Encyclopedia of Clinical Toxicology

Encyclopedia of Clinical Toxicology

A COMPREHENSIVE GUIDE AND REFERENCE TO THE TOXICOLOGY OF PRESCRIPTION AND OTC DRUGS, CHEMICALS, HERBALS, PLANTS, FUNGI, MARINE LIFE, REPTILES AND INSECT VENOMS, FOOD INGREDIENTS, CLOTHING AND ENVIRONMENTAL TOXINS

Irving S. Rossoff DVM, FACVPT

The Parthenon Publishing Group
International Publishers in Medicine, Science & Technology

A CRC PRESS COMPANY
BOCA RATON LONDON NEW YORK WASHINGTON, D.C.

Library of Congress Cataloging-in-Publication Data
Rossoff, Irving S.
Encyclopedia of clinical toxicology : a comprehensive guide and reference to the toxicology of prescription and OTC drugs, chemical, herbals, plants, fungi, marine life, reptiles, and insect venoms, food ingredients, clothing, and environmental toxins / I.S. Rossoff.
p. cm.
ISBN 1-84214-101-5 (alk. paper)
1. Toxicology–Encyclopedias. I. Title.

RA1193.R67 2001
615.9′003–dc21

2001036946

British Library Cataloguing in Publication Data
Rossoff, I.S.
Encyclopedia of clinical toxicology: a comprehensive guide and reference to the toxicology of prescription and OTC drugs, chemicals, herbals, plants, fungi, marine life, reptiles and insect venoms, food ingredients, clothing and environmental toxins
1. Toxicology – Encyclopedias
I. Title
615.9

ISBN 1-84214-101-5

Published in the USA by
The Parthenon Publishing Group
345 Park Avenue South, 10th Floor
New York
NY 10010
USA

Published in the UK and Europe by
The Parthenon Publishing Group
23–25 Blades Court
Deodar Road
London SW15 2NU
UK

First published in 2002

Typeset by Siva Math Setters, Chennai, India
Printed and bound in the USA

Contents

One who saves a life – it's as if one saves the whole world

Talmud

Sola docis facit venenum (Only the dose makes the poison)

Paracelsus

Even nectar is poison if taken in excess

Hindu Proverb

I maintain that you would often in the fifteenth century have heard the snobbish Roman say, in a would-be off-hand tone, 'I am dining with the Borgias tonight'. No Roman ever was able to say, 'I dined last night with the Borgias'.

Sir Max Beerbohm

Foreword

As practicing physicians who have participated in the planning, conceptual design, and cheerleading of this monumental project, we are pleased to see it come to fruition. Dr Irving Rossoff has been a leader of his medical community for both physicians and veterinarians for many decades. This eminent toxicologist has now produced a comprehensive and extremely usable reference in toxicology. Although it stretches for over 1500 pages, it is extremely readable and makes a very easy reference for quick and thorough review.

The genesis of this tome may be unique. Dr Rossoff has compiled an incredible bank of data in his office in Taylorville, Illinois. Drawing from over 100 000 of his personal files compiled since the administration of Franklin Roosevelt, the book mirrors Dr Rossoff's encyclopedic knowledge.

Watching this project come to fruition has been all the more gratifying for us who have served as his medical physicians. We had the great professional satisfaction to meet him and evaluate him for a complex fever of undetermined origin[1] in 1992. Dr Rossoff, ordinarily an extremely vigorous man, was felled at the knees by a prolonged febrile illness marked by a refractory anemia. An extensive evaluation led to the diagnosis of an infected prosthetic ascending aorta caused by *Cardiobacterium hominis.* After only 6 weeks of therapy with penicillin and gentamicin (both well described in this very edition), his symptoms cleared and have remained in remission for the last 10 years.

Users of this encyclopedia will be gratified by the wealth of information it contains. This remarkable guide provides useful information about the toxicology of medications, natural substances, herbals, venoms, foods, environmental toxins, and even clothing. Every week, patients bring in compounded herbal products that they take, often with numerous ingredients, and this encyclopedia has been very useful in explaining the purported mechanism of action and side-effects of the various ingredients. Our professional joy can only add to the encomiums of the scientific community.

Donald R. Graham, MD, FACP
Chief of Infectious Diseases
and
Stephen T. Randag, MD
Board Certified: Internal Medicine
Springfield Clinic
Springfield, Illinois
November, 2001

Reference

1. Petersdorf R, Beeson P. Fever of unexplained origin: report of 100 cases. *Medicine* 1961;40:1

Preface

I.S. Rossoff

This volume is intended to be an internationally used handbook of symptomatology in clinical toxicology; therefore, the names used are chosen to help readers in many countries and to facilitate better understanding of published papers. Detailed chemical names and structures can be found in other reference texts. The sentence structure is abbreviated and abbreviations are frequently used to permit greater space utilization. The text is easy to read, and concise, with historical tidbits and graphic highlighting for easy reference and retrieval of key information. Actual case reports are frequently cited. Of the more than 100 000 subject folders in my files (accumulated over 50 years), I have selected about 6000 for this first edition. Humans are the prime subjects in the text, followed by useful animal data. When I have no adequate data on human toxicity, animal data are emphasized.

Toxicology is a study of poisonous substances, i.e. agents that can destroy health, seriously endanger health, or cause death. It is an arbitrary general definition. Most data on human toxicology are anecdotal, but repeated observations by reliable observers make them 'facts'. The following may help to put things in a suitable perspective.

Oliver Wendell Holmes said, 'If the whole materia medica were dumped into the sea, it would be better for us on dry land and worse for the fish at sea.'

Galen said, 'All those that drink of this remedy will recover in a short time. Those who do not recover will die, therefore the remedy is of no use in the incurable patient.'

'All those who eat pickles will die'. *Anonymous*

'Cholesterol is poisonous
So never, never eat it.
Sugar, too, may murder you.
There is no way to beat it.
And fatty food may do you in.
Be certain to avoid it –
Some food was rich in vitamins,
But processing destroyed it.

So let your life be ordered
By each documented fact,
And die of malnutrition,
But with arteries intact.'

Anonymous

When the United States Department of Agriculture decided to approve sodium acid phosphate (SAP) to give hot dogs a rosy, red color, *Medical World News* (on 26th November, 1971) printed:

'I'm no faddist
Not gone organic,
Keep my cool and never panic.
I wash the clothes
Without an additive,
The phosphate problem
Won't drive me madditive.
I've learned to live
With BHA, MSG, EDTA;
I'll probably survive sodium nitrate,
Citratic acid and cal. propionate.
I'm hip that hot dogs labeled "meat"
Give us plenty of other things to eat;
But now I've had the final zap;
FDA said okay to SAP.'

Dr Raymond Woosley called for drug information free of the influence of the pharmaceutical industry. The purpose of the Food and Drug Administration (FDA) is to try to protect consumers from the untoward effects of drugs and other substances, but even it cannot always be successful in this respect.

For example, at a government hearing in Washington, Edward Weiss, Chairman of the Human Resources Subcommittee of the House Government Operations Committee, stated, 'The FDA has the obligation to make sure the toxicity of the drug is not such that it ends up killing people', and he then added, 'The FDA made a serious mistake when it approved Tambocor (flecainide) and Enkaid (encainide) for mild arrhythmias because the agency did not know if the drugs were safe for that use ... the FDA knew, prior to approval, that these drugs could kill people whose lives were not jeopardized by their conditions'. In this particular instance, it appeared that the FDA failed its statutory duty to protect the public from unsafe drugs. Therefore, we should perhaps bear in mind that, on some occasions, patients taking placebos are the lucky ones – they may get better without risking the side-effects of a new drug.

To live means to risk one's life occasionally. 'Perfect safety is a chimera, regulation must not strangle human activity in a search for the impossible' (Supreme Court Judge Warren Burger). Ancient Greeks used the word pharmacon to mean both poison and antidote, emphasizing the close similarity and relationship of the two. Often quoted are variations of the statements, 'Only the dose makes the poison' ('Sola docis facit venenum' – Paracelsus, 1493–1541) or 'Nothing is of itself good or evil, only the manner of usage makes it so'. Nothing is absolutely safe. Sir Derrick Dunlop, former head of Britain's drug regulatory agency, is quoted as saying, 'Show me a drug without side-effects, and I'll show you a drug without *any* effect.' Theophrastus, a disciple of Aristotle, wrote about *Datura metel* in the first published treatise on botany. 'One administers 1 drachma if the patient must only be animated and made to think well of himself; double that if he must enter delirium and see hallucinations; triple it, if he must be permanently damaged; give a quadruple dose if he is to die'.

Medication errors have reportedly caused 7000 deaths/year in the United States. Continuing self-education and formal education by physicians and other health-care workers, as well as patient education, can reduce these numbers. If more than one physician writes orders on a patient, drug interaction possibilities can increase.

Reported data are often difficult to interpret because of generalizations used, such as 'inert carrier' and 'suitable solvent'. A 'suitable solvent' may be esthetically fine for a drug, but can cause fatalities, as in the case of dimethylene glycol (q.v.) in a United States sulfanilamide (q.v.) elixir; a similar tragic situation was repeated 35 years later in South Africa, and again, after another 23 years, in Bangladesh, and in between there were problems in Europe. In the course of daily living, or in the workplace, people ordinarily inhale up to 20 000 liters of air contaminated with dusts, smoke, and aerosoled preparations, which can adversely affect the respiratory tract as irritants, aggravate pre-existing conditions, and be absorbed systemically. Unfortunately, many of these same substances settle on one's body surface, clothing, and food. Animals that preen themselves have increased gastrointestinal exposure.

Some drugs, such as the intestinal antiseptic, cloxiquine (chlorohydroxyquinoline), were used world-wide for decades, with only a few related reactions and then, when believed to be absolutely safe, caused an iatrogenic epidemic outbreak of toxicity in Japan. Dr Robert J. Cipolle put it succinctly, 'Drugs don't have doses – people have doses.' Nyburg and Thomas Jukes have enjoyed taunting regulatory agencies. Dr Jukes said, 'Every person, having the rudiments of perception, knows that you can sprinkle a little salt on food, but you can't eat a pound of salt a day.' Dr Nyburg has written about potatoes containing about 150 chemicals, including arsenic (q.v.), and solanine (q.v.). The latter two can be very toxic. In the United States, the average person is said to consume about 120 pounds of potatoes a year or 9.7 g solanine, enough to kill a horse, but who eats that many at one time? Normal people eat about 2 pounds of lima beans a year, containing about 40 mg hydrogen cyanide (human LD, 50–100 mg; LD_{L0}, 570 µg/kg), but, ingested daily in small quantities, the body easily detoxifies it.

Many over-the-counter items, including herbals, are often more toxic than magical, with frequently lead and arsenic as contaminants. Folk medicine is still used to treat about 80% of the world's population.

With the sudden surge of interest in nutraceuticals and herbals, one must remember that over 6000 or 3% of flowering plants contain alkaloids which can often be unsuspectedly toxic when ingested as such or as contaminants in grains, where they have adversely affected over 2000 people in one outbreak in Afghanistan. Millions of workers in the United States are exposed to known carcinogens in the workplace, with about 100 000 resulting deaths and over three and a half times as many suffering from disabilities due to it. Cigarette smoke is said to contain more than 6000 chemicals, and aryl hydrocarbon hydrolase in airway epithelium acting on them can produce potent neoplastic compounds.

Medical fads and patent medicines have given toxicologists a large quantity of material to evaluate. Circa 1920 and thereafter, radium treatments became cure-alls and the cause of many deaths. A well-known industrialist and amateur golf champion drank three bottles of Radiathor (triple-distilled water containing at least 1 μCi each of Ra^{226} and Ra^{228}) per day for more than 2 years, until his jaw fell off. The Nazis and Japanese conducted tragic lethal drug experiments during World War II. In the 1950s, Krebiozen was promoted and failed as a cancer cure. Ten and 20 years later, laetrile, from apricot pits, became a cancer cure fad, with resulting cyanide toxicity.

The safeguards for people of many prescribed, over-the-counter and herbal drugs, as well as numerous industrial and environmental chemicals, can be very fragile, as individual people and animals often have their own peculiar reactions to these. Readers must become aware that figures for lethal dose can vary by species, strain, age, genetics, sex, suspending agents, product purity, amount of food in the stomach, etc. Their concentration, their amount, and duration of ingestion of toxicants are important factors in evaluating toxicity. Low levels over a long period of time, such as with lead and methyl mercury, can cause serious permanent damage. Latency of 2–5 years has occurred for overt neurological damage after methyl mercury exposure from eating fish. As some drugs find newer uses, their increased availability finds them in cases of accidental ingestion by children, suicides by adults, and increased toxicity by previously unknown drug interactions. The tremendous indigenous botanical flora on various continents have given man many useful, as well as toxic, substances. In recent years, more attention has been given to toxic effects in fetal and early neonatal life. Pharmacokinetic differences occur in these age groups. With the realization that children are no longer 'little adults', hopefully, a decrease in toxic reports in them may occur.

Histopathological changes in mice receiving morphine are about three and half times as great when they are housed on pine, compared to corncob bedding, illustrating unusual factors influencing toxicity or therapeutic studies. Many reported adverse effects of psychotherapeutics, such as suicidal thoughts, may actually be a function of the original illness. Toxicity, as in the case of gentamicin, can be used to a clinician's advantage, by directly injecting it into the inner ear to decrease dizziness, by adversely affecting the balance nerve. Although this text is written to inform readers of adverse symptomatology, often not readily available (even on prescription drugs), users of the text must understand the basic handling of such emergencies, i.e. remove the patient from the source of exposure and remove the offending agent from the patient. In some cases of ingestion, induction of early vomiting may be indicated and use of an adsorbant and chemical inactivators may be required. Each person does not experience every untoward effect listed. Even Nostradamus would have had trouble predicting some of the unusual toxicities described in this text.

Today, a fruit or botanical ornament in Africa, Asia, or the Caribbean may be brought back to one's own country and, eventually, be eaten by a child, with disastrous results. Many such items are discussed in this text, bringing to the reader data collected during many years of world-wide consulting. Unless otherwise stated or implied, toxic doses are oral. It is, obviously, a 'human' toxicology text, but much experimental and other animal data are summarized, giving readers an idea of relative toxicity, particularly of interest to clinicians, especially where metabolic mechanisms may be similar to man's. Thus, some may consider this a 'comparative' toxicology text. It is long overdue for physicians, lawyers, toxicologists, researchers, pharmaceutical companies, veterinarians, the paint and chemical industries, the laity, and students. It has been put together so that the medical community can share my files and my memory. The text includes the production volume of many industrial chemicals and their uses.

The author is indebted to workers in many countries who have elected to help share their work with you, the reader, as well as to my parents and many teachers who imbued in me a thirst for knowledge and stimulated my curiosity and search for truth. A special debt is owed to my secretary, Mrs Connie Downs, who, for approximately 20 years, has helped to interpret my notes, collate and index my data, and help to edit my writings so that you can find this text interesting to read, use, and enjoy.

Disclaimer

The author and publisher assume no responsibility for and make no warranty with respect to results, uses, procedures, or dosages listed and do not necessarily endorse any such uses, procedures, or dosages. Neither the author nor the publisher shall be liable to any person whatsoever for any damage resulting from reliance on any information contained herein, whether with respect to drug identification, uses, procedures, dosages, toxicity, or equivalences, or by reason of any misstatement or error, negligent or otherwise, contained in this work.

The author recognizes that differences in inert carriers, manufacturing methods, etc., make secondary names useful only for information retrieval. Equivalent names used were selected from the literature to make the text useful to readers in all countries. The symbols ™ and ® have generally and purposely been excluded from this book because of major differences in product registry in different countries. No reference to any name or term is intended to indicate that it is not a proprietary term or trademark. Nothing herein is to be construed as a suggestion to violate any trademarks, patents, or laws. Because of constantly changing corporate structure, mergers, or buyouts, references to company names used refer to the way they were commonly known or written about at the time of the events reported. Little attempt is made to detail particular corporate divisions, etc. Further, any information given herein is supplied with the understanding that no discrimination is intended nor are any endorsements implied.

List of abbreviations

±	about, approximately, occurring after a number
>	greater than
<	less than
~	approximately
AIDS	acquired immune deficiency syndrome
BC	before Christ
BCG	bacillus Calmette-Guérin
BCME	bis (chloromethyl) ether
BTU	British thermal units
CD_{50}	convulsant dose in 50% of the population
CPR	cardiopulmonary resuscitation
CT	computerized tomography
DNCB	dinitrochlorobenzene
DOCA	desoxycorticosterone acetate
ED_{50}	effective dose in 50% of the population
FDA	Food & Drug Administration (U.S.)
GABA	gamma-aminobutyric acid
HIV	human immunodeficiency virus
IA	intra-arterial
ID	intradermal
IM	intramuscular
IP	intraperitoneal
IV	intravenous
LC_{LO}	lowest lethal concentration
LC_{50}	lethal concentration in 50% of the population
LD	lethal dose = DL
LD_{50}	lethal dose in 50% of the population = DL_{50}
LEL	lowest explosive (flammable) limit at room temperature
MDR	minimum daily requirement
mEq	milliequivalents
mppcf	millions of particles/cubic foot of air
MTD	minimum toxic dose
NCI	National Cancer Institute
ng	nanogram (a billionth of a gram; 10^{-9})
OSHA	Occupational Safety and Health Administration
PO	orally
ppb	parts per billion = $\mu g/kg$
pph	parts per hundred
ppm	parts per million
ppt	parts per trillion
psi	pounds per square inch

PTCA	percutaneous transluminal coronary angioplasty
q.v.	quantum vis, as much as needed
SC	subcutaneous
SGOT	serum glutamic oxaloacetic transaminase
SGPT	serum glutamic-pyruvic transaminase
TCCD	tetrachlorodibenzo-p-dioxin
TD_{LO}	lowest toxic dose reported
TD_{50}	average toxic dose in 50% of the population
TLV	threshold limit value (TWA with no adverse effects on repeated exposure to workers)
TNF	tumor necrosis factor
TWA	time weighted average (usually 8 h/day/40-h week)
UDP	unassisted diastolic pressure
U.S.	United States
WHO	World Health Organization

A

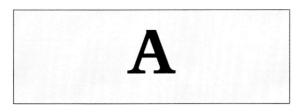

A-649
= *NSC 38,270*

Antineoplastic antibiotic. **Experimental.**

***Untoward effects*: Human**: Daily doses > 0.1 mg/kg/day, IV often associated with sudden life-threatening toxicity (severe thrombopenia, cardiovascular hemorrhage, irreversible vasomotor collapse, anuria, thrombophlebitis, nausea, vomiting, coma, confusion, psychosis, hypotension, hepatocellular necrosis).

LD_{50}, **dogs**: 0.05–0.1 mg/kg. Two out of three **dogs** died within 2 weeks with 0.24 mg/kg/day. Three **dogs** receiving 0.36 mg/kg/day died by the sixth day of therapy. IV toxic effects were produced in the liver, bone marrow, gastrointestinal tract, along with profound irritation of the vasculature. In **rats**, as in the **dog**, the toxic death rate was dose-related. Also lethal to **monkeys**.

ABACAVIR
= *Ziagen*

Antiviral.

Untoward effects: Severe hypersensitivity in ~5%, usually after 7–10 days, with gastrointestinal upsets, fever, malaise, and occasional rash. Repeat treatments lead to more rapid and more severe symptoms plus hypotension, dyspnea, and a reported death.

ABALONE

***Antiviral* (experimental polio – mice). (Paolins, in Chinese, means abalone extract.) *Antibacterial* (*Staphylococcus aureus*).**

***Untoward effects*: Human**: Poisoning from eating the viscera produces a generalized urticaria-like reaction over the entire body and gastrointestinal disturbances.

ABCIXIMAB
= *Reo Pro* = *7E3* = *Cento R* = *c7E3 Fab*

Untoward effects: Increased risk of bleeding (10–18%) at injection site; thrombocytopenia in 1–2%, ventricular tachycardia in > 1%, palpitations in 0.5% with arteriovenous fistula, incomplete atrioventricular block, atrioventricular block, nodal arrhythmia, embolism and thrombophlebitis, and gastrointestinal upsets 0.1–2%. Dizziness in 2.9%, anxiety in 1.7%, and confusion in 1.3%. All other central nervous system, respiratory, musculoskeletal, and miscellaneous symptoms < 1%, except pain (5.4%).

ABECARNIL
= *ZK-112,119*

Anxiolytic.* A *β-carboline.

Pharmacokinetics (absorption, bioavailability, and excretion) are similar in dogs, rats, mice, rabbits, cynomolgus monkeys, and baboons.

***Untoward effects*: Human**: Single oral doses of 20–40 mg, as well as multiple dosing, produced dizziness, unsteady gait, and lack of concentration.

ABIETIC ACID
= *Sylvic Acid*

A resinous acid. Extracted from Kraft paper pulp.

Use: In "sizes" in typing paper manufacture, in ester gums used in lacquers and varnishes, as well as in the manufacture of "metal resinates", soaps, surfactants, and plastics.

***Untoward effects*: Human**: Dermatitis, as a result of allergic reactions in sensitive **people**. Its 96 h LC_{50} is 0.41 mg/l in juvenile **Coho salmon**. LD_{50}, IV, **mice**: 180 mg/kg.

ABRIN
= *Agglutinin* = *Toxalbumin*

***Phytotoxin.* Both hemolytic and neurotoxic.**

An extremely toxic substance in the seeds of *Abrus precatorius* (*Rosary Bean* or *Pea*, *Jequirity Beans*, *Precatory Bean*, *Wild Licorice*, *Indian Licorice*, *Crab's Eyes*, *Lucky Bean*) q.v., a common vine in tropical countries, and southern and central Florida. Toxic principles are stable at 60°C for 30 min, but most toxicity is destroyed at 80°C for 30 min. This is not *abrine*, also isolated from the same plant. Cancerostatic, due to its strong inhibition of protein synthesis (tested in mice and rats, with many mouse and human cancers). *Abrusosides A*, *B*, *C*, and *D* in the leaves of *A. precatorius* are toxic to mice, and have sweet properties (one-half to two times sucrose) in humans.

***Untoward effects*: Human**: The toxicity is due not only to its hemagglutinating effect, but also other glycoproteins. One chewed seed reportedly fatal to an **adult**. Cases of murder by it have been reported. Severe diarrhea and shock syndrome (weakness, cold sweats, trembling of hands) are common in **man**. Symptoms may be delayed for several hours to several days after ingestion.

LD_{50}, IP, **mice**: 0.020 mg/kg. **Horses** are very susceptible; **sheep** and **goats** appear to be less affected. A penniless caste group in India illegally poisoned cattle by driving abrin soaked sticks under their skin.

Treatment: SC ***arecoline hydrobromide***, followed by an oral saline purgative, has been reported to be of some value in **ruminants**.

ABRUS PRECATORIUS

= *Abrus à chapelet* = *Crabs' Eyes* = *Gunchi* = *Indian Licorice* = *Jequirity Bean* = *Lucky Bean* = *Precatory Bean* = *Rosary Bean* = *Rosary Pea* = *Wild Licorice*

***Botanical*. The slender, high-climbing vine with feathery leaves is called *Mkandume*, *Mkangauchawi*, and *Mlazalaza* (Swahili); *Enyinmukpo* (Efik); *Mturituri* (Giriama); *Idon Zakara* (Hausa); *Anya Nwono* (Igbo); *Oju Ologbo* (Yoruba); *Muzekezeke* (Kiyanzi); and *Ngiengie* (in Kikongo). The seeds are called *Macho Ya* or *Tipitipi* (Swahili). Agglutinates red blood cells.**

Use: An antipyretic and antitussive (from ***methyl salicylate*** and ***glycyrrhizin*** content, respectively) tea is made in the Bahamas from the leaves. The shiny red and black seeds are very toxic and contain ***abrin*** (q.v.). The chief danger often is due to the fact that these seeds are used for jewelry (necklaces, earrings, belts, rosary or prayer beads, good luck charms on musical instruments, and rattles) in Latin America, and tourists may bring them back to the U.S. Some ***abrin*** exists in the licorice-flavored leaves. Has been used as an aphrodisiac, for **scorpion** bites and menstrual problems, in witchcraft, and in cancer chemotherapy.

Untoward effects: **Human** ingestion of seeds can cause intravascular hemolysis, thirst, nausea, vomiting, stomach pain, weakness, coma, hand trembling, cold sweats, cyanosis, colic, and severe diarrhea and rectal bleeding, which may persist for months. Eye irritant. Chewing one seed has been fatal in an adult and a suicide. Responsible for more plant fatalities in Florida than any other species. Unchewed seeds are less toxic. Toxicity often occurs 12–18 h, and up to 3 days after ingestion. Oliguria and uremia have been reported 812 days after poisoning. No effective antidote has been reported in **humans**.

Oil extracted from the seed has caused sterility in all **mice** and, after 150 mg, 80% of **albino rats**. Poisoned **guinea pigs** and **rats** have shown liver necrosis. **Horses** have died from eating it in neglected pastures. **Cattle** implanted with spikes 1½" long and ½" at thick end (0.6 mg/kg), or needles made from ground decorticated seeds, implanted into the throat or hindquarters of **cattle** in India and Pakistan, by unscrupulous hide dealers, die within 2–4 days, after exhibiting salivation, stiffness, incoordination, muscle spasms, convulsions, fever, and extensive, painful swelling around the implantation site. Owners attribute the death to hemorrhagic septicemia or blackleg. Aqueous extracts SC kill **cattle** within 2–4 h. **Zebu cattle** fed whole or decorticated seeds (60–70 mg/kg) stayed normal.

Chickens died within 18–96 h, after eating whole seeds or seed pulp, with mucosal hemorrhages and dehydration. Seed pods are 1½" long and occur in bunches. The pods contain 3–8 ¼" oval, green seeds, which are smaller when the pod bursts. They are then brilliant red, black at one end, and very hard. LD_{50}, **mouse**: 0.56 µg/kg.

Treatment: Experimental transfusion of hyperimmune serum to six **cattle** was effective. Since it is a caustic substance, emetics should be avoided. Cautious lavage, after adequate dilution, has been suggested. Urine should be kept alkaline (5–15 g ***sodium bicarbonate*** daily), and watch for renal failure. Treat convulsions, if necessary, with ***diazepam***. Blood transfusions and analgesics may be indicated.

ABSCISIC ACID

= *ABA* = *Abscisin* = *Dormin*

***Plant hormone*. May trigger increased formation of proteinase inhibitors, which help wound healing. *Dormin* has also been the name of an over-the-counter "tranquilizer".**

Use: As a plant defoliant, growth-inhibitor, antitranspirant, and to help fruit abscission (loosening and falling off) of citrus, olive, and pineapple.

ABSIDIA sp.

***Fungus*. Many species reported.**

Untoward effects: Acute infection of paranasal sinuses of **man** and **animals** has been reported with inflammation, local necrosis, invasion of brain tissues, frequently fatal. SC infections of all body areas have also been reported.

Ab. corymbifera: fatal in a laboratory **rabbit**, evidencing multiple tuberculoid granulomas and extensive necrotic areas in the spleen and kidney. In suppurating **human** ears, cleared by ***amphotericin B***. Associated with death, mastitis, and abortions in **cattle** and **okapis**. Also reported in systemic infections in **man**, **mink**, **mice**, and **guinea pigs**.

Ab. italina: caused systemic mycosis in **humans**.

Ab. lichtheimii: **Human**: Associated with phycomycosis of the external ear, with itching, pain, and hearing disturbances, alleviated by topical application of ***amphotericin B*** ointment.

Ab. ramosa: in rumens of **cattle** with chronic hematuria, and in the uterus, after some abortions. **Rabbits** died after 4–5 weeks, with heavy weight loss, after daily feeding of cultures. **Guinea pigs** died within 2–3 weeks (after 4 days, when pretreated with ***aspirin***). Two fatal cases of intestinal mycosis have been reported in **horses**. Apparently, nephrotoxic and a growth retardant in **pigs**. Also found in vulvovaginitis of **swine**.

ABSINTHIUM

= *Armoise* = *Wormwood*

Use: As a psychoactive (ascribed mostly to its ***thujone*** content) beverage and flavoring agent in vermouth (strong odor and acrid taste) and certain wines. Limitedly, in some countries, as a stimulant stomachic and anthelmintic. From *Artemesia absinthium*, often called ***wormwood*** or ***sage***. Contains a bitter oil, ***absinthin***, and tannins.

Untoward effects: **Human**: The volatile oil (***absinthin***), or the liqueur, ***absinthe***, has caused gastrointestinal upsets, restlessness, auditory and visual hallucinations, paranoia, acute mania, headache, hyperesthesia, apprehension, dilated pupils, vomiting, respiratory failure, nervousness, convulsions, stupor, and even death. ***Absinthe*** is a green liqueur, distilled particularly from ***wormwood***, after steeping it in ***alcohol***. In the nineteenth and early twentieth century, it was observed in France that too much imbibed caused mental deterioration. The toxic compound is ***thujone*** (q.v.). Switzerland banned its manufacture in 1908, and it was 1922 before France banned it. Its sale and use is illegal in the U.S. (In 1878, 2,000,000 gallons were imported). It was reported that Van Gogh drank ***absinthe*** in his last 4 years, and that, in his last 18 months of life, he had four or more fits, with hallucinations, that resembled those of other ***absinthe*** drinkers. Excitement and epileptoid convulsions have been reported in **animals**. Acute oral LD_{50} of the oil ranged between 0.26 and 0.48 g/kg, and the acute dermal LD_{50} in **guinea pigs** was > 5 g/kg. The oil was only slightly irritating, when applied full-strength to intact or abraded **rabbit** or **guinea pig** skin for 24 h under occlusion, but no irritation or phototoxic effects were noticed on hairless **mice** and **swine**. A 12% concentration in petrolatum produced no sensitization reactions in 24 **human** volunteers.

Treatment: **Human**: Kaye recommends IV ***pentobarbital***, to help control convulsions. Keep patient quiet and inactive. Artificial respiration, if necessary, and oxygen therapy. Oral ***glucose***, if able to swallow.

ABUTA sp.

Use: The bark of various species (*imene*, *splendida*, *yaupesensis*) is used by Colombian (South American) **Indians** in preparing potent ***curares*** (q.v.).

ACACIA

= *Gum Arabic* = *Gum Dragon* = *Gum Acacia* = *Wattle*

***Demulcent.* NOT to be confused with a number of acacia species causing toxicity in grazing livestock. The dried, gummy exudate of various African *Acacia*. Clear, sun-bleached, tasteless forms are preferred for pharmaceutical use. The pH of a 10% solution is 4.6.**

Use: In the treatment of mild diarrhea, to thicken and/or stabilize medicinal emulsions. To improve flavor stability and shelf-life of many dry pudding, dessert, and cake mixes. An ancient Egyptian concoction (1550 B.C.) used *acacia* and ***honey*** in a vaginal tampon, which is particularly interesting in that, as *acacia* ferments, it forms ***formic acid***, an active spermicidal agent in some modern contraceptives. As a binder for tablet ingredients, and extensively in jellies and confectionery products. In textile sizing, as a foam stabilizer and gush preventative in beers, stationery manufacturing, adhesives, shoe polishes, etc. As a stabilizer in parenterals at a 1% level. Over 20 million pounds ± a year are exported. African droughts have occasionally caused shortages. Chewed by **natives** in many lands, to help keep teeth clean. Hippocrates referred to it. The Ark of the Covenant and the coffins of Egyptian kings were made from the wood of *A. seyal*, the *Shittim wood* of the Bible.

Untoward effects: Contains ***tryptamine***, a vasopressor. ***Gum arabic*** dust has caused allergic rhinitis, bronchial asthma, and pneumonitis in **humans** (immediate Type 1 hypersensitivity tissue response). Kidney transplant **patients** receiving ***prednisone*** tablets on a long-term basis were sensitive to its use as a binder. The hypersensitivity manifested itself as itching, rashes, fever, and arthralgia. The fruit and pulp of *A. pennata* is piscicidal, and, in Africa, the powdered root is used as snuff for **cattle** nose maggots. *A. berlandieri* (***guajillo***) contains ***N-methyl-β-phenylethylamine***, a sympathomimetic amine that produces muscular incoordination, "rubbery" action of limbs, and prostration in **sheep** and **goats**. ***Guajillo*** is an important plant for grazing in southern Texas. It also contains ***tyramine***, ***hornedine***, and ***N-methyltyramine***. During extended droughts, **sheep** and **goats** may eat it exclusively for 6–12 months, before developing ataxia of the hind limbs ("limberleg", "***guajillo*** wobbles"). Up to 50–65% morbidity and mortality. Recumbent **animals** usually retain their appetite. Supplemental feeding will prevent the problem. Moving **animals** to a better range is advised. *A. sieberana* var. *woodii* (***Flat-topped Thorn***, ***Umbrella Thorn***, ***Paper-bark Thorn***) and *A. giraffae* may contain dangerous amounts of ***prussic acid*** (q.v. for symptoms and treatment) for **livestock**. Poisoning, especially in **cattle**, has also been reported after ingestion of

A. georginae (contains ***fluoroacetamide***, q.v.), *A. leucophloea* (contains ***hydrocyanic acid*** – affects **sheep**, **goats**, and **buffalo calves**, *A. nilotica* (in **goats**, symptoms are tachycardia, anemia, rumen atony, dyspnea, recumbency, methemglobinemia, and death – 30 g/day is lethal), and *A. salicina* (***Doolan Wattle***, ***Cooby Wattle***, ***Black Sally Wattle*** with high tannin content). *A. georginae* causes Georgina River Poisoning, in which **cattle** may die suddenly, often precipitated by exercise, fright, or by drinking water. Those less severely affected evidence cardiac and respiratory distress and nervous symptoms, which are usually aggravated by exercise. Hemorrhages in the myocardium occur, and are seen as scars in non-fatal cases. *A. constricta* contains cyanogenic constituents; *A. concinna* is used in Africa and India as an insecticide; *A. glauca* causes fibrosarcomas in **rats**. *A. aneura* (**mulga**) – **sheep** and **cattle** grazing on it during periods of drought in Australia develop "melanosis", which is actually lipofusein pigmentation of the liver. *A. cana* may concentrate dangerous amounts of ***selenium***. *A. greggii* (**Catclaw**) contains ***prussic acid*** – **ruminants** are more susceptible than **monogastric animals**. Particularly toxic when stunted, wilted, or frost-damaged. *A. nilotica* causes methemoglobinemia, abortion, dyspnea, tachycardia, rumen atony, anemia, and slight hyperglycemia with failure of normal liver and renal function in poisoned **goats**; 30 g/kg appears to be lethal, yet, 5 g/kg/day can kill over an extended period. *Acacia sp.*, in general, may accumulate high levels of ***hydrocyanic acid***.

ACAMPROSATE

= *Acamprosate 6473* = *Aotal* = *Bisacetylhomotaurine* = *Campral*

Untoward effects: Erythema multiforme with prurigenous erythematous plaques on limbs and trunk of 40-year-old **female** with alcoholic cirrhosis.

ACARBOSE

= *BAY g-5421* = *Glucobay* = *Precose*

Antidiabetic. **An α-glucidase inhibitor. For treatment of non-*insulin*-dependent diabetes mellitus.**

Untoward effects: Severe hepatotoxicity in 40-year-old **female** and 65-year-old **female** after 100 mg tid/2 months. Flatulence 77%, abdominal pain and distention 21%, and diarrhea 33%, and borborygmi due to fermentation of undigested carbohydrates. Hepatomegaly, increased aspartate aminotransferase and alanine aminotransferase after long-term therapy. Eosinophilia and generalized erythema multiforme in 58-year-old **male** after 300 mg/day/13 days.

Drug interactions: Hypoglycemia can occur when used with a ***sulfonylurea***. Can decrease ***iron*** absorption leading to anemia. Increases ***warfarin's*** anticoagulant effect and decreases ***digoxin's*** blood levels.

ACEBUTOLOL

= *M&B 17,803A* = *Neptall* = *Sectral* = *N-[3-Acetyl-4-[2-hydroxy-3-[(1-methylethyl)amino]propoxy]phenylbutanamide*

Beta$_1$-blocker, antianginal, antiarrhythmic, antihypertensive. **Large intersubject variation. Human plasma half-life is 8 h.**

Untoward effects: Occasional fatigue, cold extremities, nausea, depression, dizziness, constipation, impotence, hypotension, bradycardia, bronchospasm, arthritis, arthralgia, pulmonary complications, such as a hypersensitivity pneumonitis, and a lupus-like syndrome have been reported. Decreased elimination in renal failure. Short-term use may increase ***insulin***-induced hypoglycemia. A 15-year-old **female** ingested 7.6 g, remained conscious, but suffered from severe hypotension (60 mmHg). Electrocardiogram showed a first degree atrioventricular block and intraventricular conduction disturbances. IV ***calcium gluconate*** and plasma increased blood pressure to normal in 2 h. Use with ***verapamil*** and ***diltiazem*** may accentuate bradycardia and cause some atrioventricular blockade. High levels are found in breast milk of treated **mothers**. Their **infants** receive ~10% of a therapeutic dose (adjusted for weight). It is excreted through the liver and kidneys, and its active metabolite is excreted mainly by the kidney. Thus, it can be dangerous for **neonates** whose renal function is immature.

Treatment: ***Isoproterenol***, a β-agonist, has been suggested in severe bradycardia. ***Atropine*** will usually have little benefit. See case report above.

ACECAINIDE

= *N-Acetylprocainamide* = *NAPAh* = *N-Acetylnovocainamide*

Antiarrhythmic. **Active metabolite of *procainamide*.**

Use: For treating premature ventricular contractions.

Untoward effects: In **man**, the toxic blood level is about 1.6 mg%, compared to therapeutic levels of 0.2–1.2 mg%. An anticholinergic syndrome developed within 2 days in a 21-year-old **patient** receiving 6.3 g/day by oral and IV administration. Overdosage can cause renal failure.

Treatment: Hemoperfusion may be useful in overdoses.

ACECARBROMAL

= *Abasin* = *Acetyl Adalin* = *Acetylcarbromal* = *Carbased* = *Sedamyl*

Sedative, hypnotic.

Untoward effects: Low toxic dose, **human**: 7 mg/kg; LD_{50}, **mouse**: 1.6 g/kg.

ACECLIDINE

= *Glaucostat*

***Parasympathomimetic, miotic.* Useful topically as a 2% solution, in dropping intraocular anterior chamber eye pressures in open-angle glaucoma, especially where *pilocarpine* is not tolerated.**

***Untoward effects*:** LD_{50}, IV, **mouse**: 36 mg/kg.

ACECLOFENAC

Non-steroidal anti-inflammatory drug analgesic.

***Untoward effects*:** Leukoclastic vasculitis and hemoptysis in 68-year-old **female**.

ACEFYLLINE PIPERAZINE

= *Etaphylline*

Bronchodilator.

Untoward effects*:** Hypersensitivity to ***theophylline derivatives manifested by erythema, swelling and painful joints. Heartburn has also been reported.

ACEMETACIN

= *1-(4-Chlorobenzoyl)-5-methoxy-2-methyl-1H-indole-3-acetic Acid Carboxymethyl Ester* = *TV 1322* = *Rantudil* = *Acemix* = *Rheumibis*

***Non-steroidal anti-inflammatory agent.* A considerable amount is converted into *indomethacin* (q.v.).**

***Untoward effects*:** Acute oral toxicity LD_{50}, **rat**: 24.2 mg/kg. No teratogenic effects noted in **rat** and **rabbit** trials. Increases coronary vasodilator drug effects in **dogs**.

ACENOCOUMAROL

= *Acenocoumarin* = *Nicoumalone* = *Sintrom* = *Syncoumar*

***Anticoagulant.* A coumarin for prophylaxis and treatment of venous thrombosis, and emboli. Rapid gastrointestinal absorption. Eliminated, primarily, by renal mechanisms in man; in rats, 50% by biliary and 20% urinary excretion in 24 h. Plasma half-life in man is about 20 h, with peak oral effect within 24–48 h. After withdrawing treatment, prothrombin time approaches normal in about 30 h, and is normal in about 3–4 days. Concentrates in kidney, liver, and brain tissues.**

Untoward effects*:** Toxic serum concentration in **man** is approximately 0.1 mg/l (up to ten times upper therapeutic dose). Overdosage causes elevated and prolonged prothrombin time, hemorrhage, and, possibly, nausea, vomiting, and diarrhea. Bleeding episodes noted (cerebral, nasal, pulmonary, gastrointestinal, and ocular traumatic hematomas). Hematuria, hemoglobinuria, hemorrhage into tympanic membrane, and glandular hematomas. This tendency may increase with age. May rarely cause skin lesions (petechiae, ecchymoses, and hemorrhagic infarcts). A leukocytoclastic vasculitis has been reported, as well as cases of melena, hematemesis, skin necrosis, and asphyxiating pharyngolaryngeal hematoma. 60 mg was ingested by a 27-year-old **female**, in a suicide attempt. Five days later, **she** evidenced a severe hemorrhagic diathesis, increased menstrual bleeding, and hemorrhagic skin eruptions. ***Vitamin K treatment produced a complete recovery. A genetic resistance to its effect has been reported in **man**.

Drug interactions*:** In **humans**, it is *potentiated* by ***acetaminophen, ***alcohol***, ***amiodarone***, ***anesthetics***, ***aspirin***, ***cetirizine***, ***chloramphenicol***, ***chlorpromazine***, ***cimetidine***, ***clarithromycin***, ***clofibrate***, ***corticosteroids***, diets with a reduced saturated fat content with decrease of dietary ***vitamin K***, ***doxycycline***, ***erythromycin ethylsuccinate***, ***fluconazole***, ***indomethacin***(?), ***isoniazid***(?), ***miconazole***, ***nalidixic acid***, ***norethandrolone***, ***oral contraceptives***, ***oxyphenbutazone***, ***penicillin***, ***phenylbutazone***, ***salicylates***, ***sulfonamides***, ***tetracycline***, ***D-thyroxine***, and, possibly, ***quinine*** and ***quinidine***. Some oral antibiotics may decrease intestinal ***vitamin K*** production. Interactions with ***phenytoin*** not reported, probably due to different methods of elimination. *Antagonized* by ***aminoglutethimide***, ***antihistamines***, ***barbiturates*** (***barbital***, ***heptobarbital***, ***phenobarbital***), ***chloral hydrate***, ***corticosteroids***, ***griseofulvin***, ***menadiol***, ***meprobamate***, ***rifampin***, and ***vitamin K***.

In **rabbits** and **rats**, it is *potentiated* by ***iproniazid***, ***isocarboxid***, ***nialamide***, and ***tranylcypromine***. ***Glucagon*** *potentiates* the response in **guinea pigs**. **Guinea pigs** pretreated 4 days with ***reserpine*** or ***chlorpromazine*** had an exaggerated prothrombin response. Oral LD_{50}, **rats**: 513 mg/kg; IP, LD_{50}, **mice**: 115 mg/kg.

Treatment*:** Treatment of accidental ingestion may require gastric lavage and various supportive treatments. In the absence of hemorrhage, in overdosage cases, withhold one or more doses, until prothrombin time returns to a desired level. Oral ***vitamin K can usually control mild bleeding. In moderate or severe bleeding cases, IM or very slow IV administration of ***vitamin*** K_1 is indicated. Emergency use of fresh whole blood may also be required.

ACEPHATE

= *Orthene* = *Ortran* = *Ortho 12420* = *Acetylphosphoramidothioic Acid θ, S-Dimethyl Ester*

***Organophosphorus contact and systemic insecticide.* Used in controlling range grasshoppers. Seed protectant.**

Residual systemic activity against many plant pests for approximately 10–15 days after application. Used world-wide.

Untoward effects: Nesting **birds** in Oregon showed a 3–5 week adverse effect with 30–50% depression of brain cholinesterase. LC_{50} in **birds**, > 3000 ppm, but 90 ppm for its toxic metabolite, ***methamidophos***. **Grasshoppers** poisoned by it on western rangeland contained 10 ppm and up to 5 ppm of its toxic metabolite, ***methamidophos***. 4–6 month **male mallard ducks**: acute oral LD_{50}, 186–295 mg/kg, with death usually in 2–18 h. Symptoms were ataxia, imbalance, hopping, falling, wing spreading, and intermittent tremors, occurring within 25 min. Levels as low as 124 mg/kg produce symptoms, and recovery takes about 10 days. Exposed **mallard** embryos exhibit edema, stunted growth, scoliosis, lordosis, and malformations of bill, eye, and neck. Acute oral LD_{50}, male **rat**: 945 mg/kg; female: 866 mg/kg. LC_{50}, **quail**: 718 ppm; 871 ppm in feed was lethal to 73% by the eighth post-treatment day. Dramatic reduction in feed intake from day 1 on. Oral LD_{50}, **meadow voles**: 321 mg/kg; 96 h LC_{50} (in water), **rainbow trout**: 730 μg/l. LD_{50}, **mice**: 351–720 mg/kg (two trials); 24 h LD_{50}, **little brown bats**: > 1.5 g/kg; LD_{50}, **white-footed mice**: 380 mg/kg; LD_{LO}, **dog**: 681 mg/kg; LD_{50}, **chickens**: 852 mg/kg, and an LD_{50} of 350 mg/kg has been reported; skin LD_{50}, **rabbit**: 2 g/kg.

ACEPROMAZINE

= Acetylpromazine = Acetazine = Acetopromazine = 1-[10-[3-Dimethylamino)propyl]-10H-phenothiazin-2yl]ethanone = 10-[3-(Dimethylamino)-propyl]phenothiazin-2yl Methyl Ketone = 2-Acetyl-10-(3-dimethylaminopropyl)phenothiazine = Plegicil = 3-Acetyl-10-(3-dimethylaminopropyl) Phenothiazine = 1522 CB = 10-(3-Dimethylaminopropyl)phenothiazine-3-ethylone = Atravet = Notensil = Soprontin = Vetranquil

Tranquilizer, antiemetic.

Use: Excellent anesthetic potentiator with a wide safety margin, and with immobilizing drugs in **animals**. Controls intractable **animals**, relieves pain and controls motion sickness. Now, seldom used in **man**, where newer neuroleptics have replaced it.

Untoward effects: In **humans**, doses above 100 mg often led to hypertension and occasional convulsions. Pregnant **women** taking it are at a significantly increased risk of having a **child** with a malformation (syndactyly, microcephaly, club hands and feet, abdominal muscle aplasia, cardiac malformation, hypospadia, cleft lip, brachymesophalangy, clinodactyly, and hydrocephalus. A 19-year-old **female** attempted suicide by swallowing 25, 50 mg tablets. Sedation, lethargy, and tachycardia were the main symptoms. A case of accidental poisoning in a 41-year-old **male** from an injection of ***Immobilon***® (***etorphine***, ***acepromazine***, and ***methotrimeprazine***), a veterinary product, has been reported.

Warning: As a long-acting α-adrenergic blocker, it blocks sympathetic nervous system's ability to respond to stress. May *potentiate* the toxicity of certain organic ***phosphorus*** compounds and induces prolactin release. Do **NOT** use with ***procaine*** or ***succinylcholine***. Hypotensive!

Treatment: The 19-year-old **female** received IV fluids, an emetic (30 ml ***ipecac*** syrup), and a cathartic (30 g ***magnesium sulfate***). **She** was discharged from the hospital after 36 h. The 41-year-old **male** was successfully treated with two injections of ***nalorphine*** (10 mg, IV), 10 mg ***metoclopramide*** IV to control nausea, endotracheal intubation, intermittent positive-pressure ventilation, plus supportive measures, and was released the following day from the hospital.

Oral LD_{50}, **rats**: 130 mg/kg; oral LD_{LO}, **mouse**: 200 mg/kg; SC, LD_{50}, **mouse**: 130 mg/kg; IV, LD_{50}, **rats** and **mice**: 70 mg/kg.

ACESULFAME K

= Acesulfame Potassium = Sunette = Sweet One = Acetosulfam = 6-Methyl-1,2,3-oxathiazin-4(3H)-one 2,2-Dioxide

***Non-nutritive sweetener.* Approximately 200 times (100–300 in various trials) sweeter than *sucrose*, but also has a sour taste. Excreted unchanged in animal and human trials. It is not metabolized in the body, does not accumulate in the body after repeated administration, and has no pharmacological effect.**

Use: In chewing gum, mouth sprays, mouthwashes, toothpastes, beverage mixes, confections, canned fruit, custards, puddings, and as a table-top sweetener. Useful in baked goods, as it is stable at 400°F.

Safety tests: Feeding tests – 2-year study in **beagle dogs**, long-term **rat** studies, and carcinogenicity studies in **mice** confirmed its safety. The U.S. FDA used a 100-fold factor in calculating the maximum amount consumed by **humans**. This was less than one-hundreth the amount that caused no toxic effects in **animal** feeding trials.

ACETAL

= Diethyl Acetal = Acetaldehyde Diethyl Acetal = Ethylidene Diethyl Ether

Colorless, pungent liquid, with a fruity, pleasant odor, when diluted. A normal intermediate in the respiration of higher plants. Small amounts in tart-tasting, unripe fruits. An intermediate in the fermentation of alcoholic beverages. It can form in

wine, after exposure to air, imparting an unpleasant taste to it. Traces exist in blood, from the metabolism of sugars.

Use: Solvent found in some liquors (saki, whiskey, rum). Approximately 1000 ± lbs in use in fragrances in the U.S. annually. Concentrations are usually 0.005% in soaps and detergents; 0.0025% in creams and lotions; and 0.1% in perfumes, particularly jasmines. Rarely used as a hypnotic. Rapidly hydrolyzed in the stomach to ***acetaldehyde*** and it, in turn, is oxidized to ***acetic acid***.

Untoward effects: Acute oral LD_{50}, **rats**: approximately 4.6 g/kg. Some deaths in **rats** and **rabbits**, when given orally at 2.36–3.54 g/kg; 885 mg–1.77 g/kg, IP, caused deaths in **rats**. Acute dermal LD_{50} in **rabbits** was approximately 10 ml/kg. Fed at a 5% level to **chicks** and **rats** for 6 days without ill-effects. **Rats** can safely inhale concentrated vapors for up to 5 min; inhalation of a 4000 ppm concentration caused death in 2/6 **rats**. Caused only slight primary skin and eye irritation in **rabbits**; 10% in petrolatum caused no skin irritation or hypersensitivity reactions in **human** volunteers. High air concentrations can cause narcosis.

ACETALDEHYDE

= *Ethanal* = *"Aldehyde"* = *Acetic Aldehyde* = *Ethylaldehyde*

Use: In perfumes. ***Alcohol*** is converted in the liver to the pharmacologically active, intoxicating, and addictive metabolite, acetaldehyde. ***Antabuse***® and ***citrated calcium carbamide*** blocks this conversion, by inhibiting the production of the enzyme, aldehyde dehydrogenase, necessary for oxidation of acetaldehyde, thus increasing its body tissue levels. In the body, the latter can react with catecholamines, such as ***epinephrine***, to produce alkaloids (such as tetrahydroquinolones, among which are ***mescaline*** and ***morphine***). Demand is approximately 2 billion lbs/year in the U.S. (at least an annual average of 1.5 billion lbs). 45–50% used in the manufacture of ***acetic acid*** and ***anhydride***; 19% for ***n-butanol***; 17% for ***2-ethyl hexanol***. Flammable liquid. Commercially, as a chemical intermediate in manufacturing perfumes, polyesters, dyes, ***metaldehyde***, as a preservative for fruit and fish, as a denaturant in ***alcohol***, in fuels, in hardening gelation, as an antifungal on leather products, in silvering mirrors, as a solvent in many industries, in medicine, as an inhalant in catarrh and ozena. In a wide variety of flavor compositions in alcoholic beverages and food, including candy, baked goods, and dairy products. Granted 'generally recognized as safe' status in 1980.

Genetics: About 50% of Japanese have a reduced level of aldehyde dehydrogenase, needed for its catabolism. When they consume ***alcohol***, blood levels of acetaldehyde rise rapidly. **Patients** with severe rosacea, whose facial flushing is exacerbated by chronic ***alcohol*** consumption, may have a genetic related deficiency of aldehyde dehydrogenase (e.g. **Celtic peoples**).

Untoward effects: Found in ***tobacco*** smoke and rapidly combines with proteins in the lung, reducing longevity by stiffening connective tissues in **kangaroo** trials. It has been found in the air of **hen** houses. Increased use of, or incomplete combustion of, gasohol and ***methanol*** fuels produces airborne ***formaldehyde*** and acetaldehyde. It is also formed by the combustion of wood and cotton. The **human** fetus can oxidize ***ethanol*** to acetaldehyde; therefore, fetuses of both heavy and social ***alcohol***-drinking pregnant **women** are at risk. Fetal liver has a low capacity to convert acetaldehyde to ***acetic acid***, ***carbon dioxide***, and water. Thus, ***alcohol*** and acetaldehyde can be serious to **infants**. Appears to be selectively inhibited from entering **human** breast milk. May be produced by ionizing radiation of starch. It is also a product of reaction between ***ozone*** and the proteins of **rabbit** lung tissue. The reason for its low environmental toxicity is due to the fact that its reactivity with water and soil oxidizes it readily by chemical or biological means. TLV (threshold limit values) for **man** are 100–200 ppm. Has been found in some drinking water at levels up to 0.1 µg/l. In **humans**, eczematous contact dermatitis reported. Contamination of home-brewed lager caused erosive gastritis, abdominal pain, nausea, vomiting, and melena in 29-year-old **male**, after drinking several gallons over a 3 week period. Sweating, hepatotoxicity, neurologic shock, cardiac complications (including mycardial infarction), and disorders of ***collagen*** metabolism are also reported. Acute exposure of **humans** to vapors causes severe mucous membrane (eyes, nose) irritation, reddening of skin, coughing, pulmonary edema, and narcosis. Headache is common. Chronic exposure is associated with anemia, hallucinations, and depression, similar to symptoms seen in severe ***alcohol*** intoxication. In **man**, 0.1 mg/l is considered to be a toxic serum concentration. Said to be a cause of malformation in **mice** and chromosome damage in **humans**. It is not only a mutagen, but also a carcinogen and co-carcinogen. As blood levels build in **man**, a slight increase in blood pressure occurs, followed by a marked increase in flushing of the upper half of the body, especially the face and eyes. A tachycardia (up to 150 beats/min) may occur, with a pounding headache, some apprehension, rapid breathing, mydriasis, conjunctival congestion, and chest pains, often stimulating angina and even myocardial infarction. Occasionally, an erythematous, violaceous, non-pruritic, macular eruption occurs on the face, neck, and upper trunk. This is typical of a ***disulfiram*** response, followed after about 10 min in those cases with blood pressure drops, weakness, dizziness, pallor, nausea, vomiting, fainting, unconsciousness, and even death. If ***ethanol*** is consumed, up to several days after eating several species of mushrooms (*Coprinus* and *Clitocybe*), an "acetaldehyde storm", indistinguishable from an ***alcohol–disulfiram*** reaction can occur. (It has responded to oral dosing with 40–240 mg ***propranolol.***) Paresthesia of hands and feet has also been noted. It and ***alcohol*** exert a toxic effect on the testes of **man**, causing a decrease in ***testosterone*** production. Recovery is likely in **humans**, surviving 48 h, after acute poisoning.

Syrian golden hamsters exposed to acetaldehyde vapors of 1650–2500 ppm for 7 h/day, 5 days/week, for 52 weeks, revealed it to be a tracheobronchial irritant, as well as a respiratory tract carcinogen. 390 ppm was considered a no-toxic-effect level. High dosage (50–100 mg/kg, IP) to **rats** during gestation curtailed placental and umbilical cord growth, produced resorptions and malformations of fetuses. Similar results were produced in **mice**, by IV dosage. Vapors in air of 50 ppm/15 min causes eye irritation in **humans**, and only 25 ppm/15 min in sensitive **individuals**. 134 ppm/30 min causes respiratory tract irritation; 200 ppm/15 min causes irritation of the nose and throat. Low to moderate toxicity is reported in **shrimp** and various **fish**. **Rats** exposed to vapors 6 h/day, 5 days/week, up to 27 months of 1500–3000 ppm induced neoplastic and non-neoplastic changes in the nasal mucosa and some lesions in the larynx and trachea. **Rats** develop an aversion to ***alcohol*** as acetaldehyde levels increase. Male **mice** accumulate five times the level of acetaldehyde as females in the same time period. LD_{50}, **rats**: 1.9–4.6 g/kg; LD_{50}, **rabbit**: 2.58 g/kg; inhalation LD_{LO}, **human**: 500 mg/kg; eye irritant dose, **rabbit**: 500 mg.

Potentiators*: *Nitrofurantoin and ***chlorpropamide*** inhibit aldehyde dehydrogenase, thus preventing acetaldehyde's oxidation.

Treatment*:** One objective is to intervene in its metabolic pathways. ***Cysteine and ***ascorbic acid*** have been effective in **rats**. Street **alcoholics** appear to have higher acetaldehyde levels than **humans** in a laboratory setting; and, it appears to contribute to an increase in skin pigmentation. ***Inhalation problems***: Oxygen has been recommended in cases of pulmonary edema and diuresis, with ***furosemide***. Slow IV ***aminophylline*** for bronchoconstriction. Control restlessness with ***morphine***. *Skin contamination*: Wash with large quantities of water. *Ingestion*: Gastric lavage with water or dilute ***sodium bicarbonate*** solution. Correct electrolyte disturbances.

ACETALDEHYDE ETHYL *trans*-3-HEXENYL ACETAL

A colorless liquid, formed by combining *acetal* with *cis-3-hexenol*. Limited use in fragrances at about the same level as *acetal* (above).

***Untoward effects*:** Acute oral LD_{50}, **rats**: > 5 g/kg; acute dermal LD_{50}, **rabbits**: > 5 g/kg. Non-irritant to intact or abraded **rabbit** skin. No sensitization reactions in various tests on **human** volunteers.

ACETAMIDE

= *Acetic Acid Amide*

***Solvent*. A metabolite of *metronidazole* in the rat.**

Use*:** As a fungicide. In the early treatment of ***fluoroacetamide poisoning in **chickens**, **cattle**, and **humans**, but not in **sheep**. In the manufacture of ***penicillins*** and ***cephalosporins***. In many industries, as a solvent.

Untoward effects*:** **Animal** carcinogen. Weak carcinogen (malignant liver tumors) in the **rat**, when administered at a 2.5% level in their feed for 12 months. An equimolar amount of ***arginine glutamate (5.6%) in the same feed inhibited the carcinogenic process. It is deliquescent. In the event of eye contact, wash with copious amounts of water. Seek medical attention. For skin contact, wash thoroughly with soap and water. Copious drinking of water, followed by induction of vomiting has been suggested, in the event of ingestion.

No toxicity noted in **sheep** receiving 200 mg/kg orally. Oral, low toxic dose, **rat**: 456 g/kg; LD_{50}, IP or SC, **rat**: 10 g/kg.

5-ACETAMIDO-3-(5-NITRO-2-FURYL)-6H-1,2,4-OXADIAZINE

***A nitrofuran*.**

***Untoward effects*:** When fed to **rats**, it induces a 100% incidence in hemangioendothelial sarcomas in the mesentery, liver, and lungs. Low toxic dose, **rat**: 21 g/kg.

2-ACETAMIDO-5-NITROTHIAZOLE

= *2-Acetylamino-5-nitrothiazole* = *Acinitrazole* = *Aminitrozole* = *Enheptin-A* = *Gynofon* = *N-(5-Nitro-2-thiazolyl) acetamide* = *Pleocide* = *Trichorad* = *Trichoral* = *Tritheon*

***Protozoastat*.**

***Use*:** In prophylaxis and therapy of blackhead (histomonads) and trichomoniasis in **pigeons** and **turkeys**; also, in hemorrhagic dysentery of **swine** and hexamitiasis in **fish**.

***Untoward effects*:** Nausea and vomiting in 6/65 **women** with vaginal trichomoniasis; dark urine in 4/65, with 40% cure rate. Oral LD_{50}, **mice**: **1 g/kg**. Should not be used in **animals** or **poultry** within 7 days of slaughter, to avoid undesirable residues in food for **human** consumption. Excessive quantities to **turkeys** has caused liver necrosis, weight loss, and slight decrease in fertility. Inadequate water intake in treated **poults** can cause nephritis.

ACETAMINOPHEN

= N-(4-Hydroxyphenyl)acetamide = 4´-Hydroxy-acetanilide = p-Hydroxyacetanilide = p-Acetamido-phenol =p-Acetaminophenol = p-Acetylaminophenol = N-Acetyl-p-aminophenol = Paracetamol = Abensanil = Acamol = Acetalgin = Alpiny = Alvedon = Amadil = Anaflon = Anapap = Anhiba = Apamide = APAP = Apotel = Atasol = Ben-u-ron = Ben-u-ron Baby = Bickie-mol = Calpol = Captin = Cetadol = Crocin = Dafalgan = Dapa = Datril = Dial-a-gesic = Dirox = Disprol = Doliprane = Dolprone = Dymadon = Efferalgan = Enelfa = Eneril = Eu-Med = Exdol = Febrilix = Finimal = Gelocatil = Hedex = Homoolan = Korum = Liquiprim = Lyteca = Metalid = Momentum = Naprinol = Nobedon = Ortensan = Pacemo = Paldesic = Panadol = Panaleve = Panasorb = Panets = Panex = Panofen = Paralan = Paraspen = Parmol = Pasolind = Pasolind N = Phenaphen = Polarfen = Salzone = Tabalgin = Tapar = Temlo = Tempra = Tralgon = Tylenol = Valadol

Many combination products exist, viz. *Benalgesico, Benorylate, Campain, Coplexin, Co-Proxamol, Darth, Darvocet-N, Di-Antalvic, Distalgesic, Esgic, Excedrin, Femidol, Fiorinal-PA, FIPA, Floricet, Gerisom, Hicold, Hydratene, Mersyndol, Midrin, Nyquil, Paedo-Sed, Paidipyrin, Panadeine Co., Parafon Forte, Paramax, Percogesic-C, Phenaphen with Codeine, Predniflex, Romilar CF, Safrin, Sincarest, Sine-Off, Sinutab, Solpadeine, Talacen, Tandalgesic, Tylenol Cold & Flu, Tylenol #1,* and *Tylox.*

Analgesic, antipyretic. **It is the active metabolite of *phenacetin* (q.v.), but does not show the latter's tendency to cause hemoglobinuria or hemolysis. U.S. annual demand: > 30,000 metric tons. Excreted mainly in the urine as the glucuronide conjugate in man and dogs, but not in the cat and other felines, since they lack the liver enzymes needed for conjugation. This may be the reason for its unusual toxicity in the cat. The rat excretes it primarily as the sulfate, as do human neonates and children under 12 years of age. Passes rapidly into human breast milk, and its metabolites are then found in the urine of neonates. Carried on the Apollo 8 trip to the moon.**

Untoward effects: Agranulocytosis, plasmacytosis, and thrombocytosis, followed by a leukemoid reaction in 19-year-old **female**, after ingesting 12 g during 2 days for dental pain. Many cases of acute renal failure reported. Parents, pharmacists, and physicians should be aware of the drugs' fatal potential. It should not be considered as an innocuous substitute for ***aspirin***. Overdoses, and accidental and deliberate poisonings can easily progress to hepatic and renal failure and death. A deliberate ***cyanide*** tampering with the over-the-counter material in 1982 has led to an industry-wide re-evaluation of the need for safety seals in the food and pharmaceutical industry. Older **people** metabolize it at a slower rate than younger ones. Generally, less stomach upset and minimal ultrastructural changes on high dosage, in normal gastric mucosa than with ***aspirin***. Anorexia, nausea, and vomiting leading to night sweats occur in some **people**. Fatal toxic epidermal necrolysis is rare. Acute and chronic ingestion may deplete the body stores of ***glutathione***, which is neces-sary for the detoxification of an ***APAP*** metabolite. Accumulation of the latter can precipitate liver toxicity. A hemolytic anemia with occasional fecal blood loss has been reported. Occasional diarrhea and constipation reported. Gastrointestinal blood loss in 53-year-old **male** within 105 days, with recurring iron deficiency, after ingesting 1–2 g/day; a thrombocytopenic purpura in 63-year-old **female** within 8 h after testing, spontaneous recovery in 7 days; no bleeding time or platelet function problems in two **hemophiliacs** receiving 975 mg. In one trial on **volunteers**, 1/15 showed an increase in bleeding time; in one series, leukopenia reported in five. Small increase in methemoglobin (should always be considered in any **person** of any age) in **infants** on high dosage; hypersensitivity may be the cause of thrombopenia, gross purpura, and platelet count of 4000/mm^3 in 22-year-old **male**; possible adverse effect on bone marrow in 54-year-old **female**; a metabolite may have produced an immune thrombocytopenia; a case of fatal massive hemolysis in 37-year-old **male**, after 2 months on 1.5 g/day; generalized purpura and thrombocytopenia, after 2 weeks on 1.05 g/day in 70-year-old **male**; a case of thrombocytopenia in 29-year-old **male**, with cross-sensitivity to ***aspirin***; agranulocytosis also reported; 4–6 g qid/7 days in four **females** and one **male**, 48–71 years, related to 0.9–2.4 ml blood loss/day. Arthralgia with fever, toxic erythema, and severe headache after 12 days' self-treatment with a total of 30 tablets ***Panadol***®. In **humans**, dosing according to body weight in obesity, instead of basing it on ideal weight, can lead to toxic and lethal effects. Nephrotoxicity may be enhanced by use with ***aspirin*** and in dehydration. Hypothermia and hypotension have also been reported. Not only have deaths occurred due to hepatic failure, but also those due to cardiac arrest, asphyxia, bronchopneumonia complicating aspiration of stomach contents, cerebral anoxia, and intracerebral hemorrhage in poisoning cases. Should not be prescribed to **patients** sensitive to phenolics. In **humans**, 200 mg or more/kg causes lethargy, nausea, vomiting, and jaundice. 0.2–1.0 g/kg has been fatal. In a series of over 3000 cases, use of the drug appeared to be associated with a significant increase in melanomas (four times) and cancers of the pharynx (ten times). In 1985, the state of Illinois ordered the recall of about 377,000 bottles of ***Tylenol***® suspension, some of which contained yeasts and molds that might cause a rash in **children** allergic to ***penicillin***. Metabolic acidosis and coma in a 29-year-old **female** after ingesting 75 g.

Clinical tests: Interferes with ***uric acid***, ***theophylline***, ***glucose***, and ***chlordiazepoxide*** assays.

Eye: A congenital increase in eye cataracts has been reported, as well as conjunctivitis, dilated pupils, and even a yellow tinge to what a **person** sees.

Fetus and neonate: Despite relative safety during pregnancy, congenital cataracts reported in a **patient** whose **mother** took the drug during the tenth and twelfth weeks of gestation. Renal damage has also been reported. A 25-year-old **female**, 36 weeks pregnant, was hospitalized about 3–4 h after consuming eight cans of lager and 20 g of acetaminophen. Neither **mother** nor **child** suffered liver damage. Small amounts excreted in breast milk have caused rashes in **infants**.

Hypersensitivity reactions: May cause severe and frequent adverse rhinitis effects, and a syndrome similar to Kawasaki disease (fever, lymphadenopathy, conjunctivitis, edema, jaundice, and pneumonitis) in a 15-year-old **female**. This was confirmed by inadvertent rechallenge and a test dose of 100 mg. A third of ***aspirin***-sensitive **asthmatics** also react to it.

In pregnancy: Use during the last trimester can cause bleeding complications.

Liver: Overdosage caused necrosis in 46/57 **patients** (15–58 years), with fatal fulminant hepatic necrosis in nine. Encephalopathy, preventable only in the first 24 h, in two – no correlation between dose and degree of damage. 10–14 g is potentially life-threatening. Serum ***ferritin***, rather than transaminase, levels are more accurate predictors of pathological damage in acute cases. A single 0.5 g dose has caused a near fatality. Hepatic necrosis and encephalopathy in 11/107 intent on suicide, but unaware that 2–3 days pass before serious symptoms occur. Abdominal pain and nausea may occur by the second day, with jaundice on the third or fourth day. Chronic ingestion of therapeutic doses may produce liver necrosis and hepatitis, long after discontinuation of the drug. ***Alcohol*** consumption enhances its hepatoxicity. Its toxic intermediate, ***n-acetyl-p-benzoquinomine***, can accumulate from depletion of endogenous ***glutathione*** needed for complete metabolism leading to liver damage, even on doses of 150 mg/kg.

Pancreas: Acute attacks of pancreatitis reported in two suicide cases (31-year-old **female**, 60 g; 41-year-old **female**, 25 g).

Renal: Acute renal failure with papillary necrosis often occurs in overdoses and abuse, even without hepatic failure; frequently with tissue fragments in urine and renal colic. (Acute renal failure after ingestion in 16-year-old **male**, 26-year-old **male**; renal necrosis in 40-year-old **female** with 1 g, 12 days after ***halothane*** anesthesia; acute renal failure after 5 g in 19-year-old **female**. Her nephrotoxicity was due to contamination with ***acetic-4-chloranilide***, although denied by the manufacturer.) Hospitals should avoid dosing with it until an adequate history is taken of prior usage.

Skin: As it is a ***coal tar*** derivative, pruritus, generalized urticarial rashes, and maculopapular rashes occur infrequently; may appear on the arms and trunk, in severe poisonings. Ingestion of only two tablets by a 76-year-old **female** caused widespread bullous fixed-drug eruption. Jaundice may be noted, secondary to hepatotoxicity. Toxic epidermal necrolysis in 7-year-old **female** after three oral doses of 10 mg/kg. Confirmed by rechallenge.

Drug interactions: Slightly enhances the action of ***oral anticoagulants*** (coumarins), and, possibly, the indandiones (***anisindione***, ***dephenadione***, and ***phenindione***), ***chloramphenicol***, ***paraquat***, ***vasopressin***, and slightly decreases the plasma concentration of ***rimantadine***. Oral ***propanolol*** increases its absorption; ***albuterol***, ***desipramine***, ***propantheleine***, and ***pectin***-containing foods slow its absorption. Its hepatotoxicity and, possibly, renal toxicity, may be enhanced by ingestion of ***acetylsalicylic acid***, ***alcohol***, ***chlorpromazine***, ***chlorzoxazone***, ***isoniazid***, ***phenobarbital***, and ***probenecid***. It significantly potentiates ***warfarin*** and other coumarins (***acenocoumarol***, ***bishydroxycoumarin***, ***phenprocoumarin***), ***chloramphenicol***, and ***hydrocortisone***. Its activity may be decreased (perhaps under genetic control) by ***phenytoin***, ***oral contraceptives***, ***ethinyl estradiol***, and smoking. It may enhance ***doxorubicin's*** toxicity, and ***diazepam*** blood levels in **women**. It decreases ***digitalis*** half-life and prevents its normal absorption rate. ***Metoclopramide*** increases its rate of absorption, while ***cholestyramine resin*** binds it; ***cimetidine*** may decrease its hepatotoxicity. Lethal doses from as low as 7 g; much lower in **alcoholics**. Hepatotoxicity in 55-year-old **female** after 1300–6200 mg/day while taking ***phenytoin sodium*** 300–400 mg/day. Analgesic effect increased by 65–200 mg ***caffeine***. ***Probenecid*** inhibits its metabolization in **humans**, **mice**, and **rats**.

Low toxic dose: **Children**, **women**, **human**: 801, 260–1,200, 357 mg, respectively/kg; oral LD_{50}, **rat**, **mouse**: 2.4 g and 338 mg, respectively/kg.

Most species metabolize it by conjugation with glucuronate, and, to a small degree, with sulfate. **Cats** rapidly saturate the sulfate conjugation, and the phase I enzymes (***cytochrome-P-450***) take over, resulting in large amounts of toxic metabolites. ***Glutathione*** also metabolizes it in **cats**, but it is rapidly depleted; therefore, clinical signs of toxicosis can occur with dosages lower than 500 mg. ***Acetylcysteine*** (oral or IV) and ***sodium sulfate*** IV is antidotal (***ascorbic acid*** and ***cimetidine*** may also be useful). High oral dosage (120–500 mg/kg, respectively, in **cats** and **dogs**) produces an increase in blood methemoglobin within 4–8 h. Potentiates antitussive activity of ***dextromethorphan*** in unanesthetized **dogs**. Doses as low as 300 mg/kg, IP, consistently lead to hepatic necrosis in **mice**. Use in **cats** can cause severe anorexia, methemoglobinemia, hepato- and nephrotoxicity. It does not increase serum ***digoxin*** levels in the **dog**. Narcosis develops in **mice** at IP dosages of 300 mg/kg, in contrast to **guinea pigs**, **rabbits**, and **rats**, that show only occasional narcosis at dosages of 700–1200 mg/kg. As noted above, **felines** are relatively deficient in glucuronyl transferase, and have difficulty in metabolizing and eliminating the drug. They are deficient in the protein, ***glutathione***, found in most **mammalian** blood cells. Its role is to protect hemoglobin from conversion to methemoglobin, preventing proper transport of oxygen, and causing red blood cells to rupture. The resulting anemia can be life-threatening, and death on overdosage can occur within 3 days. In **felines**, a toxic metabolite, ***N-acetyl-p-benzoquinone***, also requires ***glutathione*** for detoxification. ***Acetylcysteine***, with a chemical structure similar to ***glutathione***, is indicated therapy. Edema of the head and extremities, hematuria, cyanosis and hemoglobinuria also occur in toxic **cats**. Within 1–2 h, **cats** may develop anorexia, salivation, and vomiting. Toxic symptoms in **dogs** (methemoglobinemia and hepatocellular necrosis) may start on dosages of 30 mg or more/kg.

Treatment: Activated charcoal, orally, is an effective adsorbant in **dogs**. **Cats**: 70–100 mg/kg ***acetylcysteine***, qid, as a 5% solution, and 50 mg/kg ***sodium sulfate***, as a 1.6% solution, orally and IV, is antidotal. Within a few hours after ingestion, 2 g/kg of ***activated charcoal*** and a cathartic of ***sodium*** sulfate (0.5 g/kg, as a 20% slurry) may be useful. ***Ascorbic acid*** (30 mg/kg, oral or SC) is also indicated in **cats** and **dogs**. ***Cimetidine*** on the first day (10 mg/kg; then, 5 mg/kg every 6 h for 48 h) is indicated in the **dog**, and, possibly, the **cat**. A protective "cocktail" has been given IP to **monkeys** (***l-cysteine***, 20 mg/kg; ***cysteamine***, 20 mg/kg; ***glutathione***, 150 mg/kg; and ***acetylcysteine***, 50 mg/kg). **Human**: Serum levels should be monitored. Hemodialysis removes the drug and its metabolites, but not effective in preventing or treating severe liver damage in poisoning cases. Charcoal hemoperfusion may be slightly better than hemodialysis. Emesis (***ipecac***) or lavage may be useful, if ingestion has been less than 24 h, and treatment should be within 12 h, to decrease liver damage and mortality to near zero. Overdosage, combined with other drugs, may delay apparent toxicity and requires a longer treatment period, regardless of apparent low serum levels. ***Acetylcysteine*** now appears to be the oral antidote of choice (140 mg/kg, diluted to a 5% solution in a carbonated beverage, ***grapefruit juice***, or water; if indicated, continue with half this level at 4–8 h intervals, for a total of at least 14 doses). For maximum effectiveness, this must be administered within 8–10 h after the overdose. A case of serum sickness-like illness has been reported on this therapy. IV infusions have also been used (usually, a 150 mg/kg bolus dose, followed by 50 mg over a 4 h period, then 100 mg over a 16 h period). Anaphylactoid reactions, asthma, cutaneous hypersensitivity, and a case of cardiac arrhythmia due to IV therapy have been reported. ***Acetylcysteine*** is a ***sulfhydryl*** donor and a ***glutathione*** precursor, which help degrade acetaminophen's toxic metabolites. Oral or IV ***methionine*** (1 g every 4 h, **child**; 2.5 g every 4 h, **adult**; for a total of 4 and 10 g, respectively) and oral ***glutathione*** are now rarely used. If **patients** have plasma levels of at least 300 µg/ml, 4 h post-ingestion, they probably already have severe liver damage. ***Cimetidine***, by virtue of its imidazole structure, can help prevent its metabolic conversion; and, in **rat** trials, significantly prevented hepatic necrosis. As with **cats** (q.v.), very early administration of large doses (30–50 g) of activated charcoal may reduce drug absorption, but may also adsorb oral ***acetylcysteine*** and ***methionine***.

ACETANILID

= *Antifebrin* = *Phenylacetamide*

Analgesic, antipyretic.

Use: To relieve common aches and reduce fevers. Occasionally as a preservative and stabilizer in ***hydrogen peroxide*** solutions; as a dyestuff intermediate; as a rubber accelerator; and in ***penicillin*** manufacturing.

Untoward effects: May cause methemoglobinemia in **animals** and **humans**. It has produced clinically significant hemolytic anemia in **humans**, particularly in those with glucose-6-phosphate dehydrogenase deficiency (within 3–4 days), and in ***primaquine***-sensitive **individuals**. Allergy and idiosyncrasy occur and are severe. Cardiac **patients** may be especially susceptible. May induce jaundice and hepatic necrosis, fixed-drug eruptions, color urine yellow to blood red, and color nails purple. Nausea, vomiting, cyanosis, weak pulse, shallow respirations, dyspnea, nephritis, stupor, and collapse have been reported in overdoses. Metabolized to ***acetaminophen*** (q.v.), but poorly metabolized in **infants**. The antibiotics ***chloramphenicol*** and ***tetracycline*** and ***alcohol*** ingestion inhibit liver microsomal activity and **potentiate** acetanilid's effect. May interfere with various ***glucose*** tests. Oral use in **dogs** (72 mg/kg/day) produced a mild anemia, but 250 mg/kg level produced a severe hemolytic anemia. A dose as low as 18 grains (1.17 g) has caused **human** death. ***Caffeine*** increases its toxicity.

Treatment: Large dose of ***vitamin C*** may hasten the remission of induced methemoglobinemia in **guinea pigs** and **humans**.

ACETARSONE

=[3-(Acetylamino)-4-hydroxyphenyl]-arsonic Acid = N Acetyl-4-hydroxy-m-arsanilic Acid = 3-Acetamido-4-hydroxyphenyl-arsonic Acid = 3-Acetamido-4-hydroxybenzenearsonic Acid = Acetarsol = Acetphenarsine = Ehrlich 594 = Fourneau 190 = 190 F = F 190 = Stovarsol = Amarsan = Arsaphen = Dynarsan = Goyl = Kharophen = Limarsol = Malagride = Gynoplix = Oralcid = Devegan = Orarsan = Osarsal = Osarsol = Osvarsan = Paroxyl = Sanogyl = Spirocid = S.V.C. = Monargan = Ginarsol = Stovarsolan = Turk-E-San

Protozoacide.

***Use*:** In spirochetosis, **turkey** blackhead, amidostomosis of **geese**, as an alterative, and topically in trichomonad vaginitis of **cattle**. For prevention and treatment of hemorrhagic dysentery in **swine**. For vaginitis in **humans**, and once a popular therapy for amebiasis and balantidiosis. Rapidly excreted in the urine.

Untoward effects*:** May be nephro- and hepatotoxic in **animals** and **man**. Fixed-drug eruptions, urticaria, dermatitis, erythema, Stevens–Johnson syndrome, and jaundice from a hypersensitivity reaction have been reported in **humans**. Gastrointestinal upsets (vomiting, etc). A 28-year-old **female** sporadically used 0.26 g pessaries over a 2 year period. Hospital treatment with 4.68 g produced vomiting; 2 days later, a 3.2 g insertion caused profuse sweating, convulsions, and anuria, followed by death, despite ***dimercaprol treatment.

Oral LD_{50}, **cat**: 150 mg/kg; LD_{LO}, **rabbit**: 125 mg/kg; LD_{50}, **mouse**: 4 mg/kg; OSHA, air, TWA: 500 μg (As)/m^3.

Treatment*:** **Human**: ***Dimercaprol, 5 mg/kg, IM, qid/2 days, then twice daily.

ACETAZOLAMIDE

= N[-5-(Aminosulfonyl)-1,3,4-thiadiazol-2-yl]acetamide = 5-Acetamido-1,3,4-thiadiazole-2-sulfonamide = 2-Acetylamino-1,3,4-thiadiazole-5-sulfonamide = Acetazoleamide = #6063 = Carbonic Anhydrase Inhibitor No. 6063 = Acetamox = Atenezol = Cidamex = Défiltran = Diacarb = Diamox = Didoc = Diluran = Diureticum-Holzinger = Diuriwas = Diutazol = Donmox = Edemox = Fonurit = Glaupax = Glupax = Natrionex = Nephramid = Vetamox

***Diuretic, antiglaucomic.* Strong carbonic anhydrase inhibitor. Developed after noting that sulfonamide produces acidosis.**

***Use*:** In laminitis, udder edema, enterotoxemia, ascites, and glaucoma in various **animal** species; and particularly for petit mal in **children**, congestive heart failure, acute mountain sickness and therapy of glaucoma in **humans**. Alkalinizes urine, but, used alone, it may produce a mild systemic acidosis.

Untoward effects*:** May produce untoward effects on blood, blood vessels, fetus, and kidney. Will produce positive results in presumptive tests for ***sulfonamides. ***Contraindicated*** in adrenal failure or low ***potassium*** and ***sodium*** syndromes. Excessive dosage may cause drowsiness, fatigue, confusion, and will not increase diuresis above optimum.

Human: Confusion and a moderate, clinically significant acidosis is not uncommon, especially in elderly **patients**. Occasional gastrointestinal disturbances, nausea, taste disturbances (particularly of carbonated beverages), tingling of toes and fingers, drowsiness, confusion (especially in **patients** with cirrhosis), anorexia, depression, paresthesia, thirst with increased water intake and polyuria, acidosis, hypokalemia, weakness, hyperuricemia, hematuria, kidney-ureteral calculi, hypersensitivity, thrombocytopenic purpura or damaged blood vessels, fatal severe bone marrow depression with leukopenia, agranulocytosis, neutropenia, pancytopenia, aplastic anemia, loss of libido, toxic epidural necrolysis, erythema multiforme, bulbous exanthema, exfoliative dermatitis, fixed-drug eruptions, abnormal thyroid function, intracellular and canalicular cholestasis of the liver with widespread patchy fatty infiltration and jaundice, and transient reversible myopia with monocular double vision. It has caused crystalluria and renal calculi, as do some ***sulfonamides*** (probably as a result of decreased urinary citrate). It may aggravate respiratory insufficiency in some **patients** unable to compensate for the acidosis. The metabolic acidosis can prolong anesthetic effects. Its use causes renal tubular reabsorption of ***amphetamine*** and prolongs the latter's effect. The same may be true for ***fenfluramine***, ***meperidine***, ***quinidine***, ***pseudoephedrine***, ***procainamide***, ***imipramine***, and ***mecamylamine***. It increases the excretion rate of ***lithium***, ***salicylates*** (but, in alkalizing the cerebrospinal fluid, it helps force ***salicylates*** into the brain, increasing its toxicity), ***phenylbutazone***, ***phenobarbital***, and ***sulfonamides***. Combined with ***phenytoin***, it has been implicated in accelerating several cases of osteomalacia.

Rat trials indicate it may enhance ***salicylate*** intoxication by increasing ***carbon dioxide*** retention in the blood. Teratogenic forelimb effects in female offspring of **rats** (300–600 mg/kg/day or twice daily for 1–3 days of gestation, days 10, 11, 12); first reported case of preferential teratogenic effect of a drug, other than a sex steroid. Diarrhea has been reported in some treated **cattle**. Forelimb defects have also been reported in offspring of treated **hamsters**, **mice**, and **monkeys**, and, possibly, **humans**.

Oral TD_{LO}, pregnant **rat**: 640 mg/kg.

Treatment: In liver disturbances, the blood ammonia level may increase. This has usually been treated with 200–500 ml of a 10% solution of ***sodium glutamate***. This was even successful in a case of coma.

ACETIC ACID

= *Ethanoic Acid* = *Methane Carboxylic Acid*

***Fungicide, acidifier*. The *diluted* form contains 6% acid. In the U.S., *vinegar* is usually standardized at 5% strength, and at 8% in some countries.**

Use: Topically, to control *Actinomyces necrophorous* and *Pseudomonas* in **animals** and *Pseudomonas* in **humans**. Topically (up to 5%), in otitis externa. Oral antidote to ***urea*** and protein poisoning in **ruminants**. Decreases enzymatic conversion of ***urea*** to ***ammonia*** within 4–10 min. In liquid feed supplements for **ruminants**, and to acidify drinking water for **poultry**. 1–2% strengths are effective against foot and mouth disease virus, and active, even in the presence of feces. 0.5–1.0% solutions have been used as a bladder irrigation against ***ammonia***-forming bacteria. As an activator in stripping epoxy resin paints. The *glacial* form (99.4% strength) is caustic, disinfectant, and antiviral, and is used to remove small warts on **animals**, and in making more dilute solutions. Several billion lbs a year are used in the U.S. It is used as a food acidulant in candies, cheeses, pickling and preserving solutions (0.7%), sauces, dressings, and relishes for **man**. After copious washing of caustics off the skin with cold water, a final rinse with a 1% solution has been used. In industry, to make colors fast on fibers, in curing leather hides, in bleaches, to acidify oil wells, and in manufacturing ***cellulose acetate***, ***vinyl acetate***, ***terephthalic acid***, ***chloroacetic acid***, ***calcium magnesium acetate*** (non-corrosive and environmentally safe alternative to rock salt as a highway de-icer), as a whitener and brightener in laundries, in boiler lime cleaners, and as a ***latex*** coagulant, etc. Often, as a pH buffer at 0.2% level in pharmaceutical liquids. Over 50% of U.S. ***vinegar*** is white vinegar, produced from ***ethyl alcohol***. Other types are made from fermented apple and grape juices, barley malt, and sugar syrup. Babylonians (c. 2700–583 B.C.) fermented grain starches and fruit juices to make beers, ciders, and wine. Vinegar was formed when these fermented. Romans dipped their ***lead*** combs in vinegar, producing ***lead acetate***, which, even today, is used by **men** as a "color restorer" to darken graying hair.

Untoward effects: Concentrated form is extremely caustic. Can cause severe skin, mouth, esophageal, and eye burns, including ulceration. Even 10% solutions are severely irritating and corrosive. Severe gastrointestinal irritation and pain, nausea, vomiting, thirst, swallowing difficulty, shallow breathing, weak pulse, twitching, convulsions, collapse, shock, and death have been caused by concentrated solutions. Death in **man** has occurred after swallowing 2 tablespoonfuls. After heating, fumes are toxic. In **humans**, exposure to air containing 2 ppm has minimal or no effect on respiration; 16 ppm is uncomfortable, with irritation of eyes, nose, throat, lungs, and skin; and ten times this level is intolerable and causes a 50% decrease in respiratory rate. Similar results are reported in **guinea pig** trials. It has been a substances of abuse (¼–½ glassful) in two **women**, causing alimentary tract burns, with development of mediastinitis, pulmonary edema, pneumothorax, and acute renal failure. 10 oz a day of vinegared gherkins caused marked changes in gastric mucosa. A case of anaphylaxis in a **woman** has been reported. ***Vinegar*** intake can aggravate acid reflux.

Two out of six **horses** who were drenched with 15 l of a 2.5% solution, as an anthelmintic, by an owner died. Survivors were seriously affected with dullness, enteritis, anorexia, and red and jaundiced mucous membranes. *Post-mortem* findings were hemorrhages in the small intestine, its mesentery, small colon and edema of the rectum.

Oral LD_{50}, **rats**: 3.53 g/kg; oral LD_{LO}, **rabbits**: 1.2 g/kg; oral LD_{50}, **mouse**: 5 g/kg.

Protective measures against injury from concentrated material include wearing rubber or neoprene clothing, full-face and head covering, as well as gas masks with white or yellow canister, or a self-contained breathing apparatus.

Caution: Do **NOT** add water to the concentrated acid; always add the concentrated acid to water. Wine vinegar may trigger ***disulfiram*** reactions. ***Clorox***® or ***sodium hypochlorite*** solutions, when mixed with vinegar, produce ***chlorine*** gas, a potentially lethal compound. Avoid mixing with ***chromic acid***, ***nitric acid***, and ***sodium peroxide***.

ACETIC ANHYDRIDE

= *Acetic Oxide* = *Acetyl Oxide*

Use: Mostly in manufacturing of ***cellulose acetate*** and in synthesis of organic compounds. 2 billion ± lbs produced annually. 2.5–10% concentrations in air are flammable.

Untoward effects: Liquid or vapor state is highly irritant to eyes and skin and causes tissue necrosis. Headaches have been reported after accidental spills. Equivalent to approximately 118% ***acetic acid*** (q.v.). It is highly flammable, like gasoline. By combining it with ***anthranilic acid***, it can be used by illicit manufacturers to make a precursor to ***methaqualone***. **Human** LD estimated at 10 g/kg and a TLV of about 5 ppm has been stated. Oral LD_{50}, **rat**: 1.78 g/kg.

ACETOACETIC ACID

= Diacetic Acid = 3-Oxobutanoic Acid = Acetylacetic Acid = Acetonecarboxylic Acid = 3-Ketobutyric Acid = 2-Ketobutyric Acid

Use: In organic synthesis.

Untoward effects: Ingestion has caused a purplish color in **human** nails.

ACETOHEXAMIDE

= N-(p-Acetylphenylsulfonyl)-N´-cyclohexylurea = Dymelor = Dimelor = Dimelin = Ordimel = Cyclamide = 4-Acetyl-N[(cyclohexylamino)-carbonyl]benzenesulfonamide = 1-[(p-Acetyl-phenyl)sulfonyl]-3cyclohexylurea = 3-Cyclohexyl-1-(p-acetyl-phenylsulfonyl)-urea = Tsiklamid

***Antidiabetic, hypoglycemic.* A sulfonylurea. Metabolized in and excreted by the kidney. Uricosuric. 6–8 h plasma half-life in humans, but biological half-life in hypoglycemic effect may be much longer.**

Untoward effects: A number of case reports exist covering obvious adverse hypoglycemic episodes, following inadvertent dispensing of it instead of ***acetazolamide***. As with any antidiabetic, hypoglycemia can occur in those not eating regularly or those who exercise without adequate caloric intake. Gastrointestinal disturbances (nausea, heartburn in approximately 3%, and headache) appear to be dose-related. Transient allergic skin manifestations with pruritus, erythema, urticaria, morbilliform and maculopapular eruptions occur in approximately 1%, and often disappear without treatment. Photosensitivity can occur. Hair loss, leukopenia, thrombocytopenia, eosinophilia, pancytopenia, agranulocytosis, and aplastic and hemolytic anemia have been reported. A very small percentage of **patients** have shown jaundice with a severe mixed hepatocellular hepatitis – possibly as a result of hypersensitivity, but cirrhosis and cholestasis have been reported. Opisthotonus and prolonged coma in 16-year-old **female**, after a suicide attempt with 12.5 g. A 39-year-old **male** had a hypoglycemic stupor with red frothing at the mouth, after taking several tablets/day/7 days, thinking they were headache pills. Increased water excretion has been reported in some.

Oral LD_{50}, **rat**: 5 g/kg.

Drug interactions: Increased hypoglycemic effect with ***phenylbutazone***, due to the latter's interference with acetohexamide's active metabolite, ***hydroxyhexamide***. ***Sulfonamides*** enhance its effect. May potentiate ***phenobarbital*** and ***allopurinol***.

Treatment of overdosage: Usually, 10–50% ***dextrose*** in water as needed. Supply other carbohydrates. Temporary use of vasopressors, respiratory assistance, and oxygen may be necessary in serious intoxications.

ACETOHYDROXAMIC ACID

= Lithostat = AHA = N-Hydroxyacetamide = N-Acetylhydroxylamine = Acetic Acid Oxime

***Antiurolith.* A bacterial enzyme (urease) blocker.**

Use: For the prevention of struvite stones in **humans**, **dogs**, and **rats**. Higher dosages cause urolith dissolution.

Untoward effects: Teratogenic in **beagle dogs** (25 mg/kg/day, oral) with anomalies of the skeletal system, heart, and ventral midline in their **pups**. Transient headache, nausea in **humans**. A reversible hemolytic anemia occurs in **dogs** and **humans** on calculolytic doses. May also be teratogenic in **rats**, after 10–33% of their lethal dose, IP.

IP, TD_{LO}, pregnant **rats**: 750 mg/kg.

ACETOIN

= Acetylmethyl Carbinol = 2-Butanolone-3 = Dimethyl Ketol = 3-Hydroxy-2-butanone = Methyl-1-hydroxyethyl Ketone = γ-Hydroxy-β-oxo-butane = 2,3-Butanolone

Fragrance, food additive.

Use: To flavor and provide an essence to beverages and foods. Also as an odorant in perfumes.

Untoward effects: Repeat IP injections of 12 or more g/kg to **mice** produce over a 50% incidence of lung tumors.

ACETONE

= Dimethyl Ketone = Propanone = β-Ketopropane = Pyroacetic Ether

***Solvent.* Volatile and highly flammable. Keep away from sparks and flames. Also called *Analar*, a term also used for *urea*. A ketone. Some is produced in the distillation of wood, by oxidation of *isopropyl alcohol*, and as a by-product of *butyl alcohol* and *phenol* manufacture. Trace amounts are found in *cigarette* smoke and cow's milk. U.S. annual production is nearly 3000 million lbs/year.**

Use: A common industrial solvent for resins, lacquers, pyroxylin (collodion), acetylene, fats, oils, topical pharmaceuticals, and paints. In nail polish, paint, varnish, and adhesive removers (usually at 66%). Also in rubber cements. In the manufacture of various methacrylates. Has demonstrated antiviral and antibacterial effects. Occasionally, along with other skin irritants, in temporary treatment of acne vulgaris. A substance of abuse, usually in **glue-sniffers**.

Untoward effects: Reported in excess waste gas emissions of Czechoslovakian plastic and footwear plants. Vapors irritate eyes, nose, and throat (at 500 ppm), often causing bronchial irritation, pulmonary congestion, weak pulse, nausea, headache, dizziness, fainting, and even unconsciousness. There may be no apparent permanent injury. Maximum allowable 8 h exposure (TLV) after recovery is 1000 ppm (2.4 g/m^3), National Institute for Occupational Safety and Health recommends 250 ppm TWA for **humans**. 23,480 ppm will cause a 50% decrease in respiratory rate. A case of non-fatal acute poisoning in 10-year-old **male** occured 9 h after inhaling fumes for an hour from an open bowl containing acetone used as a setting fluid for an immobilizing cast. The bowl was 6 ft away from his head. A very high percentage of inhaled acetone is absorbed, and may contribute to liver cirrhosis. Use as a paint remover in a very hot area contributed to a fatality. A noxious cloud-like vapor drifted over a shopping center, causing four **people** to be hospitalized and six others to be treated by paramedics. The vapors contained 95% acetone and about 2% ***mesityl oxide*** and some water. A 42-year-old **male** ingested approximately 250 ml pure acetone in a suicide attempt. **He** was stuporous on hospital admission (temperature and blood pressure normal, pulse 108/min and regular, throat red and swollen with erosions on the soft palate and entrance to the esophagus). **Patient** recovered with a hyperglycemia and decreased carbohydrate tolerance, which persisted for about 4 months. Serum levels of 20–30 mg% are very toxic and levels above 50 are apt to be fatal. **Human** oral LD has been estimated to be 370 g, and 2.6 g, IV. Other reports put the oral LD_{LO} for **humans** at 50 mg/kg. It will produce dryness (defats skin) and fissuring of skin with paresthesia of the fingers. Paresthesia may continue for several weeks after the dermatitis has subsided. One must remember that, in ***isopropyl alcohol*** poisoning, the alcohol is oxidized to acetone. It can cause delirium.

LD_{50}, **rat**: 8–13 ml/kg (less in younger **rats**). LD_{50}, **rabbits**: > 20 ml/kg, single skin application. Not a primary skin irritant to **rabbits** (no comedones); minor eye injury (temporary edema), topically – **rabbits**. Single inhalation of concentrated vapors to **rats** killed 3/6. 24 h LC_{50}, young **rainbow trout**: 6.1 g/l. May inactivate ***iodine*** disinfectants.

Treatment: Adequate ventilation. Artificial respiration and oxygen inhalation, if indicated. Flush skin or eyes with plenty of water. Remove contaminated clothing. Hemoperfusion has been effective. Induction of vomiting after ingestion.

ACETONITRILE

= *Methyl Cyanide* = *Ethanenitrile* = *Cyanomethane*

***Solvent*. Found in *tobacco* smoke and in the urine of smokers (3 days of abstinence in light smokers decreases level to that of non-smokers). A *cigarette* produces about 1 mg. Metabolized to *hydrogen cyanide*. *Ether*-like odor. Flammable liquid.**

Use: In organic chemical synthesis and in ***lithium*** batteries.

Untoward effects: Toxic by absorption through the skin, although not considered to be a skin irritant. As with other ***cyanide*** compounds, it can be a powerful poison, preventing oxygen utilization by body tissues. A number of artificial nail removers have contained up to 98% acetonitrile. One **child** died, after two **children** (16-month- and 2-year-old **males**) ingested some. 16 **painters** were accidentally exposed to its fumes (10,000–50,000 ppm) during a 15 min interruption of ventilation in the interior of a steel storage tank. One **painter** had chest pains, coughed, expectorated blood, convulsed, and became deeply comatose and died within several hours. Most had chest pains, abdominal cramps, dyspnea, nausea, bile-stained emesis, chills, headache, dysphasia, lassitude, and fatigue. Some also had pale to ashen-gray skin, shallow pulse, hypotension, transient paralysis, and tachypnea. Half had increased blood ***cyanide*** levels. A photochemical laboratory **technician** (19-year-old **male**) died, following inhalation of an unknown concentration. Nausea and vomiting occured within 4 h. Semiconciousness, confusion, sweating, hypersalivation, hematemesis, and albuminuria were noted 12 h later. **He** died 6 days after exposure.

Human TWA, 40 ppm (70 mg/m^3); oral TD_{LO}, 570 mg/kg; approximate LD_{50}, 5.6 ml, 9 ml/kg – 14 day and adult **rats** (also 3.8 g/kg reported); LD_{50}, **guinea pigs**: 177 mg/kg; LC_{50}, 2693 ppm, acute inhalation toxicity, **mice**; LC_{LO}/4 h, **rats**: 8000 ppm; LC_{LO}/4 h, **rabbit**: 4000 ppm; LC_{LO}/ 4 h, **dogs** and **guinea pigs**: 16,000 ppm.

Treatment: Multiple injections of ***sodium thiosulfate*** antagonized signs of intoxication and mortality in **mice** and **hamsters**, supporting the theoretical rationale for its use in **humans**. ***Thyroxine*** helps antagonize acetonitrile.

ACETONYL ACETONE

= *2,5-Hexanedione* = *α,β-Diacetylethane*

Use: A liquid ketone. In coated fabrics.

Untoward effects: In **man**, it is the major neurotoxic metabolite of ***hexane*** (q.v.) and ***2-hexanone*** (q.v.). The latter's hexocarbons are often "recreationally" used by **glue-sniffers** for their euphoric effects. Found in the urine of neuropathic exposed **workers**. Neuropathies consist of a gradual onset of sensory paresthesia of the limbs, with weakness and muscle atrophy after prolonged exposures. A mild, local irritant. High concentrations cause narcosis.

Oral LD_{50}, **rats**: 2.7 g/kg; inhalation LC_{LO}, **rats**: 200 ppm/4 h. **Hens** receiving 100 mg/kg/day/90 days, orally, developed ataxia in about 8–10 days and histopathology revealed severe neurotoxicity. Higher dosage produced general weakness, tremors, and dyspnea. Experimental **rats** receiving 1% in their drinking water for 6 weeks showed a distal axonopathy and testicular atrophy.

ACETOPHENAZINE

= *2-Acetyl-10[3-[-9(β-hydroxyethyl)piperazinyl]propyl]phenothiazine Dimaleate* = *10-[3-[4-(Hydroxyethyl)-1-piperazinyl] propyl]phenothiazin-2-yl Methyl Ketonedimaleate* = *1-[10-[3-[4-(2-Hydroxyethyl)-1-piperazinyl]propyl]-10H-phenothiazin-2yl] ethanone* = *1-(2-Hydroxyethyl)-4-[3-(2-acetyl-10-phenothiazyl)propyl]piperazine*

***Psychotherapeutic, mild tranquilizer.* A propylpiperazine phenothiazine. 1/4–1/16 *chlorpromazine's* dosage, usually, for the same clinical effect. Has anticholinergic activity.**

Use: As with most tranquilizers.

Untoward effects: Extrapyramidal effects are common. Dry mouth, mydriasis, cycloplegia, urinary retention, decreased gastrointestinal motility, and tachycardia may occur, as a result of its anticholinergic activity. Occasionally, weight gain, cholestatic jaundice, blood dyscrasias (leukopenia), drowsiness, lenticular deposits and opacities, blurred vision, restlessness, akathisia, convulsions, electrocardiogram abnormalities, gastritis, nausea, vomiting, dizziness, sweating, memory impairment and delirium, neuroleptic malignant syndrome, disturbed temperature regulation, erythema multiforme, and hypotension occur. Paradoxically, a rare case of hypertension has been reported. May produce a false positive or high ***bilirubin*** urine test.

Oral LD_{50}, **rats**: 433–900 mg/kg; LD_{50}, **wild birds**: 75 mg/kg.

ACETOPHENONE

= *Acetylbenzene* = *Hypone* = *1-Phenylethanone* = *Phenyl Methyl Ketone*

Fragrance, hypnotic.

Use: In perfumery, to impart an orange blossom odor, and in creating a strawberry and gardenia scent. In soaps and detergents, to impart a hawthorn odor. Permitted in foods. Rarely used as a hypnotic.

Untoward effects: Liquid may cause marked eye irritation and injury. At a concentration of 2% in petrolatum, it produced no irritation in 48 h closed-patch tests on **humans**.

Oral LD_{50}, **rats**: 2.14 ml/kg; LD_{50}, topical, **rabbits**: 15.9 ml/kg. Undiluted, it is an irritant on **rabbit** skin. 10,000 ppm in the diet of **rats** for 17 weeks produced no adverse effects. No deaths in **rats** exposed for 8 h to an atmosphere saturated with it.

Treatment: In case of contact with eyes, immediately flush eyes with plenty of water for at least 15 min. Get medical attention.

ACETOSULPHONE SODIUM

= *N-[[5-Amino-2-[(4-aminophenyl)-sulfonyl]-phenyl]sulfonyl]acetamide Monosodium Salt* = *Promacetin* = *Sulfa-diasulfone Sodium* = *Acetosulfone* = *Internal Antiseptic No. 307* = *I.A. 307* = *(N¹-Acetyl-6-sulfanilylmetanilamido) Sodium* = *Sodium-2-N-acetylsulfamyl-4,4′-diaminodiphenylsulfone* = *N-Sodium Salt*

Antibacterial, leprostatic agent.

Use: In all forms of leprosy (bacteriostatic against *Mycobacterium leprae*).

Untoward effects: In **humans**, if a lepra reaction (erythema nodoum, e. multiforme, and e. necrotans) occurs, treatment should be stopped for at least 2 weeks after all signs subside. Restart with a lower dose. A dose-related hemolysis is a common toxic effect. Methemoglobinemia, leukopenia, and a granulocytosis can occur. An exfoliative dermatitis is a serious complication of therapy. It often occurs with enlargement of the liver and may be fatal. To avoid nausea, vomiting, headaches, giddiness, and tachycardia, it is suggested that treatment be started with the lowest dose possible. Hypersensitivity-type reactions have been reported.

ACETYL ACETONE

= *2,4-Pentanedione* = *Diacetylmethane*

A ketone.

Use: As a chemical intemediate in manufacturing pharmaceuticals (viz. ***sulfamethazine***) and dyestuffs, in metal plating, in resins, in gasoline and lubricant additives, and in driers for printers' inks. Combustible – keep away from heat, sparks, and fire.

Untoward effects: Oral and inhalation toxicity reported. Vapors have a narcotic effect. Mild irritant to skin and mucous membranes.

More than one drop causes corneal burns on **rabbits**; single skin application LD_{50}, **rabbits**: 5 ml/kg; **rats**: concentrated vapors for 30 min killed none of six, and in 1 h killed 1/6. Single oral LD_{50}, **rats**: 1.12 ml/kg; inhalation LC_{LO}, **rats**: 1000 ppm/4 h.

ACETYL CARENE

Synthetic fragrance.

Use: In soaps (0.01%), detergents, creams, and lotions (0.001%), and perfumes (0.005%).

Untoward effects: Acute oral LD_{50}, **rats**: 2.1–4.1 g/kg; acute dermal LD_{50}, **rabbits**: > 5 g/kg. Only moderately irritating at full strength to intact or abraded **rabbit** skin for 24 h under occlusion. 10% concentration in petrolatum – no irritation in **human** closed-patch tests. This also produced no sensitization reactions in 25 **human** volunteers.

ACETYLCHOLINE

The bromide form = *2-(Acetyloxy)-N,N,N-trimethylethanaminium Bromide* = *Pragmoline* = *Tonocholin B*

The chloride form = *2-(Acetyloxy)-N,N,N-trimethylethanaminium Chloride* = *Arterocoline* = *Acecoline* = *Miochol* = *Ovisot*

***Cholinergic, neurotransmitter, miotic, parasympathomimetic.* Found in a number of snake venoms. Antagonized by *antihistamines*, *atropine*, and *scopolamine*.**

Untoward effects: A mediator of anaphylaxis and aquagenic urticaria. Slows heartrate, prolongs PR interval, and may cause some atrioventricular blockade. Injections in **humans** can cause cardiac arrest for 30–50 seconds. The **person** becomes pale and unconscious, with occasional twitching of facial and extremity muscles, coughing, and respiratory difficulty. Vasodilation occurs. Intraocular use in 72-year-old **male** caused a dramatic decrease in blood pressure and heart rate. It accumulates in excess in cases of organophosphorus poisonings, causing salivation, fasciculation of skeletal muscles, bradycardia, tearing, pupillary constriction, dyspnea, and involuntary defecation. Vomiting occurs in species that can vomit. Symptoms in poisonings are similar to acute poisoning with organophosphorus compounds.

Increases aggression in **cats** and **rats**.

Oral LD_{50}, **mouse**: 3 g/kg; SC, LD_{50}, **mice** and **rats**: 170 mg/kg and 250 mg/kg, respectively.

ACETYLCYSTEINE

= *N-Acetyl-L-cysteine* = *L-α-Acetamido-β-mercaptopropionic Acid* = *N-Acetyl-3-mercaptoalanine* = *Airbron* = *Broncholysin* = *Brunac* = *Fabrol* = *Fluatox* = *Fluimucil* = *Fluimucetin* = *Fluprowit* = *Inspir* = *Mucocedyl* = *Mucolator* = *Mucolyticum* = *Mucomyst* = *Muco Sanigen* = *Mucosolvin* = *Mucret* = *NAC* = *Neo-Fluimucil* = *Parvolex* = *Respaire* = *Tixair*

Mucolytic, antidote to acetaminophen poisoning, corneal vulnerary.

Use: In respiratory therapy (emphysema, bronchitis, atelectasis due to mucous obstruction, bronchiectasis, etc.). Its most common usage as a poison antidote is for ***acetaminophen*** poisoning in **humans**, **cats**, **dogs**, and **mice**. Also against toxic doses of ***doxorubicin*** (**dogs**), ***acrolein*** (a product of ***cyclophosphamide*** metabolism), ***carbon tetrachloride***, ***DENA***, ***dimethoxyphenylmorpholine propionate***, ***gold***, and ***isophosphamide***. Useful in treating deep ulcers, as it inhibits collagenase.

Untoward effects: It may irritate airways in **asthmatics** and induce bronchospasms, possibly as an allergic reaction. Nausea, vomiting, stomatitis, rhinorrhea, bronchorrhea, coughing, and respiratory arrest have been reported. Anaphylaxis (rash, angioedema, hypotension, bronchospasms, and tachycardia) has occurred. Probably inactivates ***penicillins*** and ***tetracyclines*** in sputum. Shortens ***digitalis*** half-life. Severe cutaneous excoriation at a catheter entry site, due to leakage. Since it is adsorbed by ***activated charcoal***, both should not be used at the same time orally in treating ***acetaminophen*** poisoning. Excessive dilution for IV in 3½-year-old **female** produced hyponatremia.

IP, LD_{50}, **mice**: 400 mg/kg.

ACETYL DIGITOXIN

= *Acedoxin* = *Acylanid* = *Novodigal*

Cardiotonic.

Use: As with other ***digitalis*** preparations. ***Absorption*** not impaired by co-administration with an ***aluminum hydroxide–magnesium hydroxide*** antacid tablet.

Untoward effects: Anorexia, nausea and vomiting, general malaise with mild symptoms of ***digitalis*** poisoning (including arrhythmias and circulatory failure). Pneumonia, renal and liver disorders predispose to intoxication. Visual disturbances with red-receptor damage and a "glare phenomenon" (objects appear to be covered with brown, orange, or white snow).

LD_{50}, **cat**: approximately 250 µg/kg, oral and 500 µg/kg, IV; LD_{50}, **guinea pigs**: 50 mg/kg.

ACETYLENE

= *Ethyne* = *Ethine*

***Gas, chemical intermediate. Flammable and explosive.* Approximately 400 million lbs used annually. A constituent of *coal* gas and automobile exhausts.**

Use: In welding, cutting, and soldering metals. In ***vinyl acetate***, ***vinyl chloride***, ***1,4-butanediol***, and ***acetylene black*** (in dry cell batteries). Also the fuel in flame spectrophotometry.

Untoward effects*:** Headache, urticaria, and asthma have been reported in **welders**. They have also suffered from ***cadmium, ***nickel***, ***zinc***, ***silver***, and ***chromate*** poisoning, including pulmonary edema from items they were welding with. Has been found in water supplies. OSHA recommends no exposure above 2500 ppm (10% of lowest explosive figure), to prevent narcosis and asphyxia.

Rats have developed leukopenia after exposure to it for 6 days. **Mammalian** inhalation toxicity estimated as LC_{LO}, 500,000 ppm/5 min.

ACETYLENE DICHLORIDE

= *Dioform* = *1,2-Dichloroethene* = *1,2-Dichloroethylene*

***Solvent, insecticide*. Halogenated paraffin.**

***Use*:** As a solvent for waxes, acetyl cellulose, resins, and rubber. In some antiknock gasolines, fumigants, and cleaning mixtures.

***Untoward effects*:** Absorbed from gastrointestinal tract and lungs of **humans**. May cause respiratory irritation and collapse. One fatality reported (1933) from inhalation of vapors in a small enclosure. By 1957, 31 cases of occupational poisoning were reported. Defats skin, resulting in dryness, cracking, and dermatitis. Headache, nausea, vomiting, epigastic cramps, general weakness, and tremors can occur after inhalation. Severe mucosal irritation, hemorrhagic gastroenteritis, nausea, vomiting, and abdominal pain after ingestion. Stupor, cyanosis, and death from respiratory and circulatory failure may occur within 4 days. Lesions in the liver and kidney were common. TLV, 50 ppm; TWA standards have also set a level of 200 ppm. Estimated LD by ingestion for a **human** adult: 20–30 ml; others state 500 mg/kg as an LD_{LO}.

LD_{50}, **rats**: 770 mg/kg. 1.5% vapor concentration decreases cardiac contractility in **dogs**. LD_{50}, **rats**: 1 ml/kg.

Treatment*:** Wash exposed skin thoroughly with soap and water. Apply emollients. Fresh air, warmth, rest, and fluids for inhalation problems. After ingestion, gastric lavage may be indicated. 30 g ***sodium sulfate, orally, in 250 ml water has been recommended, as well as 10 ml of 10% ***calcium gluconate***, IV, and 3–5 g/day of ***methionine***, orally or by stomach tube. ***Dimercaprol***, IM, has also been suggested, since sulphydryl compounds may be protective. If necessary, use assisted respiration or oxygen.

ACETYL ETHYL TETRAMETHYLTETRALYN

= *AETT*

***Fragrance*.**

***Use*:** In soaps, deodorants, and cosmetics, to impart a musk-like odor. In Europe, as a food additive.

***Untoward effects*:** Supposedly, no longer used in the U.S. Repeat applications to the skin of **rats** produced a blue ceroid-like discoloration of internal organs, vacuolar degeneration of the brain, and a neurological syndrome (with widespread demyelination of central nervous system and peripheral nerves). These degenerative changes were associated with hyperirritability, limb weakness, splaying, and ataxia. A contact dermatitis has been reported in **man**.

After 2–3 weeks, **rats** receiving 25–50 mg/kg/day became aggressive and were easily startled.

5-ACETYL-1,1,2,3,3,6-HEXAMETHYLINDAN

***Synthetic fragrance*. Use in the U.S. estimated at approximately 6000 lbs/year.**

***Use*:** In creams, lotions (0.02%), perfumes (0.2%), soaps (0.03%), and detergents (0.003%).

***Untoward effects*:** No irritation was noted at a 4% level in petrolatum on **volunteers** in a 48 h patch test with occlusion, and sensitization was not noted on 25 **volunteers**. Full strength applied to intact or abraded **rabbit** skin for 24 h was not irritating.

Acute oral LD_{50}, **rats**: 1.7 g/kg; acute dermal LD_{50}, **rats**: > 5 g/kg.

ACETYLIMIDAZOLE

***Laboratory reagent*.**

***Use*:** In protein analysis.

***Untoward effects*:** Care should be exercised in using the compound, as two 20, 50, and 100 mg/kg IP injections into **mice** had, after a 24 week trial, 66, 78, and 71% incidence of lung tumors, respectively.

IP, LD_{LO} in **mice** is reported at 50–250 mg/kg.

ACETYLMETHADOL

= *Methadyl Acetate* = *3-Acetoxy-6-dimethylamino-4,4-diphenylheptane* = *6-(Dimethylamino)4,4-diphenyl-3-heptanol Acetate* = *LAAM* = *β-[2-(Dimethylamino)propyl]-α-ethyl-β-phenylbenzeneethanol Acetate* = *Acemethadone* = *Amidolacetate* = *Raceacetylmethadol*

α-d Form is *DAAM*

***Narcotic analgesic*.**

Use*:** As with ***methadone, in the treatment of opiate addiction, but with longer lasting effect.

***Untoward effects*:** In **humans**, has a biphasic action with stimulation, followed by anergia and sluggishness.

Appears in amniotic fluid and breast milk of pregnant **women**. Smaller head circumference was noted in their offspring. Respiratory depression in **patients** not tolerant to opiates. Excessive sweating, constipation, decreased libido and ejaculation capability are noted.

SC and oral ED_{50} to induce corneal opacities within 3–5 days in **rats** was 4.7 and 12.6 mg/kg, respectively. LD_{50}, **rats**: 24–28.6 mg/kg; but, 8.5 mg/kg when given with ***propranolol***. LD_{50}, female **rats**: 35 mg/kg; SC, LD_{50}, **mouse**: 70 mg/kg. Treated **monkeys** (2 mg/kg, IV) had dilated pupils, droopy or closed eyelids. Some catalepsy, muscle twitching and shivering were noted. Decreased spontaneous activity in **mice**; dose-dependent acute kidney necrosis in **rats**.

ACETYLPHENYLHYDRAZINE

= *APH* = *Hydracetin* = *Pyrodine*

Untoward effects: Has caused hemolytic anemias in **man**, particularly in those with glucose-6-phosphate dehydrogenase deficiency.

Has caused Heinz bodies in erythrocytes of **cats**, and **swine**.

IP, LD_{LO}, **mouse**: 150 mg/kg.

1-ACETYL-2-PICOLINOYLHYDRAZINE

= *NSC 68,626*

Antineoplastic.

Untoward effects: Treatment of 11 **human** patients (12 mg/kg/day/14 days, IV) with lung carcinomas showed only mild toxicity (paresthesia, somnolence, and vertigo). Other reports on over 100 **patients** indicated neurotoxicity was the major dose-limiting ill-effect with nausea, ataxia, drowsiness, confusion, nystagmus, diplopia, hyperuricemia, and coma, often occurring on doses > 5–10 mg/kg.

Dog: 0.05 mg/kg/day/24 days; **monkeys**: < 0.4 mg/kg/day/28 days evidenced some toxicity. Has caused lung, ovarian, and uterine tumors in **mice** and pancreatic tumors in **rats**.

LD_{50}, **rat** and **mouse**: 670 mg/kg and 410 mg/kg, respectively. IV, LD_{50}, **rat** and **mouse**: 470 mg/kg and 255 mg/kg, respectively.

ACETYLPYRIDINE

= *Methylpyridyl Ketone* = *1-(3-Pyridinyl)ethanone*

Nicotinic acid antagonist.

Use: In experimental medicine and industry.

Untoward effects: Neurotoxic in **man**.

Causes skeletal muscle hypoplasia and teratogenicity in **chick** embryos.

IP, LD_{LO}, **mouse**: 512 mg/kg.

ACETYLSALICYLIC ACID

= *Aspirin* = *2-Acetoxybenzoic Acid* = *2-(Acetyloxy)benzoic Acid* = *Acidum Acetylsalicylicum* = *Acetilum Acidulatum* = *Acenterine* = *Aceticy* = *Acetophen* = *Acetosal* = *Acetosalic Acid* = *Acetosalin* = *Acetylin* = *Acetyl-SAL* = *Acimetten* = *Acylpyrin* = *A.S.A.* = *Asatard* = *Aspro* = *Asteric* = *Caprin* = *Claradin* = *Colfarit* = *Contrheuma Retard* = *Cosprin* = *Delgesic* = *Duramax* = *ECM* = *Ecotrin* = *Empirin* = *Encaprin* = *Endydol* = *Entrophen* = *Enterosarine* = *Helicon* = *Juvepirine* = *Levius* = *Longasa* = *Measurin* = *Neuronika* = *Platet* = *Rhodine* = *Salacetin* = *Salcetogen* = *Saletin* = *Salicylic Acid Acetate* = *Solprin* = *Solpyron* = *Xaxa*

***Analgesic, antipyretic, antithrombotic.* 20–30 million lbs are produced annually in the U.S. – most going into 5 grain tablets. Named *aspirin* by Bayer in 1898 from "a" for acetyl and "spirin" from *Spiraea*, from which *salicylic acid* was first isolated. *Aspirin* is a registered trademark in many countries.**

Use: To relieve minor aches, joint and muscular pains, and reduce body temperature. Increased acceptance in recent years of its daily or every other day use in low dosage, as an aid in **man** in treating ischemic heart disease, and in decreasing the severity or incidence of transient ischemic attacks, ischemic strokes (but can increase risk of hemorrhagic strokes), and heart attacks. This benefit may be more pronounced in **men**, rather than **women**. Numerous combination drugs are marketed. Tolerated well by **horses** and **dogs**, but not **cats**. Inhibits ***prostaglandin*** synthesis.

Contraindications: In hemophiliac **patients** or those with any bleeding state, those undergoing surgery, and those who have suffered recent trauma. Regular use during pregnancy should be avoided and it should not be used during the last trimester. Breast milk levels peak 9 h after ingestion, but are less than plasma's. Avoid use before exercise in hot weather, with ***halothane*** anesthesia, and in **children** with varicella, influenza, and Reye's syndrome. Avoid use in **asthmatics** and **cats**. Although the risk may be small, metabolic acidosis reported in a breast-fed **infant** of a **mother** ingesting large amounts to treat arthritis.

Untoward effects: Absorption is rapid and appears in the urine within 30 min; therefore, common allergic reactions of asthma, urticaria, and angioneurotic edema may make a quick occurrence. With annual production of about 140 billion 5 grain tablets and an estimated consumption of 70% of these, it is not surprising that numerous untoward effects have occurred in **people**;

mistakenly viewed as a "non-drug" by many. Abuse (a cause of abnormal electroencephalogram), suicide, asphyxia (in an 11 month **child**, following three "junior" tablets), numerous bleeding events, brain edema (in a 5-year-old **male** given large doses for 2½ years), fatal brain stem dysfunction (19-year-old **Negro** with sickle cell disease, after 2 days of 1.5 g/6 h), coronary artery abnormalities in 18/73 **children** with Kawasaki syndrome. 2 weeks after 100 mg/kg/day, 4 g/day in four **patients** exacerbated coronary artery spasm and frequency of anginal attacks; heart failure in very ill rheumatic **patients** with active carditis (probably due to increased serum ***sodium*** levels and decreased blood ***glucose***), ototoxicity (deafness: reports varying from 3 to 11/1000; degree usually increases with increased dosage; tinnitus often precedes deafness), eye problems (iridocyclitis, macular hemorrhage), possibly low teratogenicity in **humans** (and **animals**), and intracranial hemorrhage in **newborns** exposed to high dosage during the **mother's** last weeks of pregnancy (possible phocomelia and cyclopia after first trimester exposure), gastrointestinal upsets (nausea, vomiting, gastritis, bleeding, and night sweats. May color feces pink to red or black, which should be an indication of probable gastrointestinal bleeding). Hypersensitivity (especially wheezing, laryngotracheal edema, rhinorrhea, and urticaria in **asthmatics**), anaphylactoid reactions (usually 5–14 days after ingestion), death in an adult **male**, after taking two tablets (the ignored medical record had noted "***aspirin*** sensitivity"), hypoglycemia, hepatitis, bronchopneumonia (especially in **asthmatics**), renal papillary necrosis (nephropathy is probably multifactorial in nature), Reye's syndrome (especially in **children** with viral exposures – over a thousand deaths reported), skin lesions (fixed-drug eruptions, urticaria, Lyell's disease, erythema multiforme, vasculitis, toxic epidermal necrolysis, Stevens–Johnson syndrome), hair loss, gum burns (from contact blanching of mucous membranes, due to denaturation of protein in superficial epithelial cells; mouth ulceration can also occur from ***aspirin***-containing chewing gum) occur in varying degrees. May cause erroneous readings in thyroid function tests, urine ***xylose*** tests (decreased levels), blood sugar and urates (decreased levels), and many other tests. Cross-reactivity with ***FDC Yellow 5*** (***tartrazine***) is well-documented, and a possibility of similar reactivity with benzoates and non-acetylated ***salicylates*** has been reported. ***Aspirin*** poisoning in **children** under 5 years of age is very common and over ten thousand yearly, world-wide, have been noted; and has had a high mortality rate (accidental ingestion of 35 grains to a 2-year-old **male** was fatal). Death in poisoning is from overwhelming metabolic acidosis, pulmonary edema, and central nervous system depression. In addition to the drug's effect on platelet adhesiveness, thrombocytopenia, hemolytic anemia (in glucose-6-phosphate dehydrogenase deficiency), plus rare reports of agranulocytosis, and aplastic anemia (high-dose therapy) are in the literature. ***Uric acid*** clearance is decreased by 25% and serum levels increased by 6% after 1 week of 75 mg/day. At altitude, **subjects** taking ***aspirin*** develop headaches at a lower oxygen saturation than those taking a placebo. A 21-year-old **female**, after buffered ***aspirin***, suffered episodic attacks of erythema exsudativum multiforme, with a photorecall-type phenomenon, confirmed by rechallenge. A third of ***aspirin***-sensitive **asthmatics** reacted to ***acetaminophen***. Henoch–Schönlein purpura associated with its use. Excessive intake by the **elderly** can lead to disorientation.

LD, **human**: 0.07–0.1 g/kg; LD_{LO}, **human**: 50 mg/kg; LD_{LO}, **child**: 81 mg/kg.

Drug interactions: One reference lists over 70 drugs interacting with ***aspirin***. ***Alcohol*** ingestion has potentiated its gastric bleeding; ***indomethacin***, ***corticosteroids***, and ***diactyl sulfosuccinate*** enhance its ulcerogenic effect. Both adverse effects may be aggravated by stimulants of gastric acid secretion, such as appetizers or aperitifs containing bitters. ***Methocarbamol*** potentiates its analgesic effect. A synergistic antiplatelet effect has been noted with the calcium channel-blocker, ***verapamil***. It will increase ***aminosalicylic acid***, ***methotrexate***, and ***valproic acid*** toxicity; potentiates ***coumarin***, ***heparin***, ***penicillins***, and some ***sulfonamide*** activity; may increase activity of ***sulfonylureas***, such as ***tolbutamide*** and ***chlorpropamide***. Large doses may potentiate ***phenytoin***. Will increase ***ascorbic acid*** urinary excretion (three times). Pretreatment with or concurrent administration of buffered ***aspirin*** significantly increases ***indomethacin*** absorption and side-effects. Serum ***salicylate*** concentrations decrease after ***antacid*** ingestion, due to increased renal ***salicylate*** clearance in alkalinized urine; ***phenobarbital*** may decrease its analgesic effect; oral ***iron*** will chelate it, reducing its absorption; ***griseofulvin*** has dramatically decreased its serum concentration; ***caffeine*** may complex it, reducing its absorption. Reverses ***meprobamate*** analgesic effects. In some **patients**, ***benorylate*** may interfere with its binding to serum proteins. Decreases uricosuric effect of ***iopanoic acid***, ***probenecid***, and ***sulfinpyrazone***, and the diuretic effect of ***spironolactone***. Potential for transient hypoglycemia with ***glyburide***. Use with ***streptokinase*** increases number of deaths from all causes, including cerebral causes. Can decrease effect of ***diuretics***, ***β-blockers***, and **angiotension-converting enzyme** ***inhibitors*** in decreasing blood pressure. Analgesic effect is increased by 65–200 mg ***caffeine***. Use with ***ginkgo biloba*** can increase risk of bleeding and stroke.

Treatment: ***Ipecac***-induced vomiting, if in the first 3 h. Adsorbed by ***activated charcoal***. IV use of alkaline fluids, to alkalinize the urine, has been popular; hemodialysis and peritoneal dialysis provide more rapid removal of the drug.

In **animals**, gastrointestinal distress and ulceration, unsteadiness, blood dyscrasias, and fetal damage reported. Causes pronounced inhibition of ***indomethacin***, ***phenylbutazone***, and ***flufenamic acid*** in **rat** trials. This can be long-lasting, even 2 weeks later in the case of the latter. Teratogenic in the **rat**. ***Aspirin***-induced gastric lesions reported in **cats**, **dogs**, **guinea pigs**, **rats**, and **man**. In **dogs**, 50 mg twice daily leads to emesis; larger doses often cause central nervous system depression, fatal metabolic acidosis, gastric ulceration, and hematemesis. **Cats** are deficient in some metabolizing enzymes for it.

LD, **dog**: 0.2 g/kg; LD_{50}, **rat**: 1 g/kg; LD_{50}, **mouse**: 815 mg/kg; LD_{LO}, **cat**: 300 mg/kg.

ACHILLEA sp.

= *A. millefolium* = *Yarrow* = *Amarga* = *Milfoil* = *Millefoglio* = *Bloodwort*

Botanical. **Many other species exist. Related chemically and botanically to the *chamomiles*.**

Use*:** ***Yarrow is a popular source of medicine since the Trojan War (circa 1200 B.C.), when the Greek hero, Achilles, used the leaves to control bleeding of wounds in his fellow soldiers. During the U.S. Civil War, it was used for the same purpose and called ***woundwort***. Topically, also used in shampoos and as an analgesic and anti-inflammatory. Also used as a snuff, in beverages, and as a substitute for hops in beer-making.

Untoward effects*:** *A. lutea* is poisonous to **guinea pigs**, although *A. millefolium* is considered non-poisonous, the consumption of a single plant (contains varying amounts of ***nitrate) caused the rapid death of a **calf**. **People** have shown positive patch test reactions to it. Contact dermatitis and photosensitization (contains ***furocoumarins***) have also been reported in **man**. Can produce disagreeable tastes and odors in milk products and meat of **animals** grazing on it.

ACID BLUE 9

Coal tar dye. **A triphenylmethane. Diammonium and disodium salts have been used. Over 600 tons have been produced annually.**

***Use*:** Hair colorant, absorbed more strongly on damaged hair.

***Untoward effects*:** A single oral dose of 300 mg/kg has caused renal carcinoma and breast hyperplasia in **rats**.

SC, LD_{LO}, **rat**: 5.5–17 g/kg; SC, LD_{LO}, **mouse**: 4.6 g/kg.

ACITRETIN

= *Etretin* = *Neotigason* = *Ro 10-6,070* = *Soriatane*

Antipsoriatic retinoid.

***Untoward effects*:** Cheilitis in 75%, dry skin, desquamation and peeling palms and soles, and alopecia in 50–75%. Occasional hypertriglyceridemia, decreased high-density lipoprotein, retinoid hepatitis and cirrhosis, conjunctivitis, corneal erosions and opacities, iritis, decreased visual acuity, skeletal hyperostosis, phototoxicity and photosensitivity, and increase in vulvovaginal candidiasis. Possible long-term teratogenic effect for at least 3 years. Avoid blood donations during that period.

Drug interactions*:** Use with ***tetracyclines increases risk of pseudotumor cerebri. Interacts with ***ethanol*** yielding ***etretinate***. Can be additive to ***vitamin A*** toxicity. Decreases effectiveness of progestin ***oral contraceptives***.

ACOCANTHERA sp.

= *Acokanthera sp.*

Botanical. **Contains the cardiac glycosides *ouabain* (q.v.) and *acokantherin*.**

Use*:** In African (Ethiopia, Kenya, Tanzania, and Zambia) arrow poisons. Many species are found from Arabia south through East Africa and down to the Cape. The poison, ***ouabain (q.v.), is dominant in the extract used on the arrows and comes mostly from *A. schimperi*. Rapidly fatal after ingestion by **livestock**.

ACONITE

= *Monkshood* = *Wolfsbane* = *Eisenhutknollen*

Circulatory depressant, local anesthetic. **Contains potent and toxic alkaloids. In the buttercup family. All parts of the *Aconitum* plant are poisonous.**

***Use*:** As a febrifuge, and to relieve pain in neuralgia and lameness. Internal use is rare since the advent of antibiotics. 4% ointment or liniment used on sprains. In Europe, as a toxic bait for **rodents** and **insects**. In arrow poisons.

Untoward effects*:** Alcoholic, but not aqueous solutions are readily absorbed through the skin. Causes rapid death by circulatory collapse and asphyxia. Ventricular fibrillation and irregular heartbeats, diarrhea, nausea, vomiting, twitching, salivation, burning taste, numbness of mouth, throat, and, eventually, fingertips, weakness, vertigo, paralysis, and fixed-drug eruptions in **man**. Similar symptoms occur in **pets**, often after eating an aconite houseplant. After ingestion, symptoms occur rapidly. Used on iron-tipped poison arrows in China, the Himalayas, in India, and on whaling harpoons by Aleutian Indians. Also used on arrows by Ainu (Japanese) hunters. Scratches from such arrows were said even to kill a **bear**. Cases of poisoning in **man** are not infrequent in China and Hong Kong. Three fatal cases (tachyarrhythmias) from ingestion of Chinese herbal preparations by 54-year-old **male**, 35-year-old **male**, and 39-year-old **female**. Wounded flesh areas putrify rapidly. Roots, leaves, and seeds are toxic, and the roots are the most toxic. Poisoning by the wild plant is rare. *A. columbianum* grows in the northwestern U.S., and has occasionally poisoned **sheep** grazing on it. See ***aconitine below. LD_{LO}, **human**: 5 mg/kg; oral TD, **horse**: 370–400 g of root, and 5–10 g for the **dog**.

Treatment: Effectively adsorbed by ***activated charcoal***. Saline cathartics have been used. Give oxygen, IV ***glucose*** or saline, and respiratory and cardiac stimulants.

ACONITINE

Use: To produce experimental cardiac arrhythmias in **cats**, **chicks**, **dogs**, **guinea pigs**, **rabbits**, and **rats**. Strong analgesic effect against neuralgia of trifacial and sciatic nerves in **man** with a maximum oral dose of 0.2 mg. Also topically in liniments and as tinctures. In traditional Chinese, Japanese, Indian, and Alaskan medicines.

Untoward effects: Violent poison. Polish reports state 0.25 mg may cause acute poisoning and 0.5 mg may be fatal in **man**. Some reports state 0.15–5.0 mg as a fatal dose. During a 4 year period, a Hong Kong hospital reported 20 cases of typical poisoning following ingestion of *Aconitum sp.* in Chinese herbal medicines. See ***aconite*** for symptoms of poisoning.

Oral LD_{50}, **mice**, **dogs**, and **monkeys**: are reported as 2–3 mg/kg, 3–10 mg/kg, and 1.5–2.0 mg/kg, respectively. 2.4 mg has killed a **dog** in 65 min. It is estimated that 11 mg and 2 mg by injection, respectively, will kill a **horse** and a **dog**.

LD_{LO}, **rat**: IP, 75 μg/kg; SC, 100 μg/kg.

Treatment: Emetics, if early enough. Adsorbed by ***activated charcoal***. Manage the cardiac arrhythmias. Maintain blood pressure. Oxygen, if necessary.

ACREMONIUM sp.

= *Epichloe sp.*

***Fungus*. Produces *ergot*-type alkaloids.**

Untoward effects: Reproductive problems in **mares** and **cows** pastured on ***fescue***. In the Ukraine, *A. suis* is reported to be a pathogen causing atrophic rhinitis, retarded growth, and difficult breathing in **pigs**.

ACRIDINE

= *10-Azaanthracene* = *Dibenzo[b,e]pyridine*

***Coal tar* derivative.**

Use: In manufacturing dyes, chemical intermediates, acriflavine, proflavine, and reagents.

Untoward effects: Strong irritant to skin and mucous membranes in **man**. Contact dermatitis, erythema, pricking sensations and melanosis reported. A photosensitizer, phototoxic agent.

Teratogenic to **crickets** and lethal to **paramecia**, when they are exposed to sunlight.

SC, LD_{50}, **mice**: 0.4 g/kg; IV, LD_{50}, **rabbit**: 100 mg/kg.

ACRIFLAVINE

= *3,6-Diamino-10-methylacridinium Chloride mixture with 3,6-Acridinediamine* = *Euflavine* = *Gonacrine* = *Neutral Acriflavine* = *Neutroflavine* = *Trypaflavine*

Antiseptic, protozoacide, dye.

Use: Primarily bacteriostatic (bacteriocidal effect is slower) against Gram-positive organisms and some Gram-negative organisms. To irrigate wounds in trichomoniasis and genitourinary infections. To attenuate *Erysipelothrix* and various viruses. Destroys sperm. Controls fin rot and ich in **fish**.

Untoward effects: Has caused a contact dermatitis (sometimes delayed for 2–3 days) and fixed-drug eruptions in **man**. Phototoxicity (with first and second degree burns) in **man** developed after two injections of 10 cc of a 2% solution. May color urine a lemon yellow. May be found in trace amounts in milk of **cows**, after intrauterine use. LD_{LO}, IV, **human**: 1.5 mg /kg.

Oral LD_{LO}, **mouse**: 50 mg/kg; IP, LD_{LO}, 11 mg/kg; SC, LD_{50}, 14 mg/kg. LD_{LO}, **cat** IV, **dog** IV, **guinea pig** IV, and **rabbit** IV: 7.35 mg/kg, 25 mg/kg, 40 mg/kg, and 25 mg/kg, respectively.

ACROLEIN

= *Acraldehyde* = *Acrylaldehyde* = *Acrylic Aldehyde* = *Aqualin* = *2-Propenal*

***Herbicide, algicide, fungicide.* An unsaturated aldehyde. The delicious aroma of Thanksgiving turkey is, in part, caused by it. It has the typical smell of simmering or burning fat.**

Use: In manufacturing plastics, perfumes, and in organic synthesis. As a warning agent of ***methyl chloride*** refrigerant leaks, and as a lacrimating agent in military poison gases. An aquatic herbicide, useful in the control of aquatic snails and, in turn, schistosomiasis. Controls algae in irrigation canals, pulp and paper mills, cooling towers, and sewage plants.

Untoward effects: Formed from imperfect combustion of oils, wood, ***cotton***, ***wool***, ***silk***, and ***tobacco***. In ***tobacco*** smoke. Skin and mucous membrane irritant, causing contact redness, burning and swelling, large blisters, and deep tissue burns when the exposure is covered by gloves or clothing. Red conjunctiva and lacrimation from eye exposure. Hoarseness and life-threatening laryngeal edema can occur. Dyspnea, substernal pressure, hacking cough and frothy sputum indicate pulmonary edema, which may progress to cyanosis and coma. Headache, nausea, vomiting, diarrhea, dizziness, chest and abdominal pain with depressed respiratory rate are common symptoms. **Firefighters** are regularly exposed to it, and so are **people** who smoke ***tobacco***. Concentrations as low as 0.8 ppm can cause eye, mucous membrane, and respiratory symptoms. Estimated lethal concentration (10 min) for **man** is 30–100 ppm; upper TLV recommended is 0.1 ppm. Oral LD_{LO} for **humans**: 5 mg/kg. Now considered too hazardous for use in the U.S. Inhalation TC_{LO}, **human**, **rat**, **mouse**, and **guinea pig**: 1 ppm, 8 ppm/4 h, 24 mg/m^3/6 h, and 24 mg/m^3/6 h, respectively.

Oral LD_{50}, **mouse**: 40 mg/kg; oral LD_{50}, **rabbit**: 7 mg/kg; LD_{50}, **rats**: 46 mg/kg; continuous inhalation causes general body deterioration, decreased blood cholinesterase, decreased coproporphyrin urinary excretion, respiratory distress, and a 50% mortality. LD_{50}, **mallard ducks**: 6.32–13.1 mg/kg. Deaths within 32 min to several days. Toxic symptoms within 10 min and lasted up to 36 days, even in those receiving only 3.3 mg/kg. Slow response, ataxia, geotaxia, phonation, wing tremors, and myasthenia were characteristic symptoms. A small **poodle dog**, shut up for ½ h in a small unventilated kitchen with a pan of boiling fat, exhibited dyspnea, faint pulse, and cyanosis, and died within 2 days. LC_{50}, **goldfish**: < 0.08 mg/l.

Treatment: Flush eyes and skin copiously with water. Leave skin exposed to air. Get hospital attention. Resuscitate and/or administer oxygen, if respiration is depressed. Keep **patient** warm, inactive, and in a sitting or semireclining position. Combat shock. Aspirate sputum, if necessary. ***Cysteine*** and ***ascorbic acid*** have shown good protective activity in **rat** trials.

ACRYLAMIDE

= *2-Propenamide*

Use: 120 ± million lbs annually in the U.S., mostly for polyacrylamides used as flocculants in water, in pulp and paper treatments, oil recovery, as a soil, sewer and manhole grouting agent, and in the construction of dam foundations and tunnels. Has been used to prolong the action of ophthalmic preparations.

Untoward effects: Well known to be highly toxic and irritant in **man**, causing central nervous system paralysis. Absorbed through unbroken skin. Cause of mixed sensorimotor neuropathies with ataxia, desquamation of hands and soles of feet, sweating of palms and feet, paresthesias, numbness of lower limbs, tenderness to the touch, coldness, muscle weakness and tremors of the extremities, loss of body weight, lassitude, hypersomnolence, emotional changes, speech slurring, positive Romberg's sign, loss of vibration and position senses, weak or absent tendon reflexes, and footdrop. **Workers** in the synthesis and polymerization processes are at the highest risk. Residual monomer in the polymers is a cause of concern. The monomer is cumulative in its toxicity. A delayed neurotoxicity can occur. Considered to be a possible **human** carcinogen by the Environmental Protection Agency. TWA: 0.3 mg/m^3.

Rat studies indicate it is rapidly distributed throughout the body and readily metabolized by ***glutathione*** conjugation. **Rats** and **mice** receiving acrylamide by various routes initially show decreased hind limb function. **Pigtail monkeys** were dosed orally with 10 mg/kg/day/5 days a week. Toxic signs (peripheral neuropathies) developed when total dosage reached 320–450 mg/kg. **Mouse** studies indicate it causes degeneration of the epithelial cells of the seminiferous tubules in the testes. **Cats** receiving 7.5 or more mg/kg, IM/10 days developed a peripheral neuropathy.

LD_{50}, **rats** and **mice**: 170 mg/kg; LD_{LO}, **rabbit**: 126 mg/kg.

Prevention and treatment: Prevent contact with skin and eyes. ***Vitamin*** B_6 delayed the onset and decreased the severity of neurotoxicity in **rats** caused by chronic administration.

ACRYLATES and ACRYLIC ACID

Approximately 1.5 billion lbs of the acid are produced annually. Some of the published literature is vague as to which acrylate was used. See specific compounds, such as *cyanoacrylate*, *ethyl acrylate*, *methyl methacrylate*, etc.

Use: In the production of dentures, in manufacture of printing plates, fast-drying inks, fabrics, and adhesives. Acrylic resins have been used to promote bone growth, in bone cements, for film-coating sustained-action tablets, and as a thickener and suspending agents in **animal** feeds.

Untoward effects: Severe photo-onycholysis occurred after the use of ultraviolet light, to rapidly cure and set nail polish, and dermatitis venenata in industrial **workers** and dental **technicians** are well-documented. Severe contact allergen in the printing industry. A fatal inhalation abuse case was found dead in a creek bed. Acrylic resins have caused foreign-body, granulomatous, and delayed hypersensitivity reactions in **man**.

LD_{50}, **rats**: 0.19–0.58 ml/kg (340 mg/kg); LD_{50}, **mice**: 2.4 g/kg.

ACRYLONITRILE

= *Acritet* = *Cyanoethylene* = *Fumigrain* = *2-Propene-nitrile* = *Ventox* = *Vinyl Cyanide*

Explosive and flammable. Approximately 3 billion lbs produced annually in the U.S.

Use: In the manufacture of synthetic fibers, resins, plastics, nitrile, fumigants, and as a chemical intermediate.

Untoward effects: Primary toxicity is systemic. May damage the central nervous system, lungs, liver, and kidneys. Degrades slowly in the body to ***hydrocyanic acid*** (q.v.). Its slow release of ***cyanide*** (q.v.) makes it less toxic than ***hydrocyanic acid***. Excessive absorption may cause unconsciousness and respiratory arrest, without any other symptoms. Lesser amounts cause headache, nausea, constriction in the throat, apprehension, dizziness, and hyperpnea. As toxicity deepens, bradycardia, thready pulse, irregular respirations, unconsciousness, convulsions (epileptiform or tonic), opithotonus, trismus, and loss of sphincter control. This is followed by respiratory paralysis. The skin remains pink and not cyanotic. If heart action continues after respiration ceases, vigorous treatment may save a **patient**. TWA was originally 20 ppm. In 1978, Monsanto voluntarily put the exposure limit as 5 ppm, because of the **human** carcinogenic potential. OSHA later requested a 2 ppm guideline and 10 ppm for a 15 min exposure, because of an excessive incidence of lymphatic, stomach, prostate, lung, and bowel cancers in DuPont and other **workers**.

LD_{LO}, **human**: 50 mg/kg.

Inhalation and ingestion have been associated with brain, stomach, and zymbal gland cancer in **rats**.

Oral LD_{50}, **rat**: 82 mg/kg; oral LD_{50}, **mice**: 0.2 mmol/kg (27 mg/kg); IV, LD_{LO}, **dogs**: < 200 mg/kg; oral LD_{50}, **guinea pig**: 50 mg/kg; oral LD_{50}, **rabbit**: 93 mg/kg. LC_{LO}, inhalation trials, **dog**: 110 ppm/4 h; **cat**: 600 ppm/4 h; **rabbit**: 258 ppm/4 h; **guinea pig**: 576 ppm/4 h.

Prevention and treatment: Prevent skin contact. Periodic chest X-rays are advocated. The most effective treatments for ***cyanide*** poisoning include ***sodium thiosulfate*** (0.4 g/kg, slow IV, as a 20% solution), ***sodium nitrite***, ***methylene blue***, and ***hydroxycobalamin*** (50 mg/kg). ***Sodium nitrite*** rapidly forms ***methemoglobin***, which readily fixes the highly lethal ***cyanide*** ion into a relatively non-toxic cyano-hemoglobin. A simultaneous injection of ***sodium thiosulfate*** permits ***cyanide*** excretion as a harmless thiocyanate. ***Methylene blue*** serves the same purpose as ***sodium nitrite***, but is not used as often. ***Hydroxycobalamin*** has shown limited value in **dog** trials.

ACTAEA sp.

***Botanical.* Herbacious perennial. *A. spicata* = *Herb Christopher* is European and has black berries; *A. pachypoda* = *Baneberry* = *White Cohosh* = *Snakeberry* has white to reddish berries and is found in Alabama in rich woodsy areas. *A. rubra*, along the streams in Colorado and in eastern N. America, has red berries; *A. arguta* is found mostly in western N. America. Their fruits or berries LOOK edible, but should NOT be eaten, as they cause violent purgation.**

Use: *A. spicata* is used as an insecticide in Europe.

Untoward effects: May have been (*A. spicata*) one of the ingredients in Romeo's poisonous drink. An essential oil in all parts of the plant, and especially in the berries and roots, is the apparent cause of the toxicity. The root contains a local irritant, ***zimifugin*** or ***makrotin***. In **man**, tachycardia, gastroenteritis (vomiting, diarrhea), dizziness, and delirium are the chief symptoms which usually disappear in 3–4 h. Fatalities are rarely reported, but as little as six berries has caused severe symptoms. **Adults** usually spit the berries out, because of their bitter taste, but **children** often swallow them.

Cattle may also be adversely affected.

ACTINEA ODORATA

= *Bitterweed Actinea* = *Rubberweed*

***Botanical.* Grows extensively from Kansas south to Texas, then west to California. Common on overgrazed ranges and on roadsides. The genus name is also called *Hymenoxys*.**

Untoward effects: Especially toxic to **sheep** and leaves to **cattle** during drought years, whether the plant is green or dry. Anorexia, depression, rumen atony, abdominal pain, bloating, frothing at the mouth, weakness, and a green nasal discharge are usual symptoms. A pound eaten at a time will kill a **sheep** within 12–24 h. Less severe cases may recover in 3–5 days. Necropsy reveals pulmonary congestion, hemorrhages on the heart, costal pleura, lymph nodes, abomasum, and intestines.

Prevention and treatment: Move **livestock** to clean ranges and put them on supplemental feeding. Herbicides on young plants may be useful. Symptomatic treatment. No specific antidote reported successful.

ACTINIA EQUINA

An anemone that colonizes sponges.

Untoward effects: Due to a liberated toxin, ***thalassin***, which, in turn, liberates ***histamine*** and slow-reacting substance (SRS), capable of inducing severe bronchoconstriction. It is responsible for severe skin reactions and local necrosis referred to as "**sponge-diver's** disease".

ACYCLOVIR

= *Aciclovir* = *Aclovir* = *ACV* = *Acycloguanosine* = *A-655Z* = *2-Amino-1,9-dihydro-9-[(2-hydroxyethoxy)methyl]-6H-purin-6-one* = *BW248U* = *CT2336* = *9-[(2-Hydroxyethoxy) methyl]guanine* = *Vipral* = *Virorax* = *Wellcome-248U* = *Zovirax*

***Antiviral*. High selectivity.**

Use: Particularly in herpes and cytomegalus infections in **man**. Very limited **animal** use.

Untoward effects: Relatively non-toxic to host cells. Facial dermatitis, cutaneous vesicular lesions, tachycardia, fever, acute megaloblastic hemopoiesis, peripheral edema, nausea, vomiting, esophagitis, and neutropenia seizures and coma have occasionally been reported in **man**. More commonly, depression, delusions, lethargy, agitation, visual hallucinations, confusion, insomnia, hyperacusis, paranoia, headache, dysarthria, aggression, and other assorted neurological adverse effects have been noted. Some of the symptoms may possibly be interrelated to their original disease effects. Impaired renal function and failure with coma is more common. Excreted in **human** breast milk, and passes through the placenta to the fetus. Transient burning and stinging occur on topical application. High dosage should be used cautiously in bone marrow transplant **patients**. Use in pregnancy appears not to have increased the number of birth defects above norms. The intravenous form is alkaline and may cause phlebitis and cutaneous inflammation at extravasation sites. Resistance to it has occurred. Phlebitis and inflammation at injection site area. Has caused crystalline nephropathy, tremors, seizures, hallucinations, and coma.

Drug interactions: May increase serum ***lithium*** concentration. ***Probenecid*** increases its serum half-life, but it may not be clinically important. It has caused a dramatic increase in ***cyclosporine*** levels and the combination usage with ***zidovudine*** caused a rapid incapacitating drowsiness and lethargy, which was reproducible on rechallenge.

Treatment: In acute renal failure and anuria, hemodialysis has been recommended until restoration of renal function.

Testicular atrophy in **rats**. Oral LD_{50}, **mice**: > 10 g/kg; IP, 10 g.

ADAPALENE

= *CD 271* = *Differin*

Synthetic retinoid for topical treatment of acne.

Untoward effects: Erythema, burning sensation, pruritus, dryness, and scaling. Anophthalmia and agenesis of optic chiasma in **fetus** of **woman** treated with 0.1% topical gel before and during the first 13 weeks of pregnancy.

ADEFOVIR

= *Bis(pom)PMEA* = *Dipivoxil* = *GS 840* = *Piv2PMEA* = *Preveon* = *Previon*

Antiviral.

Untoward effects: Dose-related mild to moderate proximal renal tubular dysfunction, usually after ~20 weeks in > 30% of **patients**. Nausea, asthenia, and increased aminotransferases also occur.

ADENIA VOLKENSII

= *Kiliambiti Plant*

***Botanical*. Eastern Africa.**

Use: The boiled plant is used in herbal medicines.

Untoward effects: 90–95% of the plant's ***cyanide*** (q.v.) content is in the tuber. The outer bark of the tuber contains about three times the ***cyanide*** of the inner part. Another toxin, dubbed ***modeccin***, has also been isolated from *A. digitata* or *Modeca digitata*. The tuber has been used in Kenya in **homicides** and to poison **hyenas**. Also contains a phytotoxin. A ½ lb of the minced root fed to an adult **ewe** caused rapid symptoms of ***cyanide*** poisoning. Rapidly fatal to a **rabbit**.

ADENIUM sp.

***Botanical*. Five species are succulent schrubs or trees found from West Africa to Arabia, and then down southern Africa (Namibia, Botswana). Their latex and other parts contain *digitoxigenin* and *gitoxigenin*.**

Use: ***Arrow and dart poison, piscicide***, Eastern Africa.

Untoward effects: The Echuja poison of southwestern Africa comes from *A. boehmianum*. *A. honghel* = ***K'arya*** in Nigeria-Hausa, contains the cardiac glycoside ***adeniine***, a poison, whose action on the heart is similar to ***digitalin***, but also affects the central nervous system. *A. obesum* = ***Desert Rose*** = ***Mdagu Wanja*** (Swahili), ***Mwadiga*** (Duruma), ***Sarba Arba*** (Boran), is a succulent shrub, 6 ft tall, with 8" pods, borne in pairs. The large, fleshy tap root, and especially the seeds, contain cardiotoxic poison used on arrows and **fish**. In summary, theses agents cause bradycardia, arrhythmia, and fatal heart block in **man**. Very toxic to **livestock**.

ADENOSINE

= *Adenocard* = *9-β-D-Ribofuranosyl-9H-purin-6-amine* = *Adenine Riboside* = *6-Amino-9-β-D-ribofuranosyl-9H-purine* = *9-β-Ribofuranosidoadenine*

Antiarrhythmic.

Use: For IV treatment of paroxysmal supraventricular tachycardia (including Wolff–Parkinson–White syndrome in **man**). In treating arthritis, bursitis, and pruritus. In the summer of 1960, **youngsters** collected 1,200,000 **fireflies** (at 30¢/100), so that ***adenosine triphosphate*** and other lucifer materials could be extracted from their tails. Released from the brain by electrical stimulation.

Untoward effects: ***Adenosine triphosphate*** produces bronchospasm and vasoconstriction of pulmonary blood vessels, facial flushing, chest pain, slight nausea, and tendency to cough. Decreases urine volume and sodium excretion. Heart block can occur. The drug is rapidly eliminated from the extracellular areas; this helps to limit its adverse effects. After use in controlling tachycardias, arrhythmias lasting less than a minute have occurred, followed by bradycardia. Transient asystole, transient dilation of coronary arteries, transient atrial fibrillation, and premature ventricular contractions are also reported. Abdominal pain, headache, and myalgia have been reported in a few **patients** receiving the dibutyryl cyclic monophosphate form. Flushing and pain at the injection site occurred within 15 min in 12/120.

Intraventricular administration in **cats** produces sedation.

Warning: IM injections produce fewer side-effects than IVs. Use very **SLOW** IV injections with a very fine needle.

Drug interactions: ***Carbamazepine*** may increase its degree of heart block; ***dipyridamole*** may enhance its effect; and its use may require higher than normal ***theophylline*** dosage in **asthmatics**.

IV, LD_{50}, of the triphosphate, **mice**: 325 mg/kg; IP, LD_{LO}, 500 mg/kg.

S-ADENOSYLMETHIONINE

= *Ademetionine* = *Ado-met* = *Denosyl SD4* = *Gumbaral* = *SAMe*

***Anti-inflammatory, analgesic.* The active form of *methionine*.**

Use: In treating chronic liver disease, osteoarthritis, and depression. As a methylating agent in detoxification reactions.

Untoward effects: Relatively few adverse effects (nausea, vomiting, diarrhea, flatulence, anxiety, and headache) reported in **humans**. Decreased weight gain at 400 mg/kg/day and local tissue reactions in pregnant **rats** and **rabbits** at injection sites.

IV, LD_{50}, **mice**: > 6 g/kg, 560 mg/kg; IP, 2.5 g/kg.

ADINAZOLAM

= *Deracyn* = *U-41123*

Antidepressant.* A *benzodiazepine.

Untoward effects: Claimed to have induced mania in one **patient**. ***Probenecid*** decreases its clearance from the body.

ADIPONITRILE

***Organic solvent, chemical intermediate.* Made from *hydrocyanic acid*.**

Use: Neurotoxic. LC_{50}, inhalation, **rats**: 1.71 mg/l exposure caused mild salivation, irregular respiration, and some weight loss. After a number of inhalation exposures, they had increased blood glucose, urea nitrogen, creatine, and urine glucose, with decreased red blood cells, hemoglobin, and leukocytes.

Approximate LD_{50} in **rats**, **mice**, and **rabbits** is 300 mg/kg, 40 mg/kg, and 19 mg/kg, respectively. Topically, it does not appear to be a strong local irritant. The topical lethal dose for a **rabbit** is approximately 2 g/kg.

ADONIS sp.

= *A. vernalis* = *Pheasant's Eye* = *Spring Adonis* = *Ox Eye*

***Botanical.* Contains the cardiac glycosides, adonidoside and adonivernoside, with a *digitalis*-like action.**

Use: As a cardiotonic in Russia.

Untoward effects: From its major alkaloids, ***delphinine*** and ***ajacine*** and a glucoside, ***adonidin***. Toxicity in **man** may be similar to that of ***digitalis***.

Usually a severe gastroenteritis in **animals** and occasionally lethal to **horses** and **sheep**.

A. annua, although distasteful, has been poisonous to **sheep**. *A. aestivalis* has poisoned **horses**. Superpurgation and cardiac failure precede death.

ADRENOCHROME

Use: see **CARBAZOCHROME**.

Untoward effects: Bright orange-red color compound, and the cause of oxidized ***epinephrine*** (***adrenaline***) solutions turning pink. It is a psychotogenic substance, when injected into **humans**. Three elderly **patients**, within 3 months of starting topical treatment of their eyes with

2% ***epinephrine HCl*** or ***epinephryl borate*** 1%, had orange-red pigmentation in their hydrophilic lenses.

IP, LD_{LO}, **rat**: 100 mg/kg.

AESCULUS sp.

= *Buckeye* = *Venostassin* = *Eskusan* = *Vasokastan*

***A. hippocastanum* = *Horsechestnut* = *Marronnier*. Widely cultivated in Eurasia.**

***A. glabra* = *Ohio Buckeye* = *Texas Buckeye*. Trees and shrubs often found growing in canyons and valley floors, or near streams. About 25 species are native to North America.**

Use: To stimulate venous circulation, supposedly, by action of its main triterpene saponin, ***aescin***. Also in Europe, as an anticoagulant and to stupify **fish**. *A. pavia* is used in bed manufacturing, because it is **insect**-repelling. Also used by California's indigenous **peoples**.

Untoward effects: Leaves, young twigs, bark, flowers, and seeds contain the poisonous glycosides ***esculin*** and ***fraxin***, a hydroxy derivative of ***coumarin***, and a narcotic alkaloid. **Children** and **livestock** have been poisoned by eating any of the above mentioned plant parts. The tree also contains the alkaloid, ***quercitrin***. After receiving treatment with extracts of *A. hippocastanum*, 35 **male** and **female patients** aged 40 to > 60 years had cardiac palpitations, sensations of blood vessel pulsations, heartburn, and anginal pain. Occasional vomiting and bleeding disorders (hemorrhage and hemolysis) are reported. Headache, fever, thirst, mydriasis, and facial and respiratory paralysis also occur in fatal cases. Incoordination, twitching, and excitability noted with consumption of *A. pavia* (***Red Buckeye***). One Poison Control Center reported 22 cases of exposure in 1975 and four **humans** were symptomatic. Poisoning reported in **humans** consuming ***honey*** produced by **bees** consuming the toxic nectar. **Children**, in particular, are poisoned by eating the nutlike, glossy seeds. Usually 1–3 brown seeds in each leathery-looking fruit, and each seed has a tan or white scar. Death in a **child** has been reported.

Poisoned **cattle**, **sheep**, **horses**, and **swine** have a slow, staggering gait, are weak, and tremble. Their mucous membranes are congested and inflamed, and pupils are dilated. Affected **animals** go into stupor and coma, before death. A few may vomit and colic is noted in **horses**. *A. glabra* may also contain a narcotic alkaloid, and it is very detrimental to **chick** growth.

Single dose LD_{50} for seed extract of ***horsechestnuts*** to both **chicks** and **hamsters** is approximately 10.65 mg/g.

Treatment: **Livestock** are usually treated with purgatives and stimulants.

AFLATOXINS

***Fungal metabolites, mutagens, carcinogens*. Probably the best known mycotoxin. Usually produced by *Aspergillus flavus* and *A. parasiticus*. Depending on weather and storage conditions, they are usually found in varying amounts on *peanuts* (*groundnuts*) and their products, *cottonseed meal*, *corn*, dried *chilli peppers*, *coffee*, etc. *Soybeans* appear to have good general resistance to it. Has been produced on warm meats, hay, and silage. The common aflatoxins are B_1, B_2, G, G_1, M, P, and Q_1. Aflatoxin B_1 is the most potent of the group and is frequently found in food.**

Use: Experimental only.

Untoward effects: In "field" cases, aflatoxin types are usually mixed. They are among the most potent hepatocarcinogens. Long-term feeding of aflatoxin-containing diets have produced cancers in experimental **fish**, **birds**, **rats**, **mice**, **ferrets**, **tree shrews**, and **monkeys**. Its clinical significance is in poor performance. In **swine**, it impairs protein synthesis, causes hepatitis, liver necrosis, and portal cirrhosis. Clinical ill-effects are depression, anorexia, bloody diarrhea, anemia, icterus, and abortion in pregnant **animals**. Fatty infiltration, necrosis, and cirrhosis may be seen in post-mortem, but not necessarily in sub-clinical cases. **Ducklings** are the most sensitive **poultry** to its effects, followed by **turkeys**, then **chickens**. In Britain, the use of aflatoxin-contaminated ***groundnut*** meal in the feed of **turkeys** caused the loss of over 100,000 in 1960. Samples of the meal contained approximately 1–10 ppm. Averages of 7–8 ppm of aflatoxin B_1 were found in meal sold for **human** consumption in Uganda. It is estimated that the average concentration in **human** diets in the Southeast is 0.1 ppb and 0.011 ppb in the whole U.S. A positive correlation between estimates of **human** ingestion and hepatocellular cancer was identified in Uganda, Swaziland, Thailand, Kenya, Mozambique, and China. In the southeastern U.S., where aflatoxin intake is higher than the rest of the U.S., a 10% excess of hepatocellular cancer incidence has been reported. Reye's syndrome in Thailand seems to increase with seasonal increases of aflatoxin on ***rice***. A fatal case in a 3-year-old **male** followed eating 2-day-old boiled ***rice***. Assay revealed > 10 mg/kg aflatoxin B_1.

Symptoms develop rapidly in **calves** a few weeks after consuming contaminated feeds. Blindness, circling, stumbling, ear twitching, grinding teeth, followed by severe tenesmus and rectal eversion. Death follows in about 2 days. Some **cattle** are found dead without prior symptomatology. Loss of condition and slow recovery noted in non-fatal cases. Milk yields decrease. Milk contains a less toxic form, called aflatoxin M. Pale, firm, fibrosed livers, yellow kidneys, ascites, and mesenteric edema are noted in post-mortems. Centrolobular necrosis, bile duct enlargement, and catarrhal enteritis are noted. **Swine** show anorexia, jaundice, and depression. Generalized icterus, with liver color being off-white to bright orange. Blood coagulation is poor, with SC and mesenteric hemorrhages. Ascites and mesenteric edema occur occasionally. **Sheep** appear to be extremely resistant to its toxicity. **Guinea pigs**, **hamsters**, and **monkeys** rapidly develop a hepatitis. ***Ingestion*** by **ducklings**, **rainbow trout**, and **rats** is associated with liver tumor development. In addition, dietary intake in **rats** also produces an increase in carcinomas of the stomach, adenocarcinomas of the colon, and kidney tumors. Dietary concentrations in ppm and length of feeding time of mixed aflatoxins causing toxicosis have been given as 0.2–2.2 for weaning **calves** (stunting and death after 16 weeks); 0.22–0.66 for 2-year-old **steers** (liver damage – 20 weeks); 2.4 for 2 year **cows** (liver damage and clinical illness – 7 months); 0.23 for 2-week-old **pigs** (stunting – 4 days); 0.4–0.7 for 4–6 week **pigs** (stunting, liver damage – 10 weeks); 0.8 for 1 week **chicks** (stunting, liver damage – 10 weeks); 0.3 for **ducklings** (liver damage, death – 6 weeks).

LD_{LO}, **human**: 229 μg/kg; LD_{50}, **duckling**: 0.4 mg/kg; LD_{50}, **monkey**: 1.75 mg/kg. ***Aflatoxin B_1***, LD_{50}, **rat**: 7 mg/kg; LD_{LO}, **rat**: 357 μg/kg; LD_{50}, **mouse**: 9 mg/kg; LD_{50}, **monkey**: 2.2 mg/kg; LD_{50}, **cat**: 550 μg/kg; LD_{50}, **rabbit**: 300 μg/kg; LD_{50}, **pig**: 620 μg/kg; LD_{50}, **guinea pig**: 2 mg/kg; LD_{50}, **hamster**: 10 mg/kg; LD_{50}, **duck**: 335 μg/kg.

Prevention: Treating infected ***corn*** with approximately a 1% level of ***ammonia*** gas for up to 2 days destroys the bulk of aflatoxin present. An anticaking compound, hydrated ***sodium calcium aluminosilicate***, can be used in **livestock**.

Treatment: Since acidosis predisposes or aggravates the problem, IV use of alkalized saline solutions has been recommended, along with adsorbants, cardiac stimulants, and general supportive measures.

AFRAMOMUM MELEGUETA

= *Alligator Pepper*

Untoward effects: Adverse ophthalmic effects in **male Nigerians** aged 30–35 eating its seeds, leading to increased near point of convergence (17.2%) and decreased amplitude of accomodation (9.2%) without altering visual acuity or pupil size.

AGARIN

= *Muscimol* = *Pantherine*

One of the toxic substances in *Amanita* and closely related chemically to *ibotenic acid*.

Use: As a **fly** poison in Asia; for the symptomatic relief of tardive dyskinesia in **man**; has a dose-dependent inhibition of ***phencyclidine's*** effect and ***securinine***-induced convulsions in **mice**. Although structurally related to ***GABA*** (it is a ***GABA*** agonist), the brain's neurotransmitter, it appeared too toxic in **humans**. A less toxic modification, ***THIP***, is being explored.

Untoward effects: Symptoms are those of excessive ***atropine*** (q.v.) intake and can occur within 15–120 min after ingestion. Also see *Amanita*. TD_{LO}, **human**: 109 μg/kg.

SC, LD_{50}, **mice**: 3.8 mg/kg; oral, SC, IV, **rats**: 45, 3.8, and 4.5 mg/kg, respectively.

AGARITINE

= *L-Glutamic Acid 5-2-[4-(hydroxymethyl)phenyl]-hydrazide* = *β-N-[γ-L(+)-glutamyl]-4-hydroxymethylphenyl-hydrazine*

A *hydrazine* derivative, found in commercial mushrooms (*Agaricus bisporus*)(assayed at 0.04%), and in ten other species of *Agaricus*, but not in five others.

Untoward effects: Some agaritine is metabolized by the mushroom to a very potent carcinogen (single oral dose of 400 ng/g gave 30% of treated **mice** stomach tumors). Found in mushrooms in small amounts.

AGAVE

= *Century Plant*

***Botanical.* Yucca-like plants.**

Use: Ornamental. The ***maguery*** (*A. americana*) is an old antibacterial **Aztec** wound treatment, and also used as a **fish** poison. *A. cantala* of India is molluscidal. Agave is used in making Mexican beer and tequila. ***Hecogenin***, a steroidal sapogenin found in many agaves, was, for many years, an important starting agent in corticosteroid manufacturing. *A. americana* was used by **Indians** in healing ulcers.

Untoward effects: *A. americana* (***Pita***, ***Maguery Mexicana***, ***Ancaschampascera***, ***American Aloe***, ***Dalmatian Agave***), a popular desert succulent, has caused severe hemorrhagic skin rashes and systemic illness in **humans**, after contacting the sap. Severe eye irritation has occurred from exposure to the burning leaves or its ashes. Confluent papular, vesicular, and exudative dermatitis of the scalp and adjacent forehead skin, as well as the ulnar surface of the wrist in 34-year-old **male**, after using the sap topically for alopecia.

A. lechuguilla has caused severe hepatogenic photosensitivity in **animals**. White areas of **sheep**, **goats**, and occasionally **cattle** are affected. Swelling of the head and ears, serous seepage through the skin and skin sloughing, depression, and jaundice occur. Urine is dark. Emaciation, coma, and death may follow.

AGENT ORANGE

= *Trioxone*

***Defoliant*. Widely used during the Vietnam war. Has contained a highly toxic contaminant, *dioxin* (q.v.). It was actually a mixture of *2,4,5-T* and *2,4,-D* herbicides. 42.6 million liters were sprayed in Vietnam during the war. After the war, over 2 million gallons had to be disposed of. Sampling of this material indicated 0.05 to < 50 ppm *dioxin*.**

Untoward effects: Supposed cause of 58 **employee** deaths in a West Virginia plant making ***2,4,5-T***, after exposure to it and ***dioxin***. As early as 1962, the Surgeon General's office knew of this contamination and failed to notify the manufacturers, who, supposedly, manufactured the product under government requirements. ***Agent Orange***, as used, has been incriminated as a cause of chloracne, numbness of extremities, headaches, depression, suicidal attempts, violent rage, memory lapse, benign fatty tumors, increase in heart disease, nausea, stomach pains, malignant tumors, sterility, loss of libido, spontaneous abortions, birth defects, and assorted illnesses. Science and politics have become deeply intertwined in studying these problems. Many years later, the U.S. Center for Disease Control found little to support most of these claims. ***Agents Green***, ***Pink***, and ***Purple*** were also used in limited quantities (8000, 123,000, and 145,000 gallons, respectively) as ***dioxin***-tainted herbicides in Vietnam. Conflicting **mouse** data exist. See individual components in text.

AGROCLAVINE

= *8,9-Didehydro-6,8-dimethylergoline*

Ergot alkaloid.

Use: Experimental.

Untoward effects: Decreases weight gain in fetuses of **mice** fed. 250 μg/**mouse** for 2 days in the first week of pregnancy inhibited nidation. Unlike other ***ergot*** alkaloids, it is inactive when injected SC or IP. Inhibits lactation. < 300 μg/kg, IV in **rabbits** caused a 1°C increase in body temperature.

AGROSTEMMA GITHAGO

= *Corn Cockle* = *Lychnis githago* = *Nielle* = *Purple Cockle*

Very tall annual. Native to Europe and introduced into the U.S. and Canada. Flowers look like garden pinks.

Untoward effects: Contains a sapogenic glucoside, ***githagenin***, apparently identical to ***sapotoxin*** and **agrostemmic acid**, producing a soapy foam. As early as 1903, contamination of ***wheat*** was reported as a serious poison for **man**. Ingestion of these seeds, in **man**, has caused abdominal cramps, vomiting, weakness, shallow breathing, dizziness, and diarrhea. A serious contaminant in ***wheat***, causing the ***flour*** to be rejected for **human** use. Also a contaminant in ***oats***.

Seeds poisonous to **animals** (**cattle**, **horses**, and, infrequently, **hogs**). Other plant parts are also poisonous, but rarely a cause of fatalities. Especially poisonous to **poultry** (**geese**, **ducks**; single dose to **chickens** of 2.5–8.0 g of ground seeds leads to acute poisoning). Causes gastrointestinal irritation, salivation, nausea, vomiting, vertigo, muscle weakness, anorexia, diarrhea, dyspnea, and hyperexcitability. Thirty outbreaks of poisoning were recorded in Poland between 1951 and 1963, killing 132 **animals** and 381 **birds**. **Dogs** and young **animals** are particularly susceptible.

AILANTHUS ALTISSIMA

= *Frêne Puant* = *Tree of Heaven*

Untoward effects: A toxin in the bark and root, after ingestion by **man** or **animals** leads to jaundice, and gastrointestinal and urinary tract disturbances. Its tannins, resins, and oils on its leaves have caused a dermatitis on **man**.

AJMALICINE

= *Circolene* = *Hydrosarpan* = *Isoarteril* = *Lamuran* = *py-Tetrahydroserpentine* = *Ruabasine* = *Tetrahydroserpentine* = *δ-Yohimbine*

***Antihypertensive, anti-ischemic (cerebral and peripheral).* An alkaloid in some *Rauwolfia* species and extracted from *R. serpentina*.**

Use: As an arterial vasodilator, in arterial spasms, arteritis obliterans of the lower limbs, and some paresthesias.

Untoward effects: Feeling of congestion in head with facial hyperemia, feeling of heat in legs and head, secretion of tears, feeling of drunkenness and nausea lasting 20 min, decreased systolic and diastolic blood pressure, with occasional feelings of anxiety and agitation, despite early reports of no side-effects.

LD_{50}, **mouse** and **rabbit**: 400 and 500 mg/kg, respectively; IV, LD_{50}, **rat**, **mouse**, and **rabbit**: 24, 26, and 20 mg/kg, respectively.

AJMALINE

= *Aimolin* = *Ajmalan-17,21-diol* = *Arytmal* = *Cardiorhythmine* = *Gilurital* = *Gilurytmal* = *Rauwolfine* = *Ritmos* = *Tachmalin*

Antihypertensive, antiarrhythmic.* From roots of *Rauwolfia serpentina.

Use: Oral, IV, or IM treatment of cardiac arrhythmias, extra-systoles, tachycardia, auricular fibrillation, and Wolff–Parkinson–White syndrome.

Untoward effects: Acute intoxication in **humans** reported by accidental ingestion (fatal ingestion of 100 mg in **female** of 2.25 years caused death in about 90 min) or in suicide attempts. A 57-year-old **male** ingested 1 g. Unconscious, hypotensive with serious cardiac conduction disturbances. High serum and urine levels found. Forced diuresis eliminated only 4% of the ingested dose. Recovery occurred in 21 h. A 42-year-old **male** swallowed 1.05 g and suffered from a deep coma and hypotension, lasting for several hours. Supraventricular tachycardia, changes in QRS complex, and arrhythmias followed. 27-year-old **female** ingested 1.8 g and developed right bundle-branch block, followed by ventricular tachycardia and fibrillation. Shock and anuria followed. ***Lidocaine***, ***isoprenaline***, and IV ***sodium bicarbonate*** may have helped normalize the electrodiagram 30 h after the ingestion. **Child** of 17 months ingested 23 mg/kg and experienced left bundle-branch block. Recovery after treatment with gastric lavage and forced diuresis. An 18-year-old **female** ingested 1 g and had tachycardia (106/min), slight somnolence, nausea, some vomiting, and first degree atrioventricular block. Given ***pentylenetetrazol***, IM. Electrocardiogram normal in 3 days. Liver damage and jaundice, feeling of heat and facial flushing, burning sensations in the mouth, and asthenia have been reported.

ALACHLOR

= *Alanex* = *2-Chloro-N-(2,6-diethylphenyl)-N-(methoxymethyl)-acetamide* = *CP50144* = *Lasso* = *Lazo*

***Herbicide.* Annual use in the 1980s was about 80 million lbs in the U.S.**

Use: Controlling grasses and certain broad-leaf weeds in crops, especially ***corn*** and ***soybeans***. Half-life in soil is only 7–14 days. Was sold in mixtures, such as ***Bronco***® (with ***glyphosphate isopropylamine***), ***Canon***® (with ***trifluralin***), and ***Lariat***® (with ***atrazine***).

Untoward effects: After 16 years of use, some was found in tap water of northern Ohio cities and in groundwater in Iowa and Nebraska at levels ranging from 0.2 to 2.0 ppb. In the mid 1980s, Environmental Protection Agency decided **applicators** face less of a risk than they had previously believed. Concentrated solutions may cause skin irritation and a protracted dermatitis. Accidental ingestion can cause gastrointestinal upsets.

Rats fed 15 mg/kg/day for 2 years developed nasal, thyroid, stomach, and lung tumors. The lowest effect which produced a single tumor in a group of 100 **animals** was a daily dosage that was thousands of times greater than the highest potential **human** exposure. Oncogenic in **rats** and **mice**.

LD_{50}, **rats**: 1.2 g/kg; LD_{50}, **rabbit** skin: 3.5 g/kg.

Treatment: Flush chemical from the eyes with copious amounts of clean water. Severe contamination may require the attention of an ophthalmologist. Wash exposed skin areas with soap and water. Emesis might be considered, if ingestion was considerable. ***Activated charcoal***, orally, and ***sodium sulfate*** catharsis (unless diarrhea already exists) have been suggested in the event of ingestion.

ALATROFLOXACIN

= *Trovan*

Used IV and it is rapidly hydrolyzed to *trovafloxacin* (q.v.).

Untoward effects: Severe thrombocytopenia in 54-year-old **female**. Acute liver failure reported. Dizziness, rashes, and nausea occur.

ALBENDAZOLE

= *SKF 62979* = *Valbazen* = *Zental*

Anthelmintic, flukicide.

Use: In **man**, for the elimination of ascarids, hookworms, cutaneous larva migrans, cysticercosis, and capillariasis. In **animals**, to eliminate liver flukes, tapeworms, stomach and intestinal worms.

Untoward effects: In **man**, its use causes occasional diarrhea and abdominal pain, on rare occasions leukopenia, increased serum transaminase, and failure to eliminate cysticerci. Treating a 25-year-old **female** with echinococcosis with 134 mg/kg/day/5 months led to megakaryocytic thrombocytopenic purpura, fatigue, gum bleeding, and prolonged menstrual bleeding.

Warning: Very high dosage in **animals** may be embryotoxic and teratogenic.

ALBITOCIN

***A glycoside* from certain *Albizia* species.**

Use: To induce labor and/or abortion in **women** by East African doctors. This effect was confirmed by **animal** experiments.

Untoward effects: When this spasmodic is given to **women** too early, uterine ruptures are common. The **human** uterus appears more sensitive than **animal's** to its effects. Overdoses in **rabbits** produced renal tubule necrosis, and liver necrosis in **rats** and **mice**.

ED_{50} for abortion in **rats**, IP; **mice**, IP; **guinea pigs**, IP; **rabbits**, IV; and **monkeys**, IV was 0.5 mg, 3.2 mg, 0.7 mg, 1 mg, and approximately 1 mg/kg, respectively. Oral LD_{50} was 0.8, 5.9, conflicting reports, 1.8, and approximately 2.5 mg/kg, respectively.

ALBIZIA sp.

Grows as large bushes or small trees.

Use: *A. lebbek* and *A. procera* extracts have been used as arrow poisons. In East Africa, a bark extract from *A. anthelmintica* is used as an oral anthelmintic.

Untoward effects: *A. tanganyicensis* killed 8/160 **cattle** eating its leaves; 50 others were sick. The feeding of its pods to **sheep** produced within 3 h symptoms similar to poisoning by chlorinated hydrocarbons. Hyperthermia, hypersensitivity, tetanic convulsions, and dyspnea are noted. Post-mortem examination revealed petechial hemorrhages subepicardially, SC, in the diaphragm, and in some skeletal muscles. Lungs were severely congested and edematous. It adversely affects the central nervous system, including the brain. *A. versicolor* pods produced similar pathology in **cattle**, **sheep**, and **guinea pigs**. Extracts containing ***albitocin*** (q.v.) have caused poisoning and uterine ruptures in **women**. *A. odoratissima* = ***Gadad***: is used as a piscicide. *A. stipulata Boivin* = ***Ohi***: its leaves within 3–7 weeks were toxic to **rams**. Symptoms were grunting, spasmodic bleats, icteric conjunctiva, and blood-stained feces. Lesions were found in kidneys, liver, gall-bladder, spleen, and lungs. *A. lebbek*, at an oral dosage of 150 g/kg was toxic to **rats**. Saponins from its seeds prevented ovulation in 60% of treated **rabbits**. The bark of *A. petersiana* in Tanzania is used in suicides, and, when smoked, has caused unconsciousness and death in **humans**. *A. anthelmintica* bark decoctions used as an anthelmintic and purgative have caused death in **man**.

ALBUMIN

***Albumen* is Latin for egg white.**

Use: Occasionally, as a plasma volume extender. It is responsible for about 75% of plasma's osmotic pressure. A **newborn's** serum has a relatively low level of albumin; thus, any drug competing with ***bilirubin*** for binding sites on it may displace ***bilirubin***, which then crosses the blood–brain barrier, causing serious toxic central nervous system effects. In cosmetics and in microbiology as a growth medium, and in serology.

Untoward effects: Occasional hypersensitivity reactions, including angioneurotic edema. Potential source of ***aluminum***, after large amounts are given IV. Interaction and its binding of drugs may cause kernicterus. Febrile reactions, increased alkaline phosphatase levels, chills, nausea, urticaria, and variable effects on blood pressure and heart rate in **humans** also noted. A 56-year-old **female** had respiratory distress, facial flushing, and a severe decrease in responsiveness after a 50 g and a 25 g infusion. 70-year-old **male** received 1800 g IV over a 72 h period for refractory ascites, developed acute renal failure. Many drugs interfere with albumin tests.

Warning: IV usage rate should be very **slow**.

ALBUTEROL

= *Aerolin* = *AH 3365* = *Asmaven* = *Broncovaleas* = *Cetsim* = *Cobutolin* = *Ecovent* = *Proventil* = *Salbulin* = *Salbumol* = *Salbutamol* = *Salbutine* = *Salbuvent* = *Sultanol* = *Venetlin* = *Ventodisks* = *Ventolin* = *Volmax*

***Bronchodilator, tocolytic.* A β_1 and strong β_2 agonist. The sulfate is usually used.**

Use: As a bronchodilator in bronchospasm and asthma.

Untoward effects: **Patients** have reported an incidence of nervousness and tremor (20%), headache (7%), muscle cramps and pain (3%), nausea (2%), dizziness (2%), insomnia, weakness, hypokalemia, hypertension, angina, drowsiness, restlessness, irritability, vomiting, peculiar taste sensation, dry or irritated oropharnyx, facial flushing, tinnitus, chest discomfort, and difficulty in urination. Large doses have caused systemic vasodilation and compensatory tachycardia (5%), palpitations, tremors, as well as hypokalemia. Glaucoma and hypersensitivity reactions are rare. Paradoxical bronchospasm has occurred in **patients** inhaling it, apparently as a response to its irritant effect.

IV, TD_{LO}, **human**: 6 µg/kg.

May increase ocular pressure and cause myocardial ischemia, tachycardia, tachypnea, vomiting, anxiety or depression, and hyperglycemia in **dogs**; leiomyomas in **rats** (but not **mice** or **hamsters**) and teratogenic effects in **mice**. Treatment of pregnant **mice** produced cleft palates in fetuses (5/11 (45%) at 0.25 mg/kg and 10/108 (9.3%) at 2.5 mg/kg). Offspring of **rabbits** receiving 50 mg/kg had cranioschisis in 7/19 (37%).

LD_{50}, **rats** and **mice**: > 2 g/kg.

Warning: Use cautiously in **humans** with coronary insufficiency, hypertension, hyperthyroidism, diabetes mellitis, ketoacidosis, and those hypersensitive to sympathomimetic amines or taking ***monoamine oxidase inhibitors*** or ***tricyclic antidepressants***. The latter two increase albuterol's cardiovascular effects. May cause QT interval prolongation.

ALCHORNEA CORDIFOLIA

= *Christmas Bush* = *Kimvuzi-mvuzi* = *Makaya ma ngu* = *Wambunzila*

***Botanical*. Contains unidentified alkaloids of the indole type.**

Use: In many African treatments, the seed is often used for assorted ailments and as an aphrodisiac.

Untoward effects: Has a hallucinogenic principle. The leaves decrease motor activity, dilate the pupils, blanch the ears, and can cause death in **people** within 24 h.

ALCLOFENAC

= *Allopytine* = *Allopydin* = *[4-(Allyloxy)-3-chlorophenyl]acetic Acid* = *Argun* = *Epinal* = *Epnial* = *Medifenac* = *Mervan* = *Neoston* = *Prinalgin* = *Ruefenac* = *W 7320* = *Zumaril*

***Anti-inflammatory, antipyretic, analgesic*. Absorption is very variable.**

Use: In arthritis and rheumatism.

Untoward effects: Gastric discomfort, nausea, vomiting, headache, dizziness, diarrhea, transient morbilliform rash, widespread cutaneous vasculitis, and papillary necrosis have occurred.

LD_{50}, **mice** and **rats**: 1100 and 1050 mg/kg; SC, 600 and 630 mg/kg; IP, 550 and 530 mg/kg, respectively.

ALCOHOL, ALLYL

= *2-Propen-1-ol* = *1-Propenol-3* = *Vinyl Carbinol*

Clear liquid with sharp, pungent odor. Flammable and strongly reactive.

Use: In manufacture of allyl compounds, resins, plasticizers, war gases, as a weed killer in ***celery*** and ***tobacco*** seedbeds, as well as nursery plantings.

Untoward effects: TLV, **human**: skin, 2 ppm. Severe irritation to mucous membranes (eyes, nose, throat, and lungs) at 25 ppm. Absorbed through **human** skin, causing severe burns and damage to internal organs. Harmful effects may be delayed. Central nervous system depressant and causes hematuria and nephritis. ***Ingestion*** leads to gastrointestinal irritation with nausea and vomiting. Absorbed readily into **human** callus (soles of feet). Causes hepatitis and liver necrosis in **rats** and **mice**. Toxic to **chick** embryos leading to decreased hatch and corneal opacities.

LD_{LO}, **human**: 50 mg/kg; LD_{50}, **rat**, **mouse**: 64 mg/kg and 96 mg/kg, respectively; LD_{LO}, 5 mg/kg, 53 mg/kg, respectively. ***Inhalation*** LC_{LO}, **human**, **monkey**, **rabbit**: 25 ppm, 1000 ppm/4 h, and 1000 ppm, respectively. Skin LD_{50}, **rabbit**: 53 mg/kg.

Treatment: Act quickly – provide first aid – get medical assistance. Move **victim** to fresh air – give artificial respiration, if not breathing – use oxygen, if necessary. Wash exposed eyes instantly with lots of water; soap and water for skin contact. Remove contaminated clothing. If swallowed, induce vomiting.

Prevention: Use approved respirator and wear full-protective clothing, if there is danger of breathing vapors. Ventilate the area. Avoid sparks, flames, or heat.

ALCOHOL, *n*-AMYL

In human literature erroneously referred to as *Fusel Oil*, a by-product of fermentations in making *ethyl alcohol*. Commercial amyl alcohol is made by refining *fusel oil*.

Use: In organic synthesis; a solvent.

Untoward effects: More toxic than ***ethyl alcohol***. Causes an allergic contact dermatitis in **man**. It is narcotic and irritating to eyes, nose, throat, and lungs (also causes lung tumors in **mice**). Coma and dark urine due to the presence of ***methemoglobin***. Flammable. LD_{LO}, **human**: 500 mg/kg.

LD_{50}, **rats**: 3.03 g/kg; **mouse**: 200 mg/kg.

ALCOHOL, BENZYL

= *Benzenemethanol* = *Phenylcarbinol* = *Phenylmethanol* = *α-Hydroxytoluene*

Flammable. Over 6000 lbs used annually.

Use: In manufacturing; solvent and degreasing agent; in microscopy; preservative. In cough syrups, ball point inks, dermatological aerosol sprays, and ophthalmic solutions.

Untoward effects: A 5-year-old **female** treated with an IV infusion of ***diazepam*** up to 2.4 mg/kg/h/36 h for status epilepticus developed symptoms of benzyl alcohol poisoning, and died 10 days later. **Her** hypernatremia, hypotension, and metabolic acidosis were attributed to **her** receiving 5.75 mmol/kg/day of ***sodium*** and 180 mg/kg/day of benzyl alcohol. Responsible for "gasping syndrome" of premature **infants** who received parenteral solutions containing it (99 mg and 234 mg/kg daily. 0.5 ml/kg/day was considered safe in **adults**). Similar reports on a **colt**, **cats**, **dogs**, **calves**, and **monkeys**, with sudden deterioration of their condition, salivation, hyperactivity, and convulsions, usually ending in death. Contracture of **infants'** gluteus maximus muscle caused by a 2% level in ***penicillin G***. Mildly irritating in **human** patch tests. Hemolysis has also been reported. Narcotic action of low toxicity, but high exposure can depress blood pressure and the central nervous system, with death from respiratory failure. Vapors can penetrate intact skin and levels above 1 ppm cause lacrimation in **man**. Vapors above 100 ppm caused deaths in laboratory **animals**. Allergic reactions, including fever, maculopapular rash, and thrombocytopenia reported in **humans** from its presence in a sunscreen and various parenterals. Cross-sensitization to ***Peru Balsam*** noted in **humans**. **Dogs** evidenced ataxia, dyspnea, mydriasis, nystagmus, diarrhea, and urination. It is normally oxidized to ***benzoic acid*** and conjugated with ***glycine*** in the liver. Since this metabolic pathway is not fully developed in **infants**, **cats**, and the young of most species, toxic ***benzoic acid*** (q.v.) accumulates, which can cause ataxia, hyperesthesia, dyspnea, muscle fasciculations, depression, and death.

Oral LD_{LO}, **human**: 500 mg/kg.

Oral LD_{50}, **rats**, **rabbits**: 1.23 and 1.04 g/kg, respectively; **mice**: 1.58 g/kg. IV, LD_{50}, **mice**: 0.48 mg/kg; IP, LD_{50}, **rats** and **guinea pigs**: 400 and 800 mg/kg, respectively.

ALCOHOL, *n*-BUTYL

= *1-Butanol* = *n-Butanol* = *Clotol* = *Klot, Stainless*

***Solvent, antifoam.* Of little pharmacological significance, but 0.0028 ml has been found in 4 ml of *whiskey*. Flammable. 1.4 billion lbs annual demand.**

Use*:** As a solvent for many fats, gums, resins, shellacs, varnishes, nail polishes, nail polish removers, and in the manufacture of electronics, lacquers, detergents, rayons, and butyl compounds. In microscopy. Mild, systemic coagulant. Octane-enhancer in low-lead gasolines. As a denaturant (16 oz ***tert-butyl alcohol/100 gal ***ethyl alcohol***) for use in cosmetics and as a moth-proofing agent.

***Untoward effects*:** Irritant to mucous membranes (eyes, nose, and throat). Neurotoxic. Causes contact dermatitis, headache, drowsiness, blurred vision, dizziness, unconsciousness, and vomiting in **man**. Percutaneous absorption occurs. Rarely, a substance of abuse. Chronic exposure leads to pancreatic, lung, kidney, and liver damage. TLV for all butyl alcohols is 100–150 ppm (50 for skin). 127 ppm for **man** is uncomfortable, but tolerable. 1268 ppm was intolerable and caused a 50% decrease in respiratory rate.

Moderate to severe eye irritant in **rabbits**. Toxic to **chick** and **mallard duck** embryos in the form of decreased hatch, corneal opacities, cataracts, cleft palate, nerve and kidney damage.

LD_{50}, **rats**: 0.79–4.36 g/kg.

***Treatment*:** Move **victim** to fresh air. Artificial respiration and oxygen, if necessary. Wash eyes with lots of water, skin with soap and water. Induce vomiting, if **ingested**. Remove contaminated clothing. Get medical help.

***Prevention*:** Wear suitable gloves and goggles. Ventilate work areas. Wear suitable respirator in areas of spills. Eliminate all sparks, flames, and ignition sources.

ALCOHOL, CETYL

= *Cetyl Alcohol* = *Cetyl Stearyl Alcohol* = *Ethal* = *Ethol* = *1-Hexadecanol* = *Hexadecyl Alcohol* = *Palmityl Alcohol*

Made from sperm whale oil or vegetable oils.

***Use*:** As an emollient and emulsifier in cosmetics, pharmaceuticals, and in a synthetic surfactant for neonatal respiratory distress problems. In the manufacture of quaternary ammonium compounds. Promotes adhesion of face powders and adds softness and velvety feeling to cleansing and shaving creams.

***Untoward effects*:** Contact urticaria from topical medicaments is its most common adverse effect in **man**. A **woman** had an oozing dermatitis on two seperate occasions, 2 years apart, after contacting a treated green canvas sewn into tents. The cause was confirmed by patch testing. Despite early reports to the contrary, it was found to be a moderate to severe skin irritant in tests on **humans**, **rabbits**, **rats**, and **guinea pigs**.

ALCOHOL, ETHYL

= *Aqua ardens* = *Ethanol* = *Ethyl Alcohol* = *Absolute Alcohol* = *Ethyl Hydrate* = *Ethyl Hydroxide*

Solvent, antiseptic, antidote, nutrient, anesthetic, vasodilator.

Flammable. It is known that alcoholic beverages have been used for about 5000 years. Noah took *beer* and *wine* with him on the Ark. It is the most widely used drug for non-medical purposes. Annual production and consumption was over 2 billion gallons by the end of 1999, with most of it for industrial use. It is absorbed directly from the stomach (30%) and intestines (70%) within 10–90 min. 90% of it is metabolized to *acetaldehyde* to *acetic acid* to *carbon dioxide* and water. 20 ± ml of *whiskey* is metabolized per hour.

Use*:** As an antidote in ***ethylene glycol (q.v.) and ***methanol*** (q.v.) poisoning. In paint thinners, model cement, nail polishes, mouthwashes (7–27%; up to 70% in some concentrated solutions), aerosol sprays, pigment inks (0–45%), lacquers, perfumes, toiletries, detergents, plastics, hairsprays, medicinal extracts, tonics, liniments, antiseptics, and gasoline.

Untoward effects*:** **Human**: A subject for textbooks. It is a potent pharmacological agent and too little time is spent in most medical schools on its adverse effects and drug interactions. It is both a nutrient and poison. Maimonides' admonition to do everything in moderation applies well to alcohol consumption, to avoid toxicity to the liver, brain, gastrointestinal tract, hematopoietic system, pancreas, nerves, heart, fetus, etc. Large amounts of ***histamine in red ***wines*** and sherries are responsible for some of the headaches and flushing. White ***wines***, ***beer***, and port ***wine***

have intermediate amounts of ***histamine***; distilled ***whiskeys*** and cognac usually have none. Anorexia, nausea, vomiting, abdominal pain, increased levels of serum ***cholesterol***, triglycerides, and ***uric acid***, as well as hepatic and pancreatic dysfunction and hyperuricemia are reported. Common cause of diffuse burning feet. It can induce fainting spells by dilating blood vessels, interfering with normal reflex that contracts them when suddenly standing up. **Patients** with Hodgkin's disease show an increased sensitivity to alcohol. A lethal level after ingestion is about 0.4 g%. Lethality by **inhalation** would be rare, since exposure to 8000 ppm for 6 h only produced blood levels of 0.05 g%. **Inhalation** of 6000–9000 ppm has produced drowsiness. Blood alcohol levels of 0.03–0.12% usually produce mild euphoria; 0.09–0.25% leads to excitement and emotional instability; 0.18–0.3% leads to confusion and disorientation; 0.27–0.4% leads to stupor and apathy. ***Folate*** and ***thiamin*** absorption are decreased in alcoholism. Although a central nervous system depressant, it is often followed by rebound excitation, interfering with sleep.

Symptoms of alcohol intoxication in a 57-year-old **female** after ***paclitaxel*** infusions causing a blood alcohol concentration of 0.098%. A toxic neuropathy developed in 16 **patients** who drank alcohol contaminated with ***tricresyl phosphate*** (***tritolylphosphate***, q.v.). Another severe case of organophosphate poisoning (coma) occurred in a 52-year-old **male**, white, non-agricultural **worker**, who, while drunk, cleaned out the inside of a metal drum that had contained highly toxic ***dicrotophos*** in a ***turpentine*** vehicle. The **patient** survived and was discharged from the hospital 33 days later. Splenectomy may decrease tolerance to alcohol. A **pharmacist** who knowingly sold ***Sterno***® (a ***methanol*** and ***ethanol*** mixture) to **people** known to extract the alcohol for drinking purposes was held liable for their deaths. Despite obvious intoxication, emergency room personnel should be alert not only for drug interaction, but also for other undiagnosed conditions. Eye irritation in **man** from exposure to 20 ppm.

Addiction and abuse: The cause of many accidents (driving, flying, home injuries, tavern fights, etc.). Despite no definitive therapy, a wide variety of drugs has been used for treatment. Some older ones with varying degrees of success and side-effects were ***amitriptyline***, ***carbamazepine***, ***chlordiazepoxide***, ***chlorpromazine***, ***citrated calcium carbimide***, ***clomethiazole***, ***clorazepate***, ***coprin***, ***dextroamphetamine***, ***diazepam***, ***disulfiram***, ***fructose*** infusions, ***hydroxyzine***, ***lithium carbonate***, ***meprobamate***, ***metronidazole***, ***naltrexene***, ***paraldehyde***, ***pentobarbital***, ***perphenazine***, ***peyote***, ***promazine***, ***propranolol***, IV ***triiodothyronine***, and ***thiamine***. Some of them are used in withdrawal problems, along with psychotherapy and placebos.

Blood and bleeding: Significantly associated with gastrointestinal bleeding. Megaloblastic anemia, red blood cell development inhibited, due to blocking of ***folic acid*** needed in erythropoiesis. Hemolytic anemia with hepatomegaly, fatal brain hemorrhage, hemorrhagic erosive gastritis, hemorrhagic pericardial effusions, pancytopenia, thrombocytopenia, transient stomatocytosis, acanthocytosis, thrombosis, and phlebitis have occurred in chronic alcoholism. Paradoxically, a **man** with a long history of high sherry intake had a raised hemoglobin concentration and decreased plasma volume in a syndrome known as stress erythrocytosis. Vacuolation of bone marrow pronormoblasts appears to be dose-related. Hemorrhagic tendency in **alcoholics** is often secondary to folate deficiency, liver disease, and lack of ***vitamin K***-dependent factors and impaired platelet function. After ingesting 100–150 ml of ***gin***, a 38-year-old **male** Ghanian developed a severe sickle cell crisis within 12–36 h. Numerous previous crises reported with ***gin*** or palm ***wine***, but not with other ***wines*** or ***beer***. Implication is that gin's ***juniper oil*** or its metabolite was the actual causative agent. Leukopenia, acidosis, and an increase in blood triglycerides in many **alcoholics**; increase in high-density lipoprotein in moderate **drinkers**. Chronic alcoholism can also produce hypertension, and has caused orthostatic hypotension. Hemolysis from rapid IV infusions can be minimized by using a ***sodium lactate*** diluent, rather than physiological saline.

Bone: Long-term abuse of alcohol is detrimental to skeletal integrity, with decreased bone mass and mineralization, arthropathy, and necrosis of femoral and humeral heads. One report indicated social drinking was associated with increased bone mineral density in **men** and **women**.

Brain and neuropathies: One-third of hospitalized alcoholic **patients** had markedly enlarged cerebral ventricles on CT scans. 37 young, chronic **alcoholics** (usually 100–400 g/day) under the age of 35 had reductions in density of their left hemispheres. Moderate consumption over a 20–30 years period of time can damage the cerebellum. It is an important and underrecognized risk factor for strokes. Excessive alcohol intake has been associated with a reversible dementia, electroencephalogram problems, impaired gait and mental performance. Occasionally intoxication carries an increased risk of primary subarachnoid hemorrhage (2–3 times in **men**, 2–13 times in **women**). Most **alcoholics** are hyperhydrated leading to cerebral edema inside a fixed cage, causing blurred vision, throbbing headaches, nausea, and vomiting; hypoxia, due to compression on the brain's arteries and veins; and amnesia. This impairs brain function. In the early phases of a drinking binge, the body selectively excretes ***magnesium***, which causes hyperirritability of brain tissue leading to tremulousness, hallucinations, convulsions, and seizures.

Intoxication causes delayed reaction times, motor skills and coordination deteriorate, and, with central nervous system depression, stupor or coma may occur. Chronic vagal damage can occur (even vocal cord paralysis) with alcoholic polyneuropathy. Intrathecal injection has caused leg muscle paralysis, and loss of bladder and bowel sensations. Ingestion by **mothers** of breast-fed **infants** can cause the latter's impaired neurological development.

Cardiovascular*:** Arrythmias ("holiday heart syndrome"), cardiomyopathies (in addition to that caused by ***cobalt (q.v.) in ***beer***), strokes, myocardial infarction, increased heart rate and size, myocardial depression, vasodilation, Prinzmetal variant angina, alcoholic beriberi, atrial fibrillation, and development of hypertension are commonly reported in alcoholism. These adverse findings outweigh the benefits of increased high-density lipoprotein (on limited consumption). A large amount imbibed in a short period of time can cause myocarditis and weakened contractions of cardiac muscle. Sudden death reported in two markedly intoxicated **men**, probably due to potent vasodilating effect and decrease in coronary artery blood flow. Limited intake in **women** associated with decreased risk of ischemic strokes. Ethanol can cause a ***magnesium*** deficiency with its secondary cardiac effects. Overdoses can cause sinus tachycardia.

Crime*:** Involved in many serious crimes (assaults, robberies, rapes, etc.). In New York City, well-to-do **visitors** were found disoriented, agitated, with acutely distended bladders, and in various stages of undress. They had been given alcoholic drinks spiked with ***scopolamine eye drops by their "hosts" and "hostesses", before being robbed. Alcohol is involved in well over 50% of the homicides and arrests.

(Pseudo)Cushing's syndrome*:** Excessive alcohol ingestion has caused corticosteroid hypersecretion, that led to a misdiagnosis of Cushing's syndrome. ***Spironolactone therapy in an **alcoholic** also led to this false conclusion. A history of alcoholic intake should be ascertained, before making a diagnosis of Cushing's syndrome.

***Depression*:** Can be both a cause of and a result of depression.

Eye*:** May decrease intraocular pressure and mask glaucoma. This is partially due to alcohol suppressing circulating antidiuretic hormone. May be a factor in amblyopia and in sight-threatening diabetic retinopathy. Tests on **pilots** indicated it adversely affects oculomotor control, manifested as degraded vestibular nystagmus and pursuit eye movements. Chronic **alcoholics** on ***disulfiram therapy had optic neuritis that usually disappeared with discontinuance of drug therapy. In some, slight vision impairment persisted, along with some pallor of the optic disk. Miosis was noted in about ⅓ of **children** who had accidentally abused it. Substances in various distilled beverages, other than alcohol, may influence the degree of nystagmus shown. ***Injections*** of alcohol around the infraorbital nerve, to control tic douloureux, have resulted in blindness within 1 h in 60-year-old **male**, due to occlusion of central retinal artery. Total temporary corneal anesthesia with six cases of superficial keratitis in 15/111 **patients** injected 1–8 years previously, for tic douloureux. 2% reported mild trigeminal pain. Upon ***injection*** of the trigeminal ganglion in 97 **patients**, ages 30–39, 20% of them evidenced some neurotropic keratitis, and 5% had permanent corneal changes and decreased visual acuity. This can even be precipitated years later by trauma. A case of herpes zoster ophthalmicus is reported, where similar ***injection*** may have reactivated the varicella-zoster virus. Subconjunctival ***injections*** of ***framycetin sulfate*** in an 11-day-old **infant** apparently reacted with absolute alcohol, used to help remove the epithelium, and precipitated, leading to temporary corneal opacity, clearing an hour later.

Fetus and neonate*:** IV alcohol passes rapidly through the placenta and has been found in the bloodstream of the **newborn** within 1 min. It has caused neonatal depression in an **infant** born to a 30-year-old **female**, who consumed less than 50 ml of ***whiskey, 30 min before delivery. Cord levels are 20% below maternal levels, after slow IV administration. Vomiting and drowsiness occur in nursing **infants** of lactating alcoholic **mothers**. Milk ethanol levels are approximately the same as plasma's. A lactating **mother** drank 750 cc of port ***wine*** in a 24 h period, and her 8-day-old **infant** went into a deep, unrousable sleep with snoring, deep respirations, no reaction to pain, and a feeble pulse. Doses of 1 g/kg can inhibit milk ejection reflex, and doses above 2 g/kg usually completely block **suckling**-induced ***oxytocin*** release. IV alcohol to a premature **infant** associated with hypoglycemia and abnormal bone marrow morphology. Intoxication and/or abuse in pregnant **women** produced withdrawal syndromes, convulsions, hyperactivity, crying, irritability, low birth weight, reluctance to nurse in **newborns**, severe respiratory insufficiency, prenatal and postnatal growth deficits with delays in motor and intellectual development. One survey indicated a perinatal mortality of 17%. Craniofacial (short palpebral fissures, maxillary hypoplasia with prognathism) and limb defects, ophthalmic malformation, hepatic dysfunction, and cardiovascular defects are common in the fetal alcohol syndrome. Also involved are ear, nose, renogenital defects, and hernias. Pregnant alcoholic **women** have low plasma zinc levels (50.7 versus 72.2 μg/dl in controls), and may have more miscarriages. **Babies** exposed to it may fall asleep rapidly, but sleep less and for shorter periods. Breast-feeding **mothers** should allow at least 4 h between drinking and nursing. Prolonged contact (~4 h) during synaptogenesis (brain growth spurt – sixth month of gestation and for 2 years after birth) suspected of causing death of many brain cells in the **human**, based on **rat** studies.

Gastrointestinal: Increases lower esophageal sphincter pressure or reflux. May cause nausea, vomiting, gastric mucosal irritation with ulceration, hemorrhage, increased motility, diarrhea, and increased production of gastric juice and acidity. Alcohol is frequently the cause of unexplained diarrhea. Despite a 5% alcohol content, fresh palm ***wine*** from *Rafia hookeri* was not ulcerogenic.

Glucose tolerance: Significant increase in blood ***glucose*** within 1 h in young African **males**. Moderate doses significantly impaired ***glucose*** tolerance in healthy, young English **women**, attributed to impaired ***glucose*** uptake by tissues. Ethanol delays or decreases ***insulin*** secretion.

Gout: May be a symptom of alcohol abuse. Alcoholic beverage intake associated in many individuals with increased plasma ***uric acid*** levels and symptoms of gout. An exacerbation of gout was attributed to its content in IV ***nitroglycerin*** in four **patients**, 55–69 years.

Gynecomastia, libido: Gynecomastia and feminization can occur in **men** with chronic liver disease and in **male alcoholics** with only minimal liver damage. Incidence of sexual impotence about 8% (17,000 **alcoholics** studied for 37 years), and it persisted in about 50% of them, even after alcohol use was discontinued. This is probably due to alcohol's destructive effect on the neurogenic reflex that helps produce an erection. Alcohol, as a severe depressant, may also contribute to premature ejaculation and impotence, despite the belief of some **persons** that it increases their sexual desire. It depresses the central nervous system and levels of circulating testosterone, and triggers production of high levels of leutinizing hormone (LH) produced in the pituitary. The combination can produce heightened desire and decreased sexual performance.

Hallucinosis: The exact causes are unkown. One is that ***acetaldehyde***, the major metabolite formed from the metabolism of alcohol, reacts with catecholamines, such as ***epinephrine***, to produce alkaloids. *In vitro* studies show these to be related to ***mescaline***, a known hallucinogen. Alcohol withdrawal has been associated with hallucinations. Also see ***Brain and neuropathies***.

Headaches: Although associated with the intake of alcoholic beverages, it can occur for other reasons, such as ***histamine*** in red ***wine*** (2–3 µg/ml). The relationship of cerebral edema to throbbing headaches is discussed under ***Brain and neuropathies*** (q.v.).

Hearing: Lack of attention and decreased understanding of numbers, words, and facts can easily be misinterpreted as hearing failure. A temporary unilateral sensorineural hearing loss, with tinnitus in a 26-year-old **male**, after drinking 1 pt of tequila in 4 h. Spontaneous recovery after several weeks.

Hypersensitivity: Usually, in **adults** only. Anaphylactic reaction to pure ethanol in 22-year-old **female** and an immediate allergy to a test with ***acetic acid***. Other reports include acute urticaria (36-year-old **male**); edema of lips, facial flushing, itching and sensation of shrinking dentures (76-year-old **female**) on three occasions, after ***beer***, ***sherry***, and a drug with alcohol. Challenge with alcohol produced systemic reaction. A 42-year-old **female** became extremely tired and wept pathetically after a single drink. Another **patient** reacted with complete loss of memory and mood swings from silliness and restlessness to withdrawal.

Hypertension: Its induced hypertension can be suppressed by oral administration of ***dexamethasone***.

Hypoglycemia: Numerous reports in **adults**, including some who died, despite therapy with IV ***glucose***. ***Insulin***-dependent **patients** on a drinking binge are at great risk. Young **children** are not immune. Toxicity occurred after their drinking flavored mouthwashes (16% alcohol), ***wine***, ***gin***, ***sherry***, ***brandy***, etc. Acute alcohol intoxication (blood alcohol 220 mg/100 ml) by ***inhalation***, and severe hypoglycemia (blood ***glucose*** 22 mg/100 ml) in 6 month **infant**, after sponging with 750 ml over 13 h, in treating a persistent fever. Cooling with water is more comfortable and less hazardous. Fasting and treadmill running, and alcohol intake predispose to hypoglycemia.

Infections: Leukopenia may contribute to increased susceptibility. Quick alcohol preparation of a wound before a parenteral injection may not be adequate prophylaxis against infections, such as *Clostridium welchii*.

See ***Untoward effects – Respiratory system***

Lipid metabolism: One study of 3806 **patients** indicated a strong relationship between alcohol consumption and increased high-density lipoprotein. Hypertriglyceridemia occurs with light alcoholic intake.

Liver: One of the organs most seriously affected by chronic alcoholism. Its damage to liver cells causes further decrease in alcohol oxidation rate, rendering the liver vulnerable to further alcohol-induced injury. Liver disease is not always produced by alcohol alone, but often in conjunction with dietary factors and individual susceptibility. **Alcoholics** decrease their intake of protein, vitamin, and mineral-containing foods, adding to direct hepatotoxicity. Mitochondrial alterations, focal cytoplasmic degradation, steatosis, and cirrhosis are sequelae to continued use. It is claimed that, although some may be **alcoholics**, **Jews** and hairy-chested **males**, supposedly, are not particularly prone to cirrhosis. It is a cause of acute hepatic porphyrias. Surreptitious ***ingestion*** of a mouthwash preparation (25% ethanol) in 48-year-old **female** precipitated acute attacks of porphyria. **Bantus** consume alcoholic beverages with a high ***iron*** content and have hemochromatosis. ***Wines*** and commercial and illicit liquors contain ***iron*** and have been a cause of hemochromatosis. Hepatic coma, an advanced stage of alcoholism, is accompanied by portal hypertension and ascites. Alcoholic hepatitis symptoms can resemble those of liver cancer. A 48 year **patient** consumed 170 g/day ethanol from a Lysol® spray disinfectant and suffered from ascites, secondary to cirrhosis. It slows the release of liver ***glycogen***.

Malnutrition*:** After 2 weeks of intake, the intestines in about 16% of **people** go into a secretory state, flushing vitamins and minerals out of the body. Couple this with decreased dietary intake of nutrients (see ***Liver above) and disaster may result, even in the form of neuropathies. Of the vitamin deficiencies, ***folic acid***, ***pyridoxine***, ***thiamin***, and ***vitamin A***, and ***vitamin*** B_{12} are the most critical.

***Menstruation*:** Amenorrhea reported in some **women**, after heavy drinking for 3–7 months (360–540 ml hard liquor/day). Menstruation resumed after several months of sobriety.

***Muscle*:** Myopathy, especially in chronic **alcoholics**, is manifested by weakness, myoglobinuria, rhabdomyolysis, increased serum creatine phosphokinase, muscle cramps, tenderness, and wasting, and even total incapacitation. A 43-year-old **male** with history of a 26 year drinking spree had alcoholic myopathy with acute urinary retention.

Neoplasms*:** Excessive use has been associated with squamous cell carcinomas of the mouth, pharynx, larynx, esophagus, lung, and liver. *N*-nitroso group compounds are found in some alcoholic beverages, and may contribute to their carcinogenicity. Conflicting summations exist in surveys of thousands of **women**, as to the association of alcohol intake with breast cancer. Frequent use of high alcohol content mouthwashes has been associated with oral cancers. Eosinophilic leukemia was reported in an ethanol intolerant **patient**. See ***Untoward effects – Pancreas.

Pain*:** Cervical and inguinal node lymphalgia in 35-year-old **male** with early syphilis, after ***ingestion; neck pain and swelling in 18-year-old **female** with secondary syphilis noted, after consumption of any type of alcoholic beverage. Alcohol abuse by 55-year-old **male** produced a reversible, rare anterior tibial painful foot drop. Unacceptable high incidence of pain with 7–10% on IV infusions. Some **patients** with Hodgkin's disease find that their **tumors** hurt, after a sip of alcohol.

***Pancreas*:** Cause of a high incidence of pancreatitis (207/346 in one report) and has even been reported in a 15-year-old **male**. Chronic alcoholism history in 75% of pancreatic carcinomas of 83 **patients**. Increased tone of sphincter of Oddi and pancreatic duct obstruction directly related to ethanol blood levels.

Poisoning*:** Alcohol toxicity or poisoning (accidental, "recreational", or suicidal) has many symptoms similar to those in ***lead, ***phencyclidine***, and ***methaqualone*** poisoning. Chronic ***ingestion*** of untaxed ***whiskey*** ("moonshine") was a double-barreled poison – alcohol **and** ***lead*** poisoning. English upper classes, in the seventeenth and eighteenth centuries, drank large quantities of Portuguese port ***wine***, containing high levels of ***lead***. They often suffered from gout, whereas **commoners** drank mostly ***gin***, and did not suffer from gout. Transplacental passage of ***lead***, with toxicity to a **fetus** occurred from a pregnant **mother's** ingestion of moonshine ***whiskey***. Alcohol can be particularly toxic (coma, respiratory arrest, cardiovascular collapse, and even death) if drunk after fasting or ingesting it too rapidly. One pint of hard liquor (80 proof) is considered the LD_{50} for a 70 kg **adult**. A lethal dose (coma) occurs when blood ethanol levels reach 4000 mg/l. A number of over-the-counter cough medicines and mouthwashes with high alcohol content (> 25%) have caused "poor **man's** high". ***Tyramine*** is frequently found in many ***wines***. **Children** under 5 years of age are commonly involved in accidental poisoning (averaging 441–717 deaths annually in the U.S. from 1955 to 1964). Compresses or sponging with liquor or alcohol have caused fatal intoxication. Make sure alcoholic beverages are stored out of a **child's** reach, and that, after parties, all glasses are emptied. A half ounce of alcohol can be a serious poison to a **child**. If checked within 2 h of ***ingestion***, and blood levels do not exceed 50 mg/dl, ***ipecac*** is often given. ***Treatment*** varies greatly and includes some of the following: ***glucose*** IV for hypoglycemia (especially in **children**), fructose IV, hemo- and/or peritoneal dialysis, ***thiamin***, IV electrolytes, ***sodium bicarbonate*** for acidosis, ***naloxone*** for coma (reversed within 10 min in 15% of **patients** – up to 1.2 mg/injection), ***physostigmine*** for delirium tremens, etc.

Prostatitis*:** Irritant to posterior urethra and prostate. Alcohol cogeners, such as ***juniper berry, ***hops***, and ***malt***, may also be irritants and cause edema.

Renal*:** Although alcohol increases urine output, it inhibits the release of ***vasopressin. Illegal moonshine, containing lead, has caused renal pathology, as has legal ethanol.

***Respiratory system*:** Associated with respiratory depression and sleep apnea. Bronchoconstriction may be due to congeners in the alcoholic beverages, but can be due to the alcohol itself, especially in **asthmatics**. Acute pneumonia can occur in chronic alcoholism, as well as aspiration pneumonia. **Diabetics** may occasionally have alcoholic-like breath.

Skin: Porphyria cutanea tarda has been precipitated by excessive alcohol consumption. Generalized red blotchiness of skin, diaphoresis, facial flushing, purpura, increased skin and liver porphyrins, and pruritus from alcohol consumption. Scurvy, with classical perifollicular hemorrhages surrounded by corkscrew hairs, is frequently seen in **alcoholics** at emergency rooms. The cutaneous vascular reaction to topical alcohol in **Orientals** is provoked in large part by its aldehyde. Many **Japanese**, **Taiwanese**, and **Koreans** are particularly sensitive to marked facial flushing and mild to moderate symptoms of intoxication after limited alcohol ***ingestion***. ***Inhalation*** by exposure to its fumes from the manufacture of gun cotton produced typical symptoms of chronic alcoholism. Chemical burns have occurred from its use under pneumatic tourniquets in **neonates**. Severe burns from fire have occurred on a hand wrapped in gauze soaked in an alcoholic antiseptic, after the **patient** lit a ***cigarette***. Common cause of diffuse burning feet.

Sleep: Increases sleep apnea and can induce night sweats.

Temperature: Body temperature is frequently decreased by vasodilation. Severe hypothermia has occurred, due to combination of psychotropic drugs and alcohol.

Sheep and **goats** develop a taste for and a quick tolerance for alcoholic beverages, consuming 6–8 oz of ***brandy*** without serious effects. **Ducks**, **chickens**, and **parrots** will readily accept it after a few attempts. 8 oz of strong alcohol killed a **horse** within 10 min. 4–5 oz of 90 proof alcohol, if retained, kills a 20 lb **dog** in a few minutes. In general, 8 g/kg is considered potentially lethal for most **animals**. In acute poisoning, excitement is followed rapidly by collapse, coma, and death. **Animal** intoxication often occurs from excessive intake of distillery or brewery wastes, or slops from restaurants and taverns. Some **dogs** beg for ***beer***, and, like **man**, become ataxic. Neurological damage reported in very young **rats**. Induces gastric ulcers in **rats**.

Oral LD_{50}, **rat**, **rabbit**, and **guinea pig**: 14 g/kg, 6.3 g/kg, and 5.56–9.5 g/kg, respectively. Oral LD_{LO}, **human**, **men**, **women**, **dog**, and **cat**: 500 mg/kg, 1.43–50 mg/kg, 256 g/kg, 5.5 g/kg, and 6 g/kg, respectively. TWA, 1,000 ppm.

Drug interactions: Since alcohol is one of the most frequently consumed drugs by many, it is not surprising that such a pharmacologically active drug would be involved in many drug interactions. This helps complicate life for emergency room personnel. The possibility of ethanol interaction should be suspected in any drug-intoxicated **person**. It also influences gastric motility and absorption of drugs. The **elderly** are particularly susceptible to such interactions, perhaps because they use so many prescription drugs, antihistamines, analgesics, hypnotics, and tranquilizers. Confusion and altered sleep patterns are typical adverse effects. Unusual, unpredictable, and, perhaps, even humorous, drug interactions occur, such as black smoke in a **person's** mouth from a reaction between a capsule of ***Carbosylane*** (***simethicone*** and ***activated charcoal***) and ***Kisses***®, a breath deodorant containing ethyl alcohol, ***glycerin***, ***saccharin***, and ***menthol***. Alcohol, as a liver microsomal enzyme-inducer, decreases the activity of ***pentobarbital***, ***phenobarbital***, ***phenytoin***, ***tolbutamide***, and ***warfarin***.

Acetaminophen: Hepatotoxicity is enhanced by alcohol intake, especially in **alcoholics**. Alcohol may also increase the risk of non-steroidal anti-inflammatory drug-induced gastrointestinal bleeding. ***Nyquil***® ***elixir*** contains 25% alcohol.

Amitryptyline: Combination with ***amobarbital*** has caused severe hypothermia, and may adversely affect driving behavior. The potentiating effect has been responsible for deaths.

Anesthetics: **Alcoholics** may require more anesthetic than **non-alcoholics**. Autopsies on **alcoholics** have shown that, in some, myelin, the nerve coating, is destroyed. Nearly 1/3 of these have a loss of myelin in the white matter of brain, plus increased fat and ***phosphorus*** in the central nervous system. This may affect the brain's absorption of the anesthetics. The net effect of the extra anesthetics on other body organs is unknown.

Anticholinergics: See ***Crime***.

Anticoagulants: In chronic **alcoholics**, with normal liver function, the plasma half-life of ***warfarin*** was much less (26.5 h versus 41 h) than in **non-alcoholics**, perhaps due to alcohol stimulating microsomal enzyme activity. Similar results have been reported with ***heparin*** and ***heparinoids***. Moderate alcohol consumption usually has no significant adverse effects. 10–20 oz of daily mealtime ***wine*** failed to alter ***warfarin's*** anticoagulant effect. See ***Drug interactions with Aspirin***.

Antidepressants: Tricyclics seem to enhance alcohol's effects and side-effects. The combination has often been used in suicide attempts, along with hypnotics and tranquilizers. Alcohol, by hepatic enzyme induction, may decrease the clinical potency of some antidepressants.

Antidiabetics: Users of ***sulfonylureas*** may develop a ***disulfiram***-like reaction (flushing, nausea, vertigo, tachycardia, dyspnea, and severe headache) when they drink alcoholic beverages, due to inhibition of aldehyde dehydrogenase (with ***tolbutamide***, it is due to a competition for oxidative metabolic sites). ***Tolbutamide*** is metabolized about twice as fast in alcoholic **patients**. Alcohol may decrease a **patient's** need for ***insulin***, especially if there is liver damage or if large quantities of alcohol are consumed. This has caused an irreversible hypoglycemia and death in a hospitalized **patient**. ***Phenformin***, now rarely used, caused severe lactic acidosis.

Antihistamines: Combination use with ***diphenhydramine*** caused great impairment of performance through central nervous system depression. ***Chlorpheniramine*** also produced decreased psychomotor performance, when used with alcohol. Induced hypoglycemia is enhanced by ***cimetidine***, ***famotidine***, and ***ranitidine***.

Antipyrine: Its metabolism is severely impaired in **patients** with alcoholic liver cirrhosis. The combination precipitated grand mal seizures in three **Negroes** (ages 40, 47, 55) who consumed large amounts of alcohol.

Aspirin: Alcohol may precipitate asthma attacks in ***aspirin***-sensitive **individuals**. May be synergistic in breaking down the gastric mucosal acid-barrier, causing peptic ulceration and gastrointestinal bleeding. It may also prolong ***aspirin***-induced bleeding time, if taken simultaneously (but not 12 h before) or up to 12–36 h after.

Barbiturates usually interfere with the body's ability to metabolize alcohol, increasing the depressant effect. Acute ethanol intoxication delayed the elimination of ***pentobarbital*** from the blood. Deaths have occurred, due to its use as an anesthetic, without adequate history of alcohol ingestion prior to admission. Some report increased rate of ethanol elimination in **patients** without liver disease, when treated with ***phenobarbital***. Alcohol enhances the absorption of ***phenobarbital***, and the combination delays the absorption of the alcohol. As an enzyme-inducer, ethanol enhances the metabolism of ***pentobarbital***. In the U.S., ***barbital*** is frequently stolen from laboratories for purposes of abuse.

Bromocriptine: Alcohol intolerance potentiated its side-effects.

Calcium carbamide: Marked facial flushing and ***acetaldehyde*** accumulation, as with a ***disulfiram*** reaction. Potential dangers in **patients** with ischemic or myocardial disease.

Carbon tetrachloride: Fatalities associated with exposure and drinking alcohol at the same time (viz. death in 29-year-old **male alcoholic** who drank two cocktails during a 30 min period of cleaning bathroom tiles in a poorly ventilated room). Hepatic and renal toxicities also recorded.

Cephalosporins: Alcohol intolerance with ***disulfiram***-type reactions reported with ***cefamandole***, ***cefoperazone***, and ***moxalactam***. **Patients** should avoid alcohol before and for several days after therapy.

Chloral hydrate: Ethanol causes higher and more prolonged levels of ***trichlorethanol***, chloral hydrate's metabolite. Increased heart rate effect of ethanol is enhanced. Can cause a ***disulfiram***-like reaction. Additive or possible synergism in causing sedation, sleep, or respiratory failure. The combination is often called a "Mickey Finn" or "knock-out drops", but the potentiation is rarely as great as novelists portray it. Vasodilation and hypotension are common, as each drug inhibits the other's metabolism.

Chloramphenicol: ***Disulfiram***-like reaction has occurred, due to inhibition of ***acetaldehyde*** oxidation.

Chlordiazepoxide: Counters the intoxication produced by alcohol, yet, in some **patients**, it may enhance alcohol's central nervous system depressant effect.

Chloroform: Hypothrombinemia, secondary to alcoholic cirrhosis, and occupational exposure to ***chloroform*** used in cleaning spark plugs. The first clinical sign was prolonged intraoral hemorrhaging. Hepatotoxicity has also been reported.

Chlorpromazine: Depressant effect potentiated by alcohol. Bullous or erythematous skin lesions reported in **patient** with coma.

Cimetidine: Decreases the elimination of alcohol (increases blood ethanol levels, but not with ***ranitidine***). Enhances ethanol induced hypoglycemia.

Cisplatin: Long-term alcoholic intake > 80 g/day decreases vomiting and nausea in treated **patients**.

Clobazam: Serum levels increase with alcohol intake.

Clomethiazole edisylate: Action enhanced by alcohol and has been responsible for several deaths.

Clorazepate: Euphoric effects are additive.

Contraceptives, oral: Alcohol stimulation of microsomal enzyme activity may cause failure of hormonal contraception.

Demeclocycline: Toxicity may be increased by alcohol.

Diazepam: Dramatically potentiates alcohol's ability to intoxicate within 30 min–4 h. Alcohol increases its absorption rate.

Digoxin: Alcohol decreases its urinary excretion. The combination has been used in attempted suicide.

Disulfiram: The combination can produce marked facial flushing, itching, hives, nausea, palpitations, tachycardia, dyspnea, bronchospasms, severe headache, and collapse. The same reaction occurs with products containing ***thiram*** (q.v.). The **person** may only appear to be intoxicated. The reaction is utilized in treating alcoholism. Sweating and numbness in 33-year-old **male**, after ingestion of 70 ***disulfiram*** tablets (35 g) during 3 month period in food prepared by wife on doctor's prescription, without **patient's** knowledge. After ingesting 700 ml hard liquor, **patient's** condition deteriorated, with severe dyspnea, urinary and fecal incontinence, and a fever of 40°C. Death followed in a few hours. Autopsy revealed **no** alcohol in blood, hemorrhagic necrosis of left cerebral hemisphere, thalamus, and pons, thrombosis in left coronary artery, hemorrhagic pulmonary edema and focal pneumonia in right lung, tubular and glomerular patho-logy. A 60-year-old **male** died from esophageal rupture, after violent vomiting. A 25-year-old **male** developed catatonia with mutism, negativism, stereotyping, staring, combativeness on 250 mg/day. A 56-year-old **male** developed a generalized pruritic macular rash, after showering with a shampoo containing 3% alcohol. ***Disulfiram*** blocks acetaldehyde dehydrogenase, raising the level of ***acetaldehyde*** (q.v.) in the body. It is slowly eliminated from the body and may remain for 6–12 days after the last dose. Two **infants**, whose **mothers** were maintained on ***disulfiram*** during their first trimester of pregnancy, had severe limb reduction anomalies. **Users** should remember to avoid less obvious sources of alcohol, such as wine vinegars, mouthwashes, cologne, aftershave lotions, and cough medicines. Emergency medical treatment for an **adult** includes 1000 cc IV fluids, 1 g IV ***ascorbic acid*** (at 15 min intervals, if necessary; some state it has little benefit), and oxygen, when indicated.

Doxepin: Combination with alcohol intake can be very dangerous in driving.

Enalapril: Pancreatitis in 69-year-old **male**, after two 5 mg doses. Symptoms developed after taking the second dose with a bottle of beer.

Estrogens: Alcoholic cutaneous hepatic porphyria developed in **patients** receiving either ***chlorotrianisene*** or ***diethylstilbestrol*** for prostatic carcinomas.

Ethchlorvinyl: Potentiation of alcohol's effect reported. Confusion, disorientation, lethargy, unsteadiness, slurred speech, nystagmus, and weakness has been reported. Consciousness was lost in two **alcoholics**, after taking a single therapeutic dose.

Ethinamate: Additive depressant effects.

Ethionamide: Potentiates the psychotoxic effects of alcohol.

Flupentixol: Usual dosage, combined with alcohol (0.5 g/kg) seriously impaired coordination and driving skills.

Fluphenazine: Side-effects (tremor, rigidity, dystonia) were facilitated by alcohol ingestion.

Furazolidone: When given for 4–5 days, it acts as a monoamine oxidase inhibitor (q.v. in these ***Drug interactions***), potentiating alcohol's effects and inhibiting alcohol's metabolism.

Glutethimide: Overdosage effects are exaggerated by alcohol. Can increase blood alcohol levels as much as 30%. Alcohol increases its absorption.

Griseofulvin: Activity is decreased by alcohol ingestion. Flushing and tachycardia have been reported.

Haloperidol: Intense potentiation of alcohol's depressant effect.

Hexylresorcinol: Although used infrequently today, alcohol diminishes its anthelmintic activity.

Hypnotics: Effects (sedation) are worsened by concomitant alcohol intake. They potentiate each other.

Ibuprofen: In 1988, the first report of renal failure with a non-steroidal anti-inflammatory drug and alcohol occurred. Combination of these drugs accentuated memory lapses.

Imipramine: Combination with alcohol further impairs performance tasks.

Iron compounds: Increased absorption of ***ferric chloride***, when given with alcohol.

Isoniazid (INH): Alcohol may cause its rapid metabolism. This is especially true in chronic alcoholism. Alcohol–***INH*** may help trigger latent epilepsy. Driving skills deteriorated with the combination. Death occurred in several suicidal **patients**: 18-year-old **male**, 30-year-old **male**, 37-year-old **male**. Symptoms include deep coma, tonic and clonic spasms, nystagmus, severe hypertension, increased pulse rate, and irregular respiration.

Isoproterenol: Death in asthmatic 46-year-old **female**, due to cardiorespiratory failure after excessive use of an inhaler and consumption of alcohol. Ventricular arrhythmias were common.

Ketoconazole: Acute alcohol intolerance in 39-year-old **female**, after four glasses of white ***wine***. Wise to have **patients** discontinue consuming alcohol during therapy and for at least 48 h after therapy. Nausea, vomiting, and flushing have occurred.

Lithium: It may suppress the need for alcohol.

Lorazepam: Combination produced additive psychomotor impairment and increased anxiety.

Lysergic acid diethylamide (LSD): Alcohol has precipitated psychotic reactions with flash-backs, hallucinations, and delusions.

Meclizine: Potentiated by alcohol, impairing vigilance and short-term memory.

Meprobamate: Potentiates the central nervous system effects of alcohol. The combination increases driving skill impairment. Clinically significant accelerated metabolism of it, due to alcohol's stimulation of microsomal enzymes.

Methaqualone: Additive effect with alcohol.

Methotrexate: Alcohol has a major role in inducing cirrhosis in treated psoriatic **patients**. Avoid its use in alcoholic **patients**.

Methyprylon: Combination has produced abnormally deep narcotic sleep for several hours. **Patients** awakened spontaneously 3–6 h later. Many had morning-after headache, depression, disturbed coordination, heartburn, and vertigo, decreased driving competence.

Methyldopa: Increased central nervous system depression and chronic alcohol abuse helped predispose the liver to damage (including necrosis) within 1–13 weeks.

Metronidazole: It inhibits ***acetaldehyde*** oxidation, thereby potentiating alcohol's effects and producing a ***disulfiram*** (q.v.)-like reaction. This can occur with vaginal inserts. **Female** on 8 days of therapy had three drinks at a party. 2 h later, **she** was noted to be flushing and sweaty, followed by collapse and a heart rate of 170/min. Alcohol may become distasteful during treatment.

Mianserin: Additive effects with alcohol.

Monoamine oxidase inhibitors (isocarbonazid, nialamide, pargyline, phenelzine, procarbazine, tranylcypromine): Prior administration may potentiate alcohol effects. ***Procarbazine***, in particular, taken with alcohol (especially alcoholic beverages with a high ***tyramine*** content) may cause ***disulfiram***-like reactions.

Mushrooms (*Coprinus atramentarius, Coprinus erithistes, Clitocybe clavipes*, and *Amanita sp.****)***: ***Disulfiram*** (q.v. above)-like reaction with alcohol ingestion, often within a few minutes, probably due to their ***coprine*** and ***methyl hydrazine*** contents, which block ***acetaldehyde dehydrogenase*** action. Alcohol consumption up to a few days after eating cooked inky top ***mushrooms*** may cause an "***acetaldehyde*** storm".

Nalidixic acid: Its toxic effects are increased with alcohol ingestion.

Narcotics: Alcohol with ***cannabis*** often rapidly produces nausea, headache, prostration, tachycardia or bradycardia, red conjunctiva, skin pallor, sweating, behavioral impairment, and psychological distress. This can be serious with pre-existing cardiac or central nervous system disorders. Symptoms dissipated in about 4 h. Combination often produces sleepiness or euphoria. 85 cases of acute toxicosis were reported in a group of 22,000 **soldiers**. Potential teratogenic effects of the combination. ***Morphine***, ***heroin***, ***noscapine***, and ***opiates*** add to alcohol's depressant effects and always increase the danger of serious respiratory depression and even liver pathology. Drug interaction with ***morphine*** yields more than an additive effect.

Nitrofurantoin: It prevents the oxidation of alcohol's metabolite, ***acetaldehyde***.

Nitroglycerin: Causes a ***disulfiram*** (q.v. above)-like reaction. Vasodilation is potentiated and hypotension results. This can occur due to the alcohol diluent in IV ***nitroglycerin***. The condition has often been misdiagnosed as coronary insufficiency.

Paclitaxel: Has induced acute alcohol toxicity in 57-year-old **female**.

Paraldehyde: Once used frequently in the treatment of delirium tremens. One report indicated the possibility of synergistic action in explaining the deaths of nine **patients** within 30 min to several hours, after receiving 30–90 ml ***paraldehyde*** (oral, IM, or rectal) for acute alcohol intoxication.

Phenaglycodol: Additive effect of central nervous system depression.

Phenothiazines (chlorpromazine (q.v. above), ***prochlorperazine, thioridazine, etc)***: Potentiate each other, causing enhanced sedation.

Phenytoin: Chronic alcoholism can decrease its effectiveness. Hodgkin's disease in a **child** whose **mother** ingested large quantities of alcohol during her pregnancy along with 100 mg ***phenytoin***/day.

Podophyllum: Topical treatment of condylomata acuminata of the penis and anus in 25-year-old **male** associated with restlessness, irregular respiration, and cyanosis (***podophyllum*** solution contained 16.25% alcohol). Vomiting and comatose after several alcoholic beverages.

Propranolol: Seemed to protect against typical alcohol-induced behavioral changes.

Propoxyphene: Combination has been used in suicidal deaths from severe respiratory depression.

Pyrazinamide: Its toxic effects on the liver are markedly enhanced by alcohol intake.

Quinacrine: Inhibits oxidation of ***acetaldehyde***, leading to ***disulfiram***-like effects.

Reserpine and rauwolfia: Central nervous system depressant effects enhanced by alcohol intake. Incoordination, rigidity of extremities, and head drop have occurred, with only 1/10 the usual ***reserpine*** dosage.

Rifampin: Contraindicated in **alcoholics** with abnormal liver function tests. **Patients** may develop a hypersensitivity reaction with chills, nausea, myalgia, and hypertension.

Smoking: Alcohol is synergistic in the pathogenesis of oropharynx epithelial cancers. Occasional heartburn, precipitated by alcohol, in combination with heavy smoking.

Sulfonamides: Intensifies alcohol's noxious effects and may decrease the action of some ***sulfonamides***.

Testosterone: Metabolic clearance increased in most **subjects** with alcohol consumption.

Tetrachloroethylene: Potentiates central nervous system depressant effects. Avoid alcohol intake before or for 24 h after exposure.

Thirams: Typical ***disulfiram*** (q.v. above)–alcohol interaction. Rubber factory **workers** usually first report flushing of face and wrists, decreased blood pressure, increased heart rate, and "fullness in the head" after drinking alcohol.

Toluene: Alcohol increases risk of adverse effects from ***toluene*** (q.v.).

Trichlorethylene: Hepatorenal toxicity in inhalant **drug-abusers** drinking alcohol. A "**degreasers'** flush" has been described in **people** who work around degreasing vats all day, then imbibe a few after-work alcoholic drinks.

Verapamil: Inhibits ethanol elimination and prolongs intoxication.

Vitamins: Ingestion of large amounts decreases absorption of ***vitamin*** B_1, ***vitamin*** B_3, ***vitamin*** B_6, and ***vitamin A***, and can lead to their deficiency states. Alcohol also decreases ***vitamin*** B_1 storage and accelerates its destruction.

Xylene: Metabolic clearance of inhaled ***xylene*** is decreased about 50% by alcohol intake. Nausea and dizziness occur.

Zimelidine: By itself, it has no effect on memory, but it reverses alcohol's adverse effect on memory and the serotonin messenger system.

ALCOHOL, *n*-HEXYL

= *1-Hexanol* = *Amylocarbinol*

Use: In the manufacture of antiseptics and hypnotics.

Untoward effects: LD_{LO}, **human**: 500 mg/kg.

Absorbed through the skin of **mice**. ***Hexanol-1*** and ***2***, principal metabolites of the solvent ***N-hexane*** (q.v.), given by IP injections to **rats** over an 8 month period of time also led to peripheral neuropathies.

LD_{50}, **rats**: 0.7 g/kg; **mouse**: 1.95 g/kg.

ALCOHOL, HYDRATROPIC

= *Hydratropyl Alcohol* = *α-Methylphenylethyl Alcohol* = *2-Phenylpropan-1-ol*

Synthetic alcohol, artificial flavorant, and odorant. **U.S. consumption < 1000 lbs/year.**

Use: Since the 1930s, in fragrances for use in soaps, detergents, perfumes, pharmaceutical creams and lotions, and in foods.

Untoward effects: 6% in petrolatum for 48 h closed-patch test on **humans** showed no irritation. No sensitization reported in 25 **volunteers**.

Oral LD_{50}, **rats**: 2.3 g/kg; acute dermal LD_{50}, **rabbits**: > 5 g/kg.

ALCOHOL, ISOBUTYL

= *Isobutanol* = *Isopropylcarbinol* = *2-Methyl-1-propanol*

Solvent. **Flammable.**

Use: In manufacturing fruit flavorings, and in varnish removers and spray lacquers; general anesthetic for **lobsters**.

Untoward effects: Recommended by underground news media as an inhalant "orgasm extender" in **humans**. ***Inhalation*** of 182 ppm by **man** is tolerable, but uncomfortable. Very little effect is noted at 18 ppm, but 1818 ppm is intolerable, and has caused a 50% decrease in respiratory rate. Irritant to eyes, nose, and throat in **humans** leading to headache, dizziness, vertigo and vestibular disorders, drowsiness, drying and cracking of skin. Absorbed through the lungs and gastrointestinal tract. May have long-term (3 year) delayed effects in **man**. TLV, 100 ppm; 300 mg/m^3.

Found in most alcoholic beverages (292 mg/l in 86 proof distilled ***alcohol***); in gases from ***potato*** storage; in oil refinery and ***coal*** conversion emissions, and in surface water.

LD_{LO}, **human**: 500 mg/kg. LC_{LO}, **human**: 8000 ppm/4 h.

Urinary metabolites include ***acetaldehyde*** in **rabbit** studies. It is carcinogenic (liver carcinomas and sarcomas, and myeloid leukemia) in **rats**.

LD_{50}, **rats**: 2.46 g/kg; LD_{LO}, **rabbits**: 3.75 g/kg.

ALCOHOL, ISOPROPYL

= *Dimethyl Carbinol* = *Isopropanol* = *Isopropyl Alcohol* = *Petrohol* = *2-Propanol* = *2 Propyl Alcohol*

Flammable.

Use: In many hair coloring and conditioner products, nail polish and removers; and in topical disinfectants (decreases efficacy of ***Betadine***®). Effective against Gram-negative bacteria, in skin cleansers, in gas tank antifreeze (1%), and in windshield de-icer sprays; in lighter fluids, in inks, in paint thinners, and as a component of ***GB*** war gas.

Untoward effects: Accidental eye contact from splashes without immediate water rinsing or exposure to high vapor concentrations have produced ragged abrasions and patchy loss of the corneal epithelium. Eye and nose irritant at 800 ppm. Overuse as a skin disinfectant caused second and third degree burns in 700–1000 g premature **infants**. TWA, 400 ppm. LD_{LO}, **human**: 8.6 g/kg. Toxic blood levels in **man**: 340 mg/100 ml. Five deaths in **humans** in one report. Coma in 63-year-old **male *ethanol* abuser**, who ingested large quantities and survived, after a blood level of 203 mg/dl. Causes kidney damage and increase in incidence of paranasal sinus cancers from manufacturing exposures. Headaches have occurred after use in cosmetics and toiletries. Suicidal 20-year-old **male** in deep stupor after enema with 1 pt – survived after treatment. A 59-year-old **male**, after ingesting 1 l, became comatose and survived, after extracorporeal dialysis. Severe ketosis and acidosis with intoxication. Ingestion can cause a fruity breath odor. Coma, increased pulse, and shallow respirations have been produced by topical application (sponging, soaked towels) to reduce fever. TLV, skin, 400 ppm; causes an allergic contact dermatitis and erythema. ***Acetone***, a major metabolite of isopropyl alcohol, enhanced ***carbon tetrachloride*** renal and hepatic failure in 4/14 plant **workers**. Similar results have been reported in **rats**. Hyporeflexia, hypotension, and breath odor of ***acetone*** noted in lethargic or near comatose **individuals**. Exposure can be confirmed by testing for ***acetone*** in the urine. Sniffing lighter fluids has been involved in toxicosis and death. May cause red blood cell lysis and hypoglycemia. Vapors of 500 ppm are uncomfortable, but tolerable in **humans**; minimal effect at 50 ppm, but intolerable with a 50% decrease in respiratory rate at 5000 ppm.

LD_{50}, **rats**: 5.6–6.8 ml/kg. **Dogs** and **cats** may be exposed to it in antifreezes and windshield washing fluids in garages. Narcosis and death after 12,000 ppm/4 h in **rats**; 50% mortality after 8 h.

Treatment: Remove exposed **victims** to fresh air; artificial respiration, if necessary. Wash eyes with lots of water – skin with soap and water. Remove contaminated clothing. If swallowed, induce vomiting. Get medical attention!

ALCOHOL, METHYL

= *Carbinol* = *Methanol* = *Methyl Alcohol* = *Wood Alcohol* = *Wood Spirit*

***Solvent*. Flammable.**

Use: In antifreezes; an octane booster in gasolines; a fuel; a solvent in pharmaceutical manufacturing; in the manufacture of ***formaldehyde***, an intermediate in chemical and protein synthesis; and in the illegal synthesis of amphetamines. Because of its high volatility, it is used in inks for food wrappings, in the extraction of edible products, and in cleaning beer vats. Current demand is ~2 billion gallons annually.

Untoward effects: Severe poisoning occurs from ***ingestion*** (as little as 30 ml has caused death), inhalation, and percutaneous absorption in **humans**. Headache, fatigue, mydriasis, dilated pupils, impaired visual acuity, blepharospasm (even complete and permanent blindness), violent abdominal pain, nausea, vomiting, dyspnea, pulmonary edema, acidosis, areflexia, dizziness, convulsions, unconsciousness, cyanosis, circulatory and respiratory failure, followed by death (usually after ***ingestion*** of at least 100–250 ml). Rapidly absorbed after ***ingestion***. A 36-year-old **female** ingested small amounts of a blue liquid windshield wiper solution, along with **her** daily ***wine*** for 5 days causing dizziness, dull pain, ringing in an ear, nausea, and vomiting, followed by unresponsiveness and a putaminal infarct. Blindness has occurred after drinking 5 tablespoonfuls. A 39-year-old **patient** ingested approximately 500 ml (initial serum level of 920 mg/dl) and recovered uneventfully, after gastric lavage, ***sodium bicarbonate***, ***ethanol***, and hemodialysis. Often found in contaminated moonshine ***whiskey***. 46 cases of poisoning occurred at a prison, where **inmates** drank photocopier fluid. Up to 50 **students** in one school had a similar, but milder accidental exposure. Severe metabolic acidosis, central nervous system depression, and visual changes are considered the most serious manifestations. Large numbers of young **children** 1–4 years of age have ingested small amounts in gasoline. Many older young **men** did it while siphoning gasoline. Case reports are numerous, see ***Sterno***® discussion under ***Alcohol, Ethyl***. It and/or its metabolites damage the gastrointestinal tract, especially the stomach, as well as lungs, kidneys, liver, pancreas; and it selectively damages retinal cells and optic nerve (can be delayed effect). Its metabolites ***formic acid*** (q.v.) and ***formaldehyde*** (q.v.) may be the actual culprits. Half of all serious cases have some vision impairment, and even blindness. 12 **ophthalmologists** and their **wives** ingested methanol accidentally added to a hot ***wine*** concoction. The heat evaporated most of the methanol. All received liberal amounts of ***ethanol*** during the 3 h trip to the hospital, where emesis was induced, and some received gastric lavage. Blood chemistry indicated low levels of methanol and ***formic acid***. Symptoms of poisoning usually occur within 12–24 h. Acidosis usually occurs after 6–36 h. Hyperamylasemia can occur, following methanol poisoning, and, by itself, should not be equated with pancreatitis. A 985 g premature **infant** died from being cleaned with methylated spirits (95% ***ethanol*** and 3–5% methanol). 18 h later, blood levels were 259 and 26 mg/ml, respectively. Plum-colored necrotic lesions developed on the buttocks. An 8-month-old **male** died after methanol soaked pads were placed on his chest two nights in a row, for the relief of a common cold. Eczematous, as well as allergic erythematous dermatitis has been reported from skin contact (including shoe glues).

Vapors of 41,514 ppm decreased respirations by 50%. An ***inhalation*** fatality occurred from using it as a paint remover in a very hot location. Oral LD_{LO}, **human**, **mouse**, **dog**, **monkey**, and **rabbit**: 340 mg, 420 mg, 7.5 g, 7 g, and 7.5 g/kg, respectively. ***Inhalation*** TC_{LO}, **human**, **monkey**, and **cat**: 86 g/m^3, 1000 ppm, 44 g/m^3/6 h, respectively. Blood levels of 20 mg/100 ml are toxic and 89 mg or higher/100 ml has been lethal. 200 ppm (260 mg/m^3): 8 h TWA; 800 ppm/15 min. Considerable debate deferred the approval of ***aspartame*** for some time because methanol is one of its hydrolytic products.

4/40 **cattle** died from drinking it in water bled from a gas well. Methanol is used in drilling for natural gas. **Dogs** have died from drinking antifreezes and windshield washing fluids containing methanol. 8800 ppm vapor exexposure of **rats**/8 h caused lethargy; 22,500 ppm/8 h caused narcosis; 31,600 ppm/18–21 h caused 100% mortality. 4 g/kg in **pigtail monkeys** (*Macaca nemestrina*) caused marked acidosis, but not in **rhesus monkeys** (*M. mulata*). Poisoned **monkeys** develop symptoms like those in **man**. Blindness in **animals** has also occurred.

Treatment: Depends on the nature and degree of exposure. Move **patient** to fresh air. Resuscitate, if necessary. Get medical help. Many treatments have been tried and recommended. Emesis and gastric lavage, if early. Hemo- and peritoneal dialysis and ***ethyl alcohol*** (10% in 5% ***dextrose***) 1.5 ml/kg, IV, followed by 0.15 ml/kg/h, to maintain a 100–150 mg% blood level has been used very effectively. Some have recommended 0.75 mg/kg, IV, initially, followed by 0.5 mg/kg over several days or 1 ml/kg orally of a 50% ***ethanol***, followed by 0.5 ml/kg, IV or oral, for 3–4 days. If ***ethanol*** is not immediately available, 100 ± ml of ***whiskey*** is given orally. It acts as a competitive substrate for alcohol dehydrogenase, interfering with the formation of methanol's toxic metabolites. ***Sodium bicarbonate***, usually 4 mEq every 4 h as needed (5 g/h, oral, until urine is alkaline), to correct the acidosis. ***Diazepam*** for treating excitement and convulsions; oxygen in cyanosis, ***penicillin*** if comatose. High carbohydrate intake and 50 mg ***thiamin*** daily often recommended.

ALCOHOL, MYRISTYL

= *Myristyl Alcohol* = *1-Tetradecanol* = *Tetradecyl Alcohol*

Use: Emollient in cosmetic creams and lotions, soaps, detergents, and perfume.

Untoward effects: Toxic to **rats** at 5 and 10% level in their diet. **Chicks** can use it as an energy source at a dietary intake of 5%, but at 10–16%, it caused some deaths. Skin irritation and corneal injury was very low on **rabbits**, but 12% in petrolatum on **humans** showed no irritation after a 48 h closed-patch test and sensitization was demonstrated in only 2/25 **volunteers**. Slight comedonicity on **rabbit** ears.

LD_{50}, **rats**: > 5 g/kg. Acute dermal LD_{50}, **rabbits**: > 5g/kg.

ALCOHOL, OCTYL

= *Caprylic Alcohol* = *1-Octanol*

Use: Less than 10,000 lbs/year, used in fragrances. Found in many natural sources, including citrus oils. In cosmetic creams and lotions, soaps, detergents, and perfumes, with penetrating aromatic odor.

Untoward effects: A maximization test with 2% in petrolatum on 25 **volunteers** showed no sensitization.

LD_{50}, **rats**: > 3.2 g/kg; LD_{50}, **mouse**: 1.8 g/kg. Acute dermal LD_{50}, **guinea pig**: > 0.5 g/**animal**, with only slight skin irritation, and, in the **rabbit**, at > 5 g/kg.

ALCOHOL, PHENYLETHYL

= *Benzene Ethanol* = *Benzyl Carbinol* = *β-Hydroxyethylbenzene* = *Phenethyl Alcohol*

***Antiviral, antifungal, antibacterial, preservative*: Has characteristic rose-like odor.**

Use: Helps to eliminate multiple drug resistance transfer factor. 0.5% in ophthalmic solutions and parenterals.

Untoward effects: LD_{LO}, oral, **human**: 500 mg/kg. 24 h patch test at full strength produced no irritation in 20 **volunteers**. 8% in petrolatum produced no sensitization reactions in 25 **volunteers**.

LD_{50}, **rats**: 2.46 ml/kg; single skin penetration LD_{50}, **rabbits**: 0.79 ml/kg; irritant to belly skin and cornea of **rabbits**; dermal LD_{50}, **guinea pigs**: 5–10 ml/kg. 100% concentration on **miniature swine** skin and 32% in ***acetone*** on **human** skin failed to produce any irritation.

ALCOHOL, STYRALLYL

= *α-Methylbenzyl Alcohol* = *Methylphenyl Carbinol*

Use: About 10,000 lbs used annually in fragrances. In cosmetic creams and lotions, soaps, detergents, and perfumes. Also used as an artificial flavoring agent in Europe.

Untoward effects: 8% in petrolatum produced no irritation or sensitization in **human** volunteers.

LD_{50}, **rats**: 400 mg/kg.

ALCURONIUM

= *Alloferin* = *Diallylbis(nortoxiferine)* = *N,N′-Diallylnortoxiferinium* = *Diallylnortoxiferine* = *Diallyltoxiferine* = *4,4′-Didemethyl-4,4′-di-2-propenyltoxiferine* = *Ro 4-3816* = *Toxiferine*

Skeletal muscle relaxant; non-depolarizing neuromuscular blocking agent. **Over 75% eliminated unchanged in the urine.**

Use*:** In projectile syringes, for the immobilization of **wild game**. The dichloride is used. Rapid onset of action, relative lack of cardiovascular effects, and ease of antagonism has made it useful in anesthetic procedures. Reversed by ***neostigmine and ***edrophonium***.

Untoward effects*:** Hypotension and bradycardia noted in some **patients**, tachycardia in 60%. Endotracheal intubation more likely to produce blood pressure increase (20%). Delayed reversal, when used with ***streptomycin. Anaphylactic reactions have been reported. Prolonged residual paralysis, not wholly reversible by ***neostigmine***, required artificial ventilation of lungs, to overcome respiratory depression. ***Streptomycin*** enhances its effect.

Significant ***histamine*** release after use in **rats**.

ALDESLEUKIN

= *Costimulator* = *IL-2* = *Interleukin 2* = *Lymphocyte Mitogenic Factor* = *Proleukin* = *T-Cell Growth Factor* = *TCGF*

Immunoregulator. **A lymphokine. T-cells tell other cells when to increase or decrease their efforts to seek out and destroy infectious agents, tumors, or foreign tissue.**

***Use*:** In the treatment of **human** metastatic kidney cancer.

Untoward effects*:** The following adverse effects on initial treatment prohibit subsequent injections: uncontrolled cardiac arrhythmias, chest pain with electrocardiogram changes indicative of possible angina or myocardial infarction, intubation for over 72 h, pericardial tamponade, kidney dysfunction requiring more than 72 h of dialysis, coma or toxic psychosis of more than 48 h, recurrent or hard-to-manage seizures, bowel ischemia or perforation, or any gastrointestinal bleeding requiring surgery. Avoid use with any central nervous system depressants and drugs toxic to any internal organs, although ***indomethacin has been used with it to control fever. Fever, chills, shaking, pruritus, nausea, vomiting, diarrhea, fluid retention, hypotension, erythema nodosum, and bowel perforation are common untoward effects. Deterioration of kidney and liver function occurs. A potentially fatal problem called capillary leak syndrome (CLS) leads to hypotension and decreased organ perfusion, cardiac arrhythmias, myocardial infarction, respiratory insufficiency, gastrointestinal bleeding, lethargy, sleepiness, confusion, leukoencephalopathy, and agitation. Eosinophilic myocarditis developed in 44-year-old **male**. Decrease in platelet counts and secondary thrombocytopenia was common in one large series of **patients**. Its use may have triggered an acute leukoencephalopathy.

Its use has exacerbated psoriasis in three **patients** 29–62 years of age. Erythema nodosum noted in 26-year-old kidney neoplasm **patient**. It may reactivate rheumatoid arthritis.

ALDICARB

= *Alticarb* = *Ambush* = *ENT 27,093* = *2-Methyl-(methylthio)propionaldehyde-O-(methylcarbamoyl)oxime* = *2-Methyl-2-(methylthio)propanal O-[(methylamino)carbonyl]oxime* = *OMS-771* = *Temik®* = *UC 21,149*

Insecticide, acaricide, nematocide. **A carbamate.**

***Use*:** As granules, for soil application and as a seed treatment. Has been used as a **bird** repellent.

Untoward effects*:** Has caused poisoning in **man**. 17 **people** in California, and 15 in Oregon became ill within 15–30 min, after eating contaminated watermelons. Symptoms lasted for 2–4 h. A 43-year-old **male** had prolonged cholinesterase inhibition after exposure that required 5 days of intubation. IV ***atropine and ***pralidoxime*** injections are credited with his recovery. Over a period of 2 years, aldicarb was suspect in outbreaks of illness in 12 **people** from eating hydroponically grown ***cucumbers***. Illness started within 15–135 min after ***ingestion***, and was characterized by diarrhea (100%), nausea and vomiting (89 and 100%), abdominal pain (67 and 100%), dyspnea (67 and 100%), muscle fasciculation (56 and 80%), blurred vision (60 and 75%), and headache (11 and 60%). Illness ranged from 3 to 36 h, but most were ill only 3–5 h. Miosis, excessive sweating, and salivation are also noted in poisoning cases. 77 illnesses from eating possibly contaminated cantaloupes are recorded. Aldicarb is a broad-spectrum, systemic carbamate that inhibits cholinesterase (plasma and red blood cell), but it usually reverts to normal within a few hours. A leak of toxic aldicarb oxime gas at a manufacturing plant sickened at least 135 **people**. A Florida **farmer** ingested 1 teaspoonful in a suicide attempt. Accidental poisoning in 14 **people** after its mistaken use as black pepper in a salad from a mislabeled container.

LD_{50}, young **mallard ducks**, **California quail**, and **pheasants**: 3.4, 2.6–4.6, and 5.34 mg/kg, respectively. Symptoms in these **birds** were ataxia, imbalance, shakiness, wings drooped or spread, running and falling, regurgitation, hyporeactivity, tremors, convulsions, tetany, ptosis, foamy salivation, and opisthotonus. These signs appeared within 3 min, and deaths were noted within 15–40 min after dosing. 1 mg/kg was lethal to a **calf**. Has caused death of lactating **Holstein cows**. Do **NOT** use on crops intended for hay or grazing.

LD_{50}, **rats**: 0.9 mg/kg; **mallard ducks**: 3.4 mg/kg.

ALDOSTERONE

= *Aldocorten* = *Aldocortin* = *(11β)-11,21-Dihydroxy-3,20-dioxopregn-4-en-18-al* = *3,20-Diketo-11β,18-oxido-4-pregnene-18,21-diol* = *Electrocortin*

***Mineralocorticoid*. Made in the glomerular zone of the adrenal gland. *Sodium*-retaining hormone, responsible for the homeostatic control of salt balance. Produced by the adrenal cortex, after stimulation by *angiotensin II*, as well as by *potassium* and *corticotropin*.**

Use: In the prevention and treatment of cardiovascular hypotension and shock associated with ***sodium*** loss. Potentiates pressor agents in endotoxic-induced shock.

Untoward effects: ***Licorice*** ingestion can be a cause of pseudoaldosteronism in **man**. Has caused glaucoma or increased ocular pressures, as well as lens opacity. Hyperaldosteronism leads to and prolongs edema in congestive heart failure by increasing reabsorption of ***sodium*** in the kidney's proximal tubule, leading to hypertension in **people** and experimental **rats**.

ALDRIN

= *Compound 118* = *HHDN* = *Octalene*

***Insecticide*. A chlorinated hydrocarbon with a pine oil-like odor. No longer *manufactured* or *supposedly* used in the U.S. Readily converted into *dieldrin* (q.v.) in cattle, pigs, sheep, rats, and poultry. Evaporates from streams of water.**

Use: In soil treatment for insect control. To control **snakes** and **grasshoppers** (usually 2 oz/acre).

Untoward effects: A real danger in the U.S. from unused supplies that may accidentally cause exposure. Readily absorbed after ***ingestion***, ***inhalation***, or ***skin contact***. Stored in fatty tissues. Excreted slowly from the body. In 1977, Environmental Protection Agency reported 5.4 µg/l in U.S. drinking water supplies. Accidental mixing in feed has not only caused deaths of many **animals**, but also caused significant levels of ***dieldrin***, its metabolite, in meat and milk sold for **human** consumption. Occupational exposures have caused headaches, nausea, vomiting, weakness, irritability, rash, tremors, dizziness, and even death. Convulsions were preceded by myoclonic jerks – enough to project a teacup out of one's hand or make a leg suddenly shoot out. Convulsions may interfere with normal respiration and cyanosis has developed. ***Ingestion*** has been lethal to **birds**. Ten **workers** in a factory, mixing it with Fuller's earth, developed neurological disturbances, despite wearing protective clothing. It took 10 months before one **man's** electroencephalogram and ***dieldrin*** blood levels became essentially normal.

TWA, **human**: 0.15–0.25 mg/m^3. TD_{LO}, **human**: 14 mg/kg; LD_{LO}, **child**: 1.25 mg/kg.

Most **animals** show symptoms within 24 h of exposure, which usually include apprehension, belligerence, blepharospasms, fasciculations of facial and cervical muscles, clonic spasms of the neck and limbs, salivation, froth from the lips and muzzle, agitation, poor coordination, maniacal seizures, convulsions, increased temperature, coma, and death. 130–140 ppm in feed fatal to **cows** in about 30 days; 150 ppm kills **sheep** in 20 days. **Animals** with low body fat content are more susceptible to its acute toxicity. ***Phenobarbital*** and ***calcium gluconate*** have been preferred treatments in **animals**.

LD_{LO}, most farm **animals**: 20–50 mg/kg. Doses as low as 2.5 mg/kg were toxic to 1–2-week-old **calves** and fatalities occur from **animals** eating insect baits. LD_{50}, **rats**: males, 15 and females, 7 mg/kg; **birds**: 4–35 mg/kg; **mallard ducks**: 520 mg/kg; **quail**: 6.6 mg/kg; **dogs**: 65 mg/kg; and **pheasants**: 16.8 mg/kg.

Treatment: Recommendations suggest clearing the airway, aspirate secretions, and assisted pulmonary ventilation with oxygen. ***Diazepam*** is usually the anticonvulsant of choice. If aldrin was ***ingested***, and the **patient** is alert, induce emesis with ***syrup of ipecac***, followed by ***charcoal***. Intubate, if necessary. Do **NOT** give milk, cream, or fatty substances, as this will enhance systemic absorption. Do **NOT** give ***epinephrine*** or ***adrenergic drugs***, because it may aggravate aldrin's adverse cardiac effects. If indicated, remove contaminated clothing. Use soap and water to wash skin and hair. For eye exposure, flush with large amounts of water for 15 min. Seek medical consultations.

ALENDRONATE

= *Fosamax*

Untoward effects: Esophagitis, esophageal strictures and ulceration, especially if taken with less than 6 oz of water and within 30 min of lying down. Gastrointestinal upsets, decreased renal function, anemia, hypocalcemia, hypokalemia, and hypomagnesemia; leg cramps, abdominal and muscle pain, and dyspepsia (3%). Risk of acid-related upper gastrointestinal disorder increases with age and concurrent treatment with ***non-steroidal anti-inflammatory drugs***. Rarely, rash and erythema, constipation, diarrhea, dysphagia, and headache.

ALFADOLONE ACETATE

= *Alphadolone Acetate* = *GR 2/1574*

Untoward effects: Its use with ***alfaxalone (Althesin®)*** as an anesthetic has been discontinued, due to concerns over anaphlylactic reactions to its suspending agent, ***Cremophor EL (PEG-40 castor oil = Ethoxylated Castor Oil)***. Severe bronchospasms, decreased blood pressure, increased heart rate, nausea, vomiting, hiccups, occasional excitement, respiratory depression, salivation, coughing, and clonic convulsions are reported adverse effects.

ALFALFA

= *Lucerne* = *Luzerne* = *Medicago sativa* = *Methi*

Untoward effects: Ingestion of alfalfa sprouts was linked to > 20,000 **people** sickened in three salmonella outbreaks. Pancytopenia in 59-year-old **male** after intermittent ingestion of 80–160 g/day/5 months. **People** eating alfalfa sprouts should be aware that they contain assorted phytoestrogens. Ingestion causes flatulence and diarrhea in some **people**. After ingesting alfalfa tablets, two **patients** had an exacerbation of systemic lupus erythematosus, due to its *L-**canavanine*** content.

Allergic reactions to it reported in **dogs**. Hemoglobinuria reported in **cows** eating large quantities. Hepatogenous photosensitivity and skin eruptions in **cows** after ingestion. **Cattle** and **sheep** can bloat after ingesting rapidly growing plants. A diet of 20% sprouts has caused a lupus-like disease in **Macaque monkeys**. Its phytoestrogens (***coumestrol*** and ***coumestran***) have caused infertility in some **cows** and **ewes**. Its saponin content, especially ***medicagenic acid***, has decreased **poultry** growth and egg-laying.

ALFAXALONE

= *Alphaxalone*

Untoward effects: See ***Alfadolone*** above.

ALFENTANIL

= *N-[1-[2-(4-Ethyl-4,5-dihydro-5-oxo-1H-tetrazol-1-yl ethyl]-4-(methoxymethyl)-4-piperidinyl]-N-phenyl-propanamide* = *Alfenta* = *Rapifen* = *R 39,209*

***Analgesic, anesthetic.* Weak analgesic with rapid onset, but short duration of action, after IV use. Experimental use in cats, dogs, mice, monkeys, rabbits, rats, and sheep, where it may be of value for short surgical procedures, and after infusion, for longer procedures. A tetrazole derivative of *fentanyl*.**

Untoward effects: Similar to ***fentanyl*** (q.v.) and ***sulfentanil*** (q.v.). Muscle rigidity occurs in 50–88% of **patients**, chest wall rigidity (makes it difficult to ventilate a **patient**), facial movements, clenching of jaw or fist, flexion of the arm and wrist, with occasional reports of nausea, vomiting, pruritus, tachycardia, hiccups, laryngeal spasm, dysphagia, persistent erection, hypotension, dizziness, and apnea. Its metabolism is inhibited by ***erythromycin***.

IV, LD_{50}, **rats**: 47.5 mg/kg; **dogs**: 20 mg/kg.

ALGLUCERASE

= *Ceredase* = *Cerezyme* = *β-Glucocerebrosidase* = *Imiglucerase*

***Enzyme*. From human placental tissue.**

Use: To treat Type I Gaucher disease in **man**, preventing the accumulation of glycolipids in the spleen, liver, and bone marrow.

Untoward effects: Slight fever, chills, abdominal discomfort, nausea, and vomiting. These usually do not require medical intervention. A 68-year-old **female** with Gaucher's disease developed pulmonary hypertension after 2 years of treatment with it. After 1 year of treatment, **she** had progressive dyspnea during exercise.

ALITRETINOIN

= *Panretin*

Untoward effects: Erythema, rash, and pruritus. Edema and vesiculation follow additional topical use.

ALKYL-ϖ-HYDROXYPOLY(OXYETHYLENE)

= *PA-14* = *Tergitol 15-S-9*

Avian stressing agent, detergent.

Use: To control, by spraying, roosting **blackbirds**, **starlings**, **grackles**, and **cowbirds**.

Untoward effects: Keep out of reach of **children**. Corrosive. **Users** should wear protective goggles, face shield and rubber gloves. Causes eye damage and skin irritation. Avoid contamination of food. Unprotected **persons** must avoid drift areas. See ***Treatment*** below. **Fish** and other aquatic life may be killed by run-off. In Kentucky, helicopters spraying at night with it in an aqueous solution washed away the birds' protective oils, allowing at least a half million of them to die from exposure.

Treatment: Flush exposed eyes and skin with copious amounts of water for 15 min. Remove contaminated clothing. If swallowed, large amounts of milk, egg whites and gelatin solutions have been recommended. Gastric lavage may be contraindicated, if mucosal damage is suspected. Avoid alcoholic beverages. Treat circulatory shock, respiratory depression and convulsions, if indicated.

ALLAMANDA CATHARTICA

= *Golden Trumpet* = *Yellow Alamanda*

Untoward effects: The fruit and sap of this ornamental vine are poisonous leading to nausea, stomach cramps, thirst, and increased temperature. A **boy** who sucked a stem also had swollen lips.

ALLETHRIN

= *dl-2-Allyl-4-hydroxy-3-methyl-2-cyclopenten-1-one = 3-Allyl-2-methyl-4-oxocyclopent-2-enyl Chrysanthemate = ENT 17,510 = 2,2-Dimethyl-3-(2-methyl-1-propenyl)cyclopropane Carboxylic Acid,2-methyl-4-oxo-3-(2-propenyl)-2-cyclopenten-1-yl Ester = Pallethrine = Pynamin*

Insecticide, mosquito repellent, synthetic pyrethroid.

Untoward effects: Negative patch tests in **humans**. Oral LD_{LO}, **human**: 500 mg/kg.

Birds show no overt toxicity at 5000 ppm; low toxicity to **honey bees**.

LD_{50}, **rat**: 680 mg/kg, **rabbit**: 4.3 g/kg; LD_{50}, **mallard ducks**: > 2000 ppm.

ALLICIN

= *S-Allyl-2-propenethiosulfinate*

Inhibits the growth of Gram-negative bacteria and fungi.

Use: As an active ingredient in ***garlic*** (popular in herbal preparations, etc.), said to inhibit platelet aggregation and protect against atherosclerosis, coronary thrombosis, and stroke.

Untoward effects: After absorption in **man**, the body may reek of ***garlic***. A strong ***garlic*** flavor and odor can be noticed in the ***milk*** and blood of **cows** allowed to inhale the volatile substances from fresh ***wild garlic*** (*Allium vineale*) tops.

IV, LD_{50}, **mice**: 60 mg/kg; 120 mg/kg, SC.

ALLIGATOR TONGUE OIL

Use: Louisiana bayou country folk have used it for over 100 years, for relief of "the miseries", asthma, rheumatism, and arthritis. Steroidal action in the body.

Untoward effects: Causes profuse salivation.

ALLIUM SCORODOPRASUM

Use: In analgesic salves.

Untoward effects: Two reports in Japanese literature of dermatitis venenata a day after **human** application.

ALLIUM TRICOCCUM

= *Ramps = Wild Leek*

***Allium porrum* is the common cultivated leek.**

Use: As a food. Chicago gets its name from Indian languages referring to the present-day site of Chicago as "shikako" or "checagou", meaning the place of the wild garlic, leek, or onion.

Untoward effects: Contains ***oxalates***, which can cause vulva and urinary tract irritation. Leeks have caused contact allergic reactions in **housewives**, **cooks**, and a pizza **baker**. The ***milk*** of **cows** grazing on it has a ***garlic*** flavor.

ALLIUM VALIDUM, A. CANADENSE, A. CERNEUM

= *Wild Onion*

Use: Eaten by **American Indians**.

Untoward effects: ***A. canadense*** ingestion of bulbs or leaves by young **children** leads to gastrointestinal upsets.

Has caused hemolytic anemia in **horses** and **sheep** (even after consuming *A. cepi*, the cultivated onion). **Rats**, **rabbits**, **dogs**, and **poultry** fed large quantities of cooked onions became anemic. This condition develops more slowly in the dark.

ALLOBARBITAL

= *Allobarbitone = Diadol = Dial = 5,5-Diallylbarbituric Acid = Malilum*

Sedative, hypnotic.

Untoward effects: Has caused fixed-drug eruptions in **man**. Narrow margin of safety in **humans**, where 15–40 mg/l is a therapeutic serum concentration, and 40–50 may be toxic. Will block ovulation in the **rat**.

LD_{LO}, **human**: 36 mg/kg.

IP, LD_{50}, **rat**: 127 mg/kg.

ALLOPURINOL

= *Adenock = Al-100 = Allo-Puren = Allozym = Allural = Aloral = Alositol = Aluline = Anoprolin = Anzief = Apulonga = Apurin = Apurol = Bleminol = Bloxanth = BW 56158 = Caplenal = Cellidrin = Cosuric = Dabroson = 1,5-Dihydro-4H-pyrazolo[3,4-d]-pyrimidin-4-one = dura AL = Embarin = Epidropal = Foligan = Geapur = Gichtex = Hamarin = Hexanurat = HPP = 4-Hydroxypyrazolo-(3,4-d)pyrimidine = Ketanrift = Ketobun-A = Ledopur = Lopurin = Lysuron = Miniplanor = Monarch = Nektrohan = NSC 1390 = 1H-Pyrazolo[3,4-d]pyrimidin-4-ol = Remid = Riball = Suspendol = Takanarumin = Urbol = Uricemil = Uripurinol = Urobenyl = Urosin = Urtias 100 = Xanturat = Zyloprim = Zyloric*

Use: Treatment of hyperuricemia and gout.

Untoward effects: ***Eye***: Corneal opacity and possibility of cataract development demand monitoring. Intra-retinal hemorrhages and macular lesions have occurred. Minimize ultraviolet exposure, while on therapy, as it aids the development of nuclear cataracts.

Gastrointestinal: Nausea, vomiting, diarrhea, anorexia, and occasional abdominal pain, and swelling of lips and mouth occur. Long-term therapy may increase ***iron*** absorption.

Hematopoietic: Bone marrow depression has occurred in **patients** treated for malignancies; leukocytosis, leukopenia, agranulocytosis, eosinophilia, thrombocytopenia, and vasculitis have also been reported.

Hypersensitivity and skin: Pruritic maculopapular and erythematous skin eruptions, toxic epidermal necrolysis, alopecia, exfoliative and ichthyosiform dermatitis, and Stevens–Johnson syndrome occur. The hypersensitivity reactions may involve multiple organ systems. Severe forms involve renal and hepatic function and cause toxic epidermal necrolysis. Milder forms produce uncomplicated dermal lesions.

Liver: Hepatitis (cholestatic and fatty changes), massive necrosis, hepatic granulomas, and increased SGOT, SGPT, and alkaline phosphatase. Fulminant hepatic failure reported in 58-year-old **female**, due to hypersensitivity.

Renal: Exacerbates renal insufficiency. May cause nephrolithiasis, acute interstitial nephritis, and xanthine stones.

Fever, chills, metallic taste, peripheral neuritis, vertigo, headache, diffuse burning feet, bone erosions, lymphoma-like syndrome of neck lymph nodes, and bone marrow depression have also been noted.

Drug interactions: Inhibits metabolism of ***antipyrine***. Potentiates ***azathioprine***, ***cyclophosphamide***, ***cyclosporin***, ***mercaptopurine***, ***thioguanine***, ***vidarabine***, and ***vincristine*** (by enzyme inhibition), and increases their toxicity. Potentiates ***dicoumarol*** and ***warfarin*** anticoagulant effect; potentiates ***chlorpropamide***, and ***theophylline***. ***Ethacrynic acid***, ***thiazides***, and ***xanthines*** antagonize uricosuria development. ***Probenecid*** enhances its effect. Uricosuria increases with ***acetohexamide*** and ***probenecid*** which may cause ***uric acid*** crystallization in the kidney. Increases the frequency of ***ampicillin***-induced skin reactions (morbilliform rashes). ***Thiazides*** and ***ethacrynic acid*** decrease its effect.

Oral LD_{LO}, **woman**: 370 mg/kg; oral, **man**: 130 mg/kg/30 days.

ALLOXAN

= *Mesoxalylcarbamide* = *Mesoxalylurea* = *2,4,5,6 (1H,3H)-Pyrimidinetetrone* = *2,4,5,6-Tetraoxohexahydropyrimidine*

***Diabetogenic*. Related to *uric acid*. Chemically and toxicologically similar to *PNU*, a rodenticide. *Riboflavin* is a derivative.**

Use: For the induction of experimental diabetes in **cattle**, **dogs**, **guinea pigs**, **hamsters**, **mice**, **monkeys**, **owls**, **pigeons**, **rabbits**, **rats**, **sheep**, **sows**, **toads**, and **turtles**.

Untoward effects: It destroys the outer cell membrane of the β cells of the pancreas leading to diabetes and hypoglycemia. Can cause opacity of the eye's lens.

Teratogenic. **Cats** are resistant to its diabetogenic effect, but it is nephrotoxic to them.

LD_{LO}, **rat**: 4.5 g/kg; IV, LD_{50}, **rat**: 300 mg/kg; IV, LD_{LO}, **mouse** and **rabbit**: 200 and 300 mg/kg, respectively; **steers**: 110 mg/kg, IV; and **ewes**: 60 mg/kg, IV.

ALLYLAMINE

= *3-Aminopropene* = *3-Aminopropylene* = *2-Propen-1-amine*

***Unsaturated primary amine*. Annual production has been around 100 tons.**

Use: In the synthesis of mercurial diuretics, sedatives, and antiseptics; also to the dyeability of acrylonitrile polymers.

Untoward effects: Repeated IV injections cause severe necrotizing arteritis and obliterative arteriopathy. Much of it (80% ±) is metabolized to ***acrolein*** (q.v.) and may be responsible for most of its adverse effects. Monitor creatine phosphokinase serum levels in **workers** exposed to it, because studies have shown inhalation of allylamine fumes can cause myocardial damage. Strong ***ammonia*** odor causes sneezing and tears. Has a burning taste.

Inhalation TC_{LO}, **human**: 5 ppm/5 min.

Specific cardiovascular toxin, causing vascular and myocardial lesions in **rats**, **rabbits**, **dog**, **calf**, and **monkey**.

Inhalation LC_{50}, **rat**: 286 ppm/4 h; Oral LD_{50}, **rat** and **mouse**: 106 mg and 57 mg/kg, respectively. IP, LD_{50}, **mice**: 49 mg/kg.

ALLYL BROMIDE

= *Bromallylene* = *3-Bromo-1-propene* = *3-Bromopropylene*

Use: In manufacturing perfumes and allyl compounds. Used as a carminative in **man** (made by treating crushed ***garlic*** with elemental ***bromine***), after deodorizing it.

Untoward effects: Irritates eyes and respiratory tract.

Inhalation LC_{50}, **rat**: 10,000 mg/m^3/30 min; IP, LD_{50}, **mouse**: 108 mg/kg; oral LD_{50}, **guinea pig**: 30 mg/kg.

ALLYL CAPROATE

= *Allyl Hexanoate* = *2-Propenyl Hexanoate*

Use: Approximately 5000 lbs/year in fragrances for creams, lotions, perfumes, and soaps. ***Pineapple*** flavoring for candies and soft drinks.

Untoward effects: A maximization test on 25 **human** volunteers, using a 4% concentration in petrolatum, produced no reactions.

LD_{50}, **rats**: 218 mg/kg; **guinea pigs**: 280 mg/kg. Acute dermal LD_{50}, **rabbit**: 0.3 ml/kg.

ALLYL CHLORIDE
= *Chlorallylene* = *3-Chloro-1-propene* = *3-Chloro-propylene*

Chemical intermediate. **Corrosive liquid with an unpleasant, pungent odor.**

Use: In manufacturing allyl compounds.

Untoward effects: Readily absorbed through the skin. Vapors irritate eyes, nose, throat, and lungs. Acute exposures cause unconsciousness; chronic exposures cause liver, lung, and kidney pathology. High concentration of vapors can cause death. Hazardous substance, can cause severe burns and permanent injury to the eyes and skin. Occupational exposures to its vapors in **man** have caused neurotoxicity. Chronic exposure of **man** to vapor concentrations of 1 ppm to 113 ppm caused symptoms of liver damage.

Oral dosing of **mice** with 300 and 500 mg/kg for 2–17 weeks related to focal kidney damage in 70%, and the other pathological changes were restricted to the nervous system (nerve fiber degeneration in peripheral nerves, particularly motor nerves). Males were more seriously affected than females.

Inhalation LC_{LO}, **human**, **rat**, **guinea pig**: 3000 ppm, 290 ppm/8 h, and 290 ppm/4 h, respectively. Eye irritant in **rats** and **guinea pigs** at 290 ppm; IV, LD_{50}, **dog**: 7.15 mg/kg.

ALLYL CYANIDE
= *3-Butenenitrile* = *β-Butenonitrile* = *Vinylacetonitrile*

Reagent. **In some mustard oils. *Onion*-like odor.**

Untoward effects: Within a few hours of exposure to doses < 200 mg/kg to **rats**, it causes cloudy opacity of the cornea, which disappears rapidly. At a dosage > 200 mg/kg, severe and irreversible changes occurred in the cornea, as well as in the crystalline lens. The cornea became turbid and, eventually, ulcerated, with perforation or descemetocele. The lens developed a central opacity.

LD_{50}, **rats**: 115 mg/kg; LD_{LO}, **mouse**: 50 mg/kg; LC_{LO}, **rat**: 500 ppm/4 h.

ALLYL CYCLOHEXYLPROPIONATE
= *Allyl 3-Cyclopropionate* = *Allyl Hexahydrophenylpropionate*

Synthetic fragrance. **Approved for food use.**

Use: Under 30,000 lbs/year in the U.S. Europe permitted it at up to 10 ppm, as a food flavoring. In creams, lotions, perfumes, soaps, and detergents.

Untoward effects: Maximization test on 25 **human** volunteers, with 4% in petrolatum, produced no reactions. Excreted as ***hippuric acid*** in urine.

LD_{50}, **rat**: 585 mg/kg. **NO** ill-effects when fed to **rats** for 52 weeks at 2500 ppm level. LD_{50}, **guinea pig**: 380 mg/kg.

ALLYLESTRANOL
= *Gestanin* = *Gestanol* = *Gestanon* = *Gestanyn* = *Orageston* = *Organon* = *Turinal*

Progestogen.

Use: In the management of threatened abortion or premature labor. Efficacy has been questioned by some.

Untoward effects: Amenorrhea. Experimental implication of masculinization of female **fetus**.

Oral and SC, TD_{LO}, **rat** (13–21 days pregnant): 450 mg and 900 mg/kg.

ALLYL ETHER
= *Diallyl Ether* = *3,3´-Oxybis-1-propene*

Liquid. *Radish*-like odor.

Untoward effects: Local irritant. Can be absorbed through the skin.

LD_{50}, **rat**: 320 mg/kg; **rabbit**, skin: 10 mg/24 h leading to mild irritation.

ALLYL GLYCIDYL ETHER
= *AGE*

Use: Unlisted ingredient in many proprietary or trade name products. In the manufacture of epoxy resins, and as a copolymer in vulcanization of rubber.

Untoward effects: Toxic vapors and gases (including ***carbon monoxide***) can be released by fire. Odor threshold of 10 ppm. TLV of 5 ppm. Contact with strong oxidizers can cause fire or explosions. Exposure to light or air can cause the formation of explosive peroxides. Flammable. Can cause moderate irritation of the skin and severe irritation of the eyes and respiratory tract in **humans**. Chronic exposure has caused dermatitis with itching, swelling, and blisters in **man**. Skin sensitization and cross-sensitization with other glycidyl ethers can occur. National Institute of Occupational Safety and Health has specific instructions for **workers** re: respirators, clean-up, and decontamination, in plants using it.

Four IM injections of 400 mg/kg to three **rats** in a 9 day period was associated with one death and one case of testicular necrosis.

LD_{50}, **rats**: 922 mg/kg; LC_{LO}, inhalation, **rats**: 860 ppm/4 h.

2-ALLYL-2-ISOPROPYLACETAMIDE

= *AIA*

Untoward effects: Known inducer of acute hepatic porphyrias in **rats** and **chick** and **duck** embryos. Depletes liver ***glutathione***. Autocatalytic agent.

ALLYL ISOTHIOCYANATE

= *Allyl Isosulfocyanate* = *Allyl Mustard Oil* = *D4720* = *3-Isothiocyanato-1-propene* = *Isothiocyanic Acid Allyl Ester* = *Leuna* = *Oil of Mustard* = *Redskin* = *Volatile Oil of Mustard*

Isolated from *Brassica nigra* (*Black Mustard*).

Use: In manufacturing food flavors, sauce seasoning, salad dressings, and war gases. As a topical counterirritant. In nail polish removers, paint thinners, cleaning fluids, and gasoline. On shrubbery, at 0.5% as a **cat** and **dog** repellent, also as a **racoon** repellent. Used as a treatment (0.4%) of ringworm on **cattle**. The pungent chemical in ***horseradish*** and ***mustard seed***. Has been added illegally to ground up ***parsnip***, to make ersatz ***horseradish***.

Untoward effects: Vapors irritate eyes (recovery from keratitis occurs slowly), nose, and sinuses of **man**, and has been offered as an additive to deter **glue-sniffers**. Will irritate skin and, if left on it, will cause blisters (vesicular contact dermatitis in 61-year-old **male**, after using it in a household liniment). Powerful gastrointestinal irritant that induces vomiting. Estimated carcinogenic dose for **man**: 4.6 mg/day.

Toxic and lethal doses for **cattle** are 2–3 mg/kg and 5–20 mg/kg, respectively. A goitrogen in **chicks** and **rats**. Carcinogen in **rat** trials and causes chromosome aberrations in **hamsters**. Chronic administration to **rats** caused transitional cell papillomas and epithelial hyperplasia in the bladders of males.

LD_{50}, **rats** and **mice**: 92–148 mg/kg and 310 mg/kg, respectively; LD_{50}, skin, **rabbit**: 88 mg/kg; 2 mg causes eye irritation in **rabbits**.

ALLYL PHENOXYACETATE

Synthetic fragrance.

Use: In fragrances. Less than 1000 lbs/year in creams, lotions, perfumes, soaps, and detergents. Approved for food use in Europe.

Untoward effects: Mild primary irritant to **man** with 1% in petrolatum in a 48 h closed-patch test. In a repeat test, it failed to show any irritation. A 1% mixture in petrolatum in a maximization sensitization test in 26 **volunteers** gave negative results.

LD_{50}, **rats**: 0.475 ml/kg; acute dermal LD_{50}, **rabbits**: 0.82 ml/kg.

ALMITRINE BISMESYLATE

= *Duxil* = *S-2620* = *5023-SE* = *Vectarion*

***Respiratory stimulant.* A lipid-soluble *piperazine* derivative.**

Use: In sleep apnea syndromes and in **patients** with hypoxemic chronic obstructive pulmonary disease.

Untoward effects: Peripheral neuropathy reported in **humans**. Stop treatment, if paresthesia develops. Mild headache and nausea, especially if given in the fasting state. Diarrhea, abdominal pain, fatigue, weight loss, insomnia, and hyperventilation have also been reported.

IV, LD_{50}, **mice**: 210 mg/kg; IP, 390 mg/kg; oral, > 2 g/kg.

ALMONDS

Contains *salicylates*.

Use: In Near Eastern folk medicines, and as a flavoring (oils or paste) in cookies and cakes. See below for bitter almonds and sweet almonds.

Untoward effects: A common cause of allergic reactions. **People** eating mixed nuts are not necessarily allergic to it, but often, rather to ***peanut*** oil, which rubs off the ***peanuts***. Almond-sensitive **people** are often also allergic to ***apricot*** and ***peach*** seeds. Acute ***cyanide*** poisoning cases reported in the U.S., due to "almond-flavored" milkshakes, prepared from ***apricot*** kernels.

Bitter almonds (*Prunus amygdalus* var. *amara*) is found in Italy, Spain, and southern France. Used as a raw material for the production of a large number of aromatics used in perfumery and flavorings. Over-indulgence in eating bitter almonds leads to classical symptoms of ***cyanide*** poisoning (characteristic odor of bitter almonds or ***hydrocyanic acid*** eminates from the breath or stomach contents of **people** and **animals** poisoned with it). Sudden death reported in 56-year-old **female**, after ingestion of an unknown amount of bitter almonds. A survey indicated ingestion of 50–70 bitter almonds is fatal in **adults**. **Children** showed severe toxicity after 7–10. Oil of bitter almonds, *Oleum Amygdale Amarae*, contains 4% ***hydrocyanic acid***. Within 30–45 min after ingesting four or five ground up bitter almonds, a 60 kg 67-year-old **female** became light-headed, nauseated, had abdominal pain, and vomited. By the next morning **she** felt well. On the night of hospital admission, **she**, again, had consumed a slurry of 12 nuts. Within 15 min, **she** had severe abdominal cramps, went to the bathroom, and collapsed (pulse 128/min, respirations 20/min and deep). Intubation, oxygen, and aggressive treatment for ***cyanide*** (q.v.) poisoning was successfully instituted. The average ***cyanide***

content of each nut (average weight 1.32 g) was 469 mg/100 g of bitter almond. ***Caution***: ***Bitter Almond Oil*** is only for medicinal use, and it is **NOT** to be used for flavoring foods.

Sweet almonds = ***Jordan Almond*** (var. *dulcis*). Sweet almonds lack ***amygdalin***. It is used mostly in cosmetics, pharmaceuticals, and roasting oils.

Untoward effects: Comedogenic on **rabbit** ears.

Since ***beer*** in America is often made from a wide variety of fermentables, antisera to many agents, such as almonds, are available to detect use in ***beer*** manufacturing.

ALOCASIA

Untoward effects: Leaves and stems contain an irritant juice and an insoluble ***calcium oxalate***, which causes a burning sensation in the mouth and throat, and swelling of the mouth and throat, that may make breathing difficult. Nausea, vomiting, and diarrhea are rare direct systemic effects. **Children** usually shriek in pain on biting into plant parts. The rhizomes of some species are rubbed on topically for stomach aches (viz. *A. indica*) in Thailand, yet, some **people** get a terrible itch from handling it. Gastric lavage, emesis, rinsing out the mouth, applying cold packs to swollen lips and mouth, intubation, and supportive therapy have been recommended.

ALOE

= *Aloe arborescens* = *Candalabra Aloe* = *Corne de Cerf* = *Ghiu Kumari* = *Ibli* = *Jelly Leeks* = *Kir* = *Kisimamleo* = *Lu Hui* = *Msubili* = *Msubiri mwitu* = *Musabber* = *Saber* = *Zabila*

***Cathartic, wound healing agent, botanical.* Cape aloe juice has bactericidal activity. Leaves contain *emodin* and some *calcium oxalate*. Annual imports into the U.S. have occasionally been over 1 million lbs.**

Use: By **Egyptians** ca. 1000 years B.C., as a laxative. As a purgative, emmenagogue, and stomachic. In cosmetics (1–2%), to protect against sunburn (ultraviolet light absorber), and to treat skin wounds (bacteriostatic). Supposedly, brought to Egypt from South Arabia by Alexander the Great. The leaves or their mucilage have been used as a hot poultice, to relieve swellings. In India, Africa, and South America, the sap has been used as a **fly** repellent. In ***Compound Tincture of Benzoin***. The taste of the pure material is nauseating and bitter.

Untoward effects: Sufficient quantity may appear in breast milk to cause purgation in **nursing infants**. May cause intestinal injury, if used after ingestion of caustics. Excessive use in **humans** has caused a harmless pigmentation of the colon. Often present in herbal teas or tablets, causing within several hours, severe dehydration, severe diarrhea, weakness, coma, and peripheral neuropathy. Pliny the Elder commented on the Gallic ***wine*** industry adulterating the flavor and color of ***wines*** with it. Colors alkaline urine red. Allergic contact dermatitis has been reported.

Warning: Large cathartic doses may produce nephritis or colic in **animals**. ***Contraindicated*** in pregnant and debilitated **animals**. Overdosage in **animals** has caused severe gastroenteritis, collapse, and death.

ALOE VERA

= *Aloe barbadensis*

Use: Marco Polo found the **Chinese** using it to treat stomach and skin disorders. In wound and burn therapy, acting as a ***thromboxane***-blocker, allowing traumatized tissue to receive better blood circulation. ***Thromboxane*** manufactured by cells at burn or wound sites causes coagulation within injured capillaries. It is the "dried" sap or latex of *A. barbadensis*, *A. vulgaris*, *A. saponaria*, or *A. vera* Linne, but **NOT** *A. vera* Miller, and often sold to manufacturers as a gel from the mucilaginous cells of the plant. Quality is variable, but it has a moisturizing and emollient effect.

Untoward effects: Found to be cytotoxic to **human** normal and tumor cells *in vitro*. An *Aloe vera* preparation was recalled because of the presence of *Pseudomonas*, yeasts, and molds. Often on the label panel, with very little material in the product. Use of an *A. vera* jelly in 66-year-old **female** resulted in stasis dermatitis, clearing when the drug was discontinued. 21 **women** whose surgical incisions were treated with *A. vera* gel required 75% more time for adequate healing than the controls. *A. vera* Miller has caused irritant dermatitis, contact allergy, and photosensitization, after topical application in **man**. Can color alkaline urine red.

Cows that have received it IV may react with an anaphylactic reaction. 500 mg/kg twice daily to **mice** for 5 days induced hypoglycemia.

ALOIN

= *Aloinum*

Use: As the most active purgative principle of ***Aloes*** (q.v.), producing milder catharsis than ***aloes***. Intensely bitter.

Untoward effects: Thought to have an effect on over 25% of the **infant's** bowels, after **mothers** ingested it. Has caused skin reactions in **man**.

ALOSETRON

= *Lotronex*

A selective serotonin receptor antagonist for irritable bowel syndrome (IBS) with diarrhea.

Untoward effects: Commonly, constipation, occasionally severe. Occasional life-threatening acute ischemic colitis and increased aminotransferases. Use with food decreases absorption by 25%. Hospitalization, surgery, and deaths have occurred in **women**. Voluntary withdrawal of its sales in the U.S. in November 2000 by its manufacturer, Glaxo-Wellcome, Inc.

ALOXIPRIN

= *Alaprin* = *Lyman* = *Palaprin* = *Polyoxyaluminum Acetylsalicylate* = *Rumatral* = *Superpyrin*

Analgesic.

Untoward effects: Occasional gastrointestinal upsets. Avoid prolonged administration. Its use has been associated with osteomalacia.

ALOYSIA LYCIOIDES

= *Beebrush* = *Whitebrush*

Shrubby plant in the *Verbena* family. Widespread in Texas, the Southwest, and Mexico. Contains a water-soluble toxin. A *honey* plant. In Paraguay, the leaves and stems of *A. oblanceolata* (*Poleo I*) are used in a *tea* or decoction, to regulate fertility in women.

Untoward effects: When grazed heavily for 30–40 days, it causes weakness and death in **horses**, **mules**, and **burros**.

ALPHAPRODINE

= *Nisentil* = *Prisilidene*

***Narcotic, analgesic.* Related to *meperidine*, but twice as potent.**

Use: Very potent short-acting narcotic analgesic.

Untoward effects: Abuse liability is similar to ***morphine's***. Causes clinical depression in **newborns** of **mothers** receiving it. Appears in **human** milk, after use by **mothers**, producing a noticeable sedative effect on the **neonate**, which can impair its nursing and feeding. IV use in **humans** has caused severe respiratory depression, cyanosis, and loss of pulse, requiring artifical respiration. May be contraindicated in **patients** with pulmonary problems. It may aggravate pre-existing convulsive disorders. Do **NOT** use with ***diazepam*** (potentially fatal drug interaction in 1.4% of 365 **patients**). Cerebral edema and hemorrhages in lungs and brain, with death occurring 6 h after injection into buccal mucosa in 28-year-old **male**, for dental surgery. Action is enhanced by ***chlorpromazine***. **Human** LD_{LO}, at about 1 mg/kg.

IV, LD_{50}, **mouse** and **rabbit**: 54 and 19 mg/kg, respectively. IP, LD_{50}, **rat**: 22 mg/kg.

ALPRAZOLAM

= *Alplax* = *D 65MT* = *Tafil* = *Trankimazin* = *U-31889* = *Xanax* = *Xanor*

***Anxiolytic.* A benzodiazepine.**

Use: In treating panic and phobic disorders, anxiety associated with depression, and sleep disorders.

Untoward effects: In **humans**, drowsiness and ataxia, as well as rebound anxiety between doses, are the most common complaints; occasional confusion, amnesia, disinhibition, depression, dizziness, increased snoring severity, withdrawal symptoms (agitation, visual hallucinations, dysarthria, tremors, grand mal seizures, and painful myoclonus), increased salivation, hypotension, and weight decrease. Rarely, excitement, rage, hypotension, jaundice, blood dyscrasias, stuttering, dependence, and abuse, as well as dry mouth, constipation, diarrhea, nausea, vomiting, blurred vision, akathesia, headache, dermatitis, tachycardia, menstrual irregularity, incontinence or urinary retention, decreased libido, inhibition of orgasm, and delayed or no ejaculation. Hypersensitivity reactions can occur. It can exacerbate the usual gastrointestinal distress of depressed **patients**. Overdose in 28-year-old **male** leading to first degree atrioventricular block. Photosensitivity reaction reported in 65-year-old **male**. Withdrawal symptoms can occur unless dosage is slowly decreased. A 26-year-old **female** developed increased SGPT under treatment, which returned to normal levels after drug withdrawal. There is risk to breast-fed **infants** of **mothers** taking the medication. Drowsiness and withdrawal symptoms have been reported in some of these **infants**. Can cause **fetal** damage, especially if taken during a **woman's** first trimester of pregnancy.

Drug interactions: ***Erythromycin*** decreases its metabolism. ***Ketoconazole*** causes a marked decrease in its body clearance. A 40-year-old **female** who consumed 3 oz of 90 proof ***whiskey*** exhibited disinhibited behavior and caused $50,000 damage to a neighbor's home, then fell asleep. Upon awakening, **she** remembered nothing about the incident. ***Fluoxetine*** increases its half-life by about 17%. Use with ***carbamazepine*** caused its serum concentration in a **patient** to drop from 54 to 10 ng/ml. Decrease dosage by 1/3, when used with ***cimetidine***. In some **patients**, it has produced a clinically significant decreased clearance of ***digoxin***. It increases levels of ***desipramine***, ***fluoxetine***, ***fluphenazine***, ***haloperidol***, and ***sertraline***.

In **cats**, 0.1 mg/kg tid or as needed to control anxiety or aggression causes sedation and may induce ataxia.

LD_{50}, **rats**: > 2000 mg/kg; **mice**: 1020 mg/kg; IP, 610 and 540 mg/kg, respectively.

ALPRENOLOL

= *Apllobal* = *Aprobal* = *Aptine* = *Aptol Duriles* = *Betapin* = *Gubernal* = *H 56/28* = *Regletin* = *Yobir*

Antianginal, antiarrhythmic, antihypertensive, β-adrenergic blocker.

Untoward effects: Bradycardia and moderate hypotension in 90%, in one series of cases; atrioventricular block in 20%; and decreased cardiac output in 10%. Circulatory collapse, asystole, and deaths have been reported from its use. May cause 25% decrease in cardiac output, and a 12% decrease in heart rate in 18 **women** 22–25 years, undergoing ***nitrous oxide/halothane*** anesthesia. Nausea, mild sweats, skin rash, fatigue, bizarre dreams, eroticism, muscle pains, and giddiness reported in a double blind study of severe angina in 46 **males** and four **females**, ages 34–79. Two **patients** developed fever, dermatitis, eosinophilia, and renal insufficiency after several weeks of therapy. Hypersensitivity reactions can be very severe. Toxic epidermal necrolysis and eosinophilia reported in 72-year-old **male**. Poisoning in a 58-year-old **female** who ingested 10 g was successfully treated with ***glucagon***. Use with ***nifedipine*** may be contraindicated. ***Pentobarbital*** accelerates the metabolism of alprenolol.

LD_{50}, **rats**, **mice**, **rabbits**: 597, 278, and 337.3 mg/kg, respectively; IV, LD_{50}, **mouse**: 20 mg/kg.

ALPROSTADIL

= *Caverject* = *EDEX* = *Liple* = *Minprog* = *Muse* = *Palux* = *PGE₁* = *Prostaglandin E₁* = *Prostavasin* = *Prostin VR* = *Topiglan* = *U-10136*

Vasodilator. **Also used intracavernosally, topically, and intraurethrally in males.**

Use: In **infants**, to maintain patency of ductus arteriosus until surgery can be performed. Over 70% effectiveness in psychogenic impotence in **men**. Stimulates erythropoiesis. Experimentally for topical use.

Untoward effects: Testicular pain before erection has occurred, as well as priapism (1–2%), Peyronie's disease (3%), mild penile pain, hypotension, urethral burning and bleeding, and a risk of penile ischemia. Orthostatic hypotension, migraine headaches, chest pain, mild nausea, phlebitis at infusion site, shock, glaucoma, and dizziness occur. Vaginal itching and burning after use by **her** sex partner. Embryotoxic and may induce abortions. Prolonged infusions in **infants** have caused cortical hyperostosis. Throat irritation and cough reported after inhalation. Intradermal injections cause prolonged edema and erythema.

ALSTROEMERIA LIGTU

= *Lis des Incas* = *Peruvian Lily*

Untoward effects: A popular cut flower whose sap contains ***tuliposides***, as in ***tulips***, causing a contact dermatitis with erythema and blistering. The chemical can penetrate vinyl gloves.

ALTRETAMINE

= *ENT 50,852* = *Hexalen* = *Hexamethylmelamine* = *Hexastat* = *HMM* = *HXM* = *NSC 13,875*

Antineoplastic.

Use: In treating lung and ovarian cancers, acute leukemia, and many malignant neoplasms.

Untoward effects: Mild to major peripheral neuropathy (11%) with decreased tendon reflexes, confusion, disorientation, ataxia, dizziness, vertigo, depression, personality changes, rashes, alopecia, tremors, seizures, visual hallucinations, insomnia, lethargy, severe nausea and/or vomiting (68%) and anorexia, leukopenia (< 4,000), thrombocytopenia (< 100,000), gastrointestinal upsets (60%), anemia, and alopecia. Occasional reversible Parkinson-like syndrome, itching and dermatitis, and liver toxicity occur. Do **NOT** use with ***monoamine oxidase inhibitor***, as the combination may cause severe hypotension.

TD_{LO}, **human**: 8 mg/kg.

LD_{50}, **mouse**: 437 mg/kg, **chicken**: 34 mg/kg.

ALUMINUM

= *Aluminium*

The brains of some patients with Alzheimer disease contained concentrations that were toxic to cats and rabbits. These concentrations increase in the human brain with increasing age, especially in frontal and medial temporal gray matter structures. It is assumed that this is associated with neurofibrillary degeneration, but the jury is still out.

Untoward effects: Aluminum is a non-essential trace element in the body. Under normal levels of exposure, there is little evidence that it represents a risk. Persistent vaccination granulomas have been reported from its use as an adjuvant in vaccines. Boiling ***fluoride***-containing water in aluminum pots releases large amounts of aluminum. Al^{+++} causes irreversible inhibition of pancreatic lipase enzymes; constipation (especially in the **elderly**); and prolonged ingestion of aluminum-containing drugs can cause a phosphate depletion syndrome. This effect is taken advantage of clinically, by giving dialysis **patients** ***aluminum hydroxide***, to help prevent hyperphosphatemia. Aluminum intoxication has occurred in uremic **patients** on dialysis, after receiving ***aluminum hydroxide*** orally, to prevent hyperphosphatemia. Aluminum is concentrated within certain neurons in the brains of senile dementia or

Alzheimer **patients**. ***Dolomite***, frequently purchased by the **elderly** in health food stores, as a source of ***calcium*** and ***magnesium***, may contain 42–219 ppm of aluminum. ***Arsenic*** deposits on aluminum are converted to ***arsine*** (q.v.) in the presence of acids. Dermatitis from aluminum sleeve buttons has been reported. **Patients** with renal failure accumulate aluminum in their bones and this increases the risk of osteomalacia in young **children**. Encephalopathy related to aluminum-containing antacids has occurred with speech abnormalities, painful movements, myoclonus, flapping tremors, personality changes, dementia, and psychosis. Death may follow in 6–7 months. Prolonged IV feeding of preterm **infants** has a potential for neurological impairment and renal failure. Direct contact with aluminum dust has caused increased prothrombin time in **workers**. Aluminum mill **workers** have shown truncal telangiectasis, and possibly increased pulmonary emphysema, and hematopoietic and lung cancers. ***Fluoride*** poisoning by contamination of edible vegetation has been common in **humans** and **cattle** near aluminum smelters. Toxic serum level for **humans** is said to be 150–200 μg/l.

Drug interactions: ***Tetracyclines*** combine with polyvalent ions, such as aluminum, ***calcium***, etc., to form poorly absorbed complexes (a delay of 1–2 h is suggested between these drugs, when given in combination). ***Ciprofloxacin***, ***quinalone***, and ***digitalis*** activity has been decreased when combined with aluminum-containing antacids, such as ***sucralfate***. Aluminum helps degrade ***epinephrine*** solutions, decreases absorption and response to ***phenothiazines***, and decreases absorption of ***isoniazid***. Incompatibilities are noted of antineoplastic agents (***cisplatin***, ***daunorubicin***, ***doxorubicin***) with aluminum component of needles.

Treatment: ***Deferoxamine*** in renal dialysis solutions has been the main therapeutic measure.

ALUMINUM ACETATE

Astringent, antiseptic.

Use: In the form of ***Burow's Solution***, which is approximately a 5% solution on wet dressings for ulcerative conditions. Often 0.4–2% ***Burow's Solution*** with other medicaments.

Untoward effects: Strong solutions are slightly irritant to **human** eyes, as is the powder.

Warning: Homemade solutions made by mixing ***aluminum sulfate*** and ***lead acetate*** in water contain ***LEAD*** and may be potentially hazardous.

ALUMINUM ACETYLSALICYLATE

= *Alexoprine Forte* = *Aluminum Aspirin*

Analgesic, antipyretic.

Use: As ***aspirin***, but with slower absorption rate.

Untoward effects: As with ***acetylsalicylic acid*** (q.v.).

ALUMINUM AMMONIUM SULFATE

= *Burnt Ammonium Alum*

Astringent, hemostatic. **Bitter.**

Use: In purifying drinking water; in baking powders; in preventing softening of dill pickles; in making dyestuffs and pigments; as an irrigation solution in **human** bladder hemorrhage; and as a **bird** repellent.

Untoward effects: See ***Aluminum potassium sulfate***.

ALUMINUM HYDROXIDE

= *AH* = *Aluminum Hydrate* = *Aluminum Trihydrate* = *Hydrated Alumina*

Vaccine adjuvant, antacid, absorbent, adsorbent, antidiarrheal.

Use: In manufacturing glass, fire clay, pottery, printing inks, antiperspirants, and in waterproofing fabrics. As an antacid, adsorbent, and in vaccines.

Untoward effects: Encephalopathy in dialysis **patients**, primarily due to its ***phosphate*** binding. ***Aluminum***-containing phosphate binders should normally be avoided in **children**. A nondialyzed uremic 6-year-old **male** with congenital renal hypoplasia developed a severe progressive encephalopathy, after 4 years of aluminum hydroxide therapy (average 43 mg Al/day) and died. By increasing gastrointestinal pH, it decreases absorption of acidic drugs (***barbiturates***, ***dicoumarol***, ***nitrofurantoin***, ***penicillin***, ***phenylbutazone***, ***sulfonamides***, and ***warfarin***). It strongly inhibits the absorption of ***doxycycline***, ***oxytetracycline***, and ***tetracycline***; intermediate adsorption of ***troleandomycin***, ***chloramphenicol***, ***digoxin***, only slight adsorption of ***ampicillin***, and ***cloxacillin*** was not adsorbed. As a vaccine adjuvant, it has caused granulomas at injection sites. Unexpected organ uptake has interfered with bone imaging. Has caused constipation and mechanical obstruction (bezoar, even in a newborn **infant**) in the intestinal tract, especially at high or prolonged dosage. A case of large bowel perforation from ***AH*** gel intake is reported. Hypophosphatemia observed in **children** from 5 weeks old to **adults**, especially in those with renal insufficiency. This can occur even when the duration of therapy is as short as 2 weeks. Four hemodialysis **patients** who had stopped taking ***AH*** for 2 or more years stayed hypophosphatemic and developed osteomalacia. A 49-year-old **female** developed osteomalacia, after taking 20 g/day for several months. A severe proximal myopathy and hypophosphatemia developed in three **patients** with chronic renal failure ingesting 90–240 ml ***AH*** gel/day.

LD_{LO}, **child**: 122 g/kg/4 days; IP, **rat**: 150 μg/kg; SC, **rat**: 150 mg/kg.

***Drug interactions*: *Cimetidine*:** Significant decrease of bioavailability.

Ciprofloxacin and ***norfloxacin***, but not ***tosufloxacin***, absorption is decreased by it.

***Diazepam*:** Rate of absorption is decreased, but not the extent of absorption.

***Digoxin*:** Clinically significant decrease in absorption, usually by about 11%. One literature report on four **patients** said NO interference.

***Ethambutol*:** Significant decrease of absorption.

***Indomethacin*:** Absorption rate is decreased, but not the total amount of absorption.

***Naproxen*:** Decreased absorption and decreased plasma levels.

Omeprazole*:** Coadministration with ***AH granules, but not the suspension, decreases its plasma levels.

Phenytoin*:** Some reports of decreased absorption during concomitant administration of ***AH.

***Propranolol*:** Variable decrease or no difference in absorption.

Pseudoephedrine*:** ***AH increases its absorption.

Psychotherapeutics*:** Significant decrease in absorption may occur with ***fluphenazine, ***perphenazine***, ***thioridazine***, and ***trifluoperazine***.

***Rifampin*:** Decreased absorption.

***Sulpiride*:** Clinically significant decrease in availability.

Triazolam*:** Enhanced absorption in dialysis **patients** on ***AH.

ALUMINUM MONOSTEARATE

Stabilizer, suspending agent.

***Use*:** In gels, emulsions, thixotropic gels, or suspensions, vaccines, waterproofing agent in concrete products, ceramics, leather, and canvas.

Untoward effects*:** See ***Aluminum. Responsible for many of the indurations associated with ***penicillins*** in oil.

ALUMINUM NICOTINATE

= *Alunitine* = *Nicalex*

***Antilipemic, peripheral vasodilator.* A 625 mg tablet has 500 mg *nicotinic acid* activity.**

Use*:** Adjunctive therapy in treating hypercholesterolemia and hyperlipoproteinemia in **humans**. As a source of ***nicotinic acid (q.v. as ***Vitamin*** B_3).

***Untoward effects*:** Increases oral anticoagulant effects. Itching, flushing, and nausea reported. Impaired glucose tolerance and obstructive jaundice have been noted. Severe side-effects seemed almost exclusively confined to **women**. Impaired carbohydrate metabolism, intrahepatic obstructive jaundice, hyperalbuminuria, and edema in 54-year-old **male** after 3 g/day/4 years.

ALUMINUM NITRATE

Use*:** Antiperspirant, corrosion inhibitor; in leather tanning, and ***uranium extraction.

***Untoward effects*:** Severe eye irritant to **rabbits**.

LD_{50}, **rats**: 4.28 g/kg; LD_{50}, **mice**: 320 mg/kg.

ALUMINUM OXIDE

= *Alumina* = *Bauxite* = *Baymal®* = *Corundum* = *Gibbsite*

12% level in *attapulgite*.

***Use*:** Adsorbant, dessicant, abrasive, filler in paints and varnishes; in ceramics, glass, air filters, electrical insulators, cosmetics, cereals, and pharmaceuticals.

Untoward effects*:** Inhalation of ***bauxite dust has caused chronic fibrosis in **humans**. The fumes of *recycling* aluminum oxide have caused itching of **workers**' legs. Mined forms may contain contaminants (viz. ***corundum***, used in glass polishing; has also contained ***chromates***, the cause of dermatitis in sensitive **people**).

Sheep grazing near ***bauxite*** quarries developed a lymphohistiocytic reaction in the bronchial and mediastinal lymph nodes.

IP, LD_{LO}, **rat**: 90 mg/kg.

ALUMINUM POTASSIUM SULFATE

= *Alum* = *Kalinite* = *Potassium Alum*

Astringent, antiseptic, antimycotic.

***Use*:** In dye manufacturing, paper, glues and cements, baking powder, pool water purification, vaginal douches, mouth gargles, intrauterine medications, pickling solution for pickles, precipitant for phosphates and bacteria in sewage (estimated at 850,000 tons/year), tanning and photographic industries, and bladder irrigation to control hemorrhage (once called Stypteria by Dioscorides in *De Materia Medica*). 15% in purified talc, as a foot powder for **humans**. 5–10% solutions as a foot pad-toughener for **dogs**; 1% solution as a dip or spray for **sheep** with *Dermatophilus* mycotic dermatitis; topically, in "leg-tighteners" for **horses**, and as a hemostatic agent on **animal** wounds.

***Untoward effects*:** Solutions of 5–10% or more harden the epidermis of **humans** and **dogs**, but stronger solutions can cause excessive hardening. Avoid inhalation. In **swine**, 10 grains can be fatal.

ALUMINUM SILICATE

Use: In glass, paint, ceramic, colored lakes, and printing ink industries.

Untoward effects: Frequent apparent silicosis in **workers** at a ***bentonite*** (q.v.) processing plant.

IP, LD_{LO}, **rat**: 90 mg/kg.

ALUMINUM SODIUM SULFATE

= *Sodium Alum*

Astringent.

Use: Similar to ***aluminum potassium sulfate***; also, as a set accelerator in the concrete industry.

Untoward effects: Avoid inhalation. See ***Aluminum potassium sulfate***.

ALUMINUM SULFATE

Antiseptic.

Use: In foam-type fire extinguishers, styptic pencils, water purification (flocculates algae), tanning, added to canned fish, pickles, relishes, and starches. As a feed additive, it prevents ***fluorine*** toxicosis in many **animal** species.

Untoward effects: A source of controversy as a potential cause of Alzheimer disease, as it adds ***aluminum*** to potable water.

LD_{50}, **mice**: 980 mg/kg.

ALVERINE

= *Antispasmin* = *Calmabel* = *Gamatran* = *Profenil* = *Profenine* = *Prophelan* = *Proverine* = *Sestron* = *Spacolin* = *Spasmaverine*

Anticholinergic, spasmolytic.

Use: Adjunctive therapy for peptic ulcers and eneterocolitis problems in **man**.

Untoward effects: Hypotension, weakness, drowsiness, dizziness, headache, and dry mouth reported after IV use. May decrease blood pressure in hypotensive **patients**.

AMANITA sp.

Has a cup at the base of the stalk, and a loose ring around the stem, under the cup.

Untoward effects: Five species (*A. bisporigera*, *A. muscaria*, *A. phalloides*, *A. verna*, *A. virosa*, and *A. tenuifolia*) have caused serious and fatal poisonings. Other species exist. Nausea and vomiting are the first symptoms. The gastrointestinal tract, kidneys, liver, and blood can be affected with gastroenteritis, hematuria, proteinuria, and jaundice occurring. Symptoms can occur within 6–24 h (some species, within 20 min–2 h), and mortality has ranged between 50 and 90%. Adynamic ileus and small irregularly shaped kidneys are secondary to the healing process of acute tubular necrosis. Delirium, hallucinations, dry mouth, cycloplegia, and ataxia are noted. Toxic to **man** and **animals**. Avoid all fungi, unless you are positive of their identity!!

Treatment: 50–150 mg ***thioctic acid***, IV, every 6 h has been recommended. IV ***physostigmine*** is used only when indicated.

A. capensis: Poisoning in **man** reported in South Africa. Treat the same as with ***A. phalloides*** intoxication.

A. citrina contains the toxin ***bufotenine*** (q.v.).

A. crenulata: First case in the U.S. reported in New England in 1965. A 58-year-old **male** recovered, after appropriate treatment.

A. muscaria = ***Agaricus muscarius*** = ***Fly Agaric*** (so-called because it stuns or kills flies and insects). A parasympathetic stimulant. Toxins are readily absorbed from the gastrointestinal tract. Used for thousands of years in tribal religious ceremonies, and may have been the **Soma**-divine mushroom of immortality of the ancient **Aryans** (ca. 2000 B.C.) of Afghanistan, and India. This so-called food of the Gods may not be fit for **man**. The hallucinogenic effect is attributed primarily to ***ibotenic acid*** (q.v.) and ***pantherine*** (q.v.), but ***choline***, ***acetylcholine***, ***muscaridine***, ***muscarine***, and ***muscazone*** have also been isolated from the plant. Cholinergic effects include bronchial secretions, lacrimation, salivation, gastrointestinal symptoms, and pupillary constriction, and have been treated with ***atropine*** 2 mg. If anticholinergic symptoms appear, 0.5–2.0 mg ***physostigmine*** is usually given. Found in Eurasia from Portugal to Siberia, where it is orange and reddish in color; and in the U.S., where it is nearly white. Psychotomimetic effects within about 15 min after ingestion, with vomiting, tachycardia, blurred vision, excitement, and hallucinations, followed by about 2 h of sleep, marked by colored visions. After awakening, some enjoy performing extraordinary physical feats for about a 3–4 h period. The hallucinogenic principles pass into the urine and, when the plant was scarce, **people**, until recently, consumed such urine for its psycho-tomimetic effects. **Siberians** would trade one reindeer for four mushrooms brought from southern Siberia by dogsled. This helped them pass the long, dark Arctic nights. To prolong the effect, they drank their own urine, or shared it with their friends. Occasional leukocytosis, traces of hematuria, proteinuria, and cylinduria occur. Some become dehydrated with liver (jaundice after a few days) and kidney (oliguria and anuria) problems, bradycardia, miosis, and develop terminal convulsions. In the early 1960s, its juice was said to have been served in certain British cafes. Accidental ingestion reported in Uruguay. Deaths in **humans** have been rare, but one was reported in the state of Washington in the 1930s, and in 1897, when an Italian **count** ate a bowlful in Washington, D.C. A case report describes a **dog** that was poisoned by ingesting it. **Flies** that imbibed its extracts rapidly lost the use of their wings, then their legs, and, within 10–15 min, appeared dead. This effect lasted for at least 50 h.

A. ocreata should be considered toxic.

A. pantherina* = *False Blushes* = *Panther Cap*:** Strong parasympathetic stimulant. Prevalent in Asia and also found in the U.S., Europe, and South Africa. Active toxins are ***pantherine, ***ibotenic acid***, and ***muscimol***, which are rapidly absorbed from the gastrointestinal tract. Four **people** who ingested it developed neurological symptoms, followed by gastrointestinal upsets (recovery after gastric lavage, emesis, sedation, analgesia, ***atropine***, and 40% ***dextrose*** water). A **female** evidenced lightheadedness, blurred vision, and partial amnesia.

A. phalloides* = *Deadly Agaric* = *Death Angel* = *Death Cap* = *Destroying Angel (also a term used for *A. verna*) Strains growing in the states of Washington, Oregon, and California contain relatively large amounts (1.5–1.9 mg/g) of ***β-amanitin*** (q.v.). Also contains ***phalloidin***. Has caused fatty liver and centrolobular necrosis of the liver in **man**. In Switzerland, 12 **persons** were poisoned by it in six instances and one died. A Swiss review of 205 **persons** poisoned by *A. phalloides* during a 10 year period decided that age and latency between ingestion and the first clinical symptoms were of prognostic significance. Lethality was 22.4%, with 51.3% in **children** under 10 years, and 16.5% in those over 10 years. Average latency period was 10.3 h in the fatal cases and 12.6 h in those that survived. Fatal in 84% of those with thromboplastin values below 10%, while all those with values of 40% or more survived. In 1996, a 13-year-old **female** received a liver transplant after being poisoned by it in **her** spaghetti sauce. ***Penicillin***, hyperbaric ***oxygen***, and ***silybin*** improved the survival rate in **humans**. Similar benefits have been reported in **rat** and **mouse** trials. Parenteral therapy with ***thioctic acid*** (q.v.) has been reported as beneficial. Can be removed by hemoperfusion. Some writers report lethality at 30% and 75%. It is estimated that approximately 100 cases occur annually in the U.S., from early spring until frosts. As little as a teaspoonful has caused a painful, lingering death. Sometimes, a full day is symptomless, to be followed by severe abdominal cramps, diarrhea, and vomiting. Vomitus and feces should be examined for mushrooms. Vomitus may contain blood. Hepatic and renal degeneration can occur, as well as a hemorrhagic enteritis. Central nervous system lesions similar to those in ***phosphorus*** poisoning have been observed. When death occurs, it is usually within 4–6 days in **children**, and 6–8 days in **adults**, but deaths have occurred within 48 h. LD_{LO}, **human**: 500 mg/kg.

Poisoned **cattle** have painful defecation, matted feces at the base of the tail, vulva irritation with vesicles, papules, and necrotic areas. Post-mortems showed marked gastrointestinal inflammation, dark liver, and small hemorrhagic areas on the heart. **Dogs** poisoned by it evidence an extreme, decrease in red blood cells and hemoglobin, increased neutrophil leukocytes, decreased lymphocytes, monocytes, and eosinophil and basophil leukocytes. ***Penicillin G*** and ***silymarin*** prevented the abnormal increase of liver enzymes, and the decrease of clotting factors in experimental **dogs**.

A. porphyria*:** Contains ***bufotenine.

***A. preissii*:** Has caused poisoning in **humans** in Australia.

A. verna* = *Fool's Mushroom*:** Some classify it as a subspecies under *A. phalloides*. Has caused outbreaks of poisoning in **humans** and **cattle** (severe intestinal inflammation – several died). Contains ***α- and ***β-amanitines***. Large numbers of poisonings in **humans**, including fatalities, in Japan. Also found in Europe and the U.S. Hepatotoxic. Of ten **members** in one family poisoned by it, five died, the first in 72 h, and the fifth, 5 days and 20 h after the meal. A rare case in Israel reported in 1983 in a 24-year-old **male**. Vomiting and abdominal cramps appeared 10 h after ingestion. Anuria developed 48 h later and **patient** was hospitalized. Despite the late start, treatment with ***charcoal*** hemoperfusion (3 h/day/4 days), hemodialysis (4 h/day/6 days), and IV ***penicillin G*** (250 mg/kg/day), **patient** survived.

A. virosa* = *Destroying Angel (a term also used for *A. phalloides*): Contains ***amanin*** and ***α-amanatin***, acutely toxic to the liver and kidneys. An elderly **couple** ate some and, 12 h later, were admitted to the hospital with severe vomiting and diarrhea. Despite apparent improvement, the **woman** developed hepatic failure and died 2 days later.

AMANITIN

= *Isocholin*

An alkaloid from poisonous *Amanita* and *Galerina* mushrooms. Hematuria, proteinuria, gastroenteritis (abdominal pain, nausea, vomiting, and diarrhea), jaundice, and tubulonephritis are often manifested within 6–24 h, or, occasionally, 48 h after ingestion in humans. Mortality has been as high as 90% in some cases. One of the most lethal toxins known. Lethal dose for humans is about 0.1 mg/kg. *Silymarin* is an effective antidote in dogs and mice. Although considerable disagreement still exists, some still recommend *thioctic acid* and IV *glucose*

therapy. Can be removed by hemoperfusion. Apparently, does not cross the placental barrier (pregnant 21-year-old female whose maternal blood levels of *α-amanitin* were 18.5 ng/ml, had none in her amniotic fluid). *Thioctic acid* 50–150 mg, IV, every 6 h is preferred therapy by some.

α-Amanitin, IP, LD_{50}, **mice**: 0.1 mg/kg.

β-Amanitin, IP, LD_{50}, **mice**: 0.4 mg/kg.

AMANTADINE

= *1-Adamantanamine = 1-Aminoadamantane = 1 Aminodiamantane = EXP 105–1 = NSC 83,653 = Mantadix = Symmetral = Tricyclo[3.3.1.1^{3,7}]decan-1-amine*

***Antiviral. Dopamine* agonist.**

Use: Influenza control in **humans** and **animals**. IP use in **rats** decreases brain ***dopamine*** concentration and releases ***dopamine*** from storage sites in primed **dogs**. This may account for its anti-parkinsonism effect in **man**, although not anticholinergic. In treating drug-induced extrapyramidal symptoms and neuroleptic malignant syndrome (NMS).

Untoward effects: Supposedly, central nervous system or gastrointestinal disturbances of only 1–2%, at recommended dosage; incidence increases with increased dosage. Insomnia, fear, depression, confusion, grand mal seizures, and inability to work in 32% of 44 **females**. Rashes, urine retention, dizziness (33% in young **adults**), nausea, anorexia, peripheral edema, headache, dry mouth, motor restlessness, anxiety; sudden bursts of violent, aggressive behavior; spasmodic torticollis, tremors, ataxia, slurred speech, congestive heart failure, and leukopenia have regularly been reported. A 61-year-old **male** in a suicide attempt ingested 2.8 g (28 capsules), had an acute toxic psychosis, urinary retention, and acid–base disturbances. **He** was treated and recovered. A 37-year-old **female** who ingested 2.5 g in a suicide attempt, developed ventricular tachycardia, torsades de pointes, and prolonged QT interval for 36 h, and died after 10 days with respiratory distress syndrome. A 77-year-old **male** developed neuroleptic malignant syndrome, following sudden discontinuance of 100 mg/day. In treating parkinsonism, bluish red skin, due to livedo reticularis on legs and buttocks, ankle edema, visual hallucinations, and delirium, along with numerous other complaints noted. A 2½-year-old **female** ingested 600 mg, and, in 2 h, was given 15 ml ***syrup of ipecac***. Despite vomiting, **she** became very agitated, and was, possibly, hallucinating. Various treatments had little noticeable effect, but two doses of 0.5 mg ***physostigmine***, IV, 10 min apart, produced an immediate decrease of dystonic posturing and agitation. Normally excreted by the kidneys, and dosage should be decreased in renal insufficiency. Excreted in **human** breast milk. May cause vomiting, skin rashes, and urinary retention in **infants**. Overdose in 2½-year-old **female** (hallucinations, agitation, tremors, and dystonia) was reversed by IV ***physostigmine***. Teratogenicity possible in **humans**. A 29-year-old **female** consumed 100 mg/day during first trimester, to relieve parkinsonism-like movement disorder. Her **infant** was born with cardiovascular maldevelopment: single ventricle with pulmonary atresia. No adverse effect on the fetuses of pregnant **rats** and **rabbits**, in one study, but increased resorption and skeletal anomalies in another. Congestive heart failure has developed in treated **adults**. At dosage rate approximately 12 times usual **human** dose, it is teratogenic and embryocidal in **rats**, but not in **rabbits**. In **man**, a serum concentration of 1 mg/l is considered toxic.

LD_{50}, **rats**: 1275, **mice**: 700, **guinea pigs**: 360, **dogs** and **horses**: 75–500 mg/kg, respectively.

Drug interactions: A 61-year-old **male** had increased serum levels with visual hallucinations and ataxia, after taking ***Dyazide***® (***hydrochlorothiazide*** and ***triamterene***) for 1 week. Its clearance is significantly inhibited by ***quinine*** and ***quinidine***, only in **males**. May increase the effects of ***levodopa*** and ***alcohol***. When added to ***bupropion*** therapy in three nursing home **patients**, it led to neurotoxicity.

AMARANTH

= *CI Acid Red 27 = CI Food Red 9 = CI 16,185 = FD and C Red #2 = Trisodium 1-(4-sulfo-1-naphthylazo)-2-naphthol-3,6-disulfonate*

An azo dye. The term amaranth is also a synonym for *Amaranthus*.

Use: Was in foods, drugs, and cosmetics, including canned fruits, gelatins, candy bars, salad dressings, cereals, cake mixes, frankfurters, ice creams, yogurts, grape sodas, and **pet** foods. Now banned in the U.S. for the above purposes. Used in dyeing wool and silk.

Untoward effects: A 53-year-old **male** ingested 1 teaspoon of ***Elixir of Phenobarbital*** colored with it and had an attack of generalized pruritis with urticaria and angioedema of the uvula, tongue, and circumocular area. Sublingual testing with dilute amaranth solution again precipitated this syndrome. An allergic reaction to it in a **person** reported from South Africa, manifested as pigmented plaques on the shoulders and back. When baked in cookies, it decomposes, forming ***naphthionic acid***.

Of a single oral dose of 100 mg to a **rat**, only 2.8% was absorbed. Oral dosing of **cats** caused markedly red staining of urine and feces. After oral feeding of 30 mg/day/345 days, **rats** developed a fatty degeneration of their livers. **Cats** receiving 92, 187, and 264 mg/kg/day from 0 to 22 days before gestation up to 61–62nd days of gestation had NO ill-effects on implantation or fetuses. Pregnant **rats** and **rabbits** receiving 15, 50, and 150 mg/kg showed NO evidence of teratogenicity. Similar results reported with **mice**. Dietary intake of up to 30,000 ppm in one FDA study with **rats** showed early fetal deaths and another agency's tests showed an increase in malignant tumors, when it was fed to **rats** (approximately 325 mg/day/25 months – 11/450 had tumors and controls had none, which is questionable).

Oral LD_{50}, **rats**: 6 g/kg; IP, 3.5 g/kg.

AMARANTHUS

A. hypochondriacus **(Prince's Feather);** ***A. cruentas*****, and** ***A. caudatus*** **(Love-lies-bleeding, Inca Wheat)**

Use*:** The grains were basic foods for the **Aztecs**. Many species have edible leaves that taste like ***spinach, and some are grown as vegetables. It is called ***Mchicha*** in Swahili, which means ***spinach***. Probably first domesticated for their red dye pigment.

Untoward effects*:** *A. retroflexus* and some other minor species are known as ***pigweed, ***redroot***, ***carelessweed***, and ***water hemp***. Has caused pronounced abdominal distension, staggering, ataxia, sternal recumbency, and perirenal edema in **pigs** approximately 5–10 days after ingestion and which is fatal in approximately 75–80% within about 24 h after they are found ill. Since the plant is known to accumulate ***nitrates*** and ***oxalates***, it is not surprising to find that **cattle**, **sheep**, and **horses** are also occasionally poisoned by it. A common cause of ***nitrate*** poisoning in the southwestern U.S. The pollen has caused allergic reactions in **dogs**. **Rabbits** appear unaffected by it, but a single young **rabbit** may have died, after having eaten it. *A. caudatus* contains a mitogenic ***lectin***, ***β-D-galactose pyranosyl***. *A. spinosus* = ***Kata-khuturia***. The root has been inserted into a **woman's** vagina, as an abortifacient.

AMARETTO

A liquor made in Saronna, is *almond*-flavored, but not from *almonds*. Rather, the flavor is distilled from *apricot* pits. Other Italian and American amarettos owe their flavor to *amygdalin* (q.v.) in non-toxic quantities.

AMARYLLIS

Ornamental.

Untoward effects*:** The bulb contains toxic alkaloids and is a gastro-irritant, when ingested. Nausea, vomiting, purgation, dyspnea, central nervous system excitement, depression, collapse, coma, and death can occur. In **humans**, the symptoms occur rapidly and are generally limited to nausea, vomiting, abdominal cramps, shivering and occasional diarrhea. During World War II, the Dutch, short of **cattle** fodder, found it caused serious upsets when fed, due to its toxic alkaloid, ***lycorine.

AMBELANIA

***Use*:** *A. cuneata*. Its leaves and stems, when crushed and thrown onto still waters, stun **fish**, permitting their easy retrieval. The latex of *A. lopezii* is used topically on scalp ringworm. The leaves are boiled to make an arrow poison.

AMBENONIUM

= *Mysuran* = *Mytelase* = *Win 8077*

***Cholinesterase-inhibitor*. A carbamate.**

***Use*:** In the treatment of myasthenia gravis.

Untoward effects*:** As for ***Neostigmine (q.v.). May accumulate in the body, due to variations in its metabolism. Avoid use with ***atropine***, since the muscarinic effects of overdosage are then easily masked, until more serious nicotinic effects (fasciculation and paralysis of voluntary muscles) occur. Acute cholinergic crises developed in one **patient**, after 5 mg qid/19 days, and in another after 10 mg tid.

AMBLYGONOCARPUS ANDOGENSIS

The bark is used in Africa.

***Untoward effects*:** Causes pupillary dilation and death in **humans** within 24 h.

AMBRETTOLIDE

= *6-Hexadecenlactone*

***Fragrance*. Found in ambrette seed.**

***Use*:** Less than 1000 lbs/year in the U.S. Europe permitted 0.2 ppm as an artificial flavoring in foodstuffs. In the U.S., in soaps, detergents, creams, lotions, and perfumes.

***Untoward effects*:** Non-irritating at full strength to **rabbit** skin for 24 h, and 1% in petrolatum produced no irritation in a 48 h closed-patch test on **humans**, with no sensitization in 26 **volunteers**.

LD_{50}, **rats**: > 5 g/kg; acute dermal LD_{50}, **rabbits**: > 5 g/kg.

AMBUPHYLLINE

= *Bufylline* = *Butaphyllamine* = *Theophylline Aminoisobutanol*

***Bronchodilator*.**

Untoward effects: Due to its ***theophylline*** content. Adverse effects are similar to those produced by ***aminophyllline*** (q.v.) or ***theophylline*** (q.v.).

LD_{50}, **mice**: 600 mg/kg; IV, LD_{50}, **rabbits**: 163 mg/kg.

AMCINONIDE

= *Amciderm* = *CL 34,699* = *Cyclocort* = *Penticort*

Glucocorticoid, anti-inflammatory, antipruritic.

Use: In various dermatoses, including psoriasis.

Untoward effects: Burning, itching, irritation, dryness, folliculitis, acneiform eruptions, hypopigmentation, skin atrophy, striae, hypertrichosis, miliaria, and infections, especially when used topically under occlusive dressings. Avoid use during **human** pregnancies, as it has caused an increase in **fetal** abnormalities. Cushing's syndrome, hyperglycemia, and glucosuria can occur, after topical use, especially in **children**.

AMDRO

= *Amidinohydrazone* = *Hydramethylnon* = *Maxforce*

***Insecticide*. Yellow chemical.**

Use: Control of red fire **ants**.

Untoward effects: Ingestion of an unknown quantity by a **Dalmatian dog** caused vomiting and yellow diarrhea. 14 days later, it went into convulsions, coma, and died after about 24 h, despite manufacturers' claims of low **mammalian** toxicity. Nine 3–5 month **Holstein calves** were fed 113.5 g/day of the toxicant. Within 14 days, leukopenia was noted and significant eosinopenia occurred from days 21 to 49. After 25–39 days, significant decrease of body temperature was observed.

AMELANCHIER ALNIFOLIA

= *Ganháwla* = *Juneberry* = *Junebush* = *Saskatoon* = *Serviceberry*

Cyanogenic shrub.

Use: The berries or fruit are safely eaten by native **tribes** in British Columbia.

Untoward effects: **Cattle** eating very large quantities of the chopped twigs showed restlessness, scours, and dyspnea, attributed to the ***hydrocyanic acid*** in the ***prunasin*** (***mandelonitrile***). Content of the latter can be about 1.4%, but, over 2 or 3% in new spring growth, which becomes tempting forage.

AMETRYN

= *Ametrex* = *Evik* = *Gesapax*

***Herbicide*. A triazine.**

Use: For selective weed control, especially in sugar cane and ***pineapple*** plantations.

Untoward effects: NO specific antidote. Causes eye irritation. Can be absorbed through the skin. If swallowed, induce vomiting, if the **patient** is still conscious. Seek medical help.

LD_{LO}, **human**: 500 mg/kg.

Large oral doses to **rats** leading to muscle weakness, ataxia, dyspnea, salivation, and loss of righting reflex. Mild eye irritation in **rabbits** at 76 mg.

LD_{50}, **rats**: 1100–2980 mg/kg; **mice**: 965 mg/kg; **albino rats**: 1750 mg/kg. Acute dermal LD_{50}, **rabbit**: >10 g/kg.

AMEZINIUM METHYL SULFATE

= *Amezinium Metilsulfate* = *LU 1631* = *Regulton* = *Risumic* = *Supratonin*

Sympathomimetic, antihypotensive.

Untoward effects: Oral toxicity in **rats** and **mice** at doses > 4.5 mg/kg, with depression (central effect) in **mice**. **Rats** also exhibited this effect, but at higher dosage, showed slight stimulation.

LD_{50}, **rats** and **mice**: 1410 and 1630 mg/kg; IV, 45.4 and 40.4 mg/kg, respectively.

AMFEPENTOREX

= *CB 2201* = *Proligne*

Anorexigenic.

Use: In the treatment of obesity.

Untoward effects: In 54 **males** and **females**, ages 17–74, it was useful and well-tolerated, with pruritus in one, nervousness in two, mental excitement in one, heartburn in two, intoxication and discontinuance in one, and hematuric nephritis in one, also with discontinuance. In another group of 35 **males** and **females**, ages 18–60, vomiting reported in two, nausea, asthenia, and indigestion in one each; heartburn occurred, only if not taken with food.

AMICARBALIDE

= *Diamprox* = *M&B 5062A*

***Babesiacide*. The diisothionate is usually used.**

Use: To treat *Babesia* infections, particularly in **cattle** and **sheep**, and anaplasmosis in **cattle**.

Untoward effects: Potentially nephro- and hepatotoxic at the high dosage used against anaplasmosis. May cause local irritation at injection sites. A **calf** treated IV has shown ataxia for several days. **Ponies** treated with over ten times suggested dosage died within 2 days and showed necrosis of the liver, kidneys, and musculature at the injection site.

LD_{50}, SC, **mice**: 120 mg/kg.

AMIDEPHRINE MESYLATE

= *Dricol* = *Fentrinol* = *MJ 5190* = *Nalde*

Vasoconstrictor, nasal decongestant.

Untoward effects: In LD experiments, **rats** and **mice** exhibited pilomotor stimulation, ataxia, hypoactivity, decreased rate and increased depth of respiration, lacrimation, and exophthalmos. Chronic oral administration to **rats** and **rhesus monkeys** (up to 5 mg/kg/day and up to 6 mg/kg/day, respectively) revealed no adverse effects.

LD_{50}, **female rats**: 13–36 mg/kg; IP, 5–25 mg/kg; LD_{50}, **male rats**: 24–229 mg/kg; IP, 144 mg/kg. LD_{50}, **rabbits**: 12 mg/kg; IV and IP, **dogs**: 1.4 mg/kg; **mouse**: 2,284 mg/kg; IP, 780 mg/kg; IV, 190 mg/kg. Oral LD_{LO}, **cat**: 2 mg/kg.

AMIDOAZOTOLUENE

Dye.

Use: In the leather industry.

Untoward effects: Large numbers of **people** in the U.S. developed a dermatitis from their leather wrist-watch straps, due to this dye. Patch test on 47 **volunteers** confirmed the causative agent. Hopefully, the use of the dye has been discontinued.

SC injections in **rats** caused liver carcinomas.

LD_{LO}, **rat**: 1.5 g/kg; **mouse**: 800 mg/kg.

AMIFOSTINE

= *Ethiofos* = *Ethyol* = *Gamaphos* = *NSC 29,691* = *Walter Reed 2721* = *WR 2721*

***Antineoplastic.* IV.**

Untoward effects: Nausea, vomiting, flushing, anxiety, diarrhea, and urinary retention.

AMIKACIN

= *O-3-Amino-3-deoxy-α-D-glucopyranosyl-(1→6)-O-[6-amino-6-deoxy-α-D-glucopyranosyl-(1→4)]]N¹-(4-amino-2-hydroxy-1-oxobutyl)-2-deoxy-D-streptamine* = *1-N-[L(—)-4-amino-2-hydroxy-butyryl]kanamycin A* = *Amiglycin* = *Amiglyde* = *Amikin* = *BB-K8* =*Biclin* = *Biklin* = *Briclin* = *Fabianol* = *Kaminax* = *Kanbine* = *Lukadin* = *Mikavir* = *Novamin* = *Pierami*

***Aminoglycoside [illegible], antibacterial.* A *kanamycin* derivative.**

Use: Particularly active against Gram-negative organisms, *Klebsiella*, *E. coli*, *Serratia*, *Pseudomonas*, *Mycobacteria*, *Staphylococci*, *Enterobacter*, *Proteus*, *Salmonella*, *Nocardia*, *Bacteroides*, and *Bordetella*.

Untoward effects: A review of 144 published clinical trials in 10,000 **patients** summarized the incidence of nephrotoxicity as 9.4%, cochlear toxicity as 13.9%, and vestibular toxicity as 2.8%. A **patient** accidentally received 9 g IV over a 4 h period instead of ***Amicar***® (***aminocaproic acid***) as ordered, had blood levels > 309 µg/ml, and had no evidence of delayed ototoxicity or nephrotoxicity. Azotemia (uremia) reported correlated with high serum levels. Its peak **fetal** kidney concentration was six times the **fetal** serum level, and was equal to that in **maternal** serum. Milk concentrations peak within 4 h of ingestion.

Drug interactions: Increased blood levels in **infants** receiving ***indomethacin*** for patent ductus arteriosus. Of the aminoglycosides, it is probably the least adversely affected by coadministration with ***penicillin***, retaining > 75% of its activity. Synergism with ***azlocillin***, ***carbenicillin***, ***piperacillin***, and ***ticarcillin*** against clinical isolates of *Pseudomonas aeruginosa* has been demonstrated. May interfere with neuromuscular transmission, especially if given with any ***neromuscular-blocking agent***. Hemodialysis and peritoneal dialysis are effective in removing the drug.

AMILORIDE

= *Amikal* = *Amipramidin(e)* = *Amipramizide* = *Amitrid* = *Arumil* = *Colectril*® = *Guanamprazine* = *Midamor* = *MK-870* = *Modamide* = *Puritrid*

Potassium-sparing diuretic.

Use: In hypertension, edema, ascites, cirrhosis, and hyperaldosteronism.

Untoward effects: Gastrointestinal upsets (nausea, vomiting, abdominal pain, diarrhea, constipation, and anorexia), hepatic encephalopathy, taste disturbances, thirst, hyponatremia, weakness, headache, dizziness, leg cramps, photosensitivity, skin rashes, pruritus, paresthesia, impotence, decreased libido, confusion, transient visual disturbances, and hyperkalemia (an occasional problem, particularly with decreased renal function). Six **patients** treated for 7 days developed hypercalciuria. Use with caution, especially in the **elderly** with raised ***potassium*** or ***urea*** levels. A 69-year-old **female** treated with an ***amiloride/hydrochlorothiazide*** combination (***Moduretic***®) developed severe generalized muscle weakness and decreased tendon reflexes with serum ***potassium*** level at 9.3 mmol/l. Rare cases of proximal renal tubular damage and necrosis have been reported. When used in renal failure, it may interact with ***ethacrynic acid*** or ***furosemide***, to increase the risk of ototoxicity. May aggravate muscle weakness and respiratory difficulties in myasthenia gravis. Use with depolarizing skeletal muscle relaxants (e.g. ***suxamethonium***) increases the neuromuscular-blocking effect, and may induce respiratory arrest.

AMINACRINE

= *9-Acridinamine* = *9-Aminoacridine* = *5-Aminoacridine* = *Aminoacrine* = *SD-51*

The hydrochloride form is known as *Acramine Yellow* = *Monacrin*.

***Bacteriostatic agent, antiseptic.* An acridine dye. Non-staining.**

Use: Topical antibacterial agent in wound or skin disease treatments and in semen diluters. Primarily, against Gram-positive and some Gram-negative bacteria. Ineffective against spores and fungi, and, therefore, solutions should be sterilized by autoclaving before certain usage. One of the least irritating of the acridine dyes, but it can retard healing and reduce phagocytosis.

Untoward effects: One report in 1963 of a possible photosensitization reaction in 35-year-old **female**, from its use in a vaginal jelly for approximately 4 months on inflamed tissue (trichomoniasis). **Patient** used a sunlamp the day before the reaction. Can color urine a lemon-yellow color.

LD_{50}, **mice**: 78 mg/kg.

AMINEPTINE

= *Survector*

Antidepressant.

Untoward effects: Acne-like skin lesions after overdosing. Several cases in **women** who developed major macrocystic lesions after 1–9 years of therapy.

AMINOACETONITRILE

= *Cyanomethylamine* = *Glycine Nitrile* = *Glycino-nitrile*

Lathyrogen.

Untoward effects: When fed to pregnant **cows**, it caused congenital defects (excessive flexure of the limbs, malpositioning, malalignment, and rotation of forelimbs) in **calves**. Inhibits liver microsomal enzymes, collagen cross-linking, and decreases tendon elasticity.

TD_{LO}, **monkey**: 450 mg/kg (days 43–48 of pregnancy).

2-AMINOANTHRACINE

Use: Intermediate in dye production.

Untoward effects: Potent carcinogen, producing hepatomas in **mice** after SC injections, a variety of malignant and epithelial tumors in **rats**, and is a bacterial mutagen. 20 min after inhalation by **rats**, it was found in the tubinates, trachea, lungs, liver, kidneys, and gastrointestinal tract. 83% of an oral dose was absorbed. Strong potential for inhalation exposure in **humans** in smoke and refinery accidents.

TD_{LO}, **rats**: 45 mg/kg; TD_{LO}, skin, **rat**, **mouse**, **hamster**: 260, 800, and 6000 mg/kg, respectively.

2-AMINOANTHRAQUINONE

Use: As an intermediate in the synthesis of ***anthraquinone*** dyes (***CI Vat Blues 4,6,12,24***; ***Vat Yellow 1***; and ***Pigment Blue 22***). In coloring oils, fats, waxes, automotive paints, high-quality paints and enamels, plastics, rubber, printing inks, and textile dyes. Less than 1 million lbs used annually.

Untoward effects: **Human** exposure is through inhalation and skin contact. Trace amounts left in some dyes may be a cause of concern, because it has caused liver cancers and other neoplasms in male **rats**. In **mice**, it has caused liver cancer in both sexes, and cancer of blood-forming tissues (lymphomas) in females.

o-AMINOAZOTOLUENE

= *4-Amino-2′,3′-dimethylazobenzene* = *CI 11,160* = *Solvent Yellow 3* = *4-(o-Tolyazo)-o-toluidine*

Azo dye.

Use: In the manufacturing of pigments, for coloring oils, fats, and waxes (shoe polish, etc.), and as a chemical intermediate in the manufacture of ***Solvent Red 24*** and ***Acid Red 115***. Estimates vary, but approximately 400,000 lbs has been produced annually in the U.S.

Untoward effects: Caused the first known chemically induced visceral tumor, namely, hepatomas in **rats** and **mice**. Decreases liver vitamin A content. Has caused carcinomas of urinary bladder, gall bladder, and liver of **dogs**. Repeated dermal application induced liver tumors in **mice** and allergic contact dermatitis in **humans**. Dietary intake induces lung tumors and lung hemangioendotheliomas in **mice**; liver adenomas, hepatocellular and other liver carcinomas, and cholangiomas in **rats**; hepatocellular carcinomas, urinary bladder carcinomas, papilloma of the gallbladder, and mammary carcinomas in **hamsters**.

p-AMINOBENZOIC ACID

= *Amben* = *4-Aminobenzoic Acid* = *PABA* = *Pabacidum* = *Para-Aminobenzoic Acid* = *Paraminol* = *Sunbrella* = *Vitamin* B_x = *Vitamin* H_1

***Vitamin B complex factor.* A *folic acid* precursor. U.S. production is approximately 250 tons/year.**

Use: As a sunscreen, to help prevent erythema, and, possibly, the photocarcinogenic effects of ultraviolet B. Its esters are often used. As a resin modifier and in lipsticks. Antidote to ***carbarsone***.

Untoward effects: Dermatitis and skin photosensitivity with porphyrias in **humans**, as well as rare cases of photoallergy. Studies with **rhesus monkeys** and **infants** suggest it has a role in the pathogenesis of kernicterus and should be suspect in evaluating icterus in **newborns**. Fever occasionally occurs, due to hypersensitivity reaction. May cause goiter and hypothyroidism, by inhibiting ***iodine*** accumulation in the thyroid gland, and clinically significant anemia in **patients** with glucose-6-phosphate dehydrogenase deficiency. **Children** and **adults** eliminate it much more rapidly than **infants**. Will cause yellow discoloration of clothing. Can induce changes in **people's** hair color.

Adequate dietary intake makes **monkeys** more susceptible to malaria.

Drug interactions: ***Salicylate*** blood levels increase, due to its interference with the metabolism of ***salicylic acid*** to ***salicyluric acid***. ***Analgesics***, ***antipyretics***, and ***salicylates*** potentiate it. May decrease ***para-aminosalicylic acid*** activity. Cross-sensitization in **people** allergic to ***p-phenylenediamine***, ***chlorpropamide***, ***tolbutamide***, and some ***local anesthetics***. Reduces the bacteriostatic action of ***sulfonamides***, ***pyrimethamine***, ***coccidiostats***, and some ***trypanocidal drugs***.

LD_{LO}, **human**: 500 mg/kg.

LD_{50}, **rats** and **mice**: > 6 and 2.85 mg/kg; respectively; IV, LD_{50}, **rabbits**: 2 mg/kg; oral, 1.83 mg/kg. In **dogs**: 1 g/kg may produce intestinal hemorrhage and 2 g/kg may result in acute hepatic necrosis.

p-AMINOBENZYL CAFFEINE HYDROCHLORIDE

***Antihypertensive.* Experimental.**

Untoward effects: Jaundice in 66% of treated **patients** (mixed hepato-canalicular histopathology).

γ-AMINOBUTYRIC ACID

= *4-Aminobutanoic Acid* = *GABA* = *Gamarex* = *Gammalon* = *Gammasol*

Antihypertensive and inhibitory neurotransmitter.

Use: For possible experimental use in treating epilepsy, schizophrenia, and spasticity in **humans**. As an antidote for ***urea*** poisoning in **cattle**.

Untoward effects: May induce hyperprolactinemia, gastrointestinal disorders, headache, insomnia, and pyrexia in **humans**. Intracerebral LD_{LO}, **rat**: 18 mg/kg.

Σ-AMINOCAPROIC ACID

= *Afibrin* = *Amicar* = *6-Aminohexanoic Acid* = *Capralense* = *Capramol* = *Caprocid* = *CY 116* = *EACA Kabi* = *EACS* = *Epsamon* = *Epsikapron* = *Epsilcapramin* = *Epsilon-aminocaproic Acid* = *Hemocaprol* = *Hepin* = *Ipsilon*

***Antifibrinolytic agent.* Inhibitor of plasminogen activator substances, thus neutralizing plasminogen's fibrinolytic effect.**

Use: Prevention of rebleeding in subarachnoid hemorrhage. Will antagonize ***streptokinase's*** fibrinolysis. In treating **German shepherd dogs** with degenerative myelopathy.

Untoward effects: Immediately after a 2 g injection, a 71-year-old **male** developed an acute episode of delirium with auditory, visual, and kinesthetic hallucinations. Myoglobinuria in 51-year-old **female** with subarachnoid hemorrhage, after a total dose of 1.43 g over 41 days. A 20-year-old **female** with a subarachnoid hemorrhage, who received 30 g/day/7 weeks, developed acute massive muscle necrosis and acute renal failure. It took 7 months for recovery. Rhabdomyolysis and myoglobinuria in 24-year-old **female**, after 6 g every 6 h, and in a 29-year-old **female**, after 2 g every 4 h/5 weeks, for treatment of subarachnoid hemorrhage. Glomerular capillary thrombosis in 76-year-old **male** who was treated with it to control hematuria. Acute necrotic myopathy in 20-year-old **male**, after 26–36 g/day/6 weeks, and severe proximal myopathy in two **females**, after 18–30 g/day/5 weeks. Resolution occurred in both of these **patients**, after discontinuance of treatment. A 59-year-old **female** with subarachnoid hemorrhage developed a purpuric, morbilliform rash over the front and sides of the chest, after 12 days of treatment. Discontinuance of treatment resulted in the rash disappearing within 72 h. Two months later, retreatment with 6 g every 6 h caused an erythematous rash after the fourth dose, that cleared within 24 h of stopping the drug. Dyspnea, periorbital edema, urticaria and paresthesia of hands in 38-year-old **female** after two treatments for metrorrhagia. Nausea, vomiting, hypotension, dizziness, diarrhea, headaches, tinnitus, delirium, abdominal pains, tympanitis, anxiety, weakness, conjunctival suffusion, color vision impairment, anuria, metabolic acidosis, and bronchospasm are among some of the assorted, albeit often serious, side-effects. A 28-year-old **female** developed acute pericarditis, after 1 g/day/14 days for systemic sclerosis. Thrombosis and emboli are potential hazards in cases of intravascular clotting, as evidenced by thrombocytopenia and prolonged prothrombin time. Dosages as low as 8 g/24 h have, in a few cases, induced thrombotic death in **humans**.

TD_{LO}, **rat**: 150 g/kg; IV, LD_{50}, **mice**: 6.3 g/kg; IP, 7.0 g/kg; IV, LD_{50}, **rat**: 3.3 g/kg; IV, LD_{LO}, **dog**: 2.15 g/kg.

Drug interactions: A hypercoagulable state may occur, when taken with ***oral contraceptives***. Antagonizes ***coumarin*** anticoagulants.

AMINOCARB

= *A-363* = *Bay 44,646* = *4-Dimethylamino-3-toyl-N-methyl-carbamate* = *ENT 25,784* = *Matacil*

***Insecticide*. A carbamate.**

Use: Against **budworms** in spruce forests at about 1 oz/acre.

Untoward effects: Since it is cholinergic and its chemical structure resembles ***acetylcholine's***, poisoned **insects** and **mammals** exhibit violent convulsions and neuromuscular upsets. In **birds**, ataxia, tenseness, lacrimation, salivation, tachypnea, feathers fluffed or drawn tightly against the body, pilorection, diarrhea, dyspnea, tracheal congestion, convulsions, or opisthotonus are signs of intoxication. When sufficient amount is ingested, death occurs in about 1 h.

LD_{50}, **mallard ducks**: 22.5 mg/kg; **pheasants**: 42.4 mg/kg; **mule deer**: 7.5–15 mg/kg; **rabbits**: 30 mg/kg; and **rats**: 30–40 mg/kg; LD_{LO}, **guinea pig**: 50 mg/kg.

4-AMINOCATECHOL

Metabolite of phenacetin.

Untoward effects: As an abnormal metabolite of ***phenacetin*** through a genetically determined defect, a **girl** ingesting ***phenacetin*** developed cyanosis.

Extremely toxic to **dogs** (10 mg/kg caused death, after severe hematuria), but not to **rats**.

2-AMINO-5-DIETHYLAMINOTOLUENE

Use: In color photography developers, as a dye intermediate, and in diazo-type copying of blueprints.

Untoward effects: Lichenoid dermatitis (histology not that of lichen planus) and eczema, including rashes on hands and forearms in **males** and **females** 15–61 years. Eruptions cleared within 3 months after discontinuance of chemical contact. A similar reaction reported in a **worker** handling Agfa film with ***4-amino-N-diethylaniline sulfate***.

2-AMINO-3,4-DIMETHYLIMIDAZO(4,5-*F*)QUINOLINE

= *ME IQ*

Heterocyclic amine.

Untoward effects: Mutagen/carcinogen found in charred parts of fish and meat and in heated foods, that has caused neoplasms of the lung, stomach, and liver of the **mouse**, and in the mouth, liver, mammary gland, intestinal zymbal gland, clitoral gland, and skin of the **rat**. The quinoxaline form = ***ME IQx*** which causes many of the same tumors, also causes lymphomas in **mice**.

4-AMINODIPHENYL

= *4-ADP* = *4-Aminobiphenyl* = *p-Aminobiphenyl* = *p-Aminodiphenyl* = *Anilinobenzene* = *p-Biphenylamine* = *Diphenylamine* = *p-Phenylalanine* = *p-Phenylaniline* = *Xenylamine*

Use: In cancer research; as a reagent in detection of ***sulfates***; and as an antioxidant in rubber (now an obsolete use).

Untoward effects: Industrial exposure of 661 **males** was associated with development and increased incidence of clinical bladder cancer 15–35 years after their first exposure. One tumor developed 133 days after exposure. 17 year survival rate was 88.7%. Toxic exposures in **man** include inhalation, ingestion, or dermal contact (including exposure to the eyes and mucous membranes). Acute or short-term exposure of **humans** causes headache, lethargy, urinary tract burning, blood in the urine, and bluish discoloration of mucous membranes and skin (due to methemoglobinemia). Long-term exposure causes blood and pus in the urine and frequent, painful urination, in addition to bladder cancer. Urinary bladder carcinomas reported in 22 industrial **workers**. Although no longer used as a rubber antioxidant, it can be a contaminant in ***diphenylamine*** (q.v.), used in the manufacture of dyes and in a ***benzidine***-based dye. As a result, it is estimated that there is an exposure potential of ~1000 **workers** to it. Hazardous material. Toxic vapors of ***carbon monoxide*** and ***nitrogen oxides*** can be released in a fire. Research **workers** are also at risk by inhalation or percutaneous absorption. Mainstream cigarette smoke contained 4.6 ng/cigarette, while sidestream smoke contained 140 ng/cigarette. Open vessel operations are now prohibited. Protective clothing, respirator, and exhaust fans are now OSHA requirements.

Carcinogenicity, particularly of the bladder and liver, has been shown after oral administration in experimental **dogs**, **mice** and **rabbits**, as well as mammary tumors in **rats**. In **dogs**, 5 mg/kg/day/7 years was associated with the development of a urinary tract tumor after 4 years. Salivation, loss of body weight, and blood in the urine also occurs in **dogs**. In **mice**, oral and SC injections associated with bladder, liver, and mammary gland cancers.

LD_{50}, **rat**: 500 mg/kg; LD_{LO}, **dog**: 25 mg/kg.

4-AMINODIPHENYLAMINE

= 4-ADPA

Use: As a chemical intermediate in the manufacture of ***p-phenylenediamine*** antioxidants and as a component of hair dyes.

Untoward effects: Has caused contact urticaria in chemical factory **workers**. U.S. manufacturers require protective equipment for their **workers**, if airborne concentrations exceed 0.1 mg/m^3, although risk to **workers** is considered low. Teratogenic, as well as embryotoxic, in **rats**.

2-AMINODIPYRIDOL(1,2-α-3′,2′-d) IMIDAZOLE

= GLU-P-2

***Heterocyclic amine*. Isolated from cooked or charred foods.**

Untoward effects: Causes liver and blood vessel malignancies in **mice** and liver, intestine, zymbal gland, clitoral gland, and skin malignancies in **rats**.

4-(2-AMINOETHYL)-β-DIAZO-2′,4′-CYCLOHEXADIENONE

A direct mutagen from *tyramine*.

Untoward effects: Use in the drinking water of **rats** caused sqamous cell carcinomas of the oral cavity.

2-AMINO-2-ETHYL-1,3-PROPANEDIOL

An alkanolamine (aminoalcohol).

Use: In the synthesis of rubber vulcanization accelerants, surfactants, and pharmaceuticals. As an emulsifier and as an absorbant of ***carbon dioxide*** and ***hydrogen sulfide*** from industrial gases.

Untoward effects: Properly formulated into finished products dramatically decreases hazards from its use. Undiluted material can pose serious hazards for **humans** in the workplace, especially to the skin and eyes, and to the mucous membranes, if heated (low vapor pressure). Very severe eye irritant and may cause corneal burns in **man**, if not adequately washed out **immediately**.

Moderate irritant to **rabbit** skin.

LD_{50}, **mice**: 2.5 g/kg.

AMINOGLUTETHIMIDE

= 3-(4-Aminophenyl)-3-ethyl-2,6-piperidinedione = 2-(p-Aminophenyl)-2-ethylglutarimide = Cytadren = Elipten = Orimeten

***Adrenocortical hormone synthesis-blocker, aromatase inhibitor*. An amino derivative of the hypnotic *glutethimide*.**

Use: In autonomous hypercortisolism in Cushing's syndrome and other adrenal upsets; lowering estrogen levels in the treatment of metastatic breast cancer (especially in **patients** refractory to ***tamoxifen*** treatment), where it blocks the aromatization of androgens to estrogens.

Untoward effects: Are mostly dose-related, and doses of 1 g/day are said to be well-tolerated, but see case reports below. Acute toxicity – **human** leading to nausea, vomiting, abdominal cramps, diarrhea, drowsiness, and lethargy (in 41%), occasional ataxia and dizziness, rash (sometimes, severe cutaneous reactions, including hyperpigmentation and morbilliform rashes (> 30%)), and ocular toxicity. Delayed toxicity – **human** leading to bone marrow depression, fever, hypotension, masculinization and sexual precocity. Pseudohermaphroditism reported in 4-year-old **female** (associated with **mother's** drug treatment while pregnant). Cholestatic jaundice (rare), and hypothyroidism (< 5%). Serious blood dyscrasias (pancytopenia, marked leukopenia, neutropenia, and thrombocytopenia) after 3–7 weeks in 0.9% of the **women** taking it for the treatment of advanced breast cancer. The parent drug, ***glutethimide***, used as a hypnotic, has not caused these blood dyscrasias. The **mouse**, as does **man**, metabolizes the drug by hydroxylation and suffers from a reversible leukocytopenia and thrombocytopenia. The **rat** metabolizes it differently, and shows no such ill-effects. Assays of ***thyroxine-binding globulin (TBG)*** may read too low or nil by aggressive binding to the drug. Fatal thrombocytopenia reported in 49-year-old **female**, after 250 mg tid with ***prednisolone*** 10 mg orally tid for 2 weeks. Agranulocytosis in 69-year-old **female** receiving 1 g/day with ***hydrocortisone***. Similar reports in three **females** ages 64–75, treated for 5–8 weeks. Systemic lupus erythematosis in 57-year-old **female**, after 200 mg qid/6 months. Virilization including deep voice and hirsutism in 19-year-old **female** on 1 g/day/6 years. Long-term treatment of **rats** demonstrated toxic effects to the thyroid, ovaries, adrenals, and uteri of females, and atrophy and mottling of adrenals in some males.

Drug interactions: Decreases anticoagulant effect of ***warfarin*** and ***acenocoumarin***, which is clinically significant, and this has persisted for up to 2 weeks, after drug discontinuance. When used in combination therapy with ***tamoxifen***, it accelerates the latter's metabolism, as it does with ***antipyrine***, ***digitoxin***, and ***theophylline***.

p-AMINOHIPPURIC ACID
= PAH

***PAH* is also used for *Polynuclear Aromatic Hydrocarbons*.**

Use: Diagnostic agent of renal function.

Untoward effects: Causes increased levels of ***penicillin***, and its toxicity, as it blocks ***penicillin's*** renal excretion. Rapid IV infusion of its sodium salt has caused nausea, vomiting, vasomotor disturbances with flushing, tingling, abdominal cramps, and sensations of warmth. Anaphylactic shock in **female astronaut** after injection in weightlessness experiment. ***Probenecid*** decreases its excretion.

2-AMINO-6-METHYLDIPYRIDOL (1,2-α:3′,2′-*d*)IMIDAZOLE
= GLU-P-1

Heterocyclic amine isolated from cooked foods and pyrolysates of amino acids and proteins.

Untoward effects: Carcinogenic in **rats** (liver, intestines, zymbal gland, clitoral gland) and **mice** (liver, blood vessels).

2-AMINO-3-METHYLIMIDAZ(4,5-*f*) QUINOLINE
= IQ

Heterocyclic amine isolated from cooked foods and pyrolysates of amino acids and proteins.

Untoward effects: Carcinogenic in **rats** (liver, intestines, zymbal gland, clitoral gland, skin) and **mice** (liver, stomach, lung).

2-AMINO-2-METHYL-1-PROPANOL

***Corrosion inhibitor*. Solution is mildly alkaline. Combustible. An alkanolamine.**

Use: As a neutralizing amine in steam-boiler condensate return lines; an absorbent for acid gases; emulsifying agent; vulcanization accelerator.

Untoward effects: Severe irritant, if splashed in one's eye. Flush eyes **immediately** with cool water for at least 15 min. Irritant to skin – flush with water, then soap and water washing. LD_{LO}, **human**: 500 mg/kg.

LD_{50}, **rats**: 2.9 g/kg; LD_{50}, **mice**: 2.1 g/kg; IP, LD_{50}, **mice**: 0.32 g/kg.

2-AMINO-3-METHYL-9H-PYRIDO (2,3-*b*)INDOLE
= McAaC

Heterocyclic amine isolated from cooked foods and pyrolysates of amino acids and proteins.

Untoward effects: Carcinogenic (liver, blood vessels) in **mice**.

AMINONITROTHIAZOLE
= Amnizol(e) = Enheptin-T = Entramin = Histosep-S = 5-Nitro-2-thiazolamine

***Protozoacide*.**

Use: Mostly in veterinary medicine, as an antihistomonad in **turkeys** and **chickens**, for trichomonads in **pigeons**, and *Octomitus* in **tropical fish**.

Untoward effects: Three **females** received 30–40 mg/kg/day during their second month of pregnancy for *Schistosoma mansoni*. Non-viable ova shed up to 1 year after treatment. One **female** aborted, after a fall; one **infant** died; and one remained healthy. **Mothers** reported headaches, myalgia, nausea, vomiting, epigastric pain, insomnia, anorexia, adynamia, and emaciation.

Excessive amounts in **turkeys** cause liver necrosis and weight loss, with a danger of nephritis, if inadequate water supplies are available.

TD_{LO}, **rats**: 23 g/kg/46 weeks and 28 g/kg/110 weeks. IP, LD_{LO}, **mouse**: 200 mg/kg.

α-AMINO-β-OXALYLAMINOPROPIONIC ACID

***Lathyrogen*. Amino acid in *Lathyrus sativa* and *L. sylvestris*.**

Untoward effects: Causes neuro-lathyrism in **man**, characterized by ataxia and paraplegia.

AMINOOXYACETIC ACID

***Plant growth regulator, floral preservative*.**

Use: Has been explored as an antiseizure medication.

Untoward effects: Drowsiness and muscle relaxation in 20 **patients** 1–8 years of age.

Chronic administration to **rats** caused xanthurenuria, indicative of ***pyridoxine*** deficiency, yet, paradoxically, the addition of ***pyridoxine*** seemed to increase the toxicity.

6-AMINOPENICILLANIC ACID
= 6-APA = Penicin = Penin

From *Penicillium chrysogenum*.

Use: Intermediate in the manufacture of synthetic ***penicillins***.

Untoward effects: Hypersensitivity in exposed **individuals**; strong, passive cutaneous anaphylactic reactions in **guinea pigs**, and potent antigen, when injected into **rabbits**. A possible antigen in ***penicillin*** allergies.

AMINOPENTAMIDE HYDROGEN SULFATE

= *BL-139* = *Centrine* = *Valeramide OM*

Antispasmodic, anticholinergic, antiemetic.

Use: To relieve gastritis, nausea, and diarrhea in **cats** and **dogs**.

Untoward effects: Produces mydriasis. Avoid use in glaucoma.

Oral LD_{50}, **mice**: 396 mg/kg; IV, LD_{50}, **mice**: 34.7 mg/kg.

m-AMINOPHENOL

= *3-Amino-1-hydroxybenzene* = *3-Hydroxyaniline*

Use: As a dye intermediate and in the manufacture of ***p-aminosalicyclic acid***. Reactant in medium brown hair colorant.

Untoward effects: Potential as with *ortho* and *para* forms below.

LD_{50}, **rat**: 1.66 g/kg; IP, LD_{50}, **mouse**: 150 mg/kg; SC, LD_{LO}, **cat**: 70 mg/kg; Eye, **rabbit**: 100 mg/24 h leads to severe irritation; skin, **rabbit**: 12.5 mg/24 h leads to mild irritation.

o-AMINOPHENOL

= *2-Amino-1-hydroxybenzene* = *2-Aminophenol* = *2-Hydroxyaniline*

Use: In the dyeing of leather, furs, and hair; in the manufacture of azo and sulfur dyes.

Untoward effects: **Worker** exposure can occur in the dyeing industry, especially by exposure to its dust. Widespread **consumer** exposure varies, as it is in over six dozen cosmetic products.

Teratogenic, when given IP to **Syrian golden hamsters** on day 8 of gestation. Slight corneal irritation in **rabbits**.

LD_{50}, **rat**: 1.3 g/kg; IP, **mouse**: 200 mg/kg; SC, LD_{LO}, **cat**: 37 mg/kg.

p-AMINOPHENOL

= *Activol* = *4-Amino-1-hydroxybenzene* = *Azol* = *Certinol* = *Citol* = *p-Hydroxyaniline* = *Paranol* = *Rodinal* = *Unal* = *Ursol P*

Production varies, but annually has exceeded 2 million lbs world-wide.

Use: In manufacturing azo and sulfur dyes; dyeing furs, feathers, and hair; as a photographic developer; and is the key material in the manufacture of ***acetaminophen***.

Untoward effects: **Human** exposure is primarily one of occupational contact. Can cause skin sensitization (contact allergen) and dermatitis in industrial **workers** and 1–2 million **photographers** who develop their own film. Potential exposure can come from contact with dyed fur and feathers. Percutaneous absorption occurs and may cause restlessness, convulsions, and methemoglobinemia. ***Ingestion*** is rare, and causes any of the above symptoms, plus cyanosis, headaches, dizziness, hypotension, respiratory depression, and coma. Lethal dose in **man** is estimated at about 10 g. ***Inhalation*** can cause asthma and methemoglobinemia in **man**. Will cause clinically significant anemia in **patients** with glucose-6-phosphate dehydrogenase deficiency.

LD_{LO}, **human**: 50 mg/kg.

Formation of ***methemoglobin*** has been documented in 16 different **mammal** and **bird** species. Dosages as low as 0.1 mM/kg IV in **rats** caused the terminal third of the proximal convoluted kidney tubule to become necrotic. Only slightly irritant as a topical application on **rabbits'** eyes. LD_{50}, **rat**: 375 mg/kg; **mouse**: 420 mg/kg; Skin, **rabbit**: 12.5 mg/24 h leads to mild irritation; eye, **rabbit**: 100 mg leads to mild irritation.

AMINOPHYLLINE

= *Aminocardol* = *Aminodur* = *Ammophyllin* = *Cardiofilina* = *Cardiomin* = *Cardophylin* = *Cardophyllin* = *Carena* = *3,7-Dihydro-1,3-dimethyl-1H-purine-2,6-dione Compd with 1,2-ethanediamine (2:1)* = *Diophyllin* = *Etilen-Xantisan* = *Euphyllin CR* = *Genophyllin* = *Grifomin* = *Inophylline* = *Metaphyllin* = *Minaphil* = *Pecram* = *Peterphyllin* = *Phylcardin* = *Phyllindon* = *Phyllocontin* = *Somophyllin* = *Stenovasan* = *Syntophylline* = *Tefamin* = *TH 100* = *Theodrox* = *Theolamine* = *Theomin* = *Theophyldine* = *Theophyllamine* = *Theophylline Ethylenediamine* = *Variaphylline LA*

A xanthine.

Bronchodilator, ***smooth muscle relaxant***, ***antispasmodic***, ***myocardial and respiratory stimulant***, ***inotropic agent***, ***diuretic.*** Contains at least 75% ***theophylline***, but more soluble than the latter in water.

Use: In treating bronchospasms; to dilate the coronary arteries. Increases rate of urine formation, and is useful in asthma attacks. Mild, positive inotropic effect on cardiac contraction, with increased cardiac output.

Untoward effects: Nausea, vomiting, abdominal discomfort, dizziness, nervousness, restlessness, irritability, nightmares, hypotension, anorexia, dry mouth (10%), extreme thirst, insomnia, and headaches are among the commonly reported adverse effects. Pain of IM injections is long-lasting. An 83-year-old **female** inadvertently received 850 mg IV and her ***theophylline*** plasma levels were 65.1 µg/ml 1 h later, and 54.5 µg/ml 3 h later. **She** was treated successfully with hemoperfusion, ***activated***

charcoal, ***verapamil***, and ***diazepam***. Another 83-year-old **female** received 288 mg/day by continuous IV infusion for 7 days (serum concentration ranged between 6 and 10 μg/ml). On day 8, levels were 28 μg/ml and **she** experienced tremors and increased agitation. Improvement occurred after medication was discontinued. An 81-year-old **male** (48.3 kg) and a 63-year-old **male** (66 kg) developed hematemesis and gastrointestinal bleeding after treatment with continuous-release tablets (CRT). Autopsy of one revealed acute and chronic esophageal ulceration. A 31-year-old **female** developed severe esophageal ulceration, after taking a CRT without any water. This is a poignant reminder for **prescribers** and **patients**. A 19-year-old **female** developed rhabdomyolysis and acute renal failure, after swallowing 50 tablets. Will cause false positive or spuriously high readings on plasma ***catecholamine*** tests. ***Hyaluronidase*** has been successfully used for treating extravasations.

Abuse: A number of young **people** abused it in combination with ***ephedrine***, to feel "hyper".

Cardiac: Vascular collapse and death reported in cases receiving 250–500 mg in 2 h. Ventricular tachycardia in 78-year-old **male**. Acute myocardial infarction in 54-year-old **female**. Cyanosis, convulsions, and death in 68-year-old **male** after injection. Cardiac arrest in 14 critically ill **patients**. Dosages lower than those normally recommended are suggested for **patients** over 55, or those with chronic obstructive lung disease or congestive heart failure. Rapid IVs often precipitate tachycardia. May cause QT interval prolongation.

Children and neonates: Irritability, restlessness, fever, extreme thirst, and vomiting in 2-year-old **male**, after 60 mg suppository tid/3 days – then semicomatose, dehydrated, and tachycardia. Symptom-free 8 days later. **Infants** are particularly susceptible to its central nervous system effects, such as delirium and convulsions. The drug is cumulative in action and even "safe" dosages have caused vomiting and hematemesis in young **males** and **females**. Transplacental toxicity can occur in a **neonate**. The **mother** had taken the drug intermittently over a 3 year period and during her pregnancy, to control wheezing attacks. A day before delivery, the emergency room gave her 500 mg by suppository, and 100 mg orally, with 100 mg qid until 1 h before delivery. The **newborn** was irritable, vomited, and was jittery. Irritability and hematemesis occurred in a black 3-month-old **female child**, after 3.8 mg/kg IV and ***theophylline*** 2 mg/kg 2 h afterwards. **Newborn** livers are not able easily to detoxify the drug. A 1-year-old **male** vomited brown material, evidenced coma, convulsions, periods of apnea, mydriasis, hypertonus and tachycardia; **child** eventually recovered. In rectal suppositories, its absorption is very erratic in **infants** and young **children**, and has caused fever, acidosis, coma, and even death. Tachycardia, vomiting, and jitteriness in an **infant** up to 6 h after the **mother's** medication. Digit defects in **human** and **rat** fetuses have been ascribed to its maternal intake. A premature **infant** survived after a massive overdose (over 380 mg/kg during a 15–18 h period). Serum ***theophylline*** levels reached 330 μg/ml. Two other premature **infants** were accidentally overdosed in another hospital. **They** survived. Lactating **mother's** milk contains 70% of the plasma's level.

Central nervous system: A 19-year-old **female** with no history of epilepsy developed seizures, after 320 mg suppositories twice daily/3 days. Prolonged focal motor seizures developed in six previously neurologically asymptomatic **patients** (three **males** and three **females** 48–74 years) with chronic obstructive pulmonary disease. Dosage was 1500–2000 mg/day in 4–6 divided doses, with blood levels above 10–20 μg/ml. 5/6 died, with autopsy-proven focal central nervous system lesions. Brain damage in a **child** from an abnormally high dosage resulted in an award of over 2 million dollars. Delirium reported, especially on high dosage.

Gastrointestinal: Although oral doses are best absorbed on an empty stomach, it is wise to take it with food, to avoid nausea and vomiting.

Hypersensitivity: Within 10 min after infusion, shaking, chills, and myalgia in 54-year-old **female patient** with partial lipodystrophy. Reactions are usually due to the ***ethylenediamine*** component, rather than to the ***theophylline***. Hypersensitivity reaction manifests itself by itchy rash with raised plaques, feeling cold, and headaches. An 83-year-old **female**, within 10 min after a 300 mg IV loading dose, developed diffuse urticaria and hypotension. A **Negro** 12-year-old **male** had an immediate hypersensitivity reaction with urticaria and generalized pruritus after 125 mg IV. Urticaria developed on three occasions in a 62-year-old **male**, without adverse sequelae.

Poisoning: A 70-year-old **male** ingested 4.5 g and developed rhabdomyolysis, ventricular ectopic beats, and a grand mal seizure with serum concentration of 99 μg/ml. ***Charcoal*** hemoperfusion for 8 h successfully reversed the symptoms. ***Charcoal*** hemoperfusion and peritoneal dialysis was successful in four **children** (4 months–7 years), including **one** with a serum level of 180 μg/ml. An iatrogenic error in a 4-year-old **female**, who received a supposedly lethal dose (44 mg/kg with plasma concentration of 22.5 μg/ml 9 h later) instead of ***ampicillin***. **She** suffered only mild difficulty and made an uneventful recovery. A fatal case of overdosage simulated acute pancreatitis. A 48-year-old **female** consumed 50, 200 mg tablets 2 h before hospital admission. Confusion, shock, seizures, cardiac arrhythmias, with respiratory and cardiac arrests developed; but **patient** was saved by 6 h of ***charcoal***

hemoperfusion, decreasing the ***theophylline*** concentration from 170 to 20 mg/ml. One author lists the therapeutic serum range as 1–2 mg%, toxic levels as 3–4 mg%, and lethal levels of 21–25 mg%. Simple supportive and symptomatic treatment is inadequate in reversing toxic doses.

Renal: Urinary retention, albuminuria, and hematuria have been reported.

Respiratory: Occasionally aggravates an asthmatic condition and may cause rhinorrhea.

Skin: Cellulitis of perianal region, after suppository use in 54-year-old **male**. Can cross-react with the topical sensitizer ***ethylenediamine***. Dermal necrosis, graying and roughening of skin, followed by sloughing (4"×2") above injection site in 43-year-old **female**. Erythematous or papular eruptions that then become vesicular and crusty, resembling a contact allergy, but are more widespread and symmetrical, are not unusual. Eczematous eruptions are more common in the **elderly**. Also see ***Hypersensitivity*** above.

Uterus: One report indicated it decreased or abolished uterine activity in 60 **patients**.

Drug interactions: Body clearance may be decreased and serum levels increased by use with ***erythromycin***, ***cimetidine***, ***furosemide***, and chronic smoking. May cause extensor-type seizures with ***ketamine***; decreases ***diazepam's*** sedation. ***Isoproterenol*** and ***phenytoin*** may increase its requirement. ***Chloroquine*** potentiates it. Do not use with ***β-blockers***, as the latter may induce bronchial constriction and ***aminophylline*** causes β-adrenergic stimulation. May increase urinary excretion of ***lithium***. May impair absorption of oral ***prednisone***, and block ***adenosine's*** and ***dipyridamole's*** effects. Interferes with ***uric acid*** tests.

IV, TD_{LO}, **human**: 1.43 and 10 mg/kg/day; IM, TD_{LO}, **child**: 50 mg/kg; oral TD_{LO}, **child**: 8.4 mg/kg; rectal **human** TD_{LO}: 25 mg/kg.

Eight 6-month-old **pigs**, each given about 10 g, because of a manufacturer's labeling error, experienced intense excitement and incoordination, followed by collapse and thrashing. Five of the eight died. Gross post-mortem revealed a severe gastroenteritis.

Oral LD_{LO}, **cat**, **dog**, **rabbit**: 800 mg, 290 mg, and 350 mg/kg, respectively; oral LD_{50}, **mouse**: 600 mg/kg.

AMINOPROMAZINE

= Aminopropazine = [Bis(dimethylamino)-2′-3′-propyl]-10 Phenothiazine = Jenotone = Lispamol = Proquamezine = RP 3828 = Tetrameprozine = N,N,N′,N′-Tetramethyl-3-(10H-phenothiazin-10-yl)-1,2-propanediamine

***Antispasmodic*. A phenothiazine.**

Use: Smooth muscle relaxant – genitourinary, gastrointestinal, and respiratory.

Untoward effects: May potentiate ***barbiturates***, ***tranquilizers***, other ***central nervous system depressants***, and ***organophosphorus compounds***. SC, LD_{50}, **rats**: >1 g/kg; IV **rabbit**: 15 mg/kg; 50 mg/kg/day/4 weeks **rats** and **dogs** caused no toxicity.

β-AMINOPROPIONITRILE

= BAPN

Lathyrogen.

Use: In the manufacture of ***β-alanine*** and ***pantothenic acid***. Experimentally, on caustic burns and strictures. Suggested as an antiaging drug.

Untoward effects: Its poisonous effect is potentiated by ***monoamine oxidase inhibitors***. **Workers** exposed to its dimethyl form suffered from urinary retention. It is a toxic substance in sweet pea seed that blocks the formation of cross-linkage in the collagen molecule, and, thus, prevents some scarring.

Feeding 0.03% in **turkey** rations caused dissecting aortic aneurysms and ruptures; internal hemorrhages in **poults**. There are numerous reports of it causing cleft palates in the offspring of **rats** and **mice**. It is also teratogenic in **chicks** and **baboons**. Side-effects in **animals** have included weight decrease, degenerative arthritis, dissecting aneurysms, kyphoscoliosis, and hernias. Its use in **hamsters** has been associated with increased fetal death rate, retardation of uterine growth, exencephaly, encephalocele, gastroschisis, cleft palate, micrognathia, and malformed ribs, fibula, and scapula.

Oral TD_{LO}, **hamster**: 1.25 g/kg; **rat**: 2.5 g/kg; IV, TD_{LO}, **monkey**: 3 g/kg (43–48 days of pregnancy).

AMINOPTERIN

= 4-Aminofolic Acid = 4-Amino-PGA = 4-Aminopteroylglutamic Acid

***Antifolic acid; antineoplastic. Folic acid* analog.**

Use: Rodenticide. As does ***methotrexate***, it knocks out dihydrofolate reductase, essential for a leukemic cell's survival. First used in the 1940s, to induce remission in acute lymphocytic leukemia in **children**.

Untoward effects: Its use in treating psoriasis was associated with 17 cases of visceral and three cases of skin cancers. Teratogenic. Given in the first trimester to **women**, it has caused cleft lip and/or palate, hydrocephaly, meningoencephalocele, shallow supraorbital ridge, prominent eyes, anencephaly, prenatal and postnatal growth retardation, decreased ossification, micrognathia, low-set/rotated ears, cranial synostosis, broad nasal bridge, ocular hypertelorism, short forearms and hands, renal anomalies, mild retardation to normal mentality, as well as **fetal** death.

Spontaneous abortions in **women** taking it in early pregnancy have been common. A **mother** who deliberately took it in an abortion attempt, delivered a **girl** with multiple congenital anomalies, and, during a 17 year continuous follow-up, scored low–normal for mental development. A similar case in a **male child** with physical abnormalities and growth retardation, overcame **his** poor mental development by age 11.

Oral TD_{LO}, **woman**: 0.2 mg/kg (40 days pregnant); another reports 0.12 mg/kg.

Causes abnormalities of extremities and tail, cleft palate and encephaloceles in **rat** fetuses. After IP injection, toxicity is greater in male **mice** than in females. Ten *untreated* **mice**, whose litters received the drug, developed fatal bone marrow depression, apparently due to ingesting the newborns' urine. **Lambs** born to **ewes** that received the drug by SC injection developed osteoporotic fractures, as well as congenital malformations (rotational and flexing deformities of the appendicular skeleton, kyphosis, scoliosis, torticollis, aplasia of the mandible, overflexion of the carpal joint, hypermobility and dorsiflexion of the hock and stifle joints, and relaxation of the fetlock joints; aplastic ears were also noted). Doses of 0.1 mg/kg SC in **cats** at varying stages of pregnancy showed no adverse fetal effects.

Oral TD_{LO}, **rat**: 50 μg/kg (7–10 days pregnant); oral LD_{LO}, 2.5 mg/kg.

AMINOPYRIDINE

= *Avitrol* = *Compound 1861* = *DRC-1327* = *Fampridine* = *4-Pyridineamine*

Avian frightening and repellent agent.

Use: In treating Eaton–Lambert syndrome and reversing ***fentanyl***-induced postoperative respiratory depression, botulism, and myasthemia gravis in **man**, and protecting crops from **birds**. It enhances transmission at the neuromuscular junction, and, therefore, it is used experimentally and clinically, to antagonize non-depolarizing ***neuromuscular blocking agents*** and some other ***anesthetic agents***.

Untoward effects: Poisonings in **humans** have been reported from The Netherlands. Has been reported as an ingredient in some Chinese herbal remedies for rheumatism. ***Pancuronium bromide*** and artificial ventilation have been used in treating its toxicity. Therapeutic serum concentrations in **man** are reported as 0.025–0.075 mg/l.

Repellent–toxicant properties for **birds**. When ingested from broadcasted baits, it causes, within 10–20 min, poorly coordinated flight, unilateral paralysis, circling, sleepiness, panic and secondary distress calls and escape responses in nearby **birds**. Deaths in **birds** occurred in about 20 min to 2 h after ingestion. Each of two **horses** ate about 1 kg of treated ***corn***, and, within 4–6 h, showed signs of fright, backing movements, sweating, convulsions, rapid fluttering of nictitating membranes, and died within 2 h after symptoms appeared. Rigor mortis appeared rapidly, and there were no significant postmortem findings. 100 mg/ml on toxic **sheep** collars killed attacking **coyotes** within 40 min. Accidental ingestion of treated pellets killed five **white-tailed deer** and two **goats** in a zoo.

Oral LD_{50}, **pigeon**, **quail**: 8 mg/kg; **duck**: 4. 2–5. 2 g /kg; **wild birds**: 2–4 g/kg; **cowbirds**: < 1 mg/kg; **rat**: 20 mg/kg. 3 mg/kg/day was lethal to a **dog**, but lower dosage was not. LC_{50} for **catfish** and **bluegills** ranged from 2.43 to 7.56 mg/l.

AMINOPYRINE

= *Amidazophen* = *Amidofebrin* = *Amidopyrazoline* = *Amidopyrine* = *Aminophenazone* = *Anafebrina* = *Brufaneuxol* = *Dimapyrin* = *Dimethylamino-analgesine* = *4-(Dimethylamino)-antipyrine* = *Dimethylaminophenazone* = *Dipirin* = *Dipyrine* = *Febrinina* = *Itamidone* = *Mamallet-A* = *Netsusarin* = *Novamidon* = *Piridol* = *Polinalin* = *Pyradone* = *Pyramidon*

***Analgesic, antipyretic.* A pyrazolone. *Phenylbutazone* is also one, and closely related chemically.**

Use: In febrile convulsions.

Untoward effects: ***Blood***: Has caused a large number of severe agranulocytosis cases in **man**, but not reproduced in **dog** experiments. A cause of severe neutropenia and occasional leukopenia (even fatal) cases. Chinese herbal medicines sold over-the-counter in California, for the relief of arthritis and back pain, contained substantial undeclared amounts of the drug, and may have been the cause of agranulocytosis in several **people**, including the death of one. Can cause hemolytic anemia in glucose-6-phosphate dehydrogenase deficiency. Aplastic anemia has also been reported due to its use. Agranulocytosis is usually relieved within 2 weeks of drug discontinuance, but prompt recurrence is noted, even with the smallest dosage.

Clinical tests: Causes false positive or high readings on Coomb's test.

Hypersensitivity: May be involved in many agranulocytosis cases. An interesting report from Israel states aplastic anemia and agranulocytosis is rarely seen there. Some food colors (viz. ***tartrazine***) and ***sodium benzoate*** may cross-react with it. May cause anaphylaxis, and may be responsible for some of the adverse renal effects seen, as well as hemolysis of red blood cells. Urticarial wheals have been noted in the urinary bladder and intestines.

One report describes two cases of suffocation, after ingestion of therapeutic doses. See ***Skin***.

Liver: A few reports of hepatotoxicity, especially on high dosage. It is probably a hypersensitivity reaction. ***Vitamin C*** is protective in **rat** trials.

Renal: Will color urine red, due to a metabolite, ***rubazonic acid***. Hematuria and cylindruria reported on dosages > 4 g/day. Fluid and electrolyte retention also reported.

Respiratory: May cause asthma in ***aspirin***-sensitive **individuals**.

Skin: Fixed-drug eruptions, contact urticaria, toxic epidermal necrosis, acute porphyrias, and angioneurotic edema have been reported.

Drug interactions: Effect is decreased by use with ***phenylbutazone*** in **man**. Various **animal** trials have shown it can stimulate the metabolism of ***carisprodol***, ***hexobarbital***, ***meprobamate***, ***phenylbutazone***, and, probably, ***oral contraceptives***. Two days after influenza vaccination, 2/12 **people** had approximately a 50% decrease in the drug's metabolism. ***Phenobarbital*** increases the drug's elimination in **man** and **rats**.

Some countries have restricted its use, as well as its derivative, ***dipyrone***. Its use is now very limited, since the advent of newer, and, possibly, more effective, analgesics exists. Some countries also banned its use, for fear it might react with ***nitrites*** or ***nitrous oxide*** in saliva and some foods, to cause carcinogenic nitrosamines. This effect can probably be easily prevented by an increase in ***vitamin C*** intake. Considerable interindividual variation in metabolic rates can contribute to toxicity.

Convulsive in **dogs** and **rabbits** with estimated LD_{50} at 150 mg/kg. **Rabbits** receiving 0.3 g/kg and **cats** (0.2 g/kg) developed glycosuria, proteinuria, oliguria, and increased non-protein nitrogen. Pathology revealed vacuolar degeneration and necrosis of the glomerular capillaries and tubules, with an increase of fibroblasts and fibrocytes between the tubules. High dosage to **rats** causes papillary necrosis. Its metabolism, with the release of ***formaldehyde***, may cause excitement, incoordination, and muscle weakness in **cattle**.

IV, LD_{50}, **rats** and, **mice**: approximately 250–325 mg/kg; oral LD_{LO}, **rat**: 0.7–1.38 g/kg; oral LD_{50}, **mouse**: 1.85 g/kg.

p-AMINOSALICYLIC ACID

= *4-Amino-2-hydroxybenzoic Acid* = *4-Amino-salicylic Acid* = *Apacil* = *Apas* = *Deapasil* = *Hellipidyl* = *Pamacyl* = *Pamisyl* = *Para-Pas* = *Parasal* = *Parasalicil* = *Parasalindon* = *PAS* = *PAS-C* = *Pascorbic* = *Pasnodia* = *Pasolac* = *Propasa* = *Rezipas* = *Sanipirol-4*

***Antitubercular*. Bacteriostatic against *Mycobacterium tuberculosis*. Additive and synergistic with the *streptomycins* and *isoniazid*. Chemically related to *tetracaine*, which also inhibits tuberculosis.**

Use: To inhibit bacterial resistance of tuberculosis to ***streptomycin*** and ***isoniazid***.

Untoward effects: Hypokalemia (some **patients** with hypertension develop hyperkalemia), alopecia, acidosis, joint pain, nervousness, acute pancreatitis, and, rarely, ototoxicity and amblyopia have occurred. May cause hemolysis in **neonates**, due to immaturity of glucose-6-phosphate dehydrogenase enzyme system, and in glucose-6-phosphate dehydrogenase-deficient **adults**. May rarely produce a disoriented, confused state, delirium, and psychosis.

LD_{LO}, **human**: 500 mg/kg.

Cardiac: A rare cause of myocarditis.

Clinical tests: May produce increased or false positive urinary glucose (Benedict's), protein (as ***albumin***), urobilinogen, crystals, red blood cell or hemoglobin values, and discolored urine. False elevations of SGOT and altered vanillylmandelic acid assays occur.

Endocrine: Goiter, with or without myxedema. Doses of 1.5–2.0 g qid lower plasma cholesterol by 15–20%. Hypoglycemia and diabetes mellitus have been reported.

Fetus: Slight risk of ototoxicity and cerebral damage.

Gastrointestinal: This is the most common adverse effect (10%), and is manifested by bitter taste, nausea, vomiting, diarrhea, abdominal pain, steatorrhea, and blood streaked stools. A malabsorption syndrome can also occur.

Hypersensitivity: Often between the tenth and thirtieth day of treatment (7%). Fever, skin eruptions, lupus-like reactions and pruritus, Stevens–Johnson syndrome, lichenoid lesions, mononucleosis-like syndrome, lymphadenopathy, methemoglobinemia, leukopenia, eosinophilia, agranulocytosis, neutropenia, thrombocytopenia, epistaxis, sideroblastic anemia, hemolytic anemia, jaundice (1%), mixed hepatocellular hepatitis with necrosis and degeneration, pleuritis, encephalopathy (with rash that recurs on re-exposure), vasculitis, and Leoffler's syndrome. Acute hemolysis, with hemolysis and hemoglobinuria developed in a 24-year-old **female**, after a single oral dose of 6 g. Other similar fatal cases have been reported. Hemolytic anemia and jaundice have occurred after local treatment. A mortality rate of about 20% occurred when the drug was not discontinued early in the course of hepatitis. A **man** became unconscious, after swallowing 12 g. The following day, **he** evidenced bulbar paralysis (speech disturbances, limited tongue mobility, and anisocoria). Four days later, **he** showed mild jaundice, with enlarged liver and urobilinuria. It has caused acute anaphylactic shock.

Renal: Acute interstitial nephritis, dysuria, renal papillary necrosis, and a case of fatal anuria, due to acute tubular necrosis reported. Renal complications are rare.

Respiratory: Generalized systemic lupus erythematosus reaction, pulmonary eosinophilia, pleuritis, mediastinal and generalized lymphadenopathy reported.

Skin: Lichenoid eruptions, exfoliative dermatitis, erythema multiforme, and fixed eruptions. A 5+% level of skin reactions helped clinicians to switch to ***ethambutol*** in tuberculosis therapy.

Drug interactions: With high dose, ***isoniazid***, and ***vitamin*** deficiency, its use has been associated with optic neuritis. It retards the hepatic acetylation of ***isoniazid***, increases the latter's elimination half-life and duration of action in the blood, leading to an increased risk of hepatitis. Its blood levels are increased 50–200% by ***probenecid***, and its own toxicity is increased by use with ***aspirin***. It intensifies ***ethionamide's*** hypoglycemic effect. Decreases the effectiveness of sulfonamides. It decreases the absorption of ***rifampin***, ***digoxin*** (by about 20%), ***folic acid***, ***vitamin*** B_{12}, ***fats***, ***diphenhydramine***, and ***isoniazid***. It increases the toxicity of ***demeclocycline***. Potentiates ***phenytoin*** (decreases its metabolism), ***coumarin*** and ***indanedione anticoagulants***, and ***monoamine oxidase inhibitors***.

Prolonged administration to **rats** decreases microsomal metabolism of ***hexobarbital***.

LD_{50}, **mouse** and **rabbit**: 4 g and 3.65 g/kg, respectively.

5-AMINOSALICYLIC ACID

= *5-ASA* = *Asacol* = *Claversal* = *Fisalamine* = *Mesalamine* = *Mesalazine* = *Mesasal* = *Pentasa* = *Rowasa* = *Salofalk*

Anti-inflammatory agent.

Use: In treating ulcerative colitis (Crohn's disease), including **patients** unable to benefit from or tolerate ***sulfasalazine***. Has been used topically on oral and vaginal ulcers. Used in manufacturing ***sulfur***, azo dyes, and light-sensitive paper.

Untoward effects: It, like other ***salicylates***, may suppress bone marrow function, especially with the co-existence of impaired renal function. Two **patients** (17 and 37 years) reported exacerbation of pain and diarrhea. A peripheral neuropathy, confirmed by rechallenge, occurred in 38-year-old **female** on 400 mg/tid/1 month. Lupus-like syndrome in 38-year-old **female**, after 1.2 g/day/7 months. Alopecia reported in two **patients**, after ***5-ASA*** enemas. Despite the fact that little is excreted in breast milk, watery diarrhea has been noted in a breast-fed **infant** of a treated **mother**. After 17 months of treatment, a 37-year-old **female** developed Churg–Strauss syndrome with mononeuritis multiplex, rash, meningism, and intermittent blindness.

Blood: Fatal aplastic anemia in 71-year-old **male**, after 800 mg/twice daily for several months. Aplastic anemia in 48-year-old **male** on 400 mg/tid/12 months. **He** was successfully treated with ***antilymphocyte globulin***. Neutropenia in 75-year-old **male**, after 1.2 g/day/4 months; recovered only after 300 μg/day SC ***filgrastim***. Fatal thrombocytopenia in 63-year-old **female** on 800 mg/tid.

Hypersensitivity: Lung opacities, bronchospasm, and erythematous skin rash in 54-year-old **female** on 250 mg/tid/5 days. Fever, rash, arthritis, pericarditis, and pericardial effusion in 30-year-old **male** on 800 mg/day/3 weeks. Pruritic rash, fever, and arthralgia in 46-year-old **male**. A fatal myocarditis reported in a 20-year-old **female** on 1.5 g/day. Thirteen days after therapy was started, **she** reported sudden, severe chest pains and dizziness. Electrocardiogram showed left anterior hemiblock, followed by ventricular fibrillation.

Pancreas: Pancreatitis, clearly confirmed by rechallenge, in 34-year-old **female** on 250 mg every 4 h. A 32-year-old **female**, on 800 mg/tid/2 days also developed acute pancreatitis, which was confirmed by rechallenge.

Renal: Nephrotoxic with interstitial nephrotic syndrome and severe renal failure in 61-year-old **female** on 800 mg/tid/5 months.

Respiratory: Alveolitis (hypersensitivity?) in 67-year-old **male** on 1 g/day – rectally, followed by erythematous rash, dry cough, dyspnea, and progressive interstitial lung disease. Eosinophilic pneumonia in 30-year-old **female** on 800 mg/day/8 months. Pleural effusion, pulmonary infiltration, and pericardial effusion after 800 mg twice daily.

Nephrotoxicity has been demonstrated in **rat** trials. LD_{50}, **mouse**: 7.75 g/kg; IP, LD_{LO}, **mouse**: 313 mg/kg.

2-AMINO-1,3,4-THIADIAZOLE

Untoward effects: SC, LD_{50}, **rat**: 100 mg/kg in two or more doses at 17 h intervals. 75 mg/kg on days 10 and 11 of pregnancy, caused a fetal resorption rate of 20%. Viable fetuses had decreased size and weight, depressed skulls, and pointed lower jaws with absent or stunted tails; 25% had twisted or stunted hind limbs; vertebral and rib defects existed in all; and missing toes in 50%. It is hyperglycemic in the **rat**.

LD_{50}, **rat**: 200 mg/kg.

2-AMINOTHIAZOLE

= *Abadol* = *Basedol* = *2-Thiazolamine*

Use: As an intermediate in the production of ***sulfathiazole***. In treating psoriasis.

Untoward effects: Has caused contact urticaria in **humans**. LD_{50}, **rats**: 0.48 mg/kg; IP, **mouse**: 200 mg/kg.

AMINOTRIAZOLE

= *Amigol* = *3-Amino-1,2,4-triazole* = *Amitrole* = *Amizol* = *ATA* = *Cytrol* = *ENT 25,445* = *Herbizole* = *1,2,4-Triazole-3-amine* = *Weedazol*

NOT *Aminitrozole*, which is *Acetamido-5-nitrothiazole*.

***Herbicide, antithyroid agent*. Sparked the "*cranberry* crisis" of 1959.**

Untoward effects: A **farmer** used it in a ***cranberry*** bog, just prior to harvest, contrary to directions for use. Minute amounts (a few ppm) were found in a few of his samples, and the nation was almost put into panic and the industry was put into a disaster by authorities. U.S. government authorities chose not to believe science, which says everything is not black and white, and, contrary to the Delaney Amendment, there is neither an absolute "**zero**" in toxicity nor 100% safety. Its use was banned by the U.S. Environmental Protection Agency for cropland use in 1971, but it can still be used around non-food **fish**.

LD_{LO}, **human**: 5 g/kg.

Rats fed 100 ppm/68 weeks developed thyroid cancer, a natural result of its antithyroid effect. Hyperplasia of mammary glands and liver tumors have also been noted. Acute toxicity in **geese**, **cats**, **dogs**, **mice**, and a **horse** caused increased gastrointestinal motility and smooth muscle contraction in the lungs. **Sheep** that ate freshly sprayed weeds staggered, became depressed and anorexic, and died within 3–7 days. Post-mortem findings were mild pulmonary edema and congestion of the abomasal mucosa and liver. **Horses** died with a gastroenteritis. Thyroid enlargement noted in experimental chronic oral dosing of **sheep** and **chicks**. Pulmonary edema and severe gastrointestinal hemorrhages noted in **sheep**.

LD, **sheep**: 4 g/kg; LD_{50}, **rats**: 1100–2500 mg/kg; **mallard ducks**: > 2 g/kg; **quail**, **mallard ducks**, and **pheasants**: > 5000 ppm in feed.

AMIODARONE

= *Amiodar* = *Ancoron* = *Atlansil* = *Angiodarona* = *Codarone* = *Codarex* = *Codarone X* = *Miocard* = *Miodaron* = *Ortacrone* = *Ritmocardyl* = *Rhythmarone* = *SKF 33,134A* = *Trangorex*

***Antiarrhythmic, antianginal*. A benzofuran, chemically related to *thyroxine*. Contains *iodine*.**

Use: To help control some atrial and ventricular cardiac arrhythmias and block some α- and β-adrenergic effects. Effective in 50–75% of **patients** with recurrent ventricular tachycardia.

Untoward effects: Contact urticaria, delayed erythema, eczematous dermatitis, cough, pulmonary fibrosis (2–18%), pneumonitis, increased SGOT levels, increased ***thyroxine*** levels, thrombocytopenia, neutropenia, paresthesia, tremors, parkinsonism, dizziness, sinus node depression, sensitivity to sunlight with purple discoloration of the skin (***pyridoxine*** appears to be protective against this adverse effect), optic nerve degeneration, visual difficulties (yellowish-brown corneal microdeposits and halos seen around objects), nausea, vomiting, constipation, lupus-like syndrome, neuromuscular effects (20–40% – myopathy, gait disturbances, proximal muscle weakness), decreased libido, ventricular tachycardia (torsade de pointes), and only a few case reports of vasculitis; also a case report of sicca syndrome. Paradoxically, it has worsened some cases of ventricular tachycardia. Acute pancreatitis, nausea, vomiting, and constant epigastric pain 4 days after a loading dose of 800 mg/day. A 67-year-old **male** had an exacerbated congestive cardiac insufficiency culminating in acute lung edema after a 300 mg IV infusion. Two **patients** developed sinus arrest, after 600 mg and 1 g/day, respectively. A 41-year-old **male** developed gynecomastia after 200 mg/day. The cause–effect was verified by rechallenge. A 55-year-old **male** taking 400 mg/day/20 months developed bilateral epididymitis. Another 42-year-old **male** on 400 mg/twice daily had a similar reaction. Fatal liver dysfunction reported, although mild liver problems occur in up to 40% of studied **patients**. After a 1500 mg IV loading dose, caused reversible fulminant hepatitis in 69-year-old **male**. A 72-year-old **male** developed fatal hepatic encephalopathy after 100–400 mg twice daily/8 months for treatment of paroxysmal atrial flutter. A 20-year-old **female** ingested 8 g (40, 200 mg tablets) of her mother's medication in a suicide attempt. 12 h later, **she** was conscious and coherent, but perspiring profusely. Elimination half-life is 31.5 h, and this **patient** made an uneventful recovery. Depression, suicidal ideation, poor memory and concentration in 84-year-old **male**, ingesting 200 mg/day. IV use causes hypotension (15–20%), bradycardia and atrioventricular block (5%), peripheral phlebitis, plus other symptoms, as with oral use. Asymptomatic bone marrow granulomas in two **patients** on 200 mg/day. High levels can accumulate in breast milk and can be a cause of secondary toxicity (especially pulmonary) to the **infant**.

Drug interactions: Increases prothrombin time in **patients** on ***warfarin*** (often requires 50% reduction in ***warfarin*** dosage), increases ***acenocoumarol*** levels, and increases

digoxin serum levels (usually 70–100%). Does not affect ***digoxin*** kinetics in the **rabbit**. May decrease ***theophylline, phenytoin, flecainide, procainide, aprinidine, cyclosporine, clonazepam,*** and ***quinidine*** clearance, leading to toxicity in **man**. Prolongation of the QT interval noted with ***quinidine***, also noted with ***disopyramide, propafenone***, and ***mexiletine***. ***Methotrexate*** skin necrosis reported as an interaction with amiodarone. Torsades des pointes has occurred, when amiodarone is given with ***disopyramide***. ***Rifampin*** decreases its serum concentration and that of its metabolites. ***Ritonavir*** decreases its metabolism, increases its serum concentration and can increase its toxicity. It can inhibit metabolism of ***calcium channel-blockers, codeine, dextromethorphan***, and ***hydrocodone***.

IP, IV, LD_{50}, **mouse**: 254 mg and 178 mg/kg, respectively.

AMIPHENAZOLE

= *DAPT* = *Daptazile* = *Daptazole* = *2,4-Diamino-5-phenylthiazole* = *Dizol* = *Fenamizol* = *Phenamizole* = *5-Phenyl-2,4-thiazolediamine*

Narcotic and barbiturate antagonist.

Use: Has been used in ***morphine*** addiction in **man**.

Untoward effects: Lichen planus and oral ulceration in 53-year-old **male** on 100 mg qid. Many reports of lichenoid reactions in **patients**.

AMITRAZ

= *Azaform* = *BAAM* = *BTS 27419* = *ENT 27967* = *Mitaban* = *Mitac* = *Taktic* = *Topline* = *Triatox* = *U 36059*

Insecticide, acaricide.

Use: As a scabicide in **man**, in treating demodectic mange on **cats** and **dogs** and sarcoptic mange in **swine**, in treating *Acarapis* tracheal infestation in **honey bees**, and in killing **sheep** and **cattle ticks**.

Untoward effects: Poisoning in 15-month-old **female**, 30-year-old **male**, 35-year-old **male**, and 82-year-old **male** after drinking up to 100 ml (250 ppm) leading to drowsiness in 4/4; bradycardia, miosis, and/or hyperglycemia in 3/4; and hypotension and/or vomiting in 2/4.

A **dog** ate part of an amitraz-impregnated collar. Mydriasis and bradycardia were the main symptoms. Delayed gastrointestinal transit time has been demonstrated in **mice**. Transient sedation reported in approximately 8% of treated **pets**, within 2–6 h and lasting up to 72 h. Transient pruritus in 3% of the cases. Convulsions, ataxia, hypothermia, bloating, polyuria, anorexia, vomiting, diarrhea, bradycardia, edema and erythema, and several fatalities have been reported. Although a 0.25% solution appeared safe, spraying **horses** with a 0.1% solution can cause fatal colics.

LD_{50}, **dog**: 100 mg/kg; **rats** and **mice**: 400 and 1600 mg/kg, respectively.

AMITRIPTYLINE

= *Adepril* = *Amitid* = *Amitril* = *Dihydrobenzocycloheptadiene* = *Elavil®* = *Endep* = *Laroxyl* = *Lentizol* = *Redomex* = *Ro 4-1575* = *Saroten* = *Sarotex* = *Seroten* = *Triptisol* = *Triptizol* = *Tryptanol* = *Tryptizol*

***Antidepressant, anxiolytic. Doxepin* is an analog. Therefore, its topical use as an antipruritic is not surprising. *Nortriptyline* is an active metabolite. Approximately 95% metabolized in the liver, with the rest renal.**

Combined with *perphenazine = Etrafon = Mutabons = Triavil = Tritafen*.

Combined with *chlordiazepoxide = Lentizol = Limbatril = Limbitrol Forte*.

Combined with *fluphenazine = Amiperm*.

Use: In the treatment of depression, resulting from psychosis or neurosis, requiring a calming, sedative effect.

Untoward effects: Its anticholinergic activity is responsible for a number of the untoward effects (tachycardia, arrhythmias, dry mouth, blurred vision, and constipation) described below, and can often be controlled by dose reduction. Use caution and refrain from using it in glaucoma and with ***monoamine oxidase inhibitors***. The **elderly** tend to have more untoward effects.

Abuse: 86/346 **people** in a ***methadone*** maintenance program admitted abusing the drug to achieve euphoria. A non-accidental poisoning in a 2-month-old **male** expanded one author's view of "child abuse and neglect".

Blood and cardiovascular: Orthostatic hypotension, increased eosinophils in 38/100 psychiatric **outpatients** age 10–79 on 50–100 mg/day/6 months. Hypotension and/or tachycardia in about 5% of the **patients**. Leukopenia, leukocytosis, agranulocytosis, and eosinophilia occur in approximately 1% of **patients**. At lower dosage, blood pressure is usually increased, but decreased at higher dosage. Arrhythmias in **children** at dosages of 20 mg/kg or greater. Fatal cardiac insufficiency in 67-year-old **male** on less than 50 mg/day; coronary thrombosis in 66-year-old **male** on 150 mg/day/3 months; severe dyspnea in 75-year-old **female** with history of coronary, after second 25 mg dose; severe hypotension during surgery in 44-year-old **female** on 75 mg/day/4 weeks; deep coma and abnormal electrocardiogram in 34-year-old **female** after 44.2 mg/kg; unexpected deaths in 6/53 cardiac **patients** within 24 h, and no deaths in 53 controls. Thrombocytopenia, sympathetic dominance caused by this tricyclic's anticholinergic action; neutropenia also reported. ***Contra-indicated*** during acute recovery phase following myocardial infarction and in congestive heart failure.

Brain: Precipitation of seizures (3 days–3 weeks after starting treatment), including a case of tonic–clonic seizures. It decreases the seizure threshold. Central nervous system symptoms include excitation, restlessness, peripheral neuropathy and paresthesias, muscle twitching, rigidity, hyperreflexia, confusion, hypothermia, disorientation, headache, delirium, hallucinations, and coma.

Clinical tests: Chronic dosing affects thyroid function tests.

Dysarthria: A 48-year-old **female** received 25 mg tid. On day 3, dosage was increased to 50 mg tid. A day later, **she** had speech difficulty.

Eye: Blurred vision in 7/23, dizzy feelings, headache, and shakiness in 6/23, hypomania in 1/23 **females**, 47–78 years on 25–75 mg/tid/4 weeks; mydriasis, ophthalmoplegia, disturbed accommodation, acute glaucoma, and gaze paralysis also reported. Optic neuritis is rare.

Fetus: A few cases of various anomalies of the leg and foot have been reported.

Galactorrhea: Induces hyperprolactinemia.

Learning and memory: Tests on 20 healthy **students** (20–22 years), after 2 weeks on 25 mg tid, showed impairment of immediate memory and acquisition of learning tasks.

Libido: Caused impotence or delayed ejaculation in **men** and delayed orgasms in **women**.

Liver: A hypersensitivity reaction may be responsible for some of the hepatitis reported. Fatalities have occurred. Intrahepatic bile retention has caused jaundice in 0.5–1.0% of **patients**.

Miscellaneous: Hypothermia, agitation, and increased heart rate and blood pressure from apparent increased levels of free cerebral ***catecholamine*** released from binding sites, weight gain.

Neonate: Breast milk levels 15 h after a single 100 mg oral dose were approximately the same as maternal levels. None detected in nursing **infants'** serum.

Neuromuscular: Dose-related myoclonus, due to cholinergic antagonism and serotoninergic stimulation. Peripheral neuropathies (numbness, tingling, burning sensations of the extremities).

Photosensitivity: An uncommon cause.

Poisonings: Accidental and in suicide attempts. Non-dialyzable. A 19-year-old **female** survived one of the highest doses (70–75 tablets, 1750–1875 mg) reported in a poisoning case. Clonic convulsions, unconsciousness, tachycardia (100/min), blood pressure 120/80. Forced diuresis was followed by uneventful recovery. A 39-year-old **female** ceased breathing, had dilated pupils, and blood pressure 80/0, 2½ h after ingesting 800 mg (possibly with ***alcohol***). Recovered after 5280 mg ***methylphenidate*** over an 8¼ h period. Accidental ingestion of about 50, 25 mg tablets by a 15 month **child** proved fatal. In 26-year-old **female**, pulmonary consolidation of both lungs occurred within 48 h and **she** died on day 17 (the drug is concentrated in lung tissue and may inhibit sufactant production). Many treatments have been used: External cardiac massage and ***glucagon***; ***propanolol***, IV for tachycardia; ***pyridostigmine*** 0.5–1.0 mg, IM for convulsions; forced diuresis – 250 ml normal ***saline*** with 10% ***mannitol***, ***physostigmine*** 1–3 mg, IV to alleviate cardiac and decrease central nervous system toxic effects (if arrhythmias persist, ***sodium bicarbonate*** and ***potassium*** have been tried). Additionally, 1 mg ***physostigmine***, SC may be needed every h for 24 h, then 4 mg, IV every h for 24 h; then 4 mg, IV every 2 h for an additional 36 h. Hemoperfusion through ***Amberlite XAD-4*** has also been used. One **patient** reportedly survived an acute ingestion of 10 g. Hyperglycemia and self-induced water intoxication reported in some cases. Involved in hundreds of cases of lethal poisonings and suicides annually. Munchausen syndrome by proxy with cardiac arrhythmias and seizures in 5-year-old **male** after being given daily doses by his **mother**.

Psychic and central nervous system: Severe psychic upset (depersonalization–derealization response) in 27-year-old **female**, after 25 or 50 mg dosing on three separate occasions. Confusion, disorientation, visual and auditory hallucinations, slurred speech, marked paranoia, hysteria, epileptoid fits and convulsions, restless leg syndrome, and paresthesia also occur.

Renal: Urinary retention (usually managed with ***bethanechol***). Blue–green urine color.

Drug interactions: ***Alcohol***: Potentiation (deaths reported, especially in the first few days of therapy), adverse effects on driving and other skills. Amitriptyline can ehance ***alcohol–disulfiram*** reaction.

Antacids: Have caused a decrease in amitriptyline absorption.

Bethanechol: Although it is used to counter amitriptyline's anticholinergic effects, it apparently led to the development of duodenal ulcers in an obese 76-year-old **female**.

Bethanidine: This adrenergic-blocking antihypertensive agent takes up circulating ***catecholamines***, and amitriptyline is a potent inhibitor of this mechanism.

Chlordiazepoxide: High percentage of drowsiness and dizziness, inability to work, fear of responsibility, restlessness, weakness, dry mouth, constipation, nausea, impaired motor coordination, tachycardia, and blurred vision are among the symptoms in approximately 20% of the **patients** on combination therapy.

Contraceptives, oral: May decrease antidepressant effect.

Diazepam: The magnitude of the interaction varied greatly between **subjects**.

Fluoxetine: Caused decreased amitriptyline clearance and death of 40-year-old **male** when used with ***fluoxetine*** for ~6 weeks.

Furazolidone: A toxic psychosis developed in a **female**, after ***furazolidone*** (a monoamine oxidase inhibitor) was added to **her** therapeutic regime.

Guanethidine: **Patients**' hypertension was no longer controlled after the addition of amitriptyline, whose adverse effect lasted for 18 days after discontinuing amitriptyline.

Hypericum: ***St. John's Wort*** decreases its serum concentration.

Methadone: Its narcotic, analgesic action is potentiated by amitriptyline.

Methyldopa: Tachycardia, hypertension, palpitations, agitation, and hand tremors reported in 50-year-old **male**.

Methylphenidate: Inhibits the metabolism of amitriptyline and increases its blood levels.

Perphenazine: Dizziness, headache, weight gain, paresthesia of extremities, urinary retention, blurred vision, dry mouth, restlessness, nausea, drowsiness, hypotension, cholestatic jaundice, ventricular arrhythmias, decreased cardiac output, nightmares, and confusion are among the adverse effects of this combination. A 26-year-old **female** ingested 27 tablets of the combination and developed a delayed hypomania.

Phenelzine: A 38-year-old **female**, in a suicide attempt, ingested 800 mg amitriptyline and 180 mg ***phenelzine***; **she** became disoriented, somnolent, with bilateral Babinski signs, distended bladder, and difficulty in voiding. Recovered after lavage.

Protriptyline: 51 depressed **females** evidenced transient dry mouth, sweating, and decreases in both systolic and diastolic blood pressure (up to 10 mmHg).

Reserpine: Inhibited by amitriptyline.

Thioridazine: Syndrome of lethargy progressed to muteness, rigidity, drooling, cold and clammy hands, fast and thready pulse, fever, myoclonus, severe dystonia, and, eventually, coma in 44-year-old **male**. **Patient** had been on 100 mg thioridazine tid for 2 years. Ten days before this toxic crisis, 2.5 mg amitriptyline bid–tid was added to **his** treatment. **Patient** recovered after treatment. A case of life-threatening ventricular arrhythmias (tachycardia and fibrillation) responded to therapy.

Thyroid hormones: Increase receptor sensitivity to ***catecholamines*** already enhanced by amitriptyline; thus, increasing potential for cardiotoxicity.

Case reports indicate LD_{LO}, **adult human**: 500 mg; in **infants**: TD_{LO}, 50 mg/kg; in 1–2-year-old **children**: 1000 mg is usually a fatal dose. Lethal serum concentration in **humans**: 0.5–2 mg/l.

Sedation, weight increase, poor hair coat, and cholinergic effects in treated **cats**.

LD_{50}, **rats**: 320 mg/kg; LD_{50}, **mice**: 147–280 mg/kg; IV, 29 mg/kg; IV, LD, **rabbits**: 10 mg/kg; IM and SC, **dog**: 40–50 mg/kg.

AMLEXANOX

= *Aphthasol*

Topical treatment of aphthous ulcers.

Untoward effects: Local stinging and burning. Occasional nausea and diarrhea.

AMLODIPINE

= *Norvasc* = *UK 48,340*

Calcium channel-blocker.

Use: Oral treatment of hypertension, vasospastic angina, and chronic stable angina. Steady-state plasma levels after 7–8 days of daily dosing.

Untoward effects: Flushing (male: 1.5%, **female**: 4.5%); peripheral edema (dose related, **male**: 5.6%, **female**: 14.8%); dizziness; palpitations (male: 1.4%, **female**: 3.3%); headache (7.5%); fatigue (4.5%); nausea (2.9%); abdominal pain (1.6%); sleepiness (1.4%); back pain (1%). Approximately 1% had coughing, rhinitis, thirst, dry mouth, epistaxis, and tinnitus; 1–2% had dyspnea, pruritus and rashes. A 73-year-old **male** on 5 mg/day experienced very transitory moments of weakness on the sixth or seventh day. On day 9, these occurred with sensations of facial flushing and vague heat sensations in the abdomen. All of these lasted only about 2 s, with fear of falling, not fainting. The frequency increased rapidly to about 12/day in a few days. No attacks 36 h after discontinuance of medication. Rarely, increased shedding of growing hair.

Blood: A 34-year-old **female** treated with 10 mg/day/3 days developed severe thrombocytopenia with petechiae, vaginal bleeding, purpura, recurrent epistaxis, and exacerbation of acute intermittent porphyria.

Cardiovascular: Ventricular tachycardia and atrial fibrillation, bradycardia, peripheral ischemia, acute myocardial infarction and syncope, hypotension (approximately 1%). Purpura. A rare, paradoxical angina can occur.

Eye: Xerophthalmia, abnormal visual accommodation.

Gastrointestinal: Anorexia, constipation, dysphagia, diarrhea, severe abdominal pain, flatulence, and gingival hyperplasia (approximately 1%); dyspepsia (1–2%).

Libido: Decreased sexual function (1–2%).

Musculoskeletal: Arthralgia, arthrosis, and myalgia (~1%); muscle cramps (1–2%).

Poisoning: A 19-month-old **male** ingested 30 mg (approximately 2 mg/kg). Heart rate was 180/min. Treated with ***ipecac*** and recovered. An **adolescent** died after ingesting 140 mg. Therapeutic serum levels reported as 0.003–0.015 μg/ml; comatose or fatal at 0.1–0.2 μg/ml.

Renal: Frequent micturition, nocturia, polyuria.

Elderly **patients** and those with hepatic function problems may require a significant (50% ±) dosage reduction.

Drug interactions: May cause a ***sulfone*** hypersensitivity reaction when given with ***ampicillin/sulbactam***. A **patient** taking ***Lotril*** (***benazepril/amlodipine***) as a hypotensive developed a dangerously low blood pressure when taking ***Cimicifuga racemosa*** (***Black Cohosh***). Little interaction with ***grapefruit juice***. Now, it appears that it increases amlodipine bioavailability by 20%.

Single oral doses of 4 mg/kg in **dogs** led to marked peripheral vasodilation and hypotension. Induces catalepsy in **mice**. Single oral doses of 100 mg/kg in **rats** and 40 mg/kg in **mice** were fatal. After an overdose to a **dog** (36 mg/kg) it developed non-cardiogenic pulmonary edema.

AMMELINE

Use: In organic synthesis and the manufacture of synthetic resins (for utensils, on phonograph records) and in electrical insulating material.

Untoward effects: Blindness in **chicks** in western Japan was traced to this chemical added to increase the ***nitrogen*** content of fish meal. Dilation of the pupil and blindness within 2–7 days after consuming feed with 0.5% chemical content. It has a specific affinity for the retina, causing disappearance of the fuscin granules, edema, cellular degeneration, and retinal detachment. Related to ***melamine***, which does NOT cause blindness.

AMMI MAJUS

= *Bishop's Weed* = *Greater Ammi* = *Artrillal*

Use: An extract (***methoxsalen*** or ***xanthotoxin***), taken orally permits most **people** to achieve a sun tan without painful burning. Antileukodermic in Arabian medicine.

Untoward effects: A cause of phytophotodermatitis and contact dermatitis from handling the seed in **man**.

One gram of seeds fed once/day/12 days to 5-week-old **chicks** and **poults**, while exposed to 5 h sunlight/day. On the second day, all showed erythema of the comb and eyelids. Severe keratoconjunctivitis, pigmentary retinopathy, and photophobia developed. Lesions were more severe in **geese** and **ducks**, and included retinal damage. Young **geese** also developed chronic lesions of the beak and foot web. Photosensitization has also been reported in thousands of **cattle** and **sheep**. The plant may also concentrate and contain dangerous levels of ***nitrates***.

AMMI VISNAGA

= *Chellah* = *Khella* = *Toothpick Ammi*

Contains the furanochromones, *khellin* and *visnagin*, and the furocoumarins, *methoxsalen* and *imperatorin*. *Cromolyn*, an antiasthmatic, is derived from its *khellin*.

Use: **Arabs** have used it as an antiasthmatic and in the treatment of angina pectoris. Its fruiting pedicels serve as toothpicks in Egypt.

Untoward effects: Although its ***khellin*** has been used as a bronchodilator and coronary vasodilator, its use in the U.S. has been discontinued, because of its cumulative toxicity. Photosensitization, due to its ingestion by **geese** in Uruguay and **chickens**, **ducks**, and **turkeys** in Israel. Causes pronounced vasodilation and bradycardia in **rabbits**. Cause of vesicular dermatitis in **fowls**.

AMMONIA

Pungent odor; corrosive gas. Hazardous material.

Use: Annealing of stainless steel, refrigeration, chemical synthesis, water purification, sewage treatment, pH control, cleaners and detergents, direct soil application as a fertilizer, in military explosives, lacquers, dyes, rubber, in lithium batteries, in hair colorants, disinfectants; ***aflatoxin*** inactivation, and to preserve and add ***nitrogen*** to hay and silage.

Untoward effects: Ingestion of caustic ammonia products in the home is common (accidental and suicidal). Dialyzable. A 39-year-old **female** accidentally ingested "household ammonia", with resulting gastric necrosis, esophageal perforation, and duodenal and jejunal strictures – survival after three extensive surgeries. Hyperammonemia in an **infant**, due to ***ornithine transcarbamilase*** deficiency reported. Hyperammonemia also occurs in **neonates** on hyperalimentation with ***amino acid*** solutions. A 59-year-old **female** similarly developed an encephalopathy receiving an ***amino acid/dextrose*** solution

IV. In urine, causes a diaper rash in **infants**. Increased blood levels contribute to the encephalopathy of Reye's syndrome. A 46-year-old **male** accidentally swallowed some strong ammonia water. **His** breathing became rapid and stertorous, his voice husky, and **his** glottis, uvula, tonsils, lips, gums, and tongue became swollen. **He** had enlarged cervical lymphatics, headache, delirium, and coughing with bloody sputum. Serious symptoms subsided, and recovery began after about 4 days.

Acute intoxication from inhalation of it under pressure, with wheezing, laryngeal and pulmonary edema, acute asphyxia, tracheobronchitis with ulcerations, and bronchial stenosis as common symptoms; after a few days, a rapidly fatal hemorrhagic pulmonary edema. Clouds of potentially lethal leaks of anhydrous ammonia have occurred from storage tanks and terminals, injuring many, including passing **motorists**. Faulty valves on farm use tanks have also caused serious injury.

Although labels clearly warn against it, **people** often mix ***hypochlorite*** bleaches with it, releasing toxic gases. A 60-year-old **female** reported burning sensations with irritation of conjunctiva, mouth, and nasopharnyx, with flushing and uncontrollable coughing, followed by severe dyspnea, headache, and vomiting, gradually reducing in intensity after several hours. A steel cylinder fell off a **horse**-drawn wagon and released a considerable amount of gas, cauterizing the mucous membranes of the **horse's** corneas, nostrils, and mouth. Accidents have also happened in ammonia refrigeration plants.

Do **NOT** underestimate its potential severity (viz. a **person** with irreparable damage to an eye, after a bank raid, with slowly developing face, mouth, and nose burns). Early increase in intraocular pressure, mimicking acute angle-closure glaucoma, corneal edema, and semi-dilated, fixed pupils are typical. It penetrates tissues readily, and has been used in the commission of crimes. Vulnerable **subjects** should wear protective goggles or transparent shields. **Farmers** using anhydrous ammonia in the fields should always keep a 5 gallon can of water nearby, and a small bottle of water in **their** pockets, as instant flushing with water of contaminated skin and eyes is the best emergency treatment for ammonia injury. Slowly decaying **fish** in a ship's hold produced toxic quantities of ammonia and other toxic gases, leading to three deaths, and four cases of unconsciousness among commercial **fishermen**. A contact dermatitis, particularly in the paper industry, has been reported. It has a strong affinity for water. Therefore, it is very irritating to moist skin.

Inhalation – **human** threshold – maximum of 50 ppm in the workplace. 20–35 mg/m^3 appears to be safe. One report says 20 ppm is toxic and 30,000 ppm was fatal, after a 5 min exposure. Lethal dose in **humans** is about 10 ml of strong solution, although one **patient** recovered after a 60 g dose.

Ammonia can cause edema of the palpebral tissue of **birds** and nasal discharge in young **calves**. Oral ***acetic acid*** is an effective antidote for ammonia and urea poisoning in **cattle**. Livestock confinement systems can be a cause of serious, even fatal, ammonia respiratory, etc. effects on **swine**, **poultry**, and **humans**. At 100–200 ppm in the air, it induces sneezing, salivation, anorexia, and head shaking in **pigs**, which makes them more susceptible to secondary respiratory diseases. 1 oz of strong ammonia water killed a **horse** in 16 h, and 3 oz killed one in 50 min.

Inhalation LC_{LO}, **rat**: 2000 ppm/4 h; **cat** and **rabbit**: 7000 ppm/1 h.

AMMONIACAL COPPER ARSENATE

= ACA = All Weather Wood = Chemonite

Use: As a wood preservative, especially where wood is to be in direct contact with the ground.

Untoward effects: Treated lumber has been found to have 2.7–23.1 mg of ***arsenic***/sq ft of surface (90 mg on one sample). Although the ***arsenic*** does not readily penetrate **one's** skin, the wearing of gloves when handling such treated lumber is recommended. Remember, **infants** love to mouth things, and should be kept from treated lumber. Sawdust from treated lumber, in **workers** spending a lifetime sawing it, dramatically increases an ***arsenic***-induced cancer.

AMMONIUM CHLORIDE

= Amchlor = Ammonium Muriate = Darammon= Sal Ammoniac = Salmiac

Originally, from the soot after burning camels' dung. The name comes from such use in the Libyan desert, near the temple of Jupiter Ammon. Over 3 million metric tons produced annually world-wide.

Use: In fire-proofing mixtures, batteries, stain removers, inks, soaps, bread doughs and leavening agents in yeast foods as a source of ***nitrogen***, and to slow melting on ski slopes. As a systemic acidifier, expectorant, and diuretic.

Untoward effects: Use in the **mother** may cause acidosis in the **mother** and the **neonate**. Metabolic acidosis is apt to occur in liver or renal disease. May cause kernicterus in **infants** and **rhesus monkey** fetuses. Hyperchlorhydria and hypokalemia reported after injection. Two rare reports of cholinergic crisis in non-myasthenic **patients** taking small doses (31-year-old **female** on 5 mg qid/2 weeks and 81-year-old **female** on 10 mg twice daily/1½ days being treated for chronic bladder problems). One previous report of a **patient** also taking ***mecamylamine***, a ganglion blocking agent. A 54-year-old **female** developed encephalopathy and coma, after a standard IV dose for metabolic acidosis. A 58-year-old **female** with a long history of urinary tract infection and renal stones received 6 g/day, in addition to other therapy. **She** exhibited dyspnea and exhaustion in the emergency room. **Her** chronic renal disease probably triggered **her** severe metabolic acidosis.

Drug interactions: It acidifies the urine, increases the excretion of ***amphetamines***, ***magnesium***, ***chloroquine***, ***ephedrine***, ***fenfluramine***, ***procaine***, ***quinine***, ***meperidine***, ***mecamylamine***, ***levorphanol***, ***imipramine***, and some ***quinidine***. Inhibits the effects of ***tricyclic antidepressants***. The half-life of ***chlorpropamide*** is prolonged by it. Acidification of urine decreases ***salicylate*** elimination, increasing risk of ***salicylate*** toxicity, especially in the presence of large or prolonged dosage of ***salicylates***.

LD_{50}, **rat**: 1.65 g/kg; IM, LD_{50}, **rats**: 30 mg/kg.

AMMONIUM FLUORIDE

= *Ammonium Monofluoride*

Use: In ***beer*** and ***wine*** production, as an antiseptic, an inhibitor of secondary fermentation, and in decreasing cloudiness. In etching pastes for glass and for frosting electric lamp bulbs.

Untoward effects: Pustules develop at the site in 14% of **human volunteers** patch tested with 4% material/48 h. These **patients** apparently had subclinical pre-existing inflammation or edema at these sites. There is central necrosis in the large pustules, and these ***fluoride***-induced pustules result in scarring. In **rabbits**, application of 0.1–1.0% solution to skin scratches for 24 h led to a double row of sterile pustules, but not when applied to non-traumatized skin or given intradermally.

Nausea, salivation, vomiting, abdominal pain, diarrhea, hemorrhagic gastroenteritis, muscle weakness, tremors, convulsions, cardiovascular collapse, increased respirations, depression, and death can follow ingestion. Low dose chronic intake has caused mottling of dental enamel, osteosclerosis, synostoses and calcification of tendons and ligaments in **humans**. TWA, **human**: 2.5 mg as ***fluoride***/m^3.

IP, LD_{50}, **rat**: 32 mg/kg.

AMMONIUM HEXACHLOROPLATINATE

= *Ammonium Platinic Chloride*

Use: In ***platinum*** plating and making spongy ***platinum***.

Untoward effects: Industrial exposure has caused "occupational asthma" or platinosis (see ***Ammonium platinus chloride***). Inhalation LD_{LO}, **human**: 0.9 μg/m^3; TWA, air, **human**: 2 μg ***platinum***/m^3.

AMMONIUM NITRATE

Approximately 5 million metric tons produced annually world-wide.

Use: In explosives, in blasting, matches, pyrotechnics, and in making ***nitrous oxide***.

Untoward effects: As a farm-used fertilizer, with fuel oil (ANFO) it is explosive, and has replaced ***dynamite*** in many situations, and as a defoliation desiccant on grape vines. It has been involved in disastrous explosions in Texas City, Oklahoma City, and Brest (France), after being heated. The inadvertent supply and consumption of it in solution, instead of liquid whey, caused ***nitrate*** poisoning in a herd of **cattle** (75 showed clinical signs and 17 died); staggering, ataxia, shallow respirations, rapid breathing, tachycardia, very dark-colored venous blood, and jugular pulse. IV ***methylene blue*** was a rapidly effective antidote in all but one **animal**. A similar incident occurred elsewhere in **heifers** licking fertilizer off a dirt floor.

AMMONIUM OXALATE

Use: In the manufacture of explosives, in metal polishes, in dyes, in detinning ***iron***, and in assaying ***calcium***, ***lead***, and rare earth metals.

Untoward effects: **Poisonous**! 9/10 **sheep** died within 12–22 h after an oral intake of 550 mg/kg. A sharp decrease in plasma ***calcium***, an increase in urea ***nitrogen***, indicative of renal damage, plus inflammation and necrosis of the forestomachs occurred. IV use in **cattle** rapidly caused rumen atony, followed by tachycardia, skin tremors, increased respirations, polyuria, apparent blindness, ataxia, paralysis, clonic convulsions, and, eventually, death. These are also the symptoms of ***urea*** poisoning.

AMMONIUM PERSULFATE

= *Ammonium Peroxydisulfate*

Use: Oxidizer and bleacher, in dye manufacture, in electroplating and etching ***zinc***, and in cosmetics (boosting peroxide hair bleaches – usually added as a dry powder, just before use).

Untoward effects: Irritant and allergic, contact dermatitis, even with temporary hair loss, pruritus, edema, urticaria, rhinitis, asthma and syncope in **humans**. It has induced severe, life-threatening, anaphylactoid reaction in **people**. Many countries in Europe permitted it, to help whiten flour, and it caused a hand dermatitis or eczema in hundreds of **bakers**. A **woman** who baked bread in Greece acquired a severe hand dermatitis from it. **She** again had a severe reaction and flare-up of her hand dermatitis, after having **her** hair bleached in the U.S. Dermatitis of the hands has been common in **hairdressers**.

LD_{50}, **rats**: 820 mg/kg.

AMMONIUM PLATINOUS CHLORIDE

= *Ammonium Tetrachloroplatinate* = *Ammonium Chloroplatinate*

Use: In photography.

Untoward effects: One of the many compounds causing platinosis (hay-fever syndrome, scaly erythema, urticaria, eczema). Scratch or intradermal testing can be life-threatening in sensitive **people**.

Inhalation TC_{LO}, **human**: 0.9 µg/m^3.

AMMONIUM SULFAMATE

= *Amcide* = *Ammate* = *AMS*

Herbicide.

Use: In flame-proofing textiles and yarns, paper products, insulating materials, and in electroplating solutions. For exterminating ***Canada thistle***, ***poison ivy***, ***ragwood***, and hardwood trees.

Untoward effects: Irritating to skin and mucous membranes in **humans**. TWA, **human**: 15 mg/m^3.

LD_{50}, **rats**: 1.6–4 g/kg; **mouse**: 3.1 g/kg.

AMMONIUM SULFATE

= *Mascagnite*

Over 2 million metric tons produced world-wide annually.

Use: In the manufacture of ***ammonium alum***, flame-proofing fabrics and paper, as a hide-softener in the leather industry, as a fertilizer, and in holes on the top of tree stumps, to kill their roots and suckers.

Untoward effects: The most common acidic aerosol in smog. Avoid it and any ***ammonium salts*** in **patients** with liver disease.

Aerosoled material impairs breathing function in **guinea pigs**. Contamination of feedstuffs, or careless use on pastures, has caused abdominal pain, groaning, incoordination, increased pulse rate, dyspnea, hemorrhagic diarrhea with bellowing and violent struggling preceeding death in **cattle**; yet, 200 g as a drench to a 274 lb **calf** failed to produce any symptoms.

LD, **bovine**: 500 mg/kg; LD_{50}, **rat**: 3 g/kg.

AMOBARBITAL

= *Amal* = *Amasust* = *Amsearb* = *Amylobarbitone* = *Amytal* = *Barbamil* = *Barbamyl* = *Dorminal* = *Dormytal* = *5-Ethyl-5-isoamylbarbituric Acid* = *Eunoctal* = *Inmetal* = *Isomytal* = *Mylodorm* = *Pentymal* = *Sednotic* = *Somnal* = *Stadadorm*

Sedative, hypnotic, anesthetic.

Use: As above.

Untoward effects: Fatigue, headache, "hangover", and drowsiness have been common complaints. Ataxia, nausea, and vomiting have occurred. Blood pressure decrease often occurs, after IV use, as well as respiratory depression, usually returning to normal within 4 h. Necrotic ulcerations at injection sites, due to pH of 11 of a 10% solution. Addiction occurs within 30 days, and it has been ingested in attempted suicides. A single dose will interfere with the ***nalorphine*** pupil test for narcotic addiction. It concentrates and reaches equilibrium in the fetal brain and liver in 48 h, and diffuses readily across the placenta, reaching equilibrium between **mother** and **fetus**. It can appear in the milk of lactating **mothers** within 30 min. This explains why many **newborns** of **mothers** on the drug were drowsy and unable to nurse. It has been suspected of causing congenital anomalies. Has induced fixed-drug eruptions in **patients**. Poisoning with respiratory depression, coma, rigidity, and increased temperature has been successfully treated with hemoperfusion, peritoneal dialysis, artificial respiration, ***mephentermine***, IV, to increase blood pressure and urine flow, and ***physostigmine***, 2 mg IV, for comatose **patients**. It can impair driving performance and events that require similar skills. Detoxified mainly in the liver. Therapeutic serum levels are 1–5 µg/ml, toxic at 5–6 µg/ml, and lethal at ~10 µg/ml.

Contraindicated in hepatic and renal disease, ***barbiturate*** sensitivity, and in severe hypertension.

Drug interactions: Has caused severe hypothermia and hypotension, when given with ***amitriptyline***. Enhances neuromuscular blockade of certain ***antibiotics*** (***streptomycin***, ***neomycin***, ***kanamycin***, and ***colistin***), to produce apnea and muscle weakness. By stimulation of hepatic microsomal enzymes, it decrease the plasma half-life of ***coumarins***, such as ***warfarin*** and ***indandiones***. Other reports state it decreases the intestinal absorption of ***coumarins***. It decreases the plasma levels of ***imipramine*** and causes an immediate and marked decrease of ***nortriptyline*** plasma levels. ***Diazepam's*** central nervous system depressant effect is enhanced by it, resulting in respiratory depression and/or hypotension.

LD_{LO}, **human**: 43 mg/kg. Fatal cases usually have ingested at least 50 mg/kg and have blood levels above 25 mg/l, occasionally 20 mg/l.

LD_{LO}, **rat, cat**: 400 and 100 mg/kg, respectively; IV, LD_{50}, **rabbit**: 49 mg/kg. IV, LD, **dogs**: estimated at 60–75 mg/kg, depending on the rate of administration.

AMODIAQUIN

= *AQ1* = *Camoquin* = *Flavoquine* = *Miaquin* = *SN 10751*

***Antimalarial*. A quinoline.**

Use: In prevention and treatment of malaria in **humans** and in treating discoid lupus erythematosus.

Untoward effects: Serious cases of agranulocytosis (neutropenia) with some deaths, as well as thrombocytopenia and aplastic anemia. Varying degrees of eye damage, including pigmentary changes in cornea, decreased retinal function as measured by electro-oculogram. Amenorrhea with partial blindness, due to opaque corneal layers with multiple horizontal bands in a **woman**. Blue–gray macular eruption on the nose of a 41-year-old male taking the drug intermittently over a 2 year period. Pruritus has also been reported in about 1% of a large group of African **patients**. It has also induced fixed-drug eruptions. Pigmentation (bluish-gray to almost black; yellowish-brown is more likely associated with ***quinacrine***) of the hard palate and fingernail bed has occurred in **patients** on therapy longer than 2 years. Hepatitis with asthenia, jaundice, increased serum transaminases, steatosis, cholestasis and/or centrilobular or hepatocytic necrosis in many on malaria prophylaxis for 4–15 weeks. Slow resolution over 3–6 months after discontinuance of medication. A 42-year-old **female** received 780 mg/week (1.5 times recommended dosage) for malaria prevention. After 10 months, a progressive neuropathy developed with severe weakness of the legs. Muscle strength returned 2 weeks after prophylaxis was discontinued. A group of **patients** in Nigeria reported involuntary movements while on medications. Vertigo, incoordination, and spastic convulsions have also been reported. It may decrease the number of leukocytes. Gastrointestinal upsets, such as nausea, vomiting, diarrhea, and anorexia.

LD_{50}, **rat**: 495 mg/kg; **mouse**: 700 mg/kg; IV, LD_{LO}, **monkey**: 30 mg/kg.

AMOPHENONE

Untoward effects: Adverse neurological effects in **patients** with hepatic cirrhosis. Skin rashes, drowsiness, gastrointestinal upsets, methemoglobinemia, impaired liver function tests, and abnormal thyroid function have been reported with its use.

AMOPYROQUIN

= *Propoquin*

Antimalarial.

Use: In malarial prophylaxis and in the treatment of chronic discoid lupus erythematosus and rheumatoid arthritis.

Untoward effects: Similar to ***chloroquine's*** (q.v.). Insomnia, leukopenia, increased SGOT, and blurred vision in **humans**.

Albino rats and **beagle dogs** on high oral dosage developed degenerative atrophic lesions of the retinas, after 5–7 months, respectively, but **rhesus monkeys** had no lesions after a year of therapy.

AMORPHOPHALLIS

A member of the Arum family.

Untoward effects: Since all plant parts contain ***calcium oxalate***, it is not surprising to know that the curious report severe burning of mucous membranes, with swelling of the throat and tongue, nausea, salivation, vomiting, and diarrhea.

AMOTRIPHENE

= *Myordil* = *Win 5494*

Coronary vasodilator and antiarrhythmic.

Untoward effects: Occasional hypotension and tinnitus, "far-away" feeling, confusion, vomiting, and bile in the urine.

LD_{50}, **mice**: 385 mg/kg; IV, 30 mg/kg.

AMOXAPINE

= *Asendin* = *CL 67772* = *Demolox* = *Moxadil*

***Antidepressant*. A tricyclic.**

Untoward effects: Five young **women** died out of 33 **patients** who had taken overdoses. The **deceased** had seizures before death and had ingested 2–8.5 g (usual dosage is 200–300 mg/day). Irreversible central nervous system neurological damage in two overdosed **patients** (24 and 29-year-old **females**; 2 and 4 g). **Both** developed recurrent seizures, severe acidosis, and prolonged coma, as well as severe dysarthria, memory loss, poor fine motor control, and inability to walk unassisted within 1–3 months. Renal failure can occur with overdosage. Extrapyramidal adverse effects (acute dyskinesia, acute dystonia, tardive dyskinesia, and parkinsonism) can develop on normal dosage. A 46-year-old **female** taking 250 mg/day/1 week developed an akinetic syndrome with retardation of movement, speech, and thought with dysphoria. Convulsions developed in three **patients**, aged 22–58 years, who received 100–600 mg/day for 3 days–2 years. After an overdose, a 40-year-old **female**

recovered from loss of brainstem and pupillary reflexes. A 20-year-old **male** reported akathisia and cogwheel rigidity, after 300 mg during a 24 h period. Older **patients** may be particularly at risk to develop neuroleptic malignant syndrome. Acute renal failure in several overdose **patients**. Urination difficulty (1%), dry mouth, insomnia, mild hypotension, drowsiness, tachycardia, and constipation are usual anticholinergic-type side-effects (12–14%). A 42-year-old **female** showed agranulocytosis, septicemia, exfoliative dermatitis, and respiratory distress syndrome. Galactorrhea, hyperprolactinemia, amenorrhea, photosensitivity, allergic skin reactions, arrhythmias, black discoloration of the tongue, agranulocytosis, eosinophilia, testicular swelling, tinnitus, paresthesia, and visual difficulties also occur. Therapeutic serum levels are 0.18–0.6 μg/ml and toxic at 5 μg/ml.

LD_{50}, **mice**: 112 mg/kg; IP, 122 mg/kg.

AMPHETAMINE

= Actedron = Actemin = Adipan = Aktedron = Alentol = Allodene = 2-Amino-1-phenylpropane = β-Aminopropylbenzene = Amphaethamine = Benzedrine = Elastonon = Isoamyne = Isomyn = Mecodrin = dl-2-Methylphenethylamine = Momophos = Norephedrane = Novydrine = Ortédrine = Pharmamedrine = Phenamine = Phenedrine = β-Phenylisopropylamine = Profamina = Profetamine = Propisamine = Psychedrine = Psychoton = Racephen = Raphetamine = Simpamina = Simpatedrin = Sympamine = Sympatedrine

Also known by the following: *Beans, Bumblebees, Cartwheels, Chalk, Chicken Powder, Co-Pilots, Crank, Crossroads, Double Cross, Eye Openers, Hearts, Jelly Beans, Lightning, Mini-bennies, Nuggets, Oranges, Pep Pills, Purple Hearts, Roses, Thrusters, Truck Drivers, Turnabouts, Uppers, Ups, and Wake-ups*

Stimulant, analeptic.

Use: Central nervous system stimulant, anorexic; in hyperactivity of **children**, and for weight reduction. First synthesized in 1887, and in the early 1930s was marketed as a nasal decongestant by inhalation; in the mid 1930s, it was first used in treating narcolepsy in **humans**.

Untoward effects: Wakefulness and increased motor activity occur.

Addiction: Sympathomimetic stimulants are readily abused. Some say **users** are not necessarily addicted, but rather habituated. Tachycardia, dilated pupils, increased blood pressure, anorexia, hyperreflexia, nervousness, urticaria, unclear or rapid speech, muscle pain, and euphoria or dysphoria occur. Intracerebral hemorrhages in **abusers** who often have further jeopardized their health by crushing tablets for making IV solution. **Abusers** have stayed awake for days, suffer from illusions and hallucinations, then lapse into a long, deep sleep. When **they** awaken, **they** are famished and lethargic. The latter is "cured" by repeating IV usage. Dependency occurs through repeated legitimate or illegitimate use. Because of rapidly developing tolerance, chronic **abusers** may take several thousand mg daily, often with ***barbiturates*** and even ***heroin***. Very large doses, especially IV, produce a toxic psychosis, indistinguishable from paranoid schizophrenia, with frequent spells of aggressiveness and violence. Abuse of this drug has been world-wide. After World War II, it is estimated that ½–1 million **people** in Japan abused the drug, half of them taking it IV. **Some** have abused inhaler devices by using them excessively, or by extracting the drug for IV use. It was a common form of abuse by **professional ball players**. One report describes the histories of 13 **persons** who committed homicide while intoxicated with amphetamines. **Some** report imaginary bugs under **their** skin ("crankbugs"). In the early 1980s, 10% of the U.S. Armed Forces in Europe tested positive for amphetamine usage. This was, in part, due to the military prescribing it as an antidepressant and for weight control. Increased incidence of *Staphylococcus aureus* endocarditis in IV drug **abusers**. A 24-year-old **male** consumed 8–16 tablets/day for 2½ years, and was hospitalized with acute weakness after a 4 month period of illness. Post-mortem confirmed probability of acute myeloblastic leukemia with infiltrates in all major organs. Other reports indicate severe anemia, pancytopenia, megaloblastic anemia and tachycardia may also be associated with abuse of this drug. Acute congestive heart failure has been reported in a 22-year-old **female** after IV use, and in a 45-year-old **female** after high oral dosage. An acute aortic dissection reported in 27-year-old **male abuser**. It is difficult to determine its exact effect on the **fetus**, since **mothers** who abuse the drug often take others as well. Yet, a 29-year-old **female** had a stillbirth 24 h after an IV injection during her thirty-fourth week of pregnancy. Reports indicate an increase in oral clefts in **offspring** of **mothers** who ingested it during their first 2 months of pregnancy. Documentation is poor, but congenital malformations of the cardiovascular and biliary systems have been suspect. Hair loss has occurred months later, following its use in weight reduction. Dermatitis with eruptions on the face and scalp reported as early as 1939 in the treatment of narcolepsy. In general, it has an adverse effect on libido, ejaculations in **man** and orgasms in **women** – this may be dose-related. Significant amounts pass into breast milk, resulting in tremors and insomnia in **infants**. Psychotic reactions with a schizophrenic pattern in chronic **users**. These often include paranoid delusions, hallucinations, time distortion, nightmares, erectile disorders, and hyperactivity, including hypersexuality. **Some** feel these effects persist, not just for weeks or months, but even up to 2 years after drug withdrawal. It can precipitate subtle or dramatic exacerbation of previous schizophrenia. The first report of *d*-amphetamine psychoses in a **child** was in 1967. Hypersensitivity in **some** has caused hemolytic jaundice and anaphylactic shock. It has increased intraocular pressure and can cause glaucoma. If the **patient** has not reached the point of high dosage and chronic abuse, the use of ***ammonium chloride*** and ***arginine hydrochloride*** to acidify the urine will hasten the drug's excretion (4–12 times normal).

Contraindications: In agitated, psychotic hyperexcitability states, **patients** sensitive to ***sympathomimetic compounds***, and whenever ***vasoconstrictors*** (including certain nasal preparations) are contraindicated (viz. certain cardiac and cardiovascular diseases, and in **patients** with marked hypertension), in hyperthyroid **patients**, and in **those** with glaucoma or pheochromocytoma.

Poisoning: Death from cerebral hemorrhage in 16-year-old **male**, after ingesting 20 tablets. A 39-year-old **male** possibly ingested 700 mg, and was hospitalized in an unconscious state. **He** was pale, sweating, with blood pressure 65/20, dilated pupils, tachypnea and tachycardia. Slight stimuli provoked involuntary clonic movements and muscle fasciculations. Post-mortem examination 2 days later revealed subarachnoid hemorrhages and renal damage. After ingesting 2 g, a 21-year-old **male** was hospitalized with a syndrome resembling heat-stroke – delirium, coagulopathy with intramuscular hemorrhages and hyperthermia (108°F), and reversible renal failure. Soon after IV use, 4/5 **people** noted chills, fever, sweats, nausea, and abdominal cramps. Within a few hours, **they** were nauseous, vomited, had headaches, myalgias, paresthesias, and orthostasis. Cardiorespiratory arrest, accelerated bleeding, and pulmonary edema occurred in 1/5. In the first 24 h, 4–11 l of saline was required to maintain blood pressure and urine volume. All five recovered. In **humans** and **monkeys**, acute intoxications have resulted in over a 300% increase in serum ***thyroxine (T-4)***. Overuse may have been associated with an increase in Hodgkin's disease. Usual therapeutic serum levels are < 0.1 μg/ml, toxic at 0.2 μg/ml, and can be fatal at 0.5–1 μg/ml.

Drug interactions: ***Lithium carbonate*** has blocked the effects of amphetamine. ***Cocaine***, ***nialamide***, ***pargyline***, ***isocarboxazid***, ***phenelzine***, ***tranylcypromine***, ***urinary alkalinizers (sodium bicarbonate*** and ***acetazolamide)***, ***iproniazid***, and large doses of ***propoxyphene*** potentiate its action. ***Tyramine***-containing foods and ***monoamine oxidase inhibitors*** may help cause hypertensive crises (**humans**, **dogs**, and **rats**) and symptoms simulating a pheochromocytoma crisis, as can ***furazolidone***, after 4 or 5 days of treatment. ***Guanethidine's***, ***bethanidine's***, and ***debrisoquine's*** activities are inhibited by it. Amphetamine enhances the analgesic effects of ***meperidine***. IV use with ***methadone*** gives a ***heroin***-type of 24 h "high".

Treatment: Varies with the skills and familiarity of the physician. Aside from emesis or lavage as instant treatment shortly after ingestion, the following are often used: ***chlorpromazine***, to counter excitation and seizures (in 1–3 year **children** with severe poisoning – 5 mg, IV and IM; in **adults** – 0.5 mg/kg, IV and IM; occasional repeats may be necessary in 12–24 h. Halve these doses if ***barbiturates*** have also been used). Acidification of urine increases its excretion and decreases its retention. It is adsorbed by ***activated charcoal***. Dialysis may help.

LD_{LO}, **human**: 5 mg/kg; LD_{50}, **human**: < 20 mg/kg. In general, therapeutic serum levels are 2–3 μg%, 42–50 μg% is considered toxic, and 200 μg% lethal; or doses of 100–500 mg can be considered as LD_{LO} in many first-time **users**.

Has been used in the illegal "doping" of **horses** and **racing dogs**. In **animals**, half-life varies greatly: **pony**: 1.4 h; **goat**: 37 min; **pig**: 63 min; **dog**: 4.5 h; **cat**: 6.53 h. **Animal** trials indicated ***propranolol*** *antagonized* amphetamine's stimulant effects, but *potentiated* its anorexigenic effect. Stereotypy occurs in **humans**, **rhesus monkeys**, **cats**, and **rats**. LD_{LO}, **rat**: 50 mg/kg; **mouse**: 45 mg/kg.

AMPHETAMINIL

= AN-1 = Aponeuron

Psychotropic.

Use: In treating chronic schizophrenia. ***Amphetamine*** is a major metabolite.

Untoward effects: After a couple of years use in Germany, cases of abuse and dependency emerged.

LD_{50}, **rat**: 30 mg/kg; **mouse**: 59 mg/kg; IV, LD_{50}, **rat**: 22 mg/kg; SC, LD_{50}, **rat**: 48 mg/kg; SC, LD_{50}, **mouse**: 58 mg/kg; IP, LD_{50}, **mouse**: 48 mg/kg.

AMPHIACHYRIS DRAUMCULOIDES

= Annual Broomweed

Annual botanical weed with 15–30 inch woody stem under a branching top full of small, bright yellow flowers (August–October). Early **settlers** in the U.S. tied them together and used them as brooms.

Untoward effects: Toxic agent, either in the leaves or pollen, causes inflammation of the eye in **man** and **livestock**.

AMPHOMYCIN

= *Amfomycin* = *Glumamycin*

Antibiotic.

Use: Primarily topical (0.5%) against Gram-positive bacteria. Often marketed with ***neomycin***, ***kanamycin***, and ***hydrocortisone***. Effective against African Sleeping Sickness (trypanosomiasis) in **mice** and *Borrelia* in **chicks**. May induce moderate hemolysis, after parenteral use of large doses.

IV, LD_{50}, **mice**: 177.8 mg sodium salt, 120.2 mg calcium salt/kg; IP, LD_{50}, **mice**: 223–500 mg/kg.

AMPHOTERICIN B

= *Ambelcet* = *AmBisome* = *Ampho-Moronal* = *Amphotec* = *Amphozone* = *Fungilin* = *Fungizone*

From the soil of the Orinoco River region of Venezuela.

Antifungal.

Use: In treating systemic fungal infections in AIDS cases, aspergillosis, blastomycosis, candidiasis, cryptococcus, histoplasmosis, and coccidiomycosis. Occasionally, in leishmaniasis and sporotrichosis. As a preservative in vaccines.

Untoward effects: Mainly, decreased hemoglobin and red cell values, hypokalemia, hypomagnesemia, metabolic acidosis, and nephrotoxicity. During infusion, fever, increased serum ***creatinine***, nausea, vomiting, abdominal pain, muscle pain, chills, diarrhea, headache, hypotension, and tachypnea are frequent (reduced by premedication with ***chlorpromazine***, and, if needed, ***hydrocortisone*** and ***meperidine***). The first few doses produce the most severe reactions. Nephrotoxicity (including azotemia) is the most serious dose-limiting reaction. Severe hemolytic and aplastic anemias may develop after a few weeks and have been treated with ***erythropoietin***. Anorexia often occurs. Thrombophlebitis, thrombocytopenia, granulocytopenia, neutropenia, and mild leukopenia may be less common. Transient asystole, bradycardia, tachycardia, and cardiac enlargement have been reported. Severe back pain, swollen lips, hemiparesis, and dyspnea in 34-year-old **male** after intraventricular and IV injections. 4/5 **pediatric patients** (4.5 weeks–7 years) died with cardiac arrest, after receiving 4.6–40.8 mg/kg/day. Acute toxic delirium reported in 28-year-old **male**, after intrathecal therapy. Parkinsonism developed in a 26-year-old **female**, after intrathecal injection. Neurological deficits suggestive of direct spinal cord damage reported in two other **patients**, after intrathecal injections. Mild arachnoiditis may follow intrathecal use. Due to severe hypokalemia, proximal muscle weakness, rhabdomyolysis, and muscle necrosis can occur. May give falsely elevated test results (blood urea nitrogen, non-protein nitrogen, urine protein, serum ***creatine***). A 51-year-old **male** and a 29-year-old **female** developed a reversible liver toxicity, after dosage reduction; anaphylactic shock and other allergic reaction reported in a 55-year-old **male** (maculopapular eruption and eosinophilia), and in a 72-year-old **male**, which manifested itself as acute bronchospasm. Yellow discoloration of nails and photosensitivity have also been reported. It can potentiate the effects of ***digitalis*** (possibly by decreasing serum ***magnesium*** and ***potassium***) and ***non-depolarizing muscle relaxants*** (including ***aminoglycoside antibiotics***). It may increase ***cyclosporine*** toxicity. Its acute toxicity may be amplified by leukocyte transfusions leading to severe pulmonary reactions. Some degree of renal damage has been permanent. Pruritus, with burning sensations, can occur after topical use. Blurred vision, chemical meningitis, and urinary retention have followed intrathecal use. Tinnitus appears rarely. Cases of induced hypertension reported within minutes of an infusion in some **patients**. Anaphylactic reaction to liposomal form IV. Antagonizes action of ***miconazole***. Do NOT use it with saline solution, as it will cause precipitation.

Therapeutic serum levels reported as 0.1–2.5 μg/ml and toxic at 4–10 μg/ml.

IP, LD_{50}, **mice**: 88 mg/kg; IV, 4–6 mg/kg. IV, LD_{LO}, **dog**: 12 mg/kg.

AMPHOTHALIDE

= *RP 6171* = *Schistomide*

Anthelmintic.

Use: In treating schistomiasis.

Untoward effects: LD_{50}, **mice**: > 5 g/kg.

AMPRENAVIR

= *Agenerase* = *KVX* = *VX 478* = *141W94*

Protease inhibitor, antiviral.

Untoward effects: Nausea, vomiting, diarrhea, rash, and perioral paresthesias. Occasionally (~1%) severe rashes and Stevens–Johnson syndrome, hyperglycemia, increased aminotransferase, and altered distribution of body fat.

Drug interactions: Inhibits CYP3A4 metabolization of ***bepridil***, ***cisapride***, ***dihydroergotamine***, ***ergotamine***, ***midazolam***, and ***triazolam***, increasing their toxicity. Similar results can be expected with ***amiodarone***, ***tricyclic antidepressants***, IV ***lidocaine***, ***quinidine***, ***rifabutin***, ***sildenafil***, and ***statins***. ***Ritonavir*** increases and ***efavirenz*** decreases its serum concentration. A high ***fat*** meal decreases its absorption. It increases ***ketoconazole*** serum concentrations by > 40%.

AMPROLIUM

= *Amprol* = *Amprovine* = *Antococcid* = *Corid* = *Mepyrium*

Coccidiostat.

Use and untoward effects: In **livestock** feeds, to aid in the control of coccidiosis in **calves**, **chickens**, and **turkeys**. Experimentally, to produce ***thiamine*** deficiency in **horses** and polioencephalomalacia in **sheep**. To provide symptomatic relief in *Isospora belli* and *Crytosporidia* infections and diarrhea in **humans**.

AMPYRONE

= *4-Aminoantipyrine*

Metabolite of *aminopyrine* (q.v.).

Untoward effects: Shortens prothrombin time in **guinea pigs** and **man**. Experimental use in **calves** and **cows** caused excitement, incoordination, and weakness.

AMRINONE

= *5-Amino-3,4´-bipyridine-6(1H)-one* = *AWD-08-250* = *Cordemcura* = *Inocor* = *WIN 40,680* = *Wincoram*

***Cardiotonic.* Possesses positive inotropic and vasodilating properties. *Mihinone* is an analog.**

Use: In treating chronic congestive heart failure.

Untoward effects: Pancytopenia can occur after short courses of high dosage, beginning in about 2 days and reaching a nadir on the eighth day, after continuous infusion. Nausea, vomiting, abdominal pain, thrombocytopenia, tachycardia, increased liver enzymes, hypotension, and irritation at the injection site often follow IV use. Oral use has been discontinued, due to a higher incidence of adverse effects. Has a narrow therapeutic–toxic ratio. A 49-year-old **male** with arteriosclerotic heart disease and alcoholic cardiomyopathy developed hepatotoxicity after initiation of therapy. The effect was confirmed by rechallenge. A 75-year-old **patient** received 75 mg by error, when the order was for ***amiodarone***. Ventricular fibrillation followed and the **patient** was coded and recovered, with discharge 2 weeks later.

LD_{50}, **rats**, **mice**, and **rabbits**: 200–350 mg/kg; IV, **rats** and **mice**: 70–170 mg/kg.

AMSACRINE

= *Amekrin* = *m-AMSA* = *Amsidine* = *Amsidyl* = *CI-880* = *Lamasine* = *SN 11,841*

***Antineoplastic.* IV use.**

Untoward effects: Hematopoietic toxicity with neutropenia and thrombocytopenia of short duration and rapidly reversible. Mild nausea and vomiting (16%), stomatitis (80%), hyperbilirubinemia (25%), phlebitis, diarrhea, and anaphylaxis reported. Bone marrow depression, convulsions, ventricular fibrillation, congestive heart failure, alopecia, and decreased renal function are usually forms of delayed toxicity. **Patients** with liver and renal problems often require a 40% dose reduction. When mixed for IV use, it contains irritant ***lactic acid***. Extravasated material may cause severe necrotic tissue reactions. Use gloves and eye shield! After accidental contamination, wash thoroughly with water, then rinse with buffered ***phosphate*** solution.

IP, LD_{50}, **mice**: 24–60 mg/kg, methanesulfonate and hydrochloride, respectively.

AMSINCKIA INTERMEDIA

= *Fiddleneck* = *Fireweed* = *Tarweed*

Weed found primarily in northwestern U.S. ranges. The term *fireweed* also refers to *Epilobium*.

Untoward effects: Contains toxic pyrrolizidine alkaloids (***intermedine***, ***lycopsamine***, and ***echiumine***). Consumption of weedy alfalfa hay, containing this weed at a 5–10% level, after 2½–6 months, caused the deaths of **dairy calves**. Post-mortems revealed megalocytosis and fibrosis of the liver. Adult **cattle** did not seem to be affected. **Pigs** have been fatally poisoned by consuming it. It has caused "walking disease" in **horses**, so-called because affected **horses** may walk for miles in a straight line. Post-mortems reveal liver pathology. **Rats** fed extracts of the seeds developed acute central nervous system effects and liver toxicity. The plant is also a ***nitrate*** accumulator.

LD_{LO}, **rats**: 500 mg/kg. Another report is much higher.

AMYGDALIN

= *Mandelonitrile-β-D-gentiobioside*

Untoward effects: Confusion exists in the literature, because the term is also used for ***laetrile***, which is ***mandelonitrile glucuronide***. Toxic symptoms would be similar, and due to ***cyanide***.

It is a cyanogenic glycoside, widely found in plants and in their parts (viz. ***apples***, ***apricots***, ***peaches***, ***pears***, ***wild cherry***, ***hydrangeas***, some ***laurels***, and some ***lima beans***). To promote sales, it was given the name of ***Vitamin*** B_{17}. The hypoxia it causes can produce dizziness, nausea, vomiting, shock, stupor, coma, respiratory failure, and death. At subtoxic levels, the ***cyanide*** is converted to ***thiocyanate***, which is goitrogenic, and, since goitrogens are potentially carcinogenic, it is considered carcinogenic.

LD_{LO}, **infant**: 50 mg/kg.

AMYL CINNAMATE

= *Isoamyl Cinnamate*

Synthetic.

Use: In fragrances and foods.

Untoward effects: 8% in petrolatum under an occlusive patch test for 48 h on 25 **human** volunteers produced no irritation.

Acute LD_{50} in **rats** and acute dermal LD_{50} in **rabbits**: > 5 g/kg.

AMYL CINNAMIC ALDEHYDE DIETHYL ACETAL

Synthetic.

Use: In fragrances, particularly in creams and lotions (up to 0.05%) and perfumes (up to 2%).

Untoward effects: At 10% in petrolatum – no irritation after a 48 h closed-patch test in 25 **human** volunteers. 32% in ***acetone*** produced no irritation on **humans**.

Pure material showed minor irritation on **guinea pigs** and **rabbits**, but not on **miniature swine**.

LD_{50}, **rats** and acute dermal LD_{50}, **rabbits**: > 5 g/kg.

4-*tert*-AMYLCYCLOHEXANONE

Synthetic.

Use: As a fragrance in soaps, detergents, creams, lotions, and perfumes.

Untoward effects: Kligman's testing of 8% in petrolatum showed no irritation.

Acute LD_{50} **rats** and acute dermal LD_{50} **rabbits**: 4.7 g/kg.

AMYLENE HYDRATE

= *Amylene Hydroxide* = *Dimethyl Ethyl Carbinol* = *2-Methyl-2-butanol* = *tert-Pentyl Alcohol*

Hypnotic.

Use: As an ingredient with ***tribromoethanol*** in ***Avertin***®.

Untoward effects: Irritating to mucous membranes. Narcotic in high concentrations. Deep narcosis, dilated pupils, loss of corneal reflexes, decreased pulse, decreased temperature, slow, deep, and irregular breathing in **human** poisoning cases.

LD_{50}, **rats**: 1 g/kg; IP, LD_{LO}, **rats**: 1.53 g/kg; SC and rectal LD_{LO}, **rats**: 1.4 g/kg.

AMYLIN

A pancreatic hormone discovered in the 1980s.

Untoward effects: Induces ***insulin*** resistance, is involved in pathogenesis of Type II diabetes mellitus, and increase in blood renin concentrations.

AMYL NITRITE

= *Pentyl Nitrite*

***Vasodilator.* Will expand when heated to 95°C or 203°F. Slang = *Poppers* = *Snappers*. *Locker Room* and *Rush* are also used for the *butyl nitrites*.**

Use: In relieving angina pectoris by inhalation, in emergency treatment of ***cyanide*** poisoning (methemoglobin forms, to be followed by ***sodium nitrite*** and ***sodium thiosulfate***, IV).

Untoward effects: A drug of abuse by inhalation, giving a long-lasting "high", transient increase in ocular pressure, tachycardia, often causing faintness, dizziness, pain around the eyes and headache, decreased blood pressure, and giddiness. A 29-year-old **male** rapidly developed methemoglobinemia (33% concentration), after ingesting 10 ml. Some **people** have a methemoglobin reductase deficiency. Symptoms were reversed rapidly by 100 mg ***methylene blue***, IV. A 43-year-old **male** ruptured an intracranial aneurysm after abuse of the drug. A 28-year-old **male** had two episodes of hemolytic anemia associated with its inhalation. 2/23 **volunteers** showed symptoms of allergic reactions after inhalation. There may be a relationship between its abuse and Kaposi's sarcoma. It causes a warm feeling in the pelvis and genitalia and a 3–4 min "high". If used to increase sexual pleasure, it is often used during intercourse, to prolong the time to orgasm and enhance the orgasm. If hypotension is marked, loss of erection occurs. Incontinence, collapse, cyanosis, and sudden deaths have been reported. Inhalation of 5–10 drops produces a rapid, violent flushing of the face, increased heart rate, and a feeling that "one's head would burst". Corneal necrosis from eye contact has occurred.

AMYL SALICYLATE

= *Amylenol* = *Isoamyl-o-hydroxybenzoate*

Synthetic.

Use: As a fragrance in soaps, detergents, creams, lotions, perfumes, and in foods. About 30 tons/year used in the U.S. Topically, as a less odorous substitute for ***methyl salicylate*** in rheumatism.

Untoward effects: Produced no adverse reactions on 25 **volunteers** at a 10% concentration in petrolatum.

Avoid use in ***salicylate***-sensitive **individuals**. 100% concentration topically on the skin of **rabbits** and **guinea pigs** produced some irritation, but not on **miniature swine**, or with 32% in ***acetone*** on **human** skin.

Oral and IV, LD_{50}, **dogs**: 0.5–0.8 g/kg.

AMYRIS OIL ACETYLATED

Synthetic.

Use: As a fragrance in soaps, detergents, creams, lotions, perfumes, and in foods.

Untoward effects: At a 10% level in petrolatum, no irritation noted in 25 **volunteers** after a 48 h closed-patch test.

LD_{50}, **rats** and dermal LD_{50} in **rabbits**: > 5 g/kg; LD_{50}, **rats**, ***amyris oil***: 4.5 g/kg.

ANABASINE

= *Gamibasin* = *Neonicotine*

***Insecticide. Tobacco alkaloid* found in *Nicotinia glauca* and other species; a venom in a poison gland of *Aphaenogaster* ants; a toxin used by nemertine worms to paralyze prey and deter predators. A derivative, *GTS-21*, has a great potential in the treatment of Alzheimer's disease in humans.**

Use: Against **aphids**. Identical to ***neonicotine***. As an anti-smoking drug.

Untoward effects: As a drug of abuse, acute poisoning reported in 29-year-old **male** who administered it rectally. General weakness, vertigo, increased salivation, nausea, hypothermia (35.2°C), confusion, disturbed vision and hearing, photophobia, diarrhea, clonic spasms, bradycardia (42/min), hypotension (80/50), acrocyanosis, dyspnea, dilated pupils, and coma occurred, followed by recovery after treatment. A young **male** died from ingesting it. First stimulates, then decreases autonomic ganglia. LD_{LO}, **human**: 5 mg/kg.

May be teratogenic in **cows**.

LD_{LO}, **dog**: 50 mg/kg.

ANABENA FLOS-AQUAE

Algae.

Untoward effects: A highly toxic blue-green algae, common in shallow lakes in the U.S. Midwest, particularly during hot, dry weather. In poisoned **animals**, vomiting, muscle trembling, incoordination, paralysis, respiratory arrest, and death may occur within an hour after drinking it in water. **Calves**, **rats**, **ducks**, and **fish** have died from its postsynaptic depolarizing neuromuscular blockade. Other forms, such as *A. lemmermannii*, *A. torulosa*, *A. variabilis*, and *A. spiroides* have been involved in poisonings in other species, as well as a cause of dysentery in **man**.

ANACARDIACEA

The family includes *Anacardia occidentale*, *Mangifera indica*, and *Sclerocarya caffra*, which are used respectively for its seed and bark oils as an insect repellant due to its *anacardic acid*, its powdered flowers as a mosquito repellent (South America), and its fruit extracts against ticks (Africa).

Untoward effects: ***Renghas***, a collective noun for it, have a strongly irritating juice. Its wood is used in furniture manufacturing and contains a resin in its sap for lacquer, the so-called Japan lacquer. It can cause poisoning, especially a dermatitis – even from the smoke of burning its wood.

Anacardic acid and ***urushiol*** are found in the oils of the ***cashew*** nut (*Anacardium occidentale*) family and are skin irritants. ***Mango*** trees and poorly processed ***cashew*** nuts contain it. ***Poison ivy***, ***poison oak***, and ***sumac*** plants contain a similar chemical.

A. rhinocarpus stem bark is used as a **fish** poison and *M. indica* (***mango***) old leaves have caused death in **cattle**, after prolonged consumption.

ANADENANTHERA PEREGRINA

= *Piptadenia peregrina*

Untoward effects: Hallucinogenic snuffs (***cohoba***, ***paricá***, ***yopo***) are used by South American **Indians** and are made from its toasted beans. Other species are also used and they all contain ***dimethyl*** and ***hydroxytryptamines***, as well as other ***tryptamines***.

ANAGALLIS ARVENSIS

= *Mouron Rouge* = *Poison Chickweed* = *Poor Man's Weather Glass* = *Red Chickweed* = *Scarlet Pimpernel* = *Shepherd's Clock* = *Shepherd's Weather-glass*

Untoward effects: Despite its use in various parts of the world as a diuretic, laxative, resolvent or drawing agent, and antiviral in humans, it has caused fatal nephrosis (coagulative necrosis and intratubular hemorrhage in the renal cortex) when fed to sheep and has killed calves. It is used for killing fish. Oral intake of four drops in man caused intense headache and nausea for about 24 h and body pain. Fatal gastroenteritis in a dog was caused by 12 ml. Extracts have poisoned a horse. The leaves cause contact dermatitis in man, due to *primin* in its leaf "hairs". Renal congestion, staggering, and erythema have also been reported in man. Contains the poisonous glycoside, *cyclamin*.

ANAGESTONE ACETATE

Progestin.

Use: Explored in ***oral contraceptives***.

Untoward effects: Excessive breakthrough bleeding in **humans** was a common complaint, when it was used as a progestogen component with ***estrogen***.

At high dosage, it was associated with a high incidence of mammary nodules in **beagle dogs**.

ANAGRELIDE

= *Agrylin* = *BL 41,624* = *BMY 26,538*

***Antineoplastic, antithrombotic.* A quinazolin.**

Use: In treating thrombocytosis, due to its potent anti-aggregating effect on platelets. It is especially useful in **patients** with chronic myeloproliferative disease.

Untoward effects: Headache (44.5%), palpitations (27%), diarrhea (24%), asthenia (22%), edema (20%), abdominal pain (17%), nausea and dizziness (15%), vasodilation, hypotension, and possible inotropic effects. Rarely, myocardial infarction and heart failure, atrial fibrillation, pericarditis, occasional hypertension, pancreatitis, gastric and duodenal ulceration, seizures, dyspnea and flatulence (10.5%), chest pain, rash and urticaria (8%), vomiting, paresthesia, tachycardia, peripheral edema (7%), dyspepsia, back pain, anorexia, and malaise (6%).

ANAGYRINE

= *Monolupine* = *Rhombinin*

An alkaloid in *lupine* seeds and early-growth leaves and stems.

Untoward effects: Passes rapidly into **goat's** milk after ingestion, posing a great risk for birth defects in **humans** consuming such **goat's** milk during their pregnancy. In one family, a baby **boy**, a litter of **puppies**, and **goat kids** had bone birth defects, after drinking milk from goats that had foraged on lupines. Congenital crooked **calf** disease occurs in those whose **dams** consumed this alkaloid at a minimum of 1.44 g/kg, while grazing on lupines during their 40–75th days of gestation. Arthrogryposis, torticollis, scoliosis, and occasional cleft palates occur in these **calves**, similar to effects caused by ***aminoacetonitrile***. Has been found in herbal remedies containing *Caulophyllum* (***blue cohosh***), and is potentially dangerous, especially for pregnant **women**. It is related chemically to ***lobeline***, and can cause death, due to respiratory paralysis.

ANASTROZOLE

= *Arimidex* = *ICI-D 1033* = *ZD 1033*

Non-steroidal aromatase inhibitor, antineoplastic.

Untoward effects: Nausea, headache, hot flushes, abdominal pain, and diarrhea or constipation. Delayed reactions leading to asthenia, dyspnea, bone pain, back pain, chest pain, pharyngitis, peripheral edema, and anorexia.

ANCHUSA ARVENSIS

Use: In red dyes.

Untoward effects: Hepatotoxic, due to its pyrrolizidine alkaloid, ***echinatine***. *A. officinalis*, and *A. italica* have been used in Italy, Germany, and Africa in herbal sudorific combinations.

ANCROD

= *Arvin*

***Anticoagulant.* A fraction from the venom of the Malayan adder (pit viper).**

Use: In treating deep vein thrombosis and recurrent thromboembolic disease in **humans**. Successfully used in **dogs** and **ponies**. Markedly decreases circulating fibrinogen to an unstable form of fibrin that cannot be clotted by physiological thrombin.

Untoward effects: Tendency for blood to ooze from venipuncture and spontaneous signs of bruising have been noted. Localized edema and erythema at injection sites. Will cause afibrinogenemia or severe hypofibrinogenemia, if given IV. Probably weakly antigenic, and resistance to its effects reported after many repeat injections. Urticaria, hematuria, and occasional hypersensitivity reactions. As a foreign protein, it can induce antibodies, which can decrease its effect. Do **NOT** use with drugs that inhibit the reticulo-endothelial system (e.g. ***dextrans***) or the physiological fibrinolytic system (e.g. ***Σ-aminocaproic acid*** and ***tranexamic acid***).

ANDROCTONUS AUSTRALIS

An African scorpion.

Untoward effects: In Greek, androctonus means "killer of **man**".

IV, Toxin I, LD_{50}, **mice**: 19 μg/kg, Toxin II, 10 μg/kg.

ANDROMEDA sp.

A. arboreum = *Sorreltree* = *Sour Wood*

Untoward effects: In 1794, it was discovered that **bee's** honey (q.v.) from it could cause mild to serious poisoning in **people**. After ingestion (*A. floribunda*), a pregnant **goat** became weak, depressed, anorexic, vomiting, with prostration and death in 2 days.

ANDROMEDOTOXINS

= *Grayanotoxins*

In the leaves, young twigs, flowers, and nectar of *Rhododendons*, *Kalmias*, *Azaleas*, *Monotropa*, and *Leucothoe*.

Untoward effects*:** They cause hypotension, mucous nasal discharge, frothing at the mouth, vomiting, diarrhea, dyspnea, bloating, teeth grinding, projectile vomiting, convulsions, muscle paralysis, and even death (within 3–14 h after a lethal dose) in **cattle**, **sheep**, and **goats**. Poisoning in **humans** has also occurred. A 27-year-old **female** was poisoned after eating contaminated ***honey and suffered from arrhythmias, requiring a temporary pacemaker. **Humans** have been poisoned by honey made from *Kalmia* nectar, but its bitterness reduces the number of **people** poisoned. These **people** have had central nervous system suppression after transitory excitement, tearing, sialism, nasal discharge, nausea, vomiting, cardiac arrhythmias, respiratory arrest, and limb paralysis. It has a ***curare***-like effect on skeletal muscle end-plates and a direct stimulatory effect on striated muscle, followed by depression.

Type I, IP, LD_{50}, **mice**: 1.31 mg/kg; Type II, 26.1 mg/kg; Type III, 0.84 mg/kg.

ANDROSTENEDIONE

= *Androtex*

Precursor of *estrone* and *testosterone*. Intranasal or oral as an anabolic agent by athletes.

Untoward effects*:** The same as exogenous ***testosterone (q.v.). Use is banned by many athletic organizations. Hirsutism common in **women**. Can be converted to ***estrone*** in **women**. Lowers high-density lipoprotein ***cholesterol*** levels in **males** and increases coronary heart disease risk. ***Ketoconazole*** can block ***testosterone*** formation. ***Phenobarbital*** increases its metabolism.

ANDROSTERONE

Untoward effects*: *Phenobarbital*, *phenylbutazone, and ***chlorcyclizine*** stimulate its metabolism in **rats**.

TD_{LO}, **rats** (15–20 days pregnant): 400 mg/kg.

ANEMONE sp.

= *Windflowers*

***Botanical*. Native to the Midwest and the West.**

Untoward effects*:** Severe vesicular inflammation of the skin reported from the leaves of the bush anemone (*A. nemorosam*), used externally by **people** in the treatment of rheumatism. A **female** put a ***prairie crocus (*A. patens* = ***Easter Flower*** = ***Pasque Flower***) compress on her knees, to relieve osteoarthritis. Within an hour, **she** had extreme erythema, some swelling, and severe burning sensations around the knee. **She** was hospitalized 9 days later with blisters, ruptured blisters, and severe edema of the knee. The seeds and young plants contain alkaloids that increase in concentration as the plants age. Stomach pain, depression, incoordination, and fatalities in **people**, after ingesting large quantities. Fresh plants contain the alkaloid, ***anemonine***, that causes contact irritation and inflammation in **people**.

Potentially hazardous to **livestock**. **Sheep** are poisoned by eating the flowers or young plants. Although all parts of the plant are poisonous, the roots are particularly so to **pigs** and **poultry**.

ANETHOLE

= *Anise Camphor* = *Monasirup*

Use*:** Volatile oil, imparting a very sweet licorice-like flavor and odor to pharmaceuticals (carminatives, expectorants), herbal teas, **livestock** feeds, dental preparations and mouthwashes, liquors, color photography, soaps, foods, pastries, and "absinthe***"-type beverages. About 8 tons/year used in the U.S. It is estrogenic.

Untoward effects*:** It is considered to be a naturally occurring carcinogen (viz. in ***fennel), but, despite the Delaney Amendment, still permitted for **human** consumption, because actual carcinogenicity has not been demonstrated, except for lung tumors in **mice**. It is the main constituent of the essential oils of ***fennel*** and ***anise***, and is their active estrogenic principle. Related to the psychoactive chemicals, ***mescaline*** and ***asarone***. Extrapolating indicates little or no hepatotoxicity for **man**.

Repeated doses to **rats** of 695 mg/kg caused slight discoloration, mottling, and blunting of the lobes of the liver.

LD_{50}, **rats**: 2 g/kg; IP, **rats**: 93 mg/kg; oral, LD_{50}, **mice**: 3 g/kg; **guinea pig**: 2.16 g/kg. Acute dermal LD_{50}, **rabbit**: > 5 g/kg.

ANGEL HAIR

***Use*:** As a Christmas ornament.

***Untoward effects*:** It is made of spun glass and causes eye, skin, and gastrointestinal irritation.

ANGELICA sp.

***Botanical*. The root of Chinese *Angelica* (*A. sinesis*) is called *Danggui*. Dahurian *Angelica sinensis* is known as *Baizhi* or *Dong Quai*.**

***Use*:** As a carminative, diuretic, diaphoretic, and in treating psoriasis. Its root oil is used in flavoring liqueurs, such as Cointreau, "absinthe", and cigarettes. In Chinese and other herbal remedies, as well as in homeopathic medicines.

Untoward effects: A known photosensitizing plant. Many species exist and are used. When fresh, the roots of *A. atropurpurea* are poisonous, and once ingested by **Indians** of the eastern U.S., to commit suicide. Potentiates oral ***anticoagulants*** including ***warfarin***. It contains ***psoralens*** leading to photosensitivity and dermatitis. The seed oil is used in seasoning. Photocarcinogenic in laboratory **animals**. The plant resembles ***water hemlock***.

LD_{50}, **rats** and **mice**: 11 g and 2.2 g/kg, respectively.

A. officinalis (*A. archagelica*) seed oil approved for food use and in spices and seasonings. Its LD_{50}, **rats**: 11.2 g/kg; **mice**: 2.2 g; acute dermal LD_{50}, **rabbits**: > 5 g/kg, and not irritant topically on **mice** and **humans**, but only mildly irritating on abraded **rabbit** skin.

ANGIOTENSIN

***Vasoconstrictor*. Angiotensin I is converted in the blood by enzyme action to angiotensin II, the active pressor form.**

Use: A powerful elevator of blood pressure.

Untoward effects: Definite risk to life (rupture of cerebral aneurysm, etc.), when preinfusion diastolic blood pressure > 110 mmHg. A 27-year-old **female** became edematous with extreme expansion of extracellular volume. Avoid use in **patients** with aortic incompetence. By decreasing renal clearance, serum urates increase. Infusion of angiotensin II has caused mild to moderate headaches and urticarial rashes, clearing after the infusion is stopped. Infusion of the amide = ***Hypertensin*** = ***Ipertensina*** has caused general malaise, fatigue, tachycardia, bradycardia, severe headaches, flushing, anxiety, sensations of pressure in **one's** head, vomiting, and increased blood pressure.

IV, LD_{50}, **rat**: 8 mg/kg.

ANGYLOCALYX OLIGOPHYLLUS

Untoward effects: Bark extracts decrease motor activity, cause muscle relaxation, enophthalmos, followed by death in about 4 h. Native to Africa.

ANHYDROTETRACYCLINE and ANHYDRO-4-EPI-TETRACYCLINE

Degradation products of outdated *tetracycline*, or *tetracycline* stored in a hot, humid environment.

Untoward effects: Ingestion of outdated ***tetracycline*** has caused Fanconi's syndrome in **man**, with glycosuria, aminoaciduria, phosphaturia, and proteinuria, and would be particularly dangerous in **patients** with renal disease. These adverse effects have been reported world-wide. Degraded material has caused a syndrome similar to systemic lupus erythematosus. One report from Spain in a 41-year-old **female** indicated that 60% of the ***tetracycline*** tablets **she** ingested contained these two degradation products.

A large group of **calves** with respiratory disease were treated with ***tetracycline*** that proved to contain large amounts of degradation products. Post-mortems on five revealed abnormal coloration of skeletal muscles, myocardium and brain, as well as liver degeneration. The epi form seemed more toxic to experimental **dogs**.

ANILAZINE

= *B 622* = *2,4-Dichloro-6-σ-chloroanilino-s-triazine* = *Direx* = *Dyrene* = *Kemate* = *Triasyn*

***Fungicide*. For botanical (crop, vegetable, turf, and ornamental) use.**

Untoward effects: Dermatitis in field **workers**, lasting at least 2 days after exposure. Can cause irreversible eye damage. **Users** must wear protective clothing and goggles. LD_{LO}, **human**: 50 mg/kg. In **ducks**, regurgitation, ataxia, weakness, and falling, lasting for about 10 days. In chronic feeding to **rats** leads to enlarged liver, kidney, and spleen; in **dogs**, splenic enlargement.

LD_{50}, 3–4 month male **mallard ducks**: > 2 g/kg, female **rats**: 2.7 g/kg, **rabbit**: 400 mg/kg; IP, LD_{50}, **rats** and **mice**: 25 mg and 50 mg/kg, respectively. Oral LD_{50}, **cats**: > 620 mg/kg; **monkeys**: > 3.2 g/kg; **dogs**: > 7 g/kg.

ANILERIDINE

= *Alidine* = *Apodol* = *Leritine* = *MK 89* = *Nipecotan*

***Narcotic analgesic*. Synthetic opiate.**

Use: As a strong analgesic.

Untoward effects: IM use has caused vomiting, headache, pruritus, and respiratory decrease, especially at high dosage. Rapid IV use has caused apnea. Abuse liability is similar to ***morphine's***. Estimated **human** lethal dose is 500 + mg in **those** who have not developed a tolerance to it. Interferes with vanillylmandelic acid assays.

It is a useful central nervous system stimulant in the **horse** at 0.5–0.8 mg/kg.

LD_{50}, **rats**: 42–205 mg/kg; **mouse**: 128 mg/kg; SC, LD_{50}, **rats**: 163 mg/kg; **mouse**: 100 mg/kg.

ANILINE

= *Aminobenzene* = *Aminophen* = *Aniline Oil* = *Benzenamine* = *Kyanol* = *Phenylamine*

Over 1 billion lbs/year currently used.

Use: 75% in MDI, 15% in rubber-processing chemicals, 4% in herbicides, 3% in dyes, printing, and pigments, and 3% in specialty fibers, pharmaceuticals, and photographic chemicals. In the illegal production of ***fentanyls***.

Untoward effects: Causes anoxemia and central nervous system depression, usually within 1–3 h. Readily absorbed through the unbroken skin. Death or permanent injury can follow ingestion. It can cause methemoglobinemia. If heated, its fumes are highly toxic. If **workers** are exposed, remove all clothing; **they** must take a warm shower with soap and water. Rubber or neoprene gloves and full face and head covering are required for protection in work areas. **Human** LD_{LO}: 50 mg/kg. ***Aniline dyes*** have been allergens in high-speed duplicator papers. **Neonates** can easily be harmed by cutaneous application of aniline-containing disinfectant (viz. ***trichlorocarbanilide***, used as an incubator and bedding disinfectant) solutions. **Infants** can develop methemoglobinemia, after wearing diapers stamped with aniline-containing inks. A 22-year-old **male** accidentally ingested some aniline oil, intended as an octane-booster for his racing car. In about an hour, **he** was drowsy and cyanotic. **He** failed noticeably to respond to IV ***methylene blue***. After a total of 72 h, **he** passed dark red urine and showed an increase in jaundice. On day 6, **he** required blood transfusions, and was discharged on the tenth day. **Workers** in the aniline dye industry have shown an increase of bladder cancer above normal numbers. ***p-Phenylenediamine*** and ***p-amino sulfonamide*** sweetening agents (***saccharin***, ***cyclamates***, ***Sucaryl®***, and ***Sweeta®***) may cross-react with aniline dyes. May cause nephrotoxicity. Blue to violet nail pigmentation in **man** after exposure, as well as cutaneous eruptions from its dyes. This often occurred in **workers** using rubber vulcanizing accelerators. Dietary administration to **rats** has induced hemangiosarcomas, fibrosarcomas, and sarcomas. Aniline **workers** are also exposed to ***chromates***. The chemical is dialyzable. In 1981, it was an adulterant in ***olive oil***, leading to thousands of poisoning cases and several deaths. A 3-year-old **male** ingested 10–20 ml shoe polish with 27% aniline and developed cyanosis and bilateral Babinski signs. Recovery after 2 days of treatment. Suicide attempts have occurred. A 17-year-old **female** was poisoned by eating a dog-shaped deodorant. Glucose-6-phosphate dehydrogenase-deficient **people** may be unable to utilize ***methylene blue*** as an antidote. Even moderate exposure to vapors causes cyanosis, readily observed in cheeks, lips, and ears. Severe exposures cause facial flushing, irritability, drowsiness, headache, shallow respiration, decreased blood pressure, photophobia, and burning sensations on eyelids. Average lethal dose by ingestion is estimated at 4 g, but 1 g has caused death, and one **person** survived a 30 g dose. Aplastic and hemolytic anemia and leukopenia have occurred. Methemoglobin formation is enhanced by ***alcohol*** intake in **humans** and **cats**. See ***acetamide***, ***acetaminophen***, and ***phenacetin*** (aniline compounds) for other untoward effects.

Aniline dyes: Used in marking crayons, pencils, and shoe polish. Toxicity is usually manifested by apathy, dyspnea, vomiting, convulsions, and methemoglobinemia. Ingestion and some skin absorption occurred in a mentally retarded 17-year-old **female epileptic** during a 3 week period of stenciling, requiring 50 mg ***methylene blue***, IV, and repeated in 3 h. Each treatment produced nausea, vomiting, and abdominal pain. **She** recovered 2 days later. Hypersensitivity to ***aniline black*** from a dyed dress reported. A 46-year-old **male** ingested ***aniline black*** in a suicide attempt. Cyanosis, coma, hyper motor activity, sweating, shallow respirations, diarrhea, and frequent vomiting occurred. Initial therapy consisted of IV ***methylene blue*** and exchange transfusions. Recovery took 40 days. Cases of allergic reactions to the blue dyes (***rosaniline***) in **their** uniforms reported. Aniline derivatives may be associated with increased incidence of cleft lip and cleft palate. A definite increased in bladder tumors in **humans**, after 2–36 years of employment in the industry (average 12 years). ***Tartrazine*** (q.v.) is an aniline dye.

LD_{50}, **rats**: 440 and 1,070 mg/kg; **dog**: 195 mg/kg; LC_{LO}, inhalation, **rat**: 250 ppm/4 h; **cat**: 180 ppm/8 h; skin LD_{50}, **rat**: 1.4 g/kg; **cat**: 254 mg/kg; **rabbit**: 820 mg/kg; **guinea pig**: 1.3 g/kg.

ANISALDEHYDE

= *Anisic Aldehyde* = *Aubepine* = *4-Methoxybenzaldehyde*

Approximately 25 tons/year used.

Use: As a fragrance in soaps, detergents, creams, lotions, and perfumes, imparting the odor of ***coumarin***, new-mown hay, clover, and fougere.

Untoward effects: Applied full-strength to intact or abraded **rabbit** skin for 24 h under occlusion was moderately irritating, and 10% in petrolatum was not irritating to **human** volunteers, after a 48 h closed-patch trial.

LD_{50}, **rats**: 1.5 g/kg; **guinea pigs**: 1.26 g/kg. Acute dermal LD_{50} in **rabbits**: > 5 g/kg.

ANISE

= *Anny* = *Pimpinella anisum* = *Sweet Cumin*

***Flavoring agent, carminative*. China appears to be its origin.**

***Use*:** In foods (**livestock** and **human**), soaps, detergents, creams, lotions, perfumes, carbonated beverages (viz. root beer), candies, cordials and liqueurs, expectorants (increased respiratory tract fluid), and carminatives.

Untoward effects*:** Used as a flavoring agent, it has caused an allergic cheilitis in **people**. In a denture cleaning cream, it caused a dermatitis in the **user**. It is estrogenic. Has been abused by ***paregoric addicts, injecting it IV with usual problems from such abuse. Russian anise is sometimes admixed with up to 30% ***coriander***, and Italian anise usually has 2–10%, and, on occasion, even 50% ***conium*** (q.v.). A case of poisoning is on record from the accidental admixture of the fruits of ***conium***.

Anise oil*:** No irritation to the skin of 25 **human** volunteers, after a 48 h closed-patch test with 2% in petrolatum. Several cases of sensitivity to it in **man**, supposedly due to its high (80–90%) ***anethole content. The dermatitis consisted of erythema, scaling, and vesiculation.

LD_{50}, **rats**: 2.25 g/kg; acute dermal LD_{50}, **rabbit**: > 5 g/kg.

***Star anise oil*:** From the seeds of *Illicium verum*. No sensitization in **humans**, **swine**, and **mice**.

LD_{50}, **rats**: 2.57 g/kg; acute dermal LD_{50}, **rabbits**: > 5 g/kg.

ANISE ALCOHOL

= *Anisic Alcohol* = *Anisyl Alcohol* = *4-Methoxy-benzene-methanol* = *p-Methoxybenzyl Alcohol*

***Use*:** In soaps, detergents, creams, lotions, perfumes, and foods.

***Untoward effects*:** 5% in petrolatum in a 48 h closed-patch test on the skin of 25 **human volunteers** showed no irritation or sensitization.

Full-strength on intact or abraded **rabbit** skin was moderately irritating.

LD_{50}, **rats**: 1.2 g/kg; **mouse**: 1.6 g/kg; acute dermal LD_{50}, **rabbits**: 3 g/kg.

o-ANISIDINE

= *o-Methoxybenzamine*

***Aniline* derivative.**

Use*:** Approximately 2 million lbs/year. In dye manufacturing, including synthesis of ***guaiacol, azo dyes, and in hair dyes.

Untoward effects*:** Traces of the chemical may remain in finished products and OSHA, noting a potential for skin absorption, established a TWA of 0.5 mg/m^3. Inhalation and dermal contact are the chief routes of **human** exposure. It is estimated that a potential risk exists for about 2000 **workers**/year in the U.S. to its derivatives. Headache, dizziness, and cyanosis have been reported. When heated, it emits toxic fumes of ***nitrogen oxides and ***hydrochloric acid***.

Carcinogenic, when fed to **rats** and **mice** (cancer and neoplasms of the bladder, cancer of the pelvis and kidneys, and thyroid tumors in male **rats**).

IP, LD_{LO}, **mouse**: 250 mg/kg.

ANISINDIONE

= *Miradon* = *SPE 2792* = *Unidone*

Anticoagulant.

***Use*:** In **people** sensitive to other oral anticoagulant products.

***Untoward effects*:** Hematomas at injection site in some **patients**. Rectal hemorrhage reported in one **patient**. Epistaxis, hematuria, gastrointestinal and adrenal hemorrhages, and hypersensitivity reactions (rash, fever, leukopenia, and hepatitis) also reported. Use with caution, as renal and hepatic toxicity has occurred.

ANISOLE

= *Methoxybenzene*

Aromatic.

Use*:** An intermediate for aroma ingredients and a component of ***butylated hydroxyanisole (q.v.), a food additive.

***Untoward effects*:** LD_{50}, **rat**: 3.7 g/kg.

ANISOMYCIN

= *Antibiotic PA-106* = *Flagecidin*

***Antifungal* and *antiprotozoal* (trichomonads) *antibiotic*.**

***Untoward effects*:** Induces amnesia for a specific learning task in **mice**, if given shortly before or after training.

IV, LD_{50}, **mice**: 140 mg/kg; IP, 400 mg/kg; oral, 148 mg/kg.

ANISOTROPINE METHYLBROMIDE

= *Lytispasm* = *Octylatropine Bromide* = *Valpin*

Anticholinergic, antispasmodic.

***Untoward effects*:** Dry mouth, headache, constipation, and nausea has been reported. Mild mydriatic effect.

ANISYLIDENE ACETONE

Synthetic.

***Use*:** As a fragrance in soaps, detergents, creams, lotions, and perfumes.

Untoward effects: 2% in petrolatum in a 48 h closed-patch test on **humans** revealed no irritation. A maximization test on 22 **volunteers** produced two sensitization reactions.

Full-strength to intact or abraded **rabbit** skin for 24 h under occlusion was not irritating.

LD_{50}, **rats** and acute dermal LD_{50}, **rabbits**: > 5 g/kg.

ANISYL PROPIONATE

= *p-Methoxybenzyl Propionate*

Synthetic.

Use: As a fragrance in soaps, detergents, creams, lotions, perfumes, and foods:

Untoward effects: 4% in petrolatum in a 48 h closed-patch test produced no irritation and no sensitization reactions on 22 **volunteers**.

Full-strength to intact or abraded **rabbit** skin for 24 h under occlusion was not irritating.

LD_{50}, **rats**: 3.33 g/kg and acute dermal LD_{50}, **rabbits**: > 5 g/kg.

ANONA sp.

***Botanical.* Various species found in Africa, Europe, Cuba, U.S., Asia, and South America.**

Use: As a vermifuge, insecticide, and **insect** repellent.

Untoward effects: Ingestion of bark extracts of *A. senegalensis* = ***Gwandar daii*** (Hausa) = ***Uburu Ocha*** (Igbo) = ***Abo*** (Yoruba) = ***Nkoke-lo*** (Kiyanzi) = ***Niolo*** (Kikongo) caused enophthalmus in **man** within 3 h, and short-lasting hypertension in **dogs**. The roots of *A. chrysophylla* = ***Malamute*** (Boni) = ***Mbokwe*** (Digo) = ***Mkonokono mwitu*** (Swahili) = ***Wild Custard Apple*** are occasionally poisonous to **man**. *A. glabra* = ***Cajuda*** reported poisonous to **fish**. *A. squamosa* = ***Sherifa*** = ***Sugar Apple*** was abortifacient in **rats**.

ANONIDIUM MANNI

***Botanical.* From Africa.**

Untoward effects: Ingestion has caused decreased motor activity, muscle relaxation, and enophthalmus within 3 h.

ANTS

Untoward effects: Extreme **human** sensitivity to its venom and stings, with serious, even fatal reactions. Jumper ants, bull ants, and green ants are the most venomous in Australia. Their tail stings cause immediate pain, which may last for hours. It has been relieved by warm to hot water, and the pruritus eased by ***antihistamines***. Some **people** required ***epinephrine*** injections or an inhaler. **Workers** gathering ant eggs have developed an eczematous skin scaling attributed to their ***formic acid*** content. Fire ants (*Solenopsis sp.*) stings are painful and anaphylactic shock and death can occur in hypersensitive **patients**. Wheals appear 8–10 h later, which are sterile, but easily infected after rupturing. Occasionally, febrile reactions of 24–48 h duration occur with localized burning sensations, nausea, generalized urticaria, shock, and bronchial asthma. ***Epinephrine*** and ***antihistamines*** are helpful. The venom contains a necrotoxin and a hemolytic fraction. The death of a 97-year-old **female** who died in a Florida hospital, supposedly, after being bitten by red fire ants, while **she** was restrained in **her** bed, was the basis of a multimillion dollar lawsuit. **Drunks** have frequently fallen asleep on the ground near a fire ant mound. One **drunk** used such a mound as a pillow. Several hours later, **he** was brought to a hospital with approximately 5000 1–3 mm vesicles on **his** face, trunk, and extremities from the stings. They became pustular, and healed in about 2 weeks, leaving small, white scars. Thousands of fire ant-related injuries to **livestock** occur annually in the U.S. Harvester ants or desert ants (*Pogonomyrmex maricopa, P. badius,* and *P. rugosus*) bites have caused serious reactions in **people**. The ants vary in color from red to black. The venom produced averages about 12 µg/ant. Pharaohis ants (*Monomorium pharaohis*) have chewed their way through fiber glass and ***Teflon*** pads of the air vent of sterile administration sets. They have been pathogen vectors in hospitals.

Intense salivation, gagging, vomiting, and diarrhea follow toxic IP doses in the **cat** and **dog**, while similar SC doses only produce local irritation, pain, redness, burning, itching, local swelling, and swollen lymph glands.

48 h IP, LD_{50}, **mice**: 810 µg/kg; 72 h IP, LD_{50}, **cats**: 3 mg/kg. IP route is 55 times more toxic than the SC route in the **mouse**.

ANTAZOLINE

= *Antastan* = *Antasten* = *Antihistal* = *Antistin* = *Antistine* = *Azalone* = *Ben-a-hist* = *Fenazolina* = *Histazine* = *Histostab* = *Imidamine* = *5512 M*

Antihistamine.

Use: For symptomatic relief of ocular allergies.

Untoward effects: Chills, fever, lightheadedness, leg muscle spasms, drowsiness (20%), nausea (20%), dyspnea, contact dermatitis, tachycardia, cardiac arrhythmias and arrest, dry mouth, and flushed facial sensations are all reported occasionally. Acute interstitial pneumonitis and rashes reported in a 37-year-old **female**, 7 h after the second 60 mg dose. Rechallenge confirmed the cause. Its use has been associated with severe immune hemolytic anemias, neutropenia, leukopenia, agranulocytosis, thrombocytopenia, hemoglobinuria, and acute renal failure. Can be excreted in breast milk. High dosage can cause incoordination, lethargy, amnesia, confusion, hallucinations, stupor and coma, followed by tremors, hypothermia, and convulsions. Seriously poisoned **individuals** have about a 10% mortality. **Children** are more susceptible to adverse effects of overdoses. Estimated LD, **humans**: 50 mg/kg. May react with ***norepinephrine*** to enhance the latter's cardiovascular effects; with ***monoamine oxidase inhibitors*** to potentiate endogenous ***norepinephrine*** effects, and to potentiate the antihistamine effects.

IV, LD_{50}, **dog**: 30 mg/kg.

ANTHEMIS COTULA

= *Dog Chamomile* = *Dog Fennel* = *Iron Wort* = *May Weed* = *Meadow-cress*

***Weed*. Common in roadsides and farmyards. Belongs to the *aster* family.**

Untoward effects*:** Leaves, flowers, and seeds have blistered the feet, hands, and ankles of **farmers** and **others**. Contains an allergenic vesicant oil in the sap. Phototoxic. Ingestion has caused gastrointestinal upsets. May contain dangerous levels of ***hydrocyanic acid. Has produced a disagreeable taste and odor to ***milk***, milk products, and meat of **animals** that graze it.

ANTHIOLIMINE

= *Antholamine* = *Antimony Lithium Thiomalate* = *Delepine* = *Caillot* = *Lithium Antimony Thiomalate*

Anthelmintic, schistosomal, filaricide.

Untoward effects*:** Dermatitis, gastrointestinal upsets, sciatic pain and paralysis, after intragluteal injections. See ***Antimony sodium tartrate for additional adverse effects.

TL_{LO}, **human**: 11 mg/kg.

IP, LD_{50}, **rats**: 90 mg/kg; **mice**: 82 mg/kg.

ANTHODISCUS sp.

Trees and shrubs in South America.

Untoward effects*:** Fast-acting **piscicide** made from its leaves and young branches. The bark of one species is used by Columbian **Indians** in making a type of ***curare.

ANTHRACENE

***Use*:** In manufacturing dyestuffs. A coal tar derivative. Used in treating psoriasis and eczemas.

Untoward effects*:** Topical application has shown it to be a photosensitizer. A by-product of ***coke plants, in ***coal tar*** pitch, and found in the smoke of ***charcoal*** grills. Exposure has caused tumors of the nose, larynx, lung, skin, and scrotum in diesel jet **testers** and textile **weavers**.

It is a common air pollutant, lethal to **tadpoles** exposed to sunlight or ultraviolet light, but not when they are kept in the dark.

TD_{LO}, **rats**: 18 g/kg; SC, 3.3 g/kg.

***Carcinogenicity*:** Chronic oral toxicity, **rats**: 3.3 g/kg; topically, **mice**: 80 μg/kg.

2-ANTHRAMINE

= *2-Aminoanthracene* = *2-Anthracenamine*

***Untoward effects*:** Carcinogenic (mammary tumors – lobular hyperplasia and carcinoma) in **rats**.

TD_{LO}, **rat**: 45 mg/kg/30 days; skin, **rat**: 260 mg/kg/33 weeks; skin, **mouse**: 800 mg/kg/20 weeks; skin, **hamster**: 6 g/kg/55 weeks.

ANTHRAMYCIN

= *Refuin*

***Antineoplastic; chemosterilant*. The more stable methyl ether form = *AME* = *RO 5-9000*.**

Untoward effects*:** Increase in serum ***uric acid. Extravasation will cause ulceration of the skin. Lethargy, somnolence, and disorientation has been reported. Rapid IV may cause irreversible shock to the **patient**.

ANTHRAQUINONE

= *Anthradione* = *Corbit* = *Heolite* = *Morkit*

Use*:** In manufacturing inks and dyes, and as a **bird** repellent. In ***cascara sagrada, ***senna***, ***danthon***, and ***aloes***. ***Prune*** juice contains ***oxyphenisatin***, an anthraquinone. Used to break down wood chips into cellulose pulp and in improving adhesion and heating stability of rubber tire cords.

Untoward effects*:** Despite low systemic toxicity, it can cause skin irritation (dyes on bikinis, bathing suits, etc.) and sensitization, especially in dye manufacturing plants. Melanosis coli, a dark pigmentation of the colonic mucosa, has occurred in chronic **users** of anthraquinone-containing ***laxatives. A corrosive gastroenteritis can occur in some **people** eating ***rhubarb***, ***spinach***, and ***beets*** that contain a high concentration of its glycosides. Conjunctivitis, diarrhea, and coughing occur. A small amount is excreted in **mother's** milk, at times, enough to cause colic and catharsis in an **infant**. Will color alkaline urine red. Slight fire hazard, when exposed to heat or flame.

ANTHURIUM ANDRAEANUM

= *Flamingo Lily*

Untoward effects: Ingesting or sucking on the leaves or stems of this ornamental house plant leads to severe discomfort with painful swelling and blistering of the mouth and throat, hoarseness, dysphagia, and possible asphyxiation from its ***calcium oxalate*** crystal content.

ANTIARIS TOXICARIA

= *False Iroko* = *Mkunde (Swahili)* = *Mnguonguo (Digbo)* = *Ubas Tree* = *Upas Tree*

Forest tree in China, Burma, Fiji, Africa, Indonesia, and the Philippines.

Untoward effects: The latex, once used for poisoning arrows, was made famous in the poetry of Erasmus Darwin's *The Garden*, where he called it the "hydra tree of death". Used in executions. In Java in 1776, 13 of the king's **concubines** were convicted of infidelity and **their** bare breasts were pierced with an awl-like instrument dipped in the latex leading to agonal death within 5 min.

Estimated IV LD of its toxic fraction is 0.116 mg/kg in **cats**.

ANTIMONY

= *Sb*

Use: In metallurgy, battery alloys, fireworks, fire-retardants, and a stabilizer in polyvinylchloride manufacture. Its compounds are used in the treatment of schistosomiasis, leishmaniasis, and filariasis. About 80,000 tons produced annually. **Egyptians** of the Stone Age used it in cosmetics and as prophylaxis against disease of the eye. There are references to its therapeutic use in the Bible and ancient writings from China, India, and Mexico.

Untoward effects: Nausea, skin and mucous membrane irritation, and occasional cardiac complications with IV and IM use of its compounds, which usually require a long course of treatment. Symptoms and lesions resemble those of ***arsenical*** poisoning. Pneumoconiosis in a ***stibnite***-processing plant's **workers**. Toxic amblyopia and yellowing of skin and sclera have been ocular side-effects. Although enzyme changes occur, actual hepatitis is rare. Radiological examination of **workers** at an ***Sb*** ore-smelting plant showed a very high percentage had a pneumopathy, unrelated to **their** length of time employed. Acute renal tubular necrosis is reported. Gray enamelware containing ***Sb***, when in contact with acidic foods, causes leaching of ***Sb***. Within minutes to a few hours, vomiting, abdominal pains, spasms, collapse, and even death can occur. Laboratory **users** must remember that when ***Sb*** or its compounds are treated with acids, ***antimony hydride*** (***SbH_3***), a toxic gas, can be generated. Gun **shooters** may get traces of ***Sb*** on **their** hands from "blowback" of gunpowder primers. Will cause false positive blood urea nitrogen tests. Suitable respirators should be worn by **those** with potential exposure. Can be adsorbed by activated ***charcoal***. ***Dimercaprol*** (3–5 mg/kg, IM every 4 h for 2 days, every 6 h on day 3, then every 12 h for ten or more days) and ***penicillamine*** has been used in treatment. TWA, 0.5 mg/kg/8 h. See specific ***antimony compounds*** for more details.

Has caused toxicity (anorexia, anxiety, diarrhea, decreased rumination, and even death) in **animals**. Overtoasted cornflakes were fed to **mink** and all died of ***Sb*** poisoning. How it got into the food chain was never proven. Yet, commercial cornflakes off a grocery shelf, at that time, revealed traces of ***antimony***.

ANTIMONY PENTACHLORIDE

Use: Industrial catalyst.

Untoward effects: Reacts with atmospheric moisture to form ***hydrochloric acid*** (q.v.). Approximately 500–600 **people** were evacuated from **their** homes, after a tank truck leaked approximately 1,000 lbs. The site was covered with ***soda ash***, ***bicarbonate of soda***, and sand. 12 **people** were treated in area hospitals for dizziness, throat and stomach pains, and **one** with the emergency-response team with a chemical burn.

LD_{50}, **rat**: 1,115 mg/kg; **guinea pig**: 900 mg/kg.

ANTIMONY POTASSIUM TARTRATE

= *Potassium Antimonyl Tartrate* = *Tarter Emetic*

Use: Anthelmintic and ruminatoric. In treating schistosomiasis, in denaturing ***alcohol***, and, rarely, as an emetic.

Untoward effects: Has caused fixed-drug reactions, often pustular (on chronic **users**), with occasional hyperpigmented skin areas. Auricular fibrillation, bradycardia, and cardiac arrest may occur. Acute poisoning is characterized by a metallic taste, nausea, vomiting, frequent hiccups, Adams–Stokes syndrome with severe ventricular arrhythmias, burning pain in the stomach, colic, frequent stools and tenesmus, fainting, rapid, feeble pulse, cold skin, respiratory distress, cutaneous anesthesia, convulsive motions, leg cramps, prostration, and death. Occasionally vomiting and purging does not occur. Pustular eruptions have been reported. ***Coproporphyrin*** test for abnormal ***lead*** absorption may be unreliable after treatment. Death has occurred after 150 mg oral dose. 48.5 mg has been fatal to a **child**. A 25-year-old **female** took a maximal therapeutic dose of 1.5 grains (97.2 mg) without ill-effect. A smaller dose, 24 h later, caused violent purging and vomiting, with death in 36 h. Fatalities are sometimes delayed for several days, but an **adult** has died in 7 h and a **child** within 45 min. LD_{LO}, **human**: 2 mg/kg. ***BAL*** has been suggested for treating its toxic reactions.

IV use in **rabbits** causes a leukopenia. 200–400 mg is said to be toxic to **dogs**; 15–30 g for **horses**; 50 g for **cows**; and 15 g for **sheep**. Acute gastroenteritis, fatty degeneration of the liver, and dilated blood vessels are frequent post-mortem findings.

LD_{50}, **rat**: 115 mg/kg; IP, 11 mg/kg; oral, **mouse**: 600 mg/kg; oral, **rabbit**: 115 mg/kg.

ANTIMONY SODIUM GLUCONATE

= *Myostibin* = *Pentostam* = *Stibanate* = *Stibogluconate Sodium*

The pentavalent form is used in treating leishmaniasis.

Use: In treating leishmaniasis (kala-azar).

Untoward effects: IV treatment for 10 days (0.6 g/day) of 22 **soldiers** (18–31-year-old **males**) in Hong Kong with cutaneous leishmaniasis caused a reversible decrease of T wave amplitude on electrocardiogram. This can occur after about the fourth day of treatment and after a total dose of 2 g has been given. Muscle pain, joint stiffness, nausea, and vomiting are common complaints after treatment. Colic, diarrhea, rashes, pruritus, myocardial and liver damage, and bradycardia are occasional problems. Rarely, hemolytic anemia, renal tubular dysfunction, shock, and sudden death have occurred. A 32-year-old **male** inadvertently given a 6.5 g dose instead of 0.65 g dose developed increased pancreatic enzyme activity for 48 h with no adverse clinical symptoms.

ANTIMONY SODIUM TARTRATE

Use: In treating schistosomiasis.

Untoward effects: In 17 **male** (8–27 years) Egyptian farm **workers**, six became nauseous and vomited; spasmodic cough in nine, after injection; temperature increased to 102°F in four, with marked eosinophilia in two (decreased with ***antihistamines*** and postponement of injections for a few days). Local necrosis reported at injection sites. Slightly less toxic than the potassium form, when given IV. Causes changes in the electrocardiogram (flattening or inversion of the T wave), bradycardia, and cardiac arrhythmias. Collapse and sudden death appear to be due to anaphylactic-type reactions. Liver and renal damage are rare. TWA, **human**: 0.5 mg/m^3.

IP, SC, IV, LD_{50}, **mouse**: 60, 48, and 25 mg/kg, respectively.

ANTIMONY THIOANTIMONATE

Use: As an industrial lubricant additive.

Untoward effects: Mild skin and eye irritant.

Acute oral, LD, **rats**: > 5 g/kg; acute dermal LD, **rabbits**: > 2 g/kg.

ANTIMONY TRICHLORIDE

= *Butter of Antimony*

Untoward effects: Corrosive, escharotic. Seven **males** accidentally exposed to its fumes had respiratory tract irritation from the ***hydrochloric acid*** produced. Five later suffered from gastrointestinal upsets, including abdominal pain and anorexia. **Their** environmental ***Sb*** exposure had been up to 73 mg/m^3 and ***hydrochloric acid*** up to 146 mg/m^3. In eight cases of poisoning in **humans**, four fatalities occurred after each consumed 2 oz; two that ingested 1 oz recovered.

TD_{LO}, **human**: 73 mg/kg; TC_{LO}, 73 mg/m^3.

LD_{50}, **rat** and **guinea pig**: 525 and 574 mg/kg, respectively.

ANTIMONY TRIOXIDE

= *Antimony Oxide* = *Diantimony Trioxide* = *Exitelite* = *Flowers of Antimony* = *Senarmontite* = *Valentinite* = *Weisspiessglanz*

Use: As a flame-retardant on canvas, in paints, opacifier in porcelain enamels, ceramics, and in the glass industry as a mordant.

Untoward effects: Hazardous dust. Can react with acids yielding poisonous ***stibine*** (q.v.) gas. Red, itchy rashes from skin contact. Some **workers** may be hypersensitive. Oncogenic potential. Alopecia reported. Accidental spills onto bags of green coffee beans in a warehouse prompted FDA to establish a limit for it in ***coffee***. TWA, **human**: 0.5 mg/m^3.

Extreme eye irritation after 4 s contact in **rabbits**.

LD_{50}, **rats**: > 20 g/kg; acute LD_{50} dermal toxicity, **rabbits**: > 2 g/kg.

ANTIMONY TRISULFIDE

= *Antimony Sulfide* = *Needle Antimony* = *Stibnite*

Use: In pyrotechnics, "strike anywhere" matches, explosives, in ruby glass, and as a paint pigment.

Untoward effects: Benign pneumoconiosis in **workers** processing the crude ore.

TC_{LO}, **human**: 580 μg/m^3/35 weeks.

ANTIMYCIN A

= *Blastmycin* = *Fintrol*

Antifungal antibiotic, piscicide. **Possesses antifungal and miticidal properties.**

Use: In **catfish** farming, to eliminate "trash" or undesireable **fish**.

Untoward effects: Avoid eye and skin contact. Wear protective goggles and gloves. Treated waters are unsafe for any use (**livestock** and **human** consumption or crop irrigation)until fingerling **trout** or **bluegills** can survive in them for 48 h. For use with non-food **fish** only.

IP, LD, **rat**: 800 µg/kg; IV, LD_{50}, **mice**: 1 mg/kg; SC, LD_{50}, **mice**: 1.6 mg/kg; IP, LD_{50}, **mice**: 1.5–1.8 mg/kg. IP injections caused hind limb ataxia, impaired reflexes, and respiratory distress.

ANTIPYRINE

= *Analgesine* = *Anodynine* = *Dimethyloxychinizin* = *Dimethyloxyquinazine* = *Dolo-Med-Much* = *Parodyne* = *Phenazone* = *Phenylone* = *Sedatine*

Analgesic, antipyretic, local anesthetic. **Based on the Greek word *pyretos*, meaning fever. Chemically related to *phenylbutazone*.**

Use: To relieve aches and pains, reduce fever, and as a topical anesthetic and vasoconstrictor for external use in otic preparations. Often combined with other analgesics. Half-life in **man** is about 11–12 h, and less than 2 h in **dogs** and **monkeys**. In use for over 100 years, but use has decreased with the advent of safer alternatives.

Untoward effects: Has caused hemolytic anemia in ***primaquine***-sensitive and glucose-6-phosphate dehydrogenase-deficient **individuals** (but not in **blacks**), as well as fixed-drug eruptions, exfoliative dermatitis, erythema multiforme, pruritus, toxic epidermal necrolysis and Stevens–Johnson syndrome. In a group of 52 cases of dermatitis due to it, 41 were morbilliform, seven erythrometopapular, and four urticarial. Eruptions were prone to itch and desquamate. A 30-year-old **male** suffered a severe allergic reaction that included collapse and peripheral cyanosis. Another 23-year-old **male** also had a severe allergic reaction, characterized by diaphoresis, flushing, throat and upper lip swelling, vomiting, and a diffuse urticarial rash. A 32-year-old **male** had four episodes of acute renal failure, due to hypersensitivity, after ingestion of it. One hour after ingestion, a 33-year-old **male** collapsed and was semi-comatose. An hour later, **he** was able to walk with assistance, but had periods of dizziness, fainting, nausea, periodic chills, and pruritus. **He** recovered in 24 h, after ***diphenhydramine***, IV. An 87-year-old **female** died of cardiac arrest after 500 mg, IV. Agranulocytosis, urticaria of eyelids, keratoconjunctivitis, amblyopia, nystagmus, grand mal seizures, delirium, kidney papillary necrosis (especially after abuse of the drug), abdominal pain, nausea, and proteinuria have also been reported. Use may color urine yellow–red. It is effectectively adsorbed by ***activated charcoal***. Serum concentration of 100 mg/l is considered toxic to **man**. May cause inaccurate high test readings of sugar in urine (Benedict's test) and urine urobilinagen.

Drug interactions: Increased urinary excretion by acidifying agents, such as ***ammonium chloride***. Decreased urinary excretion by cigarette smoking, alkalinizing agents, such as ***sodium bicarbonate***, and ***propranolol***. Half-life increased by ***Δ^9-THC***, ***allopurinol***, ***cimetidine***, ***disulfiram***, ***flurbiprofen***, ***methyldopa***, ***nortriptyline***, and ***propoxyphene***. Half-life decreased by ***alcohol***, ***Brussels sprouts***, ***cabbage***, ***charcoal-broiled beef***, ***chlorinated hydrocarbons***, ***halothane***, ***insulin***, ***phenobarbital***, and ***quinine***. Enhances ***warfarin*** metabolism, and its own effect may be enhanced by ***oral contraceptives*** and ***ketoconazole***. Oral intake of ***mineral oil*** decreases its absorption. Increases uptake of ***radiopharmaceuticals*** by the kidney. ***Caffeine*** elimination increased by it with ~25% decrease in half-life.

Dexamethasone and ***griseofulvin*** increase antipyrine elimination, and it was decreased by ***chloramphenicol*** or ***oxytetracycline*** use in two 12-weeks-old **calves**. **Rabbits** on 30 mg/kg consistently developed a proteinuria. **Rats** on approximately 1 g/kg for up to 2 months had normal renal histology but some cells in their urine.

LD_{LO}, **human**: 50 mg/kg; **dog**: 500 mg/kg; **guinea pig**: 1.4 g/kg; **cat**: 100 mg/kg; LD_{50}, **rat**: 1.8 g/kg; **mouse**: 358 mg/kg.

ANTITHYMOCYTE GLOBULIN or SERUM

= *Atgan* = *ATS* = *Pressimmune* = *Thymoglobuline*

Immunosuppressant. **Usually from equine serum; occasionally, goat and rabbit sera have been a source.**

Use: In treatment of aplastic anemia; in preventing organ transplant rejection in **patients**.

Untoward effects: Assorted infections have been reported, due to its immunosuppression. It may also have a role in the development of neoplasms, such as multiple myeloma. Infusions in **patients** with bone marrow failure cause varying degrees of serum sickness with urticarial and morbilliform skin rashes, as well as a thin erythematous band along the sides of hands, fingers, and toes, rapidly replaced by petechiae or purpura. The use of horse serum may cause serious reactions in **people** previously sensitized to it.

APAMIN

***Neurotoxin.* In the venom of honey bees.**

***Untoward effects*:** 1 mg/kg, injected into **mice** caused incoordination, constant motion, convulsions, and death. Survivors were extremely hyperexcitable for up to 60 h.

IV, LD_{50}, **mice**: 4 mg/kg.

APAZONE

= *AHR 3018* = *Azapren* = *Azapropazone* = *Cinnamin* = *Mi 85* = *Prolixan* = *Rheumox* = *Sinnamin* = *Tolyprin*

Anti-inflammatory, analgesic.

***Use*:** In the treatment of rheumatoid arthritis.

***Untoward effects*:** Anorexia, nausea, vomiting, transient diarrhea, pruritus, petechiae, alveolitis, and hemolytic anemia (usually after 3 months–4 years), acute renal failure, bullous skin eruptions, photosensitivity, and hepatitis, due to hypersensitivity have all been reported.

Drug interactions*:** Clinically important increase of ***warfarin anticoagulation. A 77-year-old **female** developed hypoglycemia induced by its coadministration with ***tolbutamide***. Four **male volunteers** had their ***tolbutamide*** half-life of 7.7 ± 1.6 h increased to 25.2 ± 6 h by apazone usage. Similarly, apazone inhibits the metabolism of ***phenytoin***, and has caused neurological toxicity, including diplopia, nystagmus, and vertigo. ***Methotrexate*** toxicity can be precipitated by it. Chronic tests and teratogenic tests in **rats** with 100 mg/kg/day doses showed no untoward effects.

LD_{50}, **rat**: 2 g/kg; **mouse**: 1 g/kg; IV, LD_{50}, **cat**: 500 mg/kg; **rat**: 660 mg/kg.

APHOLATE

Insect chemosterilant.

***Untoward effects*:** LD_{LO}, **human**: 5 mg/kg.

Japanese quail with it in their feed at 125 ppm had only slight decrease in body weight and feed consumption, but decreased egg production, fertility, and hatchability. A few were found dead at 500 ppm, and mortality was high at 1000 ppm. Similar ill-effects were noted at lower levels in **chicks** and mature **chickens**. Both sexes had marked leukopenia, and mortality was higher in **males**. Will sterilize **house flies** that eat it or walk on it. Lethal to **sheep** (5 mg/kg, IM) with leukopenia, and, if given low doses orally over a period of time, causes decreased thrombocytes or platelets. An IM dose of 2.5 mg/kg killed 4–5 month **calves** in 5–7 days (after rumen atony, anorexia, general weakness, and depression). Marked leukocytopenia and lymphocytopenia occurred within 24 h. Possible congenital anomalies in a **lamb**, after the **ewe** and **ram** had received oral doses of it.

LD_{50}, **male rat**: 98 mg/kg; **female rat**: 113 mg/kg.

APIOLE

***Insecticide synergist, volatile oil.* Found in *parsley* and *dill*.**

Untoward effects*:** Can cause psychic and motor disturbances, because of its ***thujone (q.v.) content damaging the central nervous system. Has had a long history as an abortifacient in **women**. Numerous reports of toxicities from such use occurred in the 1930s. It was later discovered that an adulterant, ***o-tricresyl phosphate***, was the cause.

SC, LD_{LO}, **dog**: 500 mg/kg; **mouse**: 1 g/kg; **frog**: 1.5 g/kg.

APOATROPINE

= *Atropamine* = *Atropyltropeine*

Anticholinergic, antispasmodic.

Untoward effects*:** A toxic contaminant that has occurred in ***atropine solutions. Approximately 16 times ***atropine's*** toxicity. It has been formed by ***zinc*** compounds eluting from rubber stoppers into the citrate buffer of ***atropine*** solution (***Atropen*** cartridge). Death can occur with overdoses.

LD_{50}, **mice**: 160 mg/kg; IP, 14.1 mg/kg.

APOCYNUM CANNABINUM

= *American Hemp* = *Apocyn Chanvrin* = *Canadian Hemp* = *Hemp Dogbane* = *Indian Dogbane* = *Indian Hemp* = *Indian Physic* = *Rheumatism Weed*

Contains the tumor-inhibiting compounds, *apocannoside* and *cymarin*.

Use*:** In folklore medicine, for the topical treatment of condylomas and cancers. **Indians** used fibers from its bark for making rope and as a **piscicide**. As a diuretic and cardiotonic (digitalis***-like properties).

***Untoward effects*:** Despite **animals** avoiding its bitter, sticky, milk-white juice, when grazing is sparse, **cattle**, **horse**, and particularly **sheep** are poisoned by it. Signs occur 6–12 h after consumption. LD, **sheep** ½–1 oz/cwt, **cattle**, and **horses** ½–¾ oz/cwt. Leaves are poisonous, even when dry or in hay. Young, tender shoots are particularly poisonous. It increases temperature and pulse, pupils are dilated, sores and discoloration of mouth, anorexia, sweating, cold extremities, frequent bowel movements, and death can occur. Other species have similar effects. *Apocynum* in the acrid juices is responsible for the vomiting, diarrhea, and superpurgation.

APOMORPHINE

Emetic.

Use: As an emetic in conscious **animals** and **man**, and in the treatment of Parkinson disease.

Untoward effects: When used as an immediate treatment for ingested poisons, 50% showed central nervous system depression. May potentiate erections in **man**. Causes bradycardia, respiratory depression, coma, pinpoint pupils, protracted vomiting, shock, dyskinesia, faintness, tachycardia, dizziness, central nervous system depression, sleepiness, hypotension, metallic taste, and a case of progressive edema of the lower limbs reported. Very large doses can cause a period of excitement and convulsions, followed by death from asphyxia. Occasionally, induces blinking and yawning, tremors, and euphoria. Extreme caution should be used in giving it to **children**, where its efficacy may be very variable.

Overdosing in **dogs** has produced central nervous system and/or respiratory acceleration or depression, and, rarely, restlessness, excitement, tetanic convulsions, protracted vomiting, and death from asphyxia. **Horses** exhibit delirium, sweating, and dyspnea. Long-time storage in a polyethylene syringe causes discoloration of its solutions. Induces stereotypy in **mice**.

IV, LD_{LO}, **dog**: 80 mg/kg; **rat**: 40 mg/kg; IV, LD_{50}, **mouse**: 56 mg/kg; IP, 160 mg/kg.

APPLE

Untoward effects: Immediate aggravation of chronic hand dermatitis reported after handling apples. In a selected group of **atopics**, some experienced itching, tingling and/or edema of **their** lips, mouth, and tongue, and throat irritation with hoarseness after eating raw apples. **People** allergic to ***cinnamaldehyde*** in aftershaves and cosmetics have become allergic to apples. Consumption of apples can cause osmotic diarrhea and they are potent sources of gastrointestinal gas production. The fruit, and especially its seeds, contain ***amygdalin glycoside***, and it may become hydrolyzed in the gut by β-glucosidases, to release ***cyanide*** ions. A **man** who ate apples daily saved the seeds and ate them when **he** had a cupful. **He** died shortly afterward from cyanide poisoning. In less sudden poisoning from the ***cyanide*** (q.v.), nausea, vomiting, abdominal pain, diarrhea, dyspnea, muscle weakness, dizziness, stupor, and even convulsions have occurred. ***Apple pericarp oil*** used in Nigeria can be caustic and vesicant; if taken orally, it causes diarrhea and dehydration. **People** with ***chromate***-induced dermatitis should avoid apples. Dry cider from inedible apples in the French Normandy peninsula has been implicated as a possible cause of esophageal cancer. Apples contain ***estrogens***. A carcinogenic mycotoxin has been found in apple juice. The juice is a frequent cause of diarrhea in **children**. Large increase in renal calculi after 8 oz/day. *Bacillus cereus* on home-dried apples has also been reported as causing diarrhea in **adults**. All parts of the apple tree, and less in the very mature fruit, contain ***phloridzin***, which interferes with various enzyme systems and causes a glucosuria. This may occur in cases of considerable apple consumption. Can cause migraine headaches in some **people**. **People** who develop urticaria after eating apples are probably ***salicylate***-sensitive, because apples contain ***salicylates***. Other causes of the urticaria can be from residual sprays, after consuming unwashed apples. An ***Alar*** (***daminozide*** – q.v.) scare by the FDA regarding its possible contamination of apples was ridiculous in many respects. ***Cholestyramine resin*** with its bad taste and odor will turn apple juice black, when admixed without altering its activity.

Apple pomace fed to **cattle** and **sheep**, along with ***nonprotein nitrogen***, caused the latter to pass rapidly through the rumen, without proper utilization, and caused **calves** and **lambs** to have weak pasterns and crooked legs.

APRICOT

= *Prunus armeniaca*

Use: Extracts are used in flavoring beverages and cigarettes. A food.

Untoward effects: Its pits and the seeds therein contain ***cyanogenic amygdalin***, that releases ***hydrocyanic acid*** and can result in ***cyanide*** (q.v. and discussion above, under ***Apple***) poisoning. A 49-year-old **female** ingested 20–40 pits and within 30 min was nauseous, vomited, had a headache, and became weak with a pulse of 140. In the emergency room, **her** whole blood ***cyanide*** level was 3.2 mg/l (> 1 is considered highly toxic). **She** recovered after treatment. Approximately 48 seeds were ingested by a 34-year-old **male**, and within 1 h, **he** showed symptoms of ***cyanide*** poisoning. Emergency treatment with ***ipecac*** helped recovery. A 2-year-old **female** ate a large number of apricot kernels and became unconscious. Prompt gastric lavage aided recovery. After eating some cracked open kernels (15 ±), a 3½-year-old **female** rapidly developed a marked pallor, tachycardia, tachypnea, mydriasis, disorientation, confusion, vertigo, and was restless with irrational actions. **She** recovered after treatment. Some **children** have died from eating the seeds. **Adults** have also been poisoned by drinking ***almond***-flavored milkshakes that contained apricot kernels. As little as 50 kernels may kill an **adult** and seven, a **child**. As with ***apples*** (q.v. above), caution may be needed by ***salicylate***-sensitive **people**. Two over-the-counter products had potentially dangerous levels of ***hydrocyanic acid***. ***Aprikern***® and

Bee-Seventeen® packets were also dangerous. 10 mg ***hydrocyanic acid*** may be fatal to a **child** and 40 mg to an **adult** may be lethal. **People** allergic to ***almonds*** may also be allergic to apricots, and **those** with ***cobalt***-induced dermatitis should avoid apricots.

Acute toxicity has also been reported in **mice**.

APRINDINE

= *Amidonal* = *Aspenon* = *Compound 83846* = *Fibocil* = *Fiboran* = *Ritmusin*

Antiarrhythmic.

Untoward effects: Some risk of agranulocytosis and aplastic anemia reported. These may be due to an immune-related mechanism. Five cases of hepatitis within 2–3 weeks of initiating therapy. Mild disorientation, confusion, and nightmares in three at toxic plasma levels of 3.1–4.2 μg/ml (0.75–1.5 levels are therapeutic and 2–3 may be toxic). It appeared to aggravate arrhythmias in about 11% of those tested.

APROBARBITAL

= *Allypropymal* = *Alurate* = *Aprozal* = *Isonal* = *Numal* = *Sommipron*

Sedative, hypnotic.

Untoward effects: Fatal intoxications have occurred. Serum concentrations of 35–50 mg/l are considered toxic.

LD_{LO}, **human**: 5 mg/kg (one case at 286 μg/kg); **rabbit**: 160 mg/kg; IP, LD_{LO}, **rat**: 100 mg/kg; **rabbit**: 90 mg/kg.

APRONALIDE

= *Apronal* = *Isodormid* = *2-Isopropyl-4-pentenoyl Urea* = *Sedormid*

***Sedative, hypnotic.* A *urea* derivative.**

Untoward effects: Thrombocytopenia and non-thrombocytopenic purpura, gross hemorrhages into skin, and fixed-drug skin eruptions. Fatalities dramatically reduced its use.

Rats and **rabbits** also exhibited a hepatic type of porphyria.

LD_{LO}, **rat**, **mouse**, and **guinea pig**: 600 mg/kg, **dog**: 300 mg/kg; **rabbit**: 1.25 g/kg.

APROTININ

= *Antagosan* = *Antikrein* = *Apronitine* = *Bayer A-128* = *Bayer 128* = *Contrykal*® = *Fosten* = *Gordox* = *Iniprol* = *Kir Richter* = *Onquinin* = *Repulson* = *Riker 52G* = *RP 9921* = *Trasylol*® = *Trazinin* = *Zymofren*

Proteinase inhibitor.

Use: In treating pancreatitis, and as a fibrinolysis inhibitor for IV use, to decrease bleeding and decrease the need for blood transfusions in high-risk coronary artery by-pass graft **patients**.

Untoward effects: Allergic or even rapid anaphylactoid reactions occur, since it is of bovine origin. Even a death has occurred. Severe pruritus and edema of hands and knees, colic, diarrhea, and possible disseminated intravascular coagulation has forced the therapy to be aborted. Acute respiratory distress syndrome in 24-year-old **male** after its use as a hemostatic following tonsilectomy. Interaction with ***muscle relaxants***, such as ***succinylcholine***, has caused apnea of 30 min duration, possibly due to its inhibition of serum ***cholinesterase***.

APSAC

= *Anisoylated Plasminogen Streptokinase Activator Complex* = *Anistreplase* = *BRL-26921* = *Eminase*

Thrombolytic, plasminogen activator.

Use: For IV therapy of coronary thrombosis, with longer half-life (90 min), compared to ***TPA*** (5 min), ***urokinase*** (16 min), and ***streptokinase*** (23 min).

Untoward effects: Hematomas at the catheter entry site in about 1/3 of **patients**, intracranial hemorrhages (1%), allergic and anaphylactic reactions, and hypotension occur. Fever, hematuria, proteinuria, vasculitis, maculopapular rash, and arthralgia in 46-year-old **male**, after 30 Units, IV, and 162.5 mg ***aspirin*** orally. Multiple microemboli in six (57–70 years), and most died. Cerebral hemorrhage reported in 58-year-old **male**, fatal IP hemorrhage in 47-year-old **male**, and severe hypotension in three **patients** (35–69 years) with massive pulmonary embolism. ***Contraindicated***, if major internal surgery occurred within 2 weeks, known bleeding diathesis cases, active internal bleeding, central nervous system surgery within two previous months, pregnancy, intracranial neoplasms, prolonged cardiopulmonary resuscitation, diabetic retinopathy, and in the very **aged**.

AQUILIDE A

= *Ptaquiloside*

Carcinogen.

Untoward effects: In ***bracken fern*** (q.v.), which, when fed to **cows**, causes bladder cancer.

ARALIA sp.

= *Spikenard*

Use: All parts, in many folklore and herbal medicines, and the official ***Compound White Pine Syrup***.

Untoward effects: *A. spinosa* seed has caused poisoning of **livestock** in southeastern U.S. and dermatitis (inflammation and blisters) in **man**, after handling the bark and roots. *A. manchuria* leaves are more toxic than roots in **mice** (hypotensive) and normalize blood pressure in hypertensive **dogs**.

ARAMITE

= *Acaracide* = *Aracide* = *Aratron* = *Compound 88R* = *ENT 16519* = *Niagaramite* = *Ortho-Mite*

Miticide, antimicrobial.

Use: For preharvest application on fruit and nut trees. No longer produced in the U.S.

Untoward effects: Little data available on possible **human** adverse effects from manufacturing or use exposure. Skin irritant in **man**. Large doses may cause central nervous system depression.

Carcinogenic at high oral doses. When fed to **dogs** at 500 ppm, led to adenocarcinomas of the gall bladders and biliary ducts; caused liver tumors in **rats** (at 100–400 ppm), and hepatomas in **males** of one strain of **mice**.

LD_{LO}, **human**: 429 mg/kg; **dogs**: 20 g/kg/2 years; **mouse**: 130 g/kg/80 weeks. LC_{50} in feed/5 days, > 5000 ppm for **Bobwhite**, **Japanese quail**, and **pheasants**, but with a 20% mortality at 5000 ppm in **Bobwhites**. LD_{50}, **rats**: 3.9 g/kg (4–6.3 g/kg in some studies); **mice**: 1.7 g/kg.

ARDEPARIN

= *Normiflo*

Low-molecular-weight heparin.

Untoward effects: Occasionally induces thrombocytopenia.

ARECA

= *Betel Nut* = *Pinangi* = *Ping Lang* = *Sopari*

***Anthelmintic, purgative, euphoric.* Also see *catechu*.**

Untoward effects: Has parasympathomimetic effects, due to its ***arecoline*** content, which decreases heart rate and blood pressure while increasing sweating. **Human *betel nut*** chewers have had a high rate of oral and esophageal cancers, due to its condensed catechin tannin content. Imparts a red color to the saliva of nut **chewers**. Cardiac arrhythmias, increased chest discomfort, palpitations, epigastralgia, and dyspnea after chewing 1–4 nuts.

SC injection of these tannins in **rats** caused tumors. Topical application of extracts to the buccal pouch caused tumors in 38% of **hamsters**. Produces nausea and vomiting in **cats** and **dogs**. ***Contraindicated*** in pregnancy, severe colics, and very young, aged, sick, and weak **animals**. ***Atropine*** is antidotal.

SC, TD_{LO}, **rat**: 1.4 g/kg/29 weeks; **mouse**: 48 g/kg/5 weeks.

ARECOLINE

***Cholinergic, parasympathomimetic.* Enhances brain *theta* activity, and *acetylcholine*-enhanced learning, but not memory in humans and rodents.**

Use: Anthelmintic in **humans** and **animals**, and as a ruminatoric.

Untoward effects: Diaphoresis and vomiting after IV use in Alzheimer **patients**. Decreases heart rate and blood pressure, and has caused marked bronchial constriction in **man**. Central paralysis can occur with large doses. Incoordination, depression, emesis, straining, and hypothermia can occur in **animals**.

SC, LD_{50}, **dog**: 5 mg/kg; IV, LD_{50}, **mouse**: 36 mg/kg; SC, LD_{LO}, **mouse**: 65 mg/kg.

ARGEMONE MEXICANA

= *Bara sheal-kanta (Bengal)* = *Cardo Amarillo* = *Cardo Santa* = *Ekan-ekun (Yoruba)* = *Kwarko (Hausa)* = *Mexican Poppy* = *Prickly Poppy*

Widely distributed tropical weed in Africa and the Americas. Contains throughout the plant the alkaloids *arterine*, *berberine*, *protopine*, and *sanguinarine* (in the seeds).

Use: Widely used in many folklore medicines; the seed's oil is used for lighting and lubrication, and the treatment of skin diseases.

Untoward effects: Overdosage, especially of the seeds, causes trembling, visual disturbances, staggering, edema of the feet and legs (sometimes with petechiae and telangiectasis), body pain, respiratory distress, nausea, and can be life-threatening in **humans**. Some of the poisonings in Africa have resulted from contamination of grains. An epidemic of poisoning reported in 27 **patients** in Bombay, from the seed extracts, as an adulterant in edible oils. Cardiac failure in seven, with death in three. Sinus tachycardia noted. Ingestion of its oil has caused "epidemic dropsy" in **man**.

The leaves and roots are poisonous, but, because of their unpleasant taste and prickles, **animals** seldom eat them, although the prickles have caused mechanical injury. The seeds have poisoned **poultry**. The alkaloids pass into **goats'** milk, and caution should be used in drinking milk from these areas during drought periods.

ARGININE

= *1-2-Amino-5-guanidinovaleric Acid* = *Argivene* = *R-gene*

***Amino acid.* Essential for rats.**

Use: In ***ammonia*** detoxification and diagnostic tests.

Untoward effects: Most of the adverse effects reported are when combined with other drugs. Increase in serum ***insulin*** and ***human growth hormone***, hypotension, and dangerous hyperkalemia have occurred. A 10½-year-old **male** experienced a rare anaphylactic reaction to an IV of a 5% solution (8 ml/min). Within 5 min, coughing, choking, nasal obstruction, and sweating was noted. The infusion stopped itself, due to venospasm. The needle was removed after **he** had received 100 ml in 15 min. I have found reference to only two other such reports. Tissue necrosis from extravasation can be serious. A **patient** had a seizure after accidentally being given 2 g IV, due to a labeling error.

ARGON

Untoward effects: Although an inert gas, it has been used for euthanasia of **cats** and **dogs**, simply by replacing the air (***oxygen***) in a closed chamber. It was the cause of three deaths at an army hospital, where a supplier hooked up an argon tank to the operating room's ***oxygen*** line. The **patients** died of suffocation.

ARGYREIA NERVOSA

= *Baby Wood Rose* = *Hawaiian Baby Wood Rose* = *Hawaiian Wood Rose* = *Wood Rose*

In the *Convolvulaceae* (*morning glory*) family.

Untoward effects: Hallucinogenic, due to its ***ergoline*** alkaloids (3 mg/g of seeds), ***lysergamide*** (1.04 mg/seed), ***isoergine***, and ***penniclavine***. Seeds are used by poor **Hawaiians** for a "high". Nausea, severe hangover, constipation, blurred vision, and vertigo often follow. Effects from 4–8 seeds usually are noted in 1–2 h, and last about 6 h.

ARIOCARPUS RETUSUS

= *False Peyote*

A globular cactus, found in open, stony places in central and northern Mexico.

Untoward effects: Popular narcotic and hallucinogen. Contains the mildly hallucinogenic alkaloids, ***hordenine***, ***N-methyltyramine***, and others with weak stimulatory (parasympathetic) effects.

ARISAEMA sp.

Untoward effects: *A. dracontium* = ***Green Dragon***: The corm is used as an **insecticide** in India, and is poisonous to **livestock** in the U.S., due to its ***calcium oxalate*** content, which causes severe irritation of mucous membranes and salivation. *A. japonicum* in Asia, and *A. speciosum* and *A. tortusom's* corms in India are also used as an **insecticide**.

The entire plant of *A. triphyllum* = ***Arum*** = ***Bog Onion*** = ***Brown Dragon*** = ***Devil's Ear*** = ***Dragon Turnip*** = ***Indian Turnip*** = ***Jack-In-The-Pulpit*** = ***Lords and Ladies*** = ***Marsh Turnip*** = ***Pepper Turnip*** = ***Petit-prêcheur*** = ***Priest's Pintle*** = ***Starchwort*** = ***Swamp Turnip*** = ***Wakerobin*** = ***Wild Pepper*** = ***Wild Turnip***, particularly the tuberous root stock (or corm), is poisonous. It contains small needle-like crystals of ***calcium oxalate***, that cause salivation, intense irritation and burning sensations of the lips, tongue, mouth, and stomach (with colic) that may last for several days. Patency of the airway should be a concern in preventing fatalities. Contact with its watery sap (leaves or corms) causes an itching rash (**Floridians** put ***lime*** juice on it as an antidote) in some **people**. Accidental poisoning from chewing the stalks or leaves has been common in Florida **children**, where it grows naturally, and even from potted plants in New York. When eaten fresh, it is poisonous. **American Indians** found that cooking the corms gave them a safe, albeit tasteless, source of starch. May cause nephritis.

Cattle, **sheep**, **goats**, and **swine** can be poisoned by it, but rarely eat enough of it.

ARISTIDA sp.

= *Poverty Grasses* = *Wire Grasses*

Untoward effects: Awns (Pineland three-awn grass) often penetrate the mucous membranes of **dogs**, especially field-trial hunting **dogs** in the U.S. Other forms get into the wool of **sheep** and work their way into the skin and flesh. Reports from Nigeria state that they help form bezoars, impaction, and even death in **animals**.

ARISTOLOCHIA sp.

Use: In many folklore medicaments. Extracts have been used since early Greek–Roman times, as a cure for cancer.

Untoward effects: *A. petersiana* = ***Mtangwa mwiyi*** (Swahili) = ***Lunkulwe*** (Zigua), a perennial climbing shrub in Nairobi, is known to be poisonous for **man** and **animals**. A few leaves can kill a **goat**. *A. cymbifera* in large doses caused cries and excitement, followed by sleep and muscle relaxation, gastrointestinal irritation, and depression of motor and sensory nerves. After very large doses, decreased blood pressure and respiratory arrest, followed by death, with the heart still beating. *A. grandiflora* in Panama has killed **deer** and has been used by **criminals** to kill **people**. *A. fangchi* is a nephrotoxic Chinese herb (viz. in ***Mu Tong***).

Nephritis in **animals** has been reported. In **dogs**, intestinal hemorrhages, and a dramatic decrease in blood pressure, but, unlike the **rabbit**, no nephritis. Tachycardia, weak pulse, anorexia, constipation, polyuria, and leukocytosis in poisoned **horses**. Also see ***Snakeroot***.

ARISTOLOCHIC ACID

= Aristolochine

Use: An alkaloid in herbal medicines. From *Aristolochi sp.*, particularly, *indica*.

Untoward effects: Terminal renal failure in 30/70 **patients**. 70 had progressive renal failure from its content in Chinese herbals (viz. *A. fangchi*).

IV, LD_{LO}, **human**: 3 mg/kg/2 days.

In chronic studies, at various dosages, oral treatment of **rats** led to hyperplasia, papillomas or carcinomas of the transitional epithelium of the renal pelvis and urinary bladder. After 12–16 months, papillomas or squamous cell carcinomas were found in the stomach. See above for other effects in **animals**.

IV, LD_{LO}, **cat**: 40 mg/kg; LD_{50}, male and female **rats**: 203 and 184 mg/kg, respectively; male and female **mice**: 56 and 106 mg/kg, respectively.

ARNICA

= Leopard's Bane = Mountain Tobacco = Wolf's Bane

Topical counterirritant, cholinergic.

Use: The oil is used in treating sprains, etc, in cosmetics, and as an ***alcohol*** denaturant.

Untoward effects: Use mask, goggles, and gloves when handling the flowers, to help prevent contact dermatitis, due to hypersensitivity reactions to its ***arnicin*** content. Has caused severe gastroenteritis, intense muscle weakness, decreased pulse rate, giddiness, collapse, and death. Occasionally, only cerebral toxicity is noted. **Children** who have eaten it develop a severe gastroenteritis, decreased pulse rate, and collapse.

ARROWGRASS

= Goose Grass = Goose Tongue = Sour-grass = Triglochin sp. = Troscart

***Cyanogenic plant.* Common in the western U.S. in salt or alkaline marshes.**

Untoward effects: Leaves and stems are particularly poisonous to **cattle** and **sheep**, when growth is stunted by lack of water or early frost. The plant contains ***hydrocyanic acid***. Dyspnea, salivation, lacrimation, rapid pulse, staggering, frequent voiding of urine and feces, convulsions, and paralysis finally precede death in poisoned **animals**. The blood of affected **animals** is a bright, cherry red.

ARSANILIC ACID

= p-Aminobenzenearsonic Acid = AS 101 = Atoxylic Acid = Pro-Gen = R-Sonic

***Growth stimulant, coccidiostat, alterative.* In veterinary medicine.**

Untoward effects: Blindness in ***arsenic*** toxicity occurs in **humans**. **Humans** should avoid inhaling its dust. If swallowed, induction of vomiting is recommended. Wash exposed areas thoroughly. Its use in feed for **animals** is discontinued 5 days before slaughtering **swine** or **poultry** for **human** consumption, to avoid residues in edible tissues.

Posterior weakness, collapse, paralysis, and death in **swine**, when fed 16 times the recommended therapeutic levels. When fed at 1000 ppm (25–400 ppm used for growth and severe enteritis, respectively) for a month, 7/10 **pigs** developed optic disc atrophy and blindness. Caused convulsions in **lambs** when incorporated at 2000–4000 ppm in their rations.

LD_{50}, male **rats**: > 1 g/kg; **rats**: 216 mg/kg; IP, LD_{LO}, **rats**: 400 mg/kg.

ARSENAMIDE

= Caparsolate = Caparcide = Caparside = Filaramide = Filcide = TDC #970 = Thiacetarsamide

***Filaricide.* Widely used IV (*sodium salt*) treatment of dogs, to eradicate mature *Dirofilaria immitis* (heartworms).**

Untoward effects: Potentially hepato- and nephrotoxic, along with the same effect from the dying worms. Its ***benzyl alcohol*** content may also be a factor. In cases of severe toxicity, ***BAL*** offers dramatic relief.

ARSENIC

Approximately 20 million lbs in U.S. annual production.

Use: In metallurgy for hardening metals, in glass and ceramic manufacturing, in wood preservation, and in pesticides and rodenticides.

Untoward effects: Prejudice against arsenic started in 1820, when a British physician blamed it for cancer in **chimney-sweeps**. Arsenic is a cumulative poison. Old literature references were mostly for ***arsenic trioxide***. Yet, when ***Fowler's Solution*** (***potassium arsenite***) is taken internally by **humans**, multiple skin cancers characteristically develop on unexposed parts of the body and palms of the hands, along with pigmentation and hyperkeratosis. Arsenic-induced skin, lung, liver (angiosarcomas), bronchi, and genitourinary cancers have been common in **miners**, **smelters**, **tanners**, **oil refiners**, and **vintners**. It is a carcinogen in ***cigarette*** and ***coal*** smoke; lung cancers have been caused by occupational dust exposures. In one factory, lung and lymphatic cancers were 6–7 times normal in post-mortems on 27 **men**. In one group of **patients**, bronchial carcinomas had an average clinical onset of 32 years, after parenteral therapy for psoriasis. So-called "wallpaper poisoning", due it its arsenic pigments, were actually cases of aspergillosis and penicilliosis. ***Garlic***-like odors in the room were from fungal action on the arsenic, and the illness was due to the inhalation of fungal spores. Inhalation of dust has caused ulceration of nasal septum. Gastroenteritis with vomiting and abdominal pain (acute abdomen), diarrhea, or constipation occurs. Pulmonary edema, liver and renal failure can be fatal. ***Garlic*** odor to breath occurs. One of the most commonly used poisons for criminal purposes. Peripheral neuropathy has been common from arsenic in pesticides, pharmaceuticals, sprays, dusts, contaminated foods and beverages, and suicide attempts. Arsenic poisoning often misdiagnosed as Guillain–Barré syndrome. Quadriplegia and ventricular fibrillation developed in 16-year-old **male** in a suicide attempt who then had cardiac arrest 19 and 21 days later, during induction of anesthesia for surgery (gastrointestinal hemorrhage). Arsenic may not only cause permanent damage, but also increase risks during anesthesia. A 67-year-old **male**, poisoned by arsenic-containing pesticides, was first diagnosed with aplastic anemia, and, 5 months later, diagnosed with myelogenous leukemia. Nausea, vomiting, delayed liver failure, jaundice, anemia, leukopenia, thrombocytopenia, and fatal arrhythmias have also been reported. Increases dilation and permeability of capillaries. Periarteritis nodosa and allergic vasculitis are due to hypersensitivity. Some organic arsenicals have caused fever and eosinophilia. A 39-year-old **male**, treated for psoriasis with an arsenical for 3 years, developed retinal hemorrhages and atrophy of the optic nerve. Improvment after withdrawal of the drug. Inorganic arsenic can cross the placenta and cause **fetal** death. Arsenic is concentrated in **their** brains. The chemical is excreted in **human** breast milk in significant amounts. A **newborn**, due to a labeling error, was poisoned by it in a dusting powder, instead of containing ***zinc oxide***. In Iran, hundreds of **villagers** suffered from probable ***arsenic*** poisoning from **their** arsenic-contaminated wells. Bath water from the village pool containing arsenic in a depilatory salve was the source. The **inhabitants'** tongues were blue and **their** skin splotchy. An 18-year-old **male** crack **abuser** became poisoned by the use of arsenic as a diluent by **his** supplier. A 39-year-old **male** with arsenic poisoning after drinking illicit whiskey for 10 days, developed renal cortical necrosis and renal calcification with anuria and azoturia. An outbreak of poisoning occurred in about 6000 **people** and, later in Nova Scotia, from arsenic-contaminated beer and ales, as a result of using malting fuels high in ***arsenic*** content. Some Chinese herbal medicines have contained dangerous ***arsenic*** levels. ***Arsenic*** is a cumulative poison, particularly in **man** and **rats**, but limitedly in the **cow** (where it can pass into their milk). Vesicobullous lesions, erythroderma, pigmentation changes, lichenoid eruptions, exfoliative dermatitis, eczematous rashes, vasculitis, erythema multiforme, hair loss, Stevens–Johnson syndrome, and transverse white lines or grooves on nails are among **man's** many cutaneous reactions to arsenical toxicity. Ototoxicity (deafness, tinnitus, and vertigo) has also been reported. Its use may give falsely elevated blood urea nitrogen results and decreased alkaline phosphatase readings. In Bangladesh, an estimated 20–70 million **people** are at risk from tube well water and 73,000 are currently with adverse effects. It attacks proteins with ***sulfur–sulfur*** bonds leading to skin lesions, hyperpigmentation of limbs, alopecia, coughing, fever, numbness and weakness of limbs, liver and renal pathology, and various neoplasms. A four-fold risk of diabetes mellitus. TWA, **human**: 10 $\mu g/m^3$ reduces expected mortality to 8–10 of every exposed 1000 **workers** over a 45 year period from the previous allowable limit (before 1978 – U.S.) of 500 $\mu g/m^3$, which had a likely death rate of 375–465/1000 **workers**. About 10,000 **workers** in the U.S. are exposed annually. Estimated minimum lethal dose, 200 mg and 5 mg/kg. Levels in drinking water should be limited to a maximum of 5 ppb.

Animals appear to be resistant to the cancer-causing effect of ***arsenicals***. Post-mortems reveal dry and cracking skin, gastrointestinal inflammation with edema and necrosis, plus bloody mucosae. Pulmonary congestion, endocarditis, hepatitis, and nephritis in most **animals**.

TD_{LO}, pregnant **mouse**: 120 mg/kg.

ARSENIC PENTOXIDE

Contains 65% *arsenic*.

Untoward effects: LD_{LO}, **human**: 5 mg/kg.

Several deaths in **cattle** from eating wood ashes of ***arsenic***-treated lumber.

LD_{50}, **rat**: 8 mg/kg; **mouse**: 55 mg/kg.

ARSENIC TRIOXIDE

= *Arsenous or Arsenious Acid* = *Arsenous or Arsenious Oxide* = *Trisenox* = *White Arsenic*

Contains nearly 76% *arsenic*. Half-life in soil – 6½ years.

Use: In poisonous baits, **livestock** dips for external parasites, in insecticides, and in treating acute promyelocytic leukemia.

Untoward effects: Chronic toxicity in ***copper*** smelter **employees**. Fatal poisoning in 17-year-old **male** 36 h after ingesting 3 g in **rat** poison. Severe polyneuropathy in 28-year-old **male** who drank a similar **rat** poison; survived after ***BAL*** therapy. An **infant** died after a 7 month pregnant 17-year-old **female** took it in an attempted suicide. The **infant** had a high ***arsenic*** concentration in the brain, kidney, and liver. 128 **infants** died in Japan of 12,038 who were poisoned when poor quality ***sodium phosphate dibasic*** (containing ***arsenic*** contamination) was added to a dry milk **infant** powder, to improve its stability. Can cause cancer of the skin, liver, lung, and, possibly, the lymphatic system. Many poisoned in Singapore, from ingesting a Chinese herbal preparation, ***Lu Shen Wan***, containing the chemical. The tolerance to any of its ill-effects in **mountaineers**, **dogs**, and **rabbits** may be due to decreased absorption over a period of time. An unusual case followed the accidental spill of a 47% solution on a charcoal grill and the gravel under it. The gravel was removed and the grill "thoroughly" scrubbed. Chicken was grilled on it 9 h later and eaten with resulting headaches, nausea, vomiting, and abdominal pain. A protein-losing enteropathy has been described with associated abdominal pain, diarrhea, edema of the face and legs, and desquamation and scaling of palms; successfully treated with ***BAL***. Finer ground material appears to be considerably more toxic.

Use in treating leukemia initially leads to fatigue, musculoskeletal pain, and light-headedness, and may be followed by peripheral neuropathies, flaccid paralysis, dysesthesias, bone marrow depression, alopecia, and renal toxicity.

LD, **human**: 70–180 mg/kg (also a report of 7 mg/kg).

LD_{50}, **rat**: 15.1 mg/kg; **rabbits**: 8 mg/kg. LD, **cattle**: 33–55 mg/kg. Lethal doses, **cows** and **horses**: 10–45 g; **sheep** and **goats**: 3–10 g; **pigs**: 0.5–1 g (10 ± mg/kg); **dog**: 0.1–1.5 g; **poultry**: 0.05–3 g. 100 **cattle** died after overdosing topically to control lice. SC edematous swellings, petechial hemorrhages in areas of application. **Cattle** usually die 12–36 h after the onset of symptoms (depression, trembling, bloody diarrhea, restlessness, incoordination, stumbling, groaning, teeth grinding, salivation, dyspnea, and convulsions). In all **animal** species tested, application to a wound is dramatically more toxic than when given orally (viz. **horse**: 150–700 grains, oral and 30–60 grains on a wound; **dog**: 1.5–3 grains oral and 0.03 grains on a wound). Chronic poisoning in **animals** is infrequent, compared to **man**.

ARSENIC TRISELENIDE

= *Arsenious Selenide* = *Arsenous Selenide*

Untoward effects: No evidence of irritation or sensitization in **humans** handling treated disks.

LD_{50}, **rats**: > 5 g/kg. Acute dermal toxicity tests **rabbits**: slight erythema at 200–2,000 mg/kg; acute inhalation, **rats**: 2.9 g/l of atmosphere lethal to all.

ARSENOSOBENZENE

= *Arzene* = *Phenylarsenoxide* = *Phenyl Arsenoxide*

Coccidiostat, pesticide. **Rarely used.**

Untoward effects: Thrombocytopenia reported in **humans**.

LD_{50}, **mouse**: 490 μg/kg.

ARSINE

= *Arsenic Hydride*

Very poisonous gas with *garlic*-like odor. Contains 96% *arsenic*.

Use: In organic synthesis.

Untoward effects: Inhalation of escaping gas from a cylinder in a freighter's hold produced a severe toxicity within a few hours, with fever, weakness, nausea, vomiting, diarrhea, abdominal pain, hemolysis, hemoglobinuria, and intravascular hemolysis in eight crew **members**. Renal failure in four, with one totally anuric for 5 weeks; peripheral neuropathy in two, with one still severely disabled 6 months later. Headache, dyspnea, hematuria (dark-red urine), acute renal tubular necrosis, tachycardia, jaundice-like (bronzing) discoloration of the skin and mucous membranes, enlarged and tender liver and spleen, and back pain are also noted. Hemoglobin concentrations usually decrease below 10 g/100 ml with reticulocytosis and leukocytosis. Up to 0.2–0.3 mg/m^3 considered safe for **humans** but National Institute for Occupational Safety and Health set it at 2 $μg/m^3$. May form and endanger **workers** during battery recycling and **those** engaged in the ***cyanide*** extraction of ***gold***. Inhalation of 250 ppm in air is instantly fatal; 25–50 ppm for ½ h is lethal and 10 ppm for longer periods is also lethal. TWA, **human**: 0.2 mg/m^3 (0.05 ppm)/8 h. **Workers** around ***arsenic*** compounds must be warned to stay away from freshly forming ***hydrogen***. Typical of this is the case where two **workers** became ill after attempting to flush a clogged drain with 2000 gallons of water from a tank that had once held arsenical herbicides. A drain cleaner containing ***sodium hydroxide*** was used. It reacted with the traces of ***arsenic*** to give off poisonous ***arsine*** gas.

ARSPHENAMINE

= *606* = *Arsaminol* = *Arsphenolamine* = *Ehrlich 606* = *Kharsivan* = *Salvarsan* = *Sanluol*

The first drug commercially manufactured for a specific disease.

Untoward effects: Anorexia, nausea, vomiting, malaise, red blood cell aplasia, aplastic anemia, pruritus, erythrodermia, lichenoid and fixed-drug skin eruptions, contact dermatitis, headache, chemosis, hepatocanalicular jaundice, dark urine, and light stools have occurred in **man**.

Allergic reactions in **dogs** have been reported.

IV, LD_{100}, **rats**: 140 mg/kg; IV, LD_{LO}, **cat**: 36 mg/kg.

ARTEMESIA ABSINTHIUM

= *Wormwood*

See *Absinthium*.

Use*:** The oil is used in soaps, detergents, creams, lotions, and perfumes. Its ketone, ***thujone (q.v. for toxicity), imparts the odor of ***tansy***, ***cedarleaf***, and ***Dalmatian sage*** oils.

Untoward effects*:** If ***thujone-free, it has been used in foods. Orally, it has damaged the nervous system and caused mental deterioration in **man**. A **patient** aged 31 drank ~10 ml of the oil and within several hours developed acute renal failure, hypernatremia, hypokalemia, hypobicarbonatemia, sore muscles, and eventually congestive heart failure. 2% in petrolatum showed no irritation or sensitization in **human volunteers**, after a 48 h closedpatch test. Yet, later clinical reports indicate it has caused lichenified dermatitis of exposed skin areas, particularly in **men** over 40. It starts out seasonal, but may become permanent after repeated exposures.

The oil was only slightly irritating to intact or abraded **rabbit** skin after 24 h under occlusion. No phototoxic effects, when applied undiluted on hairless **mice** and **pigs**. When eaten by **cows**, it gives an undesirable flavor to milk.

Oil, LD_{50}, **rats**: 960 mg/kg.

ARTEMETHER

= *Artenam* = *Paluther*

Antimalarial.

Untoward effects*:** ***Grapefruit juice decreases its serum levels.

ARTESUNATE

Antimalarial.

***Untoward effects*:** Acute cerebellar dysfunction, ataxia, and slurred speech after 5 day treatment in **patient** aged 36.

ARTICHOKE

= *Cynara scolymus*

A member of the *Compositae* family.

Untoward effects*:** Its sesquiterpene lactones have caused a serious dermatitis, cheilitis, and chronic photosensitivity. A 53-year-old **female** had eczematous lesions, mainly on **her** hands, eyebrows, and back of the neck, which disappeared when **she** quit cleaning and selling artichokes. About 20% of the **women** in that occupation were affected. Many (60% of nearly 250 at a biologists' dinner) noticed that water (**others** have sometimes noted this effect after ***wine or ***milk***) tastes sweet after consuming it.

ARUM sp.

This plant family includes *American* and *English Ivy*, *Caladium*, *Colocasi*, *Cuckoo-pint*, *Dieffenbachia*, *Holly*, *Lords & Ladies*, *Mistletoe*, *Virginia Creeper*, and *Wake-Robin*.

Untoward effects*:** Acute gastritis, vomiting and severe diarrhea, and shock, followed by death in **humans**. Its berries are particularly toxic. The toxic agents vary and are discussed under their respective headings. Mechanical injury of the oral cavity occurs with some of these. Polish reports indicate *A. exoticum* in two **males** aged 4 and 9, after ingestion of root and leaf, developed sore throat, retrosternal and epigastric pain, dull headache, burning of mucous membranes, and transient central nervous system disturbances attributed to the glycoside, ***aroine.

A. maculatum ingestion caused toxicity in five **horses** in Hungary that grazed near the edge of a wooded area. One died and post-mortem revealed hyperemia of the gastrointestinal mucosa and kidneys, and fatty degeneration and karyolysis of the liver. It has caused excessive salivation and edema of the face, lips, and tongue in a **water buffalo**. The toxin is destroyed by boiling, heating, or thorough drying.

ARYLCARBONOCHLORIDO-THIOICATE

Chemical intermediate.

***Untoward effects*:** Manufacturing procedures and safety clothing are designed to prevent accidental **employee** exposure.

A 4 h exposure of ten **rats** to an atmosphere of 2.1 mg/l mean concentration caused labored breathing and death in all (80% mortality within 9 days).

ARYL OXIME

Chemical intermediate.

***Untoward effects*:** Preliminary results with gavage of female **rats** during gestation days 6–15 killed 8/8 at 1 g/kg/day and only 1/8 at 500 mg/kg/day. Weight gain was decreased at lower levels. At 250 mg/kg/day, fetuses exhibited rare limb defects (short legs, short and/or missing toes).

ASAFETIDA
= *Asant* = *Devil's Dung* = *Hing*

The oleo-gum-resin from the roots and rhizomes of various species of *Ferula* in Persia, India, and Afghanistan.

Use and untoward effects*:** Antispasmodic, carminative, and expectorant. Adds a characteristic, slightly ***garlic-like aroma and taste to Worcestershire sauce. Because undiluted it has an offensive odor, it has been buried under doorsteps of houses and barns, as well as worn in amulets around one's neck, to keep evil spirits away from **people** and **cattle**. As a vile smelling and tasting concentrate, it was useful in treating hysteria in **people**, presumably by olfactory and psychic reflexes; the **patient** was eager to shun further treatments.

Topically, to prevent cannabalism in **poultry**, bandage-chewing in **dogs**, and as an **animal** repellent.

ASARUM CANADENSE
= *Wild Ginger*

***Untoward effects*:** Contact with the leaves of this wildflower of eastern Canada and northeastern U.S. has caused a few reported cases of dermatitis in **people**.

ASBESTOS

***Use*:** Over 5000 uses known, including electrical insulation, roofing, siding, cement pipe and sheets, flooring, gaskets, paper products, clothing, and in fire protection. Consists of four important commercial forms: anosite, anthophlite, chrysotile, and crocidolite.

***Untoward effects*:** A subject for textbooks. Primarily by inhalation, skin contact, and ingestion. Dyspnea, decreased exercise tolerance, finger clubbing, malignant neoplasms of the larynx, bronchogenic carcinomas, peritoneal and pleural mesotheliomas, lung cancers and fibrosis, asbestosis, and possibly some gastrointestinal cancers (from millions of fibers in each liter of beers, sherries, vermouths, and drinking water). The greatest exposure has probably been in shipyard, mining, brake, insulation, and roofing **workers**. **Family members** of such **workers** have been indirectly exposed to its dust. Over 80,000 lawsuits have been filed in the U.S. on behalf of the dead and dying. Its fibers or spicules are apparently transported through the intestinal mucosa and then by lymphatics to many areas of the body, with some even eliminated in **human** urine. Various recommendations on **human** exposure levels exist, including 100,000 fibers (> 5 μm in length)/m^3/8 h. Mesothelial cancers generally develop 30 or more years after exposure.

Inhalation TC_{LO}, **human**: 1.2 fibers/cc/19 years.

Trials in numerous **animal** species confirm these ill-effects. Inhalation in **mice** leads to bone marrow alterations that may have relevance to the leukopenia reported in asbestos **workers**.

Inhalation TC_{LO}, **rat**: 12 mg/m^3/13 weeks.

ASCLEPIAS sp.
= *Milkweeds*

Untoward effects*:** Although palatability is low, **animals** can be poisoned by relatively little of the plants (2–3 oz/**sheep**) and become extremely depressed, weak, with abdominal pain, trembling, dyspnea, weak rapid pulse, increased temperatures, violent seizures, and coma. Death occurs in 1–4 days after ingestion, with congestion of liver and kidneys. **Horses** and **cattle** have also been poisoned. Ingestion of ***milkweed pods, as a food, has caused illness in **man**.

A. curassavica = ***Cancerillo*** has poisoned **cattle** in Australia and Brazil. It contains antineoplastic ***digitalis***-like (***cardenolides***) fractions.

A. eriocarpa and ***A. latifolia*** = ***Broadleaf Milkweeds*** will kill **sheep** eating 0.1% of their body weight in a day. Toxic symptoms and post-mortem lesions were similar to ***digitoxin*** poisoning in **sheep**. Also toxic to **rabbits**.

A. galioides and ***A. mexicana*** = ***Whorled Milkweed*** have caused incoordination, weak and rapid pulse, bloating, salivation, convulsions, and death in **cattle** and **sheep** eating the leaves and stems at 0.2% of their body weight in a day. **Chickens** have also been poisoned by ***A. mexicana***.

A. labriformis causes similar toxicity in **cattle** and **sheep** eating 0.05–0.25% of their body weight, again, due to its ***cardenolides***.

A. speciosa = ***Belle asclépiade*** = ***Showy Milkweed***, a broad-leaved plant native to the rangeland of western Canada and western and midwestern U.S., is distasteful to grazing **animals**, but, in droughts, is eaten and poisons **cattle** and especially **sheep**. Dyspnea, grunting, recumbency, and anorexia are common in poisoned **sheep**.

A. subverticullata = ***Horsetail Milkweed***, a ***whorled milkweed*** of Texas, New Mexico, Colorado, and Utah has killed many range **cattle** and an occasional **horse**.

A. verticullata, also called ***Eastern Whorled Milkweed*** = ***Asclépiade verticullée***, has poisoned **sheep**, **cattle**, **horses**, **chickens**, and **turkeys**, particularly in the western U.S. **Horses** show azoturia-like symptoms.

Many other species are also toxic. Some species (***A. syriaca*** = ***Asclépiade de Syrie*** = ***Common Milkweed*** = ***Cottonweed*** = ***Milkweed***; ***A. tuberosa*** = ***Butterfly Weed*** =

Pleurisy Root = ***Orange Milkweed***) were boiled and eaten by **Indians** in New England and British Columbia. ***A. syriaca*** cardiac glycosides have poisoned and killed **sheep** in the eastern U.S. during droughts.

l-ASPARAGINASE

= *l-Asparagine Amidohydrolase* = *Colaspase* = *Crasnitin* = *Elspar* = *Erwinase* = *Kidrolase* = *NSC-109,229*

Enzyme, antineoplastic.

Use: In various leukemias (by limiting the amount of ***asparagine*** necessary for neoplastic lymphocyte growth) in **man** and lymphomas in **animals**.

Untoward effects: Nausea, vomiting, hypersensitivity, anaphylaxis, urticaria, rash, bronchospasm, flushing, pruritus, acute hemorrhagic pancreatitis, abdominal pain, hypo- and one case of hyperglycemia, leading to coma, shock, high serum amylase levels, hypocalcemia, 80% decrease in serum protein (mainly ***albumin***), leukopenia, edema, stomatitis, agranulocytopenia, increased pulse, thrombocytopenia, disseminated intravascular coagulation, megaloblastic anemia, fatty changes (lipidosis) in the liver, uremia, toxic epidermal necrolysis, fever (extreme hyperthermia in 13-year-old **male** after second injection – fever responded rapidly to 10 mg ***dantrolene***, IV), headache, disturbed consciousness, personality changes, lethargy, somnolence, confusion, and 15% of **patients** evidenced hallucinations and disorientation. Electroencephalogram abnormalities are similar to those found in early hepatic coma, especially in **patients** who collapse. When encephalopathy appears on the first day of treatment, it appears to resolve itself in a few days. Chest tightness, ST segment elevation, and myocardial infarction in 21-year-old **male**. A stroke in a 44-year-old **female** appeared to be related to its use. The source of the enzyme affects the hypersensitivity in some **patients**. Delayed toxicity leads to acute hemorrhagic pancreatitis, central nervous system depression or stimulation, and thrombosis.

IM, TD_{LO}, **child**: 8,145 IU/kg/1 week.

Drug interactions: Can increase ***vincristine*** toxicity. It decreased serum levels of ***phenytoin*** in a 10-year-old **male**.

Encephalopathy and skeletal defects observed in offspring of treated **pregnant rats** and **mice**. Offspring of treated **pregnant rabbits** evidenced spina bifida, missing tails, lung defects, and abdominal extrusion.

IP, TD_{LO}, **mouse**: 10 IU/kg/4 days.

ASPARAGUS

= *Asparago* = *Asperage* = *Asperge* = *Asperjo*

Use: As a food. A decoction made from its roots is a diuretic, sedative, and used in cardiac disorders.

Untoward effects: **Many** have urine with a strong odor, due to ***S-methyl thioesters***. Some **people** develop delayed eczematous patch test results with it and dyshidrosis. **Others** with ***nickel*** dermatitis may need to avoid it.

ASPARTAME

= *Equa*® = *Canderel* = *Nutrasweet*® = *SC-18,862* = *Tri-Sweet*®

Untoward effects: Little potential toxicity, in view of the tremendous amount used (over 5 billion dollars' worth in Europe alone, and more than that in the U.S. Found in over 5000 products). One of the biggest objections raised to its FDA approval was possible danger in **those** suffering from the metabolic defect, phenylketonuria (PKU) with an estimated incidence of 1 in 16,000. Elevated ***phenylalanine*** frequently causes mental retardation. All natural protein contains about 4% ***phenylalanine***. Therefore, all **infants** who test positive for PKU must be put on a ***phenylalanine***- free intake of protein hydrolysates or ***amino acid*** mixtures. Aspartame is produced commercially from two amino acids, ***L-phenylalanine*** and ***L-aspartic acid***, by several methods, one of which I helped design. A Chinese restaurant syndrome was reported in a 32-year-old **male**. A 33-year-old **female** had typical panic attacks on 20 cans/day of aspartame-sweetened soft drinks. The attacks subsided when **her** consumption was decreased to three cans/day. Grand mal seizures and acute mania developed in a 54-year-old **female**, after excessive intake of aspartame-sweetened ***tea*** (1 gallon/day/for about 3 weeks). Symptoms subsided after substituting ***sugar***-sweetened ***tea***. Granulomatous panniculitis after drinking 1000–1200 ml daily of aspartame-sweetened drink. It recurred on rechallenge with 50 mg aspartame/day. Persistent generalized urticaria (hives on one eyelid and lower lip edema) and episodes of dyspnea in 23-year-old **female** after 14 oz of a diet soft drink containing 185 mg aspartame, diminished with abstinence, but recurred with rechallenge. A 42-year-old **female** had intermittent angioedema and urticaria within an hour after consuming an aspartame-sweetened drink. The cause was reaffirmed by rechallenge. Headaches, seizures, and aggravation of mental depression are also reported. Loses its sweetness under high temperature (cooking), and it breaks down into ***methyl alcohol*** (q.v.).

Aspartame has caused decreased blood pressure and possibly tumors in **rats** and seizures in susceptible **mice**. Oral administration to infant **mice** (1–2.5 g/kg), but not infant **monkeys** (2 g/kg), has caused neuronal necrosis.

l-ASPARTIC ACID

= *Aminosuccinic Acid* = *Asparagic Acid* = *Asparaginic Acid*

Amino acid.

Use: In hyperalimentation solutions, in organic synthesis, antioxidant and stabilizer in soaps, in treating fatigue, in decreasing blood ***ammonia*** levels, in ***aspartame*** manufacturing.

Untoward effects: Can cause nausea and vomiting. Abnormal electroencephalogram with its ***magnesium*** and ***potassium*** salts in both **male** and **female patients**.

Called an excitatory neurotransmitter, comparable to ***glutamic acid*** in **mouse** trials. **Mice** given 1 g/kg orally show hypothalamic damage (hypoplasia). Ten day-old **mice** given 1–4 g/kg developed obesity, skeletal stunting, and sterility (in **females**). Appears to be a non-toxic emetic agent in the **rhesus monkey** (IV, 200 mg/kg). Lethal doses were approximately 4–7 g/kg and leading to neurotoxicity with tonic and clonic convulsions. Emetic doses led to increased cardiac and respiratory rates. Other adverse clinical effects were salivation, facial pallor, and decreased muscle tone.

ASPERGILLIC ACID

Untoward effects: Neurotoxic for **mice** and causes gas gangrene in **guinea pigs**.

LD_{LO}, **mouse**: 200 mg/kg; IP, 150 mg/kg.

ASPERGILLUS sp.

Untoward effects: The cause of systemic mycoses of **mammals** (**man**, **bison**, **cat**, **deer**, **dog**, **goat**, **guinea pig**, **monkey**, **pig**, and **rabbit**) and many domesticated and wild **birds**. This usually occurs from ingestion of ***mycotoxin***-contaminated grains or foods over a period of several weeks, causing anorexia, listlessness, weakness, icterus, prostration, fecal blood, fatty degeneration of the liver, carcinomas, and even death. Mycotic pneumonias with respiratory distress are common in **birds**. Irreversible vision loss has occurred in intravenous **drug abusers** with ***A. flavus*** from non-sterile injections. **Physicians** should be alert to airborne infections. This has occurred from a hospital's contaminated air conditioning system.

A. flavus: One of the most common affecting **poultry**. Produces ***aflatoxins B*** and ***G***, causing diarrhea, vascular ruptures, and hepatic necrosis in **poultry**, carcinomas in **rats**; tremors in **mice**; decreased feed efficiency, anorexia, abortions, stillbirths, hemorrhagic enteritis, and death in **swine**. Thousands of **turkeys** in England died from eating contaminated South African ***groundnut*** (***peanut***) meal. The U.S. government monitors the degree of contamination in ***peanuts*** used for making ***peanut*** butter. Its toxins can be passed into a **cow's** milk. Has caused photosensitivity in **mice**. Has been found on moldy, country-cured hams. It has caused peripheral green or punctate white lesions of the nail plate of **man**, as well as a keratitis, dermatomycosis, and otomycosis. Pulmonary adenomatosis reported in a 68-year-old **male** working for 3 months on a method of sterilizing contaminated ***peanut*** meal, probably inhaled it. **Rats** fed a diet of 20% contaminated meal developed lung and liver lesions, including hepatomas within 6 months. Hepatomas are about seven times as common in **Bantus** as in U.S. **whites**, probably from ***A. flavus***-contaminated foods. A 2-month-old **female** in Singapore had a 3 day history of diarrhea and died 13 days later from disseminated ***A. flavus*** infection of the brain, lungs, pericardium, myocardium, thyroid, kidneys, small intestine, and diaphragm.

A. fumigatus: A common cause of fungal abortions and tremors in **cattle** and **sheep**. Also adversely affects **hens**, **ducks**, **turkeys**, **swans**, **canaries**, etc. It is often a cause of atopic asthma with eosinophilic infiltrations in **man**. It, and occasionally other species, grows saprophytically in air-containing spaces of the lungs and even in **man's** pleural cavity. Found in a rapidly fatal case of acute cerebral aspergillosis in a 38-year-old **male**. A 15-year-old **male** became ill the same day **he** threshed moldy ***oats*** and died 24 h later with "**farmer's** lung". ***A. fumigatus*** was isolated from **his** lung. Has been isolated from some **patients** with asthma (allergic bronchopulmonary aspergillosis) and from an aortic embolus in a 39-year-old **male**. It is one of the more common fungi isolated from ***mushroom*** **farmers** with chest pain, dyspnea, nausea, headaches, and fevers, as well as in the sputum of **workers** in poultry houses. Also in otomycosis of man, in the vaginal discharge of a pregnant **woman**, and in the granuloma of the eyelid in a 27-year-old **male**.

A. niger: Its toxic metabolite, ***malformin C***, is antibacterial and the fungus itself is a source of ***amylase***, ***cellulase***, ***lipase***, and ***pectase***, useful in baking, ***coffee*** concentrates, flavoring **cheese**, clarifying ***wine*** and fruit juices, and in ***gluconic*** and ***citric acid*** production. It was the cause of aspergillosis in a 9-year-old **male**, after open-heart surgery. Found in the toe webs of a 61-year-old **male**, in a **woman's** vaginal discharge, on the eyes and lids of three **patients**, and in aspergillomas of **man**. It has been associated with otitis externa in **dogs** and in tympanic disease of **man**, sterility and placentitis in **cows**, hepatitis in **mink**, toxicosis in **rabbits**, decreased feed efficiency and weight gain in **chicks**, uterine hypertrophy in **mice**, hyperestrogenism in **sows** and **rabbit does**, air sac and lung infections in **turkeys**, and mycotic gastritis in **cattle**.

A. ochraceus: Produces ***ochratoxin A***, a nephrotoxic substance, that causes fatty infiltration of the liver, staggers, diarrhea, and death in **ducks**; depresses growth, enteritis, and nephritis, decreases sexual maturity and egg production, and increases mortality in **poultry**; sudden death in **heifers** and **lambs** (flaccid paralysis before death); nephrotoxic and hepatotoxic in **swine**; and pulmonary aspergilloma in a 66-year-old **female**.

A. oryzae*: *Use*:** In the food industry for its amylolytic enzymes, to help ferment foods and tenderize meats. The ***alcohol and soy food industries of the Far East utilize this mold, as does the pharmaceutical and **livestock** feed industries, for its proteolytic activity. Also as a ***beer*** stabilizer and in textile desizing. Its fibrinolytic effect found useful in some cases of arterial venous shunt obstructions (***Brinase*, *Protease 1*, *CA-7***).

Untoward effects*:** Can produce the toxic metabolites, ***cyclopiazonic acid*, *β-nitropropionic acid*, *maltoryzine, and ***kojic acid***. May be cause of hepatotoxicity in **foals**. *A. oryzae* var. *microsporus* has a toxic metabolite, ***maltoryzine***, found in malt sprouts, that caused cases of feed poisoning in **dairy cattle**. Many strains produce ***aflatoxin***. A 34-year-old **female** was treated for a case of *A. oryzae* meningitis.

A. parasiticus*: *Untoward effects*:** Metabolites of its ***aflatoxins can appear within 12 h in **cows'** milk. A common cause of ***aflatoxins*** on ***peanuts***.

A. terreus*: *Untoward effects*:** Produces many toxins (citrinin*, *patulin*, *terretonin***) and their metabolites. Has been isolated from over a dozen cases of disseminated aspergillosis in **dogs**, primarily **German shepherds**. Has been isolated from an aborted **bovine** fetus, and a case of mastitis in a **cow**. Found on autopsy in an immunocompromised **patient**, in **human** ear canals, and in the middle ear of **patients** after ear surgery.

ASPHALT

= *Bitumen* = *Mineral Pitch*

***Use*:** In making or covering roofs and roads; in making walls and tanks waterproof.

Untoward effects*:** Its odor has caused migraine headaches in some **people**. Thermal injury after contact with hot pitch has caused > 50% of all injury cases and **worker's** compensation costs in the U.S. Asphalt **workers** are exposed to ***polynuclear aromatic hydrocarbons* (*PAH*)**, considered to be the cause of skin eruptions and skin cancers in them. ***Ethylenediamine is used in some asphalt-wetting agents and may be a cause of contact dermatitis in some **people**. It is an eye and respiratory irritant. A maximum of 5 mg/m^3 of air for 15 min is recommended. Prevent skin contact!!

A known cause of ***pitch*** poisoning in **hogs** that eat blacktop. Approximately 1/3 of **mice** whose skins were painted with it developed skin tumors.

ASPIDIUM

= *Farnwurzel* = *Filicis Malis* = *Filix Mas* = *Fougère Mâle* = *Helecho Maco* = *Male Fern* = *Male Shield Fern*

***Use*:** As a taeniafuge in **humans**, and particularly in **animals** and **poultry**. The oleoresin is also used.

***Untoward effects*:** A breast-fed **infant** whose **mother** received the extract during the lactation period developed optic atrophy. The association was not proven. Headache, nausea, vomiting, severe abdominal pain, diarrhea, dyspnea, glycosuria, albuminuria, bilirubinemia, poor reflexes, mydriasis, optic neuritis and atrophy, yellow vision, temporary or permanent blindness, coma, convulsions, and death from respiratory or cardiac failure reported in **man**. Death was caused by ingestion of 23 g in a **person** (normal dosage in **man** was 4 to almost 8 g). Another report stated TD_{LO}, 50 mg/kg.

61/68 **cattle** became drowsy after eating young rhizomes; 45 became blind, most recovering in a week, but eight remained blind. Has caused nausea, vomiting, and diarrhea in some treated **dogs**. In acute toxicity experiments in **mice**, concomitant use of ***ascorbic acid*, *glycine*, *glucuronic acid***, and ***cystine*** were effective detoxicants. Colic, degenerative kidney, heart, and liver changes, and death from respiratory and cardiac failure have been noted in **cats**.

ASTATINE

Radioactive halogen.

***Untoward effects*:** Experimentally, will induce mammary and pituitary tumors with a single injection. Concentrates in thyroid tissue.

ASTEMIZOLE

= *Astemisan* = *Hismanal* = *Histamen* = *Histaminos* = *Histazol* = *Kelp* = *Laridal* = *Metodik* = *MJD 30* = *Novo-Nastizol A* = *Paralergin* = *Retolen* = *R 43512* = *Waruzol*

H_1 Antihistamine

Untoward effects*:** Ventricular tachycardias from overdoses. Cardiotoxic effects in five **children** (1½–11 years) leading to prolonged corrected QT intervals, ventricular dysrhythmias, and atrioventricular heart block lasting an average of 2.5 days. Accidental poisoning in six **children** (1–3 years) with 2.5–16.7 mg/kg related to prolonged QT interval in five and ventricular arrhythmias in **one**. After 10 weeks of treatment for allergic rhinitis with 10 mg/day, a 15-year-old **female** developed episodes of torsades de pointes. Two **patients** developed paresthesia of the extremities a week after 10 mg/day for rhinitis. The paresthesia gradually disappeared after a 6 week period of no medication. Self-overdosing (up to 90 mg/day/3 days) by a 42-year-old **female** caused ventricular fibrillation and anoxic encephalopathy with a severe amnesic syndrome and a doubtful prognosis for a full recovery. May have putative action in the inner ear. A 14-year-old **female** took an overdose of 200 mg without any serious adverse effects. Some increased appetite, weight gain, and cholestatic jaundice reported. It is ***contraindicated in **patients** with liver disorders, as that can increase its serum levels.

Drug interactions: Its metabolism is decreased and its serum levels are increased by ***clarithromycin***, ***delavirdine***, ***erythromycin***, ***fluvoxamine***, ***grapefruit juice***, ***itraconazole***, ***ketoconazole***, ***nefazodone***, ***nelfinavir***, and ***ritonavir*** leading to an increase in their toxicity, which, with **some**, can result in a risk for torsades de pointe, sudden cardiac arrest, and death. Also decreases clearance of ***indocyanine green***.

Inhibits the immune reactions in the **guinea pig** and **rat**.

ASTERS

Untoward effects: **Patients** allergic to them may develop severe hypersensitivity reactions to ***chamomile***, ***goldenrod***, ***marigold***, ***yarrow***. Dermatitis of the hands and feet of **workers** harvesting *Anthemis cotula*, a member of the aster family.

The leaves and stems of the parry aster = ***Woody Aster*** in the western U.S., after ingestion (1.25 lbs/day for **sheep**) cause leg and neck muscle weakness, falling, fast pulse, then weakened pulse, increased temperature, dyspnea, bloat, increased urination, bloody froth on the nose and mouth, unconsciousness, cyanosis, and death in 4 h to 4 days. Asters can cause abortions in grazing **cattle**, due to their ***nitrate***-concentrating ability. Also concentrates ***selenium***.

ASTRAGALUS

= *Crazyweed* = *Locoweed* = *Peavine*

Untoward effects: Many species are ***selenium***-accumulators. Toxicity varies with the type of soil it grows on. I have found one reference regarding addiction to it for hallucinogenic purposes in **man**. Toxicity has occurred from the ***milk*** of affected **cows**.

Some species are coarse and not ordinarily consumed by **animals**, although they often develop a craving for other species. **Cattle**, **sheep**, **goats**, and **horses** are commonly poisoned by it. Coughing, hind leg weakness, irregular gait, nervousness, loss of sense of direction, poor vision due to cytoplasmic vacuolation of the retina (**cattle**, **sheep**, and **horses**), constipation, bloat, violent actions if disturbed, running through fences, falling down wells, drowning, paralysis, convulsions, and death can occur. Excess ***selenium*** intake has caused severe lameness, deformed hooves, kneeling, and teratogenicity. Abortion occurs in **cows**; arthrogryposis and skeletal defects in offspring of affected **cattle**, **sheep**, and **horses**. It is an additional cause of congestive right heart failure in **cattle** at altitudes above 7000 ft.

A. lentiginosus = ***Spotted Locoweed***. Its alkaloid, ***swainsonine***, causes locoism with depression and, when disturbed – excitement in **sheep** and **cattle**. Teratogenic effects occur in **lambs** and **foals**. At high altitudes, causes congestive heart failure in **cattle** ("high mountain disease"). Abortions also reported in **cows**.

ATENOLOL

= *AteHexal* = *Atenol* = *Cuxanorm* = *Ibinolo* = *ICI-66,082* = *Myocord* = *Prenormine* = *Seles Beta* = *Selobloc* = *Tenobasan* = *Tenoblock* = *Tenormin* = *Uniloc*

Antihypertensive, antiarrhythmic, antianginal, β-adrenergic-blocking agent. **The human kidney eliminates over 75% unchanged in the urine.**

Untoward effects: A breast-feeding **newborn** had toxicity from her **mother's** milk, after **she** received 100 mg for postpartum hypertension. The **infant** was clinically normal 6 h after discontinuance of breast-feeding on the eighth day. Concentration in the milk was 469 ng/ml; 48 h after breast-feeding it was 2010 ng/ml in the **infant's** serum – 24 h later, it decreased to 140 ng/ml. Profound orthostatic hypotension occurred in two **females** aged 27 and 51, after routine dosage. Has increased sinus arrhythmia in some **patients**. Bradycardia, migraine headaches, delirium in 85-year-old **female** (after dosage increase from 50 mg twice daily to 100 mg twice daily), orthostatic hypotension, dizziness, ataxia, nausea, diarrhea, agranulocytosis, bronchospasm, mesenteric thrombosis, Peyronie's disease, possible retroperitoneal fibrosis and sclerosing obstructive peritonitis, arthropathy, muscle fatigue, visual fatigue with scotoma and scintillations and possibly blindness (**patient** also received ***diazoxide*** and ***furosemide***), Raynaud's phenomenon, renal artery thrombosis, peripheral skin necrosis, acute pancreatitis, initiation or aggravation of psoriasis, watery nasal secretion, impotence, erectile disorders, and some slight adverse effects on memory have all been reported. Hepatic dysfunction developed in 78-year-old **male** after a few days of treatment. Treatment stopped upon hospital admission and liver function was almost normal after 2 weeks. A 64-year-old **female**, after 18 months of therapy, developed systemic lupus erythematosus with fever and pleuropericarditis, which resolved on discontinuing medication. Self-poisoning with 4–6.5 g in a 34-year-old **male** resulted in several episodes of ventricular asystoles, starting after about 7 h, none lasting over 40 s, but requiring cardiac massage and a temporary pacemaker. A 15-year-old **female** ingested 500 mg in a suicide attempt and developed a severe bradycardia, hypotension, and marked hyperglycemia. Gastric lavage, ***activated charcoal***, 5% ***dextrose*** IV, and ***atropine sulfate*** are credited for **her** recovery. **Patients** on ***β-blockers*** may be at increased risk from *Hymenoptera* stings. It may be prudent to reduce dosage when given with thrombolytic drugs. Rarely, coughing may occur. Usual therapeutic serum levels are 0.1–1 μg/ml, toxic at 2–3 μg/ml, and a fatality reported at 27 μg/ml.

Drug interactions: Its antihypertensive action is significantly decreased by ***indomethacin*** 50 mg twice daily. ***Ampicillin*** decreases its absorption when both are administered orally together. ***Nisoldipine*** increases its serum concentration by about 20%. ***Aspirin*** and ***allopurinol*** decrease its bioavailability to nearly half, and ***aluminum hydroxide*** decreases it by about ⅓. Complete heart block in a 44-year-old **patient** receiving 100 mg/day with ***verapamil*** 360 mg/day. Sinus arrest with ventricular escape rhythm and shock in a **diabetic** 42-year-old **male** with the same drug combination. 20% decrease in the clearance of ***disopyramide*** when given with it. Potentiates adverse effects of ***general anesthetics*** and prolongs effects of the non-depolarizing muscle relaxants, ***atracurium***, ***pancuronium***, and ***vecuronium bromide***.

LD_{50}, **rats**: 3 g/kg; **mice**: 2 g/kg; IV, LD_{50}, **rats**: 59 mg/kg; **mice**: 99 mg/kg.

ATIPAMEZOLE

= *Antisedan* = *MPV 1248*

***α_2-Adrenoreceptor antagonist.* Oral and IV in human; IM in animals.**

Untoward effects: Tremors, shivering, sialism, motor restlessness, diaphoresis (hands and feet), increased systolic and diastolic blood pressure, tachycardia, and muscle cramps. Can be absorbed from eye, skin, and mouth contamination leading to systemic effects.

Dogs may show nausea, occasional vomiting, periods of excitement, tremors, panting, scleral vasodilation, sialisis, bradycardia, bradypnea, hypothermia, and diarrhea.

ATORVASTATIN

= *CI 981* = *Lipitor* = *Sortis*

Antihyperlipoproteinimic.

Untoward effects: Toxic epidermal necrolysis, thrombocytopenia, myalgia, hepatotoxicity, gastritis, peripheral edema, chest pain, bronchitis, rhinitis, dizziness, and abdominal pain are among the more important adverse effects. Rarely, coughing can occur.

Drug interactions: ***Cyclosporine***, ***erythromycin***, ***gemfibrozil***, ***grapefruit***, ***itraconazole***, ***ketoconazole***, and possibly ***niacin*** can increase its serum concentration and induced myopathy. It can increase serum concentration of ***digoxin*** 20%. Use with recalled ***mibefradil*** could also increase its serum concentration. ***Colestipol*** and ***antacids*** can decrease its serum concentration by 25 and 35%, respectively. Can increase serum concentration of ***norethindrone*** and ***ethinyl estradiol*** by ~ 30 and 20%, respectively, in an ***oral contraceptive***.

ATOVAQUONE

= *Acuvel* = *Mepron* = *566C80*

Antiprotozoal.

Use: For oral treatment of toxoplasmosis and mild to moderate *Pneumocystis carinii* pneumonia in **patients** with AIDS, **who** cannot tolerate a ***trimethoprim–sulfamethoxazole*** regimen. It has fewer treatment-limiting side-effects than ***pentamidine***.

Untoward effects: Over 5% of **patients** report adverse effects, such as nausea, vomiting, diarrhea, headache, fever, insomnia, increased liver enzymes, weakness, abdominal pain, dizziness, fungal infections in the mouth, and pruritus with rash in about 20%. Causes a slight decrease in ***azithromycin*** serum levels. Plasma concentrations decrease 40–50% when coadministered with ***metoclopramide***, ***rifampin***, or ***tetracycline***.

ATRACTYLOSIDE

Untoward effects: A natural, highly toxic glycoside, mostly from the rhizome of *Atractylis gummifera*, producing ***strychnine***-like and hypoglycemia-like convulsions in **animals**. IP injection of 50 mg/kg causes tubular nephrosis within 3 h in **rats**.

LD_{50}, **rat**: 431 mg/kg.

ATRACURIUM

= *BW 33A* = *Tracrium* = *Wellcome 33-A-74*

Skeletal muscle relaxant during anesthesia.

Untoward effects: Bradycardia observed in 19/376 surgical **patients** at one hospital and four cases (ages 29–84 years) at another hospital. Any other ill-effects, if any, are hard to define, because it is usually administered with many other drugs. Anaphylaxis occurred in a **patient** allergic to other ***neuromuscular-blocking agents***. Can cause hypotension and asthma-like symptoms in **man**. Its muscle relaxant effects are prolonged by ***β-blockers***.

Gentamicin potentiates its effect and increases recovery time in **cats**.

ATRAZINE

***Herbicide.* Over 5000 metric tons have been used annually in the U.S. Restrictions on its use have caused a dramatic phase-out.**

Untoward effects: Under acid conditions, it can react with ***nitrites***, to form ***N-nitrosoatrazine***, a potential carcinogen. A **farmer** accidentally ingesting a solution equivalent to approximately 4 mg/kg showed no adverse symptoms and no treatment was given.

A 2-year-old **child** ingested approximately 800 mg without ill-effects. It failed to cause primary irritation or sensitization on repeated insult patch tests in 50 **volunteers**.

Sheep and **cattle** eating hay from treated areas or grazing on them have shown no clinical toxicity. Experimental use of two daily doses of 250 mg/kg has killed **cattle** and **sheep**. Not secreted in **cows'** milk. Hemorrhages in the heart, lungs, kidneys, and central nervous system on post-mortem examination. It has caused conjunctivitis, salivation, dyspnea, weakness, decreased hemoglobin, and anorexia in **rabbits**, after 6 months of feeding. **Dogs** fed it at a dietary concentration of 150 ppm for 2 years showed no ill-effects. Most **birds** and **poultry** are unaffected by 2500–5000 ppm in their rations. Weakness, ataxia, and tremors occur on toxic doses in as little as 1 h after ingestion. Various trials in **rats** have shown an LD_{50} of 1750–3080 mg/kg. Muscle weakness and ptosis have been noted. Acute oral LD_{50}, **mice**: 1750 mg/kg. Has caused tumors in male **rats** (750 ppm/126 weeks). Acute dermal LD_{50} toxicity, **rabbits**: 7.5–9.3 g/kg; only slightly irritating to skin and eyes (undiluted).

ATRIPLEX sp.

= *Arrach* = *Green Orach* = *Nutall's Salt-bush* = *Saltbush* = *Shad-scale*

Untoward effects: Toxicity due to the fact that it is a ***selenium*** (q.v.) accumulator. Also contains harmful ***oxalates***.

ATROPA BELLADONNA

= *Deadly Nightshade*

Contains *hyoscyamine* (q.v.) in all plant parts, particularly the roots, leaves, and seeds. *Atropine* and other alkaloids are also found in the roots.

Untoward effects: Narcotic and mydriatic. **Children** have been poisoned (flushed skin, dry mouth, dilated pupils, delirium, and death from respiratory failure) from eating its black berries. Its ***scopolamine*** content has helped cause hallucinations. Part of the Middle Ages magic brews (witches' brews – often used topically).

Many, but not all, **rabbits** and **guinea pigs** can thrive on belladonna leaves, due to high levels of a serum atropinesterase. Over 100 years ago, this became the first fully documented report of an inherited modification of a pharmacological response.

ATROPINE

***Anticholinergic, spasmolytic, parasympatholytic, mydriatic.* The sulfate is usually used. First isolated from the above plant.**

Use: Pharmacological antidote for poisoning by organic ***phosphorus*** and ***carbamate*** compounds.

Untoward effects: Often, dose-related and more frequent in **children**. At low dosage, dryness of mouth and nose, bradycardia, and decreased sweating. As dosage increases, thirst, slight mydriasis, and cycloplegia occurs. At 2 mg and over in **adults**, sinus tachycardia, blurred near-vision, constipation, flushed and dry skin; at about 5 mg additional, symptoms such as swallowing difficulty, slurred speech, headache, restlessness, and weakness. Increased gastric reflux may cause retrosternal pain. When dosage in **adults** reaches 10 mg or above, these symptoms become much more dramatic with ataxia, delirium, hallucinations, talkativeness, quarrelsomeness, relaxation of sphincter muscles, excitement, disorientation, and coma. 65 mg has been fatal. Urinary retention is common. Rarely, erectile disorders. Hypersensitivity, manifesting as asthma, contact dermatitis, and follicular conjunctivitis are rare complications of its use. Leukopenia reported in one **patient**. Deaths reported in **children** at 10 mg. An **adult** even recovered after a 1 g dose, indicating the wide variability in susceptibility to it. Decreases milk flow in **women** and can pass into milk, where it might produce anticholinergic effects in an **infant**. After IV injections in pregnant **women**, **fetal** plasma levels are almost half of the **mother's** with tachycardia in the **fetus** (occasionally preceded by bradycardia). The meat of **animals** poisoned by atropine can be poisonous to **man**. Symptoms of poisoning have been reported in **cats**, **dogs**, and **man** that have eaten apparently healthy **rabbits** that have consumed large quantities of ***deadly nightshade***. In Venezuela, 15 **people**, during a 14 year period, consumed **wasp *honey*** leading to atropine poisoning, apparently from **wasps** using nectar from nearby ***Datura*** plants. Also see ***Proatropine***.

Contraindications: Acute angle glaucoma, asthma, intestinal atony, prostatic hypertrophy with difficult urination, pregnancy, adhesions between the iris and lens, cardiac failure, and pyloric obstruction.

Drug interactions: Its effects are enhanced by ***desipramine***, ***imipramine***, ***isoniazid***, and ***meperidine***. Cardiac arrhythmias may develop when used with ***neostigmine***, due to competition for the same receptor site. Dangerous in glaucoma with ***isoniazid***. May cause hypotension with ***propanidid***. The combination of atropine premedication and ***propanolol*** therapy may cause a marked decrease in cardiac output with anesthetic agents like ***ether***. When given with ***diphenoxylate*** as ***Lomotil***, it can delay opiate effects and caused ataxia, lethargy, respiratory decrease, and seizures in **infants** after only one tablet.

LD_{50}, **rat**: 622 mg/kg; **rabbit**: 600 mg/kg; **guinea pig**: 1,100 mg/kg; SC, LD_{50}, **rat**: 250 mg/kg; **mouse**: 75 mg/kg; SC, LD_{LO}, **dog**: 225 mg/kg; **cat**: 140 mg/kg; **rabbit**: 500 mg/kg; **guinea pig**: 450 mg/kg; **frog**: 1 g/kg.

ATTAPULGITE

= *Minugel* = *Pharmasorb*

A rare clay, containing an *aluminum–magnesium* complex found only in Russia, Spain, Africa, and between Quincy, Florida and Attapulgus, Georgia.

Use: As a toxin-adsorbing antidiarrheal, and, preferably, as the "activated" grade. As an insecticide carrier and in drilling fluids.

Untoward effects: Adsorbs many drugs (***lincomycin***, ***methylene blue***, ***neomycin***, ***promazine***, ***quinine***, and ***strychnine***) and interferes with many drug assays. Preferably, administer other drug therapy after a suitable time interval.

AURAMINE

= *Basic Yellow 2* = *Solvent Yellow 2*

An aromatic amine. Yellow or colorless crystalline material.

Use: As a dye or dye intermediate for coloring textiles, inks, wool, paper, and leather, and in sensitizing electrophotographic papers, but **NOT** for use in foods.

Untoward effects: **Human** exposure of several thousand **workers** is primarily by absorption through the skin or by inhalation. Higher than normal incidence of benign and malignant bladder tumors in industrial **workers** after exposure. Limited dermal exposure to **consumers** is expected from the dye's migration from its dyed products.

Orally, it has caused liver tumors in **rats**, and local sarcomas after SC injection. Lesions in the kidneys and spleens of experimental **animals** also reported.

TD_{LO}, **rat**: 37 g/kg; **mouse**: 29 g/kg; SC, TD_{LO}, **mouse**: 2.6 g/kg.

AURANOFIN

= *Aktil* = *Crisinar* = *Crisofin* = *Ridaura* = *Ridauran* = *SKF 39162*

Use: An oral chrysotherapeutic in rheumatoid arthritis in **man** and occasionally in unresponsive discoid lupus erythematosis.

Untoward effects: The most common is mild–severe diarrhea (44% in 52 **patients**; 4% in 303 **adults**). Colitis and eosinophilia in 50-year-old **female** after 3 mg twice daily/3 months, then 3 mg/tid/2 months. Rapid improvement within 2 weeks after discontinuance. Proteinuria in 3% of 1283 **patients** and 39% of an 1800 **patient** group. Rash (1% in 303 **adults**), thrombocytopenia (also 1% of 303 **adults**), leukopenia (1.3%), anemia (3.1%), renal toxicity, nausea, palpitations, and presyncope also reported. Five fatal cases of thrombocytopenia reported by 1985.

Contraindications: Necrotizing enterocolitis, pulmonary fibrosis, exfoliative dermatitis, bone marrow aplasia, or severe hematological disorders.

LD_{50}, **rats**: 265 mg/kg; **mice**: 310 mg/kg.

AUROTHIOGLUCOSE

= *Aureotan* = *Aurumine* = *Gold Thioglucose* = *(1-D-Glucosylthio)gold* = *Oronol* = *Solganal* = *Solganal B*

Antirheumatic.

Use: Parenterally, to treat rheumatism in **man**, indolent lip ulcers in **cats**, pemphigus in **cats** and **dogs**, and experimentally, to produce obesity in **mice**.

Untoward effects: Toxic effects occur in up to 50% of treated **patients**, with some even fatal. Stomatitis, pruritus, rash, erythematous scaly areas on face and scalp with some alopecia, leukopenia, thrombocytopenia, and mild nephrotoxicity with proteinuria occur. Interstitial pneumonitis, aplastic anemia, eosinophilia, and enterocolitis, although all are rare, can be fatal. Contact dermatitis often occurs in those sensitive to ***nickel***. Osteoporosis, facial flushing, giddiness, photosensitivity, keratitis, corneal ulceration and subepithelial deposits of metallic gold in the cornea (chryseosis) have also occurred. Since renal excretion is slow, serious cases are often given ***dimercaprol*** therapy. It can be excreted into a **mother's** breast milk, causing ***gold*** to be found in her **infant's** serum and red blood cells, with a potential for rashes and idiosyncratic reactions in such **infants**.

IP injections in **mice** cause lesions in the ventral medial hypothalamus, causing hyperphagia and obesity. Hypothalamic changes in injected **rats** preceded gastric ulcers.

SC, TD_{LO}, **mouse**: 400 mg/kg; IM, TD_{LO}, **chicken**: 300 mg/kg; IV, LD_{50}, **chicken**: 1 g/kg.

AVOBENZONE

= *Parsol 1789*

Ultraviolet A-absorbing suncreen. An ingredient at a 3% level in ***Photoplex***® and ***Shade UVAGuard***®.

Untoward effects: Photosensitivity reactions reported. When the sunscreen is applied to the face and not near the eyes, "ocular stinging" with excessive tearing has occurred, possibly from its vapors.

AVOCADO

= *Aguate* = *Auacatyl (Aztec)* = *Avocatier* = *Persea americana* = *Zabōka*

Rich in *potassium*. The oil is an effective sunscreen.

Untoward effects*:** Contains cyanogenic glycosides and a number of pressor amines (as in ***broad beans and some ***cheeses***) capable of causing hypertensive episodes. Most **people** have an enzyme, monoamine oxidase, which rapidly breaks the amines down into harmless entities, but a few **people** have a reduced enzyme capacity or they are taking monoamine oxidase inhibitors. A 35-year-old **male** developed a severe hypertensive crisis with severe throbbing headache, chest pains, and sweating, requiring an emergency room visit after 50 mg ***tranylcypromine***, four avocados and some guacomole. Repeated injections of ***phentolamine*** were administered, to decrease the blood pressure of 206/108 to normal in about 4½ h. In addition to interaction with monoamine oxidase inhibitors, it can cause a sharp decrease in the international normalized ratio of **patients** on ***warfarin***.

Horses usually recover from poisoning after grazing in an avocado grove. Symptoms are swelling around the mouth and head, often extending to the neck and chest, anorexia, lethargy, and, sometimes, colic. **Cows** and **goats** have decreased milk flow and swollen udders. Caged **birds**, particularly **budgies**, become agitated, pull their feathers, and may die from eating the fruit (contains 23 μg vasoactive amines/g). **Cattle**, **goats**, **rabbits**, and **fish** have all been poisoned by it or its leaves or twigs. Guatemalan cultivars have caused mastitis and deaths of **rabbits**, but Mexican cultivars did not.

AYAHUASCA

= *Caapi* = *Yagé* = *Yaje*

Narcotic beverage from several species of *Banisteriopsis*, used in various rituals in the Amazon.

Untoward effects*:** Psychomimetic hallucinogen containing the β-carboline alkaloids, ***harmine (q.v.), ***harmaline***, and ***tetrahydroharmine***. These are monoamine oxidase inhibitors and produce a stimulatory effect. Intoxication usually starts with giddiness, nervousness, nausea, profuse perspiration, followed by lassitude, listlessness, and detachment. Then, bluish or purple color visions (occasionally rich red, green, or orange) mixed with lightning-like flashes of light appear. Synesthesia and deep sleep interrupted by dreams. In the Amazon, these alkaloids are often taken as a snuff with similar effects.

AZACOSTEROL

= *20,25-Diazacholestenol* = *20,25-Diazacholesterol* = *Ornitrol* = *SC 12,937*

Hypocholesteremic, avian chemosterilant.

***Untoward effects*:** Accidental ingestion of grain treated as a sterilant may cause muscle spasms (myotonis) in **man** and **rats**.

LD_{50}, **rat**: 470 mg/kg; **mouse**: 380–610 mg/kg; IP, LD_{50}, **rat**: 60 mg/kg; **mouse**: 92 mg/kg.

AZACYCLONAL

= *Ataractan* = *Calmeran* = *Frenoton* = *Frenquel* = *MER-17* = *Psychosan* = *γ-Pipradol*

Tranquilizer.

***Untoward effects*:** No side-effects were noted with oral dosing of 1.2 g/day/3 weeks in one **patient** and 120 mg/day/3–7 months in another **patient**.

LD_{LO}, **human**: 500 mg/kg.

Convulsions and death occurred in 20–240 min. **Rats** given 86 mg/kg SC showed flaccid paralysis of their hind limbs, decreased spontaneous activity, and increased response to auditory stimuli. **Cats** and **dogs** receiving oral doses of 25 and 50 mg/kg had tremors and increase in irritability. **Dogs** receiving 100 mg/kg orally had constant activity, yet, **monkeys** fed 160 mg/kg showed no behavioral changes or toxicity.

LD_{50}, **mouse**: 650 mg/kg; IP, 220 mg/kg; IV, 177 mg/kg; SC, 355 mg/kg.

5-AZACYTIDINE

= *Azacitidine* = *Ladakamycin* = *Mylosar* = *NSC-102816* = *U-18,496*

***Antineoplastic*. An analog of *cytidine* and it is incorporated into RNA and DNA.**

***Untoward effects*:** One of the most mutagenic molecules known, but rapidly (15–60 min) metabolized in **man**. Acute toxicity in **man** may be manifested by nausea, vomiting, diarrhea, fever, rash, and drowsiness. Bone marrow depression, muscle pain, weakness, hepatotoxicity, and some cardiotoxicity can result from a delayed toxicity. Leukopenia, often prolonged thrombocytopenia and hepatotoxicity, all unrelated to dosage, can occur.

IV, TD_{LO}, **woman**: 500 μg/kg, 6 mg/kg given over a 10 day period.

IP to **mice** for 52 weeks and **rats** for 34 weeks caused tumors of the liver, bone marrow, and spleen, and interstitial cell testicular tumors in **rats**. Embryotoxic in **mice**. Mild irritant to **rabbit** skin, but not to **man's**. Hepatotoxic in experimental **animals**.

IV, LD_{LO}, **dog**: 13 mg/kg; LD_{50}, **mice**: 572 mg/kg; IP, 116 mg/kg.

AZALEA

Untoward effects: Ingestion of approximately ten flowers of an indoor azalea by a 13-month-old **female** caused vomiting within 30 min. Has a ***curare***-like effect. All parts of the plant contain ***andromedotoxin*** and various cardiac glycosides. Has caused bradycardia and heart block, salivation, nausea, profuse vomiting (in species that can), abdominal pain, weakness, dyspnea, incoordination, and collapse in **man** and **animals**. ***Honey*** from azaleas can be toxic. Its toxicity was recognized in ancient Greece.

The California azalea (*A. occidentalis*) has been the cause of many fatalities in grazing **sheep** in the southern Sierras. A few hours after eating azalea, a **llama** began drooling, then vocalized, as if in pain, and had neck pain. Atropinization effected a cure. A few ounces of the plant can cause serious symptoms of poisoning in **horses**, **sheep**, **cattle**, and **goats**.

AZARIBINE

= *Azauridine Triacetate* = *AzUR* = *Ribo Azauracil* = *Triacetyl-6-Azauridine* = *Triazure®*

Antineoplastic.

Use: The triacetate, ***TA-AzUR***, has been used in the treatment of psoriasis.

Untoward effects: Mild leukopenia (white blood cells 3500–4000/m^3) and some anemia, i.e. hematocrits > 30, decreased hemoglobin, decreased ***iron*** utilization and megaloblastic changes in bone marrow, temporary erythema, pruritus, nausea, epigastric burning sensations, giddiness, mental depression, fever, headache, stiffness, dizziness, seizures, cramps and diarrhea have occurred. The FDA ordered it off the market because its use was associated with life-threatening or fatal blood clots.

It was responsible for delayed abortions in **macaque monkeys**. **Dogs** were particularly sensitive to it and evidenced a severe leukopenia. When leukopenia intensified in **man**, it was preceded by uremia.

AZASERINE

= *CI 337* = *CN 15757* = *O-Diazoacetyl-l-serine* = *P 165*

Antineoplastic, antibiotic, abortifacient.

Untoward effects: Anorexia, vomiting, hemorrhagic diarrhea, jaundice, glossitis, stomatitis, ulceration of the buccal mucosa, alopecia, dermatitis, and one **patient** died. **Patients** are often taking other drugs simultaneously.

Teratogenic in **rats** with cleft lips, palate abnormalities, and syndactyly. Degree of hepatotoxicity is dose-related in **rats** in which it also causes pancreatic tumors.

LD_{50}, **rat**: 170 mg/kg; **mouse**: 150 mg/kg; IP, LD_{50}, **rat**: and **mouse**: 100 mg/kg; IV, TD_{LO}, **dog**: 30 mg/kg.

AZATADINE DIMALEATE

= *Bonamid* = *Idulian* = *Optimine* = *Sch 10649* = *Zadine*

Antihistamine.

Untoward effects: Sedation is not uncommon in **people** taking 20 mg or greater doses, but it can occur on much lower dosage. Urticaria, rash, photosensitivity, dry mouth, nose, and throat, anaphylactic shock, dizziness, restlessness, nervousness, excitation, irritability, tremors, insomnia, euphoria, blurred vision, diplopia, vertigo, tinnitus, urinary retention, and nasal stuffiness occur in some **people**. Has anticholinergic (***atropine***-like) activity; therefore, use with caution in narrow angle glaucoma, peptic ulcers, pyloroduodenal obstruction, prostatic hypertrophy, bronchial asthma, hypertension, cardiovascular disease, and hyperthyroidism.

AZATHIOPRINE

= *Azamune* = *Azanin* = *Azoran* = *Azothioprine* = *BW 57-322* = *Imuran* = *Imurek* = *Imurel* = *NSC-39084*

***Antineoplastic, immunosuppressant, antirheumatic, antimetabolite.* A *purine* analog that interferes with *purine* biosynthesis, and it is an imidazolyl derivative of *6-mercaptopurine*, which is one of its metabolites. Metabolized in the liver.**

Use: Mostly with other immunosuppressants in preventing rejection of kidney transplants; as a cytostatic agent in the treatment of leukemia, and in the treatment of rheumatoid arthritis.

Untoward effects: Dosage used in the treatment of rheumatoid arthritis usually, but not always, produces only a mild toxicity including nausea, vomiting, abdominal pain, hepatitis, and reversible bone marrow toxicity (depression), moderate leukopenia, pancytopenia, and transient thrombocytopenia. A 19-year-old **female** with ulcerative colitis had an abrupt fall in white blood cell count on day 3 (from 6200 to 800/m^3) and, despite withdrawal of drug treatment, died on day 5 from fulminating chest infection. Anorexia, alopecia, pancytopenia, anemias, leukocytosis, bleeding, plasmacytosis, fast atrial fibrillation, chromosome aberrations in 2–20% of bone marrow cells (after long-term therapy, but reversible), subcapsular lens opacities and increased ocular pressure have all been occasaionally reported. Many reports indicate no adverse teratogenic effects when given during pregnancy for many other conditions. Possible association with retardation of growth in very young **children**. Hypersensitivity has manifested as a vasculitis, nausea, steatorrhea, muscle and joint pains, diarrhea, ulcerative colitis, pyrexia, rash, dysphagia, dysarthria, exacerbation of myasthenia gravis, and even death, usually after 1–2 weeks of treatment. Some of these symptoms are the same as seen in transplant rejection, but these **patients** had not received any transplants. As an immunosuppressant, it is not surprising that many **patients** have an increased incidence of bacterial, fungal, viral, and protozoal infections. Lymphomas and squamous cell carcinomas may be associated with its use. Although used in treating hepatitis, it can worsen the disease and, in rare instances, cause cholestatic jaundice and reactivate a viral hepatitis. Its use has been suspect in the development of many types of neoplasms, but, it appears that it is more likely with the coadministration of ***prednisone***. Yet, it is not unusual for a single immunosuppressant to be involved (viz. a 73-year-old **female** developed an acute myeloid leukemia after 3 years of treatment with 50–100 mg/day for **her** rheumatoid arthritis). It has caused acute pancreatitis in **patients**. An accidental case of poisoning in a renal transplant **patient** who ingested 7.5 g caused a prolonged sensitivity. Acute interstitial nephritis has been reported. Decreases taste acuity. Occasionally a decrease in wound healing rate, and an increase in fever and gingivitis. Fetal damage has occurred after use during pregnancy in **women**. Hematologic toxicities are usually dose related. Leukopenia, anemias (aplastic, macrocytic, and megaloblastic), neutropenia, thrombocytopenia, lymphopenia, pancytopenia, leukocytosis, eosinophilia, leukemia, and thromobocytosis, roughly listed in decreasing order of frequency, have been reported.

Drug interactions*: *Allopurinol inhibition of xanthine oxidase activity dramatically inhibits the metabolism of azathioprine and its metabolite, ***mercaptopurine***, in **man**. The combination caused a near-fatal interaction in a 40-year-old **male**. The dosage is usually decreased by 65–75% when the combination is used. May potentiate ***anticoagulant*** activity, yet decreased ***warfarin*** serum levels in a 50-year-old **female**. Potency of ***curare*** and probably other muscle relaxants, greatly decreased by it. Its bone marrow depression effects may be enhanced when given with a ***trimethoprim*** and ***sulfamethoxazole*** combination and this may cause increased susceptibility to infections that may not be controlled by the antibacterial combination. Its use with corticosteroids, such as ***prednisone***, has been suspect in the development of reticulum cell sarcoma. Mixing it with alkaline solution will produce ***mercaptopurine***.

One French report indicates it was teratogenic in **pika**, **mice**, **rats**, and **rabbits**. Increased fetal death rate, ear cancers, squamous cell carcinomas, and thymic lymphomas reported in **rats**. Use in **mice** associated with neoplasms of the lymph system, and bladder and kidney tumors. Thrombocytosis, neutropenia, and thrombocytopenia reported in treated **cats**.

Intraduodenal LD_{50}, **rat**: 630 mg/kg; **mouse**: 2.4 g/kg; IP, LD_{50}, **mouse**: 273 mg/kg.

AZAURIDINE

= 6-Azauracil Riboside = AzUR = CB 304 = Ribo-Azauracil = Triazure

***Antineoplastic, antiviral.* The triacetate is used as an antipsoriatic.**

***Untoward effects*:** Significant leukopenia and thrombocytopenia in most treated **patients**. Stomatitis, diarrhea, hyperuricemia, teratogenicity, and anemia often occur. Fever, joint pain and swelling, nausea, vomiting, exanthema, painful and rigid muscles in some **patients**. Teratogenic in pregnant **women**.

In normal **dogs**, 6–9 mg/kg/tid caused a fatal leukopenia within 10–14 days, yet, 600 mg/kg given to **man** on a similar schedule has produced no toxicity. Causes fatty liver changes resembling ***ethionine*** toxicity in **animals**.

IP, LD_{50}, **rat**: 9.4 g/kg; **cat**: 2.4 g/kg; **dog**: 3.4 g/kg.

AZELAIC ACID

= Anchoic Acid = Azelex = 1,7-Heptanedicarboxylic Acid = Lepargylic Acid = Nonanedioic Acid = Skinoren

Oxidation product of *oleic acid*.

***Use*:** In the treatment of acne and skin hyperpigmentation disorders; in alkyd resins and polyesters.

Untoward effects*:** A 20% cream with ***glycolic acid led to more peeling, burning, stinging, and skin dryness than in **patients** using ***hydroquinone***.

Minimum lethal dose, **rabbit**: eye – 3 mg; skin – 500 mg/24 h.

AZELASTINE

= A 5610 = Allergodil = Astelin = Azeptin = Radethazin

***Histamine-H_1 receptor antagonist.* Nasal spray.**

***Untoward effects*:** Somnolence (5–12%), transient bitter taste in ~10–30% (exacerbated by cold drinks), xerostomia, sore throat, nasal burning, and epistaxis.

Teratogenic in **mice** at extremely high dosage.

AZINPHOS-ETHYL

= Bayer 16259 = ENT 22014 = Ethyl Guthion = Gusathion = R 1513

Organic phosphorus insecticide, anticholinesterase.

***Untoward effects*:** 50 or 100 mg/kg was toxic for **calves**. Found in the milk for 6 days after dermal application to **Holstein cows** as a 0.5% emulsion.

LD_{50}, **rats**: 9 mg/kg; **chicken**: 34 mg/kg; dermal LD_{50}, **rabbits**: 280 mg/kg; **rats**: 250 mg/kg; IP, LD_{50}, **rat**: 7.5 mg/kg.

AZINPHOS-METHYL

= 17/147 = Bay 9,027 = Bay 17,147 = Carfene = Cotnion-methyl = DBD = ENT 23,233 = Gusathion = Guthion® = Methyl Guthion = Metiltriazotion = R 1582

Organic phosphorus insecticide, anticholinesterase.

Untoward effects: Food tolerance levels varied in different countries from 0.1–2.0 ppm and in dried citrus pulp for **livestock** feeding of up to 5 ppm. Poisoning has been reported in **man**. ***Pralidoxine*** and ***atropine*** have been usual antidotes for this type of poisoning. Can be poisonous if swallowed, inhaled, or from skin absorption. Use respirators and protective clothing when working with it. Wash exposed body areas and gloves with soap and water. Hypersalivation, gastrointestinal hypermotility, abdominal cramping, vomiting, diarrhea, sweating, dyspnea, cyanosis, miosis, blurred vision, aching eyes, muscle twitching (in extreme cases, tetany, followed by weakness and paralysis), convulsive seizures and laryngeal spasms are noted in **man**. Death usually results from hypoxia due to bronchoconstriction. Extreme bradycardia may occur in some cases.

Although 0.1% as a spray or dip is safe on **sheep**, 0.01% is toxic to **calves**, and 0.02% is lethal to **calves**. 1% levels are lethal to **sheep** and **goats**. Orally, 25 mg/kg is toxic to **sheep** and lethal to **calves**. Can be found in the milk of treated **cows**. Treated hay and grain, when harvested too early, have poisoned **calves**. In many species, symptoms of toxicity can occur within 15 min in **mallard ducks**, **bobwhite quail**, and **pheasants**. Death usually occurs between 1 and 22 h after exposure. Symptoms include regurgitation, goose-stepping ataxia, withdrawal, lethargy, salivation, tremors, hypoactivity, anorexia, wing-drop, wing spasms, tenesmus, diarrhea, myasthenia, dyspnea, prostration, terminal wing-beat convulsions or opisthotonos. Very toxic to **honey bees**. **Dogs** have tolerated up to 5 ppm in their diet for 90 days without ill-effects. 4–8 or more mg/kg has been lethal to some **fish**.

TLV, 200 μg/m^3. LD_{LO}, **human** and **mouse**: 5 mg/kg; LD_{50}, **rat**: 11 mg/kg; **guinea pig**: 80 mg/kg; **chicken**: 277 mg/kg; **wild birds**: 8 mg/kg; **mallard ducks**: 136 mg/kg; **bobwhite quail**: 60–120 mg/kg; **pheasants** and **Leghorn chicks**: 280 mg/kg; dermal LD_{50}, **rat**: 220 mg/kg.

AZITHROMYCIN

= CP-62,993 = Sumamed = XZ-450 = ZithroMax

***Antibacterial, macrolide antibiotic.* Chemically related to *erythromycin*.**

Use: In treatment of many bacterial infections and explored in treating chlamydial urethritis and cervicitis, acute sinusitis, toxoplasmosis, and *Mycobacterium avium*.

Untoward effects: Low incidence of disabling nausea, diarrhea, abdominal pain, headache, and dizziness. Dose-related hearing loss usually disappears within 2–4 weeks after medication is stopped. Tinnitus and hearing loss after 500 mg IV/day/8 days in 47-year-old **female**. A 43-year-old **male** suffered from an irreversible, biopsy-proven acute interstitial nephritis, after taking 250 mg/day/9 days for a respiratory infection. Has appeared in **human** breast milk after oral use.

Drug interactions: Increases ***warfarin's*** hypoprothrombinemic effect. Use with ***rifabutin*** leads to severe neutropenia, fever, and myalgia; rhabdomyolysis with ***lovastatin***. Large decrease in ***theophylline*** serum concentration when azithromycin therapy is withdrawn. Slight decrease in serum levels when ***atovaquone*** is added to therapy.

AZOBENZENE

= Azobenzide = Azobenzol = Azophenylenebenzene = Benzeneazobenzene = Diphenyldiazene = Diphenyldiimide = Nitrogen Benzide

Use: Acaricide, antioxidant in resin soaps for paper manufacturing, in synthetic dyestuffs and ***carbon tetrachloride*** fire extinguishers.

Untoward effects: Carcinogen. Azobenzenes have caused chloracne in chemical plant **workers**.

LD_{LO}, **human**: 50 mg/kg; LD_{50}, **rats**: 1 g/kg.

AZOSULFAMIDE

= Drometil = Neoprontosil = Prontosil S = Prontosil Soluble

Antibacterial.

Untoward effects: Tests on **animals** (**mice**, **rabbits**, and **cats**) demonstrated that azosulfamide has low toxicity even in large doses. In **man**, side-effects similar to those occurring during treatment with ***sulfanilamide*** have sometimes been observed, much less severe and less frequent. These are generally due to individual hypersensitivity, and include tinnitus, dizziness, anorexia, malaise, fever, paresthesia, acidosis, cyanosis, methemoglobinemia, sulfhemoglobinemia, cutaneous eruptions, jaundice, hemolytic anemia and agranulocytosis. A reddish tinge in the urine and skin may occur. In certain cases (for example, hemolytic anemia), blood transfusions may be indicated. Toxic manifestations, similar to those observed in **man**, have not been reported in veterinary medicine.

AZTREONAM

= Azactam = Azonam = Az-threonam = Aztreon = Dynabiotic = Nebactam = Primbactam = SQ 26776

***Antibacterial, antibiotic.* A monobactam.**

Use: Primarily against β-lactamase-producing and non-producing Gram-negative bacteria, including *Pseudomonas aeruginosa*.

Untoward effects: In one review, phlebitis and swelling (2.4%), rash and pruritus (1.6%), nausea and vomiting (0.9%), diarrhea (0.8%). Anaphylaxis, abdominal cramps, symptomless liver function abnormalities and myelosuppression (after 7–10 days of IV treatment). A 70-year-old **male** developed acute renal failure, generalized maculopapular rash, eosinophilia, eosinophiluria, and fever after receiving 1 g IV/tid/9 days. Recovered in 5 days after discontinuing the drug and receiving ***dopamine***, IV fluids, and oral ***corticosteroids***.

AZULENE

= *Cyclopentacycloheptene*

Anti-inflammatory constituent of *chamomile*. The term is also used for various derivatives, such as *guaiazulene* (q.v.).

Untoward effects: One report implies a contact dermatitis in **man** from its use in a cosmetic.

B

BACCHARIS sp.

= *Consumption Weed* = *Groundseltree* = *Silverling*

Leaves and flowering shoots contain toxic saponins and glycosides and the alkaloid, *bacharrine*. Numerous species exist and many are unpalatable, but they are eaten and can cause toxicity when consumed in hay. In the southeastern U.S., *B. halinifolia* has poisoned cattle, sheep, and poultry. *B. coridifolia* is a common cause of poisoning in Brazilian cattle and sheep. Clinical symptoms include anorexia, bloat, muscle tremors, tachycardia, and death. (In sheep eating 1–2 g/kg of the flowering plants, or 3–4 g/kg of sprouting plants; similarly, in rabbits, 1.36 g/kg and 4 g/kg caused death.)

BACITRACIN

= *Parentracin* = *Penitracin*

***Antibiotic. Cobalt Bacitracin*, *Manganese Bacitracin*, *Bacitracin Methylene Disalicylate*, and *Zinc Bacitracin* have also been used. Oral use has NO direct systemic benefits.**

Untoward effects: Renal toxicity reported for parenteral bacitracins is also a function of original renal damage and/or drug purity. Oral use may cause malabsorption of proteins and fats in **humans**. Mediastinal irrigation with it at 50,000 U/l (100 ml/h/3 days) led to potentially toxic serum level of 10.7 U/ml, as a result of the availability of a significant absorptive surface. Anaphylaxis has been reported after its topical use on a varicose ulcer. This type of reaction has been confirmed by intradermal testing. It has induced weakness with hypoactive or absent deep tendon reflexes, often fixed, dilated pupils, and prominent bulbar or respiratory muscle weakness. Parenteral administration always causes albuminuria with casts, and, if treatment continues too long, oliguria and increased blood urea. The latter problems resolve, if dosage is decreased or discontinued. May potentiate the action of neuromuscular-blocking agents (***curare***, ***succinylcholine***, ***tubocurarine***), decreasing blood pressure and respiration. Although unrelated chemically, it may "co-react" with ***neomycin***. May lead to false positive or elevated blood urea nitrogen and albumin test results. ***Decamethonium*** may enhance its effect.

BACLOFEN

= *Ba 34647* = *Baclon* = *Lioresal* = *Myospan*

Skeletal muscle relaxant, a ***γ-aminobutyric acid* (*GABA*) derivative.**

Use: In treating muscle spasticity, intractable hiccups, and neuralgias.

Untoward effects: Can cause depression in **patients**, especially in those with central nervous system disease. A hypothyroid 60-year-old **female** developed buccolingual dyskinesia and hallucinations while being treated for hemifacial spasm. Akinetic mutism in 76-year-old **male** after treatment with 10 mg tid/3 days. Some treated **patients** have experienced short-term memory loss, decreased taste acuitiy and increased (6–10 times normal) levels of SGOT and SGPT. Euphoria in 2/113 treated **patients**. Drowsiness, light-headedness, vertigo, fatigue, nausea, vomiting, rashes, mental confusion, giddiness, hypotonia, hypotension, asthenia, diarrhea, and agitation can often be overcome by dosage decrease. A reversible metabolic coma developed in a 25-year-old **male** with tetanus, overdosed with an intrathecal injection (2 mg/day). Use has aggravated bronchoconstriction in an asthmatic **patient** and induced status epilepticus 7 h after dosing in another. This can be used in the treatment of bladder emptying failure. Relaxes the bladder's internal sphincter and can cause incontinence. An 11 kg 2-year-old **male** accidentally ingested 120 mg (10.9 mg/kg). 2 h later, **he** was semi-comatose with some respiratory distress and sluggish lower limb tendon reflexes. Two hours later, **he** had respiratory arrest, requiring 18 h of positive pressure ventilation. Respiratory distress and stridor continued for an additional 24 h. **He** made an uneventful recovery with no sequelae. A 17-year-old **female** responded to IV ***atropine*** after overdosing (symptoms included hypothermia, respiratory decrease, bradycardia, and hypotension). Coma, hypothermia, bradycardia, hypertension, and hyporeflexia within 1–2 h after eight **adolescents** at a party ingested 3–30, 20 mg tablets, required a mean length of assisted ventilation of 40 h. Cardiac conduction abnormalities noted in a **patient** that ingested approximately 500 mg. Acts like an endogenous psychotogen and worsens schizophrenia 80% of the time. Its use has been associated with the loss of libido and inability to ejaculate. Sudden withdrawal has led to profound psychiatric upsets, including euphoria, visual and auditory hallucinations, paranoia, nightmares, mania, depression, anxiety, convulsions, sleeplessness, increased activity and agitation, confusion, and refusal to eat. Only an extremely small fraction of the ingested drug has been found in **human** breast milk. Renal and hepatic impairment have occurred. Usual therapeutic serum levels are 0.08–0.4 μg/ml and toxic at 1 μg/ml.

Drug interactions: Use with ***ibuprofen***, ***lithium***, and ***tricyclic antidepressants*** has increased its toxicity. ***Taurine*** decreases its absorption.

In **cats** and **dogs**, it has caused transient drowsiness, skeletal muscle weakness, and nausea. Over a 5 year period more than 80 cases of toxicity have been reported in animal **pets**. Appears to be an effective antagonist to ***phencyclidine*** in **mice**.

LD_{50}, **rats**: 145 mg/kg; **mice**: 200 mg/kg; IV, **rats**: 78 mg/kg; **mice**: 45 mg/kg; SC, **rats**: 115 mg/kg; **mice**: 103 mg/kg.

BAGASSE

Fibrous material (50% cellulose, 25% lignin, 25% pentosans, and traces of insecticides) from cane sugar manufacture.

Use: In making paper, cardboard, boards, garden mulches, ***furfural*** and ***furfuryl alcohol***.

Untoward effects: Acute pneumonitis, pulmonary edema, bronchitis and mycosis reported in bagasse **workers**, usually after exposure to overheated or moldy material. A clinical case developed in a 59-year-old **male** only 4–5 h after working on a board-making technique.

After 25 **bulls** were fed on a diet of it with ***molasses*** and a protein supplement (without extra ***vitamin E*** and ***selenium***) for 20 days, they developed typical symptoms of white muscle disease, including some deaths. Its use as a litter under **poultry** has led to a high incidence of mortality from *Aspergillus fumigatus*.

BAILEYA MULTIRADIATA

= *Desert Baileya*

Botanical.

Untoward effects: Toxic to **rabbits**, **sheep**, and **goats**, due to its ***hymenoxin*** content. Survival rate is dramatically increased by IV use of ***L-cysteine***. All parts of the green or dried plant (especially the flowers and seed heads) are poisonous. The latter are often eaten by **sheep**, even when grass is plentiful. Anorexia, emaciation, regurgitation of green, frothy material, weakness, and pneumonia with audible rales are common in **ruminants**. Some tremble, develop incoordination, opisthotonus, red urine, and a rapid, pounding heart action. Post-mortem lesions include hemorrhages on the endocardium and epicardium, ascites (amber to red), friable and petechiated livers, dark kidneys with subcapsular hemorrhage, severe hemorrhagic gastroenteritis, and even edematous brain. **Rabbits** often die within 2 h after ingesting it.

BALANITES sp.

= *Aduwa (Hausa)* = *Desert Date* = *Soapberry Tree*

***Botanical.* Various species are used.**

Untoward effects: The bark's saponin is used to kill **fish** and in arrow poisons (usually with *Strophanthus*). Apparently, non-toxic to **man** and warm-blooded **animals**.

BALSAM, CANADIAN

= *Balsam of Fir* = *Canadian Turpentine*

Use: About 2 tons a years used in the U.S. for fragrances in cosmetic creams and lotions, soaps, detergents, and perfumes. Also in microscopy, in manufacturing fine lacquers, and as a lens cement.

Untoward effects: 2% in petrolatum topically in a 48 h closed-patch test on **human** volunteers showed no irritation. No sensitization was noted on 25 **human** volunteers after application of similar material.

Full-strength topically to intact or abraded **rabbit** skin for 24 h under occlusion was not irritating.

LD_{50}, **rats**: > 5 g/kg; acute dermal LD_{50}, **rabbits**: > 5 g/kg.

BALSAM PERU

= *Black Balsam* = *China Oil* = *Indian Balsam* = *Honduras Balsam* = *Myroxylon pereirae* = *Peruvian Balsam* = *Surinam Balsam*

Use: In perfumes and chocolate flavorings; as a scabicide and miticide; and in the healing of indolent wounds and ulcers.

Untoward effects: Has caused allergic contact dermatitis and pruritus in approximately 2–6% of a **human** test population, and 10% in those with diseases of the skin. Has been an allergen in the baking industry, causing hand dermatitis. Hypersensitivity reactions can occur within hours of topical or oral use. Dental resins containing it can cause an allergic cheilitis. It cross-reacts with ***cinnamon oil*** and ***clove oil***. Certain respiratory allergies in **man** may be aggravated by its ***cinnamic aldehyde*** content.

BALSAM TOLU

= *Myroxylon balsamum* = *Opobalsam* = *Resin Tolu* = *Thomas Balsam*

Untoward effects: Sometimes adulterated with ***rosin*** and ***pine*** resins. Skin minimum lethal dose, **rabbits**: 500 mg/24 h.

BAMBUTEROL

= *Bambec* = *KWD-2183*

***Bronchodilator.* A carbamate derivative of *terbutaline*.**

Untoward effects: Prolongs the duration of ***succinylcholine*** effects in **human** surgeries.

BAMETHAN

= *Bupatol* = *Butedrin* = *Garmian* = *Rotesar* = *Vasculat* = *Vasculit* = *Vascunicol* = *Vaskulat*

Vasodilator. **The sulfate is usually used.**

Untoward effects: Tired feeling, restlessness, stimulant effects, and increased heart rate have been reported.

BAMIFYLLINE

= *AC-3810* = *BAX 2739Z* = *Briofil* = *1802CB* = *Trentadil*

Bronchodilator. **A *theophylline* derivative.**

Untoward effects: Cephalgia, perspiration, "hot flashes", palpitations, somnolence, and vertigo reported after IV use.

BAMIPINE

= *Soventol* = *Taumidrine*

Antihistamine.

Untoward effects: Giddiness, nausea, vomiting, and excessive weakness in 11/139 Parkinson **patients**. Excessive dosage aggravates parkinsonism.

BANANA

Many types exist. The common one used by most people is the ready-to-eat yellow (bitter when green) one.

Untoward effects: Environmental Protection Agency "safe" level of ***aldicarb*** (q.v.) was found on 2% of them collected from 75 markets. Some **people** find it a potent source of intestinal gas. Contains natural estrogens. The skin contains 65 μg ***tyramine***/kg, 150 μg ***serotonin***/g, ***norepinephrine*** 250 μg/banana, and ***dopamine***, the pulp only, 7 μg/kg and some ***serotonin***. The inside of the skin has been called "***mellow yellow***" by drug addicts. The consumption by **people** of four bananas barely increases ***catecholamine*** excretion, because most **people** have the endogenous monoamine oxidase enzyme. Unless very large quantities are consumed, one should not expect restrictions to be needed by **those** ingesting ***monoamine oxidase inhibitors***, and a hypertensive crisis should not occur. The effects of banana smoke are more psychologic than psychedelic for "**hippies**" and others. Banana intake can interfere with vanillylmandelic acid assays and trigger migraines in some **people**. Licorice-colored vomitus occurred in a **patient** on ***levodopa*** therapy. African **tribes** that eat large amounts of bananas have a high incidence of right-sided heart lesions, similar to **those** suffering from "carcinoid syndrome" with their production of high ***serotonin*** levels.

Monkeys fed bananas have increased urinary excretion of ***5-hydroxyindoleacetic acid***, a ***serotonin*** metabolite.

BANEBERRY

= *Actaea sp.*

Botanical.

Untoward effects: From all parts, especially the berries (red or white). Causes gastroenteritis and increased heart rate in **humans**. As few as six berries have caused these symptoms, plus vomiting, diarrhea, colic, dizziness, and delirium. The rootstock contains a violent purgative. **Children** have died from eating the berries.

BANGALA

Botanical.

Untoward effects: **Bantu men** grind the roots into a powder to be mixed and ingested with ***tea*** or ***coffee*** 2 h before a romantic encounter; causes a powerful erection, subsiding only after intercourse.

BAPTISIA sp.

= *False Indigo* = *Indigo Weed* = *Wild Indigo* = *Yellow Indigo*

Botanical. **Contains *sparteine*, *baptisin*, and *cytisine*.**

Untoward effects: Ingestion can cause anorexia and severe diarrhea in **animals**, and, because of its ***sparteine*** content, could cause overstimulation of the myometrium and cause abortions in **women** and **animals**. Particularly toxic to **horses**.

BARBAN

= *Barbamate* = *Barbane* = *Carbyne* = *Chlorinat* = *CS-847*

Herbicide. **A carbamate.**

Untoward effects: May cause skin irritation in **man**. Feeding trials for up to 10 days to **cattle**, **sheep**, and **chickens** indicated toxic doses were 25 mg/kg, 10 mg/kg, and 50 mg/kg, respectively.

LD_{50}, **rats**: 600 mg/kg.

BARBAREA VULGARIS

= *Yellow Rocket*

Untoward effects: Poisoning of a **horse** that ate large quantities of the leaves and stems. Possibly due to its glucosinolate content.

BARBASCO

Untoward effects: The ground-up roots and leaves are used to poison and stun **fish**. The *Dioscoreae* or wild yams of Mexico are a source of ***diosgenin***, a starting point for many ***corticosteroids***. The tea is a natural contraceptive.

BARBERRY

= *Berberis sp.*

Contains the alkaloids, *berberine*, *venetine*, *palmatine*, and *berbamine*.

Untoward effects: The roots, shoots, and berries have been incriminated as causing colic, diarrhea, and dermatitis in **animals** and **man**. A 33-year-old **male** developed a severe pruritus on **his** arms, legs, and thighs 1 month after trimming a hedge. The skin eruption was persistent with pustular papules and hyperpigmented nodules.

Adverse effects on the central nervous system occur in **animals**, causing paralysis of the respiratory and vasomotor centers.

BARBITAL

= *Barbitone* = *Deba* = *Diemal* = *Diethylmalonylurea* = *Diemalum* = *Dormonal* = *Embinal* = *Hypnogène* = *Malonal* = *Medinal* = *Pyrabital* = *Sédeval* = *Uronal* = *Veroletten* = *Veronal* = *Vespéral*

Sedative, hypnotic.

Untoward effects: In **humans**, therapeutic serum concentrations have been given by some as 30–60 mg/l and toxic concentrations as 50–80 mg/l, indicating a narrow margin between the two. With a half-life of 4.8 h, 65–90% of an administered dose appears in the urine unaltered. Even though it sometimes does not appear in **human** breast milk (viz. in one **patient** after a 650 mg dose), its use by a **mother** has produced sedation in a nursing **infant**. It passes through the **human** placenta readily and accumulates in **fetal** brain and liver. Physicians should be aware of possible alterations in **neonatal** microsomal enzyme activity. Morbilliform and scarlatiniform skin eruptions have occurred after its use in **man**. After withdrawal, addicted **people** may develop convulsions in 4–5 days. It is a dialyzable drug. Therapeutic serum levels are > 2 µg/ml; toxic at 20–50 µg/ml, with coma or fatalities often at 50 or more µg/ml.

LD_{LO}, **human**: 50 mg/kg.

Drug interactions: Pretreatment with it in **human** volunteers, **dogs**, **guinea pigs**, and **rats** has antagonized the anticoagulant effect of ***bishydroxycoumarin*** (***Dicoumarol***), by increasing the metabolism of the latter.

It increases the toxicity of ***methyl salicylate*** in **dogs**. Withdrawal convulsions have been noted in **rats**. ***Disulfiram*** significantly increases barbital toxicity and sleeping time in **rats**. ***Alcohol*** increases its depressing effects in **rabbit**, **rat**, **dog**, and **mouse** trials. Tolerance to it and physical dependence to it within 2 weeks has been demonstrated in **monkeys**.

LD_{LO}, **cat**: 280 mg/kg; **rabbit**: 275 mg/kg; IP, **rat**: 300 mg/kg; IP, **rabbit**: 225 mg/kg.

BARIUM

Untoward effects: Mixtures of finely divided barium and a number of ***halogenated hydrocarbons*** are explosive. Granular barium with ***monofluorotrichloromethane***, ***trichlorotrifluoroethane***, ***carbon tetrachloride***, ***trichloroethylene***, and ***tetrachloroethylene*** can detonate on impact. 1 mg/l is the maximum allowable level permitted in drinking water for **humans**, and 20 ppm is set as the maximum tolerable level in **animal** feed. Its salts are used for coloring fireplace flames. Ingestion causes intense gastrointestinal irritation with nausea, vomiting, and diarrhea. The TWA for **humans** of its soluble salts is 0.5 mg/m^3, to prevent undue irritation of eyes, nose, throat, bronchi, and skin, dry mouth, stomach pains, diarrhea, bradycardia, irregular heartbeats, increased blood pressure, tinnitus, dizziness, muscle spasms, confusion, and convulsions. Death can occur within an hour from cardiovascular failure or paralysis of respiratory muscles. Implicated in causing hypokalemia; and ***potassium*** administration has been useful in treating a case of poisoning by barium. Early administration of ***magnesium sulfate***, orally, is given to precipitate toxic salts, as the insoluble sulfate, followed by gastric lavage. Respiratory assistance and ***atropine*** may also be necessary. ***Sodium sulfate*** 300 mg/kg is used in **humans** as an antidote against unabsorbed barium salts by precipitating them, and it has also been used as a 2–5% solution in lavage. In the most severe poisoning, 10 ml of a 10% solution has been suggested for IV use in adult **humans**. Inhaled dust is retained in lung and regional lymph nodes. Clay-colored, whitish stools in **humans** after ingestion. Water- and acid-soluble salts are toxic, and the lethal **human** adult dose is between 1 and 15 g. Can interfere with ***tetracycline*** absorption.

Has caused increased absorption and toxicity of ***nitrofurazone*** in **horses**. In adult **cattle**, 250–1000 g of ***sodium*** and ***magnesium sulfates***, as a 20% solution, is recommended orally as a treatment. In **dogs**, 2–25 g are given in a similar manner. Ingestion by **poultry** causes darkening of their combs and incoordination.

BARIUM CARBONATE

= *Witherite*

Over 20,000 tons used annually in the U.S.

Use: In manufacturing chemicals, glass, ceramic glazes, brick, porcelain, and clay products, TV picture tubes, rodenticides, silverfish baits, in cements, and to impart a flat, white appearance to photographic paper.

Untoward effects: Very poisonous. Acute **human** toxicity manifested by extreme salivation, vomiting, colic, intense diarrhea, tremors, convulsions, hypertension, muscle paralysis, and gastrointestinal and renal hemorrhages.

LD_{LO}, **human**: 57 mg/kg.

LD_{50}, **rats**: 630–800 mg/kg; LD_{LO}, **dog**: 400 mg/kg; **mouse**: 200 mg/kg.

BARIUM CHLORIDE

Use: In many dyes, pesticides, leather tanning, boiler water softeners, mordants, occasionally as a ruminatoric in **cattle** and **sheep**, and as an irritant purgative in **animals**. U.S. production is estimated at about 10,000 metric tons annually.

Untoward effects: Transient dysplasia in a **woman's** cervical cytology after three applications to **her** cervix at 3–4 week intervals. Similar reports after incorporation into plastic intrauterine devices, to permit X-ray detection and maneuverability inside the uterine cavity. In **man**, respiratory tract irritation from its inhalation; gastroenteritis from its ingestion. Can cause eye irritation, mydriasis, skin burns, bradycardia, hypokalemia, and extrasystoles. Death usually occurs from respiratory failure and fibrillary contractions continue for some time. Acute renal failure in a 52-year-old **male** after ingestion of approximately 13 g.

LD_{LO}, **human**: 11.4 mg/kg; **dog**: 90 mg/kg; **rabbit**: 170 mg/kg.

BARIUM HYDROXIDE

Use: In cements, storage batteries, refining oils, inks, insecticides, depilatories, corrosion inhibitors, softening boiler water, rubber vulcanizing, making artificial pearls, ivory, and stone, and sewage treatments.

Untoward effects: Can cause alkali burns of the corneal epithelium in **workers** handling it. Irrigate as soon as possible with copious amounts of water.

IP, TD_{LO}, 200 mg/kg; LD, **chicks**: 467 mg/kg; 0.1% in the feed without ill-effects, 0.2% decreased rate of growth, 0.8% for 4 weeks killed over half of them, and 1.6% killed all.

BARIUM SULFATE

= *Actybaryte* = *Bakontal* = *Baridol* = *Baritop* = *Barosperse* = *Barotrast* = *CI 77,120* = *CI Pigment White 21-22* = *Citobaryum* = *Esophotrast* = *E-Z-HD* = *E-Z-Paque* = *Intestibar* = *Microbar* = *Micropaque* = *Microtrast* = *Neobar* = *Oratrast* = *Polybar* = *Prontobario* = *Radiopaque* = *Raybar* = *Redipaque* = *Steripaque* = *Telebar* = *Unibaryt*

Use: In gastrointestinal radiography and fluoroscopy, silicone elastomers.

Untoward effects: Extensive bacterial contamination of enema solutions has caused obvious problems. Enema material has penetrated the rectal wall; cecal and colonic perforation has occurred due to carcinoma, diverticula, etc. Barium enemas are contraindicated in severe pseudomembranous colitis. Peritonitis can follow such use. Distal small bowel may communicate with peritoneal space in some **infants**. Volvulus is possible if early evacuation does not occur. An asymptomatic, adherent, non-obstructing barium fecalith (barolith) of the cecum reported 2 years after previous barium studies. Constipation, occasionally severe, has followed its use as an enema. Interferes with abdominal echography. Retained barium has resulted in acute appendicitis 7 days after barium enema. A rectal granuloma developed after rectal injury during a barium enema. During routine enemas, potentially new arrhythmias noted in 27/58 **patients**, ages 60–98. Use as a contrast media has caused a non-fatal cardiac arrest in a **child** and bradycardia in another. The electrocardiogram changes occurred more commonly in the **elderly** and those with heart disease. Of 95 **patients**, 46 showed alterations, including ventricular bigeminy and trigeminy, premature ventricular contractions, atrioventricular block, intermittent bundle-branch blocks, and transient sinus arrest. An anaphylactic reaction to a single rectal enema reported in a 51-year-old **female** and others. A **patient's** allergic history, including any sensitivity to ***latex*** should be ascertained, before an enema is given. Ocular injury to **children's** eyes from taking golf balls apart containing it, or a mixture with ***zinc sulfide***. Contents are under high pressure (2000–2500 lb/sq in) and have been deposited in conjunctival and palpebral tissues, often requiring cosmetic surgery. In one case, a fine stream was ejected and penetrated the eyeball, causing great pain. The lesson learned is, "don't keep your eye on the ball". May give false positive readings to a number of diagnostic tests. Death due to intravasation of a barium enema into the portal venous system has occurred. One report indicated 20/175 **patients** developed a bacteremia after the enemas. Adding ***tannic acid*** at > 0.5% concentration in a barium sulfate enema appeared to be associated with hepatotoxicity.

BARIUM SULFIDE

Use: In depilatories, for leather hides and pelts, fungicides, luminous paints, fire-proofing wood, and vulcanizing rubber. Over 25,000 tons used annually in the U.S.

Untoward effects: Allergic skin reactions virtually eliminated its use as a depilatory on **humans**. A suicide attempt with it by a 42-year-old **female** produced stupor, cyanosis, slight dyspnea, hiccup, salivation, abdominal pain, muscle spasms, and dilated pupils several hours after ingestion. High QRS complex, high T-wave, supraventricular arrhythmias and other electrocardiogram changes, apparently due to hypokalemia. Rapid recovery after IV ***potassium chloride***. Has caused death in **people** when taken internally by mistake for ***barium sulfate***.

BARLEY

= *Hordeum vulgare*

Contains the vasopressor, *gramine*.

Use: As a food and in ***beer*** manufacturing.

Untoward effects: Usually, from toxic molds on it, which can cause pancytopenia and "yellow rain", as well as "**brewers'** or **maltworkers'**" lung (alveolitis). Has caused an allergic contact dermatitis in **man**, especially in ***bread*** **bakers**. Death occurred in 459 **Iraqis** of 6530 poisoned from eating ***bread*** baked from ***methylmercury***-contaminated barley. Contains significant amounts of ***gluten*** (q.v.), which can cause problems in **people** intolerant to it.

Vulvovaginitis occurred in **sows**, due to the estrogenic content of moldy barley. Acute alcoholic intoxication and hepatitis developed in **cows** eating moist, fermenting barley. Hyperkeratosis and acanthosis of the ruminal wall developed in **calves** a month after eating a diet containing 90% barley. Barley has also been a cause of ***nitrate*** poisoning. Occasionally, **dogs**, **sheep**, and **cattle** have sustained severe eye, nose, mouth, and lung injury from its awns.

BARRINGTONIA sp.

Use and untoward effects: The bark and roots as a **fish** poison in Samoa, Solomon Islands, and Africa. Some **humans** cannot eat its nuts, because **they** develop a mild, but long-lasting headache. An active cardiac poison.

BARTERIA FISTULOSA

Botanical.

Use: The powdered bark and roots are used in Africa to control head lice.

Untoward effects: Ingestion of the bark or its extracts causes enophthalmia in **humans** and death within about 4 h.

BARTHRIN

= *ENT 21,557*

Insecticide.* A *pyrethrin*-type ester of *chrysanthemumic acid.

Untoward effects: In seven adult **male human** volunteers, ten daily cutaneous applications of 0.5 ml undiluted material under plastic failed to induce any hyperemia or irritation.

Only 1/8 as toxic as ***pyrethrins*** and 1/3 as toxic as ***allethrin*** in **warm-blooded animals**. No cutaneous reactions from single application on **rabbits**, nor on repeated cumulative exposure.

LD_{50}, **rat**: 24 g/kg.

BASILIXIMAB

= *Simulect*

***Monoclonal antibody.* Blocks interleukin-2 receptor on T-lymphocytes. For IV infusion.**

Untoward effects: ***Blood***: Anemia, decreased red blood cells, hypo- or hyperkalemia, hyperuricemia, hypocalcemia, hypercholesterolemia.

Cardiovascular: Hypertension.

Central nervous system: Headache, tremors, and dizziness.

Gastrointestinal: Constipation, abdominal pain, nausea, and vomiting.

Miscellaneous: Moniliasis, acidosis, increased weight, peripheral edema, fever, asthenia, chills, dysuria, increased non-protein nitrogen, leg and/or back pain, acne, and delayed wound healing are among a long list of adverse effects.

Respiratory: Coughing, chest tightness, pharyngitis, rhinitis, and dyspnea.

BASIL OIL, SWEET

Use: In fragrances for creams, lotions, perfumes, and foods. From the leaves and flower heads of *Ocimum basillicum*.

Untoward effects: A 4% concentration in petrolatum on 25 **human** volunteers failed to produce any irritation. No phototoxic effects noted.

LD_{50}, **rats**: 1.4 ml/kg; acute dermal LD_{50}, > 5 ml/kg.

BASSIA sp.

Annual plant.

Untoward effects: **Sheep** have died after a single feeding on *B. hyssopifolia* in Utah, after weakness, incoordination, tetany and coma with hemorrhages in the rumen and occasionally kidney enlargement. ***Oxalates*** are suspect. *B. longifolia* var. *latifolia* contains a toxic saponin often contaminating ***groundnut*** (***peanut***) cake and has caused heavy mortality of **poultry** in India. Its seeds are used as a piscicide. ***Digitalis***-like effect on the heart and saponin-like effect on destroying red blood cells. A ***nitrate***-concentrating plant.

BATRACHOTOXIN

***Untoward effects*:** One of the most toxic chemicals known to **man**, from the skin of the small (approximately 1" in length), vividly colored Colombian kokoi **frog** (*Phyllobates aurotaenia*), which it uses to protect itself from predators. Causes irreversible blockade of nerve impulse transmission to muscles, producing cardiac arrhythmias and cardiac arrest. 1/50,000 mg, SC kills a **mouse**. Poisonous to **man** from eating anything exposed to the poison, only if **one** has an oral scratch or gastrointestinal ulcer. Has been used in arrow and dart poisons. Now made synthetically, for studies of neuromuscular transmission.

IV, LD_{50}, **mouse**: 2 μg/kg (***strychnine's*** LD_{50} is 500 μg/kg).

BATTERIES

Untoward effects*:** Esophageal and stomach erosions and burns from ***sodium or ***potassium hydroxide*** released after ingestion of button batteries from watches, calculators, hearing aids, and cameras. The batteries often also contain ***mercury***, ***cadmium***, ***zinc***, ***nickel***, ***silver oxide***, ***potassium hydroxide***, or ***lithium***, with their respective toxicities. Hundreds of cases are reported in **children** annually. **Dogs** have also ingested them. **Workers** and **people** living near ***zinc–cadmium*** battery factories may inhale these elements from contaminated air. ***Carbon black*** used in dry cell batteries causes **workers** to have coughing with phlegm production, chest pain, chronic rhinitis, fatigue, headache, and skin irritation. Contact dermatitis from ***nickel*** in alkaline batteries and from ***chromates*** in dry batteries. Fatal poisoning in 3-year-old **male** after ingesting ***ferrous sulfate*** in a pinch of gray mass from car battery terminals. ***Lithium–sulfur dioxide*** batteries may exhibit radioactivity. Auto **mechanics** wearing metal objects while handling batteries risk getting deep second degree burns. **People** who "jump" batteries should stand away, especially in poorly ventilated areas, as the ***hydrogen*** gas released can explode. **Children** of battery **workers** have increased ***lead*** burden, as do the **workers**. **Children** often bite into flashlight and pen-type batteries. A size D battery contains ~1/5 minimum lethal dose of ***mercuric chloride*** for a **child**.

Lead poisoning in **cattle** from licking discarded storage batteries and from eating silage transported in a trailer that had hauled broken storage batteries.

BAY LEAF

= *Laurus nobilis*

***Aromatic spice*.** From the evergreen sweet bay or laurel tree.

***Untoward effects*:** A 45-year-old **female** with a 2 day history of cramping and right-side lower abdomen pain and tenderness. Leukocyte count of 8000/mm^3 with strong left shift. Meckel's diverticulum had been perforated by a bay leaf. A 25-year-old **male** had excruciating rectal pain on defecation, due to a large, wedged bay leaf. Leaf stems have been found wedged in the mucosa of the esophagus. Remove bay leaves before eating foods cooked with it.

BCG

= *Bacillus Calmette–Guérin* = *ImmuCyst* = *TheraCys*

***Immunostimulant, antineoplastic*. Intravesical and occasionally percutaneous use. Oral and parenteral of various strains. In some countries, SC for animals.**

Untoward effects*:** Nausea, vomiting, cramping, anorexia, sepsis, fever, and bladder irritation. **Some** later develop urethral obstruction, granulomatous pyelonephritis, epididymitis, orchitis, hepatitis, renal abscesses, arthralgia, and cytopenia. Rarely, induced lupus vulgaris, Stevens–Johnson syndrome, and polyneuritis. Inadvertent use instead of ***tuberculin for skin test caused **many** to develop the expected local reaction; **those** with previous ***tuberculin*** reactions had more severe effects. Serious and fatal reactions in many immunosuppressed **patients**. Allergic reactions occur. Hypersensitivity to its ***dextran*** content has also occurred.

Drug interactions*:** Immunostimulation after BCG decreases liver metabolism of ***theophylline and probably ***oral contraceptives***.

No longer popular for **animal** use because it needs to be repeated annually and because it causes sensitivity to ***tuberculin***.

BEANS

Untoward effects*:** In general, beans produce an offensive flatus odor, due to their content of undigestible sugars, such as ***raffinose and ***stachyose***. Baked beans, and beans in general, contain ***oxalates***, which are undesirable for many **people** with urinary problems. Mycotic infections of the fingers and hands have occurred in cannery and flour mill **workers**, handling contaminated beans. Asthmatic-type of respiratory upsets reported in mill **workers**. Some **people** develop migraine headaches from eating beans, due to their ***tyramine*** content. Bean pastes can cause ***sulfite***-induced problems in some **people** from the ***sodium hyposulfite*** treatment of beans. In some **people**, dyshidrosis has been associated with its ingestion and it can increase ***uric acid*** serum levels. Bean bag infant cushions were implicated in the deaths of **infants** by suffocation.

Faba Bean = *Broad Bean* = *Fava Bean* = *Field Bean* = *Horse Bean* = *Italian Green Bean* = *Tick Bean* = *Vicia faba*

Untoward effects*:** Within a few minutes to a few hours, many **people** develop favism, a condition that includes headache, dizziness, nausea, yawning, vomiting, jaundice, hemoglobinuria, chills, pallor, malaise, and collapse. This from either consuming the bean or inhaling the blossom's pollen. It occurs primarily in the Mediterranean areas or in **individuals** of that ethnic background living in other areas. The toxins include ***trypsin-inhibitors, ***glycosides*** (***vicine*** and ***convicine***), ***lectins***, ***tyramines***, and ***levodopa***. Maternal ingestion has induced adverse hemolytic reactions in four glucose-6-phosphate dehydrogenase-deficient nursing **infants**. More common in **male** than **female children**. Its content of nitrosable mutagen precursors has been implicated, along with high ***nitrate*** and ***nitrite*** content of well water in the high incidence of stomach cancer in **Colombians**. Jaundice reported in **children** eating it. Blood dyscrasias (hemolytic anemia and eosinophilia; occasionally thrombocytopenia and methemoglobinemia) have occurred in unusually sensitive **individuals**. Even relatively small intake has caused hemolysis or hemolytic anemia in glucose-6-phosphate dehydrogenase-deficient **people** (10–30% of Negroes, 3–30% of Sardinians, 25% of Iraqis, 12% of Filipinos, and 60% of Kurdish; also in Chinese and Greeks). A **Chinese** 30-year-old **female** developed hydrops fetalis with glucose-6-phosphate dehydrogenase anemia after eating fava beans. Serious intrauterine hemolysis led to fetal death at 38 weeks of gestation. An active hemolytic principle is ***dopaquinone*** and the ill-effects are dependent on the rate of action of ***tyrosinase*** on the ***levodopa*** content. ***Divicine*** is also a hemolytic agent in the bean. The hemolytic anemia it causes may be caused by a susceptibility, genetic in nature, in addition to a glucose-6-phosphate dehydrogenase deficiency. Low blood ***glutathione*** increases a **person's** susceptibility. Approximately 100 million **people** worldwide are affected. A 6-year-old **male** developed hemolysis after a fava bean meal. This was alleviated by an injection of 500 mg ***desferrioxamine***. One fava bean meal has caused hypertensive crises in **people** on ***monoamine oxidase inhibitors***, such as ***isocarboxazid***, ***nialamide***, ***pargyline***, ***phenelzine***, and ***tranylcypromine***, often with resultant faintness, dizziness, palpitations, nausea, vomiting, torticollis, apprehension, and headaches. If ***furazolidone*** is given for over a few days, fava bean intake may cause hypertensive crisis.

Ingestion by **swine** caused abdominal pain, decreased activity, depression, flatulence, anorexia, constipation, and death after consuming very large quantities. Post-mortem showed gastrointestinal inflammation and pale yellow liver and kidneys. In **chickens** caused decreased egg production, decreased rate of growth, and decreased liver and pancreas size.

Bean, Kidney = Common Bean = French Bean = Garden Bean = Haricot Bean = Navy Bean = Phaseolus vulgaris = String Bean = Wax Bean

Untoward effects: Contains two antagonists to ***vitamin E***: one, heat stable, and one, heat labile. Half of this antagonism is eliminated by cooking. Contains ***trypsin-inhibitors***, ***estrone***, ***estriol***, ***17α-estradiol***, ***hydrocyanic glycosides*** (destroyed by normal body defense mechanisms), and ***phaseolamin***. The latter inhibits the enzyme, ***α-amylase***, that helps convert ***starch*** to ***glucose***. Causes gastrointestinal upsets and allergic reactions in some **people**. A **hemophiliac** 31-month-old **male** died from a phytobezoar.

Excessive amounts in the diet of **livestock** and **poultry** were associated with decreased growth rate, muscular dystrophy, and/or deaths from ***vitamin E*** deficiency. IP use of an extract caused abortions in **mice**. **Dogs** have been allergic to it in their diet.

Beans, Lima = Akpaka (Igbo) = *Awuje* (Yoruba) = *Java Bean = Phaseolus limensis = Phaseolus lunatus = Wake* (Hausa)

Untoward effects: Average U.S. consumption by **adults** is nearly 2 lb/year, which contains about 40 mg ***hydrogen cyanide***, easily detoxified by the body and released by cooking. Inadequately cooked lima beans contain more ***cyanides***. Contains ***trypsin-*** or ***protease-inhibitors*** and ***oxalates***. Giddiness, vomiting, colic, purging, and prostration has occurred in **man** from eating large quantities of uncooked beans. The plant has contained 0.0002% ***hydrocyanic acid***.

Fatal cases occurred in **livestock**.

Pinto Bean

Untoward effects: Allergic reactions in **people**.

Rangoon Bean

Untoward effects: Dried beans have caused ***cyanide*** poisoning in **swine**.

Winged Bean = Asparagus Bean = Botor = Calocan = Goa Bean = Kachang Botor = Kechipir = Psophocarpus tetragonolobus = Too-a-poo

Untoward effects: ***Trypsin-inhibitor*** in the seed. The ripe seeds are toxic (Burma), and the toxin can be destroyed by cooking.

BECANTHONE

= WIN 13,820

Antihistomonal.

Untoward effects: Marked central nervous system and gastrointestinal toxicity. Acute toxic effects in 14/37 **patients** with poor cure rate. Use eventually discontinued.

BECLAMIDE

= *Benzchlorpropamide* = *Chloracon* = *Chloroethylphenamide* = *Hibicon* = *Neuracen* = *Nidrane* = *Nydrane* = *Posedrine* = *Seclar*

Anticonvulsant.

Untoward effects: 2/27 **patients** developed a morbilliform rash.

LD_{50}, **rat**: 3.2 g/kg; **mouse**: 2.3 g/kg.

BECLOBRATE

= *Beclipur* = *Becloslerin* = *CHF-1511* = *Turec*

Antihyperlipoproteinemic.

Untoward effects: A fatal case of acute hepatitis in 64-year-old **male**, treated with 100 mg/day for hypercholesterolemia. Anorexic and jaundiced 6½ months after starting treatment. Admitted to a hospital 3 weeks later, and died 7 days after admission. Acute fatal liver failure in a 63-year-old **female** who received 100 mg/day/4 months for moderate hyperlipidemia. Despite drug withdrawal, the failure progressed and extensive liver necrosis was noted on autopsy.

BECLOMETHASONE

= *Aldecin* = *Anceron* = *Andion* = *Beclacin* = *Becloforte* = *Beclomet* = *Beclorhinol* = *Becloval* = *Beclovent* = *Becodisks* = *Beconase* = *Beconasol* = *Becotide* = *Clenil-A* = *Entyderma* = *Inalone O* = *Inalone R* = *Korbutone* = *Propaderm* = *Rino-Clenil* = *Sanasthmax* = *Sanasthmyl* = *Sch 18020W* = *Vancenase* = *Viarex* = *Viarox*

Corticosteroid, anti-inflammatory.

Use: Orally and intranasally in allergies, by inhalation in asthma; topically, as an anti-inflammatory. Inhaled material is poorly absorbed from the oropharyngeal and airway mucosa.

Untoward effects: Muscle cramps in a 48-year-old **female** attributed to its oral inhalation (200 µg/twice daily) for asthma control. Growth retardation and adrenal suppression reported in **children** (2½–9½ years) with asthma, inhaling 400–1000 µg/day/1.6–6.5 years. Purpura and dermal thinning occurred in **patients** inhaling 1000–2250 µg/day. Purpura reported in 54-year-old **female** who inhaled 400–1500 µg/day. 28/70 (40%) **outpatients** developed coughing after inhaling it. Tests on three **asthmatics** showed that the propellants were **NOT** the cause. A **patient** using the nasal spray (100 µg/tid/2 weeks) developed ulceration of the nasal mucosa. ***Steroid***-induced mania has been reported with its use as a nasal spray. Stinging sensations, sneezing, and hemorrhagic secretions in 5% of **patients** receiving 400 µg/day and increased incidence of intra- and paranasal infections and mucosal atrophy. A 23-year-old **female**, using it as aerosol therapy, developed esophageal candidiasis. Oropharyngeal candidiasis (thrush) and moniliasis after continuous use as an aerosoled inhalant is dose-related and more common in **women** than in **men**. This is unlikely with doses < 600 µg/day. **Some** develop sore throats with asthma-like symptoms and become hoarse and have difficulty speaking after use. Should **not** be used in **patients** with tuberculosis. A case of aspergillosis reported in a 70-year-old **male**, occurring after 2 years of inhalation therapy for asthma. Bacterial contamination of nebulizers has been found. Cases of severe chickenpox in **children** may have resulted from inhalation therapy immunosuppression. Paradoxically, a 38 month **child** developed an exacerbation of asthma 7 months after inhalation therapy started. Although the risk of cataract formation in **patients** receiving it as inhalation therapy for asthma is minimal, cases have been reported. A 30-year-old **female** receiving 100–200 µg/day/5 years developed posterior, subcapsular cataracts; three **females** (40–52 years) also developed similar cataracts, all of whom received 400 µg/day/1 year, and the third also received 1500 µg/day for an additional year. Suppression of the hypothalamic–pituitary–adrenal axis has occurred in **children** and **adults** on inhalation therapy and must be considered before recommending it. Deaths have occurred after transferring **patients** from systemic to inhalation therapy. Physicians would be wise to insist that **patients** have systemic ***corticosteroids*** to use with impending problems and carry warning cards or wear Medic-Alert® bracelets.

BEER

Commercial beer contains antioxidants, such as *sodium isoascorbate*, to protect clarity, by preventing oxidation haze; proteolytic enzymes, such as *papain*, to chill-proof it; *diethylpyrocarbonate*, to sterilize it at room temperature, as a prevention against development of yeasts and bacteria. No longer permitted for this use in the U.S. Beer, so-treated, often developed, by reaction with *ammonia*, 1–3 mg *urethan* (q.v.)/l, which is a carcinogen. *Gum arabic* and *alginates* are used as foam stabilizers. Dry beers are less sweet and often have less *alcohol* than others.

Untoward effects: Headache and flushing experienced by some **people** may, in part, be due to beer's ***histamine*** content. Since beer consumption decreases intraocular pressure, it can mask the diagnosis of glaucoma. Regular consumption increases stomach acidity, increases weight gain, and decreases blood ***sodium***. Weakness, drowsiness, extrapyramidal effects, disorientation, and attacks of epilepsy, due to hyponatremia, in **some** who drank over 4 l of beer a day. Beer and sun increase the danger of dehydration; therefore, a **sunbather** who returns from the beach should drink other fluids before drinking beer. The latter may cause **him** to feel dizzy, nauseous, and faint. A hypo-osmolarity syndrome occurred in a 66-year-old beer **drinker** (4 l daily and total fluid intake was only 5 l). Speculation still exists on its relationship to the development of stomach and large bowel cancer. The ***nitrosamine*** content, particularly in imported beers (German, Puerto Rican), was the suspected agent. It formed when ***nitric oxide*** contacted beer malt in the drying process. The FDA now has a limit on its content in beer. In the last decade, some Austrian beer was found contaminated by deliberate inclusion of ***bromoacetic acid*** as a disinfectant. Hepatic toxicity reported. Acute renal failure after drinking large quantities of beer. **One** drank 9 l of home-brew during an 8 h period, and **another** drank 17 l (time not specified). **Both** responded rapidly to peritoneal dialysis with little permanent renal damage. Large volume beer intake has caused cardiovascular beriberi. Severe heart failure usually occurs 3–4 weeks after dyspnea, tachycardia, cyanosis, and gallop rhythms appear, and death has occurred. Immediate contact urticarial reactions have been reported, possibly due to ***sulfites*** added to beer. Beer has definitely increased the aggression in some intoxicated social **drinkers**. **Those** susceptible to urinary ***oxalate*** problems should remember that beer can contain considerable amounts. Beer consumption can induce and aggravate gout attacks.

In Canada and Australia, **beer-drinkers** suffered from what was called "Quebec **beer-drinkers'** myopathy", with the main causal factor being ***cobalt***. It was also used to a lesser extent in the U.S., to help control foaming. In Omaha or Quebec, the typical **patient** was in their early 40s and drinking 3–6 l of beer/day for at least 5 years. Mortality was often about 50%.

Alcohol-free beers actually contain < 0.5% ***alcohol*** and may cause increased breath ***alcohol*** test results for up to an hour after ingestion.

Drug interactions: Use with ***methyprylon*** has resulted in euphoria, ataxia, and deep sleep. Due to its ***alcohol*** (q.v.) content, many interactions can occur. A ***disulfiram*** reaction (pruritic macular rash) in a 56-year-old **male** reported, due to use of a hair shampoo containing beer. Similar reactions with facial flushing, tachycardia, and hypertension in a number of **patients** taking ***cefoperazone***. Flushing with a florid macular eruption on the face and chest, and sweating in a **patient** drinking a can of beer 6 h after ***moxalactam***, IV. Hypertensive crises in some **patients** using ***monoamine oxidase inhibitors*** and drinking beer. The ***tyramine*** level in beers varies. Nausea, vomiting, and indigestion reported in a **beer-drinker** taking ***nifuratel*** for a chronic urinary infection. Beer appears to be diuretic and soporific.

BEES

= *Apies sp.*

Untoward effects: Hypersensitivity to their stings occurs in **humans** and **animals**. Desensitization vaccinations are occasionally used. Between 1960 and 1981, 25 bee sting-related deaths were reported in Australia. African killer bees have caused death and serious injuries in **people**, **animals**, and **turkeys**. Between 1950 and 1959, bee stings caused 124 deaths in the U.S. Many are mistaken for cardiac or respiratory deaths. Allergic reactions, such as chest pains, profuse sweating, hypotension, fainting, dyspnea, and urticaria can occur. In one Georgia study of emergency room visits, 25% of over 2000 visits were for *Hymenoptera* stings. More severe reactions than expected were reported in **patients** on long-term therapy with ***ibuprofen*** and ***diclofenac***. A 31-year-old **male** developed a severe uvulitis, after being stung twice at his uvula by a honey bee. A 38-year-old **male**, stung repeatedly by yellow jackets, showed no anaphylaxis until 16 h later, when **he** developed cardiac problems, including premature ventricular contractions and junctional rhythm. **Beekeepers** have become allergic to inhalants at beehives, and occasionally develop a dermatitis from bee ***propolis***. A case of thrombotic thrombocytopenic purpura complex with hemolytic anemia, renal failure, and fever occurred in a 44-year-old **female** after a bee sting. Diffuse hives, wheezing, laryngeal edema, rhinorrhea, headache, dizziness, pain, swelling, urticaria, chills, fever, and even shock can occur rapidly after a sting. Local allergic reactions are common in the mouth area of **dogs**, and death of a **dog** has followed being stung by a bee. Bee stings can cause leukopenia and a decrease in red blood cells. Do **NOT** rub the affected area, because it may force more venom out of the embedded stinger. Instead, try to scratch it out of the skin with a fingernail or an instrument. The venom sac takes 2–3 min to empty fully. Many treatments have been recommended: 1% ***lidocaine*** with ***epinephrine*** at and under each site; rub the site with a juicy ***onion***; or apply a paste of a ***papain***-containing meat tenderizer. ***Diphenhydramine*** and ***corticosteroids***, topically or orally, have also been used. ***Epinephrine***, IM and ***calcium***, IV have been used in the treatment of severe allergic reactions.

Bee pollen, a health food antiaging fad, is just another food source of protein, and is responsible for acute anaphylaxis, including asthma, in **people** allergic to bees or pollen. It is pollen scraped from the legs of bees, and often contains a significant amount of bee protein.

BEESWAX

= *Cera Flava* = *Yellow Wax*

Use: In cosmetics, confectionary glazes, and in topical pharmaceuticals.

Untoward effects: A known, but infrequent, skin sensitizer. Occupational dermatitis reported in a 65-year-old **beekeeper** from resins in the beeswax. An allergic reaction occurred in a **person** using cold cream containing beeswax.

BEETS

= *Beta vulgaris*

Untoward effects: Beets contain a high level of ***nitrate*** (up to 3000 ppm, or an average of 200 mg/100 g). These can be reduced to ***nitrites***, under conditions of low acidity or during storage. Carcinogenic activity is probably due to ultimate formation of ***nitrosamines***. It is suggested that **patients** with ***cobalt*** dermatitis avoid it. Consumption of beets can color urine a brilliant red, due to ***anthocyanin*** pigments. Feces may also become red. Contains ***oxalates*** and ***anthraquinone glycosides***, which can lead to a serious gastroenteritis in **humans**, **cats**, and **dogs**. Its leaves contain 0.186% ***β-sitosterol***, an estrogen. Beets have low acid content and must be cooked longer at a high heat level to render them sterile.

Numerous cases of poisoning of **cattle** and **sheep** and increase in urinary calculi, due to excessive intake of beets or their leaves as fodder are due to the ***oxalate*** and ***nitrate*** content.

BEHENIC ACID

Use: In paint removers to retard evaporation, as an antifoam in floor polishes, and in photocopy paper.

Untoward effects: Purpura.

BELLADONNA

= *Banewort* = *Deadly Nightshade* = *Death's Herb* = *Dwale* = *Poison Black Cherry*

***Anticholinergic, antispasmodic, mydriatic, cycloplegic.* A very poisonous perennial known in Egypt, India, and Mesopotamia seven thousand years B.C. The leaves contain 0.3–0.5% alkaloids (primarily, *atropine, hyoscyamine,* and *scopolamine*) and the dried roots 0.4–0.7% alkaloids.**

Untoward effects: See ***Atropine*** and ***Atropa belladonna*** for additional ***Use*** and ***Untoward effects***. Additionally, it has led to fixed-drug eruptions, Stevens–Johnson syndrome, mydriasis, and predisposes **people** to heat stroke. Its delirium-causing properties were used in witchcraft, during the Middle Ages. In the 1960s and 1970s, it became an important drug of abuse, particularly by **teenagers** and young **adults**, seeking unpublicized hallucinogens, often resulting in acute toxic psychosis. Ingestion of ***Asthmador***®, a powdered mixture of belladonna and ***stramonium*** intended for burning and inhalation of its smoke for asthma, was a frequent cause of toxicosis. Serious cases of ***anticholinergic*** poisoning from belladonna contamination of herbal teas have been reported. Pharmaceutical industry **workers** must remember to use masks, goggles, and gloves when working with it.

LD_{LO}, **human**: 50 mg/kg.

BEMEGRIDE

= *Eukraton* = *Malysol* = *Megimide* = *Methetharimide* = *Mikedimide* = *NP 13*

***Analeptic.* Not a true barbiturate antagonist, and should no longer be used as such.**

Untoward effects: Excessive doses can be convulsive and cause of cardiac dysrhythmias, confusion, restlessness, twitching of facial and finger muscles, epileptiform seizures, hallucinations, vomiting, and psychoses. Has caused acute attacks of urticaria and porphyrias in **man**. 15/50 **patients** treated with it for ***barbiturate*** poisoning became delirious and had visual hallucinations. Black specks, smoke, fire, colored patterns, and auditory hallucinations have occurred 1–4 days after use and have persisted for 2–6 days.

LD_{50}, **rats**: 93.5 mg/kg; IP, **rats**: 28 mg/kg.

BENACTYZINE

= *Actozine* = *Amitacon* = *Amizil* = *Amizyl* = *Arcadine* = *AY5406-1* = *Cafron* = *Cedad* = *Cevanol* = *CP 1215* = *Diazyl* = *2-Diethylaminoethylbenzylate* = *Finaline* = *Fobex* = *Ibiotyzil* = *Lucidil* = *Nervacton* = *Neuroleptone* = *Nutinal* = *Parasan* = *Parason* = *Parpon* = *Phobex* = *Procalm* = *Suavitil* = *Tranquillin*

Anticholinergic, antidepressant.

Untoward effects: Large doses have ***atropine***-like effects, especially dryness of the mouth and difficulty with visual accommodation. In **man**, it causes giddiness, difficulty in concentration, and a tendency to overestimate the passage of time. Dryness of mouth, loss of concentration, and palpitations on 16 mg/day/2 months. Deliriant psychoses and optic hallucinations helped cause its absence from current therapeutic armamentariums. One report stated that all suicide attempts with up to 1300 mg failed. No toxicity noted in **man** on 90 mg as a single dose, nor on 15 mg, SC six times a day/6 days. One report of LD_{LO} in **man** of 14 mg/kg.

Dryness of mouth, slight ataxia, mydriasis, and tenseness noted on 15 mg/kg/day/5 weeks in **rats**, and 5 mg/day/4½ weeks in **dogs**.

LD_{50}, **rats** and **mice**: 160 mg/kg.

BENAZEPRIL

= *CGS-14,824A* = *Lotensin*

Angiotensin-converting enzyme inhibitor, antihypertensive.

Untoward effects: It and its active forms are excreted mainly in the urine; therefore, dosage adjustment must be made when renal function is impaired. Excessive hypotension can occur with diuretics. It can cause a dry, hacking cough, that may last for months after the drug is discontinued. Angioneurotic edema and asthma-like symptoms may occur. Erectile disorder and decreased libido have occurred.

BENDAZAC

= *AF 983* = *Bendazolic Acid* = *Bindazac* = *Versus* = *Zildasac*

Anti-inflammatory.

Untoward effects: Hepatotoxicity reported.

LD_{50}, **rats** and **mice**: approximately 1200 mg/kg. IV, LD_{50}, **rats**: 304 mg/kg; **mice**: 380 mg/kg.

BENDIOCARB

Carbamate insecticide, anticholinesterase.

Untoward effects: Use in relatively confined premises may cause headache and impaired memory and concentration in **people**. Toxic to **fish** and **wildlife**. Highly toxic to **bees**.

LD_{50}, **rat**: 179 mg/kg; dermal, 1 g/kg; LD_{50}, **bobwhite quail**: 21 mg/kg.

BENDROFLUMETHIAZIDE

= *Aprinox* = *Benuron* = *Bendrofluazide* = *Benzydroflumethiazide* = *Benzylhydroflumethiazide* = *Benzyl-Rodiuran* = *Berkozide* = *Bristuric* = *Bristuron* = *Centyl* = *Flumesil* = *Naturetin* = *Naturine* = *Neo-Naclex* = *Niagaril* = *Nikion* = *Orsile* = *Pluryle* = *Plusuril* = *Poliuron* = *Relan Beta* = *Salural* = *Salures* = *Sinesalin* = *Sodiuretic* = *Urlea*

Diuretic, antihypertensive.

Untoward effects: Hyperglycemia and decrease in serum and body ***potassium***. ***Potassium*** supplementation may correct the former. By inhibiting renal tubular secretion, serum urates increase. Increases serum ***cholesterol***. Weakness is another major complaint, and, less commonly, are anorexia, tingling sensation of hands and exacerbation of arthritis, thrombocytopenia, and skin rash. Sever acute pancreatitis in a third trimester pregnant 19-year-old **female** attributed to 5 days of therapy with the drug. May cause photosensitivity, headache, dizziness, tiredness, and paraesthesia in **people**. Bone marrow depression with thrombocytopenia, pancytopenia and purpura have been reported. Impotence in **men** also noted. Heinz body hemolytic anemia in a 4-day-old **infant** of a **mother** who took the drug 14 days before delivery. Suppresses lactation in **women**. A 14-year-old **female** ingested between 150 and 200 mg and developed grand mal convulsions. May cause toxic ***lithium*** levels when given with ***lithium carbonate***.

LD_{50}, **mouse**: 395 mg/kg.

BENFLURALIN

= *Balan* = *Balfin* = *Banafine* = *Benefex* = *Benefin* = *Bethrodine* = *Binnell* = *Blulan* = *Bonalan* = *Carpidor* = *EL-110* = *Quilan*

Herbicide.

Untoward effects: Poisoned **ducks** show ataxia, weakness, weight loss, and falling after 2–14 days. Regurgitation may occur within the first 2 h.

LD_{50}, female **rats**: > 10 g/kg; **mouse**: 5 mg/kg; **mallard ducks**: > 2 g/kg.

BENOMYL

= *Benlate* = *DuPont 1991* = *F 1991*

***Anthelmintic, fungicide.* A benzimidazole.**

Untoward effects: Low **mammalian** toxicity. Contact dermatitis reported in seven females after second exposure in 6 weeks, when it was used to spray flowers in a greenhouse. Ten other **workers** were unaffected. May irritate eyes, nose, and throat.

No overt signs of toxicity in **quail** at 5000 ppm in the diet or in **rats** with 2500 ppm for 2 years. Dietary levels of 0.5% to pregnant **rats** showed no teratogenic effects. Run-offs are toxic to **fish**.

LD_{50}, **rats**: > 9.5 g/kg; inhalation, 10 g/kg.

BENORYLATE

= *Benoral* = *Benortan* = *Fenasprate* = *Quinexin* = *Salipran* = *Win 11450*

Analgesic, antipyretic, anti-inflammatory.

Untoward effects: A component is an ***aspirin*** ester; it can cause intestinal blood loss and salicylism, which is especially dangerous in the **elderly**. Indigestion, nausea, heartburn, constipation, disorientation, diarrhea, headache, skin rashes, and severe tinnitus with hearing loss has occurred. Can produce severe bronchospasm and should be avoided in **asthmatics** (25% are intolerant to ***aspirin***), since ***aspirin*** is a main metabolite. Liver necrosis in a 13-year-old **female** given therapeutic doses of it with *D-**pencillamine***. Treatment of its toxicity is based on its two components (q.v.), ***aspirin*** and ***acetaminophen***.

BENOXAPROFEN

= *Compound 90459* = *Coxigon* = *Opren* = *Oraflex* = *Uniprofen*

Analgesic, anti-inflammatory.

Untoward effects: Marketing was discontinued because of an unusual number of deaths associated with its use in the U.K. and the U.S. Hepatotoxicity with cholestasis, necrosis, and jaundice occurred. Photosensitivity, hypertrichosis, koilonychia onycholysis, gastrointestinal discomfort including hemorrhage, rashes, milia, drowsiness, confusion, palpitations, dysuria, cold sensations, and neutropenia.

Has caused liver cancer in **mice**.

BENPERIDOL

= *Anquil* = *Benzperidol* = *Frénactil* = *Frenactyl* = *Glianimon* = *McN-JR 4584* = *R 4584*

Neuroleptic.

Untoward effects: Has caused parkinsonian and hyperkinetic–dystonic syndromes. Anorexia, apathy, confusion, increased salivation, sweating, lymphocytosis with neutropenia, dry mouth, diuresis, and sedation also reported. Similar toxicity to ***haloperidol's*** (q.v.).

LD_{50}, **mouse**: 1 g/kg; SC, 210 mg/kg; IV, 27 mg/kg; **rat**, SC, 218 mg/kg; IV, 27 mg/kg.

BENSERAZIDE

= *RO 4-4602*

Use: As a ***dopa decarboxylase-*** inhibitor with ***levodopa*** in treating parkinsonism.

Untoward effects: The combination is associated with nausea, vomiting, hypotension, dizziness, sweating, somnolence, cardiac arrhythmias, and lupus-like autoimmune syndrome. Most of these adverse effects are attributed to the ***levodopa***. Alone, it caused disturbed endochondral bone ossification in the **rat**, but not in **dogs** or **humans**.

BENSULIDE

= *Betasan*

Herbicide.

Untoward effects: Cholinesterase inhibitor that can cause vomiting, diarrhea, increased salivation, miosis, incoordination, muscle trembling, dyspnea, convulsions, paralysis, and even death in **animals** and **man**. Can be absorbed through the skin. Flush eyes and exposed areas with copious amounts of water. Remove contaminated clothing and shoes. ***Atropine*** is antidotal. ***Pralidoxine chloride*** may also be useful, if administered early.

TD_{LO}, **human**: 500 mg/kg.

Egg hatchability, but not fertility, markedly decreased in **Japanese quail** fed a diet with 1000 ppm for 5 weeks.

LD_{50}, **rat**: 770 mg/kg; dermal, 4 g/kg; dermal; **rabbit**: 2 g/kg; 96 h LC_{50}, **channel catfish**: 379 μg/l.

BENTONITE

= *Wilkinite*

Use: In oil-drilling muds, as a pH buffer in **cattle** feeds, as a pharmaceutical suspending agent, as a wine clarifier, and as a bacterial toxin and ***aflatoxin*** adsorber.

Untoward effects: The reported decreased absorption of ***rifampin*** by ***p-aminosalicylic acid (PAS)*** was due to the bentonite in the ***PAS*** granules. Interferes with the assays and/or absorption of ***neomycin***, ***thiabendazole***, ***nihydrazone***, ***norfloxacin***, and ***spironolactone***. Frequent apparent silicosis in **workmen** in bentonite processing plants. Silicosis proven by mineral analysis of lung biopsy in 67-year-old **male** with 10 years of exposure. Potentially hazardous dust for **man** and **animals**, because of its free crystalline silica content. Should not be applied to wounds without prior sterilization, because it can contain bacterial spores, including tetanus.

BENZALDEHYDE

Use: As a synthetic ***bitter almond*** oil in limited use in flavorings and perfumes today. It serves mostly as a raw material in manufacturing ***cinnamic aldehyde***, dyes and medical products, including ***ampicillin***, and as a **bee** repellent to help **beekeepers** collect ***honey***.

Untoward effects: Has caused contact urticaria in **man** and **guinea pigs**. Unlike the natural oil, it is free of ***hydrocyanic acid***. Vapors have a mild narcotic effect in **man**, and skin or eye contact should be avoided. Found leached into IV solutions contained in ***PVC*** bags and from rubber stoppers. Recommended workroom air limits based on potential irritation range from 3 to 33 ppm. Oxidizes readily in air to ***benzoic acid***. Marked decrease in its concentration when in formulations with ***aspartame***.

LD_{LO}, **human**: 500 mg/kg and 50–60 g is the usual fatal ingested dose.

LD_{50}, **rat**: 1.3 g/kg; **mouse**: 28 mg/kg; **guinea pig**: 1 g/kg.

BENZALKONIUM CHLORIDE

= *A-33* = *Barquat* = *Benirol* = *Benz-all* = *BTC* = *Capitol* = *Cequartyl* = *Drapolene* = *Drapolex* = *Enuclen* = *Germinol* = *Germitol* = *Hyamine 3500* = *Osvan* = *Padicide* = *Paralkan* = *Quat 50* = *Quatricide* = *Roccal* = *Rodalon* = *Serten* = *Zephiran Chloride* = *Zephirol*

Topical cationic antiseptic, deodorant, virucide.

Untoward effects*:** Generalized corneal drying from topical application in **humans** and **rabbits**. Has caused keratoconjunctivitis with endothelial damage after use during cataract surgery in **humans**. Hypersensitivity to this cationic detergent as a result of its adsorption on contact lens. Contact dermatitis of the trachea in 30-year-old **male** from its use a disinfectant on a tracheotomy tube. In one series of tests, 3/100 **persons** had severe local skin reactions to a weak solution; 1% solution caused hair loss and inflammation of the middle ear of **guinea pigs**. Skin erythema, pruritus, urticaria, edema, and rhinorrhea in 18-year-old **female**, after one application to abraded skin; facial ulcers and rash after one application (1:1000) as a pre-op preparation in an 18-year-old **male**. Patches of erythema with some axillary plaques in 45-year-old **male** after use in a deodorant. Poisoning with severe circumoral and pharyngeal burns in a **male** and **female** 2-week-old **twins** given an 11% solution instead of a 0.05% (1:50,000) solution. Death in a 22-year-old **female** after swallowing 300 ml ***Zephirol. 10–15% solutions, if ingested or inhaled, can corrode mucous membranes of the gastrointestinal tract and pulmonary tree. Intrauterine and IV use of 5–15 mg/kg was fatal in five **patients**, and oral ingestion of 100–400 mg/kg caused death in five **others** within 3 h. Ingested material can be neutralized by anionic non-detergent bar soaps. Inadvertent administration of a 3% solution as an enema to two women caused nausea, sweating, and death. Viscera were congested and cyanosed. Another **woman** died from respiratory paralysis soon after vaginal douching with it. A 47-year-old **male** with third degree burns on arm treated with wet dressings at 1:750 before grafts developed *Staphylococcus aureus* infection under the dressing and cultures of the stock bottle showed coagulase-negative *S. aureus*. Viable *Serratia marcescens* in its solution has occurred and its use as a skin disinfectant was suspect as the nosocomial cause of meningitis in a **patient**. *Pseudomonas* resistance is also known. Has failed to decontaminate fecally contaminated hands carying the dysentery organism, *Shigella sonnei*. Its inefficiency in killing *Alcaligenes faecalis* in hemodialysis equipment has led to bacteremia with chills and fever. **Physicians** and health care **workers** must remember that it can be an important source of nosocomial infections. Compatible with glass and ***polyethylene***, but not ***PVC*** containers. Incompatible with anionics.

LD_{LO}, **human**: 50 mg/kg; minimum lethal dose, skin, **human**: 150 μg/3 days.

Has caused vomiting, salivation, and diarrhea in **dogs**; 25 mg/kg/day/several weeks is lethal. **Mice** have been killed by topical application of one drop of a 13 or 50% solution. A young **kitten** became comatose after licking a baby's dusting powder off its fur containing it. Contact with and ingestion by **cats** caused anorexia, increased salivation, depression, dehydration, nasal and ocular discharge, and ulceration of the mouth and skin.

LD_{50}, **rats**: 234–447 mg/kg; IV, 31 mg/kg; IP, 100 mg/kg. LD_{50}, **rabbits** and **dogs**: 496 mg/kg; **guinea pigs**: 200 mg/kg; **mouse**: 340 mg/kg; **frog**: 30 mg/kg; **mallard duck**: > 2 g/kg.

1,2-BENZANTHRACENE

= *Benz[a]anthracene* = *Benzanthrene* = *2,3-Benzphenanthrene* = *Naphthanthracene* = *Tetraphene*

Untoward effects*:** **Humans** are exposed through inhalation or ingestion. Carcinogen in ***cigarette smoke and ***coal tars***, automobile exhaust fumes, soot and in emissions from gas and electric plants. Microgram quantities are found in ***charcoal*** broiled, barbecued, or smoked foods, some vegetables, and coffee. In **humans**, **hamsters**, and **rats**, higher concentrations were found in tracheal and bronchial mucosa than in lung.

Mutagenic in **mouse** trials, and, after repeated oral administration to young **mice**, leading to liver cancers and lung neoplasms. Carcinogenic on **mouse** skin. Causes recurrent aphthous stomatitis in **cattle**. 1 ppm killed 87/100 **bluegills**.

TD_{LO}, **mouse**, skin, 18 mg/kg; SC, 2 mg/kg.

BENZBROMARONE

= *Azubromaron* = *Besuric* = *Desuric* = *L 2214* = *Max-Uric* = *Minuric* = *MJ 10061* = *Narcaricin* = *Normurat* = *Uricovac* = *Urinorm*

Uricosuric.

***Untoward effects*:** An increase in frequency of gout attacks in nine **people** of three **females** and nine **males** (35–71 years), during the beginning of **their** treatment. Anorexia, nausea, and vomiting in one, requiring withdrawal from treatment, and a ureteral stone in one. In some **patients**, hepatitis reported within 7 months of starting treatment.

BENZENE

= *Benzol* = *Coal Tar Naphtha* = *Cyclohexatriene* = *Phenyl Hydride*

Over 2000 millions of gallons produced annually.

Use*:** In manufacturing ***ethylbenzene, ***styrenes***, ***cumene***, ***phenol***, ***cyclohexane***, ***nitrobenzene***, ***aniline***, rubber tires,

chlorobenzenes, linoleum, explosives, solvents, glues, antiknock premium gasolines, shoes, and in printing.

Untoward effects: Literature references to its carcinogenic effects actually refer to its inducing leukemias in **man** by its adverse effects on bone marrow. Rubber **workers** reportedly had a nine-fold increase in leukemia, after long-time inhalation exposure to at least 100 ppm. Prolonged exposure to 10 ppm (0.001%) and, possibly, 1 ppm (0.0001%) can cause myeloid leukemia. Inhalation by **workers** in the rubber industry, from **shoemaker's** glues, and by **painters** have caused serious bone marrow depression. Chronic inhalation can cause irreversible encephalopathy and anemia. Addiction to inhaling benzene-containing products has often caused deaths. Death occurred in a 48-year-old **male** after spending several hours in a very hot place removing paint. Benzene was found in a large amount of ***cocaine*** seized in Florida. Mass poisoning in a shoe factory in Finland with abnormal blood counts in 100. I found the common denominator that caused approximately 40 serious poisoning cases in Italy. These were 15–70 year-olds in southern and northern Italy, and **all** worked over open benzene vats during one of the hottest summers recorded. Exposure to benzene can have a direct toxic effect on red blood cells, causing their lysis. It reportedly caused a case of agranulocytosis after several months of contact with benzene. The **patient** died of pneumonia and nephritis, after 17 days. The gradual development of aplastic anemia has regularly been reported as an ill-effect, as is long-lasting chromosomal damage. Even in the year 2000, serious anemias and other illness due to benzene were common in the rapidly expanding Chinese shoe industry. It has a desiccating effect on skin, and has caused dermatitis in **users** of rubber cement, varnishes, and cleaning fluids. Paresthesia of the fingers and fissuring from contact has occurred. Anosmia is not uncommon. Can appear in **mother's** breast milk. Deaths from inhalation of ***gasoline*** fumes may be due to ventricular fibrillation from various fractions, including benzene (usually at a 1–4% level). Fatigue and irritability reported in 6/10 **workers** chronically exposed for many years to vapors of < 25 ppm, and accidental exposure for 3 months to 85–115 ppm. Hypersegmentation of neutrophils, increased corpuscular volume, and mild anemia were noted. 9/10 recovered in 4–8 months. Post-exposure breath analysis considered rapid diagnostic index of benzene exposure. Urinary ***phenol*** excretion can be used as a measure of benzene exposure (ingestion), but only of inhalation levels of 25 ppm and over. **Human** exposure to high airborne levels (3000 ppm or 0.3%) has caused acute narcotic effects (vertigo, coma) on the central nervous system, and, if the **victim** is not immediately (within 30–60 min) removed to fresh air and given emergency resuscitation, death may occur quickly from respiratory and/or cardiac failure. This usually occurs from working in confined spaces, like storage tanks or reactor vessels. Benzene poisoning in **children** is usually a delayed event. Three denture adhesives were found to contain benzene levels of 65–191 ppm and it is found in cigarette smoke (50–150 μg/cigarette). 1 ppm in the air has now become a realistic TWA for **man**. Environmental Protection Agency has set a maximum contaminant level of 0.0005 ppm (5 ppb) in drinking water for **humans**. An oral dose of 30 ml caused headache, giddiness, bluish discoloration of the face, delirium, convulsions, and death in a **person**. Fatal doses as low as 10 ml have been reported.

In summary, benzene exposure in **humans** is, primarily, by inhalation and ingestion (dermal absorption is slow), and has caused headache, nausea, vomiting, dizziness, rapid irregular pulse, ataxia, anemias (including aplastic and hemolytic), leukopenia, epistaxis, sensitizes the heart to ***catecholamines***, restlessness, excitement, convulsions, delirium, narcosis, temporary blindness, dermatitis, facial pallor, cyanosis of the extremities, coma, and death. It is metabolized in the liver and bone marrow and excreted via the lungs and urine of **man**. Many treatments have been suggested, such as gastric lavage and/or emetics (if conscious), blood transfusions, artificial respiration, ***oxygen***, ***antibiotics***, and supportive care and therapy. Use soap and water washing for dermal exposure. Avoid use of ***sympathomimetic amines***, as they may induce ventricular fibrillation. Necropsy findings are usually those of suffocation, edema of cerebral meninges and lungs, small hemorrhages in many organs, and a tendency for the blood to remain fluid for a long time. It becomes a classic illustration of FDA's vascillation, when nit-picking over residues and outlawing potential cancer-inducing chemicals. FDA Commissioner, James Benson, is quoted as saying, in regard to benzene contamination of U.S. supplies of Perrier® water – "If I had a bottle in the refrigerator, I would drink it." Three cheers for our protector.

Tremendous amounts (over 1 billion lbs) are released into the atmosphere annually.

TD_{LO}, **human**: 50 mg/kg; inhalation, 210 ppm.

Cattle drinking petroleum-contaminated water have had benzene found in their milk. **Cats** exposed to its vapors show long-term (about 2 months) adverse effects on electroencephalogram. Has caused Zymbal gland, skin, and mouth tumors in **rats**; lymphomas, tumors of the lung, ovaries, mammary, preputid, Zymbal and Harderian glands in **mice**. Chromosome damage in **rats** exposed to 1 ppm/6 h.

LD_{50}, **rats**: 3.8 g/kg; **mouse**: 4.7 g/kg; LD_{LO}, **dog**: 2 g/kg. LC_{50}, **goldfish**: 46 mg/l.

BENZENE HEXACHLORIDE

= *BHC* = *HCH* = *Hexachlorocyclohexane* = *Hexalon*

Chlorinated hydrocarbon insecticide.

Use: As a scabicide and pediculicide.

Untoward effects: A 39-year-old **female** exposed to 2% spray each week/2 years became easily fatigued, had spontaneous skin bruises, bone marrow hypoplasia, and developed fatal aplastic anemia. Other fatal cases of aplastic anemia from prolonged exposure to its vapors have been reported. Vomiting, drowsiness, dilated pupils, bradycardia, sinus arrhythmia, decreased tendon and absent abdominal reflexes, and grand mal convulsions in an 8-year-old **male** after eating 12 ***chocolate*** biscuits sprayed with a 4% solution in rubbish the night before. Being fat-soluble, it accumulates in the fat of **human** breast milk. In the 1970s, levels up to 0.125 µg/ml were reported. In pest-control **operators**, dermal exposure was greater than inhalation exposure. Apprehension, headache, disorientation, paresthesia, and tonic and clonic convulsions (often epileptiform) are the most common symptoms of poisoning. Others include excitability, dizziness, weakness, muscle twitching, tremor, coma, and leukemia. Pallor occurs even in moderate poisoning and cyanosis is due to convulsions interfering with respiration. Treatment includes establishing a clear airway (aspirate secretions, if necessary) and oxygenation with ***oxygen***, and even mechanically assisted pulmonary ventilation. ***Diazepam*** is often used to control convulsions. Emetics are used if the **victim** is alert and respiration is not depressed. This is often followed by oral administration of ***activated charcoal***. Give supportive therapy. Avoid ***adrenergic amines***, as they aggravate BHC's myocardial irritability. The gamma isomer (***lindane***, q.v.) is its major toxic component.

LD_{LO}, **human**: 50 mg/kg.

Poisoned **birds** show signs of polydipsia, regurgitation, hyperexcitability, ataxia, ptosis, fluffed feathers, hyporeactivity, imbalance, slowness, stumbling, phonation, tenseness, shakiness, jitteriness, sitting, ataraxia, withdrawal, tremors, masseter tenseness, spasms, aggressiveness, fear–threat displays, backing, circling, asthenia, tongue protruding sideways from the bill (**mallards**), and immobility. Prolonged signs include falling, sitting, using wings for pedestrian locomotion, ataraxia, and withdrawal. Signs appeared as soon as 30 min in **mallards** and 2 h in **pheasants**, and mortalities usually occurred between 2 and 5 days in **mallards** and between 4 and 9 days after treatment in **pheasants**. Remission took up to 20 days. Most of the symptoms noted in **man** also occur in poisoned **cattle** and **sheep**.

LD_{LO}, **guinea pig**: 1.4 g/kg; **chicken**: 700 mg/kg. LD_{50}, **mallard ducks**: 1.4 g/kg; **pheasants** and **quail**: 118 mg/kg; LC_{50}, **trout**: 15–20 µg/l. LD_{50}, **rat**: 100 mg/kg; dermal, **rat**: 900 mg/kg.

BENZENESULFONYL CHLORIDE

Use: In organic synthesis and in the manufacture of illicit ***amphetamines***.

Untoward effects: LD_{50}, **rats**: 2 g/kg.

BENZETHONIUM CHLORIDE

= *Banagerm* = *Formula 144* = *Hyamine 1622* = *Inactisol* = *Phemeride* = *Phemerol* = *Phemithyn* = *Q.A.C.* = *Quatrachlor* = *Sanizol* = *Solamin*

Topical cationic antiseptic.

Untoward effects: Allergic penile and scrotal dermatitis in two **men** after sexual intercourse with **women** who had used it as part of a vaginal spray deodorant. 1:1000–1:10,000 dilutions (0.1–0.01%) are used. Occasionally up to 0.2% is used in powders. 1% solutions can damage mucous membranes, and ingestion can cause corrosive damage to the esophageal mucosa, vomiting, depression, collapse, and coma. Depolarizing muscle relaxant effects, dyspnea and cyanosis, due to paralysis of the respiratory muscles, can occur after ingestion.

LD_{LO}, **human**: 50 mg/kg.

LD_{50}, **rats**: 368–420 mg/kg; **mouse**: 338 mg/kg.

BENZIDINE

An aromatic amine. Millions of pounds once produced annually in the U.S.

Use: In dye manufacturing, leather and rubber industries, and by chemists.

Untoward effects: Although once used extensively as a test for occult blood, it is seldom used today in U.S. forensic and clinical laboratories. Anemia, hemolysis, cystitis, hepatitis, and painful urination after ingestion, inhalation, and skin contact. Can cause glycosuria, hemolysis and paralysis in heavy exposures. Plant **workers** exposed to it excreted it in **their** urine, along witih several metabolites. Has caused malignant bladder tumors in **man** up to 20 years after exposure. It was first recognized as a possible carcinogen in 1934. Occupational hazards (carcinogenesis) have been declared for a number of its dyes. It causes bladder tumors (papillomas) in **man**, **monkeys**, and **dogs**; liver tumors in **hamsters** and **mice**; acoustic neuromas, Zymbal, mammary, colon, and liver tumors in **rats**. If exposed to spills or sprays, **people** should use eye-wash fountains, soap and water showers, and remove contaminated clothing. **Workers** should **NOT** wear contact lenses.

Benzidine dyes are no longer sold for home use in the U.S., but many are still on shelves, and, although unpalatable, **children** have ingested them.

Inhalation TC_{LO}, **man**: 18 mg/m^3.

Acute oral administration of benzidine to **dogs**, **rabbits**, **rats**, and **mice** leads to weight loss, cloudy swelling and cirrhosis of the liver, degeneration of renal tubules, and hyperplasia of thymic and splenic myeloid elements and lymphoid cells.

LD_{50}, **rat**: 309 mg/kg; **mouse**: 214 mg/kg; LD_{LO}, **dog** and **rabbit**: 200 mg/kg.

BENZIODARONE

= *Algocor* = *Amplivix* = *Cardivix* = *Dilafurane* = *DilaVasal* = *L 2329* = *2329 Labaz* = *Retrangor*

***Coronary vasodilator, uricosuric.* 49% of its molecular weight is *iodine*.**

Untoward effects: Drug was withdrawn from the U.S. market, despite widespread use in Europe, because it had adverse effects on the liver, including jaundice. Hypothyroidism and hyperthyroidism, renal colic, nausea, diarrhea, and mild allergic reactions reported. Use required decreased dosage of oral ***anticoagulants*** (50% with ***warfarin***, 40% with ***diphenadione***, 25% with ***nicoumalone***, and 20% with ***ethylbiscoumacetate***) in **people**.

BENZNIDAZOLE

= *Radanil* = *RO 7–1051* = *Rochagan*

***Antitrypanocide.* A nitroimidazole.**

Untoward effects: Frequently causes an allergic rash, gastrointestinal upsets, dose-dependent polyneuropathy, and psychic disturbances.

BENZOIC ACID

= *Benzenecarboxylic Acid* = *Dracylic Acid* = *Phenylformic Acid*

***Antifungal.* U.S. demand is for approximately 250 million lbs/year.**

Use: In the manufacture of dyes, perfumes, pharmaceuticals, and benzoates. Antiseptic, antifungal preservative in foods, fats, and cosmetics.

Untoward effects: Irritation to skin and mucous membranes. May cause severe eye irritation and contact urticaria. Eliminated in urine as ***hippuric acid*** with its characteristic odor. Percutaneous absorption occurs in **humans** and **rhesus monkeys**. Ingestion with ***tartrazine***-containing substances may decrease air flow rates in **people**. Hypersensitivity to it in many ***aspirin***-sensitive **humans** and **asthmatics**. **People** are occasionally allergic to its esters, such as ***procaine*** and ***benzocaine***.

LD_{LO}, **human**: 500 mg/kg; dermal, 6 mg/kg.

The teratogenic effect of ***aspirin*** in **rodents** is enhanced by it. Causes ataxia, anorexia, salivation, hyperesthesia, muscle fasciculations, convulsions, and depression in **cats**. Poisoning noted in **cats** eating diets containing 0.2% or more or receiving 450 mg/kg. Acute lethal dose, **dogs**, **rats**, and **rabbits**: 2 g/kg; **cat**: 800 mg/kg.

LD_{50}, **rats**: 2.57 g/kg; inhalation LC_{50}, **rats**: > 12.2 mg/l/4 h; LD_{50}, **dog**: 2 g/kg.

BENZOIN GUM

= *Gum Benjamin* = *Resin Benzoin*

Use: In soaps, cosmetic creams, lotions, detergents, and perfumes; topically, to aid wound healing, inhalant for bronchitis, and orally as an expectorant.

Untoward effects: Known skin sensitizer causing a contact dermatitis (including exfoliative). Can provoke asthmatic attacks in **patients** with vasomotor-type asthma and/or rhinitis. A flash fire occurred on the abdomen of a 42-year-old **female** after compound tincture of benzoin was used and an electrocautery unit was used. A severe allergic reaction from the compound tincture noted in a **patient** undergoing rhinoplasty. Edema, itching of face, ecchymosis, and discoloration of submandibular glands occurred. Essential oils used in dentistry and ***Balsam Peru*** may cross-react with it.

LD_{LO}, **human**: 500 mg/kg.

LD_{50}, **rat**: 10 g/kg; dermal LD_{50}, **rabbit**: 9 g/kg.

BENZONATATE

= *Exangit* = *KM 56* = *Tessalon* = *Ventussin*

***Antitussive.* Chemically related to *tetracaine*.**

Untoward effects: Has caused headache, heartburn, drowsiness, nausea, nasal congestion, numbness in chest, sensation of burning in **one's** eyes, malaise, asthenia, pruritus, and mild erythema in **people**. Used in two fatalities, one a suicide, the other an accidental overdose.

BENZONITRILE

= *Cyanobenzene* = *Phenyl Cyanide*

Solvent.

Untoward effects: LD_{LO}, **rat**: 720 mg/kg; inhalation LC_{50}, male **rats**: 900 ppm/1 h.

BENZO(*g,h,i*)PERYLENE

***Polycyclic aromatic hydrocarbon.* Found in *cigarette* smoke.**

Untoward effects: Topical LD_{50}, **mice**: 0.8 mg/kg.

BENZOPHENONES

= *Uvasorbs* = *Uvinals*

Ultraviolet light (UVA and UVB) absorbers, flavoring, rose-like perfumery.

Use: For the above and in soaps, cosmetic creams, lotions, detergents, and perfumes.

Untoward effects: Has caused contact urticaria and photoallergic reactions in some **people**. 6% in petrolatum led to no reactions in 25 **human** volunteers.

Oral administration to **rats** and **mice** has caused hepatocellular hypertrophy; cholestatic and renal damage in **rats**. Application to **rabbit** ears, but not **mouse** skin, caused proliferative, benign, and malignant ear tumors.

LD_{50}, **rat**: > 10 g/kg; dermal LD_{50}, **rabbit**: 3.5 g/kg.

BENZO[α]PYRENE

= *3,4-Benzopyrene* = *BP*

Procarcinogen. A polycyclic aromatic hydrocarbon.

Use: As a research chemical, to produce experimental tumors.

Untoward effects: A product of combustion, with about 2 million lbs released into the atmosphere annually. In ***cigarette*** smoke at up to 17 μg/100 ***cigarettes***; in foods at 0.1–50 ppb; and from char-broiling meats (> 2 mg/lb), grilling bacon (129 μg/kg), etc. Readily activated by body enzymes, such as aryl hydrocarbon hydroxylase, to form carcinogenic epoxides and mutagenic metabolites. Pregnant **women** who smoke have increased concentrations of the enzyme in **their** placentas. **Human** exposure is highest in **those** working with ***coal tar pitch*** (paving operations), at incinerators, power plants, ***coke*** manufacturing, in the vicinity of volcanoes, and in research. In **man**, it has caused "***tar*** smarts", with pricking sensations, erythema, and melanosis. Has caused squamous cell carcinomas of **male** genitals, hyperplasia of bronchial epithelium, and leukemia. Occupational exposure to BP-containing substances has been associated with increased incidence of skin, respiratory tract, and upper gastrointestinal tumors. An inducer of microsomal enzyme activity that can accelerate the oxidation of some administered drugs in **fish**, **mice**, **man**, and, probably, other species.

Inhalation TC_{LO}, **human**: 70 ng/m^3.

Secreted in the milk of **rats**, less so in **rabbits**, and extremely low transfer in **sheep**. Enhances ***glucuronide*** formation in **rats**. Oral, dermal, and IP use has induced tumors in **rats**, **mice**, **guinea pigs**, **hamsters**, **rabbits**, **ducks**, and **monkeys**. SC use leads to sarcomas in **rats**, **mice**, **guinea pigs**, and **tree shrews**. Embryotoxic in **mallard ducks**. Placental transfer into **mouse** embryos occurs. Enhances the hepatotoxicity of ***carbon tetrachloride*** in **mice**. IP and oral use in **mice** increases the number of lung tumors; oral use increases gastric tumors; and topical use leads to skin cancers in them. Topical use of a non-tumorigenic dose on **mouse** skin, followed a year later by a non-tumorigenic dose of topical ***croton oil***, caused skin tumors. Pretreatment with it decreases the hypothrombinemia effect of ***warfarin***. TWA, 0.2mg/m^3/8 h.

TD_{LO}, **pregnant rat**: 1 g/kg; LD_{50}, **rats** and **mice**: 50 mg/kg.

BENZOTAMINE

= *BA 30,843* = *Tacitin*

Anxiolytic, skeletal muscle relaxant.

Untoward effects: Epigastic discomfort, nausea, dizziness, muscular weakness, blurred vision, hypotension, increased SGOT and SGPT, nervousness and aggression, nightmares, and hangover feelings have been reported in some **patients**. Drowsiness, sleepiness, and dry mouth were the most common complaints.

LD_{50}, **rat**: 600 mg/kg; **mouse**: 280 mg/kg; **rabbit**: 800 mg/kg.

BENZOTRICHLORIDE

= *Benzenyl Trichloride* = *Phenylchloroform* = *Toluene Trichloride*

Use: Chemical intermediate in dye and pigment production, pharmaceutical and antimicrobial manufacturing. About 40 million lbs used in the U.S. annually.

Untoward effects: Highly irritating to eyes, mucous membranes, and skin, and is a potential carcinogen for **man**. **Human** exposure from its manufacture and use as an intermediate.

Intragastric administration to **mice** caused stomach and lung cancers in them. Topically on **mice**, it caused skin, lung, lip, and tongue tumors and lymph system neoplasms.

LD_{50}, **rats**: 6 g/kg; inhalation LC_{LO}, 125 ppm/4 h.

BENZOYL CHLORIDE

= *Benzenecarbonyl Chloride*

Use: In the manufacture of ***benzoyl peroxide*** and dye intermediates.

Untoward effects: Eye, skin, and mucous membrane irritant in **man**. Strong lacrimator. Periodic chest X-rays and pulmonary function testing required of **workers** with it.

TWA, 5 mg/m^3 (1 ppm)/8 h. TC_{LO}, **human**: 2 ppm/1 min.

BENZOYLPAS

= *Be-Calcipas* = *Benzacyl* = *Benzapas* = *B-Paracipan* = *Therapas*

Tuberculostatic.

Untoward effects: May cause goiter, with or without myxedema, in **man**. Hypersensitivity can be manifested as fever, skin eruptions, infections, mononeucleosis-like syndrome, leukopenia, agranulocytosis, thrombocytopenia, hemolytic anemia, jaundice, encephalopathy, vasculitis, and Löffler's syndrome.

BENZOYL PEROXIDE

= *Acetoxyl* = *Acnegel* = *Benoxyl* = *Benzagel 10* = *Benz-aken* = *BZW* = *Debroxide* = *Desanden* = *Loroxide* = *Lucidol* = *Nericur* = *Oxy-5* = *Oxydex* = *Oxy-L* = *PanOxyl* = *Peroxydex* = *Persadox* = *Persa-gel* = *Sanoxit* = *Theraderm* = *Vanoxide* = *Xerac BP 5* = *Xerac BP 10*

Use: As a keratolytic, oxidizing agent, catalyst in plastic manufacturing, in bread baking, bleaching agent for flour, oils, fats, waxes, and milk in cheese production.

Untoward effects: Potent contact allergen. Photosensitization, erythema, and skin peeling reported in **man**. Has caused dermatitis in **bakers**. Fumes can cause respiratory, eye, and skin irritation in **man**. A fish-odor syndrome (trimethylaminuria), noticed only by others, after topical use has occurred.

TWA, 5 mg/m^3 (1 ppm)/8 h; TC_{LO}, **human**: 12 mg/m^3; LD_{LO}, **human**: 500 mg/kg.

Topically or orally induced skin tumors, respectively, in **rats** and **mice**.

BENZPHETAMINE

= *Didrex* = *Inapetyl*

Anorexic.

Untoward effects: Insomnia, dry mouth, and indigestion in **people**. Physical dependence may occur. In severe poisoning, bradycardia, decreased blood pressure, cardiac arrhythmias, vomiting, hematuria, convulsions, deep coma with pyramidal reflexes, extreme hypotension, and cardiac failure can occur. A depressed 16-year-old **male** took an overdose in an apparent suicide attempt during the night. When found dead in the morning, bloody fluid came from the nose and mouth. Autopsy revealed a red and inflamed trachea, generalized visceral congestion, 10 cc of blood subdurally, and gastric, duodenal, and jejunal petechiae.

IP, LD_{50}, **mice**: 102–123 mg/kg.

BENZQUINAMIDE

= *BZQ* = *Emete-Con* = *Emeticon* = *NSC 64375* = *Promecon* = *Quantril*

Antipsychotic, antiemetic.

Untoward effects: Occasional hypertension. Exercise caution in giving it to **people** who have experienced extrapyramidal reactions to ***phenothiazines***, since it may do the same. This reaction was reversed by ***diphenhydramine*** in a 28-year-old **male**. Somnolence, nasal congestion, nausea, vomiting, indigestion, tremor, mask-like facies, flushing, salivation, transitory premature ventricular contractions, weakness, fatigue, anxiety, agitation, excitement, dystonic reactions, leukopenia, hypotension and hypertension, sweating, grand mal seizures, and dizziness are among the many adverse effects occasionally reported. Cerebral thrombosis and aphasia reported in a **patient**. Delirium induced in an 18-year-old **female** was reversed by 1 mg ***physostigmine***, IV.

LD_{50}, **rats**: 990 mg/kg; **mouse**: 580 mg/kg; IP, **mice**: 376 mg/kg.

BENZTHIAZIDE

= *Aquatag* = *Benzothiazide* = *Dihydrex* = *Diucen* = *Edemex* = *ExNa* = *Exosalt* = *Fovane* = *Freeuril* = *Hy-Drine* = *Lemazide* = *Naclex* = *Proaqua* = *Urese*

Diuretic, antihypertensive.

Untoward effects: Has caused photosensitivity reactions in **man**. Diabetogenic. Crosses the **human** placenta and can cause thrombocytopenia in the latter part of pregnancy. Decreases serum ***sodium***. May occasionally cause skin rashes, anorexia, nausea, vomiting, diarrhea or constipation, epigastric pain, headache, dizziness, salivary gland inflammation, paresthesias, agranulocytosis, aplastic anemia, leukopenia, and jaundice. Hyperglycemia and glycosuria in **diabetics** and susceptible **people**. Prolonged use may cause hypokalemia, intensifying ***digitalis'*** effects. Orthostatic hypotension may be aggravated by ***alcohol***, ***barbiturates***, and ***narcotics***. Approximately a 50% decrease of dosage of ***hydralazine*** and ***ganglionic blockers*** is required when it is used.

LD_{50}, **dogs, rats, mice**: > 5 g/kg; IV, **dogs**: 200 mg/kg; IV, **mice**: 410 mg/kg.

BENZTROPINE MESYLATE

= *Cobrentin Methanesulfonate* = *Cogentin* = *Cogentinol*

Antiparkinsonian, anticholinergic.

Untoward effects*:** Has caused the anticholinergic intoxication syndrome (including dilated pupils, flushed skin, tachycardia, hallucinations, and disorientation) in **man**. Deliberate abuse for the euphoric effect of high dosage has been reported. Can cause decreased libido and impotence. Increased intraocular pressure and glaucoma reported from its use. A toxic psychosis occurs on high dosage. Dry mouth, blurred vision, dilated pupils, nervousness, tachycardia, nausea, and drowsiness reported. Akathesia, confusion, memory impairment, numbness of fingers, constipation, dysuria, urinary retention, and decreased gastric secretions also occur. Delirium and disorientation from overdosage. Occasional allergic reaction, manifesting as a skin rash. May cause tremor and profuse sweating, when given with ***monoamine oxidase inhibitors. Case of paralytic ileus reported when given with ***mesoridazine***. Prescribing error with double-dosing caused hyperpyrexia (108°F) and coma in 35-year-old **male**.

A hyperthermia was reported in **rabbits** also receiving ***monoamine oxidase inhibitors***.

BENZYDAMINE

= *Afloben* = *Andolex* = *Benaalgin* = *Benzindamine* = *Benzyrin* = *C1523* = *Difflam* = *Dorinamin* = *Enzamin* = *Epirotin* = *Imotryl* = *Indolin* = *Ririlim* = *Riripen* = *Salyzoron* = *Tamas* = *Tantum* = *Verax*

Analgesic, antipyretic, anti-inflammatory.

Untoward effects*:** Heartburn and vomiting (disappeared when administered rectally); heartburn in **others**, feeling of abdominal fullness, urobilinuria, pruritus, nervous tension, and insomnia reported. Symptoms appear to be aggravated by co-administration with ***tetracycline. Renal disease precipitated in a 57-year-old **female** after using 400 g of a 3% cream over a 4 month period for musculoskeletal pain. Restlessness, choreatic movements, increased muscle tone, mydriasis, somnolence, and pallor in 2½-year-old **female** about 2 h after ingesting 900 mg.

High dosage (5–10 mg/kg) in **rabbits**, **mice**, and **rats** leads to convulsions and prostration.

LD_{50}, **rats**: 1 g/kg; **mice**: 515 mg/kg; IP, **rats** and **mice**: 100 mg/kg.

BENZYL ACETATE

***Use*:** As a flavor ingredient in chewing gum, puddings, candy, baked goods, and soft drinks. Also in laundry detergents, soaps, and incense fragrances. Approximately 500 ± tons/year used in fragrances in the U.S.

***Untoward effects*:** Vapors are irritating to eyes and respiratory tract of **man**, triggering coughing. Gastrointestinal irritant with nausea and vomiting. Irritant and narcotic at 15 ppm.

TC_{LO}, **human**: 50 ppm.

Has been narcotic and lethal to **animals**. Mildly irritating to the skin of **rabbits**. Vapors of its esters have caused hyperemia and pulmonary edema in **mice**.

LD_{50}, **rats**: 2.49–3.69 g/kg; dermal LD_{50}, **rabbit**: > 5 g/kg.

BENZYL BENZOATE

***Use*:** As a scabicide, pediculicide, mosquito, tick, and mite repellent (Vietnam War), as a solvent for synthetic musks, and to decrease volatility for flavors used in jellies, jams, and puddings. About 250 ± tons used annually in the U.S. In fragrances for soaps, cosmetic creams, lotions, detergents, and perfumes.

Untoward effects*:** Pyoderma and burning sensations, especially in **children**. A 51-year-old **male** with scabies, applying the solution twice daily to **his** body, became flushed and had a swift downward-spreading burning facial erythema on the third day. Sudden headache, tearing, rhinorrhea, sweating, and malaise also occurred. The **patient** recovered in a few hours. Convulsions can occur. A ***disulfiram-like reaction probably occurred, as the **patient** had red ***wine*** with a meal before the symptoms occurred. Probably wise to suggest abstinence from ***alcohol*** for a couple of days after the last treatment. Metabolized to ***benzoic acid*** (q.v.).

Application to too large an area of an **animal's** skin, or too frequent use, has caused nausea, vomiting, diarrhea, decreased respiratory rate and heart function, as well as death of **dogs**. Contraindicated on **cats**, where it causes serious nervous excitability.

LD_{50}, **rats**: 2.8 g/kg; **rabbits**: 1.68 g/kg; **cats**: 2.24 g/kg; **dogs**: > 22 g/kg; dermal LD_{50}, **rabbits**: 4 ml/kg.

BENZYL CHLORIDE

= *(Chloromethyl)benzene* = *α-Chlorotoluene*

About 75 million lbs used annually.

Use*:** As a plasticizer in vinyl flooring, in manufacturing pharmaceuticals, ***aspartame, perfumes, dyes, and as a chemical intermediate.

***Untoward effects*:** Very irritating to eyes, mucous membranes, and skin. **Workers** with long-term exposure have increases in respiratory illness, dermatitis, headaches, abnormal liver function, and decreased white blood cells, as well as pulmonary edema and increase in lung cancer. High concentration leads to central nervous system depression. TWA, 1 ppm/8 h; 2 ppm is uncomfortable, but tolerable; 17 ppm leads to a 50% decrease in respiratory rate. Chest X-rays and pulmonary tests are required for **those** working with it.

Inhalation TC_{LO}, **human**: 16 ppm/1 min.

LD_{50}, male **rats**: 1.23 g/kg; SC injection in **rats** caused local sarcomas. Inhalation LC_{50}, **rats**: 150 ppm/2 h; LD_{50}, **mouse**: 1.6 g/kg; inhalation LC_{LO}, 80 ppm/2 h.

BENZYL FORMATE

Use: Solvent for cellulose esters, in soaps, cosmetic creams and lotions, detergents, and perfumes.

Untoward effects: LD_{50}, **rats**: 1.7 ml/kg; dermal LD_{50}, **rabbits**: 2 ml/kg.

BENZYL ISOBUTYRATE

= *Benzyl-2-methyl Propionate*

Use: In fragrances for soap, cosmetic creams and lotions, detergents, and perfumes. Permitted in foods.

Untoward effects: 4% in petrolatum caused no irritation in a 48 h closed-patch test on 25 **human** volunteers.

LD_{50}, **rats**: 2.85 g/kg; dermal LD_{50}, **rabbits**: > 5 ml/kg.

BENZYL ISOEUGENOL

= *Isoeugenol Ether*

Use: In fragrances for soap, cosmetic creams and lotions, detergents, and perfumes. Permitted in foods.

Untoward effects: 5% in petrolatum caused no irritation in a 48 h closed-patch test on 25 **humans**.

Only mildly irritating at full-strength to intact or abraded **rabbit** skin.

LD_{50}, **rats**: 4.9 g/kg; dermal LD_{50}, **rabbits**: > 3 g/kg.

BENZYL ISOVALERATE

Use: In fragrances for soap, cosmetic creams and lotions, detergents, and perfumes. Permitted in foods.

Untoward effects: 4% in petrolatum caused no irritation in a 48 h closed-patch test on 25 **humans**.

Slightly irritating at full-strength to intact or abraded **rabbit** skin.

LD_{50}, **rats**: > 5 g/kg; dermal LD_{50}, **rabbits**: > 5 g/kg.

BENZYL NICOTINATE

= *Nicotinic Acid Benzyl Ester*

Untoward effects: Percutaneous absorption occurs in **men** and **mice**. Topical use causes hyperemia in **people**, with or without edema, within 5–10 min. **Customers** at a hairdressing salon developed urticaria-like lesions of the skin after **their** hair was done by the owner's daughter, who, annoyed by her father's reaction to her pregnancy, applied it to his **customers'** skin. **She** found from experience that absorption from the palms was poor, and **she** could easily vent her anger on his **clients**, without **herself** reacting.

BENZYL PHENYLACETATE

= *Benzyl* α*-toluate*

Use: In fragrances for soaps, cosmetic creams and lotions, detergents, and perfumes. Permitted in foods.

Untoward effects: 2% in petrolatum caused no irritation in a 48 h closed-patch test on 25 **humans**.

LD_{50}, **rats**: > 5 g/kg; dermal LD_{50}, **rabbit**: > 10 ml/kg.

BENZYL PROPIONATE

= *Benzyl Propanoate*

Use: In fragrances for soaps, cosmetic creams and lotions, detergents, and perfumes. Permitted in foods.

Untoward effects: 4% in petrolatum caused no irritation in a 48 h closed-patch test on 25 **humans**.

LD_{50}, **rat**: 3.3 g/kg; dermal LD_{50}, **rabbit**: > 5 g/kg.

BENZYL SALICYLATE

Use: In perfumery and sunscreens, and as a plasticizer in making cellulose derivatives. In soaps, cosmetic creams and lotions, and detergents. Permitted in foods.

Untoward effects: Hypersensitivity to it can cause skin to blister and to increase pigmentation. Has caused severe pruritus and moderate to marked erythema in 6/15, when in a ***trioxalen*** lotion. Skin reactions to it are enhanced by ***methoxsalen***. Probably one of the causes of dermatitis in **patients** using ***Balsam Peru***.

LD_{50}, **rat**: 22.3 g/kg; dermal LD_{50}, **rabbit**: 14 g/kg.

BEPHENIUM

Anthelmintic. *The Embonate = Naftamen; the Hydroxynaphthoate = Alcopar = Befeniol = Locibis = Naphthamone = Nemex; the Pamoate = Frantin*

Untoward effects: Vomiting, nausea, headache, vertigo, abdominal pain, and weakness, especially in **children** under 15 years. Palpitations reported in one (15 years) with the hydroxynaphthoate. It is estimated that a dermatitis occurs from it in about 8% of **people** sensitive to perfumes.

Salivation, vomiting, and hepatotoxicity in **dogs**; the pamoate associated with slight increase in temperature, anorexia, and rumen hypotonia in **calves**.

BEPRIDIL

= *Angopril = Bepadin = Cardium = Vascor*

Antianginal, calcium channel-blocker. **A dihydropyridine.**

Untoward effects: Because of these, its use is reserved for **patients** who fail to respond to other therapy: dyspnea (12%), nausea (18%), dizziness (22%), headache, tremor (5%), asthenia (14%), diarrhea (14%), nervousness (8%), occasional agranulocytosis, QT, QTc, and RR interval prolongation, and ventricular arrhythmias (torsades de pointes – in about 1%, mostly in **women**). ***Ritonavir*** decreases its metabolism, increases its toxicity with serum levels of 0.05–0.1 μg/ml.

LD_{50}, **mice**: 2 g/kg; IV, 23.5 mg/kg.

BEQUAERTIODENDRON MAGALIESMONTANUM

Untoward effects: The bark extract can cause death in **people** within 24 h.

BERBERINE

= *Umbellatine*

Coloring agent, stimulant to healing, antifungal, antidiarrheal. **An alkaloid found in many unrelated plants.**

Untoward effects: Vomiting in three of 100 **children** 2 months–6 years. May interfere with ***morphine*** assays.

Slowed growth when fed to young **rabbits** for 4 months, and caused synostosis at epiphyses, osteoporosis, and degenerative changes in articular cartilage. Causes hypoglycemia in **mice**.

BERGAMOT OIL

Use: To mask undesirable odors and as a bacteriostat and fungistat in cosmetics, particularly in citrus colognes. Also used in foods and ***tobacco***, for its refreshing fruit notes or sweet freshness. Over 40,000 kg imported into the U.S. annually. A "tan without burn" compound. In the treatment of vitiligo.

Untoward effects: Photosensitizer. Contains ***5-methoxypsoralen***. The photosensitive agent is ***beregaptine*** (***bergapten***), a furocoumarin (as are ***psoralens***). Some oil is available where this agent has been removed. Berlock dermatitis (contact dermatitis) with erythema and hyperpigmentation reported. Can cross-react with orange peel, since it is a citrus oil. In the 1970s, was an ingredient in Chanel No 5®.

May produce skin cancers in **mice** exposed to ultraviolet A radiation.

LD_{50}, **rat**: > 10 g/kg; dermal **rabbit**: > 20 g/kg.

BERSAMIA sp.

African shrubs or trees.

Untoward effects: Ingestion of *B. abyssinia* leaves has killed **calves** and **cows**. They had gastrointestinal irritation, convulsions, and death from heart failure. In **cattle**, the IV lethal dose of a purified extract was 0.045 mg/kg. SC injection of the extract causes degenerative changes in the kidneys and liver of **rabbits**. In **cattle**, **rabbits**, and **mice**, death usually occurs within 24 h. The symptoms comprise limb weakness, paralysis, hyperesthesia, decreased temperature, irregular shallow respirations, trismus, mystagmus, opisthotonis, convulsions, and death. The leaves of *B. swynnertoni* have killed **sheep**.

BERYLLIUM

= *Glucinium*

Use: In manufacturing rockets, ceramics, appliance circuitry, neon lights, nuclear weapons and reactors, and aerospace guidance systems.

Untoward effects: Toxic to all living cells, causing a delayed skin hypersensitivity, skin ulcers, and conjunctivitis. Respiratory problems with granulomatous and fibrotic lung lesions. Thickened alveolar walls, progressive loss of respiratory function with overburdened right heart function, and death from cor pulmonare after inhalation of its dusts. The dust is easy for **workers** to carry home on **their** clothing. A metallic taste, coughing, blood-stained sputum, anorexia, and weight decrease occur in chronic cases. The average **patient** had a life expectancy of 8 years after diagnosis of berylliosis. Highly carcinogenic and a short-term poison in high concentrations. Use in second stage, but not first stage, rockets is to protect populated areas from possible exposure. Can also delay healing of open wounds and cause liver necrosis. Chronic berylliosis reported in 34 **females** and 26 **males** (median ages 30–35) in half of them after only 1–2 years, and often high intensity exposure. Main symptoms in a 33-year-old **male** were weight loss, fatigue, and muscle pain. A 35-year-old **male**, after 3 years exposure, had bronchitis, dyspnea, and lung crepitation. Many **people** living up to a half mile away from the source have been disabled by exposure to its toxic compounds. Nephrotoxic (interstitial nephritis) and causes secondary hyperuricemia, hypercalcemia, and urinary tract stones. Can be absorbed through burned, abraded, and skin wounds and is cumulative in bone. Periodic chest X-rays and pulmonary function testing required for **workers**. TWA, 2 $\mu g/m^3$/8 h; 25 $\mu g/m^3$/30 min. For eye exposure, use eye-wash fountain. Soap and water showering for skin exposure.

Osteomalacia in experimental **animals**, osteosarcoma in **rabbits** and **guinea pigs** after inhalation. **Guinea pigs** and **rats** have not shown lung changes from its inhalation. **Beryllium oxide** and **beryllium sulfate** have caused lung tumors in **monkeys** and **rats** after inhalation.

Chronic ulcers and granulomata by lymphatic spread in a 48-year-old **male**, who cut **his** finger on a silicon carbide grinding wheel contaminated with the oxide. The sulfate has caused contact irritation in **man** with a papulovesicular dermatitis.

BETAHISTINE

= *Acquamen* = *Betaserc* = *Medan* = *Melopat* = *Menitazine* = *Merislon* = *PT-9* = *Remark* = *Ribrain* = *Serc* = *Suzutolon* = *Tenyl* = *Vasomotal*

Vasodilator.

Untoward effects: Drowsiness, sleepiness, nausea, vomiting, anorexia, nervousness, abdominal cramps, epigastric burning, insomnia, dizziness, light-headedness, and aggravation of (or cause of) peptic ulcer symptoms in **people**. Antagonizes the secretion-reducing effects of ***anticholinergics***.

BETAINE

= *Cystadene* = *Glycine Betaine* = *Trimethylaminoglycine*

Lipotrope. **Decreases *homocysteine* levels.**

Untoward effects: Occasional nausea and diarrhea.

BETAMETHASONE

= *Becort* = *Betadexamethasone* = *Betasolon* = *Betnelan* = *Betnesol* = *Celestan* = *Celestene* = *Celestone* = *beta-Corlan* = *Dermabet* = *Diprolene* = *Flubenisolone* = *β-Methasone* = *NSC-39470* = *Sch 4831* = *Visu-beta*

Corticosteroid.

Untoward effects: Skin irritation, burning, itching, folliculitis, hypertrichosis, hypopigmentation, acne-like eruptions, perioral dermatitis, allergic contact dermatitis, purpura, atrophy or hives, striae, miliaria, glaucoma, increased intraocular pressure, anosmia and intracranial hypertension, growth suppression, and infection (viz. candidiasis) have occurred following topical use. After systemic use, hyperglycemia, glucosuria, Cushing syndrome, gastric distress (even bleeding), peptic ulcers, vasomotor reaction with facial mooning, osteoporosis, weight increase (fluid retention), facial flushing, diabetes mellitus, headache, nausea, dizziness, weakness, and confusion have been noted. Its use can induce neutrophilia, eosinopenia, and lymphopenia. Benefits of antenatal use in the prevention of respiratory distress syndrome in **infants** are greater than the risks, which can include a leukemoid reaction and even **fetal** death. Pulmonary edema can occur when given with ***magnesium sulfate*** or ***ritodrine*** for pre-eclampsia in **women**. Rupture of both calcaneal tendons above insertion reported in 66-year-old **male** on 3 mg/day/2 years, with marked wasting of calf and thigh muscles. In a 9½-year-old **male** with ***steroid***-dependent nephrosis, its use may have been the cause of diabetic ketoacidosis with acute hemorrhagic pancreatitis. Doses > 10 mg/day/2 or more years may increase the incidence of cataracts by 50%. Acute adrenal crisis in 87-year-old **female** after 7 mg intra-articularly. ***Magnesium*** and ***aluminum trisilicates*** may adsorb the drug.

Daily topical administration of betamethasone valerate to **dogs** can cause marked pituitary–adrenocortical suppression. Teratogenic in the **rat** and **mouse**, and decreases natural defense mechanisms, including those against parasitism. Weight decrease, anorexia, diarrhea, polydipsia, polyuria, osteoporosis, and increased wound healing time reported in **dogs** and other **animals**. TD_{LO}, SC, **rat**: 4 mg/kg (12–15 days pregnant); SC, **mouse**: 17 mg/kg (11–15 days pregnant); IM, **rabbit**: 8 mg/kg (13–16 days pregnant).

Betamethasone Acetate = *Betafluorene*

Betamethasone Adamantoate = *Betsovet*

Betamethasone Benzoate = *Bebate* = *Beben* = *Benisone* = *Euvaderm* = *Flurobate* = *Parbetan* = *Uticort* = *W 5975*

Betamethasone Dipropionate = *Diproderm* = *Diprophos* = *Diprosis* = *Diprosone* = *Maxivate* = *Rinderon-DP* = *Sch 11460*

Betamethasone Divalerate = *Betadival*

Betamethasone Phosphate Disodium Salt = *Bentalan* = *Betnesol Injectable* = *Durabetason* = *Vista-Methasone*

Betamethasone Valerate = *Bedermin* = *Betnesol-V* = *Betneval* = *Betnovate* = *Bextasol* = *Celestan-V* = *Celestoderm-V* = *Dermosol* = *Dermovaleas* = *Ecoval 70* = *Hormezon* = *Tokuderm* = *Valisone*

BETAXOLOL

= *Betoptic* = *Betoptima* = *Kerlone*

β-Adrenergic-blocking agent. **A B_1 cardio-selective agent.**

Use: For oral treatment of hypertension and topically for the treatment of glaucoma.

Untoward effects: May occasionally cause bronchospasm, dyspnea, and wheezing, even in ordinary dosage, especially in **asthmatics**. Heart failure, hypotension, and atrioventricular block can occur. Bradycardia is more common in the **elderly**. Fatigue, depression, insomnia, impotence, and antinuclear antibodies have been reported. Lupus, with weight decrease, exertional dyspnea, pleuritic chest pain, and diffuse aches and pains for several months in a **patient** after taking it for years. Cystoid macular edema and decreased visual acuity in 85-year-old **female**, and ankle edema, pulmonary rales, and shortness of breath in 80-year-old **male**, from topical use. An 81-year-old

male experienced an acute inferior myocardial infarction, after one drop topically onto one eye. Alopecia has been attributed to its topical use. Contraindicated in **patients** with obstructive airway disease, heart failure, and ***insulin-*** dependent diabetes.

LD_{50}, **mice**: 860–944 mg/kg; IV, **mice**: 37 mg/kg.

BETAZOLE

= *Histalog* = *Histimin*

Diagnostic aid, stimulant to gastric secretions. **A histamine analog.**

Untoward effects: Within 5 min of IM injection, 54-year-old **male** had severe precordial pain, relieved by ***nitroglycerin***. After recurrent episodes, an electrocardiogram revealed a subendocardial infarction (inversion of Tv2-v-4). Shock and oliguria developed within 15 min after a single injection in a 55-year-old **female**. Myocardial ischemia, atrial fibrillation, systolic hypotension, headache (3% of **patients**), rash, urticaria, nausea, wheezing, and occasional, but not serious, gastric bleeding and syncope reported in some **patients**. Flushing of skin occurs in about 20% of **patients**. A 75-year-old **patient** received 50 mg, IM, as a test for gastric acidity, and, within 15 min, developed nausea, dyspnea, and pulmonary edema; died after 40 min. Antagonizes the secretion-reducing effects of ***anticholinergics***.

BETEL

= *Piper Betel*

Untoward effects: **People** in the southwest Pacific and Asia (India, Ceylon, and Malaysia) chew the leaves and nuts as a stimulant to salivation, and for its stimulant and mind-altering effects. Its essential oil can be hypotensive, relaxant, and a cardiac and respiratory depressant. Also see ***Areca***.

BETHANECHOL

= *Duvoid* = *Mechothane* = *Mictone* = *Mictrol* = *Myocholine* = *Myotonine* = *Myotonochol* = *Urecholine* = *Uro-Carb*

Cholinergic, parasympathomimetic. **A methyl form of *carbachol*. A long-acting *muscarinic*, resistant to cholinesterase hydrolysis.**

Untoward effects: Salivation, shivering, shortness of breath, belching, borborygmi, involuntary defecation, sweating, and diarrhea from its cholinergic action. Nausea and vomiting may occur, if taken by a **patient** too soon after a meal. Miliaria, due to the profuse sweating has been reported. Appears in breast milk! Headache, facial flushing, urinary urgency (on large doses), hypotension, transient syncope with cardiac arrest, transient heart block and atrial fibrillation (in **hyperthyroids**) also occur. A 58-year-old **male** developed an esophageal rupture 2 h after a SC injection to relieve urinary retention. May produce false positive or increased serum amylase and lipase test readings and delay radiotracer clearance from the gall bladder. Contraindicated in asthma, intestinal or urinary tract obstructions, bradycardia and hypotension, epilepsy, pregnancy, parkinsonism, and in **patients** with recent myocardial infarctions. Use with ***tricyclic antidepressants*** can reverse some of the latter's adverse effects, but cause gastrointestinal distress or bleeding.

SC, TD_{LO}, **human**: 130 mg/kg.

LD_{50}, **rat**: 1.5 g/kg; **mouse**: 250 mg/kg.

BETHANIDINE

= *Bendogen* = *Benzaidin* = *Benzoxine* = *Betaling* = *Betanidine* = *Betanidol* = *BW 467C60* = *Esbatal* = *Eusmanid* = *Hypersin* = *Tenathan*

Antihypertensive, adrenergic-blocker.

Untoward effects: Depression, exercise, and orthostatic postural hypotension (especially in the **elderly**), impotence and/or absence of ejaculation, sweating, transient headache, dizziness, nasal stuffiness, diarrhea, and possible asymptomatic thrombocytopenia have occurred. Its hypotensive effects may be reversed by ***tricyclic antidepressants*** (***amitriptyline***, ***desipramine***, ***imipramine***), ***chlorpromazine***, ***amphetamines***, ***phenylpropanolamine***, and ***mazindol***. ***Methyldopa*** has an additive hypotensive effect. Related chemically and pharmacologically to ***guanethidine***, with more rapid onset, shorter duration, and less tendency to cause diarrhea.

Blocks sweating in **buffalo**, **cattle**, **goats**, **donkeys**, and **sheep**.

LD_{50}, **mouse**: 520 mg/kg; IV, LD_{50}, **mice**: 12 mg/kg; IP, 150 mg/kg; SC, 260 mg/kg.

BEXAROTENE

= *Targretin*

Use: In treatment of refractory cutaneous T-cell lymphoma.

Untoward effects: Mostly dose-dependent leading to hypercholesterolemia, hypertriglyceridemia, and decreased high-density lipoprotein. Headache, rash, photosensitivity, hypothyroidism, asthenia, leukopenia, anemia, and infections are common. Occasionally increased aminotransferases, alopecia, and even fatal cholestasis and pancreatitis.

Drug interactions: It is metabolized by the liver's CYP3A4 enzyme system. ***Gemfibrizol*** increases its plasma concentration. Since it is a ***retinoid*** analog, ***vitamin A*** may increase its toxicity.

BEZAFIBRATE

= *Befizal* = *Bezalip* = *Bezatol* = *BM 15,075* = *Cedur* = *Difaterol*

Antihyperlipoproteinemia.

Untoward effects*:** A mild peripheral neuropathy in a 52-year-old **male** attributed to 1 month of therapy with 200 mg, tid. Has induced rhabdomyolysis in 43-year-old **female** receiving 400 mg/day. A 36-year-old **female** had severe recurrent headaches after the fourth dose (200 mg tid). Apparent cause of severe reversible renal failure in a heart transplant **patient**, after nearly 10 months of therapy with 400 mg/day. Enhances ***warfarin's anticoagulant effect, reducing the latter's requirement. Can potentiate action of ***oral hypoglycemics*** and interact with ***statins*** increasing the risk of rhabdomyolysis.

BEZITRAMIDE

= *Benzitramide* = *Burgodin* = *R-4845*

Analgesic.

***Untoward effects*:** Infrequent nausea and vomiting, drowsiness, and jaundice reported.

LD_{50}, **rats**: 141 mg/kg; **mice**: 2.1 g/kg.

BIALAMICOL

= *Biallylamicol* = *Biethylamicol* = *Camoform* = *PAA 701* = *SN 6771*

Amoebicide.

***Untoward effects*:** A 27-year-old **patient** received 1 g/day for *Entamoeba histolytica*. Fever noted on the fifth day, and on the seventh day, the leukocyte count was $1200/mm^3$ with 56% polymorphs (pretreatment counts were 5200 and 70%). Nausea, vomiting, and skin rashes have occurred during treatment.

BICALUTAMIDE

= *Casodex*

Antiandrogen.

***Untoward effects*:** Hot flashes, nausea, diarrhea, increased liver enzymes, hematuria, and gynecomastia.

BICOZAMYCIN

= *Aizumycin* = *Antibiotic 5879* = *Bacfeed* = *Bacteron* = *Bicyclomycin* = *CGP 3543/E* = *WS-4545 antibiotic*

Antimicrobial.

***Untoward effects*:** On rare occasions, causes rashes in **people**.

LD_{50}, **mice**: > 4 g/kg.

BICUCULLINE

Alkaloid.

Untoward effects*:** As a ***GABA-antagonist in the central nervous system of **mammals**, it causes epilepsy and convulsions in **mice**, **rats**, **cats**, and non-human **primates**.

IP, CD_{50}, **mice**: 8 mg/kg; IP, LD_{LO}, 25 mg/kg.

BIETASERPINE

= *DL 152* = *Tensibar*

Antihypertensive.

***Untoward effects*:** Somnolence, asthenia, and vertigo reported in **patients**.

Ptosis, hypothermia, and decreased spontaneous activity in **mice**.

LD_{50}, **mice**: 620 mg/kg; IP, 236–430 mg/kg; IV, 215 mg/kg; LD_{50}, **dog**: 20 mg/kg.

BIFONAZOLE

= *Amycor* = *Azolmen* = *Bay h 4502* = *Bedriol* = *Mycospor* = *Mycosporan*

Antifungal. An imidazole.

***Untoward effects*:** After topical use, one **patient** reported burning sensations; burning and itching in two **others**. Non-toxic in topical trials on **dogs**, **rabbits**, **rats**, and **mice**, but oral doses > 10 mg/kg in **rats** had adverse effects on hemoglobin and retarded growth.

BINAPACRYL

= *Acricid* = *Ambox* = *Endosan* = *ENT 25793* = *HOE 2784* = *Morocide* = *Niagara 9044*

Fungicide (powdery mildew), miticide.

***Untoward effects*:** A cumulative poison that can be absorbed after inhalation, accidental ingestion, or from the skin. Sweating, thirst, yellow discoloration of sclera, skin, and hair occur after heavy exposure. Nausea, abdominal pain, restlessness, dyspnea, and tachycardia occur from massive exposure or ingestion. TWA, **human**: 2.5 μg/kg.

Poor skin absorption in **mice** and **rabbits**. Orally to **guinea pigs** metabolized ***dinitroisopropylphenol***, which was toxic. Moderate eye irritation in **rabbits** and 4000 ppm in **chick** diets led to cataracts.

LD_{50}, **rats**: 58–225 mg/kg; **mice**: 1.6–3.2 g/kg; **guinea pig**: 200–400 mg/kg; **chick**: > 800 mg/kg; dermal LD_{50}, **rabbit**: 0.75–1.35 g/kg; **rat**: 720 mg/kg; **dog**: 50 mg/kg.

BIPERIDEN

= *Akineton* = *Akinophyl* = *KL 373*

Anticholinergic, antiparkinsonism.

Untoward effects: Can produce the anticholinergic syndrome in **man**. Can increase intraocular pressure and either aggravate an existing glaucoma or cause it. Dryness of mucous membranes, nausea, blurred vision, depression, giddiness, headache, dizziness, balance and gait disturbances, tiredness, decreased libido, restlessness with angor animi, tremors, impaired memory and concentration, mild postural hypotension, hallucinations, and erythematous rashes reported. Abused for its euphoric effect. **NOT** advisable to use in epileptic **patients**.

LD_{50}, **mice**: 545 mg/kg; IV, 56 mg/kg.

BIRCH

= *Betula*

Use: Extracts of the bark and wood have been used topically as antiseptic wound treatments. Its ***Tar Oil*** is used in fragrances for soaps, cosmetic creams and lotions, detergents, perfumes, shampoos, and in foods.

Untoward effects: The ***Tar Oil***: 2% in petrolatum produced no irritation in a 48 h closed-patch test in 25 **human** volunteers. An occasional **person** is hypersensitive to it. It contains ***methyl salicylate*** (q.v., for additional untoward effects). Some **people** are allergic to birch pollen. A 6-year-old **male**, sensitive to ***aspirin***, developed a generalized pustular dermatitis after walking in a Pennsylvania woodland in the spring, from inhaling the pollen and chewing its twigs. ***Oil***: LD_{LO}, **human**: 170 mg/kg.

No irritation when applied undiluted on the backs of hairless **mice**, but was on **rabbits**.

LD_{50}, **rat**: > 5 g/kg; dermal, **rabbit**: > 2 g/kg; ***Oil***: LD_{50}, **rat**: 887 mg/kg; **guinea pig**: 1 g/kg.

BISABOLENE

= *Limene*

Use: In fragrances for soaps, cosmetic creams and lotions, detergents, and perfumes, and as a solubilizing agent for biliary calculi. Found in many essential oils.

Untoward effects: 10% in petrolatum produced no sensitization on 21 **volunteers**.

Only slightly irritating when applied full-strength to intact or abraded **rabbit** skin for 25 h under occlusion.

LD_{50}, **rats**: > 5 g/kg; dermal, **rabbits**: > 5 g/kg.

BISACODYL

= *Bicol* = *Broxalax* = *Contalax* = *DAMP* = *Deficol* = *Dulcolan* = *Dulcolax* = *Durolax* = *Endokolat* = *Eulaxan* = *Godalax* = *Laco* = *Laxadin* = *Laxagetten* = *Laxanin N* = *Laxorex* = *Nigalax* = *Perilax* = *Pyrilax* = *Stadalax* = *Telemin* = *Theralax* = *Ulcolax* = *VDH*

Cathartic.* Chemically related to *phenolphthalein.

Untoward effects: Vomiting and diarrhea can occur if suppositories are swallowed. A 34-year-old **female** hospitalized for chronic diarrhea and weight loss had ingested 200, 5 mg tablets/week/1½ years and developed a mild malabsorption syndrome and hypoalbuminemia. Carpopedal tetany and seizures in a 61-year-old **female** may be associated with ingestion of two tablets daily. Abdominal pain and gas are frequent complaints. A 54-year-old **male** developed nausea with vomiting, a rash and chills (within 15 min). So-called cathartic colon with chronic ulcerative colitis and bloody stools in 42-year-old **female** taking two tablets qid/4 years. Tablets are enteric-coated and should not be chewed or taken with milk or ***antacids***. Mild rectal irritation, burning sensations, and tenesmus reported with the suppository.

No lethal effects in **dogs** given 10–15 g/kg. LD_{50}, **rats**: 4.32 g/kg; **mice**: 17.5 g/kg; LD_{LO}, **rat**: 3 g/kg.

BIS(CHLOROMETHYL) ETHER

= *BCME* = *sym-Dichloromethyl Ether* = *Oxybis [chloromethane]*

Use: In the synthesis of chemicals and in the manufacture of plastics and ion exchange resins. Estimated annual production is 500 ± million lbs.

Untoward effects: **Human** exposure is mostly through inhalation and some skin absorption. In the presence of moisture, it decomposes into ***hydrogen chloride*** and ***formaldehyde***. Exposed **workers** have an increased risk of lung cancers, mostly oat-cell carcinomas. Closely related chemically to a number of inhalation anesthetics. A class-action lawsuit on behalf of exposed industrial **workers** was filed. Direct-acting alkylating carcinogen, causing squamous cell lung tumors and bronchogenic carcinomas. Irritation of eyes, skin, and respiratory mucous membranes; corneal damage, pulmonary congestion, edema, coughing, dyspnea, wheezing, and blood-stained sputum and bronchial secretions. Exposure limits of 0.005 mg/m^3 (0.001 ppm) suggested.

Rats exposed to 0.1 ppm air concentration 6 h/day/5 days per week, for a total of 101 exposures, developed high incidence of lung cancers and all were dead at the end of 659 days after exposure. **Mice** exposed to 1–2 ppm for 82 days also had an increase in lung tumors. A single

SC injection of 12.5 µl/kg into newborn **mice** induced a high incidence of lung adenomas. **Rats** exposed to vapors (100 ppb) for 10 days developed nasal neuroepitheliomas and/or pulmonary carcinomas. Topical application on **mice** caused tumors at the site. Squamous cell papillomas found after oral use in **mice**.

LD_{50}, **rats**: 240 mg/kg; dermal, **rabbits**: 3 g/kg; inhalation LC_{LO}, **rats**: 700 ppm/5 h.

BIS(2-ETHYLHEXYL)PHTHALATE

= *DEHP* = *Di(2-ethylhexyl)phthalate* = *Dioctyl Phthalate* = *DNOP* = *DOP* = *Octoil*

Plasticizer, stabilizer. **Millions of pounds produced annually. In pacifiers and teethers.**

Untoward effects: Has leached into stored blood from its plastic containers, into ***PVC*** endotracheal catheters, possibly causing tracheobronchial irritation when using ***methoxyflurane***, and into hemodialysis blood tubing. Accumulates in liver, spleen, lung, and fat of **patients**, but no overt toxicity noted from the small amounts. Eye and mucous membrane irritant. May be a cause of "**meat-wrappers**'" asthma. Heating of ***PVC*** film releases ***DEHP*** and ***hydrogen chloride***. The Environmental Protection Agency and the FDA have not considered the chemical a significant **human** health risk, although its use increased the number of liver cancers in **rats** and **mice**. 8 h TWA, 5 mg/m^3.

TD_{LO}, **human**: 143 mg/kg.

Mice ingesting 1 mg/kg on day 7 of pregnancy led to 58% of fetuses dead with absence of tail bones, radius, ulna, tibia, or fibula. Testicular atrophy reported in **rats**.

LD_{50}, **rat**: 31 g/kg; **mouse**: 30 g/kg; **rabbit**: 34 g/kg; IV, TD_{LO}, **rat**: 300 mg/kg.

BISETHYLXANTHOGEN

= *Auligen* = *Aulinogen* = *Bexide* = *Diethyl Xanthogenate* = *Dixanthogen* = *Ethylxanthic Disulfide* = *EXD* = *Herbisan* = *Lenisarin* = *Preparation K* = *Sulfasan*

Scabicide, herbicide.

Untoward effects: LD_{LO}, **human**: 500 mg/kg.

LD_{50}, **rats**: 480–603 mg/kg; **rabbit**: 620 mg/kg; **guinea pig**: 400 mg/kg; dermal LD_{LO}, **rat**: 2.1 g/kg.

BISHYDROXYCOUMARIN

= *Dicoumarol* = *Dicumarol* = *Dicumol* = *Dufalone* = *Melitoxin*

Anticoagulant. **A *coumarin* derivative. The cause of hemorrhagic sweet *clover* poisoning of cattle, sheep, and horses, isolated by Winconsin Alumni Research Foundation in 1939.**

Untoward effects: Intestinal obstruction, due to spontaneous intramural small bowel hemorrhage in a hypertensive 75-year-old **male** and in a 68-year-old **female** with basilar artery insufficiency. Large oral doses frequently cause anorexia, nausea, burning sensations of the face, epigastric cramps, pyrosis, flatulence, diarrhea, and, rarely, vomiting, all within 1–3 h. Hematuria, gluteal ecchymoses, upper gastrointestinal bleeding, hypoprothrombinemia, epistaxis, decreased ***uric acid*** assay results, and intraocular hemorrhage and blindness have also been reported in less than 10 years of its early clinical usage. 2 mg% was considered a therapeutic blood level and 7 mg% toxic in **man**. Transplacental transfer has been demonstrated and it has caused **fetal** death. If given to the **mother** shortly before delivery, postnatal hemorrhage can occur. The **fetus** is slow in metabolizing the drug. Very small amounts can be found in breast milk of treated **mothers**. Suicide and malingering attempts have occurred with it. Petechiae, ecchymosis, hemorrhagic infarcts, or gangrenous necrosis 3–6 days after administration, eventually followed by sloughing eschar, often requiring skin grafting or amputation. Urticaria, rashes, alopecia, pruritus, and angioedema have occurred. Frank hemorrhages into joints and liver necrosis have occurred.

Drug interactions: ***Antidepressants***: Its bioavailability is increased in many **subjects** by ***amitriptyline*** and ***nortriptyline***. ***Antidiabetic agents***: Increases effect of ***sulfonylureas*** (***tolbutamide***, ***chlorpropamide***), with resultant hypoglycemia. ***Barbiturates***: Its metabolism is increased by ***barbital***, ***phenobarbital***, ***heptabarbital***, and ***secobarbital*** by microsomal enzyme induction.

Carbon tetrachloride: Hypoprothrombinemia in 59-year-old **male** during treatment after accidental ingestion of approximately 0.1 ml CCl_4.

Chloral hydrate: Apparently stimulated its metabolism and, when ***chloral hydrate*** therapy was discontinued, prothrombin time became abnormally high and the **patient** suffered a fatal hemorrhage.

Chloramphenicol: Inhibits biotransformation of bishydroxycoumarin and increases its anticoagulant action.

Clofibrate: Inhibits metabolism of bishydroxycoumarin.

Contraceptives, oral: Inhibited the hypoprothrombinemic action of bishydroxycoumarin.

Miscellaneous: ***Adrenocorticotropic hormone***, ***cortisone***, ***ethchlorvynol***, ***glutethimide***, ***griseofulvin***, and ***prednisone*** antagonize the anticoagulant effect.

Penicillins: Activity is increased by protein displacement.

Phenyramidol: Potentiates the anticoagulant effect.

Phenytoin: Metabolism is dramatically inhibited by the anticoagulant (***phenytoin's*** half-life increased from 9 to 36 h). ***Sulfonamides***: Activity and potential toxicity is enhanced by the anticoagulant.

Thyroid preparations: ***Dextrothyroxine*** increases the anticoagulant's effects.

Anticoagulant effect potentiated by ***acetaminophen***, ***allopurinol*** (in some **patients**), ***aspirin***, ***cinchophen***, ***clofibric acid***, ***ethylestrenol***, ***indomethacin***, ***magnesium hydroxide***, ***methandrostenolone***, ***methyldopa***, ***methylphenidate***, ***norethandrolone***, ***oxyphenbutazone***, ***phenylbutazone*** (effects are variable), ***quinidine***, and ***tetracycline***.

All the above interactions have essentially been reported for most experimental **animals** as well. 'In the **cow**, **dog**, and **rabbit**, its ingestion has caused fetal death and hemorrhage.

LD_{50}, **rats**: 542 mg/kg; **mice**: 230 mg/kg; IV, LD_{50}, **rat**: 52 mg/kg; **mouse**: 64 mg/kg; **guinea pig**: 59 mg/kg.

BISMUTH

Use: In manufacturing of its salts, electric fuses, solders, and "silvering" mirrors.

Untoward effects: The salts (q.v. below) are usually well tolerated and darken feces, but encephalopathies and fixed-drug eruptions, alopecia, and constipation have occurred. They decrease absorption of ***tetracyclines*** and ***lincomycin***. Bismuth, especially from ***bismuth subnitrate***, can be passed into the milk of lactating **women**. Oral pigmentation can occur after the systemic absorption of bismuth can react with the ***hydrogen sulfide*** in debris between **one's** teeth, to form an insoluble ***bismuth sulfide***. This is deposited in the oral tissues as a dark, bluish-black pigmentation. This blue line or "bismuth line" may persist for years. Gingivitis and oral ulceration also reported. Garlicky breath often noted after ingestion with many bismuth preparations. Some **people** detect a metallic taste. A bismuth paste was applied topically to an open ulceration on a 13 month **child**. X-ray examination 9 weeks later showed that the metal was deposited in areas of maximal growth across metaphysical ends of **her** long bones. Bone changes resembling osteochondritis in **children** have occurred not only as a result of the **child's** therapy but also as a result of the **mother** being treated during pregnancy. Severe osteoporosis, mainly of the vertebrae and pelvic bones reported in many **patients**, after repeated bismuth injections. Acute renal tubular necrosis has followed massive accidental or parenteral administration. A toxic serum concentration is reported as 0.1 mg/l. Methemoglooinemia and hepatitis have also occurred. Bismuth has been used in some topical treatments for nipple care of lactating **women**, and its use is potentially dangerous for the nursing **infant**. Many bismuth compounds were once used in treating syphilis, and their medical use is now very limited. Toxicity has been reported.

BISMUTH AMMONIUM CITRATE

Untoward effects: Darkens stools, and occasionally tongues. Dry mouth.

BISMUTH GLYCOL ARSANILATE

= *Broxolin* = *Dysentulin* = *Glycobiarsol* = *Ig 9659A* = *Milibis* = *Viasept* = *Wintodon*

Anthelmintic, antifungal, antiprotozoan.

Untoward effects: Will color feces black in treated **patients**. Less than 2% of treated **dogs** show nausea and vomiting, and a clinical case may have had drug-related bloody stool, seizures, and renal necrosis. Suppositories have caused local irritation. The arsenical portion is most apt to give systemic toxicity.

BISMUTH SODIUM THIOGLYCOLLATE

Untoward effects: Will blacken stools. A 7-year-old **male** developed severe oliguric renal failure after an IM injection of 200 mg for warts (*verruca vulgara*). Recovered after 10 days of hemodialysis. This stresses danger of bismuth preparations in young **children**.

IP, LD_{LO}, **guinea pig** and **rat**: 26 mg/kg; IM, LD_{LO}, **rabbit**: 26 mg/kg; IV, LD_{LO}, **rabbit**: 21 mg/kg.

BISMUTH SODIUM TRIGLYCOLLAMATE

= *Bistrimate* = *Oribis*

Untoward effects: Causes albuminuria. A 15-year-old **female** ingested 7–8 tablets (75 mg ***Bi***/tablet) during one day, given to **her** by a friend for a sore throat. She began to vomit, had decreased urine output, and acute renal failure with blood urea nitrogen > 150. Reversible renal failure with proximal tubular involvement in 19-year-old **female** after 21 tablets (1.5 g ***Bi***). An 8-year-old **female** took two tablets twice daily/3 months, skipped 6 weeks ±, then ingested more for 5 weeks. **She** developed severe renal failure which, fortunately, was reversed by hemodialysis. A 19-year-old **female** ingested 21 tablets and recovered after treatment with ***dimercaprol***. Reversible renal failure in a 7-year-old after 200 mg, IM.

BISMUTH SUBGALLATE

Use: As an ***astringent***, wound protective, and in ostomy odor control.

Untoward effects*:** Neurological syndrome (confusion, tremors, clumsiness, myoclonic jerks, malaise, peculiar sensations in fingers and toes, loss of memory and power to concentrate, and difficulty in walking) after ingestion, usually by colostomy **patients** (200–400 mg, up to four times daily). It is unclear if this toxicity is due to the bismuth or the ***gallic acid content. Some of the same symptoms appeared in **patients** on long-term use of a bismuth-containing skin cream.

BISMUTH SUBNITRATE

Antidiarrheal, wound protectant.

Untoward effects*:** Causes methemoglobinemia, often occurring after cessation of therapy. Its ingestion has been serious, and often fatal, in **children**. Methemoglobinemia has even occurred after its topical use as a powder on **infants**. After ingestion, the subnitrate fraction, when reduced by intestinal bacteria, can cause fatal ***nitrate poisoning.

Absorption after oral use has caused brain dysfunction in **rats**.

BISMUTH SUBSALICYLATE

= *Bismogenol* = *Stabisol*

Untoward effects*:** Tongue and stools turn black. The popular ***Pepto-Bismol® also contains the equivalent of 6–7, 5 grain ***aspirin*** tablets/240 ml. Myoclonic encephalitis in a 45-year-old **male** after 7 days therapy with it (5.2–9.4 g/day in ***Pepto-Bismol***®). A 62-year-old **male**, taking it off and on for many years, developed muscle twitches of the face and limbs, incontinence, disorientation, and an electroencephalogram pathognomonic of bismuth encephalopathy. **His** blood bismuth level was 72 μg/l (normal values are < 5 μg/l). A 42-year-old **male** had black ***carbon***-like particles in **his** skin for 2–3 days after ingesting it every 2–4 weeks. Verified by chemical analysis. Chronic large doses in the **elderly** can cause dementia.

Drug interactions*:** Peak ***doxycycline serum levels in **man** decrease if bismuth subsalicylate is given 2 h before, but not when given 2 h after the ***antibiotic***. Absorption of the ***antibiotic*** is significantly decreased by approximately 40–50%.

Cats are particularly susceptible to its ***salicylate*** toxicity. **Cats** and **cynomolgus monkeys** frequently vomit after being given commercial suspensions, such as ***Pepto-Bismol***®.

BISOPROLOL

= *CL 297939* = *Concor* = *Detensiel* = *Emconcor* = *Emcor* = *EMD 33 512* = *Euradal* = *Isoten* = *Maintate* = *Monocor* = *Soprol* = *Zebeta*

β_1-Adrenergic-blocker, antihypertensive.

***Untoward effects*:** Preliminary trials indicate fatigue, dizziness, and diarrhea may be among the more common adverse effects. A 60-year-old **female** developed circumscribed scleroderma after 4 months of treatment. Causes bronchospasm in some asthmatic **patients**.

BISOXATIN ACETATE

= *Laxonalin* = *Maratan* = *Talsis* = *Tasis* = *WY 8138*

Cathartic.

***Untoward effects*:** Abdominal pain and gas are the most common complaints. Occasional nausea reported.

LD_{50}, **mouse**: 600 mg/kg.

BISPHENOL A

= *2,2-Bis(4-hydroxyphenyl)propane* = *4,4´-Isopropylidenediphenol* = *4,4´-(1-Methylethylidene)bisphenol*

Over a billion lbs/year used.

***Use*:** In manufacturing many resins, particularly polycarbonate and epoxy; in flame-retardants, in a coating for teeth, and in shatterproof windows.

Untoward effects*:** Hypersensitivity in **patients** from uncured epoxy resin. Dust can irritate **workers'** eyes, nose, and throat. Goggles and dust respirator or filters should be worn. Wash with lots of water, in the event of eye and skin contact. Dust clouds can explode if exposed to spark or flames. Will burn. Can leach from ***PVC containers into parenterals.

90-day inhalation of the dust caused a reversible damage to the nasal cavity of **rats**.

LD_{50}, **rat**: 4 g/kg; IP, **mouse**: 150 mg/kg; dermal, **rabbit**: 3 g/kg.

BIS(TRI-*n*-BUTYLTIN)OXIDE

Use*:** In marine antifouling *paints* (approximately 2 million lbs/year); as a ***molluscicide; and as a ***microbiocide*** in cooling-waters.

***Untoward effects*:** Causes thymus atrophy, lymphocytopenia, increases immunoglobulin M and decreases immunoglobulin G, increases release of luteinizing hormone and follicle stimulating hormone, increases platelets, increases prolactinomas and pheochromocytomas in **humans**.

In **rats**, 320 ppm in the diet is lethal at 30 days.

LD_{50}, **rat**: 87–194 mg/kg; **mouse**: 55 mg/kg; LD_{LO}, **rabbit**: 50 mg/kg.

BITHIONOL

= *Actamer* = *Bitin* = *Lorothidol* = *Thiofen* = *2,2´-Thiobis (4,6-dichlorophenol)* = *XL-7*

Anthelmintic, antiseptic, fungicide. **Chemically related to *hexachlorophene*.**

Untoward effects: Contact photodermatitis in **humans** with cross-sensitivity to **users** of ***dichlorophene***, ***hexachlorophene***, ***dibromo-*** and ***tetrachlorosalicylanilide***. Diffuse erythematous lesions in a **patient** disappeared after discontinuation; a toxic hepatitis (rare) in **another**; leukopenia (rare), gastrointestinal complaints (including anorexia, diarrhea, abdominal discomfort); dizziness and itching in **patients** after oral use in treatment of cerebral paragonimiasis. 220 **individuals** used throat lozenges (average of 8/day/25–68 days) containing 5 mg with no reports of oral or gastrointestinal irritation, sensitization, or changes in blood, urine, or liver function.

LD_{50}, **rats**: 6.6 g/kg; **mice**: 0.9 g/kg; **rabbits**: lethal dose approximately 7 g/kg; **guinea pigs**: barely tolerated 250 mg/kg.

BITOLTEROL

= *Effectin* = *Tornalate* = *Win 32,784*

Bronchodilator, β_2-adrenergic agent. **The methanesulfate (mesylate) form is commonly used.**

Untoward effects: Tachycardia, palpitations, and tremors can occur in **people**, but less often than with ***isoproterenol*** or ***metaproterenol***. High dosage can cause hypokalemia. As-needed use, rather than regular use, as an inhalant will decrease chances of hyperresponsiveness and increased mortality. In a series of 140 adult **patients** treated by inhalation, tremors (5.7%), nervousness (2.1%), headache (0.7%), dizziness (1.4%) and light-headedness (1.4%), palpitations (1.4%), chest discomfort (0.7%), and nausea (2.9%) occurred. Tachycardia and musculoskeletal tremors in **rats** and other **animal** models, when given orally or parenterally.

BLACKBERRY

= *Bramble Berry* = *Cloudberry* = *Fingerberry*

Use: As a food; as an *antidiarrheal*, and in many folklore medicines. Contains natural ***salicylates***.

Untoward effects: Excess ingestion by **people** will darken stools enough to mislead some **people** into thinking they have gastrointestinal bleeding. May also color the urine red. The plants have caused mechanical injury to hunting **dogs** and other **animals**.

BLASTICIDIN S

Antibiotic. **To inhibit growth of *rice* blast fungus.**

Untoward effects: **Human** exposure occurs.

LD_{50}, **rat**: 2.8 mg/kg; **mouse**: 40 mg/kg.

BLEOMYCIN

= *Blenoxane* = *NSC 125,066*

Antibiotic, antineoplastic. **Actually a group of basic *cytotoxic glycopeptide antibiotics*.**

Use: In treating lung, testicular, and soft tissue tumors.

Untoward effects: Ototoxic potential. Hypotension, if IV is pushed too fast. Occasional acute pneumonitis, interstitial and alveolar infiltration of lymphocytes, eosinophils, and plasma cells, fibrosis (5–15% of **patients** receiving > 100 mg/m^2), bronchiolitis obliterans, dyspnea, and hypersensitivity pneumonitis in 3–5%, and, rarely, up to 40% of a group of **patients**. Nausea, vomiting, fever, anaphylaxis, and other allergic reactions are manifestations of acute toxicity. Rash, pruritus, hyperpigmentation, hyperkeratosis, ulcerated stomatitis, alopecia, Raynaud's phenomenon, and cavitating granulomas occur in delayed toxicity. Collagenous plaques (scleroderma-like) or skin nodules (often diffuse and resembling progressive systemic sclerosis) disappear when treatment ceases. Longitudinal brown–blue pigmentation of nail plates and nail beds, cutaneous striae, soft tissue (including penile) calcification, substantial weight loss, weakness, paresthesia, back pain, hallucinations, paranoia, hyperkeratosis of fingers, palms, and feet, leukopenia, thrombocytopenia, pancytopenia, bone marrow depression (rarely), liver toxicity, pancreatitis, encephalopathy (after previous occurrence with ***ifosfamide***), chest pain, pericarditis, and ulcerations of hand and forearm from extravasation have all been reported in different **patients**. Local reactions and thrombophlebitis may follow parenteral use. **Oxygen** administration has potentiated its pulmonary toxicity. Metallic taste noted by treated **patients**.

Treated **mice** have cutaneous ***collagen*** levels 1.75 times those of untreated controls; it induces pulmonary fibrosis in **rats**, **mice**, and **dogs**. **Dog** and **monkey** trials indicate there is a dose-related skin toxicity and skin changes occur at pressure sites implying an alert to physicians to help prevent bed sores in their **patients**. Pyrogenic in **rabbits**.

IP, TD_{LO}, **rat**: 400 μg/kg (9–14 days pregnant), **mouse**: 8 mg/kg (7–12 days pregnant).

BLIGHIA SAPIDA

= *Ackee* = *Akee* = *Fisa* = *Gwanja kusa* = *Ishin* = *Okpu Ocha*

Untoward effects: **Nigerians** remove the tree's fruit's (arils) poisonous raphe (although this is now questioned as the source of toxicity) before eating it as a food, after parboiling and frying. Unripe fruit causes vomiting and even death in **humans**. It is the cause of "vomiting sickness" in Jamaica, with nausea, vomiting, abdominal cramps, collapse, and, sometimes, death occurring. The encephalopathy and hepatic steatitis mimic symptoms of Reye's syndrome. It appears to be especially toxic, and frequently fatal, to **children**. Epidemic of fatal encephalopathy in 29 West African **children** – 2–6 years. **Their** urine contained ***dicarboxylic acid*** concentrations at 4–200 times those of controls. Literature states that the ripe fruit is non-toxic. Improperly prepared fruit contains ***hypoglycin A*** (q.v.), which produces severe hypoglycemia, (with a mortality of 40–80%), depletion of liver ***glycogen***, and fatty liver degeneration in **humans**. In the last 100 years, its ingestion has caused over 5000 deaths in Jamaica. The U.S. has banned its importation, because overripe or underripe (when gathered), it is particularly poisonous to **man**. Some of the trees are grown as a curiosity or ornament in South Florida.

BLUEBELL

= *Endymion nonscriptus* = *Scilla nonscriptus*

Flowering plant.

Untoward effects*:** All parts of the plant, including bulbs, leaves, and flowers, contain glycosides similar to ***digitalis. Ingestion by **humans** after mistaking the bulbs for ***onions*** leads to abdominal pain, bradycardia, flushed skin, and diarrhea. Diarrhea in a **child** of 18 months after eating 6–10 fruit pods with seeds. Slow recovery. The plant's sap has caused skin irritation in some **people**.

Has caused poisoning in **horses**. Symptoms were abdominal pain, attempts at vomiting; cold, clammy skin; decreased temperature, bradycardia and weak pulse, bloody diarrhea, and anuria. Recovery was slow. **Cattle** became dull, reluctant to move, and constipated; they staggered and evidenced bradycardia and decreased temperature; and eventually went down, unable to move.

S. siberica = ***Siberian Scilla***. The entire plant also contains cardiac glycosides. Ingestion could cause abdominal pain, cramps, diarrhea, arrythmias, and vomiting.

BLUE VRS

= *Acid Blue 1* = *Patent Blue V* = *Sulphan Blue*

Food coloring.

***Untoward effects*:** Data indicate its use should **NOT** be permitted in any country. Carcinogenicity has been reported. Persistent fibroplastic reaction after SC injection resulted in thickened collagenous tissue. No longer permitted for food use in the U.S. and Britain. Occasionally causes nausea, allergic reactions, and asthma attacks, after IV use in **man**. Fatal allergic shock reported in a burn **victim**, given 6 ml of 10% solution, IV.

LD_{50}, **rat**: > 10 g/kg; **mouse**: > 5 g/kg.

BOIS DE ROSE, ACETYLATED

Rosewood oil from the wood of *Aniba rosaedora* is acetylated, yielding a spicy-lavender odor used in soaps, cosmetic creams and lotions, detergents, and perfumes. The oil has been approved for food use.

***Untoward effects*:** 12% in petrolatum produced no irritation in a 48 h closed-patch test in 25 human **volunteers**, nor sensitization or phototoxicity.

LD_{50}, **rat**: 4.3 g/kg; acute dermal, **rabbit**: > 5 g/kg.

BOLDENONE

= *Ba 29,038* = *Dehydrotestosterone* = *Equipoise* = *Parenabol*

***Anabolic.* The undecylenate is a long-acting injectable agent.**

***Untoward effects*:** Possible bilateral gynecomastia attributed to it in a **male**.

Not to be used in **horses** intended for **human** food consumption.

BOLETUS sp.

***A mushroom.* Contains the cholimimetic alkaloid, *muscarine*.**

Untoward effects*:** Contains gastrointestinal irritants, causing mild to severe nausea, vomiting, abdominal cramps, diarrhea, usually within ½–2 h after ingestion, and lasting several hours. The incidence and severity of these symptoms are greatest with the raw mushrooms. Rarely, a fatality has occurred. In a 40-year-old **male**, previously unaffected by eating *B. variipes*, tolerance changed adversely when receiving ***hydroxychloroquine for discoid lupus erythematosus.

BOLOGNA

Untoward effects*:** A pounding, vascular headache often occurs in **people** within minutes after eating this ***nitrate-containing food. ***Nitrate*** addition to foods helps cure it and stabilizes its color, by virtue of the food's ***salt*** content, enabling micrococci to reduce the ***nitrate*** to ***nitrite***, which improves the product's color stability and keeping qualities.

BOMYL

= *ENT 24,833* = *GC 3707* = *Swat*

Insecticide.

***Untoward effects*:** **Human** exposure has occurred.

LD_{50}, **rats**: 32 mg/kg; 12 day male **chicks**: 9.43 mg/kg.

BOÖPHONE

= *Haemanthus*

Untoward effects*:** Extract of the bulbs contains the alkaloid, ***buphanine, which is used in southern Africa as an arrow and dart poison.

BORAGO OFFICINALIS

= *Bourrache* = *Bourragi*

***Garden herb.* Contains *pyrrolizidine* alkaloids and *γ-linolenic acid.* Native to the Mediterranean region, but now cultivated in Baja California, Mexico.**

***Use*:** As a flavoring agent, diuretic, and anti-inflammatory agent. Its oil is used in cosmetics.

Untoward effects*:** Since it contains large quantitites of ***potassium nitrate, prolonged exposure to it can cause gastroenteritis, methemoglobinemia, anemia, and nephritis in **humans**. The potential toxicity of its alkaloids has not been clearly defined.

BORAX

= *Jaikin* = *Sodium Biborate* = *Sodium Borate* = *Sodium Pyroborate* = *Sodium Tetraborate*

***Weak antiseptic, larvicide, detergent, astringent.* Industry production capacity is approximately 1 million tons/year.**

***Use*:** In soaps and washing compounds, glass, insecticides, ceramics, fire-proofing, photographic developers, weed control, in pH control in drilling muds, and in wood preservatives.

Untoward effects*:** One of the first preservatives tested on **volunteers** by the late Dr Harvey W. Wiley's "Poison Squad". Of the ingested borax, 86–94% was recovered in **their** urine and traces were still detected 8 days later. As dosages increased, appetite decreases, with fullness and discomfort of the stomach along with dull, persistent headaches. Up to 3 g/day for a short period of time was tolerated, but at the 4–5 g/day level, **participants** were unable to do work of any kind. 0.5 g/day/50 days upset the **volunteers**. Its use in many foods in 1902 gave the general public about 0.25 g/day. Dr Wiley then concluded the U.S. should ban it as a food preservative. **Children** consuming 5–10 g have experienced severe vomiting, diarrhea, and death. The dust causes acute respiratory irritation in borax **workers** at 4 mg/m^3. Alopecia may occur 2 weeks after ingestion or occupational exposure. Recurrent seizures, vomiting, loose stools, irritability in 4½-month-old **male** and 9-month-old **male** given pacifiers coated with a mixture of it and ***honey (107 g borax/l or 3–9 g borax ingested over a 1 h period). Bone marrow depression and a severe normocytic, hypochromic anemia in **one**. Both recovered when the borax was withheld. A 50-year-old **male** storekeeper, who daily weighed out borax washing powder for about a year, developed nervousness, excitement, headache, fever, muscle pains, redness of face and head, and insomnia. Symptoms disappeared when **he** avoided contact with it. **He** still excreted ***boron*** 6 weeks later. 15–30 g may be lethal for **adults**.

TD_{LO}, **human**: 500 mg/kg; **infant**: 1 g/kg.

Due to accumulation in testicular germinal tissues, it has caused **male** infertility in **rats** and **dogs**. **Cattle** consumed chunks of borax at an abandoned oil-well drilling site. Five died and others were weak and staggering. The predominant post-mortem lesion was hemorrhagic gastroenteritis. **Cattle** receiving it in their drinking water equivalent to 15.3 mg boron/kg/day/30 days had decreased feed consumption, weight loss, edema, and altered blood chemistry.

LD_{50}, **rats**: 2,000–5,330 mg/kg; **mice**, **cattle**, **goats**, and **dogs**: estimated at 2–3.5 g/kg.

BORIC ACID

= *Boracic Acid* = *Borofax* = *Orthoboric Acid*

Antiseptic, antifungal, astringent, insecticide.

Untoward effects*:** See ***Borax. Unfortunately, it is still considered harmless by the general public and some health care providers. Nausea, vomiting, diarrhea, abdominal cramps, oliguria or anuria, erythematous skin and mucuous membrane lesions, toxic epidermal necrolysis, rash with desquamation, alopecia (from casual ingestion of mouthwashes), tachycardia, hematemesis, bloody stools, circulatory collapse, cyanosis, delirium, convulsions, coma, and death, after ingestion or from its topical use on abraded or burned skin. Especially dangerous to **infants** and **children**. Ingestion by **children** of *roach powder* can lead to a "boiled lobster" appearance. Fatality in 14-year-old **male** after 1% soaks twice daily–tid for burns over 12% of **his** body. This resulted in sudden anorexia, nausea, disorientation, vomiting, generalized rash, fever, hypotension, uremia (blood urea nitrogen 140), and death after 2 days. Topical poisonings are usually 70% fatal after onset of symptoms. Over 100 reported fatal cases in **infants** after its use in treating diaper rash. Inadvertent addition of 2.5% solution in **infants'** feeding formula accounted for 11 cases of poisoning, five of which were fatal. Poisonings with fatalities occurred in English **children** consuming *milk* which contained 0.7 g/l. Continuous wet dressings over 14 days on a 35-year-old **female** caused lethargy, coma, and death. **Her** blood ***boron*** level was 350 mg/100 ml. After ingestion, feces may develop a bluish-green color. Has been in ***Silly Putty***® at a 1% level. Pancreatic inclusion bodies, injection of cerebral vessels and kidneys found in some fatal cases. A dialyzable (hemo- and peritoneal) poison. Little or no adsorption by ***activated charcoal***. 10–30 g as a lethal dose for **adults**; 0.1 g for **infants** and 2–6 g for small **children**. Coincidentally, boric acid comes from Death Valley.

LD_{LO}, **human**: 214 mg/kg; **infant**: 934 mg/kg.

In **animals**, it has induced vomiting, anorexia, diarrhea, central nervous system depression with muscle weakness, ataxia, tremors, and occasional seizures. Oliguria, anuria, and occasional albuminuria, hematuria, and metabolic acidosis. Contact poison for **roaches**.

LD_{LO}, **dog**: 1.8 g/kg; **rabbit**: 4 g/kg; LD_{50}, **rat**: 2.66 g/kg; **mouse**: 3.45 g/kg. Older literature reports toxic doses for **dogs**: 2.5–3 g/kg.

BORNEOL
= *Baros Camphor* = *Borneo Camphor* = *Bornyl Alcohol* = *Camphol* = *Malayan Camphor* = *Sumatra Camphor*

Use: As a chemical intermediate, and in perfumery, soaps, cosmetic creams and lotions, detergents, perfumes, flavorings, foods, and in nasal inhalers.

Untoward effects: Nausea, vomiting, confusion, dizziness, and convulsions reported. 2% in petrolatum in a 48 h closed-patch test on 25 **human** volunteers indicated no irritation or sensitization.

LD_{LO}, **human**: 50 mg/kg; **rabbits**: 2 g/kg.

BORNYL ACETATE
= *Borneol Acetate*

Use: In soaps, cosmetic creams and lotions, detergents, perfumes, flavorings, food, and in nasal inhalers.

Untoward effects: 2% in petrolatum in a 48 h closed-patch test on 25 **human** volunteers indicated no irritation or sensitization.

LD_{50}, **rat**: > 5 g/kg; dermal, **rabbit**: > 10 ml/kg.

BORNYL ISOVALERATE
= *Bornyl 3-methyl Butyrate* = *Bornyval*

Sedative, flavoring agent, perfume.

Use: Primarily, to impart the aroma of valerian and camphor to soaps, cosmetic creams and lotions, detergents, and perfumes, and flavor to foods.

Untoward effects: 4% in petrolatum in a 48 h closed-patch test on 25 **human** volunteers indicated no irritation or sensitization.

LD_{50}, **rat** and dermal, **rabbit**: > 5 g/kg.

BORON

Use: To harden other metals, and to prevent microbial contamination in ***jet*** and other ***hydrocarbon fuels***.

Untoward effects: Vomiting and stomach pain are common symptoms. See ***Boric acid*** and ***Borax*** for further details on symptoms. Serum levels of 0.08 mg% considered normal; 4 mg%, toxic; and 5 mg%, lethal in **adults**. **Human** potable water should **not** contain over 10 mg/l (10 ppm); half this level is considered safe for **cattle** and other **livestock**. Deficiencies may be a factor in decreased alertness and increase in osteoporosis. These can be overcome by eating more ***apples***, ***broccoli***, ***carrots***, ***grapes***, and ***pears***. Gastroenteritis, central nervous system disturbances, respiratory disease and nephritis reported in **sheep** ingesting too much in their forage.

LD_{50}, **mouse**: 2 g/kg.

BORON TRIFLUORIDE

Use: As a catalyst for a variety of resins and to protect ***magnesium*** and its alloys from oxidation.

Untoward effects: An industrial toxicant, irritating to eyes and mucous membranes, and a cause of a delayed pulmonary edema. A 1 ppm (3 mg/m^3) limit on airborne exposure has been set by OSHA with pulmonary function testing of **workers** required. Epistaxis in both **man** and **animals** from inhalation.

Inhalation LC_{LO}, **rat**: 750 ppm/5.5 h; LC_{50}, inhalation, **mouse**: 3460 mg/m^3/2 h; **guinea pig**: 109 mg/m^3/4 h.

BOUGAINVILLEA

Botanical.

Untoward effects: May cause an acute contact dermatitis in **people**, and apt to be very responsive to topical ***corticosteroids***.

BOUILLON
= *Beef Tea*

Untoward effects: Can cause a hypernatremia. Two 6 month **infants** had diarrhea and generalized edema (**one** after 3000 ml of chicken broth and the **other** after "large" amounts of broth prepared from bouillon cubes); **both** recovered after the broths were discontinued and diuresis instituted.

BOUNCING BET
= *Fuller's Herb* = *Saponaria officinalis* = *Soapwort*

Perennial herb.

Untoward effects: Poisonings in grazing **animals** are usually mild and rarely cause death, because the most poisonous parts, the seeds, are usually not eaten by **animals**. Its ***saponin***, probably ***githagenin***, as well as ***saponarin*** and ***saporubin*** (the most toxic portion), is also found in the leaves and roots, causes nausea, salivation, vomiting, muscle weakness, dizziness, diarrhea, dyspnea, slow respirations, and, rarely, coma.

BRACHIARIA

= *Signal Grass* = *St. Lucia Grass* = *Tanner Grass*

Untoward effects: Its ***oxalate*** and ***nitrate*** content has caused poisoning in grazing **horses** and **cattle** in Australia, Africa, and Brazil. Photosensitization reported in grazing **sheep**; hemolytic anemia, pale mucous membranes, straining on micturition, loss of balance, salivation, hematuria, and prostration reported in affected **cattle**.

BRACKEN FERN

= *Brake Fern* = *Brake Fiddlehead* = *Fougère d'aigle* = *Hog Brake* = *Pteridium aquilinum* = *Pteris aquilina* = *Warabi*

Untoward effects: Although young shoots (after cooking) and roots have been eaten safely by **people** (see below), the plant should be recognized as poisonous. It contains a cumulative "poison", often requiring ingestion for 1–3 months, and symptoms may even occur over 2 weeks after ingestion. It contains thiaminase, destroying **thiamin**, with resulting encephalopathy and other symptoms. In Japan, where bracken shoots are a delicacy, the incidence of stomach cancer is among the highest in the world. Its carcinogens include ***ptaquiloside*** (***aquilide A***) and ***shikimic acid***. Even low doses of these toxins caused malignancies of the lungs, bladder, and lymph system in **mice**, **rats**, **quail**, and **cattle**. Its spores are the most carcinogenic part of the fern. The carcinogens can be transfered through **milk**. Also contains ***prunasin***.

In **cattle**, the key effect is degeneration of bone marrow's hematopoietic ability with thrombocytopenia and leukopenia. Hemorrhages occur in the body, especially on mucous membranes. Anorexia, depression, dyspnea, salivation, hematuria, and nasal and rectal bleeding are common symptoms. Urinary bladder tumors (papillomas and carcinomas) developed after **cattle** consumed 2 g/kg for a mean of 550 days. Hematuria developed as early as 60 days. Anemia and leukocytosis developed later. Cerebrocortical necrosis has occurred. The potential carcinogenic hazards for **humans** from its direct consumption in salads or from affected **cow's** milk should not be ignored.

Horses become weak, incoordinate, depressed with tachycardia, muscle twitiches, timidity, palsy-like lateral head movements, congested mucous membranes, constipation, convulsions, opisthotonus, paralysis, and inability to rise. **Sheep** and some **swine** have also been affected. Symptoms in **sheep** resemble those in **cattle**, and, in **swine**, those of **horses**. **Sheep** have also developed degeneration of retinal epithelium, thrombocytopenia, and leukopenia. When fed to **rats**, anemia, thrombocytopenia, decreased platelet adhesiveness, and increased tumor incidence (mammary, ileal, and bladder) developed. Embryotoxic and teratogenic in **mice**. ***Thiamin*** injection and *DL*-***butyl alcohol*** have helped bring about recovery in various species.

BRADYKININ

= *BRS 640* = *Callidin I* = *Kallidin I*

Vasodilator.

Untoward effects: Decreased blood pressure, tachycardia, sensations of facial and body warmth, headache, metallic taste, vomiting, dyspnea, and palpitations. White ***wine*** contains large amounts. Causes anaphylaxis in many **species**. Angioneurotic edema, urticaria, and asthma-like symptoms from many drugs can be due to non-specific release of vasoactive substances, such as bradykinin and ***anaphylatoxin***.

IV, TD_{LO}, pregnant **mouse**: 25 μg/kg.

BRALLOBARBITAL

= *Vesperone*

Sedative, hypnotic.

Untoward effects: Therapeutic levels are 5–15 mg/l and 20–25 mg/l levels are toxic to **man**. Other workers report levels 50% of the above.

BRASSAIA ACTINOPHYLLA

= *Australian Umbrella Tree*

Untoward effects: Erythema and contact dermatitis in **man**. Contains ***oxalates***, alkaloids, ***saponins***, and cardiac glycosides.

Vomiting, ataxia, leukopenia, and anorexia in a **dog** that ate its leaves. Gastrointestinal hemorrhage and death in **rats**.

BRASSICA sp.

Untoward effects: At least two dozen species are goitrogenic, by preventing the thyroid from accumulating inorganic ***iodine***. Many vegetables, such as ***broccoli***, ***cabbage***, ***cauliflower***, ***Chinese mustard***, ***kale***, ***kohlrabi***, ***rape***, ***rutabaga***, ***Swede***, and ***turnips*** are members of the *Brassica* family and contain ***thiocyanates***. New cultivars have decreased toxicity.

B. campestris = ***Bird Rape*** = ***Chinese Cabbage*** = ***Field Cabbage*** = ***Russian Turnip*** = ***Rutabaga*** = ***Swede*** = ***Swedish Turnip***. Decreases the rate of ***iodine*** uptake by the thyroid in **man**. Poisonings occur, mostly in **children** eating large cooked quantities. Ingestion by **cattle** of the seeds has caused acute hemorrhagic gastroenteritis from its content of ***crotonyl isosulphocyanate*** and *S*-***methylcysteine sulphoxide*** (kale anemia factor).

B. juncea* = *Indian Mustard* = *Moutarde de l'Inde*.** Its leaves and seeds contain ***glucosinolates including ***S-methyl-L-cysteine sulfoxide*** and ingestion by **cattle** causes staggering, abortion, dehydration, and death. The seeds contain ***allyl isothiocyanate*** (q.v.).

B. rapa* = *Kohlrabi seeds have had similar effects on **cattle** as described above.

BRAZIL NUTS

= *Cream Nuts*

***Untoward effects*:** A common cause of atopic allergic reactions in **man**.

BREAD

Untoward effects*:** The likelihood of it being discarded when moldy decreases the possibility of it being a health hazard. **Farmers** who ate ***rye bread infested with *Claviceps purpura* developed psychiatric and circulatory problems from ergotism (q.v.), a condition also known in the Middle Ages. It was then known as "St. Anthony's Fire". In 1789, the fungus produced on ***rye*** grain caused outbreaks of poisoning, panic, and violence that helped to incite the French Revolution. The hungry peasants' desire for bread and Queen Marie Antoinette's famous "Let them eat cake" remark helped inflame the countryside. ***Aflatoxins*** are capable of penetrating into bread from surface mycelia. Several molds on bread can cause a dramatic increase in content of ***amines***. Bread baked from ***methylmercury*** treated ***wheat*** and ***barley*** flour caused the death of 459/6530 **Iraquis**. Other reports put the number at about 1000/60,000. Accidental contamination of flour with an epoxy resin hardener in baking bread caused 84 cases of hepatotoxicity. ***Parathion*** contamination in bread caused the death of 17 **people** in Mexico. So-called "bloody bread" is due to a pigment produced by *Bacillus subtilis*. Hazardous and potentially carcinogenic is ***urethane***, a natural product of fermentation in bread-making. ***Calcium***-enriched white bread may have contributed to a case of renal ***calcium*** stone formation. ***Iodate*** addition to bread in Tasmania, to prevent goiter, helped induced thyrotoxicosis. **Patients** on ***monoamine oxidase inhibitors*** may develop adverse reactions when consuming yeast breads. **Bakers** of bread often develop contact dermatitis from the many additives in bread. ***Potassium bromate***, an antioxidant, used to strengthen bread dough, is a carcinogen, and always assumed to be safe, because it was, supposedly, converted to ***bromide*** during baking; yet, residues of 50–300 ppb have been found. Canada now prohibits its use; California placed it on its Proposition 65 list, while in late 1994, the FDA had taken no action against it.

Ethanol toxicosis in **dogs** after eating sourdough bread.

BRETYLIUM TOSYLATE

= *Bretylan* = *Bretylate* = *Bretylol* = *Darenthin* = *Ornid*

Antiarrhythmic.

***Untoward effects*:** Postural hypotension (21/63), bradycardia (after 5–10 min of tachycardia and hypertension), premature ventricular contractions, palpitations, persistent headache, rash, parotid pain and swelling, weakness, fatigue, dyspnea, increased pulmonary vascular resistance and pulmonary arterial pressure, nausea, vomiting, loose stools, blurred vision, polyuria, decreased libido, ejaculation failure, edema, nasal stuffiness, hyperthermia, dry mouth, and flatulence have all been reported. Avoid rapid IV use.

Drug interactions*:** May aggravate ***digitalis-induced arrhythmias, and, with ***quinidine***, its inotropic effect may be decreased, possibly leading to hypotension.

LD_{50}, **mice**: 50 mg/kg; IM, **rats**: 250 mg/kg.

BREVETOXINS

= *BTX*

***Untoward effects*:** Neurotoxic. Has caused extensive **fish** kills, **mollusc** poisoning, and **human** food poisoning. It is produced by the "red tide's" red algae, *Ptychodiscus brevis* (*Gymnodinium breve*). Causes arrhythmias in **guinea pig** and **rat** hearts, including ventricular tachycardia and atrioventricular blockade.

BRILLIANT GREEN

= *C.I. Basic Green* = *C.I. 42040* = *Diamond Green G* = *Emerald Green* = *Ethyl Green* = *Fast Green J* = *Malachite Green G* = *Solid Green*

Antiseptic dye.

***Untoward effects*:** Hypersensitivity contact reactions on skin and eczema occur. Hemolytic anemia in glucose-6-phosphate dehydrogenase-deficient **infants**. May cause nausea, vomiting, diarrhea, headache, and dizziness in **man**. Necrotic skin reactions in two **volunteers** by application of 1% solution to abraded skin.

LD_{LO}, **human**: 50 mg/kg.

LD_{LO}, **mouse**: 25 mg/kg; **rabbit**: 75 mg/kg. IV, LD in **mice**: 3 mg/kg.

BRIMONIDINE TARTRATE

= *AGN 190,342-LF* = *Alphagan* = *UK 14,304-18*

***Antiglaucoma, α_2-adrenergic agonist.* Topical ophthalmic use.**

***Untoward effects*:** In ~25%, eye stinging and redness, xerostomia; drowsiness and allergic reactions in ~10%.

BRINZOLAMIDE
= *Azopt*

Carbonic anhydrase inhibitor. **For topical ophthalmic use.**

***Untoward effects*:** Bitter aftertaste and transient blurred vision. Occasional headache, blepharitis, dry eye, ocular stinging or pain, and sensation of foreign body presence on eye.

BROCCOLI
= *Brassica oleracea*

Untoward effects*:** Contains five goitrogens. Allergic **individuals** often cross-react with ***cabbage, ***brussel sprouts***, and ***radishes***. Excessive intake may cause carotenemia, in which the palms, soles, and nasolabial folds turn deep yellow, and, in **women**, amenorrhea. Can cause bloating, gas, and irritable bowel syndrome in some **people**. Has been used to cut (dilute) ***cocaine*** and ***PCP***. Contains ***allyl isothiocyanates***, ***oxalates***, ***neochlorogenic acid*** 50–500 ppm (a known carcinogen, after conversion to ***caffeic acid***), and an antiestrogen compound.

Drug interactions*:** Its ***vitamin K content (200 µg/100 g) has caused ***warfarin*** antagonism in **patients** consuming 230–450 g/day.

BRODIFACOUM
= *Assault* = *Bromfenacoum* = *Havoc* = *Mouse Pruff II* = *PP 581* = *Ratak* = *Ropax* = *Talon* = *Volid* = *WBA 8119* = *WeatherBlock*

Rodenticide. **A coumarin anticoagulant.**

Untoward effects*:** A 31-year-old **female** ingested approximately 1500 g of such bait and was successfully treated with ***vitamin K and transfusions. A 65 kg 17-year-old **male** attempted suicide by ingesting 65 g, of similar bait. **He** developed flank pain and gross hematuria. ***Vitamin K*** seemed to have little benefit, and blood values took 51 days to return to normal. ***Vitamin*** K_1 is antidotal and repeat treatments are often required.

Rodents ingesting it in bait at 0.005% level begin dying after 4 days from internal hemorrhage. Has caused sudden death in **swine**. Toxic to **fish** and **wildlife**.

LD_{50}, **rats**: 270 µg/kg; LD_{50}, **dogs**: 0.25–3.6 mg/kg.

BROMACIL
= *Borea* = *Bromoacil* = *DuPont Herbicide 976* = *Hyvar* = *Krovar II* = *Uragon* = *Urox B*

Herbicide.

***Untoward effects*:** **Users** should avoid getting it on **their** eyes or skin.

LD_{LO}, **human**: 500 mg/kg.

Single oral doses of 250 mg/kg were toxic to **cattle**; **sheep** were poisoned by 50 mg/kg/day/10 days. Anorexia and depression occur. Occasionally tympanitis amd muscle incoordination are noted. When fed to **cows** at 5–30 ppm, milk levels reached 0.019 and 0.13 ppm, respectively. **Quail**: up to 5000 ppm led to no toxicity; fed to pregnant **rabbits** (day 8–day 16) at up to 150 ppm, NOT teratogenic. **Rats** showed NO adverse effects on reproduction or lactation, when fed it at up to 250 ppm; 1250 ppm/2 years decreased rate of weight gain.

LD_{50}, **rats**: 5.2 g/kg.

BROMADIOLONE
= *Bromodialone* = *Bromone* = *Contrac* = *Just One Bite* = *Lightning* = *LM-637* = *Maki* = *Ratimus* = *Super-Caid* = *Super-Rozol* = *Trax-One*

Rodenticide. **A coumarin anticoagulant.**

Untoward effects*:** Accidental poisonings in 2- and 3-year-old **children** responded to therapy with ***vitamin K_1 and ***factor IX complex*** over a period of time, because of the chemical's long half-life.

Toxic to **fish**. **Owls** have died after feeding on **rats** killed by it. Cause of sudden death in **swine**.

LD_{50}, **rat**: 1.13 mg/kg; **mouse**: 1.75 mg/kg; **dog**: 11–20 mg/kg.

BROMAZEPAM
= *Compendium* = *Creosedin* = *Durazanil* = *Lectopam* = *Lexomil* = *Lexotan* = *Lexotanil* = *Normoc*

Psychotherapeutic. **A benzodiazepine.**

***Untoward effects*:** Therapeutic serum concentrations for **man** are 0.08–0.17 mg/l and 0.25–0.5 mg/l are toxic. Over-sedation is a serious side-effect. A comatose state and fatalities can occur at > 1 mg/l.

Drug interactions*: *Fluvoxamine significantly increases its serum level and half-life.

Cardiotoxic in **mice**.

LD_{50}, **rat**: 3 g/kg; **mouse**: 2.35 g/kg; **rabbit**: 1.7 g/kg; **guinea pig**: LD, 500 mg/kg.

BROMELAIN
= *Ananase* = *Bromelin* = *Extranase* = *Inflamen* = *Traumanase*

Use*:** Actually, a mixture of several enzymes from the ***pineapple (*Ananas cosmosus*) plant, capable of digesting proteins to the ***peptide*** and ***amino acids***. It is, therefore, used in chill-proofing or clarifying ***beer***, by breaking down visible protein particles into invisible portions. Also in meat tenderizing, and as an anti-inflammatory agent and for wound debridement in **man** (experimentally in **rats**). In making protein hydrolysates and used in dissolution of a phytobezoar and esophageal obstruction with meat chunks.

Untoward effects: Can cause an irritating, itching contact dermatitis or menorrhagia. In addition, it can cause separation of **one's** superficial skin layers with increased skin and capillary permeability, looking very much like an allergic wheal reaction. Gastric discomfort in 13 **males** and nine **females**, 6–76 years, after ingestion for anti-inflammatory purposes. Nausea, vomiting, and diarrhea can occur, on rare occasions.

Drug interactions: Theoretically, it may potentiate oral ***anticoagulant*** activity in **man**.

In **rabbits**, its oral use increases the blood levels of ***ethambutol***. Inhalation of a 3% aerosoling/4 h caused pulmonary emphysema in **Syrian hamsters**. **Rats** receiving rapid IV developed a hypotension for about 10 min and heart rate increase for about 17 min.

IV, LD_{50}, **mice**: 30–35 mg/kg; IP, 36.7 mg/kg; IP, LD_{50}, **rats**: 85 mg/kg; **mouse**: 37 mg/kg.

BROMETHALIN

= *Assault* = *EL 614* = *Vengeance*

Rodenticide. NOT an anticoagulant.

Untoward effects: Neurotoxic to **rats** and **mice**, causing pulmonary and myelin edema, with increased pressure on nerves, decreasing nerve transmission leading to excitability; grand mal seizures, paralysis, and death follows after 2 days.

LD_{50}, **rats**: 2.3 mg/kg; **mouse**: 2–6.7 mg/kg; **cats**: 2 mg/kg; **dogs**: 5 mg/kg.

BROMFENAC

= *AHR 10,282* = *Duract*

Analgesic, non-steroidal anti-inflammatory drug.

Untoward effects: Gastrointestinal upsets with nausea, abdominal cramps, bleeding, and ulcers. Renal toxicity reported with long-term use. Removed from U.S. market because of liver failure, deaths, and need for eight liver transplants.

Drug interactions: Serum concentration decreased by ***phenytoin*** and increased by ***cimetidine***. May increase serum concentration of ***lithium*** and decrease those of ***antihypertensives***.

BROMHEXINE

= *Auxit* = *Bisolvon* = *Ophtosol* = *Tossimex*

Mucolytic, expectorant.

Untoward effects: Some increase in fatigue, slight gastric distress (epigastric fullness, occasionally requires withdrawal of therapy), nausea, and vertigo can occur. Gastrointestinal distress can be more serious in **those** with peptic ulcers. A transient increased serum transaminase reported. It can cross the placenta and, possibly, stimulate production of surface fluids on **fetal** lung.

Drug interactions: Given with ***oxytetracycline***, it doubles the antibiotic concentration in bronchial secretions of **man** by increasing capillary permeability. Admixture with ***gentamicin*** solution causes slow deterioration of the latter.

LD_{50}, **rabbit**: > 10 g/kg; LD_{LO}, **rat**: 1 g/kg.

BROMINDIONE

= *Fluidane* = *Halinone* = *HL 255* = *MG 2555*

Anticoagulant.

Untoward effects: Hematemesis, subdural hemorrhage, epistaxis, purpura, and hematoma at injection sites reported.

LD_{LO}, **human**: 214 μg/kg.

BROMINE

Use: Annual demand of approximately 400 million lbs with approximately ⅓ for flame retardants and 5% for water treatment. Bromide intoxication from many sedatives was often manifested as one of mental confusion, delusion, disorientation, and intellectual function impairment. Bromoderma has also occurred, even with low bromine blood levels. Has similar toxic manifestations as ***chlorine*** (q.v.). Inhalation can cause severe irritation of eyes and respiratory tract. Glottal spasms, choking, vomiting, pulmonary edema, dyspnea, asphyxia, cyanosis, congestive heart failure, and even death. Estimated safe air concentration is 1 ppm (0.1 ppm/8 h) and 1000 ppm is rapidly fatal. Skin contact can corrode tissues, and cause pain and ulcerations. Estimated lethal dose is 1 ml. It can be teratogenic in **people** (delayed bone development and microcephaly) and cause postnatal growth retardation. It can cause serious acneform eruptions, skin burns, and blisters. Inhalation of vapors can be more irritant than ***chlorine's***, but less fatal. Ingestion can also cause serious gastroenteritis and even death. Therapeutic serum levels of bromides range between 10 and 15 μg/ml, are toxic at > 500 μg/ml, and comatose states reported at 2000 μg/ml.

LD_{LO}, **human**: 14 mg/kg; inhalation LC_{LO}, **human**, 1000 ppm.

Inhalation LC_{LO}, **rabbit**: 180 ppm/6.5 h; inhalation LC_{50}, **mouse**: 750 ppm/9 h.

BROMISOVALUM

= *Alluval* = *Bromisoval* = *(α-Bromoisovaleryl)urea* = *Bromural* = *Bromuvan* = *Bromvaletone* = *Brovalurea* = *B.V.U.* = *Dormigene* = *Isobromyl* = *Somnurol*

Sedative, hypnotic.

Untoward effects: Blood serum levels of 25–40 mg/l are toxic in **man** (10–20 mg/l are therapeutic). A 16-year-old **female** ingested 114 tablets containing 11 g and became comatose. After 3 h of ***charcoal*** hemoperfusion, **her** blood levels decreased and **her** coma lightened.

LD_{LO}, **woman**: 400 mg/kg.

LD_{50}, **rat** and **rabbit**: 1.1 and 1.2 g/kg, respectively.

BROMOACETONE

= *Bromacetone* = *1-Bromo-2-propanone*

War gas.

Untoward effects: Severe lacrimator. A **chemist** suffered serious eye injuries plus lung edema and bronchitis from inhaling its fumes. Hypersensitivity to it has developed from previous exposures, as in gas drills.

Inhalation LC_{LO}, **human**: 572 ppm/10 min; **mouse**: 600 mg/m³.

BROMOBENZENE

Untoward effects: Causes hepatitis with necrosis, as well as in the bronchial epithelium and proximal convoluted kidney tubules in **rats** and **mice**.

BROMOCRIPTINE

= *CB 154* = *Parlodel* = *Pravidel* = *Serono-Bagren*

***Antiparkinsonian, dopamine receptor agonist.* The methanesulfonate (mesylate) form is usually used. An *ergot* derivative.**

Untoward effects: Reduces serum ***prolactin***. High incidence of side-effects including: nausea (49%), diarrhea, headache (19%), bad metallic taste, alopecia, hypertension, dizziness (17%), postural hypotension, syncope, incontinence, dry mouth, constipation, giddiness, confusion, nightmares, heartburn, hallucinations, delusions, paranoia, depression, anxiety, schizophrenic relapse, palpitations, erythema, edema, erythema and tenderness of ankles, burning sensation of eyes, diplopia, reversible myopia, ophthalmic trigeminal neuralgia, dyskinesia, rhinorrhea, collagenosis-like symptoms, retroperitoneal fibrosis, leukopenia, hyponatremia, cerebral thrombosis, chest pain, pleuritis, pulmonary fibrosis, pulmonary edema, seizures, erectile disorders, painful clitoral tumescence, neuroleptic malignant syndrome, myocardial infarction, strokes, psychoses including mania, digital pallor with vasospasms, and fatalities. Postpartum intracerebral hemorrhage after 2.5 mg/day/4 days.

In view of some of these, the FDA, in late 1994, advised that it no longer be used for lactation suppression, but still be used in treating Parkinson's disease, acromegaly, and dysfunctions associated with amenorrhea and galactorrhea. Excreted in breast milk. Acute poisoning developed in a 351 lb 19-year-old **female**, after ingesting 90 tablets (225 mg) plus other drugs in a suicide attempt. Fortunately, due to **her** excessive weight, it only induced a brief period of hallucinations 8 h afterwards. Chronic constrictive pericarditis in 63-year-old **male** and **male** aged 69 after 30–40 mg/day/2–4 years in treatment of parkinsonism. Two **children**, 2 and 2½ years, accidentally ingested 25 mg and 7.5 mg, respectively. **They** were successfully treated with ***activated charcoal*** and ***magnesium citrate*** by nasogastric tube. Two **males** aged 20 and 31 years ingested 50 and 75 mg, respectively; **one** experienced spontaneous vomiting, sweating, and mild hypotension; the **other**, only spontaneous vomiting.

Drug interactions: Its side-effects have been potentiated by ***alcohol*** intake, and ***thioridazine*** may interfere with its action. Enhanced dopaminergic effect when used with ***erythromycin***, requires a 67–75% decrease in its dosage. ***Caffeine*** increases its plasma concentration and ***ethanol*** increases ***dopamine*** receptor sensitivity, increases intolerance to it.

Treatment of pregnant **pony mares** caused symptoms of ***fescue*** (q.v.) toxicosis. In Europe, where it has been used to abort **dogs**, behavioral changes were noted.

TD_{LO}, **rat**: 31 g/kg (100 weeks); LD_{50}, **rabbit**: 12 mg/kg.

4-BROMO-2,5-DIMETHOXYAMPHETAMINE

= *4-Bromo-2,5-dimethoxy-m-methylphenethylamine* = *4-Bromo-2,5-DMA* = *100X*

Psychotomimetic.

Untoward effects: A hallucinogenic substance of abuse. A 21-year-old **female** was found lifeless, with pupils fixed and dilated; her 22-year-old **male** companion was found in a convulsive state and remained comatose for several weeks. A 39-year-old **male** had diffuse progressive arterial spasms, right hand pain, cold and blue feet and hands 36 h after ingestion. **He** responded to treatment with ***tolazoline*** and ***sodium nitroprusside***.

IP, LD_{50}, **rat**: 50 mg/kg; IV, **mouse**: 80 mg/kg; IV, **dog**: 6.4 mg/kg.

BROMODIPHENHYDRAMINE

= *Ambodryl* = *Bromanautine* = *Bromazine* = *Bromdiphenhydramine* = *Bromo-Benadryl* = *Deserol* = *Histabromamine*

Antihistamine.

Untoward effects: Contains 24% ***bromide*** and physicians should be aware that excessive intake can cause ***bromine***-induced neurological changes. Has a high affinity for ***antacids*** and ***adsorbents***. Sedation can occur (but less with equal amounts of ***diphenhydramine***). Has anticholinergic effects with occasional dry mouth reported. Photoallergic and phototoxic reactions have occurred in **people**. Rarely, nausea, vomiting, and diarrhea have been reported.

LD_{50}, **rat**: 602 mg/kg; **mouse**: 366 mg/kg; IV, **dog**: 21 mg/kg; IV, **rat**: 55 mg/kg; IV, **mouse**: 63 mg/kg.

9-BROMOFLUORENE

= *Diphenylenemethyl Bromide*

Untoward effects: A cause of eczematous or erythema multiforme reactions in 10% of about 250 **students** doing a lab experiment. They occurred on exposed skin areas 9–12 days after contact, and extended over most of the body, with denuding in some areas. Four **students** had increased temperature (38.3–39.4°C), but were not ill. A 22-year-old **male** exposed to it had extensive bullous eruptions. The reaction was mild at 24 h, but became severe after 9–10 days, with epidermal hyperplasia, pseudo-epitheliomatous, to some degree, and a dense cellular infiltrate in the upper dermis.

IV, LD_{50}, **mice**: 180 mg/kg.

BROMOFORM

= *Methyl Tribromide* = *Tribromomethane*

***Sedative, hypnotic, antitussive.* Contaminant in some drinking water supplies.**

Untoward effects: Use may color urine green to deep yellow color. A potent carcinogen. Irritant to eyes and respiratory tract after inhalation. Decreased central nervous system activity and hepatotoxicity after ingestion or absorption from skin in **humans**. A 4 year **child** became unconscious with dyspnea, miosis, and livid complexion after a 15 minim dose (2–5 drops is the usual **child** dose). Symptoms resemble ***chloroform*** toxicity.

Sedation, flaccid muscle tone, ataxia, piloerection, and prostration noted in toxic **rats**. Pathology showed liver and kidney congestion. When fed to **rats**, they developed an autoimmune thyroiditis.

LD_{50}, male **rats**: 1388 mg/kg; female **rats**: 1147 mg/kg; **mice**: 1.4 g/kg.

BROMOLYSERGIDE

= *BOL-148* = *2-Bromo-N,N-diethyl-D-lysergamide* = *Bromo-LSD* = *D-2-Bromolysergic Acid Diethylamide*

Serotonin antagonist.

Untoward effects: TD_{LO}, **human**: 75 µg/kg. Produces congenital malformations when injected into pregnant **hamsters**, but not orally in pregnant **rats**. As a ***tryptamine*** antagonist in increased pecking in **pigeons**.

TD_{LO}, SC, **hamster** (80 days pregnant): 2 µg/kg; IP, LD_{50}, **mouse**: 75 mg/kg; IV, **mouse**: 20 mg/kg; IV, **rabbit**: 6 mg/kg.

BROMOPHOS

= *Cela S-1942* = *Brofene* = *Brophene* = *ENT 27,162* = *Nexion* = *OMS 658* = *S1942* = *Shg 1942* = *Z51*

Organophosphorus insecticide, acaricide, anticholinesterase.

Untoward effects: Slow onset with protracted symptoms. High dosages to **rats** decreased weight gain with degenerative liver and renal changes. Males were more susceptible than females. **Dogs** have tolerated 500 mg/kg; **cattle** and **sheep**, up to 300 mg/kg; **horses**, up to 800 mg/kg. 400 mg/kg was toxic to a **Zebu cow**. 2 g/kg was the LD_{50}, topically, to abraded **rabbit** skin.

LD_{50}, **rat**: 1.6–3.7 g/kg; **mouse**: 2.8–5.8 g/kg; **guinea pig**: 1.5 g/kg; **rabbits**: 0.7 g/kg; **hens**: 9.7 g/kg; LD_{LO}, **cat**: 750 mg/kg.

BROMOPHOS-ETHYL

= *Filariol*

Organophosphorus insecticide, acaricide, anticholinesterase.

Untoward effects: Slow onset with protracted symptoms. High dosages to **rats** decreased weight gain with degenerative liver and renal changes. Males were more susceptible than females.

LD_{50}, **rat**: 52–116 mg/kg; **mouse**: 230–331 mg/kg; **dog**: 360 mg/kg; **quail**: 200 mg/kg; 3 day **chicks**: 142 mg/kg; dermal LD_{50}, **rat**: 1 g/kg; **rabbit**: 0.5–1.37 g/kg.

BROMOXYNIL

= *Bronate* = *Buctril* = *One Shot*

Pre-emergence herbicide.

Untoward effects: May cause severe allergic contact dermatitis in **handlers**. Can cause extensive, but temporary, eye injury. Use protective goggles and clothing. Toxic to **fish**.

LD_{50}, **rat**: 190 mg/kg; **mouse**: 110 mg/kg; **guinea pig**: 63 mg/kg; **rabbit**: 260 mg/kg.

BROMPHENIRAMINE

***Antihistamine.* The maleate is commonly used.**

Untoward effects: Sedation, sleepiness, dry mouth, nose, and throat, thickened bronchial secretions, dizziness, epigastric distress, and incoordination are among the most common complaints. Insomnia, anorexia, euphoria, gastrointestinal upsets (constipation or diarrhea), wheezing, rash, urticaria, blurred vision, diplopia, vertigo, tinnitus, headache, palpitations, urinary retention, photosensitivity, hemolytic anemia, thrombocytopenia, leukopenia, and agranulocytosis are also reported. Given in a cough syrup to promote a **child's** sleep led to reckless homicide charges in the death of a 1-year-old **female**. Therapeutic serum levels are 0.8–1.5 μg%.

LD_{50}, **rat**: 318 mg/kg; IP, **rat**: 76 mg/kg.

BROMPHENOL BLUE

= *Albustix* = *Albutest* = *Tetrabromol Blue*

pH indicator.

Use: In testing urinary pH, and by IV injection, where normal tissue becomes bluish, and dead tissue from burns can be differentiated by their dark blue color.

Untoward effects: In test strips, 23/53 **patients** showed false positive proteinurea results after ***cifenline***.

BROMSTYROL

= *α-Bromo-β-phenylethylene*

Use: As a fragrance, in public use for over 75 years in soaps, detergents, cosmetic creams and lotions, and perfumes. 4% in petrolatum showed no irritation or sensitization in 25 **human** volunteers. Subsequent reports indicate it can be strongly irritant, and that it should be excluded from preparations used on **infants**.

LD_{50}, **rats**: 1.25 ml/kg; acute dermal LD_{50}, **rabbit**: > 6 ml/kg.

BROMTHYMOL BLUE

= *Thybromol*

***pH indicator.* Especially popular for aquaria.**

Use: By milk testing, for determination of mastitis in **cows**, and in various test strips (viz. ***Azostix***®, ***Combistix***®) for use with **human** or **animal** urines.

Untoward effects: Often underestimates blood urea nitrogen in **patients** with acidosis, and, if alkalosis exists, a slight increase in blood urea nitrogen is not uncommon. Some test kits contain ***potassium hydroxide***, which is a potentially hazardous caustic.

BROMUS sp.

Untoward effects: *B. sphaerocarpa* has caused poisoning of domestic **animals** who grazed on it in Brazil.

BRONOPOL

= *2-Bromo-2-nitropropane-1,3-diol* = *Bronosol* = *Onyxcide*

***Preservative, antiseptic. Formaldehyde* donor.**

Use: In cosmetic shampoo, toiletries, **bovine** mastitis preparations, and air-conditioning water-cooling systems.

Untoward effects: Reported to be non-irritating and non-sensitizing at 0.1% in one series of tests, yet, it has been reported by others as a common contact allergen. Activity may be reduced by increase in alkaline pH. **People** who react to ***tris-nitro*** (in machine cutting oils) cross-react with bronopol. It may contribute to ***nitrosamine*** formation. 13.2% of 190 **humans** evidenced a contact dermatitis from it. **Not** for use in hypo-allergenic creams.

Percutaneous absorption in **rats** and **rabbits**. Decreased motor activity and respiratory rate after oral use in **rats**. Transient gastric irritation in **dogs** at 40 or more mg/kg.

LD_{50}, **rats**, for males and females, respectively: 307 and 342 mg/kg; IP, LD_{50} are approximately 1/10 the oral dosage. LD_{50}, **dog**: 250 mg/kg.

BROOMWEED

= *Gutierrezia sp.* = *Snakeweed* = *Xanthocephalum sp.*

Untoward effects: Injections of extracts cause abortion in **cows**, **goats**, **horses**, and **rabbits**, and after high oral intake (even after 1 week of ingestion). The toxic agent is a triterpinoid saponin. Dyspnea, depression, anorexia, and frequent urination (occasionally bloody). Term **calves** are often weak and eventually die (10–60% are stillborn). Retained placentas are common in **cows** that abort. Similar effects noted in affected **sheep**. Commonly found in Texas and southwestern U.S.

BROTIZOLAN

= *Lendorm* = *Lendormin* = *WE 941*

***Sedative, hypnotic.* A benzodiazepine.**

Untoward effects: ***Alcohol*** (in a social cocktail) impaired its plasma clearance in 13 healthy **male volunteers** (21–43 years).

In **cats**, it caused a marked toxic interaction with ***chloralose-urethane*** anesthesia, but little with ***pentobarbital*** anesthesia. Ataxia, salivation, and diarrhea were seen in **dogs** receiving it IV over a 4 week period; increased feed intake and muscle spasms were seen in 13 week and 12 month oral studies in **monkeys**. In **cats**, **dogs**, **monkeys**, and **rodents**, significant adverse effects when doses used were 7–10 mg/kg, i.e. about 1000 times the intended **human** dosage.

BROXALDINE

= *Brobenzoxaldine*

Use: In combination with ***broxyquinoline*** = ***Intestopan*** for treating intestinal amebiasis and leprosy.

Untoward effects: Nausea with transient erythematous and macular skin rashes, as well as diarrhea and pruritus, have been reported.

BRUCINE

= *2,3-Dimethoxystrychnidin-10-one* = *10,11-Dimethoxystrychnine*

***Alkaloid*. Less potent, but chemically and pharmacologically related to *strychnine*.**

Use: Primarily, as a denaturant in ***alcohol*** and in aerosol hairspray. Occasionally, with ***nux vomica*** as a central nervous system stimulant for **animals** and **man**. For safety, dry powder must be handled under a hood, to prevent inhalation and ***strychnine***-like seizures. Finding it in appreciable amounts in malicious poisonings indicates ***nux vomica*** or Chinese herbals were used, rather than ***strychnine***. Paralyzes peripheral nerves. Can produce a ***curare***-like paralysis with respiratory effects. Treatment, therefore, demands the maintenance of adequate respiration.

LD_{LO}, **human**: 5 mg/kg.

LD_{50}, **rat**: 1 mg/kg; **rabbit**: 4 mg/kg.

BRUGMANSIA

***Botanical. Solanaceae* family. Closely related to *Datura stramonium* (q.v.) = *Jimson Weed*.**

Untoward effects: ***B. arborea***, by accidental or deliberate ingestion, reported as an abortifacient in Uruguay.

B. sanguinea of the western Amazon and the Andes is hallucinogenic.

B. suaveolens = ***Angel's Trumpet***. Its ingestion leads to delirium, disorientation, and hallucinations.

BRUNFELSIA

Large flowering shrub. In the *nightshade* (*Solanaceae*) family.

Untoward effects: ***B. calcyina*** = ***"Yesterday, Today, and Tomorrow"*** has caused the death within 2½ h of a **dog** in Texas that ate its leaves and fruit.

Dogs that have eaten berries of ***B. bonodora*** in Australia had dose-dependent symptoms that included vomiting, enteritis, frequent urination, with occasional ataxia, salivation, muscle tremors, and convulsions.

B. hopeana dried roots = ***Manaca*** depresses central nervous system activity in **rats** (California), and is used for shamanistic purposes and as an aphrodisiac in Brazil. Although highly toxic, the berries are, fortunately, not attractive to **children**.

BRUSSELS SPROUTS

= *Brassica oleracea*

Untoward effects: Contains 110–1560 ppm ***sinigrin*** (converted to ***allyl isothiocyanate***) and 50–500 ppm ***neochlorogenic acid*** (converted to ***caffeic acid***), which are carcinogenic in **rodents**. **People** allergic to its ingestion are usually allergic to ***broccoli***, ***cabbage***, and ***radishes***. A major cause of intestinal gas production in **man**.

Drug interactions: Ingestion in **man** dramatically enhances the metabolism of ***acetaminophen*** (17%) and ***phenacetin***. Its ***vitamin K*** content can cause ***warfarin*** antagonism.

Excessive ingestion has caused hemoglobinemia in **cattle**, with symptoms similar to those produced by ***kale*** (q.v.).

BRYONIA

= *Bryony*

***Cathartic*. Found in the British Isles. A member of the *cucumber* family.**

Untoward effects: Decoctions associated with abortions. Irritant to eye, nose, and throat mucous membranes. It is estimated that ingestion of 15 berries can be fatal to a **child**. Giddiness, delirium, mydriasis, decreased temperature, weak pulse, perspiration, and collapse noted in poisonings in **man**.

Sheep and **goats**, apparently, eat the leaves without ill-effect. After eating ½ oz of the root, **dogs** become dull and die within 24 h, without significant symptoms. 2 lbs of fresh, or 6–8 oz of dried root did not cause purging, but did cause abdominal pain, dyspnea, anorexia, posterior paralysis, diuresis, and hematuria in some **horses**. The berries have been lethal to **poultry**.

BUCHU

Use: As a urinary antiseptic and diuretic in **humans** and **animals**. Gastric stimulant in herbals. From the dried leaves of various *Barosma* species. Contains a volatile oil. The **Hottentots** of South Africa used it before white men arrived there.

BUCINDOLOL

= *MJ 13,105*

β-Adrenergic-blocker.

Untoward effects: Within 2 h after ingesting 200 mg, seven **patients** exhibited toxic effects related to hypotension.

BUCKBEAN

= *Bean Trefoil* = *Bitterworm* = *Bog Bean* = *Bog Hop* = *Bog Myrtle* = *Bog Nut* = *Brook Bean* = *Marsh Clover* = *Marsh Trefoil* = *Menyanthes trifoliate* = *Moonflower* = *Water Shamrock* = *Water Trefoil*

Use: Listed in a number of European pharmacopeias for the treatment of muscular rheumatism.

Untoward effects: Large doses have caused vomiting and purgation.

BUCKWHEAT

= *Fagopyrum esculentum* = *Sarrasin Commun*

Untoward effects: Ingestion has caused anaphylactic reactions in **man**. Buckwheat ***honey*** has caused severe allergic reactions similar to hay fever.

Inflammation and swelling of eyelids, ears, and face with pruritus have been noted, particularly in **animals**. Feeding the plant has caused liver cirrhosis and neoplasms. Ingestion of *F. sagittatum* has caused severe photosensitization of white or unpigmented areas of skin, including blistering of and skin death, infections, and liver damage in **cattle**. *F. esculentum* has caused this photosensitization in **cattle**, **sheep**, **goats**, **pigs**, **horses**, and **fowl** exposed to sunlight. Acute poisoning in **sheep** is characterized by severe cerebral excitement, convulsions, meningitis, gastroenteritis, paralysis, and death. Hallucinations have been reported. Feeding the flowers, seed husks, or hay causes dermatitis in **rats**, **mice**, and **guinea pigs** exposed to ultraviolet light. Usually, ingestion of only 0.25 g/100 g body weight is enough to produce this inflammation in about 24 h. Its photosensitizing agent, ***fagopyrin***, is found in the green or dry plant and little is found in the seeds.

BUCLIZINE

= *Aphilan R* = *Buclifen* = *Buclina* = *Histabutizine* = *Histabutyzine* = *Longifene* = *Posdel* = *Postafen Tabl* = *Softran* = *UCB 4445* = *Vibazine*

Antihistamine, antiemetic.

Untoward effects: Suspect in causing **human** birth defects. Therapeutic doses usually cause mild sedation, dry mouth, jitteriness, and headache. Overdoses have caused convulsions.

TD_{LO}, **human**: 5 mg/kg.

Anophthalmia, microphthalmia, cataracts, and cleft palates reported in offspring, after use in pregnant **rabbits** and **rats**. Microstomia, micrognathia, and laryngeal swelling also reported in **rat** fetuses.

LD_{50}, **rat**: 1 g/kg; **mouse**: 2.1 g/kg; IV, **rat**: 0.5 mg/kg.

BUCLOSAMIDE

= *N-Butyl-4-chlorosalicylanide* = *4-Chloro-2-hydroxy-benzoic acid-N-n-butylamide*

Antifungal.

Untoward effects: Photoallergic contact reactions (spreading dermatitis with swelling and vesiculation) reported. The slow decrease in reactions after removal of the topically applied product indicates its retention in the skin. Cross-sensitivity with ***halogen-hydroxy-benzoics***, ***p-aminobenzoic acid***, ***sulfanilamide*** and its derivatives, ***carbutamide***, ***chlorpropamide***, ***tolbutamide***, ***chlorothiazide***, ***hydrochlorothiazide***, ***cyclopenthiazide***, and ***furosemide***.

BUDESONIDE

= *Budeson* = *Preferid* = *Pulmicort* = *Rhinocort* = *S-1320* = *Spirocort*

Glucocorticoid, anti-inflammatory.

Untoward effects: Severe bronchospasm in an asthmatic **patient** after inhalation. A 5-year-old **female asthmatic** became hyperactive, and aggressive with mania symptoms 48 h after two 50 μg puffs twice daily. Behavior improved after dosage reduced to one puff twice daily. Four young **children** (2½–3 years) showed behavioral disturbances after 400 μg puffs 2–3 times a day. A 50-year-old **female** also showed an allergic reaction to a single, two, 200 μg puffs dose from an inhaler. **Father** (39-year-old **male**) and **son** (9-year-old **male**) **asthmatics** both had allergic reactions to an inhalation of 200 and 100 μg puffs, twice daily. Adrenal insufficiency has developed with inhaled doses of 1.6–2.4 mg. Nasal irritation, epistaxis, and skin bruising occur. Six weeks treatment with it intranasally in **children** has suppressed lower leg growth in **children** with allergic rhinitis. Allergic contact dermatitis from topical use has also been reported.

Teratogenicity demonstrated in **rabbits**.

BUFENCARB

= *Bux* = *ENT 27127* = *Metalkamate* = *Ortho 5353* = *RE5353*

Insecticide, anticholinesterase. **A methylcarbamate.**

Untoward effects: Can be absorbed after ingestion, inhalation, or dermal application, causing within 12 h headache, weakness, ataxia, blurred vision, miosis, muscle twitching, and tremors. Occasional convulsions, mental confusion, nausea, cramps, bradycardia, dyspnea, sweating, incontinence, salivation, pulmonary edema, and unconsciousness.

LD_{50}, male and female **rats**: 97 and 61 mg/kg, respectively; **chickens**: 44 mg/kg; **mallard ducks**: 10.5 mg/kg; **pheasants**: 88 mg/kg with ataxia, dyspnea, wing-drop, tremors, regurgitation, ptosis, and tetany. Symptoms can appear within 15 min and deaths usually occur between 20 min and 1 h.

BUFEXAMAC

= *CP 1044 J3* = *Droxarol* = *Droxaryl* = *Feximac* = *Malipuran* = *Mofenar* = *Norfemac* = *Parfenac* = *Parfenal*

Anti-inflammatory, analgesic, antipyretic.

Untoward effects: Possible allergic reactions causing discontinuance of treatment, dyspnea in 2/24, drug rash in 1/24. Local intolerance, such as burning, irritation, redness, swelling, and itching reported. A 55-year-old **male** developed an erythema multiforme-like dermatitis after a few days of topical application of a 5% lotion.

BUFORMIN

= *Andere* = *Biforon* = *Bigunal* = *Bufonamin* = *Bulbonin* = *Butformin* = *DBV* = *Diabrin* = *Dibetos* = *Gliporal* = *Insulamin* = *Krebon* = *Panformin* = *Silubin* = *Sindiatil* = *Tidemol* = *W37* = *Ziavetine*

Untoward effects: A **patient** with diabetes mellitus developed a severe lactic acidosis after taking 600 mg/day. Nausea and diarrhea in 4–10%, pruriginous exanthema in 1% of **patients**. Increased ***bromophthalein*** retention in **some** at beginning of treatment.

LD_{50}, **rat**: 320 mg/kg; oral and IP, LD_{50}, **mice**: 380 mg/kg.

BUFOTENINE

= *5-Hydroxy-N,N-dimethyltryptamine* = *Mappine*

An indole isolated from many plants including *Amanita citrina* and *A. porphyria*, the snuff, *cohoba*, and in tropical toad (*Bufo vulgaris*) venoms.

Untoward effects: Hallucinogenic in **man** and **animals**.

IV, TD_{LO}, **human**: 57 μg/kg.

Parenteral doses > 1 mg/kg may cause acute heart failure and death in **cattle** and **sheep**. Lower dosages induced hyperexcitability, salivation, head nodding, mydriasis, poor coordination, and lateral recumbency.

IP, LD_{50}, **mouse**: 290 mg/kg.

BUMETANIDE

= *Bumex* = *Burinex* = *Fontego* = *Fordiuran* = *Lixil* = *Lunetoron* = *PF 1593* = *Ro 10-6338* = *Segurex*

Diuretic.

Untoward effects: Increases ***potassium*** and decreases ***uric acid*** excretion, increases serum ***uric acid***, and may cause ototoxicity in **humans** (also in **cats**, **dogs**, and **guinea pigs**). There is an uncertain risk of potentiation of ***aminoglycoside antibiotic*** ototoxicity. An anaphylactic reaction has been reported in a 41-year-old **male**. This may occur in **patients** allergic to ***sulfonamides***. Adverse reactions occurred in about 4% of treated **patients**, with myopathy, muscle cramps, dizziness, hypotension, headache, nausea, increased serum ***uric acid***, and encephalopathy roughly in 0.6–1.1%. Less common have been impaired hearing, skin rash, pruritus, hives, abdominal and arthritic pain, vomiting, vertigo, fatigue, sweating, dry mouth, diarrhea, nipple tenderness, gynecomastia, thrombocytopenia, granulocytopenia or leukopenia, premature ejaculation, and difficulty maintaining an erection.

Drug interactions: Can increase toxicity of ***digitalis*** preparations by causing a decrease in serum ***potassium***. ***Probenecid*** and ***indomethacin*** decrease its diuretic effect. No significant effects by use with ***aspirin*** or ***warfarin***.

Its toxicity is noted more in **rabbits** than **rats**, **dogs**, **hamsters**, **mice**, or **baboons**.

IV, LD_{50}, **mouse**: 330 mg/kg.

BUNAMIDINE

= *N,N-Dibutyl-4-(hexyloxy)-1-naphthamidine* = *Scolaban*

Anthelmintic.

Untoward effects: It is a cholinesterase inhibitor and can produce irreversible neuromuscular block (respiratory and cardiac failure with ensuing death), particularly after parenteral usage. Do NOT overdose. Avoid certain ***antibiotics***, ***organic phosphorus anticholinesterase compounds***, ***succinylcholine***, and other compounds that might potentiate these effects. May be extremely irritant to **one's** eyes.

Some vomiting and diarrhea is occasionally noted in **cats** and **dogs**. Avoid vigorous exercise for at least 48 h after use. Avoid its use in cardiac cases. Avoid its use in male **dogs** within 1 month of their breeding or in the debilitated or very young. In male **dogs**, it has interfered with spermatogenesis. In cardiac cases, it may sensitize the myocardium to endogenous ***catecholamines***.

LD_{50}, **mouse**: 540 mg/kg.

BUNAMIODYL

= *Bunaiod* = *Buniodyl* = *Orabilex* = *Orabilix*

Cholecystographic diagnostic aid.

Untoward effects: It has fallen into disuse because its use, especially at high dosage (> 4.5 g) was associated with oliguria, tubular necrosis, and acute renal failure.

LD_{50}, **mice**: 2.78 g/kg; IV, 418 mg/kg.

BUNAZOSIN

= *Detanol* = *Detantol* = *E-643* = *E1015*

Antihypertensive, α_1-adrenergic blocker.

Untoward effects: Occasional orthostatic hypotension, fatigue, and headache. Concurrent use with ***rifampin*** decreases time to maximum plasma concentration by about 20%.

BUPHANE DISTICHA

= *BRC 968* = *Haemanthus toxicarius*

Use: As an arrow nerve poison, made from its bulbs (6–8" in diameter). For bringing down small game in South Africa and the Congo.

Untoward effects: Due to its alkaloid, ***haemanthine***, found to be stable and active on poison arrows stored for over 100 years. Oral contact by **humans** with the juice of these bulbs causes dry mouth, general weakness, and delirium.

5 mg SC into a 1 kg **rabbit** increased breathing rate, and caused loss of reflexes. Within 15 min, its head drooped and the **rabbit** trembled. Spasms occurred after 25 min, followed by death. 15 mg/kg caused death within 15 min. Injections of 7–10 mg/kg to **pigeons** caused vomiting, exhaustion, and sleepiness (for at least 2 h), and catatonic-like effects, ending in death from respiratory failure.

BUPIVACAINE

= *AH-2250* = *Anekain* = *Carbostesin* = *LAC-43* = *Marcaina* = *Marcaine*

Long-acting local anesthetic.

Untoward effects: Methemoglobinemia reported with doses > 600 mg in **adults**, two with slight chills, and one with headache with peridural injection. A 36-year-old **female** had a generalized seizure after 10 ml of 0.75% solution via epidural catheter. A 40-year-old **female** inadvertently received it IV during an attempted epidural procedure, and also developed seizures. Another report indicates convulsions in four **women** in labor within 30 s after injection through an epidural catheter, probably accidentally in a vein. Mild cerebral stimulation in **others** after inadvertent IV injections. Acidosis, bradycardia, orthostatic hypotension, nausea, vomiting, and **fetal** bradycardia or tachycardia. Cardiac arrest with 16/20 **women** dead after use in obstetrical procedures. Placental passage occurs (**fetal** concentration approximately ¼ that of maternal) and it is deactivated in the liver of **adults**. Horner's syndrome with marked ptosis, miosis, and cyanosis of fingertips in 18-year-old **female** after its use for lumbar extradural blockade. Symptoms started subsiding after about 50 min. Therapeutic or normal serum levels are said to be about 0.25%, and toxicity occurs at 1.0 mg%. Toxic serum levels have also been reported as 2–4 µg/ml. Amide local anesthetics can cause acute hyperthermia with muscle rigidity, hyperkalemia, metabolic acidosis, and ventricular arrhythmias. Use may increase **patients'** susceptibility to ***opioid*** respiratory decrease. Intolerable metallic taste reported in 48-year-old **female** after 5 ml/h infusion.

TD_{LO}, **human**: 4.3 g/kg.

SC, LD_{50}, **rat**: 48 mg/kg; IV, 5.6 g/kg; SC, **mouse**: 53 mg/kg; IV, 7.3 g/kg.

BUPRENORPHINE

= *Buprenex* = *CL 112,302* = *6029M* = *NIH 8805* = *RX 6029M* = *Subutex* = *Temgesic* = *UM 952*

Untoward effects: Some nausea, vomiting, and dizziness in treated **patients** that persists longer than that following ***morphine***. Marked miosis within 10 min of IV injection. 30–40 µg/kg caused severe respiratory depression (< 8/min) in 9/12 cholecystectomy **patients**. A group of **females** (20–50 years) who received 800 µg sublingually after hysterectomy had late onset prolonged respiratory depression. Auditory hallucinations with hearing of commanding voices in 50-year-old **male** telling **him** to jump through a window after being given 200 mg sublingually for pain, and he complied. Nightmares and visual hallucinations attributed to 750–1200 µg epidurally in five postoperative **patients**, aged 38–58 years. Slight abuse potential, especially after IV usage (viz. 24-year-old **male** taking 4.5 mg/day IV for 2 months). Ulceration on the upper surface of the tongue of an 82 year **patient** after placing it on the tongue instead of sublingually. A single postsurgical injection in 65-year-old **male** led to hypertension and tachycardia. Abrupt withdrawal syndromes have been reported and sometimes do not appear until 2 weeks after discontinuing the drug. A 66-year-old **male** developed acute urinary retention after treatment with six 200 µg tablets over an 18 h period. Significant psychomotor impairment was noted to peak in about 4 h in several tests.

BUPROPION

= *Wellbatrin* = *Wellbutin* = *Zyban*

Antidepressant.

Untoward effects: Doses ⅓ greater than recommended are associated with a high risk of seizures, especially in bulimic **patients**. In the 1980s, the incidence of seizures was reported in 0.4–5.8% of treated **patients**, and were, again, apparently dose-related. Anxiety, agitation, insomnia, tremor, increased motor activity, dry mouth, blurred vision, constipation, nausea, vomiting, syncope, dizziness, headache, tachycardia, palpitations, alopecia, some sexual dysfunction, diarrhea, somnolence, insomnia, anorexia, and weight loss have been reported. Rarely, mania, tinnitus, nightmares, psychosis, hallucinations, dose-related seizures and serum sickness, and falling backward have been reported. May mimic a transient ischemic attack. Accumulates in **human** breast milk at levels higher than in **maternal** serum.

Drug interactions*: *Ritonavir decreases its metabolism and may increase its toxicity. Use with ***amantadine*** and ***fluoxetine*** may cause panic symptoms and psychotic reactions. ***Carbamazepine*** increases its metabolism and decreases its effectiveness. Addition of ***amantadine*** to its use in three nursing home **patients** caused neurotoxicity.

Paradoxical response (increased aggression) in some treated **cats** and **dogs**. Half-life is ~1.7 h in **dogs**. Nausea, vomiting, dyspnea, ataxia, salivation, arrhythmias, hypotension, tremors, and seizures occur within hours after poisoning or overdoses.

LD_{50}, **rats**: 600 mg/kg; **mice**: 575 mg/kg; IP, **rats**: 210 mg/kg; **mice**: 230 mg/kg.

BUQUINOLATE
= *Bonaid*

***Coccidiostat*. A quinoline.**

***Untoward effects*:** Drug resistance and cross-resistance has developed.

LD, **mice** and **rats**: > 7.5 g/kg. No signs of toxicosis in **dogs** receiving 450 mg/kg/day/90 days. 20 g/kg to 7-day-old **chicks** was non-toxic. Oral doses of 10 g/kg made 2-week-old **turkey poults** lethargic, with recovery after 24 h.

BUR BUTTERCUP
= *Ceratocephalus testiculatus*

Untoward effects*:** Most toxic to grazing **animals** in the early flower stage (containing the toxin, ***ranunculin (q.v.), at about 2.3% of the plant's dry weight), when they are short of supplement. About 500 g is enough to kill a 100 lb **sheep**. Weakness, watery diarrhea, dyspnea, tachycardia, anorexia, occasional increased temperature precede death. On post-mortem, they show varying degrees of inflammation and edema of the rumen wall, hemorrhage in the heart's left ventricle, congestion of lungs, liver, and kidneys, and excessive fluids in the thoracic and abdominal cavities. Crushed plants, when dry, are less toxic.

BURDOCK
= *Arctium sp.* = *Bardane* = *Bardock* = *Beggar's Buttons* = *Burr Burr* = *Clotbur* = *Cockle Button* = *Cuckold Dock* = *Cuckoo Button* = *Lappa* = *Stick Button*

Untoward effects*:** A case of poisoning in **humans** from consuming ½ a cup of its root tea (*A. minus* and *A. lappa*) has occurred. The root was purchased in a health food store. The tea caused typical anticholinergic symptoms of blurred vision, mydriasis, dry mouth, inability to urinate, bizarre speech and behavior. This ***atropine-like agent has been declared a contaminant by those unable to discover such activity in burdock. The root can also be a strong purgative.

A. lappa has caused severe itching and localized mechanical injury in Alabama **livestock**, a granular stomatitis with necrosis and tissue sloughing in **dogs**. It is the weed that Shakespeare describes Lear as crowned with. Has caused abortions in **cattle** because it has a high ***nitrate*** content.

BUSH TEAS

A poorly defined group of plant teas.

Untoward effects*: *Crotolaria and ***Senecio*** (q.v.) have been incriminated as having a role in causing liver disease, cirrhosis, and esophageal cancer in many underdeveloped countries. Their teas have also been suspect in increased incidence of esophageal cancer in **humans** in areas around Charleston, South Carolina.

BUSPIRONE
= *Bespar* = *Buspar* = *Buspinol* = *Censpar* = *Lucelan* = *MJ 9022-1* = *Travin*

***Anxiolytic*. Non-benzodiazepine.**

Untoward effects*:** Relatively free from the abuse potential, dependence, and withdrawal symptoms that exist with various ***benzodiazepines. Fatigue, dizziness, and slight headaches occur. Occasionally, nausea, diarrhea, and paresthesias are reported. Rarely, psychosis, vivid dreams, delirium, decreased libido, and mania occur. Panic attacks and mild hypertension occurred in a 40-year-old **male** on two occasions following a single 10 mg dose. A 30-year-old **male** abused it by crushing and snorting a 10 mg oral tablet. **He** reported sensations similar to a mild ***cocaine***-induced stimulation. These occurred within 10 min and lasted about 15–20 min. The abuse continued and was stopped in about a week, as **he** then experienced dysphoria. Has caused priapism.

Drug interactions*:** It interferes with ***metanephrine urinary assays. ***Grapefruit*** juice increases its serum concentration by more than twice and increases its elimination half-life.

BUSULFAN
= *Busulphan* = *CB 2041* = *GT 41* = *Mielucin* = *Misulban* = *Mitosan* = *Myeleukon* = *Myeloleukon* = *Myelosan* = *Myleran* = *Sulfabutin*

Antineoplastic, insect sterilant. **An alkylating agent.**

***Untoward effects*:** Can cause a cutaneous vasculitis with a combination of papules, palpable purpura, urticaria, nodules, ulcerations, and hemorrhagic blisters, even involving liver, kidney, brain, and joints. Acute toxicity has essentially been one of nausea and vomiting, and, rarely, diarrhea. Delayed toxicity is manifested primarily as bone marrow depression (myeloid leukemia aplasia), pulmonary infiltrates, and fibrosis (often related to duration of treatment) in approximately 12–43% of treated **patients**, dyspnea, cough, alopecia, gynecomastia, azoospermia (testicular atrophy and damage to seminiferous epithelium), and ovarian failure (amenorrhea and premature menopause), hyperpigmentation (more common in **people** with darker skin), leukemia, chromosome defects, hepatitis with jaundice and veno-occlusive disease, cataracts, seizures (with high dosages), cerebral hemorrhages (secondary to bone marrow exhaustion), decreased erythropoiesis, and hyperuricemia. Has caused abortions, teratogenicity, intrauterine and postnatal growth retardation, including decreased white blood cells, hydronephrosis, and cleft palate, brown nails, weight loss, weakness, fever, cholestasis, leukopenia, and thrombocytopenia. Treated **patients** have developed endocardial fibrosis, many cytological abnormalities, breast cancer, cancer of the external **female** genitalia, bronchiolar cell carcinoma, and, rarely, an impairment of neuromuscular conduction. Myeloschisis in a **human** embryo of a 39-year-old **female** primipara. Possible myasthenia gravis in a 70-year-old **male** after 9 months of treatment. Sideroblastic anemia in a 50-year-old **female**. Cardiac tamponade occurred in 8/400 thalassemic **patients** treated before bone marrow transplantation. Endocardial fibrosis of the left ventricle demonstrated on post-mortem of **patient** receiving the drug for 9 years. Radiation predisposes to additional pulmonary toxicity.

TD_{LO}, **women**: 798 mg/kg.

IP injections in **mice** caused lymph system neoplasms, and IV use increased the number of thymic and ovarian tumors. Offspring of **rats** given it 5–6 days before full-term birth were sterile. Teratogenic in **rats**.

IV, LD_{50}, **rats**: 1.8 mg/kg; IP, 18 mg/kg; IV, TD_{LO}, **dog** and **monkey**: 8 mg/kg.

BUTABARBITAL

= *Asturidon* = *Bubarbital* = *Butabarbitone* = *Butabarpal* = *Busodium* = *Busotran* = *Butabar* = *Butabon* = *Butak* = *Buta-Kay* = *Butalan* = *Butalix* = *Butanotic* = *Butased* = *Butatran* = *Butazem* = *Butex* = *Buticaps* = *Butisol* = *Butrate* = *Butte* = *Carrbutabarb* = *Ethnor* = *Loubarb* = *Neravan* = *Prelital* = *Sarisol* = *Secbutobarbitone*

Sedative, hypnotic. **NOT *butobarbital.***

Untoward effects*:** Sedation, respiratory depression, and even coma. "Hangover" effects are commonly reported. Induces liver microsomal enzyme activity and decreases ***warfarin, ***coumarin***, and ***indandione*** plasma levels, their effectiveness, and decreases prothrombin time. This interference peaks in about 1–2 weeks and continues for 6 weeks after cessation of butabarbital therapy. Can also decrease the therapeutic response to ***tricyclic antidepressants***, ***griseofulvin***, and ***steroids***. Excreted in **human** breast milk. It crosses the placenta and enters **fetal** tissues. Withdrawal symptoms have occurred in **infants** born to **mothers** treated with it. Some elderly **patients** react to it paradoxically, showing excitement, confusion, or depression. Can be removed by hemoperfusion. A 24-year-old **female** ingested 6 g. **Her** blood level was 8.6 mg/100 ml. **She** survived after intensive supportive therapy, moderate diuresis, and assisted respiration.

LD_{LO}, **human**: 120 mg/kg.

LD_{50}, **rat**: 78 mg/kg; **rabbit**: 194 mg/kg; IV, LD_{50}, **dog**: 90 mg/kg; **rat**: 70 mg/kg; **rabbit**: 91 mg/kg.

BUTACAINE

= *Butelline*

Topical anesthetic. **A *para-aminobenzoic acid* ester. A "caine".**

Untoward effects*:** Being much more toxic SC than ***procaine, its use is limited to topical applications. Kligman reported a 25% concentration leading to moderate skin sensitization. Contact eczema has been reported. Cross-sensitivity with ***procaine***, ***benzocaine***, and ***tetracaine***. Dermatitis of the eyelids and face followed its instillation into a **patient's** conjunctival sac. Its use is contraindicated where breaks exist in the mucuous membrane (viz. urethra), as symptoms of acute poisoning (anxiety, fainting, pallor, dyspnea, brief convulsions, and respiratory arrest) can arise from such accidental entry into the general circulation. Avoid unnecessary parenteral use (**NOT** over 0.25 or 0.5% solution), or use with ***epinephrine*** or ***morphine***. Deteriorated by light.

SC, LD_{LO}, **rat**: 150 mg/kg; **mouse**: 100 mg/kg; **cat**: 30 mg/kg; **rabbit**: 50 mg/kg; **guinea pig**: 45 mg/kg; IV, LD_{50}, **mice**: 12.4 mg/kg.

BUTACARB

= *Bassa* = *BPMC* = *3,5-Di-t-butylphenyl-N-methylcarbamate*

Carbamate pesticide.

Untoward effects*:** ***Atropine is usually antidotal. Use of ***pralidoxime*** and other ***oximes*** is not recommended.

Goats, **sheep**, and **pigs** dosed with 300 mg/kg had profuse salivation, shivering, and dyspnea. All the **pigs** vomited and all the **sheep** died within 3 h. Visceral hemorrhages were noted on those that died. Doses of 30 mg/kg led to mild salivation and very rapid breathing. Heart rate increased 3–12 h after dosing, then decreased. Atony of the gizzard was noted in affected **birds**. Respiration ceased before the heart stopped beating.

LD_{50}, **rat**: 1.8 g/kg; **mouse**: 3.2 g/kg; **dog**: 1 g/kg; 6 week **cockerels**: 1.4 g/kg; 9 week male **quail**: 240 mg/kg.

BUTACHLOR

= *Butanex* = *CP 53,619* = *Machete*

Herbicide.

Untoward effects: LD_{50}, **rats**: 1.74 g/kg; dermal, **rabbit**: 2.5 g/kg.

BUTACLAMOL

Neuroleptic, antischizophrenic.

Untoward effects: A much higher incidence of extrapyramidal signs than with ***chlorpromazine***.

1,3-BUTADIENE

= *Biethylene* = *Bivinyl* = *α,γ-Butadiene* = *Divinyl* = *Erythrene* = *Pyrrolylene* = *Vinylethylene*

Colorless, non-corrosive, flammable gas with a mild aromatic or gasoline-like odor.

Use: Chemical intermediate, in manufacturing various rubbers in tires and conveyor belts, resins, carpet backings, paper coatings, polymers for pipes, automobile and appliance parts. Oil-based chemical giving rubber its elasticity and flexibility. Demand varies, but it averages about 4–5 thousand million lbs annually.

Untoward effects: Highly toxic. Should be handled with extreme care. Highly irritating vapors to nose and throat, causing nausea and blurring of vision. Compressed liquid causes freeze burns. A great potential for fire or explosion. It is estimated that 8000–65000 U.S. **workers** are potentially exposed to it in the workplace. OSHA has recommended 1000 ppm (2200 mg/m^3)/8 h TWA concentration for occupational exposure. Various pressures are attempting to make 10, or 1 and 2 ppm the newest standards. Factory health record surveys indicate it a probable industrial cause of liver cancer and leukemia.

Female **mice** exposed to 625 and 1250 ppm showed an increase in granulomatous tumors of the ovary, cardiac hemangiosarcomas, malignant lymphomas, stomach papillomas, alveolar/bronchiolar adenomas, carcinomas of the lungs, and an increase of mammary gland tumors at 8000 ppm. **Mice** have developed lung tumors at 60 ppm exposures. **Monkeys** are less susceptible. Various teratogenic effects noted in pregnant **rats**.

Inhalation LC_{50}, **rats** and **mice**: 129,000 and 122,000 ppm, respectively; LC_{100}, **rabbits**: 250,000 ppm. Toxic effects were light anesthesia, running movements, tremors, then deep anesthesia and death. LD_{50}, **rat**: 5.5 g/kg, **mouse**: 3.2 g/kg.

BUTALBITAL

= *Alisobumal* = *Allybarbital* = *Itobarbital* = *Lotusate* = *Talbutal* = *Tetrallobarbital* = *Sandoptal*

Sedative, hypnotic.

Untoward effects: May become habit-forming (addictive). Dizziness, nausea, and grogginess or drowsiness. Has been used in suicide attempts. 10–25 mg/l of serum is considered toxic. Hemoperfusion with ***Amberlite XAD-4*** reported effective as treatment.

LD_{LO}, **human**: 3 mg/kg.

Drug interactions: May decrease ***imipramine*** serum concentration by 50%.

LD_{50}, **rats**: 57.5 mg/kg; **red-wing blackbirds**: 75 mg/kg; **starlings**: > 100 mg/kg; **pigeons**: 56 mg/kg; **mallard ducks**: 100 mg/kg.

BUTAMBEN

= *Butyl Aminobenzoate* = *Butesin* = *Butoform* = *Planoform* = *Scuroforme*

***Topical anesthetic.* A *para-aminobenzoic acid* ester.**

Untoward effects: Cyanosis and methemoglobinemia may have resulted from its topical use prior to intubation. Cases of dermatitis eczematosa reported from the use of it topically as the picrate.

BUTAMISOLE

= *CL 206,214* = *Styquin*

Anthelmintic.

Untoward effects: Do **NOT** use in heartworm-positive, debilitated, or **dogs** under 2 months of age, since it may be lethal to them. **NOT** to be given simultaneously with ***bunamidine***, which increases the former's toxicity. The latter can be given 48 h later. Localized swellings and induration have occurred at injection sites.

BUTANE

Use: As a fuel, in aerosol propellants, and in the manufacture of synthetic rubber. An aliphatic hydrocarbon.

Untoward effects: Butane lighters have exploded when left near heaters (or on a cockpit instrument panel, or a window ledge in sunlight) and if outside atmospheric pressures drop (as at high altitudes or loss of aircraft cabin pressure). An exploding butane lighter has the explosive force of approximately three sticks of ***dynamite***. A Union-Pacific **employee** had a spark from **his** welding hit the butane lighter in **his** pocket, causing an explosion, which killed **him** instantly. Another fatal accident also occurred there. Inhalation from an aerosol propellant caused ventricular fibrillation in a **2-year-old** and sudden death in an **11-** and **15-year-old** by asphyxiation. Can be narcotic in high concentrations.

Inhalation LC_{50}, **rat**: 658 g/m^3/4 h; **mouse**: 680 g/m^3/ 2 h.

BUTANILICAINE

= *Hostacain*

Local anesthetic.

Untoward effects: Fatigue and transient severe cyanosis of the arm in some **patients** (13½–75 years) receiving brachial plexus block anesthetic for surgery of arm and hand.

Degenerative spinal cord changes in four **dogs** accidentally given subarachnoid, instead of epidural, injections of 1 and 2% solutions. Two were paralyzed.

BUTANONE

= *2-Butanone* = *MEK* = *Methyl Acetone* = *Methyl Ethyl Ketone* = *2-Oxobutane*

Over 50 million lbs imported annually into the U.S.

Use: In liquid cements, adhesives, magnetic tapes, lube oil dewaxing, smokeless powders, plastic resin catalysts, solvents, in vinyl coatings, antiozonants, painting, cleaning fluids, printing inks, organic synthesis, and in manufacturing ***cocaine***.

Untoward effects: Flammable, eye, nose, throat, and skin irritant. Can cause nausea, headache, dizziness, vomiting, fainting (300–500 ppm exposure), central nervous system depression, and unconsciousness with emphysema and congestion of the liver and kidneys. Has a ***mint*** or ***acetone***-like odor. Odor and irritation prevent narcosis in **humans**. It has been an abused inhalant. Severe toxic polyneuropathy in shoe factory and coated fabrics plant **workers** and in 25 young **people** in Berlin addicted to it in solvents, required 2½–3 years for the peripheral motor defects to regress. Daily ***glue***-sniffing (mixture of ***MEK*** and ***toluene***) caused weakness of proximal and distal muscles after several years in three young **male Australians**. Peripheral nerve conduction was decreased. Percutaneous absorption occurs. Its oxime's chronic toxicity may have hematopoietic and oncogenic effects. Its *peroxide* (***MEKP***) vapors (in addition to eye, skin, nose, and throat irritation) cause coughing, dyspnea, pulmonary edema, blurred vision, blisters, abdominal pain, vomiting, and diarrhea. Dermatitis has occurred on exposed skin and exposure to vapors or liquid has caused numbness of fingers and arms of **workers**. 350 ppm irritant to **human** eyes. OSHA suggests maximum exposure at 0.2 ppm for the *peroxide*. OSHA and National Institute for Occupational Safety and Health set upper exposure limits of 200 ppm; 15-min exposure, 300 ppm; and immediate danger to life or health, 3000 ppm.

TD_{LO}, **human**: 500 mg/kg. 200 ppm TLV (500 mg/m^3).

Moderate reversible injuries to **rabbit** eyes from high vapor concentration. Extreme exposure led to opaque corneas, healing spontaneously in 8 days. Inhalation of 1000 or 3000 ppm by **rats** on days 6–15 of gestation 7 h/day was embryotoxic, fetotoxic, and teratogenic. Concentrated vapors killed 1/6 **rats** within 30 min and 5/6 in 1 h. **Guinea pigs** have survived 1 min exposure to 100,000 ppm and 1 h exposure to 10,000. The latter caused eye and nose irritation and narcosis after 4–5 h with emphysema, corneal opacities, congestion of brain, lungs, liver, and kidneys.

LD_{50}, **rats**: 3.4 ml or 2.6 g/kg, dermal; **rabbit**: 12.6 ml/kg.

BUTAPERAZINE

= *AHR 712* = *Bayer 1362* = *Butyrylperazine* = *Megalectil* = *Randolectil* = *Repoise* = *Tyrylen*

Antipsychotic. A non-halogenated propylpiperazine and phenothiazine.

Untoward effects: Low dosage (5 mg tid) has caused hypertension in some **patients**, decreasing threshold for seizures, has caused dystonias and cataracts, drowsiness, and decreased mental alertness. Avoid its use in **patients** with a history of jaundice or blood disorders, as it has caused blood dyscrasias and liver disorders. Has caused cardiac arrhythmias. If severe hypotension occurs, use ***levarterenol*** or ***phenylephrine***, but **NOT** ***epinephrine***, which, paradoxically, may cause a further decrease in blood pressure. Dermatological reactions, allergic reactions, photosensitivity, blurred vision, aggravation of glaucoma, nausea, vomiting, transitory leukopenia, headache, dry mouth, constipation, tachycardia, weight gain, ataxia, dizziness, agitation, flushing, convulsions, and extrapyrimidal symptoms; paradoxically, exacerbation of psychoses has been reported. Decreases metabolic elimination when given with ***desipramine***. Higher incidence of tardive dyskinesias when given with ***conjugated estrogens***.

Treated **pregnant mice** are more susceptible to cleft palates by its use than are **rats** and **rabbits**. As a

prolactin-inducing drug, it may have caused mammary gland tumors in **rats** and **mice**.

LD_{50}, **rat**: 264 mg/kg, **mouse**: 296 mg/kg; IV, **rat**: 63 mg/kg; IV, **mouse**: 67 mg/kg.

BUTENAFINE

= *KP 363* = *Mentax*

Antifungal.* Topical in treating *Tinea sp.

Untoward effects: Occasional burning and stinging from 1% cream.

BUTENOLIDE

= *4-Acetamido-4-hydroxy-2-butenoic Acid*

***Mycotoxin.* A toxin isolated from *Fusarium tricinctum* on toxic fescue.**

Untoward effects: Has killed **calves** within 3–4 days. Low repeated intake by **calves** has caused inflammatory encrusted skin lesions and tail necrosis. Disease syndromes from contaminated feed reported in **fish**, **rat**, **mouse**, **rabbit**, **guinea pig**, **horse**, **donkey**, **cattle**, **sheep**, **goat**, **swine**, **poultry**, **cat**, **insects**, and **man**. Vapors at 130 ppm caused eye irritation, salivation, nasal discharge, growth retardation, and increased liver weight in **hamsters**.

LD_{50}, **mice**: 275 mg/kg; IP, 44 mg/kg.

BUTETHAL

= *Butobarbital* = *Butobarbitone* = *Etoval* = *Neonal* = *Soneryl*

Sedative, hypnotic.

Untoward effects: Has caused fixed-drug eruptions in **patients**. In general, its toxic effects are similar to ***phenobarbital's***. Self-poisoning usually responded to a combination of dialysis, forced diuresis, and intensive respiratory support. Vertical gaze palsy in a 28-year-old **female** who had ingested 60, 100 mg tablets for a headache 3 days before hospital admission. Therapeutic serum concentrations are 5–15 mg/l, and toxic concentrations are 20–25 mg/l. ***Metronidazole*** and ***diloxanide*** decrease its rate of metabolism. It increases ***warfarin's*** metabolism.

LD_{LO}, **human**: 1 mg/kg; **cat**: 84 mg/kg; **rabbit**: 160 mg/kg.

BUTETHAMINE

= *Ibylcaine* = *Monocaine*

***Local anesthetic.* A *para-aminobenzoic acid* ester.**

Untoward effects: IV, LD_{50}, **rat**: 28 mg/kg; **mouse**: 43 mg/kg; IV, LD_{LO}, **rabbit**: 30 mg/kg; SC, LD_{LO}, **guinea pig**: 215 mg/kg.

BUTIROSIN

= *Ambutyrosin* = *CL 642*

Aminoglycoside antibiotic.

Untoward effects: Transient dizziness and increased liver enzymes after IM use in **man**. Cross-sensitivity with ***gentamicin*** and ***neomycin***.

BUTOCONAZOLE

= *Exelgyn* = *Femstat* = *Gynomyk* = *RS-35887*

Antifungal.

Use: Primarily, for topical treatment of vulvovaginal candidiasis.

Untoward effects: A 54-year-old **female** with a 19 year history of rheumatoid arthritis, after 1 week of topical treatment for a vaginal yeast infection, developed life-threatening thrombocytopenia. This may also be related to simultaneous use of ***methotrexate*** and ***ibuprofen***.

BUTONATE

= *F-139* = *T-113*

Organophosphorous insecticide, anthelmintic. Cholinesterase inhibitor.

Untoward effects: LD_{LO}, **human**: 500 mg/kg.

A small percentage of treated **horses** have exhibited varying degrees of anorexia, depression, diarrhea, stiffness, salivation, laminitis, sweating, and colic.

LD_{50}, **rat**: 1.1–7.5 g/kg.

BUTOPYRANOXYL

= *Indalone*

Insect repellent, rotenone solvent.

Untoward effects: LD_{LO}, **human**: 5 g/kg.

LD_{50}, **rats**: 7.4 ml/kg; **mice**: 11.6 ml/kg; **rabbit**: 5.4 g/kg; **guinea pig**: 3.2 g/kg.

BUTORPHANOL

= *Stadol* = *Torate* = *Torbugesic* = *Torbutrol*

***Narcotic analgesic, antitussive.* A synthetic opioid. Agonist–antagonist analgesic of the *nalorphine* type.**

Untoward effects: Depresses respirations, but, unlike ***morphine***, does not increase the intensity of the depression with higher dosages, only prolongs it. Rarely, causes psychotomimetic reactions, physical dependence, rashes, hives, and pruritus. By the summer of 1997, the FDA reported 44 deaths and 774 addiction-related side-effects. Few were from the injectable form, sold since 1978, but 18 deaths and 654 addiction-related reactions reported due to the use since 1991 of the nasal spray form. Some use it for the "buzz" it gives **them**, but **most** abuse it for the rebound syndrome associated with migraines. Apraxia (inability to move or speak) in a 43-year-old **female** after a single intranasal use.

In **cats** and **dogs** causes bradycardia, dysphoria, and respiratory decrease.

LD_{50}, **rats**: 570–756 mg/kg; **mice**: 395–527 mg/kg; IV, **rats**: 17–20 mg/kg; **mice**: 40–57 mg/kg.

2-BUTOXYETHANOL

= *Butyl Cellosolve* = *Ethylene Glycol Monobutyl Ether*

***Solvent.* In dry cleaning and as a solvent for nitrocellulose, resins, greases, oils, cyanoacrylates, and paints.**

***Untoward effects*:** Readily absorbed through the skin. OSHA has recommended a 5 ppm of air as a workplace limit, with a TWA of 25 ppm for air to skin exposure. Can cause anemia, macrocytosis, young granulocytes in hemolysis, hemoglobinuria, and central nervous system symptoms. Has leached into solutions from rubber stoppers.

LD_{LO}, **human**: 50 mg/kg; inhalation TC_{LO}, **human**: 195 ppm/8 h.

LD_{50}, **rats**: 1.48 g/kg; **mice**: 1.23 g/kg; **rabbit**: 320 mg/kg; **guinea pig**: 1.2 g/kg. LC_{50}, **goldfish**: 1.7 g/l. Inhalation LC_{LO}, **rat**: 500 ppm/4 h; inhalation LC_{50}, **mouse**: 700 ppm.

BUTRIPTYLINE

= *AY 62,014*

Tricyclic antidepressant.

Untoward effects*:** Produces greater electrocardiogram abnormalities than ***amitriptyline.

LD_{50}, **rat**: 700 mg/kg; IP, 150 mg/kg; LD_{50}, **mice**: 345 mg/kg; IP, 120 mg/kg.

BUTTER

Untoward effects*:** Conflicting reports have not resolved the possible adverse effect of eating large quantities of various fats and oils. It had a low incidence of focal myocarditis and fibrosis in **rats**. Coloring it with ***AB or ***OB dyes*** no longer permitted in the U.S., nor preserving it with ***boric acid***, as was done in France, where it caused indigestion and deaths in at least 40 **children**. ***Estrogen*** content varies widely from 0.8 to 28 mg/100 g. Rancid butter has caused hepatic dysfunction. When ***DDT*** was a popular insecticide on **cows**, butter made from their milk contained 65 ppm.

n-BUTYL ACETATE

***Use*:** In nail polish, nail polish removers, perfumes, paper coatings, gasoline, lacquers for wooden furniture and automotive topcoat, in making safety glass, and as an ink and adhesive solvent. About 250 million lbs used annually.

***Untoward effects*:** Poor ventilation at electronic plants caused headaches, dizziness, drowsiness, and vomiting. Rarely sensitizing, occasionally irritant, but dehydrating to skin. Vapors are irritating to eyes, nose, and throat. Weakness and even unconsciousness reported. Known to be an abused inhalant substance.

LD_{LO}, **human**: 500 mg/kg; inhalation TC_{LO}, **human**: 200 ppm; eye, **human**: 300 ppm. TLV of 150 ppm.

17,500 ppm/30 min caused narcosis and some deaths in **cats**.

LD_{50}, **rat**: 14 g/kg. Inhalation LC_{50}, **rats**: 160 ppm; LC_{50}, **bluegills**: 100 mg/l/96 h.

n-BUTYL ACRYLATE

***Use*:** In manufacturing polymers and resins in the textile, leather, and paint industries.

***Untoward effects*:** LD_{50}, **rats**: 3.73 g, 9 ml/kg; **mice**: 128 mg/kg; dermal, **rabbits**: 2 ml/kg; inhalation LC_{LO}, **rat**: 1000 ppm/4 h.

sec-BUTYLAMINE

= *2-Aminobutane* = *2-Butanamine* = *Deccotane* = *Frucote* = *Tutane*

Fungicide, fungistat.

***Use*:** On post-harvest citrus. A tolerance of 90 ppm for citrus molasses and dried pulp for **cattle** feeding.

***Untoward effects*:** Irritant to skin (flushing and burns) and mucous membranes, eyes, nose, and throat of **workers**. Causes headaches.

LD_{50}, **rats**: 152–380 mg/kg; **dog**: 225 mg/kg; skin, **rabbit**: 2.5 g/kg; LC_{50}, **bluegills**: 32 mg/l/96 h.

tert-BUTYLAMINOETHYL METHACRYLATE

***Untoward effects*:** IP, LD_{50}, **mice**: 0.19 ml/kg.

BUTYLATED HYDROXYANISOLE

= *Antrancine 12* = *BHA* = *Embanox* = *Nipantiox 1-F* = *Sustane 1-F* = *Tenox BHA*

***Antioxidant.* Over 20 million lbs used annually in foods, parenterals, in lipsticks and eye shadow, in**

plastics to help retard rancidity, and helps prevent encephalomalacia in chicks.

ntoward effects: Estrogenicity has been reported. icaria produced in some **humans**. Allergic contact ntitis to it in latex gloves has occurred. 125 mg doses exacerbation of rhinitis, asthma flare-ups, marked sis, somnolence, headaches, back pain, flushing, ion of conjunctivae in **people**.

0.1% level in the diet, it has caused pregnant to give birth to eyeless offspring. Levels h to **monkeys** for 2 years had no ill- dered safe for **man**, whose metabolic ns are very similar to **monkeys'**. Low the carcinogen seem to inhibit ng, stomach, mammary, and liver 0.5% in the diet of **rats** from ostnatal induced some develop- xicity in **rats**. High dosage liver weight in **rats**.

g.

oods.

espira- ntation ale **rats**. eight was

TOLUENE

rvuol = *Ionol CP* =

nterals, topi- oaps, paints,

d **female** herpes teroid *lene* f a g

INE

nal cell carci-

gents, cosmetic and in foods in

caused no irritation **nans** and sensitization am introduced into a um level of 300 μg/l of

$_{50}$, **rabbits**: >5 g/kg.

E

plastic industry can cause ensitivity to further exposure

p-*tert*-BUTYLCATECHOL

= *TBC*

Astringent, antioxidant, polymerization inhibitor for styrene–butadiene.

Untoward effects: Allergen in high-speed duplicator and phototypesetting papers caused focal, red, itchy, scaling lesions, resembling lichen planus. Has caused hypo-pigmentation and leukoderma in 4/75 tappet assembly **workers** using an oil with it as an antioxidant. Positive patch tests in 3/4 with 0.1% in ***acetone***; 1/4 developed depigmentation at the site, which persisted for about 2 months. 0.5% or more is irritant to **human**, **guinea pig**, and **rabbit** skin. May cross-react with ***pentadecylcatechol*** in ***poison ivy***.

n-BUTYL CHLORIDE

= *Bu-Chlorin* = *1 Chlorobutane* = *NBC*

Anthelmintic, solvent.

Untoward effects: LD_{LO}, **human**: 500 mg/kg.

LD_{50}, **rats**: 2.67 g/kg; inhalation LC_{LO}, **rat**: 8000 ppm/4 h.

BUTYL CYANOACRYLATE

= *Embucrilate* = *Histoacryl Blue* = *Nexabond* = *Vetbond*

Tissue adhesive.

Untoward effects: Neither bacteriostatic nor bacteriocidal. Allergic reactions in 1/1283 who underwent laparoscopy. Slow degradation in the body. Avoid accidental contact with eyes and stainless steel sutures. Aerosols set or polymerize in nearly 0.1 s, and drops in about 1 s. Bonding of fingers has accidentally occurred.

4-*tert*-BUTYLCYCLOHEXANOL

Use: About 5 tons used annually in fragrances, providing musty, woody odor for soaps, detergents, cosmetic ams and lotions, and perfumes.

ward effects: 4% in petrolatum produced no irritation itization in 25 **human** volunteers.

ngth for 24 h under occlusion on abraded **rabbit** d no irritation.

2 g/kg, acute dermal LD_{50}, **rabbits**: >5 g/kg; **se**: 50 mg/kg.

-2,6- LPHENOL

In rubber and petroleum

Untoward effects: It and/or its metabolites have anti-***vitamin K*** properties and have caused death from hemorrhage in male **rats**.

tert-BUTYLDIMETHYLCHLORO-SILANE

Untoward effects: Two IP injections of 200–1000 mg/kg resulted in up to a 47% increase of lung tumors in **mice**.

IP, LD_{LO}, **mouse**: 1 g/kg.

1,3-BUTYLENE DIMETHACRYLATE

Untoward effects: IP, LD_{50}, **mice**: 3.6 ml/kg

1,3-BUTYLENE GLYCOL
= *1,3-Butanediol*

***Humectant, preservative, food additive, aircraft de-icer, antilipogenic agent, cosmetic solvent, aldehyde dehydrogenase inhibitor (mice), and flavoring agent.* Experimentally, effective in treating *ethylene glycol* toxicity in dogs.**

Untoward effects: Potentiates ***carbon tetrachloride*** hepatotoxicity in **rats**.

LD_{50}, **rat**: 23 g/kg; **guinea pig**: 11 g/kg.

n-BUTYL ETHER

Untoward effects: Eye, **human**: 200 ppm/15 min, inhalation TC_{LO}, **human**: 200 ppm.

LD_{50}, **rats**: 7.4 g/kg; skin LD_{50}, **rabbit**: 10 g/kg.

n-BUTYL GLYCIDYL ETHER
= *BGE*

Untoward effects: Contact with strong oxidizing agents may cause fire or explosions. Exposure to light or air may cause the formation of explosive ***peroxides***. Contact with strong caustics may cause strong polymerization with liberation of heat, that may cause the container to burst. OSHA's exposure limit is 50 ppm (270 mg/m^3) of air. National Institute for Occupational Safety and Health recommends an exposure limit of 5.6 ppm (30 mg/m^3) and a TLV of 25 ppm (135 mg/m^3) as a TWA for an 8 h workday and 40 h work-week. Toxicity in **man** can occur after inhalation, ingestion, dermal or eye contact, and can be in the form of eye, nose, and skin irritation, sensitization, and narcosis.

Inhalation or ingestion by **rats** and **mice** caused central nervous system depression, pulmonary edema, and death. IM injection in **rats** increases white blood cells. Dermal exposure of male **mice** subsequently bred to unexposed **females** caused increased incidence of fetal deaths. Inhalation of 300 ppm (7 h/day/30 days) by **rats** caused atrophic testes in 50%. Mild irritant to skin, eyes, nose, throat, and respiratory tract. Long-term exposure causes inflammation and sensitization of skin.

LD_{50}, **rat**: 2 g/kg; **mouse**: 1.5 g/kg. Inhalation LC_{LO}, **rat**: 670 ppm.

BUTYLGLYCOL ACETATE

Solvent.

Untoward effects: Percutaneous absorption in **rabbits** is rapid (LD_{50}, 1.5 g/kg). Death occurred within 1–2 days with hemoglobinuria and/or hematuria, decreased red blood cells and hemoglobin. Low doses caused necrotizing hemorrhagic interstitial tubular nephrosis wit[h] occasional glomerular lesions in **rats** and **rabbits**.

LD_{50}, male **rats**: 3 g/kg; female **rats**: 2.4 g/kg.

tert-BUTYLHYDROQUINONE
= *TBHQ* = *Tenox*

***Antioxidant.* In oils, fats, lard, nuts, and snack f**

Untoward effects: Increase in hyperplasia of nasal tory epithelium and increase of splenic pigm at 5000 ppm in male and 10,000 ppm in fem Some hair discoloration and decrease in body w also noted.

N-BUTYL-*N*-(4-HYDROXYBUTYL)NITROSAM

Untoward effects: Induces bladder transitio nomas in **rats** and **mice**.

LD_{50}, **rat**: 1.8 g/kg.

BUTYL ISOBUTYRATE
= *n-Butyl-2-methylpropanoate*

***Fragrance.* Used in soaps, dete creams and lotions, perfumes, Europe (up to 40 ppm).**

Untoward effects: 4% in petrolatu after a 48 h closed-patch test on **hu** in only 1/25 **volunteers**. One g **cow's** rumen produced a maxim milk after 2–4 h.

LD_{50}, **rats**, and acute dermal L

BUTYL ISOCYANAT

Untoward effects: Use in th delayed reactions and hypers in **humans**. Isocyanates are potentially dangerous (see ***methyl isocyanate***).

LD_{50}, **rat**: 600 mg/kg; **mouse**: 150 mg/kg; **guinea pig**: 250 mg/kg. Inhalation LC_{50}, **rat**: 3 g/m^3; **mouse**: 680 mg/m^3.

n-BUTYL MERCAPTAN

= *1-Butanethiol* = *Thiobutyl Alcohol*

Antioxidant. **Rat repellent (skunk-like odor).**

Untoward effects: Incompatible with strong oxidizing agents and ***copper*** utensils. May release toxic vapors (***sulfur dioxide*** and ***carbon monoxide***) in a fire. In an industrial accident, a 1 h exposure of seven **workers** to 50–500 ppm caused irritation of mucous membranes, weakness, malaise, increased respiration, neck pain, drowsiness, nausea, vomiting, sweating, dizziness, confusion, and coma. OSHA suggests maximum exposure of 10 ppm (35 mg/m^3)/8 h TWA. TLV suggested at 0.5 ppm/8 h. Cyanosis, and liver and kidney damage have also been reported. A toxic metabolite of ***DEF*** and ***merphos***.

Inhalation by **mice** and **rats** leads to increased respirations, hyperactivity, staggering, weakness, cyanosis, sedation, and death.

LD_{50}, **rat**: 1.5 g/kg; inhalation LC_{50}, **rat**: 4020 ppm/4 h; **mouse**: 2500 ppm/4 h.

BUTYL NITRITE

Propellant.

Untoward effects: Methemoglobinemia from sniffing it. Closely related chemically to ***amyl nitrite***. Both dilate blood vessels, decrease blood pressure, and decrease the heart's ***oxygen*** consumption. Sold over-the-counter as a "room deodorizer" = ***Black Jack*** = ***Lockaroma*** = ***Locker Room*** = ***Poppers*** (also used for ***amyl nitrite***) = ***Rush*** = ***Snappers***, and, unfortunately, used as a drug, to provide a brief, but intense, "high" after inhalation. Ingestion is even more toxic. Headaches, dizziness, increased perspiration, skin irritation, and facial flushing occur, miscarriages, increase in tumors, increased risk of heart problems with increased heart rate, and, less commonly, nausea, vomiting, and fainting. Deaths have followed from its use (viz. fatal methemoglobinemia in a 30-year-old **male** after ingesting a room deodorizer). Severe methemoglobinemia (54%) in 28-year-old **drug abuser** after ingesting a 0.4 oz bottle of room deodorizer – responded to treatment with 100 mg ***methylene blue***, IV. Can cause hearing loss in **humans** and **rats**. It has been used in a suicide attempt.

p-tert-BUTYLPHENOL

= *1-Hydroxy-4-tert-butylbenzene*

Soap antioxidant, in motor oils, in neoprene adhesives, oil well additives, and in manufacturing lacquers and varnishes. **As a fragrance in soaps, detergents, cosmetic creams and lotions, and perfumes.**

Untoward effects: Use in hospital disinfectants has caused skin inflammation and vitiligo-like depigmentation (not always reversible). Mildly abnormal liver function may be linked to it. One of the agents responsible for off-flavor salmon caught in Lake Michigan. A common cause of shoe contact dermatitis. Leached from an improperly cured liner of a water storage tank, it caused intermittent foul taste in food and water at a new Georgia hospital. 1% in petrolatum caused no irritation after a 48 h closed-patch test on 25 **volunteers**.

Irritating to intact or abraded **rabbit** skin after 24 h occlusion.

LD_{50}, **rats**: 1.4 g/kg; acute dermal, **rabbit**: 2.5 g/kg.

BUTYL STEARATE

Solvent. **In plastics, cosmetics, textiles, and rubber.**

Untoward effects: Comedogenic in **man** and **rabbits**.

BUTYRALDEHYDE

= *Butanol*

Use: In manufacturing rubber accelerators, resins, solvents, and plasticizers.

Untoward effects: 1000 ppm is irritant; therefore, a TLV of 10–100 has been suggested for **man**. Inhalation TC_{LO}, **human**: 580 mg/m^3.

LD_{50}, **rats**: 2.5–5.89 g/kg.

BUTYROLACTONE

= *1,2-Butanolide* = *1,4-Butanolide* = *γ-Butyrolactone* = *Dihydro-2(3H)-furanone* = *GBL*

Solvent.

Untoward effects: As oral premedication before anesthesia, caused parasympathetic stimulation in the form of salivation and bradycardia during ***cyclopropane*** anesthesia. Soporific found in many ***wines***. Illegally marketed as (***Blue Nitro*** = ***Blue Nitro Vitality*** = ***Firewater*** = ***GammaG*** = ***GH Revitalizer*** = ***Renewtrient*** = ***Revivarant*** = ***Revivarant G***) a dietary supplement and, by February 1999, 55 **consumers** had seizures, vomiting, bradypnea, bradycardia, and 19 became unconscious or comatose; one died despite intubation and assisted breathing; central nervous system depression and respiratory depression in 41 ages 11–46.

LD_{50}, **rats**: 17.2 ml/kg.

n-BUTYRONITRILE

Untoward effects: TWA, 8 ppm (22 mg/m^3).

LD_{50}, **rats**: 140 mg/kg; inhalation LC_{LO}, **rat**: 400 ppm. Inhalation LC_{50}, **mice**: 249 ppm.

BUXUS sp.

= *Box* = *Boxwood*

Known in many areas as *Hedge* = *Buis* (Africa) = *Baqs* = *Shimshar* (Syria) = *Pyxos* (Greece). Probably, the tree or hedge referred to in the *Book of Isaiah* (Chap. 41:19 and 60:13 – eighth century, B.C.) as the box or plane tree, used to beautify God's sanctuary and make the wilderness flourish.

Untoward effects*: *Honey from ***B. calcarica*** has caused mild to serious poisoning in **man**.

B. sempervirens contains a volatile oil, resin, and a number of alkaloids, the most toxic of which is ***buxine***, found in all parts of the plant, especially in the dark, green leaves. Its dark fruits are often mistaken for wild ***grapes***, and cause stomach pain, diarrhea, bloody stools, and vomiting. Large amounts can cause convulsions and respiratory failure in **humans**.

Its ***cyclobuxine*** is teratogenic in **animal** trials. Ingestion of leaves has caused the death of **cattle**, **horses**, **sheep**, **camels**, and **hogs** who browse on it or eat clippings. The early symptoms are those of hemorrhagic enteritis. It is purgative and has caused central nervous system disturbances and convulsions. 750 g of leaves has caused intense abdominal pain, dysentery, convulsions, and death from asphyxia in a **horse**.

BYSSOCHLAMYS NIVEA

Fungus.

Untoward effects*:** Produces a mycotoxin, ***patulin. An injection of 20 mg/kg killed **sheep** within 5 h. The same dose, by esophageal tube, caused slight temporary intoxication. Symptoms were nasal discharge, rumen atony, weight loss and anorexia, and 50–200% increase in serum ***urea*** levels. Its spores on fruit are heat-tolerant and can withstand 3 h at 85°C.

LD_{50}, **rats**: 140 mg/kg; inhalation LC_{LO}, **rat**: 400 ppm. Inhalation LC_{50}, **mice**: 249 ppm.

BUXUS sp.

= *Box* = *Boxwood*

Known in many areas as *Hedge* = *Buis* (Africa) = *Baqs* = *Shimshar* (Syria) = *Pyxos* (Greece). Probably, the tree or hedge referred to in the *Book of Isaiah* (Chap. 41:19 and 60:13 – eighth century, B.C.) as the box or plane tree, used to beautify God's sanctuary and make the wilderness flourish.

Untoward effects*: *Honey from ***B. calcarica*** has caused mild to serious poisoning in **man**.

B. sempervirens contains a volatile oil, resin, and a number of alkaloids, the most toxic of which is ***buxine***, found in all parts of the plant, especially in the dark, green leaves. Its dark fruits are often mistaken for wild ***grapes***, and cause stomach pain, diarrhea, bloody stools, and vomiting. Large amounts can cause convulsions and respiratory failure in **humans**.

Its ***cyclobuxine*** is teratogenic in **animal** trials. Ingestion of leaves has caused the death of **cattle**, **horses**, **sheep**, **camels**, and **hogs** who browse on it or eat clippings. The early symptoms are those of hemorrhagic enteritis. It is purgative and has caused central nervous system disturbances and convulsions. 750 g of leaves has caused intense abdominal pain, dysentery, convulsions, and death from asphyxia in a **horse**.

BYSSOCHLAMYS NIVEA

Fungus.

Untoward effects: Produces a mycotoxin, ***patulin***. An injection of 20 mg/kg killed **sheep** within 5 h. The same dose, by esophageal tube, caused slight temporary intoxication. Symptoms were nasal discharge, rumen atony, weight loss and anorexia, and 50–200% increase in serum ***urea*** levels. Its spores on fruit are heat-tolerant and can withstand 3 h at 85°C.

in **humans**. Isocyanates are potentially dangerous (see ***methyl isocyanate***).

LD_{50}, **rat**: 600 mg/kg; **mouse**: 150 mg/kg; **guinea pig**: 250 mg/kg. Inhalation LC_{50}, **rat**: 3 g/m^3; **mouse**: 680 mg/m^3.

n-BUTYL MERCAPTAN

= *1-Butanethiol* = *Thiobutyl Alcohol*

Antioxidant. **Rat repellent (skunk-like odor).**

Untoward effects*:** Incompatible with strong oxidizing agents and ***copper utensils. May release toxic vapors (***sulfur dioxide*** and ***carbon monoxide***) in a fire. In an industrial accident, a 1 h exposure of seven **workers** to 50–500 ppm caused irritation of mucous membranes, weakness, malaise, increased respiration, neck pain, drowsiness, nausea, vomiting, sweating, dizziness, confusion, and coma. OSHA suggests maximum exposure of 10 ppm (35 mg/m^3)/8 h TWA. TLV suggested at 0.5 ppm/8 h. Cyanosis, and liver and kidney damage have also been reported. A toxic metabolite of ***DEF*** and ***merphos***.

Inhalation by **mice** and **rats** leads to increased respirations, hyperactivity, staggering, weakness, cyanosis, sedation, and death.

LD_{50}, **rat**: 1.5 g/kg; inhalation LC_{50}, **rat**: 4020 ppm/4 h; **mouse**: 2500 ppm/4 h.

BUTYL NITRITE

Propellant.

Untoward effects*:** Methemoglobinemia from sniffing it. Closely related chemically to ***amyl nitrite. Both dilate blood vessels, decrease blood pressure, and decrease the heart's ***oxygen*** consumption. Sold over-the-counter as a "room deodorizer" = ***Black Jack*** = ***Lockaroma*** = ***Locker Room*** = ***Poppers*** (also used for ***amyl nitrite***) = ***Rush*** = ***Snappers***, and, unfortunately, used as a drug, to provide a brief, but intense, "high" after inhalation. Ingestion is even more toxic. Headaches, dizziness, increased perspiration, skin irritation, and facial flushing occur, miscarriages, increase in tumors, increased risk of heart problems with increased heart rate, and, less commonly, nausea, vomiting, and fainting. Deaths have followed from its use (viz. fatal methemoglobinemia in a 30-year-old **male** after ingesting a room deodorizer). Severe methemoglobinemia (54%) in 28-year-old **drug abuser** after ingesting a 0.4 oz bottle of room deodorizer – responded to treatment with 100 mg ***methylene blue***, IV. Can cause hearing loss in **humans** and **rats**. It has been used in a suicide attempt.

p-tert-BUTYLPHENOL

= *1-Hydroxy-4-tert-butylbenzene*

Soap antioxidant, in motor oils, in neoprene adhesives, oil well additives, and in manufacturing lacquers and varnishes. **As a fragrance in soaps, detergents, cosmetic creams and lotions, and perfumes.**

***Untoward effects*:** Use in hospital disinfectants has caused skin inflammation and vitiligo-like depigmentation (not always reversible). Mildly abnormal liver function may be linked to it. One of the agents responsible for off-flavor salmon caught in Lake Michigan. A common cause of shoe contact dermatitis. Leached from an improperly cured liner of a water storage tank, it caused intermittent foul taste in food and water at a new Georgia hospital. 1% in petrolatum caused no irritation after a 48 h closed-patch test on 25 **volunteers**.

Irritating to intact or abraded **rabbit** skin after 24 h occlusion.

LD_{50}, **rats**: 1.4 g/kg; acute dermal, **rabbit**: 2.5 g/kg.

BUTYL STEARATE

Solvent. **In plastics, cosmetics, textiles, and rubber.**

***Untoward effects*:** Comedogenic in **man** and **rabbits**.

BUTYRALDEHYDE

= *Butanol*

***Use*:** In manufacturing rubber accelerators, resins, solvents, and plasticizers.

***Untoward effects*:** 1000 ppm is irritant; therefore, a TLV of 10–100 has been suggested for **man**. Inhalation TC_{LO}, **human**: 580 mg/m^3.

LD_{50}, **rats**: 2.5–5.89 g/kg.

BUTYROLACTONE

= *1,2-Butanolide* = *1,4-Butanolide* = *γ-Butyrolactone* = *Dihydro-2(3H)-furanone* = *GBL*

Solvent.

Untoward effects*:** As oral premedication before anesthesia, caused parasympathetic stimulation in the form of salivation and bradycardia during ***cyclopropane anesthesia. Soporific found in many ***wines***. Illegally marketed as (***Blue Nitro*** = ***Blue Nitro Vitality*** = ***Firewater*** = ***GammaG*** = ***GH Revitalizer*** = ***Renewtrient*** = ***Revivarant*** = ***Revivarant G***) a dietary supplement and, by February 1999, 55 **consumers** had seizures, vomiting, bradypnea, bradycardia, and 19 became unconscious or comatose; one died despite intubation and assisted breathing; central nervous system depression and respiratory depression in 41 ages 11–46.

LD_{50}, **rats**: 17.2 ml/kg.

n-BUTYRONITRILE

***Untoward effects*:** TWA, 8 ppm (22 mg/m^3).

plastics to help retard rancidity, and helps prevent encephalomalacia in chicks.

Untoward effects: Estrogenicity has been reported. Urticaria produced in some **humans**. Allergic contact dermatitis to it in latex gloves has occurred. 125 mg doses caused exacerbation of rhinitis, asthma flare-ups, marked diaphoresis, somnolence, headaches, back pain, flushing, and suffusion of conjunctivae in **people**.

When fed at a 0.1% level in the diet, it has caused pregnant **rats** and **mice** to give birth to eyeless offspring. Levels 200 times as high to **monkeys** for 2 years had no ill-effects. It is considered safe for **man**, whose metabolic and excretion patterns are very similar to **monkeys'**. Low doses given before the carcinogen seem to inhibit carcinogen-induced lung, stomach, mammary, and liver tumors. Feeding it up to 0.5% in the diet of **rats** from preconception to 90 days postnatal induced some developmental neurobehavioral toxicity in **rats**. High dosage causes a transitory increase in liver weight in **rats**.

LD_{50}, **rats**: 2.2 g/kg; **mice**: 2 g/kg.

BUTYLATED HYDROXYTOLUENE

= *Antrancine 8* = *BHT* = *Dalpac* = *Impruvol* = *Ionol CP* = *Sustane* = *Tenox BHT* = *Vianol*

***Antioxidant.* In foods, medicinals, parenterals, topicals, synthetic rubbers, plastics, oils, soaps, paints, and inks.**

Untoward effects: Severe gastritis in a 22-year-old **female** after ingesting 4 g for self-medication of genital herpes simplex virus. May stimulate the activity of steroid hydroxylating hormones. Leaching from a ***polyethylene*** film may have caused hypopigmentation of the feet of a **Negro** 9-year-old **male** treated for psoriasis. 125 mg caused similar rhinitis symptoms, as with ***BHA*** above. Minimum lethal dose, skin, **human**: 500 mg/48 h.

LD_{50}, **rat**: 890 mg/kg; **mice**: 1 g/kg; LD_{LO}, **cat**: 940 mg/kg.

BUTYL BENZYL PHTHALATE

= *Palatinol BB* = *Santicizer 160* = *Sicol 160* = *Unimoll BB*

***Plasticizer.* Added to polymers to give flexibility and softness (in vinyl flooring).**

Untoward effects: In 2 year feeding studies, **rats** given as high as 25,000 ppm in their feed (approximately 2.2 g/kg) for 10 weeks, developed slight decrease in weight, slight anemia, and decreased spermatozoa counts. All **rats** survived the 2 year study, but there was a slight increase in pancreatic acinar cell hyperplasia and transitional epithelial papillomas of the bladder. IP, LD_{50}, **mouse**: 3.2 g/kg.

p-tert-BUTYLCATECHOL

= *TBC*

Astringent, antioxidant, polymerization inhibitor for styrene–butadiene.

Untoward effects: Allergen in high-speed duplicator and phototypesetting papers caused focal, red, itchy, scaling lesions, resembling lichen planus. Has caused hypopigmentation and leukoderma in 4/75 tappet assembly **workers** using an oil with it as an antioxidant. Positive patch tests in 3/4 with 0.1% in ***acetone***; 1/4 developed depigmentation at the site, which persisted for about 2 months. 0.5% or more is irritant to **human**, **guinea pig**, and **rabbit** skin. May cross-react with ***pentadecylcatechol*** in ***poison ivy***.

n-BUTYL CHLORIDE

= *Bu-Chlorin* = *1 Chlorobutane* = *NBC*

Anthelmintic, solvent.

Untoward effects: LD_{LO}, **human**: 500 mg/kg.

LD_{50}, **rats**: 2.67 g/kg; inhalation LC_{LO}, **rat**: 8000 ppm/4 h.

BUTYL CYANOACRYLATE

= *Embucrilate* = *Histoacryl Blue* = *Nexabond* = *Vetbond*

Tissue adhesive.

Untoward effects: Neither bacteriostatic nor bacteriocidal. Allergic reactions in 1/1283 who underwent laparoscopy. Slow degradation in the body. Avoid accidental contact with eyes and stainless steel sutures. Aerosols set or polymerize in nearly 0.1 s, and drops in about 1 s. Bonding of fingers has accidentally occurred.

4-*tert*-BUTYLCYCLOHEXANOL

Use: About 5 tons used annually in fragrances, providing a musty, woody odor for soaps, detergents, cosmetic creams and lotions, and perfumes.

Untoward effects: 4% in petrolatum produced no irritation or sensitization in 25 **human** volunteers.

Full-strength for 24 h under occlusion on abraded **rabbit** skin showed no irritation.

LD_{50}, **rats**: 4.2 g/kg, acute dermal LD_{50}, **rabbits**: >5 g/kg; IP, LD_{LO}, **mouse**: 50 mg/kg.

4-*tert*-BUTYL-2,6-DIISOPROPYLPHENOL

***Phenolic antioxidant.* In rubber and petroleum chemicals.**

Untoward effects: It and/or its metabolites have anti-***vitamin K*** properties and have caused death from hemorrhage in male **rats**.

tert-BUTYLDIMETHYLCHLORO-SILANE

Untoward effects: Two IP injections of 200–1000 mg/kg resulted in up to a 47% increase of lung tumors in **mice**.

IP, LD_{LO}, **mouse**: 1 g/kg.

1,3-BUTYLENE DIMETHACRYLATE

Untoward effects: IP, LD_{50}, **mice**: 3.6 ml/kg

1,3-BUTYLENE GLYCOL

= *1,3-Butanediol*

***Humectant, preservative, food additive, aircraft de-icer, antilipogenic agent, cosmetic solvent, aldehyde dehydrogenase inhibitor (mice), and flavoring agent.* Experimentally, effective in treating *ethylene glycol* toxicity in dogs.**

Untoward effects: Potentiates ***carbon tetrachloride*** hepatotoxicity in **rats**.

LD_{50}, **rat**: 23 g/kg; **guinea pig**: 11 g/kg.

n-BUTYL ETHER

Untoward effects: Eye, **human**: 200 ppm/15 min, inhalation TC_{LO}, **human**: 200 ppm.

LD_{50}, **rats**: 7.4 g/kg; skin LD_{50}, **rabbit**: 10 g/kg.

n-BUTYL GLYCIDYL ETHER

= *BGE*

Untoward effects: Contact with strong oxidizing agents may cause fire or explosions. Exposure to light or air may cause the formation of explosive ***peroxides***. Contact with strong caustics may cause strong polymerization with liberation of heat, that may cause the container to burst. OSHA's exposure limit is 50 ppm (270 mg/m^3) of air. National Institute for Occupational Safety and Health recommends an exposure limit of 5.6 ppm (30 mg/m^3) and a TLV of 25 ppm (135 mg/m^3) as a TWA for an 8 h workday and 40 h work-week. Toxicity in **man** can occur after inhalation, ingestion, dermal or eye contact, and can be in the form of eye, nose, and skin irritation, sensitization, and narcosis.

Inhalation or ingestion by **rats** and **mice** caused central nervous system depression, pulmonary edema, and death. IM injection in **rats** increases white blood cells. Dermal exposure of male **mice** subsequently bred to unexposed **females** caused increased incidence of fetal deaths. Inhalation of 300 ppm (7 h/day/30 days) by **rats** caused atrophic testes in 50%. Mild irritant to skin, eyes, nose, throat, and respiratory tract. Long-term exposure causes inflammation and sensitization of skin.

LD_{50}, **rat**: 2 g/kg; **mouse**: 1.5 g/kg. Inhalation LC_{LO}, **rat**: 670 ppm.

BUTYLGLYCOL ACETATE

Solvent.

Untoward effects: Percutaneous absorption in **rabbits** is rapid (LD_{50}, 1.5 g/kg). Death occurred within 1–2 days with hemoglobinuria and/or hematuria, decreased red blood cells and hemoglobin. Low doses caused necrotizing hemorrhagic interstitial tubular nephrosis with occasional glomerular lesions in **rats** and **rabbits**.

LD_{50}, male **rats**: 3 g/kg; female **rats**: 2.4 g/kg.

tert-BUTYLHYDROQUINONE

= *TBHQ* = *Tenox*

***Antioxidant.* In oils, fats, lard, nuts, and snack foods.**

Untoward effects: Increase in hyperplasia of nasal respiratory epithelium and increase of splenic pigmentation at 5000 ppm in male and 10,000 ppm in female **rats**. Some hair discoloration and decrease in body weight was also noted.

N-BUTYL-*N*-(4-HYDROXYBUTYL)NITROSAMINE

Untoward effects: Induces bladder transitional cell carcinomas in **rats** and **mice**.

LD_{50}, **rat**: 1.8 g/kg.

BUTYL ISOBUTYRATE

= *n-Butyl-2-methylpropanoate*

***Fragrance.* Used in soaps, detergents, cosmetic creams and lotions, perfumes, and in foods in Europe (up to 40 ppm).**

Untoward effects: 4% in petrolatum caused no irritation after a 48 h closed-patch test on **humans** and sensitization in only 1/25 **volunteers**. One gram introduced into a **cow's** rumen produced a maximum level of 300 μg/l of milk after 2–4 h.

LD_{50}, **rats**, and acute dermal LD_{50}, **rabbits**: >5 g/kg.

BUTYL ISOCYANATE

Untoward effects: Use in the plastic industry can cause delayed reactions and hypersensitivity to further exposure

C

CABBAGE

= *Brassica oleracea var. capitata*

Untoward effects*: Patients** with ***cobalt dermatitis should avoid cabbage. Maternal ingestion can cause hypersensitivity reactions in nursing **infants**. A goitrogen. Consumption can decrease ***iodine*** uptake by the thyroid in **man** and **rabbits**. Cabbages accumulate ***sulfates***. Can increase irritable bowel syndrome, bloating, and flatus. Contains the carcinogens, ***allylisothiocyanate*** (from ***sinigrin***) and ***neochlorogenic acid*** (converted to ***caffeic acid***), yet, raw cabbage contains anticancer agents, such as ***indole-3-carbinol*** and ***cyanohydroxybutene***. ***Allyisothiocyanate*** is also a mutagen. Allergic cross-sensitizations with ***broccoli***, ***brussel sprouts***, and ***radishes*** can induce hypoglycemia. Increases the metabolism of ***phenacetin*** and ***acetaminophen***.

Drug interactions*:** Its ***vitamin K content can cause ***warfarin*** antagonism.

Can cause Heinz body anemias in **animals**. Hemoglobinemia in **cattle** eating large quantities. Large increase in rumen ***lactic acid*** content after over-feeding it to **ruminants**. Endemic goiter in **kids** of **goats** eating large amounts during pregnancy.

CABERGOLINE

= *Dostinex* = *FCE 21,336*

Prolactin inhibitor.

***Untoward effects*:** Nausea and vomiting (up to 35%), constipation, headache (up to 30%), dizziness and vertigo (up to 25%), and asthenia are chief complaints (~10–25%); fatigue in 7–15%, abdominal pain and somnolence in 5–15%, apathy, and occasionally polydipsia, anorexia, xerostomia, diarrhea, breast pain, hot flushes, pruritus, dry eyes, leg paresthesias, dyspnea, digital vasospasm, leg cramps, and a sense of suffocation.

CACALIA DECOMPOSITA

= *Matarique*

A member of the sunflower family.

Untoward effects*:** Matar in Spanish means to kill or murder, so its common name is appropriate. It is indigenous to northern Mexico, especially in the states of Chihuahua and Sonora. Contains a number of ***pyrrolizidine-type alkaloids, and yet, sold in Mexican-American herbal stores, often causing symptoms of cholera and cardiac problems.

CACAO

= *Chocolate*

It is not *cocoa* (the palm yielding *coconut*).

Untoward effects*:** The fruit waste products have caused death from sudden heart failure. Toxicity is primarily from ***theobromine. The beans are from the tropical plant, *Theobroma cacao*.

The shells have been fed to **livestock** and ground up, used as a diluent for **livestock** medications. Poisoning has occurred in **horses**, **calves**, **pigs**, **dogs**, **ducks**, and **chickens**. A **dog** that ate some shells sold for horticultural use became restless and excitable 8 h later, followed by thirst, diarrhea, and vomiting, with red urine. The **dog** convulsed and died 17 h after ingestion. Post-mortem revealed congested and hyperemic liver, kidneys, and spleen. Other reports of toxicity in **dogs** indicate death after about 12 h. Ingestion can have a cumulative effect in **dogs**, where ***theobromine*** absorption is faster than its excretion. ***Theobromine*** content of cocoa shells has varied from 0.19 to 2.98%. Toxicity has also occurred from ***aflatoxin*** contamination. **Calves** fed waste chocolate develop excitement, sweating, increased respiratory and pulse rates, collapse and convulsions, and even death.

CACODYLIC ACID

= *Ansar* = *Dimethylarsinic Acid* = *Phytar* = *Silvisar*

***Herbicide, alterative, defoliant.* Contains 54.3% *arsenic*. Major metabolite of orally administered *arsenic trioxide*.**

***Untoward effects*:** LD_{LO}, **human**: 500 mg/kg.

Readily crosses the placenta of pregnant **rats**. In the **mouse**, 400–600 mg/kg on days 7–16 of gestation leads to cleft palate and delayed skeletal ossification. Pregnant **rats** receiving 50–60 mg/kg/day during gestational days 6–13 leads to fetal and maternal death. Extremely toxic to **honey bees** at 100 ppm.

LD_{50}, **rats**: 0.7–1.35 g/kg; SC LD_{LO}, **dog**: 1 g/kg.

CACTINOMYCIN

= *Actinomycin C* = *HBF 386* = *Sanamycin*

Antineoplastic antibiotic.

***Untoward effects*:** Rashes, alopecia, mucocutaneous ulceration, exfoliative erythroderma, and sweat gland disturbances. Local reactions follow SC and IM use, or from extravasation after IV use.

Testicular atrophy in treated **rats**.

IV LD_{50}, **rat**: 100 mg/kg; IP LD_{LO}, **mouse**: 5 mg/kg.

CADMIUM

= *Cd*

Use: In paint pigments, batteries, TV tube phosphors, electroplating, fungicides, stabilizer in plastics, curing agent in rubber tires and other products, fertilizers, coal, oil, gasoline, solder, cigarettes and tobacco, newsprint, scotch tape, sheetrock, plaster, shellac, varnish, and jewelry fabrication. Approximately 10,000 metric tons are used annually in the U.S.

Untoward effects: Has a long half-life in **mammals**. It interferes with the functioning of ***zinc*** in enzyme systems. Cadmium toxicity has known similarities with ***zinc*** deficiency. Has caused lung cancer, emphysema, and kidney damage in **man**. ***Metallothionein*** protects against the acute toxicosis, but contributes to its persistence in the kidneys. Toxicosis after an acute oral exposure is manifested by nausea, salivation, vomiting, diarrhea, abdominal pain, myalgia, throat irritation, and a metallic taste. Chronic oral exposure signs are tachycardia, loss of smell, coughing, bronchitis, dyspnea, cyanosis, anorexia, weight loss, growth retardation, microcytic hypochromic anemia, decreased hemoglobin, lumbar pain, myalgia, yellow staining of teeth, hypertension, severe proteinuria, glycosuria, degeneration and necrosis of renal tubular epithelium, thinning of cortical osseous tissue, and spontaneous fractures. Acute pulmonary exposure causes coughing, throat dryness, bronchitis, dyspnea and pneumonitis, collapse, and even death. Chronic pulmonary exposure results in proteinuria, anemia, and emphysema. **Man** and **animals** are exposed to contamination water from galvanized pipes, especially when they join ***copper*** pipes. Has caused food poisoning outbreaks and deaths after its use in plating containers used for food, drink, or cooking utensils. Placental transfer is low. Some passes into the milk of lactating **women**, **cows**, and **rats**, where it can pose a neonatal neurotoxic risk to **infants**. Hemorrhage and abortion have occurred when **mothers** were exposed to toxic doses. Although wood ash contains low levels of cadmium, avoid using large quantities in gardens where ***lettuce*** (an accumulator) grows. ***Zinc*** smelters contribute a large amount into the air. When overheated (it is found in certain types of silver solder), it vaporizes and produces dangerous fumes of ***cadmium oxide*** (q.v.). **People** in the vicinity of, but not working with, cadmium in ***brass***, and ***iron*** foundries have an increased incidence of mortality from prostatic cancer. A **plumber** was fatally poisoned in an unventilated area from fumes while brazing a valve into ***copper*** pipe with silver solder containing about 25% ***Cd***. The 20 min exposure produced pneumonitis in 1 h and **he** died in 5 days (***Cd*** concentration in lung 0.15 mg; in liver, 0.1 mg; and 0.5 mg in kidney per 100 g of tissue). ***Cd*** is a regular contaminant of the ***zinc*** used to galvanize metal and pipes. It is readily leached out by soft acidic waters. Two outbreaks of poisoning occurred in one season in California. Within 1 h after consuming pink lemonade, which had been stored for 3½ h in a 3 gallon ***Cd***-plated war surplus container, 23/32 school **children** (5–9 years) became ill (left-over beverage assayed 21 ppm ***Cd***) and recovered in 2 days. Sudden nausea, vomiting, abdominal cramps, some dizziness and fainting in 25/100 **persons** 10–40 min after consuming an orange-flavored beverage (pH 2.6) stored in a 10 gallon 15-year-old thermos, lined with ***copper*** and plated with ***Cd***; only one **person** was hospitalized. Many states now ban ***Cd***-plated containers. Ingestion of small flat disc ***Cd*** batteries has caused poisoning in **children** and **animals**. Toxic serum levels are approximately 5 µg%. Suggested occupational exposure limits (OSHA) for its fumes are 0.1 mg/m^3 TWA, with a ceiling of 0.3 mg/m^3. Twice this level is suggested for dusts. National Institute for Occupational Safety & Health suggestions are 1/25th of these. Toxic serum levels are 100 µg/l, and the estimated lethal dose by ingestion is about 1 g. The painful "itai-itai" disease ("ouch-ouch" or "painful-painful" disease in English) of Japan, so-called because sudden pains in various parts of the body cause its **victims** to cry out loudly, has been attributed to a mining company polluting the Jintsh River, a source of water for drinking and watering rice fields, with cadmium wastes. Mortality was about 50% in about 200 **people** affected. Symptoms also include waddling gait, aminoaciduria, glycosuria, severe osteomalacia, and multiple fractures. Poisoning occurred in five **workmen** cutting steel bolts plated with ***Cd*** (to prevent corrosion). Within a few hours, all had malaise, shivering, sweating, chest pain, an irritating cough, and dyspnea. Later, pulmonary edema and pneumonitis developed and one died on day 5. ***Cadmium oxide*** fumes and ***Cd*** dust may have been the chief causes. Possible allergic sensitization where cotton fabrics in a hospital were treated with it as a bactericidal agent. 70% of the ***Cd*** in tobacco is found in its smoke. **Patients** dying of chronic bronchitis and/or emphysema in one British trial had mean ***Cd*** levels three times those of **patients** dying from other diseases. Since many ***lead***-based paints contain ***Cd***, **children** poisoned by ***lead*** may actually have had a synergistic type of poisoning.

Most ***Cd*** is excreted in the feces and some is stored in the kidneys, liver, and spleen. There is little transfer into milk. Excessive ***Cd*** is not secreted into milk, but is bound by mammary gland tissue. Market milk contains 0.017–0.03 ppm and lower after hand milking (0.001 ppm), and higher after machine milking (up to 0.1 ppm), indicating milk can be contaminated by the milking equipment. Residues in foods for U.S. **adults** ranged between 0.002 and 0.05 ppm.

Inhalation TC_{LO}, **human**: 39 mg/m^3/20 min; 88 µg/m^3/8.6 years. Decreased resistance of **rats** to ***endotoxin***. Has shown immunosuppression in **mice**, **rats**, and **rabbits**. **Calves** fed 640–2560 ppm in the diet looked unthrifty, with rough hair coat, hair loss, dry scaly skin, mouth lesions, were dehydrated, had poor vision, edematous and often shrunken and scaly scrotums, and enlarged and painful joints. All those fed the higher level died within 2–8 weeks. Causes testicular atrophy, decreased longevity, hypertension, toxemia of pregnancy, and thinning of cortical bone in **rats**. Dietary ***zinc***, ***selenium***, and ***ascorbic acid*** have some protective effect against ***Cd*** toxicity. Growth retardant in **chicks**, **ducklings**, and **quail**. ***Cd*** intake produces an enteropathy in **quail** similar to that noted in **human** malabsorption syndrome. Ingestion causes testicular hypoplasia, severe anemia, bone marrow hyperplasia, and cardiac hypertrophy in **Japanese quail**.

Oral LD_{LO}, **rabbit**: 70 mg/kg; SC, 6 mg/kg.

CADMIUM ACETATE

Use: In electroplating and producing iridescence on porcelains, pottery, and textiles.

Untoward effects: Emits ***cadmium oxide*** when heated. Wear protective clothing, National Institute for Occupational Safety & Health-approved respirator, safety glasses, and avoid ingestion, inhalation, or contact with skin or eyes. May cause renal tubular dysfunction and "itai-itai" disease in **man**.

Accumulates in the liver and intestines of the **turtle** and the liver, kidneys, and gastrointestinal tract in **rats**.

CADMIUM CARBONATE

= *Otavite*

Use: In fungicides for lawns and as a chemical intermediate.

Untoward effects: Poor solubility and may appear in waters from run-off.

CADMIUM CHLORIDE

= *Cadmium Dichloride* = *Dichlorocadmium*

Use: In dyes, photography, turf fungicides, and vacuum tubes.

Untoward effects: Ingestion of small amounts causes severe, painful, and occasionally fatal, results. High solubility causes it to be found in run-off waters. Inhalation can cause lung damage in **man** and **animals** (**guinea pigs**, **rats**, etc.), including a dose-dependent incidence of malignant lung tumors and centrilobular emphysema in **rats**.
Mallard ducks fed it at 200 ppm for 90 days had testicular atrophy, with cessation of the spermatogenic process. **Rabbits** given 300 ppm in their drinking water for 70 days had decreased antibody response to antigens. Although not lethal when given alone to **rat** fetuses or dams, its lethality was 55–100% when given with ***sodium nitrilotriacetate*** (***NTA***). ***NTA*** also enhances cadmium chloride's teratogenicity (micrognathia, cleft palate, club foot, and small lungs) in **rats** and **mice**. SC injections in **rats** caused a marked decrease in testicular weight. Intratesticular injections in **dogs** are painful and cause fibrosis and atrophy by the 60th day. Same results in **goats** and **boars** or by SC injections in **gerbils**; by intra-ovarian injection to sterilize scrub **cows**. 15–25 mg is emetic in a **cat** and produces nephritis. 62 ppm in a **rat's** ration causes a severe anemia with increased reticulocytes, rapid bleaching of the incisor teeth, cardiac hypertrophy, and hyperplasia of the bone marrow. 1000 ppm is lethal within a few days in affected **rats**. There have been serious losses in wild **animals** drinking road-side slush when it has been used as a road de-icer.

LD_{50}, **rat**: 88 mg/kg; **mouse**: 175 mg/kg; **guinea pig**: 63 mg/kg; inhalation LC_{90}, **dog**: 420 mg/m^3/30 min.

CADMIUM FLUORIDE

= *Cadmium Difluoride* = *Cadmium Fluorure*

Contains 74.74% *cadmium* and 25.26% *fluoride*.

Use: In manufacturing of nuclear reactor controls, phosphors, and glass and laser crystals.

Untoward effects: OSHA standard, 200 µg as ***Cd***/m^3/ 8 h in air and 600 µg limit as dust/15 min.

SC LD_{LO}, **frog**: 200 mg/kg.

CADMIUM NITRATE

= *Cadmium Dinitrate*

Use: In photographic emulsions, coloring glass and porcelain, manufacturing light-sensitive paper, and in the production of ***nickel-cadmium*** batteries.

Untoward effects: Demands use of skin protection and filter mask. Strong irritant and highly toxic when ingested or inhaled. 4 mg dust in a **person's** lungs can be fatal. Ingestion of 14.5 mg causes nausea and vomiting and 8.9 g has been fatal. Will emit toxic vapors when heated. OSHA standard for **human** exposure to the dust is 0.2 mg ***Cd***/m^3/8 h, with a ceiling limit of 0.6 mg/m^3/15 min. National Institute for Occupational Safety & Health recommends a TWA of 40 µg/***Cd***/m^3/10 h.

LD_{50}, **rat**: 300 mg/kg.

CADMIUM OXIDE

Use: In storage battery electrodes, silver alloys, rubber acceleration, electroplating, ceramic glazes, and now, rarely, as an anthelmintic in **swine**.

Untoward effects: Acute exposure in **man** causes a unique clinical picture. Initially, there are no symptoms, so **workers** (often **welders**) do not know **they** are being poisoned until after 4–12 h, when feelings of chest constriction, substernal pain, coughing, dyspnea, wheezing, or hemoptysis. Occasionally, shaking chills, aching back and limb muscles, disturbed liver function, and, eventually, pulmonary edema. Severe pulmonary edema has persisted for weeks, with recovery taking several months. In severe respiratory problems, death often occurs from pulmonary edema 5–7 days after exposure (14% of cases). Some evidence of increased incidence of prostatic cancer in **workers** exposed to it for a minimum period of 1 year. Acute renal cortical necrosis in a fatal case. Acute toxicity developed within 30 min to 5 h in **subjects** in New Jersey, eating candy beads that, unfortunately, contained large amounts of ***Cd***. Short-lived symptoms were salivation, nausea, vomiting, abdominal pain, diarrhea (often with weakness), prostration, and myalgias. The exact form of the ***Cd*** was uncertain. An excess of lung cancer mortality has been reported in exposed **workers**. **People** are exposed from the melting of steel scrap, waste incineration, and burning of fossil fuels. TWA, 40 μg/m^3.

Inhalation TC_{LO}, **human**: 500 μg/m^3/over 5 year period.

Intratracheal instillation increases the incidence of mammary tumors and that of tumors at multiple sites in male **rats**.

Inhalation TC_{LO}, **rat**: 10 mg/m^3; **monkey**: 250 mg/m^3; LD_{50}, **rat**: 72 mg/kg.

CADMIUM SULFATE

Use: In pigments, phosphors, cathode ray tube screens, seborrhea treatments, watch and instrument dials, and plastics. U.S. consumption is in millions of lbs. Found in surface waters near ***zinc*** and ***zinc-lead*** oil refineries.

Untoward effects: Suspect carcinogen. Teratogenic in **hamsters**; decreases water consumption and growth rate in **mice**; increases incidence of interstitial cell testicular tumors in **rats** and **mice**.

LD_{LO}, **dog**: 105 mg/kg; SC LD_{LO}, **dog**: 27 mg/kg.

CADMIUM SULFIDE

= *Cadmium Yellow* = *CI 77,199* = *CI Pigment Yellow 37* = *Jaune Brilliant* = *Radiant Yellow*

Use: Coloring paper, glass, textiles, inks, and tattoos.

Untoward effects: Photoallergic reactions in tattoos have been reported; varying from a lymphohistiocytic infiltrate to lichenoid reactions, simulating lichen planus, to sarcoidal granulomas in nature. Nephropathy and weight loss reported in **man**.

SC TD_{LO}, **rat**: 90–135 mg/kg; IM TD_{LO}, **rat**: 150–250 mg/kg.

CAFFEINE

= *Guaramine* = *Methyl Theobromine* = *Thein*

Caffeine is the only *methylxanthine* present in *coffee*. In *tea*, it occurs in lower concentration with *theobromine* and *theophylline*. In *cocoa* and *chocolate* products, it exists at about 1/8th the amount of the predominant *theobromine* and with traces (up to 0.001%) of *theophylline*. In 1892, it, fruit syrup, and an extract of *coca* leaf (*cocaine*) was called *Coca-Cola*®. It is one of the most widely used alkaloid drugs world-wide. Instant *coffees* contain about 3–4% and decaffeinated ones < 0.3%.

Use: Generally, the pharmaceutical industry uses synthetic caffeine and the beverage industry uses natural. As a flavoring agent in ***cola*** beverages, usually from the ***cola*** nut and ***guarana***. Therapeutic uses are based on its central nervous system and cardiovascular stimulation, diuretic, and muscle relaxant effects.

Untoward effects: Caffeine in various beverages can cause a transient increase in blood pressure for at least 30 min. Despite wide-spread availability, acute poisonings are relatively few. Fatal results have been reported in **children** with doses of 5.3 g and, in **adults** with 6.5 g (average of 10 g, yet some have survived after 30 g). **Children** have died after estimated doses of 3–18 g. Surviving **infants** have ingested as much as 1.5 g–78 mg/kg. Some **children** sensitive to it have had toxic symptoms at 15 mg/kg. Insomnia, restlessness, irritability, excitement, mild delirium, nausea, vomiting, increased urinary frequency, tachycardia, and premature atrial and ventricular contractions can occur with large doses. Fever, headache, muscle tremors, diuresis, tinnitus, photophobia, scintillating scotoma, increased gastric secretions, gastric ulcerations, and hematemesis have also been reported. One of the common causes of recurrent headaches. Seizures rarely occur (150–200 mg/kg has been reported as a fatal dose in **man** with convulsions, emesis, and coma; 57 mg/kg, IV). Increased lipolysis, glycogenolysis, and gluconeogenesis increase serum ***glucose*** and lipids, which may be particularly significant in **diabetics**. Doses of 85–250 mg in **man** decrease fatigue and drowsiness, increase sustained intellectual thought, yet, decrease actual reaction time. Daily doses of 350 mg can produce physical dependence in **adults**

and > 600 mg/day has caused other health problems (i.e. 4–6 cups of ***coffee***, respectively). Low doses, by vagal stimulation, decrease heart rate, while large doses, by direct stimulation of the myocardium, lead to tachycardia. Death from cardiovascular collapse has been reported. **Fetal** arrhythmias have been reported in three **females** (22–26 years), consuming large quantities of caffeine. When exposure was stopped, sinus rhythm returned to normal. Causes vasodilation of the peripheral vasculature, yet it induces vasoconstriction by sympathetic stimulation, with the net result of a slight increase in blood pressure. Strong cerebral vasoconstriction has made it useful in treating headaches, although it has caused headaches on rare occasions. Its use for treating preterm **infants** for apnea is dangerous, because its half-life in them is 36–144 h, compared to 3–3½ h in **adults**. Many adverse caffeine effects reported are biased (viz. teratogenicity, due to easy placental transfer into the **fetus**, abortion, and fertility problems in **women**), and fail to consider other factors, such as smoking and ***alcohol*** consumption. A term **child** born to a **mother** who consumed 35–40 cups of ***coffee*** daily was born jittery for the first few days of its life. **Mother's** milk: plasma ratios have been reported as 0.5:1–1:1. Milk levels decrease rapidly by 4 h after ingestion. Doses equivalent to that found in eight cups of ***coffee*** daily produce increased fear, nervousness, anxiety, nausea, and restlessness in **patients** with panic disorders. Some **people** are so sensitive to it, that a single cup of ***coffee*** produces near-toxic effects, while **others** may only develop heartburn. Within a period of 7 months in Illinois, three deaths occurred in **students** from ingesting so-called ***"Legal Speed"*** (a mixture of caffeine and ***phenylpropanolamine***. Abdominal distress reported in **some**, as a result of central nervous system stimulation, and it has been incriminated in mastodynia, nipple leakage, and fibrocystic breast disease and negative ***calcium*** balance with resulting increases in hip fractures. Can occasionally cause small transient increases in diuresis and in intraocular pressure in both normal and glaucomatous eyes. Rhabdomyolysis, hallucinations, mania, depression, and acute renal failure in 21-year-old **male** after ingesting 3.57 g in a suicide attempt. May cause negative or false low ***bilirubin*** serum readings, and false positive or high ***vanillylmandelic acid*** test readings. A syndrome has been described (1886) of chronic caffeine poisoning in "excessive" **tea-drinkers** (five cups or 8 gr caffeine), with dyspepsia-epigastric uneasiness after meals, reduced mental and physical tone, followed by restlessness, excitability, tremors, disturbed sleep, and then, anorexia, headache, vertigo and confusion, palpitations, and dyspnea. A **father** deliberately poisoned his 5-week-old infant **son**, causing persistent tachycardia, agitation, crying, vigorous limb movements, subarachnoid hemorrhages, and death. Withdrawal symptoms of lethargy, headaches, incapacitation, anxiety, and muscle tension occur.

Therapeutic serum levels reported as 2–10 mg/l, toxic at 15–20 mg/l, and fatal at > 80 mg/l.

LD_{LO}, **human**: 192 mg/kg; **child**: 320 mg/kg.

Drug interactions: ***Acetaminophen***: Caffeine induces a significant decrease in its clearance.

Antipyrine: Increases caffeine elimination.

Bromocriptine: Caffeine increases its plasma concentration, but its clinical significance is uncertain.

Cimetidine: Decreases metabolism and elimination of caffeine.

Ciprofloxacin: Decreased clearance by caffeine.

Clozapine: Two cups of ***coffee*** in a 39-year-old **male** caused exacerbation of **his** psychotic symptoms.

Contraceptives, Oral: Decreases metabolism and elimination of caffeine.

Dipyridamole: Caffeine prevents its induction of chest pain and ST depressions on the electrocardiogram. **Patients** must avoid caffeine products for 24 h before Tl-201/***dipyridamole*** testing.

Disulfiram: Decreases caffeine elimination (increases half-life by about 30%).

Grapefruit Juice: Decreases metabolism of caffeine (increases its half-life by 31%).

Monoamine oxidase Inhibitors: Hypertensive crisis may be precipitated by use with caffeine.

Mexiletine: May impair caffeine's use in liver function tests and increase retention of caffeine.

Phenylpropanolamine: A combination increases the toxicity of each.

Propoxyphene: Use together increases central nervous system stimulation.

Salicylates: Decreases ***aspirin*** absorption and enhances its ulcerogenic effect.

Smoking: Shortens half-life by about 50% and leads to increased ***coffee*** drinking by **smokers**.

Sulfamethazine: 100 mg decreased urinary excretion in 4/5 **volunteers**. Its bioavailability was increased in **some** and decreased in **others**.

Theophylline: Both compete for ***adenosine*** receptors and asthma **patients** may require an increase in **their *theophylline*** dosage.

Doses of 65–200 mg can also enhance the analgesic effects of ***aspirin*** and ***ibuprofen***.

Its oral use increases body temperature in **humans** and **cats**, and decreases body temperature in the **mouse** and **rabbit**. Toxic syndrome in **guinea pigs** and **rats** consists of excitement, convulsions, gastroenteritis, edema, congestion of liver, heart, lungs, spleen, and adrenal glands, and respiratory failure. Large doses have caused birth defects and reproductive problems in **rats**, **mice**, and **rabbits**. Accidental ingestion of 50 gr (3.24 g) was fatal to a 7 kg **dog**, who first showed epileptiform, followed by tonic convulsions. **Roosters** fed 0.1% in their ration had decreased fertility, with decreased sperm counts. This was reversible when caffeine was withheld. Its teratogenic effects in **mice** were more serious after a SC dose, compared to IP dosage. Poisoning in **dogs** is often after ingestion of an **owner's** so-called "pep" pills, or in **horses** heavily "doped" (half-life of approximately 20 h). Leading to hyperexcitability, panting (in **dogs**), salivation, increased heart rate and force of cardiac contractions, yet, an 8-year-old 24 kg **Boxer dog** trembled, vomited, had tachycardia, and was breathless 1 day after eating 2 kg of ground coffee. Renal tubular degeneration is noted in poisoned **dogs**, but not in most experimental species.

LD_{50}, **rat**: 192–200 mg/kg; **mouse**: 127 mg/kg; **guinea pig** and **hamster**: 230 mg/kg; **rabbit**: 246 mg/kg; LD_{LO}, **cat**: 100 mg/kg; **dog**: 140 mg/kg.

Caffeine Sodium Benzoate: LD_{50}, **mouse**: 878 mg/kg.

CAJEPUT

= *Australian Tea Tree* = *Melaleuca sp.* = *Paper Bark* = *Punk Tree*

Use: Its oil is used internally as a carminative, and topically as a rubefacient for muscle aches, and in udder ointments, primarily in veterinary medicine. Occasionally, as a topical antiseptic and antifungal on cuts, burns, athlete's foot, in vaginal douches, or by inhalation, for the treatment of sinus and throat infections in **humans**, as well as by injection, in bronchitis.

Untoward effects: Dermatitis has been reported after use in the nasal area. Common in Australia and South Florida. **Asthmatics** and **people** with respiratory or cardiac problems find themselves in misery, with difficult breathing, nasal and sinus congestion, sneezing, throat irritation, coughing, swollen eyelids, burning of the eyes, facial irritation, and rashes, not from exposure to its pollen, but to an odiferous and irritant substance dispersed by a tree in bloom. A 17-month-old **male** drank < 10 ml of the oil leading to ataxia and drowsiness lasting several hours.

LD_{50}, **rat**: 387 mg oil/kg.

CALADIUM

= *Elephant's Ear* = *Exposition* = *Mother-in-law Plant* = *Pink Cloud* = *Sea Gull* = *Stoplight* = *Texas Wonder* = *Xanthosoma*

Untoward effects: All parts of the plant are toxic to **man** and **animals**. It contains sharp ***calcium oxalate*** crystals that cause intense burning and swelling of the mouth's mucous membranes. The throat, tongue, and respiratory passages can become swollen enough to cause death by asphyxia. Gastroenteritis has occurred with salivation, nausea, vomiting, and diarrhea in **some**. **Children** that chew on it usually shreik in pain. A 27-year-old **male**, intrigued by its pleasant odor, sucked on a small piece for a few seconds and instantly developed swelling of the lips and tongue, severe lacrimation, copious salivation, nausea, and vomiting. Pain in the oral cavity lasted about 5 days.

Similar symptoms have been reported in **animals**, particularly, **dogs** and, occasionally, **cats**.

CALAMINE

Use: As a protective, antiseptic, astringent, composed of ***zinc carbonate***, with traces of ***zinc*** and ***ferric oxides***. The lotions contain about 8%.

Untoward effects: A 9-year-old **male** develop a psychosis and delirium from topical use of ***Caladryl***® (calamine with ***diphenhydramine***), apparently, from the antihistamine. Generously applied, it has caused interference with X-ray interpretation, as it is radiopaque.

CALAMUS

= *Acorus calamus* = *Acorus spurius* = *Bach* = *Boja* = *Bojho* = *Muskrat Root* = *Rat Root* = *Sweet Cane* = *Sweet Cinnamon* = *Sweet Flag* = *Sweet Grass* = *Sweet Myrtle* = *Sweet Root* = *Sweet Rush* = *Sweet Sedge* = *Wye*

Use: A commonly used folklore herb. Contains 1–3% of a volatile oil, ***acorin***, and minute amounts of ***choline***. In India and China, as an insecticide (vapors sterilize male **houseflies**). Roman **soldiers** carried the root as standard equipment, to disinfect questionable water supplies on forced marches and during campaigns. In Exodus 30, Moses tells how God directed him to formulate a sacred oil containing it, to anoint items of ritual importance. Also, as an anthelmintic, carminative, in herbal teas, and in perfumery. North American material contains **no** ***β-asarone***, whereas the European material contains nearly 10% in the oil and up to 0.3% in the dried plant. ***β-asarone*** is considered to be the carcinogenic principle.

Untoward effects: A carcinogen in **rat** trials, and now, no longer permitted in the U.S. in foods or medicinals. Some say it has psychedelic properties. It is used with *Artemesia* *absinthium* in ***absinthe***. Masks, goggles, and gloves should be used when working with it.

The oil potentiates ***barbiturate*** and ***ethanol*** hypnosis in **mice**. An 18 week feeding trial caused mortality, decreased growth, liver and heart abnormalities, and serous abdominal effusions in **rats**. In a 2-year feeding study in **rats**, livers developed blunted edges and discoloration, and histopathology showed degenerative changes in the liver, and slight or moderate focal or diffuse myocardial degeneration. After 59 weeks, malignant tumors were found in the duodenal region of those fed 500–5,000 ppm.

LD_{50}, **rat**: 777 mg/kg; IP, **rats**: 221 mg/kg.

CALATHEA VEITCHIANA

= *Pulma*

Untoward effects: Peruvian Amazonian **Indians** mix it with ***ayahuasca***, to see visions.

CALCIPOTRIOL

= *Calcipotriene* = *Dovonex* = *MC 903* = *Psorcutan*

Use: Topical treatment of psoriasis. A ***vitamin*** D_3 analog.

Untoward effects: Skin irritation with burning, pruritus, skin dryness or atrophy, folliculitis, worsening of psoriasis, and hyperpigmentation in up to 15% of **patients**. Systemic absorption also occurs and has caused hypercalcemia. It, therefore, is not recommended for use in **children**, during pregnancy, or for nursing **mothers**.

In **rabbits** ≥ 12 μg/kg/day, and 54 μg/kg/day in **rats**, were teratogenic. Progressive lethargy, nausea, depression, anorexia, and recumbency in **dogs** from 1–3 days after ingestion. Acute renal failure and soft tissue mineralizaion occurred.

~LD_{50}, **dog**: 100–150 μg/kg.

CALCITONIN

= *Calcimar* = *Calcitare* = *Calsyn* = *Calsynar* = *Cibacalcin* = *Miacalcic* = *Prontocalcin* = *Salcatonin* = *Salmotonin* = *Thyrocalcitonin* = *Tomocalcin*

A salmon, swine, and synthetic human *calcium*-regulating hormone, lowering blood *calcium* levels by inhibiting bone resorption.

Untoward effects: Phosphaturia, epistaxis (in seven **females**, 51–65 years, taking 50–125 IU for post-menopausal osteoporosis), mild gastrointestinal upsets in 3/9, nausea, flushing, tingling, strange taste sensations with pain, erythema and pruritus at injection site. Antibody formation may occur with non-synthetic forms. The serotonin antagonist, ***pizotifen*** (5 mg tid), may abolish most of these side-effects, except flushing. Do **NOT** use concurrently with ***mithramycin***. Intranasal form eliminates nausea, but nasal dryness, irritation, and occasionally epistaxis are common complaints.

SC use in **cats** and **dogs** may induce nausea and vomiting.

CALCIUM

Untoward effects: Complexes ***tetracyclines***, ***kanamycin***, and ***ciprofloxacin***, and decreases their availability. Ionized forms, IV, help produce heart block and potentiate ***digitalis*** and ***cardiac glycoside*** toxicity. Prolonged hypercalcemia normally causes a ***calcitonin*** release, with a decrease in serum calcium. Hypercalcemia in **man** causes severe abdominal pain and pancreatitis, can lead to kidney stones, and make **one** sluggish and comatose. Hypercalcemia due to IV infusions in **cows** is associated with bradycardia, sinus arrest, and ectopic beats. Calcium infusions in **man** have also been associated with hypertension. Hypertension and renal symptoms occur in fatal arterial calcification in **infants** that may, in some way, be associated with ***vitamin D*** intoxication. Renal calcium stone formation from hypercalciuria due to excessive oral intake in foods and supplements. Excessive intake has caused constipation in **man** and **animals**, and pulmonary calcification in **patients** with chronic renal failure. Hypercalcemia has been associated with ***aluminum*** in dialysis fluid and ***vitamin A***, as well as from excess calcium in dialysis solutions. Hypercalcemia leads to weakness, fatigue, pollakiuria, constipation, anorexia, and disordered mental thought and perception.

Drug interactions: Can decrease ***phenytoin*** availability.

CALCIUM ACETATE

Use: As a food stabilizer, in dyeing, curing skins and leather tanning, in corrosion inhibitors and lubricants, and in ***aluminum acetate*** solution.

Untoward effects: A cause of dermatitis in **bakers**.

IV LD_{LO}, **rat**: 147 mg/kg; IV LD_{50}, **mouse**: 52 mg/kg.

CALCIUM ACETYLSALICYLATE

= *Ascal* = *Cal-Aspirin* = *Calcium Aspirin* = *Dispril* = *Disprin* = *Kalmopyrin* = *Kalsetal* = *Solaspin* = *Solprin* = *Tylcalsin*

Untoward effects: Similar to ***acetylsalicylic acid*** (q.v.). In a double-blind cross-over study on 54 **patients**, severe epigastric pain was noted in ten, flatulence in two, with 20% complaining of assorted side-effects.

CALCIUM ARSENATE

= *Pencal* = *Tricalcium Arsenate*

Use: Insecticide, molluscicide, herbicide, and as a fruit tree spray.

***Untoward effects*:** Weakness, gastrointestinal disturbances, and peripheral neuropathy reported after inhalation; skin absorption and/or ingestion has caused hyperpigmentation, palmar and plantar hyperkeratoses. **Handlers** should remember that it is poisonous.

LD_{LO}, **human**: 5 mg/kg.

Vomiting (garlic odor), restlessness, severe abdominal pain, dysphagia, diarrhea (occasionally bloody), cyanosis, weak pulse, shock, and collapse reported in poisoned **dogs**. ***BAL*** is antidotal. Cancerigenic in lung, lymphatic systems, and skin. Lung tumors in **hamsters** after intratracheal injections. Single oral doses of 23 mg/kg are fatal within 3 days. Liver damage in poisoned **animals**.

LD_{50}, **rat**: 20 mg/kg; **mouse**: 794 mg/kg; **dog**: 38 mg/kg; **hares, partridges**: < 100 mg/kg; LD_{LO}, **rabbit**: 50 mg/kg.

CALCIUM BENZOYLPAS

= *Benzacyl* = *Benzapas* = *B-Paracipan* = *Therapas*

Antibacterial, tuberculostatic.

***Untoward effects*:** Fever, muscle pain, pruritus, dermatitis, epigastric pain, followed by jaundice, hepatosplenomegaly, and lymphadenopathy.

CALCIUM CARBIDE

= *Acetyleneogen* = *Miner's Carbide*

Use*:** Primarily, in the production of ***acetylene, neoprene, prochloroethylene, and ***vinyl acetate***.

***Untoward effects*:** Flammable. Dangerous when wet.

CALCIUM CARBIMIDE

= *Calcium Cyanamide* = *Nitrolime*

The citrated form = *Abstem* = *Colme* = *Dipsan* = *Temposil*

Use*:** Foliant, herbicide, defoliant, soil sterilant, molluscicide, pesticide, and in the manufacture of ***calcium cyanide, melamine, and in ***iron*** refining. The citrated form has been used, as is ***disulfiram***, to discourage ***alcohol*** consumption. This sensitizing action to ***alcohol*** occurs within about 20–30 min, and lasts 3–12 h. Has shown less side-effects than ***disulfiram***, and, in **some**, has worked where the latter hasn't. Can be hazardous, especially in **alcoholics** with ischemic and other forms of myocardial disease. Flushing, palpitations, tachycardia, dyspnea, hypotension, headache, nausea, and vomiting do occur, and is the basis for its use as an aversion therapy. Pounding headache, increased blood pressure, dizziness, precordial pain, sense of apprehension, weakness, and faintness are also noted. Inhalation or ingestion of large quantities has caused immediate unconsciousness, convulsions, and death within 15 min. Its circulatory changes with ***alcohol*** are usually less severe than with ***disulfiram***. A peripheral neuropathy occurred in a 50-year-old **male** after treatment with it. Caused increased urinary excretion of ***oxalates*** in 42-year-old **male**; skin rash and pyrexia, with abrupt deterioration of renal function in 28-year-old **female**. A rare fatal case in a 43-year-old **male** receiving 100 mg occurred following an ***alcohol*** test. Severe skin irritation with ulceration and systemic effects have followed skin contact.

Untoward effects*:** Causes headache, tremor, dyspnea, tachycardia, facial flushing, lassitude, and prostration. These symptoms are intensified, if ***alcohol is consumed. Dermatitis, skin ulceration (due to its corrosive action), with pigmentation, especially if the **person** exposed is perspiring, as with fruit **cultivators** and ***iron*** **workers**. Dermatitis due to hypersensitivity also occurs. When moist air contacts it, ***phosphine, ammonia,*** and ***hydrogen*** may be released.

LD_{LO}, **human**: 571 mg/kg.

Animals have been poisoned by it when applied or stored as fertilizer.

LD_{50}, **rat** and **rabbit**: 1.4 g/kg.

CALCIUM CARBONATE

= *Aeromatt* = *Albacar* = *Calcichew* = *Calcidia* = *Chalk* = *Citrical* = *Iceland Spar* = *Os-Cal* = *Prepared Chalk* = *Precipitated Chalk* = *Purecal* = *Spar* = *Tums®*

Antacid, protective, antidiarrheal, mineral nutrient.* Contains 62.5% *calcium.

Untoward effects*:** Acute hypercalcemia and nephrolithiasis in **people** regularly ingesting large quantities over many years. Chronic ingestion of ***Tums® (estimated daily ingestion of 6 g ***calcium***) by a 65-year-old **female** led to hypercalcemia. Hypercalcemia, azotemia, alkalosis, anorexia, nausea, vomiting, and epigastric pain in 40 **male** peptic ulcer **patients**. A 40-year-old **female** ingested approximately 16 g/day, along with other ***antacids*** for 9 years and developed classic ***milk-alkali*** syndrome with acidosis. ***Milk-alkali*** syndrome can follow ingestion of 4–60 g/day/2–60 days. Care must be used in hemodialysis **patients** where it is used as a ***phosphate***-binding antacid. Severe hypercalcemia developed in 3/10 hemodialysis **patients** receiving 3.2–6.4 g/day/4–8 weeks. A 23-year-old **male** hemodialysis **patient** developed nausea, vomiting, muscle weakness, personality changes, and subconjunctival calcification. It has been a major cause of constipation, even concretions and diarrhea, especially in chronic **users**, and is **NOT** recommended in pregnant **women** with a history of constipation, hypertension, or renal impairment. After effective early neutralization of gastric secretions, it then stimulates gastric acid secretions for 2 h with an increase in serum ***gastrin*** levels for

30–60 min. Hypercalciuria and reduced renal function can develop. Bloating can occur from gas released by acid neutralization. Contact dermatitis from chalk has been reported. A chalk impregnated with ***cypermethrin*** or ***deltamethrin*** that kills **ants**, has been illegally imported from China into the U.S. and was been eaten by a **child**, who required hospitalization. One trade name has been ***Pretty Baby Chalk***®.

LD_{LO}, **human**: 5 g/kg.

Drug interactions: Most significant is its complexing and reducing the availabiliy of ***tetracyclines*** and ***ciprofloxacin*** (40% decrease in maximum plasma concentration). Use with ***hydrochlorothiazide*** can increase the chance of hypercalcemia. It can decrease the availability of ***phenytoin*** for the first few days of treatment. Varying decreases in the effectiveness of ***nalidixic acid***, ***nitrofurantoin***, ***anticoagulants***, ***penicillin G***, ***phenylbutazone***, and ***sulfonamides***.

CALCIUM CHLORIDE

Use: In roadway de-icing, antifreezes, fire extinguishers, oil field drilling muds, tire ballast, fire-proofing fabrics, and in concrete. Therapeutically, as a diuretic, urinary acidifier, hemostatic, and, intravenously, as a ***calcium*** source, and in treating "brown blood disease" in **catfish**.

Untoward effects: May produce thrombophlebitis on rapid IV injection and spasms of coronary or cerebral vasculature leading to decreased oxygenation of vital tissues. Caused hypercalcemia, cardiac arrhythmias, and death in a 2-month-old **female** during open-heart surgery. IM injections are painful and may cause necrosis of the site. Over 2 days oral treatment may cause metabolic acidosis in **infants**. Severe superficial tissue necrosis with pain after perivenous extravasation. Dry, scaling, and erythematous eruption on scalp and forehead at sites of contact with electroencephalogram electrodes in a 5-year-old **female**. Symptoms recurred 3 years later on a subsequent electroencephalogram. Paste contained calcium chloride. Healing usually takes about 6 weeks and leaves an atrophic scar. Other similar cases reported. Skin lesions on the feet of ice and popsicle plant **workers** are caused by dripping solutions. The first symptom is erythema. Agricultural **workers** can also develop a dermatitis from its use in fertilizers. The dry powder is not irritating to skin, but a dermatitis develops within 10 min after contact with a paste, and in 3–4 min if it is rubbed onto the skin. Calcification of superficial scalp veins in a 10-day-old **male** followed its use via a scalp vein.

Drug interactions: Incompatible in solution with ***methicillin***, ***amphotericin B***, ***cephalothin***, ***chlorpheniramine***, ***ciprofloxacin***, ***folic acid***, ***nitrofurantoin***, ***sodium bicarbonate***, and ***tetracyclines***.

An accidental exposure of a **dog's** skin to a commercial hygroscopic landscaping product containing nearly 80% calcium chloride caused papules and ulcerations within 24 h. IV use in **dogs** and **horses** will produce, in many, ventricular fibrillation, arrhythmias, extrasystoles, and tachycardia. Do **NOT** use IM, as it will cause severe necrosis of tissue.

LD_{50}, **rat**: 1 g/kg; LD_{LO}, **rabbit**: 1.4 g/kg; IV LD_{LO}, **dog**: 274 mg/kg; **cat**: 249 mg/kg; **rabbit**: 274 mg/kg.

CALCIUM CHROMATE

= *Calcium Chrome Yellow* = *C.I. 77223* = *C.I. Pigment Yellow* = *Gelbin* = *Yellow Ultramarine*

Use: As a pigment, corrosion inhibitor, and in batteries.

Untoward effects: A cause of lung cancer in pigment **workers**.

Carcinogenic in **rats** by various routes. Chronic inhalation of the dust by **mice** causes a marked increase in pulmonary adenomas.

LD_{50}, **rat**: 327 mg/kg.

CALCIUM CYCLAMATE

= *Calcium N-Cyclohexylsulfamate* = *Cyclan* = *Sucaryl Calcium*

Non-nutritive sweetener. **Approximately 10 million lbs consumed annually in the mid-60s in the U.S.**

Untoward effects: Removed from the U.S. and Great Britain markets in 1969, because its metabolite, ***cyclohexylamine***, produced by bacterial action in the gut, was suspected to be a bladder carcinogen in **rats**. A metabolite in some, but not all, **humans**. It is also a starting chemical in manufacturing. A ***sulfonamide*** with a potential for cross-sensitivity. Photosensitization reported in **users** with severe eczema and acute renal tubular acidosis (also in **rats**). Consumption of over 5 g/day can have a laxative effect. A 25-year-old **female** airline **stewardess** ingested 12 g/day and develop erythema of the face and neck, as well as patchy areas on the trunk, along with severe leg and muscle weakness, and decreased blood pressure. This was reversed with oral ***sodium bicarbonate***. **She** was left with a microcytic anemia. Simultaneous use with ***lincomycin*** can decrease absorption of the latter by about 75%. Chromosome breakage reported in **human** *in vitro* cell studies.

Large oral doses to **hamsters** caused a 75% mortality with calcifying lesions in muscle, myocardium, and kidneys.

LD_{LO}, **rat** and **mouse**: 10 mg/kg.

CALCIUM DISODIUM EDETATE

= *Antallin* = *Calcitetracemate Disodium* = *Calcium Disodium Versenate* = *Ledclair* = *Mosatil* = *EDTA Calcium* = *Edathamil Calcium Disodium* = *Sormetal* = *Versene CA*

Untoward effects: Principal toxic effect is renal tubular necrosis. Dosage schedules must not be exceeded, to prevent toxic and potentially fatal effects. Do **NOT** give during periods of anuria. Localized thrombophlebitis at injection site. Nausea, diarrhea, muscle cramps and pain, fever, and headache can occur. May aggravate symptoms of systemic lupus erythematosus.

Drug interactions: Physically incompatible in ***lactated Ringer's solution***.

LD_{50}, **dog**: 12 g/kg; **rabbit**: 7 g/kg.

CALCIUM EDETATE

= *Calcium Ethylenediaminetetraacetate* = *Calcium Versenate*

Use: In treating ***lead, manganese, copper***, and radioactive metals poisoning by chelation.

Untoward effects: **Rat** experiments demonstrated an induced enteropathy with inhibition of DNA synthesis, and pathology in the proximal renal tubules. Increased **rat's** sensitivity to ***pentobarbital*** anesthesia. Its use was associated with many pyknotic nuclei in the duodenum of **horses**.

CALCIUM FLUORIDE

In *fluorspar*.

Use: For remineralization of teeth, and fluoridation of drinking water.

Untoward effects: When fed to **dairy heifers**, definite mottling of tooth enamel, with staining occurred; **guinea pigs** lost weight, slobbered, had dental deformities, and defects in bone calcification; pregnant **mice** had fetal resorption, and even small doses in late pregnancy caused defects in their progeny's teeth and jaw bones.

LD_{50}, **mouse** 11–21 days pregnant: 28 mg/kg; LD, **guinea pig**: > 5 g/kg.

See ***Fluorine***.

CALCIUM GLUCEPTATE

= *Calcium Glucoheptonate*

Use: For the treatment of hypocalcemia, or in its prevention during exchange transfusions.

Untoward effects: Local reactions at IM injection sites. Rapid IV injections cause "tingling sensations, a ***calcium*** taste, and a sense of oppression or heat waves". Use caution in **patients** on ***digitalis*** or ***digitalis***-like preparations. Perivenous infiltration has caused local necrosis.

Drug interactions: Physically incompatible with ***cephalothin, tetracycline***, or ***aminophylline*** solutions.

CALCIUM GLUCONATE

= *Calcium Glyconate*

Use: In treating hypocalcemias, sewage purification, and as an anticaking agent in ***coffee*** powders. In prevention of **poultry** egg shell breakage, to antagonize the cardiotoxic effects of hyperkalemia, and the toxins of **toads, lizards**, and various **spiders** and other **insects**, topically, to localize ***fluoride*** ions, after ***hydrofluoric acid*** burns, IV in treating ***chlorinated hydrocarbon*** toxicity, and in treating neuromuscular blockade after ***antibiotic*** therapy.

Untoward effects: Bradycardia and heart block can occur with too rapid or too concentrated solutions. SC and venous calcifications reported after IV use in **man**. It is irritating and can cause necrosis and soft tissue calcification after extravasation. Use dilute solutions! *Avoid* intra-arterial injections! Intramuscular injections have been associated with necrosis. The addition of a 10% solution to a hyperalimentation formulation caused paresthesia in a 24-year-old **female**. The reaction was traced to ***disodium edetate*** as a stabilizer, that chelated trace elements in the formulation. Iatrogenic hypercalcemia developed in a 4-year-old **male** after receiving 800 mg via parenteral hyperalimentation. Hypercalcemia and hypertension in a 35-year-old **female** after receiving 15 g/day. Calcinosis cutis in an 11-year-old **male** developed within 3 weeks after IV injections for hypocalcemia associated with rhabdomyolysis. Hypomagnesemia developed in 12/50 **infants** with chronic diarrhea. A 35-year-old **female** developed mania after two 3 g IV injections (30 min for each) in a 24 h period, to correct hypocalcemia, secondary to short bowel syndrome.

IM TD_{LO}, **infant**: 143 mg/kg.

Drug interactions: 30% of **patients** receiving it had impaired absorption of ***vitamin*** B_{12}. May adversely affect imaging studies using ***gallium***^{67} ***citrate***. Physically incompatible with ***amphotericin B, ampicillin, cefazolin, cephalothin, digoxin, kanamycin, sodium bicarbonate***, and ***tobramycin***. Incompatible in solution with ***aminophylline, folic acid, sodium iodide***, and ***prednisolone sodium succinate***.

Large doses or rapid IV in **cattle** can cause cyanosis, cardiac arrest or ventricular fibrillation, due to the direct effect of ***calcium*** ions on cardiac muscle. Toxic doses may cause polycythemia.

CALCIUM HYDROXIDE

= *Hydrated Lime* = *Slaked Lime*

Use: As an antacid, as an antidote against ***tannins*** (viz. in shin post or sand shinnery **oak** ingestion in **rabbits**, **cattle**, and **horses**), and in ***gossypol*** poisoning in **swine**; helps forestall formation of oxalate kidney stones in **humans**, in cements, in disinfectant solutions or powders, in depilatories (even for loosening hair on leather), and in lime water (a 0.14% saturated aqueous solution).

Untoward effects: Relatively insoluble antacid tablets have caused partial obstruction of the terminal ileum in a 26-year-old **male**. A dermatitis has been reported from its use topically (viz. in depilatories) in **man**. Has caused extensive edema and necrosis when used to cap **human** dental pulp after exposure. May slowly cause corneal damage – wash promptly with water. Particles on the body should be removed with cotton, etc., as water makes them slick and stick better to tissue. Skin contact with wet cement for prolonged periods has caused severe burns with thick, tenacious eschars.

CALCIUM HYPOCHLORITE

= *Bleaching Powder* = *Calcium Oxychloride* = *Chloride of Lime* = *Chlorinated Lime*

Use: As a deodorant, germicide, bleach, oxidizing agent, swimming pool disinfectant, in food processing, sugar refining, shrimp farming, sanitizing wells and water mains, as well as controlling algae. U.S. consumption is approximately 100,000 tons/year.

Untoward effects: Mildly caustic to skin and an eye and nose irritant. Dangerous if heated to decomposition, or if in contact with acid or acid fumes. It will then emit toxic, corrosive ***chlorine*** fumes (as it does if contacted by water or steam) in an explosion.

Frogs kept in water containing 4 ppm ***chlorine*** from it develop petechiation and ulceration of the skin, and died within 12 h.

CALCIUM OXALATE

Use: In water-proof and acid-proof cements, and the manufacture of candies and confections.

Untoward effects: Severe burning of the mouth, tongue, skin, etc., occurs after ingesting or chewing on various plants (q.v. ***Caladium, skunk cabbage, tulips, hyacinths, narcissus, oxalis;*** *Araceae*, including ***Cutleaf Philodendron*** (*Monstera sp.*); *Nyctaginaceae*, ***green dragon*** (*Arisaemia dracontium*) ***fishtail palm*** (*Caryota mitis*), ***Dumb Cane*** (*Dieffenbachia*), ***autumn crocus*** (*Colchicum*), ***Lily of the Valley*** bulbs, ***rhubarb*** leaves, and ***Calla Lily***). Some drugs also precipitate the problem, such as ***piridoxilate*** and ***nafronyl oxalate***. Heavy crystalluria has been reported in **people** drinking ***ethylene glycol***. In **man**, hyperoxaluria usually appears before actual nephrolithiasis. A calcium oxalate stone developed in a 21-year-old **male** taking 1 g ***vitamin C***/day/several months; permanent kidney failure from intratubular and intrarenal stone deposits reported in a 70-year-old **male** who received 2.5 g ***vitamin C***, IV over a 5 h period. A 58-year-old **female** who received a single IV dose of 45 g ***vitamin C*** for primary amyloidosis developed renal failure, secondary to tubular obstruction with calcium oxalate crystals. Chronic inhalation or ingestion of ***oxalic acid*** and soluble oxalates precipitate in the kidneys as insoluble calcium oxalate. Chronic absorption through the skin has occurred from hand contact with ***oxalic acid*** or soluble oxalate cleaning solutions. Pain and bluish discoloration of the nails develop and continued contact may cause gangrene.There is a genetic disorder that results in accumulation of these crystals in various parts of the body. It is suspect in cases of interstitial cystitis in **women**. ***Oxalic acid*** or ***oxalates*** absorbed are exreted as calcium oxalate, which can inflame the bladder and block the urethra.

Crystal deposits in the kidneys of **dogs** are common after ingestion of ***ethylene glycol***, and have caused death.

CALCIUM OXIDE

= *Burnt Lime* = *Calx* = *Chaux Vive* = *Gebraunter Kalk* = *Lime* = *Quick Lime*

Untoward effects: Becomes ***calcium hydroxide*** (q.v.) when wet. A caustic irritant to eyes and respiratory tract, and has caused perforation of the nasal septum. Will produce burns on wet or sweaty skin. Contact dermatitis is common, ranging from localized necrosis to serious burns over large areas. Erythema, vesicles, and ulcers appear more commonly in the summer. Hair of exposed **workers** becomes a grayish-red and **their** nails become thickened and ridged. Rock wool may contain 10% calcium oxide, and the latter may be a main cause of its ill effects on exposed skin. Leukoedema on the side of cheek mucosa, in which a quid of ***cocaine*** (in ***coca*** leaf) and lime was placed, reported in 35 (76%) **Bolivian Indians** (17–75 years of age). NO buccal cancer found on biopsy specimens. A **patient** developed bronchospasm from inhalation of ***soda lime*** (***calcium oxalate*** and ***sodium hydroxide***), due to its contamination of the breathing circuit of an anesthetic machine.

It is often scattered in bedding of **cows** and on concrete floors, where it tends to dry the skin and hooves of **animals**, sometimes, causing hoof cracks, leading to "foot rot". Ingestion by **animals** may cause abdominal pain, bloody vomitus, and erythema. Skin erosions have occurred on **frogs** contacting lime from concrete blocks.

CALLA LILY

= *Zantedeschia sp.*

Calla palustris* = *Water Dragon* = *Wild Arum* = *Wild Calla

Untoward effects: All parts of the plant (an aroid, a member of the *Araceae* or *Arum* family) are poisonous. **Children** are often poisoned by the berries of the wild calla. The watery sap is very irritant on contact, causing an itching rash. Biting or chewing on the stems or leaves has caused severe burning of the mucous membranes, with inflammation and swelling of the tongue and throat (including blisters), and swollen lips. Nausea, salivation, vomiting, and diarrhea reported. The tongue can swell enough to block the airway. Older **children**, as a prank, have induced **youngsters** to take a bite, because of its instant mouth burning effect. The effects are due to ***calcium oxalate*** (q.v.) crystals, and a possibly unidentified agent.

Cats and **dogs**, when left alone, have chewed on the plant with the same distressing symptoms.

CALLILEPIS LAUREOLA

An African plant. Extracts are used to kill maggots in wounds. Also in herbals.

Untoward effects: A case of hepatitis and liver necrosis is reported from its use as a herbal.

CALLIONYMUS CALAUROPOMUS

= *Common Stinkfish*

Untoward effects: It has a bitter flavor when eaten, and has caused nausea in **man**.

CALOTROPIN

Untoward effects: A cytotoxic extract of *Apocynum cannabinum* and *Asclepias curassavica* used in folk medicine, to treat cancer and warts in many parts of the world. Chemical structure is similar to cardiac glycosides. An extremely toxic African arrow poison in the milky sap of *Calotropis procera* (***Crown flower***, ***Giant Milkweed***, ***Tumfafiya*** [Hausa], ***Bomubomu*** [Yoruba], ***Aak*** or ***Madar*** [India], ***Bohr*** [Somali]). Minute amounts cause death, and it has been used in tropical America for murder and suicide. Caustic to skin and mucous membranes.

The **horse** is quite susceptible to its poison; **goats** and **sheep** can safely eat the withered, but not the fresh, leaves. When the latter eat the fresh leaves, they die within 33–83 days. Anorexia, diarrhea, dyspnea, and alopecia precede death. Hepatocellular necrosis, portal fibrosis, catarrhal enteritis, splenic hemosiderosis, pulmonary congestion and edema, cardiac hemorrhage, degeneration or necrosis in the convoluted renal tubules, and a normocytic anemia are noted.

LD, **cats**: 0.12 mg/kg; LD_{LO}, **rat**: 120 mg/kg.

CALTHA PALUSTRIS

= *Cowslip* = *King-cup* = *Marsh Marigold*

Untoward effects: All parts are poisonous, especially the seeds and young plants. Paresthesia, burning sensations of the skin and in the mouth, nausea, vomiting, weak pulse, hypotension, and convulsions have been reported in **man** from its irritant oils. It causes an unpleasant taste and odor from milk, milk products, and meat after **animals** have grazed it.

Despite its acrid taste, **cattle**, **sheep**, and **horses** have been poisoned by eating the fresh tops. The alkaloid, ***jervine***, and the glucoside, ***helleborin***, are the toxic principles. The dried plant is, supposedly, harmless.

CALUSTERONE

= *7β,17α-Dimethyltestosterone* = *Methosarb* = *NSC 88,536* = *U 22,550*

Use: In oral treatment of advanced breast cancer, and in stimulating platelet formation.

Untoward effects: Hepatotoxic and virilizing androgenic effects. Hirsutism, acne, nausea, moderate ***sulfobromophthalein*** retention, increased oiliness of skin, mild loss of scalp hair with loss or deepening of voice, fluid retention, and one case of increased libido reported.

CALYCANTHUS FLORIDUS

= *Allspice* = *Spicebush* = *Sweetshrub*

Shrub or tree 10 ft tall; occasionally cultivated as an ornamental. Found in southern U.S. in moist, rich woods, hillsides, and streambanks. When crushed, it has the odor of strawberries. *C. fertilis* is more toxic.

Untoward effects: Violent convulsions, paralysis, and cardiac depression are caused by its alkaloid, ***calycanthine***, found in its seeds.

Poisoning in **cattle** and **sheep** occurs mostly in the fall. It causes severe muscle tremors and convulsions, similar to ***strychnine***. It is a cardiac depressant.

CAMAZEPAM

= *Albego* = *SB 5833*

Anxiolytic. A benzodiazepinone.

Untoward effects: Generalized morbilliform or urticarial skin rashes. Therapeutic serum levels are 0.1–0.6 and toxic ones are 2 mg/l.

LD_{50}, **rats**: > 4 g/kg; **mice**: 970 mg/kg.

CAMBENDAZOLE

= *Bonlam* = *Bovicam* = *Cambenzole* = *Cambet* = *Equiben* = *MK-905* = *Novazole* = *Noviben*

A *benzimidazole*.

***Untoward effects*:** Doses well above those recommended have been reported to cause inappetence, listlessness, pleural effusion, lymphadenopathy, pulmonary edema, hepatomegaly, and even death in **animals**. Embryotoxic and teratogenic in **rats**. Doses 20–100% above recommendations to **ewes** in early pregnancy have been associated with malformed **lambs**. It has been a suspect teratogen in **mares** treated in early pregnancy.

CAMPHENE

***Use*:** In fragrances for cosmetic creams and lotions, soaps, detergents, and perfumes, and in foods. Found in the essential oils of many plants and conifers, which are often damaged by field **animals**.

***Untoward effects*:** 4% in petrolatum produced no irritation after a 48 h closed-patch test on **humans**, and no sensitization in 25 **human** volunteers.

Full-strength to abraded or intact **rabbit** skin for 24 h under occlusion was slightly irritating.

LD_{50}, **rat**: > 5 g/kg; and acute dermal LD_{50}, **rabbit**: > 2.5 g/kg.

CAMPHOR

= *Alcanfor* = *2-Bornanone* = *Gum Camphor*

***Analeptic, counterirritant, mild antiseptic, insecticide. Camphorated Oil* or *Camphor Liniment* is 20% camphor in cottonseed oil.**

***Camphor Spirits* is a 10% *alcohol* solution.**

***Camphor, Oil of* is a volatile oil (natural or synthetic).**

Use*:** In explosives, lacquers and varnishes, pyrotechnics, embalming fluids, and antipruritic mixture. As a moth repellent. Topically, with massage, it can cause vasodilation. Parenterally, in veterinary medicine, as a circulatory and respiratory stimulant. It is a component of ***paregoric.

Untoward effects*:** Eczema in **workers** handling the synthetic material which is often made from ***turpentine. These **workers** had positive patch tests for ***turpentine***, but not for camphor. Contact urticaria is also reported. A comedonal eruption in **infants** and **children** often occurred when it was a popular therapeutic. In **man**, ingestion or injection has been associated with nausea, vomiting, diarrhea, hysteria, fever, mental confusion, dizziness, vertigo, excitement, delirium, hallucinations, clonic epileptiform convulsions, anuria, tinnitus, coma, respiratory failure, and even death. Orally, 1 teaspoonful of camphorated oil in a 3-year-old **male** caused convulsions. Often, mistakenly taken instead of ***castor oil*** and 2 ounces has caused the death of an **adult**. It can cross the placental barrier and, 30 min after delivery, death occurred in an **infant** whose **mother** had accidentally ingested 12 g 36 h before. Another pregnant **woman** accidentally ingested nearly 2 oz of the ***camphorated oil***; 20 min later **she** became nauseous and had three grand mal seizures. Spontaneous labor began 24 h later (10 days before the estimated due date). Although the **baby** was not in distress, the amniotic fluid and the **baby's** skin and breath had the distinct camphor odor. In the U.S., ***camphorated oil*** poisoning cases during 1973–1975 were nearly 200 annually, and 494 cases of camphor poisonings occurred in 1993. Through the efforts of a pharmacist, Carmine Verano, the ***camphorated oil*** is no longer an over-the-counter product in the U.S. Camphor is *rapidly* absorbed from the gastrointestinal tract. A small portion is exhaled with the expired air. The fumes can irritate the eyes and throat, and even cause some of the symptoms described above for ingestion. Less than 1 g of camphor has been reported as fatal to **children**. 20 ml of the ***camphorated oil*** has been lethal to **adults**, with characteristic odor on breath and in urine. Various therapies have been reported successful, including emesis, ***olive oil*** stomach lavage, ***activated charcoal***, ***amberlite resin*** hemoperfusion, and ***diazepam*** for convulsions. Convalescence may take several days or weeks.

LD_{LO}, **human**: 50 mg/kg; **rabbit**: 2 g/kg.

In the **dog**: 5–15 g has been fatal; 2–4 oz in **cattle** and **horses**: 10–15 g to **sheep** increases respirations and, occasionally, causes convulsions.

CAMPSIS RADICANS

= *Trumpet Creeper*

***Untoward effects*:** A severe contact dermatitis has been reported in **people** from Alabama to Illinois. The poisonous principle has not been identified.

CAMPTOTHECIN

***Antineoplastic* extracted from the fruit, flowers, bark, wood and twigs of *Camptotheca acuminata*, a tree native to mainland China.**

***Untoward effects*:** Bone marrow depression in 9/12, alopecia in 4/12, vomiting in 4/12, and hemorrhagic inflammation of the bladder in 8% of the **patients** with advanced intestinal cancer. Leukopenia in 14/18 (78%), thrombopenia in 50%, maculopapular reactions in 2/18. Diarrhea in 3/15, stomatitis, and anemia.

IV LD_{LO}, **human**: 2.5 mg/kg/7 days.

Clinical signs of toxicity in treated **dogs** and **monkeys** were emesis, hemorrhagic diarrhea, dehydration, coma, and death. Histopathology in **dogs** showed intestinal epithelial metaplasia, and occasionally necrotic cholecystitis. **Monkeys** showed hypocellularity of the bone marrow and lymphoid tissues, and focal necrosis of the liver. In **rats** and **mice**, oral administration was more toxic than IV.

LD_{50}, **rat**: 66–87 mg/kg; **mice**: 27 mg/kg; IV LD_{LO}, **dog**: 80 mg/kg and 0.625 mg/kg/day/5 consecutive days.

CANARY GRASS

= *Phalaris sp.*

Untoward effects: In the early 60s, "*Phalaris* staggers" in Californian **cattle** was described after ingestion of *P. minor*. Similar cases occasionally occurred in **sheep**. In Australia, ***tryptamine*** alkaloids were isolated from *P. tuberosa*. Parenteral administration of these alkaloids caused central nervous system disturbances and "staggers". Low dosage caused hyperexcitability, tremors, salivation, head bobbing, dilated pupils, stiff hopping gait, incoordination, and lateral recumbency.

CANAVANINE

An insecticidal *arginine* analog, found in some leguminous seeds and sprouts.

Untoward effects: Causes a systemic lupus erythematous-like disease in **monkeys** when ingested.

CANDESARTAN CILEXETIL

= *Atacand* = *TCV 116*

Angiotensin II receptor antagonist, antihypertensive.

Untoward effects: Initial trials reveal little difference from placebo. Overdosage would probably be manifested by hypotension, dizziness, and tachycardia. Vagal stimulation can cause bradycardia. Angioedema can occur.

CANDICIDIN

= *Candeptin* = *Candimon* = *Levorin* = *Vanobid*

Antifungal antibiotic.

Untoward effects: Irritation and increased vaginal discharge in 34/154 **patients**. Sensitivity rarely noted.

LD_{50}, **mouse**: 90 mg/kg; IP LD_{50}, **mice**: 2.1 mg/kg.

CANNABIDIOL

= *CBD*

Untoward effects: **Male mice** ingesting it had a decrease in impregnation of **female mice** with an increase in prenatal and postnatal deaths after the impregnation. It prolongs ***pentobarbital*** sleeping time in **mice** and 0.32 mg/kg orally did not modify ***alcohol's*** adverse effects in **students**.

CANNABIS

= *Acapulco Gold* = *Bhang* = *Black Russian* = *"Broccoli"* = *Bush* = *Canamo Indiano* = *Cannabis sativa* = *Chanvre Indien* = *Charas* = *Chares* = *Daccha* = *Dagga* = *Diamba* (Kikongo) = *Dry High* = *Esrar* = *Gage* = *Ganga* = *Ganja* = *Goof Butts* = *Grass* = *Great White Shark* = *Griffo* = *Hanfkraut* = *Hash* = *Hashish* = *Hasis* = *"Hay"* = *Hemp* = *Herb* = *Indian Hay* = *J* = *Jane* = *Jay* = *Joints* = *Kannabis* = *Kenevir* = *Kif* = *Kynja* = *Lebake* = *Ma* = *Machony* = *Maconha* = *Marihuana* = *Marijuana* = *Mary Jane* = *Mary Walkers* = *Maui Waui* = *Mota* = *Mutah* = *Njemu* = *Nkaya diamba* (Kiyanni) = *Northern Lights* = *Panama Red* = *Peace-maker* = *Pod* = *Pot* = *Quarter Moon* = *Reefers* = *Sativa* = *Sawi* = *Shesha* = *Skunk* = *Smoke* = *Soles* = *Stick* = *Stuff* = *Suma* = *Tea* = *13* = *Vongory* = *Weed*

Potency in plants varies widely. Known in India, Egypt, and Mesopotamia several thousand years B.C.

Use: As an analgesic, ataractic, antiemetic in **humans**. In treating glaucoma, multiple sclerosis, and various neurological diseases. About 2,000 years ago, Chinese surgeons used it as an anesthetic. Commonly smoked, but has been used as a beverage, for oral ingestion (dried), and intravenously. As an excellent hypnotic for severe intestinal pain or before surgery in **horses**. Decreased activity, induces catatonia and ptosis in **mice**, **monkeys**, and **rats**, increased vocalization, and emotional defecation in **rats**, tames **monkeys**, and induces electroencephalogram changes in **monkeys**, **rabbits**, and **cats**. Tolerance to it has been noted, after chronic use in **pigeons** and **rats**. Gram-positive bacteria are very sensitive to its essential oils. Enhances depressant effects of numerous ***sedatives*** and hypnotic ***anesthetics*** in **rats**, **mice**, and **dogs**.

Untoward effects: Narcotic, intoxicant, and hallucinogen (*C. sativa*). Behavioral changes, increased pulse rate, decreased blood pressure, muscle weakness, conjunctival injection (bloodshot), depression, stupor, ataxia, hypothermia, and vomiting occur in **humans** and **dogs**. Catatonia, narcotic plastic rigidity, or tendency to maintain abnormal positions reported in **man** after excessive usage. When the dosage is high enough, it becomes a psychedelic agent, and no longer just a mild euphoriant. Effects can be unpredictable and vary with **individuals**, and on the same **individual** at various times and circumstances. Although it may open the airways in **asthmatics**, its use for that is not recommended because it may induce bronchitis and dryness and irritation of the mouth and throat. Heavy cannabis **smokers** may experience pharyngitis and bronchitis. Its euphoria and a distorted perception of time are interrelated. Smoking it has caused decreased driving performance, increased heart rate and blood pressure, increased sensitivity to intermittent light, increased hunger and thirst, decreased self-confidence, and may cause crying, indifference or apathy, headaches, restlessness, jittery feelings, aggressiveness (including criminal assaults), clumsiness, suicidal tendencies, poor memory, and even the "hearing of voices". Yet, a Harvard University report stated its use does not constitute a significant influence on non-violent or delinquent conduct. Adverse consequences, in terms of performance and psychological states are relatively infrequent, and it has often decreased some hostility responses. But, **pilots**, who had smoked it 24 h before, were unaware that, in flight simulators, **they** had made major deviations from the proper angle of descent, etc. Its use has been implicated as a cause of a commercial air crash and in train wrecks. In regard to the plane crash, the National Transportation Safety Board stated "when the smoke ... is inhaled, the drug reaches the brain. The psychological and cardiovascular effects are evident within a few *seconds* .... When smoked, ***THC*** (tetrahydrocannabinol) is absorbed by the blood in minutes. ... there is long-term retention of ***THC*** in the fatty tissues of the body. Marijuana may be active in the nervous system long after it is no longer detectable in the blood." Passive **inhalers** of smoke showed no cannabinoid blood levels up to 3 h after exposure, in contrast to significant urinary levels of its metabolites up to 6 h after exposure. These **people** cannot pass cannabis tests required for employment. A manuscript, *The Slavic Chronicles*, written in the 12th century, stated "***Hemp*** raises them to a state of ecstasy or folly or intoxicates them. Then, sorcerers draw near and exhibit pleasures and amusements to the sleeper's phantasms. They then promise that these delights will become perpetual, if the orders given them are executed with the daggers provided." **Hashish-eaters** were known as *Hashishins* or *Hashshäsins*, who were called "Assassins" by Englishmen unable to pronounce the former names. These men, under the influence of the drug, became hired *Hashshäsins* or assassins. *Salmonella muenchen* contaminated marijuana was responsible for a number of outbreaks of the infection in several states. Minor adverse pulmonary effects were reported in **people** who smoked ***paraquat***-treated plants. Bacterial and fungal contaminants on the plant are a potential risk factor, especially in an immunocompromised **patient**. A **patient** smoked illicitly obtained material during combination antineoplastic therapy for small cell lung cancer and developed invasive pulmonary aspergillosis. An allergic reaction to it occurred in a 29-year-old **female** first time **user**. Heavy use has been associated with a fatality. It impairs the immune response and the production of sperm and ***testosterone*** levels, as well as interfering with ovulation and prenatal development. Several cases of recurrent genital herpes each time the **person** smoked a reefer have occurred. Byssinosis in 39% of 102 non-smoking **female** hemp carding and spinning textile **workers** inhaling hemp dust. Its stimulation of ***β-adrenergic receptors*** causes a loss of body heat via increase muscle blood flow, and is dangerous for use by **skiers**. A few cases of urticaria have been reported after its use by some U.S. **soldiers** in Germany. Progressive obliterative arteritis of the lower extremities reported in 29 young heavy smoking **Moroccan males**. **Children** who accidentally ingested cannabis experienced ataxia, tremor, pallor, and mood alterations, as do **adults**. Serious symptoms of poisoning occurred in a **child** eating its unripe fruits or so-called seeds, which are more narcotic.

A study without controls indicated mental illness in many chronic pot **smokers** was associated with cerebral atrophy or irreversible brain deterioration. Their symptoms included headaches, loss of recent memory, poor concentration, depression, inefficiency, paranoid psychosis, and hallucinations. Other studies showed no cerebral atrophy in heavy **smokers**. Mutism, plus disorientation and confusion in four **males**, aged 19–31 years, after each smoked a pipeful. It appears to exacerbate symptoms in schizophrenics. Chromosomal breakage has been reported, as well as decreased ***uric acid*** assay results. Visual disturbances, including nystagmus, double vision, poor accomodation, and conjunctivitis have been reported in **abusers**. A **patient** developed bilateral burns of the central retina with central scotomas, loss of visual acuity, and retinal edema from sun-gazing. Significant, but subtle, neonatal effects (nervous symptoms and visual responses affected in a dose-related manner) from **mother's** use of cannabis. Breast milk exposure can delay motor development of **infants**. **Babies** born to cannabis-smoking **mothers** may be prone to leukemia. Intrauterine growth retardation, small head, epicanthus, posteriorly rotated ears, and long philtrum noted in **human neonates** and **animals**. Lactation is inhibited by large doses of cannabis, and it is excreted in breast milk. ***THC*** is absorbed by the **baby** from the milk. An author referred to an unusual case of "pot belly" in a 32-year-old **male**, who had swallowed 18 home-made sachets filled with approximately 55 g hashish oil in each and suffered from blood-stained vomitus. Surgery removed the offending sachets. Regular cannabis **smokers** have had clinical gynecomastia and raised plasma ***prolactin*** levels. It has been involved in fatal intoxications and a suicide attempt. During the Vietnamese War, the U.S. Army's medical teams were more concerned with cannabis than venereal disease. Vietnamese marijuana had more resin content than U.S. varieties, and was often contamination with ***opium*** and other agents. The number of resulting psychotic reactions was much higher than with U.S. plant material. Variations in ***THC*** content appears to be the key factor. Various types of non-dose-dependent, even long-lasting paranoia have been reported, even in **those** who had no past history of emotional problems. Schizophrenic-like psychosis also reported. Koro, an acute anxiety state, characterized by the perception of penile retraction into the abdomen, with fear of impending death, precipitated during the first experience of cannabis smoking has been reported from India. Bronchitis, coughing, dyspnea, and irreversible pulmonary function loss is greater than with ordinary cigarette smoking. A 50-year-old **male**, on several occasions, had urinary retention after ingesting cannabis. Recurrent herpes (type 2 simplex isolated) in a 30-year-old **male** on shaft of penis started 2–3 days after smoking cannabis, with attacks lasting 5–8 days. Attacks stopped with discontinuance of smoking and recurred when restarted. A drop or two of hashish oil on a regular cigarette is equal in psychoactive effect to an entire marijuana cigarette. Estimates of the amount smuggled annually into the U.S. range up to 15,000 metric tons, mostly from Mexico, Jamaica, and Columbia. Drug effects from smoked material are generally two to three times those of orally ingested material. 22 **persons** attending an office party became ill with muscular incoordination, dizziness, concentration difficulties, confusion, dysarthria, dry mouth, dysphagia, blurred vision, and vomiting. A bundt cake was suspect. Illness began within 15–120 min, and most symptoms resolved within a few hours, but two evidenced extreme excitability and paronia, lasting about 3 weeks. Assay of a few crumbs revealed ***THC*** and urine analysis 3 days later confirmed the probability of marijuana contamination of the cake. Reported to be the most common cause of poisoning deaths in South Africa. Toxic effects have been produced by 0.6 g, and death has occurred after 1.5–3 g, yet, **some** have recovered from larger doses. Hepatotoxicity is rare. Adverse effects of intravenous injections of its broth or extract have been reported leading to acute illness with severe vomiting, diarrhea, abdominal cramps, hypotension, tachycardia, peripheral vasodilation, moderate azotemia and oliguria, leukocytosis with left shift, thrombocytopenia, increased partial thromboplastin time, increased fibrin degradation products, and a positive ***protamine sulfate*** test. Also noted was hyperventilation, hypoxemia, pulmonary edema and rhabdomyolysis, with all the abnormalities reversed in 3–7 days, with no significant sequelae.

Drug interactions: An increase in adverse effects reported with use of ***alcohol***, ***barbiturates***, ***dextroamphetamine***, ***dimethyltryptamine***, ***lysergic acid diethylamide***, ***opium***, ***phencyclidine***, and ***scopolamine***. It may be found adulterated with ***oregano***, ***parsley***, and ***phenytoin***. Toxic symptoms resembling botulism in two 22-year-old **males**, after ingesting ***phenytoin***-adulterated material (approximately 1 g cannabis), and smoking 0.5 g 2 days later. Symptoms included vomiting, diarrhea, muscle stiffness and pain, dry mouth, dysphagia, blurred vision and dysphonia, with recovery after 5 days. Its use increased the metabolism of ***theophylline***.

Cattle and **horses** are frequently adversely affected after ingestion of the more palatable seedlings and young plants with symptoms of excitation, dyspnea, muscle spasms, decreased temperature, salivation, sweating, weak heart action, and death, often within 30 min after ingestion. Atrophy of spermatogenic elements noted after chronic use in **monkeys**. **Rhesus monkeys**, after the first dose, did not care for repeated self-administration. After being given the injections automatically, discontinuance caused abstinence signs of irritability, biting and licking fingers, hair-pulling, tremors, shaking, and hallucinations. **Dogs** and **monkeys** may show central nervous system stimulation, including symptoms of delirium and mania, as in **man**, prior to depressant effects. As a vasodilator (especially conjunctiva), hypothermia may be noted. Fetal resorption in **mice** and both it and teratogenicity in **hamsters**, **rabbits**, and **rats**. 19 of 37 adolescent **users** made their pet **dogs** (12) or **cats** (13) "high", by blowing an average of five puffs of smoke into the **animal's** nostrils on 1–15 occasions. The **animals** evidenced somnolence (87%), "glassy" eyes (72%), ataxia (72%), bumping into furniture (66%), biting or aggressive behavior (28%), and vomiting (5%). Intoxication lasted over an hour. Interferes with visual discrimination in **monkeys** and **pigeons**; **cats** and **monkeys** just gaze into space.

Usually, ataxia in fox-terrier **dogs** is used for assay. Although normally non-fatal for **animals**, 15 **horses** and **mules** died rapidly after ingesting it where it was grown illegally for hashish production. Normal toxic symptoms in **animals** include narcosis, stupor, decreased respiration, and bradycardia. A **puppy** that ate some marijuana cigarettes salivated, had muscle weakness, a pounding pulse, and, alternately, was sound asleep or wide awake for about a 6 h period. Hyperactivity, staggering, staring, somnolence, and vomiting in a **dog** after eating hashish brownies. Caused sneezing and ataxia in a pet **ferret** before it became comatose.

CANRENONE

= *Phanurane* = *RP 11,641* = *SC 9376*

***Aldosterone antagonist, diuretic*. The major active metabolite of *spironolactone*.**

Untoward effects: Interferes with accurate plasma ***11-deoxycorticosterone*** assays, usually, giving raised results. Hepatic coma and death possible in **patients** with severe hepatic insufficiency.

CANTHARIDES

= *Blistering Beetle* = *Blistering Fly* = *Russian Fly* = *Spanish Fly*

Use: In vesicant ointments, due to its ***cantharidin*** (0.5–0.9%) content.

Untoward effects: Cramps and abdominal pain, hematemesis, increased urinary frequency, dysuria, and hematuria within 2 h, followed by diarrhea (occasionally bloody), genitalia blisters plus inflamed bullae on penis and scrotum in a 20-year-old **male** after ingesting one capsule. Although not an aphrodisiac, it can cause persistent abnormal, and occasionally painful, penile erection from irritation and elimination via the urinary tract. Also causes clitoral erectile tissue congestion. **Recipients** interpret this as a sign of increased sexual desire. Vomiting, bloody stools, abdominal pain, salivation, fetid breath, increased respiratory rate, increased pulse, burning thirst, delirium, tetany, convulsions, and death have occurred from its ingestion. Abortion and menorrhagia have occurred in **women**. A fatality followed the ingestion of 1.5 g of the powder. Post-mortem examinations reveal severe and acute inflammation of the kidney, intestines, and, sometimes, the spleen.

TD_{LO}, **human**: 5 mg/kg.

Frogs can eat them sometimes without ill-effects, and, in Sudan, the beetles are often drowned during the late winter flooding. A **man** who ate an apparently healthy **frog** for supper had extreme gastric pain and thirst, bloody urine and priapism during the night. The muscles of a normal-looking experimentally treated **frog** were fed to two **guinea pigs**. They died with nephritis and cystitis. It can be adsorbed by ***charcoal***. Consumption of hay or feed containing the beetles has caused severe and even fatal disease in **horses**, **mules**, **cattle**, and **sheep**. In Africa, **cattle** have been poisoned by drinking water contaminated with the beetles, often showing mild uneasiness and abdominal pain to profound shock and sudden death. **Goats**, **rabbits**, **hedgehogs**, **rats**, **mice**, and **dogs** have also been poisoned by cantharides and ***cantharidin***. Poisonous doses for **horses** and **cattle** are > 15 g; **sheep**: 3–4 g and **dogs**: 1–2 g. A **pony** died after consuming 1.4 g of the ground dry beetles.

LD, **horses**: 4–6 g dried beetles.

CANTHARIDIN

= *Cantharides Camphor*

Use: In treating warts. The blistering agent in ***cantharides*** is usually at a 0.6–0.9% level. 0.7% in ***Cantharon*** and 0.4% in ***Liquor Epispasticus***, and in ***Cantharoplast***. See ***Cantharides***.

Untoward effects: Use can be associated with Guillain-Barré syndrome. **People** who ingested it require cardiac evaluation and follow-up. A **patient** had oral ulceration after ingesting 1.5–2.0 ml of a 0.7% solution. This caused sinus tachycardia and T wave inversion, followed in several days by sinus bradycardia. A 42-year-old **male** had frequent painful urination, hematuria, abdominal discomfort, diarrhea, and priapism after drinking 5 ml of ***Liquor epispasticus*** (blistering liquid) to increase libido. The following day, **he** had blistering and ulceration of **his** cheeks, tongue, and palate, with protein in **his** urine. Death has occurred after oral use, and topical application can be followed by local blistering, myocardial damage, denudation of the gastrointestinal tract, central nervous system toxicity, and renal damage. Death occurred within 24 h after two young **women** ingested ***coconut*** ice deliberately contaminated by a **male** friend, who had blisters on **his** face from the drug. Vomiting of blood, chest and upper abdominal pain preceeded **their** collapse and early death. Hemorrhagic necrosis of gastrointestinal and urinary tracts noted on autopsies. Effectively adsorbed by ***charcoal***. Ingestion of 15–30 mg may be fatal in **man**.

LD_{LO}, **human**: 428 μg/kg.

Hay, especially ***alfalfa*** hay, particularly in southwestern U.S., is often contaminated by blister beetles. Male beetles contain an abundance of the toxin. Colic, depression, anorexia, frequent urination, hematuria, dysuria, oral erosions, drooling, pyrexia, tachycardia, tachypnea, diarrhea, dehydration, hypocalcemia, muscle fasciculations, abnormal gait, dysphagia, disorientation, and aggressiveness have been reported in **horses**. **Cattle**, **goats**, **sheep**, **emus**, and **chickens** have been adversely affected.

CANTHAXANTHIN

= *Carobronze* = *Carophyll Red* = *Carotaben plus* = *β,β-Carotene-4,4′-dione* = *C.I. 40850* = *4,4′-Dioxo-β-carotene* = *Food Orange 8* = *Orobronze* = *Roxanthin Red 10*

Use: Coloring agent for food and drugs. A carotenoid that has been used by **some** as an oral "tanning agent from within". As little as 3 mg/quart gives a tomato hue to spaghetti sauce, tomato soup, pizza sauce, and salad dressings.

Untoward effects: Its oral use tends to cause an orange–red tan. A blood **donor's** blood was a bright orange, due to its oral use. A 20-year-old **female**, after a 4 month course of treatment, developed a fatal aplastic anemia. Transient gastrointestinal symptoms have occurred. Elderly **people** and **patients** with retinopathy should not use it, if **they** have pre-existing epitheliopathy or ocular hypertension.

CAPECITABINE

= *Ro 09-1978* = *Xeloda*

***Antineoplastic*. Pro-drug of *5-fluorouracil*.**

Untoward effects: Diarrhea, nausea, vomiting, stomatitis, abdominal pain, constipation, dyspepsia; tingling, pain, swelling, numbness, and erythema of soles and palms; dermatitis, fatigue, pyrexia, paresthesias, headache, dizziness, anorexia, neutropenia, thrombocytopenia, leukopenia, anemia, and hyperbilirubinemia. Many of these can be severe. Marked ocular irritation, corneal deposits, and reduced vision were reported in two **patients**, resolving 6–8 weeks after treatment withdrawal. Confirmed by rechallenge in one **patient**. Possibly teratogenic.

CAPERS

Use: To impart a pungent taste to sauces.

Untoward effects: The plants can produce an irritant, and, possibly an allergic, contact dermatitis. The flower buds are the seeds and add the spicy and peppery flavor; and the top shoots, young leaves, and roots of the plant contain a rubefacient and vesicant principle, along with ***mustard oil*** (***isothiocyanates***).

CAPPARIS

Untoward effects: ***Andal*** = ***C. tomentosa*** = ***Haujeri*** (Hausa) contains a ***sulfur*** oil poisonous to **camels**, **sheep**, **cattle**, and **humans**. Hind limb weakness, swaying, inappetence, and recumbency occur. Necrosis of the cardiac fat, edema, congestion of many organs, and peritoneal and pericardial effusions are reported.

CAPREOMYCIN

= *Capastat* = *Caprolin* = *Capromycin* = *Ogostal*

Antitubercular aminoglycoside antibiotic. **Chemically related to, but less toxic than,** ***viomycin.***

Untoward effects: Nephro- and ototoxic in a small number of cases, and may cause hypocalcemia and cataracts in **man**. Nausea, vomiting, fever, increased blood urea nitrogen, temporary eosinophilia, allergic and hypersensitivity skin rashes, pain with induration at injection site, hypokalemia, headache, and raised serum urates reported as adverse effects. May cause increased or false positive urinary protein or albumin results, and occult urine cast values. Cross-resistance with ***viomycin*** is common.

CAPROLACTAM

= *Aminocaproic Lactam*

Use: In the manufacture of synthetic nylon fibers. Millions of **workers** are exposed to it.

Untoward effects: A 22-year-old **male**, after 3 days of occupational exposure, developed grand mal seizures, nausea, vomiting, slight fever, and dermatitis of the hands and feet. **He** had inhaled caprolactam dust and vapors. **His** clothing, as well as exposed areas of **his** skin, were coated with the dust. Stiffness of fingers, thickening and a shiny appearance of the skin on **his** hands, as well as bluish patches on the dorsal surface of **his** feet, buttocks, and thighs developed during the first day or so. Severely affected skin eventually desquamates. Other **workers** in the area with less exposure had similar skin changes of the hands. Neurological symptoms in other exposed **workers** were minor complaints of irritability, apprehension, and some confusion, which resolved after being in fresh air.

Neurological toxicity in experimental **animals** is characterized by tremor and apprehension, and, after large doses, convulsions.

CAPRYLIC ACID

= *Octanoic Acid*

Use: In manufacturing of dyes and esters for perfumery.

Untoward effects: Relatively small quantities decrease ***glucose*** tolerance in **people** in the presence of increased serum ***insulin*** levels.

LD_{50}, **rat**: 10 g/kg.

CAPSAICIN

= *Axsain* = *Finalgon* = *Mioton* = *Zostrix*

Topical analgesic, irritant pepper alkaloid.

Use: As a rubefacient and counterirritant in **human** arthritis; jetted by mailmen as a defense against **dogs** (***Halt***®) with a telltale orange face area identifying the unruly **dog**. In **cow** teat preparations, to discourage self-licking and on bandages or devices, to discourage chewing; intravesical administration in the treatment of neurogenic bladder; and in relieving the pain of postherpetic neuralgia and various dermatologic and peripheral pain disorders.

Untoward effects: Causes "Hunan Hand", a superficial burning of the skin in **people** handling hot peppers. Highly irritating at 1 : 1,000,000. Inhalation leads to coughing, sneezing, and airway resistance. Burning and erythema are common at the application site. Warm water and sweating can intensify these effects, which, in some rare cases, last up to a month. Nausea, internal pain, and fainting occurred in an 18-year-old **female** using a topical cream with 0.12% concentration. Skin testing confirmed the hypersensitivity. Unfortunately, jalapeno peppers have been used in a case of **child** abuse, with its potential for neurologic, gastrointestinal, pulmonary, dermatologic, and adverse ocular effects. Very irritant to mucous membranes and the eye. Avoid use on cracked skin. A review of 81 emergency room **patients**, representing 10% of those sprayed by police officers demonstrated ocular burning and redness; corneal abrasions were noted in seven and respiratory symptoms in six. A healthy 4-week-old **infant** accidentally sprayed it from his **mother's** key chain container leading to immediate life-threatening respiratory distress, epistaxis, apnea, and cyanosis, with eventual recovery after heroic treatments.

SC produces long-lasting neural desensitization in neonatal **rats**; after ~1 month, lung atelectasis, yellowish fur, thickened skin, hematuria, urine retention, and death.

CAPSELLA BURSA-PASTORIS

= *Case Weed* = *Mother's Heart* = *Pick Purse* = *Shepherd's Purse*

Untoward effects: Contains a goitrogenic substance that may pass into the milk of **cows** that consume it.

CAPSICUM

= *Cayenne Pepper* = *Chillies* = *Pfeffer* = *Red Pepper*

Use: Externally, as a counterirritant, and internally, as a carminative, and in some ginger ales.

Untoward effects: A cause of delayed eczematous reactions in **man**. See ***Capsaicin*** and ***Pepper*** for additional untoward effects. Poison control centers have reported its accidental use as having caused many cases of severe eye irritation and lacrimation.

Rats fed it at a 10% level in their diet had an increase in liver tumors. Not a true rubefacient, as it does not produce reddening of the skin.

CAPTAFOL

***Fungicide, pesticide.* A chlorinated compound.**

Untoward effects: TLV, air, **human**: 100 μg/m^3. Voluntarily removed from use on fruits and crops (4.5–5 million lbs annually in the U.S.) by the manufacturer, after evidence indicated it caused birth defects and was oncogenic in **rats** and **mice**.

LD_{50}, **rat**: 2.5 g/kg.

CAPTAN

= *ENT 26,538* = *Glyodex 37–22* = *Merpan* = *Orthocide 406* = *SR 406* = *Vancide 89* = *Vangard 45* = *Vondecaptan*

***Fungicide, bacteriostat, bird repellent.* A chlorinated compound.**

Use: In hair shampoos, insecticide mixtures, on equipment, and for direct application on **animals**, and on seed, to repel **pheasants**. As a spray on vegetable grain crops and on fruits. In the U.S., ~10 million lbs used annually.

Untoward effects: Contact dermatitis with urticaria and eye irritation reported in **people**. Estimated lifetime cancer risk of < 1/million. Ingestion can cause vomiting and diarrhea. Allergic sensitization reactions occur in **man**. National Institute for Occupational Safety & Health sets upper **human** exposure limit as 5 mg/m^3.

LD_{LO}, **human**: 1 g/kg.

High doses to **mice** in lifetime studies, but not **rats**, showed an increase in intestinal tumors. Fatal to **sheep** at 250 mg/kg. They show increased respiration, anorexia, and tend to stand alone. Excess fluid, and even blood, may be found in the thoracic and abdominal cavities. Petechial hemorrhages may occur on the liver; the gallbladder may be distended and the gastrointestinal tract inflamed. Use has caused nervous system and skeletal defects, especially phocomelia and amelia in **chicks**, but not in **hamsters**, **rabbits**, or **rats**.

LC_{50} on feed, **Bobwhite quail**: > 2,400 ppm; **Japanese quail**, **pheasants**, and **mallards**: > 5,000 ppm; LD_{50}, **rat**: 480 mg/kg or 9 g/kg (both values quoted in literature); **rabbits**: 2 g/kg; **ruminants**: 250–500 mg/kg.

CAPTODIAMINE

= *Captodiam* = *Captodramin* = *Covatin(e)* = *Suvren*

Anxiolytic.

Untoward effects: In one study, it caused nausea and increased restlessness. A metallic or bitter taste reported by some **patients**.

Inhibited sexual desire in male **dogs**. Excitation, convulsions, and exhaustion before death in **mice**.

LD_{50}, **rat**: 3.8 g/kg; **mouse**: 1.63 g/kg; CD_{50}, **mice**: 247 mg/kg.

CAPTOPRIL

= *Acediur* = *Aceplus* = *Acepress* = *Acepril* = *Acezide* = *Alopresin* = *Capoten* = *Captolane* = *Captoril* = *Cesplon* = *Dilabar* = *Garranil* = *Hypertil* = *Lopirin* = *Lopril* = *SQ 14225* = *Tensobon* = *Tensoprel*

Angiotensin-converting enzyme inhibitor.

Untoward effects: Dry, persistent cough and angioneurotic edema (can be fatal, if it involves the tongue, glottis, or larynx). Occasionally nausea, vomiting, diarrhea, anorexia (possibly, related to dysgeusia), constipation, headache, dizziness, malaise, dyspnea, insomnia, alopecia, and paresthesias are reported by **patients**. Bone marrow aplasia, hemolytic anemia, neutropenia, myeloid hypoplasia, leukopenia, agranulocytosis, eosinophilia, granulocytopenia, and pancytopenia (occasionally fatal) occur. A strong first dose hypotensive effect has been seen, especially when given with ***thiazide*** diuretics. A case of pericarditis developed in a 39-year-old **male** after 7 days of treatment. Encephalopathy, Guillain-Barré syndrome, cholestatic jaundice and hepatotoxicity, alterations of taste, stomatitis, ulcerations of the tongue and mouth, hyponatremia, nephrotic syndrome, maculopapular skin rashes and pityriasis rosea-like and lichenoid rashes, and, rarely, an exfoliative dermatitis or photosensitivity reaction. Toxic epidermal necrolysis, pemphigus erythematous, and separation and crumbling of fingernails also occur. Tachycardia, chest pain and palpitation occur in about 1% of **patients**; angina, myocardial infarction, and Raynaud's phenomenon, if related, occur in about 0.25% of treated **patients**. Fatal and non-fatal strokes have occurred. Deliberate self-poisoning with doses of 150–750 mg have not been fatal, although frightening. A successful suicide in a 75-year-old **male** after ingesting ~1.1 g. **His** plasma level was 60.4 mg/l. Several cases of mania were possibly precipitated by captopril dosing. Has caused **fetal** and **neonatal** morbidity and death, when given to pregnant **women**. Hypotension, neonatal skull hypoplasia, acute interstitial nephritis, anuria, renal failure, fetal limb contractures, cranio-facial deformation, and hypoplastic lung development have been reported. Its use has been suspect in intrauterine growth retardation and patent ductus arteriosus. Ingestion of 7.5 g leads to mild hypotension. Therapeutic serum levels are reported as 0.05–0.5 mg/l; toxic at 6 mg/l; and fatal at 60 mg/l.

Drug interactions: Acebutolol: Its uncertain if captopril (75 mg/day) alone or the combination with ***acebutolol*** (200 mg/day) resulted in Kaposi's sarcoma of the left arm in a 70-year-old **male**.

Allopurinol: The combination caused a fatal case of Stevens-Johnson syndrome in a 71-year-old hypertensive **patient** with chronic renal failure.

Aspirin: Potential serious interactions can occur. The combination should be avoided.

Chlorpromazine: The combination in a 49-year-old **male** with severe hypertension and chronic schizophrenia, resulted in marked hypotension with postural syncope.

Cimetidine: Neuropathies have developed on the combination, due to an allergic reaction or autoimmune processes.

Digoxin: 37.5–150 mg/day of captopril added to a stable dose of ***digoxin*** increase serum ***digoxin*** levels from 1.3 to 1.7 mmol/l, due to decreased renal clearance in 20 **patients** with congestive heart failure.

Diuretics: Some **patients** receiving the first dose of captopril while on ***diuretics***, especially ***thiazides***, have experienced (usually within 1 hour) a dangerous precipitous decrease in blood pressure. This effect has also been demonstrated in **rats** treated with ***furosemide***. Also see ***Potassium*** below.

Erythropoietin: In ten healthy **volunteers** (19–48 years) treated with captopril for 28 days, a significant decrease in plasma levels of ***erythropoietin*** occurred.

Indomethacin: It can block the immunosuppressive effect of captopril.

Metolazone: The combination was suspect in renal failure in a 65-year-old **female**. When both drugs were discontinued, renal function returned to normal.

Potassium: The combination of captopril with ***potassium*** supplements or ***potassium***-sparing diuretics can cause a significant hyperkalemia.

Probenecid: Decreased renal excretion of captopril.

Doses > 2 mg/kg have caused renal failure in **dogs**.

LD_{50}, **mice**: 6 g/kg, IV, 1 g/kg.

CAPURIDE

= *McN X-94* = *Pacinox*

Hypnotic, anxiolytic.

Untoward effects: Nausea, lethargy, vertigo, orthostatic hypotension, dizziness, depression, stupor, headache, and blurred vision reported on oral use in **patients**.

CARAMEL

= *Burnt Sugar*

Use: In coloring foods, confectioneries, cosmetics, beer, soft drinks, and pharmaceuticals.

Untoward effects: Has contained significant levels of ***sulfites*** with its potential for allergic reactions in **people**. One batch was found to be contaminated with ***leptophos*** and condemned.

CARAMIPHEN

The ethanedisulfonate or edisylate = *Alcopon* = *Taoryl* = *Toryn*, a weak antitussive, and the hydrochloride = *Panpanit* = *Panparnit*, an anticholinergic = *Pentaphen* in some countries, but the term is used elsewhere for *p-tert-pentylphenol*.

Untoward effects: Although the anticholinergic effects are weak, use it with caution in **patients** with glaucoma or urinary retention due to prostatic hypertrophy.

LD_{50}, **cat**: 390 mg/kg; IV LD_{50}, **mice**: 39.5 mg/kg; IP, **mouse**: 220 mg/kg.

CARAPA PROCERA

Untoward effects: Employed in the Congo as a stimulant, but others report it to depress the central nervous system with ptosis, and used "to calm **fools**". Has anticholinergic effects.

LD in **rats**: 0.25 g/kg to 1.25 g/kg.

CARAWAY

Use: As a cooking and baking spice; as a flavoring in pharmaceuticals and liquors. The dried ripe fruit of *Carum carvi*. The oil is frequently used and is also made synthetically. U.S. imports of the oil are approximately 40,000 lbs annually for flavorings and as an anti-nausea substance. It is bacteriostatic and fungistatic.

Untoward effects: A common cause of allergic dermatitis and cheilitis, often erroneously attributed to the food or substance it has been added to. Will cross-react with ***orange***-based products. Violent vomiting, abdominal pain, and loss of consciousness with eventual recovery followed the ingestion of 3.5 oz. The **patient's** urine contained ***acetone*** and ***albumin***.

Oil LD_{50}, **rat**: 3.5 g/kg.

CARAZOLOL

= *BM 51,052* = *Conducton* = *Suacron*

β-Adrenergic blocker.

Use: Weak antihypertensive in **man**; prevents adverse effects of stress in transporting **swine** and to reduce tachycardia and tachypnea in farrowing **sows**, and in preventing heart attacks in racing **hounds**.

Untoward effects: IV LD_{50}, **mouse**: 14 mg/kg; **rabbit**: 5 mg/kg.

CARBACHOL

= *Carbamylcholine Chloride* = *Carbocholine* = *Carcholin* = *Choline Chloride Carbamate* = *Coletyl* = *Doryl* = *Lentin* = *Miostat* = *Moryl* = *Enterotonin (also refers to a proprietary preparation in Europe of chloramphenicol and furazolidone)*

Cholinergic, miotic, parasympathomimetic, emetic, cathartic, ruminatoric.

Untoward effects: Urge to urinate and defecate, abdominal cramping, salivation, nausea, and sweating occurs. Dosage must be carefully titrated or small doses should be given every 30 min, if necessary. Can cause decreased blood pressure, syncope, and cardiac arrhythmias in **man**. SC injection in a 58-year-old **male** to relieve urinary retention caused a fatal esophageal rupture. A 60-year-old **female** with glaucoma suffered toxic side-effects from the use of a 3% solution as eye drops in error, instead of 3% ***pilocarpine***. Within 48 h after drug withdrawal, full remission occurred. Routine use **NOT** advised during pregnancy. Can produce hypotension, headache, bronchoconstriction, and dyspnea. Avoid use in **asthmatics**, in known cardiac disease, or in hot environments.

Experimental intrahypothalamic injections in the **cat** cause severe emotional states of aggression or fear, characterized by growling, crouching, then, hissing, spitting, rage, and attack; followed by vicious attacks on the experimenter. Can cause duodenal ulcers in the **rat**.

LD_{50}, **rats**: 40 mg/kg; **mice**: 15 mg/kg; IV, 300 μg/kg. LD_{LO}, **dog**: 1 mg/kg.

CARBADOX

= *Fortigro* = *GS-6244* = *Mecadox*

Antimicrobial.

Use: Extensively used in the U.S., primarily in the prevention and treatment of *Treponema* infections in **swine**.

Untoward effects: The U.K. banned it in the mid-80s, since it was declared a genotoxic carcinogen, and its use would give unacceptable risks to **mill operators** and **workers** handling it. Yet, despite the Delaney Ammendment, its use continued in the U.S.

Above recommended doses in **swine** it causes hepatomegaly and splenomegaly. It also causes pathological changes in the adrenal gland of **swine**. Toxicity is potentiated by ***furazolidone***.

CARBAMAZEPINE

= *Biston* = *Calepsin* = *Carbelan* = *Epitol* = *Finlepsin* = *G 32883* = *Sirtal* = *Stazepine* = *Tegretal* = *Tegretol* = *Telesmin* = *Timonil*

***Anticonvulsant, analgesic.* An *iminostilbene* derivative, chemically related to *tricyclic antidepressants*.**

Untoward effects: One early British summary on 501 **patients** with trigeminal neuralgia in 1969 indicated nervous disorders in 176 **patients** (35.1%), cutaneous effects in 19 (3.7%), gastrointestinal upsets, xerostomia, aplastic anemia (3/4 fatal), and biliary problems. It caused the onset of Tourette's syndrome or exacerbation of tics and vocalizations in three **patients** (7–12 years) treated for the control of seizures. Reducing its dosage improved the appetite of a 27-year-old **female** who suffered from severe anorexia on 1 g/day. A pseudotoxic-shock syndrome (fever, exfoliative dermatitis, renal, hepatic, and gastrointestinal upsets) occurred in a 13-year-old **female** after 1 month of therapy. Mania, hypomania, irritability, and aggression with abnormal electroencephalogram readings in psychiatric cases. Dystonia and opisthotomus in three brain-damaged **children** after 2–3 weeks. The problems subsided after drug withdrawal. In **children** with complex partial seizures, treatment may exacerbate them.

Reduced dosage may be required in the **elderly**. Significant absorption of undiluted, but not diluted, drug occurs into ***polyvinyl*** nasogastric tubes. It can lose at least 1/3 of its effectiveness, if stored under humid conditions. Both rapid and delayed leukopenia, leukocytosis, thrombocytopenia, pulmonary infiltrates, eosinophilia, severe rashes, mild dermatitis and pruritus, diarrhea, psychosis, transient disturbances of accommodation, nystagmus, drowsiness, dizziness, vertigo, nausea, vomiting, giddiness, restlessness, hand tremors, apathy, syncope, ataxia, palpitation, impotence, pancreatitis, mild paresthesia of face, and (in 25%, unless taken with meals) a feeling of gastric fullness occurs if dosage is increased too rapidly. Blood dyscrasias including fatal aplastic anemia (often in **patients** with trigeminal neuralgia) are usually preceded by rashes, which can be suppressed by ***antihistamines***, but, if treatment continues, the results can be fatal. Bone marrow aplasia, neutropenia, pancytopenia, hemolytic anemia, purpura, and hyponatremia, as well as decreased free ***testosterone*** levels in plasma occur. Produces acute dose-related adverse effects, which may disappear in a few hours, because it is rapidly absorbed. Increasing the dosing frequency to at least three times/day reduces some of these effects. Leukopenia (white blood cells < 4,000/μl) usually appears within 2–3 months in **children**; approximately 1.5% of **those** treated have a moderate leukopenia (white blood cells 3,000–3,500/μl). Cell counts usually return to normal during continuing treatment; < 1% of pediatric **patients** have white blood cells as low as 2,000/μl. Cases of erythroid arrest or hypoplasia were quickly resolved following cessation of therapy. Hypertension, congestive heart failure, and arrhythmias have been reported within 24–48 h after treatment is instituted. Hemorrhagic diathesis with bleeding gums after 24 months of treatment of 44-year-old **male**. Antiretroviral therapy failure in an HIV-positive 48-year-old **male** given ***indinavir*** (decreased plasma levels). **Fetal** exposure to it has been associated with cardiac, oral cleft, and neural tube congenital anomalies.

TD_{LO}, **human**: 43 mg/kg.

Bone: Osteoporosis of metaphyseal areas of long bone reported in **patients** aged 4–23 years, treated for > 3 years with high dosage.

Brain: Has caused aseptic meningitis, a rare drug reaction, and transient mental confusion.

Cardiovascular: Congestive heart failure reported in a 33-year-old **male epileptic** after its use at 200–600 mg/day/13 days. A 56-year-old **female** developed heart block, which resolved itself, after discontinuation of medication. Many additional cases are in the literature. Bradycardia also occurs, as did a failure of a functioning pacemaker 5 days after starting carbamazepine therapy for mania. It may increase the degree of heart block caused by ***adenosine***. Eosinophilic myocarditis in a 13-year-old **male** given 200–800 mg/day to control aggressive and impulsive behavior caused fever, chest pain, tachypnea, tachycardia, conjunctivitis, hepatomegaly, severe uncontrolled arrhythmias, and death.

Dystonia: A number of reports of drug-induced dystonia occur, especially at doses of 10–25 mg/kg/day. It may be due to carbamazepine's antagonist effect on ***dopamine***. Worsening of multiple sclerosis symptoms in many **patients**.

Ear: Tinnitus has been reported.

Eye: Lenticular opacities, due to exposure to ultraviolet or sunlight. Sunglasses should be recommended, to help avoid photosensitivity. Diplopia, ophthalmoplegia, retinopathy, mydriasis, and nystagmus have been reported. Exhibits mild anticholinergic effects – use cautiously in **patients** with increased ocular pressure.

Fetus and newborn: In some studies, congenital malformations have been reported in 2/3 pregnancies. Closed spina bifida in a **female infant** of a 27-year-old **female** who had taken 600 mg/day throughout **her** pregnancy. A drug-related case of myelomeningocele has also been reported. May increase drug metabolizing capacity in the **neonate**. Insignificant amounts excreted in breast milk and **infant's** plasma level is equal to the milk level after the third day. In another report, the milk:serum ratio in three **women** was 0.6, other workers had ratios of 0.9. Drowsiness and somnolence are, theoretically, possible in the nursing **neonate**.

Hallucinations: Visual hallucinations in a 76-year-old **female** with refractory trigeminal neuralgia (40–120 mg qid/6 years). Hallucinations stopped when drug discontinued, and reappeared within 2 days, when rechallenged with 60 mg qid.

Hypersensitivity: Hypersensitivity reaction or syndrome in an 81-year-old **male** led to bone marrow failure, erythroderma, severe pruritus, and fever after ingesting 200 mg tid/50 days.

Liver: Raised alkaline phosphatase is caused by liver hypersensitivity. Granulomatous hepatitis in three **patients** after 200–1200 mg/day for less than 1 month. Symptoms disappeared after drug withdrawal. Similarly, in 54-year-old **male** and 52-year-old **male**, after 200 mg tid/3 weeks. Also reported are cases of acute cholestatic hepatitis within 1 or 3 weeks after therapy. Cholestasis often occurs without hepatitis. Centrilobular or midzonal hepatic necrosis has been common and has also been fatal. Transient hepatic dysfunction noted in an **infant** of an epileptic **mother** treated with it during pregnancy and breast feeding.

Poisonings: One report covers 33 cases of overdosing (1.6–45 g ingested; 51% of the **patients** also took other drugs). Reduced consciousness, mydriasis, abnormal muscle tone and tendon reflexes, ataxia, nystagmus, ophthalmoplegia, hyperglycemia and hypokalemia (both on higher intake), hepatic dysfunction, hepatic dysfunction in 50%, and hyponatremia in 12% occurred. In pediatric (1–17 years) studies, **patients** with serum levels of about 100 μmol/l (23.5 mg/l) required close observation, and levels > 150 (35 mg/l) require intensive life support, and may be lethal. ***Activated charcoal*** and ***magnesium sulfate*** via nasogastric tube, after suctioning and ***physostigmine***, to control dystonias have been used in treatment. Serum levels of 15–25 mg/l are considered toxic, but so have levels as low as 12. Encephalopathy and profound stupor occurred in a 16-year-old **male** after a massive overdose (5.8 g). A 6-year-old **male** ingested 10 g producing vomiting, bradypnea, and coma; he recovered in 24 h after IV fluid. A massive overdose of ~20 g by a 56-year-old **male** led to hospitalization in a comatose state with renal failure; he recovered in 3 weeks. Usual therapeutic serum concentrations are 2–8 mg/l, toxic at 10 mg/l, and fatal at 20 mg/l.

Pyrexia: A number of cases of fever have been associated with its use. It can involve a hypersensitivity reaction.

Renal: Syndrome of inappropriate antidiuretic hormone secretion, acute interstitial nephritis, acute renal failure, and dysuria reported.

Respiratory: Pulmonary eosinophilia and asthma reported in a 52-year-old **male** taking 200 mg tid/9 months. Acute pulmonary hypersensitivity occurred in a 55-year-old **female** after 5 weeks of 100–200 mg twice daily. It precipitated coughing, shortness of breath, and rashes (arms and thighs, then, spreading to **her** trunk). Chest X-ray showed diffuse nodular and reticular pattern in both lungs, with prominence of hila, suggesting adenopathy. Marked improvement within 3 days after drug withdrawal. Hypersensitivity-induced pneumonitis is rare.

Skin: Its use has been associated with systemic lupus erythematosus and Stevens-Johnson syndrome. Pruritus, eczematous dermatitis, alopecia, erythema multiforme, toxic epidermal necrolysis, exanthema, lichen planus, facial edema, and prurigo have been reported. In a near-fatal case a 12-year-old **female** receiving 200 mg twice daily developed an exfoliative dermatitis; 1 month later, a morbilliform rash spread over **her** entire body during a 2 week period. A nodular lympho-proliferative disorder in a 63-year-old **male** receiving 400 mg/day for a painful leg neuropathy, resolved after 3 weeks with discontinuance of treatment. Other **patients** have developed maculopapular rashes with urticaria and acute porphyria syndrome.

Taste: Use has given **patients** reduced taste acuity or a bitter taste.

Water intoxication: Edema, fluid retention, and hyponatremia occur, due to its antidiuretic effect.

Drug interactions: It stimulates its own metabolism and that of ***clonazepam***, ***haloperidol*** (up to 60%), ***oral contraceptives***, ***phenytoin***, and ***warfarin*** and reduces their blood levels. ***Didanosine*** inhibits its absorption. Inhibits ***dexamethasone***-induced suppression of urinary corticosteroids, and can yield falsely raised ***cortisol*** test levels. Toxic ***serotonin*** syndrome (tremor, hyperactivity, and salivation), leukopenia, and thrombocytopenia develop with ***fluoxetine***. The reactions resolved with discontinuance of the ***fluoxetine***. ***Cimetidine***, but not ***ranitidine***, potentiates it by decreasing its hepatic metabolism, thus, increasing its blood levels. ***Macrolide antibiotics***, such as ***erythromycin***, potentiate its action and adverse effects. ***Isoniazid*** and ***fluoxetine*** also inhibit its metabolism. ***Monoamine oxidases*** may potentiate its adverse effects, and its use should be avoided within 14 days of ***monoamine oxidase*** ingestion. ***Ethyl alcohol*** increases its metabolization. ***Alprazolam*** serum concentration decreased from 54 to 10 ng/ml when carbamazepine was added. A dramatic worsening of the **patient's** clinical status occurred. It may cause neurotoxicity in **patients** on ***lithium***. Neurotoxicity, due to the addition of ***diltiazem*** therapy in a 78-year-old **male** dramatically reduced carbamazepine's elimination, requiring a 62% reduction of the latter to resolve the problem. The addition of ***clozapine*** to its therapy resulted in muscle rigidity, hyperpyrexia, tachycardia, sweating, and somnolence within 3 days. After ***clozapine*** was discontinued, the symptoms abated. Cross-hypersensitivity has been reported with ***phenytoin*** and ***phenobarbital***. Displaces ***thyroxine*** and ***liothyronine*** from serum binding proteins. An apparent dangerous interaction occurred in a 59-year-old **male** when ***viloxazine*** was added to treatment leading to a dramatic increase in ***viloxazine*** serum concentration and a decrease in carbamazepine serum levels. ***Grapefruit juice*** inhibits gut and liver metabolic enzymes, increasing its bioavailability. It increases ***valproic acid*** clearance by 10%; decreases ***topiramate*** serum levels by 40%. Use causes a ***carnitine*** deficiency. ***Fluvoxamine*** and ***sertraline*** increase its serum concentration. ***Fluconazole*** (150 mg/day/3 days) induced stupor and increased carbamazepine serum concentration in a 33-year-old **male**. ***Trazodone*** increased carbamazepine concentration in a 53-year-old **patient**. Decreases plasma concentration of ***mianserin***.

Numerous teratogenic effects have been noted in offspring of treated **mice**. ***Clarithromycin*** decreases its serum concentration by 28.5% in **rabbits**.

LD_{50}, **rat**: 4 g/kg; **mouse**: 936–3750 mg/kg; **rabbit**: 1500–2680 mg/kg; **dog**: 5620 mg/kg; **guinea pig**: 920 mg/kg.

CARBANOLATE

= *Banol* = *SOK Improved*

Use: As a methylcarbamate insecticide, to control **fleas**, **ticks**, and **lice**.

Untoward effects: A fast-acting cholinesterase inhibitor that, after topical application, causes some **pets** to salivate, lacrimate, have pupillary constriction, dyspnea, muscle tremors, and even convulsions. ***Atropine*** should be used as an antidote to its cholinergic effects. At a 1% concentration in topically applied powder form, it has been toxic, and even fatal, to some **cats**. Avoid use with any ***phenothiazines***.

LD_{50}, **rat**: 30 mg/kg; **pigeon**: 4.2 mg/kg; **duck**: 2.4 mg/kg; **birds**: 2 mg/kg.

CARBARSONE

= *Amabevan* = *Ameban* = *Amebarsone* = *Amibiarson* = *Aminarson(e)* = *Arsambide* = *p-Arsonophenylurea* = *Carbazon* = *Carb-O-Sep* = *Fenarsone* = *Histocarb* = *Kutan* = *Leucarsone* = *p-Ureidobenzenearsonic Acid*

Amebicide, histomonadicide.* Contains 28.8% *arsenic.

Untoward effects: Has caused nausea (27%), vomiting and diarrhea (76%), exfoliative dermatitis, chills, fever, and prostration, apparently, due to a hypersensitivity (13%), in **man**. Other adverse effects reported have been epigastric burning, right upper quadrant pain and tenderness, central nervous system disturbances (encephalitis after a 10 day course with suppositories), liver dysfunction (hepatocanicular with jaundice), polyuria, local edema, headache (26%), anorexia (21%), pruritus ani (21%), dizziness (20%), eosinophilia, leukocytosis, and three **patients** died from ingestion of out-dated tablets, apparently, due to some chemical change in them. Avoid use with any other ***arsenic***-containing products. Intolerance reported in one in every 330 **patients**. Since its ***arsenic*** is eliminate slowly in the urine, interrupted therapy has been suggested.

Discontinue use at least 5 days before slaughtering food-producing **birds** or **swine**, to eliminate ***arsenic*** tissue residues.

LD_{LO}, **rat**: 1 g/kg; **cat**: 250 mg/kg; **rabbit** and **guinea pig**: 200 mg/kg.

CARBARYL

= *Arylam* = *Arylan* = *Carylderm* = *Clinicide* = *Derbac* = *Dicarbam* = *ENT 23969* = *1-Naphthyl-N-methylcarbamate* = *OMS 29* = *Ravyon* = *Seffein* = *7* = *7744* = *Sevin* = *UC 7744*

Use: An N-methylcarbamate insecticide for use on plants, trees, **animals**, and premises.

Untoward effects: Relatively low **mammalian** toxicity, yet, in **man**, its depression of acetylcholinesterase activity has led to salivation, gastrointestinal hypermotility, abdominal cramping, nausea, diarrhea, sweating, miosis, tearing, blurred vision, headache, dizziness, ataxia, bradycardia, dyspnea, cyanosis, and muscle twitching or tremors. In extreme cases, tetany, mental confusion, incontinence, weakness, collapse, paralysis, convulsive seizures, and even death, can occur. Can cause eye and skin irritation, reduces normal development of **human** sperm, and may reduce fertility. Intubation as a treatment is not without risk, as it often contains ***petroleum distillates***, and the risk that the solvent might be aspirated is real. Usual therapy of poisoning includes ***atropine*** (after ***oxygen***, to prevent ventricular fibrillation), open airways, and ***oxygen***. Use **NO *oximes***, as they may be harmful. ***Activated charcoal*** can be used to bind toxicant in the gastrointestinal tract. Watch for vomitus. ***Diazepam*** has been suggested for convulsions. Deliberate ingestion of 250 mg has caused a "moderately severe poisoning". Avoid skin and eye contact and use near pregnant **women**.

Do NOT use on food-producing **animals** or **birds** within 1 week of slaughter. Do NOT use on premises housing food **animals** or **birds** within 1 week of their slaughter. Avoid spraying or dusting eggs or egg-laying nests. Do NOT treat **kittens** less than 4 weeks of age. Do NOT retreat **birds** for 4 weeks or **cats** and **dogs** for 7–10 days. May decrease blood cholinesterase levels, especially in Brahma or Brahma-cross **cattle**, and the drug can be recovered in the milk of lactating treated **animals** (1 tablespoonful of 5 or 50% dust after each milking was without milk levels in one trial). Avoid feed contamination of milking **animals**, pregnant **dogs** (teratogenicity in **pups** has been noted), or ingestion by **hamsters** (300 mg/kg was teratogenic). Some workers suggest a 3 week withdrawal time for **swine**. Teratogenic in **dogs**. **Swine** are affected by 750 mg/kg. Severe poisoning in **horses** at 50 mg/kg, and in **cattle** at > 250 mg/kg, but 50 mg/kg produced symptoms in a 5-month-old **steer** within 30–40 min. Has caused vertebral defects in off-spring of treated **guinea pigs**. Death in **sheep** within 1–48 h after ≥ 300 mg/kg. Delayed neurotoxicity noted in treated **hens**. In general, most **mammalian** LD_{50} toxicity is around 500 mg/kg, most **birds**: 2,000 mg/kg, and LC_{50} for tested **fish**, about 2,000 μg/l. Toxicity symptoms in **birds** resemble that of **man**.

LD_{50}, **rat**: 250–500 mg/kg; **mice**: 200 mg/kg; **chicks**: 500 mg/kg.

CARBAZOCHROME SALICYLATE

= *Adenogen* = *Adrenosem* = *Adrestat-F* = *Hemostop* = *Statimo*

Systemic hemostat.

Untoward effects: Nine psychotic and four neurastheniform reactions occurred in one trial on 15 **subjects**.

IP LD_{LO}, **mouse**: 750 mg/kg.

CARBENOXOLONE

= *Biogastrone* = *Bioplex* = *Bioral* = *Duogastrone* = *Gastrausil* = *Liquiriton* = *Neogel* = *Pyrogastrone* = *Sanodin* = *Ulcus-Tablinen*

Anti-ulcerative. **Increases mucous secretion in the stomach and duodenum, and may have a local anti-inflammatory action. Related to *enoxalone*. Prepared from *licorice* root, one of the oldest remedies known to man.**

Untoward effects: Has caused severe and prolonged hypokalemia, resulting in a myopathy, with focal muscle necrosis and myoglobinuria, increases ***sodium, chloride,*** and ***water*** retention. This may be a contraindication for its use in elderly **patients** with hypertension, ischemic heart disease, or renal parenchymal disease. It is hazardous with ***digitalis*** preparations, because of the increased ***digitalis*** toxicity, and with non-depolarizing ***muscle relaxants*** associated with hypokalemia. Nephropathy and impaired ***glucose*** tolerance also occur. In the **elderly**, it can precipitate cardiac failure, hypertension, and signs mimicking neurologic disease. One of the main adverse effects is edema (ankle, etc.), due to its anti-diuretic effect. Liver function tests were abnormal in a significant number of **patients**. Acute renal failure in a 38-year-old **female** treated with 200 mg/day/5 years. Orthopnea, dyspnea, headaches, increased blood pressure, heartburn, mild abdominal pain, flushing, metabolic alkalosis, muscle weakness, weight gain, hematemesis, ascites, systemic lupus erythematosis, pulmonary eosinophilia, vision impairment and papilledema, left ventricular failure, and cardiac failure (especially in the **elderly**) have also been reported. Use with ***spironolactone*** abolishes its ulcer-healing properties.

TD_{LO}, **human**: 120 mg/kg/56 days.

LD_{50}, **male rats**: 3.2 g/kg; IV LD_{50}, **mice**: 198 mg/kg; IP, **mice**: 120 mg/kg.

CARBETAPENTANE CITRATE

= *Antees* = *Aslos* = *Calnathal* = *Carbetane* = *Cossym* = *Fustpentane* = *Germapect* = *Pencal* = *Sedotussin* = *Toclase* = *Tosnone* = *Tuclase* = *UCB 2543*

Antitussive. **Mild, *atropine*-like properties.**

Untoward effects: Slight transient decrease in blood pressure, nausea, restlessness, and reduced cardiac function. Sensations of heat and facial hyperemia reported with doses of 200 mg or more.

LD_{50}, **rat**: 1.48 g/kg; **mouse**: 625 mg/kg.

CARBIDOPA

= *HMD* = *Lodosin* = *Lodosyn* = *MK-486*

Use: In treating parkinsonism, usually, with ***levodopa***, where it reduces enzyme degradation of ***levodopa***. It is a dopa decarboxylase inhibitor, decreasing ***levodopa*** requirement by at least 75%.

Untoward effects: May increase dyskinesias of **patients** receiving ***levodopa***. Nausea, tremors, and psychosis develop over a period of time. Its addition to ***levodopa*** therapy caused a reversible autoimmune hemolytic anemia in a 65-year-old **male** and Henoch-Schönlein syndrome in 68-year-old **male**. Decreases urinary ***dopamine*** and urinary ***sodium*** excretion.

LD_{50}, **rat**: 4.8 g/kg; **mouse**: 1.75 g/kg.

CARBIMAZOLE

= *Athyromazole* = *CG-1* = *Neomercazole* = *Neothyreostat*

Antithyroid agent, goitrogen.

Untoward effects: A 30-year-old **female** (10 mg, tid) developed a blotchy pruritic rash, conjunctivitis, and throat discomfort after 19 days, and, on day 22, developed arthalgia of the first carpometacarpal joints, shoulders, and knees. Muscle pain reported in three **patients** (18–39 years) with hyperthyroidism. Agranulocytosis, aplastic anemia, leukopenia, eosinophilia, leukocytosis, pancytopenia, thrombocytopenia, keratitis, neutropenia, bone marrow aplasia, and purpuras can occur. It has induced cholestatic hepatitis and jaundice. Acute confusional psychosis and paranoid delusions have been precipitated by the hypothyroidism produced in some cases. Loss of taste and, occasionally, smell, occur while taking it. Taking it during pregnancy may cause goitre and mental retardation in the **fetus** and **neonate**. Maculopapular and urticarial rashes with pruritus are common, but alopecia 2–6 weeks after starting therapy is not too common. Hearing loss and tinnitus reported in a 28-year-old **female**. Advisable not to breast feed while taking it. A single dose decreases peak ***digoxin*** serum levels by ~20%.

CARBINOXAMINE MALEATE

= *Allergefon* = *Ciberon* = *Clistin* = *Hislosine* = *Lergefin* = *Polistin* = *T-Caps*

Antihistamine.

Untoward effects: Photosensitivity reactions (exanthemas) on the skin. Most reactions reported are due to a combination with ***pseudoephedrine***, yet, a 5-year-old **female**, apparently, hallucinated due to it. Infrequent gastrointestinal symptoms and sedation have been reported. Significant anticholinergic effects. Adverse effects are the same as those for ***antihistamines***, in general.

LD_{50}, **mouse**: 162 mg/kg; **guinea pig**: 411 mg/kg.

CARBIPHENE

= *Bandol* = *Etymide* = *Jubalon* = *SQ 10,269*

Analgesic.

Untoward effects: Nausea (31.6%), drowsiness (36.8%), headache (21%), dizziness (15.7%), and vomiting (5.2%) in 19 **patients**. Constipation and vertigo have also been reported in **others**. ***Hexobarbital*** sleeping times increase in **rats**, **dogs**, and **rhesus monkeys**.

CARBOFURAN

= *BAY 70143* = *Brifur* = *Curaterr* = *D1221* = *ENT 27,164* = *FMC 10,242* = *Furadan* = *NIA 10242* = *OMS 864*

Insecticide, nematocide. A carbamate.

Untoward effects: Considered highly toxic for **man** with symptoms and signs of poisoning typical of all ***carbamates***. Headache, weakness, dizziness, ataxia, miosis, twitching, tremors, nausea, bradycardia, sweating, salivation, and decreased respiratory rate are key symptoms. Other symptoms include blurred or dark vision, mental confusion, convulsions, incontinence, coma, abdominal cramps, vomiting, diarrhea, wheezing, rhinorrhea, lacrimation, coughing, and unconsciousness.

LD_{LO}, **human**: 11 mg/kg.

Has caused field cases of death in **cattle**. Initially, affected **cows** quivered, then weaved, collapsed, were dyspneic, convulsed, and died. Some had paralysis of the tongue. **English sparrows** that roosted above their contaminated hay also died. Toxic dose in **cattle** appears to be about 4 mg/kg. Experimental oral and topical use was toxic to **sheep** and **cattle**. **Quail** had reduced weight gain, egg production, fertility, and hatchability when fed it at 200 ppm. Has caused large-scale deaths in **waterfowl**, from misuse in rice fields. Toxic to most **birds** and **mammals** at about 10 mg/kg. In October 1999, ~27,000 **birds** died after eating treated ***wheat*** spread in a field.

LD_{50}, **rat**: 5 mg/kg; **mouse**: 2 mg/kg; **dog**: 19 mg/kg; **mallard ducks**: 0.4–0.6 mg/kg; **pheasants**: 4 mg/kg; **chickens**: 6 mg/kg; LC_{50}, **quail**: 746+ ppm; **fish**: 2 mg/l.

CARBOMYCIN

= *Antibiotic M-4209* = *Elcarmycin* = *Magnamycin*

Antibiotic.

Untoward effects: Cross-resistance shown with ***oleandomycin*** and ***spiramycin***. Oral or IM to **dogs** at 200 mg/kg for 20 weeks without ill-effects.

LD_{50}, **rats**: > 5 g/kg; **mice**: < 3.5 g/kg.

CARBONS

Untoward effects: Well-known as the causative agent in soot, the cause of "Chimney Sweep's Cancer". Ground coke material, with oily material from ***tar*** or ***pitch*** is molded into a desired shape and baked. **Workers** in the molding department, due to lack of immaculate hygiene, develop a scrotal irritation and epitheliomas. **They** are also at risk for ***polycyclic aromatic hydrocarbon*** (q.v.) toxicity. Inhalation of carbon dust over a period of many years led to cases of anthracofibrosis in **workers**. In 1982, the first cases of black lung diagnosed outside a ***coal***-mining region were reported in two petrochemical **workers** at a Texas plant making ***petroleum coke***. A **chemist** producing ***carbon black*** in **his** dusty laboratory with an experimental furnace developed a carcinoma of Stenson's duct and black material was found in the non-cancerous parotid gland. Follicular coniosis noted on ***carbon black*** **worker's** skin. After long-term use of mascara and eyeliner containing ***carbon black***, particles penetrate the conjunctival stroma, eventually causing a follicular–papillary reaction. Suspended dust particles in a dry cell battery plant ranged from 25–34 mg/m^3, and **workers** had considerable adverse pulmonary effects (cough with phlegm, chest pain, tiredness, rhinitis, headache) and skin irritation. Three **men** were suffocating and **one** of them died, after entering a ***carbon black*** storage tank, probably, from ***carbon monoxide***. A mutagenic contaminant was, apparently, responsible for mutagenicity associated with xerographic copies and toners. Oral use will color stools black. ***Charcoal***-broiled meat may be carcinogenic. A number of ***polycyclic aromatic hydrocarbons*** have been produced in the process, and particularly, ***benzo(α)pyrene***, are on the meat and in the smoke. **"Grill-cooks"** run a risk of inhaling carcinogens, as well as smoke. It can significantly alter the utilization of many drugs (see ***Activated Carbon***).

Inhalation of carbon dust reduces the cell-mediated immune response in **mice**.

Activated Carbon:

Untoward effects: Its use in hemoperfusion equipment in a severe case of ***phenobarbital*** poisoning was associated with lowered arterial blood pressure with a transient feeling of unrest, inefficient respiration, facial flushing and a burning sensation in the throat, urethra, and anus. This was attributed to ***sulfur*** compounds liberated from the ***charcoal***. Prescribing of activated carbon for treating flatulence or other gastrointestinal problems should be done carefully for **patients** on ***antidepressants, digitalis,***

phenytoin, and ***theophylline***, where it could easily prevent absorption of the drug. Aspiration of activated carbon with gastric contents has been reported. ***Ice cream*** causes inhibition of activated carbon's adsorptive effects.

Briquettes:

Untoward effects: In the U.S., over 30 deaths reported from burning in confined spaces (indoors or in tents, trailers, campers, automobiles, and boats), due to the large amount of ***carbon monoxide*** produced and the depletion of ***oxygen***. Burning 17 briquettes in an 8 × 10 × 8 ft kitchen releases 1,920 ppm ***carbon monoxide***, enough to cause unconsciousness, and even death in 20 min.

Carbon Black:

Untoward effects: No longer approved in the U.S. as a food coloring in such items as licorice and black jelly beans for **people**, because its use may be linked to tumors in experimental **animals**. Long-term inhalation exposure of **monkeys** failed to produce any malignancies, but the topical application of extracts caused fibrosarcomas on **mice**.

Channel Black = Furnace Black = Gas Black

Untoward effects: Has contained ***benzpyrenes***. Can be used in ink pigments, if there is NO direct contact with foods. Has been found in lung tissue, probably, from soot, and not necessarily from industrial exposure. In several plants with dust levels >10 mg/m^3, 9/52 long-term **workers** had pneumoconiosis.

CARBON DIOXIDE

= *Carbonic Acid Gas* = *Carbonic Anhydride* = *Carboxide*

Use: In food freezing, carbonation of many beverages, fire extinguishers, manufacturing ***carbonates***, baking, dry ice, in aerosols, anesthetizing equipment in slaughter-houses and for euthanasia, in effervescent vaginal and rectal suppositories, and in diluting ***ethylene oxide*** gas to prevent explosions. Room air contains about 0.04%.

Untoward effects: Levels of > 5,000 ppm in the atmosphere have caused drowsiness, asphyxiation, coma, paralysis, increased respiration, and cyanosis in **man**. 100,000 ppm has caused dizziness and unconsciousness; death usually occurs within a few hours after exposure to 250,000 ppm or more. Caution must be used in entering buildings where it may be present, remembering that it is colorless, odorless, and heavier than air. Headache, dizziness, shortness of breath, muscular weakness, drowsiness and tinnitus are warnings to get into fresh air immediately. A **farmer** opened his barn door on a still morning and saw some of his **chickens** run in ahead of him and drop dead. Carbon dioxide spilling down the silo chute collected on the barn floor. Concentrations in enclosed buildings, such as confinement hog houses, can rise rapidly, if manure is stirred or ventilation equipment fails. Illness and death have occurred from entering silos during or shortly after filling, as it and ***nitrogen dioxide*** (q.v.) are released from the fermentation. Three **men** died from asphyxiation as each went into a 6 ft deep drainage pit, to recover a fallen lid. ***Oxygen*** levels were about 20% at the top, and 3% at the bottom; carbon dioxide levels were 22% at the bottom. The accepted lethal level for **man** is 10–11% where ataxia is followed by unconsciousness in about 10 min. At about 8%, dyspnea, raised blood pressure, and congestion occurs. Rectal suppositories have caused nausea, vomiting, rectal discomfort, abdominal cramping, sweating, and even collapse. Post-treatment swelling after cryotherapy of keratoses often occurs. Retinal abnormalities were noted in an asphyxiated **worker**. Cardiac arrest and death following insufflation of the Fallopian tubes in a 35-year-old **female**. Increased ocular pressure, and glaucoma, as well as headache, tinnitus, tachycardia, and increased blood pressure also reported. Coughing spells during its use in vizualizing the inferior vena cava, hepatic and renal veins. A case of toxic shock syndrome reported in a non-menstruating 24-year-old **female** after laser treatment for genital tract condylomas. Discharge of a carbon dioxide fire extinguisher in a tightly-closed place, such as in an aircraft, can be extremely hazardous, just as can transporting a large amount of dry ice in a small aircraft. Welding releases carbon dioxide, as well as other potentially dangerous fumes. Simultaneous exposure to ***carbon monoxide*** makes carbon dioxide more lethal. In 1986, a volcanic eruption in the Cameroons spilled carbon dioxide into a lake bottom, sending 75–250 ft-high waves through a valley (estimated at 50 million gallons), killing 1,746 **people** and affecting 10,000 more. Use protective gloves when handling dry ice.

TWA, **human**: 5,000 ppm/8 h (9,000 mg/m^3) and 30,000 ppm/10 min. Inhalation LC_{LO}, **human**: 100,000 ppm/1 min.

Inhalation LC_{LO}, **rats**: 657,190 ppm/15 min. Only 200–400 ppm/4 min is anesthetic for **fish**. **Hogs** can tolerate levels 50–100% greater than **man**.

CARBON DISULFIDE

= *Carbon Bisulfide* = *Dithiocarbonic Anhydride*

Use: In manufacturing rayon, rubber chemicals, cellophane and regenerated cellulosis, agricultural chemicals, and as a fumigant. Over 300 million lbs used annually in the U.S.

Untoward effects: Primarily, a central nervous system poison. Can cause behavioral and temporary or permanent mental problems. Toxicity is normally from inhalation. **Workers** exposed to the fumes are affected with garlic breath, headaches, sleeplessness, vertigo, incoherent singing, laughter or weeping, weakness, cachexia, decreased sexual desire, memory impairment, dullness of sight and hearing, and paralytic symptoms.

It is volatile at room temperature. Although major exposure can cause unconsciousness, industrial ill-effects are, essentially, from long-term exposure to low concentrations. Irritant to the eye and mucous membranes. A TLV of 20 ppm (60 mg/m^3) has been established. Can be absorbed through the skin, and such contact usually causes local irritation as well as systemic effects. Ingestion is a rare cause of its toxicity. After swallowing ½ oz, a **man** became unconscious within 30 min, had a rapid and feeble pulse, decreased respirations and dyspnea, clammy skin, and was dead in about 2 h. Carbon disulfide, a major ***disulfiram*** metabolite, is probably responsible for the latter's behavioral and neurological adverse effects. **Workers** exposed to carbon disulfide have experienced ***disulfiram*** reactions after ***alcohol*** consumption. Neurotoxicity has included demyelination of cerebellum, pyrmidal tracts, and anterior spinal columns in **man**; increased ***dopamine*** and decreased ***norepinephrine*** in **rat** brain; inhibition of ***dopamine-β-hydroxylase*** in the adrenal gland of **steers**; and swelling and degeneration of peripheral nerves in the **dog**. Inhalation of 100–200 mg/m^3 for weeks or months (8 h/day) causes weakness, hyperreflexia, positive Babinski, ataxia, motor incoordination, optic neuritis, nystagmus, ascending peripheral neuropathy, tremor, and extrapyramidal signs; recovery was rapid. Several months exposure of **rats**, **cats**, **dogs**, and **man** to atmospheric levels of 400–2250 mg/m^3 were associated with hypertrophy of coronary arteries, myocardial hemorrhage, vacuolar and fatty degeneration. Similar levels in **rats**, **rabbits**, and **man** for several months caused microhematuria, renal atherosclerosis, glomerulonephrosis, and chronic interstitial nephritis. Exposure of **workers** to 30–60 mg/m^3 for 1–5 years caused increased total cholesterol, anorexia, anemia, fatigue, amnesia, chronic gastritis, and frequent achlorhydria. Between 1933 and 1962, 223 viscose ***rayon*** **workers** in Wales and England died of coronary heart disease. Other countries have also reported similar **workers** developing atherosclerosis. Thrombotoxic and arrhythmogenic. Parkinsonian-like effect is infrequent, but hallucinations, auditory and visual disturbances, tremors, anosmia, paresthesia, and depression have been reported. Abortions, chronic metritis, sterility, amenorrhea, and abnormal menstrual cycles occur in exposed **women**, and decreased libido in both sexes. A high percentage (viz. 50/79) of adrenal insufficiency occurs in poisoned **patients**. In **man**, 14 mg/kg or 10 ml reported as an oral LD; by inhalation, 3,212 ppm/1 h and 2,000 ppm/5 min have been LDs. When capsules were used orally for treating **horses** for bot infestations, breakage in the mouth caused severe coughing and weakness. It is highly flammable and explosive and can ignite, if it contacts hot steam lines or a light bulb, and it can be ignited by static electricity from clothing. High-purity material has an ***ether***-like odor. TWA, **human**: 20 ppm/8 h.

LD_{LO}, **human**: 14 mg/kg; inhalation LC_{LO}, **human**: 4,000 ppm/30 min.

Dogs exposed to 400 ppm/8 h/day/51 days developed symptoms similar to those in ***rubber*** **workers** with chronic low-level inhalation. **Rabbits** exposed to 200 ppm for 10–20 min/day for up to 8 months developed thickening of their coronary vessel endothelium, sclerosis of the intima, and extramural hemorrhages with hyalinization.

CARBON MONOXIDE

= *CO* = *Oxyde De Carbone*

Untoward effects*:** A tasteless, colorless, odorless, nonirritating gas product of incomplete combustion of ***hydrocarbons, mostly from internal combustion engines (especially dangerous in confined spaces; 6–10% level in idling automobile exhausts), explosions, fires, improperly ventilated stoves, grills, and furnaces, and ***tobacco*** smoking (levels in **smokers** are usually 3–8%, and **chain smokers** up to 15%). It has no direct effect on the lungs, but, since its affinity for hemoglobin is about 210–250 times that of ***oxygen***, it combines with the Fe^{++} in hemoglobin to form ***carboxyhemoglobin*** (***CBH***). When these levels reach 3–6%, neurological defects, such as restlessness, irritability, delays in reaction time, and lack of attention to tasks, such as driving occur. As levels approach 10%, headache, confusion, incoordination, and tunnel vision are reported. At 20%, palpitations, bradycardia, metabolic acidosis, and generalized weakness are common. Nausea and vomiting often occur at a 30% level. Death often occurs as low as at a 35% level. Levels above this cause electroencephalogram abnormalities, ischemia, and loss of consciousness. Coma, convulsions, and death are frequent at 50–70%. ***CBH*** levels are not reliable in predicting outcome. May cause angina or myocardial infarction in **patients** with coronary artery disease. A dead **girl** found in an unlocked room of a partially burned house had a 55% ***CBH*** level; **her** ***diazepam*** serum level may explain **her** inability to get to safety. CO levels of 3,000 ppm have been reported at some fires. **Firefighters**, especially those involved in final clean-up work after interior fires, are especially vulnerable, and should wear protective breathing equipment with proper mask seal. In one fire, a level of 4,800 ppm CO was discovered. If a **mother** and **fetus** survive CO intoxication, the **fetus** may later show growth and neurological defects, including mental retardation. It can be teratogenic. A case of fatal **fetal** poisoning, after accidental non-lethal **maternal** CO intoxication has been reported. Disorders of menses, pre-eclampsia and eclampsia incidence are more common in **smokers** than in **non-smokers**. It is an insidious killer. With car windows closed, a **mother** discovered her two young **sons** dead in the back seat, from CO that had leaked from the car's exhaust. Rain forced a **woman** into a tool shed, and **she** continued cooking with **her** hibachi grill. When found later, **she** was dead. Four members of a family were found dead after using a gas space heater to heat **their** tightly sealed home. Gas ranges, ovens, and ***charcoal*** grills give off large amounts of CO (see ***Carbon Briquettes*** above). Symptoms of CO poisoning often mimic influenza or gastroenteritis (including food poisoning), and misdiagnoses have not been uncommon. Vague, flu-like illnesses demand an adequate history, to eliminate the possibility of CO toxicity, which may also include edema of pharyngeal mucosa. The expected cherry-red color of the skin is often absent. Unintentional CO poisoning has occurred from using a ***gasoline***-powered pressure washer in cleaning a farrowing barn; use of alternative heating during a power failure (after 9 h, 38 **patients** sought emergency room help in the next 12 h – 41% of the **patients** were Asian and 20% Hispanic, despite over 2,000 fire department distributed door-to-door warnings in ENGLISH); 63/148 high school **students** reported acute respiratory distress for 1–32 h after attending an indoor ice hockey tournament that relied upon a ***gasoline***-powered ice resurfacing machine (a simulated test showed 150 ppm CO or five times the recommended limits). High levels (52–69 ppm) in a dental office drifted into adjoining businesses. Two **workers** fainted 4 h after arriving for work, after the floor was burnished by a ***propane***-powered unit. A ***propane***-powered fork lift in a 20,000 ft^2 garment manufacturing plant and a suspended heating unit were responsible for illness in all 12 **workers**; levels up to 250 ppm were recorded in an indoor tractor-pull event; exhaust fumes killed three **children** riding in the back of a camper-truck. Fumes from a defective heater in a bowling alley sent 15 **people** to an emergency room, and underground garages or tunnels can easily build up toxic levels. Some automobile **drivers** thought by the **police** to be drunk may be suffering from CO poisoning from exhaust fumes. It has been estimated that 10,000 **people** annually in the U.S. seek medical attention for CO intoxication; at least 1,500 die; and 2,000 commit suicide with it. Akinetic mutism from widespread brain damage in a 42-year-old **female** who attempted suicide by it. Pink to cherry body color, especially the lunula, often occurs. On post-mortem, normal tissues lose their red color within a few hours, but cherry-red colored tissue from an acute CO poisoning case will retain its color for a few days in ***formaldehyde*** (a useful qualitative test). Leukocytosis, albuminuria, leukonychia, dilated pupils, retinal hemorrhages, blindness, and hearing loss (prolonged exposure, combined with loud noises), and even a case of thrombotic thrombocytopenic purpura have been reported. Retrobulbar neuritis and neuroretinal edema have been delayed manifestations of CO poisoning. Not only does it mimic flu, but it also confuses other diagnoses, as it can cause cerebellar ataxia, renal failure, and acute rhabdomyonecrosis. In severely poisoned **patients**, an early increase in glutamate transaminase, as well as malate and sorbitol dehydrogenase occur with a relatively rapid drop towards normal. This, undoubtedly, represents hepatic parechymal damage. The possibility of ***methylene chloride*** (in ***paint removers***, q.v.) exposure must be determined, since it is metabolized to CO, and since the former is stored in the tissues, the continued slow release of CO can produce a syndrome far worse than originally expected. Necrosis of sweat glands, intra- and subepidermal vesicles, bulla, intracellular edema, muscle necrosis, hemolytic anemia, polycythemia, and alopecia are also reported sequelae. A TWA of 35–50 ppm/8 h for **man**. Artificial respiration is often indicated in serious cases. Pure ***oxygen*** inhalation can dramatically decreases its half-life from 240–320 min in room air to 80 min, and hyperbaric ***oxygen*** to 25 min. Toxic at concentrations of 25–30% and fatal at 50–60%.

Inhal LC_{LO}, **human**: 4,000 ppm/30 min and TC_{LO}, 650 ppm/45 min.

When **sows** are in a 50–100 ppm environment, their **pigs** often have low birth weights and are less vigorous. Late-term abortions, increases in stillborn **pigs**, and decreased growth rate of their **pigs** occur at 150–350 ppm exposure, but without obvious ill-effects on the **sows**. After exposure of pregnant **rats** to CO, their **pups** were born with cardiac hypertrophy and reduced activity. Bright red blood and pink to red tissues noted in sudden death of **cattle** from CO. In **chickens**, exposure to 100 ppm reduced hatching and higher levels caused embryonic mortality and decreased growth rate. It has been used for euthanasia of many **animals** at a 6–8% level.

Inhalation LC_{LO}, **dog**: 4,000 ppm/46 min, **cat**: 9,000 ppm/35 h, **rabbit**: 4,000 ppm. Inhalation LC_{50}, **rats**: 14,200 ppm.

CARBON PAPER

Untoward effects: Occupational dermatitis from exposure to many such papers has been rare, but I found two cases in the literature.

CARBON TETRABROMIDE

Untoward effects: TLV, **human**, 100 ppb.

Hepato- and nephrotoxic in **rats**.

LD_{LO}, **rat**: 1 g/kg.

CARBON TETRACHLORIDE

= *Benzinoform* = *CCl_4* = *Necatorina* = *Perchloromethane* = *Tetrachloromethane*

***Solvent.* In industrial cleansers, home spot removers and lacquers, and in the manufacture of *chlorofluorocarbons.* Use only in well-ventilated areas. Atmospheric half-life of 30–100 years.**

Untoward effects*:** Central nervous system depression. Cardiac dysrhythmias and coagulation disorders occur. Highly nephrotoxic (acute tubular necrosis) and hepatotoxic (proximal centrolobular zone). Can cause permanent damage to these organs, often after a 2 week delay. Hepatomas have been reported from its use. ***Alcohols and ***acetone*** potentiate the toxicity. A 21-year-old **female** placing ***isopropyl alcohol*** on a packaging line (2–4 pm) 60 ft from an area using it became ill at approximately 7:30 pm. **She** had headache, backache, nausea, vomiting, and was drowsy. Nausea and vomiting persisted for 3 days. **Her** abdomen was tender, and **her** face edematous. Intestinal cramps, diarrhea, liver enlargement, jaundice, azotemia with hypertension, pulmonary edema, convulsions, and anemias are also reported. **Human** exposure is by the inhalation of fumes or by skin absorption, and, occasionally, by contaminated drinking water and soil. It has been an inhalant substance of abuse, even from fire extinguishers. A 32-year-old **male** died following use of it to put out a fire in a closed room. A 2-year-old **male** playing on a rug cleaned with it develop hematuria, oliguria, and hematemesis after about 10 days. Deliberate inhalation has caused hallucinations. Increased or decreased temperature, gastrointestinal disturbances, enlarged liver, increased SGOT, and abnormal ***glucose*** tolerance reported in **many** from air pollution. Death usually follows uremic convulsions. Can occasionally cause dermatitis and severe skin necrosis. A study of laundry and dry-cleaning **workers** suggested it may also be a cause of increased respiratory and liver tumors and leukemias. Sale for use by the general U.S. public has been banned. Use with ***coumarin*** may cause hypoprothrombinemia. As little as 3 ml has been fatal to **adults** by inhalation or ingestion. A dialyzable poison and hemoperfusion has been used in treating ***CCl_4*** poisoning. When used to put out electrical fires, ***phosgene*** (q.v.) may be formed. TWA, 10 ppm/8 h; 200 ppm maximum/5 min in 4 h. Proposed safety levels, **human**: 5 ppb in air.

LD_{LO}, **human**: 43 mg/kg. Inhalation TC_{LO}, **human**: 317 ppm/30 min.

Carcinogenic (livers) in **rats** and **mice**, and, for many years, a standard experimental procedure for causing liver toxicity in **rats**. Occasionally, 0.2 ml/kg has been fatal to **sheep** within 12 h. Although some **dogs** tolerate 3 ml/kg, others develop serious liver pathology at 0.25 ml/kg.

LD_{LO}, **dog**: 1 g/kg; LD_{50}, **rat**: 2.8 g/kg; **mouse**: 12.8 g/kg; **rabbit**: 6.38 g/kg.

CARBOPHENOTHION

= *Acarithion* = *Carbothion* = *Dagadip* = *ENT 23,708* = *Garrathion* = *Hexathion* = *R-1303* = *Stauffer R-1303* = *Trithion*

***Organophosphorous insecticide.* Cholinesterase inhibitor.**

Untoward effects*:** One of the most toxic of this class. Salivation, tremors and muscle fasciculations, ataxia, convulsions, and miosis, followed by mydriasis, are common signs of toxicity. **Spraymen** have received an average of 1.12% of a **human** toxic dose per hour of exposure. Minimal to very severe cholinergic crisis in seven family **members** (six hospitalized, four comatose), after eating tortillas made from ***flour contaminated with it. **All** recovered. The ***flour*** contained 3,220 ppm (0.3%), possibly from a spill in the storage area. Insecticides should **NOT** be stored near foods. In general, 2 weeks has been established as a safety net for consuming fruit sprayed with it.

TD_{LO}, **human**: 5 mg/kg.

Dairy calves sprayed with it at 0.05% in water were poisoned by it, but older **cattle** could tolerate 0.05–0.1%. **Sheep** were poisoned by 25 mg/kg, but not at 10 mg/kg. **Cattle** sprayed accidentally with a 1% solution had profuse diarrhea, salivation, and a stiff gait; nearly 50% died. Topically, a 1% powder was fatal to a **cat**. Affected **birds**, within 1/2 h, show rubber-legged goose-stepping, ataxia, wing drop, hyperexcitability, falling, nutation, dyspnea, and convulsions. Deaths may occur within 1–3 h. Signs may persist in recovering **pheasants** for 34 days, and 14 days for **mallards**. U.S. regulations do not now permit its use on food-producing **animals**. Simultaneous ***malathion*** exposure has tripled its toxicity.

LD_{50}, **ducks** and **grouse**: approximately 120 mg/kg; **pheasants**: 270 mg/kg; **rats**: 10–30 mg/kg; **mice**: 218 mg/kg; **rabbit**: 1250 mg/kg; **chicken**: 57 mg/kg; **wild birds**: 6 mg/ kg.

CARBOPLATIN

= *CBDCA* = *JM-8* = *NSC-241240* = *Paraplatin*

***Antineoplastic.* A *cisplatin* analog with reduced toxicity.**

***Untoward effects*:** Nausea and vomiting represent the acute toxicity. Bone marrow depression, hearing loss, hemolytic anemia, transient renal toxicity, thombocytopenia, transient cortical blindness, and, less commonly, a peripheral neuropathy and ototoxicity, are more delayed toxic symptoms. A 38-year-old **male** developed acute promyelocytic leukemia after four courses of treatment for seminoma. Hypersensitivity reactions (including erythroderma, chest tightness, wheezing, dyspnea, tachycardia, facial swelling, and increased or decreased blood pressure) in ~12% of 205 **patients**, ages 35–72.

IV or SC in **rats** causes thrombocytopenia. Hypersensitivity reactions with erythema, urticaria, pruritus, and bronchospasm are rare in **cats** and **dogs**. Rarely, vomiting or diarrhea has occurred in them. Strong myelo depression reported in **cats** within 2–3 weeks and in **dogs** within 2 weeks.

CARBOPROST

= *Dinoprost Methyl* = *Methyl Dinoprost* = *15 ME-PGF$_{2\alpha}$* = *(15S)-15-Methyl PGF$_{2\alpha}$* = *Prostin/15M* = *U-32921*

Oxytocic, abortifacient. **Induces menstruation.**

Untoward effects: Vomiting (95%) and diarrhea (100%) are the most commonly reported adverse effects. These are reduced to about 26% in **those** receiving ***antiemetics*** and ***antidiarrheal*** pretreatment. Pyrexia, headache, flushing, retrosternal tightness and breast tenderness occasionally occur. Uterine rupture has also been reported.

CARBOPROST METHYL

= *U-36,384*

Untoward effects: Over 50% had diarrhea and 39% vomited. Pyrexia, shaking and chills, and vaginal bleeding occur.

CARBOPROST TROMETHAMINE

= *Carboprost Trometamol* = *Hemabate* = *Prostin/15M* = *U 32,921E*

Abortifacient.

Untoward effects: 12/236 **patients** required surgical intervention to control postpartum hemorrhage. Pyrexia, nausea, vomiting, diarrhea, flushing, chills, shivering, headache, increased diastolic pressure, coughing, hiccough, and endocarditis occur, and usually disappear, when therapy ceases. *Contraindicated* in **patients** with cardiac, pulmonary, renal, or hepatic disease.

CARBOXIN

= *D735* = *DCMO* = *Vitavax*

Systemic plant fungicide. **Used as a seed dressing before crops are planted.**

Untoward effects: Toxic to **humans**. Avoid contact with skin, eyes, or clothing. **Users** should wash hands and face thoroughly with soap and water after use.

LD_{50} in **poultry** is about 24 g/kg. Feeding 240 mg/kg/day/6 weeks caused some deaths. **Birds** should not be slaughtered earlier than 2 weeks after consuming any treated grains, to avoid residues for **man**. In Canada, **cattle** died after eating ***canola*** seed contaminated with both ***lindane*** and ***carboxin***. Toxic to **fish**.

LD_{50}, **rats**: 430 mg/kg; **mice**: 3.2 g/kg.

CARBOXYPOLYMETHYLENE

= *Carbomer* = *Carbopol* = *Carboxyvinyl Polymer*

Use: In cosmetics, pharmaceuticals, foods, printing inks, waxes, and paints as a thickening, suspending, or emulsifying agent. ***Carbopol 934*** is commonly used.

Untoward effects: Neither an irritant nor a sensitizer in long-term trials on nearly 200 **people**.

Carbopol 934: LD_{50}, **rats**: 4.1 g/kg; **mice**: 4.6 g/kg; **guinea pig**: 2.5 g/kg. There was no toxicity from dosages up to 1 g/kg in the **dog** and 3.36 g/kg (5% of the diet) in **rats** for up to 13 months in **dogs** and 14 months in **rats**.

CARBROMAL

= *Adalin* = *Addisomnol* = *Bromadal* = *Diacid* = *Nyctal* =*Planadalin* = *Tildin* = *Uradal*

Sedative, hypnotic. **A bromide.**

Untoward effects: See ***Bromine***. Chronic intoxication has been reported in **patients**; **their** symptoms include stupor, giddiness, ataxia, clumsiness, irritability, dysarthria, memory disturbances, tremor and fainting. Apathy and depression often followed, but, occasionally, euphoria occurred. Impotence and significant increases in abnormal sperm, as well as suicide attempts reported. Additional neurological symptoms include brisk reflexes, slurring of speech, circumoral tremors, facial paresis, nystagmus, inequality of pupil size, pale optic discs, and scotomas. Addiction and cataract formation have been reported. A few cases of non-thrombocytopenic purpuric dermatitis (usually on the ankles, progressing to the upper legs, buttocks, and lower abdomen), papular exanthema and itching have been reported. Narcosis and death have occurred with excessive doses. Thrombocytopenia and hemolytic anemia have been associated with its use. It can be removed by hemoperfusion. A withdrawal psychosis, resembling delirium tremens, can occur. Cross-sensitivity with ***meprobamate*** has been reported. Therapeutic concentrations are 2–10 mg/l, toxic at 20 mg/l, and fatal at 40 mg/l.

LD_{50}, **human**: 10–15 g and 50 mg/kg.

LD_{50}, **rat**: 316 mg/kg; **dog**: 450 mg/kg; LD_{LO}, **cat**: 350 mg/kg; **rabbit**: 600 mg/kg.

CARBUTAMIDE

= *Alentin* = *Bucarban* = *Bucrol* = *Bukarban* = *BZ 55* = *Cidoral* = *Emedan* = *Glucidoral* = *Glucofren* = *Invenol* = *Nadisan* = *Norboral* = *Oranil* = *Orasulin* = *U 6987*

Antibacterial, antidiabetic. **A sulfonamide; a sulfonylurea.**

Use*:** Originally, as an antibacterial, until Franke and Fuchs, in 1955, discovered that it lowered blood sugar in **patients** treated for infectious dieases. This led to the development of ***tolbutamide (q.v.), although not antibacterial, it is less toxic as an antidiabetic. Used in Europe and elsewhere after U.S. sales were discontinued.

Untoward effects*:** 5% of **patients** may experience side-effects similar to those of other sulfonamides. Hyperlipemia in 26% of 918 **patients** (20–40 years). Hypoglycemia (viz. in a 70-year-old **male** hypertensive on 1 g/day), and deaths have occurred, particularly in older **patients** (65–75 years) and **those** with uremia. Skin allergies (141/3936), urticaria, bullous dermatitis, gastrointestinal upsets (63/3936), drug fever (5/3936), and blood dyscrasias (5/3936), oliguria, conjunctivitis, cheilitis, pharyngitis, glossitis, and facial edema also noted. Maculopapular skin rashes (53/1078), usually disappearing with continued use, but exfoliative dermatitis requires discontinuance of treatment. Fever, extensive erythematous nodular and pustular exanthema of face, neck, and arms in a 63-year-old **female** on the 8th day of treatment. Hepatocanalicular jaundice (1%) and diffuse hepatic degeneration and necrosis, photoallergic dermatitis, bone marrow aplasia and aplastic anemia, cataracts, transient leukopenia, thrombocytopenia, pancytopenia, agranulocytosis (fatal in a 77-year-old **female**, and, in one report, 8/15 died), lymphocytosis, hyperplasia, and neoformation of islets of Longerhans, toxic epidermal necrolysis, pyrexia, subendocardial hemorrhages, nausea, vomiting, hypokalemia, nephrotic syndrome with minimal glomerulonephritis, retinal hemorrhages and hemorrhages into the vitreous, cataracts, and antithyroid effects have been reported. Cholestatic hepatitis has been fatal. Various degrees of liver dysfunction occur in > 25% of **chronic users**. A **patient** developed an exanthema resembling scarlet fever after 3 weeks of treatment. Pain and paresthesia continued, followed by flaccid paralysis of arms and legs, with some loss of tactile sensation. A 40-year-old **female** took approximately 150 g from the 3rd to 7th month of pregnancy and was hospitalized in pre-coma. Her **child** was stillborn (**fetal** liver contained 1.8 mg/100 g). Drug resistance was reported in 7–8%, and **patients** were switched to ***insulin. May result in intolerance to ***alcohol*** and false positive or raised urine ***albumin*** test results. Profound hypoglycemia in a **patient** on ***phenylbutazone***.

Treatment of pregnant **rats** caused a high percentage of fetal deaths and malformations. 1–2% in the diet of **rats** resulted, respectively, in a 24% and 34% incidence of corneal and lenticular opacities, with some deaths, in a 50-week study.

LD_{50}, **rats**: 10.3 g/kg; **mice**: 3.46 g/kg.

CARBUTEROL

= *Bronsecur* = *Pirem* = *SKF 40,383*

β-Adrenergic bronchodilator.

***Untoward effects*:** Decreased esophageal sphincter tone.

LD_{50}, **mice**: 3 g/kg; IV, 32.8 mg/kg; IV LD_{50}, **rats**: 72.2 mg/kg.

CARDAMON

Carminative, flavoring agent in foods, pickles, meats, sauces, curry powder, and in perfumes. **The seeds are used.**

Untoward effects*:** Allergic contact dermatitis in **man** from its ***borneol (q.v.) content. Used in foods.

***Oil*:** Non-irritant to the backs of hairless **mice**, and 4% in petrolatum showed no irritation, photosensitivity, or sensitization in **human** volunteers.

LD_{50}, **rats**: >5 g/kg; acute dermal LD_{50}, **rabbits**: >5 g/kg.

Δ^3-CARENE

= *Isodiprene*

Use*:** In the manufacture of synthetic ***menthol and in cosmetic creams and lotions, soaps, detergents, and perfumes. A constituent of ***turpentine***.

Untoward effects*:** An occasional skin irritant. Its oxidation products are the main actual eczematogen in ***turpentine.

Undiluted to the skin of **guinea pigs** it caused skin reactions (wide-spread erythematous papules with infiltration and desquamation) within 24–48 h.

LD_{50}, **rats**: 4.8 g/kg; acute dermal LD_{50}, **rabbits**: > 5 g/kg.

CAREX sp.

Untoward effects*:** Its aeroallergens have caused allergic rhinitis, bronchial asthma and/or hypersensitivity pneumonitis. Despite the fact that *C. crus-corvi's* stems and roots have been eaten by Louisiana **natives** without ill-effects, this grass-like genus has been a ***hydrocyanic acid accumulator.

Several incidents of acute poisoning by *C. vulpina*, *C. buckii*, and *C. contigua* reported in 90% of the **cattle** that had grazed them (114–158 mg ***hydrocyanic acid***/100 g).

CARFENTANIL

= *R 33,799*

Synthetic opioid. **IM use.**

Untoward effects: Even though ***diprenorphine*** and ***naloxone*** reverse its narcotic effects, they have a relatively short half-life, and renarcotization occurs. It is 20–40 times as potent as ***fentanyl*** and several thousand times more potent than ***morphine***.

Catatonia, hypertension, bradycardia, and bradypnea in treated **goats**.

CARISPRODOL

= *Apesan* = *Arusal* = *Caprodat* = *Carisoma* = *Carisoprodate* = *Domarax* = *Flexal* = *Flexartal* = *Isobamate* = *Miolisodal* = *Mioril* = *Rela* = *Relasom* = *Sanoma* = *Soma* = *Somadril* = *Somalgit*

Tranquilizer, skeletal muscle relaxant. **A *propanediol* derivative. Metabolized to *meprobamate* (q.v.).**

Untoward effects: Sleepiness (12/107, 27/50), gastralgia, nausea (3/50, 6/44), headache (2/50), diarrhea, dizziness, vertigo (10/127), constipation, pruritus of hands (1/50), numbness of fingers or tongue (1/50), severe rash in a **female** with previous ***penicillin*** allergy, apnea (1–250 mg in a 34-year-old **female**) with loss of consciousness, peripheral neuropathy with loss of sensation in both legs of a 19-year-old **female** 4 days after starting medication (1 tablet, twice daily), sweating (7/44), fixed-drug reactions, acute attacks of porphyria, and lobular hepatitis reported. In two poisoning cases (27 and 24 tablets), some of the central nervous system symptoms above were noted, along with amnesia. Nervousness and agitation in 77-year-old **female** after 350 mg up to qid/3 days. Has an abuse potential. An addicted 46-year-old **female** recovered after ingesting approximately 100–350 mg tablets. Therapeutic serum levels are 1–4 mg%, and toxic levels are 6 mg%. It stimulates the metabolization of ***meprobamate***, by stimulating liver microsomal enzyme activity. Being structurally related to ***meprobamate***, cross-allergenicity (skin manifestations) between the two have occurred. ***Alcohol, central nervous system depressants,*** and ***psychotropic drug*** effects may be addictive. It is excreted in **human** breast milk with levels equal to or up to four times the plasma levels. Central nervous system depression and gastrointestinal upsets may occur in the **infant**. Therapeutic serum levels are 10–30 mg/l, toxic at 40 mg/l, and fatal at 50 mg/l.

Its metabolization varies greatly in different species (duration of effect 6 min/**mouse** and 10 h/**cat**), but plasma levels are almost identical on recovery of their righting reflex.

LD_{50}, **rats**: 1.3 g/kg; **mice**: 2.34 g/kg.

CARISSA

Untoward effects: In the Congo, a toxic principle (***Bushegwe***) is obtained from the tree, *C. oppositifolia* (***Umushegwe***) and used as an arrow poison. SC use in a **goat** caused abdominal pain, dull staring eyes, cardiac problems, and death within 10 min.

CARMINE

Aztec Indians of Mexico cultivated the cochineal bug for its coloring value. European women used it in rouge, lipstick, and eye make-up. By the end of the 16th century, 500,000 lbs of the bugs were shipped annually from Mexico to Spain. The dried insects contain approximately 10% *carminic acid* (*C.I. Natural Red 4*), which is precipitated onto an *aluminum hydrate*; the resulting lake is called carmine. It is used in various countries as needed in foods, drugs, and cosmetics.

Untoward effects: Hospital outbreaks of *Salmonella cubana*, particularly in **infants**, were reported from its use, primarily as a gastrointestinal transit time marker. Severe diarrhea, meningitis, and septicemia and even an **infant** death, were reported. High heat sterilization of the raw material should be used. An extrinsic allergic alveolitis in a **person** who experienced recurrent episodes of spasmodic coughing, then a flu-like febrile syndrome, whenever food additives were prepared with it has been described. This, despite the fact that the **patient's** office was only located near the room where the food additives were prepared. A 34-year-old **female** had a severe anaphylactic reaction 15 min after drinking ***Campari-Orange***®, which contained the dye, carmine.

CARMUSTINE

= *BCNU* = *BiCNU* = *Becenun* = *Carmubris* = *Nitrumon* = *NSC-409962*

Antineoplastic. **A nitrosourea.**

Use: Primarily, in treating Hodgkin's lymphoma, multiple myeloma, and brain tumors.

Untoward effects: Most common acute effects are nausea, vomiting (in nearly 40%), and local phlebitis; delayed effects are, primarily, bone marrow depression, leukopenia, and even a prolonged thrombocytopenia, pneumonitis, pulmonary fibrosis (may take years to show and not always reversible), renal damage, reversible liver damage, leukemia, and myocardial ischemia. Lung toxicity reported in different series at 1.5–20% in man and also in **animals**. Optic neuritis reported. Gynecomastia in **men** associated with its use. Concurrent use with ***streptozocin*** greatly enhances bone marrow toxicity and thrombocytopenia. Proper work practices and containment equipment must be used by **health professionals**. ***Cimetidine*** is additive to its bone marrow decreasing effects. It is carcinogenic, and, in **man**, it has been suspect in causing mammary tumors, kidney and lung carcinomas, lymphosarcomas, brain, lung, intra-abdominal, and skin tumors.

Toxicity was greater in tumor-bearing **mice**. Hepatotoxicity in treated **rats**, and vasculitis and blindness after intracarotid injection in **dogs**, but only minor ill-effects in **rhesus monkeys**. LD_{50}, **dogs**: 5 mg/kg; **rats**: 3 mg/kg.

CARNAUBA WAX

= *Brazil Wax*

Use: In furniture, automobile, and floor waxes, ointments and suppositories, depilatories and deodorant sticks, in the final stage of tablet coating, and glazing confectionaries. Millions of lbs are imported into the U.S. annually.

Untoward effects: A skin sensitizer, causing dermatitis of **workers'** hands.

CARNE SECA

Untoward effects: Salmonellosis in **man** reported from eating it (sliced, partially thawed ***beef***, salted, seasoned, and dried for 2–3 days in a passive solar drying room [New Mexico] after marinating with a chile marinade for 24 h). *S. cerro* was incriminated and a previous outbreak involving 41 **people** was due to *S. thompson* in carne seca.

CAROB

= *Algarroba* = *Arobon* = *Johannisbrotmehl* = *Locust Bean Gum* = *St. John's Bread*

Use: In pharmaceuticals, tobaccos, wines, foods, and candies, as a flavoring agent, chocolate substitute, and stabilizer. The seeds were once the **goldsmith's** measure of weight being equal to 1 carat (nearly 3.2 g).

Untoward effects: An eczematous contact dermatitis reaction to ***benzoin*** occurred in a **student** who showed cross-sensitivity to carob.

β-CAROTENE

Use: ***Vitamin A*** precursor (except **cats**) and nutritional supplement.

Untoward effects: In **humans** and **mice**, long-term oral use decreases plasma concentrations of ***α-tocopherol***. Ingestion in **humans** from tanning capsules may cause orange plasma. Chronic dosing with > 30 mg/day occasionally leads to carotenodermia (yellow pigmentation of skin) and pigmentation may last for 2–6 weeks after discontinuance of supplementation. Loose stools occur in some **individuals**. Abnormal yellow nail pigmentation also reported. Its ingestion can give false positive or high serum ***albumin*** test readings. High daily doses may increase the risk of lung cancer. More deaths occurred from it than from placebo in trials with 1,862 **men** who had previous myocardial infarctions. A thirsty 5-year-old **female** drank > 1½ quarts of ***Sunny Delight***® (two times recommended daily intake for **adults**) leading to yellow skin.

Cats lack the ability to convert it into ***vitamin A***.

CARPERONE

= *AL1021*

Use: A ***butyrophenone carbamate*** for treating schizophrenia.

Untoward effects: Skin rash, central nervous system effects, dyskinesia, akathisia, weight gain, parkinsonism, tremor, postural hypotension, rigidity, tachycardia, dry mouth, headache, and nausea reported. Excessive side-effects led to its lack of popularity.

CARPHENAZINE

= *Proketazine* = *WY 2445*

Antipsychotic. A propylpiperazine.

Untoward effects: Transient mild hepatotoxicity in 4/12, mild normochromic normocytic anemia with narrowing of neutrophile/lymphocyte ratio in 4/12, drowsiness (8/457), hypotension (3/457), extrapyramidal symptoms (27/457), postural hypotension, leukopenia, photophobia, lenticular opacities (8/31 after 3–9 years of therapy), mild transient rigidity, stiffness, tremor, muscular twitching, epileptic seizures, gastrointestinal upsets (including vomiting), allergic purpura with swollen eyelids, dystonia, skin rashes, dry mouth, and tardive dyskinesia noted. These are, essentially, typical of most ***phenothiazines*** in **man**. Non-thrombocytopenic purpura and a polymorphonuclear leucocytosis with increased white blood cells (13,400), edematous eyelids, and muscle twitching in a 67-year-old **female**, after ingesting 175 mg/day/55 days. This was then followed by a catatonic state.

CARPOBROTUS CHILENSIS

= *Garra de Léon*

Untoward effects: A 16-month-old **female** ate some of the fruits of this wild vine-like creeping plant common on the sand dunes of Uruguay, and was hospitalized with renal failure. **Her** condition was characterized by hemorrhagic gastroenteritis, generalized edema, hypertension, and seizures, apparently, due to the fruit's high content of ***oxalates***.

CARPROFEN

= *C 5720* = *Imadyl* = *Rimadyl* = *Ro 20-5720/000*

Anti-inflammatory.* An *non-steroidal anti-inflammatory drug.

Untoward effects*: *Probenecid inhibits its clearance in **man**.

In **dogs**, vomiting, lethargy, constipation, diarrhea, gastrointestinal ulceration and bleeding, pancreatitis, hematemesis, anorexia, jaundice, raised liver enzymes, ataxia, paralysis, seizures, disorientation, aggressiveness, hematuria, polyuria, urinary incontinence, azotemia, glucosuria, acute tubular necrosis and renal failure, thrombocytopenia, epistaxis, anemia, ecchymosis, pruritus, alopecia, and hives. Death has been reported. Thousands of adverse reactions have been reported to the FDA.

CARRAGEENAN

= *Chondrus Extract* = *Hens Dulse* = *Irish Moss* = *Killeen* = *Pearl Moss* = *Pigwrack* = *Rocksalt Moss* = *Salt Rock Moss* = *Viscarin*

A dried, sun-bleached, red seaweed (*Chondrus crispus* or *Gigartina mamilosa*, and, occasionally, *Euchema sp.*), really a perennial algae, commercially collected by scraping it off rocks on the coasts of Ireland, Portugal, France, and the U.S.

Use: As a gelling and stabilizing agent in foods, toothpastes, pharmaceuticals, and cosmetics. 5 lbs in every 1,000 lbs of meat for McDonalds's – McLean®.

Untoward effects: Injected into the subplantar area of **rat** paws to produce edema, or IP to produce pleurisy and granulomatous tumors, and in the studies of anti-inflammatory agents. Toxic to macrophages and suppresses the immune response. Hepatotoxic and causes sarcomas by SC injection. Since high molecular weight carrageenans are not absorbed from the gastrointestinal tract of **primates**, they are presumed to be safe for addition to **human** food. Hazard as a potential antipeptic ulcer agent, and its anticoagulant property would contraindicate its use in contact with hemorrhagic lesions. It has caused urticaria in **humans** and colitis in **laboratory animals**, and it may be carinogenic by injection in the latter.

MRC outbred **rats** fed it for 110 days developed adenocarcinoma of the testes.

Calcium and ***Sodium Salts***: LD_{50}, **rats**: 5.5 g/kg; **mouse**: 9 g/kg; **hamster**: 7 g/kg.

CARROTS

= *Daucus carota*

Untoward effects: ***Estrogen*** content of 3.6–10.5 mg/100 g. **Mothers** having eaten 2–3 lbs/week excrete carotenoids in **their** milk, which caused yellowish discoloration of the nursing **infant's** skin. A 2-week-old **male Negro infant** fed only 500 ml of carrot juice containing 525 ppm ***nitrate*** and 775 ppm ***nitrite*** in 24 h, developed methemoglobinemia. ***Methemoglobin*** was 9 g/100 ml – 60% of the total hemoglobin. His **twin**, receiving only a modified milk formula, was asymptomatic. So-called "organically grown" ***carrots*** contain more ***nitrates*** (320 ppm [220–405]) than conventionally-grown ones (72 ppm [39–108]). Six **children** (five < 1 year of age) had "benign" carotenemia, especially of the palms, soles, and noso-labial folds. Five ate a partially homogenized diet, and yellow pigmentation of the skin decreased as the ***carotene*** intake was decreased. Particle size influences absorption with the homogenized or pureed forms allowing more than chopped carrots. Serious and fatal cases in **infants** from ***sodium nitrate*** in carrots. Ingestion may color urine a carrot color. A selected group of **atopics** experienced itching, tingling and/or edema of lips, mouth, and tongue, throat irritation and hoarseness, after ingesting carrots; **some** had dermatitis of the hands. May cause photosensitization, due to its ***furocoumarins*** and positive delayed eczematous patch test. Many **patients** have shown fissured contact dermatitis, including cheilitis, from contact with them, and **patients** with ***nickel*** dermatitis should avoid them because carrots may exacerbate their vesicular hand eczemas. Carrots can contain naturally occurring cholinesterase inhibitors, and even some gastrogenic activity, as well as ***carototoxin***, a potent neurotoxic agent, and ***myristicin*** (q.v.), an hallucinogen. A **faddist** died of liver cirrhosis after drinking gallons of carrot juice to wash down **his** usual ***vitamin A*** intake. **One** should be alert to the fact that carrots contain ***oxalates***. Intake has helped cause ***vitamin A*** intoxication (pseudotumor cerebri, skeletal pain, desquamating rashes, and hepatitis), and can cause tests for ***albumin*** to be abnormally high.

Some **dogs** have shown a skin allergy to it.

CARSALAM

= *Beaprine* = *CSA*

Analgesic, antipyretic.

Untoward effects: At 50 mg/kg, decreases body temperature of experimental **rabbits**. At eight times this level, they have marked diuresis and are limp for a few hours.

LD_{50}, **rats**: 750 mg/kg; **mice**: 1–1.5 g/kg.

CARTEOLOL HYDROCHLORIDE

= *Abbott 43326* = *Arteoptic* = *Caltidren* = *Carteol* = *Cartrol* = *Endak* = *Mikelan* = *OPC 1085* = *Optipress* = *Tenalet* = *Tenalin* = *Teoptic*

β-adrenergic blocker, sympatholytic.

Untoward effects: Adverse effects in 8/30 **patients** with angina, and 4/378 of another group, treated for cardiac arrhythmias, discontinued therapy because of adverse effects.

CARVACROL

= *Isopropyl-σ-cresol* = *Isothymol*

Antiseptic.

Untoward effects: In ***marjoram*** (q.v.). Suspect carcinogen.

LD_{LO}, **human**: 50 mg/kg; **cat** and **rabbit**: 100 mg/kg; LD_{50}, **rat**: 810 mg/kg.

CARVEDILOL

= *BM 14190* = *Coreg* = *Dilatrend* = *Dimitone* = *DQ 2466* = *Eucardic* = *Kredex* = *Querto*

Antihypertensive, β-adrenergic-blocker.

Untoward effects: Bradycardia, atrioventricular-block, orthostatic hypotension, hepatotoxic, edema/fluid retention, bronchospasm (death from status asthmaticus was reported in two **persons** after single doses), diarrhea, nausea, hyperglycemia, increased blood urea nitrogen and non-protein nitrogen, arthralgia, abnormal vision, and Stevens-Johnson syndrome.

Drug interactions: ***Cimetidine*** increases its serum concentration ~30%; ***rifampin*** decreases its serum concentration ~70%. ***Fluoxetine, paroxetine, propafenone,*** and ***quinidine*** may also increase its serum levels ~15%.

Propantheline decreases its rate of absorption in **rats**.

l-CARVEOL

Use: In fragrances for cosmetic creams, lotions, soaps, detergents, and perfumes. Found in ***spearmint oil*** and imparting a ***spearmint***-like odor to products. Approved for food and flavoring uses.

Untoward effects: No irritation or sensitization at 4% in petrolatum on **human** volunteers.

Produces irritation when applied full-strength to intact or abraded **rabbit** skin.

LD_{50}, **rat**: 3 g/kg; acute dermal LD_{50}, **rabbit**: > 0.5 g/kg.

CARYOCAR sp.

Untoward effects: *C. gracile*, when fed to **dogs**, causes their slow death within a week. Found in Central and South America, where the fruit of various species is mashed with mud; when dumped into a stream, the **fish** come to the surface and are caught by hand.

CASANTHRANOL

= *Cantralax* = *Casanol* = *Parenterin* = *Perilaxin* = *Peristin* = *Peristaltin*

Laxative, cathartic.

Untoward effects: Avoid its use in pregnant **women** and **nursing women**, as its anthranol glycosides may be excreted in breast milk, causing purgation in **infants**.

CASCARA SAGRADA

= *Bearberry Bark* = *Bearwood* = *Chittem Bark* = *Chittim Bark* = *Persian Bark* = *Purshiana Bark* = *Sacred Bark*

Laxative, cathartic.

Untoward effects: A colon irritant. Diarrhea in ***laxative-abusers***. May enter breast milk in significant amounts, and cause purgation in approximately 50% of **infants**. Its use can cause harmless pigmentation of the colon's mucosa (melanosis coli, after chronic use) and, in urine, a red tint (brown in alkaline urine; black, after urine stands for a while), due to its conversion to ***anthraquinones***. Chronic **users** have suffered from hypokalemia with resultant weakness, nausea, anorexia, and severe diarrhea.

CASEIN

Use: As a nutrient, in adhesives, inks, coatings, glass lamination, and plastics.

Untoward effects: One of the many antigens, but the major protein in cows' milk, causing allergic reactions in **humans**. Increases urinary ***calcium*** and ***magnesium***, while decreasing ***potassium*** excretion in **man** and **animals**. Severe anaphylaxis in an 11-month **infant** from casein in ingested milk, and after application to the perineal area of a diaper ointment containing casein.

Cows, under certain conditions, resorb their own milk and develop an allergic reaction to its casein. Aggravates ***safrole***-induced hepatic adenomas in **rats**.

CASHEW

= *Anacardium occidentale* = *Kaju* (Yoruba) = *Kanju* (Hausa)

A bushy, spreading tree, 35–40 ft high with evergreen 8-inch long, leathery leaves. Has very juicy pear-shaped false fruit, or "cashew apple". Native to Brazil and Venezuela, and found in sandy areas of East Africa, India, Southern Florida.

Untoward effects: The corrosive oil from the skin of the fruit (pericarp) contains ***cardol*** and is used in Nairobi to produce decorative scars on the arms and hands of **natives**. It is caustic and a vesicant to the skin, causing inflammation, irritation, blisters, itching, burning, and open sores. When the oil is taken orally, it causes diarrhea and dehydration. **People** and **children** who have bitten into the raw nuts (inside the fruit) have suffered severe swellings of the face and lips, followed by blistering, and later, blackening and peeling. A **female** had cut open and eaten the raw cashews, displayed by an importer in a supermarket, with severe injuries. It contains ***oxalates*** and ***oxalic acid***. Use *only* commercially "roasted" (boiled) nuts. The smoke from open-fire roasting is very irritating, and handling or opening the nuts after such roasting can also be very irritating. Even after skillful processing, cashew kernels may still be contaminated by ***CNSL*** (cashew shell nut liquid) causing oral and anal itching in **people** who have bought and eaten so-called "raw" kernels in health food stores. Commercial roasting decontaminates them. ***CNSL*** has been used in industry for its high heat and friction-resisting attributes; and, when first used, it caused extensive dermatitis in factory **workers**, until efficient means of detoxification were adopted. Improper processing of the nuts that left small shell remnants on them produced a ***poison ivy***-like dermatitis (bright erythematous and maculopapular eruptions, mostly, on the extremities [97%], trunk [66%], groin [45%], axillas [34%], and buttocks [21%], and perianal itching in four and mouth blisters in three) in 54/274. The rash developed 1–8 days after consumption of the nuts, and lasted 5–21 days; the **patients'** ages were 9–72 years. A similar ***poison ivy***-like dermatitis developed only in **boys** in an elementary class handling cashew nut toys from Puerto Rico. The face of a **female** exposed to the fumes of the roasting nuts became so swollen that no features of **her** face were clearly discernable. From external contact, a **boy's** tongue, face, neck, hands, forearms, and scrotum became extremely swollen, red, and very painful. Cross-reacts in **people** sensitive to ***poison ivy, poison oak,*** and ***poison sumac***.

Animals can also have toxic reactions, as does **man**, from eating the fruit.

CASSAINE

***Untoward effects*:** A cardiotoxic principle used in arrow and dart poisons.

Causes convulsions in **frogs** and **cats**.

IV LD_{50}, **guinea pigs**: 2.6 mg/kg.

CASSAVA

= *Akpu* (Igbo) = *Casabe* = *Casave* = *Cazabe* = *Gbaguda* (Yoruba) = *Mandioca* = *Manihot esculenta* = *Manioc* = *Rogor* (Hausa)

A major food plant for approximately 300 million people. Of American origin, and, now, at least 90 million tons are produced annually in the tropics. In the last 30 years, it has been used as an animal and poultry feed component in Europe. Edible varieties are sweet, and those that contain excessive amounts of *hydrocyanic acid* are bitter. Tapioca is heated cassava starch.

Untoward effects*:** Under tropical conditions, *Aspergillus flavus* can grow on it and create secondary problems. Contains variable quantities of the cyanogenic glucosides, ***linamarin (up to 90%) and ***lotaustralin***, which, upon hydrolysis, yield ***hydrocyanic acid*** (15–400 mg/kg of fresh material). **Children** have been poisoned and **some** have died after eating cassava beans. Chronic toxicity in **man** is characterized by blindness, goiter (55% incidence in northern areas and 5% in southern areas, which may also be a factor of the amount consumed and ***iodine*** consumed), and an ataxic neuropathy is less prevalent in Latin America than Africa, where **people** in the former have learned how to detoxify it. The sweet varieties cultivated in the northern regions of Africa have less ***cyanide*** than those in southern areas. Poisoning from eating the raw root or peelings from the tubers has the symptoms of dyspnea, loss of voice, spasms, twitching, shaking, and coma. Fatalities have occurred in **children** who chewed twigs, its seeds, or tea made from it. When improperly prepared or large quantities have been eaten raw, nausea, vomiting, abdominal distention, dyspnea, and collapse have been reported. Cassava roots must be soaked in water, for retting out the ***cyanide***. Degenerative neuropathy is not uncommon in **those** poorly nourished and in **those** consuming the bitter variety with its higher levels of cyanogenic material in Nigeria.

Water buffaloes in Indonesia were poisoned by eating the plant tops (***prussic acid*** 39 mg/kg); in Zimbabwe, extensive deaths in **goats** and a **bull**, after eating the leaves. **Cows** will tend to refuse it when the ***hydrocyanic acid*** level is very high. **Animals** have been poisoned by eating its roots or drinking water in which the roots have been soaked.

CASSIA

Contains *cinnamaldehyde* and is used for its *cinnamon*-like flavor.

Untoward effects*:** Many species contain ***anthraquinones and are laxative.

C. alata IP LD_{50}, **mice**: 905 mg/kg.

C. absus contains a toxalbumin, ***absin***, which is similar to ***abrin***.

C. acutifolia is ***Senna*** (q.v.).

C. fasiculata = ***Partridge Pea*** causes severe purgation in **livestock** that consume it.

C. obtusifolia = ***Sicklepod*** = ***Indigo*** = ***Sicklepod Senna***, is less toxic than ***C. occidentalis***, but has caused severe purgation and some of the same symptoms. After ingestion, **cows** are alert, but unable to rise. Early symptoms are diarrhea, anorexia, weakness, incoordination, a swaying gait, and muscle trembling. All parts of the plant, and possibly its seeds, can be toxic to **chicks** (decreases growth rate).

C. occidentalis = ***Rai 'dore*** and ***Rairai*** (Hausa) = ***Rere*** (Yoruba) = ***Kunde Nyika*** (Zanzibar) = ***Mnuka Uvundo*** and ***Mrambazi*** (Swahili) = ***Msalafu*** (Duruma) = ***Ogbunmo*** (Ibo) = ***Spider Pea*** = ***Stinking Weed*** = ***Stypicweed*** = ***Taperyva Hu*** (Paraguay) = ***Coffee Senna*** has caused poisoning in **sheep**, **goats**, **swine**, **horses**, **rabbits**, and **chickens**, and myodegenerative disease in **cattle**. Affected species usually have dark red or coffee-colored urine, decreased weight, are weak, occasionally, ataxic before recumbency and death within 24 h after signs of acute illness. Liver and kidney pathology may also be found. Toxicity is due, primarily, to its ***chrysarobin*** or ***chrysophanic acid*** content.

C. roemeriana, after ingestion, caused skeletal myopathy and some liver damage in **cattle** and **goats**.

C. siamea has caused rapid death in **pigs**, but, apparently, non-poisonous to **cattle** and **sheep**.

The oil causes a contact dermatitis in **man**.

CASTANEA sp.

= *Chestnuts*

***Use*:** As a food and as a timber crop.

***Untoward effects*:** Both the wood and the nuts have caused an allergic dermatitis in **man**.

CASTOR BEAN

= The castor plant, *Ricinus communis* = *Kurū'a* (Arabic) = *Lara* (Yoruba) = *mu Pfuta* (Shona) = *Palma Christi* = *Tartago* = *Ugba* (Igbo) = *um Fude* (Ndebele) = *Zurma* (Hausa)

Native to India, now, extensively grown in the West Indies and southern U.S. states.

Untoward effects*:** After expressing its oil, it leaves a cake residue or pomace whose dust has caused sensitivity reactions (rhinitis, asthma, dermatitis) in **man**. The beans represent the seeds of the plant, which contain water-soluble ***ricin (not oil-soluble and, therefore, not found in expressed ***castor oil***). After a latent period of several hours, a poisoning characterized by salivation, abdominal pain, nausea, vomiting, diarrhea, violent purging, intravascular hemolysis, and oliguria after ingestion of the seeds develops in **man**. The ingestion of one to six well-chewed seeds has been fatal in **humans**. The hemolysis may cause secondary renal failure with uremia, hemolytic anemia, and convulsions. Severe thirst, cold sweats, and collapse may precede death. The leaves are much less toxic. When seeds are not chewed, the hard seed coat prevents the absorption and poisoning associated with the toxalbumin, ***ricin*** (q.v). When the seeds are fermented for several days, they have been used safely as a condiment in soup. Ingestion of unrefined plant parts has caused burning in the mouth. The flowers' odor sometimes causes headache and nausea, as does handling or wearing necklaces of the seeds, especially, if a seed is broken. Some sensitive **people** react to unseen seed particles in the air. A 57-year-old **male**, in shock, was hospitalized after chewing a single bean and recovered $2\frac{1}{2}$ h later, after symptomatic treatment. One or two beans, popular in decorative gourds and Mexican and Southwestern jewelry, can be fatal to small **children**. **They** often have upset stomachs, nausea, mouth and throat burning, jaundice, dizziness, and convulsions. There has been a high frequency of sensitivity to it in **farmers** harvesting it. An epidemic (57 suspected cases) of allergy to the beans occurred near a factory expressing the oil; symptoms included asthma, conjunctivitis, lacrimation, edema of the eyelids, rhinitis, skin rashes, and pharyngolaryngitis. The ***ricin*** has been used in suicides and assasinations.

LD_{LO}, **child**: 500 μg/kg.

Can cause severe gastroenteritis and purging in **pets** that chew the beans. Dullness, violent purging, tetanic spasms, and strong heartbeats. **Cattle** have been poisoned by contamination of their feed and symptoms (weakness, salivation, excessive eructation, abdominal pain, incoordination, abortions, and bloody diarrhea) developed 12–48 h later. Post-mortem examination revealed intense inflammation of the gastrointestinal tract and mesenteric lymph nodes, edematous enlarged kidneys and liver, and petechiae on the heart. **Goats** have been poisoned by eating the pressed cake. **Horses**, **pigs**, **sheep**, and **poultry** have also been poisoned by the beans. **Pigs** may vomit, have non-bloody diarrhea, and abdominal pain with some ataxia and convulsions; their ears, flanks, and hams appear cyanotic. Affected **pullets** appeared dull, had diarrhea, droopy wings, ruffled feathers, gray-colored wattles and combs, and have died. Older **hens** were less seriously affected. There is a wide individual variation in toxic doses for **animals**.

LD, **cow**: 2 g/kg; **horse**: 0.1 g/kg; **sheep**: 1.25 g/kg; **pig**: 1.4 g/kg; **goats**: 5.5 g/kg; **rabbit**: 1 g/kg; **goose**: 0.4 g/ kg; **hens**: 14 g/kg.

CASTOR OIL

= *Cosmetol* = *Koli* = *Neoloid* = *Oil of Palma Christi* = *Phorbyol* = *Ricinus Oil* = *Tangantangan Oil*

In addition to domestic production, the U.S. imports approximately 40,000 tons annually for use

in coatings, lubricants, hydralic fluids, paints, varnishes, inks, and in candy manufacturing.

Untoward effects: Superpurgation can happen with excessive use. Causes pelvic congestion and should not be used during pregnancy or menstruation, or in fecal impactions, especially in the **elderly**. *Contraindicated* in ***phosphorus*** or ***chlorinated hydrocarbon*** poisoning, where it can increase absorption of the toxic substances. As a parenteral vehicle, it causes moderate irritation, since it is poorly absorbed. Long term use can lead to malabsorption of nutrients, diarrhea, and damage to duodenal and jejunal mucosa. Hypersensitivity to it (viz. as a solvent in ***paclitaxel***) can induce asthma-like symptoms.

May cause an acute lethal toxicosis with pulmonary thromboembolism and central nervous system upsets in **cattle**.

CASTOREUM

From the oil glands of the beaver (*Castor fiber*) a by-product of the fur industry.

Use: In fragrances for soaps, creams, lotions, and perfumes.

Untoward effects: 4% in petrolatum showed no irritation in 48 h closed-patch test or sensitization in 25 **human** volunteers.

LD_{50}, **rat**: > 5 g/kg; acute dermal LD_{50}, **rabbit**: > 5 g/kg.

CATECHOL

= *1,2-Dihydroxybenzene*

The most abundant *phenol* in *tobacco* smoke (0.08–0.28 mg/cigarette). An *estradiol* metabolite. An intermediate for a number of cosmetic, perfume, and aroma ingredients.

Untoward effects: Increases the invasiveness of tumor cells by spurring the production of ***oxidants***.

A strong cocarcinogen on **mouse** skin applied with ***benzo(α)pyrene***. Rapidly absorbed from cigarette smoke, redistributed, and excreted in the urine (91%, 120 min after exposure) in **mice**.

CD_{50}, in **mice:** 0.38 μM/kg.

CATERPILLARS

Untoward effects: Some stinging species (***Io***, ***Puss***, and ***Saddleback***) are common in Florida and elsewhere.

Puss Caterpillar = ***Megalopyge opercularis*** = ***Tree Asp*** is the most toxic. It is described as beige to gray or chocolate-colored, and is ~1 inch long when mature. Clusters of venomous spines are distibuted on its back, along with fine hairs. It is particularly abundant in October and November, but also found in the spring. It is a vegetarian, eating leaves of bushes and trees, and is often seen on exterior walls of houses and buildings, and on fences and gates. An immediate, burning sensation is followed by redness, swelling, vesiculation, numbness, and a severe, radiating, stabbing pain within 10 min to the axillary or inguinal region. Lymphadenitis is common and a dermatitis, often in a grid-like pattern, at the sting site can persist for many days. In severe cases, systemic reactions, including seizures and lymphadenopathy, occur. Occasionally irritability, restlessness, and excitement are reported. Rarely, paralysis has occurred. Stinging hairs are usually imbedded in the areas of the sting. Poison Control Center reported > 2,000 caterpillar bites in 1994.

CATHA EDULIS

= *Cafta* = *Chat* = *Ciat* = *Flower of Paradise* = *Gat* = *Kat* = *Khat* = *Miraa* = *Murungu* (Kiswahili) = *Qât* (Yemen) = *Quat* = *Tschat*

A little-known narcotic drug plant (contains *cathine* = *D-norisoephedrine*, *celastrin*, and *cathinone*). Has a sweet licorice-like odor and taste.

Untoward effects: Chewing its leaves or branches for about 10 min is common in Africa and the Arabian peninsula. It has recently become popular in Europe. It inhibits amine oxidases, giving a sense of euphoria and excitation, irritability, thirst, hyperthermia, and increased respiration. The effects are mostly from ***cathinone***, an alkaloid chemically similar to ***amphetamine***. A 23-year-old **male** North African college **student** in Ohio had manic psychosis, and increased sympathetic activity after chewing about 2 dozen leaves. **He** showed hyperactivity, pacing, shouting, rapid speech, dizziness, thirst, and demanded sexual activity from **his** girlfriend. ***Phenylpropanolamine*** was isolated from **his** urine. **His** symptoms abated in about 5 h. Chronic **users** may have decreased libido, and even impotence. It decreases gastric juice secretions, is, apparently, habituating, with a loss of sense of reality, causes parotitis, intestinal hypotonia, and anorexia. In 1999, Custom and U.S. Drug Enforcement Administration agents estimated ~2,500 lbs smuggled annually into Los Angeles.

Daily feedings to **guinea pigs** during the last 3 weeks of pregnancy decreases maternal feed intake and fetal birth weights.

CATNIP

= *Catmint* = *Catnep* = *Catrup* = *Catwort* = *Field Balm* = *Nep* = *Nepeta cataria* = *Nip*

Untoward effects: The oil in the plant contains about 70–90% ***nepetalactone***. Ingesting the concentrated oil by **humans** has caused central nervous system toxicity. It has been smoked and used in teas for hallucinogenic effects, while adding a mint-like taste and odor. An **infant** may have had a convulsion from an overdose. In the 60s, it was frequently smoked alone or used as a filler in ***marijuana*** cigarettes. Inhalation caused altered consciousness for up to 3 days in one **patient**; headache and malaise in **another**. Ingestion of large amounts of tea has caused emesis in some. A 19-month-old **male**, after consuming an unknown amount of its tea or tea bag had central nervous system inhibition, and was obtunded and hypotonic with slow recovery (~4 days).

Cats often respond to it sniffing, licking, head shaking, chin and cheek rubbing; occasionally, spontaneous vocalization and estrus-type rolling. **Cats** response to it is controlled by an autosomal dominant trait. Two **cats** ingesting excessive amounts were depressed for several days to a week.

CAULIFLOWER
= *Brassica oleracea*

Untoward effects: Contains the carcinogen, ***sinigrin***, and ***thiocyanate***, a goitrogen, which stimulates the rate of nitrosation in the stomach acid environment. If consumed raw, contains a significant amount of ***vitamin C***, a protective antioxidant, but over $\frac{1}{3}$ of it is lost on freezing, blanching, steaming, or boiling. Also contains ***indoles***, protective against some carcinogens. Coliforms and other bacteria often contaminate the raw vegetable. Bloating and intestinal gas formation (> 2 quarts/day) can follow its ingestion in many **people**. Its high ***purine*** content can raise ***uric acid*** blood levels causing gout-like attacks.

CAULOPHYLLINE
= *N-Methylcytosine*

Untoward effects: One of the main toxic substances in ***laburnum*** (q.v.) and *Caulophyllum thalictroides* (***Blue Cohosh***). Handling the latter's powdered root can irritate mucous membranes. Has caused mouth irritation.

Its alkaloids cause teratogenicity and embryotoxicity in **rats**.

CAVIAR

Untoward effects: Contains approximately 680 µg ***tyramine***/g, which can cause hypertension in **patients** on ***monoamine oxidase inhibitors***.

CEDAR, WESTERN RED
= *Thuja plicata*

Untoward effects: Inhalation of the dust causes a delayed onset of occupational asthma, rhinitis, and, in chronic exposure over a period of years to its dust, an irreversible pulmonary airway obstruction with bronchospasms in **lumber workers** often triggered by minute quantities of the antigen. Contact dermatitis and photodermatitis has been reported.

Male mice on cedar chip bedding had their ***pentobarbital*** or ***hexobarbital*** sleeping time reduced by 50% on the second day and it decreased the threshold for ***pentylenetetrazol***-induced clonic seizures in **mice**.

CEDRENOL

Use: As a fragrance from various conifers and used in foods, cosmetic creams and lotions, soaps, detergents, and perfumes.

Untoward effects: An 8% level in petrolatum on the skin of 25 **volunteers** showed no sensitization reactions, nor did full-strength for 24 h on the backs of intact or abraded **rabbit** skin show any irritation.

LD_{50}, **rats** and acute dermal LD_{50}, **rabbits**: > 5g/kg.

CEDRENYL ACETATE

Use: In fragrances for cosmetic creams and lotions, soaps, detergents, and perfumes.

Untoward effects: 8% in petrolatum in a 48 h closed-patch irritation test on **humans** was negative, as was its results on sensitization in 25 **human** volunteers.

Applied full-strength for 24 h to intact or abraded **rabbit** skin revealed it to be an irritant.

LD_{50}, **rats** and acute dermal LD_{50}, **rabbits**: > 5 g/kg.

CEDROL
= *Cedarwood Oil Alcohols*

Use: From various cedarwood conifers, particularly, cypresses and cedars, for use as a fragrance in foods, soaps, detergents, cosmetic creams and lotions, and perfumes. Over 20 ton/year used in the U.S.

Untoward effects: 8% in petrolatum produced sensitization reactions in 2/25 test **subjects**, and, subsequently, no reactions on the same **subjects**. Similarly, no irritation after a 48 h closed-patch test on **humans**.

Decreased ***hexobarbital*** and ***pentobarbital*** sleeping time in **mice** when this was sprayed on their bedding.

Acute dermal LD_{50}, **rabbits**: > 5 g/kg.

CEDRYL ACETATE

Use: Approximately 50 tons used annually in the U.S. as a fragrance in cosmetic creams and lotions, soaps, detergents, and perfumes.

Untoward effects: 8% in petrolatum, after a 48 h closed-patch test, was negative for irritation on **humans** and the same mixture produced no sensitization reactions on 26 **human** volunteers.

LD_{50}, **rats**: 44.75 g/kg; acute dermal LD_{50}, **rabbits**: > 5/kg.

CEFACLOR

= *Alfatil* = *Ceclor* = *Compound 99638* = *Distaclor* = *Lilly 99,638* = *Panacef* = *Panoral*

Second generation semi-synthetic oral cephalosporin β-lactam antibiotic, in clinical use for about 20 years.

Untoward effects: Its rate of absorption, but not necessarily the total amount, is decreased by ingestion with food. Preferably, it should be taken 1 h before or 2 h after eating. Approximately 50–55% (43–85%) is eliminated unchanged in the urine. Short bowel syndrome also decreases absorption. Gastrointestinal upsets (nausea, vomiting, diarrhea) are the most common adverse effects. Others include morbilliform eruptions, urticaria, pruritus, positive Coombs test, eosinophilia, leukocytosis, vaginitis (often, *Candida*), transiently increased SGOT and SGPT enzyme levels, serum sickness-like reactions in **children** (with acute polyarthritis and urticarial rash), and anaphylaxis (five cases in one reported, and three cases of hematological disorders in eight **Japanese** [aged 27–73 years]). Hundreds of reports of arthralgias and erythematous rashes. A **child**, nearly 4 years old had cholestatic jaundice, hematuria, rash, abdominal pain, vomiting, anorexia, and emotional disturbances after 40 mg/kg, tid/8 days. A 36-year-old **male** showed acute interstitial nephritis and non-oliguric renal failure after 500 mg, tid/2 days. A 19-year-old **male**, after treatment for Henoch-Schönlein purpura syndrome, had the condition worsened by cefaclor. Avoid use in **patients** allergic to any ***cephalosporin*** or ***penicillins***, as a similar starting chemical, ***7-ACA*** (***7-aminocephalosporanic acid***), is used.

CEFADROXIL

= *Baxan* = *Bidocef* = *BL-S 578* = *Cefa-Drops* = *Cefamox* = *Cefa-Tabs* = *Ceforal* = *Cefroxil* = *Cephos* = *Duracef* = *Duricef* = *Kefroxil* = *MJF 11567-3* = *Moxacef* = *Oracéfal* = *Sedral* = *Ultracef*

A first generation semi-synthetic oral cephalosporin.

Untoward effects: Nausea is the most common adverse effect, and can be decreased by administration with food. Angioneurotic edema, rashes, urticaria, pemphigus, diarrhea, dysuria, genital pruritus and moniliasis, vaginitis, moderate transient neutropenia, and positive direct Coombs tests also noted. Indications and toxicity are, essentially, the same as for ***cephalexin*** (q.v.). Contraindicated in **patients** allergic to ***cephalosporins*** or ***penicillins***, as a similar starting chemical is used.

No teratogenicity or antifertility effects noted in usual **rat** and **mouse** studies. **Cats** receiving ten times recommended dosage/21 days had decreased food intake, diarrhea, and vomiting.

LD_{50}, **dog**: > 0.5 g/kg.

CEFAMANDOLE

The nafate is usually used = *Bergacef* = *Cedol* = *Cefam* = *Cefiran* = *Cemado* = *Cemandil* = *Fado* = *Kefadol* = *Kefandol* = *Lampomandol* = *Mandokef* = *Mandol* = *Mandolsan* = *Neocefal* = *Pavecef*

A second generation semi-synthetic cephalosporin for parenteral use. In use about 20 years.

Untoward effects: Maculopapular rash, urticaria, eosinophilia, and drug fever have been reported as hypersensitivity reactions, especially in **those** allergic to other ***cephalosporins*** or ***penicillin***. Occasionally pain on injection and thrombophlebitis noted. Less common are neutropenia, leukopenia, thrombocytopenia, positive direct Coombs tests, transient increases in SGOT, SGPT, and alkaline phosphatase levels, decreased ***creatine*** clearance (in **patients** with prior renal problems), transitory increased blood urea nitrogen, especially in older **patients**, and a mild increase in serum ***creatinine***. Some **patients** develop severe upper gastrointestinal bleeding on receiving it after surgery. It may give a false positive for ***glucose*** if the latter is tested for with Benedict's solution or ***Clinitest***® tablets, but not ***Tes-Tape***® or ***Diastix***® testing. A 30-year-old **male** developed jaundice 5 days after starting therapy. Immune hemolytic anemia has been reported, as well as a bad taste in the mouth of a 22-year-old **female**.

Disulfiram-like reactions with it and ***alcohol*** occur occasionally. Can potentiate ***warfarin's*** anticoagulant action by decreasing clotting factor synthesis and impairing platelet function. Prolongation of prothrombin time, although rare, has caused serious bleeding in some **patients**. Hypoprothrombinemia has been reported in other cases. ***Probenecid*** prolongs its blood levels.

CEFAZOLIN SODIUM

= *Acef* = *Ancef* = *Atirin* = *Biazolina* = *Bor-Cefazol* = *Cefacidal* = *Cefamedin* = *Cefamezin* = *Cefazil* = *Cefazina* = *Elzogram* = *Firmacef* = *Gramaxin* = *Kefzol* = *Lampocef* = *Lividina* = *SKF 41558* = *Sodium CEZ* = *Totacef* = *Zolicef*

A first generation semi-synthetic cephalosporin for parenteral use.

Untoward effects: Eosinophilia is the most common blood dyscrasia, followed by a positive direct or indirect Coomb's test and anemias. Occasionally, thrombocytosis, thrombocytopenia, eosinophilia, neutropenia, leukopenia, and phlebitis at IV injection site. Increased thrombin, prothrombin, and thromboplastin times were noted in a **patient**, 12 days after receiving 0.5 g/tid. Immune hemolytic anemia, acute confusional state, nausea, vomiting, anorexia, drug fever, skin rashes, and acute interstitial nephritis noted. Generalized convulsions developed in a uremic 52-year-old **male** receiving it after dialysis. Several cases of generalized seizures noted in **patients** with high cerebrospinal fluid concentrations; some, due to renal dysfunction, multiple dosing, or overdosage. Because of its poor passage through the blood–brain barrier, it should not be used when meningitis exists or possible. Crystalluria associated with its therapy. A 70-year-old **male** developed chills and fever after 1 g tid/2 days, IM. Hypersensitivity may occur in **patients** allergic to ***penicillin*** and other ***cephalosporins***. Diarrhea and pseudomembranous colitis reported. **Fetal** serum levels are about 10% of **maternal** levels, and only low concentrations appear in breast milk. Determination of glycosuria is unpredictable with ***Clinitest***®, but normal with ***Diastix***® and ***Tes-Tape***®. May precipitate in IV solutions containing ***cimetidine***. Enhances ***warfarin's*** anticoagulant activity by ~60% in post-operative cardiac valve **patients**, possibly due to an independent hypothrombinemic effect.

Nephrotoxic in **animals**, with focal necrosis of the proximal tubules in **rats**, **rabbits**, and **guinea pigs**.

SC LD_{50}, **rat**: 10 g/kg; **mouse**: 7.6 g/kg; IV LD_{50}, **rat**: 3 g/kg; **mouse**: 5 g/kg.

CEFDINIR

= *Omnicef*

***Antibiotic.* A third generation cephalosporin. Oral.**

Untoward effects: Toxic in 16–18% of **patients**. Diarrhea and vaginal candidiasis are common. Hemorrhagic colitis, rash, nausea, headache, vomiting, and abdominal pain also reported. Some **children** have found the suspension unpalatable.

Drug interactions: Its blood levels are decreased if taken within 2 h after ***antacids*** or ***iron***-containing compounds.

CEFEPIME

= *BMY 28,142* = *Maxipime*

Semi-synthetic fourth generation parenteral cephalosporin. IM and IV.

Untoward effects: **Patients**, aged 42 and 48, developed neutropenia after 2 g IV twice daily/1 month. Headache, nausea, fever, pruritus, rash, and diarrhea in 1–4% of **patients**. False positive urinary ***glucose*** reaction with ***Clinitest***® tablets. Local reactions after IV cause phlebitis, pain, and inflammation.

Drug interactions: Antagonism expected with ***imipenem*** or ***polymyxin B***.

CEFIXIME

= *Cefspan* = *CL 284635* = *FK-027* = *FR 17027* = *Oroken* = *Supprax*

A third generation semi-synthetic cephalosporin for oral use.

Untoward effects: Diarrhea and stool changes in approximately 15–20%, headaches (10%), nausea (8%), abdominal pain (6%), flatulence (2%), dyspepsia and dizziness (3%), and hypersensitivity (rash, pruritus, urticaria). Hypersensitivity reactions can occur in **those** allergic to other ***cephalosporins*** or ***penicillin***.

Drug interactions: ***Ambroxol*** increases its lung concentration.

CEFMETAZOLE

= *CS 1170* = *SKF 83088* = *Zefazone*

A second generation semi-synthetic cephalosporin for parenteral use.

Untoward effects: Contains a ***methylthiotetrazole*** group that is related to its decrease of prothrombin levels, and to bleeding problems, and also to ***disulfiram***-like reactions in **those** drinking ***alcohol***. After IV use, 70–85% is eliminated in the urine within 12 h. ***Probenecid*** or renal failure prolongs its elimination half-life. Nausea, vomiting, chills, fever, headache, rash, vaginal candidiasis, increased blood urea nitrogen and aminotransferase, pain at the injection site, and pseudomembranous colitis reported. Hypersensitivity reactions can occur in **those** allergic to other ***cephalosporins*** or ***penicillin***.

Drug interactions: Dramatic increase in international normalized ratio when added to ***warfarin*** therapy.

IV LD_{50}, **rats**: > 5 g/kg.

CEFONICID

A second generation semi-synthetic cephalosporin for parenteral use.

Untoward effects: Hypersensitivity reactions can occur in **those** allergic to other ***cephalosporins*** and ***penicillin***. Pseudomembranous colitis in a 29-year-old **female** after one dose, and seizures in a 14-year-old **female** after two IV injections. Has given falsely raised tests results for ***glucose*** with ***Clinitest***®.

CEFOPERAZONE

= *Cefobid* = *Cefobine* = *Cefobis* = *Cefosint* = *Farecef* = *Tomabef*

A third generation semi-synthetic cephalosporin for parenteral use.

Untoward effects*:** Hypersensitivity reactions can occur in **those** allergic to other ***cephalosporins and ***penicillin***. ***Disulfiram***-like reactions (tachycardia, facial flushing and dyspnea, headache) reported in **patients** drinking ***alcohol***. Skin rash, diarrhea, nausea, vomiting, fever, urticaria, neutropenia, transient eosinophilia (10% of **patients**), and increased serum transaminases (5–10%), phlebitis, and pain at injection site reported. Contains a ***methylthiotetrazole*** group (as does ***cefmetazole*** above), which is related to its reduction in prothrombin levels (hypoprothrombinemia) and increase in bleeding. Investigators reported diarrhea in 5–48% of **patients**; decreased neutrophils (2% of **patients**), and decreased hemoglobin and decreased hematocrit (5%).

CEFORANIDE

= *BL-S786* = *Precef*

A second generation semi-synthetic cephalosporin for parenteral use.

Untoward effects*:** Hypersensitivity reactions can occur in **those** allergic to other ***cephalosporins and ***penicillin***. May cause false raising of ***creatinine*** test results.

CEFOTAXIME

= *Cefotax* = *Chemcef* = *Claforan* = *HR-756* = *Makrocef* = *Pretor* = *Primafen* = *Ru-24756* = *Tolycar*

A third generation semi-synthetic cephalosporin for parenteral use.

Untoward effects*:** Dose-related side-effects were fatigue, diarrhea, and night sweats. Pain at injection sites, thrombophlebitis, and minor changes in laboratory parameters of liver and blood. Hypersensitivity reactions can occur in **those** allergic to other ***cephalosporins and ***penicillin***. These reactions (2%) involve fever, rash, and pruritus. Gastrointestinal upsets (2%) include nausea, vomiting, colitis, diarrhea, and pseudomembranous colitis. Nearly 5% reported pain, induration, and tenderness with IM injection, or phlebitis after IV injection. Granulocytopenia, thrombocytosis, transient leukopenia, eosinophilia, and neutropenia, as well as positive direct Coomb's tests (3%), moniliasis, vaginitis, headache, fever, nephrotoxicity, transient increases in SGOT, SGPT, serum lactate dehydrogenase and serum alkaline phosphatase levels, and blood urea nitrogen. Coagulopathy, due to hypoprothrombinemia reported in about 2% of **patients**. Results with ***Clinitest***® tablet ***glucose*** determinations are generally unreliable. It interferes with Guthrie ***galactose*** test. Very small levels peak in breast milk about 2 h after **maternal** use.

Drug interactions*:** ***Probenecid may increase plasma concentration 200%.

CEFOTETAN

= *Apatef* = *Cefotan* = *Cepan* = *CTN* = *Darvilen* = *ICI 156834* = *Yamatetan* = *YM 09330*

A second generation semi-synthetic cephalosporin for parenteral use.

Untoward effects*:** Hypersensitivity reactions can occur in **those** allergic to other ***cephalosporins and ***penicillin***. Colitis, pseudomembranous colitis, diarrhea, decreased prothrombin levels and bleeding, ***disulfiram***-like reactions (flushing, sweating, dyspnea, headache, tachycardia) with ***alcohol***, and thrombocytopenia reported. A single 2 g IV injection in three **females** (26–50 years) given it for surgical prophylaxis, caused anaphylaxis. Its use caused clinically important false increases of ***creatinine*** concentrations and causes unpredictable errors in urine ***glucose*** tests with ***Clinitest***®. A 46-year-old **female** developed fulminant hemolytic anemia after a second exposure to it for postoperative prophylaxis. Respiratory and acute renal failure occurred, and, despite temporary improvement with plasmapharesis, **she** died on the 25th postoperative day. Severe hemolysis and hemoglobin reduction from 15.2 g/dl to 5.9 g/dl in a 46-year-old **male** after four IV doses of 1 g every 12 h.

It has caused adverse testicular effects on prepubertal **rats**.

LD_{50}, **rats** and male **mice**: > 10 g/kg; IV, **rats**: 8.5 g/kg; **mice**: 8.12 g/kg.

CEFOXITIN SODIUM

= *Betacef* = *Cenomycin* = *Farmoxin* = *Mefoxin* = *Mefoxitin* = *Merxin* = *MK-306*

A second generation semi-synthetic cephalosporin for parenteral use.

Untoward effects*:** Serum sickness (fever, malaise, skin rashes, arthralgias, gastrointestinal upsets, lymphadenopathy), hemolytic anemia, pancytopenia, acute interstitial nephritis, pseudomembranous enterocolitis, exfoliative dermatitis, acute reversible leukopenia, nonthrombocytopenic purpura and coagulopathy (clinical bleeding, increased prothrombin time) have often been reported. Interferes with most serum ***creatinine tests, and, at high drug concentrations, may give false positive or raised ***glucose*** results with ***Clinitest***® on urine. Hypersensitivity reactions can occur in **those** allergic to other ***cephalosporins*** and ***penicillins***.

Use is associated with a lethal enterocolitis (*Clostridium difficile*) in **hamsters**.

LD_{50}, **rats**: 9 g/kg; **mice**: 5 mg/kg; IV, **dogs**: > 10 g/kg.

CEFPROZIL

= *BMY 281-03-800* = *Cefzil*

A second generation semi-synthetic cephalosporin for oral use. 95% absorbed from the gastrointestinal tract and unaffected by the presence of food.

Untoward effects: Serum sickness-like reactions, rash, diarrhea, nausea, vomiting, and abdominal discomfort reported. Hypersensitivity reactions can occur in **those** allergic to other ***cephalosporins*** and ***penicillin***.

CEFSULODIN SODIUM

= *Abbott-46811* = *Cefomonil* = *CGP-7174/E* = *Monaspor* = *Pseudocef* = *Pseudomonil* = *Pyocefal* = *SCE-129* = *Sulcephalosporin* = *Takesulin* = *Tilmapor* = *Ulfaret*

A third generation semi-synthetic cephalosporin for parenteral use against *Pseudomonas aeruginosa*.

Untoward effects: Nausea and vomiting increase with increased infusion rate, diarrhea (> 8%), rash and fever (6%), and itching (3%) are the common adverse effects. Hypersensitivity reactions can occur in **those** allergic to other ***cephalosporins*** and ***penicillin***.

IP LD_{50}, **mice**: > 4 g/kg.

CEFTAZIDIMES

= *Ceftim* = *Ceptaz* = *Fortaz* = *Fortum* = *Glazidim* = *GR 2063* = *Kefazim* = *Panzid* = *Pentacef* = *Spectrum* = *Tazicef* = *Tazidime*

Third generation semi-synthetic cephalosporins for parenteral use.

Untoward effects: Nephrotoxicity, especially in **patients** with pre-existing renal impairment. Neurotoxicity in two **males** (71 and 75) with renal dysfunction after 2–3 g IV/twice daily. Symptoms were confusion, disorientation, agitation, weakness, and myoclonus. Skin rash, lowered white blood cells, transient increases in serum transaminases, photosensitivity in a 26-year-old **female** with cystic fibrosis, visual and auditory hallucinations in a 62-year-old **female** after 2 g/twice daily/3 days (possibly, due to prolonged half-life from impaired renal clearance), neutropenia, agranulocytosis, and thrombocytopenia reported. Irritation at the injection site occurs. Encephalopathy and severe multifocal myoclonic-like jerking in an 80-year-old **male** with osteomyelitis and renal insufficiency treated with 2 g, IV/tid/25 days. Improved with discontinuance of treatment, but returned on rechallenge. Epileptiform effects have occurrred with high dosage. Hypersensitivity reactions can occur in **those** allergic to other ***cephalosporins*** and ***penicillins***. The sodium carbonate form releases ***carbon dioxide*** when dissolved in water. Psychonosema during a period of regular hemodialysis reported.

Drug interactions: Has increased ***digoxin*** toxicity. Fever, rash, and reversible pancytopenia in 43-year-old **female** after administration of ***vancomycin*** with it.

CEFTIBUTIN

= *Cedax* = *Isocef* = *Keimax* = *Sch 39,720*

Semi-synthetic cephalosporin antibiotic for oral administration.

Untoward effects: Diarrhea is common; nausea, headache, dyspepsia, abdominal pain, vomiting, and dizziness also reported.

CEFTIOFUR

= *Excenel* = *Naxcel*

A third generation semi-synthetic cephalosporin for IM use in animals.

Untoward effects: Pain at injection site is common. Often ineffective unless higher than recommended doses are used.

CEFTIZOXIME

= *Cefizox* = *Ceftix* = *Epocelin* = *Eposerin* = *FK-749* = *FR-13,479* = *SKF-88373*

A third generation semi-synthetic cephalosporin for parenteral use.

Untoward effects: Hypersensitivity (fever, rash, pruritus), transient increases in SGOT, SGPT, and alkaline phosphatase, eosinophilia, thrombocytosis, and, occasionally, positive Coomb's test and injection site problems (phlebitis, cellulitis, pain, burning sensations, induration, tenderness, and paresthesia are among the more common reactions. Less common are transient increases in blood urea nitrogen and ***creatinine***, nausea, vomiting, diarrhea, and pseudomembranous colitis. Rarely, neutropenia, leukopenia, thrombocytopenia, hypoprothrombinemia, and vaginitis were reported. ***Glucose*** testing results are unpredictable with ***Clinitest***®. Hypersensitivity reactions can occur in **those** allergic to other ***cephalosporins*** and ***penicillin***.

IV LD_{50}, **rats** and **mice**: approximately 6 g/kg.

CEFTRIAXONE

= *Cefatriaxone* = *Ro-13-9904/001* = *Rocefin* = *Rocephin(e)*

A third generation semi-synthetic cephalosporin for parenteral use.

Untoward effects: Pain, induration, and tenderness at the site of IM injection, and, occasionally, phlebitis at the IV site. Hypersensitivity (rash, some pruritus, and fever), eosinophilia (5–23%), thrombocytosis (7%), leukopenia (> 8%), diarrhea (10–44%), thrombocytopenia (1%), nausea (3%), vomiting, dysgeusia, neutropenia, headache, dizziness, increased SGOT (> 3%), increased SGPT (> 3%), increased alkaline phosphatase, increased ***bilirubin***, cholethiasis and biliary colic, *Clostridium difficile* enterocolitis, hypoprothrombinemia and bleeding, reduction in ***factor VII*** activity, acute pancreatitis, pemphigus, moniliasis or vaginitis, and flushing have all been reported. A case of pseudolithiasis and intractable hiccups in a 10-year-old **male** receiving 50 mg/kg/day, IV. Inadvertent rapid IV (2 g in 5 min) in a 21-year-old **female** was associated with shivering, diaphoresis, dilated pupils, and palpitations. Psychonosema reported during a period of regular hemodialysis. Hypersensitivity reactions can occur in **those** allergic to other ***cephalosporins*** and ***penicillins***. May precipitate acute ***verapamil*** toxicity.

IV LD_{50}, **rats** and **mice**: > 10 g/kg.

CEFUROXIME

Sodium forms = *Anaptivan* = *Biociclin* = *Biofurex* = *Bioxima* = *Cefamar* = *Cefoprim* = *Cefossim* = *Cefumax* = *Cefurex* = *Cefurin* = *Curocef* = *Curoxim* = *Duxima* = *Gibicef* = *Ipacef* = *Kefurox* = *Kesint* = *Lamsporin* = *Medoxim* = *Polixima* = *Ultroxim* = *Zinacef*

Axetil forms = *Axoril* = *CCI 15641* = *Ceftin* = *Cepazine* = *Ximos* = *Zinnat*

A second generation semi-synthetic cephalosporin for oral use. The sodium form is used parenterally.

Untoward effects: Fever noted in **patients** 5–7 days after initiation of therapy. Nausea in ~4%. Catheter-related failure due to IV-induced phlebitis. It caused acute renal failure (allergic nephritis) in a 58-year-old **male** after 1.5 g IV, tid/8 days. Hypersensitivity reactions can occur in **those** allergic to other ***cephalosporins*** and ***penicillin***.

Drug interactions: ***Probenecid*** will block its tubular excretion almost completely.

CELASTRUS SCANDENS

= *Bittersweet, American*

A common woody-stemmed vine with bright clusters of orange and scarlet fruits, found mostly in eastern North America, often found along fence rows and in dried house bouquets. This is not the European *bittersweet*, *Solanum dulcamara*.

Untoward effects: All parts of the plant are mildly poisonous and gastroirritants. Physicians should be aware that especially the fruits can cause vomiting and diarrhea. Some claim the toxin is ***euonymin***.

Rarely fatal to **animals** that have ingested the leaves. It has caused mild to severe purgation in **horses** and may have ***digitalis***-like ill-effects.

CELECOXIB

= *Celebra* = *Celebrex* = *SC 58,635* = *YM 177*

***Analgesic, anti-arthritic, anti-inflammatory non-steroidal anti-inflammatory drug.* A selective COX-2 inhibitor.**

Untoward effects: Severe erosive gastropathy with petechial hemorrhages and small erosions of the body and antrum. Fluid retention and edema reported. Abdominal pain, flatulence, diarrhea, dyspepsia, palpitations, tachycardia, chest pain, fever, nausea, back pain, sinusitis, hot flushes, epistaxis, nervousness, tinnitus, rash, anemia, blurred vision, and renal toxicity. May cause allergic reactions in ***sulfonamide***-sensitive **people**. Sudden death may have occurred due to its use.

Drug interactions: Can decrease effectiveness of ***angiotensin-converting enzyme inhibitors***, ***furosemide***, and ***thiazides***. ***Fluconazole*** at 200 mg twice daily causes a doubling in celecoxib serum levels. Increases ***warfarin*** anticoagulant effect.

CELERY

= *Apium graveolens* = *Kinchai*

***Vegetable.* Wide application in foods and tobaccos.**

Untoward effects: Contains a high amount of ***nitrates***. May be a photosensitizer and cause cutaneous eruptions in some **people**, due to its ***5-methoxypsoralen*** content. It can also become infected with a fungus that produces phototoxicity. Its β-glucosidases (similar to those found in ***almonds***), when simultaneously consumed with ***laetrile*** tablets, can release ***hydrocyanic acid***, and, possibly, cause poisoning. Anaphylactic reactions from ingestion and positive delayed eczematous patch test results reported, as well as "celery dermatitis" in celery **pickers**, **handlers**, or celery **washers** from the fungus called "pink rot", after **their** exposure to sunlight. The latter causes the celery to synthesize large amounts of ***furocoumarins*** and ***psoralens*** (concentrations on an infected plant can reach hundreds of thousands of ppm) as ***phytoalexins***. Maximum photoreactivity occurs in 2–4 h and disappears about 8 h after ingestion. Exercise after consuming it has induced anaphylaxis and coma. Contains ***oxalic acid***, ***caffeic acid*** (50–200 ppm), and ***3-n-butylphthalide***, a vasodilator. May cross-react with ***orange*** peels. Its seed oil is used in fragrances for cosmetic creams and lotions, soaps, detergents, and perfumes. ***Latex***-sensitive **people** can show hypersensitivity to it.

Excessive ingestion of celery grown on heavily fertilized fields caused the death of 47 southern California **feedlot calves** from ***nitrate*** poisoning. A number of **cows** died of ***cyanide*** poisoning after eating frost-damaged celery tops. Its aqueous extract, IV, has caused a large drop in blood pressure in **rabbits**. Experimentally, it can induce hypoglycemia.

CELIPROLOL

= *Carden* = *Celectol* = *Corliprol* = *REV 5320A* = *Selecor* = *Selectol* = *ST 1396*

β-Adrenergic blocker, antihypertensive, antianginal.

Untoward effects: It shows relatively β_1 selective blockade and partial β_1 and β_2 agonist activities. A 61-year-old **male** developed severe diarrhea after 2 weeks on 200 mg/day for angina pectoris.

LD_{50}, **rats**: 3.8 g/kg; **mice**: 1.8 g/kg; IV LD_{50}, **rats**: 68 mg/kg; **mice**: 56 mg/kg.

CELLASENE

***Herbal.* Supposedly, eliminates cellulite.**

Untoward effects: Usual dose contains five times **human** MDR of ***iodine*** leading to thyroid problems.

CELLULOSE

= *Avicel* = *Elcema* = *Emcocel* = *PH 105* = *Sulfafloc*

Many forms exist: *Cellosize* = *Cellulose, Ethyl Hydroxyethyl Ether*

Hemopek = *Isopto Tears* = *Methocel* = *Methylcellulose* = *Ophthakote* = *Pharmacoat 606* = *Oxycel* = *Oxycellulose* = *Surgicel* = *Tabotamp* = *Tearisol* = Cellulose, Oxidized

Sodium Carboxymethylcellulose = *Carbose D* = *Carmethose* = *Cellolax* = *Cel-O-Brandt* = *Cethylose* = *CMC* = *Glykocellon* = *Polycell* = *Thylose* = *Tylose MGA* = *Xylo-Mucine*

Untoward effects: As an adsorbable hemostatic in internal, wound, and bone surgery, its sterility must be maintained to avoid infection. The same applies to its use as a topical suspending agent in ophthalmic preparations. When used in ***Rely***® tampons, its use was associated with severe Staphylococcic septic shock reactions. It was, apparently, broken down into ***glucose*** by bacteria and/or yeasts in the vagina. Staphylococci trapped in the ***CMC*** gel used the ***glucose*** to produce toxin. Anaphylaxis in a 63-year-old **female** was induced by its use as an excipient in a ***barium sulfate*** suspension. Older dialysis equipment uses cellulose acetate fibers in their filters. As they age, their use has been associated with severe post-treatment reactions, including intense headaches, and virtual blindness, and deafness.

CEMENT

Untoward effects: Regular Portland Cement (named after the isle of Portland in Dorset, England) contains ***calcium oxide*** (64% [***lime***]), ***silicon dioxide*** (21%), ***di-*** and ***tricalcium silicates*** with varying amounts of ***aluminum oxide*** (5.8%), ***tricalcium aluminate, iron oxide*** (2.9%), and traces of ***chromates***. Not only is the mixture highly alkaline, but also the ***calcium oxide*** reacts with water and with sweat as well, liberating large amounts of heat, known to cause severe burns on the skin areas of **workers**. It even causes necrotic burns and ulcers on **workers** kneeling on wet cement. ***Sulfur dioxide*** (q.v.) is released in crushing and milling it, especially under damp and humid conditions, which, in turn, has caused burning of eyes, sore throats, and chest tightness. Fine particles may stick to the cornea and inside the eyelids, causing irritation. Sharp silica fragments can be abrasive. Many cement **workers** become sensitized by ***chrome*** (especially the hexavalent form) in cement with recurrent dermatitis. ***Thallium***-containing atmospheric dust was found in the vicinity of a cement processing plant in Germany.

CENTALLA ASIATICA

= *Centasium* = *Centelase* = *Emdecassol* = *Gotu Kola* = *Hydrocotyle* = *Indian Pennywort* = *Madecassol* = *Marsh Penny* = *White Rot*

***Herbal.* Unlike *cola* or *kola* nuts, it does not contain *caffeine*.**

Untoward effects: Contact dermatitis.

Antifertility effects in **mice**.

CENTAUREA sp.

Untoward effects: ***Honey*** from **bees** pollinating ***C. scabiosa*** has caused poisoning in **man**.
C. repens = ***Centaurée de Russie*** = ***Knapweed*** = ***Russian Knapweed*** is a cause of fatal nigropallidal encephalomalacia in California **horses** after its chronic ingestion. After about 3–11 weeks, they are unable to eat or drink, muscles around the mouth are tense, and lips twitch with involuntary chewing motions. Plants are also found in Turkestan, Australia, and South Africa. **Sheep** may also be affected.
C. solstitialis = ***Yellow Star Thistle*** causes the same symptoms in **horses**. Its spines can cause mechanical injury in the mouth of **those** ingesting it.

CENTAURIUM sp.

= *Centaury* = *Mountain Pink*

***Botanical.* Found in the mid-western and southern U.S., Europe, and Australia.**

Untoward effects: In the U.S., ingestion has caused anorexia, abdominal pain, and diarrhea in **cattle**, **goats**, and **sheep**. Death has also occurred, and, on post-mortem, severe gastroenteritis with ulcers and hemorrhages, as well as congestion of the liver and kidneys was noted.

CENTOXIN

= *HA-1A* = *Nebacumab*

A monoclonal IgM antibody against endotoxin from Gram-negative bacteria.

Untoward effects: Hypersensitivity-type reactions (erythema, flushing, edema, hypoxia) and hypotension are uncommon adverse effects. Extremely expensive!!

CEPHADRINE

= *Anspor* = *Cefradrex* = *Cefrag* = *Cefro* = *Celex* = *Cesporan* = *Dimacef* = *Ecosporina* = *Eskacef* = *Forticef* = *Lenzacef* = *Lisacef* = *Medicef* = *Megacef* = *Samedrin* = *Sefril* = *SQ 22022* = *Velocef* = *Velosef*

A first generation semi-synthetic cephalosporin for oral and parenteral use.

Untoward effects: Gastrointestinal upsets (glossitis, nausea, vomiting, diarrhea, tenesmus, abdominal pain, pseudomembranous colitis), pain at injection sites, urticaria, skin rashes, erythema, pruritus, edema, joint pain, acute interstitial nephritis, drug fever, transient eosinophilia, leukopenia, occasionally increased SGOT and SGPT levels, hepatomegaly, increased bilirubin, increased alkaline phosphatase, increased lactate dehydrogenase, increased blood urea nitrogen, headache, dizziness, dyspnea, paresthesia, candidiasis (particularly, vaginal), and thrombophlebitis reported, especially in young **children** and **patients** > 50. Will give a false positive urinary ***glucose*** test with ***Clinitest***® tablets. Hypersensitivity reactions can occur in **those** allergic to other ***cephalosporins*** and ***penicillin***. Incompatible with ***Lactated Ringer's*** solution, due to the ***sodium bicarbonate*** and ***calcium*** ions in it. Withdrawn from the Scandinavian market in 1997.

Drug interactions: Use with ***probenecid*** will increase its plasma concentration 96% at 60 min, and increase its half-life by 88%. IV use increases ***aztreonam's*** binding to serum protein by about 5%.

CEPHALANTHUS OCCIDENTALIS

= *Buttonbush*

Tall shrub found in North America. Unpalatable and found in low areas, particularly near creeks, swamps, and ponds.

Untoward effects: During summer and fall dry periods, **cattle** have consumed the toxic leaves (other parts are less toxic). The poisonous principles are ***glycosides*** (***cephalanthin*** and ***cephalin***), which cause vomiting, paralysis, spasms, hemolysis, and, infrequently, death. **Indians** have made a tea from it, for its emetic and laxative effects.

The bark contains ***cephalanthin***, a poison, toxic to both cold- and warm-blooded **animals**, which destroys red blood cells, converts ***oxyhemoglobin*** to ***methemoglobin***, causes violent vomiting, convulsions, and paralysis in **animals**.

CEPHALEXIN

= *Alfaspoven* = *Ausocef* = *Cefadros* = *Cefa-Iskia* = *Cefaloto* = *Cefibacter* = *Ceporex* = *Ceporexine* = *Cex* = *Derantel* = *Efalexin* = *Farexin* = *Fergon 500* = *Garasin* = *Ibilex* = *Iwalexin* = *Keflet* = *Keflex* = *Keforal* = *Keftab* = *Larixin* = *Lexibiotico* = *Llonexina* = *LY061188* = *Madlexin* = *Mamalexin* = *Mecilex* = *Neolexina* = *Ohlexin* = *Oracef* = *Oracocin* = *Ortisporina* = *Rinesal* = *Sartosona* = *Sencephalin* = *Sintolexyn* = *Syncl* = *Taicelexin* = *Tokiolexin* = *Xahl*

A first generation semi-synthetic cephalosporin for oral use.

Untoward effects: Acute interstitial nephritis reported. Crystalluria has followed overdosing with 2.5 g in a 3-year-old **male**. A **diabetic** 70-year-old **male** received 500 mg qid for a UTI. **He** became dehydrated, with raised blood urea nitrogen (129 mg/100 ml), ***creatinine*** (5.6 mg/100 ml), and ***bilirubin*** (2.4 mg/100 ml). Generalized skin rashes (usually, maculopapular), itching, pemphigus, urticaria, angioneurotic edema, toxic epidermal necrolysis, and Stevens-Johnson syndrome affecting the entire skin and mucosal areas of the body have occurred. Hematuria, eosinophilia, transient neutropenia, lowered white blood cells, hemolytic anemia, leukopenia, leukocytosis, thrombocytopenia, and, in a **hemophiliac**, acute intravascular hemolysis noted with its use. Granulomatous hepatitis, pseudomembranous colitis, nausea, vomiting, diarrhea, heartburn, abdominal cramping, flatulence, dry mouth and stomatitis, dizziness, headache, swelling of the tongue, vulvovaginitis, and swollen ankles in a ***penicillin***-sensitive **individual**, dyspnea and asthma (occupational exposure, as well) occur. A uremic 46-year-old **female** developed symptoms of euphoria, agitation, and grand mal seizures after 5 g over a 3-day period. This was probably due to accumulation of the drug, due to **her** renal impairment. Urinary ***glucose*** determinations with ***Clinitest***® are unreliable. Hypersensitivity reactions can occur in **those** allergic to other ***cephalosporins*** and ***penicillin***.

Drug interactions: Can decrease the effectiveness of ***oral contraceptives***. Its absorption is decreased by simultaneous use of ***cholestyramine***.

Anorexia, vomiting, diarrhea, and cutaneous reactions have been reported in treated **cats**.

LD_{50}, **mouse**: 1.6 g/kg.

CEPHALOGLYCIN

= *Kafocin*

Untoward effects: Rarely used now, except as an antibacterial in urinary tract infections. Mild abdominal cramps, itching, nausea, generalized rash, headache, transient increases in serum transaminase and alkaline phosphatase, diarrhea, gastrointestinal bleeding, monilial vaginitis, eosinophilia (> 10%), and stomatitis have all been reported. Hypersensitivity can occur in **those** allergic to other ***cephalosporins*** and ***penicillin***.

CEPHALORIDINE

= *Aliporina* = *Ampligram* = *Cefaloridin* = *Ceflorin* = *Cepaloridin* = *Cepalorin* = *Ceporan* = *Ceporin* = *Cer* = *Cilifor* = *Deflorin* = *Faredina* = *Floridin* = *Intrasporin* = *Keflodin* = *Kefspor* = *Lloncefal* = *Loridine* = *Sefacin*

Untoward effects: Shedding of finger and toe nails, angioneurotic edema, burning of skin and eyes, respiratory problems (wheezing, sneezing paroxysms, cyanosis), nervousness, urticaria, skin rash (maculopapular or erythematous), itching, eosinophilia, hemolytic anemia, neutropenia, drug fever, coagulopathy, encephalopathy, epileptic convulsions after injection into cerebral ventricles, hallucinations, nystagmus, parkinsonism syndrome, false positive Coomb's test, leukopenia, nausea, vomiting, pain in **some** with IM use, and, rarely, phlebitis, after IV use. May increase SGOT. Has increased prothrombin time to 29 s. Nephrotoxicity with significant proximal tubule damage (also noted in **mice**, **guinea pigs**, and **rabbits**), including anuria and severe acute renal failure and death, especially in **patients** on high dosage. Hypersensitivity can occur in **those** allergic to other ***cephalosporins*** and ***penicillin***.

Drug interactions: Nephrotoxicity is potentiated by ***furosemide***, which decreases the renal excretion of the drug by ***aminoglycoside antibiotics***.

High dosage causes renal tubular necrosis in **rats**, **mice**, **guinea pigs**, **rabbits**, and **monkeys**.

Oral and SC LD_{50}, 2.5 g/kg; IV, 1.4 g/kg.

CEPHALOTHIN

= *Averon-1* = *Cefalotin* = *Cephation* = *Ceporacin* = *Cepovenin* = *Chephalotin* = *Coaxin* = *Keflin* = *Lospoven* = *Microtin* = *Synclotin* = *Toricelocin*

A first generation semi-synthetic cephalosporin for parenteral use.

Untoward effects: Pruritus, flushing, wheezing, asthma, tachypnea, hypotension, allergic skin rashes, urticaria, and some drug fever reactions, such as nausea, chills, myalgia, and hypertension. Some increases in SGOT. Metabolized by the liver, therefore, reduce dosage in the presence of renal–hepatic disease. Painful on IM administration, and phlebitis, and even necrosis, can occur after IV use. Nephrotoxic, including acute interstitial nephritis, tubular necrosis, and renal failure when given alone. This is aggravated by prior renal disease or use with ***aminoglycosides*** and ***furosemide***. Thrombophlebitis (reduced by using IV "piggy-back" system and scalp needles), thrombocytopenia, thrombocytosis, erythroblastopenia, leukopenia, eosinophilia, pancytopenia, occasionally agranulocytosis, high incidence of Coomb's positive hemolytic anemias, increases in prothrombin time, pseudomembranous colitis, diarrhea, tachycardia, and hemolysis, and even a cardiac arrest. Anaphylactic shock reactions and death and hepatotoxicity with jaundice also reported. A toxic paranoid reaction in a 57-year-old **female**; after treatment, **she** complained of headache and nausea, followed the next day by jitteriness, limb jerks, confusion, and suspected people were going to kill **her**. Hypersensitivity can occur in **those** allergic to other ***cephalosporins*** and ***penicillin***. Incompatible physically in solution with ***aminophylline, diphenhydramine, erythromycin, hydrocortisone, kanamycin, oxytetracycline, polymyxin B, saline solutions, sodium bicarbonate, tetracycline,*** and ***theophylline***. Falsely elevates ***creatinine*** and urinary ***glucose*** (***Clinitest***®) test levels.

Drug interactions: ***Probenecid*** slows its tubular excretion and increases its half-life by 90%, and increases its blood levels (by 343% at 60 min).

LD_{50}, **rats**: > 10 g/kg; **mice**: > 20 g/kg; IV LD_{50}, **mice**: 5 g/kg; **guinea pigs**: 150 mg/kg; IP LD_{50}, **rats**: 7.7 g/kg; **mice**: 5.7 g/kg.

CEPHAPIRIN

= *BL-P 1322* = *Ambrocef* = *Brisfirina* = *Bristocef* = *Cefadyl* = *CefaLak* = *Cefatrexyl* = *Piricef* = *ToDay*

A first generation semi-synthetic cephalosporin for parenteral use.

Untoward effects: Mild phlebitis (grade 1) develops in approximately 50% of the IV injection sites, and 25% of 12 **volunteers** developed moderate (grade 2) phlebitis. Acute interstitial nephritis, pain on IM injection, increases in SGOT and alkaline phosphatase levels, eosinophilia, positive direct Coomb's test, jaundice, transient increases in blood urea nitrogen, and neutropenia reported in **patients**. Interferes with urine ***glucose*** determination using ***Clinitest***® tablets. Hypersensitivity reactions can occur in **those** allergic to other ***cephalosporins*** and ***penicillins***.

IM TD_{LO}, **man**: 350 mg/kg/6 days.

Produces extensive acute tubular necrosis in **rats**, even on low dosage when given with ***furosemide***.

CERCOCARPUS MONTANUS

= *Mountain Mahogany*

Untoward effects: **Ruminants** are more susceptible than monogastric **animals** to its ***cyanogenic glycosides***, which are particularly toxic after freezing, wilting, or crushing of the plants. Death in **cattle** may occur within minutes after ingestion. Symptoms before death are panting, gasping, rapid breathing, colic, tremors, excess tearing, mydriasis, and convulsions.

CERIUM

Rare earth.

Use*:** The nitrate, on burns unresponsive to ***silver sulfadiazine; the oxalate, as a gastric protectant and antiemetic; in decolorizing glass.

***Untoward effects*:** Increases blood clotting time and lowers platelets.

Has caused chromosomal aberrations in **mouse** bone marrow cells. Passes into the milk of **cows**.

LD_{50} of the nitrate, **mice**: 470 mg/kg.

CERIVASTATIN

= *Baycol* = *BAY-W 6228* = *Lipobay* = *Rivastatin*

***Antihyperlipoproteinemic*. Synthetic HMG-COA reductase inhibitor.**

***Untoward effects*:** Arthralgia, myalgia, asthenia, leg pain, rhabdomyolysis, and increases in serum aminotransferases and creatine kinase.

Drug interactions*: *Cholestyramine decreases its absorption if taken within 2 h. ***Erythromycin*** increases its serum levels. ***Cyclosporine*** increases its plasma levels three to five times.

CERULETIDE

= *Caerulein* = *Caerulin* = *Ceosunin* = *FI 6934* = *Takus*

***Use*:** Gastric secretory stimulant, diagnostic aid in cholecystography in **man**, and in production of pancreatitis in **rats** and **mice**.

***Untoward effects*:** Transient heat sensations on face, nausea, and abdominal discomfort reported in **patients**.

Long-term SC trials in **rats** and **dogs** with high doses cause pancreatic hypertrophy. Higher doses cause acinar cell damage with atrophy of exocrine pancreatic tissue. Decreased blood pressure in **rabbits** and **dogs**, and a hypertensive or biphasic response in **rats**.

IV LD_{50}, **mice**: 1 g/kg.

CERULOPLASMIN

= *Caeruloplasmin* = *Ferroxidase*

Untoward effects*:** It can cause plasma to look green, particularly, when it is raised, due to the ***estrogen content of ***oral contraceptives***, or in rheumatoid arthritis **patients** who have a decrease of yellow pigments (***hem*** and ***carotenoids***) in **their** urine.

CESTRUM

Shrub.

Untoward effects*: *C. aurantiacum fruit (berries) and green leaves are toxic, and have killed **cattle** and **goats** in Rhodesia, Kenya, and Guatemala. Affected **cattle** become irritable and hypersensitive to stimuli, then, a progressive posterior paralysis. Pathology described includes a necrotic hemorrhagic gastroenteritis, and degenerative changes in the liver, kidney, and brain, and extensive hemorrhages SC in the neck, shoulder, and rump. **Sheep** eating 200–500 g of leaves/day are often found dead on the fourth day.

C. diurnum = ***Chinese Inkberry*** = ***Day-blooming Jessamine*** = ***Day Cestrum*** = ***Day Jessamine*** = ***Wild Jasmin*** has caused hypercalcemia and calcinosis in **horses** and **pigs**. As an ornamental shrub in Florida, it caused toxicosis (***atropine***-like) in **people** and **pets** eating the ripe berries. Symptoms initially including salivation, followed by labored respirations, tachycardia, pyrexia, partial paralysis, hallucinations, and, occasionally, mydriasis. It contains a ***1,25-dihydroxycholecalciferol***-like compound that, upon ingestion, causes the hypercalcemia and excess deposits of ***calcium*** in the heart, liver, kidneys, and other organs, and painful death in **cattle** and **horses** in Florida. Florida **children** have ingested these fruits, but, fortunately, **their** stomachs have been emptied in the emergency room.

C. Humboldtii's purplish-black fruit of the Amazon is considered to be poisonous by the **Kamsá Indians**.

C. lanatum of Central Africa is powdered and the dust sprinkled on **chickens** to kill lice.

C. laevigatum's leaves and berries have caused anorexia, excitement, aggressive behavior, incoordination, and death of **cattle** in Brazil.

C. loretense is considered toxic by **Tikuna Indians** of the Amazons. Contains the alkaloids, ***parquine*** and ***solasodine***.

C. nocturnum (***Night-blooming Jessamine***), native to the West Indies, Mexico, and Central America, and transplanted to Florida, has caused dyspnea, depression, restlessness, severe headache, nausea, dizziness, sneezing, throat irritation, and other ***anticholinergic*** effects in **people**. It blooms several times a year and its far-reaching sweet-scented odor causes respiratory distress and illness in many **humans** and some **animal** pets.

C. ochraceum (***Sauco Blanco***) of Colombia is poisonous. Indian **medicine men** in the Sibundoy Valley make a "therapeutic" tea from it, to provoke intense sweating. If too much is taken, delirium occurs.

C. parquii of Argentina, Australia, and Italy has killed **sheep**, **cattle**, **pigs**, and **horses**. It and ***C. laevigatum*** contain hallucinogenic properties.

C. reflexum's leaves and roots are reported to be violently toxic by the **Witoto Indians** of the Amazon.

CETIRIZINE

= *Alerlisin* = *Formistin* = *P-071* = *Reactine* = *Virlix* = *Zirtek* = *Zyrtec* = *Zyrlex*

Second generation antihistamine.

Untoward effects: More sedation and impairment of driving performance than other second generation antihistamines. Xerostomia, pharyngitis, dizziness, urticaria, drowsiness, and fatigue also reported.

Drug interactions: Its clearance is decreased 16% when given ***theophylline*** 400 mg/day. Strong potentiation of ***acenocoumarol*** anticoagulant effect.

CETRIMONIUM BROMIDE

= *Bromat* = *Cetab* = *Cetavlon* = *Cetylamine* = *C.T.A.B.* = *Lissolamine V* = *Micol* = *Quamonium*

Antimicrobial, preservative, sterilizing agent.

Use: In pharmaceuticals, cosmetics, and in equipment and premise sterilization.

Untoward effects: Irrigation with 0.1% solution after a cyst excision in a 45-year-old **female** caused severe methemoglobinemia. Injection of a 5% solution into hydatid cysts and 0.5% as a peritoneal washout in three **patients** caused a sterile chemical peritonitis. It has caused a contact dermatitis in **man**. A 20-month-old **male** had an allergic reaction to a 12% solution in a shampoo that spilled accidentally on **his** chest. Rechallenge with a 12% solution to a small area produced similar lesions. It has a potential for ***bromide*** toxicity after ingestion.

LD_{50}, **rats**: 500–760 mg/kg; **rabbits**: 150 mg/kg; SC LD_{50}, **guinea pigs** and **rabbits**: 125 mg/kg; IV LD_{50}, **rats**: 44 mg/kg.

CEVIMELINE

= *Evoxac*

Use: To treat xerostomia.

Untoward effects: Nausea and diaphoresis. Occasionally, diarrhea, rhinitis, and vision problems (especially at night).

CHAEROPHYLLUM sp.

= *Chervils*

Untoward effects: ***C. sylvestre*** = ***Asses' Parsley*** = ***Wild Chervil*** reported to cause enteritis, mydriasis, anorexia, and paralysis in **pigs** in Germany, yet, **animals** eating fresh material in Britain seemed unaffected. Post-mortem reveals acute gastrointestinal inflammation.

C. temulum = ***Rough Chervil*** causes diarrhea, weakness, stupor, incoordination, mydriasis, tremors, and central nervous system depression. Species affected include **cattle**, **pigs**, and **nutria**. **Horses** are particularly sensitive to its toxins.

CHAETOMIUM sp.

Untoward effects: A fungus often found on ***corn*** in the U.S. and Europe. It has been toxic to **rats** and **cockerels**.

CHAMAESYA sp.

Untoward effects: **People** in Florida, pulling up this weed from **their** lawns without gloves, developed rashes between **their** fingers and even over **their** bodies.

C. hyssopifolia is used as an abortifacient in Cuba and Brazil.

CHAMOMILE

= *German Chamomile* = *Horse Gowan* = *Manzanila* = *Matricaria chamomilla* = *Wild Chamomile*

Untoward effects: A localized vesicular dermatitis appeared on the back of a **person's** hands after picking the plant. The skin reaction was negative with the leaves, but positive for the flowers. After oral challenge, the **patient** suffered from an anaphylactic attack with severe symptoms (leukopenia, asthma). A **gardener** developed a severe dermatitis of the face and hands after chamomile dressings were applied to **his** eczema. Another **individual** developed an allergic dermatitis from its presence in a shampoo. It has also been a photosensitizer, after topical application. The tea from its flower heads has caused a contact dermatitis, anaphylaxis, and hypersensitivity cross-reactions with ***ragweed***, ***goldenrod***, ***asters***, and ***chrysanthemums***. A 54-year-old **female** drank a cup of its tea and two ***aspirin*** tablets on arising, to help relieve a headache. Within 30 min **she** developed a severe anaphylactic reaction, with generalized hives and pharyngeal edema. **She** had no previous sensitivity to ***aspirin***, nor to rechallenge with it. A 35-year-old **female** suffered a severe allergic reaction after a few sips of its tea.

Chamomile Oil, German = ***Blue Chamomile Oil*** = ***Hungarian Chamomile Oil***. Used in foods, soaps, detergents, cosmetic creams and lotions, and perfumes. 4% in petrolatum in 25 **volunteers** showed no sensitization or irritation (48 h closed-patch test).

LD_{50}, **rats** and acute dermal LD_{50}, **rabbits**: > 5 g/kg.

Chamomile Oil, Roman = ***Camomile Oil*** = ***English Chamomile Oil***. Used in foods, soaps, detergents, cosmetic creams and lotions, and perfumes. 4% in petrolatum in 25 **volunteers** showed no sensitization or irritation (48 h closed-patch test).

Moderately irritating, when applied for 24 h under occlusion, to intact or abraded **rabbit** skin.

LD_{50}, **rats** and acute dermal LD_{50}, **rabbits**: > 5 g/kg.

CHEESES

Untoward effects: They have caused immediate urticarial reactions to intact or scratched skin of **people**. In normal **individuals**, ***tyramine*** is broken down into its non-toxic derivative, ***p-hydroxyphenylacetic acid***, in the intestine by ***monoamine oxidase*** enzyme. Any cheese, but especially strong aged, blue, brie, boursin, Camembert, cheddar, Cheshire, Gouda, Gruyére, and Swiss cheeses containing ***tyramine*** react with ***monoamine oxidase inhibitors***, such as ***debrisoquine, iproniazid, isocarboxide, isoniazid, mebanazine, nialamide, pargyline, phenelzine, procarbazine, prochlorperazine, tranylcypromine,*** and ***trifluoperazine***, to cause headaches, increase blood pressure and hypertensive crises, neck stiffness, chills, sweating, nausea, vomiting, flushing, palpitations, and retrosternal pain. 20 g New York cheddar provoked a pressor response in two ***monoamine oxidase***-treated **patients**. Cottage cheese contains about 0.2 µg/g; cheddar has up to 1.4 mg/g; Leiderkrantz, up to 1.68 mg/g; Gruyére, up to 516 µg/g, and Swiss, up to 434 µg/g. ***Aflatoxin*** has been found on the surface of cheddar cheese. Red ***wines*** contain 2–3 µg of ***histamine***/ml, plus other ***amines*** and are often consumed at the same time. Urticaria, itching, palpitations, headache, and diarrhea have occurred, due to ***histamine*** in various cheeses, including Gouda. Not all such untoward effects of cheeses are due to ***monoamine oxidase inhibitors*** or allergic reactions. Some are due to their intrinsic pharmacologically active substances, such as vasoactive ***amines***. These can precipitate migraine headaches by increasing circulating ***epinephrine*** and ***catecholamines***, increasing platelet aggregation, and ***serotonin*** release. A type III hypersensitivity pneumonitis, known as **cheese worker's** lung or disease, exists in cheese **washers** from inhalation of the cheese mold, *Penicillium casei*. **Patients** on ***furazolidone*** should avoid cheeses. A rare case of myonecrosis in a 27-year-old **female** during a hypertensive crisis associated with eating cheese and ***tranylcypromine*** use. Several cases of contact dermatitis were due to cheese powder in potato chip factory **workers**. ***Nitrosamines*** have been found in cheeses. ***Penitrem A***, a tremorgen, has been found on moldy cream cheese. Cheese can cause loose stools in ***lactose***-deficient **people**. European studies indicated ***uranium*** concentration in cheese is approximately eight times that of milk. The FDA has found ***strontium***^{90}, ***potassium***^{40}, ***cesium***^{137}, and ***ruthenium***^{106} in imported cheeses. Hypersensitivity reported in 39-year-old **female**, due to *Penicillium* mold content in blue cheese, manifested by recurrent erythema annulare centrifugum. Imported Camembert or brie cheese from France caused illness, due to their *E. coli* and *Citrobacter freundii* content. A group C β-hemolytic streptococcus caused the illness of many New Mexico **Hispanics** after eating homemade cheese. *Listeria monocytogenes* in Mexican-style soft cheese sickened at least 243 **people** in California and 39 deaths were linked to its ingestion. Several brands of imported French brie cheese were also incriminated as carriers of this organism. Several hundred cases of salmonellosis (*S. typhimurium*) in Canada reported from ingestion of various contaminated cheeses. In England, seven **people** developed *Brucella melitensis* infection after eating Pecorino cheese from Italy, made from unpasteurized sheep's milk. Storage of such cheese for 3 months usually makes it free of viable organisms. A 19-year-old **female** had severe itching and urticaria after eating cheese containing ***sodium benzoate***. A week later, **she** experienced a life-threatening anaphylactic reaction (flushing, angioneurotic edema, dyspnea, and severe hypertension) after eating cheeses and ***mustard*** with the same preservative. Pruritic vesiculopapular eruptions in a food parcel **inspector** 1–2 days after opening packages of homemade cheeses from southern Italy. The dust on the cheese was swarming with mites. In the 18th century, a text details a poisoning in a **man** in England **who** ate cheese whose color was intensified by adulteration with ***red lead***.

CHEILANTHEA SIEBERI

= *Mulfa Fern* = *Rock Fern*

Untoward effects: Ingestion has caused hematuria and death in **cattle** and **sheep** in Australia.

CHELIDONIUM MAJUS

= *Celandine* = *Celandine Poppy* = *Celidonia* = *Cockfoot* = *Devil's Milk* = *Erbe da fossi* = *Felonwort* = *Jacob's Ladder* = *Killwort* = *Kina* = *Otompy-Kina* = *Swallow Wort* = *Wart Flower* = *Wartweed* = *Wartwort* = *Wretweed*

Untoward effects: ***Morphine***-like central nervous system depressant in poisonings and used in arrow poisons for that purpose. Acrid taste prevents most **animal** and **human** poisonings, yet, **people** have mistakenly eaten the leaves, thinking they were ***parsley***, and there are reports of deaths in Europe from it. Contains various ***poppy*** alkaloids (***chelidonine, sanguinarine, berberine, protopine***, etc.). Its latex irritates and blisters skin, and irritates mucous membranes of the mouth, pharynx, and stomach in **man**. Ingestion of latex or leaves leads to headache, somnolence, vomiting, diarrhea, collapse, circulatory failure, and death. Postmortem reveals severe inflammation of the colon. A 28-year-old **female** ingested ***celadine*** (***Panchelidon***) for

4 months, which is extracted from it causing thickening of gallbladder wall, hepatotoxicity, jaundice, itching, abdominal pain, diarrhea, decolorized feces, and brown-colored urine.

Similar adverse symptoms have been reported in **animals**. Toxic effects reported in **cattle** and **horses** after ingesting ~500 g. In Britain, **cattle** died after ingesting its ripe fruit. Symptoms include staggering, salivation, drowsiness, and convulsions before death.

CHENODEOXYCHOLIC ACID

= *CD* = *CDC* = *CDCA* = *Chendol* = *Chenix* = *Chenocedon* = *Chenocol* = *Chenodex* = *Chenodiol* = *Chenofalk* = *Chenosäure* = *Chenossil* = *Cholanorm* = *Fluibil* = *Hekbilin* = *Kebilis* = *Ulmenide*

Bile acid.

***Untoward effects*:** Liver toxin (3%) production resembles cholestasis. Diarrhea (50%), mild cramping, nausea, decreased appetite in the obese and increased serum bile acid levels, atopic dermatitis-like eruption in a 30-year-old **female** treated for 2 weeks, increased serum transaminases, acute cholecystitis and obstructive jaundice, possibly, from partially dissolved gallstones into the bile ducts, and pruritus reported. Avoid use in pregnancy, severe liver disease, jaundice, pancreatitis, and cholangitis.

Long-term (6 months) therapy with high doses (40–100 mg/kg/day) in **rhesus monkeys** led to bile duct proliferation, portal tract inflammation and fibrosis, bile canalicular bleb formation, and hypertrophy of endoplasmic reticulum noted with increases in SGOT, SGPT, and leucine aminopeptidase. Offspring of **monkeys** treated during pregnancy had congenital hepatic, renal, and adrenal abnormalities.

CHENOPODIUM

Untoward effects*: *C. album or ***Chénopode Blanc*** = ***Fat Hen*** = ***Lamb's Quarter*** = ***White Goosefoot*** ingestion has caused toxic nephrosis (perirenal edema) in **swine**. A ***nitrate*** and soluble ***oxalate*** accumulator. Can cause bradypnea, diarrhea, yellowing of skin and mucous membranes, recumbency, unconsciousness, abortion, and perirenal disease in **cattle**. Fatal to **sheep** eating large quantities, with symptoms resembling hypocalcemia. The dry herbage assayed 1.26% ***nitrates*** and 7.97% ***oxalic acid***. The pollen is highly allergenic to **people**, but less than ***ragweed***.

C. ambrosioides var ***Anthelminticum*** = ***American Wormseed*** = ***Epasote*** = ***Epazote de Comer*** = ***Hipazote*** = ***Jerusalem Oak*** = ***Pasote*** = ***Pazote*** = ***Quenopodium*** yields an unpleasant poisonous volatile oil with a bitter, burning taste = ***Chenopodium Oil*** = ***Wormseed Oil***. Has caused toxic epidermal necrolysis and ototoxicity. In the Caribbean, Mexico, and Latin America, it has been used as a vermifuge, insecticide, abortifacient, and food flavoring or spice. British **soldiers** in 1797 mistook it for ***Datura***, for use in greens, with at least one fatality; **children** in Argentina have been poisoned by the oil. The chief symptoms of poisoning by the oil are gastroenteritis and central nervous system disturbances, if much is absorbed from the gastrointestinal tract. The latter is manifested by muscle spasms, convulsions, paralysis, coma, and, eventually, death. Severe poisoning in **man** after two doses of 0.6 cc, 4 h apart, and death after two treatments with 3 cc within a week. Facial flushing, decreased respiration, extreme dizziness, headache, stupor, disturbed hearing, eye muscle paralysis, and albuminuria occur. About 70% of the severe cases are fatal, preceded by tachycardia, convulsions, and coma. Its ***ascaridol*** content varies between 45 and 75%; this may be a major factor in its varying toxicity.

LD_{LO}, **human**: 50 mg/kg.

Poultry and **geese** have been poisoned by eating the seeds. **Pigs** and **lambs** have been poisoned by 4 ml, and **dogs** by 1 ml.

LD_{50}, **rat**: 255 mg/kg.

CHERRY

= *Cerisier*

Untoward effects*: *Prunus emarginata = ***Bitter*** or ***Wild Cherry***. The twigs, leaves, bark, and the fruit's stones contain a compound that releases ***cyanide*** upon ingestion. Within minutes or an hour, **man** or **beast** may show dyspnea, excitement, vertigo, weak cardiac action, spasms, convulsions, prostration, cyanosis, coma, and death. Ingestion of the fruit's stones can cause a polyneuropathy in **children**.

Consumption of the dry, wilted, or frosted leaves is often fatal to **cows** and **sheep**. Low-level chronic intake may cause **cows** to have chronic diarrhea and discontinue milk production.

Prunus laurocerasus = ***Cherry Laurel***. Contains ***amygdalin*** in its stony seeds, bark, and leaves.

Prunus pensylvanica = ***Pin Cherry***. Its ***amygdalin*** and ***pru***-nasin content has caused **moose** in Newfoundland and Alberta, Canada that browse it to vomit a gray, chalky, pasty substance. The leaves also contain a very high level of ***nitrogen*** (average of 91 mg/100 g).

Prunus serotina = ***Rum Cherry*** = ***Wild Black Cherry***. Its small, white flowers appear in the spring, and their odor causes headaches and, sometimes, nausea in **people**.

Animals often eat the leaves where the trees or shrubs grow along fence rows, but eating from the branches that have been cut or broken, or the leaves that have

been bruised, wilted, frosted, or dried in hay releases ***hydrocyanic acid*** from its cyanogenetic glucoside, ***amygdalin***, is often fatal to them. Symptoms are similar to those described above under ***P. emarginata***. Consumption by **sows** has caused malformations in their **pigs** (arthrogryposis, cleft palate, brachygnathia, skull deformities, and atresia ani).
Prunus sphaerocarpa, a cyanogenic plant that has caused losses of **cattle** and **goats** in Brazil.
Prunus virginiana = ***Chokecherry***. **Sheep**, **cattle**, **horses**, and **deer** have been poisoned by eating the leaves as noted above. **Children** have been poisoned and have died from eating the seeds or pits. **Livestock** have even been poisoned from eating young twigs and bark. 2–4 oz eaten at once may kill a 100 lb **sheep**. Symptoms are the same as described under ***P. emarginata*** above. Nausea, vomiting, abdominal pain, diarrhea, dyspnea, muscle weakness, stupor, and convulsions occur in **man**. Two small **girls** playing had a tea party including crushed ***chokecherry*** seeds. **They** were admitted to a hospital emergency room with ***cyanide*** poisoning. The cherrries themselves are edible. A **dog** that enjoyed jumping up and pulling the leaves off the tree made the mistake of swallowing some, causing its death.

A 65-year-old **male** was allergic to all fruit with large pits, including cherries. Familial relationships in these allergies to cherries include ***almonds, apricots, peaches,*** and ***plums***. Caution should be used in drinking or eating cherry-flavored items, such as colas, as well as maraschino cherries, which have contained ***FDC Red 4 (Ponceau SX)***, ***Red 40***, and ***FDC Red 3***. ***Apricot*** and ***cherry*** pit ingestion are common causes of poisoning in **people**, due to their ***amygdalin*** and, eventually, ***cyanide*** content. Edible sour cherries contain significant amounts of ***oxalates*** and ***salicylates***. The ***oxalates*** may contribute to urinary problems and the ***salicylates*** to cross-reactions in **people** hypersensitive to ***aspirin***. Cherries may contain 50–100 ppm ***caffeic acid***.

Pet **birds** (**budgies** and **canaries**) have been poisoned (fluffed feathers and dark-purple droppings) by the leaves, bark, and twigs.

CHICKEN

Untoward effects: Contact urticaria reported in **man**. A case of hand and finger dermatitis in a **man** employed in killing and cleaning chickens was caused by a hypersensitivity to chicken blood. Chicken livers have caused migraine headaches in **people**. The aromatic vapors produced by grilling chicken can expose **humans** to carcinogenic ***heterocyclic amines***. Cancer of the esophagus in a local area of China may be connected to a similar condition in chickens. Allergic reactions from ingestion reported in **man**, **cats**, and **dogs**. The potential for bacterial contamination is always present.

CHICORY

= *Cichorium intybus* = *Coffee Weed* = *Blue Sailors' Succory*

Untoward effects: Allergic contact dermatitis occurs in **man**.

Young **shoats** became ill with anorexia, dullness, pyrexia, and clonic convulsions after eating the leaves and ***oats*** over a week's time. After eating chicory roots (18 kg/day/5 days), 40, 2–3-year-old **heifers** became ill. They scoured, salivated profusely, and became anorexic; one became ataxic and two died. Post-mortem revealed inflammation of the gastrointestinal tract, endocardial petechiae, and congestion of the liver, lungs, and kidneys. Consumption by **cows** causes an undesirable, bitter taste in their milk.

CHILLIES

Also see *Peppers*. U.S. imports are mostly from India, China, Egypt, and Hungary. About 40,000 metric tons used annually.

Untoward effects: Suspect carcinogen, due to its ***capsaicin*** content. Thousands of **women** in the Budapest area developed "**chilli splitter's** lung" from a toxomycosis due to *Penicillium glaucum* and *Aspergillus* sp. Contamination of chili powder used in making curry powder with ***lead chromate*** caused ***lead*** poisoning in Gurka **soldiers** in Hong Kong.

CHINESE TALLOW TREE

= *Sapium sebiferum*

Untoward effects: Dermatitis and gastroenteritis reported in Alabama from its irritant latex.

Severe diarrhea, weakness, and dehydration in **steers** within 12–14 h after ingestion. One died after 5 days of feeding it and 2/4 had to be euthanized after 10–14 feedings during a 3 week period. A **goat** and two **ewes** were also adversely affected in Texas trials.

CHINIOFON

= *Quiniofon* = *Quinoxyl* = *Sefona* = *Yatren* = *Yochinol*

A mixture of 80% *8-hydroxy-7-iodoquinoline-5-sulfonic acid* and 20% *sodium bicarbonate*.

Antiamebic.

Untoward effects: Slight nausea, vomiting, diarrhea, and pruritus from this now rarely used treatment. Large doses cause liver damage and are contraindicated in **patients** intolerant to ***iodine***.

IV LD_{LO}, **rat**: 500 mg/kg; IM, **rat**: 1 g/kg.

CHIRACANTHIUM PUNCTORIUM

***Untoward effects*:** A venomous **spider** (q.v.) of Europe.

CHIRONEX FLEKERI

***Untoward effects*:** One of the most venomous marine **cnidarians**; its sting has caused death in less than 3 min. The lethal dose is estimated at 0.005 ml/kg.

CHIRONIA PALUSTRIS subsp. TRANSVAALENSIS

= *Wild Gentian*

***Untoward effects*:** Ingestion of stems and leaves caused death of **sheep** and **rabbits**. Post-mortem revealed general cyanosis, hemorrhages of the suprapharyngeal and submaxillary lymph glands, hyperemia of the lungs, and degeneration of the myocardium. Edema of the rumen in **sheep** has been reported regularly. Hypocalcemia developed a few days before death. It is not a common field problem, because it isn't too palatable.

CHIVES

= *Allium schoenoprasum* = *Ciboulette* = *Ezonegi*

***Untoward effects*:** A common allergen causing a protein contact dermatitis and positive delayed eczematous patch-test results in **man**.

A hemolytic anemia and jaundice with degeneration and necrosis in parenchymatous organs described in Japanese **horses** consuming it.

CHLORACETYL CHLORIDE

***Use*:** In organic synthetics.

***Untoward effects*:** Acute toxicity of heated material in an industrial accident may have caused burning sensation of the skin, followed by respiratory difficulty, grand mal seizure, and cardiopulmonary arrest requires cardiopulmonary resuscitation. Transient pancytopenia development on the 5th day of hospitalization. Eye and mucous membrane irritant. Decomposes with water contact.

CHLORACIZINE

= *Chloracysin* = *Chlorocizin* = *Chlorocyzin* = *G-020* = *Khlorasizin*

Coronary vasodilator.

Untoward effects*:** Decreases free ***heparin level and thrombin time, increases prothrombin index and fibrinolysis time.

CHLORAL BETAINE

= *Beta-Chlor* = *MJ 5107* = *Somilan*

Untoward effects*:** Many toxic effects are similar to those listed under ***chloral hydrate (q.v.).

Drug interactions*:** Metabolism increases and the half-life and anticoagulant effect of ***warfarin and ***coumarin anticoagulants*** are decreased.

CHLORAL HYDRATE

= *Chloraldurat* = *Chloratex* = *Escre* = *Lorinal* = *Noctec* = *Nycton* = *Somnos* = *Somnox*

Hypnotic, sedative, anesthetic.

Untoward effects*:** Gastrointestinal irritation, dryness of oral mucosa, unpleasant after-taste, respiratory depression, miosis, eyelid edema, conjunctivitis, transient amaurosis, diplopia, idiosyncratic hyperactivity, sinus bradycardia, and rage reactions reported. Mydriasis may occur with overdosage. Possible relationship to a few cases of cancer in the mouth and prostate. Can be harmful to esophageal mucosa, if trapped there. The vapors irritate the eyes and nose, and strong solutions irritate the skin. **One's** breath has a ***banana-like odor. Pulmonary edema can follow coma and lead to bronchial pneumonia, renal damage, cerebral edema, and, eventually, death. Several weeks of use can cause dependence on it, and was referred to in the literature as "chloral tippling". Leukopenia has been reported with its use. Acute intoxication leading to central nervous system depression, hypotension, hypoventilation, hypothermia, cyanosis, hallucinations, delusions, gastric necrosis, incoordination, ataxia, dizziness, sleepiness, coma, respiratory arrest, and cardiac irregularities, including premature ventricular beats and tachycardia. Hemodialysis has been successful in treatment. Has caused fixed-drug reactions (hypersensitivity with erythematous, bullous, urticarial and/or pruritic plaques, sometimes pustular or hemorrhagic), toxic epidermal necrolysis, flushing of face, neck, chest, and extensor surfaces of the extremities (usually, after 4–6 days of dosing). Rectal suppositories with 1.33 g given to nursing **mothers** led to levels of it and its metabolites in the maternal blood and milk within 15 min, and up to 24 h. Maximum milk levels were 10 mg/100 ml. The minimum amount for **infant** sedation is thought to be 10 or more mg. A 35-year-old **female** died shortly after ingesting about 35 g. The **human** LD is approximately 10 g, although 4 g was fatal in one case, and an **adult** survived a 30 g dose. Tolerance can develop with continued use. Habituation and physical dependence can happen. A **dependent** 45-year-old **female** ingested up to 10 g daily without any distressing symptoms. It crosses the placenta and intrauterine **fetal** death can occur with large doses. Use should be limited or *contraindicated* in hepatic disease, because it is metabolized in the liver. Therapeutic serum levels are reported as 1.5–15 mg/l, toxic at 50 mg/l, and fatal at 100 mg/l.

Drug interactions: Inhibition of liver microsomal enzymes enhances and prolongs sedation when given with ***antidepressants, monoamine oxidase inhibitors,*** and ***tranquilizers***. Its action is potentiated by ***furazolidone***, which acts as an ***monoamine oxidase inhibitor*** when given for 4 or more days. Potentiates ***ethyl alcohol's*** central nervous system depression. Pretreatment with it for 7 days before consuming ***ethyl alcohol*** led to marked vasodilation, flushing, palpitations, and headache, due to production of ***trichloroethanol***, a metabolite. The combination should be avoided in cardiovascular **patients**. This combination of drugs has caused ***disulfiram***-like reaction in some **people**. When added surreptitiously to ***alcohol***, it has been called ***"Knock-out drops"*** or ***"Mickey-Finns"***. **Patients** receiving it orally for 24 h or less prior to IV ***furosemide***, had rapid onset of vasodilation, hot flashes, hypertension, diaphoresis, flushing and/or tachycardia. Chloral hydrate significantly decreases ***phenytoin, oral contraceptive, analgesic, antihistamine,*** and ***anti-inflammatory agent*** levels. Has reduced prothrombin time and ***anticoagulant*** effectiveness, with decrease in ***coumarin*** half-life, thus increasing the risk of thrombus formation. As an acidic drug, it or its metabolite, displaces ***warfarin*** from its ***albumin***-binding sites, thus, increasing the available ***warfarin*** and increasing anticoagulation. Cross-sensitization may occur with ***chlorbutanol*** leading to skin eruptions. May interfere with fluorimetric tests for ***catecholemines*** where it causes increased levels, and it may lead to false positive test results for renal malfunction and ***uric acid*** readings.

Markedly decreases half-life of ***bishydroxycoumarin*** (***dicoumarol***) in **man** and **dogs**, but not in **rats**, so much that sudden withdrawal has caused fatal hemorrhage. If too shallow anesthesia is used, excitement or pain may trigger sudden cardiac arrest. Death from it in **animals** is usually preceded by gasping, muscle spasms, and vocalization.

IV LD_{50}, **mouse**: 500 mg/kg; LD, **horse**: 100–150 g.

CHLORALOSE

= *Alphakil* = $C_8H_{11}Cl_3O_6$ = *Chloralosane* = *Dorcalm* = *Glucochloral* = *α-D-Glucochoralose* = *Somio*

A condensation product of *glucose* and *chloral hydrate*.

Use: As a sedative, hypnotic, anesthetic, rodenticide, and **avian** repellent.

Untoward effects: Ingestion of it in a **rat** poison caused five **people** in France to experience generalized convulsions, which were controlled by IV ***diazepam***. A 62-year-old **male**, in an apparent suicide attempt with it, developed miosis, areflexia, generalized hypotonia, and myoclonic jerking. Deep hypnosis, stupor, hypothermia, and unconsciousness with myoclonic movements are common symptoms of poisoning, and can lead to death. Respiratory failure, loss of reflexes, and electroencephalogram changes are usually suggestive of very large dosage. A **patient** with uremia died from cerebral anoxia and hemorrhagic diathesis.

Has caused coma, convulsions, excitement, and incoordination in various **animals** and salivation in **dogs**; strong respiratory depressant in **cats**, **dogs**, and **rats**.

LD_{50}, **rat**: 300–400 mg/kg; **mice**: 400 mg/kg; **cats** and **dogs**: 400–600 mg/kg; **mallard ducks**: 33.73 mg/kg.

CHLORAMBEN

= *ACPM-629* = *ACPM-728* = *Amchem-65-81B* = *Amchem 66-206* = *Amiben* = *Amoben* = *Vegaben* = *Vegiben* = *Verbigen*

***Pre-emergence herbicide*. Various salts, esters, etc. are used.**

Untoward effects: Harmful if swallowed. Avoid contact with skin or eyes or inhaling mists.

At 10 ppm in water, < 10% mortality in fingerling channel **catfish** in 48 h. Non-toxic to newly-emerged **honey bees** at levels up to 1,000 ppm in a 60% ***sucrose*** syrup.

LD_{50}, **rats**: 3500–5620 mg/kg; LD_{50}, acute dermal, **rabbit**: 3136 mg/kg.

CHLORAMBUCIL

= *Amboclorin* = *CB 1348* = *Chloraminophene* = *Chloroambucil* = *Leukeran*

***Antineoplastic alkylating agent*. A *nitrogen mustard* derivative.**

Untoward effects: Mutagenic, oncogenic, and teratogenic. Low emetogenic potential. Myelosuppression and significant association of leukemia development with its use. One study indicated chromosome breakage frequency of 2–3%. A 65-year-old **male** receiving 2 mg twice daily/ 5 years as prophylaxis for non-Hodgkin's lymphoma developed reduced visual acuity and optic atrophy. Diplopia and retinal hemorrhages reported. Given to **mothers** in the first trimester, spontaneous abortions, congenital anomalies (80%) and low birth weights (40%) can be expected. It is a ***folic acid*** antagonist. Hypersensitivity-type of allergic reactions have been pruritus, skin rash, and even anaphylaxis. A potential exists for mycobacterial infections during therapy. Erectile dysfunction and decreased libido may be from the therapy and/or many other factors. Oligospermia, azoospermia, and aspermia have also followed its use, and the severity appears to be dose-related. Stomatitis and oral ulceration noted, as with other ***antineoplastics***. Its action mimics ionizing radiation, carcinogenic to many organs. The radiogenic tumor with the shortest latent period is leukemia, and leukemia has been reported as a result of treatment of polycythemia vera, Wegener's granulomatosis, non-Hodgkin's lymphoma, and ovarian cancer. Two **patients** treated for 7½–10 years with it for chronic cold hemaglutinin disease developed malignant lymphomas. Papillomas also occur, gastrointestinal upsets are uncommon, except with dosages > 15 mg. Peripheral neuropathy in a 49-year-old **male**, and seizures during or after treatment with it in 7/91 **children** demands careful observation of **patients**. Poisoning occurred in a 2-year-old **female** who ingested a ten times overdose leading to vomiting, central nervous system irritation, and moderate pancytopenia, from which there was complete recovery. Pulmonary fibrosis, pneumonitis, and pyrexia have been reported from its use. Acute renal failure, various degrees of nephropathy, seizures, lethargy, stupor, and coma have developed with it. In treatment of ovarian cancer, mild to severe vomiting, urticaria, fever, edema of the face and extremities, increased risk of developing acute non-lymphocytic leukemia, and even hallucinations have been noted. Nausea, abdominal pain, jaundice, (a low incidence of indirect hepatic toxicity or metabolic idiosyncracy with fatty metamorphosis and necrosis), decreased white blood cells, vertigo, herpes zoster, viral pneumonias, thrombocytopenia, leukopenia, hemolytic anemia, hematuria, cystitis, oliguria, anuria, alopecia, drug fever, pruritus, painful morbilliform exanthema with bullae causing loss of nails on the hands, toxic epidermal necrolysis, amenorrhea, and skin pigmentation are among the many adverse effects attributed to it when used alone. Neutropenia can develop up to 10 days after treatment. May potentiate the action of neuromuscular-blocking agents. **Human** exposure is primarily in **workers** involved in manufacturing, and in **those** administering the finished form, as well as in **patients** receiving the drug. Myoclonus in an 81-year-old **male** and a 75-year-old **female** after treatment for 3 and 5 days, respectively. When **male** rechallenged, symptoms returned.

TD_{LO}, **women**: 13 mg/kg/1–18 weeks pregnant.

When given to pregnant **rats** on the 10th day of gestation (normal pregnancy is 22 days), it caused, as in **humans**, a high incidence of renal defects (including absence of kidneys or ureters). It also causes defects of the nervous and skeletal systems, as well as the soft palate in **rats** and **mice**. Active in producing lymphosarcomas and ovarian tumors in **mice**. Also carcinogenic in **rats**, causing lymphomas.

SC TD_{LO}, **rat**: 32 mg/kg; IP LD_{50}, **rat**: 14 mg/kg.

CHLORAMPHENICOL

= Ak-Chlor = Alficetyn = Amphicol = Anacetin = Aquamycetin = Austracol = Azramycin = Bemachol = C.A.F. = Chemicetina = Chlomycol = Chloramex = Chloramfilin = Chloramsaar = Chlorasol = Chloricol = Chlorocaps = Chlorocid = Chloromycetin = Chloronitrin = Chloroptic = Chronicin = Cidocetine = Ciplamycetin = Cloramfen = Cloramficin = Cloramicol = Clorocyn = Cloromisan = Cylphenicol = Detreomycin = Duphenicol = Embacetin = Enicol = Enteromycetin = Farmicetina = Fenicol = Globenicol = Interomycetine = Intramycetin = Juvamycetin = Kamaver = Kemicetine = Klorita = Leukomycin = Levomycetina = Levomycetin = Loromisin = Mastiphen = Medichol = Micloretin = Micoclorina = Microcetina = Mychel = Mycinol = Novomycetin = NSC 3096 = Opclor = Ophthochlor = Pantovernil = Paraxin = Quemicetina = Ronfenil = Septicol = Sintomicetina = Sintomycin = Sno Phenicol = Stanomycetin = Synthomycetine = Synthomycin = Tega-Cetin = Tevcocin = Tifomycine = Treomicetina = Unimycetin = Vetcytine = Veticol = Viceton = Zoomycetin

Antibiotic.

Untoward effects: Introduced in 1948, and, in 1967, a record of 42,000 kg was certified by the FDA. In the 15 years prior to this, most U.S. physicians had an unbroken series of excellent results with it and hated to part with such a useful tool. By then, the average physician, in his lifetime, had given < 5,000 courses of therapy, and, statistically, 20,000–40,000 courses must be given before a death occurred due to its use. Slowly, the trend in its use reversed. **Human** carriers of the Mediterranean mutant of glucose-6-phosphate dehydrogenase deficiency develop hemolytic anemia from its ingestion. The hemolysis may also be one of a drug–disease synergism. Aplastic anemia reported, due to its use. This also occurred in identical **twins**, indicating that genetic factors may have predisposed to such an anemia. Aplastic anemia has also been associated with its topical ophthalmic use. An inherited hypersensitivity was suggested in one case. Bone marrow depression 50% in **patients** with liver disease and 30% in **those** with renal disease. Agranulocytosis, thrombocytopenia, erythroid hypoplasia, and neutropenia associated with aplastic anemia has been responsible for many deaths. One of the drugs most often associated with pancytopenia. Myeloid and myeloblastic leukemia in a 47-year-old **female** (40 g/3 months) and a 38-year-old **female** (84 g/6 weeks) and **others**; **both** died. One review gives an incidence of about eight aplastic anemia cases/million. An interesting report in December 1988 about its use in Hong Kong (11–442 times that of several western countries and Australia) indicated the death rate from aplastic anemia was only 0.4/1,000 deaths, compared to 1.0/1,000 in England and Wales. The incidence of aplastic anemia in the Near East has been very low, even with high usage of chloramphenicol. Administration to **prematures** and **newborns**, especially with excessive doses, has caused the "Gray **Baby** Syndrome" (cardiovascular collapse, left ventricular dysfunction) and deaths because of immature glucuronide conjugation. **Fetal** circulatory collapse, meteorism, cyanosis, and death have also been attributed to it. Cleft lips and cleft palate in the **fetus** has been attributed to its use in pregnant **women**. Has adverse effects on the **neonate's** cardiovascular system. **Patients** receiving high dosage have shown chromosome damage. May give falsely raised results for urine ***glucose***, serum ***iron*** or ***iron***-binding capacity, while giving falsely lowered results of ***urobilinogen*** tests. Encephalopathy with acute toxic delirium occurs. Neurotoxicity, such as optic neuritis, with marked visual impairments in 19/95 **patients** on long-term therapy. A 4-year-old **female** had permanent loss of vision after receiving 425 g during 69 weeks. Neuroretinitis with long-term treatment led to reduced vision, paralysis of accomadation, and problems with color vision reported, especially in young **children**. Colitis and pseudomembranous colitis, as well as dry mouth, urticarial rashes, nausea, vomiting, diarrhea, stomatitis, rectal and vaginal irritation, moniliasis, headache, clouding of consciousness, mental depression, and fatigue have occasionally been associated with its use. Has been suspect in some deafness cases. Hypersensitivity may have caused circulatory collapse, hemorrhage, shock, dermatitis, prurigo, thrombopenia, purpura, and psychomotor excitement. Bronchospasm and respiratory failure within minutes after injection of 250 mg occurred in a 12-year-old **female**. **She** became unconscious, cyanotic, with arrhythmias, with mydriasis, dyspnea, bradycardia, incipent pulmonary edema, sialorrhea, and convulsions. **Workers** exposed to ***DNCB*** may become sensitized to chloramphenicol. Intestinal moniliasis with *Candida* has followed its use. Jaundice reported in 50-day-old **male who** received 60 mg/kg/day/16 days. Mammary excretion occurs in **women**, **cows**, and **sows**. ***Charcoal*** hemoperfusion has been effective in poisonings by it (viz: in accidental intoxication in a 12-day-old **male** it decreased serum levels from 98 μg/ml to 13.5 μg/ml in 3 h). A misinterpretation of new labeling on vials of ***chloramphenicol sodium succinate*** had five **patients** inadvertently receiving 10 times the intended dose. Plasma levels were 5–20 times normal therapeutic levels. Skin reactions to it include urticaria, diffuse pruritus, morbilliform rashes, exacerbation of eczema, vasculitis, Stevens Johnson syndrome, porphyrias, photo-onycholysis, pyoderma from topical use, erythematous papules and nodules that may ulcerate and become gangrenous, contact eczema and urticaria. Cases of black, hairy tongue reported. Inhibition of ***bilirubin*** destruction, leads to jaundice in **newborns**. Jaundice in the **newborn** nursling occurs, due to the inhibition of the immature hepatic conjugating enzyme system (UDP-glucuronyl transferase) in the liver. Breast feeding is probably contraindicated when the **mother** is taking the drug. It interferes with enzymes in protein synthesis. Erythropoietic depression is increased when jaundice is present. Due to its decreased metabolization in some elderly **patients**, confusion and delirium may be induced. Therapeutic serum concentrations range from 10–20 mg/l and toxic concentrations start at about 25 mg/l.

TD_{LO}, **infant**: 440 mg/kg; **woman**: 400 mg/kg.

Drug interactions: It potentiates ***warfarin*** and other ***coumarins*** by inhibiting its enzyme metabolization; also potentiates ***tolbutamide*** and ***chlorpropamide*** leading to hypoglycemia. Doses of ***phenytoin*** and ***phenobarbital*** may need to be reduced approximately 50 and 60%, respectively, during therapy with chloramphenicol. Simultaneous use causes a 2–3 fold increase in ***nortriptyline*** blood levels, inhibits the metabolization of ***acetanilide, aminopyrine, chlorpropamide, codeine, cyclophosphamide, cyclosporine***, and ***tolbutamide***. Acts, as does ***disulfiram*** in some **patients** drinking ***alcohol***. Used topically as ophthalmic drops by a 83-year-old **female** it caused ~four times increased international normalized ratio after 11 days of treatment.

Risk of synergistic bone marrow suppression and aplastic anemia with ***cimetidine***. ***Rifampin*** may increase chloramphenicol's metabolization. It increases ***demeclocycline*** toxicity. May inhibit the bactericidal action of ***penicillins***.

Administer it cautiously in **animals** with impaired kidney or liver function. Avoid parenteral administration, within at least 2 h before or after ***barbiturate*** anesthesia in **monkeys**, **rabbits**, **rats**, **mice**, **cats**, and **dogs**, as it may prolong surgical anesthesia by at least 100%. Potentiates ***monensin*** toxicity in **turkeys**. Ototoxic in **guinea pigs**. May potentiate epileptic problems in **dogs**. Long-term (26 days) use in **cats** may cause myelosuppression. Has induced memory impairment in **mice**. Recommended doses have not been toxic to **monkeys**, **calves**, or **sheep**, but can be to **dogs** and **foals**. Toxicity in **calves** noted at 50 mg/kg. Acute ***digoxin*** toxicity in **dogs** receiving it.

LD_{50}, **rat**: 3.4 g/kg; **mouse**: 2.64 g/kg; **dog**: 300 mg/kg; IV LD_{50}, **rat**: 171 mg/kg; **rabbit**: 117 mg/kg; **dog**: 150 mg/kg.

CHLORANIL

= *Spurgon* = *Vulklor*

Fungicide.

Untoward effects: Skin and mucous membrane irritant.

LD_{LO}, **human**: 500 mg/kg.

LD_{50}, **rat**: 4 g/kg.

CHLORAZANIL

= *ASA-226* = *Chlorazinil* = *Daquin* = *Diurazine* = *Doclizid-T* = *Neo-Urofort* = *Neurofort* = *Orpidan* = *Orpizin* = *Triazurol*

Diuretic.

***Untoward effects*:** Contraindicated in hepatic coma, hypokalemia, and renal insufficiency with anuria.

Alfalfa meal protects against toxicity in **rats**.

CHLORAZEPATE

= *Tranxene*

***Anticonvulsant.* A long-acting benzodiazepine prodrug for *desmethyldiazepam*.**

***Untoward effects*:** Irritates esophageal mucosa. Transfers into **mother's** milk.

Drug interactions*: *Cimetidine, erythromycin, isoniazid, ketaconazole, propranolol, and ***valproic acid*** increase its anticonvulsant activity.

LD_{50}, **mouse**: 700 mg/kg; IP LD_{50}, **mouse**: 290 mg/kg.

CHLORBENSIDE

= *Chlorocide* = *Chlorparacide* = *Chlorsulphacide* = *Mitox*

Acaricide.

***Untoward effects*:** Skin irritant.

Kidney and liver pathology in experimental **animals**.

LD_{50}, **rat**: 20 g/kg.

CHLORCYCLIZINE

= *Alergicide* = *Chlorocyclizine* = *Compound 47-282* = *Di-Paralene* = *Perazyl* = *Trihistan*

Antihistamine.

***Untoward effects*:** Drowsiness, light-headedness, nausea, and dry mouth. Contact eczematous dermatitis, lichenoid reactions, and central nervous system depression reported with its use. Contraindicated in pregnant **women**, because of possible teratogenic effects.

Drug interactions*:** Antagonizes ***coumarin and ***indandione*** anticoagulant response. Inhibited by ***desoxycorticosterone, estradiol, progesterone,*** and ***oral contraceptives***.

By enzyme induction, enhances the metabolization of ***barbiturates*** in **rats**. Teratogenic (cleft palates, anophthalmia, micropthalmia, cataracts, hydrocephalus, hydronephrosis, undescended testicles, microstomia, micrognathia, and micromelia) in **rats, mice**, and **rabbits**.

IP LD_{50}, **mice**: 137 mg/kg.

CHLORDANE

= *Belt* = *CD-68* = *Chlordan* = *Corodane* = *Endane* = *Niran* = *Octachlor* = *Octaklor* = *Ortho-Klor* = *Synklor* = *Toxichlor* = *Velsicol 1068*

***Insecticide.* A *chlorinated hydrocarbon*, readily absorbed through cutaneous, respiratory, and gastrointestinal surfaces.**

Untoward effects*: Exterminators** using it had reduced ***antipyrine half-lives (7.7 ± 2.6 h), compared to office **personnel** in the same company (13.1 ± 7.5 h). It is strongly lipophilic and stored in fatty tissues. Long-term **human** occupational exposures lead to skin rash, headache, fatigue, dizziness, diarrhea, cough, dyspnea, sore throat, nausea, anorexia, diaphoresis, vertigo, lacrimation, dermatitis of hands with bleeding, and degenerative liver pathology. Acute poisonings in **humans** may, initially, cause central nervous system stimulation, followed by dizziness, nausea, vomiting, malaise, fever, diarrhea, headache, apprehension, aplastic anemia, blurred vision, watery eyes, sore throat, convulsions, delirium, and encephalopathy with paresthesias, central nervous system excitation, tremors, seizures, and respiratory failure. One swallow of a 70% solution can be lethal. Thrombocytopenia, leukopenia, neutropenia, leukocytosis, and erythroid hypoplasia also reported. A 35-year-old **male** was successfully treated with ***charcoal*** hemoperfusion 4 h after ingesting about 90 g. High concentrations can be found in **human** mother's milk. The fatal oral dose for **man** is estimated to be 6–60 g, with symptoms starting in about 45 min, but a **patient** died after only 100 mg. A 54-year-old **male** attempted suicide with rectally administered chlordane (57–205 mg) leading to tachycardia, fasciculations, and tremors, followed by large amounts of liquid stool. **He** was discharged after 15 days of hospitalization. An anecdotal report of **maternal** exposure to it and subsequent development of neuroblastoma 2 years later is in the literature. Persists in the environment for at least 20–30 years, sticking strongly to soil particles, and not likely to enter ground water. Therefore, **human** exposure can easily come from touching soil near homes treated with it for **termites**. Eating **fish**, **birds**, or **mammals** storing it in their tissues, or eating root crops grown on treated soil are other current sources of **human** exposure.

LD_{LO}, **human**: 40 mg/kg.

Drug interactions*:** It increases the catabolism of steroids, such as ***estradiol and ***testosterone***. ***Adrenergic amines*** or ***atropine*** are contraindicated with it, because they increase myocardial irritability. It decreases the effectiveness of oral ***anticoagulants*** and ***phenylbutazone***.

Liver microsomal enzyme inducer in **rats**, increases metabolization and decreases effectiveness of ***androsterone, cyclophosphamide, DOCA, estradiol, estrone, phenylbutazone, and progesterone***. Similar interacting effects [illegible] with ***bishydroxycoumarin*** in **dogs**; ***hexobarbital*** and ***zoxazolamine*** in **rabbits**: ***aminopyrine, antipyrine, hexobarbital,*** and ***zoxazolamine*** in **monkeys**. Given to pregnant

rabbit dose 4 days before term increases liver microsomal activity in their newborn, as well as in the adults. Lung, liver, and renal damage described in poisoned **animals**. Lethal doses in **animals** range from 200–300 mg/kg. **Sheep** are very sensitive (70 mg/kg) to it, as are young **calves** (55 mg/kg).

LD_{50}, **rats**: 283–590 mg/kg; **mouse**: 430 mg/kg; **rabbit**: 100 mg/kg; **hamster**: 1.7 g/kg; **chicken**: 220 mg/kg; topical LD_{50}, **rat**: 700 mg/kg; **rabbit**: 780 mg/kg.

CHLORDANTOIN

= *Sporostacin*

Fungicide.

Untoward effects: Allergic vulvar contact dermatitis from vaginal cream.

LD_{50}, **rats**: > 1.1 g/kg; **monkeys**: > 400 mg/kg.

CHLORDIAZEPOXIDE

= *Ansiacal* = *Bent* = *Benzodiapin* = *Calmoden* = *Cebrum* = *Chlordiazachel* = *Clopoxide* = *Corax* = *Decacil* = *Disarim* = *Droxol* = *Eden-psich* = *Elenium* = *Equibral* = *Helogaphen* = *J-Liberty* = *Kalmocaps* = *Labican* = *Lentotran* = *Libritabs* = *Librium* = *Menrium* = *Mesural* = *Mildmen* = *Multum* = *Napoton* = *Novosed* = *O.C.M.* = *Psichial* = *Psicosan* = *Psicoterina* = *Reliberan* = *Risolid* = *Seren Vita* = *Silibrin* = *SK-Lygen* = *Sonimen* = *Sophiamin* = *Tensinyl* = *Timosin* = *Trakipeal* = *Tranimul* = *Tropium* = *Viansin* = *Viopsicol* = *Zeisin* = *Zetran*

Anxiolytic, tranquilizer.

Untoward effects: A basic chemical, unstable in solution and when exposed to ultraviolet light. Parenteral injections must be prepared immediately prior to use. Habituation (44-year-old **female** ingesting average of 140 mg/day/2 months led to weight loss, inability to concentrate, memory lapse, euphoria, fine tremor of hands, and mydriasis – depression upon withdrawal) or addiction has occurred. Avoid ataxia and over-sedation in its use. Passes into **mother's** milk. Has caused lactation suppression in a 41-year-old **female**. Hiccups have followed a single oral dose of 10 mg in 19-year-old **male**. Sudden withdrawal has caused convulsions and delirium in **some**. Metabolied in the liver. Cholestatic jaundice, hepatomegaly, and increased SGOT occur. Renal clearance decreases with age and **patients** > 70 years should only receive a maximum of 50% of the usual dose; **those** with cirrhosis, only 33%. 100% of the drug passes the placental barrier and congenital anomalies, particularly cardiovascular, are found in **children**, especially if **maternal** ingestion is during the first 42 days of pregnancy. Some clinicians found increased **fetal** death, mental deficiency, spastic diplegia, deafness, cleft palate, cleft lip, microcephaly with retardation, duodenal atresia, and Meckel's diverticulum with its use. Drowsiness, dry mouth, constipation, blurred vision, dizziness, diaphoresis, bloating, slurred speech, vertigo, euphoria, paresthesias, thrombocytopenia, neutropenia, pancytopenia, leukocytosis, and hemolytic and aplastic anemia noted. Occasionally, vivid dreams, impotence and failure to ejaculate, tremor, dyspnea, confusion, and nasal congestion reported. Paradoxically, it can increase aggressive behavior and induce rage reactions. Fixed-drug reactions (erythematous, bulbous, urticarial and/or pruritic plaques) and porphyria occur, and often leave hyperpigmented areas. May cause hypoplastic and hemolytic anemias, leukopenia, neutropenia, agranulocytosis, and photosensitivity reactions. Frequently tried in suicide attempts. Drug overdosage (425 mg) in 11-year-old **male** caused coma, weak pulse, decreased blood pressure, fever, and miosis. Recovered in 3 days. 22 other cases described in the same report. Respiratory failure and coma are common in poisonings. It is dialyzable. **Suicides** usually ingest other drugs as well. Usual oral dose is 10–100 mg, and **people** have survived single oral doses > 2,000 mg. Strong potential for psychological dependence after about 60 days. Blood levels > 20 μg/ml were associated with grade 2 or 3 central nervous system depression. Therapeutic levels are 0.5–2.0 μg/ml or 0.5–2 mg/l, and toxic levels are 3.5–10 mg/l.

Estimated **human** IV LD is 24 mg/kg with blood concentration of 26 mg/l. TD_{LO}, **women**: 4 mg/kg.

Drug interactions: Additive sedation and enhanced ***atropine***-like effects when given with ***tricyclic antidepressants***. ***Disulfiram*** may alter its plasma clearance, leading to drowsiness and excess sedation. Effectiveness decreased with smoking. ***Cimetidine***, by decreasing its metabolism, impairs the elimination of it and its metabolite. ***Heparin***, within 90 s, causes a 150–250% increase in free-circulating fraction, which decreases to baseline in 30–45 min. May induce liver enzymes to cause failure of ***oral contraceptives***. Also can interfere with thyroid function tests. Absorption rate decreases with ***aluminum hydroxide–magnesium hydroxide antacid*** combination. Decreases ***warfarin's*** effectiveness, but decreases ***phenytoin's*** metabolization, increasing its potential toxicity.

LD_{50}, **mice**: 820 mg/kg; **rat**: 548 mg/kg; **rabbit**: 590 mg/kg.

CHLORDIMEFORM

= *C 8514* = *CDM* = *Chlorphenamidine* = *Ciba 8514* = *Fundal* = *Galecron* = *Schering 36,268* = *Spanon*

***Insecticide, acaricide.* A formamidine.**

Untoward effects: Hematuria in 9/22 **workers** in a separate area for packaging in a chemical plant. **Those** severely ill had abdominal pain, dysuria, and urgency to void urine. Four additional **workers** who had packaged the chemical the previous year had similar histories. The illness lasted 1 week to 2 months, with complete recovery. Ventilation was poor and **they** wore no protective garb. Hemorrhagic cystitis developed in a **woman** who had romantic contact with one of the **men** and rode home with them in a car. SC injury produced similar, but less severe, symptoms in 2/3 **cats**. The chemical's major metabolite is ***2-methyl-4-chloroaniline(4-chloro-orthotoluidine)***, well-known for its toxicity. Six physicians saw one or more of these **workers** without suspecting a work-related toxicant.

Long-term feeding of high doses to some strains of **mice** developed malignant tumors. This was about 3,000 times the exposure to **workers** in a Louisiana manufacturing plant.

LC_{50}, **quail**: 5,079 ppm; LD_{50}, **rats**: 170 mg/kg; **mouse**: 160 mg/kg; **rabbit**: 625 mg/kg; IP LD_{50}, **rats**: 238 mg/kg; topical LD_{50}, **rat**: 4 g/kg; **rabbit**: 640 mg/kg.

CHLORETHYLBENZENE

Untoward effects: Causes chloracne. Formed during the manufacture of ***chlorinated diphenyl*** from ***benzene***. On contact with the skin, ***hydrochloric acid*** is liberated.

CHLORFENVINPHOS

= *Birlane* = *Compound 4072* = *CVP* = *Dermaton* = *ENT 24,969* = *Gen. Chem 4072* = *Sapecron* = *SD 7859* = *Shell Compound 4072* = *Steladone* = *Supona*

Organo-phosphorus insecticide, acaricide.

Untoward effects: Each of two **children** (1½ and 5 years) received a teaspoon of a 24.5% solution, instead of a cough medicine. Vomiting and muscle weakness occurred about 4 h later; muscarinic symptoms with coma developed in another 2–6 h. Slow onset of symptoms delayed appropriate therapy. A 43-year-old **female** dog **groomer** sponged 8–12 dogs/day/3 years on their flea-infested areas. For years, **she** had periodic dizziness, fatigue, black-outs, blurred vision, chest pain, sweating, chills, and, during these episodes, extreme miosis and decreased blood cell cholinesterase levels. This was a typical case among many with this compound or ***phosmet***. With therapy, **they** eventually recovered. One must read labels!!

Topical LD_{LO}, **human**: 10 mg/kg.

Use of levels above recommendations in **cattle** caused expected symptoms of salivation, dyspnea, muscular weakness, and diarrhea. ***Atropine*** reversed the toxicity. No toxicity noted in **dogs** at 12 g/kg.

LD_{50}, **rats**: 12 mg/kg; **mice**: 100–200 mg/kg; **chicken**: 29 mg/kg; **mallard ducks**: 85.5 mg/kg; **bobwhite quail**: 80–160 mg/kg; **pheasants**: 63.5 mg/kg; dermal LD_{50}, **rabbit**: 108 mg/kg.

CHLORHEXIDINE

= *10040* = *Bacticlens* = *Chlorasept 2000* = *Corsodyl* = *Hibiclens* = *Hibidil* = *Hibiscrub* = *Hibitane* = *Klorhexidin* = *Lisium* = *Nolvasan* = *Orahexal* = *Peridex* = *pHiso-Med* = *Plac Out* = *Plurexid* = *Rotersept* = *Sterilon* = *Unisept*

Antiseptic.

Untoward effects: Instillation into the bladder of a 0.02% solution caused hematuria. Two **men** had penile and scrotal contact dermatitis after sexual intercourse with **women** who had used it in a feminine hygiene spray. Hospital solutions have been infected with *Pseudomonas maltophilia*. Urticaria, dyspnea, and anaphylactic shock in six Japanese **patients** (9–48 years), due to the topical application of the gluconate form. Severe anaphylaxis in 49-year-old **male** during whole body bath with 0.05% chlorhexidine gluconate solution. Accidental application of a 4% gluconate solution in four **females** (9 months to 83 years) during surgical preparation caused irreversible corneal injuries and opacification. Chlorhexidine can damage plastic lenses. Can alter **one's** taste, and, topically, it can be ototoxic (total irreversible and permanent deafness). Inactivated by ***tannic acid*** and ***tannins***, which can leach out of cork-stoppered containers. A case of attempted suicide by swallowing the equivalent of 30 g of the pure chemical (400 mg/kg or 150 ml ***Hibitane*** solution) caused pharyngeal edema, necrotic lesions in the esophagus, and increased aminotransferase levels (30 times normal 5 days later), indicating liver necrosis. A 89-year-old **female** drank 30 ml of ***Hibiclens*** solution (chlorhexidine gluconate). The only symptoms observed were giddiness and unusual laughter.

Mild irritation of **rabbit's** eyes that were fitted with plastic lenses (0.005% of the acetate for overnight storage), and after 3 weeks, some corneal opacities developed.

LD_{50}, **rat**: 1.8 g/kg; IV LD_{50}, **rat**: 22 mg/kg.

CHLORINATED DIPHENYL

= *Chlorodiphenyl*

Use: In synthetic waxes.

Untoward effects: An acne-like contact dermatitis after 2 years exposure in some **individuals** with a predilection for the face and extensor surfaces of the body. A form of chloracne, as it is manufactured from ***benzene***. Liver damage noted.

CHLORINATED LIME

= *Bleaching Powder* = *Carrel-Dakin's Solution* = *Dakin's Solution*

Disinfectant, topical antiseptic.

Untoward effects: Strong solutions or paste are caustic skin irritants. Many eczemas in European dye factories were traced to use in hand-washing. Inhalation can cause throat and pulmonary irritation with pulmonary edema and even death. Ingestion has caused severe mouth, esophageal, and gastric irritation.

Animals have been poisoned by its use in error in limewash.

CHLORINATED NAPHTHALENES

= *Halowax*

Use: In synthetic waxes in electrical insulating materials, in wood preservation, lubricating oils, and varnishes.

Untoward effects: Causes a contact dermatitis called "***Halowax*** acne" or "cable rash" or "**electrician's** rash". A few exposed **workers** develop yellow atrophy of the liver.

After dermal contact, it causes **bovine** hyperkeratosis ("X-disease"), and can be excreted in their milk. Listlessness, depression, lacrimation, salivation, watery nasal discharge, intermittent diarrhea, swollen red areas in the mouth and on the muzzle, polyuria, weight loss and abortions are common in chronic exposures. The skin of the scrotum, udder, and anterior $\frac{1}{3}$ (neck and withers) of the **animal** thickens, becomes dry, wrinkled, and no longer elastic. Toxic doses for **calves** are 11–22 mg/kg, with symptoms often occurring by the third day and death within 8 weeks. Postmortem reveals metaplasia of the salivary glands.

CHLORINE

Green-yellow gas. Disagreeable, suffocating odor may be noted at about 0.2–0.4 ppm in some people, but at 3.5 ppm in most others. Supplied under pressure as a liquid for commercial use.

Use: The gas is used as a biocide in water disinfection (5%), in plastic manufacturing (60%), in pulp paper bleaching (15%), and in chemical manufacturing. Approximately 25 million lbs used annually in the U.S. Density is greater than air, thus, leaks keep the gas near the ground.

Untoward effects: Many are from secondary compounds (viz. chlorine bleaching of wood pulp generates approximately 250 different chlorinated chemical compounds. Extreme respiratory and eye irritant. Can cause glottal spasms, pulmonary edema, cyanosis, and death. Has caused skin burns. Throat irritation occurs at 15 ppm; coughing at 30 ppm; pneumonitis and pulmonary edema at 40–60 ppm; and a few deep breaths at 100 ppm can be fatal. Inhalation of 430 ppm/30 min is usually considered fatal, and 1,000 ppm is rapidly fatal. Inhalation of low levels can trigger asthma attacks, substernal pain, mouth irritation, and syncope. Reacts with substances in drinking water to form ***trihalomethanes***, a cause of cancer in laboratory **animals**, and estimated to increase the risk of gastrointestinal cancer 50–100% over a **person's** lifetime. Free chlorine concentration of 0.4–0.6 ppm provides adequate swimming pool disinfection, and irritation of the eyes is very noticeable at 0.75 ppm. The gas was used by the **French** against the **Germans**, and the **Germans** against the **British** and the **French** in the early days of World War I. This attack in Belgium killed about 5,000 **persons** and claimed some 20,000 Allied **casualties**. Accidental exposure of 150 **longshoremen** to the gas decreased respiratory function for 2–3 years. Large scale gas release and evacuations of **people** have occurred in various industrial areas. Death from severe hemorrhagic pulmonary edema in 2/22 within 24–72 h, after exposure from a leaking storage tank. Some **workers** still had reduced pulmonary air flow a year later. Headaches are not unusual. A **worker** exposed to the gas developed chloracne. Slight increases in the chlorine level in swimming pool water, plus a decrease in water pH, apparently, caused corneal epithelial denudation, erosions, and edema. Most of these **subjects** were in the pool about 30 min, and symptoms disappeared within 24 h. Halos or rainbows around light sources continued for about 30 min in 34 (68%) of the **swimmers**. A follicular-type of dermatitis and urticaria reported from being in a swimming pool. Erosion of dental enamel was reported in competitive **swimmers** where swimming pools used chlorine gas instead of hypochlorites, and the water pH was below 6. An 8 h TWA is 1 ppm and 3 ppm/15 min. Do **not** mix hypochlorite solutions with ***vinegar***, as it releases dangerous chlorine gas. Do **not** mix chlorine bleaches with toilet bowl cleaners, strong acids, or ***ammonia*** for the same reason. Forms explosive mixtures with ***turpentine, ammonia,*** and ***ether***.

Accidental gas release in Alberta, Canada over a 10 sq mile area exposed **cattle**, **pigs**, **goats**, **horses**, **rabbits**, and **poultry**. Vegetation in the area was damaged and two **cattle** died with pulmonary lesions. The escape of 15 tons of chlorine from a chemical works that drifted downwind covered 10 sq miles. **Cattle**, the most severely affected, showed dyspnea, lacrimation, profuse nasal discharge, and depression; some died within a day or 2. Salivation, lacrimation, coughing, vomiting, and anorexia occurred in **pigs**. Frequent urination and dyspnea were characteristic signs in **horses**. Excess chlorine in the water can be very toxic to **fish** and **frogs**. It damages the gills of **fish** and causes spasms. The lethal concentration for **fish** is 0.8 mg/l. Repeated exposures of **rats** to 1, 3, and 9 ppm/6 h a day 5 days/week/6 weeks caused extensive respiratory tract inflammation and hyperplasia. Hepatic and renal pathology noted at 3 and 9 ppm in this trial. **Dog** trials suggest volume depletion and increased blood viscosity, rather than just hypoxia, as the cause of death.

LC_{50}, **rats**: 293 ppm.

CHLORINE DIOXIDE

= *Anthium Dioxide* = *Chlorine Peroxide* = *Oxine*

***Use*:** To remove obnoxious tastes and odors from plant effluents and in smelly lagoons; to inhibit bacterial growth. A bleaching agent for paper pulp, flour, and textiles.

Untoward effects*:** Does not form ***triethanolamines with drinking water, as does ***chlorine***. Odor may not be noticed before 17 ppm, but 5 ppm is irritating to mucous membranes and eyes. May cause pulmonary edema, coughing, wheezing, and chronic bronchitis. Heating it will cause release of toxic chlorine gas. Shock can cause the product to explode. Powerful oxidizing agent, and can react violently with reducing agents and organic material.

Inhalation LC_{LO}, **rat**: 500 ppm/15 min.

CHLORIS sp.

Untoward effects*:** ***C. trumcata = ***Windmill Grass*** of Australia is a useful fodder plant, but suspect as a cause of "yellow bighead" and "bighead jaundice" in **sheep**, due to ***phyloerythin***, probably generated in the rumen, and causing photosensitization. Some species contain dangerous levels of ***hydrocyanic acid***.

CHLORISONDAMINE

= *Ecolid* = *Ecolid Chloride* = *SU 3088*

***Antihypertensive*. Ganglionic-blocking agent.**

***Untoward effects*:** 20/28 **patients** experienced constipation, nausea, blurred vision, dry mouth, dizziness, weakness, and syncope. Orthostatic hypotension and syncope are decreased by taking the medicine after eating.

LD_{50}, **mouse**: 480 mg/kg; IV LD_{50}, **rat** and **mouse**: 28 mg/kg.

CHLORMADINONE

= *CAP* = *Cero* = *Clogestone* = *Delmadinone* = *Gestfortin* = *Lormin* = *Luteran* = *Lutoral* = *Matrol* = *Normenon* = *Menstridyl* = *Prostal* = *Skedule* = *Traslan* = *Verton*

Progestogen, antineoplastic, estrus regulator.

***Untoward effects*:** Long-term oral use has caused acceleration of platelet aggregation. Accelerated lung metastases of breast carcinomas; break-through bleeding when used in **female**/day for contraception, venous thrombosis, weight gain, decreases protein in **mother's** milk, and a relatively high pregnancy rate made its use undesireable.

High oral dosage in beagle **dogs** (0.6 mg/kg/day/1 year) has caused renal abnormalities similar to those in **human** diabetes. Non-carcinomatous breast nodules reported in female **dogs** on long-term toxicity study. Similar nodules have been reported in **women**. May lead to slight clitoral and ovarian enlargement; occasionally, mammary hypertrophy and false pregnancy; and, near the expiration of its effects, depression and some sexual behavior changes in **bitches**.

TD_{LO}, **dog**: 18 mg/kg, 2 year study.

CHLORMAPHAZINE

= *CB 1048* = *Chloronaftina* = *Erysan* = *R48*

***Antineoplastic*. A *nitrogen mustard* compound.**

***Untoward effects*:** Carcinogen in **man**, where it has caused bladder cancer in 10/61 **patients** treated for polycythemia. Once used in Europe.

IP injection has caused lung tumors in **mice** and local sarcomas after SC use in **rats**.

CHLORMEQUAT CHLORIDE

= *CCC* = *Chlorocholine* = *Cycocel* = *TUR*

***Plant growth regulator*. Acts as a retardant, permitting early flower development, compact plants, and increased yield.**

***Untoward effects*:** Neuromuscular-blocking effect order in the **cat** > **man** > **rabbit** > **monkey** > **mouse** > **rat**. In **ducks**, miosis, falling, walking on wing tips, tremors, tetanic seizures, opisthotonus within 5 min (remission in some in up to 3 days), and deaths within about 15–40 min. Excreted in the urine and milk of **cows**. 200 mg/kg lethal to **sheep**.

LD_{50}, 670–750 mg/kg; **mice**: 540 mg/kg; **mallard ducks**: 265 mg/kg; **guinea pigs**: 615 mg/kg; dermal LD_{50}, **rabbit**: 440 mg/kg; IM LD_{50}, **hens**: 30 mg/kg.

CHLORMERODRIN

= *Mercuroxyl* = *Neohydrin*

***Oral diuretic*. Contains 54.6% *mercury*.**

***Untoward effects*:** Nausea, headache, pruritus, dermatitis, stomatitis, and abdominal pain in some **patients**. At high dosage, some **patients** report gastric irritation and diarrhea. Nephrotoxic.

Radioactive chlormerodrin is not concentrated in **dog's** normal myocardium, but concentrates in infarcts.

LD_{LO}, **rats**: 82 mg/kg; LD_{50}, **mice**: 560 mg/kg.

CHLORMEZANONE

= *Alinam* = *Banabin-Sintyal* = *Chlormethazone* = *Fenarol* = *Lobak* = *Mio-Sed* = *Rexan* = *Rilansyl* = *Rilaquil* =

Rilassol = Supotran = Suprotan = Tanafol = Trancopal = Trancote = Transanate

Anxiolytic, skeletal muscle relaxant.

Untoward effects*:** Mainly, drowsiness, with transient nausea, colic, dizziness, headache, difficulty in urination, skin rash, flushing of skin, dry mouth, jaundice, weakness, and increased anxiety less common. Decreases taste acuity. A Korean 26-year-old **male** had a fixed-drug reaction, confirmed by oral provocation and patch tests. Thrombocytopenia in a 36-year-old **patient**, after 300 mg/day/10 days. Hepatitis in a 70-year-old **female**, based on positive lymphocyte stimulation test. Cholestatic hepatitis has also been reported. Alternating coma and excitement with hypertonicity after 32-year-old **patient** overdosed with 11 g orally. A 36-year-old **female** overdosed by ingesting 7 g and became comatose and hypotensive. Consciousness returned after 15 h. Acute renal failure in a 28-year-old **male** with glucose-6-phosphate dehydrogenase deficiency after ingesting 8 g with 40 g ***acetominophen. Recovery over a 7-week period with peritoneal dialysis. Overdosing causes hypotension and coma. Therapeutic serum levels are generally 5–9 mg/l, toxic at ~20 mg/l, and fatal at > 50 mg/l. Hepatitis and serious toxic epidermal necrolysis reactions led to its withdrawal from the marketplace.

LD_{LO}, **human**: 500 mg/kg.

LD_{50}, **rat**: 840 mg/kg; **mouse**: 1380 mg/kg; **dog**: 500 mg/kg.

CHLOROACETALDEHYDE

= *2-Chloro-1-ethanal = Monochloroacetaldehyde*

Untoward effects*:** Extremely irritating and corrosive to skin, mucous membranes, eyes, and respiratory tract. Has caused pulmonary edema. Acrid odor. An oxidation product of ***2-chlorethanol. May have causative role in neurotoxicity. TLV for **man** is 1 ppm.

Inhalation by **rats** (LC_{50} ~0.65 g/m^3) leads to pulmonary edema, usually, with hydrothorax, and, occasionally, pulmonary atelectasis and a cardiac thrombus.

LD_{50}, **rat**: 23 mg/kg; **mouse**: 21 mg/kg; IP LD_{50}, **rats** and **mice**: 2 mg/kg; **rabbit**: 1.4 mg/kg; **guinea pig**: 636 µg/kg.

CHLOROACETAMIDE

***Use*:** As a preservative in cosmetics, pharmaceuticals, paints, glues, coolant oils, as a finish for nylon thread, in the leather and tannery industry, and on woods.

***Untoward effects*:** Strong skin sensitizer. A cause of a 6 month recurrent pruritic facial allergic contact dermatitis from a 0.2% solution in a topical spray astringent. Causes hepatotoxicity in **rats**.

CHLOROACETIC ACID

= *Chloroethanoic Acid = Monochloroacetic Acid = MCA*

***Herbicide.* Approximately 125 million lbs used annually in the U.S.**

Use*:** Manufacturing of dyes and organic chemicals; food and beverage stabilizer; its carboxymethyl cellulose is used in drilling muds, detergents, and pharmaceuticals, and its ***thioglycollic acid is used in ***polyvinyl chloride*** stabilizers.

***Untoward effects*:** Its local caustic effects have been used to destroy warts and other cutaneous growths.

LD_{LO}, **human**: 50 mg/kg.

LD_{50}, **rats**: 76 mg/kg; **mice**: 165–255 mg/kg; **guinea pigs**: 80 mg/kg.

CHLOROACETONE

Lacrimator. Pungent odor.

***Untoward effects*:** LC_{LO}, inhalation **human**: 605 ppm/10 min.

After 1 h inhalation exposure, **rats** became restless, sat in hump-back position, rubbed their snouts, and salivated.

14 day LD_{50}, **rats**: 100 mg/kg; **mice**: 127 mg/kg; 1 h inhalation LC_{50}, **rats**: 262 ppm.

CHLOROACETOPHENONE

= *Chemical Mace = CN = Mace*

***Lacrimator.* Pungent odor.**

***Use*:** As a tear gas component at about a 0.9% concentration as a spray for police or military use in grenades.

***Untoward effects*:** Potent eye, throat, nose, and skin irritant. May cause corneal damage, especially, if it is exposed directly to **CN** in liquid form. Speed or direction of spray important factors. Total permanent blindness from corneal damage, as well as conjunctival ischemia and chemosis has occurred. Residual effects may expose **nurses** and **physicians** from the **victim's** clothing. Pulmonary edema, allergic contact dermatitis, and sensitivity reported. A **patient** suffered acute inflammation and swelling of **her** face, after viewing tear gas in a closed glass container. Lacrimation occurs at 0.3–0.5 mg/m^3, nasal irritation at 1 mg/m^3, vision impossible at 2 mg/m^3, and **man** cannot tolerate 4–5 mg/m^3 for more than 1 min. Exploding tear gas pens can cause permanent nerve damage in **one's** hands.

TC_{LO}, **human**: 119 mg/m^3/2 min; LD_{LO}, 159 mg/kg/20 min; TLV, **human**: 0.05 ppm.

Ocular injury with conjunctival inflammation, corneal edema, and epithelial loss in experimental **rabbits**.

LD_{50}, **rat**: 52 mg/kg; **mice**: 139 mg/kg; inhalation LD_{50}, **rat**: 14 mg/kg; **mice**: 59 mg/kg; **guinea pigs**: 1.1 mg/kg.

N-(3-CHLOROALLYL)HEXAMINIUM CHLORIDE

= *Dowicide Q* = *Dowicil 75* = *Dowicil 200* = *Quaternium 15*

Bactericide, preservative.

Untoward effects: Decomposes rapidly near 100°C leading to heat and toxic flammable vapors. Not a primary irritant or skin sensitizer up to 1.0% of ***Dowicil 75***. A common allergen in cosmetics (moisturizing lotions and shampoos).

Has caused transient eye irritation, lasting about 3 days, in **animals**. Prolonged or repeated exposure can cause skin irritation.

LD_{50}, skin absorption, **rabbits**: 2877 mg/kg dry powder, and 600 mg/kg for strong solutions. Repeated oral doses in **rats** of 15–60 mg/day has caused birth defects.

LD_{50}, **rat**: 1.2 g/kg; **mallard duck**: > 2.5 g/kg.

σ-CHLOROANILINE

= *σ-Aminochlorobenzene*

Dye intermediate.

Untoward effects: Vapors or skin contact lead to cyanosis and anemia.

CHLOROBENZENE

= *Benzene Chloride* = *Monochlorobenzene* = *Phenyl Chloride*

Use: In manufacturing ***phenol*** (now, ***cumine*** is preferred), ***nitrochlorobenzene***, and ***aniline***; as a solvent. Demand has decreased to about 250 million lbs annually.

Untoward effects: Increases ***taurine*** excretion. Neurotoxic. Has caused lens opacities and cataracts, skin irritation, drowsiness, incoordination, headaches, syncope, methemoglobinemia, and anuria. Pallor, cyanosis, fibrillary twitching, and collapse after long exposure. Case reports of acne and comedone formation in **workers** exposed for 8 weeks to 3 years. TLV: 75 ppm.

The 2-chloro form (***2-CNB***) has caused acute toxicity in **man**, manifested as anemia and cyanosis. Occupational exposure has become low, despite use of 20 million lbs/year in the U.S. Suggested TLV: 1 mg/m^3.

Causes necrosis of pulmonary epithelium and of proximal convoluted kidney tubules in **mice**. Liver toxicity noted. Has produced adrenal and pituitary tumors, thyroid adenocarcinoma, lymphosarcoma, cholangiosarcoma of the liver, and SC fibroma in **rats**; hepatocellular carcinomas in **mice**.

LD_{50}, **rat**: 2.9 g/kg; **rabbit**: 2.8 g/kg. Others state LD_{50}, **rats** and **mice**: 110–400 mg/kg.

Over half of the 100 million lbs produced annually in the U.S. of the 4-chloro form was used to make ***4-nitrophenol***, used in the manufacture of ***parathion***. The 4-chloro form causes anemia with decreased erythrocyte counts, hemoglobin levels, and hematocrit values, along with dizziness, cyanosis, headache, and nausea. TLV of 1 mg/m^3. Vascular type tumors in **mice**.

LD_{50}, **rats**: 400–850 mg/kg.

CHLOROBENZILATE

= *Acaraben* = *Akar* = *Chlorbenzilat* = *Compound 338* = *Folbex* = *G 23922*

***Acaricide.* Chlorinated hydrocarbon.**

Untoward effects: In 56 **human** volunteers, a 1% solution in patch tests did not produce primary irritation or sensitization.

LD_{LO}, **human**: 500 mg/kg.

Low toxicity to **honeybees**. Liver and renal damage in **animals**. **Mice** given 2–4 g/kg had depression, labored breathing, ataxia, hypnosis, tremors, and lacrimation. Poisoning in **rats** led to depression, salivation, lacrimation, diarrhea, and deep and rapid respiration. Hemorrhagic lungs on post-mortem.

LD_{50}, **rats**: 700–3200 mg/kg; **mice**: 12.8 g/kg; acute dermal LD_{50}, **rabbit**: > 10 g/kg.

2-(5-CHLORO-2H-BENZOTRIAZOL-2-YL)-4,6-BIS(1,1-DIMETHYLETHYL)-PHENOL

= *Tinuvin 327*

***Ultraviolet light-absorber.* For stabilizing plastics and polymer coatings.**

Untoward effects: In **rat** trials, a no adverse toxic effect level was < 50 ppm in the ration. 100 or 200 ppm/3 months led to increased liver, kidney, and thyroid weights in **males**.

σ-CHLOROBENZYLIDINEMALONITRILE

= *Chemshield* = *CS*

Lacrimator, chemical warf[illegible]gent.

Untoward effects: Inte[illegible]nful eye, ear, nose, throat, and skin (erythema) [illegible]nt. More potent, but less incapacitating and slower acting, than ***CN*** as a riot control agent. As with ***CN***, washing or showering with water will increase stinging sensations, but will help prevent first degree burns. ***Cyanide*** formation is an important part of its toxicity. Causes blepharospasm, coughing, nostril pain, bronchospasm, chest pain, vomiting, and dyspnea. Pneumonitis in a 4-month-old **infant** after prolonged (2–3 h) exposure to ***CS***. Persistent leukocytosis (white blood cells 20,000–30,000/m^3) with slow resolution. ***CS*** grenades (1 lb with 115 g ***CS***) have a 12 s delay fuse delivered by grenade-launcher or by air-drop. The U.S. Army bought 7,000 tons for use in Vietnam. Contact dermatitis in 25/28 (89%) in chemical plant **workers** when air levels exceeded Department of Labor standards. TLV, **human**: 0.05 ppm. Minimum lethal dose, skin, **human**: 10 mg/1 h.

Inhalation TC_{LO}, **human**: 1.5 mg/m^3/90 min.

High air concentration (3600 mg minimum/m^3) caused death in some **rats**. **Rats**, **mice**, **guinea pigs**, **rabbits**, and **pigeons** exposed to it by injection or aerosoling lacrimate, salivate, become lethargic and dyspneic, after an initial response of excitability and hyperactivity. **Dogs** are rather resistant to its ill-effects.

LD_{50}, **rat**: 178 mg/kg; **mouse**: 282 mg/kg; **rabbit**: 143 mg/kg; **guinea pig**: 212 mg/kg; inhalation LC_{LO}, **rat**: 1.8 g/m^3/45 min; **mouse**: 2.75 g/m^3/20 min; **rabbit**: 1.8 g/m^3/10 min; **guinea pig**: 2.3 g/m^3/10 min.

CHLOROBROMOMETHANE

Use: In fire-extinguishers, especially, in aircraft.

Untoward effects: Has caused superficial eye irritation, disorientation, dizziness, skin irritation, pulmonary edema, and gastrointestinal upsets in **people**. Kidney and liver pathology reported. Three **firemen** extinguishing an aircraft fire inhaled high concentrations for short periods of time. **One** developed apnea and **another**, generalized convulsions. **Both** went into deep coma and needed supportive therapy. TLV: 200 ppm.

TD_{LO}, **human**: 50 mg/kg.

LD_{50}, **rat**: 5 g/kg; **mouse**: 4.3 g/kg; inhalation LC_{50}, **rat**: 465 ppm/15 min; **mouse**: 2300 ppm/72 h.

CHLOROBUTANOL

= *Acetone Chloroform* = *Chlorbutol* = *Chloretone* = *Coliquifilm* = *Methaform* = *Sedaform*

***Analgesic, sedative, antiseptic, preservative.* Also, as an anesthetic for frogs and narcosis in fish.**

Untoward effects: A **patient** in whom it had a long half-life (13.2 days) developed toxicity and dependence on it, when used as a sedative. Delayed hypersensitivity to it in 52-year-old **male** receiving ***thiamine*** injections for ***alcohol*** withdrawal syndrome. A 19-year-old cancer **patient** receiving high-dose ***morphine*** infusion (6.8 mg/kg/h) not only had pain relief from the ***morphine***, but also induced somnolence, because the chlorobutanol exceeded the sedative dose. A 26-year-old **female** developed hypersensitivity to it in ***heparin*** solution. This was confirmed by subsequent rechallenge. A **diabetic** 54-year-old **female** developed pruritus from it in a ***desmopressin*** intranasal spray. A **patient** developed a contact dermatitis from it in an ear wax solvent, and a 7-day-old **infant** may have developed methemoglobinemia and death from its use (0.5%) in 30 ml of a bladder infusion for hemorrhagic cystitis. Eye irritation has followed ophthalmic use (in artificial tears and contact lens solutions). Acute poisoning leads to deep stupor, hypotension, dyspnea, and cyanosis. Enhances ***vitamin C*** metabolize and excretion. May cause increased blood urea nitrogen readings. Diffuses through or adsorbed by ***polyethylene*** and ***polypropylene***. Reacts with ***rubber*** closures, rapidly losing its effectiveness. The FDA makes a zero residue mandatory in milk for **human** consumption.

When used for motion sickness control in **cats**, it can cause respiratory depression and death. Do **NOT** use in **animals** with liver or kidney pathology.

4-CHLORO-*m*-CRESOL

Untoward effects: Has caused hypersensitivity reactions after IV use, as in ***heparin*** solutions. Pallor, sweating, hypotension, and tachycardia in 35-year-old **female** receiving ***heparin*** for treatment of deep vein thrombosis and pulmonary embolism. A 55-year-old **male** developed nasal congestion, profuse sweating, and generalized urticaria. Six myocardial infarct **patients** also suffered reactions from it in parenterals. Has been a sensitizing agent in ***corticosteroid*** creams. It and ***lye*** were involved in 11 homicides and 16 attempted homicides of senile, demented hospital **patients** by a staff **employee**. After 1 week at 50°C, 85% is adsorbed by ***nylon*** syringes. *Pseudomonas aeruginosa* has been resistant to it at 0.25% concentration. *Warning*: Has separated from emulsions into the oily phase, leaving the water phase unprotected. Physically inactivated by ***nylon*** devices (syringes, etc.). Avoid use in eyes – has produced corneal opacities. ***Antioxidants*** in ***polyethylene*** plastic containers inactivate it.

Use in eye drops has caused corneal opacities in **rabbits**.

LD_{LO}, **rat**: 500 mg/kg; IP LD_{LO}, **mouse**: 30 mg/kg; SC LD_{LO}, **mouse**: 200 mg/kg; SC LD_{50}, **rat**: 400 mg/kg.

N-(2-CHLOROETHYL) DIBENZYLAMINE

= *Dibenamine* = *Sympatholytin*

α-Adrenergic-blocker.

Untoward effects: Miosis in poisoned **children**. Inhibits platelet function in **man** and **animals**.

IV TD_{LO}, **human**: 60 μg/kg.

Blocks ovulation in **rats**, **rabbits**, and **hens**. SC LD_{50}, **mice**: 800 mg/kg.

CHLOROFORM

= *Trichloromethane*

Use: Approximately 650 million lbs used annually in manufacture of fluorocarbons, refrigerants, fluoropolymers, and solvents. In the past, significant usage was for anesthesia of **humans** and **horses**.

Untoward effects: As with ***ether***, it had a history of recreational use, before it was used as an anesthetic. As an anesthetic in **humans**, nausea, vomiting, tachypnea, tachycardia, hypotension, arrythmias, and bradycardia were often noted. Sniffing it often associated with health care **workers**. A 21-year-old **male** died after habitual inhalation of it. Its ingestion by a 19-year-old **male** at a chloroform party led to **his** death. Acute respiratory distress and acute hemolysis observed in a **patient** who abused it IV. Mild jaundice developed 3 days postoperative in a 54-year-old **male**. Severe liver damage in 10-year-old **male** after inhaling large amounts at a chloroform party. Excreted in the milk of **humans** and **goats**. Readily enters the fetal circulation, causing deep sleep in the unborn **infant**. A new **mother** receiving it for afterpains, caused her **baby** to go into a deep sleep for 8 h. Central nervous system depression occurs with toxic doses. Blood concentrations of 7–25 mg% are considered toxic, and 29 mg% or above become lethal. It increases myocardial irritability and impairs contractile strength. Inhalation of large quantities may cause death, by inducing ventricular fibrillation. Malignant hyperpyrexia and myopathies occurred with its use as a **human** anesthetic. The limited amounts found in drinking water supplies may not be as great a danger as some say. In some countries, it is allowed in drug preparations up to 60 ppm. Considered potentially carcinogenic for **man**, as it has caused liver cancers in **mice**, kidney cancers in male **rats**, and thyroid tumors in male **mice**. Dilated pupils, irregular respirations, acidosis, apnea, cold clammy skin, slow weak irregular pulse, cyanosis, coma, hypotension, cardiac failure, ventricular fibrillation, oliguria, anuria, fatty infiltration and degenerative liver changes can occur before death. Accidental overdose or suicidal attempts cause pain in the throat and stomach, with or without vomiting. Some vapors are inhaled and the balance is rapidly absorbed. This can cause a short period of excitement or stupor. Diarrhea and jaundice may occur during recovery. Prolonged skin contact causes erythema, pain, and severe blisters. Although non-explosive, its use in the presence of air or daylight, or a flame can result in the production of ***phosgene gas***, ***chlorine***, and ***hydrochloric acid***, which causes severe coughing, dyspnea, and collapse. In 1902, Gerlinger reported fatalities in surgical **spectators** from this reaction. It causes a characteristic sweet breath in **those** poisoned by it. Its use, medically, is, now, almost non-existent. Only **those** occupationally exposed to it must notify **their** physicians if taking ***propranolol***. TLV 10 ppm; minimum lethal dose, **human**: 10–15 ml.

LD_{LO}, **human**: 140 mg/kg; inhalation TC_{LO}, **human**: 1 g/m³/1 year and 5 g/m³/7 min.

When liver damage (fatty degeneration) occurs in **horses**, it happens about 24 or more hours after use. Oral use in **donkeys** has caused centrilobular necrosis of the liver.

CHLOROGUANIDE

= *Chlorguanide* = *Diguanyl* = *Drinupal* = *Guanatol* = *M 4888* = *Paludrine* = *Palusil* = *Proguanil* = *RP 3359* = *SN 12837* = *Tirian*

***Antimalarial.* A *folate* antagonist.**

Untoward effects: Occasionally vomiting, abdominal pain, oral ulceration, alopecia, fixed-drug eruptions, scaling of palms of the hands and soles of the feet, thrombocytopenia, and, with large doses, diarrhea, and, rarely, hematuria. One writer denies the association of alopecia and scaling of the skin, based on 5,000 African **patients**. May cause occult urinary casts and hemoglobinuria. Best given after meals.

Drug interactions: Coadministration with ***cimetidine*** significantly increases its blood levels and decreases its body clearance.

LD_{50}, **rats**: 200 mg/kg; **rabbits**: 150 mg/kg; LD_{LO}, **mouse**: 50 mg/kg; **chicken**: 400 mg/kg.

α-CHLOROHYDRIN

= *3-Chloro-1,2-propanediol* = *Epibloc*

Use: In dye intermediates, decreases ***dynamite's*** freezing point, and as a **rodent** control agent.

Untoward effects: By its action on epididymal vasculature with decreased oxygenation of the area, as well as it damaging the plasma membrane of sperm, it induces reversible sterility in **rats** and **monkeys**.

LD_{50}, **rat**: 150 mg/kg; **mouse**: 160 mg/kg; inhalation LC_{LO}, **rat**: 125 ppm/4 h.

CHLOROMETHYL METHYL ETHER

= *CMME*

Use: In organic synthesis.

Untoward effects: Its volatile vapors are a direct-acting alkylating carcinogen for **man**. Eye, mucous membrane, and skin irritant for **man**. Causes pulmonary congestion and edema, coughing, wheezing, pneumonia, and blood-stained sputum. Strong concentrations cause burning on contact. The technical grade is almost always contaminated with ***BCME***, leading to local sarcomas after SC injections in **mice**, and may be an initiator of skin and oat cell lung tumors, and bronchitis in **man**.

Carcinogenic by inhalation in **mice** and SC in **rats**.

Inhalation LC_{50}, **rat**: 55 ppm/7 h; **hamster**: 65 ppm/ 7 h.

CHLOROPHACINONE

= *Afnor* = *Caid* = *Drat* = *LM 91* = *Liphadione* = *Microzul* = *Mr. Rat Guard II* = *Muriol* = *Parapel* = *Quick* = *Ramucide* = *Ratindon* = *Raviac* = *Redentin* = *Rotomet* = *Rozol* = *Topitox*

Anticoagulant, rodenticide.

Use: In baits at a 0.005% concentration, to kill **voles, rats, and mice**.

Untoward effects: Risk to **humans** is from accidental, malicious, or suicidal exposure. ***Vitamin*** $\boldsymbol{K_1}$ is antidotal.

LD_{50}, **rats**: 2.1 mg/kg; **mice**: 1 mg/kg; **rabbit**, 200 mg/kg; **duck**: 100 mg/kg.

CHLOROPHENOL

Untoward effects: Traces of ***dioxin*** (***TCDD***) were found in the fat of six **workers** contaminated during a 1953 explosion in West Germany. Approximately 30 years later, one of these **workers** still had ***dioxin*** levels in **his** fat 10–20 times the general population's level. In 1979, quick action by authorities, and the benefit of cold weather prevented a disaster in a town of 800 when 20,000 gallons of the σ form spilled from a rail car.

In **mice**, the convulsant action of $p > m > \sigma$.

m, LD_{50}, **rats**: 570 mg/kg; σ, LD_{50}, **rats** and **mice**: 670 mg/kg.

4-(4-CHLOROPHENOXY)-BUTYRIC ACID

= *4-CPB*

Herbicide.

Untoward effects: Anorexia, rumen stasis, weakness, ataxia, muscle spasms, gastrointestinal inflammation, lung and kidney congestion reported in exposed **ruminants**.

p-CHLOROPHENYLALANINE

= *Fenclonine* = *PCPA*

Antimetabolite.

Untoward effects: Injection into **man** has caused marked hypothermia. Allergic reaction reported in 5/11 within 12 days to 8 weeks. Depression, anxiety, restlessness, irritability, crying, and agitation in 5/7. Tiredness, dizziness, nausea, uneasiness, paresthesias, headache, and constipation in six prison **inmate volunteers** on 1 g/day. Mental confusion, nightmares and insomnia, abdominal pain, alopecia, transient eosinophilia, hyperalgesia, and lens opacities have been reported.

As a potent inhibitor of ***tryptophan hydroxylase***, it causes a decrease in brain ***serotonin*** synthesis, causing hypersexuality in **animals**. Causes ***bananas*** to turn brown. Aphrodisiac in **cats**, **rabbits**, and **rats** and increases aggressiveness in **cats**. After dosing, **rats** have an aversion to ***alcohol***.

4-CHLORO-σ-PHENYLENEDIAMINE

= *Ursol Olive 6G*

Use: In dyes, including hair dyes.

Untoward effects: Carcinogenic; urinary bladder and stomach tumors in **rats**; hepatocellular carcinomas in **mice**.

CHLORO-2-PHENYLPHENOL

Antiseptic, fungicide.

Untoward effects: Photoallergic contact dermatitis. Acneform eruptions have occurred up to several months after exposure. A number of related "***Dowicides***", such as ***G, 4, 31*** and ***32*** powders were taken off the market.

31-Sodium Salt, LD_{50}, **rat**: 3.5 g/kg.

CHLOROPHYLL

Use: Touted as an air (***Airwick***®) and breath deodorant, yet,

Why reeks the **goat** on yonder hill
That seems to dote on ***chlorophyll***?

Untoward effects: It is a catalyst for dermatitis in **bathers**. Has been suspect as a cause of hypersensitivity in some cosmetic products, causing berlock dermatitis. In inherited Refsum's disease, ***phytannic acid***, a derivative of chlorophyll, accumulates. Some of its features are reversible, if the **patient** is maintained on a chlorophyll-free diet. Once used in lymphography, where it caused granular eosinophilic material to accumulate in the lymph nodes. Chlorophyll is even present in white ***potatoes***.

IP LD_{50}, **mouse**: 400 mg/kg; IV LD_{50}, **mouse**: 285 mg/kg; IV LD_{LO}, **guinea pig**: 80 mg/kg.

CHLOROPHYLLUM MOLYBDITES

= *Lepiota morgani*

Untoward effects: This ***mushroom***'s toxic effect after ingestion is occasionally reported. Symptoms occur after about 1–2 h, occasionally, up to 6 h, and gastrointestinal symptoms (nausea, vomiting, abdominal pain, diarrhea [occasionally, bloody]), fever, headache, and diaphoresis are common, and more severe when the raw ***mushrooms*** are eaten. Tachycardia, postural hypotension, thirst, and giddiness are also reported. Some **people** who ingest it do not suffer any toxic effects, but, in rare cases, it can be fatal (viz. a 2-year-old **infant** died with convulsions 17 h after eating it).

2-CHLORO-3-PHYTYL-1,4-NAPHTHOQUINONE

= *2-Chloro-vitamin K*

Anticoagulant.

Untoward effects: Orally, its effect is more pronounced than ***warfarin*** in **wild rats**.

CHLOROPICRIN

= *Acquinite* = *Chloro-pic* = *Dolochlor* = *G25* = *Larvacide 100* = *Microlysin* = *Nitrochloroform* = *Pic-clor* = *Picfume* = *Picride* = *Telone* = *Trichloronitromethane*

Fumigant.

Untoward effects: Skin burns in **workers** after 2–4 h exposure. A riot control agent that causes severe irritation of the upper respiratory tract and skin. Vomiting often accompanies profuse lacrimation. A lung irritant (intermediate between ***chlorine*** and ***phosgene***) in gas warfare, where it was used in shells. Known as "***vomiting gas***", forcing **troops** to remove **their** gas masks, making **them** victims of other, more destructive gases, used at the same time. Causes profuse lacrimation and frontal headaches.

LD_{LO}, **human**: 5 mg/kg; inhalation LC_{LO}, **human**: 2.4 g/m^3/min.

Its fumes are toxic and preferred over ***hydrocyanic acid*** in France for killing **foxes**. Vapors killed 200/6,000 broiler **chickens** housed 3 meters from its field use.

Inhalation LC_{LO}, **cat**, **rabbit**, and **guinea pig**: 800 mg/m^3/20 min.

CHLOROPRENE

= *Neoprene*

Use: In the manufacture of neoprene elastomers, cable sheaths, hoses, and rubber items.

Untoward effects: Neurotoxic. Fertility reduction, spontaneous abortions, mutagenesis, and suspect in lung and skin cancers in **man**. Chest X-ray and pulmonary function testing required for U.S. **workers**. Pregnant **women** need counseling, concerning continued exposure. TWA, 25 ppm.

Carcinogenic in **mice**.

LD_{LO}, **rat**: 1.6 g/kg.

CHLOROPROCAINE

= *Nesacaine*

***Local anesthetic.* A *p*-aminobenzoic ester.**

Untoward effects: IV regional anesthesia has caused generalized twitching, sensations of warmth, tingling, dizziness, sleepiness, sinking feeling, anxiety and fearfulness, slurring of speech, and, rarely, thrombophlebitis. Overdoses rapidly cause nausea, vomiting, syncope, bradycardia, ventricular arrhythmias, talkativeness, and convulsions. In addition to some of the above symptoms, allergic reactions can include urticaria, pruritus, angioneurotic edema, sneezing, and diaphoresis. ***Oxygen***, lowering the **patient's** head, and an injection to control convulsions has been suggested. Less toxic than ***procaine*** (q.v.). Some solutions contain ***epinephrine*** or ***norepinephrine***, and may cause prolonged hypotension or hypertension in **patients** on ***monoamine oxidase inhibitors***. May inhibit the therapeutic benefits of ***sulfonamides***.

In **dogs**, its relative myocardial toxicity was 2.2 times that of ***procaine*** and ⅓ that of ***lidocaine***.

CHLOROPROPHAM

= *Chloro-IPC* = *Chlorpropham* = *CIPC* = *Sprout Nip* = *Spud-Nic* = *Taterpex* = *Triherbide-CIPC* = *Y-3*

Herbicide, plant growth regulator. A carbamate. Commercial mixtures, such as Sprout Nip® also contain methanol. Workers entering storage areas must wear protective clothing and respirators.

Untoward effects: Can cause anorexia, diarrhea, bloat, lung congestion, enlarged lymph nodes, and gastrointestinal inflammation with hemorrhage in affected **animals**.

LD_{LO}, **human**: 5 g/kg; **mouse**: 600 mg/kg; LD_{50}, **rats**: 1.2–8 g/kg; **rabbit**: 5 g/kg; **mallard ducks**: > 2 g/ Kg.

CHLOROPROPYLATE

= *Acaralate* = *Chlormite* = *Gesakur* = *Isopropyl-4,4′-dichlorobenzilate* = *Rospan* = *Rospin*

***Miticide.* Used on fruit trees. Structurally related to dichloro-diphenyl-trichloroethane (*DDT*). A *chlorinated hydrocarbon*.**

Untoward effects: **Human** toxicity would be similar to ***DDT***'s (q.v.). Repeated insult patch test of a 1% emulsion on 50 **human** volunteers failed to show any irritation or sensitivity.

Rats receiving 4.6–10.2 g/kg of a 25.2% emulsion in ***xylene*** and a ***petroleum*** derivative evidenced ptosis, salivation, hypoactivity, dyspnea, and muscle weakness.

Of a 40% emulsion, LD_{50}, **rats**: 8 g/kg; acute dermal, **rabbits**: > 10.2 g/kg.

6-CHLOROPURINE

Untoward effects: Bone marrow depression, thrombocytopenia, anorexia, vomiting, rashes, and hyperuricemia have been reported. 16/36 **patients**, after treatment for acute leukemia, developed jaundice. Renal blockage with ***uric acid*** crystals in one case.

Causes decreased embryonic growth and increase fetal anomalies (aplasia-limbs) in **rats**.

IP LD_{50}, **rat**: 400 mg/kg; IP TD_{LO}, pregnant **rats**: 100 mg/kg.

CHLOROPYRAMINE

= *Halopyramine* = *Suprastin* = *Synopen* = *Synpen*

Untoward effects: Colic-like syndrome, cyanosis, irritability, weight loss, hyperhydrosis, respiratory distress, urinary retention, and **fetal** anomalies (craniofacial, central nervous system, and skeletal).

CHLOROQUINE

= *Amokin* = *Aralen* = *Arechin* = *Arequine* = *Athrochin* = *Atrichin* = *Avlochlor* = *Avloclor* = *Bemaco* = *Bemaphate* = *Bemasulph* = *Benaquin* = *Bipiquin* = *Capquin* = *Chemochin* = *Chingamin* = *Chloraquine* = *Chlorochin* = *Chlorquin* = *Chloroquina* = *Chloroquinium* = *Cidanchin* = *Clorochina* = *Cocartrit* = *Delagil* = *Dichinalex* = *Elestol* = *Gontochin* = *Heliopar* = *Imagon* = *Iroquine* = *Klorokin* = *Lapaquin* = *Malaquin* = *Malaren* = *Malarex* = *Mesylith* = *Neochin* = *Nivachine* = *Nivaquine* = *Nivaquine B* = *Quinachlor* = *Quinagamin* = *Quinagamine* = *Quinercyl* = *Quingamine* = *Quinilon* = *Quinoscan* = *Resochen* = *Resochin* = *Resoquina* = *Résoquine* = *Reumachlor* = *Reumaquin* = *Roquine* = *RP 3377* = *Sanoquin* = *Serviquin* = *Silbesan* = *Siragan* = *SN 6718* = *SN 7618* = *Solprina* = *Sopaquin* = *Tanakan* = *Tankan* = *Tresochin* = *Trochin* = *W 7618* = *WIN 244*

Protozoacide, antimalarial, lupus suppressant.

Untoward effects: ***Cardiac and hematologic***: Aplastic anemia, agranulocytosis, thrombocytopenia, leukopenia, eosinopenia, idiopathic thrombocytopenic purpura, prolonged QT interval, and heart block. Anemia and hemolysis, especially, in glucose-6-phosphate dehydrogenase-deficient **individuals**. Methemoglobinemia and cyanosis reported. IV use may cause cardiotoxicity, collapse, and death.

Central nervous system: Excitation, irritability, fatigue, convulsions, equilibrium upsets, confusion, delirium, delusions, mania, headache, and depression.
Ear: Tinnitus, progressive irreversible deafness that can follow cochlear and vestibular toxicity, congenital deafness.
Eye: Blurring of vision; visual halos; and decreased speed of accomodation, reversible on withdrawal; irreversible retinopathy, which starts as a non-specific mottling of the macula, developing into a typical pathognomonic "bull's eye" pigmentation of the ventral region, surrounded by a zone of depigmentation. Lenticular and corneal haze (reversible on discontinuance), corneal edema, loss of central visual acuity and color vision, retinal artery constriction, photophobia (irreversible), increases in dark adaptation time, ocular palsies, ptosis, whitening of eyelashes, and decreased corneal sensitivity. **Patients** receiving the drug for at least 3 months deposit some of it in the cornea without any symptoms. Anesthesia of the cornea may occur, leading to embedment of foreign bodies, without the **patient's** knowledge. This is especially dangerous in **those** wearing contact lenses.
Gastrointestinal: Nausea, vomiting, diarrhea, anorexia, abdominal distress, and, rarely, hematemesis.
Miscellaneous: Giddiness, psychosis, porphyrinuria, rust yellow or brown urine color, weight loss and liver necrosis due to hypersensitivity, and abortions. Avoid use in hepatic disease, because it concentrates in the liver. Readily crosses the placenta and appears in breast milk. Suspect teratogen in **humans** and inhibits DNA repair. Has induced changes in hair coloring.
Neuromuscular: Tremors; myalgia; muscle rigidity, weakness, and atrophy; pre-synaptic local anesthetic effect, decreases the release of neurotransmitter. Extrapyramidal and dystonic symptoms are more apt to occur in combination with other drugs. Peripheral neuropathy (both motor and sensory), and convulsive seizures reported.
Poisoning: Cardiac arrest and persistent hypotension in a 20-month-old **female** after ingesting an estimated 800 mg is typical of poisoning problems in **infants**. 2 g or more proved fatal to more than one 26-year-old **female**. Blood levels in overdose deaths, in one review, ranged between 1.1 and 10.3 mg/100 ml. Urinary excretion can be increased by urinary acidification. Hypokalemia after acute ingestion.
Skin: Pruritus, dry skin, desquamation, exacerbates psoriasis, bleaching of hair (3 months delay), fixed-drug reactions, rashes on exposure to sunlight, pigmentation of skin, mucous membranes, palate, and under nails (bluish color). Ultraviolet (UV) A and UVB sensitivity may take 2–3 months to appear and a year to resolve after discontinuance.

LD_{LO}, **woman**: 20 mg/kg.

Drug interactions: Dystonic reactions can occur in the presence of ***metronidazole***. As a weak ***base***, its urinary excretion is increased by acidifying agents, such as ***ammonium chloride*** and decreased by alkalinizing agents, such as ***sodium bicarbonate***. Potentiates ***folic acid*** antagonists, like ***methotrexate***. Metabolic degradation in the liver is inhibited by ***monoamine oxidase inhibitors***, increaseing toxic effects and retinal damage. May react with SGOT to give false positive test results. ***Cimetidine***, but not ***ranitidine***, decreases its absorption, and it decreases the bioavailability of ***praziquantel*** in **humans**. Do **NOT** use with glass syringes or containers (decreases concentration up to 40%). It may take months or years for the body to eliminate all of it from the tissues. Can increase thyroid stimulating hormone (***TSH)*** levels.

LD_{50}, **rat**: 1 g/kg; **mouse**: 400 mg/kg.

Warning: Its use has been associated with retinopathy and cardiotoxicity in **rabbits**, **rats**, and **man** (especially, pigmented eyes). Do **NOT** use in lactating **animals** producing milk for **human** consumption. Its use has been associated with central nervous system and other pathology in **swine** and **dogs** and various myopathies on high continuous dosage (40 mg/kg/day/4–7 weeks) in **rats**. The U.S. has withdrawn official approval for veterinary use of this drug.

5-CHLORO-8-QUINOLINOL

= *Cloxiquine* = *Cloxyquin* = *Dermofungin A*

Antifungal, antibacterial.

Untoward effects: Epigastric discomfort, contact dermatitis, and neuropathy. Its use in Japan in aged, chronically ill **patients** or after major surgery caused epidemic neuropathies. It can be reproduced in **dogs** receiving excessive doses.

CHLOROSULFONIC ACID

Untoward effects: Will fume in moist atmospheres. Contact with moisture causes it to decompose producing ***hydrochloric*** and ***sulfuric acids***. This can occur in the moist passages of the respiratory tract. Inhalation results are the same as inhalation of the strong acid fumes. Severe irritation of the nose, mouth, and throat can occur and, if the exposure is massive, pulmonary edema. The latter is followed by fluid in the lungs, dyspnea, and chest pains. Industrial exposures are of the acute type and are due to equipment failure, careless work habits, or failure to use or maintain safety equipment. Skin and eye burns are more common than inhalation problems. Thorough flushing of affected areas with copious amounts of cool water should be used instantly. ***Hydrogen*** gas release can be an end result of contact of it with moisture and metals. **Handlers** should use eye protection and rubber gloves.

CHLOROTHALONIL

= *Bravo* = *DAC-2787* = *Daconil 2787* = *Exotherm Termil* = *Forturf* = *Termil*

***Fungicide, bactericide, nematocide.* Used in latex paints and for turf treatments.**

Untoward effects: A Navy **lieutenant** died 2 weeks after **he** spent 3 consecutive days playing golf at a country club. **He** had a habit of licking **his** golf balls, for good luck. The pathologist attributed **his** death to the unusual heavy exposure to this pesticide. The golf course was treated weekly with it. **Turf applicators** are cautioned about temporary allergic effects. Causes eye irritation by direct contact, or as an allergic reaction. Allergic reactions can also including mild bronchial irritation and erythema on exposed skin areas.

Possibly, carcinogenic in **rats**. Chronic feeding (175 mg/kg/day/91 days) to **rats** produced gastric ulcers, hyperplasia, and hyperkeratosis. Within the first week, degeneration of the proximal tubular epithelium was noted.

LD_{50}, **rats**: > 10 g/kg.

CHLOROTHEN

= *Tagathen* = *Thenclor*

Antihistamine.* In *Achrocidin.

Untoward effects: Unconsciousness and convulsions developed in a 15-month-old **female** after ingesting 30 **adult** doses along with ***phenacetin, caffeine,*** and ***salicylamide***. Treated with peritoneal dialysis. No indications of any permanent physical or mental damage 7 months after ingestion.

LD_{LO}, **human**: 50 mg/kg.

IP LD_{50}, **mice**: 105 mg/kg.

8-CHLOROTHEOPHYLLINE

Untoward effects: ***Dimenhydrinate*** (***Dramamine***®) is the 8-chlorotheophylline salt of ***diphenhydramine*** (***Benadryl***®). A fixed-drug skin eruption occurred with the former and flared again when this compound was ingested.

CHLOROTHIAZIDE

= *Alurene* = *Chlorosal* = *Chlorurit* = *Chlotride* = *Clotride* = *Diuresal* = *Diuril* = *Diurilix* = *Diurite* = *Exuril* = *Flumen* = *Minzil* = *Neo-Dema* = *Ro-Chlorozide* = *Salisan* = *Salunil* = *Saluretil* = *Saluric* = *Salutrid* = *Urinex* = *Warduzide* = *Yadulan*

Diuretic, antihypertensive.

Untoward effects: Sialadenitis, alkalosis, lethargy, nasal stuffiness, decreased libido. Hyperparathyroidism after long-term use. Hypersensitivity, muscle spasms or cramps, fever, pneumonitis, respiratory distress, necrotizing angitis, vasculitis, and anaphylaxis reported. Anorexia, weakness, headache, dizziness, and acute pancreatitis also occur. Acute non-cardiogenic pulmonary edema reported twice in a 66-year-old **female**. Long-term therapy may cause a ***magnesium*** deficiency. Diabetogenic in large dosage (usually prevented by ***adrenergic blockers*** and ***potassium*** supplementation).

Blood: Hypersensitivity manifested as non-thrombocytopenic purpura of Henoch-Schönlein type; decreased in blood volume, hypercalcemia, hyperuricemia, hemolysis, bone marrow depression and neutropenia, agranulocytosis, aplastic anemia, leukopenia, eosinophilia, immuno-thrombocytopenia, pancytopenia, and thrombocytopenia. A 49-year-old **male** developed severe hypokalemia, eventually associated with acute myoglobinuric renal failure. Can exacerbate asymptomatic hyponatremia, and, as a result, can cause a potential danger in aldosteronism. Can cause false positive or increased ***uric acid*** in blood tests.

Cardiovascular: Hypotension, orthostatic hypotension – aggravated by concommitant use of ***alcohol, barbiturates, antihypertensive drugs,*** and ***narcotics***.

Central nervous system: Dizziness, vertigo, headache, paresthesia, weakness, muscle spasms, confusion and disorientation, nervousness, seizures, and insomnia.

Eye: Xanthopsia, transient myopia, and retinal edema. Use in a 62-year-old **female** with severe glaucoma abruptly lowered systemic blood pressure causing severe loss of visual field and acuity.

Fetus and neonate: Readily crosses the placenta and coagulation defects have occurred (thrombocytopenia and hemorrhage). Has caused **maternal** and **fetal** acidosis. In **neonates**, hypoelectrolytemia, profound anemia, convulsions, respiratory distress, and death reported. A **newborn** had hemolytic anemia after its **mother** had received 500 mg/day plus 3 g ***sulfamethazole***/day for 2 weeks before delivery. Increased risk of hypertension at maturity.

Gastrointestinal: Gastric irritation, nausea, vomiting, cramping, epigastric pain, diarrhea, constipation, anorexia, dyspepsia, gastrointestinal ulcerations, and stricture of ileum.

Glycosuria, gout, and diabetes: Hyperglycemia, asymptomatic hyperuricemia, occasionally acute gout, coma, and diabetes reported.

Liver: Jaundice (1–3% after 1–4 week exposure in 90% of **patients**, but has occurred after only one dose). Cholestasis and increased alkaline phosphatase due to hypersensitivity occur. Reports of liver dysfunction occur in up to 50% of **patients**. Different opinions exist regarding its causing acute cholecystitis. Hepatic encephalopathy in **patients** with cirrhosis. Hepatic coma in gravida 36-year-old **female** with cirrhosis after receiving 500 mg twice daily/5 days. After a 24 h recovery, **she** then went into spontaneous labor and delivered a stillborn **infant**.

Photosensitivity and skin: Photosensitivity dermatitis with eruptions similar to lichen planus, occasionally with residual hyperpigmentation (fixed-drug eruptions). Toxic epidermal necrolysis and alopecia. Erythema multiforme, Stevens-Johnson syndrome, and purpura with necrotic blisters (primarily, on the legs). Reactions occur in a wavelength range of 2,750–3,100 angstroms. A 53-year-old **male**, after 2 months of therapy, had a near-fatal drug reaction along with pruritus, fever, exfoliative dermatitis, lymphadenopathy, splenomegaly, eosinophilia, and shock.

Poisoning: Acute intoxication in a 14-year-old **female** and a 2½-year-old **female** after accidental ingestion of 15 g (30 tablets) leading to lethargy after 3.5 h increasing to deep coma in the next 9 h (lasting about 3–4 h) with acute gaseous abdominal distention and flatus. Recovery after 2 days. Coma and death following hyponatremia, hypokalemia, and hypochloremic alkalosis in a pregnant 33-year-old **female** ingesting over 80 g over a 3 week period. Pneumonia, cerebral hypoxia, and ***salt*** depletion were probable cause of death. Has also been implicated in a suicide.

Renal: Tubular degeneration due to acute ***potassium*** depletion reported. A hypersensitive patient developed fatal tubular necrosis after 1 g/day/7 months. Anuria, oliguria, and acute renal failure also noted.

Drug interactions: By virtue of its ***potassium***-depleting effect, it enhances the activity and toxicity of ***digitalis glycosides, curare, d-tubocurarine,*** and ***gallamine***. It decreases the renal clearance of ***ganglion-blocking agents***. Its effect on the ***curariform drugs*** increases the danger of postoperative respiratory depression. Potentiates ***demeclocycline, hydralazine, lithium, mecamylamine,*** and ***pargyline***. Caution should be used in prescribing it with ***angiotensin-converting enzyme inhibitors*** in **patients** with renal vascular disease. ***Probenecid*** will displace it in active transport in the renal tubules, prolongs its diuretic effects, and blocks its ability to eliminate ***uric acid***. Effect is reversed by ***allopurinol***. Immunochemically related to *p*-amino compounds, such as ***benzocaine*** and ***p-phenylenediamine***, and cross-sensitivity exists with resultant skin eruptions. Its induced orthostatic hypotension can be aggravated by ***alcohol, barbiturates,*** or ***narcotics***. It may impair the control of diabetes by ***chlorpropamide***, or other oral ***hypoglycemic agents*** and ***insulin***.

Because ***cholestyramine*** interferes with its absorption, it should be given 1 h before or 4–6 h after the administration of ***cholestyramine***. Its bioavailability is decreased by concurrent administration of ***colestipol***. Interferes with radiolabeling of red blood cells with ***Tc 99m pertechnetate***.

Fluid and electrolyte imbalances can occur with its use in **animals**, just as it occurs in **humans**. Hyperglycemic in **rats**. Phototoxicity, pancreatitis, and pancreatic necrosis in **mice**. I have suggested a 3 day withdrawal period in **animals** before their meat or milk can be consumed by **humans**.

IV LD_{50}, **dog** and **mouse**: 1 g/kg; **rat**: 1.39 g/kg.

3-CHLORO-*p*-TOLUIDINE

= *DRC 1339* = *Starlicide*

Avicide.

Untoward effects: Highly toxic, slow-acting avicide for **birds** (**starlings**, **seagulls**, **ducks**, **blackbirds**, and **poultry**) and is eaten readily by them in their feed. Causes renal failure, regurgitation, polydipsia, ataxia, and tremor. Tail fanning, ataraxia, falling, ptosis, dyspnea, tachypnea, and immobility are common in **birds**. Some signs appear in 10 min, often peaking in 6–12 h, with death usually occurring within 2–3 days. ***Starlicide***® is marketed as a 0.1% concentration in feed pellets. Low **mammalian** toxicity. Secondary poisoning of **carnivorous mammals** and **birds of prey** is low (LD_{50} for the **African bulbul** was 6.74 mg/kg. A **mongoose** fed 21 of them over a period of 6 days showed no toxicity). IP injections in **mice** and **rats** caused methemoglobinemia, decreased pulmonary ventilation, hind limb paralysis, reduced temperature, cyanosis, and coma. Causes corrosive effects on the eyes and skin of **rabbits**, and induces dermal sensitization in **guinea pigs**. **People** must avoid accidental ingestion of treated grains, unnecessary dermal and eye contact, and inhalation.

LD_{50} to most **passerines**, **columbids**, and **corvids** is < 10 mg/kg, but > 100 mg to most **raptors** and **mammals**. LD_{50}, **rats**: 300 mg/kg; **pigeon**: 13 mg/kg; dermal LD_{50}, **rabbits**: > 2 g/kg.

CHLOROTRIANISENE

= *Hormonisene* = *Merbentul* = *Tace*

***Pro-estrogen.* Metabolizes into an active form, stored in body fat and released slowly. Possesses unique anti-*estrogen* properties.**

Untoward effects: Excreted in breast milk. Cardiovascular and thromboembolic complications have occurred with its use. Appears to cause a ***pyridoxine*** deficiency. Vascular headaches and mental confusion in 61-year-old **male** and hyperlipemia and thrombosis in 76-year-old **male**.

SC TD_{LO}, **mouse**: 180 mg/kg/89 weeks.

CHLOROTRIFLUOROETHYLENE

Use: An intermediate in plastic manufacturing, fire extinguishing chemicals, and pharmaceuticals, such as ***halothane***.

Untoward effects: Industrial use of this fluorocarbon monomer gives it a potential for **human** exposure.

Nephrotoxic to **rats** and **mice**. Hepatotoxicity and cardiotoxicity in **mice**.

Inhalation LC_{50}, **rat**: 1,000 ppm/4 h, 1,000 ppm/24 h.

CHLOROXURON

= *C-1983* = *Chloroxifenidim* = *Norex* = *Tenoran*

Herbicide.

Untoward effects: In **ducks** ataxia, weakness, sideways walking, and falling noted about 3 days after oral administration. Symptoms persist for up to 2 weeks.

LD_{50}, **rats**: 3.1 g/kg; **mallard ducks**: > 2 g/kg; **dog**: 10 mg/kg.

CHLOROXYLENOL

= *Benzyltol* = *Dettol* = *Ottasept* = *PCMX*

***Antiseptic, fungicide, preservative.* A halogenated phenolic. The term *Dettol* was also used, at one time, for a chloroxylenol product that contained additional *terpineol* and *alcohol*.**

Untoward effects: Contact eczematous dermatitis. *Serratia marcesens* has been found in a 3% solution. Ulceration of vocal cords in a 22-month-old **female** noted 40 min after drinking 125 ml. A schizophrenic 56-year-old **female** became addicted to a 4.8% solution in ***isopropylalcohol***. **She** admitted to drinking over 250 ml/days for over 10 years. A 70-year-old **female** attempted suicide by ingesting 350 ml of a 4.8% solution that also contained ***terpineol*** and ***ethyl alcohol***. Toxicity similar to, but less serious than, ***phenol***. **She** rapidly developed a strong reduction of central nervous system and cardiovascular function and coma, but recovered after treatment with gastric lavage, ***dopamine***, and IV ***verapamil***.

Abnormal position of wings and legs in 2% of **chick** embryos whose eggs had been dipped in a 10% solution before incubation.

LD_{50}, **rats**: 3.83 g/kg.

CHLOROZOTOCIN
= *DCNU* = *NSC 178,248*

Antineoplastic.

Untoward effects: Cholestasis without hepatitis. Causes moderate reversible increases of SGOT and SGPT, which peak at about 4 weeks of therapy. Occasionally nausea and vomiting. Leukopenia and thrombocytopenia are common delayed reactions. Chronic anemia developed in 45-year-old **female** before renal failure.

In vitro mutagenesis.

IV LD_{50}, **rat**: 40 mg/kg; IV LD_{10}, **mouse**: 20 mg/kg.

CHLORPHENESIN
= *Mycil*

Topical antifungal.

Untoward effects: Dermatitis reported.

CHLORPHENESIN CARBAMATE
= *Maolate* = *Rinlaxer*

Skeletal muscle relaxant.

Untoward effects: Drowsiness (30/50, 5/30, and 5/37), dizziness (4/30 and 22/194), weakness (3/30), headaches (10/50), gastrointestinal upsets and nausea (4/50 and 3/37), insomnia, distorted vision (3/30), pruritus, erythematous and submaxillary gland swelling, fixed-drug reactions, and paradoxic stimulation and nervousness. Urticaria, diarrhea, muscle tremors, and mild depression in **humans** and **animals**. Avoid use in **patients** or **animals** with liver pathology. Rare cases of leukopenia, thrombocytopenia, agranulocytosis, pancytopenia, and anaphylactic reactions can occur.

Adverse effects in **dogs**, although infrequent, include urticaria, tremors, vomiting, and diarrhea.

LD_{50}, **rats**: 748 mg/kg.

CHLORPHENIRAMINE
= *Haynon*

The maleate = *Allerclor* = *Allergisan* = *Alunex* = *Antagonate* = *Chlo-amine* = *Chlormene* = *Chlor-Trimeton* = *Chlor-Tripolon* = *Cloropiril* = *C-Meton* = *d-Chlorpheniramine* = *Dexchlorpheniramine* = *Fortamine* = *Histadur* = *Histalen* = *Histaspan* = *Isomerine* = *Lorphen* = *M.P. Chlorcaps T.D.* = *Phenamin* = *Phendextro* = *Piriton* = *Polaramine* = *Polaronil* = *Pyridamal-100* = *Sensidyn* = *Teldrin*

The tannate = *Duraband*

Antihistamine.

Untoward effects: Drowsiness (34%), dry mouth (7%), blurred vision (1%), itchy palate, sweating, diarrhea, flatulence, anorexia, anosomia, ataxia, urticaria, rash, chills, tachycardia, extra systoles, palpitation, hypotension, dizziness, disturbed coordination, tremor, euphoria, insomnia, neuritis, convulsions, urinary retention, nasal stuffiness, wheezing, headache, nausea, vertigo, weak pulse, decreased blood pressure, impaired concentration and performance, abnormal thyroid function, and even anaphylactic shock have occurred. Facial dyskinesia in a **patient** on a daily intake of 12 mg timed-release for over 10 years. Symptoms disappeared with cessation of self-medication. Overdosage can cause delirium. Induced thrombocytopenia in a 53-year-old **male**. Aplastic anemia in 51-year-old **male** after ingesting 6 mg/day for 3–4 day/week/10 years. Bone marrow suppression with neutropenia and pancytopenia reported. Fatal agranulocytosis in 48-year-old **female**. Giddiness reported on rare occasions. A 64-year-old **male** was hospitalized with a severe systemic hypersensitivity reaction that included lip swelling, chest pain, and body rashes after only two 4 mg doses. Adsorbed by ***activated charcoal***. Marked restlessness and slight drowsiness occurred in 14-year-old **female** after swallowing eight 8 mg capsules in a failed suicide attempt. A fatal case of pulmonary edema in a young **man** with blood concentration 65 times normal after consuming an unknown amount with ***alcohol***. Therapeutic concentrations are up to 5 μg%. Avoid use in **patients** with bronchial asthma, increased intraocular pressure, hyperthyroidism, cardiovascular disease, hypertension, or diabetes. Therapeutic serum concentration range 0.003–0.017 mg/l and toxic at 20–30 mg/l.

Drug interactions: It increases ***phenytoin's*** concentration and anticonvulsant activity. Enhances the cardiovascular effects of ***norepinephrine***. ***Antacids*** may alter its dissolution rate and adsorption may occur. On high dosage, its anticholinergic effect can reduce urination ability.

Half-life in **dogs** is 1.17 h, compared to 16–39 h in **man**.

LD_{50}, **mouse**: 142 mg/kg; **guinea pig**: 198 mg/kg; IV LD_{LO}, **dog**: 98 mg/kg.

CHLORPHENOXAMINE
= *Clorevan* = *Contristamine* = *Phenoxene* = *Phenoxine* = *Systral*

Anticholinergic.

Untoward effects: Nausea, vomiting, ataxia, dizziness, urinary retention, tremors, dry mouth, delirium, delusions, irritability, insomnia, and visual and auditory hallucinations on high dosage. Overdoses are convulsant. The toxic effects are similar to ***atropine's*** (q.v.).

LD_{50}, **rat**: 1 g/kg; **mouse**: 345 mg/kg.

CHLORPHENTERMINE

= *Avicol* = *Avipron* = *Clorfentermina* = *Lucofen* = *Teramine*

***Anorexic. Avicol* is also a name for *quintozene*.**

Untoward effects*:** Mydriasis, nausea, thirst, rash, drowsiness, dry mouth, constipation, sweating, headache, urticaria, and difficulty in urinating occur occasionally. Nervousness, nightmares, increased blood pressure, and insomnia have also been reported. It releases ***norepinephrine from peripheral adrenergic neurons and depletes the brain of ***serotonin***. Concurrent use of chlorphentermine, a parasympathetic amine (***amphetamine***-like) and an ***monoamine oxidase inhibitor***, such as ***phenelzine***, causes severe hypertension and cerebral hemorrhage.

Causes a phospholipid storage disorder in **rat** lungs.

LD_{50}, **rats**: 230 mg/kg; **mice**: 270 mg/kg.

CHLORPROMAZINE

= *2601 A* = *Aminazine* = *Ampliactil* = *Amplictil* = *Chloractil* = *Chlorazin* = *Chlorderazin* = *Chlorpromados* = *Chlor-Promanyl* = *Chlor-PZ* = *Contomin* = *Cromedazine* = *Elmarin* = *Esmind* = *Fenactil* = *Hebanil* = *Hibanil* = *Hibernal* = *HL 5746* = *Klorpromex* = *Largactil* = *Largaktyl* = *Marazine* = *Megaphen* = *Novomazina* = *Plegomazin* = *Proma* = *Promacid* = *Promactil* = *Promazil* = *Propaphenin* = *Prozil* = *RP 4560* = *Sanopron* = *SKF 2601-A* = *Sonazine* = *Taroctyl* = *Thorazine* = *Torazina* = *Wintermin*

***Halogenated tranquilizer, α-adrenergic blocking agent.* Possesses significant anticholinergic activity. Originally, intended for use as an antihistamine, its use as a tranquilizer was discovered by a serendipitous accident, as was *promazine*, a degradation product of chlorpromazine, after exposure to sunlight.**

***Untoward effects*:** Ataxia and taste disorders are among the miscellaneous adverse effects. Possibility of it being a bladder carcinogen in **patients** treated for polycythemia. Significant weight gain and drug-induced asthma reported with its use.

***Blood*:** Agranulocytosis frequently reported after a long latency period, probably, due to bone marrow interference with leukocyte development, although it can occur suddenly. Decreased erythropoiesis, eosinophilia, leukopenia, thrombocytopenic purpura, pancytopenia, monocytosis, and aplastic anemia reported with its use. Hemolysis reported, due to formation of a stable photoproduct or by an immunologic response and it may be difficult to distinguish from true aplastic anemia. Hemolytic anemia, usually of the autoimmune Coomb's positive-type, occurs. Purpura also reported. Agranulocytosis often occurs with jaundice, especially in fatal cases. Hemolysis and clinically significant anemia has occurred in **patients** with glucose-6-phosphate dehydrogenase deficiency. It may cause damage to platelets and other coagulation factors. Has hypercholesteremic properties and reduces serum urates.

***Cardiovascular*:** Hypotension often develops as it blocks the sympathetic nervous system at a number of sites, especially peripheral ones and also by depression of the hypothalmic centers. It usually occurs as orthostatic hypotension, especially in older **people**, even at low dosage. Has caused electrocardiographic abnormalities, ventricular arrhythmias, myocardial ischemia, and tachycardia. A schizophrenic 30-year-old **female** treated with 75 mg tid/ 10 years developed transient heart block with right bundle branch block. Hypothermia occurs and a **schizophrenic** on 300 mg/days collapsed in a swimming pool.

***Chromosomes*:** Chromosome breaks and other abnormalities reported.

Clinical tests*:** False positive phenylketonuria test with ***Phenistix®. Hospital **personnel** must be made aware of its use by a **patient**, to better interpret blood counts, liver function, circulatory and digestive problems. Has caused false pregnancy tests on urine, especially those with breast engorgement, and also on ***bilirubin*** tests. Causes false positive on ***metyrapone*** tests, and an increase in ***uric acid*** levels.

***Central nervous system*:** In a Veterans Administration series, there was a 12% incidence of convulsions and 7% with delirium tremens in the treatment of alcoholism. It has aggravated or unmasked cases of myasthemia gravis. Disturbances of consciousness ranged from drowsiness to coma after 1–3 weeks of treatment. Acute dystonia developed within 6–36 h in 3% (5/165) ***narcotic* addicts**. It has strong epileptogenic potential. Catatonia developed within 24 h in a 40-year-old **female** with a 9 year history of recurrent mania after 200 mg, IM and 100 mg every 6 h afterward. Simlar cases also reported. It caused slowing of electroencephalogram activity and was most pronounced in **patients** with a history of encephalopathy. Dyskinesias, dystonia, psychotic reactions, assorted extrapyramidal symptoms, trismus, and tremors reported. Agitation or depression also noted occasionally. It will aggravate parkinsonism. A tardive dyskinesia, often irreversible, has occurred. Although it's recommended that its use be discontinued 24 h before anesthesia, one 67-year-old **male** discontinued **his** treatment 7 days before surgery and became uncontrollable in the operating room. Two **patients** who followed abstinence recommendation suffered grand mal seizures 30 min after **their** surgery was completed. A New York State Anesthetic Study Committee then recommended that ½ **their** oral dose be given parenterally, when the surgery is completed, despite the inherent risk of severe circulatory depression. A potentially lethal neuroleptic malignant syndrome has

occurred with chlorpromazine, due to its ability to block ***dopamine*** receptors.
Eye: Brown, granular (occasionally, whitish) pigmentation of the skin, cornea, lens, and retina in those **patients** on high dose therapy for 3–5 years. Ocular effects first reported 10 years after its introduction. Anterior cortical lens hypersensitivity opacities, posterior corneal opacities, and optic atrophy reported. It may, on occasions, increase intraocular pressure, cause temporary cycloplegia, temporary toxic amblyopia and miosis, oculogyric crises, and phototoxicity.
Fetus and neonate: It crosses the placental barrier, where it competes with ***albumin***-binding sites, may cause hypoglycemia, goiter, increased **fetal** mortality and teratogenicity (omphalocele, ectromelia, syndactyly, clinodactyly, brachymesophalangy, club hands, club feet, microcephaly, endocardial fibroelastosis, and stillbirths). Chromosomal abnormalities, involuntary muscle movements, jaundice, and possible pigmentary retinal disease reported in the **fetus**. Part of its ill-effects in the **newborn** is due to the immature detoxification processes in the **neonate**. Levels in breast milk of treated mothers are extremely variable – from physiologically insignificant amounts to higher than **maternal** plasma levels leading to drowsiness, extrapyramidal syndrome, respiratory distress, jaundice, and lethargy in the **baby**. A **neonate** had fever and cyanotic spells as possible signs of withdrawal, following **maternal** exposure during pregnancy.
Gastrointestinal: Fatal paralytic ileus in a 19-year-old **female**. It appears to have unmasked a case of adult-onset Hirschsprungs disease in a 53-year-old **male**. Suspected as a cause of violet-red patches on loops of small intestines noted during surgery on a 37-year-old **male**. Unusual deposits of ***melanin*** in many internal organs in a number of fatal cases. Constipating, due to its anticholinergic effects.
Hyperglycemia, diabetes: Increases blood ***glucose*** and also some antagonism to the hypoglycemic effects of ***insulin*** and ***oral hypoglycemics***. Causes a so-called "***phenothiazine*** diabetes" in 15–27% of treated **patients**. After drug withdrawal, remission occurs in only 25% of these **patients**.
Immunity: Use is associated with increased ***IgM*** after long-term (> 2½ years) therapy.
Libido or sexual dysfunction: Libido decreases, impotence, delay or failure to ejaculate, and priapism. It stimulates ***prolactin*** production, which may cause the impotence, galactorrhea (5%) in **females** and gynecomastia in **men**.
Liver: Cholestatic hepatocanalicular jaundice (1–8% – occasionally fatal and usually occurring in the second to fourth weeks, although it has occurred after a single dose), and increases alkaline phosphatase. Prodromal symptoms are fever, chills, and abdominal distress. Do NOT administer to **patients** with known liver disease. Eosinophilia and dark urine also reported.
Poisoning: Fatal for a 3-year-old **male** accidentally ingesting 800 mg. Death in 19-year-old **male**, 15½ h after ingesting 250 mg. Tachycardia noticed 4 h after ingestion. **He** received stomach lavage. At 10½ h after ingestion, temperature was 103 °F, then 104 °F, tachycardia and decreased blood pressure, symptoms progressed into irreversible unconsciousness. Extreme hypotension blood pressure 85/50 and 60/0 in 48-year-old **female** and 47-year-old **female**. Both recovered after lavage and ***amphetamine***. A 16-year-old **female** swallowed 4.3 g in a suicide attempt and survived. Respiratory distress is a common early symptom. Single drug fatal blood levels reported as 5.7–52.5 mg/l.
Pyrexia: Hyperpyrexia reported in many toxic cases. The drug alters normal thermoregulation. **Patients** should avoid being in the hot sun or in cold conditions. Hypothermia also reported in some older **people**. Neuroleptic malignant syndrome also occurs.
Renal: May cause transient reduction in renal plasma flow. Use in renal insufficiency has caused death in 47-year-old **male** who received 780 mg/day/5 days.
Skin: Photosensitivity allergic reactions (exaggerated sunburn-type of reaction with rashes, hyperpigmentation of purple or slate-gray tones, delayed erythema, urticaria, maculopapular eruptions, lichenoid, bullous and eczematous reactions). Can occur months after termination of exposure. May cause toxic epidermal necrolysis. Fixed-drug reactions, as well as contact dermatitis and urticaria noted. Cross-reacts with ***phenergan*** and ***tripellenamine***. Black escharing of skin over injection site in 57-year-old **male**. Oral dryness, ulcers, and moniliasis, as well as lip swellings, reported. **Physicians** and **nurses** who handle it are at risk. Repeated exposure has caused severe dermatitis in **nurses**. It is found in the urine of treated **patients**.
Systemic lupus erythematosus: Lupus-like illness (low-grade fever, polyserositis, dyspnea, raised antinuclear antibodies, bilateral pleural effusions, etc.).

Usual therapeutic serum levels are 0.03–0.1 mg/l, toxic at 1–2 mg/l, and fatal at 4 mg/l.

LD_{LO}, **human**: 50 mg/kg; **woman**: 200 μg/kg.

Drug interactions: Potentiated by ***ethyl alcohol*** with further impairment of muscular coordination and judgement. Serum levels of ***ethanol*** and ***acetaldehyde*** are often increased. Antagonist to ***amphetamine***, and this effect is used clinically. Antagonist the antihypertensive effect of ***guanethidine***. ***Antacids***, particularly ***aluminum*** and ***magnesium hydoxides*** or ***trisilicates***, by adsorption, impair its absorption. ***Anticholinergics*** also inhibit its absorption and decrease its plasma levels. Can produce extrapyramidal reactions and hypertension with ***monoamine oxidase inhibitors***; orthostatic hypotension with ***antihypertensives***. Prolongs ***antipyrine*** half-life. Poten-tiates ***pargyline*** and ***reserpine*** effects. It potentiates the sedative effects of ***alphaprodine*** and ***barbiturates***, yet, ***phenobarbital*** can increase its metabolization and excretion by 11–69%, due to enzyme induction. Its use increases the toxicity of ***demeclocycline***. Drowsiness produced by it is less in cigarette **smokers**. Possible failure of ***oral contraceptives*** by its stimulation of liver microsomal enzymes. Increased neuro-muscular blockade occurs when used with ***gallamine***. Metabolization and excretion of ***imipramine*** reduced by it. With ***lithium***, either one can decrease the serum level of the other. The same may be true for ***propranolol***. Plasma concentration increases when used with ***nortriptyline***. May have caused deaths when therapy combined with ***isoproterenol*** in **asthmatics**. It antagonizes the behavioral effects of lysergic acid diethylamide (***LSD***), but does not modify ***LSD***-induced increases in brain ***serotonin***; it decreases the mydriasis and unusual visual experiences caused by ***psilocybin*** and can aggravate ***PCP***-induced abnormal behavior. Antagonizes effects of ***phenmetrazine***. Inhibits ***phenytoin*** metabolization and increases its toxicity. ***Orphenadrine*** increases liver microsomal enzymes to decrease its blood level. Potentiated by ***salicylates***. Convulsions developed in a **child** given it a few days after treatment with ***piperazine***. ***Piperazine***, apparently, exaggerates its extrapyramidal effects. The results were confirmed in six **dogs** and nine **goats**. A 27-year-old **female** developed cardiac arrest after ingesting 8 g and 150 mg ***flurazepam***. **She** was successfully resuscitated and **her** ventricular tachycardia was controlled with IV ***lidocaine***. Use of ***epinephrine*** with it may enhance the hypotension.

In **horses**, 2–4 mg/kg causes tachycardia, hypotension, and depression. **Calves** receiving 50% more than usual dosage (1 mg/kg) develop incoordination. 75–250 mg to **sows** promotes lactation. Daily injections to **dogs** cause non-fatal kidney and lung pathology. Absorption from the gastrointestinal tract is rapid and toxicity in **animals** is, essentially, the same for oral or SC use. High dosage leads to fatal hypotension and jaundice, eosinophilia, and agranulocytosis. Use in pregnant **goats** can cause fetal tachycardia and degenerative liver pathology.

LD_{50}, **rat**: 141 mg/kg; **rabbit**: 20 mg/kg; IV LD_{50}, **dog**: 37 mg/kg; **mouse**: 23 mg/kg; **rabbit**: 16 mg/kg.

CHLORPROPAMIDE

= *Adiaben* = *Asucrol* = *Catanil* = *Chloronase* = *Diabechlor* = *Diabenal* = *Diabetoral* = *Diabinese* = *Diabitex* = *Insulase* = *Melitase* = *Millinese* = *Oradian* = *P-607* = *Stabinol*

Antidiabetic.

Untoward effects: ***Blood***: Thrombocytopenia (hemorrhagic diathesis [gums, nose, generalized petechiae and ecchymoses] and splenomegaly also reported), agranulocytosis (fatal in 70-year-old **female** after short-term therapy with 100 mg qid), leukopenia, leukocytosis, purpura, neutropenia, pancytopenia, eosinophilia, aplastic anemia, hemolytic anemia, red blood cell aplasia, hyponatremia, and water intoxication.
Cardiac: Atrial dysrhythmia in a diabetic 63-year-old **female**. Since it causes fluid retention, it, apparently, caused nocturnal angina in a 76-year-old **male** receiving 500 mg/day. Has caused decreased blood pressure and serious hypotension.
Clinical tests: Higher alkaline phosphatase and sulfobromophthalein retention results occur.
Eye: Blurred vision, corneal opacity, color disturbances, central scotoma, mydriasis, diplopia, toxic amblyopia, change in refractive error, visual hallucinations, and exudative conjunctivitis.
Fetus and neonate: Not recommended for use during pregnancy. Has caused **fetal** mortality, microencephalopathy and spastic quadriplegia, apneic spells, and prolonged **neonatal** hypoglycemia. Most of these occurred with doses of 500 mg/day to the mother.
Hypoglycemia: Fatal hypoglycemic coma with brain damage reported. It has occurred as a result of assorted dispensing errors (℞ instead of ***chlorpromazine, quinidine***, or ***aluminum hydroxide***). Aphasia and confusion has preceded a case of severe hypoglycemia. Intensive treatment increases risk of hypoglycemia.
Liver: Hypersensitivity a cause of cholestatic hepatocanalicular jaundice, often appearing after 2–6 weeks with an incidence of 0.4%. Alkaline phosphatase levels increase in about 25% of **patients**. Occasionally occurs 1–2 years after usage. A case of anicteric hepatitis also reported.
Miscellaneous: Cranial neuropathies are rare and have affected the eighth cranial nerve, as well as the eye. Peripheral neuropathies have caused numbness, tingling, or burning sensations of the extremities. Mild adverse effects may including nausea, vomiting, epigastric pain, dizziness, weakness, headache, paresthesia, fever, and facial flushing.
Poisoning: Fatal in a 11-year-old **female** intentionally ingesting 7.5 g. Hospitalized in a coma. Irreversible neurological damage in 3-year-old **male** ingesting it with ***aspirin***. ***Aspirin*** may compete with it for renal excretion. Serum level was 22.6 mg/100 ml (toxic levels have ranged between 20 and 75 mg/100 ml. Therapeutic levels are 3–14 mg/100 ml. Factitious hypoglycemia in **nurse** who self-administered because of a family history of diabetes. Self-poisoning cases, including one whose blood sugar levels could not be raised by IV ***glucose*** until ***glucagon*** was administered. A pregnant 22-year-old **female** misread instructions and took 250 mg every 2 h, instead of every 12 h. **She** suffered from irritability, headache, and, on the third day, diplopia, nausea,

tremor, sensation of alternating heat and cold, sweating, somnolence, occasionally myoclonia, mental confusion, and stupor. **She** had severe reversible neurological sequelae 6 months later.
Renal: Antidiuretic effect.
Skin: Photosensitivity reactions; eczematous photodermatitis. Erythema multiforme and Stevens-Johnson syndrome, toxic epidermal necrolysis, maculopapular rash, morbilliform rash, porphyrias, exfoliative dermatitis – lichen planus-like (often on the lips), and purpura. Fatal case of toxic erythema in 71-year-old **female** after approximately 2 weeks of therapy. **Users** may become sensitized to widely used para-amino compounds, such as hair dyes, ***p-phenylenediamine***, sunscreens containing ***p-aminobenzoic acid*** or its esters, ***benzoic acid***, local anesthetics, such as ***benzocaine***, and ***p-aminosulfonamide*** compounds.
Thyroid: Clinically significant antithyroid activity in some **patients**.

Usual therapeutic serum levels are 30–150 mg/l; toxic at 700 mg/l.

LD_{LO}, **woman**: 300 mg/kg.

Drug and food interactions: ***Alcohol, ethyl***: Causes an additive hypoglycemia and a ***disulfiram***-like intolerance to ***alcohol*** with nausea, vomiting, flushing, headache, dyspnea, tachycardia, weakness, and dizziness, and an incidence rate of 33%. Genetically, this is an autosomal dominant effect in 30% of **Caucasians**. ***Alcohol*** inhibits its metabolization by about 40%. An **alcoholic** 43-year-old **male** developed angina when **he** also drank ***ethanol***. Prolonged fasting or malnutrition increase its effect.
Antidepressants: ***Tricyclics*** potentiate its hypoglycemic effect.
Anticoagulants: ***Dicoumarol*** has doubled its half-life and caused hypoglycemic coma, especially in elderly **patients**.
Barbiturates, hypnotics, and sedatives: Their hypnotic and sedative effects may be prolonged.
Chloramphenicol: Causes an increased half-life of chlorpropamide, which can result in severe hypoglycemia.
Chlorothiazide: May cause loss of diabetes control.
Clofibrate: Slightly increases its half-life, requires a slight reduction in dosage.
Demeclocycline: Its toxicity is increased by chlorpropamide administration.
Monoamine oxidase inhibitor: Interaction may require reduction in chlorpropamide dosage.
Miscellaneous: ***Karela***, a curry ingredient from *Momordica charantis*, increases its hypoglycemic effect. Also potentiated by ***halofenate, oxyphenbutazone***, and ***phenylbutazone***. Causes a small increase in ***phenytoin*** plasma concentration.
Orphenadine: A case of hypoglycemia in a **patient** deserves attention, because they are often frequently prescribed together.
Phenylbutazone: Potentiates chlorpropamide with increased hypoglycemic effect.
Phenyramidol: Causes an enhanced hypoglycemic effect.
Propranolol: A 59-year-old **male** developed hyperglycemia after its co-administration. Can also cause enhanced hypoglycemia.
Rifampicin: Decreases the effectiveness of chlorpropamide.
Salicylates: ***Aspirin*** or ***sodium salicylate*** potentiates its hypoglycemia.
Sulfonamides: ***Sulfisoxazole, sulfamethazine, sulfamethoxazole***, and ***sulfaphenazole*** increase its hypoglycemic effects.
Vasopressin: Action is potentiated by it.

Has produced corneal opacities in **rats**. Potentially toxic to **dogs**.

IP LD_{50}, **rats**: 580 mg/kg.

CHLORPROPHAM

= Chlor-IFC = Chloropropham = CIPC = Furloe = Sprout-Nip

Herbicide, plant growth regulator.

Untoward effects: LD_{LO}, **human**: 5 g/kg.

LD_{50}, **rats**: 1.2–8 g/kg; **rabbit**: 5 g/kg. Inhalation LD_{50}, **rat**: 3.35 g/kg.

CHLORPROTHIXENE

= N-714 = Taractan = Tarasan = Truxal = Truxaletten

Antipsychotic, ataractic, central nervous system depressant, tranquilizer.

Untoward effects: ***Blood***: Increases thrombocytes, decreases prothrombin time, neutropenia, and a lupus erythematosus syndrome.
Cardiac: Electrocardiogram changes, sinus tachycardia, decreases blood pressure.
Extrapyramidal: Occasional rigidity, inability to close mouth, tremor, dystonia, dyskinesia, spasmodic torticollis, dizziness, convulsions, and akathesia on high dosage.
Eyes: Lenticular deposits and opacities occur rarely. **Patients** should protect their eyes from exposure to sunlight or ultraviolet light by wearing sunglasses. Miosis reported.
Gastrointestinal: Rarely, nausea, vomiting, gastritis, and dry mouth.
Liver: Obstructive jaundice reported.
Miscellaneous: Galactorrhea (potentiated by ***oral contraceptives***), increases libido, decreases ejaculation, impotence, poor temperature regulation (pyrexia), skin pigmentation, lupus-like syndrome, decreases serum urates after 2–3 days, respiratory depression, oliguria, hematuria, drowsiness, and false increases in urine

bilirubin tests. Sudden withdrawal may precipitate nausea and vomiting.
Poisoning: One Danish poison center reported 26 cases in 5 years (6/26 were comatose; 5/26 had brainstem seizures; three had transient neurological signs; one fatality after 6 g causing violent seizures, respiratory arrest, and laryngospasm). A 13 month **infant** died after ingesting 1.4 g.

Therapeutic blood levels are approximately 0.5–1 mg/100 ml; toxic and lethal levels are > 1.5 mg/100 ml. Others list therapeutic serum levels as 0.02–0.2 mg/l; toxic at 0.4 mg/l; and lethal as 0.8 mg/l.

LD_{LO}, **human**: 50 mg/kg.

Causes anal sphincter relaxation in many **animal** species.

LD_{50}, **rat**: 380 mg/kg; **rabbit**: 182 mg/kg.

CHLORPYRIFOS

= *Dowco 179* = *Dursban* = *ENT 27311* = *Lorsban* = *Pyrinex* = *Sect-A-Chlor*

Organophosphorus insecticide, miticide.

Untoward effects: Harmful and even fatal if swallowed. Avoid inhaling sprays or mists or contact with eyes, skin, or clothing. Wash thoroughly after handling. Avoid contamination of feed or feeding utensils. May cause irreversible eye damage. Do NOT re-enter treated areas in less than 24 h, unless suitable protective gear is worn. A 23-year-old **female** was found in a stupor next to an empty container with episodic grimacing, choreoathetotic movements of all **her** limbs, and very pronounced in upper limbs, occurring every 3 min. A 34-year-old **female secretary**, unaware that **her** building was sprayed with it to control **cockroaches**, had a 5 year disability with rashes, weight gain, and arthritis. In the U.S., use is now being limited by law to professional **applicators**.

Highly toxic to **honeybees**. Toxicity in **animals** regularly reported. Death in 7/7 newborn **pigs** after applying a $2\frac{1}{2}$% solution to their tails and umbilicus. Symptoms including weakness, lethargy, ataxia, lateral recumbency, limb paddling, tremors, salivation, and diarrhea. Chronic toxicosis with anorexia and weakness reported in a **cat**, where it was used every 3 weeks as a premise spray for flea control. **Llamas** exposed by topical application (25 mg/kg) had dramatic decrease of pseudocholinesterase levels (requiring 5–7 weeks for return to near normal levels). Miosis, hyperglycemia, metabolic acidosis, salivation, and inability to stand occurred. Embryotoxic and fetotoxic to **mice**. **Bulls** have been particularly susceptible to its adverse effects. Symptoms develop slowly over 3–10 days and last up to 10 weeks, if death doesn't occur.

CHLORPYRIFOS-METHYL

= *Dowco 214* = *ENT 27520* = *OMS-1155* = *Reldan*

Organophosphorus insecticide.

Untoward effects: Cholinesterase inhibitor. A pesticide **applicator** tainted millions of bushels of ***oats*** with it and caused a class action suit against General Mills for using it in cereals. Your protectors, government food regulators, declared that the millions of boxes did not pose a **human** health hazard.

Symptoms in affected **quail** and **pheasants** are ataxia, ataraxis, falling, ptosis, catatonia, bradypnea, and dyspnea.

LD_{50}, **rat**: 941–2140 mg/kg; **mouse**: 1122 mg/kg; **rabbit**: 2 g/kg; **mallard ducks** and **pheasants**: > 2 g/kg. LC_{50}, **quail**: > 5,000 ppm.

CHLORQUINALDOL

= *Afungil* = *Chloroquinaldol* = *Gyno-Sterosan* = *Gynotherax* = *Quesil* = *Saprosan* = *Siogène* = *Siosteran* = *Sterosan* = *Steroxin*

Bacteriostatic and fungistatic agent.

Untoward effects: Irritates mucous membranes and eyes. Restlessness, depression, and gastritis can follow ingestion. Dermatitis in 5% of **patients** in patch testing with a petrolatum vehicle. Can cause serious permanent neuropathy. Other toxicity may be similar to that of ***halquinol*** (q.v.). Causes renal damage in **hamsters**.

LD_{50}, **rats**: 660 mg/kg; **rabbits**: 160.4 mg/kg; **dogs**: approximately 2.25 g/kg.

CHLORTETRACYCLINE

= *Acronize* = *Algromix* = *Aureocina* = *Aureocycline* = *Aureomycin* = *Aurofac* = *Biomitsin* = *Biomycin* = *Biovetin* = *Biovit* = *Chlorachel* = *CLTC* = *Chrysomykine* = *Duomycin* = *Factor A-377* = *Fermycin* = *Pfichlor* = *ViMycin*

Antibiotic.

Untoward effects: ***Blood and cardiovascular***: Coagulation time reduced in **man**, **rabbits**, and **cats**. Two cases of increased clotting time in **man**. Agranulocytosis and ulcerative necrotizing stomatitis in 22-year-old **female** after 8 days of treatment with ***penicillin*** as well. Phlebitis with long golden-colored thrombus extricated from 8-year-old **female**, 12 days after successful treatment via catheter for an infection resistant to ***penicillin*** and ***chloramphenicol***. Few details available re: thrombocytopenia in 14 **patients**, neutropenia in seven, pancytopenia in three, non-thrombocytopenic purpura in two, leukopenia in one, and aplastic anemia in ten.
Clinical tests: Turns urine a bright yellow, which can cause errors in jaundice tests.

Central nervous system: Vertigo and vestibular dysfunction.
Dental and bone: Gray-brown discoloration of teeth in **children** < 10 years receiving it. Defective enamel development after fourth months of pregnancy. May decrease **fetal** skeletal growth. See ***Fetus*** below. Taken up by newly formed bone and calcifying cartilage, where it fluoresces under ultraviolet light. Bulging fontanels in treated young **infants**, disappearing rapidly with drug discontinuance.
Eye: Has reduced color vision perception.
Fetus: May be hepatotoxic. Hypoplasia with yellow-brown staining of teeth in nearly 50% exposed from 29th week until term. If given close to term, the crowns of permanent teeth may be stained. Use in the first trimester not apt to cause such defects. Fetal levels are 25–75% of maternal levels. Micromelia and syndactyly reported. Possibly, a cause of cataracts and cleft palate. See ***Dental and bone*** above.
Gastrointestinal: Minimal adverse effects include diarrhea, nausea, vomiting, anorexia, abdominal pain, intestinal hemorrhage, enterocolitis, fungal overgrowth (colon, rectum, or body orifices), glossitis, and dysphagia.
Infection: May activate *Monilia* and *Candida* species.
Liver: Frequently produces a fine fatty vacuolization of hepatic parenchyma. Parenteral doses of 1.5 g or more and high oral dosages are hepatotoxic, especially in pregnant **women** with complicating pyelonephritis or in **anyone** with renal insufficiency. Causes cholestatic hepatitis and centrilobular necrosis. Abnormal liver function tests, include increased SGOT, increased SGPT, increased alkaline phosphatase, increased thyroid turbidity, and increased sulfobromophthalein retention.
Miscellaneous: Hypotension, lethargy, coma, anaphylactoid reactions and shock, local moniliasis, anosmia, pericarditis, and brown-black microscopic discoloration of thyroid.
Renal: Azotemia without increases in ***creatinine***. Degradation products of outdated materials or those stored in a hot environment have caused a Lignac-Fanconi syndrome with tubular dysfunction, glucosuria, proteinuria, aminoaciduria, and metabolic acidosis. Increased blood urea nitrogen.
Skin: Fixed-drug eruptions (hypersensitivity with erythematous, bullous, urticarial and/or pruritic plaques). Hyperpigmented areas may remain. Photosensitivity reactions in 10/63 **patients** receiving 15 mg/kg. Occasionally Gram-negative folliculitis. Angioneurotic edema and exacerbation of systemic lupus erythematosus.

Some quote therapeutic serum levels as 1–5 mg/l and occasionally 10 mg/l; toxic at 30 mg/l.

Drug interactions: ***Antacids***: Those containing ***calcium***, ***magnesium***, and ***aluminum*** decrease its gastrointestinal absorption. Dairy products, ***iron***, and ***zinc*** behave similarly.
Anticoagulants: ***Bishydroxycoumarin*** use requirements have been doubled in some **patients** with its use. Effect may be due to decrease in intestinal bacterial synthesis of ***vitamin K***. Monitor prothrombin time levels with any ***coumarin*** or ***indandione anticoagulants***.
Lithium: Avoid its use. Acneiform eruptions due to ***lithium***, as it increases ***lithium*** toxicity and produces a nephrotoxic effect.
Penicillin G: Significantly decreases clinical effectiveness of ***penicillin***. **Patients** with pneumococcal meningitis had a significant increase in mortality, compared to either antibiotic used alone.
Riboflavin: Its loss of biological potency with ***vitamin B complex*** is attributed to the latter's ***riboflavin*** content.

Individual **pigs** have shown sensitivity to large doses with ear and foot swellings, and anal and nasal hemorrhages after chronic administration of large doses. Of the expiramental **laboratory animals**, the **guinea pig** is particularly susceptible to its indirect toxicity resulting from the rapid kill of Gram-positive organisms and the overgrowth of Gram-negative toxin producing organisms. Care must, therefore, be used in its administration to **chinchillas**, **rabbits**, and **rodents**. Excessive oral administration of large dosage to **ruminants** may be toxic, due to secondary effects from sudden alteration of the rumen flora. **Cats**, **dogs**, and **pigs** have occasionally shown nausea and diarrhea from high dosage. To avoid undesirable tissue residues, do not use tissues from **animals** treated within 48 h of slaughter for **human** consumption (4 days after in-water treatment of baby **calves**).

Diarrhea after its oral use in **horses** may not occur until several weeks after administer. Bile flow decrease *in situ* in **guinea pigs**, **monkeys**, **cats**, and **dogs**. After IV use in **cattle**, ptyalism, dyspnea, lung edema, and even death have been reported.

LD_{50}, **rats**: 10.3 g/kg; IV, 118–150 mg/kg.

CHLORTHALIDONE

= *Chlorphthalidolone* = *G 33182* = *Hydro-long* = *Hydroton* = *Hygroton* = *Phthalamodine* = *Thalitone*

Diuretic, antihypertensive. **Action is more prolonged than that of the *thiazides*.**

Untoward effects: ***Blood***: Leukopenia, agranulocytosis, thrombocytopenia, neutropenia, aplastic anemia, hyperlipidemia, hypokalemia, hyponatremia, increased blood urea nitrogen. Rarely, hypercalcemia.
Cardiovascular: Orthostatic hypotension, necrotizing angiitis, and tachycardia. Decrease of extracellular fluid and electrolytes can lead to cardiovascular collapse.
Central nervous system: Headache, dizziness, vertigo, paresthesias, muscle spasm, weakness, restlessness, drowsiness, convulsions, and, eventually, coma.

Eye: Transient myopia, retinal and periorbital edema, uveitis, optic atrophy, blurred vision, xanthopsia, conjunctivitis, and eye pain.
Fetus and neonate: Possibility of congenital anomalies in 5/291. Appears in maternal milk, with half-life of 60 h.
Gastrointestinal: Gastric irritation, nausea, vomiting, cramping, diarrhea, and pancreatitis.
Glucose: Hyperglycemia and glycosuria. May cause the need for increases in ***insulin*** or ***sulfonylurea*** dosage. A case of hyperosmolar hyperglycemic coma in 43-year-old **male** with essential hypertension 14 days after 100 mg/day.
Liver: Cholestatic jaundice. Hepatic coma in 45-year-old **female** after 100 mg qid/5 days.
Miscellaneous: Fever and malaise, fatigue and sweating. Pancreatitis and necrotizing vasculitis are rare.
Renal: Hyperuricemia, may precipitate acute gout attacks. Acute interstitial nephritis.
Sexuality: Impotence, decreased libido, and decreased vaginal secretions.
Skin: Purpura, photosensitivity, urticaria, rash, toxic epidermal necrolysis, and cutaneous vasculitis.

Drug interactions: ***Atenolol***: Cold extremities in nearly 30% of **patients**. Potentiates ***atenolol's*** antihypertensive effect.
Digitalis: Its effect on electrolyte excretion can induce or aggravate ***digitalis*** and ***digoxin*** toxicity.
Lithium: Can induce ***lithium*** toxicity.
Miscellaneous: ***Epinephrine*** decreases blood pressure when given with ***alcohol***, ***narcotics***, or ***barbiturates***. Enhances the effect of ***gallamine***.
Probenecid: Blocks its ability to inhibit the renal tubular secretion of ***uric acid***.
Sulfinpyrazone: Blocks its ability to inhibit renal tubular excretion of ***uric acid***.
Tubocurarine: Possibility of increasing toxicity of ***tubocurarine***.

CHLORTHION

= *Chlorothion* = *Compound 22/190*

***Organophosphorus insecticide.* A cholinesterase-inhibitor.**

Untoward effects: Salivation, nausea, vomiting, diarrhea, weakness, miosis, muscle twitching, headache, dizziness, incontinence, convulsions, coma, and respiratory failure can occur.

LD_{LO}, **man**: 500 mg/kg.

50 mg/kg, but not 25 mg/kg is lethal to **cattle** > 6 months.

LD_{50}, **rats**: 625–1500 mg/kg; **mouse**: 1250 mg/kg, acute dermal, **rabbits**: 1.5–4.5 g/kg.

CHLORTHIOPHOS

= *Celamerk S 2957* = *Celathion* = *OMS 1342*

***Organophosphorus insecticide.* A cholinesterase-inhibitor.**

Untoward effects: In **animals**, symptoms of poisoning were tremor, clonic convulsions, exophthalmos, lacrimation, salivation, dyspnea, and paralysis.

LD_{50}, **rat**: 8 mg/kg; **mouse**: 141 mg/kg; **guinea pig**: 58 mg/kg; **rabbit**: 20 mg/kg; **quail** and **hen**: 45 mg/kg; acute dermal LD_{50}, **rat**: 58 mg/kg, **rabbit**: 48 mg/kg.

CHLORZOXAZONE

= *Biomioran* = *Chlorozoxazone* = *Paraflex* = *Solaxin*

Skeletal muscle relaxant, anxiolytic.

Untoward effects: Gastrointestinal disturbances (nausea, vomiting, heartburn, abdominal discomfort), hypothrombinemia, headache, drowsiness, dizziness, dry mouth, spasmodic torticolis, and hepatic toxicity. An allergen-type of skin reaction (petechiae or echymoses) and anaphylaxis is rare – extremely rare. Will production an orange to purplish-red color of urine. Hepatotoxicity can be fatal. Early symptoms are usually fever, rash, anorexia, nausea, vomiting, fatigue, upper right abdominal quadrant pain, jaundice, and dark urine.

Drug interactions: ***Alcohol*** will enhance its drowsiness effect. Serious idiosyncratic centrilobular hepatic necrosis in 55-year-old **female** after therapeutic doses for several months combined with ***APAP***. Jaundice recurred within 5 h of rechallenge with the combination, but 1 week of treatment with ***APAP*** alone had no ill-effect. A single ingestion of 50 g ***watercress*** inhibits its metabolization.

Can cause paralysis in **chicks**.

LD_{50}, **rat**: 763 mg/kg; **mouse**: 658 mg/kg; **hamster**: 662 mg/kg.

CHOCOLATE

= *Cioccolata* = *Schokolade*

Made from 30 different types of *cacao* beans. Many brands contain *vanillin* (q.v.). Often contains sugar, starch, and flavors. Approximately 1 million tons consumed annually. Contains *theobromine* and *caffeine*. Its source, the plant *Theobroma cacao*, was brought from Mexico by Hernando Cortes, who learned about it from the Aztecs. Converted from a drink to a solid form in 1847 by the British. The Swiss introduced milk chocolate ~30 years later.

Untoward effects: Contains vasoactive ***tyramine*** and can precipitate migraine headache attacks. This pressor amine, in combination with ***monoamine oxidase inhibitors***, can cause a hypertensive crisis by stimulating the release of ***norepinephrine*** and ***serotonin*** (particularly in the brain). Has caused facial flushing. Ingestion of ***cacao*** waste products has caused ***theobromine*** poisoning, with ensuing death from sudden cardiac failure. Hypersensitivity manifested by ***gastrointestinal***: abdominal pain, itching of mouth, vomiting; ***miscellaneous***: aversion, chills; and ***respiratory symptoms***: clogging of nose, sneezing, itching, cough, and wheezing. A 23-year-old **female** with a chocolate allergy presented to an emergency room after sneezing every 5–10 s for 2 h. Prior to the sneezing, an inebriated dinner companion shoved some chocolate-topped ***ice cream*** into **her** nose. Washing **her** nasal passages did not help. Local application of ***cocaine*** stopped the sneezing within 5 min. ***Skin***: Eczematous eruptions, pruritus, morbilliform or scarlatinaform eruptions, urticarial eruptions, circumoral erythema, and redness of ears. Stevens-Johnson syndrome. Five chocolate **dippers** in industry exhibited skin eruptions on **their** fingers, thumbs, and the distal half of the dorsum of one hand.

Coumarin, which tastes and smells like ***vanilla***, was often added to chocolate. In the 1940s, I complained to the FDA that it caused nosebleeds. They said I was wrong, and 10 years later, this use was withdrawn.

Maternal intake has been considered as a cause of rashes, eczemas, and allergic manifestations in nursing **infants**. Its ***theobromine*** concentration in **mother's** milk is > 80% maternal serum levels 2–3 h after ingestion of chocolate. Vesicular hand eczema in **people** with ***chromate***, ***cobalt***, and ***nickel*** dermatitis may find **they** can cross-react with chocolate. *Salmonella* has been found to survive in milk chocolate. Avoid its use by **patients** on ***lysine*** therapy for herpes simplex infection. Contains high levels of ***oxalates*** and should be avoided by **patients** prone to ***calcium oxalate*** urinary stones. Also contains a high level of ***copper*** and should be avoided in Wilson's disease. Avoid its use in **patients** with esophageal reflux, as it aggravates acid reflux. Colors feces dull gray–dark red–chocolate brown. Can increase pain in interstitial nephritis cases. Dark chocolate contains 21 mg ***caffeine***/oz and 132 mg ***theobromine***/oz; milk chocolate contains 6 and 55 mg/oz, respectively; and white chocolate contains 850 and 250–450 μg/oz, respectively.

1–3 oz can be lethal to a 20 lb **dog** (usually 1 oz/kg, but deaths reported at 115 mg/kg). **Dogs** cannot readily excrete its ***theobromine*** and ***caffeine*** content. Baker's chocolate is particularly toxic, less toxicity with milk chocolate or sweetened chocolate (~4–60 mg/oz). Symptoms are usually vomiting, diarrhea, diuresis, hyperactivity or restlessness, tachycardia, tachypnea, cyanosis, ataxia, and seizures. These occur within 4–5 h after ingestion. Excitement, sweating, increased respiratory rate, tachycardia, convulsions, and collapse occur in **calves** after ingesting it.

CHOLESTEROL

Use: As an emulsifying agent primarily in topical pharmaceuticals.

The predominant sterol in higher animals. Found in all body tissues, especially fats, oils, brain, and spinal cord. Its esters are found in wool fat (***lanolins***). A basic structure in sex hormones.

Untoward effects: Low serum levels in **people** may be associated with excess cholesterol excretion, and it has been suggested that this could be a promoter of colon cancer.

Drug interactions: In **humans**, its plasma levels have been increased in varying degrees by ***phenytoin***, ***oral contraceptives***, ***estrogens***, ***vitamins C*** and ***E***, ***hydrochlorthiazide***, and ***chlorthalidone***. The ***vitamins*** have also been reported to decrease its levels.

Increased levels potentiate ***thiopental*** anesthesia in **dogs** and **rabbits**. Raised blood levels may often be found in acanthosis nigricans in **dogs**. Much of the confusion over its role in heart attacks was based on early reports with **rabbits**, who are vegetarians. **Rat** studies are also poor for extrapolation. **Rats** do not have the normal high concentration of cholesterol that **humans** have, and they lack the low-density lipoprotein fractions closely linked to **human** atherosclerosis. **Rats** are also hardly affected by dietary cholesterol and are very resistant to atherosclerosis. Bile salts are produced from it. Infant **rat** studies indicate it is essential for development of the protective myelin sheath, and surrounding nerve cells. **Cats** fed diets containing 2% levels developed marked cholesterolemia, while **cats** fed at 0.5% levels do not. An important constituent of the plasma membrane of **mammalian** cells, and it helps regulate the membrane's fluidity and secondary effects on cell membrane receptors. **Shrimp** are unable to synthesize it and they must be supplemented. Young **rabbits** fed it at a 0.25% level in their diet for 6–10 months developed xanthomas.

CHOLESTIN®

***Over-the-counter dietary supplement.* A Chinese condiment of red *yeast* (*Monasus purpureus Went*), fermented on *rice*, containing at least ten compounds, including *lovastatin* (q.v.), all of which inhibit *HMG-CoA reductase*, an enzyme that contributes to the endogenous synthesis of *cholesterol*.**

Untoward effects: Although used in the production of ***rice wine*** and as a spice in China, it is not standardized in its content or type of statins. Dependence on it for lowering ***cholesterol*** can lead to a false sense of security, especially in **those** who attempt to treat themselves without making necessary changes in dietary intake.

CHOLESTYRAMINE RESIN

= *Cholybar* = *Colestran* = *Colestyramin* = *Colyar* = *Cuemid* = *Dowex 1-X2-C1* = *MK-135* = *Quantalan* = *Questran*

Anionic ion exchange resin.

Use: Antihyperlipoproteinemic, bile salt complexer, in the treatment of pruritus associated with obstructive jaundice, as *E. coli* and *C. difficile* toxin adsorbers, and to bind overdosage of ***digitalis***. In ***kepone*** poisoning.

Untoward effects: Steatorrhea, anorexia, unpleasant taste, heartburn, constipation, nausea, abdominal distention and cramping, belching, night blindness, uveitis, hyperchloremic metabolic acidosis, reduced absorption of fat-soluble ***vitamins***, hypoprothrombinemia, possible bone thinning, or florid osteomalacia, xanthomatosis, irritation of skin, tongue and perianal areas. Hiccups, sour taste, pancreatitis, headache, fatigue, tinnitus, vertigo, dizziness, urticaria, pancreatitis, abnormal liver function tests, backache, muscle and joint pain, and arthritis are among the many adverse effects reported. Implicated as a cause of bezoar formation.

Drug interactions: Binds ***digitoxin*** and ***digoxin***, requires an increase in their dosage of up to 30%. By absorption, interferes with thyroid replacement therapy, but little effect after a 4–5 h interval between dosages. The same recommendation is made for ***non-steroidal anti-inflammatory drugs, phenylbutazone, tetracyclines,*** and ***penicillin***. Interferes with intestinal absorption of ***vitamin K, folic acid, B vitamins, vitamin D, phenylbutazone, coumadin,*** and other ***oral anticoagulants***. Resultant ***vitamin K*** deficiency may cause hypoprothrombinemia with associated bleeding tendencies. A 17-year-old **female** with erythropoietic protoporphyria received 12 g/day/4 months and developed ***iron***-deficiency anemia. Although the cause was not proven, it does inhibit ***iron*** absorption in **rats**. A 88-year-old **female** developed an unexplained hypoprothrombinemia after one packet qid. **She** was also being treated with ***ciprofloxacin*** and ***vancomycin***. Bioavailability of ***calcium carbonate, diclofenac, doxepin, hydrochlorothiazide, imipramine,*** and ***ketoprofen*** is decreased by it. Binds ***acetaminophen***, only if given shortly after ingestion. Decreased absorption of ***valproic acid***, of ***troglitazone*** by 70% and ***raloxifene*** by 60%.

CHOLINE

***Lipotrope.* A B-complex vitamin.**

Untoward effects: A common health food whose intake can lead to depression. Degraded by intestinal bacteria to ***trimethylamine***, which has a pungent, fishy, ammoniacal odor. The liver converts it to the odorless ***trimethylamine oxide***. When the liver's oxidase system is defective, the circulating ***trimethylamine*** is excreted in the urine, breath, and skin, with its characteristic rotten fish odor.

IP LD_{50}, **rat**: 400 mg/kg.

CHOLINE MAGNESIUM TRISALICYLATE

= *Trilisate*

Analgesic, antipyretic.

Untoward effects: As with all ***salicylates***. Tinnitus, fever, and hepatotoxicity may occur.

CHOLINE SALICYLATE

= *Actasal* = *Arret* = *Arthropan* = *Atrobione* = *Audax* = *Mundisal*

Analgesic, antipyretic.

Untoward effects: Unpleasant taste, some gastrointestinal upsets, vomiting, and headache in 1/7 **children** receiving 0.07–0.09 mg/kg, headache in 3/18 **children**, and vomiting in 1/18 **children**.

LD_{50}, **rat**: 1.53 g/kg.

CHOLINE THEOPHYLLINATE

= *Choledyl* = *Cholinophylline* = *Filoral* = *Oxtriphylline* = *Oxytrimethylline* = *Sabidal* = *Soliphylline* = *Teofilcolina* = *Teokolin* = *Theoxylline*

Bronchodilator.

Untoward effects: Gastrointestinal upsets (nausea, vomiting, stomach pain, diarrhea and cramping), restlessness, headache, dizziness, and insomnia. A 21-year-old **female** was hospitalized with nausea, vomiting, and diarrhea 7 h after ingesting 50 200 mg tablets. Seizures and cardiac arrest followed. **She** became comatose and died a day later.

Drug interactions: Possible clinically important interactions between it and ***rifampin*** caused its increased elimination, but this was decreased with ***erythromycin***.

LD_{50}, **mouse**: 770 mg/kg; **guinea pig**: 210 mg/kg.

CHONDROITIN

Untoward effects: Has caused nausea in **dogs** at recommended doses. Its unregulated **bovine** source could be a rare source of bovine spongiform encephalopathy and unwanted products.

CHRISTI

House plant.

Untoward effects: See ***Castor Oil***.

CHRISTMAS FACTOR

= *Antihemophilic Factor B* = *Autoprothrombin II* = *Blood Coagulation Factor IX* = *Plasma Thromboplastin Component* = *PTC*

Its complex is known as *Konyne* and *Proplex*.

Untoward effects: Hepatitis, hypersensitivity (leg itching progressing to generalized pruritus), unconsciousness without detectible blood pressure, and sporadic respiration in 9-year-old **male**. Myocardial infarction, hypotension, and ventricular tachycardia in a 51-year-old **male hemophiliac** treated before oral surgery; some cases of fatal viral hepatitis reported. Limiting use of it in **hemophiliacs** should decrease the incidence of fatal myocardial infarctions. Irritation at the site of injection occurs, as does venous thrombosis. Flushing, chills, headache, and tingling, if administered too rapidly. May cause increased bleeding tendency, due to temporary interference with platelet function.

Drug interactions: Use of ***oral contraceptives*** may cause a false elevation of it in laboratory tests. Synthesis is reduced by oral ***anticoagulants***.

CHRISTMAS PEPPER

= *Capsicum annum var. onoides*

House plant.

Untoward effects: Poisonous to **animals** after ingestion. One of most violently "hot" peppers.

CHROMATED COPPER ARSENATE

= *Boliden CCA* = *Boliden Salt K-33* = *CAC* = *CCA* = *Chrom-Ar-Cu* = *Greensalt* = *Langwood* = *Osmose K-33*

Wood preservative.

Untoward effects: Occupational exposure by inhalation was common before it became commercially supplied as a water solution. Poorly absorbed through the skin, but **infants** and **animals** must be kept away from treated lumber as they like chewing on things. Since it is used to pressure treat lumber (yellow-green to brown color), its scraps must NOT be burned in a wood stove or fireplace, to avoid inhalation of arsenical fumes. It persists in such treated lumber for decades. Burning wood treated with this complex has produced systemic symptoms which were, primarily, due to the ***arsenic*** content. It was involved in recurring seasonal (winter) alopecia in eight **family members**, a significant clue that ***arsenic*** was the culprit. Skin rashes, bleeding, gastrointestinal upsets, and severe respiratory distress that even led to a tracheostomy in the youngest **child** and blackouts. The ***arsenic*** was in the *penta*, a less lethal, form. The burned wood air-ash and dust contained ***arsenic-V***, ***copper***, and ***chrome***.

CHROMIC CHLORIDE

Use: In chromizing, corrosion inhibitors, tanning, as a textile mordant, and as a ***chromium*** source in **infants** with impaired ***glucose*** tolerance. Cr^{51} is also used in radioactive studies.

Untoward effects: See ***Chromium*** for potential toxicity.

LD_{50}, **rat**: 1.87 g/kg.

CHROMIC NITRATE

Use: In corrosion inhibitors and textile dyeing.

Untoward effects: See ***Chromium*** for potential toxicity.

LD_{50}, **rats**: 1540 mg/kg.

CHROMIC OXIDE

= *Anadonis Green* = *Chrome Green* = *Chrome Ocher* = *Chrome Oxide Green* = *Chromia* = *Chromium Sesquioxide* = *C.I. 77288* = *C.I. Pigment Green 17* = *Green Cinnabar* = *Green Oxide of Chromium* = *Green Rouge* = *Leaf Green* = *Oil Green* = *Ultramarine Green*

Use: In paints, pigments, Portland cement, bricks, abrasives, and leather goods. In coloring glass.

Untoward effects: Allergic cutaneous reactions reported. See ***Chromium*** for additional potential toxicity.

CHROMIC POTASSIUM SULFATE

= *Potassium Chrome*

Use: In photography, ink manufacturing, leather tanning, chrome plating, paints, and as a mordant in textile dyeing.

Untoward effects: See ***Chromium*** for potential toxicity.

LC_{50}, **quail**: > 5,000 ppm.

CHROMIC SULFATE

Use: In photography, ink manufacturing, leather tanning, chrome plating, paints, porcelain glazes, green varnishes, and as a mordant in textile dyeing.

Untoward effects: Hypersensitivity to it reported in ***chromate***-sensitive **subjects**.

Broilers eating hydrolyzed leather meal had raised ***chromium*** levels only in their kidneys.

IV LD_{LO}, **mice**: 85–247 mg/kg; LC_{50}, **quail**: > 5,000 ppm.

CHROMIUM

Use: In wood protective finishes, and in activating ***insulin*** for maximum utilization of ***glucose***. Adds corrosion resistance and hardness to steel. Used in treating hot parts of airplane engines, petroleum refining, cutting tools, drill bits, tanning of leather, and in chromic catgut. Common in green Christmas wrapping paper. Traces are required in **human** and **animal** nutrition.

Untoward effects: Hexavalent forms are carcinogenic (squamous cell cancers of the lung) to **man** at high levels. First described in 1827.

Chromates in the ash of treated lumber may be a hazard of its burning. One batch of such ashes analyzed 9,600 ppm or 9.6 g/kg.

Contact dermatitis, especially in industrial and cement **workers** and **lithographers**, and by contact with book matches, certain detergents, and bleaching agents. This is usually by trivalent forms. Even found in chrome tanned leather sweat bands of hats and welding fumes. Hexavalent forms produce mucodermal ulcerations from irritation or allergic manifestations. It also causes respiratory ailments. This occurs frequently in **chrome-platers**. Burning and itching of arms, face, and hands noted due to sensitization in an **artist**. A **baker** in Germany developed eczema of **his** hands after contact with flour containing traces of chromium (estimated between 0.01 and 0.09 mg/100 ml). Vertigo, thirst, vomiting, diarrhea, abdominal pain, oliguria, anuria, shock, convulsions, coma, and enlarged liver reported. Uremia may lead to death. Rhinitis, perforation of nasal septa, conjunctivitis, stomach cancer, ulceration of the bucal cavity, tonsils, and pharynx have also been reported. Acute tubular necrosis due to massive accidental exposure noted. Inhalation or ingestion may have caused pulmonary fibrosis. Inflamed and hypertrophic squamous oral mucosa from dentures containing chromium. **Patients** with chromated joint prosthesis have shown hypersensitivity. Blood levels of 5.9 mg/l in one fatal poisoning case. 0.1 mg/m^3 considered upper safe air limit for **workers** near plating tanks.

Livestock feeding trials indicated it is, apparently, safe and beneficial in feeds at 100 ppm. Potentially hazardous quantities may exist in some water supplies. Levels of 0.05 mg/l are set as upper limits for quality water to **livestock** and **poultry** and 5 mg/l is regularly used in irrigation waters.

CHROMIUM PICOLINATE

= *Cromax 2*

Use: By health food **fadists** and **athletes**, as a supposedly safe form of oral ***chromium***. Usual dose has been 1.6 mg/day which contains 200 μg elemental ***chromium***. Currently promoted at 200 μg/kg of **swine** feed to produce leaner pork. The FDA considers it an over-the-counter item for **man** and feed supplement for **hogs**, not requiring their surveillance, despite the following reports.

Untoward effects: A 35-year-old **male** experienced disruption of and slowed down thinking on 1.6–3.2 mg/day. A 49-year-old **female** ingesting 600 mg/day/6 weeks for weight loss developed chronic renal failure with blood urea nitrogen 74 mg/dl and ***creatinine*** of 5.9 mg/dl. A 33-year-old **female** ingested 1.2–2.4 mg/day/4–5 months to enhance weight loss leading to anemia, thrombocytopenia, hemolysis, liver dysfunction, weight loss, and renal failure. A 24-year-old **female** during 48 h ingested 1.2 mg leading to rhabdomyolysis with diffuse muscle weakness, pain, and bilateral leg muscle cramping.

Drug interactions: It can react with ***vitamin C*** and other ***antioxidants*** within the cells leading to a reduced form of ***chromium*** capable of producing DNA mutations.

CHROMIUM TRIOXIDE

= *Chromic Acid* = *Chromic Anhydride*

Use: In chromium plating, corrosion inhibitors, wood preserving, photography, oil, and acetylene purification. 50–60 thousand tons produced annually in the U.S. Rarely, as a topical astringent. A hexavalent form of chromium.

Untoward effects: Present in the mist above chrome plating tanks. Leukocytosis, leukopenia, monocytosis, and eosinophilia. Erythema, eczema, and ulceration of the skin and nasal mucous membrane are common and follow local irritation. A cause of shoe leather contact dermatitis, resembling fungal infections. "Chrome ulcers" or "chrome holes" are due to contact with the dust or concentrated solution. Inhalation toxicity ranges from minor tissue corrosion in the nose to fatal lung disorders. If swallowed, the fatal dose is < 6 g. Abraded skin, such as **workers'** knuckles, easily becomes ulcerated. Renal damage is a common sequelae to systemic absorption. Acute poisoning occurred in a 9-year-old **male** after **he** accidentally fell into a drum of the solution (both limbs were immersed for about 5 min). **He** developed anuria and recovered after treatment, being discharged from the hospital 50 days after the accident. Severe sensorineural deafness in a 25-year-old **male** after it was applied to a tympanic membrane that had been perforated. Immediate vaginal burning in a 26-year-old

female and a 28-year-old **female** after a midwife, in cauterizing cervical erosions, applied tampons soaked in a 50% solution. Pain in the throat, pharyngitis, headache, dizziness, intense vomiting, tachycardia, general weakness followed by numbness of lower limbs; vaginal necrosis with greenish-yellow membrane, abdominal pain, conjunctivitis, and even high fever (39 °C) were some of the many symptoms. OSHA has established a ceiling of 1 mg/10 m^3 for it.

LD_{LO}, **human**: 50 mg/kg; TC_{LO}, **human**: 110 µg/m^3.

Teratogenic in **hamsters** at relatively low dosage.

CHROMOMYCIN A

= *Aburamycin B* = *NSC 58,514* = *Toyomycin*

Antineoplastic.

Untoward effects: Alopecia, skin rashes, sweat gland disturbances, and mucocutaneous ulceration. Mild nausea, transient leukopenia, increased SGOT and alkaline phosphatase occur. Acute renal tubular necrosis in one **patient** receiving 1 mg/m^2/day/5 days. Extravasation causes pain, induration, and necrosis.

Emesis, hemorrhagic diarrhea, fever, dehydration, intestinal and lymphoid tissue necrosis in treated **dogs** and **monkeys**.

IV LD_{50}, **mice**: 603 µg/kg/day/5 days. IV LD_{LO}, **dogs**: 200 µg/kg; **monkeys**: 330 µg/kg.

CHROMONAR

= *A27,053* = *AG-3* = *Antiangor* = *Carbochromen* = *Carbocromen(e)* = *Cassella 4489* = *Intenkordin* = *Intensain®*

Coronary vasodilator, anti-anginal.

Untoward effects: Pruritus (2/30), tachycardia (2/52), arthralgia, back pain, nausea, vomiting, and headache.

LD_{50}, **mice**: 6.3 g/kg.

CHRYSANTHEMUM

= *Juhua*

Many species exist, one of which is a common source of *pyrethrum* (q.v.).

Untoward effects: **People** with chronic photosensitivity have shown an allergic response to extracts of chrysanthemum. **They** often cross-react with ***Achillea, chamomile, ragweed, dandelions, daisies,*** and ***pyrethrum***. Hypersensitivity contact dermatitis with eczema reported in a **gardener**. Contact dermatitis in **women** picking and packing flowers for market. ***C. leucantheum*** or (***Wild Daisy, Moon Daisy, Ox-Eye, Marguerite***) used in flower arrangements is also phototoxic. It is a common contaminant in *Matricaria Chamomilla* of commerce. The sensitizing agent is ***arteglasin A***. ***Ox-Eye Daisy*** causes a disagreeable taste and odor in milk, milk products, and meat of **animals** grazing it.

Also, See ***Pyrethrum***.

CHRYSAORA

Untoward effects: ***C. quinquecinha***, a **jellyfish** occurring in swarms along the eastern coast of the U.S. have caused swelling, edema, and pain, after stinging many **people**.

CHRYSAROBIN

= *Purified Goa Powder* = *Purified Araroba*

Untoward effects: Excessive use or high concentrations (> 10%) can cause an orange-brown skin discoloration, brown nails, conjunctivitis, keratitis, and skin destruction. Will color urine red if significant systemic absorption occurs.

CHRYSENE

= *1,2-Benzphenanthrene*

Use: Occurs in ***coal tar*** and used in organic synthesis. Formed during ***coal*** distillation and pyrolysis of fats and oils. Found in gasoline and diesel engine exhausts, crude oil, ***cigarette*** and wood smoke, and even foods. A polycyclic aromatic hydrocarbon.

Untoward effects: Potential carcinogenicity in **man**. 0.2mg/m^3/8 h TWA as OSHA standard.

Carcinogenic (liver and skin) in **animals**. Embryotoxic in **ducks**.

CHRYSOPHAMIC ACID

= *Chrysophanol*

In ***cascara sagrada***, ***senna***, and ***rhubarb***.

Untoward effects: Will color urine orange.

CHUIFONG TOUKUWAN

Use: In self-medication of arthritis. Usually, imported from Hong Kong since 1974.

Untoward effects: In 1982, 13 rheumatoid arthritis **patients** developed flare-ups, lesions, Cushingoid appearance, arrhythmias, and depression over a period of 1 year. Tested samples contained 1 µg ***lead*** per unit. In 1988, the same product was ingested by 93 **people** in Texas with harmful results. The product was found to contain ***diazepam***, ***indomethacin***, ***hydrochlorothiazide***, ***mefenamic acid***, ***dexamethasone***, ***lead***, and ***cadmium***.

CHYMOPAPAIN

= *BAX 1526* = *Chymodiactin* = *Discase*

Proteolytic enzyme. Nucleolysis agent.

Untoward effects: Severe allergic reactions (1–2%), often life-threatening, in treated **patients**. This has included deaths and subsequent neurological problems that have including paraplegia, cerebral hemorrhage (especially if injected into the subarachnoid space), drowsiness, confusion, and headache. Itching, urticaria, involuntary muscle spasms of the back, back pain, stiffness, soreness, hematuria, and urinary retention.

Drug interactions: ***Benzyl alcohol*** may inactivate it.

CHYMOTRYPSIN

= *Avazyme* = *Catarase* = *Chymar* = *Chymetin* = *Chymolase* = *Enzeon* = *Kymo-trypure* = *Quimar* = *Quimoral* = *Quimotrase* = *Quimotripsina* = *Zolyse*

A major proteolytic enzyme in pancreatic juice. Anti-inflammatory.

Untoward effects: Has caused severe generalized dermatitis in a **patient** and a localized inflamatory reaction in **another**. A severe anaphylaxis after the tenth daily IM injection and a high incidence of increased intraocular pressure after 2–5 days when used for zonolysis. The tension then decreased rapidly. Extremely toxic to the retina and uveitis has been reported.

Drug interactions: Concomittant oral use has increased absorption of ***phenethicillin***, and, perhaps, ***tetracycline***. Potentiation of ***oral anticoagulants*** may occur.

CIBENZOLINE

= *Cifenline* = *Cipralan* = *Ritmalan* = *RO 22-7796* = *UP 33-901*

Anti-arrhythmic.

Untoward effects: Induces hypoglycemia occasionally with central nervous system depression, hyperkalemia, respiratory distress, and electrocardiogram abnormalities. A 65-year-old **female** who received 6 mg/kg/day/2 months developed cardiac insufficiency and renal failure. Cardiogenic shock developed on rechallenge.

Therapeutic serum levels are generally 0.2–0.4 mg/l; toxic at 1 mg/l.

Drug interactions: False positive proteinuria reactions in 23/53 **whose** urine was tested with ***bromphenol blue*** reagent strips. ***Cimetidine***, but not ***ranitidine***, increases its plasma levels and decreases its clearance by depression of hepatic oxidative mechanisms.

CICLOPIROX

= *Bactrafen* = *Ciclopiroxolamin* = *HOE 296* = *Loprox* = *Penlac*

Antifungal.

Untoward effects: Occasionally pruritus and edema at application site, initially, and even burning and worsening of lesions.

CICUTA sp.

= *Water Hemlock*

Untoward effects: The most neurotoxic plant (at least eight species identified) in the U.S., with the primary manifestation being convulsions. They are commonly found in wet swampy areas and wet highway ditches, bogs, and streams, where **children** and **adults** have often mistaken it for ***wild parsnip***. Within 15–60 min after ingestion, salivation, nausea, emesis, and tremors occur, followed quickly by grand mal seizures. When death occurs, it is usually secondary to prolonged anoxia during violent tonic contractions. Control of the convulsions usually permits complete recovery within 24 h with no adverse residual effects. All of the parts of the plant are poisonous, particularly the small succulent area above the roots.

A walnut-size piece of root will kill a **cow**. Although **pigs** may be more resistant, the minimum lethal dose for its toxin, ***cicutoxin***, in most domestic **animals** is ~50–110 mg/kg. Toxicity is retained in the dry plant, but appears absent when in hay. **Cattle** have been poisoned by eating very small (walnut-size) portions of the tuberous root, and by drinking from a stream after crushing the tubers as they stepped on them. When forage is scarce in the spring, its roots are easily pulled up by grazing **cattle**.

C. douglassi = ***Cicutaire pourpe*** = ***Western Water Hemlock*** is one of the more common poisonous species and common in western U.S. A 50 lb 8-year-old **male** died a few hours after eating a small piece of the root. **He** had severe tonic–clonic convulsions, temperature 105 °F, and undetectable pulse when hospitalized. Its primary toxic substance is ***cicutoxin***, which has a carrot-like or raw parsnip odor. **Cattle** and, sometimes, **horses** and **sheep** are also poisoned by it. Muscle twitching, tachycardia, tachypnea, increased salivation, mydriasis, frothing at the mouth, tremors, convulsions, and coma usually occur within 1–6 h. **Pigs** are more resistant to it, but have been poisoned by it.

C. maculata = ***Cowbane*** = ***Musquash Root*** = ***Poison Parsnip*** = ***Spotted Water Hemlock***. One of the most

poisonous of the water hemlocks. A single mouthful has killed a **man** within 15 min. Typical symptoms including salivation, severe stomach pain, vomiting, mental excitement, frenzied action, mydriasis, tremors, spasms, violent convulsions, delirium, opisthotonia, coma, and even death. The tuberous root has even been eaten after confusing it with a turnip. The roots, reportedly, were eaten by American Indian **women** to commit suicide. Approximately 2 oz of root will kill a **sheep**; 10–12 oz, a **cow**; and 8 oz, a **horse**. Contains ***cicutoxin*** and ***cicutal***.
C. occidentalis is a potent toxic species for **man** and **animals** in southwestern U.S.
C. vagans also occurs in the U.S. and has poisoned **livestock**.
C. virosa is the poisonous water hemlock referred to by Shakespeare. It has killed large numbers of **cattle** in Poland and Russia. Used on poison arrows in ancient Japan.

CIDER

Untoward effects: Often due to sulfiting agents added as an antioxidant in some ciders, to help prevent spoilage and browning. See specific ***sulfites***. Usually contains 4–10% ***alcohol***. Its acidic nature can cause leaching of heavy metals from certain containers. On rare occasions, it has contained ***arsenic*** from a fruit spray. *E. coli* can live in high acid foods and has been responsible for three outbreaks in the U.S. from cider. The ***apples*** used had not been washed and were not taken from the tree, but rather from the ground.

CIDOFOVIR

= *GS 504* = *GS 0504* = *Vistide*

***Antiviral.* For IV use.**

Untoward effects: Severe nephrotoxicity, iritis, ocular hypotonia, hearing loss, metabolic acidosis, Fanconi syndrome, neutropenia, and nephrogenic diabetes insipidous. Local irritation and ulceration after topical use, especially as concentrations approach 3%. Uveitis in two HIV-positive **patients** with retinitis.

CIGUATERA

An algal toxin concentrated in the flesh of certain fish (grouper, red snapper, barracuda, amberjack, surgeon fish, parrot fish, sea bass, and moray eels).

Untoward effects: The poisoning in **man** occurs after ingesting certain tropical marine reef **fish** that occasionally become toxic from eating smaller marine organisms containing the toxin. The syndrome was first described by Spanish explorers in the Caribbean area. Also occurs in the Indo-Pacific areas. Although a non-reportable disease, Hawaii and Florida report the greatest number of cases. The syndrome usually occurs within 1–36 h after eating affected **fish** and involves both gastrointestinal and paresthetic neurological symptoms. Rarely fatal, but tingling (mouth, nose, lips, tongue, hands, and feet), numbness, ataxia, dizziness, temporary paralysis, myalgia, headache, diaphoresis, chills, pruritus, dysesthesia, respiratory failure, shock, coma, reversal of hot and cold sensations, "aching" teeth, blurred vision, nausea, vomiting, diarrhea, and abdominal cramping may occur. Most **people** recover in ~3 weeks. **Some** are ill for months, and a few suffer for years. The toxin is heat-stable.

Cats have been poisoned by eating affected **fish**. They salivate, become anorexic and stagger.

CILASTATIN

= *L 642,957* = *MK 0791* = *Tienam*

Use: IV with ***impenem***, a ***thienamycin*** derivative = ***Primaxin*** = ***Zienam***. Prevents renal enzyme metabolism of carbapenem β-lactam antibiotics.

Untoward effects: Severe vomiting, hypokalemia, neurotoxicity, increased liver enzymes, phlebitis, rash, and thrombophlebitis after IV with ***impenem***. Occasionally seizures, increased prothrombin time, hypotension, increases in urinary red blood cells, vertigo, aplastic anemia, pseudomembranous colitis, pleural effusion, and hiccups.

CILOSTAZOL

= *Pletal* = *OPC 13,013*

***Phosphorodiesterase III inhibitor, antithrombotic.* A *quinoline* derivative.**

Untoward effects: Severe headache, palpitations, dizziness, and diarrhea. Occasionally peripheral edema and dyspepsia. Adverse effects in < 9%.

Drug interactions: ***Diltiazem***, ***erythromycin***, and ***CYP3A4 inhibitors*** increase its serum concentration by ~50%. ***Omeprazole*** increases its serum concentration by ~70%. ***Grapefruit juice***, ***itraconazole***, and ***ketoconazole*** may also increase its serum concentration.

Use in pregnant experimental **animals** causes decreased fetal weights; increases in cardiovascular, renal, and skeletal defects in **neonates**; and increased number of stillbirths.

CIMETIDINE

= *Acibilin* = *Aciloc* = *Acinil* = *Biomag* = *Brumetidina* = *Cimal* = *Cimet* = *Cimetag* = *Cimetum* = *CRC 1820*

= *Duractin* = *Dyspamet* = *Edalene* = *Eureceptor* = *Gastromet* = *Metracin* = *Notul* = *Peptol* = *SKF 92334* = *Tagagel* = *Tagamet* = *Tratul* =*Ulcedin* = *Ulcidine* = *Ulcerfen* = *Ulcimet* = *Ulcofalk* = *Ulcomedina* = *Ulcomet* = *Ulhys* = *Valmagen* = *Venopex*

***H_2 receptor antagonist*. Production of stomach acid decreased by about 90% for 5–7 h.**

Untoward effects: TD_{LO}, **human**: 80 mg/kg/8 days and 1300 mg/kg/12 weeks.
Blood: Agranulocytosis, thrombocytopenia, neutropenia, leukopenia, and, rarely, pancytopenia; hemolytic and aplastic anemias, and prolonged prothrombin time.
Cardiovascular: Bradycardia, tachycardia, and atrioventricular heart block have occurred on rare occasions. Leukoclastic vasculitis (palpable purpura) has occurred. Life-threatening arrhythmias in two **patients** receiving 400 mg qid. One report described three **patients** who had cardiac arrest following bolus IV injections of 150–400 mg, suggesting IV infusion should be over at least a 30 min period. Other reports confirm this. Has caused hot flashes after 1 month of use in a 31-year-old **male** dialysis **patient**.
Clinical tests: False positive hemoccult tests and Fe-Cult guaiac tests reported.
Central nervous system: Headache, dizziness, mental confusion, mania, agitation, depression, stupor, coma, and hallucinations. Delirium, psychosis, and aggressiveness often occur in older **patients** (> 50–60 years), because plasma clearance of the drug is usually < 50% of younger **patients**.
Eye: Can increase intraocular pressure and exacerbate existing glaucoma.
Fetus and neonate: In breast milk, suppressing gastric acidity, decreasing drug metabolism, and increasing central nervous system stimulation.
Gastrointestinal: Diarrhea in about 1% of **patients**. Causes esophagitis. Cases of resistance to its effect.
Liver: Increases serum transaminase with increased dosage. Hepatic fibrosis in one **patient**. Occasionally, acute hepatitis and jaundice, possibly, due to a hypersensitivity.
Miscellaneous: Induction of hyperprolactinemia, breast tenderness in **women**, gynecomastia in **men**, galactorrhea, and occasionally bronchoconstriction, especially in **asthmatics**. Use has caused delays in diagnosing gastric cancers and ulcers. Hypersensitivity induced laryngospasm, Quincke's facial edema (angioneurotic edema), pruritus, urticaria, and anaphylaxis after an IV injection, but not after the same **patient** received it orally. It has enhanced reactions to a tuberculin skin test. Lymphadenopathy and, possibly, increased risk of gastric adenocarcinomas. The latter may be a result of increased concentrations of ***N-nitrosamine***. Rapid IV has even caused severe pain in a **patient**. Parathyroid hormone markedly decreased in some **patients** by its use. Cross-sensitivity with ***famotidine***.
Musculoskeletal: Cramps, myokymia, weakness, reversible arthralgia and myalgia, and rhabdomyolysis.
Pancreas: Pancreatitis has occurred after 2–12 weeks of treatment.
Renal: Nephrotoxic with interstitial nephritis. A 56-year-old **female** who received 1.2 g/day/8 days had greatly decreased renal function (80% increase in serum ***creatinine***). Uremia occurs.
Sexuality: Decreased libido in **male** and **female**, impotence in **man**, decreased sperm count without affecting fertility.
Skin: Alopecia, exacerbates cutaneous lupus erythematous, toxic epidermal necrolysis, decreases sebum excretion rate with xerosis and asteototic dermatitis, rashes, exfoliative dermatitis, Stevens-Johnson syndrome, and psoriasis.

Usual therapeutic serum levels are 0.25–3 mg/l; toxic at 30–50 mg/l; and fatal at 110 mg/l.

Drug interactions: ***Alcohol***: Peak plasma concentrations increased 10% by cimetidine.
Alfentanil: Serum levels were significantly increased by cimetidine.
Antacids: Impair its absorption (3–48%) in some **patients**.
Antibiotics: ***Tetracycline*** absorption is significantly reduced with single dose. It increases absorption of oral ***penicillin G***.
Antipyrine: Its metabolization is decreased and lower dosage may be required.
Caffeine: Cimetidine decreases its clearance by about 31% in **smokers** and 42% in **non-smokers**, due to inhibition of microsomal metabolism.
Captopril: A neuropathy may develop.
Carbamazepine: Toxicity increased by cimetidine.
Carmustine: May add to its bone marrow suppression.
Chloramphenicol: Its aplastic anemia potential may be potentiated by cimetidine.
Chlordiazepoxide: Decreases elimination due to decreased demethylation.
Chloroguanidine: Cimetidine significantly increases ***chloroguanidine*** blood levels and decreases ***chloroguanidine*** body clearance.
Chloroquine: Absorption decreased by cimetidine.
Citaloprim: Cimetidine increases its serum levels.
Clarithromycin: Absorption is prolonged by cimetidine.
Cyclosporine: Increases blood levels.
Diazepam: Its clearance decreases by about 40% (its half-life increases from 30 h to 51 h) when cimetidine is given. Inhibition may also occur with ***clorazepate***, ***flurazepam***, ***halazepam***, and ***prazepam***.
Doxepin: ***Doxepin*** blood levels increase by nearly 70%.
Erythromycin: Otic toxicity increased by it.

Glucose and insulin: ***Insulin*** metabolism of ***glucose*** is decreased by about 20%.
Imipramine: May precipitate a psychosis.
Indomethacin: Absorption increased.
Iron: Achlorhydria will decrease ***iron*** absorption and requires supplementation.
Ketoconazole: Absorption decreased.
Labetalol: Cimetidine increases its bioavailability.
Lidocaine: Its hepatic metabolization is decreased because cimetidine decreases liver blood flow. Clearance of ***lidocaine*** was decreased by about 16%. This may not occur with slow constant infusion.
Mebendazole: Peak serum concentrations increase by nearly 50%.
Meperidine: Its oxidation is decreased and is a clinically significant interaction.
Metformin: ***Metformin*** serum levels increase.
Methadone: Potentiated by cimetidine.
Metoclopramide: Cimetidine's oral bioavailability decreased by 25–30%.
Metoprolol: Blood level is increased.
Metronidazole: Plasma clearance decreased.
Morphine: The combination may pose a potentially lethal interaction. 150 mg twice daily with 15 mg IM ***morphine*** caused mental confusion and apnea.
Nortriptylline: Serum levels increase by about 40% with potential for significant toxicity.
Penicillin G: Absorption may be increased due to decreases in stomach acidity.
Phenobarbital: May require increased cimetidine to offset the increased metabolization of the latter.
Phenytoin: Its plasma concentrations increase by 13–33%, and possibly, even more in one case. This has caused ataxia and confusion.
Praziquantel: Cimetidine increases its serum concentration.
Procainamide: Significant increase in serum levels, probably, due to decreased renal tubular clearance.
Propranolol: Serum levels increase by up to 80%.
Quinidine: Oral absorption increased.
Quinine: Clearance decreased.
Succinylcholine: Duration of effect increased by about 100%.
Tacrine: It increases ***tacrine*** serum levels by ~30%.
Theophylline: Its clearance is decreased and its half-life is increased.
Thiobarbiturates and benzodiazepines: It slows the metabolization of injected ***thiobarbiturates*** and ***benzodiazepines*** thus increasing their duration of action and requires a 10–25% decrease in their dosage.
Triazolam: Cimetidine decreases its body clearance, increases intensity and duration of its action.
Verapamil: Elimination is significantly decreased.
Vitamin B_{12}: Long-term cimetidine therapy will require ***B_{12}*** supplementation.
Warfarin: Action is increased by decreasing liver metabolism and decreasing renal clearance by about 36%.
Zidovudine: Its renal clearance is decreased.
Zolmitriptan: Cimetidine decreases its metabolism.

Surprisingly, the U.S. FDA, in 1996, approved cimetidine for over-the-counter sales.

Male progeny of treated female **rats** had decreased sexual motivation, testicle weights, and serum levels of ***testosterone***. In **poultry**, it has potentiated the bronchoconstriction of ***histamine***. Causes a high incidence of gastric carcinoids in **rats**. Can cause thrombocytopenia in **cats** and **dogs**. Potentiates and prolongs hypoglycemic effect of ***glyburide*** in **rabbits**.

LD_{50}, **rat**: 5 g/kg; **mouse**: 2.6 g/kg.

CIMICIFUGA RACEMOSA

= Black Cohosh = Black Snakeroot = Bugbane = Bugwort = Rattleroot = Rattleweed = Remifemin = Richweed = Squaw Root

Untoward effects: Ingestion of leaves and roots may induce nausea and vomiting. Occasionally dizziness, limb pain, hypotension, and headache; less commonly, intolerance of contact lenses, changes in corneal shape, and blood clots in the back of the eye.

Drug interactions: Can potentiate hypotensive effects of ***antihypertensive drugs***.

CINASERIN

= SQ 10,643

Antiserotonin, tranquilizer.

Untoward effects: In chronic **schizophrenics**, facial flushing, dizziness, weakness, akathisia, dystonia, tremors, rigidity, parkinsonoid appearance, agitation, mild transient decreased white blood cells, and increased SGPT reported.

Liver tumors in **rats**, convulsions in **mice**, and electroencephalogram stimulation in **cats**.

CINCHONA

= Calisaya Bark = Jesuit's Bark = Peruvian Bark

Antimalarial, antipyretic. **The bark contains about 7% alkaloids, of which 50–90% is *quinine*. It has been used in treating atrial fibrillation.**

Untoward effects: See ***Quinidine*** and ***Quinine***. Even small doses may be a stomach irritant, causing nausea and gastric pain. Potentiates ***anticoagulant*** action.

LD_{LO}, **human**: 50 mg/kg.

CINCHOPHEN

= *Agotan* = *Alutyl* = *Artam* = *Atocin* = *Atophan* = *Cinconal* = *Mylofanol* = *Phenylcinchoninic Acid* = *Phenoquin* = *Polyphlogin* = *Quinofen* = *Quinophan* = *Rhematan* = *Rheumin* = *Sovcain* = *Tophol* = *Tophosan* = *Vantyl* = *Viophan*

Anti-arthritic, analgesic.

Untoward effects: Fixed-drug eruptions, anorexia, nausea, vomiting, diarrhea, hypotension, bradycardia, bradypnea, apnea, agranulocytosis, shock, anaphylaxis, increased alkaline phosphatase, and decreased serum urates. Hypersensitivity hepatitis with hepatocellular jaundice in 0.1%, with a mortality of nearly 50%, that discouraged its clinical use. Several hundred cases of hepatitis, including postnecrotic cirrhosis, have been reported.

LD_{LO}, **human**: 214 mg/kg.

Drug interactions: Direct hypothrombinemic action that has potentiated oral ***anticoagulants***. Toxicity increases when used with ***demeclocycline***. Has caused false positive or spuriously high ***glucose*** test readings.

Experimental ulcerogenic agent in **animals**.

LD_{LO}, **rat**: 500 mg/kg.

CINNAMALDEHYDE

= *Cinnamal* = *Phenylacrolein* = *3-Phenyl-2-propenal*

***Flavoring and perfumery agent. Antifungal.* In *cinnamon* oils. Applied to lambs, it repels coyotes.**

Untoward effects: An allergen causing a contact dermatitis, including urticaria. Suspect carcinogen.

LD_{50}, **rats**: 2.2 g/kg.

CINNAMEDRINE

= *N-Cinnamylephedrine*

***Antispasmodic, smooth muscle relaxant.* With *caffeine* and *aspirin* in *Midol*®.**

Untoward effects: Potential for abuse. Psychological dependence reported.

CINNAMIC ACID

= *β-Phenylacrylic Acid*

In ultraviolet B sunscreens. In *Balsam of Peru*. Used in manufacturing *1-phenylalanine* and perfumes.

Untoward effects: Causes an allergic contact urticaria.

IV in **rabbits** initially decreases, then increases blood platelets and causes leukocytosis.

CINNAMIC ALCOHOL

= *Cinnamyl Alcohol*

Approximately 150,000 lbs/years in U.S. fragrances. Approved for food use. Odor of *hyacinth*.

Untoward effects: Contact dermatitis, especially when used in toilet paper, sanitary napkins, and facial tissues. Cross-reacts in **people** sensitive to ***Balsam of Peru***.

LD_{50}, **rats**: 2 g/kg; acute dermal LD_{50}, **rabbits**: > 5 g/kg.

CINNAMIC ALDEHYDE

Untoward effects: Contact urticaria dermatitis from cosmetics, toothpastes (Aim®, regular Close-Up®, Crests®, Gleem®), shaving cream, ***chocolate***, toilet paper, sanitary napkins, and facial tissues. Also, a cause of allergic contact reactions with ***hyacinths***, ***myrrh***, ***Bulgarian rose***, ***patchouli***, and ***Balsam of Peru***. A cause of dermal hypopigmentation.

LD_{50}, **rat**: 2.22 g/kg; **guinea pig**: 1.16 g/kg.

CINNAMON

***Carminative, flavoring agent, spice.* From evergreen trees. The Old Testament makes reference to it. Over 15 million lbs imported into the U.S. annually. Its oil is also known as *Cassia Oil*.**

Untoward effects: Has shown delayed positive patch tests. Has caused a contact urticarial dermatitis. The causative agent of contact dermatitis in many **bakers**. One **baker** had his hand dermatitis flare up when **he** drank ***vermouth*** containing it. This **baker** also developed a cheilitis from a toothpaste containing cinnamon. Has been a cause of lipstick-induced cheilitis. **Candy makers** also have had dermatis from exposure to it. Has triggered migraine headaches in some **people**. A cause of food allergies. A suspect carcinogen. Gingivostomatitis and cheilitis from its use in toothpaste. It may cross-react with ***Balsam of Peru***.

A couple of tons of Ceylon leaf oil is used in the U.S. annually in soaps, detergents, cosmetic creams and lotions, and perfumes. Its LD_{50}, **rats**: is 2.65 g/kg and acute dermal LD_{50}, **rabbits**: is > 5 g/kg. Low-level phototoxicity has been demonstrated for the Ceylon bark oil.

LD_{50}, **rats**: 3.4 ml/kg; acute dermal LD_{50}, **rabbits**: 0.69 ml/kg.

CINNAMYL ACETATE

***Fragrance.* Approximately 2 tons annually in the U.S. for soaps, detergents, cosmetic creams and lotions, and perfumes. Approved for food use.**

Untoward effects: Mildly irritating to some **humans**.

LD_{50}, **rats**: 3.3 g/kg; acute dermal LD_{50}, **rabbit**: > 5 g/kg.

CINNAMYL ANTHRANILATE

Use: As an imitation grape and cherry flavor, and as a fragrance in soaps, detergents, cosmetic creams and lotions, and in perfumes.

Untoward effects: Has caused primary lung and liver tumors in a strain of **mice**, and pancreatic and renal tumors in **rats**.

LD_{50}, **rats**: > 5 g/kg.

CINNAMYL CINNAMATE

Limited use in fragrances for soaps, detergents, cosmetic creams and lotions, and perfumes. In *styrax*.

Untoward effects: LD_{50}, **rats**: 4.2 g/kg; acute dermal LD_{50}, **rabbits**: > 5 g/kg.

CINNAMYL ISOVALERATE

Limited use in fragrances for soaps, detergents, cosmetic creams and lotions, and perfumes.

Untoward effects: LD_{50}, **rats**: > 5 g/kg; acute dermal LD_{50}, **rabbits**: > 5 g/kg.

CINNARIZINE

= *Aplactan* = *Aplexal* = *Apotomin* = *Artate* = *Carecin* = *Cerebolan* = *Cerepar* = *Cinaperazine* = *Cinazyn* = *Cinnacet* = *Cinnageron* = *Cinnipirine* = *Corathiem* = *Denapol* = *Dimitron* = *Eglen* = *Folcodal* = *Giganten* = *Glanil* = *Hilactan* = *Ixertol* = *Izaberizin* = *Katoseran* = *Labyrin* = *MD-516* = *Midronal* = *Mitronal* = *Olamin* = *Processine* = *R-516* = *Rinomar* = *Sedatromin* = *Sepan* = *Siptazin* = *Spaderizine* = *Stugeron* = *Stutgeron* = *Stutgin* = *Toliman*

***Antihistamine, vasodilator, antimotion sickness drug.* A *piperazine* derivative. A lipophilic class III *calcium channel-blocker*.**

Untoward effects: Disorientation and mania in **patients** with arteriosclerotic disease, lethargy, drowsiness, in about 50%, occasionally weight gain in **women**, impotence in a 55-year-old **male**, parkinsonism, tremor, depression, extrapyramidal symptoms, and cholestasis reported.

CINOBUFAGIN

A bufandienolide.

Untoward effects: An hallucinogen found in some Chinese herbal, anesthetic, and cardiac medicines. Related to ***cinobufaginol***, also found in Chinese herbals, and ***cinobufotalin***, a **toad** venom.

IV LD_{LO}, **cat**: 230 µg/kg.

CINOXACIN

= *Cinobac* = *Compound 64716* = *Noxigram* = *Uronorm*

***Antimicrobial.* A quinolone.**

Untoward effects: Bullous skin eruptions after 500 mg twice daily/2 weeks and exposure to a sunlamp producing ultraviolet A radiation.

LD_{50}, **rats**: 4.16 g/kg.

CINOXATE

= *Deep Tan* = *EEMC* = *2-Ethoxyethyl-p-methoxycinnamate* = *Giv Tan F* = *RVPaque* = *Sunbloc*

Sunscreen.

Untoward effects: Acute allergic contact dermatitis in 60-year-old **female**, 3 days after application. Within 1 h after sunlight exposure, raised red areas appeared, and within 24 h, progressed to closely spaced large vesicles. 3 weeks after treatment, skin became normal. Other reports refer to a contact hypersensitivity dermatitis and photodermatitis.

CIODRIN

= *Akrodeks* = *Crotoxyphos* = *ENT 24,717* = *Kemdrin* = *SD 4294* = *Shell 4294* = *Simax*

***Organic phosphorus insecticide.* Cholinesterase inhibitor.**

Untoward effects: Symptoms of acute poisoning develop during exposure or within 12 h, and include headache, dizziness, extreme weakness, ataxia, miosis, muscle twitching, tremors, nausea, bradycardia, pulmonary edema, bradypnea, toxic psychosis, and sweating. Other symptoms, such as salivation, vomiting, abdominal cramps, diarrhea, wheezing, anorexia, convulsions, and blurred vision also occur as with other ***organophosphate cholinesterase-inhibiting pesticides***. May potentiate this effect by other insecticides, anthelmintics, phenothiazine tranquilizers, and anesthetic agents. Avoid unnecessary exposure to spray operator or contamination of feedstuffs.

Brahman cattle are particularly susceptible to toxicity from it (even at 0.1% concentration). Do NOT treat **calves** under 6 months of age. NO withdrawal period before milking or slaughtering **cattle** after aqueous sprays or dusts. Keep away from **children**. Typical adverse effects in **mammals**. **Mallard ducks** have shown ataxia, leg weakness, wings crossed over their backs, opisthotonus, LD_{50}, 790 mg/kg. Moderately toxic to **bees**. LD_{50}, **chicks**: 111 mg/kg; **rat**: 74 mg/kg; **mouse**: 90 mg/kg; LD_{50}, skin, **rabbit**: 385 mg/kg; **rat**: 202 mg/kg.

CIPROFLOXACIN

= *Bay σ 9867* = *Ciflox* = *Ciprobay* = *Ciproxan* = *Ciproxin* = *Velmonit*

Fluorinated quinolone antibiotic.

Untoward effects: ***Cardiovascular***: Palpitations, atrial flutter, syncope, angina pectoris, hypertension, myocardial infarction, and cerebral thrombosis.
Clinical test interference: May produce false positive glycosuria results.
Central nervous system: Dizziness, anxiety, tremors, muscle twitching, paresthesia, difficulty in walking, seizures, ataxia, hemiparesis with facial nerve involvement, exacerbation of myasthenia gravis, delirium with nausea and vomiting.
Eye: Blurred vision, poor color perception, diplopia, eye pain, and overbrightness of lights.
Gastrointestinal: 3–6%. Painful oral mucosa and candidiasis, intestinal perforation and bleeding, dysphagia, and *Clostridium difficile* disease.
Hypersensitivity, anaphylaxis: Generalized itching, hives, shortness of breath, facial edema, laryngeal edema, serum sickness-like reaction with polyarthralgia, myalgia and generalized urticarial rash. A single 500 mg oral dose to a 32-year-old **male** precipitated an acute anaphylaxis, which, despite indicated treatment, caused cardiac arrest and death within 24 h. Photosensitivity reported.
Liver: Fulminant hepatic failure in a 66-year-old **male** with a history of drug allergies received 500 mg twice daily for urinary tract infection. After three doses, **he** was hypoglycemic, lethargic, unresponsive to oral and tactile stimuli, had raised hepatic enzymes, and died within a week. A 92-year-old **male** died with hepatic failure after 2 days of IV treatment for a suspected urinary tract infection with 200 mg twice daily. Jaundice and pruritus in a 44-year-old **female** after 500 mg twice daily/7 days.
Miscellaneous: Serious arthropathies have been induced in young **patients** with cystic fibrosis. Transitory von Willebrand disease, Henoch-Schonlein purpura, pseudomembranous colitis, increased bleeding time in 19-year-old **male**, treated for 14 days. Tinnitus, tendinitis, and tendon ruptures (particularly of the Achilles tendon) also reported.
Renal: Acute interstitial nephritis, polyuria, urinary retention, urethral bleeding, and renal failure.
Respiratory: Dyspnea, hiccough, laryngeal and pulmonary edema, bronchospasm, pulmonary embolism, epistaxis, and hemoptysis.
Skin: 0.5–2.0%. Toxic pustuloderma after 250 mg twice daily/2 days. Cutaneous vasculitis, toxic epidermal necrolysis, lobular panniculitis, pruritus, urticaria, edema of the face, neck, lips, conjunctiva, and hands, hyperpigmentation. Sensitizes skin to sunlight.

Therapeutic serum levels are 2.5–4 mg/l and toxic at 11.5 mg/l.

Drug interactions: ***Antacids, Dairy Products, and Sucralate***: Serum peak concentrations in **those** taking ***aluminum***-containing ***antacids*** were 14–50% of **those** taking ciprofloxacin without ***antacids***. Its absorption is decreased by ***antacids, milk, sucralate, milk, yogurt,*** and ***magnesium citrate***. A single dose of ***calcium carbonate*** given 2 h before a single oral dose did not alter its absorption.
Caffeine: When taken at the same time, blood levels of ***caffeine*** become increased longer than normal.
Cyclosporin: The combination appears to act synergistically to produce nephrotoxicity.
Digoxin: Its serum concentrations are increased by ciprofloxacin.
Foscarnet: The combination therapy resulted in tonic–clonic seizures in two **patients**.
Iron: Even when ferrous ***iron*** is given orally a ferric ion–ciprofloxacin complex significantly decreases ciprofloxacin absorption.
Miscellaneous: ***Zinc*** had the same effect as ***iron***.
Phenytoin: Its metabolization is inhibited and ***phenytoin*** toxicity began 2 days after ciprofloxacin treatment was initiated. (By the 5th day, serum levels of 44.3 μg/ml, compared to previous levels of 12–13 μg/ml). Unexpected low concentrations in 78-year-old **female** after ciprofloxacin addition.
Probenecid: Ciprofloxacin serum levels increased by it.
Quinidine: Its serum levels are sometimes increased by ciprofloxacin.
Theophylline: Ciprofloxacin decreases its metabolic clearance, increases ***theophylline*** blood levels and half-life, causing seizures, necessitating reduced ***theophylline*** dosage by up to 67%, to prevent toxicity.
Warfarin: Significant interaction occurs increasing prothrombin time by ~25–65% in various reports.

Interaction with ***warfarin*** in **rats** can cause serious or fatal increased prothrombin time. ***Antacids*** cause decreased availability in **dog** trials. Oral ***Taraxacum mongolicum*** (***dandelion***) increases its elimination half-life in **rats**.

CIRSIUM ARVENSE

= *Canada Thistle* = *Cnicus*

Untoward effects: Accumulates ***nitrates***. Rarely eaten by **livestock**, but has caused dyspnea, cyanosis, and chocolate-brown discoloration of their blood. Emetic and an emmenagogue in **humans**.

CISAPRIDE

= *Acenalin* = *Prepulsid* = *Propulsid* = *R 51619* = *Risamal*

***Prokinetic gastrointestinal stimulant.* Enhances and promotes gastrointestinal motility.**

Untoward effects: Urinary tract disorders, central nervous system ill-effects (headache and rhinitis) reported and gastrointestinal effects (abdominal cramps, diarrhea, flatulence, borborygimi, nausea, and vomiting). Bronchospasm in four **patients** (33–56 years, on 10–30 mg/day), three of whom had a history of asthma. Rare cases of aplastic anemia, granulocytopenia, leukopenia, thrombocytopenia, tachycardia, torsades de pointes, and QT prolongation reported. Secreted in **human** breast milk. Arrhythmias and deaths after its use have diminished its prescribing, especially for **infants** and **premature babies**. Tremors, stiffness, and numbness in Japanese 77-year-old **female** after 15 mg/day/6 months. Tremors reported in other **patients**. Buttock rash and anal excoriation in 2-month-old **male**. Serious arrhythmias and deaths reported. An 8-month-old **female** was given 8 ml of cisapride suspension, instead of 0.8 ml leading to glassy-eyes and loss of facial expression within 5 min, followed by vomiting, abnormal behavior, tachycardia, hypertension, hyperactive bowel sounds, and thrombocytosis. As of December 31, 1999, the FDA cited 80 deaths and 341 assorted heartbeat abnormalities associated with its use. An additional 23 deaths due to it were reported in the first couple of months of year 2000. From March, 2000, Johnson & Johnson only market it to **patients** who can't take other medications for the same purpose.

Drug interactions: May increase effects of ***anticoagulants*** (such as ***warfarin*** and ***acenocoumarol***), ***alcohol***, and ***diazepam*** in some **patients**. Accelerated gastric emptying and intestinal motility may alter absorption of certain drugs. ***Anticholinergics*** will decrease its effects. ***Itraconazole*** increases its serum concentration and may precipitate cardiac arrhythmias and death. ***Ketoconazole, miconazole,*** and ***troleandomycin*** can also cause a marked increase in its plasma levels. ***Metronidazole*** will probably produce the same effect. Use with ***clarithromycin*** causes a three times increase is cisapride concentration, torsades de pointes, and deaths. ***Diltiazem*** with it prolongs QT interval; ***erythromycin, fluconazole, ketoconazole,*** and ***metronidazole*** should have the same interaction. ***Cimetidine, delavirdine, indinavir, nelfinavir,*** and ***ritonavir*** decrease its metabolization, increase the potential for increased toxicity. Considerable interindividual variation in effects of ***grapefruit juice*** on it.

Overdoses in **dogs** at > 15 mg/kg cause diarrhea, ataxia, disorientation, tremors, dyspnea, hyperthermia, and hyperactivity.

LD, **dog**: 640 mg/kg.

CISMETHRIN

***Insecticide.* A pyrethroid.**

Untoward effects: Contact with the pure form can cause a reversible neuropathy in **man**.

IV use in **rats** rapidly produces fine tremors.

CISPLATIN

= *Briplatin* = *CACP* = *cis-DDP* = *Cismaplat* = *cis-Platinum II* = *Cisplatyl* = *Citoplatino* = *CPDC* = *DDP* = *Lederplatin* = *Neoplatin* = *NSC-119875* = *Platamine* = *Platinex* = *Platiblastin* = *Platinol* = *Platinoxin* = *Platistin* = *Platosin* = *Randa*

***Antineoplastic.* IV use.**

Untoward effects: Blood and bone marrow: Myelotoxic; cause of severe immune hemolytic anemia, often after several courses of treatment, and almost complete tumor regression. Severe anemia has also resulted from a decrease in erythroid cells and precursors, not from hemolysis or decreased ***erythropoietin***. Leukopenia, neutropenia, and thrombocytopenia reported. Hypomagnesemia is common. Hypocalcemia, hypokalemia, hyperuricemia, hyponatremia, hypophosphatemia, and decreased white blood cells also occur.
Cardiovascular: Hypertension after infusion, often lasting for many months after discontinuation of therapy. Acute vascular ischemia, angina, bradycardia, thrombotic microangiopathy, atrial fibrillation, and Raynaud's disease reported. Fatal cerebrovascular event and probable myocardial infarction reported.
Central nervous system: Neurotoxicity, convulsions, transient cortical blindness, focal encephalopathy, associated with peripheral neuropathies, even presenting as a myasthenic syndrome with paresthesia, absent ankle jerks and loss of vibration sense, tetany or seizures, aphasia, confusion, agitation, Lhermitte's sign, and coma occur. Some neuropathies have, infrequently, persisted for years after therapy.
Ear: High-frequency hearing loss in nearly 90% of 24 pediatric **patients** with the amount of hearing loss inversely related to age. It affects the inner ear. In older **patients**, significant hearing loss in 36%. ***Ifosfamide*** exacerbates the hearing loss.
Eye: Optic neuritis, papilledema and cerebral blindness have been infrequently reported.
Gastrointestinal: Early manifestations of acute toxicity are nausea, vomiting, anorexia, diarrhea including *Clostridium*

difficile diarrhea. These symptoms can often be very severe. Treated **patients** often reported a metallic taste.
Liver and pancreas: Hepatotoxicity is an uncommon adverse effect and acute pancreatitis has been reported.
Miscellaneous: Anaphylactic-like reactions include facial edema, wheezing, tachycardia, hypotension, pruritus, urticaria, and skin rash. Photosensitivity, as well as hypersensitivity, reactions reported. Overdoses have led to headaches, hyperpnea, numbness of all limbs, confusion followed by unconsciousness, and some hearing impairment. A 50-year-old **female** was successfully treated on day 1–5 every 3 weeks for a nasopharynx cancer with metastases resulting in an acral toxic erythema of **her** hands. A 36-year-old **male** inadvertently received 40 mg/m^2 tid/4 days, instead of 40 mg/m^2/day divided into three equal 8 h infusions. **He** developed myelosuppression, nephrotoxicity, neurotoxicity, and irreversible ototoxicity. The other toxic effects were partially or totally reversed. A 2-year-old **male**, in another hospital incident, received 100 mg/m^2/day/5 day, instead of the dosage over a 5 day period. The **patient** died from a septicemia caused by extensive drug-induced intestinal ulceration. Alopecia and sterility can be sequelae. Hyperthermia, fatal respiratory failure, osteonecrosis, optic disc swelling, and, possibly, a delayed acute leukemia occur. A 10-month-old **child** died after receiving 204 mg instead of 20.4 mg, due to the omission of a decimal point by the **prescriber**. Carcinogenic, mutagenic, and teratogenic potential for careless **handlers**. Use with stainless steel needles because ***aluminum*** needles will cause its precipitation.

Drug interactions: ***Cephalosporins***: Enhanced nephrotoxicity of cisplatin.
Etoposide: Systemic clearance decreases when given shortly after cisplatin.
Gentamicin: Its elimination is decreased, requires monitoring and nephrotoxicity is enhanced.
Methotreate: Its elimination is significantly decreased and may require decreased dosage and escalated ***leucovorin*** dosage to prevent toxicity.
Phenytoin: Bloods levels were decreased by 63% in a 36-year-old **female**.
Sodium bisulfite: May inactivate cisplatin.

In **animals**, nausea, vomiting, and loss of hearing may occur, and, if not severe, reverses on withdrawal of treatment. The drug can cross the blood–brain barrier. Cerebellar toxicity in **cats** may occur from ***fluorocitrate***, a degraded form of the drug's major metabolite. In **dogs** and **monkeys**, as in **humans**, a major adverse effect is nephrotoxicity. Causes severe pulmonary edema in **cats**. Use gloves and mask in clean-up of accidental spills. May be embryotoxic. Causes an increase in lung adenomas in female **mice**. Multiple injections have caused leukemia in **rats**. At 3 mg/kg, it will cause emesis in **dogs**.

CITALOPRAM

= *Celexa* = *Cipramil* = *LU 10-171* = *Seropram*

Psychotherapeutic; selective serotonin reuptake inhibitor anti-depressant.

Untoward effects: Acute overdoses in six suicides (23–56 years) with postmortem concentrations in femoral vein of 5.2–49 μg/ml. Doses > 600 mg cause mild nausea, dizziness, tachycardia, tremor, rash, headache, postural hypotension, xerostomia, delayed ejaculation, paresthesias, fever, syndrome of inappropriate antidiuretic hormone secretion, drowsiness, and somnolence. Doses of 600 mg–1.9 g produce convulsions in 18% of **patients** and in 47% of **patients** at 1.9–5.2 g. Electrocardiogram aberrations in 25%; rarely, rhabdomyolysis, hypokalemia, aspiration pneumonia, and transient bundle branch block. Treatment dosage is 20–40 mg. Extrapyramidal symptoms and attempted suicides reported. ***Serotonin*** syndrome with QT prolongation, tremors, convulsions, coma, and even a few fatalities after overdoses. Secreted in breast milk leading to somnolence, weight loss, and decreased feeding in **infants**. Alopecia reported with its use.

Drug interactions: ***Cimetidine*** increases its serum levels. Its use can increase serum concentrations of ***imipramine*** and ***metoprolol***. Use with ***monoamine oxidase inhibitors*** can precipitate a ***serotonin*** syndrome.

CITRAL

= *Geranial* = *Lemarone*

***Flavoring and perfumery agent.* Strong lemon flavor and odor used in the food industry. Has been used in *vitamin A* synthesis. In oranges and lemongrass oil.**

Untoward effects: Skin sensitizer and irritant in **humans**, **miniature swine**, **guinea pigs**, and **rabbits** at high or 100% concentration.

In **animals**, oral doses can damage the lining of blood vessels, the optic nerve, and, clinically, is a ***vitamin A*** antagonist.

LD_{50}, **rat**: 5 g/kg; **mice**: 6 g/kg. IP LD_{50}, **mice**: 450 mg/kg.

CITRIC ACID

Use: In effervescent preparations, hair rinse, flavorings, drugs, antioxidant enhancers for use with ***vitamins***, food and pharmaceutical acidulant, in beverages, candies, jellies, ***wines***, canned fruits and vegetables. As a chelator and buffer in detergents, hair spray, boiler water, liquid fertilizers, animal feeds, metal finishing and cleaning, on ***Kleenex***® as an antiviral agent, in ***beer***, to maintain the color of fresh pork cuts, and in ***acid-citrate-dextrose*** (***A-C-D***) solutions to extend storage time of blood. Over 450 million lbs made annually in the U.S.

Untoward effects: Excessive IV use can cause hypocalcemia and ***A-C-D*** solutions can increase prothrombin time. Inhalation causes coughing. May cause chronic urticaria. Frequent or excessive ingestion can cause erosion of the teeth and gingivitis. Use to help solubilize ***tetracycline*** may actually decompose the latter to the more toxic epianhydro derivatives. Use in ***benzocaine*** throat lozenges causes a dramatic decrease in ***benzocaine*** assay.

LD_{50}, **rats**: 11.7 g/kg; **mice**: 5 g/kg; LD_{LO}, **rabbit**: 7 g/kg; IV LD_{50}, **mouse**: 42 mg/kg; **rabbit**: 330 mg/kg.

CITRININ

= *Antimycin*

***Antibiotic, mycotoxin.* Produced by *Penicillium* and *Aspergillus* sp. Found in many stored grains.**

Untoward effects: Nephrotoxic in **dogs**, **poultry**, **rats**, and **pigs**. Causes increased water consumption, decreased growth rate, and severe diarrhea in young **poultry**. Perirenal edema and hepatotoxic (fatty infiltration and necrosis) in **swine**. Platelet count and hematocrit lowered; respiratory and cardiac problems in **mice**. Suspected cause of death over a 2 year period of several hundred **steers** of approximately 9,600. Pneumonia, esophageal and small intestineal ulcers on post-mortem and anorexia, fever, and diarrhea in clinically ill. Citrinin was isolated from their moldy ***corn*** diet. Affected **animals** recovered and were marketable 2 weeks after the discontinuance of the contaminated feed. Kidneys may be residue risk for **humans**.

LD_{50}, **mouse**: 112 mg/kg; IP LD_{50}, **rats**: 67 mg/kg; **mice**: 35 mg/kg.

CITROBACTER

Untoward effects: Nosocomial meningitis, caused by ***C. diversus***, has been reported in **neonates** in Florida and Connecticut, as well as ***C. freundii*** in Louisiana. Has been associated with outbreaks of food poisoning.

C. freundii causes hunched backs, rough hair coats, rectal prolapse, and dehydration in young **mice**. Mucosal hyperplasia of their colons found at necropsy. Various species can cause the so-called "red-leg" syndrome or bacterial septicemia in **amphibians**. ***C. freundii*** causes a cutaneous ulcerative disease in **turtles**.

CITRON OIL

Untoward effects: Causes photosensitivity reactions in **humans**.

CITRONELLA OIL

Approximately 1.5–2 million lbs still imported into the U.S. annually.

Use: As an insect repellant, in fragrances for cheap soaps, and in foods.

Untoward effects: Causes an allergic contact dermatitis. Three cases of eczematous contact dermatitis. Acneform-type folliculitis and papulovesicular eczema of the hands, fingers, and forearms has been confirmed by patch testing. Several **workers** have reported it as a primary irritant and sensitizer in perfumes, although Kligman did not.

LD_{LO}, **human**: 500 mg/kg.

LD_{50}, **rats**: > 5 g/kg; acute dermal LD_{50}, **rabbits**: 4.7 ml/kg.

CITRONELLAL

Use: Insect repellent, and in foods. In soaps, detergents, cosmetic creams and lotions, and in perfumes.

Untoward effects: The allergen in ***citronella oil***, causing the eczematous contact-type hypersensitivity.

Moderately irritating to **rabbit** skin after 24 h under occlusion.

LD_{50}, **rats**: > 5 g/kg; acute dermal LD_{50}, **rabbits**: > 2.5 g/kg.

CITRONELLOL

= *Cephrol*

Use: Insect repellent, and in foods. In soaps, detergents, cosmetic creams and lotions, and in perfumes.

Untoward effects: A positive reaction to a 1% concentration in ***acetone*** in **people** sensitive to ***citronella oil***. A 32% concentration was severely irritating in a **human** patch test.

Moderately irritating to **guinea pigs** and **miniature swine**. Moderately irritating to intact or abraded **rabbit** skin for 24 h under occlusion.

LD_{50}, **rats**: 3.45 g/kg; acute dermal LD_{50}, **rabbit**: 2.65 g/kg.

CITRONELLYLS

Use: The acetate, butyrate, formate, oxyacetaldehyde, and propionate forms in foods, soaps, detergents, cosmetic creams and lotions, and perfumes.

Untoward effects: The acetate and formate at 4% in petrolatum caused mild irritation on **humans**.

LD_{50} in **rats** for the various forms noted above are 6.8 g/kg, 5 g/kg, 8.4 g/kg, > 5 g/kg, and > 5 g/kg, respectively; acute dermal LD_{50}, **rabbit**: > 2 g/kg, > 5 g/kg, > 2 g/kg, > 5 g/kg, and > 5 g/kg, respectively.

CITRUS

Untoward effects: Photosensitivity reactions after topical exposure to citrus rind oils and contact skin allergic reactions to citrus oil. Ingestion of citrus oils alkalinize urine, which may be undesirable. The National Cancer Institute survey indicates an increase of lung cancer in **women** with increased consumption of citrus fruit and juices. In Singapore, a 49-year-old **male** developed extensive photodermatitis after applying ***Citrus hystrix*** juice, an old folk remedy, topically to deter biting insects. Sun exposure through a thin shirt triggered **his** reaction. See ***Grapefruit***, ***Lemon***, ***Lime***, and ***Orange***.

Repeated application of the oils to the skin of **mice** has caused epidermic hyperplasia and, eventually, tumors. Citrus pulp fed to lactating **cows** may cause a false positive spot test for ***penicillin*** residues in milk. There has been a debate over the safety for **cats** of citrus oil or extract-based insecticides.

CITRUS RED 2

= *CI 12,156* = *CI Solvent Red 80*

Untoward effects: This oil-soluble color for use on the skin of mature ***oranges*** is not to exceed 2 ppm in the whole ***orange***. The safe level for **man** is estimated at 18 mg/day.

Fed to **dogs** at 50 and 200 mg/kg/day/6 days/week for 76 or 104 weeks, only the higher level caused a decrease in red blood cells and hemoglobin, and increased white blood cells. In **rat** and **mouse** (40 of each species) 2 year feeding trials at 0.05 and 0.25% levels, hyperplasia of the bladder wall occurred in ten **rats** and six **mice**, bladder papilloma in two **rats** and papillary carcinoma in one **rat**.

LD_{50}, **rat** and **mouse**: > 4 g/kg.

CIVET

= *Zibeth* = *Zibethum*

Use: In fragrances for food, soaps and detergents, cosmetic creams and lotions, and perfumes.

Untoward effects: Pungent odor in pure form.

Undiluted, it is mildly irritating to skin of **mice**.

LD_{50}, **rats** and acute dermal LD_{50}, **rabbits**: > 5 g/kg; skin, minimum lethal dose, **mouse**: 500 mg.

CLADRIBINE

= *2-CdA* = *2-Chlorodeoxyadenosine* = *Leustatin*

In treating hairy cell leukemia.

Untoward effects: Acute toxicity presents as fever and bone marrow depression represents a delayed toxicity. Severe pancytopenia and thrombocytopenia developed in six **patients** (52–77 years). Fatigue, nausea, headache, and rashes are common adverse effects. Diarrhea, petechiae, epistaxis, abdominal pain, and myalgia occur, to a lesser extent. A 69-year-old **female** developed a large fatal lymphoma after treatment with it. Has caused a delayed reactivation of hepatitis B infection. Peripheral neuropathy has developed with high dosage.

CLAMS

Untoward effects: Raw clams, or those steamed for too short a time, have been incriminated as a source of hepatitis A. Consumption of raw clams has also been associated with numerous outbreaks of gastroenteritis with diarrhea, abdominal cramps, nausea, vomiting, and fever, which have occurred up to 2–3 days later. *Vibrio vulnificus* infections reported in raw **clam eaters**. Clams and other mollusks (mussels and scallops) may concentrate a neurotoxin, highly poisonous to **man** and **animals**, causing a condition known as paralytic shellfish poisoning (PSP). The toxin is a dinaflagellate, *Gonyaulax catenella*, a plankton that mollusks feed on. These plankton multiply rapidly in warm weather. Their rapid proliferation often causes water to turn red, frequently called "Red Tide". In **man**, the PSP toxin has caused disturbances of the central nervous system within a few minutes or hours after ingestion. Tingling and numbness of lips, tongue, and fingertips are the first symptoms, followed by poor muscle coordination and balance, slurred speech, swallowing difficulties, and even complete muscular paralysis with death from asphyxiation. The toxin is more concentrated in the dark meat (digestive organs), and is not destroyed by cooking. Clams accumulate lead in their shells and tissues (0.5–1.0 ppm). ***Saxitoxin*** has also been found in clams. Clam extracts and raw clams inactivate up to 50% of ***thiamin*** in **people**. Anaphylactic shock, especially if exertion occurs before or after eating clams, has been reported.

Alaskan Butter Clam = *Saxidomus giganticus* toxicity has been associated with "Red Tide" in Alaskan waters, causing a neuromuscular blockade by the ***saxitoxin*** produced. It interferes with ***sodium*** permeability of the nerve. Symptoms usually occur within a few hours after ingestion and cause skeletal muscle weakness with respiratory distress. Mortality has varied between 1 and 10%.

CLARITHROMYCIN

= *A 56,268* = *Biaxin* = *Klaricid* = *6-Methoxy-eryth-romycin* = *TE 031*

Antiobiotic.* A macrolide related to *erythromycin.

Untoward effects: Nausea (3–5%), diarrhea (3–6%), and abdominal pain (2%) are less common than with ***erythromycin***. Headache (2%) and dizziness, urticaria, skin eruptions, dermatitis, glossitis, oral moniliasis, vomiting, insomnia, cholestasis, altered taste (6%), and rare cases of Stevens-Johnson syndrome and anaphylaxis also occur. High dosage has caused reversible hearing loss. A 68-year-old **male**, after two doses of 250 mg, developed leukocytoclastic vasculitis with 5 mm areas of purpuric rash on **his** arms and legs. A 26-month-old **female** developed pseudo-membranous enterocolitis while being treated for otitis media. A 66-year-old **female** who received 500 mg twice daily/4 days developed ventricular tachycardia and torsades de pointes. In a series of 3,768 **patients**, the most common symptoms were nausea, diarrhea, abdominal pain, and headache, although not serious. Pancreatitis in 58-year-old **female** after 500 mg twice daily/5 days. Prolonged QT syndrome in 70-year-old **female** after 250 mg twice daily/4 days. Hypersensitivity swelling of lips, jaw, tongue, mouth, and face in 90-year-old **female** after two 500 mg doses. Mania reported in two **patients**. Some adverse effects are noted in 50% of treated **patients**.

Drug interactions: ***Acenocoumarol***: Significantly increases serum concentrations of ***acenocoumarol***.
Astemizole: Significantly increases serum concentrations of ***asmizole***.
Carbamazepine: May increase serum concentrations of ***carbamazepine*** when given concurrently. ***Carbamazepine*** toxicity has been induced by clarithromycin in a 35-year-old **female epileptic**.
Cimetidine: Prolongs its absorption.
Cisapride: Use with ***cisapride*** leads to syncopal episodes, QT prolongation, torsades de pointes, three times increase in ***cisapride*** serum concentration leading to death.
Cyclosporine: It inhibits the metabolization of ***cyclosporine***.
Delavirdine: ***Delavirdine*** decreases its metabolization and increases its toxicity.
Digoxin: It induces ***digoxin*** toxicity, causing nausea and confusion.
Disopyramide: Ventricular fibrillation requiring resuscitation in 74-year-old **female** 6 days after starting it at 250 mg twice daily after 7 years on ***disopyramide*** 200 mg twice daily.
Fluoxetine: Clarithromycin inhibits the metabolization of ***fluoxetine***. The increased blood levels induce delirium.
Itraconazole: Has caused increased concentration of clarithromycin in 38-year-old **female**.
Lansoprazole: Glossitis, stomatitis and/or black tongue when combined with ***lansoprazole***.
Lovastatin: Significantly increases peak concentrations of ***lovastatin***. Addition to ***lovastatin*** therapy can induce rhabdomyolysis.
Midazolam: Prolongs ***midazolam***-induced sleep by decreasing its metabolization.
Omeprazole: Clarithromycin inhibits its metabolization.
Pimozide: May increase serum concentrations of ***pimozide*** when given concurrently. Use with ***pimozide*** leads to prolonged QT interval, cardiac toxicity, and death may occur if > 10 mg ***pimozide***/day is used.
Rifabutin: Significantly increases peak concentrations of ***rifabutin***. Severe neutropenia and uveitis with prophylactic doses of ***rifabutin***.
Rifampin: ***Rifampin*** significantly decreases its serum concentration.
Terfenadine: Significantly increases peak concentrations of ***terfenadine***.
Simvastatin: Significantly increases peak concentrations of ***simvastatin***.
Theophylline: May increase serum concentrations of ***theophylline*** when given concurrently.
Verapamil: Addition to ***verapamil*** therapy leads to bradycardia.
Warfarin: Significantly increases peak concentrations of ***warfarin***.

Renal tubular degeneration in **rats** at two times **human** dosage (**monkeys**: eight times, **dogs**: 12 times). Testicular atrophy in **rats** at seven times **human** dosage (**monkeys**: eight times, **dogs**: three times). Corneal opacity in **dogs** at 12 times **human** dosage (**monkeys**: eight times). Dramatically increases ***carbamazepine*** serum levels in **rabbits**.

CLAVICEPS

C. purpurea*'s dried sclerotium is known as *Ergot.

Untoward effects: A parasitic fungus blackening grains and grasses, fortunately, making them unpalatable. Ergotism caused by alkaloids produced by it were involved in "St. Anthony's Fire" and "Holy Fire" that swept through Europe. Unfortunately, ***rye*** infested with it was ground into bread flour, baked, and eaten by **humans**. See ***Bread*** for its historical significance during the French Revolution. Ingestion has caused visual hallucinations, possibly, from ***lysergic acid diethylamide*** content. Acute poisoning is associated with nausea, vomiting, weak pulse, confusion, numbness and tingling of extremities, and unconsciousness. Its ***ergotamines*** have caused hyperexcitability and lameness with gangrene. More chronic symptoms including circulatory disturbances with thrombi formation and vasoconstriction, coldness of skin, severe muscle pain, dry peripheral gangrene, angina, tachycardia or bradycardia, hypo- or hypertension, headache, dizziness, diarrhea, leg weakness, and miosis.

Ergotism caused by ***C. purpurea*** has caused decreased rate of gain, anorexia, abortions, agalactia, cold extremities, and gangrene in **swine**; lameness, gangrene of the extremities, and occasionally inflammation of the digestive tract in **cattle**, **sheep**, and **horses**; and causes a vesicular dermatitis in **chickens**.

C. paspali has caused a neurological disorder in **cattle** and **horses** called "Paspalum Staggers".

LD_{50}, **sheep**: 4.6 mg/kg.

CLAVULANIC ACID

= *Claventin* = *MM 14,151*

Usually used with *amoxicillin* = *Amoksiklav* = *Augmentin* = *Clavamox* = *Co-amoxiclav* = *Stacillin* = *Synulox or with ticarcillin* = *Betabactyl* = *Timentin*

Antibacterial. A β-lactamase inhibitor.

Synergistic with other *penicillins*.

Untoward effects*:** Since its antibacterial activity is poor but it enhances the activity of some other antibiotics, particularly, ***penicillins, adverse effects are reported in terms of the two most popular combinations. May cause false positive results for leukocytes in urine dipstick tests.

Drug interactions*:** Hepatitis and jaundice reported when used with ***amoxicillin in many **patients**. These symptoms usually appear about 1 month after commencement of treatment. Gastrointestinal upsets reported in up to 40% of **patients** treated with this combination. Erythema multiforme also reported. Thrombocytosis and hyperkalemia associated with its use with ***ticarcillin***. The combination may cause slightly decreased ***digoxin*** serum levels.

In **cats**, use with ***ticarcillin*** may have been associated with increased fevers (up to 106 °F) and systemic vasculitis in one case.

CLAY PIGEON TARGETS

Untoward effects*:** Pieces and powder of these on land used by skeet clubs or similar sportsmen's organizations are very toxic to **hogs** after ingestion. ***Coal-tar pitch appears to be the key toxic ingredient, and is used on stone chips for road repairs or tarred walls. Centrolobular necrosis and hemorrhage in their livers are the key pathology. Presents a potential hazard to **children**, who are prone to eat or chew on most things.

CLEBOPRIDE

The malate = *Amicos* = *Clanzol* = *Clast*= *Cleboril* = *Cleprid* = *Motilex*

Antispasmodic, anti-emetic.

***Untoward effects*:** The most common, and often severe, ill-effect is drowsiness. Reported to accentuate the symptoms of Parkinson's disease.

CLEMASTINE

= *Aloginan* = *Alphamin* = *Anhistan* = *Clemanil* = *Fuluminol* = *HS-592* = *Inbestan* = *Kinotomin* = *Lacretin* = *Lecasol* = *Maikohi* = *Mallermin-F* = *Marsthine* = *Masletine* = *Meclastine* = *Mecloprodin* = *Piloral* = *Reconin* = *Tavegil* = *Tavegyl* = *Tavist* =*Telgin-G* = *Trabest* = *Xolamin*

Antihistamine.

***Untoward effects*:** Drowsiness, fatigue, sedation, and impaired coordination and performance commonly noted. Dry mouth, light-headedness, dizziness, vertigo, giddiness, skin irritation, urticaria, eczema, photosensitivity, prurigo, malaise, nausea, headache, "hangover" sensation, "creeping scalp", and anosmia reported occasionally. In breast milk, it has caused drowsiness in a nursing **child**. In overdoses, also mydriasis, flushing, and hypotension.

Drug interactions*: *Monoamine oxidase-inhibitors prolong and intensify its anticholinergic effects.

CLEMATIS sp.

***Untoward effects*:** Contains a vesicant sap that causes gastroenteritis and irritation of skin and mucous membranes. Its alkaloids, in large quantities, have caused incoordination and death. Leaf and flower secretions are irritant and have an acrid burning taste and can cause sneezing. When bruised, they irritate the eyes and throat, causing tearing and coughing. Inflammation and vesication led to its old name, *Flammula jovis*. Its glycosides account for its toxicity from Chinese arrow poisons.

0.5 g has caused vomiting in **budgies** and **canaries**. Used by **Sioux Indians** as an irritant and stimulant for tired **horses** by putting small pieces of raw root containing ***anemonin*** in their nostrils. Poisoned **horses** have shown tachycardia, weak pulse, decreased appetite, constipation, and polyuria.

C. flammula's alkaloid, ***clematine*** (2 mg), causes copious and frequent urination, tremors, respiratory distress, bradycardia, and, within a few minutes, death in **guinea pigs**.
C. hirsuta = ***Yamanza*** (Hausa) = ***Adapopo*** (Yoruba) is very poisonous to **cattle** and other **livestock** in Nigeria.
C. vitalla = ***Old Man's Beard*** = ***Traveler's Joy*** has caused dyspnea, conjunctivitis, recumbency, abdominal pain, loss of muscle tone, muzzle ulceration, dysentery, and death within 24 h in a **heifer**. Externally, it is a strong irritant and vesicant.

CLEMIZOLE

= *Allercur* = *Histacuran* = *Klemidox* = *Reactrol*

Antihistamine.

Untoward effects: Mild drowsiness in some **patients**. Rarely, weakness, diplopia, nausea, dry nose, and burning of the chest occur. Supraventricular and ventricular ectopic beats, ventricular tachycardia, and fibrillation reported in a **patient** receiving 4.5 mg/kg.

CLENBUTEROL

= *Monores* = *NAB 365* = *Planipart* = *Spiropent* = *Ventipulmin*

Anti-asthmatic, bronchodilator, tocolytic, β-sympathomimetic, β-adrenergic agonist.

Angel Dust is its popular name in Ireland. The same name is used for ***phencyclidine***. It is a substituted catecholamine with a chemical and pharmacological relationship to ***epinephrine***, causing increased gains and decreased fat deposits in **sheep**, **cattle**, **swine**, and **poultry**.

Untoward effects: Residues in the meat (concentrated in the liver) of treated **animals** have caused serious problems in **humans** consuming beef livers. Symptoms often appear suddenly after ingesting it, and are tachycardia, muscle tremors, nervousness, headache, dizziness, nausea, chills, and fever. These reports after its illegal use in **animals** were common in Ireland, France, Spain, and Italy. Consumption of residues in clandestine-slaughtered veal **calves** by 62 **people** (7–65 years) led to tachycardia, palpitations, dizziness, nervousness, tremors, nausea, and headache within 10 min–3 h. **Athletes** have abused it. Some of these same symptoms develop in **people** given high dosages for therapeutic reasons.

Drug interactions: In **patients** with alcoholism, it decreases ***chlorzoxazone*** clearance.

Animals have shown nervousness, sweating, tremors, weakness, and even vomiting (in species that can) at high dosage.

CLIBADIUM SYLVESTRE

Untoward effects: Extracts of its leaves are used by South American **Indians** as a **fish** poison.

CLIDANAC

= *Britai* = *Indanal* = *TAI-284*

Antipyretic, anti-inflammatory.

Untoward effects: LD_{50}, **rats**: 35 mg/kg.

CLIDINIUM BROMIDE

= *Quarzan* = *Ro 2-3773*

Anticholinergic.

Untoward effects: At one time, an analog was blamed for enhanced anticholinergic effects, such as blurred vision, mydriasis, mouth dryness, constipation, difficulty in urination, headaches, impotence, and allergic reactions.

CLINDAMYCIN

= *Antirobe* = *Cleocin* = *Dalacin C* = *Dalacin T* = *Dalactine* = *Klimicin* = *MJ 1986* = *Sobelin* = *U-21251*

***Antibacterial.* A faster absorbing form of *lincomycin*, with a broader spectrum of activity.**

Untoward effects: A small amount is secreted in breast milk.
Arthritis: Acute migratory polyarthritis and pseudomembranous colitis in 55-year-old **male** after therapy.
Blood: Reversible (after discontinuance of therapy) granulocytopenia, leukopenia (chiefly neutropenia), leukocytosis, anemia, antinuclear antibody, thrombocytic purpura, and eosinophilia in a limited number of cases.
Gastrointestinal: Numerous cases of treatment-associated nausea, diarrhea, vomiting, abdominal pain, flatulence, and pseudomembranous colitis. Centrilobular or midzonal hepatic necrosis. Increased SGOT, SGPT, and alkaline phosphatase in many **patients**. Colitis has also been reported from its use in a topical skin cream. Constipation, heartburn, a "barnyard taste", and esophageal ulceration (probably, from prolonged mucosal contact) have occurred.
Hypersensitivity: Most common have been pruritus, rash, and urticaria, up to 10% in some series. Angioedema, lip and nasal edema, exfoliative dermatitis, serum sickness, Stevens-Johnson syndrome, leukocytoclastic angiitis and anaphylactic reactions are rare.
Miscellaneous: Hypoalbuminemia, ascites, pleural effusion, vertigo, eyes burning, sore throat, frequent uri nation, fatigue, headache, thrombophlebitis from IV use, vaginal candidiasis with long-term usage, and nephrotoxicity.

Potential for serious reactions in nursing **infants**, as it appears in breast milk. Undiluted solutions IV may cause cardiac arrest.

Drug interactions: Solutions are chemically incompatible with ***ampicillin, barbiturates, aminophylline, calcium gluconate, magnesium sulfate***, and ***phenytoin***.
Anesthesia: Postoperative respiratory decrease occurs. This has also unpredictably occurred during surgery, and, apparently, was not reversible with ***anticholinesterase agents*** or ***calcium***. It exhibits a ***curare***-like effect and presynaptic local anesthetic-like effect at high concentrations. It has prolonged the action of muscle relaxants, such as ***pancuronium***, in some **patients**.
Contraceptives, oral: May increase incidence of breakthrough bleeding and spotting. Greater risk of pregnancy.
Contrast media: Diffuse kidney and/or colon uptake when given with ***gallium***^{67} ***citrate*** imaging.
Cyclosporin: Requirements are increased when used with it.
Digoxin: Clindamycin may increase its toxicity.
Verapamil: Acute toxicity to it followed clindamycin administration.

Enhanced non-depolarizing neuromuscular blockade and enterocolitis occur in **guinea pigs**. It has caused enterocolitis in **hamsters** and neuromuscular paralysis in **mice**. Fatal colitis may develop in **horses**. **Neonates** have limited capacity to metabolize it.

CLITOCYBE sp.

Untoward effects: This ***mushroom*** is one of many toxic ones causing muscarinic effects (salivation, lacrimation, rhinitis, nausea, vomiting, bronchospasm, dyspnea, diarrhea, abdominal pain, and headache) within 15–30 min, and sometimes, up to 2 h, after ingestion. Bradycardia, hypotension, muscle paralysis, shock, and even hallucinations also occur. Death is infrequent. ***Atropine*** is usually an effective antidote for its neurotoxic agent, ***muscarine***.
C. acromegala in Japan is toxic.
C. cerussata is toxic.
C. claviceps in Japan contains a ***disulfiram***-like compound, ***coprine***.
C. dealbata contains ***muscarine***.
C. olearia is a common cause of toadstool poisoning in Yugoslavia and Mediterranean areas. Symptoms develop within a few minutes to 3 h. General symptoms are those listed above plus occasionally vertigo, metallic taste in the mouth, somnolence and confusion, darkened vision, double vision, difficulty in swallowing, and enlarged livers. A 86-year-old **female**, mistakenly thinking it was ***coffee***, drank the blackish water some had been boiled in, and was poisoned. Contains some ***muscarine***.
C. toxica of South Africa contains a toxic factor, apparently, destroyed by boiling for a least 5 min. Toxic orally to **guinea pigs**, **mice**, and **rats**, and its extracts, parenterally, to **cats**, **frogs**, **mice**, **rabbits**, and **rats**.
C. trunicola has toxic muscarinic effects in **humans**.

CLOBAZAM

= *Frisium* = *H-4723* = *HR 376* = *LM 2717* = *Urbadan* = *Urbanyl*

Anxiolytic. **A benzodiazepine.**

Untoward effects: Conflicting reports concerning its adverse effects on driving and braking abilities. Since hepatic metabolism may be impaired with increasing age, its clearance can be decreased and serum levels increased.

Sedation with behavioral depression, increased muscle relaxation, hyporeflexia, and ataxia in **cats**, **dogs**, and **mice**. Potentiates drug-induced anesthesia in **mice**. Long-term dosing in **dogs** and **monkeys** has shown withdrawal effects (notably, convulsions leading to death) similar to those that occur with other benzodiazepines.

LD_{50}, **dog**: 100 mg/kg; **rat**: 6 g/kg.

CLIVIA MINIATA

= *Kaffir Lily*

An African lily grown elsewhere as an ornamental house plant for its flowers.

Untoward effects: Contains a toxic alkaloid, ***lycorine*** (0.43% – dry weight in the bulb). Excessive intake by **children** and **pets** leads to salivation, vomiting, diarrhea, collapse, and paralysis.

CLOBETASOL

Glucocorticoid. The propionate = *Clobesol* = *Dermoval* = *Dermovate* = *Dermoxin* = *Dermoxinale* = *Temovate*

Untoward effects: A 34-year-old **male** presented with arthralgias, easy bruising, truncal obesity, moon facies, and violaceous striae. To treat **his** psoriasis, **he** used approximately 30 g/week, mainly on **his** face, axillae, and groin for approximately 5 years. A 55-year-old **female** developed systemic side-effects after weekly topical application of 25 g. These cases put the lie to statements that levels under 50 g/week are without systemic effects. Cushingoid features and symptoms of adrenocortical insufficiency in three **patients** noted after withdrawal of therapy. ***Steroid***-induced skin blanching and milia can occur. Allergic contact dermatitis in some **patients**. Topical use has caused stinging sensations, itching, skin cracking, erythema, folliculitis, numbness of fingers, and telangiectasis. Can pass into **mother's** milk. Epidermal thinning of skin in **mice**.

CLOCORTOLONE PIVALATE

= *CL 68* = *Cloderm* = *Purantix* = *SH863*

Glucocorticoid.

Untoward effects: Skin irritation, burning, itching, stinging, leg blisters, dryness, folliculitis, hypertrichosis, hypopigmentation, perioral dermatitis, allergic contact dermatitis, skin atrophy, striae and milaria.

CLOFAZIMINE

= *B 663* = *G 30320* = *Lampren(e)*

Antibacterial, leprostatic.

Untoward effects: Slight pink tinge in skin and urine. Hyperpigmentation and red-brown pigmentation of the skin commonly reported. This may also involve the conjunctiva. May have induced a facial granuloma in a **patient**. Rash, dermatitis, skin dryness, icthyosis, and photosensitivity in up to 28% of **patients**; rash and pruritus in 1–3%. Enteropathy with splenic infarction, abdominal discomfort, mild diarrhea, nausea, vomiting, anorexia, and weight decrease in ~50%. Red tinting of the urine, sputum, and sweat have been reported. Transient increase in aspartate transaminase. Nausea, vomiting, and diarrhea are uncommon. Dizziness, drowsiness, fatigue, headache, taste disorders, and neuralgia also noted. Corneal changes in the form of fine linear brownish subepithelial opacifications and retinal pigment epithelial degeneration reported. **Mothers** who have ingested it have had **infants** pigmented at birth or after drinking their mother's milk.

Drug interactions: ***Isoniazid*** decreases its tissue concentration.

Impaired fertility in **rats** at high dosage.

CLOFIBRATE

= *Amotril* = *Anparton* = *Apolan* = *Arterioflexin* = *Arteriosan* = *Arterosol* = *Artevil* = *Ateculon* = *Atheromide* = *Atheropront* = *Atromidin* = *Atromid-S* = *Bioscleran* = *Claripex* = *Cloberat* = *Clobren-SF* = *Clofinit* = *CPIB* = *Hyclorate* = *Lipavlon* = *Liprinal* = *Neo-Atromid* = *Normet* = *Normolipol* = *Recolip* = *Regelan* = *Serotinex* = *Sklerolip* = *Skleromexe* = *Sklero-Tablinen* = *Ticlobran* = *Xyduril*

Antihyperlipoproteinemic.

Untoward effects: LD_{LO}, **woman**: 90 mg/kg; **man**: 200 mg/kg.

Blood: A 49-year-old **male** had a reversible neutropenia and leukopenia after 500 mg qid/5 weeks. Agranulocytosis in a 57-year-old **female** may have been an idiosyncracy reaction. Serum alkaline phosphatase decreased in 24/26 receiving 2 g/day. Increased fibrinolytic activity and decreased plasma fibrinogen. Inhibits platelet function and adhesiveness in **man** and **animals**. Leukopenia and neutropenia have been common, while thrombocyto-penia, eosinophilia, antinuclear antibody, and coagulation defects occur in **humans** less often.
Cardiovascular: Ventricular bigeminal rhythm within 24 h and lasted up to 6 days after discontinuance in 58-year-old **female**. Extrasystole, arrhythmia, and leg pain in 54-year-old **male**. Increased intermittent claudication, angina, and thromboembolic events reported.
Gallbladder and gastrointestinal: Marked increase in hypercholesterolemia and intrahepatic gallstones, secondary pruritus, jaundice, nausea, vomiting, epigastric discomfort, heartburn, flatulence, and diarrhea.
Libido: Libido decreased in **women**, impotence in **men** estimated at 3–4%. Improvement occurs 3–4 weeks after discontinuance of therapy.
Miscellaneous: Has antidiuretic-like action, but may not cause clinical hyponatremia. May decrease ***carbohydrate*** absorption by inactivation of enzymes and cause hypoglycemia. Has caused a lupus erythematous syndrome and pancreatitis. Headache, drowsiness, fatigue, hypogeusia, asthenia, flu-like symptoms, abnormal liver function, and ventricular ectopy reported. May harm **fetus**. Strict birth control is recommended before, during, and after taking the drug. Exacerbation of gouty arthritis
Muscle: Acute muscular or necrotizing myopathy syndrome, including pain, weakness, tenderness, myokymia, and cramps, can follow its use. Occurs frequently in **patients** with renal failure, nephrotic syndrome, or hypothyroidism, and may be dose-dependent.
Neoplasm: WHO indicated an association with increases in colon cancer.
Poisoning and suicide: A 15-year-old **male** attempted suicide by ingesting 49 capsules, but was asymptomatic.
Pyrexia: Temperature increased after 500 mg qid 1 month.
Renal: Renal plasma flow decreased and renal failure and acute interstitial fibrosis can be precipitated by its use.
Skin and hair: Blotchy purpuric eruptions with macules and papules in 54-year-old **male** on 1 g twice daily/ 1 week. Alopecia is rare. Photosensitivity reactions occur. Allergic reactions including urticaria and sweating.
Taste: Taste acuity decreased and unpleasant or altered taste. Smell may also be adversely affected.

Drug interactions: ***Acetylsalicylic acid***: Prothrombin time increases.

***Androsterone*:** Postprandial discomfort, agranulocytosis and, rarely, erythematous facial rash, transient weight increase, nausea, and muscle ache.
Anticoagulants*:** (acenocoumarol, anisindione, bishydroxy-coumarin, coumarins, diphenadione, ethylbiscoumacetate, phenindione, phenprocoumon, warfarin***) increases peak plasma concentrations and prolongs their half-life, causing a significant decrease in ***anticoagulant*** required in 15–70% of anticoagulated **patients**. Generally, requires a 33% decrease in ***warfarin*** dosage.
***Chlorpropamide*:** Its half-life is prolonged by clofibrate.
Contraceptives, oral*:** A 30-year-old **female** had increased serum cholesterol concentrations with an ***estrogen–progesterone combination, although it was previously controlled by clofibrate.
***Nalidixic acid*:** Action is decreased.
***Neomycin*:** Clofibrate action is increased.
***Nitrofurantoin*:** Action is decreased.
***Penicillins*:** Effect is increased.
***Probenecid*:** Renal and metabolic clearance of clofibrate decreased by it.
***Rifampin*:** Being a strong liver microsomal enzyme-inducer it decreases clofibrate levels by nearly 50%.
Thyroxine*:** Rate of metabolization decreases. *D-**thyroxine decreases anticoagulant dosage by $\frac{1}{3}$–$\frac{1}{2}$.
***Thyroxin-binding globulin*:** May be rendered unavailable by clofibrate.
***Tolbutamide*:** Its hypoglycemic activity is increased by a dramatic increase in its half-life.
***Vasopressins*:** A significant prolongation of action of both IV and intranasal forms.
Vitamin C*:** Daily doses of 1 g/day increase serum ***cholesterol.

Liver enlargement, and both malignant and benign tumors in **rats** and **mice**. Potentiates ***warfarin's*** activity in **dogs**. Liver glycogen is decreased in the **dog**, **monkey**, and **rat**. Reduces viral replication in **chick** embryos.

LD_{50}, **rat**: 1.65 g/kg; **mouse**: 1.28 g/kg.

CLOGESTONE ACETATE
= *AY 11,440*

***Progestogen*. Explored as a treatment for migraine.**

***Untoward effects*:** Polymenorrhea (19/22), intermenstrual bleeding (14/22), depression (4/22), amenorrhea (3/22), mastalgia (2/22), nausea (2/22), abdominal swelling (1/22), and acne (1/22).

CLOMACRON
= *SKF 14,336*

***Psychotherapeutic*. An acridan compound.**

***Untoward effects*:** Akathisia (6/79), dystonic episodes (3/79), tremors (1/79), dry mouth (7/79), constipation (5/79), dermatitis (1/79), and weight loss (1/79). Headaches are common; occasionally drowsiness, dizzy spells, facial and hand edema, and anemia noted. In a group of 36 **patients**, extrapyramidial symptoms in 16, pruritus in three, salivation in two, drowsiness in two, dry mouth in two, dizziness, hyperalertness, increased SGOT in one each reported. Other reactions noted have been pigment deposits in eye lenses, neutropenia, and photosensitivity.

IP LD_{50}, **mouse**: 140 mg/kg.

CLOMETHIAZOLES
= *Chlorethiazol* = *Chlormethiazole* = *Distraneurin* = *Hemineurin* = *Heminevrin* = *SCTZ*

Sedative, hypnotic, and anticonvulsant.

Untoward effects*:** Drug abuse and dependency, drowsiness, sneezing reflex, decreased respiration and blood pressure, tingling sensations in nose and sneezing, apnea, euphoria, headache, and rhinorrhea were commonly reported. Cerebral depression and severe asphyxia in a newborn **female** whose **mother** took it, along with other drugs, 22 h before delivery. Orally in late pregnancy, it caused alopecia in four **infants**. Allergic reactions, poisonings (especially with ***alcohol), coma from overdoses, and parotitis also reported. Therapeutic serum levels reported as 0.7–2 mg/ml; toxic at 4–15 mg/ml; and fatal at 50 mg/ml.

Drug interactions*:** Use with ***diazoxide in pregnant **mothers** resulted in 21 depressed **infants**, and neonatal hypotonia, hypoventilation or apnea. Plasticization of the polymers by it occurs, as well as softening of plastics. ***Cimetidine*** potentiates it and increases its half-life by decreasing its clearance. Its elimination is slowed during ***alcohol*** withdrawal and fatalities have occurred. Other central nervous system depressants can enhance its effects.

Mouse studies indicated ***phenobarbital*** decreased its sleeping time.

CLOMIPHENE CITRATE
= *Chloramiphene* = *Clomid* = *Clomifene* = *Clomivid* = *Clomphid* = *Clostilbegyt Dyneric* = *Enclomiphene* = *Ikaclamin* = *Ikaclomine* = *MRL-41* = *Pergotime* = *Serophene* = *Zuclomiphene*

***Ovulation inducer*. Estrogen and anti-estrogen activity.**

***Untoward effects*: *Cardiovascular*:** Vasomotor flushing (> 10%).

Central nervous system: Nervousness and insomnia (~2%), headache (1%), dizziness or light-headedness (1%), fatigue, depression, and possible psychosis.
Eye: Blurred vision (13%), spots or flashes seen (scintillating scotomata), electroretinographic changes, cataracts or lens opacities and multiple images, diplopia.
Fetus: Neural tube defects if given just prior to or at time of conception, meningomyelocele, spina bifida, Down's syndrome, club foot, talipes equinovarus, tibial torsion, blocked tear duct, hemangioma, and anencephaly.
Gastrointestinal: Abdominal pain or discomfort (> 7%), nausea and vomiting (> 2%).
Genitourinary: Ovarian cysts and enlargements (10–32%), hot flashes (8–52%), uterine bleeding, increased urination, exacerbation of endometriosis, mittelschmerz, multiple births and leukorrhea, intra- and extrauterine pregnancies, tubal pregnancy, abnormal karyotypes in abortions after clomiphene-induced ovulation, and endometrial cysts reported. Use in 37-year-old **female** for 6 months led to grade 2 immature teratoma.
Miscellaneous: Breast tenderness (> 2%), galactorrhea, weight gain, moniliasis, leg cramps, ascites, hydatidiform mole, and toxemia.

Injections into neonatal **rats** cause extensive reproductive tract abnormalities in adult **rats** and effects are also seen in masculinized female **rats**. High dosage induces ovarian cysts in **rats**, but not in **dogs**. Some alopecia noted in **rats** of both sexes, but not in **dogs**. Lenticular changes noted in 4/29 **rats**, but none in **dogs**. Teratogenic in **rabbits** and **rats**.

LD_{50}, **rats**: 5.75 g/kg; IP, 530 mg/kg.

CLOMIPRAMINE

= *Anafranil* = *Chlorimipramine* = *Clomicalm* = *G34,586*

***Tricyclic antidepressant.* Inhibits *serotonin* and *norepinephrine* re-uptake. High incidence of untoward effects (~20%).**

Untoward effects: True toxic effects are rare. Most are unwanted pharmacological effects.
Anticholinergic: Dry mouth, decreased salivation, mild to moderate constipation reported. Blurred vision and urinary retention are less common.
Blood: Adverse effects are rare, but agranulocytosis, pancytopenia, and hyponatremia reported.
Cardiovascular: Hypotension, tachycardia, venous thrombosis, and cardiac arrest.
Central nervous system: Tremor, myoclonus, rigidity, sedation, depression, vertigo, agitation, confusion, disorientation, headache, ataxia, seizures (< 1%) (often these follow withdrawal of treatment and in **neonates** born to treated **mothers**), mania, and clumsiness. Intolerance or overdosage may lead to delirium.
Fetus and neonate: **Babies** born to **mothers** treated in pregnancy may be lethargic, tachypneic, and cyanotic with some metabolic acidosis. This has persisted for 5–16 days. See ***Central nervous system*** above for discussion of seizures.
Gastrointestinal: Primarily, nausea and vomiting. Occasionally, slight anorexia, dysgeusia, gingivitis, glossitis, salivation, and dental caries. Rarely, pharyngeal edema and paralytic ileus.
Genitourinary: Increased serum ***prolactin***, uncomfortable breast engorgement, lactation, and irregular or absent menstruation.
Libido: Sexual dysfunction that includes decreased libido in both sexes, partial or total anorgasmia, delayed ejaculation, and, less commonly, orgasm while yawning.
Miscellaneous: Weight gain, dyskinesia, candidiasis, decreased rapid eye movement sleep. Metallic taste, insomnia, tinnitus, yawning, coma, and death.
Skin: Photosensitivity, rash, and sweating.

Ingestion of 10–20 mg/kg can cause moderate to severe adverse effects.

Therapeutic serum concentrations are 0.09–0.25 mg/l; toxic at 0.4–0.6 mg/l; and lethal at 1 mg/l.

Drug interactions: Hyperreflexia, tremors, rigidity, and clonus in 3/6 after a 1 month washout, when given an ***monoamine oxidase inhibitor*** (***clorgyline***) and seizures occurred when given after discontinuance of ***phenelzine*** (an ***monoamine oxidase inhibitor***). With ***lithium***, a 59-year-old **male** developed a ***serotonin*** syndrome with myoclonus, tremors, shivering, incoordination, and euphoria. Similar effects from a combination with ***moclobemide***. ***Tryptophan*** potentiates its antidepressant action. Severe encephalopathy and death reported with ***meperidine***. Use with ***fluvoxamine*** increases plasma levels of clomipramine and its metabolites ten times.

Hyperthermia noted in **rats** and **mice**. Significant vomiting, lethargy, diarrhea, polydipsia, and seizures reported in treated **dogs**.

CLOMOCYCLINE

= *Chlormethylenecycline* = *Megaclor*

Antibacterial.

Untoward effects: Nausea, vomiting, diarrhea, indigestion, epigastric pain, rash, and pruritus ani. Possible teratogen in 34-year-old **female** on 170–340 mg/day during **her** first 8 weeks of pregnancy.

Phototoxic in **mice**. Has caused decreased weight and malformation in newborn **chicks** of treated **hens**.

LD_{50}, **mouse**: 564, 150, and 30 mg/kg; oral, IP, and IV, respectively.

CLONAZEPAM

= *Clonopin* = *Iktorivil* = *Klonopin* = *Landsen* = *Rivotril* = *Ro 5-4023*

***Anticonvulsant.* A benzodiazepine.**

Untoward effects*: *Blood: Thrombocytopenia, anemia, leukopenia, and eosinophilia (1%), and agranulocytosis (1.3%).
Cardiovascular: Hypotension and palpitations.
Central nervous system: Drowsiness, fatigue, ataxia, muscular hypotonia, depression, confusion, amnesia, slurred speech, paradoxical rage, agitation, excitement and/or mania, dizziness, tremors, and vertigo. Overdoses have caused cyclic coma.
Eye: Nystagmus and diplopia.
Fetus and newborn: Apnea, hypotension, and lethargy in premature **newborn** after maternal use.
Gastrointestinal: Nausea, anorexia.
Genitourinary: Dysuria, nocturia, enuresis, and urinary retention.
Libido: Intermittent vaginal discharge and development of pubic hair in 15-month-old **female** receiving 0.5 mg twice daily. Symptoms subsided after medicine discontinued.
Liver: Jaundice, centrolobular necrosis, and increased hepatic enzymes and ***bilirubin***.
Miscellaneous: Behavioral changes. Withdrawal may precipitate delirium, generalized seizures (5%) and status epilepticus, drooling (especially in **infants** and **children**), allergic reactions, hyperglycemia, and fever. Often, tolerance after 3 months. 3 months high dosage therapy for headache in 43-year-old **male** taking ***insulin*** led to severe hypoglycemia.
Skin: Rashes, alopecia, hirsutism, ankle and facial edema.

Therapeutic serum levels reported as 0.01–0.08 mg/l; toxic at 0.1 mg/l.

Drug interactions*: *Alcohol and other ***depressants*** may exaggerate its ill effects. Its use with ***amiodarone*** in 78-year-old **male** caused development of slurred speech, disturbed sleep, difficulty in walking, dry mouth, urinary incontinence, and confusion. **He** improved considerably within 5 days after discontinuance of clonazepam. Its metabolization is increased by ***carbamazepine***, decreasing its serum levels by 19–37%. In normal **subjects**, ***phenytoin*** pretreatment decreases clonazepam by 46–58% and ***phenobarbital*** pretreatment increases clonazepam clearance by 19–24%. A similar ***phenytoin*** relationship exists in epileptic **patients**. Rarely does it decrease ***phenytoin*** levels. Metabolization of ***valproate sodium*** is increased by use with clonazepam. The ***carbamazepine*** effect also occurs in **monkeys**. **Animal** trials indicate that ***cimetidine, erythromycin, isoniazid, ketoconazole, propranolol,*** and ***valproic acid*** may increase clonazepam's anticonvulsant action.

LD_{50}, **rat**: 3 g/kg; **mice**: 4 g/kg; **rabbits**: 2 g/ kg.

CLONIDINE

= *Catapres* = *Catapresan* = *Clofelin* = *Clonistada* = *Dixarit* = *Duraclon* = *Haemiton* = *Ipotensium* = *Isoglaucon* = *TensoTimelets*

***Antihypertensive.* In shaving soaps. Central α-adrenergic stimulant.**

Untoward effects*: *Cardiovascular: Severe bradycardia in 62-year-old **male** after 0.2 mg tid. Arrhythmia in 10-year-old **male** given two doses of 0.05 mg in 8 h with explosive disorder. A 0.3 mg dose was associated with severe hypotension, bradycardia, and coma in 21-month-old **female**. Paradoxical hypertension reported on high dosage, or after IV use. A 59-year-old **male** developed cardiac arrest after abrupt withdrawal of treatment. Abrupt withdrawal has caused a syndrome associated with sudden high increase in blood pressure, tachycardia, arrhythmias, sweating, anxiety, restlessness, headache, nausea, and vomiting. Avoid prescribing to non-compliant **patients**. This rebound hypertension may be worse in the presence of a ***β-blocker***. Raynaud's disease (16%).
Eye: Can decrease intraocular pressure and/or augment such effects by other agents. Postmenopausal **women** tend to require higher dosage and are more prone to blurred vision, as well as drowsiness and insomnia. Causes miosis and unreactive pupils.
Fetus and neonate: Avoid during pregnancy. Can induce changes in an **infant's *glucose*** metabolism. Approximately 8% of the **maternal** dose appears in breast milk. This may be more than a therapeutic dose for **infants** up to 8 months of age. Has caused a withdrawal syndrome.
Gastrointestinal: Constipation is common (strongly decreased distal colon motility in 27/30). Pseudo-obstruction of large bowel in 26-year-old **male** after a kidney transplant and in other **patients**. Diarrhea reported in 44-year-old **male** after a total of 5 mg. Xerostoma, as well as dry mouth, nausea, and vomiting, occurs.
Hyperglycemia*: *Glucose intolerance associated, particularly with increased dosage.
Hypersensitivity: Contact dermatitis reported with clonidine transdermal patches. Symptoms disappeared within a week after treatment discontinued.
Libido: Up to 50% impotence, delayed or retrograde ejaculation in **men** and decreased libido and orgasms in **women**.
Miscellaneous: Transient fatigue, sedation, depression, dry mouth (46%), dizziness, nasal congestion, headache, anxiety, impotence and difficult urination, asthenia,

bradypnea, lassitude, vertigo, somnolence, hemiparesis, and facial paresis noted. Collapse, sweating, and disorientation in 74-year-old **male** who took twice **his** usual dosage. Gynecomastia in **men** and exacerbation of psoriasis. Insomnia and vivid dreams, hypernatremia, and transient weight gain occur on occasions. Fluid retention, rashes, itching, thinning of hair, and burning and itching eyes. Hypotonia, hyporeflexia, irritability, and even seizures, also reported. A 63-year-old **male diabetic** whose blood sugar was controlled found it deteriorating when **his** clonidine dosage was increased.
***Poisoning*:** Primary symptoms of overdosage are central nervous system depression, bradycardia, hypotension, miosis, hypotonia, hyperglycemia, syncope, respiratory decrease, and, occasionally, seizures. Severe poisoning in 21-month-old **female** after only 0.3 mg, with resultant drowsiness, severe hypotension, bradycardia, recurrent apnea, and unconsciousness. Toxic in **children** with doses as low as 0.1 mg. Some **adults** have survived ingestion of up to 18.8 mg.
***Psychosis*:** Dementias, delirium, depression, and hallucinations occur. Case reports in the **elderly**.

Therapeutic serum levels reported as 0.001–0.004 mg/l and toxic at 0.01–0.05 mg/l.

TD_{LO}, **human**: 69 μg/kg.

Drug interactions*:** Sedation may increase when given with ***alcohol, sedatives, tranquilizers, antidepressants, and ***narcotics***.
Desipramine*:** Decreases antihypertensive effect. This may also be true of other ***tricyclic antidepressants.
***Fluphenazine*:** Delirium.
Imipramine*:** As with ***desipramine.
***Levodopa*:** Its antiparkinsonism effect is decreased by clonidine.
Metoprolol*:** See ***Propranolol. With both, if necessary, clonidine withdrawal must be done very slowly to avoid rebound hypertension.
***Mirtazepine*:** Hypertensive crisis in 20-year-old **male** when it was added to clonidine therapy.
***Nitroprusside*:** **Patients** who have received this may be predisposed to severe, sudden hypotension.
***Propranolol*:** Withdraw treatment with it very slowly, to avoid rebound hypertension.
***Sotalol*:** Blood pressure increases within 1 week.
***Verapamil*:** The combination caused atrioventricular block in two **patients** and hypotension in one.

Clonidine potentiates ***apomorphine*** effects and delays intestinal transit in **rats**. Reported to be embryotoxic in experimental **animals**.

CLONITRATE

= *Dylate*

Coronary vasodilator.

***Untoward effects*:** Palpitations, pounding headache, dizziness, and faintness reported.

CLOPAMIDE

= *Adurix* = *Aquex* = *Brinaldix* = *Chlosumdimeprimyl* = *DT-327*

Antihypertensive, diuretic.

Untoward effects*:** Dizziness (6–9%) and vertigo (5%). Often used with ***pindolol. Tachycardia, weakness, occasionally nausea, abdominal pain, and headaches, slight hyperglycemia (average increase of 24%), and glycosuria in 3/12 with diabetes, and slight decrease in serum ***potassium***.

CLOPENTHIXOL

= *AY 62021* = *Ciatyl* = *Cisordinol* = *Clopixol* = *N-746* = *Sordenac* = *Sordinal* = *Zuclopenthixol*

Psychotherapeutic, tranquilizer.

Untoward effects*:** Extrapyramidal disorders (5–10% more frequent in **female**) including tremors, akathisia, dyskinesia, vertigo, parkinsonism, severe drowsiness (usually decreasing after 1 week), tachycardia, nausea, diarrhea, headache, dry mouth, dermatitis, urinary incontinence, increased SGPT and ***bilirubin, decreased libido, hyperglycemia, blurred vision, photophobia, sweating, anxiety and excitation, mania, delirium, and increased temperature. Thrombotic episodes have followed prolonged immobilization. Blood levels of 0.01–0.1 mg/l are considered therapeutic and 0.3 mg/l toxic.

LD_{50}, **rat**: 660 mg/kg; IV LD_{50}, male **mice**: 105 mg/kg; **rat**: 125 mg/kg.

CLOPIDOGREL BISULFATE

= *Plavix* = *sustained release 25,990C*

Antithrombotic, platelet aggregation inhibitor.

Untoward effects*:** Rash, dyspepsia, abdominal pain, and diarrhea in ~4–6%; gastrointestinal bleeding in 2%, acute arthritis, reversible ageusia, intracranial bleeding in 0.4%, and severe neutropenia in 0.04%. Rare cases of thrombotic thrombocytopenic purpura (TTP) occur, often associated with anemia, renal failure, abnormal heartbeats, neurological problems mimicking strokes, and death. In **volunteers** taking ***naproxen, occult gastrointestinal blood loss.

CLOPIDOL

= *Clopindol* = *Coyden* = *Meticlopindol* = *Rigecoccin*

Coccidiostat.

***Untoward effects*:** The FDA set tolerances for its residues at 5 and 15 ppm, respectively, in muscle, and in liver and kidney of **chickens**; at 0.2, 1.5, and 3 ppm, respectively, in uncooked muscle, liver, and kidney of **cattle**, **goats**, and **sheep**; at 0.2 ppm in uncooked edible tissues of **swine**; at 0.02 ppm in milk; and at 0.2 ppm in cereal grains, vegetables, and fruits. Withdraw 227 g level medication 5 days before slaughter or decrease to lower indicated level. Fairly stable in the environment, and, apparently, non-deletorious to non-target organisms.

CLOPROSTENOL

= *Estrumate* = *Heifex* = *ICI 80,996* = *Planate*

***Synthetic prostaglandin analog*. Estrus synchronizer related to *dinoprost*.**

***Untoward effects*:** Keep away from pregnant **women**, or those of child-bearing age.

Induces abortions in **cattle** and **swine**. At extremely high dosage, uneasiness and frothing at the mouth reported in **cattle**. Withhold milk of treated **animals** for 12 h and must not be slaughtered for food use for 48 h.

CLORAZEPATE DIPOTASSIUM

= *Abbott 35,616* = *CB 4306* = *Belseren* = *Mendon* = *Tranxilène* = *Tranxène*

***Psychotherapeutic, anxiolytic*. A benzodiazepine and a precursor of *nordiazepam*.**

***Untoward effects*:** Sexual asthenia and somnolence, dizziness, diplopia, headache, bad taste, nausea, rash, gastric upsets, dry mouth, mental confusion, insomnia, slurred speech, tremors, ataxia, weakness, pharyngitis, coughing, dysuria. Jaundice, paradoxical rage reactions and destructive behavior occur on rare occasions. Overdoses cause diarrhea, vomiting, and increased blood pressure. Uneventful recovery after a **patient** ingested 187.5 g (25 capsules). Multiple congenital abnormalities in a **newborn** whose 17-year-old **mother** ingested 23 dose units during the first trimester of pregnancy. After 24 h, the **infant** died of pneumothorax and progressive respiratory failure. Placental transfer blamed for hypotonia, decreased reactivity, and early apnea in some **neonates**. Its active metabolites are excreted in breast milk, where they may cause drowsiness and decreased sucking. Therapeutic levels are reported as 0.5–0.8 mg/l and toxic levels at 1.5 mg/l.

Drug interactions*: *Magnesium*-containing *antacids decrease its absorption. Hypokalemia, hyperglycemia, and metabolic acidosis followed a clinical interaction in a 24-year-old **male** taking ***theophylline***.

LD_{50}, **mice**: 700 mg/kg; IP, 290 mg/kg.

CLOREXOLONE

= *Flonatril* = *M & B 8430* = *Nefrolan* = *RP 12833*

Diuretic.

Untoward effects*:** Nausea, diarrhea, asthenia, vertigo, anorexia, profuse perspiration, and, occasionally, constipation, belching, and raised blood ***glucose. May cause hyperuricemia and precipitate gout attacks.

CLORGYLINE

= *Clorgiline* = *M & B 9302*

Monoamine oxidase inhibitor.

***Untoward effects*:** Dry mouth (6/40), headache (4/40), nausea (4/40), constipation (3/40), and vertigo (2/40). Ingestion of 2–3 mg/kg can be life-threatening.

LD_{50}, **rat**: 210 mg/kg; **mouse**: 430 mg/kg.

CLORINDIONE

= *Chlorathrombon* = *Chlorindione* = *G 25,766* = *Indaliton*

Anticoagulant.

***Untoward effects*:** Abdominal pain, nausea, vomiting, hematuria, skin necrosis, ecchymosis, alopecia, pruritus, and bleeding, including epistaxis.

LD_{50}, **rat**: 290 mg/kg; **mouse**: 180 mg/kg.

CLORPRENALINE

= *Broncon* = *Clopinerin* = *Conselt* = *Fusca* = *Isoprofenamine* = *Isoprophenamine* = *Kalutein* = *Lilly 20,025* = *Pentadoll* = *Restanolon*

Bronchodilator.

***Untoward effects*:** Central nervous system stimulation and tachycardia reported.

CLORTERMINE

= *Chlortermine* = *Su 10,568* = *Voranil*

Anorexic.

***Untoward effects*:** Headache is prime complaint; dry mouth in some (4%), transient increased SGPT, constipation (2%), insomnia (3%), and chronic intoxication can lead to psychosis indistinguishable from schizophrenia.

LD_{50}, **rats**: 332 mg/kg.

CLOTHIAPINE

= *Dibenzothiazepine* = *Entumine* = *Etumine* = *HF 2159* = *LW 2159*

Psychotherapeutic.

Untoward effects: Tiredness, hypokinesia, and tremor. Degree of ill-effects is also dose-related. Tardive dyskinesia in **some** in 36th week of treatment. Tachycardia (120–130/min) is very common, trismus, torticollis, disturbed speech, sweating, and salivation also reported.

LD_{50}, **guinea pig**: 150 mg/kg.

CLOTIAZEPAM

= *Clozan* = *Rise* = *Rize* = *Rizen* = *Tienor* = *Trecalmo* = *Veratran* = *Y-6047*

***Anxiolytic.* A benzodiazepam.**

Untoward effects: Potential to accumulate, due to reduced hepatic metabolism with age.

CLOTRIMAZOLE

= *Bay b 5097* = *Canesten* = *Canifug* = *Empecid* = *Gyne-Lotrimin* = *Lotrimin* = *Mono-Baycuten* = *Mycelex-G* = *Mycofug* = *Mycosporin* = *Pedisafe* = *Rimazole* = *Tibatin* = *Trimysten* = *Veltrim*

Antifungal.

Untoward effects: Itching and irritation of vulva and vaginal area in < 1%. Contact allergic dermatitis (2%). Oral burning or pricking sensation in local treatment of stomatitis. Erythema, skin peeling, burning, pruritus, urticaria, and blistering also occur after local treatment. Oral use has caused gastrointestinal upsets (abdominal cramps, bloating), and induction of microsomal drug metabolism. Slight increase in urinary frequency and even burning or irritation in the sexual **partner** has followed topical use. Depression, drowsiness, disorientation, and visual hallucinations have followed systemic use.

LD_{50}, male **rats**: 708 mg/kg; male **mice**: 923 mg/kg.

CLOVERS

= *Tréfle*

Untoward effects: ***Alsike Clover*** (*Trifolium hybridum*): Photosensitization (Trifoliosis or "bighead") developed in light-colored **animals** (usually **cattle**, **horses**, and **sheep**) grazing when it is wet. Anorexia, drooling, swollen lips and tongue, and peeling of skin are common in **cattle**. Occasionally dyspnea. staggering, and cyanosis. Swollen eyelids, ears, nose, and mouth, and intense itching are common in **sheep**. Progressive loss of condition, hepatic failure (enzootic hypertrophic cirrhosis), nephrosis, and neurological impairment, including blindness and staggering, also occur in **horses**.

Ladino Clover: (*T. repens*): also known as ***White Clover***. Contains cyanogens, phytoestrogens, and bloating agents. Has caused laminitis in **horses**.
Lucerne Clover: Associated with bloating in South Africa and Australia.
Red Clover: (*T. pratense*): Also known as ***Crimson Clover*** and ***Meadow Clover***. Causes bloat, salivation, and decreased milk production in **cattle**, and its phytoestrogenic content decreases the incidence of multiple pregnancies during spring and autumn. Contains ***slaframine*** and is cyanogenic in winter. High intake by **cows** has been associated with dwarfism, joint laxity, and superior brachygnatha in their **calves**.
Subterranean Clover (*T. subterraneum*): Estrogenic content decreases fertility in **ewes** and **cows**, and has been associated with urinary calculi in **wethers**, lactation in **rams**, hypertrophy of the clitoris in **ewes**, and photosensitization in **turkeys**.
Sweet Clover = *Melilotus sp.* = ***White Sweet Clover*** = ***Yellow Sweet Clover***. May have caused eye problems after ingestion by **herbal users**. Despite its use as a valuable and nutritious forage for **livestock** and soil improvement legume, it causes serious problems. Its ***coumarins*** are converted into ***dicoumarol***, a fungal metabolite, which is the toxic agent in or on it, normally, at a level of 1.5–2.0%. When the clover is spoiled, the ***dicoumarol*** is converted to ***bishydroxycoumarin***, a cause of serious hemorrhage. Understanding this led, in 1941, to Dr Link's discovery of ***warfarin*** (***coumadin***). Its hay or silage (especially if damaged or spoiled) has caused extensive internal hemorrhages in **cattle** and **sheep**, if fed continuously for over 2 weeks. Bloody milk has been common. **Sheep** have become ill and have died after grazing on it. Abortions, anemia, hemorrhage, and tachycardia in **cattle**; anemia and hemorrhages reported in **horses**. Marked swellings occur anywhere on the body and contain blood. **Rabbits** are particularly susceptible to the toxicity and are often used to test for the poisonous properties. Also see ***Melilot***.

CLOVES

= *Carophyllus*

Untoward effects: In addition to the normal ill-effects of ***cigarettes***, the smoking of ***cigarettes*** with cloves has caused hemoptysis, spasmodic lung contractions, pulmonary edema, and asthma (due to its ***eugenol*** content), often within hours. Nausea, angina, chronic cough, and respiratory infections noted. **Patients** with ***cobalt*** dermatitis should avoid cloves. **Bakers** have developed allergic contact dermatitis to it. It gives positive delayed eczematous patch test.

Clove oil: Millions of lbs are used annually in the U.S. (80% is leaf and stem oil, bud oil is 20%). Some is used as a fragrance in soaps, detergents, cosmetic creams and lotions, and in perfumes. Phototoxicity and contact dermatitis occur. False positive ***phenytoin*** serum assays reported.

LD_{50}, **rats**: 2.02 and 2.65 g/kg, for stem and bud oil, respectively. The acute dermal LD_{50}, **rabbits** was > 5 g/kg and 5 g/kg, respectively.

CLOZAPINE

= *Clozaril* = *HF-1854* = *Leponex*

Psychotherapeutic.

Untoward effects: Most common are drowsiness (21%), fatigue, nausea and vomiting (10%), dizziness (14%), headache (10%), postural hypotension (13%), raised temperature (13%), salivation, weight increase, anticholinergic toxicity (confusion, urinary retention, agitation, dysarthria, delirium with visual and auditory hallucinations, convulsions, tachycardia [140–150/min] [17%], and electrocardiogram changes [arrhythmias]), hypersalivation (13%), and increased ***triglycerides***. Other ill-effects are constipation (16%), paradoxic hypertension (12%), granulocytopenia, agranulocytosis (1–2%), neutropenia, eosinophilia, leukocytosis, and seizures. Incidence of epilepsy reported in 11.5% of 1,303 **patients**. Thromboembolism in 12 **patients** (26–59 years) with doses up to 500 mg/day for up to 2 years. A chronic paranoid 37-year-old **female schizophrenic** had pleural effusion, and maculopapular rash 1 week after starting treatment with it. Rarely, pancreatitis, hepatitis, diarrhea, stuttering, dystonia, akathisia, priapism, neuroleptic malignant syndrome, extrapyramidal reactions, diaphoresis, myalgia, arthralgia, urticarial plaques, swollen ankles and eyes, gastric outlet obstruction, hyponatremia, cardiorespiratory failure, and death. Acute interstitial nephritis in 38-year-old **female** after 125 mg twice daily/11 days. Overdoses were associated with aspiration pneumonia, hypotension, renal failure, seizures, electrocardiogram changes, coma, and death. Ingestion of 3 g in a suicide attempt by a 29-year-old **male** led to hypotonic coma, respiratory decrease, inhalation pneumonia, vascular collapse, and peak serum concentration of 5200 ng/ml. Survived with no sequelae. Some **adults** have survived ingestion of up to 4 or more g, but fatalities reported at 2.5 g.

Therapeutic serum concentrations reported as 0.3–0.6 mg/l; toxic at 0.6 mg/l; and usually fatal at 2 mg/l.

Drug interactions: A 79-year-old **male** experienced acute exacerbation of psychotic symptoms when taking ***caffeine*** in ***coffee*** or ***Diet Coke***®. Seizure followed a 10 day course of ***erythromycin*** (250 mg qid). Two **women** reported seizures after receiving ***lithium carbonate*** (900 mg/day for 4 and 6 days). Its blood levels have been increased by simultaneous use of ***fluoxetine, fluvoxamine, paroxetine,*** and ***sertraline***. Syncope developed in two, 32-year-old **male** and 36-year-old, when given ***enalapril***. ***Phenytoin*** may decrease its blood levels. ***Phenobarbital*** and ***ritonavir*** decrease its metabolization and may increase its toxicity. After sudden cessation of smoking by a **schizophrenic** 35-year-old **male**, it caused toxicity with tonic – clonic seizures, stupor, and coma, requiring a 400% decrease in dosage.

LD_{50}, **mouse**: 210 mg/kg; IV LD_{50}, **rats**: 58 mg/kg; **mice**: 61 mg/kg.

CNIDARIANS

Untoward effects: Approximately 70 of ~9,000 species cause injuries to **man**. These include **jellyfish**, **hydroids**, **sea anemones**, and **corals**. They cause solitary wheals fringed with erythema or papular linear eruptions, and may become pustular. Their stinging units are the nematocysts. Only some are capable of penetrating the skin and poisoning **humans**. After storms or rough seas, tentacles often break free of the organism and drift, but are still capable of firing their nematocysts. Intense pain can be followed by weakness, nausea, headache, perspiration, vertigo, respiratory distress, and cyanosis. In severe stingings, pain is instantaneous and severe, often followed by collapse and violent twitching. Cardiac arrest and death may follow.

COAL

Untoward effects: In one study, pneumoconiosis (Black Lung Disease, anthracosis) was 20 times more prevalent in coal **miners**. A pernicious vascular disease was found in a large number of soft coal **miners**. Chronic fibrosis, bronchitis, and emphysema can occur from coal dust inhalation. "Black lung" has also been diagnosed in a couple of ***petrochemical*** **workers**.

In the vicinity of a Czechoslovakian brown coal (high ***sulfur*** and ***arsenic*** content) burning power station, not only does the water, soil, and vegetation have a high ***sulfur*** and ***arsenic*** content, but local **inhabitants** also have a high ***arsenic*** level in **their** hair and nails, and a high incidence of premature deliveries and monsters. **Bees** in the area died out. **Calves** and **pigs** have also died and abortions were common. Coal-burning plants also contaminate the environment with ***thallium***, ***arsenic***, ***mercury***, and ***selenium dioxide*** in their air emissions. Some coals contain up to 54 ppm ***lead*** and in the vicinity of ***coke*** ovens, herbage was found to contain 25–46 ppm, which is sufficient to poison grazing **cattle** and **sheep**. Grains or vegetables grown on these soils present a significant risk to **humans**.

Unsanitary conditions in coal mining, particularly in the days before mechanization, led to a high incidence of dermatitis. ***Anthracite*** coal dust (90–95% ***carbon***) in a factory accumulated in the hair follicles, causing folliculitis, furunculosis, first and second degree burns, rhagades, and thickened palms. ***Anthracite*** coal **workers** have a significant occupational exposure to ***silica***. In the U.S., coal-burning utilities must limit their ***sulfur dioxide*** emissions to 1.2 lb/million British Thermal Units.

Workers producing coal or coal products are at high risk of ***polycyclic aromatic hydrocarbons*** exposure.

Coal soot was a common cause of scrotal cancer in **chimney sweeps** in England, but not in Scandanavian countries where employers provided bathing facilities. These slightly malignant epitheliomas didn't occur until the **patient** was 30 or 40 years of age, often, long after **he** had outgrown the trade. Cancer of the penis and scrotum occurred in a circumcised **man** who had for 18 years shoveled coal on a locomotive. This is extremely rare in circumcised **men**. Coal dust is explosive (***aspirin*** powder has ten times the explosive capacity).

COAL TAR

= *Clinitar* = *Pix Lantharis*

Untoward effects: Photodermatitis, skin irritation, erythema, melanosis, hyperpigmentation, carcinogenicity (papillomas, and, eventually, epitheliomas), chronic eczematous contact dermatitis, comedonicity; conjunctivitis and keratitis (often seen in **roofers**); and Stevens-Johnson syndrome. Rarely, blond, bleached, tinted, white or gray hair may be temporarily stained green. At one time, **fishermen** developed a high incidence of lip cancer from holding a tar-smeared needle in **their** mouths while repairing tarred nets. Inhalation of volatile fumes can cause bronchial irritation.

In general, it and its derivatives can induce chronic liver damage with icterus, anemia, ascites, and even death.

Comedogenicity noted in **rabbits** and **hamsters**. Carcinogenic in **mice**. Photophobia and keratoconjunctivitis in **rabbits**.

COBALT

Untoward effects: Occupational cancer among **workers** in the chemical industry working with cobalt or its compounds. Aerosols of finely powdered cobalt may cause severe lung abnormalities including severe coughing and pulmonary fibrosis in **man** and **guinea pigs**. Upper respiratory irritation with symptoms of shortness of breath, coughing, and, occasionally, permanent disability and death in **people** with exposure to 1–2 mg/m^3 in air. Folliculitis, erythema, and activation of acne, as well as growth disturbances in **children** can occur. Sensitivity to it has occurred due to various metal prostheses or dentures. Granulomas and uveitis reported after light blue tattooing. Cobalt contact dermatitis in many **workers**, particularly in areas subject to friction. Cross-sensitivity with ***nickel*** is debateable. Large oral doses lead to anorexia, nausea, vomiting, diarrhea, tinnitus, neurogenic deafness, facial flushing, and skin rashes. Causes goiter by blocking ***iodine*** uptake (oxidation), especially in **children**. Toxicity usually occurs as a result of excessive intake or frequent exposure. It enhances the proliferation of erythropoietic cells in bone marrow and cells of the thyroid. In 1960, it was used in Canada to stabilize the foam in ***beer***. When ***beer*** is poured into inadequately rinsed glasses, detergent residues caused the foam layer to disappear. Usually the addition of 1–5 ppm cobalt prevents this. In one area of Canada, 20 ***beer***-related deaths occurred with cardiac hypertrophy and degenerative changes in the myocardium. These heavy ***beer***-drinking **patients** were ingesting 5–10 mg cobalt/day. Their cardiomyopathy was usually accompanied by pericardial effusion and, occasionally, thyroid hyperplasia. 11/28 with similar symptoms died in Omaha, Nebraska. Other fatal cases appeared in Minneapolis, Minnesota, western Australia, and in Leuven, Belgium. Many poisoned cases had polycythemia. The epidemic ceased when cobalt was no longer added to ***beer***. Intermittent claudication and angina pectoris also reported in toxic cases. Inflammatory gastrointestinal lesions and central hemorrhagic necrosis of the liver also occurred. At that time, western Australian ***beer*** had 0.4–0.8 ppm cobalt, while most other Australian ***beers*** had only 0.003 ppm. In general, **people** ingesting 6–8 mg/day do not develop any of the above symptomology. The answer to the obvious dilema is that these **beer-drinkers** had significant nutritional deficiencies. Fatal cardiomyopathy reported in a **metal-worker**, as a result of industrial exposure. Toxicity due to excess ***vitamin*** B_{12} has occurred in **people** from direct exposure to cobalt in mines. Although it contains no ***iodine***, cobalt alters the results of protein-bound iodine tests.

8 h TWA, 0.1 mg/m^3.

In **livestock**, cobalt decreases the availability of ***copper, iron,*** and ***manganese***. Maximum safe dietary levels are 20, 50, 10, and 4 ppm, respectively, for **cattle**, **sheep**, **swine**, and **poultry**. 24 ppm level has been considered toxic to **cattle** and symptoms resemble those of cobalt deficiency. 6/15 **cattle** died after eating hay harvested near a metal works, contaminated with it and other metals. Other poisoned **cattle** were in poor condition and anorexic, had stiffened gait, lacrimation, salivation, dyspnea, poor hair color, and diarrhea. Causes a polycythemia in **rabbits**, **ducks**, **dogs**, young **rats**, and **dogs**.

LD_{LO}, **rat**: 1.5 g/kg; **rabbit**: 20 mg/kg.

Radioactive Cobalt (***Cobalt 60***): It is an intense emitter of highly penetrating *gamma* or X-rays, and can damage living cells, especially hematopoietic tissues. Its half-life is approximately 5.3 years. Heavy exposure decreases the body's ability to produce antibodies. Diarrhea, vomiting, dry skin with erythema, cancer, and even fatal bone marrow aplasia and pancytopenia has resulted from its use. The Chinese have done extensive tests with it in **monkeys** to destroy sperm production, apparently, exploring methods of population control.

COBALT ARSENATE

= *Cobalt Violet*

Untoward effects: As an artist color, it is, obviously, potentially dangerous as a source of ***cobalt*** and ***arsenic***.

COBALT CHLORIDE

= *Cobaltous Chloride*

Untoward effects: Has caused mild acneiform eruptions, nausea, vomiting, and cataracts or lens opacity after oral intake. Cardiomyopathy in a 17-year-old **female** on hemodialysis given 25 mg twice daily/7 months for **her** anemia. A 35-year-old **female** received 25 mg qid/6 months for anemia associated with chronic nephritis and developed severe damage to vestibular and cochlear parts of the eighth nerve, paresthesia of limbs, and peripheral neuritis. It causes a contact urticaria (within 5 min in a 20-year-old **male**). A 6-year-old **male** ingested about 2.5 g from a toy crystal growing set. Abdominal pain, vomiting, and neutropenia resulted.

TD_{LO}, **child**: 48–1500 mg/kg.

LD_{50}, **rat** and **mouse**: 80 mg/kg; **guinea pig**: 55 mg/kg.

As ***Cobalt 60***, see ***radioactive cobalt*** above.

COBALT NAPHTHENATE

Use: To accelerate oxidation and drying of varnishes, paints, and inks.

Untoward effects: Dermatitis has resulted in **workers** having frequent contact with it in inks and varnishes.

COBALT OLEATE

Untoward effects: Dermatitis in **workers** where it is used as a drying accelerator in printing inks.

COBALTOUS ALUMINATE

Untoward effects: Photosensitivity of skin in blue tattooed areas.

COBALT SULFATE

= *Cobaltous Sulfate*

Untoward effects: Allergic contact dermatitis from paper pulp manufacturing. Formerly used as an antifoaming agent in beverages and ***beer*** (see ***Cobalt*** above) with some adverse effects.

Drug interactions: It breaks down ***hydrocortisone acetate*** in ointments.

Experimentally, to produce cardiomyopathy in **dogs**. Decreases immune response, inhibits the action of ***interferon***, and has caused encephalitis and carditis in **mice**.

For **sheep**, the LD is approximately 330mg/kg.

COCA

= *Brain Tabasco* = *Cuca* = *Dandruff of the Gods* = *Erythroxylon* = *Hayp* = *Ipado* = *Toot*

***Central nervous system stimulant*. Utilized for centuries in Peru in religious ritualism that stressed its personal ecstasy as a means of contacting the supernatural. Its use is documented from as early as 2500 B.C. It was considered to be the property of the Incan royal family. It was so revered that its leaves were put into the graves of noblemen, to give them a sense of well-being and freedom from hunger and fatigue in the hereafter. Inca workers toiled tirelessly in the high Andes, chewing slow-acting coca leaf. Contains large amounts of *oxalates* and cocaine.**

Untoward effects: Quids of coca leaf and ***lime*** containing about 45 mg ***cocaine*** are chewed and have caused leukodema of cheek mucosa, as well as many of ***cocaine's*** (q.v.) adverse effects. **Users** often show fatigue, decreased hunger and malnutrition, uncertain steps, general apathy, trembling lips, green and crusted teeth, sunken eyes, fetid breath, and blackened corners of the mouth. Extreme cases evidence dropsy and marasmus. The fruit's waste products can cause ***theobromine*** (q.v.) poisoning, with death from sudden heart failure. In 1885, a mixture from coca leaves and ***kola*** nuts (a source of ***caffeine***) was made and sold in 1886 as a quick "pick-up" and "brain tonic" – "to cure neuralgia, hysteria, and melancholy", and called ***Coca-Cola***®. The formula was eventually changed and, since 1904, is no longer a dangerous beverage.

COCAINE

= *Benzoylmethylecgonine* = *Bernice* = *Bernies* = *Big C* = *Blow* = *C* = *Cecil* = *Charlie* = *Coke* = *Dream* = *Flake* = *Gift of the Sun God* = *Girl* = *Gold Dust* = *Happy Dust* = *Happy Trails* = *Heaven Dust* = *Heaven Leaf* = *Her* = *Jam* = *La Dama Blanca* = *Lady* = *Nose Candy* = *Paradise* = *Rock*

= *Snow* = *Snowbirds* = *Stardust* = *The Pimp's Drug* = *The Rich Man's Drug* = *The Star Spangled Powder* = *White(s)*

Brain Tabasco, *Dandruff of the Gods*, and *Toot* are also used, as with *Coca*.

Untoward effects: Its stimulation of the brain causes excitement, restlessness, euphoria, decreases normal inhibitions, paranoid delusions, and garrulousness. IV use leads to ecstasy, increased activity, and lack of fatigue and hunger within 2 min. These effects last about 15 min. Repeat doses increase deep tendon reflexes, tremors, muscle spasms, convulsions, pulmonary edema and hemorrhage. Toxic psychosis, cyanosis, and addiction also occur. Blood level of 90 μg% is considered toxic; 0.1–0.2 mg% is usually fatal. Cocaine is usually snorted (nasal inhalation), but ***Crack***, a modification, is usually smoked. Since the drug is a vasoconstrictor, **snorters** have slow drug absorption (20–60 min to reach peak blood level), although necrosis of the nasal septum (osteolytic sinusitis) can occur. When smoked, peak blood levels occur within 2–5 min. In the late 1980s, ***Basuco*** (Spanish for ***cocaine*** paste or ***cocaine sulfate***) = ***Bazooka*** = ***Coke Paste*** = ***Diesel*** = ***Little Devil*** made a prominent appearance in the U.S., especially in New Jersey. Unfortunately, it was particularly harmful when smoked because it had high levels of ***lead***. A report cited that 78% of $1.00 bills in Miami contained up to 1 mg cocaine on each. British notes held less.
Addiction: Sallow or ashen complexion and pinpoint-size pupils noted. The terms abuse or habituation are preferred by some. A previously unrecognized risk of cocaine inhalation was the development of granulomas in the lungs from inhalation ***cellulose*** fibers used as a filler. Thrombocytopenia in 70 cocaine IV **abusers** (mean of 10 years). Abuse has resulted in raised temperature, tachycardia, premature ventricular contractions, tremors, hallucinations, rhabdomyolysis, cerebral vasculitis, *Clostridium botulinum* sinusitis, sudden generalized seizures and death. High ambient temperature (> 88 °F or 31.1 °C) increases mortality of overdosage.
Cardiovascular: These can occur after relatively small doses in **persons** who have an idiosyncrasy against it. Nausea, pallor, vertigo, and unconsciousness may occur. Tachycardia, premature ventricular contractions, myocardial ischemia and infarction, coronary artery spasm and cerebral vasoconstriction leading to sudden death, increased blood pressure, cardiomyopathy, and coronary artery dissection. Blocks platelet adhesion. Severe thrombocytopenia in six **males** aged 30–45 years abusing cocaine. Strong association between use of crack cocaine with ischemia and cerebrovascular complications.
Ear: Severe vertigo after 10% solution applied to middle ear. Hearing loss of considerable magnitude after use as a local anesthetic on the round window. Similar ill-effects in **cats**.
Eye: Local eye toxicity due to inhibition of mitosis and corneal healing may, in part, be due to sensitivity to ***epinephrine***, usually used with it. Slight vasoconstriction, mydriasis, occasionally decreased of intraocular pressure, partial cycloplegia, slight exophthalmia, and narrow angle glaucoma occurs in **man** from topical opthalmic use. Keratitis from its use in cataract surgery. Extensive topical use may cause addiction and cocainism, including anxiety, agitation, psychosis, and convulsions.
Fetus and neonate: Acute IV and intranasal use associated with abruptio placentae in 21 weeks and 33 weeks pregnant **females** leading to stillbirth and premature delivery, respectively. Its use by 70 pregnant **women** led to lower gestational age at delivery, increase in preterm labor and delivery, and small **infants**. Not only does its abuse influence the outcome of pregnancy, but it can also adversely affect the neurological behavior of the **newborn** (language delay, autism), may be associated with Turner's syndrome, absent toes, respiratory distress, pulmonary hypertension, increased perinatal mortality, intestinal atresia, cranial abnormalities, urinary tract anomalies, increased patent ductus arteriosus surgical intervention, seizures, reduced cerebrospinal fluid ***homovanillic acid*** levels, increased incidence of necrotizing enterocolitis, transient myocardial ischemia, and even **fetal** death due generali-zed cerebral ischemia and necrosis from a 28-year-old **female** who abused the drug late in pregnancy. A pregnant 26-year-old **female** cocaine **abuser** and her **fetus** died. The **mother** had severe bradycardia when admitted to the emergency room. A primigravid 19-year-old **female** at 32nd week of gestation had a rupture of an unscarred uterus. Hyponatremia in a **male neonate** of a 27-year-old **female mother**. Even full-term **infants** of ***crack***-using **mothers** have increased blood pressure and cerebral blood flow velocity, increasing **their** risk of intracranial hemorrhage. This drug abuse epidemic has resulted in the unnecessary exposure of nursing **infants** with seizure, hypertonia, and withdrawal problems. Passes into breast milk of addicted **mothers**.
Hypersensitivity: Relatively low level of sensitivity from topical use on the eye.
Liver: Minimal increase of liver enzymes in non-parenteral cocaine-**abusers**.
Miscellaneous: Intestinal infarction of a 38-year-old **male** cocaine-**abuser** after 4 g, IV. *Pseudomonas cepacia* has contaminated anesthetic atomizers and caused an outbreak of 18 nosocomial respiratory tract infections that lasted about a year. Convulsions followed spraying of the throat in a 42-year-old **female** before bronchoscopy. Acute intoxication in a 14-month-old **male** after topical use of 30 mg during bronchoscopy caused hyperactivity, increased blood pressure, and tachycardia, lasting about 18 h. Toxicity often follows ingestion to avoid detection by police. Muscle and skin infarction in 20-year-old

female after smoking ***crack***. Acute porphyria variegata in 36-year-old **male** after snorting 1.5 g cocaine 12 h previously. Hemoptysis in 20-year-old **female** after smoking it. Cases of dissolution of dental enamel and dentin from oral use or nasal insufflation. Severe or life-threatening exacerbation of asthma provoked by cocaine inhalation. ***Crack*** is often used with ***heroin*** to stave off suicidal depression after a burst of euphoria. Cocaine abuse unmasked and exacerbated myasthenia gravis in a 24-year-old **female**. Priapism reported in cocaine **abusers** that even required surgery, but, more commonly, decreased libido and erectile function in chronic **abusers**. In **some**, early use can increase libido. A 34-year-old **male** was hospitalized with a 3 day history of priapism after self-administering intraurethral cocaine to enhance **his** sexual performance. Muscle and skin infarction occurred after inhalation cocaine from a base pipe. Spontaneous pneumomediastinum in a 21-year-old **male** after oral cocaine inhalation. A **pilot** of a Continental Express commuter plane was found, by blood and urine tests, to have used cocaine prior to a fatal crash. Diaphoresis occurs.
Poisoning: Overdosing has been common, especially in view of the variations and degree of adulteration. Route of administration can play a key role. Severe poisonings usually do well at the emergency room, but sudden deaths are common outside the hospitals. A 36-year-old **male** developed fatal pulmonary edema after "free-base" cocaine IV. The emergency room finds severe hypertension, severe tachycardia, supraventricular tachyarrhythmias, hallucinations, tremors, and seizures are the poison symptoms that demand prompt attention. Potentially fatal cases have blood levels of 0.6–11.8 mg/l. It has been used in suicides. **People** with a cocaine idiosyncrasy have died of cardiac failure after doses of only 20 mg. Smuggling by swallowing up to 75–102 cocaine-filled condoms has often caused death, due to rupturing of the condom (usually containing 5 g each). Some reports list therapeutic serum levels as 0.1–0.3 mg/l; toxic at 0.5 mg/l; and lethal at 2–4 mg/l.
Temperature: Hyperpyrexia reported in 26-year-old **male** after nasal insufflation. Known as "cocaine fever". Causes a decrease in temperature in **guinea pigs**.

Drug interactions: ***Alcohol, ethyl***: Taken with cocaine, it increases the formation of ***norcocaine***, shown in **rats** to be more potent and to prolong the effects on heart rate and QRS interval.
Amphetamines: Potentiate the central nervous system stimulation.
Buprenorphine: It increases response to intranasal cocaine. Prolonged use can give a reverse effect.
Catecholamines: Deaths, hypertensive crises when combined for local anesthesia. Use of cocaine and ***epinephrine*** paste for intranasal anesthesia has caused life-threatening cardiac arrhythmias. ***Amphetamine, chlorpromazine, guanethidine, imipramine, methyldopa, phentolamine, propranolol,*** and ***reserpine*** may cause similar interactions.
Furazolidone: When given for 4–5 days, it acts as an ***monoamine oxidase inhibitor*** and potentiates cocaine.
Monoamine oxidase inhibitor: Potentiates cocaine.

TD_{LO}, **human**: 714 mg/kg.

Addictive in **monkeys** whose reactions are similar to **man's** (excitement, visual and tactile hallucinations). Absorbed from mucous membranes and wounds with secondary systemic effects. Used by freedom-loving **Swedes** in World War II to foil Nazi war **dogs** searching for refugees. Some **rabbits** lack cocaine esterase. In **horses**, 4–6 g causes restlessness, excitement, salivation, mydriasis, and acute mania. In **dogs**, 15–20 mg/kg leads to anxiety, fear, then weakness, convulsions, stupor, dyspnea, respiratory and then cardiac failure.

LD_{LO}, **rabbit**: 126 mg/kg; IV LD_{50}, **rat**: 18 mg/kg; **mouse**: 30 mg/kg; **cat**: 15 mg/kg.

COCARBOXYLASE

= *Berolase* = *Bivitasi* = *Cocalose* = *Cocarvit* = *Nutrase* = *Pyrolase* = *TDP* = *Thiamine Diphosphate* = *TPP*

Untoward effects: Pain at injection sites. Like ***thiamine***, may be a topical skin sensitizer with eruption.

In **sheep** parenteral dose of 50 mg/kg was toxic leading to irregular pulse, deep intermittent breathing, diarrhea, weakness, and inability to stand. Even 20–25 mg/kg caused adverse symptoms.

COCCULUS sp.

Untoward effects: ***C. carolinas = Red Moonseed = Coral-beads***. Widely distributed in the southeastern U.S., where its roots are used to make a tea with "blood-clarifying" properties. Unfortunately, it contains an alkaloid particularly toxic to **children** who are attracted by the colorful berries, which can cause gastroenteritis, paralysis, and even death.
C. filipendula has caused dermatitis, itching, and rashes.
C. indicus = Fish Berry = Indian Berry used in India as a pediculicide on **people** and to stupify **fish**. Also known as ***"Knockout Drops"***, because it was once used by robbers to disable their victims. It is a dangerous and poisonous drug. Its chief toxic principle is ***picrotoxin***, found in its seeds at a 1.5–5% concentration. See ***Picrotoxin*** for **animal** toxicity.

COCHINEAL

Use: It is the dried dead **female** insects of *Coccus cacti*, a plant louse native to Mexico and Peru, feeding on cactus branches. Its ***carminic acid*** content gives foods and inks a red or pink color. It is also used in chemical analysis, where its solutions are yellow in the presence of acids and violet to bright red in alkaline solutions.

Untoward effects*:** Pigmented plaques on shoulders and back a manifestation of an allergic reaction. Has occasionally been adulterated by mixing with ***talc, ***lead carbonate***, and ***barium sulfate***.

TD_{LO}, **rat**: 524 g/kg.

COCKLES

Untoward effects*: *Corn Cockle* = *Agrostemma githago* = *Corn Rose* = *Corn Campion* = *Crown of the Field* = *Purple Cockle, in grain or screenings has been lethal, particularly to **poultry**. Cracked or ground seed is more lethal than unbroken seed containing the same amount of poisons, a glucoside, ***githagen***, and a saponin, ***agrostemmic acid***, both of which are absorbed from the gastrointestinal tract. **Sheep**, **cattle**, and especially **hogs** can graze enough to poison them. It is also a ***nitrate***-accumulator. **Human** poisoning is now rare, due to improved methods of cleaning wheat seed, but occurred after eating ***bread*** baked from contaminated ***wheat*** flour. Vomiting, colic, frothy foul-smelling diarrhea, and inability to stand is noted in **pigs**, followed by spasms and death. **Cattle** exhibit nervousness, slobbering, and teeth grinding, followed by excitement, colic, and coughing, which lasts 5–8 h. This is followed by recumbancy, fetid diarrhea, rapid and noisy breathing, tachycardia, coma, and death. Slobbering, yawning, colic, tachycardia, dyspnea, coma, and death occur in affected **horses**.
Cow Cockle* = *Saponaria vaccaria found as a weed in spring ***wheat***. All parts of the plant, especially the seeds, are poisonous, due to its ***saponin*** content. Dizziness, vomiting, and diarrhea are common ill-effects.

COCKLEBURS

= Ban-okra = Bhangra = Burweed = Clotbur = Noogoora Bur = Sheep Bur = Xanthum sp.

Untoward effects*:** Allergic contact dermatitis, especially from its pollen, in **humans**, and has poisoned **children** in Argentina. ***Hydroquinone carboxyatractyloside and ***xanthostrumarin*** are its poisonous principles, found particularly in young plants and seeds. May also contain ***cyanogenic glycosides***.

Tender seedlings, especially in the two-leaf stage, are often consumed with other forage and are very poisonous to **livestock** and can cause death in 12–24 h. In **swine**, symptoms are anorexia, nausea, vomiting, depression, methemoglobinemia, decreased temperature, weakness, and prostration. Occasionally, nervous symptoms occur. At necropsy, gastrointestinal irritation, hemorrhagic nephritis, and hepatitis are noted. **Chickens** are also commonly affected. After spring and fall rains, **cattle** often eat the sprouting seedlings and rapidly develop nervousness, excitement, trembling, stilted gait, apparent abdominal pain, and death. Subserous hemorrhages, straw-colored fluid in the abdomen, and hemorrhages on the gall bladder and intestines noted. **Sheep**, **horses**, **swine**, and **goslings** have also been poisoned by the seeds. The spiny burs may cause intestinal obstructions.

Cockleburs are caught in the fur of **animals**, on the clothing and even bodies of **people**. They are not only annoying, but also irritating. The burrs have caused a granular stomatitis in **dogs**, progressing to papules on the tongue, which eventually became ulcerated and necrotic.

COCOA

= Cacao

A breakfast drink from the tree, *Theobroma cacao*, contains 2% *theobromine* (q.v.) and 1% *caffeine* (q.v.). After drying and fermenting, the beans are made into *chocolate* (q.v.). There are about 30 different kinds of cocoa beans, each with a distinct flavor.

Untoward effects*:** Contains ***oxalates and ***oxalic acid***. Ingestion in excessive amounts reported to cause diuresis, and inflammation of the neck of the bladder, prostate, and urethra. Its relationship with fibrocystic breast disease in **women** is still debatable. May color feces a light gray. Contains ***catecholamines***. Reported to be a teratogen. **People** with ***cobalt*** or ***chromate***-induced dermatitis should avoid ***chocolate***.

Deaths began 5 days after **pullets** ate a ration containing 10–30% of undecorticated cocoa cake meal. Ordinary household cocoa powder can be fatal to **poultry**. The toxicity is presumed to be due to ***theobromine*** (q.v.) which causes nervousness in **poultry**. **Calves** fed waste ***chocolate*** can show excitement, sweating, tachycardia, increased respiratory rate, and, eventually, convulsions and death.

COCOA BEAN SHELLS or HUSKS

Use*:** To give a ***chocolate taste to inexpensive and dietetic ***chocolates***, and, occasionally, fed to **horses**.

Untoward effects*:** Fed to **lambs** at about a 15% dietary level it reduced food intake and weight gain. Fed to race **horses**, it has caused a number of inadvertent and embarassing drug test positives for ***caffeine and ***theobromine***.

COCOBOLO

Untoward effects: An exotic Brazilian wood that has caused dermatitis of the face, scalp, neck, arms, and extensor surface of the knees of **those** working with it. A **female** had edematous swelling of **her** lips from contact with **her** flute's cocobolo wood mouthpiece. Also reported was a **patient** with difficulty in swallowing, diarrhea, and acute eczema from exposure to the wood.

COCONUT

= *Copra* = *Naryal*

Untoward effects: Diarrhea due to its oil content may occur. An allergen, especially a contact allergen, causing a dermatitis, in some **people**. Its oil contains over 80% saturated fats.

CODEINE

= *Methyl Morphine* = *Morphine-3-methyl-ether*

An opioid.

Untoward effects: ***Addiction***: Not uncommon in **people** using cough syrups with codeine, **who** develop a mild euphoria from its use. Withdrawal has precipitated irritability, insomnia, lacrimation, diarrhea, and muscle aching in **some**. Tolerance to it develops. The late Howard Hughes, supposedly, received 35 grains (4.27 g) parenterally/day. ***Loads***, a combination of the sedative ***glutethimide*** and codeine is particularly addictive. Ineffective in 10% of **Caucasians** and 2% of **Chinese** because they lack the liver microsomal enzyme CYP2D6 that converts it to ***morphine*** (q.v.), its active form.

Behavior, driving skills, performance: Significantly impairs performance and causes depression in **some** after initial exhilaration.

Blood: Thrombocytopenia.

Clinical tests: May give a false positive or increased amylase and lipase, serum transaminase, and SGOT readings.

Eye: Miosis, near-sightedness, and blurred vision reported.

Fetus and neonate: Increased incidence of cleft lip and cleft palate. After 8 h, a little is excreted in breast milk, but exercise caution in the case of **addicts** and in non-addicted **mothers** who ingest some prior to delivery. Their **infants** may show withdrawal symptoms.

Gastrointestinal: If taken within 24 h of a blood test, it may cause false positive or increased amylase or lipase values, usually indicative of a pancreatitis. Constipation, occasionally epigastric pain, and ageusia have occurred.

Hypersensitivity: Urticaria and tachycardia, due to ***histamine*** release and anaphylactic action.

Miscellaneous: The same as with most ***narcotics***, ie., nausea, vomiting, sedation, constipation, confusion, slurred speech, tremors, excitement, hyperactivity, facial and neck flushing, dizziness, respiratory decrease, anorexia, and headache. Large doses can also cause semiconsciousness, delirium, convulsions, tachycardia, circulatory failure, dyspnea, respiratory collapse, and coma.

Poisoning: Many overdose cases and fatalities reported. Toxic and lethal doses generally lead to blood levels of 0.2–0.5 mg/l. 1.5 g is an estimated fatal dose for **man**. Of 234 **children** who had taken > 5 mg/kg, respiratory arrest requiring intubation and artificial ventilation occurred in **eight** and **two** of them died. Some of the following symptoms occurred in the **others**: somnolence, ataxia, miosis, vomiting, and rash, swellings, and itching of the skin. Adverse symptoms in some **children** with doses > 1 mg/kg. LD_{LO}, **human**: 5 mg/kg. Fatal cases reported from overdosing with various marketed drug combinations. Collapsed **patients** with pinpoint pupils, regardless of **their** age, may need an immediate therapeutic trial with ***naloxone***. Pulmonary edema and convulsions also occur. Doses of 200 mg may be dangerous, and 800–1000 mg has been fatal. Therapeutic serum levels reported as 0.03–0.25 mg/l; toxic at 1 mg/l; and fatal at 1.8 mg/l.

Skin: Fixed-drug eruptions, erythema multiforme, itching of the nostrils, non-allergic urticaria, and exfoliative and contact dermatis reported.

Withdrawal: In the **neonate**, these symptoms may includes irritability, tremors, high-pitched crying, hyperactivity, diarrhea, respiratory alkalosis, lacrimation, disorganized sucking reflex, yawning, sneezing, hiccups, and, in severe cases, seizures.

Drug interactions: ***Alcohol***: It increases drowsiness. Potentially dangerous, especially due to respiratory depression.

Analgesics: Their action can potentiate codeine's analgesic effect.

Antibiotics: ***Chloramphenicol*** and ***tetracyclines*** inhibit liver microsomal enzymes, resulting in codeine potentiation. Cross-sensitivity with ***tetracyclines*** increases risk of photosensitivity.

Chlordiazepoxide: Coma.

Heparin: Within 5 min, respiratory arrest, cyanosis, dizziness, loss of consciousness, and nausea reported.

Kaolin: Adsorbs codeine in the gastrointestinal tract.

Monoamine oxidase inhibitor: Can cause hypotension, respiratory arrest, coma, and shock.

Nalorphine, a narcotic antagonist, may still synergize or potentiate codeine's antitussive effects. Heavy dosage for narcosis may lead to restlessness or excitement during recovery. Warnings against the use of codeine in **cats** are frequently repeated by those who have never used it for them.

LD_{50}, **rat**: 600 mg/kg; SC LD, **dog**: 100 mg/kg; **cat**: 60–90 mg/kg; SC LD_{50}, **mice**: 231 mg/kg; **rat**: 500 mg/kg.

COD LIVER OIL

= *Gaduol* = *Oleum Jecoris* = *Oleum Morrhuae* = *Tunol*

Untoward effects: Severe angioedemas of the larynx in a 5-year-old **female** after eating two bites of fish. **She** had unknowingly become sensitized, after taking cod liver oil when younger. Has caused acneiform eruptions on the face, chest, and back, as well as a contact dermatitis. Prolongs bleeding time and platelet aggregation, platelet count, and platelet ***thromboxane*** B_2 decrease. ***Mercury*** and ***selenium*** occurs in it. When oxidized, can decrease reticulo-endothelial system. A 63-year-old **female** who smoked regularly, developed lipoid pneumonia after regular ingestion of the capsules.

Excessive intake has caused muscular dystrophy in **calves** and encephalomalacia in **poultry** due to its high level of unsaturated fats, precipitating a ***vitamin E*** deficiency.

COELENTERATES

Including fire coral (q.v.), jellyfish (q.v.), Portugese Man-of-War (q.v.), sea anemones (q.v.), and sting-rays (q.v.).

Untoward effects: All contain stinging capsules which, upon contact, release a stinging barb, injecting a small amount of a potent toxin. This causes a bradycardia and decreases nerve transmission. Rapid, severe, and often life-threatening reactions are probably anaphylactic. Severe pain, erythema, and edema occur. Itching is usually mild. The coelenterates can release antigens that can sensitize some **swimmers** and **divers** without **their** being stung. Sensitized **divers** may develop edema of the throat and larynx and extreme weakness. ***Alcohol*** neutralizes the nematocysts (high-proof liquors and ***baking soda*** in *sea* water – NOT fresh water, have also been used).

COFFEE

= *Coffea*

Contains ~400 distinct chemical entities, including *caffeine* (q.v.). Discovered in Ethiopia in the 9th century by a goatherder who saw his goats acting frisky. He tried some of the berries they had been eating off an evergreen bush, and he, too, felt a sense of exhilaration. From there, coffee bushes were cultivated in southern Arabia, then, its use spread to Turkey, Venice, and the U.S.A. When the great explorer, Sir Henry Morton Stanley, found Dr David Livingston, he is quoted as saying, "Dr Livingston, I presume". Few are aware of the doctor's equally casual reply, "Just in time for coffee, Stanley".

Untoward effects: Drinking coffee with 200 mg ***caffeine*** increases blood pressure by an average of 10/7 mmHg for ½–2 h. Drinking five or more cups/day is said to increase a **person's** chances of developing lung cancer, bladder and pancreatic cancer; and result in a negative ***calcium*** balance and increased risk of hip fractures. Another study links drinking > 2 cups/day by sedentary and mildly active **men** to increased serum ***cholesterol*** and heart disease. A British study indicated **patients** drinking > 6 cups/day had a 2½-fold risk of myocardial infarction. Excessive consumption can cause hypomania. Burning sensations in the pit of the stomach (heartburn), indigestion, nausea, abdominal cramps, anxiety, dizziness, apprehension, frequent diarrhea, insomnia, and nervousness are often related to coffee intake. High intake may be teratogenic. It may trigger migraine headaches, esophageal reflux, release of prostaglandins in the stomach, and can cause an allergic contact cheilitis and general allergic reactions. Drinking > 6 cups/day appears to increase chances of miscarriages in **women** and may increase benign fibrocystic mammary disease. A safe level has not been established for pregnant **women**. Has caused a brownish discoloration of the nail plate. A common cause of urinary frequency, bladder cancer, recurring prostatitis, and increased urinary ***oxalates***. The popularity of coffee enemas as therapy for cancer has, fortunately, waned, after causing many deaths due to extreme hyponatremia and hypokalemia, followed by bronchopneumonia and cerebral hypoxia. The FDA has seized coffee containing over 2 ppm ***antimony oxide***. It inhibits the absorption of ***non-heme iron***. May produce spuriously elevated vanillylmandelic acid assays. Has contained ***ochratoxin A***, insects, rodent droppings, ***cesium-137***, ***strontium-90***, ***potassium-40***, and pesticides. Contains a considerable amount of burned material, including the mutagenic pyrolysis product, ***methylglyoxal***. May cause malabsorption of neuroleptic drugs. Coffee should be avoided in **patients** with ***cobalt*** dermatitis. "Coffee-worker's lung", a hypersensitivity pneumonitis, results from inhalation of the dust from hulls and chaff of green coffee beans. Rhinitis, asthma, and dermatitis can follow the inhalation of its hapten, ***chlorogenic acid***. After one cup of coffee, a 56-year-old **female** had ataxia, paresthesias of legs, nausea, loss of limb motor control, gaze nystagmus, and difficulty speaking clearly. ***Alcohol*** and ***aspirin*** can also trigger such attacks. Netherland **workers** found a 10% increase in ***homocysteine*** serum levels in **those** who drank six cups of unfiltered coffee/day. Drinking it or colas with ***cimetidine***, ***ciprofloxacin***, ***famotidine***, or ***ranitidine*** can increase ***caffeine*** serum concentrations leading to stomach irritation and jitters.

Coffee grounds have killed **hogs** eating them.

Decaffeinated Coffee (Decaf)

Untoward effects: In the U.S., the ***caffeine*** was removed by ***trichloroethylene*** (q.v.), a potent liver carcinogen. The FDA allowed its residues, up to 10 ppm, in instant decaffeinated coffee, and 25 ppm in ground decaffeinated coffee. U.S. manufacturers voluntarily quit using it in 1975. This coffee is a stimulant to both gastric juice and the hormone, ***gastrin***. Later, ***methylene chloride*** (q.v.), another liver carcinogen, came into use as the decaffeinating agent. In some studies, this coffee causes more acid to be formed in the stomach than by ordinary coffee. Strangely enough, these decaffeinating agents put the FDA, your protectors, in violation of the Delaney Clause that they are supposed to enforce. In recent years, ***ethyl acetate*** (q.v.) and a steam method are used by some manufacturers. ***Decaf* drinkers** have shown ~6% increase in low-density lipoprotein. This increases the risk of coronary artery disease by about 10%, compared to no changes in **those** drinking the same amount of regular coffee. **Maternal** consumption of both regular and decaf coffee increases **fetal** breathing activity and decreases **fetal** heart rates. Limiting the consumption of ***decaf*** to 2 cups/day decreases premenstrual tension in **women**.

COGNAC

Ten casks of white *wine* make one cask of brandy (cognac).

Untoward effects: Even though it is distilled, it contains no ***histamine***, but its ***alcohol*** content can produce the ill-effects of ***alcohol*** (q.v.). A cognac oil is distilled from ***grape*** leaves used in ***wine***-making or by steam distillation of the ***yeast*** and other sediment in wine lees. Used as a fragrance in foods, soaps, detergents, cosmetic creams and lotions, and in perfumes. In **human** trials, no sensitization or phototoxic effects noted.

Oil, LD_{50}, **mice** and acute dermal LD_{50} in **guinea pigs**: > 5 g/kg.

COHOBA

= *Yopo*

Earliest reports of its use were from Columbus' second voyage to the Caribbean where the Tainos of Santo Domingo extracted it from *Anadenthera peregrina*.

Untoward effects: A hallucinogenic snuff of northern South America and the West Indies, containing ***bufotenine*** (q.v.), ***N,N-dimethyltryptamine*** (q.v.), and some lesser indoles.

TD_{LO}, **human**: 57 µg/kg.

COKE

Untoward effects: Coke oven **workers** have an increased risk of malignant neoplasms of the kidney, urinary tract, and lungs. Released in the emissions are an extensive group of potential carcinogens and co-carcinogens, including ~3% ***polycyclic aromatic hydrocarbons*** (q.v.). **Fetuses** can be at risk from ***polycyclic aromatic hydrocarbons*** exposure.

OSHA standard is for a maximum exposure of 150 µg/m^3, 8 h TWA.

COLAS

= *Kolas*

Untoward effects: Contain acids that can damage tooth enamel, but saliva tends to neutralize this effect. Their flavors tend to stimulate salivation. Cola drinks are considered fairly high in ***oxalates*** (q.v.), and should be avoided by **those** prone to urinary ***oxalate*** stones. ***Pepsi-Cola***® reportedly contains 16 times ***Coca-Cola's***® ***sodium*** content (recently only 1.4 : 7) and only 1/17th the ***potassium***. Their methylxanthines can stimulate the central nervous system and produce abdominal distress. For many years, ***Coca-Cola***® was used as a pre- and postcoital contraceptive. The precoital effect may have been due to the low pH. Postcoital effectiveness was probably poor because sperm move rapidly through the cervix. See ***Coca*** for early history of ***Coca-Cola***®. Most diet colas have contained artificial sweeteners, such as ***saccharin*** (q.v.) and ***aspartame*** (q.v.). Drinking it or ***coffee*** with ***cimetidine, ciprofloxacin, famotidine,*** or ***ranitidine*** can increase ***caffeine*** serum concentrations leading to stomach irritation and jitters.

Cola nuts contain about 2.13% ***caffeine***, and have been used for "doping" **horses** in so-called "speed balls".

COLCHICINE

= *Colcemid* = *Colchicin* = *Colchicinum*

Use: Gout suppressant, treatment of familial Mediterranean Fever in **humans**, and as an antineoplastic, antimitotic, and analgesic in **animals**. From *Colchicum autumnale*. Also see *Gloriosa*. Used by plant-breeders to produce mutations.

Untoward effects: Blood: Several cases of aplastic anemia reported. Bone marrow depression and agranulocytosis with prolonged use. Occasionally cases of leukopenia, neutropenia, pancytopenia, and thrombocytopenia.
Deaths: Fatal uremia in a 35-year-old **female** who received 12.5 mg over 12 days for gouty arthritis. Of six overdosed (12.5–45 mg) **patients** (15–57 years), four died of multiorgan failure. Five of these were young and healthy and took someone else's medicine. A 57-year-old **male** unintentionally overdosed **himself** by taking ten 1 mg tablets in 1 h, instead of over a 4 day period.

Autopsy revealed bilateral adrenal hemorrhage with cortex degeneration, and a large fatty liver, which may have decreased **his** ability to metabolize the drug. An 18-year-old **female** had a drug-induced respiratory distress syndrome and died after ingesting 150 mg of the drug.
***Fetus*: Human** teratogen.
Malabsorption: Decreases ***cyanocobalamin*** (***vitamin*** B_{12}) absorption, often accompanied by steatorrhea and diarrhea. Usually reversible after ileal receptor sites in the intestinal mucosa heal.
Miscellaneous: A rare acute anaphylactic reaction in a 61-year-old **male** after ingesting one tablet with ***probenecid***. Rash, pruritus, diarrhea from injury to intestinal mucosa, malaise, nausea, vomiting, abdominal pain, hypotension, fever, weakness, confusion, muscle cramps, rhabdomyolysis or myoneuropathy, tonic–clonic seizures, alopecia, pain and ecchymosis, paralytic ileus, loss of deep tendon reflexes, megaloblastic anemia, leukopenia, thrombocytopenia, granulocytopenia, hepatitis, hemolysis in a glucose-6-phosphate dehydrogenase deficient **patient**, gastrointestinal bleeding, purpura, hematuria, decreased glomerular filtration rate, renal failure, altered taste and smell, and complete azoospermia in a previously fertile 36-year-old **male** taking the drug intermittently for ~3 years. Suspect as a cause of glaucoma and cataracts or lens opacities. May decrease the effectiveness of ***antihypertensive*** and ***anticoagulant drugs***, and delay the intestinal absorption of ***vitamin*** B_{12}. Excreted in the milk of **women**, **cows**, and **goats**.
Suicide: A 38-year-old **male** who tried to commit suicide by ingesting 80 mg exhibited symptoms of a toxic gastroenteritis, followed by stomatitis, pharyngitis, epithelial damage on the genitalia, and some bone marrow and hematological damage. A 16-year-old **female** ingested 35 mg, trying to commit suicide. After major hematopoietic, gastrointestinal, respiratory, and central nervous system ill-effects, **she** died from the latter on day 13. An 18-year-old **female** was successful in **her** suicide attempt within 42 h after ingesting 150 mg in **her** tea. Mild epigastric pain and nausea developed 4 h later, followed by dyspnea, tachycardia, diffuse pulmonary râles, and death. These are typical of the numerous cases in the literature. As little as 6 mg has been fatal. Death is usually from cardiovascular collapse.

Therapeutic serum levels reported as 0.0003–0.0025 mg/l; toxic from 0.005 mg/l; and fatal at 0.024 mg/l.

LD, **human**: 1–5 mg/kg; IM LD, **human**: 30 mg.

In **animals**, excessive oral dosage may produce nausea, vomiting, and diarrhea. Oral doses of 4 mg/day for 14 days to **dogs** has produced a severe malabsorption syndrome. Parenteral doses may produce similar untoward effects plus weakness, depression, and anorexia. Potential risk with its use during pregnancy. Has demonstrated teratogenic effects in **laboratory animals**. Fetal morbidity, skeletal deformities, encephalopathy, and eye defects reported in **hamsters**. In **cattle**, and to a lesser extent in **sheep** and **goats**, resistance to its effects has occurred.

Average LD in most **animals** is ~1 mg/kg. LD_{LO}, **rabbit**: 50 mg/kg; LD_{50}, **hamster**: 600 mg/kg; male **quail**: 42.1 mg/kg.

COLCHICUM AUTUMNALE

= *Autumn Crocus* = *Colchique d'automme* = *Fall Crocus* = *Meadow Saffron*

Fall-blooming, as is *Crocus sativa*, a safe garden type.

Untoward effects: Its toxicity has been known to **man** since the beginning of the Christian era. Contains ***colchicine***, a toxic alkaloid. All parts of the plant are toxic. Ingestion of the bulbs and seeds from the flower garden, where it resembles the common crocus, causes nausea, diarrhea, vomiting, excitement, burning of the mouth, cerebral depression, dyspnea, renal failure, thirst, temporary alopecia, hemoglobinuria, circulatory collapse, coma, and death. The corms (bulbs) are often mistaken by **people** for ***onions*** or ***garlic***. Symptoms of poisoning occur after 3–6 h. Recovery is slow and relapses can occur. Its extract has been used for denaturing ***rubbing alcohol***. *C. luteum*, found in India, is a common toxic plant.

Animals are killed by eating its young, succulent leaves in the spring, or in the fall from eating its flowers growing wild in the pasture. **Cattle** show abdominal pain, violent purgation with fetid green or black feces, teeth grinding, incoordination, and collapse; they may take several days to die of respiratory failure. Its ***colchicine*** is excreted primarily in the milk and urine. **Rats** and **mice** are readily killed by consuming the seeds, but **hamsters** are relatively resistant to its ill-effects. Lethal to **cattle**, **goats**, **lambs**, **swine**, and **horses**. Pregnant **cows** may give birth to **calves** with skeletal defects.

Leaves: LD_{50}, **guinea pigs**: 12 g/kg; **young lambs**: 6.4 g/kg.

COLESEVELAM

= *Welchol*

Cholesterol sequestrant.

Untoward effects: Flatulence and constipation. Either ***atorvastatin***, ***lovastatin***, or ***simvastatin*** with it decreased low-density lipoprotein ***cholesterol*** more than with these statins alone. Decreases bioavailability of ***verapamil***.

COLESTIPOL

= *Cholestabyl* = *Colestid* = *Lestid* = *U-26597A*

***Antihyperlipoproteinemic.* A bile acid-binding anion-exchange resin.**

Untoward effects: Similar to ***cholestyramine*** (q.v.). Hot flashes, constipation, intestinal discomfort, belching, bloating, nausea, unpleasant taste, steatorrhea, and joint pain reported.

Drug interactions: Will bind ***bile acids, digoxin*** and ***hydrocortisone***. May interfere with fat-soluble ***vitamins***.

LD_{50}, **rat**: > 1 g/kg.

COLEUS

Untoward effects: ***C. pumila*** and ***C. blumei*** introduced from southeastern Asia have been used for their hallucinogenic properties.

C. forschkohlii, from India, is moderately toxic (IP LD_{50}, **rat**: 92 mg/kg) and a source of ***forskolein*** (***colforsin***), a possible inotropic and hypotensive agent. It is the only *Coleus* decribed in ancient Hindu medical texts and Ayurvedic materia medica.

Many asymptomatic cases of ingestion by young **children** reported to Poison Control Centers. Common as an hallucinogen in Mexico.

COLISTIN

= *Colimycin* = *Colisticina* = *Coly-Mycin* = *Kolimitsin* =*Polymyxin E* = *Totazina*

Antibiotic. The sodium methanesulfonate form = *Alficetin* = *Colistimethate Sodium* = *Methacolimycin*. The sulfate form = *Malimyxin* = *Multimycine*

Untoward effects: The sodium salt appears to be less toxic than the sulfate. Toxicity is often unrelated to dosage used.
Blood: Hemolytic anemia in a sickle cell anemia **patient**. May decrease bone marrow effects.
Clinical tests: False positive or increased readings in urine ***albumin*** and blood urea nitrogen tests.
Fetus: It crosses the placenta.
Hypersensitivity: An allergic reaction from the sulfate form in ophthalmic drops.
Liver: Fatal nephro- and hepatotoxicity in 64-year-old **male** on 150 mg tid/less than 2 weeks.
Miscellaneous: Drug fever, dermatitis, visual disturbances, headache, and ototoxicity.
Neonate: It increases blood urea nitrogen; nephrotoxicity. Breast milk contains about 17% the **maternal** serum level.
Neuromuscular: Postoperative blockade with persistent apnea reported. An unusual case of myasthenic syndrome occurred in a 48-year-old **female**. Perioral numbness and tingling with tinnitus after 150 mg, IM in a **patient**; numbness of limbs and ataxia in **another**; respiratory arrest, cardiac arrest, and death in a 53-year-old **male**, 1 h after 150 mg IV for treatment of *Klebsiella* pneumonia. Weakness, dizziness, vertigo, circumoral paresthesia, and slurred speech regularly reported.
Poisoning: Peritoneal dialysis appeared to be helpful.
Renal: Nephrotoxicity (20%), azotemia, acute renal failure, anuria, and renal tubular necrosis in **some**.

Drug interactions: Like ***kanamycin***, ***neomycin***, and ***streptomycin*** prolonged neuromuscular blockade and apnea can occur when used with muscle-relaxing ***anesthetics*** or peripheral-acting ***skeletal muscle relaxants***. This problem occurs when used with a polarizing ***muscle relaxant***, such as ***curare***, and depolarizing types, such as ***decamethonium, edrophonium***, and ***succinylcholine***. ***Neostigmine***-resistant ***curare*** reversal occurred in **patients** also receiving colistin. Interaction in the body to enhance the effects of ***neostigmine*** and *d-****tubocurarine***.

Nephrotoxicity is enhanced by ***furosemide*** in **rat** trials. May depress cardiac function in **cats** and **dogs**. Curariform toxicity is possible on high dosage.

LD_{LO}, **mouse**: 27 mg/kg; LD_{50}, **mice**: 720 mg/kg; **dog**: 42 mg/kg.

In France, Russia, and some Russian satellite countries, the word ***Colimycin*** frequently refers to the antibiotic, ***Neomycin***, and *not* ***Colistin***. Elsewhere, ***Colimycin*** *is* ***Colistin***.

COLLAGEN

= *Aviderm* = *Avitene* = *Beristypt* = *Collasol* = *Helistat* = *Instat* = *Ossein* = *Zyderm*

Comprises 70–79% of the dry weight of human skin.

Untoward effects: Has elicited an immediate Type I hypersensitivity reaction. Severe allergic reactions can occur in **patients** with connective tissue diseases, such as scleroderma and rheumatoid arthritis. It induces the release of ***antiheparin factor***. Vaginal contraceptive collagen sponges can become suitable substrates for the growth of bacteria. Collagen implants have caused foreign body granulomas and an allergic reaction (0.2–0.3%) in **some**. One **patient** developed an eczematous reaction to its use as an implant. ~3% of **patients** have a positive skin test response to implantable bovine collagen.

In **dogs**, topical use of excessive amounts has caused adverse foreign body reactions.

COLOCASIA sp.

Untoward effects: The stems and leaves of ***C. antiquorum*** = ***Dasheen*** = ***Dashin*** = ***Elephant Ear***, a broad-leafed ornamental plant, contain sharp ***calcium oxalate*** crystals. These can penetrate lips, tongue, palate, and oral tissues, causing severe pain, burning sensations, and swelling. If air passageways are blocked by swelling, it can be fatal. Gastroenteritis has also occurred. Contains ***sapotoxin***. In 1975, 116 **children** under 5 years of age presented at Poison Control Center after exposure. 51 had important symptoms and two required hospitalization.

C. esculenta and other species have poisoned **cattle** and **water buffalo**, causing edema of the face, lips, and tongue. Irregular erosions on their tongues and increased salivation reported.

COLOCYNTH

= *Bitter Apple* = *Bitter Cucumber* = *Bitter Gourd* = *Citrullus colycynthus* = *Colocynthis* = *Coloquinte* = *Ground Gourd* = *Handal*

***Cathartic.* A tar extracted from the seeds is used for treating mange of camels. In Arab lands, the phrase "bitter as *handal*" is a substitute for the western world's "bitter as gall".**

***Untoward effects*:** Bitter, drastic cathartic, causing mucosal irritation, abdominal pain, diarrhea with blood-stained watery stools, nausea, vomiting, delirium, and prostration. Intravaginal use is a common practice in India to induce abortions of dead **infants**.

LD_{LO}, **human**: 50 mg/kg.

COLOGNES

Untoward effects*:** Contact dermatitis, even after a **spouse** has used it. Some ***alcohol-based colognes have induced ***disulfiram***-like reactions. Often contain ***musk, bergamot,*** and ***citrus oils,*** and ***furocoumarins***, which have caused erythema multiforme, toxic epidermal necrolysis, and photosensitivity.

COLTSFOOT

= *Carogna* = *Coughwort* = *Drouya* = *Farfara* = *Mieas* = *Prepouli* = *Tussilago farfara*

Untoward effects*:** Contains hepatotoxic pyrrolizidine alkaloids, including ***senkirkine (2.8–4.1 ppm in the leaves and 2.4 ppm in the flowers). In herbal teas; contains ***petasitenine***, a carcinogenic agent.

COLUBRINA

= *Hogplum*

A shrub of the buckthorn family (*Rhamnaceae*) commonly found from Texas to Colorado.

Untoward effects*:** Contains a hepatotoxin causing poisoning in **sheep** similar to that from ***Lechuguilla (q.v.).

COLUMBINE

= *Aquilegia*

***Untoward effects*:** The whole plant is poisonous. In Italy, it is used as a poultice; the leaves for an antiseptic decoction, for healing skin sores, etc., so it is not surprising that it might be a poison, especially for **children**. Seeds have been reported to be fatal to **children**.

When ingested by pastured **animals** in the U.S., it has caused numbness, paralysis, and convulsions. Symptoms resemble those of ***aconite***. **Animals** generally do not eat the plant.

COMBRETUM sp.

= *muBupu* (Shona) = *mu Rhka* = *Red Wings*

Untoward effects*:** Poisonous plants in Rhodesia. In Brazil, the red flowers of ***C. cacoucia are, reportedly, toxic.

Two **human** adults became seriously ill with stomach pains and vomited after eating some ***C. oatesii*** seeds. Rhodesia warns **children** of the danger of eating the seeds. The seeds of ***C. oatesii*** and ***C. platypetalum*** are highly toxic to **pigs'** central nervous system, although nausea and vomiting occur in the early stages.

Pulmonary edema, liver degeneration, increased temperature (112 °F), and increased pulse (240/min) are noted in **swine**. Cerebral blood vessels of **pigs** are markedly injected and numerous small hemorrhages are found inside the heart on post-mortem.

COMFREY

= *Blackwort* = *Symphytum*

Approximately 25 species.

Untoward effects*:** Despite its popularity as a "cure-all" in herbal teas, it contains hepatotoxic and carcinogenic ***pyrrolizidine (q.v.) alkaloids, such as ***symphytine***. ***S. asperum*** = ***Consoude âpre*** = ***Prickly Comfrey*** contains at least five toxic ***pyrrolizidine*** alkaloids. A 47-year-old **female** consuming up to 10 cups of the tea daily plus tablets for over a year developed hepatic veno-occlusive disease. Biopsy revealed a dense fibrosis of portal tracts that contained proliferating bile ductules. A 13-year-old Crohn disease **patient** who received several months of leaf tea developed thrombosis of the hepatic veins. A ***nitrate***-accumulator. When fed green to **swine** caused ***nitrate*** (q.v.) poisoning.

Russian Comfrey = ***S. X uplandicum*** and causes chronic hepatotoxicity and hepatocellular adenomas in **rats**.

Several **people** have suffered and even died from ***digitalis*** (q.v.) poisoning after mistaking ***foxglove*** for comfrey.

Estrogen content is the probable cause of its antigonadotropic activity in **mice**.

CONCRETE

Untoward effects: Allergic dermatitis from ***potassium dichromate*** and other ***chromium*** salts in it that are often aggravated by alkali in the cement. See ***Cement***. A number of concrete treatments or sealants, such as ***coal tars*** (q.v.) and ***polychlorinated biphenyls*** (q.v.) have caused problems.

Erosion of the pars esophagus in **pigs** ingesting ***silica*** crystals while licking food off a concrete floor. **Frogs** have developed skin erosions from the ***lime*** of concrete blocks.

CONE SHELL FISH

= *Cone Fish* = *Cone Shell* = *Conus textile*

Untoward effects: One of the most dangerous marine **snails** that suddenly extrudes a small, myotoxic, venom-soaked "harpoon" from its proboscis. Severe reactions can lead to paresthesia, paresis, dysphagia, chest tightness, blurred vision, and, eventually, cardiovascular collapse. Deaths follows respiratory paralysis. Their ***serotonin***-like toxin, ***GVIA***, is virulent enough to kill a small **child**. Mild cases resemble **bee** or **wasp** sting effects. Found in warm waters, particularly, in the Indian and Pacific Oceans, and have killed many **people** near Queensland, Australia. *C. geographus,* in the area of the Phillipines, produces similar symptoms and causes ~50% mortality. Localized ischemia and numbness occur around the wound and spread rapidly over the body. Death can occur in about 30 min with either species of harpooning. 18 of ~400 species of the genus have been implicated in **human** envenomation, often, resulting from careless handling of live specimens or from rummaging through sand or debris about **corals**, into which they have burrowed.

CONGO RED

Untoward effects: Pyrexia, respiratory paralysis, collapse, and death in 38-year-old **male** 3 min after IV of 10cc solution. Small blood vessel thrombosis and irreversible brain damage on autopsy. A number of reactions reported in the literature, including of couple of deaths. Excessive dosage and rapid rate of administration may lead to thromboplastic action, intravascular clotting, and death. Skin-staining may occur from repeated injections.

Teratogenic IP in **rats**.

IV LD_{50}, **rats**: 190 mg/kg.

CONIINE

= *Cicutine* = *Conicine*

Untoward effects: The poisonous volatile alkaloid in ***Conium maculatum*** (q.v.), with a burning taste and causing a mouse-like odor to breath and urine. Symptoms occur rapidly (20–30 min). Drowsiness, colic, paresthesias, weakness, ataxia, headache, profuse salivation, nausea, initially bradycardia, and, eventually tachycardia, mydriasis, double vision, amblyopia, decreased hearing, and decreased temperature reported. Various literature reports give 2, 130, and 350 mg as estimated **human** fatal doses. Lesser amounts produce a respiratory paralysis.

LD, **dog**: 50 mg/kg; SC LD_{LO}, **rabbit**: 90 mg/kg.

CONIUM MACULATUM

= *Baldiran* = *California Fern* = *Cigüe maculée* = *Deadly Hemlock* = *European Hemlock* = *Fool's Parsley* = *Nebraska Fern* = *Poison Hemlock* = *Poison Parsley* = *Spotted Cowbane* = *Spotted Hemlock* = *St. Bennet's Herb*

Untoward effects: In general, salivation, trembling, lack of coordination, mydriasis, bloating, bradycardia followed by tachycardia, weak pulse, bluish mucous membranes in the mouth, cold extremities, coma, respiratory paralysis, and death in **humans** and all classes of **livestock**. **Humans**, **cattle**, **sheep**, **horses**, **goats**, **poultry**, **nutria**, **rabbits**, **swine**, and **zoo animals** have been affected. Its general central nervous system depressant effects are due to its 0.5–1.5% ***coniine*** (q.v.) alkaloid content. All parts of the plant are poisonous, particularly, the seeds and roots. During sunny seasons, the alkaloid content may be twice that found in wet seasons. A native of Europe and Asia, it was brought to the U.S. as a showy garden plant and is now widespread. Leaves are an especially toxic source in early spring. Other poisonous alkaloids in the plant are ***conic acid, conhydrine, ethylpiperidine, Lambda-coniceine, n-methyl coniine,*** and ***pseudoconhydrine***.

In the rural areas of Turkey, it is a significant cause of poisoning in 2–11-year-old **children**. Its toxicity was recognized in the B.C. era. It was used in ancient Greece to kill condemned **prisoners** and **Socrates** met **his** death in 400 B.C. by a "cup of hemlock". **Royalty** found it a respectable method of commiting suicide or disposing of one's adversaries. Since it is a member of the ***parsley*** family, it is not surprising that **people** and **children** have been poisoned by it when using it for salads and from eating the roots. Convulsions after consuming it are rare, but common with ***water hemlock*** consumption. Poisoning often occurs when **victims** confuse its roots with ***carrots*** or ***wild parsnips***, its seeds with ***anise***, and its leaves with ***parsley***. Whistles, peashooters, and telescopes made from its hollow stems have caused death in **children**. Its stems usually have small purple spots, hairless, and hollow, while those of the ***wild carrot*** are hairy. In literature, it is commonly referred to as ***Hemlock***, NOT the tree. After crushing leaves or after ingestion, **one** may notice a distinctive mouse or cat urine odor to the breath and urine. Shortly after ingestion by **humans**, **they** report irritation of the mouth, increased salivation, nausea and vomiting, headache, mydriasis, thirst, sweating, dizziness, and general discomfort. Severe cases progress to coma and, rarely, convulsions. Despite intense and rapid symptomology, mortality is often low with early ***ipecac*** and ***activated charcoal*** treatment. Has caused dermatitis in **man**. Causes a disagreeable taste and odor in ***milk***, milk products, and meat of **animals** that graze on it. See ***Coniine*** above.

Has caused congenital skeletal malformations (arthrogryposis [twisted or bowed limbs], scoliosis and kyphosis [twisted or bowed spine], torticollis, and cleft palates) in **calves** and **pigs**. Colic, salivation, bloating, mydriasis, tachycardia, weak pulse, decreased milk production, muscular weakness, and respiratory failure occur in **cattle**. Abortions have also occurred. Reports of death in **horses** and **pigs**. The alkaloids are destroyed by drying, and, thus, cause no problems from **livestock** eating it in hay. Acute toxicity in **cows** within 1.5–2 h after 3.3 mg/kg; within 30–40 min in **mares** after 15.5 mg/kg; and within 1.5–2 h in **ewes** after 44 mg/kg.

LD (of fresh leaves), **ducks**: 50–70 g; **sheep**: 800 g; **horses**: 2 kg.

CONVALLARIA MAJALIS

= *Lily of the Valley* = *May Lily* = *Muguet* = *Sneezeweed*

Untoward effects: Contains > 20 ***digitalis***-like glycosides that increase the contractile force and decrease the rate of the heart beat and can cause cardiac arrhythmias. Also increase diuresis. In excessive dosage, mouth irritation, headache, dizziness, mydriasis, diarrhea, abdominal pain, and vomiting occur. Gastrointestinal symptoms are due, primarily, to its content of ***saponins*** and other irritants. Has caused poisoning in Greek **soldiers** about 400 B.C. Often causes sneezing and often found in "sneeze powders". Digestive upsets, dyspnea, nervous symptoms, mental confusion, and even death have followed ingestion of the leaves or flowers. **Children** have been poisoned by drinking the water from vases that contained the plants. Contains the cardiac glycosides, ***convallamarin*** and ***convallanin***. The former acts like ***digitalin*** and the latter is a purgative. Also contains ***convallatoxin*** with ***digitalis***-like activity. 100 mg is roughly equal in potency to 12 ***digitalis*** units. It has been brewed as a tea and has caused serious illness in **people**.

Often, **dogs** that are bored dig in the shade and chew on the plant with sometimes toxic results. **Poultry** have died from eating the plants.

CONVALLATOXIN

Extracted from the blossoms of *Convallaria majalis*. In all parts of the plant.

Untoward effects: Due to its ***digitalis***-like action, the plant source can cause serious or fatal ill-effects.

IV LD_{50}, **rats**: 16 mg/kg; **mice**: 1 mg/kg; **guinea pig**: 180 μg/kg; **cat**: 80 μg/kg; IP LD_{50}, **mice**: 10 mg/kg.

CONVICINE

Untoward effects: A toxic glycoside in *Vicia faba* (***Broad*** or ***Fava Bean***) at a level of about 2% of its dry weight, and is associated with low ***glutathione*** concentration leading to marked sensitivity to agents that cause oxidative damage. Sensitive **individuals** or glucose-6-phosphate dehydrogenase-deficient **people** may develop severe hemolytic anemia.

CONYZA COULTERI

= *Coulter Conyza*

Found in the western U.S. and Mexico.

Untoward effects: **Sheep**, **goats**, and **cattle** have been poisoned after eating it during periods of drought, when it grows near watering holes. Has caused serious losses of **cattle** in Texas.

COPAIBA

= *Balsam Capivi* = *Balsam Copaiba* = *Jesuit's Balsam*

Use: The oil is used in fingerprinting inks, in paints, varnish and lacquers, color printing on textiles, and in pharmaceuticals and foods.

Untoward effects: Dermatitis, usually, maculopapular, sometimes, scarlatinoid or morbilliform. Eruptions are not uncommon. Fixed-drug eruptions. Gastrointestinal irritant. Cross-reacts with ***Balsam of Peru*** (q.v.). Copaiba oil is steam distilled from the balsam and used as a fragrance in foods, soaps, detergents, cosmetic creams and lotions, and perfumes.

Oil, LD_{50}, **rats** and acute dermal **rabbits**: > 5 g/kg.

Full-strength to intact or abraded **rabbit** skin was irritating.

COPPER
= *Cu*

Untoward effects: In **man**, excess intake adversely affects the gastrointestinal tract, erythrocytes, and the liver. Syringes with metal plungers, used by **dentists** for local anesthetics, have caused swellings lasting for hours or days, from ***Cu*** ions leached into the solution. Can be excreted in **mother's** milk. Has caused anemia, hematemesis, and bloody stools in **infants** treated with ***Cu*** salts. Ingestion of fireplace ***Cu*** coloring salts has caused intense gastrointestinal irritation. ***Cu***-induced severe dystonia, secondary to cholestatic liver disease in a 19-year-old **female**. High degree of ***Cu*** ionization in acidic solutions, when used in hemodialysis with copper pipes.
Blood: Hemolysis and methemoglobinemia from ***Cu*** in water pipe, acute hemolytic anemia after hemodialysis from ***Cu***-contaminated dialysate, some cases of methemglobinemia, and rare cases of leukocytosis.
Cardiovascular: Atherosclerosis and increased frequency of coronary heart disease reported with increased levels of ***Cu***.
Clinical test interference: Test levels may be increased by ***oral contraceptives***.
Eye: Deposits of ***Cu*** in the cornea reported in Wilson's disease. Irritant. Spasms of the lids and ulceration occur.
Fetus: Small amounts of ***Cu*** from an intrauterine device can prevent embryogenesis.
Liver: ***Cu*** may have a role in the etiology of liver cirrhosis, and the ***Cu*** content of liver in primary biliary cirrhosis is 30 times normal level.
Neoplasms: High rate of lung cancer in **people** living near copper smelters or involved in copper mining.
Poisoning: ***Cu*** in the water supply is a probable cause of ascites and severe micronodular cirrhosis with biochemical evidence of Wilson's disease. This in a 14-month-old **male** drinking water with pH of 4.4 and a ***Cu*** level of 675 μg/100 ml, after passing through new pipes. Nausea in all and vomiting and/or diarrhea in 11/20 **workmen** who drank tea left in a copper-lined hot water geyser. ***Cu*** contamination of dialysis solutions has caused hemolysis. ***Cu*** salts has caused drowsiness, convulsions, neuropathy, and coma. Behavior changes, diarrhea and progressive emaciation, hypotonia, photophobia, peripheral edema, and red extremities (Pink Disease) in 15-month-old **male** after ingesting ***Cu*** in hot and cold water for 3 months (79 and 35 μg/ml, respectively). Conjunctivitis and pharyngitis after exposure to a copper-containing insecticide dust. A serious exposure comes from consuming any acidic food or drink (especially carbonated), stored in a ***Cu*** container or food laid out on ***Cu*** serving dishes. ***Cu*** salts are especially popular as pigments for blue coloring of yuletide wrapping chewed on by **children** and **pets**.
Renal: Nephrotoxic. Acute tubular necrosis has followed massive exposure.
Schizophrenia, extrapyramidal symptoms: These may, in some cases, be related to excessive amounts of ***Cu*** in the body. Increased ***Cu*** levels occur in some **patients** taking ***phenytoin***, **who** develop toxic effects (to which entity?). Psychiatric symptoms have followed ***Cu*** contamination in hemodialysis.
Skin and hair: Oral lichen planus has been a reaction to ***Cu***-containing dental alloys. A rash in **mother** and **infant** was attributed to **their** soft water supply passing through 475 meters of ***Cu*** pipe. The rashes disappeared after the pipe was replaced with ***polyethylene***. Eczematous contact dermatitis 2 weeks after insertion of a ***Cu***-containing intrauterine device, or from eating unwashed apples treated with a ***Cu*** spray. ***Cu*** in swimming pool water can be absorbed into hair and turns it green.
Wilson's disease: A rare hereditary ailment is characterized by the body's inability to eliminate ***Cu*** and can cause hepatolenticular degeneration. These **patients** avoid ***chocolate*** and ***shellfish***, which have a high ***Cu*** content.

OSHA's **human** TWA for its dusts or mists is 1 mg/m^3. First symptoms would be eye and skin irritation and a flu-like illness.

TD_{LO}, **human**: 120 μg/kg.

Suggested levels on a dry feed basis for **cattle** are 8 ppm, and toxic levels are about 95 ppm. Icterus, hemoglobinuria, listlessness, and depression are common toxic symptoms, and bellowing, cramping, and dyspnea are noted in severe poisonings, before coma and death. In acute poisoning of **calves**, anorexia, ascites, and renal tubular necrosis are also reported. **Cattle** have suffered from ***Cu*** poisoning after eating wood ashes from preservative-treated woods, from eating chicken litter, and from grazing near copper smelters. **Bedlington terriers** and, occasionally, individuals in other **dog** breeds, suffer from an autosomal recessive genetic deficiency in their ability to store and metabolize ***Cu***, causing a hepatotoxicosis resembling Wilson's disease in **man**. Sublethal ***Cu*** exposure of **trout** increases their susceptibility to virus diseases. Mass death of ~34,000 kg of **fish** has followed an industrial waste discharge into a reservoir of a very fine suspension of ***Cu*** particles. ***Cu*** in coated tanks will cause a fatal reaction in **fish** anesthetized with ***MS 222***. Teratogenicity after IV use in a **hamster**. Growth decreases and decreases feed efficiency in **poultry** consuming 400 ppm in their diet. As in **man**, excess ***Cu*** is stored in the **rat's** liver parenchymal cells, and the excess reduces liver aniline hydroxylase activity in **rats**. Fatal jaundice of **sheep** in vineyards and orchards from spraying of copper salts or from grazing pastures similarly sprayed to kill **snails**. Has caused necrotic pancreatitis in **sheep**. Hemoglobinuria, hemolysis, methemoglobin formation, weight loss, and muscle damage occur in poisoned **sheep**. Of all the domestic **animals**, **sheep** are the most susceptible to ***Cu*** toxicosis. In **swine**, addition of 250 ppm ***Cu*** to the ration is usually toxic, decreasing hemoglobin and hematocrit (100 ppm ***zinc*** will negate this effect). High dietary levels in **pigs** cause ulceration of the pars esophagea. Jaundice, dullness, weakness, decreased growth rate, anemia, trembling, and respiratory distress are the first symptoms noticed in poisoned **pigs**. The bone marrow is red and the feces are hard and dark. Death may follow. **Poultry** may develop diarrhea and listlessness.

COPPER ARSENITE

= *Cupric Arsenite* = *Scheele's Green* = *Swedish Green*

Use: As a wood preservative, insecticide, fungicide, rodenticide, and as a pigment.

Untoward effects: Very poisonous. In the 1850s, Dr Hassall, an English physician, wrote a series of articles in *The Lancet* on food adulteration, including the use of this poisonous substance to color confectionery products, candies, and tea. Contains 34% ***Cu*** and 44% ***arsenic***.

The autoarsenite form (***Emerald Green*** = ***Paris Green***) was once used as an inexpensive green coloring for wallpaper and artificial flowers. Its use for those purposes was discontinued, because it caused many **human** deaths. It is still used in **slug** baits.

LD_{LO}, **human**: 5 mg/kg.

COPPER CARBONATE

= *Cupric Carbonate* = *Bremen Blue* = *Bremen Green*

Untoward effects: Contains 50–57% ***copper***. In the 1800s, ***tea*** in England was adulterated with it; one batch seized contained 35% by weight. A Russian report describes skin lesions from its penetration into skin after use as a dust for disinfecting dry ***corn***.

COPPER CHLORIDE

= *Cupric Chloride*

Untoward effects: Once used for coloring confectionery products. Contains 47% ***Cu***. OSHA sets limits at 1 mg ***Cu***/m^3.

LD_{50}, **rat**: 140 mg/kg; **mouse**: 190 mg/kg; **guinea pig**: 31 mg/kg.

COPPER CYANIDE

= *Cupricin* = *Cuprous Cyanide*

Untoward effects: Found in industrial waste effluents in the petrochemical, electroplating, steel, and coking industries. Also a residue from use in insecticides, fungicides, and antifouling agents in paints. Contains 71% ***Cu***.

COPPER NAPHTHENATE

= *Cuprinol* = *Cupri-Nox* = *Kopertox* = *Thrush-XX*

Use: Fungicide for wood, cordage, canvas, sandbags, ship hulls, piers, wharves, piles, rope, for treating "foot rot" of **cattle**, thrush on **horses** and **ponies**, ringworm treatment and foot pad-toughener for **dogs** and **pigs**.

Untoward effects: The risk to **carpenters** whose saws send the dust flying has not been reported. Avoid eye contact. Contact can cause hair loss.

Isolated reports of skin burns and blistering on **horses** and **dogs**. Kills **termites** and **parasites** on **chicken** perches.

LD_{LO}, **mouse**: 110 mg/kg.

COPPER NITRATE

= *Cupric Nitrate*

Use: In light-sensitive copy papers, textile dyeing and coloring, herbicides, fungicides, pyrotechnics, and as a catalyst in rocket fuels.

Untoward effects: Irritating to skin and mucous membranes. A 2-year-old **child** died with gastrointestinal, respiratory, cardiac, liver, kidney, and acid–base balance problems after ingesting 15 ml of ***Gun Blue***® solution. The ***selenious acid*** content, rather than the ***Cu*** nitrate was considered to be the prime cause of death.

Extremely irritant (pH 2) to the cornea of **rabbits**.

LD_{50}, **rats**: 940 mg/kg; 96 h LC_{50}, **rainbow trout**: 253 μg/l.

COPPER OXALATE

= *Crow Chex*

Untoward effects: As a **bird** and **rodent** repellent on seeds because of its bitter taste. ***Oxalates*** are rapidly absorbed from the gastrointestinal tract and moderately toxic to **birds** and **mammals**. The ***oxalates*** combine in the body to form ***calcium oxalate***, a central nervous system stimulant. Death can be caused by it and/or renal obstruction with ***oxalate*** crystals.

COPPER OXIDE, RED

= *Caocobra* = *Copper Pentoxide* = *Cuprous Oxide*

Use: Molluscicide, fungicide, on fish nets, in marine paints, in photoelectric cells, and as a red pigment for glass and ceramic glazes.

Untoward effects: Used as a molluscicide at 1–2 ppm and potentially dangerous to **children** because of its attractive bright red color.

COPPER OXYCHLORIDE

Fungicide.

Untoward effects: Severe diarrhea and large increases in liver weights, testicular atrophy, spermatogenic arrest, and increased lipid accumulation in Leydig cells, and seminiferous tubules of **poultry**.

LD_{50}, **rats**: 700 mg/kg; **hens**: 2 g/kg; **roosters**: 1.26 g/kg.

COPPER SULFATE

= *Algimycin* = *Bluestone* = *Blue Vitriol* = *Cupric Sulfate* = *Roman Vitriol* = *Salzburg Vitriol*

Use: Fungicide, bactericide, algicide, mordant in textile dyeing, in electroplating, in pyrotechnics, in hide preservation and leather tanning, in wood preservation, as a veterinary anthelmintic, and as a ***Cu*** source in **livestock** feeds. U.S. use is ~50,000 metric tons annually. Some has even been used as a clarifying agent for wine, is an emetic, and to neutralize ***white phosphorus*** (particularly in the military), and in the treatment of molybdenosis in **cattle** and **sheep**.

Untoward effects: Most outbreaks of poisoning are due to negligence and/or ignorance of the methods of storage, application, and disposal. In the 1800s, used in England for coloring pickles, bottled fruit, and confectioneries. In 1902, Dr. Harvey W. Wiley established a "Poison Squad" that identified industry's use of ***$CuSO_4$*** to color canned peas and other green vegetables. Oral use delays gastric emptying and leads to nausea and vomiting by its action on peripheral gastrointestinal receptors.

A 18-month-old **male** drank a solution containing ~3 g and serum ***Cu*** reached 1.65 mg%. The **child** recovered after over a month of treatment and, during that time, **he** suffered from hemolytic anemia and renal tubule damage. A 27-year-old **male** developed methemoglobinemia after ingesting > 50 g, and was unsuccessfully treated. Acute poisoning with vomiting and diarrhea in 20 **workmen** who drank tea made with copper sulfate-contaminated water from an unserviced corroding gas hot water heater. A 23-year-old **patient** ingested 15 g and, despite adequate therapy, had persistent hemolytic anemia and gastric ulceration, when discharged 13 days later. Accidental ingestion of 10–15 mg can lead to hepatic damage with jaundice and hemolysis within 3 days. Hemolysis may occur due to glucose-6-phosphate dehydrogenase inhibition, making glucose-6-phosphate dehydrogenase-deficient **patients** more susceptible. A 2-year-old **male**, apparently, drank about 30 ml of a saturated solution used for growing crystals. **He** vomited immediately, developed diarrhea, hemolytic anemia, renal failure, cardiac arrhythmia, and central nervous system depression. **He** did not developed fatal renal failure or jaundice, and recovered after treatment. Fatal poisoning occurred in four **patients** (21–50 years) after ingesting it in spiritual water. In New Delhi, India, 32 **males** and 16 **females** were admitted to Irwin Hospital during 1 year, and 7/48 died. Ingestion of ***$CuSO_4$*** is a favorite method of attempting suicide in India, but infrequent in the U.S., where most poisonings are from drinking water from contaminated pipes, ***copper***-containing creams applied to eczematous skin, ***$CuSO_4$*** irrigation of fistulas, applying it to skin burns, or exposure from ***$CuSO_4$***-containing dialysis equipment. Severe abdominal pain, gastrointestinal bleeding, ulceration and perforation, hemolysis, vomiting, and diarrhea reported. In severe cases of intoxication, a metallic taste, burning sensations in the epigastrium, eructations, repeated vomiting of greenish material, followed by bloody diarrhea, shock, smoky urine, oliguria, anuria, uremia, enlarged tender liver, jaundice, coma, and even death. In 1982, 200 **people** reported stomach cramps, nausea, numbness, dizziness, and tingling of the extremities. Some cases may have resulted from mass hysteria, after enjoying soft drinks, probably caused by ***$CuSO_4$*** contamination of the dispenser at one site in the stadium. Use as an emetic, although effective in ~85% of the cases, should now be considered too dangerous for general use. It can form a weak complex with hair, turning it green, cause a contact dermatitis, and turn nails green.

TD_{LO}, **child**: 200 mg/kg; **human**: 11 and 50 mg/kg (both values quoted in literature).

In a dairy herd, **cows** received 11 g/day and chronic ***Cu*** poisoning developed in milk-producers fed 22 g/day. After 6 months, three became ill and died. Their symptoms were anorexia, decreased milk production, frequent recumbency, and jaundice. Death followed in 4 days. Excessive intake will give milk an oxidized off-flavor. Splenic congestion with hemosiderosis, enteritis, and nephritis have been reported in poisoned **calves** and **cattle**. Levels that were toxic as a liquid drench (12 g/day) were toxic and fatal after 2 months, but the same level was not toxic to the **beef steers**. High levels (> 4 ppm) as an algicide can rapidly be toxic to **fish**. Can be fatal to **goats** after oral dose of 20 mg/kg. ***Cu*** at 500 ppm (as ***$CuSO_4$***) decreases growth rate and feed efficiency in **chicks**. Levels above this to **chickens** or **turkeys** decrease feed efficiency and egg production. Watery diarrhea, listlessness, burns or erosions on the gizzard, and a greenish seromucous intestinal exudate found in toxic or lethal doses to poultry. **Birds** that die of toxic doses within 24 h, show acute tubular necrosis and necrotic changes in the intestinal epithelium. Those that die within 2–7 days exhibit mostly liver lesions and advanced degenerative renal changes. In **turkeys**, its use may interfere with ***arsenicals*** used in controlling blackhead, and decrease the benefits of ***penicillin-streptomycin*** (1 : 3) in their feed. **Sheep** have died with fatal jaundice after grazing vineyards and orchards that were sprayed with ***$CuSO_4$***, or on pastures where it was used as a molluscicide spray at 1%. Poisoned **sheep** have shown hemolysis, liver, kidney, spleen, and brain damage. **Sheep** are particularly susceptible, as seen by the LD_{50}s. As levels approach and exceed 50 ppm in **pig** feed, toxicity begins to develop with decreased hemoglobin, decreased weight gains, decreased hematocrit, jaundice, dullness, trembling, and respiratory distress. This is aggravated by low ***zinc*** and ***vitamin E*** levels. Undesirable taste in water, forcing treated **birds** to seek non-medicated sources. Incompatable with ***iodides***. Long, hard rains can lower pond alkalinity and increase ***copper*** toxicity. ***Copper***-containing feces can be very corrosive to metal.

LD_{50}, **rats**: 300 mg/kg; **mink**: 7.5 mg/kg; **cattle**: 400 mg/kg; **sheep**: 25–50 mg/kg. LD_{LO}, **duck**: 600 mg/kg; LD, **chickens**: 693 mg/kg; **dog**: 2.5–4 g; **cow**: 200 mg/kg; **sheep**: 130 mg/kg; **lambs**: 25–50 mg/kg.

COPPER SULFIDE

Untoward effects*:** Common in ***copper ores and oxidized to ***$CuSO_4$*** by moist air. It induces vomiting. **Miners** must avoid inhalation and other contamination. May become caustic on sweating skin.

COPPER THIOCYANATE

= *Cuprous Thiocyanate*

***Use*:** In marine antifouling paints and in primers for explosives.

Untoward effects*:** Interferes with thyroid function tests. Thyroid uptake of ***iodide strikingly impaired. Radio-uptake tests invalid in one **female** and three **males** (21–36 years) taking it.

COPPER TRIETHANOLAMINE

= *Cutrine* = *Swimtrine*

Algicide.

***Untoward effects*:** Concentrate will irritate skin and eyes.

Toxic to **fish** at concentrations > 4 ppm.

COPPERWEED

= *Oxytenia acerosa*

***Untoward effects*:** Causes anorexia, depression, weakness, some struggling when down, coma, and, eventually, death, especially in **cattle** who die within 24–48 h after eating about 3 lb. Poisoned **sheep** may linger 1–3 weeks before dying. Found in southwestern U.S. along old stream beds and gullies. Contains a toxic alkaloid which peaks at the bush's maturity. Hepatotoxic. Leaves and stems are equally toxic. **Sheep** generally eat the fallen leaves.

COPRINE

Untoward effects*:** A toxin in the inky capsule mushroom (*Coprinus atramentarius*, *C. erethistes*, and *Clitocybe claviceps*) with properties similar to ***disulfiram, blocking the action of the enzyme, acetaldehyde dehydrogenase. After ***alcohol*** ingestion, ***acetaldehyde*** (q.v.) accumulates in the body. Coprine toxicity only occurs if ***alcohol*** is taken with or shortly after (occasionally, up to 4–5 days) ingesting the ***mushrooms***. Flushing, neck vein distention, paresthesia, and swelling of the hands and feet, metallic taste, palpitations, and tachycardia are early (often within 20 min) symptoms. Later, nausea, vomiting, sweating, and, occasionally, diarrhea occur. Severe cases may show weakness, visual disturbances, mydriasis, vertigo, hypotension, arrythmias, dyspnea, and coma. Generally, the toxicity is of early onset and short duration, with a favorable prognosis.

Causes severe changes in testicular germ cells of **rats** and **dogs**.

COPY PAPER AND COPYING MACHINES

Untoward effects*:** Carbonless copy paper has caused a toxic dermatitis and irritation of the eyes and mucous membranes, and solvents and dyes, such as ***triarylmethanes and ***triphenylmethanes*** are suspect. Respiratory irritation; migraine headaches; burning or erythema of the face; dryness of the eyes, nose, mouth, and throat; and dermatitis reported from use of activators, toners, or solvents in old-style copying machines. Necrotizing skin vasculitis has occurred after inhalation of photocopier fumes. Photocopy paper has contained the allergen, ***ethyl butythiourea***.

CORALS

***Untoward effects*:** Dermatosis in coral **workers** is generally caused by mechanical injury, and not sensitization. Stony or true corals and soft corals do not envenomate. Soft corals are toxic to **fish**. The hydroid, stinging, or fire corals are not true corals (*Millepora sp.*), and are found near coral reefs. A **person** brushing lightly against it will immediately feel a burning or numbness of the area, which can progress into urticarial-like dermatitis with erythematous, papular, or patchy eruptions.

CORIAL YELLOW

***Untoward effects*:** As a textile dye sprayed on silk through a stencil, 29 **women** in Europe were poisoned, with **their** illness lasting 16–20 weeks.

CORIANDER

One of the oldest flavoring essences known to man. References have been made to it in the Bible.

***Untoward effects*:** Contact dermatitis in **man**.

Poisoning and death of **horses** and **cattle** when incorporated into their feed at over 3% for **dairy cows** and 5% for fattening **cattle**. Antifertility effect in **rats**. Hypoglycemic.

CORIARIA

Untoward effects*:** Two circus **elephants** showed very strong central nervous system stimulation, including convulsions, after eating the foliage of ***C. arborea* (*Tutu*)**. They recovered after ~1 day. Two **dogs** who fed on ruminal contents from a **lamb** that had died after eating the leaves and flowers, showed central nervous system disturbances. ***Tutin is also its toxic alkaloid.

C. rusifolia has caused poisoning in **humans** due to the toxic alkaloid, ***tutin***. **They** exhibited giddiness, delirium, convulsions, stupor, and coma.

C. sarmentosa and ***C. arborea*** seeds and shoots are poisonous to **cattle** and **sheep** in New Zealand. Symptoms include excitement, vomiting, convulsions, exhaustion, and death. Secondary toxicity in another **dog** from eating offal of a poisoned **lamb**.

C. thymifolia fruits have poisoned **children**.

CORN

= *Maize* = *Zea mays*

Untoward effects*:** A major cause of food allergies in **people**. Symptoms have included gastrointestinal upsets, headache, dyspnea, fatigue, runny nose, itchy eyes, bronchial asthma, and hypersensitivity pneumonitis. Contamination with *Aspergillus flavus* and ***aflatoxins occurs and is suspect as a cause of hepatomas in **Bantus** (14/100,000) versus an incidence of 1.7/100,000 in U.S. **whites**. A pregnant 23-year-old **female** developed orofacial swelling, dyspnea, hypotension, arrhythmias, voice hoarseness, and flushing within 8 min of receiving an IV dose of corn-derived ***dextrose***. Ingestion of corn has been associated with migraines. 18/43 **patients** in an Illinois nursing home developed facial flushing and/or erythematous macular rash on the upper arms and face within 15–30 min after eating breakfast. **All** had eaten cornmeal mush. ***Niacin*** is a common additive (16–24 mg/lb), and samples of the meal contained > 1 g ***niacin***/lb. Corn dermatitis (corn itch) is an irritant occupational eruption. Excoriations and secondary infection has occurred. Corn oil is comedogenic. It is suggested that **patients** with ***nickel*** dermatitis avoid it. Corn contains ***cyanogenic glycosides***. ***Antimony*** has been found in corn flakes. This author forced the FDA to stop the shipment of several ***mercury***-contaminated trainloads of corn destined for a corn flakes manufacturer. Several thousand **patients**, including 170 **children** (16 months–14 years), in Spain developed a toxic syndrome from ingesting an oil mixture contaminated by a ***rape oil*** denatured with ***aniline***. Mydriasis in corn field **workers** called **"corn-pickers** pupils", apparently, occurs from corn field exposure to ***jimson weed* (*Datura stramonium*,** q.v.), and will clear spontaneously within a few days after leaving the site of exposure. Corn oil often contains significant amounts of ***peroxides***. Cornmeal and ***tranylcypromine*** may cause undesirable effects.

Swine, **horses**, and other **livestock** have been seriously affected by various ***mycotoxins***, ***aflatoxins***, and ***fuminoisin*** from *Fusarium* contamination on moldy corn. Stalks tend to accumulate ***nitrates*** and may contain a ***cyanogenic glycoside***. High incidence of dietary cirrhosis in **rats** fed corn oil.

CORNMINT OIL

= *Japanese Mint Oil* = *Olei Mentha arvensis*

***Use*:** As a fragrance in soaps, detergents, cosmetic creams and lotions, and in perfumes.

***Untoward effects*:** No sensitization reactions in 22 **human volunteers**. No phototoxic reactions on **swine** and hairless **mice**. Full-strength to the backs of **swine** and hairless **mice**, under 24 h occlusion on **rabbits**, and 8% in petrolatum on **humans** all produced no irritation.

LD_{50}, **rats**: 1.24 g/kg; acute dermal LD_{50}, **rabbits**: > 5 g/kg.

CORN SYRUP

= *Glucosum* = *Liquid Glucose* = *Syrupy Glucose*

***Untoward effects*:** The New York State Health Dept. has warned **mothers** not to feed corn syrup to **babies** < 1 year of age, because of the risk of botulism. For unknown reasons, the intestinal tract of susceptible **infants** < 1 year of age is ideal for the germination of botulism spores. Oral use of concentrated material may cause nausea and vomiting.

CORONILLA sp.

= *Crownvetch*

Untoward effects: Toxic to **non-ruminants**. **Ruminants** are not adversely affected because rumen bacteria detoxify the compounds. ***Digitalis***-like activity and can lead to ventricular spasms and diastolic failure.

CORONOPUS sp.

Untoward effects: ***C. squamatus*** and ***C. didymus*** are, possibly, goitrogenic in Australian reports on **children** drinking milk from **cows** that have eaten it. ***C. didymus*** (***Bitter Cress*** = ***Mastuergo Hembra*** = ***Quimpe***), after ingestion, not only taints **cows'** ***milk***, but also their meat from its ***benzyl isothiocyanate***, ***benzylmethylsulfide***, ***benzylmercaptan***, and ***benzylcyanide*** content.

CORTICOTROPIN

= *Acethropan* = *Acortan* = *Acorto* = *ACTH* = *Acthar* = *Acton* = *Actonar* = *Adrenocorticotropic Hormone* = *Adrenocorticotrop(h)in* = *Adrenomone* = *Alfatrofin* = *Cibacthen* = *Corstiline* = *Cortiphyson* = *Cortrosyn* = *Cortrophin* = *Isactid* = *Reacthin* = *Solacthyl* = *Tricortan* = *Tubex*

Adrenocorticotropic hormone.

Untoward effects: Weight increase and Cushingoid syndrome common with prolonged use.

Allergy, anaphylaxis: Most reactions occurred due to foreign proteins (viz. pork) in early commercial preparations, manifesting symptoms were anaphylaxis with shock, asthma, dyspnea, and urticaria. Abdominal cramps, headache, vomiting, syncope, and cold sweats can occur. Occasionally deaths have been reported. Local reactions with depot form are probably due to the ***zinc hydroxide*** contained in it. Synthetic forms have less reactivity.
Blood: Hyponatremia, hypokalemia, anemia, eosinopenia, leukocytosis, non-thrombocytopenic purpura, low-grade gastrointestinal bleeding or gastric ulceration, and hyperprolactinemia noted.
Bone: Osteoporosis, aseptic necrosis of femoral head.
Cardiovascular: Altered T waves, prolonged QT segments, and uncontrolled atrial flutter-fibrillation reported.
Children and neonates: Vascular spasms as an allergic reaction in 10-year-old **male** led to death from renal cortical necrosis and necrotic foci in lungs. Marked growth retardation reported in **some**. Transient brain shrinkage in **infants** after treatment for infantile spasms. Enlarged thyroid in **neonate** after ***ACTH*** on first day of life for thymic enlargement. The **infant** was thyrotoxic at birth. Said to increase ***potassium*** and decrease ***sodium*** in **mother's** milk.
Clinical test interference: Interferes with uptake of ***radioiodine*** by the thyroid for about 1 week. Will cause increased or false positive ***glucose***, decreased or false negative ***potassium***, decreased chemical ***estriol***, and increased serum protein test levels.
Eye: Posterior subcapsular cataracts, macular lesions, and transient myopia. Occasionally edema, twitching, or ptosis of the eyelids, corneal edema, mydriasis, increased intraocular pressure, slowed healing of ocular tissue, and exophthalmos.
Infection: Increases susceptibility, as with ***corticosteroids***.
Miscellaneous: May aggravate diabetes mellitus, cause dependence, cause a brown discoloration of the skin, and electrolyte disturbances.
Muscle: Myasthenia and myopathy.
Parathyroids: Can produce hyperplasia and hypertrophy of the glands.
Psychosis: Malignant euphoria, mania, emotional problems, depression, and confusion reported.
Skin: Photosensitivity (2–3%), pigmentation (melanoderma), urticaria, erythema, acneform lesions, and hirsutism.
Wound healing: Delayed healing.

Drug interactions: May increase ***anticoagulant*** requirement and increase ***salicylate*** clearance rate. Interaction in the body decreases the effectiveness of vaccines.

Do NOT use in **animals** with marked adrenal insufficiency. Failure to respond to corticotropin therapy is a frequent indication of such failure. Prolonged or excessive use may cause hyperplasia or hypertrophy of the adrenal cortex and requires ***potassium chloride*** supplementation, especially if muscle weakness is noted. May accent osteoporosis in affected **animals**, and repeated use may increase **animal's** susceptibility to infections.

CORTINARIUS sp.

Untoward effects: The largest genus of ***mushrooms*** in Europe, and most of the ~250 British species are toxic. Renal failure and death in **humans** have been reported. The symptoms may become apparent after a latent period of 3–20 days after ingestion and are manifested by intense burning thirst, vomiting, and headache.

Goats and **snails** eat large quantities of these ***mushrooms*** without ill-effects, but, when **goat** milk and meat, as well as **snails**, are consumed, the potential for secondary toxicity exists. **Sheep** have become ill and died after ingesting them.

CORTISONE

= *Adrenalex* = *Adreson* = *Compound E* = *Corlin* = *Cortadren* = *Cortelan* = *Cortisal* = *Cortisate* = *Cortistab* = *Cortisyl* = *Cortogen* = *Cortone* = *Incortin* = *Irisone* = *Kendall's Compound E* = *Reichstein's substance Fa* = *Ricortex* = *Scherosone* = *Winter-steiner's compound F*

Adrenal cortex hormone, anti-inflammatory, glucocorticoid. **The acetate forms are commonly used. Africa's Bantu tribesmen knew of its value long before we did. I had the pleasure of visiting with the witch-doctor who arranged for the shipment to the U.S. of the first cortisone-containing African plants.**

Untoward effects: ***Addiction***: Not uncommon in **patients** with polyarthritic conditions. Hypercortisonism may be responsible for anorexia, chills and fever with myositis and vasculitis, euphoria, depression, and nocturnal agitation noted on withdrawal.
Anesthesia: Requires pretreatment before surgery in **patients** who have been receiving it. IV ***corticosteroids*** should be immediately available in the operating room.
Behavior: Pathological excitement, euphoria, psychosis, depression and mania reported.
Blood and cardiovascular: Hypokalemic alkalosis, hypernatremia, hyperglycemia, hypertension, fluid retention, and congestive heart failure noted.
Bone and growth: May adversely affect final stature by stimulation of epiphyseal maturation. Can increase osteoporosis and necrosis of the femoral head.
Clinical tests: False positive and increased or false negative and decreased ***cholesterol*** results. ***Estriol*** measurements to assess placental viability in high-risk pregnancies may be increased. ***Thyroxine***-binding globulin concentrations are decreased by it. Clinical tests must be interpreted with ***sodium*** retention, ***potassium*** loss, and hyperglycemia in mind.
Cushing's syndrome: May result from normal doses or overdoses. A 19-year-old **female** gained 21 lbs while being treated for an ***indomethacin***-induced rash.
Eye: Glaucoma and cataracts, even from topical application on the eye. Loss of vision has occurred, due to a dendritic ulcer in a **patient** applying it for a herpes eye infection. Papilledema and benign intracranial hypertension reported.
Fetus: Although it frequently caused fetal malformations in **laboratory animals**, it rarely has in **humans**. Estimated cleft palates in < 1% of **children** exposed *in utero*, if the **mother** received large doses before the 14th week of gestation. One report of transient adrenal atrophy. Its use in pregnancy should be reserved only for serious diseases. ~1% is excreted in breast milk.
Gastrointestinal: Impairs the production of natural mucous barriers that protect the stomach's mucosa.
Infection: Can exacerbate respiratory problems and increase susceptibility to infections by decreasing body defenses. This occurs by reducing protein synthesis, antibody formation, ***interferon*** levels, and phagocytic activity of the reticulo endothelial system. Avoid its use during early stages of any infectious disease unless intensive antibacterial and/or antiviral therapy is also used. Decreases anterior pituitary trophins and ***adrenocorticotropin***. May protect against endotoxin lethality.
Liver: Therapeutic doses have caused fatty infiltration.
Miscellaneous: Arms and legs become thinner, mostly from loss of fat, and upset menstrual cycles and accumulation of fat on the upper back (buffalo hump) often occur with its overproduction, due to pituitary or adrenal tumors.
Muscle: Muscle weakness and ***steroid*** myopathy with loss of muscle mass.
Renal: Lesions resembling diabetic glomerulosclerosis in ~50% of nephrosis cases treated with it.
Skin: Fragmentation and disappearance of elastic fibers and small transient yellow spots after local infiltration. Hyperpigmentation noted around an implanted pellet of cortisone acetate. May cause exacerbation of some previous skin conditions, thinning of skin, and some atrophy. Acne, purple stretch marks on skin, and easy bruising reported.
Wound healing: Delayed, but topical ***vitamin A*** antagonizes cortisone's effect in the wound. Yet, systemic ***vitamin A*** may reactivate the basic disease process. May delay healing in certain ligament injuries.

Drug interactions: Conflicting reports on its raising and lowering effects on various ***anticoagulants***. Increases ***steroid*** dose required, due to enhanced ***steroid*** metabolization by ***rifampin***. May require an increase in ***tolbutamide*** dosage.

Hyperglycemia produced only during treatment in the **rat**, **rabbit**, **guinea pig**, and **dog**. Fetal anomalies when pregnant **rabbits'** eyes were treated topically. Can induce diabetes mellitus in **rabbits** and **cats**. May contribute to parathyroid insufficiency, if given to **calves** at 1 week of age. Teratogenic in the **mouse**, **rabbit**, and **rat**. Causes fatty livers, delayed healing, and eosinopenia in **rats**. Activity of ***tetanus toxin*** increases and decreases effectiveness of ***tetanus antitoxin*** in **mice**.

IM TD_{LO}, **rabbit**: 25 mg/kg.

CORYDALIS

= *Bicuculla* = *Squirrel Corn* = *Turkey Corn*

Untoward effects: Trembling, visual disturbances, dyspnea, nausea, and possibly life-threatening convulsions, especially in **children** and **pets**. Contains ***isoquinoline*** alkaloids (***apomorphine, protoberberine, protopine***), which are more poisonous to **cattle** than **sheep**. ***C. caseana*** = ***Fitweed*** is the usual cause.

C. cava contains the convulsant alkaloid, ***bicuculline***, as potent as ***strychnine***. ***Bulbocapnine*** (***Corydaline***), ***corydine***, ***corycavine***, and ***corydine*** are also found in Corydalis.

COSTUS ROOT OIL

Use: In fragrances in cosmetic creams, lotions, and perfumes, and foods.

Untoward effects: 4% in petrolatum produced sensitization in 17 **human volunteers**. Initial irritation was so severe that an additional eight **volunteers** were not challenged.

LD_{50}, **rats**: 3.4 g/kg; acute dermal LD_{50}, **rabbits**: > 5 g/kg.

COTONEASTER

Untoward effects: Exposed **people** have been common asymptomatic **visitors** to Poison Control Centers. ***Cyanogenic glycosides*** were highest in the leaves and least in the bark and fruit in eight species. Feeding the dried fruit to **rats**, **cats**, and **dogs** was without incidence, and it is the fruit that is more likely to be ingested by **children**.

COTTON

= *Abduga* (Hausa) = *Gossypium sp.* = *Owu* (Yoruba) = *Owubi* (Igbo)

Use: In Mexico and South Carolina, **natives** use a root decoction of *G. hirsutism* (***Upland Cotton***) or *G. barbadense*, as an emmenagogue, to ease delivery, as a contraceptive, and as an abortifacient. It was once officially recognized in the U.S. as a vasoconstrictor and hemostat. Its greatest use is as yarn and clothing.

Untoward effects: Byssinosis is common in long-term **employees** in carding areas of cotton mills. Symptoms including dyspnea, chest tightness, bronchitis, wheezing, cough, and decreased pulmonary function. In the early stages, these occur, especially at the beginning of each work week. Oral and pharyngeal cancers may have an increased incidence in cotton textile **workers**. Years ago, the skin cancers, including scrotal cancers on **laborers** in cotton mule-spinning mills were traced to **their** prolonged contact with ***mineral oils*** used for lubricating machinery. Bilateral symmetrical buccal carcinomas reported from years of packing both posterior buccal areas with wads of plain sterile cotton, to improve appearance of sunken cheeks. Contact dermatitis in **workers** from abrasiveness of cotton sheeting. Some reactions from using cotton have been traced to chemicals (***dichlorophen, cadmium,*** etc.) added to prevent fungal and bacterial deterioration, or to impart smoulder resistance (***boric acid***). Fatal disseminated coccidioidomycosis in 56-year-old **male** opening cotton bales for 42 years. Chemicals used on the plants may remain on cotton. Cotton fibers have been found in the lungs of **drug-abusers** after being drawn into **their** syringes from cottons used as a filter. OSHA list 1 mg/m^3 as maximum exposure for cotton dust or waste processing exposure; 200 μg/m^3 for yarn manufacturing; 750 μg/m^3 for textile slashing and weaving, and 500 μg/m^3 for all other operations. See ***Gossypol***.

Dust, TC_{LO}, **human**: 1 g/m^3/10 years.

Cotton pellets SC have been used to produce granulomas in **rats**.

COTTONSEED MEAL

Untoward effects: Contains ~1% ***gossypol*** (q.v.), a yellow pigment in some ***cotton*** plants. A ***gossypol***-free cottonseed has also been grown. ***Aflatoxin*** contamination can pass into the milk of **dairy cows**. Over ~20 years ago, other contaminants including ***mercury***, pentachloronitrobenzene, ***disulfoton***, and a solvent, ***n-octylamine***. The meal's *Aspergillus flavus* ***aflatoxin*** has been a causative factor of hepatomas in **fish**. A liver cancer cluster involved three **brothers** who lived near a large cottonseed mill and several grain mills while **they** were growing up.

Causes discoloration of **poultry** egg yolks. Toxicity is variable. The amount in feed that is toxic to **pigs** is not to **chicks** or **rats**. Young or pregnant **animals** are more susceptible to its ill-effects, which develop in about 10–30 days. Symptoms include anorexia, dyspnea, restlessness, and spasms. **Calves** from cottonseed-fed **cows** may suffer from congenital defects, become weak, and die.

COTTONSEED OIL

= *Aceite de Algodon* = *Acid Oil* = *Oleo de Algodoeiro* = *Oleum Gossypi Seminis*

Untoward effects: In one series, 14.3% of **patients** (4/28) receiving 284 infusions (***Lipiphysan***®, also containing ***lecithin*** and ***vitamin E***) IV had severe reactions, including headache, nausea, vomiting, alternating flushing and pallor, shivering, and hyper- or hypotension. In other trials, pain in the chest, back, and abdomen; dyspnea, and cyanosis in 3/341 infusions of ***Inponutrol***®.

Hepatomas in **trout** fed it for 5 months. **Rats** receiving up to 20 ml qid died within 3–4 days after signs of pallor, listlessness, diarrhea, cyanosis, epistaxis, ataxia, dyspnea, anorexia, weight decrease, hypothermia, oligodypsia, oliguria, and aciduria.

COTYLEDON ORBICULATA

Untoward effects: Causes plant poisoning in **sheep**, **goats**, and **rabbits** in South Africa. The syndrome is called Krimpsiekte, and includes salivation, torticollis, colic, tachycardia, tachypnea, convulsions, collapse, coma, and death. About 1 g of the dry plant can be fatal to **sheep**.

COUMACHLOR

= *G 23,133* = *Ratilan* = *Tomorin*

***Rodenticide.* An anticoagulant coumarin.**

***Untoward effects*:** Highly toxic.

In **dogs**, single oral doses of 5 mg/kg were lethal in 6–13 days. ***Warfarin*** at 100 mg/kg was relatively innocuous. For **pigs**, single doses of 0.5–30 mg were fatal in half of them. Symptoms were exhaustion and coma; occasionally, joint swellings and hematomas of the neck, as well as paralysis.

LD, **baby pig**: 1 mg/kg; LD_{50}, **rat**: 900 mg/kg or 100 mg/kg/day/5 days.

COUMAFURYL

= *Fumarin* = *Fumisol* = *Krumkil* = *Lurat* = *Prolin* = *Rat-A-Way* = *Tomarin* = *Warficide*

***Rodenticide.* An anticoagulant coumarin.**

Untoward effects*:** Gastrointestinal absorption is efficient. Symptoms are as with ***coumachlor. Accidental ingestion by **humans** and **pets** must be avoided.

LD_{50}, **rat**: 25 mg/kg.

COUMAPHOS

= *Asuntol* = *Bayer 21/199* = *Baymix* = *Co-Ral* = *Meldane* = *Muscatox* = *Resistox* = *Umbethion*

***Organic phosphorus insecticide, anthelmintic.* Cholinesterase inhibitor.**

Untoward effects*:** Do not use in conjunction with or within a few days of (before *and/or* after) any other cholinesterase-inhibiting compound. (Avoid use with ***phenothiazine, phenothiazine tranquilizers). Three weeks appears to be a safer interval. High levels of ***vitamin A*** may increase toxicity. Caution should be exercised in using it on **Brahmas**, due to their generally recognized increased sensitivity to organic phosphorus anticholinesterase agents. Layers of colored **poultry** breeds should not be treated when in egg production, as they appear unusually sensitive to the drug's toxic effects, compared to the white breeds. Do NOT contaminate feed or water, except as in use directions. Keep from **children**. Do NOT repeat treatment for **cattle** within 45 days or use within 45 days of their slaughter. Salivation, dyspnea, and stiff gait after about 1 day. **Horses**, **sheep**, and **goats** often develop a diarrhea, and **horses** also become restless and roll, as with colic.

LD, **calves**: 10–40 mg/kg; **sheep**: 8–25 mg/kg; LD_{50}, male **rats**: 41 mg/kg; female **rats**: 16 mg/kg; **quela** and **redwinged blackbirds**: 3.2 mg/kg; **sparrows**: 10 mg/kg.

COUMARIN

= *Benzopyrone* = *Cis-Ortho-Hydroxycinnamic Acid Lactone* = *Lodema* = *Lysodem* = *Tonka Bean Camphor*

***Flavoring agent.* Odor of new-mown hay. Structurally related to *aflatoxin* and *coumarin anticoagulants*. Tastes and smells like *vanilla*.**

***Use*:** In fragrances and in electroplating, to decrease porosity and increase brightness. Flavoring agent for **cattle** and, unfortunately, in candies for **humans**. I finally convinced the FDA to prohibit its use in candy bars and foodstuffs.

Untoward effects*:** A natural toxin in some feedstuffs. An occasionally illegal adulterant in ***vanilla products. During 1990–1998, France's pharmavigilence system reported 33 cases of hepatotoxicity in **people**. A case of lung cancer when used in ***cigarettes***. Can and has caused interference with ***vitamin K*** transport, with resulting hemorrhage. In **man**, 80% is excreted as ***7-hydroxycoumarin***; in **rats**, only 1%.

TD_{LO}, **human**: 50 mg/kg causing vomiting and weakness.

Should no longer be used for **livestock**! **Mice** fed 0.25% levels on days 6–17 of pregnancy had an increase in stillbirths and delayed ossification; increases mortality at 0.05–0.25% levels.

LD_{50}, **rats**: 293–680 mg/kg; **mice**: 196 mg/kg; **guinea pigs**: 202 mg/kg.

COUMATETRALYL

= *Bay 25,634* = *Endox* = *Racumin*

***Rodenticide.* Used on *warfarin*-resistant rats.**

***Untoward effects*:** Outbreak of poisoning and deaths in a **piggery** where it was estimated that a total of 1–2 mg/kg taken over 7–12 days was lethal.

LD, **rat**: 16.5 mg/kg; **cat**: 20–50 mg/kg; LD_{50}, **rat**: 300 µg/kg/day/5 days.

COUSSAPOA CINNAMOMEA

Piscicide.

***Untoward effects*:** A minor **fish** poison (leaves and fruit pounded and mixed with mud). Used in the Amazon.

COUTOUBEA RAMOSA

***Untoward effects*:** Kills grazing **animals** in Columbia.

COWHAGE

= *Cowage* = *Mucuna*

A leguminous climbing plant found in the West Indies and tropical countries of both hemispheres.

Untoward effects: Its hairs contain an enzyme, ***mucanain***, which, in **man**, causes unbearable itching of the fingers within 5 min and lasts up to 30 min after handling.

COXIELLA BURNETII

= *Rickettsia burnetii*

Untoward effects: The cause of Q fever acquired by **man** from inhalation of contaminated dusts or aerosols, and by direct contact with **animals** through their raw milk, hair, feces, or infected placentas. Causes an interstitial inflammation and infiltrate in the lung with swollen and desquamating alveolar cells. Symptoms in **man** include fever, chills, myalgia, severe headache, malaise, and, after 2–4 weeks, a non-productive cough. Chronic endocarditis has been an occasional sequelae. **Veterinarians**, slaughterhouse and research **workers** are particularly at risk.

A California study noted evidence of previous exposures in **coyotes** (78%), **foxes** (55%), **brush rabbits** (53%), **grey squirrels** (6%), **wood rats** (3%), **kangaroo rats** (2%), **brush mice** (31%), **pinyon mice** (9%), as well as many other small **mammals**. **Marsupials** and **birds** have also been infected.

COYOTILLO

= *Karwinskia humboldtiana* = *Tullidora*

Untoward effects: Its berry-shaped black fruit contains myotoxic anthraquinones and can adversely affect all domestic **animals** and **humans** with weakness, muscular incoordination, paralysis, and death (usually from paralysis of respiratory muscles). Post-mortems reveal a mild gastroenteritis, widespread cardiac and skeletal muscle degeneration, and hemorrhages on the heart. In **man**, symptoms have occurred several days to several weeks after ingestion of the berries.

Up to 4 months between ingestion and onset of symptoms in **cattle** and **sheep**.

CRABS

Untoward effects: A common cause of food allergies. **Sand** and **horseshoe crabs** contain a ***saxitoxin***-like poison that has powerful hemagglutinating properties. Symptoms from exposure consist of weakness, dizziness, and bradycardia. Severe poisonings lead to hypersalivation, paralysis, coma, and death within 16 h. An asthma-like illness has occurred in crab-processing **workers**. Over a period of years, in the Phillipines, the crab, *Lophozozymus pictor*, has caused death after **people** have ingested its soup or muscles. Fatality rate in **people** is ~80% and 100% in **mice**. Within 30 min after ingestion, **people** have complained of abdominal pain, dizziness, profuse vomiting, numbness of extremities, back pain, spasms of the hands and face, carpopedal spasms, dyspnea, coma, and death. A Filipino 4-year-old **female** developed pulmonary paragonimiasis and hemoptysis, apparently, from eating raw crab meat.

CRANBERRY

= *Vaccinium oxycoccus*

Untoward effects: Contains ***benzoic acid*** and ***oxalates*** and, in quantity, it acidifies the urine.

CRATAEVA BENTHMII

= *Tamara*

Untoward effects: Fruit or bark in Peru is poisonous for **man**. In Brazil, it is employed as a tonic.

CRAYONS

Untoward effects: Those bearing the C.P or A.P. designation are, supposedly, non-toxic. In the U.S., **children's** wax and chalk crayons are normally required to be non-toxic. Red and orange wax crayons used to be harmful, and may still be, due to ***aniline*** dye or ***chromate*** colors. Methemglobinemia followed accidental ingestion of para red, ***p-nitroaniline*** (q.v.), etc., especially in **children.**

CREAM

Untoward effects: Has triggered migraine headaches. Avoid large quantities (contains some ***tyramines***) with ***monoamine oxidase inhibitors***, which can produce hypertension and severe headaches. Often contains ***phosphate*** stabilizers, and whipping creams often contain monoglycerides and ***algin***. Contact urticaria reported, probably, due to a ***polysorbate*** additive.

CREATINE

Untoward effects: Occasional reports of vomiting, diarrhea, fatigue, migraine, rash, dyspnea, nervousness, anxiety, increased blood urea nitrogen, myopathy, polymyositis, atrial fibrillation, and seizures. Renal failure reported in a **patient**. The chief ingredient in ***Krebiozen***, a 1950 "cancer cure".

CREMOPHOR EL

= *Polyethoxylated Castor Oil* = *Polyoxyethylated Castor Oil*

Untoward effects: As a non-ionic surfactant or solvent for many IV preparations, it can cause its own side-effects, such as fat embolism, severe bronchospasm, anaphylaxis (sweating, pallor, nausea, and flushing).

A histamine-releaser causing allergic reactions in the **cat** and **dog**, but not in the **rabbit**, **pig**, or **man**.

CRENOTHRIX POLYSPORA

Untoward effects: This "***iron*** bacterium" has a thick, gelatinous capsule, usually heavily encrusted with a deposit of red ***ferric oxide***. Outbreaks in drinking water have occurred and it is not poisonous. In England, it caused a school swimming pool to become blood red.

CREOLIN

***Antiseptic, disinfectant, parasiticide, oral antiferment.* The safest one for internal use is *Pearson's Creolin = VCP*.**

Untoward effects: ***Pearson's Creolin*** contains ~20% less toxic, high-boiling homologs of ***phenol***. Avoid unnecessary use on young **animals** or those prone to lick excessive amounts off their bodies, such as **cats**. ***Milk*** may pick up the odor from excessive use. A creolin-induced stomatitis has occurred in **cows** from gelatin capsules that break or separate in their mouths.

CREOSOTE

***Wood preservative, antiseptic, expectorant.* From *coal tar*. *Beechwood Creosote = Wood Creosote* is the only one for internal use.**

Untoward effects: Because of potential **human** health risks, the U.S. Environmental Protection Agency now limits its use to killing **gypsy moth** egg masses and the preservation of wood (~80 million gallons annually in the U.S.). Used as an expectorant, it colors the urine a dark green. **Workers** in the creosote industry have higher-than-normal incidence of some cancers. Cancers of the lip reported in **fishermen** after repairing **their** nets with tarred twine and holding the bone or wooden needle between **their** lips. Gangrene of the fingers has developed in a railway **worker** who handled creosoted material without gloves. Squamous papilloma, as well as warts on hands, forearms, nose, and thighs of a **worker** handling creosoted lumber. An acute toxic polioencephalitis with severe neurological disturbances in 48-year-old **male** painting fence posts in a poorly ventilated garage for 1 h. Symptom free after 2 weeks. Readily absorbed through the gastrointestinal tract and skin. Salivation, headache, vomiting, dyspnea, vertigo, decreased pupillary reflexes occur in systemic poisoning. Irritation of skin, eyes, and mucous membranes manifested by burning and itching sensations, keratoconjunctivitis, erythema, and papular and vesicular skin lesions from contact. ***Milk*** may pick up the odor.

LD_{LO}, **human**: 50 mg.

LD, **sheep**: 4–6 g/kg; **calves**: 4 g/kg; LD_{LO}, **cat**, **dog**, and **rabbit**: 600 mg/kg; LD_{50}, **rat**: 725 mg/kg.

m-CRESIDINE

Dye intermediate.

Untoward effects: Causes bladder cancer in **rats**.

CRESOL

= *Cresylic Acid* = *Cresylol* = *Tricresol*

Antiseptic, disinfectant, and preservative.

Untoward effects: Poisonous. Ill-effects on skin, liver, kidney, and pancreas. Cresols can cause allergic hypersensitivity, abdominal pain, nausea, vomiting, gastric hemorrhage, diarrhea, esophageal strictures, shock, and respiratory decrease, if large amounts are ingested. Reported to induce corneal opacities. Absorption by rubber, plastics, and fabrics may cause burns when these later come in contact with skin. TWA, 2.3 ppm (10 mg/m^3); for skin, 5 ppm (22 mg/m^3). Odor threshold appears to be 1 ppm. Will color urine brown-black.

TD_{LO}, **human**: 50 mg/kg.

LD_{50}, **rat**: 1.5 g/kg; **mouse**: 861 mg/kg.

m-Cresol is incompatible with ***chlorpromazine***. Cleft palates in **rats** after mothers received 150 mg/kg/day during their pregnancy. Upper limb defects at 300 mg/kg level for *m* and *p* forms.

LD_{50}, **rat**: 242 mg/kg; **mouse**: 828 mg/kg.

p-Cresol* = *4-Hydroxytoluene* = *1-Methyl-4-hydroxybenzene has caused hypopigmentation of the skin.

LD_{50}, **rat**: 207 mg to 1.8 g/kg; **mouse**: 344 mg/kg; acute dermal LD_{50}, **rabbits**: 3.6 g/kg.

CRESOL, SAPONATED SOLUTION of

= *Lysol®* = *Odorit* = *Tekresol*

Untoward effects: Has caused hemolytic anemia, methemoglobinemia, erythrocytes' Heinz bodies, drowsiness, confusion, respiratory decrease, and collapse. In the 40s and 50s, its use as a vaginal douche was strongly linked to an increase in cervical cancers of **women**. A 9-month-old **female** developed a methemoglobin level of 56% from inhalation of its fumes and was successfully treated with IV ***methylene blue***. 3 days after ingestion, a **person** developed massive intravascular hemolysis, due to a direct oxidant action on red blood cells by its metabolite, ***hydroquinone***.

CRESS

= *Coronopus sp.*

Untoward effects: *C. didymus* = ***Bitter Cress*** = ***Land Cress*** when eaten by **dairy cows**, gives their ***milk*** a distinctive off-flavor, due to its ***benzyl isothiocyanate*** and ***benzyl thiocyanate*** content.

m-CRESYL ACETATE

= *Cresatin* = *Kresatin* = *Metacresol Acetate* = *Metacresylacetate*

Antifungal, antiseptic.

Untoward effects: LD_{LO}, **human**: 50 mg/kg.

LD_{50}, **rat**: 1.9 g/kg; skin, **rabbit**: 2.1 g/kg.

p-CRESYL ACETATE

= *p-Tolyl Acetate*

Use: In fragrances for soaps, detergents, cosmetic creams and lotions, perfumes, and in foods.

Untoward effects: No irritation or sensitization noted in **human** volunteers.

LD_{50}, **rats**: 1.9 g/kg; acute dermal LD_{50}, **rabbits**: 2.1 g/kg.

CRIMIDINE

= *Castrix*

Rodenticide.

Untoward effects: LD_{LO}, **human**: 5 mg/kg.

Violent convulsions, salivation, restlessness, fright, and muscular trembling noted in poisoned **dogs**. Poisoned carcasses are innocuous to scavengers.

LD_{50}, **rats**: 1 mg/kg; female **mice**: 1.2 mg/kg; **hens**: 10 mg/kg; **rabbit**: 5 mg/kg; **guinea pig**: 2.7 mg/kg; **birds**: 5 mg/kg.

CROCUS SATIVUS

= *Saffron Crocus*

Not to be confused with common garden crocuses or *Colchicum autumnale*, the *Meadow Saffron*. *Saffron*, the dried stigmas and styles, from this yellow crocus is used as a food dye. The name comes from the Arabic word Za'faran or Z'afarrân, meaning yellow. During the 14th century, after introduction into England, the term was anglicized to *saffron*. Because of its scarcity, the Saffron War of 1374 was fought and the blossoms became part of Basle, Switzerland's coat of arms.

CROMOLYN SODIUM

= *Aarane* = *Alercrom* = *Alerion* = *Allergocrom* = *Colimune* = *Cromovet* = *Disodium Cromoglycate* = *DSCG* = *Fivent* = *FPL-670* = *Gastrocrom* = *Gastrofrenal* = *Inostral* = *Intal* = *Introl* = *Irtan* = *Lomudal* = *Lomupren* = *Lomusol* = *Lomuspray* = *Nalcrom* = *Nalcron* = *Nasalcrom* = *Nasmil* = *Opticrom* = *Opticron* = *Rynacrom* = *Sodium Cromoglycate* = *Sofro* = *Vicrom* = *Vividrin*

***Antiasthmatic, antiallergic.* In prevention, not treatment.**

Untoward effects: Despite different routes of administration, adverse reactions, especially after repeated treatments, involve primarily the respiratory tract. Bronchospasm and constriction, cough, dryness or irritation of the throat, nasal congestion, pharyngeal irritation, wheezing, tightness of chest, exacerbation of asthma, eosinopenia, eosinophilia, and, rarely, laryngeal edema occur. Less frequent are angioedema, dizziness, sneezing, esophagitis, dysuria, joint swelling and pain, lacrimation, nausea, headache, maculopapular rash, urticaria, hypotension, swollen parotid glands, and anaphylaxis. Rare and suspect reactions include anemia, exfoliative dermatitis, hemoptysis, hoarseness, myalgia, myositis, nephrosis, neuritis, pericarditis, photodermatitis, granulomatous hepatitis, pulmonary infiltrate with eosinophilia, vasculitis, and vertigo. Capsules may help produce a pharyngitis from local irritant effects. ***Intal***® capsules have contained ***tartrazine*** (q.v.). Elimination half-life is 81 min. Cromolyn eyedrops have caused transient stinging sensations and, rarely, itching, redness, and chemosis.

LD_{LO}, **human**: 34 mg/kg/4 weeks.

LD_{50}, **rats** and **mice**: > 8 g/kg.

CROPROPAMIDE

Analgesic.

Untoward effects: Flushing, paresthesia, restlessness, and dyspnea reported.

CROTACTIN

Untoward effects: A neurotoxin in ***crotoxin*** from the venom of *Crotalus durissus terrificus*, a South American **rattlesnake**. See ***Snakes***.

CROTALARIA

Untoward effects: In many countries, most species contain ***pyrrolizidine alkaloids*** and are hepatotoxic for **man**, **animals**, and **poultry**. There are ~600 species, most of which are in Africa. The FDA has seized many contaminated carloads with 1.5–351 seeds/lb. They haven't seized grains with less than 1 seed/lb.

C. aegyptia contains ***monocrotaline*** and ***crosemperine***.

C. albida (***C. montana***) contains ***croalbidine***.

C. angulata, ***C. mucronata***, ***C. purshii***, ***C. sagittalis***, and ***C. spectabilis*** are toxic species in Alabama.

C.aridicola in Queensland was the cause of "Chillagoe **horse** disease". Affected ***horses*** cannot swallow and become emaciated. They have esophageal mucosa erosions, and, occasionally, some erosions in the stomach.
C. burkeana in South Africa caused "stiff sickness" ("stywesiekte") in **cattle**. Laminitis is noted in acute poisonings and overgrown and deformed hooves with a stiff, staggering gait is noted in chronic cases. Post-mortems revealed a gastroenteritis with hemorrhagic mucosa and parenchymatous visceral degeneration.
C. dissitifolia in Australia has caused mortality in **horses**.

In South Africa, ***C. dura*** and ***C. globifera*** cause "jaagsiekte" in **horses**, due to its ***dicrotaline*** content, with its dyspnea, increased temperature, rapid and weak pulse, collapse and death.
C. fulva, the presumptive cause of veno-occlusive disease of **man** in Jamaica causes liver lesions in **rats**.
C. geminiflora in Zimbabwe causes an acute laminitis of all four feet in **cattle**, which requires about 1 year for recovery.
C. goreensis seeds have caused stunting and mortality of **chicks** and **poultry** in northern Queensland.
C. intermedia in Nigeria is poisonous to **cattle**.
C. retusa = ***birana*** or ***chika saura*** (Hausa) = ***Koropo*** (Yoruba) = **Akidìnmo** (Igbo) and ***C. crispata*** are causes of "walk-about" or "Kimberly **horse** disease" in Northern Australia. It causes anorexia, dysentery, and, eventually, liver cirrhosis and death. The stems, leaves, roots, and seeds are toxic and lethal to **pigs** and **poultry**.
C. mucronata has proven to be toxic to **sheep** and ***C. novae-hollandiae*** to **horses** in Australia.
C. sagittalais = ***Arrow Crotalaria*** = ***Rattlebox*** = ***Showy Crotalaria*** is found mostly in southeastern U.S., where it may cause liver cirrhosis in **horses**. All parts of the plant, green or dry, are poisonous to **livestock**, particularly **cattle** and **horses**. Seeds have a higher concentration of the toxic alkaloid, ***monocrotaline***. Symptoms are anorexia, depression, dullness, and weight decrease. Poison effects are slow and death may occur after weeks or months. Some die within 10 days. As early as 1884, it was incriminated as the cause of "Missouri River bottom" disease. **Deer** browsing in a Louisiana wildlife refuge showed signs of hepatoencephalopathy, with apparent blindness, incoordination, icterus, hemoglobinuria, emaciation, anemia, and no fear of **man** a few months after consuming the plant.

In the U.S., ***C. spectabilis*** may kill **poultry** within a few hours. **Rats** die within a week from ***monocrotaline*** toxicity, and, on post-mortem, show hemorrhagic hepatic necrosis, ascites, pulmonary edema, and pleural effusions. In **hogs**, liver necrosis and cirrhosis, liver hypertrophy, ascites, esophageal and gastric ulcers, interstitial kidney fibrosis, glomerular sclerosis, and tubular necrosis occurs. Hemorrhagic necrosis of the liver is produced in **turkeys**. Its seeds at 0.25–1.0% in the feed of **rhesus monkeys** caused hemorrhagic hepatic necrosis, pulmonary edema, and leukocytic infiltrate of the pulmonary arteries. Acute toxicity in **horses** caused gastric irritation, anorexia, tenesmus, and bloody feces; chronic toxicity caused emaciation and depression. 1% level to **guinea pigs** was without ill-effects, but 5% level caused problems, but was less toxic to **pigs**, **sheep**, and **cattle**. In the U.S., ***C. striata*** is also very toxic to young **chickens** and **rats**.
C. trifoliastrum has caused symptoms and lesions similar to those of "Chillagoe **horse** disease".

CROTAMITON
= *Crotamitex* = *Eurax* = *Euraxil* = *Veteusan*

Scabicide.

Untoward effects: Allergic contact dermatitis in some **patients**. Severe itching and burning of ear and adjacent skin in 24-year-old **male** 96 h after application. Patch-testing confirmed delayed reaction and none to ointment base. Similar reactions all over body of 14-year-old **female** 5 days after treatment, and in a 35-year-old **male** 10 days after treatment. Conjunctivitis has been reported. Not to be applied to acutely inflamed or raw surfaces. Avoid use by the eye or mouth. Do NOT use to cover extensive areas on **infants**. If ingested, it will cause a burning sensation in the mouth, abdominal pain, and vomiting.

LD_{50}, **rats**: 1.5 g/kg; **mice**: 1.6 g/kg.

CROTETHAMIDE
= *Crotamide*

Analgesic.

Untoward effects: Suspect as the cause of flushing, paresthesias, restlessness, and dyspnea, yet, these may have been due to a combination with ***cropropamide***.

CROTON
= *Codiaeum variegatum*

Untoward effects: Many species exist and ***Croton tiglium*** is the commercial source of croton oil. The oil is a gastrointestinal irritant and a drastic cathartic, causing nausea and vomiting, headache, dizziness, general prostration, weak pulse, cold clammy skin, decreased temperature, decreased blood pressure, collapse and death in excessive dosage. Topical application has caused skin blistering. Application, even a year after a non-tumorigenic dose of an initiator, such as ***benzopyrene***, can lead to a high incidence of tumors in **mice**. Contains toxalbumins. Chronic administration of ***C. penduliflorus*** to **mice** causes abortions in late pregnancy and 100% fetal mortality, in addition to purgation. The oil is irritant, caused tachypnea, decreased motor activity, and some **mice** convulsed before dying. Its toxic agent is ***5-deoxyingenol***. ***C. texensis*** = ***Texas Croton***. Its oil or milky juice is a strong irritant and blistering agent on **man**. Pustulation can lead to sloughing and even permanent scarring. Contact with eyes can cause conjunctival and corneal damage. Fatal to **dogs** after ingesting 10 drops of the oil. **Cattle** have been poisoned by eating it in hay. ***Honey*** from its nectar has poisoned young **bees** and **humans**. It has been used experimentally to induce granulomas in **rats**.

LD_{50}, **mice**: 543 mg/kg; **frogs**: 49 mg/kg.

Oil, LD_{LO}, **human**: 5 mg/kg.

CRUSTACEANS

Untoward effects: A common cause of food-induced allergies in **people**. See ***Crabs***, ***Lobsters***, and ***Shrimp***.

CRYPTENAMINE TANNATE

= *Unitensin Tannate*

Antihypertensive.

Untoward effects: Nausea, vomiting, and apnea reported.

CRYPTOSTEGIA GRANDIFLORA

= *Madagascar Rubber Vine* = *Rubber Vine*

Untoward effects: Has caused deaths in **humans** in India.

Highly toxic to **cattle**, **horses**, **goats**, and **pigs**. Grows profusely along Queensland's coastal waters. Usual annual mortality in **cattle** reported as 5/800 and up to 30/1,000 when it is eaten after a previous-year burning. Some **cattle** drop dead after a short gallop or trot, with little or no struggling. Cyanosis and congestion, particularly, of cerebral and myocardial blood vessels noted. Experimentally fed **horses** had profuse sweating, dyspnea, muscle twitching, ataxia, and abdominal pain, with epicardial hemorrhage noted on post-mortem. Evidence of toxic cardiac glycoside action in **guinea pigs**.

Leaves, LD, **cattle** and **sheep**: < 1g/kg.

CRYPTOSTEMMA sp.

= *Capeweed*

Untoward effects: After ingestion, its high ***nitrate*** content killed ~500/5000 **sheep**.

CRYPTOSTROMA CORTICALE

Untoward effects: A mold on ***maple*** tree bark that has caused ***maple*** **barkstripper's** lung in **people** and **woodworkers**.

CUBE ROOT

Untoward effects: Due to its ***rotenone*** (q.v.) content. Low order of **mammalian** toxicity. Excitement and convulsions are easily produced in **fish**, **frogs**, **guinea pigs**, and **rats**. In Peru, for paralyzing **fish** for **human** food.

CUCUMBERS

Untoward effects: Allergic reactions are often familial. Often gives positive delayed eczematous patch-test results. Contains moderate amounts of ***oxalates***. Molds within brined cucumbers, not the brine, cause softening of the cucumbers. A small % of cucumbers sampled harbored *Listeria monocytogenes*, a known cause of illness and death in **humans**. Particularly vulnerable are **fetuses**, **newborns**, **people** with depressed immune systems, and some **elderly**.

In South Africa, wild cucumbers contain highly toxic ***cucurbitacins***, and they are poisonous to **sheep**.

CUMENE

= *Cumol* = *Isopropyl Benzene*

Use: In making ***phenol***, ***acetone***, and other chemicals. Annual demand is for ~5 billion lbs.

Untoward effects: Eye and mucous membrane irritation, headaches, dermatitis, narcosis, and coma reported. The narcotic effects develop slowly, but are of long duration. Coma can occur. Acute toxicity is > ***benzene*** (q.v.) or ***toluene*** (q.v.). 50 ppm, TWA, 8 h.

Inhalation TC_{LO}, **human**: 200 ppm.

LD_{50}, **rats**: 1.4–2.91 g/kg.

CUMIN, BLACK

= *Awassida* = *Carum nigrum* = *Cuminum cyminum* = *Cyminii* = *Habba Sûdâ* = *Habbat al-barakah*

Condiment, carminative.

Untoward effects: Phototoxicity of oil. No irritation or sensitization reported on **human** volunteers.

Aqueous extracts of the seeds were hepatotoxic to **rats**. Blood ***glucose*** levels decrease in normal or ***alloxan***-diabetic **rabbits**.

Oil, LD_{50}, **rats**: 2.5 ml/kg; acute dermal LD_{50}, **rabbits**: 3.6 ml/kg.

CUMINALDEHYDE

= *Cuminic Aldehyde*

Use: In fragrances for soaps, detergents, cosmetic creams and lotions, perfumes, and foods. Found in at least 50 essential oils.

Untoward effects: No irritation, sensitization, or phototoxicity reported in **humans**.

LD_{50}, **rats**: 1.9 g/kg; acute dermal LD_{50}, **rabbits**: 2.8 g/kg.

CUPFERON

***Reagent and industrial chemical.* Used for separating *tin* from *zinc* and *copper* and *iron* from other metals. ~25 tons produced annually.**

Untoward effects: **Human** exposure is primarily through ingestion or inhalation of the dry dust. Skin exposure is also likely. Chief exposures are to manufacturing **workers** (~4,000), and **those** in analytical or research work. When heated, it emits toxic fumes of ***ammonia*** and ***nitrogen oxides***.

In **rats** of both sexes, it causes hemangiosarcomas, hepatocellular carcinomas, and squamous cell carcinomas of the forestomach; carcinomas of the auditory sebaceous gland in **females**. In male **mice**, causes hemangiosarcomas; in females, hepatocellular carcinomas, carcinomas of the auditory sebaceous gland, hemangiomas, hemangiosarcomas and adenomas of the Harderian gland.

TD_{LO}, **rat**: 250 mg/kg.

CUPRIC ACETOARSENITE

= *C.I. Pigment Green 21* = *C.I. 77410* = *Copper Acetate Arsenite* = *Emerald Green* = *French Green* = *Imperial Green* = *Mineral Green* = *Mitis Green* = *Paris Green* = *Parrot Green* = *Schweinfurt Green* = *Vienna Green*

Untoward effects: Dermatitis and a bluish discoloration of the skin, resembling argyria, in a ***cotton***-field **worker**, due to inhalation of it. Gastrointestinal upsets, muscle cramps, tremors, and peripheral neuritis, primarily, from the 44% ***arsenic*** content. Vomiting (***garlic*** odor), abdominal pain, restlessness, dysphagia, diarrhea (often bloody), weak pulse, cyanosis, shock, and collapse in poisoned **dogs**.

LC_{50} in feed, **Bobwhite quail**: 480 ppm; **pheasants**: 1,043 ppm; **mallard ducks**: > 5,000 ppm; LD_{50}, female **rats**: 100 mg/kg; **rabbits**: < 100 mg/kg.

CUPRIZONE

Copper chelating agent.

Untoward effects: Neurotoxic.

At 0.5% level in feed for 150 days, causes toxicosis, spongiform encephalopathy with extracellular vacuolation, gliosis, and astrocyte hypertrophy in **mice**. **Syrian hamsters** fed a 3% level in their feed for 20–70 days developed widespread lesions in the brain, consisting of vacuolation and astrocyte hypertrophy. Lesions in both species are similar to those scrapie-induced and enlarged glial mitochondria resembling those of Alzheimer's disease in **man**. Produces giant mitochondria in hepatocytes of **mice** and is a strong inhibitor of **bovine** hepatic mitochondria.

CURARE

= *Ourari* = *Urari* = *Woorari*

***Skeletal muscle relaxant, parasympatholytic.* Contains *tubocurarine* (q.v.) and over 60 other alkaloids. Unitage usually standardized by rabbit "head drop" test. Extracted from *Chondodendum tomentosum*. Tube curare, bamboo curare, pot curare, and gourd curare are curare, and only refer to the type of container it is in, although, occasionally, abstracted from other plants.**

Untoward effects: Used in arrow poisons, by the **Indians** of the Amazon and Orinoco Valleys, and in the Guianas. Action is due primarily to its ***tubocurarine*** (q.v.) content. Blocks nerve conduction with resulting paralysis, by competing with ***acetylcholine*** for the nerve ending receptors (muscle endplate). Acts rapidly, even after many years of storage in the dry state. Urticaria and angioedema have occurred, due to hypersensitivity to the drug. Respiratory failure and generalized muscle weakness in 53-year-old **male** sensitive to it. Crosses the placenta poorly, yet, prolonged use or high **maternal** concentrations have caused paralysis *in utero* with position deformities. Fatal apnea when used with some ***antibiotics*** (***bacitracin, colistin, gramicidin, kanamycin, neomycin***, and ***streptomycin***). ***Thiazide diuretics*** and ***monoamine oxidase inhibitors***, by decreasing ***potassium*** levels, enhance the muscle paralyzing effects of curare. Dosage of curare requirement was two to four times anticipated requirement in **patients** receiving ***azathioprine***, alone or with ***guanethidine***. Anticholinesterases, such as ***neostigmine*** and ***edrophonium***, antagonize curare's action. ***Succinylcholine*** and ***decamethonium*** prolong curare's muscle relaxation. Use with ***quinidine*** can decrease neuromuscular transmission for up to 24 h and lead to apnea and bizarre cardiac arrhythmias. ***Propranolol*** and ***diazepam*** may also have prolonged postoperative curarization. Use caution with other ***β-blockers***. A **physician** was accused of murdering **patients** with lethal doses of curare, to discredit other **surgeons**. A **patient** survived an amazing rise in temperature from 97 to 103.5 °F within a few minutes after receiving it as the sole relaxing agent. Elimination half-life in **man** is 76–372 min. Slow IV is essential. Avoid use with ***ether*** or ***atropine*** in **humans** and **animals**, and in **those** recently exposed to ***organic phosphorus*** compounds, ***phenothiazine***, or ***phenothiazine tranquilizers***. Vigorous manipulation or stimulation of the pharynx while under curarization may precipitate cardiac arrest. Death is due to paralysis of respiratory muscles. Partial paralysis, head drop, mydriasis, frontal headache, dizziness, euphoria, blurred or double vision, salivation, decreased blood pressure, ***histamine*** release, and occasionally glycosuria in a conscious **patient** are characteristic responses. Oral doses are relatively non-toxic.

LD_{50}, **mice**: 140 µg/kg.

CURVULARIA GENICULATA

***Untoward effects*:** A fungus on Bermuda grass that is hepatotoxic to **cattle** and, possibly, **man**. A cause of chromomycetomas of the skin and SC tissues of **dog**, **cat**, and **horse** limbs, and nasal mucosa of **cattle**. The ulcerated, suppurative, fibrotic granulomas contain small, irregular black granules.

CUSCUTA sp.

= *Dodder*

***Untoward effects*:** A parasitic weed suspect in causing gastrointestinal upsets and colic in Ohio **horses** and scours in Connecticut **cattle**. Poisoning occurs when level in hay is > 50%.

C. breviflora in Russia has, after 12–18 months, caused up to 50% loss of weight and colic in **horses**.

C. campestris has caused poisoning in **cattle**, **horses**, and **rabbits**, characterized by extreme nervousness, anorexia, staggering, wobbling, circling, salivation, increased peristalsis, followed by atony, diarrhea, and vomiting (in species that can). Endocardial and interstitial hemorrhagic lesions noted.

C. europaea in Armenia has caused similar problems in **horses**.

C. reflexa* = *Akashilota* = *Amarbel causes hypotension and bradycardia in **rats** in India.

CUSSONIA CORBISIEN

***Untoward effects*:** In herbal preparations, causing shock, hypotension, acidosis, dehydration, paralytic ileus, perforation of the intestines, and severe penile burns.

CUTTING OILS

Untoward effects*:** Folliculitis, pustule formation, comedones, keratoses, and acne are the most common untoward effects on the skin, attributed to ***chlorine and ***sulfur*** compounds in them, followed by new growths and, rarely, eczemas. Malignant scrotal cancer has followed regular saturation of clothing. Some dermatitis may be due to various germicidal agents used in them (***carbamates, p-chloro-m-cresol, cresols, dichlorophen, mercaptobenzothiazole, σ-phenylphenol, tris-nitro, etc.***). Cutting oil mists can be potentially toxic to **workers**. Inhalation can cause mucous membrane irritation. Metal shavings in the oils can cause skin wounds.

N-Nitrosodiethanolamine has been found at 0.02–3% levels in some synthetic cutting oils and is carcinogenic to livers in **rats**. Causes a high incidence of fatty infiltration of **duck** livers after ingestion.

CYANAMIDE

= *Alzogur* = *Amidocyanogen* = *Carbimide* = *Carbodiimide* = *Hydrogen Cyanamide*

The term "cyanamide" is also used for *calcium cyanamide*.

Untoward effects*:** Irritating and caustic, especially on moist skin. Inhalation causes irritated mucous membranes. Transitory redness of face, headache, vertigo, tachycardia, tachypnea, and hypotension can follow inhalation or ingestion. Can cause a ***disulfiram-like reaction with ***alcohol*** ingestion.

LD, **pig**: 200 mg/kg; LD_{50}, **rats**: 125 mg/kg.

CYANATES

***Potassium* and *sodium cyanate* are used. Genetically, the sickle-cell trait helps protect against malaria and is common in Africa. Yet, there is very little sickle-cell anemia in Africa, probably, due to African diets containing at least 1 g of cyanates/day from *yam*, *sorghum*, *millet*, and *cassava*. The average U.S. diet contains 25 mg/day.**

Untoward effects*:** ***K, IP LD_{50}, **mice**: 320 mg/kg; ***Na***, **mice**: 360 mg/kg.

Has caused development of cataracts in **guinea pigs**.

CYANAZINE

= *Bladex* = *DW-3418* = *Fortrol* = *Payze* = *SD-15418* = *WL-19805*

***Untoward effects*:** DuPont decided to phase out its U.S. sales of this triazine, effective 12/31/99, since it is suspect as a **human** carcinogen. Possible risk to pregnant **women**.

Poisoned **birds** show a goose-stepping ataxia, ruffled feathers, wing tremors and crossed high over their backs. Despite the high LD_{50} for **ducks**, they showed clinical symptoms at 150 mg/kg. Particularly poisonous to **cattle**.

LD_{50}, **rats**: 149 mg/kg; **mouse**: 380 mg/kg; **quail**: 445 mg/kg; **mallard ducks**: > 2.4 mg/kg; **rabbit**: 141 mg/kg.

CYANIDES

Use: ***Hydrogen cyanide*** and inorganic cyanides are used as pesticides, fumigants, rodenticides, in electroplating, in hardening steel, and in extracting ***gold*** and ***silver*** from ores.

Untoward effects: Powerful poisons that prevent body tissues' utilization of oxygen by enzyme (cytochrome oxidase) inhibition, preventing transfer of oxygen from blood to tissues. Oxygen remains bound in the blood and venous blood becomes the same bright red color as arterial blood. Asphyxiation and failure of all body functions follow. Cyanides may be inhaled, ingested, or absorbed through intact skin. **Victims** may collapse within seconds after large doses, with tachypnea, followed by irregular breathing and gasping. Convulsions and death may follow. Dizziness, weakness, headache, bitter almond odor on breath, anxiety, confusion, ataxia, giddiness, burning taste, lower jaw stiffness, and feeling of constriction or numbness in throat can follow low-level exposures; salivation, nausea, and vomiting may occur in ingestion cases. If the **patient** survives low exposures, breathing becomes normal and consciousness returns. Full recovery may take several days. The cyanide in the blood, especially during treatment, is converted to the less toxic thiocyanate. Low-dose chronic exposure permits the body to continuously convert absorbed cyanide to thiocyanate. **Patients** with moderate levels of toxicity report hypernoia, dyspnea, tachypnea – then slowed short inspiration and prolonged expiration and tachycardia. High exposure levels cause loss of consciousness, lactic acidosis, cardiac arrhythmias, hypotension, opisthotonus, convulsions, involuntary micturation and defecation, sweating, paralysis, mydriasis, no eye reflexes, bloody foam around the mouth, bradycardia, and respiratory arrest precede death that occurs within 2–10 min of exposure. Fatal cases with intermediate exposures occur in about 4 h. Toxic serum concentrations are usually between 0.1–1.0 mg/l of blood. Acute exposures are more common and **some** that survive severe non-fatal poisoning may suffer from permanent damage (encephalopathy, frank goiter, and ultrastructural changes in heart muscle). A 23-year-old **male** medical **student** saw his **puppy** collapse suddenly and gave cardio pulmonary resuscitation and mouth to nose ventilation. Within a few minutes, the **dog** died and the **student** became nauseous, vomited, and lost consciousness. An alert emergency room **doctor** smelled an odor of bitter almond and treated the **patient** for cyanide poisoning, apparently from inhalation exposure trying to save his **puppy**. Cases of self-poisoning have been common. Bradycardia, absence of cyanosis, pink mucous membranes or even skin in an unconscious **patient** suggest cyanide poisoning, while waiting for confirmatory tests. As severe hypoxia sets in, the **victim** turns cyanotic. Murder convictions in 1983 stemmed from the death of an **employee** who inhaled cyanide fumes in a plant recovering ***silver*** from used X-ray film. Cyanide was the apparent cause of death in ~900 **people** in a mass suicide in Guyana. In 1991, three **people** had acute cyanide poisoning after taking contaminated ***Sudafed***® capsules. Two died. During 1982–1986, 12 **people** died of cyanide poisoning from criminal product tampering, murder, or suicide. Cyanide exposure also occurred in Tokyo subways in 1995. Nearly 1,500 **people** were asphyxiated near a lake in the Cameroons from a volcanic discharge of ***hydrogen cyanide*** and a tidal wave. About 20–50% of the general population lack the ability to taste or smell cyanide. Some **pathologists** can determine its presence by their sense of smell. Use of ***sodium nitroprusside*** by **cardiologists** has a potential for cyanide to accumulate in the body. Metal **workers** using cyanides may develop a diffuse dark blue pigmentation of their nails. Silver polish containing cyanide can accidentally contaminate foods. Fatal methemoglobinemia has happened from overdosing with the antidotes ***sodium nitrite*** and ***sodium thiosulfate***, especially in **children**. Of 75 illicit ***phencyclidine*** samples in Virginia, about 33% contained a ***nitrile*** contaminant that generated ***hydrogen cyanide*** while being smoked. Visual deterioration (***tobacco*** amblyopia) may occur when **smokers** fail to detoxify cyanide in their cigarette smoke. Cyanide toxicity can occur from domestic fires or Laetriles. Cyanide content of about 150 species of cyanogenic plants varies with climate, season, stage of growth, rainfall, and fertilization. ***Apple*** (q.v.) seeds, ***apricot*** (q.v.), ***choke cherry*** (q.v.), and ***peach*** (q.v.) pits contain cyanide, as do ***cassava*** (q.v.) and bitter ***almonds***. In 1989, two ***grapes*** from Chile were found to have been injected with cyanide. ~800 plants, such as ***arrowgrass***, ***cassava*** (***manioc***), ***elderberry***, ***hydrangea***, ***jackbeans***, ***Johnson grass***, ***kidney beans***, ***Kiliambiti*** (*Adenia*), ***lablab bean***, ***Lathyrus*** and ***Vicia peas***, ***lima beans***, ***linseed***, ***maize***, ***mesquite***, ***millet***, ***milo***, ***peanuts***, ***precatory bean***, ***sugar cane***, ***sorghum***, ***Sudan grass***, ***sweet potatoes***, ***velvet grass***, ***yam***, etc. are sources of cyanides. **Maggots** can thrive on the flesh of cyanide-poisoned **victims**. Some describe the breath of poisoned **victims** as that of silver polish; most describe it as ***almond***-like. TWA, 5 mg/m^3, 8 h. Minimum lethal dose, **human**: ~0.5 mg/kg, others say 50 mg/**patient**.

A massive cyanide spill over a dam at a Romanian ***gold*** mine led to a tremendous **fish** kill in Romania, Hungary, and Yugoslavia, due to water levels as high as 0.13 mg/l. **Farmers** near earlier such spills reported their **livestock** died or became blind from the exposure.

Cyanide poisoning has been reported in **pigeons**, **wildlife** (especially from leachates of ***gold*** and ***silver*** extraction from ores), **owls**, **eagles**, **ravens**, **vultures**, **wild turkeys**, **cats**, **dogs**, **rabbits**, **rats**, **mice**, **sheep**, **fish**, **chickens**, **geese**, **coyotes**, and **cattle** (mostly from young, frosted, or drought-raised plants, such as ***arrow-grass, corn, Johnson Grass, sorghum, Sudan grass, wild cherry,*** and pesticides. Cyanide is liberated from ***amygdalin***, present in ***apricot, cherry, peach,*** and ***plum*** pits, as well as the seeds of bitter ***almonds***. **Cows** have been poisoned by eating frost-damaged ***celery*** tops. Has caused polioencephalomalacia in **dogs**; excitement, muscle tremors, dyspnea, mydriasis, increased salivation, lacrimation, depression, convulsions, and death in **cats**. **Lambs** born of **ewes** ingesting cyanogenic plants often have goiters as a result of liver generated thiocyanate impairing thyroid function. This can be offset by increasing ***iodine*** intake to the **ewes**. The toxic dose of cyanide for most **animals** is ~2 mg/kg.

Some of the plants which, under certain conditions, can contain dangerous levels are: *Acacia sp., Adenia digitata, Ageratum sp., Andrachne decaisnei, Anthemis sp., Aquilegia vulgaris, Bahia oppositifolia, Brachyachne sp., Bridelia ovata, Calotis scapigera, Canthium vaccinifolium, Cassia sp., Cercocarpus sp., Chenopodium sp., Chloris distichophylla, Cotoneaster, Crescentia sp., Cynodon sp., Dactyloctenium radulans, Digitaria sanguinalis, Dimorphotheca sp., Dolichos sp., Eleusine indica, Eremophila maculata, Eriobotrya japonica, Eschscholtzia sp., Eucalyptus sp., Euphorpia sp., Florestina tripteris, Glyceria striata, Goodia lotifolia, Grevillea sp., Gyrostemon ramulosus, Hakea sp., Haloragis sp., Heterodendron oleifolium, Holcus sp., Hydrangea sp., Indigofera australis, Lambertia sp., Linum sp., Lotonis laxa, Lotus sp., Loudonia roei, Macadamia ternifolia, Malu sp., Manihot esculenta, Nerium oleander, Panicum sp., Passiflora sp., Phaseolus lunatus, Phyllanthus gasstroemii, Poranthera microphylla, Prunus sp., Pyrus sp., Rhodotypos sp., Sambucus nigra, Sorghum sp., Stillingia treculeana, Suckleya suckleyana, Trifolium repens, Triglochin sp., Trigonella sp., Vicia sativa, Ximenia sp., Xylomelum angustifolium, Zea mays,* and *Zieria sp.*

CYANOACRYLATES

= *Eastman 910* = *Histacryl* = *Instant Magic* = *Krazy Glue* = *Miracle Glue* = *Nail Glue* = *Nexaband* = *Oneida Instant Weld* = *Permabond 102* = *Rapid Set* = *Superbonder* = *Super Glue* = *Super Three Cement* = *Zip Bond* = *Zip Grip*

Alkyl-, butyl-, ethyl-, heptyl-, isobutyl-, and methyl-forms are in common use.

Untoward effects*:** May cause weeping and double vision while the polymer is clearing from the eyeball. Caution must be used because of their rapid, strong adhesive properties. **Users'** fingers can easily be glued together so strongly that, if one tries to pull them apart, skin will be ripped. Fingers should be soaked in warm, soapy water for several minutes, before peeling or rolling them apart. An ***acetone-like liquid is sold for removing the hardened glue and ***rubbing alcohol*** works in some cases. Do NOT use either of the last two near the eyes. Will adhere to stainless steel sutures, making their removal difficult. Will permanently stain clothing. Nail adhesives have inadvertently been used as eye drops in six **patients** (23–47 years), resulting in abrasions in four and punctate epitheliopathy in two. Treatment was successful. A similar report covers 34 more **patients**.

CYANOGEN

= *Dicyan* = *Ethanedinitrile* = *Oxalic Acid Dinitrile*

Untoward effects*:** Idling car engines roughly emit 0.2–0.7 mg/m^3 of car emissions, and accelerating engines emit 0.9–4.5 mg/m^3 of this poisonous gas with an ***almond-like odor. Toxicity is similar to ***hydrogen cyanide*** (q.v.). An important systemic poison for **man**. 16 ppm/6 min is irritant to the **human** eye.

Inhalation TC_{LO}, **human**: 16 ppm; inhalation LC_{50}, **rat**: 350 ppm/1 h.

CYANOGEN CHLORIDE

***Use*:** In chemical synthesis and as a military poison gas.

Untoward effects*:** Toxicity is described above under ***Cyanogen (q.v.). Produces cachexia on long, continued exposure. Inhaled vapors cause bronchoconstriction and irritation of respiratory tract.

Inhalation TC_{LO}, **human**: 10 mg/m^3.

Inhalation LC_{LO}, **mouse**: 780 mg/m^3/7.5 min; **dog**: 800 mg/m^3/7.5 min; **rabbit**: 3,000 ppm/2 min; SC LD, **rabbit**: 20 mg/kg.

CYCADALES

Woody, fern-like plants (members of *Cycas, Bowenia,* and *Macrozamia* genera).

***Untoward effects*:** Causes toxic reactions (carcinogenic, hepatotoxic, neurotoxic, and teratogenic) in **humans** and **animals**. High incidence of neurotoxic problems in the Caribbean and Australia.

Called "*Zamia* rickets" or "wobbles" in Australian **cattle** that ultimately lose control of their hind limbs. The toxic agent, ***methylazoxymethanol***, is teratogenic. **Golden hamster** offspring had hydrocephalus, microcephalus, cranioschisis, rachischisis, anophthalmia, and microphthalia. Microencephaly occurred in **rats**.

CYCADS

= *Cycas sp.*

Untoward effects*:** The seeds contain an excitotoxic amino acid that may cause neuronal degeneration associated with a high incidence of amyotrophic lateral sclerosis with features of parkinsonism and an Alzheimer-type dementia in Chamorro **people** in Guam. An amino acid, ***β-N-methylamino-L-alanine) in the seed of the food plant, *C. circinalis*, has been implicated as the specific cause of this disease in **man** and a similar neurodegenerative syndrome in non-human **primates**. Hepatotoxic in **man**. **Natives** use the kernel for food and the seed's outer hull as a confection, and have developed a progressive, incurable paralysis. Recognizing that the kernels are toxic, **they** have learned to eliminate the toxicity, by soaking the kernels in water for 7–10 days. It is then dried and powdered into a flour. The husk is about 1/6th as toxic as the kernel. *Aspergillus sp.* and other fungi often contaminate the seeds, husks, and flour, adding to the toxicity.

Seeds have caused liver, kidney, pulmonary, and pancreatic cancers in **rats**, and liver tumors with centrolobular necrosis in **guinea pigs**. Flour made from *C. circinalis* nuts (seeds), when fed at 1–2% level in the ration for 3–4 months, led to portal fibrosis with bile duct proliferation in **cattle** and **horses**; portal and centrolobular fibrosis in **pigs**; hepatic cell degeneration, chronic interstitial nephritis, and peritoneal and pleural effusions in **cattle**, **horses**, **pigs**, and **rats**. Similar lesions were found in fetuses and the young dams fed the flour. **Chickens** eating the flour for 6 months had degenerative liver lesions with hemorrhages and edema. Posterior weakness and ataxia have been reported in **cattle**. *C. media* has caused demyelination of the spinal cord in **goats** and **cattle**, with resulting posterior weakness and ataxia. Cephalic anomalies reported in offspring of **golden hamster** that ate various species of *Cycas*.

CYCASIN

Toxic substance in the seeds, roots, and leaves of *Cycad* (q.v.) plants.

Untoward effects*:** It or its metabolite, ***methylazoxymethanol, is carcinogenic in **rats**, **guinea pigs**, **mice**, **hamsters**, **fish**, and **rhesus**, **cynomolgus**, and **African green monkeys**. The waste water from soaking ***cycads*** to leach out the cycasin content represents a potential source of exposure. Neurological disorders and locomotor difficulties have been produced experimentally in **mice**, **ferrets**, and **rats**. Oral intake leads to extensive biliary proliferation and fatty degeneration of **ducklings'** livers.

LD_{50}, **rats**: 270 mg/kg; **mice**: 500 mg/kg; **guinea pigs**: < 20 mg/kg; **hamsters**: < 250 mg/kg; **rabbits**: ~30 mg/kg.

CYCLAMATES

These sweeteners were banned for food use in the U.S. and Great Britain in 1969, after discovery that they caused cancer in rats. See *Calcium Cyclamate*, *Sodium Cyclamate*, and *Cyclohexylamine*. Causes a marked decrease in *lincomycin* absorption. At least 1–2 h should occur between each dosage.

CYCLAMEN

= *Alpine Violet* = *Persian Violet* = *Sowbread*

Untoward effects*:** Contains toxic triterpinoid saponins, the most prevalent one being ***cyclamin, in its acrid tuberous rhizome, a fungitoxic defense mechanism for the plant. The saponins have a local irritant effect, causing nausea, vomiting, and diarrhea. After absorption, they have caused convulsions and diarrhea. Ingestion is apt to be at a minimum, because cyclamen, especially the rhizomes of the plant, has an objectionable, rancid taste. Its toxicity has been known since the late 1700s. The juice of ***C. hederifolium*** crushed tubers is used as an abortifacient in Italy and can cause a fatal gastroenteritis in **humans**.

CYCLAMEN ALDEHYDE

Use*:** Approximately 75 tons have been used annually in the U.S. as a fragrance in soaps, detergents, cosmetic creams and lotions, perfumes, and foods. Has a ***lily-of-the-valley aroma.

***Untoward effects*:** 3% in petrolatum led to mild irritation in 25 **human** volunteers after a 48 h closed-patch test.

LD_{50}, **rats**: 3.81 g/kg.

CYCLANDELATE

= *BS-572* = *Cyclergine* = *Cyclobral* = *Cyclolyt* = *Cyclomandol* = *Cyclospasmol* = *Natil* = *Novodil* = *Perebral* = *Spasmocyclon*

Vasodilator.

Untoward effects*:** Heartburn, eructation, nausea, flushing, headache, sweating, weakness, and tachycardia are the most common adverse reactions. Other symptoms reported are stuffy nose, unsteadiness, and sensations of giddiness. ***Nicotine in ***tobacco*** may interfere with its effectiveness.

CYCLAZOCINE

Narcotic antagonist.

Untoward effects*:** Headache, blurred vision, hallucinations, agitation, anxiety, euphoria, hyperactivity, drowsiness, dizziness, dry mouth, constipation, dysphoria, increased libido, and insomnia. Circulatory effects similar to ***morphine's (q.v.). High incidence of unpleasant psychic effects after 500 μg IV.

SC LD_{50}, **rat**: 310 mg/kg; **mouse**: 155 mg/kg; IV LD_{50}, **rat** and **mouse**: 30 mg/kg.

CYCLIZINE

= *Compd 47-83* = *Marezine* = *Marzine* = *Nautazine* = *Neo-Devomit* = *Valoid* = *Wellcome prepn 47–83*

Antiemetic.

Untoward effects: Hypersensitivity hepatitis, malaise, drowsiness, sedation, blurred vision, dry mouth, anorexia, nausea, vomiting, and vertigo. A 26-year-old **male** had a sore and inflamed patch on penile skin in < 12 h. A 13-year-old **male** had auditory and visual hallucinations after an 800 mg overdose. Euphoria, tachycardia, hypertension, exhilaration, disorientation, reduced judgement, slurred speech, and ataxia in three, 17-year-old **students**, after **each** ingested 750 mg "for kicks". A 16-year-old **female** ingested 1 g and became delirious. Congenital abnormalities in **fetuses** of **mothers** taking cyclizine during early pregnancy. Large doses interfere with ***heparin's*** effect in the body.

LD_{50}, **mice**: 147 mg/kg.

CYCLOBARBITAL

= *Cavonyl* = *Cyclobarbitone* = *Cyclodorm* = *Cyklodorm* = *Fanodormo* = *Hexemal* = *Irifan* = *Itridal* = *Kollerdormfix* = *Namuron* = *Palinum* = *Phanodorm* = *Phanodorn* = *Philodorm* = *Prälumin* = *Pronox* = *Pro-Sonil* = *Sonaform* = *Tetrahydrophenobarbital*

Anesthetic, sedative, hypnotic.

Untoward effects: Hypersensitivity reactions have been reported. Toxic serum concentrations are > 20 mg/l. Other untoward effects are similar to those of ***phenobarbital*** (q.v.). Therapeutic serum levels are 2–6 mg/l; toxic at 10 mg/l; and fatal at 20 mg/l.

LD_{LO}, **human**: 50 mg/kg.

Alcohol potentiates its effects in **rabbits**, **rats**, **dogs**, and **mice**.

LD_{LO}, **rat**: 300 mg/kg; **cat**: 120 mg/kg; **rabbit**: 450 mg/kg; IP LD_{LO}, **rat**: 195 mg/kg; **rabbit**: 450 mg/kg.

CYCLOBENZAPRINE

= *Flexeril* = *Flexiban* = *MK-130* = *Proheptatriene* = *Ro-4-1577* = *RP-9715*

***Skeletal muscle relaxant.* A tricyclic antidepressant with anticholinergic effects.**

Untoward effects: Drowsiness (40%), dry mouth (28%), dizziness (11%), tachycardia, weakness, fatigue, dyspepsia, nausea, paresthesia, blurred vision, unpleasant taste, and insomnia occur. Rare adverse effects include sweating, myalgia, tremors, dyspnea, dysarthria, jaundice, abdominal pain, depression, hallucinations, urinary retention, skin rash, urticaria, alopecia, and facial and lingual edema. Specifically:
Allergic: Photosensitization.
Blood and bone marrow: Agranulocytosis, leukopenia, eosinophilia, thrombocytopenia, purpura, and bone marrow decrease.
Cardiovascular: Palpitations, hypo- or hypertension, angina, myocardial infarction, heart block, and stroke.
Central nervous system and neuromuscular: Excitement, delusions, anxiety, restlessness, nightmares, vertigo, peripheral neuropathy, incoordination, seizures, digital tremors, tinnitus, electroencephalogram changes, and mania. Confusion and delirium reported in some elderly **patients**.
Endocrine: Testicular swelling and gynecomastia in **male**, galactorrhea and breast enlargement in **female**.
Gastrointestinal: Paralytic ileus, constipation, nausea, vomiting, coated or black tongue, diarrhea, and parotid swelling.
Libido and sexual dysfunction: Occasionally reports of increased or decreased libido and erectile problems.

Ingestion of doses > 100 mg can produce many of the above adverse effects.

Drug interactions: May block the antihypertensive effect of ***guanethidine***. Do NOT use with or within 2 weeks of any ***monoamine oxidase inhibitor***. May potentiate ***valerian***.

CYCLOCHLOROTINE

A hepatotoxin produced by *Penicillium islandicum* growing on yellow *rice*.

Untoward effects: Suspect carcinogen.

CYCLOCOUMAROL

= *BL5* = *Coumopyran* = *Coumopyrin*

Anticoagulant.

Untoward effects: Its effectiveness is decreased by ***butabarbital, chloral hydrate, glutethimide, griseofulvin*** and ***phenobarbital***, thus, increasing the risk of thrombus formation. It is potentiated by ***chloramphenicol, indomethacin, oxyphenbutazone, penicillin, phenylbutazone, salicylates,*** and ***tetracyclines***, thus, increasing chances of hemorrhagic complications.

CYCLOGUANYL PAMOATE
= *Camolar* = *CI501* = *Cycloguanil Embonate*

Antimalarial.

Untoward effects: Pain and swelling at injection sites, headache, vertigo, fever, and lymphangitis. No serious sequelae in > 15,000 **patients**. Urticaria and skin rashes have followed second injections. Prolonged use causes slight decrease in hemopoiesis due to interference with ***folic acid*** metabolism.

Rat studies indicated changes consistent with ***folic acid*** deficiency.

IP LD_{LO}, **rat**: 400 mg/kg.

CYCLOHEXAMINE
= *9* = *CI 400* = *PCE*

An *N*-ethyl analog of *PCP*.

Untoward effects: First appeared in the illicit street drug market in 1969, in the Los Angeles, California area. Potency and effects appear to be slightly greater than ***PCP*** (q.v.) in **humans** and **animals**.

CYCLOHEXANE
= *Benzene Hexahydride* = *Hexahydrobenzene* = *Hexamethylene* = *Hexanaphthene*

Approximately 375 million gallons used/year.

Use: Primarily, for manufacturing ***adipic acid*** and ***caprolactam*** for nylons, and as a solvent.

Untoward effects: Irritant to eyes and respiratory tract. Dermatitis, drowsiness, dizziness, nausea, narcosis, and coma reported. Chronic exposure causes neurotoxicity, oncogenicity, and teratogenicity. Many of its toxic effects are due to ***benzene*** (q.v.) as a contaminant.

Eye irritation, **human**: 5 ppm; LD_{LO}, **human**: 500 mg/kg; TWA, 300 ppm.

LD_{50}, **rats, 2 weeks**: 8 ml/kg; **young adults:** 39 ml/kg; **older rats**: 16.5 ml/kg; **mouse**: 1.3 g/kg.

CYCLOHEXANOL
= *Hexahydrophenol* = *Hexalin* = *Hydralin*

Solvent.

Untoward effects: Prolonged low-level industrial exposure leads to headache and conjunctival irritation. Permitted as a fragrance in soaps, detergents, cosmetic creams and lotions, and foods. Eye, nose, throat, and skin irritant from over-exposure. Eye irritant for **man** at 100 ppm.

LD_{LO}, **human**: 500 mg/kg. TLV, 50 ppm.

Vapors cause conjunctival and nasal irritation in **guinea pigs**. High concentrations cause narcosis, coma, and death in them.

LD_{50}, **rats**: 2 g/kg; IM LD_{50}, **mice**: 1 g/kg; IV LD_{50}, **mice**: 272 mg/kg; **rabbits**: 2.2 g/kg.

CYCLOHEXANONE
= *Anone* = *Hytrol O* = *Ketohexamethylene* = *Nadone* = *Pimelic Ketone*

Solvent for lacquers, *nitrocellulose*, *polyvinyl chloride*, in cleaning leather and textiles, in crude rubber, insecticides, epoxy resins, and in illicit drug manufacturing.

Untoward effects: Irritant to eyes, nose, throat, mucous membranes, and skin. Causes dermatitis, headache, dizziness, narcosis, coma, and liver and kidney degeneration. Ingestion causes gastrointestinal irritation and narcosis. In waste gas emissions of plastic and footwear manufacturing plants. Abused by solvent **inhalers**. Keep away from heat, sparks, and fire. Vapors at 50 ppm cause mild throat irritation. TWA, 50 ppm, 8 h.

Inhalation TC_{LO}, **human**: 75 ppm.

IV use in **beagle dogs** caused vocalization, lacrimation, salivation, mydriasis, urination, defecation, restlessness, stupor, and ataxia. Occasionally, convulsive movements, hyperpnea and/or dyspnea. Cutaneous or SC use for several days caused cataracts in **guinea pigs**. Inhalation by **rabbits** and **monkeys** has caused central nervous system depression and liver and kidney degeneration. 1 g/kg will kill a **rabbit**.

LD_{50}, **rats**: 1.6–3.46 g/kg; skin, **rabbits**: 1 ml/kg.

CYCLOHEXENES
= *Tetrahydrobenzenes*

Untoward effects: Irritant to eyes, nose, throat, and skin. Causes drowsiness on inhalation. Found to be a cause of ***phenolic*** contamination with foul taste and odor of potable water, due to leaching from a ***phenolic*** resin liner of a solar water tank.

CYCLOHEXIMIDE
= *Actidione* = *Naramycin A* = *NSC-185* = *U-4527*

Fungicidal antibiotic.

Use: To eliminate yeasts and molds in bacterial culturing, and as a protein synthesis inhibitor.

Untoward effects: In poisoned **rats**, **dogs**, and **monkeys**, salivation and diarrhea are common. Excitement, followed by depression is noted in **dogs** and **rats**. Toxic

birds show goose-stepping ataxia, wing-drop, tremors, hypoactivity, and prostration. **Mallard ducks** also exhibit polydipsia and regurgitation. Symptoms appear within 10 min and **birds** die in ~12 h. Up to 1 month for remission in those less seriously ill.

LD_{50}, **rats**: 1.8 mg/kg; **pheasants**: 9.38 mg/kg; **mallard ducks**: 82.5 mg/kg; **mice**: 133 mg/kg; **dogs**: 65 mg/kg; **monkeys**: 60 mg/kg; **guinea pigs**: 65 mg/kg.

CYCLOHEXYLAMINE

Use: In organic synthesis, boiler water corrosion-inhibiting treatments, and rubber accelerator chemicals. Approximately 18 million lbs used annually.

Untoward effects: Irritation and sensitization. Nausea and narcosis after heavy exposure. 77/121 **workers** became ill from a boiler steam humidification. Symptoms were headache, nausea, vomiting, dizziness; eye, nose, and throat irritation. A musty, pungent, ***ammonia***-like, or radiator-like odor was noted in the workplace. In 1966, it was discovered that ***cyclamates*** are sometimes converted (varying in **individuals** from 1–50% of the total ***cyclamate*** dose ingested) into a far more toxic substance, cyclohexylamine. TWA, **Human**: skin, 10 ppm.

LD_{LO}, **human**: 50 mg/kg.

Suspect as a bladder carcinogen and 60 mg/kg causes severe liver and kidney damage in **rats**.

LD_{50}, 300 mg/kg.

2-CYCLOHEXYL CYCLOHEXANONE

Use: Over 15 ton/year used in the U.S. as a fragrance in soaps, detergents, cosmetic creams and lotions, and perfumes.

Untoward effects: No sensitization or irritation reactions in **human volunteers**.

LD_{50}, **rat** and acute dermal, **rabbit**: > 5 g/kg.

CYCLOHEXYL ETHYLS

The *acetate* form is used in soaps, detergents, cosmetic creams and lotions, perfumes, and foods as a fragrance.

Untoward effects: No irritation or sensitization in **human volunteers**.

LD_{50}, **rats**: 3.2 g/kg, acute dermal, **rabbit**: 5 g/kg.

The *alcohol* is used in soaps, detergents, cosmetic creams and lotions, and perfumes as a fragrance.

Untoward effects: No irritation or sensitization in **human volunteers**.

LD_{50}, **rats**: 940 mg/kg, acute dermal, **rabbit**: 1.22 g/kg.

CYCLOMETHICONE

A slowly volatile silicone for cosmetic use. Also referred to as a *cyclic dimethyl polysiloxane*.

Untoward effects: A 10% concentration is non-irritating to **rabbit** skin; pure material is only slightly irritating. No adverse effects at 1% level in diet of **rats** and **rabbits** for 1 year and 8 months, respectively. Slight transitory conjunctival irritation and discomfort after full-strength application to eye. Neither **rats** nor **Cynomolgus monkeys** showed adverse effects from repeated aerosol inhalation.

LD_{50}, **rat**: 35 g/kg.

CYCLOMETHYCAINE

= *Surfacaine* = *Surfathesin* = *Topocaine*

Topical anesthetic.

Untoward effects: As a "caine", it cross-reacts with other "caines", leading to local skin reactions. May cause transitory stinging or burning. Allergic reactions in 5/413 with previous contact dermatoses. A **baby** developed cardiac arrhythmias after topical application of a 0.5% and ***methapyriline HCl*** 2% ointment for seborrheic dermatitis.

CYCLONITE

= C_4 = *Hexogen* = *RDX* = T_4

***Explosive, rodenticide.* RDX designation stands for Royal Demolition Explosive.**

Untoward effects: A number of **soldiers** ingested ~25–180 g of C_4 plastic explosive containing 91% cyclonite. **They** developed grand mal seizures, hematuria, nausea, vomiting, muscle twitching, and amnesia. Inhalation of its vapors while cooking with it caused some of the same symptoms. Epileptic attacks in **man** as a manifestation of industrial exposure first reported in 1949.

Caused liver tumors in **mice** and decreased size of **rat** off-spring.

LD_{50}, **rat**: 200 mg/kg; LD_{LO}, **mouse** and **rabbit**: 500 mg/kg; **cat**: 100 mg/kg.

CYCLOOCTYLAMINE

= *SKF 23,880-A*

Antiviral.

Untoward effects: Transient, mild nasopharyngeal irritation, bitter taste, and headache from intranasal use.

CYCLOPAMINE

= *11-Deoxojervine*

Untoward effects: Its effect on fetal **lambs** was a first in medical history to reveal a cause of major birth defects in **mammals**. This steroidal alkaloid occurs in *Veratrum californicum* (***False Hellebore***). **Lambs** are born with a cyclops eye and monkey face when the **ewe** eats it on the 14th day of pregnancy. Eaten later during gestation, their **lambs** develop leg deformities, including loose knees and hock joints. In **rabbits**, ingestion between the sixth and nineth days of pregnancy causes cyclopia and other cephalic malformations in their offspring. When applied to developing **chick** embryos at 14–28 h of incubation, a high frequency of head abnormalities is reported; trunk and limb deformities, if used at 30–40 h of incubation. Stomach acids can induce conversion to non-teratogenic ***veratremine*** in **non-ruminants**.

TD_{LO}, **rat**: 960 mg/kg (6–9 days pregnant), **rabbit**: 180 mg/kg (6–9 days pregnant), **hamster**: 170 mg/kg (7 days pregnant).

CYCLOPENTADECANOLIDE

= *Pentadecanolide*

From *angelica root oil*, imparting a musk-like odor for soaps, detergents, cosmetic creams and lotions, perfumes, and food.

Untoward effects: No irritation or sensitization in **human volunteers**.

LD_{50}, **rats** and acute dermal **rabbits**: > 5 g/kg.

CYCLOPENTAMINE

= *Clopane* = *Cyclonarol* = *Cyclopentadrine* = *Cyklosal* = *Sinos*

Vasoconstrictor, nasal decongestant.

Untoward effects: Essentially, the same as for ***ephedrine*** (q.v.).

LD_{50}, **rat**: 169 mg/kg.

CYCLOPENTANONE

= *Adipic Ketone* = *Ketocyclopentane* = *Ketopentamethylene*

Found in wood and *cigarette* smoke, and in glucose pyrolysates.

Untoward effects: Severe eye irritant for **rabbits**.

LC, 640 ppm for **Syrian golden hamsters** and, in lower dosage, it caused slight growth retardation.

CYCLOPENTHIAZIDE

= *Cyclomethiazide* = *Navidrex* = *Navidrix* = *Salimid* = *Su-8341* = *Tsiklometiazid*

Diuretic, antihypertensive.

Untoward effects: Hypochloremia, metabolic acidosis, hypokalemia, increased blood ***uric acid***, hyponatremia, and, possibly, acute pancreatitis in some **patients**. As a ***sulfanilamide*** derivative, certain photoallergic reactions may occur with other drugs. Interference with ***pyrimidine*** metabolism may occur when given with ***allopurinol***. Reports of decreased efficacy with ***oral contraceptives*** may be due to the latter's induced ***sodium*** and fluid retention.

IV LD_{50}, **rat**: 142 mg/kg; **mouse**: 232 mg/kg.

CYCLOPENTOLATE

= *Ak-Pentolate* = *Alnide* = *Cyclogyl* = *Mydplegic* = *Mydrilate* = *Zyklolat*

Mydriatic, cycloplegic.

Untoward effects: Has increased intraocular pressure and caused lens opacities and narrow-angle glaucoma in **man**. Transient psychotic episodes with dizziness, disorientation, purposeless movements, memory loss, incoherent speech, and inability to stand. Has also caused ***atropine***-like toxicity with vomiting, distention, and ileus. Central nervous system toxicity in a 82-year-old **male** causing delirium after ophthalmic instillation of four drops of a 0.2% solution. **He** failed to recover and had to be institutionalized. Blood pressure decreases in **infants** with it alone. A 7-year-old **male** developed a neurotoxic reaction with incoherence, visual hallucinations, slurred speech, and ataxia, with all symptoms disappearing in about 6 h. A 6-year-old **female** developed hallucinations and ataxia following 1% eye drops with 0.25% ***scopolamine***. Two recent reports concerned systemic complications of drug abuse by an 18-year-old **female** and a 30-year-old **female** with 100–400 drops/day over a long period of time.

SC TD_{LO}, **child**: 40 μg/kg.

Conjunctival edema noted in a **dog** after its use.

CYCLOPENTOPHENANTHRENE

Untoward effects: Oral dosing caused Grade III lobular carcinoma of the breast and Grade II mixed-type columnar cell carcinoma of the kidney in **rats**.

CYCLOPHOSPHAMIDE

= *B-518* = *Cycloblastin* = *Cyclophosphane* = *Cyclostin* = *Cytophosphane* = *Cytoxan* = *Endoxan* = *Endoxana* = *Genoxal* = *Neosar* = *NSC 26,271* = *Procytox* = *Sendoxan*

***Antineoplastic*. An alkylating agent related to *nitrogen mustard*. Over a thousand lbs used annually in the U.S. (oral and parenteral) on several hundred thousand patients.**

Untoward effects: LD_{LO}, **woman**: 45 mg/kg; **human**: 20 mg/kg.

Blood and bone marrow: Bone marrow depressed, hyponatremia. Water retention, leukopenia, thrombocytopenia, neutropenia, hemolytic anemia. Myelosuppression peaks at about seventh day. Case reports of increased and decreased prothrombin time.
Cardiovascular: Extremely high dosage has caused cardiac hemorrhagic necrosis and edema and electrocardiogram changes with symptoms resembling congestive heart failure. Neutropenia in an **infant** who was breast fed while the **mother** was receiving the drug. Significant amounts pass into breast milk. A hypotensive episode and cardiac arrest immediately followed 1 g IV dose in a 66-year -old **female**. Dose >180 mg/kg can cause cardiac failure through direct endothelial damage.
Chromosomes: Aberrations, particularly in peripheral lymphocytes. Mutagenic.
Ear and eye: Corneal opacity and cataracts, visual blurring and burning. Bilateral optic atrophy and blindness in 18-year-old **male** after five courses of therapy with 2 mg IV + oral of 600 mg/day/5 days. Myopia in a 15-year-old **female**. Ototoxicity reported.
Fetus and neonate: Abnormalities (digital defects, flattened nasal bridge, ecrodactyly, palatal anomalies, and single coronary artery) in first and second trimester, and abortion reported. Found in breast milk and breast feeding should be avoided in treated **mothers**.
Gastrointestinal: Nausea, vomiting, anorexia, and abdominal pain. Metallic taste. Pigmentation of teeth in a 3-year-old **male**.
Genitourinary: Sterility (often temporary), azoospermia, oligospermia, testicular atrophy, amenorrhea, and decreased libido. Long-term (> 8 weeks) use in prepubertal **boys** can result in gonadal damage and sterility. Ovulation in **women** permanently impaired in 30–50%. Azoospermia reported.
Infections: Due to immunosuppression.
Liver: Hepatic necrosis and fibrosis.
Miscellaneous: Anaphylaxis and delayed wound healing. Possibility of pancreatitis and thyrotoxicosis. Has caused pyrexia and inflammation mucous membranes, and decreased plasma cholinesterase, which can increase duration of ***suxamethonium*** action. Has induced acute type 1 diabetes.
Neoplasms: Cancer of the bladder, leukemia, lymphoma, especially after long-term usage.
Renal: Syndrome of inappropriate antidiuretic homone secretion, bladder carcinoma, hemorrhagic cystitis, and bladder fibrosis. Must be used with extreme caution in **patients** with poor renal function.
Respiratory system: Pulmonary infiltrates, fibrosis, alveolitis, edema, and hypersensitivity pneumonia.
Skin: Alopecia (~30%), erythrodysesthesia syndrome, urticaria, pruritus, hyperpigmentation, longitudinal brown pigmented band on nails, brown-black nail plate and nail bed. Facial burning with IV use. Cutaneous and SC necrosis after IV.

Drug interactions: Its serum metabolite levels increase with ***allopurinol***. ***Chloramphenicol*** inhibits its metabolization. ***Grapefruit juice*** up to 90 min before or after the drug increases its peak concentration in the blood. ***Methotrexate*** may enhance its hepato- and nephrotoxicity. ***Oral contraceptives*** stimulate its metabolization. ***Phenobarbital*** increases its conversion to its active metabolite. ***Prednisolone*** increases its activity. ***Proadifen***, a competitive inhibitor of microsomal drug metabolism, increases its toxicity. It increases ***vasopressin*** excretion.

Carcinogenic in **rats** (oral and IV) and testicular and lung tumors in **mice** (SC). A cause of plasmacytoma in **hamsters**. Damaged bladder mucosa and hemorrhagic cystitis in **dogs**, but rarely in **cats**. Some **cats** treated for 2 months–1 year have developed cystitis with hematuria, stranguria, and incontinence. Transitional cell bladder carcinoma reported in **dogs** and **humans**. Alopecia may occur in **poodles** and **Old English Sheepdogs**. Has caused a toxic nephrosis in **dogs** and **rats** and pulmonary edema in **dogs**. **Dogs** are suitable as an experimental model of **human** cyclical neutropenia by daily injections of the drug. **Dogs** do not tolerate the drug well. Only suckling **rats** (< 2 weeks) develop internal hemorrhages on toxic doses. Causes hemopoietic decreases in **rats**, **mice**, **dogs**, **rhesus monkeys**, and a high percentage of fetal deformities in **rats**, **mice**, and **rabbits** whose mothers received the drug during pregnancy. Fetal death rate increased, abortions, and intrauterine and postnatal growth retardation in **rats**. High doses to **rats** a few hours before mating doubled the rate of preimplantation embryo losses. The wool of **sheep** falls off within a few weeks after being sprayed with it or given it orally or by injection. High fevers may be noted at defleecing time. Approximately 4 weeks needed for its elimination from edible tissues of treated **sheep**. Induced neurogenic sarcomas on peripheral nerves of **rats**. SC use has led to mammary carcinomas, lymphomas and lymphoreticular neoplasms, pulmonary adenomas and sarcomas, squamous cell carcinomas, and ovarian carcinomas in **mice**.

LD_{50}, **rats**: 94 mg/kg; **mice**: 137 mg/kg; **dog**: 44 mg/kg.

CYCLOPIAZONIC ACID

= *CPA*

A toxic indole tetramic acid, first isolated from *Penicillium cyclopium* and found in some strains of *Aspergillus flavus*, *A. oryzae*, *A. tamari*, *A. versicolor*, *P. camemberti*, *P. crustosum*, *P. griseofulum*, *P. puberculum*, and *P. viridicatum*. Often found associated with *aflatoxins*.

Untoward effects: To avoid potential risk to **humans** consuming ***peanuts*** and their products, farmer's stock ***peanuts*** are examined for the presence of ***aflatoxins***. These are removed from the food supply. Levels of 6.5–32 ppb have been found in U.S. ***peanuts***. Many of these species were isolated from stored ***corn***, other grains, ***cheese***, ***walnuts***, ***sausage***, and ***ham***. Cultures of *P. camemberti* and *A. oryzae*, used in the feed industry, are controlled, to prevent production of ***CPA***.

Rats given oral doses of 30–82.6 mg/kg died in 1–5 days with degeneration and necrosis of liver, spleen, pancreas, kidney, salivary glands, myocardium, and skeletal muscle. **Chickens** given 100 ppm in their feed developed ulcerative proventriculitis, ulcerative ventriculitis, chronic necrotizing hepatitis and splenitis, decreased weight gain, high mortality rate, and poor feed conversion. **Pigs** given orally 1 and 10 mg/kg developed necrotizing gastritis and villous blunting in the jejunum and ileum. **Pigs** on the higher dosage also developed focal hepatocellular necrosis, hepatic peripheral lobular fatty changes, and focal renal tubular nephrosis with focal suppurative tubulointerstitial nephritis. **Swine** have developed hemorrhage and gastrointestinal ulceration after oral dosing. All **dogs** that received 0.25–1.0 mg/kg/day became toxic in 2–44 days. Anorexia, vomiting, diarrhea, dehydration, pyrexia, weight decrease, and central nervous system depression noted. Renal infarcts, necrotizing epididymitis, ulcerative dermatitis, leukocytosis, lymphopenia, monocytosis, and increased serum alkaline phosphatase occur. All **dogs** receiving 0.5–1.0 mg/kg/day died. More toxic than ***aflatoxin*** to **channel catfish**. May be a cause of a Theilers-like disease in **foals** and **horses**.

LD_{50}, male **rats**: 36 mg/kg; female, 63 mg/kg. IP LD_{50}, **fingerling channel catfish**: 2.82 mg/kg; **rats**: 2 mg/kg.

CYCLOPROPANE

= *Cyclopane* = *Trimethylene*

Inhalation anesthetic. **Flammable. Do NOT use near open flames, electrocautery devices, or anything capable of producing a spark or static electricity!!**

Untoward effects: Malignant hyperpyrexia with muscular rigidity, acidosis, hyperkalemia, disseminated intravascular coagulation, and renal failure reported. Duration of administration to the **mother** influences degree of depression in the **human fetus**. This drug is highly lipid-soluble and easily enters the **fetal** circulation. Can increase airway resistance and cardiac arrhythmias. Adverse hepatic effects are related to initial dysfunction. Therapeutic or normal anesthetic serum levels are reported as 13 mg%. Postoperative headaches are common and it can increase hemorrhage.

Drug interactions: Cardiodepression can occur when given with ***β-adrenergic-blockers***, such as ***propranolol***. Use with ***anterior pituitary hormone*** increases risk of hypotension and cardiac arrhythmias. The latter are also precipitated by exogenous pressor amines, such as ***dopamine***, ***ephedrine***, ***epinephrine***, ***metaraminol***, and ***norepinephrine***.

Chick embryos exposed to it have shown decreased growth rate, increased fetal death, and congenital anomalies. **Dogs** and **cats** show similar adverse effects to exogenous pressor amines, as does **man**. High concentrations have caused hepatic injury in experimental **animals**, and it increased ***morphine*** bradycardia in **dogs**.

CYCLOSERINE

= *106-7* = *Closina* = *Farmiserina* = *Micoserina* = *Orientomycin* = *Oxamycin* = *PA-94* = *Seromycin*

Antibiotic.

Untoward effects: At therapeutic dosage, ~30% of **patients** experience untoward effects. Dosage should be decreased in **patients** with renal insufficiency, as up to 60% is normally excreted unchanged in the urine. 5–10% of **patients** receiving it as an antitubercular develop a disoriented confused state, psychosis, and convulsions; thus, its use is limited. Hypersensitivity reactions manifest as fever and/or skin rashes. Muscle twitching (6%), severe headaches (4%), disturbances of equilibrium and drowsiness (8%), tremors (4%), paresthesias (4%), and excitement (18%) reported in a group of 51 **patients**. Dizziness, torpor, pains in the extremities, depression, memory loss, decreased visual acuity, anxiety, and aggression have also been noted. Breast milk levels are usually 6–19 μg/ml, with 34 μg/ml the plasma level. Others reported milk levels at 50–75% plasma levels. It is structurally related to the active isoxazoles in *Amanita muscaria* (q.v.), and, in **humans**, they share similar side-effects: mental confusion, acute psychotic episodes, convulsions, and other abnormal behavior. These adverse effects are common in the **elderly**. Sideroblastic anemia, due to induced ***pyridoxine*** deficiency, responds to ***pyridoxine*** therapy. A **patient**, in a suicide attempt, ingested 3 g, and was successfully treated with peritoneal dialysis. A good reduction in plasma levels and toxic symptoms was noted after 21 h. Megaloblastic anemia is due to its impairment of ***folate*** absorption and/or utilization. Probably, a cause of neutropenia.

LD_{50}, **rats** and **mice**: 5.3 g/kg; **dogs**: 2 g/kg.

SC LD_{50}, **mouse**: 2.8 g/kg.

CYCLOSPORINE

= *Ciclosporin* = *Cyclosporin A* = *Neoplanta* = *Neoral* = *Sandimmune* = *Sang Cya* = *Sang Stat*

Immunosuppressant.

Untoward effects: Anaphylaxis: After IV infusions. May be due to the emulsifying agent, ***Cremophor EL*** (q.v.). Severe headaches, shortness of breath, wheezing, and diffuse pruritic urticaria reported. Angioedema noted on oral therapy.
Blood: Hyperbilirubinemia, hypertriglyceridemia, hyperkalemia, hyperuricemia with gouty arthritis, thrombocytosis, and bone marrow toxicity reported. Hypomagnesemia also noted.
Cardiovascular: Induces sympathetic activation and hypertension. It may also follow nephrotoxicity. Raynaud's phenomenon reported. **Patients** have developed vitreous hemorrhage and severely decreased vision. A **patient** on normal dosage developed colitis with extensive, vascular thromboses in the colonic submucosa, and died. Has caused prolongation of QT interval. Hypertension reported in six treated **patients**. Atrial fibrillation in 61-year-old **male** after accidental overdose of 1,000 mg.
Central nervous system: Grand mal seizures, transient aphasia, cerebellar ataxia with tremor and depression, confusion, disorientation, and a case of mania. Encephalopathy reported, due to fat embolism from the emulsifying agent in an infusion.
Hyperplasia and neoplasms: Gingival hyperplasia, benign tumors, carcinoma of the corneoscleral limbus, malignant lymphomas, warts, squamous cell carcinoma, and premalignant keratoses reported.
Miscellaneous: Myopathy, ingrown toenails, hypertrichosis, hepatotoxicity, and gastrointestinal intolerance. Osseous malformation in an **infant** whose **mother** had been on therapy for 9 months before **her** pregnancy. Topical use for severe alopecia areata has caused transient folliculitis and pruritus. May induce gout. Has caused pustular psoriasis and cervical lymphadenopathy in 32-year-old **female**. Fever in 31-year-old **female** within 2 h after each 250 mg/day dose. Mee's lines on nails and rhabdomyolysis have been reported. Will pass into breast milk.
Poisoning: A 23-year-old **female**, treated for Hodgkin's disesase, received it by mistake, instead of ***tranexamic acid***, IV. Caused anxiety, diarrhea, vomiting, and perspiration, with weak and irregular pulse and renal insufficiency for ~36 h. Several accidental overdoses of up to 25 g during an 8 day period recovered. One report lists seven cases where the drug was given in error, instead of another drug.
Renal: Nephrotoxicity, *including* hemolytic-uremic syndrome, decreased ***potassium*** excretion, glomerular microthrombi, drug deposits in renal transplants, hemorrhagic cystitis, tubular atrophy, interstitial fibrosis, acute tubular necrosis, decreased glomerular filtration rate, and decreased ***creatine*** and ***urea*** clearances. Adverse renal effects responsible for **Baby Fae's** death, after receiving a **baboon** heart.

Drug interactions: It has a propensity for many drug interactions and interactions with immunostimulants, such as vaccines and toxoids.

Ganciclovir, Hypericum, phenobarbital, phenytoin, probucol, and ***rifampin*** increase its metabolization and have precipitated rejection episodes.

Acyclovir, allopurinol, amiodarone, amphotericin B, cimetidine, clarithromycin, dalfopristin, danazol, diclofenac, diltiazem, erythromycin, ethinyl estradiol with ***levonorgestrel, fluconazole, fluoxetine, glyburide, grapefruit juice, griseofulvin, itraconazole, josamycin, ketoconazole, methylprednisolone, methyl testosterone, metoclopramide, metronidazole, nicardipine, nifedipine, pristinamycin, quinupristin, sulfamethazine, sulindac, trimethoprim,*** and ***verapamil*** decrease its metabolization.

The interactions (if any) between ***ciprofloxacin, fluconazole,*** and ***cyclosporine*** are controversial. It decreases clearance of ***digoxin, prednisolone*** and ***simvastatin***. When used with ***Iodine***^{131} ***Iodohippuric acid***, it can cause a nephrotoxicity indistinguishable from ischemic acute tubular necrosis. IV solutions contain surfactants which cause leaching of relatively large amounts of ***diethylhexyl phthalate*** (q.v.) in 24 h from ***polyvinyl chloride*** containers and tubing. Although ***grapefruit juice*** decreases its absorption, it can increase its blood levels. ***Aluminum hydroxide*** gel decreases its blood levels in **patients** and **rats**. ***Chloramphenicol*** may increase its serum concentration. Use with ***nifedipine*** increases blood urea nitrogen and gingival hyperplasia. It can increase ***cervistatin*** serum levels three to five times. After several months of treatment, ***mifepristone*** causes a dramatic increase in cyclosporin serum levels, requiring a 70% decrease in cyclosporin dosage. ***Amphotericin*** can potentiate its nephrotoxicity. Use in a 76-year-old **female** stabilized on ***diltiazem*** (q.v.) led to aplastic anemia and thrombocytopenia. ***Ciclopidine*** and ***rifampin*** have decreased cyclosporin serum levels. ***Hypericum*** (***St. John's Wort***) also decreases its plasma concentration and has led to heart transplant rejection.

In **dogs**, the use of a 0.2% ophthalmic ointment caused ocular and periocular irritation in ~13%. Prolongs ***barbiturate*** anesthesia in **rats**; and causes mild hepatic necrosis in **pigs**. High dosage caused nephro- and hepatotoxicity in **rats** and **cynomolgus monkeys**, but not in other experimental **animals**. In **rats**, use with ***simvastatin*** may cause hepatic toxicity and decrease renal function.

LD_{50}, **rats**: 1.5 g/kg; **mice**: 2.33 g/kg; **rabbits**: >1 g/kg. IV LD_{50}, **rats**: 25 mg/kg; **mice**: 107 mg/kg; **rabbits**: > 10 mg/kg.

CYCLOTHIAZIDE

= *Anhydron* = *Aquirel* = *Doburil* = *Fluidil* = *Lilly 35483* = *Renazide*

Diuretic, antihypertensive.

Untoward effects: Inappropriate antidiuretic secretion syndrome with stupor and hyponatremia. Hyperuricemia, hypokalemia (potentiating ***digitalis*** toxicity), decreased arterial response to pressor amines, and muscle cramps. Occasionally, nausea, dizziness, vertigo, headaches, hot flushes, and increased photosensitivity. Enhances ***tubocurarine*** effect, requires discontinuance several days before elective surgery.

CYMARIN

= *Alvonal MR* = *K-strophanthin-*α

***Cardiotonic.* Isolated from *Strophanthus kombé*, *Adonis sp.*, and other plants.**

Untoward effects: Nausea, vomiting, diarrhea, and extrasystoles. Palpitations, hot flushes, anxiety, anorexia, hypertension, decreased pulse, decrease of ST and U waves in electrocardiogram, pallor, and myocardial or coronary insufficiency with overdosage. A strong vasoconstrictor.

It is one of the poisonous glucosides in ***Apocynum cannabinum*** (q.v.) that can adversely affect **cattle**, **horses**, and **sheep** ingesting the plant.

LD_{LO}, **cat**: 130 μg/kg.

p-CYMENE

= *Dolcymene*

Fragrance with ***kerosene***-like odor for use in soaps, detergents, cosmetic creams and lotions, perfumes, and foods. Found in ~100 volatile oils. Approximately 4 tons used in the U.S. annually.

Untoward effects: Absorbed through the skin, and, undiluted, is a primary skin irritant to **man**, leading to erythema, dryness, and defatting. 4% in petrolatum was not irritating after a 48 h closed-patch test in 25 **volunteers**.

Full-strength was moderately irritating to **rabbit** skin.

LD_{50}, **rats**: 4.75 g/kg; acute dermal LD_{50}, **rabbits**: > 5 g/kg. IP LD, **guinea pigs**: 2.162 g/kg.

CYMOPTERUS sp.

Untoward effects: ***C. watsonii*** (***Spring Parsley***) contains photosensitizing furocoumarins that cause lesions after skin contact or ingestion.

A photosensitizer in **sheep**. Feeding it or ***C. longipes*** seed or whole plants to 2-week-old **broiler chicks** or **turkey poults** caused severe photosensitivity, particularly in the **chicks**. Photophobia, red discoloration of beak, comb, and feet; loss of feathers in the periorbital area; dried serous fluid on combs and edge of beak; keratoconjunctivitis; and foot and leg lesions were common signs of toxicity. Mortality was high in **turkeys** fed ***C. longipes*** seed. ***Isoimperitorin*** and ***oxypencedanin*** were two isolated phototoxic furocoumarins.

CYNANCHUM

Untoward effects: In Asia and Northern Pacific region, ***C. caudatum*** = ***Enup*** = ***Ikema***, as a hunting arrow poison and to immobilize **birds**.
C. vincetoxicum = ***White Swallow-wort***, found in Europe and in the grasslands of the northeastern U.S., contains a poisonous glucoside, ***vincetoxin***, which can cause vomiting in **people**, and, in large quantities, dangerous, even fatal, stomach inflammation.

Several species in Africa have convulsed and poisoned **cattle**, **sheep**, and **horses**.

CYNODON

Untoward effects: ***C. dactylon*** = ***Bermuda Grass*** = ***Dhub*** = ***Dubo*** = ***Durba***. Nervousness, prostration, posterior paralysis, tremors or muscle twitching in flanks, and death, especially in white-skinned **cattle**. Fungus sporulating on its dead leaves 3–5 weeks after a frost may be associated with photosensitization in **cattle** consuming the plant. Paralyzed **cattle** have drowned in shallow ponds. In the fall of 1971, this "***Bermuda grass*** tremor" syndrome affected and killed nearly 25,000 **cattle** in Louisiana. Contains ***hydrocyanic acid*** under some environmental conditions. A **dog** was reported to be allergic to its inhaled pollen, and it can cause allergic rhinitis, bronchial asthma, and hypersensitivity pneumonia in **man**.
C. plectostachyus = ***Giant Star Grass*** = ***Star Grass***. As with other species, it contains ***prussic acid*** and has been the cause of neonatal goiters and skeletal deformities in autumn-born **lambs** in Zimbabwe.

CYNOGLOSSUM OFFICIANALE

= *Hound's Tongue*

Untoward effects: Contains hepatotoxic ***pyrrolizidine*** alkaloids (0.6–2.1% of dry matter), mostly ***heliosupine*** and ***echinatine***, causing thirst, nervousness, and liver damage (portal edema, hepatocellular necrosis, bile duct proliferation, and enlarged gallbladder) in **sheep**, **cattle**, and **horses**. Photosensitization and dermatitis has also been reported. Extracts paralyze motor nerves of **vertebrates**, due to a poisonous alkaloid, ***cynoglossine***, with ***curare***-like action. Death has usually occurred within 24 h after ingestion.

CYPERMETHRIN
= *Agrothrin* = *Ammo* = *Arrivo* = *Barricade* = *Cymbush* = *Cynoff* = *Cypercare* = *Cyperkill* = *Cypersect* = *Demon* = *Dysect* = *Ectomin* = *Ectopor* = *Fastac* = *Flectron* = *FMC-30980* = *NRDC 149* = *Nurelle* = *Polytrin* = *PP-383* = *Ripcord* = *Rycopel* = *Sherpa* = *Topclip Parasol*

Insecticide, synthetic pyrethroid.

Untoward effects: Some **workers** handling it develop symptoms within 30 min–3 h, which last 30 min–8 h. Symptoms include a feeling as if a cold wind was hitting the face; feeling of a ***nettle***-like rash or sunburn; and hot or cold facial sensations. Allergic reactions from skin contact or inhalation are common in **man**. The U.S. Environmental Protection Agency declared it a weak oncogen with tolerances of 0.05 ppm in **sheep** fat and milk. The highest level then found in ***lanolin*** was 13.6 ppm. Extrapolating this to breast-fed **infants** in a worst-case scenario yielded a daily dose of 0.003 mg/kg.

The highest no-effect level in **rats** is 7.5 mg/kg, and 1 mg/kg in **dogs**. A 45-year-old **male** ingested it with symptoms of nausea, vomiting, colic, tenesmus, diarrhea, respiratory paralysis and coma.

CYPRIPEDIUM sp.
= *Lady's Slippers*

Untoward effects: Contact with the leaves or stems by **people** leads to dermatitis and weeping blisters from their ***cypripedin*** content.

CYPROHEPTADINE
= *Anarexol* = *Antegan* = *Cipractin* = *Ifrasarl* = *Nuran* = *Periactin* = *Peritol* = *Vimicon*

Antihistamine, antipruritic.

Untoward effects: Drowsiness, sleepiness, tiredness, insomnia, restlessness, confusion, anorexia, increased appetite, weight gain, dry mucous membranes, blurred vision, increased ocular pressure, cholestasis, pruritus, and lichenoid skin eruptions. Rarely, central nervous system stimulation and photosensitivity reactions. Birth deformity (shortened arm) in **child** born to **mother** treated with it during pregnancy. Can appear in breast milk. May cause jaundice when given with ***imipramine pamoate***. May falsely imply drug-induced pancreatitis, due to increased serum amylase and lipases. Central anticholinergic syndrome with agitation and hallucinations in 8-year-old **male** taking 8 mg/day for migraine prophylaxis.

Occasionally, causes pupillary dilation in **dogs**. Although useful as an appetite stimulation in anorexic **cats**, it should not be a substitute for adequate diagnostic work. It has caused tachycardia, hyperactivity, or drowsiness in some **cats**. Fetal resorption occurs in treated **rats** and **mice**.

LD_{50}, **rat**: 295 mg/kg; **mouse**: 123 mg/kg; LD_{LO}, **dog**: 50 mg/kg.

CYPROLIDOL
= *IN 1060*

Psychotherapeutic.

Untoward effects: Motor restlessness, insomnia, burning sensation in epigastrium, and diarrhea.

IV LD_{50}, **mice**: 63 mg/kg.

CYPROTERONE
= *Androcur* = *CPA* = *Cyprostat* = *Diane 35* = *Dianette* = *SH-714* = *SH-881* = *SH-80881* = *Sinevir*

***Antiandrogen, progestogen.* The acetate is commonly used.**

Untoward effects: Gynecomastia (occasionally, painful), alopecia, decreased libido, profound decrease in sperm count, transient testicular pain, depression, and edema in **males**. Irregular menstruation, gynecomastia, skin dryness, porphyria cutanea tarda, and decreased libido in **females**. Nausea, weight decrease, fatigue, vertigo, headache, liver damage, and pulmonary embolism have also occurred. A case of cyproterone-induced hyperosmolar hyperglycemia reported in a 84-year-old **male**. Approximately 2% of the administered dose is transferred to **newborns** via breast milk.

Male fetuses of treated **dog bitches** showed feminization of their genital organs. Decreased weight of testicles, spermatogenesis, and diameter of tubules in treated **boars**. Adrenal suppression in **rats** and **humans**, but not **monkeys**. Has caused fetal anomalies in **mice**. Decreased uterine motility in **rats**.

IM TD_{LO}, **rat**: 500 mg/kg.

CYROMAZINE
= *CGA-72662* = *Citation* = *Larvadex* = *Trigard* = *Vetrazin*

Insecticide.

Untoward effects*: *Melamine (q.v.), its metabolite, is a potential carcinogen for **man**, since it has caused tumors in **rats**.

Treatment of egg-laying **hens** must cease 3 days before being slaughtered for **human** food use.

CYSTE ABSOLUTE
= *Ciste Absolute*

A greenish alcoholic extract of *Cistus ladaniferus* or *C. incanus*, posessing a sweet amber-like odor. Used

as a fragrance in soaps, detergents, cosmetic creams and lotions, perfumes, and in foods.

***Untoward effects*:** 4% in petrolatum caused no irritation in a 48 h closed patch test on 25 **human volunteers**. No sensitization reported in **humans**.

LD_{50}, **rats** and acute dermal LD_{50} in **rabbits**: > 5 g/kg.

CYSTEAMINE

= *Becaptan* = *L 1573* = *Lambratene* = *MEA* = *Mercamine* = *Mercaptamine* = *2-Mercaptoethylamine*

Experimental radioprotective agent, acetaminophen antidote, and cystinosis treatment.

***Untoward effects*:** Nausea, anorexia, and diarrhea. Rarely, skin rashes, encephalopathy, seizures, reversible leukopenia, and abnormal liver function tests.

Causes duodenal ulcers and decreases ***prolactin*** production in **rats** within a few hours after oral or SC use.

LD_{50}, **mouse**: 625 mg/kg; IP LD_{50}, **rat**: 232 mg/kg, **mouse**: 305 mg/kg; IV, **rabbit**: 150 mg/kg.

Acetate: IP LD_{50}, **mouse**: 3.2 g/kg.

CYSTEINE

= *Thioserine*

***Amino acid, dough conditioner, radioprotective, antioxidant, flavor-enhancer, glutathione precursor, detoxicant, and reducing agent in permanent wave solutions.* Approximately 60 tons used annually in the U.S.**

***Untoward effects*:** Careless over-the-counter use of this chelating agent can leach out essential, as well as toxic, minerals.

High dosages in **infant mice** cause brain damage with end stage neuronal necrosis in 2–3 h; lower dosage caused more extensive damage in 4–6 h. Increases the mortality of **rats** poisoned by ***cadmium***.

CYSTINE

Amino acid.

Untoward effects*:** A recessively inherited disease, cystinosis, in **humans** deposits cystine crystals in cornea, bone marrow, and reticulo-endothelial cells. Cystine stones occur in the urinary tract of **people** with a tendency to gout, urate stone formation, and cytinuria, especially with excessive acidification of urine with > 4 g ***ascorbic acid, etc. Poor solubility in urine. Cystinuria occurs in **man**, **dogs**, and **mink**.

With excessive dietary intake by **rats** fed low levels of ***Cu*** and high levels of ***Mo***, anemia, diarrhea, and deaths occur. Spontaneous aortic rupture potentiated in **turkeys**, when a 3% level is in their feed. Intestinal bacteria may be a factor in producing ***β-mercaptoethylamine*** (***cysteamine***) from the cystine in their gastrointestinal tract. Dietary cystine aggravates foot pad dermatitis in **poults** and ***Se-vitamin E*** deficiency in **ducklings**.

CYTARABINE

= *Alexan* = *Arabinoside* = *Arabinosyl Cytosine* = *Arabitin* = *Ara-C* = *Aracytidine* = *Aracytine* = *CHX-3311* = *Cytar* = *Cytarbel* = *Cytosar* = *Cytosine* = *Erpalfa* = *Iretin* = *NSC 63,878* = *U-19920*

Antineoplastic, antiviral.

Untoward effects*:** Nausea, vomiting, diarrhea, headache, anaphylaxis, an acute cerebellar syndrome characterized by dysarthria and ataxia, blanching between finger joints, and dusky red bands near the articulations. Blanched areas become bullous and cultures are sterile. Over the next few weeks, desquamation and healing occur. Delayed toxicity includes bone marrow decrease with leukopenia, thrombocytopenia, granulocytopenia, reticulocytopenia, anemia, megaloblastosis, stomatitis, hepatic dysfunction and peliosis, thrombophlebitis, pseudotumor cerebri, inflamed seborrheic dermatitis, severe parkinsonism with associated autonomic dysfunction, and fever. Conjunctivitis, oral ulceration, pulmonary edema, pneumonitis, rhabdomyolysis, peripheral neurotoxicity, hearing loss, and hyperuricemia may develop, due to lysis of neoplastic cells. High dosage can induce confusion. White blood cells decrease and platelet decrease of 30–40% may limit its use, unless transfusions are available. Bone marrow aplasia may be fatal. Reversible corneal toxicity with opacities in lower epithelial layers with punctate staining in later stages, sometimes, with corneal ulcers, pain, and iritis. Chromosomal and congenital abnormalities in some **infants** born to treated **mothers**. The former occurred when used with ***thioguanine. Peritonitis, pericarditis, tachypnea, maculopapular rash, palmar and plantar skin toxicity with bullae, increased liver enzymes, alopecia, and herpes zoster have followed therapy. Pancreatitis reported when used with ***asparaginase***. Intrauterine growth retardation, teratogenicity, and **fetal** death reported. Therapeutic serum levels are 10–15 mg/l and 30–40 mg/l is toxic.

Causes abnormalities and severe growth retardation in **chicks**. Suppressed antibody responses in **hamsters** and **rabbits**. Increased fetal mortality and numerous fetal abnormalities in treated **rats**. **Rabbit** corneas treated with it develop central megalocytes in the basal layers of the corneal epithelium. Treatment of young **rats** consistently produced alopecia. Suppresses immune response and antibody production in many species.

CYTHIOATE

= *CL 26,691* = *Cyflee* = *ENT 25,640* = *Proban*

***Organo-phophorus insecticide.* The term *Proban* is also used for a fire retardant on *cotton* textiles.**

***Untoward effects*:** Cholinesterase inhibitor. Can cause mild eye irritation in **man** and cause a sense of tightness in the chest, stomach pains, miosis, vomiting, sweating, and diarrhea. Prolonged exposure may cause flu-like symptoms with weakness, anorexia, and malaise. Avoid concurrent use with other anticholinesterase agents, phenothiazines, and muscle relaxant drugs. Use with other anticholinesterases increases the risk of toxicity. Dermal absorption occurs. In **dogs**, 175 mg/kg as a single oral dose leads to emesis, tremors, salivation, and diarrhea in 4–6 h.

LD_{50}, **rats**: 160 mg/kg; **mouse**: 38 mg/kg; acute dermal LD_{50}, **rabbit**: > 2.5 g/kg.

CYTISINE

= *Baptitoxine* = *Cytiton* = *Laburnine* = *Sophorine* = *Tabex* = *Ulexine*

Untoward effects*:** A very toxic alkaloid in all parts of ***Cytisus canariensis (q.v.), ***C. laburnum*** (q.v.), ***mescal beans*** (q.v.)., and ***Kentucky coffee tree*** (q.v.) sprouts. A variable no-effect period follows ingestion, then, excessive salivation, burning of mouth and esophagus, thirst and nausea. Violent and persistent vomiting, sometimes hemorrhagic. Severe abdominal pain, meteorism, and diarrhea. Skin cold and clammy, pulse at first rapid, then slow and irregular, respiration dyspneic and sometimes Cheyne-Stokes type. Headache, staggering gait, vertigo, somnolence. Muscular twitchings and cramps. Aphasia, visual disturbances, delirium, and hallucinations. Unconsciousness. Death occurs by respiratory paralysis, or later, from uremia, following oliguria and anuria. **Children** who shell and eat the seeds from the pea-like pods of the ***Golden Chain Tree*** are often poisoned by it. Toxic action is similar to ***nicotine***, but is more potent to the sympathetic ganglia.

Excreted in **cow's *milk***, giving it a bitter taste.

LD, **cats** and **dogs**: 2–3 mg/kg; **goats**: ~70 mg/kg; **chickens** and **pigeons**: 7–9 mg/kg; LD_{50}, **mouse**: 101 mg/kg.

CYTISUS (GENISTA) CANARIENSIS

Untoward effects*:** Native to the Canary Islands and used by Yaqui **medicine men** in northern Mexico as an hallucinogenic plant. Contains ***cytisine, the psychoactive principal of ***mescal beans***. Contains small amounts of ***sparteine***, an abortifacient and oxytocic.

CYTISUS LABURNUM

= *Golden Chain Tree* = *Laburnum* = *Laburnum anagyroides*

Untoward effects*:** From its toxic alkaloid, ***cytisine (q.v.). Symptoms include excitement, vomiting, incoordination, convulsions, coma, and, rarely, death. **Children** often eat seeds, pods, and flowers; symptoms are vomiting and restlessness. More rare, but violent symptoms are abdominal pain, dizziness, and convulsion. All occur within a short time after ingestion and subside in about 12 h. Mild to serious poisoning reported from its contribution to ***honey*** production. All parts of the plant are probably poisonous. 58 **boys** were poisoned simultaneously by eating the roots with symptoms of extreme sleepiness, vomiting, convulsive movements (usually tetanic), coma, some frothing at the mouth, and unequally dilated pupils. **Some** had severe diarrhea, loss of pupillary reflex, increased temperature, delirium, and cyanosis. Extreme cerebral hyperemia, nephritis, and mucous membrane erosions in the colon were noted in the **dead**.

Toxic symptoms in **animals** (**cattle**, **dogs**, **pigs**, and **sheep**) are similar to those in **man**. Deaths have occurred. Affected **horses** exhibit general and sexual excitement, muscle tremors, incoordination, convulsions, and coma.

Seeds, LD_{50}, **dog**: 6 g/kg; **horse**: 0.5 g/kg.

D

DACARBAZINE

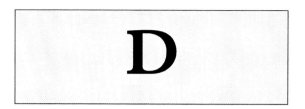

= Dacatic = Deticene = DIC = DTIC = DTIC-Dome = NSC-45388

***Antineoplastic.* IV use.**

***Untoward effects*:** Nausea and vomiting (usually within 1–3 h, lasting up to 12 h in 90% of treated **patients**), diarrhea, and local pain are acute toxic reactions. May cause thrombophlebitis and extravasation; causes severe local necrotic reactions. May induce or potentiate parkinsonism. Delayed toxicity consists of bone marrow depression, with leukopenia (30–50%), thrombocytopenia, alopecia, and peripheral eosinophilia, as well as a flu-like syndrome, consisting of fever, myalgia, and malaise. Mild to moderate SGOT increase with centrilobular or midzonal hepatic necrosis. A case of jaundice reported associated with its use. May induce photosensitivity with occasional rashes and hyperpigmentation, and is a local skin irritant. A local cutaneous reaction histologically resembling a fixed-drug reaction in a 19-year-old **male**. Facial flushing, paresthesia, and photosensitivity reported. Anaphylactic shock in 3-year-old **female** on third IV course; during **her** second course, **she** exhibited restlessness, coughing, and crying. Potency can be decreased 50% in 4 h at room temperature from exposure to light. During infusion, a **patient** may show auricular fibrillation, followed by paroxysmal tachycardia.

Teratogenic, mutagenic, and carcinogenic in experimental **animals**. Oral or IP use in **rats** may lead to mammary carcinomas (100%), sarcomas of thymus (80%); spleen (20%) and brain (20%) tumors in about 18 weeks. IP use in **mice** may lead to lung, hematopoietic tissue, and uterine tumors.

DACLIZUMAB

= Daclximab = Zenapax

***Monoclonal antibody.* Blocks interleukin-2 receptor on T-lymphocytes. For IV infusion.**

Untoward effects*:** Reactions similar to those of ***basiliximab (q.v.) occur (i.e. chills, fever, wheezing, chest tightness, dyspnea, pulmonary edema, tremors, nausea, vomiting, seizures, cerebral edema, and encephalopathy).

DACTINOMYCIN

= *Actinomycin* A_{IV} = *Actinomycin* C_1 = *Actinomycin D* = *Actinomycin* I_1 = *Actinomycin IV* = *Actinomycin X* = *Cosmegen* = *Meractinomycin* = *NSC-3053*

***Antineoplastic, antibiotic.* The first antibiotic for which antitumor activity was reported in the early 1950s in Germany. For IV use.**

***Untoward effects*:** Acute toxicity within a few hours including nausea, abdominal pain, vomiting, fever, diarrhea, and local irritation (thrombophlebitis and cellulitis) after extravasation. Delayed toxicity including bone marrow suppression, with thrombocytopenia, leukopenia, agranulocytosis, pancytopenia, reticulopenia, and anemia within 1–2 weeks; alopecia, oral and gastrointestinal mucositis and ulceration, glossitis, proctitis, cheilitis, and pneumonitis. Acneiform eruptions, folliculitis, cutaneous erythema, desquamation, hyperpigmentation, anaphylactic reactions, and a fatal case of agranulocytosis have been reported. Has caused pericarditis in **man**. Hepatomegaly has occurred. Death in a newborn **infant** treated with it has caused it to be contraindicated in **newborns**. Sweat gland disturbances, folliculitis of the face, back and chest, and Stevens–Johnson syndrome reported. Most symptoms enhanced by concurrent radiation therapy.

IV, TD_{LO}, **human**: 40 μg/kg/4 days.

Teratogenic and carcinogenic in **animals**. Dosing **rats** on days 7–9 of gestation led to malformations (eyes, central nervous system, viscera, cardiovascular, and skeletal) in up to 50% of surviving fetuses. Bone marrow suppression, stomatitis, and alopecia in **dogs** noted 2–4 days after a course of therapy, and may not be maximal before 1–2 weeks.

LD_{50}, **rat**: 7.2 mg/kg; **mouse**: 13 mg/kg; IV, LD_{50}, **rat**: 460 μg/kg.

DACTYLIS GLOMERATA

= Cocksfoot = Orchard Grass

Untoward effects*:** After ingestion, **lambs** fail to grow and exhibit stiffness. Often contaminated with ***ergot (q.v.).

DACTYLOCTENIUM AEGYPTIUM

A fodder for fattening and milk production in West Bengal livestock.

Untoward effects*:** Releases ***hydrocyanic acid in hot dry periods, poisoning **livestock**.

DAFFODIL

= Jonquille = Narcissus pseudonarcissus

Ornamental plant.

Untoward effects: Pretty, but dangerous. All parts, especially the bulbs, are poisonous. One bite of the bulb by a **child** can cause severe gastrointestinal irritation with nausea and vomiting. Abdominal cramps, diarrhea, shivering, and dyspnea occur, and, less commonly, central nervous system excitement, depression, collapse, coma, and even death. The alkaloid, ***narcissine***, is responsible. A **woman** used a daffodil bulb instead of an ***onion*** in stewing beef. Vomiting and shivering occurred in five **adults** eating the meal. Severe diarrhea also occurred in two. **All** recovered in 2 h. A contact dermatitis reported in **growers** and **those** who regularly handle the flowers.

DALFOPRISTIN

Use: In a 70 : 30 combination with ***quinupristin***, known as ***Synercid***, for IV use against life-threatening ***vancomycin***-resistant *Enterococcus faecium* infections. Both are streptogramin antibacterials.

Untoward effects: Pain, inflammation and edema at injection site, and thrombophlebitis reported. Arthralgias, myalgias, and increased conjugated ***bilirubin*** occur frequently.

Drug interactions: May increase serum concentrations of ***cyclosporine, midazolam,*** and ***nifedipine*** and potentiate ***cisapride***, increasing QTc interval.

DALTEPARIN

= *Boxol* = *FR860* = *Fragmin* = *Kabi 2165* = *Tedelparin*

***Antithrombotic.* A low molecular weight fragment of *heparin*.**

Untoward effects: Hematomas and pain at injection site and minor bleeding complications. Increased serum transaminase in ~3%, thrombocytopenia in < 1%. Skin necrosis and allergic reactions (pruritus, rash, fever, bullous eruptions) and anaphylaxis reported.

DAMAR

= *Dammar* = *Gum Damar* = *Resin Damar*

Use: In adhesive plasters, varnishes, lacquers, and in microscopy.

Untoward effects: A **person** was sensitive to it in adhesive plaster, and an **artist** had a positive patch test to it, after suffering from dermatitis.

DAMINOZIDE

= *Alar* = *B-9* = *B-995* = *B-Nine* = *Kylar*

Plant growth regulator.

Untoward effects: Carcinogenic in **rats** and **mice**. Some breaks down in the presence of water and under acidic pH levels into ***unsymmetrical 1,1-dimethylhydrazine*** (***UDMH***), also a carcinogen for **rats** and **mice**, when treated ***apples*** are cooked. In 1989, the Environmental Protection Agency's proposed banning of its use on foods caused havoc in the ***apple, peanut,*** and ***grapefruit*** markets.

LD_{50}, **rat**: 8.4 g/kg; **mouse**: 6.3 g/kg.

DANAPAROID

= *Orgaron* = *Org 10,172* = *Lamoparin* = *Lomoparan*

***Heparinoid.* A partially depolymerized mixture of porcine *chondroitin sulfate, dermatan sulfate,* and *heparan sulfate*. Used SC.**

Untoward effects: Occasional bleeding complications. Fever, nausea, constipation, injection site pain, rash, peripheral edema, insomnia, joint problems, urinary tract infection, edema, asthenia, dizziness, and urinary retention reported in decreasing frequency. Use with spinal or epidural anesthetics can induce hematomas in those areas that can result in long-term paralysis. Recurrence of deep vein thrombosis has occurred, but less than with ***heparin***.

DANAZOL

= *Bonzol* = *Chronogyn* = *Cyclomen* = *Danocrine* = *Danol* = *Danoval* = *Ladogal* = *Win-17757* = *Winobanin*

Antigonadotropin.

Untoward effects: Due to its weak androgenic and anabolic activity, adverse reactions, such as weight gain, edema, acne, oily skin, decreased breast size, mild virilism, hirsutism (~10%), and increase or decrease in libido can occur. Flushing (68%), sweating, vaginitis, and clitoral hypertrophy reported in **women**. Irregular menses or amenorrhea and atrophy of endometrial tissue are common, especially on large doses. Nausea, vomiting, headache, nervousness, backache and muscle pain, and, paradoxically, rare cases of alopecia. Has caused abnormal liver function tests and cholestatic jaundice. A **female infant** whose **mother** inadvertently received it (800 mg/day) in early pregnancy developed pseudohermaphroditism. Many cases of virilization of **female fetuses** after their **mothers** received 800 mg/day dosage. Clitoral hypertrophy has occurred in **children** treated for precocious puberty. Thrombocytopenia developed in four **females** of 22–49 years, after 400–800 mg/day/7–10 days. Polycythemia in an anephric **patient** on 400 mg/day/6 months. Voice changes reported in several **women**. The longer the therapy, the less it may be reversible on withdrawal. Hypertension developed in a 28-year-old **male** 3 months after 200 mg twice daily. Hepatic peliosis in a 74-year-old **female** on 600 mg/day/3 months. A 26-year-old **female** developed ***insulin***-dependent diabetes mellitus after 400 mg twice daily/8 weeks. Diabetes resolved after drug withdrawal. Erythema multiforme in a 49-year-old **male** and a

67-year-old **female** on 600 mg/day and exacerbation of lupus-like illness in a 70-year-old **male**. Occasionally severe skin rashes in **women** attributed to the gelatin capsule dyes (***D&C Yellow, 10, FDC Red #3,*** and ***FDC Yellow #6***), as the contents alone caused no such reaction. Alopecia, microscopic hematuria, headaches, myalgia, myokymia, weakness, increased or decreased libido, pseudotumor cerebri after withdrawal, hepatocellular adenoma in a 34-year-old **female** after 600 mg or 400 mg/day/4 years. Hepatocellular carcinoma in a 49-year-old **female** after 200 mg/2 years, and possible cause of uterine leiomyomas in a 32-year-old **female** after 200–400 mg/day/6 months.

Drug interactions: May require ~50% decrease in ***warfarin*** and ***carbamazepine*** dosage. Can cause clinically serious increase in serum concentration of ***cyclosporine*** and ***tacrolimus***.

DANILEUKIN DIFTITOX
= *Ontak*

Antineoplastic.

Untoward effects: Dyspnea, rash, chest tightness, tachycardia, hypotension, dysphagia, back pain, headache, chills, fever, pruritus, diarrhea, nausea, and vomiting. Delayed reactions include anemia, hypoalbuminemia, anorexia, asthenia, hypocalcemia, increased aminotransferases, and edema.

DANSYL CHLORIDE

Untoward effects: Hypersensitivity reactions, skin sensitizer.

IV, LD_{50}, **mouse**: 56 mg/kg.

DANTHRON
= *Altan* = *Antrapurol* = *Chrysazin* = *Dantron* = *Diaquone* = *Dorbane* = *Istizin* = *Modane*

***Irritant cathartic.* An anthraquinone.**

Untoward effects: Excreted in **human** breast milk and large doses will affect bowel movements of **infants**. May cross the placenta into **fetus**. May color urine red (reddish-violet to yellow-brown, depending on pH). A benign melanosis coli occurs and will gradually disappear after drug withdrawal. A periportal hepatitis and a perianal exanthema, especially in **infants** and **children**, resembling erysipelas, has been reported. The lesions are sharply demarked and dark red. After a few days, they become brown and livid, and then disappear. Occasionally skin eruptions. If given 6–10 h before a ***PSP*** (***phenolsulphonphthalein***) test, the red dye will give a false high ***PSP*** value. Given with surfactants, such as ***dioctyl calcium*** or ***sodium sulfosuccinate***, it may cause hepatitis, due to increased absorption.

Prolonged dosage to **rats** and **mice** induced intestinal and liver tumors, prompting the FDA, in 1987, to withdraw its approval of danthron products for **man**.

IP, LD_{LO}, **mouse**: 500 mg/kg.

DANTROLENE
= *Dantamacrin* = *Dantrium* = *F-440*

Skeletal muscle relaxant.

Untoward effects: Drowsiness, dizziness, nausea, vomiting, salivation, weakness, fatigue, headache, nervousness, insomnia, depression, and diarrhea, often disappearing during treatment. Anorexia, pruritus, and acne-like skin eruptions are common during long-term therapy. Has produced a dose-related fatal hepatocellular syndrome (0.1–0.2%) in susceptible **patients**. This is common in **women** over 30 years of age ingesting > 300 mg/day/2–6 months and taking ***estrogens***. Transient elevations of SGOT and SGPT occur. Symptomatic hepatitis in 0.35–0.5% of **patients**. Some decrease in spermatogenesis and motility noted. Malignant liver tumors reported after long-term use. Rarely, pleuritis, pleural effusions, pneumonitis, visual disturbances, constipation, hallucinations, and incontinence reported. Bilateral sensorineural hearing loss in 19-year-old **female** after 50 mg. In decreasing order, leukocytosis, leukopenia, and thrombocytopenia reported. Tachycardia, phlebitis, erratic blood pressure, and heart failure occur. May color urine orange to red.

TD_{LO}, **human**: 320 mg/kg; LD_{LO}, **woman**: 600 mg/kg.

Drug interactions: Prolonged recovery time from ***vecuronium*** neuromuscular block.

Hepatotox, sedation, and skeletal muscle weakness reported in treated **cats** and **dogs**. **Rat** trials indicate synergistic action with ***theophylline***, requires decrease in dantrolene dosage.

DAPHNE

The genus of ~70 species occurs world-wide and is a common ornamental.

Untoward effects: Its bark, leaves, and berries contain virulent poisons, particularly ***daphnetoxin*** or ***daphnin***. Ingestion of a few attractive red berries or a small amount of bark has been fatal (~30%). Causes gastrointestinal burning or ulceration, nausea, salivation, bloody vomitus, and diarrhea. Abdominal pain, convulsions, hallucinations, stupor, renal failure, internal bleeding, and coma precede death. Skin vesicant. **Children** are more commonly affected by the red berries. ***Honey*** from *Daphne sp.* has caused mild to serious poisoning in **man**.

D. Mezereum = ***February Daphne*** = ***Daphné jolibois*** = ***Dwarf Bay*** = ***Mezereon*** = ***Spurge Flax*** = ***Spurge Laurel*** = ***Spurge Olive*** = ***Wood Laurel*** is a common offender, and, in Europe, its berries have been used in bait to kill **wolves** and **foxes**. Its toxic chemicals are ***daphnetoxin, dihydroxycoumarin,*** and ***mezerein***. A drastic irritant and purgative. Severe burning of lips, mouth, and throat, followed by inflammation in **man**. In severe cases, inflammation of stomach and intestines, colic, vomiting, bloody diarrhea, nephritis, nervousness, weakness, and convulsions. Ingestion of 12 berries has been fatal to a **human**. 1 oz of the bark has killed a **horse**, and 12 g has killed a **dog**.

D. Kamtschatica is a **walrus** arrow and dart hunting poison in the North Pacific region.

IP, LD_{LO}, **mouse**: 500 mg/kg.

DAPSONE

= *Avlosulfon* = *Croysulfone* = *DADPS* = *DDS* = *Diaphenyl-sulfone* = *Diphenasone* = *Disulone* = *Dumitone* = *Eporal* = *1358F* = *Novophone* = *Sulfona-Mae* = *Sulphadione* = *Udolac*

***Antibacterial, coccidiostat.* A sulfone.**

Untoward effects: TD_{LO}, **human**: 18 g/kg/15 years, **woman**: 168 mg/kg/14 days.

Blood and cardiovascular: Hemolytic anemias in glucose-6-phosphate dehydrogenase-deficient **patients** and in normal **individuals**. Leukopenia, leukocytosis, eosinophilia, neutropenia, agranulocytosis, methemoglobinemia, Heinz-body anemia, and tachycardia can occur. Cyanosis, severe methemoglobinemia, and delayed sulfhemoglobinemia after ingesting 3 g. Severe and even fatal hypoalbuminemia and bone marrow depression reported. Agranulocytosis in 16 U.S. **soldiers** given daily doses for 3–13 weeks, as prophylaxis against falciparum malaria. Within 9 days, eight **men** died, due to sepsis. Tachycardia, secondary to methemoglobinemia and hemolysis, reported.

Fetus and neonate: Passes readily into breast milk and has caused hemolytic anemia in **infants** breast feeding on treated **mothers**. Neonatal hyperbilirubinemia after **mother's** use during pregnancy.

Miscellaneous: Hepatitis, depression, psychosis, hematuria, cyanosis of lips, pale-gray complexion, anorexia, nausea, vomiting, dizziness, progressive weakness, facial edema, cervical lymphadenopathy, exudative tonsillitis, nephrotic syndrome, fever, malaise, manic-depressive psychosis, pulmonary eosinophilia, and death due to sepsis in U.S. **soldiers** in Vietnam, taking 25 mg/day/3 weeks–3 months as prophylaxis for falciparum malaria. A 17-year-old **male** with leprosy had a fatal reaction to 100 mg/day/3 weeks. Dapsone-resistant *Mycobacterium leprae* is increased.

Neuropathy: Sensory and motor peripheral neuropathies and paresthesia may, in part, be dose-related. A 20-year-old **male** ingesting 600 mg/day/10 days developed severe neuropathic leg weakness and blindness. Neither event improved after 6 months. Agitation, insomnia, depression, headache, and hallucinations have been reported. Large doses may cause a psychosis.

Poisoning: A 23-month-old **female** swallowed 1.5 g and a 3-year-old **male** (unknown quantity ingested) developed tachycardia, exertional dyspnea (decreased hemoglobin and leukocytes in **male**); methemoglobin level of 130%, early vomiting, purpuric spots, and cyanosis in **female**. IV ***methylene blue*** 2 mg/kg gave dramatic results. Two, **male** 25-day-old **twins** each received 100 mg in error leading to methemoglobinemia. **One** responded to IV ***ascorbic acid***; the **other** required ***methylene blue***. In an 18-month-old **child**, ***activated charcoal***, 10 g every 6 h for 64 h to reverse methemoglobinemia. Similar treatment for 7 days saved a 45-year-old **male** who swallowed ~10 g. Its half-life is ~88 h and 67 h for its metabolite, ***monoacetyldapsone***. A 31-year-old **male** ingested 7.5 g in a suicide attempt and had permanent retinal damage. Therapeutic serum concentrations are 10–15 mg/l and toxic levels are generally 30–40 mg/l. Other workers reported these numbers as 0.5–2 and 10, respectively.

Skin: Rashes, exfoliative dermatitis, urticaria, fixed eruptions, porphyrsia, toxic epidermal necrolysis, dermatitis herpetiformis, morbilliform rashes, and alopecia noted. Severe photosensitivity, later, developing discoid lupus erythematous lesions.

Drug interactions: Clearance is decreased by ***probenecid*** and increased by ***rifampin***. May decrease effectiveness of some ***oral contraceptives***. ***Didanosine*** decreases its absorption.

Withhold milk from treated quarters for at least 96 h after last treatment, to avoid undesirable residues in food for **human** consumption. Nausea, vomiting, and liver pathology has occurred in treated **dogs**. Caused tumors in **mice** and mesenchymal tumors of spleen and peritoneum of **rats**. Causes methemoglobinemia in **horses**, **pigs**, and **dogs**, but not in **sheep** and **cattle**. Central nervous system damage in **sheep**, and 200 mg/kg may be fatal. **Rats** fed a diet containing 0.3% for 2 years developed intestinal, splenic, and thyroid neoplasms.

LD_{LO}, **rat**: 1 g/kg; LD_{50}, **mouse**: 496 mg/kg.

DATURA

A number of species, such as *D. candida* and *D. molis*, are called *Angel's Trumpets*, but, in the U.S., it

usually refers to *D. suaveolens* (q.v.) or *D. metaloides* (q.v.), all of which cause similar toxicity. *Burgmansia* is considered to be a subgenus under *Datura*.

Untoward effects: It has played an important role in shamanism and religious rituals, due to its hallucinogenic, stimulant, and narcotic effects, mainly from its alkaloids, ***scopolamine*** (***hyoscine*** (q.v.)), ***hyoscyamine*** (q.v.), and ***atropine*** (q.v.). Literature often fails to identify the species. Six **people** in England, 17–34 years old, ate the leaves and brewed a tea from them, leading to mydriasis, hallucinations, disorientation, and delirium. School **children** eating the leaves staggered, were confused, showed mydriasis and blurred vision, and increased creatine phosphokinase, lactate dehydrogenase, and SGOT. The **wives** and **slaves** of deceased **Indian chieftains** in the Bogota area were given potions of it to induce stupor before being buried alive with their departed **chieftain**.

D. discolor = ***Kamóstim*** ("plant make-squint") = ***Karóokkoot*** ("plant make-crazy") is used by **Seri Indians** of Mexico.

D. fastuosa = ***Mondzo*** seeds are used in Africa as a rodenticide with stored grain, and in fertility rites with **Tsonga girls**.

D. metaloides = ***Angel's Trumpet*** = ***Moonflower*** = ***Sacred Datura***. Occasionally called ***Locoweed***, a term also used for ***Astragulus*** and ***Oxytropis sp***. The leaves, flowers, nectar, and seeds are poisonous. Symptoms are mydriasis, dry mouth, increased temperature, intense thirst, confusion, delirium, hallucinations, rapid pulse, and even death.

D. metel = ***Apikan*** = ***Hairy Thornapple*** = ***Haukata*** = ***Hindu Datura*** = ***Man T'o Lo*** = ***Mnanaha*** = ***Yaro*** = ***Zakami*** is often mistaken for *Veronia amagdalina*, which is edible and used for chewsticks, with resultant toxicity. Its seeds and young leaves are especially poisonous. **Children** are often poisoned by eating the flowers or sucking the nectar. A **family** of four were poisoned after 3 oz of its seeds were added to soup instead of lentils. Within 10–15 min, all four vomited, showed incoordination, hallucinations, hyperactivity, confusion, flushed faces, dilated and fixed pupils, increased blood pressure, increased extensor plantar reflexes, and dry mouth (with some salivation). The **family** recovered in 12 h. **Animals** consume it when starved for forage, or when it is accidentally in their hay. Tachycardia, tachypnea, mydriasis, diarrhea, and stiff gait are characteristic symptoms in affected **cattle**.

D. rosei poisoning from flower stamens in 2-year-old **male** and 3-year-old **female** in New Zealand. Symptoms were of acute ***atropine*** poisoning and responded to ***physostigmine*** (IV pushes can cause cardiac arrhythmias) as an antidote. Discharged from the hospital after 36 h.

D. strammonium = ***Angel's Trumpet*** = ***Apple of Peru*** = ***Chamico*** = ***Devil's Apple*** = ***Devil's Trumpet*** = ***Dewtry*** = ***Fireweed*** = ***Green Dragon*** = ***Jamestown Lily*** = ***Jamestown Weed*** = ***Jimson Weed*** = ***Mad Apple*** = ***Mezerbae*** = ***Stinkblaar*** = ***Stinkweed*** = ***Stinkwort*** = ***Thorn Apple*** = ***Tolguacha*** = ***Tonco-tonco***. All parts of the plant are toxic, especially the seed in its hard prickly fruit and the nectar in its flowers. Used criminally by **Peruvian Indians** who feel they've been wronged, or to solve love affair jealousies. Particularly common in the western U.S. In 1676, **British soldiers** in Jamestown, Virginia mistook it for ***lamb's quarters*** (*Chenopodium album*) used as salad greens. At least one died. The mass poisoning of the British **troops** with comical, unmilitary behavior, was followed by amnesia and hallucinations. One of the first recorded cases of its toxicity was in 37–38 B.C., when Marc Antony's **troops** ate it. **Those** who did acted "silly", confused, and stupified, and even death occurred. Poisoning has been caused by smoking the seeds or leaves or drinking decoctions of them. Its sap was once used as knock-out drops to shanghai **sailors**. In India, **criminals** use the seeds in rail and road robberies, to stupefy their **victims**. An increasingly common substance of abuse in some areas. In Ohio, 6/10 **people** were hospitalized with acute anticholinergic syndrome, after deliberately ingesting some seeds. Key symptoms are the same as for other species, namely, insatiable thirst, delirium, decrease or loss of motor control, tachycardia, vertigo, urinary retention, mydriasis, increased temperature, convulsions, coma, and death. Large numbers of intoxications in **adolescents** reported in Baltimore, Maryland in the 1970s and 1980s. In the 1980s and 1990s, Kentucky Poison Control Center reported 29 cases in those aged 16–39. An additional 29 adolescent **patients** were hospitalized in El Paso, Texas in 1979 and five required catheterization, due to urinary retention. Disorientation and hallucination occurred in a 28-year-old **male** after smoking *Datura*-containing cigarettes. In 1980, many U.S. **soldiers** who had used it for "highs" were warned not to eat the German variety, which often caused dramatic increase in blood pressure and heart rates, fever, failure of **their** eyes to respond to light, and a rash spreading over **their** chests and shoulders. **GIs** in Europe often became ill after picking and devouring the poisonous *Datura* berries, thinking they were ***blueberries***. A condition known as **"cornpickers'** pupil" consists of dilated pupils in **farmers** exposed to its pollen while harvesting grain. It can cause a contact dermatitis and it cross-reacts with ***chrysanthemums, marigolds, sunflowers,*** and ***zinnias***. Its alkaloids are found in the nectar and have contaminated ***honey***. Total alkaloid content of the plant has been 0.25–0.7%.

Animals normally avoid eating it, due to its unpleasant taste and odor, but eat it when pressed for forage or when it is in hay. Mydriasis, tachypnea, tachycardia, restlessness, nervousness, frequent urination, and muscle twitching occur. Convulsions or coma precede death. **Cattle** have tolerated up to 7 g seed/day; **hogs**, up to 5 g/day, and **hens**, up to 15 g/day. **Birds**, **cats**, and **dogs** are more sensitive to its toxicity. Some claim the toxic dose of seed for **cattle** is 600–900 mg. An affected **horse** was unable to rise, mouth partially open, very dry tongue, tachycardia, congested mucous membranes, and dilated pupils. **Ostrich** chicks that had eaten the seeds staggered, had spasmodic jerking of the neck, unnatural contortions, followed by stupor, coma, and death. A lethal dose for **cattle** is ~ ½–1 lb of the green plant. **Goats**, **sheep**, and **swine** have also been poisoned by it.

D. suaveolens = ***Angel's Trumpet***. A **woman**, thinking the seeds were seasoning, used them for hamburgers that **she** and her **husband** ate. Within an hour **both** began hallucinating and had tachycardia and diarrhea. Hospital discharges after 3 days. Used as a "legal" hallucinogen in southern U.S. by ingestion of its flowers or tea, leading to anticholinergic syndrome with fever, delirium, agitation, and memory disturbances. Excessive intake leading to flaccid paralysis, convulsions, cardiac arrhythmias, stupor, coma, and death. Touching one eye after handling the plant will dilate that eye only. The heavy, sweet aroma emanating from the flowers has caused violent headaches, nausea, dizziness, and even stupor. Additional symptoms from ingestion including dry, burning sensations in the mouth, difficulty in swallowing, cold extremities, dry skin, and muscular and respiratory paralysis. Occasionally rashes develop, particularly on the neck.

DAUBENTONIA PUMICEA

= *False Poinciana* = *Purple Rattlebox* = *Purple Sesbania* = *Rattlebox* = *Sesbane*

Untoward effects: The saponin-containing seeds are poisonous to **humans** (especially **children**), **cattle**, **goats**, **pigeons**, **poultry**, **cats**, and **sheep**, causing gastrointestinal irritation, vomiting, diarrhea, syncope, tremor, urinary incontinence, slowed breathing, visceral vasodilation, excitement, and, in severe cases, coma.

D. drummondii = ***Poison Bean*** found on the coastal plains of Florida and Texas can cause similar symptoms.

DAUCUS CAROTA

= *Queen Anne's Lace*

Untoward effects: Erythema and a contact dermatitis, apparently due to irritation by the plant's hairs, especially when the leaves are wet. Nervousness and photosensitization also reported.

Poisoning in **horses** characterized by medullary depression; in **cattle**, by depression and convulsions. Plant has contained dangerous concentrations of ***nitrates***. Suspected as a cause of lymphoid leukosis in Swedish **cattle** grazing neglected pastures overrun with it. Bitter tasting milk from **cows** that ingest it.

DAUNORUBICIN

= *Cerubidin* = *Cérubidine* = *Daunoblastina* = *Daunomycin* = *DaunoXome* = *Leukaemomycin C* = *NSC 83,142* = *Ondena* = *RP-13057* = *Rubidomycin*

***Antineoplastic*. IV use.**

Untoward effects: ***Anaphylaxis***: Such reactions are rare.

Blood and bone marrow: Myelosuppression and aplasia (nadir in ~ 2 weeks) with leukopenia (50%), decreased hemoglobin, thrombocytopenia, decreased reticulocytes; granulocytopenia, pancytopenia, hemorrhage, and anemia. A high percentage (23) of fatalities in some series.

Cardiovascular: 4–10% cardiomyopathy, congestive heart failure, tachycardia, gallop rhythm, electrocardiogram changes (S-T and T wave abnormalities), cardiomegaly, atrial ectopic beats and other supraventricular arrhythmias, especially in **those** with previous cardiac disease, cardiac vasoconstriction, hypotension, pericarditis, pleural effusion, and pulmonary edema. Has been fatal.

Gastrointestinal: Anorexia, nausea, vomiting, mucositis (stomatitis, esophagitis, etc.) and ulceration can be severe, abdominal pain, and diarrhea.

Hypersensitivity: Angioneurotic edema of lips and eyes and urticaria on trunk and extremities.

Liver: Hepatotoxicity. Veno-occlusive disease.

Miscellaneous: Malaise, vaginal ulceration, cellulitis at injection site, myelophthisis, fever, chills, high percentage of infections, transient red urine (not hematuria). Rarely, renal tubular damages and convulsions occur.

Skin: Alopecia (80%) with hair regrowth in 2–5 months after cessation of therapy. Photosensitivity reported. Erythematous hyperpigmentation on face, neck, arms, chest, and back, lightening and disappearing within 6 weeks. Transverse brown–black bands have occurred on nails. It is a vesicant and extravasation can produce local pain and necrosis. Causes dermatitis in previously irradiated areas. Contact dermatitis reported in a laboratory **technician**. Thick ***polyvinyl chloride*** gloves are permeable to it. Avoid any device, such as needles, that contain ***aluminum***, which reacts with it.

LD_{LO}, **human**: 6 mg/kg.

Long-term therapy in **dogs** led to cardiac toxicity with degenerative and necrotic lesions on the heart, decreased coronary blood flow, and cardiomyopathy. Causes a cumulative dose-related cardiomyopathy in **rabbits**. IV dosage to **rabbits** led to necrosis and fatty degeneration of cardiac muscle fibers within 2 months. Similar results occur in the **hamster**. Mammary and renal tumors, as well as teratogenicity, reported in **rats**. Causes fatal respiratory depression in **guinea pigs** at 75–175 mg/kg, yet doses of 300 mg/kg have no such effect on **rats**; 50 mg/kg led to acute ventricular arrhythmias in **rhesus monkeys**. Testicular atrophy noted in **dogs**, and fibrosarcomas at injection sites in **mice**.

IV, LD_{50}, **mice**: 20 mg/kg; **rabbits**: 4 mg/kg; **dogs**: 4, 6 mg/kg; **rats**: 13 mg/kg; **guinea pigs**: 4 mg/kg. SC, LD_{50}, **mice**: 16, 47 mg/kg; IP, LD_{50}, **mice**: 2.5, 5.6 mg/kg.

DAURICINE

Untoward effects: An alkaloid in ***moonseed*** (*Menispermum canadense* (q.v.)) and *M. dauricum* of Asia, with ***curare***-like action. Has been fatal to **children** eating the plant's fruit.

DAZOMET

= *Biocide N-521* = *Cosan S* = *Crag 974* = *DMTT* = *Mylone* = *N-521* = *Tiazon*

Herbicide, soil fungicide, nematocide.

Untoward effects: LD_{LO}, **human**: 500 mg/kg.

LD_{50}, **rats**: 363 mg/kg; **mouse**: 180 mg/kg; **rabbit**: 120 mg/kg.

DDT

= *Agritan* = *Chlorophenothane* = *Citox* = *Clofenotane* = *p,p′-DDT* = *Dicophane* = *Gesapon* = *Gesarex* = *Gesarol* = *Guesapon* = *Neocid* = *Pentachlorin*

Insecticide.

Use: Restricted in many countries, due to its persistence in the environment.

Untoward effects: Poisoning can occur due to ingestion or by absorption through the skin or respiratory tract. Deaths have occurred. Accidental and/or suicidal ingestion has caused hepatitis with jaundice and kidney damage. Hypersensitivity reactions have been associated with periarteritis nodosa or allergic vasculitis. A **farm-hand** at the U.S. National Museum chewed a plug of ***tobacco*** contaminated with DDT. After 2 h, **his** jaws stiffened, **he** vomited, became giddy, and developed an anxiety complex. **He** recovered.

Blood: Aplastic anemia, resulting in fatalities in some **patients**. Thrombocytopenia reported.

Central nervous system: Visual disturbances; headache; perceptual abnormalities; muscle weakness; ataxia; coarse tremors; slowed mental and motor activities; memory failures; motor polyneuropathy; slurred speech; excitement; nervousness; irritability; anxiety; euphoria; paralysis; paresthesia of extremities, tongue, lips, and face; albumino-cytologic dissociation, resembling Guillain–Barré syndrome; delirium, convulsions; coma; and death from respiratory failure.

Fetus and neonate: Levels up to 5 mg% found in the milk of exposed **mothers**.

Miscellaneous: Nausea, vomiting, sore throat, and pulmonary edema reported. A 63-year-old **female** exposed to it while planting ***watermelons*** developed painful erosive lesions of the mouth's mucous membranes; and blisters on **her** neck, shoulders, and sacral area.

Drug interactions: By enzyme induction, increase in the catabolism of steroids (***estradiol, progesterone,*** and ***testosterone***), ***antipyrene, bilirubin, diphenhydramine, phenobarbital,*** and ***phenylbutazone***.

Phenytoin and ***phenobarbital*** help in removing DDT from the body by increasing its metabolism. TWA, 8 h; skin, 1 mg/m^3.

LD_{LO}, **human**: 50 mg/kg; **infant**: 150 mg/kg; TD_{LO}, **human**: 6 mg/kg.

Bats are the most susceptible **mammals** to DDT toxicity. Over 0.1 ppm may be toxic to **fish**. So-called thin egg shell problems in **birds** may be secondary to a decrease in ***calcium*** absorption. Pretreatment of **rats** with DDT decreased toxicity and increased excretion of metabolites of ***phenylbutazone*** and ***dieldrin***; shortened the duration of action of ***androsterone, chlordiazepoxide, deoxycorticosterone, meprobamate, methprylon,*** and ***progesterone***. In many of these, the increased metabolism was noted for at least 12 weeks. The increase in metabolic enzyme stimulation is, apparently, dose-dependent, and the maximum induction usually occurs within the first 3 weeks. ***Phenytoin*** in the **rat** decreased DDT levels in the fat, and increased liver microsomal enzyme activity. Pretreatment of **pigeons** with DDT also increased liver metabolism of ***progesterone*** and ***testosterone***. WARNING: Oil-based solutions are more dangerous for **animals** than water-based emulsions, in terms of percutaneous absorption. Do NOT use on **cats**, milk **cows**, or **goats**. Obey control recommendations on its use. Highly publicized mystery of arctic water contamination is less surprising, in view of its lesser known uses, such as control of warble fly on Russian **reindeer** herds by spraying with a 2% DDT emulsion up to 20 times a year. Central nervous system adverse effects are similar to **man's**. Induces hepatomas in **mice** and **rats**; lymphomas, lung carcinomas, and adenomas in **mice**. Can alter male **fish** to lay eggs that produce young.

Minimum lethal dose, young **calves**: 250 mg/kg; **cattle**: 0.5–2 g/kg; **sheep**: < 0.5 g/kg; LD_{50}, **rats**: 115 mg/kg; **mouse**: 135 mg/kg; **guinea pig**: 150 mg/kg; LD_{LO}, **dog** and **chicken**: 300 mg/kg; **cat**: 250 mg/kg.

DEANOL

= *Dimethylaminoethanol* = *DMAE*

***Central nervous system stimulant.* Central cholinergic antidepressant.**

Aceglumate*: *Clérégil* = *Otrun* = *Ristarun

Acetamidobenzoate*: *Deaner* = *Pabenol* = *Riker 546

Bitartrate*: *Liparon* = *Paxanol

Hemisuccinate*: *Rischiaril* = *Tonibral

***Untoward effects*:** Petulance, insomnia, dry mouth, dizziness, drowsiness, jitteriness, fatigue, dyspnea, mild heartburn, apathy, vomiting, headaches, and blurred vision occasionally reported. Palpitations in a 10-year-old **male** on 300 mg/day.

LD_{50}, **rat**: 2.3 g/kg.

DEBRISOQUIN

= *Declinax* = *Isocaramidine* = *Ro-5-3307/1* = *Tendor*

Antihypertensive.

Untoward effects*: *Cardiovascular*:** IV use may produce a transient hypertensive effect. May predispose to cerebral and myocardial infarction. Associated with total ischemia in a **patient** receiving a SC injection of ***lidocaine and ***epinephrine***.

***Miscellaneous*:** Dizziness, weakness, muscle pain, nasal stuffiness, diarrhea, constipation, cutaneous reactions (pruritic rashes), sweating, impotence, ejaculation difficulty, headache, blurred vision, and wheezing. Crosses the placenta. ~8% of **Caucasians** and 30% of Hong Kong **Chinese** are unable to metabolize the drug. Postural hypotension is particularly of concern in the **elderly**.

Drug interactions*:** Potentiates circulatory effects of ***phenylephrine and mydriatic response to ***epinephrine*** and ***phenylephrine***. Its hypotensive effects are antagonized by ***tricyclic antidepressants, chlorpromazine***, and ***sympathomimetics*** (***amphetamines, phenylpropanolamine***) and, possibly, by Gruyère and other aged ***cheeses***.

LD_{50}, **mouse**: 235 mg/kg.

DECABORANE

= *Decaboron Tetradechydride*

***Use*:** In rocket propellents.

***Untoward effects*:** **Workers** exposed to this and other boranes become toxic. Both depression and excitement occur. Nausea, dizziness, headache, light-headedness, fatigue, incoordination, muscle spasms, and convulsions noted. Liver and kidney lesions.

Animals evidence dyspnea, weakness, liver, and kidney damage.

LD_{50}, **rat**: 64 mg/kg; **mouse**: 40 mg/kg; inhalation LC_{50}, **rat**: 46 ppm/4 h; **mouse**: 12 ppm/4 h.

DECADIBROMODIPHENYL OXIDE

= *FR 300*

Fire-retardant additive for thermoplastics.

***Untoward effects*:** Not a primary skin irritant or skin sensitizer in **man** and only mildly irritating to eyes.

DECAHYDRONAPHTHALENE

= *DEC* = *Decalin* = *Dekalin* = *Naphthalane* = *Naphthane*

***Solvent.* In motor fuels, lubricants, lacquers, floor waxes and shoe polish.**

***Untoward effects*:** Causes cataracts.

Inhalation TC_{LO}, **human**: 100 ppm.

LD_{50}, **rats**: 4.2 g/kg; inhalation TC_{LO}, **rats**: 500 ppm/2 h.

DECAHYDRO-β-NAPHTHOL

= *2-Decalol*

***Use*:** As a fragrance in soaps, detergents, cosmetic lotions and creams, and perfumes for ~70 years.

***Untoward effects*:** No irritation or sensitization in **human** trials.

LD_{50}, **rats** and acute dermal LD_{50}, **rabbits**: > 5 g/kg.

DECAMETHONIUM BROMIDE

= *C-10* = *Syncurine*

***Skeletal muscle relaxant, neuromusclar blocking agent.* Synthetic *curare*.**

***Untoward effects*:** Bradycardia, increased intraocular pressure (10%), muscle fasciculations, marked hypotension, and delayed anaphylactic reaction.

Drug interactions*:** Many ***antibiotics, anticholinesterases, dexpanthenol, fluorine anesthetics, lidocaine, procaine, and ***quinidine*** prolong the muscle relaxing effect of this depolarizing muscle relaxant.

LD_{50}, **rats**: 2.9 mg/kg; IV, LD_{50}, **mouse**: 750 μg/kg.

DECAMETHRIN

= *Decis* = *NRDC 161*

A synthetic pyrethroid.

Untoward effects: One of the most toxic pyrethroids to **mammals**. When adequate brain concentrations (~0.5 ppm) are reached, salivation, ataxia, choreoathetoid movements, and clonic seizures occur in various **animal** species. Extremely effective against **mosquito** larvae at only 0.05–0.09 ppb. Toxic to **birds**.

LD_{50}, **rat**: 125 mg/kg; male **mice**: 5.54 g/kg; female **mice**: 3.45 g/kg; **houseflies**: 0.17 μg/g.

DECOQUINATE

= *Deccox* = *Dekoks* = *M & B 15497*

***Coccidiostat*. Used in many animal species and poultry.**

Untoward effects: Do NOT feed to laying **chickens**. Withdraw from feeding it 3–5 days before slaughtering, to avoid undesirable tissue residues for **humans** eating the **poultry**. No withdrawal time required in Canada. The FDA has established a maximum tolerance of 1 ppm for its residues in edible skeletal tissues and 2 ppm in other tissues of **chickens**. Not for breeding **animals** or **cows** producing ***milk*** for **human** consumption.

DECYL ACETATE

= *Acetate C-10*

***Fragrance*. Used in soaps, detergents, cosmetic creams and lotions, perfumes, and foods. Occurs in *citronella* and *mandarin* oils.**

Untoward effects: No sensitization reported in **human volunteers**. Fed to **chicks**, it has caused encephalomalacia. Mild erythema when applied undiluted to **rabbit** skin.

LD_{50}, **rats** and acute dermal LD_{50}, **rabbits**: > 5 g/kg.

DECYLENIC ALCOHOL

= *9-Decenol-1*

***Fragrance*. ~10 tons used annually in soaps, detergents, cosmetic creams and lotions, and perfumes.**

Untoward effects: No irritation or sensitization reported in trials with **human volunteers**. Moderately irritating when applied full-strength to intact or abraded **rabbit** skin.

LD_{50}, **rats** and acute dermal LD_{50}, **rabbits**: > 5 g/kg.

DECYL OLEATE

= *Cetaphyl 140*

Untoward effects: Common in **pre-teen** facial creams, and is comedogenic, probably by interfering with normal exfoliation within the pores' ducts. Although **rabbits** tolerate a 15% concentration of this fatty acid ester, full strength causes acute ocular and primary cutaneous irritation.

DECYLTRIMETHYLAMMONIUM BROMIDE

Untoward effects: This detergent quaternary ammonium antiseptic surfactant is significantly irritating or injurious to **animal** eyes at concentrations of 0.1–1.0%.

DEER LEATHER

Untoward effects: Has caused a contact dermatitis after its use as a covering on an artificial leg. Patch tests were positive.

DEFERIPRONE

= *CGP 37,391* = *CP 020* = *1,2-Dimethyl-3-hydroxypyrid-4-one* = *L1*

Iron chelator.

Untoward effects: Joint and musculoskeletal pain, as well as serious decrease in immunity, resulting in systemic lupus erythematosus in 18-year-old **male**. Agranulocytosis, thrombocytopenia, and neutropenia can occur.

DEFEROXAMINES

= *Ba 29,837* = *Ba-33,112* = *Desferal* = *Desferrioxamines* = *DFOM*

Iron chelators.

Untoward effects: Gastrointestinal discomfort after oral administration and hypotension after large rapid (even 15 min) IV administration. Urine may become red from ***ferrioxamine*** excretion. It enhances the growth and virulence of *Yersinia enterocolitica*, with complicating gastrointestinal or pulmonary symptoms, fever, or diarrhea. Mild skin itching, immediate wheal and flare reactions, urticaria, nausea, and vomiting 30 min after 800 mg. Tachycardia, abdominal pain, leg cramps, fever, anaphylaxis, thrombocytopenia, mucormycosis, including a fatal cerebral case, due to *Rhizopus* sp.; pseudo deep vein thrombosis, leg cramps, acute hepatitis, and growth retardation have been reported. Neurotoxicity is dose-dependent. Permanent sensorineural hearing loss and optic neuropathy with vision loss occur. Reversible retinal toxicity (night blindness, visual field defects), eye irritation, and lenticular opacities reported. Dysuria also occurs.

Prolonged treatment of **cats**, **dogs**, and **rats** with high doses leads to irreversible lens opacities and severe gastroenteritis. As in **man**, experiments with **mice** led to a *Yersinia enterocolitica* septicemia.

IV, LD_{50}, **rats**: 329 mg/kg; **mice**: 287 mg/kg.

DEHYDROACETIC ACID

Use: It and its salts are used as antibacterials and antimycotics in toothpastes, wrapping paper for foods, cosmetics and pharmaceuticals, and as a plasticizer.

Untoward effects: Although it can be absorbed through the skin, usual concentrations of 0.05–0.2% have not been irritant to **man**.

LD_{LO}, **human**: 500 mg/kg. LD_{50}, **rats**: 500 mg/kg; LD_{LO}, **dog**: 400 mg/kg.

DEHYDROCHOLIC ACID

= *Acolen* = *Bilidren* = *Bilostat* = *Cholagon* = *Cholan-DH* = *Cholepatin* = *Chologon* = *Decholin* = *Dehychol* = *Deidrocolico Vita* = *Didrocolo* = *Erebile* = *Felacrinos* = *Procholon*

The sodium salt = *Carachol* = *Dycholium* = *Sodium Dehydrocholate* = *Suprachol*

A synthetic derivative of *cholic acid*, acting as a laxative, due to its hydrocholeretic action.

Untoward effects: > 600 million tablets sold in the U.S. between 1929 and 1965, and, in court, the FDA was unable to show a single case of harm to any **person** taking it. The FDA contended the product was unsafe for lay use and was misbranded. In 1967, the Federal judge disagreed with them. Since then, after IV use, an acute asthmatic attack occurred in a 61-year-old **male**; and an anaphylactic reaction with hypotension, clammy skin, and convulsions was reported after IV use in a 79-year-old **male**.

Sodium: IV, LD_{50}, **rat**: 887 mg/kg; **mouse**: 1.1 g/kg.

DEHYDROEMETINE

= *BT 436* = *Dametin* = *Mebadin* = *Ro1-9334*

Amebicide.

Untoward effects: Gastrointestinal intolerance with diarrhea, nausea and vomiting, colic, hypotension, tachycardia, dyspnea, burning sensation of face and moderate asthenia, anorexia, transitory itching, papular rash, pyrexia, T-wave inversion and widened QRS 10–20 days after beginning of treatment, psychosis, and blurred vision reported. Use in the presence of heart and kidney disease is contraindicated, due to its cardiotoxicity. Hypotension and bradycardia in **cats**.

IP, LD_{50}, **mice**: 37 mg/kg.

DELAVIRDINE MESYLATE

= *Rescriptor*

Untoward effects: Rash is common. Can cause neutropenia.

Drug interactions: Decreased metabolism of ***astemizole, benzodiazepines, cisapride, clarithromycin, indinavir, nelfinavir, rifabutin, saquinavir,*** and ***terfinadine***. ***Rifampin*** induces its metabolism. Use with ***didanosine*** decreases metabolism of both drugs.

DELPHININE

Untoward effects: Toxic alkaloid in *Delphinium* (see **Larkspur**), *Anemone*, *Adonis vernalis*, *Ranunculus*, and *Helleborus niger*, especially in the seeds and leaves of young plants, causing stomach pain, nausea, vomiting, diarrhea, bradycardia, depression, and incoordination in **people**. Occasionally fatal. Toxicity of the plant decreases with its aging.

LD, **human**: 2–5 mg.

Horses may show trembling, collapse, prostration, and convulsions; rarely, salivation and constipation. **Sheep** and **goats** appear to be resistant to serious ill-effects from it; 10–30 mg causes severe symptoms in **cats** and **dogs**.

DEMECARIUM BROMIDE

Cholinergic, miotic, anticholinesterase (irreversible).

Untoward effects: Cataractogenic, increased intraocular pressure (glaucoma), increased conjunctival congestion, has caused uveitis with posterior synechiae, eye ache, blurred vision, headache, and, rarely, retinal detachment. Avoid use with other anticholinesterase inhibitors, such as ***organophosphates*** and ***phenothiazine tranquilizers***.

DEMECLOCYCLINE

= *Biotericlin* = *Clortetrin* = *Declomycin* = *Deganol* = *Demetraciclina* = *Detravis* = *Ledermycin* = *Meciclin* = *Mexocine* = *Periciclina* = *RP-10192*

Untoward effects: ***Anaphylaxis***: Mild to severe reported. A 34-year-old **male** ingested 1 tablet from a previous prescription (taken without reaction), developed severe anaphylactic shock with severe headache, throat tightness, general flushing, dyspnea, limb and facial swelling, rash, and abnormal electrocardiogram.

Blood: Hemolytic anemia, eosinophilia, neutropenia, and thrombocytopenia.

Dental: Given to **children** < 8 years of age causes a dark yellow tooth discoloration.

Gastrointestinal: Some irritation (7.5–20%; anorexia, nausea, vomiting, abdominal cramps, glossitis, and dysphagia) reported.

Miscellaneous: Acute, reversible liver toxicity, vaginal candidiasis, light-headedness, headache, pancytopenia, aplastic anemia, hypocellular bone marrow, increased intracranial pressure, tinnitus, and brown–black thyroid discoloration. Interferes with assay procedures used to diagnose pheochromocytomas.

Neonate: Doses > 600 mg/day leading to breast milk levels 70% of serum's. Detectable for up to 3 days after last dose. Bulging fontanels in young **infants** disappear when the drug is discontinued.

Photosensitization: Of all the ***tetracyclines***, it is the most potent phototoxic agent. Affects about 10% of **patients**. Exaggerated in sunburn in ~50%. A phototoxic persistent light reaction has lasted for at least 3 years in a **patient**. One **patient** experienced a photoreaction through a pane of glass. Photo-onycholysis, followed in 5 weeks by onychodynia, plus black discoloration of nails.

Renal: Dose-dependent, reversible decrease in urinary concentrating ability by inhibition of antidiuretic homone (nephrogenic diabetes insipidus). Polydipsia and polyuria. A 91-year-old **male** developed these last two, severe dysphagia, black hairy tongue, pronounced weight loss and fatigue, dehydration, transient hypotension, uremia (296 mg/100 ml), and anemia.

Skin: Acneform eruptions reported and 3/200 develop a lichenoid eruption. Erythema of ear and erythematous eruptions over hands and legs. Persistent erythema nodosum in a 55-year-old **female**. Severe facial hyperpigmentation potentiated by X-radiation. Monilial overgrowth in the anogenital region occurs.

Drug interaction: Use with ***aluminum hydroxide, calcium*** salts, ***milk***, or ***milk of magnesia*** decreases its serum levels ~70%.

TD_{LO}, **woman** (20–24 weeks pregnant): 240 mg/kg; **human**: 10 mg/kg.

Experimental phototoxicity to it has been demonstrated in albino **guinea pigs** and **mice**. Use in **pine voles** led to a bacterial superinfection and subsequent lethal enteritis. IV use may be fatal to **horses**.

LD_{50}, **rats**: > 4 g/kg; **mice**: 4 g/kg.

DEMECOLCINE

= *Colcemid* = *Colchamine* = *Demecolin* = *Omaine* = *Santavy's substance F*

***Antineoplastic.* Isolated from *Colchicum autumnale* (*Autumn Crocus*) tubers.**

Untoward effects: Effective dosage caused severe nausea, vomiting, dysphagia, burning sensations in the mouth and throat, choking sensations, chromosome aberrations, decreased white blood cell count, transient hair loss, and bone marrow aplasia (occasionally severe and fatal). Used throughout one pregnancy with other drugs, it may have caused a premature birth without malformations, but, if used after the first trimester, it appears to be without ill-effects. Pruritus, maculopapular rash, agranulocytosis, severe transient leukopenia with decrease of erythroid bone marrow cells, intrahepatic cholestatic jaundice, phlebitis at injection site, and inhibition of spermatogenesis are assorted symptoms reported in several **patients**.

It has caused fetal hypoplasia and damaged ovarian tissue in **rats**. In one test series at 1.5 mg/kg IP or SC, it caused a high incidence of teratogenesis in **mice**, a few adverse effects in **rabbits**, and no ill-effects in **monkeys**. Leukocytopenia in **mice**.

IV, TD_{LO}, **rat**: 5.7 mg/kg; SC, TD_{LO}, **rabbit** (9 days pregnant): 0.1 mg/kg; IP, TD_{LO}, **mouse**: 75 mg/kg.

DEMETON

= *BAY 8173* = *BAY 10,756* = *Bayer 8169* = *Bayer 19,639* = *Demox* = *E-1059* = *Isosystox* = *Mercaptophos* = *Septox* = *Systemox* = *Systox*

***Organophosphorus insecticide.* Strong cholinesterase inhibitor at motor end-plates.**

Untoward effects: Minimal toxicity in **man** at 6.75–7.125 mg/day, symptoms usually appear within 1 h of exposure and peak within 2–8 h. Miosis, sweating, abdominal pain, aching eyes, "tightness" in chest, headache, rhinorrhea, dyspnea, laryngeal spasms, nausea, vomiting, diarrhea, muscle fasciculations, giddiness, confusion, ataxia, decreased blood pressure, arrhythmias, convulsions, coma, cyanosis, pulmonary edema, and death reported. **Florists** exposed to flowers containing up to 22 ppm evidenced neurological symptoms, including diplopia, blurred vision, headaches, paresthesia, muscle fasciculations and weakness. A 3-year-old **male** developed a staggering gait, ptosis, colicky abdominal pain with transient vomiting, and diarrhea, and, eventually, stupor from exposure to demeton-contamination ***fuel oil***. Estimated LD for **humans** is ~5 g.

LD_{50}, **rats**: 1.4 mg/kg; **mice**: 3.9 mg/kg; **mallard ducks**: 7.19 mg/kg; **pigeons** and **quail**: 9 mg/kg; **pheasants**: 8 mg/kg; **goats**: 8–18 mg/kg. In general, the toxic symptoms in **animals** are similar to those in **man**. **Birds** also evidence polydipsia, lacrimation, goose-stepping ataxia, tachypnea, and opisthotonus. Moderately toxic to **bees**.

FDA regulations permit a tolerance of up to 5 ppm for residues in dehydrated ***beet*** pulp for **livestock** feed only as a result of its necessary agricultural usage. Zero tolerance in meat and poultry.

DEMETON-METHYL

= *Bayer 21/116* = *Meta-isosystox* = *Metasystox* = *Methyl Demeton* = *Methylmercaptophos*

Organophosphorus insecticide. Cholinesterase inhibitor.

Untoward effects: Nausea and vomiting and atrial fibrillation 6–8 h after inhalation of sprayed material. Nausea, vomiting, diarrhea (74), sweating, salivation, lacrimation (15), hyperreflexia (7) and fasciculations (3); other symptoms persisted for several days leading to abdominal pain and dyspepsia, dizziness (29), weakness and fatigue (23), dyspnea (20), headache (19), tremors and ataxia (5), hiccough (3), drowsiness (3), and coma (1) of 673 **spraymen**. Decreased serum cholinesterase correlated with symptomatology. A number of suicide attempts with it are reported. Dermal absorption occurs.

LD_{50}, **rats**: 44 mg/kg.

DENATONIUM BROMIDE

= *Bitrex* = *Lignocaine Benzyl Benzoate* = *THS 839*

Untoward effects: One of the most bitter substances known to **man**. Used to denature ***ethyl alcohol*** and as an additive (58 mg/gal) to deter ingestion of toxic substances. A liquid containing only 10 ppm will cause a **child** or an **adult** to stop drinking it, pucker up their faces, and induce them to spit out the liquid.

DENDROCHIUM TOXICUM

Untoward effects: A saprophytic contamination of **animal** fodder and straw, producing the highly toxic ***dendrodochin***, killing **horses**, **sheep**, **pigs**, **rabbits**, **rats**, and **mice**, with disturbances of the cardiovascular and central nervous system in natural and experimental outbreaks. In a natural outbreak, 30% of 170 **pigs** died. Nervous symptoms, indigestion; cardiac, circulatory, and respiratory disorders; hemorrhagic diathesis, thrombocytopenia, leukocytosis, increased hemoglobin, and necrotic, ulcerative and degenerative visceral lesions noted. The toxin is thermo-stable.

DENZIMOL

= *Rec 15–1533*

Anticonvulsant.

Untoward effects: 100 mg/kg to **dogs** led to head tremors, clonic-tonic convulsions, and ataxia.

LD_{50}, **rats**: > 2.7 g/kg; male and female **mice**: 434 and 463 mg/kg, respectively.

DEOXYCHOLIC ACID

= *Degalol* = *Desoxycholic Acid*

Choleretic.

Untoward effects: Oral intake by **humans** and **rats** reduced food intake. Possibly, toxic to small intestine mucosa of **man**, and causes gastric mucosal bleeding in **rats** after ***aspirin*** ingestion. It increases the blepharoptotic effect of ***reserpine***, because it increases the latter's solubility.

DEOXYCORTICOSTERONE

= *Deoxycortone* = *Desoxycorticosterone*

Acetate: *Cortate* = *Cortiron* = *DCA* = *Decosteron* = *DOCA* = *Dorcostrin* = *Percorten* = *Syncortyl*

Pivalate: *DOCP* = *Percorten Pivalate*

Adrenocorticosteroid, mineralocorticoid.

Untoward effects: ***Sodium*** and water retention, ***potassium*** loss, muscle weakness, paralysis, hypertension, edema, gynecomastia, cardiac enlargement, and congestive heart failure. Has caused increased intraocular pressure and lens opacities or cataracts. Overdosage has caused hypertensive encephalopathy with edema and permanent brain damage in a 3-week-old **male**. Calcified auricular cartilages in Addison's disease may, in part, be due to prolonged usage. Joint pain and pyrexia have occurred with prolonged treatment. ***Chlorcyclizine, phenobarbital,*** and ***phenylbutazone*** increase its metabolism by enzyme induction. Increased renal clearance of ***radiopharmaceuticals*** used in cardiac imaging.

Easy to produce hypertension in **rats** with its use. Increases the incidence of rheumatoid-like joint lesions in **guinea pigs** and the hydropericardium syndrome in **poultry**. Overdosage produces edema, cardiac necrosis, and cardiac failure in **animals**. Small oral ***salt*** supplementation must accompany therapy.

IP, LD_{50}, **rat**: 326 mg/kg.

2-DEOXY-D-GLUCOSE

= *Ba 2758* = *2-DG*

Antiviral, appetite stimulant.

Untoward effects: Increases hunger in **humans**, **goats**, **sheep**, **mice**, **monkeys**, **rabbits**, and **rats**.

IP, LD_{50}, **rat**: 2 g/kg.

DEOXYNIVALENOL

= *Vomitoxin*

***Mycotoxin.* From *Fusarium sp.* Commonly on moldy *corn.* A trichothecene toxin.**

Untoward effects: Implicated in the lethal "***Yellow Rain***" Russian canisters introduced in 1967 to Vietnam, which, supposedly, suppressed antibiotic efficacy, and caused lethargy, and deaths. 0.7–2.0 ppm in feed causes feed refusal and vomiting in **swine**. **Poultry** tolerate about two times the level **swine** do and three times that of **dairy cattle**. Causes **pigeons** to vomit.

LD_{50}, **mice**: 46 mg/kg; SC, LD_{50}, **dog**: 27 mg/kg.

4-DEOXYPYRIDOXINE

= *4-Desoxypyridoxine* = *4-DOP*

Untoward effects: A ***vitamin*** B_6 antagonist, used to produce experimental deficiency in studies of ***pyridoxine***–drug interactions.

DEOXYRIBONUCLEIC ACID

= *Desoxiribon* = *Desoxyribonucleic Acid* = *DNA* = *Eucytol*

Untoward effects: Autosensitization in 39-year-old **male** with hemorrhagic cutaneous anaphylaxis. Cutaneous hypersensitivity of delayed type only on extremities in 50-year-old **male** with painful SC hemorrhages and hysteria or schizophrenia.

4-DEOXYSCIRPENOL

Untoward effects: One of the trichothecene mycotoxins identified in the chemical warfare agent "***Yellow Rain***". Definitely a dermal irritant.

DEOXYTHIOGUANOSINES

Antineoplastics.

Untoward effects: The α-2' form has relatively low toxicity. Emesis, diarrhea, anorexia, anemia, and leukopenia are early symptoms in experimental **dogs** and **monkeys**. Later, neutropenia and thrombocytopenia develop, especially at high dosage.

The β-2' form (***B-TGdR***) after a year of IP injections to **rats** was associated with ear canal cancer in male **rats**.

IV, LD_{LO}, **dog**: 142 mg/kg.

DEPRENYL

= *Anipryl* = *Deprenil* = *Déprényl* = *E-250* = *Eldéprine* = *Eldepryl* = *Jumex* = *Movergan* = *Plurimen* = *Selegiline*

Antiparkinsonian agent. A monoamine oxidase inhibitor.

Untoward effects: Has caused nausea, dizziness, excitation and restlessness, insomnia or disturbed sleep, decreased blood pressure, tachycardia, palpitations, hyperhydrosis, abdominal pain, dyskinesia, headache, asthma-like symptoms, and suicide attempts.

Drug interactions: Taken with ***fluoxetine***, a toxic serotonin syndrome develops with myoclonic jerking, acute delirium, convulsive movements, and akathesia. Enhances the effect of and decreases the dosage needed for ***levodopa***. Use with ***levodopa*** in 821 Parkinson **patients** led to 16% increase in relative risk of death, compared to 809 controls.

Gastrointestinal upsets were noted in some **dogs** and use with other ***monoamine oxidase inhibitors***, ***tricyclic antidepressants***, ***selective serotonin reuptake inhibitors*** (viz., ***fluoxetine*** and ***sertraline***), ***narcotics***, and ***amitraz*** (in flea collars or dips) should be avoided.

DEPTROPINE

= *Brontine* = *BS-6987* = *Dibenzheptropine*

Anticholinergic, antihistamine.

Untoward effects: Dry mouth (12%), headache, mydriasis (10%), urticaria and nasal pruritus (6.6%) in 30 **male** and **female asthmatics**. Drowsiness in 34% in another trial. Serious psychic adverse effects of large doses in pediatric **patients** (1–10 years; 0.6–3 mg/day), with hallucinations, aggressive behavior, agitation, and ataxia reported.

LD_{50}, **rats**: 445 mg/kg; **dogs**: 75 mg/kg.

DEQUALINIUMS

The chloride = *BAQD-10* = *Decamine* = *Decatylen* = *Dekadin* = *Dekamin* = *Dequadin Chloride* = *Dequafungan* = *Dequavagyn* = *Dequavet* = *Eriosept* = *Evazol* = *Groceme* = *Labosept* = *Optipect* = *Phylletten* = *Polycidine* = *Sorot*

***Antimicrobial, bacteriostat, fungicide.* Rapidly penetrates bacterial cells; their *nucleic acid* components are the probable targets.**

Untoward effects: Exacerbation of ulcerating balanitis. A manufacturer has warned against using it in balanitis, especially when the skin is occluded. Do NOT use in genital region and intertriginous areas. A case of eye irritation has been reported. A 5% incidence of positive skin reactions in patch testing in over 600 **patients**.

SC, LD_{50}, **mouse**: 70 mg/kg.

DERRIS ROOT

= *Akar Root* = *Deguelia Root* = *Toeba Root* = *Toefa Root* = *Tuba Root*

Insecticide, rotenone and rotenoid source.

Untoward effects: Gastrointestinal upsets including diarrhea, nausea, and vomiting. Depression, stupor, and convulsions after heavy inhalation exposure. The **Malays** make an arrow poison called ***Ipoh*** with it. Eye and mucous membrane irritant.

As a piscicide to kill undesirable **fish**. In South America, it's used at low dosage to stun edible **fish**. **Cats** and **dogs** vomit after small amounts. **Pigs** and **guinea pigs** are very susceptible. **Poultry** are rather resistant to its adverse effects.

Dried root LD_{50}, **mice**: 350 mg/kg; **rabbits**: 3 g/kg; **guinea pigs**: 50–200 mg/kg; LD_{LO}, **dog** and **rat**: 100 mg/kg.

DESAMINOOXYTOCIN

= *Deaminooxytocin* = *Demoxytocin* = *ODA 914* = *Sandopart*

Transbuccal labor-inducer.

Untoward effects: Avoid use in high-risk cases because of its rapid effect. Has caused hypoglycemia and increase in plasma non-esterified ***fatty acids***, which can lead to ketosis and disturbed acid–base balance in the blood.

DESCURAINIA PINNATA

= *Tansy Mustard*

Untoward effects: In southwestern United States, central and western Canada, poisonous by ingestion of leaves and seeds to **cattle**, **sheep**, and **goats**. The first symptoms noted are partial or complete blindness, inability to use the tongue for swallowing (natives call it "paralyzed tongue" disease). **Animals** either wander aimlessly until exhausted or stand pushing for hours against a solid object in their path. They become thin and weak, and many die. Contains glucosinolates (***isothiocyanate, nitrites***, and ***thiocyanates***) in fresh or dried seeds and vegetative parts.

DESERPIDINE

= *Canescine* = *Decaserpyl* = *11-Desmethoxyreserpine* = *Harmonyl* = *Methoserpidine* = *10-Methoxydeserpidine* = *Raunormine* = *Recanescine*

***Antihypertensive*. Alkaloid from *Rauwolfia canescens*.**

Untoward effects: Depression, increased body weight, sedation, increased dreams, nasal stuffiness, and edema. Less common are constipation, postural dizziness, and gastrointestinal upset.

LD_{LO}, **human**: 50 mg/kg.

Relatively low toxicity to **rodents** and **dogs**. Miosis in **dogs** at ~30 mg/kg, along with dyspnea, diarrhea, and salivation.

LD_{50}, **mice**: 513 mg/kg; IV, LD_{50}, **rat**: 15 mg/kg; IP, LD_{50}, **mice**: 60 mg/kg.

DESFLURANE

= *1653* = *Suprane*

Inhalation anesthetic.

Untoward effects: Nausea (27%), vomiting (16%), apnea, coughing (34%), laryngospasm (3–10%), and breath-holding. As with ***isoflurane***, it reduces myocardial and respiratory activity and is a vasodilator (peripheral, coronary, and cerebral). Malignant hyperthermia in **humans** and **animals**.

DESIPRAMINE

= *Desmethylimipramine* = *G-35020* = *JB-8181* = *Norimipramine* = *Norpramin* = *Nortimil* = *NSC-114901* = *Pertofran* = *Pertofrane*

Tricyclic antidepressant.

Untoward effects: May cause an anticholinergic intoxication syndrome. Depression, delirium, agranulocytosis, thrombocytopenia, neutropenia, cyanosis, irritability, increased intraocular pressure or glaucoma, blurred vision, tachypnea, tachycardia, profuse sweating, pruritus, urticaria, skin rash, alopecia, gynecomastia, decreased libido, impotence, painful orgasm, inhibition of ejaculation, dry mouth, urinary retention, pruritic morbilliform rash, photosensitivity, jaundice, centrilobular or mid-zonal hepatic necrosis, convulsions (avoid use in **epileptics**, as it decreases the convulsive threshhold), mania, tremors, and confusion. Cases of acute toxicity reported are usually from overdosage leading to cardiac toxicity with hypotension within 1–2 h, alterations in cardiac rhythm, due to conduction defects (first-degree atrioventricular block, bundle branch block, and intraventricular disturbances of conduction. Several reports of cardiac conduction defects, eosinophilia, purpura, and hemolytic anemia on clinical doses. Poisoning and suicide following somnolence, nystagmus, nausea, ataxia, cyanosis, irregular slow pulse, decreased blood pressure, tachycardia, ventricular extrasystoles, and QRS disturbances. Even healthy **subjects** have up to five times variation in half-lives or steady-state plasma concentrations. Therapeutic blood levels are usually 15–30 μg%, and toxic levels, 50 μg%. Despite treatment, a 23-month-old **child** died after ingesting ~2 tablets. Although a 24.3 lb 19-month-old **male** ingested 28 25-mg tablets, **he** survived after major transient toxic effects. Very limited transfer into breast milk, but readily crosses the placenta. Withdrawal symptoms manifested in **neonates** including breathlessness, tachypnea, tachycardia, and profuse sweating. Adverse central nervous system effects often occur in elderly **patients**.

Therapeutic serum concentrations are 0.01–0.5 mg/l and fatal above 3 mg/l.

Drug interactions: Use with ***monoamine oxidase inhibitor*** (viz. ***isocarboxide, nialamide, pargyline, phenelzine***, and ***tranylcypromine***) is contraindicated, as it often produces excitation, hyperpyrexia, convulsions, and, occasionally, coma and death. When given with ***trihexyphenidyl***, three elderly **patients** developed excitement, confusion, and hallucinations. ***Oral contraceptives*** and ***phenobarbital*** increase its metabolism. ***Fluoxetine*** causes dangerous increase in its blood levels, and its metabolism is decreased by ***methylphenidate***. Appears to potentiate central nervous system effects of ***amphetamines*** and ***adrenergic compounds*** and reduces absorption of ***phenylbutazone***. Antagonizes the antihypertensive effect of ***bethanidine, guanethidine,*** and ***clonidine***. Delays the absorption of ***ethyl alcohol*** and ***acetaminophen***. Multiple doses of ***sertraline*** decrease its plasma clearance, increase its plasma concentration and prolong its half-life. ***Alprazolam*** increases its serum concentration.

DESLANOSIDE

= *Cedilanid-D* = *Desace* = *Desacetyldigilanide C* = *Desaci* = *Lanimerck* = *Purpurea Glycoside C*

Untoward effects: As for ***digitalis***. Arrhythmias and conduction disturbances, anorexia, nausea, vomiting, blurred vision with yellow and green flickering, and disorientation. Occasionally pain at IM injection site. Average half-life of 33 h.

LD_{LO}, **human**: 5 mg/kg.

DESMANTHUS VIRGATUS

Untoward effects: In Cuba has killed **goats** grazing on it.

DESMOPRESSIN

= *Adiuretin SD* = *DAV Ritter* = *DDAVP* = *Desmospray* = *Minirin* = *Octostim*

Antidiuretic.* A synthetic analog of *vasopressin.

Untoward effects: Transient headache, conjunctivitis, rhinitis, chills, dizziness, epistaxis, cough, nausea, mild abdominal cramps, and vulval and nostril pain. Facial flushing occurs. Injections have caused local erythema, swelling or burning pain, slight increase in blood pressure, water intoxication, hyponatremia, convulsions, and coma for **some** who drink more than needed to satisfy thirst. Increased ***Factors VII*** and ***VIII***. Myocardial infarctions and death have been reported. A 21-month-old **male** with tetralogy of Fallot developed marked cyanosis and dyspnea 30 min after infusion. **Patient** recovered. Hyponatremic convulsion in a 10-year-old **male** appeared related to the co-administration of ***imipramine***.

DESOGESTREL

= *Org-2969*

Progestogen.* With *ethinyl estradiol* as an *oral contraceptive.

Untoward effects: The combination increased apolipoprotein, triglycerides, bound globulin values, and venous thrombosis. Use with ***rifampin*** may decrease ***oral contraceptive*** efficacy, and increase breakthrough bleeding and menstrual irregularities. To a lesser extent, the same may occur when taken with ***ampicillin, barbiturates, carbamazepine, griseofulvin, phenylbutazone,*** and ***tetracyclines***.

DESONIDE

= *Apolar* = *D-2083* = *Desowen* = *Locapred* = *Prednacinolone* = *Sterax* = *Steroderm* = *Topifug* = *Tridesilon*

Anti-inflammatory non-fluorinated corticosteroid.

Untoward effects: Topical use can cause irritation, erythema, itching, burning sensations, dryness, folliculitis, hypertrichosis, telangiectasia, acneform eruptions, skin blanching (380 times ***hydrocortisone's*** vasoconstrictor effects), and hypopigmentation. Secondary infection, skin atrophy, striae and milaria can occur under occlusive dressings. Avoid use in pregnancy.

SC, LD_{50}, **rat**: 93 mg/kg.

DESOXIMETASONE

= *A-41-304* = *Desoxymethasone* = *Esperson* = *HOE-304* = *Ibaril* = *R-2113* = *Stiedex* = *Topicorte* = *Topisolon*

Topical fluorinated corticosteroid.

Untoward effects: See ***Desonide***. May cause photosensitivity reactions. A 45-year-old **female** with psoriasis applying 30 g/day of a 0.25% cream/5 years developed Cushing's syndrome with hirsutism.

DESOXYLAPACHOL

Untoward effects: Strong sensitizer in ***teakwood***, causing an allergic contact dermatitis in **man**.

DETOMIDINE

= *Domosedan*

Sedative, α_2-adrenoreceptor agonist.

Untoward effects: Hypertension in **horses**, followed rapidly by bradycardia. Occassionally, they salivate, sweat, show mild muscle tremors, piloerection, and even transient penile prolapse. Partial atrioventricular and sinoauricular blocks with decreased heart and respiratory rates, and a few minutes of incoordination and staggering may occur.

DEW GRASS

= *Knaeulgras*

Untoward effects: Contact urticaria with positive patch tests.

DEXAMETHASONE

= *Aeroseb-Dex* = *Azium* = *Corson* = *Corson-P* = *Cortisumman* = *Decacort* = *Decaderm* = *Decadron* = *Decalix* = *Decasone* = *Dekacort* = *Deltafluorene* = *Deronil* = *Desametasone* = *Deseronil* = *Dexacortal* = *Dexacortin* =

Dexafarma = *Dexa-Mamallet* = *Dexameth* = *Dexameth-A-Vet* = *Dexamonozon* = *Dexapos* = *Dexa-sine* = *Dexasone* = *Dex-A-Vet* = *Dexinoral* = *Dinormon* = *9α-Fluoro, 16α-methylprednisolone* = *Fluormone* = *Gammacorten* = *Hexadecadrol* = *Hostadex* = *Isopto-Dex* = *Lokalison F* = *Loverine* = *Luxazone* = *Maxidex* = *Millicorten* = *Oradexon* = *Pet-Derm III* = *Zonometh*

Adrenocortical steroid hormone, glucocorticoid, anti-inflammatory.

***Untoward effects*: *Central nervous system*:** Headache, vertigo, increased intracranial pressure with papilledema, convulsions, mania, and euphoria.

***Dermatologic*:** Thin fragile skin and reduced wound healing. Occasionally, erythema, petechiae, ecchymoses, allergic dermatitis and urticaria, angioneurotic edema, and diaphoresis. A known recurrent urticaria and bronchial asthma sufferer, a 23-year-old **female**, developed severe pruritus, itching of tongue and uvula, and wheezing within 15 min after ingesting a 0.5 mg tablet.

Endocrine*:** Cushingoid state, suppressed growth in **children**, hirsutism, adrenal atrophy, increased ***insulin or hypoglycemic needs, and reduced response to stress. Cushingoid syndrome developed in a 45-year-old **female** who abused a nasal spray for 2 years.

***Eye*:** Cataracts, increased intraocular pressure, glaucoma, and exophthalmos. Cataracts have also developed following long-term topical use on the eye.

Fluid and electrolytes*:** Increased fluid and ***sodium retention, decreased ***potassium***, hypertension, and congestive heart failure.

***Gastrointestinal*:** Esophageal, peptic, and intestinal ulceration. Abdominal distention and pancreatitis.

Miscellaneous*:** Nausea, malaise, increased appetite, increased weight, eosinopenia, hiccups, ventricular arrhythmias, thromboembolism, and hypersensitivity (especially with ***sulfites in parenteral material). Bradycardia reported in **infants** after oral or IV use. Use in 118 low-birth-weight **infants** to decrease ventilator dependency led to higher percentage with cerebral palsy and abnormal neurologic findings.

***Musculoskeletal*:** Muscle weakness, myopathy, osteoporosis, vertebral compression fractures, fractures of long bones, tendon ruptures, and aseptic necrosis of femoral and humeral heads.

***Neonate*:** Subclinical adrenal insufficiency after **mother** received high dosage during pregnancy. Passes into breast milk. Induced adrenal hyperplasia has virilizing effects.

Drug interactions*: *Cimetidine, phenobarbital, phenytoin, and ***rifampin*** decrease its concentration in blood. Most reports indicate that it also dramatically increases serum ***phenytoin*** levels. One report indicated the opposite effect.

In **animals**, use during the last trimester of pregnancy may shorten time to parturition, with dystocia, retained placenta, metritis, and, occasionally, fetal death. Use in otic preparations has caused adrenocortical suppression. Can cause laminitis in **horses**. It is contraindicated in most viral infections, chronic nephritis, or tuberculosis. Slow IV use!! Solutions have a pH of approximately 4, and secondary problems can occur at the injection site, also from acid material reaching the brain in too great a concentration. Avoid excessive use of ophthalmic preparations during pregnancy, due to systemic absorption.

LD_{50}, female **mice**: 6.5 g/kg.

DEXBROMPHENIRAMINE

***Antihistamine.* Rarely used alone.**

Untoward effects*:** Contains about 25% ***bromine (q.v.). A morbilliform rash in a 25-year-old **male** after ingesting 6 mg. An 18-month-old **female** developed transient involuntary facial and lingual movements after ingesting 24 mg. Hodgkin's disease positively associated with its use.

DEXCHLORPHENIRAMINE

= *Fortamine* = *Isomerine* = *Phenamin* = *Phendextro* = *Polamin* = *Polaramine* = *Polaronil* = *Sensidyn*

Antihistamine.

***Untoward effects*:** Malaise, nausea, vomiting, and vertigo in 2/30 postoperative **patients**. Has exacerbated the symptoms of Tourette's disorder. Some sedation, but, more frequently, central nervous system stimulation.

DEXFENFLURAMINE

= *Adifax* = *Dextrofenfluramine* = *DFF* = *Glypolix* = *Isomeride* = *Redux* = *S-5614*

Active isomer of *fenfluramine*, used as an appetite suppressant.

Untoward effects*:** Drowsiness, fatigue, dry mouth, diarrhea, and polyuria. Myocardial infarction in 31-year-old **female** after 15 mg tid/8 days. Lung pathology reported and possible association with diplopia in a 38-year-old **female** and subacute ischemia of lower limb in another **patient**. Ischemic necrosis of fingers required amputation after 15 mg twice daily/3 months led to increased cardiac valve disorders, including aortic regurgitation (~9%) and often fatal pulmonary hypertension. May cause cardiac damage in **infants** when taken by **mothers** during pregnancy. Can cause serotonin syndrome when taken with ***monoamine oxidase inhibitor*s** or other ***selective

serotonin reuptake inhibitors within 2–3 weeks after taking it for weight loss. A 54-year-old **female** with a 30 year history of bipolar disorder developed mania. Abrupt withdrawal can precipitate a serotonin withdrawal syndrome.

Has caused damage to the serotonergic neurons in the brain of **squirrel monkeys** after injection for 4 days with 5 times recommended **human** dose.

DEXPANTHENOL

= *Alcopan-250* = *Bepanthen* = *Cozyme* = *Ilopan* = *Intrapan* = *Motilyn* = *Pantenyl* = *Panthoderm* = *Pantothenol* = *Pantothenyl Alcohol* = *Urupan*

Cholinergic, surfactant, vitamin B_5 source.

Untoward effects: Prolonged muscle relaxation and respiratory depression reported in a **patient**, when given with ***succinylcholine***, the depolarizing muscle relaxant. Also enhances the action of ***decamethonium*** and ***anticholinesterases***.

DEXRAZOXANE

= *Zinecard*

Untoward effects: Leukopenia, granulocytopenia, and thrombocytopenia. Bone marrow suppression also in **patients** with tumors other than breast cancer. Occasionally abnormal hepatic and renal tests.

DEXTRAN

= *Dextraven* = *Gentran* = *Hemodex* = *Intradex* = *Macrose* = *Onkotin* = *Plavolex* = *Polyglucin* = *Promit*

Plasma volume expander.

Untoward effects: Antigenicity and coagulation defect with increased bleeding is more common with mw > 80,000. Cardiac arrest, fatal allergy, severe fatal hypotension, raised blood urea nitrogen, urticaria, and alopecia reported. May potentiate ***anticoagulants***. Brittleness and transverse ridging of nails may occur 3 months after infusion. Anaphylactoid reactions occur at an estimated frequency of 3/10,000. Even traces in ***BCG*** vaccine caused a hypersensitivity reaction in a 9-month-old **female**. Severe bleeding can occur in defibrinated **patients** receiving ***ancrod***.

Dextran 40 = *Eudextran* = *Gentran 40* = *LMD* = *LMWD* = *LVD* = *Rheomacrodex* = *Rheotran*

Average mw of this low mw product is 40,000.

Untoward effects: Anuria or oliguria, pulmonary edema, fever, flushing, dyspnea, hypotension, cardiac arrhythmias, nausea, vomiting, rashes, edema of eyelids, congestive heart failure, increased SGOT, anxiety, tremors, shock, headache, and occasional increase in clotting time. In rare cases, deaths have followed within a few minutes of the infusion of less than 10 ml. Interferes with photocolorimetric determination of blood urea.

Dextran 70 = *Hyskon* = *Macrodex*

Average mw of this high mw *dextran* is 70,000–75,000, with a range of 50,000–120,000.

Untoward effects: Marked hypotension in severe, rapid, and fatal anaphylactoid reactions. Urticaria, pruritus, macular rash, wheezing, nasal stuffiness, flushing, pulmonary edema, ascites, severe apprehension, nausea, vomiting, diarrhea, convulsions, increased clotting time, and acute renal failure reported.

IV, TD_{LO}, **woman**: 6 mg/kg.

DEXTROAMPHETAMINE SULFATE

= *Afatin* = *Albemap* = *d-Amfetasul* = *Amsustain* = *Ardex* = *Betafedrina* = *d-Betafedrine* = *Dadex* = *Dexalone* = *Dexampex* = *Dexamphetamine* = *Dexedrine Sulfate* = *Dexten* = *Diocurb* = *Domafate* = *Maxiton* = *Obesedrin* = *Simpamina-D* = *Sympamin*

Untoward effects: ***Cardiovascular***: Palpitation, tachycardia, increased blood pressure; occasionally, cardiomyopathy with chronic use.

Central nervous system: Restlessness, dizziness, euphoria, dyskinesia, dysphoria, tremors, headaches, insomnia, frequent lip-licking, tics, Tourette's syndrome, and psychosis (paranoid and schizophreniform) reported.

Endocrine: Altered libido and impotence.

Fetus and neonate: May be related to significant increase in congenital heart disease, especially transposition of great vessels. Passes into breast milk.

Gastrointestinal: Dry mouth, diarrhea, constipation, anorexia, weight decrease, and lowered threshold for bitter taste.

Miscellaneous: Addiction, lower growth rate in **children**; hypersensitivity reactions including facial swelling, malaise, weakness, fixed-drug eruptions, thrombocytopenia, and euphoria reported.

Overdosage and poisoning: Excitement, agitation, hypertension, tachycardia, tachypnea, cyanosis, mydriasis, tremors, ataxia, slurred speech, chills, hyperreflexia, fever, headache, paranoid delusions, auditory and visual hallucinations, schizophreniform psychosis, collapse, coma, vomiting, incontinence of urine and feces, and death.

Activated charcoal is an effective adsorbant. Estimated LD for an **adult** is 200 mg and for a 2-year-old **child** is 20 mg.

Respiratory: Allergic-type bronchial asthma has been due to ***tartrazine*** as a capsule coloring agent.

LD_{LO}, **human**: 5 mg/kg.

Drug interactions: ***Acidifiers***: ***Ascorbic acid, fruit juices, glutamic acid hydrochloride, guanethidine***, and ***reserpine*** decrease its absorption. Urinary acidifiers, such as ***ammonium chloride, methionine***, and ***sodium acid phosphate*** increase its urinary excretion.

Adrenergic blockers: Action inhibited by it.

Alkalinizers: ***Sodium bicarbonate*** and other gastrointestinal alkalinizers increase its absorption. Urinary alkalinizers, such as ***acetazolamide*** and some ***thiazides***, decrease its urinary excretion. Both of these groups increase its blood concentration and potentiate its actions.

Antidepressants, monoamine oxidase inhibitor: Potentiates its effects. Has caused hypertensive crisis, headaches, malignant hyperpyrexia, and death (viz. ***phenelzine, tranylcypromine***). A 41-year-old **female** developed 109.4°F temperature, sweating, agitation, coma, hyperkinesis, opisthotonus, and convulsions within 3+ h after 10 mg ***tranylcypromine*** and ***dexamyl***. ***Monoamine oxidase inhibitor*** effect may last for 2 weeks after withdrawal.

Antidepressants, tricyclic: Increase its concentration in the brain.

Antihistamines: Their sedative effect is neutralized.

Antihypertensives: Their (***guanethidine, veratrum alkaloids***, etc.) effect may be antagonistic.

Chlorpromazine: Inhibits its toxicity by blocking ***dopamine*** and ***norepinephrine*** reuptake. It and ***thioridazine*** reduce its anorexic effect.

Ethosuximide: Its rate of absorption is decreased by amphetamines.

Furazolidone: Acts as a ***monoamine oxidase inhibitor*** (see above).

Haloperidol: Acts as ***Chlorpromazine*** (see above).

Lithium carbonate: Can inhibit its stimulant and anti-obesity effects.

Meperidine: Its analgesic effect is potentiated.

Norepinephrine: Its adrenergic effects are enhanced.

Phenobarbital: Its absorption may be delayed.

Phenytoin: May delay its absorption and potentiate its anticonvulsant effect.

Propoxyphene: Overdosage may increase amphetamine's central nervous system stimulation and cause fatal convulsions.

Congenital cardiac malformations and stunted fetuses in **mice**. Toxicity in **rats** and **mice** manifested by enhanced motor activity and excitement, followed by depression. **Rhesus monkeys** have shown addiction to it. Hyperpyrexia noted in **rabbits**.

LD_{50}, **rats**: 32 mg/kg; **mice**: 9 mg/kg; **dogs**: 10 mg/kg.

DEXTROMETHORPHAN HYDROBROMIDE

= *Benylin DM* = *Canfodion* = *Cosylan* = *Delsym* = *Demorphan Hydrobromide* = *Hihustan M* = *D-Methorphan* = *Ro-1-5470/5* = *Romilar HBr* = *Silentium*

Antitussive.

Untoward effects: Miosis. A drug of abuse, particularly in cough syrups. Fixed-drug eruptions in a 45-year-old **female** after ingesting 30 mg. Drowsiness in sensitive **individuals** and with extremely high doses. Rarely, nausea, vomiting, indigestion, dizziness, and unconsciousness. High dosage may decrease respiration. Toxic psychosis in a 23-year-old **male drug addict** after ingesting 20 tablets, characterized by hyperactive behavior and marked visual and auditory hallucinations. A possible lethal reaction, due to ***phenelizine*** with it in a 26-year-old **female**. Excitation and hyperpyrexia have occurred when given with a ***monoamine oxidase inhibitor***, as it can cause a serotonin syndrome. A 32-year-old **female** receiving 30 mg/day ***isocarboxid*** developed persistent myoclonic leg jerks after a single 15 mg dose. Avoid use with ***monoamine oxidase inhibitors***. A 2-year-old **male** was given 1½ teaspoonsful every 6 h of an over-the-counter cough and cold medicine that also contained ***pseudoephedrine***. **He** awoke after 1 h showing hyperexcitability, irritability, babbling, and poor balance control. In general, its toxic effects can be synergistic with ***monoamine oxidase inhibitors, selective serotoin reuptake inhibitors***, and ***tricyclic antidepressants*** in **humans** and **animals**.

Therapeutic serum levels are 0.01–0.04 mg/l; toxic at 0.1 mg/l; and comatose or fatal at > 3 mg/l.

TD_{LO}, **human**: 50 mg/kg; **child**: 30 mg/kg.

Addiction demonstrated in **dogs**.

LD_{50}, **rat**: 350 mg/kg; **mouse**: 165 mg/kg.

DEXTROMORAMIDE

= *Dimorlin* = *Jetrium* = *d-Moramid(e)* = *Palfium* = *Palphium* = *Pyrrolamidol* = *R-875* = *SKF-5137*

Narcotic, analgesic.

Untoward effects: Addiction, nausea, vomiting (4%), respiratoy depression (3/107) and occasionally apnea, sleepiness, vertigo, dizziness, sweating, muscle twitching and weakness. Miosis, itching of the nose, and tachycardia reported. Induced euphoria ~three times that of ***heroin***.

Estimated LD, **adult**: 500 mg.

DEXTROTHYROXINE SODIUM

= *Biotirmone* = *Choloxin* = *Dethyrona* = *Detyroxin* = *Dextroid* = DT_4 = *Dynothel* = *Eulipos*

Cholesteropenic and lipopenic.

Untoward effects*:** Generally, mimics effects of hyperthyroidism, such as weight decrease, nervousness, irritability, light-headedness, insomnia, tremors, anxiety, increased severity of angina, palpitations, tachycardia and arrhythmias, tinnitus, dizziness, ptosis, visual disturbances, hyperhydrosis, and menstrual irregularity. Occasionally, altered taste, vertigo, constipation, and diarrhea during the first 1–2 months of therapy. Alopecia, rash, and pruritus can develop in **patients** hypersensitive to ***iodine. May increase serum protein-bound ***iodine***. Progressive deterioration of diabetic control, necessitating increased ***insulin*** or ***oral hypoglycemic*** dosages. Suspected cause of increased deaths in **patients** with heart disease.

Drug interactions*:** May increase anticoagulant effect of various drugs, such as ***bishydroxycoumarin and ***warfarin***, necessitating a 30–50% decrease in ***anticoagulant*** dosage.

DEZOCINE

= *Dalgen* = *WY 16,225*

Narcotic, analgesic. **Synthetic opioid agonist/antagonist.**

***Untoward effects*:** As with other opioids, nausea, vomiting, and sedation. Anxiety, dizziness, vertigo, blurred vision, diplopia, slurred speech, urinary retention, tinnitus, hiccups, disorientation, hallucinations, sweating, chills, flushing, decreased hemoglobin, edema, and tachycardia. Skin reactions at IM injection site and after IV, and acute, even fatal, respiratory depression. Use with caution in **cardiac patients**.

LD_{50}, **rats**: 232 mg/kg; **mice**: 313 mg/kg. IM, LD_{50}, **rats**: 270 mg/kg; **mice**: 129 mg/kg.

DHURRIN

Untoward effects*:** The cyanogenetic glucoside (yields ***hydrocyanic acid on hydrolysis) in ***sorghums***.

DIACETAZOTOL

= *Diacetotoluide* = *Dermagan* = *Dimazon* = *Pellidol*

Epithelization stimulant.

***Untoward effects*:** Scarlet-red color.

N, N′-DI(ACETOACETYL)-σ-TOLIDINE

Chemical intermediate in the synthesis of many dyestuffs and pigments.

Untoward effects*:** Cancer. Metabolized to ***σ-tolidine, a known bladder carcinogen.

DIACETOXYSCIRPENOL

= *Anguidine*

A trichothecene mycotoxin.

***Untoward effects*:** Hemorrhagic bowel lesions after IV, but not oral, administration to **swine**. Emesis, lethargy, extreme hunger or refusal to eat with decreased weight gains, frequent defecation, and posterior paresis noted in **swine**. Causes severe necrotic oral lesions in **chicks** and **pigs**. Yellowish plaque-type lesions develop on the beak, tongue, and angle of the mouth in **chicks**, and their weight gains decrease.

LD_{50}, **rats** and **mice**: 7.3 mg/kg; **guinea pigs** and **chickens**: 4 mg/kg; IV, LD_{50}, **swine**: 376 μg/kg.

DIALIFOR

= *Dialifos* = *Hercules 14503* = *Torak*

Insecticide, acaricide. **An organophosphorus compound.**

***Untoward effects*:** Teratogenic in **hamsters**. Feed refusal by **dogs** at dietary level of 500 ppm.

LD_{50}, **rats**, male: 24 mg/kg; female: 6 mg/kg; **mouse**: 39 mg/kg; **dog**: 94 mg/kg; **rabbit**: 35 mg/kg.

DIALLATE

= *Avadex* = *CP-15336* = *DATC*

Herbicide.

***Untoward effects*:** LD_{LO}, **human**: 50 mg/kg.

Sheep became toxic after a total of 180 g was ingested. Symptoms in **cattle** and **sheep** including bloating, salivation, ataxia, muscle spasms, alopecia, dyspnea, ascites, and pulmonary congestion. May be carcinogenic in **mice**. In **chickens**, anorexia, decreased weight gains, and prostration are noted.

LD_{50}, **rats**: 395 mg/kg; **dog**: 510 mg/kg; dermal LD_{50}, **rabbits**: 2–2.5 g/kg.

N, N-DIALLYL-2-CHLOROACETAMIDE

= *Allidochlor* = *CDAA* = *CP 6343* = *Randox*

Herbicide.

***Untoward effects*:** LD_{LO}, **human**: 500 mg/kg.

LD_{50}, **rats**: 700 mg/kg; skin, **rat**: 360 mg/kg.

DIALLYL DIGLYCOL CARBONATE

= *CR 39 monomer*

Untoward effects: Repeated skin contact has caused severe irritation with blistering and secondary infection. **Some** develop hives. Burns in the mouth, gastrointestinal tract, and even illness and death might be expected after ingestion. **Human** experience has shown eye exposure caused moderate to severe irritation, tearing, increased sensitivity to light, burning, and pain.

Topical application of 1 ml/kg to 6–18-day pregnant **rabbits** was fatal to some. This dose, as well as 0.5 ml, decreased weight gains and led to liver pathology. Abortions increased and lens opacities and small lenses occurred in the fetuses.

2,4-DIAMINOANISOLE

= *C.I. 76050* = *C.I. Oxidation Base 12* = *2,4-DAA* = *4-MMPD*

Use: In hair and fur dyes.

Untoward effects: At high dosage, skin, lymph system, and thyroid cancers in **rats** and **mice**. Mammary, clitoral, and pituitary tumors in female **rats**. Percutaneous absorption occurs in **man** and **monkeys**.

LD_{LO}, **rat**: 151 g/kg/72 weeks; **mouse**: 157 g/kg/78 weeks.

2,4-DIAMINOAZOBENZENE

= *Azoangin* = *p-Azoaniline* = *Azohel* = *Basic Orange 2* = *Chrysoidine Hydrochloride Citrate* = *Solvent Orange 3*

Azo dye.

Untoward effects: Possibly, bladder tumors in **men**, due to handling fishing bait dyed with it. Eye irritant.

Liver tumors, leukemias, and reticulum cell sarcomas in **mice**.

LD_{50}, **rat**: 1650 mg/kg; LD_{LO}, **mouse**: 1670 mg/kg.

2,4-DIAMINOBUTYRIC ACID

Untoward effects: Neurotoxic amino acid in ***Lathyrus*** (q.v.) seeds.

4,4´-DIAMINODIPHENYLAMINE

Antineoplastic, fur dye, hydrocyanic acid diagnostic agent.

Untoward effects: 20% incidence of cholestatic-hepatocanalicular damage with bile casts.

2,6-DIAMINOPURINE

Untoward effects: ***Purine*** antagonist. Causes limb aplasia in **rats**.

IP, LD_{50}, **rats**: 250 mg/kg.

DIAMPHIDIA

= *Nga*

Untoward effects: The live larvae contain a very poisonous juice causing general paralysis. **Bushmen** of northern Namibia and Botswana put about 100 drops on a poison arrow. The juice also has a hemolytic action. Non-toxic orally.

DIAMTHAZOLE

= *Amicazole* = *Asterol* = *Atelor* = *Ro 2-2453*

Antifungal. **The dihydrochloride is used.**

Untoward effects: Convulsions developed in several **infants** after topical therapy. Tremors, ataxia, hallucinations, and behavioral changes have also been reported. A photosensitizer.

LD_{LO}, **human**: 50 mg/kg.

LD_{50}, **mouse**: 500 mg/kg; SC, LD_{50}, **mouse**: 375 mg/kg.

2,5-DI-*tert*-AMYLQUINONE

Untoward effects: In thermally activated price labels affixed to meat wrappings. One of the causes of "**meatwrappers'** asthma".

σ-DIANISIDINE

= *3,3´-Dimethoxybenzidine* = *DMOB*

Use: In numerous azo dyes. Estimated U.S. annual production varies between 3,000 and 5,000 tons.

Untoward effects: Suspected **human** carcinogen.

Causes bladder, stomach, acoustic duct, and mammary gland tumors in **rats** after stomach tubing with it.

LD_{50}, **rat**: 1920 mg/kg; LD_{LO}, **dog**: 600 mg/kg.

DIAPAMIDE

= *CI 456* = *Thiamizide* = *Vectren*

Diuretic.

Untoward effects: Headache, gastrointestinal upsets, epistaxis, nasal congestion, and increase in serum urates.

LD_{50}, **rat**: 1.4 g/kg; **mouse**: 2.6 g/kg; IP, LD_{50}, **mouse**: 520 mg/kg.

DIATOMACEOUS EARTHS

= *Diatomite* = *Dicalite* = *Infusorial Earths* = *Siliceous Earth*

Adsorbent. **Siliceous earthy rock composed of the skeletons of diatoms.**

Use: As a toxin adsorbing carrier for oral ***antibiotics*** in diarrhea control, and as a feed additive. United States regulations limit it to a maximum of 2% by weight of the total ration.

Untoward effects: Dust may temporarily irritate eyes, throat, nasal passageways and upper respiratory tract. Its small amount (3%) of crystalline ***silica*** makes it a possible carcinogen and a cause of silicosis, a non-cancerous lung disease.

LD_{50}, **rat**: 3160 mg/kg.

DIATRIZOATE SODIUM

= *Hypaque*

Contains 59.8% *iodine*.

Untoward effects: Nausea, vomiting, sensations of heat, weakness, dizziness, flushing, dyspnea, bad taste, transient apnea, salivary gland enlargement, headache, thirst, coughing, sneezing, itching, pallor, tachycardia, hypotension, and mild diarrhea occur. Convulsions, ventricular fibrillations, pulmonary edema, bradycardia, circulatory failure, cardiac arrest, complement-induced fibrinogenopenia, and pancytopenia are rare complications. Anaphylactic shock with dyspnea, edema of the face and glottis, and urticaria also reported. Pain may occur at the injection site, especially with extravasation. This may be followed by phlebitis, thrombosis, and embolism. Renal failure also reported. After cerebral arteriography leads to paresthesia, aphasia, hemiplegia, and encephalopathy. After carotid angiography leads to photophobia, blepharospasm, periorbital and retinal hemorrhage, retinal edema, and transient blindness. A 74-year-old **female** suffered a fatal ***radiographic contrast medium***-induced pulmonary granulocyte aggregation (leukostasis) during an intravenous pyelogram procedure.

High IV doses have caused pulmonary edema in **rats**.

IV, LD_{50}, **cat**: 11.3 g/kg; **dog**: 13.2 g/kg; **rabbit**: 12.2 g/kg; **mouse**: 1.4 g/kg.

DIAZEPAM

= *Aliseum* = *Ansiolin* = *Aposepam* = *Apaurin* = *Atensine* = *Atilen* = *Bialzepam* = *Calmpose* = *Ceregulart* = *Diacepin* = *Dialar* = *Diazemuls* = *Dipam* = *Dizac* = *Ducene* = *Eridan* = *Eurosan* = *Evacalm* = *Faustan* = *Gewacalm* = *Horizon* = *Lamra* = *Lembrol* = *Levium* = *Mandrozep* = *Neurolytril* = *Noan* = *Nocu* = *Novazam* = *Paceum* = *Pacitran* = *Paxate* = *Pro-Pam* = *Q-Pam* = *Relanium* = *Sedapam* = *Seduxen* = *Servizepam* = *Setonil* = *Sibazon* = *Solis* = *Stesolid* = *Tranimal* = *Tranquase* = *Tranquo-Puren* = *Tranquo-Tablinen* = *Unisedil* = *Valaxona* = *Valiquid* = *Valium* = *Valrelease* = *Vival* = *Vivol*

A benzodiazepine.

Untoward effects: ***Addiction***: Dependence potential. Slurred speech, ataxia, dry mouth, and mydriasis in **some**. Toxic psychosis and transient dysmnestic syndrome after drug withdrawal. Has led to suicide.

Anesthesia: As a preoperative agent, it has caused transient respiratory depression, moderate nausea, and slight shivering.

Behavior: Paradoxical aggression and excitement is rare and has followed abuse of the drug.

Blood and bone marrow: Hypoplastic bone marrow, agranulocytosis, pancytopenia, thrombocytopenia, and lactic acidosis.

Cardiovascular: Cyanosis, apnea, and unconsciousness within a few seconds after an IV; pain and thrombophlebitis at injection site, venous thrombosis, decreased blood pressure, bradycardia, ventricular arrhythmias and fibrillation, and hypotension. Irritant due to extravasation. Apnea and cardiac arrest have been reported after IV use. A case of transitory leukopenia has occurred.

Chromosomes: A **patient** had a significant (15.3%) breakage increase in leukocytes.

Clinical tests: Transient decrease in ***insulin*** and ***p-aminohippuric acid*** clearance. Decreased thyroid function is debatable and may not be clinically significant.

Central nervous system: Sedation, "hangover", dizziness, amnesia, tremors, insomnia, depression, confusion, headache, vertigo, dysarthria, cerebral ischemia, paranoid delusions, hallucinations, grand mal seizures, spasmodic torticollis, hypnosis, and transient memory loss. Infrequent paradoxical excitement, rage, or stimulation. A 2-year-old **male** ingested 20–24 tablets and 2 h later was unable to sit or stand, was ataxic, weak, and hyporeflexive. Uneventful recovery.

Eye: Blurred vision, diplopia, nystagmus, glaucoma, cataracts, conjunctivitis, and decreased corneal reflex.

Fetus and newborn: Decreased respiration in **newborn**, drowsiness with decreased suckling and weight loss ("floppy **infant** syndrome"). Levels in breast milk are about one-tenth of maternal plasma's. One report indicates umbilical cord concentration equal to **mother's** serum level. **Infant's** plasma levels on day 4 are nearly 40% of maternal plasma levels, decreasing to about 12% on day 6. Cleft lip and/or cleft palate, limb and digit malformations (agenesis of thumbs, syndactyly, limb reduction defects, hypoplastic femur, dislocated hips), ventricular septal defects, defective anterior abdominal wall, diaphragmatic defects, omphalocele, hypertelorism, and low-set ears attributed to its use during the first trimester. Three **infants** experienced withdrawal symptoms after prolonged uterine exposure. Displaces ***bilirubin*** from ***albumin*** binding sites, causing kernicterus in premature **infants** and **neonates**.

Gynecomastia: Particularly after high dosage in **men**.

Hypersensitivity: Anaphylactic reactions have been reported after IM and IV injections. Immediate asthmatic reaction in a 9-year-old **male** after 5 mg IV. ***Propylene glycol***, a potent contact sensitizer, occurs in the parenteral form.

Libido: Impotence or inhibited ejaculation in **men** and decreased libido in **women**. This may be due to the drug's sedative effect or the state of anxiety for which the **patient** is under treatment.

Liver: Increased alkaline phosphatase is probably a manifestation of hypersensitivity. Granulomatous hepatitis and cholestatic jaundice have rarely been associated with its use. Hepatotoxicity may also be due to ***sodium benzoate*** and ***benzoic acid*** content in parenteral products.

Miscellaneous: Gout, nausea, and constipation. Altered driving skills and reduced reaction time. A report of hypothermia.

Poisoning: A 4½-year-old **male** collapsed and became unconscious after ingestion of 35 mg (7 tablets). Central nervous system disturbances. Recovery only after 12 days. A 62-year-old **female** received, in error, 50 mg IV, instead of 5 mg. **She** was somnolent for 4 h. Death occurred within 30 min in a 46-year-old **female** after 5 mg IM to control agitation due to unrecognized respiratory failure. Cardiac arrest in a 2-year-old **female** 3 h after ingesting 70 mg, with four more episodes in the next 30 min, before cardiac rate stabilized. Alone or with other drugs in suicide attempts. In suicide attempts, a 61-year-old **female** ingested 450–500 mg and a 28-year-old **male** ingested 2000 mg. With treatment, both recovered in 48 h. Near-fatal poisoning in a 46-year-old **male alcoholic** due to the ***propylene glycol*** (q.v.) solvent in treating ***alcohol*** withdrawal symptoms (~3 g during 24 h). Therapeutic blood levels are stated as 0.05–0.25 mg% (0.25–0.75 mg/l), and toxic levels as 0.5–2.0 mg% (1.5–3.0 mg/l).

Renal: Gross hematuria reported from large (80–180 mg/4 days in 35-year-old **male**) IV dosage. Allergic interstitial nephritis after use for 2 months. Urinary incontinence and/or retention may occur.

Respiratory: Pulmonary complications reported in 2/1000 **patients**. Respiratory arrest, even after slow IV of 10 mg, in a 79-year-old **male**, a 56-year-old **male**, a 47-year-old **male**, an 18-year-old **female**, and **others**. Apnea occurred in a 20-year-old **female** after 2.5 mg IV, requiring 20 min of artificial ventilation before spontaneous breathing resumed.

Skin: Urticarial skin rashes in 7% of **psychoneurotics**; fixed-drug eruptions and photosensitization occur.

Withdrawal: Tremors, tachycardia, and hypertension in a 31-year-old **female** who took 100–120 mg/day/2 years. Grand mal convulsions in a 23-year-old **male** who had been on 10 mg/tid. Psychosis in a 49-year-old **male** after 7 years on 15–30 mg/day. Trembling, agitation, fearfulness, panic attacks, major depression, stomach cramps, nausea, vomiting, paresthesia, tinnitus, and sweating reported in **others**. Avoid abrupt withdrawal of treatment.

Drug interactions: ***Alcohol, Ethyl***: Potentiation of both, but mostly of diazepam, with greatest effect at about 30 min. Diazepam absorption is increased with concomitant ***alcohol*** ingestion.

Anticoagulants: One case describes ***bishydroxycoumarin*** potentiation by diazepam, but others, such as ***warfarin***, appear to have no significant interaction.

Barbiturates: Hypotension, dizziness, and disorientation have occurred. In **animals**, ***phenobarbital*** has increased diazepam metabolism. Similarly, ***pentobarbital*** has affected diazepam in **humans** and **rats**.

Cimetidine: Decreases diazepam metabolism, potentiates its action, with increased incidence of sedation, especially in the **elderly** and **those** with liver dysfunction.

Meperidine: Respiratory failure and cardiovascular collapse.

Miscellaneous: Raises plasma half-life of ***ketamine*** (30–40%) and ***digoxin***. ***Isoniazid, erythromycin, ketoconazole, propranolol, estrogen-containing oral contraceptives, fluvoxamine,*** and ***propoxyphene*** increase its elimination half-life and decrease its total clearance. Profound hypothermia in a 38-year-old **female** on three occasions after also taking ***lithium carbonate***. Increased drowsiness with ***scopolamine, nortriptyline,*** and ***amitriptyline***; galactorrhea with ***phenelzine***. ***Disulfiram*** decreases its plasma clearance, increases its drowsiness and sedation. Anorexia, vomiting, and headache with ***niridazole***. Respiratory depression from use with ***alphaprodine*** can be potentially fatal. Rate of absorption, but not the amount of absorption, decreased by ***antacids***. ***Δ^9-Tetrahydrocannabinol*** augments the central nervous system and respiratory decrease caused by diazepam. Although it may help control the agitation, it may unexpectedly prolong ***PCP*** psychosis, by inhibiting the metabolic breakdown of ***PCP***. Slight increase in its clearance, and its metabolism is occasionally increased when used with ***sertraline***. ***Fluvoxamine*** inhibits its metabolism.

Muscle relaxants: *d-**Tubocurarine*** effect enhanced by it. Profound potentiation (three times) of ***gallamine triethiodide***. ***Succinylcholine***-induced neuromuscular block decreased by 20–30%.

LD_{LO}, **human**: 50 mg/kg; IV, LD_{LO}, **human**: 143 µg/kg.

Used alone in **cats** and **dogs**, it can produce excitement. IV use in **dogs** will cause excitement, disorientation, and seriously aggravate ***digitalis***-induced arrhythmias. Can increase appetite in **cats** and cause hepatotoxicity. Has an effect on **harbor seals**, but NO effect on **California sea lions**. Crosses the placental barrier in **hamsters**, **mice**, and **monkeys**.

LD_{50}, **rat**: 710 mg/kg; **mouse**: 535 mg/kg.

DIAZINON

= *Alfa-Tox* = *Antigal* = *Basudin* = *Chlormethiuron* = *Dazzel* = *Diazajet* = *Diazide* = *Diazitol* = *Diazol* = *Dimpylate* = *Dipofene* = *Dipophene* = *D-z-n* = *ENT 19,507* = *Escort* = *Ferti-Lome* = *G-24480* = *Garden Tox* = *HelfaCat* = *HelfaDog* = *Kayazinon* = *Neocidol* = *Nucidol* = *Parasitex* = *Sarolex* = *Spectracide* = *Taberdog* = *Tabergat* = *Terminator* = *Tik-20* = *Timpylate*

Pesticide, insecticide, nematocide. **An organophosphorus compound, whose metabolite, *diazoxon*, has even more potent anticholinesterase activity.**

Untoward effects*:** Causes rapid and severe reduction in serum cholinesterase, demanding early treatment, before advanced symptomology. After spraying in a central laboratory for **cockroach** control, **workers** and laboratory **rats** and **monkeys** still had decreased serum cholinesterase 3 months later. Similar use in various premises has caused acute gastroenteritis, nausea, vomiting, tightness in chest, pulmonary edema, bradycardia, epigastric pain, giddiness, headache, diarrhea, cyanosis, edema, miosis, blurred vision, tearing, papilledema, coma, loss of reflexes and sphincter control, myalgias, nervousness, muscle twitching, dyspnea, weakness, sweating, lethargy, and encephalopathy. Acute short-term confusional psychosis in a 43-year-old **male** after intermittent exposure after spraying **sheep** without washing of hands or changing clothes. Nephrotoxicity manifested as crystalluria in an exposed 26-year-old **patient**. After a **minister** put 4 oz ***Spectracide® into a gallon milk jug half full of water, making a milky suspension, the **wife**, seeing it on the kitchen counter, put it in the refrigerator and used it for making pancakes. All her four **children** were treated in the emergency room. Its greatest reported ill-effects were from suicide attempts.

The estimated LD for **man** is 10–25 g.

LD_{LO}, **human**: 50 mg/kg. On post-mortem, a 75 kg 54-year-old **female** had petechial hemorrhages in the stomach, gastric mucosa, and brain, after ingesting a fatal dose. Five cases of intentional ingestion of 60–180 ml of a 25% solution survived after prompt treatment.

A **farmer** killed 30 **cows** and a herd **bull** as a result of spraying them with a mixture made for a premises spray. Salivation, tremors, muscle fasciculations, ataxia, convulsions, miosis followed by mydriasis, diarrhea, and vomiting (in species that can) are common symptoms in poisoned **animals**. May induce pancreatitis and pancreatic pseudocysts in **dogs**. Causes skeletal defects in **chick** embryos. **Bird** mortalities from it have been associated with their feeding on treated grasses, such as golf courses. Very toxic to **goslings**, **ducks**, **pheasants**, and **turkeys**. Goose-stepping ataxia, wing spasms, wing-drop, hunched back, dyspnea, tenesmus, diarrhea, salivation, lacrimation, ptosis, prostration, opisthotonus, and convulsions in **birds**. Highly toxic to **bees**. **Dorset Down sheep** may be genetically more susceptible to its toxic effects than other **sheep**. Persian and Himalayan **cats** usually lack adequate enzyme levels to metabolize it. Fatal poisoning in young **calves** occurs at 0.88 mg/kg; adult **cattle** by 22 mg/kg; **sheep** by 26 mg/kg; **hens** 50 mg/kg; and **ducks** 14 mg/kg.

LD_{50}, **rats**: 100 mg/kg; acute dermal: 500–900 mg/kg; LD_{50}, **cattle** and **swine**: 100 mg/kg; **turkeys**: 2 mg/kg; **birds**: 4 mg/kg; **mallard ducks**: 3.54 mg/kg; **pheasants**: 4.33 mg/kg; **bullfrogs**: > 2 g/kg; **mouse**: 85 mg/kg; **guinea pig**: 240 mg/kg; **rabbit**: 130 mg/kg.

DIAZOACETYL GLYCINE HYDRAZIDE

= *DGI* = *NSC 58,404*

Antineoplastic. **Experimental.**

***Untoward effects*:** Nausea and vomiting.

Necrotizing nephrosis and lympoid decrease in **mice**, **rabbit**, **rat**, and **dog** fatal poisonings. Granulocytopenia was most severe in affected **dogs**.

IP, LD_{50}, **rats**: 1.34 g/kg; **mice**: 2.58 g/kg.

DIAZOLIDINYL UREA

= *Germall II*

Preservative.

***Untoward effects*:** Positive allergic responses in 1.76% of **patients**.

DIAZOMETHANE

= *Azimethylene*

***Untoward effects*:** Explosive. Very toxic yellow gas. Brief inhalation exposure in a 40-year-old **male** led to general weakness, severe headache, flushing, cough, and tracheobronchial irritation. Other cases also reported fever, bradypnea, asthma, chest pain, pulmonary edema, pneumonitis, and conjunctivitis. TWA 0.4 mg/m^3 (0.2 ppm); 2 ppm – danger to life (within 30 min).

6-DIAZO-5-OXO-L-NORLEUCINE

= DON = NSC 7,635

Antineoplastic.

Untoward effects: Oral mucosa is red and ulcerated. Abdominal pain, vomiting, and diarrhea occur. Alopecia is common and related to severity of gastrointestinal upsets and bone marrow suppression. Tongue may become bright red after 7–10 days and requires decreased dosage.

Given at 5 mg/kg to **rats** 24 h before palatal closure, it causes an incidence of > 90% cleft palate fetuses, in addition to some aplastic long bones, syndactyly, and fetal deaths.

LD_{50}, **rat**: 80 mg/kg; IP, LD_{LO}, **rat**: 5 mg/kg (15 days' pregnancy).

DIAZOXIDE

= Eudemine = Hyperstat = Hypertonalum = Mutabase = Proglicem = Proglycem = SRG-95213

Antihypertensive, vasodilator, antidiuretic.

Untoward effects: Major adverse effects are ***sodium*** and ***water*** retention (explored for preslaughter shipment of **livestock**), hyperglycemia, and hyperuricemia. Occasionally, nausea, vomiting, muscle cramps, unpleasant taste, and anorexia. Temporary interuption of labor reported. Local burning with extravasation and route of injection. Acute pancreatitis, gout, galactorrhea, and extrapyramidal symptoms occur. Therapeutic serum levels reported as 5–25 mg/l and toxic concentration as 30–40 mg/l.

Blood: Granulocytopenia, eosinophilia, leukopenia, neutropenia, thrombocytopenia, hemolytic anemia, and hyperuricemia.

Cardiovascular: Has caused large and almost irreversible decrease in blood pressure in 3% of **patients**. One **patient** developed severe hypotension, anginal syndrome, cerebral ischemia and edema, and right hemiplegia. **Another** developed a myocardial infarction. Palpitations, flushing, and headache also noted. Peripheral resistance decreased with dilation of arterioles and cardiac output increased. Has caused heart block in a 6-month-old **child**.

Clinical tests: False-negative results in IV ***glucagon***-induced ***insulin*** stimulation test.

Central nervous system: Cerebral ischemia and infarction leading to unconsciousness, confusion, convulsions, and paralysis. Numbness of hands, orthostatic hypotension, sweating, flushing, tinnitus, weakness or euphoria also noted.

Eye: Rapid fall of blood pressure after IV usage in 14-year-old **male** may have precipitated a case of blindness. Similar results in a 30-year-old **female** and a 32-year-old **female** treated for malignant hypertension. Oculogyric crisis. Has caused lens opacities, cataracts, subconjunctival hemorrhages, diplopia, and blurred vision.

Fetus and neonate: Crosses the placenta and can cause hyperbilirubinemia, thrombocytopenia, and altered carbohydrate metabolism in an **infant**. Orally, in late pregnancy, it can cause alopecia or hypertrichosis in **infants**. Decreased IgG has had no clinical significance. Advanced bone age has been reported. See ***Glucose*** and ***Hair*** below.

Glucose: Diabetogenic on long-term use. Hyperglycemia with blurred vision, weakness, hunger, and thirst in a 53-year-old **male** after 4 weeks on 400 mg/day. May cause direct inhibition of Islets of Langerhans. Ketosis, hyperglycemia, increase in body hair, vomiting, facial edema, and facial hair. Death of one **child**. **All** were treated for idiopathic hypoglycemia. Impaired ***glucose*** tolerance in two **offspring** of diabetic **mothers** given it orally during pregnancy. Abnormal hair growth in them, even after 5–12 months. Acute diabetic ketoacidosis in a 29-year-old **female**. Polyuria has also been reported.

Hair: Hirsutism and hypertrichosis lanuginosa reported in young and old. See ***Glucose*** above.

Hypersensitivity: Rash, leukopenia, and fever are uncommon.

Skin: Alopecia noted above under ***Glucose*** and ***Hair***. Thrombocytopenia and maculopapular rash on trunk and lower limbs. Diffuse petechial eruptions, pruritus, and paresthesia also occur.

Therapeutic serum levels are 10–20 mg/l and toxic at 50–100 mg/l.

Drug interactions: ***Anticoagulants***: Displaces ***warfarin*** from its binding to ***albumin***, causing potentiation of ***warfarin's*** effect.

Chlorpromazine: Will enhance the hyperglycemic effect of diazoxide. A 2-year-old **child** developed diabetic pre-coma after a single dose while receiving long-term diazoxide therapy.

Clomethiazole: Used orally in late pregnancy with diazoxide it can cause **neonatal** depression, hypotonia, hypoventilation, and apnea.

Hydralazine: The combination has been associated with angina, ischemia, electrocardiogram changes, tachycardia, myocardial infraction, and stroke.

Phenytoin: Markedly reduced therapeutic levels and seizure control.

Hyperglycemia, glycosuria, and ketonuria reported in treated **dogs**. Fatal hyperglycemia (215) and glycosuria (4+) in pregnant **dog** after 3 weeks on 200 mg/kg/day. Diazoxide (5 mg/kg) IV did not affect the blood ***glucose*** levels of pregnant (110 days) Hampshire **sheep** and Angora **goats**, but 75% of their **lambs** and **kids** were born diabetic. Hyperglycemia induced in **mice**, **rats**, **monkeys**, and **rabbits**. Sinus tachycardia, arrhythmias, 15-min exposure segment decrease and myocardial necrosis in treated **dogs**. Lenticular aberrations in beagle **dogs**. Treated **pigs** are more susceptible to the toxic effects of ***benzene hexachloride*** and ***carbaryl***.

LD_{50}, **rats**, **mice**, **guinea pigs**: 150–200 mg/kg.

DIBEKACIN

= *Debecacin* = *Débékacyl* = *3´,4´-Dideoxykanamycin B* = *DKB* = *Icacine* = *Kappabi* = *Orbicin* = *Panamicin* = *Panimycin* = *Tokocin*

Aminoglycoside antibiotic.

Untoward effects: Nephrotoxicity reported in **humans**, **rats**, and **rabbits**. Cochlear ototoxicity in **guinea pigs** less than ½ that of ***gentamicin***.

IV, LD_{50}, **rat**: 140 mg/kg; **mouse**: 63 mg/kg.

DIBENZACRIDINE

A heterocyclic compound used in research.

Untoward effects: Carcinogen found in ***cigarette*** smoke, automobile exhausts, and polluted air from ***coal*** combustion, petroleum refineries, and incinerators. 8 h TWA 0.2 mg/m^3.

Exposed **mice** have reduced pregnancies. SC to **mice** leads to local sarcomas and lung tumors; IV leads to increase in multiple lung tumors; topically leads to skin tumors.

SC, TD_{LO}, **mice**: 40 mg/kg.

1,2 : 5,6-DIBENZANTHRACENE

= *DBA* = *Dibenz[a,h]anthracene*

Widespread in the environment as a product of incomplete combustion.

Untoward effects: Reported in airborne soot at 64–705 mg/1000 m^3, and in ***coal tar*** at 230 mg/kg. OSHA exposure limits for **people** are, essentially, the same as for ***dibenzacridine*** (q.v.).

In **mice**, increased abortions, resorptions, and stillbirths. Over 80% with primary lung tumors at 5 months. Topically leads to papillomas and epitheliomas. Leukemia in IV treated **guinea pigs** and local tumors after SC injection. Also, tumors by different routes of administration in **rats**, **frogs**, **pigeons**, and **chickens**.

IV, LD_{LO}, **mouse**: 10 mg/kg; **guinea pig**: 30 mg/kg; **frog**: 8 mg/kg.

DIBENZEPIN

= *HF-1927* = *Neodalit* = *Noveril*

Antidepressant.

Untoward effects: Fatigue (26%), malaise (15%), blurred vision (11%), anorexia (7%), nausea (7%), drowsiness (7%), and hot flushes (7%). Headache, vertigo, collapse, skin rash, dry mouth, sweating, and constipation have also been reported. Coma in a 59-year-old **female** 2 h after ingestion of 2 g (50 tablets). Convulsions 1 h later, systolic blood pressure 60 mmHg, irregular respiration, cyanosis, and cardiac arrhythmia. Two **children** died 5 and 10 h, respectively, after ingesting 160 mg/kg (23-month-old **male**) and 35 mg/kg (3½-year-old **female**). Severe circulatory collapse with hypotension and respiratory depression in both. The **boy** had cardiac rhythm problems with atrioventricular block; the **girl** had frequency seizures. A 39-year-old **female** ingested eight sustained release-240 mg tablets leading to pulmonary edema. Avoid use with ***monoamine oxidase inhibitors***. Therapeutic serum levels are 0.1–0.3 mg/l; toxic at 3 mg/l; and fatal at 18 mg/l.

LD_{LO}, **child**: 160 mg/kg.

LD_{50}, **rat** and **mouse**: 220 mg/kg; **guinea pig**: 110 mg/kg.

7H-DIBENZO(C,G)CARBAZOLE

Untoward effects: Carcinogenic in the **mouse**, **rat**, **hamster**, and **dog**. Oral use causes stomach tumors and hepatomas in **mice**; intratracheal administration leads to respiratory tract tumors in **hamsters**; intravesicular use leads to epithelial neoplasms in the urinary bladder of **dogs**.

IP, TD_{LO}, **mouse**: 13 mg/kg; SC, TD_{LO}, **mouse**: 8 mg/kg.

DIBENZODIOXINS

A chemical contaminant in *pentachlorophenol* (q.v.)-treated lumber. The halogenated forms persist in the environment. Found in municipal fly ash.

Untoward effects: Sensitivity to its lethal effects are great in **guinea pigs** and **chickens**; **rats** are intermediate in sensitivity, and **mice** evidence less. Causes **monkeys** to progressively lose $\frac{1}{3}$ of their body weight and **chick** edema disease in **broilers**. A report of high mortality in affected **horses**, **birds**, **cats**, **dogs**, **rodents**, and **humans** from soil contamination (riding arena treated with waste oil).

DIBENZO(a,h)PYRENE

A polycyclic aromatic hydrocarbon found in engine exhausts, *coal tar* pitch, and *cigarette* tars.

Untoward effects: Carcinogenic after repeated topical use on **mice**, and after injections into **rats** and **mice**. The (a, i) form causes local sarcomas in the **hamster** and **mouse** after SC use.

DIBORANE

= *Boroethane*

Use: In rocket propellents and ***rubber*** vulcanizing. Flammable gas with sickly-sweet odor.

Untoward effects: Precordial or chest pain, dyspnea, cough, nausea, headache, vertigo, chills, fever, fatigue, weakness, tremor, and muscle fasciculations. TWA 0.1 mg/m^3 (0.1 ppm); 40 ppm – immediate danger to life (within 30 min).

Inhalation LC_{50}, **rat**: 50 ppm/4 h; **mouse**: 29 ppm/4 h.

DIBROM

= *Alvora* = *Arthrodibrom* = *Bromchlophos* = *Bromex* = *Dibromfos* = *ENT-24988* = *Naled* = *OMS-75* = *Ortho-Dibrom* = *RE-4355*

***Organophosphorus insecticide*. Cholinesterase inhibitor.**

Untoward effects: Miosis, lacrimation, headache, chest tightness, wheezing, laryngeal spasm, salivation, anorexia, cyanosis, nausea, vomiting, abdominal cramps, diarrhea, weakness, twitching, paralysis, giddiness, ataxia, convulsions, decreased blood pressure, cardiac arrhythmias, and eye and skin irritation. TWA 0.1 mg/m^3; 40 ppm leading to immediate danger to life (within 30 min).

LD_{LO}, **human**: 50 mg/kg.

Highly toxic to **honeybees**. Topical sprays have caused skin irritation to **cattle** and were toxic to **ducks**. See general cautions under ***diazinon***. Avoid use on **chickens** under 6 weeks of age and **turkeys** under 3 months of age. Tissue residues in **poultry** are usually gone by 72 h post-treatment.

Inhalation LD_{50}, **rat**: 7.7 mg/kg; **mouse**: 156 mg/kg; LD_{50}, **rat**: 56 mg/kg; **rabbits**: ~7.5 mg/kg; **mice**: ~200 mg/kg; **birds**: 120 mg/kg. 24 h, LC_{50}, **Atlantic salmon**: 165 μg/l; 96 h, **trout**: 195 μg/l.

DIBROMOCHLOROPROPANE

= *BBC 12* = *3-Chloro-1,2-dibromopropane* = *DBCP* = *Fumazone* = *Nemafume* = *Nemagon* = *OS-1897*

Soil fumigant, nematocide.

Untoward effects: Drowsiness, nausea, vomiting, pulmonary edema, and eye, skin, nose, and throat irritant. 10 weeks exposure caused many manufacturing plant **workers**, **farmers**, and crop-spraying **workers** to be sterile and have reduced sperm counts. Carcinogen and mutagen. Renal and liver toxicity reported. OSHA 8 h TWA 1 ppb, and a limit of 10 ppb for a 15 min period.

LD_{LO}, **human**: 5 mg/kg.

Causes stomach cancer in **rats** and **mice**, decreased sperm counts and sterility in **rats**, and mammary tumors in female **rats**. Inhalation led to nasal cavity tumors and tumors of the tongue in **rats**, and adrenal cortical adenomas in female **rats**. Nasal cavity and lung tumors in **mice**. Ataxia, hyporeactivity, slowness, falling, ptosis, tremors, lethargy, asthenia, myasthenia, ataraxia, and somersaulting onto back or side developments within 15 min in exposed **mallard ducks** and **pheasants**. Deaths occur within 1–9 days.

LD_{50}, **rats**: 170 mg/kg; **mice**: 260 mg/kg; **mallard ducks**: 66.8 mg/kg; **pheasants**: 156 mg/kg; 96 h, LC_{50}, **bluegills**: 320 mg/l.

5,7-DIBROMO-8-QUINOLINOL

= *Brodiar* = *Broxykinolin* = *Broxyquinoline* = *Colepur* = *Fenilor* = *Intensopan*

Antiseptic, disinfectant.

Untoward effects: Subacute myelo-optic neuropathy. Permanent blindness reported in **children**. Single large doses led to cerebral dysfunction with disorientation and retrograde amnesia.

TD_{LO}, **child**: 1000 mg/kg/27 days.

Symptoms in **dogs** within 3–12 h. A single dose of 10 mg/kg has been fatal to some **dogs**. Renal lesions in treated **rats**. Symptoms were nervousness, restlessness, excitation, and salivation, followed by muscle twitches of the ears, mouth corners, shoulders, and thighs. Postmortem revealed nephritis; acute non-purulent lymphohistiocytic meningoencephalitis; acute myocardial degeneration; necrosis and fatty infiltration of the liver.

DIBROMSALAN

= *4′,5-Dibromsalicylanilide*

Antiseptic.

Untoward effects*:** Primary photosensitizing agent in sensitive **individuals**, and it induces cross-reactions with ***tribromsalan and ***impregon***.

DIBUCAINE

= *Benzolin* = *Cincaine* = *Cinchocaine* = *Nupercaine* = *Percaine* = *Sovcaine*

Local anesthetic, diagnostic agent.

Untoward effects*:** Frequently used to stop itching, but it sensitizes the **individual** so that later use causes an allergic reaction and pruritus. This a real hazard from its indiscriminate use in first-aid sprays. Possible toxic cauda equina syndrome in 35-year-old **male** receiving spinal anesthesia with 12 mg prior to lumbar laminectomy. Seizures, dysarrhythmias, and death reported. An over-the-counter item at 0.5–1.0% in ointments or creams. Toxicity estimated at ten times that of ***lidocaine and 20 times that of ***procaine***.

LD_{LO}, **child**: 50 mg/kg.

Avoid IV use in **humans** and **animals**. Highly cardiotoxic, leading to ventricular arrhythmias. Will antagonize antibacterial effects of ***sulfonamides***. Salivation, vomiting, hypothermia, tremors, bradycardia, weakness, and seizures reported in **cats** and **dogs**.

SC, LD_{50}, **rabbit**: 8.5 mg/kg; SC, LD_{LO}, **guinea pig**: 112 mg/kg.

DIBUTYL PHTHALATE

= *n-Butyl Phthalate* = *DBP*

Insect repellent in clothing, plasticizer, solvent, antifoam. In microcapsules of some carbonless papers.

Untoward effects*:** Mild dermatitis and severe irritation of upper respiratory tract and stomach. Can cause coma if ingested. Identified in ***alcohol sponges. May be estrogenic. TWA 5 mg/m^3; 9300 mg/m^3 – danger to life (within 30 min).

TD_{LO}, **human**: 140 mg/kg.

Cats often bite into glow-in-the-dark jewelry containing it leading to sudden profuse salivation. Transient hyperactivity and aggressive behavior follow in ~50%. High dietary intake causes severe testicular atrophy in **rats**.

LD_{50}, **mice**: 3.57 g/kg; **mammal**: ~5–15 g/kg.

DIBUTYL SQUARATE

***Untoward effects*:** Transient leukoderma after topical treatment of alopecia areata. Potent skin sensitizer and contact allergen.

DIBUTYL TIN DILAURATE

= *Butynorate* = *Davainex* = *DBTD* = *Tinostat*

***Coccidiostat, anthelmintic, stabilizer for polyvinyl chloride resins.* Contains 18.8% *tin*.**

Untoward effects*:** A feed-mill mixing error caused over 1000 **cattle** to become clinically ill. 171 died and 287 were slaughtered in a cachectic condition. Young **calves** often died suddenly, occasionally with colic, collapse, and visible paralysis. In those dying up to 3 days after symptoms, colic, watery diarrhea, weakness, tremors, and convulsions occurred. In older **calves** that died between 4 and 14 days hemorrhagic diarrhea was also noted. Polydypsia, severe weight decrease, tremors, and weakness were also noted. ***Milk yields of **cows** fell about 80% and abortions occurred. 50–5000 ppm ***tin*** (equivalent to 250–25,000 ppm ***DBTD***) was found on different affected farms. **Palm doves** and **mink** were also poisoned.

LD_{50}, **rat**: 243 mg/kg.

DICAMBA

= *Banex* = *Banvel D* = *Dianat* = *Mediben* = *Mondak* = *Vel 58-CS-11* = *Velsicol Compound R* = *Velsicol-58-CS-11*

***Herbicide.* Metabolized to *salicylic acid*.**

Untoward effects*:** A **female** intentionally ingested 100 ml of a mixture of it and ***2,4-D (probably, ~2 g with 20 g of the latter salt) and survived after treatment.

Fed at 50 ppm (1.25 mg/kg/day) to **dogs** for 2 years, it revealed no clinical or laboratory ill-effects. It requires 250 ppm (6.25 mg/kg/day) to produce one adverse finding (slight yellowish cast to the liver) in one **dog**. No toxicity to **quail** up to 5000 ppm. Estimated toxicity for **cattle** and **sheep** is ~300 mg/kg. Symptoms are salivation, depression, and tympanites.

LD_{50}, **rat**: 1.04 g/kg; **mouse**: 1.2 g/kg; **rabbit**: 2 g/kg; **guinea pig**: 3 g/kg.

DICAPTHON

= *ACC 4124* = *Dicaptan* = *Isochlorothion*

***Insecticide.* Organophosphorus cholinesterase inhibitor. *Paranitrophenol* is a urinary metabolite.**

***Untoward effects*:** LD, **calves**: 25 mg/kg. LD_{50}, male **rats**: 400 mg/kg; female **rats**: 330 mg/kg; dermal: 790 and 1250 mg/kg, respectively.

DICENTRA CANADENSIS

= *Squirrel Corn* = *Stagger Weed* = *Turkey Corn*

Found in rich moist soils in eastern North America.

Untoward effects: Contains the alkaloids ***apomorphine*** and ***protopine***, causing restlessness in **cattle** that ingest its leaves or tubers.

DICENTRA CUCULLARIA

= *Bleeding Heart* = *Dutchman's Breeches* = *Little Blue Stagger*

Untoward effects: The leaves, stems, and roots are poisonous to **man**. Contains ***isoquinoline***-type alkaloids (***apomorphine, corydaline, cryptopin, cucullarine, corydine, isocorydine, bicuculine, bulbocapnine, protoberberine,*** and ***protopine***). Tremors, ataxia, dyspnea, salivation, nausea, visual disturbances, and convulsions. Its ***furocoumarin*** content causes photosensitization reactions after ingestion or skin contact. Other species of *Dicentra* are also poisonous.

Has caused trembling, staggering, glassy-eyed appearance, dyspnea, frothing of mouth, opisthotonus, projectile vomiting, convulsions, and death of **cattle**. Occasionally, **horses** are poisoned, but rarely **sheep**, after grazing it in wooded pastures. Ingestion of leaves has poisoned **budgies**. Ingestion by **cats** and **dogs** leads to trembling, salivation, vomiting, hyperexcitability, recumbency, and (within 10 min) opisthotonus and convulsions, but rarely fatal in them.

DICHAPETALUM sp.

Untoward effects: ***D. cymosum*** = ***Gifblaar*** = ***Ndebele*** = ***Poison Leaf*** = ***Umkauzaan***, an extremely poisonous South African plant, contains ***fluoroacetic acid***. ***Fluoroacetamide***, a simple derivative, is a potent rodenticide. The plant has killed many grazing **sheep**, **goats**, and **cattle** with hyperemic lungs, liver, and kidneys; ascites, enteritis, and diarrhea. Cardiotoxic. 2–3 oz of young leaves can kill **sheep** or a **bovine**.

D. barteri = ***akwuosa*** = ***ngbu-ewi*** (Igbo meaning "**goat** killer") has killed **mice**, **rabbits**, and **goats** in Nigeria. Convulsions, dyspnea, cyanosis, and paralysis occur.

LD_{LO}, **mouse**: 200 mg/kg; **rabbit**: 800 mg/kg.

D. toxicarium = ***Ratsbane***, like other species, may be more toxic if the **animal** drinks water after ingesting the plant.

Most other species are toxic and accumulate the monofluoroacetate ion. Pulp of the fruit of many species is edible, but the seeds are poisonous.

DICHLOFENTHION

= *Mobilawn* = *Nemacide* = *VC-13* = *VC 1-13*

Organophosphorus insecticide. **Cholinesterase inhibitor.**

Untoward effects: Ingestion of a 75% solution with suicidal intent by four **males** and one **female** led to mild miosis, sialism, vomiting, fecal and urinary incontinence, diarrhea, sweating, cramps, and weakness. Severe cholinergic symptoms persisted 5–48 days in **three**; almost total inhibition of cholinesterase enzymes for 66 days in **one**; **two** died. A senile **female** died after inadvertently drinking some. Non-fatal cases usually involve agricultural and lawn-spray **workers**.

LD_{LO}, **human**: 50 mg/kg.

In the feed of **hens** at 800 ppm it gave their eggs an undesirable flavor. Residues were found in the liver, muscle, fat, and egg yolk of treated **hens**. Do NOT use in conjunction with or within a few days of (before and/or after) any other cholinesterase-inhibiting compound. (Avoid use with ***phenothiazine*** or ***phenothiazine tranquilizers***). A 3 week interval is safer. **Brahman cattle** often appear to have increased sensitivity to organic phosphorus cholinesterase inhibitors. Do NOT contaminate feed or water. Do NOT use within 2 weeks of slaughter, to help prevent undesirable residues in food for **human** consumption. Use on **livestock** is not permitted in some areas. Keep away from **children**.

LD_{50}, **rat**: 250 mg/kg; **chicken**: 148 mg/kg.

DICHLONE

= *Algistat* = *Compound 604* = *ENT 3776* = *Phygon* = *Sanquinon*

Fungicide, algicide, herbicide.

Untoward effects: For non-food **fish** only; not for potable water. Treat only ⅓ of a pond at a time; repeat in 3 days, after algae has decomposed.

LD_{LO}, **human**: 500 mg/kg.

LD_{50}, **rat**: 1.3 g/kg; **bird**: > 5000 ppm in feed; 24 h, LC_{50}, **bluegill**: 0.04 mg/l.

DICHLORALANTIPYRINE

= *Bihypnal* = *Bonadorm* = *Dichloralphenazone* = *Dormwell* = *Sedor* = *Sominat* = *Welldorm*

Sedative, hypnotic. **A complex of *chloral hydrate* and *phenazone*.**

Untoward effects: Dizziness, weakness, and palpitations. General erythematous rash with body temperature increase to 40°C (104°F) in a 60-year-old **female** after ingesting 650 mg. Nausea, vomiting, itching red skin, and aching limbs on two occasions in a 60-year-old **female** after 650 mg. **Others** reported urticarial rashes, giddiness, tremors, and sweating. Anaphylaxis has been reported on a number of occasions and has included periorbital edema. Life-threatening arrhythmia after suicide attempt with 6.5 g. Sudden withdrawal may precipitate fits.

Drug interactions: Increases ***warfarin*** metabolism and decreases its effect.

Rat studies showed increased absence of fifth sternovertebra and some inhibition of ossification in T_{13}; no effect on offspring mortality, litter size, weight, or pregnancy rate.

DICHLORISONE

= *Astroderm* = *Diloderm* = *Disoderm*

Topical anti-inflammatory, antiallergic steroid.

Untoward effects: Cataracts, insomnia, mild facial hirsutism, transient amenorrhea, increased physical capacity, occasional gastrointestinal upsets, rash, increased appetite and weight gain, and increased serum ***uric acid*** level. An episode of gout in a 51-year-old **male** with a history of gout.

DICHLORISOPROTERENOL

= *DCI* = *Dichlorisoprenaline*

β-Adrenergic-blocker. **Early experimental agent before the synthesis of *propranolol*.**

Untoward effects: Suspect carcinogen.

DICHLOROACETIC ACID

= *Bichloracetic Acid* = *Dichlorethanoic Acid* = *DCA*

Keratolytic, caustic.

Untoward effects: Local pain and occasional circumferential erythema after skin lesion is sloughed off. Apply only to the lesion being treated and not to the surrounding area. Neurotoxic. Has been found in some ***pangamic acid*** preparations, which, when taken internally, can cause renal ***oxalic acid*** stones and dysfunction.

LD_{LO}, **human**: 5 g/kg.

Carcinogenic in **rats**.

LD_{50}, **rat**: 2.8 g/kg; **mouse**: 5.5 g/kg; dermal, **rabbit**: 510 mg/kg.

1,3-DICHLOROACETONE

= *Fyrol FR2*

Untoward effects: Lacrimator and vesicant. A potent metabolite is a probable carcinogen and immunosuppressant.

3,4-DICHLOROANILINE

= *3,4-Dichlorobenzeneamine*

Use: In manufacturing the herbicide and nematocide, ***propanil***; and various deodorant cosmetics, solvents, and dyestuffs.

Untoward effects: Chloracne was found in 38% of the above **workers. Many** were treated for acute methemoglobinemia, with the possibility of anemia and cyanosis. ***TCAB*** and ***TCAOB*** were contaminants in the manufacture of both compounds and also cause chloracne. In 1977, a leak from a railroad tank car sent its toxic (in high concentration), dense, dark, acrid fumes over 4 miles from the plant. Several **people** were hospitalized.

LD_{50}, **rat**: 648 mg/kg; **mouse**: 740 mg/kg; skin, **cat**: 700 mg/kg.

DICHLOROBENZALKONIUM CHLORIDE

= *Botran* = *Dichloran* = *Dimethyl Dichlorobenzyl Ammonium Chloride*

Cationic germicide. **For skin, premises, and surgical instrument use.**

Untoward effects: Concentrations of 0.1% or less in water are non-irritating to normal **human** skin.

LD_{50}, **rats**: 700 mg/kg; **guinea pigs**: 300 mg/kg.

σ-DICHLOROBENZENE

= *DCB* = *1,2-Dichlorobenzene* = *MPC*

Solvent, insecticide and dye intermediate. **Approximately 55 million lbs used annually.**

Untoward effects: Central nervous system depression, liver and renal toxicity, and severe eye, nose, and throat irritation from inhalation. Death and/or renal toxicity, as well as other symptoms, as with *para* form below from ingestion. Pain, blistering, and brown pigmentation from skin contact. Dermatitis in a **glazier** caused by dipping window sashes in a solution, and in another **worker** from contact with a synthetic varnish. Conjunctival irritation on eye contact. One author reports it $2\frac{1}{2}$ times as toxic as CCl_4. 180 ppm exposure is expected to decrease respiratory rate by 50%. TWA, 50 ppm.

Approximate LD in **adults**: 15 g; LD_{LO}, 500 mg/kg.

Nausea, vomiting, depression, and hemolytic signs occur in **animals** and **humans**.

LD_{50}, **rat** and **rabbit**: 500 mg/kg; 96 h LC_{50}, **bluegills**: 27 mg/l.

p-DICHLOROBENZENE

= *Di-chloricide* = *Paracide* = *Paradichlorobenzene* = *Paradow* = *Paramoth* = *Para-zene* = *PDB* = *PDCB*

Fumigant, room deodorants. **Kills and repels moths. Annual demand is for about 100 million lbs annually.**

Untoward effects: Anaphylactoid purpura and acute glomerulonephritis in a 69-year-old **male**. Cataracts in **humans** and **animals** have followed inhalation of vapors, as well as headache, vertigo, weakness, mild anemia, ***alcohol***-type excitement, eye irritation, and dyspnea. Nausea, vomiting, diarrhea, abdominal cramps, anorexia, rhinitis, jaundice, renal damage, methemoglobinemia, and cyanosis may follow ingestion. Absorption is enhanced by ***milk*** or ***oils***. Occasional dermal sensitization. Recommended maximum levels in drinking water set at 0.75 mg/l and 75 ppm (450 mg/m^3) in air.

LD_{LO}, **human**: 500 mg/kg; eye, **human**: 80 ppm.

Extremely high dosage in laboratory **animals** has caused tremors and liver injury. Induces carcinoma and adenomas in **mice** and adenocarcinomas in male **rats**.

LD_{50}, **rat**: 500 mg/kg; **mouse**: 2950 mg/kg.

3,3´-DICHLOROBENZIDINE

Use: In many azo dyes, ***benzidine*** pigments, and in manufacturing, the dihydrochloride salt.

Untoward effects: **Human** exposure is, primarily, by inhalation and skin contact (including mucous membranes and eyes) during industrial procedures. It is a potential **human** carcinogen, since ***benzidine*** is a **human** bladder carcinogen. Dermatitis, caustic burns, headache, dizziness, frequent urination, dysuria, hematuria, gastrointestinal upsets, upper respiratory irritation, and secondary infections occur.

Acute inhalation by **rats** led to moderate pulmonary congestion, and chronic oral administration led to gastrointestinal congestion and hemorrhage, granulocytic leukemia, and cancers of the intestines, mammary, sebaceous, and Zymbal glands. Chronic oral administration led to bladder cancers in **hamsters** and female **dogs**, liver cancers in **mice**, and liver and bladder cancers in female **dogs**. Given to pregnant **mice**, it led to lymphoid leukemias in their offspring.

LD_{50}, **rat**: 4.7 g/kg; **dog**: 16 g/kg.

2,4-DICHLOROBENZYL ALCOHOL

= *2,4-DCBA* = *Dybenal* = *Myacide*

***Antiseptic, fungicide.* In pharmaceuticals, cosmetics, and mouthwashes.**

Untoward effects: No irritation by 2.5% cream to **humans** and no irritation or sensitization noted in **guinea pigs**.

LD_{50}, **rats**: > 3 g/kg; **mice**: 2.3 g/kg.

1,1-DICHLORO-2,2-BIS(*P*-CHLOROPHENYL)ETHANE

= *DDD* = *p,p´-DDD* = *ENT 4225* = *ME 1700* = *Rhothane* = *TDE* = *p,p´-TDE* = *Tetrachlorodiphenylethane*

***Insecticide*. A break-down product of *DDT*.**

Untoward effects: Nausea, anorexia, abdominal cramps, ataxia, weight decrease, and lethargy. Slightly irritating to skin. Symptoms often similar to ***DDT's*** (q.v.). Chronic exposure leads to liver damage and atrophy of adrenal cortex. Body fat is the major storage depot in **man** and **animals**. Estimated fatal LD_{LO} oral dose is ~5 g/kg.

Low toxicity to **bees**. Various **birds** have been poisoned by it. **Mallard ducks**, **California quail**, and **pheasants** (LD_{50} > 2 g/kg, > 760 mg/kg, and 386 mg/kg, respectively) when intoxicated showed ataxia, imbalance, hyperexcitability, tenseness, jerkiness, shakiness, slowness, hypoactivity, ptosis, tremors, and ataraxia within 2 or more hours. Deaths occur between 1 and 5 days post-treatment. Remission took up to 11 days. **Chickens** are also poisoned by it. In the feed of lactating **cows**, it appears in their milk, and ***butter*** and ***cheese*** made from it. LD_{LO}, 1–2 week **calves**: 250 mg/kg. Coordination impaired in exposed **fiddler crabs** (*Uca pugnax*). Causes gross atrophy of adrenal glands and it increases ***pentobarbital*** half-life in the **dog**. In **goats**, 50 mg/day caused adrenal gland fatty infiltration, atrophy, and connective tissue replacement; 100 mg/day led to renal and liver pathology.

LD_{50}, **golden hamsters**: > 5 g/kg; **rats**: 113 mg/kg.

DICHLORODIFLUOROMETHANE

= *Arcton 12* = *Difluorodichloromethane* = *F12* = *Freon 12* = *Frigen 12* = *Genetron 12* = *Halon* = *Isotron 2* = *Pyroforane*

***Aerosol propellant, refrigerant.* In deodorant, air freshner, cooking lubricant, and television tuner fluids.**

Untoward effects: Dizziness and tremors; abuse also incurs hallucinations, narcosis, unconsciousness, cardiac arrhythmias, cardiac arrest, and death. Acute fatal bronchitis and pneumonia in a 68-year-old **male**, 96 h after exposure. Possible inhalation of degradation products. Hypopigmentation, followed by hyperpigmentation in combination with ***Freon 11***. Contraindicated in **those** with pigmented skin (viz. **Negroes**, **Puerto Ricans**). Ventricular tachy- and bradyarrythmias (sinus bradycardia arrest and atrioventricular block), acute heart failure, arterial hypotension, asphyxia and sudden death in young inhalation **abusers**. May increase the "greenhouse" effect in the atmosphere. Exposure limits of 1000 ppm or 4950 mg/m^3 suggested.

Induces cardiac arrhythmias at inhaled concentrations as low as 0.25% in **dogs**. Tachycardia, spontaneous arrhythmias, decreased contractility, and bronchospasm at 10% in **monkeys**. Cardiac arrhythmias, apnea, and even cardiac arrest in **rats** and **mice**, following aerosolized material. After 90 days of exposure, liver damage noted in **dogs**, **guinea pigs**, **monkeys**, **rabbits**, and **rats**.

3,3´-DICHLORO-5,5´-DINITRO-σ, σ´-DIPHENOL

= *Distolon* = *Menichlopholan* = *Niclofolan*

Fasciolicide.

Untoward effects: Headache, weakness, lumbar pain, and profuse sweating, followed by dyspnea, oliguria, fever, tachycardia, leukocytosis, and coma in an 83-year-old **male** after ingestion of 2 tablets (400 mg), mistaken for ***aspirin***. Died 36 h later, and necropsy showed hemorrhagic nephritis, cerebral edema, myocardial and liver degeneration, and focal necrosis in adrenals. Usual dose in **cattle** is 1 tablet/50 kg.

1,1-DICHLOROETHANE

= *Ethylidene Chloride* = *1,1-Ethylidene Dichloride*

Halogenated hydrocarbon. U.S. annual consumption is over 4 billion lbs, as a component of leaded fuels, a solvent, and fumigant.

Untoward effects: With other drugs in suicide attempts, leads to sudden dyspnea, shock, coma, and cardiac arrest. Necropsy showed acute yellow atrophy of liver, focal myocardial necrosis, and multiple hemorrhages. ***Chloroform***-like odor. Vapors cause lacrimation, corneal clouding, dizziness, headache, nausea, vomiting, weakness, narcosis, pulmonary congestion, epigastric cramps, tremors, rhinitis, and vertigo (due to central nervous system depression). Dermatitis from skin contact. Ingestion leads to liver, lung, and kidney damage, leukocytosis, hemorrhagic gastroenteritis, cerebral hemorrhages, dizziness, nausea, vomiting, abdominal pain, stupor, and cyanosis, and death can occur from respiratory or circulatory failure within 4 days. LD by ingestion in **adults** is ~20–30 ml. Minimize exposure, as it is structurally related to carcinogenic ***chloroethanes***. TWA 8 h exposure limit of 100 ppm. Immediate danger to life or health, 3000 ppm.

LD_{LO}, **human**: 500 mg/kg.

Increased incidence of hepatocellular carcinomas in male **mice**, mammary gland adenocarcinomas and uterine neoplasms in female **mice**, and lung adenomas in **mice** of both sexes. Hemorrhagic necrosis of the adrenal cortex occurs in many **animals**. **Dogs** and **foxes** have developed bilateral swelling and turbidity of the cornea, that usually clears after the exposure ceases.

LD_{50}, **rat**: 725 mg/kg; **mouse**: 625 mg/kg.

sym-DICHLOROETHYL ETHER

= *Bis(2-chloroethyl)ether* = *Chlorex* = *DCEE*

Solvent, soil fumigant.

Untoward effects: ***1,4-Dioxane*** has been a contaminant. Nose, throat, and respiratory tract irritation; lacrimation, nausea, vomiting, and coughing. Skin and eye irritation from vapors and direct contact. Ingestion or percutaneous absorption leads to liver and kidney damage. Exposure limits set at 5 ppm (30 mg/m^3).

LD_{LO}, **human**: 50 mg/kg.

Carcinogenic in **mice**. Inhalation of 1000 ppm lethal to **guinea pigs**.

LD_{50}, **rats**: 75 mg/kg; LC_{LO}, inhalation, **rat**: 1000 ppm/45 min.

DICHLOROMETHYLENE DIPHOSPHONATE

= *Bonefos* = *Clasteon* = *Cl$_2$MDP* = *Clodronate* = *Difosfonal* = *DMDP* = *Loron* = *Mebonat* = *Ossiten* = *Ostac*

Calcium regulator.

Untoward effects: This useful drug normally lacks clinically significant side-effects. Bronchoconstriction reported in ***aspirin***-sensitive **asthmatics**. Reduce dosage in **patients** with impaired renal function.

2,6-DICHLORO-4-NITROANILINE

= *Allisan* = *Botran* = *DCNA* = *Dichloran* = *Ditranil*

Fungicide. **For vegetable, fruit, and soil treatment.**

Untoward effects: Corneal deposits and cataracts.

Corneal opacities in **dogs** and **pigs** after exposure to sunlight. 160 mg/kg/day lethal to **monkeys** within 3 months. More toxic to female than male **monkeys**. Orange-colored urine in **rats**, but not **monkeys**, after treatment. Causes liver enlargement in **rats** and centrolobular fatty infiltration in the liver of **monkeys**.

LD_{50}, **rats, mice**, and **guinea pigs**: 1.5 g/kg.

DICHLOROPHEN(E)

= *Antiphen* = *Dicestal* = *Didroxane* = *Di-phenthane-70* = *G-4* = *Nuophene* = *Parabis* = *Plath-Lyse* = *Preventol G-D* = *Teniathane* = *Teniatol* = *Wespuril*

Use: Germicide in soaps, shampoos, cosmetics, cutting oils, water cooling systems, and textiles. Cestocide and fungicide. A chlorinated phenol.

Untoward effects: Allergic contact dermatitis. Cross-sensitivity may exist with ***bithional, dibromsalicylanilides,***

hexachlorophene, and ***tetrachlorosalicylanilide***. Nausea, vomiting, abdominal cramps, diarrhea, urticaria, and jaundice from oral anthelmintic use.

LD_{LO}, **human**: 500 mg/kg.

Vomiting and diarrhea may occur on therapeutic doses in **dogs**. Overdosage or dosage of weak **animals** may also cause tremors, depression, and anorexia. Similar symptoms in **sheep**.

LD_{50}, **rats**: 1.5 g/kg; **mice**: 1 g/kg; **dogs**: 2 g/kg.

2,4-DICHLOROPHENOXYACETIC ACID

= *2,4-D* = *Esteron 44* = *Lironox* = *Trinoxol* = *Weedone*

***Herbicide.* With *2,4,5-T* in *Agent Orange*. *Dioxin* has been a frequent contaminant in the manufacturing process. Various salts and esters are commonly used, and these metabolize to the acid form in the body and in the environment. There have been ~1500 products in the marketplace used to kill weeds.**

Untoward effects: An **infant female** was born at 40 weeks' gestation to a primiparous 20-year-old **mother** and a 35 year-old **father**. **Both** worked in forestry, spraying with ***2,4-D*** 6 days/week, 7 h/day from 6 months prior to conception until 5 weeks after the **mother's** last menstrual period. No protective clothing was worn and ***2,4-D*** was frequently inhaled. Mental retardation, frontal bossing, low-set and posteriorly rotated auricles, saddle-shaped nasal bridge, hypoplastic philtrum, and abnormal hair growth pattern occurred in the **child**. Careless **users** face an increased risk of lymphomas. In 1951, Britain sprayed areas of Malaysia with it, and its embassies said it was "harmless to **human** and **animal** life". Pyrexia, tachycardia, hyperventilation, sweating, coma, muscle twitching, myopathy, convulsions, and hypoxia have also been reported after severe exposures. A 46-year-old **male** swallowed about 30 ml of a concentrate (with ***EPTC*** = ***Knoxweed***®) in error, thinking it was iced tea. **His** tongue and throat felt burnt. An hour later, **his** stomach was washed out and he was sweating profusely, had burning of mouth, chest, and lower abdomen, and flushing. After 18 h, **he** was cyanotic and had a fever with fibrillary twitching and paralysis of intercostal muscles. Hemoglobinuria, myoglobinuria, increased glutamic oxalacetic transaminase, lactic dehydrogenase, aldolase, and creatine phosphokinase then developed. Weakness, tiredness, and insomnia persisted for about 2 weeks, and sexual potency was gone for about 4 months. A peripheral neuropathy occurred in a 39-year-old **male** 4 days after spraying weeds on a windy day. **He** repeatedly used **his** bare hands to unplug the sprayer. **He** still had mild hand numbness 1 year later. A fatal suicide occurred after ingestion of the diethylamine salt. Toxic serum concentrations are listed as 200–400 mg/l. Ingestion has caused vomiting, abdominal cramps, diarrhea, anorexia, myotonia, and increased salivation. Ataxia, mental confusion, hypotension, and tachycardia have also occurred. Although death has occurred with ingestion of 6 g, a fatality was reported with less. Contact dermatitis and four cases of aplastic anemia reported. Toxic serum levels are ~100 mg/l; comatose and fatal levels reported as 320 and 720 mg/l, respectively.

LD_{50}, **human**: 80 mg/kg.

Large doses to experimental **animals** leads to vomiting, diarrhea, anorexia, weight loss, pulmonary congestion, mouth and pharyngeal ulcers; liver, kidney, and central nervous system toxicity. Some species develop hind limb stiffness and incoordination. Treated plants have an increased level of ***nitrates***. Intoxicated **birds** show polydipsia, ataxia, slowness, ruffled feathers, tachypnea, tremors, prostration, ptosis, and salivation. 0.1 ppm permitted in potable water and 0.1 ppm in **fish** and **shellfish**. High doses disturb cerebral electrical activity in **cats** and **rats**. **Cattle** grazing on treated pastures have not become ill, but can be poisoned by drinking contaminated water. Liver damage may be pronounced in affected **dogs**. 30% increased incidence of malignant lymphomas in **dogs** whose owners used it on their lawns. Reproductive disorders in **cattle** commenced 1–6 weeks after consuming contaminated feedstuffs.

LD_{50}, **rat** and **mouse**: 310 mg/kg; **guinea pig**: 400 mg/kg; **chuckars**: 200 mg/kg; **doves**: 668 mg/kg; **ducks**, **quails**, and **pheasants**: > 5 g/kg; **chickens**: 541 mg/kg; **dogs**: ~100 mg/kg.

1,3-DICHLOROPROPENE

= *Telone II*

Soil fumigant, nematocide, and pesticide intermediate.

Untoward effects: Inhalation leading to skin, eye, nose, and throat irritation or burns; coughing, nausea, headache, and fatigue. In 1975, a highway accident caused about 80 **people** in California to be exposed to its vapors. Acute symptoms were headache, mucous membrane irritation, dizziness, chest discomfort, nausea, and vomiting. **Three** lost consciousness. **Some** had persistent dyspnea and coughing. May cause allergic dermatitis. Suspect carcinogen for **man**.

Oncogenic in **rats** and **mice** (stomach, bladder, and lungs) after ingestion. High dosage also caused walking difficulties and liver and lung pathology.

LD_{50}, male **rats**: 715 mg/kg; female **rats**: 470 mg/kg; acute dermal, **rabbits**: 504 mg/kg.

1,2-DICHLORO-1,1,2,2-TETRACHLOROETHANE

= *Arcton 114* = *Cryofluorane* = *Freon 114* = *Frigen 114* = *Frigiderm*

Refrigerant, aerosol propellent.

Untoward effects: Inhalation leads to ventricular tachy- and bradyarrhythmias (sinus bradycardia and arrest, and atrioventricular block), acute heart failure, arterial hypotension, respiratory tract irritation, and asphyxia. Frequently involved in aerosol-sniffing deaths.

In **dogs**, inhalation of a 20% concentration led to hypotension, tachycardia, and increased pulmonary resistance. Bronchoconstriction and cardiac arrhythmias also occurred in the **mouse**, and apnea and cardiac arrest in **rats**.

Inhalation LC_{50}, **mouse**: 700,000 ppm/30 min.

DICHLORPHENAMIDE

= *Antidrasi* = *Daranide* = *Oratrol*

Carbonic anhydrase inhibitor.

Untoward effects: Hyperchloremic metabolic acidosis with sickle cell anemia in a number of **patients**. After 90 days of treatment, one **patient** developed an urticaria severe enough to discontinue treatment. Headache and twitching of fingers with chronic respiratory insufficiency, paresthesia of fingers, breathlessness on exertion, apathy, feelings of unreality, short temper, poor concentration, irritation of gastric mucosa, decreased libido, and impotence have occurred. Decrease of arterial pH can be hazardous in **patients** with respiratory acidosis. Transient myopia, retinal edema and loss of accommodation reported. Ataxia, tinnitus, tremors, weakness, nervousness, globus hystericus, depression, confusion, dizziness, constipation, weight decrease, hyperuricemia, pruritus, skin eruptions, fever, agranulocytosis, thrombocytopenia, increased urination, renal calculi, and phosphaturia are among the many adverse effects that have occurred.

Drug interactions: May potentiate ***salicylate*** toxicity, possibly by increasing cerebral blood flow. Severe hypokalemia may occur when given with ***hydrochlorothiazide*** and other ***thiazides***.

Fetal limb malformations (primarily ulna and/or digits) in **rats**.

LD_{50}, **rat**: 2.6 g/kg; **mouse**: 1.7 g/kg; IV, **dog**: 200 mg/kg.

DICHLORVOS

= *Astrobot* = *Atgard* = *Canogard* = *DDVP* = *Dedevap* = *Dichlorman* = *Dichlorovos* = *Divipan* = *Equigard* = *Equi-gel* = *Estrosol* = *Herkol* = *Nogos* = *Nuvan* = *SD-1750* = *Task* = *Vapona* = *Verdisol*

***Organophosphorus insecticide and anthelmintic.* Cholinesterase inhibitor.**

Untoward effects: Nervousness, progressive dyspnea, jerking of extremities, muscle twitching, hyperhidrosis, lacrimation, nausea, vomiting, rigidity, severe convulsions, and coma in a 33-year-old **male** after a failed suicide attempt. Mild to moderate plasma cholinesterase decrease in **handlers** of dispensers used for malaria eradication. A 54-year-old **male** pest control **worker** developed an acute contact dermatitis with extensive areas of erythema and bullae and systemic toxicity, probably from leaky equipment carried on **his** back. Primary irritant contact dermatitis from handling **dog's** dichlorvos-impregnated flea collars. Wetting (rain or baths) the collars increases the amount of chemical released. Miosis, headache, chest tightening, wheezing, laryngeal spasms, decreased blood pressure, cardiac irregularities, cyanosis, anorexia, paralysis, giddiness, and ataxia have also been reported symptoms. A **woman** in an area hospital exposed to numerous ***DDVP*** fly strips in **her** kitchen remained on artificial ventilation for several days after receiving ***succinylcholine*** in the operating room. ***Oils*** increase its absorption.

LD_{LO}, **human**: 50 mg/kg.

Dogs and **cats** have also had a contact dermatitis from the collars. An ataxia-depression has also been reported frequency in **cats** wearing these flea control collars. Highly toxic to **bees**, **canaries**, and **finches**. 10 mg/kg to **calves** and 25 mg/kg to **horses**, **sheep**, and **dogs** have been toxic. 100% lethal to **guppies** at 10 mg/l/24 h. Malicious fatal poisoning of 27,000 month-old **chicks** by water contamination caused staggering, frothing from the beak, paralysis, and tonic convulsions. Goose-stepping ataxia, tremors, and convulsions occur in poisoned **ducks** and **pheasants**. Very toxic to ***Peromyscus* mice**. Minimum lethal dose for **rabbits** is 10 mg/kg. Intraruminal doses > 70 mg/kg can kill **reindeer**. **Peafowl** (*Pavo* sp.) and **guineafowl** (*Numida* sp.) have been killed from foraging in feces from treated **zebras**. Being easily metabolized and excreted, symptoms of poisoning are usually of short duration, but dyspnea, increased salivation, stiffness, colic and diarrhea (especially in **horses**), and convulsions may be followed by death. Readily passes into milk. In **animals** and **man**, do NOT use in conjunction with or within a few days of (before and/or after) any other cholinesterase inhibitors and avoid use with ***arsenicals, purgatives, phenothiazine, phenothiazine tranquilizers, succinylcholine***, or drugs producing purgation as a side-effect. **Brahman cattle**, **greyhound** and **whippet dogs** are generally recognized as often having increased sensitivity to this class of compound. Some have disputed this. Possibility of adverse reactions in heartworm-infested **dogs** has been noted.

LD_{50}, **rat**: 56 mg/kg; **mouse**: 135 mg/kg; **dog**: 1.1 g/kg; **chicken**: 15 mg/kg; **mallard ducks**: 7.8 mg/kg; **pheasants**: 11.3 mg/kg.

DICLOFENAC

= *Allvoran* = *Assaren* = *Benfofen* = *Cataflam* = *Delphimix* = *Dichronic* = *Diclobenin* = *Diclo-Phlogont* = *Diclo-Puren* = *Diclord* = *Dicloreum* = *Dolobasan* = *Duravolten* = *Ecofenac* = *Effekton* = *GP-45840* = *Kriplex* = *Neriodin* = *Novapirina* = *Primofenac* = *Prophenatin* = *Rhumalgan* = *Tsudohmin* = *Valetan* = *Voldal* = *Voltaren* = *Voltarol* = *Voveran* = *Xenid*

***Anti-inflammatory.* Non-steroidal, with analgesic and antipyretic activity.**

Untoward effects*: *Anaphylaxis: An 80-year-old arthritic **patient** developed a shock-like syndrome on three occasions after a change-over from ***ibuprofen***. May increase sensitivity to allergic reactions from **bee** or **wasp** stings.

Blood: Spontaneous platelet aggregation, hemolytic anemia, eosinophilia, agranulocytosis, thrombocytopenia, and neutropenia. Prolonged bleeding time and bruising occurs. Spontaneous hematoma of thigh after 100 mg/day/2 years in a 73-year-old **female**.

Cardiovascular: Hypertension and congestive heart failure. Increased blood pressure in **some**, apparently due to increase in circulating fluids.

Central nervous system: Headache, dizziness, paresthesia, memory upsets, tremors, tics, convulsions, disorientation, vertigo, and psychotic reactions. Impotence.

Eyes and ears: Dry and irritated in a 51-year-old **female** after 25 mg tid. Tinnitus, amblyopia, and night blindness.

Gastrointestinal: Gastric lesions, melena, gastrointestinal bleeding in 2%, colonic ulcerations, abdominal cramping and distension. Ileal perforation in an 80-year-old **female** on 75 mg, IM/twice daily. Rarely, dry mouth and mucous membranes, aphthous stomatitis, vomiting, and pancreatitis.

Liver: Hepatitis developed within several weeks in seven **patients**, 45–69 years, six of whom received 75 mg twice daily, and one, 25 mg, tid. Generally, resolved 4–6 weeks after withdrawal of treatment.

***Miscellaneous*: *Non-steroidal anti-inflammatory drug*s** are not recommended in **women** trying to get pregnant since they interfere with implantation of fertilized eggs in **animals**.

Muscle: Nicolau syndrome with isolated muscle necrosis in a 35-year-old **male** after 75 mg, IM. IM use can result in severe tissue necrosis or necrotizing fasciitis.

Renal: Papillary necrosis after 50 mg tid/6 years. Nephrotic syndrome may be due to a hypersensitivity reaction. Inappropriate antidiuretic homone syndrome in an 82-year-old **female** receiving 100 mg/day. Lipoid nephrosis in 56- and 70-year-old **females** on 100 mg/day/3 months or 150 mg/day for two periods of 13 and 20 days. Azotemia, nocturia, and hematuria.

Respiratory: Dyspnea, cough, and diffuse pulmonary infiltrate after 50 mg twice daily.

Skin: Rashes, pruritus, erythema multiforme, Stevens–Johnson syndrome, and hyperhidrosis.

Therapeutic serum levels reported as 0.5–3 mg/l; toxic at 60 mg/l.

Drug interactions*: *Anticoagulants: True interactions have not occurred, but note below possible effects on platelets.

Aspirin: Decreases its plasma concentration over 50%. Asthmatic attacks can occur in ***aspirin***-sensitive **patients**.

Cyclosporine: Co-administration significantly increased diclofenac plasma concentration and may increase nephrotoxicity.

Digoxin: Serum concentration increased by diclofenac. ***Digitoxin*** had no effect.

Lithium: Trials indicated ***lithium's*** renal clearance decreases by 23% and plasma levels increase by 26%.

Methotrexate: Serum concentration increased by diclofenac and may have been involved in a fatal reaction.

Sucralate: Possible increase in bioavailability. Potentiates the hypokalemic effect of ***hydrochlorothiazide*** in **rats**.

LD_{50}, **rats**: 150 mg/kg; **mice**: ~390 mg/kg.

DICLOFOP-METHYL

= *HOE 23,408* = *Hoe-grass* = *Hoelon* = *Illoxan*

Herbicide.

Untoward effects: Wear goggles and respirator when using it, as it causes skin and eye damage in **people**.

Toxic to **fish**.

LD_{50}, **rat**: 580 mg/kg; LC_{50}, **quail**: > 5000 ppm.

DICOFOL

= *Acarin* = *DTMC* = *ENT-23648* = *FW-293* = *Kelthane* = *Mitigan*

***Acaricide.* Was extensively used on *grapes*, *almonds*, and other crops and trees. Annual U.S. consumption of this chlorinated hydrocarbon was 2–3 million lbs.**

Untoward effects: Tolerance levels on foodstuffs vary from 10 ppm in the U.S., down to 3.5 ppm in West Germany. Originally tainted in manufacture with ***DDT*** or its analogs, including ***DDE***. Harmful to eyes and skin. Wear goggles, mask, and chemical-resistant gloves.

LD_{LO}, **human**: 500 mg/kg.

Will pass into the milk of **cows**. Causes liver cancer in male **mice**, but not in females or **rats**. Low toxicity to **bees**. Lethality in the **cat** was due to cardiovascular failure. Poisoned **pheasants** lose their balance, show ptosis, fluffed feathers, tremors, wing-beat convulsions, emprosthotonus, and ataxia. Deaths occur 1–8 days after treatment. Toxic to **fish**.

LD_{50}, **rats**: 575 mg/kg; **dog**: > 4 g/kg; **pheasants**: 265 mg/kg; LC_{50}, **Japanese quail**: 905 mg/kg for 1 day olds up to 1746 for 21 days olds; **mallard ducks**: 1651 ppm; **bobwhite quail**: 3010 ppm; acute dermal, **rabbits**: 1000 mg/kg.

DICROTOPHOS

= Bidrin = C-709 = Carbicron = Ektafos = ENT-24482 = SD-3562

Organophosphorus insecticide. **Cholinesterase inhibitor.**

Untoward effects: Environmental Protection Agency established a 48 h waiting period before **workers** can re-enter treated areas. Acute poisoning in a 41-year-old **male** who sprayed inside of house day/2–3 weeks against **mosquitoes**. Recovery after hospitalization for 22 days. Weakness, headache, tightness in chest, blurred vision, extreme miosis, salivation, sweating, nausea, vomiting, abdominal cramps, and diarrhea are classic symptoms.

LD_{LO}, **human**: 5 mg/kg.

Goose-stepping ataxia, asthenia, miosis, salivation, lacrimation, tonic spasms, diarrhea, tachypnea, anorexia, prostration, tetany with wings extended, and convulsions in **birds**. 10% of a lethal dose produces toxic symptoms.

LD_{50}, **rats**: 16 mg/kg; **mice**: 11 mg/kg; male **mallard ducks**: 4.24 mg/kg; **pheasants**: 3.21 mg/kg; **chicks**: 8 mg/kg; **chukars**: 9.63 mg/kg; **ducks**: 4 mg/kg; **pigeon**: 2 mg/kg; **California quail**: 1.89 mg/kg; **Japanese quail**: 4.32 mg/kg; **bullfrogs**: 2 g/kg; **mule deer**: 12.5–25 mg/kg; **fish**: 6300 µg/l.

DICTAMNUS ALBUS

= Burning Bush = Dictame = Dittany = Fraxinella = Gas Plant

Untoward effects: Itchy eruption, resembling a ***poison ivy*** dermatitis and a phytodermatitis that can cause severe burns, erythema, and bullae, especially from the seed pods and plant juices. Staining occurs and may persist for months, as with ***berlock*** dermatitis. A quote in a dermatology journal states that "when in flower, … bring on intoxication in **gardeners** who touch them, as if the **gardeners** had drunk ***wine***".

DICYCLOHEXYL PHTHALATE

Plasticizer.

Untoward effects: Recurrent episodes of severe asthma in **meat-wrappers** from price labels affixed to wrapped meat, by activation of the adhesive with a heating element.

DICYCLOMINE

= Autumin = Benacol = Bentomine = Bentyl = Bentylol = Dicycloverin = Diocyl = Di-Syntramine = Dyspas = Mamiesan = Merbentyl = Procyclomin = Wyovin

Anticholinergic.

Untoward effects: Dry mouth, mild blurring of vision, dizziness, drowsiness, and psychic effects are the key complaints. A 5-week-old **infant** developed apnea, when given it with ***dimethicone*** (***Ovol***). Two **infants**, 5 and 6 weeks, developed apneic episodes after dosing with ***Merbentyl***. Respiratory disorders developed in three **infants**, 7–10 weeks, who received 10 mg, twice the recommended dosage. Two **children**, 18 and 15 months, swallowed 23 and 30 enteric-coated ***Debendox*** tablets, respectively. The first **child** died, despite treatment, and the second survived after treatment. Lowers esophageal sphincter tone and increases the risk of thyroid cancer. Avoid ***antacids*** and ***antidiarrheals*** within 1 h; ***alcohol*** may potentiate its drowsiness effect. Avoid use in obstructive disease of the gastrointestinal and urinary tracts. May cause constipation, tachycardia, palpitations, mydriasis, urinary retention, altered taste, headache, mild nausea, nervousness, transient vertigo, impotence, decreased lactation, cycloplegia, decreased sweating, and, rarely, allergic reactions. Hot, dry skin, central nervous system stimulation, confusion, and accenting of other symptoms occur with overdosage or in the **elderly**. In high concentrations, it and ***doxylamine*** may cause death. The **maternal** use of the combination (***Bendectin***) may be associated with pyloric stenosis in **infants**. Taken during the first trimester, most evidence shows it is not causally associated with major birth defects.

LD_{50}, **mouse**: 625 mg/kg; IV, **rabbit**: 35 mg/kg.

DIDANOSINE

= ddI = ddIno = Dideoxyinosine = 2′,3′-Dideoxyinosine = NSC-612049 = Videx

Antiviral.

Untoward effects*:** Painful peripheral neuropathy (usually after 2–6 months, with distal numbness, and tingling, painful hands and feet), abdominal cramps, diarrhea, impaired ***glucose tolerance, acute pancreatitis (sometimes fatal), hepatic failure, headache, insomnia, nausea, vomiting, constipation, fever, rash, confusion, dry mouth, increased ***uric acid***, increased aminotransferase, hypokalemia, hypocalcemia, hypomagnesemia, thrombocytopenia, eosinophilia, cardiomyopathy, severe lactic acidosis, Stevens–Johnson syndrome, optic neuritis, atrophy of retinal-pigment epithelium, mania, and secondary respiratory failure leading to death. Has induced asthma-like symptoms in **adults** (1–4%) and **children** (21%). Latent dysbetalipoproteinemia precipitated by it and other HIV-protease inhibitors. Placental transfer occurs.

Drug interactions*:** Gastrointestinal absorption of ***ciprofloxacin, delavirdine, doxycycline, indinavir, itraconazole, ketoconazole, and, possibly, ***dapsone***, is decreased by it. Food decreases its absorption by up to 50%, and ***ranitidine*** increases its bioavailability. ***Ganciclovir*** decreases its renal excretion and increases the risk of toxicity.

Excreted in the milk of lactating **rats**.

DIDEOXYCYTIDINE

= *ddCyd* = *2′,3′-Dideoxycytidine* = *Hivid* = *Zalcitabine*

Antiviral.

***Untoward effects*:** Rash, aphthous stomatitis, fever, painful sensorimotor neuropathy with paresthesia, esophageal ulceration, pancreatitis, ototoxicity, and, possibly, cardiomyopathy reported.

Less toxic than ***AZT*** in **animal** trials.

DIEFFENBACHIA sp.

= *Apior* = *Dumb Cane*

Untoward effects*: *D. maculata (formerly, ***D. picata***) = ***Spotted Dumbcane*** = ***Tropic Snow*** has succulent, thick stems, and variegated leaves up to 15 inches long with very acrid sap, and a frequent cause of toxicity. It is a vigorous tall-growing plant that **people** make cuttings from. The juicy sap causes severe inflammation of the hands and arms, and, when conveyed to the mouth, as in smoking, causes acute pain, swelling, and respiratory distress that can last for several days. The sap is released under pressure in invisible jets, when the plant stem is cut, and causes respiratory irritation. ***D. exotica*** appears to be less toxic.

D. obliqua in Ecuador is poisonous to touch, causing severe itching.

D. seguine = ***Mother-in-law Plant***, a more common houseplant, causes rapid mouth and tongue irritation and swelling and salivation. The latter symptoms often leave the **victim** speechless, and give the plant its common name, ***Dumb Cane***. Water-insoluble ***oxalate*** crystals and a water-soluble toxin may be the cause. Esophagitis and gastritis can also occur. Contains glycosides that can adversely affect the heart. If the swelling blocks the throat, death can occur. The sap may cause a severe skin irritation and eye injury. Used by **Indians** of the Upper Amazon with ***curare*** and ***strychnine***, as an arrow poison.

Burning of the mouth and throat has occurred in **toddlers** chewing on the plant. Excluding ***mushrooms*** and ***toadstools***, ~4% of plant poisonings reported to Poison Control Center are due to it, with ~20% showing symptoms and, occasionally, **some** requiring hospitalization. Even the leaves can actively fire slender, needle-like crystals. A **child's** shrieks of pain as soon as **he** takes a bite of the plant is enough to alert a parent. Nausea and vomiting often occur. Used by **Amazon natives** as a narcotic, for arrow-tipping, and to cause temporary infertility in **men**.

Like small **children**, **puppies** and **kittens** chew or attempt to eat its leaves, and suffer similar symptoms, including incoordination and listlessness. Renal damage can occur. Has seriously, adversely affected **canaries** and other pet **birds**.

DIELDRIN

= *Compound 497* = *ENT-16225* = *HEOD* = *Insecticide No. 497* = *Octalox*

***Insecticide.* Now banned for use in most countries. A chlorinated hydrocarbon, with some anticholinesterase properties.**

Untoward effects*:** Acute toxicity manifested as irritability within 1–4 h, then miosis, limb jerks, convulsions, central nervous system depression, and death within 24 h. In more chronic exposures, headache, dizziness, giddiness, muscle twitching, nausea, vomiting, diarrhea, dyspnea, thrombocytopenia, and liver and renal damage have occurred. May have caused aplastic anemia and contact dermatitis. Percutaneous absorption occurs readily. Dermatitis in 288/1209 **men**, after wearing socks impregnated with it for moth-proofing. Convulsions reported in insecticide plant **employees**. Several young **males** had vomiting and diarrhea within 9 h of eating restaurant sandwiches, accidentally sprayed by a **child** of the restaurant **staff**. **All** recovered. A **schoolmaster** mistook the powder for ***cocoa. Two **boys** became dizzy in about 30 min, vomited, and went into a coma with repeated convulsions and decreased temperature. Consciousness returned in about 24 h. **One** had disturbed sleep rhythm and was manic for a week. **Both** had palpitations and epigastric pains for weeks. Involved in a suicide attempt. Hemolytic anemia, leukopenia, thrombocytopenia, and splenomegaly have been reported. Has been found in **human** breast milk. Serum levels of 15–30 μg are considered toxic for **man**. Guideline limit for airborne exposure to **man** is 1 μg/m^3, for exposures not exceeding 3 years. TWA, 8 h, skin: 0.25 mg/m^3.

LD_{LO}, **human**: 5 mg/kg.

Highly toxic to **bees**, **birds**, **cats**, and **fish**. Passes into the milk of **cows** and **deer**. Anorexia, weight decrease, depression, immobility, irritability, and severe liver and kidney damage in poisoned **puppies**. Intermittent convulsions noted in older **dogs**, followed by cyanosis, coma, and death. Abortions occurred in two out of three female **guinea pigs** given it at 100 ppm in feed. Placental transfer occurs. No carcinogenic action has been demonstrated in long-term feeding trials with **rats**, **dogs**, or **monkeys**. Residues in African game returned to pre-aerial spraying levels within 2 years. Feeding it to **rabbits** has caused blindness.

Minimum lethal dose, **cattle**, **horses**, and **sheep**: 25 mg/kg; **pigs**: ~50 mg/kg; LD_{50}, **rats**: 45 mg/kg; **mice**: 38 mg/kg; **dog**: 65 mg/kg; **guinea pig**: 49 mg/kg; **chicken**: 20 mg/kg; **pigeon**: 27 mg/kg; **quail**: 70 mg/kg; **duck**: 381 mg/kg.

DIENESTROL

= *Cycladiene* = *Dienoestrol* = *Dienol* = *Estrodienol* = *Estroral* = *Gynefollin* = *Hormofemin* = *Lipamone* = *Oestrasid* = *Oestroral* = *Restrol* = *Retalon* = *Retalon-Oral* = *Synestrol*

***Untoward effects*:** Use during pregnancy to prevent vaginal spotting and abortions associated with vaginal adenocarcinomas in at least five **females**, 18–20 years. Nausea and vomiting has occurred on 20 mg/day. Gynecomastia, impotence, testicular atrophy, and porphyria cutanea with blisters and erosions on back of hands 1 year after treatment in a 47-year-old **male**. A 70-year-old **male** developed gynecomastia after 6 weeks' exposure two or three times a week to a 0.1% vaginal cream his **wife** with atrophic vaginitis used to facilitate intercourse. Skin reactions can include chloasma, rashes, urticaria, erythema multiforme, and either hair loss or hirsutism. Breakthrough bleeding, increased size of uterine fibromas, cyst-like syndrome, dysmenorrhea, amenorrhea, bloating, intolerance to contact lenses from steepening of corneal curvature, headaches, dizziness, and edema have all been reported. Cardiovascular defects reported in **newborns** of treated **mothers**. Large doses may cause premature closure of epiphyses.

Young **male calves** given it parenterally have developed fibrosis and muscular hypertrophy of the accessory genital system. An abortion outbreak in **beef cows** traced to their consumption of **chicken** manure containing it. Female **mink** fed 5 μg/day during breeding season decreased their litter size, and 20 μg/day prevented conception or nidation. Feeding it to **turkeys** causes a high incidence of aortic ruptures.

DIESEL FUEL

***Use*:** As a fuel and pesticide carrier.

Untoward effects*:** Reports on toxicity usually refer to #2 grade; #1 is more volatile. Diesel exhaust can contain various carcinogens, including ***polycyclic aromatic hydrocarbons. Diesel engine exhaust is called a potential carcinogen by National Institute for Occupational Safety & Health. Possible cause of scleroderma in **man** after repeated skin contact.

LD_{LO}, **human**: 500 mg/kg.

Chronic inhalation exposure of **rats** suppresses alveolar macrophages. **Mallard ducks** consuming it at levels greater than they might ingest in nature develop nervous disorders. Weakness, diarrhea, and regurgitation occur. The LD_{50} for them is > 20 ml/kg. A **ewe** developed inappetence, depression, conjunctival congestion, and mucopurulent nasal discharge after drinking it. #1 diesel fuel may have caused a contact dermatitis to a **cow's** legs. **Cattle** will often drink it, with disastrous results, although it seems unpalatable to us.

DIETHANOLAMINE

= *Diolamine*

***Use*:** In the manufacture of detergents, as a softening, moisturizing, and dispersing agent. In gas purification.

***Untoward effects*:** May cause nausea, vomiting, and abdominal pain after ingestion. Skin irritant, and will cause skin burns if exposed area is covered. Severe eye irritant. Vapors may cause coughing, headache, nausea, and vomiting. Can form nitrosamines.

Implicated in **dog** and **cat** clinical cases. Nausea, vomiting; liver, renal, and pancreatic pathology; epistaxis, convulsions, and anemia.

LD_{50}, **rat**: ~0.7 g/kg; **mice**: 2.3 g/kg.

DIETHAZINE

= *Antipar* = *Aparkazin* = *Casantin* = *Deparkin* = *Dinezin* = *Diparcol* = *Dolisina* = *Eazaminum* = *Ethylemin* = *Latibon* = *Parkazin* = *RP-2987* = *Thiantan* = *Thiotan*

Anticholinergic, antiparkinsonian, tranquilizer, hypnotic.

Untoward effects: Has caused cataracts, agranulocytosis, burning taste, and transient tongue numbness.

LD_{50}, **mice**: 450 mg/kg; IP, **mouse**: 225 mg/kg.

2-DIETHYLAMINOETHANOL

A component in *procaine*.

Untoward effects: Nausea and vomiting within 5 min after 30 s accidental exposure to 100–200 ppm; **others** in area complained of nauseating ***ammonia***-like odor. Skin, eye, and respiratory tract irritant. **Some** feel warm and dizzy, and may complain of a bitter taste. OSHA exposure limit of 100 ppm (860 mg/m^3); 10 ppm on skin.

Rats exposed to 500 ppm vapors 6 h/day/5 days showed severe weight loss, high mortality, eye and nasal irritation on the first day, increasing to marked corneal opacity on day 3; at 200 ppm/1 month, 7/50 died of bronchopneumonia. Recovery from decreased growth rate after 6 months.

LD_{50}, **rats**: 1.3 g/kg.

DIETHYLCARBAMAZINE

= *Banocide* = *Carbam* = *Carbamazine* = *Carbilizine* = *Caricide* = *Caritrol* = *Cypip* = *Dec* = *Difil* = *Dioform* = *Dirocide* = *Ditrazin* = *Ethodryl* = *Filaribits* = *Filazine* = *Filban* = *Franoside* = *Heterosan* = *Hetrazan* = *84L* = *Longicid* = *Notézine* = *RP-3799* = *Spatonin* = *Superil*

Anthelmintic.

Untoward effects: Nausea, vomiting, headache, malaise, weakness, dizziness, drowsiness, tachypnea, tachycardia, hypotension, lymphadenopathy, fever for 2 days, edematous papular rash and bullae, pruritus, coughing, proteinuria, chorioretinitis, and abdominal discomfort. The severe allergic and febrile reactions often coincide with the death of microfilaria and include pedal edema, dermatitis, pruritus, fever, colic, lymphadenitis, and, occasionally, tachycardia. Rarely, an encephalopathy and loss of vision in onchocerciasis is reported. Eye, skin, and/or respiratory tract irritations can occur from contact with the dust.

To prevent acute nephritis in **dogs** with established **heartworm** infestations, it should not be given until they become negative. Large doses can make **rats** and **dogs** more sensitive to loud noises for about 2 h. Vomiting, behavioral changes, rash, anorexia, and shock-like reactions rarely reported in **dogs**.

LD_{50}, **rats**: 1.38 g/kg; **mouse**: 660 mg/kg.

DIETHYLENE GLYCOL

= *Diglycol* = *Polyglycol*

Use: In antifreezes; ***glycol, polyester,*** and ***morpholine*** manufacturing; solvents, ***tobacco*** humectant, printing inks, aircraft de-icing, cloth finishing, and dehydrating natural gas in pipelines.

Untoward effects: Ingestion leads to nausea, vomiting, abdominal cramps, diarrhea, lumbar pain, dizziness, drowsiness, anuria and/or oliguria within 2–5 days, tachypnea, jaundice, liver degeneration, facial edema, bradycardia, hypothermia, cyanosis, and death may occur within 2–3 weeks. ***Oxalic acid*** is a metabolite and crystals may form in the urine. A 39-year-old **male** had an eczematous rash on **his** left index and middle fingers and the left side of **his** upper lip from it as a humectant in the cigarette ***tobacco*** **he** smoked. On September 4, 1937, the S.E. Massengil Company distributed the first batches of a ***sulfanilamide*** elixir (72% ***diethylene glycol***, 10% ***sulfanilamide***, and about 15% ***water***) with an aromatic resembling ***raspberry*** and ***anise***, with a sweet taste. Over 100 **people**, mostly **children**, died from consuming it. Renal failure was the known cause of death. This prompted changes in the FDA's responsibility to include pre-market approval. Seizure of the balance prevented an estimated 4000 deaths. The fine was 26,000 dollars for "misbranding". It was sold as an elixir, which falsely implied that it contained ***alcohol***, which it did not. The LD for **adults** is ~1 ml/kg. That should have ended similar problems, but, in 1972, seven **people**, 6–31 years, in South Africa, died after receiving a sedative mixture (***Pronap*** or ***Plaxin***), where, in error, it was substituted for ***propylene glycol***. Symptoms included a prodromal febrile illness, vomiting, oliguria and anuria, dehydration, irritability, depression of consciousness, acidotic breathing, and hepatomegaly. A **few** also had diarrhea and enlarged kidneys. In 1985, a number of Austrian and Italian wines, some of which were shipped by a West German firm, had traces of it, and most was destroyed. In 1995, a sudden increase in fatal renal failure in Bangladesh was traced to it in an ***acetaminophen*** elixir, where over 300 **children** were affected. In Haiti during 1995 and 1996, 109 **children** 3 months to 18 years, ingested a similar ***acetaminophen*** syrup leading to acute anuric renal failure; fatalities in 85 due to contamination of its ***glycerin*** ingredient by diethylene glycol. It has also been involved in successful suicide attempts. It can catch fire. Keep it away from open flames.

Bone marrow cytotoxicity occurred in **rats** after a single IP injection of 100 mg/kg. **Rats** died within 2–5 days after two doses of 2 ml with terminal anuria and coma, as in **man**. Prolonged feeding of **rats** with 4% in their diet led to stones and fibrosarcomas of the bladder. Expect high mortality in **cats** and **dogs**. Causes **dogs** to vomit.

LD_{50}, **rat**: 14.8 g/kg; **mouse**: 23.7 g/kg; **dog**: 9 g/kg; **cat**: 3.3 g/kg.

1,2-DIETHYLHYDRAZINE

Untoward effects: SC to **rats** leads to lymphoreticular, nerve tissue, hepatic, and ethmoturbinal tumors. Causes transplacental induction of carcinogenesis in **rats**, **mice**, and **hamsters**.

IV, TD_{LO}, **rat** (15 days pregnant): 10 mg/kg.

DIETHYLMERCURIC SULFIDE

Untoward effects: **Cats** treated during pregnancy gave birth to young with ataxia, decreased growth of cerebellum, and disarrangement of Purkinje and other brain cells.

DIETHYL-*p*-NITROPHENYLPHOSPHATE

= *E-600* = *Ester 25* = *Eticol* = *Fosfakol* = *Mintacol* = *Miotisal A* = *Paraoxon* = *Phosphacol* = *Soluglaucit*

Miotic, insecticide, anticholinesterase.

Untoward effects: Cataractogenic. An epidemic of illness in 186 ***peach*** orchard **workers** was due to its residue in ***parathion*** used as a spray. Oxidation of ***parathion*** forms it, and it becomes 10–100 times more toxic to **human** red blood cells. At least 29 separate poisoning incidents in field **workers** after citrus sprayings. Toxic serum concentration in **man** is 5 µg/l.

TD_{LO}, **human**: 5 mg/kg.

Causes respiratory failure and pressor effects in **rats**. Prolongs ***hexobarbital*** sleeping time in **mice**.

LD_{50}, **rat**: 1.8 mg/kg; **mouse**: 19 mg/kg.

DIETHYLPHTHALATE

= *DEP* = *Ethyl Phthalate*

Use: In making celluloid and in making plastics more flexible; found in toothbrushes, tools, toys, automobile parts, and food packaging.

Untoward effects: Easily elutes from the above products into food products. **People** ingestion it from water wells near landfills or waste sites and by eating exposed **fish** or **shellfish**. An oily liquid with a bitter taste. Inhalation leads to eye, skin, nose, and throat irritation; headache, dizziness, nausea, and lacrimation. Ingestion has caused weakness, numbness, and leg and arm spasms. National Institute for Occupational Safety & Health recommends maximum exposure limit of 5 mg/m^3.

Daily application/2 years to the skin of **mice** led to liver tumors. Continuous feeding to laboratory **animals** led to decreased litter size; high doses by injection into pregnant **rats** caused an extra rib in their **newborns**.

IP, LD_{50}, **mouse**: 3.22 g/kg.

DIETHYLPROPION

= *Amfepramone* = *Anfamon* = *Anorex* = *Apisate* = *α-Benzoyltriethylamine* = *Danylen* = *Dobesin* = *Frekentine* = *Keramik* = *Keramin* = *Magrene* = *Marvate* = *Moderatan* = *Modulor* = *o.b.c.t.* = *Parabolin* = *Prefamon* = *Regenon* = *Ro-Diet* = *Tenuate* = *Tenuate Dospan* = *Tepanil* = *Tylinal*

Sympathomimetic, anorexogenic.* Related chemically to *amphetamine.

Untoward effects: Psychotic reactions. A 22-year-old **female** who took over 200 tablets/week showed extreme restlessness, hyperhidrosis, tachycardia, perplexity, incoherent thinking, and severe visual and auditory hallucinosis. Central nervous system stimulation and euphoriant effects, particularly in abuse cases. A 14-year-old **female** took an overdose leading to tachyarrhythmias and hemodynamic deterioration that failed to respond to DC shock, but converted to normal sinus rhythm after 2 mg ***propranolol***, IV. Dry mouth, depression, insomnia, constipation, vertigo, nausea, dizziness, headache, polyuria, and short periods of double vision reported. Pruriginous dermatosis, rashes, and gynecomastia occur. Can react with ***sodium nitrite*** in the stomach to form ***N-nitrosodiethylamine***.

LD_{50}, **mouse**: 160 mg/kg.

DIETHYL PYROCARBONATE

= *Baycovin* = *DEP* = *DEPC*

***Antimicrobial.* In fruit juices, *wine*, and beer. *DEP* is also used for *Diethyl Phosphate*.**

Untoward effects: Foods treated with it can contain the carcinogen, ***urethan***, if pH and ***ammonia*** content are optimum. Therefore, no longer approved for food use in the U.S. and some other countries.

LD_{50}, **rat**: 850 mg/kg; **dog**: 500 mg/kg; LD_{LO}, **cat**: 100 mg/kg.

DIETHYLSTILBESTROL

= *Antigestil* = *Bufon* = *Clinestrol* = *Cyren A* = *Cyren B* = *Domestrol* = *Estilben* = *Estrobene* = *Estrobene DP* = *Estrosyn* = *Euvestin* = *Fonatol* = *Grafestrol* = *Idroestril* = *Makarol* = *Micrest* = *Milestrol* = *Momozol* = *Neo-Oestranol I and II* = *Oestrogine* = *Oestromenin* = *Oestromensyl* = *Oestromon* = *Orestol* = *Pabestrol D* = *Palestrol* = *Serral* = *Sexocretin* = *Sibol* = *Sinestrol* = *Stilb[illegible]* = *Stilbetin* = *Stilboefral* = *Stilboestroform* = *Stilb[illegible] DP* = *Stilbofax* = *Stilkap* = *Stilpalmitate* = *Stilronate* = *Synestrin* = *Syntestrine* = *Synthoestrin* = *Vagestrol* = *Willestrol*

A popular treatment during 1941–1971 (especially, 1945–1955) to prevent miscarriage in women.

Untoward effects: Vaginal adenosis in 30–90% of post-pubertal **females** whose **mothers** received it during the 6–36th weeks of their pregnancy. Vaginal adenocarcinomas have also been seen in many whose **mothers** all received ***DES***, often at high dosage, 11–25 years previously. Similarly, cervical and endometrial carcinomas, uterine pseudopolyps, and cervical abnormalities have occurred. Non-malignant epithelial changes in vaginal and cervical epithelium was common in **females** whose **mothers** took ***DES*** during their pregnancy. These effects are, essentially, an effect of the drug's estrogenic potency, although it has been like a time bomb, creating nightmares for thousands of **daughters** of **mothers** who took ***DES*** during pregnancy. **Many** subsequently also had pregnancy problems, ectopic pregnancies, miscarriages, or premature deliveries, and infertility. The **mothers** often had nausea, vomiting, headache, breast tenderness and enlargement, and a modest but significant increase of breast cancer. Has caused virilization of **female fetuses**. Steepening of corneal curvature and intolerance to contact lenses reported. Risk of epididymal cysts, testicular cancer (seminomas), and cryptorchidism, as well as increased abnormalities of penile urethra and infertile sperm may occur from prenatal exposure of **males** to ***DES***. Hypogonadotropic hypogonadism with anosmia in a 17-year-old **male** whose **mother** received large doses of ***DES*** throughout **her** pregnancy. Metastatic prostatic cancer in the breast of a 72-year-old **male** receiving 1 mg/tid/2½ years for prostatic disease. A 77-year-old **male** with prostate cancer receiving up to 30 mg/day/6 years developed cancer of the nipple. Daily doses to **men** of 5 mg led to gynecomastia, nausea, anorexia, testicular atrophy, decreased libido, and increase in fatal and non-fatal cardiovascular complications. In 1972, one **male worker** had surgery to remove **his** enlarged breasts, due to ***DES*** exposure in a manufacturing plant.

Menopausal **patients** with metastic breast cancer and osteolytic bone metastases developed hypercalcemia within 24–48 h after treatment with it, and **patients** with Turner's syndrome developed endometrial adenocarcinoma. Thrombocytopenia, anemia, pancytopenia, increased bone marrow myeloblasts, increased in ***factor IX***, increase in peurperal thrombosis and pulmonary embolism, and increased congestive heart failure have occurred. A 12-year-old **female** with malignant ovarian teratoma reported, due to prenatal ***DES*** exposure. In large doses, it has been associated with **neonatal** bleeding. Exposure can result from inhalation, percutaneous absorption or ingestion. Gynecomastia in a 44-month-old **male** and a 23-month-old **female** and twice monthly menstrual periods in the **mother** who injected paste ***DES*** implants into **chickens**, and it occasionally splattered on **her** clothes. Her **children** often played with the tablets. I recall a U.S. firm I consulted to where, despite total mechanization or hands off manufacturing, the **men** in the plant had decreased libido and gynecomastia, and the **women** had excessive menstrual bleeding. By using one-way mirrors for observation, it was discovered that wrenches were used to loosen pipe unions, setting up air surges and vibrating the material so it moved faster through the pipes and **people** finished work sooner. Skin contact with it and inhalation of the dust occurred. These **people** were replaced with more adult postmenopausal **women**. It wasn't long before **they** too learned how to speed up production and go home earlier. The only problem was that **they** all began menstruating again, from inhaling or contacting the ***DES*** dust. When ***DES*** was used for prevention of lactation, **women** often had breast hardness, leakage, and pain. Jaundice, peliosis hepatitis, hepatocellular carcinoma, and cutaneous hepatic porphyria reported after ***DES*** treatments. Fixed-drug eruptions, acanthosis nigricans, and porphyria tarda occur from treatment. ***Alcohol*** intake may precipitate the latter in **some**. Hematometra and uterine bleeding noted after treatment withdrawal. There is a possibility that a 13-year-old **female** who menstruated at age 9 and subsequently died of cancer showed results of her **grandmother's** taking ***DES*** when pregnant. Some **animal** studies also indicated ***DES*** effects can pass through more than one generation.

TD_{LO}, **woman** (2–17 week pregnant): 35 mg/kg.

Carcinogenic in **rats**, **mice**, **hamsters**, **squirrel monkeys**, and **frogs**, leading to tumors, principally in estrogen-responsive tissues. ***DES***-treated baits have been used to help prevent pregnancy in **coyotes** and **deer**. When it was used for caponizing **chickens**, the implanted necks with ***DES*** residues were fed to **mink** and caused serious disruption of normal reproduction, even failure to produce **kits**, as well as dealing a severe blow to the fur industry. Excessive use or dosage in **dogs** led to bone marrow suppression, transient leukocytosis, anemia, and leukopenia.

LD_{LO}, **mouse**: 2.5 g/kg.

DIETHYL SULFATE

Millions of lbs used annually.

Untoward effects: Suspected as an increasing cause of cancers of the upper respiratory tract, particularly laryngeal, associated with its release during ***alcohol*** production.

Prenatal exposure of **rats** leads to tumors of the nervous system; weekly doses by gastric gavage lead to gastric tumors, and local sarcomas follow SC use.

LD_{50}, **mouse**: 647 mg/kg; LD_{LO}, **rat**: 750 mg/kg; inhalation LC_{LO}, **rat**: 250 ppm/4 h.

N,N,-DIETHYL-*m*-TOLUAMIDE

= *Autan* = *Deet* = *m-Delphene* = *m-DETA* = *Detamide* = *Dieltamide* = *ENT-20218* = *Flypel* = *M-Det* = *Metadelphene* = *Off*® = *Repel*

Insect repellent, resin solvent.

Untoward effects: Toxic encephalopathy in a 3-year-old **female** after 180 ml of ***Off***® had been sprayed on **her**, her bedding, and nightclothes over a 2 week period, which allowed for skin absorption and ingestion. Symptoms were disorientation, staggering, slurred speech, crying out, stiffening into a sitting position, extending extremities, flexing fingers, and dorsiflexing the toes. Hypersensitivity was suspected in this isolated case, but, then, more cases were reported. An 18½-month-old **female** sprayed daily for 3 months developed a toxic encephalopathy manifested with coarse tremors, opisthotonus, and nystagmus. Symptoms slowly disappeared over a 3 month period. Accidental or deliberate poisoning by ingestion in five **patients** 1–33 years. **Two** died. Seizures occurred in four **males** 3–7 years and a 29-year-old **male** after its topical use. A 30-year-old **male** applied a 70% solution over a 3 week period and developed an acute manic psychosis. A 61-year-old **female** developed cardiovascular toxicity after topical use. **She** recovered within several hours. Allergic reactions have been reported. Contact urticaria and erythema noted. Military **personnel** in Vietnam suffered bullous eruptions on antecubital fossae with subsequent severe necrosis that appeared to be a response to irritation, rather than an allergic reaction. Can persist in the skin of **humans** and **animals**. Hypotension, bradycardia, and coma also reported. Do NOT apply near eyes, groin, mucous membranes, antecubital or popliteal areas, as irritation or blistering may occur. Excessive or prolonged use leads to mania and hallucinations. Persian Gulf War **veterans** showed arthro-myo-neuropathy increase with increasing quantity used and increasing frequency of exposure.

LD_{LO}, **human**: 500 mg/kg; skin, TD_{LO}, **human**: 35 mg/kg/5 days.

Seizures, tremors, incoordination, salivation, and vomiting occurs in **cats** after excessive application. Use of 15% or more has caused a dermatitis and steatosis in **horses**.

LD_{50}, **rats**: 1.95 g/kg; skin, **rat**: 5 g/kg; skin, **rabbit**: 3.2 g/kg.

DIETHYLTRYPTAMINE

= *DET*

Untoward effects: Hallucinogenic indole drug of abuse, and a synthetic derivative of ***tryptamine***. Properties are similar to ***LSD*** (q.v.) and ***psilocybin*** (q.v.), potent, but short-lived.

TD_{LO}, **human**: 200 μg/kg.

DIFENACOUM

= *Neosorex* = *PP580* = *Ratak* = *WBA 107*

Rodenticide. **Anticoagulant rodenticide inhibiting production of prothrombin, inhibiting clotting mechanism.**

Untoward effects: Pale mucous membranes, anemia, lameness, vomiting, and diarrhea occur in **animals** accidentally ingesting it. Sublethal hemorrhaging occurred in **owls** that fed on **rats** killed by it.

DIFLORASONE

= *Dermaflor* = *Diacort* = *Difulal* = *Florone* = *Maxiflor* = *Psorcon* = *Soriflor*

Corticosteroid.

Untoward effects: Extensive topical use has caused adrenal suppression. In general, it and topical corticosteroids can cause irritation, burning, striae, skin atrophy, secondary infections, dryness, folliculitis hypertrichosis, acneiform eruptions, allergic contact dermatitis, perioral dermatitis, maceration of the skin, milaria, and hypopigmentation. Some of these may occur more frequently under occlusive dressings. Percutaneous absorption was greater in the **Cynomolgus monkey** (88.2%) than in **man**, and comparable to that in the **rat** (49.6%). Avoid use in pregnancy, since it might be teratogenic.

DIFLOXACIN

= *A 56,619* = *Abbott 56.619* = *Dicural*

Antibiotic. **A fluoroquinolone.**

Untoward effects: In **dogs** with a history of central nervous system disorders, convulsive seizures can occur. Induces arthropathy in young **animals**, especially **dogs**. Anorexia, emesis, and diarrhea reported in some **dogs**.

DIFLUNISAL

= *Adomal* = *Difludol* = *Dolisal* = *Dolobid* = *Dolobis* = *Flovacil* = *Fluniget* = *Fluodonil* = *Flustar* = *MK-647*

Anti-inflammatory, analgesic. **A synthetic analog of *salicyclic acid*, but not metabolizable to it.**

Untoward effects: Gastrointestinal upsets, such as nausea, dyspepsia, gastrointestinal pain, diarrhea, constipation, and flatulence have a variable incidence in different studies, ranging from 1 to 10%. Gastrointestinal bleeding and ulceration occurs in ~1%. Dizziness, weakness, vertigo, tinnitus, fatigue, somnolence, insomnia, headache, rash, pruritus, erythematous eruptions, sweating, photosensitivity reactions, Stevens–Johnson syndrome, visual disturbances, peripheral edema, stomatitis, decreased serum ***uric acid*** levels and increased ***uric acid*** renal clearance, acute interstitial nephritis and oliguric renal failure, cholestatic jaundice, and neutropenia reported. A 23-year-old **male** ingested 6 tablets/24 h for a minor injury and developed a life-threatening necrotizing fasciitis and severe suppression of lymphocyte function and lymphopenia. In lactating **women**, the concentration in milk is 2–7% of plasma's. Generalized hypersensitivity reactions have been reported. Usual therapeutic serum levels reported as 40–100 mg/l and toxicity at 500 mg/l.

TD_{LO}, **human**: 71 mg/kg/5 days.

Drug interactions: Clinically significant interactions may occur with ***acenocoumarol*** increasing prothrombin time, but not ***warfarin***. Anticoagulation effects may decrease with ***warfarin*** treatment when diflunisal is discontinued. ***Oxazepam*** renal clearance is reduced by it. ***Aluminum-containing antacids*** decrease its absorption. Increases plasma levels of ***hydrochlorothiazide***, but decreases its hyperuricemic effects. Decreases the hyperuricemic effect of ***furosemide***. Increases the plasma levels of ***acetaminophen, indomethacin,*** and ***naproxen***.

Toxic doses cause pyrexia in **rabbits**. More ulcerogenic and lethal to **rats** than ***aspirin***.

LD_{50}, **rat**: 710 mg/kg; female **mice**: 439 mg/kg; **rabbit**: 603 mg/kg.

DIFLUOROTETRACHLORETHANE

= *Freon 112*

Anthelmintic, flukicide, refrigerant, solvent.

Untoward effects: Flammable. Considerable variation in isomer content may be the cause of erratic results and varying toxicity. Apparently safe and effective in **beef cattle**, but it has been lethal at the same mg/kg in some lactating **dairy cattle**.

DIGALLOYL TRIOLEATE

Sunscreen.

Untoward effects: Photosensitizing agent on rare occasions; possibly allergenic.

DIGITALIN

= *Digitalinue German* = *Merck's Digitalin*

***Cardiotonic.* Standardized to 10 times *digitalis* powder. It is NOT *digitoxin* (*unfortunately, digitoxin is also known as digitalin[e]*). May not possess the cumulative properties of *digitalis, digitoxin,* or *digoxin*.**

Untoward effects: As for ***digitalis***. See below.

LD, **horse**: ~130 mg; **cat**: ~11 mg; **dog**: ~22 mg.

DIGITALIS

= *Digifortis* = *Digitora* = *Fairy Gloves* = *Foxglove* = *Neodigitalis* = *Pil-Digis* = *Purple Foxglove* = *Yüksükotu*

***Cardiotonic, diuretic.* Introduced into clinical medicine in 1775 by Dr. William Withering, after obtaining knowledge about its use by Mrs. Hutton, a Shropshire woman, to treat dropsy.**

Untoward effects: In 1785, Withering noted that "when given in large and quickly-repeated doses, occasions sickness, vomiting, giddiness, confused vision, and objects appearing green and yellow". In acute overdoses, hyperkalemia is common. Anorexia, nausea (81%), vomiting (40%), and abdominal pain (65%) are common gastrointestinal upsets. Diarrhea, fatal intestinal necrosis and cachexia occur less frequently. Neurotoxicity is usually in the form of fatigue (95%), drowsiness, weakness (82%), lethargy, depression, decreased libido, restlessness, nightmares, headache (45%), dizziness (59%), vertigo, confusion, delirium, and personality changes. Psychotic reactions (65%) and visual hallucinations are rare. Digitalis affects the retinal receptor cells decreasing visual acuity with bilateral central scotomata. Mydriasis, conjunctivitis, decreased intraocular pressure, photophobia, ocular muscle palsy, retrobulbar neuritis, color perception changes (usually in yellows – xanthopsia), and visions of flashing or flickering lights and halos. **Vincent Van Gogh's** favoring the use of yellows and the use of halos in his paintings is attributed to **his** being treated with digitalis. **He** twice painted a portrait of his **physician** holding a ***foxglove*** plant. Visual complaints in 95% of **patients**. Rarely, temporary blindness, thrombocytopenia, agranulocytosis, pancytopenia, aplastic anemia, sweating, hypersensitivity reactions (urticaria, eosinophilia), and gynecomastia (especially in older **men**) reported.

Any type of arrhythmia may occur, especially multifocal ventricular premature complexes and non-paroxysmal atrioventricular junctional tachycardia. This requires dosage adjustments in at least 20% of **patients**. The latter occurs more commonly if there was pre-existing atrial fibrillation. Sinus bradycardia, sinus arrest, sinoatrial block, atrial tachycardia, atrioventricular blocks, premature ventricular contraction, and ventricular tachycardias. Hypokalemia can aggravate digitalis toxicity.

Its ***estrogen***-like activity may interfere with some diagnostic tests. Decreases gastric retention. It decreases myocardial concentration of ***thallium*** radiotracer. It has induced Stevens–Johnson syndrome; fixed-drug eruptions; scarlatiniform, maculopapular urticaria; anaphylactoid or serum-sickness-like reaction; and allergic dermatitis. Chronic administration may cause ***thiamin*** deficiency. Due to slow elimination, it can be a cumulative poison.

Digitalis toxicity can readily be precipitated by renal disease or by nephrotoxicity induced by other drugs, such as ***gentamicin***, or by a hypothyroid state. **Children** often eat the seeds or flowers and suck on the flowers with severe nausea, vomiting, purging, headache, arrhythmias, convulsions, and sudden death. Accidental or intentional (suicide) and overdosage poisoning has occurred with exacerbation of all the above symptoms, and, occasionally, with coma and death. Accidental poisoning in **children** has occurred from access to poorly secured medications or consuming large amounts of ***D. purpurea*** leaves or seeds (or even the water the cut flowers were in) leading to cardiac arrhythmias, gastrointestinal upsets, mental confusion, and death. Poisoning has also occurred from ingesting multicolored digitalis weight-reducing pills. Poisonings have been reported from drinking ***herbal teas*** intentionally made with it, or when made in error instead of using ***comfrey***. An elderly **couple** in Washington State died within 24 h after drinking a **herbal tea** made in error from ***foxglove*** instead of ***comfrey***. An 85-year-old **male** inadvertently made a similar mistake and became toxic.

LD, ~2 g. A Unit is the LD_{50} for a 3 lb **cat**.

Drug interactions: Bradycardia with ***guanethidine***, ***halothane***, ***quinidine***, and ***reserpine***. ***Amphotericin B***, ***bretyllium***, IV ***calcium***, ***diuretics***, ***ethacrynic acid***, ***gelatin polymers*** (***Haemaccel***), ***gentamicin***, hypomagnesemia, ***isoproterenol***, ***kanamycin***, decreased ***potassium***, ***propranolol***, ***succinylcholine***, ***thiazides***, ***verapamil***, and ***vitamin E*** may increase its toxicity. ***Phenobarbital*** may increase the dose required. Effects are cumulative and the heart becomes sensitive to lower dosage after digitalization (maximum effect short of digitalis toxicity) is reached or approached. Avoid use with ***hematinics*** containing ***ferric salts*** (physical incompatability).

Even low therapeutic doses can induce vomiting in **cats**. The lethal dose for a **cat** is ~60% that of a **dog**. ***Foxglove*** poisoning has caused deaths and unthriftiness in **red deer**. Toxicity in mature **dogs** and **guinea pigs** less than in the young and aggravated by hypomagnesemia in **dogs**. **Dogs** and **cats** can be poisoned by drinking water from a vase containing the flower. In the BIO 14.6 strain of **Syrian hamsters** with hereditary cardiac disease, the adverse effects of digitalis can be abolished by ***KCl*** or ***$MgCl_2$***. Digitalis glycoside data from **rat** and **rabbit** studies cannot be readily applied to **man** because of different modes of biotransformation and excretion. The ration of 10–50 kg **pigs** was inadvertently contaminated with 50–100 g of dried ***D. lanata*** leaves (source of ***digoxin***) and five died within 24 h. A post-mortem revealed gastroenteritis, splenic border infarcts, and severe fatty degeneration of the bundle of His. **Horses** that have eaten it show mydriasis, swollen eyelids, sleepiness, conjunctivitis, submaxillary and head swellings, and raised temperature. Dyspnea, pulmonary edema, and death reported in **dairy cows** that ate the plants. Toxicity in **dogs** has caused junctional tachycardia (blood concentrations usually > 4.5 ng/ml), central nervous system suppression, anorexia, vomiting, diarrhea, and arrhythmias. Ingestion of the plant by **deer** and **sheep** led to unthriftiness and deaths. **Canaries** consuming ~240 mg showed trembling and died in ~2.5 h. **Turkeys**, after eating its leaves, showed mydriasis, drowsiness, anorexia, and convulsions before death.

Dried leaves LD, **horses**: ~25 g; **dog**: 5 g; fresh tops LD, **horse**: ~100–200 g.

DIGITOXIN

= *Cardigin* = *Carditoxin* = *Coramedan* = *Cristapurat* = *Crystodigin* = *DAP* = *Digitalin, crystalline* = *Digitaline* (in France) = *Digitaline Nativelle* = *Digicor* = *Digilong* = *Digimed* = *Digimerck* = *Digipural* = *Digisidin* = *Digitophyllin* = *Foxalin* = *Lanatoxin* = *Myodigin* = *Purodigin* = *Purpurid* = *Tardigal* = *Unidigin*

***Cardiotonic*. Most active glycoside of *Digitalis pupurea*. 1 mg is approximately therapeutically equivalent to 1 g of *Digitalis* leaf.**

Untoward effects: Has a 5–9 days half-life in **man**. Essentially, the same adverse reactions as with ***digitalis*** above. A 67-year-old **female** receiving 0.2 mg/day/20 days orally reported seeing green and blue lines while looking at white objects. Acute esophagitis and thrombocytopenia have occurred. **Fetal** heart levels may exceed **maternal**. A **fetal** death reported from a **maternal** overdose. Fatal suicidal and accidental ingestion reported. Chronic administration may cause a ***thiamin*** deficiency. Toxic serum concentration > 0.03 mg/l and fatal at 0.04 mg/l.

TD_{LO}, **human**: 5 mg/kg; **infant**: 150 μg/kg.

Drug interactions: ***Phenobarbital*** reduces its half-life. ***Cholestyramine***, ***colestipol*** (9.3 days to 2.75 days), ***phenylbutazone***, ***phenytoin***, and ***rifampin*** act similarly. ***Quinidine*** and ***spironolactone*** increase its serum levels and half-life. It is theorized that its metabolism may be inhibited by ***cimetidine***.

Markedly increased depressive effect of ***reserpine*** in **mice**. In the **dog**, half-life is not significantly altered by ***phenobarbital***. Absorption is very variable in the **dog**, unless the tincture is used. Placental transfer in **guinea pigs** and **rats**. ***Chlordane*** raises its metabolism in the **monkey**.

LD_{50}, **rat**: 56 mg/kg; **cats**: 180 μg/kg; **guinea pigs**: 60 mg/kg.

DIGLYCIDYL ETHER

= *DGE*

Untoward effects: Strong irritating odor (threshold ~5 ppm) and eye irritation level of 10 ppm. Contact with strong oxidizing agents leads to fires and explosions with release of toxic vapors and ***CO***. Will deteriorate some plastics and rubber. OSHA air exposure limit of 0.5 ppm (2.8 mg/m^3). TWA/8 h/40 h work-week is 0.1 ppm (0.5 mg/m^3). Acute short-term exposure leads to skin burns and severe irritation to skin, eyes, and respiratory tract. Chronic long-term exposures cause skin irritation and dermatitis. Potential **human** carcinogen. 25 ppm in air is immediately dangerous to life.

Acute inhalation or oral intake in **rats** and **mice** leads to central nervous system suppression, incoordination, dyspnea, respiratory failure, and death. In **rats** and **rabbits**, chronic inhalation and percutaneous absorption leads to weight loss, corneal opacities, testicular degeneration and necrosis, thymus atrophy, and decreased white blood cell and bone marrow cell counts. Topical application leads to tumors in **mice**. Lung, liver, and kidney damage also reported from skin and eye exposure.

DIGOXIN

= *Acygoxin* = *Cardiox* = *Cardoxin* = *Coragoxine* = *Cordioxil* = *Davoxin* = *Digacin* = *Dilanacin* = *Dixina* = *Dokim* = *Dynamos* = *Eudigox* = *Lanacordin* = *Lanicor* = *Lanoxin* = *LenoxiCaps* = *Lenoxin* = *Longdigox* = *Neo-Dioxanin* = *Rougoxin* = *Stillacor* = *Vanoxin*

Cardiotonic.

Untoward effects: Essentially, the same as with ***digitalis*** above. Non-thrombocytopenic purpura also reported. Readily crosses the **human** and **rat**, but not **ovine**, placenta. Very rare is a hypersensitivity reaction with wheezing. Toxicity increases with increasing **patient** age. Toxicity from suicidal intent and accidental ingestion or overdosage occurs with serum levels > 0.2 μg%. Usual therapeutic serum levels are 0.0005–0.002 mg/l; toxic at 0.0025–0.003 mg/l; and fatal at > 0.005 mg/l.

LD, **human**: ~10 mg.

Drug interactions: Serum levels reduced by ***acarbose, cholestyramine, carbimazole, cimetidine, colestipol, grapefruit juice, metoclopramide, penicillamine, saccharin***, and ***sulfasalazine***. ***Antacids, metoclopramide***, and ***propantheline*** may interfere with its absorption. Chronic use can cause a ***thiamin*** deficiency. A psoriasiform eruption was induced by it and confirmed by re-exposure. Has caused acute esophagitis. May physiologically decrease ***testosterone*** levels, decrease libido, and occasionally lead to gynecomastia in **men**. Toxic serum concentrations are usually > 0.002 mg/l. Initially, ***ibuprofen*** increases its serum level 59% within 7 days, but essentially the same by the 28th day. ***Alprazolam, amiodarone, diazepam, diltiazem, itraconazole, quinidine, quinine, nifedipine, propafenone, spironolactone***, and ***verapamil*** (~60%) increase its serum levels and potentiate its toxicity. Cardiotoxicity increases by decrease in intracellular ***potassium***, as occurs from ***diuretics***. Nephrotoxic ***aminoglycoside antibiotics***, such as ***gentamicin*** and ***kanamycin***, can precipitate toxicity, because ~70% clears the body via the kidneys. Other ***antibiotics***, such as ***clarithromycin*** and ***erythromycin***, may alter intestinal bacteria, requiring a decrease in digoxin intake. ***Rifampin*** is a potent inducer of its metabolism. A case report indicated ***lithium*** increases its toxicity (confusion and bradycardia), as will ***hydroxychloroquin*** and ***itraconazole***. ***Trimethoprim*** has increased its serum levels by 22%, due to its decrease of digoxin renal tubular secretion. Given with ***brandy*** (40% ***alcohol***) it reduces digoxin urinary excretion. Use with ***methyldopa*** causes sinus node dysfunction and symptomatic bradycardia. ***Anticholinergics*** and ***opiates***, by decreasing intestinal motility, increase its absorption. Visual disturbances following a drug interaction of ***metildigoxin*** and ***verapamil*** in a 65-year-old **female**. After 900 mg ***Hypericum extract***/day/ 10 days, digoxin plasma levels were decreased by 25%. After 5 days, use with ***valspodar*** increased its serum levels by 211%. ***Diazepam*** increases its serum levels.

Digoxin toxicosis in treated **dogs** does not correlate well with the amounts given IV. Analgesic drugs frequently given to **dogs** for arthritic syndromes raise digoxin levels ~twice. Enteric absorption was lower in **miniature pigs** and **guinea pigs** than **cats**, and **cats** were 3–6 times more sensitive than **guinea pigs** to IV injections. Currently, the more popular cardiotonic for **dogs** with depression, anorexia, vomiting, diarrhea, severe weakness, dehydration, and cardiac arrhythmias (heart block, premature ventricular contraction, tachycardias, and fibrillation) as toxic symptoms. Half-time of elimination in **horses** is 23 h. **Horses** are similar to **humans** in sensitivity to digoxin. Toxicity in **animals** varies with degree of sensitivity and protein binding in different species and within individuals, decreased clearance rate in congestive heart failure and renal failure, and bioavailability. In the **horse**, the latter varies between 15 and

25% after oral dosing. Depression and anorexia are usually the early signs of toxicity in many **animal** species. ***Cisapride*** significantly decreased its half-life in **rabbits**, but ***amiodarone*** had no effect. ***Quinidine*** increased its serum levels in the **rat**.

LD_{50}, **rats**: 30 mg/kg; **cat**: 200 μg/kg; **guinea pigs**: 3.5 mg/kg; LD_{LO}, **dog**: 300 μg/kg.

DIHYDRALAZINE

= *Depressan* = *Dihyzin* = *Nepresol* = *Pressalin* = *Pressunic*

Vasodilator, antihypertensive.

***Untoward effects*:** Orthostatic collapse, dizziness, nausea, hot flashes, nasal mucosa swelling, headache, drowsiness, heart pain, and palpitations noted.

Drug interactions*:** With ***reserpine leads to excessive sedation and gastrointestinal bleeding.

IP, LD_{50}, **rats**: 1084 μmol/kg.

DIHYDROABIETYL ALCOHOL

= *Abitol*

***Untoward effects*:** Contact dermatitis, sensitizer in mascara.

DIHYDROANETHOLE

= *p-Propyl Anisole*

Limited amount used in fragrances for use in soaps, detergents, cosmetic creams and lotions, perfumes, and in flavoring foods.

***Untoward effects*:** No sensitization or irritation reactions to 10% in petrolatum on **human** volunteers.

Levels of 1000+ ppm fed to **rats** led to osteoporosis.

LD_{50}, **rat**: 4.4 g/kg; **mouse**: 7.3 g/kg; acute dermal LD_{50}, **rabbits**: > 5 g/kg.

DIHYDROCODEINE

= *Codhydrine* = *Dehacodin* = *DF 118* = *DH-codeine* = *Didrate* = *Dihydrin* = *Dihydroneopine* = *Drocode* = *Hydrocodin* = *Nadeine* = *Novicodin* = *Paracodin* = *Parzone* = *Rapacodin*

Antitussive, analgesic.

Untoward effects*:** Similar to ***morphine (q.v.), but less pronounced. Nausea and dizziness in 55% of **those** receiving 30 mg of the tartrate and in 33% after placebo. As dosage was increased to 60 mg, vertigo was also reported. A 51-year-old **male** on 60 mg qid orally developed perceptual problems. Psychotic states occur. Abuse potential. Has caused clinical depression in the **newborn** after use during labor. Dihydrocodeine-related deaths had blood levels of 0.8–12 μg/ml.

TD_{LO}, **human**: 6.5 g/kg/26 weeks. Estimated LD_{LO}, **infants**: 2.5 mg.

In **mice**, depression and muscle weakness noted with clonic convulsions at 75% of the fatal dose. **Rabbits** showed initial depression and muscle weakness, followed by increased muscle tone, twitching, and trembling with convulsions at 75% of a fatal dose.

LD_{50}, **rat**: 240 mg/kg; SC, LD, **mice**: 225 mg/kg; **rabbits**: 129 mg/kg.

DIHYDROERGOTAMINE METHANESULFONATE (MESYLATE)

= *Agit* = *Angionorm* = *Dergotamine* = *DET MS* = *D.H.E. 45* = *Diergo* = *Dihydergot* = *Dirgotarl* = *Endophleban* = *Ergomimet* = *Ergont* = *Ergotonin* = *Ikaran* = *Migranal* = *Morena* = *Orstanorm* = *Séglor* = *Tonopres* = *Verladyn*

***Sympatholytic, vasoconstrictor.* In treating migraine headaches.**

Untoward effects*:** Can cause diarrhea and muscle cramps. Ergotism with vascular spasms, especially in combination with ***heparin, nicotine, and ***macrolide antibiotics***, such as ***erythromycin***. May interact with and decrease efficacy of ***oral contraceptives*** by liver enzyme induction. Symptoms similar to ***ergot*** (q.v.).

IV, LD_{50}, **rat**: 110 mg/kg.

DIHYDROSAFROLE

***Untoward effects*:** When fed at dietary levels of 2500 ppm and over, it caused esophageal cancer in **rats**, which, under the terms of the Delaney Amendment, precluded its use as a food flavoring agent.

LD_{LO}, **human**: 500 mg/kg.

LD_{50}, **rat**: 2260 mg/kg; **mouse**: 3.7 g/kg.

DIHYDROSTREPTOMYCIN

= *Abiocine* = *DHSM* = *DST* = *Vibriomycin*

***Antibiotic.* Used in human and veterinary medicine, and for botanical purposes.**

Untoward effects*:** A few **nurses** were reported to have developed positive sensitivity reactions to it, but no cross-sensitivity to ***streptomycin. Paralysis, and even deaths, reported with it or other ***antibiotics*** with ***anesthetics*** or ***muscle relaxants***, such as ***decamethonium*** and ***succinylcholine***. Curariform action has caused apnea and potentiated apnea produced by ***neuromuscular-blocking agents***. Allergic reactions reported at an incidence of 1.7%, much less than for ***penicillin***. Renal damage, decreased color perception, and numerous reports of hearing loss due to eighth cranial nerve cochlear damage. The latter may develop after a delay of several years. Crosses the placenta and has caused ototoxicity in **children** whose **mothers** received it during pregnancy. Cerebral damage to the **fetus** also noted.

In **animals**, large doses have caused nephritis, dizziness, hypotension, loss of balance, and hearing loss. SC injections may lead to inflammatory reactions. **Cats** may salivate and become ataxic and depressed. Respiratory and ambulatory problems are, possibly, due to inhibition of ***calcium*** uptake at cellular membrane storage sites. Has induced hypocalcemia in **cows** and respiratory depression in **monkeys** during ***pentobarbital*** anesthesia. Within minutes, induces fatal toxicity in **gerbils**.

IV, LD_{50}, **rat**: 200 mg/kg; IM, LD_{50}, **mouse**: 350 mg/kg.

DIHYDROTACHYSTEROL

= *Antitanil* = *Anti-tetany substance 10* = *AT 10* = *Calcamine* = *Dichystrolum* = *Dihydral* = *Dygratyl* = *Hytakerol* = *Parterol* = *Tachyrol*

Calcium regulator.

Untoward effects: "Calcium Gout" and nephrocalcinosis caused by prolonged use. Calcifications of shoulder joints, hips, and fingers. Excreted in **human** breast milk. Avoid its use in pediatrics.

Hans Selye reported it to cause calciphylaxis and skin shedding in **rats**.

DIHYDROTELEOCIDIN B

= *DHTB*

Untoward effects: The strongest tumor-promoter known is a breakdown product of ***teleocidin***, and has caused severe irritation and eruptive vesications on **human** skin.

Highly toxic to **Japanese killifish** within 1 h at 0.01 μg/ml.

DIHYDRO-α-TERPINEOL

= *p-Methan-8-ol*

Has a stronger woody pine-like odor than *terpineol*.

Use: In soaps, detergents, cosmetic creams and lotions, and in perfumes.

Untoward effects: 10% concentration in petrolatum led to no irritation or sensitization on **human** volunteers.

Moderate irritation when applied full-strength on intact or abraded **rabbit** skin.

LD_{50}, **rats** and acute dermal LD_{50} **rabbits**: > 5 g/kg.

5-α-DIHYDROTESTOSTERONE

= *5-α-DHT*

A *testosterone* metabolite with androgenic properties.

Untoward effects: Hyperplastic prostate glands have a higher than normal level. Associated with idiopathic hirsutism of pubic and scalp hairs in **males** and **females**. Its production by the skin of acne **patients** is abnormally high (2–20 times).

DIHYDROTERPINYL ACETATE

= *p-Methan-8-yl Acetate*

Use: ~5 ton used annually in the U.S. for its herbaceous-like odor in soaps, detergents, cosmetic creams and lotions, and in perfumes.

Untoward effects: No irritation or sensitization of 12% in petrolatum in **human** volunteers.

Acute dermal LD_{50} in **rabbits**: > 5 g/kg.

1,2-DIHYDRO-2,2,4-TRIMETHYLQUINOLINE

= *Acetone Anil* = *Flectol* = *Methylquinoline* = *Vulkanox*

Use: Antioxidant in rubber and latex.

Untoward effects: Renal tubule adenomas or carcinomas in female **rats** and male and female **mice** after 2 year dermal studies, as well as acanthosis in male and hyperkeratosis in female **rats**.

LD_{50}, **rat**: 2 g/kg; **mouse**: 1.45 g/kg.

DIHYDROXYACETONE

= *1,3-Dihydroxydimethyl Ketone* = *1,3-Dihydroxy-2-propanone* = *Ketochromin* = *Protosol*

***Skin tanning agent, catalyst in plastic polymerization.* Sunlight exposure is not required.**

Untoward effects: Severe allergic contact dermatitis. The temporary tan staining of skin may be difficult to remove. Does not increase ***melanin*** production.

DIISOBUTYL KETONE

= *DIBK* = *Isovalerone* = *Valerone*

***Solvent*. Slow evaporation makes it useful for nitrocellulose and vinyl coatings.**

Untoward effects: Combustible liquid. Irritant to eyes, nose, throat, and skin. Headache, dizziness, drowsiness, and unconsciousness, as well as nephritis and hepatitis reported.

TWA 25 ppm (150 mg/m^3). 2000 ppm is immediately dangerous to life. Minimum lethal dose for the **human** eye was 25 ppm/15 min.

LD_{50}, **rat**: 6.16 ml/kg; single dermal, **rabbit**: 20 ml/kg.

DIISOBUTYL PHTHALATE

= *DIBC*

Untoward effects: May leach in ***polyvinyl chloride*** blood bags and, potentially, into tissues of **patients** receiving extensive transfusions. Has caused severe atrophy of the testes in **rats**.

IP, LD_{50}, **mice**: 4 g/kg; **rat**: 3.7 g/kg.

2,4-DIISOCYANATE

Untoward effects: Allergenic occupational hazard and irritant to skin and mucous membranes. Skin sensitization can occur from exposure to the liquid or vapors, and asthma from inhalation. See ***Toluene 2,4-Diisocyanate***.

DIISOPROPANOLNITROSAMINE

Untoward effects: Injections in **hamsters** led to ductal pancreatic cancer similar to that seen in **man**.

DIISOPROPYLAMINE

= *DIPA*

Use: In organic synthesis.

Untoward effects: Gastrointestinal and pulmonary irritation, nausea, vomiting, headache, and visual disturbances after ingestion. Skin or eye contact is irritating. TWA for skin exposure 5 ppm (20 mg/m^3). 1000 ppm is immediately dangerous to **human** life.

LD_{50}, **rat**: 700 mg/kg; inhalation LC_{LO}, **rat**: 1000 ppm/4 h; **cat**: 2207 ppm/72 min.

DIISOPROPYL FLUOROPHOSPHATE

= *DFP* = *Diflupyl* = *Dyflos* = *Dyphlos* = *Floropryl* = *Fluostigmine* = *Fluropryl* = *Isofluorphate* = *Isoflurophate*

***Cholinesterase inhibitor*. Miotic agent in the treatment of glaucoma and strabismus.**

Untoward effects: Use in glaucoma occasionally causes cholinergic systemic effects leading to nausea, diarrhea, bradycardia, pain, salivation, dim vision, and central nervous system excitation. Cataractogenic. In a study with 121 **patients** for the treatment of esotrobia, 12% had amblyopia, 16% iris cysts, 6% eye or lid irritation, 3% blurred vision, and 3% discomfort or nausea. Poisoning in a 12-month-old **male** treated with 1 drop/day in both eyes of 1% solution for 2 months without occluding the nasolacrimal duct led to abdominal cramps and otitis media, apneic spells with 1–2 min seizures, miotic unreactive pupils, profuse nasal discharge, some motor weakness, and temperature of 103°F. Strong anticholinesterase agent. Avoid use with drugs potentiating such effects (***organophosphorus compounds, phenothiazines, muscle relaxants***, etc.). Useful when other agents fail. Regular use may lead to unnecessary congestion, even in glaucoma. Avoid skin contact and inhalation. Potential war "nerve gas". Lacrimation, salivation, flushing, head throbbing, sweating, hypotension, tremors, convulsions, miosis, cyanosis, collapse, respiratory arrest, and death also reported.

LD ~12 mg; LD_{LO}, **human**: 0.5 mg/kg.

Delayed neuropathy has definitely been demonstrated in **cats**, **dogs**, **ducks**, and **pheasants**. Clonic convulsions in 50% of **mice** by 3 mg/kg SC and 1.6 mg/kg leading to apnea. Probably detoxified by the liver. IV or SC use in **mice** is more toxic than IP. May paralyze spontaneous respiration in **rats** and **rabbits**.

LD_{50}, **rats**: 6 mg/kg; **mice**: 36.8 mg/kg; **rabbits**: 9.8 mg/kg; SC, LD_{50}, **mice**: 3.71 mg/kg; **rats** and **dogs**: 3 mg/kg; **rabbit**: 1 mg/kg.

DILAURYL SUCCINATE

Untoward effects: In feed use, causes encephalomalacia in **chicks**, decreased fertility of **roosters**, increased fragility of erythrocytes in **rats**, **chicks**, and **hens**, and prevented by ***antioxidants***.

DILL

= *Anethum graveolens* = *Aneto* = *Dilta* = *Eneldo* = *Peucedanum graveolens*

Untoward effects: Contains ***myristicin*** (q.v.), and has been mixed with ***MSG*** and smoked, which has been known to **abusers** as ***ZNA*** leading to central nervous system effect. Contains furocoumarins, natural photosensitizers, and has caused cutaneous eruptions. May cross-react with ***orange*** peel.

Oil, LD_{50}, **rat**: 4 g/kg.

DILOXANIDE FUROATE

= *Diloxanide 2-Furoic Acid Ester* = *Furamide* = *Histomibal* = *Miforon*

Antiamebic.

Untoward effects: Flatulence and abdominal distension are common. Occasionally, nausea, vomiting, diarrhea, and albuminuria; rarely, pruritus, urticaria, dizziness, and diplopia.

DILTIAZEM

= *Adizem* = *Altiazem* = *Anginyl* = *Angizem* = *Britiazem* = *Bruzem* = *Calcicard* = *Cardizem* = *Citizem* = *Cormax* = *Deltazen* = *Dilacor* = *Diladel* = *Dilpral* = *Dilrene* = *Dilzem* = *Dilzene* = *Herbesser* = *Masdil* = *Tildiem*

***Calcium channel-blocker, antihypertensive, antianginal, and antiarrhythmic.* Usually given orally. Has been given IV.**

Untoward effects: Frequently, nausea (1.9–2.7%), edema (2.4%), headache (2%), arrhythmia (2%), rash (~1.5%), asthenia (1.2%), and transient atrioventricular block (1.1%). Other adverse effects usually < 1% are flushing, congestive heart failure, hypotension, bradycardia, palpitations, syncope, drowsiness, paresthesia, dizziness, nervousness, akathisia, tremors, insomnia, depression, confusion, amnesia, hallucinations, acute psychosis, onset of mania labile mood, pressured speech, restlessness, hyperactivity, paranoid ideations, delusions of grandeur, threatening behavior, vomiting, diarrhea, constipation, dyspepsia, indigestion, alterations of taste and smell, petechiae, urticaria, nonthrombocytopenic purpura, myocardial infarction, leg swellings, arthralgia, Stevens–Johnson syndrome, cutaneous vasculitis, exanthematous pustular dermatitis, erythematous macular skin eruptions, photosensitivity, decreased plasma ***insulin***, acute interstitial nephritis, polyuria, nocturia, osteoarticular pain; it induces parkinsonism (or aggravates pre-existing mild Parkinson disease) and immune thrombocytopenia, agranulocytosis, gingival hyperplasia, gynecomastia, hypersensitivity manifested as generalized lymphadenopathy, and fatal liver and kidney failure. May stimulate ***prostacyclin*** generation, which, in turn, may increase circulatory ***calcium*** by mobilizing it from the bones. Has been involved in suicides. Excreted in **human** breast milk. Steatorrhea in a 77-year-old **female** taking a slow-release form. IV use in a 52-year-old **male** led to serious tetany responsive to 1 g ***calcium chloride*** IV. Fatalities reported after ingestion of 0.7–2.88 g, yet **some** survived doses up to 12 g. Usual therapeutic serum levels are 0.03–0.25 mg/l; toxic at 0.8–1 mg/l; comatose or fatal at > 2 mg/l.

Drug interactions: Increased bioavailability of ***encainide*** by diltiazem. Time to reach maximum concentrations increases from 2.9 h to 3.5 h when taken with ***propranolol*** and can induce bradycardia with ***β-blockers***, including ***pindolol***. Increases ***digoxin*** serum concentration. Impairs hepatic oxidation of ***aminophylline***, but not enough to cause clinically significant increase in ***theophylline*** concentration. May inhibit ***cyclosporine*** metabolism. ***Carbamazepine*** elimination decreases, serum levels increase, and neurotoxicity develops with the addition of diltiazem. Caused 6–31% increase in ***digitoxin*** levels in 5/10 **patients**. Increases ***nifedipine's*** and ***midzolam's*** elimination half-life. May cause transient increase in chest pain, increased heart rate and blood pressure with ***ibopamine***, dizziness and fine tremors with development of ***phenytoin*** toxicity. A 52-year-old **male** demonstrated a dermatologic cross-sensitivity with ***amlodipine***. **His** pruritic maculopapular rash followed one dose of ***amlodipine***, after being discontinued from diltiazem for the same adverse effect.

Enhances absorption and bioavailablity of ***diazepam***, ***dipyridamole***, and ***furosemide*** in **dogs**, but not in **humans**. Catalepsy, hypotension, bradycardia, and heart block in **cats**.

LD_{50}, **rat**: 585 mg/kg; **mouse**: 690 mg/kg.

DIMEFLINE

= *Remeflin*

Respiratory stimulant.

Untoward effects: In nearly 1000 **patients**, IV use led to tremor and vertigo in 2%. Agitation, nausea, vomiting, urticaria, pruritus, and slight tachycardia also reported.

LD_{50}, **rat**: 40 mg/kg; **mouse**: 12 mg/kg; **hamster**: 25 mg/kg; IV, LD_{50}, **rat**: 1.8 mg/kg; **mouse**: 1.5 mg/kg.

DIMEFOX

= *Hanane* = *S-14* = *Terrasytam*

Organophosphorus insecticide, acaricide.

Untoward effects: Highly toxic anticholinesterase agent. Symptoms in **man** similar to ***parathion*** (q.v.).

LD_{LO}, **human**: 5 mg/kg.

LD_{50}, **rat**: 1 mg/kg; **mice** and **cats**: 2 mg/kg; **rabbits**: 3 mg/kg.

DIMENHYDRINATE

= *Amosyt* = *Anautine* = *Andramine* = *Antemin* = *Diamarin* = *Dimate* = *Dramamine* = *Dramarin* = *Dramocen* = *Dramyl* = *Emedyl* = *Emes* = *Epha* = *Faston* = *Gravol* = *Menhydrinate* = *Reidamine* = *Removine* = *Travamine* = *Travel-Gum* = *Travelin* = *Travelmin* = *Vomex A* = *Xamamina*

Antiemetic.

Untoward effects: Has anticholinergic activity and can simulate anticholinergic ill-effects or amplify the adverse effects of anticholinergics. Drowsiness, occipital headaches, vertigo, fixed-drug eruptions, acute porphyria attacks, convulsions, coma, cyanosis, and death reported. **Children** ($6\frac{1}{2}$–14 years) developed extrapyramidal symptoms, including torticollis, anxiety, tremor, dysarthria, neck pain, akathisia, and walking difficulties after antiemetic drugs, including dimenhydrinate. A 22-year-old **male** and a 20-year-old **male** each ingested 800 mg and developed hallucinogenic experiences. Abuse with dependence and withdrawal symptoms reported in three **females** (30–41 years) taking 1250–3750 mg/day. Withdrawal symptoms include hallucinations, nausea, and vomiting. 15 min increase in blood pressure after IV. Can, unfortunately, mask symptoms of ***streptomycin*** toxicity until they have progressed to a dangerous level.

Large doses in **animals** lead to sedation, incoordination, blurred vision, and xerostomia.

DIMERCAPROL

= *BAL* = *British Anti-Lewisite* = *Dicaptol* = *Dikantol* = *1,2-Dithioglycerol* = *Sulfactin*

Heavy metal chelating agent.

Untoward effects: Pain at injection site; occasionally nausea and vomiting; headache; burning sensation, edema, and ulceration of the lips, mouth and throat; constriction or pain in the throat, chest, or hands; conjunctivitis, lacrimation, rhinorrhea, and salivation; burning or tingling paresthesias of the hands or penis, dermatitis with papular or urticarial rash after skin contact; sweating and abdominal pain; increased blood pressure, tachycardia, and unpleasant breath odor. Adverse effects peak in about 20 min and disappear within ~2–4 h. **Children** may have raised temperature during treatment. Forms a complex with any administered ***iron***. Ill-effects diminish after 4 h. Similar cautions suggested for ***zinc***, because its use after ***BAL*** therapy may cause severe gastrointestinal symptoms and lethargy. Nephrotoxic. Hemolysis and methemoglobinemia in **patients** with glucose-6-phosphate dehydrogenase deficiency.

IM, TD_{LO}, **human**: 3 mg/kg.

Lameness, stiff gait, deepened respirations, salivation, nasal discharge, recumbency, convulsions, and death reported in **sheep** receiving large doses. Interindividual variations in toxicity attributed to ***1,2,3-mercaptopropane***, an impurity.

IM, LD_{50}, **rats**: 87 mg/kg; **rabbit**: 99 mg/kg.

DIMERCAPTOPROPANE SULPHONATE

= *Dimaval* = *DMPS* = *Unithiol* = *Unitiol*

Heavy metal chelating agent. The sodium salt is used. Used more in Europe and Asia than in the U.S.

Untoward effects: Expected to be similar to ***dimercaprol***. Some reports indicate it to be less toxic.

IP, LD_{50}, **mice**: 5.22 mmol/kg.

DIMETHADIONE

= *AC-1198* = *BAX-1400Z* = *DMO* = *Eupractone* = *NSC-30152*

Anticonvulsant.

Untoward effects: Photophobia with itching eyes in 5–16 year **patients**. Causes a high incidence of decreased fetal weight, fused and wavy ribs, and retarded sternebrae in **rats**.

TD_{LO}, **rats** (6–15 days pregnant): 540 mg/kg; IV, LD_{50}, **mouse**: 450 mg/kg.

DIMETHICONE

= *Dimethylpolysiloxane*

Antiflatulent. Usually, mixed with 4–4.5% *silicon dioxide (silicaaerogel)* = *Antifoam A* = *Baros* = *Colicon* = *Covicone* = *Dimethicone* = *Dimeticone* = *Endo-Paractol* = *Gas-X* = *Infacol* = *Lefax* = *Mylicon* = *Phasil* = *Phazyme* = *sab simplex* = *Sicol* = *Silain* = *Silastic 200* = *Silicote* = *Silizaz* = *Simethicone, activated* = *Tympanol*

Use: As a defoaming agent in gelatin desserts, antibiotics, corticosteroids, cosmetics, and implants.

Untoward effects: Death due to bilateral pulmonary edema after injection. Indurated areas at injection sites, malaise, increased SGOT, granulomatous hepatitis, enlarged liver, hypopigmentation, autoimmune disorders from implants and the material can migrate in the body. Illicit injections in 34-year-old **female** breast soft tissue 5 years previously and local injections over a 3 year period led to severe granulomas. Siliconized syringes may react adversely with ***insulin***. Impaired absorption of ***warfarin*** and ***phenindione*** was discovered after several **patients** had eaten ***potato*** chips prepared with cooking oil containing it as an additive.

DIMETHINDENE

= *Dimetindene* = *Fenistil* = *Fenostil* = *Forhistal*

Antihistamine. **The maleate is used.**

Untoward effects: Some sedation or drowsiness. < 1% report dryness of mouth, gastrointestinal discomfort, nausea, diarrhea, central nervous system stimulation, irritability, insomnia, dizziness, headache, bladder discomfort, and nocturia.

LD_{50}, **rat**: 618 mg/kg; **guinea pig**: 888 mg/kg; IV, LD_{50}, **rat**: 27 mg/kg; **dog**: 45 mg/kg.

DIMETHISTERONE

= *P 5048* = *Secrosteron*

Progeston.

Untoward effects: Potentially, masculinizing to **female fetus**. One case of hypospadia reported from prenatal use. Large doses lead to pelvic pain, as in dysmenorrhea, breast swelling, nausea, and vertigo.

LD_{50}, **mouse**: 7.65 g/kg.

DIMETHOATE

= *AC-18682* = *American Cyanamid 12880* = *BI 58* = *CL 12,880* = *Cygon* = *Daphene* = *De-Fend* = *Demos-L40* = *Dimetate* = *Dimethogen* = *EI 12880* = *ENT 24650* = *Equigard* = *Ferkethion* = *Fosfamid* = *Fostion MM* = *L395* = *Le-Kuo* = *NC 262* = *Perfekthion* = *Rebelate* = *Rogor* = *Roxion* = *Trimetion*

Organophosphorus insecticide. **Anticholinesterase.**

Untoward effects: Poisonings have occurred from suicidal intent or spraying accidents. Acute cholinergic crisis, followed by intermediate syndrome with paralysis of proximal limb muscles, neck flexors, motor cranial nerves, and respiratory muscles, ending with a delayed neuropathy. A 48-year-old **male** ingested an unknown quantity leading to nausea, vomiting, excitement, and coma. Direct hemoperfusion led to return of consciousness within 4 h. A 34-year-old **female** ingested ~10 g, 30 min before hospital admission. **Her** symptoms were nausea, vomiting, diarrhea, intense diaphoresis, and increased salivation. Atropinization, gastric lavage, ***activated charcoal*** orally, hemoperfusion, and hemodialysis for 24 h was effective in saving **her**. Percutaneous absorption can occur.

LD_{50}, **human**: 30 mg/kg.

Mice receiving ~10 mg/kg/day in their drinking water had lower pregnancy rate, decreased survival and growth weight of their pups. Fetuses of **rats** given 24 mg/kg had waxy-rib anomalies. **Cats** receiving 12 mg/kg/day had 8/39 fetuses with polydactyly. Older material was more toxic to **sheep** than fresh material. Highly toxic to **bees**. LD, adult **cattle**: 22 mg/kg. Young **calves** can tolerate a higher level. **Horses** are poisoned by 3–4 times that level. **Sheep** have died from 100 mg/kg, and some were poisoned by 50 mg/kg. Chronic poisoning in **sheep** leads to depression, anorexia, salivation, and diarrhea.

LD_{50}, **rats** and **cats**: 150 mg/kg; **mouse**: 60 mg/kg; **dog**: 400 mg/kg; **guinea pig**: 350 mg/kg; male **chicks**: 36.6 mg/kg; LC_{50}, **fish**: 6.2 mg/l; feed LC_{50}, **quail**: 496 ppm.

DIMETHOXANE

= *Acetomethoxane* = *Giv-Gard DXN*

Biocide. **Preservative in cutting oils, inks, and gasoline.**

Untoward effects: Long-term feeding to **rats** leads to tumors of liver, kidney, skin, SC and lymphoid tissues, as well as leukemias in ~20%.

2,5-DIMETHOXY-4-ETHYLAMPHETAMINE

Psychotropic.

Untoward effects: Euphoria and feeling "talkative", as with ***amphetamine***. 8/10 **subjects** had difficulty in concentrating. 8/10 had prominent pupillary dilation in about 4 h. **Some** felt "light-headed", and only two reported visual disturbances. Despite a chemical resemblance to ***amphetamine***, its effects are different from those of ***amphetamine***.

DIMETHOXYETHYLPHTHALATE

= *DMEP*

Use: Up to a million or more lbs annually used as a defoaming agent in paper manufacturing and in lubricating oils, and as a plasticizer in ***polyvinyl chloride*** blood storage containers.

Untoward effects: **Human** TLV of 5 mg/m^3.

In **mice**, decrease in implantations and spermatogenesis, especially at $\frac{1}{3}$ of the LD_{50}. Decreased liver enzymes in **rats** and increased liver and kidney weights of **guinea pigs** and **dogs**. Skeletal abnormalities after IP injections in **rats**.

LD_{50}, male **mice**: 3.6 ml/kg; **guinea pigs**: 1.6 g/kg; LD_{LO}, **rat**: 2.75 g/kg.

DIMETHOXYMETHYLAMPHETAMINE

= *DOM* = *STP*

Hallucinogen.

Untoward effects: Hallucinogenic effects estimated to be 5–100 times that of ***mescaline***. In 1967, 12 **users** in California were hospitalized with 3 day mania, euphoria, depression, irrational behavior, dry mouth, blurred vision, mydriasis, preconvulsive jerks, swallowing or breathing difficulties, raised blood pressure, photophobia, tachycardia, fear, and confusion. A single dose of the drug has produced these results, and the effects are, unexpectedly, aggravated by ***chlorpromazine***, used to control psychotic reactions to ***LSD***. Severe shock and death has occurred. Doses > 3 mg lead to hallucinations for ~8 h. ***STP*** stands for serenity, tranquility, and peace, yet its use meant anything but that. It is not the automobile oil additive of the same name. Symptoms were of a powerful anticholinergic agent, and death occurs due to respiratory paralysis. Analysis of samples indicated variations in its chemistry. By 1970, it was a contaminant in some ***LSD*** sold or substituted for ***mescaline***.

Also increased systolic blood pressure in **rats**. Piloerection in **rats** at high dosage. Elicits rage-like behavior in **cats**.

LD_{50}, **mice**: 330 mg/kg.

N,N-DIMETHYLACETAMIDE

= *DMA* = *DMAC*

Use: Excellent solvent for ***antibiotics*** (***chloramphenicol*** and ***tetracyclines***), pharmaceuticals, polymers, and resins.

Untoward effects: Enhances skin permeability of various compounds. Inhalation leads to jaundice, and liver damage; skin absorption leads to irritation, depression, lethargy, and hallucinations; ingestion leads to delusions. TWA, skin contact, 10 ppm. 400 ppm can cause immediate risk to health.

LD_{LO}, **human**: 500 mg/kg; inhalation TC_{LO}, **human**: 20 ppm.

Marked embryotoxicity from topical application on **rabbits**. Depression, coma, and some trembling occurred in **mice** before death.

LD_{50}, **rats**: 3.5 g/kg; **mice**: 3.2 g/kg.

N,N-DIMETHYLAMINE

Use: Rubber vulcanizing accelerator, hide depilatory in tanneries, and in the manufacture of ***antihistamines*** and some rocket fuels.

Untoward effects: Nose and throat irritation from inhalation; sneezing, coughing, and dyspnea from ingestion; pulmonary edema, conjunctivitis, dermatitis, and burns of the skin and mucous membranes from contact. TWA, 10 ppm (18 mg/m^3). 2000 ppm is immediately dangerous to life and health. Found in some samples of ***ampicillin***.

Dimethylamine, present naturally in ***fish***, reacts with ***nitrites*** added to ***fish*** to preserve it before dehydrating and processing, to form ***nitrosodimethylamine***, a toxic substance, causing the deaths of many **mink** and **foxes** on farms in Norway and Britain, because of the incorporation of herring meal into their diet. Has caused hepatic necrosis in **rats** and **mice** in the presence of **nitrite**. This is inhibited by ***ascorbic acid*** or ***vitamin E***. At 770 mg/m^3, **rabbits** and **dogs** had nose and eye irritation and **rats** and **guinea pigs** had inflammation of the lungs. Continuous exposure to 9 mg/m^3 led to inflammatory lung changes in all the **animals** tested. Concomitant daily oral ingestion of it with ***sodium nitrite*** led to lung and soft tissue carcinogenesis in **rats**; hepatic necrosis, weight loss, and mortality in **mice**. Eye irritation in **rabbits** at 50 mg/5 min. 75% of it is volatilized in about 40 min in landfarming.

LD_{50}, **rat**: 698 mg/kg; **mouse**: 316 mg/kg; **guinea pig** and **rabbit**: 240 mg/kg.

p-DIMETHYLAMINOBENZENE

= *Butter Yellow* = *C.I. 11020* = *C.I.Solvent Yellow 2* = *DAB* = *Methyl Yellow*

pH indicator.

Untoward effects: Once used to color butter or margarine. Now banned for food use in the U.S. **Human** exposure can now occur by inhalation or percutaneous absorption. Chronic exposures lead to contact dermatitis, weakness, dizziness, euphoria, dyspnea, methemoglobinemia with bluish discoloration of skin and mucous membranes, coughing, bloody sputum, bronchial secretions, frequent urination, hematuria, and dysuria.

Hepatocellular carcinomas and degeneration and liver cirrhosis in **rats** after gavage, dietary inclusion or SC. Bladder tumors in **dogs**. SC to **mice** leads to local and liver neoplasms. Newborn **mice** develop liver and lung tumors after treatment. Topical use caused skin tumors in **rats**, but not **mice**. Single doses to male **mice** led to abnormal sperm, and to pregnant **mice** led to skeletal defects in pups.

LD_{50}, **rat**: 200 mg/kg; **mouse**: 300 mg/kg.

2′,3-DIMETHYL-4-AMINOBIPHENYL

Untoward effects: This aromatic amine produces colonic tumors in **rodents**, but not in germ-free **rats**.

SC, TD_{LO}, **rat**: 680 mg/kg.

2-DIMETHYLAMINO-3′,4-DIHYDROXYACETOPHENONE HCl

= *NSC 62,512*

Antineoplastic.

Untoward effects: Bone marrow depression, vomiting, and hepatotoxicity in **humans**. The latter two also occur in **dogs** and **monkeys**.

N,N-DIMETHYLAMINO-ETHOXYETHANOL

Untoward effects: Causes skin irritation; prolonged or repeated skin contact can cause chemical burns. In the presence of ***nitrites***, ***nitrosamines*** may be formed.

LD_{50}, **rats**: 2.46 ml/kg; skin, **rabbits**: 1.41 ml/kg.

4-DIMETHYLAMINOPHENOL

= *DMAP*

Cyanide antidote.

Untoward effects: This is one of the agents that induces methemoglobinemia in **humans**, **cattle**, **mice**, **cats**, and **dogs**, making it an effective antidote in ***cyanide*** poisoning.

Oxidizes hemoglobin to ferrihemoglobin, and at twice the LD_{50} leads to renal tubular necrosis in **rats**.

LD_{50}, male **rats**: 780 mg/kg; female **rats**: 689 mg/kg; female **mice**: 946 mg/kg; IV, LD_{50}, **rats**: 57 mg/kg; female **mice**: 70 mg/kg; IP, LD_{50}, **rat**: 90 mg/kg.

DIMETHYLAMINOPROPIONITRILE

= *DMAPN*

Use: A catalyst in the manufacture of flexible ***polyurethane*** foam.

Untoward effects: Neurological bladder dysfunction with urinary retention and frequency, painful urination in exposed **workers**. Impotence and dysesthesias (pins and needles sensations in hands and feet), muscle weakness, lassitude, nausea and vomiting have occasionally been reported. **Human** exposure occurs by inhalation, skin absorption, and ingestion.

LD_{50}, **rat**: 2.6 g/kg; IV, **mouse**: 180 mg/kg; dermal, **rabbit**: 1.4 g/kg.

3-DIMETHYLAMINOPROPYLAMINE

= *3-DMAPA*

Use: Epoxy resin hardener, pharmaceutical manufacturing, and petroleum processing.

Untoward effects: As an epoxy resin hardener, it or the resin can cause skin irritations and sensitization in a high percentage of exposed **workers**. **Workers** exposed to air levels of 0.41–1.38 ppm in a snow-ski manufacturing plant showed significant decrease in pulmonary function. It is suggested that exposures be limited to a maximum of 0.55 ppm. Has produced corneal edema and staining in 11 **males** (20–54 years) after chemical plant exposure to 30 ppm vapors. Mydriasis in 1/11, blurred sight in 11, glare in 10/11, "blue sight" in 7/11, halos and pain in 4/11, photophobia in 2/11, inability to focus in 1/11, severe headache in 5/11, nausea in 4/11, vomiting in 2/11, salivation in 1/11, and deafness in 1/11.

Instilled onto **rabbit** eyes it led to corneal staining and mydriasis. Low IV dosage caused mydriasis and high dosage led to miosis.

DIMETHYLAMINOPROPYL-METHACRYLAMIDE

= *BM 611* = *DMAPMA*

Untoward effects: Moderately irritating to skin (severe if exposure is prolonged) and severe irritation to eyes.

LD_{50}, **rat**: 3.54 ml/kg; acute dermal, **rabbit**: 2.5 ml/kg.

4-DIMETHYLAMINOSTILBENE

Untoward effects: **Rats** consuming ~500 μg/day of this aromatic amine in their feed, after about 8 months (total of 112 mg/**rat**), developed ear duct carcinomas.

4-(*p*-DIMETHYLAMINOSTYRYL) QUINOLINE

= *4M20*

Radiomimetic.

Untoward effects: Hepatotoxic in **mice** within an hour with some still present after 17 days. Cataractogenic in **mice** in ~3 months after 75–100 μg/kg, IV; effect similar to those by X-radiation.

IV, LD_{LO}, **mouse**: 100 mg/kg.

N,N-DIMETHYLANILINE

Use: In the manufacture of paint and allied coatings.

Untoward effects: Inhalation leads to anoxia and cyanosis. Topical exposure leads to drowsiness, weakness, ataxia, headache, nausea, and vomiting. Exposure limits of 5 ppm and 100 ppm, respectively, present an immediate danger to health and life by damaging liver, kidneys, and cardiovascular systems.

LD_{LO}, **human**: 50 mg/kg.

LD_{LO}, **rats**: 1.41 ml/kg.

DIMETHYL ANTHRANILATE

= *Methyl Methylaminobenzoate* = *Methyl-n-methylanthranilate*

A major constituent in *Oil of Mandarin Leaves* and a minor constituent in oils of *Hyacinth, Mandarin, Petitgrain,* and *Rue*. As a fragrance for soaps, detergents, cosmetic creams and lotions, perfumes, and in foods.

Untoward effects: No irritation when 10% in petrolatum tested on **human** volunteers. In a similar test for sensitization on 25 **human** volunteers, two had possible reactions. No reactions on a new group of 25 **volunteers**. Estimated daily intake in food for a 70 kg **man** is ~14 mg or 100 times the no-effect level in **rat** 90 day feeding trials.

A **bird** repellent feed additive for **starlings** and **geese**.

LD_{50}, **rats**: 3.7 ml/kg. In another trial, it was > 2.25 g/kg and < 3.38 g/kg (100% mortality); acute dermal LD_{50}, **rabbits**: > 5 g/kg.

9,10-DIMETHYL-1,2-BENZANTHRACENE

= *7,12-Dimethylbenz[a]anthracene* = *DMBA*

Synthetic polycyclic aromatic hydrocarbon, widely accepted as a cancer-inducer in research.

Untoward effects: A potent carcinogen. Although it only initiates hepatocytes, it both initiates and promotes malignant skin cells. Can induce squamous cell carcinomas of the buccal pouches of **hamsters**; melanomas, morphologically similar to blue nevi of **man**. Oral feeding leads to mammary tumors in **rats**; inhalation causes tracheal hyperplasia. Initiator of skin tumors and papillomas on **mice** and mammary tumors, stillbirths, and fetal resorption. Embryotoxic to **mallard ducks** when applied to the surface of eggs. Topical use led to benign and malignant ear tumors on **rabbits**. Single SC injections to **guinea pigs** after birth led to tumors at the injection site, and tumors of the lungs, uterus, ovaries, breasts, skin, etc. Induces dermal hyperpigmentation and blue nevus-like tumors in **gerbils**.

LD_{50}, **rat**: 327 mg/kg; **mouse**: 340 mg/kg; LD_{LO}, **rat**: 15 mg/kg; **mouse**: 4 mg/kg.

DIMETHYLBENZYL CARBINOL

Use: As a fragrance in soaps, detergents, cosmetic creams and lotions, perfumes, and in food.

Untoward effects: No irritation or sensitization at 8% in petrolatum on 25 **human** volunteers.

No adverse effects when fed to **rats** at 10,000 ppm/16 weeks or 1000 ppm/28 weeks in their diet.

LD_{50}, **rats**: 1.35 g/kg; **guinea pigs**: 988 mg/kg; acute dermal, **rabbits**: > 5 g/kg.

DIMETHYLBENZYL CARBINYL ACETATE

Use: Up to 25 tons/year used in fragrances in the U.S. for soaps, detergents, cosmetic creams and lotions, perfumes, and foods.

Untoward effects: No sensitization of 4% in petrolatum, and only mild irritation by 4% in petrolatum after a 48 h closed-patch test in groups of 25 **human** volunteers.

Moderate irritation full-strength on intact or abraded **rabbit** skin for 24 h under occlusion.

LD_{50}, **rats**: 3.3 g/kg; acute dermal, **rabbits**: > 3 g/kg.

DIMETHYLCARBAMOYL CHLORIDE

= *DMCC*

Use: Chemical intermediate in the manufacture of ***neostigmines***, ***pyridostigmine***, ***dyes***, and ***insecticides***.

Untoward effects: One of the most potent carcinogens known. N.Y.U. estimated < 200 **persons** potentially exposed to this direct-acting alkylating compound. Dermal exposure and inhalation are probable sources for **man** in addition to contacting residues in dyed materials.

Within 7 months, 89/93 **rats** exposed to a concentration of 1 ppm in the air developed squamous cell carcinomas of the nose. This also occurred in **male hamsters**.

LD_{50}, **rat**: 1 g/kg.

trans-4,4′-DIMETHYL-α-α′-DIETHYLSTILBENE

= *DMES*

Estrogen. **Experimental.**

Untoward effects: Daily oral use led to papillomatous growths on internal genitalia of female **dogs**. Alopecia and vulvar edema also occurred. Slow regression.

DIMETHYL DISULFIDE

An attractant pheromone in **hamster** vaginal secretions. Used in sour gas wells for presulfiding petroleum catalysts, in manufacturing polyurethanes, and in synthesizing drugs and pesticides.

Untoward effects: Causes hemolytic anemia, protein inactivation, and decreased appetite in **sheep** and **goats**, similar to that in ***kale*** poisoning, in which it is a metabolite.

Inhalation LC_{50}, **rat**: 805 ppm.

DIMETHYLDODECLAMINE OXIDE

Untoward effects: Chronic oral intake by **rats** with ***sodium nitrite*** led to liver tumors, but not with only the amine.

LD_{LO}, **rat**: 1 g/kg.

7,10-DIMETHYLELLIPTICINE

= *NSC 102,726*

Antineoplastic.

Untoward effects: IV use in anesthetized **rhesus monkeys** led to immediate and marked hemolysis.

DIMETHYLFORMAMIDE

= DMF = DMFO

Use: A universal type of organic solvent.

Untoward effects: National Institute for Occupational Safety & Health estimated > 100,000 U.S. **workers** may be exposed to it, primarily from poorly ventilated work areas. Readily absorbed through the skin; leads to hepatitis, hepatomegaly, ***disulfiram***-like reactions to ***alcohol*** intake, increased blood pressure, facial flushing, anorexia, nausea, vomiting, and dermatitis. Promotes increased permeability to many compounds. Ingestion can cause similar symptoms. Clusters of testicular cancer may have been due to ***DMF*** exposure in leather **tanners**. For about 4 h, a 34-year-old **male** maintenance **fitter** repaired a pipe under a ***DMF*** reaction vessel. **He** noticed an unusual smell and the surrounding atmospheric concentration was 30 ppm. Normal exposure limits are 10 ppm. For several days, **he** developed dyspnea, tightness of the chest, and generalized red blotchiness of the skin after ***alcohol*** consumption. Acute pancreatitis was suspected in two factory **workers** after ***DMF*** exposure. **They** experienced abdominal pain, nausea, vomiting, and erythema of exposed skin areas. Kidney and liver damage reported.

LD_{LO}, **human**: 500 mg/kg.

Mice become depressed and comatose during the first day post-treatment and tremble before dying. Ototoxic with torticollis in **guinea pigs** after two topical doses of 0.1 ml. Embryotoxic in **rats** and **rabbits** after cutaneous application. Jitteriness, slowness, ataxia, geotaxia, and falling in **ducks**, **quail**, and **gulls**, with recovery in 24 h.

LD_{50}, **rats**: 2.8 g/kg; **mice**: 3.75 g/kg; **mallard ducks**: > 2 g/kg; IP, LD_{50}, **rat**: 1.4 g/kg; **mice**: 650 mg/kg; **cat**: 500 mg/kg; **rabbit**: 1 g/kg; IV, LD_{50}, **rat**: 2 g/kg; **mouse**: 3.5 g/kg; **dog**: 470 mg/kg; **rabbit**: 1.8 g/kg; **guinea pig**: 1 g/kg.

DIMETHYLHEPTENAL

A melon-like fragrance approved for use in soaps, detergents, cosmetic creams and lotions, perfumes, and foods.

Untoward effects: 4% in petrolatum produced no irritation or sensitization in **human** volunteers.

LD_{50}, **rats** and acute dermal, **rabbits**: > 5 g/kg.

1,1-DIMETHYLHYDRAZINE

= Dimazine = DMH = UDMH

Use: Fuel for certain rocket engines.

Untoward effects: Potential **human** carcinogen. ***Ammonia*** or **fish**-like odor. Irritant to eyes, skin, and respiratory tract. Weakness and lethargy are first toxic symptoms. Percutaneous absorption occurs. **Human** exposure can be through inhalation, ingestion, and dermal or eye contact. Short-term exposures lead to choking sensation, chest pain, dyspnea, lethargy, nausea, vomiting, and mucous membrane irritation. Chronic exposure also leads to gastrointestinal irritation, tremors, hypoglycemia, fatty infiltration of the liver, raised SGPT, hemolysis, increased blood pressure, and anoxia. Exposure limits at 0.06–0.5 ppm (0.15–1.0 mg/m^3) – the lower figures for inhalation, and the upper, for skin. In the presence of water and an acidic pH, it's a breakdown product of ***daminozide*** (q.v.). National Institute for Occupational Safety & Health recommends exposure limit of 0.06 ppm (150 $\mu g/m^3$). A Titan II missile silo leaked an ***Arozene 5*** fuel (50% ***UDMH***).

By gavage to female **mice** led to lung tumors. In drinking water led to angiosarcomas and tumors of kidneys, lungs, and liver in male and female **mice**, as well as liver and colon carcinomas in **rats**. Significant decrease in performance of **macaque monkeys** in about 2 h after IP injection. Inhalation led to central nervous system stimulation, dyspnea, convulsions, and deaths in **rat**, **mouse**, and **dog** trials. Chronic exposure of **dogs** to its vapors led to central nervous system depression, salivation, vomiting, diarrhea, ataxia, convulsions, and hemolytic anemia. Chronic oral gavage in **mice** led to blood vessel, lung, kidney, and liver tumors.

Inhalation LC_{50}, **rat**: 252 ppm/4 h; **mouse**: 172 ppm/4 h; **dog**: 3580 ppm/15 min; **hamster**: 392 ppm/4 h.

1,2-DIMETHYLHYDRAZINE

= SDMH

Untoward effects: Similar to above.

Powerful carcinogen for the colon in some **rats** and **mice**. Blood vessel and lung tumors in **mice**; liver, stomach, intestine, and blood vessel tumors in **hamsters**.

DIMETHYL MERCURY

Reagent.

Untoward effects: Said to be biologically inert, but it can be converted to the poisonous ***monomethyl mercury***. Has caused neurotoxic poisoning in **man**. In 1997, it killed a **female** research **chemist** when a drop seeped into **her** ***latex*** glove.

LD_{50}, **rat**: 100 mg/kg.

DIMETHYL METHYL PHOSPHONATE

= DMMP

Organophosphorus chemical and biological warfare nerve gas simulant. **Lacks neurotoxic properties of other anticholinesterase agents. Also used as a flame retardant.**

***Untoward effects*:** Decreased spermatogenesis or degeneration and necrosis of spermatogenic tubules in **rats**.

DIMETHYLOL DIMETHYL HYDANTOIN

= DMDM Hydantoin = Glydant

Use*:** As a preservative in cosmetics and soaps, and, infrequently, as a ***formaldehyde (q.v.)-releaser in permanent press products.

***Untoward effects*:** Contact allergen in **people**.

DIMETHYL PHOSPHATE

Embargoed by the U.S. for export to Iran, Iraq, and Syria; now subject to global export licensing, because of its use in chemical and biological warfare.

DIMETHYL PHTHALATE

= Avolin = DMP = Fermine = Lupersol = Methyl Phthalate = Mipax = Palatinol M

Insect repellent, solvent, and plasticizer.

***Untoward effects*:** Inhalation irritates the upper respiratory tract and ingestion is highly toxic. Ingestion leads to irritation of buccal mucosa, stomach pain, nausea, vomiting, vertigo, decreased blood pressure, and coma. Contact dermatitis reported. OSHA limits exposure to 5 mg/m^3.

LD_{LO}, **human**: 5 g/kg.

Causes gross skeletal fetal anomalies in **rabbits**. LD_{50}, **rat**: 6.9 g/kg; **mouse**: 4 g/kg; **rabbit**: 4.4 g/kg; **guinea pig**: 2.4 g/kg; **chicken**: 8.5 g/kg; Inhalation LC_{LO}, **cat**: 10 g/m^3.

DIMETHYLSAFRANINE

***Untoward effects*:** This dye in lipstick has caused cheilitis.

DIMETHYL-*p*-STYRYLANILINE

***Untoward effects*:** A single oral dose induced lobular carcinomas and diffuse fibrous hyperplasia of **rat** mammary glands.

LD_{LO}, **rat**: 50 mg/kg.

DIMETHYL SULFATE

= DMS = Methyl Sulfate

Corrosive to metals.

***Use*:** Methylating agent in organic chemical and perfume manufacturing. War gases.

Untoward effects*:** Poisonous. Decomposes in water to ***sulfuric acid. Overexposure to vapors or skin contamination leads to eye, nose, throat, lung, and skin irritation or burns, severe rhinitis, headache, giddiness, photophobia, blepharospasm, impaired color vision, periorbital edema, dysphonia, aphonia, dysphagia, productive cough, chest pain, dyspnea, vomiting, diarrhea, dysuria, hematuria, albuminuria, icterus, and raised temperature. Caustic. Neurotoxicity (nystagmus, convulsions, and respiratory failure during coma, leading to death) reported. An accidental spill in a laboratory led to first degree burns on dorsum of feet, conjunctivitis, nasopharyngitis, and laryngitis. Lung signs cleared within 24 h, except for the burns; the rest improved by day 3. A case of choroidal melanoma attributed to it in a **male** after 6 year exposure. Death occurs from pulmonary edema and circulatory collapse. TLV, 1 ppm (5 mg/m^3). Estimated lethal dose 1–5 g.

Inhalation LC_{LO}, **human**: 97 ppm/10 min.

SC to **rats** leading to local sarcomas and vapor inhalation at 3 ppm/1 h/5 times a week led to squamous carcinomas of the nasal cavity and neurogenic tumors in 8/27.

LD_{50}, **rats**: 440 mg/kg; inhalation LC_{LO}, **rat**: 32 ppm/4 h; **mouse**: 75 ppm/17 min; **guinea pig**: 75 ppm/24 min.

DIMETHYL SULFOXIDE

= Deltan = Demeso = Dermasorb = DMSO = Dolicur = Domoso = Dromisol = Gamasol 90 = Hyadur = Infiltrine = Kemsol = Methyl Sulfoxide = Rimso-50 = Sclerosol = Somipront = SQ 9453 = Sulfinylbismethane = Syntexan

***Use*:** Solvents, antifreeze, skin penetration-enhancer, analgesic, and anti-inflammatory agent. Millions of lbs are produced annually.

Untoward effects*:** A ***garlic-like or ***oyster***-like taste, breath, and skin odor often occurs within minutes of topical or IV use, due to ***dimethyl sulfide***, a minor metabolite. It may last up to 3 days. Suprapubic discomfort, headache, nausea, and lethargy can occur. Topically it leads to urticaria, itching, erythema, occasional vesiculation, photosensitivity reactions, cracked skin, sloughing, burning, allergic shortening of breath, anaphylactoid facial swelling, gastrointestinal upsets (including nausea and diarrhea), lethargy, dizziness, conjunctivitis, photophobia, and transient disturbances in color vision. Fatigue, cyanosis, and dyspnea within 24 h after topical application of 4 oz to the lower abdomen of a 43-year-old **female** due to sulfhemoglobinemia (6.2%) formation. Strong solutions into the bladder lead to transient cystitis and contact dermatitis. Scaling of skin, contact dermatitis, and hyperpigmentation have followed chronic use. When used as a penetration-enhancer, it can increase the toxicity of the drug mixed with it. It can activate a latent intracellular virus infection. Severe partial thickness burn in an 18-year-old **male** after overnight wrapping of a finger with a cloth soaked in 90% ***DMSO***. An 18-year-old **female**, after endodontic treatment of a mandibular molar, had severe postoperative pain. A friend gave **her** some ***DMSO*** to apply topically, and **she** also ingested 1 oz/day/4 days, which gave **her** relief, but **she** experienced tissue destruction at the site of application, ***oyster***-like breath, glassy-eyed stare, and raised alkaline phosphatase. ~25% of 65 **patients** treated with it topically on the eye reported burning or stinging sensations when the drops were applied. IV use to treat arthritis of the knees in an **elderly couple** caused raised aspartate transaminase, hydroxybutyrate dehydrogenase, creatine phosphokinase, and hematemesis in **one**. An 82-year-old **male** received IV treatment 5 days/week/3 weeks and showed encephalopathy with agitation and poor cooperation. Another **two** received 100 g of a 20% solution IV, tid on two separate occasions. After the second series, **one** became drowsy and vomited blood. During the next 2 days, the **patient** showed jaundice, flapping tremor, oliguria, and increased liver and muscle enzymes. Milder, but less severe symptoms developed in the other **patient**. In a similar trial with 14 **patients**, a dose-dependent hemolysis and hemoglobinuria developed without renal toxicity. Growth inhibition, precocious puberty induced, decreased testicular size in prepubertal ages, ***testosterone*** inhibited during puberty; decreased testicular size and libido in **adults**. Hepatotoxic in **humans** and experimental **animals**. Will penetrate rubber gloves, except ***neoprene***, taking with it any dissolved drugs. Do NOT use it in plastic syringes or IV sets, because it will release toxic materials from them.

It inhibits the metabolism of ***sulindac*** (***Clinoril***) into its active form. A 68-year-old **male** developed a progressive sensorimotor peripheral neuropathy after using 150 mg orally twice daily with topical ***DMSO***. **His** disabilities gradually improved when the latter was discontinued.

Guinea pigs treated topically developed erythema; thickened, leathery texture, and dryness and sloughing of skin, as noted in **humans** and **rabbits**. Its use (route?) may have been a cause of hepatitis and death in a **horse**. Hepatotoxic in **rats**, **dogs**, **monkeys**, and **humans**. In the **dog**, IV injection of 100 mg/kg led to increased cardiac output, increased stroke volume, increased heart rate, and increased arterial and pulse pressure. Orally, decreased lucency of lens cortex in **dogs** and **monkeys**. Causes nuclear sclerosis (whitish growth in the lens) of **dogs**. IP to **hamsters** (8 days pregnant) teratogenic and embryocidal. Has caused vascular disturbances in **rabbits**, neurolysis in **rats**, and anemia and hematuria in **dogs**. 0.6 g/kg/60 days topically in **horses** led to refractive index lens changes, leading to myopia. Intravesical instillation in **cats** can be irritating.

LD_{50}, **rat**: ~18 g/kg; **mouse**: 21 g/kg; **chicken**: 12.5 g/kg; IV, LD_{50}, **rat**: 5.4 g/kg; **mouse**: ~4 g/kg; **dog**: 2.5 g/kg.

DIMETHYL TEREPHTHALATE

= *DMT*

Use: In manufacturing ***polyester*** textiles, ***dacron***, and surgical prostheses.

Untoward effects: Little or NO migration from beverage bottles. Cataractogenic in experimental **animals**.

LD_{50}, **rat**: 4.4 g/kg.

N,N-DIMETHYLTRYPTAMINE

= *DMT*

Untoward effects: Hallucinogen, similar in pharmacological properties to ***lysergide*** (***LSD***). Found in a number of South American and West Indies hallucinatory snuffs. Inactive by mouth, but produces a rapid intoxication when smoked, injected, or used as a snuff. Effects last 30–60 min. In addition to hallucinations, it can cause tremors, anxiety, nausea, hypertension, mydriasis, incoherent speech, cold hands and feet, laughing, crying, and panic reactions. It can be made synthetically, and was once rather popular, being known as ***Businessman's Special***. Flashbacks or distortions of perception may occur afterwards. Some of the manifestations of the illness in ~70% of the **schizophrenics** have resulted from a large conversion in **their** brain cells of ***tryptophan*** to ***DMT***. Abnormally high levels of ***DMT*** have been found in **their** blood and urine. IM doses of 0.7–1.0 mg/kg appeared optimum for psychotic effects within 3–4 min.

IM, TD_{LO}, **human**: 1 mg/kg.

Cattle consuming *Phalaris* sp. developed central nervous system disturbances, including classical staggers that can lead to cardiovascular failure. ***DMT*** is one of the plant's constituents. Lower dosage leads to salivation, hyperexcitability, head-nodding, mydriasis, incoordination, and lateral recumbency.

IV, LD_{50}, **mouse**: 32 mg/kg.

DIMETHYL TUBOCURARINE IODIDE

= *Metocurine Iodide* = *Metubine Iodide*

Muscle relaxant. **Non-depolarizing neuromuscular blockade by competing with the cholinergic receptor sites at the motor end-plates.**

Untoward effects: Large IV doses may cause ***histamine*** release with peripheral vasodilation, hypotension, and bronchospasm. Although not readily crossing the placenta, repeated large doses can cause **fetal** paralysis.

Drug interactions: ***Enflurane*** and ***isoflurane*** can potentiate or prolong its action at the neuromuscular junction. ***Methoxyflurane*** affects it less and ***halothane*** little. This requires significant decrease in dosage during anesthesia. Many ***antibiotics***, such as ***amikacin, bacitracin, clindamycin, colistin, gentamicin, kanamycin, neomycin, polymyxin B,*** and ***streptomycins*** enhance the neuromuscular blockade. The same has been reported for ***magnesium sulfate, quinidine,*** and ***trimethaphan***.

In **animals**, causes respiratory paralysis and regurgitation, due to relaxation of esophageal muscle and stomach sphincters. ***Neostigmine*** has been an effective antidote.

IV, LD_{50}, **rat**: 35 μg/kg; **mouse**: 238 μg/kg; **rabbit**: 32 μg/kg; **guinea pig**: 50 mg/kg.

DIMETHYLVINYL CHLORIDE

Use: Intermediate in organic chemical synthesis.

Untoward effects: Extremely volatile and flammable at room temperature. Emits toxic fumes of ***hydrochloric acid*** and other chlorinated compounds when heated. **Human** exposure most probably by inhalation.

By gavage to **rats** led to carcinomas and adenocarcinomas of the nasal cavity; squamous cell papillomas and carcinomas of the mouth, esophagus, and stomach. By gavage to **mice** led to squamous cell carcinomas of the stomach (male and female) and preputial gland in male.

Inhalation LC_{LO}, **mouse**: 181 g/m^3/10 min.

DIMETILAN

= *G-22870* = *GS-13332* = *Snip*

Insecticide. **A carbamate.**

Untoward effects: Cholinergic. I have not seen a report on poisoning in **people**, but, if it has occurred, symptoms would be those of a cholinesterase inhibitor leading to salivation, blurred vision, miosis, tearing, weakness, headache, giddiness, nervousness, diarrhea, cramps, sweating, chest discomfort, pulmonary edema, uncontrollable muscle twitching, papilledema, convulsions, cyanosis, coma, and loss of reflexes and sphincter control, responsive to large doses of ***atropine***.

LD_{LO}, **human**: 50 mg/kg.

Moderate transient eye irritation in **rabbits**. **Dairy calves** 1–2 weeks became toxic within 2 h when sprayed with 0.1% solution or within 40 min when given 0.5 mg/kg orally.

LD_{50}, **rats**: 25 mg/kg; **mouse**: 60 mg/kg; **guinea pig**: 63 mg/kg; acute dermal, **rabbits**: > 3 g/kg.

DIMETRIDAZOLE

= *DMZ* = *Emtryl* = *RP-8595* = *Unizole*

Antiprotozoal.

Untoward effects: None reported in **man**, but inhalation may be irritating and cause slight coughing.

Despite the fact that it has been very effective against *Giardia* in **puppies** and **calves**, *Trichomonas* in **bulls**, *Treponema* in **swine** and sewage, *Campylobacter* and *Histomonas* in **turkeys** without ill-effects, the U.S. FDA has said all imidazoles are potentially carcinogenic, and ordered it off the U.S. market, while giving other imidazoles (that have never had such detailed experimental and clinical studies done) carte blanche. In the industry, this was considered to be selective enforcement against a company less in their favor. The particular director involved, knowing this, allowed its use when he was with the "Directorate" in Canada. It is still considered safe and effective there and in other countries, where it is withdrawn 5 days before slaughtering, to avoid undesirable residues. Do NOT feed to **birds** laying eggs for **human** consumption. **Bulls** may have transient diarrhea 2–3 days after the start of treatment.

DIMINAZENE ACETURATE

= *Azidin* = *Berenil* = *Ganasag*

Antiprotozoal.

In treating babesiosis and trypanosomiasis.

Untoward effects: Painful induration from IM injections. Cardiotoxic and implicated as a cause of acute idiopathic polyneuritis (Landry–Guillain–Barré syndrome) in one **patient**. Other adverse effects similar to ***pentamidine*** (q.v.). Brain damage in **dogs**, toxicity in **camels** and **horses**, but few adverse effects in **cattle**, **rats**, and **mice**.

m-DINITROBENZENE

Untoward effects: Methemoglobinemia, with smoky urine a few hours after handling. Acute poisoning leads to headache, vertigo, vomiting, exhaustion, leg numbness, staggering, somnolence, and loss of consciousness.

Secondary anemia in subacute or chronic poisoning. TLV, 1 ppm.

Dermal TD_{LO}, **human**: 4 mg/kg/2 days.

TD_{LO}, **cat**: 27 mg/kg; LD_{50}, **birds**: 42 mg/kg.

p-DINITROBENZENE

An aromatic nitro compound.

Untoward effects: Five **steam-press operators** in an Ohio ***rubber*** plant became ill. Symptoms were yellow discoloration of hands, blue discoloration of lips and nail beds, headache, nausea, chest pain, dizziness, confusion, and difficulty in concentrating. **One** had a seizure. ***Methemoglobin*** (***Met Hb***) levels ranged from 3.8 to 41.2% (normal is ≤ 1%). The **workers** were bonding metal studs into ***rubber*** strips for auto bumpers. ***DNCB*** appeared to be a contamination in an old lot of adhesive and is readily absorbed through the skin. When ***Met Hb*** is > 15%, cyanosis and headache appear first; followed by dizziness and fatigue (> 40% ***Met Hb***), then ataxia, shortness of breath, tachycardia, nausea, vomiting, and drowsiness; when levels reach > 70%, stupor, coma, and even death. Jaundice, enlargement with tenderness of the liver, anemia, diarrhea with colorless stools, and dark brown urine also reported. TWA, 1 $mg/m^3/8$ h.

An inflammatory reaction, histologically analogous to allergic contact dermatitis, was produced in colons of skinsensitized **guinea pigs**, suggesting that tissues other than skin have the ability to react to contact allergens.

LD_{LO}, **cats**: 29.4 mg/kg.

2,4-DINITROCHLOROBENZENE

= *DNCB*

Reagent. **Used to treat recalcitrant verrucae in man and in the manufacture of azo dyes, fungicides, rubber chemicals, and explosives. Hundreds of tons produced annually.**

Untoward effects: Potent skin irritant and sensitizing agent leading to a contact-type hypersensitivity. **Women** using nail-hardeners complained about browning of **their** nails and tissue damage under **their** nails, resulting in its removal from the market for that purpose. Chemical laboratory **workers** in contact with it develop urticarial skin eruptions. May contain a very toxic contaminant, ***nitrochlorobenzene***.

Irritation, **human** skin: 30 µg; **rabbit**: 100 µg. LD_{50}, **rat**: 1 g/kg; skin LD_{50}, **rabbit**: 130 mg/kg.

DINITRO-σ-CRESOL

= *Antinonnin* = *Decrysil* = *Dekrysil* = *Detal* = *Dinitrol* = *Ditrosol* = *DN* = *DNC* = *DNOC* = *Effusan* = *Elgetol* = *ENT 154* = *K III* = *K IV* = *Lipan* = *Nitrador* = *Prokarbol* = *Selinon* = *Sinox* = *Trifoside*

Herbicide, insecticide.

Untoward effects: Euphoria, headache, fever, lassitude, excessive thirst, hyperhidrosis, tachycardia, hyperpnea, shortness of breath, chest pain, coughing, anxiety, contact dermatitis; yellow staining of skin, hair, and nails; jaundice, weight loss, peripheral neuritis, liver and kidney damage, conjunctivitis, green-yellow urine, insomnia and cataracts, depending on the degree and duration of exposure. A cumulative poison in **man**. Exposure to 5 ppm for several weeks may be fatal. Said to be 3–4 times more toxic than ***dinitrophenol***. OSHA TWA, 0.2 mg/m^3. This coal dye was once sold for ***saffron*** in Berlin with fatal results. Blood levels > 20–40 µg/ml are toxic and 7.5 mg% is fatal. TLV, 0.2 ppm.

TC_{LO}, **human**: 1 mg/m^3; LD_{LO}, **human**: 5 mg/kg.

Intoxication in **birds** led to ataxia, wings crossed high over the back, tail tremors, tachypnea, dyspnea, hyporeactivity, ptosis, and opisthonus within 15 min and deaths within 1–21 h. Lens opacities in **ducklings**. Has caused abortions, stillbirths, and decreased growth rate in **swine**. Poisoning in **foals**, **dogs**, and **calves** led to salivation, vomiting, thirst, fever, listlessness, sweating, dyspnea, weakness, tremors, convulsions, coma, terminal hyperpyrexia, and death. **Poultry** and **bees** have also been poisoned by it.

LD_{50}, **rats**: 10 mg/kg; **ducks** and **pheasants**: ~30 mg/kg; **mouse**: 47 mg/kg; inhalation LC_{LO}, **cat**: 40 mg/m^3.

2,4-DINITRO-3-METHYL-6-*tert*-BUTYLANISOLE

= *Musk Ambrette*

Use: A fragrance in cosmetics, particularly, **men's** colognes and after-shaves.

Untoward effects: Has caused phototoxicity and photoallergic reactions. One dermatology clinic reported on 34 **patients** with contact or photocontact allergy to it. Neurotoxicity has also occurred.

2,6-DINITRONAPHTHOL

= *Martius Yellow*

Untoward effects: Toxicity noted in 1885, when it was used as a yellow food coloring. Increases body temperature. (See ***dinitrophenol***).

IP, LD_{50}, **rats**: 47.5 mg/kg; **mice**: 55 mg/kg.

2,4-DINITROPHENOL

= *Aldifen* = *a-Dinitrophenol* = *DNP*

***Use*:** Wood preservative, insecticide, fungicide; in manufacturing dyes, chemicals, and munitions; and as a pH indicator.

Untoward effects*:** Its dangers as a reducing agent were colorfully emphasized in an article entitled ***DNP** is **TNT. In 1935, several cases were reported of **women** developing cataracts from such use. Metabolic rate increased 40% by 3–5 mg/kg/day. Cumulative. Absorbed from skin contact or inhalation of vapors. Corrosive to skin and mucous membranes. Hyperhidrosis, hyperpyrexia (up to 110°F), increased thirst, nausea, vomiting, dyspnea, ventricular arrhythmias, acute hepatitis, cyanosis, collapse, anoxia, and death occur in acute poisoning with almost immediate rigor mortis. Chronic poisoning leads to dermatitis, exfoliative dermatitis; yellow staining of skin, hair, sclerae, and urine; typical ***phenol*** smell, anorexia, decreased weight, jaundice, polyneuropathy, cataracts (often with latent period of months or years), acidosis, myelotoxicity, neutropenia, pancytopenia, thrombocytopenia, agranulocytosis, purpuric anemia, dehydration, occasional pulmonary edema, weakness, nausea, and vomiting. A case of methemoglobinemia in **man** has been reported. Most **human** exposure reports were from the French munitions industry in World War I. Fatal oral dose is 1–3 g, and 3 g has been fatal, even in divided doses over 5 days.

LD_{LO}, **human**: 4.3 mg/kg.

Cataractogenic in **chickens**, **ducklings**, **male Macaca phillipiniensis monkeys** and young **rabbits**. Hyperpyrexia (**cat**, **dog**, **guinea pig**, **mouse**, **rabbit**, and **rat**), nervousness, convulsions, coma, and death in **animals**. They usually show a yellow area around the mouth and other parts of the body and acute swelling of the liver and kidneys has been reported. Partially excreted by the kidneys. Has caused toxicity in young **pigs** chewing on treated wood. Toxicity increases with increased environmental temperatures.

LD_{50}, **rat** and **rabbit**: 30 mg/kg; **mouse**: 45 mg/kg; **guinea pig**: 81 mg/kg.

DINITROTOLUENE
= Dinitrotoluol = DNT = Methyldinitrobenzene

Use*:** In manufacture of ***polyurethanes and explosives.

***Untoward effects*:** Exposed manufacturing plant **workers** had a 65% decrease in sperm counts. Anoxia, cyanosis, methemoglobinemia, headaches, irritability, dizziness, weakness, nausea, vomiting, dyspnea, drowsiness, anemia (after repeated or prolonged exposure), jaundice, unconsciousness, and even death reported. Exposure limit of 1.5 mg/m^3/8 h for skin.

LD_{LO}, **human**: 50 mg/kg.

Potent hepatocarcinogen in **rats**. Decreased sperm counts and testicular atrophy in **rats**, **mice**, and **dogs**. Its 2,6-isomer in the technical material is responsible for its carcinogenic effect noted in **rat** and **mouse** studies.

LD_{50}, **rats**: 1.1 g/kg.

DINOBUTON
= Acrex = Dessin = Systasol

Miticide.

Untoward effects*:** Orchard **sprayers** may develop a yellow nail pigmentation. Systemic toxicity may resemble that of ***dinitrophenol (q.v.) or ***dinoseb*** (q.v.).

LD_{50}, **rat**: 59 mg/kg; **mouse:** 170 mg/kg; **chicken**: 150 mg/kg; acute dermal, **rabbit**: 2.5 g/kg.

DINOCAP
= Arathane = CR-1639 = Crotothane = DNOCP = ENT-24727 = Isocothane = Karathane = Mildex

Acaricide, fungicide.

***Untoward effects*:** A fatal poisoning by it reported. Skin irritant.

LD_{LO}, **human**: 500 mg/kg.

Causes severe irritation and necrosis of **rabbit** skin. Teratogenic in the **rabbit**. In **dogs**, dietary concentration of 250 ppm/1 year was lethal, with necrosis of the pancreas, liver, and zona glomerulosa of the adrenals.

LD_{50}, **rat**: 1 g/kg; **dogs**: ~100 mg/kg; **rabbits**: 2 g/kg; acute dermal, **rabbit**: > 9 g/kg; in feed, LC_{50}, **quail**: 790 ppm.

DINOPROST
= Dinolytic = Enzaprost F = Glandin = ICI 79,939 = Lutalyse = $PGF_{2\alpha}$ = Prostaglandin $F_{2\alpha}$ = Prostarmon F = U-14583

Oxytoxic, abortifacient.

Tromethamine salt = Prostamate = Prostin $F_{2\alpha}$

***Untoward effects*:** Nausea, vomiting, and diarrhea are common. Hypogastric pain, headache, cough, shivering, flushing, phlebitis, tachycardia, cardiac arrhythmias, raised temperature, dizziness, increased white blood cells, hiccups, dysuria, bronchoconstriction, itching, and maculopapular rashes have also occurred; rarely, convulsions. Failure to complete the abortion procedure due to vomiting resulted in multiple congenital anomalies. NOT to be handled by pregnant **women**, **asthmatics**, or **individuals** with respiratory problems, yet **cows** produce about 2 μg daily in their milk. Readily absorbed through the skin, and can cause abortions or bronchospasms in

humans. Accidental spills on **men** or **women** should immediately be washed off with soap and water. Erythema and pain occasionally at injection site.

Abortifacient in **rats**, **hamsters**, **guinea pigs**, **monkeys**, and **rabbits**. Low levels cause luteal regression in the **cow**, **ewe**, **goat**, **mare**, **sow**, **hamster**, and **guinea pig**; higher levels are needed in **primates**, **bitches**, and **rats**. Sweating, colic, and ataxia lasting a few minutes to an hour have been noted in **mares** on a rare occasion. IV usage is not without risk. **Bitches** occasionally become restless, hypersalivate, vomit, defecate, and pant. May be toxic to **dogs**, when used as an abortifaciient. Hypothermia may be noted in treated **dogs**. Parturition induction in **swine** may be accompanied by abdominal discomfort for an hour or two, as well as occasionally defecation, shivering, and increased respiratory rate. Severe swelling at injection site and deaths reported in **cattle**.

LD_{50}, **rat**: 1.2 g/kg; **mouse**: 1.3 g/kg; IM, LD_{50}, **rabbits**: 2.5–5.0 mg/kg; **rat**: 112 mg/kg; **mouse**: 152 mg/kg; IV, LD_{50}, **rabbits**: 2.5–5.0 mg/kg.

DINOPROSTONE

= *Cervidil* = *Cerviprost* = *Minprostin* E_2 = *PGE*$_2$ = *Prepidil* = *Propess* = *Prostaglandin* E_2 = *Prostarmon-E* = *Prostin* E_2 = *U-12062*

Oxytocic and abortifacient.

Untoward effects: Erythema and/or phlebitis at injection site, gastrointestinal disturbances including vomiting (65%) and diarrhea (50%) are the most common. Headache, pyrexia, chills, decreased diastolic pressure, tachycardia, arrhythmias, blurred vision, uterine rupture, hypotension, and collapse reported. May lower esophageal sphincter tone. Retinal hemorrhages are more frequent in **neonates** after labor induced by oral dosing (40%) than by IV ***oxytocin*** (28%). **Fetal** heart rate abnormalities after prolonged uterine contractions are < 1%. Neonatal aspiration syndrome by an **infant** of undissolved remnants of an intravaginal tablet stresses the preference of a viscous gel which dissappears more quickly from the vagina. Causes sickling of red blood cells. A 38-year-old **female** developed myocardial infarction 30 min after a 20 mg vaginal suppository for termination of **her** pregnancy. Nausea, vomiting, fever, diarrhea and abdominal pain, headache, shivering, and chills in < 1% of **patients**, except when used as an abortifacient, when their incidence is much higher.

Produces coronary vasodilation in the **rat**. IV in **cows** for milk let-down. Central nervous system depressant effect, with rapid onset and short duration in the **rhesus monkey**, at 2–8 mg/kg, SC. Hypotensive in **hens** (0.1 µg, IV) and **calves**.

DINOSEB

= *Aretit* = *Basanite* = *2-sec-Butyl-4,6-dinitrophenol* = *Caldon* = *Chemox DN* = *Chemox PE* = *Chemox Selective* = *4,6-Dinitro-2-sec-butylphenol* = *DN-289* = *DNBP* = *DNOSBP* = *Dow General* = *Dow Selective* = *Elgetol 318* = *ENT-1122* = *Gebutox* = *HOE-2904* = *Ivosit* = *Kiloseb* = *Notropone C* = *Premerge* = *Sinox W* = *Subitex* = *WSX-8365*

***Herbicide, insecticide, miticide.* Up to 11 million lbs sprayed annually in the 1980s in the U.S.**

Untoward effects: Absorption occurs through gastrointestinal, skin, or lung exposures. Environmental Protection Agency banned it after considering it a very serious risk of birth defects in pregnant **women**. Death in a 50-year-old **male** after delibrate repeated ingestion of unknown amount. Autopsy revealed yellow discoloration of hands and conjunctiva. Cause of death was not stated.

Carcinogenic, embryotoxic, and teratogenic in **mice**. Teratogenic in **rabbits**. Causes sterility in male **rats** and **mice**. Mistaken for an ***iodine*** solution, 200 ml was applied to the face and neck of each of four **dairy heifers**, which was fatal in less than 1 day. Intraruminal in **sheep** leading to methemoglobinemia lasting 2 or more days, serious hemolysis, dyspnea, and hyperthermia. Rapid deaths of two **dogs** whose coats were yellow-stained. Their symptoms were tachypnea, hyperexcitability, and ataxia. Has caused yellow-stained hair coats on **rabbits**. Most exposed **animals** show tachypnea, tachycardia, raised temperature, central nervous system disturbances, convulsions, coma, and death. Poisoned **mallard ducks** and **pheasants** show slow reactions, tachypnea, ataxia, goose-stepping, falling, and sitting within 20–60 min after exposure. Toxic to **fish**.

LD_{50}, **rats**: 25 mg/kg; **mice** and **guinea pigs**: 20 mg/kg; **mallard ducks**: 27 mg/kg; **pheasants**: 26.4 mg/kg; **chickens**: 40 mg/kg; acute dermal, **rabbit:** 80 mg/kg.

DIOSCOREA

= *Yams*

Edible roots and a source of sapogenins, useful in *steroid* production, especially for *oral contraceptives*. Mexican and Central American tubers contain about 4–5% *diosgenin*. > 600 species. They are an important source of food, but require extensive washing and/or boiling, to remove their poisonous saponins. Also see *Barbasco*.

Untoward effects: Contains large amounts of ***prussic acid*** (***hydrocyanic acid***) and ***discorine***, an alkaloid with ***picrotoxin*** (q.v.)-like action. **Africans** eat ~1 g cyanates/day.

D. dumetorium in Nigeria = ***Bitter Yam*** = ***K'osain rogo*** (Hausa) = ***Esuru gudugudu*** (Yoruba) = ***Ji ona*** and ***Ighu*** (Igbo), used as a food after special preparation, has caused convulsions, paralysis, and death.

D. quartiniana has been suspected as a cause of poisoning in northern Zimbabwe.

Extracted ***discorans***, IP cause hypoglycemia in **mice**. Weanling **rats** on a yam diet for 100 days had a high incidence (10/13) of liver necrosis, cholangiofibrosis, and nodular hypoplasia.

DIOSPYRUS sp.

Many species of these trees are used for their wood (Indian ebony, West African ebony, etc.). Their fruit is eaten and extracts are used for tribal medicine cures (human and veterinary).

Untoward effects: ***D. mollis***, used for centuries in Thailand for treating worm infestations, causes blindness in **people**. Other species used in Nigeria are suspect.

Oil of ***D. tomentosa*** seeds decreased blood pressure and increased respirations in **cats** and was a central nervous system suppressant in **mice**.

DIOXANE

= *1,4-Diethylene Dioxide*

Solvent. **Flammable and can become explosive. Sympatholytic.**

Untoward effects: Vapors cause irritation to skin and mucous membranes of eyes and upper respiratory tract. Inhalation leads to nausea, vomiting, drowsiness, narcosis, headache, and lacrimation. Ingestion leads to severe gastrointestinal upsets, and liver and kidney dysfunction. Large doses led to incoordination, coma, and death (from hemorrhagic nephritis, uremia, and central necrosis of the liver) reported in five factory **workers** who died from exposure. Exposure limit of 1 ppm (3.6 mg/m^3)/30 min and TWA 25 ppm (90 mg/m^3) to skin suggested. Skin exposure may cause acne. Once found in many food products, cosmetics, and surfactants. Use of the contraceptive ***Today Sponge***, made of ***polyurethane*** and dioxane, was associated with a high incidence of vaginal irritation and toxic shock syndrome.

LD_{LO}, **human**: 500 mg/kg; inhalation TC_{LO}, **human**: 470 ppm.

Orally in **rats** leads to squamous cell carcinomas and hepatocellular carcinomas; hepatomas and gallbladder carcinomas in male **guinea pigs**; hepatocellular carcinomas in **mice** and as a promoter, increasing their susceptibility to papillomas, squamous cell carcinomas and sarcomas. IP to male **mice** causes lung tumors. **Dogs** develop severe gastrointestinal upsets and weakness.

LD_{50}, **rats**: 4.2 g/kg; **mouse**: 5.7 g/kg; **cat** and **rabbit**: 2 g/kg; **guinea pig**: 3.2 g/kg.

DIOXATHION

= *Co-Nav* = *Delnatex* = *Delnav* = *Deltox* = *ENT-22879* = *Hercules 528* = *Kavadel* = *Navadel* = *Polythion*

Organophosphorus insecticide, acaricide. **Cholinesterase inhibitor.**

Untoward effects: In an unusual incident, **cattle** sprayed at 0.15% developed a ***selenium/alpha tocopherol*** deficiency, as did the **sprayer operator**.

LD_{LO}, **human**: 50 mg/kg.

Adult **cattle** and 1–2 week **dairy calves** were killed by topical emulsions of 0.8% and 0.5%, respectively. Even sprays as low as 0.1% have been toxic to young **calves**. Excessive salivation, dyspnea, stiff gait, weakness, prostration, and death occur. Ataxia, weakness, slowness, immobility, wing-beat convulsions, and tetany within 30 min in **ducks** and **pheasants**. Do NOT contaminate **animal** or **human** feed or water supplies. Do NOT repeat applications for at least 2 weeks. Keep it away from **children**. Sprays 0.184% or greater are apt to be toxic.

Do NOT treat **animals** under 3 months of age. No withdrawal time in **sheep**, **swine**, or **cattle**. Very toxic to **fish**, **wildlife**, and **birds**.

LD_{50}, male **rats**: 43 mg/kg; female **rats**: 23 mg/kg; **mouse**: 50 mg/kg; **guinea pig**: 40 mg/kg; **chicks** and **chickens**: 170 mg/kg; **mallard ducks**: 277 mg/kg; **pheasants**: 240 mg/kg; **dogs**: 10 mg/kg; acute dermal, male **rats**: 235 mg/kg; female **rats**: 63 mg/kg. LC_{50} in feed, **ducks**, **pheasants**, **quail**: 3600–7000 ppm.

DIOXIN

= *TCDBD* = *TCDD* = *2,3,7,8-Tetrachlorodibenzo-p-dioxin*

One of the most toxic chemicals known to man and an inevitable contaminant in the manufacture of *2,4,5-T*. Levels were originally up to 30 ppm, and now < 0.5 ppm. Persists unchanged for years in the environment. Refuse incinerators, fossil-fueled powerhouses, gasoline-powered vehicles, fireplaces, charcoal grills, and cigarettes discharge it into the atmosphere. Paper products contain up to 13 ppt produced during the *chlorine* bleaching of wood pulp.

Untoward effects: A complex issue, compounded by emotions and media hype. Causes skin eruptions of chloracne, soft tissue sarcoma, and porphyria cutanea tarda associated with liver dysfunction, especially after ***Agent Orange*** (q.v.) exposure. The FDA stated that the dioxin levels (0.02–0.62 ppt) in milk from contaminated paper cartons might pose a lifetime cancer risk of < 1/million. Chemical **workers** exposed to it for over a year had a 46% higher mortality rate from cancer after a 20 year latency period. The Environmental Protection Agency currently considers it as a "probable" **human** carcinogen. A 1976 chemical plant explosion at Seveso, Italy released large amounts of dioxin. Over 700 **people** were evacuated. Thousands were probably exposed. The cloud of noxious vapors that hung over **them** was a metaphor, also describing the persistent uncertainties in **their** future. 6 years after the incident, 152 **people** still had chloracne. Hepatomegaly and slight increase in SGPT and γ-glutamyl transpeptidase in potentially exposed Seveso **children**. **People** living near the plant have increased rates of leukemia, myelomas, liver cancers, and lymphomas. Some of the **victims** had a mild form of porphyria cutanea tarda 2–4 years later, decreased velocity of nerve conduction, numbness and pain in **their** extremities, muscle weakness, and disturbances of vision. In the U.S., homes have been abandoned in Times Beach, Missouri; New Jersey, and Niagara Falls, New York. Dioxin exposures from accidents or explosions have occurred in a number of other countries. Air exposure to 2.8–2.9 ppm occurred after a major transformer fire. Concentrated in the body's fatty tissues and, eventually, it was identified as the "**chick**-edema factor" that caused deaths and tremendous numbness of **chicks** from feeding fat wastes from the tanning industry. Dangerous to **people** at 1 ppb. Some exposed **individuals** have complained of fatigue, decreased weight, myalgias, nausea, insomnia, irritability, decreased libido, increased serum lipids, increased prothrombin time, and headache. Half-life in **human** serum and adipose tissue is 7.1 years, and can be over 10 years in soil. Dioxins have been found in **human** breast milk. An English study found immunosuppression in **human** subjects 10 years after exposure. Limited evidence indicated it might have caused four deaths from soft-tissue sarcoma among 5000 chemical plant **workers**. Over a thousand exposed veteran **workers** in a Hamburg, Germany chemical manufacturing plant appeared to have > 3 times the risk of dying from cancer and 2½ times the risk of dying from ischemic heart disease. This has not been corroborated in other studies. ***Agents Orange, Pink, Green, Purple,*** and ***Orange II***, as well as ***2,4,5-T*** and ***2,4-D***, have all been contaminated with dioxin.

Higher than normal rates of birth defects and stillbirths in **rats** and **mice**, but not in **humans**, according to the Vietnamese experience. **Rats** and **mice** exposed prenatally may have cleft palates, gastrointestinal hemorrhage, kidney abnormalities and malfunction, immunosuppression, and atrophy of the thymus in many species. Atrophy of the thymus in **guinea pigs**, a species most susceptible to adverse effects of dioxin. Their liver Kupffer cells may be pigmented yellow-brown. Causes hepatocellular carcinomas and squamous cell carcinomas in **rats**, but only female **rats** develop porphyria and the liver tumors. When fed to pregnant **rats**, it had a permanent feminizing effect on the livers of their male pups. Tumors of the skin when painted on the backs of **mice**. There are 40% fewer pregnancies in **mice** in contaminated areas, compared to **mice** in uncontaminated areas. In the Missouri case, some of the contaminated waste oil was used to spray four stable areas and rural roads and appeared to cause the death of many **birds**, **cats**, **dogs**, and at least 60 **horses**. Many **mares** had serious breeding problems, spontaneous abortions, stillbirths, and deformed foals. Several **people** became ill, including a 6-year-old **female** who played on a treated horse arena floor. **She** was hospitalized with nephritis and a hemorrhaging bladder. Exposed **monkeys** often failed to conceive, had abortions, stillbirths, and smaller than normal offspring. Poisoned **rhesus monkeys** had reduced weight, blepharitis, loss of fingernails and eyelashes, facial alopecia with acneform eruptions, mild anemia, neutrophilia, and lymphopenia; male **monkeys** had reduced spermatogenesis. **Rhesus monkeys** receiving 2–3 μg/kg over a 9 month period died. Moderate to severe ileitis and peritonitis in **hamsters**. **Fish** and **crabs** concentrate large amounts in their tissues. The FDA recommends they should not be eaten with levels > 50 ppt. Toxicity in **fish** leads to fin necrosis, erratic swimming, hemorrhaging from anus and lower jaw, growth retardation, edema, and death. After exposure of **bird** eggs to it, it caused asymmetry of brain, twisted beaks, anophthalmia, and clubbed feet.

LD_{50}, male **rats**: 22 μg/kg; female **rats**: 45 μg/kg; **mice**: 0.3 μg/kg; male **hamsters**: 1.2 mg/kg; **guinea pigs**: 0.5 μg/kg; **chickens**: 25–50 μg/kg; **dogs**: 100 μg/kg; **rabbits**: 115 μg/kg; female **rhesus monkeys**: < 70 mg/kg; acute dermal, **rabbit**: 275 μg/kg.

DIOXYBENZONE

= *Benzophenone 8* = *Spectra-Sorb UV24*

UV light absorber, sunscreen.

Untoward effects: Contact dermatitis reported.

DIPENTAMETHYLENE THIURAM DISULFIDE

Rubber accelerator.

Untoward effects: A **male patient** became sensitized to it after using a ***rubber*** condom. A study of **patients** with contact dermatitis from manufactured ***rubber*** articles revealed 17/72 with positive patch tests to it.

DIPHEMANIL METHYLSULFATE

= *Demotil* = *Diphenatil* = *Nivelona* = *Prantal* = *Variton*

Anticholinergic.

Untoward effects: Erythema and skin irritation from topical use. Mild mydriasis noted. Slight tachycardia, especially after IV use. Constipation, decreased peristalsis, esophageal spasms, drowsiness, headache, and blurred vision. May cause urinary retention after prolonged use. Severe adverse cardiac effects after 1–18 days of treatment in premature **babies**.

LD_{50}, **rat**: 1.1 g/kg; **mouse**: 64 mg/kg; **guinea pig**: 404 mg/kg.

DIPHENADIONE

= *Diaphenadione* = *Didandin* = *Dipaxin* = *Diphacin* = *Diphacinone* = *2-Diphenylacetyl-1,3-indandione* = *Oragulant* = *Ramik* = *Solvan* = *U 1363*

Oral anticoagulant, rodenticide, baticide.

Untoward effects: Bleeding diathesis, poor coagulation, epistaxis, hematuria, melena, ecchymotic hemorrhages of trunk, and 6.4 g/100 ml hemoglobin in 5-year-old **male** after unauthorized ingestion of several tablets. Renal and hepatic damage, rash, fever, pruritus, and eosinophilia have also occurred. Decrease dosage by about 40% when given with ***benziodarone***.

LD_{LO}, **human**: 5 mg/kg.

Secondary toxicity occurs. Tissue residues in **nutria** killed with the drug proved toxic when their meat was fed to **mink**. Similarly, meat from killed **coyotes** (containing ~0.5 ppm) killed **rats** and **eagles**, and they evidenced toxicity after eating **sheep** meat containing 2.7 ppm. **Owls** died of hemorrhaging after feeding on **rats** killed by it. Has caused sudden or slow death from internal hemorrhaging in **swine** and **dogs**, usually from eating **rat** baits. As part of a **vampire bat** control program, adult **cattle** tolerated 1 mg/kg intraruminal doses, but it killed 9/14 young **calves**.

LD_{50}, **rat**: 1.5 mg/kg; **mouse**: 141 mg/kg; **cat**: 15 mg/kg; **dog**: 3 mg/kg; **pig**: 150 mg/kg.

DIPHENAMID

= *Dymid* = *Enide* = *L-34314*

Untoward effects: Large amounts have caused muscular incoordination in **steers** and weight loss in **chickens**.

LD_{50}, **rat**: 700 mg/kg; **mouse**: 600 mg/kg; **dog** and **monkey**: 1 g/kg; **rabbit**: 1.5 g/kg; **cattle**: 440 mg/kg; **sheep**: 22 mg/kg; acute dermal **rabbit**: >2 g/kg.

DIPHENAZINE

= *Quietidin*

Tranquilizer, neuroplegic.

Untoward effects: Somnolence. Cortical cataract formation reported after long-term use.

SC, LD_{50}, **mouse**: 193 mg/kg; IV, 30.5 mg/kg.

DIPHENCYPRONE

= *DCP* = *DPC*

Untoward effects: Persistent vitiligo in a 37-year-old **male** after topical application. A 55-year-old **wife** developed patchy vitiligo while her **husband** was treated topically with it for alopecia universalis. The lesions resolved when her **husband's** treatment ceased.

DIPHENHYDRAMINE

= *A 254* = *Alledryl* = *Allergina* = *Amidryl* = *Bagodryl* = *Bax* = *Benadryl* = *Bendylaten* = *Benocten* = *Benodine* = *Benzantin* = *Benzhydramine* = *Dabylen* = *Dibondrin* = *Dihydral* = *Diphantine* = *Dimedrol* = *Dolestan* = *Fenylhist* = *Halbmond* = *Histacyl* = *Noctomin* = *S 8* = *Sedopretten* = *Sekundal-D* = *Syntedril* = *Valdrene* = *Wehydryl*

Antihistamine, local anesthetic, parasympatholytic.

Untoward effects: Sedation, confusion, delusions, nervousness, restlessness, tremors, nausea, vomiting, diarrhea, constipation, tightness of chest, hyperhidrosis, wheezing; dryness of mouth, nose, and throat; tingling or weakness in hands, thickened bronchial secretions, vertigo, nasal stuffiness, tinnitus, headache, palpitations, insomnia, hypotension, anaphylactic shock, tachycardia, extrasystoles, urinary retention, spontaneous facial flushing, penile erectile problems, decreased libido, early menses, infrequent gastrointestinal symptoms, "restless legs" syndrome, thrombocytopenia, agranulocytosis, seizures, hallucinations, blurred vision, double vision, anisocoria, mydriasis, reduced tolerance to contact lenses, and nystagmus. Has potent or significant anticholinergic effects. Paradoxic excitation in a 46-year-old **female** after 50 mg IV. Severe anaphylactoid reaction in a 32-year-old **female** after receiving 25 mg IV for generalized urticaria. **Patient** had it orally several years previously after a ***penicillin*** injection. Drug abuse by self-injections with ***butorphanol*** caused confusion and stupor. In many poisoning cases, 6–40 times the therapeutic dose was ingested. Bizarre behavior reported in several **children** treated with it orally or topically on varicella-zoster lesions leading to visual and auditory hallucinations. A 6-month-old **male** died within 29 h after hospital admission from its topical application with ***calamine*** in treating chickenpox lesions. Toxic encephalopathy (delirium and auditory, tactile, and visual hallucinations) after repeated whole body application of a 1% lotion with ***calamine*** on a 9-year-old **male** during a 48 h period. ~20 **children** under 5 years have been hospitalized annually in the U.S. from ingesting it. Phototoxic and photoallergic reactions, photosensitivity reactions, fixed-drug eruptions, eczematous reactions, urticaria, papules, edema, and patchy erythema in light-exposed areas. Pink diphenhydramine cream contains ***red*** and ***yellow ferric oxides*** which act as effective sun-blocking agents. Allergic contact dermatitis in the white cream has been due to ***propylene glycol*** (q.v.). Has caused an acute episode of cutaneous allergic vasculitis with papulopurpuric eruption and dystonia in a 28-year-old **female** and a 66-year-old **female** after a 50 mg dose (oral and IV, respectively), and in a **child** after 2–25 mg doses. Trismus in a 19-year-old **female** after an IM injection. Hemolytic anemia, especially in glucose-6-phosphate dehydrogenase-deficient **patients**. Aplastic anemia, neutropenia, thrombocytopenia, purpura, and leukocytosis also reported. Danger of bleeding, due to impairment of platelet function (even in the **fetus**) suggests caution is needed in combination with ***anticoagulants***. Large doses may cause delirium and central nervous system depression, mydriasis, tachycardia, hypertension, followed by coma and death. Delirium may follow sudden withdrawal. Potential toxic dose for a 2-year-old **child** is 100–180 mg. Severe central nervous system symptoms, including depression, may occur. In breast milk. In pregnant **women** near term, it may cause withdrawal symptoms. Finnish reports cite an association of its intake during the first trimester with cleft palate. Therapeutic serum concentrations are 0.025–0.1 mg/l; toxic at 1–2 mg/l; and comatose or dead at >75 mg/l. A 28-year-old **male** was hyperreactive to noise, developed hyperpyrexia and ventricular tachycardia, and died with a serum concentration of 5 mg/l. Often fatal in **adults** after 25 mg/kg; less in **children**.

LD_{LO}, **human**: 5 mg/kg.

Drug interactions: Potentiates cardiovascular effect of ***noradrenaline*** with possible hypertensive episodes when combined with ***isocarboxide, nialamide, phenelzine,*** and ***tranylcypromine***. Its anticholinergic effects are additive to other drugs, such as ***imipramine*** and ***trihexyphenidyl***, causing ***atropine***-like complications, such as dental caries and loss of teeth from prolonged drug-induced xerostoma. It, itself, in overdose, can produce the anticholinergic syndrome. Nasal bleeding, swollen tongue, dry mouth, and disorientation reported in ten **patients** receiving ***Mandrax*** (250 mg ***methaqualone*** and 25 mg ***diphenhydramine***), when taken with ***thioridazine*** or a ***tricyclic antidepressant***. Apnea and severe muscle relaxation with ***Mandrax–diazepam*** combination. Three fatal and five severe cases of adynamic ileus in the same report. Decreases absorption rate of ***p-aminosalicylic acid***. Enhances ***alcohol's*** impairment of performance. Large doses decrease effectiveness of ***heparin***.

Rat studies indicate it increases the metabolism of many drugs metabolized by hepatic microsomal enzymes. Causes a fatal hyperpyrexia in **rabbits** pretreated with ***monoamine oxidase inhibitor*** from ***5-hydroxytryptamine*** potentiation. Drowsiness and depression may occur on high doses or in sensitive individuals. Potentiates ***sedatives*** and ***hypnotics***.

LD_{50}, **rat**: 390 mg/kg; **mouse**: 114 mg/kg; **guinea pigs**: 284 mg/kg.

DIPHENIDOL

= *Ansmin* = *Cefadol* = *Celmidol* = *Cephadrol* = *Defenidol* = *Difenidolin* = *Maniol* = *Mecalmin* = *Pineroro* = *Satanolon* = *SKF 478* = *SKF 478-A* = *Tenesdol* = *Vontril* = *Vontrol* = *Wansar*

***Antiemetic*. In treating vertigo. Central anticholinergic action.**

Untoward effects: Auditory and visual hallucinations, disorientation, confusion, skin rash, slight dizziness, dry mouth, nausea, heartburn, headache, blurred vision, palpitations, drowsiness, and acute toxic encephalopathy with abnormal electroencephalogram and delirium. Transient hypotension in some **patients**. Contraindicated in renal impairment as > 90% eliminated through the kidneys.

LD_{50}, **rat**: 815 mg/kg; **mouse**: 450 mg/kg.

DIPHENOXYLATE

= *R 1132*

Antidiarrheal. Structurally related to *meperidine*. With *atropine sulfate* = *Diarsed* = *Lomotil* = *Lowquel* = *Reasec*

Untoward effects: Faintness or dizziness (6%), headache (5%) in double-blind study with 244 **patients**. An increased risk of thyroid cancer and limited addiction potential reported. Appears to have adverse effects in **patients** with pseudomembranous colitis from antibiotic therapy.

With ***atropine***: Addiction potential, ***morphine***-like euphoria, drowsiness, dizziness, depression, ileus, abdominal distension, pancreatitis, nausea, lethargy, respiratory depression, flushing, pruritus, gum swelling, angioneurotic edema, anaphylaxis, coma, hypotonic reflexes, miosis, nystagmus, and tachycardia. Many of these symptoms are actually those of diphenoxylate, especially in high dosage. Respiratory depression with apnea in **children** on overdoses. A 2-year-old **male** died after accidental ingestion of 12 tablets, and a 2-year-old **female** survived respiratory depression and cardiac arrest after ingesting 6 tablets.

Treatment was with ***nalorphine***. A 2-year-old **male** ingested 30–40 tablets and a 27-month-old **female** ingested 40 tablets. Sudden high fever lasting 2–3 h developed from the ***atropine***, followed by central nervous system depression. Shallow respirations and possible seizures responded to resuscitative measures and ***nalorphine***. Numerous other cases reported, perhaps making its use contraindicated in **children**. These cases illustrate a disadvantage of polypharmacy, i.e. dual toxicity. ***Naloxone*** has also been effective in some of these cases.

LD_{LO}, **human**, **child** and **adult**: 5 mg/kg.

DIPHENYL

= *Bibenzene* = *1,1′-Biphenyl* = *Phenyl Benzene*

Use: Deodorant and fungistat for ***orange*** containers or wrappers. In organic synthesis, and as a heat transfer agent.

Untoward effects: Nephrotoxic. Eye and throat irritant, headache, nausea, vomiting, abdominal pain, liver damage, fatigue, and numbness or aching of limbs reported. 33 fruit paper **workers** were exposed to more than the threshold limit of 1 mg/m^3 (0.2 ppm). **One** died from acute yellow liver atrophy; histological evidence of liver damage in three **others**. **Some** showed deterioration when seen 1 year later.

Inhalation TC_{LO}, **human**: 4.4 mg/m^3.

Weight decrease in **monkeys**, **mice**, and **chickens**, after chronic exposure. Intubation of **rats** with 1 g/kg was required before fetal and maternal toxicity was noted.

LD_{50}, **rat**: 3.3 g/kg; **rabbit**: 2.4 g/kg.

DIPHENYLAMINE

= *DPA*

Use: In manufacturing dyes, explosives, and rubber tires; as a ***nitrate*** and ***nitrite*** diagnostic reagent, antioxidant, and to prevent apples from “scalding” (turning brown) during shipment.

Untoward effects: ***4-Aminodiphenyl***, a carcinogenic impurity, has been found in material used in the tire industry. Contact dermatitis, eczema, tachycardia, hypertension, methemoglobinemia, and bladder cancer in **man**.

LD_{LO}, **human**: 500 mg/kg.

Causes anemia in **rats** and **mice**, and hypoglycemia in **rabbits**. Cystic dilation of renal tubules with interstitial inflammation noted in two feeding trials with **rats**.

LD_{LO}, **rat**: 3 g/kg; LD_{50}, **guinea pig**: 300 mg/kg.

DIPHENYLMETHANE

= *Benzylbenzene*

Use: Fragrance used in soaps, detergents, cosmetic lotions and creams, and in perfumes. Has an odor of oranges.

Untoward effects: No irritation or sensitization at 8% in petrolatum in 25 **volunteers**.

LD_{50}, **rats**: 2.25 g/kg; acute dermal, **rabbits**: > 5 g/kg.

p,p-DIPHENYLMETHANE DIISOCYANATE

= *MDI*

Less volatile than, and often substituted for, toluene 2,4-di-isocyanate in manufacturing.

Untoward effects: Respiratory tract irritation and sensitization after repeat exposures, causing occupational asthma. Body temperature increased to 101°F after inhalation exposure to low concentration of 0.13 ppm/30 min.

N,N′-DIPHENYL-*p*-PHENYLENEDIAMINE

= *DPPD*

Antioxidant.

Untoward effects: The first antioxidant used on a large scale in **livestock** feeds. Now banned in the U.S. because it was detrimental to reproduction in the **rat** and hazardous to the fetus.

LD_{50}, **rat**: 2.4 g/kg; **mouse**: 18 g/kg.

DIPIPANONE

= *Amidone* = *378C48* = *Fenpidon* = *Hoechst 10805* = *Pamedon(e)* = *Phenylpiperone* = *Pipadone* = *Piperidyl*

***Analgesic.* Controlled substance. Synthetic narcotic.**

Untoward effects: Especially with overdoses: drowsiness; coma; convulsions can occur in **children**; miosis, except in hypoxia, when mydriasis occurs; respiratory decrease, and pulmonary edema, which can be severe and fatal. A 50-year-old **female** had visual hallucinations and paranoid delusions after doses up to 60 mg/day/8 days for metastatic bone pain. **She** returned to normal 48 h after drug withdrawal. Other untoward effects are similar to ***methadone's*** (q.v.), and include nausea, vomiting, and dizziness (especially when standing erect after recumbency). IV injection recommendations were ceased, due to dangerous hypotension produced. Drug of abuse, especially with ***cyclizine*** = ***Diconal***, which has been withdrawn from some markets in 1971.

DIPIVEFRIN

= *Diopine* = *Diphemin* = *d Epifrin* = *Pivalephrine* = *Propine*

Use: α- and β-Adrenergic used in treating glaucoma. A pro-drug of ***epinephrine***.

Untoward effects: Tolerated by many **patients** who cannot tolerate ***epinephrine*** (q.v.). Slight burning sensations and irritation, allergic conjunctivitis during long-term therapy. Mydriasis and, on rare occasions, tachycardia, premature ventricular contraction, hypertension, headache, tremors, sweating, and even disorientation.

DIPLODIA sp.

Fungal pathogen on plants.

Untoward effects: Its toxicity to **humans** is still unknown. ***D. maydis*** (***D. zeae***) is responsible for dry rot or cob rot of ***corn***.

Toxigenic strains cause growth retardation of **mice** and **hamsters**, death of **rats**, **chicks**, and **ducklings**, and general weakness, salivation, incoordination, paralysis, and death of **cattle** and **sheep**. Myocarditis, enteritis, focal renal tubular necrosis, degeneration and peripheral necrosis of the islets of Langerhans, and generalized venous congestion have been reported.

DIPLOPTERYS CABRERANA

Untoward effects: Its leaves contain ***tryptamines***, psychoactive alkaloids that prolong the effect of ***serotonin*** in the body, producing hallucinations at high dosage, and used to increase the duration and intensity of ***ayahuasca***, a hallucinogenic drink of the **Amazon Indians**.

DIPRENORPHINE

= *M 50 - 50* = *Revivon*

Controlled substance. Reverses the neuroleptanalgesia of *etorphine*.

Untoward effects: Undesireable hyperexcitability and cardiac stimulation can often occur in some species. Said to have caused the death of a **horse**, and, possibly some other **animals**.

SC, LD_{50}, **mouse**: ~300 mg/kg.

DIPROPYLENE GLYCOL MONOMETHYL ETHER

= *DPGME* = *DPM*

***Solvent.* In enamels, hydraulic fluids, inks, and cosmetics.**

Untoward effects: Vapors: Nose and eye irritant, headache, and lightheadedness. National Institute for Occupational Safety & Health/OSHA exposure limits are 100 ppm (600 mg/m^3); skin, 150 ppm.

LD_{LO}, **rat**: 4.9 g/kg; LD_{50}, **dog**: 7.5 g/kg.

DIPYRIDAMOLE

= *Anginal* = *Cardoxin* = *Cleridium* = *Corodil* = *Coronarine* = *Curantyl* = *Dipyridan* = *Gulliostin* = *Natyl* = *NSC-515776* = *Peridamol* = *Persantine* = *Piroan* = *Prandiol* = *Protangix* = *RA-8*

***Vasodilator, antianginal, antiplatelet agent.* As a vasodilator, it inhibits platelet release induced by *collagen*, *epinephrine*, and *thrombin*. It inhibits phosphodiesterse activity in platelets, increasing their cyclic AMP levels, which helps inhibit platelet function.**

Untoward effects: Eosinopenia and neutropenia occur. A 72-year-old **male** reported dysgeusia after short-term use of 150 mg qid; hemorrhoidal rectal bleeding in 13 **patients** on 225–300 mg/day; severe pharyngeal hemorrhage in a diabetic 63-year-old **female** on 225 mg/day/8 months; severe headaches in a 52-year-old **female** on 100 mg/day/10 days; vascular-type headaches in 36/169 (42%) after 400 mg/day/14 months; severe myalgic syndrome in a 52-year-old **male** 2 days after 300 mg ID; and severe epistaxis in a 70-year-old **male** on 225 mg/day/4 months reported. Less common symptoms have been transient blurred vision, decreased β-thromboglobulin, myocardial ischemia, pseudomyalgia rheumatica, and agranulocytosis. Overdoses lead to hypotension, vertigo, headache, dizziness, flushing, weakness, fainting, and occasionally rashes, pruritus, nausea, and gastrointestinal upsets. On rare occasions, it has aggravated angina, especially at the beginning of therapy. Gallstones containing 70% and 15%, respectively, in an 80-year-old **female** and an 83-year-old **male** taking 225 mg/day for 15 and 10 years, respectively.

LD_{50}, **rats**: 8.4 g/kg; **dogs**: ~400 mg/kg.

Diltiazem enhances its absorption in **dogs**. Low-dose ***aspirin*** potentiates its antithrombotic effect in **rabbits** and **man**. At high dosage, the reverse occurs for the first day or two in **rabbits**, and then the effect is potentiated. Dose-dependent decrease in heart rate and blood pressure in **rabbits** and **rats**.

α,α´-DIPYRIDYL

= *2,2´-Bipyridine*

Chelating reagent for ferrous iron determinations.

Untoward effects: A respiratory irritant, where it can bind to and injure epithelial tissues. When smoked, it is converted from ~60–70% of the ***paraquat*** in contaminated ***marijuana***. See ***Diquat*** and ***Paraquat***, other dipyridyls.

Decreases gastric secretory volumes and ***pepsin*** output in **rats**.

LD_{LO}, **rat**: 250 mg/kg.

DIPYRONE

= *Alginodia* = *Algocalmin* = *Analate* = *Analgin* = *Bonpyrin* = *Conmel* = *Dimethone* = *Dioprin* = *Divarine* = *Dolazon* = *D-Pron* = *Dya-Tron* = *Espyre* = *Farmolisina* = *Feverall* = *Fevonil* = *Keypyrone* = *Metamisole* = *Metamizol* = *Methampyrone* = *Metilon* = *Minalgin* = *Mypron* = *Narone* = *Nartate* = *Nevralgina* = *Nolotil* = *Novacid* = *Novaldin* = *Novalgin* = *Novemina* = *Novil* = *Novin* = *Paralgin* = *Pidyrone* = *Pydirone* = *Pyralgin* = *Pyrilgin* = *Pyrojec* = *Sulpirine* = *Sulpyrin* = *Sulpyrine* = *Tega-Pyrone* = *Unagen*

Analgesic, antipyretic, antispasmodic. **An enolic acid derivative of *aminopyrine*.**

Untoward effects*:** Associated with many cases of agranulocytosis, some fatal, possibly due to a hypersensitivity reaction. One case was associated with thrombopenic hemorrhagic diathesis. Implicated as a cause of immune hemolytic anemia, other anemias, cold sweats, hypotension, vertigo, tongue edema, tingling over arms and body, bone marrow hypoplasia, neutropenia, thrombocytopenia, asthma, and fixed-drug eruptions. Marketing and importing have been discontinued in some countries. Cancer **patients** taking ***amygdalin (***Laetrile***) from Mexico ran an increased potential risk, because one imported shipment contained 100% dipyrone.

Drug interactions*:** Severe hypothermia can occur with ***chlorpromazine. ***Barbiturates*** and ***glutethimide***, by enzyme induction, have raised its metabolism and decreased its activity in **man**. Oral doses potentiated ***ethylbiscoumacetate*** with maximum effect in ~4 h and lasted 7½ h.

Found in the milk of treated **animals** up to 4 days. Has caused nausea in some **animals**. May be toxic to **cats**. Enhances myocardial concentration of radiopharmaceuticals.

SC, LD_{50}, **mouse**: 69 mg/kg.

DIQUAT DIBROMIDE

= *Aquacide* = *Dextrone* = *FB/2* = *Preeglone* = *Reglone*

Herbicide. **Variable gastrointestinal tract absorption. Excreted unchanged in the urine.**

Untoward effects*:** Ingestion of limited amounts led to diarrhea, dysphagia, gastrointestinal irritation, and tongue and pharynx ulceration. Large amounts led to ulcerations of alimentary tract mucosa, paralytic ileus, proximal tubular degeneration of the kidney, focal degeneration of the liver, degeneration of testicular seminiferous epithelium, vomiting, hallucinations, delirium, pulmonary edema, irreversible fibrosing pneumonitis, cerebral hemorrhage, epistaxis, poor nail growth, brownish nail discoloration, loss of nails, and cataracts. Deaths have been due to suicid[illegible]nd vocational exposure, including inhalation. An [illegible]ar-old **male** accidentally ingested part of a mouthful leading to diarrhea. Excreted in **his** urine until day 11. A 16-year-old **female** ingested ~50 ml ***Reglone® (20% diquat) in a suicide attempt. Hospitalized with abdomnial pain, vomiting, and diarrhea, and died 24 h after the ingestion. **Her** serum level was 4.5 ppm (0.45 mg/100 ml). A 60-year-old **female** ingested 20 ml ***Reglone***®. Despite treatment, **she** developed respiratory arrest, ventricular arrhythmias, and died 5 days after the ingestion. Decreased wound healing reported. Corneal scarring from accidental splashing on eye, even after immediate washing with water.

LD_{LO}, **human**: 50 mg/kg.

Cattle have died 4 years after exposure to an 'empty' previously discarded used container. In poisoned **quail**, symptoms were rubbery-legged and tense-legged ataxia, wing-drop, shivers, and nutation. Airborne application to enhance drying of ***rapeseed***-contaminated fodder and **livestock** enclosures led to diarrhea, abdominal pain, depression, anorexia, incoordination, recumbency, and dyspnea, and vomiting occurred in 75% of 520 **ewes**, 90% of 32 **dairy cows**, and in **piglets** suckled by **sows** given the fodder. Post-mortems revealed gastroenteritis, pulmonary edema, and liver congestion. Proteinuria, glucosuria, degeneration of proximal convoluted renal tubules, and gastrointestinal distension after oral administration to **rats**, and fetal toxicity in **rats** and **mice**. Occasionally bilateral cataracts in **rats**. Oral dosing of **cynomolgus monkeys** led to diarrhea, coma, and death.

LD_{50}, **rat**: 231 mg/kg; **mice**: 125 mg/kg; **guinea pig**: 101 mg/kg; **rabbit**: 100 mg/kg; **dog**: 150 mg/kg; **cattle**: 20–40 mg/kg; **cat**: 35 mg/kg; **mallard ducks**: 564 mg/kg; **quail**: 1337 ppm in feed.

DIRCA PALUSTRIS

= *Leatherwood* = *Moosewood* = *Wicopy*

A shrub of northeastern North America.

***Untoward effects*:** Chewing the bark has caused severe irritation and blistering of the mouth and skin. Dermatitis from contact with its flowers and fruit.

DIRECT BLACK 38

***Use*:** Dye for textiles, inks, plastics, wood, leather, typewriter ribbons, and biological stains. Although U.S. consumption was > 70 tons in 1980, it is slowly being replaced by non-benzidine black dyes.

***Untoward effects*:** No longer used in hair dyes for **humans**. Mutagenic in Ames tests.

Carcinogenic in **rats**, but not in **mice**. Liver carcinomas in **rats**, hepatocellular carcinomas in male **rats**, papillomas and carcinomas of the bladder, and adenocarcinomas of the colon in **rats**.

TD_{LO}, **rats**: 6.8 g/kg/13 weeks.

DIRECT BLUE 6

A benzidine dye. No longer used in hair dyes for humans. Some uses are the same as for *Direct Black 38*.

Untoward effects: Carcinogenic in **rats**, but not in **mice**. Hepatocellular carcinomas in **rats**.

TD_{LO}, **rats**: 6.8 g/kg/13 weeks.

DIRECT BROWN 95

A benzidine dye.

Untoward effects: Apparently, carcinogenic only to female **rats**. High dosage was lethal.

TD_{LO}, **rat**: 6.8 g/kg/6 weeks.

DIRITHROMYCIN

= *AS-E 136* = *Dynabac* = *LY-237216* = *Noriclan* = *Nortron* = *Valodin*

Macrolide antibiotic.

Untoward effects: Gastrointestinal symptoms, mostly nausea and abdominal pain, in about 10% of **patients**. Symptoms may resemble ***erythromycin's***. Until more is known, avoid use with ***terfenadine***.

LD_{50}, **mice**: > 1 g/kg.

DISODIUM EDETATE

= *Chealamide* = *Chelaplex III* = *Chelaton III* = *Edathamil Disodium* = *Edetate Disodium* = *EDTA Disodium* = *Endrate Disodium* = *Sodium Versenate* = *Tetracemate* = *Titriplex III*

***Chelating agent, anticoagulant.* Present in practically all contact lens solutions.**

Untoward effects: IV use has caused burning at infusion site and thrombophlebitis (unless given slowly and well-diluted), ***potassium*** diuresis, gastrointestinal irritation with nausea and diarrhea, fever with shivering, arthralgia and diffuse myalgia (including leg cramps), hypocalcemic tetany, and temporary decreased prothrombin time (returning to normal in about 12 h). Small amounts in a 10% ***calcium gluconate*** solution, added to a hyperalimentation formulation, led to paresthesias in a 24-year-old **female**. Albuminuria, oliguria, and renal failure has occurred, but clears up in a few days after treatment ceases. A dermatitis similar to that of ***pyridoxine*** deficiency has occurred after long-term therapy. Used topically on the eye, it has caused transient stinging and chemosis.

LD_{50}, **rat**: 2 g/kg; **mouse**: 21 mg/kg; **rabbit**: 2.3 g/kg.

DISODIUM METHANEARSONATE

= *Ansar 8100* = *Arrhenal* = *Arsinyl* = *Clout* = *Crab-E-Rad* = *Dal-E-Rad* = *Disodium Monomethanearsonate* = *DSMA* = *Sodar*

***Herbicide.* Pentavalent arsenical.**

Untoward effects: Acute inhalation toxicity reported in **man**. Symptoms are those of ***arsenic*** (q.v.) toxicity.

Extremely toxic to **honeybees**.

DISODIUM METHYLARSONATE

***Herbicide.* Pentavalent arsenical.**

Untoward effects: Near-fatal case in a **client** due to inhalation, reported to the FDA.

LD_{LO}, **human**: 50 mg/kg.

Embryotoxic when 1–2 mg injected into a **chicken** egg.

LD_{50}, **rat**: 1.8 g/kg.

DISOPHENOL

= *Ancylol* = *Iodofen* = *Syngamix*

Anthelmintic.

Untoward effects: Respiratory distress, hyperthermia, depression, recumbency, hypoglycemia (especially in young **pups** and **toy breeds**), local irritation, and temporary yellow staining at injection site in **dogs**. Cataractogenic in **pups** up to ~4 months of age. May elevate ***thyroxine*** levels in **dogs** with respiratory problems. Will stain **bird** feathers yellow at injection site.

LD_{LO}, **human**: 5 mg/kg.

LD_{50}, **rat**: 170 mg/kg, **mouse**: 212 mg/kg.

DISOPYRAMIDE

= *Dicorantil* = *Dirythmin SA* = *Diso-Duriles* = *Durbid* = *Durbis* = *H-3292* = *Isorythm* = *Lispine* = *Norpace* = *Ritmodan* = *Ritmoforine* = *Ritnodar* = *Rhythmodan* = *Rythmodul* = *SC-7031* = *SC-13957*

Antiarrhythmic.

Untoward effects: Anticholinergic effects (dry mouth, nose, and eyes; blurred vision, confusion, constipation, and urinary retention) are common, and a frequent cause of discontinuing the medication. Rarely, a cause of acute-angle closure glaucoma, delirium, hallucinations, paranoia, panic, depression, and psychosis. Nausea, vomiting, slight anorexia, feeling of gastric fullness, gastric pain, and diarrhea have occurred. Agranulocytosis, hypotension, heart block, tachyarrhythmias, torsades de pointes, and aggravation of existing heart failure reported. Slow rates of infusion (> 10 min) led to less myocardial depression than those given over 2–5 min. Less common are headache, dizziness, nervousness, fatigue, muscle weakness, hypoglycemia, intrahepatic cholestasis, urticaria, rash, photosensitivity reactions, anaphylactoid reactions, decreased libido, orthostatic hypotension, epileptiform convulsions, and peripheral neuropathy (burning in both feet). Rarely, gynecomastia and pruritus noted. May induce uterine contractions and delivery. Found in **human** breast milk at similar levels to **mother's** serum levels. A number of fatal deliberate overdose cases reported. Symptoms were early loss of consciousness after apneic episodes. Necropsy revealed pulmonary congestion, secondary to left ventricular failure. Its induced xerostomia caused a failure of dissolution of a sublingual ***isosorbide dinitrate*** tablet in a 75-year-old **male**. Therapeutic serum levels are 0.2–0.4 mg%, and toxic at 0.7 mg%.

Drug interactions*: *Alcohol reduces its clearance by 20%. ***Practolol*** caused sinus bradycardia in one **patient**, and death due to asystole in **another**. ***Phenytoin*** and ***rifampin*** increase its hepatic metabolism. Life-threatening interaction may occur in **patients** taking ***erythromycin***. May potentiate other drugs' anticholinergic effects. Should not be given within 48 h before or 24 h after ***nifedipine***, and probably ***verapamil***. Addition of ***clarithromycin*** to treatment with disopyramide in a 74-year-old **female** led to QTc prolongation and hypokalemia requiring resuscitation. Coadministration with ***ritonavir*** led to adverse cardiac and neurological effects.

Anticholinergic effects noted in **dogs** and **rabbits**. Has induced left ventricular dysfunction in **dogs**. Prolonged QT interval in **rats** enhanced by coadministration with ***erythromycin***.

DISPERSE NUMBERS

Untoward effects*:** Many of these cause contact dermatitis problems for **patients**. ***Disperse Orange 3 and ***Disperse Yellow 3***, azo dyes used in coloring ***nylon*** (as in pantyhose), and dermatologists use them in patch-test allergen kits. ***Disperse Blue 35*** caused a photocontact dermatitis in 16 **male employees** of a manufacturing plant.

DISTILLATE

= *Coal Oil* = *Diesel Fuel* = *Kerosene* = *Petroleum Distillate*

***Use*:** For heating, motor fuels, and vehicle for pesticides and paint thinners.

Untoward effects*:** Estimated LD, **adults**: 3–4 oz. Aspiration of as little as 4 ml can be fatal. Potential hazard for **humans** and **animals**, when used in insecticide aerosols, as well as pneumonitis from induced vomiting. Absorbed after oral use or via the respiratory tract. Dermal absorption is not apt to be significant. After ingestion, central nervous system depression with liver and renal failure in severe cases. Aerosoled material in poorly ventilated places leads to headache, blurred vision, dizziness, nausea, deep drowsiness, and ataxia. Massive exposure leads to collapse, nervous twitching, and coma. Ingestion often leads to gagging, coughing leading to aspiration, and, in serious cases, bronchopneumonia can occur in 24–36 h. Accidental or suicidal ingestion is common. **Abusers** have also used it IV and SC, causing systemic and local toxicity, including pneumonitis. ***Toluene (q.v.) and ***xylene*** (q.v.) can be significant contaminants.

Farmers drenching **cattle** with it to treat bloat have often gotten some down into the trachea and lungs leading to coughing, pneumonia, and even death. Topically, some distillates have induced skin tumors on **mice**.

DISULFIRAM

= *Abstensil* = *Abstinyl* = *Antabuse* = *Antadix* = *Antalcol* = *Anthetyl* = *Anticol* = *Antietanol* = *Contralin* = *Cronetal* = *Disulphuram* = *DSF* = *Esperal* = *Etabus* = *Ethyl Thiurad* = *Ethyl Tuads* = *Exhoran* = *Noxal* = *Stopetyl* = *TETD* = *Tetradine* = *Tetraetil* = *Teturamin* = *Thiuranide* = *TTD*

***Alcohol deterrent, antioxidant in rubber acceleration, metal chelator, fungicide.* In 1948, two Danish physician researchers testing it for anthelmintic efficacy tasted some and then went to a party. They became violently ill after drinking some liquor. Serendipity and repeated stories did the rest. It stays in the body for up to 2 weeks.**

Untoward effects*:** Without ***alcohol: Drowsiness, fatigue, mental dullness, decreased libido in **females**, impotence in **males**, headache, acute hypertension, and occasionally anorexia and nausea. Chronic administration may increase the incidence of arteriosclerotic cardiovascular disease, as in cases of chronic exposure to ***carbon disulfide*** (q.v.), its major metabolite. Rarely, allergic dermatitis, itching, redness, fixed-drug eruptions, acneiform rashes (resembles ***rubber***-induced dermatitis), decreased libido, anxiety states and psychoses (severe depression, schizophrenia, mania, catatonia, delirium, and organic encephalopathy). Polyneuropathy, peripheral neuritis with demyelination of motor and sensory fibers (nerve conduction velocity decrease), and, on very rare occasions (usually reversible), optic neuropathy and grand mal seizures, possibly due to ***carbon disulfide***, a metabolite. A metallic or ***garlic***-like aftertaste and body odor is often reported. Symptoms often decrease after 1–2 weeks of therapy. A 5-year-old **female** ingested 4.75 g leading to basal ganglion infarction with persistence of dystonic movements for 4 months. Hepatotoxicity (lobular), although rare, may be very severe and occur 3–25 weeks after ingestion. Suicidal ingestion reported. Estimated LD, 30 g.

With ***alcohol***: It interferes with aldehyde dehydrogenase increasing blood levels of ***acetaldehyde*** (q.v.), which causes uncomfortable symptoms leading to flushing, dyspnea, hyperventilation, pounding headache, nausea, vomiting, thirst, chest pain, fixed-drug eruptions, palpitation, peripheral neuritis, vertigo, tachycardia, hyperhidrosis, hypotension, syncope, confusion, cardiac arrhythmias, pulmonary edema, and even convulsions and congestive heart failure, leading to a possible fatality. Many fatal outcomes are reported in the literature, including one on fatal shock. Estimated LD, 1 g. The ***mushroom*** toxin, ***coprine***, a disulfiram-like substance, can cause severe symptoms up to 5 days after ***mushrooms*** and ***alcohol*** are consumed together. A 56-year-old **male**, after ingesting 250 mg disulfiram/day developed a generalized pruritic macular rash after showering with a ***beer***-containing shampoo with 3% ***alcohol***. Spontaneous esophageal rupture in a **patient** who consumed a large amount of ***alcohol*** 1 month after disulfiram therapy. The implied threat of illness may be a more powerful influence to avoid intake of ***alcohol*** than the pharmacological action itself. An unpleasant odor from a colostomy noted in a 40-year-old **male alcoholic** while on disulfiram. **Users** must avoid colognes, after-shave lotions, mouth washes, wine vinegars, cooking ***wines***, and many cough medicines.

The drug has been implicated in causing pancreatitis; causes a marked, but short-lived, increase in blood ***acetone*** levels, and implicated in severe limb reduction anomalies (radial aplasia, vertebral fusion, tracheo-esophageal fistula, and phocomelia of the lower extremities) in two **infants** of **mothers** who took it during their first trimester of pregnancy. Its use has interfered with vanillylmandelic acid assay (it increases vanillylmandelic acid urinary excretion), and increases the incidence of cardiovascular disease. **People** sensitive to it may also be sensitive to ***thiram***(q.v.) and ***tetramethylthiuram disulfide*** (q.v.). Usual therapeutic serum levels reported as 0.05–0.4 mg/l; toxic at 5 mg/l; and can be fatal at 8 mg/l. LD_{LO}, **human**: 160 mg/kg.

Drug interactions: In most studies, it increases ***warfarin*** anticoagulant effect by decreasing its metabolism. It inhibits the inactivation of ***ethotoin, mesantoin,*** and ***phenytoin,*** by the liver. It increases ***phenytoin*** serum concentration by 100–500% by the same mechanism. Its use inhibits the metabolism of ***isoniazid*** leading to coordination difficulties and behavioral changes; with ***metronidazole***, in about 15–20% of **patients** leads to ataxia; acute psychosis in 6/29 **males** (confusion and disorientation in the six, paranoid delusions in five, and visual and auditory hallucinations in three). It increases half-life of ***antipyrine***, decreases elimination of ***caffeine***, and has precipitated ***temazepam*** toxicity. Increases peak plasma levels and elimination half-life of the tricyclic antidepressants ***desipramine*** and ***imipramine***. Can potentiate the actions of ***oral contraceptives*** by decreasing the hepatic metabolism of their ***estrogen*** and ***progestogen*** components leading to increased fluid retention, hypertension, increased risk of thromboembolic disorders, and having diabetogenic activity. By inhibiting the metabolism of ***chlordiazepoxide, clorazepate, diazepam, flurazepam, halazepam, prazepam, rifampin,*** and ***terfenadine***, their toxicity can also be increased, and dosage adjustment or discontinuation of their use is necessary. Lowers levels of ***perphenazine***. Disulfiram-like reactions also occur in **patients** ingesting ***alcoholic*** beverages and receiving ***moxalactam, sulfonylureas***, (viz. ***chlorpropamide, glipizide***), ***cephalosporins (cefoperazone, cefotetan), chloram-phenicol, griseofulvin, metronidazole***, and ***trimethoprim-sulfamethoxizole***.

Its use with ***amphetamine, barbiturates, meperidine,*** and ***morphine*** led to morbidity in experimental **animals**. A toxic interaction between inhaled ***ethylene dibromide*** and ingested disulfiram led to high death rates in **rats**, and this is the basis for National Institute for Occupational Safety & Health recommendations that no **workers** should be simultaneously exposed to both. The minimum blood ***alcohol*** levels and disulfiram doses in **rats** which caused neuronal degeneration in the brain are close to those described in a clinical case of abuse of disulfiram in **man**. It has induced myocardial and skeletal muscle degeneration in **rats**. Has produced slight corneal opacities in **rabbits**. Causes a three times increase in ***hexobarbital*** sleeping time in **rats**. Oral administration to **cats** dramatically increased ***dopamine*** concentration in the iris and nictating membrane. Intensifies the ***ethanol***-induced hypotension and cardiac suppression in **dogs**.

LD_{50}, **rats**: 1.3 g/kg; **rabbit**: 2 g/kg. IP, LD_{50}, **rats**: 335 mg/kg; **mice**: 75 mg/kg.

DISULFOTON

= *Bay 19639* = *Bayer 28/63* = *Bug Dart Plus* = *Dimaz* = *Disipton* = *Disyston* = *Disystox* = *Dithiodemeton* = *Dithioseptox* = *Dithiosystox* = *ENT-23347* = *Frumin AL* = *Frumin G* = *M74* = *S276* = *Solvirex* = *Thiodemeton*

***Organophosphorus insecticide and acaricide.* Highly toxic cholinesterase inhibitor.**

Untoward effects: In **humans**, ingestion and percutaneous absorption leads to miosis, drooling, vomiting, diarrhea, dyspnea, sweating, stomach pains, tightness in chest, arrhythmias, tremors, convulsions, and even death. Limited skin contact has caused fatigue, weakness, and local irritation. Rhinitis has followed inhalation of fumes. Environmental Protection Agency suggests a maximum of 10 ppb in **children's** drinking water for a maximum of 10 days. For longer periods, they suggest 3 ppb maximum, and 9 ppb for **adults**. National Institute for Occupational

Safety & Health recommends an exposure limit of 0.1 mg/m^3/10 h, day/40 h week.

LD_{LO}, **human**: 5 mg/kg.

Animals exposed for long periods of time may, like **humans**, become near-sighted and have blurred vision. Ingestion by **animals** during pregnancy led to newborns with undeveloped bones and testes, and damaged livers and kidneys. Poisoned **birds** show typical symptoms of organophosphorus poisoning in **birds**: goose-stepping ataxia, wing-drop, tremors, wing-beat convulsions, tetany, visual disturbances, and even death.

LD_{50}, **rats**: 2.1 mg/kg; **mouse**: 5.5 mg/kg; **mallard ducks**: 6.54 mg/kg; **pheasants**: 11.9 mg/kg; **mule deer**: 2.5–5.0 mg/kg; male **guinea pigs**: 10.8 mg/kg. LC_{50} in feed/5 days, young birds, **Japanese quail**: 334 ppm, **mallard ducks**: 510 ppm, **pheasants**: 634 ppm, **bob-white quail**: 715 ppm.

DITHIAZANINE IODIDE

= *Abminthic* = *Anelmid* = *Anguifugan* = *Dejo* = *Delvex* = *Déselmine* = *Dilombrin* = *Dizan* = *Nectocyd* = *Partel* = *Telmicid* = *Telmid*

Anthelmintic, sensitizer for photographic emulsions.

Untoward effects: Lactic acidosis from systemic absorption, nausea, vomiting, diarrhea, shock, pulmonary edema, and fatalities reported with its use in **humans**. Gives false positive protein-bound iodine values and has caused albuminuria.

Will stain **animals** and surroundings. Do NOT use in nephritis.

2,4-DITHIOBIURET

= *Thioimidodicarbonic Diamide*

Untoward effects: Several daily doses to **rats** led to progressive skeletal muscle paralysis by central action.

LD_{50}, **rat**: 5 mg/kg.

1,4-DITHIOTHREITOL

= *Cleland's reagent* = *DTT*

Mucolytic.

Untoward effects: Although very effective in laboratory work to reduce mucus viscosity, it is too toxic for routine *in vivo* use. Severe nausea and vomiting at 25 mg/kg in a 10-year-old **male cystinotic** in terminal uremia. Tolerated 15 mg/kg tid.

DITHIZONE

= *Diphenylthiocarbazone*

Metal chelating agent. **Used to treat *thallium* poisoning.**

Untoward effects: There is considerable disagreement as to its safety for **man**, even on a risk versus benefit scale.

Extremely large doses led to diabetes in **rabbits**, and daily dosage for 10 days is goitrogenic for young **rats**. IV use has caused irreversible retinal damage in **dogs** and **rabbits**, probably due to tapetal ***zinc*** chelation.

DITHRANOL

= *Anthra-Derm* = *Anthralin* = *Antra-derm* = *Batidrol* = *Cignolin* = *Cigthranol* = *1,8-Dihydroxyanthrone* = *Dithrocream* = *Psoradrate* = *Psoriacide*

Psoriasis treatment.

Untoward effects: Normally, an absence of systemic effects, but very irritating to eyes, and causes temporary staining of skin and light hair. Fabrics may be stained permanently. ***Salicylic acid*** is often added to decrease skin staining and increase its stability and effectiveness of older preparations. Hands must be washed after application, because it is cathartic. Percutaneous absorption occurs, and, although it has not been reported, the possibility of renal toxicity can occur in **people** with impaired renal function. It is a photosensitizer and can cause erythema in normal skin adjacent to treated areas. Contact dermatitis, bullous pemphigoid, fixed-drug eruptions, and nail disorders occur. **Psoriatics** are more tolerant of its local inflammatory effects than **non-psoriatics**. Not genotoxic or carcinogenic, but can enhance the effect of previously direct-acting carcinogens or pro-carcinogens.

DIURON

= *Dailon* = *DCMU* = *Dichlorofenidim* = *Di-On* = *Diurex* = *DMU* =*Duran* = *Dynax* = *Herbatox* = *Karmex* = *Marmer* = *Urox D* = *Vonduron*

Pre-emergent herbicide. **Non-ionic.**

Untoward effects: Eye, skin, and nose irritant from vapors. Throat irritant, if ingested. If ingested in large quantities, causes nausea, vomiting, abdominal distress, and diarrhea. National Institute for Occupational Safety & Health suggests limit of exposure at 10 mg/m^3.

LD_{LO}, **human**: 500 mg/kg.

Repeated oral administration to **rats** leads to anemia and methemoglobinemia. Proliferative endocarditis in **carp** after exposure. < 10% mortality in **catfish** exposed to 10 ppm in their water for 48 h. Poisoned **mallard ducks** developed ataxia, which persisted for 11 days. Incoordination noted in poisoned **sheep**; **cattle, sheep**, and **chickens** developed anorexia, dyspnea, and prostration. **Cattle**, although poisoned by 100 mg/kg/day/ 10 days, survived; the same was true for **sheep** treated for 2 days, but **chickens** were killed by the same dose for

10 days. Single oral doses of 220–440 mg/kg are toxic to **cattle** and **sheep**. Persists in soil for 3–6 months.

LD_{50}, **rats**: 437 mg/kg; **mallard ducks**: > 2 g/kg; feed LC_{50}, 12 days **Japanese quail**: > 5000 ppm (2/14 died at concentrations of 2500 and 5000 ppm), **bobwhite quail**: 1730 ppm.

DIVINYLBENZENE

= ***Diethylbenzene*** = ***DVB*** = ***Vinylstyrene***

Untoward effects: Inhalation causes eye, skin, and respiratory tract irritation.

Central nervous system depression in experimental **animals**.

LD_{50}, **rats**: 4 g/kg.

DIXYRAZINE

= *Esocalm* = *Esucos* = *Eucepha* = *UCB-3412*

Neuroleptic, antipsychotic.

Untoward effects: Tiredness. Rigidity of masticatory muscles in **patient** on 100 mg/day. Sedative effect troubled **some** at 60 mg/day. Ichthyosis-like dermatosis, hair loss, depigmentation of hair, and blepharoconjunctivitis reported.

Mammotropic effect in ***estrogen***-primed female **rats**.

LD_{50}, **rat**: 400 mg/kg; IV, 3.75 mg/kg.

DJENKOLIC BEAN

= *Pithecolobium lobatum*

Food, nutrient.

Untoward effects: Although a desirable food item in Java, despite its foul smell, it frequently causes toxic effects, due to its ***sulfur***-containing amino acid, ***djenkolic acid***. ***Djenkolic acid*** is the cysteine thioacetal of ***formaldehyde***. Its toxic properties are caused by its insolubility, similar to that of ***cystine***. When, and if, crystallization occurs in the genitourinary tract, strangury renal colic, hematuria, proteinuria, oliguria, anuria, and even fatalities reported.

DOBUTAMINE

= *Compound 81929* = *Dobutrex*

***Sympathomimetic, vasopressor, β-adrenergic with weak α-adrenergic properties. A dopamine* derivative with direct inotropic action.**

Untoward effects: Tachycardia and hypertension are the most common. Occasionally ventricular arrhythmias, premature ventricular contraction, anginal pain, nausea, palpitations, and dyspnea. Rarely, thrombocytopenia or eosinophilia. Transmural myocardial ischemia in a 69-year-old **female** during dobutamine (40 µg/kg/min and 0.2 mg ***atropine***) stress echocardiography. It can alter the distribution of coronary blood flow. Hemodynamic tolerance to it has occurred. A hypersensitivity reaction with dermal cellulitis in a 64-year-old **female** within 4 h after start of 8 µg/kg/min infusion. Redness, swelling, itching, and warmth developed over an extensive area surrounding the infusion site. Severe necrotic reactions and phlebitis have followed extravasation. Pruritus of the scalp in a 60-year-old **male** during infusion of 10 µg/kg/min. A 71-year-old **female** with congestive heart failure developed fever after three separate courses of 3–5 µg/kg/min. An erroneous overdose of 130 µg/kg/min, instead of an antibiotic infusion, led to hypertension, tachycardia, tachypnea, chest pain, urinary incontinence, and paresthesia. Symptoms subsided after 2 h. Hypotension in a 54-year-old **male** after low dosage IV. Its half-life is only a few minutes and is eliminated from the body in ~12 h.

Drug interactions: Bilateral retinal infarction in a 30-year-old **female** after dobutamine 9–54 µg/kg/min and ***dopamine***, 14–115 µg/kg/ min for 5 days. Ineffective, if ***β-blockers*** have recently been administered. ***Insulin*** requirements in **diabetics** may be increased by it. It has been implicated in severe drug interactions with ***monoamine oxidase inhibitors***.

IV, LD_{50}, **mouse**: 73 mg/kg.

DOCOSAHEXAENOIC ACID

An essential fatty acid.

Untoward effects: A 68-year-old **female** developed lymphadenitis with chills, myalgia, and fever after ingesting ***Super-EPA***®, a combination of it and ***eicosapentaenoic acid***. Symptoms resolved when medication was withdrawn, and recurred when **she** was rechallenged with it. The combination is not recommended to lower plasma ***cholesterol*** in middle-aged **men** with hypercholesterolemia, and can increase fasting ***glucose*** levels.

DOCUSATE SODIUM

= *Aerosol OT* = *Colace* = *Comfolax* = *Coprola* = *Dioctyl* = *Dioctylal* = *Dioctyl Sodium Sulfosuccinate* = *Diotilan* = *Disonate* = *Disposaject* = *Doxinate* = *Doxol* = *DSS* = *Dulcivac* = *Jamylène* = *Molatoc* = *Molcer* = *Molofac* = *Nevax* = *Regutol* = *Sodium Dioctyl Sulfosuccinate* = *Soliwax* = *Velmol* = *Waxsol*

Stool softener, anionic surfactant in pharmaceutical, food, and industrial applications, such as sugar manufacturing.

Untoward effects: Occasionally epigastric cramps and flatulence. **Maternal** overuse throughout pregnancy caused diarrhea and the loss of ***magnesium***, with resulting hypomagnesemia in **mother** and **infant**. Acute vesicular eczema reported in six **patients** from its occurence as a constituent in orthopedic ***wool*** applied under a cast. Bitter taste, nausea, and throat irritation reported with its use as a syrup. It has been suggested that liver toxicity can occur with it alone.

Drug interactions*:** It has increased the absorption and toxicity of drugs such as ***danthron, oxyphenisatin, quinidine, and ***salicylates***. It increases the absorption of ***mineral oil***. It decreases the bioavailability and absorption of ***tetracycline***.

Use in water causes a significant increase of ***phenobarbital*** absorption in **goldfish**, apparently by altering membrane permeability. Doses 3–5 times recommended amounts lead to severe diarrhea, rapid dehydration, and death in many **horses** and **guinea pigs**. Toxicity has also been demonstrated in **dogs**, **monkeys**, **rabbits**, and **mice** after topical or oral use in high concentrations.

LD_{50}, **rats**: 1.8 g/kg; **guinea pig**: 0.65 g/kg.

n-DODECANE

Solvent.

***Untoward effects*:** A co-carcinogen in **mice**. Severe reddening, induration, and crust formation when applied topically to **miniature swine** for 48 h closed-patch method.

1-DODECANETHIOL

= *Dodecyl Mercaptan*

***Solvent*. Oily liquid with a mild skunk-like odor.**

***Untoward effects*:** Vapors cause eye, skin, and respiratory tract irritation. Ingestion leads to weakness, dizziness, dyspnea, and coughing. Skin contact leads to sensitization, confusion, cyanosis, abdominal pain and nausea.

1-DODECANOL

= *Dodecyl Alcohol* = *Lauryl Alcohol*

***Use*:** Up to 10 ton/years used as a fragrance in soaps, detergents, cosmetic creams and lotions, and foods.

***Untoward effects*:** A 4% concentration in petrolatum demonstrated no cases of sensitization in 25 **volunteers**.

Mildly comedogenic when applied to **rabbit** ears.

LD_{50}, **rats**: 12.8 g/kg; skin LD_{50}, **guinea pig**: > 36 ml/kg.

DOFETILIDE

= *Tikosyn*

Antiarrhythmic.

***Untoward effects*:** Dose-dependent torsades de pointes (< 1%) in early trials; > 3.0% in **patients** with left ventricular dysfunction, clinical heart failure, and two deaths.

Drug interactions*: *Cimetidine, ketoconazole, megestrol, prochlorperazine, trimethoprim, and ***verapamil*** can increase its plasma concentration. Small increase in plasma concentration may occur with macrolide antibiotics, other azole fungicides, selective serotonin reuptake inhibitors, ***grapefruit juice***, and protease inhibitors.

DOGWOOD

= *Cornus sp.* = *Piscidia sp.*

Untoward effects*:** The Jamaican variety has caused ataxia, respiratory paralysis, convulsions, and death. It is narcotic to **man**, **rabbits**, and **frogs**. Although extracts are used in various herbal ***teas, it is suggested that **people** working with the bark of the Jamaican varieties use mask, goggles, and gloves to avoid toxicity (salivation, mydriasis, tetanoid state, bradycardia). In the Bahamas, it is used as a piscicide, but the government discourages its use for this, because of lingering toxic effects. Contains low levels of ***nitrates***.

DOLASETRON

= *Anzemet* = *MDL 73,147*

***Antiemetic, antimigraine, serotonin receptor antagonist.* IV or oral.**

***Untoward effects*:** Headache (3.8%), dizziness, and hot flashes after IV; headache, abdominal cramps or pain, constipation (0.8%), and diarrhea (0.4%) after oral use. Infrequently, anorexia, taste perversion, tremors, diaphoresis, rash, and agitation. Rarely, pancreatitis, hematuria, epistaxis, anemia, thrombocytopenia, facial edema, urticaria, anaphylaxis, arthralgia, ataxia, dyspnea, and thrombophlebitis. Venous irritation (~20%) reduced by slow (15 min) infusion and/or dilution with normal saline.

DOLICHOS LALAB

= *Bean-Vine* = *Hyacinth Bean*

***Untoward effects*:** Contains cyanogenic glycosides, leading to respiratory difficulties, loss of voice, spasms, shaking, twitching, and coma. All parts of the plant are poisonous. Boiling the fruit (pods and seeds) make it safe to eat. Also contains saponins.

Various species in Africa are suspected of causing ***cyanide*** poisoning of **livestock**.

DOMIPHEN BROMIDE

= *Bradosol Bromide* = *Modibromide* = *Modicare* = *Neo-Bradoral* = *NSC-39415* = *Oradol* = *PDDB* = *Phenododecinium Bromide*

Topical antifungal, bactericide, and detergent.

***Untoward effects*:** Has caused dermatitis.

DOMOIC ACID

A potent *glutamate* analog.

Untoward effects: In December, 1987, ~150 **Canadians** got sick after eating blue **mussels** that contained a high level of it, a neuroexcitatory ***amino acid*** also found in a red alga, *Chondria armata*. Four of these died. Twelve **survivors** had permanent memory loss similar to that of Alzheimer's disease. Autopsies on the **dead** revealed severe damage to the neurons in the hippocampus, a brain structure associated with memory. Domoic acid has also sickened **animals** who have eaten contaminated seafood.

In 1961, a **seabird** attack in Monterey Bay, California inspired Alfred Hitchcock's chilling movie, "The Birds". Dr Garrison of the University of California later suspected the **birds** had suffered from a phytoplankton-(*pseudo-Nitrizia australis*) spawned food poisoning, due to their domoic acid content. **Birds** appeared out of the fog and descended on shore communities, where they were "hanging off of lampposts, running into police cars, and being chased by **cats**". *The Santa Cruz County Sentinel* reported eight **people** nipped by the **birds**. This toxin affected **anchovies**, **sardines**, and **krill**, that can affect the food chain. It was responsible for killing > 50 **sea lions** in Monterey Bay, California in 1998.

DOMPERIDONE

= *Eucitin* = *Evoxin* = *Gastronorm* = *Mod* = *Motilium* = *Nauzelin* = *Peridon* = *Peridys* = *R-33812*

***Antiemetic. Dopamine* antagonist.**

Untoward effects: Has induced galactorrhea in **patients**. Facial flushing, headache, some somnolence, and dry mouth occur. A 4-month-old **infant** developed an opisthotonic dystonic reaction after 0.33 mg/kg every 6–8 h/2 days. Improvement followed discontinuance. Grand mal seizures in four cancer **patients** on high IV dosage. Ventricular fibrillation, followed by death occurred in a 69-year-old **male** after a bolus IV injection. Four fatal cases of cardiac arrest in cancer **patients** (33–53 years) after 25–50 mg IV. Its IV use is now restricted to a named-**patient** basis by the manufacturer.

Drug interactions: May occasionally increase bioavailability and absorption of ***L-dopa***.

DONEPEZIL

= *Aricept* = *E2020*

***Acetylcholinesterase inhibitor.* Slow-working (up to several months) aid to correct mild symptoms of Alzheimer's disease. A *piperidine*.**

Untoward effects: Nausea, vomiting, diarrhea, and muscle cramps occur in ~5% of **patients**. Occasionally, fatigue, insomnia, anorexia, depression, pollakiuria, weight loss, abnormal dreams, and purpuric rash reported. May worsen Parkinson's disease symptoms. Purpuric rash after 5 mg/day/4 days – confirmed by rechallenge. Chest pain, toothache, hypo- or hypertension, atrial fibrillation, hot flashes, bloating, epigastric pain, gastrointestinal bleeding, delusions, tremors, vertigo, ataxia, increased libido, aggression, dyspnea, bronchitis, pruritus, and fecal and urinary incontinence also observed.

Drug interactions: Potential exists for interactions with ***ketoconazole, quinidine, risperidone,*** and ***succinylcholine***. Drowsiness and shuffling when given to a **patient** receiving ***risperidone***. ***Ketoconazole*** and ***quinidine*** decrease its metabolism *in vitro*. Use caution with ***succinylcholine***, etc.

In experimental **animals**, overdoses cause ataxia, clonic convulsions, respiratory depression, salivation, lacrimation, miosis, tremors, fasciculations, spontaneous movements, and decreased body surface temperature.

DOPA

= *3-Hydroxytyrosine*

Untoward effects: Hypertensive crises have developed in ***monoamine oxidase***-inhibited **patients** after eating ***broad beans*** rich in it and ***dopamine***. Has a chemical structure very similar to ***mimosine***, a toxic amino acid. Transitory nausea and vomiting has occurred in **many**, orthostatic faintness in **some**, transient decrease of granulocytes in a **few**, and reversible dose-dependent abnormal head movements in 10/26 with parkinsonism. Feeling of pressure in head on high dosage (100 mg). Augments tremor of parkinsonism and that induced by ***tremorine***. Converted into ***dopamine*** in the body.

L-DOPA

= *Bendopa* = *Bio-Dopa* = *Brocadopa* = *Cidandopa* = *Deadopa* = *Dopaflex* = *Dopaidan* = *Dopal* = *Dopalina* = *Dopar* = *Doparkine* = *Doparl* = *Dopasol* = *Dopaston* = *Dopastral* = *Doprin* = *Eldopal* = *Eldopar* = *Eldopatec* = *Emeldopa* = *Eurodopa* = *3-Hydroxy-L-tyrosine* = *Laradopa* = *Larodopa* = *Ledopa* = *Levodopa* = *Levopa* = *Maipedopa* = *Parda* = *Prodopa* = *Syndopa* = *Veldopa*

***Antiparkinsonism.* An *amino acid* intermediate in the formation of *norepinephrine*.**

Untoward effects: ***Cardiovascular***: Postural hypotension, faintness, dizziness, palpitations, flushing, hyperhidrosis, cardiac dysrhythmias with atrial and ventricular ectopic beats, and occasionally hypertension. May rarely cause recurrence of angina. May be undesireable to use prior to inhalation anesthetics to prevent increase in ***norepinephrine*** adverse effects.

Central nervous system: Headache, weakness, ataxia, and, rarely, paresthesias and convulsions. Psychiatric symptoms including agitation, anxiety, drowsiness, depression, insomnia, euphoria, increased libido, aggression, mania, panic attacks, confusion, paranoid delusions, auditory and visual hallucinations, and suicidal behavior. May unmask dementia. Involuntary movements, starting particularly in the mouth, jaws, and tongue are common. After several months of treatment, abnormal limb and severe choreoathetoid movements have been reported. Accentuates latent tremors and can exacerbate schizophrenia.

Eye: Has caused recurrence of oculogyric crises, and, occasionally, muscle twitching and blepharospasm. Rarely, mydriasis, aniscoria, diplopia, and glaucoma. Opisthotonus, terror, rage, and catalepsy have occasionally accompanied the crises. It has been suggested that it not be used in suspicious cases of or in **those** with a history of melanomas, since, in 1982, a choroidal melanoma was diagnosed in a treated **patient**.

Gastrointestinal: Nausea and anorexia are common. Vomiting during initial treatment and with high dosage. Abdominal pain, constipation, diarrhea, dysphagia, reduced taste acuity, and gastrointestinal bleeding have been reported.

Genitourinary: Polyuria, incontinence, urinary retention, priapism, and increased libido. Colors urine brown.

Miscellaneous: Increased levels of liver enzymes and blood urea nitrogen, tachypnea, bradypnea, alopecia, skin rashes, pemphigus, lichen planus-like eruptions, pruritus, urticaria, possible systemic lupus erythematosus syndrome, agranulocytosis, thrombocytopenia, hemolytic anemia; severe erosions of some teeth, probably due to abnormal dystonia-type movements (bruxism); akinesia paradoxica (start hesitation), positive direct Coombs' test, "burning feet", adrenal suppression, increased lactate dehydrogenase and bone fractures in **patients** resuming activities too rapidly after treatment. Lowers gastric emptying rate and increases first-pass metabolism in the gut wall. May cause false positive ketonuria results with many dipsticks, increase in colorimetric ***uric acid*** and ***creatinine*** results; false negative glucose oxidase tests. Changes in hair color reported.

TD_{LO}, **human**: 13 g/kg/1 year, 156 g/kg/10 years.

Drug interactions: In pink capsules colored with ***FD & C Red 3*** (***erythrosine***) containing free ***iodine*** has produced falsely elevated protein-bound iodine values in **patients**. Its effectiveness is reduced with ***chlordiazepoxide***. ***Pyridoxine*** 5 mg or more/day can decrease or abolish its effectiveness in parkinsonism. ***Clonidine*** antagonizes its antiparkinsonian effect. ***Propranolol*** and ***metaprolol*** may antagonize its hypotensive and positive inotropic effects. ***Phenothiazines*** block ***dopamine*** receptors and block the effect of L-dopa. ***Pargyline***, ***phenelzine***, and other ***monoamine oxidase inhibitor***s may cause an undesirable increase of tissue catecholamines, and increase blood pressure and central nervous system activity. ***Isoniazid*** can decrease its induced dyskinesia and increase parkinsonian symptoms. ***Droperidol*** causes sleep disturbance, orthostatic hypotension, sexual dysfunction, weight gain, increased rigidity and reappearance of symptoms in L-dopa-treated **patients**. The dose of L-dopa usually needs to be decreased if the **patient** is taking ***amantadine***. This potentiation of effect can also be induced by ***antacids*** and ***anticholinergics*** (***parasympatholytics***). It inhibits ***metoclopramide***-induced rise in esophageal sphincter pressure and ***metoclopramide*** increases its availability by delivering it more rapidly to the site of absorption. ***Carbidopa*** decreases the frequency of its adverse effects and permits a dramatic decrease in L-dopa dosage. High protein intake decreases its therapeutic and cerebral side-effects and low protein diets potentiate and stabilize its therapeutic effects. ***Ferrous sulfate*** can decrease its efficacy by decreasing its gastrointestinal absorption. ***Methyldopa*** may antagonize its effect. Addition of ***orphenadrine*** led to neuropsychiatric side-effects in a few days in 7/12, confusion in 6/12, facial dyskinesia or chorea in 3/12, suggesting a synergistic effect. ***Papaverine*** (100 mg/day) antagonized L-dopa therapeutic benefits in a 71-year-old **female**. This effect has been observed in **others**, confirming this iatrogenic cause of treatment failure. ***Phenytoin*** can also decrease its therapeutic effects. ***Licorice***-colored vomitus has occurred in **patients** due to enzyme systems in certain fruits (***bananas, grapes,*** and ***plums***), degrading the drug. ***Reserpine*** and ***tetrabenazine*** can decrease its dopaminergic function by decreasing presynaptic ***dopamine*** stores. In 821 clinical **patients**, an apparent increase of 16% in the risk of death when ***selegiline*** was added to the treatment. Vitiligo reported in a 50-year-old **male** after 375 mg/day and ***tolcapone*** 300 mg/day.

Produces a state of intense behavioral excitement in **cats**, induces gastric erosions and hypertension in **rats**, and demonstrates *in vitro* enhanced toxicity to melanoma cells and antitumor activity in **mice** against neuroblastomas.

LD_{50}, **rats**: 1.8 g/kg; **mice**: 2.4 g/kg; **rabbit**: 609 mg/kg.

DOPAMINE

= *ASL-279* = *Cardiosteril* = *Dopastat* = *Dynatra* = *3-Hydroxytyramine* = *Inovan* = *Inotropin*

***Sympathetic nervous system neurotransmitter, antihypotensive.* IV. It should be diluted before use.**

Untoward effects: TD_{LO}, **woman** (1–7 weeks pregnant): 300 μg/kg/day; **child**: 13 mg/kg.

Cardiovascular: Ventricular and supraventricular tachycardia, anginal pain, decreased blood pressure (although used to treat hypotension), and dysrhythmias from stimulating ***β-adrenergic receptors***. A 30-year-old **female** experienced an unexpected decrease in heart rate after receiving 25 μg/kg/min.

Eye: Bilateral retinal infarction in a 30-year-old **female** after 14–115 μg/kg/min and ***dobutamine*** 9–54 μg/kg/min/5 days. Mild transient impairment of accommodation, and, rarely, extreme mydriasis reported.

Gangrene and ischemia: Vasoconstriction, followed by ischemic necrosis and necrosis of extremities and digits (often requiring amputation) reported frequently. Severe necrotic reactions have also followed extravasation into surrounding tissues. Use with caution in **patients** with previous vascular damage.

Miscellaneous: Polyuria, decreased esophageal sphincter tone, nausea, vomiting, headache, hypersensitivity with increased cerebral blood flow in some migraines, interference with glucose oxidase reaction in urine led to false positive ketonuria, its increase in cerebrospinal fluid may contribute to the restless leg syndrome, and wheezing may be due to ***sulfites*** as preservatives. Acute diabetes insipidus induced by 12 μg/kg/min in an 18-year-old **female**.

Neurological syndromes: Choreiform disorders, induced parkinsonism, mania, and tardive dyskinesia occur. It depletes ***pyridoxine*** and can cause a peripheral neuropathy.

Drug interactions: ***Disulfiram*** reduces its metabolism. Antagonism by ***droperidol*** may increase rigidity. Its excessive excretions in manic-depressive disorders return to normal with ***lithium carbonate*** treatment. ***Monoamine oxidase inhibitors***, combined with foods or ***wines*** containing pressor amines, increase its serum levels and can increase blood pressure with an acute hypertensive crisis. This requires a 90% decrease in dopamine dosage. ***Monoamine oxidase*** inactivates dopamine. Use with ***methysergide*** can increase the risks of arterial spasms and occlusions. ***Metoclopramide*** may impair its vascular effects. IV ***phenytoin*** used with it has caused a dramatic decrease in blood pressure, usually with bradycardia. Cardiac arrest reported in a **patient** given it after a small dose of ***tolazoline***.

In **cats**, its concentration is increased 4–5 times by oral ***disulfiram***. Causes atrial and ventricular arrhythmias in **dogs**. Large doses of ***ascorbic acid*** increase brain dopamine levels in the **guinea pig**. Induces hypoglycemia in **mice**.

IV, LD_{50}, **rats**: 20 mg/kg; **mice**: 32 mg/kg.

DOPAN

= *Elderfield Pyrimidine Mustard* = *NSC 23,436* = *Ypenyl*

Antineoplastic.

Untoward effects: Gastroenteropathy, anorexia, nausea, vomiting, and bone marrow suppression.

IP, LD_{LO}, **mouse**: 3 mg/kg.

DORONICUM sp.

Untoward effects: ***Doronicum* sp.** in Colorado = ***Arnica*** = ***Leopardsbane***, whose leaves' ***arnicin*** content causes a skin rash in some **people**.

D. caucasicum has killed **cattle**, **horses**, and **sheep** in Germany. Minimum lethal dose of the dry leaves was 0.5 g, 0.75 g, and 0.1 g, respectively. **Sheep** develop toxic and clonic spasms; **cattle** and **horses** become depressed, somnolent, and comatose. Enlargement and degeneration of the liver and kidneys were noted on post-mortem.

D. hungaricum (***Wild Sunflower***), in Bulgaria, causes **cattle** and **sheep** to suddenly fall over about 16 h after ingestion. Convulsions, apathy, and paralysis follow. Some die.

DORZOLAMIDE

= *MK-507* = *Trusopt*

Antiglaucoma agent.

Untoward effects: Bitter taste and ocular stinging immediately after use. Blurred vision, tearing, dry eye, and photophobia occur. Superficial punctate keratitis, conjunctivitis, and allergic lid reactions in 10–15% of **patients**. Infrequently, headache, nausea, asthenia, fatigue, skin rashes, urolithiasis, and iridocyclitis.

DOTHIEPIN

= *Altepin* = *Depresyn* = *Dosulepine* = *Prothiaden(e)*

***Tricyclic antidepressant.* An *amitriptyline* analog with less anticholinergic activity.**

Untoward effects: Blurred vision, orthostatic hypotension, slurred speech, shakes (hand), incoordination, tachycardia and extrasystoles, agranulocytosis, and fatal fibrosing alveolitis. May trigger epileptoid symptoms in the **elderly**. A depressed 39-year-old **male** with pancytopenia and portal hypertension became drowsy, disoriented, and confused, had persistent hyponatremia and raised ***arginine-vasopressin*** levels after dosing with 75 mg/day and then up to 150 mg/day. A 72-year-old **female** developed very noisy and frequent eructations 3 days after 75 mg/day. A 55-year-old **female** developed the same after 20 mg/day. Tachycardia developed in the **fetus** of a 26-year-old **female** treated with doses of 25–75 mg/day. Arrhythmia resolved after drug withdrawal, and the **infant**, born at term, was normal. Numerous fatal overdoses were reported in Great Britain and Australia. So-called **child**-resistant containers were of little use in preventing one **child** from biting through the bottom and gaining access to an unknown quantity. About 0.2% of the **maternal** dose appears in **mother's** milk without any apparent effect on breast-feeding **infants**. Usual therapeutic serum levels are 0.02–0.1 mg/l; toxic at 0.8 mg/l; and a comatose state develops at 1 mg/l.

TD_{LO}, **woman**: 4.5 mg/kg.

IV use in **rabbits** leads to ataxia, hind limb relaxation, and electroencephalogram slowed frequency and increased amplitude, with occasional bursts of rapid activity.

DOXACURIUM CHLORIDE

= *BW-A938U* = *Nuromax*

Skeletal muscle relaxant. **Non-depolarizing neuromuscular-blocking agent.**

Untoward effects: It may have a prolonged effect beyond its need during surgery. It may result in skeletal muscle weakness, and even profound and prolonged skeletal muscle paralysis, with respiratory insufficiency and apnea that may require manual or mechanical ventilation. Small decrease in heart rate and mean arterial pressure, hypotension, flushing, ventricular fibrillation, myocardial infarction, wheezing, bronchospasms, urticaria, reactions at injection site, diplopia, and fever reported in less than 1% of the **patients**.

DOXAPRAM

= *AHR-619* = *Dopram* = *Doxapril* = *Stimulexin*

Respiratory stimulant.

Untoward effects: Faintness, nausea, vomiting, restlessness, shivering, tremors, twitching, rigidity, coughing, laryngospasm, bronchospasm, dyspnea, hiccups, salivation, anxiety, confusion, fever, hypertension, flushing, cardiac arrhythmias, tachycardia, pruritus, pelvic warmth, diaphoresis, ringing in ears and head, pounding heart, sour taste, mydriasis, alkalosis, and convulsions have been reported in **people** after normal or high IV dosage. Use may cause increase or false positive blood urea nitrogen results. Agitation and hallucinations are rare. Thrombophlebitis has followed extravasation.

Amphetamine in certain proportions may antagonize the drug's beneficial effects in **animals**. Excessive doses may lead to hyperventilation, systemic alkalosis, muscle spasms, abnormal posture, and even aggressive behavior.

LD_{50}, **rats**: 261 mg/kg; **mouse**: 270 mg/kg; **dog**: 150 mg/kg; IV, LD_{LO}, **dog**: 40 mg/kg.

DOXAZOSIN MESYLATE

= *Alfadil* = *Cardenalin* = *Cardular* = *Cardura* = *Carduran* = *Diblocin* = *Normothen* = *Supressin* = *UK-33274-27*

Antihypertensive. ***α_1-Adrenergic blocker.***

Untoward effects: Orthostatic hypotension with syncope, dizziness, vertigo, arrhythmias, arthralgia, arthritis, myalgia, and muscle weakness. Has caused stress incontinence in **females** and an increase in postural hypotension in older **people**. Psychosis in a 71-year-old **female** after dosage increased from 8 to 16 mg/day. Has induced a high incidence of congestive heart failure.

DOXEPIN

= *Adapin* = *Aponal* = *Curatin* = *MF 10* = *MF 110* = *P 3693A* = *Quitaxon* = *Sinequan*

Tricyclic antidepressant. **An analog of *amitriptyline*.**

Untoward effects: Anticholinergic effects (dry mouth, blurred vision, constipation, urinary retention), drowsiness, postural hypotension, weight increase, and tachycardia are the most common. Psychosis, mania, heart block and other electrocardiogram abnormalities, rash, sweating, confusion, respiratory depression, insomnia, anorgasm, decreased or increased libido, ejaculatory dysfunction and erectile problems, glossitis, aphthous stomatitis, gingivitis, dental caries, and withdrawal symptoms occur occasionally. Rarely, gynecomastia in **males**, breast enlargement and galactorrhea in **females**, tinnitus, bone marrow depression, eosinophilia, purpura, anemia, neutrophilia, lymphopenia, agranulocytosis, thrombocytopenia, photosensitivity dysarthria, stuttering, nausea, tremors, increased ocular pressure, and hepatitis. Confusion, memory impairment, and delirium can occur in the **elderly**. Although little passes into **mother's** milk, its **maternal** use for depression during lactation leads to a pale, unresponsive **infant**. Poisoning from accidental ingestion and suicide attempts has been reported. These usually cause respiratory depression, unconsciousness, and convulsions. Therapeutic serum concentrations are 0.1–0.25 mg/l, and toxic levels are 0.4–0.5 mg/l. **Patients** are comatose at levels of 2–4 mg/l.

LD_{LO}, **human**: 60 mg/kg.

Drug interactions: Cimetidine 600 mg twice daily and 50 mg doxepin/day increase doxepin plasma levels by ~65% and its half-life by ~50%. ***Propoxyphene*** increased doxepin blood levels > 50% in an 89-year-old **male**. ***Oral contraceptives*** usually decrease its efficacy. Antagonizes ***guanethidine***. Increases effects of ***alcohol***. Discontinue ***monoamine oxidase inhibitor*** at least 2 weeks before ingesting it or using it as a topical cream.

LD_{50}, **rat**: 147, 681 mg/kg; **mouse**: 135, 275 mg/kg.

DOXORUBICIN

= *Adriacin* = *Adriamycin* = *Adriblastina* = *Doxil* = *FI 106* = *1,4-Hydroxydaunomycin* = *Lip-Dox* = *NSC 123,127* = *Rubex*

Antineoplastic. **An anthracyline antibiotic used IV. Isolated from a fungus.**

Untoward effects: Acute toxicity: Anaphylaxis (histamine-mediated), nausea, vomiting, diarrhea, transient electrocardiogram changes, ventricular arrhythmias (premature ventricular contractions, sinus tachycardia), increase in heart size, chest pain, cardiomyopathy, shortness of breath, rales, chromosome aberrations, and red-colored urine (not hematuria) noted. Potent vesicant after extravasation.

Chronic toxicity: Myelodepression, thrombocytopenia, leukopenia, anemia, agranulocytosis, degeneration of cardiac muscle and death from congestive heart failure, cardiomyopathy, mucositis (stomatitis, vulvitis, esophagitis) with ulceration, fever, chills, urticaria, angioneurotic edema, palmar–plantar erythrodysesthesia syndrome, alopecia, onycholysis, hyperpigmentation of nails and nail beds, melanin pigmentation of phalangeal and dermal creases, acral erythema, photosensitivity, jaundice, and liver and renal toxicity.

Its use has an inherent risk to the **fetus**, especially during the first trimester, when liver concentrations were ~10 times **maternal** plasma levels. Extravasation has caused severe local irritation and necrosis. Skin cancer in a 58-year-old **female** at the site of extravasation after 10 years. A serious case of bone marrow toxicity due to ***Adria*** followed a **doctor's** poor handwriting, prescription calling for ***Aredia*** (***pamidronate sodium***). Recall of skin reactions from prior radiotherapy has occurred.

IV, TD_{LO}, **human**: 15 mg/kg/day; another reported 400 μg/kg.

Drug interactions: Its cardiotoxic effects may be potentiated by ***cyclophosphamide***. ***Acetaminophen*** may also enhance its toxicity, as does hyperthermia and radiation therapy. It alters uptake of ***radiopharmaceuticals***.

Its use in **cats** has shown it to be detrimental to wound-healing. Gastrointestinal upsets, anaphylaxis, myelosuppression, alopecia, and cardiomyopathy occur in **dogs** and other **animals**. ***Theobromine*** with it increases its inhibition of DNA biosynthesis in **mice**. IP in **rats** of 16 mg/kg leads to severe myocardial damage. Ingestion of ***tea*** polyphenols decreases the damage. Cardiotoxic and hepatotoxic in **mice**; nephrotoxic in **cats**. Cumulative cardiotoxicity in **dogs** and **rats**.

IV, LD_{LO}, **dog** and **rabbit**: 250 mg/kg.

DOXYCYCLINE

= *D 100* = *DOOTC* = *Doxychel* = *GS 3065* = *Jenacycline* = *Liomycin* = *Periostat* = *Spanor* = *Supracyclin* = *Vibradox* = *Vibramicina* = *Vibramycin* = *Vibraveineuse*

Antibiotic. **Synthetic analog of *oxytetracycline*. Serum half-life is ~18 h+, permitting daily or twice daily dosing; it is eliminated by extrarenal mechanisms.**

Untoward effects: Gastrointestinal upsets (heartburn, anorexia, stomach ache, constipation, nausea, vomiting; diarrhea, 10%; ulcerative esophagitis); with or without adequate water intake after swallowing capsules, yellow urine that can cause misreading of tests for jaundice; agranulocytosis, eosinophilia, photosensitivity (10%), papular rashes, and onycholysis. Rarely, permanent discoloration of deciduous and permanent teeth. Generally, avoid use in pregnant **women** and young **children**. Anaphylactoid reaction from IV use during general anesthesia and ***β-blockade*** therapy reported. 100 mg twice daily to a 15-year-old **male** with recent history of Lyme disease led to painful discoloration of fingernails, resolving 10 days after discontinuation of treatment. Extravasation of IV material has caused phlebitis and induration or burning at infusion sites, and severe local necrotic reactions. Vaginitis, glossitis, stomatitis, proctitis (monilial overgrowth in the anogenital region has occurred), urticaria, rashes, allergic dermatitis, anaphylactoid reactions, photo-onycholysis, lactic acidosis, and photohemolytic effects have also been reported. Since ***calcium*** decreases its availability, avoid dairy products and ***calcium***-containing supplements during treatment. May aggravate azotemia in **patients** with severe renal failure. Usual therapeutic serum levels are 1–5 mg/l; toxic at 30 mg/l.

Drug interactions: Half-life decreased 50% by ***carbamazepine***, ***phenobarbital***, and ***phenytoin***. ***Bismuth subsalicylate*** decreases its bioavailability by 37–51%; waiting for 2 h after doxycycline before taking it, caused little or no decrease in bioavailability. Giving ***aluminum phosphate*** before doxycycline significantly decreased (52%) its absorption and only 26% if given 2 h after. Other ***antacids*** have had similar effects. Food only decreases its absorption slightly. Even small doses of ***ferrous sulfate*** decrease its absorption by up to 50%. Enhanced anticoagulant effect has been reported when given with ***acenocoumarol*** and ***warfarin***. Use in a 17-year-old **female** caused a dramatic increase in ***methotrexate*** levels leading to hematological and gastrointestinal toxicity.

High IV dosage to **dogs** reduced cardiac function. Toxic to **bovine** spermatozoa at 125 μg/ml. Characteristic staining effects in bone, teeth, and thyroids in chronic **rat**, **dog**, and **monkey** trials. Histaminergic effect causes a peripheral vasodilation in **dogs** and vasoconstriction in **rabbits**. Phototoxic to the **mouse** orally or IP. Hepatotoxic in **mice**, especially females. Non-teratogenic in **rat**, **rabbit**, and **monkey** trials.

LD_{50}, **rat**: > 2 g/kg; **mouse**: 1.9 g/kg; **dog**: 500 mg/kg.

DOXYLAMINE SUCCINATE

= *Alsadorm* = *Decapryn Succinate* = *Gittalun* =*Hoggar N* = *Mereprine* = *Sedaplus* = *Unisom*

Antihistamine, hypnotic. **After November, 1976, *dicyclomine* was removed from the U.S. *Bendectin*. In the year 2000, many physicians prescribed its ingredients separately.**

Untoward effects: High dosage has hazardous sedative effects. Despite numerous reports indicating it, in combination with ***pyridoxine*** (***Bendectin***), was not teratogenic, public opinion forced the combination off the market. Any evidence that they were involved in congenital heart disease was slight and involved use in the first trimester, when ***aspirin*** is associated with a two times increase in septation of truncus arteriosis. Pros and cons reported on the association of the combination with pyloric stenosis and limb deformities in **infants**. Fatal and non-fatal cases in very young **children** ingesting unknown quantities of it with other drugs. Rhabdomyolysis reported in seven **patients** (18–31) who had self-overdosed with 500–3000 mg. It has been theorized that maternal ***tobacco*** usage may have a synergistic effect with it, causing genital tract abnormalities in **infants**. Therapeutic serum levels are 0.05–0.2 mg/l; toxic at 1–2 mg/l; serious comatose state at 5 mg/l.

LD_{LO}, **human**: 50 mg/kg.

In **animals**, therapeutic doses can cause central nervous system depression and incoordination, along with occasional gastrointestinal upsets. Overdosage has caused excitement, ataxia, and convulsions. Use has been associated with diaphragm and cardiac defects in their offspring.

LD_{50}, **mice**: 470 mg/kg; **rabbits**: 250 mg/kg.

DROMOSTANOLONE PROPIONATE

= *Drolban* = *Drostanolone Propionate* = *Emdisterone* = *Masterid* = *Masteril* = *Masterone* = *Metormon* = *Metholone* = *NSC-12198* = *Permastril*

Antineoplastic.

Untoward effects: Masculinization in ~70% of **females** treated for > 3 months. Vocal changes, hirsutism, alopecia, fluid retention, mild pustular rashes; clinical jaundice (26%), coarser skin, and increased facial hair. Rarely, hepatocellular neoplasms associated with long-term use. Contra-indicated in carcinoma of **male** breast, in premenopausal **females**, in cardiorenal disease, in hypercalcemia, and in pregnant or nursing **mothers**.

DROPERIDOL

= *Dehydrobenzperidol* = *Dridol* = *Droleptan* = *Inapsine* = *McN-J-R 4729* = *R-4749*

Tranquilizer, antipsychotic, antiemetic. **Often used with *fentanyl*.**

Untoward effects: Incidence of nausea, vomiting, and pain at injection site is low. Oculogyric crises in two young **children** (a 2-year-old **female** and a 6-year-old **female** on 7.5 mg qid/2 weeks). Anxiety, tachycardia, blurred vision, and acute dyskinesias and dystonias. Extreme increase in blood pressure in a 13-year-old **male** with pheochromocytoma after 1.25 mg IV. Given with ***L-dopa***, it increases rigidity in parkinsonian **patients**. Causes hypotension when given with ***perphenazine*** and ***phenelzine***. Recommended injection dose has been decreased from 10 mg to 5 mg to decrease the risk of severe cardiac arrhythmias and sudden death.

Droperidol IV increases body temperature in **rabbits**, decreases body temperature after SC in **rats**, produces adrenergic blockade in **dogs**, and brief slight hypotension in **cats** and **dogs**.

Its use in **animals** (**armadillos**, **agoutis**, **aoudads**, **cats**, **dogs**, **ground squirrels**, **guinea pigs**, **marmots**, **mice**, **monkeys**, **opossums**, **rabbits**, **rats**, and **swine**) has, primarily, been with ***fentanyl***. Seizures, respiratory depression, apnea, excitability, aggressive behavior, hyperthermia, prolonged depression, and death from them has occurred in ~1% of **dogs**. Heart block in them can be prevented by the addition of ***atropine***.

LD_{50}, **rats**: 700 mg/kg; 40 mg/kg, SC; male **mice**: 125 mg/kg; SC and 43 mg/kg, SC.

DROTAVERINE

= *Dihydroethaverine* = *No-Spa* = *Nospanum*

Coronary vasodilator, antispasmodic.

Untoward effects: Allergic dermatitis in **patients**.

DRYMARIA sp.

Untoward effects: ***D. arenaroides*** = ***Alfombrilla***. Found growing in Mexico and New Mexico in acid soils; poisonous to **cattle**, **sheep**, and **goats**. After ingestion, symptoms develop rapidly: normal or decreased body temperature, trembling, muscle spasms, salivation, dyspnea, reluctance to move, collapse, struggling, and death within 2–24 h after the first symptoms. Lung congestion with petechial hemorrhages on the lung and heart. Liver, spleen, and kidney are congested and friable.

D. pachyphylla = ***Inkberry*** = ***Inkweed*** = ***Thickleaf Drymary***, in Texas, New Mexico, Arizona, and Mexico is a glabrous annual that also affects **cattle**, **sheep**, and **goats**. Anorexia, diarrhea, depression, weakness, incoordination, and red urine are characteristic after consumption of small amounts. Needs to be differentiated from anthrax. Coma and death can occur. Gastroenteritis with congestion of liver, kidneys, and spleen, and petechial hemorrhages on the heart noted on post-mortem.

DUBOISIA MYOPOROIDES

Untoward effects*:** In Australia, has caused poisoning in **cattle**. In October, contains almost pure ***hyoscyamine (q.v.), and, in April, almost pure ***hyoscine*** (***scopolamine***, q.v.). ***D. hopwoodii*** in Australia is also poisonous to **livestock**. It contains the alkaloid ***piturine*** (***D-nornicotine***) with actions identical to that of ***nicotine*** (q.v.).

DUGALDIN

Untoward effects*:** The poisonous glucoside principle in ***Helenium Amarum and ***H. autumnale***, ***Bitter Sneeze Weed*** = ***Bitterweed***, common weeds in Alabama that cause salivation, weakness, bloating, dyspnea, and terminal spasms in **cattle**, **sheep**, and **horses** grazing on it. ***H. autumnale*** in the western U.S. is also known as ***Owl's Claws*** and ***Yellow Weed***.

DULCAMARIN

Untoward effects*:** A poisonous alkaloid in ***Solanum dulcamara (q.v.), unfortunately found in some herbal ***teas***.

DULCIN

= *Sucrol* = *Valzin*

Non-nutritive sweetener.

***Untoward effects*:** Suspected carcinogen for **man**. Sale banned in the U.S.

TD_{LO}, **human**: 600 mg/kg; **child**: 400 mg/kg.

LD_{50}, **rat**: 1.9 g/kg; TD_{LO}, **dog**: 1 g/kg.

DUROIA

Untoward effects*:** In Brazil, Venezuela, and Columbia, the seeds of the tree ***D. saccifera, and, in Venezuela, ***D. sprucei***, are toxic when ingested. In Columbia, the seeds of ***D. petiolaris*** are also poisonous.

DUSTS

Untoward effects*:** Adverse effects can be from specific chemicals discussed elsewhere in the text, or be from assorted industrial, house, feed, or farm dusts. These have carried irritants, live bacteria, virus, fungal spores, and mites. Some have caused anosmia, asthma, byssinosis, nasal adenocarcinomas (particularly in the boot and shoe industry), chronic bronchitis, emphysema, and pulmonary fibrosis. Traumatic dust tattooing in a **worker** repairing a gas pipe that exploded. Grain bin explosions attributed to dust have been common. ***Kaolin, ***silica***, and ***talc*** dusts are particularly fibrogenic. Sensitization is not uncommon in **people** exposed to peptide hydrolases (***Bacillus subtilis***, etc.) and ***antibiotics***. A 6-year-old **male** had severe recurrent pneumonitis and hay fever while living in a city near a live chicken market. Improvement followed moving from the area. The symptoms are similar to many dust problems in **man**, called **farmer's** lung, **pigeon-breeder's** disease, **coffee-worker's** lung, sequoiosis, **cork-worker's** lung, etc. House dusts often contain irritating allergens: decaying food particles, animal and insect danders, mites, and lint particles. Dust-removing machines, particularly those using high voltage, give off toxic amounts of ***ozone***.

Animals have suffered from many of the same ill-effects. **Turkeys** have had an increased incidence and severity of air sac lesions from high dust concentration. **Swine** have had an increased incidence and severity of atrophic rhinitis, coughing, and pneumonias. **Farmers** working in confined **swine** and **poultry** operations have developed serious respiratory problems.

DYCLONINE

= *Dyclone* = *Tanaclone*

Topical anesthetic.

Untoward effects*:** Severe bronchospasm, acute wheezing, dyspnea, grand mal seizures, excitement, cardiac arrest, and death have occurred after use in preparation for bronchoscopy and gastroscopy. Euphoria, confusion, dizziness, tinnitus, blurred or double vision, vomiting, hot and cold sensations, twitching, tremors, drowsiness, convulsions, bradycardia, and hypotension also reported. Will precipitate the ***iodine in pyelographic ***contrast media***.

In **animal** trials, 3–4% has irritated the nictitating membrane and 10% has caused corneal opacification.

LD_{LO}, **mouse**: 100 mg/kg; **dog**: 40 mg/kg; **rabbit**: 200 mg/kg.

DYDROGESTERONE

= *Dufaston* = *Duphaston* = *Gestatron* = *Gynorest* = *Prodel* = *Retrone*

Progestogen.

***Untoward effects*:** Potentially masculinizing and teratogenic to **female fetus**. Nausea, vomiting, and breakthrough bleeding.

DYMANTHINE
= *Dimantine* = *GS 1339*

Anthelmintic.

Untoward effects: Nausea (40%), giddiness (30%), epigastric pain (30%), vomiting (20%), diarrhea (20%), headache and disturbed sleep (10%).

DYPHYLLINE
= *AFI-phyllin* = *Astmamasit* = *Asthmolysin* = *Astrophyllin* = *Circair* = *Coronarin* = *Cor-Theophylline* = *Dilor* = *Diprophylline* = *Glyfyllin* = *Glyphylline* = *Hiphyllin* = *Hyphylline* = *Lufyllin* = *Neostenovasan* = *Neothylline* = *Neotilina* = *Neo-Vasophylline* = *Neutrafil* = *Neutraphylline* = *Prophyllen* = *Silbephylline* = *Solufilin* = *Solufyllin* = *Theal* = *Thefylan*

Bronchodilator.

Untoward effects: Similar to ***theophylline*** (q.v.), but nervousness, dizziness, and palpitations may be less severe. Headache, nausea, dyspnea, and a feeling of "heaviness" reported by **some**. ***Probenecid*** inhibits its renal excretion in **man**, **rat**, and **chicken**.

LD_{50}, **mice**: 3.4 g/kg.

E

EBASTINE
= *Bastel* = *Ebastel* = *Kestine* = *LAS 90* = *LAS W90* = *LAS-W-090* = *W 090*

***Antihistamine* (H_1).**

***Untoward effects*:** Headache and dry mouth are common. Can increase QT interval.

EBONY
= *Dalbergia*

Untoward effects*:** Eczema has been reported from Jamaican or West Indian ebony = ***Granadilla; hypersensitivity to an ebony wood bracelet has occurred. Ebony **workers** have developed a dark yellow or blackish discoloration in the nail bed region. Root concoctions of ***D. nitidula*** and ***D. stuhlmannii*** used in eastern African herbal medicines has caused death in many **patients**.

EBROTIDINE
= *FI 3,542*

***Antihistamine* (H_2).**

***Untoward effects*:** Erectile dysfunction and impotence in a 34-year-old **male**.

ECHINACEA

Untoward effects*:** Can cause allergic reactions, particulary in ***ragweed-allergy **sufferers**.

ECHINOCHLOA CRUS-GALLI
= *Japanese Millet*

Untoward effects*:** Swollen pendulous ears on **lambs** in Australia after grazing on it. Photophobia, photosensitization, and irritation and intense yellowing of skin reported. May contain dangerous concentrations of ***nitrates.

ECHINOMYCIN
= *Levomycin* = *Quinomycin A*

Antineoplastic antibiotic.

***Untoward effects*:** Nausea, vomiting, allergic reactions, thrombocytopenia, and reversible liver enzyme abnormalities.

Beagle dogs had gastrointestinal and liver upsets, as well as toxicity to their lymphoreticular systems.

IP, LD_{50}, **mouse**: 400 μg/kg.

ECHIUM PLANTAGINEUM
= *Paterson's Curse* = *Purple Bugloss* = *Salvation Jane*

Untoward effects*:** An annual garden plant that has escaped into Australian pastures, causing liver and renal toxicity in **sheep**, **horses**, **cattle**, and **swine**, due to its hepatotoxic ***pyrrolizidine (q.v.) content. ***Honey*** derived from the plant's nectar has contained up to 1 ppm of the alkaloid. A **colt**, after grazing on it for 4 months, became blind, had abdominal pain, and died within 5 h after symptoms were noticed. ***E. vulgare*** = ***Blueweed*** = ***Blue Devil*** = ***Blue Thistle*** = ***Vipérine*** = ***Viper's Bugloss*** is found in northeastern North America. The plant's hairs (bristles when mature) on its stems and leaves cause a severe dermatitis with inflammation and pruritus, especially after being imbedded in the skin. Chronic use of *Echium sp.* in ***tea*** can lead to veno-occlusive liver disease with thrombosis of the hepatic vein leading to cirrhosis.

ECHOTHIOPHATE IODIDE
= *217 MI* = *Phospholine Iodide*

***Cholinergic.* An organophosphorus compound. Used topically and orally.**

***Untoward effects*:** Cholinesterase inhibitor with prolonged effects. Progressive visual impairment, due to lens opacities in 50% of **patients** given 0.06–0.25% topically twice daily. Iris pigment epithelial cyst formation occurred in many (especially **children**) on high concentrations. Miosis-induced edema, periorbital headache, eye ache, transient increased intraocular pressure, conjunctival irritation, conjunctival and ciliary congestion, mild anterior uveitis with formation of posterior synechiae, blurred vision, poor night vision, and retinal detachment occur. Acute cholinergic crises have been reported. Crosses the placental barrier in pregnancy. Pseudocholinesterase levels decreased in **infant** after **mother** treated with it for glaucoma during first 32 weeks of pregnancy.

Systemic effects are unusual, but have occurred after 0.125 or 0.25%, topically, daily or twice daily leading to increase parasympathomimetic activity with nausea, anorexia, abdominal cramps, diarrhea, salivation, bradycardia, confusion, dizziness, headache, anxiety, paresthesia, weakness, fatigue, sweating, wheezing, and dysphagia. Systemic effects can follow drip down the tear duct.

Drug interactions*:** Use while exposed to ***anticholinesterase insecticides can cause serious adverse systemic effects. If used for over 2 weeks, **patients** risk severe prolonged apnea from ***succinylcholine*** or electroshock therapy.

In **animals**, its cholinergic effects may include nausea, diarrhea, salivation, and bradycardia. Avoid its use with ***organophosphorus insecticides***.

IP, and SC, LD_{50}, **mouse**: 135 µg/kg; ocular LD_{50}, **rabbit**: 250 µg/kg.

ECONAZOLE NITRATE

= *Ecostatin* = *Epi-Pevaryl* = *Gyno-Pevaryl* = *Ifenec* = *Micofugal* = *Micogin* = *Palavale* = *Pargin* = *Pevaryl* = *R-14827* = *Spectazole*

Antifungal. **A topical imidazole.**

Untoward effects: Local irritation, especially after vaginal use (> 10%), including itching, burning, stinging, erythema, and, rarely, a rash.

Prolongs gestation and excreted in the milk of **rats** after oral use.

LD_{50}, **rats**: 668 mg/kg; **mice**: 462 mg/kg; **guinea pigs**: 272 mg/kg; **dogs**: >160 mg/kg.

ECTYLUREA

= *Astyn* = *Cronil* = *Ektyl* = *Levanil* = *Neuroprocin* = *Nostal* = *Nostyn* = *Pacetyn*

Sedative, hypnotic.

Untoward effects: Low incidence of drowsiness. Occasionally, nausea, rashes, dizziness, nervousness, and cholestatic jaundice. Sleep, slight ataxia and brief periods of dysarthria in suicide attempts with ~30 tablets (9 g).

LD_{LO}, **human**: 500 mg/kg.

LD_{50}, **rats**: 2.5 g/kg; **mice**: 4.5 g/kg; **dogs**: 2.7 g/kg; **rabbit**: 1.2 g/kg.

EDETIC ACID

= *EDTA*

Chelating agent. **The most common forms used are *Calcium Sodium Edetate* and *Sodium Edetate*.**

Untoward effects: Renal irritation has been the most serious. As an IV antidote for ***lead*** poisoning, it has precipitated the death of some **patients**, causing albuminuria, uremia, tubular nephrosis, kidney necrosis, oliguria, renal failure, and arteriosclerotic changes. A high proportion of **those** receiving single IV dose develop the "excessive chelation syndrome", apparently related to amount and rate of injection. Symptoms include light-headedness, dizziness, vertigo, sneezing, nasal congestion, nausea, vomiting, abdominal cramps, fever, and headache with chest, muscle, and joint pains. Pain at the injection site and hypotension have occurred. Occasionally, glycosuria and hypocalcemic tetany. Rarely, angioneurotic edema and mucocutaneous lesions have followed prolonged usage. Dermatitis lesions resemble those of ***pyridoxine*** deficiency. Its use in ophthalmic solutions leads to acute allergic conjunctivitis and/or periorbital dermatitis. Its dust leads to skin, eye, and respiratory tract irritation. Spermicidal *in vitro*.

Drug interactions: Enhances neuromuscular blockade caused by many ***antibiotics***, leading to apnea and increased muscle weakness. ***Glucocorticoids*** have enhanced its toxicity.

IV use leads to hypocalcemia in **cattle**. **Rat** studies confirmed destruction of proximal tubular cells. The lesions are dose- and treatment duration-related.

EDIFENPHOS

= *BAY 78,418* = *Ediphenphos* = *Hinosan* = *SRA 7847*

Fungicide. **Used in *rice* and *cotton* fields.**

Untoward effects: Rapidly metabolized (within a few days) in **goats**, **dogs**, and **rats**. Oral dosing in **mallard ducks** (LD_{50}, > 2.8 g/kg) and **pheasants** (LD_{50}, 500 mg/kg) led to polydipsia, goose-stepping ataxia, geotaxia, imbalance, asthenia, falling, hypoactivity, ptsosis, and death.

LD_{50}, **rat**: 150 mg/kg; **mouse**: 218 mg/kg; **guinea pig** and **rabbit**: 350 mg/kg.

EDROPHONIUM

= *Antirex* = *Enlon* = *Reversol* = *Tensilon*

Anticholinesterase, cholinergic, diagnostic aid, curariform antagonist.

Untoward effects: Occasionally, cholinergic side-effects increase vagal activity leading to miosis, sweating, blurred vision, diplopia, lacrimation, increased salivary and bronchial secretions, laryngeal spasms, respiratory paralysis, weakness, incontinence, and gastrointestinal upsets. Has caused sinus bradycardia, arrhythmias, ventricular asystole, decreased blood pressure, dizziness, atrioventricular block, and, rarely, cardiac arrest, particularly in the **elderly**, in **those** receiving ***digitalis*** preparations, and in **those** with a history of cardiac disease. Therapeutic serum levels are < 0.1 mg/l and usually toxic at 0.15 mg/l.

LD, **human**: ~100 mg.

Drug interactions: Large doses can aggravate circulatory collapse during muscular paralysis, due to ***skeletal neuromuscular-blocking agents***. Cholinergic effects may be antagonized by ***quinidine***. Its actions are enhanced by ***colistin*** and it reduces the effects of ***kanamycin***.

LD_{50}, **mouse**: 600 mg/kg.

EEL, MORAY

= *Muraena helena*

Untoward effects*:** Rarely attack unless provoked. May grow to 10 ft long and 1 ft in diameter. They have powerful jaws with tearing, crushing teeth. **Swimmers** in temperate zones should not explore crevices and holes with **their** fingers. Many tropical species are poisonous when eaten by **man**, some, from the **eel's** consumption of the algal toxin, ***ciguatera (q.v.), and then concentrated in their flesh.

EFAVIRENZ

= *DMP 266 = EFV = L 743,726 = Sustiva*

Antiviral.

Untoward effects*:** Dizziness, insomnia, impaired concentration, abnormal dreams, drowsiness, headache, and severe skin rashes ($\frac{1}{3}$ of **patients**). Occasionally, nightmares, hallucinations, vivid dreams, diarrhea, hypersensitivity, hepatotoxicity, and neutropenia. Adverse central nervous system effects often occur 1–3 h after each dose and generally cease within a few days or weeks. A 34-year-old **male** with HIV stabilized on ***methadone 40 mg/day showed withdrawal symptoms after efavirenz added to **his** therapy, despite raising ***methadone*** dosage to 80 mg/day. Fetal anomalies after dosing pregnant **monkeys**.

EFLORNITHINE

= *DFMO = Difluoromethylornithine = Ornidyl = RMI 71,782 = Vaniqua*

Antiprotozoal.

***Untoward effects*:** Anemia and leukopenia. Occasionally, thrombocytopenia, diarrhea, nausea, vomiting, and seizures. Rarely, hearing loss. Its slowing of facial hair growth may become commercially useful. When used topically to prevent hair growth, it has caused some skin irritation and "razor bumps".

Prevents embryonic growth in **rats** and **rabbits**.

EGGS

Untoward effects*:** 1 mg ***avidin, a constituent in raw or unheated egg whites, inactivates 13.8 µg ***biotin***, and the deficiency has caused diarrhea and dermatitis in **infants** and **mink** consuming large quantities. Eggs have inhibited ***vitamin*** $\boldsymbol{B_{12}}$ absorption. Although rich in ***iron***, 95% of it occurs in the yolk, tightly bound to a protein, ***phosvitin***. Theoretically, it can further bind other ***iron*** in the digestive tract, and it does inhibit ***iron*** absorption from foods cooked with eggs. The inference that eggs cause increased ***cholesterol*** levels in **man** is questionable, since a large chicken egg contains about 5.1 g of fat. This consists of 1.7 g ***saturated fats***, 0.7 g ***polyunsaturated fatty acids***, and 2.7 g ***monounsaturated fatty acids***. The latter two are considered to be ***cholesterol***-lowering and the ratio in the egg of ***unsaturated*** to ***saturated fatty acids*** is 2 : 1. **Romanians** eat 3 eggs/day and have less coronary artery disease than **Americans**. Eggs have triggered migraines, and the whites have cross-reacted with ***gliadin*** fractions in ***gluten*** and caused toxicity to celiac **patients**. Hypersensitivity to it has caused anaphylactic reactions, contact urticaria, nausea, vomiting, laryngeal edema, bronchospasm, skin rashes, diarrhea, shock, and collapse. Reactions have occurred from its use in the manufacture of ***measles***, ***mumps***, ***rubella***, and ***rabies vaccines***. Transmission of egg proteins into breast milk in allergenic levels can occur. It is estimated that 10% of **children** are sensitive to egg protein in vaccines. Egg white is used to make the froth in ***root beer***. Cooking can alter a food's chemistry, so that some **people** allergic to eggs can safely consume hard-boiled eggs. Eggs from different areas have contained 1–40 ppb ***mercury***. WHO recommends **mercury** food intake to be <50 ppb. A rare farm had levels up to 420 ppb. Rotten eggs contain ***hydrogen sulfide***, a colorless, poisonous gas, and toxicity to **man** from an egg source is unlikely. The most highly publicized toxicity is from the increased incidence of *Salmonella enteritidis*, from the consumption of raw or undercooked eggs from certain sources. In schools and nursing homes, scrambled eggs was the most common culprit. It was estimated that 224,000 cases of food poisoning (salmonellosis) occurred in over 3 million **people** who ate ~1 million gallons of ***Schwan's Ice Cream***®, as a result of tanker trucks that were used to haul both ***ice cream*** mix and raw eggs. Enterotoxigenic staphylococcal contamination of intact, hard boiled eggs caused ~300/850 **children** at an Easter egg hunt to become ill. An infected **cook** appeared to be the source. Heating and then cooling dyed eggs, to facilitate peeling, permitted the absorption of ~2 ml contaminated cooling water into each egg. Fishy-tasting eggs are not from fish meals fed to **chickens**, but rather from feeding too much ***rapeseed*** meal, with more ***trimethylamine*** than the **hen** can metabolize. Occupational asthma occurs in **workers** inhaling egg protein in dusts of powdered eggs. Easter egg dyes (***Handy Easter Egg Color***®) caused a 7-year-old **male** to develop a severe circumocular and oral edema, and swelling of **his** uvula, palms of **his** hands, and soles of **his** feet within ~12 min.

A carcinogenic substance for **mice** has been found in egg yolks. **Dogs** have had allergic reactions due to egg proteins in various ***vaccines***. Experimentally, it has been used to cause ***biotin*** deficiency in **cats**, **dogs**, **foxes**, **hamsters**, **mink**, **poultry**, and **rats**. Egg yolks fed to **rabbits** causes ***cholesterol*** to deposit in their arteries.

EGGPLANT

= *Solanum melogena*

Untoward effects: Can occur from infrequent, occasionally high levels of ***solanine*** (q.v.), a glycoalkaloid in it, which causes gastrointestinal irritation with nausea, cramping, constipation, diarrhea, and cholinesterase inhibition.

Contains ***caffeic acid***, a carcinogen in **rabbits**.

ELDERBERRY

= *American Elder* = *Black Elder* = *Sambucus canadensis*

Use: The berries, fresh, cooked, or dried, have been used in making jelly, jams, ***wine***, soup, pies, and dyes; the blossoms have been steeped for ***tea*** or cooked. The ***tea*** is a mild laxative.

Untoward effects: Under certain conditions, the roots, leaves, bark, and young shoots (but not the purple-black, juicy berries) concentrate ***nitrates***, ***alkaloids*** (particularly ***sambucine***), and ***cyanogenic glycosides***. The latter are particularly abundant in shrubs that have been stunted, wilted, or frost-damaged. The ***cyanide*** can cause initial excitement and muscle tremors, polypnea, dyspnea, tachycardia, salivation, lacrimation, loss of voice, nausea, diarrhea, clonic convulsions, and coma. Blood is bright, cherry red. Nausea and digestive upsets are the most common symptoms in **adults** and **children**. **Children** have been poisoned by using pieces of the pithy stems for blowguns or pea-shooters, and from chewing or sucking on the bark.

S. mexicana = ***Marsh Elder***, is found in southwestern U.S., Mexico, Central America, Peru, and Venezuela. Eight **Californians** were poisoned consuming a beverage 2 days after crushing the berries with their leaves and branches into a pot. **They** were flown by helicopter to a hospital. Within 15 min after drinking the juice, 11/25 became nauseous and vomited. Abdominal cramps, weakness, dizziness, and numbness reported in the **eight**. **One** of those became stuporous. **All** recovered. **Patients** sensitive to ***ragweed*** usually test positive to it.

S. nigra = ***Sambuco*** = ***Sammuco*** has caused an acute papulovascular dermatitis in a **female** after contact exposure to its branches.

S. racemosa = ***Red or Scarlet Elderberry***. Ingestion of its leaves, shoots, or bark has caused nausea, vomiting, and diarrhea.

Cattle, **hogs**, **sheep**, and **horses** have been poisoned from eating the roots, leaves, and young shoots.

ELEDOISIN

= *ELD 950*

A peptide chemical from the posterior salivary gland of two species of octopus found in the Bay of Naples, Italy.

Untoward effects: Powerful hypotensive and vasodilator in **man**, **dog**, and **rabbit**, but not in **sheep**. IA in **man** led to transient feelings of heat, headache, paresthesias, vertigo, and facial hyperemia.

ELM

= *Ulmus*

Untoward effects: Hypersensitivity leading to eczema in **people** and its pollen has caused an allergic rhinitis in **some**.

ELYMOCLAVINE

An ergot alkaloid from *Claviceps purpurea* fungi, parasitic on *Elymus* and *Pennisetum*.

Untoward effects: Elicits an excitation syndrome in **animals** and has been misused by **drug-abusers**.

Embryocidal and teratogenic in **mice**.

IP, LD_{50}, **mouse**: 67 mg/kg.

ELYMUS sp.

Untoward effects: From the ***elymoclavine*** (q.v. above) on it.

The ***Medusa-head Rye*** has coarsely bearded heads that cause mechanical injury to the mouth, eyes, and noses of **sheep** and **cattle** grazing on it.

EMEPRONIUM

= *Cetiprin* = *Restenacht* = *Ripirin* = *Uro-Ripirin*

Antispasmodic, ganglion-blocking agent.

Untoward effects: Esophageal ulceration, strictures, and inflammation, as well as buccal mucosal ulceration were common, especially, when taken with insufficient water. Dry mouth, nausea, blurred vision, dizziness, abdominal pain, retrosternal pain, headache, painful micturation, and dysphagia have also occurred.

TD_{LO}, **human**: 175 mg/kg/21 days.

EMETINE

= *Hemometina* = *NSC 33,669*

Untoward effects: Occur in 50–75% of treated **patients**. Cardiovascular reactions are the most serious leading to atrial tachycardia, premature ventricular contractions, pericarditis, hypotension, epigastric pain, dyspnea, gallop rhythm, right ventricular dilation, congestive heart failure, myocardial necrosis, heart stops in diastole, and death. Prolonged P-R and QT intervals, ST segment changes, inversion or flattening of the T wave, and premature beats noted on electrocardiogram. Oral doses are often vomited, so, in the U.S., it is given IM or SC.

Nausea, vomiting, fatigue, skeletal muscle weakness, diaphoresis, headache, peripheral neuropathy, wrist drop, vertigo, necrotizing myopathy, convulsions, nephritis, cellulitis at injection site, urticaria, fixed-drug eruptions, and variable leukonychia or white coloration of nails. It is the principle active ingredient of **Syrup of Ipecac**.

LD_{LO}, **human**: 5 mg/kg.

Toxic dose for **cats** ~20 mg, **dogs**, 30–500 mg. **Sheep** have died from 4.5 ml of 1% solution IM. Symptoms are violent vomiting, increased secretions in respiratory and alimentary tracts, gastrointestinal inflammation, coma, and death. Therapeutic and toxic levels in **monkeys** are very close, and 80% of those treated show some toxicity.

LD_{LO}, **guinea pigs**: 20 mg/kg.

EMODIN

= *Archin* = *Frangulic Acid*

***Cathartic.* In *rhubarb* root, *alder buckthorn*, and *cascara*.**

***Untoward effects*:** In large doses, sufficient amounts may be excreted in **mother's** milk to affect the **infant**. Causes a pink to red to red-brown color in alkaline urine, and yellow-brown in acid urine. It is a diarrheagenic toxin extracted from cultures of *Aspergillus wentii*.

EMYLCAMATE

= *Kabi 925* = *JD-91* = *Nuncital* = *Restetal* = *Statran* = *Striatran*

***Anxiolytic.* A carbamate.**

Untoward effects*:** Similar to ***meprobamate (q.v.). Gastrointestinal irritation (5%), drowsiness (3%), headache, dry mouth, dizziness, palpitations, paresthesia, insomnia, increased anxiety, irritability, and skin rash reported. Severe allergic reactions can occur.

LD_{LO}, **human**: 500 mg/kg.

LD_{50}, **mouse**: 760 mg/kg.

ENALAPRIL

= *Amprace* = *Bitensil* = *Cardiovet* = *Enacard* = *Enaloc* = *Enapren* = *Glioten* = *Hipoartel* = *Innovace* = *Lotrial* = *MK-421* = *Olivin* = *Pres* = *Renitec* = *Reniten* = *Renivace* = *Vasotec* = *Xanef*

Antiotensin-converting enzyme inhibitor, antihypertensive.

***Untoward effects*:** Headache and dizziness reported in 4–5% of **patients**; fatigue, diarrhea, rash, hypotension, cough, nausea, and orthostatic effects in 1–2%.

***Allergy*:** Numerous case reports of angioneurotic edema, often with pruritus; tongue edema and edema of the submaxillary area. Laryngeal edema can be fatal.

Blood*:** Two cases of induced agranulocytosis in the literature. A case of eosinophilia in a 66-year-old **male** has also occurred. Rarely, bone marrow suppression, thrombocytopenia, and neutropenia. Decreased hemoglobin and hematocrit with anemia occurs in rare cases. Slight increase in blood urea nitrogen and ***creatinine can occur.

***Cardiovascular*:** Syncope, orthostatic hypotension, palpitations, facial flushing, and chest pain reported.

***Gastrointestinal*:** Abdominal pain, vomiting, and dyspepsia.

***Gynecomastia*:** Reported in a 58-year-old **male** and a 72-year-old **male**.

***Liver and pancreas*:** Hepatotoxicity (usually, cholestatic) and pancreatitis occur. May decrease liver plasma flow.

***Miscellaneous*:** Intermittent diplopia and gait disturbance in a 63-year-old **male** 1 day after 5 mg twice daily. Hyperkalemia, fever, nasal blockage and rhinorrhea, mucosal and submucosal edema of the small intestine. Infrequent erectile disorders.

***Mouth*:** A scalded sensation of the oral mucosa and altered taste reported. A 63-year-old **male** develop pemphigus of the larynx and esophagus.

***Myopathy and neuropathy*:** Myalgia, arthralgia, and peripheral neuropathy (decreased motor and sensory nerve conduction).

***Poisoning and suicide*:** Failed suicide attempts reported in **patients** after 300–440 mg by ingestion, and 600 mg IV.

***Psychosis*:** A 41-year-old **male** developed an acute psychosis 4 weeks after receiving 10 mg/day. Agitation, panic, extreme depression, and insomnia occurred.

***Renal*:** Decreased renal function, acute renal failure, oliguria, and anuria. Thrombosis of the renal artery in a renal transplant **patient**, after a single 5 mg dose. Probable cause of early renal artery occlusion in a 71-year-old **female** and anuria in an **infant** whose **mother** was given the drug during pregnancy.

***Respiratory*:** Numerous reports of an induced, dry, nonproductive cough, and occasionally bronchitis and wheezing exist. Unfortunately, many clinicians are unaware that this can persist for at least 6 months (particularly, in **women**), necessitating unnecessary and costly diagnostic procedures. Induced bronchospasm is rare.

***Skin*:** Pemphigus, vulvovaginal pruritus, bullous eruptions, and Schönlein-Henoch purpura reported.

Drug interactions*:** May increase ***imipramine blood levels, especially with ***fluoxetine***. ***Rifampin*** may decrease its hypotensive effect.

LD_{50}, **rats** and **mice**: 2 g/kg.

ENAMEL

Untoward effects: Dermatitis reported. Some contain ***bromates***, ***chromates*** (q.v.), ***lead*** (q.v.), and ***nickel oxides*** (q.v.) in green enamels. Art and craft **enamelers** need to be aware of possible reproductive system damage from those containing ***glycol ethers***.

ENCAINIDE

= *Encaid* = *Enkade* = *MJ 9067*

Antiarrhythmic.

Untoward effects: Light-headedness, dizziness, headache, nausea, diarrhea, constipation, tremors, ataxia, hyperglycemia, bradycardia, heart-block, sustained tachycardia, new ventricular fibrillation, heart failure, prolonged PR and QRS on electrocardiogram, dermatitis, and visual disturbances. Because of the cardiac problems and some increased mortality due to induced arrhythmias and shock, Bristol-Myers Company withdrew its sales ~3 years after its release by the FDA.

LD_{50}, **mice**: 86 mg/kg; **dogs**: 43 mg/kg; IV, **mice**: 16 mg/kg; **dogs**: 17 mg/kg.

ENDIVE

Untoward effects: Has caused an immediate urticarial reaction to intact or scratched skin, particularly in agricultural **workers**, **chefs**, and **housewives**.

ENDOSULFAN

= *Benzoepin* = *BIO 5462* = *Chlorthiepin* = *Cyclodan* = *ENT 23,979* = *FMC 5462* = *HOE 2671* = *Insectophene* = *Kop-Thiodan* = *Malix* = *NIA 5462* = *Thimul* = *Thiodan* = *Thiofor* = *Thionex*

***Insecticide.* A *chlorinated hydrocarbon* compound.**

Untoward effects: Skin irritant. Ingestion leads to nausea, confusion, agitation, flushing, dry mouth, headache, tremors, convulsions, and hepatitis. Acute poisoning in an industrial **worker** who cleaned vats containing residues of it. Malaise, fainting, and a convulsive fit were **his** first symptoms. After recovery, **he** became disoriented and agitated. Then, 2 years later, came cognitive and emotional deterioration, severe memory impairment, and inability to do many tasks. Ingestion, inhalation, and dermal exposure can be fatal. Causes eye damage. Do NOT enter treated areas without protective clothing for 24 h. Environmental Protection Agency limits content in foods at 0.1–2.0 ppm. FDA establishes a maximum limit of 24 ppm in dried ***tea***.

Moderately toxic to **bees**. Some steel drums of it dumped into the Rhine River near Bingen, West Germany in 1967 rusted, and, in 1969, caused the death of ~40 million **fish** along 200 miles of the river before the ebb tide carried the poison into the North Sea. Rotterdam and Amsterdam closed the water conduits supplying them with drinking water for their **people**. **Calves** have died from contamination of their feed bunks. **Cattle** accidentally sprayed with a 0.1% emulsion became listless, staggered, became restless and hyperexcitable, goose-stepped, and had muscle spasms and violent convulsions. 50/250 became ill, and 11 of those died. ~8 g killed **cattle** within 15 h. Will pass into **cow's** milk and eggs of laying **hens**. In poisoned **ducks** and **pheasants** they develop ataxia, goose-stepping, jerkiness, dyspnea, tremors, and falling.

LD_{50}, **rat**: 18 mg/kg; **mouse**: 75 mg/kg; **cat**: 2 mg/kg; **hamster**: 118 mg/kg; **dog**: 77 mg/kg; **mallard ducks**: 33 mg/kg; **pheasants**: 80–190 mg/kg.

ENDOTHALL

= *Accelerate* = *Aquathol* = *Des-I-Cate* = *Herbicide 282* = *Herbicide 283* = *Hydout* = *Hydrothol 47* = *Hydrothol 191* = *Niagrathal* = *Tri-Endothol*

Herbicide, defoliant.

Untoward effects: Irritant to skin and mucous membranes. Ingestion leads to gastrointestinal tract necrosis, vomiting, hematemesis, diarrhea, melena, arrhythmias, central nervous system depression, collapse, and coma. Corrosive.

LD, **human**: 1–3 g.

Very toxic to **honeybees**. Do NOT use with food **fish** for at least 3 days after water treatment. Poisoned **ducks** and **pheasants** show dyspnea, tachypnea, ataxia, goose-stepping, falling, regurgitation, jerkiness, and lethargy. **Fish** are easily killed by concentrations > 0.3 ppm.

LD_{50}, **rats**: 51 mg/kg; **guinea pig**: 250 mg/kg; **mallard ducks**: 229 mg/kg; **pheasants**: < 198 mg/kg.

ENDRIN

= *Compound 269* = *Endox* = *Endrex* = *ENT-17251* = *Hexadrin* = *Mendrin* = *Nendrin* = *SD3419*

***Insecticide.* Chlorinated hydrocarbon.**

Untoward effects: One of the most toxic chlorinated hydrocarbons; related to ***aldrin*** and ***dieldrin***. All uses now prohibited in the U.S. Blood concentration of 3 μg% toxic in **man**. Acute poisoning with sudden convulsions in three **Egyptians** eating ***bread*** made with contaminated ***flour***. 6/13 **children** (2 months to 16 years) in Israel died after signs of weakness, vertigo, headache, blurred vision, abdominal pain, nausea, vomiting, convulsions, and coma. Autopsy revealed congestion of brain and meninges, edematous white matter, pulmonary congestion and edema, pleural hemorrhages, and liver congestion. India reported 60 fatal cases in a 5 year period, of which 41 were suicides and five murders. U.S. Center for Disease Control reported 3/4 with hyperexcitability and sudden convulsions after eating ***bread*** baked from ***flour*** stored in contaminated sacks. Of 874 **people** hospitalized after four poisoning outbreaks in Qatar and Hofuf from accidentally contaminated ***flour***, 26 died. In Pakistan, 19/192 died. Sudden collapse, bilateral jerking of upper extremities followed by generalized tonic and clonic contractions, frothing of the mouth or vomiting. **Some** had headaches and nausea 30 min before collapsing. Contaminated ***sugar*** was the apparent source. The first reported outbreak was in Great Britain, where bags of ***flour*** had been in a railroad boxcar a month after an endrin spill. In 1988, three **family members** in California became ill after eating store-bought taquitos (**corn** tortillas wrapped around a meat filling). Other **people** also became ill. The source was not located. It is also hepatotoxic.

LD_{LO}, **human**: 5 mg/kg.

Toxic to **honeybees** and **fish**. An avicide. 0.2–0.4% sprayed on pasture grasses toxic to **calves**. Caused hypertension in **dogs** and major **fish** kills in the Mississippi River. **Goats** eliminated it in their milk for 22 days after a 1 month feed exposure. Carcinogenic in **rats**. Used as a rodenticide. Apprehension, belligerence, salivation, hyperesthesia, convulsions, and death are common in **cats**. Poisoned **cattle** become restless, twitch, stagger, slobber, and convulse.

LD_{50}, **rats**: 3 mg/kg; **mouse**: 1.4 mg/kg; **cats**: 3–6 mg/kg; **hamster**: 10 mg/kg; **rabbit**: 7 mg/kg; **goats**: 25–50 mg/kg; **pheasants**: 3–4 mg/kg; **mallard ducks**: 5.6 mg/kg; **birds**: < 2 mg/kg; **grouse** and **quail**: ~1 mg/kg; **pigeon**: 5.6 mg/kg; LC_{50}, 96 h; **fish**: <1 µg/l.

ENFLURANE

= *Alyrane* = *Compound 347* = *Efrane* = *Ethrane* = *Methylflurether* = *NSC-115944*

Inhalation anesthetic.

Untoward effects: Involuntary muscle movements and seizures with deep levels of anesthesia; hypocapnia with light levels. Hypotension, respiratory depression severe bronchospasm, nausea, vomiting, shivering, arrhythmias, increased white blood cells, hepatitis, hiccups, delayed asthmatic response, and malignant hyperthermia have occurred. Risk of convulsions in **people** with a history of epilepsy.

Drug interactions: Use with ***β-adrenergics*** and drugs that increase ***norepinephrine*** levels (***tricyclic antidepressants***, ***monoamine oxidase inhibitor***, ***L-dopa***) can precipitate dangerous and fatal arrhythmias. Can prolong ***mivacurium*** neuromuscular blockade.

Inhalation TC_{LO}, **human**: 1%/6 h.

Malignant hyperthermia reported in **dogs**. Central nervous system stimulation, muscle twitching, and even convulsions have been reported during recovery, therefore premedication is indicated.

ENOXACIN

= *Abenox* = *AT-2266* = *Bactidan* = *Comprecin* = *Enoxen* = *Enoxor* = *Flumark* = *Gyramid* = *Penetrex*

***Antibacterial.* A *nalidixic* analog.**

Untoward effects: Achilles tendinitis and tendon rupture in an 85-year-old **male** after 400 mg twice daily/7 days. By mid-1996, over 150 cases of induced tendinitis were reported world-wide on quinolones. Photosensitivity reactions reported. Headache, dizziness, nausea, abdominal pain, diarrhea, anorexia, insomnia, seizures, rashes, vasculitis, serum sickness-like reactions, and anaphylaxis occur. Reversible leukocytopenia, eosinophilia, increased hepatic enzymes, and acute hepatitis also reported. Avoid use in **people** < 18 years, and pregnant or nursing **women**.

Drug interactions: Inhibits ***theophylline*** clearance, increasing its plasma levels and potential toxicity. Causes some decrease in ***warfarin*** clearance. A six times increased ***caffeine*** elimination time. Absorption is reduced by ***ranitidine*** and ***antacids***.

Cartilage erosions in young **dogs**. Interactions with ***theophylline*** in **rats** are similar to those in **man**.

LD_{50}, **rats** and **mice**: > 5 g/kg.

ENOXAPARIN

= *Clexane* = *Klexane* = *Lovenox* = *PK-10169* = *RP-54563* = *Ultraparin*

***Antithrombotic.* A low molecular weight *heparin*.**

Untoward effects: Hematomas, urticaria; skin irritation, erythema, and necrosis at injection site; and thrombocytopenia reported. Bleeding complications appear to be less than with standard ***heparin***. **Patients** with renal insufficiency have a high (30%) risk of bleeding. A hospitalized 69-year-old **male** suffered a fatal retroperitoneal hematoma 4 days after 80 mg twice daily.

ENOXIMONE

Inotropic agent, phosphodiasterase inhibitor.

Untoward effects: A 66-year-old **male** developed convulsions and diarrhea after 6 µg/kg/min IV.

ENROFLOXACIN

= *Baytril* = *BAY VP-2674* = *CFPQ*

***Antibacterial.* A quinolone.**

Untoward effects: Has caused gastrointestinal upsets including nausea and vomiting in **dogs**. In **dogs**, doses of 125 mg/kg/day/11 days has been lethal. **Dogs** up to 7 months have developed an erosive arthropathy with carpal and hind limb weakness. Not recommended in larger breeds < 12–18 months of age. Has a negative effect on Marek vaccine titer when mixed together. Oral or IV led to infrequent temporary blindness, partial blindness, blindness, and mydriasis after extra-label use in **cats** at high dosage (>5 mg/kg/day).

LD_{50}, **mice**: ~5 g/kg.

ENTACAPONE

= *Comtan*

A catechol-σ-methyltransferase (COMT) inhibitor.

Untoward effects: Possibly rhabdomyolysis and neuroleptic malignant syndrome, increases ***L-dopa*** adverse effects (anorexia, dyskinesia, hallucinations, insomnia, nausea, syncope, and vomiting).

ENTEROLOBIUM

Untoward effects: South American **Indians** use the crushed bark (***keenimowe***) as a piscicide.

In **cattle**, ingestion of ***E. contortisiliquum*** (***timbauba***) fruit leads to anorexia, lassitude, foul-smelling diarrhea, sunken eyeballs, and death within hours to a few days. Liver degeneration and nephritis occur.

The sawdust of the tree ***E. cyclocarpum*** kills **fish** and **cattle** when it contaminates streams.

The pods of ***E. gummiferum*** (***tamboril de campo***) in Brazil have caused hepatogenic photosensitization in **cattle**.

ENTOLOMA

= *Jaunet* = *Livid Agaric* = *Perfide* = *Rhodophyllus* = *Videau*

Untoward effects: A ***mushroom*** that has caused severe and fatal poisonings in Europe from spring to late fall in **humans**. Violent symptoms within ½–6 h after ingestion included epigastric cramping, hemorrhagic vomiting, weakness, diarrhea, dyspnea, bradycardia, liver pathology, and uremia. Death occurs during uremic coma.

ENTOMOPHTHORA CORONATA

Untoward effects: In Africa, inhalation of this fungus led to granulomatous upper respiratory lesions in **man** and **horses**. In Nigeria, **people** have had soft-tissue swellings over the nose and cheek, unilateral nasal obstructions, opacity of the antra and ethnoid cells on the same side, but no bone destruction shown on X-ray.

EOSIN(E)

= *Acid Red 87* = *CI 45,380* = *CI Acid Red 87* = *D&C Red #22* = *Eosine Y* = *Eosine Yellowish - (YS)*

A *coal tar* dye.

Untoward effects: Used in lipsticks in the 1920s in the U.S., where it caused photosensitization and sore lips in thousands of **women**. Systemic lupus erythematosis may have been caused in **females** by retention of it in the body after inadvertent ingestion in **their** lipstick, where it can act as a persistent immunological trigger of autoimmunity. Use in nail enamels can cause onycholysis. Fixed-drug eruptions have occurred.

Photosensitization in **mice**.

LD_{50}, **rabbits**: 10 g/kg.

EPHEDRINE

= *Eciphin* = *Fedrin* = *Isofedrol* = *Mandrin* = *Sanedrine* =*Zephrol*

Sympathomimetic.

Untoward effects: Even necessary therapeutic doses have caused nervousness, insomnia, tremor, vertigo, headache, tachycardia, arrhythmias, cardiomyopathy, palpitations, increased blood pressure, fixed-drug eruptions, eosinopenia, hypertension, mydriasis, sweating, anorexia, nausea, vomiting, and feeling of warmth. Chronic use has caused anxiety states, dyshidrosis, and dermatitis. Overdoses have led to chills, cyanosis, fever, irritability, nausea, vomiting, mydriasis, aggravated glaucoma (lesser response in **Negroes**), blurred vision, tachycardia, urinary retention, spasms, opisthotonus, and convulsions. Dyspnea, decreased blood pressure, and coma may follow. Prolonged topical use in the nostrils can cause a paradoxical nasal congestion. A toxic psychosis developed in a 65-year-old **male** and a 54-year-old **female** after ingesting excessive doses. **He** took up to 200 60 mg tablets every week for its stimulating effect while driving a truck; **she** took up to 15 30 mg tablets/5 times a day during exacerbations of asthma. **They** had clear consciousness, but had persecution delusions and auditory and visual hallucinations. Several case reports of arterial thrombosis, myocardial infarction, or seizures after ingesting weight loss ephedrine-containing dietary supplements at recommended dosage for 1–30 days.

Ephedra = ***Brigham Tea*** = ***Brigham Young Weed*** = ***Ma-Huang*** = ***Mexican Tea*** = ***Mormon Tea*** = ***Whore-house Tea*** used in China for ~5000 years. Its chief active ingredient is ephedrine, rediscovered in 1924. Use of it in herbal remedies has induced manic and hypomanic reactions, as well as tachycardia, palpitations, nervousness, increased blood pressure, insomnia, renal calculi, and possibility of myocardial infarction, stroke, and death. Of 20 herbal samples analyzed, over 50% had > 20% content discrepancies with their labeling. Ephedrine is excreted by the kidneys, mostly unchanged or as ***norephedrine***. A small amount can be reduced to ***desoxyephedrine*** (***methamphetamine***) and cause an injustice by misinterpretation of urine analysis without adequate history of medications used. Abuse for "kicks", by ingesting tablets or dissolving them for IV use ("main-lining"), plus the fact that about 15 **people** have died and about 400 have suffered seizures, cardiac problems, and psychoses from over-the-counter sales, has forced the FDA to curtail their use and sales.

LD, 2-year-old **child**: ~200 mg; **adult**: ~2 g; LD_{LO}, **human**: 50 mg/kg.

Drug interactions: Cardiovascular effects are attenuated by ***reserpine***. Use with ***monoamine oxidase inhibitors*** leads to acute hypertensive crisis, possibly, with intracranial hemorrhage, hyperthermia, convulsions, and even death, in some cases. It may cause a need for an increased dosage of ***digitalis***-type preparations. Will antagonize ***guanethidine*** and decrease its effect. Use with ***halothane*** leads to cardiac arhythmias. ***Reserpine*** reduces its effect. It increases ***dexamethasone*** metabolism.

Frequent secondary atrioventricular blocks in **horses**. Central nervous system stimulation, anxiety, tachycardia, cardiac arrhythmias, hypertension, and urinary retention in **cats** and **dogs**.

LD_{50}, **rat**: 600 mg/kg; **mouse**: 400 mg/kg; IV, LD_{LO}, **rat**: 130 mg/kg; **cat**: 60 mg/kg; **dog**: 70 mg/kg; **rabbit**: 50 mg/kg.

EPICHLOHYDRIN

= *ECH* = *Epi*

***Solvent*. Nearly 700 million lbs used annually in the U.S.**

Untoward effects: Causes poisoning by swallowing, inhalation, or by absorption through the skin. Painful irritation of eyes, nasal passageways, and skin; nausea, vomiting, abdominal pain, cough, pulmonary edema, liver and renal problems, decreased fertility, dyspnea, and cyanosis. Probably carcinogenic (respiratory tract) and increased chromosome abnormalities in **humans**. Gives off ***hydrochloric acid*** in a humid atmosphere. OSHA limits exposure to 5 ppm (19 mg/m^3)/8 h TWA. 75 ppm is considered an immediate danger to life and health.

LD_{LO}, **human**: 50 mg/kg; inhalation TC_{LO}, **human**: 20 ppm.

Causes nasal carcinomas, papillomas, and stomach cancers in **rats**, and infertility in male **rats**; causes local sarcomas in female **mice**. 100 mg/kg to **rats** causes aneuresis and death within 36 h.

LD_{50}, **rats**: 90 mg/kg; inhalation TC_{LO}, **rat**: 250 ppm/4 h.

EPINEPHRINE

= *Adrenalin* = *Asthmanefrin* = *Bronkaid* = *Epifrin* =*Epiglaufrin* = *Epinal* = *Epi-Pen* = *Epitrate* = *Eppy* = *Glaucon* = *Glauposine* = *Levorenin* = *Primatene Mist* = *Simplene* = *Sus-phrine* = *Suprarenalin* = *Suprarenin* = *Vaponefrin*

***Vasoconstrictor, adrenergic, hormone, bronchodilator, hemostatic, mydriatic*. For systemic or topical use; rarely, used orally.**

Untoward effects: From parenteral use, usually IM, and occasionally SC; rarely IV leads to anxiety, apprehension, tremors, restlessness, raised blood sugar, mydriasis, asthenia, dizziness, throbbing headache, pallor, eosinopenia, palpitations, tachycardia, and arrhythmias. Systemic absorption can occur after topical use. Intranasally, can result in rebound congestion, and rebound bronchoconstriction has followed inhalation. Inhalation can cause irritation of pharyngeal and bronchial mucosa. Habituation or psychic dependence reported. Can cause marked hypokalemia within 30 s of SC use. May increase fibrinolysis. Endogenous material causes the "Fight or Flight" response in **people**. Many reports of hospital or **physician** errors exist, including 30 ml of 1 : 1000 or 30 mg in 1 min (the **man** survived after hypertension, pulmonary edema, hematemesis, shock, and renal shutdown); 16 mg (use of 1 : 1000, instead of 1 : 200,000) infiltrated for a tympanoplasty in a 23-year-old **female** led to hypertension (systolic > 300 mmHg), tachycardia, vasoconstriction and peripheral circulatory failure, followed by hypotension for 2 days and an uneventful recovery. A **patient** survived 110 mg SC. ~30 deaths, due to overdoses, reported in the literature. Acidified solutions intracardially have been considered potentially lethal. A 29-year-old **female drug-abuser** self-injected 82.5–124 mg **she** extracted from a ***Primatene Mist***® inhaler and suffered myocardial infarction, atrial fibrillation, and acute renal failure.

Clinical tests: May increase fasting ***glucose*** test readings, decrease inorganic ***phosphate*** test readings, and interfere with ***uric acid*** assays, increase urinary excretion of ***vanillylmandelic acid***, increase blood ***glucose*** by stimulating its production by the liver, and decrease its peripheral tissue uptake.

Eye: Black conjunctival and corneal lesions after topical use. It can undergo oxidative changes and cause ocular prostheses to turn black. Has caused irritation, tearing, conjunctival hyperemia, increased intraocular pressure, and aggravated or induced narrow angle glaucoma, blurred vision, and madarosis.

Fetus and neonates*:** Causes greater cardiac acceleration in the **fetus** at mid-gestation than in mature **fetuses**. A severe illness mimicking an epidemic of neonatal sepsis in nine premature **infants**, due to **their** receiving 0.5 ml (11.25 mg) of ***Vaponefrin® inhalation solution, instead of 0.5 ml (25 IU) of ***vitamin E***. **Fetal** bradycardia noted after a maternal paracervical block with 1% ***lidocaine*** and epinephrine.

Hypersensitivity*:** Fixed-drug eruptions and allergic reactions, including syncope, wheezing, pruritus, and dermatitis. ***Sulfite content may be a factor.

***Infections*:** Due to its vasoconstrictive effects and poor topical disinfection, infections, including gas gangrene, have followed some injections.

Respiratory*:** Sudden-onset pulmonary edema in a 44-year-old **male** after accidental infiltration of 5 ml (5 mg) 1 : 1000 instead of 1 : 100,000 of incision site for open reduction of fractured mandible and 1 ml (1 mg) instead of 0.4 mg ***atropine to decrease heart rate in a 57-year-old **male**. Failure of response has caused panic in **patients** who develop tolerance to it.

Drug interactions*: *Alcohol consumption increases its excretion. ***Protriptyline*** and ***desipramine*** potentiate its ill-effects 3-fold. Deaths of asthma **patients** using it IV with ***isoproterenol*** by inhalation. ***Phenothiazine*** derivatives are potent ***adrenergic-blockers***, thus, antagonize epinephrine's pressor effect, and can cause a deepened comatose state. Even small amounts in dentists' ***local anesthetics*** can react (but often do not, clinically) with ***monoamine oxidase inhibitors***. ***Guanethidine*** potentiates its mydriatic action. ***Entacapone*** potentiates the chronotropic and arrhythmogenic effects of exogenously administered epinephrine. Hypertensive crises and death can occur when used with ***cocaine***. Tachycardia of epinephrine is potentiated by ***cannabis***. Use with ***propranolol*** and ***metaprolol*** has caused marked bradycardia, atrioventricular block, and increased blood pressure; decreases blood pressure with ***haloperidol***. If its solution is mixed with ***penicillin*** or ***streptomycin***, body levels of it are increased, but not with ***neomycin***. It prolongs the irritation produced by ***polymyxin***. Use with various inhalation anesthetics, such as ***chloroform***, ***cyclopropane***, ***enflurane***, ***fluroxene***, ***halothane***, ***isoflurane***, and ***methoxyflurane***, has caused increase in cardiac arrhythmias. ***Hydrocortisone*** potentiates its response on topical application on the eye, by increasing epinephrine affinity for ***adrenergic receptors***. Causes ***polyethylene*** devices to discolor, due to degradation products, after storage for 1 month at 50°C.

IV to **buffalo calves** (25 μg/kg) led to increase in blood ***glucose***, ***lactic acid***, ***potassium***, packed-cell volume, and decrease in phospholipids. Avoid IV ***lidocaine*** with epinephrine in **animals**. As in **man**, it causes ventricular arrhythmias in **cats** and **dogs**. Causes hyperglycemia and raised hematocrit in **cattle**. IV use (0.25 μg/kg/min) in **sheep** and **goats**, but not in **cows** or **horses**, causes a 10% decrease in serum ***calcium***. Inhibits milk ejection in **cows** and **women**, and raises blood ***glucose***, ***cholesterol***, and other ***lipids*** in **dogs**, and promotes platelet aggregation and myocardial necrosis. It increases incidence of gastrointestinal ulcers and, with ***nicotine*** increase, it triggers ventricular fibrillation in **dogs**. Use with ***cyclopropane***, ***halothane***, ***methohexital***, ***methoxyflurane***, ***thiamylal***, ***thiopental***, ***trichloroethylene***, and various other ***fluorinated hydrocarbon propellants*** has induced ventricular fibrillation and death in **cats**, **dogs**, and **swine**. IV dosage of 25 μg/kg in **goats** caused a few minutes of bradycardia. Causes ectopic heart beats in **horses** after IV use. Use with ***local anesthetics*** in thin-skinned areas, particularly in **Arabian horses**, or in emergency use, alone, in **cattle**, causes vasoconstriction with possible infarcts, skin scarring, and white hair regrowth. Cataractogenic in **rats** and **mice**. Hyperglycemic and teratogenic in **rats**. Hyperglycemic in **swine**. Induces hemoconcentration in **hens**; hemorrhage, edema, and necrosis of distal extremities in **rabbits**.

LD_{LO}, **rat** and **rabbit**: 30 mg/kg; SC, LD_{50}, **mouse**: 2.8 mg/kg; IV, LD_{50}, **rat**: 980 μg/kg; **mouse**: 217 μg/kg; IV, LD_{LO}, **cat**: 700 μg/kg; **dog**: 150 μg/kg; **guinea pig**: 150 mg/kg; **rabbit**: 500 μg/kg.

EPIRUBICIN

= Ellence = 4′-Epidoxorubicin = Epidx = Farmorubicin = IMI 28 = Pharmorubicin − Pidorubicin

Antineoplastic.

Untoward effects*:** About 40% less cardiotoxic than ***doxorubicin (q.v.). Bone marrow suppression, leukemia, paresthesias, anaphylaxis, decreased diastolic performance, and has induced cardiomyopathy. Delayed congestive heart failure after withdrawal. Alopecia, mucositis, nausea, red urine, and vomiting.

Drug interactions*: *Cimetidine can cause dramatic increase (up to 50%) in its serum concentration.

Embryotoxic and teratogenic in laboratory **animals**.

EPN

***Insecticide, acaricide.* Organophosphorus cholinesterase inhibitor.**

***Untoward effects*:** Exposure is by inhalation, skin contact with absorption, and ingestion. Eye and skin irritation, lacrimation, miosis, rhinitis, wheezing, laryngeal spasm, salivation, nausea, abdominal cramps, diarrhea, anorexia, cyanosis, decreased blood pressure, arrhythmias, tightness in chest, paralysis, and convulsions. Causes a delayed neuropathy in **man** and experimental **animals**. OSHA exposure limit for skin is 0.5 mg/m^3 and exposure to 5 mg/m^3 is considered potentially lethal to **man**.

LD_{LO}, **human**: 5 mg/kg.

Ingestion of 1 mg/kg is toxic to **cattle** and 20 mg/kg is lethal to **sheep**. Spray or dip of > 0.05% is toxic and 0.25% is lethal to **calves**. Highly toxic to **honeybees**. Delayed neurotoxicity to **chickens** and **mallard ducks**. Exposed **rats** become ataxic. Teratogenic in **birds**. Symptoms of toxicity can develop within 30 min. **Birds** become ataxic, hyperexcitable, asthenic, and dyspneic; they regurgitate, fall, and salivate; they also have wing-drop, penile extrusion, tetany, and convulsions. Toxicity to female **rats** is greater than in males.

LD_{50}, **weanling rats**: 7.7 mg/kg; **adults**: 33 mg/kg; **mice**: 23.9 mg/kg; **dogs**: 20 mg/kg; **mallard ducks**: 3 mg/kg; **Japanese quail**: 5.25 mg/kg; **pheasants**: 51.9 mg/kg; **pigeon**: 6 mg/kg.

EPOXY RESINS

Untoward effects: Local skin irritation from splashes, due to its many amines. These can also cause severe conjunctivitis and facial rashes. Symptoms can occur within minutes or several weeks. The hardened or cured epoxys have little or no toxicity, but the curing or hardening agent often causes a contact dermatitis, with pruritus, erythema, and erythroedema. Allergic papular eczematous areas around the knees of **children** have appeared after wearing worn, older jeans with reinforced knee patches. The glue has, since, supposedly, been replaced with less irritating material. Similar reactions have followed contact with plastic gloves, artificial pearls, water-proof adhesive tapes, baby pants, and sheets of plastic. 1/1000 **people** not employed in the resin industry showed hypersensitivity to it, especially, those with ***glycidyl ether*** bases. Persistent photosensitivity developed in a number of **men** after occupational exposure to hot epoxy resin fumes. Pneumonitis reported after inhalation of epoxy pipe coatings. Two **males** exposed to the resin powder, containing ***trimellitic anhydride*** as a curing agent, developed hemolytic anemia and repeated hemoptyses. Model airplane **builders** working in confined areas have developed pulmonary inflammation and coronary problems.

EPTC

= *Eptam* = *FDA-1541* = *R-1608*

Untoward effects: Depression, anorexia, vomiting, salivation, muscle spasms, and dyspnea, can occur in **man** and **animals**. Avoid contact with eyes, skin, and clothing. Avoid breathing dust.

LD_{LO}, **human**: 500 mg/kg.

Run-off is toxic to **shrimp**.

LD_{50}, **rats**: 1.6 g/kg; **mouse**: 750 mg/kg; **cat**: 112 mg/kg; **birds**: 100 mg/kg.

EPTIFIBATIDE

= *Integrilin*

***Platelet glycoprotein II b/III a receptor antagonist.* Prevents platelet aggregation. IV use.**

Untoward effects: Increases risk of bleeding, especially at access site.

Drug interactions: Use with ***alteplase*** led to coma and death in a 47-year-old **male** and an 85-year-old **female**.

EQUISETUM

= *Horsetail* = *Mares Tail* = *Scouring-rush*

Untoward effects: Causes a disagreeable taste and odor for **people** in ***milk***, milk products, and meat of **animals** that have grazed it. **Horses**, and, occasionally, **cattle** and **sheep**, are poisoned by grazing on its tops in wet meadows, or eating it in hay. It causes weakness, nervousness, trembling, diarrhea, ataxia, weight loss, tachycardia, and an apparent craving for the plant. Occasionally, constipation. In serious cases, diarrhea, decreased milk production (**cattle** and **sheep**), convulsions, and coma usually precede death. Contains thiaminase, ***aconitic acid***, ***equisetin***, ***nicotine***, and ***silica***.

E. arvense* = *Field Horsetail* = *Horsetail Fern* = *Meadow Pine* = *Prêle des Champs is found in the U.S., Canada, France, Italy, and Russia. It is the cause of most of the ***Equisetum*** poisonings, especially, in **cattle** and **sheep**, in the U.S. from this genus.

E. fluviatile* = *Swamp Horsetail has poisoned **horses** in the U.S.

E. giganteum* = *Cola de caballo, a giant horsetail in the Andes, from Chile and Argentina northward, is 1⅓ in thick and 30 ft tall.

E. palustre* = *Marsh Horsetail* = *Prêle des Marais in Europe, Asia, and North America has poisoned **horses**, decreased or stopped milk production and caused diarrhea in **cows**, and caused patchy hyperhidrosis in **calves**. Diarrhea, muscle weakness, and hyperhydrosis reported in affected **sheep**. Contains ***palustrine***, a toxic alkaloid.

ERBON

= *Baron* = *Novon*

Herbicide.

Untoward effects: Dermatitis and conjunctivitis.

LD_{LO}, **human**: 500 mg/kg.

LD_{50}, **rat**: 650 mg/kg; **mouse**: 912 mg/kg; **rabbit**: 2.2 g/kg; **chicken**: 3.2 g/kg.

EREMOCARPUS SETIGERUS

= *Dove Weed* = *Turkey Mullein*

Untoward effects: **Sheep** and **rabbits** on the U.S. Pacific Coast have been killed by eating it, possibly from mechanical injury in the digestive tract by its leaves and stems that are covered with a mat of hairs. Indigenous Californian **people** used it to poison, stun, or stupefy **fish**.

EREMOPHILA MACULATA

= *Fuchsia*

Untoward effects: High concentration of ***hydrocyanic acid*** makes it dangerous for **livestock** grazing on it. Found in Australia.

ERGOLOID MESYLATES

= *CCK-179* = *Circanol* = *Coristin* = *Dacoren* = *DCCK* = *Deapril-ST* = *Decril* = *Dulcion* = *Ergodesit* = *Ergohydrin* = *Ergoplus* = *Hydergine* = *Lysergin* = *Novofluen* = *Orphol* = *Pérénan* = *Progeril* = *Redergin* = *Sponsin* = *Trigot*

***α-Adrenergic-blocker.* A mixture of equal parts of *dihydroergocornine*, *dihydroergocristine*, and *dihydroergocryptine mesylates*.**

Untoward effects: Transient nausea, vomiting, and gastrointestinal upsets. Sublingual irritation from sublingual tablets. Ergotism can develop. High dosage usually leads to bradycardia, listlessness, general apathy, withdrawal, and refusal to eat.

Induces vomiting in **dogs** and bradycardia in **rats**.

ERGONOVINE

= *Ergobasine* = *Ergoklinine* = *Ergometrine* = *Ergonil* = *Ergostetrine* = *Ergotocine* = *Ergotrate* = *Syntometrine*

Oxytocic.

Untoward effects: Increases blood pressure and slight headache on IV; causes emesis on IM; severe gangrene of legs in a 20-year-old **female**. Within 20 min after 0.2 mg IM in a 22-year-old **female**, myocardial infarction developed, possibly secondary to induced coronary artery spasm. IM on second neonatal day led to ecchymoses and cyanotic erythematous lesions with hemorrhagic blebs on extremities, groin, and genitals in five **newborns**. A 41-year-old **female** given 9 mg in 5 days to control hemorrhage after premature birth developed mental problems, insomnia, diarrhea, renal failure, and pain in legs, followed by gangrene, requiring amputation of both legs. Death after $3\frac{1}{2}$ months. Post-mortem revealed enlarged liver with fatty infiltration, lung and kidney infarcts, and gangrene of hands and limb stumps. An accidental 0.2 mg IM in a **female newborn**, instead of ***vitamin K***, led within 3 h to tonic/clonic seizures of all extremities and peripheral necrosis. Has caused marked increase in systolic and diastolic blood pressure by 20 mmHg, and, when given with other oxytocics, such as ***oxytocin*** or ***sparteine***, it may cause a rupture of cerebral blood vessels. Avoid use with other vasoconstrictor/pressor drugs. Passes into **human** milk, and 90% of **babies** who used this milk had signs of ergotism (diarrhea, vomiting, irritability, convulsions, weak pulse, and drowsiness). Can suppress ***prolactin*** secretion in breast-feeding **women**. Use with ***methoxamine***, an α-adrenergic agonist, led to hypertension and severe headaches.

IV, LD_{LO}, **guinea pig**: 80 mg/kg; **rabbit**: 7.5 mg/kg.

ERGOT

= *Secale cornutum*

***Vasoconstrictor, oxytoxic.* From the dried sclerotia of the fungus *Claviceps purpurea* that develops in the heads of grasses and cereal grains. Contains many alkaloids.**

Untoward effects: ***Acute***: Nausea, vomiting, hematemesis, thirst, colic, diarrhea, weakness, salivation, chest pains, dysphagia, tachycardia, fainting, dyspnea, miosis, tinnitus, confusion, agitation, headache, and coma.

Chronic: Convulsions, poor digital circulation, impairment of speech and vision, anuria, hallucinations, and gangrene.

Confusion and hallucinations are probably due to its ***LSD*** fractions. Can induce acute attacks of porphyrias; cataract formation, fixed-drug eruptions, pain and paresthesia of limbs, retinal edema, psychiatric states (St. Anthony's fire, see ***Bread***), renal failure in poisonings, hypertension, angina, myocardial infarction, dysesthesia, and teratogenicity. A healthy pregnant **female** ingested 12.5 grains/day/2 days leading to tingling in hands and feet, bluish tinge of nose, headache, breast rigor and pain, dyspnea and anorexia, followed by blue-black discoloration of feet and nose, septic abortion, and **her** death on the fourth day. Its alkaloids can pass into **mother's** milk (see ***Ergonovine***), with serious consequences in **infants**. Can suppress lactation. Its use has been associated with abortions and stillbirths. Most of France's ***rye*** crop, in 1789, was affected with it, probably causing hallucinations and panic (see ***Bread***). It was also called ***la Grande Peur*** or the ***Great Fear of 1789***. Some historians think it responsible for the Salem, Massachusetts witchcraft episodes of 1602. Now, the U.S. forbids the sale of grain for **human** consumption containing > 0.3% ergot.

Outbreaks of ergotism in **cattle** result in necrosis of feet, tails, and ears, with occasional sloughing; hyperexcitability, muscular incoordination or "staggers", edema of hind limbs, epilation, areas of skin necrosis, diarrhea, and abortion. Causes complete cessation of lactation in **sows**, lower pregnancy rate, and can cause gangrene. Weakens and kills baby **pigs**. Has caused ergotism in **guinea pigs**, **buffalo**, **horses**, **sheep**, **pullets**, and **ducks**. Reduces the number of **ewes** lambing and has caused fetal deaths. Convulsions, staggering, muscle spasms, and temporary paralysis reported in **sheep** and **horses**. Embryocidal and teratogenic in **mice**. Affected **poultry** develop dry, blackened combs. **Pigeons** are very meticulous about peeling the ergot off of grain, and a company I consulted to used them in its ergot production.

ERGOTAMINE TARTRATE

= *Ergate* = *Ergomar* = *Ergostat* = *Ergotartrat* = *Exmigra* = *Femergin* = *Gynergen* = *Lingraine* = *Lingran*

***Vasoconstrictor. Serotonin*-antagonist. Used SC, IM, orally, and by suppository.**

Untoward effects*:** This is the cause of almost all the adverse effects of ***ergot (q.v.). Arteriospasms, vasospasms, arterial aneurysms, acute peripheral ischemia, peripheral neuropathy, gangrene, intermittent claudication, nausea, vomiting, paresthesias, angina pectoris, ventricular fibrillation, polydipsia, renal failure, tongue cyanosis and partial necrosis, intractable headaches, hypersensitivity with calf and knee pain, retroperitoneal fibrosis, and addiction have occurred. Induced psychic disturbances can be serious with paranoid schizophrenia reactions, inability to concentrate, confusion, anxiety, nightmares, and hallucinations. Adverse reactions are more common when plasma concentration is > 1.8 ng/ml. After a 35-year-old **female** received 1 ml ***Gynergen***® for a migraine headache, **she** convulsed and died within 30 min from a subarachnoid hemorrhage. A lactation inhibitor and excreted in **human** breast milk. Poisoning with cyanosis and severe intermittent claudication with decreased pulse amplitudes in oscillogram after unauthorized overdosage of suppositories (5/week/4 years, then, 5/10–14 days/3 years) against migraine in a multiparous 33-year-old **female** in fourth month of pregnancy. **She** recovered. A 62-year-old **female** developed chronic ergotism after ~20 year treatment for migraine. Successfully treated by periarterial sympathectomy. A 40-year-old **female** with intermittent hypertension and cluster headaches received a total of 10 mg in five suppositories during 60 h, and developed renal arterial spasms. Accidental poisonings in **infants**: in a 14-month-old **male**, fatal after ingesting 12 mg in ***Cafergot***® tablets; a 13-month-old **female** eventually recovered after ingesting ~15 mg in 50 ***Bellergal***® tablets; a 14-year-old **male** failed to respond to treatment after developing severe dyspnea, pyrexia, hemorrhagic gastritis, and cerebral edema.

Drug interactions*:** A 61-year-old **male** treated with ***Cafergot® suppository twice daily/6 years after ***propranolol*** 30 mg/day developed purple, painful feet. Recovered with treatment. Ergot alkaloids are potent vasoconstrictors and one natural pathway for vasodilation is blocked by ***β-blockers***. ***Troleandomycin*** reduces hepatic enzyme metabolism of ergotamine, leading to toxicity. Ergotism reported when used with ***indinavir*** or ***ritonavir***.

ERUCIC ACID

***Use*:** In manufacturing steel and rubber.

Untoward effects*:** Incriminated as the toxic substance in ***rapeseed and ***canola oil***, causing growth decrease, hydropericardium, ascites, and fatty changes in the heart, skeletal muscles, spleen, and kidneys of **chickens**, **ducks**, and **turkeys**. Also, present in the remaining plant fibers after oil extraction. New varieties have extremely low levels of it.

ERYSIMUM CHEIRANTHOIDES

= *Vélar Fausse Giroflée* = *Worseed Mustard*

Untoward effects*:** Young plant ingestion may be harmless, but, after eating large quantities of its seeds, their ***glucosinolates become hydrolyzed to ***allylisothiocyanate***. If the seeds are moistened with cold water before feeding, enzymes are liberated that help produce blistering ***oil of mustard***. Boiling water inactivates the enzymes. **Cattle** develop colic, depression, hemorrhagic diarrhea, enteritis, and cardiac and respiratory failure. Death has occurred. **Horses** and **swine** are also affected with similar symptoms.

ERYTHRINA sp.

Untoward effects*:** The seeds of some species may act as do ***mescal beans, and are used as hallucinogens in Mexico, where they are called ***colorines***.

E. herbacea* = *Eastern Coralbean* = *Cardinalspear seeds are poisonous and contain saponins.

E. vespertilis is poisonous to **animals**. Various species are commonly called ***Coral Bean*** and contain alkaloids and ***curare***-like action, if the bark or seeds are ingested. Crushed stems are used as a piscicide. Some species are native in some of the warmer parts of the U.S. Their attractive seeds make them particularly hazardous for **children**.

ERYTHRITOL ANHYDRIDE

= *Bioxiran* = *Butanedione* = *Diepoxybutane*

***Use*:** In polymer and textile manufacturing.

***Untoward effects*:** Primarily, from inhalation and dermal contact. Also, in the manufacturing and handling of fabrics, pharmaceuticals, and polymers where it has been used. Carcinogenicity in **man** is suspect.

Carcinogenic in experimental **animals**. Dermal application led to skin papillomas and squamous cell carcinomas in **mice**. Local fibromas in female **rats** and **mice** after SC. IP leads to lung tumors in **mice**.

Its diacetyl form is a fermentation product in ***wine*** and other foods, and is responsible for the aroma of ***butter***.

LD_{50}, **rats**: 78 mg/kg.

ERYTHRITYL TETRANITRATE

= *Cardilate* = *Cardiloid* = *Erythrol Tetranitrate*

***Vasodilator*.** **Orally, sublingually, and buccally.**

Untoward effects: The crystals are explosive on percussion, unless mixed with an inert substance, such as ***lactose*** or ***sodium bicarbonate***. Headache, flushing, and dizziness; occasionally, orthostatic hypotension and reflex tachycardia. A 71-year-old **male** ingested 500 $\frac{1}{4}$ gr tablets/1 year leading to severe refractory macrocytic anemia. Similar results occurred with high dosage in a **cat**.

ERYTHROMYCIN

= *Abomacetin* = *Ak-Mycin* = *Aknin* = *E-Base* = *EMU* = *E-mycin* = *Eritrocina* = *ERYC* = *Erycen* = *Erycin* = *Erycinum* = *Ery Derm* = *Erymax* = *Erymysin* = *Ery-Tab* = *Erythromast 36* = *Erythromid* = *Erythromycin A* = *Erytromycin* = *Gallimycin* = *Ilotycin* = *Inderm* = *Kenmycin* = *Proterytrin* = *Retcin* = *Robimycin* = *Staticin* = *Stiemycin* = *Torlamicina*

Untoward effects: ***Blood***: Agranulocytosis, eosinophilia, leukopenia, clotting factor X decrease, uremia, severe immune hemolysis, and anemia reported.

Cardiac: Use is associated with development of torsades de pointes and QT interval prolongation. Palpitations. **Women** may be twice as susceptible as **men**.

Clinical test interference: Causes false or increased ***bilirubin***, alkaline phosphatase, SGOT, and SGPT results; errors in fluorimetric determination of urinary catecholamines; interferes with *Lactobacillus casei* assay for serum ***folate***.

Ear: Many reports of ototoxicity, some, reversible. Usually cochlear damage.

Fetus and neonate: Crosses the placenta and **fetal** concentrations are usually < $\frac{1}{4}$ of **maternal**. In **mother's** milk 3–6 times **her** plasma levels. Has caused icterus in **newborns**. Oral treatment in **neonates** leads to increased risk of infantile hypertrophic pyloric stenosis.

Gastrointestinal: Gastrointestinal upsets usually mild (nausea, vomiting, heartburn, diarrhea, abdominal pain, indigestion, flatus, constipation, and esophagitis; rarely, pseudomembranous colitis).

Liver: Cholestatic jaundice estimated at ~3.6/100,000. Cases of acute hepatitis require hospitalization.

Miscellaneous: Headache, dizziness, pyrexia, arthralgia, hypothermia, pancreatitis, nightmares, hypovolemic shock with difficulty in rousing, paleness, weak pulse, decreased blood pressure, wheezing, hallucinations, seizures, and vertigo. A topical for treatment of acne was withdrawn from the French market because it contained 36% ***ethoxyethanol*** (q.v.).

Neuromuscular: May interfere with transmission.

Renal: Has caused interstitial nephritis and acute renal failure.

Skin: Urticaria with pruritus and local pain, due to hypersensitivity. Comedogenic. Acneform eruptions, rashes, and eosinophilic subcorneal pustular dermatosis reported. Photoallergenic responses and Stevens–Johnson syndrome occur.

Teeth: Excessive high doses implicated in development of dental enamel hypoplasia and discoloration of teeth in **neonates**.

Use of IV forms has caused hypotension and local tissue necrosis after extravasation. Has increased renal uptake of ***radiopharmaceuticals***. See below for specific reactions to some.

Drug interactions: As an inhibitor of CYP3A4, the metabolism of the anxiolytic ***alprazolam***, the sedative hypnotic ***triazolam***, the antihistamine ***astemizole***, the antiparkinsonian drugs ***bromocriptine*** and probably ***lisuride*** and ***pergolide***, and the antihypertensive ***felodipine*** is decreased. Prolonged QT intervals, torsades de pointes, and deaths reported due to decreased metabolism of ***cisapride***. Similar cardiac effects reported with ***quinidine***. Bradycardia reported when taken with ***verapamil***. Serotonin syndrome 4 weeks after adding erythromycin – 400 mg/day in a 12-year-old **male** taking ***sertraline*** 37.5 mg/day.

Anticoagulants: By decreasing ***warfarin*** and ***phenindione*** clearance, it potentiates the anticoagulant-induced hypoprothrombinemia.

Anticonvulsants: Anxiety, confusion, and dysarthria in a 38-year-old **female** taking 3500 mg ***valproic acid*** daily with 250 mg qid/1 week. ***Valproic acid*** serum levels of 260.4 μg/ml reported. ***Anticonvulsant*** (***carbamazepine***, ***chlorazepate***, ***clonazepam***, and ***diazepam***) serum levels and toxicity are increased by it.

Clarithromycin: Life-threatening interaction reported with ***clarithromycin***.

Contraceptives, oral: Some **women** lose birth control protection when it is taken with ***oral contraceptives***.

Cyclosporine: Its absorption is enhanced, its metabolism is decreased, raising its serum concentration by 2–5 times.

Digoxin: By altering enteric flora, it helps prevent ***digoxin's*** breakdown and increases its serum level.

Lovastatin: Rhabdomyolysis with weakness and myalgia occurred when erythromycin inactivated liver enzymes and increased ***lovastatin*** serum concentrations up to 8 times usual expectations. Also raises ***simvastin*** serum levels.

Miscellaneous: Decreases metabolism and elimination of ***alfentanil*** and ***clozapine*** (increases serum levels ~2 times, causing tonic-clonic seizures). A 43-year-old **female** maintained on 10 mg/day ***felodipine*** extended release developed palpitations, flushing, and ankle edema 2–3 days after erythromycin 250 mg/twice daily. By stimulating gastric motility, it decreases the rate of ***glyburide*** absorption. When given together, it increases plasma concentration of ***loratadine***. It increases the risk of arterial spasms and occlusion by ***methysergide***, increases and prolongs the effect of ***midazolam***, and increases ***methylprednisolone*** and ***tacrolimus*** serum levels. With ***lincomycin***, the benefits of both are reduced. Decreased effectiveness of ***penicillins***.

Terfenadine: Its clearance and that of other ***antihistamines*** are decreased by erythromycin, leading to syncope, ventricular arrhythmias, prolonged QT interval, torsades de pointes, and death.

Theophylline: Its clearance is decreased and toxicity increased by erythromycin.

In **rats**, QT prolongation by ***disopyramide*** is enhanced by coadministration with erythromycin.

LD_{50}, **mice**: ~1.8 g/kg.

Various ***erythromycin*** forms have had specific problems described in the literature, and these are listed below.

ERYTHROMYCIN ESTOLATE

= *Eriscel* = *Eritroger* = *Eromycin* = *Estomicina* = *Eupragin* = *Ilosone* = *Lauromicina* = *Marcoeritrex* = *Neo-Erycinum* = *PELS* = *Propiocine Enfant* = *Propionyl Erythromycin Lauryl Sulfate* = *Roxomicina* = *Stellamicina* = *Togiren*

Untoward effects: Vomiting (4%), abdominal cramps (4%), diarrhea (2%), rash (2%), sinus tachycardia, tachypnea, fever, increased white blood cells, dyspepsia, glossitis, anorexia, Stevens–Johnson syndrome, and hypovolemic shock. Accumulates in the liver more than other forms, and has up to a 15% incidence of liver reactions (intrahepatic cholestasis, jaundice, fever, eosinophilia, increased serum transaminase and ***bilirubin*** levels). Contraindicated in pregnancy.

ERYTHROMYCIN ETHYL SUCCINATE

= *Anamycin* = *Arpimycin* = *Durapaediat* = *E.E.S.* = *E-Mycin E* = *Eryliquid* = *Eryped* = *Erythrocin* = *Erythro ES* = *Erythro-Holz* = *Erythroped* = *Esinol* = *Monomycin* = *Paediathrocin* = *Pediamycin* = *Refkas* = *Sigapedil* = *Wyamycin E*

Untoward effects: Liver reactions with fever and abdominal pain, nausea, heartburn, mild glossitis, oral ulceration, and skin reactions. ***Theophylline*** clearance decreased by it 7–44% in asthmatic **children**. Increases ***cyclosporine*** serum levels.

ERYTHROMYCIN GLUCEPTATE

= *Erythromycin Glucoheptonate*

Untoward effects: Reversible ototoxicity and deafness in a 53-year-old **male**.

ERYTHROMYCIN LACTOBIONATE

Untoward effects: Ototoxicity, cardiac toxicity, torsades de pointes, QT interval prolongation, hypotension, nausea, and vomiting. IV use causes vein irritation and phlebitis, acute pancreatitis, thrombophlebitis from a too concentrated and too rapid infusion, inhibition of ***quinidine*** metabolism, causing ventricular tachycardia (torsades de pointes), and a potentially fatal interaction with ***disopyramide*** leads to increased QT prolongation, polymorphic ventricular tachycardia, and increased levels of ***disopyramide***. Markedly potentiates ***warfarin*** anticoagulation by ~2 times.

ERYTHROMYCIN PROPIONATE

= *Propiocine*

Untoward effects: Transitory diarrhea, nausea, vomiting, epigastric pain, and urticaria.

ERYTHROMYCIN STEARATE

= *Abboticine* = *Bristamycin* = *Dowmycin E* =*Eratrex* = *Erypar* = *Eryprim* = *Erythro S* = *Erythrocin* = *Ethril* = *Ethryn* = *Gallimycin* = *Meberyt* = *Pantomicina* = *Pfizer-E* = *SK- Erythromycin* = *Wemid*

Untoward effects: Abdominal pain, nausea, vomiting, flatulence, diarrhea, vaginal pruritus, jaundice, malaise, vertigo, heartburn, rashes, and allergic respiratory distress. Confusion, fear, feeling of impending loss of consciousness and of being drugged in a 45-year-old **male**. Use with ***warfarin*** markedly increases prothrombin time (76.3 s). Retroperitoneal hematoma after 250 mg qid/ 7 days, then, twice daily/7 days. Inhibits ***theophylline*** clearance by close to 50%.

The ethyl carbamate form in a 4-year-old **male** was associated with uncontrolled weeping and hysterical laughing. The tetraoleate form has caused hepatotoxicity in **man** after prolonged administration.

Mammary excretion occurs in the milk of **cows**, **dogs**, **goats**, and **women**. Readily lethal to **hamsters** (100% at 5 mg tid/5 days). Kernicterus in **rhesus monkey** fetuses and **infants**. Administration to **rabbits** causes an overgrowth of Gram-negative bacteria, diarrhea, and even death. SC or IM in **swine** leads to severe local reactions, necrosis, hemorrhages, and fibrosis.

ERYTHRONIUM

= *Dogtooth Violet* = *Easter Lily* = *Fawn Lily* = *Rattlesnake Violet* = *Scrofula Root* = *Yellow Adder's Tongue* = *Yellow Snowdrop*

Untoward effects: ***E. americanum***, ***E. dens-canis***, and ***E. revoltum*** have caused a contact dermatitis from ***α-methylene-γ-butyrolactone***, a glycoside in its bulbs, stems, leaves, and petals. Ingestion causes emesis.

The bulbs are known to poison **poultry** that have eaten some.

ERYTHROPHLEUM

= *Ordeal Tree* = *Akpa (Efic)* = *Erun (Yoruba)* = *Gwaska (Hausa)* = *Inyi (Igbo)* = *Red Water Tree*

Untoward effects: ***E. africanum*** = ***Samberu*** (Hausa) = ***Sungwoi*** (Fulani) contains ***erythrophleine*** and kills **livestock** within minutes in acute cases. Chronic cases increase dyspnea, salivation, nasal discharge, sunken eyes, and bloated appearance. Blood-stained watery diarrhea, in some cases.

E. chlorostachys = ***Ironwood*** in Australia has similar effects and is extremely toxic to **camels**, **cattle**, **sheep**, **goats**, and **horses**. It has also been used in poisonous baits to kill **animals**. Secondary toxicity from eating internal organs of such killed **animals**.

The leaves and bark of ***E. guineense*** are poisonous for **horses** and **cattle**. Also used in Africa in arrow and dart poisons.

Bark extracts of ***E. suaveolens*** are very poisonous, and were once used by African **witch doctors** in trials of ordeal for **those** suspected of witchcraft or serious crimes. Vomiting, diarrhea, and dyspnea with death from cardiac and respiratory arrest. These adverse effects are from cardiotonic ***digitalis***-like action from its alkaloids, ***cassaidine***, ***cassaine***, ***erythrophlamine***, and ***erythrophleine***.

ERYTHROPOIETIN

= *Ep* = *Epo* = *Epoade* = *Epoetin alfa* = *Epogen* = *Eprex* = *Erypo* = *Erytropoetine* = *ESF* = *Espo* = *Hemax* = *Procrit*

Hematopoietic. **Usually, given SC; less commonly, IV, and, rarely, IP. The recombinant human *Epo* is also known as *rh EPO*.**

Untoward effects: Resistance to it has been reported, especially in **patients** with severe secondary hyperparathyroidism, osteitis fibrosa, pyruvate kinase deficiency, ***valproic acid*** toxicity, and ***aluminum*** intoxication in dialysis **patients**. ***Iron*** deficiency is a common cause of resistance to it. Infectious diseases, malignancies, hematological diseases (thalassemia and myelodysplasia), ***folic acid*** or ***vitamin*** B_{12} deficiencies, hemolysis, and ***aluminum*** intoxication may also prevent a suitable response. Hypertension is a severe complication of its use. Skin rashes, momentary pain at injection site, and urticaria occur. Splenic infarction in an 8-year-old **male** with aplastic anemia, several months after 200 IU/kg/three times a week. Hypertension and grand mal convulsions in a 56-year-old **female** on hemodialysis, due to 50 IU/kg/three times a week. Apparent causal relationship to an early thrombosis in a renal transplant **patient**. Has caused spontaneous platelet aggregation in hemodialysis **patients**. Pseudoporphyria cutanea tarda in two **children** (7 and 12 years) on peritoneal dialysis, receiving 500–1500 U/three times a week and 2500 U/twice a week, respectively. Skin eruptions appeared 2 months and 2 weeks, respectively, after sunlight exposure. Peripheral subretinal neovascularizaion and Coat's-type retinitis pigmentosa associated with its use in a 37-year nephrotic **patient**. Aggravation of splenomegaly reported in two Japanese **patients** (59 and 65 years), after IV use. Headache and arthralgia in 4–5% of **patients**. Rarely, shortness of breath. ***Theophylline*** attenuates the endogenous production of erythropoietin. A 51-year-old **male** with renal cell carcinoma developed adrenal gland neoplasms after long-term ***EPO*** treatment. An anemic **patient** developed antibodies against it. Concurrent treatment with ***ketanserin*** delayed its effect on bleeding time. Headache in renal failure **patients** may be due to induced hypertension. Mucocutaneous myeloid metaplasia at SC injection site in a 74-year-old **male** with previous polycythemic myeloid metaplasia.

Some **animals** produce antibodies to it, which may result in ***iron*** deficiency, hyperkalemia, hypertension, polycythemia, and life-threatening erythroid hypoplasia. Use in **horses** to improve stamina and performance has induced an immune mediated anemia and deaths of some.

EPOETIN *beta* = ***Epogin*** = ***Marogen*** = ***Recormon.*** Pain at the injection site is said to be less with this form.

ERYTHROSINE

= *C.I. 45430* = *C.I. Acid Red 51* = *C.I. Food Red 14* = *Erythrosine B* = *Erythrosine BS* = *FD & C Red No. 3*

Use: In foods, pharmaceuticals, cosmetics, and dog food. Its safety has been controversial.

Untoward effects: Its ***iodine*** content may falsely increase protein-bound iodine test results of thyroid function. Has caused fixed-drug eruptions, and possibly, an allergic bronchoconstriction or dyspnea in perennial **asthmatics**. It is a photosensitizer. In the presence of ***acids***, ***iron***, and ***tin***, it breaks down into ***fluorescein***.

Mongolian gerbils fed it at 1–4% in their feed developed reticulocytopenia, leukopenia, anemia, decreased hematocrit, and follicular hyperplasia of the thyroid. Male **rats** developed precancerous thyroid lesions when fed it, and, at high dosage, decreased red blood cells, decreased hemoglobin, decreased ***cholesterol***, and showed diarrhea, and growth retardation. It's lethal, often within minutes, to 85% of **house flies** in all stages of their growth.

LD_{50}, **rat** and **mouse**: > 7.5 g/kg; LD_{LO}, **mouse**: 2.5 g/kg.

ESCHSCHOLTZIA CALIFORNICA

= *California Poppy*

Untoward effects: It is a member of the ***opium poppy*** family, but contains NO narcotic alkaloids. Its alkaloids (***coptisine***, ***protopine***, and ***sanguinarine***) have no physiological activity. Alleged abuse from smoking it is actually due to chemicals or plant materials added by **man**.

ESCIN

= *Aescin* = *Aescusan* = *Reparil*

A bioflavonoid glycoside.

Untoward effects: See ***Aesculus***. The sodium form is usually used IV. Occasional nausea, fatigue, and restlessness on high oral dosage. IV leads to thickening of vein without inflammated in 2/264. Vomiting, diarrhea, inflamed mucous membranes, fever, stupor, muscle weakness and twitching, respiratory paralysis, and hemolysis in oral poisonings in **man**.

IP, LD_{50}, **rat**: 17 mg/kg.

ESMOLOL

= *ASL 8052* = *Brevibloc*

***β-Adrenergic-blocker, antiarrhythmic.* Ultrashort-acting. Cardioselective for IV use.**

Untoward effects: Generalized seizures in an 89-year-old **male** within 14 min of starting a normal infusion rate. Has caused bradycardia, hypotension, heart block, heart failure, bronchospasm, dermatological reactions, and pain at infusion site. A **patient** coded after receiving a portion of a concentrated 2.5 g solution, designed to be diluted in a large volume of IV solution. A number of such errors have occurred.

Drug interactions: Has caused an increase in ***digoxin*** levels. Can prolong effects of non-depolarizing muscle relaxants, such as ***atraconium***, ***pancuronium***, and ***vecuronium***, as well as potentiating the adverse effects of general anesthesia.

ESTAZOLAM

= *Bay K-4200* = *Cannoc* = *D-40TA* = *Esilgan* = *Eurodin* = *Julodin* = *Nemurel* = *Nuctalon* = *ProSom* = *Somnatrol*

***Sedative, hypnotic.* A benzodiazepine.**

Untoward effects: Drowsiness, sedation, dizziness, hypokinesia, incoordination, rebound insomnia, constipation, dry mouth, chills, fever, neck pain, flushing, palpitations, anorexia, gastritis, flatulence, euphoria, and hostility; rarely, edema, jaw pain, breast swelling, arrhythmias, syncope, leukopenia, purpura, thyroid nodules, decreased libido, hallucinations, nystagmus, tremor, arthralgia, neuritis, melena, mouth ulcerations, "hang-over", and, possibly, dose-dependent impairment or recall. Plasma concentration peaks 1.5–2 h after swallowing. Metabolized by liver microsomal oxidation, therefore, increases in age, cirrhosis, or ***cimetidine***, an enzyme inhibitor, prolongs its half-life. Potential for **fetal** damage. Marketed in an over-the-counter dietary supplement, ***Sleeping Buddha***.

Cimetidine, but not ***ranitidine***, increases its depressing activity in **rats**.

LD_{50}, **rats**: 3.2 g/kg; male **mice**: 740 mg/kg; **rabbits**: 300 mg/kg.

ESTIL

= *Acetamidoeugenol* = *Detrovel* = *G-29505* = *2-M-4-A*

***Anesthetic for IV use.* A *eugenol* derivative.**

Untoward effects: Vomiting and nausea during first 6 h. Occasionally involuntary muscle movement and thrombophlebitis in **humans**.

Hyperpnea, limb twitches, miosis, and incontinence in **cats**.

ESTRADIOL

= *Aquadiol* = *Compudose* = *Dihydrofollicular Hormone* = *Dihydrofolliculin* = *Dihydrotheelin* = *Dihidroxyestrin* = *Dimenformon* = *Diogyn* = *3,17-Epidihydroxyestratriene* = *Estrace* = *Estraderm* = *β-Estradiol* = *cis-Estradiol* = *Estratrienediol* = *Estroclim* = *Evorel* = *Gynoestryl* = *Macrodiol* = *Monorest* = *Oestrogel* = *Ovocyclin* = *Ovocylin* = *Profoliol B* = *Progynon* = *Systen* = *Vagifem* = *Zumenon*

Untoward effects: Dose-related nausea and vomiting, ***sodium*** retention and edema, weight increase, breast tenderness and enlargement, jaundice, depression, headache, dizziness, withdrawal bleeding, amenorrhea, chloasma, rashes, urticaria, endometrial carcinomas and breast cancers on prolonged therapy, premature closure of the epiphyses with large doses, "barnyard" aftertaste, urinary incontinence, vaginitis, nipple pigmentation, and increased intraocular tension. Passes into breast milk. Skin irritation and dermatitis from transdermal patches. Gynecomastia in **males**. Anaphylactic shock reported with ***estradiol benzoate***. Unclear if the antigen was the hormone itself, an impurity, or an additive. 500 g of meat from treated **cattle** contains ~15,000 times less than the average daily production rate in **men**, and several million times less than the production rate/day in **women**. An 8 oz steak from an untreated **steer** has 93 ng of estradiol, and 95 ng from an implanted **steer**. ***Grapefruit*** juice decreases its metabolism in ovariectomized **women**. ***Phenobarbital*** decreases its effectiveness.

Estradiol-17β in adult and newborn male and female **mice** and female **rats** caused adenocarcinomas, papillary carcinomas, and anaplastic carcinomas, pituitary adenomas in male **rats**, fibromyomas of the uterus, mesentery, and abdomen in female **guinea pigs**, and malignant renal tumors in **hamsters**.

ESTRAGOLE

= *p-Allylanisole* = *Esdragole* = *Methyl Chavicol*

Use: In soaps, detergents, cosmetic lotions and creams, perfumes, and foods, as a flavoring agent. Found in ***basil***, ***fennel***, and ***tarragon***. ~2 ton/year used in U.S.

Untoward effects: Suspected carcinogen for **man**. Europe restricts its use in foodstuffs to 40 ppm. No sensitization reactions in 25 **human** volunteers at 3% in petrolatum. Daily intake for **man** estimated at 0.4 mg.

Rats receiving 605 mg/kg/day/4 days developed discoloration, mottling, and blunting of liver lobe edges. Carcinogenic in **mice** (50 mg/kg) and **rats**.

LD_{50}, **rats**: 1.23 g/kg; **mice**: 1.25 g/kg.

ESTRAMUSTINE

= *Emcyt* = *Estracyst* = *NSC 89,199* = *Ro 21-8837*

***Antineoplastic. Estradiol–nitrogen mustard* complex for IV and oral use.**

Untoward effects: Nausea, vomiting, diarrhea, local reactions at injection site, thrombophlebitis, myocardial infarction, congestive heart failure, mild gynecomastia, esophagitis, edema, dyspnea; pulmonary infiltrates, fibrosis and emboli; leukemia, hypertension, and decreased ***glucose*** tolerance in diabetes. Very irritant – do not contact skin or mucous membranes. Long-term use, as with other ***estrogens***, has potential for breast and liver carcinomas.

ESTRIOL

= *Aacifemine* = *Colpogyn* = *Destriol* = *Estro-Dequavagn* = *Follicular Hormone Hydrate* = *Gynäsan* = *Hormomed* = *16 α- Hydroxy-estradiol* = *Klimax E* = *Klimoral* = *Oekolp* = *Oestriol* = *Ortho-Gynest* = *Ovesterin* = *Ovestin* = *Ovo-Vinces* = *Theelol* = *Tridestrin* = *Trihydroxyestrin* = *Triovex*

Untoward effects: Decreases hepatic sulfobromophthalein excretion in postmenopausal **women**. Increases hepatic metabolism of ***estrogens*** leading to decreased levels in postmenopausal **women smokers**, probably contributes to **their** osteoporosis. **Pregnant women** have decreased serum concentraton after receiving ***phenobarbital***. Urinary levels decreased after oral ***aspirin***, ***erythromycin***, ***methenamine mandelate***, and ***penicillin V***.

ESTROGENS, CONJUGATED

= *Co-Estro* = *Conest* = *Equigyne* = *Estroate* = *Femest* = *Fem-H* = *Genisis* = *Kestrin* = *Menogen* = *Menotab* = *Menrium* = *Ovest* = *Palopause* = *Premarin* = *Presomin* = *Sodestrin-H*

The synthetic *Conestin* contains nine of the ten fractions found in *Premarin*.

Untoward effects: Nausea. Rarely, vomiting. Withdrawal bleeding, climacteric complaints, dysmenorrhea, occasional fatigue, decreased libido, gynecomastia, breast tenderness, recurrent cerebrovascular accidents, myocardial infarction, pulmonary emboli, nervousness, burning and/or dry mouth, jaw pain, palate soreness, increased nipple pigmentation, increased diastolic blood pressure, darkening of skin and face, chloasma, pruritic blisters on arms after sun exposure, reddish urine, hair loss, hypertrichosis on forehead and temples, porphyrea cutanea tarda, vaginal candidiasis, red nodules on skin of calves, weight gain, increased fasting ***glucose***, tremors, lethargy, headaches, "barnyard" aftertaste, hemolytic uremic syndrome. Causes enlargement of uterine fibroid tumors and increases incidence of endometrial cancer, breast and liver hyperplasia and cancers, and gallstones. Asthma may be due to ***tartrazine*** coloring. Dr. Jukes has pointed out that certain substances essential to life (viz. estrogens produced in the body and ***chromium***) are carcinogenic, depending on the dosage level. Leg swelling has been reported.

Drug interactions: High dosage, apparently, induced extreme lethargy, nausea, constant headaches, and hypotension, when given with ***imipramine***. ***Phenytoin*** impairs its efficacy, apparently by inducing hepatic microsomal enzyme action. Increased maintenance levels of ***butaperazine*** are required when given with it. Can cause shock reactions in ***aspirin***-sensitive **patients**.

ESTRONE

= *Cristallovar* = *Crinovaryl* = *Destrone* = *Disynformon* = *Endofolliculina* = *Estrin* = *Estrol* = *Estrugenone* = *Estrusol* = *Femestrone* = *Femidyn* = *Folikrin* = *Folipex* = *Folisan* = *Follestrine* = *Follicular Hormone* = *Folliculin* = *Follicunodis* = *Follidrin* = *Glandubolin* = *Hiestrone* = *Hormofollin* = *Hormovarine* = *Ketodestrin* = *Ketohydroxyestrin* = *Kolpon* = *Menformon* = *Oestrin* = *Oestrone* = *Ovifollin* = *Perlatan* = *Theelin* = *Thelykinin* = *Tokokin*= *Wynestron*

Oral, parenteral, and topical.

Untoward effects: Increases risk of endometrial carcinomas, nausea, vomiting, nipple pigmentation, vaginitis, ***sodium*** and water retention, and anemia. See ***estradiol*** for additional untoward effects. ***Conjugated estrogens*** consist mainly of ***sodium estrone sulfate***.

ESTRONE PIPERAZINE SULFATE

= *Estropipate* = *Harmogen* = *Ogen* = *Piperazine Estrone Sulfate* = *Sulsetrex Piperazine*

Untoward effects: In addition to the above, loss of scalp hair and increase in facial hair have been uncommon complaints.

Cow's milk contains < 10 ng/100 ml; their colostrum contains 141–199 ng/100 ml. Found in many plants. In **rats**, ***phenobarbital*** inhibits its uterotrophic activity and accelerates its metabolism. Oral, topical, or SC in **mice** led to increase in mammary tumors; SC in **rats** led to pituitary, adrenal, mammary, and bladder tumors, as well as bladder calculi; SC in **hamsters** led to renal tumors and pituitary tumors in castrated males.

ETANERCEPT

= *Enbrel* = *rh Tumor Necrosis Factor Receptor* = *THFR:Fc* = *TNR 001*

Antiarthritic. **For SC use. Experimentally by infusion.**

Untoward effects: Mild local inflammation reactions at injection site. Occasionally, abdominal pain, nausea, and vomiting in Crohn's disease **patients**.

ETHACRIDINE LACTATE

= *Acrinol* = *Acrolactine* = *Ethodin* = *Metifex* =*Rimaon* = *Rivanol* = *Vucine*

Antiseptic, abortifacient. **In Triple Dye.**

Untoward effects: Shivering, nausea, vomiting, and strong epigastric pain in 10%, when used by injection for transcervical abortion; occasional decrease in blood pressure and bleeding. Topically, it caused an allergic contact eczema and brownish nail plate pigmentation.

SC, LD_{50}, **mice**: 120 mg/kg; SC, LD_{LO}, **rabbit**: 100 mg/kg; IV, LD_{LO}, **rabbit**: 30 mg/kg.

ETHACRYNIC ACID

= *Crinuryl* = *Edecril* = *Edecrin* = *Endecril* = *Endercin* = *Hydromedin* = *MK-595* = *Reomax* = *Taladren* = *Uregit*

Diuretic.

Untoward effects: Pain and thrombophlebitis, diaphoresis, and generalized body pain (burning) after IV.

TD_{LO}, **man**: 3 mg/kg; **women**: 4 mg/kg; IV, **women**: 3 mg/kg.

Blood and cardiovascular: Agranulocytosis, thrombocytopenia, transient aplastic anemia in **one**, anemia, neutropenia, hypokalemia, hyponatremia, decreased serum ***chloride***, increased serum ***bicarbonate***, hypoglycemia, hypotension, shock, collapse, and decreased pulmonary artery systolic and diastolic pressures.

Clinical test interference: Decreased or false negative ***sodium***, increased or false positive blood urea nitrogen and fasting ***glucose***. Occasionally decreased or false negative fasting ***glucose***, followed by elevation.

Gastrointestinal: Nausea, abdominal pain, and diarrhea are common; occasional gastrointestinal bleeding, dry mouth, thirst, anorexia, and heartburn.

Gout: Hyperuricemia, an asymptomatic side-effect, may lead to gout.

Hearing: Ototoxicity with hearing loss (transient and permanent) often within a few minutes, especially after large dosage infusions; occasionally, associated with tinnitus and vertigo. Has occurred after oral dosing.

Liver: Hepatocanicular jaundice, focal necrosis, hepatic coma and encephalopathy, and death.

Miscellaneous: Diabetogenic, acute Henoch–Schönlein purpura, pancreatitis, nystagmus, fever, chills, wheezing, hematuria, maculopapular rash, headache, and deaths. Marked dehydration can impair mental function, leading to acute brain syndrome with confusion, hallucinations, inappropriate behavior, and maniacal manifestations in elderly **patients**; and reproducible when administered again.

Drug interactions: Low doses of ***aminoglycoside antibiotics*** (***gentamycin***, ***kanamycin***, and ***streptomycin***) with it can cause permanent hearing loss. It has a potential for increasing ***warfarin's*** anticoagulant effect. Antagonizes ***allopurinol's*** uricosuria development and decreases ***probenecid's*** uricosuric effect. Decreases the effect of ***nalidixic acid*** and ***nitrofurantoin***. Potentiates ***sulfonylurea's*** antidiabetic effect, ***digitalis***, and ***hypotensive agents***.

Increases serum ***lithium*** levels with possible ***lithium*** toxicity. Increases urinary excretion of ***chloramphenicol*** and its metabolites. ***Gallamine***, ***hydralazine***, ***mecamylamine***, ***pentolinium***, ***d-tubocurarine***, and ***veratrum*** alkaloid effects are potentiated by it.

Potentiates hypoprothrombinemic effects of ***ethylbiscoumacetate*** and ***warfarin*** in **rats**. Rapid ototoxicity in **cats**, but not **guinea pigs**. Do not administer in cases of decreased renal function.

LD_{LO}, **mouse**: 600 mg/kg; IV, **mouse**: 221 mg/kg.

ETHAMBUTOL

= *CL 40,881* = *Dadibutol* = *Dexambutol* = *Ebutol* = *EMB* = *Etapiam* = *Etibi* = *Myambutol* = *Mycobutol* = *Sural* = *Tibutol*

Untoward effects: Optic neuritis and atrophy of optic nerve more frequent with dosage > 25 mg/kg/day. Reduced visual acuity or constriction of peripheral visual fields, slowly develops over a period of a month, with blurred vision and color blindness. Discontinuance of treatment usually avoids permanent injury, but results may take several months. Dermatitis, pruritus, toxic epidermal necrolysis, lichenoid eruptions, paresthesias, anaphylactic reactions, joint pain, fever, nausea, anorexia, dizziness, malaise, headache, dyspnea, fever, twitching of limbs, increased transaminase, increased blood urea nitrogen, cholestatic jaundice, hyperuricemia, peripheral neuropathy (primarily sensory), shock, metallic taste, thrombocytopenia, leukopenia, neutropenia, aplastic anemia, purpura, melena, and pancreatitis have been reported. Slight risk of ototoxicity and cerebral damage to **fetus**. Nephrotoxic. May cause a false positive ***phentolamine*** test for pheochromocytoma. ~80% excreted unchanged in urine. Decrease dosage in **patients** with renal impairment. Therapeutic serum concentrations are 2–8 mg/l; toxic at 10 mg/l.

TD_{LO}, **man**: 600 mg/kg.

Drug interactions: ***Aluminum hydroxide***, ***glycopyrrolate***, and ***metoclopramide*** retard its gastrointestinal absorption. Interaction with ***rifampin*** leads to Stevens–Johnson syndrome with increase in the latter's toxic effects.

Teratogenicity reported in **animals**, but not in **man**.

ETHAMIVAN

= *Cardiovanil* = *Emivan* = *Vandid* = *Vanillic Acid Diethylamide* = *Vanillic Diethylamide*

Central nervous system and respiratory stimulant. Oral or IV.

Untoward effects: Restlessness, irritability, sneezing, gasping-type of respiration, laryngospasm, coughing, pruritus, hiccups, blotchy erythematous rash lasting 30 min, shivering, mild frontal headaches, color hallucinations, and substernal chest pain. High IV dosages cause rapid return to consciousness, ventricular extrasystoles, nausea, vomiting, flashes before eyes, generalized body warmth, pain at injection site, and grand mal seizures or convulsions. Its effect is enhanced by ***monoamine oxidase inhibitors***.

LD_{50}, **mouse**: 67 mg/kg; **dog**: 300 mg/kg; IV, **mouse**: 15 mg/kg; **dog**: 30 mg/kg.

ETHAMOXYTRIPHETOL

= *MER 25*

Antiestrogen.

Untoward effects: Experimentally effective antifertility agent, by preventing estrus, and, specifically, conception in **rats**, **mice**, **monkeys**, **chickens**, **rabbits**, **dogs**, and **sheep**. Limited published work indicated it may cause developmental abnormalities, when given to pregnant **animals**.

LD_{LO}, **rat** (13–19 days pregnant): 175 mg/kg.

ETHAMSYLATE

= *Aglumin* = *Altodor* = *Biosinon* = *Cyclonamine* = *Dicynene* = *Dicynone* = *E-141* = *Etamsylate* = *MD-141*

***Hemostatic.* IV or IM.**

Untoward effects: Nausea, abdominal and gastric pain, diarrhea, transient hypotension, headache, skin rash, and some leg vein thrombosis, when given IV before 42 vaginal surgeries.

IV, LD_{50}, **rats**: 1.35 g/kg; **mice**: 0.8 g/kg.

ETHANE

= *Bimethyl* = *Dimethyl* = *Ethyl Hydride* = *Methylmethane*

Use: Fuel for lighter-than-air aircraft. Has practically the same density as air and does not appreciably alter the weight of the ship. In "bottled gas" with 90% ***propane***.

Untoward effects: A simple inert asphyxiant. High concentrations cause narcosis.

Sensitizes the **dog's** heart to ventricular fibrillation effects of ***epinephrine***.

ETHANETHIOL

= *Ethyl Mercaptan* = *Ethyl Sulfhydrate* = *Mercaptoethane* = *Thioethyl Alcohol* = *Uromitexan*

Use: Odorant (earthy, sulfidy, leek-like) in natural gas, detectable at 0.001 ppm, in manufacturing plastics, insecticides, and antioxidants.

Untoward effects: Headache, nausea, and mucous membrane irritation. **People** exposed to low concentration

(5–10 days) show altered taste to sweet and bitter substances and increased olfactory threshold. TLV, 0.5 ppm/TWA. OSHA ceiling is 10 ppm.

LD_{LO}, **human**: 420 mg/kg; inhalation TC_{LO}, **human**: 200 ppm.

Acute inhalation by **rats** and **mice** led to mucous membrane irritation, incoordination, weakness, staggering, partial skeletal muscle paralysis, cyanosis, respiratory depression, coma, and death. Chronic inhalation by **rats** and **rabbits** led to cardiovascular disorders, excitability, and decreased red blood cells.

LD_{50}, **rats**: 1.7 g/kg; inhalation LC_{LO}, **dog**: 76,000 ppm, **rabbit**: 106,000 ppm.

ETHANOLAMINE

= *2-Aminoethanol* = *β-Aminoethyl Alcohol* = *Colamine* = *Ethamolin* = *Ethylolamine* = *2-Hydroxyethylamine* = *β-Hydroxyethylamine* = *Monoethanolamine* = *Tindanol*

Use: Surfactant in parenterals and agricultural chemicals, hair-waving solutions, cleansers and detergents, natural gas conditioning, petroleums, ***carbon dioxide*** absorption, manufacturing ***antihistamines***, and as a sclerosing agent in medicine. ~600 million tons/year used.

Untoward effects: Necrosis at injection site with persistent scar when used as a sclerosing agent; local pigmentation. Dermatitis when used as an amine "curing" agent or "hardener" in epoxy resins. Nausea, vomiting, abdominal pain, and even death, if swallowed. Can be absorbed through the skin. Vapors are very irritating leading to coughing and nose, throat, and chest discomfort. Contact with eyes causes burns. Airborne TLV, 3 ppm/8 h.

LD_{50}, **rat**: 2.14–2.74 g/kg; dermal, **rabbit**: 1 g/kg; severe eye irritation, **rabbit**: 763 μg.

ETHCHLORVYNOL

= *Arvynol* = *Ethclorvynol* = *Normoson* = *Placidyl* = *Roeridorm* = *Serenesil*

Sedative, hypnotic. **Oral and IV use.**

Untoward effects: Gastric discomfort, nausea, vomiting, unpleasant aftertaste, decreased blood pressure, skin rashes, fixed-drug reactions, pruritus, porphyrias, blurred vision, facial numbness, drowsiness, dizziness, fatigue, headache, ataxia, confusion, visual and auditory hallucinations, giddiness, nystagmus, addiction, respiratory distress, raised alkaline phophatase and hepatotoxicity, nightmares, apnea, coma, shock, and cardiac arrest. Episodes of recurring fulminant thrombopenia in a 33-year-old **female**, the last one resulting in death. Dependence and abuse reported. Rarely, prolonged hypnosis, cholestatic jaundice, pancytopenia, and thrombocytopenia. IV use has induced pulmonary edema, hypotension, and decreased cardiac output in **man** and **dogs**. High risk of toxicity to **children**. Therapeutic serum range is 0.05–0.5 mg/%; toxic levels 2–10 mg/%. Others report 0.5–83 μg/ml. Deaths have been reported after ingestion of 5–49.5 g, yet **others** have recovered after 7.5–100 g in suicide attempts. In these cases, unconsciousness and respiratory depression may be prolonged; pancytopenia and hemolysis have been reported in other cases. "Delirium tremens" on withdrawal in a 66-year-old **female** on 2000–3000 mg/day. A 67-year-old **male** on 1 g/day increased dosage to at least 1.25 g/day/5 months. **His** withdrawal symptoms included convulsions, hypertonicity, and hyperreflexia. Withdrawal symptoms also occurred in an **infant** whose **mother** had taken the drug during **her** last 3 months of pregnancy. It was detected in the **infant's** urine for 4 days. Very variable half-life (21–105 h).

TD_{LO}, **woman**: 10 mg/kg.

Drug interactions: Effects enhanced by ***alcohol*** and ***barbiturates*** with some fatalities. Consciousness lost in two **alcoholics** after taking a single therapeutic dose. Can antagnize hypothrombinemic action of ***bishydroxycoumarin*** and ***warfarin*** in **man**, but not in **rats** or **dogs**. Can decrease the effectiveness of ***oral contraceptives***.

LD_{50}, **mice**: 290 mg/kg; **wild birds**: 42 mg/kg.

ETHEPHON

= *Camposan* = *CEPA* = *CEPHA* = *Cerone* = *Ethrel* = *Florel*

Use: Plant growth regulator hastens the maturation of numerous fruits (***cherries, figs, grapes, peppers, pineapple, plums, tomatoes, walnuts***, etc.) and used to defoliate ***beans*** and ***cotton***, by supplying ***ethylene*** (q.v.), a naturally occurring ripening agent found in most fruits.

Untoward effects: Formulations for spraying are very acidic (pH ~1). Very irritating to eyes and skin, and to respiratory tract mucous membranes if inhaled. May cause irreversible eye damage.

Long-term suppression of plasma cholinesterase in **rats** and **mice** after oral intake.

LD_{50}, **rat**: 4.2 g/kg; **mouse**: 2.9 g/kg.

ETHER, ETHYL

= *Anesthetic Ether* = *Diethyl Ether* = *Diethyl Oxide* = *Ether* = *Ethoxyethane* = *Ethyl Oxide* = *1,1′-Oxybisethane* = *Sulfuric Ether*

Over a hundred million lbs produced annually in the U.S.

***Anesthetic.* A term coined by Oliver Wendell Holmes for this revolutionary discovery and use of it by Dr. Crawford W. Long, a physician, and, shortly thereafter, by William T. Morton, a dentist.**

***Use*:** Surface antiseptic and cleansing agent, coolant, in pharmaceutical extractions and liniments, solvent, as a cold weather ignition initiator, and in manufacturing synthetic textiles, smokeless gun powders, and explosives.

Untoward effects*:** Irritant to respiratory tract mucous membranes, stimulates salivation, and, often, dramatically increases bronchial secretions. Causes vasodilation decrease in blood pressure (occasionally, severe), hyperglycemia, cystitis, decreased renal blood flow and function, increased capillary bleeding and occasional increase in prothrombin time, increased polymorphonuclear leukocytes, decreased respiration, and rectal (but not skin) temperature may increase. Overexposure leads to mydriasis, acidosis, headache, dizziness, drowsiness, nausea, vomiting, narcosis, ventricular fibrillation and heart block, and eye irritation. Occasional laryngeal spasms, and, if blood pressure increases, cerebral hemorrhage can occur. Convulsions (particularly in young **children** under deep anesthesia), malignant hyperpyrexia, and nephrotoxicity occur, and overdoses lead to respiratory failure and cardiac arrest. Some reports of decerebration-type syndromes in anesthetized **children**. Breast milk levels for ~8–10 h are about equal to **maternal** blood levels. Crosses the placenta rapidly, depressing the **infant** by direct narcotic effect without interfering with oxygenation. In the early 1800s, its first uses were as intoxicants. Its use for recreational purposes continued in Europe and America throughout the nineteenth century and exhibited an occasional resurgence in the 1920s and 1940s. The drinking of ether (liquid at cool room temperatures) may have become popular as a substitute for ***alcohol, as it left no hangover symptoms. The U.S. Drug Enforcement Administration has clamped down on the sale of ether, as it is used to convert impure ***cocaine*** paste into the crystalline form used for snorting or injecting. Also used in the illegal synthesis of ***amphetamines***. Addiction to it has occurred. Sniffing has been reported in **health care workers**. Some tolerance to the fumes may occur with daily use. Temporary stimulation may occur before central nervous system suppression. Airborne limit for continuous exposure is 2.1 mg/l. Others state it at 100–400 ppm. Oral LD, **human**: ~30 ml. Continuous inhalation of 6% in air has stopped respirations in 10 min.

LD_{LO}, **human**: 420 mg/kg; TC_{LO}, inhalation, **human**: 200 ppm.

Bite down hard on a wintergreen-flavored ***Certs***® or ***Life-Saver***® and you will generate a spark. Avoid their use in explosive atmospheres (operating rooms, ***oxygen*** tents, and spacecrafts). Explosive in air at 1.85–100% concentration.

Drug interactions*:** Synergistic toxicity (decreases neuromuscular transmission) with ***colistin, curare, decamethonium, kanamycin, neomycin, streptomycin, and ***succinylcholine***. This can cause respiratory suppression, apnea, and paralysis. Cardiodepressant (bradycardia) with ***propranolol*** by enhancing the latter's action.

Has been used as an inhalation anesthetic in ~20 species of **animals** and **fish**. A very narrow margin of safety between therapeutic and lethal doses in **birds**. Mammary excretion into breast milk also occurs in **goats**. Flammable vapors from **animals** euthanized with it can become a hazard if the **animal** is cremated. In older **dogs**, it may induce temporary albuminuria, oliguria, and even anuria.

Inhalation LC_{LO}, **dog**: 76,000 ppm; **rabbit**: 106,000 ppm.

ETHER, VINYL

= *Divinyl Ether* = *Divinyl Oxide* = *Ethydan* = *Vinethine* = *Vinesthine* = *Vinydan* = *Vinyl Ether*

***Anesthetic.* For inhalation.**

***Untoward effects*:** Convulsions in eight, cyanosis in one, bradycardia in one, and conjunctivitis in one of 4196 **females** anesthetized for vaginal deliveries. Five cases of acute liver atrophy reported in **man**. Use in 1418 **infants** and **children** led to 1.6% with jerky, violent twitchings, and 1% with convulsions.

ETHINAMATE

= *Ethynylcyclohexyl Carbamate* = *Valamin* = *Valmid* = *Valmidate*

Sedative hypnotic.

Untoward effects*:** Habituation. Confusion, agitation, disorientation, and unsteady gait on 2–15 g/day in **male pharmacist**. Convulsions, syncopal episodes, tremors, and hallucinations 12–24 h after withdrawal. Hospitalized for 26 days. Central nervous system effects and aplastic anemia. Enhances ***alcohol's decreased attention and depressant effects. Gastrointestinal upsets and skin rashes. Excitement in **children**. Rarely, thrombocytopenic purpura, fever, and hypersensitivity reactions. Fatality after ingestion of 15 g, yet survival in one case after 28 g. Interferes with various steroid test readings. A wide range reported for therapeutic serum levels of 1.5–10 mg/l; usually toxic at 100 mg/l; comatose or fatal at > 200 mg/l.

TD_{LO}, **human**: 7 mg/kg.

LD_{50}, **rabbit**: 1 g/kg; **guinea pig**: 250 mg/kg.

ETHINYL ESTRADIOL

= *Diogyn E* = *Estigyn* = *Estinyl* = *17-Ethinylestradiol* = *Ethy 11* = *Ethynylestradiol* = *Eticylol* = *Eticyclin* = *Etivex* = *Feminone* = *Gynolett* = *Kolpolyn* = *Linoral* = *Lynoral* = *Oradiol* = *Orestralyn* = *Primogyn C* = *Progynon C* = *Progynon M*

Untoward effects: See ***Estradiol***. Cardiovascular defects, hypercalcemia, decreased plasma ***calcium*** in **patients** with osteoporosis, slight decrease in body temperature, hepatic necrosis and hepatitis, increased acute intermittent porphyria, headache, anorexia, feminization in **men**; thrombophlebitis, thrombosis, and pulmonary infarction in **men** and **women**; increased blood pressure, decreased libido in **males** and **females**, recurrent severe attacks of urticaria and angioedema in two **sisters**, decreased ***insulin*** sensitivity, erythema multiforme, Stevens–Johnson syndrome, plasma : milk ratio of 10 : 1 and carcinogenicity occur. The acronym, **VACTREL**, describes some of the adverse effects when used with various progestogens. (**V**-vertebral; **A**-anal; **C**-cardiac; **T**-tracheal; **R**-renal; **E**-esophageal; **L**-limb)

TD_{LO}, **woman**: 21 mg/kg/21 days.

Drug interactions: Antibiotics, such as ***ampicillin***, ***neomycin***, ***penicillin V***, and ***rifampin***, increase its hepatic metabolism. This may cause breakthrough bleeding or contraception failures in **patients** on ***oral contraceptives***.

Causes pituitary tumors and malignant mammary tumors in **mice**; endometrial hyperplasia and malignant uterine and cervical tumors in mice; benign gonadal tumors in male **mice**; increase in neoplastic liver nodules and pituitary adenomas in **rats**; induces hepatocellular carcinomas in **hamsters**; and its enterohepatic circulation in **rats** is decreased by ***ampicillin*** and ***neomycin***.

LD_{50}, **rat**: 3 g/kg; **mouse**: 1.7 g/kg.

ETHIODIZED OIL

= *Ethiodol* = *Lipiodol* = *Neohydriol* = *Oleum Iodatum* = *Poppyseed Oil, Iodized*

Use: In lymphography, and is antineoplastic when its ***iodine*** is $^{131}\boldsymbol{I}$.

Untoward effects: Lipid embolization in kidney and brain has occurred. Residual material in the lungs associated with granulomas. Inadvertent IV has caused pulmonary oil embolism. Hypersensitivity reactions (dyspnea, dermatitis, tachycardia, increased temperature, nausea, and vomiting) reported. May interfere with thyroid function tests. As $^{131}\boldsymbol{I}$, extravasation necrosis, lymphangitis, hypothyroidism, hepatic and pulmonary insufficiency, hemolytic anemia, and amenorrhea have occurred.

ETHION

= *Diethion* = *ENT-24105* = *FMC-1240* = *Niagara 1240* = *Nialate* = *RP 8167*

Insecticide, acaricide. **Anticholinesterase.**

Untoward effects: Toxic exposure by inhalation, dermal absorption, and ingestion. Irritant to eyes and skin; nausea, vomiting, abdominal cramps, diarrhea, salivation, headache, giddiness, vertigo, weakness, tightness in chest, miosis, blurred vision, cardiac arrhythmias, tremors, dyspnea, and decreased blood cholinesterase.

TD_{LO}, **human**: 100 μg/kg.

Paresis, myoclonic spasms, increased salivation, trembling, arched backs, exophthalmos, and mucosal hemorrhages in poisoned **rats**, **mice**, **rabbits**, **dogs**, and **fowl**. Poisoned **birds** rapidly exhibit ataxia, wing-drop, dyspnea, and falling.

LD_{50}, male **rats**: 65 mg/kg; female **rats**: 27 mg/kg; **mallard ducks**: ~2.6 g/kg; **pheasants**: 1.3 g/kg.

ETHIONAMIDE

= *Aetina* = *Amidazine* = *Bayer 5312* = *Ethimide* = *Ethioniamide* = *2-Ethylisothionicotinomide* = *Iridocin* = *Nisotin* = *1314-Th* = *Tio-Mid* = *Trecator* = *Trescatyl*

Antitubercular.

Untoward effects: Hepatocellular incidence of 3–5% and acute hepatitis in ~5–10%, particularly in **diabetics**, alopecia with long-term therapy, gastric irritation (nausea, vomiting, abdominal pain, heartburn, salivation, metallic taste, swallowing problems, and anorexia) in 50% receiving 1 g/day, leading to poor acceptance of this therapy, immediate Type 1 hypersensitivity with asthma, arthritis and arthralgia (sometimes with joint pain), insomnia, malaise, hypoglycemia, peripheral neuropathies, slight deafness, diplopia, thrombocytopenia, postural hypotension, depression, hepatotoxicity, severe allergic skin rash, purpura, amenorrhea, stomatitis, decreased libido, gynecomastia, confusion, and psychosis. May cause false decrease in alkaline phosphatase tests and increase or false positive SGOT and SGPT values. Probably teratogenic, based on limited usage in **women**. Suicide attempts led to hyperglycemia.

LD_{LO}, **human**: 7 mg/kg.

Drug interactions: Potentiates the psychotoxic effects of ***alcohol*** ingestion. Exaggerated toxicity of both when given with ***isoniazid*** (q.v.).

Teratogenic in **rats** and **rabbits**. Causes fatty livers in **rats**.

LD_{50}, **rabbit**: 1 g/kg; **guinea pig**: 250 mg/kg.

ETHIONINE

***Amino acid antagonist. Methionine* analog.**

Untoward effects*:** Use led to a better understanding of chemical carcinogenesis. Blocks utilization of ***methionine by **rats**; induces fatty liver, pancreatitis, muscle degeneration, fetal resorption, hepatomas, and cholangiomas in **rats** and **pigs**; and pancreatic necrosis in **mice**. Decreased rate of gain in **chicks** and **rats**, bile duct hyperplasia in **ducks**; liver lesions in **guinea pigs**; potentiates ***endotoxin*** toxicity in **mice**, **rats**, and **guinea pigs**; fatty degeneration, hepatic necrosis, bile retention, and proliferation of bile ducts in **dogs** and **rabbits**; diffuse and moderate breast tissue hyperplasia in **rats**; and fatty livers in **calves**.

ETHISTERONE

= *Anhydrohydroxyprogesterone* = *Ethinyltestosterone* = *17α-Ethynyltestosterone* = *Lutocyclin* = *Lutocylol* = *Ora-Lutin* = *Pranone* = *Pregneninolone* = *Progestoral* = *Syngestrotabs* = *Trosinone*

Progestogen.

***Untoward effects*:** Androgen effects may predominate. Masculinization of external genitalia, lumbosacral fusion, and clitoral enlargement in ~15% of exposed **newborns** of **mothers** who had taken it, usually prior to their 16th week of pregnancy. Traces appear in breast milk. Masculinizing effects of antenatal administration has also been shown in **rabbits**.

ETHOGLUCID

= *AY 62,013* = *Epodyl* = *Etoglucid* = *ICI-32865* = *TDE* = *Triethylene Glycol Diglycidyl Ether*

Antineoplastic.

***Untoward effects*:** Nausea and vomiting for 24 h; transient dizziness during IV; maximum decrease in leukocytes and platelets between days 10–17. Severe facial edema, alopecia, rashes, mucocutaneous ulceration, leukopenia, and thrombocytopenia reported. IA has caused tissue necrosis, alopecia, and brain and nerve damage. Reacts with plastics. Use glass syringes.

Neurotoxic, bone marrow suppression, infection, and widespread hemorrhage in **dogs**. Testicular atrophy and decreased spermatogenesis after IP in **mice**.

ETHOHEPTAZINE

= *Ethyl Heptazine* = *WY 401* = *Zactane*

***Analgesic.* Analog of *meperidine*.**

***Untoward effects*:** Nausea, drowsiness, dizziness, epigastric distress, rashes, and pruritus. Above normal dosage also leads to headache, nervousness, visual and auditory hallucinations, and syncope. Also reported, thrombotic thrombocytopenic purpura with fatality in a **female** after therapy for headache.

Extreme hyperpyrexia and death in **rabbits** pretreated with ***monoamine oxidase inhibitor***.

LD_{50}, **rat**: 355 mg/kg; **mouse**: 318 mg/kg.

ETHOHEXADIOL

= *612* = *Ethylhexanediol* = *Octylene Glycol* = *Rutgers 612*

Insect repellent.

***Untoward effects*:** Irritant to eyes and mucuous membranes, but not to skin. Safety to **pregnant women** has been questioned.

LD_{LO}, **human**: 500 mg/kg.

Repeated skin applications on **mice** led to inflammation and ulceration. Oral administration (2–4 g/kg/day) to pregnant **rats** led to maternal toxicity and lethality. Fetal developmental errors occurred even with levels of 500 mg/kg/day to **dams**. Severe irritation by 5 mg to eye of **rabbit**.

LD_{50}, **rat**: 1.4 g/kg; **mouse**: 1.9 g/kg; **rabbit**: 2.6 g/kg; **chicken**: 1.4 g/kg.

ETHOPROP

= *ENT 27,318* = *Ethoprophos* = *Jolt* = *Mocap* = *Phosethoprop* = *Prophos* = *VC9-104*

Insecticide, nematocide. Organophosphate anticholinesterase agent.

***Untoward effects*:** Undue exposure of **man** can be expected to cause usual symptoms of this class of chemicals, i.e. headache, dizziness, weakness, ataxia, miosis, muscle twitching (especially tongue and eyelids), confusion, drowsiness, disorientation, nausea, vomiting, abdominal cramps, increased salivation and tearing, dyspnea, cyanosis, pulmonary edema, bradycardia, sweating, convulsions, and even death. Rapidly absorbed through the skin.

5 mg/kg to **calves** decreased blood cholinesterase 91%, and is lethal in ~6 h. 25 mg/kg to **sheep** caused diarrhea in 48 h. More toxic to **mallard ducks** by skin absorption than by mouth. Hypoactivity, ataxia, stumbling, dyspnea, spasms, lacrimation, asthenia, and death in poisoned **birds**.

LD_{50}, **rat**: 34 mg/kg; **chicks**: 5.5 mg/kg; **mallard ducks**: 12.6 mg/kg; **pheasants**: 4.2 mg/kg.

ETHOPROPAZINE

= *Dibutil* = *Isothazine* = *Isothiazine* = *Lysivane* = *Pardisol* = *Parkin* = *Parphezein* = *Parphezin* = *Parsidol* = *Parsitan* = *Parsotil* = *Phenopropazine* = *Profenamine* = *Rodipal* = *RP-3356* = *W-483*

Anticholinergic, antiparkinsonian, tranquilizer. **A phenothiazine.**

***Untoward effects*:** Nausea, drowsiness, muscle cramps, paresthesia, blurred vision, corneal opacity, dizziness, palpitations, dryness of mouth and throat, frequent micturation, and impaired memory and concentration. Less frequently, confusion, angor animi, decreased libido, face and extremity pain, paresthesia, insomnia, and incontinence.

LD_{50}, **mouse**: 485 mg/kg; SC, LD_{50}, **rat** and **rabbit**: 200 mg/kg; **mouse**: 500 mg/kg.

ETHOSUXIMIDE

= *Atysmal* = *Capitus* = *Emeside* = *Epileo Petitmal* = *Ethymal* = *Mesentol* = *Pemal* = *Peptinimid* = *Petinimid* = *Petnidan* = *Pyknolepsinum* = *Simatin* = *Succimal* = *Suxilep* = *Suximal* = *Suxinutin* = *Zarontin*

Anticonvulsant.

***Untoward effects*:** Frequently, nausea, vomiting, epigastric pain, and anorexia. Headache, vertigo, hiccups, drowsiness, fatigue, ataxia, and even euphoria. Occasionally, parkinsonism, photophobia, blurred vision, personality changes, psychotic behavior, skin rashes, and increased sore throats. Agranulocytosis in a 7-year-old **male**. **He** died of massive thrombocytopenic hemorrhages. Hemolytic anemia, nephrosis, and lupus-like syndrome in a 4½-year-old **male** after 6 months of therapy. Lupus erythematosus reported in many **adults**. Fatal complete bone marrow aplasia in an 8-year-old **female** treated for petit mal epilepsy for 4 years. Leukopenia, eosinophilia (10%), pancytopenia, hypochromic anemia, aplastic anemia, granulopenia, and thrombocytopenia occur. Breast-fed **infants** whose **mothers** were on the drug have suffered from hemorrhagic disorders. Erythema multiforme, Stevens–Johnson syndrome, and acute intermittent porphyria reported. Average milk : **maternal** plasma ratio is 0.8. Limited data indicate **maternal** use can cause developmental errors (mongoloid facies, short neck, simian crease, accessory nipple, etc.) in their **infants**. Effective blood levels are > 40 μg/ml, and toxic levels are > 100 μg/ml.

Drug interactions*:** Its metabolism is induced by ***carbamazepine; it decreases ***furosemide's*** diuretic effect; its serum concentration is dramatically increased by ***valproic acid***. May cause protective failure of ***oral contraceptives***, and give false positive or spuriously high urine ***albumin*** readings.

Crosses the placenta in the **rat**. Its absorption in **mice** is delayed by ***amphetamine***. Teratogenic (cleft palates) in **mice**.

LD_{50}, **mice**: 1.75 g/kg; LD_{LO}, **rats**: 1820 mg/kg.

ETHOTOIN

= *Peganone*

Anticonvulsant.

Untoward effects*:** Skin rashes, exanthema, anorexia, nausea, vomiting, epigastric burning, drowsiness, nervousness, diplopia, nystagmus, headache, and mental depression. Rarely, ataxia, gingival hyperplasia, hirsutism, lymph node and malignant lymphoma-like syndrome, macular exanthema, fever, thrombocytopenia, eosinophilia, and anemias (aplastic and megaloblastic). May be teratogenic in **humans**. Its action may be enhanced for some time by ***disulfiram.

LD_{LO}, **rat**: 1500 mg/kg; **mice**: 1750 mg/kg.

ETHOXAZENE

= *Carmurit* = *Cystural* = *p-Ethoxychrysoidine* = *Serenium* = *SN-612*

***Analgesic, pH indicator. Serenium*®** **is also used for *medazepam*.**

Untoward effects*:** Its metabolites can color urine pink, red, or reddish-orange, and give false readings for many ***bilirubin, ***porphyrin***, and ***phenolsulfonphthalein*** urine tests.

2-ETHOXYETHANOL

= *Cellosolve* = *Dowanol EE* = *Ethylene Glycol Monoethyl Ether* = *Glycol Ether EE* = *MIL-127,686E* = *Oxitol*

***Use*:** Solvent for nitrocellulose, lacquers, enamels, and varnish removers, and as a plasticizer in cosmetics.

***Untoward effects*:** Liquid and vapors are irritating to skin and eyes. Inhalation, ingestion, or absorption through skin can cause liver, kidney, and lung pathology. Suspect as a cause of testicular dysfunction in shipyard **painters** and leather tanning **workers**. Marked increase in gynecological disorders, and birth defects probably due to it. National Institute for Occupational Safety & Health limits exposure to 0.5 ppm (2.7 mg/m^3) and OSHA to 100 ppm (540 mg/m^3) for skin exposures. 500 ppm presents an immediate life-threatening situation.

LD_{LO}, **human**: 500 mg/kg.

Teratogenic and increased embryonic deaths in **rats** and **rabbits** at 200 and 160 ppm, respectively. Anemia, liver and lung pathology, and reproductive failures in **animals**. Testicular atrophy in **rats**, **mice**, and **dogs**.

LD_{50}, **rat**: 3 g/kg; **mouse**: 4.3 g/kg; **rabbit**: 3.1 g/kg.

2-(2-ETHOXYETHOXY)ETHANOL

= *Carbitol* = *Diethylene Glycol Monoethyl Ether* = *Dowanol DE* = *2-EEE* = *Ethyl Diethylene Glycol* =*Ethyl Digol* = *Transcutal*

Solvent.

Use: Similar to ***2-Ethoxyethanol*** and in soaps, detergents, cosmetic creams and lotions, and perfumes, to impart a faint ethereal odor. Over 2 tons/year in the U.S.

Untoward effects: 20% in petrolatum showed no irritation or sensitization in 25 **human volunteers**.

LD_{LO}, **human**: 500 mg/kg.

Not irritant to **rabbit** skin (dermal LD_{50}, 10.3 g/kg). Dyspnea, somnolence, and mild ataxia in a 4 week feeding study to **rats** (0.2–0.4 l/kg/day).

LD_{50}, **rats**: 5.4 g/kg; **mice**: 6.58 g/kg; **guinea pigs**: 3.67 g/kg; **rabbits**: 3.62 g/kg.

2-ETHOXYETHYL ACETATE

= *Cellosolve Acetate* = *EEAC* = *EGEEA* =*Ethylene Glycol Monoethyl Ether Acetate*

Use: Retards evaporation and imparts high gloss to automobile lacquers.

Untoward effects: Irritant to nose and eyes. Nausea, vomiting, dyspnea, renal damage, and paralysis. National Institute for Occupational Safety & Health limit at 0.5 ppm (2.7 mg/m^3) and OSHA at 100 ppm (540 mg/m^3) for skin exposures. 500 ppm presents an immediate life-threatening situation.

LD_{LO}, **human**: 500 mg/kg.

Inhalation by **rats** leads to fetal resorptions, decreased fetal weight, and teratogenicity at 130–600 ppm.

LD_{50}, **rat**: 5.1 g/kg; inhalation: 1500 ppm/8 h; LD_{50}, **guinea pigs**: 1.9 g/kg.

1-ETHOXYETHYLIDENE PROPANEDINITRILE

Use: Chemical intermediate (Eastman Kodak).

Untoward effects: All **rats** died after single gavage with 625 mg/kg. Post-mortem revealed hemorrhagic thymus and gastric edema, necrosis and hemorrhage in addition to some liver, splenic, and kidney lesions. May cause contact skin irritation in **man**.

2-ETHOXYNAPHTHALENE

= *Bromelia* = *β-Naphthyl Ethyl Ether* = *Nerolin*

Use: ~5 tons/year in the U.S., as a fragrance in soaps, detergents, cosmetic creams and lotions, and perfumes. In Europe, in foods.

Untoward effects: No sensitization in tests on 25–50 **volunteers** at 2% in petrolatum and no irritation in 48 h closed-patch tests on **humans**.

LD_{50}, **rats**: 3.1 g/kg; acute dermal, **rabbit**: > 5 g/ kg.

ETHOXYQUINE

= *EMQ* = *Santoflex* = *Santoquin*

Use: Antioxidant in foods and rubber.

Untoward effects: Direct skin contact is irritating to some **people**. Although safe when used by themselves, the combination with ***BHT***, as hay preservatives leads to ***hydrogen cyanide*** (q.v.) gas. This occurs when temperatures in a fire reach 240°F and adversely affected ~80% of **firemen** and **aides** at a barn fire. Symptoms were weakness, nausea, headache, dyspnea, and tachycardia.

LD_{LO}, **human**: 500 mg/kg.

Carcinogenic potential in **rats** at levels 33 times the recommended maximum in feeds. SC in neonatal **mice** caused hepatomas.

LD_{50}, **rat**: 800 mg/kg.

ETHOXZOLAMIDE

= *Cardrase* = *Ethamide* = *Ethoxyzolamide* = *Glaucotensil* = *Mingoral* = *Redupresin*

Diuretic.

Untoward effects: Nausea, vertigo; numbness and paresthesia of fingers and toes, or at mucocutaneous junctions of lips and/or anus; intoxicated feeling, light-headedness, anorexia, drowsiness, fatigue, headache, dry mouth, thirst, depression, hypokalemic acidosis, and decreased libido.

TD_{LO}, **rat** (pregnant): 150 mg/kg.

ETHYBENZTROPINE

= *Ponalid* = *UK 738*

Anticholinergic.

Untoward effects: Similar to ***atropine*** (q.v.). Excessive mouth and throat dryness, blurred vision, gastrointestinal upsets, dizziness, tiredness, rigidity, tremor, headache, confusion, impaired concentration, oliguria, paresthesias of hands and feet, and short hallucinatory episodes.

LD_{LO}, **rat**: 560 mg/kg; **mouse**: 66 mg/kg; **rabbit**: 215 g/kg.

ETHYL ACETATE

= *Acetic Ether* = *Vinegar Naphtha*

Use: In adhesives, paint thinners and enamel removers, nail enamels, as a solvent in making decaffeinated ***coffee***, in chemical synthesis as a solvent and solvent-extender, and as a fragrance for soaps, detergents, cosmetic creams and lotions, and in foods. 200–300 million lbs used annually in the U.S.

Untoward effects: Eye, nose, throat, and skin irritant. Weakness, drowsiness, and unconsciousness occur. Cutaneous eruptions in industrial **workers** (adhesive **cementers** in the shoe industry, etc.). National Institute for Occupational Safety & Health and OSHA suggest upper exposure limit TWA of 400 ppm (1400 mg/m^3). Exposure to 2000 ppm may be life-threatening, with acute pulmonary edema and congestion of liver and lungs. Chronic exposure leads to anemia and leukocytosis.

LD_{LO}, **human**: 500 mg/kg; TC_{LO}, **human**: 400 ppm; TLV, 400 ppm/8 h.

12,000 ppm/5 h causes narcosis in **cats**.

LD_{50}, **rats**: 5.6 g/kg; **rabbit**: 4.9 g/kg; inhalation LC_{50}, **rats**: 1600 ppm/8 h.

ETHYL ACRYLATE

= *Ethyl Propenoate*

Millions of lbs produced annually.

Use: In water emulsion paint vehicles; in adhesives, in fragrances in soaps, detergents, cosmetic creams and lotions, and foods. Permitted in foods at 1 ppm.

Untoward effects: Irritant to eyes, skin, and respiratory tract, and would probably cause exposed **individuals** to retreat from highly contaminated areas. OSHA limit of 25 ppm (100 mg/m^3) TWA. 300 ppm may be life-threatening. 4% in petrolatum led to sensitization in 10/24 **volunteers**. Suspect as a **human** carcinogen in 1989 and delisted 11 years later because ingestion by **humans** is unlikely.

Decrease food intake and weight in **rats**. Emesis in **dogs**. Saturated vapors killed **rats** in 15 min. Gavage in **rats** and **mice** caused gastric squamous cell carcinomas or papillomas.

LD_{50}, **rats**: 1 g/kg; **rabbits**: 400 mg/kg; inhalation LC_{50}, **rats**: < 1000 ppm/4 h. Inhalation LC_{LO}, **guinea pigs** and **rabbits**: 1204 ppm/7 h.

ETHYLAMINE

= *Aminoethane* = *Ethanamine*

Untoward effects: Eye and respiratory irritant, skin burns, and dermatitis. National Institute for Occupational Safety & Health/OSHA limits exposure to 10 ppm (18 mg/m^3) TWA. 600 ppm can be life-threatening.

LD_{LO}, **rat**: 400 mg/kg.

ETHYL-*p*-AMINOBENZOATE

= *Aethoform* = *Americaine* = *Anesthesin* = *Anesthone* = *Benzocaine* = *Ethoforme* = *Hurricane* = *Orthesin* = *Parathesin*

***Topical anesthetic.* Found in hundreds of over-the-counter products.**

Untoward effects: 3–5% of **patients** with dermatitis or eczema are positive for it on patch testing. Contact eczema and photosensitivity reported. Methemoglobinemia in susceptible **infants** (even from rectal absorption) and young **children**. Has also been reported in **adults** up to 83 years of age. Continuous use can induce sensitization with itching and urticaria. Methemoglobinemia and marked cyanosis in a 77-year-old **female** after a 1 s exposure of a 20% spray to the back of **her** throat. Toxicosis leads to vasodilation, hypotension, sedation, respiratory depression, cardiac arrhythmias, tremors, seizures, and occasionally death. Cross-sensitivity with ***butacaine***, ***p-aminobenzoate*** sunscreens, ***butacaine, procaine, sulfonylureas***, such as ***chlorpropamide*** and ***tolbutamide***, and ***tetracaine*** with severe edema, erythema, and pruritus. Severe conjunctival and corneal injury can occur after accidental eye contact. Can antagnize benefits of ***sulfonamides***.

Methemoglobinemia also reported in **cats**, **dogs**, **ferrets**, **mice**, **rats**, **monkeys**, **rabbits**, **miniature pigs**, and **sheep**. Enough can be absorbed from inflammatory skin to production toxicosis. 5 days' withdrawal required after use in food **fish**. Sensitizer in **guinea pigs**.

ETHYL *sec*-AMYL KETONE

= *EAK* = *5-Methyl-3-heptanone*

Use: Solvent in nitrocellulose and vinyl based resins in paint and coating products.

Untoward effects: Eye, nose, throat, and skin irritant; headache, dizziness, nausea, narcosis, and unconsciousness. OSHA suggests exposure limits of 25 ppm TWA. Combustible.

LD_{50}, **rat**: 3.5 g/kg; **mouse**: 3.8 g/kg; **guinea pig**: 2.5 g/kg; inhalation LC_{LO}, **rats** and **mice**: 3484 ppm/4 h.

ETHYLBENZENE

= *Ethyl Benzol* = *Phenylethane*

Use: Nearly 100% for ***styrene*** manufacturing. Also, as a solvent in detergents and perfumes. ~13 million lbs produced annually.

Untoward effects: Eye, nose, throat, and skin irritant; dermatitis, headache, weakness, dizziness, drowsiness, narcosis, and coma. National Institute for Occupational Safety & Health and OSHA TWA exposure limits of 100 ppm (545 mg/m^3). 800 ppm is life-threatening. Flammable.

TC_{LO}, **human**: inhalation 100 ppm/8 h.

Rats given 1 ml/kg/14 days, unlike with ***benzene***, gain weight, have normal or increased leukocyte count, thymus and spleen weight.

LD_{50}, **rats**: 3.5 g/kg.

ETHYL BENZOATE

Use: 2–3 tons used annually as fragrance in soaps, detergents, cosmetic creams and lotions, and in foods.

Untoward effects: 8% in petrolatum produced no irritation or sensitization in 25 **human volunteers**.

LD_{50}, **rats**: 2.1 g/kg; **rabbit**: 2.6 g/kg; acute dermal, **rabbits**: 1.94 g/kg.

ETHYL BISCOUMACETATE

= *B.O.E.A.* = *Dicumacyl* = *Ethyldicoumarol Acetate* = *Neodicoumarol* = *Pelentan* = *Stabilene* = *Tromexan* = *Tromexan Ethyl Acetate*

Anticoagulant.

Untoward effects: Nausea, abdominal cramps, vomiting, diarrhea, hematuria, hemoptysis, melena, petechiae, ecchymoses, epistaxis, hematomas, aneurysms, ready bruising, joint hemorrhages, cardiovascular collapse, and death. Metabolized at widely variable rates (half-life ~0.7–14 h, usually 2.4 h). Variable excretion in breast milk. Severe hemorrhagic manifestations in **infants** born to **mothers** treated with it during pregnancy. Acute poisoning in a 17-year-old **female** after ingesting 50 tablets (15 g) over 4 days in an abortion attempt leading to weakness, dizziness, and bloody stools. Metabolism increased and its effect decreased by ***ACTH***, ***barbiturates***, ***chloral hydrate***, ***cortisone***, ***glutethimide***, ***griseofulvin***, and ***prednisone***; half-life increased by ***cinchophen***, ***clofibrate***, ***dipyrone***, ***methylphenidate***, ***oxyphenbutazone***, ***phytonadione***; and increases activity of long-acting ***sulfonamides***, such as ***sulfamethoxypyridazine***.

Mammary excretion also occurs in the **rat** and **dog**. Causes kernicterus in **monkeys**. High doses of ***iproniazid***, ***isocarboxazid***, ***nialamide***, and ***tranylcypromine*** potentiate its hypoprothrombinemic action in **rats** and **rabbits**. High doses of ***ethacrynic acid*** and ***hydrochlorothiazide*** enhance its effects in **rats**. Increases activity of ***sulfonamides*** in **rats**.

LD_{50}, **rats**: 260 mg/kg; **mice**: 880 mg/kg; **rabbit**: 1.1 g/kg.

ETHYL BROMIDE

= *Bromic Ether* = *Bromoethane* = *Hydrobromic Ether* = *Monobromoethane*

Use: Refrigerant, solvent, and in organic synthesis.

Untoward effects: Vapors are irritant to eyes, skin, and respiratory tract mucous membranes. $\frac{2}{3}$ is eliminated unchanged by the lungs and $\frac{1}{3}$ by metabolism into ***acetic acid***, inorganic ***bromide***, and ***ethanol***. Ingestion leads to narcosis, pulmonary edema, liver and kidney dysfunction, and cadiac arrhythmias and arrest. OSHA limits exposure to 200 ppm TWA.

Inhalation LC_{50}, **rat**: 149,000 ppm/15 min, **mouse**: 16,230 ppm/1 h, **guinea pig**: 3200 ppm/9 h.

ETHYL BROMOACETATE

= *EBA*

Untoward effects: A tear gas/war gas with a fruity smell and high vapor pressure.

ETHYL BUTYL KETONE

= *Butyl Ethyl Ketone* = *3-Heptanone*

Characteristic fruity odor.

Untoward effects: Irritant to eyes and mucous membranes. Especially irritant to skin, with resulting dermatitis. Headache, narcosis, and coma. National Institute for Occupational Safety & Health/OSHA exposure limits at 50 ppm (230 mg/m^3). 1000 ppm is life-threatening.

LD_{50}, **rats**: 2.76 g/kg; dermal, **rabbits**: > 20 ml/kg; inhalation LC_{LO}, **rat**: 2000 ppm/4 h.

ETHYL BUTYL THIOUREA

Untoward effects: Contact dermatitis of plantar surface of the foot from the boot inserts and innersoles of athletic shoes. Allergen found in some ***neoprene*** or ***polychloroprene rubber*** products (diver's wet suit, backing on adhesive tape, etc.) and in photocopy paper where it serves as an anti-oxidant to prevent its yellow discoloration.

ETHYL BUTYRATE

= *Butyric Ether* = *Ethyl-n-butanoate*

Use: ~ 25 ton/year in U.S. as a fruity, ***pineapple*** fragrance in soap, detergent, cosmetic creams and lotions, and foods.

Untoward effects: No sensitization or irritation at 5% in petrolatum on 25 **human volunteers**. Full-strength under occlusion or intact or abraded **rabbit** skin for 24 h indicated moderate irritation.

LD_{50}, **rats**: 13 g/kg; **rabbit**: 5.2 g/kg; acute dermal, **rabbits**: > 2 g/kg.

ETHYL CHLORIDE

= *Aethylis Chloridum* = *Anodynon* = *Chelen* = *Chlorethyl* = *Chloroethane* = *Chloryl Anesthetic* = *Ether Chloratus* = *Ether*

Hydrochloric = Ether Muriatic = Kelene = Monochlorethane = Narcotile

Use: Refrigerant, solvent, in manufacturing ***tetraethyl lead***, and as an anesthetic (topical and by inhalation).

Untoward effects: Flammable gas, ethereal odor, burning taste. Nausea, vomiting, abdominal cramps, cardiovascular suppression, bradycardia, fibrillation, arrythmias, cardiac arrest, respiratory depression and irritation, occasionally mydriasis, salivation, laryngeal spasms, headache, dizziness, hangovers, unconsciousness, and liver and kidney pathology. Sensitizes the heart to ***epinephrine***. Pigmentation changes can occur in skin after freezing. Occupational exposure limits have varied between 200 and 1000 ppm TWA. 3800 ppm is considered life-threatening. Sudden death was not an unusual occurrence in its early use during anesthesia, even with relatively low dosage. 40 mg% can be lethal. May cause tissue trauma undesirable for biopsy. Probably carcinogenic in **humans**.

TC_{LO}, **human**: 13,000 ppm.

ETHYL CINNAMATE

= *Ethyl-β-phenylacrylate = Ethyl-3-phenylpropenoate*

Use: Limited amounts used as a fragrance in soaps, detergents, cosmetic creams and lotions, and foods, and in glass prisms and lenses.

Untoward effects: No sensitization at 4% in petrolatum in 25 **human volunteers**. Full-strength after a 48 h closed-patch test on the forearms of **volunteers** led to irritation in 1/22, but not on **rabbit** skin.

LD_{LO}, **human**: 500 mg/kg.

LD_{50}, **rats**: 7.8 g/kg; acute dermal, **rabbits**: > 5 g/kg.

a-ETHYL CYANOACRYLATE

***Surgical adhesive.* For vascular and bone use.**

Untoward effects: A 32-year-old **male** developed asthma and rhinitis from inhaling its vapors after using the adhesive on balsa wood used in building remote-control model airplanes.

5-ETHYL DIISOBUTYLTHIO-CARBAMATE

= *Buylate = Diisocarb = Genate = R-1910 = Sutan*

Herbicide.

Untoward effects: Poisoned **mallard ducks** leading to wide stance, stumbling, goose-stepping ataxia, and regurgitation within 10 min, persisting up to 10 days.

LD_{50}, **rats**: 4 g/kg; **mallard ducks**: > 2 g/kg.

ETHYLENE

= *Elayl = Ethene = Olefiant Gas*

Use: Although we have used it to ripen fruit for less than a century, the **Chinese** have used it for thousands of years to ripen ***persimmons*** and citrus fruits. In chemical synthesis and now, rarely, as an anesthetic. > 40 billion lbs used annually in the U.S.

Untoward effects: Flammable and explosive anesthetic agent, stored in violet-colored tanks. Decreases **fetal** oxygenation, but less than ***nitrous oxide***.

Daily to **animals** leads to leukopenia.

LC_{50}, **mice**: 95%.

ETHYLENE BISDITHIOCARBAMATES

= *EBDC's*

***Fungicide.* See *Maneb* and *Zineb*.**

Untoward effects: Used on fruits, nuts, and vegetables, but, when warmed, can degrade to ***Ethylene Thiourea*** (q.v.) with major health risks. Cross-sensitivity with thiurams.

LD_{LO}, **human**: 50 mg/kg.

LD_{50}, **rat**: 395 mg/kg; **mouse**: 580 mg/kg.

ETHYLENEBISTETRA-BROMOPHTHALIC IMIDE

Untoward effects: An industrial **employee** suffered respiratory distress while working in a production area for this chemical and required hospitalization. A second **employee** was removed from the same work area with cough and respiratory complaints.

ETHYLENE BRASSYLATE

= *Ethylene Undecane Dicarboxylate*

Use: As a fragrance in soaps, detergents, cosmetic creams and lotions, and foods. < 2 million lbs used annually in the U.S.

Untoward effects: No irritation with 30% in petrolatum after a 48 h closed-patch test in 25 **human volunteers**. Full-strength was moderately irritating to intact or abraded **rabbit** skin for 24 h under occlusion. Dermal irritation in a 20 day **rabbit** study at 30, 70, and 700 mg/kg/day. At the highest level, enlargement of regional lymph nodes was noted.

LD_{50}, **rats** and acute dermal, **rabbits**: > 5 g/kg.

ETHYLENE CHLOROHYDRIN

= *2-Chloroethanol = 2-Chloroethyl Alcohol = Cinecol = Glycol Chlorohydrin*

Use: As a solvent, in preplanting treatment of ***sweet potatoes***, in manufacturing insecticides and dyes, and in chemical synthesis.

Untoward effects: Potent irritant to skin and mucous membranes, nausea, vomiting, vertigo, incoordination, numbness, blurred vision, headache, thirst, delirium, decreased blood pressure, liver and kidney pathology, stupor, collapse, shock, coma, and death from respiratory failure. Fatal within 12 h after ingestion of ~2 ml by a 23-month-old **male**. National Institute for Occupational Safety & Health and OSHA TWA upper exposure limits for skin 1 and 5 ppm, respectively. 5–7 ppm can be life-threatening. Measurable amounts found in materials sterilized with ***ethylene oxide*** (q.v.). Store out of sunlight and in glass containers in the laboratory. Explosive reactions with ***aluminum*** containers.

LD_{LO}, **human**: 50 mg/kg.

Irritant to **rabbit** eyes. LD_{50} for **rats**, **mice**, **rabbits**, and **guinea pigs** is between 58 and 98 mg/kg. Liver, kidney, and lung pathology noted in these **animals**. Toxicity in **rabbits** greatest when applied topically.

ETHYLENEDIAMINE

= *1,2-Diaminoethane* = *1,2-Ethanediamine*

Use: In chelating agents, such as ***EDTA***, carbamate fungicides, lube and oil additives, asphalt-wetting agents, stabilizing rubber latex, and as a solubilizer.

Untoward effects: Irritant to nose and respiratory system. Asthma and allergic contact dermatitis (18/100 allergic **patients**) and itching. Perlèche dermatitis on cheeks of a 50-year-old **female** from its use as a stabilizer in ***Mycolog***® cream. Allergic contact dermatitis in several **infants** after use of ***Mycolog***® cream in the diaper areas. It prevents the incompatibility in the aqueous phase between ***neomycin***, ***nystatin***, and the ***corticosteroid***. One of the more common skin sensitizers (Type I hypersensitivity) – an ingredient in some hair spray and nail preparations. ***Antazoline***, ***chlorothen***, ***diphenylpyraline***, ***hydroxyzine***, ***methapyrilene***, ***pyrilamine***, ***thenyldiamine hydrochloride***, ***tripelennamine***, and ***zolamine*** are ethylenediamines. Cross-reactivity with ***EDTA***, ***hydroxyzine***, ***piperazine citrate***, and ***promethazine***, as well as polyamines, such as ***diethylenetetramine*** and ***triethylenetetramine***, and other industrial exposures, such as resins, dyes, fungicides, rubber accelerators, shellacs, and synthetic waxes. A stabilizer in ***Renografin***® (***meglumine diatrizoate***) where it can cause systemic reactions. National Institute for Occupational Safety & Health/OSHA upper limits of permitted exposure, 10 ppm; TWA. A zero tolerance is established for its residues in milk for **human** consumption.

TC_{LO}, **human**: 200 ppm.

Drug interactions: With ***theophylline***, it is ***aminophylline***, and has caused allergic and severe disabling contact dermatitis with generalized exfoliative dermatitis and asthma.

LD_{50}, **rat**: 76 mg/kg; **guinea pig**: 470 mg/kg.

ETHYLENE DIBROMIDE

= *1,2-Dibromoethane* = *sym-Dibromoethane* = *Dowfume W 85* = *EDB* = *Ethylene Bromide* = *FC-3* = *Glycol Dibromide*

Use: Leaded gasoline octane-enhancing additive, fumigant, solvent, and chemical intermediate.

Untoward effects: Immediate irritation to eyes and respiratory tract. Severe skin irritant with vesiculation. Inhalation leads to headache, depression, and anorexia, in addition to severe eye and throat irritation. Two fatalities following acute occupational exposure after **their** climbing into a tank that had ***EDB*** fumes. **They** had severe acidosis, acute liver and kidney failure, and localized areas of tissue death in skeletal muscle and various organs. In the 1980s, hysteria gripped the news media and legislatures failed to act based on science. The FDA allowed peanut butter to be sold with ***aflatoxin***, a potent carcinogen, at levels of 15 ppb, while demanding ***EDB*** restrictions with only evidence of carcinogenicity in **rats** and **mice**. Although suspect, there is insufficient evidence to indicate carcinogenicity or reproductive problems (sperm damage) in **people** due to ***EDB***. Suggestions of possible liver, kidney, and heart damage are also based on **animal** studies. Exposure to 10,000 ppm for a few minutes can be fatal. OSHA suggests limits of 20 ppm/TWA. 100 ppm can cause a serious threat to health.

Oral LD_{LO}, **human**: 5 ml (50 mg/kg).

Gastric squamous cell carcinomas induced in **rats** and **mice**; blood vessel cancers in male **rats**; liver cancers in female **rats**; and lung cancers in male and female **mice**. A serious toxic interaction between inhaled ***EDB*** in **rats** and ingested ***disulfiram***. Inhalation of vapors by **rats** led to central nervous system suppression, decreased sperm count in males, pulmonary irritation, liver and kidney pathology, and it caused death in **guinea pigs**. They had kidney degeneration and pancreas, liver, spleen, heart, and adrenal pathology. Vapor-exposed **rats** and **guinea pigs** had pale and swollen spleens, while dermally exposed ones had highly congested and swollen spleens. Oral doses to **bulls** reduced sperm count and motility. Teratospermia in **rams**.

In **hens**, it decreased egg production, size, and fertility.

LD_{50}, **rats**: 108 mg/kg; inhalation, 200 ppm.

ETHYLENE DICHLORIDE

= *Brocide* = *1,2-Dichloroethane* = *sym-Di-chloroethane* = *Dutch Liquid* = *EDC* = *Ethylene Chloride*

Use: Solvent, fumigant, and in chemical synthesis. Annual use ~20 billion lbs.

Untoward effects: Irritant to eyes leading to corneal opacity; nausea, vomiting, stomach pain, dizziness, "drunkenness", dermatitis, liver and kidney pathology, internal bleeding, bluish-purple mucous membranes and skin (cyanosis), rapid weak pulse, unconsciousness, and death from circulatory and respiratory failure. Acute tubular necrosis can follow its use in industrial cleansers and home spot-removers in unventilated areas. **Human** exposure by inhalation, dermal contact, and ingestion. Leukocytosis in three industrial **workers** exposed to it for 4 h while cleaning yarn. **They** recovered in 1 week. A 63-year-old **male** development leukocytosis (11,800/mm^3 with 86% neutrophils) 8 h after ingesting 2 oz and died 22 h after ingestion. Long-term exposures lead to anorexia, neurological problems, gastrointestinal upsets, mucous membrane irritation, liver and kidney damage, and death. In **human** milk and in the exhaled breath of nursing **mothers** exposed to it. Their nursing **infants** are at risk. A 14-year-old **male** drank ~15 ml to get "high". Had few initial symptoms, but died in 5 days. Hypoglycemia and hypercalcemia are prominent symptoms. Postmortem indicated florid liver and renal tubular necrosis, and focal adrenal degeneration and necrosis. After 4 h of intermittent hand immersion into it, they appeared red, raw, and scalded. Later, the hands became dry and shiny. It has been abused as an inhalant. National Institute for Occupational Safety & Health limits exposure to 1 ppm/TWA; OSHA to 50 ppm, TWA, which can be life-threatening. May be carcinogenic. When used as a grain fumigant, enzymatic action on it can release dangerous ***phosgene*** gas.

TD_{LO}, **human**: 428 mg/kg; TC_{LO}, inhalation, **human**: 4000 ppm/h.

In male **rats**, squamous cell carcinomas of the stomach, hemangiosarcomas and fibromas in SC tissues; adenocarcinomas of the mammary glands of female **rats** and **mice**; in female **mice**, stromal polyps and sarcomas of the endometrium; in male and female **mice**, adenomas of the alveoli and bronchioli. Crosses the placenta in the **rat** leading to abnormal fetal development. Hemorrhagic necrosis of adrenal cortex common in poisoned **animals**. Minimal concentration of 0.5% in inhaled vapors by anesthetized **dogs** decreases myocardial contractility.

LD_{LO}, **mouse**: 600 mg/kg; **dog**: 2 g/kg; **rabbit**: 860 mg/kg.

ETHYLENE GLYCOL

= *1,2-Ethanediol*

Use: In antifreeze, manufacturing of polyesters, solvent in inks and coatings, humectant for ***tobacco***, and in brake fluids.

Untoward effects: Ingestion leads to severe vomiting, abdominal pain, restlessness, ataxia, "drunk"-like state, convulsions, drowsiness, oliguria, anuria, dizziness, stupor, central nervous system damage (may be permanent); visible CT findings of brain demyelination and edema; decreased blood pressure (occasionally, increased blood pressure), increased serum osmolality, tachycardia, noisy shallow respirations, coma, and death due to renal failure with uremia, crystalline deposits of ***calcium oxalate*** in lumina of renal tubules, acidosis, pulmonary edema, and circulatory failure, usually within 2 days. ***Oxalic acid*** is a toxic metabolite and IV ***ethanol*** is an early emergency antidote, because it inhibits the biotransformation, by acting as a competitive substrate for alcohol dehydrogenase, permitting more to be excreted unchanged in the urine. Unusual complications including severe hypotension of such ***alcohol*** therapy in a 36-year-old **male** on ***disulfiram***. Bilateral optic atrophy in a 58-year-old **male** after drinking ~250 ml/48 h. After 5 months, visual acuity decreased 4/60 in the right eye and counting fingers in the left. A 33-year-old **female** ingested 2 l in a suicide attempt and survived after prompt treatment. Prolonged or repeated skin contact has caused a dermatitis in some **people**. Irritant to eyes, skin, nose, and throat. An area hospital was sued by families of two **patients** who died after a faulty valve permitted ethylene glycol to leak into water for dialysis solutions. A mouthful in a small **child** caused systemic acidosis and renal toxicity with ***calcium oxalate*** crystals within 1–3 days. A summertime cluster of intentional ingestion by 22 **patients** ~16–19 years within a 3 month period reported in a Chicago area. After ingestion by a 31-year-old **female** it led to cranial neuropathy, metabolic acidosis, and renal failure.

LD, **humans**: ~1.4 ml/kg or 100 ml/**adult**. Toxic levels estimated at 150 mg% and 200–300 mg/l. TD_{LO}, **child**: 7.4 g/kg; **human**: 710 mg/kg; TC_{LO}, inhalation, **human**: 10,000 mg/m^3.

Because of its sweet taste, **cats**, **dogs**, and **poultry** readily drink it and exhibit many symptoms similar to **man's**: depression, decreased temperature, vomiting, dyspnea, ataxia, convulsions, acidosis, coma, and death within 12–36 h. Chronic ingestion leads to renal failure, oliguria, anuria, uremia, polydypsia, increased thirst, behavior changes, occasionally blindness, hypocalcemia, oxalates in urine, and convulsions. **Poultry** and **swine** have been poisoned by fluid leaking from water pipes (viz. radiant heating systems). **Swine** and **cattle** have been poisoned by drinking material drained from radiators. **Mink** have had seizures from it. **Pigeons** and **geese** appear to be more resistant than other species.

LD_{50}, **rats**: 5.84 g/kg; **mice**: 7.5 g/kg; **guinea pigs**: 6.6 g/kg; **cat**: 1.68 g/kg; **dog**: 7.4 g/kg; **poultry**: ~8.5 g/kg.

Flash-fires can occur in aircraft when it is used as a deicing agent and contacts ***silver***-coated electrical circuitry, wiring, switches, or circuit breakers that use positive direct electrical current.

ETHYLENE GLYCOL DINITRATE

= *EGDN* = *1,2-Ethanediol Dinitrate* = *Ethylene Dinitrate* = *Ethylene Nitrate* = *Glycol Dinitrate* = *Nitroglycol*

Use: Additive to dynamite and explosives in mining and fuel industries. Presence in ambient air as evidence of hidden bombs.

Untoward effects: Angina and sudden death in exposed **workers**. Acute exposure leads to severe throbbing headache, dizziness, nausea, heart palpitations, hypotension, and flushing. Chronic exposure leads to ***alcohol*** intolerance, chest pain during brief periods away from the workplace, and reported cases were first evidence of an occupational toxic factor in the etiology of ischemic heart disease. Methemoglobinemia, delirium, central nervous system suppression, and skin irritation on direct contact reported. Poor circulation in **workers** with repeated short exposures of 3 ppm. OSHA exposure limits for skin, 0.2 ppm (1 mg/ m^3)/8 h.

LD_{LO}, **human**: 5 mg/kg.

In **animals**, anemia and liver and kidney damage occur.

LD_{50}, **rat**: 616 mg/kg; SC, LD_{LO}, **cat**: 50 mg/kg; **rabbit**: 300 mg/kg.

ETHYLENE GLYCOL MONOBUTYL ETHER

= *Butyl Cellosolve®* = *2-Butoxyethanol* = *EGBE*

Use: Solvent. Dry cleaning.

Untoward effects: Skin, eye, nose, and throat irritant. Taste disturbances, erythropenia, hemolysis, hemoglobinuria, central nervous system depression, headache, nausea, vomiting, and kidney and liver pathology. Failed suicide attempts reported with 250–500 ml of a cleaning solution (12% ***EGBE***) in a **female** and in another **female** ingesting a window cleaning solution. Severe metabolic acidosis followed. Treatments were successful. Readily absorbed through the skin. National Institute for Occupational Safety & Health upper limit of exposure, 5 ppm (33 mg/m^3).

Animal experiments indicate effects on central nervous system (inactivity, weakness, dyspnea), liver, and kidneys occur at higher exposure levels than for hematotoxicity.

LD_{50}, **rat**: 2450 mg/kg; **rabbit**: 320 mg/kg; **mouse** and **guinea pig**: 1200 mg/kg; inhalation LC_{50}, **rat**: 450 and 486 ppm/4 h, **mouse**: 700 ppm/7 h.

ETHYLENE GLYCOL MONOBUTYL ETHER ACETATE

= *EGBEA*

Use and untoward effects: Similar to above.

LD_{50}, male **rat**: 3000 and 7000 mg/kg; female **rat**: 2400 mg/kg; dermal, **rat**: 1580 mg/kg.

ETHYLENEIMINE

= *Aminoethylene* = *Azacyclopropane* = *Aziridine* = *Dimethylenimine* = *EI* = *Ethylenimine* = *Ethylimine* = *Etilenimina* = *Vinylamine*

Use: Chemical intermediate; in vaccines to inactivate foot and mouth, pseudorabies, rabies, African swine fever, and parvo viruses; in manufacturing paper, textiles, rocket and jet fuels, resins, lacquers, varnishes, ion-exchange resins, cosmetics, and surfactants.

Untoward effects: Acute exposure leads to severe irritation and inflammation of eyes, nose, and throat; delayed onset headache, dizziness, nausea and vomiting, pulmonary edema, skin sensitization, liver pathology, and necrosis of kidney tubular epithelium. Acute inhalation or dermal exposure in laboratory **workers** led to central nervous system disturbances, fluid in lungs, liver and kidney damage, and even death. In an effort to dislodge five college **students** from a locked, poorly ventilated room, it was poured under a door and through a broken window. It is strongly alkaline and has an intense ***ammonia***-like odor. A **man** who handled it developed a severe necrotizing, painless burn of **his** hand. The **occupants** in 3–7½ h suffered from severe eye irritation, severe sore throat, coughing, and vomiting. Conjunctivitis, respiratory tract inflammation, transitory polycythemia, leukocytosis, eosinophilia, and albuminuria were noted. **Two** were unable, because of eye troubles, to work for a full term. **Three** fully recovered, except for some eye trouble in **one**, after 3 months. Considered potentially carcinogenic for **humans**.

Instillation on the eyes of **cats** and **rabbits** led to blindness and death. Orally to **rats** caused decreased white blood cells, and degeneration of liver, kidney, and cardiac tissues. Subchronic SC in **rats** led to injection site skin cancers; chronic oral administration to **mice** caused cancers of the liver and lungs. Embryotoxic and teratogenic in **rats**.

LD_{50}, **rats**: 15 mg/kg.

ETHYLENE OXIDE

= *Anprolene* = *Dimethylene Oxide* = *1,2-Epoxyethane* = *Eto* = *Freoxide* = *Oxirane*

Use: Foodstuffs and textile fumigant, surgical instrument sterilization, fungicide, and in manufacturing ***ethylene glycol***, ***acrylonitrile***, and non-ionic surfactants.

***Untoward effects*:** **Human** exposure by inhalation, ingestion, and dermal contact. Highly explosive and reactive. Uncontrolled industrial emissions occur during loading and unloading transports, sampling procedures, and equipment maintenance. Use of this powerful alkylating agent for sterilization and fumigation accounts for only 2% of its use, but for most of the exposures to **hospital personnel**. Acute exposures lead to nausea, headache, weakness, vomiting, drowsiness, paleness, light-headedness, infrequent muscle twitching, incoordination, collapse, and irritation to the eyes, nose, throat, and lungs. Dyspnea, apnea, and pulmonary edema have often followed. Early deaths are due to respiratory failure; later ones, to pulmonary edema, and liver and kidney pathology. Chronic exposure has caused anemia, peripheral neuropathy, chromosomal damage in white blood cells, skin sensitization, respiratory infections, and anosmia. Inhalation by 41 factory **workers** caused repeated vomiting and severe headaches. Occasional circulatory collapse from electrolyte changes. Transient conjunctivitis occurred. When **their** shoe leather was contaminated, localized edema, followed by blistering, reported.

Skin contact has caused blisters, edema, burns, frostbite, and severe dermatitis. It is sufficiently absorbed by plastics and rubber to be irritant to skin, mucous membranes, and exposed tissues. Hand burns reported from use of sterilized surgical gloves; facial irritation, swelling, and erythema from use of anesthesia masks; burns from prepackaged sterile dressings, laryngeal edema from sterilized nasogastric tubes, and postoperative burns of buttocks and backs from contact with reusable surgical gowns and drapes containing 3600–10,800 ppm residual ethylene oxide, which was 16–54 times the safe level (200 ppm) for skin contact. Anaphylaxis has been reported, including in a **patient** on long-term hemodialysis with materials sterilized by ***Eto***. **She** developed acute urticaria, hypotension, and dyspnea. Cutaneous eruptions and eosinophilia in a 32-week-old **male** on continuous ambulatory peritoneal dialysis, using ***Eto***-sterilized tubing. An eye drug was suspected of causing blindness from the traces of ***Eto*** in it. Spontaneous abortions may have occurred in exposed hospital **nurses**. OSHA limits exposure to 1 ppm/TWA, 8 h. Russia limits it to 0.5 ppm.

Inhalation TC_{LO}, **human**: 12,500 ppm/10 s.

Acute inhalation by **rats** and **guinea pigs** led to pulmonary edema, paralysis, and corneal edema. Subchronic inhalation in **rats** led to decrease in pregnancy and litter size. Chronic inhalation by **rats** caused leukemia, brain tumors, and peritoneal malignancies. Testicular damage in **rats** and **guinea pigs**. Chromosomal and sperm damage in **monkeys**. Chronic oral intake by **rats** caused stomach cancers.

Inhalation LC_{50}, **rat**: 1462 ppm/4 h, **mouse**: 836 ppm/4 h, **dog**: 960 ppm/4 h.

ETHYLESTRENOL

= *Bolenol* = *Durabolin-O* = *Grabolin* = *Maxibolic* = *Orabolin* = *Orgabolin* = *Orgaboral*

Anabolic.

Untoward effects*:** An abused steroid. Acne, liver upsets, liver tumors, masculinization of **females**, decreased growth in **children**, amenorrhea, changes in SGOT and sulfobromophthalein in **some**. Potentiates ***dicumarol and ***warfarin*** effects, and decreases ***insulin*** requirements in **diabetics**.

Rat studies confirm shortenting of ***barbiturate***-induced anesthesia. Abortifacient effect in **rats**.

ETHYL FORMATE

***Use*:** Flavoring agent (0.01–0.05%) in lemonades and artificial rum, solvent, larvicide, and in organic synthesis. An aldehyde.

Untoward effects*:** Systemic toxicity can occur from its degradation to ***formic acid (q.v.). Inhalation leads to irritation to eyes and upper respiratory tract. OSHA upper exposure limits 100 ppm, TWA/8 h. 1500 ppm may be life-threatening.

Inhalation TC_{LO}, **human**: 330 ppm. LD_{LO}, **human**: 500 mg/kg.

Can cause narcosis in **animals**. Causes tissue necrosis after IM in **chickens**.

LD_{50}, **rats**: 1850 mg/kg; **guinea pig**: 1110 mg/kg; **rabbit**: 2.1 g/kg.

2-ETHYL-1-HEXANOL

= *2EH* = *2-Ethylhexyl Alcohol*

Solvent, plasticizer. **For dyes, resins, and oils. Variable quantities used, now averaging about 750 million lbs annually.**

***Untoward effects*:** Vapors cause headache, dizziness, nausea, and unconsciousness. Liquid is slightly irritating to eyes and skin. Combustible.

LD_{50}, **rats**: 12.46 ml/kg; dermal, **rabbit**: 2380 mg/kg.

ETHYLIDENE GYROMITRIN

The main poisonous compound in the *mushroom False Morel* or *Lorchel* (*Gyromitra esculenta* - q.v.).

***Untoward effects*:** Large number of **human** poisonings by it in Europe and North America. Contact causes skin and eye irritation. After ingestion it leads to severe cramps, nausea, violent vomiting for up to 24 h, watery diarrhea that may last several days, headache, debilitation, jaundice, tender liver and spleen, fatty degeneration of liver, arrhythmias, tachycardia, dyspnea, cold and clammy skin, delirium, and convulsions.

ETHYLIDENE NORBORNENE

= *ENB*

Use: In rubber manufacturing.

Untoward effects: Stored under ***nitrogen*** gas, to reduce its reactivity with ***oxygen***. Eye, nose, throat, and skin irritant; headache, coughing, dyspnea, nausea, vomiting, olfactory and taste changes, and aspiration pneumonia.

Inhalation TC_{LO}, **human**: 6 ppm/30 min.

In **animals**, liver and kidney pathology, cyanosis, and narcosis.

4 h inhalation LC_{50}, **rat**: 4000 ppm; **mouse**: 732 ppm; **rabbit**: 3104 ppm; **guinea pig**: 2896 ppm.

ETHYL ISOBUTRAZINE

= *Diquel* = *Ethotrimeprazine* = *Etymemazine* = *Nuital* = *RP-6484* = *Sergetyl*

***Tranquilizer, antihistamine.* A *phenothiazine* derivative.**

Untoward effects: Occasionally salivation and sweating. Do NOT use in conjunction with agents possessing anticholinesterase-like action, such as ***organophosphates*** or large amounts of ***procaine*** in **man** and **animals**, since they can potentiate each other's toxicity and anti-cholinesterase effects.

LD_{50}, **mouse**: IP, 110 mg/kg; IV, 70 mg/kg.

ETHYL ISOTHIOCYANATE

= *Ethyl Mustard Oil*

Use: Chemical and biological warfare agent.

Untoward effects: Vapors irritate eyes and blister skin.

ETHYL LAURATE

= *Ethyl Dodecanoate* = *Ethyl Dodecylate*

Limited use in fragrances for soaps, detergents, cosmetic creams and lotions, and in foods.

Untoward effects: 12% in petrolatum showed no irritation or sensitization in 25 **human volunteers**. Full-strength to intact or abraded **rabbit** skin led to no irritation.

LD_{50}, **rats** and acute dermal in **rabbits**: > 5 g/kg.

ETHYL LOFLAZEPATE

= *CM 6912* = *Meilax* = *Victan*

Anxiolytic.

Untoward effects: Many poisoning cases reported (~25% in **children** < 16 years). Sleepiness, agitation, and ataxia were common symptoms. Hypotonia in very severe cases.

ETHYL MALEATE

= *Diethyl Maleate*

Use: In organic synthesis and limited amounts in fragrances for soaps, detergents, cosmetic creams and lotions, and perfumes (0.0033–0.4%).

Untoward effects: 4% in petrolatum was a sensitizer in all 25 **human volunteers**. Proven to be a sensitizer in patch tests performed on four **men** working with unsaturated ***polyester resins***. The same concentration showed NO irritation in **volunteers**.

Moderately irritating when applied full-strength under occlusion to intact or abraded **rabbit** skin for 24 h. No deaths in **rats** exposed to saturated vapors.

LD_{50}, dermal, **rabbit**: 4 g/kg.

ETHYL MALTOL

= *3-Hydroxy-2-ethyl-4-pyrone* = *Veltol Plus*

Use: In fragrances for soaps, detergents, cosmetic creams and lotions, and foods.

Untoward effects: No irritation or sensitization as 10% in petrolatum on 25 **human volunteers**.

Non-irritating when applied full-strength to intact or abraded **rabbit** skin. In 90 day studies, some kidney lesions in **rats** on 10 g/kg/day. In 90 day studies on **dogs** on 500 mg/kg/day, mild hemolytic anemia noted. No adverse effects in 2 year feeding studies with **rats** and **dogs** on 200 mg/kg/day.

LD_{50}, **rats**: 1.2 g/kg; **mouse**: 780 mg/kg; **chicken**: 1.3 g/kg.

ETHYLMERCURIC CHLORIDE

= *Granosan*

Seed fungicide.

Untoward effects: Symptoms of poisonings in **man** from grain treated with it consisted of central nervous system disturbances, spasticity, cerebellar ataxia, hearing impairment and deafness; narrowing of the visual field, followed by blindness, mental confusion, and exaggerated stretch reflexes. Absorbed through the skin and has caused skin burns. Ingestion of ***bread*** from seed ***wheat*** treated with 50 g/100 kg led to 24 **people** who developed sore mouths, metallic taste, blue gum line, vomiting, diarrhea, polyuria, fever, insomnia, hand tremors, dysarthria, ataxia, hyperreflexia, Babinski reflex, hyporeflexia, limb atrophy, and abnormal electrocardiograms. Ingestion by **pregnant mothers** led to mental retardation, decreased muscle tone, and decreased weight in their **infants**.

LD_{LO}, **human**: 5 mg/kg.

Ingestion of 0.38 mg/kg/day by **swine** caused anorexia, decreased weight, diarrhea, weakness, and paralysis in < 2 weeks. Poisoning has also occurred in **calves**, **sheep**, and **poultry**.

LD_{50}, **rat**: 40 mg/kg.

N-(ETHYLMERCURI)-*p*-TOLUENE-SULFONANILIDE

= *Ceresan M*

***Fungicide*. Pungent, *garlic*-like odor.**

***Untoward effects*:** Deafness, ataxia, dysarthria, visual deterioration, dysphagia, incontinence, confusion, stupor, and death.

LD_{LO}, **human**: 50 mg/kg.

Toxic effects and deaths reported in **cattle**, **sheep**, **quail**, **mallard ducks**, and **pheasants**.

LD_{50}, **rat**: 100 mg/kg.

ETHYL METHANESULFONATE

= *EMS*

***Use*:** Experimentally, as a mutagen, teratogen, and brain carcinogen.

***Untoward effects*:** Lung adenomas or adenocarcinomas in **mice** after SC. Lung carcinomas in **rats** and renal carcinomas in female **rats** after three IP injections. In drinking water, caused mammary adenocarcinomas in **rats**.

TD_{LO}, **rat**: 300 mg/kg; **mouse**: 1000 mg/5 days; SC, **mouse**: 83 mg/kg; IP, **mouse**: 200 mg/kg.

ETHYL METHYLPHENYLGLYCIDATE

= *Aldehyde C-16* = *Strawberry Aldehyde*

Use*:** Major ingredient in imitation ***strawberry flavors and ***strawberry*** scents in soaps, detergents, cosmetic creams and lotions, and perfumes.

***Untoward effects*:** No irritation or sensitization in 25 **human volunteers**. A 16 week feeding trial at 10,000 ppm in **rats** led to decreased growth, especially in males, and marked testicular atrophy in male **rats**. 2500 ppm had no adverse effects. In a 2 year feeding study in **rats** on 5000 ppm it caused paralysis of hind limbs and demyelinating degenerative changes of the sciatic nerve.

LD_{50}, **rat**: 5.47 g/kg; **guinea pig**: 4.05 g/kg.

ETHYLMORPHINE HYDROCHLORIDE

= *Codethyline* = *Dionin*

Analgesic, antitussive, mydriatic.

Untoward effects*:** Similar to ***morphine (q.v.), but less of a respiratory depression. A 57-year-old **female** and two (22 and 24 years) of her four **children** had very painful attacks of biliary colic shortly after swallowing it in a cough mixture.

Has caused congenital malformations, including cranioschisis in **hamsters**.

LD_{50}, **rats**: 810 mg/kg; **mice**: 771 mg/kg; SC, **mouse**: 200 mg/kg.

N-ETHYLMORPHOLINE

***Untoward effects*: *Ammonia*-**like odor. Eye, nose, and throat irritant. Has caused corneal edema, and blue-gray vision with colored haloes. OSHA sets maximum exposure limits at 20 ppm to skin.

Inhalation TC_{LO}, **human**: 100 ppm/2.5 min; **rat**: 2000 ppm/4 h.

ETHYL PARATHION

***Insecticide*. Organophosphorus anticholinesterase.**

Untoward effects*:** Similar to ***parathion (q.v.). An 18-month-old **male** accidentally drank a "small amount". **He** was taken to a hospital immediately and arrived with fixed, dilated pupils, and was without spontaneous respiration. Recovered after resuscitation, ***atropine*** and ***pralidoxime***. Little safety factor between its insecticidal activity and dermal toxicity for **man**. In 1975, 68 cases of **human** toxicity reported in California, where 921,000 lbs were used. A number of suicides and one murder reported with it in Belgium.

LD_{LO}, **human**: 240 µg/kg.

LD_{50}, **rat**: 2 mg/kg; **mouse**: 6 mg/kg; **dog**: 3 mg/kg; **rabbit**: 10 mg/kg; **guinea pig**: 8 mg/kg; **pigeon**: 3 mg/kg; **quail**: 6 mg/kg; **duck**: 2.3 mg/kg.

ETHYL PHENYLEPHRINE

= *Effortil* = *Ethyl Adrianol* = *Etiladrianol* =*Etilefrine* = *K 30,052* = *MI 36* = *Phetanol*

Sympathomimetic.

***Untoward effects*:** Tachypnea after 30 s – normal after 4 min; tachycardia after 30–50 s with bradycardia after 80–120 s, and normal after 4 min. Transient tonic-clonic unilateral convulsions in an **infant** with birth trauma of the brain after 1 ml. Hypertensive after long-term oral therapy.

Depressive effect in **rats** and **mice** on high dosage.

ETHYL PHENYLGLYCIDATE

***Use*:** A synthetic fragrance in soaps, detergents, cosmetic creams and lotions, and foods.

***Untoward effects*:** 4% in petrolatum led to no irritation or sensitization in tests on 25 **volunteers**. Non-irritating full-strength on intact or abraded **rabbit** skin.

LD_{50}, **rats**: 2.3 ml/kg; acute dermal, **rabbits**: > 5 g/kg.

ETHYL SILICATE

= *Tetraethyl Silicate*

***Use*:** Weatherproofing stone and cements.

Untoward effects*:** Sharp, ***alcohol-like odor – 250 ppm in air is unpleasant for **humans**. Eye, nose, and throat irritant at 3000 ppm. OSHA sets maximum exposure limit at 100 ppm/8 h TWA.

Slight injury topically on **rabbit** eyes. Breathing concentrated vapors killed all six **rats** in 4 h, but none in 2 h. Exposed **animals** show dyspnea, lacrimation, pulmonary edema, tremors, anemia, and liver and kidney pathology.

LD_{LO}, **rat**: 1 g/kg; inhalation LC_{LO}, **rat**: 1000 ppm/4 h.

ETHYL VANILLIN

= *Bourbonal* = *Ethavan* = *Ethovan* = *Ethylprotal* = *Vanillal*

Use*:** A synthetic fragrance in soaps, detergents, cosmetic creams and lotions, and foods. ~15 ton/year used as a substitute for ***vanilla.

***Untoward effects*:** 2% in petrolatum caused mild irritation, but no sensitization in groups of 25 **human volunteers**. Minimum lethal dose, **human**: skin, 10 mg/48 h occlusion.

20 mg/kg/day/18 weeks to **rats** had no adverse effects; 64 mg/kg/day/10 weeks decreased growth rate and caused myocardial, renal, hepatic, lung, spleen, and stomach pathology.

LD_{50}, **rats**: > 2 g/kg; **rabbits**: ~3 g/kg.

ETHYNODIOL DIACETATE

= *Femulen* = *Luteonorm* = *Luto-Metrodiol* = *SC-11800*

Progestogen.

Untoward effects*:** Similar to ***progesterone (q.v.). Occasionally prurigo, abdominal pain, and delayed menses. Teratogenic in **mice**.

ETIDOCAINE HYDROCHLORIDE

= *Duranest* = *W 19,053*

Local anesthetic.

***Untoward effects*:** Occasionally hypersensitivity. Central nervous system reactions (excitatory and/or depressant) leading to nervousness, dizziness, tremors, blurred or double vision, muscle weakness, drowsiness, tinnitus, vomiting, convulsions, unconsciousness, and respiratory arrest. Bradycardia, hypotension, urticaria, edema, anaphylactoid reactions, and trismus also reported.

Therapeutic serum levels are stated as 1–1.5 mg/l with a narrow safety margin; toxic at 1.6–2 mg/l.

In the **dog**, central nervous system toxicity leading to mydriasis, nystagmus, and head tremors.

IV, LD_{LO}, **dog**: 10 mg/kg.

ETODOLAC

= *AY-24236* = *Edolan* = *Etodolic Acid* = *Etogesic* = *Lodine* = *Ramodar* = *Tedolan* = *Ultradol* = *Zedolac*

Non-steroidal anti-inflammatory drug, analgesic.

Untoward effects*:** Agranulocytosis in a 72-year-old **female** on 300 mg twice daily for osteoarthritis. A 62-year-old **male** on 200 mg/day developed Schönlein-Henoch vascular purpura. Hepatitis reported with its use. Some gastrointestinal microbleeding. Nausea, abdominal pain, diarrhea, flatulence, and dizziness in > 2% of **patients**. ***Non-steroidal anti-inflammatory drugs are not recommended in **women** trying to become pregnant since they interfere with implantation of fertilized eggs in **animals**.

In **dogs**, after 8 days of treatment it caused vomiting, lethargy, anorexia, loose stools, hypoproteinemia, urticaria, and behavioral changes, such as urinating in the house.

ETOFENAMATE

= *B-577* = *Bayrogel* = *Glasel* = *Rheumon Gel* = *Traumon Gel* = *TV-485*

Anti-inflammatory.

Untoward effects*:** Its metabolite, ***flufenamic acid (q.v.), appears in the milk of treated **women** and **goats**.

LD_{LO}, **rat**: 292 mg/kg; **mouse**: 743 mg/kg.

ETOMIDATE

= *Amidate* = *Hypnomidate* = *R-16,659*

Hypnotic, IV anesthetic induction agent.

Untoward effects*:** Reversible suppressive adrenocortical function, especially with repeated doses or long-term therapy. Epileptiform seizures reported after prolonged infusions. Myocardial infarction and asystole in a 42-year-old **male** and a 55-year-old **female**; lactic acidosis and hyperosmolality in a 70-year-old **female**; all probably due to the ***propylene glycol content.

IV, TD_{LO}, **human**: 300 μg/kg.

Induces excitement, myoclonus, and pain on injection; vomiting and apnea during anesthetic induction in **dogs**. Adrenocortical function suppressed 2–3 h after use in **canine** surgical patients.

ETOPOSIDE

= *EPE* = *EPEG* = *Epidophyllin* = *Epidophyllotoxin* = *EPT* = *Lastet* = *NSC-141540* = *PTG* = *Vepesid* = *VP-16-213*

Antineoplastic. Similar to ***teniposide***. Both are semi-synthetic derivatives of ***podophyllotoxin*** (q.v.). Oral and IV use.

Untoward effects: Acute: Nausea, vomiting (33–71%), red urine (not hematuria), diarrhea, fever, hypotension (especially with rapid infusion), anaphylactoid reactions (bronchospasm, facial flushing, facial edema, urticaria, apnea), phlebitis at infusion site and thrombosis, myocardial infarctions, and arrhythmias. Some anaphylactic reactions after IV treatment may have been due to surfactant. Accidental needlestick injury reported in a **nurse**. Delayed: Bone marrow depression, encephalopathy, rashes, alopecia (13–66%), urticaria, pruritus, peripheral neuropathy, and leukemia; with high dosage, mucositis and liver pathology; facial and tongue swelling, coughing, laryngospasm, aftertaste, pigmentation, constipation, transient cortical blindness, and optical neuritis.

Acute dystonia in an 11-year-old **male** and possible transient loss of consciousness in a 70-year-old **male**. Hematological toxicity, non-dose-related: leukopenia (~40%), thrombocytopenia and pancytopenia (~29%). Life-threatening vasospastic phenomena (viz. angina) and increased risk of myelodysplasia and leukemia. Dramatic increase in ***warfarin*** anticoagulation (nearly eight times increase in international normalized ratio) after 8 days of treatment with etoposide and ***carboplatin***.

Mildly irritating to **rabbit** skin on contact. Causes hair loss in **mice**.

ETORPHINE

Made from *thebaine*; ~1000–80,000 times the analgesic, sedative, and respiratory depressing effects of *morphine*. *Diprenorphine* counteracts its effects.

Untoward effects: Its use in **humans** has been limited because of its extremely high potency and inherent danger of overdosage.

Effects on central nervous system vary greatly with **animal** species. Always decreases respiration and increases bradycardia. Rapid effect and long-acting. Profound depression or excitation in different species. Addictive in **rhesus monkeys**. With ***acepromazine*** in **horses** leads to muscle tremors and rigidity, mydriasis, sweating, tachycardia, hypertension, respiratory depression, anorexia, constipation, and counterclockwise circling. **Cattle** and **sheep** may show increased blood pressure, respiratory depression, muscle tremors, and hyperpyrexia. In addition, **cattle** may bleat, bellow, bloat and grind teeth. **Camels** and **giraffes** may regurgitate and show opisthotonus. ***Azaperone*** and some ***tranquilizers*** actually counteract the respiratory depression of etorphine. **Monkeys** may be very sensitive to it and results are extremely variable – respiratory arrest may occur, without any noticeable anesthesia.

LD_{50}, **rat**: 72 mg/kg; **mouse**: 1.9 g/kg; SC, LD_{50}, **rat**: 53 mg/kg; **mouse**: 425 mg/kg.

ETOXADROL HYDROCHLORIDE

= *CL 1,848C* = *Thoxan*

IV anesthetic.

Untoward effects: Depersonalization, transient ocular myopathy, panic feelings, post-arousal self-image distortion.

ETRETINATE

= *RO 10-9359* = *Tegison* = *Tigason*

Antipsoriatic. **Synthetic *vitamin A* derivative. An aromatic retinoid.**

Untoward effects: Dry mucous membranes, chapped lips, cheilitis, alopecia, bone and joint pain; peeling skin, including desquamation of palms; pruritus, eye irritation and eyeball pain, pseudotumor cerebri, hepatitis, hyperostosis, premature epiphyseal closure, erection difficulties, decreased libido, may cause pathological increase in serum triglycerides, papular dermatitis, inflammation of urethral meatus, headaches, giddiness, vomiting, nosebleeds, abdominal pain, muscle cramps, and photosensitivity reactions. Remains in **human** tissue for up to 2 years. Possibly, major teratogenic effects in **pregnant women**.

Embryotoxic and teratogenic in **rats** and **rabbits** on high dosage. Therapy may have to be discontinued, due to hypersensitivity reactions, and symptoms indicative of ***vitamin A*** toxicity.

LD_{50}, **rats**: > 4 g/kg; **mice** and **rabbits**: > 2 g/kg.

EUCALYPTOL

= *Cineole* = *Cajeputol*

Flavoring agent. **In pharmaceuticals. ~2 ton/year used in the U.S. in soaps, detergents, cosmetic creams and lotions, and foods.**

Untoward effects: At 16% in petrolatum, it failed to cause irritation or sensitization in 25 **volunteers**. In 1935, a case of eczema and sensitivity was reported after it contacted a **patient's** nose. Has caused death of ~⅓ of **those** who ingested 10–30 ml. Large doses cause excitement, tachycardia, nausea, vomiting, mydriasis, and death.

LD_{LO}, **human**: 500 mg/kg.

LD_{50}, **rats**: 2.48 g/kg; acute dermal, **rabbits**: > 5 g/kg.

EUCALYPTUS GLOBULUS

= *Australian Fever Tree* = *Blue-gum Tree* = *Gum Wood* = *Malee*

Untoward effects: Contains 1–3% volatile oil, ***Eucalyptus Oil*** = ***Neat Oil***, that causes fixed-drug eruptions. A 3-year-old **male** ingested 25 ml of the oil leading to emesis within 15 min; 30 min later neurological responses deteriorated. Exposure to the pure oil in 14 **patients** (7 months to 28 years) by inhalation or skin exposure led to no toxicity, but ingestion led to gastrointestinal upsets, central nervous system depression, and one **child** developed apnea. **Individuals** poisoned by the oil exhibit gastroenteritis, coma, seizures, kidney irritation, and occasionally miosis. These usually occur in < 30 min and the coma may last 2–3 days. Over 15 ton/year used in the U.S. as a fragrance in soaps, detergents, cosmetic creams and lotions, and foods. No photosensitivity reactions in **man** and 10% in petrolatum failed to produce any irritation or sensitization in 25 **volunteers**.

Rumors of no ill-effects when ingested by **koala bears** are false, as young shoots and leaves have killed many of them with ***prussic acid*** poisoning.

Oil, LD_{50}, **rat**: 2.5 g/kg.

EUGENOL

= *Allylguaiacol* = *Caryophyllic Acid* = *Eugenic Acid*

50 ton/year, mostly in soaps, detergents, cosmetic creams and lotions, and foods.

Untoward effects: Mild inflammation or allergic reactions (cheilitis and stomatitis) when used in periodontal surgical dressings. Allergen causing dermatitis from miscellaneous use as a fragrance and in **bakers**. Severe pulmonary toxicity when smoking cigarettes containing it. 8% in petrolatum caused mild irritation and no sensitization reactions in 25 **volunteers**. Suspect as a carcinogen. As an oily inhalant, potentially dangerous (foreign body pneumonias).

Dogs given single oral doses of 250 and 500 mg/kg vomited, and 50% at the higher level died. Topically for 5 min on the ventral surface of the tongue caused erythema and occasional ulcers in **dogs**. In **dogs**, IV of diluted eugenol led to transient decrease in arterial blood pressure and myocardial contractile force, impaired motor activity, and increased salivation. Large doses caused pulmonary edema in some **dogs**. Tumor-promoting on skin of **mice**.

LD_{50}, **rats**: 2.68 g/kg; **mice**: 3 g/kg; **guinea pig**: 2.13 g/kg.

EUGLENA SANGUINEA

= *Red Tide*

Untoward effects: In the Bible (Exodus) there is the first written account of an algal bloom: "...waters in the river were turned to blood. The fish in the river died. The river stank and the Egyptians could not drink the river water." This may have been due to an explosive population of these algae.

EUONYMUS

Untoward effects: ***E. americanus*** = ***Burning Bush*** = ***Fusain*** = ***Strawberry Plant***. Found in southeastern U.S. pastures. **Animals** poisoned by it develop diarrhea, weakness, and coma. ***E. europaeus*** = ***Spindle Tree*** = ***Spindle Wood***. Ingestion of fruits by **children** led to mucoid and watery diarrhea within 10–12 h. Persistent vomiting, fever, hallucinations, somnolence, unconsciousness, convulsions, and violent purging. Death has followed in **some**. Less toxic after fruits ripen. **Sheep**, **goats**, and **horses** have also been poisoned by it. Contains a ***digitalis***-like toxic glycoside, ***evonoside***, that causes bradycardia. Leaves and fruits are purgative.

EUPATORIUM sp.

Untoward effects: Many species contain ***pyrrolizidine*** alkaloids and have a potential for hepatotoxicity, despite their use world-wide in herbal ***teas***. They may also contain dangerous concentrations of ***nitrates***.

E. adenophorum = ***Crofton Weed***. Ingestion caused coughing, dyspnea, and decreased exercise tolerance in **horses** in Queensland, Australia. Chronic cases lead to fibrosis, alveolar epithelial proliferation, pulmonary edema, neutrophil infiltration, and abscessation. Occasionally vascular thrombosis and infarction of the lungs. Similar pathology in experimental **rabbits**, but not in **sheep** or **rats**. The cause of "Numimbah **horse** sickness" of New South Wales.

E. oderatum. Entire plant is piscicidal. In Nigeria, suspected of causing milk sickness in **man**.

E. ogeratoides and ***E. urticelfolium*** = ***White Sanicle*** may also cause milk sickness in **humans**.

E. rugosum = ***Eupatoire Rugueuse*** = ***Richweed*** = White ***Snakeroot*** (q.v.) contains a toxic, fat-soluble principle, ***tremetol***, the cause of "milk sickness" in **man**. Called "trembles" in **animals**. Symptoms in **humans** include weakness, dizziness, anorexia, nausea, persistent vomiting, abdominal pain, thirstiness, swollen and coated tongue, dry skin, bradypnea, decreased temperature, weakness, collapse, delirium, and coma from drinking the ***milk*** of **cows** that have consumed it. Mortality ranged from 10 to 25% in some villages. Affected **cows** and **sheep**

tremble, especially in the nose, shoulders, and limb muscles. **Cattle** evidence dyspnea, anorexia, vomiting, constipation, ataxia, nasal discharge, recumbency, trembling, prostration, and dark urine preceding death. **Horses** perspire profusely, are dyspneic, constipated, stiff-gaited, tremble, have nasal discharge, mydriasis, and can become prostrate before death. After eating it, **rabbits** and **guinea pigs** show "trembles". **Calves**, **cats**, and **dogs** drinking ***milk*** from affected **cows** show "milk sickness".

EUPHORBIA sp.

Most are known as *Spurges* = *Sütleğen* (Turkey)

Untoward effects: Accidental or intentional ingestion can cause severe vomiting and catharsis (viz. ***ipecac*** q.v.). The resinous substance in their latex helps the adhesion of arrow and dart poisons, and their highly irritant ingenol esters (in most species) promote the absorption of toxic materials. ***Honey*** made by **bees** visiting their flowers can be poisonous for **man**. Dermatitis and blistering is common in **people** contacting their white latex. Extracts and latex have been used as insecticides. Many species may contain dangerous levels of ***nitrates***. Causes a disagreeable taste and odor in ***milk***, milk products, and meat of **animals** that graze it.

E. corollata = ***Flowering Spurge*** = ***White-Flowered Milkweed***. Contains ***euphorbin***, a poisonous resinoid and other poisonous compounds in the milky sap, causing severe gastrointestinal irritation, including diarrhea and vomiting (in species that can), contact dermatitis, alopecia, emaciation, collapse, and death in **humans**, **cattle**, **sheep**, and **horses**, primarily in southeastern and central U.S.

E. cotinifolia = ***Red Spurge***. A shrub or small tree, native to northern South America and Central America, well-known as a cause of dermatitis in **people** and many blisters in the mouths of grazing **animals**. Introduced into Florida nurseries in 1959, it has since caused skin rashes and blisters, as well as eye inflammation, depending on the amount of contact. The seeds are violently purgative.

E. cyparissias = ***Cypress Spurge*** = ***Graveyard Weed***. The same effects and areas as with ***E. corollata***. Its key toxic agent is ***5-deoxyingenol***.

E. dendroides = ***Tassu***. The sap is used as a piscicide.

E. drumondii. Causes deaths of **cattle** and **sheep** in Australia.

E. eremophila. Ingestion associated with death of **livestock** in Australia.

E. esula = ***Leafy Spurge***. Introduced into the U.S. and Canada from Europe during the early 1800s. **Cattle** grazing it develop severe dermatitis of the mouth and diarrhea. Has caused severe blistering and loss of hair on the ankles of **horses** working in grain fields in eastern New York. Skin irritant for **humans**. The toxic agent is ***5-deoxyingenol***.

E. fortissima. Six skin irritants for **man** in it have been isolated.

E. genistoides = ***Piss-goed*** = ***Piss-grass***. Causes a severe urethritis in **camels**, **oxen**, and **geldings**. Contains an acrid, irritating juice.

E. helioscopia = ***Sun Spurge***. Contact with it has caused an itching generalized eruption of the face, neck, and forearms of **people**. A **child** died after ingesting it. A common herbaceous weed in the British Isles and parts of Europe. Causes severe inflammation of the mucous membranes and eyes of exposed **livestock**, especially **sheep**, from a variety of ***deoxyphorbols***. Common in northeastern U.S. and eastern Canada.

E. heterophylla = ***Fire-On-The-Mountain*** = ***Painted Leaf*** = ***Summer Poinsettia*** = ***Wild Poinsettia***. Causes irritation of the mucous membranes in Alabama **livestock** and causes interdigital rashes and even body rashes of **people** in Florida weeding **their** gardens.

E. hirta = ***Asthma Plant*** = ***Khar***. Causes rashes between the fingers and on the body of **people** in Florida from weeding their gardens. Its extract used in herbal teas reduces cardiac and respiratory function. Causes death in **ducklings** ingesting it.

E. kamerunica = ***K'erana*** (Hausa) = ***Oro Onigum Meta*** (Yoruba). Although a strong purgative for **livestock**, it is so poisonous that, in Nigeria, it is primarily an ordeal poison for **people**. It is caustic and large quantities cause **their** deaths within a few minutes.

E. kansui. Intra-amniotic injection of a root extract caused mid-term abortions (99.5% success rate).

E. lactea = ***Candelabra Cactus*** = ***False Cactus***. Direct contact dermatitis. One drop of sap in the eye of a 42-year-old **male** caused severe burning and temporary blindness (keratoconjunctivitis and uveitis). Ingestion caused nausea, vomiting, and severe, persistent diarrhea. Corneal opacity in **dogs** after eye contact. Native to the East Indies. Now widespread in the West Indies and U.S. as popular houseplants.

E. lathyris = ***Cagarrino*** = ***Caper Spurge*** = ***Garden Spurge*** = ***Gopher Plant*** = ***Mole Plant***. Its juice, or latex, and seed cases are highly poisonous to **livestock** and **man**. **Children** have died from eating the plant, common in Europe. **People** have become seriously ill (mouth burning, abdominal pains, diarrhea, vomiting), mistaking its seeds as ***capers*** for **their** dill pickles.

E. leonensis = ***Oro*** (Yoruba). A drastic purgative. Acrid poison leads to blistering of the mouth. Fatal symptoms of pericarditis and dropsy.

E. maculata = ***Eyebane*** = ***Milk Purslane*** = ***Spotted Spurge***. A toxic plant found in Florida and Alabama, westward to Oregon. Symptoms and affected species same as with ***E. corollata***, above. Photosensitization in **animals** has caused blistering and death of skin and hepatitis.

E. marginata = ***Snow-On-The-Mountain***. A common ornamental garden and house plant, growing wild in the midwest and western U.S. Its milky sap causes skin irritation on contact. Ingestion causes severe gastrointestinal upsets. **Cattle** find it unpalatable, and eat it only during drought or if in hay. May cause skin inflammation and they may die from its cumulative effects. **Kittens**, **puppies**, and small **children** may eat it, with poisonous consequences. Most *Euphorbia* poisonings in **people** are from this variety.

E. mauritanica. Toxic to **sheep** with symptoms of enterotoxemia (groaning, salivation, ruminal paresis, tachycardia, shivering, and stiff and unsteady gait).

E. mili = ***Crown of Thorns***. Despite its thorns, **people** contact its sap leading to severe dermatitis; thorn punctures are so painful that injured **persons** have often sought relief in emergency rooms. All parts are gastrointestinal irritants when ingested and may be fatal. Its toxic agent is ***5-deoxyingenol***.

E. myrsinites = ***Creeping Spurge*** = ***Donkey Tail***. Many cases of dermatitis and, at a Utah Poisin Control Center, 16/17 cases were in **patients** < 6 years. Severity varies from mild erythema to vasculation and crusting. The degree of swelling and vesicle formation is related to the amount of sap contacted and duration of exposure. Appears 2–8 h after exposure, increasing in severity during the next 12 h. Mild conjunctival irritation from eye contact.

E. peplus = ***Euphorbe des jardins*** = ***Petty Spurge*** common in the eastrn U.S. Its latex contains ***ingenol*** and a toxic diterpene, ***5-deoxyingenol***, that is caustic and irritant leading to mouth irritation and eye discharge in **humans**; salivation and blood-stained feces in **cattle**.

E. poissonii = ***Candle Plant*** = ***Tinya*** (Hausa) = ***Oro Adete*** (Yoruba). Latex is a strong mucous membrane irritant and a homicidal poison (Nigeria), added to food, drinking water, or ***kola*** nuts. The latex causes burning of the mucous membranes, and even blindness in **humans** and **cattle**. Also used as an arrow poison and insecticide.

E. pulcherrima. See ***Poinsettia***.

E. tirucalli = ***African Milk Bush*** = ***Euphorbe effilée*** = ***Finger Euphorbia*** = ***Malabar Tree*** = ***Milk Bush*** = ***Monkey Fiddle*** = ***Pencil Tree*** = ***Spurge Tree***. Often used as foundation plants in Florida. They frequently become too large and have to be cut back. Then, or when accidentally broken, it exudes a milky sap, bitter and toxic (severe irritation of mouth and gastrointestinal tract) if taken internally. Inflames or blisters skin after contact. After eye contact, it causes intense pain, acute keratoconjunctivitis and temporary blindness. Eye contact can come from rubbing or fine airborne particles. These are common occurences in Florida. This plant or bushy tree is now sold in other states. A **man** moving a potted specimen in a Galveston, Texas library had a twig strike **his** eye. After it broke and exuded sap, it caused a severe eye inflammation, requiring emergency treatment. The latex has caused **human** fatalities in East Africa, due to its toxic ***ingenol*** and ***5-deoxyingenol***. In Africa, it is an effective barrier against small **animals** and **humans**, but not **rhinos**. Contact has caused corneal opacity in **dogs**.

E. variegata = ***Japanese Edelweiss***. Gathering and binding it has caused marked swelling, erythema, and vesiculation in **people**.

EUPHORBIN

Untoward effects: A deadly resinoid in ***poinsettia*** (q.v.) leaves and various ***Euphorbias***.

EUPROCIN

= *Eucupine*

Local anesthetic, antiseptic.

Untoward effects: Serious toxic symptoms (vision problems, amblyopia, ptosis, blindness) have followed injections.

SC, LD, **mice**: 300 mg/kg; IV, **rabbits**: 13 mg/kg.

EVANS BLUE

= *Azovan Blue* = *C.I. 23,860* = *T-1824*

Use: Diagnostic aid for blood volume determinations and as a contrast medium in lymphangiography.

Untoward effects: Severe pruritus, erythema, and marked edema of the dorsum of the feet and toes.

Has caused liver tumors and teratogenicity in **rats**.

EXEMESTANE

= *Aromasin*

Use: Irreversible aromatase inactivator for treatment of advanced breast cancer in postmenopausal **women** after ***tamoxifen*** failure.

Untoward effects: Hot flashes and nausea in 20%; weight increase and fatigue in ~10%; diaphoresis and increased appetite in 3–4%.

Drug interactions: A potent CYP3A4 enzyme-inducer did not increase its serum concentration, but other CYP3A4 inducers, such as ***rifampin***, may.

F

FACE TISSUES

Untoward effects: Results of analysis of nine brands in hospital use showed them to be a potential source of infection, due to bacterial contamination. **Formaldehyde** has been used in its manufacture to improve its wet strength.

FADOGIA sp.

Untoward effects: One of several genera in South Africa causing gousiekte (sudden death) in **cattle**, **sheep**, and **goats** after ingestion. Affected **animals** lose their appetite, have tachypnea, prostration, and death from heart failure, due to chronic interstitial myocarditis.

FAGUS SYLVATICA

= *Beech*

Untoward effects: The nuts (beech mast), but not the expressed oil, can cause poisoning. Ingestion of up to 50 nuts has caused headache, abdominal pain, vomiting, diarrhea, vertigo, and fever, followed occasionally by syncope, fatigue, pallor, and lassitude within an hour or less and lasting $\frac{1}{2}$–6 h. Occasionally, induces allergic rhinitis, bronchial asthma, and a hypersensitivity pneumonitis.

Poisoning has also been reported in **animals** (***F. sylvatica*** in Europe and ***F. grandiflora*** in North America). Gastrointestinal distress, including abdominal pain, due to a saponin glycoside, which is irritant and corrosive, and not destroyed by cooking. Pain may become severe; tetanic convulsions and death may follow. Poisoning has followed the feeding of "cake" left after the extraction of the oil.

FAMCICLOVIR

= *BRL-42810* = *Famvir* = *FCV*

***Antiviral*. A pro-drug for *penciclovir* (q.v.).**

Untoward effects: Even at recommended dosages, it may precipitate new, or aggravate existing central nervous system disorders (confusion and bradycardia in an 80-year-old **female** on 500 mg twice daily/7 days), particularly in elderly **patients**. Headache, nausea, and diarrhea occasionally reported.

In female **rats**, $1\frac{1}{2}$ times the **human** daily dose/kg for 2 years led to increase in mammary adenocarcinomas, and within 10 weeks after ~2 times the **human** daily dose/kg to male **rats** testicular changes and decreased fertility was noted.

FAMOTIDINE

= *Amfamox* = *Bancolon* = *Dispromil* = *Famodil* = *Famodine* = *Famosan* = *Famoxal* = *Fanosin* = *Fibonel* = *Ganor* = *Gaster* = *Gastridin* = *Gastrion* = *Gastropen* = *Ifada* = *Lecedil* = *MK-208* = *Motiax* = *Muclox* = *Nulcerin* = *Pepcid* = *Pepcidina* = *Pepcidine* = *Pepdine* = *Pepdul* = *Peptan* = *Ulcetrax* = *Ulfamid* = *Ulfinol* = *YM-11170*

H_2-receptor antagonist.

Untoward effects: Has induced neutropenia, thrombocytopenia, hyperprolactinemia, galactorrhea, headache, dizziness, diarrhea, constipation, palpitations, sinus bradycardia, negative inotropic cardiac effects, impotence (without recorded increase of ***prolactin***), hepatitis, hepatocellular jaundice, pyrexia, Stevens–Johnson syndrome, alopecia, flushing, toxic epidermal necrolysis, seizures, mental confusion, and hallucinations after oral therapy, and, in elderly **patients** with mild renal insufficiency, after IV therapy. A 40-year-old **male** with normal sinus rhythm, 89 beats/min and no conductive problems, complained of dizziness and malaise after 40 μg/day/3 weeks. Electrocardiogram showed sinus bradycardia (38/min) and second degree atrioventricular block. 6 months later, **he** was retreated with it and electrocardiogram results were similar with 32 beats/min. 48 h after treatment withdrawal, **he** reverted back to normal sinus rhythm. Rarely, agranulocytosis, pancytopenia, leukopenia, thrombocytopenia, decreased libido, and gynecomastia. A case of hypergastrinemia in a 29-year-old **male** after continuous treatment with ***cimetidine***, then, ***ranitidine***, and, finally, famotidine reported. Nausea, vomiting, anorexia, dry mouth, anaphylaxis, angioneurotic edema, facial edema, rash, urticaria, and muscle and joint pain. **Non-responders** to therapy have been reported.

Drug interactions: Has induced statistically significant slowing of ***theophylline*** elimination. Concomitant use with ***diclofenac*** has caused a more rapid onset of action by the latter. In aldehyde dehydrogenase-deficient **individuals**, its co-administration with ***alcohol*** leads to an inhibition of ***acetaldehyde*** metabolism. ***Probenecid*** may decrease its renal elimination. Can increase ***caffeine*** serum levels if taken with ***coffee*** or ***colas*** leading to stomach irritation or jitters. It reduces absorption of ***itraconazole*** and ***ketoconazole***, as they require gastric acidity for absorption.

Reduced clearance of ***ciprofloxacin*** in treated **rats**, probably due to famotidine inhibition of its renal tubular secretion. Repeated administration led to atrophy of seminiferous tubules, decreased sperm count and motility, and reduced fertility in **rats**, **mice**, and **dogs**. Excreted in the breast milk of lactating **rats**.

IV, LD_{50}, **mice**: 244.4 mg/kg.

FAMPHUR

= *American Cyanamid 38023* = *Bo-Ana* = *Dovip* = *ENT-25644* = *Famophos* = *Warbex*

***Organophosphorus insecticide*. Cholinesterase inhibitor.**

Untoward effects: Do not treat stressed **animals** or **calves** under 3 months of age. A limited study indicated that a group of **Brahman cattle**, including bulls, tolerated without ill-effects the recommended pour-on dosage of 40 mg/kg, as well as twice this dosage. 45 day withdrawal period is suggested for use of **reindeer** meat and 35 days for **cattle** (pour-on). Discontinue use 3 weeks before freshening of **dairy cattle**. Occasionally scurfing and temporary hair loss may be noted in treated **calves**. Avoid run-off water exposure for **hogs**, **birds**, and **fish**. Do NOT treat **cattle** within 6 weeks of grub emergence. **Brahman heifers** and **Holstein calves** show greater cholinesterase depression than **Hereford heifers** after treatment. **Holstein calves** (40–190 kg) require 5–6 weeks before cholinesterase levels return to normal.

LD_{50}, male **rats**: 35 mg/kg; female **rats**: 62 mg/kg; **mouse**: 27 mg/kg; **sheep**: 400 mg/kg; **starlings**: 4.2 mg/kg; **mallard ducks**: 9.87 mg/kg; **red-winged blackbird**: 1.8 mg/kg; dermal, **rabbits**: 1.5 g/kg.

FATS

Fatty acid complexes are used by the body as an energy source, and supplied as animal or vegetable oils. When metabolized, they yield ~9 calories/g. Many are essential to health and are precursors of prostaglandins. An excess of saturated fats may predispose a **person** to, or aggravate, atherosclerosis. Type II hyperlipoproteinemia is common and usually detectable in early childhood. Type IV, also common, appears after age 20. Adverse effects of treatments for hyperlipidemia including gastrointestinal upsets, myopathy, hepatic injury, gallstones, ventricular arrhythmias, and drug interactions. Habitual use of vegetable oils has caused diarrhea and flatulence. A diet high in polyunsaturated vegetable oil (***unsaturated fatty acid***) may increase the incidence of malignant melanomas and squamous-cell carcinomas, as well as increasing the frequency of thrombophlebitis. It doubles the incidence of gall bladder disease and may increase the effect of ***coumarin anticoagulants***. Preterm **infants**, especially **those** on high polyunsaturated fats, can develop a ***vitamin E*** deficiency, characterized by hemolytic anemia. Edema, skin lesions, increased platelets and morphological changes in **infants** from high polyunsaturated fat diets. Fatty foods tend to decrease lower esophageal sphincter tone and cause reflux esophagitis. High dietary fat intake has been associated with increased mammary adenocarcinomas, and colon and prostrate cancer. In the fatal Tay–Sachs disease, a genetic defect preventing the enzymatic metabolism of brain lipids causes cerebral sphingolipidosis. Lipemia can cause increase of false positive sulfobromophthalein retention and SGPT test results. ***Cholestyramine*** can bind and decrease fat absorption. ***Neomycin*** and ***p-aminosalicyclic acid*** decrease fat absorption. Fats enhance the absorption of ***griseofulvin***. Some hyperlipidemia develops during ***isotretinoin*** administration, but rapidly reverts to normal with cessation of treatment.

Accidental IV injections of the lipoid vehicles in parenterals, such as ***penicillin*** suspensions, have caused pulmonary embolism. Fat emboli from ***alcohol***-induced fatty livers have caused idiopathic avascular necrosis of bone. ***Fatty acids*** (0.5%) and ***olive oil*** (0.5%) have been emulsified with 99% liquid ***silicone*** in the famous Sakurai formula for breast implants. IV fat emulsions can lead to acute decreased blood pressure, occasionally hypertension, tachycardia, sweating, cyanosis, transitory coma, chills, headache, nausea, vomiting, fever, tachypnea, dyspnea, flushing, cough, hemoptysis, hematuria, chest and back pain, abdominal pain and tenderness, transient increase in liver enzymes, hepatosplenomegaly, azotemia, anemia (usually normocytic with hematocrit decrease of 4–18%, thrombophlebitis and hypercoagulability, thrombopenia, leukopenia, peculiar taste, priapism, anorexia, somnolence, icterus, hyperlipemia, and fat embolism, among the many symptoms reported with different emulsions (***Intralipid***, ***Infonutrol***, ***Lipofundin***, ***Lipomul***, ***Lipiphysan***, etc.). Bacterial contamination has occasionally been reported. Sepsis, due to contamination, has occurred with IV therapy procedures. Has caused increased heart rate in the **fetus** of treated **mothers**. Oils used as industrial lubricants have caused acne, folliculitis, furunculosis, and, rarely, keratoses, squamous cell skin cancers, and allergic reactions. These are often due to various contaminants.

Fish require ***omega-3-fatty acids*** and **crustaceans** require ***cholesterol*** and ***lecithin*** in their diets. Cyclopropenoid ***fatty acids*** are carcinogens in **trout**, probably due to the fact that they are easily metabolized to ***peroxides*** and ***free radicals***. Excessive intake of rancid ***unsaturated fatty acid*** causes a steatitis (yellow fat disease) and death in **mink**, as well as myocarditis and death in **mice**.

FAT, TOXIC
= *Chick Edema Factor*

Untoward effects: Causes a hydropericardium syndrome (including SC edema, ascites, hydrothorax, decreased red blood cells and white blood cells, and myocardial hypertrophy and edema) in **chicks**, **monkeys**, and to some **rats**.

FAVEIRA TREE
= *Diamorphandra mollis*

Untoward effects: In Brazil, **cattle** consuming large amounts of the pods in 1 day died within 14 days. They developed anorexia and blood-streaked feces. Postmortem revealed serous fluid in the abdomen, mesenteric edema, hemorrhagic cecum and large intestines, coagulation necrosis of the proximal convoluted renal tubules, and hyaline casts elsewhere in the tubules.

FDC BLUE #1
= *Acid Blue 9 = Brilliant Blue FCF = CI 671 = CI 42,090 = CI Food Blue 2 = Erioglancine = Patent Blue AE = Schultz 770*

Untoward effects: FDA-approved for food, drug, and cosmetic uses. In February 1994, the FDA finally approved its use for coloring external cosmetics and drugs in the eye area. Considered a suspect carcinogen by some. A hospitalized 11-year-old **female** received enteral nutrition with it with this blue food coloring leading to clinical cyanotic appearance. Pediatric code response team found normal respiratory rate and normal oxygen saturation. Non-colored enteral nutrition was continued and cyanotic appearance resolved within 24 h.

In **rats**, it has led to an intense macrophage necrosis and fibroplastic proliferation.

SC, LD_{50}, **mice**: 4.6 g/kg.

FDC GREEN #3
= *CI 42,053 = CI Food Green = Fast Green FCF*

Untoward effects: FDA-approved for food, drugs, and cosmetics, excluding use in the eye area. Considered a suspect carcinogen by some.

In **rats**, it has led to an intense macrophage necrosis and fibroplastic proliferation.

LD_{50}, **rats**: > 2 g/kg.

FDC RED #40
= *Allura Red AC = CI 16,035 = Food Red 17*

Untoward effects: Now, one of the most widely used dyes for foods, drugs, and cosmetics. When fed in large amounts to **mice**, it was suspect as a cause of malignant lymph tumors and death. A subsequent test failed to demonstrate this. ***p-Cresidine***, an intermediate in its manufacture, is a known carcinogen.

FDC VIOLET #1
= *Acid Violet 49 = Violet 6B*

Use: Primarily, for meat inspectors' rubber stamps, and in candies and pet foods.

Untoward effects: Banned in various countries and by the FDA in 1973, based on unpublished data that it was carcinogenic.

TD_{LO}, **rat**: 800 g/kg/46 weeks.

FDC YELLOW #6
= *C.I. 15985 = C.I. Food Yellow 3 = DFG29 = E 110 = Orange RGL = Sunset Yellow FCF = Yellow Orange S*

Use: Provisionally, accepted for use in foods, drugs, hair rinses, and cosmetics. An azo dye.

Untoward effects: Allergic problems (including anaphylactic shock, that may have been precipitated by its use in an enema, as well as episodes of Quincke's edema and gastroenteritis) and mutagenic effects reported. 5 mg/kg/day considered maximum acceptable intake for **man** (2.5 mg/kg/day, in France). The FDA permitted up to 300 mg/day intake. Nausea, leukopenia, and rashes reported in 3/10 **patients** with AIDS.

LD_{50}, **rat**: > 10 g/kg; **mouse**: > 6 g/kg. IP, LD_{50}, **rat**: 3.8 g/kg; **mouse**: 4.6 g/kg.

FELBAMATE
= *ADD-03055 = Felbamyl = Felbatol = Taloxa = W-554*

Anticonvulsant.

Untoward effects: Despite usual mild and transient adverse effects, it has caused severe hematological and hepatic toxicity. Aplastic anemia and acute liver failure are the most serious adverse effects. A 33-year-old **female** developed toxic epidermal necrolysis after 3600 mg/day/16 days. Anaphylactoid reaction in a 46-year-old **female** on 300 mg twice daily for refractory trigeminal neuralgia. Has caused intermenstrual bleeding in a young **woman** during treatment.

Drug interactions: Decreased ***phenobarbital*** hydroxylation, causing a dramatic increase in ***phenobarbital*** plasma concentration. Increased blood levels of ***phenytoin*** and ***valproic acid***, and decreased ***carbamazepine*** concentration. Dramatic increase in ***warfarin*** concentration also reported.

IP, LD_{50}, **mice**: 4 g/kg.

FELDSPAR

Untoward effects: Inhalation of its dust implicated as a cause of respiratory disease in **workers**.

FELODIPINE

= *Agon* = *Feloday* = *Flodil* = *H-154/82* = *Hydac* = *Munobal* = *Plendil* = *Prevex* = *Splendil*

Antihypertensive, antianginal, calcium channel-blocker.

Untoward effects: Usually, causes an increased heart rate, especially during the first week of therapy. Occasionally, precipitates significant hypotension, myocardial infarction, and reflex tachycardia that, in turn, can precipitate angina. Rarely, syncope. Peripheral edema, headache, chest pain, facial edema, abdominal pain, flatulence, vomiting, dry mouth, muscle cramps, leg and arm pain, reduced libido, depression, anxiety, urticaria, erythema, dyspnea, bronchitis, visual disturbances, urinary problems (urgency, dysuria, and polyuria), and gingival hyperplasia reported in < 0.5–1.5%. In the **elderly** and in **patients** with impaired liver function, there is a potential for increased plasma levels.

Drug interactions: Its serum concentrations are increased by ***itraconazole***. In healthy **men**, the mean bioavailability of 5 mg with 250 ml ***grapefruit juice*** was 284% of that with water. A 43-year-old **female**, after 2–3 days of treatment with ***erythromycin***, experienced palpitations, flushing, and ankle edema. Adverse symptoms subsided completely a few days after completing antibiotic therapy. Its bioavailability is reduced by ***anticonvulsants***, such as ***carbamazepine*** (94%) and ***oxcarbazepine*** (~30%).

FELYPRESSIN

= *Octapressin* = *PLV 2*

Vasoconstrictor.

Untoward effects: In general, similar to ***Vasopressin*** (q.v.). Nausea, abdominal pain, fainting, headache, skin pallor, chest pain, dyspnea, tachyphylaxis, slight bradycardia, urge to micturate or defecate. Rarely, some diarrhea and vomiting. Unlike ***epinephrine***, it is compatible with ***halogenated hydrocarbon*** and ***cyclopropane anesthesia***.

FENAMIPHOS

= *Bay 68,138* = *ENT 27,572* = *Nemacur* = *Phenamiphos*

Nematocide.

Untoward effects: Nausea, vomiting, abdominal cramps, diarrhea, salivation, headache, giddiness, vertigo, weakness, rhinitis, chest tightness, sweating, blurred vision, miosis, arrhythmias, muscle fasciculation, and dyspnea. Has caused irreversible eye damage. May be fatal if excess quantities are absorbed through the skin. National Institute for Occupational Safety & Health has established a limit of 0.1 mg/m^3 for skin exposure.

Goose-stepping ataxia, tachypnea, salivation, tremors, and opisthotonus within 5 min in **birds**. Most deaths within 15 min–3½ h.

LD_{50}, **rat**: 8 mg/kg; **mouse**: 23 mg/kg; **dog** and **cat**: 10 mg/kg; **guinea pig**: 75 mg/kg; **mallard duck**: 1.68 mg/kg; **California quail**: 1.83 mg/kg; **pheasants**: 0.5–1.0 mg/kg.

FENBUFEN

= *Bufemid* = *Cinopal* = *Cinopol* = *CL-82204* = *Lederfen*

Anti-inflammatory.

Untoward effects: Heartburn is common. Hemolytic anemia in a 66-year-old **male** on 300 mg twice daily/10 days. A case of fatal aplastic anemia in a 65-year-old **female** a few years after a suspected case of aplastic anemia in **her** due to ***sulindac***. Dermatoses and pulmonary eosinophilia reported in some **patients** (68–85 years). Erythema multiforme in a 65-year-old **female** 3 days after 900 mg/day. Severe vasculitis in an elderly **female** on 900 mg/day. Visual hallucinations in a 75-year-old **patient** after two doses of 300 mg. Generalized hypersensitivity reported in a 41-year-old **female** after 900 mg/day/10 days. Collagenous colitis may have been associated with its use in a 55-year-old **female**. Less gastric microbleeding than that produced by ***aspirin***.

Increased the serum, brain, and cerebrospinal fluid levels of ***nalidixic acid*** in **rats**.

FENCAMFAMINE

= *Euvitol* = *H 610* = *W 1206*

Central nervous system stimulant.

Untoward effects: Increased irritability, restlessness, increased motor activity, and mild transient urticarial rashes.

LD_{50}, **rat**: 83 mg/kg; **mouse**: 135 mg/kg; **dog**: 30 mg/kg; **cat**: 34 mg/kg; IV, **rat**: 24 mg/kg; **mouse**: 16 mg/kg; **dog**: 15 mg/kg.

FENCLOFENAC

= *Flenac* = *RX 67,408*

Anti-inflammatory.

Untoward effects: Regularly severe; the manufacturer has ceased selling it. Skin rashes (~7%), gastrointestinal upsets and blood loss (~5%), central nervous system problems (1.3%) including headache and dizziness, acute interstitial nephritis, profound decrease in circulating thyroid hormone, and eye irritation.

FENCLOZIC ACID

= *ICI 54,540* = *Myalex*

Analgesic, anti-inflammatory.

Untoward effects: Use discontinued because of serious hepatotoxicity.

LD_{50}, **rat**: 850 mg/kg; **mouse**: 1 g/kg.

FENFLURAMINE

= *Acino* = *Adipomin* = *AHR 965* = *EMPT* = *Ganal* = *Obedrex* = *Pesos* = *Ponderal* = *Ponderax* = *Ponderex* = *Pondimin* = *Rotondin* = *S 768*

***Anorexiant.* Has sympathomimetic amine activity.**

Untoward effects: High incidence. Dry mouth, dizziness, urinary frequency, drowsiness, diarrhea, flatulence, abdominal pain, impotence in **men** and decreased libido in **females** not associated with its sedative effects, lethargy, headache, vertigo, giddiness, sleeplessness, nightmares, depression, palpitations, shivering, jaw tremors, sensations of heat and cold, temporary mood depression on sudden withdrawal, metrorrhagia, hemolytic anemia, bruxism, dyskinesia, psychosis, pulmonary hypertension, angina, and interference with ***growth hormone*** release. Often induces confusion in the **elderly**. Coma, convulsions, and death from overdoses. It became a regular substance for abuse. A severe cardiac depressant. Therapeutic levels are considered to be 10–12 μg%, and toxic levels 20–90 μg%; 0.6–1.5 mg% has been lethal. 0.3–2.0 g causes toxicity. Other workers quote therapeutic serum levels as 0.04–0.3 mg/l; toxic at 0.5–0.7 mg/l; and fatal at 6 mg/l. Ten of 53 cases reported in Germany were lethal after 28.7–70 mg/kg. Three young **females** ingested 400, 600, and 800 mg, respectively, leading to dilated, unreactive pupils; rotary nystagmus, hyperreflexia, jaw tremor, and feelings of heat. A 2½-year-old **child** who ingested 160–200 mg developed convulsions, pyrexia, and tachycardia; recovered. Another **child** (2½-year-old **female**) ingested 440 mg leading to semiconsciousness, sweating, agitation, mydriasis, cyanotic feet, pulse 200/min, and generalized convulsions. Recovered after treatment. A 17-year-old **female** ingested 1.6 g (80 tablets) and presented at an emergency room in a diaphoretic state with tachycardia. Seizures developed 90 min later, with shallow respirations and lack of responsiveness. Fatality in a 5-year-old **female** after ingesting ~20 60 mg slow-release capsules. Cardiac, neurologic, and respiratory symptoms predominated. Malignant hyperpyrexia (rectal, 109°F) in a 34-year-old **female** who probably ingested 100 tablets. **She** also exhibited coma, hyperventilation, tachycardia, sweating, rigidity, flexing of arms and legs, decreased blood pressure, and rolling eye movements. Despite treatment, death occurred ~13 h after admission. Renal excretion increases in acid urine (e.g. after ***ammonium chloride***) and decreases in alkaline urine (e.g. after ***sodium bicarbonate*** and ***acetazolamide***).

Although the combination was never approved by the FDA, it is estimated that 6 million **Americans** in 1996 used it in combination with ***phentermine*** (q.v.) leading to heart valve problems and deaths, endocardial fibrosis, and, possibly, ischemic colitis. Abrupt withdrawal of ***phentermine-dexfenfluramine*** combination can cause ***serotonin*** withdrawal syndrome.

Drug interactions: May react with ***monoamine oxidase inhibitors***, leading to hypertension, hyperpyrexia, headache, subarachnoid hemorrhage, coma, and potentiation of ***hypoglycemics*** and ***antihypertensives***. Hypotension in **patients** on ***methyldopa***, ***reserpine***, or ***Rauwolfia***.

LD_{50}, **rat**: 130 mg/kg; **dog**: 100 mg/kg; **cat**: 60 mg/kg; **rabbit**: 50 mg/kg.

FENNEL

The dried fruit of *Foeniculum vulgare*. Contains *anethole* (q.v.). In *absinthe*, some *breads*, and as a food spice.

Untoward effects: Contains furocoumarins, which are naturally occurring photosensitizers. Many **people** who contact the plant develop phytodermatitis. Stomatitis and dermatitis of the hand attributed to its presence in a denture-cleaning cream. Some estrogenic activity in its oil.

Oil, LD_{50}, **rat**: 3.1 g/kg; skin irritation, **mouse** and **rabbit**: 500 mg/kg/24 h.

FENOFIBRATE

= *Ankebin* = *Elasterin* = *Fenobrate* = *Fenotard* = *LF-178* = *Lipanthyl* = *Lipantil* = *Lipidil* = *Lipoclar* = *Lipofene* = *Liposit* = *Lipsin* = *Nolipax* = *Procetofen(e)* = *Procetoken* = *Protolipan* = *Secalip* = *Tricor*

Antihyperlipoproteinemic.

Untoward effects: Skin rashes and pruritus are the most common complaints, followed by gastrointestinal upsets and eye irritation. Cholelithiasis, hepatitis, myositis, and increased serum ***creatinine*** also reported. Uricosuric.

Drug interactions: Use with ***warfarin*** increases the latter's anticoagulant effect and has caused epigastric discomfort, nausea, and ***tea***-colored urine.

High dosage in **rats** was teratogenic and increased incidence of liver and pancreatic cancer.

FENOLDOPAM MESYLATE

= *Corlopam* = *SKF-82526-J*

***Antihypertensive.* IV use.**

***Untoward effects*:** Hypotension, flushing, dizziness, headache, tachycardia, increased intraocular pressure, hypokalemia, and nausea.

Drug interactions*:** Use with ***acetaminophen increases its peak serum concentration.

Arterial lesions (medial necrosis and hemorrhages) in **rats** after IV infusions, but not in **dogs** or **mice**. Oral dosage has induced polyarteritis nodosa in **rats**.

FENOPROFEN CALCIUM

= *Fenopron* = *Fepron* = *Feprona* = *Lilly 69323* = *Nalfon* = *Nalgesic* = *Progesic*

Anti-inflammatory, analgesic. non-steroidal anti-inflammatory drug.

Untoward effects*:** Nephrotic syndrome and renal failure reported in arthritic **patients**. Causes proteinuria and renal papillary necrosis after long-term therapy. Nausea, vomiting, constipation, diarrhea, dry mouth, epigastric pain, gastrointestinal bleeding, prolonged bleeding time, tachycardia, palpitations, weakness, dyspnea, headache, hepatotoxicity, thrombocytopenia, agranulocytosis, aplastic anemia, fever, toxic epidermal necrolysis, sweating, rashes, urticaria, pruritus, erythema multiforme, Stevens–Johnson syndrome, bronchospasms, confusion, tremors, tinnitus, blurred vision, insomnia, and dizziness reported. Intentional overdose with 60–72 g in a 41-year-old **female** caused non-oliguric renal failure and hypotension. When discharged 25 days after ingestion, **her** renal function was stabilized, but **her** mental concentration was not normal. ***Non-steroidal anti-inflammatory drug*s** are not recommended in **women** trying to become pregnant since they interfere with implantation of fertilized eggs in **animals**. Cross-allergenicity to ***aspirin (75–100%). Food delays and decreases its absorption.

Drug interactions*:** Can increase ***warfarin's anticoagulant effect and ***warfarin*** can accentuate gastrointestinal bleeding.

LD_{50}, **mouse**: 800 mg/kg.

FENOTEROL HYDROBROMIDE

= *Airum* = *Berotec* = *Dosberotec* = *Partusisten* = *TH-1165a*

***Bronchodilator, tocolytic, sympathomimetic, β-adrenergic agonist.* Given by inhalation, oral, SC, IV, or rectal.**

***Untoward effects*:** Hypokalemia, increased chances of cardiac dysrhythmias, and finger tremors. Increases risk of death or near-death with regular inhalant use (particularly with self-administration). Aerosol forms are unlikely to affect the **fetus**, but IV or oral use can cause **fetal** tachycardia. **Neonatal** hypoglycemia after **mother's** tocolysis. Pulmonary edema in a **patient** during continuous tocolysis. Inhalation not apt to cause adverse cardiovascular effects, as might larger oral or SC doses (viz. tachycardia). IV increases heart rate.

Rats treated by gavage for 35 days had dose-dependent increase in cardiac weights and scarring. Newborn **rats** died from gastrointestinal disorders or cachexia.

LD_{50}, **mice**: 2 g/kg; SC, 1 g/kg.

FENOVERINE

= *Spasmopriv*

Antispasmodic.

***Untoward effects*:** Dry mouth and a case of acute rhabdomyolysis.

LD_{50}, **mice**: ~1.5 g/kg.

FENOXEDIL

= *Suplexedil*

***Untoward effects*:** May cause QT interval prolongation.

LD_{50}, **rats**: 2.4 g/kg; **mice**: 750 mg/kg; **rabbit**: 850 mg/kg.

FENRETINIDE

Antineoplastic.

***Untoward effects*:** Diminished dark adaptation, alterations of dark adaptometry have been common. Mild and transient dermatoses also reported.

FENSULFOTHION

= *Bay 25,141* = *Dasanit* = *ENT 24,945* = *S 767* = *Terracur P*

***Insecticide, nematocide.* Organophosphate anticholinesterase.**

Untoward effects*:** A 65-year-old **male** accidentally ingested an unkown quantity and became unconscious. A total of 182.5 g of ***pralidoxime over 47 days was needed to stabilize the **patient**, who was finally discharged 69 days after hospitalization, with neurological impairment, characterized by confusion and vagueness. Skin irritant. Induces nausea, vomiting, abdominal cramps, diarrhea, salivation, headache, giddiness, vertigo, weakness, coughing, rhinitis, chest tightness, blurred vision, muscle fasciculations, dyspnea, and arrhythmias. Skin exposure limit of 1 mg/m^3.

It is highly toxic to **bees** and **earthworms**. Toxicity in affected **birds** includes regurgitation, ataxia, wing and tail tremors, dyspnea, and opisthotonus. Symptoms appear within 20 min after exposure and death within $\frac{1}{2}$–3 h.

LD_{50}, male **rats**: 6 mg/kg; female: 2.5 mg/kg; **ewe lambs**: 3.5 mg/kg; **mallard ducks**: 747 μg/kg; male and female **quail**: 1.68 and 1.19 mg/kg, respectively; **pheasants**: 1.34 mg/kg; **chicken**: 1 mg/kg.

FENTANYL

= *Actiq* = *Durogesic* = *Fentanest* = *Leptanal* = *McN-J-R 4263* = *POTFC* = *Pentanyl* = *Phentanyl* = *R4263* = *Sublimaze*

Analgesic, narcotic, tranquilizer. **Parenteral, transdermal, and transmucosal. Usually IM.**

Untoward effects*:** Respiratory decrease has often been delayed (3–5 h after surgical anesthesia with moderate doses of 55–75 μg/kg). It is usually of short duration, but has been prolonged and recurrent, and can lead to apnea. Rigidity is associated with increased muscle tone and occasionally chest wall spasms, and exacerbated by $\mathbf{N_2O}$. Hypotension, bradycardia, peripheral vasodilation, headache, gastrointestinal upsets, amnesia, hallucinations, pruritus, rash, amblyopia, wheezing (histamine release), and urinary problems. Significant long-term deterioration of cognitive function reported. In an 11-year-old **patient**, a 5 μg/kg IV bolus caused an abrupt increase in intracranial pressure. Decreased ***radiotracer uptake by the myocardium, due to decreased myocardial perfusion. Drug dependence reported in **instructors** and anesthesia **residents**. A 33-year-old **male**, a physician and military officer, was found dead, after apparently self-injecting it. Transdermal fentanyl leads to confusion in 10% of long-term **users**. Many so-called "designer drugs" are among the hundreds of fentanyl modifications **chemists** have made. Some of these analogs, such as ***3-methylfentanyl*** or ***α-methylfentanyl*** (sold on the street as ***China White***), have ***heroin***-like effects and have caused over 100 deaths and thousands of emergency room episodes. Some **purchasers** of ***benzy-lamine, methyl acrylate, phenethylamine, propionic anhydride***, and other chemicals have used these to manufacture illicit fentanyls.

TD_{LO}, **human**: 2 μg/kg.

Decreases respiratory frequency in **cats** and, occasionally, **rabbits** and **rats**; decreases myocardial blood flow in **dogs**. IV or IM causes pacing and rapid motor stimulation in **horses**.

LD_{50}, **rat**: 18 mg/kg; SC, 12 mg/kg; IV, 6 mg//kg; **mice**: IV, 3 mg/kg; SC, 62 mg/kg.

FENTHION

= *Baycid* = *Bayer 29493* = *Baytex* = *ENT-25540* = *Entex* = *Lebaycid* = *Mercaptophos* = *Pro-Spot* = *Queletox* = *S-1752* = *Spotton* = *Talodex* = *Tiguvon*

Insecticide, nematocide. **Organophosphorus cholinesterase inhibitor.**

Untoward effects*:** When used as a residual spray in Nigeria for malaria eradication, 400 **natives** showed a moderate decrease in plasma (but not red blood cells) cholinesterase for 5 weeks after spraying. In India, two **males** and one **female** in suicide attempts showed salivation, bronchial constriction, vomiting, and pulmonary edema. A young **male**, in a suicide attempt, developed shock, blurred vision, vomiting, diarrhea, slurred speech, and respiratory paralysis, requiring artificial ventilation for 7 days. A suicidal 43-year-old **male**, ingesting ~18 g, within 20 min showed nausea, biliary vomiting and abundant, fluidy diarrhea for ~12 h. Hospitalized the next morning, ~22 h after ingestion, with atrial tachycardia, sporadic ventricular extrasystoles, systolic-diastolic hypertension (220/110), persistent nausea and vomiting, sialorrhea and bronchorrhea. 9 h later, intense generalized fasciculations, and marked tracheo-bronchorrhea with bronchospasm (requiring the use of a respirator). 70 h after ingestion, symptoms worsened leading to generalized fasiculations, hypersudation, sialorrhea, bronchorrhea, bronchospasm, miosis, bradycardia (40/min), obtundation, and repetitive convulsive episodes, requiring 18 days of hospitalization. A 26-year-old **male** spilled it on **his** clothing and may have ingested some. **He** developed periumbilical abdominal pain and vomiting 8 h later, continuing for 18 h before becoming lethargic. 42 h after termination of exposures to contaminated clothing, **he** developed respiratory arrest, requiring mechanical ventilation. **Others** with suicidal intent showed cutaneous paresthesia; and paralysis of proximal limb muscles, neck flexors, motor cranial nerves, and respiratory muscles. Veterinary hospital **employees** developed toxicity, after using it topically for **flea** control. Assorted symptoms are the same as summarized above, under ***fensulfothion.

LD_{LO}, **human**: 50 mg/kg.

Ducklings have died from eating exposed **tadpoles**. Very toxic to **birds**, where it has been used as a perch toxicant. Topical treatment is very toxic to young **kittens**. **Sheep** are killed by 50 mg/kg. **Cattle** show toxicity after 25 mg/kg. As with other **birds**, it is extremely toxic to **poultry**. Topically, in **dogs**, it has caused muscular tremors, salivation, vomiting, diarrhea, and weakness in **puppies** and in older **dogs**, from overdosing. Found in the milk of lactating **cows** for a week after topical use on them. **Angora goats** were poisoned by 0.25% topically (**sheep**, at 0.5%, were not) leading to salivation and weakness. IM poisoned **cattle** at 5 or 10 mg/kg.

LD_{50}, **rats**: 178 mg/kg; **rabbit**: 150 mg/kg; **redwinged blackbirds** and **pigeons**: 1.8 mg/kg; **starlings**: 3 mg/kg; **chickens**: 28 mg/kg; **quail**: 11 mg/kg; **ducks**: 6 mg/kg.

FENTICLOR

= *D 25* = *Novex* = *Oksid* = *Ovitrol* = *Oxid*

***Fungicide.* Chemically related to *bithional* (q.v.).**

Untoward effects*:** A 60-year-old **male** developed a photo-contact dermatitis of the exposed skin of **his** face and neck, after using a hair cream containing it. The sensitivity persisted. Present in some soaps. **Patient** reactors cross-react with ***hexachlorophene.

FENTICONAZOLE

= *Falvin* = *Fentiderm* = *Fentigyn* = *Lomexin* = *Rec-15–1476*

***Antifungal.* An imidazole.**

***Untoward effects*:** Mild sensitization reported in a couple of **patients**.

FENUGREEK

= *Trigonella foenum graecum* = *'ULBA (Yemenites)* = *Fenugrène*

Use*:** Seeds in making curry and as an aromatic. The seeds are called ***chilbe by Yemenites and used as an antidiabetic folk medicine. Confirmed in **rat** trials. In Egypt, its use associated with 160–900% increase in **mother's** milk. Used in making imitation maple syrup and mixed spices for **people**.

***Untoward effects*:** Large intake causes hypoglycemia in **humans** and **dogs**.

Causes myopathy in **cattle** and **sheep** after long-term feeding of its straw. **Calves** exhibited chronic lameness (starting, first, in the hind limbs), abduction of limbs, extensor weakness, dropped fetlocks, overgrown toes and stumbling gait over a period of 1–2 months. Cardiac arrhythmias also noted, as well as skeletal and cardiac muscle degeneration determined on post-mortem. Affected **sheep** evidenced similar symptoms. Post-mortems revealed ascites, SC edema, hydropericardium, and mild parenchymatous degeneration of the liver.

FENURON

= *Dybar* = *Fenidrim* = *Fenulon* = *PDU* = *Uro b*

Herbicide.

Untoward effects*:** Hydrolyzed to ***aniline (q.v.), which can cause methemoglobinemia, anorexia, and central nervous system depression.

LD_{LO}, **human**: 5 g/kg.

Cattle were poisoned by two oral doses of 500 mg/kg and **sheep** were killed by five such daily doses. Depression, anorexia, and incoordination were characteristic symptoms. One **heifer** aborted. Recovery is very slow. Pulmonary congestion and large hemorrhages on the heart noted on post-mortem.

LD_{50}, **rats**: 6.4 g/kg; **quail**, **pheasants**, and **mallard ducks**, LC_{50} in feed: > 5000 ppm.

FENVALERATE

= *Agmatrin* = *Belmark* = *Ectrin* = *Fenkill* = *Phenvalerate* = *Pydrin* = *Pyridin* = *S-5602* = *Sanmarton* = *SD-43775* = *Sumicidin* = *Sumifly* = *Sumipower* = *Sumitox* = *Tirade* = *WL-43775*

***Insecticide.* A synthetic pyrethroid. Nearly 20 tons used in 1980 in the U.S.**

***Untoward effects*:** Irritant to skin, eyes, and respiratory tract. Varying degrees of dysesthesia in agricultural **applicators**. Stinging or burning sensation progresses to numbness in ~$\frac{1}{3}$ of **those** exposed. This usually occurs several hours after contact, peaking in the evening, and rarely, present in the morning. Increased nasal secretions and sneezing in 19% of conifer seedling **handlers**; 6% had increased coughing, 4% had dyspnea, and 10% had eye irritation. Itching and tingling of facial skin has been common after contact. May be fatal if swallowed.

Despite its efficacy, use with ***deet*** for **flea** control on **cats** and **dogs**, for unknown reasons, seemed linked to illnesses and deaths in hundreds of them.

LD_{50}, **rats**: 3.2 g/kg.

FEPRAZONE

= *Analud* = *DA-2370* = *Methrazone* = *Phenylprenazone* = *Prenazone* = *Zepelin*

Anti-inflammatory.

***Untoward effects*:** Slight heartburn, occasionally nausea, erythematous rash, headache, cramps, and metallic taste reported. Severe thrombocytopenia and immune hemolytic anemia has caused its withdrawal from usage.

FERRIC AMMONIUM CITRATE

= *Iron Ammonium Citrate* = *Minadex*

Untoward effects*:** ***Orange-flavored ***Minadex***®, at a dosage 64 times normal, caused a fatality in a 2-year-old **male**. Some **people** have gastrointestinal upsets with it, especially if taken without food. See ***Ferrous Sulfate*** and ***Iron***. Harmlessly, turns stools dark green or black. Excessive quantities or concentrations may cause diarrhea. Incompatible with ***acacia*** preparations, ***iodides***, and ***tannin*** preparations.

Although effective parenterally against baby **pig** anemia, it is occasionally lethal.

LD_{50}, **mouse**: 5 g/kg; **guinea pig**: 1.75 g/kg; **rabbit**: 2.8 g/kg.

FERRIC CHLORIDE

= *Flores martis* = *Iron Trichloride*

Use: 70% in sewage and waste water treatment. ~300,000 tons used annually in the U.S.

Untoward effects: Permanent pigmentation of the skin reported from its use in the treatment of ***poison ivy*** dermatitis. See ***Ferrous Sulfate*** and ***Iron***. ***Whiskey*** or ***brandy*** increase its absorption. Incompatible with ***acacia*** preparations, ***iodides***, and ***tannins***.

LD_{50}, **mouse**: 1.5 g/kg; **guinea pig**: 600 mg/kg; **rabbit**: 1.2 g/kg.

FERRIC FERROCYANIDE

= *Berlin Blue* = *Chinese Blue* = *CI 77,510* = *CI Pigment Blue 27* = *Paris Blue* = *Prussian Blue*

Use: Pigment in coloring inks, paints, resins, linoleum, and typewriter ribbons. Also, in the early treatment of ***thallium*** poisoning, in diagnosing ***cyanide*** poisoning, and in removing ***H_2S*** from gas pipelines and storage tanks.

Untoward effects: Caused the death of 11 **cattle** and blindness in a twelfth, due to contamination of a pasture from a gas line leak.

FERRIC HYDROXIDE

= *Hydrated Oxide of Iron* = *Limonite*

Use: Arsenical antidote, a pigment, and in water purification.

Untoward effects: High intake from bore water caused scouring, decreased weight and decrease in milk and butterfat production by **dairy cows**.

FERRIC OXIDE

= *Blood Stone* = *Hematite* = *Red Iron Oxide* = *Rouge*

***Coloring agent.* Worthless as a source of oral *iron* for man and animals. Over 2 billion lbs used annually in the U.S.**

Untoward effects: Overexposure to its dust caused benign pneumoconiosis (siderosis), with X-ray shadows identical to those of fibrotic pneumoconiosis. Suspect as a lung carcinogen from dust exposures, especially in underground mining operations. **Miners** often had an increase in emphysema and bronchitis. Aircraft **workers** contacting it have developed dermatoses and **housewives** have shown cheilitis from its use as a rouge. OSHA exposure limits set at 10 mg/m^3, TWA.

SC, TD_{LO}, **rat**: 135 mg/kg.

FERROUS CARBONATE MASS

= *Blaud's Mass*

Untoward effects: Despite poor bioavailability, a 60-year-old **female** developed hemochromatosis ingesting ~12,000 g (including some ***iron gluconate***) over a 27 year period. Rarely, nausea, diarrhea and/or constipation reported.

FERROUS FUMARATE

= *Cpiron* = *Erco-Fer* = *Ferrofume* = *Ferronat* = *Ferrone* = *Ferrotemp* = *Ferrum* = *Fersamal* = *Firon* = *Fumafer* = *Fumar F* = *Fumiron* = *Galfer* = *Heferol* = *Ircon* = *Meterfer* = *One-Iron* = *Toleron* = *Tolferain* = *Tolifer*

Untoward effects: Nausea and gastrointestinal upsets reported.

FERROUS GLUCONATE

= *Fergon* = *Ferlucon* = *Ferronicum* = *Iromon* = *Irox* = *Nionate*

Use: Hematinic and as a food colorant with ***olives***, etc.

Untoward effects: In one group of pregnant **women**, 3/141 were nauseous, 8/141 vomited, 10/141 had gastrointestinal disorders, 3/141 had diarrhea, and 7/141 showed "intolerance". Accidental overdosing by **children** 11 months and 2½ years reported.

TD_{LO}, **child**: 162 mg/kg.

LD_{50}, **mice**: 3.7 g/kg.

FERROUS SUCCINATE

= *Cerevon* = *Ferromyn*

Untoward effects: A 19-year-old **female** hospitalized 2½ h after ingesting 60 tablets (2 g ***Fe^{++}*** or 42 mg ***Fe^{++}***/kg) was drowsy and had epigastric pain. **She** was treated and recovered, except for bloody stool the next day. A 26-year-old **female** hospitalized 3 h after ingesting 150 tablets (11 g ***Fe^{++}*** or 180 mg ***Fe^{++}***/kg) was warm, conscious, hypertensive, and felt ill. Vomitus contained blood. **She** became hypovolemic with tachycardia and decreased blood pressure, pale, and sweaty, and had increased SGOT, SGPT, and ***bilirubin***. Treatment continued for 1 week. To be taken with or immediately after meals, but not with ***antacids*** that complex it and decrease its absorption.

FERROUS SULFATE

= *Feosol* = *Fesofor*

Use: In **animal** feeds, as a herbicide, in preventing chlorosis in plants, in fertilizers, in treating hazardous wastes and soils contaminated with ***chromium***, in manufacturing, and as an ***iron*** supplement for **man**. > 90,000 tons of dried material, plus tonnage of moist material, used annually in the U.S.

Untoward effects: Gastrointestinal upsets (dyspepsia, nausea, vomiting, colic, anorexia, heartburn, cramping, constipation, diarrhea, metallic taste, pyloric stenosis) are even common with therapeutic doses, especially during pregnancy. One of the more common causes of poisoning in young **children**, due to easy access, poor packaging, or attractive coloring. **Some** have out-witted the designers of **child**-resistant containers. In addition to common **adult** symptoms noted above, **many** have become lethargic with decreased muscle tone, acidosis, gastrointestinal bleeding, shock, acute hepatic failure, hypoglycemia, pallor, cyanosis, melena, cardiovascular collapse, increased serum ***iron*** and *Yersinia enterocolitica* septicemia; death has occurred in a high percentage of poisonings in **children**. The latter can occur in as little as 1–4 h. Occasionally, fatal collapse can follow apparent recovery. As few as 10–15 5-grain (324 mg) tablets have been lethal to **children**, despite treatment. An unusual fatal poisoning in a 3-year-old **male** 1 day after ingesting a pinch of gray mass from car battery terminals. Numerous cases in the literature of deliberate (apparently, suicidal) poisonings in **adults**. Esophagitis, including ulceration and swelling of the hypopharynx, in an 89-year-old **female**, caused by a tablet lodging there. Rarely, skin eruptions reported. Taken orally, it can turn stools greenish-black, but they are ***guaiac*** and ***benzidine*** tests negative for occult blood. 200 mg tid in a rheumatoid arthritic 49-year-old **female** caused a clinical flare-up after 3 days' therapy. Gangrene of Meckel's diverticulum reported. May cause a yellow-brown discoloration of the eyes. Large doses can cause severe liver pathology and shock. In fatal cases, post-mortem reveals periportal necrosis, gastric edema and ulceration, and brown-staining of the mucous membranes.

Minimum lethal dose, **adult**: 50 g; LD_{LO}, **child**: 150 mg/kg; **woman**: 60 mg/kg.

Drug interactions: In normal **volunteers**, 200 mg tablet with 500 mg ***tetracycline*** decreased the latter's peak serum concentration 40–50%; with ***oxytetracycline***, a decrease of 50–60%; with 200 mg ***doxycycline***, a decrease of 80–90%; with ***methacycline*** 300 mg, a decrease of 80–85%. Limited data indicate that it may complex and decrease the availability of ***carbidopa***, ***levodopa***, and ***methyldopa***. Its absorption is decreased by ***antacids*** and it may decrease the efficacy of ***thyroxine*** in some **patients**. May alter distribution of ***radiopharmaceuticals***. ***Allopurinol*** and ***ascorbic acid*** increase its absorption.

LD_{50}, **rat**: 319 mg/kg; **mouse**: 980 mg/kg; **guinea pig**: 1.2 g/kg; **rabbit**: 3 g/kg; **cat**: > 500 mg/kg; **dog**: 800 mg/kg.

FERROVANADIUM

An alloy containing 50–80% *vanadium*.

Untoward effects: The dust can irritate the eyes and respiratory tract. A non-combustible solid, but the dust can be an explosion hazard. OSHA limits exposure to 1 mg/m^3, TWA.

Bronchitis and pneumonitis in experimental **animals** exposed to the dust.

FERULA COMMUNIS

= *Giant Fennel*

Untoward effects: Poisoning in **sheep**, **pigs**, **rabbits**, **guinea pigs**, and **dogs** from its ingestion. Found in hilly Mediterranean areas. Contains a ***coumarin*** anticoagulant that has caused a large number of fatalities in **sheep** and **swine** in Spain and Algeria, and a severe hemorrhagic diathesis in grazing **ewes** and their **lambs** *in utero* in Israel.

FESCUE, TALL

= *Festuca arundinacea*

Untoward effects: Commonly referred to as ***reed***, ***giant fescue***, ***ditch-bank fescue***, and ***Williams grass***. Its toxicity reported in the U.S., Italy, Australia, and New Zealand. Its grains or seeds are often parasitized by species of *Claviceps*, whose sclerotia are known as ***ergot*** (q.v.), and cause a poisoning commonly known as ergotism, that has caused losses of **cattle** and **sheep**. *Fusarium tricinctum* isolated from the plant produces a toxin, ***butenolide***, as well as a ***T-2*** toxin. IM or oral administration to **cattle** of these has caused tail tip necrosis and death, with massive internal hemorrhage and fatty degeneration of the liver. An endophytic fungus. *Acremonium coenophialum* on it appears to be the cause of fescue foot or lameness in **cattle**. Symptoms in **cattle** and **horses** can include fever, tachypnea, rough hair coat, sialism, and agalactia. Weak **foals** and thickened placenta reported in affected **horses**. It is a ***nitrate***-accumulating plant.

FEVERFEW

= *Chrysanthemum parthenium* = *Featherfew* = *Medieval Aspirin* = *Midsummer Daisy* = *Nose Bleed* = *Tanacetum parthenium*

Untoward effects: Mild tranquilizing and sedative effects. Its airborne pollen is contamination with ***sesquilactones*** and has caused itchy, runny nose and watery eyes. Adverse effects may develop after 1–8 weeks in ~18% leading to mouth ulcers, abdominal pain, and unpleasant taste. Nervousness, insomnia, tension, and joint stiffness and pain often occur after withdrawal. Skin sensitization and contact dermatitis reported. The seeds were inadvertently shipped to India along with ***wheat*** and the plants spread, causing severe contact dermatitis and some fatalities. Its ingestion can increase risk of bleeding in **patients** taking ***aspirin*** or ***warfarin***.

FEXOFENADINE

= *Allegra* = *MDL 16,455A* = *Telfast*

***Antihistamine*. The active metabolite of *terfenadine*, with less adverse effects. An H_1-receptor blocker.**

Untoward effects: Occasionally, dysmenorrhea, dyspepsia, drowsiness, and fatigue. QT lengthening and life-threatening arrhythmias, ventricular fibrillation, and syncope in a 67-year-old **male** after 180 mg/day/1 month.

Drug interactions: Coadministration of ***erythromycin*** and ***ketoconazole*** for 1 week may increase its peak plasma concentration. ***Grapefruit juice*** may inhibit its absorption.

FIBERGLASS

= *Fibrous Glass*

Made from sand, limestone, and soda.

Untoward effects: Allergic or irritant dermatitis. Almost all glass fibers are coated with binders, lubricants, or coupling agents. Sensitization to these resins is rare, because they are fully cured prior to **human** exposure. Fine particles or dust can readily penetrate the skin leading to milarial eruption with small erythematous follicular papules. Purpura, telangiectasis, folliculitis, urticaria, and linear erosions in skin creases (particularly in **those** with fair complexions or dry skin) reported. Penetration of glass spicules beneath nailfold may cause paronychia. Pneumoconiosis, coughing, and eye irritation are commonly reported from dust exposure. Suspect lung carcinogen in **man**. An anaphylactic attack in a 33-year-old **female** reported. **She** had an asthma attack and marked acidosis.

Lung tumors can be produced with it in **rats**.

FIBRINOLYSIN

= *Plasmin* = *Thrombolysin*

Untoward effects: Fever, hypotension, tachycardia, and shock have been reported within 3 h after injection. Nausea, vomiting, epigastric pain, chills, dizziness, and urticaria can occur. Echymoses, SC hematomas, and excessive bleeding reported. May augment hypoprothrombinemia of ***anticoagulants***. Associated with deaths in respiratory distress syndrome in **infants**. As a foreign protein, it can be immunogenic.

FICIN

***Proteolytic enzyme*. From *fig* (*Ficus*) latex.**

Use: In chill-proofing ***beer***, as a meat-tenderizer, and as a ***rennin*** substitute for coagulating ***milk*** in the ***cheese*** industry.

Untoward effects: Very irritating to eyes, nose, and skin.

LD_{50}, **rat**, **rabbit**, and **guinea pig**: 5 g/kg; 10 min LC_{LO}/m^3, **mouse**, **cat**, **rabbit**, **guinea pig**: 290 ppm.

FIG

= *Ficus aurea* = *Ficus carica*

Untoward effects: Contains psoralens and can induce phototoxic reactions, usually peaking within 2–4 h. These are manifested as a dermatitis. A cheilitis from figs also occurs. Canned figs contain ***tyramine*** and have caused hypertension in **patients** taking ***monoamine oxidase inhibitor*** drugs.

FILGRASTIM

= *Neupogen*

A recombinant *granulocyte colony-stimulating factor* for IV or SC use.

Untoward effects: Acute anaphylaxis in a 48-year-old **female** after 10 μg/kg/day leading to itching of trunk and extremities, hoarseness, tightness in chest, and a "lump" in **her** throat within 10 min of third dose. Rash and edema in a 19-year-old **female** after third daily dose of 300 μg SC. Mild to moderately severe medullary bone pain reported in 24% of **patients**. The following adverse reactions, in decreasing frequency, are nausea/vomiting, skeletal pain, alopecia, diarrhea, neutropenic fever, mucositis, fever, fatigue, anorexia, dyspnea, headache, cough, skin rash, chest pain, weakness, sore throat, stomatitis, and constipation. Reversible increase in ***uric acid***, lactate dehydrogenase, and alkaline phosphatase. It induced a neurological syndrome with asthenia, blurred vision, and headache due to leukocytosis in a 34-year-old **male**.

Drug interactions: Potentiates the pulmonary toxicity of ***cyclophosphamide***.

FINASTERIDE

= *Andozac* = *Chibro-Proscar* = *Finastid* = *MK-906* = *Proscar* = *Prostide*

Use: For treatment of benign prostatic hyperplasia by inhibiting steroid 5-α-reductase, an enzyme that catalyzes conversion of ***testosterone*** to ***dihydrotestosterone***, a chemical that helps develop and maintain prostatic hypertrophy. Also used in treating idiopathic hirsutism in **women**. A side-effect helps regrowth of hair and/or prevents the loss of more hair.

Untoward effects: Gynecomastia and intraductal breast cancer reported in two treated **men**. Concentrations of serum ***prostate specific antigen*** may decrease. Occasionally reports of decreased libido, impotence, decreased ejaculate volume, and severe myopathy in **men**. **Women** who are or may become pregnant are warned not to handle crushed or broken tablets.

Drug interactions: Increases ***theophylline*** clearance by 7% and decreases its half-life by 10% after IV ***aminophylline***. Probably, no clinical significance.

Teratogenic in **monkeys**.

FIPRONIL

= *Frontline* = *MB 46,030*

Pesticide. **Flammable. For topical use.**

Untoward effects: Manufacturer states **applicators** must wear latex gloves and avoid contamination of feed and water. Harmful to **humans** and **animals** if swallowed. Expected symptoms can be lethargy, muscle tremors, and, in extreme cases, convulsions. Will irritate eyes on contact.

May be harmful to young **puppies**, and debilitated, aged, pregnant, or nursing **animals**. Can induce thyroid tumors in **rats**. Toxic in **rabbits**.

LD_{50}, **rat**: 100 mg/kg.

FIRE COLORING SALTS

Untoward effects: These are potentially dangerous, especially for **children**. ***Copper sulfate***, for green flames; ***calcium chloride***, for orange; ***copper chloride***, for blue; ***lithium chloride***, for rich reds; and ***potassium chloride***, for purple.

FIRE CORAL

= *Millipora complanata*

Not a true coral, but a coelenterate hydroid, living in shallow, sunlit waters of coral reefs in Florida, the Caribbean, Hawaii, and tropical Pacific waters.

Untoward effects: When **bathers** brush against the branches of this dull reddish or mustard-colored, finger-like tree, **they** receive severe, painful burning sensations, numbness; and (within 1–10 h) non-linear, erythematous, papular, patchy lesions. Its nematocysts are in the mucus it secretes. The lesions still remain for 3 days after treatment. Rarely, sytemic effects occur leading to anaphylaxis, shock, nausea, vomiting, diarrhea, headache, muscle spasms, chest or abdominal pain, and chills.

FIRECRACKERS

Untoward effects: ***Jumping Jack*** and ***Totes***, etc., after ingestion, have caused large numbers of hospitalizations, coma, hepatic encephalopathy, and fatalities from ***yellow phosphorus*** (q.v.) content. Numerous injuries, including burns and eye injuries, reported.

FIRE EXTINGUISHERS

Untoward effects: Some in use contain toxic chemicals (***carbon dioxide***, ***carbon tetrachloride***, ***chromates***, and ***methyl bromide***).

FISH

Untoward effects: There are > 700 poisonous and venomous species and over 220 of these are venomous. Poisonous vertebrate fish causing **human** illness are generally classified according to the location of their toxins. *Ichthyosarcotoxic* ones have the toxin in their musculature, viscera, skin, or mucus, and are the most common offenders. *Ichthyotoxic* fish contain the toxin in their gonads. *Ichthyohemotoxic* fish contain the toxin in their blood and are rarely a cause of **human** poisonings. ***Ciguatera*** and **scombroid** fish poisonings account for ~99% of vertebrate fish-poisoning outbreaks.

Ciguatera fish poisoning is due to ***ciguatoxin*** (lipid-soluble and relatively heat-stable) found in > 400 fish species that are, primarily, bottom-dwelling shore fish caught near reefs. The toxin is probably produced by a marine dinoflagellate, *Gambierdiscus toxicus*, growing with red algae. These are then eaten by herbivorous fish, who pass it up the food chain. Then, it is concentrated in large, predacious fish, such as **barracuda**, **red snapper**, **amberjack**, **parrot fish**, and **grouper**. The liver is the most toxic part, followed by the intestines, testes or ovaries, and muscle. After ingestion, symptoms usually develop within a few minutes to 48 h (usually, 1–7 h) and generally start with abdominal cramps, nausea, vomiting, and watery diarrhea. Some **people** also develop numbness and paresthesia of the lips, tongue, and throat. Paresthesia of the extremities may follow. Malaise, metallic taste, dry mouth, myalgia, arthralgia, itching, blurred vision, photophobia, transient blindness, sensations of looseness or pain in the teeth; and sharp, shooting pains in the extremities often reported. In very severe cases, early symptoms include sinus bradycardia, hypotension, cranial nerve palsies, alterations in hot and cold sensations, and respiratory paralysis. Fatalities occur. Duration of illness is usually a few days, but symptoms can, occasionally, persist for a few months. The toxin is not destroyed by cooking. Attacks often occur after **travelers** return home from areas where these fish abound. *Gymnothorax* poisoning from tropical reef **eels** (**moray eel**, **mullet moray**) is now classified as ***ciguatera*** poisoning.

Scombroid fish poisoning is due to action by ***histamine*** and ***saurine***, produced by bacteria on the flesh of certain

Scombroideae fish, such as **bonito**, **butterfly kingfish**, **mackerel**, **saury**, **seerfish**, **skipjack**, and **tuna**, and non-scombroid fish, such as **Pacific dolphins** (***mahi-mahi***) and **bluefish** (caught off the coast of Connecticut). Optimal ***histamine*** produced by enzymatic action on ***histidine*** occurs at 20–30°C (68–86°F). Symptoms are, essentially, those of a ***histamine*** reaction and often occur within minutes. May occur several hours after ingesting the fish. Usual symptoms are itching, urticaria, an acute and stinging erythema, primarily of the face (flushing), neck, and upper part of the trunk, resembling a sunburn, dizziness, abdominal cramps, nausea, vomiting, diarrhea, and burning of the mouth and throat. Most symptoms abate within hours. Bronchospasm, respiratory distress, and hypotension, in severe cases. Fatalities are extremely rare. The toxin is heat-stable. A common illness in large metropolitan areas, from serving fish improperly refrigerated or gutted. **People** often complain afterwards that the fish **they** ate had a hot, spicy, or "Cajun" taste. Although these fish are well-distributed throughout temperate and tropical marine waters, some are occasionally found in polar waters.

Puffer poisoning (***fugu***, ***tetradon*** poisoning, tetradotoxication) from **people** ingesting the viscera, skin, or gonads of **puffers** (***toado***, ***toadfish***, ***globefish***, ***swellfish***), **porcupine fishes**, and **ocean sunfishes**. Symptoms include ataxia, orofacial paresthesias, dizziness, pallor, numbness and tingling of extremities with central spreading, variable gastrointestinal upsets, skin desquamation, and shock. In Japan, **puffers** are considered a delicacy and have caused fatalities from improper preparation. Japan now only permits special, approved restaurants with licensed ***fugu*** **cooks** to prepare the fish. ***Tetrodotoxin*** is heat-stable. Cases are extremely rare in the U.S. **Captain Cook** was twice poisoned by eating toxic Pacific fishes – once, by **puffers** and once, by a ciguatoxic fish.

Venomous fish include **stingrays**, **scorpion fishes** (***lionfishes***, ***hogfishes***, ***turkeyfishes***, ***stonefishes***, ***zebrafishes***), **weevers**, **stargazers**, **ratfishes**, **catfishes**, **surgeon fishes**, and some **sharks**. The venoms are not very stable and usually used as defense mechanisms against other fishes. The more hazardous ones for **man** frequent tropical waters. Stings usually occur when **people** thrust **their** hands into dark underwater crevices, inadvertently stepping on a fish partially buried in the sand, or **divers** attempting to catch some of the attractive ones. Some **people** are stung from handling some of these fish in home salt-water aquariums. Pain is instantaneous, intense, throbbing, and often radiating to other body areas. Pain from the hand can go to shoulder, neck, and head; pain from foot stings radiate into the abdomen. Wound areas become red, swollen, hot, and cyanotic. Numbness, necrosis, tissue sloughing, and gangrene may occur later. Secondary infections and tetanus are possible from the puncture wounds. Nausea, vomiting, shock, delirium, convulsions, and cardiac failure can occur. The ***stonefish's*** sting is most painful and some **victims** thrash about, scream in agony, and even lose consciousness. After the pain subsides (hours to days after the sting) numbness and paralysis of the entire limb has occurred. Watery blisters develop on the affected limb. Very severe systemic effects with hypotension, peripheral vasodilation; skeletal, cardiac, involuntary and diaphramatic muscle weakness can progress to paralysis, and even death from hypotensive shock and cardiac and/or respiratory failure. The venom of **scorpion fishes** is heat-labile and accounts for the usefulness of immediately putting the wound in the hottest water tolerable for 30–90 min.

Methylmercury toxicity in **swordfish** and **tuna** became a serious public health problem, especially in **pregnant women** on a high fish diet, as the **fetus** concentrates ***mercury*** up to 30 times the **mother's** level. In Japan, ***methylmercury*** was responsible for central nervous system damage and deaths (Minamata disease) in **people** after consuming fish that had fed on waste by-products from a plastic factory. Some have contained ***arsenic, DDT, dioxins, furans, iodides, lead, polychlorinated biphenyls***, and ***selenium*** in more than normal levels. ***Selenium*** has blunted the ill-effects of ***mercury*** in **swordfish**. Diphyllobothriasis (fish tapeworm disease) from eating raw fish, such as Japanese ***sashimi*** and ***sushi***, Latin American ***ceviche***, and Dutch green **herring**. The tasting of **gefilte fish** for flavor, while chopping **pickerel** and **northern pike**, has also been a cause of *Diphyllobothrium latum* infestation. The parasite is destroyed by cooking and lengthy freezing. Anisakiasis, a roundworm infestation, is acquired from eating raw marine fish (usually, **herring**) or uncooked seafood. It causes gastrointestinal upsets and a tingling throat syndrome (from worms wiggling in the mouth after migrating up the esophagus); also killed by cooking and adequate freezing. Within a matter of days in 1982, **physicians** in operating rooms in a Baltimore, Maryland hospital saw **patients** with bright red worms (*Eustrongylides*) wriggling through the intestinal wall into the abdominal cavity. The worm lives in fish-eating **birds** and the eggs are passed in bird droppings. The droppings with the eggs are ingested by the fish who will be eaten by other **birds**. The **patients** were **fishermen** who engaged in a not-too-uncommon pastime of swallowing live bait **minnows**, when the fish weren't biting. The worms grow to nearly 30 cm in length. Slowly decaying fish in ships' holds deplete the ***oxygen*** and lead to toxic levels of ***hydrogen sulfide*** (q.v.) and ***ammonia*** (q.v.) that have caused many incidents of asphyxia and death in **fishermen**. Excessive ingestion of **anchovies** can induce or aggravate gout attacks.

Hypersensitivity to fish has been reported. Contact dermatitis in kitchen **workers**, including urticaria; on the hands it may be vesicular; angioedema, laryngeal edema, bronchospasm, vomiting, diarrhea, shock, and even collapse reported. Severe angioedema of the larynx in a 5-year-old **female** after eating two bites of fish. **She** had unknowingly been sensitized after taking ***cod liver oil*** capsules. Inhaling the odor of cooking fish has caused a hypersensitivity reaction. Many clinical reports of interaction with ***monoamine oxidase inhibitors*** leading to palpitations, skin and facial flushing, conjunctival irritation, headache, nausea, vomiting, and itching. Less common were giddiness, urticaria, diarrhea, sweating, wheezing, and nasal stuffiness. Most of these reports were with the consumption of ***scombroid*** fishes and smoked or pickled fish (mostly **herring** – ***tyramine*** content of 3 mg/g; Russian ***caviar*** contains ~0.7 mg/g). ***Histamine*** poisoning has occurred in **patients** stabilized on ***isoniazid*** (a potent ***histaminase***-inhibitor) after ingestion of **tuna** fish (rich in ***histamine***). These **patients** are also at risk from eating **skipjack** or **sardinella**.

Haff disease in U.S. due to an unidentified toxin in **buffalo fish** led to a paroxysm of rhabdomyolysis, muscle tenderness, rigidity, and dark brown urine. Uncommon fish-borne illnesses include *Clostridium botulinum* poisoning from the home preparation of smoked fish and from eating **kapchunka**, an ungutted, dried and salted **whitefish** that is not cooked before eating. The latter is distributed to New York City delicatessens. A Russian immigrant **husband** and **wife** died within a few days after consuming its ***Type E toxin***. Acute hepatitis, renal failure, abdominal pain, nausea, vomiting, and diarrhea after ingesting raw **carp** gallbladders has been common in Asia and in **immigrants** to the U.S. Long-time consumption of salted, particularly rotted fish, common to the **Chinese** diet, is the apparent cause of nasopharyngeal carcinoma, a leading cancer in China. ***Clupeotoxin*** from eating a marinated raw fish, **tambon** (*Sardinella* sp.). After nausea, vomiting, stomach pain, headache, and weakness, ~10% died in Phillipine outbreaks. Fish-roe poisoning has been reported from ingestion in Europe; poisoning from ingestion of blood from a number of European **eels**; hallucinatory fish poisoning in Norfolk Island, Hawaii, Mauritins, and South Africa from consumption of **rock cods**, **grey mullets**, **drummers**, **rudderfish**, **damselfish** (***sergeant-majors***), **surgeonfishes** (***unicorn fishes***), and **goatfishes**. Symptoms also include dizziness, ataxia, depression, and nightmares within minutes to 2 h after ingestion. Heat-stable toxin. Erysipeloid and *Leptospira* infections in **fishermen**, **fish-handlers** and **-cutters**, from punctures by fish spines, etc. A 55-year-old **male** diagnosed with *Mycobacterium marinum* infection, following trauma through fish-handling. A 39-year-old **male rice-grower** in Cambodia was stung in a foot. 21 months later *Pyrenochaeta romeroi* was isolated from a tumor that developed at the site. Numerous pathogenic bacteria isolated from aquaria. Infectious **human** viruses have been found in the guts of fish caught near sewer outlets into rivers. ***Nitrates*** and ***sodium nitrite***-forming ***nitrosamines*** and fungi on smoked or salted fish can cause digestive disorders, and, possibly, cancers. Use of fish flour can, possibly, cause mottling of teeth, due to its ***fluoride*** content. Fish, as do leafy green vegetables, contain ***vitamin K*** or its analogs, and can antagonize ***anti-coagulants***. **Children** rarely outgrow **their** allergy to fish.

Steatitis in **cats** fed canned red **tuna**. Beriberi reported in a **queen** and her **kittens** feeding exclusively on fresh fish. As little as 10% raw fish in the diet can cause acute **mammalian** ***thiamine*** deficiency. **Atlantic herring**, **whiting**, **Pacific mackerel**, **clams**, and ~50% of all fresh water fish have heat-sensitive anti-***thiamine*** activity. ***Ciguatera*** poisoning reported in Brisbane, Australia in **cats** after eating some of the same batch of **narrow-barred Spanish mackerel** that caused poisoning in 33 **people**. Partial limb paralysis, inappetance, and sialism occurred, lasting up to 2 weeks. Numerous elements, ***DDE***, and ***polychlorinated biphenyls***, as well as ***mercury***, have been found in **cat** foods. Central nervous system damage observed in **kittens** fed **tuna** containing 0.5 ppm ***mercury***. Raw fish fed to **dogs** has helped cause a ***thiamine*** deficiency, and, occasionally, tapeworms. Allergic reactions reported in some **dogs**. "Cotton-fur" in fish-fed **mink**, due to ***iron*** deficiency. Raw fish also causes ***thiamine*** deficiency (Chastek paralysis) in **mink** **Poultry** develop a paralysis, decreased egg production, diarrhea, cyanosis, thirst, hepato-nephrosis; and degeneration of spleen, ovaries, and testicles, after being fed a fish paste prepared from **anchovies**.

FISH OILS

> 700,000 tons produced annually by Japan, Peru, and the U.S.

Untoward effects: Despite various beneficial claims touting it as a magical elixir, with little scientific basis, its use as a source of highly unsaturated ***omega-3-fatty acids*** in capsules for **man** has reached tremendous volume. There are cases where **people** consuming it have suffered a precipitous decrease in platelets. Because some **people** consuming it have platelets less likely to stick or aggregate together, **they** have an increased tendency toward skin bruises. Care should be used with it in **patients** on ***anti-coagulants***. Fish oil covering the ***duraluminum*** in the air-craft industry causes an occupational dermatitis (***dural*** poisoning), apparently due to high levels of bacterial growth on it. Impure fish oil on ***jute*** threads has caused skin irritation. It has decreased the immune system's response to simulated infection. Fish oil is more vulnerable to oxidation than most other fats.

Feeding it has caused steatitis in **mink**, muscle degeneration and liver necrosis in **pigs**, muscular dystrophy in the **lambs** of **ewes** fed it, feather-picking in **ducks**; enlarged hocks in **turkey poults**; gizzard erosion in **broilers**; and liver dystrophy and false positive ***tuberculin*** reactions in **chickens**.

FLAIRA

= *African Coffee* = *Akinoboglo (Igbo)* = *Bellyache Bush* = *Botuje Pupa (Yoruba)* = *Jatropha gossypifolia* = *Tua Tua*

Untoward effects: Contains an irritant toxalbumin, ***curcin***. Leaves, roots, and stems are poisonous. The seeds are purgative, and poisoning by curious **children** ingesting them is common in Nigeria leading to burning sensations in the mouth and throat, nausea, abdominal pain, vomiting, diarrhea, and collapse.

FLAVOXATE HCL

= *Bladderon* = *DW-61* = *Genurin* = *Patricin* = *Rec-7-0040* = *Spasuret* = *Urispas*

Antispasmodic, antimuscarinic.

Untoward effects: Lowered esophageal sphincter tone, urinary retention by decreased bladder contractility, blurred vision, tachycardia, constipation, nausea, vomiting, dry mouth, nervousness, vertigo, confusion, delirium, decreased sweating, increased intraocular tension, abdominal pain, urticaria, fever, eosinophilia, one case of reversible leukopenia, and headache. Use in a 65-year-old **female** with juvenile rheumatoid arthritis, treated for cystitis, caused high fever and the drug was immediately withdrawn. Retreatment in about 6 months led to high spiking fever and death from acute respiratory distress, apparently due to a hypersensitivity reaction.

In the **dog**, nausea, vomiting, tachycardia, eosinophilia, and increased intraocular pressure.

LD_{50}, **rat**: 2.8 g/kg; IV, LD_{50}, **rat**: 27.4 mg/kg.

FLAX

= *Linum usitatissimum*

Flaxseed is known as *linseed*.

Untoward effects: Byssinosis with acute effects precipitated by inhalation of flax dust during the processing of its fibers. Allergic reactions, chest tightness, and coughing were commonly reported symptoms. The plant contains ***linamarin***, a cyanogenic glycoside. Argentinian reports of **children** poisoned by it after ingestion of ***teas***, or using cataplasms or clysters containing it.

Dry, immature flaxseed has caused ***cyanide*** poisoning in **swine**. ***Linseed cake***, green flax in the fall, and chaff, when eaten, can yield ***prussic acid***; poisoned **cattle**, **swine**, and **sheep** in North Dakota. Goitrogenic when fed to **lambs**. Hyperesthesia and terminal vomiting noted in **sheep**. The plant is a ***nitrate***-accumulator. Other species have caused poisoning in Europe. ***Cucurbitacins*** in *Linum* sp. may also account for some of the plant's toxicity.

FLECAINIDE

= *Almarytm* = *Apocard* = *Ecrinal* = *R-818* = *Tambocor*

***Antiarrhythmic.* Class 1C.**

Untoward effects: Blurred vision, corneal deposits, and dizziness are the commonest, followed by nausea, headache, fatigue, tremors, nervousness, paresthesias, dermatologic reactions, interstitial pneumonitis, immune-mediated granulocytopenia, sensations of warmth, hypotension, bradycardia, heart block, ventricular arrhythmias, and tachycardia. Proarrhythmia reported in 12% with Holter monitoring and 9% with electrophysiologic testing; congestive heart failure in ~5%; sinus node dysfunction in 1.2%; second- and third-degree atrioventricular-block in < 1%. 2–3 times greater risk of cardiac arrest or death, compared to placebo, after average treatment of 10 months. FDA suggests use only for immediate life-threatening arrhythmias. Therapeutic serum levels reported as 0.2 to a maximum of 0.8 mg/l; toxic at 1–2 mg/l; and life-threatening or fatal at 2.6–13 mg/l.

Drug interactions: Reduce dosage about ⅓ when given with ***amiodarone***. Use may cause slight increase in serum ***digoxin*** levels. Additive negative inotropic and chronotropic effects with ***verapamil*** may help cause asystole and cardiogenic shock. Levels of both may increase when given with ***propranolol***. Increase in plasma levels and urinary excretion can be caused by ***cimetidine*** maintenance therapy. ***Ritonavir***, as does the combined ***lopinavir/ritonavir***, decrease its metabolism and increase its toxicity.

Simultaneous administration with ***talipexole*** increases its absorption and decrease its metabolism in **rats**.

FLEROXACIN

= *AM-833* = *Megalocin* = *Megalone* = *Quinodis* = *Ro-23-6240*

***Antibacterial.* Long-acting *fluoroquinolone*.**

Untoward effects: High rate of adverse effects. Breast milk levels in **women** were 62% of plasma's. Since ***quinolones*** can cause arthropathy in young **animals**, it should not be administered to breast-feeding **women**. Inhibits ***theophylline*** metabolism.

May induce convulsions in **mice**.

FLOCTAFENINE

= *Diralgan* = *Idalon* = *Idarac* = *Novodolan* = *R-4318* = *Ru-15750*

Analgesic.

Untoward effects: Occasional drowsiness and dizziness. Coma induced in a 53-year-old **alcoholic** with cirrhosis. Avoid use in **patients** with severely reduced liver function. Nephrotoxic (acute interstitial nephritis and tubular cell toxicity). Anaphylaxis reported.

LD_{50}, **rats**: 960 mg/kg; **mice**: 2.25 g/kg; **rabbit**: 700 mg/kg.

FLORFENICOL

= *Aquafen* = *Florphenicol* = *Nuflor* = *Sch 25,298*

***Antibiotic.* For IM use only.**

Untoward effects: Commercial formula can cause skin and eye irritation in **man**.

In **cattle**, anorexia, decreased water intake, and transient diarrhea. IM injections have caused local tissue reactions and necrosis (with large amounts at one site). Seminiferous tubular degeneration and atrophy after oral treatment are dose- and frequency-dependent in **dogs** and **rats**. Liver hypertrophy, central nervous system vacuolation and decreased hematopoiesis occur in **dogs**.

FLOSEQUINAN

= *BTS-49465* = *Flosequinon* = *Manoplax*

Antihypertensive, vasodilator.

Untoward effects: Headache, arrhythmias, hypotension, asthenia, dizziness, palpitations, decrease in serum ***potassium***, syncope, nausea, vomiting, increased aminotransferases, taste disturbances, and tachycardia. **Patients** taking the recommended dose of 100 mg/day had a mortality rate 1.51 times higher than **those** on placebo, causing its marketing withdrawal (still available for special cases). **Patients** with liver or renal failure require decreased dosage. Can cause renal failure.

Drug interactions: ***Cimetidine*** increases its plasma levels. Flosequinan can increase the effects of ***anticoagulants***, such as ***warfarin***.

FLOUR

Untoward effects: For ~30 years, **millers** used ***nitrogen trichloride*** (q.v.) to induce its aging, until it was recognized as a cause of "running fits" in **dogs**. ***Chlorine dioxde*** was then used as a substitute and as a bleaching agent. Many other additives, such ***acetone peroxide***, ***ammonium persulfate***, ***azodicarbonamide, benzoyl peroxide***, and ***hydrogen peroxide*** followed. The addition of ***iron*** proved a possible threat to **patients** with hemochromatosis, liver cirrhosis, and ***iron*** storage diseases. Excessive amounts of ***potassium bromate*** (q.v.) have also caused poisoning. Toxic chemical contamination in trucks and ships' holds has also caused toxicity. Renal ***calcium*** stone formation from excessive consumption of ***calcium***-enriched white ***bread***. Allergic respiratory response (rhinitis and asthma) from exposure to the flour dust. After 1 year, 8% of 880 **apprentices** developed skin reactions to ***rye*** and ***wheat*** flour. Allergic contact urticaria. Dermatitis increased in **bakers** followed the increased use of chemicals with flour. ***B-complex*** addition to flour caused a dermatitis in a 50-year-old **female** who previously reacted to oral intake of ***B-complex vitamin*** capsules. Some **people** have a sensitivity (sprue) to the ***gluten*** in flour. Toxicity reported from contamination by ***Lolium temulentum*** (q.v.) flour, known since ancient times ("tares" of the Bible), and ***Saponaria*** (q.v.) saponins. Cases of ***lead*** poisoning reported after ingestion of ***lead***-contaminated flour (from pieces of ***lead*** castings that came off the stone mill used for grinding grain.

FLOURENSIA CERNUA

= *Blackbrush* = *Tarbush*

Leafy shrub found in Texas, New Mexico, Arizona, and Mexico.

Untoward effects: Poisonous to **sheep**, **goats**, **cattle**, and **rabbits** after ingestion of blossoms, buds, and immature and ripe fruit. Anorexia, listlessness, muscular twitching, abdominal pain, groaning, bruxism, nostrils clogged with mucus, and ataxia occur. On post-mortem, severe gastroenteritis with some hemorrhage in the abomasum (**ruminants**) and first part of the small intestines; marked liver and renal congestion.

FLOXURIDINE

= *2′-Deoxy-5-fluorouridine* = *5 FDU* = *FUDR* = *NSC-27640*

Antiviral, antineoplastic.

Untoward effects: Nausea, vomiting, enteritis, diarrhea, and localized erythema. Delayed toxicity includes stomatitis, oral and gastrointestinal ulceration, bone marrow depression; also anorexia, leukopenia, thrombocytopenia, alopecia, dermatitis, rash, fatal liver cirrhosis with continous hepatic artery infusion, myocardial ischemia, and decreased colonic mucosal cell mitosis and renewal. Fever, weakness, and malaise also reported. Potentially teratogenic.

Use in pregnant **rats** caused syndactly and tibial aplasia.

LD_{50}, **rat**: 215 mg/kg; **mouse**: 147 mg/kg; **rabbit**: 94 mg/kg; **dog**: 157 mg/kg.

FLUANISONE

= *Haloanisone* = *R-2028* = *Sedalande*

Antipsychotic, tranquilizer.

Untoward effects: Extrapyramidal symptoms in > 50% of **patients**. Transient drowsiness, ataxia, profuse perspiration, postural hypotension, mild hypotension, mild eosinophilia, slight increase in SGOT and alkaline phosphatase, and dyspnea occur.

May cause slight pyrexia in **animals**. Some **dog** breeds are particularly sensitive to its effects, requiring up to 50% decrease in dosage.

IP, LD_{50}, **mice**: 200 mg/kg.

FLUBENDAZOLE

= *Flubenol* = *Flumoxal* = *Flumoxane* = *Fluvermal* = *R-17889*

Anthelmintic.

Untoward effects: Occasionally diarrhea and abdominal pain. Rarely, leukopenia, agranulocytosis, and hypospermia. Contraindicated in pregnancy, based on **rat** trials with ***mebendazole***.

LD_{50}, **rats, mice, guinea pigs**: > 2.5 g/kg.

FLUCONAZOLE

= *Biozolene* = *Diflucan* = *Elazor* = *Fluconal* = *Triflucan* = *UK-49858* = *Zoltec*

***Antifungal.* A systemic triazole.**

Untoward effects: Gastrointestinal upsets ~15% (nausea, abdominal pain, vomiting, diarrhea) are common. Headache ~2%, skin rash ~2% (occasionally, allergic), alopecia, eosinophilia, fixed-drug eruption, Stevens–Johnson syndrome, toxic epitermal necrolysis, jaundice, hepatic necrosis, hypokalemia, anemia, leukopenia, neutropenia, agranulocytosis, and thrombocytopenia. Anaphylaxis, angioedema with buccal ulceration, swollen tongue, orbital and buccal swelling, and paresthesia of mouth and perineum. QT prolongation and torsades de pointes in a 59-year-old **female** after 400–800 mg/day/5 weeks IV, then 150 mg/day IP. Therapeutic serum levels are ~1–5 mg/l and toxic at 95 mg/l. Contact dermatitis, vulvar irritation, and edema can occur after topical use.

Drug interactions: Potentiates ***cyclosporine, nortriptyline, phenytoin, warfarin***, and some ***oral hypoglycemics***. ***Rifampin*** may decrease its serum concentration and ***hydrochlorthiazide*** may increase it. Causes a variable increase or decrease of ***ethinyl estradiol*** and ***levonorgestrel*** serum levels. Inhibits CYP3A4 and CYP2C9 metabolism of certain drugs including ***astemizole*** and ***terfenadine***. It increases ***theophylline*** serum concentrations. Use with ***astemizole*** and ***cisapride*** led to torsades de pointes, prolonged QT intervals, and deaths. Potentiates ***diclofenac, glyburide, ibuprofen, indinavir, losartan, piroxicam, progesterone, rifabutin, tacrolimus, testosterone***, and ***zidovudine***. Addition of 150 mg/day/3 days for a 33-year-old **male** taking ***carbamazepine*** 1200 mg/day/5 years led to stupor and doubling of **his** ***carbamazepine*** levels. Inhibits metabolism of ***cisapride***.

FLUCYTOSINE

Antifungal.

Untoward effects: Serious or life-threatening leukopenia, thrombocytopenia, eosinophilia, granulocytopenia, anemia, and pancytopenia. Hepatitis, rash (< 5%), false increase in serum ***creatinine***, gastrointestinal intolerance (6–18% with diarrhea, anorexia, nausea, vomiting, diffuse abdominal pain), crystalluria, and photosensitivity. Dry mouth, gastrointestinal hemorrhage, jaundice, raised ***bilirubin***, chest pain, dyspnea, and cardiac and respiratory arrest reported. Rarely, confusion, hallucinations, psychosis, sedation, headache, ataxia, vertigo, hearing loss, paresthesia, parkinsonism, peripheral neuropathy, and pyrexia. Its bone marrow suppression is potentially lethal in azotemic **patients**. Serum concentrations of 25–50 mg/l are therapeutic and 100 mg/l is considered toxic.

Resistance to it develops rapidly in **animals**. Teratogenic in **rats**. Hematologic toxicity may be due to some of its conversion to ***fluorouracil***. Nephrotoxic.

LD_{50}, **rats**: 8 g/kg; **mice**: > 2 g/kg; IV, LD_{50}, **rats**: 100 mg/kg; **mice**: 500 mg/kg.

FLUDARABINE PHOSPHATE

= *2-F-ara-AMP* = *Fludara* = *NSC-312887* = *NSC-328002*

Antineoplastic.

Untoward effects: Myelosuppression (with neutropenia, thrombocytopenia, and anemia), fever, infection, edema, tumor lysis syndrome, nausea, vomiting, stomatitis, gastrointestinal bleeding, rash, weakness, anorexia, paresthesias, peripheral neuropathy, encephalopathy, visual disturbances, agitation, confusion, hemorrhagic cystitis, hypersensitivity pneumonitis, and cytopenia.

FLUDROCORTISONE

= *Alflorone* = *F-Cortef* = *Florinef*

Mineralcorticoid.

Untoward effects: A 49-year-old **female** developed steroid diabetes and progressive muscular paralysis after large doses. Improvement followed drug withdrawal. Has induced glaucoma, exophthalmus, and corneal opacity. Arterial hypertension, even with encephalopathy, in **children** with congenital adrenal hypoplasia. Hypokalemia, cardiomegaly, metabolic alkalosis, reversible myopathy, ***sodium*** and water retention, congestive heart failure, osteoporosis, aseptic necrosis of femoral and humeral heads, spontaneous fractures, delayed wound-healing; thin, fragile skin; facial erythema, petechiae and echymoses, increased sweating, SC fat atrophy, skin and nail hyperpigmentation, hirsutism, acneiform eruptions, urticaria, skin rashes, vertigo, headache, convulsions, decreased growth rate, hyperglycemia, necrotizing angiitis, thrombophlebitis, and anaphylactoid reactions.

FLUFENAMIC ACID

= *Achless* = *Ansatin* = *Arlef* = *CI-440* = *Fullsafe* = *INF-1837* = *Meralen* = *Paraflu* = *Parlef* = *Ristogen* = *Sastridex* = *Surika* = *Tecramine*

Analgesic, anti-inflammatory.

Untoward effects: Indigestion, upset stomach, heartburn, nausea, diarrhea, headache, drowsiness, dizziness, and bronchoconstriction. Less common are depression, acute porphyrias, skin rash, tinnitus, raised serum transaminases, abdominal pain, and, rarely, hemolytic anemia, and leukopenia. Can retard wound-healing. Excreted in **human** breast milk.

LD_{50}, **rat**: 272 mg/kg; **mouse**: 2 g/kg; IV, **rat**: 98 mg/kg; **mouse**: 158 mg/kg.

FLUMAZENIL

= *Anexate* = *Flumazepil* = *Lanexate* = *Mazicon* = *Ro-15-1788* = *Romazicon*

Benzodiazepine antagonist.

Untoward effects: Ventricular arrhythmias, fibrillation and tachycardia, cardiac arrest, and death reported. These followed use as an antidote to other drugs taken with benzodiazepines. Some **patients** treated after benzodiazepine overdosage suddenly become alert and are predisposed to a ***catecholamine*** rush leading to hypertension and, rarely, cardiac arrhythmias, such as supraventricular tachycardia and premature ventricular contraction. Avoid its use in **patients** with a known history of seizures or suspected of ingesting ***tricyclic antidepressants***. Use with the latter has resulted in fatalities. Some **patients** report pain at the injection site, mild agitation, dizziness, headache, nausea, vomiting, abnormal vision, transient hearing impairment, tinnitus, confusion, convulsions, shivering, hiccups, and dysphonia. Usual therapeutic serum levels are 0.02–0.1 mg/l and toxic at > 5 mg/l.

LD_{50}, **rats**: 6 g/kg; **mice**: 4.3 g/kg.

FLUMETHASONE

= *Anaprime* = *Aniprime* = *Asistar* = *Cortexilar* = *Flucort* = *Flumetasone* = *6α-Fluorodexamethasone* = *Fluvert* = *Locacorten* = *Locorten* = *Lorinden* = *Losalen* = *Methagon*

Glucocorticoid.

Untoward effects: Prolonged use of the pivalate form over large body areas or under occlusion has caused adrenocortical suppression. Topical use can cause burning sensations, itching, dryness, folliculitis, hypertrichosis, acneiform eruptions, and hypopigmentation. Under occlusive dressings, secondary infection, skin atrophy, striae, milaria, and maceration of skin also reported.

Do NOT treat over 5–10% of a pregnant **dog's** skin surface. Parenteral use during the last trimester of pregnancy in **cattle** and, possibly, other species, has caused premature parturition and retained placenta.

FLUNARIZINE HCL

= *Dinaplex* = *Flugeral* = *Flunagen* = *Flunarl* = *Fluxarten* = *Gradient* = *Issium* = *Mondus* = *R 14,950* = *Sibelium*

***Vasodilator, calcium channel-blocker.* A *piperazine* derivative.**

Untoward effects: Depression, parkinsonism, akathisia, tardive dyskinesia, tremors, and bradykinesia in **geriatric patients**. Therapeutic serum levels are 0.025–0.2 mg/l and toxic at 0.3 mg/l.

Induces catalepsy in **mice**.

FLUNISOLIDE

= *Aerobid* = *Bronalide* = *Lunis* = *Nasalide* = *Rhinalar* = *RS-3999* = *Synaclyn* = *Syntaris*

***Glucocoticoid, anti-asthmatic.* For oral inhalation only.**

Untoward effects: Diarrhea (10%), nausea and/or vomiting (25%), upset stomach (10%), flu-like (10%), cold symptoms (15%), nasal congestion (15%), upper respiratory infection (25%), sore throat (20%), unpleasant taste (10%), and headache (25%). An incidence of 3–9% for palpitations, heartburn, abdominal pain, chest pain, decreased appetite, edema, fever, *Candida* infection, irritability, nervousness, shakiness, dizziness, menstrual upsets, chest congestion, cough, hoarseness, rhinitis, sinusitis, sneezing, wheezing, eczema, pruritus, loss of smell or taste, and ear infection; 1–3% for chills, increased appetite, weight increase, malaise, sweating, weakness, hypertension, tachycardia, dyspepsia, flatus, capillary fragility, enlarged lymph nodes, glossitis, dry throat and irritation, pharyngitis, anxiety, depression, fatigue, insomnia, numbness, vertigo, bronchitis, dyspnea, epistaxis, laryngitis, pleurisy, acne, hives, urticaria, increased intraocular pressure, blurred vision, and earache. May inhibit wound-healing.

FLUNITRAZEPAM

= *Narcozep* = *Ro-5-4200* = *Rohypnol* = *Roipnol* = *"Roofies"*

***Hypnotic.* A benzodiazepine. A common date rape drug.**

Untoward effects*:** Postoperative respiratory depression, apnea, "hangover", temporary amnesia, and drowsiness. Rebound insomnia reported. Transient erythema and occasional nausea and vomiting. Found in breast milk at levels apparently too low to be of clinical significance. Placental transfer occurs. Unlike ***diazepam, is virtually free of post-injection thromboses. Snorting has an abuse potential, due to rapid absorption. Poisonings reported in France. Therapeutic serum concentrations reported as 0.005–0.015 mg/l, and toxic concentrations of 0.03 mg/l or more.

Use in pregnant **rats** led to high level of neonatal loss at birth and during the next couple of days.

LD_{50}, **rat**: 485 mg/kg.

FLUNIXIN MEGLUMINE

= *Meflosyl* = *Sch-14714*

***Anti-inflammatory, analgesic, antipyretic.* Inhibits *prostaglandin* synthesis. Used IM, IV, and SC.**

***Untoward effects*:** After IM use in **horses**, caused localized swellings, sweating, and stiffness. Anaphylactoid reactions, including fatalities, after IV. Significantly reduced tensile strength of sutured wounds in **animals**. Interferes with platelet cohesiveness and may aggravate bleeding. May cause renal damage.

FLUOCINOLONE ACETONIDE

= *Coriphate* = *Cortiplastol* = *Dermalar* = *Fluonid* = *Fluovitef* = *Fluvean* = *Fluzon* = *Jellin* = *Localyn* = *Synalar* = *Synamol* = *Synandone* = *Synemol* = *Synotic* = *Synsac*

***Topical adrenocorticoid.* A very potent fluorinated corticosteroid for topical use.**

Untoward effects*:** See ***Flumethasone. Necrosis of leg tissues, probably due to its vasoconstrictive properties. Atrophy of distal phalanx of index finger in a 14-year-old **male** after 1 month of treatment including occlusive dressings. Diabetogenic action. Folliculitis and other skin inflammation can also be due to the vehicle. Has induced rosacea and transient increase in intraocular pressure. Strong inhibitory effect on wound-healing.

FLUOCINONIDE

= *Bicosal* = *Dermaplus* = *Ledemol* = *Lidex* = *Metosyn* = *Straderm* = *Topsym* = *Topsymin* = *Topsyn* = *Topsyne*

***Topical glucocorticoid, anti-inflammatory.* A very potent fluorinated corticosteroid for topical use.**

***Untoward effects*:** Burning, itching, irritation, dryness, folliculitis, hypertrichosis, acneiform folliculitis, hypopigmentation, perioral and contact dermatitis, skin maceration, skin atrophy, striae, milaria, rosacea, and secondary infection.

Daily topical use in **dogs** has caused marked and rapid pituitary–adrenocortical suppression.

FLUOMETURON

= *Ciba 2059* = *Cotoran* = *Cottonex* = *Lanex*

***Herbicide.* A phenylurea.**

***Untoward effects*:** Causes irritation. Harmful if swallowed, inhaled, or absorbed through the skin.

Ataxia, wing-drop, fluffed feathers, hyperexcitability, phonation, and falling within 15 min and lasting up to a week in poisoned **mallard ducks**.

LD_{50}, **rats**: 89 mg/kg; **mice**: 900 mg/kg; **mallard ducks**: > 2 g/kg.

2,7-FLUORENEDIAMINE

Laboratory reagent for detection of *cadmium, cobalt, copper, zinc, bromides, chlorides, nitrates,* and *persulfates.*

***Untoward effects*:** Causes lobular carcinoma of the mammary glands of **rats**. Single dose of 100 mg/kg led to 20% mortality in **rats** within 30 days.

N-2-FLUORENYLACETAMIDE

= *2-AAF* = *2-Acetylaminofluorene* = *2-FAA*

***Untoward effects*:** By inhalation and dermal contact. May decrease function of liver, kidney, and pancreas. It is estimated that < 1000 **people** in 200 laboratories are potentially at risk.

It is a carcinogen in every species that forms its *N*-hydroxy derivative, a more potent carcinogen than the parent compound. **Man**, **monkeys**, **hamsters**, and **dogs** form this metabolite. In **rats** it causes carcinomas of bladder, kidneys, liver, pancreas, salivary glands, eyes, ears, sebaceous glands, and face; testicular mesotheliomas, mammary tumors, and Zymbal gland tumors. In **mice** it causes carcinomas of liver and bladder, and bladder cancer in the **dog**. Given to **mice** on the 8–11th day of gestation leads to teratogenicity, and decreased growth or lethal to the fetus; to **rats** on the 8–12th day of gestation leads to fetal hydrocephalus. The **guinea pig** and **steppe lemming** are very resistant to its carcinogenicity. Oral ***chloramphenicol*** to the **rat** overcomes its toxicity and carcinogenicity. Colon cancer in **rodents**.

FLUORESCEIN

= *D&C Yellow #7*

The sodium form = Ak-Fluor = D&C Yellow #8 = Fluorescite = Ful-Glo = Fundescein = Irescein = Uranine

Untoward effects: Photosensitivity reactions, yellow pigmentation of nail plate, colors urine yellow-orange, nausea, coughing; and, during angiography, cardiac arrest. The latter may, in part, be due to an anaphylactoid reaction leading to nausea, pruritus, dyspnea, shock, and death. Panophthalmitis and eye infections have occurred from *Pseudomonas aeruginosa* contaminated solutions. Diffuse yellowing of the skin and sclera lasting 36–48 h. Longer discoloration has marked the insidious onset of hepatitis. Chemical meningitis with status epilepticus in a 69-year-old **male** after intrathecal injection.

After fluorescein angiography in a 35-year-old **male** led to psoriasiform drug eruption.

FLUORIDES and FLUORINE

Untoward effects: Fluorine gas is pale yellow or greenish, with a pungent odor causing eye, nose, and respiratory irritation; nausea, vomiting, bronchospasms, pulmonary edema, and skin and eye burns in **man**. Universally distributed in the earth's crust, primarily in ***fluorite*** and ***cryolite*** minerals. Found in most plant leaves at 2–20 ppm and in *Camellia* and ***teas*** at ~2000–8000 ppm. Many plants contain toxic fluoroacetates. Water can be an important source and 2 mg/l is considered the upper safe limit for **livestock** and **poultry**. So-called safety standards are not necessarily valid. Fluoride emissions from an ***aluminum*** plant on the south bank of the St. Lawrence River reportedly emitted 34 kg or 0.812 metric tons or more for many years. This rained down on Canada's Cornwall Isle. **Cows** became sick and had stunted growth and dental fluorosis, interfering with drinking and mastication; some died. The **bees** and most **partridges** left, ***pine*** trees started dying, and several thousand **residents** were exposed. **Some** had excessive fluoride levels in **their** blood and **some** had respiratory, endocrinological and neurological problems. ***Phosphate*** (including ***superphosphate***) manufacturers have emitted, into the air and water, significant amounts to damage vegetation, **animals**, and **human** health. Non-skeletal phase of fluoride intoxication was first described in ***cryolite* workers**. It has also occurred in **people** from errors in fluoridation of drinking water, air and edible vegetation contaminated near ***aluminum***, steel, other metal smelters, and fertilizer plants. In 1960, it was stated that, in the manufacture of ***superphosphate*** fertilizers, > 4400 tons of fluorine are discharged annually into the atmosphere of the British Isles. In November 1979, one physician described, in detail, the symptoms in 21 of his **patients** affected by fumes from an enamel smelter. **They** had respiratory symptoms after inhalation of fumes. Similar symptoms were noted in **people** in Annapolis, Maryland (after a water plant **worker's** error) in **those** who drank water with up to 36 ppm fluorine. **They** developed ulcers in **their** mouths, had subcostal pains, urticaria, and skeletal fluorosis (osteosclerosis, spondylosis, osteopetrosis). > 200 cases of bone disease in South African **children** drinking subterranean waters containing 3.6–3.8 ppm. Goiter and dental problems have also been reported. Skeletal fluorosis also reported in 20 **American Indians** (22–85 years) in the Southwest ingesting well water with 0.5–18 ppm. Serious bone disease in ten **patients** receiving hemodialysis for up to 31 months with fluorinated water. Neurological symptoms in 70 **patients** receiving naturally fluorinated water containing 1.2–11.8 ppm. A death reported in the U.S. from drinking water containing 4 ppm. In some areas of Morocco, fluorosis in **man** and **sheep** has been recognized for centuries. It is called darmous, from the Berber words Dr'Ghamas, meaning tooth abnormality. It also causes debility in **people**. "***Wine*** fluorosis" (fluorosis vinica) reported from its fraudulent addition to prevent wine fermentation. A number of **patients** regularly ingest 7.6–72.6 mg/l, and presented with deforming periostitis, osteosclerosis, or osteoperosis-osteomalacia. Dental fluorosis with mottling and white or yellow-brown flecks in the enamel after young **children** drinking water with 2 ppm; coarse trabecular bone structure from 5–10 mg/day. Besides staining of **their** teeth, **some** develop stomatitis on high intake. In a survey of 662 ***tea***-drinking English **children** (5–15 years), **they** averaged 410 ml/day or 1.26 mg F (0.52 mg F/cup). Perioral rosacea-like dermatitis and intra-oral lesions from fluorides in dentifrices. Suspect teratogen. Will pass into breast milk and **mothers** should not drink water with > 1 ppm. No final conclusion on its possible carcinogenicity for **man**. Fluoride ion, produced by catabolism of ***methoxyflurane***, may be responsible for postoperative kidney toxicity and renal disease. Despite so-called adequate ventilation systems, exposed hospital **personnel** can have urinary fluoride levels. Fluorine compounds have been used as a preservative in wood, and an exposure can easily occur from the dust. **People** firing ceramics can be exposed to fluoride fumes. Optic neuritis, chronic gastroenteritis, arthritis of the spine, acneiform eruptions (often rapid, pustular, and inflammatory), brown bands in the nail plate and matrix, leukonychia, and even brain damage reported from excessive drinking water exposures. Illness reversible upon elimination of the water intake and recurs upon resumption. Decreased ***radiopharmaceutical*** distribution in thyroid imaging and by methodologic interference, decreased alkaline phosphatase results. Can lead to increased serum amylase and lipase values if given within 24 h of testing. Organic fluoride preparations have caused cardiac problems in **man**. Breast milk levels have been reported at 0.05–0.13 ppm, depending on water source, and is insufficient to prevent caries in an **infant's** teeth. Higher than recommended (1 ppm) water levels may decrease senile dementia from ***aluminum*** toxicity. OSHA limits air exposure to 2.5 mg F/m^3/8 h TWA. See ***hydrofluosilicic acid***, ***sodium fluoride***, and ***stannous fluoride*** for additional dental problems.

Depression, inappetence, incoordination, and lameness have been reported in many **animal** species, including **cats**, **dogs**, **cattle**, **horses**, **sheep**, and **swine**. Maximum suggested safe dietary levels of fluorine are: **cattle**: 30 ppm, slaughter **cattle**: 100 ppm, **sheep** and **horses**: 60 ppm, fattening **lambs**, **pigs**, and **sows**: 150 ppm, and **poultry**: 300 ppm. Decreased reproduction in exposed **owls**. Stained, eroded and irregularly worn, pitted, and mottled teeth; exostoses of the mandibles, and emaciation in **deer** grazing inside an industrial complex. Fluorine toxicosis noted in **bison** and **elk** near Yellowstone National Park hot springs containing high levels of fluorine. Chronic fluorosis in certain parts of India in **cattle** and **water buffalo**. **Cattle** are particularly susceptible to excessive fluorine exposure. Minor exposure leads to discoloration of teeth, and, with increasing exposures, leads to softening of teeth and premature wearing; then, skeletal changes, lameness, exostoses (primarily on the shafts of long bones), enlarged joints, difficulty in limb extension, hip arthritis, hypothyroidism, anemia, abnormal heart sounds, recumbency, and even death. Dental enamel defects also reported in **dogs** and their aborted fetuses. High oral feed intake has caused exostoses in **dogs**. High fluoride content of rock ***phosphate*** in the feed for **guinea pigs** led to chronic fluorosis with weight loss, inappetence, "slobbers", abnormal growth of molars, and death. Their feed contained 130–400 ppm. **Horses** with severe fluorosis developed lameness, molar abrasions; and exostoses on the ribs, mandibles, metatarsus, and metacarpals. **Moose** have shown retarded growth rate, infertility, and anemia. Decreased food intake and weight gain in **chicks**, and decreased **hen's** egg shell breaking strength. **Japanese quail** tolerate 200 ppm in their drinking water, but 500 ppm was lethal. Young **poults**, from hatching to 8 weeks of age, fed 800 ppm developed severe leg deformities. High water fluorine content (> 3 mg/l) from artesian wells led to fluorosis with exostoses on the shafts of long bones in **rabbits** within 13–17 months. Induces hyperglycemia and diuresis (at high doses has caused bone cancer) in **rats**. **Rats** inhaling the gas for a short interval develop lung damage. Repeated short-interval exposures also caused liver and kidney damage. Passes readily through the placenta of **mice** and **dogs**. Excess intake by **sheep** caused severe dental pathology, exostoses on the periosteal surface of long bones, and decreased health and wool production. Toxic symptoms and dental lesions reported in domestic **animals** in Iceland since volcanic eruptions in 1694. In 1970, the volcano, Hekla, erupted and samples of ash had up to 2000 ppm, which decreased rapidly within weeks. ~3% of the adult **sheep** and 8–9% of the area **lambs** died from acute fluorosis. Dental lesions found a years later in surviving **lambs**. In **pigs**, excess intake resembles rickets, with periosteal hyperostosis, and is very severe on distal limb bones, the pelvic girdle, ribs, mandible, and lumbar vertebrae, particularly, at sites of tendinous and fascial muscle insertions, without damage to the articular surfaces. Anorexia, decreased body weight, gastroenteritis, muscle weakness, and respiratory problems.

FLUORINE MONOXIDE

= *Fluorine Oxide* = *Oxygen Fluoride*

Untoward effects: Irritant to eyes, skin, and mucous membranes; severe headache, pulmonary edema, eye and skin burns (especially from contact with the gas under pressure) OSHA limits exposure to 0.05 ppm (0.1 mg/m^3).

Inhalation LC_{50}, **rats**: 2.6 ppm, **mice**: 1.5 ppm.

FLUORINE NITRATE

= *Nitrogen Trioxyfluoride* = *Nitroxy Fluoride* = *Nitryl Hypofluorite*

Use: A war gas and oxidizing agent in rocket propellants.

Untoward effects: Powerful oxidizing agent and irritant.

FLUOROACETAMIDE

= *1081* = *FAA* = *Fluorakil 100* = *Fluoroacetic Acid Amide* = *Monofluoroacetamide* = *Fussol*

Rodenticide and insecticide.

Untoward effects: Estimated LD for **man** is 13–14 mg/kg, Absorbed after ingestion, through the skin, and after inhalation. Onset of toxicity is slower, but symptoms are the same as for ***fluoroacetic acid*** (q.v.).

The LD_{50} in **chickens** is 4.25 mg/kg, and reported as 0.06–0.2 mg/kg in **dogs**. Therefore, it is not surprising that apparently healthy chicken meat as dog food could be lethal for them. One **cat** and ~70 **dogs** died after being fed frozen minced chicken meat sold in 1 kg lots. In England, ~100 **cats** and **dogs** died within 21 days, after the meat of a dead **pony** sold for **animal** consumption was ingested. After 3–7 days, **dogs** develop erythema, necrosis, and sloughing of skin. **Cattle** and **sheep** have been accidentally poisoned by consuming grains contaminated years previously. They developed erosions and necrosis of the muzzle, lips, tongue, esophagus, and necrotizing duodenitis and jejunitis. Later, patches of alopecia developed and loss of wool occurred in **sheep**. Has been lethal to **golden jackals**, **foxes**, wild **cats**, **mongooses**, and wild **birds**.

LD_{50}, **rat**: 5.75 mg/kg; **mouse**: 52 mg/kg; **male quail**: 13.3 mg/kg.

FLUOROACETIC ACID

= *FAC* = *Fluoroethanoic Acid* = *Gifblaar* = *Monofluoroacetic Acid*

The sodium salt = *1080* = *Compound 1080* = *Fratol*

Rodenticide.

Untoward effects: Extremely toxic after a latent period of 30–120 min. Central nervous system stimulation, hypoglycemia, tingling of the nose, facial numbness, paresthesias of the arms and legs, muscle spasms, epileptiform convulsions; then, central nervous system depression and ventricular fibrillation. Early cardiac arrest is usual in **children** and death usually occurs within a few hours after onset of symptoms. Nausea and vomiting may follow the muscle spasms. A British **professor**, Lord Adrian, ingested some of the sodium salt and, afterwards, produced urine lethal to **guinea pigs**. An 8-year-old **male** who chewed on some treated ***wheat*** was hospitalized with generalized convulsions, vomiting, respiratory depression and tachycardia (160/min); then, cardiac irregularities and asystole, and was resuscitated. Consolidation and collapse of left lung, oliguria, convulsions, and coma. Slow recovery after 2 weeks of intensive therapy. Moderate leg paresis and severe mental defect considered permanent. Delayed generalized seizures in an 8-month-old **female**, after chewing on a contaminated cup. **She** recovered. A 15-year-old **female** ingested some in a suicide attempt, leading to grand mal seizures, psychomotor agitation, deteriorated consciousness, and diffuse brain atrophy. End-stage cerebellar ataxia evident 18 months later. Acute renal failure in five **patients** who ingested 8–40 ml of a 1% solution of the sodium salt. Recovered after treatment. The acid and sodium salt become extremely toxic after conversion into ***fluorocitrate*** by enzymatic action, which then interferes with the Krebs cycle, especially in cardiac and central nervous system cells. Estimated LD, **adults**: 50–150 mg. A lethal case reported at 2 mg/kg.

Found in many poisonous plants (viz. ***Acacia***, ***Dichapetalum***, ***Gastrolobium***) described elsewhere in the text. Convulsant in the **cockroach** and **mouse**. This is followed by prostration and tremors. **Cattle** show anorexia, weakness, and incoordination, and may die suddenly; **dogs**: nausea, vomiting, dyspnea, twitching, and convulsions; **cats** and **pigs**: central nervous system stimulation, convulsions, vomiting, diarrhea, and cardiac arrhythmias; **goats** and **horses**: cardiac arrhythmias and ventricular fibrillation; **sheep**: anorexia, weakness, tachycardia, dyspnea, hyperexcitability, mania, and muscle spasms; **monkeys**: ventricular fibrillation and arrhythmias; **guinea pig**, **hamster**, and **rats**: convulsions. **Rats** poisoned by it were eaten by **dogs** who vomited portions of the **rats** before dying. Farm **chickens** ate the vomitus and, in turn, died.

LD, **dogs**: 50 μg/kg. **Cats** and **rabbits** are as sensitive as the **dog**. The **mouse** responds like some **monkeys** (LD, 50 mg/kg). A leak from a British factory caused deaths in farm **animals** months later; contaminated soil was dumped at sea. The meat and organs of **deer** poisoned by it contained up to 920 μg/100 g, considered insufficient to adversely affect **humans** that might consume it. Its use now banned in most countries, but supplies still exist.

~LD in mg/kg, **horse**: 0.5–1.75, **sheep**: 0.25–0.5, **goat**: 0.3–0.7, **guinea pig** and **pig**: 0.3–0.4, **cattle** and **cats**: 0.2–0.5, **dog**: 0.06–0.2, **poultry**: 6–30, **rabbits**: 0.3, **rat**: 0.22, **monkey**: 5–50; **frog**: 1–2 g/kg.

FLUOROMETHOLONE

= *Cortilet* = *Delmeson* = *Efflumidex* = *Fluaton* = *Flumetholon* = *Fluolon* = *Fluormetholon* = *FML* = *Loticort* = *NSC 33,001* = *Oxylone* = *Ursnon*

Glucocorticoid, anti-inflammatory.

Untoward effects: Increased intraocular pressure, but less than ***betamethasone*** or ***dexamethasone***. Lens opacity, weight increase, Cushing's syndrome, irritability, and hypertension.

FLUOROSULFONIC ACID

= *Flusulfonic Acid*

Use: Fluorinating agent. In manufacturing of petroleum products. Colorless liquid. Fumes in moist air.

Untoward effects: Highly irritating to skin and mucous membranes. A leak from a derailed train forced the evacuation of ~1500 **residents** of Bridgman, Michigan. Some **residents** were treated at area hospitals for eye irritation.

FLUOROURACIL

= *Adrucil* = *Arumel* = *Efudex* = *Efudix* = *Fluoroplex* = *Fluracil* = *Fluril* = *Fluroblastin* = *Fluro Uracil* = *5-FU* = *NSC- 19893* = *Ro-2-9757* = *Timazin*

Antineoplastic, antimetabolite. **Oral, topical, IA, IV, intrathecal, and intralesional use.**

Untoward effects: Non-hematological toxicity often attributed to it may be due to underlying disease. Acute adverse effects from parenteral use including nausea, vomiting, diarrhea, stomatitis, and, occasionally, a hypersensitivity reaction. Later, oral and gastrointestinal ulcers, bone marrow depression, leukopenia, thrombocytopenia, eosinophilia, leukocytosis, and an acute reversible cerebellar syndrome (1–2%) with ataxia, slurred speech, confusion, aphasia, dizziness, dysmetria, poor coordination, tremors, parkinsonism, tonic-clonic seizures, nystagmus, increased lacrimation, decreased vision, conjunctivitis, nasal irritation, insomnia, decreased cognitive function, and memory loss. It is detoxified in the liver and 80% excreted via the lungs. Cardiac arrhythmias, angina pectoris, cardiomyopathy, tachycardia, hypotension, atrial and ventricular fibrillation, myocardial infarction, cardiac arrest, palmar–plantar erythrodyesthesia, alopecia, skin rashes, telangiectasia, urticaria, ichthyosis, hyperpigmentation, brown–blue coloration of the nail plate, bullous pemphigoid lesions, maculopapular eruptions, toxic epitermal necrolysis, and photosensitivity also occur. Topical use often causes red sores, blisters, and scabs. A few cases of squamous cell carcinoma reported after topical use. A normal **infant** was born to a **mother** who was treated an estimated 4–6½ months during **her** pregnancy, although the **infant** showed some toxic symptoms. An abortion and birth defects have been reported. Apparently, a **human** teratogen when given during the first trimester of pregnancy. Myelotoxicity is rare with IV use. Severe, localized hemorrhagic scrotal lesion on an 81-year-old **male** who inadvertently scratched **his** scrotum before washing off a 5% cream on **his** hands.

TD_{LO}, **human**: 450 mg/kg/30 days, IV, 480 mg/kg/32 days, IV, 6 mg/kg/3 days.

Drug interactions: Interferes with the hepatic metabolism of ***warfarin*** leading to epistaxis, hematuria, and increased prothrombin time. ***Cimetidine*** decreases its hepatic metabolism, but its clinical significance is unknown. Use with α_{2b}***-interferon*** increases its blood levels. ***Thiazides*** may potentiate its myelosuppression.

Cats, as do **humans**, suffer from acute cerebellar ***5-FU*** toxicity. It is, apparently, caused by the metabolite, ***fluorocitrate***, which inhibits a key Krebs cycle enzyme. It also tends to cause lethal convulsions in **cats**, as well as in **dogs**, at **human** equivalent therapeutic doses. Topical use of a 5% cream has caused irreversible neurotoxicosis in **cats**. IV use has caused diarrhea, bone marrow suppression, and stomatitis in **dogs** and **monkeys**. Teratogen (central nervous system, palate, and skeletal defects) in **rats**, **mice**, and **monkeys**. Co-carcinogen and carcinogen in **rats**. **Calves** are twice as sensitive to its toxicity as **dogs**, and show decreased hematopoiesis, diarrhea, and death. **Rabbits** show much less skin irritation than **man** from topical use.

FLUOSILICIC ACID

= *Hydrofluosilic Acid*

Use: As a 1–2% solution for sterilizing equipment in brewing and bottling plants; in hardening ***cement***, in electroplating, in removing ***lime*** from tanning hides, as a lumber preservative, and as a source of ***fluorine***.

Untoward effects: Extremely caustic to skin and mucous membranes. In 1979, in Annapolis, Maryland, a water treatment plant with cross-connections allowed an accidental spill of 3800 l of a 22% solution to directly enter the drinking water supply. The water was used in mixing ***dialysate*** and eight **patients** became ill; one died. Symptoms were nausea, hypotension, substernal pain or pressure, diarrhea, itching, vomiting, malaise, dyspnea, flushing, localized numbness, diaphoresis, and headache. Undoubtedly, there was unreported widespread mild ***fluorine*** intoxication in the community.

FLUOXETINE

= *Adofen* = *Fluctin* = *Fluoxeren* = *Fontex* = *Foxetin* = *LY-110140* = *Prozac* = *Reneuron* = *Sarafem*

Antidepressant. A selective serotonin reuptake inhibitor. **Inhibits reuptake and increases availability of *serotonin* in the central nervous system.**

Untoward effects: Nausea, vomiting, anxiety, nervousness, headache, drowsiness, fatigue, insomnia (11.7%); and sexual dysfunction (10%) including anorgasmia, delayed orgasms, spontaneous orgasm, ejaculation difficulties, and decreased libido are the most common adverse effects. Akathisia, asthenia, severe neutropenia, rash, alopecia, pruritus, toxic epitermal necrolysis, arthralgia, mania, panic attacks, anger attacks, increased aminotransferase, and suicidal ideation are less common. Uncommon are other extrapyramidal reactions, seizures, dystonia with involuntary movements (can be long-lasting after drug discontinuance), blepharospasm, and lip tremors, aplastic anemia, bruising, weight decrease (weight increase after prolonged use), sweating, decreased platelet aggregation, syncope, confusion, acute cognitive impairment, vaginal and vulval anesthesia, penile anesthesia, hyponatremia with induced inappropriate antidiuretic homone syndrome, bradycardia, tachycardia, respiratory distress, anorexia, psoriasis, serum sickness, stomatitis, hepatitis, and IV abuse. A **newborn** whose 17-year-old **mother** supposedly took 20 mg/day during the onset of **her** pregnancy showed jitteriness, tachypnea, fluctuating temperature, restlessness, stuffy nose, and poor sucking ability. Symptoms disappeared by the fourth day. Taken during the first trimester of pregnancy increases chances of an **infant** having more than two minor malformations, such as fused toes; taken during the third trimester leads to premature deliveries, poor neonatal adaptation, dyspnea, jitteriness, and decreased birth weight and length. Tachycardia and drowsiness may be seen with overdoses. No toxicity noted in **children** taking < 3.6 mg/kg. Small amounts in breast milk of treated **mothers** leads to colic in **infants**.

Drug interactions: A 45-year-old **male** apparently developed heat intolerance when also taking ***lithium***. A 30-year-old **female** developed a ***serotonin*** syndrome from the same combination. A **female** also developed a toxic ***serotonin*** syndrome (tremors, hyperactivity, and salivation), plus leukopenia and thrombocytopenia, after taking 20 mg/day/14 days while on ***carbamazepine*** therapy. The same toxic syndrome suspected in a 37-year-old **male** also taking ***buspirone***. Use with ***tricyclic antidepressants*** (***amitriptyline***, ***desipramine***, ***imipramine***, ***nortriptyline***, and ***protrityline***) has increased the latter's serum levels. Cardiac toxicity reported with ***terfenedine***. Acute delirium and toxic ***serotonin*** syndrome in a 72-year-old **male** given it with ***selegiline*** for 9 weeks. It has increased blood levels of ***cyclosporine***, increased ***warfarin's*** international normalized ratio, increased ***haloperidol*** serum levels (~20%), increased ***clozapine*** serum levels, and inhibits ***phenytoin*** metabolism. The clinical significance of its slight increase in serum ***digitoxin*** and ***diazepam*** is unknown. Probably, related to fatalities in **patients** on ***monoamine oxidase inhibitors***, such as ***isocarboxid***, ***phenelzine***, and ***tranylcypromine***. Bradycardia in a 54-year-old **male**, 2 days after adding 20 mg/day to **his** ***metoprolol*** regimen. ***Sotalolol*** did not induce bradycardia in **him**. ***Haloperidol***, ***lithium***, and ***tryptophan*** (as with ***carbamazepine***) can cause adverse behavioral and neurological effects. A 53-year-old **male** on ***propranolol*** 80 mg/day developed complete heart-block a week after adding 20 mg fluoxetine/day. Noisy and frequent eructations in a **female** 3 days after adding ***dothiepin HCl*** to **her** therapy. ***Clarithromycin*** (250 mg/twice daily) inhibited its metabolism in a 53-year-old **male** and caused delirium and psychosis. Light-headedness, anxiety, nausea, and paresthesia in a **patient** on ***pentazocine*** and it antagonized the anxiolytic effect of ***buspirone***. Causes a slight increase in ***methadone*** levels. Serotonin syndrome induced by use with ***venla-faxine***. Phototoxic when used with ***alprazolam***. Inhibits ***benzodiazepine*** metabolism including ***alprazolam***, ***diazepam***, ***midazolam***, and ***triazolam***. ***Alprazolam*** serum concentrations are increased ~30%. Use with ***ritonavir*** led to cardiac and central nervous system problems and a 19% increase in ***ritonavir*** serum levels. Action has been potentiated by ***Hypericum*** (***St. John's Wort***). ***Propafenone*** metabolism significantly decreased after fluoxetine metabolism.

Increases serum ***prolactin*** in **male rats**, increases ***carbamazepine*** plasma levels in **rats**, and decreases food intake and body weight in **mice**. Lethargy, polydipsia (17%), ~20% had increased appetite and ~20% had decreased appetite, seizures, and diarrhea (4.5%). Jump-starts reproductive behavior in **clams** and **mussels**.

Minimum lethal dose, **dog**: > 100 mg/kg.

FLUOXYMESTERONE

= *Androfluorene* = *Androfluorone* = *Androsterolo* = *Halotestin* = *Oratestin* = *Ora-Testryl* = *Testoral* = *Ultandren*

Androgen, anabolic agent.

Untoward effects: Oily skin (36%), alopecia (28%), hoarseness (66%), fluid retention and edema (33%), nausea (15%), vomiting, hypercalcemia (7%, especially in immobile **patients** and **those** with metastatic breast cancer), deepening of voice, acne, growth of facial hair, seborrhea, increased libido and clitoral enlargement in **some**, insomnia, excitation, fever, postmenopausal bleeding, increased sulfobromophthalein retention, increased SGOT, increased hemoglobin and red blood cells in **some**, amenorrhea or metrorrhagia, and cholestasis with jaundice; rarely, peliosis hepatitis and hepatocellular neoplasms, and anaphylactoid reactions. Priapism, gynecomastia, and decreased ejaculatory volume reported in some **males**. Contraindicated in **patients** with liver pathology.

TD_{LO}, **human**: 400 μg/kg.

Drug interactions: May potentiate action of ***anticoagulants***, oral ***hypoglycemics***, or ***oxyphenbutazone***.

Overdoses in **animals** will interfere with bone growth.

FLUPENTIXOL

= *Flupenthixol* = *LC-44* = *N-7009*

The decanoate = *Depixol* = *Fluanxol Dépot* = *LU-5-110* = *Viscoleo*.

The dihydrochloride = *Emergil* = *Fluanxol* = *Metamin* = *Siplarol*

Antipsychotic, neuroleptic.

Untoward effects: Restlessness, insomnia, extrapyramidal symptoms, oropharyngeal muscle dyskinesias, disorientation, increased platelet aggregation, urinary incontinence, and dizziness; transient hypomania with the chloride form. A 3½-year-old **female** ingested 36–40 of these

0.5 mg tablets. Attempts with emesis failed. Gastric lavage failed to remove any tablets. Extrapyramidal side-effects of neck extension and catalepsy noted after 23 h. Then, **she** alternated between consciousness and relapses; recovered in 48 h. Hair loss, extrapyramidal symptoms, akathisia, intractable involuntary movements; paradoxical excitement, restlessness, and aggression; transient hypomania, anergia, and weight increase with the decanoate.

Drug interactions: Possible cause of neuroleptic malignant syndrome associated with inappropriate antidiuresis and hyponatremia in a 41-year-old **schizophrenic** on the decanoate, when oral ***orphenadine*** was added. A 48-year-old **female** developed reversible parkinsonian symptoms when 75 mg decanoate was added. A 31-year-old **male volunteer** died suddenly after ***eproxindine*** infusion/5 min. It was discovered that earlier that day **he** received 40 mg of the decanoate IM. Dyskinesias probably associated with a treated **patient** chewing ***betel nuts***.

FLUPHENAZINE DECANOATE

= *Dapotum D* = *Lyogen Depot* = *Modecate* = *Prolixin Decanoate* = *QD-10733* = *Siqualine* = *SQ-10733*

Dihydrochloride = *Anatensol* = *Dapotum* = *Lyogen* = *Moditen* = *Omca* = *Pacinol* = *Permitil* = *Prolixin* = *Siqualone* = *Tensofin* = *Valamina*

Enanthate = *Moditen Enanthate* = *Prolixin Enanthate* = *SQ-16144*

Antipsychotic, neuroleptic. A piperazine (halogenated propyl piperazine).

Untoward effects: High potency with low sedative effects, but very apt to produce extrapyramidal symptoms, especially dyskinetic reactions in **children**. Neuroleptic malignant syndrome, a combination of extrapyramidal symptoms, hyperthermia, autonomic dysfunction, hypertension, and coma has been reported. Erection difficulties, decreased libido, ejaculation inhibition, priapism, hypersensitivity cholestatic jaundice, corneal opacity, oculogyric crises, tardive dyskinesia, false results on pregnancy test, itching, erythema, interference with blood urea nitrogen tests, pancytopenia, phosensitivity, increased ***prolactin***, dystonia, akinesia, diaphoresis, leukopenia, agranulocytosis, leukocytosis, neutropenia, lymphocytosis, thrombocytopenia, nervousness, insomnia, vomiting, parkinsonism, slight tremors and stiffness, retinopathy with lenticular deposits and opacities, psychotic reactions, anemia, tongue swelling, trismus, swallowing difficulties, opisthotonus, catatonia, urinary incontinence, rash, anorexia, increased weight, convulsions, amenorrhea, and galactorrhea. Pallor and dyspnea occur in some **patients** on ***neuroleptics***.

Side-effects with the decanoate including hair loss, movement disorders, parkinsonism, akathisia, hypothalamic syndrome, upset temperature regulation, profuse salivation, extrapyramidal rigidity in all limbs, catatonia, urinary incontinence, impaired learning, dizziness, blurred vision, thrombocytopenia, and paradoxical responses, such as excitement, restlessness, and aggression dominate the published case reports. The use of the decanoate throughout the pregnancy of a 32-year-old **female** led to minor extrapyramidal symptoms in the **infant** 4 weeks after delivery. Administration to a nursing **mother** may have caused similar symptoms in her **infant**.

The enanthate has caused most of the same reactions, plus hypertension or hypotension, hallucinations, headache, decreased blood pressure, and spasms of the mouth and tongue. The hydrochloride's adverse effects are also the same, plus reports of ballismus, nuchal spasms, transient arrhythmias, and increased SGPT and SGOT. Respiratory distress with severe rhinorrhea and vomiting in a **newborn** exposed prenatally.

Therapeutic serum levels are 0.001–0.01 mg/l, and toxic serum levels are 0.1 mg/l.

Decanoate TD_{LO}, **human**: 179 μg/kg.

Enanthate TD_{LO}, SC, **human**: 357 μg/kg.

Drug interactions: ***Fluphenazine–clonidine*** therapy now associated with delirium (acute organic brain syndrome). Use with ***hydrochlorothiazide*** was associated with a compulsive water drinking (~96 glasses/day) leading to heart failure and acute water intoxication on two occasions, resulting in death from the second. Effectiveness decreased by many ***antacids*** and ***antidiarrheal*** preparations. ***Ascorbic acid*** decreases its blood level. ***Methaqualone*** with it leads to epistaxis and disturbed menstruation. Neurotoxicity reported with ***lithium***. Smoking ***tobacco*** decreases its blood levels.

An adult **male** *Macaca fascicularis* (crab-eating **monkey**) tolerated 0.07 mg of the enanthate/kg IM well, while an adult **male** *M. arctoides* developed a severe and prolonged parkinsonian neurological reaction, requiring 4 days of treatment and 8 days of hand-feeding. A **horse** became progressively agitated and restless, and made unusual repetitive motions after 50 mg enanthate IM.

LD_{50}, **mouse**: 220 mg/kg.

FLUPREDNISOLONE

= *Alphadrol* = *Etadrol* = *NSC-47439* = *U-7800*

Glucocorticoid, anti-inflammatory.

Untoward effects: As with other ***glucocorticoids***: ***sodium*** and fluid retention, ***potassium*** loss and related alkalosis, hypertension, congestive heart failure, decreased muscle mass and weakness, osteoporosis, vertebral compression fractures, aseptic necrosis of femoral and humeral heads, fractures of long bones, abdominal distension, peptic and esophageal ulcers, pancreatitis, diaphoresis; thin, fragile skin with petechiea; ecchymoses and impaired wound-healing, vertigo, headaches, decreased growth in **children**, Cushingoid state, menstrual irregularities, and posterior subcapsular cataracts and glaucoma.

FLUPROSTENOL

= *Equimate* = *ICI 81,008*

***Luteolytic agent.* A synthetic *prostaglandin* analog, related to *Dinoprost* (q.v.) [*prostaglandin* F_{2a}].**

Untoward effects: Can be absorbed through the skin. Pregnant **women** or **those** of child-bearing age, **asthmatics**, or **people** with respiratory problems should not handle the product. If **they** do, **they** should wear disposable gloves.

IM in **horses** has occasionally caused transient diaphoresis and diarrhea and may aggravate respiratory problems.

FLURANDRENOLIDE

= *Cordran* = *Drenison* = *Drocort* = *Fludroxycortide* = *Fluorandrenolone* = *Flurandrenolone* = *Haelan* = *Lilly 33,379* = *Sermaka*

Topical corticosteroid, glucocorticoid, anti-inflammatory.

Untoward effects: Increased frequency of use and application under occlusive dressings can aggravate the following adverse effects, listed in ~decreasing frequency: burning, itching, irritation, dryness, folliculitis, hypertrichosis, acneiform eruptions, hypopigmentation, perioral dermatitis, and allergic contact dermatitis. The following are more common under occlusive dressings: skin maceration, secondary infection, skin atrophy, striae and milaria. Increased intraocular pressure in the eyes of an 18-year-old **male** after prolonged topical use on the eyes. After a few weeks, spontaneous remission occurred in one eye; the other eye continued to develop further glaucomatous field defect and loss over 4 years since withdrawal of treatment. That eye became myopic. See ***Fluprednisolone*** above for other possible adverse effects.

FLURAZEPAM

= *Dalmane* = *Dalmadorm* = *Felmane* = *Noctosom* = *Ro-5-6901* = *Stauroderm*

Sedative, hypnotic.

Untoward effects: Dizziness, drowsiness, staggering, ataxia, disorientation, coma, headache, heartburn, nausea, vomiting, diarrhea, constipation, stomach pains, nervousness, talkativeness, irritability, chest pains, and body and joint pains. Less common are leukopenia, granulocytopenia, diaphoresis, flushing, blurred vision, burning eyes, hypotension, dyspnea, pruritus, rashes, cholestasis, dry mouth, bitter taste, salivation, impaired renal function, anorexia, confusion, hallucinations, and increased SGOT, SGPT, ***bilirubin***, and alkaline phosphatase. Rarely, paradoxical excitement, rage, hostility, hyperactivity, and euphoria. Allergic reaction (tongue swelling after 4 days) in a 65-year-old **male** on 30 mg/day. ***Lactose*** (q.v.) filler in a capsule was a cause of cramps and diarrhea in a **patient**. **Neonatal** depression has followed the **maternal** use of this drug prior to delivery. Excreted in **mother's** milk. Despite its supposed safety, it was inevitable that many fatalities occurred from its use. It has a 1–4 h plasma elimination half-life, but its pharmacologically active metabolite, ***desalkylflurazepam***, has a half-life of 45–300 h. Therapeutic serum levels are 0.05–0.15 mg% and toxic levels at 0.25 mg% (measured by its major metabolite).

Drug interactions: ***Cimetidine*** slows the elimination of its ***desalkylflurazepam*** metabolite. ***Narcotic analgesics***, ***antide pressants***, ***antipsychotics***, and acute ingestion of ***alcohol*** increase its central nervous system suppressive effects.

TD_{LO}, **human**: 430 μg/kg.

LD_{50}, **rats**: 1.2 g/kg; **mice**: 870 mg/kg, **rabbits**: 568 mg/kg.

FLURBIPROFEN

= *Adfeed* = *Ansaid* = *Antadys* = *BTS-18322* = *Cebutid* = *Froben* = *Flurofen* = *Ocufen* = *Stayban* = *U-27182* = *Zepolas*

Analgesic, anti-inflammatory.

Untoward effects: ~9–10% discontinue treatment because of serious adverse reactions.

Central nervous system: Headache in 5%±; 1–3% have nervousness, anxiety, insomnia, tremors, enhanced reflexes, vertigo, amnesia, asthenia, dizziness, somnolence, malaise, and depression.

Gastrointestinal: Dyspepsia, diarrhea, nausea, and abdominal pain have a 5%± incidence. 1–3% have constipation, gastrointestinal bleeding, vomiting, flatulence, anorexia, cholestatic jaundice, and increased liver enzymes.

Miscellaneous: Ototoxicity, renal papillary necrosis, acute parkinsonism, acute flank pain, thrombocytopenia, neutropenia, lichen planus, edema, rash, rhinitis, tinnitus, and numerous uncertain causal relationships.

Drug interactions: Prolongs plasma half-life of ***antipyrine***, reduces the diuretic effect of ***furosemide***, and attenuates the hypotensive effect of ***propranolol***, but not ***atenolol***.

Decreased body temperature in **goats**, **rabbits**, and **rats**.

FLUROTHYL

= *Hexafluorodiethyl Ether* = *Indoklon*

Central nervous system stimulant.

***Untoward effects*:** Inhalation leads to occasional nausea and vomiting; headache, dizziness, amnesia, disorientation, confusion, convulsions, esophagitis, and tracheitis. Hypertension (200/180) in two **patients**, persisting for ~2 weeks; connection to this drug therapy not proven.

Inhalation by **hamsters** caused myoclonic jerk, clonic convulsions, catatonia, and depression.

IP, LD_{50}, **rat**: 1.26 g/kg; IV, **mouse**: 46 mg/kg.

FLUROXENE

= *Fluoromar*

Inhalation anesthetic.

***Untoward effects*:** Eye irritation, central nervous system and respiratory depression, and seizure. Explosion hazard has decreased its popularity for clinical use.

FLUSPIRILINE

= *Imap* = *Redeptin*

Psychotherapeutic.

***Untoward effects*:** Extrapyramidal effects (50–75%) including akathesia, tremors, dysarthria, dyskinesia, oculogyric crises, gastrointestinal upsets (5%), insomnia (16%), transient fatigue and apathy (33%), dysphagia, drowsiness (25%), and local irritation after IM dosing.

IM, LD_{50}, **rats**: 146 mg/kg; **mouse**: 125 mg/kg.

FLUTAMIDE

= *Drogenil* = *Euflex* = *Eulexin* = *Flucinom* = *Fugerel* = *Niftolid* = *Niphtholide* = *Sch 13,521*

Antiandrogen, antineoplastic.

***Untoward effects*:** Nausea, diarrhea, hot flashes, gynecomastia (~50% in **men**), nipple pain, hepatotoxicity (fatal and non-fatal), decreased beard growth in **men**, constipation, cholestatic jaundice, hepatic encephalopathy, hepatic necrosis, photosensitivity (including erythema, ulceration, bullous eruptions, and epidermal necrolysis), probable malignant breast neoplasm, drowsiness, insomnia, anxiety, headache, diaphoresis, depression, skin rash, hypotension, transient blurred vision, methemoglobinemia, hemolytic and macrocytic anemia (6%), leukopenia (3%), and thrombocytopenia (1%).

FLUTICASONE PROPIONATE

= *CCI-18781* = *Cutivate* = *Flixonase* = *Flixotide* = *Flovent* = *Flunase*

***Anti-inflammatory, anti-allergic, glucocorticoid.* Topical (0.05%) in dermatology and as a nasal spray.**

Untoward effects*:** Usually, self-limiting and mild in dermatology leading to pruritus (2.9%), dryness (1.2%), numbness of fingers (1%), burning sensations (0.6%), hypertrichosis, erythema, hives, and light-headedness. After intranasal use, leads to nasal burning (3–6%), nasal irritation (1–3%), epistaxis, pharyngitis, and headache (1–3%). Occasionally reports of glycosuria, sneezing, runny nose, nasal dryness, sinusitis, bronchitis, nasal ulcers and septal excoriation, dizziness, unpleasant taste, dry mouth, nausea, vomiting, urticaria, rashes, facial–lingual edema and bronchospasm. High dosage ($\geq$ 1000 μg/day) by inhalation in asthmatic **children** may cause adrenal suppression and growth retardation. Invasive pulmonary aspergillosis in an asthmatic 44-year-old **male** after inhalation of 1.76 mg/day along with 20 mg/day ***zafirlukast.

FLUVASTATIN

= *Fluindostatin* = *Lescol* = *XU 62-320*

Antihypertensive.

***Untoward effects*:** ~1% of **patients** had serum aminotransferases > 3 times normal. A **patient** developed myopathy with creatine kinase 10 times normal after exercise. Upper respiratory tract infections, pharyngitis, rhinitis, bronchitis, sinusitis, coughing, dyspepsia, diarrhea, abdominal pain, nausea, constipation, flatulence, dizziness, insomnia, headache, fatigue, back pain, arthropathy, and rashes are among the many other adverse effects. A fatal lupus-like syndrome with respiratory distress after 20 mg/day in a 67-year-old **male**.

Drug interactions*:** Its serum concentrations are increased by ***cimetidine, ***omeprazole***, and ***ranitidine***; its serum concentration is decreased by ***rifampin***. Bound up by bile acid-binding resins. Induces clinically significant increase in ***warfarin*** anticoagulation. Inhibits CYP1A2 and CYP2C19 ***cytochrome P-450*** metabolizing enzymes.

FLUVOXAMINE MALEATE

= *DU-23000* = *Dumirox* = *Faverin* = *Fevarin* = *Floxyfral* = *Luvox* = *Maveral* = *MK-264*

***Antidepressant. Serotonin* reuptake inhibitor.**

***Untoward effects*:** Nausea, vomiting, headache, nervousness, insomnia, alopecia, asthenia, dyspepsia, drowsiness, delayed or failed orgasm in **males** and **females** (< 0.01%), failed erection, toxic epitermal necrolysis, galactorrhea in a **female**, induced or unmasked manic behavior, inappropriate secretion of antidiuretic homone leading to hyponatremia and mental confusion. Suicidal ideation, accidental and intentional overdoses, convulsions, and deaths also reported. Excreted in breast milk (0.09 mg/l; **mother's** plasma, 0.31 mg/l).

Drug interactions: Increases serum concentration of ***alprazolam***, ***astemizole***, ***bromazepam***, ***caffeine***, ***carbamazepine***, ***clomipramine***, ***clozapine***, ***diazepam***, ***haloperidol***, ***imipramine***, ***lansoprazole***, ***mephenytoin***, ***methadone***, ***nelfinavir***, ***omeprazole***, ***terfenadine***, ***theophylline*** (~3%), ***triazolam***, and ***warfarin*** (~98%). Seizures have been reported when used with ***lithium***. Fatal reactions are possible if taken within 2–3 weeks of ***monoamine oxidase inhibitors***. Decreased serum concentration in ***cigarette*** **smokers**. Electrocardiogram changes with high doses of it and ***pipamperone***. Acute dystonia in an obese 14-year-old **male** when added to ***metoclopramide*** therapy.

In **dogs**, 10 mg/kg leads to depression and tremors. Overdoses in **dogs** cause bradycardia, lethargy, tremors, and vomiting. Half-life was 15 h.

FLY ASH

Residue from incinerators or *coal* burning. Contains *arsenic*, *cadmium*, *mercury*, *selenium*, precious metals, and some radioactive elements.

Untoward effects: Suspect as an airborne carcinogen. Influenza virus growth was enhanced *in vitro* in the presence of lignite fly ash and suppressed ***interferon*** production by cells, suggesting decreased **human** host resistance to viral infections.

Intratracheal application in **hamsters** caused pulmonary toxicity.

FOLIC ACID

= *Factor U* = *Folacin* = *Folvite* = = *Lactobacillus casei growth factor* = *PGA* = *Pteroylglutamic Acid* = *Vitamin* B_c = *Vitamin M*

***Vitamin, hematinic.* Given the name *Vitamin* B_c to honor the thousands of chicks used in the experimental studies of this B-complex fraction. The *Vitamin M* designation honors the monkeys that developed a sprue-like syndrome when deprived of it.**

Untoward effects: A possible allergic reaction with erythema, urticaria, pruritus, fever, and bronchospasm in **patients** and a 9-month-old **child**. 15 mg/day to healthy **volunteers** led to anorexia, nausea, abdominal distension and discomfort, flatulence, sleep disturbances, vivid dreams, malaise, irritability, excitability, and overactivity. **Many** quit therapy after a month, due to these adverse effects. May precipitate acute neurological disorders in **patients** with unsuspected early pernicious anemia, and may mask ***vitamin*** B_{12} deficiency states. Apparently, has potentiated ***vitamin*** B_{12} deficiency syndrome of psychosis with paranoid delusions and auditory hallucinations. **Mother's** plasma levels are ~40 times breast milk levels. A **patient** had degeneration of the spinal cord after extensive resection of the ileum. Previous use of folic acid was suspect in the acute onset of symptoms. Recovery took place rapidly after folic acid was withdrawn and ***vitamin*** B_{12} was administed. Acute lethal dose > 200 mg/kg. Toxic levels in **infants** are ~20 mg/day; ~40 mg/day in **adults**.

Drug interactions: ***Anticonvulsants*** (***phenytoin***, ***phenobarbital***, and ***primidone***) inhibit the absorption of dietary folic acid leading to megaloblastic anemia, gingival hyperplasia, increased seizure frequency, schizophrenia-like psychoses of epilepsy, macrocytosis, dizziness, ataxia, stomatoglossitis, maculopapular rash, and epilepsy neuropathies. It increases the metabolic conversion of ***phenytoin*** to its inactive derivative, ***p-hydroxyphenytoin***, decreasing ***anticonvulsant*** control. ***Estrogen***-containing ***oral contraceptives*** decrease its plasma concentration and have caused megaloblastic anemia in some **women**. Adsorbents, such as ***cholestyramine*** and ***magnesium trisilicate***, decrease its availability. ***Methotrexate*** increases its urinary excretion and may decrease its synthesis. It may inhibit the activity of ***doxycycline hyclate***. ***Metformin***, ***pyrimethamine***, ***sulfamethoxazole***, ***sulfasalazine***, and ***trimethoprim*** effects are decreased by it. Intestinal absorption is interfered with by ***aspirin***, ***alcohol***, and ***p-aminosalicylic acid***.

IP, LD_{50}, **mouse**: 100 mg/kg.

FOLINIC ACID

= *CF* = *Citrovorum Factor* = *Leucovorin*

Use: Antidote to ***folic acid*** antagonists.

Untoward effects: Local tissue reaction, nausea, and diarrhea after IV. Hepatic toxicity reported. ***Methotrexate***, ***phenytoin***, ***pyrimethamine***, and ***sulfonamide*** interactions similar to those with ***folic acid*** (q.v.).

FOLPET

= *Advacide TMP* = *ENT 26,539* = *Folpan* = *Fungitrol 11* = *Phaltan*

Fungicide.

Untoward effects: Mucous membrane irritant.

LD_{LO}, **human**: 5 g/kg.

Mild ataxia and slight weight decrease for about 18 days in **mallard ducks**. Teratogenic potential in **hamsters**.

LD_{50}, **mallard ducks**: > 2 g/kg; dermal, **rabbits**: > 10 g/kg; TD_{LO}, **rat**: 500 mg/kg/5 days.

FOMIVERSEN SODIUM

= *Vitravene*

An antisense oligonucleotide for intravitreal injection.

Untoward effects: Iritis, vitreoitis, increased intraocular pressure, and vision changes are common.

Drug interactions: The manufacturer suggests it not be used if ***cidofovir*** was used within 2–3 weeks.

FONAZINE MESYLATE

= *Banistyl* = *Bonpac* = *Calsekin* = *Dimethothiazine Mesylate* = *Fusaben* = *Migristène* = *Neomestine* = *Promaquid* = *Yoristen*

***Antimigraine.* A *phenothiazine* derivative.**

Untoward effects: Vertigo, nausea, somnolence, diarrhea, itchy palate, and dry mouth.

FONOPHOS

= *Dyfonate* = *ENT 25,796* = *Fonofos* = *N-2790*

***Organophosphorus insecticide.* Cholinesterase inhibitor.**

***Untoward effects*: Human** toxicity from inhalation, ingestion, absorption through the skin, and by skin contact leading to nausea, vomiting, abdominal cramps, diarrhea, salivation, diaphoresis, headache, giddiness, weakness, vertigo, rhinitis, tightness of the chest, miosis, blurred vision, muscle fasciculations, dyspnea, and arrhythmias.

Estimated accidental intake of 67 mg by five 250 lb **calves** (~600 μg/kg) after eating contaminated grain led to clinical symptoms of staggering and increased lacrimation within 36–40 h. Treated with ***atropine***/day/3 days. Recovered. Hyperexcitability, wide stance, ataxia, goose-stepping, falling, tremors, wing-drop, and convulsions in **mallard ducks** within 25 min. Death in ~1–3 h. Up to 4 days for remission in others.

LD_{50}, **rat**: 3 mg/kg; **dog**: 3.5 mg/kg; **mallard duck**: 16.9 mg/kg; **cattle**: 1.3 mg/kg. LC_{50}, **quail**: 284 ppm.

FOODS

Untoward effects: Natural doesn't mean it is safe. Natural carcinogens, such as ***hydrazines***, are in ***mushrooms***; ***tannins*** in ***coffee***, ***tea***, and red ***wines***; ***ethyl carbamate*** in ***beer***, ***wine***, soy sauce, ***yogurt***, and ***bread***; ***allyl isothiocyanate*** in ***broccoli***, ***cabbage***, ***horseradish***, and ***mustard***; ***benzopyrenes*** on grilled meats; ***aflatoxin*** on grains and ***peanuts***, etc. Numerous other toxic substances are found in foods for **humans**. Cholinesterase inhibitors are found in ***radishes***, ***carrots***, ***celery***, and ***potatoes***. ***Potatoes*** and ***tomatoes*** contain ***solanine***, an alkaloid, used as an insecticide; lima ***beans***, ***almonds***, ***cassava***, ***yams***, ***maize***, ***sugar*** cane, ***sorghum***, ***linseed***, ***cycads***, and ***apricot*** kernels contain ***hydrogen cyanide*** (***prussic acid***); ***tea***, ***cocoa***, ***spinach***, ***rhubarb***, ***cashews***, and ***almonds*** contain ***oxalates*** and ***oxalic acid***. ***Caffeine***, a known mutagen, is in ***coffee***, ***tea***, and ***cocoa***; ***nutmeg*** contains ***myristicin***, a hallucinogen, and ***safrole***, a liver carcinogen; ***licorice*** induces hypertension; uncooked broad or fava ***beans*** (rich in ***levodopa***) cause hypertension and hemolytic anemia; high intake of ***calcium***-enriched white ***bread*** and dairy products leads to renal ***calcium*** stones; high dietary intake of ***lettuce***, ***broccoli***, and ***turnip*** greens' ***vitamin K*** content has interfered with ***warfarin*** prophylaxis and caused myocardial infarction. Numerous foods trigger migraines in some **people**; ***betel nut***-chewing has been associated with increase in oral carcinomas; many foods increase lower esophageal sphincter reflux; and fatal and near-fatal anaphylactic reactions to different foods have occurred. Migraineous neuralgia and hypertension from ***tyramine***, ***histamine***, and other pressor amines in red ***wines***, strong ***beers***, aged ***cheeses***, ***liver***, pickled or kippered ***herring***, ***chocolate***, ***oranges***, and ***bananas***. ***Parsley*** contains ***psoralen***, a photosensitizer; ***celery*** **harvesters** develop phototoxic bullae; and ***broccoli*** contains five goitrogens. Some contain ***lead***, ***arsenic***, ***mercury***, etc. Hypersensitivity reported in **adults** and **children** (usually, ***egg*** white, ***milk***, ***wheat***, ***beef***, etc.). *Salmonella*, *Staphylococcus*, *Listeria*, and *Clostridium* (including botulism) infections occur from time to time. ***Bananas***, citrus fruits, ***tea***, ***coffee***, ***chocolate***, ***vanilla***, and ***nuts*** often produce spuriously increased urinary vanillylmandelic acid levels. A wide range of contaminants have precipitated problems and confused diagnoses (viz. ***molds***, pesticides, ***polychlorinated biphenyls***, ***methyl alcohol***, etc.). ***Nitrates*** and ***nitrites*** in food can help form ***nitrosamines***, which are potentially carcinogenic. Food–drug interactions are numerous and, sometimes, life-threatening (viz. ***monoamine oxidase inhibitors*** with aged ***cheeses***, etc.). It is said in the U.S., the chief food-related hazard for **man** is over-eating.

FORMALDEHYDE

= *Formic Aldehyde* = *Methanal* = *Methyl Aldehyde* = *Methylene Oxide* = *Oxomethane* = *Oxymethylene*

***Disinfectant, antiseptic, astringent, embalming fluid, and in resins.* A gas.**

Its water solution, 37% w/w or 40% w/v (usually stabilized with 10–15% *methyl alcohol*) = *Formalin* = *Formol* = *Morbicid* = *Veracur*

Untoward effects: Ingestion causes immediate intense pain in the mouth, pharynx, and stomach; nausea, vomiting, hematemesis, diarrhea, shock with pale clammy skin, nephritis, dysuria, oliguria, hematuria, anuria, vertigo, stupor, convulsions, coma, and death can follow respiratory failure. Less than 1 oz has been lethal. It is rapidly metabolized to ***formic acid***. A **female** drank ~120 ml of 10% solution causing shock and severe abdominal pain. Frequent episodes of epigastric distress, nausea, vomiting, and anorexia, increase in severity over a 3 month period.

Total gastrectomy required after exploratory surgery revealed diffuse gastric ulceration, fibrosis, and almost complete obstruction. Inhalation and contact leads to eye, nose, and throat irritation, lacrimation, coughing, bronchospasm, palpitations, pulmonary edema, and skin burns. **Pathologists** and medical **students** have had itchy eyes, rhinitis; browning, hardening, or blistering and local anesthesia of the skin, urticaria, coughing, and respiratory distress from continuous exposure. Repeated application of strong solutions leads to superficial necrosis of the skin and nails and persistent eczema. Topical formaldehyde–***gelatin*** sponges have been ototoxic. ***Urea***–formaldehyde resin foams in insulation release formaldehyde over a long period of time; and ***phenol***–formaldehyde resins (used as adhesives in wood products, such as particle board, fiber board, and plywoods extensively used in housing and furniture construction) have been major sources of **human** exposure. The formaldehyde-based glue on new furniture causes nausea and possibility of fatal multiple myeloma. Solutions splashed or dropped on the eyes lead to transient injury or discomfort, as well as severe corneal opacification and loss of vision. Exposure to the gas form causes a protective reflex eye closure. Airborne levels < 1 ppm often cause direct irritation of skin, eyes, nose, throat, and lungs. **Employees** in dialysis hospital units have had similar experience from attempting to control bacterial contamination in water distribution systems and in the dialysis fluid pathways of artificial kidney machines. Its use in nail-hardeners (now banned in the U.S.) led to leukonychia with variable brown–azure discoloration of the nail plate surface. Skin irritation, sensitization, and allergic contact dermatitis (urticaria, purpura) have occurred from low concentrations, such as those used as cosmetic preservatives, or from methenamines in oral urinary antiseptics. Allergic contact dermatitis reported from formaldehyde resins in permanent-press fabrics. Also used to improve wet-strength of facial tissue, toilet tissue, and paper towels; in cutting oils as an antimicrobial, ***phenol***–formaldehyde resin on typewriter correction paper as a binder for the powdery coating; and in newspapers and many paper products, such as plates, cups, and examination table coverings. After 25 years of low-level occupational exposure in the textile finishing industry, a 57-year-old **male** developed a squamous cell carcinoma of the nasal cavity. A number of poisoning cases are in the literature. They range from a 68-year-old **male** who developed mucosal damage and ulceration, with bleeding of the oral cavity and tonsils, after inhalation or gargling with it in a deliberate suicide attempt. A 7-year-old **male** was accidentally given it rectally and recovered. Hospital clinic pharmacies inadvertently put ***formalin*** into distilled water containers, and used the latter to reconstitute ***antibiotic*** solutions. 39 **children** in one hospital received it and became sick with upset stomachs and burning mouths; 15 **children** in the second hospital. **All** recovered. OSHA now suggests an upper airborne exposure limit of 0.75 ppm/8 h TWA and 2 ppm/15 min exposure. National Institute for Occupational Safety & Health sets the levels at 0.016 ppm and 0.1 ppm, respectively. Now considered a *probable* **human** carcinogen.

LD_{LO}, **woman**: 36 mg/kg; inhalation TC_{LO}, **human**: 17 mg/m^3/30 min.

A high incidence of nasal squamous cell carcinomas in **rats** exposed to 15 ppm of the vapors for 18–24 months. Severe synovitis reported in two **horses**, due to the use of electrolyte solutions (containing formaldehyde and ***methanol*** as preservatives) for irrigating tissues and in arthroscopy. Intra-articular ***formalin*** is used to create experimental degenerative joint disease. **Monkeys** and **man**, but not **rats**, metabolize ***formalin*** to ***formic acid*** leading to metabolic acidosis, central nervous system depression, retinal edema, and nephritis. Its vapors are used as an incubator fumigant on **bird** eggs. Higher than recommended concentrations led to high mortality during the first 3 days of incubation. Avoid ***zinc*** equipment when using formaldehyde with ***malachite green*** (increases the toxicity of the latter).

LD_{50}, **rat**: 800 mg/kg; **guinea pig**: 260 mg/kg; inhalation LC_{LO}, **rat**: 250 ppm/4 h; **mouse**: 900 mg/m^3/2 h, **cat**: 820 mg/m^3/8 h.

FORMAMIDE

= *Carbamaldehyde* = *Methanamide*

Use: Solvent; softener for ***paper***, animal glues, and water-soluble gums.

Untoward effects: Drowsiness, fatigue, nausea, and acidosis from ingestion. Skin eruptions and eye and mucous membrane irritation from contact. Industrial grades have a slight ***ammonia***-like odor.

IP, LD_{50}, **rats**: 5.7 g/kg; **mice**: 4.6 g/kg.

FORMIC ACID

= *Methanoic Acid*

Use: In leather tanning, coagulating rubber latex, textile dyeing and finishing, a fumigant, and as a silage preservative.

Untoward effects: Eye irritation, lacrimation, throat irritation, coughing, rhinitis, and dyspnea from inhalation; skin and eye burns and dermatitis from contact; and nausea from ingestion. It is a rapidly formed toxic metabolite of ***methanol*** and ***formaldehyde*** (q.v.) leading to severe metabolic acidosis, central nervous system depression and nephritis. A 2-year-old **male** who swallowed some kettle descaler (60% formic acid) developed hematemesis, 12 h oliguria, shock, blood pressure 70/50, slight liver damage, severe laryngeal stridor (requiring tracheostomy), and, eventually, an esophageal stricture. One of the poisonous agents in *Taxus* (***yew***) poisonings, and an airborne irritant from combustion of ***wood***, ***cotton***, and ***newspapers***. Dermatitis in **man** from contact with it in the sting-hairs of ***nettles***. **Workers** gathering **ant** eggs develop an erythematous, scaly dermatitis from their formic acid content. National Institute for Occupational Safety & Health/OSHA limit of 5 ppm TWA; 30 ppm presents immediate danger to a **person's** health. Estimated lethal **human** dose: 30 g. Lethal levels reported in blood, 9–68 mg%; in urine, 216–785 mg%; and liver, 60–99 mg% after ingestion by **man**. Formic acid near the outer shell of the tropical **fire coral**, a stinging **coral**, causes the burning lesions upon contact. Similar lesions from formaldehyde released by warm sterilized rubber gloves. ***Glyoxylic acid***, a toxic metabolite of ***ethylene glycol***, is further oxidized to formic acid.

Has induced mammary, kidney, liver, intestinal, and ear canal carcinomas in **animals**.

LD_{50}, **rats**: 1.2 g/kg; **mice**: 1.1 g/kg; **dog**: 3 g/kg.

FORMOTEROL
= *Atock* = *BD-40A* = *Foradil*

Antiasthmatic.

Untoward effects: Hypokalemia and tachyphylaxis reported.

FOSCARNET SODIUM
= *A 29,622* = *Foscavir* = *Trisodium Phosphonformate*

Antiviral.

Untoward effects: Nephrotoxicity that can progress to renal failure, especially in dehydrated **patients** or **those** receiving other potentially nephrotoxic drugs, such as ***amphotericin B***, ***aminoglycoside antibiotics***, and ***pentamidine***. Anemia, decreased hemoglobin, decreased hematocrit, increased liver enzymes, nausea, vomiting, fatigue, mild headache, anxiety, seizures, hypomagnesemia, hypocalcemia, hypophosphatemia (occasionally hypermagnesemia, calcemia, and phosphatemia), proteinuria, leukopenia, thrombophlebitis and painful erosive ulcerations of the penis, uvula, and esophagus, which appear to be forms of fixed-drug eruptions. Deposited and retained in bone matrix of **humans** and **animals**.

Drug interactions: Concurrent use with ***pentamidine*** increases the risk of hypocalcemia-related fatalities and increases propensity with ***ciprofloxacin*** for seizures. Can increase risk of ***zidovudine***-induced anemia.

FOSFINOPRIL SODIUM
= *Accor* = *Monopril* = *Secorvas* = *Staril*

Antiotensin-converting enzyme inhibitor.

Untoward effects: Cough, bronchospasm, and angioneurotic edema. Weakness, diaphoresis, chest pain, nausea, vomiting, angina/myocardial infarction, palpitations, arrhythmias, hypotension, flushing, fainting, claudication, urticaria, rash, pruritus, photosensitivity, pancreatitis, hepatitis, abdominal pain and distension, dysphagia, constipation, xerostomia, heartburn, lymphadenopathy, arthralgia, muscle pain and cramps, drowsiness, vertigo, memory and sleep disturbances, tremors, confusion, paresthesia, pharyngitis, sinusitis, rhinitis, epistaxis, tinnitus, vision and taste disturbances, increased urinary frequency, neutropenia, leucopenia, eosinophilia; and increased transaminases, lactate dehydrogenase, alkaline phosphatase, and serum ***bilirubin*** are among the many adverse effects reported by some **patients**.

FOSFOMYCIN TROMETHAMINE
= *Monurol*

Antibiotic. **Used parenterally in Europe, orally in the U.S.**

Untoward effects: Diarrhea and headache (~10%), vaginitis (7.6%), nausea (5.2%), rhinitis (4.5%), back pain (3%); dysmenorrhea, pharyngitis, dizziness, abdominal pain (> 2%), dyspepsia, asthenia, rash, pruritus, flatulence, xerostomia, dysuria, hematuria, toxic megacolon, and increased SGPT. Rarely, aplastic anemia and angioedema. Placental transfer occurs.

Drug interactions: ***Metoclopramide*** decreases its serum concentration.

Excessive dosage in **dogs** and **rats** leads to diarrhea and anorexia.

FRAMYCETIN
= *Actilin* = *Enterfram* = *Framomycin* = *Neomycin B* = *Soframycin* = *Sofra-Tulle*

Antibiotic.

Untoward effects: Limited oral absorption. Parenteral and topical use has caused ototoxicity (primarily, cochlear damage; some vestibular damage). Contact dermatitis from topical use. Cross-sensitivity with ***gentamicin***, ***kanamycin***, ***neomycin***, and ***streptomycin***. Nephrotoxic.

Drug interactions: Parenteral use may enhance the activity of ***neuromuscular blocking agents***. May decrease serum levels of ***penicillin V***.

SC, LD_{50}, **mouse**: 220 mg/kg, IV, 65 mg/kg.

FRANGULA

= *Alder Buckthorn* = *Arrow Wood* = *Berry Alder* = *Black Dogwood* = *Buckthorn Bark* = *Nerprun Bourdain* = *Persian Berries* = *Rhamnus frangula*

Untoward effects: The leaves, bark, and fruit (berries) contain ***franulin***, a strong purgative. Has induced fixed-drug eruptions, as well as *in vitro* genotoxic effects, with potential for tumor promotion. Its ***anthraquinones*** may cause colon pigmentation. Fluid and electrolyte depletions have followed the diarrhea in **children** who eat the berries (green to brilliant red, then black), despite their unpleasant taste.

If **they** eat at least 20 berries, **they** often vomit, but can still develop renal damage, convulsions, abdominal pain, hemorrhage, dyspnea, and collapse.

Rhamnus cathartica = ***European Buckthorn*** is a related shrub. Although severe poisoning is rare, it contains ***emodine*** and ***anthraquinones***, and **children** have often eaten its black berries and chewed its twigs leading to transient vomiting, abdominal pain, and diarrhea. If a quantity is consumed, more severe symptoms develop, as with Frangula.

FRANKFURTERS

= *Hot Dogs*

Untoward effects: Concentrations of up to 100 ppb ***nitrosamines*** in some. ***Sodium nitrate*** and ***sodium nitrite*** added to help color and preserve some brands leading to pounding, vascular, migraine-type headaches. Excessive intake by **children** has been cited as increasing **their** risk for brain tumors.

FRAXINUS sp.

= *Ash*

Untoward effects: The pollen has caused allergic rhinitis, bronchial asthma, and hypersensitivity pneumonitis, and its oleoresins have caused a dermatitis.

Cows that had eaten the leaves and fruit of ***F. excelsior*** developed SC purple-colored edema (udder, perineal region, ribs, flanks, and hips), incoordination, abdominal pain, and collapse.

FREE RADICAL

Contains an unpaired electron, that makes it highly reactive.

Untoward effects: Although a natural product of body metabolism, it can overwhelm a body's natural defense against it, as well as exogenous sources from ***cigarette*** smoke or other environmental pollutants leading to arthritis and inflammation disease states, damage to cells with breakage or mutation of their ***DNA*** (potential inducer of cancer), destruction of endothelial cells of blood vessels, hastens the aging process and lung injury. Said to be involved in Duchenne's disease (childhood muscular dystrophy) and Becker's disease (cardiomyopathy with congestive heart failure).

FREEZE BALLS

Untoward effects: Infectious hepatitis in all ten **members** of a family (age 2–40) exposed to some leaking plastic ***ice balls*** imported from Hong Kong. Cultures indicated probability of fecal contamination of their water content.

FREONS

= *Arctons* = *Frigens* = *Ledons*

Untoward effects: Inhalation of 20% vapors of ***dichlorofluoromethane*** led to wheezing, analgesia and confusion, but not unconsciousness. Tremors and convulsions on high concentrations. Non-flammable, but, when heated, produce ***phosgene***, ***chlorine***, and a volatile acid.

Animals survive 20% vapors for 120 h; 80% for only 15 min, but die by 30 min.

Check individual freons in the index.

FRITILLARIA MELEAGRIS

= *Chequered Daffodil* = *Drooping Tulip* = *Fritillary* = *Snake's Head*

Untoward effects: Contains a cardiac depressant alkaloid, ***imperialine***, and has caused poisoning of **people** in Europe.

FROGS

Untoward effects: ***Batrachotoxin*** (q.v.), one of the most toxic chemicals known to **man**, is found in the vividly colorful skin of the Columbian kokoi **frog**, *Phyllobates aurotaenia*, and *P. bicolor*, and used in arrow and dart poisons. A less potent, but also lethal toxin, ***atelopidtoxin***, is found in the golden arrow **frog**, *Atelopus zeteki* of Panama and other species in Costa Rica and Colombia. The skin of *Dendrobates* sp. is rubbed on blow darts by **Indians** of Ecuador and Colombia for the lethal effect of its toxins.

FRUCTOSE

= D-*Fructose* = *β*-D-*Fructose* = *Fructosteril* = *Fruit Sugar* = *Hexose* = *Laevoral* = *Laevosan* = *Levugen* = *Levulose*

Untoward effects: Epigastric pain, sweating, and flushing have occurred after IV of large volumes of 40% solution or by rapid infusions. In some **patients**, oral intake led to mild nausea, due to sweetness. Severe lactic acidosis has followed IV infusion of large amounts, therefore not recommended as a sweetener for **diabetics**. Its IV use has induced discomfort, hyperuricemia, increased urinary ***calcium*** and ***magnesium***, and decreased ***potassium*** excretion. Ingestion from **infant** foods in those < 6 months led to vomiting, diarrhea, anorexia, failure to thrive, hepatorenal dysfunction, hypoglycemic convulsions, and transient bleeding, usually from a deficiency of fructose-1-phosphate aldolase. Some cases have been fatal. **People** can have a hereditary fructose intolerance with vomiting and severe, prolonged hypoglycemia after ingestion or infusion of small amounts. An IV dose of only 250 mg/kg (17.5 g/kg) can harm these **patients**. IV solutions can cause a local thrombophlebitis. High dietary levels can exacerbate the effects of ***copper*** deficiency. In **pigs**, this has been demonstrated to cause very enlarged hearts within 10 weeks.

Newborn **pigs** and **calves** up to 1 month of age cannot utilize it, nor can **rats**.

TD_{LO}, **mouse**: 5 g/kg.

FRUITS and FRUIT JUICES

Untoward effects: A number of sulfiting agents are effective antioxidants, and are, therefore, added to fruit juices, dried fruits, and cut up raw fruits at restaurants. They can cause allergic reactions. Some **people** are allergic to certain fruits (***oranges, lemons, tomatoes***) and can develop a contact uticaria. A 65-year-old **male** was allergic to all fruits with large pits in the ***plum*** family (***almonds, apricots, avocados, cherries, nectarines***) and ***avocados*** in the ***laurel*** family (also including ***bay leaves*** and ***cinnamon***). **Others** react to the ***apple*** family (***apples, pears***, and ***quinces***). These families do not cross-react. ***Apples, grapefruit, lemons, limes***, and ***oranges*** can trigger migraines in some **people**. ***Akee*** fruit can help lead to hypoglyeemia. Fruit pits can cause intestinal obstruction with abdominal pain and cramping. ***Apricot, choke cherry, peach*** pits, and ***apple*** seeds contain ***cyanide*** (q.v.). Radioactive ***strontium*** and ***potassium*** have been found in imported fresh and canned fruits. ***Lead*** poisoning in two **children** who drank fruit juice stored in an earthenware jug; one died. In a survey of 264 contemporary glazed surfaces 50% were found unsafe for table use, with ***lead*** > 7 ppm; 10% of imported and commercial earthenware released > 100 ppm ***lead***. Fruit punch, prepared the evening before, was served to a home economics class after storage in containers that had galvanized metal linings with large areas of corrosion. Onset of illness ranged from 5 min to 2 h and intensity varied with amount consumed. In decreasing order, symptoms were nausea, abdominal cramps, metallic taste, headache, dizziness, vomiting, and chills. In other similar outbreaks, fever and diarrhea have also been reported, and ***zinc*** levels > 1000 ppm. ***Erythromycin*** is a basic substance and rapidly hydrolyzed by acidic juices. ***Erythromycin stearate*** bioavailability was not altered by ***orange*** juice.

FUELS, OIL

Untoward effects: Exposure to fuel oils may cause a rapidly progressing glomerulonephritis. After ingestion (especially in **cattle**) chlorinated chemicals, ***sulfur***, and ***lead*** contamination in used oils can add to its toxicity. Also, see ***kerosene***. ***Diesel***: Acute renal failure reported after its use as a hair shampoo. Diesel engines give off practically no ***carbon monoxide***, but give off 1000 times as much ***nitric oxide*** as is considered safe for **man** to breathe. Feedlot **calves** on the bottom deck of short-stacked diesels hauling them did poorer than those on the top deck. Contact dermatitis reported in **cows**.

Jet: Airport **workers** exposed to its vapors daily/5 years developed increasing symptoms of neurasthenia, psychasthenia, and polyneuropathy. Hyperactivity and polydipsia in **rats** given it by gavage.

TD_{LO}, **human**: 500 mg/kg.

FULLER'S EARTH

Untoward effects: It has induced hypercalcemia and distension of the bowel in a 39-year-old **male** treated for ***paraquat*** poisoning.

FUMAGILLIN

= *Amebacilin* = *Fugillin* = *Fumadil B* = *Fumidil*

Antibiotic, antiprotozoal.

Untoward effects: May cause headache, insomnia, nausea, vomiting, intestinal spasms, diarrhea, urticaria, and maculopapular and vesicular skin eruptions.

Photosensitivity in **mice** after SC injection. Toxic to some **dogs**. DO NOT use on **bees** to control Nosema disease during or immediately before ***honey*** flow, to prevent secondary exposure of **humans**.

SC, LD_{50}, **mice**: 800 mg/kg.

FUMARIA sp.

Untoward effects: ***F. officinalis*** extract, ***Oddibil***, is used in treating migraines and biliary diseases. Dry mouth, metallic taste, nausea, palpebral edema, slight vertigo, and slight epigastric pain have occasionally been reported. Contains ***fumaric acid***.

F. parvifolia led to hypoglycemia in **rabbits**. Apparently, safe for **man** at 3–6 g/kg.

FUMARIC ACID

= *Allomaleic Acid* = *Boletic Acid*

Use: Paper size resins (40%), food acidulant (20%), polyester and alkyd resins (20%), and in oils, inks, and lacquers.

Untoward effects: Gastrointestinal complaints, disturbed liver function, acute interstitial nephritis, and tubular cell toxicity.

LD_{50}, **rats**: 10.7 g/kg.

FUMITREMORGIN B

Untoward effects: A tremorgenic mycotoxin produced by *Aspergillus fumigatus*, *A. caespitosus*, and *Penicillium piscarium* growing on plants, apparently, causing staggers in **sheep** and neurological disorders in **cattle**, after grazing on some of the dead vegetative material.

FUMONISIN

A toxin from *corn* molds, *Fusarium moniliforme* and *F. proliferatur*.

Untoward effects: Suspect as a carcinogen for **man**. 10 ppm has killed > 100 **horses** with leukoencephalomacia (ELEM); 30 ppm led to pulmonary edema, reproductive problems, and neonatal deaths in **swine**; deaths in **chicks** at 10 or 30 days of age; 50 ppm led to liver cancer in **rats**; 100 ppm led to hepatotoxicity and decreased gains in **cattle**; 20 ppm led to decreased rate of gain and increased susceptibility to bacterial infection in **catfish**.

FUR

Untoward effects: Dermatitis, asthma, "furrier's lung", and anaphylactoid reactions from contact with the skins themselves, or the acids, chemicals, and dyes used on them. Anthrax infections have been a problem occasionally.

FURALTADONE

= *Altabactina* = *Altafur* = *Furasol* = *Furazolin* = *Furmethonol* = *Ibifur* = *Medifuran* = *NF-260* = *Nitraldone* = *Nitrofurmethone* = *Otifuril* = *Sepsinol* = *Ultrafur* = *Unifur* = *Valsyn*

Antibacterial.

Untoward effects: Retinal toxicity with adverse effects on color vision by selective action on cones, as well as ocular palsies, blurred vision, diplopia, and nystagmus. Sensory and motor changes in peripheral nerve function occur. A case of cerebral nerve paralysis and four of atrial fibrillation. Has caused hemolytic anemia in **patients** with glucose-6-phosphate dehydrogenase deficiency. **Patients** on it have had an intolerance to ***ethyl alcohol***.

FURAZOLIDONE

= *Colifuran* = *Furovag* = *Furoxane* = *Furoxone* = *Giardil* = *Giarlam* = *Medaron* = *Neftin* = *NF-180* = *Nicolen* = *Nifulidone* = *Ortazol* = *Roptazol* = *Tikofuran* = *Topazone* = *Tricofuran*

Antibacterial, antiprotozoal.

Untoward effects: Nausea and vomiting occur frequently. Agranulocytosis, headache, anorexia, fever, leukocytosis, eosinophilia, hypoglycemia, vesicular rash, urticaria, restlessness, insomnia, orthostatic collapse, acute pulmonary hypersensitivity with shortness of breath and pleuritic chest pain. Hemolytic anemia in glucose-6-phosphate dehydrogenase-deficient **patients**. Usually causes a harmless brownish discoloration of the urine.

LD_{LO}, **human**: 500 mg/kg.

Drug and food interactions: It is a ***monoamine oxidase inhibitor*** when given for 4 or more days, therefore interacts with many foods and drugs (***alcohol***, ***amphetamines***, ***tricyclic antidepressants***, ***antiparkinson drugs***, ***barbiturates***, ***catecholamines***, ***choral hydrate***, ***cocaine***, ***insulin***, ***meperidine***, ***phenothiazines***, ***propranolol***, ***reserpine***, and foods with ***sympathomimetics*** or ***tyramines*** enhancing their effects. Reverses the effect of ***α-methyldopa***. ***Disulfiram***-like reaction in many **patients** consuming ***alcohol***, and hypertensive crises (such as agitation, tremors, pyrexia, subarachnoid bleeding, and even death can happen from intracerebral bleeding) have occurred with ***sympathomimetics***. May enhance the effects of ***oral hypoglycemics*** and ***chloral hydrate***.

Overdoses can cause neurotoxic symptoms in **animals**. The FDA has established a zero tolerance for it in uncooked edible tissues of treated **animals** and **poultry**. **Poults** fed ~2.25 times maximum recommended dose for 5 weeks continuously had a 77% incidence of round heart disease; **chicks** and **ducklings** fed ~3 times maximum recommended dosage have increased deaths from cardiomyopathy; sexual maturity delayed indefinitely in male **birds**; induces mammary neoplasia in **rats**; and **calves** consuming 4 mg/kg/day develop diarrhea, hematuria, and fatal hemorrhagic diathesis in < 8 weeks.

LD_{50}, **rat**: 2.34 g/kg; LD_{LO}, **mouse**: 793 mg/kg; **chicken**: 150 mg/kg.

FURAZOLIUM CHLORIDE

= *Dermafur* = *NF 963* = *Novofur*

Antibacterial. A nitrofuran.

Untoward effects: Skin reactions from topical use in **children**. Interferes with ***creatinine*** urine test. Will color urine pink–purple–rust shades.

Oral use caused skin tumors in **rats**, leading to its withdrawal from the marketplace.

FURFENDREX CYCLAMATE

= *E-106E* = *Frugalan* = *Furfenorex*

Anorexic.

***Untoward effects*:** Acute facial eczema, general urticaria, nervousness, insomnia, vertigo, nausea, and epigastric pain in limited reports.

FURFURAL

= *Fural* = *Furfuraldehyde* = *Furfurol*

***Untoward effects*:** Irritant to eyes, upper respiratory tract, and, particularly, skin. Headache, dermatitis, unconsciousness, loss of sense of taste, and numbness of tongue reported from exposures.

Death in **rabbits** and **mice** from pulmonary congestion after inhalation. Liver, kidney, blood, and coagulation tests more adversely affected by SC than inhalation route.

Inhalation LD_{50}, **mice**: 222.7 mg/kg (1 day); **rabbit**: 500 mg/ kg (30–50 min).

FURFURYL ALCOHOL

= *2-Furylmethanol*

***Untoward effects*: *Acute*:** Excitement, drowsiness, nausea, vomiting, salivation, diarrhea, dizziness, decreased temperature, shortness of breath, irregular breathing, and diuresis. Eye contact leads to serious irritation, redness, tearing, and corneal opacity. Irritant to mucous membranes and skin.

***Chronic*:** Headache, dermatitis, and eye irritation. OSHA sets exposure limits at 50 ppm (200 mg/m^3). 75 ppm is immediate danger to life or health.

Daily exposure of **mice**, **rats**, and **rabbits** leads to tachypnea, respiratory tract irritation, and vascular congestion, followed by bradypnea, decreased blood pressure, weak cardiac contractions, sensory nerve paralysis, and death.

LD_{50}, **rat**: 275 mg/kg; **mouse**: 160 mg/kg; inhalation LC_{50}, **rat**: 233 ppm/4 h.

FUROSEMIDE

= *Aisemide* = *Beronald* = *Desdemin* = *Disal* = *Discoid* = *Diural* = *Dryptal* = *Durafurid* = *Errolon* = *Eutensin* = *Frusemide* = *Frusetic* = *Frusid* = *Fulsix* = *Fuluvamide* = *Furesis* = *Furo-Puren* = *Furosedon* = *Fursemide* = *Hydrorapid* = *Impugan* = *Katlex* = *Lasilix* = *Lasix* = *LB-502* = *Lowpston* = *Macasirool* = *Mirfat* = *Nicorol* = *Odemase* = *Oedemex* = *Profemin* = *Rosemide* = *Rusyde* = *Trofurit* = *Urex*

Diuretic, antihypertensive.

Untoward effects*: *Blood*:** Decreased renal clearance of urates, slight increase in β-lipoproteins, thrombocytopenia in an 87-year-old **male** treated > 5 years; hyperglycemia, and increased or false positive blood urea nitrogen and plasma catecholamines. Rapid and high dosage IV leads to hypokalemia, decreased blood volume, dehydration, shock, orthostatic hypotension, collapse, hyponatremia, hypochloremia, hyperuricemia, ototoxicity, transient deafness and tinnitus (especially in **patients** with renal disease or **those** taking ***aminoglycoside antibiotics), interstitial nephritis, vertigo, anaphylactoid reactions, burning paresthesia, and hepatotoxicity (rare). Delirium has followed overdosage or intolerance. Vascular thrombosis and embolism can occur, especially in the **elderly** or **debilitated**.

***Oral*:** Paradoxically, several cases of induced edema on large doses. Pancreatitis, rash, photosensitivity, erythema multiforme, epidermolysis bullosa (including bullous pemphigoid and bullous hemorrhagic lesions), pruritic rashes, pseudoporphyria cutanea tarda, hypercalciuria, hypermagnesuria, hypokalemia (occasionally hyperkalemia). Also anorexia, cramping diarrhea, altered color vision (rare), thrombocytopenia, purpura, neutropenia, leukopenia, eosinophilia, agranulocytosis, hemolytic anemia, aplastic anemia, macrocytic anemia, and possible association with increase in lung cancer. As a drug of abuse without significant side-effects in **patients** with a neurotic compulsion to lose weight. Contraindicated in anuria and hepatic coma. Aggressive use can lead to metabolic alkalosis. Wheezing also reported.

***Fetus and neonate*:** Crosses the placenta leading to increased **fetal** urine production. Increased incidence of patent ductus arteriosus in premature **infants** with respiratory distress syndrome, as well as secondary hyperparathyroidism, bone disease, and calcification in the heart and kidney.

IV, TD_{LO}, **man**: 29 mg/kg.

Drug interactions*:** Chronic ***anticonvulsant therapy may reduce its gastrointestinal absorption and decrease its diuretic action. By increasing blood sugar, it can antagonize the hypoglycemic effect of ***insulin*** or oral ***sulfonylureas***, such as ***acetohexamide***, ***chlorpropamide***, and ***tolbutamide***. ***Acetaminophen***, ***aspirin***, ***flurbiprofen***, ***ibuprofen***, ***indomethacin***, ***flurbiprofen***, and ***naproxen*** significantly reduce its diuretic effect, and the concentration and potential toxicity of this group may increase. Nephrotoxicity, especially in the **elderly**, occurs with ***cephaloridine***. Use with ***chloral hydrate*** leads to diaphoresis, hot flashes, hypertension, and uneasiness apparently by inducing increased metabolic level with increased ***thyroxine***. Given > once/day with ***digitalis***, ***digoxin***, and other cardiotonic drugs increases ***potassium*** loss and has caused arrhythmias. ***Probenecid*** dramatically increases its serum levels. Given to edematous **patients** with ***aminoglycoside antibiotics, kanamycin,*** or ***tobramycin*** may increase chances of oto- and nephrotoxicity. Given with ***monoamine oxidase inhibitors***, it can augment hypotensive effect (including shock). Non-depolarizing (***gallamine*** and ***d-tubocurarine***) and depolarizing (***decamethonium*** and ***succinylcholine***) skeletal muscle relaxants with it can prolong paralysis of respiratory muscles. Increases plasma levels and β-blocking effects of ***propranolol***. With ***aminophylline*** infusions, it increases ***theophylline*** urinary excretion. May increase ***lithium*** serum levels. It can precipitate renal failure with ***cisplatin*** and cause toxic reactions with ***methotrexate***. ***Hydralazine*** decreases its half-life. The addition of ***piroxicam*** to therapy with it increases symptoms of edema and congestive heart failure. Enhances the myocardial uptake of ***radiopharmaceuticals***. Delayed absorption if taken with foods. Wernicke's encephalopathy in an 85-year-old **female** taking it with 40 mg ***amiloride*** (***Co-amilofruse***) for 3 years. Use with ***acetazolamide*** in preterm **infants** with periventricular dilation and hemorrhage worsened the poor prognosis. In preterm **infants** receiving ***vancomycin*** with ***aminophylline*** or ***theophylline***, addition of furosemide decreases ***vancomycin*** serum levels.

Residues in food-producing **animals** or in their milk is prohibited for **human** consumption in the U.S. Enhances ***cephaloridine*** and ***cephalothin***-induced nephrotoxicity in **rats** and **mice**. Causes a dose-dependent hepatic necrosis in **mice**, and dehydration and electrolyte imbalance in any species.

LD_{50}, male **rats**: 2.8 g/kg; female **rats**: 2.6 g/kg; **mice**: 4.6 g/kg; **dogs**: > 1 g/kg; IV, LD_{50}, **mice**: 308 mg/kg.

FUSEL OIL

Untoward effects: Along with ***alcohol***, it is responsible for hangovers after consuming ***brandy***.

In tests on **rats**, it was more hepatotoxic than ***alcohol***.

TD_{LO}, **mouse**: 12.5 g/kg/5 days.

FUSIDIC ACID

= *Fucithalmic* = *SQ 16,603*

Fusidate, Sodium = *Fucidin* = *Fucidina* = *Fucidine* = *Fucidin Intertulle* = *SQ 16,360*

Antibacterial.

Untoward effects: Mild gastrointestinal upsets are common in **children** 1–3 years with staphylococcal disease or salmonellosis (abdominal pain 6/120, diarrhea 3/120, vomiting 2/120), as well as transient exanthema with eosinophilia 3/120, hypoproteinemia and increased SGPT, indigestion, and anorexia. Fatal cases of bleeding peptic ulcers, possibly due to the fact that it has a ***cyclopentophenanthrene*** ring, as do anti-inflammation steroids. Granulocytopenia reported in several cases when given with ***floxacillin***. Occasional reports of jaundice and skin reactions. Venospasms and thrombophlebitis with the IV diethanolamine form. Can interfere with plasma ***hydrocortisone*** assays.

Cholestyramine interferes with its absorption in **rats**.
LD_{50}, **mouse**: 1.2 g/kg.

G

GABAPENTIN

= *CI-945* = *Gö-3450* = *GOE-3450* = *Neurontin*

Anticonvulsant. **Also used in pain control of peripheral neuropathy.**

Untoward effects: Somnolence (1.2%), dizziness (0.6%), ataxia (0.8%), peripheral edema, aggressive behavior in **children**, mouth sores, cytopenia, fatigue (0.6%), nystagmus, nausea and/or vomiting (0.6%), alopecia (rare), and rash (rare). Acute overdoses up to 49 g reported leading to double vision, slurred speech, drowsiness, lethargy, and diarrhea. **All** recovered with supportive care. ***Antacids*** (***Maalox***®) reduces bioavailability by ~20%, but only a 5% decrease if given 2 h apart. Weight increase (> 3 kg) in 11 **patients** (18–56 years) receiving 1.8–3.6 g/day/mean of $3\frac{1}{2}$ months. Anorgasmia reported.

Given to **mice** at 200–2000 mg/kg/day, and to **rats** at 250–2000 mg/kg/day for 2 years associated with increased non-metastasizing pancreatic acinar cell adenomas and carcinomas in male **rats** at the highest dose. Increased incidence of postimplantation fetal loss in **rabbits** on 60–1500 mg/kg/day. Fetotoxic in **mice** given 1–3 g/ kg/day during organogenesis. Acute toxicity in experimental **animals** leading to ataxia, dyspnea, ptosis, sedation, hypoactivity, and excitement.

GABEXATE MESYLATE

= *FOY* = *Megacert*

Protease inhibitor. **Parenteral treatment of pancreatitis and disseminated intravascular coagulation.**

Untoward effects: Several case reports of anaphylaxis after IV use.

LD_{50}, **mouse**: 8 g/kg; IV, 25 mg/kg.

GADODIAMIDE

= *Omniscan*

Magnetic resonance imaging contrast agent.

Untoward effects: Acute pancreatitis in a 58-year-old **male** 3 h after 16 ml IV led immediate fatigue, nausea, vomiting, and sweating; followed later by epigastric tenderness and intense pain radiating to the chest.

GADOTERIDOL

= *ProHance*

Non-ionic contrast medium for magnetic resonance imaging.

Untoward effects: Nausea and taste perversion in 1.4% of **patients**. < 1% had facial edema, pain at injection site, tachycardia, dry mouth, anxiety, dizziness, dyspnea, rash, and tinnitus. Avoid its use or use caution in **patients** with renal or liver disease, and **those** with a history of grand mal seizures.

GAILLARDIA

= *Blanket Flower*

Untoward effects: Several cases of dermatitis venenata reported.

GALACTOFLAVIN

Untoward effects: ***Riboflavin*** antagonism and symptoms of the resulting deficiency reported in **humans** and **animals**. Teratogenic in **mice** with aplasia of the tibia and fibula; and malformed limbs in **rat** fetuses.

GALACTOSAMINE

Untoward effects: Causes hepatotoxicity and liver necrosis at 400 mg/kg IP in **dogs** and **rats**.

TD_{LO}, **rat**: 135 g/kg/72 weeks.

GALACTOSE

= *Cerebrose*

Untoward effects: Causes cataracts in **humans** with galactosemia (autosomal recessive disorder); in young **rats**; in **guinea pigs** on a scorbutic diet; and in young **kangaroos**. It may be the cause of increased incidence of **human** cataracts in India, where ***yogurt*** is a large part of the diet. Galactose is a major component of ***lactose*** in ***yogurt***. It is a dietary risk factor for ovarian cancer in **women**. It increases urinary ***calcium*** and ***magnesium***, and decreases ***potassium*** excretion. Bacterial infections are common in **infants** with galctosemia, apparently by impairing white blood cell function. Anaphylactic shock reported in a 27-year-old **male** after receiving 67 ml of 30% over 3–4 min as a liver function test.

Causes neurotoxicity in **chicks**. The **goat's** placenta is freely permeable to it. SC to **mice** causes occasional sarcomas. High dietary intake in **non-ruminants** leads to diarrhea and death. **Chicks** fed a diet with 55% level develop violent wing and leg spasms, and die within a few days.

TD_{LO}, **rat**: 50 g/kg (8–10 days pregnant).

GALALITH

Untoward effects: An **organist** and a **pianist** developed an occupational dermatitis from galalith, a synthetic plastic used in the manufacture of the keys for **their** instruments.

GALANTHAMINE

= *Galantamine* = *Galantone* = *Jilkon* = *Lycoremine* = *Nivalin*

***Cholinesterase inhibitor.* An alkaloid from *Caucasian snowdrops* (*Galanthus woronowii*).**

Untoward effects: Psychomotor excitation. Moderate toxicity and slight sensitivity to it reported in 7/36 **patients** with simple chronic glaucoma.

GALANTHUS NIVALIS

= *Perce-neige* = *Snowdrop*

Untoward effects: The bulb contains ***lycorine*** (***narcissine***) and the symptoms are the same as for ***Hyacinth*** (q.v.). The bulbs were served as emergency food in Holland during World War II; ingestion of large quantities caused nausea, vomiting, and diarrhea.

The leaves are not palatable to **livestock**. Native to Great Britain, southern Europe, and western Asia.

GALBANUM GUM

From Iranian, Indian, Afghanistan, and Asian *Ferula* sp., and used for its antimicrobial action.

Untoward effects: Eczematous contact dermatitis with cross-sensitivity to tincture of ***benzoin***. Contains ***asafoetida*** (q.v.).

GALEGA OFFICINALIS

= *French Honeysuckle* = *Goat's Rue* = *Pro-Forma*

Galegine is probably its key toxic agent.

Untoward effects: Lactation, disruption of menstrual cycle, fever, stomach spasms, and dysmenorrhea reported in a **woman**.

Ingestion by **cows**, but not in **dogs** or **goats**, increases milk secretion. Poisoning and deaths in **sheep** eating it in hay or freshly cut fodder, reported from France and Hungary. Symptoms develop after 18–24 h and lead to death in a few hours from severe progressive dyspnea with pulmonary congestion or edema and voluminous hydrothorax. 400–500 g of the green plant or 3 g of the dry plant/kg will kill a mature **sheep**. Hypoglycemic in **rats**. Plant has an undesireable odor and is unpalatable.

Galegine sulfate - LD_{50}, **mice**: 122 mg/kg.

GALENA

Untoward effects: These rocks contain ***lead*** and should not be used in aquaria.

LD_{LO}, **guinea pig**: 10 mg/kg.

GALEOPSIS

= *Hempnettle*

Untoward effects: Feed containing 1.1–2.5% of its seeds killed 4/18 **horses** and 79/150 **pigs**, after they had ingested it for about a week. Symptoms were anorexia and coma. Found throughout Europe and Siberia.

GALERINA sp.

(autumnalis, marginata, and venenata)

Commonly found on lawns, meadows, and decaying woods.

Untoward effects: A toxic ***mushroom***, frequently collected by **amateurs** in the U.S., unaware that it can cause serious illness. Contains ***amanitins***, and symptoms in **man** from ingestion are the same as for ***Amanita phalloides*** (q.v.), despite the fact that it contains no ***phalloidin***. These include an explosive type of violent abdominal pain, nausea, vomiting, diarrhea, hematuria, fever, tachycardia, hypotension, and jaundice. Convulsions and death can follow. Symptoms usually occur 6–24 h after ingestion.

GALITOXIN

Untoward effects: The key resinoid toxic principle in ***milkweeds, Asclepias* sp.** (q.v.), adversely affecting **sheep**, **cattle**, and **horses**.

GALIUM sp.

Untoward effects*: *G. triflorum* = *Sweet-Scented Bedstraw*.** Its ***coumarin content has caused hemorrhages in **cattle**. The roots, when eaten by **animals**, color their bones red.

G. vercum roots color the bones yellow. The dye, ***pseudo-purpurin***, is found in ***Natural Red 8*, *9***, and ***14***.

GALLAMINE TRIETHIODIDE

= *Benzcurine Iodide* = *F-2559* = *Flaxedil* = *Relaxan* = *Retensin* = *RP-3697* = *Tricuran*

***Skeletal muscle relaxant.* A synthetic non-depolarizing type used IV and IM. Antagonist of *acetylcholine* that acts by blocking *choline*-sensitive receptors on motor end plates.**

Untoward effects: As for ***tubocurarine*** (q.v.) or ***curare*** (q.v.). Tachycardia, hypertension, and increased cardiac output, due to release of ***catecholamines***. This can increase when ***anticholinergics, phenothiazines***, and ***tricyclic antidepressants*** are taken. Gallamine itself can also have a central anticholinergic action. Affects all skeletal muscles, including those of the diaphragm, therefore apnea is common. Causes bronchoconstriction due to histamine release. Has caused urticarial wheals and angioneurotic edema, as well as anaphylactoid reactions. Eliminated by the kidneys, and renal failure can prolong its induced muscle paralysis, as well as causing some **patients** to experience postoperative recurarization that has lasted for 48 h, with flaccid paralysis and severe respiratory paralysis. May cause erroneously high protein-bound iodine readings.

Drug interactions: ***Antibiotics***, such as ***kanamycin, neomycin, polymyxin B***, and ***streptomycin*** can cause serious potentiation. ***Acetazolamide, chlorpromazine, diazepam, chlorthalidone, ethacrynic acid, furosemide***, and ***quinethazone*** can also cause profound potentiation of duration and intensity of its neuromuscular block. ***Thiazide diuretics***, such as ***chlorothiazide***, enhance its activity by ***potassium*** depletion. Ventricular arrhythmogenicity IV, LD, **human**: ~1.1–2.9 mg/kg, when used with ***cyclopropane***. Synergism, rather than a simple additive effect, when used with ***tubocurarine***. ***Cincona*** alkaloids, ***ether***, and ***quinidine*** can potentiate its respiratory depression. Can cross the placenta, but, in usual doses, ill-effects on the **fetus** have not been reported. Be prepared to administer artificial respiration and ***neostigmine***. When the latter is metabolized, artificial respiration may have to be re-instituted.

Increased heart rate and decreased respiratory rate in **sheep**, **goats**, **dogs**, and **buffalo calves**.

LD_{50}, **mouse**: 425 mg/kg; **rabbit**: 100 mg/kg. IV, LD_{50}, **rat**: 5.5 mg/kg; **mouse**: 1.8 mg/kg; **rabbit**: 0.4 mg/kg.

GALLIC ACID

***Antioxidant, astringent.* Chemical intermediate. In manufacturing permanent inks; in engraving and lithography; as a developer in photography; in fur dyeing; and in dyes, wallboard, and pyrotechnics.**

Untoward effects: A ***tannic acid*** metabolite that causes ***egg*** yolk mottling, when excesses occur in feed for **hens**.

LD_{LO}, **human**: 500 mg/kg.

LD_{50}, **rats** and **rabbits**: 5 g/kg; IV, **mice**: 320 mg/kg.

GALLIUM

= *Ga*

***Ga 67* has an affinity for human and animal neoplasms, and some infected tissues, when injected IV.**

Untoward effects: Excreted in **human** breast milk, and may be bound to ***lactoferrin***. Can cause skin rashes and a metallic taste. The citrate form of ***Ga 67*** has accumulated in inflamed mucosa in pseudomembranous colitis, and has had an abnormal accumulation in the lung after ***bleomycin*** therapy. ***Ga 67 Nitrate = Ganite = NSC 15,200***. Renal tubular plugs can lead to renal failure, unless adequate hydration of the **patient** is maintained. Hypocalcemia reported with acute toxicity and hypophosphatemia, nephrotoxicity, anemia, optic neuritis, and bone marrow depression with chronic toxicity. Contraindicated with nephrotoxic drugs, including ***aminoglycoside antibiotics***.

Ga 67 has been taken up by infarcts in **dogs**. The sulfate form of ***Ga 67*** has been teratogenic in **hamsters**.

Nitrate LD_{50}, **mice**: 80 mg/kg.

GALLOPAMIL

= *Algocor* = *D-600* = *Methoxyverapamil* = *Procorum*

Antianginal, calcium channel-blocker.

Untoward effects: 16% increase in ***digoxin*** palsma concentration.

Occasionally leads to spontaneous tonic-clonic seizures in **mice**.

GALVANIZED METALS

Untoward effects: ***Chromium*** dermatitis in **man** from ***chromates*** used to prevent rusting on the surface of ***iron*** sheets galvanized with ***zinc***. Chronic alkali eczema in **galvanizers** after prolonged contact with alkalis. ***Cadmium*** dissolved from galvanized pipes suspect as a cause of hypertension and sudden death.

GAMBIERDISCUS TOXICUS

Untoward effects: This marine dinoflagellate produces ***maitotoxin*** (***MTX***) which is passed up the food chain; when concentrated in certain **fish** (q.v.), causes **ciguatera fish** poisoning.

GAMBOGE

= *Gambogia*

A gum-resin from *Garcinia hanburyi* in the East Indies.

Untoward effects: Violent purgative for **man** leading to severe vomiting and diarrhea, cardiovascular collapse, and death. Once, used as a water-color pigment and in large **animal** purgatives at 4–15.5 g/dose. 4 g has been reported as fatal for **man**. Topical cross-sensitivity with tincture of ***benzoin***.

LD_{LO}, **human**: 50 mg/kg.

GANCICLOVIR

= BIOLF-62 = BW-759 = BW-B759U = BW-759U = Cymevan = Cymevene = Cytovene = Denosine = DHPG = 2'NDG = RS-21592

***Antiviral.* IV. *Vitrasert* = Intraocular implant.**

Untoward effects: Neutropenia (13–67%) is the most frequent one. Adverse hematological effects increase (9/10) when therapy is combined with ***zidovudine***. Pancytopenia, leukopenia, granulocytopenia, thrombocytopenia, increased liver transaminases, phlebitis, fever, rash, confusion, seizures, and psychosis reported. Vitiligo in a 44-year-old **male** 3 weeks after 5 mg/kg twice daily/5 days.

Drug interactions: May decrease ***cyclosporine*** serum levels and decrease ***didanosine*** renal excretion.

Teratogenic, carcinogenic, mutagenic, and aspermatogenic in various experimental **animals**.

GARCINIA sp.

Untoward effects: The resinous bark exudates of ***G. hanburyi***, ***G. lanessonii***, and ***G. morella*** of southeastern Asia and ***G. polyantha*** of Africa used as drastic purgatives, in large doses, cause nausea, vomiting, griping diarrhea, and even death (within 24 h).

G. kola contains potent antioxidant bioflavanoids and, in Africa, its nuts are chewed for enhancing the immune system, and for treating hepatitis, arthritis, and viral diseases. **Natives** use it as an antidote for *Strophanthus gratus* (***ouabain***) poisoning. Also see ***Gamboge***.

GARDONA

= Appex = Dietreen = Endrol = ENT-25841 = Rabon = Rabond = ROL = SD-8447 = Stirofos = Tetrachlorvinphos

***Insecticide, acaricide.* Organophosphorus anticholinesterase.**

Untoward effects: Highly toxic to **bees**. Oral doses of 94 mg/kg/13 days to **hens** had no ill-effects; 188–752 mg/kg/13 days led to inactivity and lethargy by day 3; by day 14, 60% on 752 mg/kg had died. 1 g/kg to **fowls** leads to central nervous system disturbance, incoordination, depression, paresis, catarrhal hemorrhagic enteritis, and decreased blood acetylcholinesterase. 800 ppm to laying **hens** caused decreased body weight and decreased egg production; at 400 ppm, body weight was maintained, but ***egg*** flavor deteriorated. No residues detected in the ***eggs***, but tissue residues were 0.02–0.03 ppm. **Pheasants** show convulsions, tremors, and prostration in ~2–3 days. Limited evidence that it may be carcinogenic in **rats** and **mice**. Has caused severe keratitis in a **horse**. Avoid contamination of food and water supplies. Avoid inhaling mists or skin contact. Contrary to FDA news releases, residues can appear in **cows'** milk. Use only as directed.

LD_{50}, **rats**: 1.1 g/kg; **mouse**: 1.6 g/kg, 10–12-day-old **chicks**: 2.53 g/kg; **chukars**, **pheasants**, **mallard ducks**: ~2 g/kg. Minimum lethal dose **wild birds** and **swine**: 100 mg/kg.

GARLIC

= Ail = Ajo = Allium sativum = Kwai = Sapec

Untoward effects: A field **worker's** hands, forearms, legs, and face developed an erythematous and vesicular dermatitis 1 day after pulling up garlic. Severe toxic contact dermatitis in a 42-year-old **female** 10 h after handling fresh-crushed garlic cloves. Patch tests with garlic juice were positive on the **patient** and negative on two **controls**. **Volunteers** who ate 10 g of fresh garlic without any other food had inhibition of ***serotonin***-induced platelet aggregation. ***Diallyldisulfide*** is the main allergen in garlic, and a cause of "house-wife eczema". Its ***allyldisulfides*** have caused rashes, hypotension, and leukocytosis, and fatalities have been reported when garlic has been overused in **children**. Ingestion may trigger migraine headaches in some **people**. It contains ***allyl isothiocyanate***, a known carcinogen for **man**. Many **patients** with fissured contact dermatitis from vegetables were most commonly positive to patch tests for garlic. A 29-year-old **female** developed a painful, erythematous, blistering rash on **her** chest and upper abdomen. It was originally misdiagnosed as herpes zoster, until it was determined it was a burn from the application of a compress of crushed garlic, wrapped in cloth, to these areas for 18 h. In 1594, Thomas Nagle wrote in *The Unfortunate Traveler* that "Garlick makes a **man** winke, drinke, and stinke". Enhances bleeding in **patients** taking oral ***anticoagulants***. Occupational asthmatic reactions to garlic dust reported.

Horses, **cattle**, and **sheep** have died from overeating it. ***Wheat*** is often contaminated with it and fed to **cows**, where its strong flavor and odor can readily be detected in their milk. Wild garlic, *Allium vineale*, in pastures is often the culprit. **Cows** can inhale the volatile substances in garlic and eliminate them in their milk. Intravascular hemolysis has been reported from it in **cats** and **dogs**.

GASOLINE

= Benzin = Petrol = Petroleum Ether

Untoward effects: ***Contact***: Skin and eye irritation. Defatting of skin; dermatitis and chemical burns and ulcers with repeated exposures. Possible cataract due to splashing gasoline on the eye. Occupational dermatitis from ***alcohol*** denatured with gasoline.

Ingestion: Vomiting, inebriation, drowsiness, vertigo, confusion, and fever. Excessive amounts also cause some excitation, unconsciousness, convulsions, cyanosis, and pneumonitis, within 24 h. Capillary hemorrhaging in lungs and internal organs, and death from circulatory failure. A death reported after ingestion of $\frac{1}{2}$ oz and $\sim\frac{1}{3}$ oz in **children**.

Inhalation*:** Irritation and burning of nose, mucous membranes of the mouth and lungs, mydriasis, dizziness, systemic fat embolism, anesthesia, asphyxiation, and, eventually, bronchopneumonia. 15–20 deep breaths of the vapors caused euphoria, ataxia, and disorientation lasting 5–6 h. Repeated exposures led to nausea, vomiting, diarrhea, headache, dizziness, insomnia, peripheral neuritis (probably from ***triorthocresyl phosphate component), acute encephalopathy, hallucinations, euphoria, abnormal behavior, homicidal tendencies, loss of sense of direction or pain (analgesia), myoclonus, addiction, dyspnea, anorexia, glomerulonephritis, thrombocytopenia, neutropenia, hypochromia, and anemia. Autopsies on adolescent **gasoline-sniffers** found severe arteriosclerotic damage to the coronary vessels. Very high levels led to central nervous system suppression, ataxia, coma, cardiac arrhythmias, convulsions, hepatotoxicity and sudden death, and hypothrombinemia in **people** on ***warfarin***.

***Injection*:** A 38-year-old **male** with a long history of drug abuse injected 5 ml IV into **his** right forearm. Some may have gone IM and SC, with severe swelling of hand and forearm from damage to **his** right median, ulnar, and radial nerves.

Dermatitis after repeated handwashing with gasoline containing ***tetraethyl lead***. ***Lead*** encephalopathy reported in **gasoline-sniffers**, four **workers** cleaning a tank/6 days that had held leaded gasoline, and in the inexperienced siphoning gasoline (especially during the energy crunch of the 1970s), with unintentional ingestion and aspiration. Numerous other additives can cause their own adverse effects.

Has been implicated in the etiology of various occupational neoplasms, including those of the nose, larynx, and pancreas. ***Benzene*** (q.v.) content can aid carcinogenicity and hematotoxicity.

LD_{LO}, **human**: 500 mg/kg; TC_{LO}, inhalation, 900 ppm/1 h. **Human** eye: vapor irritation, 140 ppm/8 h.

Abomasal displacement after four **heifers** drank 3 qts. Increased numbers of male **rats** with kidney cancer as vapor levels increase and duration of exposure increases. Exposing **cats** and **dogs** to its vapors sensitizes the heart to the action of catecholamines, resulting in cardiac arrhythmias, vagal paralysis, and, occasionally, sudden death. Visceral congestion and petechial hemorrhages, particularly in the lungs of **animals** after death by inhalation.

Guinea pig and **rabbit**, skin irritant: 375 mg/24 h.

GASTROLOBIUM sp.

Untoward effects*:** Palatable plants on Australian ranges containing varying amounts of ***fluoroacetates and ***fluoroacetamides*** that kill **sheep** ingesting them. Symptoms are weakness, tachycardia, dyspnea, hyperexcitability, mania, muscular spasms, dejected appearance, and collapse.

G. callistachys = ***Rock Poison***.

G. calycinum = ***York Road Poison***.

G. floribundum = ***Wodjii Poison***.

G. grandiflorum = ***Heart Leaf*** poisoned 2000 **sheep** at one location and caused the deaths of 150–300 **cattle** in another location.

G. laytonic = ***Breelya Poison*** = ***Kite-leaf Poison***.

GATIFLOXACIN

= *AM 1155* = *Tequin*

Untoward effects*:** Gastrointestinal upsets, skin rashes, dizziness, nervousness, restlessness, tremors, palpitations, confusion, hallucinations, insomnia, and seizures are apt to be the most common adverse effects of this relatively new ***fluoroquinolone. Eventually, tendinitis will probably be reported, especially in young **patients**.

GAULTHERIA ANASTOMOSANA

***Untoward effects*:** Glycosides from this low shrub in Colombia poison **cattle** and **sheep**.

GEFARNATE

= *71,285* = *Alsanate* = *Arsanyl* = *DA-688* = *Dixnalate* = *Gefanil* = *Gefarnil* = *Gefarnyl* = *Gefulcer* = *Osteol* = *Salanil* = *Zackal*

***Antiulcerative*.**

***Untoward effects*:** Urticarial rashes.

LD_{50}, **mice**: > 8 g/kg; IV, 2.8 g/kg.

GEIGERIA sp.

= *Vermeerbos*

Untoward effects*:** Ingestion of ***G. africana or ***G. aspera*** causes "vomiting sickness" or "vermeersiekte" in South African **sheep** and **goats**. Symptoms are debility, stiffness, salivation, paralysis, and vomiting. Only the paralytic effects are noted in **cattle**. Mortality is about 60% and the disease resembles that of ***sneezeweed*** (***Helenium***, q.v.). The toxic principle is ***vermeeric acid***.

GELATIN

= *Gelfoam* = *Puragel*

A natural product derived by the partial hydrolysis of *collagen* from skin, tendons, ligaments, and bone.

Untoward effects*:** Immediate hypersensitivity of lung tissues reported from its use in aerosols. IV use of 250 ml of an ***oxypolygelatin (***Gelifundol***® – a 5.6% solution in normal saline) caused severe anaphylactic shock within 15 min with bronchospasms and cardiac arrest. **Patient** was resuscitated and survived. **He** was sensitized before, while working in a food store. Of 11 pregnant **women** receiving an infusion of ***polygelatin*** (***Haemaccel***®), six showed allergic reactions (angioneurotic edema, urticaria, bronchoconstriction, and hypotension), possibly from ***histamine*** release. A 67-year-old **male** surgical **patient** received 2 l of ***Gelofusine***® and developed acute renal failure. Cardiac arrhythmias developed in fully digitalized **patients** receiving ***Haemaccel***® containing 12.5 mEq ***calcium*** ions. Use of absorbable gelatin sponges (***Gelfoam***®) for hemostasis is dangerous in closed or rigid spaces (e.g. after craniotomy). Increased protein-bound iodine to hyperthyroid levels (9–11 μg/100 ml) in a **patient**, due to ***erythrosine*** as the pink coloring agent of a ***lithium*** capsule. See ***tartrazine*** for adverse effects from its use in coloring ***secobarbital*** and ***tetracycline*** capsules. ***Formaldehyde*** used to harden gelatin capsules, so they can bypass the stomach, has continued its reaction and made the capsules impervious to moisture, allowing them to be excreted intact. When ***calcium sulfate hemihydrate*** (***plaster of Paris***) was substituted as a filler, instead of the ***dihydrate*** (***terra alba***), it extracted moisture from the capsule wall on aging and formed a cement and unavailable bullet.

GELSEMIUM SEMPERVIRENS

= *Carolina Jassaminen* = *Carolina Wild Woodbine* = *Evening Trumpetflower* = *Yellow Jasmine* = *Yellow Jessamine*

Contains alkaloids (*gelsemine, gelseminine,* and *gelsemoidine*).

Untoward effects*:** The entire plant, especially the roots and the flowers with their nectar, has poisoned **children** and various **livestock** species, paralyzing motor nerve endings, causing dizziness, ptosis, muscle weakness, decreased temperature, decreased blood pressure, double vision, dimness of vision, mydriasis, tremors, dropping of the jaw; tachycardia, then bradycardia; bradypnea, dyspnea, prostration, convulsions, respiratory arrest, and death. 12 drops of the fluid extract was fatal to a 3-year-old **male**. ***Honey from the nectar is poisonous.

Gelsemine has the same action as ***nicotine*** (q.v.). LD_{LO}, **human**: 5 mg/kg. Extensive losses in **turkeys** after becoming lethargic, sleepy, and staggery, with loss of control of neck muscles. Hypodermically to **cats**, **rabbits**, and **frogs** led to prostration, mydriasis, convulsions, and asphyxia.

Cestrum nocturnum (***Night-Blooming Jessamine***) in the tropics and up to Florida, Texas, and the Carolinas, has poisoned **children**, **pets**, and **poultry**.

In tropical parts of Asia, ***G. rigens*** is used in committing suicide.

GEMCITABINE

= *dFdC* = *dFdCyd* = *Gemzar* = *LY-188011*

Antineoplastic.

***Untoward effects*:** Initially, fatigue, nausea, vomiting, skin reactions, pruritus, and, eventually, bone marrow depression with thrombocytopenia, neutropenia, granulocytopenia, and leukopenia. Acute respiratory insufficiency in a 69-year-old **female** after treatment with it. Autonomic neuropathy reported in a 55-year-old **male** with lung cancer.

Drug interactions*:** Increases ***warfarin's anticoagulant effect.

GEMEPROST

= *Cergem* = *Cervagem(e)* = *Preglandin*

***Abortifacient, oxytocic.* A *prostaglandin*.**

***Untoward effects*:** Occasional vomiting and diarrhea. Uterine rupture in one case where cervix failed to dilate and severe uterine contractions occurred.

GEMFIBROZIL

= *CI-719* = *Decrelip* = *Genlip* = *Gevilon* = *Lipozid* = *Lipur* = *Lopid*

Antihyperlipoproteinemic.

***Untoward effects*: *Blood*:** Eosinophilia, leukopenia, anemia, and hypokalemia.

***Central nervous system*:** Headache, dizziness, blurred vison, nystagmus, and rombergism.

Gastrointestinal*:** Abdominal pain (6%), epigastric pain (4.9%), diarrhea (4.8%), nausea (4%), vomiting (1.6%), flatulence (1.1%), dry mouth, anorexia, colitis, constipation, and dyspepsia. May increase biliary ***cholesterol saturation leading to increased formation of gallstones.

Miscellaneous*:** Decreased libido and impotence in **men**. Decreased fasting and oral ***glucose tolerance test values and increased SGOT, SGPT, lactate dehydrogenase, creatine phosphokinase, and alkaline phosphatase values. Rhabdomyolysis and acute renal failure in a 63-year-old **male** also receiving ***lovastatin***. Middle-aged **men** with heart disease taking the drug had an increased risk of dying from heart disease and all causes.

Musculoskeletal: Back pain, muscle cramps, myalgia, rhabdomyolysis, arthralgia, and swollen joints.

Skin: Dermatitis, erythema, rash, photosensitivity, pruritus and urticaria.

Drug interactions: Initial increase in prothrombin time that may require significant decrease in ***warfarin*** dosage. This can be followed by a biphasic interaction to previous dose requirements. Can cause dramatic increase in ***lovastatin*** and ***simvastatin*** serum concentrations, increasing the risk of myopathy.

Rats develop liver cancer on 1.3 times the **human** dosage.

GEMTUZIMAB OZOGAMICIN

= *CDP 771* = *CMA 676* = *Gemtuzimab Zogamicin* = *Mylotarg* = *WAY-CMA 676*

Use: Recombinant **human** monoclonal antibody bound to a cytotoxic derivative of ***calicheamicin***, an antibiotic for the treatment, by infusion, of acute myeloid leukemias.

Untoward effects: Within several hours, fever, nausea, hypotension, shortness of breath, and chills, eventually followed by severe neutropenia and thrombocytopenia in most **patients**. Bilirubinemia, increased aminotransferases, and a fatal hepatic failure also reported.

GENTAMICIN

= *Cidomycin* = *Garamycin* = *Garasol* = *Gentabac* = *Genta-Gobens* = *Gentalline* = *Gentalyn* = *Gentamina* = *Genticin* = *Gentamycin* = *Gentavet* = *Gentavetina* = *Gentocin* = *Gevramycin* = *Glevomicina* = *Refobacin* = *Ribomicin* = *Septigen* = *Sulmycin*

***Antibiotic.* An *aminoglycoside.* IV or IM use in humans. The sulfate form is used.**

Untoward effects: Nephrotoxicity is the most common adverse effect (proteinuria, albuminuria, cylinduria, increased blood urea nitrogen, increased ***creatinine***, and azotemia) in up to 55% of **patients**. Although usually reversible with discontinuance of therapy, months may pass before the **patient** returns to normal. A rule of thumb suggests reduced dosage of 50% for blood urea nitrogen of 50–100 mg/ml or serum ***creatinine*** of 2.0–2.2 mg%. Acute tubular necrosis is its more common renal pathology.

Ototoxicity is the next most serious adverse effect (vestibular and cochlear neural degeneration). This can also be induced by topical use in the ear canal. Dizziness and sensorineural hearing loss in **adults** and **children**. Therapeutic serum concentrations are 4–10 mg/l, but can be toxic at 12 mg/l.

Blood: Neutropenia, anemia, agranulocytosis, thrombocytopenia, hypokalemia, and hypomagnesemia. Often increases ***creatinine*** by at least 0.5 mg/dl in older **patients**.

Eye: Conjunctival hyperemia and edema, exacerbation of photophobia, and short-lasting stinging sensation from topical use. Acute ischemic retinopathy and blindness in **patients** erroneously given it as an intraocular injection.

Miscellaneous: Given to pregnant **mothers**, may cause a toxic effect on the eighth cranial nerve (vestibulo cochlear) of the **fetus**. ***Sulfite*** sensitivity from ***sodium metabisulfite*** preservative leads to pain, cyanosis, wheezing, dyspnea, tachypnea, tachycardia, and hypoxemia. Topical use may decrease sense of smell when applied to nasal epithelium. It may cause erroneous increase in SGPT tests. A 21-year-old **female**, 1 h after 450 mg injection, showed shakes, chills, and rigors. Once-daily injections of > 2.5 mg/kg caused similar ***endotoxin***-like reactions in 6/52 **patients**.

Neuromuscular blockade: Myasthenic syndrome. Profound transitory muscle weakness in a few hours after 80 mg tid, then 45 mg tid for postoperative fever in a 64-year-old **male**, including flaccidity and inability to perform voluntary muscular movements. Has caused respiratory muscle paralysis (curariform toxicity with high dosages and certain drugs, including ***anesthetics***). Respiratory arrest reported in an **infant**.

Psychosis: A 40-year-old **female** received 60 mg tid, then 40 mg tid after genitourinary surgery. Lethargy, delusions, vomiting, grand mal seizures, and unresponsiveness, and **she** became mute. A rare occurrence.

Skin: Alopecia, progressive loss of eyebrows, vesicular eruptions, maculopapular rash, and contact urticaria. Cross-reacts with ***neomycin***. Cutaneous necrosis reported after SC use.

Drug interactions: Use with other nephrotoxic or ototoxic ***antibiotics*** (***amphotericin***, ***bacitracin***, ***cephalexin***, ***cephaloridine***, ***cephalothin***, ***colistin***, ***kanamycin***, ***neomycin***, ***polymyxin B***, and ***streptomycin***) should be avoided. Enhances the nephrotoxicity of ***cisplatin***. ***Carbenicillin***, ***piperacillin***, and ***ticarcillin*** may cause a profound decrease in its blood levels if mixed together. Many reports of anuria and acute tubular necrosis and a death, when given with ***cephalothin***. Rapid and irreversible deafness reported with ***ethacrynic acid***, and use with ***furosemide*** reduces gentamicin's urinary clearance. A report of a **patient** with extremely high serum levels while also receiving ***cloxacillin***. ***Indomethacin*** significantly increases its serum levels. Physically incompatible with ***heparin***.

Stinging sensations when used on the eyes of **animals**. Has caused ototoxicity, neuromuscular block, pruritus, and skin lesions in treated **cats**. Nephrotoxicity reported after topical use in a **cat**. Nephrotoxic in the **dog**, **horse**, **sheep**, **cougar**, **rat**, **mouse**, **snakes**, and **turtles**. Causes ototoxicity in **squirrel monkeys** and **guinea pigs**. Neuromuscular blockade and death in **hawks** (20 mg/kg twice daily, IV), **cats**, and **monkeys**. Dose-dependent cardiovascular depression in **monkeys**. Polyuria, polydipsia, and nephrotoxicosis in **psittacine** and **raptorial birds**. IM use causes myositis in many species.

Biological half-life in **snakes** is 82 h, 1½–3 h in **dogs**, and 2–3 h in **man**.

IV, LD_{50} **rats**: 108 mg/kg; **mice**: 57 mg/kg.

GENTIAN VIOLET

= *Adergon* = *Aniline Violet* = *Axuris* = *Badil* = *C.I. 42555* = *C.I. Basic Violet 3* = *Crystal Violet* = *Dye-Gen* = *Gentiaverm* = *GV-Eleven* = *Hexamethyl-p-rosaniline* = *Inter-Gen-16* = *Meroxyl* = *Meroxylan* = *Methylrosaniline* = *Pyoktan(n)in* = *Vianin* = *Viocid* = *Violeta de metilo*

Antimicrobial, anthelmintic.

Untoward effects: Treatment of **infants** with thrush (oral candidiasis) has caused oral ulceration, including grayish, gelatinous-like lesions under the tongue and on the gingiva. Hematuria and hemorrhagic cystitis in a 32-year-old **female** who accidentally injected it into the bladder, instead of the vagina. Irritation and sensitivity from its topical (including vaginal) use in various mixtures, such as ***Triple Dye***. The use of this dye mixture to prevent umbilical sepsis was responsible for severe hemolytic jaundice in glucose-6-phosphate dehydrogenase-deficient **infants**. *Mycobacterium chelonae* infections in eight **patients** after face-lifts and augmentation-mammoplasty, from its use as a surgical skin-marking solution. Very rarely, shock reactions from its use. Its topical use on nails leads to variable purple discoloration of the nail plate.

LD_{LO}, **human**: 50 mg/kg.

Hepatocellular carcinomas at 300 and 600 ppm in the feed after 24 months in male **mice** and 18–24 months in female **mice**. Bile salts neutralize its anti-candidal effects. IV use NOT recommended, as it can cause lung emboli with secondary thrombosis and infarction. Toxic to spermatozoa, and intrauterine use may interfere with conception. May cause undesirable violet staining, although this is deliberately added to some medicaments to demark treated areas.

LD_{50}, **rats** and **dogs**: 1 g/kg; **mice**: 1.2 g/kg; LD_{LO}, **cat** and **guinea pig**: 100 mg/kg; **rabbit**: 125 mg/kg.

GENTISIC ACID

***Analgesic, anti-inflammatory, antipyretic.* Found in *gentian*. The active metabolite of *acetylsalicylic acid* that produces its antipyretic effect due to its accumulation in the pituitary, in proximity to the heat control center in the hypothalamus.**

Untoward effects: Interferes with ***uric acid*** assay and the glucose oxidase reaction. It has caused a negative glucose oxidase test in the presence of known glucosuria. It is an inhibitor of glucose-6-phosphate dehydrogenase.

LD_{50}, **rat**: 800 mg/kg.

GEORGINA GIDYEA

Untoward effects: Although it grows in only a small area of Australia's Northern Territory, it causes the deaths of > 2000 **cattle** annually.

GEOSMIN

Produced by species of blue-green algae and some actinomycetes, both of which flourish in heavily fed fish ponds and muddy river waters.

Untoward effects: Contributes a potent, earthy odor to **fish**, ***beans***, and water supplies. It is absorbed through the gills and digestive tract of **fish**. Oxidation of drinking water decreases its odor. **Human** threshold taste concentration is 0.1 ppb.

GERANIOL

= *Lemonol*

Found in many essential oils and *muscat wines*. A million ± lbs used annually in the U.S., to impart the odor of roses to perfumes, soaps, detergents, cosmetic creams and lotions, and has been used in *vitamin A* production.

Untoward effects: Contact irritant in **man**; confirmed by patch-testing. Hypersensitivity reported in some **individuals**. May cross-react in **people** sensitive to ***balsams of Peru***.

LD_{LO}, **human**: 500 mg/kg.

Skin irritant to **rabbits** and **guinea pigs**, but not to **miniature swine**.

LD_{50}, **rats**: 3.6 g/kg.

GERANIUM

Geranium Oil Bourbon = *Oil Geranium Reunion* = *Oil Rose-geranium from Pelargonium graveolens* = *Geranio*

~50 tons used in perfumery, in soaps, and in tooth and dusting powders in the U.S. annually. Contact

with the leaves has caused a vesicular dermatitis, and cosmetics containing the oil may cause a dermatitis in hypersensitive individuals. No sensitization reactions in 25 volunteers. Approved in Europe and the U.S. for food use.

LD_{50}, **rats**: > 5 g/kg; acute dermal, **rabbit**: 2.5 g/kg.

Geranium Oil - East Indian = Indian Grass Oil = Palmarosa Oil = Rusa Oil

Found in the grass, *Cymbopogon martini*. ~15 ton used in perfumery in the U.S. annually. Approved for food use. No sensitization reported in 26 volunteers.

LD_{50}, **rats** and acute dermal, **rabbits**: > 5 g/kg.

G. maculatum = Wild Geranium = Crane's Bill = Crow Foot = Dove's Foot = Spotted Geranium = Wild Alum Root

Has caused a high mortality in Australian sheep grazing it, due to its high *oxalate* content.

GERBERA

G. discolor = Wild Yellow Barberton Daisy

G. jarnessonii = Barbeton Daisy = Transvaal Daisy

Untoward effects: Native to Zimbabwe; contains ***prussic acid***, poisonous to **livestock**.

GERMANDER, WILD

= *Teucrium chamaedys*

Untoward effects: Numerous cases of hepatitis in **humans** reported in Europe from its use in herbals and ***teas***, as a diuretic, diaphoretic, and emmenagogue. Now outlawed in France.

GERMANIUM

= *Germane*

A trace element, found in certain plants and water supplies.

Untoward effects: It and its compounds are purchased in health food stores. Chronic use has caused persistent renal dysfunction and myopathies. Inhalation of the tetrahydride form leads to malaise, headache, giddiness, fainting, dyspnea, nausea, vomiting, renal dysfunction, and hemolysis. Pungent odor. National Institute for Occupational Safety & Health sets upper exposure limits of 0.2 ppm (0.6 mg/m^3).

The trioxide is teratogenic in **hamsters**.

GERMINE DIACETATE

Use: A ***veratrum*** alkaloid explored in the IM and IV treatment of myasthenia gravis.

Untoward effects: Skin tingling, cool sensations in the mouth, ***menthol***-like or metallic taste, and some nausea reported.

GESTONORONE CAPROATE

= *Depostat = SH 582*

Progestogen.

Untoward effects: Hyperthermia and occasionally fatigue, vomiting, depression, and uterine hemorrhage.

GF

= *Cyclohexyl Methylphosphonofluoridate*

***Chemical and biological warfare "nerve agent".* Liquid at ambient temperatures.**

Untoward effects: More potent than organophospahte insecticides. It inhibits acetylcholinesterase at cholinergic receptor sites. The resulting build-up of ***acetylcholine*** leads to hyperactivity at other receptors and stimulates the central nervous system, with eventual paralysis and respiratory failure. Absorbed through the skin leading to sweating, fasciculations, nausea, salivation, rhinorrhea, dyspnea, miosis, and abdominal cramps. Symptoms may not appear for 12–18 h after light exposures. Symptoms of sudden unconsciousness, convulsions, paralysis, and apnea can occur within a few minutes of heavy exposure.

Severe neuropathy in **hens** after 300 µg/kg parenterally.

GIBBERELLA ZEAE

A fungus causing *corn* root, stalk, and ear rot, as well as scab in *barley* and *wheat*. Particularly prevalent during cool, wet weather.

Untoward effects: Its toxin, usually ***F-2 toxin*** (***zearalenon***), causes **hogs** to refuse to eat the grain and, sometimes, vomit and have diarrhea. **Cattle** and **poultry** are not particularly bothered by it, but it causes emesis in **pigeons**. The mold also has estrogenic activity and is associated with abortions and infertility in **swine**, and vulvovaginitis and edema in young **pigs** and **gilts**. Estrogenism in some **cattle** and **sheep**, and reduced egg production in **hens**.

GIDYEA

= *Acacia georginae*

Untoward effects: Ingestion of the tree's leaves and pods by **cattle**, **sheep**, and **goats** led to Georgina River poisoning in Australia, with sudden deaths due to acute myocardial damage from ***fluoroacetate*** ion. It is precipitated by fright or driving the herd. Myocardial hemorrhages are noted on autopsy and scarring in survivors.

GILA MONSTER

= *Heloderma suspectum*

Untoward effects: Their bites produce weakness, pain, swelling, and cyanosis around the wound. Hypotension, nausea, vomiting, and tachycardia are common sequelae. Serious bites may requires intensive care hospitalization and antivenin.

GIN

In 1650, a medicine with diuretic properties was made in Holland by distilling *juniper* berries with spirits. In French, *juniper* is *genievre* – the English finally corrupted the pronunciation to *gin* and the name became synonymous with the resulting alcoholic drink.

Untoward effects: In eighteenth century London, it became known as "gin mania" or addiction because the government did not tax it as they did ***beer*** or ***ale***. In 1736, there were > 7000 gin houses in London, and the drink was blamed for ~75% of the deaths of **children** < 5 years with fetal ***alcohol*** (q.v.) syndrome. In 1971, ingestion of 100–150 ml by a **Ghanaian** 38-year-old **male** in London caused a severe sickle cell crisis within 12–36 h. Gin and tonic has provoked insulinemia and a hyperglycemic response, but, after ~2 h, hypoglycemia occurred.

GINGER, JAMAICAN

= *Zingiber officinale* = *Atale* (Yoruba) = *Chittar* (Hausa) = *Shengjiang* (Chinese) = *Zinjabil* (Arabic)

It is the rhizome of a reed-like perennial herb of the tropics, especially Jamaica, India, and Africa. > 12 million lbs used annually in the U.S.

Untoward effects: Can increase bleeding risk with oral ***anticoagulants***. Most of the pungency is due to ***gingerol***, a cardiotoxic agent with positive inotropic effects that stimulate the ***calcium*** pump from skeletal and cardiac muscles. It has earned its place in a toxicology text because it became involved in a case illustrating the result of relaxed vigilance in monitoring of adulteration of foods and drugs. In 1930, in the U.S., an estimated 50,000 **people** developed muscle pain, weakness of all extremities, sensory impairment, and gait impairment (sometimes permanent) = "Jakeleg", Jakewalk", "Jake Paralysis", "Jake Leg Paralysis", or "Jamaican Ginger Paralysis", caused by the consumption of adulterated Jamaican Ginger Extract (up to 70–80% ***alcohol***), or due to a new additive, ***tri-σ-cresylphosphate*** (q.v.), that caused only partially reversible spinal cord and peripheral nerve damage and peripheral nerve spasticity. The same additive in cooking oils caused ~10,000 cases of paralysis in 1959 in Morocco. It was substituted for the more expensive ***castor oil***. ***Essence of Ginger***, an inexpensive alcoholic preparation of ginger, was once sold as an intoxicant. Cases of amblyopia and blindness resulted from incorporation of ***methyl alcohol***.

Ginger oil, LD_{50}, **rats**, and acute dermal, **rabbits**: > 5 g/kg.

GINGER ALE

Untoward effects: In a fruit punch (pH 3.8) stored overnight in corroded galvanized containers caused nausea, abdominal cramps, headache, metallic taste, dizziness, chills, and vomiting from heavy ***zinc*** concentration (443 ppm).

GINKGO BILOBA

= *Maidenhair* = *Rokas* = *Tanakan* = *Tebonin*

Use: IV or orally to increase cerebral and peripheral blood flow, particularly in the **elderly** in the Orient and Europe.

Untoward effects: Oral or IV ill-effects appear limited to headache and gastrointestinal upsets. Acute dermatitis in a 25-year-old **female** working with the fruit and a case of dermatitis venenata 48 h after contact with the berries. Sidewalks become slippery with the decaying pulp and **people** who touch the juices often get ***poison ivy***-type skin reactions. Inhalation of the leaf extract has triggered occasional bronchospastic crises in some **patients** with a history of respiratory problems. Large doses can induce restlessness and retard clot formation (increasing surgical risk). The crude extract, ***gin-nan***, a popular folk medicine in China and Japan, if consumed in excess, leads to convulsions and death. Spontaneous hyphema in a 70-year-old **male** taking 40 mg concentrated extract/twice daily/7 days in addition to ***aspirin*** 325 mg/day/3 years. A 61-year-old **male** taking 40 mg tid or qid < 6 months suffered subarachnoid hemorrhage. Use with ***aspirin*** or ***warfarin*** increases risk of bleeding. Surgeons should request its use be discontinued 2 weeks before surgery. It has caused eye problems, due to its adverse effects on platelets.

GINSENG

= *Ninjin* = *Panax* = *Renshenlutou* = *Shinseng Korean Ginseng* = *Ginsana* = *G 115*

Untoward effects: Although a popular health food remedy, its small amount of ***estrogen*** content has caused swollen and painful breasts and diffuse mammary nodularity and vaginal bleeding in postmenopausal **women**, even with normal ***prolactin*** levels. A postmenopausal 44-year-old **female** developed vaginal bleeding after use of a ginseng face cream (***Fang Fang***) from China. It was confirmed by rechallenge. Maternal use throughout pregnancy with 1.3 g/day in a 30-year-old **female** caused neonatal androgenization with pubic and head hair growth and enlarged testes in her **male infant**. **She** also experienced increased hair growth. Many cases of induced hypertension, following long-term use or abuse (average of 3 g/day), probably due to its ***corticosteroid***-like properties. A 27-year-old **male** reportedly developed Stevens–Johnson syndrome from its use. Some sold on the American market has been adulterated with ***mandrake*** root (containing ***hyoscine***) and ***Rauwolfia serpentina***. Cerebral arteritis in a 28-year-old **female** ingesting large quantities. Central nervous system stimulation with insomnia, nervousness, dizziness, hypertension, edema, eye problems, euphoria, and inability to concentrate, particularly noted in **those** with abuse syndrome. ***Warfarin*** anticoagulant effect decreased ~50% by it.

GLADIOLA

***Untoward effects*:** The bulbs cause gastrointestinal irritation and purgation, and can adversely affect **children** and **animals**.

GLAFENINE

= *Glaphenine* = *Glifanan* = *Privadol* = *R 1707*

***Analgesic*. A *non-steroidal anti-inflammatory drug*.**

Untoward effects*:** Nausea, somnolence, anorexia, lumbar and abdominal pain, heartburn, sweating, dry mouth, anaphylaxis, and dizziness. After high dosage (2–3 g/day), chills, hemolysis, severe jaundice, oliguria, albuminuria, and acute renal failure. Acute hemolysis reported in a 43-year-old **male** with glucose-6-phosphate dehydrogenase deficiency. Withdrawn from the market in France in 1992. ***Non-steroidal anti-inflammatory drugs are not recommended in **women** trying to become pregnant since they interfere with the implantation of fertilized eggs in **animals**.

GLASS

Untoward effects*:** Glass colorers, polishers, and glaze **workers** are exposed to ***chromates (q.v.) and dust. Dermatitis in **workers** also occurs from ***alkalis*** and ***arsenic*** in glass manufacturing. A high percentage of **glassblowers** develop erosions of the mucous membranes and cheeks and leukoplakia, as well as hyperkeratosis of the palm and sides of the fingers. ***Angel hair***, made of spun glass, has caused eye, skin, and gastrointestinal tract irritation. Glass particles have been found in parenteral solutions, especially those drawn from ampules. Alkaline substances leach from some glass containers and many drugs are adsorbed onto them. ***Lead*** and ***cadmium*** leaching into foods can be a serious problem, as can drinking from a ***lead***-lined glass. See ***Fiber Glass***. Glass chillers are sniffed for their intoxicating effects.

Ingestion of ground glass has been blamed for the deaths of **animals**. Other than some gastroenteritis, most of it was eliminated in the stool within 48 h after feeding 5 g to each of 16 **dogs**. No gross or histological changes were noted in **broilers** fed ground glass.

GLATIRAMER

= *COP 1* = *Copaxone* = *Copolymer 1*

***Immunomodulator*. Synthetic *polypeptides* for SC treatment of multiple sclerosis.**

***Untoward effects*:** Pain, erythema, pruritus, and induration at injection site; flushing, dyspnea, anxiety, chest tightness, and occasionally palpitations.

GLAUCARUBIN

= *Glaumeba*

***Antiamebic*.**

***Untoward effects*:** Decreased prothrombin in 7/10 on 3 mg/day/5–7 days. Anorexia, nausea, and diarrhea may develop. One **patient** developed a transient leukopenia.

GLIADIN

Present in *wheat, oats, rye, barley,* and *rice*.

***Untoward effects*:** Damage to the small intestine's mucosa in celiac disease of **man** by the α, but not the γ-form.

GLICLAZIDE

= *Diamicron* = *Glimicron* = *Nordialex* = *S-1702*

***Antidiabetic*. A *sulfonylurea*.**

***Untoward effects*:** Unrelated to dose leading to severe hypoglycemia, and deaths can occur.

Fibrinolytic effect and decreased platelet adhesiveness in experimental **animals**.

LD_{50}, **mice**: > 3 g/kg.

GLIMEPIRIDE

= *Amaryl* = *HOE-490*

***Antidiabetic*. A *sulfonylurea*.**

***Untoward effects*:** < 2% of **patients** had dizziness, asthenia, headache, and nausea. Hypoglycemia can occur, especially in a **child** ingesting a single tablet. The effect may be delayed for up to a day and gradually increase in intensity. Thrombocytopenic purpura in a 68-year-old **male** after 1 mg/day.

GLINUS sp.

Untoward effects*: *G. oppositifolius roots, used in West Bengal, are irritant, cathartic, and abortifacient.

GLIOCLADIUM sp.

Untoward effects: The ***epoxytricothene***'s toxins produced by these molds found on ***corn*** and other cereal grains have caused severe outbreaks of alimentary toxic aleukia in **man** in the U.S.S.R., leukopenia, increased prothrombin time, and cardiovascular collapse.

Suspect as one of the causes of gastric ulcers in **pigs**, digestive disorders in **cattle**, and hemorrhagic syndromes in **poultry**.

GLIPIZIDE

= *Glibenese* = *Glucotrol* = *Glydiazinamide* = *K-4024* = *Mindiab* = *Minidiab*

Antidiabetic. A sulfonylurea.

Untoward effects: A high risk of toxicity in **children**. Hypoglycemia can occur, especially in a **child** ingesting a single tablet. The effect may be delayed for up to a day and gradually increase in intensity.

Blood: Hypoglycemia (3%), agranulocytosis, thrombocytopenia, pancytopenia, and hemolytic and aplastic anemia.

Clinical tests: Increased SGOT, lactate dehydrogenase, alkaline phosphatase, blood urea nitrogen, and ***creatinine*** results are usually transient.

Gastrointestinal: Nausea, vomiting, and diarrhea (1.4%); gastralgia and constipation (1%).

Miscellaneous: Dizziness, drowsiness, fever, sore throat, dark urine, and headache (2%). Empty shell of the extended release tablet can be found in the feces.

Skin: Allergic reactions (1.4%) such as erythema, morbilliform or maculopapular eruptions, urticaria, pruritus, pigmented purpuric dermatosis, and eczema. Possible photosensitivity reactions.

Drug interactions: An ***alcohol***-induced ***disulfiram***-like reaction in up to 15% of **patients** while on it. Its hypoglycemic effect may be dramatically increased by ***cimetidine*** and ***ranitidine***. Prolonged hypoglycemia in **patients** receiving ***heparin*** SC. A hypoglycemic coma was, apparently, induced by ***fluconazole***. Oral ***miconazole*** can possibly cause a similar reaction. Results have varied, but increased hypoglycemia has occurred with ***trimethoprim–sulfa*** combination. Half-life is increased by ***indobufen***. ***Activated charcoal*** and ***cholestyramine*** decreases its absorption. Hypoglycemic effect may also be potentiated by ***β-adrenergic-blockers, chloramphenicol, coumarins, furazolidone, monoamine oxidase inhibitors***, ***non-steroidal anti-inflammatory drugs***, ***salicylates***, and ***sulfonamides***. ***Calcium channel-blockers, corticosteroids, oral contraceptives***; ***diuretics,*** such as the ***thiazides; estrogens, isoniazid, nicotinic acid, phenothiazines, phenytoin, sympathomimetics,*** and ***thyroid preparations***, by increasing blood ***glucose***, may negate its effects.

Vomiting, hypoglycemia, and increased serum hepatic enzymes in diabetic **cats** receiving 5 mg twice daily.

GLOCOTRICHIA ECHINALATA

Untoward effects: A freshwater blue-green alga, implicated in poisonings.

GLORIOSA

G. simplex = *Glorieuse du Malabar* = *Gloriosa Lily* = *Homa* (Boran) = *Mkalamu and Msufari* (Swahili)

Untoward effects: Widely spread in East Africa, where its powdered rhizomes are used for **criminal** poisoning. Now a cultivated ornamental in Europe and the U.S.

A dangerous house plant for pet **animals**.

G. superba = *African Climbing Lily* = *Glory Lily* = *Baurere* (Hausa) = *Mora* (Yoruba) = *Uguele* (Igbo)

Untoward effects: Contains ***colchicine*** (q.v.), highly toxic to **man** and **animals** in Africa. In Ceylon, a 21-year-old **female** ingested ~125 g of the tubers (containing ~350 mg ***colchicine***), mistaking it for a ***sweet potato***, which it resembles. Initial symptoms were vomiting and profuse, watery diarrhea, then a massive alopecia involving all body hair within 11 days; and menorrhagia continuing for 20 additional days after end of **her** menstrual period on day of ingestion. Eventual recovery. Tingling, numbing of the lips, tongue, and throat; intense nausea, vomiting, diarrhea, and narcosis reported in **people**. Used for suicidal purposes in Burma and India. Also contains another toxic alkaloid, ***gloriosine***.

The root-stock, stem, and leaves are poisonous to grazing **livestock**. **Cattle** become uneasy, salivate, and have abdominal pain and violent purging with green–black feces.

GLOTTIDIUM VESICARIUM

= *Bagpod* = *Bladderpod* = *Castle Bean* = *Coffee Bean Weed* = *Coffee Weed*

Untoward effects: The beans are particularly poisonous after frost. **Cattle** develop tachycardia, tachypnea, and diarrhea, and become comatose before death.

GLOVES

Untoward effects: One of the more serious adverse effects of use is from starches used on surgical gloves, that has caused serious granulomatous peritonitis, epididymitis, periorchitis, etc. ***Talc*** caused even more problems. Dermatitis from ***rubber*** gloves in ***thiram***- (q.v.) or

mercaptobenzothiazole- (q.v.) sensitive **people** from their use as ***rubber*** accelerators. ***Antineoplastics*** and ***methylmethacrylate*** can penetrate ***polyvinyl chloride*** and ***rubber*** gloves. **Surgeons** doing hip replacements and denture-repairing **technicians** are at risk from dermatitis from the latter. Severe anaphylactic reactions from latex gloves reported in a **patient** and **obstetricians**. Loss of insulating quality of surgical gloves noted after use in a wet environment, allowing serious risk of electrical injury. Seam failures and small punctures make for potential infection of **surgeon** or **patient**. Renal petechiae and focal glomerulitis from starch powder on gloves contaminating catheters, syringes, or ***contrast media*** while preparing the equipment for angiography. Contact dermatitis reported in industrial **workers** from ***chromium*** in leather gloves.

GLUCAGON

Antihypoglycemic.

Untoward effects*:** Nausea and vomiting after parenteral use. Hypokalemia, light-headedness, premature ventricular contractions, diarrhea, papular skin rash, decreased lower esophageal sphincter pressure, abdominal cramping, and loss of consciousness; hypersensitivity reactions are less common. May provoke severe hypertension in **patients** with pheochromocytoma. Myocardial ischemia in a 74-year-old **male** after 0.5 mg SC. Fatal cardiopulmonary arrest after 0.5 mg IV during a ***barium enema in a 62-year-old **female** attributed to a hypersensitivity reaction. Inhibits platelet aggregation. Has caused ventricular tachycardia and fibrillation. High dosage can enhance ***warfarin's*** effect.

IM, TD_{LO}, **human**: 28 μg/kg.

Potentiates hypothrombinemic response to ***acenocoumarol*** in **guinea pigs**. Increases ***hexobarbital*** sleeping time in **rats**. Given to pregnant **rats**, it increases fetal deaths.

GLUCOSAMINE

= *Arth-X-Plus*

***Antiarthritic.* Human blood contains ~80 mg/100 ml. They have tolerated 20 g/day.**

***Untoward effects*:** Poorly absorbed from the gastrointestinal tract. Mild gastrointestinal upsets (heartburn, epigastric pain, nausea, diarrhea, and slight to moderate constipation) reported.

Increases serum levels of oral ***oxytetracycline*** in **dogs**.

LD_{50}, **mice**: 14 g/kg; IV, **mice**: > 1 g/kg.

GLUCOSE

= *Blood Sugar* = *Cerelose* = *Corn Sugar* = *Dextropur* = *Dextrose* = *Dextrosol* = *Glucolin* = *D-Glucose* = *Grape Sugar* = *Lucozade* = *Traubenzucker*

Untoward effects*:** Premature and low-birth-weight **infants** cannot handle high quantities of sugars in **their** first 48 h, and easily become hyperglycemic when given 10%, instead of 5%, solutions. **Maternal** IV use has caused **neonatal** hypoglycemia and jaundice. Headache, continuous seizures, and respiratory arrest in a 6½-year-old **female** accidentally given 300 ml of a 5% solution IV. **She** was resuscitated and survived with treatment. May be dangerous when given to a diabetic **patient**. ***Thiazides may increase blood glucose and may not be significant, except in **diabetics**. IV use can cause hypokalemia, with resulting ventricular arrhythmias, and is reported to have caused an anaphylactic reaction. Paradoxical hyperglycemia reported in a **diabetic**. When large amounts are used in peritoneal or hemodialysis, hyperglycemia may develop. 50% solutions are highly irritating and sclerosing to veins, and, after dilution, should be given slowly IV. Extravasated material can be irritating, due to its acidic pH. IV glucose to conscious healthy **men** or anesthetized **dogs** causes a rapid drop in free ***fatty acids***. Electrocardiogram alterations, especially rate, 15-min exposure level, and T-wave amplitude in almost all 117 **female**s tested after excess oral glucose. Ingestion of 100 g may cause periodontal pathology. Cataractogenic. Suspect teratogen, if given during embryogenesis. May give falsely increased ***creatinine*** and decreased blood urea nitrogen readings. Intra-amniotic injection of hypertonic solution has caused a **maternal** death. Use ***cimetidine*** cautiously in **patients** predisposed to hyperglycemia. Results have been conflicting. ***Arginine***, ***chlorpromazine***, some ***oral contraceptives***, ***corticosteroids***, ***dextrothyroxine***, ***diazoxide***, ***ephedrine***, ***epinephrine***, ***furosemide***, ***isoproterenol***, ***nicotinic acid***, ***phenytoin***, and ***thiazides*** usually cause varying degrees of hyperglycemia. ***Ethacrynic acid*** and ***thiabendazole*** have caused an increase or decrease in blood sugar. ***Acetaminophen***, ***androgens***, ***haloperidol***, ***monoamine oxidase inhibitors***, and ***propranolol*** cause a decrease in blood sugar. ***Salicylates*** can falsely decrease fasting glucose levels and falsely increase oral glucose test results.

Concentrated solutions SC to **mice** and **rats** may cause sarcomas. High dietary concentration to **pigs** causes a hemorrhagic syndrome with increased prothrombin time, as well as producing cardiac lesions. IV use potentiates and prolongs ***barbiturate*** anesthesia in **dogs**. IV use induces diuresis in **animals**.

LD_{50}, **rat**: 25.8 g/kg; LD_{LO}, **dog**: 8 g/kg; **rabbit**: 20 g/kg.

GLUCOSINOLATES

Include *epiprogoitrin*, *sinalbin*, and *sinigrin*. *Cabbage*, *canola*, *crambe*, *kale*, *mustard*, and *turnips* are rich in them.

Untoward effects: They and their derivatives [***goitrin*** (q.v.), ***isothiocyanates***, and ***nitrites***] are implicated in the development of hepatic hemorrhage in **chickens** and **hens**. Feeds with high concentrations of ***crambe*** or ***rapeseed*** decrease feed intake, decrease growth rate, and cause enlargement of liver, kidney, and adrenal glands in **rats**, **mice**, **chickens**, and **swine**. Decreases growth rate in ***canola***-fed **trout**. Induced goiter in many species. Although ***canola***, unlike ***rapeseed***, is low in ***erucic acid***, its glucosinates have discouraged its use as a food for **humans**.

GLUCOSULFONE SODIUM

= *Angeli's Sulfone* = *501-P* = *Promanide* = *Promin* = *Protomin*

Antibacterial.

Untoward effects: Dose-related hemolysis, including many cases of hemolytic anemia (in **patients** with or without glucose-6-phosphate dehydrogenase deficiency), hemolytic jaundice (hypersensitivity), enlarged liver, methemoglobinemia, leukopenia, agranulocytosis, severe dermatitis (herpetiformis, exfoliative, and lepra reaction), and psychosis. Less common if dosage is slowly increased are nausea, vomiting, headaches, giddiness, and tachycardia.

LD_{50}, **mice**: 3.93 g/kg.

GLUE

Untoward effects: "Glue-sniffing" usually involves the deliberate inhalation of its volatile ***hydrocarbons***. A psychological dependence, intoxication, hallucination, hypoxia, dizziness, vertigo, nausea, vomiting, neuropathies, myocardial infarction, cardiomyopathy, ventricular fibrillation (due to sensitization of the heart to ***catecholamines***), Fanconi syndrome, renal tubular acidosis, other drug vices, behavioral problems, aplastic anemia, eosinophilia, neutropenia, and numerous deaths and suicides. The solvents usually were ***acetone, alcohols, benzene, hexane, methylethyl ketone, toluene***, and ***xylene***. Liver, pulmonary, and cerebral damage on post-mortem. Superglues or ***cyanoacrylates*** (q.v.) have caused serious adverse effects. Shoe and book-binding glues have caused dermatitis in **workers**. Glues often contain antimicrobials causing hypersensitivity reactions.

Airplane glue containing ***acetone***, ***phorone***, and ***toluene*** caused cardiac arrhythmias in **dogs** and heart-block in **mice** after inhalation.

GLUTARALDEHYDE

= *Glutaral* = *1,5-Pentanedial* = *Sonacide* = *Ucaricide* = *Verucasep* = *Wavicide*

Use: Disinfection of instruments, plastics, and rubber. Buffered to an alkaline pH = ***Cidex***.

Untoward effects: Use in rapid-process X-ray film emulsions has caused dermatitis in **radiologists** and **X-ray technicians**. Contact dermatitis in **nurses** handling instruments sterilized with it. Renal dialysis and inhalation therapy **personnel**, as well as **those** tanning leather or handling embalming fluids, can develop a dermatitis from it. Use of a 25% solution for hyperhidrosis of the soles of the feet showed no reactions, but severe dermatitis reported from a 2.5% solution on the antecubital fossa and axilla. Deep ulceration and scarring reported, when used under plastic occlusion for treatment of warts. Golden brown discoloration of the nail plate reported. Pungent odor. Eye and nose irritation noted at 0.3 ppm. Skin staining, irritation, and sensitization occurs. Burning sensations if it penetrates into fissures. Allergic contact dermatitis in a 24-year-old **female** dental **assistant**. Inhalation of vapors leads to symptoms similar to those caused by ***formaldehyde*** (q.v.) fumes. Ingestion has caused nausea, vomiting, and other symptoms, similar to ***formaldehyde*** ingestion. The mislabeling of a glutaraldehyde solution as "Spinal Fluid", when it was intended to be used as a fixative for a cancerous eye, led to its accidental use in the spine, resulting in brain death.

LD_{LO}, **human**: 500 mg/kg.

Topically, 0.5% solution caused severe eye injury in **rabbits**. Fetotoxic in **mice**.

LD_{50}, **rats**: 600 mg/kg; acute dermal, **rabbits**: 2.56 mg/kg.

GLUTEN

Constitutes 90% of *wheat* protein and is composed of *glutenin* and *gliadin*. Also found in most other grains.

Untoward effects: When Dutch **children** went through the famine of World War II, **those** with celiac disease did well, but became ill as cereals were reintroduced into **their** post-war diet. **They** then became pale, anorexic, lost weight, and had diarrhea and increased steatorrhea. Enteropathy and dermatitis herpetiformis are manifestations of hypersensitivity to gluten. Enteropathy causes malabsorption of ***calcium***, ***iron***, and ***vitamins***. Cutaneous eruptions reported in industrial **workers** who handle hollow castings and **their** molds treated with a gluten preparation. **Patients** with gluten-sensitive enteropathy fail to absorb ***penicillin V***.

LD_{LO}, **human**: 50 mg/kg.

A cause of allergic skin reactions in **dogs**.

LD_{50}, **rat**: 600 mg/kg; **mouse**: 360 mg/kg; **dog**: 500 mg/kg; **rabbit**: 600 mg/kg.

GLUTETHIMIDE

= *Ciba 11,511* = *Doriden* = *Doriden-Sed* = *Elrodorm* = *Glimid* = *Noxyron*

Sedative, hypnotic. **Slang terms for it by abusers =** ***C.D.*** **=** ***Cibas***

***Untoward effects*:** Physical dependence, mydriasis, often unequal fixed pupils with cyclic pattern dry mouth, headache, vertigo, cardiovascular depression, blurred vision, cerebral depression, focal neurologic abnormalities, slurred speech, disorientation, tachycardia, hypotension, bradypnea, sudden apnea, laryngeal spasm, asterixis, coma and agranulocytosis, acute intermittent porphyria, thrombocytopenic purpura, neutropenia, leukopenia, aplastic anemia, megaloblastic anemia, urticaria, bullae, exfoliative dermatitis, erythema multiforme, Stevens–Johnson syndrome, diffuse burning feet, depigmentation (one case report); erythematous, oozing lesion on the penis (on three occasions) in a 23-year-old **male** after 500 mg doses; increased hepatic microsomal enzymes, and osteomalacia reported. The coma produced is of long duration, due to its active hydroxy metabolite (Rotarod tests in **mice** indicated the ED_{50} for the metabolite was 24 mg/kg and 47 mg/kg for glutethimide). A 49-year-old **male** experienced a shock reaction after ingesting a 250 mg tablet. A 39-year-old **female** increased dosage over a 5 year period to 3 g causing ataxic, slurred speech and limb paresthesias. Peripheral neuropathy persisted after drug withdrawal. Metabolized in the liver, therefore avoid use in **people** with liver impairment. In poisoning or suicides, fever or decreased temperature, tetanic spasticity, and gastric concretions. Anxiety, depression, delirium, convulsions, and, occasionally, euphoria may follow withdrawal. Withdrawal symptoms have been noted in the **neonate**, including severe respiratory depression. Therapeutic serum levels reported as 0.2–5 mg/l; toxic at 10 mg/l; shock and death at 20–50 mg/l. Deaths, despite adequate treatment, have been reported in suicides with as little as 9 g.

Drug interactions*:** Because it is a potent microsomal enzyme-inducer, it enhances ***anticoagulant (***acencoumarol, bishydroxycoumarin, cyclocoumarol, ethylbiscoumacetate***, and ***warfarin***) ***antipyrine, dipyrone***, and, possibly, ***phenytoin*** metabolism, requiring up to twice usual ***warfarin*** dosage. Increases blood ***alcohol*** concentration by ~11% within 2 h, and its hypnotic effects are potentiated by ***phenothiazine tranquilizers***. Its absorption is increased in **smokers**. Hypotension, respiratory arrest, coma, and shock reported, when taken with ***monoamine oxidase inhibitors***.

Increased fetal resorption rate in **rat, mouse**, and **rabbit** trials. Accelerates metabolism of ***carisprodol, meprobamate***, and ***pentobarbital*** in **rats**, and ***warfarin*** in **dogs**. **Dogs** develop seizures on withdrawal after chronic use.

GLYBURIDE

= *Adiab* = *Azuglucon* = *Bastiverit* = *Daonil* = *Dia-basan* = *Diabeta* = *Duraglucon* = *Euglucon* = *Gilemal* = *Glibenclamide* = *Gliben-Puren N* = *Glidiabet* = *Glimidstada* = *Glubate* = *Glucoremed* = *Gluco-Tablinen* = *Glybenzcyclamide* = *Glycolande* = *HB-419* = *Lederglib* = *Libanil* = *Lisaglucon* = *Malix* = *Maninil* = *Micronase* = *Praeciglucon* = *RO6-4563* = *U-26452*

Antidiabetic.

Untoward effects*:** Severe hypoglycemia has followed inadvertent administration (5 mg to a **non-diabetic** 79-year-old **female**) or overdoses in **adults** and **children**. Hypoglycemia can occur, especially in a **child** ingesting a single 2.5 mg tablet. The effect may be delayed for up to a day and gradually increase in intensity. Brain damage in a **patient** when a pharmacist dispensed it in error for ***amoxicillin (***Daonil*** for ***Amoxil***). Optic nerve atrophy developed in a healthy **infant** after accidental ingestion and resulting hypoglycemia. Nausea, epigastric fullness, anorexia, and heartburn in < 2%; allergic skin reactions: pruritus, erythema, urticaria, photosensitization, porphyria cutanea tarda, morbilliform or maculopapular eruptions in < 2%; dizziness, headache, hot flushes, weight gain, tinnitus, thrombocytopenia, leukopenia, hemolytic anemia, eosinophilia, constipation, diarrhea, transient myopia, water retention with dilutional hyponatremia, fatigue, dark urine, light-colored stools, and increased hepatic aminotransferases; rarely, cholestatic jaundice, impotence, neuropathy, and a ***disulfiram***-like reaction. Usual therapeutic serum levels are 0.1–0.2 mg/l and toxic at 0.6 mg/l.

Drug interactions*:** Use with ***cimetidine and ***ranitidine*** has, in some cases, led to severe hypoglycemia. Apparent increased risk of hepatotoxicity with ***fluconazole***. Hypoglycemia potentiated by ***dicoumarol, phenylbutazone, maprotiline***, and ***sulfaphenazole***. Flushing reaction in about 50% of **patients** ingesting ***ethyl alcohol***. It causes a 57% increase in ***cyclosporine*** serum levels. Use with ***aspirin*** can induce a transient hypoglycemia.

LD_{50}, **rats** and **mice**: > 20 g/kg.

GLYCERIA sp.

Untoward effects*:** **G.** or ***Poa aquatica and **G. *maxima*** = ***Reed Sweet-grass*** caused rapid ***cyanide*** (q.v.) poisoning in **cattle**, **horses**, and **sheep** in New Zealand, Russia, and Romania after ingestion, with usual symptoms of dyspnea, shivering, restlessness, staring, coma, and death.

G. grandis = ***Tall Mannagrass*** has also caused ***cyanide***-induced deaths of **cattle** in British Columbia.

G. striata = ***Fowl Mannagrass*** in Alabama and Maryland also contains a cyanogenic glycoside that has killed **cattle**.

GLYCERIN

= *Bulbold* = *Cristal* = *Glyceol* = *Glycerine* = *Glycerol* = *IFP* = *Incorporation Factor* = *Ophthalgan* = *1,2,3-Propanetriol* = *Trihydroxypropane*

Untoward effects: Hyperemia from enemas; nausea, headache, some light-headedness, and mild diuresis after treatment for glaucoma; nausea (17%), emesis, severe headache and diarrhea when used after cataract surgery; hemolysis, hemoglobinuria, and renal failure after IV to reduce intracranial pressure. Nasogastric use for cerebral edema associated with non-ketotic hyperosmolar hyperglycemia; occipital headache, shaking right arm, quivering of eyes, and nausea in an 82-year-old **female** after 200 ml of 50% for glaucoma; and danger of explosion if ***hydrogen peroxide*** used within a few minutes of glycerin-containing mouthwash. The mist can irritate eyes, skin, and respiratory tract.

Large doses parenterally in **dogs** have caused hemolysis, hypotension, and central nervous system disturbances; experimental renal failure in **rats**.

LD_{50}, **rat**: 1.26 g/kg; **guinea pig**: 750 mg/kg.

GLYCERYL GUAIACOLATE

= *Actifed-C* = *Calmipan* = *Colrex* = *Equicol* = *GGE* = *Gecolate* = *Glycodex* = *Guaiacol Glyceryl Ether* = *Guaiacuran* = *Guaiamar* = *Guaiphenesin* = *Guayanesin* = *Miocaina* = *MY-301* = *Myocaine* = *Myoscain* = *Oresol* = *Oreson* = *Relaxil G* = *Resyl* = *Robitussin*

Expectorant, muscle relaxant.

Untoward effects: Nausea and decreased platelet adhesiveness and may exacerbate a bleeding tendency. Inadvertent paravenous injection caused localized aseptic necrosis. A metabolic by-product causes a false positive ***5-hydroxyindoleacetic acid*** (***5-HIAA***) test reading and may cause interfering color reactions during ***vanillylmandelic acid*** (***VMA***) tests. High continuous oral dosing can cause hypouricemia.

LD_{LO}, **human**: 5 g/kg.

LD_{50}, **mouse**: 690 mg/kg; IP, 495 mg/kg.

GLYCIDOL

= *2,3-Epoxy-1-propanol* = *Epoxypropyl Alcohol* = *Glycide*

Use: In manufacturing vinyl polymers, hydraulic fluids, and some epoxy resins. > 10 million lbs of glycidyl compounds used annually in the U.S.

Untoward effects: Irritation of eyes, nose, throat, and skin; narcosis. National Institute for Occupational Safety & Health and OSHA suggest exposure limits of 25 and 50 ppm, respectively; 150 ppm immediate danger to life or health.

Gavage for 2 years to **mice** and **rats** led to mesotheliomas of tunica vaginalis, fibroadenomas of mammary glands, gliomas, neoplasms of forestomach, intestine, skin, and zymbal and thyroid glands in male **rats**; fibroadenomas and adenocarcinomas of mammary glands, gliomas, leukemia, and neoplasms of oral mucosa, forestomach, and clitoral and thyroid glands in female **rats**; harderian, uterine, and mammary gland tumors; as well as SC and on-skin tumors in female **mice**; harderian gland, forestomach, skin, liver, and lung tumors in male **mice**.

LD_{50}, **rat**: 850 mg/kg; **mouse**: 450 mg/kg; inhalation LC_{50}, **rat**: 580 ppm/8 h; **mouse**: 450 ppm/4 h. LD_{50}, skin, **rabbit**: 558 mg/3 days.

GLYCIDYL METHACRYLATE

Untoward effects: Contact dermatitis in **workers** handling ***Sta-Loc*** used in joining metal surfaces.

LD_{50}, **rat**: 770 mg/kg.

GLYCOLONITRILE

= *Cyanomethanol*

Untoward effects: Eye, skin, respiratory tract irritant; headache, dizziness, weakness, giddiness, confusion, convulsions, dyspnea, abdominal pain, nausea, and vomiting. Forms ***cyanide*** in the body.

GLYCOPYRROLATE

= *AHR-504* = *Glycopyrronium Bromide* = *Nodapton* = *Robanul* = *Robinul* = *Tarodyl* = *Tarodyn*

Anticholinergic.

Untoward effects: Occasional dryness of mouth and pharynx, disturbance of accommodation (blurred vision), constipation, nausea, vomiting, bloated feeling, urticaria, confusion, headache, and loss of libido. A 39-year-old **female** stabilized on ***ritordine*** 0.3 mg/min IV for premature labor, abruptly developed severe supraventricular tachycardia after 0.2 mg IV in preparation for emergency Cesarean section. Slow placental passage.

In **dogs**, mild mydriasis, xerostomia, and occasionally tachycardia.

LD_{50}, **rat**: 1280 mg/kg (some **rat** strains: 107, 196, and 1150 mg/kg); **mouse**: 550 mg/kg; IV, **rat**: 14.6 mg/kg; **mouse**: 14.7 mg/kg.

GLYCYRRHETINIC ACID

= *Enoxolone* = *Glycyrrhetic acid*

Untoward effects: Causes retention of ***chlorine***, ***sodium***, and ***water***. Causes hypertension by mimicking the hormone ***aldosterone***.

GLYCYRRHIZA GLABRA

= *Antesite* = *Gancoa* = *Licorice* = *Saila* = (*Sûs* = *Aûd* = *Assûs* - Arabic)

Untoward effects: Due to its glycyrrhizinic content, 20 g/day causes water and ***sodium*** retention, hypertension, hypokalemia, alkalosis, periodic weakness, decreased or absent reflexes, and cardiac arrest. A 62-year-old **male** ingested 100 g during Ramadan led to severe hypokalemic rhabdomyolysis. Flaccid quadriplegia, due to severe hypokalemia, and myoglobinuria in a 70-year-old **female** ingesting it intermittently in a laxative preparation. Fulminant congestive heart failure developed over a week in a healthy 53-year-old **male** after ingesting 700 g ***licorice*** candy in 9 days. **He** had symptoms of pseudoaldosteronism, tetany, paresis, and myoglobinuria. Transient hypertensive encephalopathy in a 2½-year-old **female** 39 h after ingesting ¼ lb candy (~1 g ***licorice*** extract or 0.05–0.3 g ***glycyrrhizic acid***), despite vomiting soon after ingestion. Some symptoms continued for 2 weeks. Headaches and hypophosphatemia also occur. Certain chewing ***tobaccos*** contain large amounts of pure ***licorice***. An 85-year-old **male** was hospitalized with profound muscle weakness, hypertension, and marked hypokalemic alkalosis. **He** chewed 8–12, 3 oz bags of chewing ***tobacco***/day and swallowed, rather than spit out, the saliva. **His** intake of ***glycyrrhizic acid*** was 0.88–1.33 g/day/50 years.

GLYHEXAMIDE

= *SQ-15,860* = *Subose*

Antidiabetic.

Untoward effects: Heartburn, epigastric discomfort, thrombocytopenia, leukopenia, and basophilia.

LD_{50}, **mice**: > 8 g/kg.

GLYMIDINE

= *Glycodiathine* = *Glycodiazine* = *Glyconormal* = *Gondafon* = *Lycanol* = *Redul* = *SH-717*

Antidiabetic. **The sodium form is usually used.**

Untoward effects: Malaise and gastrointestinal symptoms in ~2%; leukopenia < 1%, jaundice and abnormal liver function in ~2%. Dizziness, allergic dermatitis ~2%, acute eczema, buccal aphthae, purpura, pruritus, giddiness, anorexia, shooting pains in hands and arms, pancytopenia, and euphoria also occur. Plasma concentration increase by ***oxyphenbutazone***, but not ***phenylbutazone***. Efficacy is decreased in ***rifampin***-treated **diabetics**.

LD_{50}, **rats**: 2.85 g/kg; **mice**: 5.3 g; IV, **rats**: 2 g/kg; **mice**: 1.48 g/kg.

GLYODIN

Agricultural fungicide.

Untoward effects: LD_{LO}, **human**: 50 mg/kg.

LD_{50}, **rats**: 4.6 g/kg.

GLYOXAL

= *Biformyl* = *Diformyl* = *Ethanedial* = *Oxalaldehyde*

Disinfectant, preservative. **In glues, textiles and ophthalmic solutions.**

Untoward effects: Moderately irritating to skin and mucous membranes. Strong contact sensitizer in **man**.

LD_{50}, **rats**: 2 g/kg; **guinea pigs**: 760 mg/kg.

GLYOXYLIC ACID

Untoward effects: Irritant and corrosive. A toxic intermediate of ***ethylene glycol*** (q.v.) metabolism.

IM, LD_{LO}, **rat**: 25 mg/kg.

GLYPHOSPHATE

= *CP 67,573* = *CP 70,139* = *Deploy* = *Honcho* = *MON 39* = *MON 2,139* = *Ranger* = *Rodeo* = *Roundup*

Herbicide.

Untoward effects: Self-poisoning in two **adults** caused no serious toxic effects. May cause eye and skin irritation.

LD_{50}, **rat**: 4.32 g/kg.

GNIDIA KRAUSSIANA

= *Tururibi* (Hausa) = *Yellow Heads*

Untoward effects: Roots and leaves used to kill **people**, and deaths in **Africans** using it for medicinal purposes (purgative). Extracts used for poison arrows and to kill **fish**. Poisoned ***fish*** cause diarrhea, if eaten. Hemorrhagic gastroenteritis in grazing **animals**. 350 g of the young leaves and flowering tips rapidly kill a **horse** or **cow**.

GOAT FISH

Untoward effects: A toxin isolated from it, common in north Queensland waters, is the cause of hallucinations and nightmares reported from its ingestion.

GOLD

Renal excretion is very slow and it remains in the body for months, requiring chelating agents to accelerate its removal and decrease its toxicity.

***Untoward effects*: *Blood and bone marrow*:** After excessive dosage or in hypersensitive **individuals**: eosinophilia, leukopenia, thrombocytopenia, agranulocytosis, pancytopenia, neutropenia, leukocytosis, aplastic anemia, bone marrow aplasia, and hemolytic anemia.

***Chromosomes*:** Intra-articular injection of radioactive gold caused 8.5% white cell chromosome damage.

***Clinical tests*:** Has caused false decrease of protein-bound iodine in colorimetric assays.

***Eye*:** Reversible subepithelial metallic gold deposits in the cornea (chryseosis) and keratoconjunctivitis is common. Allergic intolerance is manifested by inflammation or ulcers.

***Liver*:** Has induced mixed types of jaundice and acute intrahepatic cholestasis. Considered to be a hypersensitivity reaction.

Miscellaneous*:** Muscle palsies, polyarteritis, systemic lupus erythematosus; sudden nausea, vomiting, and weakness after injections; entercolitis, interstitial lung disease, flushing, dizziness, sweating, and paralysis preceded by pain; rarely, proctitis and vaginitis. Excreted in breast milk. Traces found in **infant's** serum and red blood cells. A potential cause of rashes and idiosyncratic reactions in an **infant**. ***Mercury poisoning has occurred from inhalation of vapors from home gold ore processing. ***Angiotensin-converting enzyme inhibitors*** exacerbate gold-induced vasomotor reactions of flushing, weakness, palpitations, and dyspnea.

***Mouth*:** A frequent complication of IM therapy of rheumatoid arthritis is glossitis and stomatitis. Allergic contact stomatitis with gingival mucosa sloughing, lichen planus-like stomatitis under the tongue, and reactivation of previous dermatitis from jewelry after a gold crown was placed on a tooth. Metallic taste.

***Renal*:** Nephropathy with proteinuria and albuminuria; hematuria.

Skin*:** Exfoliative and lichenoid dermatitis. Fixed-drug eruptions, toxic epidermal necrolysis, and photosensitivity. Variable brown nail pigmentation has occurred. Contact dermatitis with erythema, blisters, and hyperkeratosis from contamination of gold with decaying ***radon. A squamous cell carcinoma required amputation after chronic gold exposure. Gold has been found in the skin and hair. Some golds contain ***nickel*** (q.v.). Has caused hair loss in **some**. A 33-year-old **female** with known sensitivity to gold drank 90–120 ml ***Goldschlager***, a gold-containing liquor, leading to erythematous pruritic rash.

Radioactive gold caused cystitis, ureteritis, and hydronephrosis when used in uterine carcinomas; bone marrow aplasia and severe injury to liver and kidneys when used in myelomas.

Gold is extracted from soil by plants, and after these are ingested, it has been found in the horns and hides of **deer**, in **ants**, in **cockchafers**, and in **human** blood and hair. **Rat** trials show that gold concentrates in the kidney's proximal tubule cells, proportional to the amount injected. Accumulation destroys the mitochondria. Tubules can return to normal after cessation of treatment. Has caused cataracts in **rats** and **rabbits**. Can cause thrombocytopenia in **cats** and **dogs**.

GOLDENROD

Approximately 100 species.

Untoward effects*:** Causes allergic rhinitis, bronchial asthma, and/or pneumonitis in some **people** exposed to its flowers. Its ***diterpenes are toxic to **sheep**. It accumulates ***nitrates***. ***Aplopappus heterophylus*** = ***Rayless Goldenrod*** = ***Alkaliweed*** = ***Jimmy Weed*** = ***Isocoma heterophylus*** or ***wrightii***. During colonial times, it was recognized as a cause of "milk sickness" in North Carolina, then it appeared in the Midwest (Ohio, Indiana, and Illinois), and, finally, it spread to southwestern U.S. and Mexico. Its ***tremetol*** (q.v.) content in green or dry material is poisonous to **cattle**, **sheep**, **goats**, **horses**, and **swine** who eat it, and to their nursing young. Symptoms are marked weakness, trembling or "jimmies" (especially after exercise), humped-up stance, stiff gait (pronounced in front legs), inability to stand, constipation, vomiting in **cattle** and **sheep**, nearly continuous urine dribbling, and dyspnea with prolonged inspiration. **Horses** sweat profusely and have partial throat paralysis. Death usually follows. **Humans** have developed "milk sickness" or "trembles" after drinking milk from **cows** ingesting it. Some **people** have become delirious, comatose, and died (10–25% mortality).

A. fruticosus also contained ***tremetol***.

Solidago mollis causes weight loss in **calves**, **sheep**, and **laboratory animals** fed it experimentally, possibly due to fungal rust or corrosive resin on it. **Sheep** have also shown dyspnea, nausea, vomiting, and death.

Solidago virgaurea = ***European Goldenrod***. Herbal ***teas*** made from its floral heads have caused severe hypersensitivity reactions in **people** allergic to ***asters, chamomile, chrysanthemums***, and ***ragweed***.

GOLD SODIUM THIOMALATE

= *Kidon* = *Myochrysine* = *Myocrisin* = *Shiosol* = *Sodium Aurothiomalate* = *Tauredon*

Untoward effects**: Infant** dosage from breast milk was 5.5% of **mother's**. Numerous deposits found in placenta, and less in **fetal** liver and kidney, with no obvious **fetal** abnormalities after pregnancy terminated in **women** who received a total of 570 mg. Pruritus, dermatitis, glossitis, gingivitis, stomatitis, metallic taste, cholestasis, vasomotor rhinitis, flushing, weakness, tachycardia, photosensitivity, black–brown nail dyschromy, corneal deposits, hematuria, proteinuria, edema, nausea, vomiting, diarrhea, peripheral neuropathy, confusion, hallucinations, seizures, bronchitis, interstitial pneumonitis and fibrosis, alopecia, intestinal hemorrhage, non-bacterial thrombotic endocarditis, Guillain–Barré syndrome, neuropathies, myokymia, bone marrow aplasia, aplastic anemia, thrombocytopenia, neutropenia, agranulocytosis, pancytopenia, exanthematous rash, grayish skin discoloration, increased alkaline phosphatase, herpes zoster, pleuritic pain, dyspnea, contact dermatitis after IM, and fever reported.

Oral doses of 8 mg/kg to **rats** caused gastric hemorrhage.

IM, LD_{50}, **rats**: > 400 mg/kg; **mice**: ~800 mg/kg.

GOLD SODIUM THIOSULFATE

= *Auricidine* = *Aurocidin* = *Aurolin* = *Auropex* = *Auropin* = *Aurosan* = *Aurothion* = *Crisalbine* = *Crytion* = *Novacrysin* = *Sanochrysine* = *Sodium Aurothiosulfate* = *Solfocrisol* = *Thiochrysine*

Untoward effects**:** Similar to ***gold*** or the thiomalate form above, especially stomatitis, nephritis, dermatitis, gastrointestinal upsets and hemorrhagic enteritis, erythrodermas, fever, dizziness, decreased white blood cells, hemoglobinuria, proteinuria, and hematologic reactions.

LD_{LO}, **man**: 5.5 mg/kg; IM, TD_{LO}, **woman**: 182 µg/kg, 20 mg/kg/1 year; **man**: 5 mg/kg/13 weeks.

IV, LD, **rabbits**: 100 mg/kg.

GOLF BALLS

Untoward effects**:** Most serious is ocular injury, particularly from **children** taking them apart and having the contents (usually a mixture of ***barium sulfate, corn syrup, gelatin, silicone, sulfuric acid***, and ***zinc sulfide***) under high pressure hit the eye and, in some cases, penetrate the skin without grossly breaking the skin surface. "Golf-course dermatitis" is due to use of ***thiram*** as a fungicide on the grass.

GOMPHRENA CELOSIOIDES

Untoward effects**: Animals** in Australia, other than **ruminants**, became ataxic after eating it. Severe cases become recumbent and die.

GONADOTROPINS, CHORIONIC

Measurable levels throughout pregnancy in women, gorillas, orangutans, and chimpanzees, but only for a brief period in early pregnancy in rhesus monkeys and baboons.

Chorionic, Human = *Ambinon* = *Antuitrin S* = *A.P.L.* = *Apoidine* = *Choragon* = *Chorex* = *Chorigon* = *Chorgon* = *Choriogonin* = *Choron* = *Coriantin* = *Coriovis* = *Corulon* = *Ekluton* = *Endocorion* = *Follutein* = *Glukor* = *Gonamone* = *Gonadotraphon L.H.* = *Gonic* = *HCG* = *Libigen* = *Lutormone*= *Luteogonin B* = *Physex* = *Predalon* = *Pregnancy Urine Extract* = *Pregnesin* = *Pregnyl* = *Profasi* = *Profasi HP* = *Primogonyl* = *Progon* = *Suigonan*

Untoward effects**:** Headache, irritability, restlessness, depression, fatigue, edema, precocious puberty, gynecomastia, pain at injection site, ovarian hyperstimulation, hematospermia, virilization, increased thromboembolism, ovarian cyst formation and rupture, penile enlargements and frequent erections, vomiting, blurred vision, incessant menstrual flow, and decreased weight. Slipped capital femoral epiphysis 3 months after injections in a 17-year-old **male**.

GONYAULAX

A dinoflagellate plankton (algae) that is known as "red tide" when its numbers exceed 20,000/ml. From 1698 to 1972, ~250 deaths reported and > 1000 poisoning cases world-wide in humans.

Untoward effects**: *G. catenella*** render **shellfish** (**clams**, **mussels**, **scallops**, and **oysters**) unfit for **human** consumption, due to a neurotoxin (***saxitoxin***) that is concentrated in them. Causes deaths in **humans** and **animals**. Symptoms of this "paralytic shellfish poisoning" in **man** include numbness and trembling of the lips, muscle weakness, nausea, vomiting, and stomach aches. Toxicity is ~50 times that of ***curare*** and is about 10,000 times as toxic as ***cyanide***. It has killed in as little as 12 h. **Mussels** can become toxic when they are only 100–200/ml of ocean water. They have been poisonous to **whales**, **dolphins**, **eels**, **crabs**, and, possibly, to **geese** and **mackerel**. Weanlings and young **rats**, but not newborns, have convulsions before death.

Also of significance are other "red tides":

G. breva, reportedly, has not killed **humans**, but has caused severe asthma-like attacks and neurotoxic shellfish poisoning by a toxin, ***brevetoxin B*** from the almost identical ***Gymnodium breve*** or ***Ptychodiscus brevis***.

G. polyhedra kills **shellfish** and **fish**, but not **rats** or **humans**.

G. tamarensis var. ***excavata*** has caused **human** deaths along the Florida coasts, and its cysts have been found off New England. In May, 1968, **sea birds**, **eels**, and **mollusks** died and 80 **people** off the northeast coast of England had paralytic shellfish poisoning.

GOPHACIDE

= *BAY 38,819* = *DRC 714* = *Phosacetin*

Organophosphorus rodenticide.

Untoward effects: Causes myasthenia, arched back, anorexia, ataxia, salivation, miosis, ptosis, dyspnea, diarrhea, tremors, loss of righting reflex, opisthotonus, and tetanic seizures in **birds**. Symptoms may be delayed up to 4 days.

LD_{50}, **rats**: 13 mg/kg; **mice**: 12 mg/kg; **dog**: 23 mg/kg; **guinea pig**: 20 mg/kg; **mallard ducks**: 24 mg/kg; **golden eagle**: 2.5–5.0 mg/kg; **pheasants**: 161 mg/kg; **chukars**: 322 mg/kg; **starlings**: 18 mg/kg; **red-winged blackbird**: 4.2 mg/kg; **pigeon**: 15 mg/kg.

GOSSYPOL

A yellow pigment extracted from the seeds of the *cotton* plant (*Gossypium* sp.). ~1% level in *cottonseed*.

Untoward effects: Severe decrease in sperm motility, maturation, oligospermia or azoospermia in Chinese and non-Chinese **men**, without decrease in libido. A survey in the 1950s of common diseases of **people** in the rural communities of Hunan and Hubei provinces of China revealed subnormal fertility. This was linked to the extensive use of crude ***cottonseed oil***. By the 1960s, gossypol was identified as the cause of **male** sterility. In > 8800 clinical trials, it was considered 99.89% effective. **Some** require up to 2 years for recovery after withdrawal of therapy. Occasionally irreversible changes. Hypokalemia is the most serious side-effect. Some gastrointestinal irritation, fatigue, and skin rashes also occur. In 67 **women** taking up to 20 mg for up to 3 months, atrophy of the endometrium was the main morphologic change, requiring up to 6 months for full recovery.

Toxicity reported in **pigs**, **poultry**, young **calves**, **lambs**, **goats**, and **dogs**. In **pigs**, ~200 mg/kg leads to muscle weakness, dyspnea, pulmonary edema, hydrothorax, hydropericardium, ascites, myocarditis, hepatitis, and nephritis before death. In **poultry**, decreased growth, yolk discoloration, cardiac edema, dyspnea, weakness, and anorexia. **Ruminants** can ingest more than monogastric **animals** before any toxicity develops. Gossypol can pass into the milk of apparently healthy **cows**, that could be consumed by **humans**, causing concern for medical investigators. **Calves** evidenced poor growth, dyspnea, cardiac arrest, pleural and peritoneal effusions, and death. Congestive heart failure reported in poisoned **dogs**. It is a potent initiator and promoter of carcinogenesis in skin-painting studies on **mice**.

LD_{50}, **rat**: 2.3 g/kg; **pig**: 550 mg/kg.

GRACILLARIA

An edible seaweed eaten by people in Japan. *G. lichenoides* is the source of *agar*.

Untoward effects: Nausea, vomiting, abdominal pain, general malaise, shock, convulsions, and cyanosis, with some fatalities. This is attributed to an enzymatic release of ***dinoprostone*** during its preparation.

GRAMICIDIN

***Antibiotic.* For topical use.**

Untoward effects: Alleged allergic skin reactions with its use in dermatologic creams was shown to be due to ***ethylenediamine*** stabilizer hypersensitivity, or to that of its other ingredients. Use with ***tyrothricin*** in an intranasal production may have induced anosmia and/or parosmia. Use with ***succinylcholine*** or ***decamethonium*** can prolong muscle relaxation.

No oral toxicity to **mice** or **rats** at up to 1 g/kg. Can cause a delayed hemolysis *in vitro*, which is prevented by ***glucose***. In **guinea pigs**, topical use causes moderate hair loss and severe inflammation of the middle ear mucosa and organ of Corti. Avoid excessive use on nose areas of guard or hunting **dogs**. Do NOT use parenterally.

Drug interactions: It is inactivated, or its action is interfered with, by ***bisulfites***, ***cephalin***, ***lecithin***, ***phospholipids***, ***"Spans"***, ***"Tweens"***, and ***sulfites***.

GRANISETRON

= *BRL 43,694* = *Kytril*

***Antiemetic.* For IV use.**

Untoward effects: Headache (3.2%), weakness, somnolence, constipation (2.3%), diarrhea (3.6%), sinus bradycardia, atrial fibrillation, electrocardiogram changes, agitation, anxiety, insomnia, fever, taste disorders, and skin rashes occur. Rarely, extrapyramidal and hypersensitivity reactions.

GRANULOCYTE COLONY STIMULATING FACTOR

= *G-CSF* = *Filgrastin* (q.v.) = *KRN 8601* = *Nartograstin* = *Neupogen*

Hematopoietic, antineutropenic.

Untoward effects: Acute neutrophilic dermatitis, increased lactate dehydrogenase, increased alkaline phosphatase, bone pain, splenomegaly, and increased risk of myelodysplasia and leukemia in **patients** with severe aplastic anemia. Nausea, hyperbilirubinemia, linear bullous dermatosis, and Sweet's disease occasionally reported.

GRANULOCYTE MACROPHAGE COLONY STIMULATING FACTOR

= *Colony Stimulating Factor 2* = *CSF-2* = *Ecogramostatin* = *GM-CSF* = *Leucomax* = *Leukine* = *Molgramostatin* = *Sargramostim* = *Sch 39,300*

Hematopoietic, antineutropenic.

Untoward effects: Pleural and pericardial effusions, transient increased serum ***creatinine*** and aminotransferase, transient fever, bone pain, diarrhea, asthenia, rash, malaise, severe hypokalemia, hypersensitivity vasculitis, severe hypocalcemia and hypomagnesemia, necrotizing vasculitis, and hypoalbuminemia, nausea, and hyperbilirubinemia.

GRAPE

= *Vitis sp.*

Untoward effects: Dermatitis from washing grapes. Patch-tests on the **patient** with the inside of the grape skin were positive only after eating grapes. **Wine-makers** often have developed itching and burning sensations on arms and legs within 20 min of contacting the pulp and stems of ***V. labrussa*** and ***V. vinifera***. This may be due to the ***methyl-anthranilate*** content. Erythema follows, then maculopapular lesions appear, which may persist for days. This does not occur with Concord grapes. Dermatitis also occurs as a result of contact with some insecticides used on the grapes. Grapes and raisins also contain ***caffeic acid*** (a carcinogen in **rodents**), and natural ***oxalates*** and ***salicylates***. Asthma has occurred after drinking Catawba grape juice. Grape juice is a potent cause of flatulence in some **people**. Grape flavors in sodas, etc. may contain ***amaranth***, ***brilliant blue*** and ***tartrazine*** dyes, ***cinnamyl anthranilate*** and ***EDTA***. Grape seeds have caused an obstruction of the sigmoid colon in a 66-year-old **male**. May cause degradation of ***levodopa***. Small grapes contain more ***tannin*** than larger ones, and more in their vintage.

GRAPEFRUIT

= *Citrus paradisi*

Untoward effects: May trigger migraine headaches in some **people**. Large quantities cause an alkaline urine and increasing ***quinidine*** serum concentration by 44% as well as increasing risk of renal calculi (***calcium oxalate*** stones). Large quantities (8 oz) of the juice increase renal calculi incidence by 44%.

Drug interactions: 200 ml increases plasma concentration of ***felodipine*** ~200%. Increases concentrations of ***amlodipine***, ***atorvastatin***, ***buspirone***, ***carbamazepine*** (40%), ***cisapride***, ***clarithromycin***, ***diltiazem***, ***β-estradiol***, ***felodipine***, ***lovastatin***, ***nifedipine***, ***nimodipine*** (50%), ***progesterone***, ***saquinavir***, ***simvastatin***, and ***testosterone***; increases concentration of ***nitrendipine*** ~100%; increases concentration of ***caffeine*** (~30%), ***cyclosporine*** (20–60%), oral ***midazolam***, ***nisoldipine***, ***terfenadine***, ***triazolam*** and ***verapamil***. These interactions are apparently due to its suppression of the drug-degrading enzyme CYP3A4. Inhibits absorption of ***digoxin*** and ***fexofenadine***. These effects can last for a day after a single drink of only 250 ml of the juice. In some countries, such as New Zealand, grapefruit interaction warning labels are required on ***benzodiazepines*** (***alprazolam***, ***midazolam***, ***triazolam***), ***caffeine***, ***calcium antagonists*** (***amlodipine***, ***diltiazem***, ***felodipine***, ***isradipine***, ***nifedipine***, ***nimodipine***, ***verapamil***), ***cisapride***, ***clomipramine***, ***cyclosporin***, ***ethinyl estradiol***, ***HMG CoA reductase inhibitors*** (***fluvastatin***, ***pravastatin***, ***simvastatin***), non-sedating ***antihistamines*** (***astemizole***, ***terfenadine***), ***perhexiline***, ***quinidine***, and ***tacrolimus***.

Can aggravate ***valsartan***-induced flatulence. Its juice decreases systemic availability of ***itraconazole*** and decreases plasma levels of ***artemether*** ~65% after 5 days of treatment.

Grapefruit oil is approved as a fragrance in soaps, detergents, cosmetic creams and lotions, and perfumes. It produces no sensitization or irritation in **human** volunteers.

It significantly increases ***nifedipine*** bioavailability in **rats**.

LD_{50} in **rats** and acute dermal in **rabbits**: > 5 g/kg.

GRAPHITE

= *Black Lead* = *Mineral Carbon* = *Plumbago*

Untoward effects: Inhalation by **workers** has caused coughing, dyspnea, black sputum, reduced pulmonary function, and lung fibrosis. Increased risk of ***polycyclic aromatic hydrocarbon*** exposure. No longer permitted in cosmetics in the U.S. George I of England forbade its use as an adulterant in ***tea***.

GRASSES

Untoward effects: Allergic asthmatic reactions caused by lawn-mowing are usually due to disturbed pollen, rather than molds. Hypersensitivity pneumonitis in **people** from the dried grass and leaves (thatched roof disease, Papuan or New Guinea lung). Dermatitis also occurs in **people** from contact. Some grasses contain ***cyanide*** and ***oxalates***.

Dogs often show allergic reactions from contact or inhalation of pollen.

GREAT WEEVER
= *Trachinus draco*

Untoward effects: An 18 inch **fish** in the eastern Atlantic, from Norway to North Africa and the Mediterranean, often living in the sand or mud of bays, emerging quickly, to strike at **divers** with venomous spines from erect dorsal and cheek spines. Pain from the sting is excruciating. Also see ***Fish***.

GREPAFLOXACIN
= *OPC 17,116* = *Raxar*

Antibiotic. A fluoroquinolone.

Untoward effects: Nausea, taste perversion, and headache. Also dizziness, rash, diarrhea, prolongation of QTc interval, tendinitis, tendon rupture, and, infrequently, photosensitivity.

Drug interactions: It decreases ***theophylline*** clearance. Absorption decreases if taken within 4 h of taking ***iron*** or ***zinc*** compounds, ***sucralfate***, or ***antacids*** containing ***aluminum, calcium***, or ***magnesium***.

GRETA
= *Alarcon* = *Azarcon* = *Coral* = *Liga* = *Maria Luisa* = *Rueda*

An orange-yellow powder containing up to 90% *lead oxide* (q.v.), used as a folk remedy for digestive disorders in infants and young children. Popular in Mexican-Hispanic communities and migrant workers in Texas, Florida, California, New Mexico, and Colorado.

Untoward effects: Anorexia, vomiting, abdominal pain, diarrhea, headache, irritability, and muscle soreness.

GRISEOFULVIN
= *Amudane* = *Curling Factor* = *Fulcin* = *Fulvicin* = *Grifulvin* = *Grisactin* = *Griséfuline* = *Grisovin* = *Grisowen* = *Gris-PEG* = *Grysio* = *Lamoryl* = *Likuden* = *Neo-Fulcin* = *Polygris* = *Poncyl-FP* = *Spirofulvin* = *Sporostatin*

Antifungal, antibiotic.

Untoward effects: Headaches are common (~15%), but usually transient. Dysgeusia, xerostomia, nausea, vomiting, diarrhea, flatulence, arthralgia, systemic lupus erythematosus, angioneurotic edema, vertigo, dizziness, fever, fatigue, drowsiness, leukopenia, persistent anemia, leukocytosis, eosinophilia, lupus erythematosus-like syndrome, neutropenia, pancytopenia, and decreased serum ***uric acid*** occur. Has caused growth retardation and weight loss due to appetite suppression, and asthma, due to hypersensitivity. Large doses in **children** can cause gynecomastia. A fatal case of pancytopenia in a 50-year-old **male** after 2½ years of therapy. Syncope, diplopia, strabismus, horizontal nystagmus, irritability, hallucinations, confusion, severe psychosis, neuropathy, dyspnea, cyanosis, urticaria, cholestatic jaundice, increased alkaline phosphatase, increased SGOT, increased lactate dehydrogenase, nephrotoxicity, exfoliative dermatitis, vesiculobullous eruptions, macular edema, photosensitivity, pharyngeal edema, porphyria, hairy tongue, glossodynia, erythema multiforme-like purpura, toxic epidermal necrolysis, Stevens–Johnson syndrome, and fixed-drug eruptions are infrequent adverse effects.

Drug interactions: ***Phenobarbital*** impairs its absorption, possibly by liver microsomal enzyme stimulating an increase in its metabolism. ***Foods***, and especially ***fats***, increase its intestinal absorption. Increases effects of ***alcohol***. The combination has caused ***disulfiram***-like reactions in some **patients**. Stimulating the metabolism of ***warfarin*** in most **patients** requires increasing dosage ~40% of the latter. Use with ***oral contraceptives*** has led to ***contraceptive*** failure and intermenstrual bleeding. Has decreased serum ***salicylates*** in an 8-year-old **male** taking ***aspirin***. Potentiated by ***phenytoin*** and decreased effect by ***chloral hydrate, chlorcyclizine, phenylbutazone, diphenhydramine, glutethimide, orphenadrine***, and ***primidone***. May increase blood levels and nephrotoxicity of ***cyclosporine***.

Digestive upsets may occur and pruritus, scurfiness, and malaise have been reported as untoward effects in **cats**. Its metabolism is accelerated by ***phenobarbital*** and ***primidone*** in **rats** and **man**. Concurrent use of ***phenothiazines*** (including tranquilizers) has produced acute porphyria lesions in **animals** (a genetic factor may be involved). Parenteral use causes undesirable inhibition of mitosis in bone marrow and seminal epithelium. It is a teratogenic agent in **cats**, and doses of 125–1500 mg/kg decrease survival rates and lead to teratogenesis in treated **rats**. Administration to pregnant **cats** in the first half of gestation is associated with stillborns and weak neonates. Prolonged administration of 1% level in the feed of **mice** increases liver size, hepatic ***protoporphyrin***, and hepatomas, especially in **males**. May interfere with the metabolism of drugs metabolized by the liver. High dosage in **cattle** decreases spermatogenesis, and is associated with liver damage in pregnant **cows**. Causes a marked increase in mortality in **mice** when given with ***colchicine***, and this combination should be pursued cautiously in **man**. Can induce thrombocytopenia in **cats** and **dogs**.

TD_{LO}, **rat** (pregnant 6–15 days): 2.5 g/kg; **mouse**: 730 g/kg/52 weeks; **cat** (pregnant 1–22 days): 2 g/kg.

GROUND CHERRY
= *Husk-tomato* = *Physalis sp.*

Untoward effects: Death preceded by dyspnea, blindness, and coma in a 6-year-old **male** shortly after ingestion of the "berries" in a Philadelphia public park, due to anticholinergic effects of its ***hyoscyamine*** (q.v.) and ***scopolamine*** (q.v.) content.

Concentrates ***nitrates*** in the plants and **sheep** grazing on the sprouts, leaves, and cherries have had dual poisoning effect by ***nitrates*** and the alkaloids. Potential abortifacient for **cattle** and **sheep**.

GROWTH HORMONES

= *Asellacrin* = *BGH* = *BST* = *CB-311* = *CL-291894* = *Crescormon* = *EL-349* = *Genotonorm* = *Genotropin* = *GH* = *Grolean* = *Grorm* = *Humatrope* = *Hypophyseal Growth Hormone* = *Leanstar* = *LY-177837* = *met-HGH* = *Nanonorm* = *Norditropin* = *Nutropin* = *Optiflex* = *pGH* = *Phyone* = *Pituitary Growth Hormone* = *Protropin* = *PST* = *Quest* = *Saizen* = *Somacton* = *Somagrebove* = *Somalapor* = *Somato-gen* = *Somatonorm* = *Somatotrophin* = *Somatotropic Hormone* = *Somatotropin* = *Somatrem* = *Somavubove* = *Somenopor* = *Sometribove* = *Sometripor* = *Somfasepor* = *Somidobove* = *STH* = *Umatrope*

Untoward effects*:** A number of cases of Creutzfeldt–Jacob disease reported in **patients** receiving **human** GH (hGH***). After receiving hCG, a **patient** died after sudden convulsions and deep coma, reportedly not due to a pathogenic virus. Cases of benign intracranial hypertension (pseudotumor cerebri) and papilledema. Diabetes, hypertension, acromegaly, osteosarcomas, and diabetic retinopathy are potential adverse effects. ***hGH*** use stimulates increased growth of melanocytic nevi. Menstrual bleeding after IM use in a 4-year-old **female**; hyperglycemia and increased glycohemoglobin in a 12-year-old **female**; acute pancreatitis in a 12-year-old **male**; acute renal graft rejection in a $4\frac{1}{2}$-year-old **male** and an 8-year-old **male**; and trace contamination with ***vasopressin*** (***antidiuretic homone***) caused marked antidiuresis. Arthralgia, carpal tunnel compression, and fluid retention in 12 **patients** (63–76 years). Abdominal lipohypertrophy after long-term 0.05 mg/kg SC in abdominal area of a $12\frac{1}{2}$-year-old **male**. A 15-year-old **male** given 0.07 units/kg/day/~$4\frac{1}{2}$ years developed a testicular tumor. High dosage to **adults** in intensive care unit led to two times increase in mortality rate; **survivors** had increased duration of mechanical ventilation and increased time in an intensive care unit. It can increase the propagation of pre-existing cancer cells. A 10-year-old **male** developed acanthosis nigricans after 3–4 units/week SC/7 years.

Drug interactions*: *Oral contraceptives increase laboratory assays; ***propranolol*** with it may lead to hypoglycemia; ***clonidine*** increases its serum levels and ***pergolide*** decreases its serum levels.

Various forms are diabetogenic; young **puppies** and **kittens** are resistant to this effect, while adult **animals** are not. **Rodents** and **lagomorphs** show only a transient glycosuria. Potentiates ***erythropoietin***-stimulated erythropoiesis.

GUAIAC

= *Gum Guaiac*

***Use*:** Clinical reagent; preservative for food fats and oils.

***Untoward effects*:** Skin sensitizer.

LD_{50}, **guinea pigs**: 1120 mg/kg.

GUAIACOL

= *Anastil* = *2-Methoxyphenol*

Use*:** As an expectorant; in ***inks as an antioxidant; and in manufacturing ***vanillin*** and ***glyceryl guaiacolate***.

***Untoward effects*:** Repeated or prolonged contact causes skin or eye irritation.

LD_{LO}, **human**: 50 mg/kg.

LD_{50}, **rat**: 725 mg/kg.

GUAIJILLO

= *Acacia berlandieri*

Untoward effects*:** Poisoning of **sheep** and **goats** in Texas and Mexico by it is negligible during periods of normal rain, but, during prolonged droughts, its exclusive ingestion causes ataxia ("guaijillo wobbles" or "limberleg"), due to hind limb weakness, lateral bending of the hocks, and stumbling; and mortality up to 65%. Nervousness, tachypnea, and convulsions can also occur in affected **sheep**. The toxic substance in the plant is a sympathomimetic amine, *N-**methyl-β-phenylethylamine (q.v.).

GUANABENZ ACETATE

= *Rexitene* = *Tenelid* = *WY 8678* = *Wytensin*

Antihypertensive.

***Untoward effects*:** Sedation, xerostomia, weakness, orthostatic hypotension, dizziness, sexual dysfunction, headache, palpitations, nasal congestion, blurred vision, frequent urination, nausea, vomiting, constipation, gynecomastia, anxiety, depression, insomnia, arrhythmias, pruritus, rashes, and rebound hypertension upon sudden withdrawal of treatment have all been reported.

GUANACLINE SULFATE

= *Cyclazenine* = *Leron*

Antihypertensive.

***Untoward effects*:** Severe postural hypotension; dizziness, vertigo, asthenia, nausea, headache, myalgia; parotid and, sometimes, submandibular pain in ~50%.

GUANADREL SULFATE

= *Anorel* = *Hylorel*

Antihypertensive.

Untoward effects: Similar to ***guanethidine*** (q.v.), but causes less orthostatic hypotension, diarrhea, decreased libido, erectile disorders, and inhibition of ejaculation.

GUANETHIDINE

= *Abapresin* = *Dopom* = *Esimil* = *Eutensol* = *Guethine* = *Iporal* = *Ismelin* = *Isobarin* = *Octadine* = *Oktadin* = *Oktatensin* = *Oktatensine* = *Oktatenzin* = *Sanotensin* = *Su-5864*

Antihypertensive.

Untoward effects: ***Cardiovascular***: Average increase in blood volume of 10% associated with ***salt*** and water retention (can lead to congestive heart failure in **patients** with limited cardiac reserve, unless used with a diuretic), polyarteritis nodosa, bradycardia, transient tachycardia, orthostatic and exertional hypotension, syncope, angina-like chest pains, hemolytic anemia, neutropenia, atrioventricular block, and vasculitis occur.

Digestive: Diarrhea, nausea, vomiting, xerostomia, and parotid tenderness.

Eye: Burning sensation for up to 30 s, dilation of conjunctival blood vessels, ptosis, keratitis, miosis, and blurred vision. A case of occlusion of central retinal artery and Bell's palsy attributed to it.

Fetus and neonate: May cause **neonatal** ileus if given to the **mother** in the third trimester of pregnancy.

Genitourinary: Urinary incontinence, increased blood urea nitrogen, nocturia, ejaculation failure, erection problems, azotemia, and decreased libido in **males** and **females**.

Miscellaneous: Myalgia, muscle weakness, giddiness, and weight increase. Since it prevents the release of catecholamines at their storage sites, as protective measures during surgery, caution must be used.

Neurologic: Syncope, dizziness, muscle tremors, shivering, and depression with suicidal intention.

Respiratory: Dyspnea; occasional asthma and nasal congestion.

Skin: Rashes, urticaria, erythematous papules and nodules that may ulcerate and become gangrenous, pruritus, and alopecia.

Drug interactions: ***Amphetamines, dextroamphetamines, methamphetamines, amytriptyline, cocaine, desipramine, diethylpropion, doxepin, ephedrine, haloperidol, imipramine, monoamine oxidase inhibitors, methylphenidate, nialamide, nortriptyline, phenothiazines, phenylbutazone, pipradol, protriptyline, pyrilamine***, and ***tripelennamine*** antagonize its hypotensive effect. ***Anesthetics*** with it enhance its actions and can cause cardiovascular collapse. Has also decreased the effect of ***imipramine***. Both its effects and that of ***antidepressants*** may be diminished by their combined therapy. Withdrawal of treatment led to striking rise in ***blood sugar*** levels and ***insulin*** requirement in a **patient**. Increases cardiac sensitivity to ***norepinephrine***-induced arrhythmias and hypertension after guanethidine treatment. Use with thiazides enhances its effects, often increasing nausea, headache, and angina. Use with ***digitalis*** can cause increased susceptibility to vagal effects, causing excessive atrioventricular block (conduction failures), and intolerable bradycardia. ***Alcohol*** potentiates its postural hypotension effect. Decreased effect with ***oral contraceptives*** is probably due to the latter's induced ***sodium*** and fluid retention.

Guanethidine completely eliminates ***morphine's*** analgesic effect in **animals**.

LD_{50}, **mouse**: 3 g/kg; IV, LD_{LO}, **cat** and **rabbit**: 50 mg/kg.

GUANFACINE

= *BS 100–141* = *Entulic* = *Estulic* = *LON 798* = *Tenex*

Antihypertensive, α-sympathomimetic.

Untoward effects: Xerostomia, tiredness, impotence, bradycardia, orthostatic hypotension, dizziness, constipation, and asthenia. A 2-year-old **male** ingested 4 mg (~30 times normal **adult** dose), became sleepy, and had decreased blood pressure. **He** was given gastric lavage and ***activated charcoal***. Discharged 24 h later. Sudden treatment withdrawal led to increased blood pressure and heart rate.

GUANIDINE

= *Carbamidine* = *Iminourea*

Untoward effects: Shock, cardiac arrhythmias, atrial fibrillation, hypotension, diarrhea, weakness, occasional tingling of mouth and tongue, gastric discomfort, nausea, distal paresthesia, faciculations, tremors, ataxia, nervousness, anemia, leukopenia, aplastic anemia, increased blood ***creatinine***, and dryness and scaling of the skin have occurred.

Injection in **rabbits** aggravates ***histamine*** intoxication and, like ***histamine***, causes **cats** to vomit.

LD_{50}, **mice**: 600 mg/kg; LD, **rabbits**: 500 mg/kg.

GUANOXAN

Antihypertensive.

Untoward effects: Hepatotoxicity with frank jaundice, severe abnormalities in liver function tests, and death caused the product's withdrawal from the marketplace. Systemic lupus erythematosus-like syndrome, gastrointestinal upsets, asthenia, hypotension, arthralgia, headache, vertigo, and numerous other adverse effects reported.

GUARANA

The dried paste from the seeds of *Paullinia cupana*, a South American shrub. Contains ~4% *caffeine*, also traces of *theobromine* and *tannic acid*, and used as a stimulant. Especially popular in Brazil, where it is used in medicines and as a beverage, as others use *coffee* or *tea*. This non-alcoholic, fruity-flavored beverage is sold in Europe, North America, and the Orient.

Untoward effects*:** Sold in the U.S. as ***Zoom, cheaper than ***coffee***, and providing an "upper" feeling, like ***"speed"*** (***amphetamine***), and a "buzz", like ***cocaine***. Overdosing, especially in **children**, is possible and the symptoms are essentially those from ***caffeine*** (q.v.).

Stems, leaves, and roots are used as a piscicide. **Fish** so obtained can be used as food for **humans**.

GUAREA

Some species are used as emetics, expectorants, and purgatives.

G. thompsonii. In Africa. Bark extracts have been lethal in an hour.

G. trichilloids* = *Camboatá Tree. In Brazil. The fruits are poisonous. The leaves and stems have caused abortions and deaths in **cattle** and deaths in **guinea pigs**. In **man**, it is a powerful topical irritant and the powder from the fruit causes edema of the eyes and, if inhaled, leads to nasal bleeding. The fruit has been used as an abortifacient in **women**. Extracts given orally to **rats** caused death in 2–8 h.

GUAR GUM

= *Cyamopsis tetragonolobus* = *Jaguar*

Untoward effects*:** Hypersensitivity reaction in occupational exposure to its dust led to rhinitis and asthma. A single case reported after oral ingestion. In 1990, the FDA banned weight loss products, such as ***Cal-Ban 3000, because as it swells in the throat, ~50 **people** were injured by it and one **person** died from an acute gastric obstruction.

GURANIA

Untoward effects*:** Many species are used medicinally by Amazon **natives**. Unfortunately, ***G. rufipila stems, mistaken for the others, are poisonous for **man**.

G. guaransenia contains stinging hairs that cause irritant dermatitis, and often harbor a **butterfly** larva with similar devices.

GUTIERREZIA MICROCEPHALA

= *Broomweed* = *Slinkweed* = *Snakeweed* = *Turpentine Weed* = *Xanthocepthalum microcephala*

***Untoward effects*:** **Cattle**, **sheep**, **goats**, and **swine** in southwestern U.S. are poisoned by its toxic triterpinoid saponin. Acute poisoning leads to listlessness, diarrhea/constipation, hematuria, weight loss, and anorexia. Abortion, stillbirths, retained placentas, and weak **calves** are common in **cattle**.

GUTTA SIAC

***Untoward effects*:** Causes a dermatitis when used in some adhesive plasters.

GVIA

Untoward effects*:** An ***ϖ-conotoxin in the **fish**-hunting cone **snail**, *Conus geographus*, neurotoxic to vertebrate nervous systems, causing paralysis and death.

GYPSY MOTH

= *Lymantria dispar*

***Untoward effects*:** Almost ⅓ of 620 **students** in two Pennsylvania schools developed rashes on the exposed parts of **their** bodies (75% on arms, 23% on neck and 21% on legs) from contact with the highly mobile and airborne larval stages (**caterpillars**).

GYROMITRA ESCULENTA

= *False Morel* = *Fruhjahrslorchel* = *Helvela suspecta* = *Lorchel*

Untoward effects*:** Although commonly used as a wild, edible ***mushroom in Europe and North America, it has caused many cases of fatal poisoning in **man**. Toxicity noted in food-processing plant **workers** sensitized to ***mushroom*** vapors. The main toxic agent in the ***mushroom*** is ***ethylidene gyromitrin*** (q.v. for symptoms), which is metabolized to ***methylhydrazine*** (q.v.). Ingestion by five **women** (14–57 years) after a latent period of 6–10 h caused vomiting, epigastric pains, and jaundice. Hemolysis and central nervous system effects have been reported. Some **people** are resistant to its ill-effects. Mortality estimated at 2–4%. Severe hepatic damage reported in a 47-year-old **female**. Symptoms are usually less severe than with ***amanitin*** (q.v.).

A fatal hemolytic episode in a 10-week-old **dog**.

H

HAFNIUM
= *Celtium*

Untoward effects: Explosive in dry form. National Institute for Occupational Safety & Health/OSHA limit exposure to 0.5 mg/m^3; immediate danger to life or health, 50 mg/m^3.

Animal trials indicate it is an irritant to eyes, mucous membranes, and skin and causes liver pathology.

HAIR

Untoward effects: Trichobezoars occur most frequently in **children**, mentally disturbed **persons**, and in postgastrectomy **patients**. Occupational dermatitis in **workers** handling hair. Permanent wave solutions cause numerous cases of contact eczematous or papular eruptions from some of their ingredients (viz. ***glyceryl monothioglycolate*** and other ***thioglycolates***). These **people** usually cross-react with ***"caines"***, ***PABA*** sunscreens, ***sulfonamides***, hair dyes, and ***azo*** and ***aniline dyes***. Acute renal failure after ingestion by a **child** of solution containing ***sodium bromate*** (q.v.). ***Potassium bromate*** has also been ingested. Contact urticaria and thesaurosis from exposure to or inhalation of hair sprays, with dyspnea, coughing, granulomatous pneumonia, blood dyscrasias, aplastic anemia, hypersensitivity pneumonias, and migraines. Severe chemical burns with ulceration of labia minora, after mistakenly using it as a vaginal deodorant; 3 weeks healing time and 6 weeks dyspareunia. Hair creams have caused photocontact dermatitis from ***fenticlor*** (q.v.); sensitization in **children** from ***sodium hydroxide*** (q.v.) in a cream "hair-relaxer". Comedonal acne occurs in **children** and **adults** using certain brialliantines or hair-straightening pomades. See individual ingredients. Shampoo ingestion usually causes nausea, vomiting, and, later, diarrhea. Eye contact usually causes a conjunctival erythema. ***Sodium lauryl sulfate*** (q.v.) is the chief offender. A generalized pruritic eruption after showering or shampooing in a **patient** on ***disulfiram*** was traced to the use of a ***beer***-containing shampoo. Hair wave dermatitis is often caused by ***karaya gum*** (q.v.). **Patients** have been burned by heat-activated solutions. ***Ammonium persulfate*** (q.v.) in hair bleaches has caused eczematous contact dermatitis, severe urticaria, and anaphylactic shock.

Many symptoms reproduced experimentally in **dogs**, **rats**, and **guinea pigs**. Trichobezoars have occurred in **cattle**.

HALAZEPAM
= *Paxipam* = *SCH 12,041*

Anxiolytic. **A *benzodiazepam*.**

Untoward effects: Drowsiness (~30%), headache, apathy, confusion, disorientation, euphoria, gastrointestinal upsets, skin rashes, pruritus, abdominal distress, xerostomia, increased serum transaminases, dysarthria, depression, syncope, dizziness (8%), ataxia (5%), fatigue (4%), and paradoxical agitation or rage (1%). Crosses the placenta and appears in breast milk. Delirium and convulsions can occur with sudden withdrawal of therapy. Low potential for abuse.

Drug interactions: Decreased metabolism is to be expected when used with ***cimetidine***.

LD_{50}, **mice**: > 4 g/kg.

HALOFANTRINE
= *Halfan* = *SKF 102,886* = *WR 171,669*

Antimalarial.

Untoward effects: Diarrhea, abdominal pain, pruritus, arrythmia, prolonged QTc and PR intervals, severe intravascular hemolysis, and acute renal failure (black water fever). Resistance to it can develop rapidly and there is cross-resistance with ***mefloquine***.

Drug interactions: Use with ***magnesium carbonate*** decreases its serum levels; ***ketoconazole*** decreases its metabolism.

Significant increase in oral bioavailability after stimulation of gastric acid secretion by ***pentagastrin*** noted in **beagle dogs**.

HALOFENATE
= *MK 185*

Hypolipidemic, uricosuric.

Untoward effects: Nausea and abdominal pain (~10%) and increased serum ***thyroxine***.

Drug interactions: Decreases absorption and β-blocking activity of ***propranolol*** and rebound effect on withdrawal; decreases plasma protein binding of ***phenytoin*** and potentiates ***sulfonylureas'*** hypoglycemic effects.

Can decrease ***warfarin*** effect in **dogs**, when administered prior to ***warfarin***.

HALOGETON

Poisonous plant related to the Russian thistle, growing best on roadsides and abandoned land.

Untoward effects: Due to ***sodium*** and ***potassium oxalates*** in the above ground parts, increase in concentration as the plant grows, and when it is frozen and dry. **Sheep** are easily killed by 12–18 oz, with symptoms appearing in 2–6 h. Dullness, anorexia, colic, lowered head, and reluctance to follow the herd are early symptoms, followed by drooling, frothing about the mouth, nasal discharge, weakness, dyspnea, coma, and death, usually due to renal failure. ~1200 **sheep** died in one incident in Utah. **Cattle** are less commonly affected and their deaths are infrequent. Usually, they become stiff and recumbent. **Calves** have developed convulsions and hematuria.

Estimated LD_{LO}, **sheep**: 1 g/kg.

HALOPERIDOL

= *Aloperidin* = *Bioperidolo* = *Brotopon* = *Dozic* = *Einalon S* = *Eukystol* = *Haldol* = *Halosten* = *Keselan* = *Linton* = *McN JR 1625* = *Peluces* = *R-1625* = *Serenace* = *Serenase* = *Sigaperidol*

Psychotherapeutic.* A *butyrphenone.

Untoward effects: Primarily, extrapyramidal motor symptoms in > 75% of **patients**, accompanied by decrease of ***dopamine*** in the caudate nucleus and increased excretion of ***dopamine's*** principle metabolite, ***homovanillic acid***. This results in Parkinson-like symptoms and, less frequently, motor restlessness, dystonia, akathesia, hyperreflexia, opisthotonus, oculogyric crises, tardive dyskinesia, and seizures. Occasionally hypotension, bradycardia (one fatality, due to cardiovascular collapse), tachycardia, depressive syndromes, increased or decreased libido, painful ejaculation, xerostomia, hyposalivation, rhabdomyolysis, hair loss, decreased liver function, and jaundice. Rarely, a neuroleptic malignant syndrome (muscle rigidity, fever, autonomic hyperactivity, and altered consciousness) in ~1%, lasting at least 5–10 days after drug withdrawal (lasting longer after IM use). Loss of voice and sudden death are also rare occurrences. Photosensitivity, maculopapular and acneiform eruptions, corneal opacity and blurred vision, insomnia, lethargy, headache, vertigo, confusion, hallucinations, catatonia, sedation, gynecomastia, increased serum ***prolactin***, lactation, mastodynia, menstrual irregularities, hyper- or hypoglycemia, anorexia, diarrhea, constipation, increased weight, neurotoxicity in hyperthyroid **patients**, platelet aggregation, torsades de pointes, cardiac arrest and acute fatal pulmonary edema in one case, diaphoresis, urinary retention, syndrome of inappropriate antidiuretic hormone secretion, decreased thirst, and dehydration. **Human** breast milk levels are ~0.5% of **maternal** levels leading to **infant** maximum intake of 0.0075 ng/day. Two case reports of severe limb malformations in **infants** following early **maternal** use during the first trimester of pregnancy. One of these **women** had also been taking ***phenytoin*** and ***methylphenidate*** during this period. Overdosing has been common (24 **children** in one hospital in Japan). Accidental overdosing in one case, seeking a "high" and mistaking it for ***diazepam***. Withdrawal symptoms include mutism, staring, posturing, catatonia, and refusing food and drink.

TD_{LO}, **child**: 13 mg/kg; **woman** (1–7 weeks pregnant), 300 μg/kg/day.

Drug interactions: Hypotension and cardiopulmonary arrest with ***propranolol***. A case report of decreased effect of ***phenindione*** with it. Another case report of intense potentiation of depressant effects of ***alcohol***. Increased blood levels by ***fluoxetine*** (~20%), ***fluvoxamine***, ***nortriptyline***, and ***alprazolam***; decreased levels with ***tobacco*** use and when put in ***tea*** (90% decrease) or ***coffee*** (10% decrease); decreased by ~60% within 2–3 weeks after use with ***carbamazepine*** and 22% decrease in serum levels with concomitant ***phenobarbital*** use. ***Epinephrine*** may interact with ***α-adrenergic antagonist***, such as haloperidol. It can potentiate the analgesia and respiratory depression of ***phenothiazine tranquilizers***. Decreases metabolism and excretion of ***imipramine***. Use with ***α-methyl tyrosine*** caused severe extrapyramidal symptoms within 24 h. Effects potentiated by ***methyldopa***. ***Indomethacin*** increases its induced drowsiness and confusion. Some of its benefits decreased by ***benztropine mesylate***. Presumed neurotoxicity from interaction with ***lithium*** is, probably, ***lithium's*** own toxicity.

Haloperidol has usually decreased body temperature after parenteral use in the **cat**, **goat**, **mouse**, **rabbit**, and **rat**. Dyskinesias in **squirrel** and **cebus monkeys**, decreased seizure threshold in laboratory **animals**, and behavioral changes in newborn **rabbits** whose mothers had received the drug. In **rats** and **mice**, cataleptic state, delayed embryo implantation, prolonged gestation, and intrauterine growth retardation; teratogenic in **mice**; and decreased blood pressure and heart rate in **rats**.

LD_{50}, **rat**: 165 mg/kg; **mouse**: 145 mg/kg.

HALOPROGIN

= *Halotex* = *M-1028* = *Mycanden* = *Mycilan* = *Polik*

***Antifungal.* Topical.**

Untoward effects: Local irritation, slight erythema, allergic contact dermatitis, burning sensation, vesicle formation, increased maceration, pruritus and exacerbation of pre-existing lesions. Avoid eye contact.

HALOPROPANE

= *FHD-3* = *Tebron*

Anesthetic. **For inhalation.**

Untoward effects: ~20% incidence of ventricular arrhythmias in **humans** and **cats**. Tends to decrease blood pressure and effective ventilation in **humans** with frequent bigeminal cardiac rhythms and slow induction and waking.

TC_{LO}, **human**: 4000 ppm/10 min.

HALOTHANE

= *Fluothane* = *Fluotic* = *Somnothane*

Anesthetic. **For inhalation.**

Untoward effects: ***Cardiovascular***: Hypotension (~25%), bradycardia (20–25%), peripheral vasodilation, asystole, partial or total atrioventricular block, extrasystole, abnormal electrocardiogram, ventricular arrhythmias, especially after ***atropine*** or ***epinephrine***, bigeminy, and occasionally paradoxical hypertension. Various equipment failures and topical ***epinephrine*** during halothane anesthesia have caused fatal cardiac arrests. Significantly decreases ventricular function in **children**. Readily enters **fetal** circulation. Increases sensitivity of cardiac muscle to ***β-adrenergic*** activity.

Eye: Nystagmus, diplopia, transient vision loss, and decreased intraocular pressure reported. Ophthalmic hypersensitivity in **anesthetist**. Irritation was progressive and uncomfortable.

Liver: Mild hepatitis, jaundice, and even massive liver necrosis (0.00001%) have occurred. May be an allergic reaction, as it is accompanied by fever and eosinophilia.

Miscellaneous: Increased cerebrospinal pressure, headache, respiratory acidosis, hypoxemia with seizures, shivering, porphyrias, acneiform eruptions ("haloderma"), and urticaria. An operating room **nurse** developed acne from halothane exposure. Contact later with post-surgical **patients** in intensive care unit caused recurrence of widespread acneiform eruptions. Nausea, mild vomiting, dizziness, and suppression of renal tubular transport occur. A fatal case among hospital **personnel** abusing it. Many deaths in **abusers**. Unfortunate adulteration or contamination reported. Suicides and accidental IV use have also caused fatalities.

Drug interactions: Malignant hyperpyrexia and increased serum creatine phosphokinase with ***succinylcholine***; neuromuscular block with ***gallamine***, ***gentamicin***, ***neomycin***, ***streptomycin***, and ***tubocurarine***; nearly fatal shock-producing hepatic reaction to ***rifampin*** given immediately after anesthetic induction; increased jaundice and liver necrosis in combination with radiotherapy; sinus bradycardia, transient atrioventricular block, hypotension, and asthma when used with ***propranolol***; with ***nitrous oxide***, leading to transient nodal rhythm (17%), premature ventricular contractions (7%), premature atrial contraction (0.5%), bigeminy (0.2%), and headaches. ***Methohexital sodium*** IV or ***hexafluorenium*** increase the arrhythmias; decreased hepatic metabolism and increased toxicity of ***phenytoin***. Cross-sensitivity with ***methoxyflurane*** frequency reported; ventricular arrhythmias with ***tricyclic antidepressants***, ***anticholinesterases***, and ***pancuronium*** in experimental **animals**, and, to a lesser extent, in **man**; and increase in renal and hepatic damage after ***acetaminophen***. Cardiac arrhythmias with ***ephedrine***, ***epinephrine***, and ***methamphetamine***.

Epinephrine (even for hemostasis) and ***β-adrenergics*** will produce arrhythmias in **animals**. Hypothermia is the rule in **animals**, but acute, highly fatal hyperthermia has occurred in **swine** and **man**. Tachycardia is common in anesthetized **cattle**. **Horse** and **dog** myocardium is sensitized by ***xylazine*** leading to occasional fibrillation. Permanent learning deficits in young **rats** exposed to it during early development. Cardiac arrhythmias, fibrillation, and bigeminy in **Pekin ducks**.

LD_{LO}, **human**: 140 mg/kg; inhalation LC_{LO}, 7000 ppm/3 h.

TC_{LO}, **rat** (8 days pregnant): 8000 ppm/12 h.

HALOXON

= *Eustidil* = *Galloxon* = *Galoxone* = *96-H-60* = *Halox* = *Helmirone* = *Loxon* = *Luxon*

Anthelmintic. **Organic *phosphorus* compound.**

Untoward effects: Many **sheep** can tolerate 3–4 times the recommended dose, yet some **sheep**, particularly **Suffolks**, lack a hydrolyzing esterase and develop a delayed neurotoxicity leading to rear leg weakness, ataxia, posterior paralysis, and death. **Pigeon** dose may be lethal to **geese** and toxicicity has to **ducks** and **quail**. Delayed neurotoxicity has also been reported in **swine** and **horses**.

LD_{50}, **rats**: 900 mg/kg.

HALQUINOL

= *Chlorquinol* = *CHQ* = *Halquivet* = *Quinolor* = *Quixalin* = *Quixalud* = *SQ-16401* = *Tarquinor*

Antimicrobial.

Untoward effects: Ingestion leads to skin rashes, pruritus, nausea, vomiting, headache, dizziness, and dysuria. Repeated topical use leads to skin sensitization, contact allergy, and irritation. Cross-reacts with ***chlorquinadol*** and ***iodochlorhydroxyquin***.

HALVA(H)

= *Tahneeya*

Untoward effects*:** **Ants** avoid this Near Eastern candy delicacy made of crushed ***sesame seeds and ***sugar*** because ***sesame oil*** contains the lignans ***sesamin*** and ***sesmolin***, which are powerful natural insecticides and insecticidal synergists.

HAM

Untoward effects*:** Contains 30–150 ppm ***nitrites and smoked ham also contains ***polycyclic hydrocarbons***, including the procarcinogen ***benzopyrene***.

HAMAMELIS VIRGINIANA

= *Amamelide* = *Witch Hazel*

Untoward effects*:** Leaves contain up to 9.5% ***tannins. Ingestion causes a severe gastritis. 1 g has caused nausea, vomiting, and constipation. ***Tannins*** are poorly absorbed, yet can cause liver and kidney damage.

LD_{LO}, **human**: 5 g/kg.

HAMBURGERS

= *Beef Patties*

Untoward effects*:** Incomplete cooking has caused severe outbreaks of diarrhea, cramps, and even hemolytic uremic syndrome from ***E. coli. In 1986, 140 **people** in Minnesota, South Dakota, and Iowa became ill from eating "gullet-trimmed meat" in **their** hamburgers. Gullet trimmings include **animal** thyroid glands. This increases metabolism to discomforting levels. United States Department of Agriculture now forbids "gullet trimmings". Inhalation of fumes from the grilling process increases potential risk from cancer-causing heterocyclic amines.

HAMYCIN

= *Primamycin*

Antifungal antibiotic.

***Untoward effects*:** Nausea, vomiting, diarrhea, and abdominal pain in **people**.

Caused diarrhea and nephritis in **dogs**.

HARMALINE

= *Harmidine*

An alkaloid in *Peganum harmala*, *Banisteriopsis* sp., and *Passiflora jorullensi*.

Untoward effects*:** Nausea, dizziness, malaise; paresthesia of hands, feet, and face, followed by numbness. Oral doses lead to tremors and clonic convulsions, motor paralysis, central nervous system depression, decreased body temperature, decreased blood pressure, and respiratory paralysis. Hallucinogenic in **humans**. Twice the toxicity of ***harmine.

Parenteral use decreases body temperature in **rats**; and causes bradycardia, salivation, and increased pulse pressure in **dogs**.

HARMALOL

An alkaloid in *Peganum harmala*, *Banisteriopsis* sp., and *Passiflora jorullensi*.

***Untoward effects*:** Hallucinogenic at doses > 4 mg. Bradycardia, increased pulse pressure, and increased myocardial contractile force.

SC, LD_{LO}, **cat**, **rat**, and **rabbit**: ~100 mg/kg.

HARMAN

= *Aribine* = *Loturine* = *Passiflorin*

An alkaloid in *Peganum harmala*, *Passiflora jorullensi*, and in cured commercial *tobacco* and smoke condensate of *Nicotiana rustica*.

***Untoward effects*:** Hallucinogenic.

HARMINE

= *Banisterine* = *Yageine*

An alkaloid in *Peganum harmala*, *Banisteriopsis* sp., and *Passiflora jorullensis*. In some South American snuffs.

Untoward effects*:** Hallucinogenic (150–200 mg IV) and definite psychotic symptoms, nausea, ataxia, and hallucinations at 300 mg orally. Euphoria after 25–75 mg SC. A ***monoamine oxidase inhibitor.

IM, TD_{LO}, **human**: 3 mg/kg.

Parenteral use reduces body temperature in **rats** and decreases blood pressure in **dogs**.

HARMOL

An alkaloid in *Peganum harmala* and *Passiflora jorullensis*.

Untoward effects*:** Increases ***epinephrine activity at low doses and cardiac arrest with high dosage.

Monoamine oxidase inhibitor in **guinea pigs** and **rats**.

HARPAGOPHYTUM

= *Devil's Claw* = *Grapple Plant* = *Wood Spider*

***Untoward effects*:** May cause allergic reactions. Contraindicated in **patients** with gastrointestinal ulcers, as it increases gastric acid secretion.

HARUNGANA MADAGASCARIENSIS

Untoward effects: Stem bark or roots as an abortifacient in Nigeria. The orange berries are edible.

Initially, hyperactivity in **rats**, followed by inability to move. Purgation, vaginal bleeding and discharge, abortions, and deaths within 5 h in experimental **rats** and **mice**.

HATS

Untoward effects: "Mad as a hatter" became a more common expression after the 1930s, when it became known that **workers** who cut fur for felting hats developed central nervous system disturbances from ***mercuric nitrate*** treatment of the fur. Callosities on palms of felt **hat-formers** from dipping the felt in a solution of ***sulfuric acid*** and ***mercuric nitrate***. ***Chrome***-tanned leather sweat bands in hats have caused a dermatitis. Dermatitis occurred in felt **hat-makers**. A hat **saleswoman** developed a severe purpuric eruption of exposed areas of the face, neck, and arms, due to a sensitivity to ***p-phenylenediamine*** (q.v.) used in fur dyeing.

HAWTHORN

= *Crataegus sp.* = *Haws* = *Whitehorn*

Untoward effects: Often used by young **women** to alleviate palpitations, but may cause hypotension.

Ingestion of fermented fruit has caused **birds** to fall off rooftops. Post-mortems revealed crop ***alcohol*** content of 380 ppm and 238–989 ppm in their livers.

HAZEL NUTS

= *Corylus americana*

Untoward effects: A common allergen for **people**, often cross-reacting with ***apples*** and ***potatoes***. In Germany, antiserum is used to detect its use in ***beer*** manufacturing. Its use is illegal there, but not in the U.S.

HEDEOMA sp.

Untoward effects: ***H. oblongifolia = Nana = Pennyroyal = Poleo Chino***. In New Mexico, an emmenagogue and abortifacient in a ***tea***.

H. pulegioides = American Pennyroyal = Squaw Mint = Tickweed and its oil (***Pennyroyal Oil***) as a stronger abortifacient and emmenagogue in New Mexico. Many poisoning cases reported to US Center for Disease Control. After ingesting 1 oz of the oil, one **patient** died following abdominal pain, nausea, vomiting, lethargy, and agitation. The **patient** developed massive hepatic necrosis and intravascular coagulation, and died of cardio-pulmonary arrest after 7 days.

HEDERA HELIX

= *Curré* = *English Ivy* = *Lierre Commun*

Untoward effects: **Children** have been poisoned by eating the acidy, pungent berries that contain ***hederin*** and ***hederogenin***. Symptoms include nausea, vomiting, diarrhea, purgation, dyspnea, depression, or excitement. A contact dermatitis developed in some sensitive **people** with slow-healing weeping lesions and blisters. Disagreeable taste and unpleasant odor in ***milk***, milk products, and meat of **animals** that graze on it.

The leaves have an ***atropine***-like effect, causing some **cattle** to become recumbent in a milk fever-like condition after browsing on them or after eating them in clippings from the vine. They usually recover. **Dogs** show vomiting, diarrhea, muscle spasms, and paralysis.

HELENIUM sp.

Untoward effects: ***H. amarum = H. tenuifólium = Bitter Sneezeweed***. Its poisonous principle, ***dugaldin***, when eaten by **livestock** during droughts, causes weakness, dyspnea, and spasms before death. In southeastern U.S., Cuba, Yucatan, and Puerto Rico. Disagreeable taste and unpleasant odor in ***milk***, milk products, and meat of **animals** that graze on it.

H. autumnale = Autumn Sneezeweed. Leaves and flowers can cause violent sneezing in **people**. Spasms with delirium and loss of consciousness reported in four **people**.

All parts of the plant, especially the blossoms, are poisonous, due to its ***helenalin*** content, and, despite its bitterness, individual **cattle** and **sheep** develop a craving for it. Found in the U.S. and Canadian midwest. Symptoms are tachycardia, restlessness, dyspnea, convulsions, severe gastroenteritis, ataxia, and staggering. Can be fatal. Causes **cows** to give bitter milk.

H. flexuosum = Helénie nudiflore = Naled-flowered Sneezeweed. After ingesting its leaves and stems containing ***sesquiterpene lactones***, ***flexuosins A and B***, **horses** and **sheep** develop weakness, dyspnea, and convulsions.

H. hoopesii = Orange Sneezeweed. Common in western U.S. at elevations > 5000 ft; contains ***dugaldin***, a poisonous glucoside and cumulative poison. **Sheep** are particularly susceptible. Their symptoms are profuse vomiting ("spewing sickness"), sialism, coughing, bloating, weakness, belching, borborygmi, arrhythmias, and tachycardia. Thousands have died annually. Some **sheep** die within a few days after symptoms are noted (usually, 2 lbs leaves/day/10 days). **Cattle** are occasionally affected.

H. integrifolium has caused large losses in Mexican **sheep** (altitudes > 7500 ft).

***H. microcephalum* = *Smallhead Sneezeweed*.** In its flowering stage, as little as 2.5 g/kg causes acute poisoning and death in Texas **sheep** and **goats**. Nasal discharge, hydrothorax, hydropericardium, cardiac hemorrhages, gastroenteritis, and occasionally death in **cattle**.

H. nudiflorum* = *Purple-headed Sneezeweed*.** Except for its color, its effects are the same as for ***H. autumnale. Also in the Midwest.

HELIOTRINE

A *pyrrolizidine* alkaloid from *Senecio jacobaea* (q.v.).

***Untoward effects*:** Has been incriminated in **human** liver disease in Asia.

Injection during the second week of gestation in **rats** led to malformed ribs, vertebrae, mandibles, palate, limbs, and tail. Oral doses are hepatotoxic to **rats**. Liver necrosis in **rats** 20 h after IP injection and large doses caused muscle weakness, dyspnea, and death within 30 min, apparently from neuromuscular block.

LD_{LO}, **rats**: 50 mg/kg.

HELIOTROPIN

= *Piperonal*

***Use*:** As a fragrance in soaps, detergents, cosmetic lotions and creams, and foods. ~75 ton used annually in the U.S.

***Untoward effects*:** No sensitization and limited irritation (full-strength) in **man**.

LD_{LO}, **human**: 500 mg/kg.

Rats fed 1000 ppm/28 weeks or 10,000 ppm/15 weeks in their diet showed no ill-effects.

LD_{50}, **rat**: 2.7 g/kg.

HELIOTROPIUM sp.

= *Heliotrope*

Not to be confused with *garden heliotrope* (*Valeriana officinalis*).

Untoward effects*:** Contains hepatotoxic ***pyrrolizidine alkaloids. Used in herbal medicines and "***bush teas***", for **people**, in many parts of the world, especially in Asia, Africa, and the West Indies, where it appears to be associated with the high incidence of liver disease in **people** in those areas.

H. europaeum* = *Wild Heliotrope*,** introduced in the early days of settlement in New South Wales and South Australia. Less poisonous and lethal to **sheep** than **cattle** (after a 1–2 month period grazing on it). Predisposes **sheep** to liver ***copper accumulation and the hemolytic crisis of chronic ***copper*** poisoning leads to hematogenous jaundice and hemoglobinuria, and is usually fatal. **Sheep** are relatively resistant to the ill-effects of grazing a pure stand of the plant, but, if they graze it for two consecutive summers, death rates of 50–70% reported.

H. indicum* = *Hatisura* = *Kalkashin (Hausa) = ***Korama***(Hausa) Flowers are abortifacient for **women** in large doses and an emmenagogue in small doses. A ***nitrate***-accumulator, causing methemoglobinemia in **calves**. Contains pyrrolizidine alkaloids ***indicine***, ***heliotrine***, and ***lasiocarpine***. In 1946, consumption of certain cereal grains contaminated by seeds of ***H. lasiocarpum*** was shown to be associated with liver disease in central Asia. **Livestock** in the same region fed on grains and offal contaminated with the same seed also suffered from similar liver disease. Although these alkaloids primarily produce cell damage in the liver, they also adversely effect lungs and other organs, and, in some cases, are carcinogenic. Used in herbal ***teas***, where its excessive intake can cause veno-occlusive liver disease with hepatic vein thrombosis.

Other species have also been reported as toxic.

HELIUM

Untoward effects*:** Donald Duck-type voice distortion and increased body heat loss when **aquanauts** breathe it with ***oxygen, as a substitute for ***nitrogen*** in air. Progressive reduction of speech intelligibility with increased depth.

Inhalation also causes decreased temperature in **guinea pigs**, **hamsters**, **rabbits**, **mice**, and **rats**.

HELLEBORUS NIGER

= *Christmas Rose*

Untoward effects*:** Rootstocks and leaves contain a number of toxic cardiac glycosides, ***germidine, ***germitrine***, ***hellebrin***, ***helleborein***, and ***helleborin***. Known to be poisonous since the ancient Grecian era, causing violent purging, abdominal pain, diarrhea, cramps in calf muscles, palpitations, bradycardia, hypotension, sedation, delirium, headache, vertigo, tinnitus, mydriasis, photophobia, convulsions, and death from respiratory failure. The glycosides have ***digitalis***-like action leading to increased blood pressure in small doses, but, after large doses, tachycardia, followed by cardiac arrest in systole. ***Helleborin*** is cardiotoxic and narcotic and ***helleborein*** is a local irritant and cardiotoxin. Touching the plant has caused diarrhea and uprooting it can cause vomiting in some **people**. The juice causes skin inflammation and numbing of the mouth; conjunctivitis, redness and eyelid swelling, and sneezing after eye or nose contact.

Pets and **livestock** are also adversely affected. Bradypnea, bradycardia, mydriasis, and anesthetic-like symptoms occur. LD for a **horse** is ~1 kg of green leaves. **Horses**

and **cows** poisoned by eating 8–10 g of the roots; **goats** and **sheep** by 4–12 g. **Pigs** and **dogs** may vomit after eating 0.3–1 g of the roots. Secreted into **animals'** milk and found in their meat, making consumption of such products from affected **animals** dangerous for **man**. SC ***H. purpurescens*** extract causes leukoblastosis in **horses**, **cattle**, **pigs**, and **dogs**. Other hellebore species have also been found toxic.

HEMATIN

= *Hydroxyhemin* = *Phenodin*

Use: In the treatment of acute intermittent porphyria and the neurological crisis of tyrosinemia.

Untoward effects: Hemolysis in a 32-year-old **patient** after receiving 4 mg/kg/12 h IV for 36 h. Hematological values returned to normal 72 h after cessation of therapy. Another **patient** developed coagulopathy (markedly increased prothrombin time, partial thromboplastin time, thrombocytopenia, mild hypofibrinogenemia, mild increase of fibrin split products, and 10% decrease in hematocrit), 1 day after 196 mg IV every 12 h for acute intermittent porphyria.

IV, LD_{50}, **rat**: 43.2 mg/kg.

HEMATOPORPHYRIN

= *Photodyn*

A pigment from the decomposition of blood.

Untoward effects: Cataracts. Produces a burgundy color to urine in ***sulfonmethane***, ***barbital***, and, occasionally, in ***lead*** poisoning.

HEMIN

= *Panhemin*

Use: In treatment of acute intermittent porphyria.

Untoward effects: Circulatory collapse in a 42-year-old **female** after 4 mg/kg/15 min infusion. Phlebitis can occur. Induced hemolysis in **women**, **mice**, and **rabbits** enhanced by ***chlorpromazine***. Avoid use with ***barbiturates***, ***estrogens***, and ***steroids***.

HENNA

Untoward effects: Most of the literature states that it is non-allergenic and non-irritating, but there are case reports of immediate-type hypersensitivity contact dermatitis, and variable brown nail pigmentation; neutral henna has caused excessive brittleness and breakage of **people's** hair.

Alcoholic extracts of the *Lawsonia infernis* roots are abortifacient in **rats**, **mice**, and **guinea pigs**.

HEPARIN

= *Arteven* = *Bioheprin* = *Calciparine* = *Cutheparine* = *CY216* = *Ecasolv* = *Fluxum* = *Fragmin* = *Fraxiparine* = *Heprinar* = *Hepsal* = *Kabi 2165* = *Leparan* = *Lipohepin* = *Liquaemin* = *Liquémin* = *LipoHepinette* = *Logiparin* = *Longheparin* = *Lovenox* = *Minihep* = *Monoheparin* = *Panheprin* = *PK 10,169* = *Pularin* = *Thrombohepin* = *Thromboliquine* = *Thrombophob* = *Unihep*

Anticoagulant.

Untoward effects: ***Adrenal insufficiency***: Acute syndrome mimicking septic shock and cardiovascular shock, abdominal pain radiating to flanks, and fever.

Blood: Thrombocytopenia. **Patients** bleeding while receiving the drug may not be bleeding from over anticoagulation, but rather from induced thrombocytopenia (in ~3% after 5 days). Hematomas, hemothorax, retroperitoneal hematoma; gastrointestinal, genitourinary, respiratory tract, and intracerebral bleeding. Ischemic cerebrovascular events, cerebral venous thromboses, and transient confusional states have also occurred.

Bone: Osteoporosis, bone pain, and multiple spontaneous fractures after long-term therapy (usually, 2–12 months) with high dosage.

Cardiovascular: Paradoxical thromboembolism, disseminated intravascular coagulation; myocardial infarction, phlebitis, cardiovascular collapse with bronchial spasm, dyspnea, and fatal cardiac arrest.

Clinical test interference: Decreases protein-bound iodine values, increases free ***fatty acids*** in serum of **diabetics**, spuriously increases erythrocyte sedimentation rate. May give false positive or spuriously high readings with sulfobromophthalein retention and serum ***phosphate*** and ***cholesterol*** levels; false negative or spuriously low readings with serum ***sodium*** and ***calcium*** levels, and ***5-hydroxyindoleacetic acid tests***.

Fetus: Transfer across the placenta is extremely retarded and only low levels are found in the fetal circulation.

Hypersensitivity: Anaphylactoid. Urticaria and edema and papules at site of previous injections. Cutaneous reactions can occur due to sensitivity to the animal of its origin. Preservatives have often triggered reactions. Severe bronchospasm reported occasionally.

Miscellaneous: Femoral neuropathy secondary to retroperitoneal bleeding, pyrexia, renal failure, purpura; pain, erythema, induration and skin necrosis at injection site. Hyperkalemia, priapism, and limb gangrene reported.

Skin and hair: Transient alopecia, ranging from very rare to 40% of **those** on high dosage for 1–15 years, occasionally within 1 month. Exfoliative dermatitis reported.

Drug interactions: Effect antagonized by ***penicillin*** and ***gentamicin***. It reduces ***neomycin's*** antibacterial activity. Non-cardiogenic pulmonary edema due to potentiation of IV ***dextran*** effects. Gastrointestinal bleeding and melena with ***ethacrynic acid***. Synergized by ***dihydroergotamine***. Heparin's anticoagulant effect is markedly decreased by interaction with ***nitroglycerin***. It prolongs hypoglycemia caused by ***glipizide***. Heparin's half-life is decreased in **smokers**. Decreases ***radiopharmaceutical*** uptake by thyroid. ***Protamine sulfate*** (q.v.) is used to reverse or antagonize its anticoagulant effect. Its anticoagulant effect is decreased by large doses of ***chlorpheniramine***, ***cyclizine***, ***diphenhydramine***, ***hydroxyzine***, ***perphenazine***, ***promazine***, and ***promethazine***.

Some **horses** develop a moderately severe anemia after 4–7 days of continuous therapy. Do NOT drop or bruise treated **animals**.

HEPATICA sp.

Untoward effects: ***H. americana*** = ***Liverleaf*** = ***Liverwort*** can cause photosensitivity reactions.

Although ***H. nobilis*** is used in many herbal preparations and as a poultice in Europe, its leaves are said to be poisonous and should not be used during pregnancy.

HEPTABARBITAL

= *Heptabarb* = *Heptadorm* = *Medomin*

Sedative, hypnotic.

Untoward effects: Similar to ***phenobarbital*** (q.v.). In 55 male and **female** (15 – > 71 years) treated for insomnia, 12% had "hangover" effect, and in 71% when treatment continued for > 3 months. As with other barbiturates, overdoses lead to deep sleep or coma, central nervous system depression occasionally preceded by periods of excitement and delirium; slow or rapid and shallow respiration; cyanosis; and Cheyne–Stokes breathing may occur. Therapeutic serum levels in **man** are 2–5 mg/l; toxic concentrations are reported as 15–20 mg/l; and coma or death at 50 mg/l.

TD_{LO}, **human**: 700 μg/kg; LD_{LO}, **human**: 50 mg/kg.

Drug interactions: By liver microsomal enzyme induction, it increases metabolism of and decreases plasma level and half-life of ***acenocoumarol***, ***dicumarol***, ***ethyl biscoumacetate***, and ***warfarin***. Also decreases ***dicumarol*** and ***warfarin*** absorption.

In **rats**, oral use causes a dramatic increase in ***dieldrin*** metabolism, decreasing storage in adipose tissue by 58–74%. As in **man**, it decreases ***dicumarol's*** anticoagulant effect in **guinea pigs** and **dogs**.

LD_{50}, **rats** and **mice**: ~4.5 g/kg.

HEPTACHLOR

= *Drinox* = *E-3314* = *ENT 15,152* = *Heptamul* = *Velsicol 104*

Insecticide. A chlorinated hydrocarbon.

Untoward effects: Can accumulate in **human** fat (breast milk contains 1–4% fat, and a maximum of 2.05 mg/kg found in milk samples from 1460 **women**). Exposure can be after ingestion, skin contamination, and inhalation. Nausea, lacrimation, sore throat, and chest discomfort have occurred. Apprehension, excitability, dizziness, headache, disorientation, weakness, paresthesia, muscle twitching, tremors, tonic and clonic convulsions, coma, pallor, and cyanosis are also to be expected before death. A case of Coombs' positive hemolytic anemia with leukopenia and thrombocytopenia may have occurred in a **farmer** after regular exposure. Relatively persistent in the environment. *In vitro* trials with **human** breast cells showed estrogenicity. The U.S. has suspended its use in most previously used situations.

LD_{LO}, **human**: 50 mg/kg.

It and its metabolites are cumulative poisons. Convulsions occur rapidly in **cats**. Its epoxide metabolite is more acutely toxic than heptachlor. Hepatocellular carcinomas and hepatotoxic in **mice** and **rats**. Minimum toxic dose for baby **calves** is 20 mg/kg and 50 mg/kg for adult **sheep**. Deaths reported in **kestrels**, **minnows**, **ospreys**, **snakes**, **geese**, **quail**, **woodcocks**, **pigeons**, **grasshoppers**, **bees**, **cats**, and **fish**. FDA legal limits are < 0.1 ppm in **cow's** milk and 0.3 ppm in meats for **human** consumption. Increases metabolism of ***hexobarbital*** in **rats**. Ataxia, excessive swallowing and nutation before death in **mallard ducks**. Ataxia, spasms, and opisthotonus in **pigeons**.

LD_{50}, **rats**: 40 mg/kg; **mice**: 68 mg/kg; **hamsters**: 100 mg/kg; **mallard ducks**: ~2 g/kg; **pigeons**: ~50 mg/kg; **guinea pig**: 116 mg/kg; **chicken**: 62 mg/kg; **birds**: 50–100 mg/kg. LC_{50}, **fish**: 96 h, 7–8 μg/l; in feed LC_{50}, **quail**: ~95 ppm; **pheasants**: 224 ppm; **mallard ducks**: 480 ppm.

n-HEPTANE

***Solvent.* Flammable.**

Use: In tire industry, adhesives, rubber cement, gasoline, and paint thinners.

Untoward effects: Vapors irritate eyes, nose, and throat; light-headedness, dizziness, nausea, and anorexia. Aspiration can cause a chemical pneumonia and unconsciousness. Dermatitis from skin contact. National Institute for Occupational Safety & Health suggests 85 ppm TWA and 510 ppm ceiling; OSHA suggests 500 ppm TWA limit and 750 ppm is considered immediate danger to life or health.

Inhalation TC_{LO}, 1000 ppm/6 min.

HEPTANETHIOL
= *Heptylmercaptan*

Untoward effects: Inhalation leads to eye, skin, nose, and throat irritation. Ingestion leads to weakness, cyanosis, tachypnea, nausea, vomiting, drowsiness, and headache. Can be absorbed through the skin. National Institute for Occupational Safety & Health suggests 0.5 ppm/15 min as an exposure ceiling.

1-HEPTANOL
= *Alcohol C-7 = n-Heptyl Alcohol*

Found in essential oils of *hyacinth*, *violet* leaves, etc.

Use: As a fragrance in soaps, detergents, cosmetic creams and lotions, and foods.

Untoward effects: No sensitization reactions to 1% in petrolatum and no irritation after a 48 h closed-patch test in 20 **human volunteers**.

Subacute inhalation exposure in **rabbits** and **rats** led to conjunctivitis and decreased blood cholinesterase. Moderately irritating when applied full-strength for 24 h under occlusion to abraded or intact **rabbit** skin.

LD_{50}, **rats**: 3.25 g/kg; **mice**: 1.5 g/kg; **rabbit**: 750 mg/kg; acute inhalation LC_{50}, **mice**: 6.6 mg/l.

2-HEPTANONE
= *Amyl Methyl Ketone = Methyl n-Amyl Ketone*

***Solvent. Banana*-like fruity odor. A natural pheromone of ants. Found in *clove* and *Ceylon cinnamon oil*.**

Use: A fragrance in soaps, detergents, cosmetic creams and lotions, and foods. Used in artificial carnation oils.

Untoward effects: Inhalation causes eye, nose, and throat irritation, headache, dizziness, narcosis, coma, and liver and renal toxicity; dermatitis from skin contact. No sensitization in 26 **human volunteers** from 4% in petrolatum. National Institute for Occupational Safety & Health and OSHA TWA exposure limits are 100 ppm.

None of six **rats** survived 4 h exposure to 4000 ppm. Strong irritant to **guinea pigs'** mucous membranes and 4800 ppm/4–8 h leads to narcosis and death.

LD_{50}, **rats**: 1.67 g/kg; **mice**: 730 mg/kg.

HEPTYLCYANOACRYLATE

Untoward effects: Moderate engorgement of conjunctival blood vessels for 1–2 weeks, when used as an adhesive for the closure of corneal perforations in **man**.

2-*n*-HEPTYL CYCLOPENTANONE

Use: In fragrances for soaps, detergents, cosmetic creams and lotions, and perfume.

Untoward effects: No sensitization to 10% in petrolatum on 25 **human volunteers**, nor any irritation after a 48 h closed-patch test.

Full-strength to intact or abraded **rabbit** skin for 24 h under occlusion was irritating.

LD_{50}, **rats**: > 5 g/kg; acute dermal LD_{50}, **rabbits**: 5 g/kg.

HEPTYL HYDRAZINE

Untoward effects: 5 or 6 mg/kg/day of this experimental ***monoamine oxidase inhibitor*** caused gliosis in the cerebellum and one case of profound neuronal vacuolation in **dogs**.

IP, LD_{50}, **mouse**: 150 mg/kg.

HERACLEUM sp.
= *Cow Parsnip = Hogweeds*

Untoward effects*: *H. lanatum, also called ***Cow Cabbage*** or ***Indian Celery*** in the western U.S. Some cases of contact dermatitis and erythema reported. **Cattle** are poisoned from eating the leaves only when they have little else to eat.

H. mantegazzianum = Giant Hogweed, native to the Caucasus and often cultivated in European gardens. It is juicier and more irritant than the former. Some **people** evidence photosensitivity to bright sunlight after contact with its leaves and stems, and develop erythematous patches or painful bullous eruptions on the exposed skin due to the ***furocoumarins*** (***psoralens***) present. After healing, brown pigmentation may last for years. Malaise and headache have also been reported following development of skin erythema and blistering. Contact dermatitis also reported in a **dog**.

H. spondylum = Common Hogweed = Keck. Occasionally eaten as a vegetable, but can cause gastrointestinal irritation and photosensitivity, due to its ***psoralen*** (***bergapten, imperatorin, trioxsalen***, and ***xanthotoxin***) content.

HEROIN
= *Baifen = Diacetylmorphine = Diamorphine = Heroish*

Narcotic, analgesic, antitussive.* Semi-synthetic *opiate*. Common street names are *Ack-Ack*, *Bag*, *China

***White*, *Crap*, *Dragon Chasing*, *H*, *Horse*, *Junk*, and *Smack*. The term *China White* has also been used for a form of *fentanyl* (q.v.).**

Untoward effects*:** Miosis, euphoria, anxiolytic, decreased activity and mental concentration, deep sleep, diaphoresis, anorexia, decreased sense of time and space, pruritus, xerostomia, Cheyne–Stokes type of respiratory depression, bradycardia, nausea, vomiting, dizziness, constipation, decreased and occasionally increased libido, menstrual abnormalities (especially amenorrhea), infertility, eosinophilia, lymphocytosis, leukopenia, hemolytic anemia, delirium, cyanosis, anoxia, coma, respiratory failure, and death have occurred. Death from overdosage can occur within 2 h, but usually within 12 h; occasionally after several days. Seizures, arrhythmias, and pulmonary edema can occur from overdosage. Pupils may dilate prior to death. Rhabdomyolysis, frank myoglobinuria, renal failure, confusion, suspiciousness, hostility, belligerence, pneumonias, acute polyneuritis (Guillain–Barré syndrome), anisocoria, occasionally acute or subacute muscle pains, decreased appetite, increased ***norepinephrine levels, difficulty in urination, and suicides also reported. Several hundred thousand **addicts** exist in the world. For a number of years, over a half million **addicts** existed in the U.S. ~5.5 metric tons were smuggled into the U.S. in 1977. Mexico was the source of 50%; Southeast Asia 30%; and the Middle East 20%. Decreased birth weights, decreased neurological development, respiratory depression, and increased congenital defects in **infants** of addicted **mothers**. Placental transfer into breast milk occurs. **Neonatal** narcotic dependence from such breast milk. Withdrawal phenomena are seen within 4 days in ~70% of the addicted mother's **infants**. This occurs in < 12 h to 96 h after birth. These often consist of hyperactivity, irritability, trembling, twitchings; shrill, high-pitched, prolonged crying; convulsions, wakefulness, fever, muscle rigidity, vomiting, hyperhidrosis, jaundice, diarrhea, disorganized sucking reflex, lacrimation, respiratory alkalosis, hiccups, sneezing, yawning, and increased synthesis of ***serotonin***. Sudden withdrawal in the **mother** can lead to premature labor and intrauterine death. Due to its induced amenorrhea, the pregnancy rate in heroin **addicts** is low.

In the 1980s, use in the U.S. spread rapidly of an unusually more potent form from Mexico (***Black Tar***, because it resembled roofing tar in color and consistency) leading to marked increase in hospitalizations and deaths. ***Black Tar*** contained 60–90% heroin, whereas conventional Mexican material assayed only 2–6%. Atypical signs of wound botulism after SC or IM.

IV use ("mainlining") often leaves scars, usually on the inner surfaces of the arms and elbows; *Candida* infections, ophthalmitis, chorioretinitis, anterior uveitis, leukodystrophy, cerebral arteritis, cerebral infarction, thrombocytopenia, septic arthritis, jaundice, hepatitis, spinal osteomyelitis, thrombophlebitis, endocarditis, septicemia, malaria, and serious problems from various adulterants (***baking soda, lactose, magnesium silicate [talc]***, ***mannitol, procaine, quinine, sucrose***, and the neurotoxic ***MPTP***, that caused Parkinson's disease symptoms and paralysis). Pseudogynecomastia in **male addicts** after repeated injections into breasts. Necrotizing cellulitis of the scrotum and medial thigh area after IA of the left femoral artery in a 36-years-old **male addict**. A deep necrotic ulcer after extravasation, following an unsucessful attempt at IV injection of the dorsal vein of the penis in a 25-year-old **male addict**. Sudden death in **addicts** reported after IV use. "**Skin-poppers**" (intradermal use) may develop multiple indolent cutaneous ulcers, *Candida* infections, renal failure, and tetanus. Intranasal use ("snorting") also produces myoglobinuria, tongue pigmentation, and many of its other adverse effects, as it is absorbed through the mucous membranes. This snuff is commonly called "***snow***" by **addicts**. A 17-year-old **male** snorted an unknown quantity leading to respiratory failure, shock, seizures, and a stroke. Hair of dead **addicts** assayed 1.15 ng/mg, 6.07 ng/mg in active **addicts**, and 0.74 ng/mg in formerly active **addicts** in a survey of 37 deaths. Inhalation of vapors ("chasing the dragon") leads to progressive spongiform leukoencephalopathy.

~5 times the toxicity of ***morphine***. MLD ~60 mg.

Drug interactions*: *Propranolol and ***ethanol*** may cause toxic interactions and prolong withdrawal symptoms. May decrease absorption of ***acetaminophen*** and ***mexilitine***.

Congenital malformations of the central nervous system increased in **hamster** offspring, increasing with dosage. They also had decreased growth rate and low birth weight. Causing excitability in **horses**, it has been used in "***speed balls***" for doping. SC injection of 350 mg has caused a **horse's** death in 6 h. Given to pregnant **monkeys**, it reaches the fetus quicker than ***morphine***. In pregnant **rabbits**, it causes intrauterine fetal growth retardation and accelerates lung maturation. See ***morphine*** for its ill-effects in **cats**. Symptoms in affected **dogs** are drowsiness, ataxia, and decreased reaction to pain, progressing to miosis, delirium, convulsions, bradypnea, coma, severe hypotension, and death by respiratory arrest within 12 h.

HERRING

Untoward effects*:** **Patients** taking ***monoamine oxidase inhibitors (***isocarboxide, nialamide, phenelzine, tranylcypromine***, etc.) or drugs with such action (viz. ***furazolidone***) for > 4–5 days' therapy must avoid ingesting smoked or pickled herring, to avoid precipitating a ***tyramine***-induced hypertensive crisis. An occasional cause of dyshidrosis. Herring may precipitate eczemas in **people** who have experienced a ***nickel*** dermatitis. Can raise ***uric acid*** blood levels and precipitate gout attacks.

Domestic **animals** (**cattle**, **sheep**, **foxes**, and **mink**) have developed hepatic disorders from ***dimethylnitrosamine*** after being fed ***nitrite***-treated herring meal. ***Sodium nitrite*** was used by **fishermen** in Norway to prevent bacterial spoilage of the **fish** caught, but, when some spoilage occurs, ***trimethylamine*** is formed, which then reacts with the ***nitrite***, forming the toxic ***nitrosamine***.

HETEROMELES ARBUTIFOLIA

= *Christmas Berry* = *Photinia arbutifolia* = *Toyon*

Untoward effects: A shrub-like evergreen tree in California whose leaves contain ***hydrocyanic acid*** (q.v.), which can be dangerous if **animals** are forced to eat them during droughts. Symptoms include dyspnea, stupor, hemolysis, reddening and congestion of mucous membranes, mydriasis, weakness, and convulsions.

HETERONIUM BROMIDE

= *Hetrum Bromide* = *Lilly 31,814*

Anticholinergic.

Untoward effects: Xerostomia, headache, dysuria, nausea, constipation, and abdominal pain in some **people**.

LD_{50}, **rat**: 3.6 g/kg.

HETEROPHYLLEA PUSTULATA

= *Cegadera*

Untoward effects: Dermatitis, keratoconjunctivitis, and blindness due to photosensitization in South American **cattle**, **sheep**, **rabbits**, and **guinea pigs** after ingestion of the shrub's leaves, flowers, and fruits.

HEXACHLOROBENZENE

= *Anticarie* = *Bunt-cure* = *Bunt-no-more* = *HCB* = *Hexachlorobenzol* = *Julin's Carbon Chloride* = *Perchlorobenzene*

Use: Formerly, in extensive use as a fungicide on seeds, soil, and in industry for manufacturing dyes, organic chemicals, and wood preservatives. It is a ***chlorinated hydrocarbon*** by-product of chemical and pesticide manufacturing.

Untoward effects: In 1955, during a famine, > 4000 **Turks** were poisoned after eating ***bread*** made from treated ***wheat*** seed. Although the treated ***wheat*** had been dyed to identify it, washing failed to remove a lot of the poison. **They** developed a condition similar to porphyria cutanea tarda, known as "porphyria turcica", "kara yara", or "black sore". Roughly 10% of **those** who ate the ***bread*** died. **Children**, in particular, developed porphyrias that caused neurological damage and disfiguring sensitivity to light, destroying the previously held belief that all porphyrias were genetic in origin. The syndrome has persisted for decades. Many **children** < 12 years died of "Pembe Yara" or "pink sore" caused by ***HCB*** in the placenta or **mother's** breast milk. Excessive intake by nursing **mothers** can lead to vomiting, diarrhea, fever, papular skin lesions, anemia, dystrophy, and even death in their **infants**. These **children** died in convulsions, after weakness and a toxic annulaerythema. In many Turkish villages in 1955–1960, no **children** remained between the ages of 2 and 5 years. In **others**, general weakness was the most prominent symptom, followed by severe scarring (86%), hyperpigmentation (81%), hypertrichosis (54%), painless arthritis/small hands (69%), short stature (47%), pinched-in faces (46%), sensory shading (60%), enlarged thyroid (40%), myotonia (42%), cogwheeling (28%), enlarged livers (5%), and port-***wine*** colored urine. Many of the **children** were called "**monkey children**" by peasants, because of their dark pigmentation and fine layer of dark hair on **their** face, trunk, and extremities. Severe blistering of **their** skin and severe photosensitivity caused **their** skin to rub off easily, with poor healing, ulcer formation, and infection that led to the scarring. Chloracne was not seen in this outbreak. Chloracne and comedone formation reported after chronic exposure in **workers**. Its metabolites have been found in seminal fluids. Stored in body fats.

LD_{LO}, **human**: 500 mg/kg.

Algae, **daphnids**, **snails**, and **mosquito fish** concentrate ***HCB*** water levels 60–2000 times; **catfish** biomagnify the levels 600–1600 times the water level. In 1972, ~1 million kg became air or waterborne during manufacture of ***carbon tetrachloride***, ***perchloroethylene***, ***trichloroethylene***, and other ***chlorinated hydrocarbons***, including pesticides. Residues have been found in many forms of **wildlife**. Levels found in Louisiana **cattle** were due to airborne industrial emissions, while those in Texas and California **sheep** were due to contaminated pesticides. The Environmental Protection Agency set tolerance limits for edible meat products for **human** consumption. Found in the milk of exposed **cows**. Anorexia, decreased weight, neutrophilia, serositis, steatitis, hyperplasia of gastric lymphoid tissue, vasculitis, and mortality in experimentally exposed **dogs**. **Japanese quail** developed tremors, reddish fluorescence of tissues, hepatitis, and nephritis when fed at 5–80 ppm. Chronic ingestion has caused porphyria in **rats** and **swine**. Has caused liver and other tumors in laboratory **mice**, as well as immature fetuses or fetal death; liver tumors in **rats**; hepatomas, hemangioendotheliomas, and thyroid adenomas in **hamsters**. Exposed **cats** develop muscle tremors, ataxia, weakness, paralysis, and often die.

LD_{50}, **cat**: 1.7 g/kg; **rat**: 3.5 g/kg; **guinea pig**: 3 g/kg; **rabbit**: 2.6 g/kg.

HEXACHLOROBUTADIENE

= *HCBD* = *Perchlorobutadiene*

Use: A by-product of ***tetrachloroethylene*** and ***trichloroethylene*** manufacturing, used as a solvent, in gyroscopes, as a heat transfer and hydraulic fluids, and in making ***rubber*** products.

Untoward effects: At 20 mg/kg/day/2 years, led to renal tubular adenocarcinomas in **rats**. Nephrotoxic in **rats**, **mice**, and **rabbits**. Fumes can irritate eyes, skin, and respiratory tract of **animals**.

LD_{50}, **rat**: 90 mg/kg; **mouse**: 110 mg/kg; **guinea pig**: 90 mg/kg; inhalation LC_{LO}, **mouse**: 235 ppm/4 h.

HEXACHLOROCYCLOPENTADIENE

= *HCCP* = *HCCPD*

Use: In the manufacture of chlorinated pesticides, fire retardants, heat-resistant and shock-proof plastics, resins, and dyes.

Untoward effects: Tonnage was discharged from pesticide plants in Memphis, Tennessee; Louisville, Kentucky; and Michigan into sewage and rivers, and was lethal to **fish**. The Kentucky plant was closed for several months after **employees** became ill from inhaling it. Symptoms included irritation of eyes, skin, and respiratory tract, with eye and skin burns, lacrimation, sneezing, coughing, dyspnea, salivation, and pulmonary edema. TLV, 10 ppb.

Diarrhea, lethargy, and bradypnea were toxicity symptoms after oral ingestion in **rats** and **rabbits**, but a single inhalation exposure of **mice** and **rabbits** to 1.5 ppm/7 h caused hepatitis and nephritis, and was lethal. 75 mg/kg/day/12 days was toxic to **rabbit does**.

LD_{50}, **rat**: 113 mg/kg; LD_{LO}, **rabbit**: 420 mg/kg; dermal LD_{50}, **rabbit**: 430 mg/kg; skin irritation, **monkey**: 10 mg/kg; **guinea pig**: 20 mg/kg; severe eye irritation, **rabbit**: 20 mg/24 h.

HEXACHLORODIBENZO-*p*-DIOXINS

= *HCDD*

Present in *fly ash* from industrial heating facilities and municipal incinerators.

Untoward effects: Causes chloracne in **mice**, **chick** edema, and hepatocellular carcinomas in **rats** and **mice**.

LD_{50}, **mouse**: 1.25 mg/kg; **guinea pig**: 70 μg/kg; LD_{LO}, **rat**: 100 mg/kg; 6–15 days pregnant: 100 μg/kg; moderate eye irritation, **rabbit**: 2 mg/kg.

HEXACHLOROETHANE

= *Avlothane* = *Carbon Hexachloride* = *Chlorethane* = *Ethylene Hexachloride* = *Fasciolin* = *Hexachlorethane* = *Perchloroethane*

Use: In metallurgy, hair styling agents, pyrotechnics, grenades for generating smoke, flame-proofing and fire extinguishers, and in vulcanizing and manufacture of synthetic diamonds. Veterinary flukicide for **cattle**, **sheep**, **goats**, **swine**, and **nutria**.

Untoward effects: Because of potential carcinogenicity, OSHA limits skin exposure to 1 ppm (10 mg/m^3/8 h TWA. Inhalation, although rare, can cause irritation to mucous membranes. Eye and skin irritant.

LD_{LO}, **human**: 50 mg/kg.

Kidney tumors and pheochromocytomas in male **rats** and hepatocellular carcinomas in **mice**. Decreased safety in debilitated **livestock**, which evidence weakness, staggering, and death.

LD_{50}, **rat**: 6 g/kg; IV, LD_{LO}, **dogs**: 325 mg/kg.

HEXACHLORONAPHTHALENE

In lubricants on farm machinery, and in pelleting equipment for animal feeds.

Untoward effects: Chloracne, primarily of the face and extensor surfaces of the body, develop in some **individuals** after 1–15 months' skin exposure. Can cause nausea, confusion, jaundice, and coma following inhalation or skin absorption.

Affected **cattle** had central nervous system depression, anorexia, weight loss, lacrimation, and increased salivation. Has caused decreased weight gain, enlarged and darkened livers, and deaths of **poults**.

LD_{LO}, **cattle**: 11 mg/kg.

HEXACHLORONORBORNADIENE

= *HEX-BCH*

Untoward effects: As a chemical intermediate, its use in millions of lbs annually has decreased since ***endrin*** and ***isodrin*** production in the U.S. has ceased.

Neuromuscular toxicity in **rats**. Lethal to **fish** at 0.42–1.4 ppm in water. In **fish** > 0.3 ppm leads to loss of equilibrium.

LD_{50}, **rat**: 776 mg/kg.

HEXACHLOROPHENE

= *AT-7* = *Bilevon* = *Dermadex* = *Exofene* = *G-11* = *HCP* = *Hexachlorophane* = *Hexosan* = *pHisohex* = *Surgi-Cen* = *Surofene*

Untoward effects: Despite apparent safety in large numbers of **people** in a wide variety of products for

> 20 years, the FDA has, now, greatly restricted its use for **man** and **animals**. A 6-year-old **female** died 9 h after ingesting ~250 mg/kg as 4–5 oz ***pHisohex***, despite its undesirable taste. A 7-year-old **male** accidentally ingested ~1 g, leading to blindness, lethargy, and a convulsive episode. A 46-year-old **female** ingested 200 ml ***pHisohex*** (3% hexachlorophene), leading to vomiting, lethargy, confusion, coma, and death. Topically, after 4 days of treatment in a **newborn**, it led to buttock and face excoriations; slight twitching of arms, legs, and face; followed by convulsions and nystagmus. Two weeks later, peeling of affected skin, followed by recovery. Absorbed through the skin from cosmetics, vaginal sprays, deodorants, and skin cleansers, leading to ~1 ppb in blood. Heavy use leads to 0.38 ppm (⅓ the blood level causing gross brain damage in **rats**). Premature **infants** develop spongiform myelinopathy after repeated exposures. Application to burned skin caused 15 fatalities in the U.S. by 1973. In France, 39 **infants** died after use on them of a dusting powder (***Bébé***), containing ~6% level. Can penetrate the placenta, and caused an increase in severe congenital malformations in **infants** born to **mothers** who used 10–60 times/day soaps and creams containing it. Surgical **personnel** scrubbing with ***pHisohex*** had mean blood levels of 0.22 ppm, compared to 0.07 ppm in **those** using ***Septisol***. Skin sensitizer, photosensitization, and cross-sensitization with ***bithionol*** and ***salicylanilides***. Allergic contact dermatitis, erythema, and transitory pruritus. Hypersensitivity after 5 years of industrial exposure leading to asthma.

LD_{LO}, **child**: 250 mg/kg; **human**: 50 mg/kg; TD_{LO}, **infant**: 257 mg/kg/7 days; **woman**: 600 mg/kg; mild irritation, skin, **human**: 3 mg/3 days.

Topically, 1.21 ppm blood level caused brain damage in **rats**. Cerebral edema and cystic spaces in **rats** given it orally. Total body baths of newborn **monkeys** with 3% led to spongy brain lesions (plasma levels 2.3 μg/ml). High oral or repeated oral dosage may cause atrophy of the seminal epithelium and brain damage in **rats**, and brain damage (status spongiosis) and blindness in **sheep**. Also, a polyneuropathy paralysis associated with limb wasting and weakness in **pigs**, **rabbits**, **rats**, and **cats**. Four daily intravaginal doses of 300 mg/kg on days 7–10 of gestation in **rats** caused 40% malformed fetuses, versus 4% in the control group. Surfactants are often antagositic to it. As a ***phenol*** derivative, its use is associated with depression and paralysis in **cats**, especially in **kittens**. First aid use on burns prevents ***collagenase*** (q.v.) treatment and increased healing time and scarring. High topical dosage to abraded epithelium can also cause brain damage in **rats**. The FDA requires a zero tolerance in edible tissues and milk designed for **human** consumption.

LD_{50}, **rats**: 60 mg/kg; **mice**: 67–290 mg/kg; **guinea pigs**: 280 mg/kg; **sheep**: 70 mg/kg; skin LD_{LO}, **rat**: 600 mg/kg.

1-HEXADECANETHIOL

= *Cetyl Mercaptan* = *Hexadecylmercaptan*

Untoward effects: Skin, eye, and respiratory tract irritant; headache, dizziness, weakness, cyanosis, nausea, and convulsions. National Institute for Occupational Safety & Health suggests exposure limit as 0.5 ppm (5.3 mg/m^3)/ 15 min.

HEXADECYLDIMETHYLAMINE

= *ADMA 6*

Untoward effects: Corrosive to corneas. Skin irritant and can cause a delayed dermatitis. Prolonged skin contact can cause severe irritation and a delayed burn within 48 h.

Slight oral toxicity in **rats** and slightly toxic dermally in **rabbits**.

LD_{50}, **rat**: 600 mg/kg; dermal, **rabbits**: > 2 g/kg.

HEXADIMETHRINE BROMIDE

= *Polybrene*

Heparin antagonist.

Untoward effects: Can be additive to ***curare*** and ***ganglionic-blocking agents***. Adverse reactions in 14/30 **patients**. Dizziness, facial flushing, burning in the chest, dyspnea, coughing, lumbar pain, hidrosis, fatigue, numbness in the arms and spasms in the legs, starting within 2–3 min after IV. Tachycardia, tachypnea, and thrombocytopenia also occur; rarely, bradycardia and hypertension. Dosages over 4–5 mg/kg in **humans** and **dogs** can cause renal failure with marked tubular vacuolization and necrosis.

IV, LD_{50}, **rats**: 20 mg/kg; **mice**: 25 mg/kg; **dogs**: 15 mg/kg; **rabbits**: 10 mg/kg.

HEXAETHYLTETRAPHOSPHATE

= *Ethyl Tetraphosphate* = *HETP*

Organophosphorus insecticide, parasympathomimetic. **Cholinesterase inhibitor.**

Untoward effects: Anorexia, nausea, vomiting, diarrhea, increased salivation, miosis, loss of depth perception, headache, giddiness, chest pains, dyspnea, cyanosis, hidrosis, convulsions, coma, and death reported. Death can occur within 30 min to 4 h.

LD_{LO}, **human**: ~350 mg (5 mg/kg).

In **dogs**, salivation, muscle fasciculations, tremors, vomiting, diarrhea, ataxia, and convulsions occur. Initially, miosis, followed by mydriasis. IV use in the **cat** (0.8–2 mg/kg) causes violent convulsions.

LD_{50}, **rat**: 5 mg/kg; **mouse**: 56 mg/kg.

HEXAFLUORENIUM BROMIDE

= *Hexafluronium Bromide* = *Mylaxen*

Skeletal muscle relaxant.

Untoward effects: Cardiac arrhythmias when used with ***halothane***. A selective inhibitor of pseudocholinesterase, it prolongs the neuromuscular blockade of ***succinylcholine*** and ***decamethonium***. Death, after a short episode of bronchospasm and ventricular fibrillation in a 27-year-old **male** with a history of heavy smoking and regular use of bronchodilator drugs.

LD_{50}, **mouse**: 280 mg/kg; SC, 240 mg/kg; IV, 1.8 mg/kg.

HEXAFLUOROACETONE

= *HFA* = *Perfluoroacetone*

Colorless gas with musty odor, shipped as a liquified compressed gas.

Untoward effects: Severe skin, eye, mucous membrane, and respiratory tract irritant. DuPont will not permit **women** to work with it, for fear that it could damage a **fetus**. Can cause pulmonary edema and frostbite. TLV, air: 100 ppb.

Teratogenic in **rats**, retarded ossification, and increased resorptions at 7 ppm in air; 30 ppm caused deaths in 4/6 pregnant **rats**; decreased fetal weights noted at 0.1 ppm.

HEXAFLURATE

= *Nopalmate* = *Nopalumate* = *Potassium Hexafluoroarsenate* = *TD-480*

Herbicide.* Used, primarily, against *prickly pear cactus.

Untoward effects: Regurgitation, circling, tremors, tenseness, spasms, dyspnea, tachypnea, polydipsia, opisthotonus, convulsions, and tetany in **birds** within 10 min after exposure. Deaths within 1–3 h.

LD_{50}, **rat**: 1.2 g/kg; **mallard ducks**: 193 mg/kg; **California quail**: 229 mg/kg; **gray partridges**: 142 mg/kg.

HEXAMARIUM BROMIDE

= *BC 51* = *Distigmine Bromide* = *Ubretid*

Cholinesterase inhibitor.

Untoward effects: Colic, palpitations, salivation, and bronchial spasms.

HEXAMETAPOL

= *Dorcol* = *ENT 50,882* = *HEMPA* = *Hexamethylphosphoramide* = *Hexamethylphosphoric Triamide* = *HMPA* = *HMPT*

Use: Solvent, de-icing additive in jet fuels, and **insect** chemosterilant.

Untoward effects: Eye, skin, respiratory tract irritant. Dyspnea and abdominal pain reported.

Inhalation of 400 and 4000 ppb by **rats** led to nasal epidermoid carcinomas, papillomas, and other carcinomas after 8 months, as well as degenerative changes in convoluted kidney tubules. Potential carcinogen for **man**. Oral treatment of **rats** caused testicular atrophy and aspermia; nephritis, bronchiectasis, bronchopneumonia with squamous metaplasia, and fibrosis in lungs. Oral use inhibits testicular development in **cockerels**. Repeated topical applications on **rabbits** led to weight decrease, and gastrointestinal and central nervous system dysfunction.

LD_{50}, **rat**: 2.65 g/kg; **chicken**: 835 mg/kg.

HEXAMETHONIUM

Bistrium = *Hexamethone* = *Hexathonide* = *Hexonium* = *Methium* = *Vegolysen*

***Antihypertensive, ganglion-blocking agent.* The bromide and chloride forms are generally used.**

Untoward effects: Mydriasis, tachycardia, hypertension, dry mouth and throat, decreased peristalsis, myopia, cycloplegia, optic nerve neuropathy, increased intraocular pressure, decreased lacrimation, hypotension, urinary retention, ileus, constipation, dyspnea, ataxia, decreased libido, dry skin, flushing, pulmonary alveolitis and fibrosis, dry mouth, postural syncope, cerebral and coronary thrombosis, dissecting aneurisms, collapse, and death reported. Crosses the placenta, accumulating in amniotic fluid leading to **neonatal** ileus, pneumonia, and **fetal** death, when ingested throughout pregnancy.

LD_{50}, **mouse**: 1 g/kg.

HEXAMETHYLENE DIISOCYANATE

= *1,6-Diisocyanatohexane* = *HDI*

***Polyurethane* activator.**

Untoward effects: Eye, skin, and respiratory tract irritant. Sharp, pungent odor. Coughing, dyspnea, bronchitis, wheezing, pulmonary edema, asthma, corneal damage, and skin blisters reported. National Institute for Occupational Safety & Health limits exposure to a maximum of 0.005 ppm (35 $\mu g/m^3$) TWA.

Similar symptoms in **rats**, **rabbits**, **guinea pigs**, **mice**, and **dogs**. Convulsions and hyperplasia of anterior nasal areas in **rats**. Gastroenteritis and cyanosis in **rats** after oral dosing.

LD_{50}, **rat**: 710 mg/kg; inhalation LC_{LO}, **rat**: 60 mg/m^3/4 h; **mouse**: 1570 mg/m^3/10 min.

HEXAMETHYLENE GLYCOL

= *2,5-Hexanediol*

Use: In ***formaldehyde***-free durable press finishing of ***cotton***, in nylon manufacture, in ***gasoline*** refining, as a plasticizer, and in manufacture of polyesters and ***polyurethane***.

Untoward effects: Peripheral neuropathy in **rats** and **hens**, probably from metabolic conversion to ***2,5-hexanedione***. Low eye irritation and low hemolytic activity.

LD_{50}, **rat**: 2.24 g/kg.

n-HEXANE

Use: Solvent for extraction of ***soybean, linseed, peanut, cottonseed***, and ***corn*** oils. In rubber cements and adhesives, shoe manufacturing, artificial leather finishes, cleaning precision instruments, and colored as a substitute for ***mercury*** in thermometers.

Untoward effects: Peripheral polyneuropathy with decreased motor nerve conduction velocities, hallucinations, unconsciousness, dysesthesia, muscle weakness and atrophy, ataxia, sensory loss, and impaired visual acuity after inhalation. Cabinet **workers** in a room with 650–1300 ppm became ill before adequate ventilation was installed. Symptoms have mimicked encephalitis and multiple sclerosis. Vapors irritate eyes, nose, and throat at 1500 ppm. Affected **people** often feel lightheaded, nauseous, and headachey, and develop a dermatitis from contact; aspiration pneumonia has occurred. Giddiness and dizziness occur at 5,00 ppm. Metabolized, primarily, to ***2-heptanol*** and ***2,5-hexanedione***. Two **workers** using it as a glue solvent in manufacturing sandals became quadriplegic. Sewers in Louisville, Kentucky blew, when a car passing over a manhole ignited vapors.

TLV, 500 ppm; National Institute for Occupational Safety & Health, 100 ppm, TWA. Inhalation TC_{LO}, **human**: 5000 ppm/10 min.

Cats, **dogs**, **mice**, **hens**, and **rats** also developed peripheral neuropathies, after exposure to its fumes. Comedogenic in **rabbit** ears. Potent in elicitating arrhythmias in **dogs** with ***epinephrine***.

LD_{50}, **rat**: 30 g/kg. 40,000 ppm is lethal for **mice**.

LC_{50}, **rats**: 3366 ppm.

HEXANETHIOL

= *Hexylmercaptan* = *n-Hexylthiol*

Untoward effects: Eye, skin, nose, and throat irritant. Tachypnea, weakness, cyanosis, nausea, vomiting, headache, and drowsiness reported. Flammable when wet.

National Institute for Occupational Safety & Health, 0.5 ppm (2.7 mg/m^3)/15 min.

LD_{50}, **rat**: 1.3 g/kg; inhalation, **mouse**: 528 mg/kg.

HEXAPROPYMATE

= *Hexopropymate* = *L-2103* = *Lunamin* = *Merinax* = *Modirax*

Sedative, hypnotic.

Untoward effects: Gastric irritation can cause nausea. A 31-year-old **nurse** addicted to it displayed physical and psychological dependence, habituation, and tolerance to it, taking as much as 16 g/24 h. A 28-year-old **male** took 40 400 mg tablets (16 g) leading to deep coma, hypotension, supraventricular tachycardia, hypothermia, respiratory depression, inhalation pneumonia, and lactic acidosis. Resuscitation and treatment resulted in recovery. Therapeutic concentrations are 5 mg/l and toxic levels are 10–20 mg/l.

TD_{LO}, **human**: 228 mg/kg.

LD_{50}, **mouse**: 900 mg/kg.

HEXESTROL

= *Cycloestrol* = *Dihydrodiethylstilbestrol* = *Hexanoestrol* = *Hexevan* = *Hexoestrol* = *Hormoestrol* = *Sintofolin* = *Synthovo* = *Syntrogène*

Estrogen, antineoplastic.

Untoward effects: Primary toxicity, when used as an antineoplastic, is a hypersensitivity with chills and fever. Gastrointestinal upsets with nausea and vomiting (~13% severe, mild to moderate in ~52%), nipple and areolar tenderness, skin pigmentation, hypercalcemia, and gynecomastia in **men**. Uterine cell proliferation, postmenopausal bleeding, jaundice, ***sodium*** retention and edema, weight increase, chloasma, rashes, and urticaria can also occur.

In **guinea pigs**, it decreases blood ***glucose*** and potentiates ***tolbutamide***-induced hypoglycemia.

HEXOBARBITAL

= *Citodon* = *Citopan* = *Cyclonal* = *Dorico* = *Enhexymal* = *Hexanastab Oral* = *Hexobarbitone* = *Noctivane* = *Sombucaps* = *Sombulex* = *Somnalert*

Sedative, hypnotic. **The sodium form has been used IV.**

Untoward effects*:** Bradycardia and hypotension after IV. ***p-Aminobenzoic acid, chloramphenicol, imipramine, and ***iproniazid*** enhance its effects. ***Phenobarbital, phenytoin***, and ***rifampin*** induce its enzymatic metabolism.

TD_{LO}, **human**: 4 μg/kg.

In **rats**, **rabbits**, **mice**, **dogs**, and **man**, ***alcohol*** potentiates its depressive effect. In the **rat**, its metabolism is enhanced by ***chlorcyclizine***. In **mice**, ***chlorpromazine, hydroxyzine, propranolol***, and ***reserpine*** increase its hypnotic effect. ***L-Thyroxine*** pretreatment increases its sleeping time in **rats**, as does ***disulfiram***. ***Chloramphenicol*** prolongs the duration of anesthesia in **rats**, **mice**, and **monkeys**.

LD_{50}, **rat** and **mouse**: 468 mg/kg; **rabbit**: 1200 mg/kg.

HEXOBENDINE

= *Andiamine* = *Hexabendin* = *Reoxyl* = *ST-7090* = *Ustimon*

Coronary vasodilator.

***Untoward effects*:** Headache, probably due to vasodilation of cerebral blood vessels after IV use. Occasionallly, weakness, vertigo, epigastric discomfort, diarrhea, epistaxis, allergic skin reaction, alopecia, insomnia, and tendency to collapse have also occurred.

HEXOCYCLIUM METHYLSULFATE

= *Tral* = *Tralin*

Anticholinergic.

***Untoward effects*:** Slight xerostomia, decreased salivation, and visual disturbances reported. Toxic doses can produce depolarizing neuromuscular blocking effects.

Drug interactions*:** Enhanced mydriasis with ***sympathomimetics. Decreases ***iron*** absorption from ***ferrous sulfate*** and ***ferrous citrate***.

HEXOL

Untoward effects*:** ***Turpentine-related ***pine oil*** distillate, used as a household disinfectant, caused acute renal failure within 24 h after intrauterine use in a 31-year-old **female** trying to induce an abortion. Uremia slowly cleared; anemia persisted for 6 months; peripheral neuropathy developed 4 weeks after abortion, and slowly cleared. Renal biopsy revealed focal fibrosis and tubular atrophy.

HEXOPRENALINE

= *Bronalin* = *BYK 1512* = *Delaprem* = *Etoscol* = *Gynipral* = *Ipradol* = *Leanol*

Bronchodialtor, tocolytic.

***Untoward effects*:** Arrhythmias (6%), increased heart rate of 18–26% with 5 or 10 μg, decreased blood pressure with 10 μg, tremors (12%), palpitations (6%); sweating, nausea, and anxiety (2% for each) after IV use. Continuous IV infusion, as a tocolytic, in a 20-year-old **female** led to atrial fibrillation and abrupt pulse increase to 160/min after $1\frac{1}{2}$ h.

HEXYL ACETATES

The *n*-hexyl form is used in limited quantities for its fruity aroma in soaps, detergents, cosmetic creams and lotions, and foods.

***Untoward effects*:** 4% in petrolatum caused no irritation or sensitization in trials on **human** volunteers.

LD_{50}, **rats**: 6.16 g/kg; acute dermal, **rabbit**: > 5 g/kg.

The *sec*-hexyl form can cause headache; and eye, nose, throat, and skin irritation in **humans**. Fruity aroma with National Institute for Occupational Safety & Health/OSHA limits of 50 ppm (300 mg/m^3), TWA; immediate danger to life or health: 500 ppm.

HEXYLCAINE

= *Cyclaine*

Local anesthetic.

***Untoward effects*:** Convulsions reported in 5/200 after extradural use; neuritis and paralysis of 2–4 weeks in 3/100 given it caudally for vaginal delivery. Edema of the glans penis with urethral discharge reported in four **patients**, after use of a 5% gel. Tissue irritation, burning, swelling, tissue necrosis with sloughing, after topical and inadvertent parenteral use.

Respiratory failure reported in three **puppies** after its SC use for tail docking.

SC, LD_{50}, **mouse**: 1080 mg/kg; **rabbit**: 164 mg/kg; IV, LD_{50}, **mouse**: 30 mg/kg; **rabbit**: 14 mg/kg.

HEXYLENE GLYCOL

Use*:** In hydraulic brake fluids and latex paints, as a ***gasoline additive, and de-icing agent.

***Untoward effects*:** After prolonged contact, leads to skin and eye irritation. Use in burn dressing gauze led to coma and death in ~3–6% of treated **children**. Headache, dizziness, nausea, incoordination, central nervous system depression, and dermatitis also reported after various exposures. National Institute for Occupational Safety & Health, 25 ppm, TLV–TWA.

LD_{LO}, **human**: 500 mg/kg; inhalation TC_{LO}, 50 ppm/10 min.

LD_{50}, **rats**: 4 g/kg; **mouse**: 3.86 g/kg; **rabbit**: 3.2 g/kg; **guinea pig**: 2.8 g/kg.

HEXYL METHYL KETONE

= *Methyl Hexyl Ketone* = *2-Octanone*

Use: ***Apple***-like odor in fragrances for soaps, detergents, cosmetic creams and lotions, perfumes, and foods.

Untoward effects: 4% in petrolatum caused no sensitization or irritation in **human** volunteers. Immediate eye and nose irritation of saturated atmosphere (1300 ppm) in **guinea pigs**. After 1 h, weakness, narcosis, and, eventually, coma reported.

LD_{50}, **rat** and acute dermal, **rabbit**: > 5 g/kg.

HEXYLRESORCINOL

= *Ascaryl* = *Caprokol* = *Crystoids* = *Gelovermin* = *Sorunex* = *ST-37* = *Sucrets* = *Viraspray* = *Worm-Agen*

Anthelmintic, topical antiseptic.

Untoward effects: Repeated oral doses caused severe gastrointestinal irritation, necrosis of the small intestine, and cardiac and liver upsets. Acquired sensitivity after 5 months of topical application. A single, brief contact with the drug 26 months later caused a severe dermatitis. Oral use with ***alcohol*** reduces its anthelmintic activity. Respiratory tract and skin irritant. ***Alcoholic*** solutions are vesicant.

Drug interactions: It can cross-react with ***hydroquinone***, ***phloroglucinol***, ***pyrocatechol***, ***pyrogallol***, and ***resorcinol***.

LD_{LO}, **human**: 500 mg/kg.

LD_{50}, **rat**: 550 mg/kg; LD_{LO}, **guinea pig**: 400 mg/kg.

HEXYL SALICYLATE

= *n-Hexyl-σ-hydroxybenzoate*

Use: Fungistatic agent in soaps, detergents, cosmetic creams and lotions, and perfumes.

Untoward effects: No sensitization or irritation at 3% in petrolatum on **human volunteers**.

Full-strength to intact or abraded **rabbit** skin was moderately irritating, but not irritating when applied to the backs of hairless **mice** or **swine**.

LD_{50}, **rats** and acute dermal, **rabbits**: > 5 g/kg.

HIBISCUS ROSA-SINENSIS

Untoward effects: Extracts of the flowers have an anti-implantation effect in **rats** and **mice**.

HICKORY

= *Carya orata*

Untoward effects: In North America, its nuts and pollen have been common causes of allergic rhinitis, bronchial asthma, and/or a hypersensitivity pneumonitis in **people**.

HIMATANTHUS sp.

Untoward effects: The roots of ***H. phagedoenicus*** are reported to be poisonous. The roots of ***H. sucumba*** are very poisonous for **man**. A decoction of its stem bark is used in Brazil for gastritis and hemorrhoids of **people**, and has low reproductive toxicity and teratogenic potential in **rats**.

HIRUDINS

= *Exhirud* = *Exhirudine* = *Hirudex*

Recombinant form = *CGP 39,393* = *HBW 023* = *Rec-hirudin* = *Revasc*

***Anticoagulants.* From the salivary gland of the medicinal leech, *Hirudo medicinalis*.**

Untoward effects: Intracerebral bleeding is a risk. Nausea has occurred in **patients** undergoing PTCA. Allergic reactions are rare. At the turn of the twentieth century, highly toxic reactions were due to putrefaction and poor quality control.

HISTAMINE

A pressor amine. Usually, not very active orally.

Untoward effects: Capillary dilation, constricts bronchioles and uterus, and relaxes arteriolar tone. Causes intravascular fluids to migrate into tissues, leading to hypovolemia and tissue edema. Ultraviolet light onto the skin leads to dilation of arterioles and capillaries, known as sunburn. Headache and flushing of face and neck sometimes experienced with ***alcoholic*** drinks are, in part, due to it (large amounts in red ***wines*** and ***sherries***; intermediate amounts in white ***wine***, ***beer***, and ***port***; none in ***whiskey*** or ***cognac***). Injections often cause transient nausea, increased gastric acidity, dyspnea, decreased appetite, increased lower esophageal sphincter, hyperglycemia, decreased blood pressure, tachycardia, and pancreatitis. Drugs causing its sudden release cause burning sensation of the skin, pruritus, urticaria, erythema, and decreased blood pressure. High histamine contents in various scombroid ***fish*** (q.v.) interact with ***isoniazid*** and other drugs, leading to severe headaches, redness and itching of eyes and face, chills, palpitations, loose stools, and variations in pulse rate. The same has been reported due to the high histamine content in strong ***cheeses*** (Gouda, Cheshire, and Swiss). **Patients** on ***monoamine oxidase inhibitors*** may develop hypertensive crises with histamine, as in ***yeast*** extracts and ***sauerkraut***. Can cause unpredictable results with ***d-tubocurarine***, as the latter can often cause histamine release and hypotension. Poisonous plants, such as ***stinging nettles***, inject some into skin, causing capillary dilation. This produces a wheal, which aids the rapid absorption of its second component, ***acetylcholine***, in sufficient concentrations to affect a **person's** nerve endings, causing the stinging sensation. Extreme acute sensitivity to cold and cold products causes a massive release of histamine leading to angioedema with swelling of the lips, throat, and tongue. Exercise also increases histamine release.

It is the main constituent in **honeybee** venom. A single SC injection of 250–350 mg/**rat** causes duodenal ulcers. In **swine**, 5 mg/kg SC causes gastroesophageal ulcers. It plays a major role in anaphylaxis in the **guinea pig** and **dog**. Exposure to cold causes a dramatic increase in its excretion in **rats**. Some **mouse** strains are many times more sensitive to it than others. Often, hypertensive in **guinea pigs** and **rats**, but not in **hamsters**. May increase sensitivity to endotoxic shock. Avoid rapid IV.

IV, LD_{50}, **monkey**: 50 mg/kg; **guinea pig**: 180 μg/kg; IV, LD_{LO}, **dog**: 30 mg/kg; IP, LD_{50}, **mouse**: 2 g/kg.

HISTIDINE

Essential amino acid.

Untoward effects: High doses can cause dysgeusia, hypogeusia, dysosmia, mental and cerebellar dysfunction, increased urinary ***zinc*** excretion, and decreased appetite. A bacterium that multiplies rapidly with increased temperatures in **scombroid fish** converts histidine to ***histamine*** (q.v.). Histidinemia in **man** occurs due to a metabolizing enzyme defect leading to retardation and neurological manifestations.

HISTRELIN

= *Hydron* = *ORF 17,070* = *RWJ 17,070* = *Supprelin*

Use: The acetate form is injected daily to treat central precocious puberty.

Untoward effects: Serious hypersensitivity reactions (angioneurotic edema and urticaria) and 45% of treated **children** showed redness, swelling and itching at the injection sites; **girls** often had slight vaginal bleeding, starting ~6 weeks after beginning therapy.

HISTRIONICOTOXIN

= *Gephyrotoxin*

Untoward effects: A lethal neurotoxin in the skin of a **frog** (q.v.) *Dendrobates* sp., that is rubbed on blow-darts by **Indians** of Ecuador and Colombia.

HOGWORT

= *Croton capitatus*

Untoward effects: Its disagreeable taste prevents most **animals** from grazing it, but, where it grows abundantly, it can be cut with hay, and, then, can poison **animals**. Contains ***croton oil*** (q.v.), a drastic purgative and skin irritant. Affected **animals** have diarrhea, and colic, and are hyperexcitable. Rarely fatal.

HOLCUS LANATUS

= *Mesquitegrass* = *Velvetgrass* = *Yorkshire Fog*

Untoward effects: Poisonous to **swine**, due to its ***hydrocyanic acid*** (q.v.) content when eaten in the fresh or wilted condition. Often infected with ***ergot*** (q.v.), adding to its toxicity.

HO LEAF OIL

Use: Tonnage, as a fragrance in soaps, detergents, cosmetic creams and lotions, and perfumes. ***Linalool*** (q.v.) is its chief constituent.

Untoward effects: No irritation or sensitization of 10% in petrolatum on **human volunteers**. Moderately irritating when applied full-strength for 24 h under occlusion on intact or abraded **rabbit** skin.

LD_{50}, **rats**: 3.27 g/kg; acute dermal, **rabbit**: > 5 g/kg.

HOLLY

= *Ilex sp.*

Untoward effects: The berries contain complex ***phenols***, and are particularly toxic, causing severe vomiting, abdominal pain, diarrhea, and central nervous system depression (stupor), when ingested. **Children** < 5 years have been particularly affected (in 1977, 333 reports to U.S. Poison Control Centers; 19 showed symptoms, and one required hospitalization) after eating the attractive red or black berries. One 2-year-old **female** experienced 2 h of nausea after eating only two berries. ***Ilicin***, a bitter principle, is the suspected toxic agent.

I. aquifolium = ***English Holly***. In Europe in 1889, 20–30 of its berries were reported to have caused a **human** fatality. Since then, European and American reports only note vomiting and diarrhea from ingestion of the berries, which contain ***ilicin*** and a cyanogenic glycoside, ***dihydromandelonitrile***.

I. vomitoria = ***Yaupon*** used by **Indians** in southwestern U.S. in making **"*Black Drink*"** (a ***tea***) containing ***caffeine*** and, in large quantities, was purgative. The leaves of all hollies are also poisonous, but not eaten. Also see ***Ilex***.

HOLOCALYX sp.

= *Alecrim*

Untoward effects: ***H. balansae*** can be acutely toxic to **rabbits**.

H. glaziovii = ***Garden Rosemary*** in South America is an important photosensitizing plant.

HOLOTHURIN

Untoward effects: A highly toxic substance produced by **sea cucumbers** (a marine animal) can kill **fish** within 30 min and cause developmental anomalies in **sea urchins'** eggs, after only a 5 min exposure to 1 part per 100 million.

HOMATROPINE

= *Arkitropin* = *Homapin* = *Malcotran* = *Mesopin* = *Novatrin* = *Novatropine* = *Sethyl*

Parasympatholytic, anticholinergic. **The methylbromide is commonly used.**

Untoward effects: Oral use in 130 **patients** with parkinsonian phenomena led to vertigo (four), xerostomia (11); **some** with headache, restlessness, drowsiness, tenseness, impaired accommodation, weakness, disorientation, and optic hallucinations. In **others**, hallucinations can also be auditory or tactile. Increased intraocular pressure, acute narrow-angle glaucoma, mydriasis, amblyopia, and cycloplegia. An 84-year-old **female** developed tachycardia within 1–2 h after topical ophthalmic use. Severe psychomotor agitation with confusion, tachycardia, incontinence, and mydriasis in a 12-year-old **child** after ophthalmic instillation of 12 drops of 1% solution three times in 16 h. Several case reports of induced psychosis in other **children** after use of ophthalmic drops. Restlessness can be a prominent symptom with periods of hyperactivity, alternating with periods of quiet, with the **patient** just sitting and staring while talking endlessly. After 2 drops instilled onto the left eye of a 33-year-old **female**, toxicity developed 2 h after the last treatment. Combative behavior has also followed 3–10 one drop applications of 2% into each eye. SC doses < 4 mg usually induced bradycardia, and 4 mg usually induces tachycardia.

LD_{LO}, **human**: 50 mg/kg.

LD_{50}, **mice**: 1.4 g/kg; IP, **mice**: 60 mg/kg; IP, LD_{LO}, **rat**: 180 mg/kg.

HOMERIA sp.

Untoward effects: ***H. collina*** = a ***Cape Tulip*** (also including the genus *Moraea*) causes "***tulip***" poisoning, with heavy losses in grazing **cattle**.

H. glauca = ***Natal Yellow Tulip***, contains a toxic glycoside, ***epoxyscillirosidin***. Reports of it causing fatalities in **humans** and **animals** in New Zealand have been documented. ***Epoxyscillirosidin***, its main toxic principle (0.4 mg/kg in **guinea pigs** leading to ***curare***-like paralysis with death from respiratory failure) also caused similar symptoms in **rabbits** and convulsions in **mice**.

H. pallida, another South African Cape ***tulip***, is the most toxic (100–200 g to kill an **ox**). The bulb and leaves are poisonous leading to colic; often, blood-stained diarrhea; frequent urination, tympany, weakness, collapse, and, eventually, death of **animals**.

HOMOBATRACHOTOXIN

Untoward effects: One of the most toxic chemicals from **frogs** in Colombia, as is ***batrachotoxin*** (q.v.), known to **man**. It is estimated that only 170 μg would kill a 68 kg **man**. It causes hypotension, anoxia, loss of equilibrium, arrhythmias, dyspnea, and convulsions in **animals**.

SC, LD_{50}, **mouse**: 3 μg/kg.

HOMOCHLORCYCLIZINE

= *Curosajin* = *Homoclomin* = *Homoginin* = *Homorestar* = *SA-97*

Serotonin antagonist.

Untoward effects: Dizziness, itching, weakness, syncope, poor coordination, and depression.

LD_{50}, **rat**: 490 mg/kg; **mouse** and **guinea pig**: 390 mg/kg; IV, LD_{50}, **rat**: 36 mg/kg; **mouse**: 47 mg/kg; **guinea pig**: 19 mg/kg; **dog**: 20 mg/kg.

HOMOCYSTEINE

Untoward effects: The death of a 2-month-old **male** from arteriosclerosis in 1968 and **he** was related to an 8-year-old **male** who died from arteriosclerosis in 1933. Both **children** had increased blood levels of this ***amino acid*** product by the liver from ***methionine***. **Men** and **women** with high concentrations are 13–14 times more likely to have cardiac disease. Netherland researchers found an increase of 10% above controls in **those** who drank 6 cups/day of unfiltered ***coffee***.

It induces atherosclerosis in ***vitamin*** B_6-deficient **pigs** and **monkeys** and its ***thiolactone*** (present in animal fats) fed to **rabbits** causes arteriosclerotic patches or plaques in 3 weeks.

IP, LD_{LO}, **mouse**: 500 mg/kg.

HOMOCYSTINE

Almost identical, chemically, to *homocysteine*.

Untoward effects: Helps cause platelet aggregation *in vitro*. Homocystinuria is an inherited disorder of ***amino acid***

metabolism, often associated with brain damage, mental retardation, myointimal hyperplasia, pulmonary embolism, and bone abnormalities.

HOMOMENTHYL SALICYLATE

= *Heliophan* = *Homosalate*

***Ultraviolet-absorber, suncreen, skin-tanning agent.* In *Coppertones, Filtrosol A, Heliophan*, etc.**

Untoward effects: Contact dermatitis and photosensitivity reported.

HONEY

= *Mel*

Untoward effects: Despite its great popularity as a condiment since antiquity (> 200 million lbs used annually in the U.S.), its use is not without problems. Color and taste vary with the flowers the **bees** visit. Honey has occasionally contained significant quantities (up to 1 ppm) of ***pyrrolizidine alkaloids***, which can be carcinogenic, mutagenic, and teratogenic. Ingestion of toxic honey produced by **bees** collecting nectar from ***Rhododendron sp.*** caused the hospitalization of 11 **people** in Turkey with ***andromedotoxin*** (q.v.) poisoning (1983–1988). Similar reports from Brazil, British Columbia, and North America. **People** often buy and bring such honey to other countries. ***Kalmia*** and other nectars can also prove toxic. In 400 B.C., Greek **soldiers** were poisoned by the toxin, and many case reports exist up through present times. In New Jersey, in 1794, a **botanist** reported similar mild to serious poisonings. ***Euphorbias*** in South Africa cause similar problems. Poisonings in **man** also reported, due to honey from ***Aesculus sp.*** (q.v.), ***Gelsemium sempervirens*** (q.v.), ***Datura stramonium*** (q.v.), ***Cytisus laburnum*** (q.v.), and ***Coriaria*** (q.v.). ***Tutin*** 2 mg and 15 mg ***mellitoxin*** found in a poisonous sample from ***Coriaria***-derived honey. Reactions from ***buckwheat*** honey are allergic in nature. Its ***fructose*** (q.v.) content has occasionally caused problems after excessive ingestion. *Clostridium botulinum* spores, but not toxin, have been found in honey fed to **infants**. It is suggested that neither honey nor any raw agricultural products be given to **infants** < 1 year.

HONEYSUCKLE

= *Lonicera sp.*

Untoward effects: Ingestion of seeds and berries has caused nausea, vomiting, and purgation. Most cases reported were asymptomatic.

L. tatarica berries have caused mild toxicity in European **children**.

L. xylosteum* = *Fly Honeysuckle is reported as having caused serious poisoning in **man**.

HOPS

= *Houblon* = *Lucro* = *Lupulo*

The dried seed cones of *Humulus lupulus*.

Untoward effects: ***Sulfur dioxide*** (q.v.) is sometimes added to preserve its color. Sedative or hypnotic properties discovered when **hop-pickers** were observed to tire easily. A dermatitis, "hop-rash", occurs in **pickers**. Hairs on the leaves have also caused mechanical abrasions of the skin. Erythema, blistering, and conjunctivitis have also been reported. Additive to ***alcohol, ethyl's*** prostatic irritant effects.

HORDEUM JUBATUM

= *Foxtail* = *Squirreltail Grass* = *Wild Barley*

Untoward effects: Its posteriorly directed barbs cause it to migrate in a unidirectional manner in tissue, penetrating the eyes, nose, mediastinum, bronchi, lungs, pericardium, spinal cord, joints, and interdigital areas of a **dog**, often causing abscess formation. It has caused mechanical injury of the mouth and eyes, as well as death of **sheep**. Related to ***barley*** (q.v.) [***Hordeum vulgare***], ***H. pusillum* = *Little Barley*** of western North America causes similar problems in grazing **sheep**.

HORNETS

Untoward effects: Painful stings. Allergic reactions can occur and extreme sensitivity has resulted in death. Local swelling, urticaria, fever, chills, and hypotension also reported. Hornet's stinger apparatus is not barbed and they can sting their **victims** many times. Major cross-reactivity with venom of **yellow jackets**.

HORSERADISH

= *Armoracia rusticana* = *Raifort*

Untoward effects: Contains **allyl isothiocyanate** (q.v.), a pungent chemical. Can cause contact eczema in **people**. If the roots are ingested in excess, gastrointestinal distress, diarrhea, and even bloody vomiting can occur.

Fatal to **livestock** feeding on tops and roots. Dead **cattle** have a frothy bloat and rumen contents are a fluorescent shade of green. Despite its bitterness, **pigs** and **horses** have been poisoned by it. Acute gastric erythema, pain, collapse, and death in **pigs** ingesting roots equal to 1% of their body weight.

HOYA sp.

= *Wax Flower* = *Wax Plant*

Untoward effects: Many species are poisonous. **Canaries** may develop a whitish diarrhea after pecking at the leaves.

HURA CREPITANS

= *Sandbox*

A tall tree native to the West Indies, Costa Rica, and Bolivia.

Untoward effects: In Florida, seed ingestion has caused abdominal pain, violent vomiting, and diarrhea. The fruit is a pumpkin-shaped, woody, and ribbed pod, 2–4 inches wide. It explodes when ripe, scattering out thin, flat, circular ¾ inch brown-speckled seeds, containing a purgative oil and the toxic proteins ***humin*** and ***crepitan***. These extracts and ***huratoxin*** have been used on poison arrows and darts. The juice is extremely acrid, causing erythema and pustular eruptions after skin contact.

Its toxic proteins are used in poison baits to kill **coyotes**, and an extract used as a piscicide.

HYACINTH

= *Hyacinthus orientalis*

Untoward effects: Nausea, salivation, vomiting, abdominal pain, and diarrhea in ~15% after ingestion of very small quantities of a bulb. Fatalities suspected. Needle-sharp ***calcium oxalate*** crystals in the bulbs' outer layers have caused severe itching and wheals. Dermatitis from handling the bulbs and from hyacinth oil.

The bulbs have caused severe purgation in **cattle**.

HYALURONIDASE

= *Alidase* = *Apertase* = *Diffusin* = *Diffusing Factor* = *Enzodase* = *Haglodase* = *Harodase* = *Hyalase* = *Hyalidase* = *Hyalozima* = *Hyasmonta* = *Hyason* = *Hyazyme* = *Hylase* = *Infiltrase* = *Invasin* = *Jalovis* = *Kinaden* = *Kinetin* = *Luronase* = *Pendase* = *Permease* = *Rondase* = *Ronidase* = *Spreading Factor* = *Thiomucase* = *Unidasa* = *Wydase*

Untoward effects: Generalized urticaria closing both eyes and pitting edema of the dorsum of the hand; a hypersensitivity reaction; and a case of severe anaphylactic shock 20 min after injection reported. Although used in medicine to depolymerize ***hyaluronic acid*** that glues cells together, permitting ready diffusion of injected substances, the same effect permits the faster spread, absorption, and toxicity of ***bee***, ***snake***, and ***spider*** venom, and even ***lidocaine***.

HYCANTHONE

= *WIN 24,933*

The mesylate or methanesulfonate form = *Etrenol*.

Untoward effects: Nausea, 35–50%; vomiting, 15–47%; abdominal pain, 32%; anorexia, 11%; headache, 7%; dizziness, 8%; orange-yellow urine, weakness, hepatitis, hepatic necrosis, and pain and tenderness at injection site reported.

High dosage was teratogenic, mutagenic, and, possibly, carcinogenic in **mice** and **hamsters**.

LD_{50}, **rat**: 980 mg/kg; **mouse**: 1120 mg/kg.

HYDRALAZINE

= *Apresoline* = *C-5968* = *Ciba 5968* = *1-Hydrazinophthalazine* = *Hipoftalin* = *Hypophthalin* = *Lopress* = *Präparat*

Antihypertensive, calcium channel-blocker, arteriole vasodilator.

Untoward effects: ***Blood***: Asymptomatic anemia and immune hemolysis in a 63-year-old **male** receiving 25 mg/day/3 years. Thrombocytopenia in two **neonates** whose **mothers** received the drug during their last trimesters. Leukopenia, agranulocytosis, pancytopenia, eosinophilia, decreased hemoglobin, hyperuricemia, and hyperglycemia reported.

Cardiovascular: Weakness, hypotension, palpitations, tachycardia, flushing, headache, polyarteritis nodosa, and angina occur. Myocardial ischemia after parenteral use can cause focal subendocardial necrosis. Has caused increased pulmonary artery pressure and clinical deterioration in **patients** with primary pulmonary hypertension. Pericarditis may be a manifestation of hydralazine-induced systemic lupus erythematosus. May aggravate congestive heart failure.

Fetus and neonate: Several cases of **neonatal** thrombocytopenia after **maternal** use during the third trimester, with recovery in a few weeks. Premature atrial contractions in a 36-week-old **fetus** of a 33-year-old **female** on 25 mg twice daily. Withdrawal of the drug led to resolution of the arrhythmia, and normal delivery occurred.

Liver: Granulomatous hepatitis in a 51-year-old **female** after 25 mg twice daily. Centrilobular zonal necrosis in a 68-year-old **male** after 75–150 mg/day/3 years. Hepatotoxicity confirmed by rechallenge. Drug sensitivity (50 mg/day) in a 39-year-old **female** induced acute hepatitis with necrosis on several occasions. Confirmed by rechallenge.

Miscellaneous: Fever, dyspnea, and nervousness are common symptoms. Lymphadenopathy, splenomegaly, constipation, tingling of hands and feet, muscle cramps, tremors, diaphoresis, fluid retention, pulmonary edema, depression, chills, gastric erosions, swollen feet, lacrimation, myopia, blurred vision, headache, and aching joints. Peripheral neuropathy in a 55-year-old **male drycleaner** on 50 mg qid and in many other **patients**. Many side-effects are due to ***histamine***, because the drug inactivates histaminase. Decreased libido, as well as priapism, in some **patients**. Increases antinuclear antibody titers, leading to arthralgias, arthritis, anemias, glomerulonephritis, anorexia, and nasal congestion. Can interfere with ***uric acid*** assays. Nausea, dizziness, vomiting, diarrhea, and nasal congestion are common with overdosage.

Renal: Proliferative necrotizing glomerulonephritis reported in four **females** (35–65 years) receiving 150–300 mg/day/6 months to 7 years. Another eight **patients** rapidly developed proliferative necrotizing glomerulonephritis after treatment. Hematuria and difficulty in urination is also reported.

Skin: A common cause of diffuse burning feet. It has induced a necrotizing vasculitis with widespread, itchy, maculopapular, and purpuric eruptions after 50 mg tid. Resolved after treatment ceased. Has caused Stevens–Johnson syndrome, erythema multiforme, fatal toxic epidermal necrolysis, and fixed-drug eruptions. Rashes may be associated with photosensitivity and systemic lupus erythematosus syndrome. Dermatitis after previous exposure to ***hydrazine*** (q.v.). Purpura, urticaria, and pruritus also reported.

Systemic lupus erythematosus: Not only associated with causing systemic lupus erythematosus (especially in **slow-acetylators**), but also may often unmask an underlying lupus diathesis. More common in **Caucasians** and **slow-acetylators** of the drug. A 54-year-old **male** received an emergency operation for violent abdominal pains after treatment with hydralazine. Necrotic areas of the jejunum and mesentery were noted. Some weeks later, **he** was hospitalized again and died. Post-mortem findings and diagnosis were post-therapeutic lupus erythematosus, necrotic arteriolitis, thrombophlebitis of right leg, and paralysis agitans, with the drug as the suspect antigen. Bilateral retinal vasculitis in **patients** with induced lupus syndromes.

Drug interactions: Since it can cause gastric erosions, it can increase the risk of hemorrhage in **patients** on ***anticoagulants***. May increase systemic availability of ***propranolol***. Addition to ***furosemide*** and ***propranolol*** therapy caused a **patient** to be unable to maintain an erection. Confirmed on rechallenge. ***Ethanol*** ingested will add to its hypotensive effect. Food can increase its bioavailablity by 2–3 times. May increase mutual toxicity with ***monoamine oxidase inhibitors***. Potentiated by ***thiazide diuretics*** and catastrophic hypotension is possible with ***diazoxide***. Effectiveness decreases with ***sympathomimetics*** and ***amphetamines***. Increased effect with ***triamterene***.

In **cats**, hypotension, weakness, and tachycardia; in **rats**, myocardial necrosis and cardiac hypertrophy; and induced antinuclear antibodies in **dogs**. Chronic use in **dogs** and **rats** has caused various ***collagen*** diseases and systemic lupus erythematosus, and may lead to reflex tachycardia, hypotension, and gastrointestinal upsets, with occasional vomiting and/or anorexia.

LD_{50}, **rat**: 90 mg/kg; **mouse**: 122 mg/kg.

HYDRANGEA

= *Hortensia*

Untoward effects: Contains the glycoside, ***hydrangin***, in the leaves and buds, which is hydrolyzed in the gut by β-glucosidases to ***cyanide*** (q.v.), poisonous to **man** and **livestock**. Early symptoms are abdominal pain, nausea, vomiting, diarrhea, dyspnea, weakness, dizziness, stupor, and convulsions. Smoked or used in ***teas*** for euphoric, stimulant, or hallucinogenic effects. Phytodermatitis with pruritus, rash, and vesicular eruptions reported in **florists**.

Cattle and **horses** have been poisoned, with typical ***cyanide*** symptoms, by eating it, especially in the spring. The bulbous roots have caused poisoning in **pups** playing with the stored roots.

HYDRARGAPHEN

= *Conotrane* = *Fibrotan* = *Hydraphen* = *Octrane* = *Optrane* = *Penotrane* = *Phenyl Mercuric Fixtan* = *P.M.F.* = *Septotan* = *Versotrane*

Bactericide.

Untoward effects: Hypersensitivity reactions reported. Eye irritation in ~40% after ophthalmic use. Local pruritus and swelling, causing discontinuance of vaginal suppositories in 3/1457.

LD_{100}, **mice**: 80 mg/kg.

HYDRASTIS CANADENSIS

= *Golden Seal* = *Indian Dye* = *Indian Tumeric* = *Orange Root* = *Yellow Puccoon* = *Yellow Root*

Contains 2–4% *hydrastine*, 2–3% *berberine* (q.v.), and *canadine*.

Untoward effects: Ingestion causes mucosal inflammation, ulceration, stomach pain, poor coordination, and, in large quantities, leads to vomiting, diarrhea, convulsions, paralysis, respiratory failure, and death.

LD_{LO}, **human**: 500 mg/kg.

HYDRAULIC FLUIDS

Untoward effects: Neurological disturbances from ***tri-o-cresylphosphate*** (q.v.) used in airplane hydraulic fluids (also see ***Ginger, Jamaican***) and many ***triaryl phosphate*** additives. Other additives, such as ***diethylene glycol*** (q.v.), ***polychlorinated biphenyls***, and ***ethylene glycol*** (q.v.) also caused toxicity. A **technician** noticed a red, discolored area on a 3000 psi air craft hydraulic line. A fine stream squirting out of a pin-hole penetrated **his** finger, which had to be amputated.

HYDRAZINE

= *Diamide* = *Diamine*

Millions of lbs produced annually.

Use: In rocket fuels; in manufacturing organic chemicals, spandex fibers, and foam rubber; in ***tobacco*** and ***tobacco*** smoke, in cutting oils, in welding, and as a soldering flux.

Untoward effects: Eye, nose, throat, and lung irritant; skin and eye burns, temporary blindness, dizziness, nausea, dermatitis, and systemic lupus erythematosus. Contact dermatitis with sensitization to its derivatives, ***hydralazine***, ***isoniazid***, and ***phenelzine***. The latter two often contain significant amounts of free hydrazine. Accidental ingestion of a mouthful by a 24-year-old **male** led to delayed sensory polyneuropathy. Hydrazine, potentially hepatotoxic, is formed in significant concentration during ***isoniazid*** metabolism in young **children**. It is a ***monoamine oxidase inhibitor***. Hydrazine hydrobromide solder flux leading to contact dermatitis with mild dermatitis or erythema to severe vesiculation and edema in many **workers**. If large quantities are inhaled or absorbed, tremors, convulsions, lethargy, coma, hepatitis, and nephritis can occur. Delayed pulmonary edema reported. A **worker** exposed to it once/week/6 months developed pleural effusions and pulmonary edema, bronchitis, enlarged liver and kidneys, liver necrosis, and intestinal hemorrhage, and he died. Abnormal liver function in 17/140 **persons** handling missile fuel. Fatty liver degeneration noted in two of them. In 1980, after a Titan II missile silo exploded, 200 **residents** of a small town 5 miles away complained of nausea, burning sensations in **their** noses, throats, and lungs; and dry, salty lips, after exposure to the "fog" and debris that fell. The propellant was a mixture of ***hydrazine*** and ***dimethylhydrazine***.

Caused lung and liver tumors in **mice** and **rats**; liver necrosis in **rats**; myeloid leukemias in **mice**; and, topically, it caused convulsions in **mice** and **rats**.

Inhalation LC_{50}, **rat**: 570 ppm/4 h; **mouse**: 252 ppm/4 h; skin LD_{LO}, **dog**: 90 mg/kg.

HYDRAZOBENZENE

Use: In dye manufacture; in manufacture of ***benzidine*** (q.v.), ***sulfinpyrazone*** (q.v.), and ***phenylbutazone***; and as an antisludging additive in motor oils.

Untoward effects: Induced hepatocellular and squamous cell carcinomas, squamous cell papillomas of the Zymbal gland, ear canal, and skin of the ear in male **rats**; mammary adenomas in female **rats** and hepatocellular carcinomas in female **mice**. SC and topical use in solvents also caused various neoplasms in mice and **rats**.

LD_{50}, **rat**: 301 mg/kg.

HYDRAZOIC ACID

Use: In manufacturing shell-detonators.

Untoward effects: Vapors irritate skin and mucous membranes, leading to conjunctivitis and coughing; inhalation of high concentrations leads to hypotension; facial flushing, followed by pallor; sweating, nausea, fatigue, vertigo, and collapse. Arthralgia, splenic enlargement, and nephritis also reported.

LD_{LO}, **human**: 5 mg/kg; inhalation TC_{LO}, **human**: 300 ppb.

Inhalation LC_{LO}, **rat**: 1100 ppm/1 h.

HYDRIODIC ACID

= *Hydrogen Iodide*

Use: In manufacturing ***methamphetamine*** and pharmaceuticals.

Untoward effects: Severe lung irritation by fumes. Rapidly caustic on skin contact.

HYDROABIETYL ALCOHOL

Several tons used annually in the U.S. as a fragrance in soaps, detergents, cosmetic creams, lotions, and perfumes.

Untoward effects: No irritation or sensitization of 10% in petrolatum on **human volunteers**.

Not irritating at full-strength on intact or abraded **rabbit** skin for 24 h under occlusion.

LD_{50}, **rats** and acute dermal, **rabbits**: > 5 g/kg.

HYDROCHLORIC ACID

= *Muriatic Acid* = *Spirit of Salt*

Use: Metal cleaner, in soldering, in removing ***calcium*** scale, and acidifying oil wells.

Untoward effects: Corrosive with mouth, lips, eyes, and skin burns from fumes or contact. Ingestion causes severe gastrointestinal upsets, nausea, violent vomiting, abdominal pain, ulceration, gastric hemorrhage, and peritonitis. Thirst, pulmonary edema, laryngeal spasms, hiccups, pneumonia, restlessness, shock, collapse, and death also reported. Contact with the eye has caused conjunctival and corneal necrosis, perforation, scarring, and loss of vision. In Australia and North Carolina, hospital **pharmacists** made fatal mathematic errors, leading to solutions 10–20 times prescribed strength. Skin burns are dark brown, which later blacken. The frequent vomiting of bulimic **persons** causes tooth enamel erosion and exposes the dentin. Numerous cases of self-poisoning from ingestion of the acid itself or in toilet bowl-cleaners and soldering fluxes. Maximum allowable exposure is 5 ppm/ 8 h/day; 50 ppm cannot be tolerated for > 1 h; and 1500–2000 ppm are lethal to **humans** in a few minutes.

Inhalation LC_{LO}, **human**: 1,300 ppm/30 min.

Drug interactions: Mixing with ***bleaches*** can release dangerous fumes.

Inhalation LC_{50}, **rat**: 3124 ppm/1 h; **mouse**: 2142 ppm/30 min. Oral, LD_{50}, **rabbit**: 900 mg/kg.

HYDROCHLOROTHIAZIDE

= *Aquarius* = *Bremil* = *Chlorosulthiadil* = *Cidrex* = *Dichlorosal* = *Dichlotride* = *Diclotride* = *Direma* = *Diumelusin* = *Disalunil* = *Esidrex* = *Esidrix* = *Fluvin* = *Hydresis* = *Hydrid* = *Hydril* = *Hydro-Aquil* = *Hydroazide* = *Hydrodiuretic* = *Hydro-Diuril* = *Hydroronol* = *Hydrosaluric* = *Hydrothide* = *Hydrozide* = *Hypothiazide* = *Ivaugan* = *Manuril* = *Maschitt* = *Nefrix* = *Neo- Codema* = *Neoflumen* = *Novohydrazide* = *Oretic* = *Panurin* = *Thiaretic* = *Thiuretic* = *Urodiazin* = *Urozide* = *Vetidrex*

Diuretic, antihypertensive.

Untoward effects: ***Blood***: Agranulocytosis, neutropenia, leukopenia, thrombocytopenia, purpura, granulocytopenia, increase in ***uric acid*** content or red blood cells, immune hemolytic anemia, hypovolemia, hyponatremia, hyperuricemia, hyperglycemia, hypercalcemia (decreased urinary ***calcium*** loss), hypokalemia with associated weakness and vertigo, hypochloremia, increased ***cholesterol*** and ***triglycerides***, and increased ***zinc*** excretion.

Bone: Long-term use associated with increased bone mineral content and decreased spinal fracture incidence.

Cardiovascular: Decreased serum electrolytes may induce arrhythmias and palpitations. Hypotension and orthostatic hypotension usually associated with increased dosage and other drugs.

Diabetogenic: Due to its hyperglycemic effect, especially on large or long-term dosage, it may aggravate existing diabetes, necessitating increase in ***insulin*** dosage.

Eye: Transient myopia, blurred vision, and xanthopsia. Permanent blindness in a **patient** also receiving oral ***propranolol*** and ***amiloride***.

Fetus and neonate: Limited data indicate intrauterine growth retardation, decreased placental perfusion, **fetal** resorption, thrombocytopenia, purpura, and **neonatal** death.

Gastrointestinal: Rarely, cholestasis, severe necrotizing and hemorrhagic pancreatitis, nausea, vomiting, diarrhea, sialadenitis, and anorexia.

Hypersensitivity: Severe allergic pneumonitis after ingesting a single 50 mg tablet. Confirmed on rechallenge. Pulmonary edema, fever, rashes, urticaria, necrotizing angiitis including cutaneous vasculitis, photosensitivity (closely related, chemically, to ***sulfonamides***) to ultraviolet light. A 50-year-old **female** developed acute pulmonary edema, leukopenia, and thrombocytopenia after a single 50 mg tablet. A single 25 mg dose led to acute nausea, vomiting, diarrhea, headache, and severe rigors in a 61-year-old **female** with psoriasis, diabetes mellitus, alcoholic liver disease, and hypertension. Reaction recurred on two rechallenges. Acute pulmonary edema in a 46-year-old **female** 30 min after 25 mg.

Miscellaneous: Hair loss, gout associated with hyperuricemia. Water intoxication (stupor, lethargy, somnolence, disorientation, and convulsion in **one**) in two compulsive **waterdrinkers**, due to acute hyponatremic encephalopathy, leg cramps, increased thirst, decreased libido, anorgasmia, depression, and hyperparathyroidism associated with induced hypercalcemia and hypophosphatemia. A psychiatrically disturbed 28-year-old **female patient** abusing the drug developed severe hypokalemic hypochloremic alkalosis and trigeminal rhythm. Excreted in **mother's** milk and may decrease volume of milk secreted. Increases urinary ***estriol*** assays.

Renal: Interstitial nephritis and urinary calculi reported.

Skin: Hypersensitivity reactions, purpura with pemphigoid necrotic blisters (primarily on the legs), lichenoid eruptions, photodynamic exanthemas, erythematous or papular eruptions that can become vesicular and tend to crust, cutaneous lupus erythematosus, toxic epidermal necrolysis, and Stevens–Johnson syndrome.

Drug interactions: Plasma levels are raised and its hyperuricemic effect decreases with ***diflunisal***. Combined with ***guanethidine*** permits decrease in dosage of the latter, and the latter's side-effects, particularly orthostatic hypotension. Severe hyperkalemia, hyponatremia, and metabolic acidosis after combined treatment with ***amiloride***. Severe hypokalemia and hypophosphatemia with chronic ingestion of large quantities of ***licorice***. ***Methyldopa*** potentiates its hypotensive effects and dry mouth, sleepiness, headache, increase of uremia, decreased blood ***cholesterol***, and orthostatic dizziness. Fixed combinations with ***potassium chloride*** have increased incidence of gastric, ileal, and jejunal ulceration with secondary problems. Acute interstitial nephritis and urolithiasis in fixed combination with ***triamterene***. An 80-year-old **female** with well-controlled mild hypertension on 25 mg/day developed a severe hypertensive crisis with the first dose of ***naproxen*** 250 mg and some subsequent twice daily doses over 18 days. No crises developed after ***naproxen*** discontinuance. Reactions with ***digoxin***, ***digitalis***, and ***digitoxin*** preparations (due to depletion of intracellular ***potassium***, leading to extrasystoles, tachycardia, and fibrillation) have been common. Use with ***ibuprofen*** causes a small but statistically significant increase in blood pressure. ***Probenecid*** can decrease its hyperuremic effect (common in 40–50% of **male hypertensives**). Increases arrhythmias with ***ketanserin***. Absorption decreased by ***colestipol***. Can decrease ***lithium*** excretion.

Milk from **cows** treated with it must be withheld from **human** consumption for 3 days after their last treatment.

LD_{50}, **mouse**: 3 g/kg; IV, LD_{50}, **mouse**: 590 mg/kg; **rabbit**: 461 mg/kg; **dog**: 250 mg/kg.

HYDROCODONE BITARTRATE

= *Bicotussin* = *Calmodid* = *Codinovo* = *Duodin* = *Dihydrocodeineone Bitartartrate* = *Hycodan* = *Hydrokon* = *Kolikodal* = *Mercodinone* = *Norgan* = *Orthoxycol* = *Synkonin*

Antitussive, analgesic, narcotic.

Untoward effects: Nausea, dizziness, drowsiness, and constipation are common. Occasionally, tightness in the chest and dryness of the pharynx. Dependence liability may be more than ***codeine's***. Poisoning with large doses led to central nervous system stimulation; convulsions, vomiting, drowsiness, respiratory depression, cyanosis, and coma. Rarely, skin rashes. Use in a resin-complex cough mixture, ***Tussionex***®, that also contained ***phenyltoloxamine***, was associated with three deaths in young **males**. Therapeutic serum levels in **man** are 0.01–0.03 mg/l and 0.1 mg/l is toxic.

Rat studies showed the resin complex 3.8 times as toxic as the drug in solution.

HYDROCORTISONE

= *Aeroseb-HC* = *Ala-Cort* = *Anflam* = *Cetacort* = *Cortaid* = *Cort-Dome* = *Cortef* = *Cortenema* = *Cortisol* = *Cortril* = *Dermacort* = *Dermocortal* = *Dermolate* = *Dioderm* = *Efcortelan* = *Evacort* = *Ficortril* = *Hydracort* = *Hydro-Adreson* = *Hydrocort* = *17-Hydroxycorticosterone* = *Hydrocortisyl* = *Hydrocortone* = *Hytone* = *Kendall's compound F* = *Lacticare-HC* = *Liefcort* = *Medicort* = *Mildison* = *Nutracort* = *Penecort* = *Proctocort* = *Reichstein's substance M* = *Scheroson F* = *Synacort* = *Texacort* = *Timocort* = *Zenoxone*

The acetate form = *Colifoam* = *Colofoam* = *Cortaid* = *Cordes* = *Cortifoam* = *Efcorlin* = *Hc45* = *Hydrocal* = *Hydrocortistab* = *Hydrocortone Acetate* = *Lanacort* = *Lenirit* = *Sigmacort* = *Sintotrat* = *Velopural*

The butyrate form = *Alfason* = *Laticort* = *Locoid* = *Plancol*

The sodium phosphate form = *Cleiton* = *Efcortesol* = *Hydrocortisone Sodium Phosphate* = *Hydrocortone Phosphate*

The sodium succinate = *A-hydroCort* = *Buccalsone* = *Corlan* = *Efcortelan* = *Hydrocortisone Hemisuccinate Sodium Salt* = *Saxizon* = *Solu-Cortef* = *Solu-Glyc*

The valerate form = *Hydrocortisone Valerate* = *Westcort*

Corticosteroid.

Untoward effects: Depending on the dosage, type, frequency of use, and route of administration, these can vary considerably in severity.

Blood: Lymphopenia (4–8 h after a dose), eosinopenia, leukocytosis, non-thrombocytopenic purpura, and, rarely, thrombocytopenia, and anemias (including that from gastrointestinal bleeding).

Cardiovascular: Hypertension.

Central nervous system: Headache, psychic disturbances, vertigo, pseudotumor cerebri, and convulsions. Delirium following overdosage or intolerance. Increased anxiety-proneness.

Endocrine: Cushingoid state, decreases growth rate in **children**, can aggravate latent diabetes and increase need for antidiabetic medications, decrease ***carbohydrate*** tolerance; menstrual irregularities, hirsutism, and slow adrenal and pituitary response to trauma, surgery, or illness.

Eye: Increased intraocular pressure, glaucoma, posterior subcapsular cataracts, and exophthalmus.

Fluids and electrolytes: ***Sodium*** and fluid retention, ***potassium*** loss, and alkalosis.

Gastrointestinal: Peptic ulcers, intestinal ulcers, esophageal ulcers, abdominal distension, pancreatitis.

Miscellaneous: Induced protein catabolism, causes negative ***nitrogen*** balance, increased weight, increased appetite, nausea, malaise, euphoria, thromboembolism, hypersensitivity, hearing loss, and increased susceptibility to infection. A case report of induced mania. Given to premature **infants** in their first day of life has been associated with decreased motor skills at 1 year of age, and decreased T-lymphocytes, coinciding with increased infections at 5 years of age. IV use in preterm **infants** increases their risk for disseminated candidal infections. A **child** given large doses of it and ***prednisone***/11 weeks suffered marked fatty infiltration of the liver with massive fat embolism and sudden death. A 7-year-old **male** with chronic eczema treated with 1% ointment (< 10 g/day/3 years) developed benign intracranial hypertension.

Musculoskeletal: Weakness, loss of muscle mass, osteoporosis, vertebral compression fractures, fractures of long bones, aseptic necrosis of femoral and humeral heads, decreased tendon tensile strength and their predisposition to rupture, and acute painful myopathy in asthma **patients** treated with large IV doses.

Skin: Thin, fragile skin; delayed wound-healing, erythema, ecchymoses, hypopigmentation, milaria, folliculitis, pruritus, perioral dermatitis, hypertrichosis, acneiform eruptions, increased diaphoresis, decreased response to some skin tests, SC and cutaneous atrophy, fat necrosis, sterile abscesses, and arthropathies after injections, and, paradoxically, allergic contact; dermatitis, urticaria, and angioneurotic edema. The butyrate, being very potent topically, has caused allergic contact sensitivity.

Drug interactions: ***Oral contraceptives*** may decrease test levels; IV use has led to pruritus and exacerbated bronchospasms in an **asthmatic**, due to ***paraben*** content, and ***diphenhydramine, butabarbital, phenobarbital, phenylbutazone***, and ***phenytoin*** may decrease its systemic levels. ***Creatinine*** is a solubilizing agent in parenteral ***sodium phosphate***, and can lead to false readings. It potentiates the response to catecholamines (***epinephrine, metaraminol, norepinephrine***, and ***phenylephrine***) on the eye.

FDA regulations allow a tolerance of 10 ppb for this drug in milk for **human** consumption. After maternal injection of 10–30 mg/2 kg/day/3 days in pregnant **rabbits**, many fetuses were born dead; live ones were smaller than controls, and without accelerated lung maturation. Topical ophthalmic use of the *sodium succinate* form on pregnant **mice** caused cleft palates in their offspring. Lymphopenia develops rapidly in **dogs** and **guinea pigs**. High dosage led to decreased fetal growth and increased incidence of cleft palates in **rats**. ***Phenobarbital*** and ***phenytoin*** increased its metabolism in the **guinea pig** and ***desipramine*** and ***imipramine*** increased its metabolism in the **rat**. Use during the third trimester of pregnancy may precipitate premature parturition with associated dystocia, fetal death, metritis, and retained placenta.

SC, LD_{50}, **rat**: 566 mg/kg; IM, TD_{LO}, ~11 days pregnant **mouse** and **hamster**: 400 and 125 mg/kg, respectively.

HYDROFLUMETHIAZIDE

= *Bristab* = *Bristurin* = *Di-Ademil* = *Dihydroflumethiazide* = *Diucardin* = *Elodrine* = *Finuret* = *Hydol* = *Hydrenox* = *Leodrine* = *NaClex* = *Olmagran* = *Rodiuran* = *Rontyl* = *Saluron* = *Sisuril* = *Vergonil*

Untoward effects: ***Blood***: Leukopenia, agranulocytosis, thrombocytopenia, pancytopenia, and aplastic anemia.

Cardiovascular: Orthostatic hypotension (may be potentiated by ***alcohol, barbiturates***, and ***narcotics***).

Clinical tests: Hyperglycemia, glycosuria, decreased protein-bound iodine, increased serum ***nitrogen***, hyperuricemia, and hypercalcemia.

Eye: Increased intraocular pressure.

Gastrointestinal: Gastric irritation, nausea, vomiting, abdominal pain, diarrhea, constipation, and acute pancreatitis.

Hypersensitivity: Photosensitivity, rash, urticaria, purpura, necrotizing angiitis, glomerulonephritis, exfoliative dermatitis, and cholestatic hepatitis, and it is immunochemically related to topical skin sensitizers containing ***PABA***, such as *p-****phenylenediamine*** and ***benzocaine***.

Musculoskeletal: Weakness and muscle cramps.

Drug interactions: Has caused a 24% decrease in ***lithium*** clearance.

LD_{50}, **mice**: > 8 g/kg; IV, 750 mg/kg.

HYDROFLUORIC ACID

= *Fluohydric Acid* = *HF* = *Hydrogen Fluoride*

Approximately 300,000 tons used annually in the U.S.

Use: In manufacturing fluorides, polishing and etching glass, rust and stain removal, steel pickling, aluminum manufacture, in acidifying oil wells, and in improving octane rating of gasoline.

Untoward effects: Strong solutions cause severe, painful, deep, and slow-healing burns. The intense pain can be delayed for several hours, so that a **person** is unaware until then that **they** have been burned. 2–5% solutions have little or no effect on **human** skin. A **worker**, accidentally burned by spillage from breakage of a vat, died 10 h after admission to a hospital. Post-mortem revealed severe tracheobronchitis and hemorrhagic pulmonary edema. Etching of the teeth has occurred in **workers** exposed to it. Toxicity reported from low-grade chronic exposure. Fluorocarbon aerosol propellents can break down into ***HF*** and ***phosgene*** on contact with hot surfaces. In 1986, a tank ruptured at a ***uranium*** processing plant in Oklahoma, forcing the closure of Interstate 40, due to release of a cloud of ***uranium hexafluoride*** gas, which breaks down in the atmosphere into ***HF*** and ***uranyl fluoride***. One **worker** died and eight were injured. Exposure to the gas caused eye, skin, and respiratory tract irritation. Coughing, choking, and chills have lasted up to 2 h, with a latent period of up to 2 days before fever, tightness in chest, bronchial pneumonia, and pulmonary edema with cyanosis. Accidental overdose ingestion by **children** of ***sodium fluoride*** < 5 mg/kg caused nausea and vomiting, due to the corrosive action of ***HF*** formed in the stomach. A number of suicidal ingestions, as well as a murder, have been reported in Dade County, Florida. Rust-removers are a common cause and many contain 6–11% ***HF***. Deliberate ingestion of ~3–4 oz ***Erusticator***® caused the death of a 33-year-old **male** in an estimated 45–60 min. **He** was acting in a bizarre manner, wandering aimlessly, periodically crying out for water, and clutching **his** stomach. **His** breathing became shallow and large amounts of mucus dripped from his nose and mouth before convulsing and dying. Post-mortem revealed erosions of the buccal mucosa, edematous lungs, large amounts of clotted blood in the stomach, and pancreatic necrosis. Topically, it can damage the nail plate leading to yellow or black discoloration and cause dermal swelling and/or redness, blistering, and pain. A ***cocaine***-intoxicated **addict** accidentally injected an automotive rust-remover into **his** rectum, and, in ~36 h, developed abdominal and perianal pain, eventually requiring surgical resolution of the resulting sigmoid colon perforation and other problems. TLV, 3 ppm.

Inhalation TC_{LO}, **Man**: 110 ppm/1 min; LC_{LO}, **human**: 50 ppm/30 min.

Many **cows** have been poisoned, due to pasture contaminated by ***aluminum*** factory emissions. Symptoms are lameness, emaciation, apathy, decreased milk production, exostoses of bone, especially on ribs, dental erosions, and reproductive disorders. Chronic exposure also causes teeth and bone abnormalities in **hamsters**.

HYDROGEN

Untoward effects: Odorless, lethal asphyxiant that can kill by excluding ***oxygen***. Lighter than air and it can accumulate from and above electric storage batteries, causing an explosion when ignited. Forms an explosive mixture with ***nitrous oxide*** (q.v.). ***Tritium*** (q.v.), its radioactive isomer, sometimes used in nuclear weapons, readily coats any moist surface, and easily enters the body.

HYDROGEN BROMIDE

= *Anhydrous Hydrobromic Acid*

Untoward effects: The vapors will irritate skin, eyes, nose, and throat. Solutions or liquid will cause skin and eye burns and frostbite. National Institute for Occupational Safety & Health and OSHA limit exposure to 3 ppm (10 mg/m^3).

HYDROGEN CHLORIDE

= *Anhydrous Hydrochloric Acid*

Untoward effects: Formed during decomposition and pyrolysis of ***polyvinyl chloride*** and ***polyester*** resins, causing severe irritation of larynx, nose, throat, and eyes; pulmonary edema with coughing and choking. Fumes have precipitated asthma attacks (acute bronchospasm, exertional dyspnea, and coughing) in **meat-wrappers** from electrically heated wires cutting ***polyvinyl chloride*** soft-wrap and from heating adhesive price labels. Fumes are a frequent cause of respiratory problems and deaths in **fire-fighters**. In solution, it causes dermatitis and eye and skin burns; the liquid has caused frostbite symptoms. Highly irritating and corrosive when swallowed leading to hiccups, violent retching, agonizing stomach pain, and increased thirst. OSHA: 5 ppm; TWA: 8 h; immediate danger to life or health: 50 ppm. Inhalation LC_{LO}, **human**: 1000 ppm/1 min.

Laryngeal spasms and pulmonary edema in exposed **animals**.

Inhalation LC_{50}, **rat**: 4701 ppm/30 min; **mouse**: 2644 ppm/30 min; inhalation LC_{LO}, **rabbit** and **guinea pig**: 4416 ppm/30 min.

Also see ***Hydrochloric Acid***.

HYDROGEN CYANIDE

= *Formonitrile* = *HCN* = *Hydrocyanic Acid* = *Prussic Acid* = *Zyklon*

Nearly 1.5 billion lbs used annually in the U.S.

Use: Fumigant for insect and rodent control in buildings; in manufacturing nylon and chemicals.

Untoward effects: Inhalation causes weakness, headache, confusion, nausea, vomiting, tachypnea, dyspnea, and a few tenths of 1% will cause death within minutes. This is used by some states for gas-chamber executions. Pyrolysis of ***urea–formaldehyde resins***, ***polyurethane*** foam, ***polyacronitrile*** (in clothing, carpeting, and blankets – ***Acrilon***, ***Creslan***, and ***Orlon***), ***acrylonitrile styrene***, ***acrylonitrile butadiene styrene***, ***asphalt***, ***silk***, ***wood***, and hay preserved with ***ethoxyquine*** and ***BHT*** release hydrocyanic acid. Lake Monoun, in the Cameroons, near Lake Nyos, described under ***cyanide*** (q.v.), also emitted a cloud of hydrogen cyanide gas, causing red skin spots, chemical burns, bleeding from the mouths and noses of **victims**, and deaths. Hydrocyanic acid has been sold in health food stores as ***Bee-Seventeen*** and ***Aprikern*** capsules. The latter contain 2 mg each; five are estimated to be fatal to a **child** and 20 for an **adult**. Present in ***tobacco*** smoke. Self-poisonings, some fatal, have been reported. After ingestion led to abnormal odor from breath, photophobia, tachypnea, dyspnea, vertigo, tinnitus, cyanohemoglobin, cherry-red colored blood, coma, and death. ***Lima beans***, containing 200 ppm, can be safely eaten in reasonable amounts, but the Environmental Protection Agency says sewers must not contain over 0.65 ppm. 20% of the **population** can neither taste it nor smell it. It was discovered in 1782 by Scheele, **who** died from accidentally inhaling it, thus demonstrating its lethal effects.

OSHA limits skin exposure to 10 ppm (11 mg/m^3). LD, **human**: 50–100 mg.

Many plants and foods consumed by **people** and **animals** cause hydrocyanic acid toxicity, under certain conditions (e.g. ***apple***, ***apricot***, ***arrow grass***, ***bitter almond***, ***Brachyachne convergens***, ***Cassava*** [***manioc***], ***castor bean***, ***catclaw***, ***cherry***, ***cherry laurel***, ***choke cherry***, ***cocklebur***, ***corn***,

cycads, ***cynodon***, ***flax***, ***fuchsia***, ***Heteromeles sp.***, ***Johnson grass***, ***Lathyrus***, ***linseed***, ***peach***, ***pear***, ***plum***, ***reed canary-grass***, ***sorghum***, ***sudan***, ***sugar cane***, ***Vicia peas***, ***Viquiera***, and ***yams***).

LD_{LO}, **human**, oral, 570 μg/kg; inhalation LC_{LO}, 120 mg/m^3/1 h, 200 mg/m^3/10 min; IV, LD_{50}, **human**: 1 mg/kg; SC, LD_{LO}, **human**: 1 mg/kg.

In **animals**, as in **people**, symptoms will vary according to the amount of and duration of exposure. **Ruminants**, particularly **cattle**, are more susceptible to its toxicity than monogastric **animals**. After large doses, muscle tremors and cyanosis develop rapidly, and death can occur within minutes. Lower doses also cause sudden onset with salivation, tachypnea, and dyspnea within 15 min, followed by muscle fasciculations, spasms, incoordination, collapse, convulsions, and death within 45 min. Those that live for a few hours recover after treatment. 600 mg will kill a 600 lb **cow** and 2.3 mg/kg is lethal for **sheep**. Although it causes rapid death when used for euthanasia, it causes undesirable convulsive seizures and endangers **personnel**.

Inhalation LC_{50}, **rats**: 540 ppm/5 min; **mice**: 169 ppm/30 min; **dogs**: 300 ppm/3 min; LD_{LO}, **dog** and **rabbit**: 4 mg/kg; LD_{100}, **mice**: 3–10 mg/kg.

HYDROGEN PEROXIDE

= *Albone* = *Hioxyl* = *H_2O_2* = *Hydrogen Dioxide* = *Hydroperoxide* = *Lensan A* = *Mirasept* = *Oxysept* = *Pegasyl* = *Perhydrol*

Over a billion lbs used annually in the U.S.

Use: Large amounts in the pulp and paper industry, as a ***bread*** dough conditioner, bleaching agent, germicide, sporicide, cosmetic use, and water treatment.

Untoward effects: Intestinal gangrene from gas embolism during colonic lavage in a 36-hour-old **male**. Gas embolism has been common with 0.75%–1.25% solutions in ***saline***. Experiments with **puppies** and **dogs** indicated bowel lavage with 0.75% was regularly associated with gas embolism. Acute ulcerative colitis after use of 3% in a number of **adults**. A 2-year-old **male** ingested an unknown quantity of 3%, leading to portal venous embolism. Methemoglibinemia can result from a rare condition, acatalasia. After a 56-year-old **female** ingested 15–20 ml 30% solution, within an hour she became confused and developed paralysis of **her** extremities. Hyperoxemia led to cerebral hypoxia and ischemia, due to vasoconstriction of cerebral veins. Recovered after 3 weeks. **Children** < 5 years are often hospitalized after ingesting it. Fatal hemolysis and methemoglobinemia in a 29-year-old **male** after suicidal ingestion of a glassful of 30%. Use in certain mouthwashes can cause the tongue to become red and sore and irritate buccal mucosa. Not to be used in the body where its gases cannot easily escape into the atmosphere. Can induce skin blanching and there is potential for eye injury from concentrations > 3%. Severe skin burns from contact with 70% solution. Inhalation of vapors or mists can cause extreme nose and throat irritation and inflammation. Will not burn, but is capable of causing spontaneous combustion after contacting easily oxidized materials. An explosion can occur if added to mouthwashes containing ***glycerin***. Equal parts of ***povidone-iodine*** (***Betadine***®) solution with 3% ***H_2O_2*** exploded shortly after mixing. ***Cigarette*-smokers** exposed to ***asbestos*** are more likely to die from lung cancer than **non-smokers**. Cells exposed to ***asbestos*** liberate hydrogen peroxide, which acts on ***benzo(α)pyrene*** in ***cigarette*** smoke and changes it from a precarcinogen into a potent carcinogen.

HYDROGEN SELENIDE

= *Selenium Dihydride* = *Selenium Hydride*

Untoward effects: Eye, nose, and throat irritant after inhalation of vapors. Nausea, vomiting, diarrhea, ***garlic***-like breath, metallic taste, dizziness, fatigue, and frostbite effects after eye and skin contact. National Institute for Occupational Safety & Health and OSHA: TWA limits of 0.05 ppm (0.2 mg/m^3). Immediate danger to life or health of 1 ppm.

Inhalation TC_{LO}, **human**: 0.2 ppm.

Pneumonitis and hepatitis in experimental **animals**.

Inhalation LC_{LO}, **guinea pig**: 280 mg/m^3/10 min.

HYDROGEN SULFIDE

= *H_2S* = *Hydrosulfuric Acid*

Use: In rayon manufacturing.

Untoward effects: Toxic, colorless gas with odor of rotten eggs. Common in "stink bombs". Inhalation of high concentrations can produce immediate unconsciousness and death by vagal reflex. Symptoms due to lower concentrations are eye and respiratory system irritation, pharyngitis, bronchitis, pulmonary edema, nausea, vomiting, sulfhemoglobinemia, headache, dizziness, palpitations, apoplectic collapse (rarely, with convulsions), cyanosis, greenish skin discoloration, weak pulse, and apnea. With chronic poisoning, there can also be coughing, sneezing, coryza, anorexia, confusion, insomnia, abdominal cramps, muscle weakness, anemia, and dermatitis. It is flammable, heavier than air, and has a **human** odor threshold of 0.13 ppm, followed by olfactory fatigue. Eye irritant at 10 ppm; lung and mucous membrane irritant at 20 ppm. Severe headaches, dizziness, and excitement after 500 ppm/30 min. Collapse and death in a few seconds after one or two inspirations of pure ***H_2S***. Accidental and sudden death in three **males** from acute intoxication by concentration of 750–1000+ ppm, which caused olfactory paralysis, that precluded a warning of peril. Agitation of stored manure has produced dangerous and lethal levels of ***H_2S***. Blood levels of 0.092 mg% are lethal. The gas was released from a tanker truck during water washing after emptying a load of refinery waste, and killed the **driver** and **dogs** and **birds** in the area. Five **onlookers** and **rescuers** were hospitalized. ***H_2S*** gas from a reaction of a strong acid in an industrial cleaner with ***plaster of Paris*** sludge in an acute-care hospital resulted in serious injury to the **workman** and less serious injury to three other **workers** and a **physician**. Slowly decaying **fish** in a ship's hold depleted it of ***oxygen*** and produced toxic quantities of ***H_2S*** and, during a 5-year period, led to three deaths and four cases of unconsciousness. Can cause disturbances in color vision. Lethality was noted in the area of volcanic emissions. Postmortem results are similar to those from ***carbon dioxide***, except for the strong odor of ***H_2S*** from the entire corpse. National Institute for Occupational Safety & Health states 10 ppm/10 min as maximum exposure. OSHA states 50 ppm/10 min. LC_{LO}, **human**: 600 ppm/30 min.

Pigs are frequently poisoned by it. Continuous exposure to 20 ppm results in fear of light, anorexia, and nervous ness; 50–100 ppm leads to nausea, vomiting, and diarrhea. Deaths occur at 400 ppm and above. Convulsions are occasionally noted. **Cattle** and **workmen** emptying or cleaning slurry tanks have developed similar symptoms. Its industrial discharge into a river from a glass factory caused sudden decrease of pH and death of **fish**. Exposure to 20 ppm reduced growth rate in **poultry** and significant drop in egg production at 80 ppm. **Rats** should be limited to 5 ppm exposure.

Inhalation LC_{50}, **rat**: 444 ppm; **mouse**: 673 ppm/1 h.

HYDROIDS

A hydrozoan found in tropical and subtropical oceans.

Untoward effects: Their stings produce mild, transient discomfort at the areas of contact, except for those of the order *Leptomedusea*, that causes severe contact dermatitis and urticaria within a few minutes and up to 2 h, or papular, hemorrhagic, or zosteriform reaction 4–12 h later. Erythema multiforme with morbilliform, vesicular, or desquamative eruptions have also developed. Systemic reactions lead to malaise, apprehension, muscle spasms, rigor, severe abdominal pain, diarrhea, chills, and fever. Sensitization and anaphylaxis have followed repeated exposure.

HYDROMORPHONE

= *Dihydromorphinone* = *Dilaudid* = *Dimorphone* = *Hymorphan* = *Laudicon* = *Lords* = *Novolaudon*

Untoward effects: Respiratory depression, apnea, drowsiness, irritability, anxiety, euphoria, depression, confusion, panic, tremors, dry mouth, pruritus, sweating, dysphoria, slurred speech, dizziness, constipation, nausea, and abuse. Secondary amenorrhea for 10 years, despite imprisonment and cessation of ***narcotic*** addiction. A **nurse** erroneously administered 0.5 mg, instead of ***morphine***, and caused the death of a 5-year-old **patient**. A **physician** caused respiratory arrest in an 18-year-old **male** by an erroneous order for 25 mg, IM, when **he** meant it for ***meperidine***. Psychic and physical dependence occurs. Orthostatic hypotension, ureteral spasms, and urinary retention reported. Use by a **mother** during pregnancy caused withdrawal symptoms (tremors, hyperactivity, crying, diarrhea, and vomiting) in her **neonate**.

LD_{LO}, human: 1428 μg/kg. Serum levels of 10–30 μg% are lethal.

IV, LD_{50}, **mice**: 61 mg/kg; SC, 84 mg/kg. Teratogenic in the **hamster**.

HYDROQUINIDINE

= *Dihydroquinidine* = *Hydroconchinine*

Antiarrhythmic.

Untoward effects: Syncope, cyanosis, cardiac and respiratory failure in 11 **females** and four **males**, 12–62 years. Overdose in a 16-year-old **female** caused headache, vertigo, sleepiness, photophobia, and sinus tachycardia.

IV, LD_{50}, **mouse**: 76 mg/kg.

HYDROQUINONE

= *Aida* = *Artra* = *1,4-Benzenediol* = *Black and White Bleaching Cream* = *Derma-Blanch* = *p-Dihydroxybenzene* = *Eldopaque* = *Eldoquin* = *Hydrochinone* = *Idrochinone* = *NSC-09247* = *Tecquinol*

Use: In photography as a reducing agent (developer); bleaching hyperpigmented skin areas; as an antioxidant for acrylic, polyester, and other resins, and for rubber, greases, and oils. An ***oxygen*** scavenger in water boilers.

Untoward effects: Occasionally localized contact dermatitis reported. Tingling, burning sensations and erythema observed. Idiopathic vitiligo, orange–brown finger nails (plates). It is a polymerization-inhibitor and a cause of dermatitis in **dental technicians**, and in **surgeons** doing hip replacements, as it is found in ***methylmethacrylate monomer***. It readily passes through surgical gloves. Cross-reacts with ***phenolics*** and ***resorcinols*** (including

hexylresorcinol, pyrocatechol, and ***phloroglucinol***). Ochronosis and colloid milia with jelly-like substance over the cheekbones reported in South African Bantu **women**, after latency period of 3–4 years, usually from prolonged use of 3.5–7.5% concentration. In industrial poisonings, a peculiar brown coloration of the corneal layers occurs in the areas of the palpebral fissure, as well as the conjunctiva. After many years of no abnormalities reported from such exposure, corneal applanation and accompanying severe irregular astigmatism did occur. A reported contact allergen in some high-speed duplicator papers. A photosensitizer. Acute inhalation exposure leads to headache, dizziness, nausea, vomiting, tachypnea, dyspnea, sensation of suffocation, tinnitus, pale and bluish skin discoloration, and green or brownish-green urine (usually darkens on standing). In 1977, 544 **crewmen** aboard a large U.S. Navy ship developed an acute onset of nausea, vomiting, abdominal cramps, and diarrhea after drinking potable water contaminated by a make-shift cross-connection from an automatic photographic developer tank. Heavier exposure has caused convulsions, cardiovascular collapse, pulmonary edema, and systemic acidosis. Rarely, may be involved in methemoglobinemia and renal and hepatic failure. OSHA standard recommendation of 2 mg/m^3/8 h TWA and National Institute for Occupational Safety & Health suggests this as a 15 min ceiling.

LD_{LO}, **human**: 29 mg/kg.

Ingestion leads to convulsions in **cats**; changed a treated **cat's** hair from black to gray after it walked ¼ mile in the sunlight; implicated as a cause of depigmentation of planum nasale in **dogs** from eating out of a rubber bowl treated with it as an antioxidant. **Livestock**, especially **hogs** and **chickens**, are poisoned by eating the young sprouts of ***cockleburs*** (q.v.) or its seeds accidentally ground up in their feed. Vomiting, weak heart, and convulsions are common symptoms, due to the burs' hydroquinone complex.

LD_{50}, **rat**: 320 mg/kg; **mouse**: 400 mg/kg; **dog**: 200 mg/kg; **cat**: 70 mg/kg; **rabbit** and **guinea pig**: 550 mg/kg.

HYDROXYAMPHETAMINE

= *Paredrine* = *Paredrinex* = *Pulsoton*

Mydriatic.

Untoward effects: Increased intraocular pressure. Blue-eyed **people** are more sensitive than brown-eyed **people** to its effects. Protracted treatment with locally applied ***guanethidine*** abolishes its mydriatic effect.

TD_{LO}, **human**: 300 μg/kg.

Drug interactions: Toxicity increases with ***monoamine oxidase inhibitors***.

p-HYDROXYBENZOIC ACID

= *p-Oxybenzoic Acid*

Use: In fungicides, dyes, preservatives, and organic synthesis.

Untoward effects: Its esters (see ***Parabens***) occasionally cause skin irritation and allergic sensitivity similar to those caused by ***aspirin*** or ***tartrazine***.

γ-HYDROXYBUTYRATE

= *GHB* = *Grievous Bodily Harm* = *4-Hydroxybutyrate* = *Liquid Ecstasy* = *Liquid G*

***Anesthetic.* IV use.**

Untoward effects: Cardiorespiratory arrest and death in a 21-year-old **female** abusing it by ingestion. A drug of abuse by **athletes**. Hallucinogenic. Nasal bleeding, central nervous system depression, coma, and brain damage reported. Used in "date rapes". Only a few drops (odorless and tasteless) can cause calmness, euphoria, quick "buzz", and then loss of consciousness in ~20 min with no memory of events during the following period. Causes akinesia and rigidity in **mice**.

HYDROXYCHLOROQUINE

= *Ercoquin* = *Oxichlorochine* = *Oxychloroquine* = *Plaquenil* = *Quensyl*

Antimalarial, antirheumatic, and lupus erythematosus treatment.

Untoward effects: Reactions may be more serious in young **children**.

Blood: Agranulocytosis, decreased hemoglobin, decreased platelet aggregation, neutropenia, aplastic anemia, slow decrease in blood viscosity. Fatal hemolytic anemia can develop in **patients** with glucose-6-phosphate dehydrogenase deficiency.

Eye: Usually irreversible retinopathy, especially with more than usual therapeutic doses. May follow gastrointestinal upsets, malaise, and depression. Lack of color perception. Corneal lesions of scattered opaque white dots, limbal aggravates, or curvilinear white subpapillary streaks. Perimacular fluorescence. Retinopathy may be delayed at least a year.

Miscellaneous: Malaise, nausea, vomiting, fever, mild headache, diarrhea, burning in the mouth and epigastrium, dizziness, severe vertigo, cardiomyopathy (especially in **children** < 5 years. May increase an **individual's** toxic reactions to ***mushrooms*** previously tolerated. Auditory nerve damage and hearing loss. Cranial nerve palsies are rare. Toxic myopathy and neuropathy reported. About 3.3% of **maternal** dose is secreted in **human** breast milk.

Poisoning: Fatal suicide in a 16-year-old **male** after ingesting > 54 200 mg tablets. A **man** survived swallowing 36 tablets after treatment. **He** was drowsy, had slurred speech, mydriasis, decreased blood pressure, bradypnea, and irregular respirations. Toxic serum levels are 0.5–0.8 mg/l and often fatal at 4 mg/l.

Skin and hair: Pigmentation of soft tissues in mouth and maculopapular eruptions from photosensitivity. Local vesicular eruptions around intralesional injections.

TD_{LO}, **human**: 429 mg/kg/25 days; LD_{LO}, **human**: 50 mg/kg.

Drug interactions: Increases plasma ***digoxin*** concentrations, potentially increases ***digoxin*** toxicity.

LD_{50}, **mouse**: 3.1 g/kg.

HYDROXYCITRONELLAL

Use: ~500,000 lbs/year in fragrances for soaps, detergents, cosmetic lotions and creams, perfumes, and foods.

Untoward effects: No sensitization reactions at 5% and 12% in petrolatum on **human** volunteers. Another report indicated a high percentage of sensitivity at 12% level. Contact dermatitis reported from its use in perfumes.

Pure material was moderately irritating on **rabbit** and **guinea pig**, but not on **miniature swine** skin.

LD_{50}, **rats**: > 5 g/kg; acute dermal LD_{50}, **rabbits**: > 2 g/kg; another report states 500 mg.

HYDROXYDIONE SODIUM

= *Presuren* = *Viadril*

***Anesthetic.* For IV use.**

Untoward effects: Pain and burning at IV site, venospasm, thrombophlebitis, hypotension, respiratory depression, tachycardia, and endothelial damage. The latter led to its discontinued use. Accidental IA injection led to thrombosis and amputation of the affected arm.

Venous thrombosis and respiratory decrease in some **animal** species.

HYDROXYETHYL STARCH

= *Hespan* = *Hetastarch* = *Plasmasteril* = *Volex*

***Plasma volume-expander.* 6% colloidal solutions are commonly used.**

Untoward effects: Although rare, allergic reactions occur (estimated at 8/10,000). Serum amylase levels increase up to 200%, peaking in 24 h, and may persist for 3–4 days, preventing its use as a diagnostic test for pancreatitis; erythrocyte sedimentation rate may increase 2–3 times for 48 h. Coagulopathy reported and occasionally frank bleeding. Pruritus, chills, increased temperature, tachycardia, hypertension, headache, and hyperkalemia have also been reported. Use cautiously in **patients** with bleeding disorders, congestive heart failure, or neurosurgery. Use in potential **organ donors** leads to subsequent osmotic-nephrosis-like lesions in kidney transplant **recipients**.

Dogs given > 25% of blood volume have overt bleeding.

N-HYDROXY-*N*-2-FLUORENYLACETAMIDE

***Carcinogen.* A metabolite of *N-2-fluorenylacetamide* (q.v.).**

Untoward effects: IP in female **rats** caused malignant liver tumors. Orally, to induce experimental hepatomas in male **mice**. Topical application to **rat** left thoracic glands led to (within 3–6 months) mammary adenocarcinomas at the site.

HYDROXYLAMINE

= *Oxammonium*

Antiviral agent.

Untoward effects: Can cause methemoglobinemia, sulfhemoglobinemia, hypotension, cyanosis, convulsions, and coma. Skin irritant on topical application. A 10% solution is extremely irritating to skin leading to intense redness, violent burning, sweating, and occasionally vesication. The skin of some **people** cannot tolerate a 1% solution.

LD_{LO}, **human**: 50 mg/kg.

High dosage causes convulsions in **monkeys** and **rats**. Experimentally used to produce anemias and blood dyscrasias in **animals**.

HYDROXYPHENAMATE

= *AL 0361* = *Listica*

Anxiolytic.

Untoward effects: Drowsiness and ataxia.

Transient hypotension in **cats**. High dosage causes some ataxia in **cats** and **dogs**.

LD_{50}, **rats**: 607 mg/kg; **mice**: 830 mg/kg.

17-α-HYDROXYPROGESTERONE

= *17α-Acetoxyprogesterone* = *Gestageno Acetate* = *Prodox*.

Caproate = *Delalutin* = *Hyproval P.A.* = *Lentogest* = *Neolutin* = *Pharlon* = *Primolut* = *Proge* = *Prolutin* = *Teralutin*.

Progestogen.

***Untoward effects*: *Caproate*:** Rashes, localized nodules at injection site, intraductal papilloma, gluteal abscesses, mild allergic reactions, postpartum dysfunctional uterine bleeding, and impotence in 25% of **males** treated for prostatic hypertrophy. Congenital defects in **infants** of **mothers** treated for threatened abortion. Agenesis of both thumbs and metacarpal bones, dislocation of radius or hips, masculinization of **female fetus** and clitoral enlargement.

Avoid breeding treated **animals** until the second post-treatment estrus cycle. Severe pyometra-like syndromes have occurred in treated **animals** deprived of normal estrus stimulating effects on uterine endometrium.

3-HYDROXY-4(1H)PYRIDONE

= *DHP*

Untoward effects*:** Produced from ***mimosine (q.v.) by enzymes in the leaves of ***Leucaena*** (q.v.), a shrub poisonous to **cattle**, but not to **goats**.

8-HYDROXYQUINOLINE

= *Mycantine* = *Ossidrochinone* = *Oxine* = *8-Oxychinolin* = *Oxyquinoline* = *8-Oxyquinolinol* = *Ozichinolini* = *8-Quinolinol*

Antiseptic, fungistat, chelator.

Untoward effects*:** Potentially neurotoxic (myelo-opticoneuropathy). Methemoglobinemia in a 4-week-old **female** may have been due to it or ***benzocaine in a rectal suppository. May cause false positive reaction in test for ***ketones***.

LD_{LO}, **human**: 500 mg/kg.

LD_{50}, **rat**: 1.2 g/kg.

HYDROXYSTILBAMIDINE

Used in IV treatment of blastomycosis and leishmaniasis.

***Untoward effects*:** Hypotension, tachycardia, anorexia, nausea, vomiting, headache, and malaise. Occasionally, short-lived dizziness; flushing, hyperhidrosis, dyspnea, salivation, syncope, paresthesias, fecal and urinary incontinence, formication, and edema of the face and eyelids. Hepatotoxicity and trigeminal neuropathy also reported.

5-HYDROXYTRYPTOPHAN

= *5 HTP* = *Oxitriptan*

The metabolic precursor of *serotonin*.

***Untoward effects*:** Diarrhea and hypertension. Increased REM sleep in a 7-year-old **male** and a 9-year-old **male**. Worsening of akinesia and rigidity in **parkinsonian patients**.

Deterioration in performance in **cats** and **monkeys**, and pyrexia in **rabbits**.

HYDROXYUREA

= *Hydroxycarbamide* = *Hydrea* = *Litalir* = *NSC 32,065* = *SQ 1089*

Antineoplastic.

***Untoward effects*: *Blood and bone marrow*:** Myelosuppression (thrombocytopenia, leukopenia, anemias, megaloblastosis, reticulopenia, agranulocytosis, macrocytosis) is the major toxic effect, and usually rapidly reversible.

Miscellaneous*:** Cardiac arrhythmias, nausea, vomiting, diarrhea, abdominal pain, stomatitis, anorexia, pyrexia, mental confusion with hallucinations, hyperuricemia, alopecia, acute pancreatitis, vasculitis, decreased uptake of ***iron by red blood cells, liver toxicity, headache, dizziness, disorientation, convulsions, pulmonary infiltrates; occasional hypersensitivity reactions and, rarely, dysuria. Hypersensitivity can occur to older formulations containing ***tartrazine***.

***Skin and hair*:** Skin rashes, facial erythema, pruritus, alopecia, lichenoid eruptions, and browning of nails. Leg ulcerations, particularly over the malleoli. Squamous cell carcinoma in a 59-year-old **male** after 1 g/day/> 8 years.

TD_{LO}, **human**: 80 mg/kg; IV, 86 mg/kg.

Drug interactions*:** Decreased cellular uptake of ***methotrexate.

Teratogenic in **hamsters**, **rats**, **dogs**, and **chick embryos**. Microphthalmia, hydrocephaly, defects of palate and skeleton, and decreased postnatal learning in offspring of treated **rats**.

TD_{LO}, **rat**: 750 mg/kg (13 days pregnant); **cat**: 100 mg/kg (13–14 days pregnant).

HYDROXYZINE

= *Alamon* = *Atarax* = *Aterax* = *Durrax* = *Orgatrax* = *Quiess* = *Tran-Q* = *Tranquizine* = *UCB-4492* = *Vistaril Parenteral*

Pamoate = *Equipose* = *Masmoran* = *Paxistil* = *Vistaril Pamoate*

Untoward effects*:** Drowsiness, xerostomia, nausea, vomiting, diarrhea, dizziness, weakness, hypotension, headache, mild skin rash, pruritus, cough, chest tightness, tachycardia, and altered electrocardiogram T-waves. A 13-year-old **female** ingested 20–25 200 mg tablets leading to generalized seizures, tachycardia, mydriasis, apnea, and peripheral vasodilation. A **neonate** whose **mother** received 600 mg/day and ***phenobarbital 60 mg/day during pregnancy was jittery, tachypneic, and irritable, and had clonic movements, shrill cries, and poor sucking reflex, which was typical of ***narcotic*** withdrawal. Pain at IM injection sites. Microscopic and gross hematuria reported after single injections in young **children**. Accidental IA injections have caused gangrene and necessitated amputations. Tremors and convulsions are rare, but can occur on high dosage. Severe exfoliative dermatitis, Stevens–Johnson syndrome, and hyperpigmentation of the palm have occurred. Found in breast milk.

LD_{LO}, **human**: 5 mg/kg.

Drug interactions: Acute dystonic reaction with opisthotonus, anxiety, hypertension, and fever after ***metoclopramide*** IV. Increased plasma concentration when given with ***cimetidine***. Additive effects from ***alcohol, barbiturates, narcotics***, and ***pentazocine***. May decrease dosage requirements of ***oral anticoagulants***. May lead to false elevations of ***theophylline*** levels and interfere with response to skin test antigens. Decreases effectiveness of ***phenothiazine psychotropics*** and ***phenytoin***. Contact dermatitis can occur with other ***ethylenediamines***. Recovery time prolonged by 30–40% when used as a premedicant with ***ketamine***.

Teratogenic in the **rat** and **mouse**, and causes abortions in the **rhesus monkey** and **sows**.

LD_{50}, **rat**: 840 mg/kg; **mouse**: 480 mg/kg.

HYGROMYCIN B

= *Hygramix* = *Hygromix* = *Hygrovetin* = *Gigromitsina B* = *Gigrovetina* = *Hyanthel*

Antibiotic, anthelmintic.

Untoward effects: Discontinue its use for **swine** 15 days, and for **chickens** 3 days before slaughter. The FDA has established a zero tolerance for its residues in or on eggs or uncooked edible tissues of **swine** and **poultry**. **Users** should avoid direct contact of this antibiotic with their skin or eyes. Keep it away from farm **dogs** who frequently eat medicated feeds. Labels, unfortunately, do NOT carry this warning, despite the fact that cataract formation has been attributed to its ingestion by **dogs**. Deafness in **dogs** and **swine** eating medicated food has also been noted. Its use in **swine** has also been suspected as causing an encephalitis under some conditions (larvae abscesses, etc.). Has interfered with weight gain in **sheep** trials.

HYMENOCALLIS OCCIDENTALIS

= *Spider Lily*

Untoward effects: A member of the ***Amaryllis*** (q.v.) family (*Amaryllidaceae*), many of whose species are poisonous. Contains toxic alkaloids, including ***lycorine***, and, in the spring, the bulbs have been poisonous to **cattle** and potentially dangerous for **children**.

HYMENOPTERA

Untoward effects: Thousands of bites and stings by this group (**bees**, **hornets**, **yellow jackets**, **wasps** - q.v.). Venoms of the last three are probably twice as allergenic as that of the **honey bee**. Most stings are relatively mild; 10% can involve an entire extremity with swelling and erythema; and 1% develops an anaphylactoid (and even life-threatening) reaction. Death has occurred in > 400 cases in the U.S. during a 10-year period. Guillaine–Barré syndrome and seizures have followed 3–10 days after stings. ***β-Blockers*** may worsen the anaphylactic reactions.

Dogs have been particularly susceptible to sting toxicity.

HYMENOXYS sp.

H. odorata = *Bitter Rubberweed* = *Bitterweed* = *Western Bitterweed*

H. richardsoni = *Actinea richardsoni* = *Colorado Rubberweed* = *Pingue*

Untoward effects: Serious economic losses in **sheep**, **goats**, and occasionally **cattle** from their alkaloids, ***hymenoxon*** and ***hymenovin***, leading to anorexia, depression, abdominal pain, green regurgitated material about their mouths and noses, coma, and death. Post-mortem revealed gastroenteritis, cardiac hemorrhages, and congestion of lungs, liver, and kidneys.

l-HYOSCYAMINE

= *Bellafolina* = *Cystospaz* = *Daturine* = *Duboisine* = *Egacin* = *Egazil* = *Levsin* = *Peptard*

Anticholinergic.* It is a *belladonna* alkaloid found in *belladonna, duboisia, Datura stramonium, deadly nightshade, Hyoscyamus,* and *mandragora.* The *dl* form is *atropine.

Untoward effects: Xerostomia, mydriasis, diplopia, blurred vision, flushing, constipation, decreased libido, central nervous system stimulation with irritability, disorientation, confusion, hallucinations, memory impairment, and delirium, followed by coma and even death from respiratory paralysis. Found in some arrow and dart poisons. Has become a drug of abuse (viz. from ingestion or smoking excessive amounts of ***Asthmador cigarettes***). Mydriasis, increased intraocular pressure, and tachycardia reported.

LD_{LO}, **human**: 5 mg/kg.

HYOSCYAMUS

= *Banotu* = *Beleño* = *Devil's Eye* = *Hog's Bean* = *Insane Root* = *Khurasani ajwain* = *Poison Tobacco* = *Sakaraan*

***Anticholinergic, parasympatholytic, antispasmodic.* Contains 0.04–1.5% alkaloids (primarily, *atropine* [q.v.], *l-hyoscyamine* [q.v.], and *scopolamine* [q.v.]) in the leaves.**

Untoward effects: Long known in the Mediterranean as a poison and ***narcotic*** capable of causing permanent insanity. Used in European sorcery in the Middle Ages in conjuration of demons, prophecy, and soothsaying, along with fantastic visual hallucinations.

H. muticus is smoked as an inebriant in India.

Four ***H. niger*** = ***Black Henbane*** = ***Jusquiame Noire*** flowers were deliberately ingested by a 20-year-old **male** seeking euphoria. Instead, **he** became agitated, restless, had dry mouth, mydriasis, tachycardia (120/min), and hot dry skin. The symptoms dissipated over 48 h. Ingestion causes the anticholinergic syndrome, which also includes hypertension, fever, urinary retention, gastric hypomotility, confusion, incoordination, agitation, delirium, and hallucinations. Rarely, seizures, coma, and cardiovascular collapse have occurred. **Humans** have been poisoned by mistakenly eating the roots as ***parsnips*** or in the hope of producing euphoria. Smoked in the Himalayas as an hallucinogen. Unintentional ingestion of seeds and roots by young (2–11 years) **children** in Turkey has caused poisonings. Supposedly, caused the death of Hamlet's father.

Poultry have died from eating the seeds; **pigs** from eating the fleshy roots; and **cattle** avoid grazing on the foliage. The alkaloids are retained after drying. If drought forces **cattle** to eat it, dyspnea, bloat, tachycardia, mydriasis, restlessness, ataxia, cyanosis, and convulsions can occur. Its alkaloids can pass into a **cow's** milk.

HYPERICIN

= *Hypericum Red*

Antidepressant.

Untoward effects: A photosensitizer in ***Hypericum perforatum*** (q.v.) and cannot be used as a beverage flavor in the U.S. Levels in the plant range from 0.03 to 0.38%. Causes dermatitis, skin blistering, and loss of hair in unpigmented skin areas of **cattle**, **horses**, and **sheep** exposed to bright sunlight.

HYPERICUM

Untoward effects: ***H. crispum*** has caused photosensitivity reactions in **cattle**, **sheep**, and **horses** in Turkey.

H. perforatum = ***Amber*** = ***Goat Weed*** = ***Herba*** = ***Klamath Weed*** = ***Rosin-rose*** = ***STE1-300*** = ***St. John's Wort*** = ***Tipton Weed***. A delayed hypersensitivity or photodermatitis reported in **people** after ingestion of ***tea*** made from its leaves. Acute neuropathy in a 35-year-old **female** ingesting 500 mg/day after sun exposure. Some constituents lead to ***monoamine oxidase inhibitor*** activity, xerostomia, dizziness, constipation, gastrointestinal upsets, and confusion in some **people**. Hypomania in a 47-year-old **female**. Possible manic reactions also reported. Use of 300 mg/day in a 42-year-old **female** decreased ***theophylline*** serum levels, requiring a 2.7 times increase in ***theophylline*** dosage. It also decreases ***cyclosporine*** and ***indinavir*** effectiveness. Its ***cyclosporine*** interaction led to cases of acute heart transplant rejection. Has decreased effectiveness of ***antihypertensives***. Effect can be additive to ***fluoxetine***. After ~10 days of use, its extract caused a 25% decrease in ***digoxin*** plasma levels. Disagreeable taste and odor in ***milk***, milk products, and meat of **animals** that graze it. Sampling of commercial productions in the U.S. revealed ***hypericin*** (q.v.) concentrations 47–165% of labeled content.

Skin blistering, scabbiness, itching, and loss of hair on non-pigmented areas of **cats**, **dogs**, **cattle**, **horses**, **rabbits**, **sheep**, and **swine** from sun exposure. Mature plants are usually not eaten. Affected **animals** are restless and have dyspnea, diarrhea, pyrexia, tachycardia, depression, mydriasis, tremors, and mental confusion. Contains a volatile oil (***red oil***) and two fluorescent compounds, ***hypericin*** and ***hypericum red***. ***H. triquetrifolium*** = ***Dirnach***. Photosensitization reactions in **sheep**, **goats**, and **cows** in Israel. Uneasiness, hyperirritation, severe pruritus and dermatitis, ophthalmia, lacrimation, corneal opacity, blindness, edema of lower jaw and abdomen, hematuria, and hemorrhagic diarrhea reported. No increased temperature.

HYPOCHLOROUS ACID

Untoward effects: It is now known that the irritating effect of ***chlorine*** gas on the pulmonary, gastrointestinal, ophthalmic, and cutaneous systems are also due to this, and not just ***hydrochloric acid***.

HYPOGLYCIN A

A toxic *amino acid* in the arils of unripe *Ackee* (*Blighia sapida* - q.v.). The *B* form is less toxic.

Untoward effects: Profuse vomiting, collapse, hypothermia, hypoglycemia, and fatalities in **man** and **animals**.

IP, TD_{LO}, **rat** (1–6 days pregnant): 30 mg/kg.

HYSSOP OIL

Untoward effects: An ingredient of the psychoactive beverage ***absinthe***. Has caused poisoning in **humans**, characterized by tonic-clonic or plain clonic convulsions.

I

IBOGAINE

Untoward effects: Central nervous system stimulation. Large doses are hallucinogenic in **man**, **swine**, **porcupines**, and **gorillas**. General paralysis then occurs in **man** with death due to respiratory paralysis. Controlled drug in the U.S. The chief active indole alkaloid in ***Tabernanthe iboga*** found in Gabon and the Congo. Ingestion reduced cravings in **people** with various drug addictions. Fantasies of floating in space or a fast review of **one's** past life occur after 4–8 h, followed by another 8 h of mulling over the meaning of these visions, including an insight into the cause of **their** addictions. **Patients** rarely sleep during the first 36 h of the ordeal.

Rat studies found some brain cell damage. **Dogs** have developed seizures and cardiac arrest.

IV, LD_{50}, **mouse**: 56 mg/kg.

IBOPAMINE

= *Inopamil* = *N-Methyldopamine Diisobutyrate* = *SB 7505* = *Scandine* = *SKF 100,168-A*

Cardiotonic.

Untoward effects: Dyspnea.

IBOTENIC ACID

Untoward effects: A key compound causing psychotropic poisoning by ***Amanita muscaria*** (q.v.). Nausea, vomiting, tachycardia, blurred vision, excitement, and hallucinations; then, deep sleep for up to 12 h. Deaths have occurred in **people** with cardiac problems.

LD_{50}, **rat**: 129 mg/kg; **mouse**: 38 mg/kg.

IBUFENAC

= *Dytransin* = *Ibunac*

Analgesic, anti-inflammatory.

Untoward effects: Jaundice, malaise, lassitude, anorexia, vomiting, flatulence, itching, giddiness, urinary frequency, sleep disturbances, and fecal blood loss. Deafness and tinnitus also reported. An ***aspirin***-sensitive **asthmatic** developed urticarial rash, dyspnea, laryngeal edema, and tightness of the chest.

LD_{50}, **mouse**: 1.8 g/kg.

IBUPROFEN

= *Adran* = *Advil* = *Amibufen* = *Anco* = *Anflagen* = *Apsifen* = *Ardinex* = *Artril 300* = *Bluton* = *Brufen* = *Brufort* = *Buburone* = *Butylenin* = *Dansida* = *Dentigoa* = *Dolgin* = *Dolgirid* = *Dolgit* = *Dolocyl* = *Dolo-Dolgit* = *Ebufac* = *Emodin* = *Epobron* = *Femadon* = *Fenbid* = *Gynofug* = *Halprin* = *Haltran* = *Ibu-Attritin* = *Ibumetin* = *Ibuprocin* = *Ibutad* = *Ibutid* = *Ibutop* = *Inabrin* = *Inoven* = *Lamidon* = *Lebrufen* = *Liptan* = *Lobufen* = *Medipren* = *Motricit* = *Motrin* = *Mynosedin* = *Napacetin* = *Nerofen* = *Nobfen* = *Nobgen* = *Novogent N* = *Nuprin* = *Nurofen* = *Opturem* = *Orafen* = *Pediaprofen* = *Proflex* = *Prontalgin* = *RD-13621* = *Recidol* = *Roidenin* = *Rufen* = *Seclodin* = *Spedifen* = *Suspren* = *Tabalon* = *Trendar* = *Urem*

Anti-inflammatory. **World consumption is ~7000 metric tons annually, and > 3000 tons in the U.S.**

Untoward effects: ***Anaphylactoid***: Can occur in **patients** hypersensitive to other ***non-steroidal anti-inflammatory drugs***, such as ***aspirin***.

Blood: Neutropenia, agranulocytosis, aplastic and hemolytic anemia, thrombocytopenia, purpura, eosinophilia, and decreased hemoglobin and hematocrit. Inhibits platelet aggregation and increases bleeding time.

Central nervous system: Dizziness (3–9%), headache and nervousness (1–3%), depression, insomnia, confusion, hallucinations, somnolence, aseptic meningitis with fever and coma (especially in **patients** with systemic lupus erythematosus or connective tissue disease).

Cardiovascular: Edema and fluid retention (1–3%), increased blood pressure, palpitations, and possible congestive heart failure.

Clinical tests: May cause false positive test results for ***cannabinoids***.

Ear: Tinnitus (1–3%) and hearing loss.

Eye: Amblyopia, blurred vision, scotomata, diplopia, uveitis, conjunctivitis, and changes in color vision.

Fetus and neonate: Can cause premature closure of ductus arteriosus and pulmonary hypertension. < 0.6% appears in breast milk.

Gastrointestinal: Nausea, epigastric pain, and heartburn are common (3–9%); diarrhea, abdominal distress (cramps or pain), indigestion, vomiting, bloating, and flatulence in 1–3%; gastric or duodenal ulcers with bleeding or perforation, melena, hepatitis, jaundice, constipation, and abnormal liver function tests (increased SGOT and SGPT). Ingesting it with food does not necessarily stop the adverse gastrointestinal effects. After 1 g/day/30 days, a

12-year-old **female** with cystic fibrosis showed pyloric channel stricture.

Hypersensitivity: Bronchospasms and asthma. A 65-year-old **female** developed severe fatal asthma after ingesting 400 mg. Cross-allergenicity to ***aspirin*** is > 75%. Cross-reactions can also exist with ***tartrazine*** and ***sodium benzoate***.

Miscellaneous: Dry eyes and mouth, rhinitis, and gingival ulcers. A febrile reaction in a 36-year-old **female** with lupus erythematosus after 400 mg/5 times a day/1 week. Night sweats and stiff neck reported. Can cause swollen ankles, feet, or legs.

Poisoning: Shock, metabolic acidosis, and coma in a 6-year-old **male** after ingesting 30 200 mg tablets. A 13 kg 2½-year-old **male** ingested 30 400 mg tablets leading to deep sedation and vomiting. Estimates were that **he** only retained one 400 mg tablet. A 70-year-old **male** ingested 30 400 mg tablets, leading to unconsciousness and hypotension. **All** survived after treatment. Hypercalcemia, hypomagnesemia, and seizures in a 21-year-old **male** after an 8 g dose. Therapeutic serum concentrations are listed between 10 and 50 mg/l and toxicity at 100–500 mg/l.

Renal: Decreased ***creatinine*** clearance, acute interstitial nephritis, polyuria, azotemia, cystitis, and hematuria. Avoid use in **patients** with renal impairment. Renal papillary necrosis in a 70-year-old **male** on 400 mg qid/3 weeks. Oliguria in 5/74 **infants** (24–32 weeks) given IV 10 mg/kg, followed by 5 mg/day/2 days to treat patent ductus arteriosis.

Skin: Photosensitivity, rashes (3–9%), pruritus, urticaria, erythema multiforme, alopecia, and Stevens–Johnson syndrome.

Drug interactions: Can increase effects of ***anticoagulant drugs*** and ***digoxin*** serum levels. Others report no changes on both drugs. Heavy ***alcohol*** intake with it can induce gastric bleeding and renal failure. May require decrease in ***lithium*** and ***phenytoin*** dosage. Transient renal failure while simultaneously receiving ***gentamicin*** and ***tobramycin***. Rhabdomyolysis and renal failure in a 29-year-old **male** when added to ***ciprofibrate*** therapy. Has increased ***raloxifene*** serum concentration by 20–30%. Use while taking ***zidovudine*** prolongs bleeding time. Can decrease antihypertensive effect of ***angiotensin-converting enzyme inhibitors***. Analgesic effect can be potentiated by 65–200 mg ***caffeine***.

TD_{LO}, **woman**: 240 mg/kg/10 days or 4 g/kg/32 weeks.

May markedly increase serum ***digoxin*** levels in **dogs**. In general, **dogs** and **cats** given single doses of 100 mg and 50 mg/kg are asymptomatic. Gastrointestinal irritation and hemorrhage after repeated or large doses, in addition to vomiting, hematemesis, anorexia, melena, and some central nervous system depression.

LD_{50}, **rat** and **mouse**: 1 g/kg.

IBUTILIDE FUMARATE

= *Corvert*

***Antiarrhythmic - Class III.* IV use.**

Untoward effects: Ventricular asystoles, tachycardia (3%), headache, nausea, hypotension, and bundle branch block. Palpitations and acute renal failure in a 52-year-old **male** after two doses of 0.87 mg. Accidental spillage on hands led to erythematous bullous lesions.

ICACINA SENEGALENSIS

= *Ibiala* (Igbo)

Untoward effects: Tubers are eaten during periods of food scarcity. Unless sliced and steeped in water for a long time, they are poisonous to **man**. Death has occurred within a few hours if not steeped properly. During a siege of Kumasi in Ghana, it caused deaths within a few days. Colic, dysentery, and swelling of the face can precede death.

ICE

Untoward effects: In ***freeze balls*** (q.v.), it has caused infectious hepatitis in **man**. In 1981, its use in beverages at a fund-raising luncheon in Jackson, Michigan was the source of typhoid fever. Acute respiratory illness in hockey **players** and **spectators** occurred from too high a concentration of ***nitrogen dioxide*** used to refreeze the ice. An outbreak of *Enterobacter cloacae* and *Pseudomonas aeruginosa* in a cardiothoracic surgery unit was traced to the use of a poorly installed ice machine.

ICE CREAM

Untoward effects: A case report indicates probable inhibition of ***warfarin*** effect in a **patient** from ice cream ingestion. Pruritic, erythematous rash, and severe headache from ice cream made from milk containing ***penicillin*** 10 U/ml. The **patient** had a previous history of ***penicillin*** sensitivity. Adding ice cream to ***activated charcoal*** decreases its adsorptive capacity. Can cause migraine in some **people**. Sudden, severe, steady pain in the posterior bridge of the nose, frontal and retro-orbital areas often occurs after swallowing large amounts or other cold foods, due to irritation of the V, IX, and X cranial nerves. Contamination with *Listeria monocytogenes* caused a major recall of ice cream bars, and in 1994, > 200,000 **people** were poisoned as a result of tanker trucks hauling ice cream mix after hauling raw eggs, without sanitizing between loads. A **person** with cold sensitivity developed angioedema (throat constriction, lip and tongue swelling) after eating it, due to sudden massive release of ***histamine***.

ICELANDITOXIN

Untoward effects: A fungal toxin produced by ***Penicillium icelandicum*** (q.v.) that causes liver damage and hemorrhage at low doses in **mice** and carcinomas at higher dosage (oral, 6.5 mg/kg).

ICODEXTRIN

= *Extraneal* = *Icodial*

Use: As a peritoneal dialysis fluid.

Untoward effects: Exfoliative and blistering skin reactions in ~15% of 102 **patients**, serious in ~5%.

IDARUBICIN

= *DMDR* = *Idamycin* = *IMI-30* = *NSC-256439* = *Zavedos*

***Antineoplastic, anthracycline.* A synthetic analog of *daunorubicin*.**

Untoward effects: Nausea and vomiting as acute toxicity. Delayed toxicity leads to bone marrow depression, alopecia, headache, stomatitis, myocardial toxicity, decreased left ventricular ejection fractions, cardiomyopathy, electrocardiogram changes, arrhythmias, diarrhea, and abdominal pain. Liver and renal injury has been reported. It is vesicant and direct contact can cause skin reactions. Extravasation has caused severe local tissue necrosis.

IDOXURIDINE

= *Dendrid* = *Emanil* = *Herpes-Gel* = *Herplex* = *Idexur* = *Idoxene* = *IDU* = *Idulea* = *IDUR* = *Iduridin* = *Iduviran* = *5-Iodo-2′-deoxyuridine* = *IUDR* = *Kerecid* = *NSC 39,661* = *Ophthalmadine* = *SKF 14,287* = *Stoxil* = *Virudox*

***Antiviral.* An analog of *thymidine*.**

Untoward effects: Used systemically, causes host toxicity, therefore it is limited to topical use, where it has caused pain, pruritus, and inflammation and edema of the eyelid or eye. Alopecia is transient and hair regrows within 1 year after treatment withdrawal. Cholestatic jaundice reported in one **patient**; hyperbilirubinemia and SGOT elevation reported in **another**. Allergic reactions occasionally reported. Contact dermatitis, punctate epithelial keratopathy, follicular conjunctivitis, and lacrimation after topical therapy for herpes keratitis. Systemic toxicity is due to its inhibition of rapidly proliferating tissues [hematopoietic system, oral mucosa, dermal appendages (hair follicles, nails)]. This inhibition leads to leukopenia, thrombocytopenia, neutropenia, pancytopenia, stomatitis, alopecia, and transverse ridging of nails.

Teratogenic in **rabbits** (exophthalmus, clubbing of forelegs) and **rats** (eye, cerebellum, and kidney). *Contraindicated* for parenteral use during pregnancy, as it interferes with DNA formation in dividing cells. Will cause fatal leukopenia in **dogs** within 2 weeks on the same schedule that shows no toxicity in **man** at 70–100 times **dog** dosage on a mg/kg basis.

IFOSFAMIDE

= *A-4942* = *Asta Z-4942* = *Holoxan* = *Ifex* = *Ifomide* = *Iphosphamid(e)* = *Isoendoxan* = *Isophosphamide* = *Mitoxana* = *MJF-9325* = *NSC-109724* = *Z-4942*

***Antineoplastic.* IV use. Alkylating agent.**

Untoward effects: Leukopenia, cystitis, hematuria, dysuria, urinary incontinence, blurred vision, confusion, hallucinations, somnolence, nausea, vomiting, alopecia, bone marrow depression, encephalopathy, cerebellar ataxia, electroencephalogram abnormalities, twitching of the mouth and neck, non-convulsive status epilepticus, hypokalemia (fatal in one case), Fanconi syndrome, nephrotoxicity, hypophosphatemic rickets in **children**, pulmonary embolism, coma, congestive heart failure, and, occasionally, hypersensitivity reactions. Extravasation has caused pain and inflammation.

TD_{LO}, **human**: 100 mg/kg; IV, TD_{LO}, **human**: 2.3 g/kg/3 days.

Drug interactions: Exacerbates ***cisplatin***-induced hearing loss. Several **patients**, after previous treatment with ***cisplatin***, developed nephrotoxicity from it.

Dogs given it IV at dosages causing electrocardiogram abnormalities in **man** died with myocardial damage within 4½–18 h. Carcinogenic and nephrotoxic in **rats**. Embryotoxic and teratogenic in **mice**, **rats**, and **rabbits**. Uterine leiomyosarcomas in **rats** and malignant lymphomas of the hematopoietic system in **mice**.

IP, LD_{50}, **rat**: 150 mg/kg.

ILEX sp.

= *Holly*

I. paraguariensis* = *Jesuit's Tea* = *Maté* = *South American Holly* = *Yerba mate

Untoward effects: The dried leaves are used to make a stimulating ***tea***, ***Paraguay Tea***, due to its ***xanthine*** content (***caffeine*** 0.56%, ***theobromine*** 0.03%, and ***theophylline*** 0.02%). ***Caffeine*** content in the ***tea*** is ~2%. South American **natives** drinking such ***tea*** during a 1–3 h period usually consume 80–120 mg of ***caffeine*** (q.v.). Intake of large quantities of such ***tea*** leads to purging and vomiting. Quality control of exported material has been poor and a number of emergency cases of anticholinergic syndrome poisoning were reported by New York City emergency rooms in 1994 from contamination with ***belladonna alkaloids***. Liver damage and death of a **female** who consumed large amounts of this ***tea*** for several years, due to its ***pyrrolizidine*** content. Also, see ***Holly***.

ILLICIUM ANISATUM

= *Japanese Star Anise* = *Shikimi* = *Skimmi*

Untoward effects: Due to its ***sesquiterpene, anisatin***, as well as ***myristicin*** content in its fruits. Causes epiletiform convulsions (analogous to ***picrotoxin's***), mydriasis, and cyanosis, especially in **children**.

I. parviflorum in the hills of U.S. Georgia and the Carolinas is also considered to be the same, and a decoction of its seeds caused violent gastrointestinal irritation, followed by motor and sensory paralysis, with convulsions and death on high dosage.

The same as ***I. religiosum*** in India.

ILOPROST

= *Ciloprost* = *Ilomedin* = *ZK 36,374*

***Peripheral vasodilator, antithrombotic.* A stable analog of *prostacyclin*.**

Untoward effects: Had a variable effect on platelet aggregability and can cause hypotension and a sudden, unheralded sinus bradycardia. Adverse effects are usually dose-related, and include mild coughing, jaw pain, and slight headaches. Confusion and psychosis have been common in the **elderly**.

IMICLOPAZINE

= *Chlorimpiphenine* = *P-4241* = *Ponsital*

***Psychotherapeutic.* Oral and IM.**

Untoward effects: Dyskinesia (35), parkinson syndrome (32), insomnia (17), hyperhidrosis (11), sialorrhea (10), tachycardia (7), tremor (6), orthostatic collapse (4), hypotension (3), urinary retention (3), respiratory complaints (2), and swallowing difficulty (2) in 55 **males** and 19 **females**.

IMIDAN

= *Decemthion* = *ENT 25,705* = *Ftalofos* = *GX 118* = *Paramite* = *Phosmet* = *Prolate* = *R 1504*

***Insecticide, acaricide.* Cholinesterase-inhibitor.**

Untoward effects: LD_{LO}, **human**: 50 mg/kg; inhalation TD_{LO}, **human**: 2 mg/m^3/day.

Highly toxic to **honeybees**. Regurgitation, ataxia, excessive preening, hyperactivity, falling, ptosis, salivation, prostration, dyspnea, tremors, convulsions, and tetanic seizures within 10 min, and deaths in **mallard ducks** and **pheasants** within $\frac{1}{2}$–17 h after treatment. **Mallards** also show polydipsia. The FDA has a zero tolerance for it in milk and eggs and a tolerance of 1 ppm in meat, fat, and meat by-products of **livestock** and **poultry**. Improper use or use with other drugs caused hypersalivation, ataxia, tremors, and death in a large number of **cattle**.

LD_{50}, **rats**: 113–160 mg/kg; **mice**: 26 mg/kg; male **chicks**: 707 mg/kg; **guinea pig**: 200 mg/kg.

IMIDAZOLIDINYL UREA

= *A Biol* = *Germall 115* = *IDZU* = *Imidazoline Urea* = *Imidurea*

***Preservative, biocide.* Widely used in cosmetics, often in combination with *parabens* to control *Pseudomonas* contamination.**

Untoward effects: Allergic contact sensitivity occurs. It was identified as the cause of chronic facial dermatitis in a 26-year-old **female** due to its presence in **her** facial cream.

IMIDOCARB

= *4A65* = *Imizol*

***Protozoacide.* Used IM for babesiosis in equines and other species.**

Untoward effects: Suspect carcinogen. Occasionally local irritation at injection site. Serous nasal discharge, ptyalism, diarrhea, dyspnea, tremors, incoordination, and polyuria in **cattle** at 10 mg/kg. Avoid use in **horses** less than 1 year of age, or in near-term pregnancy. Do NOT use with or 10 days before or after treatment with, or exposure to, any cholinesterase inhibiting drugs. NOT for use in **horses** for **human** consumption. Drug reactions in **horses** are usually minimal, until dosage exceeds 2 mg/kg/day. **Donkeys** have died receiving as little as a single dose of 1 mg/kg. Nearly 6 months of residues in the liver and kidneys of treated **cattle**. Dosage of 10 mg/kg in **cattle** can cause death.

IMIPENEM

= *N-Formimidoyl Thienamycin* = *Imipemide* = *MK 787*

Combined with Cilastin Sodium = *Imipem* = *Primaxin* = *Tenacid* = *Tienam* = *Tracix* = *Zienam*

***Antibiotic.* A *β*-lactam. Eliminated in the urine, where it is metabolized by an enzyme in the renal tubular cells. *Cilastin* is often used simultaneously, to inhibit this inactivation.**

Untoward effects: A syndrome of nausea and hypotension (especially after rapid infusion) and a predisposition to seizures (estimated at < 1–7.5% with the combination). In 1984, some deaths were associated with the combination, and U.S. marketing approval for the combination was temporarily rescinded. Skin rashes, diarrhea, vertigo, hiccups, false low ***glucose*** measurements with the ***Clinitest***®, and, possibly, a case of aplastic anemia also reported with the combination. False positive results for urine leukocytes, and three cases of suspected induced dental staining (13-, 33-, and 52-year-old **females**), after IV imipenem reported.

IMIPRAMINE

= *Antipress* = *Berkomine* = *Chrytemin* = *Deprinol* = *Efuranol* = *Feinalmin* = *G 22,355* = *Imavate* = *Imidol* = *Imilanyle* = *Imiprin* = *Imisine* = *Imizin* = *'Iramil* = *Janimine* = *Melipramine* = *Norfanil* = *Norpramine* = *Pramine* = *Presamine* = *Pryleugan* = *Tofranil* = *W.D.D.*

***Antidepressant.* A tricyclic. The HCl is usually used. Potent anticholinesterase inhibitor.**

Untoward effects: ***Blood***: Agranulocytosis and aplastic anemia, inhibition of platelet function and aggregation, asymptomatic eosinophilia, increased lactate dehydrogenase, leukocytosis, leukopenia, lymphocytosis, neutropenia, purpura, and thrombocytopenia.

Cardiovascular: Cardiac arrest, tachycardia, arrhythmias, ventricular and auricular fibrillation, hypotension, postural hypotension, palpitations, heart-block, and cardiomyopathy reported. Even therapeutic doses have shown cardiotoxic effects leading to deaths, especially in elderly **patients** or **those** receiving drugs that deplete cardiac ***catecholamines***, such as ***guanethidine***.

Central nervous system: Neurological or brain syndromes. Convulsive coma and respiratory depression, followed by dysarthria and ataxia. Akathesia, confusion, hypomania, psychic reactions, and delirium. IV can provoke epileptic features on the electroencephalogram.

Eye: Increased intraocular pressure and narrow-angle glaucoma, blurred vision, lens opacity, and cataracts; mydriasis due to its peripheral anticholinergic effects. Cycloplegia and diplopia can also occur.

Fetus and neonate: A few scattered reports, primarily of amelia and limb defects, after **mothers** ingested it during the first trimester. Despite many surveys, a true causal relationship has not been established. Small quantities excreted in breast milk.

Liver: Jaundice with nausea and vomiting; even fatal cases of massive hepatic necrosis. Increased alkaline phosphatase. An 11-year-old **male** developed hepatic failure and massive necrosis, requiring a liver transplant after treatment with 25 mg at bedtime for enuresis.

Miscellaneous: Visual hallucinations, agitation, aggressiveness, restlessness, insomnia, seizures, hyperpyrexia, tremors, ataxia, roaring in ears, temporary retardation of growth in **children** with hyperkinetic behavior, hyperprolactinemia, breast enlargement and galactorrhea in the **female**, gynecomastia and erectile impotence in the **male**, delayed orgasm in **women**, thyrotoxicosis, pulmonary eosinophilia, peroneal nerve palsy; Löffler's syndrome, sweating, dry mouth, glossitis, dysarthria, drowsiness, syncope, obstipation, nausea, vomiting, diarrhea, transient headache, dizziness, anorexia, altered taste, and tearfulness occur. Severe withdrawal symptoms reported, especially in **children**. Increased age is correlated with increased serum levels.

Poisoning: Hyperpyrexia, areflexia, hypotension, cyanosis, cardiac arrhythmias (ventricular flutter or tachycardia; atrial fibrillation or tachycardia), electrocardiogram evidence of impaired conduction (atrioventricular or intraventricular block), signs of congestive heart failure, convulsions, respiratory depression, and cardiac arrest in accidental and suicidal overdoses. In overdosage, symptoms can also include drowsiness, stupor, tachycardia, ataxia, vomiting, shock, restlessness, agitation, severe perspiration, hyperactive reflexes, muscle rigidity, anthetoid movements, and mydriasis. Coma may ensue. Therapeutic serum levels are generally 0.05–0.3 mg/l; toxic at > 1 mg/l; and coma or death at 1.5–2 mg/l. Doses of 10 mg/kg can be very toxic in **some**.

Renal: Urinary retention and damage (blood urea nitrogen 80 mg/100 ml; blood ***creatinine*** 2.5 mg/ml). Syndrome of inappropriate antidiuretic hormone secretion, decreased bladder contractility, and increased urethral pressure. 30% decrease in urinary ***vanillylmandelic acid*** excretion.

Skin: Hyperpigmentation, bullous lesions, dermatitis, erythematous maculopapular rash; pruritus, photosensitivity, porphyria, and alopecia reported.

TD_{LO}, **woman**: 50 mg/kg; **child**: 30 mg/kg.

Drug interactions: More severe hepatic necrosis if used within 3 weeks of ***acetaminophen*** overdose. Additive sedation with ***alcohol***. ***Amobarbital*** and ***phenobarbital*** decrease its plasma levels. Antagonizes the hypotensive effect of ***guanethidine*** and increases its own toxicity. Collapse, unconsciousness, and death has occurred in several **patients** after use with ***chlorpromazine***. Speech impairment, generalized tremor, hyperreflexia, bilateral clonus, and equivocal Babinski reflexes in a 38-year-old **female** receiving 75 mg of each. ***Chlorpromazine, haloperidol, perphenazine, sertraline***, and ***thioridazine*** decrease its excretion. Can increase serum ***phenytoin*** levels to the point of intoxication. ***Estrogens*** and ***oral contraceptives*** may interfere with its effect. Convulsions have followed its ingestion with old ***cheese***. Coma, hyperpyrexia, and deaths in **patients** also ingesting ***monoamine oxidase inhibitors*** (a 2 week interval is suggested), due to its potentiation of imipramine effects. These include ***isocarboxazid, pargyline, phenelzine***, and ***tranylcypromine***. It slows the absorption of ***levodopa*** and ***phenylbutazone***. Leukopenia with ***protriptyline***. ***Benzodiazepines, meprobamates***, and ***phenothiazines*** potentiate its effects. Also potentiates ***meprobamate*** effects. ***Methylphenidate*** decreases its metabolism. Use with ***clonidine*** interferes with the latter's antihypertensive effects. ***Cigarette*** smoking increases its clearance rate leading to plasma levels only 55% of **non-smokers**. Psychosis in a 38-year-old **female** after its use with ***cimetidine***. ***Cimetidine*** and ***fluoxetine*** decrease its metabolism and increase its serum concentration. Several case reports indicate ***fluvoxamine*** dramatically increases its blood levels. It antagonizes the panicogenic effects of exogenous ***cholecystokinin tetrapeptide***. ***Cholestyramine*** decreases its plasma levels by ~23%. ***Butalbital*** also decreases its serum levels. Potentiation of pressor effects of ***phenylephrine*** (2–3 times), ***norepinephrine*** (4–8 times), and ***epinephrine*** (2–4 times). Convulsions due to hyponatremia when given with ***desmopressin***. Mania-type reactions with ***reserpine***. Enhances the effects of ***anticholinergics, atropine, carisoprodol, hexobarbital, pentobarbital***, and ***thiopental***. Use with ***aspirin*** and ***salicylates*** has been fatal. Can dramtically alter blood pressure control by ***bethanidine***. Adverse reactions can occur if given with ***thyroid*** preparations and ***anticholinergic*** agents. Use with ***diphenhydramine*** may prolong xerostoma. In a 54-year-old **male**, it potentiated the antispastic effect of ***baclofen*** and caused loss of muscle tone. ***Alprazolam*** increases its serum concentration.

In **animals**, it is potentiated by ***monoamine oxidase inhibitors***, producing coma and alterations of body temperature-regulating mechanisms. 25 mg/kg daily from the 3rd to 20th day after the mating of female **rabbits** caused fetal abnormalities. High lipid solubility accounts for very rapid maternal–fetal equilibrium noted in **rat** studies. Has caused skeletal anomalies in **rabbits** and **hamsters**. In **rats**, increases toxicity of IV ***digoxin*** and decreases the oral absorption of ***lamotrigine***.

LD_{50}, **rat**: 625 mg/kg; **dog**: 100 mg/kg; **mice**: 352 mg/kg; LD_{LO}, **cat**: 100 mg/kg; IV, LD_{50}, **rat**: 25 mg/kg.

IMIQUIMOD

= *Aldara* = *R-837* = *S-26,308*

Antiviral, immunomodulator.

Untoward effects: Frequent topical use causes local inflammation reactions, erosions, excoriation, edema, scabbing, and flaking. Occasionally itching, burning, or pain.

IMOLAMINE

= *Angolon* = *Irrigor* = *LA-1211*

Antianginal.

Untoward effects: Headache, vomiting, palpitations, flushing.

IMPATIENTS sp.

= *Balsam Weed* = *Jewel Flower or Weed* = *Speckled Jewels* = *Touch Me Not*

Untoward effects: Causes vomiting in **livestock** grazing on it.

INDALPINE

= *LM 5008* = *Upstene*

Antidepressant, selective serotonin uptake inhibitor.

Untoward effects: Mydriasis. Increased duration of QRS interval in **patients** without cardiopathy.

INDAPAMIDE

= *Bajaten* = *Damide* = *Fludex* = *Indaflex* = *Indamol* = *Ipamix* = *Lozol* = *Natrilix* = *Noranat* = *Pressural* = *RHC 2555* = *S-1520* = *SE-1520* = *Tandix* = *Veroxil*

Antihypertensive, vasodilator, diuretic.

Untoward effects: Severe hyponatremia, hypokalemia, hyperglycemia, interstitial nephritis, ventricular fibrillation, skin reactions (rashes), fever, toxic epidermal necrolysis, erythema multiforme, angioedema, decreased libido and impotence, headache, dizziness, rhinitis, blurred vision, vasculitis, tingling of extremities, and dry mouth.

LD_{50}, **rats, mice, guinea pigs**: > 3 g/kg; IP, LD_{50}, **rats**: 393 mg/kg; **mice**: 410 mg/kg; **guinea pigs**: 347 mg/kg.

INDELOXAZINE

= *CI 874* = *Elen* = *Noin* = *YM-08054*

Antidepressant.

Untoward effects: Several case reports of induced parkinsonism.

INDENE

= *Indonaphthene*

Found in tars from *coal*, lignite, and petroleum.

Untoward effects: In **animal** trials caused eye, skin, mucous membrane irritation; aspiration pneumonia; hepatitis, nephritis, and splenic pathology after injection.

SC, LD_{LO}, **rat**: 1 g/kg.

INDENO (1,2,3-cd) PYRENE

A *polycyclic aromatic hydrocarbon (PAH)*, widely found as a by-product of poorly controlled combustion systems (*coal* burning, refuse burning, *diesel*-powered vehicles, forest fires and *coke* ovens). 15 ng/kg in soot, 7300–8300 ng/kg in *coal tar pitch*, and 1 ng/kg in petroleum *asphalt*. In *cigarette* smoke (0.4–2.0 μg/*cigarette*), fresh *sausages* (0.3 μg/kg), and some in edible oils (0.9–1.6 μg/kg).

Over 200,000 workers exposed to it.

Untoward effects*:** Is a carcinogen and an initiator of skin carcinogenesis (local sarcomas after SC) in the **mouse**. Lower carcinogenicity than ***benzo(α)pyrene (q.v.).

INDIGO

= *D&C Blue 6* = *Indigo Blue*

***Untoward effects*:** When used as a food coloring, it caused some allergic reactions. May cause green or blue urine. Still approved for use on sutures (not > 0.5% by weight).

INDIGOCARMINE

= *Acid Blue 74* = *CI 1180* = *C.I. 73015* = *C.I. Acid Blue 74* = *C.I. Food Blue 1* = *DFG 105* = *E 132* = *FD & C Blue No. 2* = *Schultz 1309*

Said to be the coloring agent the Lord told Moses to use on the cord for each corner of their garments.

***Untoward effects*:** Green or blue urine. Despite FDA approval of its use in foods and drugs, Health Research Group alleges it increases incidence of tumors in experimental **animals**.

LD_{50}, **rat**: > 2 g/kg.

INDIGOFERA sp.

Untoward effects*: *I. dominii ingestion has caused poisoning (Birdsville Disease) of **horses** in the Northern Territory of Australia.

I. spicata (formerly known as ***I. endecaphylla***) = ***Birdsville Indigo*** = ***Creeping Indigo*** = ***Baba*** (Hausa) has also caused Birdsville Disease of **horses** in Australia, poisoning of **poultry** and **cattle** in Brazil, abortion and retained placentas in Fiji **cows**. Hepatotoxicity with cirrhosis reported in **rabbits**, **cattle**, and **sheep** in South Pacific islands, cleft palates and intrauterine deaths in **rats** caused by its ***indospicine*** (***arginine*** antagonist and protein synthesis inhibitor) content. Cirrhosis and hepatomegaly in **rats**, poisoning in **chicks** from its ***3-nitropropanoic acid***, dead or weak **calves**, abortions, and vulvar swelling noted in **cows** and **ewes**. Two forms of poisoning occur in **horses** and **mules**. The first involves poor coordination of limbs and hind hoof dragging. The second involves dyspnea and tetany. After improvement, they are never safe to ride, as exercise may cause symptoms to redevelop.

I. suffruticosa, native to the West Indies, and from Mexico to Uruguay. Despite the medical uses of the root or leaf decoctions, it can be severly purgative. The fumes from the fermented plants were believed to be poisonous to the **slaves** who trampled or stirred it in the dye vats. The crushed plant is a piscicide and the roots are insecticidal.

INDINAVIR

= *Crixivan* = *L 735,524* = *MK 639*

***Antiviral*. A protease inhibitor.**

Untoward effects*:** Early reports indicate it has caused nephrolithiasis. Alopecia and maculopapular eruptions reported in HIV **patients**. Latent dysbetalipoproteinemia precipitated by it and other protease inhibitors in HIV **patient** after 1 month of treatment. ***Hypericum perforatum use can decrease its serum level ~50%. Can induce renal colic and the formation of urinary stones (4%) and acute interstitial nephritis. Crystalluria, dysuria, back and flank pain in ~8%. Hemolytic anemia, increased bilirubinemia and hepatitis, focal mycobacterial lymphadenitis, hypertrophic paronychia, ingrown toenails, and pyogenic granuloma of the great toes in many **patients**. Rash, pruritus, dry skin, and alopecia usually within 2 weeks. Hyperglycemia and diabetes reported. Many case reports of increased cervicodorsal tissue (buffalo humps) or lipodystrophy, including increased abdominal girth due to fat accumulation.

Drug interactions*:** Toxicity of ***astemizole, delavirdine, ritonavir, saquinavir and ***terfenadine*** may be increased because indinavir decreases their metabolism. ***Rifampin, carbamazepine, Hypericum perforatum***, and ***nevirapine*** decrease indinavir's effect by increasing its metabolism. ***Rifabutin*** increases its metabolism, decreases its effect and the combination may increase ***rifabutin*** toxicity. Use with ***nelfinavir*** may increase the toxicity of both drugs. ***Didanosine*** decreases its absorption. ***Fluconazole*** causes slight increase or decrease in its serum levels. ***Ketoconazole*** may increase its toxicity. Can potentiate ***ergotamine***. It increases serum concentrations of ***amiodarone***.

INDIUM

Untoward effects*:** The metal is irritant to eyes, skin, and respiratory system. May cause adverse effects on liver, kidney, heart, and blood, and cause pulmonary edema after inhalation. Radioactive ^{111}In*** emmits gamma rays and is useful in many organ imaging and diagnostic procedures. ^{113}***In*** is not used much any more. Indium chloride and sulfate are poisonous.

LD_{LO}, **human**: 500 mg/kg.

Indium chloride and ***sulfate***, LD_{LO}, **human**: 500 mg/kg.

Indium nitrate, IP, LD_{50}, **mice**: 7.95 mg/kg.

INDOBUFEN

= *Ibustrin* = *K 3920*

Antithrombotic.

Untoward effects: The most common complaints are gastrointestinal, often minor and transient. It interacts with the ***sulfonylurea glipizide***, leading to enhanced blood ***glucose*** decrease.

INDOCYANINE GREEN

= *Cardio-Green* = *Fox Green* = *Ujoviridin* = *Wofaverdin*

***IV diagnostic aid* (cardiac and hepatic function, blood volume).**

Untoward effects: Contains traces of ***sodium iodide***, requiring caution in administration to **people** hypersensitive to ***iodine***. Diaphoresis, pruritus, and headache in two **patients**, and a severe anaphylactoid reaction in another three uremic **patients**. Anaphylactoid symptoms reported in three other **patients**, 49–62-year-old **males**. Possible tendency to leukopenia after catheterization in nine **males**.

Hypothermia has a dramatic effect in decreasing hepatic uptake (40%, with 1–2°C drop) in **cats**. A considerable difference has been noted between "normal" **mongrels** and purebred **beagles** in clearance times.

IP, LD_{50}, **rat**: 700 mg/kg; **mouse**: 400 mg/kg.

INDOLEACETIC ACID

= *Heteroauxin* = *IAA*

Plant growth regulator.

Untoward effects: Intraruminal, but not IV, administration can cause interstitial pulmonary edema and emphysema in **cattle**.

INDOMETHACIN

= *Amuno* = *Argun* = *Artracin* = *Atrinovo* = *Bonidon* = *Catlep* = *Chibro-Amuno* = *Chrono-Indocid* = *Confortid* = *Dolcidium* = *Dsmogit/Amuno Gits* = *Durametacin* = *Elmetacin* = *Idomethine* = *Imbrilon* = *Inacid* = *Indacin* = *Indocid* = *Indocin* = *Indomed* = *Indomee* = *Indomethine* = *Indometin* = *Indometacin* = *Indomod* = *Indo-Phlogont* = *Indoptic* = *Indoptol* = *Indorektal* = *Indos* = *Indosmos* = *Indo-Tablinen* = *Indotard* = *Indoxen* = *Inflazon* = *Infrocin* = *Inmetsin* = *Inteban SP* = *Lausit* = *Metindol* = *Mezolin* = *Mikametan* = *Mobilan* = *Osmosin* = *Rheumacin* = *Tannex* = *Vonum*

Anti-inflammatory, analgesic, antipyretic.

Untoward effects: Blood and bone marrow: Fatal aplastic anemia, eosinophilia, decreased platelet aggregation and prolonged bleeding time, hyperkalemia, thrombocytopenia, leukopenia, neutropenia, pancytopenia, agranulocytosis, leukocytosis, and secondary anemia associated with gastrointestinal bleeding and ulcers.

Cardiovascular: Salt and water retention, hypertension, tachycardia, chest pain, and aggravation of anginal pain. Occasionally congestive heart failure; palpitations, arrhythmias, and extrasystoles. Myocardial infarction in **patients** with angina.

Clinical test interference: Increased or false positive blood urea nitrogen, amylase, alkaline phosphatase, SGPT, fasting ***glucose***, and ***bilirubin***.

Eye: Corneal opacities with decreased visual acuity, altered dark adaptation, diplopia, nystagmus, and retrolental fibroplasia. Intraocular hemorrhage reported in an overdose case.

Fetus and neonate: May increase incidence of cleft lip (with or without cleft palate). *In utero* exposure has caused renal failure and irreversible renal damage in the **fetus**. Can displace ***bilirubin*** bound to serum ***albumin***. Given to pregnant **women** threatened with premature labor, it can increase risk of causing premature closing of the ductus and primary pulmonary hypertension or bronchopulmonary dysplasia in their **infants**, when delivery occurs early. Intestinal perforations have been reported in **infants** treated with it for patent ductus arteriosus. Gastrointestinal hemorrhage, oliguric renal failure, edema, hydrops, and acute pneumoperitoneum from localized ileal perforations early in the **infant's** life, after their **mothers** received chronic tocolysis with rectal indomethacin (200–300 mg/day, total dose of 500–5600 mg). General seizures occurred in a 4 day breast-fed **infant**, following the **mother's** ingestion of 3 mg/kg/2 days.

Liver and pancreas: Fatal toxic hepatitis, preceded by rectal bleeding and scleral icterus, in a 12-year-old **male** after 2 mg/kg/day: 6 months. Hepatitis with biliverdinemia and green skin in a 46-year-old **male**, 2 weeks after taking 75 mg/day. Increased SGOT and SGPT and cholestatic jaundice. Acute pancreatitis case report in a 69-year-old **male** after 25 mg/tid/~3 months has been questioned by another M.D.

Miscellaneous: High incidence of headache (~20%) and dizziness. Somnolence, vertigo (~20%), ataxia, nausea, vomiting, depression (especially in the **elderly**), gastrointestinal bleeding, hematemesis, masking of febrile illnesses, tinnitus, diarrhea, heartburn, anorexia, melena, disorientation, nightmares, insomnia, feeling of impending death, oral ulceration, esophagitis, menstrual disturbances (~20%), hypomania, breast enlargement, gynecomastia, decreased libido, and impotence. Pseudotumor cerebri in a 10-year-old **female** with Bartter's syndrome. Successfully treated with 5 mg/kg/day, and an association with its use and stomach cancer was reported in 1985. A frequent contaminant in Chinese herbals. May color feces and urine green. Poisoning reported in a 15-year-old **female** and a 29-year-old **female** after ingesting 0.9 and 0.5 g, respectively. Acute bronchial asthma, often shortly after one dose, and frequently in ***aspirin***-sensitive **individuals**. Severe malaise, headache, nausea, enlarged lymph nodes, and local discomfort in a 60-year-old **male** who received smallpox vaccine while treated with the drug, which, apparently, altered **his** body's response to the vaccination. Peripheral neuropathy (primarily motor) reported. An 80-year-old **female** receiving 25 mg/tid within 8 h developed a paranoid psychosis with hallucinations. The causal relationship is uncertain. Use in **children** < 14 years and in the **elderly** has not been recommended. The **elderly** frequently develop confusion, cognitive impairment, delirium, and amnesia from it. Rarely, may cause swollen ankles, feet, or legs.

Renal: Acute oliguric renal failure, acute reversible deterioration of renal function with hyperkalemia, and acute interstitial and glomerulonephritis. A 61-year-old **male** developed acute oliguric renal failure; swelling of **his** face, hands, and legs; proteinuria, nausea, and azotemia after 25 mg qid/10 days. Oliguria in 14/74 **infants** (24–32 weeks) given three IV 0.2 mg/kg doses at 12 h intervals. Can cause a significant, but transient, decrease in glomerular filtration rate in premature **infants**.

Skin: Macropapular rashes, urticaria, erythematous lesions, petechiae, ecchymosis, fixed-drug eruptions, indoderma (xanthomas and nodules), melanic pigmentation, pruritus, toxic epidermal necrolysis, photosensitivity reactions, and non-thrombocytopenic purpura. Can trigger or aggravate psoriasis. **Patients** sensitive to ***tartrazine*** may have similar reactions to it.

Therapeutic serum levels are 0.3–1 mg/l; toxic at 5 mg/l.

Drug interactions: ***Aluminum*** and ***magnesium antacids***, ***aspirin***, and ***cimetidine*** decrease its absorption and effect. Acute renal failure and fatalities, due to enhanced ***methotrexate*** cell toxicity, when given with it. Decreases ***furosemide***-induced urine and ***sodium*** output, which has increased congestive heart failure symptoms in a **patient**. Severe hypertension and headache after ingestion of ***phenylpropanolamine***, an appetite suppressant, while taking it. Can stop ***bumetanide***-induced diuresis and natriuresis. Acute renal failure and hyperkalemia with ***triamterene***. Intensifies the drowsiness and confusion produced by ***haloperidol***. May increase serum concentrations of ***amikacin*** and ***gentamicin*** in **infants**. 64% increase in plasma concentration given with ***probenecid***. Significantly enhances sensitivity to ***norepinephrine's*** pressor activity. Increases serum levels of ***digoxin*** and increases ***penicillamine*** plasma levels by 34%. Causes 20–59% increase in circulating ***lithium*** concentration. Reports differ on the degree of ***warfarin*** and ***anticoagulant*** potentiation by it. Caution should be used in giving ***anticoagulants*** with an ulcerogenic drug, such as indomethacin. Causes slight increase in ***penicillin*** half-life. Inhibits antihypertensive response of the ***β-blockers atenolol, captopril, labetalol, metoprolol, pindolol***, and ***propranolol***, as well as that of ***angiotensin-converting enzyme inhibitors, thiazide diuretics, furosemide, hydralazine***, and ***vasodilators***. Did not attenuate ***timolol's*** ocular hypotensive effect. Avoid use with ***urokinase***, a clot-dissolving drug. Renal clearance is decreased by ***diflunisal***.

Administration during gestation in **mice** and **rats** has caused a high incidence of fetal resorptions and growth retardation. More toxic in **dogs** than in **humans**. Except for experimental use, it is normally contraindicated in **dogs**, as 0.5 mg/kg/day is the approximate minimum lethal dose for ulceration. Single oral doses of 10 mg/kg have been tolerated by **dogs**. Will decrease renal excretion of ***digoxin*** in **dogs**. Ulcerogenic in **cats**.

LD_{50}, **rats**: 12 mg/kg; **mouse**: 72 mg/kg.

INDOPROFEN

= *Bor-Ind* = *Flosin* = *Flosint* = *IPP* = *Isindone* = *K-4277* = *Praxis* = *Reumofene*

Analgesic, anti-inflammatory.

Untoward effects: Aplastic anemia, possibly due to an allergic reaction. Breast milk levels up to 488 ng/ml (~$\frac{1}{4}$ **maternal** plasma levels).

INDORAMIN

= *Baratol* = *Doralese* = *Vidora* = *Wy-21901* = *Wydora* = *Wypres* = *Wypresin*

Antihypertensive.

Untoward effects: Drowsiness (50%), dizziness, depression, constipation, headache, palpitations, dry mouth, increased appetite, weight gain, ejaculation failure, and 50% decrease in sperm motility.

LD_{50}, **rat**: 673 mg/kg; **mouse**: 410 mg/kg.

INFLIXIMAB

= *cA2* = *Remicade*

Monoclonal antibody. **Inhibits TNF and used IV in treating Crohn's disease and rheumatoid arthritis.**

Untoward effects: Nausea in > 15%. Facial flushing, headache, fever, sepsis, chills, dyspnea, hypotension, urticaria, and chest pain also occur. A lupus-like syndrome reported in a treated **patient**.

INGENOL

Untoward effects: A diterpene isolated from the roots and aerial parts of many ***Euphorbia*** species. Its fatty acid esters in the plant's latex are known irritants, carcinogenic and cocarcinogenic in **mice**.

INKS

Untoward effects: **Infants** have developed methemoglobinemia after wearing diapers stamped with ***aniline*** dyes. Marking and indelible inks often contain ***aniline*** dyes or toxic solvents. Inks have been a source of contact dermatitis from their ***glycol ether*** solvents. Photosensitivity from ***anthraquinone*** dyes in printers' inks used on bikinis. A bank **employee** developed recurrent eczema from use of a stamp pad using ***p-phenylenediamine*** (q.v.) as a coloring agent. ***p-Phenylenediamine*** cross-reactors, ***nigrosines*** and ***indulines***, are widely used to color quick-drying inks, waxes, and polishes. Inks have often caused discolored nails by coloring the nail plate. Colored ink formulations can include diazo compounds, some of which are known or suspect **human** carcinogens (viz. ***benzidine yellow***). In early 1980, Eastern Airlines received 190 reports of episodes of red spots on the skin of **flight attendants**. After extensive epidemiological investigation, they were discovered to be due to red ink in the fabric of demonstration life vests. Some printing inks used for coloring magazine pages used to contain ***lead*** and were an obvious hazard for **children** who chewed on them. Odors from some inks have precipitated migraines in some **people**. Dermatitis from inks is common in **printers**, especially **those** using fast-drying inks containing ***acrylates*** that are polymerized by ultraviolet light. Close proximity to the manufacturing of some printing inks was correlated to liver cancer. ***Iron gallate*** and ***ethylene glycol*** are often found in inks, and ***silver nitrate*** in indelible ink. Ink removers often contain ***oxalic*** or ***acetic acids***.

5/22 printing inks given SC to **mice** caused skin epitheliomas at the injection site, with lung metastases and lymphomas.

INOCYBE sp.

Untoward effects: Of over a hundred species of this ***mushroom***, at least 40 contain significant quantities of ***muscarine***.

The most dangerous is ***I. patouillardii***. Symptoms after ingestion are profuse sweating, salivation, and lacrimation within 15–30 min. This is followed by abdominal cramps, nausea, vomiting, diarrhea, blurred vision, miosis, confusion, giddiness, dizziness, decreased blood pressure, bradycardia, dyspnea, and coma. It has been fatal and treatment with ***atropine*** is essential. Hallucinations are possible.

I. fastigiata and ***I.geophylla*** are also commonly involved in poisonings. They are highly toxic, as well, with similar symptoms.

INOSINE

= *Atorel* = *EU 2200* = *Hypoxanthosine* = *Inosie* = *Oxiamine* = *Ribonosine* = *Riboxin* = *Trophicardyl*

Untoward effects: Abdominal pain and diarrhea necessitated discontinuance by 2/47 **patients**. Anorexia and nausea are rare.

INOSITOL NIACINATE

= *Dilcit* = *Dilexpal* = *Esantene* = *Hämovannid* = *Hexanicit* = *Hexanicotol* = *Hexopal* = *Linodil* = *Mesonex* = *Mesotal* = *Palohex* = *WIN 9154*

Peripheral vasodilator.

Untoward effects: Occasional epigastric pain or burning and nausea. Flushing, dizziness, slight swelling of face, and hypotension.

INPROQUONE

= *Bayer E-39* = *E-39* = *Iminobenzoquinone* = *RP 6870*

Cytostatic.

Untoward effects: Leukopenia and thrombocytopenia have been common. Thrombocyte counts have fallen suddenly. Fatalities include a **patient** who died from apoplexy and **another** who developed skin hemorrhages, severe jaundice, and pulmonary edema.

SC injections in **animals** may cause severe local necrosis.

INSULIN

Bovine Insulin = *Hypurin*

Recombinant Human Insulin = *Huminsulin* = *Humulin* = *Humulina Semi-Synthetic Human Insulin* = *Biohulin* = *Novolin* = *Orgasuline Insulin Glargine* = *Lantus*

Isophane Insulin = *NPH Insulin* = *Neutral Protein Hagedorn Insulin*

Porcine Insulin = *Iletin II* = *Velosulin*

Protamine Zinc Insulin = *PZI Insulin*

Zinc Insulin *(Semi-Lente or Prompt, Lente, and Ultra-Lente)*

Untoward effects: ***Allergic***: Although they are rare, local reactions are ~10 times the frequency of systemic ones. Local reactions have decreased with improvements in purity. Erythematous, indurated areas at the injection site have occurred within minutes. Early reaction usually indicates previous sensitization to ***beef*** or ***pork***. Generalized reactions including anaphylaxis, angioedema, nausea, vomiting, diarrhea, dyspnea, asthma, urticaria, and pruritus. Allergic reactions have occurred not only from insulin of **animal** origin, but also from purified **human** insulin. This can be a reaction to the ***zinc*** component in some preparations.

***Blood and cardiovascular*:** Increased heart rate, and in an 80-year-old **female**, cardiac failure was attributed to an overdose. Atherosclerosis, arrhythmias, tachycardia, Arthus phenomenon and lupus erythematous cells due to hypersensitivity, immune hemolytic anemia, palpitations, flushing, thrombocytopenia, and eosinophilia. A 40-year-old **male** died of periarteritis nodosa after 13 years of treatment.

***Eye*:** Retinopathy, increased intraocular pressure, glaucoma, retinitis proliferans after long-term use, refractive media changes, mydriasis, and diplopia.

***Fetus and neonate*:** May be associated with macrosomy of the **fetus**. Hypoglycemia with apnea, bradycardia, and poor muscle tone in **newborn** of diabetic **mother**. Possible causal effect of two spontaneous abortions, two macerated **fetuses** at term, and developmental defects in offspring of 19 **females** receiving it alone or with electroshock. **Fetal** death due to ketoacidosis from an accidental interruption of a continuous SC injection in a **diabetic** 27-year-old **female**. Of 128 consecutive pregnancies in diabetic **patients** on insulin prior to conception and throughout pregnancy, 39% of their **neonates** had significant problems (polycythemia, 10%; respiratory distress syndrome, 7.7%; hypocalcemia, 4.6%). Found in breast milk, but destroyed in the **infant's** digestive tract.

Hypoglycemia*:** See ***Drug interactions. Overdosage often causes it with secondary decrease in serum ***potassium***. Unrecognized nocturnal hypoglycemia occurs in many **patients** with poorly controlled diabetes. **Some** also showed lethargy, depression, night sweats, and morning headaches. In 1 year, 204 **patients** were admitted to an emergency room with similar symptoms, and 200 of these were identified as insulin-treated. Hypoglycemia reported from its absorption after it was sprayed on decubital ulcers of a 59-year-old **female**. Rarely fatal. Factitious hypoglycemia due to surreptitious treatment studied in 10 **patients**. **Two**, eventually, committed suicide. A 3½-year-old **female** was hospitalized on two occasions with severe hypoglycemia and transient hepatomegaly, after its malicious administration.

***Lipodystrophy*:** Both atrophy and hypertrophy can occur, especially after repeated use of the same injection site.

***Miscellaneous*:** A common cause of diffuse burning feet with erythema and pruritus. Hypokalemia, seizures, convulsions (3.9/1000), tachycardia, twitching, vertigo, abuse by **drug addicts**, postural hypotension, fever, sweating, dyspnea, tremors, cramps, sensation of hunger, fatigue, shock, enlarged parotid gland, and transient headache. Rarely, permanent central nervous system damage (e.g. hemiplegia or aphasia). Fatalities reported from using insulin pumps.

Poisoning and suicide*:** See case reports under ***Hypoglycemia. Drowsiness and coma have been common, often requiring heroic measures, including excision of injection site to remove depot material, artificial pancreas, and IV ***glucose***. **Human** insulin can be injected surreptitiously or for criminal purposes. Permanent brain damage in some **diabetics** after injections of 800–3200 Un.

Renal*:** Increased ***calcium and ***magnesium*** excretion and decreased ***potassium*** excretion.

***Skin*:** Arthus reactions. Brown, hyperkeratotic papules after repeated injections in the same area. SC atrophy, urticaria, exanthematous eruptions, necrobiosis lipoidica and scleroderma-like lesions on the back of the hands.

Drug interactions*:** ***Glucose-lowering action of ***alcohol*** added to it can induce severe hypoglycemia with tremors, coma, and irreversible neurological changes. ***Propranolol*** may potentiate its hypoglycemic effect by antagonizing the hyperglycemic effects of ***catecholamines***, and has been a cause of hypertension. ***Atenolol*** did not prolong its hypoglycemic effect, as did ***propranolol***. ***Chlorpromazine, chlorthalidone, cigarette smoking, corticosteroids, dextrothyroxine, diazoxide, epinephrine, ethacrynic acid, furosemide, interferon α-2α, isoniazid, naltrexone***, large doses of ***nicotinic acid, somatostatin, thiazide diuretics***, and ***triamterene*** antagonize its action, while ***angiotensin-converting enzyme inhibitors*** increase its effect. ***Oral contraceptives*** increase ***somatotropin*** levels and these antagonize insulin action, leading to increased blood ***glucose***. ***Anabolic steroids, aspirin, chloramphenicol, fenfluramine, furazolidone, glucagon, guaunethidine, oxphenbutazone, phenylbutazone, phenypyramidol, procarbazine, sulfonamides***, and ***tetracyclines*** can increase its hypoglycemic effect. ***Monoamine oxidase inhibitors*** (***isocarboxazid, nialamide, pargyline, phenelzine, tranylcypromine***) enhance and/or prolong its hypoglycemic effect. Oral administration of ***L-leucine*** to markedly obese **children** increases blood insulin concentration after cardiac catheterization. Following ***protamine*** administration, 4/15 **patients** maintained on ***NPH Insulin*** developed a major adverse reaction, and one died. It increases the velocity of penetration and levels of ***chlorpromazine*** in the brain. Hyperthyroidism and acidosis can increase its therapeutic requirement levels. Dramatic (38%) decrease in ***antipyrine*** half-life. ***Phenytoin*** can, on rare occasions, cause hyperglycemia and, thus, interfere with insulin effects. Insulin is adsorbed onto glass or plastic surfaces. Can make urine ***sugar***-testing by ***Clinitest***® unreliable.

TD_{LO}, ***Lente*, woman**: 2.1 g/kg; 9.5 years.

May have been used illicitly to destroy non-performing **racehorses**, in order to file fraudulent insurance claims. Indications of toxicity are frequent in the form of anxiety, restlessness, sweating, tremors, hypoglycemia, and, in severe cases, coma and death. Use in **sheep** during estrus decreases conception rate. Overdoses in **cows** lead to listlessness, hypoglycemia, unwillingness to move, or inability to rise. Teratogenic in the **mouse** (exencephaly, rib and vertebral defects, growth retardation). Great variations in insulin effects on various strains of **mice**. Limb and skeletal anomalies and cleft palate in **chicks** and **rats**. Also crosses the placenta in **monkeys**.

PZI LD_{50}, female **rat**: 330 μg/kg.

INTERFERONS

= *IFNs*

Interferon α is leukocyte-derived; β is fibroblast-derived; and γ is produced by T-lymphocytes. Used IM or SC.

Untoward effects: Tinnitus and sudden hearing loss, interference with measle virus immunizations, and is suspect in some asthma attacks. Injections to cancer **patients** often cause alopecia, bone marrow depression, erythematous reactions, sudden fever (~102°F), myalgia, chest pain, numbness, paresthesias of the hands and/or toes, loss of fingertip sensory perception, fatigue, malaise, muscle weakness, headache, nausea, vomiting, dizziness, photophobia, anorexia, weight loss, diarrhea, chills, and confusion.

α-Interferon* = *Canferon* = IFNα = *Roferon-A. Early reports from France indicated a high frequency of cardiac toxicity, not substantiated in later tests elsewhere. Some **patients** had atrial arrhythmias (paroxysmal atrial tachycardia and atrial fibrillation) and a **few** have had ventricular arrhythmias (primarily premature ventricular contractions). Acute toxicity leading to fever, chills, fatigue, myalgias, arthralgia, headache, nausea, malaise, and hypotension. Chronic toxicity leading to bone marrow depression, anemia, neutropenia, reversible leukopenia, granulocytopenia, thrombocytopenia, anorexia, weight loss, depression, confusion, facial and peripheral edema, rhabdomyolysis, renal toxicity, cardiac arrhythmias, and possible hepatic toxicity. Has induced hair color changes. Depression is often associated with decreased cognitive and motor function. Confusion and forgetfulness occur in the **elderly**.

Delirium also reported. Can exacerbate ***glucose*** intolerance. A 66-year-old **male** with diabetes mellitus developed ketoacidosis from 5 MU three times a week. Renal insufficiency and nephrotoxic syndrome in a 52-year-old **female**. Acute non-lymphoblastic leukemia treated with 3 MU/day/1 month. Hypothyroidism, alopecia, seizures, psoriasis, severe hypertriglyceridemia, loss of visual acuity, and respiratory distress. Can enhance ***melphalan*** and ***zidovudine*** toxicity and potentiate ***warfarin***. Decreases ***theophylline*** clearance. Most common adverse effects for ***α-2α*** have been fever (16/36), chills (11/36), fatigue and lethargy (7/36), and anorexia (5/36). Polyarthritis, reactivation of hepatitis B infection, enhancement of ***doxorubicin*** toxicity and reversible congestive cardiomyopathy. Nine **females** and 16 **males** experienced Raynaud's disease after ***α-2β* = *Cibian* = *IFN-α2* = *Intron A* = *Introna* = *Sch-30,500* = *Viraferon* = *YM-14,090***. Hyperthyroidism, amenorrhea, fatigue, edema, leukoderma, melanoma, lichen planus-like eruption with ***α-2β*** in addition to acute and chronic toxicity noted above. Exacerbation of multiple sclerosis and chronic hepatitis B infection, liver failure, taste alterations, vertigo, cramps, paresthesias, polyarthritis, ***insulin***-dependent diabetes mellitus, acute vanishing bile duct syndrome, inhibition of metabolism by the liver of some drugs. Myopathy, acute rhabdomyolysis, and blurred vision also reported for ***α-2β***. It decreases the clearance of ***epirubicin***. Antibodies can develop against it. May precipitate adrenal insufficiency. Nephrotic syndrome reported in a 42-year-old **female** after ***α-N1* = *Agriferon***. ***Alpha-N3* = *Alferon*** has acute and chronic toxicity similar to ***α-interferon***.

β-Interferon* = *Betaseron* = *Frone* =*IFNβ. Adverse effects include local reactions at injection sites and flu-like symptoms (including fever, myalgia, and malaise). ***β-1a* = *Avonex* = *Rebif*** may cause exacerbation of multiple sclerosis. Erythematous patches and necrosis at injection site, and severe vaginal bleeding reported with use of ***β-1b***. As with the ***alpha*** form, suicides have been associated with induced depression.

γ-Interferon* = *Actimmune. The major side-effects were dose-related pyrexia with rigors and pulmonary edema. Acute tubular necrosis in a 12-year-old **male**. Fever, headache, rash, chills, erythema or tenderness at injection site, fatigue, diarrhea, nausea and vomiting, weight loss, myalgia, anorexia, and arthralgia, in decreasing order. ***γ-1b* = *Imukin*** leading to fever, chills, and muscle pain, usually subsiding by the 9–12th week.

INTERLEUKINS

Lymphokines. Used parenterally.

***Interleukin 1* = *IL-1* = *LAF* = *TRF*. A well-known pyrogen. Hypotension and arthritis can occur from its use.**

Interleukin 2* = *Aldesleukin* = *IL-2* = *Proleukin* = *T-Cell Growth Factor* = *TCGF

Untoward effects: Occasional pulmonary edema by increasing capillary permeability; cutaneous eruptions, fluid retention, hypotension, increased body weight, renal insufficiency, ischemia, cardiac arrhythmias, myocardial infarction, myocarditis, thrombocytopenia, disorientation, hallucinations, coma, hypothyroidism and hyperthyroidism. Severe malaise, fever, nausea, vomiting, diarrhea, anemia, and hyperbilirubinemia are common. Fatty infiltration of the liver and hypersensitivity reactions reported.

Interleukin 11 = rhIL 11 = Neumega = Oprelvekin = YM 294

Use: Recombinant material increases proliferation of stem cells and platelets after SC use.

Untoward effects: Some edema due to ***sodium*** retention in 60% of **patients**, increased plasma volume with a 10% decrease in hemoglobin and hematocrit, atrial fibrillation or flutter, tachycardia, conjunctival injection, and dyspnea. In **children**, ~45–50% develop tachycardia and conjunctival injection. Experimental **animals** develop thickening of their bone growth plates.

Interleukin 12 = IL-12. Use may be associated with chronic inflammatory bowel disease, hypotension, leakage of blood out of capillaries, and other shock-like symptoms.

INTERMEDINE

= *β-Melanocyte-stimulating Hormone = Melanophore-Stimulating Hormone = MSH*

Untoward effects: Transient abdominal cramps and diarrhea. Increased skin pigmentation, especially on arms and face.

INULA sp.

Untoward effects: **Cattle** fed hay contaminated with flowering ***I. conyza*** had digestive upsets, weakness, hemolysis, and death. Post-mortem showed parenchymatous degeneration of liver, heart, and kidneys, with centrolobular necrosis of the liver. Also lethal to **pigs**.

I. graveolens = Khakiweed = Stinkwort causes liver toxicity in **sheep** and **cattle** in Australia. **Humans** have developed dermatitis from handling it.

I. helenium = Elecampane. Steam distillation of it produces ***Elecampane Oil = Alantroot Oil = Inula Oil*** used as a fragrance in soaps, detergents, cosmetic creams and lotions, perfumes, and in ***alcoholic*** beverages. No irritation after 48 h closed-patch tests in **humans** with 4% in petrolatum. Severe allergic reactions in 23/25 **volunteers** after second induction application. Cross-sensitization with ***costus root oil***.

INVERT SUGAR

= *Calorose = Insubeta = Invesol = Nulomoline = Travert*

Untoward effects: IV infusion caused thrombophlebitis (~28%), due to high acidity (pH 3–4.5) and acute intravascular hemolysis in two anesthetized hypothermic **patients**.

IOBENZAMIC ACID

= *Bilibyk = Osbil = Osvil = ST 5066*

***Diagnostic aid.* In cholecystography.**

Untoward effects: Cramps (13/78), gas (10/78), nausea (23/78), vomiting (5/78), bowel movements (27/78), diarrhea (5/78), increased urinary frequency (8/78), nocturia (14/78), headache (19/78), pruritus (8/78), and burning in chest or throat (9/78) after 3 g orally. 40/78 had > 25% increase in serum ***bilirubin***; 4/78 had > 25% increase in alkaline phosphatase, 15/78 in SGPT, and 7/78 in blood urea nitrogen. Pseudoproteinuria from it or its degradation products that causes a precipitate in acid urine.

LD_{50}, **mice**: 2.87 g/kg.

IOCARMATE MEGLUMINE

= *Dimeray = Dimer X = LM 280 = Meglumine Iocarmate*

Radiopaque diagnostic aid.

Untoward effects: Vasovagal attacks with hypotension, thready pulse, sweating, nausea, vomiting, and headache in 10%, after intrathecal use. Use in cerebrospinal fluid associated with hip dislocation due to severe clonic spasms. Adhesive arachnoiditis has been frequent.

IOCETAMIC ACID

= *Cholebrine = Cholimil = DRC 1201 = MP 620*

Radiopaque diagnostic agent.

Untoward effects: Severe skin reactions in 0.18% (4/2200). Severe thrombocytopenia in a 32-year-old **male**, 2 h after 3 g orally.

LD_{50}, **rats**: 2.2 g/kg; IV, **rat**: 700 mg/kg; **mouse**: 410 mg/kg.

IOCHROMA sp.

Untoward effects: An hallucinogenic and intoxicating ***narcotic*** used by **natives** in South America.

IODAMIDE

= *Isteropac E.R. = Jodamid = Jodomiron = Opacist E.R. = Renovue-65 = Renovue-DIP = Uromiro*

***Radiopaque diagnostic aid.* Injected for urography, angiography, and myography.**

Untoward effects: Nausea, vomiting, exanthema, urticaria, dizziness, and feeling of warmth.

IV, LD_{50}, **rats**: 11.4 g/kg; **mice**: 9 g/kg; **rabbits**: 13.2 g/kg.

IODINATED I-131 SERUM ALBUMIN, AGGREGATED

= *Iodalbin*

Untoward effects: Acute pulmonary hypertension and embolism, faintness, cyanosis, agitation, diaphoresis, dyspnea, tachypnea, hypotension, engorged neck veins, aseptic meningitis, and even death reported.

IODINATED GLYCEROL

= *Iophen* = *Organidin*

Expectorant.

Untoward effects: Hypothyroidism (dyspnea, sleeplessness, decreased appetite, and cold intolerance). Chronic ***iodine*** poisoning in a 56-year-old **female** with chronic obstructive pulmonary disease receiving 60 mg qid/20 years. **She** developed hypothyroidism, migraines, gastrointestinal upsets, skin rashes, and peeling blisters on the palms of **her** hands.

IODINE

Untoward effects: ***Blood***: Thrombocytopenia, purpura; may decrease hemoglobin and leukocytes.

Clinical tests: Dyes such as ***erythrosine*** (***FD & C Red 3***), in tablets, capsules, foods, and cosmetics, contain iodine and may increase protein-bound iodine test results of thyroid function. It is a photosensitizer and has caused fixed-drug eruptions, and, possibly, an allergic bronchoconstriction or dyspnea in **asthmatics**. The same protein-bound iodine influence can occur from iodides used in parasitic diseases, ***amiodarone, contrast medias, expectorants***, skin antiseptics, and iodide shampoos. Organic-bound iodine may alter protein-bound iodine results for 4–6 weeks.

Eye: Keratitis, hypopyon, iritis, retinal hemorrhages, and bullous corneal lesions. Even vapors can be a severe irritant leading to conjunctivitis, lacrimation, and eyelid swelling.

Fetus and neonate: Use in the treatment of asthma in pregnant **women** caused respiratory distress, cyanosis, and fatal goiters (usually by suffocation) in **babies**, thyroid hyperplasia, subarachnoid hemorrhage, mental retardation, and cretinism. Iodine passes into breast milk and can alter an **infant's** thyroid function. Neonatal goiter can make labor difficult. **Maternal** administration of ***radioactive iodine*** can lead to **fetal** hypothyroidism. Nursing should be discontinued for 48 h after ^{125}***I*** and 36 h after a test dose of ^{131}***I***, and 2–3 weeks after a treatment with ^{131}***I***. ^{125}***I*** leads to slight dysphagia and thyroid tenderness, high hypothyroid incidence; ~5% in **mother's** breast milk, half-life of 60 days so may be in saliva for > 1 year, and possibly associated with laryngeal, tracheal, and thyroid malignancies, as well as leukemia. ^{131}***I*** leads to atrial fibrillation, cardiomyopathy, cerebral embolism, congenital iodide goiter in **newborn** of treated 42-year-old mother, cataracts in **fetus**, cretinism, exophthalmos, and arrested brain development in the **fetus**. Hypoparathyroidism and hypothyroidism reported. Acute toxic psychosis and vocal cord paresis has occurred. **Victims** of accidental fallout from the 1954 Bikini H-bomb test suffered thyroid damage from ingested ^{131}***I***. Myxedema, azoospermia, humeroscapula periarthritis, scleroderma parotitis and exacerbation of thyrotoxicosis (palpitation, headache, dyspnea, sweating, insomnia, painful thyroid), and transient hypocalcemia from ^{131}***I***. A breast-fed **infant** – 7½ months – had grossly elevated serum and urine iodine levels, and an iodine body odor after the **mother** used a ***povidone-iodine*** (***Betadine***®) vaginal gel once daily/6 days.

Goiter and glandular swelling: Especially with self-medication. "Iodide mumps" or enlarged submaxillary and parotid glands from repeated exposures to iodinated ***contrast media*** and other iodine-containing drugs or excessive topical applications. Excessive use of iodized ***salt*** in Tasmania was associated with a mild epidemic of thyrotoxicosis. **People** in certain northern Japanese coastal areas have enlarged thyroid glands, resulting from excessive intake of iodine-rich ***kelp*** seaweed and ***fish***. Usually, > 300 μg/day required for goiter development.

Hypersensitivity: Allergic reactions, dyspnea, and shock of a potentially fatal nature have occurred, especially from iodine-containing ***contrast media***.

Miscellaneous: Abdominal pain, nausea, vomiting, diarrhea (occasionally hemorrhagic), dehydration, laryngeal edema; yellow-brown staining after application to skin, mouth, and teeth; lip swelling, pulmonary edema, rhinitis; throat irritation and headaches from vapors; metallic taste, sore gums, burning sensation in mouth, sleeplessness, impotence, and anuria. Rarely, testicular wasting reported after long-continued use. A 50-year-old **male** with a history of hyperthyroidism developed signs of parkinsonism on three occasions after extended high iodine dosage. 2–3 g is usually fatal. National Institute for Occupational Safety & Health/OSHA exposure limit of 0.1 ppm. Immediate danger to life or health of 2 ppm.

Poisoning and suicide: Frequent premature ventricular contraction, ventricular bigeminy, ectopic atrial beats, and tachycardia from inadvertent consumption of ~15 g in a ***potassium iodide*** solution. Facial, neck, and mouth swelling noted. Death has been reported in ½ h–52 days after consumption. ***Tincture of iodine*** is in many homes. Most of its ill-effects are gastrointestinal (abdominal pain, vomiting, diarrhea, and prostration with depression and collapse). Systemic effects are minimal, as it is absorbed as relatively non-toxic iodide. **Some** may die in 24–48 h from shock, circulatory collapse, and pneumonia after gastrointestinal corrosion. Ingestion of as little as 4 ml of the (7%) tincture has been fatal, and recovery has followed the ingestion of 30 ml. Extensive topical application has also been fatal.

LD_{LO}, **human**: 5 mg/kg.

Drug interactions: ***p-Aminosalicylic acid*** decreases its accumulation in the thyroid. May potentiate ***anticoagulant*** action. ***Radishes*** interfere with use of iodine and are goitrogenic. May have synergistic hypothyroid effect with ***lithium carbonate***.

Oral or parenteral iodides and iodine complexes may increase the narcosis produced by ***phenobarbital*** in **rats**. Mammary excretion has been documented in **cows**, **dogs**, **goats**, **guinea pigs**, **humans**, **rabbits**, **rats**, and **ewes**. **Dairy cattle** fed 50, 250, and 1250 mg daily showed decreased phagocytosis, and the highest level reduced antibody response to brucellosis or leptospirosis. Excessive iodine, apparently, causes higher susceptibility to ***selenium*** poisoning. Excessive intake may be associated with respiratory distress and increased nasal discharge in **cows**. Dietary levels of 800 mg/day in growing **pigs** decrease feed intake and rate of gain, decrease hemoglobin, and cause eye lesions. Clinical signs of poisoning in **cats** may include nausea, vomiting, hypothermia, muscle cramps, cardiac damage, coma, and death. ^{131}***I*** can destroy the thyroid of **cockerels**. Thyroid gland ^{131}***I*** levels in Australian and New Zealand **sheep** rose sharply 2–4 weeks after nuclear bomb tests in the Pacific, peaking at 6–18 weeks.

IODIZED OILS

Use: In lymphography, bronchography, and myelography.

Untoward effects: A number of forms exist. The most serious ill-effect after injection has been fatal lipid embolism to the brain.

IODOCHLORHYDROXY-QUINOLINE

= *Amebil* = *Alchloquin* = *Amoenol* = *Bactol* = *Barquinol* = *Budoform* = *Chinoform* = *Clioquinol* = *Colepur* = *Eczecidin* = *Enteroquinol* = *Entero-Septol* = *Entero-Vioform* = *Enterozol* = *Entrokin* = *Entrokinol* = *Hi-Enterol* = *Iodochlorhydroxyquin* = *Iodochlorohydroxyquinoline* = *Iodochloroxyquinoline* = *Iodoenterol* = *Nioform* = *Quinambicide* = *Quiniodochlor* = *Quinoform* = *Rheaform* = *Rometin* = *Trichomonex* = *Vioform* = *Vioformio*

Antifungal, antibacterial, protozoacidal agent. **Contains 41.5%** ***iodine***.

Untoward effects: ***Clinical tests***: ***Iodine*** content increases protein-bound iodine test results.

Central nervous system: Partially reversible amnesia in a 24-year-old **female** after ingesting 4 g for treating gastroenteritis. A 24-year-old **male** ingested 30–150 mg tablets, leading to acute cerebral symptoms with confusional state, hallucinations, precoma, and then retrograde amnesia for 4–5 days. Optic neuritis, visual acuity failure, and bilateral optic atrophy, independently or associated with subacute myelo-optic-neuropathy. In the 1950s, epidemics of subacute myelo-optic-neuropathy occurred in Japan on normal or high doses of the drug in ~10% of **patients** using it for > 4 weeks. It was sold over-the-counter for diarrhea and amebiasis treatment. 11,000 cases in Japan between 1955 and 1970, after which it was removed from the Japanese market. Pharmacogenetics and high dosage may have added to the toxic nature of the drug in Japan. During the same period, it was a prescription item in the U.S., yet several cases of optic atrophy and damage to spinal cord nerves occurred. Abdominal discomfort, gastrointestinal upsets, paresthesias in the legs, and even paraplegia in some **patients**, loss of visual acuity (sometimes becoming irreversible blindness), dysesthesias, green urine, green feces, and green tongue "fur" are part of subacute myelo-optic-neuropathy symptoms. Occasional fatal. Pyrexia, severe toxemia, occasional transient diarrhea, faintness and dizziness (3%), headache (2%), myxedema, anal pruritus, malaise, nausea, abdominal distension, decreased uptake of ***radiopharmaceuticals*** used in thyroid imaging, and brownish coloring of the nail plate. Occasionally allergic dermatitis and urticaria produced by oral doses, as well as from topical use. May cross-react with ***chlorquinaldol*** and ***halquinol***.

Subacute myelo-optic-neuropathy also occurred in **animals**. The blood concentrations in subacute myelo-optic-neuropathy **patients** and **dogs** showing neurological problems (paresis of hind limbs) were similar. The same syndrome occurs in **cats** on oral dosage. **Monkeys**, **poultry**, and **mice** evidence mild symptoms from peripheral and spinal nerve damage. A single dose of 250 mg/kg is lethal to **rabbits** and less is hepatotoxic. Avoid its use in the eye or in rare cases of hypersensitivity. Overdoses cause central nervous system stimulation and neuron necrosis in **mice** and **dogs**. Strain differences have been noted. May be *contraindicated* as sole therapy in *Shigella* and *Salmonella* dysenteries and, although effective on occasion against *E. coli*, it has permitted their overgrowth in experimental **rats**. Crosses placental barrier in

mouse trials. Topical application of 1% solution or suspension caused hair cell loss and inflammatory response with thick, mucoid secretion of middle ear mucosa in **guinea pigs**.

LD_{50}, **guinea pig**: 175 mg/kg; **kittens**: 400 mg/kg.

IODOFENPHOS

= *Alfacron* = *C 9491* = *Ciba 9491* = *Jodenphos* = *Jodfenphos* = *Nivano 1-N* = *Nuvanol N*

Organophosphate, anticholinesterase, insecticide.

Untoward effects: Reported in **human** poisonings.

LD_{50}, 2.1 g/kg; **mouse** and **dog**: 3 g/kg; **rabbit**: 2 g/kg.

IODOFORM

= *Triiodomethane*

Contains 96.7% *iodine* (q.v.).

Untoward effects: Allergic reactions with erythematous rashes reported. Disagreeable odor. Eye and skin irritant. Light-headedness, dizziness, nausea, diuresis, incoordination, headache, confusion, central nervous system depression, dyspnea, hallucinations, and visual disturbances. May cause liver, kidney, and cardiac (sensitizes the myocardium to ***epinephrine***) damage, especially after internal use or ingestion. Occasionally dermatitis. Fatal poisoning is usually preceded by headache, somnolence, delirium; and rapid, feeble pulse. Fatal result after local application of 4 g powder on a **patient**. Reactions from ***BIPP*** (***bismuth*** and ***iodoform***) are usually due to precipitation of metallic ***bismuth*** in wounds.

Milk readily absorbs its characteristic odor. Avoid its use on or in milking **animals** or **animals** stabled in milking areas. Although relatively insoluble in water, it appears to be readily absorbed from treated raw wounds, mucous membranes, or from iodoform-treated gauze. It may have a toxicity of its own, distinct from iodism. Mild ***acetic acid*** or ***vinegar*** will help remove its odor from the **user's** hands. Its neurological toxicity after daily dosing in **mongrel dogs** is 3–6 times that for purebred **beagle dogs** (350–450 mg/kg).

LD_{50}, **dog**: 1 g/kg.

IODOPHORS

***Iodine antiseptic.* The term does not represent a single compound, but rather a class of compounds known as "*iodine*-bearers", "*iodine*-carriers", or "iodophors". They result from the combination of *iodine* with a solubilizing or complexing agent, which slowly liberates free *iodine* when diluted with water. The term was first used by Shelanski, when he complexed *iodine* with *polyvinylpyrrolidone* (q.v.) and other surfactants. Commercially available forms are discussed individually in this text, and some in solution solubilize up to almost 30% of their weight as *iodine*, but subsequently only release ("available *iodine*") 70–80% of this. Various *acids*, such as *phosphoric*, *hydrochloric*, and some complex organic ones, are added primarily as anticorrosive agents and for certain pH buffering. A given single iodophor is often marketed with varying percentages of these acids for different purposes. NOT necessarily interchangeable for all medical uses. Thus, the term "*Tamed Iodine*", often used with an iodophor complex, may not be so "tame", as far as tissues are concerned when incorporated in certain solutions, yet very "tame" in others. Except for pH differences, skin irritation effects, and detergency, all iodophors are, essentially, equivalent bactericides in terms of their "available *iodine*" (amounts in aqueous solutions). Their antibacterial, antiviral, and sporicidal effects are due to release of diatomic *iodine*.**

Untoward effects: Irritation or sensitivity reactions reported in **man**. Bullous pemphigoid reported in a 74-year-old **male** after its application during an orthopedic procedure. Although often useful, topical application on burns or wounds of > 20% of the body's surface, or in the presence of renal failure, may cause acidosis and renal damage.

IODOPYRACET

= *Cardiotrast* = *Diatrast* = *Diodone* = *Diodrast* = *Iopyracil* = *Neo-Methiodol* = *Neo-Skiodan* = *Neo-Tenebryl* = *Nosydrast* = *Oparenol* = *Per-Abrodil* = *Per-Radiographol* = *Pyelosil* = *Pylumbrin* = *RP 3203* = *Savac* = *Umbradil* = *Uriodone* = *Vasiodone* = *Xumbradil*

Radiopaque diagnostic agent.

Untoward effects: Occasionally deaths reported from anaphylaxis. Cerebral infarction in a 49-year-old **female** undergoing cerebral angiography (50 ml of 35% solution); **she** died 8 days later from bronchopneumonia. Has caused focal irritation of pulmonary parenchyma at site of aspiration after esophagography. Severe allergic reaction in a 56-year-old **female** after 2 ml 35% solution IV, despite no reaction to test dose. Delayed clearing from lungs reported. Increased serum urates. Pyelography in myelomas is hazardous. Avoid extravasation. Systemic reactions commonly include flushing and feelings of warmth. Nausea, vomiting, erythematous eruptions, urticaria, dyspnea, lacrimation, salivation, coughing paroxysms, choking sensations, and cyanosis are less frequent. Occasionally hypotension in **man** and **animals**, lasting about 2 h. Spuriously high protein-bound iodine readings.

IV, LD_{50}, **rat**: 6 g/kg; **mouse**: 6.4 g/kg; **cat**: 2.8 g/kg; **rabbit**: 4.7 g; IV, LD_{LO}, **dog**: 2 g/kg.

IODOQUINOL

= *Diiodohydroxyquin* = *Diiodohydroxyquinoline* = *Dinoleine* = *Diodoquin* = *Diodoxylin* = *Diolene* = *Direxiode* = *Disoquin* = *Dyodin* = *Embequin* = *Enterosept* = *Floraquin* = *Ioquin* = *Rafamebin* = *Searlequin* = *SS-578* = *Stanquinate* = *Yodoxin* = *Zoaquin*

Antimicrobial, amebicide. Iodine **~64%.**

Untoward effects: Irritation, sensitivity reactions, rash, acne, urticaria, pruritus, alopecia, coryza, sore eyes, thyroid hypertrophy, nausea, vomiting, abdominal cramps, diarrhea, anal pruritus, decreased ***radiopharmaceutical*** uptake in thyroid imaging, and protein-bound iodine values above normal. Optic neuritis and atrophy, vision loss, peripheral neuropathy, as well as the entire subacute myelo-optic-neuropathy syndrome are infrequent and may occur especially after long-term usage (3 or more weeks) at high doses. Has been used in hair shampoos for **human** or veterinary use, and may turn white hair yellow-brown. Contraindicated in **patients** with liver disease.

IOGLYCAMIC ACID

= *Biligram* = *Bilivistan* = *SH 419*

Radiopaque diagnostic aid. **In cholecystography.**

Untoward effects: 20–30% had intense nausea, flushing, subconjunctival edema, and abdominal cramps. Intense headaches in 10% of **patients**. Increases SGOT. Intravascular precipitation after IV use in a **patient**, due to a reaction with **his** monoclonal macroglobulin, causing sudden death.

IOHEXOL

= *Compd 545* = *Exypaque* = *Omnipaque* = *Win-39424*

Radiopaque diagnostic aid. Iodine **content of 46.36%.**

Untoward effects: Prolonged paraplegia reported after cervical myelography with 15 ml intrathecally in a 51-year-old **female**. Nephrotoxic with decreased renal function. Allergic reaction with extensive erythematous rash after IV in a 27-year-old **male**. On a weight adjusted basis, breast-fed **infants** of treated **mothers** receive < 1% of a standard dose. Aseptic meningitis in a 74-year-old **female** 18 h after a 12 ml lumbar myelography leading to headache, fever, sweating, shivering, neck stiffness, nausea, and mild confusion.

IV, LD_{50}, **rats**: 6.9 g/kg; **mice**: 11 g/kg.

IONONE

= *Irisone*

A fragrance used for soaps, detergents, cosmetic creams and lotions, perfumes, and foods. Both α and β isomers are approved, with α having a slightly sweeter, violet odor. The first synthesis of *vitamin A* was from the β-form.

Untoward effects: Full-strength as a 24 h patch test caused no irritation reactions in 11 **volunteers**. No sensitization reactions on 25 **volunteers** with 8% in petrolatum. May cause allergic reactions.

Fed to **rats** at 1000, 2500, and 10,000 ppm in their diet/17 weeks led to slight microscopic liver pathology with slight parenchymal cell swelling at the lowest level, to moderate swelling at the highest. **Rats** fed 13–115 mg of α and β/5–9 days showed fatty infiltration of liver parenchymal cells. Pure α topically is severely irritating to **rabbit** skin, moderately irritating on **guinea pigs**, and without ill-effects on **miniature swine**.

LD_{50}, **rats**: 4.59 g/kg; IP, **mice**: 1.8 g/kg.

IOPAMIDOL

= *B 15000* = *Iomapidol* = *Iopamiro* = *Iopamiron* = *Isovue* = *Jopamiro* = *Niopam* = *Solutrast* = *SQ-13396*

Radiopaque diagnostic aid. Iodine **content of 49%.**

Untoward effects: Generalized tonic-clonic seizures in 3/629 **patients** after myelography. Lateral rectus palsy in a 39-year-old **female** and a 46-year-old **female** after myelography, requiring 6 months to resolve. Major acute cardiovascular complications of 0.2% in 1144 **patients**. Very prolonged paraplegia in a 57-year-old **female** after 14 ml intrathecally. Aseptic meningoencephalitis and decreased renal function also reported. Incidence in Japan of adverse effects is < 30%.

IV, LD_{50}, **rat**: 6.8 g/kg; **mouse**: 10.8 g/kg; **rabbit**: 4.7 g/kg; **dog**: 8.4 g/kg.

IOPANOIC ACID

= *Cistobil* = *Colepax* = *Jopagnost* = *Telepaque* = *Teletrast*

Radiopaque diagnostic aid. **For cholecystography. Oral use. *Iodine* content of 66.7%.**

Untoward effects: Nausea, diarrhea, dysuria, and acute renal failure with large doses. After 3 g orally, abdominal cramps in 14/80, vomiting in 4/80, diarrhea in 31/80, frequent urination in 9/80, nocturia in 13/80, headache in 7/80, pruritus in 10/80, and burning in the chest or throat in 10/80. Orthostatic hypotension has been rare. Abnormal sulfobromophthalein test readings and increased urine ***albumin*** readings, apathy, skin rashes, urticaria, and strong uricosuric effect, peaking in 7 h and lasting up to 6 days. Cases of severe thrombocytopenia after ingestion of 3 g. Liver hypersensitivity to it with increased alkaline phosphatase. Peak ***iodine*** levels in breast milk after 5–19 h with 3–11 mg/**infant's** feeding (7% of a standard **maternal** dose).

LD_{50}, **rat**: 2.87 g/kg; **mouse**: 1.54 g/kg; **rabbit**: 1.38 g/kg.

IOPHENDYLATE

= *Ethiodan* = *Mulsopaque* = *Myodil* = *Neurotrast* = *Pantopaque*

***Radiopaque diagnostic aid. Iodine* content 30.48%.**

Untoward effects: After myelography, marked increase in protein-bound iodine, case reports of eosinophilic meningitis, pulmonary embolism after venous intravasation during myelography, mild lymphocytosis (lasting nearly 3 months), chronic severe urticaria and episodes of anaphylaxis in a 55-year-old **male**, and arachnoiditis in **humans** and **animals**. When severe, it can cause hydrocephalus, cranial nerve palsies, and focal epileptic seizures. Dry cough, slight circulatory collapse with skin pallor, hypotension, eosinophilic pneumonia, and hyperthyroidism have also been noted.

IP, LD_{50}, **rats**: 1.13 g/kg; **mice**: 4.6 g/kg.

IOPHENOXIC ACID

= *Teridax*

***Radiopaque diagnostic aid.* In cholecystography. Contains 66.57% *iodine*.**

Untoward effects: Can cause hypothyroidism and increased protein-bound iodine (up to 10 years after administration). Transplacental passage increased protein-bound iodine in an 8-year-old **female** 7 months after the **mother** received it during pregnancy. This prolonged retention led to its discontinued use.

IV, LD_{50}, **mice**: 374 mg/kg.

IOTHALMATE SODIUM

= *Angio-Conray* = *Angio-Contrix "48"* = *Conray-400* = *Medio-Contrix "38"*

***Radiopaque diagnostic aid.* Contains ~62% *iodine*. Parenteral use in angiography and renal arteriography.**

Untoward effects: Flushing; transient bradycardia, tachycardia, and seizures; laryngeal stridor, cyanosis, headache, ventricular fibrillation, chills, fever (up to 110°F), vomiting, sneezing, hiccups, hypotension, urticaria, coughing, dyspnea, unpleasant taste, renal failure, hyperthyroidism, and electrocardiogram changes as in myocardial infarction. Sudden hemiplegia and death from aortography with unintentional catheter passage into a brachiocephalic artery. Respiratory arrest and death in a 58-year-old **male** 48 h after injection. High ***protein-bound iodine*** readings. Severe and even fatal parenchymal necrosis, fibrosis, and calcification have occurred in the kidneys of **dogs** receiving 1 ml/kg of undiluted 80% material in their renal arteries.

IPECAC

= *Brazil Root* = *Hippo* = *Ipecacuanha*

***Antitussive, emetic, ruminatoric, protozoastat.* Brazilian material contains about 2% alkaloids, 85–95% of which consists of *emetine* and some *cephaeline*. The "syrup" is commonly used and is made from the fluid extract; the syrup is, in a sense, a 7% strength fluid extract. It is one of the oldest medicinals known.**

Untoward effects: Can cause violent emesis, fixed-drug eruptions, muscle tenderness, transaminasemia, toxic epidermal necrolysis, hypoalbuminemia, leukocytosis, granulocytosis, neutropenia, diarrhea, drowsiness, and cardiotoxicity. Allergic reactions and asthma from inhalation of the dried pulverized root once an occupational hazard for **pharmacists** and **doctors**. Poisoning occurred in a 3-year-old **male** after taking one tablet powdered ipecac that a **pharmacist** dispensed instead of syrup. Münchausen syndrome by proxy due to **parents** inducing intractable vomiting in a 3–4-month-old **child**. Confirmed by finding ***emetine*** in **child's** serum and urine. Death has also followed the inadvertent use of the fluid extract, which is 14 times the strength of the syrup. Use mask, goggles, and gloves when handling ipecac root.

Syrup of ipecac: Adsorbed and inactivated by ***charcoal***. A **child** given 3–6 times the correct dose, as treatment for a ***phenothiazine*** antiemetic overdose, developed a severe toxicity, but survived. Mallory–Weiss syndrome and pneumomediastinum has followed excessive vomiting. Aspiration can occur. Cardiotoxicity and peripheral myopathy have occurred after repeated use by anorexic and bulimic **patients**. Intentional poisoning and cardiomyopathy in Münchausen syndrome by proxy induced by their **mothers** in several **children**. Avoid its use after ingestion of caustics, petroleum products, and ***strychnine*** and in **patients** with central nervous system depression or seizures. In 1970, it was causing dizzy spells, due to contamination with ***ephedrine***. At other times, some on the market was mislabeled, due to the fact that other products were in the container.

LD_{LO}, syrup, **human**: 70 mg/kg.

In **animals**, overdoses can cause weakness, violent vomiting, and diarrhea. The powder is a strong irritant to the nostrils and skin.

IPIL-IPIL

A leaf meal of a leguminous tree in the tropics.

Untoward effects: Its ***mimosine*** (q.v.) content inhibits egg production by **hens**.

IPODATE

Sodium salt = *Biloptin* = *Oragrafin*

Ethyl ester = *Myelographin* = *SH 617L*

***Radiopaque diagnostic aid.* In cholecystography. Calcium and sodium salts, and an ethyl ester are used.**

Untoward effects: Hyperthyroidism, nausea, vomiting, abdominal pain, diarrhea, urticaria, acute renal failure with acute interstitial nephritis, decreased serum urate, dysuria, oliguria, and increased ***uric acid*** excretion. Hypersensitivity reactions reported.

LD_{50}, **rat**: 2.5 g/kg; **mouse**: 855 mg/kg.

IPOMEAMARONE

A toxin in *sweet potatoes* (q.v.) infected with *Ceratocystis fimbriata* (black rot).

Untoward effects: Causes liver, spleen, and kidney toxicity, as well as lung edema and death within 8–24 h in **mice**. Has caused the death of **cattle** and, probably, other **livestock**.

IPOMEA sp.

See *Sweet Potatoes* for *I. batatas* and *Morning Glories* for *I. hederacea*, *I. purpura*, and *I. violacea*.

Untoward effects: ***I. asarifolia*** = ***Salsa*** and ***I. fistulosa*** = ***Canudo*** in South America causes anorexia and weakness in **cattle**, **sheep**, and **goats**.

I. cairica in Nigeria is an abortifacient in **mice** and **rats**. Purgation, vaginal bleeding, and retained placentas in **rats**.

I. calobra = ***Weir Vine*** causes incoordination, stiffness, polyuria, and loss of vision in Australian **sheep**.

I. carnea ingestion by **sheep** and **zebu calves** led to staggering, hind limb weakness, anemia, and leukopenia, as well as anemia, tremors, ataxia, and death of **goats** in the Sudan.

I. digitata extracts led to vasoconstriction, bronchoconstriction, and oxytocic and peristaltic effects in **cats** and **dogs**.

I. muelleri has caused locomotion difficulties, profound weakness, leukopenia, behavioral disturbances of the ***LSD***-type, and death in **sheep** in Western Australia.

IPRATROPIUM BROMIDE

= *Atem* = *Atrovent* = *Bitrop* = *Itrop* = *Narilet* = *Rinatec* = *Sch-1000*

Bronchodilator, antiarrhythmic, anticholinergic.

Untoward effects: Nebulized intranasal use has been associated with transient mouth dryness and scratching of the throat in ~15%, bitter taste in 20–30%, nasal dryness in 5%, and mild nose-bleeds in ~10%. Oral inhalation can cause asthma, buccal ulceration, tremors, dry mouth, pharyngeal irritation, palpitations, tachycardia, urinary retention, constipation, anisocoria, and blurred vision. Can increase intraocular pressure in glaucoma. Bacterial contamination of the aerosol home nebulizers is common.

LD_{50}, **rats**: 1.67 g/kg; male **mice**: 1 g/kg; female **mice**: 1.08 g/kg; **dog**: 1.3 g/kg.

IPRINDOLE

= *Galatur* = *Prondol* = *Tertran* = *Wy-3263*

Tricyclic antidepressant.

Untoward effects: Dry mouth, constipation, blurred vision, urinary retention, liver toxicity with jaundice, malaise, sweating, nausea and vomiting, photophobia, tinnitus, and postural dizziness.

Cardiac arrhythmias noted in **animal** experiments.

LD_{50}, **rat**: 484 mg/kg; **mouse**: 775 mg/kg.

IPRONIAZID

= *Iprazid* = *Ipronid* = *Marsilid*

Antidepressant, monoamine oxidase inhibitor.

Untoward effects: Ankle edema, urination difficulty, nocturia, polyuria, impotence, orthostatic hypotension (~50% on 100–150 mg/day), viral-like hepatitis, jaundice (8%), bone marrow suppression, and vasculitis. Nausea, vomiting, coma, and death in a 39-year-old **female** after receiving 6 g over a 63 day period. Post-mortem indicated severe necrotic hepatitis, which, eventually, led to the end of its U.S. marketing. It decreases the liver metabolism of many drugs, causing increased duration of sleeping time and paralysis, excitation, giddiness, convulsions, manic psychosis in 5/34, hyper- or hypotension, and dyspnea and hallucinations with ***meperidine***. In the early 1950s, it was given to hospitalized tuberculosis **patients** (previously lethargic) in whom it induced a state of euphoria to, literally, dance in the halls of the hospital. This led to a new era of medical research, when the effect was traced to its ***monoamine oxidase inhibitor*** properties.

High dosage to **mice** led to muscle twitching, convulsions, respiratory failure, and death. Fed to **rats** it caused decreased hemoglobin, engorged spleen, red blood cell destruction, and bone marrow hyperplasia. **Dogs** evidenced similar effects. Orally, it is also a tumorigenic ***hydrazine*** compound for **mouse** lungs.

LD_{50}, **rat**: 365 mg/kg; **mouse**: 681 mg/kg; **dog**: 95 mg/kg; **rabbit**: 150 mg/kg.

IRIDIUM

Untoward effects: Mucosal and/or bone necrosis in **patients** with carcinomas of the floor of the mouth treated with ***iridium***-192 implants. In Los Angeles, an ***iridium***-192 source became detached from its shielded camera and was left on the floor. A **worker**, unfamiliar with the danger, put it in the hip pocket of **his** coveralls for about 45 min. Five other **workers** handled it, and five were exposed by standing around it. 6 h after exposure, the **man** who had put it in **his** pocket developed a burning sensation and erythema over **his** right buttock. Eventually, the area ulcerated.

Chloride, IV, LD_{LO}, **dog**: 778 mg/kg.

IRIGENIN

Untoward effects: A ***glucoside***, found in all parts of ***Iris*** (q.v.), that causes acute gastrointestinal distress.

IRINOTECAN

= Campto = Camptosar = CPT 11 = DQ 2,805

***Antineoplastic. A camptothecin* derivative for infusions.**

Untoward effects: Diarrhea, early, severe, and prolonged; nausea, vomiting, abdominal pain, bone marrow depression, diaphoresis, flushing, stomatitis, anorexia, asthenia, and alopecia. Several fatalities from accidental overdoses when the entire contents of a 5 ml (100 mg) vial was given instead of 1 ml or 20 mg. Soft tissue irritation and toxicity from extravasation.

IRIS

= Flag

Untoward effects: The bulbs and leaves have caused severe gastrointestinal irritation with nausea, vomiting, diarrhea, dyspnea, and hepatitis. ***Iridin*** in its rhizomes and juices has been implicated. The plant juices have caused a dermatitis in sensitive **people**.

Ingestion by **cattle** leads to blistering and irritation of mouth, abdominal pains, diarrhea, salivation, recumbency, and death. Diarrhea, abortions, and death reported in **swine**.

IRON

Untoward effects: ***Anaphylaxis***: Severe reactions and death have occurred after IV use; local and inflammatory reactions were frequency with IM. Injections may cause hemochromatosis-like syndrome and exacerbation of arthritis.

Blood: Hemolytic crises and hemoglobinuria in **patients** with paroxysmal nocturnal hemoglobinuria; increased red blood cell hemolysis with ***vitamin E*** deficiency; hemochromatosis from excess intake, and Sjögren's syndrome, due to iron overload after repeated blood transfusions. Neutropenia may occur.

Cardiovascular: Can induce cardiac toxicity. Tachycardia after overload. Cardiac pigmentation occurs in familial hemochromatosis.

Clinical tests: Oral intake of ferrous salts may give false positive stool occult blood results. Increased serum ***sodium*** results.

Dental: May stain tooth enamel, especially if liquid iron preparations are taken. Tooth abrasion noted in **employees** of iron works from grinding effect of ore dust.

Gastrointestinal: Nausea, vomiting, abdominal pain, constipation, blackening of feces, irritation, and necrosis. Silver stools are due to oral iron and acholic stools. Its salts have corrosive action, leading to strictures of the stomach, intestines, and esophagus; and hemorrhage. Distal areas of the bowel can be damaged while proximal portions are spared after taking enteric-coated tablets.

Infection: Apparent potentiation of ***Yersinia enterocolitica*** in young **children** after accidental oral overdoses. Deaths reported in at least 13 kwashiorkor **children** who had increase in free-circulating serum iron.

Liver: Liver pathology and necrosis, and hypoglycemia in acute iron poisoning. Liver fibrosis and pigmentation occurs in familial hemochromatosis.

Miscellaneous: ***Tea*** from China was once adulterated with iron filings. Hemosiderosis, hemochromatosis with secondary fibrosis and liver cirrhosis, and an inert reaction in the lungs from occupational dust inhalation. An excess in the joints can cause arthritis; diabetes if deposited in the pancreas.

Photosensitivity: Severe reaction with increase in red blood cell protoporphyrin to 1024 μg/100 ml cells after 3 months of treatment in an anemic 26-year-old **female** with prior history of sensitivity to sun. After stopping treatment, photosensitivity disappeared in 6–8 weeks and protoporphyrin decreased to 314 μg. Restarting iron after 6 months, increased numbers to 714 in 2 weeks, and 953 in 6 weeks. In familial hemochromatosis, most skin pigmentation occurs in areas exposed to light.

Poisoning: There is no physiological mechanism for iron excretion, but it is lost via sweat, bile, desquamation of skin and mucosal surfaces, and blood loss (especially in menses) in amounts of 1.5+ mg/day. Lethal dose is ~300 mg/kg elemental iron and 150 mg/kg can be dangerous. Hemochromatosis after long-term use in pharmacological amounts by normal **persons** in excess of need can cause damage. Iron-storage disease in South African **Bantus**, probably consuming 100–200 mg/day from rusty cooking pots, along with **their** malnutrition and ***alcohol*** consumption, had increased absorption. A 37-year-old **female** with average intake of 274 mg/day/13 years for hemolytic anemia had hemochromatosis and died. Transfusions can be an important source of excess iron. Poisoning is often characterized by bloody vomitus, gastrointestinal bleeding, tarry diarrhea, hypotension, liver failure, acidosis, pulmonary edema, and, eventually, shock and death. A 19-year-old **female** ingested 60 ***ferrous succinate*** *tablets* (42 mg Fe^{++}/kg) and became ill, drowsy, had epigastric pain, tachycardia, and hypertension. A 26-year-old **female** ingested 150 ***ferrous succinate*** tablets (180 mg Fe^{++}/kg). Three hours later, **she** was ill and hypertensive. After being forced to vomit (mixed with blood), **she** became hypovolemic with tachycardia, hypotensive, pale, sweaty, had increased SGOT, increased SGPT, and increased ***bilirubin***. A pregnant 17-year-old **female** attempted suicide by ingesting 90–325 mg ***ferrous sulfate*** capsules. Many cases of iron overload reported from consumption of home-brewed ***beer***. A case report revealed iron toxicity, due to excess intake in a hyperalimentation regimen. Fatal intoxication secondary to aspiration, ingestion, and percutaneous absorption in an 18-year-old **male**, after a fall into a vat of saturated ***ferrous chloride*** in dilute ***HCl***.

Since the first report of iron poisoning in 1947, cases have increased and **children** seem to be the largest group affected. Symptoms are, essentially, similar to those in **adults**. Lethargy, vomiting, tarry stools, weak and fast pulse, and hypotension are common. Coma often occurs within 30–60 min. After 6–24 h, cyanosis, pulmonary edema, circulatory failure, and death have been reported. The most common offender is ***ferrous sulfate*** (q.v.) in "candy"-coated or attractively colored tablets and flavored syrups. Toxicity in **children** after ingestion 20–60 mg/kg.

Renal: Nephrotoxic.

Skin: IM use has often caused staining of the skin. Incidence of diaper rash significantly increased in **infants** with iron-fortified formulas. A rare case of allergic contact dermatitis in a 66-year-old **male toolmaker**; moderately erythematous papular and papulovesicular lesions; itchy eruptions on arms and legs, spreading to upper parts of legs, hands, arms, and trunk. "Permanent" pigmentation of skin from some iron salts used in treating ***poison ivy*** lesions. Hemochromatosis can impart a bronze color to the skin.

Drug interactions: ***Alcohol*** increases absorbtion of ***ferric chloride***. ***Wines*** have high iron content, and ***wine***-drinking **alcoholics** can consume enough over 20 years to result in hemochromatosis. See ***Poisoning*** above for excessive ***alcohol*** intake and iron-storage disease in **Bantus**. Enhanced iron absorption up to 15–20% with ***allopurinol***. ***Atropine*** decreases its absorption. ***Oral contraceptives*** and ***vitamin C*** increase serum iron. ***Antacid, coffee, egg*** and ***isoniazid*** ingestion decrease iron absorption. Chronic use of ***antihistamines*** can decrease its absorption, as well. ***Ferrous gluconate*** or ***ferrous sulfate*** significantly decrease bioavailability of ***ciprofloxacin***. Use with ***tetracyclines*** causes decreased absorption of both. ***Cimetidine*** decreases gastric acid secretion leading to decreased iron absorption. DPT immunization decreases iron absorption in malnourished **infants** and young **children**.

After a high polyunsaturated ***fatty acid***–low ***vitamin E*** diet is fed to **sows**, their **piglets** are sensitive to iron and have become dyspneic, comatose, and died. Post-mortem showed muscle degeneration, hydropericardium, and hydrothorax.

IRON DEXTRANS

= *Armidexan* = *Fedex* = *Fenate* = *Ferrextran* = *Ferrimicrolex* = *Ferrobalt* = *Fervatol* = *Haemalift* = *Imferon* = *Imposil* = *Infed* = *Irondex* = *Myofer* = *NoNemic* = *Pharmatinic* = *Pidgex* = *Prolongal*

A complex of colloidal *ferric chloride* and low molecular weight dextran, used parenterally.

The hydrogenated form = *Brodex* = *Hyferdex* = *Iron Hy-Dex* = *Life Iron* = *PVL* = *Rubrafer*

Untoward effects: Anaphylactic shock, acute circulatory failure, apnea, and cardiac arrest are among the most serious. Others include local tissue staining (which can persist for years), acute exacerbation of arthritis, chronic synovitis, risk of hepatosplenic siderosis, headache, and neck stiffness in a 29-year-old **female** with high cerebrospinal fluid iron level of 36 µmol/l after 32 ml IV of ***Imferon***, increase in splenomegally and malarial parasites in treated **infants**; fever; erythema, muscle necrosis, and skin ulcers due to IM injections, increased liver uptake and decreased abscess uptake with ***gallium-67 citrate*** imaging, chest pain, dyspnea, stridor, nausea, vomiting, lower abdominal pain, thrombophlebitis, myalgia, arrhythmias, tachycardia, sweating, syncope, convulsions, seizures, disorientation, dizziness, facial edema, pruritus, pleocytosis, calciphylaxis, inguinal and cervical lymphadenopathy, sarcomas at site of IM injection, anuria, and hematuria with acute renal failure.

IM and SC injections induced sarcomas in **rats**, **mice**, **hamsters**, and **rabbits** at the injection site after a long latent period of 1–4 years. Exogenous iron may decrease resistance to bacterial invasion by decreasing RES phagocytosis. Injected baby **pigs** may stagger, fall over, and die, especially if ***vitamin E*** deficient (see ***Iron***, above). Ataxia and muscle rigidity have occurred in **rabbits**. Overdosing 1-day-old **pigs** can occasionally cause shock and death. IM use after 2–4 weeks of age in **pigs** can cause undesirable staining of future cuts of meat.

IRON DEXTRINS

= *Felac* = *Iroject* = *Tefferol*

Untoward effects: Although it has been used in **humans** parenterally, adverse effects have occurred in **pigs** and other species, as with ***Iron Dextran***.

IM, TD_{LO}, **women**: 20 mg/kg; 3 years.

IRON POLYSACCHARIDE COMPLEX

= *Dexin* = *Feraject* = *Feraplex* = *Ferropal* = *Hytinic* = *Iron Amylose Complex* = *Niferex* = *Pig Iron* = *Polyject* = *Ranch Iron Complex*

Untoward effects: Of 810 potentially toxic exposures, most were in **children** < 6 years and ~87% were unintentional leading to nausea, vomiting, abdominal pain, diarrhea, lethargy, and drowsiness.

Some of these occasionally cause an unusual mortality and irritation in **pigs** treated parenterally with them, and produce conspicuous deposition, especially in fats.

IRON SORBITEX

= *Ferastral* = *Iron Hexanehexol* = *Iron Sorbitol* = *Jectofer*

Untoward effects: IM injection leads to local pain, flushing, metallic taste, dizziness, nausea, loss of consciousness, ***folic acid*** deficiency, aches, skin discoloration, anaphylaxis, second-degree atrioventricular block, malaise, urticaria, palpitations, sweating, headache, decreased blood pressure, blackening or browning of urine, hematuria, frequent urination, and a case report of localized sarcoma. Anaphylaxis, cyanosis, collapse, and death in a 20-month-old **female** within 1 h of 1 ml ***Jectofer*** IM (50 mg ***iron***) for anemia. Post-mortem found pulmonary edema. Deaths from ventricular tachycardia and fibrillation in two **patients** and collapse in **five** after IM in malabsorption syndromes.

IRONWOOD

Untoward effects: ***Erythrophloeum chlorostachys*** is a tree in tropical Australia whose leaves, after ingestion, cause serious poisoning in **cattle**, **sheep**, **goats**, **horses**, and **camels**. They develop anorexia, staring-eyed demeanor, poor vision, frequent contractions of abdominal muscles, pale mucous membranes, increased heart sounds, and, eventually, respiratory embarrassment. **Horses** sweat profusely and have periodic extrusion of the upper and lower lips. < 2 oz of dried leaves killed a **horse** within 24 h. In 1960, ~10,000 **sheep** died from eating the leaves. Its alkaloid, ***erythrophleine***, is the toxic agent.

ISEPAMICIN

= *Exacin* = *Isepacin* = *SCH 21,420*

***Antibiotic. An aminoglycoside.* IV or IM.**

Untoward effects: Nephrotoxic in 2% and ototoxic in 1% of pediatric **patients**.

ISOAMYL ACETATE

= *Amyl Acetate* = *β-Methyl Butyl Acetate*

A technical form = *Banana Oil* = *Pear Oil*

Use: As a fragrance in soaps, detergents, cosmetic lotions and creams, perfumes, and foods.

Untoward effects: An abused solvent by inhalation. Can irritate eyes, skin, nose, and throat. No irritation or sensitization reactions in 25 **volunteers**, but dermatitis has been reported. TLV, 100 ppm. Immediate danger to life or health, 1000 ppm.

LD_{LO}, **human**: 500 mg/kg; inhalation TC_{LO}, **human**: 200 ppm.

Inhalation causes narcosis in **animals**.

LD_{50}, **rats** and acute dermal, **rabbits**: > 5 g/kg; inhalation LC_{LO}, **cat**: 35,000 ppm/m^3.

ISOAMYL ALCOHOL

= *Fusel Oil* = *Isobutyl Carbinol* = *Isopentyl Alcohol* = *3-Methyl-1-butanol*

***Industrial solvent, and in microscopy.* 2 ml/l of *whiskey*.**

Untoward effects: Skin, eye, nose, and throat irritant from inhalation. Ingestion leads to nausea, vomiting, diarrhea, headache, dyspnea, coughing, and dizziness. Contact may cause skin cracking. High concentrations cause central nervous system suppression and narcosis; intermediate concentrations cause headache and dizziness. Moderately irritating (TLV 100 ppm) to mucous membranes in **man**. TWA, 100 ppm.

LD_{LO}, **human**: 50 mg/kg; inhalation TC_{LO}, **human**: 150 ppm.

Narcotic in **animals**.

LD_{50}, **rat**: 1.3 g, 7 ml/kg; skin, **rabbit**: 4 g/kg.

Also see ***Fusel Oil*** above.

ISOAMYL PROPIONATE

= *Isopentyl Propionate*

Use: As a fragrance in soaps, detergents, cosmetic creams and lotions, perfumes, and foods.

***Untoward effects*:** 4% in petrolatum produced no irritation and sensitization in **human** volunteers.

No irritation at full-strength on intact or abraded **rabbit** skin. Narcotic to **rabbits** at 5.18 g/kg and to **tadpoles** at 0.72 g/l of water.

LD_{50}, **rat** and acute dermal, **rabbits**: > 5 ml/kg; LD_{50}, **rabbits**: 6.91 g/kg.

ISOBENZAN

= CP 14,957 = ENT 25,545X = Omtan = R-6700 = SD-4402 = Telodrin

Insecticide.

***Untoward effects*:** Has caused poisoning in **man** leading to nervousness, hyperexcitability, weakness, tremors, and convulsions.

Causes ataxia, fasciculation, tenseness, swimming backwards, tail high and fanned, loss of righting reflex, circling, and opisthotonus in **mallard ducks** within 30 min, and deaths occur after ~2 h. In **cattle**, symptoms are those of ***cyclodrine's***. It is excreted in **cow's** milk. Poisoned **cockerels** develop cyanosis of combs and wattles, tachypnea, nervousness, circular motion, and convulsions. Post-mortem revealed liver and kidney degeneration, with hemorrhagic changes; and distended gall bladder.

LD_{50}, **rats**: 5–10 mg/kg; **cockerels**: 3.85 mg/kg.

ISOBORNYL ACETATE

= 2-Camphanyl Acetate

A fragrance used in soaps, detergents, cosmetic creams and lotions, and foods. Tonnage is used.

***Untoward effects*:** No sensitization to 10% in petrolatum in 25 **human volunteers**.

Full-strength to intact or abraded **rabbit** skin for 24 h under occlusion was only moderately irritating. Daily doses of 90 or 270 mg/kg to **male rats** caused nephrotoxicity. The no-effect level was 15 mg/kg, which is 100 times the calculated maximum intake by **man**.

LD_{50}, **rats**: > 10 g/kg.

ISOBORNYL PROPIONATE

A fragrance used in soaps, detergents, cosmetic creams and lotions, and foods.

***Untoward effects*:** No sensitization reactions of 10% in petrolatum in a maximization test on 25 **human volunteers**, and no irritation after a 48 h closed-patch test.

Full-strength application to intact or abraded **rabbit** skin under occlusion was not irritating.

LD_{50}, **rats** and acute dermal, **rabbits**: > 5 g/kg.

ISOBORNYL THIOCYANOACETATE

= Barc = ENT 92 = Terpinyl Thiocyanoacetate = Thanite

Insecticide, piscicide.

***Untoward effects*:** Polydipsia, regurgitation, goose-stepping ataxia, jerkiness, falling, emaciation, and asthenia 1 day after exposure in **mallard ducks** and 7 days in **pheasants**.

LD_{LO}, **human**: 500 mg/kg.

LD_{50}, **rats**: 200 mg/kg; **ducks** and **pheasants**: > 2 g/kg; **guinea pig**: 551 mg/kg; **rabbit**: 630 mg/kg; skin, **rabbit**: 6 g/kg.

ISOBUTANE

Aerosol propellant.

Untoward effects*:** Inhalation exposure in a **2-year old** from a commercial deodorant led to ventricular tachycardia. Can cause drowsiness, narcosis, and asphyxiation after inhalation. Found in ***gasoline. Associated with death in an **inhalant-abuser**. Prolonged or repeated contact can be irritant to **human** skin. The liquid can cause frostbite effects on the skin. Flammable! National Institute for Occupational Safety & Health exposure limit of 800 ppm, TWA.

In **dogs**, it increases pulmonary resistance, decreases respiratory minute volume, increases heart rate, and, as with many other ***aliphatic hydrocarbons***, it sensitizes the heart to ***epinephrine***. Causes apnea and cardiac arrest in **rats**.

ISOBUTYL ACETATE

***Flavoring agent, solvent.* In *lacquer* thinners. Fruity, floral odor.**

***Untoward effects*:** Inhalation causes eye and throat irritation, headache, drowsiness, and anesthesia. Skin irritant. National Institute for Occupational Safety & Health and OSHA upper exposure limit of 150 ppm, TWA. Immediate danger to life or health, 1300 ppm; flammable and explosive.

Narcosis in **animals**.

LD_{50}, **rat**: 15 g/kg; **rabbit**: 4.76 g/kg; inhalation LC_{LO}, **rat**: 8000 ppm/4 h.

ISOBUTYL ALCOHOL

***Solvent.* In varnish removers. Sweet, musty odor.**

***Untoward effects*:** Inhalation leads to eye and throat irritation, headache, and drowsiness. Skin irritant, with skin cracking after contact. OSHA upper exposure limit of 100 ppm.

LD_{LO}, **human**: 500 mg/kg.

LD_{50}, **rat**: 2.46 g/kg; LD_{LO}, **rabbit**: 3.75 g/kg.

ISOBUTYL BENZOATE

Fragrance in soaps, detergents, cosmetic creams and lotions, perfumes, and foods.

Untoward effects: No sensitization or irritation in tests of 2% in petrolatum on **human volunteers**.

Non-irritating at full-strength on intact or abraded **rabbit** skin for 24 h under occlusion.

LD_{50}, **rats**: 3.7 mg/kg; acute dermal, **rabbits**: > 5 ml/kg.

ISOBUTYL NITRITE

Use: As an industrial solvent and in ***Butyl, Heart On, Locker Room, Rush***, and ***Turn-On*** as room deodorizers.

Untoward effects: In the late 1970s, reported as an inhalation drug of abuse in **humans**, leading to hypotension, hemolytic anemia, and methemoglobinemia. In the nation-wide "drug culture", it is promoted as an aphrodisiac and "orgasm-extender". Also found in ***Ban Apple Gas, Bullet, Jack Aroma, Joc Aroma, Kick***, and ***Vaporole***. Causes vasodilation, light-headedness, dizziness, syncope, hypotension, increased intraocular pressure, tachycardia, and flushing, lasting 1–2 min that is called a "rush". Headache and nausea can later last for hours.

ISOBUTYL PHENYLACETATE

= *Isobutyl α-toluate*

Fragrance in soaps, detergents, cosmetic creams and lotions, perfumes, and foods.

Untoward effects: A maximization test of 4% in petrolatum on 25 **volunteers** produced no sensitization and no irritation reactions were reported.

Full-strength to intact or abraded **rabbit** skin for 24 h under occlusion was mildly irritating.

LD_{50}, **rats** and acute dermal, **rabbits**: > 5 g/kg.

ISOBUTYL QUINOLINE

A fragrance in soaps, detergents, cosmetic creams and lotions, and perfume.

Untoward effects: No sensitization or irritation of 2% in petrolatum on **human volunteers**.

Full-strength to intact or abraded **rabbit** skin for 24 h under occlusion was not irritating.

LD_{50}, **rats**: 1.02 g/kg; acute dermal, **rabbits**: > 5 g/kg.

ISOBUTYL SALICYLATE

A fragrance in soaps, detergents, cosmetic creams and lotions, perfume, and foods.

Untoward effects: No sensitization or irritation of 10% in petrolatum on **human volunteers**.

Full-strength to intact or abraded **rabbit** skin for 24 h under occlusion was slightly irritating.

LD_{50}, **rats**: 1.56 g/kg; acute dermal, **rabbits**: > 5 g/kg.

ISOBUTYRONITRILE

= *2-Cyanopropane* = *Isopropyl Cyanide* = *2-Methylpropanenitrile*

Untoward effects: ***Almond***-like odor; forms ***cyanide*** in the body. Inhalation leads to eye, skin, nose, and throat irritation; headache, dizziness, weakness, giddiness, confusion, dyspnea, and convulsions. Ingestion causes abdominal pain, nausea, and vomiting, and some of the above symptoms, after absorption.

LD_{50}, **rat**: 102 mg/kg; inhalation LC_{LO}, **rat**: 1000 ppm/4 h.

ISOCARBOXAZID

= *Marplan* = *RO 5–0831*

Antidepressant. A hydrazine.

Untoward effects: An ***monoamine oxidase inhibitor***, causing hypotension, insomnia, restlessness, and daytime sleepiness.

Occasional: Dry mouth, nausea, constipation, diarrhea, anorexia, urinary retention, tremors, mania, paresthesias, sexual disturbances (impotence, delayed ejaculation, and, in **women**, anorgasmia), edema, weight gain, slight vertigo, postural hypotension, headaches, visual hallucinations, palpitations, and blurred vision.

Uncommon: Hepatitis and hepatocanalicular jaundice, skin rashes, lupus-like reactions, leukopenia, hypertension, and hyperthermia. Mid-brain lesion in comatose 48-year-old **female**, 1 day after hospitalization following ingestion of 30–10 mg tablets. Hypersensitivity is the apparent cause of massive generalized pitting edema of face, arms, trunk, and legs.

Human LD estimated at 200–300 mg.

Drug interactions: Increases the toxicity of ***amphetamine*** and ***phenylephrine***; severe headaches and hypertension with ***methamphetamine***; gives spuriously high sulfobromophthalein retention results and spuriously low ***glucose*** tolerance results. ***Alcohol*** intake decreases its effect. It decreases metabolism of ***amphetamines***, other ***sympathomimetics***, and ***tyramine***, a pressor amine in ***broad beans,*** gorgonzola, cheddar, and Camembert ***cheeses***; some ***chocolates***, canned ***figs***, pickled ***herring***, chicken ***livers***, and chianti ***wine*** and ***sherry*** leading to hypertensive crisis, severe occipital headache, vomiting, hyperpyrexia, chest pain, and even fatal intracranial hemorrhage. A 38-year-old **female** developed an apparent ***serotonin*** syndrome 2 days after discontinuing isocarboxazid therapy. A 32-year-old **female** developed persistent myoclonic leg jerks after a single 15 mg dose of ***dextromethorphan***. ***Venlafaxine*** in a 43-year-old **male** caused agitation, hypomania, diaphoresis, shivering, and mydriasis.

Potentiates the hypoprothrombinemic effect of ***acenocoumarol, ethylbiscoumacetate***, and ***warfarin*** in **rats** and **rabbits**. Potentiates the hypoglycemic effect of ***tolbutamide*** in **rabbits**.

LD_{50}, **rat**: 280 mg/kg; **mouse**: 193 mg/kg; **monkey**: 160 mg/kg; **cat**: 56 mg/kg; LD_{LO}, **dog**: 40 mg/kg.

ISOCETYL STEARATE

Untoward effects: A comedogenic ***fatty acid*** that should be avoided if used in cosmetics.

ISOCYCLOCITRAL

Fragrance used in soaps, detergents, cosmetic creams and lotions, and in perfumes.

Untoward effects: No irritation or sensitization of 4% in petrolatum on **human volunteers**.

Full-strength to intact or abraded **rabbit** skin for 24 h under occlusion was mildly irritating.

LD_{50}, **rats**: 4.5 ml/kg; acute dermal, **rabbits**: > 5 ml/kg.

ISODRIN

= *Compound 711*

Insecticide.

Untoward effects: LD_{LO}, **human**: 5 mg/kg.

LD_{50}, male **rats**: 15.5 mg/kg; female **rats**: 7 mg/kg; **mice**: 8.8 mg/kg; **7-day-old chicks**: 2.7 mg/kg; topically, female **housefly**: 0.06 μg/**fly**.

ISOETHARINE

= *Asthmalitan* = *Dilabron* = *Neoisuprel* = *Numotac* = *WIN 3046*

Bronchodilator. β-2 adrenergic **with some *β*-1 activity.**

Untoward effects: Palpitations, tremors, nausea, headache, tachycardia, arrhythmias, hypotension, weakness, vertigo, insomnia, nervousness, and angina. Slow development of tolerance. Hypersensitivity may trigger a severe and prolonged asthma attack. Paradoxical bronchospasm in an **asthma patient** was due to its ***sodium bisulfite*** content. Can cause myalgia, cramps, myokymia, and weakness. A case report of induced mania reported. Avoid concurrent use with ***sympathomimetics***. Overdosage may cause paradoxical airway resistance.

LD_{50}, **mice**: 1.63–2.0 g/kg.

ISOEUGENOL

Use: In manufacturing of ***vanillin*** and in perfumery.

Untoward effects: Common allergen in perfumes and fragrances, causing a moderately irritating dermatitis in **man**. It is also a minor hallucinogen (6 mg/20 g ***nutmeg***).

Severely irritating on **rabbit** and **guinea pig** skin.

LD_{50}, **rat**: 1.56 g/kg; **guinea pig**: 1.4 g/kg.

ISOFENPHOS

= *Amaze* = *Bay 92114* = *Isophenphos* = *Oftanol* = *SRA-12869*

Insecticide. **For control of grubs and mole crickets.**

Untoward effects: Anticholinesterase. **Dogs** receiving 25 mg/kg showed significant decrease in plasma and erythrocyte cholinesterase, 78% and 60%, respectively. Signs of poisoning in 1/4.

LD_{50}, **rat**: 28 mg/kg; **mice**: 91.3 mg/kg; **bobwhite quail**: 13 mg/kg; skin, **rat**: 188 mg/kg.

ISOFLUPREDONE

= *9-Fluoroprednisolone*

Anti-inflammatory. **The acetate =** ***Predef*** **=** ***U 6013***. **For oral or IM use.**

Untoward effects: In **animals**, as with ***cortisone***. Avoid its use during last trimester of pregnancy in **cows** and other species, where it may precipitate premature parturition with associated dystocia, fetal death, metritis, and retained placenta. Administration to pregnant **dogs**, **rabbits**, and **rodents** led to cleft palate in offspring. Other congenital anomalies may occur in **dogs**.

ISOFLURANE

= *Aerrane* = *Comp 469* = *Forane* = *Forene*

Inhalation anesthetic. **Slight pungent, musty, etheral odor.**

Untoward effects: During surgery, blood ***glucose*** is increased; after surgery, ***bilirubin*** and lactic dehydrogenase levels may increase. Respiratory acidosis and depression have occurred. Increased bleeding in therapeutic suction abortion. It may enhance the dysrhythmic effect of sympathetic amines, requiring dosage decrease of the latter. Potentiates the effect of ***non-depolarizing neuromuscular-blocking agents***, such as ***curare***. Post-operative nausea, vomiting, and excitation are uncommon. Decreases renal plasma flow, glomerular filtration rate, and urine output.

In **animals**, all commonly used ***muscle relaxants***, especially the non-depolarizing types, become markedly potentiated by it. High induction concentration can cause respiratory depression. Causes arteriolar vasodilation, leading to decreased vascular resistance and decreased arterial blood pressure in **birds**. Reduces renal blood flow in **dogs**. Hypotension, respiratory suppression, and arrhythmias reported in **horses**.

ISOLAN

= *G 23,611*

Carbamate insecticide.

Untoward effects: Has caused poisoning in **man**.

LD_{LO}, **human**: 5 mg/kg.

Dermal toxicity in **rats**: 5.6 g/kg and ranges in **rabbits** from 35 to 60 mg/kg.

LD_{50}, **rat**: 12 mg/kg; male **chicks**: 3.32 mg/kg.

ISOMETAMIDIUM

= *M&B 4180A* = *Samorin* = *Trypamidium*

Trypanocide.

Untoward effects: Painful local reactions at IM injection sites in **cattle**, **donkeys**, **goats**, **sheep**, **swine**, **camels**, and **dogs**. **Horses** develop very severe reactions at injection site.

ISOMENTHONE

In fragrances for soaps, detergents, cosmetic creams and lotions, and perfume.

Untoward effects: No sensitization or irritation from 8% in petrolatum on **humans**.

Full-strength to intact or abraded **rabbit** skin for 24 h under occlusion was mildly irritating.

LD_{50}, **rat**: 2.18 g/kg; Acute dermal LD_{50} in **rabbits**: > 5 ml/kg.

ISONIAZID

= *Cotinazin* = *Dinacrin* = *Ditubin* = *FSR-3* = *Hycozid* = *Hydrazid* = *INH* = *Iscotin* = *Isobicina* = *Isocid* = *Isolyn* = *Isonex* = *Isonizida* = *Isozid* = *Laniazid* = *Mybasan* = *Neoteben* = *Nicizina* = *Niconyl* = *Nicotibina* = *Nydrazid* = *Pycazide* = *Pyricidin* = *Rimifon* = *Rimitsid* = *RP-5015* = *Tibinide* = *Tubazid* = *Tubilysin* = *Tyvid*

Tuberculostatic.

Untoward effects: ***Behavior***: Causes mood changes, increased excretion of ***vitamin*** B_6, and depression in the **elderly**. A 60-year-old **female** receiving it for tuberculosis prophylaxis became hyperactive, restless, unable to sleep, and evidenced aggressiveness, mania (elation), and psychosis. Disorientation and violence after 6 weeks on 400 mg with 6 mg ***pyridoxine***/day in a **slow-inactivator**.

Blood: Severe hemolytic anemia, aplastic anemia, megaloblastic anemia, sideroblastic anemia, eosinophilia, leukopenia, neutropenia, thrombocytopenia and non-thrombocytopenic purpura, agranulocytosis, red blood cell aplasia, and increased serum ***bilirubin*** in 10%.

Carcinogenicity: Increase in lung cancer in **tuberculosis patients**, cause and effect debated.

Cardiovascular: Disseminated intravascular coagulation in a 58-year-old **female** after 300 mg/day/4 weeks. Pericarditis, vasculitis, and cardiac tamponade.

Central nervous system: Encephalopathy, peripheral neuropathy (primarily in **slow-acetylators**), grand mal seizures, dizziness, ataxia, delirium, and memory loss. Myoclonic movements of extremities and face; stammering, followed by repeated general seizures and death in a uremic, tubercular 67-year-old **male**. Mental confusion; visual, auditory, and olfactory hallucinations in a 57-year-old **female** on 600 mg/day/18 days. Paresthesia of lips and mouth in a 22-year-old **female**, after 1 g IM/4 days. Foot-drop in a 32-year-old **male** on 300 mg/day/21 days. Severe aseptic, purulent meningoencephalitis in a 27-year-old **male** on 300 mg/day. A 34-year-old **male** on 300 mg/day/4 years developed extreme lethargy and ataxia, progressing to a semi-comatose state over several weeks. Induced convulsions also reported.

Eye: Transient myopia, optic neuritis (especially with ***ethambutol***), iridoplegia, cycloplegia, toxic amblyopia, subconjunctival hemorrhage, keratitis, and swollen eyelids have occurred.

Fetus and neonate: Readily passes through the placenta and found in **fetal** blood and in amniotic fluid. Reaches the **fetus** within 15 min of oral **maternal** intake and **fetal** levels may exceed **maternal** and be toxic. Cases of retarded psychomotor activity and development reported. Some risk of ototoxicity and even cerebral damage to the **fetus**. Microcephaly, convulsions, myoclonia, and subcortical epilepsy in some **children** whose **mothers** took it during various stages of pregnancy. Mammary excretion in **women**, **cows**, **dogs**, and **rabbits**. An 11-month-old **infant** had ***isoniazid-rifampin*** (10 mg/kg/day of each) – induced fulminant liver disease. A **9-year-old** whose **mother** took it during **her** pregnancy developed a mesothelioma.

Liver: Increased SGOT and SGPT in ~10% of **patients** and aminotransferases in ~20%. Severe, and sometimes fatal, cases of hepatitis. These are age-related (< 20 years, 0–2/1000; 20–34 years, 3/1000; 35–49 years, 12/1000; 50–64 years, 23/1000; > 65 years, 8/1000). Prodromal signs include fatigue, weakness, malaise, anorexia, nausea, and vomiting. Hepatic necrosis, hepatocellular jaundice, and liver enlargement. Isoniazid-induced hepatitis may be more common in **rapid-** rather than **slow-acetylators**, because of exposure to greater concentrations of the toxic metabolites ***acetylhydrazine*** and ***acetylisoniazid***.

Miscellaneous: Fever, chills, abdominal pain, metabolic acidosis, myelopathy, acute arthritis, periorbital edema, increased bone marrow ***iron***, sore throat, peripheral adenopathy, lymphocytosis, pulmonary infiltrates, pleuritis, wheezing, anaphylaxis, lymphadenitis, and systemic lupus erythematosus. The more pigmented a **patient** is, the greater **their** chance of its rapid inactivation. Pellegra, gynecomastia, diabetes mellitus, hyperfunction of the adrenal cortex, dry mouth, and ***niacin*** deficiency have occurred. Local irritation after IM use.

Poisoning: Often due to abuse and attempted suicide. A 37-year-old **male** ingested 20 g, leading to severe acidosis, convulsions, respiratory and circulatory failure, and coma. Peripheral neuritis of lower extremities during first post-dialysis day and disappeared on the fifth day. Death after 4 days in suicide with 40 g in a 33-year-old **female**. Convulsions 2 h after 35 tablets (3.5 g) in suicide attempt by a 19-year-old **female**, plus slight mydriasis, inhibited reflex to light, hyporeflexia, tonic-clonic convulsions, moderate cyanosis, and urinary incontinence. During the second day, nystagmus and diffuse muscle pain. Accidental overdose of 5 g in a 3-year-old **male**. In a 15 year period, 60 cases of abuse in Alaska, since general knowledge that 3–5 g gives a "good trip", similar to that by ***LSD*** ingestion. Therapeutic serum concentrations have ranged between 0.5 and 10 mg/l; 20–30 mg/l has been reported toxic and 100 mg/l has been fatal.

Renal: Hemoglobinuria, acute interstitial nephritis, and urinary retention (1%).

Skin: Non-thrombopenic purpura, rash, erythroderma, pruritus, urticaria, morbilliform exanthema, exfoliative dermatitis, maculopapular lesions, lichenoid and acneiform eruptions, conjunctivitis, keratotic erythema and edema of face and hands, contact dermatitis, "burning feet", and Stevens–Johnson syndrome.

Drug interactions: A life-threatening case of renal and hepatic toxicity in a 19-year-old **female who** ingested not over 11.5 g ***acetaminophen*** during a suicide attempt, while taking isoniazid 300 mg/day/5 months. It is theorized that it may have caused kidney and renal enzyme induction, resulting in rapid metabolism of ***acetaminophen*** to its toxic metabolites. ***Alcohol*** ingestion has increased its toxicity and fatalities. Its activity and toxicity increased with ***p-aminosalicylic acid***, as both compete for the same excretion pathway. ***p-Aminosalicylic acid*** may decrease its absorption. Large doses of isoniazid with it and ***vitamin*** deficiency have caused optic neuritis. A 31-year-old **male** showed flushing, palpitations, and hypertension, after ingesting ***Swiss cheese***. ***Cheshire*** and other ***cheeses*** have also been incriminated in these reactions. ***Monoamine oxidase inhibitor*** inhibition by isoniazid. ***Antacids***, ***carbohydrates***, and food decrease its absorption. Use with ***adrenergics*** increases central nervous system stimulation. Given with oral ***antidiabetic drugs*** or ***insulin***, it may antagonize their effect and increase blood ***sugar***. Epileptic **patients** stabilized on ***carbamazepine*** became disoriented, listless, lethargic, and aggressive when given isoniazid, a potent hepatic enzyme inhibitor. This clinical effect can occur in a few days and is potentiated by ***cimetidine***. An additive anticholinergic effect may occur with ***atropine*** or ***meperidine***. It decreases protection of ***oral contraceptives***, apparently by impairing enterohepatic recirculation of its hormones. Megaloblastic anemia, due to ***folate*** deficiency, when given with ***cycloserine***; mood disturbances can also occur with this drug combination. It prolongs ***diazepam's*** half-life. Use with ***disulfiram***, in a number of cases, led to coordination difficulties, behavior disturbances, and neuropsychiatric symptoms. Causes increase in ***ethosuximide*** serum levels. It decreases serum concentration of ***itraconazole***. It inhibits non-microsomal metabolism of ***phenytoin***, especially in **slow-metabolizers** of the drug and increases the potential of ***phenytoin*** toxicity. Extra ***pyridoxine*** decreases its toxicity. Isoniazid can increase its excretion, leading to peripheral neuropathy, generalized convulsions (particularly in **infants**), and anemia. ***Rifampin*** increases its adverse liver effects. ***Theophylline*** toxicity is increased by its chronic administration. Use with ***thiacetazone***, possibly, caused diffuse hypertrophy of breasts with ulceration in an **adolescent female**. ***p-Aminosalicylic acid***, ***pyrazinamide***, and ***thiacetozone***, retard its hepatic acetylation and increase its elimination half-life, increasing the risk of hepatitis. Hyperuricemia occurs when it is used with ***pyrazinamide***. Severe neurotoxicity reported in an 85-year-old **female** on isoniazid after a single 2 mg dose of ***vincristine***. Dramatically attenuates ***vitamin D*** metabolism to its most active metabolite. Potentiates the anticoagulant effect of ***warfarin***. Will decrease uptake of ***radiopharmaceuticals*** used in thyroid imaging. Can cause potentiation of ***phenobarbital*** in **epileptics**, leading to ataxia and sedation. It enhances the effects of ***anesthetics***, ***antihypertensives***, ***antiparkinson drugs***,

narcotics, sedatives, and ***tricyclic antidepressants***. Causes spuriously high Benedict's urine ***glucose*** results. It increases ***demeclocycline*** toxicity. A drug-induced dermatitis reaction can occur in **workers** allergic to ***hydrazines***. **Plumbers** and **welders** using soldering or other fluxes, or **others** working with ***corrosion inhibitors, explosives, foam rubber, fuels, fungicides***, and ***insecticides*** can be involved in this reaction. Although it decreases ***L-dopa***-induced dyskinesia, it can increase parkinsonian signs when added to ***L-dopa*** treatment. It doubles ***chlorzoxazone*** elimination half-life.

LD_{LO}, **human**: 100 mg/kg; TD_{LO}, **human**: 430 mg/kg.

Excreted in the milk of **cows**, **dogs**, and **rabbits**. High dosage has caused carcinomas and pulmonary adenomas in **mice** and is teratogenic in **rats** and **poultry**. Its metabolites are hepatotoxic. Except for experiments, avoid continuous use on **monkey** colonies, since it can change tuberculin positive tests to negative.

LD_{50}, **rat**: 650 mg/kg; **dog**: 100 mg/kg; **rabbit**: 250 mg/kg; LD_{LO}, **mouse**: 300 mg/kg.

ISOOCTANE

Solvent.

Untoward effects: In 1992, its use with various polymers in a newly formulated ***Wilson's Leather Protector*** led to pneumonitis, chest tightness, headache, shivering, fever, weakness, and shortness of breath.

TWA, air, 350 mg/m^3; 1800 mg/m^3, 15 min.

ISOOCTYL ALCOHOL

= *Isooctanol* = *Oxooctyl Alcohol*

Untoward effects: Inhalation causes eye, skin, nose, and throat irritation. Can burn skin. National Institute for Occupational Safety & Health sets upper exposure limits at 50 ppm.

LD_{50}, **rat**: 1.5 g/kg.

ISOPHORONE

= *Isoacetophorone*

***Solvent.* Use estimated at ~100,000 lbs/year.**

Untoward effects: Acute symptoms after inhalation are anorexia, headache, dizziness, malaise, faintness, fatigue, nausea, and diarrhea. Irritant to skin, eyes, nose, and respiratory tract. Chronic exposure leads to decreased body weight, dry skin, and skin inflammation. OSHA air exposure limit of 25 ppm TWA. American Conference of Governmental Industrial Hygienists suggests 5 ppm, and National Institute for Occupational Safety & Health says 4 ppm. 27.8 ppm exposure leads to 50% decrease in respiratory rate.

In **guinea pigs** and **rats**, adverse effects on liver and kidneys.

LD_{50}, **rat**: 2.3 g/kg; inhalation LC_{LO}, 1840 ppm/4 h.

ISOPIMPINELLIN

Untoward effects: A photosensitizer in ***lemon*** juice, which may cause reactions from its use in ***hollandaise sauce***, etc.

ISOPRINOSINE

= *Aviral* = *Delimmun* = *Imunoviral* = *Inosine Pranobex* = *Inosiplex* = *Methisoprinol* = *Modimmunal* = *NP-113* = *NPT-10381* = *Pranosine* = *Viruxan*

Antiviral, immunomodulator.

Untoward effects: Gastrointestinal upsets, primarily nausea and vomiting. ***Uric acid*** is a metabolic end-product.

ISOPROPAMIDE IODIDE

= *Darbid* = *Priamide* = *R-79* = *Tyrimide*

Anticholinergic.

Untoward effects: As for ***atropine*** (q.v.). Interferes with thyroid function tests. May cause reactions in **patients** sensitive to ***iodine***. Toxic doses can cause non-depolarizing neuromuscular blocking effects. Rarely used alone.

In **dogs** it can cause gastric retention of foods, ileus, tachycardia, and increased intraocular pressure.

2-ISOPROPOXYETHANOL

= *Isopropyl Glycol*

Untoward effects: Inhalation leads to eye, skin, and nose irritation and pulmonary edema in **animals**; ingestion leads to hemoglobinuria and anemia; skin irritant.

LD_{50}, **rat**: 5.7 g/kg; inhalation LC_{LO}, **rat**: 4000 ppm/4 h, inhalation LC_{50}, **mouse**: 1930 ppm.

ISOPROPYL ACETATE

= *2-Propyl Acetate*

***Solvent.* In printing *inks*.**

Untoward effects: Inhalation causes eye, nose, and throat irritation; weakness, drowsiness, narcosis, and unconsciousness. Dermatitis from skin contact. OSHA air exposure limit of 250 ppm.

Inhalation TC_{LO}, **human**: 200 ppm; immediate danger to life or health, 2000 ppm; LD_{LO}, **human**: 500 mg/kg.

LD_{50}, **rat**: 3 g/kg; **rabbit**: 6.95 g/kg; inhalation LC_{LO}, **rat**: 32,000 ppm/4 h.

N-ISOPROPYL ACRYLAMIDE

Untoward effects: In **rats**, ataxia, weakness, and occasional dragging of hind limbs. Urinary incontinence in severe clinical cases. Testicular atrophy in **mice**. An analog of ***acrylamide*** (q.v.) leading to peripheral neuropathies in **humans** and experimental **animals**.

ISOPROPYLAMINE

= *2-Aminopropane* = *2-Propylamine*

***Ammonia*-like odor. Flammable. A gas above 91°F. A component in *sarin* (q.v.).**

Untoward effects: Irritant to eyes, skin, nose, and throat; pulmonary edema, dermatitis, eye and skin burns, and visual disturbances. OSHA sets exposure limit of 5 ppm. Immediate danger to life or health, 750 ppm.

LD_{50}, **rats**: 820 mg/kg; inhalation LC_{LO}, **rat**: 800 ppm/8 h; **mouse**: 7000 ppm/40 min.

ISOPROPYLAMINODIPHENY-LAMINE

***Antioxidant*. Used in *rubber* tires and other *rubber* products.**

Untoward effects: Causes an allergic lichenoid contact dermatitis in **workers**. Some cases developed a wide-spread purpuric capillaritis similar to sensitivity reactions due to ***cabromal*** and ***meprobamate***. Case reports of **men** who had a reaction from it in ***rubber*** boots.

LD_{50}, **rat**: 555 mg/kg.

ISOPROPYLANILINE

Untoward effects: Inhalation causes eye and skin irritation, headache, weakness, dizziness, cyanosis, ataxia, dyspnea on effort, methemoglobinemia, and tachycardia. National Institute for Occupational Safety & Health sets upper exposure limit of 2 ppm.

LD_{50}, **rat**: 1.2 g/kg.

ISOPROPYLDIBENZOYLMETHANE

***Ultraviolet-A light-absorber*. In sunscreens.**

Untoward effects: Photoallergic reactions reported.

ISOPROPYL ETHER

= *Diisopropyl Ether* = *Diisopropyl Oxide* = *DIPE* = *IPE* = *2- Isopropoxypropane* = *2,2´-Oxybispropane*

Solvent, fuel additive.

Untoward effects: Inhalation causes eye, skin, nose, and respiratory tract irritation. Dermatitis on contact. Reacts with air to form unstable ***peroxides***.

Inhalation TC_{LO}, **human**: 800 ppm.

Animal exposure to fumes led to drowsiness, unsteadiness, narcosis, and unconsciousness.

LD_{50}, 2-week-old, **rats**: 6.4 ml/kg; adult **rats**: 16 ml/kg; LC_{50}, **goldfish**: 380 mg/l.

ISOPROPYL GLYCIDYL ETHER

= *IGE*

Untoward effects: May cause adverse health effects after inhalation, ingestion, or dermal and eye contact. Acute inhalation exposure led to mental confusion; and eye, skin, and respiratory tract irritation. Chronic exposure led to dermatitis and skin sensitization. OSHA sets exposure limits of 50 ppm, TWA.

Central nervous system depression in **mice**, **rabbits**, and **rats**. Subchronic inhalation by **rats** led to decreased weight, pneumonias, and respiratory distress.

LD_{50}, **rat**: 4.2 g/kg; **mouse**: 1.3 g/kg; inhalation LC_{50}, **rat**: 1100 ppm/8 h; **mouse**: 1500 ppm/4 h; topical LD_{50}, **rabbit**: 9.65 g/kg.

ISOPROPYL LANOLATE

Cosmetic plasticizer.

Untoward effects: Strongly comedogenic in **rabbit** trials.

ISOPROPYL LINEOLATE

= *Ceraphyl 1 PI*

Cosmetic plasticizer.

Untoward effects: Extremely comedogenic in **rabbit** trials.

ISOPROPYLMETHYLHEXENES

Used in tonnage for fragrances in soaps, detergents, cosmetic creams and lotions, and in perfumes.

Untoward effects: Both the aldehyde and alcohol forms at a 10% concentration in petrolatum produced no sensitization or irritation on **volunteers**.

Acute oral LD_{50} in **rats** and acute dermal LD_{50} in **rabbits**: > 5 g/kg.

ISOPROPYL MYRISTATE

Untoward effects: Comedogenic after topical use on **humans**, **guinea pigs**, and **rabbits**. Mildly irritating to **human** skin at 28 mg/day/3 days. Topical use may also be associated with hyperplasia of **rabbit** and **dog** skin. **Mouse** skin developed erythema and, later, lichenification with fissures.

LD_{50}, **mice**: > 100 ml/kg; acute dermal, **rabbits**: > 5 g/kg.

ISOPROPYL NEOPENTANATE

Untoward effects: Use in facial cosmetics is contraindicated in acne-prone **patients**.

ISOPROPYL PALMITATE

Untoward effects: A sensitizing agent in ***corticosteroid*** cream vehicles and in ***lipstick***. Comedogenic. Mildly irritating to **human** skin at 28 mg/day/3 days.

N-ISOPROPYL-*N*-PHENYL-*p*-PHENYLENEDIAMINE

= *IPPD*

***Antioxidant.* In *rubber* products and as a stabilizer in an insecticide.**

Untoward effects: Recurrent vesicular dermatitis of hands in a 41-year-old **male** after washing **his** Volkswagen. **He** evidenced bullous reactions to occlusive patches of Volkswagen tire and bumper guard shavings, and none to Chevrolet tire shavings. Its concentration in ***rubber*** varies from 0.4 to 3.5%. 80% of Volkswagens had ***rubber*** tires containing it. Itching petechial and purpuric eruptions at contact sites from ***rubber*** diving suit, elasticized shorts, and rubberized leg support bandages that contained it.

ISOPROPYL QUINOLINE

Fragrance in soaps, detergents, cosmetic creams and lotions, and perfume.

Untoward effects: No irritation or sensitization at 2% in petrolatum in **human volunteers**.

Applied full-strength to intact or abraded **rabbit** skin for 24 h under occlusion was slightly irritating.

LD_{50}, **rats**: 940 mg/kg; acute dermal, **rabbit**: 160 mg/kg.

ISOPROTERENOL

= *A-21* = *Aerolone* = *Aerotrol* = *Aleudrin* = *Aludrin(e)* = *Asiprenol* = *Asmalar* = *Assiprenol* = *Bellasthman* = *Euspiran* = *Euspirin* = *Isadrine* = *Isomenyl* = *Isomist* = *Isonorin* = *Isoprenaline* = *Isopropydrin* = *Isopropylnorepinephrine* = *Isorenin* = *Isovon* = *Isuprel* = *Izadrin* = *Mistarel* = *Neodrenal* = *Neo-Epinine* = *Norisodrine* = *Novodrin* = *Propal* = *Proternol* = *Respifral* = *Saventrine* = *Suscardia* = *Vapo-N-Iso*

***Sympathomimetic, bronchodilator.* Usually used as an inhalant (sol or powder) or sublingual tablets.**

Untoward effects: Abuse by a 41-year-old **female** led to 15 months' history of nausea, anxiety, headache, vertigo, palpitations, pallor, tachycardia, and hypertension suggestive of pheochromocytoma. One attack was accompanied by cardiac arrest. After an "attack", urinary ***catecholamines*** were increased. Be aware that surreptitious SC self-administration of it can mimic pheochromocytoma. An 11-year-old **abuser** was, apparently, intentionally abusing the ***fluorocarbon*** propellent.

Blood and cardiovascular: During IV infusion, a progressive decrease in platelet count to 82% after 15 min, returning to normal 20 min after cessation of infusion. One of its principle ill-effects is on the myocardium. A 43-year-old **male** on an IV infusion (1–2 μg/min) developed coronary artery spasm, bradycardia, hypotension, and near-syncope. Many case reports of induced cardiomyopathy on therapeutic doses. Marked tachycardia and fall in stroke volume with aggravation of pleuritic pain in a 68-year-old **female**. Anginal pain and increased myocardial production of ***lactate***, indicative of myocardial ischemia in **patients** treated for circulatory shock with it. Hypoxemia, hypercapnea, hypotension, palpitations, tachycardia, fibrillation, facial flushing, lymphocytosis, thrombocytopenia, eosinopenia, precordial pounding, nervousness, chest pain, premature ventricular contractions, and ventricular arrhythmias (due to accelerated depolarization of pacemaker cells) often occur. Large doses can cause myocardial necrosis (due to ischemia) and ventricular fibrillation. A high death rate in **asthmatics** in Great Britain was associated with high-dose inhalers purchased over-the-counter. It was estimated that > 3500 **asthmatics** died in Britain from overuse of inhalers during a 6 year period. **Patients** often state the disease is getting worse, requiring more frequent use of the inhaled drug. Paradoxically, the more it is inhaled, the worse the disease becomes, with increase in **patient** fears, anxiety, and confusion.

Central nervous system: Another principle ill-effect. Tremors, nervousness, weakness, dizziness, headache, and delirium (from overdosage or intolerance).

Miscellaneous: Bronchospasm, hypersensitivity, diaphoresis, inhibition of vesical detrusor action, slight hyperglycemia, nausea, vomiting, pharyngeal edema, hoarseness and chronically, ulceration and hyperplastic squamous epithelium of the throat and larynx, coughing, dryness of the throat and bronchi, clammy hands, tongue ulceration, blurred vision, decreased lower esophageal sphincter, and increased myocardial concentration of ***thallous chloride radiotracer***. Occasionally induces vasodilation and congestion of the nasal passage and eustachian tube. A 12-year-old **female**, after 6 years' treatment with varying doses tid, had discoloration and destruction of dental enamel, due to acidic nature of the drug.

IM, TD_{LO}, **human**: 14 μg/kg.

Drug interactions*:** Avoid use with most ***β-adrenergic-blockers, which will negate its effects. It and other ***sympathomimetics*** with ***monoamine oxidase inhibitors*** led to acute hypertensive crisis. Cardiac arrests and deaths reported from ***epinephrine*** injections in **patients** already taking isoproterenol. Death reported in a 46-year-old **female asthmatic** after excessive isoprenaline inhalation and ***ethyl alcohol*** consumption. ***Halothane*** can sensitize the heart to it. With ***antidepressants***, the effects of both drugs may be increased.

Corticosteroid pretreatment, by inhibiting intracellular ***calcium*** function, increases its toxicity in **animals** and **man**. Inhaled aerosol sprays cause arrhythmias, tachycardia, and myocardial lesions in **dogs**. **Greyhounds** are particularly sensitive and could become a useful **animal** model. Its normal effect of increased heart rate, ***oxygen*** consumption, and increased blood pressure and peripheral vascular resistance causes a steady decrease in cardiac output, and experiments in **cats** indicate its toxicity increases 15-fold when administered to a **patient** with an already overloaded heart. ***Epinephrine*** is *contraindicated* as it may precipitate death by increasing cardiac load through increased venous return of blood. Ischemia, anoxia, and severity of isoproterenol-induced myocardial infarction can be biologically titrated on a dose:body weight ratio. **Dogs** pretreated with it by aerosol have a high incidence of myocardial necrosis after 0.5 mg/kg SC/day/2 days. Control overdosage with ***sotalol***. 200 mg/kg, IP to **ducklings** (8 weeks) induced cardiotoxicosis. It has also been reported in **frogs**, **pigeons**, **quail**, **mice**, **rats**, and **turtles**. **Dogs** are more sensitive to it than **rats**. Single SC LD in **rats** is 680 mg/kg; in **dogs** it is < 10 mg/kg. In **dogs**, 2.5 mg/kg SC on 2 successive days led to massive myocardial necrosis. **Mice**, **rabbits**, **guinea pigs**, **hamsters**, and **reptiles** are suitable for myocardial-necrosing experiments. Teratogenic in **animals**.

LD_{50}, **rat**: 2.2 g/kg; **mouse**: 450 mg/kg.

ISOPULEGOL

= *p-Menth-8-en-3-ol*

Fragrance in soaps, detergents, cosmetic creams and lotions, perfume, and food.

***Untoward effects*:** No sensitization or irritation of 8% in petrolatum in **human volunteers**.

Full-strength to intact or abraded **rabbit** skin for 24 h under occlusion was severely irritating.

LD_{50}, **rat**: 1 ml/kg; acute dermal, **rabbit**: 3 ml/kg.

ISOSAFROLE

A fragrance in limited use for soaps, detergents, cosmetic creams and lotions, and perfume.

Untoward effects*:** No longer permitted in foods in the U.S. Was used with ***methyl salicylate in ***root beer*** and ***sarsaparilla*** drinks. No irritation or sensitization by 8% in petrolatum on **human volunteers**.

LD_{LO}, **human**: 500 mg/kg.

Full-strength to intact or abraded **rabbit** skin for 24 h under occlusion was moderately irritating. **Rats** fed 10,000 ppm in their diet had growth retardation and macroscopic and microscopic liver changes, and were dead by 11 weeks of age. In a 2 year feeding study on **rats** fed it at 1000; 2500; and 5000 ppm caused slight liver damage, as well as slight growth retardation in **females**. At the highest level, primary liver tumors, increased interstitial-cell tumors of the testes, and increased chronic nephritis noted. Both upper levels led to slight thyroid damage.

LD_{50}, **rat**: 1.34 g/kg; **mouse**: 2.47 g/kg.

ISOSORBIDE

= *AT-101* = *Hydronol* = *Ismotic* = *Isobide* = *NSC-40725*

Diuretic.

***Untoward effects*:** Mild nausea, feeling of stomach fullness, watery stool, thirst, increased blood urea nitrogen, headache (25%), dizziness, giddiness, backache, chest pain, eyeache, disagreeable taste, vomiting, anorexia, increased or decreased blood pressure, cardiac failure or arrhythmias, neuritis, tinnitus, increased SGOT and SGPT, thrombocytopenia, and rashes.

ISOSORBIDE DINITRATE

= *Astridine* = *Cardio 10* = *Cardis* = *Carvanil* = *Carvasin* = *Cedocard* = *Corovliss* = *Dignonitrat* = *Dilatrate* = *Diniket* = *Disorlon* = *Duranitrat* = *EureCor* = *FDS* = *Flindix* = *Frandol* = *Glentonin* = *IBD* = *Imtack* = *Isdin* = *ISDN* = *Iso-Bid* = *Isocard* = *Isoket* = *Iso-Mack* = *Isonate* = *Iso-Puren* = *Isorbid* = *Isordil* = *Isostenase* = *Isotrate* = *Langoran* = *Laserdil* = *Maycor* = *Myorexon* = *Nitorol* = *Nitrol* = *Nitrosorbon* = *Nosim* = *Rifloc Retard* = *Rigedal* = *Risordan* = *Soni-Slo* = *Sorbangil* = *Sorbichew* = *Sorbidilat* = *Sorbid SA* = *Sorbid TD* = *Sorbitrate* = *Sorbonit* = *Sorquad* = *Vascardin* = *Vasorbate* = *Vasotrate*

Antianginal, vasodilator. **Used sublingually, transdermally, IV, and orally.**

***Untoward effects*:** Headache, flushing, increased T-wave abnormality, palpitations, restlessness of arms, epigastric distress, transient weakness and dizziness, nausea, vomiting, diarrhea, drug rash, exfoliative dermatitis, slight methemoglobin increase, laryngeal swelling, nasal congestion, and potential for decreased sensitivity or response to exercise. Pallor, perspiration, postural hypotension, and collapse have occurred on therapeutic dosage. Sinus bradycardia and ventricular asystole in a 71-year-old **female**. Accidental transmittal from the chest of a 60-year-old **male** to his **wife** during coitis precipitated adverse effects in **her**. May decrease intraocular pressure. Hemolytic anemia in two glucose-6-phosphate dehydrogenase deficient **patients** after 10–20 mg PO tid/2–4 days.

Drug interactions: Concurrent administer of ***disopyramide phosphate*** induced xerostomia in 75-year-old **male**. Some **patients** have reactions when taking ***aspirin***. May decrease response to sublingual ***nitroglycerin***.

In **cats** and **dogs**, hypotension is the most serious adverse effects. **Owners** must be advised to wear gloves when applying the cream. A continuous rise in alkaline phosphatase in **dogs** after 3 weeks of treatment.

ISOSORBIDE MONONITRATE

= *Corangin* = *Elan* = *Elantan* = *Imdur* = *Ismo*= *Isomonat* = *Monicor* = *Monit* = *MonoCedocard* = *Monoclair* = *Monoket* = *Mono Mack* = *Monosorb* = *Olicard* = *Pentacard*

***Antianginal, vasodilator.* It is the active metabolite of the dinitrate form above.**

Untoward effects: Similar to the above. Headache is the worst adverse effect. By crushing a sustained release tablet and giving it in water to a 78-year-old **male**, led to chest pain, nausea, and ischemic changes.

ISOSPARTEINE

Untoward effects: A toxic alkaloid in ***Cytisus sp.*** (q.v.) = ***Scotch Broom***. ***Sparteine*** was once used as an oxytocic.

ISOSULFAN BLUE

= *Lymphazurin* = *P 1888* = *P 4125*

***Diagnostic aid.* In lymphangiography.**

Untoward effects: A severe, immediate life-threatening anaphylaxis after 1% SC in a 51-year-old **female**.

ISOTHIPENDYL

= *Andantol* = *Andanton* = *D-201* = *Daitop* = *Nilergex* = *Theruhistin* = *Thipendyl*

Antihistamine.

Untoward effects: Photosensitizer in **man**. Somnolence (8%) and nausea (4%).

LD_{50}, **mouse**: 222 mg/kg.

ISOTOMA LONGIFLORA

Untoward effects: Has psychoactive principles and, added to a Peruvian drink, to enhance hallucinogenic effects of other ingredients. Contains ***isotomin***, which can cause cardiac arrest.

Poisonous to **livestock**.

ISOTRETINOIN

= *Accutane* = *Isotrex* = *13-cis-Retinoic Acid* = *Roaccutane*

***Keratolytic.* Synthetic retinoid used in treating cystic acne.**

Untoward effects: Fatigue, headache, nausea, vomiting, increase in serum triglycerides, decrease in high-density lipoprotein, rash, pruritus, ejaculatory failure, xerostomia, anorexia, conjunctivitis, eye irritation, bone and joint pain, acute aseptic arthritis of the knee, pseudotumor cerebri, interference with blood ***glucose*** control in **diabetics**, exercise-induced asthma, depression, galactorrhea, xerosis, intracranial hypertension, papilledema, retinal hemorrhages, dry eyes, tinnitus, neutropenia, hepatitis, skin fragility, regional ileitis, hyperuricemia, corneal opacities, skin peeling, dry skin, hair thinning, phototoxicity, vasculitis, pancreatitis, alopecia, granulocytopenia, cheilitis, dry mucous membranes, amenorrhea, leukocytosis, thrombosis, decreased night vision, canaliform nail dystrophy, eosinophilic pleural effusion, delayed wound healing, keloid formation; erythema nodosum and irreversible skeletal toxicity in several ichthyosis **patients**; nosebleeds, prolonged state of confusion, loss of consciousness, as well as a generalized seizure in a 24-year-old **male** on 80 mg/day are among the many adverse effects reported in varying degrees. Dermatitis often simulates acute pityriasis rosea. Muscle pain and weakness in two **patients** (16 and 20 years) after 40 mg/day/4 weeks. *Staphylococcus aureus* nasal colonization, acute proctosigmoiditis in a 17-year-old **male** after 40 mg/day/several days. A 19-year-old **male** developed hypercalcemia with bone pain, nausea, and lethargy after a dosage increased to 1.6 mg/kg/day. The condition resolved after decreasing dosage to 1.2 mg/kg/day. A 9-year-old **male** treated with 5 mg/kg/day/5 months for fibrodysplasia ossificans progressiva developed dense metaphyseal bands and growth arrest. A 19-year-old **female** received 40 mg twice daily for the first 18 weeks of pregnancy. Her **infant** was born with abnormalities of the craniofacial features, cardiovascular, respiratory and central nervous system. A 16-year-old **female** took 40 mg/day/first 15 weeks of gestation. Her **baby** had multiple malformations (fetal hydrocephalus and ear anomalies) and psychomotor retardation. Hydrocephalus, micrognathia, cardiac abnormalities, and poorly positioned ears in a 19-week aborted **fetus** of a 21-year-old **female** who had received 40 mg/qid. These are typical of numerous case reports. Other cases have revealed microphthalmia, cleft palate, hypertelorism, and small mouths. By 1988, 66 cases of embryopathy and 90 abortions had been reported to the FDA.

Poisoning: Facial flushing, mild tachycardia, tachypnea, and hypertension in a 17.7 kg 21-month-old **female**, after accidental ingestion of ~1.12 g. **She** recovered within 24 h. Depression, psychosis and rarely suicide ideation, attempts, and successful suicides.

Drug interactions: Levels of ***carbamazepine*** (1 mg/kg/day) clearance increase with isotretinoin.

Most signs of toxicity in **animals** of various retinoids are similar and include decreased food intake, decreased weight gain, erythema, alopecia, mucosal changes, long bone fractures, increased liver weight, pale and/or mottled liver, and decreased testicular weight. Despite decrease of most tumors in **rats**, doses of 8 and 32 mg/kg/day/ 105 weeks increased the incidence of pheocromocytomas.

LD_{50}, **rat**: > 4 g/kg; **mouse**: 3.4 g/kg; **rabbit**: ~2 g/kg.

ISOTROPIS

Untoward effects: All species seem to be very palatable to **cattle** and **sheep** in Australia, yet poisonous. Symptoms are weakness, dullness, staggering, and collapse. Post-mortem shows kidney and liver necrosis, and congestion and ecchymoses throughout the body.

I. atropurpurea = ***Poison Sage*** causes similar symptoms, and gastroenteritis is the predominant post-mortem finding.

2-ISOVALERYL-1,3-INDANEDIONE

= *PMP* = *Rat Murder* = *Valone*

***Rodenticide, insecticide*. Usually at 0.05% in rodenticides.**

Untoward effects: Meat from anticoagulant-killed **nutria** caused secondary poisoning in 15/16 **mink** and 9/17 **dogs** after they ate it.

LD_{LO}, **human**: 50 mg/kg.

LD_{LO}, **rat**: 250 mg/kg.

ISOXICAM

= *Floxicam* = *Maxicam* = *Pacyl* = *Vectren* = *W 8495*

***Anti-inflammatory*. Non-steroidal.**

Untoward effects: In France, a 28-year-old **female** on 200 mg/day developed fatal epidermal necrolysis. Reports of induced Stevens–Johnson syndrome also occurred in France. There were 13 cases of Lyell's syndrome (toxic epidermal necrolysis) reported in France, of which four were fatal. These skin hypersensitivity problems were attributed to a substance, identified as **"*553*"**, only in material made in Spain for French consumption. **Patients** stabilized on ***warfarin*** received 200 mg/day/~6 weeks. **They** all requires a decrease in **their *warfarin*** dosage, ~20% in the last 3 weeks.

ISOXSUPRINE

= *Dilavase* = *Duphaspasmin* = *Duvadilan* = *Duviculine* = *Isolait* = *Navilox* = *Suprilent* = *Vadosilan* = *Vasodilan* = *Vasoplex* = *Vasoprine* = *Vasotran*

***Vasodilator*. Used orally, IV, and IM.**

Untoward effects: Palpitations, flushing, postural hypotension, dizziness, vertigo, tachycardia, chest pain, occasionally severe rash, inhibition of vesical detrusor action, nausea, vomiting, pulmonary edema, diplopia, transient increase in intraocular pressure, alcoholic odor on breath, gastralgia, cramps, and shock.

LD_{50}, **rat**: 1750 mg/kg; **mouse**: 1.1 g/kg; IV, LD_{50}, **mouse**: 48 mg/kg; **dog**: 57 mg/kg.

ISRADIPINE

= *Clivoten* = *DynaCirc* = *Dynacrine* = *Esradin* = *Lomir* = *PN-200–110* = *Prescal* = *Rebriden*

***Calcium channel-blocker, antihypertensive*. A *dihydropyridine*, chemically related to *nicardipine* and *nifedipine*.**

Untoward effects: Flushing, headache, tachycardia, and dizziness. Less commonly, ankle edema, palpitations, leukopenia, nausea, and tinnitus. These effects are often dose-dependent. Cardiovascular problems (atrial and ventricular fibrillation, myocardial infarction, and transient ischemic attacks) reported. A 2-year-old **male** accidentally ingested 2.5 mg, leading to blood pressure 68/40, heart rate 112/min, and respirations 24/min. Usual therapeutic serum levels are 0.0005–0.002 mg/l and toxic at 0.01 mg/l.

Drug interactions: May increase serum concentration of ***propranolol*** (up to 28%).

ITRACONAZOLE

= *Itrizole* = *Oriconazole* = *R-51211* = *Sempera* = *Siros* = *Sporanos* = *Sporanox* = *Triasporin*

***Antifungal*. Orally active *triazole*. Also used topically.**

Untoward effects: A 51-year-old **patient** on 200 mg/day/ 10 weeks developed reversible alopecia; a 49-year-old **female**, after 200 mg twice daily/1 week for aspergillosis, developed pericardial effusion and pericarditis. The infection recurred after a 6 week interuption in therapy. Retreatment was initiated and, after 2 weeks, led to peripheral lung edema and cardiac hypertrophy. Dose-related nausea, vomiting, diarrhea, and abdominal discomfort reported. Edema, allergic rashes, hypokalemia, rhabdomyolysis, headache, dizziness, drowsiness, sleeplessness, tinnitus, hypertension, impotence, albuminuria, and hepatitis have also occurred. Use by 166 pregnant **women** during the first trimester of **their** pregnancies led to ~10% more major **fetal** malformations than controls.

Drug interactions: Low concentrations can cause extensive inhibition of CYP3A4 drug-metabolizing enzyme. ***Food*** and gastric acidity increase absorption from capsules, but decrease it from oral suspensions. ***Antacids, antihistamines, lansoprazole***, and ***omeprazole*** decrease its absorption. Serum concentrations are decreased by ***carbamazepine, didanosine, grapefruit juice, isoniazid, phenobarbital***, and ***rifampin***. It increases serum concentration of ***alprazolam, astemizole, cisapride, cyclosporine, digoxin, felodipine, lovastatin, methyl prednisolone, midazolam, quinidine, rifabutin, ritonavir, simvastatin, tacrolimus, terfenadine, triazolam***, and ***warfarin***, increasing their toxicity. It inhibits CYP3A4 metabolism of the macrolide antibiotics ***clarithromycin, erythromycin,*** and ***troleandomycin***, increasing their serum concentration. Cardiac arrhythmias and deaths when taken with ***cisapride*** and ***terfenadine***. Increases neurotoxicity of ***vincristine***. Rhabdomyolysis in a **patient** taking it with ***niacin*** and ***lovastatin***. Possible interaction with it and ***fluoxetine*** 20 mg/day in a 34-year-old **male**, leading to anorexia. Unlike ***ketaconazole***, it does not have a potential to interfere with ***steroid hormones*** in **patients**, but can still decrease the effectiveness of some ***oral contraceptives***. Severe hypoglycemia with oral ***hypoglycemic agents***.

Chronic exposure increases serum ***cholesterol*** in **rats**. IP itraconazole in **guinea pigs** increases ***digoxin*** serum levels. Vasculitis reported in **dogs**.

LD_{50}, **rats** and **mice**: > 320 mg/kg; **dogs**: > 200 mg/kg.

ITRAMIN TOSYLATE

= *Cardisan* = *Nilatil* = *NLT* = *Tostram*

Vasodilator.

Untoward effects: Severe gastrointestinal distress, drowsiness, headache, and chest pain in ~3% of anginal **patients**.

In **dogs**, IV use leads to hypotension, peaking in 29 min and lasting 1.5 h.

IVA sp.

Untoward effects: ***I. augustifolia = Narrowleaf Sumpweed***, a common range plant in the southern U.S. that causes abortions in **cows** consuming it during their second and third trimesters.

I. axillaris are made poisonous for range **livestock** by ***selenium*** absorbed from the soil.

I. xanthifolia = False Ragweed = Fausse Herb à Poux = Marsh Elder (q.v.) in the U.S., Canada, and Mexico, causes undesirable bitter flavors in ***milk*** or milk products after consumption by **cows**. Some **people** develop a dermatitis from contact with the plant or the oleoresins in its pollen. Its air-borne pollen in the autumn is highly allergenic and a common cause of hay fever in many **persons**.

IVERMECTIN

= *Acarexx* = *Avermectin* B_1s = *Cardomec* = *Cardotek-30* = *Eqvalan* = *Heartguard 30* = *Ivomec* = *Mectizan* = *Panomec* = *Phoenectin* = *Stromectol* = *Zimecterin*

Anthelmintic, insecticide, acaridicde.

Untoward effects: Adverse reactions in 28/87 (32%) of **patients** in Sierra Leone, treated for onchocerciasis (river blindness), a disease common to Africa, the Middle East, and Latin America. Occasionally, rash, pruritus, fever, tender and enlarged lymph nodes, joint and bone pain, and headache reported. Rarely, hypotension and insomnia. Deaths probably associated with its use in treating scabies in elderly **patients** with at least 0.2 mg/kg in reports from Canada, The Netherlands, and the U.S.

In **animals**, local tissue swellings and host–parasite reactions can occur. Do NOT treat **cattle** or **swine** within 3 weeks of slaughter. Adverse reactions, central nervous system dysfunction, and death have been reported in some **shelties** and **collies**. **Australian shepherds** and **Old English sheepdogs** may also be sensitive to treatment with it. Treatment of **dogs** with heartworms often causes tremors, ataxia, seizures, depression, mydriasis, dehydration, and possibly death.

IVY

Untoward effects: ***American Ivy = Parthenocissus quinquefolia = Psedera quinquefolia = Vigne vierge = Virginia Creeper = Wilder Wein***, a climbing vine in the eastern U.S. with small blue berries. The berries and leaves contain ***oxalic acid*** and can be gastrointestinal corrosive, and cause nausea, abdominal pain, vomiting, diarrhea, and headache after ingestion. Death has occurred in **animals** and **man**. ***Oxalate*** ions precipitate ***calcium*** ions as ***calcium oxalate*** (q.v.) crystals causing kidney tubule damage. The resulting hypocalcemia can adversely affect the heart, neuromuscular junction, and the central nervous system. Gavage with autumn-growth material in **budgies** led to depression and intermittent vomiting within an hour after the second dosing and 3–4 h after the first gavage. Symptoms disappeared after about 5 h. Repeated gavage with spring-growth led to no adverse effects.

English Ivy = Hedera helix (q.v.). The berries and leaves are poisonous, due to ***saponins*** causing gastrointestinal irritation, nausea, vomiting, salivation, diarrhea, weakness, thirst, bradypnea, syncope, headache, tremor, incontinence, sweating, excitement, visceral vasodilation, and, in severe cases, ataxia and coma. Has caused allergic contact dermatitis and photosensitization in some **people**. If its steroidal ***saponin, hederagenin***, is absorbed systemically, hemolysis occurs. A **capuchin monkey** escaped from captivity and hid in an area heavily covered with the plant. It ate the fruit and died. **Cattle** have been poisoned by leaf clippings.

Ground Ivy = Ale Hoof = Creeping Charlie = Creeping Jenny = Gill-Over-The-Ground = Glechoma hederacea = Nepeta hereacea = Turnhoof. Poisoning in **animals** is rare, as it is bitter. **Horses** are poisoned after eating large quantities, as in hay. Symptoms are sweating, salivation, mydriasis, pulmonary edema, dyspnea, and tachypnea. Rarely fatal. Contains a volatile irritating aromatic oil. Also see ***Nepeta***.

J

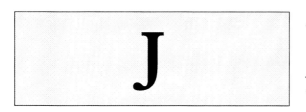

JACARANDA ENVACO

Untoward effects: In the northern part of South Africa, **carpenters** working for many years with its wood have developed a contact dermatitis.

JACK BEAN

= *Canavalia ensiformis*

Untoward effects: A common North American and West Indies legume plant, containing the lectin ***Concanavalin A***, a phytohemagglutinin. It grows well in acid soils. Contains a protease inhibitor and ***nitrogen***.

Although unpalatable to **cattle**, many in Zimbabwean ranges have died from eating it. Symptoms are severe diarrhea, decreased milk production, weakness, inability to eat or drink, and hind leg stiffness. Post-mortem showed lymph gland hemorrhages, liver necrosis, necrotic ulceration of gall bladder, emphysema, and enteritis. The seeds and foliage are poisonous.

Meal LD, **cattle**: 40 g/kg.

JACOBINE

The major *pyrrolizidine* alkaloid found in *Senecio jacobea* (q.v.), that is metabolized in the liver to toxic *pyrroles*.

Untoward effects: Hepatotoxic. Found in ***honey*** made from ***S. jacobea*** nectar, especially in the northwestern U.S. **Horses**, **cattle**, **rats**, and **chickens** are very susceptible to its toxicity.

IV, LD_{50}, **mouse**: 77 mg/kg.

JALAP

= *High John Conqueror* = *High John Root* = *High John the Conqueror* = *John Conqueror* = *St. John the Conqueror Root*

From the dried roots of *Exagonium purga*, *Ipomoea jalap*, *Ipomoea purga*, and *Exagonium jalapa*.

Untoward effects: Powerful, drastic cathartic. Can cause cramping, tenesmus, and hypercatharsis. **Workers** should wear mask, goggles, and gloves when working with it.

JANITHREMS

Untoward effects: Consists of three new tremorgenic mycotoxins, ***Janithrem A, B***, and ***C***, from ***Penicillium janthinellum***, obtained from ***ryegrass*** pastures, where ***ryegrass*** staggers outbreaks occurred in New Zealand **sheep**.

JASMINE ABSOLUTE

Chief constituent is *benzyl acetate* (q.v.). Used in perfume, cosmetic creams and lotions, soaps, detergents, and foods.

Untoward effects: No irritation on **human volunteers** with 3% in petrolatum. The same concentration caused sensitization in 2/25 **volunteers**. Another report indicates it was a frequent cause of allergic reactions in perfumes.

Undiluted material was non-irritating on the backs of **hairless mice** and **swine** or to intact or abraded **rabbit** skin for 24 h under occlusion. No phototoxic effects of undiluted material on hairless **mice** and **swine**. Ingestion by **cats** led to vomiting, diarrhea, mydriasis, increased temperature, and collapse.

LD_{50}, **rats** and acute dermal LD_{50}, **rabbits**: > 5 g/ kg.

JATROPHA sp.

= *Bini da zugu (Hausa)* = *Botu je (Yoruba)* = *Oluluidu (Igbo)*

Untoward effects*: *J. curcas* = *Barbados Nut* = *Perchnut* = *Physic Nut* = *Pignon d'Inde* = *Purging Nut* = *Ratan Jot* = *Udukaju (Thai) = ***Zeit*** (Yemen), ***J. gossypilfolia* = *Flaira*** (q.v.) and ***J. multifidia*** in Africa are highly irritant leading to burning sensation in the mouth and throat, nausea, vomiting, diarrhea, abdominal pain, weakness, collapse, and sloughing of intestinal mucosa.

J. curcas is a tall tree native to the West Indies and tropical America. The seeds contain a purgative oil more drastic than ***castor oil***; a toxic protein (***jatrophin*** or ***curcin***), and a toxic resin. Many cases of violent vomiting and diarrhea reported in Florida where it has been sold as ***pistachio*** and ***Chinese Peanut Tree***. The seeds are reported to be carcinogenic and the tree's sticky yellow sap (latex) has caused skin and eye inflammation. It is also toxic internally. The leaves, roots, and stems are poisonous as well. **Children** mix the latex with water to blow bubbles, but, unfortunately, some occasionally gets into **their** eyes or mouths, with irritant effects. During a 3 month period, eight cases of acute poisonings from the seeds occurred in **children** (ages 2–9 years) that were treated in the emergency room of a Pretoria, South Africa medical school. **Livestock** only browse it during periods of drought. The cake left after extraction of the oil (used in lamps) has caused purgation when fed. **Sheep** and **goats**

poisoned by the ***J. curcas*** showed diarrhea, dehydration, anorexia, leukocytosis, sunken eyes, and death. Post-mortem showed rumen, reticulum, lung, kidney, and myocardial hemorrhage; catarrhal or hemorrhagic enteritis, fatty livers, pulmonary congestion, edema, and straw-colored fluid in serous cavities.

In Texas, ***J. dioica*** = ***J. spathulata*** = ***Leatherstem*** and ***J. cathartica*** = ***Berlandier Nettlespurge*** have caused toxicity in **animals**. ***J. cathartica*** has poisoned **sheep**, **goats**, and **rabbits**. **Rabbits** have died in convulsions within 4 h after eating it at 2% of their body weight.

J. dioica led to progressive anemia and death in a **goat** that ate it at 3.7% of its body weight.

J. multifida = ***Còral Plant*** = ***"Nutmeg Tree"***, a common ornamental, tall, bushy tree in Florida, whose seeds are tasty, has caused (within 7–8 h after ingestion) violent vomiting and diarrhea in **children** and **adults**.

J. urens is extremely poisonous, as contact of a young plant with the wrist of a **gardener** in Kew led to such serious poisoning that, for 5 min, **he** was assumed to be dead. Its bark is known as ***pini-pini***, which is caustic and a blistering agent.

JELLYFISH

Consists of a number of dangerous species of cnidarian coelenterates, e.g. *Chironex*, *Chiropsalmus*, *Chrysaora*, and *Cyanea*. They are found in both fresh and salt water. Some are transparent, and others multi-colored, and range in size from a few millimeters to > 6 ft in diameter. They (~60 species) are free-swimming, but often ride the water currents. After storms at sea, their tentacles often break free and drift as small, transparent items, still capable of firing their nematocysts. On the U.S. east coast, they are found from Cape Cod down to Florida. Tentacles of large jellyfish can reach 50–100 ft. It is estimated that at least 1000 people are stung annually. The amount of venom in a nematocyst is very small, but, in a "sting", hundreds of thousands or millions of nematocysts may be involved.

For *Portugese Man of War*, see *Physalia pelagica*.

Untoward effects: Stings cause a linear papular urticaria with erythema and edema lasting for hours up to several days in most **people**. In some of the more serious stings described below, lead to muscle cramps, nausea, vomiting, abdominal rigidity, choking sensations, weakness, respiratory paralysis, and cardiovascular shock.

Chironex fleckeri = ***Tropical Jellyfish*** has caused dangerous and fatal injuries to **swimmers**, due to its harpoon-like stillets that pierce **human** skin. Found near the Solomon Islands, the Phillipines, Malaya, and the South China Sea.

Chiropsalmus quadrigatus = ***Sea Wasp***, a free-swimmer, has killed many **people**, near Queensland. Its **victims** die in agony within a few seconds. Found in shallow waters.

Chrysaora quiquecinha, a **jellyfish** occurring in swarms along the eastern coast of the United States and British coastal waters, has caused swelling, edema, and pain after stinging many **people**.

Cyanea capillata = ***Hairy Stinger*** = ***Lion's Mane***, 3–7 ft in diameter (yellow-pink or pale orange) with short yellow-red tentacles, and less prominent orange-pink tentacles extending down to 75 ft. A free-swimmer, found in British coastal waters drifting near the shore at depths of 2–10 ft.

Physalia utriculus = ***Bluebottle Jellyfish***, closely related to ***Portugese Man of War***. A 16-year-old **female**, ocean-swimming near Hawaii, developed classic symptoms after being stung. The welts lasted 2 days and by the tenth week had eruptions diagnosed and confirmed by biopsies of granuloma annulare. The ***Sea Nettle*** is a venomous **jellyfish** found in the Chesapeake Bay and other mid-Atlantic coastlines, where its nematocysts cause painful skin wounds.

Stomolophus nomurai = ***Chinese Jellyfish*** has caused **human** deaths.

JET FUELS

Complex mixtures of *hydrocarbons*. *JP-4* is *shale*-derived and the commonly used form in the Air Force. *JP-7* is often used in reconnaisance and supersonic aircraft; *JP-8* is *kerosene*-like and used by NATO; *JP-10* is a synthetic for use in cruise missiles. Space shuttles and Titan missiles use *hydrazines*. *Jet A* is *kerosene* (q.v.).

Untoward effects: Acute exposure to ***JP-4*** leads to central nervous system suppression, dizziness, nausea, vomiting, and drowsiness. A **pilot**, exposed to a fuel leak while in flight, later had staggering gait, muscle weakness, and decreased sensitivity to pain lasting up to 36 h. Chronic exposure to jet fuels in 30 **workers** in jet manufacturing for 4–32 years had an increase in neuropsychiatric symptoms, including anxiety, depression, fatigue, personality changes, attention deficits, and electroencephalogram abnormalities. Chronic jet fuel exposures reported in other studies leading to headache, dizziness, sleep disturbances, depression, anxiety, irritability, memory impairment, and nausea. ***JP-4*** and ***JP-7*** contain ***n-hexane*** (q.v.), a cause of peripheral neuropathy after chronic exposure; jet fuel exposures in aircraft factories produced similar symptoms. Inhalation of jet fuel vapors leads to irritation of the respiratory tract, central nervous system depression, headache, nausea, dizziness, palpitations, and chest pressure at 500 ppm. Bacteria can live in crude oil, and are probably responsible for its formation. Therefore, it is no surprise that they can proliferate in aircraft jet fuels. They then multiply in water vapor droplets (– 45°C or over at high altitudes) and interfere with fuel flow. Various bactericidal additives are used to inhibit bacterial microflora in jet fuels.

Weight decrease in male, but not female, **rats** on ***JP-8***; inhalation leads to decreased weight of fetuses. ***JP-10*** causes renal carcinomas in male **rats**, enhances nephrosis with tubular hyaline degeneration and plugging in old male **rats**.

LD_{50}, **mouse**: 500 mg/kg.

JEWELRY

Untoward effects: Like many beautiful things, it has its dangerous side. Skin reactions from ***gold*** (q.v.) in the past have been from the radioactivity (β-rays) of ***radon***-222 used in the 1930s, stored in ***gold*** tubes. Dermatitis has also been reported from ***nickel, chromium***, and ***pearls***. Anaphylaxis in a 21-year-old **female** from a necklace, due to ***castor bean*** hypersensitivity and poisoning from ***Abrus = Rosary Peas***, with their glossy scarlet and black seeds, found on Caribbean jewelry.

JIN BU HUAN

A Chinese herbal anodyne preparation.

Untoward effects: Unintentional ingestion by **children** 13, 23, and 36 months led to lethargy, bradycardia, respiratory depression, and hypotension. Acute hepatitis in **adults**, 24–66 years, within 7–52 weeks (mean 20 weeks) taking 2–4 tablets/day tid or 1–3 tablets at night. Symptoms include fever, fatigue, nausea, pruritus, abdominal pain, jaundice, and hepatomegaly. Banned for importation into the U.S. by the FDA, but supplies do come in, innocently or otherwise.

JOJOBA BEAN

= *Goat Nut* = *Ho-Ho-Ba* = *Simmondsia chinensis*

In the *Buxaceai* (*Boxwood*) family. It is a substitute for sperm oil. Has been used in *penicillin* manufacture, to decrease foaming in the vats and increase yield.

Untoward effects: The oil used in cosmetics has caused contact dermatitis and allergic reactions in some **patients**. Erythema and vesicles on **their** forearms after covered patch tests with 20% concentration on 5/6 **volunteers**. A jojoba wax is also used.

JONQUIL

= *Narcissus jonquili*

***Ornamental flowering plant*. Also see *Daffodils*.**

Untoward effects: Poisoning from ingestion of the bulbs develops rapidly. Symptoms are abdominal cramps, shivering, light-headedness, diarrhea, and occasionally vomiting. Handling the bulbs leads to "bulb fingers", a contact dermatitis in sensitive **individuals**, with wheals, probably due to ***histamine*** release. Central nervous system excitement and depression with collapse and coma have occurred in the very **young** or **elderly** from related plants, if severe vomiting has not occurred.

JOSAMYCIN

= *Alplucine* = *EN 141* = *Isalide* = *Jomybel* = *Josacine* = *Josamina* = *Josamy* = *Josaxin* = *Leucomycin* A_3 = *Wilprafen*

***Antibiotic*. A *macrolide*.**

Untoward effects: Pseudomembranous colitis in a 65-year-old **female** after 2 g/day/10 days. Pedal edema in a 56-year-old **male** after 1 g twice daily/3 days. Confirmed by rechallenge. Subcorneal pustular dermatosis after a 15 day course of treatment in a 52-year-old **male**.

Drug interactions: Significant increase in ***cyclosporine*** serum levels after high dosage (1 g tid) in three heart transplant **patients**.

JUANULLOA OCHRACEA

Climbing epiphytic shrub ~10 ft tall in Colombia.

Untoward effects: Contains the alkaloid ***parquine***, and the shrub is referred to as ***ayahausca*** (q.v.) by the **natives**. This is the name given to an intoxicating hallucinogenic *Banisteriopsis* drink, and Juanulla is used as an additve to the drink.

JUI

A traditional Chinese herbal medicine often used as a *tea*.

Untoward effects: A 51-year-old Japanese **female** ingesting it developed thrombocytopenia with gingival bleeding and numerous petechiae on legs and forearms. Confirmed by rechallenge.

JUNIPER

Untoward effects: Its berries (~1 million lbs imported annually from Europe) distilled with spirits producing ***gin*** (q.v.) that caused many deaths, especially in **children**. Its berries and plant parts have been smoked or used in ***teas*** leading to euphoria. Ingestion of a few berries has produced few symptoms other than nausea. The berries contain up to 1.5% of volatile oil and 10% resin. It is mentioned in the Bible in the Book of Kings. The berries are stored for at least a year, to ensure they lose their initial undesirable ***terbinthinate*** odor. Juniper oil, often used as a diuretic, has caused renal and gastrointestinal irritation and central nervous system toxicity. Some **people** working with the tree's wood have developed a contact dermatitis. No irritation or sensitization of the oil at 8% in petrolatum on **human volunteers**. A patch test using full-strength berry oil for 24 h caused irritation in 2/20 **volunteers**. Approximately a ton/year used in the U.S. for detergents, soaps, cosmetic creams and lotions, and perfumes. Patch tests with ***juniper tar*** (***Oil of Cade***) gave positive results in verifying the causative agent of a severe eczematous rash in the perianal, intergluteal, perineal, and scrotal areas of a 23-year-old **male**. Sensitivity to some typing papers was associated with sensitivity to ***juniper tar***. No irritation or sensitization by 2% tar in petrolatum on **human volunteers**.

On occasions, **cows** lacking desirable foliage have eaten large quantities of the needles and aborted or had weak **calves**. Undiluted oil led to no irritation on the backs of **swine** and hairless **mice**. Applied to intact or abraded **rabbit** skin for 24 h under occlusion revealed it to be moderately irritating. No phototoxic effects noted on **swine** or hairless **mice**.

Tar and ***Oil***, LD_{50}, **rats**: ~8 g/kg; acute dermal, **rabbits**: > 5 g/kg.

JUSSIAEA

Untoward effects: ***J. peruviana*** causes chronic poisoning in Colombian **cattle**. Symptoms are severe nephritis, cardiac and locomotor problems, edema, anorexia, and constipation or profuse diarrhea. Post-mortem showed generalized edema, hemorrhagic gastrointestinal tract, severe nephritis and hepatitis, generalized ecchymoses and petechiae (especially cardiac), endocarditis, and muscle degenration. **Sheep** and **horses** are also susceptible to its toxicity.

In Brazil, ***J. suffruticosa*** = ***Cruz de Malta*** = ***Jusseia*** is toxic to laboratory **animals**, **poultry**, and **cattle**.

JUSTICA PECTORALIS var STENOPHYLLA

= *Mashihiri* = *Masho-hara* = *Ya-Ko-Yoó*

Untoward effects: Its powdered leaves are added to ***Virola***, an hallucinogenic snuff in Colombia, Brazil, and Venezuela; presumed to also contain *N,N*-***dimethyltryptamine***, or it may only improve the aroma of the mixture.

JUTE

Untoward effects: Occasional byssinosis in **workers**. Sensitivity in **people** handling jute in the form of burlap or carpeting. Skin irritation, redness and swelling, dry eczema and papillary hypertrophy with occasional suppuration in **people** handling jute thread.

K

KACHASU

Untoward effects*:** In Zambia, there are clusters of esophageal cancer in areas where kachasu, a local fermented ***corn liquor is sold in stores. It is, apparently, due to its *N-**nitro-sodimethylamine*** (q.v.) content.

KAEMPFERIA GALANGA

***Untoward effects*:** Rhizome is, supposedly, used as a hallucinogen in New Guinea.

KAFFIRCORN

= *Broom Corn* = *Guinea Corn*

Untoward effects*:** **Cattle** losses from grazing it reported in Rhodesia. As with other ***Sorghum sp., it can, under some conditions, contain large amounts of ***prussic acid*** (q.v.), ***durrhine***, and ***nitrates***. Imported into Czechoslavakia for **livestock** feeding, it has caused symptoms of restlessness, asthma-like spells, rumen atony, slobbering, and paralysis.

KAINIC ACID

Derived from *Digenea simplex*, a red algae; once used as an oral anthelmintic for man.

Untoward effects*:** An ***amino acid excitotoxin that causes learning and memory impairment in **rats**, and neurotoxicity after injections into **mammals** and **amphibia**. Induces epilepsy after IP injections in **rats**.

KALANCHOE

Untoward effects*: *K. lanceolata* = *Gadalin Kura (Hausa), in Nigeria causes acute gastroenteritis when fed to **cattle**.

K. paniculata and ***K. thyrsiflorae*** have poisoned South African **sheep**. Post-mortem evidenced hemorrhagic diarrhea; bronchial, tracheal, and pulmonary hemorrhages; hyperemia of small intestine, emphysema, and edema.

K. prolifera has caused decreased hemoglobin and deaths in Rhodesian **cattle** and **sheep** when eaten.

~200 species exist, and five species tested developed to limb tremors, depression, incoordination, paralysis, collapse, and death in **chicks**. **Children**, **cats**, **dogs**, **mice**, and **rabbits** can be affected by its cardiac glycosides, including the ***bufadienolide daigremontianin***. **Cats** evidence dyspnea, convulsions, and paralysis; **chicks** show dyspnea, depression, ataxia, trembling, convulsions, paralysis, and death; **mice** show muscle spasms and paralysis; **rabbits** evidence dyspnea, teeth-grinding, opisthotonus, and paralysis.

KALE

A variety of *Brassica oleracea*.

Untoward effects*:** In a 3 year survey of goiter in Australian **schoolchildren**, goitrogenic substances in the milk of **cows** fed on kale (and other *Brassicae*, such as ***rape) appeared to be the cause. ***Thallium*** in German kale has caused illness in **people**. Eaten raw and in excess (roots, seeds, or flowers), it can cause gastrointestinal distress and hematemesis. **Children** are most apt to be affected.

The primary toxin in kale poisoning is *S-**methylcysteine sulfoxide*** (***SMCO***) and is known as the ***kale anemia factor*** that causes intravascular hemolysis. Fermentation of ***SMCO*** yields ***dimethyl sulfoxide***, the active hemolysin. Kale also accumulates ***nitrates*** and ***oxalates***. Some kales contain appreciable amounts of ***thiocyanates***. **Cattle** and **goats** are more susceptible than **sheep** to its toxicity. Causes rumen atony, dyspnea, constipation or fetid diarrhea, hemoglobinuria, icterus, hyperchromic and macrocytic anemia, weakness, ataxia, blindness, and pulmonary emphysema. May cause goiter in **animals** in ***iodine***-deficient areas. It has an estrogenic effect on **rats** and **mice**. The effect of 1 kg of kale is equal to that of 24 μg ***estradiol benzoate***. Kale contains a natural carcinogen, ***neochlorogenic acid***, which is metabolized to ***caffeic acid***. **Rabbits** and **guinea pigs** do not become anemic when fed kale.

KALLSTROEMIA

= *Carpet Weed*

Untoward effects*: *K. hirsutissima* = *Hairy Caltrop and ***K. parviflora* = *Warty Caltrop***, after ingestion, cause knuckling over of hind limb fetlocks; **animals** usually lie down, develop posterior paralysis, and some convulse before dying. Death losses occur in **cattle**, **sheep**, and **goats**. Affected **rabbits** may show similar symptoms. **Sheep** often walk on their knees before convulsions develop. Post-mortem reveals congestion and/or hemorrhage in lungs, heart, kidneys, stomach, and intestines.

KALMIA

= *Laurels*

Untoward effects*: *K. augustifolia* = *Calfkill* = *Kalmia à Feuilles Etroites* = *Lambkill* = *Narrow-Leafed Laurel* = *Sheep Laurel is extremely poisonous and found primarily in the eastern U.S. Contains the resinoid, ***andromedotoxin*** (q.v.) or ***grayanotoxin***. ***Honey*** from the nectar of its flowers has been toxic to **people** leading to bradycardia, vomiting, hypotension, convulsions, incoordination, arrhythmias, progressive paralysis of arms and legs, bradypnea, coma, and even death.

Salivation, weakness, vomiting, incoordination, and collapse noted in affected **cattle**, **sheep**, and **goats**. Suspected as a cause of death in **zebras** in a New Jersey safari park. **Rats** have died after eating a leaf extract.

K. latifolia* = *Calico Bush* = *Ivybush* = *Mountain Ivy* = *Mountain Laurel* = *Poison Laurel*.** **Humans** are not apt to be poisoned by ***honey made from its nectar because it is bitter. All parts of the shrub-like tree are poisonous. Symptoms include gastroenteritis, vomiting; watering of eyes, nose, and mouth; depressed cardiac (bradycardia, hypotension), central nervous (incoordination and leg paralysis), and respiratory systems, and death. In eastern North America. **Sheep**, **goats**, and **cattle** are usually affected. Symptoms include salivation and frothing at the mouth, nausea, vomiting, eyes watering, nasal discharge, bradycardia, spasms, incoordination, muscular paralysis, respiratory failure, and death. **Deer** and **grouse** may not be affected. Bradycardia in **frogs**.

K. occidentalis* = *Bog Laurel* = *Sagaang Kawhlaa* = *Small Kalmia* = *Xilkagan. Known by Alaskan **Indians** as the ***Poisonous Hudson's Bay Tea***. It will "make you drunk".

KAMALA

***Anthelmintic.* Once, a "safe favorite".**

Untoward effects: Overdosage (18–20 g) was lethal to most of 91 **sheep** and the rest had to be euthanized.

KANAMYCIN

= *Amforal* = *Cantrex* = *Cristalomicina* = *Enterokanacin* = *Kamycin* = *Kamynex* = *Kanabristol* = *Kanacedin* = *Kanamytrex* = *Kanaqua* = *Kanasig* = *Kanatrol* = *Kanescin* = *Kanicin* = *Kannasyn* = *Kano* = *Kantrex* = *Kantrexil* = *Kantrim* = *Kantrox* = *Klebcil* = *Otokalixin* = *Resistomycin* = *Ophthalmokalixan*

***Antibiotic. Aminoglycoside.* Usually IM, except orally for bowel sterilization.**

Untoward effects: ***Cardiac***: Cardiac and respiratory arrest within 25 min after 500 mg IV in 16-year-old **female**. Occasional decreased blood pressure.

Clinical tests: May lead to increased or false positive blood urea nitrogen, ***creatinine***, and urinary ***albumin*** values.

Ear: Ototoxicity with cochlear damage and permanent bilateral and nearly complete eighth cranial nerve deafness. Tinnitus, dizziness, light-headedness, and vertigo may be early warnings, demanding cessation of therapy. Vestibular damage has also been reported. Duration of therapy is not as important as the total dosage in causing ototoxicity.

Eye: Topically, can cause conjunctivitis and blepharitis.

Hypersensitivity: Allergic reactions may include low-grade fever after initial good response, eosinophilia (8–10%) with thrombocytopenic purpura and adverse skin responses.

Miscellaneous: Headache, stomatitis, enterocolitis, and proctitis (after oral use). Occasionally pain and sterile abscesses at IM sites. Nausea, vomiting, abdominal pain, anorexia, decreased fat absorption, and diarrhea occasionally after oral use. Slight anticoagulant effect at high dosage. **Patients** > 60 years have about $2\frac{1}{2}$ times the adverse effects reported in **those** < 60 years.

Neonate: Very little (~0.2 mg/100 ml) passes into breast milk. Poorly documented reports of ototoxicity after treatment of pregnant **mothers**. Premature **infants** (50) given audiometric examinations when **they** were 4–10 years showed no difference from 52 matched controls, if treatment did not exceed 17 days or a total of 280 mg.

Neurotoxicity: Motor and sensory neuropathy in 67-year-old **male** after 500 mg/2 cc water into extradural space before wound closure, after surgery for herniated nucleus pulposus. Immediately on recovery from anethesia, **he** had numbness and paralysis of both legs. Kanamycin has weak neuromuscular-blocking effect on combined presynaptic and post-synaptic action and it should be used cautiously in myasthenic **patients**. This curariform effect occurs, especially if dosages are too high, in cases of renal failure, and if ***acetylcholine*** body levels are low from other drugs or ***insecticides***. The possibility of postoperative respiratory depression demands caution in its use. Paresthesias reported.

Renal: Variable nephrotoxicity, increased blood urea nitrogen, azotemia, hematuria, and transient cylinduria, albuminuria, oliguria, and proteinuria in some **patients**. Kidney damage has ranged from cloudy swelling to severe proximal and moderate distal tubular necrosis in anuric cases.

Skin: Dermatitis, rashes, and pruritus. There may be cross-reactions in **patients** sensitive to ***neomycin***.

Usual therapeutic serum levels are 8–10 μg/ml and toxicity at 30 μg/ml.

Drug interactions: Large numbers of case reports implying reactions, including apnea, paralysis, and death from interactions with ***anesthetics*** and/or ***muscle relaxants***

(***curare***, ***decamethonium***, ***gallamine***, ***succinylcholine***, and ***tubocurarine***). Nephrotoxicity may be increased by use with ***methoxyflurane*** anesthesia, ***cephaloridine***, and ***cephalothin***. Use with ***ethacrynic acid*** and ***furosemide*** has caused irreversible deafness, even at usual therapeutic doses. Ototoxic effect is additive with ***dihydrostreptomycin***, ***neomycin***, and ***streptomycin***. ***Procainamide*** and ***quinidine*** add to its neuromuscular-blocking effect leading to apnea and muscle weakness. ***Calcium*** and ***neostigmine*** antagonize its neuromuscular blockade. It potentiated ***warfarin's*** anticoagulant effect. May increase ***digitalis*** activity. ***Edrophonium*** decreases its activity.

IV use in **dogs**, **owls**, **squirrels**, **rhesus monkeys**, **dog-faced baboons**, and **cats** has caused hypotension and bradycardia. ***Diuretics***, particularly after parenteral use, can precipitate an irreversible deafness. Neuromuscular blockade occurs, if given soon after anesthesia or use of ***muscle relaxants***. **Cats** and **guinea pigs** easily develop ototoxicity after its use. Subtoxic doses of ***furosemide*** or ***ethacrynic acid*** with it at subtoxic doses lead to ototoxicity. Nephrotoxicity reported in **cats**, **dogs**, **rats**, and **mice** at levels > 100 mg/kg.

LD_{50}, **mice**: 20.7 g/kg; IP, 816 mg/kg; IV, **mice**: 290 mg/kg; **rats**: 300–600 mg/kg; **rabbits**: 150–300 mg/kg; SC, **mice**: 1.65 g/kg; IM, **mice**: 1.9 g/kg.

KAOLIN

= *Argilla* = *Bolus Alba* = *China Clay* = *Pharmolin* = *Porcelain Clay* = *Vanclay* = *White Bole*

Adsorbent, poultice. **Mostly, hydrous *aluminum silicate* (q.v.).**

Use: In manufacturing porcelain, pottery, paper, bricks, Portland cement, and paints, and in liquid filtration.

Untoward effects: Pneumoconiosis (kaolinosis) in 48/533 ***China clay* workers** exposed to the dust for > 5 years. Massive fibrosis in four. There is an aspiration hazard from its incorporation into ***baby powder***. Stupor and generalized convulsions within 15 min of kaolin adulteration of IV ***paregoric***. Disseminated intravascular coagulation with encephalopathy and severe bleeding in **female** with a history of drug abuse. Another 34-year-old former ***heroin* addict** developed nosebleeds after 30 ml IV of ***Donnagel PG***, containing kaolin. The latter initiated intrinsic coagulation. Hypertension, hypokalemia, and gastrointestinal dilation in 24-year-old **female** after 6 years of chronic ingestion of a ***kaolin*** and ***morphine*** mixture. National Institute for Occupational Safety and Health limits of 10 mg/m^3 and 5 mg/m^3 for air.

Drug interactions: It delays the absorption of ***digoxin*** by up to 62%; decreases ***acetohexamide***, ***ampicillin***, ***codeine***, ***erythromycin***, ***lincomycin*** (up to 90%), ***phenytoin***, ***quinidine***, ***tolazamide***, and ***tolbutamide*** absorption. It decreases the antibacterial effectiveness of ***benzalkonium chloride*** and ***cetylpyridinium chloride***.

Despite controlling fluid losses in diarrhea, it may exacerbate ***sodium*** and ***potassium*** loss in **rats**. Corneal irritation on **rabbit** eyes. Has caused neck and mediastinal granulomas in **macaque monkeys** by faulty intubation with metal cannulas.

KAPOK

A natural fiber from the kapok tree, used in chair stuffing, throw pillows, and stuffed toys.

Untoward effects: **Humans**, **cats**, **dogs**, **guinea pigs**, **hamsters**, and **rats** have frequently shown allergic (respiratory) reactions to its dust.

KARAYA

> 300 tons used annually as a stabilizer, thickening agent, and emulsifier in foods, denture adhesives, hair-waving lotions, furniture polishes, in paper manufacturing, etc.

Untoward effects: A rare skin sensitizer leading to urticaria and fixed-drug eruptions. Use as a pastry filler and stabilizer of meringue caused a **baker's** hand dermatitis when **he** prepared pies with it. The reaction was confirmed by patch testing. A common allergen in the baking industry. A **nurse** aged 27, after using it as a skin barrier in ostomy care, developed acute respiratory symptoms (chest tightness, wheezing, coughing, and nasal congestion).

LD_{LO}, **rat**: 30 g/kg.

KASARAKOFFER YELLOW

The dye contains 13.6% *lead* (q.v.).

Untoward effects: A large number of women **workers** in a textile factory spraying it upon ***silk*** material through stencils became clinically ill after 6–8 weeks. Average duration of illness was 16–20 weeks.

KAVA

= *Ava-ava* = *HOI* = *Kava-kava* = *Kawa* = *Methysticum* = *Narcotic Pepper* = *Yangona*

The dried rhizome and roots of *Piper methysticum*, containing about 5% resin and the prime alkaloid, *kavaine*. Grows on many South Sea Islands from Hawaii to the East Indies. The Western World first discovered it in 1768, when Captain Cook was in the South Seas. It is a species of black *pepper*.

Untoward effects: The resins have a strong sedative effect on the central nervous system. When kava is chewed, it has an initial sweet taste, then, a pungency and burning sensation. Once used extensively in medicine, but, now, rarely used as a diuretic and local anesthetic. **Natives** of the Sandwich Islands, Samoa, and New Hebridian Islands make a fermented brew from the roots, which is used in religious ceremonies, to put **worshippers** in a trance. It is unlike ***alcohol***, and puts its **beneficiaries** and/or **victims** into a state of drowsiness, stupor, and silence within an hour, accompanied by almost complete loss of skeletal muscle control, as well as incoherent dreams and, possibly, suicidal ideation. Closely related chemically to ***mephenesin*** (q.v.). Chronic use may cause discoloration of the skin from two yellow pigments in the roots. Has caused mydriasis, accommodation problems, and gastrointestinal complaints.

Drug interactions: A 54-year-old **male** in the U.S. became semi-comatose from some purchased at a health food store and taken with ***alprazolam***. Additive central nervous system depression reported with ***lorazepam***.

KEBUZONE

= *Chebutan* = *Chepirol* = *Chetazolidin* = *Chetil* = *Copirene* = *Ketason* = *Ketazone* = *Ketophenylbutazone* = *KPB* = *Pecnon* = *Phloguron* = *Recheton*

Antirheumatic. A prazolidine, as is phenylbutazone.

Untoward effects: In 12/75 **patients** on 250 mg tid, diarrhea, nausea, lethargy, headache, dyspepsia, flatulence, and buccal ulcers, which caused them to withdraw from the clinical trial. It is hepatotoxic leading to jaundice and increased enzyme levels in 20/75. Water and ***sodium*** retention are serious side-effects and palpitations have also been reported.

LD_{50}, **mouse**: 650 mg/kg.

KELP

Contains 1–4 g *iodine*/kg, i.e. 1000–4000 ppm.

Untoward effects: Pustular eruptions or exacerbation of acne from its high ***iodine*** content. Health food supplements containing kelp were responsible for ***arsenic*** toxicity in a 45-year-old **female** and a 74-year-old **female**, with resulting peripheral neuropathy. Testing of kelp from various areas revealed ***arsenic*** levels of up to 58 μg/g, which appear to be readily absorbed.

Rat and **chick** trials demonstrated its goitrogenic effect.

KENTUCKY COFFEE TREE

= *Gymnocladus dioica*

Rough-barked 60–80 ft tree, native to eastern North America; now, common elsewhere in the U.S.

Untoward effects: Due to the alkaloid, ***cystine***, and ***saponins*** in the seeds and fruit pulp. Ingestion leads to gastrointestinal irritation, diarrhea, sweating, vomiting, and irregular pulse. Ingestion of large quantities causes life-threatening convulsions and coma. It seeds were once used as a ***coffee*** substitute, with varying poisonous effects. The heat used in roasting the seeds inactivated most of the alkaloid.

During drought, **cattle** and **sheep** eat the leaves and fruit leading to severe, colicky pain and profuse diarrhea and straining. Post-mortem shows white, sticky mucus in and severe congestion of the lungs and the small intestines and foamy, dark gray fecal material in the large intestine.

KEPONE

= *Chlordecone* = *Compound 1189* = *ENT 16,391* = *GC-1189*

Insecticide, fungicide. A chlorinated hydrocarbon.

Untoward effects: Was highly effective against **fire ants**, but its approval for all uses has been cancelled by the Environmental Protection Agency, and its sale is now banned in the U.S. As with many other drugs and chemicals, some supplies may still exist. It is one of those toxic agents that has often been transmitted by contaminated work clothes having been taken home. It is neurotoxic to **man** leading to ataxia, tremors, opsoclonus, headaches, stuttering, faulty memory, loss of short-term memory, irritability, convulsions, hallucinations, and slurred speech. Most **men** with high blood levels produced few or no sperm. Liver disorders (including hepatomas), nephrotoxicity, skin erythema, and chest pain were also reported. National Institute for Occupational Safety and Health suggests exposure limit of TWA, 1 $\mu g/m^3$. Of the > 440 tons once produced annually in the U.S., > 99% was exported.

LD_{LO}, **human**: 50 mg/kg.

Hepatobiliary dysfunction and hepatocellular carcinomas in **rats** and **mice**. Causes constant estrus in **mice** and decreases the size and number of litters. Has estrogenic effect and hepatotoxicity in **Japanese quail**. Jerky gait, miosis, tremors, and myasthenia in **mallard ducks** within an hour and deaths 1–8 days after treatment. **Fish** caught for commercial sale in Chesapeake Bay contained 1 ppm (ten times the FDA's recommended upper limit). After feeding it to **cows** at 5 ppm/60 days, it took 83 days to find no measurable amount in the milk.

LD_{50}, **rats**: 95 mg/kg; **mallard ducks**: 167 mg/kg; **dogs**: 250 mg/kg; **rabbit**: 65 mg/kg; **chicken**: 480 mg/kg.

KEROSENE

= *Coal Oil* = *Fuel Oil No. 1* = *Lamp Oil* = *Range Oil*

Untoward effects*:** Inhalation of vapors of high concentration leads to eye, skin, nose, and throat irritation; burning sensation in chest, headache, nausea, weakness, restlessness, incoordination, confusion, drowsiness, and coma. Repeated exposures lead to anorexia, coughing, and central nervous system symptoms. Ingestion causes gastrointestinal upsets, including vomiting and diarrhea. Serious pneumonitis, abdominal tenderness, drowsiness, fever, cyanosis, vomiting, diarrhea, bronchitis, pulmonary edema, convulsions, and death may follow aspiration of vomited material and induced vomiting is contraindicated. Bacterial invasion and pneumonia follow. From 1975–1976, 154 **children** < 5 years of 699 cases reported were hospitalized in the U.S. after ingestion. Hematotoxic. Defatting of skin from contact. Incomplete combustion in home heaters yields ***dinitropyrene emissions (some of which are probably carcinogenic), as well as ***carbon monoxide***. Kerosene is often an additive in pesticide solutions sprayed indoors. Aplastic anemia in 50-year-old **female** after rubbing **her** legs with it for several months. **She** died 5 months later. A 26-year-old **male** used it and ***carbon tetrachloride*** to clean guns for ~3 years. **His** bone marrow became markedly hypoplastic and **he** died 4 years after **his** last exposure. A 58-year-old **male** washed airplane parts in a kerosene-based solvent for ~3 years and died 1½ months later from hypoplastic anemia. Leukocytosis, methemoglobinemia, and neutropenia occur after ingestion. National Institute for Occupational Safety and Health suggests exposure limit of 100 mg/m^3.

LD_{LO}, **human**: 500 mg/kg.

A pipeline ruptured near Manassas, Virginia, spilling > 200,000 gallons into a creek and river. Many **fish** and small numbers of **beavers**, **muskrats**, and **waterfowl** were killed. **Rats**, **rabbits**, and **chickens** tolerate large doses by gastric intubation, SC, or IP, without any pulmonary damage, yet, a fraction of a milliliter instilled into the trachea or aspirated from the mouth into the lungs leads to pulmonary edema, hemorrhage, and death in a few minutes in **rats**. Fumes from kerosene heaters can be toxic to pet **birds**, while **people** may be unaffected. Given experimentally to **calves**, it has caused aspiration pneumonia and adverse central nervous system effects. Accidental ingestion by **cats** leads to salivation, nausea, vomiting, diarrhea, muscle twitching, ataxia, convulsions, and coma. Similar symptoms reported in **dogs**. Aspiration pneumonia has occurred in **monkeys** and **baboons**.

LD_{LO}, **rat**: 28 g/kg; intratracheal, 800 mg/kg; LD_{50}, **rabbit**: 28 g/kg; IV, 180 mg/kg; **guinea pig**: 20 g/kg.

KETAMINE

= *CI-581* = *Green* = *K* = *Ketaject* = *Ketalar* = *Ketanarkon* = *Ketanest* = *Ketaset* = *Ketavet* = *Jet* = *Mauve* = *Purple* = *Special K* = *Special LA Coke* = *Super C* = *Vetalar*

***Anesthetic.* Parenteral.**

Untoward effects*: *Addiction*:** In the U.S. in the 1970s and 1980s, it became an important drug of abuse, with symptoms similar to those of ***phencyclidine (q.v.), a related chemical. In England in the 1990s, it was abused by intranasal and oral use, where four young **males** became paralyzed shortly after ingesting illicit capsules; and where a **woman** had a "bad trip" for 3–4 days after ingesting some.

Cardiovascular*:** Hypertension, tachycardia, increased cardiac output, up to 25% increased arterial pressure due to stimulation of central sympathetic nervous system and inhibition of released ***norepinephrine reuptake. Thus, caution has been advised in **patients** with aneurysms, angina, congestive heart failure, cerebral trauma, or thyrotoxicosis. Cardiac stimulation has been blocked by ***diazepam***, ***enflurane***, and ***halothane***. Hypotension has occurred in severely ill **patients** (e.g. septic shock). Phlebitis reported from IV.

***Cerebrospinal fluid pressure (CSFP) and intracranial pressure (ICP)*:** Rapid increase in CSFP in 66–100% of **patients** and abrupt increase in ICP in **children** and **adults**.

***Central nervous system*:** With increased use, complications not originally noted arose, e.g. emergence reactions, such as delusional states, confusion, and frank hallucinations, often of terrifying proportions. Hallucinations may persist for months after use. Dreaming in ~45% (unpleasant in 13–17% and terrifying in 6%), increased muscle tone and movements, excitation (7%), convulsions (rare), prolonged disorientation and signs of psychotic behavior, visual hallucinations, euphoria; impaired attention, learning, and memory; catalepsy, ataxia, dizziness, and electroencephalogram changes similar to those of epilepsy have been reported in different **patients**.

***Eye*:** Transient increase in intraocular pressure noted in many **patients**. Diplopia and blurred vision recorded during recovery. Nystagmus in 60% of 100 **children**.

***Fetus and neonate*:** Lipid-soluble and readily crosses the placenta. Opinions differ, but doses of 0.23 mg/lb or 0.50 mg/kg to the **mother** have been safest for the **infant**. Respiratory arrest has occurred in **infants**.

***Miscellaneous*:** Lacrimation, profuse sweating, < 5% vomiting and nausea on recovery, temporary apnea with rapid IV, temporary respiratory depression, coughing, laryngospasm, salivation, cyanosis, aspiration of vomitus in a

6-year-old **child**, shivering, urinary retention, acidosis, fever, and fasciculations. A "date rape" drug.

Skin: In **some**, transient erythematous and morbilliform rashes and urticaria at injection site.

Some **clinicians** and **anesthesiologists** consider that its use alone has an unacceptable high incidence of adverse effects. Usual therapeutic serum levels are < 6 µg/ml and toxic at > 7 µg/ml.

IV, TD_{LO}, **human**: 2 mg/kg.

Drug interactions: Extensor-type seizures in **patients** receiving ***aminophylline*** after induction of anesthesia. Use of ***physostigmine*** has shortened anesthetic time. Possible interaction with ***thyroid hormones*** leading to extreme hypertension and supraventricular tachycardia. Excessive tachycardia with ***pancuronium***. Premedication with ***diazepam***, ***hydroxyzine***, or ***secobarbital*** prolong recovery time by ~35%.

Pedal and corneal reflexes may remain during surgical anesthesia of **cats**. Do NOT administer to **cats** and subhuman **primates** with impaired liver and renal function, as the drug is detoxified in the liver and eliminated via the kidneys. Yet, it is very useful in most cystitis cases. Excessive salivation and tachycardia may occur 10–60 min after administration to **cats**. Causes convulsions and barking fits in **dogs**, which do not occur when it is added to other psychotropic drugs. Use with caution in **cats** with circulatory problems. 10–30% of **cats** may show salivation (use with ***atropine***), and 10% may show tonic–clonic spasms. Young **kittens** frequently require 10–50% higher dosage. Rare cases show little anesthetic effect. Its metabolism in **rats** with chronic renal failure is impaired, probably due to the organelle damage within the liver. Use ***atropine*** with **ruminants**. High dosage used alone in some species (viz. **dogs**) may produce excitement, some convulsions, and muscle rigor. **Animals** may show random motor activity and increased intraocular pressure.

IP, LD_{50}, **rat**: 224 mg/kg; IV, LD_{50}, **mouse**: 180 mg/kg.

KETANSERIN

= *Ketensin* = *R-41468* = *Serefrex* = *Taseron*

Antihypertensive.

Untoward effects: Sedation, dizziness, tiredness, edema, weight increase, and decreased salivation. May cause a dose-related QT interval prolongation. Ventricular arrhythmias and syncope can also occur. Therapeutic doses may impair clearance of oral ***propranolol***.

KETAZOLAM

= *Anseren* = *Ansieten* = *Anxon* = *Contamex* = *Loftran* = *Solatran* = *U-28774* = *Unakalm*

Anxiolytic. A benzodiazepine.

Untoward effects: Drowsiness and suppression of rapid eye movement density.

KETENE

= *Carbomethane* = *Ethenone* = *Keto-ethylene*

Untoward effects: Strong penetrating odor demands use of a hood and exhaust system. Present in ***tobacco*** smoke and is one of the causative agents of **smoker's** bronchial symptoms. Also found in bonfire smoke. Fumes are irritating to skin, eyes, throat, and lungs. Pulmonary edema can occur. National Institute for Occupational Safety and Health and OSHA exposure limits are 0.5 ppm or 0.9 mg/m^3. 15-min exposure is 1.5 ppm.

LD_{50}, **rat**: 1.3 g/kg; inhalation LC_{50}, **monkey**: 200 ppm/10 min, **cat**: 750 ppm/10 min.

KETHOXAL

= *Bis(thiosemicarbazone)* = *BW 356-C61* = *Contrapar* = *KTS* = *NSC 82,116*

Anaplasmastat.

Untoward effects: Paresthesia or myalgia in 12/34; motor weakness in 8/34; nausea and vomiting in 6/34; hematological problems in 6/34; diarrhea, esophagitis, gastritis, proctitis, and skin rash in one each, psychoses in two. Large doses to **calves** for extended periods of time have caused axonal and myelin degeneration of the vagus nerve in some. In **laboratory animals**, delayed hepatic, myocardial, and adrenal toxicity noted.

KETOCONAZOLE

= *Fungarest* = *Fungoral* = *Ketoderm* = *Ketoisidin* = *Nizoral* = *Orifungal M* = *Panfungol* = *R-41400*

***Antifungal. An imidazole–piperazine* compound.**

Untoward effects: Hepatitis, increased liver enzymes, enlarged portal tracts, mononuclear cell infiltrates, and even fatal massive hepatic necrosis with 200–400 mg/day/11–110 days. Anorexia and dark urine may be early signs. Pruritus, alopecia, exfoliative erythroderma, morbilliform eruptions, headache, dizziness, somnolence, fever, chills, photophobia, diarrhea, hypothyroidism, immune hemolytic anemia, hypoglycemia, increased ***estradiol–testosterone*** ratio, decreased plasma ***testosterone***, impotence, decreased libido, decreased sperm count, gynecomastia, menstrual irregularities, epistaxis, insomnia, hyperactivity, eosinophilia, thrombocytopenia, and arthritis. On high dosage (400 mg or more/day), anorexia, nausea, vomiting, abdominal pain, dyspnea, confusion, hypertension, adrenal insufficiency or failure with decreased steroidogenesis and plasma ***cortisol***, and oligospermia. Anaphylactic reactions to 200 mg reported in two **patients**. **Patients**

may cross-react with ***miconazole*** and, possibly, other ***imidazoles***. Topical use leads to severe irritation, pruritus, and stinging sensations in ~5% of **patients**.

Drug interactions: Cimetidine, famotidine, lansoprazole, nixatidine, omeprazole, ranitidine, and ***sucralfate*** decrease gastric acidity, decrease its absorption, as do ***antacids*** taken near the same time. Toxicity and serum levels of ***cyclosporine*** and ***corticosteroids*** (particularly ***methylprednisolone***) are increased by it. Can increase ***warfarin*** effect. Use with ***rifampin*** decreases the effect of both drugs. Has increased ***clarithromycin, erythromycin, lovastatin, phenytoin, simvastatin***, and ***troleandomycin*** blood concentration. ***Disulfiram***-like reaction with ***ethyl alcohol***. Ventricular arrhythmias and torsades de pointes with ***astemizole, cisapride, loratidine***, and ***terfenadine***, due to their increased serum concentration. Can decrease clearance of ***alprazolam, antipyrine, chlorazepate, chlordiazepoxide*** (by up to 38%), ***clonazepam, diazepam, imipramine, midazolam, quinidine, tolbutamide***, and ***triazolam***. ***Amprenavir*** increases its serum concentration > 40%. Can induce ***indinavir*** toxicity. Used with ***didanosine***, decreases the absorption of both drugs. Inhibits CYP3A4 enzyme metabolism of many drugs, increasing their serum concentration. Prolongs the half-life of ***robexitine***.

In **rabbits**, it decreases hepatic metabolism of ***tolbutamide***. Long-term treatment of **rats** decreases sperm numbers and motility, and increases numbers of abnormal sperm. Teratogenic and hepatotoxic with decreased steroidogenesis in **rats**. It decreases the metabolism of ***halofantrine*** in **dogs**. Can be hepatotoxic with increased alanine aminotransferase and increased ALP in **dogs**.

LD_{50}, **rats**: 227 mg/kg; **mice**: 702 mg/kg; **guinea pigs**: 202 mg/kg; **dogs**: 780 mg/kg; IV, LD_{50}, **rats**: 86 mg/kg; **mice**: 44 mg/kg; **guinea pigs**: 28 mg/kg; **dogs**: 49 mg/kg.

KETOPROFEN

= *Alrheumat* = *Alrheumun* = *Capisten* = *Dexal* = *Epatec* = *Fastum* = *Iso-K* = *Kefenid* = *Ketopron* = *Ketum* = *Lertus* = *Menamin* = *Meprofen* = *Orudis* = *Orugesic* = *Oruvail* = *Oscorel* = *Profenid* = *RP-19583* = *Toprec* = *Toprek*

Anti-inflammatory, analgesic. A non-steroidal anti-inflammatory drug.

Untoward effects: Dyspepsia, nausea, abdominal pain, flatulence, heartburn, diarrhea, constipation, decreased renal function, increased blood urea nitrogen, edema, headache, giddiness, insomnia, and nervousness are the most common. Anorexia, stomatitis, vomiting, dizziness, malaise, depression, tinnitus, visual disturbances, rashes, urticaria, urinary tract irritation, conjunctivitis, pseudotumor cerebri, anaphylaxis, life-threatening asthma, angioedema, esophageal irritation, upper gastrointestinal hemorrhage, phototoxicity, and renal failure also occur. Its longer-acting lysine form (***Atrosilene***) is little-used and gastrointestinal upsets are its most common adverse effects. Asthmatic cross-reactions may occur with ***tartrazine*** or ***sodium benzoate***. ***Non-steroidal anti-inflammatory drug***s have not been recommended in **women** trying to become pregant, since they interfere with implantation of fertilized eggs in **animals**.

Drug interactions: Decreases renal elimination and increases serum levels of ***lithium*** (20–60%) that can cause renal damage and be life-threatening. Use with ***probenecid*** has at 6 h increased ketoprofen plasma level 400%, by inhibiting its clearance. Can precipitate increased ***methotrexate*** serum levels and toxicity. ***Tea*** consumption with it decreases urine output. May enhance ***warfarin*** anticoagulant effect, yet, in a controlled study on healthy **volunteers**, there was no such effect. ***Metoclopramide*** decreases its absorption.

IV overdoses to **horses** lead to inappetance, depression, and icterus. Post-mortem showed mild gastritis, hepatitis, and nephritis. ***Ketoprofen–probenecid*** interaction also reported in **rats**.

LD_{50}, **rat**: 101 mg/kg.

KETOROLAC

= *RS 37,619*

The tromethamine form = *Acular* = *Dolac* = *Exodol* = *Ketorolac* = *Lixidol* = *Tarasyn* = *Toradol* = *Toratex* = *Trometamol*

***Anti-inflammatory, analgesic.* Non-steroidal. Used orally and parenterally. Topically for ophthalmic use.**

Untoward effects: Drowsiness, gastrointestinal ulceration and bleeding, nausea, vomiting, dyspepsia, weight increase, rash, pruritus, purpura, palpitations, anemia, eosinophilia, thrombocytopenia, leukopenia, abnormal dreams, euphoria, depression, rhinitis, coughing, aseptic meningitis, bronchospasm, hemolytic uremic syndrome, acute renal failure, oliguria, dyspnea, hyperkalemia, seizures, decreased platelet aggregation, and anaphylactoid reactions (facial swelling, shortness of breath, and chest tightness). Occasional dizziness, diarrhea, edema, flatulence, headache, sweating, blue lips and nails, and syncope. Rarely, scaly peeling skin, hypertension, heartburn, painful glands, continuous thirst, white spots on lips and mouth, and tongue swelling. Some pain at injection site. Transient ocular irritation (stinging, burning, and keratitis) from topical ophthalmic use.

Drug interactions: Significantly increases ***lithium*** plasma concentrations by ~20–60%.

LD_{50}, **mice**: ~200 mg/kg.

KETOTIFEN FUMARATE

= *Allerkif* = *HC-20511* = *Totifen* = *Zaditen* = *Zaditor* = *Zasten*

Antiasthmatic.

Untoward effects: Dizziness and drowsiness are the main ill-effects. Ophthalmic use leads to conjunctival injection, headache, and rhinitis. Stinging, eye pain, irritation, and photophobia in 5%.

KETUHAS

= *Naniwazu* (Japanese)

An Ainu arrow poison, probably from a shrub, *Daphne* (q.v.) *kamtschatica*.

Untoward effects: The boiled-down juice of the shrub is painted on the head of an arrow. A single arrow can kill a large **walrus**.

KEVLAR®

Untoward effects: Inhalation of fibers by **rats** led to granulomatous lesions in bronchioles and alveolar ducts, which changed to patchy, fibrotic thickenings.

KHELLIN

= *Amicardine* = *Ammicardine* = *Ammipuran* = *Ammivin* = *Ammivisnagen* = *Benecardin* = *Cardio-Khellin* = *Corafurone* = *Coronin* = *Eskel* = *Gynokhellan* = *Kelamin* = *Kelicor* = *Kelicorin* = *Kellin* = *Keloid* = *Khelfren* = *Lynamine* = *Methafrone* = *Norkel* = *Simeskellina* = *Vasokellina* = *Visammin* = *Viscadan* = *Visnagalin* = *Visnagen*

Vasodilator. **From the fruit of *Ammi visnaga* (q.v.).**

Untoward effects: Nausea and vomiting. Cumulative toxicity in **man**. A photosensitizer.

Emesis in **monkeys**.

LD_{50}, **rat**: 80 mg/kg.

KIWI

= *Chinese Gooseberry*

Untoward effects: Shortly after peeling and slicing the fruit, a 53-year-old **female** developed moderately severe symptoms of immediate hypersensitivity. **She** proved to be allergic to both the skin and pulp of the fruit. Successive anaphylactic reactions reported from handling the fruit by a **person** with known atopic diathesis.

KOCHIA

= *Bluebush* = *Burning Bush* = *Fireweed* = *Tumbleweed*

Untoward effects: Highly allergenic, causing rhinitis, bronchial asthma, and pneumonitis in **people**.

Poisonings in grazing **sheep** from its high ***nitrate*** and ***sodium*** content.

K. scoparia = ***Belvedere*** = ***Fireball*** = ***Schrad*** = ***Summer Cypress***. Native to Eurasia, but now found in Australia, South America, Canada, Texas, and Iowa. A ***nitrate***-accumulator and photosensitizer. Photosensitization in **cattle**, **sheep**, and **horses**. Jaundice, enlarged livers, sore noses and eyes, and tubular nephritis reported. **Cattle** grazing it develop ataxia, incoordination, muscle spasms, recumbency, and death. Contains high ***oxalate*** (up to 11%) level and may contain thiaminase.

KOLA

= *Bichy Nuts* = *Bissy Nuts* = *Cola* = *Gooroo Nuts* = *Gotu* = *Guru Nuts*

Analeptic. **Dried cotyledons of *Cola nitida* and *C. acuminata*, containing up to 3.5% *caffeine* and 1% *theobromine*.**

Untoward effects: Ingestion of it or its extracts lead to central nervous system stimulation and adverse effects, primarily from its ***caffeine*** (q.v.) content. This is often due to prolonged chronic or excessive use of kola-containing beverages. Reported to cause hypersensitivity reactions, increased premenstrual symptoms, exacerbation of prostatitis symptoms, migraine headaches, and increased benign fibrocystic breast lesions in some **people**.

Drug interactions: A 33-year-old **female** maintained on ***phenelzine***, a ***monoamine oxidase inhibitor***, developed a severe headache and slight increase in blood pressure after drinking 12 oz of a kola-containing beverage. ***Alcohol*** and ***kola*** intake increases the risk of bladder cancer in **male smokers**.

Increases gastric acid secretion in **cats**. Use is easily detected in **horses** by saliva testing. Decreases brain cholinesterase levels.

KOMBUCHA

Untoward effects: This ***mushroom*** is incubated in sweet black ***tea***. Consumption for ~2 months led to one death and severe unexplained illness. Symptoms included pulmonary edema and metabolic acidosis in a **survivor**.

KOONG YICK HUNG FAR OIL

Untoward effects: This topical analgesic is sold in the Orient in 60 ml bottles. If the contents of the bottle were ingested, it would be equivalent to 184 **adult** 300 mg ***aspirin*** tablets, and 6 ml can kill a **child**.

KRAMERIA

= *Rhatany*

Astringent botanical.

Untoward effects*:** Severe dermatitis (eczematous rash) in the perianal, intergluteal, perineal, and scrotal areas in 23-year-old **male**, after its use in a hemorrhoidal ointment. 4 years later, **he** still showed a sensitivity to it, and not to other ingredients in the ointment. May cross-react with ***Balsam of Peru.

KRATOM

= *Ithang* = *Kakuan* = *Mitragyna speciosa* = *Thom*

A large tropical tree in Thailand.

Untoward effects*:** Addiction and psychosis from chewing, smoking, and ingesting the leaves with their ***narcotic-like effects. Long-term **addicts** become thin, **their** skin darkens, **they** are constipated with black stools and urinate frequently. **They** show hostility, aggression, lacrimation, jerky limb movements, and aching muscles as withdrawal symptoms. Large doses are said to be hallucinogenic. Contains ***mitragnine***, a ***harmine*** analog.

KUSHTAS

Untoward effects*:** A traditional medicine, used in Indo-Pakistan, containing 12–72.8% w/w ***lead (q.v.), as well as other heavy metals with potential adverse effects.

K-Y LUBRICATING JELLY®

Untoward effects*:** Allergic contact dermatitis has resulted from its ***propylene glycol (q.v.) content. A 29-year-old **female** developed a sensitivity to ***propylene glycol*** in several topical creams prescribed for chronic vaginitis, and subsequently developed a severe allergic vulvitis after its use in a gynecological examination. After eating a salad dressing containing ***propylene glycol***, **she** developed a dermatitis of the abdomen, vulva, and rectal areas. A **veterinarian** developed a severe pruritic dermatitis of the hand after using it before performing rectal examinations in **horses** and **cattle**. A 27-year-old **female** used it as an electrolyte jelly on a nerve-stimulation device for shoulder pain. In all three **patients**, the dermatitis cleared after using preparations that did not contain ***propylene glycol***.

L

LABDANUM OIL

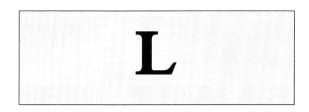

From the gum of the shrub, *Cistus ladaniferus*. > 2 tons/year used in the U.S. annually as a fragrance in soaps, detergents, cosmetic creams and lotions, perfumes, and foods.

***Untoward effects*:** No irritation or sensitization in **human** volunteers of 8% in petrolatum. Moderately irritating when applied full-strength to intact or abraded **rabbit** skin for 24 h.

LD_{50}, **rats**: 8.98 g/kg; acute dermal, **rabbits**: > 5 g/kg.

LABETALOL

= *AH-5158A* = *Albetal* = *Amipress* = *Ibidomide* = *Ipolab* = *Labelol* = *Labrocol* = *Normodyne* = *Presdate* = *Pressalolo* = *Sch-15719W* = *Trandate*

***β-Adrenergic-blocker, antihypertensive, antiarrhythmic, sympatholytic.* Has some α-blocking activity.**

***Untoward effects*:** Postural hypotension, dizziness, vertigo, headache, muscle cramps, fatigue, vasodilation, and fluid retention were common. Occasional nausea, abdominal distention, diarrhea, constipation, skin rashes, tingling of scalp, initiation or aggravation of psoriasis, maculopapular erythematous rash, urticaria, proximal myopathy, priapism, impotence, decreased libido, delayed or no ejaculation, significant decrease in vaginal lubrication, yellowing of skin or eyes, nasal stuffiness, dyspnea, and bronchospasm; and strangury and urinary retention in **men**. Rarely, circumoral paresthesia, micturition, irritability, fatal hepatotoxicity, vision problems, pruritus, atypical lichen planus and bullous lichen planus, vivid dreams, systemic lupus erythematosus, Raynaud's phenomenon, claudication, myositus, acute renal failure, and heart failure. Fever reported in **patients** that may be due to an immunoallergic reaction. Left hemiparesis in 48-year-old **female** after a single infusion of 35 mg. A 47-year-old **female** developed acute onset of urticaria, angioedema, and hypotension after a single oral dose of 100 mg. A 58-year-old **female**, after receiving 600–800 mg/day/~1½ years, developed lupus diathesis with synovitis, leukopenia, possible vasculitic edema, and antinuclear antibodies. Hypertensive response in **patients** with pheochromocytoma. Apparently, a paradoxical hypertensive effect in some **patients**. Half-life in **neonates'** plasma after oral administration for the **mother's** pregnancy-induced hypertension is ~24 h, considerably longer than in an **adult**. Only 0.07% of **maternal** dose excreted in breast milk. Usual therapeutic serum levels are 0.08–0.65 μg/ml and toxic at > 1 μg/ml.

Drug interactions*:** ***Cimetidine potentiates its actions by delaying its clearance and increasing its bioavailability and half-life. ***Epinephrine*** interacting with ***α-adrenergic-blockers*** can make management of anaphylactic reactions more difficult. ***Food*** causes a 38% increase in its mean systemic bioavailability. It increases ***antipyrine*** half-life by decreasing its clearance. Short-term administration of ***indomethacin*** or ***sulindac*** leads to clinically significant increase in systolic blood pressure in **patients** taking ***labetalol***.

In the **dog**, **cat**, and **rabbit**, 50 mg/kg causes ptosis, decreased motor activity, and occasional vomiting.

LD_{50}, **rats** and **mice**: > 2 g/kg; IV, ~50–60 mg/kg.

LACIDIPINE

= *Caldine* = *GR-43659X* = *GX-1048* = *Lacipil* = *Lacirex* = *Motens*

Antihypertensive, calcium channel-blocker.

***Untoward effects*:** Significant exercise-induced tachycardia. Most adverse effects are due to its induced vasodilation and occur during the first week of treatment, and it usually lasts 4–6 days.

LACQUER

From trees' resin of the *Anacardiaceae* family (*Rhus vernicifera*, *R. succedanea*, and *Melanorrhoes laccifera*).

***Untoward effects*:** Usually, caused by hypersensitivity to its vegetable oils and volatile solvents. Far-Eastern, Chinese, Japanese, and Indo-Chinese laquers have caused occupational dermatoses with localized edema, desquamation, erythematous and hyperemic foci, papulovesicular rashes, burning sensations, pruritus, larges blisters, nodules, erosions, and pigmentation. Nausea, vertigo, eosinophilia, and proteinuria also reported. The trees are also often used for manufacturing furniture. In addition to the resin, the trees and wood also contain a strongly irritating juice, causing Mah-Jong dermatitis. Mucous membranes are less sensitive than skin. Chronic lichenoid eczema of the scalp has been caused by lacquered steel hairpins. The odor causes migraine headaches in some **people**. Inhalation of some of its thinners in **abusers** leads to severe peripheral motor neuropathy with muscle weakness, hyperalgesia, hyperesthesia, paresthesia, respiratory distress with decreased vital capacity, asthma-like symptoms, dry cough, and hepatitis.

β-LACTAMASE

= *Cephalosporinase* = *Neutrapen* = *Penicillinase*

Untoward effects: Severe anaphylactic reactions, angioneurotic edema, erythema, and urticaria.

LACTARIUS sp.

A group of *mushrooms*.

Untoward effects: ***L. torminosis* = *Woolly Milk Capsule*** or ***L. glaucescens*** can cause a mild or severe gastroenteritis and even death in **people** who ingest it.

In Mexico, ***L. deliciosus*** is considered edible.

LACTIC ACID

Use: As an acidulant in brewing, ***yeast*** manufacturing, hide preparation, foods (jams, jellies, green ***olives***, bakery products, frozen desserts, ***pickles***, ***cheeses***, ***sauerkraut***, mincemeat, mayonnaise, and poultry), some beverages, and in medicines.

Untoward effects: Acidosis can be precipitated by a number of drugs. Has caused adverse reactions when used in some **infant** formulas. Photosensitizing and keratolytic effect after topical use on some **patients**. Twice daily topical use of gels containing it at 8.8% with 1% ***ascorbic acid*** had a skin-whitening effect during a 12 week study.

LD_{LO}, **human**: 50 mg/kg.

Sudden death syndrome ("flip-overs") in **poultry** appear to be due to lactic acidosis. Injected into a **calf's** testicle causes its degeneration.

LD_{50}, **rat**: 3.7 g/kg; **mouse**: 4.9 g/kg; **guinea pig**: 1.8 g/kg.

LACTOSE

= *Milk Sugar* = *Saccharum lactis*

Untoward effects: Intolerance to it leads to flatulence, cramps, bloating, diarrhea, and indigestion due to decreased lactase product. It has been most common in **Middle Easterners**, **Negroes**, **Orientals**, and **American Indians**. **Physicians** should be aware that it is a common diluent in medicinals. Fixed-drug eruption (inflamed plaque with subsequent blistering and crusting) on tip of nose in 64-year-old **female** ~4 weeks after each injection of ***botulism toxin A*** containing lactose as an excipient. Hydrolyzed in the body to ***glucose*** and ***galactose***. Some **children** cannot utilize the latter and are put on a ***milk***-free diet to prevent irreversible mental retardation. **Neonatal** intestinal epithelium, initially, cannot hydrolyze lactose, but develop this ability after exposure to it. ***Neomycin*** has caused lactose malabsorption. ***Milk***-sensitive **patients** had bronchospasms aggravating **their** asthma immediately after using ***Aerohalor***®, containing ***isoproterenol*** in a lactose base. In Australia, after a company changed a ***phenytoin*** diluent to lactose, caused ataxia, double vision, plus a few with nausea, pyramidal signs, and dystonia with 3–5 times increased **phenytoin** serum levels. ***Aminopterin***, ***colchicine***, and ***methotrexate*** decrease its absorption by damaging intestinal mucosal cells. Will give a false positive ***glucose*** reading in ***copper*** reducing tests. This does not occur if one uses ***glucose*** dip tests, such as ***Clinistix***®.

Skim ***milk*** feeding of **rats** leads to cataract formation, due to its lactose content. Can cause diarrhea in **foals**. It is toxic to **sea lions** and they develop cataracts by their seventh month of age. Lactose intolerance has also been reported in a **stumptail macaque**.

LACTULOSE

= *Bifiteral* = *Cephulac* = *Chronulac* = *Generlac* = *Lactuflor* = *Laevilac* = *Normase*

***Duphalac Syrup* is 50% *lactulose* with 5–8% *lactose* and 6–10% *galactose*.**

***Laxative*. < 2% of ingested material is absorbed.**

Untoward effects: Diarrhea is common, especially with high dosage. Occasionally nausea, colic, bloating, flatulence, borborygmus, hypokalemia, dehydration, and anorexia. Severe lactic acidosis in 31-year-old **male**. Hypernatremia in many **patients** with hepatic encephalopathy.

LD_{50}, **rat**: 43 ml/kg; **guinea pig**: 15 ml/kg.

LAETRILE

= *Mandelonitrile Glucuronide*

The literature often considers it synonymous with *amygdalin* (q.v., for its many adverse effects). It has frequently been touted as a cancer cure.

Untoward effects: Deaths have been reported from its ingestion or instillation by enema. It has induced agranulocytosis. Also see ***Cyanide*** for other adverse effects.

LD_{LO}, **women**: 198 mg/kg.

Orally to pregnant **hamsters** leads to skeletal malformations in the offspring.

LAGOCEPHALUS sp.

See *Fish, puffers*.

Untoward effects: Due to ***tetrodotoxin*** (q.v.). Neurological toxicity in ~50 **patients** in Italy who had eaten frozen fillets of **burbot** from Taiwan that also contained parts of

L. lunaris, a **fish** of the *Tetraodontidae* family. A few similar cases occurred in France, and one **patient** died there. ***L. laevigatus*** = ***Smooth Puffer*** in the Gulf of Mexico, has also been implicated in outbreaks of tetrodotoxicity.

LAGOCHILLUS INEBRIANS

Untoward effects: Turkestan **tribesmen** make a ***tea*** with its roasted leaves for a narcotic, intoxicating drink with hallucinogenic effects in high dosage.

LAMB

Untoward effects: Immediate reactions include urticaria aggravating chronic hand dermatitis reported from handling. A strong ***potato*** aroma, due to *Pseudomonas taetrolens* has caused rejection of carcasses.

LAMINARIA

Untoward effects: The blades of the algal plant ***L. digitata*** = ***Laminaria Tents*** are used effectively (self-lubricating) for cervical dilation and misused by self-administration, to cause abortions, with sometimes disastrous results. Contains various amounts of ***iodine***.

LAMIUM AMPLEXICAULE

= *Henbit*

Untoward effects: Ingestion causes staggers in **cattle**, **horses**, and **sheep**, especially in the spring. The toxic principle has not been identified.

LAMIVUDINE

= *Epivir* = *GR 109,714X* = *3TC*

Antiviral.

Untoward effects: A 49-year-old **male** with HIV developed anaphylaxis, angioedema, and urticaria 30 min after a first oral dose of 150 mg. Within 2 months after receiving 150–300 mg twice daily, 5/16 **male** and **female patients** (29–57 years with HIV) complained of hair loss. Other adverse symptoms were abdominal cramps, nausea, vomiting, headaches, peripheral neuritis, insomnia, pruritus, and worsening anemia and leukopenia. A 33-year-old **male** with HIV, after 300–600 mg/day/3 weeks, had exacerbation of peripheral neuropathy with painful burning paresthesia in **his** feet, extending up **his** legs. Symptoms recurred when rechallenged. Hepatitis B reactivation 5 months after completion of therapy in 29-year-old **male**. Pancreatitis, especially in **children**. Lipomatosis (buffalo hump and increased abdominal girth) has occurred. Induced paronychia in 12 HIV **patients**. In combination with ***zidovudine*** in 144 **patients**, headache (35%), nausea (33%), and pancreatitis (14.6%); > 10% showed malaise, fatigue, nasal symptoms, diarrhea, and neuropathy.

LAMOTRIGINE

= *BW 430C* = *Lamictal* = *LTG*

Anticonvulsant.

Untoward effects: Dizziness, ataxia, somnolence, headache, diplopia, nystagmus, blurred vision, nausea, vomiting, and rashes are common. Toxic epidermal necrolysis in a 56-year-old **male** receiving 200 mg/day and in **others** with doses as low as 50 mg/day. **He** had pruritus, facial swelling, SC edema of **his** limbs, and hyperkeratosis after 11 days. Unconscious 3 weeks after starting therapy. Pronounced disseminating erythematous exanthema over ~80% of **his** body, blisters, and peeling skin; **he** died 2 weeks later. Stevens–Johnson syndrome and intravascular coagulation has also been reported. A 22-year-old **female** died after 3 weeks of therapy with 50–100 mg twice daily. **She** had developed cerebral edema, a massive pulmonary embolus, and a fulminant hepatic failure. It completely inhibited erythropoiesis in a **patient** with β-thalassemia, by interfering with ***folic acid*** metabolism. Due to a pharmacy's dispensing error, a **patient** was given 150 mg twice daily instead of ***lamivudine***, and developed facial swelling, conjunctivitis and rigors, probably due to an allergic reaction. Leukopenia due to 25 mg/day/7 days and 50 mg/day/3 days in a 35-year-old **female** who also developed an erythematous rash, nausea, vomiting, dizziness, and a sore throat. A 12-year-old **female** given 50 mg/day/2 weeks developed maculopapular rash, some pruritus, abdominal discomfort, leukopenia, and increased liver enzymes. Agranulocytosis on day 3. Loss of aura in several epileptic **patients**. A case of self-poisoning in 26-year-old **male** after ingesting 1350 mg. An hour later, **he** was flushed, had vertical and horizontal nystagmus, and was hypertonic, followed the next day by ataxia and electrocardiogram changes. A 49-year-old **male** inadvertently given 2.7 g/day/4 days developed anticonvulsant hypersensitivity syndrome. Tourette syndrome symptoms with repetitive tic-like movements and inappropriate vocalization in **male**s and **female**s 7–12 years after ~2 weeks of therapy.

Drug interactions: Its half-life is decreased from ~24 h to ~15 h and its serum level is greatly decreased by ***carbamazepine***, ***phenobarbital***, ***phenytoin***, and ***primidone***. Toxicity reported in a **patient** taking ***carbamazepine***. Its serum levels are increased by ***sodium valproate*** as well as its half-life (to ~60 h). Combined treatment has caused disabling tremors in some **patients**. ***Acetaminophen*** increases its clearance. ***Imipramine*** significantly decreases its oral absorption. It increased ***amiodarone*** serum levels in 38-year-old **male**.

LAMPS

Untoward effects: Possible increase in breast and gynecological cancers in **female employees** in the coiling and wire-drawing areas of a General Electric Company Canandian subsidiary. Pyrexia (axillary temperature 39.6°C or 103.3°F), despite unwarmed IV infusions during 2 h filming of major head surgery with two 1000 W photo-flood lamps 2 m from the **patient**. Radiant heat third degree burns and tissue necrosis possible from surgical operating theater without heat filters. Some ultraviolet radiation from fluorescent lamps. Increased invasion of choroid and sclera by retinoblastoma reported after light coagulation with a Zeiss xenon lamp. Unshielded ***mercury***-vapor lamps in public gymnasiums led to conjunctivitis, eye pain, swollen eyelids, increased lacrimation, injected sclerae, skin erythema, and headache from ultraviolet radiation in ~100 **people** and **basketball players**. Symptoms lasted a few hours to several days.

Fluorescent lamp exposure of **hamster** cells has caused chromosome mutations.

LAMPTEROMYCES JAPONICUS

A *mushroom* that glows in the dark. Grows on rotten *beech* trees in Japan in October.

Untoward effects: Its appearance, like an edible ***mushroom***, has led to **human** fatalities, due to its ***lampterol*** content.

LANATOSIDE C

= *Allocor* = *Cedilanid* = *Ceglunat* = *Digilanide C* = *Celadigal* = *Cetosanol* = *Isolanid* = *Lanimerck*

***Cardiotonic. A glycoside* from *Digitalis laneta*.**

Untoward effects: Supraventricular arrythmias, ventricular extrasystoles and bigeminy, tachycardia, pallor, and weakness occurred. Pediatric poisonings leading to sleepiness, stupor, electrocardiogram changes, and renal insufficiency have been reported.

LD_{LO}, **human**: 5 mg/kg.

Vomiting in **birds**, **cats**, **dogs**, and **monkeys** after IV dosage. ½–⅓ the potency of ***digoxin*** in **dogs**.

LANDRIN

= *ENT 25,843* = *OMS 597* = *SD 8530* = *SD 8786*

Use: To establish conditioned taste aversion in certain predator **birds**, to protect **condor** eggs.

Untoward effects: Anticholinesterase carbamate pesticide. Ingestion by **humans** would be expected to cause headache, dizziness, weakness, ataxia, miosis, muscle twitching, tremors, nausea, bradycardia, salivation, and respiratory depression. Occasionally convulsions, mental confusion, incontinence, unconsciousness, vomiting, abdominal cramps, diarrhea, tightness in chest, wheezing, productive coughing, pulmonary edema, sweating, rhinorrhea, and tearing. Do NOT give ***pralidoxime***, as it has no value in carbamate poisonings. Atropinization (tachycardia, flushing, dry mouth, and mydriasis) for 2–12 h, depending on degree of exposure, is preferable.

Causes vomiting in **vultures**.

LD_{50}, **mallard ducks**: 22.4 mg/kg; **pheasants**: 51 mg/kg; **chukars**: 60 mg/kg; **Coturnix quail**: 70.8 mg/kg; **pigeons**: 168 mg/kg; **sparrows**: 46.3 mg/kg.

LANOLIN

= *Adeps Lanae* = *Agnin* = *Agnolin* = *Alapurin* = *Graisse de Suint Purifée* = *Lanain* = *Lanalin* = *Lanesin* = *Lanichol* = *Laniol* = *Lanum* = *Oesipos* = *Wollfett* = *Wool Fat*

Emollient, emulsifier.

Untoward effects: Sensitivity (dermatitis – usually, eczema) in some **people** from topical use of various ointments. Slightly > 100 cases documented in ~70 years. Estimated to be < 10/million **people**. Possibility of ***pesticide*** contamination from treatment of **sheep** caused concern over its use as an emollient in treating cracked nipples of nursing **mothers** or in ***lipsticks***. Contact urticaria reported. Highly comedogenic and should especially be avoided by **teens**. ***Acetylated*** or ***ethoxylated lanolins*** = ***Solans*** are also comedogenic. Hypersensitivity and dermatitis also reported from ***lanolin alcohols*** = ***Lanethyl*** = ***Polychols***.

Moderately comedogenic in tests on external ear canal of **rabbits**.

LANSOPRAZOLE

= *A-65006* = *AG-1749* = *Agopton* = *Lanzor* = *Ogast* = *Prevacid* = *Takepron* = *Zoton*

***Antiulcerative.* Gastric *proton-pump inhibitor*. A *benzimidazole*. Active against *Heliobacter pylori*.**

Use: In treatment of hypersecretory gastric conditions, erosive reflux esophagitis, and Zollinger–Ellison syndrome.

Untoward effects: Possible immune reaction in 48-year-old **male** leading to irreversible impaired vision from optic neuritis (papillitis) after 1 week of therapy. **He** previously had 6 weeks of therapy without reactions. Abdominal pain, nausea, constipation, flatulence, and diarrhea have occurred. Occasionally myopathy, arthralgia, severe rashes, exfoliative dermatitis, and acute interstitial nephritis. Less than 1% for each of many other possible ill-effects. Rarely, gynecomastia. Its bioavailability is decreased by ~50% when taken with ***food***.

A ***sulfone***-containing drug and it should be used cautiously in **patients** with a ***sulbactam*** allergy.

Drug interactions: Glossitis, stomatitis, and black tongue when ***clarithromycin*** was added to therapy. ***Sucralfate***

decreases its absorption. It can decrease absorption of ***itraconazole*** and ***ketoconazole***; also, possibly, ***ampicillin***, ***digoxin***, and ***iron*** compounds.

High dosage to **rats** for 2 years caused dose-related increase in gastric carcinoid tumors, probably associated with prolonged hypochlorhydria and secondary hypergastrinemia.

LANTANA sp.

= *Bunchberry* = *Hen and Chicks* = *Sages*

Many species are used world-wide for medicinal purposes without ill-effects. 3–10 ft wild perennial or cultivated ornamental. Low-growing species also exist.

Untoward effects*: *L. camana* = *Big Leaf Sage* = *Big Sage* = *Cherrypie* = *Common Lantana* = *Ewon Agogo (Yoruba) = ***Largeleaf Lantana* = *Red Sage* = *Tickberry* = *Wild Sage*.** All parts, especially the green berries are toxic, due to ***lantadene A*** and other triterpenoids, cause vomiting, diarrhea, muscle weakness, ataxia, photophobia, mydriasis followed by miosis, lethargy, decreased deep tendon reflexes, proctitis, bradypnea, labored breathing, circulatory collapse, cyanosis, and coma. Chronic intake prevents ***bilirubin*** conjugation with retention jaundice and photosensitivity. Contains ***icterogenin***, hepatotoxic to **man**. Accidental ingestion and poisoning of **children** leads to ***belladonna*** alkaloid-like syndrome, vomiting, etc., as noted above in four acute toxicity cases of 17 **children** (mean age 3 years) at one Poison Control Center; three recovered and one died with neurocirculatory collapse and acute pulmonary edema.

Constipation, tenesmus, anorexia, decreased temperature, bradycardia, bradypnea, erythema, and edematous swelling of ears, scrotum, prepuce, and perineal region in **buffalo calves**. In terminal stages, icterus, gastroenteritis, weakness, decreased Hb, decreased red blood cells, and leukocytosis. Poisoned **buffalo** have also shown lacrimation, emaciation, red streaks on the inside of their thighs, bile-stained urine, and death in 5–13 days. Post-mortem revealed hemorrhagic organs, atrophied kidneys and liver, liver necrosis, emphysema, and partial atelectasis of lungs. In India, 400 **buffalo** and **cattle** died after consuming it during a famine. Severe gastroenteritis, bloody feces, icterus; cracked, ulcerated skin on muzzle; dermatitis, itching, photosensitization, and death within a few days common in **cattle**. In Australia, 93/600 **steers** died within 3 weeks from hepatotoxicity from consuming it along the wayside on a **cattle** drive. As with **cattle**, the acute form of intoxication in **sheep** leads to cholestasis and constipation. The chronic syndrome in both leads to alimentary stasis and renal failure. Post-mortem reveals large, dilated gallbladder. **Horses** are also occasionally affected. Frosting enhances its toxicity.

***L. aculeata*, *L. horrida*,** and ***L. montevidensis* (*Creeping Lantana*)** are also poisonous to **livestock**.

LANTERNS

Untoward effects*:** The mantle of a gas lantern contains ~600 μg of ***beryllium (q.v.), which becomes volatilized and airborne during the first 15 min of using a new mantle. Its inhalation is hazardous.

LANTHANUM

***Untoward effects*:** Prolonged clotting time and decreased platelets.

LAPACHOL

= *Greenhartin* = *Lapachic Acid* = *NSC-11905* = *Taiguic Acid* = *Tecomin*

Antineoplastic.

***Untoward effects*:** Nausea and vomiting after doses > 1.5 g. Nasty taste.

LAPORTEA sp.

= *Wood Nettle*

Untoward effects*:** Dermatitis and immediate, intense pain from its stinging hairs, apparently due to their ***formic acid (q.v.) and ***histamine***-like substances in bladders within the leaf. A **human** death reported in New Guinea after severe exposure. Use of water on the affected areas intensifies the pain. ***L. gigas*** of eastern Australia is said frequently to kill **horses**.

LARKSPUR

= *Crowfoot* = *Delphinium* = *Estafiate* = *Knight's Spur* = *Pied d'alouette* = *Staggerweed*

Untoward effects*:** Ingestion of young plants or seeds containing ***delphinine can be fatal within 6 h. Symptoms include upset stomach, depression, dizziness, nervous excitement, and inflammation of mouth, lips, and tongue; followed by numbness and paresthesia starting in the extremities and extending to the entire body. Salivation, followed by dry mouth, nausea, vomiting, tinnitus, confusion, severe headache, labored breathing, pulmonary edema, bradycardia, arrhythmias, and sweating (despite cold and clammy skin). The seeds are the most toxic portion. In 1950, "Larkspur Lotion" made from the seeds was official in the U.S. Poisoning from percutaneous absorption after its topical use as a pediculicide. The "lotion" was readily absorbed through abraded skin. Toxicity was mainly due to its alkaloids, including ***delphinine*** (q.v.) and ***methyllycaconitine***.

Plains Larkspur* = *Delphinium virscens is toxic to **cattle**, especially if eaten during late spring or early summer.

Green leaves, its flowers, and roots are the most toxic portions of the plant. One of the greatest causes of poisonous plant mortality in U.S. **cattle**, as they will graze it, even when other feed is available. In wet weather, they easily pull up the entire plant and eat the roots as well. This results in bloat, salivation, constipation, weakness, paralysis, and death.

Tall Larkspur = ***D. barbeyi*** grows at elevations above 6000 ft and causes nervousness, frequent swallowing, twitching, tremors, bloating, staggers, arrhythmias, and respiratory paralysis in **cattle**. ***D. exaltatum*** in Alabama is also called ***Tall Larkspur*** and causes similar toxicity, as does ***D. glaucom*** in Canada, often poisonous to **cattle** and occasionally to **horses** and **sheep**.

D. tricorne = ***Dwarf Larkspur*** poisons **cattle** and occasionally **horses** in Alabama and Illinois.

D. ajacis = ***Rocket Larkspur***, ***D. carolinianum*** = ***Azure Larkspur***.

D. bicolor = ***Little Larkspur*** = ***Low Larkspur*** in the western U.S., is usually toxic in May and June.

Sheep are much less susceptible than **cattle** to larkspur poisoning. Affected **horses** show muscle trembling, collapse, prostration, convulsions, constipation, bloat, sialism. In Russia, ***D. confusum*** has shown ***curare***-like activity and ***D. cashmirianum*** is arrhythmogenic in **rabbits**.

LARREA sp.

= *Chaparral* = *Creosote Bush*

A common evergreen shrub in southwestern U.S. and northern Mexico.

Untoward effects: Erythematous and vesicular dermatitis on the face, neck, and hands a few hours after picking branches of the bush. Confirmed by patch-testing.

L. divaricata or ***L. tridentata*** leaves are used in making ***chaparral tea*** or sold as a powder in capsules. Cases of acute hepatitis and liver failure after ingesting this herbal remedy have been reported in the U.S. and Australia.

LASALOCID

= *Avatec* = *Bovatec* = *RO 2–2985* = *X 537A*

***Coccidiostat.* An *ionophore*.**

Untoward effects: High death losses in **calves** after overdosing. Symptoms include anorexia, watery diarrhea; cardiac, central nervous system, and electrolyte disturbances; and dyspnea.

LD_{50}, **rat**: 146 mg/kg.

LASERS

Operate continuously or in pulses.

Untoward effects: The FDA has been particularly concerned about the risk of gas or air embolism during therapeutic procedures, where they are routinely used for cooling. Laser light in wavelengths of 0.4–1.4 μm can easily damage the retinas, causing burns and blind spots. Skin burns are less of a hazard, but can be caused by high-powered lasers. High-power lasers are now in use for industrial applications. Electric shock, potentially lethal, can occur. Dangerous vapors have been produced from certain target materials. ***Cryogenic fluids*** that cool lasers can cause serious burns. Protective eyeware has been shattered by some laser beams.

LASIOCARPINE

A *pyrrolizidine* alkaloid.

Untoward effects: Hepatotoxic, teratogenic, and carcinogenic in **animals**. Deaths in **rats** from the toxic effect on the liver. Hyperacute toxicity of large doses block impulses across neuromuscular junctions, leading to progressive muscle weakness, respiratory distress, and death within 30 min in **rats**. Injections have caused acute gastrointestinal disease and intestinal atrophy in **sheep**, **rats**, and **mice**.

LD_{50}, **rat**: 150 mg/kg.

LASIOSIPHON sp.

Untoward effects: ***L. burchelli*** poisons **sheep**, leading to diarrhea, dyspnea, and rumen atony; 1.8 g/kg of the dried plant is lethal to **sheep**.

L. kraussianus = ***Pmatiniya*** (Gbari) = ***Sungoje*** (Fulani) = ***Tururubi*** (Hausa). In Nigeria, used criminally in food for **humans**. Causes toxicity and deaths in Nigerian and East African **donkeys**, **goats**, and **cows**. **Sheep** appear to be more resistant to its fatal effects. Post-mortem reveals severe hemorrhagic, ulcerative gastroenteritis; hemorrhages, congestion, and edema in the brain, liver, and myocardium. Listlessness, anorexia, nasal and ocular discharges, diarrhea, lymphopenia, eosinopenia, and increased blood urea nitrogen occur.

LASIOSPERMUM BIPINNATUM

Untoward effects: Causes abdominal pain and hepatogenous photosensitization in South African **sheep** and **calves**. Post-mortem shows necrotic livers.

LATANOPROST

***Antiglaucoma agent.* Topical.**

Untoward effects: After a year of its topical use, it was noted that > 15% of **patient's** blue, hazel, or yellow-colored eyes turned blue or brown. Hypertension in

70-year-old **male** and 71-year-old **female**. Confirmed by rechallenge in the **female**. Angina and tachycardia also reported.

LATEX

From the Brazilian *rubber* plant, *Hevea brasiliensis*. Other latexes exist. Latex contains > 250 proteins and > 30 are known human allergens.

Untoward effects: Itching, burning, rashes, hives, welts, wheezing, and even life-threatening anaphylactic shock reported in ~17 million **people** sensitive to it. Natural ***rubber*** latex is used in ~40,000 products, 300 of them with medical uses. **Nurses** who had previously handled latex gloves on a daily basis, developed hypersensitivity reactions to airborne latex allergens. A 44-year-old **female** had an anaphylactic reaction after hepatitis B vaccination. It was traced to the syringe needle's penetration and picking up traces of the vial's ***rubber*** seal. One brand of latex-cuffed enema tips was taken off the market after several **patients** died of severe anaphylactic reaction during ***barium*** enema procedures. In most cases, the reaction occurred after the enema tip was inserted and before the ***barium*** was introduced. The FDA estimates reactions to it in 6–7% of surgical **personnel** and up to 40% of spina bifida **patient's**; > 600 reports of reactions, and at least 16 deaths. **Patients** have suffered sensitivity reactions from latex eluted from obstetrician's gloves. Surgical staff should be alerted to possibility of **patient** reactions. Severe anaphylaxis in 31-year-old **female**; once from use of latex glove during a vaginal examination, and again when **her son** directed air flow from a deflating joke cushion at **her** face. **People** allergic to latex may often be allergic to ***avocados***, ***bananas***, ***celery***, ***chestnuts***, ***melons***, and ***tomatoes***. Latex gloves are permeable to ***fluorouracil*** and ***methotrexate***. Powder-free Chinese chlorinated latex gloves can burst into flame from an exothermic reaction under hot storage conditions.

A spermicidal effect has been noted from their use during **boar** semen collection.

LATHYRUS sp.

Untoward effects: Photosensitization and abortions in **cattle**. Osteolathyrism and neurolathyrism occur. In the U.S., the former is most common, leading to degeneration and enlargement in cartilage plate junctions, multiple exostoses, giant callus formation at fracture sites, and numerous hemorrhages. Neurotoxic in **man**.

L. aphaca = ***Yellow Vetchling*** contains a cyanogenic glycoside.

L. hirsutus = ***Caley Pea*** = ***Wild Winter Pea*** contains nitriles and amines as its poisonous principles leading to lameness in **livestock**.

L. latifolius causes nervous symptoms in **rats**.

L. missolia = ***Grass Vetchling*** can, apparently, be eaten safely by **horses** for 10 days, but, when exercised, they become ataxic, collapse, and struggle to rise. If the hay is discontinued, they recover completely in a few days.

L. odoratus = ***Pois de Senteur*** = ***Sweet Pea*** contains ***β-aminopropionitrile*** (q.v.) in the seeds and peas, leading to teratogenicity (cleft palates) and disturbed chondrogenetic process in **rats** and **mice**, aortic aneurysms in **rats**, and asymmetrical beak development; swollen, red, and painful joints, and fragile tibia in **chicks**. In **people** and **children**, it leads to bradycardia, weak pulse, bradypnea, skeletal deformities, paralysis, and convulsions. Seeds, eaten in quantity, can cause symptoms of and death from its ***hydrogen cyanide*** (q.v.) or ***prussic acid*** content. In **turkeys**, paralysis, pericardial hemorrhage, aortic rupture, hock deformities, and crooked toes are common. **Horses** are particularly susceptible to it, with skeletal deformity, kyphosis, scoliosis, osteoporosis, long bone curvature, poor connective tissue development, and aortic ruptures occurring.

TD_{LO}, **rats** (10–16 days pregnant): 175 g/kg; **mice** (9–18 days pregnant): 600 g/kg.

L. pusillus = ***Singletary Pea*** causes osteolathyrism, due to its ***β-aminopropionitrile*** content.

L. sativus = ***Chickling Pea*** = ***Chickling Vetch*** = ***Chick Pea*** = ***Grass Pea*** = ***Green Vetch*** = ***Indian Pea*** = ***Jarosse*** = ***Kasari*** = ***Khesari*** = ***Lentil d'Espagne*** causes neurolathyrism poisoning in **man**, chiefly in Asia, India, and Africa during periods of famine. The plant is drought-resistant. Ingestion of the peas by daily feeding to **people** in German forced-labor concentration camps caused central nervous system disturbances and the majority of affected **persons** became **invalids** with spastic paraparesis. This spastic, ataxic paraplegia of the legs developed after consuming > 300 g/day/~3 months. The pea contains a natural excitotoxin, ***β-oxaloamino-alanine*** or ***L-3-oxalylamino-2-aminopropionic acid*** that kills neurons in certain parts of the brain. Causes neurolathyrism in **monkeys**. In contrast to **humans**, it causes osteolathyrism in **rats**. **Horses** develop sudden and transient paralysis of the larynx and have difficulty breathing; it is occasionally fatal in them.

LATUA PUBIFLORA

Untoward effects: Extract of the shrub is used by **Chilean Indians** as a virulent narcotic poison, causing delirium and hallucinations.

LAUREL

= *Laurus nobilis*

Food spice.

***Untoward effects*:** Delayed eczematous contact dermatitis. Contains cardiac glycosides. The leaf oil is a fragrance used in soaps, detergents, cosmetic creams and lotions, perfumes, and foods. No irritation or sensitization of 2% in petrolatum on **human** volunteers. Others have reported the oil has caused hyperemia, severe inflammation, and allergic skin reactions.

Some irritation reported in **animal** trials. Abdominal pain, nausea, and salivation reported in pet **animals** that ingest the plant.

***Oil*:** LD_{50}, **rats**: 3.95 g/kg; acute dermal, **rabbit**: > 5 g/kg.

LAURIC ACID
= *Dodecanoic Acid*

From *coconut* and *palm kernel oils*.

***Untoward effects*:** Allergic contact dermatitis from its use in washing compounds.

LD_{50}, **rat**: 12 g/kg.

LAURYL GALLATE

Untoward effects*:** Contact dermatitis reported in a **baker** from its presence in ***margarine as an antioxidant.

LD_{50}, **rat**: 6.5 g/kg; **mouse**: 1.6 g/kg.

LAURYL ISOQUINOLINIUM BROMIDE
= *Isothan Q15*

***Cationic detergent*. A cosmetic and toiletry preservative.**

***Untoward effects*:** Irritating and injurious to **animal** eyes at 0.1–1.0% concentration.

LD_{LO}, **human**: 50 mg/kg.

LD_{50}, **rat**: 230 mg/kg; **guinea pig**: 200 mg/kg.

LAURYL OXYPROPYLAMINOBUTYRIC ACID

***Antiseptic, fungicide*. In *Sterlane*.**

***Untoward effects*:** Eczema. Skin irritation, if used more than twice daily for long periods of time. No benefit on *Tinea*.

LAVANDIN OIL

The main constituent is *linalool*. As much as 500,000 lbs/year used in the U.S. as a fragrance in soaps, detergents, cosmetic creams and lotions, perfumes, and foods. ~1000 tons lavandin produced annually.

***Untoward effects*:** No irritation or sensitization of 5% in petrolatum on **human** volunteers. Full-strength under occlusion on intact or abraded **rabbit** skin/24 h leads to slight irritation.

LD_{50}, **rat** and acute dermal, **rabbit**: > 5 g/kg.

LAVENDER
= *Lavendula sp.*

***Untoward effects*:** Has caused photosensitivity and contact allergic reactions in some **people**. Berlock dermatitis, an irregular discoloration of the skin, appears occasionally after sun exposure.

Lavender absolute*, *spike lavender oil*,** and ***lavender oil are all approved as a fragrance in soaps, detergents, cosmetic lotions and creams, and perfume. ***Lavender oil*** is also approved for food use. They produced no irritation or sensitization with **human** volunteers at 10%, 8%, and 16%, respectively, in petrolatum.

LD_{50}, **rats**: 4.25 g, 3.8 g, and > 5 g/kg, respectively; acute dermal, **guinea pigs**: > 5 g; **rabbits**: > 2 g and **rabbits**: > 5 g/kg, respectively. When applied to abraded or intact **rabbit** skin for 24 h, is slightly, moderately, and slightly irritating, respectively.

LEAD
= *Plumbum* = *Saturn*

Widely distributed in the earth's crust, in the ocean, old *paints*; in plants and water supplies, especially significant near *copper*, *lead*, and *zinc* smelters.

Untoward effects*: *Batteries*, *gasoline*,** ceramics, lead-based paints, and pigments are the major sources of **human** exposure. Lead is a cumulative poison, and atmospheric lead from vehicles burning leaded gas is the primary source. Small amounts have been found in **cow**, **goat**, and **human** milk. Toxicity can occur from oral, inhaled, or skin contact. The latter is mostly from ***tetraethyl lead. Symptoms include weakness, abdominal pain, anorexia, sleep disturbances, irritability, occasional nausea and anemia. In severe cases, a metallic astringent taste, increased abdominal pain, nausea, vomiting, muscle weakness and tenderness, diarrhea, paresthesias, encephalopathy, alopecia, nail pigmentation, decreased fertility, hyperuricemia, ototoxicity, neuritis, partial collapse, wrist or foot drop, arthralgia; hepatic, renal, and pulmonary damage; coma, and death. A dark blue line can appear on the gum margins. Erythrocyte stippling is common. Lead tastes sweet and **children** < 6 years absorb 30–75% of the lead **they** ingest, whereas **adults** only absorb 10%. Blood levels as low as 10 μg/dl associated with decreased intelligence (usually a decrease of 2 IQ points/10 μg/dl increase); impaired healing, and decreased growth. A study involving thousands of **children** aged 5–17 years found that for every increase of 5 μg/dl in blood lead levels tooth decay risk increased 80%. A 53-year-old **male** occupationally exposed to lead for 30 years had blood levels of 60–70 μg/dl, leading to optic atrophy. Environmental levels were > 0.15 mg/m^3/14 years. Poisoned **children** frequently show encephalitis, but convulsions are rare in **adults**. Highest affinity for **fetal** bone and liver is in the first trimester. Many case reports of lead poisoning with spasticity and other neurological defects, decreased intrauterine growth, and anemia in **newborns** of **mothers** who drank moonshine ***whiskey*** contaminated with lead from an old car radiator. Hemorrhage in the placenta has been followed by abortion. Mexican, Laotian, Hong Kong, and other ethnic herbal and folk-remedies have been a source of toxic lead levels, as have indoor gun-firing ranges. Some of the traditional folk medicines and cosmetics that contain lead are ***alarcon***, ***alkohl***, ***azarcon***, ***bali goli***, ***coral***, ***deshi***, ***glissard***, ***greta***, ***hai gen fen*** (***clamshell*** powder), ***kohl***, ***Koo-Sar*** pills, ***Koo-So*** pills, ***liga***, ***pay-loo-ah***, ***rueda***, and ***surma***. Lead toxicity can occur in **children** from chronic sniffing of ***gasoline***. A 41-year-old **male** using a high-pressure spray gun on the hull of a ship injected the palm of **his** left hand. Despite careful dissection and removal of ~90% of the lead-based ***paint***, **he** developed dizziness, nausea, abdominal discomfort, weakness, and insomnia over a 10 day period. A 43-year-old **female art conservator** received heavy lead exposure while restoring a Peruvian tapestry from a tomb that contained ***cinnabar*** with about 1% lead. Over a 2 month period, **she** developed neurological, gastrointestinal, and diffuse muscular symptoms, as well as severe anemia with busophilic stippling of red blood cells. **Her** blood levels were 130 μg/dl. **Infants** born to ***tobacco***-smoking **mothers** had above average concentrations of ***cadmium***, ***lead***, and ***thiocyanate***. Ingestion of beverages stored in ceramic containers and lead salts on a ***wine*** bottle's rim have been sources of exposure. In 1991, only ~3% of U.S.-produced food cans had been lead-soldered, compared to ~90% in the 1970s. Canned fruit juices should be put in non-metallic containers after opening. Lead poisoning has occurred from a retained bullet, and in **children** from swallowing lead curtain weights. Allergic contact dermatitis from lead is rare. Lead exposure reported from various countries after eating from lead-glazed pottery; and from the use of cosmetics, eye shadows, hair dyes, canned tuna, bonemeal, and ***dolomite*** supplements. Gout in aging has been attributed by some to excessive childhood exposure to lead. IV poisoning has been associated with contaminated ***methamphetamine***. Thoman's guide to lead poisoning is said to be as simple as A–B–C–D–E:

Anorexia, Apathy, Anemia

Behavioral disturbances (hyperkinesis, etc.)

Clumsiness

Developmental skill deterioration

Emesis (sporadic vomiting and colic)

For the onset of encephalopathy, use the mnemonic PAINT:

Persistent and forceful vomiting

Ataxia

Intermittent stupor and lucid intervals

Neurological intractable convulsions and coma

Tired/lethargic

Lead poisoning is said to have contributed to the fall of Roman civilization. Production in the Empire peaked at about 80,000 tons/year, where it was used to make water conduit pipes, pots, glazes, and as a preservative for ***wines*** and fruit juices. Drinking such ***wines*** led to mental incompetence and decreased birth rate. Pliny, in the first century A.D., identified palsy as a manifestation of exposure to lead dust. Roman **men** dipped their lead combs in ***vinegar*** to darken their graying hair. "Normal" or average levels are ~0.1 μg/ml and toxic at > 0.6 μg/ml. Analysis of hair samples indicated **Beethoven**, as an **adult**, was exposed to lead.

Lead poisoning has occurred in **zoo animals** (**primates**, **parrots**, **bats**) from previous use of lead ***paint*** on their cages. The **primates** develop seizures, blindness, increased blood and tissue lead levels, and acid-fast intranuclear inclusion bodies. Staggering mortality has been reported in **ducks**, **geese**, and **swans** from eating lead shot. Many other **bird** species have been affected. **Cattle** are poisoned, mostly from licking lead ***paint***, storage ***batteries***, and, occasionally, from ingesting lead salts in crankcase oil or from grazing contaminated pastures. **Dogs**, **cats**, **fish**, **gerbils**, **goats**, **guinea pigs**, **hamsters**, **horses**, **poultry**, **rabbits**, **rats** and **mice**, **sheep**, **squirrels**, **swine**, and other **animals** have also been poisoned by lead. **Pups** from lead-exposed pregnant **rats** developed 40% more dental cavities and 30% less saliva than those who were not.

LEAD ACETATE

= *Sugar of Lead*

Use: Drier in ***paints*** and varnishes; in explosives, waterproofing, insecticides, ***inks***, ***cotton*** dyes, hair dyes, pharmaceutical astringents, special glass, and as a stabilizer in plastics.

Untoward effects: It is not known for certain if its use by **adults** to darken gray hair or if **children** can also absorb this lead from contaminated hands or treated hair. A case report of acute renal insufficiency from poisoning by it. A 25-year-old **male drug addict** developed abdominal pain, vomiting, and constipation after IV of a melted down suppository containing ***lead acetate***, ***opium***, and ***cocoa butter***. Allergic contact dermatitis from it in a **patient** who had frequent contact with truck ***battery*** terminals, and in a 70-year-old **male** from its use in a hair dye. A stomach irritant in overdoses, with nausea and vomiting within 30 min, metallic taste, abdominal cramps, burning sensations in the throat, constipation, black stools from ***lead sulfide***, central nervous system upsets with paralysis, and, in fatal cases, collapse usually developing in about 3–4 days.

In **rats**, leads to renal adenomas and carcinomas; cerebral gliomas. Immunosuppressive effects in **rats**, **mice**, **swine**, and **rabbits**. Acute encephalopathy in **calves** receiving 20 mg/kg/day. **Trout** absorb ~25% in the first few hours. Anorexia, diarrhea, basophilic stippling in 6/8 **goats**, and death after 100 g or more for 10–52 days. Weight decrease, anemia, convulsions, peripheral nerve degeneration and demyelination, and occasionally mild posterior paralysis in some **guinea pigs**. **Horses** appear to be more tolerant to toxic doses. Causes neck abnormalities and retarded growth in **chick** embryos; retarded growth and anemia in **quail**. Dramatically increases endotoxin hypersensitivity in **rats**.

LEAD ARSENATE

= *NuRexform*

Insecticide, herbicide, teniacide.

Untoward effects: Occupational dermatitis in orchard **workers**. In **people** living near or working in vineyards, a number of case reports of hyperkeratoses of palms and soles, with wart-like proliferations; melanosis in the lumbar, inguinal, and axillary regions, plus poor nail growth and systemic upsets. Contains ***arsenic***. When ingested in large amounts or in high concentration, it kills many of the cells it contacts, causing intestinal ulceration. In small amounts or diluted and ingested chronically, it slowly destroys the delicate parenchymal cells of the liver.

LD_{LO}, **human**: 5 mg/kg.

Accidental deaths in **cattle** after ingestion of insecticide spray material.

LD_{50}, **rat**: 100 mg/kg; **rabbit**: 125 mg/kg; **chicken**: 450 mg/kg.

LEAD CARBONATE

= *Basic Lead Carbonate* = *White Lead*

Use: In oil ***paints***, ***cements***, putty, and in processing parchment.

Untoward effects: **Painters** develop chronic poisoning from spilling it on **their** skin for many years. **Children** have been poisoned by eating flaking lead-based ***paints***.

Sheep have been poisoned by grazing contaminated pastures near open-pit mines. It was used as a filler in linoleum manufacturing, and a **dog** was poisoned by it after eating the linoleum lining his kennel. **Livestock** and **dogs** were poisoned by gnawing on painted boards and stalls.

LD, **cattle**: 250 mg/kg; **pig**: 800 mg/kg; **horse**, **dog**, and **guinea pig**: 1 g/kg.

LEAD CHLORIDE

Untoward effects: Allergic contact dermatitis in a **patient** who had frequent contact with truck ***battery*** terminals.

LD_{LO}, **guinea pig**: 2 g/kg.

LEAD CHROMATE

= *Chrome Yellow* = *CI 77,600* = *King's Yellow* = *Leipzig Yellow* = *Paris Yellow*

Untoward effects: Poisoning in 121 Gurkha **soldiers** in Hong Kong, due to contamination of ***chilli powder***, a constituent of ***curry powder*** with it (10,800 ppm ***lead***) leading to epigastric pain, constipation, headache, some arthralgia, blue gum line in 19%, bilateral gaze limitation in 60%, and anemia and basophilic stippling in most. Can cause anemia, paralysis, abortion, and mental retardation. Inhalation of the dust is suspect as a cause of lung cancer.

LD_{LO}, **human**: 50 mg/kg.

LEAD OXIDES

Untoward effects: Used by **Romans** as a bactericide in ***wine***, with resultant chronic toxicity. Acute lead poisoning from ***Azacon*** (***Lead Tetroxide*** = ***Red Lead***) and ***Greta*** (***Lead Monoxide*** = ***Lead Protoxide*** = ***Litharge*** = ***Massicot***) used by **Hispanics** for empacho (indigestion), has caused serious problems in **people**. Similar products are ***Alarcon***, ***Coral***, ***Liga***, ***Maria Luisa***, and ***Rueda***. Demolition **workers** have had subacute effects from its fumes.

Cattle in polluted industrial areas have been poisoned by them, leading to emaciation, enteritis, bloat, colic, muscle tremors, arthralgia, joint swellings, and nervous symptoms.

LEATHER

Untoward effects: Contact dermatitis from proteases to soften it and from other chemical treatments (***chromates***, ***formaldehyde***, etc). ***Chrome***-treated shoe leather often causes symptoms that mimic tinea pedis. Leather dust is a suspected carcinogen, especially in the nose, sinuses, and bladder. Polyneuropathies reported from leather cement solvents used in Italian shoe factories.

LECITHIN

= *Actiflo* = *Alcolec* = *Clearate* = *Gliddophil* = *Granulestin* = *Kelecin* = *Lecithol* = *Maxicholine* = *Phosphatidyl Choline* = *Vitellin*

~100,000 tons produced annually.

Untoward effects*:** Despite its use as a "health food", it can induce depression. High-lecithin diet has enhanced palmar sweating in 10/11 **subjects**. May complex dietary ***potassium. IV use leads to bradycardia, hypotension, and ventricular extrasystoles. Choking and gagging sensations reported after ***soya*** lecithin was substituted for ***sorbitan trioleate*** in ***Alupent***®.

In the **cat**, IV use causes the same symptoms as in **man**, as well as contraction of the nictitating membrane. Doses > normal can produce defects that retard the development of **rat pups**. IV use in **dogs** increases blood viscosity and decreases red blood cell sedimentation rate.

LECYTHIS

Grows in northern South America.

Untoward effects*:** Hair and nail loss after eating ***Sapucaia nuts (***L. elliptica***). Nausea, vomiting, and even death have followed their ingestion.

L. ollaria = ***Coco de Mono***. Ingestion of 70–80 ***almond***-like seeds by 54-year-old **male** led to nervousness, anxiety, anorexia, asthenia, increased diuresis, and arthralgia. Hair loss (body and scalp) on the eighth day after ingestion. Another **person** experienced nausea, anxiety, and vertigo 30 min after seed ingestion.

Monkeys have died after eating its seeds. Causes loss of hair, weight decrease, and death of **mice**.

LEDUM sp.

Members of the *Ericaceae*, or *Heath* family.

Untoward effects*: *L. columbianum* = *Pacific Labrador Tea and ***L. glandulosum*** = ***Black Laurel*** = ***Western Labrador Tea***. Though not palatable, eating of young shoots along trails or when pasture is scanty, causes poisoning of **sheep** in western U.S. from their ***andromedotoxin*** (q.v.) content. ***L. palustre*** = ***Hospa*** = ***Tomamas***. After grazing on it, symptoms in **livestock** are similar to ***andromedotoxin's*** (q.v.).

LEFLUNAMIDE

= *Arava* = *HWA 486* = *SU 101*

***Immunomodulator, anti-inflammatory*. A prodrug in treatment of rheumatoid arthritis.**

Untoward effects*:** Diarrhea, rash, nausea, vomiting, fever, and alopecia. Occasional hepatotoxicity, increase in aminotransferases, leukocytoclastic vasculitis, and anaphylaxis. Serum levels still noted at 6 months. Contraindicated in **women** who plan to be **mothers**, or in **men** planning to be **fathers**, because it can be in **their** serum up to 2 years unless **they** are treated with ***cholestyramine for nearly 2 weeks.

Drug interactions*:** Induces a 13–50% increase in ***diclofenac, ***ibuprofen***, and ***tolbutamide*** serum levels. Use with ***rifampin*** increases its active metabolite, ***A77 1726***, serum levels 40%. ***Cholestyramine*** and ***activated charcoal*** rapidly decrease its serum levels. Can increase ***methotrexate*** hepatotoxicity.

Teratogenic and carcinogenic in experimental **animals**.

LEMON

= *Citrus limon* = *Citrus limonum*

Untoward effects*:** Contains pressor amines; may trigger migraines, and caused allergic contact cheilitis. Lemon-scented perfumes cause a temperature-dependent primary irritant dermatitis. Burning, stinging sensations occurred when hands were placed in lemon-scented detergent solution. Accidental use in iced ***tea of detergent samples with a picture of a lemon (***Sunlight***®) on them resulted in 85 calls to a Poison Control Center during a 6 week period. Increases urinary pH. *Candida* contaminated lemon juice as a diluent for IV ***heroin*** caused many cases of candidiasis. Lemons have been a cause of food allergies, leading to skin rashes and abdominal distress. The oil in lemon peels is a primary irritant. The main constituent in lemon oil is ***d-limonene***, and it is used in soaps, detergents, cosmetic creams and lotions, and in perfume. No sensitization or irritation of 10% in petrolatum on **human volunteers**, but showed phototoxic effects in other tests. Lemonade has caused ***zinc*** toxicity from storage in galvanized or ***cadmium***-plated containers.

Full-strength on intact or abraded **rabbit** skin for 24 h under occlusion was moderately irritating and mildly irritating on the backs of hairless **mice**.

LD_{50}, **rats**: 2.8 g/kg; acute dermal, **rabbits**: > 5 g/kg.

LENTILS

= *Adas* = *Ervum lens* = *Lens culinaris* = *Lens esculenta* = *Masoor dhal*

Cultivated since about 8000 B.C.

***Untoward effects*:** Eaten in large quantities, it has a narcotic effect, and some of the symptoms of lathyrism in

people. When fed in excess to **horses**, it seems to make them lethargic.

E. ervilia = ***Bastard Lentil*** = ***Bitter Vetch*** is poisonous to **pigs** and **poultry** in Europe. All parts of the plant are poisonous, especially the "seeds". **Pigs** become weak, incoordinated, they vomit, and become comatose. A mitogen. **Ruminants** and **humans** appear to be resistant to its ill-effects.

LEONOTIS NEPETIFOLIA

= *Hollowstalk* = *Jim Barawo* (Hausa) = *Mastranto* = *Molinollo* = *Mota Shool*

Untoward effects: A **mint** whose aeroallergens may cause an allergic rhinitis, bronchial asthma, and hypersensitivity pneumonitis after an **individual's** accidental contact with its flowers. The leaves have short hairs, causing an irritation or burning rash to the skin.

The leaves are toxic to **chicks**.

LEONURUS CARDIACA

= *Agripaume cardiaque* = *Lion's Ear* = *Lion's Tail* = *Motherwort* = *Throw Wort*

A weed in the *mint* family found in Eurasia and now naturalized in North America.

Untoward effects: As a tonic for cardiac problems and hysteria, the leaves or extract, when ingested, caused photosensitization, dermatitis, and sedation in some **people** from its ***lemon***-scented oil.

LEPIDIUM sp.

= *Pepperweeds*

Untoward effects: Contain ***mustard oils*** (with various ***isothiocyanates***) and are goitrogenic.

L. campestre = ***Pepper Grass*** in the southeastern U.S. pastures contain a glucoside that causes gastroenteritis, excitement, and death in **livestock**.

LEPIOTA sp.

Untoward effects: Poisonous ***mushrooms***.

L. brunneo-incarnata has caused serious poisonings like those of ***Amanita phalloides*** (q.v.). A case report describes the poisoning of 46 **children** in Israel.

L. dolichaulus suspected of poisoning **cattle** in New South Wales.

L. helveola causes poisoning in **man** similar to, but less serious than, that of ***A. phalloides***.

L. morgani causes gastrointestinal toxicity in **children**.

L. rachodes, although considered safe for consumption, this wild ***mushroom*** has caused toxicity. A 4-year-old **male** developed severe abdominal cramps, vomiting, and diarrhea in < 1 h post-ingestion. An **adult female** vomited, with diaphoresis, also in < 1 h. Her **husband** ate some as well, without ill-effects.

LEPIRUDIN

= *CGP 39,393* = *Desirudin* = *r-Hirudin* = *Refludan* = *Revasc*

***Anticoagulant*. A recombinant form of *hirudin* (q.v.), used IV, blocking thrombin's thrombogenic action.**

Untoward effects: Despite a half-life of 1.3 h, serious bleeding, anemia, and hematomas can occur.

LEPISOSTEUS SPATULA

= *Alligator Gar*

May reach 8–12 ft in length and weigh several hundred pounds.

Untoward effects: This large **fish** with multiple, sharp teeth has inflicted serious bite wounds on **animals** and **man**. Its flesh contains ***tetrodotoxins*** (q.v.) and is poisonous.

LEPTOMEDUSAE

Consists of four species of feather hydroids.

Untoward effects: One of the many **swimmer's** dermatoses. The venom causes, within a few minutes or up to 2 h, an urticarial reaction or papular, hemorrhagic, or zosteriform lesions after 4–12 h. With both, reaction bands of dermatitis (20 cm or 8" wide) may develop with erythema multiforme, morbilliform, vesicular, and desquamative eruptions. Systemic reactions can occur, leading to severe abdominal spasms, pain, diarrhea, rigor, chills, fever, and great apprehension. Repeated exposure can lead to sensitization and anaphylaxis.

LEPTOPHOS

= *Abar* = *K62-105* = *Lepton* = *MBCP* = *NK 711* = *Oleophosvel* = *Phosvel* = *VCS-506*

Organophosphorus insecticide.

Untoward effects: Paralysis, loss of motor coordination, blurred vision, dizziness, and choking sensations in **workers** exposed in a plant manufacturing it from 1971 to 1976.

Delayed neurotoxicity in the adult **hen** is similar to the effect in **man** and the **hen** is the experimental subject in demonstrating and testing this effect. **Ducks**, **pigeons**, **dogs**, **sheep**, and **water buffalo** also develop delayed neuropathy. Slowly metabolized, except in **rats**, **mice**, and some **non-human primates**. In Egypt, after it contaminated forages while ***cotton*** was sprayed with it, ~1300 **water buffalo** became paralyzed and died from delayed neurotoxicity. In the milk of **cows** up to 1 week after exposure.

LD_{50}, **rats**: 19 mg/kg; **mice**: 71 mg/kg; **hens**: > 1.5 g/kg; **mallard ducks**: 1.3 g/kg; **bobwhite quail**: 228 mg/kg; **California quail**: 29.2 mg/kg; **pheasants**: 62.8 mg/kg; LC_{50}, **trout**: 20 μg/l/96 h.

LETHANES

***Insecticide.* Both #60 and #384 are *thiocyanates*.**

***Untoward effects*:** Irritating to skin and mucous membranes. Narcotic in high concentration and may be absorbed through the skin. #384 is more toxic than #60.

LD_{LO}, **human**: 50 mg/kg.

LD_{50}, #60, **rats**: 500 mg/kg; **rabbits**: 200 mg/kg; **dogs**: 250 mg/kg; LD_{50}, #384, **rats**: 90 mg/kg; **dogs**: 30 mg/kg; **rabbit**: 35 mg/kg; **guinea pig**: 250 mg/kg; minimum lethal dose, **rat**: 0.5 cc/kg; **cat**: 0.2 cc/kg; **rabbit**: 0.13 cc/kg; **guinea pig**: 0.25 cc/kg.

LETROZOLE
= CGS 20,267 = Femara

***Antineoplastic.* Aromatase inhibitor decreases *estrogen* biosynthesis.**

***Untoward effects*:** Nausea, musculoskeletal pain, arthralgia, hot flashes, headache, vomiting, fatigue, dyspnea, coughing, diarrhea, constipation, anorexia, peripheral edema, weight gain, thromboembolic events, and vaginal bleeding.

Embryotox and fetotox in experimental **animals**.

LETTUCE
= French Lactucarium = Garden Lettuce = Lactuca sativa

Untoward effects*:** Known to cause an immediate urticarial reaction on contact to intact or scratched **human** skin. A ***cadmium, ***lead***, and ***nitrate*** (0.1–0.2%) accumulator. The latter can form carcinogenic ***nitrosamines***. It also contains ***caffeic acid***, a carcinogen in **rodents**. Chlorinated lettuce leaves have been smoked by **drug addicts** for alleged psychedelic effects. A 35-year-old **female** taking ***warfarin sodium*** because of a prosthetic aortic valve, began a diet of ***lettuce***, ***turnip greens***, and ***broccoli*** to lose weight. After 5 weeks, **she** developed a myocardial infarction. Because it contains β-glucosidase enzymes, it should not be eaten while taking ***laetrile***, to avoid ***hydrogen cyanide*** release. Iceberg lettuce contains moderate amounts of ***oxalates***. Contains traces of ***hyoscyamine***. ***Sodium disulfide***, a rare allergen, is commonly used in restaurants, to prevent browning of lettuce. The ***vitamin K*** in it decreases ***warfarin's*** anticoagulant effect.

Some **dogs** are allergic to it. It contains a depressant component for **toads**, interfering with neuromuscular transmission and contraction of smooth muscle.

L. scariola or ***L. virosa*** = ***Compass Plant*** = ***German Lactucarium*** = ***Laitue Scariole*** = ***Lettuce Opium*** = ***Poison Lettuce*** = ***Prickly Lettuce*** = ***Wild Lettuce*** = ***Wild Opium***. Dried juice used as a sedative for **humans**. Contains ***hyoscyamine*** and pharmacologically depressant agents, and may adversely affect **people**. Bitter sesquiterpene lactones are the major toxins in its white sap. In accidental poisoning cases leads to bradycardia, narcosis, muscle weakness, and tachypnea. After IV of an aqueous extract by three **drug addicts**, caused abdominal, back, and flank pain; fever, chills, neck stiffness, headache, leukocytosis, and faulty liver function.

Livestock often refuse to graze this plant. Has low toxicity, but in cases of excess intake muscle weakness, tachypnea, and narcosis occur. Also reported as a cause of pulmonary emphysema in **livestock**.

LEUCAENA sp.

Untoward effects*:** Hawaiian **cattle** and **goats** thrive on it, but Australian **cattle** stop gaining weight, lose hair, and develop ulcers and goiter, due to the toxic ***3–hydroxy-4-(1H)pyridone (***DHP***) broken down from the ***mimosine*** (q.v.) in the leaves. Rumen microbes from Hawaiian **goats** introduced into the rumen of Australian **cattle** permit them to consume it safely at 30% of their diet.

L. glauca = ***Browse Tree*** = ***Koa Haole*** = ***Jumbay*** = ***Jumbie Bean*** = ***L. leucocephala*** = ***Tan Tan***. **Women** eating the seeds, Australian **buffalo** that eat the leaves, and **mice** fed extracts of the seeds lose their hair. **Sheep** fed on it lose their fleece and develop hemorrhagic cystitis. **Horses** lose the long hairs on their manes and tails. Toxicity in **cattle** leads to slight constipation, some alopecia, salivation, progressive inappetence, dyspnea, tachycardia, decreased temperature, lacrimation, slight loosening of hooves, swelling of cheeks, and depression. Causes embryonic deaths when fed to pregnant **mice**. Causes fetal resorption in **sows**.

LEUCINE
= Leu = L-Leucine

Untoward effects*:** Induces hypoglycemia and can potentiate ***sulfonylurea-induced hypoglycemia. Sensitivity to it accounts for 1/3 of all cases of idiopathic hypoglycemia in **infants**; **male infants** are affected more than **females**. Its ingestion in a formula for 3½-year-old **female** was fatal after aggravation of an existing hypoglycemia. Dietary leucine may be a factor in the pathogenesis of pellegra.

Induces hypoglycemia in **dogs**. Possible bladder tumor-promoter in **rats**.

LEUCOTHOE

= *Fetterbush*

Untoward effects: ***L. davisiae* = *Black Laurel* = *Sierra Laurel*** causes salivation, vomiting, bloat, weakness, teeth-grinding, hypotension, respiratory depression, sneezing, and a brief period of excitement, followed by depression after ingestion by **sheep** of 3–4 oz, due to its ***andromedotoxin*** (q.v.) content. **Goats** and **cattle** are similarly poisoned.

LEUPROLIDE

= *A 43,818* = *Abbott 43,818* = *Carcinil* = *Enantone* = *Leuplin* = *Leuprorelin* = *Lucrin* = *Lupron* = *Prostap* = *TAP 144*

***Antineoplastic.* In treatment of metastatic prostate cancer, endometriosis, precocious puberty, and polycystic ovary syndrome.**

Untoward effects: Transient increase in bone pain, ureteral obstruction, weakness, and paresthesia. Hot flashes (3%). Delayed toxicity leading to impotence, testicular atrophy; gynecomastia and peripheral edema (13%). A 43-year-old **female** developed angina and myocardial infarction 1 month after receiving 3.75 mg of the acetate depot form. Another 43-year-old **female** with leiomyomata uteri, treated monthly with 3.75 mg of the acetate depot form developed leukopenia. A 33-year-old **male**, during treatment with it for sterility, developed pseudotumor cerebri. Use also associated with grand mal seizures, edema, flushing, migraine, alopecia, palpitations, memory loss, and suicidal thoughts.

LEVALBUTEROL

= *Xopenex*

***Bronchodilator.* For nebulizer use.**

Untoward effects: Tremors, nervousness, headache, dizziness, anxiety, chest pain, palpitations, and leg cramps. Occasionally dose-related increase in serum ***glucose*** and heart rate and decrease in serum ***potassium***.

Drug interactions: Can decrease serum ***digoxin*** levels.

LEVALLORPHAN TARTRATE

= *Lorfan*

Narcotic antagonist.

Use: For treatment IV, IM, and SC of ***narcotic***-induced respiratory depression, euphoria, pseudoptosis, and miotic effects, while maintaining most of the ***narcotic's*** analgesic effect.

Untoward effects: Dizziness, euphoria, respiratory depression, drowsiness, miosis, and occasional nausea. Bradycardia, hypotension, sweating, pallor, irritability, hot and cold flushes, and possibly temporary psychotic effects occasionally reported, especially with high dosage.

LEVAMISOLE

= *Ascaridil* = *Citarin L* = *Decaris* = *Ergamisole* = *Ketrax* = *Levacide* = *Levadin* = *Levasole* = *L-tetramisole* = *Meglum* = *Nemicide* = *Nilverm* = *Ripercol* = *Solaskil* = *Spartakon* = *Tramisole*

***Anthelmintic, immunoregulator.* Has structure similar to *thymopoietin*.**

Untoward effects: Nausea, vomiting, diarrhea, metallic taste, changes in smell, dizziness, headache, influenza-type symptoms, anaphylaxis, chills, and febrile reaction. Occasionally neutropenia, agranulocytosis, leukopenia, thrombocytopenia, encephalopathy, peripheral neuropathy, hyperlipidemia, fatigue, pancreatitis, abdominal pain, memory loss, confusion, generalized urticaria; nephropathy with granular deposits of immunoglobulin and complement in the glomeruli and dermal epidermal junction; induced arthritis and arthralgia, transient lichen planus, rashes, and fixed-drug eruptions; delayed healing of varicose ulcer, and suppresses skin reactivity in tuberculin testing. High incidence of adverse effects in **patients** with Sjögrens syndrome. Purpura of the ears in 6–11-year-old **male**s and 16-year-old **female** with nephrotic syndrome after 1.7–2.5 mg/kg/16–44 months. Reactions also include leukocytoclastic thrombotic vasculitis and thrombotic vasculitis.

TD_{LO}, **woman**: 180 mg/kg, 36 days.

Drug interactions: Major bleeding episodes and increased international normalized ratio with ***warfarin***. Concomitant use of ***fluorouracil*** reported in two cases of drug interactions; neural toxicity (multifocal cerebral demyelination) in 57-year-old **male** after adding it to ***fluorouracil*** treatment. Use with ***alcohol*** increases incidence of dizziness and headaches. Possible interaction with ***phenytoin***.

Boosting of cellular immunity appears only when cell-mediated immunity is depressed. This was the discovery of Dr Gerald Renoux, a French veterinarian, in evaluating an unusual response in *Brucella*-vaccinated **cattle**. Do NOT use orally within 3 days (7 days for parenteral and 9 days for pour-on or spray-on use) of slaughtering for **human** food use. The FDA allowed up to a 0.1 ppm residue in edible tissues of treated **cattle**, **sheep**, and **swine**. Parenteral dosages are presently experimental only. In feed treatment, may cause muzzle foam on **cattle** or **swine**. Do NOT use in adult **dairy cattle**. Post-treatment vomiting and coughing in **swine** is usually due to expulsion of lungworms. Levamisole toxicity in **pigs** is enhanced by ***nicotine***-like compounds (viz. ***pyrantel***). Overdoses have killed experimental **animals** within 15 min. Some state that one should not treat **dairy cattle** of breeding age. Occasionally salivation, vomiting, and post-treatment lassitude, tremors, anorexia, thrombocytopenia, and ataxia have been reported (**cats** and **dogs**). Aqueous solutions often produce serious irritation and tissue inflammation. This has been decreased by citrate buffering. Do NOT inject **cattle** or **pigs** within 7 days of slaughter; **goats** and **sows** within 12–14 days. Lactating **cattle** are apt to be free of the drug within 48 h. Some pour-on preparations on **cattle** have been temporarily irritating.

LEVAMPHETAMINE

= *Ad-Nil* = *Amphedrine-M* = *Levabo* = *Levoamphetamine* = *Levonor*

Untoward effects: Worsening of pre-existing and new stereotypes in nine, decreased appetite in seven, weight decrease in six, pallor in six, circles under eyes in six, increased irritability in four, and excessive sedation in four among 11 **patients** treated with it.

LEVETIRACETAM

= *Keppra*

Anticonvulsant, antiepileptic.

Untoward effects: Agitation, anxiety, psychosis, depression, somnolence, and asthenia. Rarely, small decrease in white blood cells, red blood cells, hemoglobin, and hematocrit.

LEVOBUNOLOL

= *Betagon* = *l-Bunolol* = *Gotensin* = *Vistagen* = *W 6421A* = *W 7000A*

***β-Adrenergic blocker (both β-1 and β-2).* For topical ophthalmic use.**

Untoward effects: Although initial clinical trials reported no adverse effects, pulmonary, cardiovascular, and central nervous system adverse effects reported since are, apparently, due to its absorption through the nasolacrimal areas into the systemic circulation. Ocular burning and stinging in ~40% of **patients**. Alopecia attributed to it after 1–24 months, with recovery 4–8 months after treatment withdrawal.

LEVOCARNITINE

= *Cardiogen* = *Carnicor* = *Carnitine* = *Carnitor* = *Carnum* = *Carrier* = *Miocor* = *Miotonal* = *Vitacarn*

***Antihyperlipoproteinemic.* Oral and IV.**

Untoward effects: Transient nausea, vomiting, abdominal cramps, diarrhea, and body odor.

LD_{50}, **mouse**: 19.2 g/kg.

LEVOFLOXACIN

= *Cravit* = *DR 3355* = *Levaquin* = *S-(-)-Ofloxacin* = *Tavanic*

A broad-spectrum *fluorinated quinolone*. A stereoisomer of *ofloxacin*. Oral and IV.

Untoward effects: Nausea, diarrhea (1.2%), vaginitis (0.8%), rash, abdominal pain, genital moniliasis, dizziness; dyspepsia and insomnia (0.3%) and < 0.3% with taste perversion, vomiting, anorexia, anxiety, constipation, edema, fatigue, headache, diaphoresis, leukorrhea, malaise, nervousness, insomnia, tremors, and urticaria. Injection site irritation occurs as well as many of the above ill-effects after IV. Phototoxicity, tendinitis, and tendon ruptures can be expected. Eosinophilia in 84-year-old **male** after 250 mg IV/day/6 days.

Drug interactions: Potentiates ***warfarin's*** anticoagulant effect.

Acute oral toxicity in **rats** and **mice** leading to decreased locomotor activity, ptosis, tremors, tonic convulsions, and respiratory decrease. **Rats** show increased number of fecal pellets, salivation, decreased neutrophils, and enlarged ceca. No adverse effects on fertility or teratogenicity in **rats**, until doses as high as 810 mg/kg were given. **Rat** fetuses then had decreased weight, retarded ossification, and increased mortality. Some ear swelling in **mice** at 800 mg/kg. Decreased weight gain in **rabbits** at 120 mg/kg; but not 30 mg/kg/10 days. Soft feces and vomiting in **monkeys**.

LEVONANTRADOL

Antiemetic.

Untoward effects: Dysphoria (16%), somnolence (48%), dry mouth (32%), decreased vigilance and reaction time, dizziness, weird dreams, mild hallucinations, apprehension, and confusion.

LEVONORGESTREL

= *Microlat* = *Microval* = *Norgeston* = *Norplant*

***Progestin.* Often used with *estrogens* or *androgens*.**

Untoward effects: Use as a SC implantation in silicone ***rubber*** capsules for effective (< 1% failure rate) contraception, has caused menstrual irregularities, breast pain, breast discharges, headache, acne, fluid retention, weight increase, occasional skin damage, increased hair growth, loss of scalp hair, and keloid formation at implantation site. Found in **human** breast milk. May interfere with the wearing of contact lenses. ***Phenobarbital*** use for seizure control may have prevented its oral contraceptive's effectiveness.

LEVORPHANOL TARTRATE

= *Dromoran* = *Levo-Dromoran* = *Levorphan* = *RO 1-5431/7*

***Narcotic analgesic, synthetic opiate.* A *morphine* analog.**

Untoward effects*:** Addictive. Similar to that of ***morphine (q.v.), yet some case reports claim less nausea, vomiting, and constipation. Use during labor has been associated with clinical depression of the **newborn**. Skin rashes and urticaria are manifestations of allergic reactions. Occasionally reports of hypotension, respiratory depression, cardiac arrhythmias, and urinary retention. Excretion is increased in acid urine and decreased in alkaline urine.

LD, **human**: ~60 mg.

Can cause lens opacity in **gerbils** and **mice**. Hepatotoxic in **mice**.

IV, LD_{50}, **dog**: 46 mg/kg; **mice**: 33 mg/kg.

LEWISITE

= *M-1*

***Vesicant/blistering war gas.* Has a *rose-geranium* odor.**

Untoward effects*:** A violent blistering agent that can destroy all tissues. ***Rubber is not protective. 0.3–0.5 ml left on the skin can cause severe respiratory and systemic effects and 1.4 ml may cause death in 3 h–5 days. Severely damaging to the eyes and respiratory tract during inhalation, and to the gastrointestinal tract after ingestion. Absorption leads to acute hepatitis and a less severe nephritis. Necrosis of the bile ducts and gallbladder occurs as the body excretes it by this route. Shock and death have developed a few days after exposure.

LC_{LO}, inhalation, **human**: 6 ppm/30 min; LD_{LO}, skin, **human**: 20 µg/kg.

LD_{50}, skin, **rat**: 24 mg/kg; **dog**: 15 mg/kg; **rabbit**: 6 mg/kg; **guinea pig**: 12 mg/kg.

LH-RH

= *Luteininzing Hormone-Releasing Hormone* = *LRF* = *LRH*

Buserelin = *Hoe 766* = *Receptal* = *Suprefact*

Untoward effects*:** Increases plasma ***cholesterol after 6 months of treatment, slight decrease in bone density after intranasal inhalation of 400 µg/tid/6 months in endometriosis **patients**. Hypertension may have been associated in a **patient**.

Gonadal atrophy in **dogs** after 4 weeks with 50–200 µg/day.

Gonadorelin = *A 41,070* = *AY24,031* = *Cryptocur* = *Cystorelin* = *Fertiral* = *GnRH* = *Hoe 471* = *Luliberin* = *Lutal* = *Relisorm L* = *Ru 19,847* = *Stimu-LH*

***Untoward effects*:** Anaphylaxis; transient amaurosis and headache reported. Hemorrhagic infarct of a pituitary macroadenoma after IV of 100 µg in 41-year-old **female** with 5 year history of amenorrhea.

A possible case of agalactia in a treated **cow**.

LD_{50}, **rats** and **mice**: > 60 mg/kg; **dogs**: > 0.6 mg/kg.

LICHENS

A lichen is unique in that it is composed of a microscopic blue–green algae and a colorless fungus.

Untoward effects*:** A photosensitizer and contact dermatitis-inducer, due to its ***atranorin content. Bullous, hypertrophic, follicular, annular, and ulcerative lesions.

LIDOCAINE

= *Cuivasil* = *Duncaine* = *Leostesin* = *Lidothesin* = *Lignocaine* = *Rucaina* = *Xylocaine* = *Xylocitin* = *Xylotox*

***Local anesthetic, antiarrhythmic.* Has anticholinesterase activity.**

***Untoward effects*:** The more serious are due to its central nervous system effects, leading to drowsiness, slurred speech, paresthesias, muscle twitching, agitation, convulsions, disorientation, depression, headache, hypotension, tinnitus, sweating, laryngospasms, blurred vision, numbness of tongue, yawning, restlessness, nausea, vomiting, coma, and respiratory depression with prolonged apnea. Allergic reactions reported. IV dosing at > 5 mg/kg can induce seizures. Exfoliative dermatitis in a **patient** after its use as an injectable dental anesthetic. A 42-year-old **female** developed severe thrombocytopenia 3 days after 3.6 ml of 2% solution for dental extractions. A 20-year-old **male** had an immediate fatal anaphylactic reaction from 0.8 ml of 2% solution for a dental procedure. Use of a lidocaine oral gel has caused seizures in **infants** and young **children** after they ingest it. Increase in **neonatal** jaundice, bradycardia, anemia, and depression associated with its epidural use during labor. It was implicated as a cause of a stillbirth 15 min after a 19-year-old **female** primipara was given 10 ml as a paracervical block. Passes rapidly across the placenta after IV use in the **mother**, and found in umbilical venous blood within 2–3 min.

Has caused **fetal** bradycardia and death. Large doses can depress myocardial contractility and atrioventricular conduction, leading to bradycardia, arrhythmias, and cardiac arrest, especially in cardiac disease or liver failure. Infusions for > 24 h have been associated with bradycardia, asystole, and death, due to decreased drug clearance. Normally, 70% is metabolized on the first pass through the liver. Half-life is ~100 min increased to over 24 h after administration for several days. Peripheral vasodilation and tachycardia also occur. Vasoconstriction reported after IV use in angiography. A 28-year-old **male** had two episodes of toxic methemoglobinemia after topical use of ***benzocaine*** and ***lidocaine***. A 16 day **infant** developed cyanosis and methemoglobinemia after application of a 2% gel and 20% ***benzocaine*** spray for a surgical procedure. Topical use on the round window of the ear has caused hearing loss of 15 decibels. Accidental contact with the middle ear causes vertigo and vomiting. Ophthalmic use can cause true sensitivities (keratoconjunctivitis medicamentora). Sloughing of anterior soft palate in 69-year-old **male**, 3 weeks after application of 2% gel. Complete iridoplegia in 70-year-old **male** after retrobulbar injection of 3 ml 2% with three drops 1 : 1000 ***epinephrine***. Burning sensations in the nose and numbness in or around the eyes, along with an unpleasant taste, after intranasal use. Urinary tract infections (and one death) in four, 32–68-year-old **males**, due to contamination of lidocaine jelly with *Pseudomonas aeruginosa*, after its application to endotracheal tubes used during anesthesia. In general, safe plasma levels are 0.15–0.5 mg%; toxic levels are 0.9–1.4 mg%; and lethal levels are > 2.4 mg%. Tachyphylaxis can occur with repeated epidural injections. Generally, toxicity is more common in older **patients** than younger ones due to differences in half-lives. Sensory, motor, and autonomic paralysis; and hypotension from accidental intrathecal injection while performing an epidural injection. Allergic reaction with skin wheals, itching, erythematous rash, and anaphylactic shock. Deaths have been reported after its topical use as a spray in bronchoscopy. Some liposuction deaths have been attributed to overdosing with it – peak blood levels are delayed for 8–10 h after surgery. Toxicity in 30-year-old **female** after use of 40% topical cream under plastic wraps following laser treatment for stretch marks leading to dizziness, headache, light-headedness, and confusion. Therapeutic serum levels are 1.5–5 μg/ml; toxic at 6–7 μg/ml; and can be fatal at 10 μg/ml.

Drug interactions: Clearance decreased by ***propranolol***, which decreases cardiac output and hepatic blood flow. ***Cimetidine*** decreases its clearance, due to decrease in hepatic blood flow and decreased microsomal enzyme activity. ***Glucagon*** and ***isoproterenol*** increase hepatic blood flow and increase its clearance. ***Epinephrine*** delays absorption rate and increases its duration of action. Use with ***phenytoin*** has caused sinoatrial arrest. Prolonged muscle relaxation or apnea with ***succinylcholine***, ***d-tubocurarine***, and ***decamethonium***. Sinoatrial arrest and atrioventricular junctional escape rhythm from use in a **patient** receiving ***quinidine***. **Patients** receiving sublingual ***nitrates*** and ***prenylamine*** develop atrioventricular block after bolus doses of lidocaine. Use with ***epinephrine*** can cause a dose-dependent increase in the risk of infection, due to ***epinephrine's*** vasoconstrictive effect. Hypothermia with ***diazepam***. Decreases ***chlorpromazine*** absorption.

Cardiac failure reported in **dogs**. ***Tetracaine*** not only synergizes lidocaine's anesthetic effect, but also dramatically increases its toxicity in **mice**. 0.5 ml or more of a 1% solution is usually fatal to a **parakeet**. Central nervous system disturbances, such as convulsions or excitement, can occur with inadvertent IV administration. Excesses can depress the heart. "Sensitivity" reactions are apt to be due to the paraben preservatives, rather than the drug itself. Do NOT use formulas with ***epinephrine*** for IV use. Aqueous solutions have a pH of about 4.5, and corrode all metals except stainless steel. Osmotically inactive and should always be in isotonic saline. Excessive heating causes a yellow or pink discoloration, but this apparently does not increase its toxicity. Crosses placental barrier in **sheep** (bradycardia and increased cerebral blood flow) and in **humans**. Progressive hypotension from continued absorption of excessive dosage from depots (viz. epidural) are best countered with a peripheral vasoconstrictor with positive inotropic action, such as ***ephedrine***. **Cats** are particularly sensitive to its adverse central nervous system effects, including seizures.

IV, LD_{50}, **mouse**: 25 mg/kg; LD_{LO}, **rat**: 25 mg/kg; **rabbit**: 41 mg/kg; **guinea pig**: 65 mg/kg. SC, LD_{50}, **rat**: 335 mg/kg; **mouse**: 265 mg/kg.

LIDOFLAZINE

= *Angex* = *Clinium* = *Corflazine* = *Klinium* = *McN-JR-7094* = *Ordiflazine* = *R-7094*

Calcium channel-blocker, coronary vasodilator.

Untoward effects: Gastrointestinal discomfort, dizziness, and tinnitus are common, transient, and eliminated by dosage decrease. Occasional hot flushes, loss of libido, and persistent fatigue. Minor changes in liver ultrastructure in **humans** and **dogs**, without impairment in liver function after long-term use. May induce ventricular ectopic beats, ventricular tachycardia, and fibrillation. Sudden death attributed to it has also been reported.

LIGUSTRUM VULGARE

= *Privet*

Untoward effects: The entire plant, especially the leaves and berries, is toxic. The attractive berries of this common hedge, unfortunately, are attractive to and eaten by **children**, leading to nausea, vomiting, abdominal pain, diarrhea, and death. Berries stay on the shrub during the winter when leaves fall off, and often attract the attention of **children**. Nervous symptoms develop and have been attributed to a ***nicotine***-like alkaloid. The red dye in the berries was used in Germany to color ***wine***.

Dogs and **cats** are often poisoned by eating the berries. **Horses** are poisoned by eating the leaves. Fluffed feathers, dark orange runny feces, depression, gasping, wing-drop, and reluctance to move in **canaries** that ingested 120 mg.

LILIES

= *Lilium sp.*

Untoward effects: Pollen of the ***Tiger Lily*** (***L. tigrinum***) has caused vomiting, purging, and drowsiness in a little **girl**.

L. longiflorium (***Easter Lily***). The leaves and flowers have been chewed by **cats**, leading to depression, vomiting, renal failure, and death within 2–5 days, due to ***oxalate*** content. Many Poison Control Centers, unfortunately, have told **inquirers** that it is non-poisonous. The juice of large lilies can be very irritant to **man** or **beast**. **Dogs** have vomited after eating the bulbs, apparently due to their cardiac glycoside content. ***Spathiphyllum sp.*** (***Peace Lilies***) and ***Zantedeochia sp.*** (***Calla Lilies***) contain ***oxalates*** and are toxic to **cats**.

LIME

= *Citrus aurantifolia* = *Lemu* (Hausa)

Untoward effects: ***C. aurantifolia*** (***Key Lime***) and ***C. latifolia*** (***Persian Lime***; the common ***Florida Lime***) has caused Berlock dermatitis from its use in perfumes and colognes. ***Lime oil*** in after-shave lotions causes hypersensitivity reactions to sunlight, due to its ***d-limonene*** (q.v.) and ***furocoumarin*** content. Maximum photoreactivity is induced in 2–4 h after ingestion, and drops nearly to zero after 8 h. A skin rash on the thighs of a young **woman** who had balanced and rolled limes on **her** lap during a game on a Caribbean Club Med vacation. It can trigger migraines in some **people**. It is worthy of mention that, contrary to popular opinion, limes are less antiscorbutic than ***lemons***. In 1795, the British Navy was ordered by the Admiralty to issue ***lemons*** at the insistence of a physician, Sir Gilbert Blane. A severe outbreak of scurvy occurred in 1875 on two ships of an Arctic expedition because the commander took along limes instead of ***lemons***. If people had not been confused by their apparent similarity, English **sailors** would have been called "Lemonies", instead of "Limies".

Some evidence of phototoxicity in **mice** and **pigs**.

LIMESTONE

A natural source of *calcium carbonate*.

Untoward effects: **Quarry workers**, especially **those** mining ***wollastonite***, can develop chronic bronchitis and pulmonary problems. Limestone dust has caused eye, skin, and mucous membrane irritation. Coughing, sneezing, rhinitis, and lacrimation noted. ***Wollastonite*** contains fibers resembling those of ***asbestos***. ***Dolomite***, a popular health food store mineral supplement, can cause hypercalcemia and hypermagnesemia, as well as being a source of ***aluminum***, ***antimony***, ***arsenic***, ***barium***, ***cadmium***, ***lead***, ***mercury***, and ***nickel***.

d-LIMONENE

A bacteriostatic terpene. The principle component of *citrus peel oils*.

Untoward effects: It and its derivatives are causes of occupational dermatitis on ***caraway***, ***celery***, ***citrus***, and ***dill*** **handlers**.

Carcinogenic in **rodents**. As a vasodilator, it may be the cause of reported hypothermia, ataxia, and death of **kittens** on whom it was used to kill fleas.

TD_{LO}, **mouse**: 67 g/kg, 40 weeks.

LIMU-MAKE-O-HANA

Untoward effects: This deadly seaweed is said to be found in only one locality on the island of Maui. This "poison moss" was smeared on the tips of spears in the nineteenth century, to make them lethal. Its active principle, ***palytoxin***, is one of the most poisonous used on weapons. It has recently been rediscovered in a few pools off the coasts of Oahu and Maui and is considered to be a zoanthid coelenterate, *Polythoaca toxica*.

LINALOOL

A terpene alcohol commonly found in hundreds of plants.

Untoward effects: Cases of dermatitis in perfume-bottling plant **workers** from it. Mildly irritating in **human** patch-testing.

Weakly carcinogenic in **mice**; acanthogenic in **guinea pigs**; moderately irritating on **rabbit** skin. Toxic **rats** become ataxic and usually die in 4–18 h.

LD_{50}, **rat**: 2.8 g/kg.

LINALYLS

Use: The *anthranilate*, *benzoate*, *cinnamate*, *formate*, *isobutyrate*, *phenylacetate*, and *propionate* forms are used in soap, detergent, cosmetic creams and lotions, perfumes, and foods.

Untoward effects: At 4–10% concentration in petrolatum, these produced no irritation or sensitization in **human volunteers**.

Acute oral LD_{50} in **rats** was over 4 g, and acute dermal LD_{50} in **rabbits** was > 5 g/kg. Topically, the *acetate* form is severely irritating to **rabbits** and moderately irritating to **guinea pigs**.

LINAMARIN

= *Phaseolunatin*

Untoward effects: A cause of ***cyanide*** poisoning in **people** in Africa consuming ***cassava*** (q.v.), a cyanogenic glycoside (***glucose*** and ***hydrocyanic acid*** in combination). After continued ingestion of significant amounts, a tropical neuropathy gradually developed, leading to optic atrophy, nerve deafness, ataxia (due to spinal nerve damage), stomatitis, glossitis, and scrotal dermatitis.

Has caused toxicity in **sheep** and **cattle** from ***linseed*** (q.v.) in western U.S. and Europe.

LD_{LO}, **rat**: 500 mg/kg.

LINARIA VULGARIS

= *Butter and Eggs* = *Pennywort* = *Yellow Toadflax*

Common in North America.

Untoward effects: Contains the cyanogenic glucosides, ***prunasin***, and ***antirrinoside***. Hay containing large amounts of the plant can cause ***cyanide*** (q.v.) poisoning.

LINCOMYCIN

= *Albiotic* = *Cillimycin* = *Frademicina* = *Lincocin* = *Lincolcina* = *Lincolnesin* = *Lincomix* = *Mycivin* = *NSC-70731* = *U-10149* = *Waynecomycin*

Untoward effects: Diarrhea, abdominal pain, and enterocolitis are frequent after parenteral use. Large doses IV with undiluted lincomycin have precipitated hypotension, bradycardia, and cardiac arrest. Occasional pain and erythema at IM injection site; phlebitis at IV site. Sideroblastic anemia, malaise, diarrhea, and pseudomembranous colitis in 58-year-old **female** after 1.2 g/day/8 days. Some cases of anemia with leukocytosis and a case of aplastic anemia reported. Other gastrointestinal problems are numerous, leading to esophagitis, glossitis, stomatitis, abdominal discomfort, tenesmus, colitis, nausea, vomiting, and pruritus ani. Diarrhea reported in ~20% of **patients** and pseudomembranous colitis in ~10%. Neutropenia, pancytopenia, leukopenia, granulocytopenia, aplastic anemia, agranulocytosis, thrombocytopenic purpura, eosinophilia, and skin reactions may be due to hypersensitivity. Heartburn, headache, drowsiness, generalized maculopapular rashes, urticaria, jaundice, coated tongue, ketonuria, vaginal candidiasis, tinnitus, and vertigo have occasionally been reported. May cause increased or falsely positive SGPT, SGOT, and akaline phosphate results. Can interfere with serum folate tests. Hypogeusia and dysgeusia reported. Half-life is doubled in liver dysfunction. **Fetal** serum concentration is ~25% of **maternal**. Present in **infants'** blood for 52 h after delivery. Only about 0.025% of the **mother's** plasma level found in breast milk.

Drug interactions: Has a ***curare***-like action and potentiates the action of ***muscle relaxants***, such as ***diazepam*** and ***pancuronium***. ***Kaolin***, as in ***Kaopectate***®, causes ~90% decrease in bioavailability of oral lincomycin. ***Food*** also causes a major decrease in its absorption. Beverages sweetened with ***sodium*** or ***calcium cyclamate*** also decrease its absorption. Cross-resistance can develop when used with ***erythromycin*** and it may be antagonistic.

Loose stools may occur in **swine** (6 days withdrawal time). Vomiting has been reported in treated **cats**. A single SC dose of 10 mg/kg is fatal to most **hamsters**. Mammary excretion occurs in **cattle**. Treated **animals** should not be used for food purposes for at least several days after receiving the drug. The FDA has established 0.15 and 0.1 ppm as minimum tolerances, respectively, for its negligible residues in milk and edible tissues of treated **swine** and **poultry**. At least 2 and 6 days withdrawal time in **swine**, IM and orally, respectively. Toxic to **rabbits**. 7–10 days withdrawal after water medication of **swine**. **Rabbits**, **hamsters**, **guinea pigs**, **horses**, **cattle**, and other **ruminants** have shown severe gastrointestinal upsets, including diarrhea, after access to treated feeds (metabolic upsets, diarrhea, and ketosis in **cattle**), and may be due to overgrowth with anaerobic *Clostridium difficile*. Fatal anaphylactic reactions have been reported in **monkeys**.

SC, LD_{50}, **rat**: 9.8 g/kg.

LINDANE

= *Aparasin* = *Aphtiria* = *Benason* = *Ben-Hex* = *γ-Benzene Hexachloride* = *γ-BHC* = *Emulpan* = *ENT-7796* = *Entomoxan* = *Esoderm* = *Exagamer* = *Fenoform* = *Forlin* = *Gamene* = *gamma Hexachlor* = *Gammalin* = *Gammatin* = *Gammexane* = *Gammopaz* = *Gexane* = *γHCH* = *Hexachloran* = *Isotox* = *Jacutin* = *Kwell* = *Lindflor* = *Lindagan* = *Lindatox* = *Lintox* = *Lorexane* = *Novigam* = *Quellada* = *Scabene* = *Silvanol* = *Streunex* = *Tri-6* = *Viton*

***Chlorinated hydrocarbon insecticide.* The chief active component of *benzene hexachloride* (q.v.).**

Untoward effects: Poisoning has occurred from ingestion, inhalation, and percutaneous absorption. Acute toxicity leads to headache, confusion, dizziness, seizures, mydriasis, salivation, nausea, vomiting, tremors, irritability, muscle necrosis, contact urticaria, weakness, diarrhea, convulsions, acute renal insufficiency, Henoch–Schönlein purpura, dyspnea, cyanosis, and circulatory failure. Inhalation toxicity leads to eye, nose, and throat irritation. Skin irritation is rare, except from repeated applications, and manifested as eczematous eruptions. Chronic toxicity leads to aplastic anemia, jaundice, thrombocytopenia, leukopenia, erythroid hypoplasia, transient visual field deficits, dysarthria, and syncope. ***Acetone*** in shampoos increases its systemic absorption. Erythropoietic hypoplasia in 2-year-old **male** traced to the family **dog**, dipped weekly in lindane solution. The **child**, apparently, absorbed the chemical by inhalation of vapors, skin contact, and possibly, oral ingestion from hand contamination. A 2½-year-old **female** developed irritability and grand mal seizures within 1–2 h after ingesting 1.56 g in two pellets for use in a vaporizer. **Infants** and **children** appear to absorb more than **adults** (skin surface : body weight; thinness of skin, etc.). Found in breast milk. Toxic serum levels are ~50 μg%. OSHA exposure limit of 0.5 mg/m^3, TWA. Immediate danger to life or health, 50 mg/m^3. A 43-year-old **female** intentionally ingested 8 oz of a 20% solution and developed disseminated intravascular coagulation. **She** died 11 days after the ingestion. It induces microsomal enzymes in **rats**; therefore, it is not surprising that, after using it on his **sheep**, a **rancher** failed to get the benefit from **his** administered ***warfarin*** and developed femoral thrombophlebitis.

TD_{LO}, **child**: 111 mg/kg; LD_{LO}, **human**: 840 mg/kg; **child**: 180 mg/kg.

Drug interactions: ***Antipyrine*** and ***phenylbutazone*** half-lives in plasma are decreased ~20% by exposure to it. Decreases ***warfarin*** effectiveness.

Cats are very sensitive to its ill-effects developing apprehension, belligerence, hyperesthesia, hypersalivation, convulsions, and death (from minutes to weeks post-exposure). Do NOT use on very young **animals**, sick **animals**, lactating **animals**, **cats**, **dogs**, and **exotic pets**, except in emergencies, and only after using suitable precautions to prevent self-licking, inhalation of vapors and dusts, or skin penetration. Avoid contamination of **animal** and **human** water supply, foodstuffs, and equipment used in their handling. Even where its **livestock** use is permitted, avoid slaughtering **animals** for food within 30 days of spraying or within 60 days of dipping. **Fish** are easily killed by it. Avoid contamination of rivers and streams. Check legal status before using. ***Phenobarbital*** decreases its early toxic manifestations in **dogs**, but does not ameliorate its lethal effects. Prevent **birds** from self-licking. Avoid use on **pigs** under 8 weeks of age. Not recommended in many countries for **dairy cattle**. May cause local irritation. Avoid use on **sows** within 30 days of farrowing. Benign hepatomas and fatty liver degeneration in **cattle** and **dogs**.

LD_{50}, **rat**: 76 mg/kg; **mouse**: 86 mg/kg; **guinea pig**: 127 mg/kg; **hamster**: 360 mg/kg; wild **birds**: 100 mg/kg; **rabbit**: 60 mg/kg; **dog**: 40 mg/kg. 96 h LC_{50}, **fish**: 20–36 μg/l.

LINEZOLID

= *PNU 100,766* = *U 100,766* = *Zyvox*

***Antibiotic*. An *oxazolidinone*. Contains *phenylalanine*.**

Untoward effects: Nausea, vomiting, and diarrhea. Prolonged treatment may lead to reversible thrombocytopenia. Leukopenia and increased serum liver enzymes have occurred.

Drug interactions: Presence of food delays usual 1–2 h to peak plasma concentration. Weak inhibitor of ***monoamine oxidase***. ***Tyramine***-rich foods with it lead to severe hypertension.

LINOLEIC ACID

= *Linodoxine* = *Linolic Acid*

Untoward effects: Feeding it at high levels, or with oxidized ***fats***, leads to encephalomalacia in **chickens**.

LINSEED

= *Flaxseed* = *Linum sp.*

Untoward effects: Dermatitis in **workers** manufacturing the oil from the sharp points of the linseed itself, bites by parasites in the linseed, cuts from the filter cloths made from **human** hair, or hypersensitivity to the oil itself. Chronic bronchitis and emphysema can occur from inhalation of its dust. The oil is comedogenic.

Contains ***cyanogenic glycosides***, such as ***linamarin*** (q.v.), which can be dangerous to starving **livestock** eating large quantities rapidly. Immature seed pods are the most dangerous. The plants are also ***nitrate***-accumulators. ***Boiled linseed oil***, a drying agent in paints, contains ***lead monoxide*** (***letharge***).

LINURON

= *Afalon* = *Du Pont Herbicide 326* = *HOE-2810* = *Linurex* = *Lorox* = *Methoxydiuron*

***Herbicide*. A derivative of *urea*.**

Untoward effects: Teratogenic trials in **rats** resulted in no defects with one manufacturer's material and delayed ossification with another. Benign tumors in exposed **rats** and **mice**.

LD_{50}, 1500 mg/kg.

LIOTHYRONINE

= *Cynomel* = *Cyomel* = *Cytobin* = *Cytomel* = *Cytomine* = *Tertroxin* = *Thybon* = *3,5,3′-Triiodothyronine* = *Trionine* = *Triostat* = *Triothyrone*

Thyroid hormone.

Untoward effects: Overdosage can cause headache, nervousness, sweating, tremors, tachycardia, arrhythmias, angina, increased peristalsis, menstrual irregularities, hypotension, palpitations, myocardial infarction, congestive heart failure, hostility, hallucinations, heat intolerance, manic reaction, doubly increased urinary loss of ***calcium***, diarrhea, alopecia, rashes, and glycosuria. Significant placental transfer from **mother** to **fetus** in at least 50% of cases with high dosage. False negative or spuriously low ***protein-bound iodine*** test results.

Drug interactions: Enhances therapeutic effects and adverse effects of ***tricyclic antidepressants*** and ***xanthine bronchodilators*** (***theophylline***). Decreases effectiveness of ***propranolol***. ***Cholestyramine*** decreases its oral absorption. In thyroid imaging, it decreases uptake of ***radiopharmaceuticals***. ***Salicylates***, in high dosage, give increased levels on triiodothyronine uptake tests. **Patients** on oral ***anticoagulants*** may need decreased levels of ***anticoagulants*** to stabilize their international normalized ratio. May cause an increase in ***insulin*** or ***oral hypoglycemic*** dosage. It can potentiate ***digitalis*** toxicity or even cause the need for increased ***digitalis*** dosage, due to increased metabolism of the latter. Use with ***ketamine*** can precipitate hypertension and tachycardia.

Avoid excessive dosage in **animals** with weak hearts. Readjust dosage to the symptomatology. Vomiting and excitement may occur with overdosage in **dogs**. *Contraindicated* in adrenal insufficiency, unless ***corticosteroid*** therapy is initiated first. Thyroid preparations enhance ***catecholamines'*** (***epinephrine***, etc.) adverse cardiovascular effects. Its use may potentiate the effects of ***coumarin anticoagulants***, and their dosage may have to be decreased.

LIPASE

= *Cotazym* = *Lipolase* = *Steapsin*

Untoward effects: Several case reports on 2–11 year pediatric **patients** with cystic fibrosis **who** developed abdominal pain, loose stools, large-bowel stricture, or pathological colon changes after high dosage. Contains amylase and protease in significant quantities. Direct inhibition by ***aluminum*** ions, ***neomycin***, and ***tetracyclines***. Activity is enhanced by ***vitamin C***.

LIPPIA REHMANNI

Untoward effects: Its root bark is hepatotoxic to **sheep** and **rats**, due to its ***icterogenin*** and ***rehmannic***. A photosensitizer.

LIPSTICK

Untoward effects: Allergic contact dermatitis and possible systemic lupus erythematosus from inadvertent ingestion of ***eosin*** (q.v.) and other derivatives of ***fluorescein***; blepharitis from rubbing eyes with fingers contaminated with ***silicones*** in water-repellent lipsticks. Various perfumes, such as ***methyl heptine carbonate***, with violet-like odor, has caused dermatitis. ***Dimethylsafranine***, ***tolu-safranine***, and ***carmine*** dyes in lipstick have caused a cheilitis, probably an allergic hypersensitivity. ***Lanolin*** in glossy lipsticks, ***alcohol***, ***antioxidants*** (viz. ***propyl gallate***), ***cinnamon*** flavors, ***perfumes***, and ***rose bengal*** have also caused a cheilitis. Pigmentation from photosensitizing agents in lipstick. Allergic pigmented lip dermatitis from ***D&C Red 7*** (***Lithol Rubine BCA***). FDA unapproved colors have been found in lipsticks from Taiwan. Perioral dermatitis from the ***propylene glycol*** in ***Lip Smackers***® for **children's** lips. ***Nickel***-sensitive **individuals** have developed a contact dermatitis from the holders held between **their** lips.

LISINOPRIL

= *Acerbon* = *Alapril* = *Carace* = *Coric* = *Enalaprilat Lysinate* = *MK-521* = *Novatec* = *Prinil* = *Prinivil* = *Tensopril* = *Vivatec* = *Zestril*

Antihypertensive, angiotensin-converting enzyme inhibitor.

Untoward effects: Dizziness (> 6%), headache (> 5%), fatigue, diarrhea, and upper respiratory symptoms (> 3%); cough (3–20%), particularly in **women** – may last 6 months after withdrawal; in 1–3% – nausea, hypotension, rash, orthostatic hypotension, asthenia, chest pain, vomiting, dyspnea, and dyspepsia; < 1% – paresthesia, impotence, muscle cramps, back pain, nasal congestion, decreased libido, vertigo, fatigue, fever, flushing, tachycardia, atrial fibrillation, premature ventricular contractions, bradycardia, palpitations, peripheral edema (face, lips, throat, and legs – has been fatal from difficulty in breathing), cholestatic hepatitis, mild insomnia, anorexia, flatulence, gout, joint and shoulder pain, oliguria, azotemia, renal failure, bronchospasms, acute pancreatitis, rash, proteinuria, increased blood urea nitrogen and ***creatinine***, glycosuria, blurred vision, vasculitis of legs, and angioneurotic edema (more in **black Americans**). Penile angioedema in 74-year-old **male** after 5 mg/day/6 days. Fatal aplastic anemia and liver dysfunction in 64-year-old **female** on 5 mg/day/5 days. A 64-year-old **female** on a very low-calorie diet presented to the emergency room with weakness and intermittent vomiting, due to hyperkalemia (9.7 mmol/l) after 10 mg/day. Induced a scalded mouth syndrome (burning sensation of lips and buccal mucosa) in 74-year-old **female** after treatment with 10 mg/day. Onycholysis, with loss of several toenails and discoloration of others in 7-year-old **male** after 5 mg/day/2 months. Nails were negative for fungus. Overdoses of up to 500 mg reported with fatal consequences after severe hypotension in 45-year-old **male**. Syndrome of inappropriate antidiuretic hormone secretion in 76-year-old **female** after 10–40 mg/day.

Drug interactions: Has induced ***lithium*** toxicity in **patients** taking ***lithium carbonate*** 900–1500 mg/day. Severe hyperkalemia in 33-year-old **male** taking ***lovastatin*** (20–40 mg/day) and lisinopril (50 mg/day). ***Food*** can decrease absorption by ~35%.

LISURIDE

= *Lysuride* = *Methylergol Carbamide*

Dopaminergic agonist.

Untoward effects: After IV, nausea (55–64%), hypotension (30%), faintness and collapse (10%), sedation (45%); chorea, dizziness, dystonia, and orofacial dyskinesias (30–50%) occur. Can induce mania and its psychiatric side-effects (including delusions and hallucinations) can make this product too hazardous for routine use.

LITCHI

= *Litchu chinensis* = *Lychee*

Untoward effects: **α-Methylene cyclopropylglycine**, a toxic homolog of ***hypoglycin*** (q.v.), occurs in its seeds, but not in the edible pericarp; therefore, the nut can be eaten safely by **man**.

LITHIUM

Use: In ***aluminum*** alloys, armor plate, cardiac pacemakers, ***batteries***, glass, glass ceramics, greases, and pharmaceuticals. The "***penicillin***" of psychiatry.

Untoward effects: Some side-effects in at least 35% of treated **patients**. Intentional overdosage has occurred. Dieting, ***diuretics***, and use of saunas may precipitate intoxication by altering lithium's blood concentration. Therapeutic plasma levels are 0.42–0.83 mg% (0.6–1.2 mEq/l), with toxicity at 1.39 mg% (2 mEq/l), and lethality at 3.4 mg% (5 mEq/l). After treatment is withdrawn, levels can drop 50% in 24 h, yet toxicity symptoms can still develop, due to high concentrations in the brain, red blood cells, and other tissues.

Blood and cardiovascular: Leukocytosis, ~90% inhibition of ***choline*** transport system in erythrocytes, decreases alkaline phosphatase results by methodologic interference, enhances granulopoiesis, therefore leading to neutrophilia; inhibits erythropoiesis, causes hyperparathyroidism and hypercalcemia, thrombocytopenia, chronic myeloid leukemia, acute hyperkalemia; and sinus node dysfunction, leading to bradycardia, arrhythmias, and T-wave depression. Neutropenia and pancytopenia also occur.

Central nervous system: Can cause or accentuate tremors. Delirium, convulsions, and seizures can occur, especially with high serum levels. Chronic cluster headaches reported and recurred after rechallenge. Creutzfeldt–Jakob-like syndrome in 72-year-old **female** and 69-year-old **female** after 250 mg tid, 32 years and 800 mg/day, 10 years, respectively. Recovery 2–3 weeks after treatment ceased. Manic episodes reported in **patients** with serum levels of 1.4–2.19 mEq/l. Tardive dyskinesia, extrapyramidal symptoms, ataxia, dysarthria, tremors, drowsiness, lethargy, coma, electroencephalogram abnormalities, neuromuscular blockade with aggravation of or unmasking of myasthenia gravis, exophthalmos, difficulty in writing, slurred speech, staggering, disorientation, difficulty in accommodation, crying, excitability, trembling of hands, muscle weakness, tinnitus, nystagmus, organic brain syndrome in the **elderly**, and photophobia reported.

Diabetes: Has induced diabetes mellitus and insipidus.

Fetus and neonate: Conflicting evidence on possible teratogenicity. Indications are that **infants** of treated **mothers** may have an increased incidence of congenital heart disease, especially Ebstein's anomaly (downward displacement of tricuspid valve into the right ventricle, coarctation of the aorta, dextrocardia, ventricular septal defects, mitral and tricuspid atresia, rudimentary left ventricle with aorta and pulmonary artery arising from the right ventricle, patent ductus arteriosus, and left superior vena cava). Also hydrocephalus, microtus, spina bifida–meningomyelocele, talipes, and maxilla hypoplasia. **Newborns** of treated **mothers** may be cyanotic, lethargic, and have shallow respiration, hypothermia, bradycardia, and poor sucking reflex. Found in breast milk at approximately 1/3–1/2 the nursing **woman's** serum level. Use during the first trimester of pregnancy is contraindicated.

Gastrointestinal: Induces nausea, vomiting, diarrhea, and dry mouth in only a limited number of **patients**.

Miscellaneous: Nephropathy, azotemia, glycosuria, exacerbation of psoriasis, acneiform eruptions, lichen planus, folliculitis, erythema multiforme, alopecia, goitrogenicity, hypothyroidism (4–30%), hyperparathyroidism, papillary cell carcinoma of the thyroid, weight increase and edema, flu-like syndrome, polyuria, polydipsia, anorexia, reddened shins or calf tenderness, mild muscle pain and cramps, metallic taste, altered smell, xerostomia, erectile impotence, decreased libido, and increased liver enzymes. By decreasing neuromuscular transmission, it can aggravate or unmask myasthenia gravis, especially in **those** receiving anticholinesterase agents. Transient scotoma.

Hyperthyroidism is rare. Can cause positive results on an antinuclear antibody test. Elemental lithium ignites spontaneously when exposed to water or water vapor and has caused serious burns on **people**, especially when attempts have been made to irrigate the exposed area. High lithium content in the food and water consumed by **Pima Indians** may account for their low incidence of hypertension, atherosclerosis, and myocardial infarctions. ***Wolfberries***, used for making jelly, contain 1120 ppm, an extraordinarily high amount of lithium. Toxicity in 71-year-old **female** after urinary diversion with ileal conduit.

Drug interactions: **Patients** receiving lithium can experience further impaired psychomotor performance after ingesting ***alcoholic*** beverages. ***Salt*** depletion and the use of ***diuretics***, such as ***bendroflumethiazide***, ***chlorothiazide***, ***ethacrynic acid***, ***furosemide***, ***hydrochlorothiazide***, ***triamterene***, and ***xanthines*** increase lithium levels and potential toxicity. Some reports state ***furosemide*** does the opposite. Prolongs neuromuscular-blocking activity of ***pancuronium*** and ***succinylcholine***. ***Sodium*** intake decreases lithium effectiveness in treatment. ***Aminophylline***, ***acetazolamide***, ***sodium salts***, ***theophylline***, and some ***antacids*** (bicarbonates and other alkalis) increase its excretion. ***Cisplatin*** caused a transient decrease in its serum concentration. ***Carbamazepine*** (400 mg/day) has precipitated a tardive-dyskinesia-like syndrome in 36-year-old **female** on 900 mg lithium/day and the combination was associated with potentiation of sinus node dysfunction in four mental **patients**. ***Iodides*** and ***potassium iodide*** may be synergistic with it in, leading to hypothyroidism. Use with ***acyclovir***, ***lisinopril***, ***methyldopa***, ***lisinopril***, ***phenytoin***, and ***tetracycline*** increases its toxicity. ***Spectinomycin*** decreased urine output and interfered with normal lithium excretion, leading to toxicity. ***Phenothiazine tranquilizers*** may be additive in leading to hyperglycemia and drowsiness. Neuroleptics, such as ***alprazolam***, ***chlorpromazine***, ***fluphenazine***, ***haloperidol***, ***perphenazine***, and ***thioridazine*** with it have caused neurotoxicity. Neurotoxicity in 42-year-old **female** on ***verapamil*** 80 mg tid and lithium 900 mg/day. Confirmed by rechallenge. Use with ***anticholinergics***, ***foods***, and ***opiates*** increases its absorption and decreases absorption with ***plantago*** (***psyllium***) ***seed***. Lithium prolongs and/or intensifies the effect of ***baclofen***, ***succinylcholine***, and ***tubocurarine***; increases the risk of ***digitalis*** cardiotoxicity by decreasing serum ***potassium***. ***Amiloride*** increases its clearance. ***Serotonin*** syndrome in 59-year-old **male** also given ***clomipramine*** and in a 36-year-old **female** given ***fluoxetine***. ***Metronidazole*** may increase its nephrotoxicity. ***Mazindol*** reported to increase its toxicity. Use with some ***non-steroidal anti-inflammatory drugs*** (***bromfenac***, ***diclofenac***, ***indomethacin***, ***ibuprofen***, ***ketoprofen***, ***mefenamic acid***, ***naproxen***, ***oxyphenbutazone***, ***phenylbutazone*** (possibly 100%), and ***piroxicam***) has increased its plasma concentration by 16–59%, due to inhibition of its renal clearance, but 1 g ***aspirin*** qid had no effect and ***sulindac*** may have little or no effect. Use with ***monoamine oxidase inhibitors*** can cause a serotonin syndrome.

Large doses may cause central nervous system disturbances in **animals**, since it has a ***potassium***-like effect on muscle. Use cautiously during pregnancy, since it has caused defects in unborn **chicks**, **amphibia**, and **mice**. Chronic administration can cause goiter in **rats** and **humans**. ***Sodium*** depletion can play an important role in increasing lithium poisoning. Ingestion by **cattle** of ***tacky lithium grease*** containing 5% lithium in oil, where it had been dumped in a landfill, caused illness and death in 18 head. Severe depression, diarrhea, and ataxia were noted before death.

LITHIUM ACETATE

= *Quilonorm* = *Quilonum*

Use: In the treatment of mania.

Untoward effects: A strumectomized 61-year-old **female**, treated for 21 months developed a progressive and diffuse scleromyxedematoid disorder involving legs, trunk, and arms, along with severe induced hypothyroidism.

LITHIUM CARBONATE

= *Camcolit* = *Carbolith* = *Carbolithium* = *Ceglution* = *CP 15,467-61* = *Eskalith* = *Hypnorex* = *Limas* = *Liskonum* = *Lithane* = *Lithicarb* = *Lithobid* = *Lithonate* = *Lithotabs* = *Phasal* = *Plenur* = *Priadel* = *Quilonorm-retard* = *Quilonum-retard* = *Téralithe*

Use: In the production of ***aluminum***, glass ceramics, electrical porcelain, and in medicine for treating mania. > 10 million lbs produced annually.

Untoward effects: Essentially, the same in **people** as detailed in ***Lithium*** above.

Blood: Leukocytosis, leukopenia, granulocytosis, aplastic anemia, Felty's syndrome (rheumatoid arthritis with granulocytopenia) and increased creatine phosphokinase activity.

Cardiovascular: Various ventrical and atrial arrhythmias, symptomatic sinus node abnormalities, first degree atrioventricular block, T-wave abnormalities, bradycardia, premature ventricular contractions, myocarditis, and congestive heart failure.

Central nervous system: Epileptogenic; tremors, memory loss, tinnitus, neuromuscular blockade, ataxia, slurred speech, dizziness, incontinence, hallucinations, stupor, confusion, and pseudotumor cerebri.

Clinical test interference: Pink gelatin capsules contain ***erythrosine*** (190 μg ***iodine*** each) and upset protein-bound iodine thyroid tests for up to 4 months; increased ***glucose*** tolerance results.

Diabetes: It has induced nephrogenic diabetes insipidus and hyperosmolar coma.

Fetus and neonate: Crosses the placenta and found in breast milk (0.07–0.4 mg/100 ml) of **mothers** whose plasma or serum levels were 0.2–1.1 mg/100 ml. **Newborns'** plasma levels are 1/3–1/2 those of **maternal** levels, and, during the first 2 weeks postpartum, **their** levels are equal to milk levels. Rare cardiovascular abnormal-ities after exposure of the **fetus** reported in the 1970s. **Newborns** with congenital goiter and nephrogenic diabetes insipidus have had serum lithium levels of ~1 mEq/l. Blood levels in **newborns** above that have been associated with temporary hypothermia, bradycardia, cyanosis, decreased respirations, decreased sucking reflexes, hypotonia, T-wave changes, Moro reflexes, hepatomegaly, cardiac murmurs, and arrhythmias. Use by **mothers** during **their** pregnancies has also been suspected as a cause of club feet, meningocele, stillbirths, and port wine stains in their **infants**.

Gastrointestinal: Nausea, vomiting, contact stomatitis with erosions, and diarrhea (occasionally, from ***lactose*** diluent).

Miscellaneous: Increase in chromosome breaks, edema, paroxysmal electroencephalogram abnormalities, impotence or inhibited ejaculation in **males**, decreased libido in **females**, altered taste and smell, unpleasant taste with certain ***foods***, and mild intermittent thirst.

Poisoning and suicide: A 3½ year **child** with dystonia associated with ingestion of 300 mg. Treated with ***ipecac syrup***. Lithium carbonate is considered one of the top ten drugs carrying a high risk of toxicity in **children**. A 55-year-old **female** ingested 12 g in an attempted suicide and was saved by peritoneal dialysis. A 48-year-old **female** intentionally overdosed **herself** with 29,100 mg orally and 24,000 mg intravaginally, and was treated successfully. Shock and hypoxemia can occur in overdoses. Dieting and the use of saunas can induce intoxication. A 29-year-old **female** with end-stage renal disease entered the emergency room with nausea, weakness, chills, and a change in mental status after 1800 mg/day, 2 weeks, due to a **pharmacist's** error. **Her** prescription was for ***calcium carbonate***.

Renal: Polyuria, polydipsia, and nephrotic syndrome. Acute renal failure and death reported.

Skin: Alopecia, rashes, exfoliative dermatitis with pruritus, acneiform reactions; hyperkeratotic, erythematous, and papular folliculitis.

Thyroid: Significant hypofunction (3.6%) in **some** (myxedema). Hyperparathyroidism and parathyroid adenoma reported in 51-year-old **male** after 750 mg/day/ > 10 years.

Drug interactions: ***Alcohol*** intake with it can further impair psychomotor performance. Many ***diuretics*** (see ***Lithium***) can increase serum lithium levels and toxicity, and carbonic anhydrase inhibitors (***acetazolamide*** and ***ethoxzolamide***) can increase lithium excretion, decreasing its effect. See ***Lithium – Drug interactions*** for others.

TD_{LO}, **human**: 4.1 g/kg; **woman** (pregnant): 4.5 g/kg; **woman**: 279 mg/kg, 19 days.

Incidences of cleft palates in fetuses of **mice** increase with increased doses to their mothers. High dosage (3 g/kg) in the diet of **sows** leading to weight decrease, mummies, stillbirths, and smaller and lighter litters.

LD_{LO}, **rat**: 710 mg/kg; LD_{50}, **mouse**: 531 mg/kg; **dog**: 500 mg/kg.

LITHIUM CHLORIDE

Adds a rich, red color to fireplace flames.

Untoward effects: In 1949, it killed at least seven **people** and injured many **others** in the U.S., when used in food as a ***salt*** substitute. Wet-strength paper for disposable diapers contains it and organic plasticizers.

Treated **lamb** patties are put out to nauseate **coyotes** and cause them to develop an aversion to **lambs**. Hypertension, head twitches, generalized tremors, hind leg weakness, and increased weight of heart and adrenals in **rats**. Serum levels in pregnant **rats** are similar to those in pregnant **women** on ***lithium carbonate*** associated with eye malformations in 62%, cleft palates in 39%, and external ear defects in 45%. Another **worker** failed to duplicate these results. Absence of hind limb attachment to the truncus also reported. Offspring of pregnant **mice** had increased weight of kidneys, liver, and spleen.

LITHIUM CITRATE

= *Litarex* = *Lithonate S*

Untoward effects: See ***Lithium***. Its availability as a ***sugar***-free, sweet, palatable syrup led, unfortunately, to its use in so-called lithiated beverages or soft drinks, which could add to a **patient's *lithium*** intake.

LITHIUM HYDRIDE

Use: In making missile fuels, poison gas, and possibly nuclear weapons.

Untoward effects: May ignite spontaneously on exposure to moisture vapors. Severe burns of eyes, larynx, nose, esophagus, and trachea in 29-year-old **male** following inhalation and swallowing of small amount of dust after an enclosing cylinder exploded. Severe stenosis and strictures of the larynx and esophagus resulted. Nausea, muscle twitches, mental confusion, and blurred vision have been reported in other cases. National Institute for Occupational Safety and Health and OSHA exposure limits of 0.025 mg/m^3 and immediate danger to life or health of 0.5 mg/m^3.

Irritant to **rabbit** and **guinea pig** eyes at 5 mg/m^3.

LITHIUM HYDROXIDE

= *Lithium Hydrate*

***Untoward effects*:** A strong alkali that causes tissue burns. Serious corneal damage in **animal** experiments.

LITHIUM SULFATE

= *Lithionit* = *Lithiophor*

***Untoward effects*:** Uremia in 54-year-old **male** on 3.3 g/day, 3 years. Progression ceased when treatment was withdrawn.

LD_{50}, **mouse**: 1.2 g/kg.

LITHOCHOLIC ACID

Untoward effects*:** Hemolysis, cytolysis, and hepatic damage attributed to it in a review article. Its liver toxicity suggests its etiological role in intrahepatic cholestasis of pregnancy. Feeding to **chicks** led to decreased food intake and weight gain, increased serum and liver ***cholesterol, bile duct proliferation, and liver fibrosis. Bile duct hyperplasia when fed to **ducklings**.

LD_{50}, **mouse**: 3.9 g/kg.

LITHOSPERMUM RUDERALE

= *Gray Millet* = *Gromwell* = *Stone Seed*

***Untoward effects*:** Extracts have strong inhibitory effects on ovulation, and used by Navajo and Shoshone Indian **women** as an antifertility brew.

Fed to **female mice**, it prolongs diestrus. Inhibits ***oxytocin*** effects, is hypoglycemic, and decreases blood pressure in **mice**. Injected into **roosters**, it prevents growth of testes and androgen secretion; stops egg-laying in **hens**.

LIVER

Untoward effects*:** Erythema nodosum, urticarial exanthemata, brownish discoloration of the skin at the injection site, local hypersensitivity or pain, allergic (flushing, tachycardia, urticaria) and anaphylactic (bronchospasms and circulatory failure) reactions reported after injections of liver extracts. ***Monoamine oxidase inhibitor-type of interaction if ***furazolidone*** (q.v.) is given for > 4 days, due to ***tyramine*** content of liver. As a food, it can be a source of ***uric acid***, detrimental to **patients** prone to ***uric acid*** renal calculi. Ingestion can also negate some of ***warfarin's*** anticoagulant effect. ***Vitamin A*** (q.v.) toxicity has occurred from ingestion of **chicken**, **polar bear**, **Artic huskies**, **foxes**, **wolves**; some **seals**, **shark**, and **whale** livers.

LOBELIA

Several hundred species.

Untoward effects*: *L. berlandieri in southern Texas and Mexico has killed thousands of **cattle**, **goats**, and **sheep**. **Animals** collapse, refuse to eat or drink, salivate profusely, leg muscles atrophy, and pupils are dilated. If force-fed and watered, 50% survive after about 3 weeks. **Deer** have also been affected. The dried plant contains ~ 0.08% ***lobeline***.

L. cardinalis* = *Cardinal Flower. Leaves, stems, and fruit are poisonous, with initial stimulation and, then, depression of the autonomic ganglion from its ***nicotine***-like action. Causes vomiting, respiratory depression, diarrhea, sweating, headache, dizziness, disturbed vision, paralysis, and convulsions in **humans**. Contains 0.13–0.63% ***lobeline*** (q.v.) and related alkaloids. Nausea, vomiting, mydriasis, tachycardia, stupor, exhaustion, coma, and convulsions in **animals** that eat it when feed is scant.

L. inflata* = *Asthma Weed* = *Emetic Weed* = *Indian Tobacco* = *Wild Tobacco. Poisonous plant for **humans** due to its ***lobeline*** (q.v.) and other ***pyridine*** alkaloids. All parts are poisonous. Overdoses lead to vomiting, pain, sweating, hypothermia, rapid weak pulse, collapse, miosis followed by mydriasis, muscle weakness, tremors, nervousness, paralysis, coma, and death. In herbal ***teas*** where it has caused toxicity. Toxicity reported from as little as 50 mg of the herb and 1 ml of the tincture. Has an acrid taste.

L. siphilitica* = *Blue Cardinal Flower* = *Great Lobelia of wet swampy pastures of eastern U.S., North Central U.S. states, and southern Ontario, Canada. Ingestion of its leaves and stems causes ***lobeline*** toxicity, as with ***L. inflata***.

LOBELINE

Untoward effects*:** Sweating, nausea, vomiting, hypothermia, weakness, coma, and death in large doses. See ***Lobelia above.

LOBSTER

***Untoward effects*:** Allergic reactions with urticaria in some **people**.

LOCHNERA PUSILLA

***Untoward effects*:** **Cattle** and **sheep** grazing it develop incoordination, lateral flexion of the neck, convulsions, and death.

LOCUST, BLACK
= *False Acacia* = *Robinia pseudoacacia*

Untoward effects: Bark, foliage, young sprouts, and seeds are toxic. **Children** have chewed on the bark and seeds, leading to nausea, weakness, vomiting, abdominal pain, flushed face, dry mouth, mydriasis, stupor, depression, diarrhea, weak pulse, and clamminess of arms and legs. Often fatal from its phytotoxin (***robin***) and a glycoside (***robitin***, ***robitinin***, or ***robinine***) and a heat-labile agglutinin, ***phasin***.

Diarrhea, colic, nausea, muscle weakness, posterior paresis, mydriasis, hypothermia, gastrointestinal upsets, irregular and weak pulse, and dyspnea in **cattle**, **horses**, **sheep**, and **poultry** from ingesting it. **Cattle** may appear dizzy and nervous. Ingestion of powdered bark equal to ~0.4% of a **horse's** body weight causes symptoms. **Canaries** and **budgies** develop polyuria, fluffed feathers, dyspnea, vomiting, depression, coughing, and sneezing.

LODOXAMIDE TROMETHAMINE
= *Alomide* = *U 42,585E*

Untoward effects: After topical use on the eyes, transient burning and stinging in ~15% of **patients**; ocular itching, blurred vision, tearing or dry eyes, hyperemia, crystalline deposits, and foreign body sensation in 1–5%.

Lodoxamide ethyl = ***U 42,718***. Orally effective anticholinergic for asthma.

Untoward effects: Diaphoresis, nausea, vomiting, abdominal cramps, generalized warmth or heat sensations, and hypertension common in overdose cases.

LOFENTANIL

Untoward effects: Long-acting ***narcotic analgesic*** which, in **rats**, leads to hypothermia, respiratory depression, hypoxia, hypercarbia, acidosis, and delays hepatic elimination of drugs for up to 6 h.

LOFEPRAMINE
= *Amplit* = *Gamanil* = *Gamonil* = *Leo 640* = *Lopramine* = *Timelit* = *Tymelyt*

Antidepressant.

Untoward effects: Dry mouth (6.3%), constipation (4.9%), and nausea (4.5%). An 84-year-old **female** on 70 mg twice daily had dosage increased to tid, 1 month and developed a peripheral neuropathy that resolved itself after withdrawal of medication.

LD_{50}, **rats**: > 1 g/kg; **mice**: > 2.5 g/kg.

LOFEXIDINE
= *Ba-168* = *Britlofex* = *Lofetensin* = *Loxacor* = *MDL-14042A*

***Antihypertensive. Clonidine* analog effective in *opiate* withdrawal.**

Untoward effects: Sedation, orthostatic dizziness, headache, sexual dysfunction, backache, chest pain; dry mouth, nose, and throat; blurred vision, and ear and neck pain.

High dosage, 3 mg/kg/day and 15 mg/kg/day, respectively, to **rats** and **rabbits** decreased fetal weights, increased post-implantation losses, decreased survival and growth rates in neonates, but showed no evidence of teratogenicity.

LD_{50}, **rats, mice, dogs**: ~100–125 mg/kg.

LOLIUM TEMULENTUM
= *Darnel* = *Tars*

Untoward effects: May be due to a parasitic fungus or to its alkaloids, ***temuline*** and ***loliine*** on its seeds, leading to vertigo, dizziness, headache, and somnolence, in **man**, **dogs**, **sheep**, and **horses**, but not in **hogs**, **cows**, or **poultry**. Grows in ***wheat*** fields in many countries, but not yet in the U.S. Respiratory allergens are found in ***L. perenne*** (***rye grass***).

LOLLIGUNCULA BREVIS
= *Brief Squid*

Nocturnal, up to 9 in long and 4 in wide, with eight short arms and two longer arm-like tentacles, and sharp jaws. Found in bays and inlets.

Untoward effects: Profuse bleeding from bites, possibly due to an anticoagulant. Swelling, erythema, and warmth at the wound site.

LOMEFLOXACIN
= *Bareon* = *Chimono* = *Lomebact* = *Maxaquin* = *NY-198* = *SC-47111* = *Uniquin*

Antibacterial. A fluoroquinolone.

Untoward effects: As with other ***fluoroquinolones***, ruptures of shoulder, hand, and Achilles tendons with pain and inflammation has followed its use. Photosensitivity reactions occur, and evening dosage may decrease this risk. Nausea, headache, dizziness, diarrhea, dry mouth, flushing, hyperhidrosis, fatigue, chest and back pains, allergic reactions with facial edema, syncope, bradycardia, arrhythmias, extrasystoles, cyanosis, angina, myocardial infarction, cardiomyopathy, tremors, vertigo, paresthesias, coma, convulsions, abdominal pain, vomiting, flatulence, gastrointestinal bleeding, tongue discoloration, earache, tinnitus, thrombocytopenia, purpura, lymphadenopathy, leg cramps, arthralgia and myalgia, eye pain, conjuctivitis, vaginitis, orchitis, dyspnea, coughing, epistaxis, pruritus, rash, eczema, hematuria, dysuria, stranguria, anuria, and taste alterations.

Drug interactions: As with other ***quinolones***, ***antacids*** are contraindicated. ***Food*** decreases its absorption, but, eventually, peak plasma concentrations are not altered. Use with ***ferrous sulfate*** decreases peak plasma levels by 26%. ***Sucralate*** and ***antacids*** containing ***magnesium*** or ***aluminum*** should be avoided for 2 h after dosing, to avoid decreased absorption. Can increase ***warfarin's*** anticoagulant effect and may increase serum levels of ***cyclosporine***. ***Probenecid*** causes a dramatic decrease in its renal elimination.

LD_{50}, **mice**: > 4 g/kg.

LOMUSTINE

= *Belustine* = *CCNU* = *Cecenu* = *CeeNU* = *CiNU* = *NSC-79037* = *RB-1509*

***Antineoplastic.* Oral use. An alkylating agent.**

Untoward effects: Leukemia in one **patient**. Nausea, vomiting, anorexia, bone marrow depression with delayed (up to 6 weeks) leukopenia, thrombocytopenia, and anemia. Alopecia, stomatitis, pulmonary fibrosis (rare), testicular dysfunction, and renal and hepatic toxicity. Disorientation in the **elderly**. Contact dermatitis. Avoid contact with skin or mucous membranes.

TD_{LO}, **human**: 3 mg/kg.

Injections cause lung cancers and diffuse fibrous mammary hyperplasia in **rats**, lymph system neoplasms in **mice**. Injections are teratogenic in **rats** and cause abortions in **rabbits**.

LD_{50}, **mice**: 38 mg/kg; **rat**: 70 mg/kg; LD_{LO}, **dog**: 10 mg/kg.

LONCHOCARPUS sp.

Untoward effects: The toxicity is due to the rotenoid content and is used as a piscicide in Africa, South America, and India.

LONICERA CAPRIFOLEUM

= *Honeysuckle* = *Chévrefeuille*

Untoward effects: Contact dermatitis. The fruit is cathartic.

Japanese Honeysuckle* (*L. japonica*)**, when grazed by **animals** leads to, for **man**, a disagreeable taste and unpleasant odor to ***milk, milk productions, and meat.

L. periclymenum = ***Woodbine***, a climbing ornamental ***honeysuckle***, can also be poisonous.

L. tartarica = ***Tartarian Honeysuckle***. **Children** in Europe have been poisoned by ingesting its ripe fruit.

L. xylosteum = ***Fly Honeysuckle*** has caused serious poisoning of **children** after **they** ingested its berries leading to abdominal pain, vomiting, and diarrhea. Mild diarrhea in **rabbits**. Injection of its fluid extract into **mice** leads to drowsiness and death.

LOPERAMIDE

= *Arret* = *Blox* = *Brek* = *Dissenten* = *Fortasec* = *Imodium* = *Imosec* = *Imossel* = *Lopemid* = *Lopemin* = *Loperyl* = *Suprasec* = *Tebloc*

***Antidiarrheal.* A derivative of *haloperidol*.**

Untoward effects: Has precipitated toxic megacolon in **patients** with acute colitis. Perforations can ocur in **patients** with bacterial (*E. coli*, *Salmonella*, and *Shigella*) or parasitic infection of the intestinal wall. Stomach pain, bloat, constipation, drowsiness, dizziness, headache, nausea, vomiting, dry mouth, fatigue, and skin rash. Severe respiratory insufficiency and increase of depression in 8-year-old **male** reversed by ***naloxone***. Pancreatitis in 18-year-old **female** after a suicide attempt with an overdose. Coma, bradypnea, and miosis in 10-year-old **female**, due to inadvertent overdose. A 35-year-old **male physician** developed appendicitis after excessive daily self-administration of doses for 7 days. A 3-year-old **male** received 3.5 ml every 4 h; after the fourth dose (total intake of 3 mg), **he** became lethargic, was disorientated, and had hallucinations. A 15-month-old **female**, after a single 3 mg oral dose, developed respiratory depression and coma. Acute central nervous system depression in another 15-month-old **female** after a single 1 mg oral dose. An 8 day **neonate** received an unknown amount to treat diarrhea, leading to severe respiratory insufficiency, central nervous system depression, cyanosis, miosis, hypotonia, and abdominal distention. **Children**, aged 23–34 months, given 0.5 mg orally tid became drowsy and irritable with unacceptable personality changes and behavior.

Drug interactions: ***Cholestyramine*** interferes with its absorption and effectiveness.

Sedation, constipation, nausea, vomiting, abdominal pain, and ataxia from its use in **cats** and **dogs**.

LD_{50}, **rat**: 185 mg/kg; **mouse**: 105 mg/kg.

LOPHOGORGIA sp.

= *Sea Fans* = *Sea Whips*

Horny corals or gorgonians lacking skeletal protection and found in tropical and subtropical waters.

Untoward effects: Contain a neuromuscular toxin, ***lophotoxin*** (***LTX***), which is ichythyotoxic. Causes an irreversible postsynaptic blockade at the neuromuscular junction, resembling ***bungarotoxin***, a **snake** venom.

LOPHOPETALUM

Untoward effects: A South-East Asian dart-poison plant with ***digitalis***-type effects.

LOPINAVIR

= *ABT 378* = *Kaletra*

***Protease inhibitor.* To increase concentration of other *protease inhibitors* used to treat HIV.**

Untoward effects: Nausea, fatigue, diarrhea, weakness, and headache. Hyperglycemia, hyperlipidemia, increased aminotransferases, altered body fat distribution, and fatal pancreatitis can occur.

LORATADINE

= *Claritin* = *Clarityn* = *Lisino* = *Sch-29851*

Antihistamine.

Untoward effects: For a product highly touted by TV and magazine commercials, it has, by Schering's admission, a nearly unending list of adverse effects: "Altered lacrimation, altered salivation, flushing, hypoesthesia, impotence, increased sweating, thirst, angioneurotic edema, asthenia, back pain, blurred vision, chest pain, conjunctivitis, earache, eye pain, fever, leg cramps, malaise, rigors, tinnitus, upper respiratory infection, weight increase, hypertension, hypotension, palpitations, syncope, tachycardia, blepharospasm, dizziness, dysphonia, hyperkinesia, migraine, paresthesia, tremor, vertigo, abdominal distress, altered taste, anorexia, constipation, diarrhea, dyspepsia, flatulence, gastritis, increased appetite, nausea, stomatitis, toothache, vomiting, arthralgia, myalgia, agitation, amnesia, anxiety, confusion, decreased libido, depression, impaired concentration, insomnia, nervousness, paroniria, breast pain, dysmenorrhea, menorrhagia, vaginitis, bronchitis, bronchospasm, coughing, dyspnea, epistaxis, hemoptysis, laryngitis, nasal congestion, nasal dryness, pharyngitis, sinusitis, sneezing, dermatitis, dry hair, dry skin, photosensitivity reaction, pruritus, purpura, rash, urticaria, altered micturition, urinary discoloration. In addition, the following spontaneous adverse events have been reported rarely during the marketing of loratadine: abnormal hepatic function, including jaundice, hepatitis, and hepatic necrosis; alopecia, anaphylaxis; breast enlargement; erythema multiforme; peripheral edema; seizures; and supraventricular tachyarrhythmias." Yet, the seller advertises on TV "there's a low incidence of side-effects". Headache (6–11%), somnolence (4%), fatigue (2%), xerostomia, diaphoresis, syncope also noted in other reports. Loratadine and its metabolite is excreted in breast milk. Use caution when giving it to nursing **women**. Severe necroinflammatory and subfulminant liver failure reported in 42-year-old **female** and 33-year-old **male**, required a liver transplant in the **female**.

Drug interactions: Use with ***erythromycin*** increases plasma concentration of loratadine by 40%, ***descarboethoxyloratadine*** by 46%, and decreases ***erythromycin*** plasma levels by ~15%; with ***ketoconazole***, it increases serum concentration of loratadine over 300%, which can cause ventricular arrhythmias, including torsades de pointes. ***Cimetidine*** increases loratadine serum concentration by over 100%.

Rats and **mice** given high dosage led to hepatocellular adenomas and carcinomas.

LORAZEPAM

= *Ativan* = *Emotival* = *Lorax* = *Lorsilan* = *Pro Dorm* = *Psicopax* = *Punktyl* = *Quait* = *Securit* = *Sedatival* = *Sedazin* = *Somagerol* = *Tavor* = *Temesta* = *Wypax*

***Anxiolytic.* A *benzodiazepine*.**

Untoward effects: Sedation (~16%), confusion (~9%), dizziness (~7%), weakness and fatigue (> 4%), ataxia (3.4%), which usually decrease with continued medication. Leukopenia, increased lactic dehydrogenase, disorientation, hallucinations, depression, decreased appetite, nausea, muscle twitching, headache, agitation (~12%), respiratory depression, anterograde amnesia for up to 6 h, seizures, blurred vision, impaired hand-eye coordination, and decreased libido. Rarely, blood dyscrasias, pancytopenia, jaundice, inappropriate antidiuretic homone syndrome, diaphoresis, hyponatremia, and possible exacerbation of panic disorder. Thromboses occasionally 7–10 days after IV use. After overdoses, somnolence, confusion, weakness, dizziness, diplopia, dysarthria, ataxia, coma, and respiratory and cardiovascular depression. Acute tubular necrosis after long-term IV use, due to its ***polyethylene glycol 400*** content. A 15-year-old **male** on an infusion of 2–25 mg/h/42 days had a toxic reaction ascribed to ***propylene glycol*** (q.v.) 2.25 ml/h/4 days. After abrupt withdrawal, **he** developed tonic–clonic seizures, sleep disturbances, psychosis, and weight decrease. In breast milk at > 2% of **maternal** dosage. A 25-day-old **female** and a 10-day-old **male** whose **mothers** had taken 2–3 mg/day throughout **their** pregnancies, led to aspiration pneumonia in the **girl** and the **boy** had a swaying head movement and nystagmus. IV use near or during delivery leads to "floppy **baby** syndrome", poor muscle tone, lethargy, and respiratory depression, especially if the **neonate** is premature. Therapeutic serum levels range between 0.02 and 0.2 mg/l and toxic levels reported at 0.3 mg/l.

TD_{LO}, **child**: 71 μg/kg.

Drug interactions: Concomitant use of ***alcohol*** leads to additive psychomotor impairment and increased anxiety. Causes slight increase in prothrombin time, but not necessarily clinically important with ***warfarin***. Metabolism was increased after use with combined ***oral contraceptives***. Additive central nervous system depression with ***Kava***. ***Valproate*** decreases its clearance, increases its plasma concentration. ***Probenecid*** decreases its metabolism and increases its half-life.

LD_{50}, **rat**: 4.5 g/kg; **mouse**: 3.17 g/kg.

LORCAINIDE

= *Lopantrol* = *Lorivox* = *R-15889* = *Remivox* = *Ro-13-1042*

Antiarrhythmic.

Untoward effects: Insomnia, chills, sweating, hot flashes, nightmares, vivid dreams, headache, metallic taste, dizziness, vomiting, feeling of warmth, and slight paresthesia around lips or fingertips after IV use. Aggravation of arrhythmias in 8–24% of **patients**.

LD_{50}, **rats**: 395–435 mg/kg; **mice**: 483 mg/kg. IV, LD_{50}, **rats** and **mice**: 19 mg/kg.

LOSARTAN POTASSIUM

= *Cozaar* = *DUP-753* = *Du Pont 753* = *MK-954*

Antihypertensive.

Untoward effects: Some coughing (less than with ***angiotensin-converting enzyme inhibitors***). Severe migraine in 50-year-old **female** on 50 mg/day – confirmed by rechallenge. A diabetic kidney transplant **recipient** 46-year-old **male** given 50–100 mg/day leading to renal insufficiency with increased serum ***creatinine*** concentration and uremia. The **patient** previously had renal impairment from ***enalapril***. Angioedema within 30 min (scratchy throat, facial flushing, swollen lips and right side of face, and feeling as if something stuck in throat in 52-year-old **male**. Dizziness, leg pain, upper respiratory infections, anemia, allergic rhinitis, pruritus, rash, gastrointestinal upsets, hepatotoxicity, pancreatitis, dysgeusia, increased serum ***potassium***, and insomnia also reported.

Drug interactions: ***Grapefruit*** juice inhibits its absorption. ***Rifampin*** decreases its serum levels by 35–40%.

LOTRAFIBAN

Untoward effects: An anticoagulant combination of it with ***aspirin*** was tested by SmithKline Beecham to prevent strokes and heart attacks in over 9000 **patients**; 122 **patients** repotedly died of bleeding and testing ceased.

LOTRIFEN

= *Canocenta* = *DL-717-IT* = *L-12717* = *Privaprol*

Abortifacient.

Untoward effects: Although effective in **animals**, it was withdrawn from the European market because of a high frequency of metritis in **bitches** after use.

LOTUS CORNICULATUS

= *Birdsfoot Trefoil*

Untoward effects: Occasionally contains toxic quantities of a cyanogenic glycoside for **cattle** and other grazing **animals** in the U.S. and Europe.

LOVASTATIN

= *Lovalip* = *6α-Methylcompactin* = *Mevacor* = *Mevinolin* = *Mevlor* = *MK-803* = *Monacolin A* = *Sivlor*

Antihypercholesterolemic.

Untoward effects: Dr George Mann of Vanderbilt University is quoted as saying, "I don't think it's a safe drug ... the Japanese who discovered it abandoned it because they found it was causing serious side-effects." Mild transient gastrointestinal upsets, headache, sleep disturbances (~18%), rashes, ichthyosis, thinning of hair, lupus erythematus, decreased appetite, rhabdomyolysis with myoglobinemia and myoglobinuria, changes in refraction and incipient lens opacity, cholestatic jaundice, peripheral neuropathies, impotence, and hypospermia. Use during pregnancy led to fetal abnormalities in **humans** and **animals**.

Drug interactions: Myopathy with or without rhabdomyolysis in **patients** also receiving ***cyclosporine***, ***erythromycin***, ***gemfibrozil***, ***itraconazole***, and ***niacin*** (increased lovastatin serum levels). Also increased serum levels with ***clarithromycin***, ***erythromycin***, ***gemfibrozil***, ***ketoconazole***, and ***nefazodone***. Hypothrombinemia and bleeding reported with ***warfarin*** in some **patients**. ***Propranolol*** decreases its serum concentration. ***Lisinopril*** with it led to severe hyperkalemia in 33-year-old **male**. Increase in low-density lipoprotein in **patient** on high-fiber diet. Hypothyroidism in a **patient** stabilized on ***thyroxine***. Induced cholestatic jaundice in 72-year-old **female**, possibly from use with ***cloxacillin***. ***Grapefruit*** juice causes a 12–15 times increase in its serum levels and that of its active metabolite, ***mevinolinic acid***. Use with ***protease inhibitors*** can decrease their metabolism and increase their toxicity.

Blood levels 3–4 times that in **humans** lead to liver cancer in **mice** and **rats**. Associated with neoplasms in **rats** and gastric papillomas in **mice**. Optic nerve degeneration in **dogs** at plasma levels > 30 times **human** therapeutic levels.

LD_{50}, **mice**: > 10 g/kg.

LOXAPINE

= *CL-62362* = *Oxilapine* = *S-805* = *SUM-3170*

Loxapine Succinate = *CL-71,563* = *Loxapac* = *Loxitane*

***Anxiolytic.* Has an *α-adrenergic*-blocking activity.**

Untoward effects*:** Sedation, extrapyramidal reactions (dystonia, akathisia, and parkinsonism appear early and tardive dyskinesia may appear months or years after treatment), decreased seizure threshold, rhabdomyolysis, tremors, rigidity, sinus tachycardia, oculogyric crisis, skin rash, pyrexia, edema, paresthesias, restlessness, gynecomastia, dizziness, neuroleptic malignant syndrome, anticholinergic effects (dry mouth, mydriasis, urinary retention, headache, fatigue, lethargy, tiredness, diaphoresis, delirium in high dosage, decreased gastrointestinal motility, cycloplegia, and memory impairment) and acute renal failure. An 8-year-old **male** inadvertently given 15 ml instead of the prescribed dose of 0.6 ml developed central nervous system depression. Within 30 min, the emergency room gave **him** 50 g ***activated charcoal – was asleep; aroused with difficulty for ~4 h; miosis for 7 h.

Drug interactions*:** May stimulate ***phenytoin metabolism. Non-ketotic hyperglycemia in 49-year-old **female** following ***amoxapine*** dosage.

Potential for inducing mammary cancers in **rats** and **mice**.

LD_{50}, **mouse**: 40 mg/kg.

LPG

= *Bottled Gas* = *Liquified Petroleum Gas*

A mixture of *butanes*, *butylenes*, *propane*, and *propylene*.

***Untoward effects*:** Inhalation has caused light-headedness, drowsiness, and asphyxia. Contact with the liquid can cause serious frost-bite to skin and eyes. Flammable. National Institute for Occupational Safety and Health/OSHA air exposure limits of 1000 ppm (1.8 g/m^3); immediate danger to life or health and LEL of 2000 ppm.

LUCANTHOE HYDROCHLORIDE

= *Miracil D* = *Miracol* = *MS-752* = *Nilodin* = *RP-3735* = *Tixantone*

Antischistosomal.

***Untoward effects*:** Vertigo, restlessness, confusion, convulsions, hallucinations, nausea, vomiting (76%), burning sensations in stomach, orange–yellow urine, and yellowing of skin.

LD_{50}, **mice**: 500 mg/kg.

LUFENURON

= *CGA 184,699* = *Fluphenacur* = *Program*

***Ectoparasiticide.* Accumulates in adipose tissues and then is slowly released over a period of weeks.**

***Untoward effects*:** Rarely, vomiting, depression/lethargy, pruritus, urticaria, skin congestion, diarrhea, anorexia, and dyspnea. At 20 times the recommended dose to 12 **dogs**, lacrimation, decreased appetite, and diarrhea occurred in several and one died on day 5 with cystitis and edematous kidneys. Passes into the milk of nursing **dogs** without ill-effects to the **pups**.

LD_{50}, **rats**: > 2 g/kg.

LUMBANG NUT

= *Aleurites trisperma*

Untoward effects*:** **Children** and **adults** have been poisoned from eating only part of the ***chestnut-like seeds in the fruit of this Philippine tree. Its diterpene esters cause burning sensation from mouth to stomach and, within 30–60 min, severe vomitng and diarrhea, lasting up to 8 h. Cultivated as a shade tree in southern Florida.

LUNANIA PARVIFLORA

A tree of the Amazon.

***Untoward effects*:** The **Tikuna Indians** powdered the root and added it to cooked food to kill **enemies** or unwanted **visitors** from other tribes.

LUPINES

= *Beanweed* = *Blubonnets* = *Lupins* = *Lupinus sp.*

Untoward effects*:** Leaves and seeds contain about 20 ***quinolizidine alkaloids (including ***lupinine*** and ***piperidine***) that are hepatotoxic (cirrhosis) to **man**, **sheep**, **goats**, **cattle**, **pigs**, and **horses**, as well as wild **animals**. The most poisonous ones reported are ***silky lupine*** (***L. sericeus***), ***tail cup lupine*** (***L. caudatus***), ***silvery lupine*** (***L. argentius***), and ***velvet lupine*** (***L. leucophyllus***). Photosensitization is seen in acute cases. The **animals** have decreased appetite, jaundice, decreased weight, and show dullness. Post-mortem shows very swollen, bright yellow or orange livers; and enlarged gallbladders with dark green bile. **Sheep** are most commonly poisoned. The seeds are toxic to them at 0.25–1.5% of their body weight. Affected **animals** are dyspneic, depressed, often with a lowered head pressed against an object, and, eventually, comatose before death. **Cattle** evidence weakness, trembling, walk stiff-legged, have dry noses and hard feces; rough, dry hair coats; dyspnea, twitching leg muscles, nervousness, frothing at the mouth, abortions, convulsions, and coma. Maternal ingestion is a cause of "crooked **calf** disease".

L. latifolius contains a teratogen (arthrogryposis, scoliosis, torticollis, and cleft palate), ***anagyrine*** (q.v.). A **baby** was born with severe bone deformities in **his** arms and legs, including radial aplasia and absent thumbs. The **mother** raised **goats** who, apparently, grazed on it, and who gave birth to **kids** with similar deformities. Her **dog** gave birth to deformed **pups**. Both the boy's **mother** and **dog** drank milk from these **goats**. **Horses** become nervous, lose muscle control, froth at the mouth, and have developed convulsions. Lupine toxicity is often increased after colonization of it by the fungus *Phomopsis leptostromiformis*. **Canaries** have died in < 5 min after eating 120 mg. Disagreeable taste and odor for **man** in ***milk***, milk products, and meat of **animals** that graze on them. Unfortunately, edible seeds of ***L. sericeus*** are sold in health food stores. Mild dizziness and incoordination in a **woman** who failed to soak and boil the seeds in several changes of water to remove its toxins, ***anagrine***, ***lupanine***, and ***sparteine***.

Cattle and **horses** do not eat lupines as readily as **sheep**.

TD_{LO}, **cows** (40–75 days pregnant): 60 g dried plant/kg.

LU-SHEN-WAN

A Chinese herbal preparation.

Untoward effects: Like many Chinese herbals, it contains ***arsenic trioxide***, causing chronic poisoning in many **people** in Singapore.

LUTEOSKYRIN

Untoward effects: A potent hepatotoxic pigment and ***antibiotic*** from *Penicillium islandicum* and other molds on toxic "yellowed" ***rice***.

Caused cirrhosis and hepatomas in **rats** and **mice**.

LD_{50}, **mice**: 2.21 mg/10 g; SC, 1.47 mg/10 g.

LUTEOTROPIN

= *Ferolactan* = *Galactin* = *Lactogen* = *Leuteotropic Hormone* = *LTH* = *Mammotropin* = *Pituitary Lactogenic Hormone* = *Prolactin*

Untoward effects: Excess has been suspected as a promoting factor in **human** mammary adenocarcinomas. Excessive levels have caused impotence and galactorrhea in **man**. Endogenous excess often associated with pituitary tumors. It has caused mammary cancers in **rats** and **mice**. Mildly diabetogenic. Hyperprolactinemia is associated with amenorrhea, anovulation, hirsutism, and increased adrenal androgen production in **women**.

At pathological levels (viz. with pituitary tumors and renal failure), **animal** studies show that ***prolactin*** desensitizes the testes to the action of luteinizing hormone and causes testicular regression, due to decreased ***testosterone*** levels.

LYCIUM HALIMIFOLIUM

= *Box Thorn* = *Matrimony Vine*

The name *Matrimony Vine* is also used for *Solanum dulcamara*.

Untoward effects: Native to Eurasia and now an ornamental vine in the U.S. The attractive orange–red berries are appealing to **children** and contain an alkaloid similar to ***hyoscyamine***. Can cause an ***anticholinergic*** intoxication syndrome.

The leaves and young shoots have poisoned **sheep** and **cattle** that have eaten them.

LYCOPERDON

= *Puffball*

Untoward effects: In New England, inhalation of the spores of ***L. gemmatum*** and ***L. pyriforme*** is a folk-medicine treatment for nosebleeds that has caused pulmonary lycoperdonosis, a hypersensitivity pneumonitis, in **children**.

In Mexico, ***L. marginatum*** and ***L. mixtecorum*** are used as narcotic intoxicants with auditory hallucinations.

The smoke from burning the puffballs will stupefy **bees**.

LYMECYCLINE

= *Armyl* = *Cyclolysal* = *Lumecycline* = *Mucomycin* = *Tetralisal* = *Tetramyl* = *Tetralysal*

Untoward effects: Sore mouth and lips (~7%), sore tongue (11%); discolored tongue, thrush, pruritus ani and skin rash (2%); unpleasant taste, vomiting, flatulence, and heartburn (~4%), diarrhea (9%), and retrosternal burning sensation.

LYNESTRENOL

= *Ethinylestrenol* = *Exluton(a)* = *Exlutena* = *Orgametril* = *Orgametil*

***Progestogen*. Used with *estrogens* (*ethinyl estradiol* and *mestranol*) as *oral contraceptives*.**

Untoward effects: Increased low-density lipoprotein (19%) and decreased high-density lipoprotein (32%), inhibition of sperm migration and motility, break-through bleeding (30%), fatigue (~6%), nausea (~4.6%), headache (~6%), insomnia (~6%), sweating (~6%), depression (2%), acne (~6%), decreased libido (~14%), increased libido (~4.6%), weight increase, and endometriosis. In breast milk, up to 0.5 μg/kg/day to an **infant**. Possible association with hepatocellular carcinoma.

Pretreatment of **mice** decreases anticonvulsant effects of ***hexobarbital***, ***phenobarbital***, and ***phenytoin***. Embryotoxic and teratogenic effect in **rabbits**.

LYNGBYA sp.

A freshwater blue–green algae.

Untoward effects: **Fish** that eat them are often unaffected, but they can cause dysentery and become highly toxic to **man**, in whom ~5% of the cases are, eventually, fatal. **Swimmers** develop a highly inflammatory and vesicatory contact dermatitis, known as "seaweed dermatitis". ***Lyngbyatoxin A*** is the anticholinesterase phytotoxin.

Cats have also been poisoned from such **fish**, with symptoms of ataxia, anorexia, and increased salivation. *Poecilia vittata* (***baitfish***) in Hawaii can be killed by its toxin within 30 min in seawater containing 0.15 µg/ml.

LYONIA sp.

Untoward effects: ***L. ligustrina*** = ***Fetterbush*** = ***Lyonia*** = ***Male Blueberry*** is poisonous to **sheep** that browse it during dry spells from its ***andromedotoxin*** (q.v.) content.

L. lucida and ***L. mariana***, both similar to the above, may also be poisonous.

LYOPHYLLUM CONGOBATUM

Untoward effects: A so-called edible ***mushroom*** that has caused toxicity in two **people**.

LYPRESSIN

= *Diapid* = *Postacton* = *Syntopressin* = *Vasopressin*

***Antidiuretic, vasopressor.* In the treatment of diabetes insipidus.**

Untoward effects: After inhalation, it leads to local irritation, mucous membrane blanching, rhinorrhea, nasal congestion and local pruritus, conjunctivitis, coughing, transient dyspnea, heartburn (due to drip into pharynx), headache, dizziness and tenesmus (10%), slight nausea (7%), vomiting (1%), premature menstruation, palpitations, giddiness, anxiety, and asthma attacks (~30%) in **asthmatic patients**.

LYSERGAMIDE

= *Ergine* = *Lysergic Acid Amide*

Untoward effects: An active alkaloid in the seeds of ***Argyreia nervosa*** (q.v.), ***Ipomoea violacea***, and ***Rivea corymbosa***, that causes hallucinogenic effects and sleepiness in **man**.

TD_{LO}, **human**: 14 µg/kg.

LYSERGIC ACID DIETHYLAMIDE

= *Delysid* = *LSD* = *Lysergide* = *Ololiuhqui*

Street names or slang names: *Acid* = *Beast* = *Big D* = *Blue Cheer* = *Blue Heaven* = *Blue Mist* = *Brown Dots* = *California Sunshine* = *Chocolate Chips* = *Coffee* = *Contact Lens* = *Cubes* = *Cupcakes* = *Haze* = *Mellow Yellows* = *Microdots* = *Orange Mushrooms* = *Orange Wedges* = *Owsley* = *Paper Acid* = *Royal Blue* = *Strawberry Fields* = *Sugar* = *Sunshine* = *The Hawk* = *Wedges* = *White Lightning* = *Window Pane* = *Yellows*

***Psychopharmaceutical.* A derivative of *ergot* alkaloids. Odorless, tasteless, and colorless. An indole-containing hallucinogen. See *Bread* and *Ergot*. Manufacturer's cost was ~1 cent/dose.**

Untoward effects: A single gram makes 10,000 doses, and one dose can produce hallucinations lasting from a few hours to weeks. A "normal" ***acid*** "trip" usually lasts 12–16 h.

Chromosome: Acute leukemia with abnormal bone marrow chromosomes (up to 25%) after doses of 100–250 µg. **Users** of adulterated ***LSD*** have more widespread chromosome damage than **controls** given pure material.

Eye: Abnormal visual illusions, mydriasis, blurred vision, perception distortion, known as "trailing effect" (seeing a moving object in serial, momentarily stationary positions) has persisted up to 1 year. Perceptions noted in the peripheral field, flashes of color, intensified colors, macropsia, and micropsia. Ocular malformations in an **infant** associated with **maternal** use.

Fetus and neonate: Limb defects (more severe on the right side). In a **mother** who took one dose at about the time of conception caused spondylothoracic dysplasia, clubfeet, longer than normal fingers, neurogenic bladder malfunction, horseshoe-shaped kidney, a single midline adrenal gland, and hydrocephalus in her **neonate**. Hirschspring's disease, head development more on the right side, permanent mental damage, increased incidence of chromosomal abnormalities, and stunting. Incidence of spontaneous abortions above average in **users**. **Neonates** had transient hearing loss, ventricular septal defects, cortical blindness, congenital neuroblastoma, tracheoesophageal fistula, spastic diplegia, and seizure disorders that may also be associated with **maternal** use.

Hallucinations: Visual hallucinations and flashbacks. The number of "bad trips" is probably higher than indicated by hospital admissions, as "**acid-heads**" try to hide their casualties to protect the ***LSD*** image. Often referred to as "***looney gas***".

Miscellaneous: Nausea, sleepiness, dizziness, vertigo, chilliness, trembling, tension, increased friendliness, aggression, panic reactions, facial flushing, chest pains, grand mal seizures, urinary incontinence, palpitations, hepatitis, hyperthermia (up to 106.4°F), dyspnea, hypotension, vomiting, creatinine phosphokinase in spinal fluid, increased deep tendon reflexes, synesthesia, hypertension, and carotid artery occlusion after ***LSD*** ingestion. A number of locations report sale of ***LSD***-laced lick-on cartoon transfer decals. Two **patients** (18 and 25 years) after ***LSD***-induced violence, were placed in strait jackets and developed rhabdomyolysis and acute renal failure.

Neoplasms: Leukemia, lymphoma, and choriocarcinoma of the testes reportedly may follow its abuse.

Poisoning, accidental: Acute psychosis in 5-year-old **female** after 100 μg led to abnormal electroencephalogram and disorganized visual-motor functions, lasting several months. **Children** have accidentally consumed drug-saturated ***sugar*** cubes. A 25-month-old **male** ingested two tablets (***purple microdots*** – 50–150 μg/tablet) leading to hallucinations, panic reactions, and hyperkinesia. Toxic serum levels are 0.1–0.4 μg%.

Poisoning, suicide, and homicide: Agitation, hallucinations, mydriasis, dry skin and mucosa, fever, and bizarre behavior (e.g. jumping out windows, attempting to fly, or walking in front of automobiles). Suicides of 9/1000 cases where psychosis lasts > 48 h. No relationship to dosage. By 1967, Canadian **writers** reported 19 attempted (11 successful) suicides and four attempted homicides (one successful). The Central Intelligence Agency (U.S.) tested this drug on thousands of **people**.

Psychosis: Prolonged psychotic reactions, severe depressive and anxiety states, and intensified sociopathic behavior. These may be delayed for weeks after a single dose. **Victims** are often unaware of **their** altered behavior and time distortion. **Some** experience euphoria or mania, acute paranoia, panic, and delirium.

Drug interactions: Psychosis, and violent, uncontrollable behavior after initial visual distortions; subsequent amnesia, flashbacks, and delusions with ***alcohol*** ingestion. Extreme anxiety after ***chlorpromazine*** IV as an antidote. Possible synergism with ***succinylcholine***. Cross-tolerance between it and ***psilocybin***.

TD_{LO}, **human**: 0.7 μg/kg.

Teratogenicity in **hamsters**, **mice**, **rabbits**, and **rats** has been demonstrated. Hyperexcitability and hyperthermia can occur in treated **animals**. Although usually explored in new species at 0.1 mg/kg, 1/10 this amount should be used initially in experiments with large **mammals**. The higher level has killed a young **elephant**. A serotoninergic agonist in the **rabbit**, leads to extreme behavioral excitation, with spontaneous locomotor activity, as well as a marked hyperthermic response. ***LSD*** in a **cat** made it afraid of a **mouse** during a "bad trip". A seeing-eye **dog** ingested a capsule of it, which, apparently, was the **owner's**. It was removed with an ***apomorphine*** injection. **Fish** develop a vibrating behavior after exposure and and continue to "swim" when they reach the end of the tank, unaware that they are not making any progress. Impairs performance and increases chromosomal anomalies in **rhesus monkeys**. Treated **rabbits** develop hyperexcitability, increased peripheral sympathetic activity, and hyperthermia. Half-lives in **mice**: 6–7 min; **cats**: 130 min, and **monkeys**: 100 min.

TD_{LO}, **rat** (pregnant 6–15 days): 90 μg/kg; LD_{50}, **wild birds**: 2 mg/kg.

LYSINE

Untoward effects: In congestive heart failure **patients** it causes mild azotemia with minor increase in ***creatinine***. Abdominal cramps and transient diarrhea occasionally reported.

Excessive supplementation in **poultry** and **rat** diets has created an ***arginine*** deficiency.

LD_{50}, **rat**: 10 g/kg.

LYSINE ACETYLSALICYLATE

= *Aspergesic* = *Aspidol* = *Aspirin Lysine Salt* = *Aspisol* = *Delgesic* = *Flectadol* = *Inyesprin* = *LAS* = *Lysal* = *Quinvet* = *Venopirin* = *Vetalgina*

Analgesic, antipyretic, anti-inflammatory.

Untoward effects: Occur in ~30% of **patients** and include gastrointestinal upsets and microscopic blood loss (~15% that of ***aspirin***). Doses > 64 g/day lead to abdominal cramps and diarrhea.

LYSINOALANINE

Untoward effects: A nephrotoxic ***amino acid*** for **rats** in protein of home-cooked and commercial foods and ingredients. Occurs when heating takes place under non-alkaline conditions.

LYSOZYME

= *Antalzyme* = *Globulin* G_1 = *Muramidase*

Untoward effects: Two fatalities after first IM injection of 25 mg to a 10-month-old and a 20-month-old, apparently from anaphylaxis. After aerosol use, 2/29 **patients** (18–74 years) with chronic inflammatory bronchitis had blood in **their** sputum, and required treatment interruption for 2–3 days.

LYTECHINUS VARIEGATUS

= *White Sea Urchin*

A bottom dweller with moveable solid white spines.

Untoward effects: It is both injurious and venomous. The needle-sharp spines, after puncturing **man's** skin, can produce an immediate and burning sensation, followed by redness, swelling, and aching. Occasionally numbness, muscular paralysis, and secondary infections occur.

LYUTIK

A botanical, possibly, an anemone species.

Untoward effects: Its juice was used to poison the tips of arrows. Many Russian Cossack **troops** were killed by the **Ainu's** poisoned arrows in 1770–71 on the island of Urup.

M

MACADAMIA NUTS

***Untoward effects*:** **Dogs** eating the nuts develop a back leg paralysis that lasts about 48 h.

MACHINE OILS

Untoward effects*:** Acneform and other dematoses from various lubricants (jojoba oil*, *paraffin*, *soybean oil*, *trichloronaphthalene*, *vaseline***) and allergic reactions from biocides in them.

MACKEREL

Untoward effects*:** A **fish** (q.v.) whose species are occasionally involved in scombroid poisoning. King mackerel are also involved in ***ciguatera poisonings in the Gulf of Mexico. **Americans** are told to decrease their intake of ***cholesterol*** by eating ***fish***, but 3 oz of cooked trimmed ***beef*** contains 72 mg; 3 oz of ***turkey*** has 71 mg; and 3 oz of ***mackerel*** has 86 mg. Ingestion can increase ***uric acid*** serum levels.

MACROZAMIA sp.

Untoward effects*:** Ingestion of the fruits of these palm-like plants by **animals** in Australia causes a severe fatty degeneration and fibrosis of the liver and, to a lesser extent, by the leaves. Ingestion of the leaves also causes an ataxia and posterior weakness. **Ovine** hepatosis after eating *M. reidlei* nuts and hepatic necrosis in **calves** eating *M. communis* nuts. Has caused losses in flocks of up to 9%. Post-mortem found degenerative lesions of the nerve fibers in the fasciculus gracilis. **Dogs** and **cats** have developed abdominal pain, vomiting, diarrhea, coma, and death within hours to days. Some develop hepatotoxicosis, cirrhosis, and, rarely, a hind limb paralysis. The seeds can be safely eaten, if their harmful azoxy glycosides (macrozamine*** is the major azoglycoside) are washed out.

The seeds of ***M. spiralis*** are carcinogenic.

MAERUA ANGOLENSIS

= *Galagacha (Boran)* = *Hamoloshi (Somali)* = *Mtunguru (Swahili)*

***Untoward effects*:** Although root infusions are used to treat toothaches and to relieve cold symptoms, the fruit is poisonous to **man**.

MAESA sp.

***Untoward effects*:** Leaf extracts are piscicidal and molluscicidal.

MAFENIDE

= *Ambamide* = *Emilene* = *Homonal* = *Homosulfamine* = *Marfanil* = *Mesudin* = *Mesudrin* = *Neofamid* = *Paramenyl* = *Septicid* = *Sulfamylon*

Antibacterial.

***Untoward effects*:** Pain related to hypertonicity of the burn cream, large numbers of allergic skin reactions and rashes in ~5% 10–14 days after start of treatment; erythema multiforme, slow healing, itching, facial edema, eosinophilia, dyspnea, tachypnea, hyperventilation, and deaths from metabolic acidosis occur. The deaths are due to inhibition of carbonic anhydrase by it and its principle metabolite. Death after an acute hemolytic reaction, disseminated intravascular coagulation and renal failure in 24-year-old **male** with undiagnosed glucose-6-phosphate dehydrogenase deficiency. When applied to large areas of burned skin on **infants**, methemoglobinemia can develop. A case of severe bone marrow decrease and an acute attack of porphyria reported.

MAGNESIUM

Untoward effects*:** Ingestion of large amounts of it and its compounds leads to vomiting, abdominal pain, and watery diarrhea; injection leads to decreased cardiac muscle function, central peripheral and nervous system depression, respiratory paralysis, analgesia, and anesthesia; rectal administration leads to thirst, coma, respiratory decrease, hypotension, flaccid paralysis, respiratory paralysis, and death if enough is absorbed. Deaths of **adults**, **children**, and **infants** from its use in enemas. The kidneys of **newborns** are less able to excrete it than older **people**. Danger is increased in cases of impaired renal function. An accidental overdose in a **neonate** caused ventricular fibrillation, neuromuscular block, and drowsiness. Excessive intake in ***antacids leads to diarrhea, dehydration, ***vitamin*** and electrolyte depletion, neuromuscular dysfunction, and toxic in **people** taking ***digitalis*** preparations. High concentrations (2.4 mg/dl) in the dialysate associated with pruritus in a 59-year-old chronic hemodialysis **patient**; symptoms disappeared when the concentration was decreased to 0.48 mg/dl. Can decrease laboratory test values of ***calcium***. Burning fragments of magnesium are very injurious and must be removed to prevent the small necrotic ulcers they've caused from enlarging into extensive lesions.

Drug interactions*:** Parenteral use enhances the actions of neuromuscular blockade caused by ***muscle relaxants.

Gastrointestinal absorption and effectiveness of ***clorazepate, indomethacin***, and ***tetracycline*** decreased by its presence in ***antacids***. ***Bacitracin, oleandomycin***, and ***penicillin*** activity is slightly inhibited by it. Magnesium-containing ***antacids*** with ***aluminum hydroxide*** have a laxative effect, leading to diarrhea, loss of electrolytes, osteomalacia, weakness, muscle cramps, atraumatic rib fractures, and leg and rib pain. About 20% of the magnesium is absorbed and rapidly cleared by the kidneys in normal **adults**, but, in **patients** with renal impairment and as **people** get over 50 years of age, serum magnesium rises to toxic levels. Leads to central nervous system depression, symptoms of senility, hypotension, muscle weakness, and electrocardiogram changes. Magnesium laxatives are, therefore, contraindicated in renal failure. Neuromuscular weakness and lethargy in **infants** and **neonates** from use of similar ***antacids*** and ***laxatives***. **Infants** can also have delayed respiration, even after a **mother** received it IV during delivery.

In **dogs** and **cats**, high dietary intake predisposes to formation of ***struvite*** (***magnesium-ammonium-phosphate***) uroliths. Its presence in milk has an antagonistic effect on ***tetracyclines*** used in intramammary therapy. Magnesium ions enhance blood coagulation when the ***calcium*** ion concentration is low, and may inhibit clotting time when the ***calcium*** ion concentration is increase.

LD_{50}, **dog**: 230 mg/kg.

MAGNESIUM BROMOGLUTAMATE

= *Psycho-Soma* = *Psychoverlan*

Untoward effects: Transient somnolence, decreased white blood cells, and obstipation. Increased agitation in an **infant**.

MAGNESIUM CARBONATE

= *Magnesia Alba* = *Magnesite*

Untoward effects: The dust can irritate eyes, skin, and respiratory system (induces coughing). National Institute for Occupational Safety and Health and OSHA limit air exposure to 5 mg/m^3.

Drug interactions: Gastrointestinal absorption and effectiveness of ***clorazepate, indomethacin***, and ***tetracycline*** decreased by its presence in ***antacids***. Probably decreases ***digoxin*** absorption by ~15% and increases ***halofantrine*** adsorption by up to 83%. Can prolong serum levels of ***amphetamine, fenfluramine***, and ***quinidine***.

In feed use, decreases bone ***calcium*** in **chicks** and **rats**.

MAGNESIUM CHLORIDE

Untoward effects: Traces of it in a huge, rusting, steel evaporator tank took the ***oxygen*** content in it to ~1% after 24 h, and led to the death (from cardiac arrest) of one maintenance **man** and the near-death of **another** working in it. A 65-year-old **male** with metastatic small-cell lung cancer developed gastric retention of 21 ***Slow-Mag***® tablets over a 4-day period, leading to nausea, vomiting, and gastric distress. The **patient** refused to allow removal of the tablets and died one week later. Extravasation after IV use causes severe local irritation and phlebitis. Its use in dietary supplements is to be avoided, as considerable heat is released by its solubilization.

LD_{LO}, **human**: 500 mg/kg.

Calcium renal excretion is increased after ingestion by **dogs** and **swine**. Excessive intake in **dogs** leads to hypotension, hypophosphatemia, atrioventricular block, bradycardia, prolonged QT intervals, weakness, and impaired neuromuscular transmission.

LD_{50}, **rats**: 2.8 g/kg.

MAGNESIUM CITRATE

Untoward effects: Although used as a cathartic in certain poisoning cases, it should NOT be used, as has been recommended, with ***activated charcoal***, because the citrate ion binds on the active adsorption sites on the ***charcoal***. Some **clinicians** disagree. Slight abdominal cramping, flatulence, dry mouth, and drowsiness after its use. Lethargy and refractory hypotension in 62-year-old **female** after consumption of "large quantities".

Drug interactions: Significantly decreases absorption of ***ciprofloxacin***.

MAGNESIUM HYDROXIDE

= *Polysan*

Milk of Magnesia = *Cream of Magnesia* = *Magma Magnesii* = *Mistura Magnesii Hydroxidi* = *Mixture of Magnesium Hydroxide*

Untoward effects: Diarrhea, loss of body fluids and electrolytes, and cramps are common adverse effects. It reacts with ***hydrochloric acid*** in the stomach, forming ***magnesium chloride***, some of which is absorbed. Systemic toxicity occurs if there is renal insufficiency, leading to central nervous system depression. Inhibits peptic activity and has induced fixed-drug reactions. A 1-month-old **female** who received 2 tsp/day of ***Milk of Magnesia*** for 25 days nursed poorly and had a distended abdomen, was flaccid, and had no deep tendon reflexes, required ventilatory support, ***calcium gluconate***, and aggressive fluid therapy, to treat hypermagnesemia. Excessive doses decrease ***phosphorus*** absorption, increase bone resorption, and cause osteomalacia.

LD_{LO}, **human**: 5 g/kg.

Drug interactions: Absorption rate and peak plasma levels of ***chlordiazepoxide, chlorpromazine, cimetidine***, and ***penicillamine*** are decreased. Its decreased absorption of and response to ***phenothiazines*** can result in recurrence of psychotic symptoms controlled by these drugs. Has increased the absorption of ***dicoumarol*** and peak plasma levels by 75%, but ***warfarin*** absorption was not affected. Markedly decreases serum levels of ***salicylates***. Many other case reports refer to a combination with ***aluminum hydroxide***.

MAGNESIUM OXIDE

= *Calcined Magnesite* = *Magcal* = *Maglite* = *Magnesia* = *Magox* = *Stan-Mag*

Untoward effects: Diarrhea at 600 mg twice daily in some ***calcium oxalate*** **stoneformers**. A 27-year-old **female** with myasthemia gravis on 420 mg/day developed weakness in both legs after 2 weeks. When ***magnesium*** is burned, thermally cut, or welded, the fumes become a cause of metal fume fever (coughing, chest pain, flulike fever, eye and nose irritation), due to the fine, white, airborne particles.

Inhalation TC_{LO}, **human**: 400 mg/m^3.

Drug interactions: Decreases the adsorptive capacity of ***activated charcoal***, where, with ***tannic acid***, they are in a 1 : 2 : 1 ratio in the so-called universal antidote. Due to its increasing gastrointestinal pH, it decreases the absorption of acidic drugs, such as ***barbiturates, dicoumarol, naproxen, nitrofurantoin***, many ***penicillins, phenylbutazone, sulfonamides, trichlormethiazide*** and ***warfarin***. It increases the rate of absorption of some basic drugs (***amphetamine, meperidine, quinidine***, and ***theophylline***), as well as increasing the absorption of ***penicillin G***, which is acid labile. It can cause disintegration of some enteric-coated tablets and cause degradation of ***prednisolone***.

MAGNESIUM PEROXIDE

= *Magnesium Perhydrol* = *Ozovit*

Disinfectant for seed and grain, antacid, antifermentive, and deodorant in dentifrices.

Untoward effects: Loses its ***peroxide*** when heated. Simultaneous use as an ***antacid*** decreases the bioavailability of ***digoxin***.

MAGNESIUM SILICATE

= *Ambosol* = *Florisil* = *Sclerosal* = *Sepiolite* = *Talc*

Untoward effects: Its odor has triggered migraines in some **people**. Mesotheliomas and granulomatous pneumonitis, peritonitis, epididymitis, and periorchitis, due to ***talc*** particles from the use of ***talc***-treated surgical gloves. Intestinal obstruction has been secondary to it. ***Talc*** particles deeply imbedded in 10/13 ovarian and 12/21 cervical cancers examined. High levels of bacteria and fungi in 7/27 samples.

Ingestion: It is frequently contaminated with ***asbestos*** and its addition to ***rice*** may leading to high incidence of gastric cancer in **Japanese**. Some cosmetic grade ***talc*** has contained up to 20% ***asbestos***. Excessive dusting on **babies** and **children** entails some risks from ingestion, as well as inhalation.

Inhalation leads to respiratory distress, choking, tachycardia, and cyanosis; intercostal retraction noted as well in three fatal cases. Similar symptoms within 1 h in **child** 1–2 years old, after some spilled on its face. Pneumoconiosis progressing to massive fibrosis, and, eventually, death, after years of heavy usage. Pneumoconiosis, due to chronic ***talc*** aerosol inhalation in 31-year-old **female**, while a quality control **inspector** in a cosmetics company for 9 years. It can cause a delayed hypersensitivity reaction. Excessive dusting of **children** and **babies** entails inhalation risks. A review of 34 cases of baby powder aspiration occurred when **infants** had access to its container while **their** diapers were being changed. In 50% of the aspirations, symptoms including coughing, dyspnea, sneezing, vomiting, and in one case, cyanosis. Chest X-rays on ten revealed aspiration pneumonia in one. In addition, wheezing and choking reported in another group of 64 **youngsters**.

Parenteral leads to **drug addicts** who mainline ***"blue velvet"***, a mixture of ***paregoric*** and ***tripilennamine***, developing thrombosis of pulmonary blood vessels from the ***talc*** in the tablets. Pulmonary granulomatosis, vasculitis, and pulmonary blood vessel thrombosis from similar use with ***methylphenidate*** tablets. A review details retinopathy in 17 **addicts** from such use of ***methylphenidate*** for stimulation effects. A drug **abuser** sustained ***talc*** microembolization to the brain and retina as a consequence of IV injection of pulverized ***meperidine*** tablets. Death in a 35-year-old **male addict** with pulmonary lesions and cor pulmonale after similar IV use of ***methadone***. ***Pentazocine*** tablets have also been so abused. ***Talc*** foreign bodies have been found after implantation of ***megestrol acetate*** in ***silastic*** (***dimethicone***) capsules. Some **drug addicts** have deliberately injected ***talc*** IV for its sudden "rush". Acute failure of right side of heart also reported.

Topical leads to many **infant** deaths from or caused by 6% ***hexachlorophene*** (q.v.) in baby powders. The overzealous use of it cosmetically caused talcosis and death in 39-year-old **male**. Chronic axillary granulomas have followed its use after mild abrasion of shaving. Umbilical granulomas reported from its use on raw umbilical stumps. Use as a vaginal dusting powder may be a cause of ovarian cancer. Doctors assume that ***talc*** granules migrate to the ovaries and, by their irritation, initiate the cancer.

Drug interactions: Traces of ***iron*** in some ***talc*** powders degrades ***chloramphenicol*** and ***tetracycline HCl***. ***Phenothiazine*** derivatives are adsorbed by it, as are ***benzalkonium*** and ***cetylpyridinium***, whose antibacterial activity is decreased by the adsorption.

Peritoneal adhesions in **dogs** and **rabbits** caused by the sterile powder dusting of the peritoneum. Use in **dogs** with ***dicoumarol*** decreases its plasma levels, IV use in **guinea pigs** leads to cor pulmonale. Multiple inguinal granulomas after its use to dust castration wounds in **pigs**.

MAGNESIUM STEARATE

Tablet lubricant, and in baby dusting powders.

Untoward effects: A lubricant at 0.25–1.0%. Above 1%, it increases tablet disintegration time, because it is hydrophobic. It degrades ***aspirin*** and complexes ***chloramphenicol*** and ***tetracycline***.

MAGNESIUM SULFATE

Use: In the treatment of hypomagnesemia and in its prevention during hyperalimentation. In treating and preventing eclampsia, especially near parturition. As a saline laxative for **man** and **animals**, in IV anesthesia for **animals**, in early treatment of ***lead*** ingestion, as a topical anti-inflammatory, and as a lymphogogue.

Untoward effects: Rapid IV use leads to transient nausea, headache, and palpitations, due to its peripheral vasodilative effects. Perspiration, flushing, hypotension, central nervous system and cardiac depression, nystagmus, dizziness, and dry mouth have also been reported. An early sign of impending toxicity is decreased patellar reflex, and, when magnesium concentrations reach 10 mEq/l and over, respiratory and cardiac decrease also occur. Additional early warning symptoms consist of "feeling hot all over" and extreme thirst. This demands immediate therapy because, in these cases, death can occur within 10 min–2 h. **Neonatal** drowsiness, flaccidity, weak cry, delayed onset of respiration, and neuromuscular weakness occasionally after **maternal** use. The **neonates** of **mothers** who have received continuous infusion for > 24 h may have hypermagnesemia with low Apgar scores (biological half-life is 43 h in the **newborn**), hypotonia, and respiratory insufficiency requiring temporary ventilatory support. Prolonged infusion, especially if started during the second trimester, may cause **fetal** parathyroid gland suppression and ricket-like abnormalities. Hypocalcemia has been induced. In **patients** with arterial disease, very slow IV injection of 4–6 ml of 50% solution may eliminate or decrease symptoms of intense feeling of heat throughout the body; occasionally nausea, giddiness, and slight drowsiness. Case reports of hypersensitivity in **women** with sudden onset of urticaria after its use for tocolysis of preterm labor. **Maternal** hypothermia followed its IV use in 30-year-old **female**. **Fetal** and **maternal** bradycardia occurred.

Deaths in **adult patients** with liver disease and renal failure after receiving rectal enemas with it twice daily for 2½ days. Death and extensive brain destruction after use of 100 ml 50% solution as an enema in **infants** with hyaline membrane disease. Hypermagnesemia can develop in impaired renal function, which is common in elderly **people**.

Hypertonic solutions are irritating to gastrointestinal mucosa and have caused vomiting. Sixty ml of a 50% solution can, theoretically, pull 600 ml of fluids into the intestinal lumen, and some of the magnesium cations can be absorbed, especially in cases of intestinal obstructions. Dehydration and electrolyte upsets can easily follow. A 32-year-old **female**, post-surgery, ingested 70 g and, within 10 min, had a sensation of internal heat, followed by profound prostration. Many case reports of oral or IV overdoses reported.

Drug interactions: Potentiates the neuromuscular-blocking effects of ***succinylcholine*** and ***tubocurarine***. Use with ***nifedipine*** has caused profound hypotension. Respiratory arrest in a **neonate** with hypermagnesemia (from its **maternal** use) after 2.5 mg/kg IM ***gentamicin***. Decreased lung uptake of radioactive ***technetium*** in pulmonary imaging. Chemically incompatible with ***aminophylline, clindamycin, novobiocin, procaine HCl, sodium bicarbonate***, and ***sodium iodide*** solutions.

Five **lambs** given enemas with 10 cc/lb of a 50% solution led to lost reflexes and became anesthetized, followed by cardiorespiratory decrease, and died in ~½–¾ h. Extravasation during IV can cause tissue sloughing. Excessive IV use in the face of ***antibiotics*** and/or ***anesthetic*** neuromuscular blockade can be fatal. **Cows** given only a pound orally can evidence recumbency and a "downer-cow" syndrome.

LD_{LO}, **mouse**: 5 g/kg, **rabbit**: 3 g/kg; IV, **dog**: 750 mg/kg.

MAGNESIUM TRISILICATE

Antacid.

Untoward effects: Chronic use can cause silica kidney stones, diarrhea, and acute renal failure. Can produce fatal pulmonary lesions and osteomalacia.

Drug interactions: Has a high binding affinity for ***digoxin*** and varying degrees of affinity for ***atropine, chlorguanide, chlorpheniramine, chlorpromazine, oral contraceptives, corticosteroids, folic acid, hyoscyamine, iron compounds, methantheline bromide, metoclopramide, oxyphenonium bromide, phenytoin, propantheline, rifampin,***

and ***thiamine***. Can antagonize the bactericidal activity of many preservatives.

MAGNOLIAS

Untoward effects: ***M. grandiflora*** – the odor of its flowers leads to headaches and nausea in some **people**. A case of hypersensitivity to its leaves, with eczematous eruptions, has been reported.

M. obovata bark extracts inhibit the laryngeal reflex in **cats**; inhibit locomotor activity in **rats**, and prolong ***hexobarbital*** sleeping time in **mice**.

M. officinalis – in Chinese weight loss herbals, where it has been associated with fibrosing interstitial nephritis and renal failure. Extracts have also been used in traditional Chinese and Japanese medicine for neurosis and gastrointestinal upsets, leading to sedation, ataxia, muscle relaxation, and loss of righting reflexes in **animals**.

M. sprengeri contains ***taxiphyllin***, a cyanogenic principle.

MAHONIA AQUIFOLIUM

= *California Barberry* = *Oregon Grape* = *Rocky Mountain Grape*

Untoward effects: In western North America, **children** have frequently ingested the "grapes", often with few symptoms, but symptomatic cases had nausea and (after ~9 h) vomiting and severe purgation induced by its alkaloid, ***berberine*** (q.v.).

MALATHION

= *AC 4049* = *Carbofos* = *Chemathion* = *Cythion* = *Derbac-M* = *Emmatos* = *ENT-17034* = *For-Mal* = *Fyfanon* = *Insecticide no. 4049 Karbofos* = *Kopthion* = *Kypfos* = *Malamar* = *Malaphos* = *Malaspray* = *Malatrol* = *Mercaptothion* = *MLT* = *Organoderm* = *Phosphothion* = *Prioderm* = *Suleo-M* = *Sumitox* = *Zythiol*

Untoward effects: Touted as the safe organo-phosphorus anticholinesterase insecticide. I was called regarding its toxicity by a local Medical Examiner, when a young area **farmer** poured the concentrate on the floor of a grain bin and accidentally soaked **his** pants with it. **He** drove to a nearby farm-house, stepped out of the truck, and dropped dead. The **pathologist** confirmed the cause was due to transdermal absorption. Possibly some oral intake after topical use in 3-year-old **male**, leading to vomiting, pallor, sialorrhea, miosis, tachycardia, fasciculations, hyporeflexia, convulsions, and coma 90 min after exposure. Recovered in 48 h after ***atropine*** and supportive treatment. Coma in 46-year-old **female** after ingestion of 50 ml of 50%; near-fatal poisoning of 42-year-old **female** with suicidal intent, after swallowing 120 ml of 50% (60 g or 1 g/kg); abdominal cramps, nausea, vomiting, sialism, respiratory distress, hypertension, leukocytosis, muscle weakness, fasciculations, cyanosis, pulmonary edema, bronchopneumonia, profuse sweating, unreactive pupils, absent deep tendon reflexes, and coma in 14-year-old **male** 20 min after swallowing 120 ml in a suicide attempt. In British Guiana, > 30 **natives** attempted suicide with it during an 11-weeks period. Abuse and accidental poisonings also reported. In 1976, ~2800 **people** were poisoned by it in Pakistan during 1 month of a malaria control program of a total of 7500 **workers** exposed to percutaneous absorption; five died. Asphyxiation is the main cause of death. Harvard Medical School neurophysiologists have felt that it can alter a brain's activity for more than a year. A 24-year-old **male** even tried abusing it IV. Impurities can potentiate its toxicity. Contact dermatitis from it may be due to ***dimethyl fumarate***, a contaminant in some batches. Inhalation of vapors can be irritant and toxic. Acute renal failure and massive proteinuria in a 65-year-old **patient**, 3 weeks after exposure. Since it inhibits cholinesterase, caution is requires in **those** receiving certain ***aminoglycosides***, such as ***edrophonium***; parenteral ***neomycin; neostigmine, phenothiazines, physostigmine, procaine, succinylcholine***, or any ***curare***-like acting drugs. OSHA upper limit of skin exposure is 15 mg/m^3 – TWA; immediate danger to life or health – 250 mg/m^3.

Avoid use on young **animals** (viz. **calves** < 1 month of age), on **dairy animals** within 2 weeks of freshening, on **animals** during milking time or during the 5-h period before milking. (United States residue tolerance is ZERO in milk and eggs, and not over 4 ppm in **cattle**, **swine**, and **poultry** meat and their by-products). Avoid contamination of food and food-handling equipment. Keep from **children**. Use of 0.5–1.0% sprays daily for over 3 days on **buffalo calves** (95–100 kg) causes serious decrease of cholinesterase. Witholding times before slaughtering treated **animals** have varied. Avoid use on nursing **animals**. Its rapid metabolism in **mammals**, compared to **insects**, helped give the impression that it is very safe. Exposure of **mallard embryos** led to stunted **birds**, bill defects, and scoliosis. Poisoned **dogs** show muscarinic effects: salivation, defecation, vomiting, and diarrhea within 5–30 min. Occasionally miosis reported. Nicotinic effects: muscle twitches, clonic convulsions, and polypnea within 30–60 min, followed by general paralysis of the skeletal and respiratory muscles.

LD_{50}, **rat**: 1.38 g/kg; **dog**: 1.5 g/kg; **cattle** and **sheep**: 100 mg/kg.

MALEIC ANHYDRIDE

= *2,5-Furanedione* = *Toxilic Anhydride*

Untoward effects: Inhalation led to irritation of nose and upper respiratory system; has also caused bronchial asthma. Conjunctivitis, photophobia, double vision, and dermatitis reported. National Institute for Occupational Safety and Health/OSHA exposure limits of 0.25 ppm (1 mg/m^3); immediate danger to life or health – 10 mg/m^3.

LD_{50}, **rat**: 481 mg/kg; **mouse**: 465 mg/kg.

MALEIC HYDRAZIDE

= *MH*

Plant growth-regulator.

Untoward effects: **Human** exposure from its use on ***potatoes*** and ***onions*** to inhibit their sprouting. Its permitted use has been in direct conflict with the FDA's Delaney Amendment, since it is a mutagen and carcinogen in cell cultures and in **animals**. Tolerances on foods have been 15 ppm on ***onions***, 50 ppm on ***potatoes***, and 160 ppm on ***potato chips***. Technical grades contain free ***hydrazine*** (q.v.).

LD_{LO}, **human**: 500 mg/kg.

LD_{50}, **rat**: 3.8 g/kg; dermal, **rabbit**: 4 g/kg.

MALONALDEHYDE

= *Malonic Aldehyde* = *Propanedial*

Untoward effects: A mutagen and carcinogen. A major end-product of rancidity, believed to be a cause of aging and potential carcinogen in **animals** and **man**. A single injection into **mice** is lethal to most of them by the thirtieth day. This lethality was dramatically decreased by 1 day pretreatment of the **mice** with ***vitamin E***; no evidence of malonaldehyde was seen in this group. High levels of malonaldehyde in heart muscle of **mice** given toxic doses of ***doxorubicin***. **Rats** poisoned with ***carbon tetrachloride*** have high levels in their livers.

LD_{50}, **rat**: 632 mg/kg.

MALONONITRILE

= *Cyanoacetonitrile*

Use: Solvent. An additive in detergents, ***polyurethane prepolymers***, and used in organic sythesis.

Untoward effects: Its vapors can cause eye, skin, nose, and throat irritation. Dermal absorption or ingesting solutions can cause headache, dizziness, weakness, giddiness, confusion, convulsions, dyspnea, abdominal pain, nausea, and vomiting. Releases ***cyanide***. OSHA limit – 8 mg/m^3 – TWA.

LD_{50}, **rat**: 61 mg/kg; **mouse**: 19 mg/kg.

MALOUETIA sp.

Untoward effects: Some in Amazonia are used on curarizing poison arrows.

M. nitida is said to be very poisonous.

M. tamaquarina is poisonous, but its crushed leaves are used as an hallucinogen.

MALT

Untoward effects: "**Malt-workers** lung" is an extrinsic allergic alveolitis, due to ***Aspergillus*** organisms present on moldy ***barley*** found on brewery floors. During the drying of the sprouted ***barley*** malt, ***nitrosamines*** are formed, which appear in ***beer*** and ***ale***. Malt can cause an immediate urticarial reaction to intact or scratched skin.

"Malt grain poisoning" reported in **cattle** of East Germany and Bulgaria. Outbreaks in eight herds totalling 1050 **cattle** had 346 with symptoms of hyperesthesia, tachypnea, dys-pnea, salivation, ataxia, and muscle spasms; 28% died or were euthanized, and 10% had sequelae (incoordination and dribbling of urine). *Aspergillus clavatus* on the malt sprouts was the cause.

MANCHINEEL TREE

= *Hippomane mancinella* = *Manzanillo*

An evergreen tree, native to Central America, the West Indies, the Virgin Islands, and South Florida.

Untoward effects: Eating the thin flesh of the ***crab-apple***-like fruit has caused sore tongue and mouth, abdominal pain, diarrhea, and death. Ingestion caused fatalities in early Spanish **explorers**. **Conquistadors**, victims of the latex's use on poisoned arrows, became "mad" with pain, and usually died within 3 days. Ingestion of 20 drops causes a fatal gastroenteritis. Contact with the latex or sap causes a severe dermatitis with blisters and is a co-carcinogen in susceptible **people**, and eye contact has caused an acute, painful keratoconjunctivitis. Blindness has occurred. ***Huratoxin*** is one of the highly irritant factors in the latex.

MANDEVILLA sp.

Some species are used medicinally on warts and sores, and as a depilatory and insecticide.

Untoward effects: ***M. cuneifolia***, in Columbia, contains alkaloids and is said to be poisonous.

M. Steyermarkii – its latex is very poisonous and caustic.

MANDRAGORA OFFICINARUM

= *European Mandrake* = *Mandrake*

***American Mandrake* is *Podophyllum*.**

Untoward effects: All parts of the plant (a poisonous ***narcotic***) contain the alkaloids ***atropine*** (q.v.), ***l-hyoscyamine*** (q.v.), ***mandrogorine***, and ***scopolamine*** (q.v.). In so-called "magic brews" with mysticism and hallucinogenic effects in early European history and folklore. Anticholinergic syndrome and central nervous system depression reported from ingestion. **Ancients** used it as an anesthetic before surgery and in a ***wine*** (***morion*** or ***death wine***) before torturing **people**.

MANEB

= *Manganese EBDC* = *Manganese Ethylenebisdithiocarbamate* = *Mnazate*

Fungicide. A carbamate.

Untoward effects: Several cases of contact dermatitis reported in **people**. Communities with endemic goiter may have a potential danger from its agricultural use on foods. A 42-year-old **male** sprayed ***Manzidan***® (a combination of ***Maneb*** and ***Zineb*** (q.v.)) on a ***cucumber*** plantation twice during a week. After the first application, **he** develop behavioral changes, and, after the second application, developed loss of consciousness, convulsions, right hemiparesis, and slow rhythm in the electroencephalogram. Symptoms cleared spontaneously in a few days.

LD_{LO}, **human**: 500 mg/kg.

At a dietary level of 2500 ppm, thyroid pathology developed in **rats**. **Dogs** fed 75 mg/kg/day had no thyroid hyperplasia, but did show some neuromuscular changes. Chronic oral exposure of **rats** decreased their resistance to staphylococcal infection. Doses of 5–30 mg/kg to **rats** inhibits spermatogenesis. Teratogenic in **rats**, due to a suspected metabolite, ***ethylenethiourea***.

LD_{50}, **rat**: 1 g/kg; **mouse**: 4 g/kg; **guinea pig**: 6.4 g/kg.

MANGANESE

Use: Primarily, in the steel industry, and ~5% in other uses, including ***battery*** manufacturing and welding.

Untoward effects: Chronic poisoning from inhalation is an occupational hazard in mining and industry, which causes a slow onset of parkinsonism, asthenia, insomnia, confusion; metal fume fever (dry throat, coughing, chest tightness, rales, fatigue, flu-like symptoms, and fever); lower back pain, vomiting, malaise, and renal pathology. Daily intake in **humans** from food, water, and air averages 2.2–5 g. Neurological effects (manganic madness) in **man** occur when intake is well over 5 g/day. A toxicologist stated so-called normal serum levels are 0.08–0.26 μg%, and toxicity levels at 460 μg% or above. It is absorbed from the lung and poisoned **individuals** may act as if drunk (slurred speech, slow response, poor memory, ataxia, tremors, bradykinesia), and evidence anxiety and irritability. Toxicity, often with cholestasis, reported from long-term (as little as 28 days in **infants**) hyperalimentation. Delayed hypersensitivity reactions to it in tattoos. The development of ***l-dopa*** in the 1960s as an antiparkinson drug followed its successful use in treating **Chilean miners** poisoned by manganese inhalation. Mood and personality changes from high levels of exposure lead to psychoses and suicidal tendencies.

OSHA exposure limit of 5 mg/m^3.

Cattle, **sheep**, and **poultry** tolerate up to 1000 ppm in their diets. Excessive dietary intake in **pigs** decreases gains and decreases ***iron*** absorption.

MANGANESE CYCLOPENTADIENYLTRICARBONYL

Use: An anti-knock additive in ***gasoline***.

Untoward effects: **Animal** studies indicate it can cause skin irritation, pulmonary edema, renal pathology, and decreased resistance to infection.

LD_{LO}, **rat**: 20 mg/kg; LD_{50}, **mouse**: 150 mg/kg; LC_{LO}, inhalation, **rat**: 120 mg/m^3/2 h.

MANGANESE DIOXIDE

= *Manganese Peroxide* = *Psilomelane*

Occurs in nature as the mineral *pyrolusite*. Strong oxidizing agent.

Untoward effects: **Miners** have exhibited a Parkinson's syndrome from chronic inhalation. In match heads and may be a cause of otitis externa.

MANGANESE OXIDE

Untoward effects: Inhalation, ingestion, and contact with solutions lead to asthenia, insomnia, confusion, low-back pain, vomiting, malaise, fatigue, renal damage, and pneumonitis.

MANGEL

= *Mangold*

Untoward effects: Fresh, unwilted tops fed to **sheep** or **cattle** have caused ***oxalic acid*** poisoning. Under certain conditions, the ***nitrates*** in the roots are converted to ***nitrites*** and have killed **cattle**, **sheep**, and **pigs**.

MANGO

= *Mangifera indica* = *Manguier*

Untoward effects: Acute "mango dermatitis" due to contact with the sap in young branches and the peel of the fruit. Skin irritation, primarily of the hands, neck and face, eyelids and lips, from contact with the resin in the sap, which usually lasts for only a few days and occurs in **people** of all races. Hypersensitive **individuals** have even developed dermatitis from eating the flesh of the fruit. **Others** developed respiratory distress, itching eyes, cheilitis, and facial rash from proximity to the blooming tree. Cross-reacts with ***Rhus (sumacs)***, ***cashew*** shells, and is related to ***poison ivy***. Its oleoresins cause a pollen-induced dermatitis.

Prolonged grazing on its leaves has caused illness and death in **cattle**.

MANIDIPINE

= *Calslot* = *CV 4093* = *Franidipine*

Antihypertensive. Calcium channel-blocker.

Untoward effects: Has induced parkinsonism in **humans**.

MANNITOL

= *Cordycepic Acid* = *Diosmol* = *Manicol* = *Manna Sugar* = *Mannidex* = *Mannite* = *Osmitrol* = *Osmofundin* = *Osmosal* = *Resectisol*

Use: Osmotic diuretic.

Untoward effects: Infusion leads to headache, nausea, vomiting, chills, dizziness, polydipsia, confusion, convulsions, edema, blurred vision, urinary retention, acidosis, electrolyte imbalances, nasal congestion, tachycardia, hyponatremia, urticaria, thirst, hives, rash, hematuria, asthma, lethargy, and pain or constriction in the chest. Large doses have been fatal and very rapid infusions pull fluids from intracellular spaces into extracellular spaces, and have caused thrombophlebitis, hypotension, anginal pain, tachycardia, congestive heart failure, premature ventricular contractions, and pulmonary edema. It increases cerebral blood flow and a risk of bleeding exists in neurosurgical **patients** post-operatively. Erythrocyte agglutination and crenation have occurred as well as hemolysis. Intraocular hemorrhages can be due to rapid decrease in intraocular pressure. Acute renal failure also reported. Has caused increased or false positive ***sodium*** and ***creatinine*** test results. Orally, it has caused diarrhea and hypokalemia. Anaphylactoid reactions have occurred rarely. May augment ***tubocurarine***-induced neuromuscular blockade.

LD_{50}, **rat**: 17 g/kg; **mouse**: 22 g/kg; IV, **mouse**: 17 g/kg.

MANNOMUSTINE

= *BCM* = *Degranol* = *Mannitol Nitrogen Mustard* = *Mannitol of Mustard*

Antineoplastic. A ***nitrogen mustard*** **derivative.**

Untoward effects: Nausea, vomiting, granulocytopenia, leukopenia, asthenia, bleeding tendency, diarrhea, and anorexia after IV use.

IV, LD_{50}, **rat**: 56 mg/kg; **mouse**: 90 mg/kg; **dog** and **rabbit**: 50 mg/kg.

MANSONIA ALTISSIMA

= *African Redwood*

Untoward effects: Extremely toxic. Used in the Ivory Coast and Upper Volta as an arrow poison due to its potent cardiac glycosides.

Used in furniture manufacturing; the shavings have been used as litter or bedding for **animals** where > 650 **pigs** died on eight farms during a 2 years period. Younger **pigs** were more seriously affected, showing weakness, prostration, and death within 24 h. Older **pigs** had reddening of the skin and necrotic areas where it had been in contact with the shavings. Sudden death in some after a few days or weeks exposure. Postmortem lesions including SC edema, liver congestion, gall bladder edema, necrotic foci in the mouth, and lymph node enlargement. **Dogs** and **calves** have developed similar symptoms. Electrocardiograms on **dogs** were similar to those during ***digitalis*** poisoning.

MANTIS SHRIMP

= *Gonodactylus bredini*

Untoward effects: A **lobster**-like **shrimp**, reaching up to 8 inches in length, with jack-knife claws and sharp spines projecting from its sturdy tail fin. It will attack **anyone** bothering it. The lacerations caused by its claws and tail have been severe, even cracking the face masks of **divers**. The wounds have developed anaerobic, granulomatous lesions from *Mycobacterium marinum*.

MANZANILLO

A tropical tree growing by the water's edge in the Caribbean.

Untoward effects: A milky sap in the leaves and fruit causes an eczematoid, blistering rash, and keratoconjunctivitis on contact. Ingestion of the fruit leads to painful blistering rash near and in the oral cavity and pharynx, and has caused a fatal gastroenteritis.

MAPLE

= *Acer sp.*

Untoward effects: Its pollen has caused allergic rhinitis, bronchial asthma, and hypersensitivity pneumonitis. **Loggers** have often developed a hypersensitivity pneumonitis from *Cryptostroma corticale* on rotting logs or bark (maple **bark-stripper's** disease). This condition, as well as skin reactions, also reported from contact with the wood, its dust, or its pollen.

Ingestion of ***red maple*** (***Acer rubrum***) leaves has caused jaundice, depression, hemoglobinuria (coffee-colored urine), anemia, methemoglobinemia, intravascular hemolytic anemia, brownish mucous membranes, cyanosis, scleral petechiation, polypnea, and tachycardia in **horses** and **ponies**. In the fall, these leaves have also become toxic and lethal to **cows**. Wilted or dry leaves are toxic, causing symptoms 3–4 days after ingestion. Dried leaves remain toxic for at least 30 days. A **pony** died after being fed 3 g/kg. High mortality (320/3900) in 18 day-old **ducklings** from eating young maple seedlings.

MAPROTILINE

= *Ba-34276* = *Deprilept* = *Dibencycladine* = *Ludiomil* = *Psymion*

***Antidepressant.* A *tetracyclic* with anticholinergic effects.**

Untoward effects: Drowsiness (16%), rashes, tremor (3%), unsteady gait, dizziness (8%), delirium, memory impairment, constipation (6%), dry mouth (22%), mydriasis, cycloplegia (4%), urinary retention, decreased gastro-intestinal motility, nausea (2%), abdominal cramps, headache (4%), agitation, hallucinations, decreased libido, erectile disorders, and jaundice. Transient decrease in white blood cells and increase in serum transaminases. In **patients** with no previous history of seizures, the incidence of seizures was ~15%, compared to ~2% in **those** treated with ***tricyclics***. Postural hypotension, arrhythmias, tachycardia, palpitations, heart-block, syncope, and increased PR interval after several weeks of treatment. A fatal case of toxicity associated with torsades de pointes. Skin rashes (3%), urticaria, and neutropenia after 2 weeks of treatment. Hypoglycemia in a **patient** stabilized on ***glyburide***. Cutaneous vasculitis, pulmonary alveolitis, exercise-induced bronchospasm, and photosensitivity reactions reported. Self-poisoning has increased the intensity of the above symptoms, and hallucinations have also been reported. May be teratogenic, if taken during first trimester of pregnancy. Passes into breast milk. Toxic serum concentrations are 0.5 mg or more/l.

TD_{LO}, **child**: 26 mg/kg.

Drug interactions: Acute additive effects with drugs that depress central nervous system. Concurrent use with ***monoamine oxidase inhibitor***s, or within 2 weeks of such use, may cause severe hypertension, hyperthermia, seizures, and even death.

Conscious **rabbits** receiving 50 mg/kg developed acute cardiovascular toxicity. Neurotoxicity due to IP use in **mice**.

LD_{50}, **rats**: 760–900 mg/kg; **mice**: 600–750 mg/kg; **rabbits**: > 1 g/kg, **cats**: > 300 mg/kg; **dogs**: > 30 mg/kg.

MAQUIRA sp.

Untoward effects: The latex of ***M. calophylla*** tree of the Columbian Amazon is caustic and poisonous.

M. sclerophylla is used as an hallucinogenic snuff in the central Amazon of Brazil.

MARAH sp.

= *Manroot*

Untoward effects: Used to poison, stun, or stupefy **fish** by California's early indigenous **people**.

MARASMUS OREADES

Untoward effects: Although an edible ***mushroom***, its ingestion, in one case, was a cause of poisoning.

MARBLE

A form of *calcium carbonate*.

Untoward effects: The dust has irritated eyes, skin, and mucous membranes and induced coughing, sneezing, rhinitis, and lacrimation. Marble **workers** have been exposed to ***chromates*** in coloring marble, and **workers** polishing marble with ***oxalic acid*** have developed an erythematous, vesiculo-pustular dermatosis on exposed body parts, particularly on **their** hands. The latter is now rare, because mechanical polishers have replaced hand work.

LD_{LO}, **human**: 5 g/kg; TLV, air, 10 mg/m^3.

MARBOFLOXACINE

= *Zeniquin*

***Antibiotic.* A *fluoroquinolone*.**

Untoward effects: Eye irritation and contact dermatitis in **man**. Sun exposure can induce photosensitization.

Anorexia, decreased activity, vomiting, and occasionally soft stools, increased thirst, shivering, tremors, and ataxia in treated **dogs**.

MARGARINE

= *Oleomargarine*

Untoward effects: Can result from many additives permitted in diferent countries, such as ***potassium benzoate*** (q.v.) and ***sodium benzoate*** (q.v.) preservatives, which are common allergens. ***Antioxidants***, such as ***butylated hydroxyanisole*** (q.v.), ***butylated hydroxytoluene*** (q.v.), ***lauryl gallate*** (q.v.), and ***propyl gallate*** (q.v.) have caused allergic dermatitis. Certain margarines were contaminated with ***nitrosamine*** (***n-nitrosomorpholine***) from their paper wrappers. Margarine has been implicated in causing Stevens–Johnson syndrome. Its partially hydrogenated fats have probably increased the risk of heart disease. Reports of its value in lowering blood ***cholesterol*** and heart disease do not reflect the best informed and most expert scientific opinions, but rather reflect selfish interests of industrialists and some scientists.

MARIGOLD

= *Calendula officinalis*

Untoward effects*:** Contact allergy, irritant dermatitis, and photosensitization. **Patients** allergic to members of the *Compositae* family should avoid ***teas made from its floral heads. Cross-sensitivity among species of this family as with ***sunflowers***, ***daisies***, and ***black-eyed Susans***.

MARJORAM

= *Origanum*

Untoward effects*:** Although an edible spice, it may cause genetic damage and is a suspect carcinogen, due to its ***carvacrol (q.v.) content.

MARSH ELDER

= *False Ragweed* = *Iva xanthifolia*

***Untoward effects*:** Recurrent seasonal contact dermatitis from the leaves reported in **people**. Wind-borne pollen from the flowers is a common cause of autumn hay fever in susceptible **people**.

Ingestion of leaves by **cows** leads to bitter flavor in their ***milk***.

MARSILEA DRUMMONDII

= *Nardoo Fern*

***Untoward effects*:** In 1 year, 2200/57,000 **sheep** in Australia died after grazing on almost pure stands of it. Some became blind, shook their heads; recumbency after 24–48 h, and death within 2–4 days, apparently due to high levels of thiaminase in the ferns.

MASCAGNIA sp.

Untoward effects*: *M. castenea*:** **natives** along the uppermost Rio Negro in the Brazilian Amazon used it in preparing ***curare.

M. concinna is a ***nitrate***-accumulator and has poisoned grazing **cattle** in Columbia.

M. pubiflora* = *Cipó de prata* = *Corona* = *Garland* = *Silver Liana. Toxic for **cattle** in Brazil, with tremors and convulsions before death. Apparently, not toxic to **horses**. The toxic principle is destroyed by drying the plant. ***M. rigida*** also poisons **cattle** in northeastern Brazil.

MASCARA

Over 50 million units are sold annually in the U.S.

Untoward effects*:** Even "hypoallergenic" varieties have contained ***parabens. ***Dihydroabeityl alcohol*** (***Abitol***) has been a contact sensitizer in mascara. Other eye reactions have followed flakes getting into the eyes, material getting onto the conjunctiva, contamination product introducing *Pseudomonas* onto the cornea, with resulting ulcers, and accidental trauma to the eye, especially if applied in a moving vehicle. The Maybelline® Division of Schering-Plough, aware of eye area cosmetic hazards, held a symposium on these, with special reference to mascara. Asian **children** have applied ***surma*** with varying amounts of ***lead sulfide*** to their eyes, increasing their body load of ***lead***.

MASOPROCOL

= *Actinex* = *CHX 100* = *NDGA*

Topical antineoplastic.

***Untoward effects*:** > 20% of **patients** discontinued its use due to erythema, itching, flaking, burning sensations, edema, and local soreness. May be excreted in **human** breast milk. Bleeding, crusting, oozing, and eye irritation reported in 1–5% of **patients**. It will stain clothing and other fabrics.

MASTIC

= *Mastisol*

***Untoward effects*:** Allergic contact dermatitis from the use of this tree resin, used to attach surgical dressings, and in dental cements.

MATCHES

The industry uses about 100 tons of chemicals/year. Half of this is *potassium chlorate* (q.v.).

Untoward effects*:** Contact dermatitis has been reported. "Strike anywhere" matches contain ***antimony trisulfide (q.v.), ***phosphorus***, and ***potassium chlorate***. Match factory **workers** are also exposed to ***chromates***.

MAYTANSINE

Antineoplastic, antileukemic. **Isolated from the East African shrub, *Maytenus ovatus*.**

***Untoward effects*:** Gastrointestinal toxicity (nausea, vomiting, abdominal cramps, and diarrhea), with severe dehydration within a few hours. Incapacitating neurologic toxicity, including diffuse, predominantly sensory, distal polyneuropathy and an insidious psychomotor depression with weakness, lethargy, dysphoria, and insomnia. A 24-h infusion led to severe hepatotoxicity and myelosuppression.

TD_{LO}, **human**: 190 mg/kg; 5 days.

After IP use in **mice**, decreased maternal and fetal weight gain and increased number of dead fetuses and fetal malformations.

MAZINDOL

= *Magrilon* = *Mazanor* = *Mazildene* = *Sanorex* = *Terenac* = *Teronac*

Anorexic, central nervous system stimulant.

Untoward effects: 6–12% of **patients** develop dry mouth, insomnia, constipation, nervousness, drowsiness, dizziness, nausea, headaches, flushed face, tremors, weakness, tachycardia, palpitations, chills, left chest pain, testicular pain, spontaneous ejaculation, impotence, aphrodisia, depression, rash, diaphoresis, dysphoria, unpleasant taste, and progressive rise in insulinemia with decreased fasting blood ***glucose*** and ***triglyceride*** levels.

Drug interactions: Increases ***lithium*** toxicity and increases the pressor effect of exogenous ***catecholamines***. ***Guanethidine's*** hypotensive effect may be decreased.

Corneal opacities in **dogs** given high dosage.

LD_{50}, **rat**: 180–320 mg/kg; **mouse**: 106 mg/kg; **rabbit**: 98 mg/kg; **dog**: 9–20 mg/kg.

MEBENDAZOLE

= *Bantenol* = *Equivurm Plus* = *Lomper* = *Mebendacin* = *Mebenvet* = *Mebutar* = *Nemasole* = *Noverme* = *Ovitelmin* = *Pantelmin* = *R-17635* = *Telmin* = *Telmintic* = *Vermicidin* = *Vermirax* = *Vermox*

***Anthelmintic.* An oral *benzimidazole*.**

Untoward effects: Transient abdominal pain and diarrhea in some **patients**. Only 5–10% is absorbed from the gastrointestinal tract, and this can be increased by administering with fats or oils. Severe reversible neutropenia, leukopenia, agranulocytosis, suppression of platelets and white blood cells, as well as hypospermia have also been reported with high dosage. It is suggested that it not be given to pregnant **patients** or **children** < 2 years of age. Can cause *Ascaris* to migrate aberrantly.

Drug interactions: Co-administration with ***phenytoin*** and ***carbamazepine*** may lower its plasma levels, due to induction of microsomal enzymes.

Soft stools, diarrhea, vomiting, and hepatic dysfunction reported in some treated **dogs** and **rats**. Teratogenic in **rats** and **mice**. High dosage to **rats** decreases spermatogenesis and damaged seminiferous tubules. Retreatment of **dogs** has been associated, in some cases, with icterus, emesis, anorexia, diarrhea, lethargy, and death. Postmortem revealed either liver necrosis or cirrhosis. Cross-resistance with other ***benzimidazoles*** occurs.

LD_{50}, **sheep**: > 80 mg/kg; LD_{LO}, **rat**: 320 mg/kg; **mouse**: 160 mg/kg; **guinea pig**: 1260 mg/kg.

MECAMYLAMINE

= *Inversine* = *Mervasine*

***Antihypertensive, ganglionic-blocking agent, nicotinic antagonist.* Urinary excretion increase in acid and decrease in alkaline urine.**

Untoward effects: Blocks autonomic ganglia transmission, leading to dry mouth, glossitis, mydriasis, blurred vision, urinary retention, constipation, occasionally diarrhea, impotence, decreased libido, tremors, slurred speech, decreased sweating, confusion, vertigo, delirium, hallucinations, euphoria, nausea and vomiting, paralytic ileus, mild tachycardia, hypotension (orthostatic and postural), and decreased cardiac output. Interstitial pneumonitis, progressing to pulmonary alveolitis and fibrosis reported. Transplacental passage may cause **fetal** and **neonatal** ileus.

Drug interactions: Potentiation by ***acetazolamide, ethacrynic acid, methamphetamine, sodium bicarbonate, thiazide diuretics, triamterene***, and ***urinary alkalinizers***. Has slight neuromuscular-blocking action and can cause prolonged apnea with ***antibiotics***. It potentiates ***sympathomimetics***.

Enhances ***ketamine*** catalepsy and respiratory apneusis, as well as lightening, but not shortening, anesthesia in **dogs**.

LD_{50}, **mouse**: 92 mg/kg.

MECHLORETHAMINE

= *Carolysine* = *Caryolysine* = *Chlormethine* = *Cloramin* = *Dichloren* = *Embichen* = *Embikhine* = *Erasol* = *HN2* = *MBA* = *Mustargen* = *Mustine* = *Nitrogen Mustard* = *Nitrogranulogen* = *NSC 762* = *Stickstofflost*

Antineoplastic, alkylating agent.

Untoward effects: Often, severe vomiting for ~8 h and nausea for up to 24 h. Some anorexia, weakness, and diarrhea. Serious bone marrow depression with lymphopenia (within 24 h), granulocytopenia and thrombocytopenia (within 1–3 weeks), and occasionally hemolytic anemia. Increased serum urates. Spontaneous abortion reported when given during the first trimester. May abet a delayed dissemination of cancer to unaffected parts of the body. Contact hypersensitivity and allergic dermatitis with urticaria and pruritus in ~50–80% of **patients**. Sweating, hyperpigmentation, superficial ulceration, and burning and prickling sensations reported.

IV: Toxic encephalopathy with pyrexia, coma, convulsions, local inflammation, cellulitis, thrombophlebitis, alopecia, diarrhea, oral ulcers, leukemia, amenorrhea, decreased spermatogenesis, hyperuricemia, sterility, hemiplegia, anaphylaxis, angioneurotic edema, ototoxicity (vestibular symptoms, deafness, and damage to the eighth cranial nerve), cataracts or lens deposits, necrotizing uveitis, vasculitis of choroidal vessels, delayed wound-healing, vesicant after extravasation, sinus tachycardia, oral thrush, erythema multiforme, Stevens–Johnson syndrome, confusion, disorientation, headaches, hallucinations, lethargy, tremors, seizures, brown nail plate and nail bed, vertigo, and death have been reported.

IA: Decreased white blood cells, anemia, physical and mental asthenia, coma, drowsiness, restlessness, vomiting, diarrhea, nausea, mild neurotoxic effects, hyperesthesia, hemiplegia, and hepatic failure and necrosis.

Drug interactions: Use with ***muscle relaxants*** may cause prolonged apnea. Can penetrate through ***rubber*** gloves.

Mutagenicity, carcinogenicity, and teratogenicity occur. Has caused thrombocytopenia, leukopenia, visible hemorrhages, and death in treated **calves**. Causes leukopenia, anorexia, bloody diarrhea, vomiting, hematuria, and death in **dogs**; leukopenia in **rabbits**. Teratogenic in **rats**, **mice**, **rabbits**, and **chickens**. Decreased wound-healing strength in **rat** and **rabbit** trials. Hemorrhagic encephalopathy produced in **rats**. IV, SC, or IP causes lung tumors and mutations in **mice**. IV use in **rats** leads to malignant tumors.

IV, LD_{50}, HCl, **rats**: 1.1 mg/kg; SC, 1.9 mg/kg; **mouse**: 2 mg/kg; IV, LD_{LO}, **dog**: 6 mg/kg; **monkey**: 11 mg/ kg.

MECLIZINE

= *Ancolan* = *Antivert* = *Bonadette* = *Bonamine* = *Bonine* = *Calmonal* = *Diadril* = *Histametizine* = *Meclozine* = *Navicalm* =*Neo-Istafene* = *Parachloramine* = *Peremesin* = *Postafene* = *Ru-Vert-M* = *Sabari* = *Sea-Legs* = *UCB-5062* = *Veritab*

Antiemetic.

Untoward effects: Vertigo, drowsiness, fatigue, impaired mental performance, headache, nervousness; dry mouth, nose, and throat; and blurred vision. Teratogenic effects have been debated. Use in the 5–6th weeks of pregnancy has caused an incidence of congenital defects in 12%, and only 3–4% when used in the 11–12th weeks. A 2–3 times increase in expected number of **infants** with cleft lip and/or palate after *in utero* exposure. It has also been implicated in causing omphalocele, ectromelia, and **fetal** death. Some reports claim no adverse effects on the **fetus** or **neonate**. ***Ethyl alcohol*** potentiates its sedating effects. Use with ***nicotinic acid*** (***vitamin*** B_3, q.v.), sold as ***Antivert***®, adds untoward effects of the latter.

Anophthalmia, microphthalmia, cataracts, and cleft palates in offspring of treated **rats** and **rabbits**.

LD_{50}, **rat**: 1.75 g/kg; **mouse**: 1.65 g/kg.

MECLOCYCLINES

= *GS-2989* = *Meclan* = *Mecloderm* = *Meclosorb* = *Meclutin* = *NSC-78502* = *Traumatociclina*

***Antibiotics. Tetracycline* derivative.**

Untoward effects: Contact dermatitis, with edema of the lower eyelid and conjunctivitis has occurred from topical use.

MECLOFENAMIC ACID

= *Arquel* = *CI 583* = *INF 4668*

Anti-inflammatory. Non-steroidal anti-inflammatory drug.

Untoward effects: Severe diarrhea (10–15%) and occasionally gastrointestinal distress and ulceration (33%) in **humans**. A case of thrombocytopenia reported.

Anorexia, buccal erosions, mild colic, decreased hematocrit, and occult blood in feces of treated **horses**. Aplastic anemia noted in some treated **dogs**.

Meclofenamate Sodium = *CI 583a* = *Lenidolor* = *Meclodol* = *Meclomen* = *Movens*

***Anti-inflammatory, antipyretic.* Chemically related to *mefenamic acid*.**

Untoward effects: Diarrhea in many **patients**, requires discontinuation in ~ 1/3 of these **patients**. Nausea (10%), some vomiting, flatulence, abdominal pain (< 10%), gastric or duodenal ulcers (< 1%), and little gastrointestinal bleeding. Rashes (4%), urticaria, edema, tinnitus, headache, and dizziness. Decreased hematocrit and hemoglobin in ~10% of **patients**. Occasionally increased SGOT, SGPT, ***creatinine***, and blood urea nitrogen. A case of thrombocytopenia verified by rechallenge occurred in a 81-year-old **female**. Exacerbation of psoriasis three times in a **patient**.

Drug interactions: ***Warfarin*** serum levels are increased by it. ***Aspirin*** decreases its serum levels by decreasing its absorption. Can antagonize effectiveness of ***diuretics***.

At dosage 3–5 times recommended **human** doses was only slightly embryocidal in **rabbits** and **rats**, causing only a slight increase in fetal mortality.

MECLOFENOXATE

= *Acephen* = *Analux* = *ANP-235* = *Brenal* = *Cellative* = *Centrophenoxine* = *Cetrexin* = *Clocete* = *Clofenoxate* = *Clophenoxate* = *EN 1627* = *Helfergin* = *Lucidril* = *Marucotol* = *Meclofenoxane* = *Methoxynal* = *Proserout* = *Proseryl*

Untoward effects: Heartburn, increased white blood cells and, after overdosage, light psychomotor or sexual agitation. Excitation in an **alcoholic**, requires discontinuance.

LD_{50}, **rats**: 2.6 g/kg; **mice**: 1.75 g/kg; IV, 330 mg/kg; IP, 845 mg/kg; IV, **rabbit**: 150 mg/kg.

MECLOQUALONE

= *Nubarene* = *W 4744*

Hypnotic, sedative.

Untoward effects: Jitters, nervousness, headaches, and grogginess.

LD_{50}, **mouse**: 1.1 g/kg.

MECOPROP

= *Astix CMPP* = *CMPP* = *Compitox* = *Compitox Plus* = *Iso-Cornox* = *MCPP* = *Proponex Plus* = *RD-4593*

Herbicide.

Untoward effects: Skin and eye irritant.

Dogs given 64 mg/kg developed conjunctivitis, decreased weight gains, and anemia.

LD_{50}, **rat** and **mouse**: 650 mg/kg; acute dermal LD_{50}, **rat**: 4 g/kg; **rabbit**: 900 mg/kg; LC_{50}, **quail**: > 5,000 ppm in feed.

MECRYLATE

= *Eastman 910 Adhesive* = *Methyl-2-cyanoacrylate*

Use: Tissue adhesive and in manufacturing of polymers and adhesives.

Untoward effects: Histotoxicity, necrosis, and inflammation of tissues. Eye, skin, and nose irritant; blurred vision, lacrimation, and rhinitis. Butyl and isobutyl esters are preferred for medical use.

Death of 100% of 24 **mice** given 0.2 cc intrahepatically; 90–100% in 30 given 0.1–0.3 cc IP; and 5.4% of 37 given 0.2 cc SC.

MEDAZEPAM

= *Ansilan* = *Diepin* = *Medazepol* = *Megasedan* = *Narsis* = *Nobrium* = *Psiquium* = *Resmit* = *RO 5-4556* = *Rudotel* = *Tranquilax*

Anxiolytic.

Untoward effects: Drowsiness, slight initial fatigue, ataxia, and dizziness. Occasionally muscle weakness, nausea, insomnia, dry mouth, headache, and gastric pain. Therapeutic serum concentrations are 0.05–0.5 mg/l, and toxic levels are reported as 0.6–1 mg/l.

LD_{50}, **rats**: 900 mg/kg; **mice**: 820 mg/kg; IP, **mice**: 360 mg/kg.

MEDETOMIDINE HCL

= *Domitor* = *MPV 785*

Analgesic, sedative, α_2-adrenergic agonist, anxiolytic.

Untoward effects: Dose-dependent hypotension, bradycardia, decreased cardiac output, and sedation in **humans**.

Bradycardia, apnea, cyanosis, prolonged sedation, vomiting, urination, and hypersensitivity, including excitement, in **dogs**. Deaths due to circulatory failure and congestion of lungs, liver, and kidneys have occurred.

MEDIBAZINE

= *4105S* = *Vialibran*

Coronary vasodilator, bronchodilator.

Untoward effects: Headache and allergic dermatitis.

MEDROGESTONE

= *AY-62022* = *Colpro* = *Colprone* = *Metrogestone* =*Prothil*

***Progestogen.* For oral use.**

Untoward effects: Decreased sexual function in **men** with prostatic hypertrophy. Increase in blood ***glucose***. Several **women** reported increased and irregular withdrawal bleeding. Nervousness in one **patient** that disappeared with decrease in dosage.

MEDROXYPROGESTERONE ACETATE

= *Amen* = *Clinovir* = *Curretab* = *Cycrin* = *Depo-Clinovir* = *Depo-Provera* = *DMPA* = *Farlutal* = *Gestapuran* = *G-Farlutal* = *Hysron* = *Lutoral* = *MPA* = *Nadigest* = *Nidaxin* = *Oragest* = *Perlutex* = *Prodasone* = *Provera* = *Repromix* = *Sodelut G* = *U 8839* = *Veramax*

Progestogen, antineoplastic.

Untoward effects: ***Fetus and neonate***: Transient clitoral hypertrophy. ***Salt***-losing congenital adrenal hyperplasia in 17-days-old **male**, possibly associated with **mother** taking 10 mg/day throughout pregnancy for intermittent spotting. Medicine may have decreased production of gluco- and mineralcorticoids and/or led to overproduction of androgens. Cyclopia and pseudohermaphroditism reported. Temporary masculinization may occur if given to the **mother** before the twelfth week of gestation. Increased aggression may be noted in offspring of treated **mothers**. Congenital hypospadia may have been related to its use by a **mother** during pregnancy.

Gastrointestinal: Nausea, bloating, and stomach discomfort. Ischemic colitis, crampy abdominal pain, and bloody diarrhea reported after two injections of 150 mg.

Miscellaneous*:** Weight changes (increase or decrease), edema, leg cramps, acne, urticaria, backache, flushing, fever, alopecia, decreased hair growth, occasionally hirsutism, pyrexia, insomnia, pelvic pain, decreased bone density, cholestatic jaundice, gynecomastia, breast tenderness or pain, prolonged lactation, increased nipple pigmentation, thrombophlebitis, and thromboembolism. Increased ***glucose and plasma ***insulin*** in most of 49 **females** given 400 mg IM every 6 months as a ***contraceptive***. Acromegaly also reported with this type of prophylaxis. Hypercalcemia after 50–200 mg/day in treatment of metastatic mammary carcinoma. A 26-year-old **female** developed tinnitus, vertigo, and right ear deafness, possibly from cerebrovascular occlusion after 150 mg IM. IM use every 3 months/1 year led to leg thrombosis in 26-year-old **female**.

***Nervous system*:** Headache, nervousness, irritability, depression, insomnia, weakness, and dizziness.

***Reproductive system*:** Menstrual abnormalities, with amenorrhea, spotting, bleeding, etc; decreased libido, anorgasmy, vaginal discharges, vaginal infections, sterility, cervical polyps or cysts, and pregnancy. IM, but not oral doses, increase CYPA4 activity.

Its parenteral use (particularly repeated use) in **cats** and **dogs** has been associated with a high incidence of cystic endometrial hyperplasia and pyometra, apparently due to interference with normal protective physiological hormonal cycling. **Animals** should not be bred until the second normal post-treatment heat period. This is essential in minimizing the possibility of fetal deaths, mummification, pyometra, etc. The original marketing agent has attempted to discourage its parenteral use in **cats** and **dogs**, but it is still widely used. High dosage may cause mammary swellings or lactation in **greyhounds**, and chronic high dosage has caused breast nodules in **beagles**. It may stimulate endometrial growth and secretions, and is, therefore, *contraindicated* in the presence of uterine infection or discharges, or in pseudopregnancy. If given during estrus, accidental breeding must be prevented to avoid conception and future hormonal interference with labor. United States regulations establish a zero tolerance for the drug in edible **animal** tissues, by-products, or milk. 21 days are required to reach this tolerance, but an arbitrary 40 days withholding period has been suggested by the FDA. May increase appetite, thirst, and urination, but often effective in **cats** in decreasing inappropriate urination. In **dogs**, it increases serum growth hormone levels, which are involved in mammary tumorogenesis, and, with large doses in **women**, it usually causes a decrease in those levels. High doses may rarely cause diabetes mellitus and breast enlargement in **cats** and **dogs**. Has caused decreased fetal weight, resorption, and cleft palates after SC treatment of pregnant **mice**. **Monkeys** treated with levels 50 times suggested **human** dosage developed uterine cancer in 2/12.

MEDRYSONE

= *HMS* = *Hydroxymesterone* = *Medrocort* = *Ophtocortin* = *Spectamedryn* = *U-8471*

***Glucocorticoid*. Topical.**

***Untoward effects*:** Lens opacities, cataracts, and increased intraocular pressure. Hypersensitivity with erythema and edema of skin around eyes, chemosis and erythema of conjunctiva and severe conjunctival itching in 44-year-old **female** treated for inflammed pinguecula.

LD_{LO}, **rat**: 338 mg/kg.

MEFENAMIC ACID

= *Bafhameritin-M* = *Bonabol* = *CI-473* = *Coslan* = *INF-3355* = *Lysalgo* = *Mefenacid* = *Namphen* = *Parkemed* = *Ponalar* = *Ponstan* = *Ponstel* = *Ponstil* = *Ponstyl* = *Pontal* = *Tanston* = *Vialidon*

***Anti-inflammatory*.**

Untoward effects*:** Gastrointestinal upsets with diarrhea are common. Gastrointestinal ulceration, nausea, abdominal pain, vomiting, hematemesis, sweating, constipation, blurred vision, reversible loss of color vision, insomnia, headache, drowsiness, dizziness, and vertigo occur. Also reported have been rashes, facial edema, urticaria, increased blood urea nitrogen; hemolytic, megaloblastic, and aplastic anemia; pancytopenia, agranulocytosis, neutropenia, leukopenia, bone-marrow aplasia, thrombocytopenic purpura, ear pain, and eye irritation. Pseudoporphyria on hands and forearms of 22-year-old **female** ingesting 500 mg tid/4 days of each menstrual cycle. Coma and even convulsions with cardiopulmonary arrest have occurred with overdoses. Potentially nephrotoxic (interstitial nephritis) and anemia has been reported with long-term use. May cause false positive urine test for blood and high urine ***albumin readings. Use has caused malabsorption syndromes, dysuria, hematuria, and, rarely, palpitations. Some passes into breast milk (0.017 mg/100 ml, with plasma level of 95 mg/100 ml). Usual therapeutic serum levels are 2–10 mg/ml and toxic at > 25 mg/ml.

Drug interactions*:** Can exacerbate asthma and can cause severe bronchospasm and other asthmatic reactions in ***aspirin-sensitive **individuals**. Use with ***anticoagulants*** and ***warfarin*** may increase prothrombin time, but its clinical significance has not been ascertained. Decreases effect of ***nalidixic acid*** and ***nitrofurantoin***, and increases effect of ***penicillins***. Cross-sensitization with ***tartrazine***. Decreases renal elimination of ***lithium*** and increases its serum concentration 20–60%.

Causes a severe diarrhea in treated **dogs**. Retards skin wound-healing in **rats**.

LD_{50}, **rats**: 790 mg/kg; **mice**: 600 mg/kg.

MEFENOREX
= *Incital* = *Pondinil* = *Pondinol* = *Ro-4-5282* = *Rondimen*

Anorexic.

Untoward effects: Dry throat, thirst, insomnia, headache, palpitations, giddiness, nystagmus, constipation, nausea, sweating, gastric cramps, and increased eosinophils have been reported.

MEFEXAMIDE
= *ANP 297* = *Mefexadyne* = *Mexephenamide* = *NP-297* = *Perneuron* = *Timodyne*

Central nervous system stimulant.

Untoward effects: Headache, tremor, paresthesia, sensation of heat, vertigo, and somnolence reported.

Causes hyperexcitability in treated **mice**.

IV, LD_{50}, **mice**: 168 mg/kg; **rabbits**: 135 mg/kg.

MEFLOQUINE HCL
= *Lariam* = *Ro-21-5998* = *WR-142490*

Antimalarial.

Untoward effects: Gastrointestinal upsets, including nausea, visual disturbances, dizziness, vertigo, lightheadedness, insomnia, pruritus, anorexia, nightmares, and headache. Rarely, acute fatty liver, neuropsychiatric reactions (0.7%) including visual and auditory hallucinations, thoughts of jumping from a tall building, depression, induced epilepsy, agitation, progressive delirium, generalized rigors and convulsions. Confusion is occasionally noted. Also reported have been hair loss, petechial rash, Stevens–Johnson syndrome, tinnitus, muscle aches, fever, cutaneous vasculitis, agranulocytosis, paresthesias and painful dyesthesias, anesthesia emergence delirium, and hypotension. Cross-resistance between it and ***halofantrine*** has occurred. Avoid use in **people**, such as **pilots**, that requires fine coordination and spatial determination. Usual therapeutic serum levels are 0.4–1 µg/ml and toxic at 1.5–2 µg/ml.

Drug interactions: There is a potential for inducing arrhythmias in **patients** taking ***β-blockers, chloroquin, quinidine***, and ***quinine***.

A methanesulfonate form (***WR 142***) has caused acute cardio-respiratory problems, hemoglobinuria, and pulmonary edema when given IV at high dosage. The magnitude of these effects appears to be dependent on the rate of infusion, rather than on the total dose given.

MEFRUSIDE
= *Bay 1500* = *Baycaron* = *Mephruside*

Diuretic.

Untoward effects: Rashes (> in **female** than **male**), increased serum ***urates***, decreased serum ***potassium***, and some gastric upsets. Rarely, impaired ***glucose*** metabolization, leading to hyperglycemia. Asthmatic **patients** are usually intolerant to it.

MEGESTROL ACETATE
= *Maygace* = *Megace* = *Megestat* = *Nia* = *Niagestin* = *Ovaban* = *Ovarid* = *SC 10,363* = *Synchrorin*

Progestogen.

Untoward effects: Nausea, headache, vomiting, breast tenderness, and back or stomach pain. Breakthrough bleeding, spotting, and pregnancies in 4, 14, and 15, respectively, of 415 treatment cycles with 0.25 mg tablets orally; 19, 67, and 15, respectively, in 1290 cycles with 0.5 mg tablets for contraception. Headache, depression, and lassitude in two **patients** on 10 mg/day. Occasionally hypersensitivity reaction similar to that from ***tartrazine***. Nausea, vomiting, and diarrhea are usual acute toxicities. Hot flashes, thrombophlebitis, thromboembolism, fluid retention, edema, and weight increase are often delayed adverse effects. Hyperglycemia induced by 160 mg twice daily/5 days in a 28-year-old **male** AIDS **patient**. Another 48-year-old **male** AIDS **patient** receiving 160 mg tid developed sensory loss and paresthesias on left side of face with blurred vision. Impotence in an AIDS **patient** after 12 week of treatment. A 32-year-old **male** AIDS **patient** may have developed adrenal suppression after 4 years of treatment with it. A 52-year-old **female** treated with 160 mg/day/6 months developed adrenal insufficiency. Dyspnea reported in a 62-year-old **male** treated with it for anorexia associated with lung adenocarcinoma. May be associated with increase in ectopic pregnancy. Breast milk can contain 80% of **maternal** plasma levels. ***Talc*** foreign-bodies have been identified in **patients** who received it in ***dimethicone*** implants.

Drug interactions: Use with ***flutamide*** increases prostate specific antigen levels.

Can cause lethargy, hyperglycemia (transitory diabetes mellitus), weight increase, mammary hypertrophy, endo metritis, pyometra, and testicular atrophy and infertility in **males**, on long-term oral use in **cats**. Avoid use in stressed **cats**, where it may help cause adrenocortical suppression and diabetic retinopathy, after long-term use. In **cats**, may also cause temperament changes (aggression, depression, increased affection), polyphagia with weight increase, polyuria, polydypsia, benign hypertrophy, occasionally pyometra, and diabetes mellitis.

TD_{LO}, **dog**: 183 mg/kg during 2 years.

MEGLITOL

= *Bay m 1099* = *Glyset*

Anti type 2 diabetes, α-glucosidase inhibitor.

Untoward effects: Initially and decreasing with time, abdominal pain, flatulence, and diarrhea because of unabsorbed ***carbohydrates*** and bacterial fermentation. Inhibits hydrolysis of ***sucrose***.

Drug interactions: Use with ***glyburide*** or ***metformin*** decreases their serum levels ~10–15%; decreases ***ranitidine*** and ***propranolol*** bioavailability by 60% and 40%, respectively.

MEGLUMINE DIATRIZOATE

= *Angiografin* = *Cardiografin* = *Cystografin* = *Hypaque Cysto* = *Hypaque Meglumine* = *Meglumine Amidotrizoate* = *Renografin* = *Reno M* = *Urovist*

Radiopaque, diagnostic agent.* Contains 47% *iodine.

Untoward effects: Cutaneous vasculitis and acute nephritis in **patients** after 12–22 h following IV. Weakness, dizziness, flushing, pruritus, wheezing, coughing, sneezing, dyspnea, pulmonary edema, anaphylactic shock, anaphylactoid reactions, nausea, vomiting, sialism, pallor, sweating, muscle twitching, headache, choking sensations, paresthesia, hypertension, and syncope reported. Use as an enema in newborn **infants** has caused bowel necrosis, perforation, peritonitis, and death. In addition, in ***Angiography*** it leads to cardiac arrhythmias, ventricular fibrillation; renal failure, fever, sensation of warmth, hematomas at injection site, hemiplegia, increased blood pressure, bradycardia, oliguria, urge to generalized seizures, hemorrhagic necrosis of the bowel, and even deaths have occurred. Cerebral angiography leads to transient hemiparesis, speech disorder, petechial hemorrhages in artery, temporary apnea, and visual field defects.

Arthrography: Painful spasms of back, legs, knee, and hip joints.

Gastrography: May precipitate in the stomach, causing gastric bleeding, hematemesis, and impaction. Hypotension, bowel ischemia and necrosis in 41-year-old **female** after instilling 450 ml through a Cantor tube. Delayed transport in some **patients** with adynamic ileus. Possible fluid and electrolyte problems in **infants**.

Intrauterine: **Fetal** death after induced arrhythmias, due to inadvertent introduction of catheter and ***contrast medium*** into fetal pericardial space.

Urography: Shock, asthma, anaphylactic shock, grand mal seizures, nausea, vomiting, mild cutaneous and bronchial reactions, flushing, itching, nasal stuffiness, sneezing, coughing, collapse, dizziness, palpitations, headache, myocardial infarction, cyanosis, renal failure, face and larynx edema, numbness and tingling of extremities, ventricular fibrillation, cardiac and respiratory arrest, and oliguria.

Nausea, emesis, pain, and dyspnea in **dogs**; sinus bradycardia and arterial hypotension within 5–10 s after coronary artery injections in **dogs**; and shock and death within 5 min in a **cat**. IV injection may produce a burning or stinging sensation associated with vasospasm and anaphylaxis (rhinitis, angioneurotic edema, and bronchial and laryngeal spasms).

MEGLUMINE IODIPAMIDE

= *Biligrafin* = *Cholegrafin* = *Endocistobil* = *Endografin* = *Intrabilix* = *Transbilix*

Radiopaque, IV diagnostic agent.* Contains *iodine.

Untoward effects: Significantly increase ***uric acid*** excretion for only 1–7 h, and the effect is absent after 24 h. Use has been associated with centrilobular or midzonal hepatic necrosis. Nausea, vomiting, abdominal pain, feeling of faintness, urticaria, itching, flushing, chills, asthma-like symptoms, acute oliguric renal failure, and even a fatal case of anaphylactic shock with cyanosis and convulsions. Restlessness, sensations of warmth, dizziness, headache, tremors, fever, hypotension, cyanosis, perspiration, salivation, sneezing, swollen eyelids, laryngospasm, and arthralgia have occasionally been reported. A recall of ~8000 20 ml vials, due to pyrogen content, has occurred.

MEGLUMINE IOTHALAMATE

= *Conray* = *Contrix "28"* = *Cysto-Conray*

Radiopaque, diagnostic agent.* Contains *iodine.

Untoward effects: Permanent hemiparesis and aphasia, transient hypotension, pain, hives, transient nausea and vomiting, occasionally severe facial flushing, amaurosis, headache, increased temperature, confusion, acute renal failure, coughing, sneezing, chest tightness, wheezing, laryngeal spasms, angioneurotic edema, transient cerebral blindness, trembling, bradycardia, thrombophlebitis, cardiac arrhythmias, fibrillations, palpitations, bitter taste, transient eosinophilia, myasthenia, and convulsions reported. Fatalities have occurred in **adults** and **infants**. Adverse effects reported in 7–17%.

Its epileptogenic potential has been confirmed in **cats** and **rhesus monkeys**. Seizures have been reported in treated **dogs**. Hypotension with vasodilation in **cats**.

IV, LD_{50}, **rat**: 3.4 g/kg.

MELANTHIUM
= Bunchflowers

Untoward effects: This up to 6 ft herbaceous perennial of the eastern U.S. has poisoned **cattle**, **horses**, and **sheep** after eating leaves, flowering stems, or seeds. Symptoms including tachycardia, weak pulse, dyspnea, salivation, muscle weakness, anorexia, diaphoresis, and stupor, which are probably caused by an unidentified alkaloid.

MELARSONYL POTASSIUM
= Mel W = RP 9955 = Trimelarsan

***Trypanocide, dracunculocide*. A water-soluble derivative of *melarsoprol* (q.v.).**

Untoward effects: Can produce a fatal arsenical encephalopathy. Mild allergic reactions reported.

SC to **dogs** at 5 mg/kg appeared to be safe, but 10 or 25 mg/kg killed 5/6. Non-toxic to trypanosome-infected **ponies** at dosages up to 50 mg/kg.

MELARSOPROL
= Arsobal = Mel B

Trypanocide.

Untoward effects: Albuminuria, hypertension, myocardial damage, colic, vomiting, pyrexial allergic Herxheimer-type reaction, fatal encephalopathy (12%), peripheral neuropathy, and, rarely, shock. Severe hemolytic reactions in glucose-6-phosphate dehydrogenase-deficient **patients**.

MELATONIN
= Melapure = Regulin

Pineal hormone, skin-lightening agent.

Untoward effects: Transitory sedation. Cutaneous flushing, abdominal cramps, diarrhea, scotoma lucidum, and headache in occasional **patients**. Use decreases body temperature, libido, and fertility. Acute psychotic behavior and delusional state in 73-year-old **female**. Some **children** with neurological disabilities have increased seizure activity after 5 mg at bedtime. Quality control and purity of some material has been questioned.

Granuloma-like lesions have occurred after SC use in **dogs**. Oral use in **dogs** can cause drowsiness, decreased serum ***estradiol-17β*** in **male** and **female**, decreased ***testosterone*** and ***prasterone*** in **female**, and decreased ***17-α-hydroxyprogesterone*** in **male**. Allergic reactions can occur if of **bovine** origin. Significant decrease in ***luteotropin*** in treated **mink**.

MELIA sp.

Untoward effects: ***M. azedarach*** = ***Afoforo oyibo*** (Yoruba) = ***Bakenu*** = ***Cape Lilac*** = ***Chinaball Tree*** = ***Chinaberry*** = ***Indian Lilac Tree*** = ***Kurnan masar*** (Hausa) = ***Marimara*** = ***Paraiso*** = ***Pride of China*** = ***Pride of India*** = ***Syringa Berry*** = ***Umbrella Tree*** = ***White Cedar***. Abundant in Africa, India, Asia, and in eastern North America. All parts of the tree, especially the ½ inch fleshy yellow fruit, are poisonous. Causes central nervous system stimulation with convulsions and paralysis after consuming large amounts. Nausea, vomiting, mydriasis, sweating, stomatitis, diarrhea (occasionally, bloody), syncope, ataxia, and even narcosis in some **people**. Grown as an ornamental in southern U.S. and it has escaped into the wild. It is an easy tree to climb, which often leads to **children** eating the berries. As few as 6 berries have killed a **child**. Nectar and ***tea*** made from its leaves have also caused poisonings in **people**. Hematemesis may precede death from it.Toxicity in **man**, **livestock**, and **fish** is due to a toxic resinous fraction.

Toxicity can vary in different localities. A 100 lb **sheep** died in < 20 h after consuming 28 oz of the dried fruit; 3 oz killed a 100 lb **pig** in < 4 h; and a 107 lb **pig** was killed by 1.5 oz of flesh and skin of the fruit. When forced to get up, poisoned **pigs** had severe muscle spasms 85 min after eating the fruit. They were dyspneic within 15–30 min after ingestion. Natural exposure causes illness in **hogs** in ~4 h, and they became stiff, uncoordinated, anorexic, constipated with blood-stained feces, and weak. Some recover and death usually occurs within 24 h. Toxicity is less in **goats**, **chickens**, and **ducks**. Post-mortem shows fatty degeneration of the liver and kidneys. **Robins** are narcotized after eating the fruit or bark. **Frogs** and **rabbits** appear to be immune to its ill-effects.

MELICA sp.

Untoward effects: Has caused ***hydrocyanic acid*** poisoning after ingestion by **cattle** and **buffalo**.

MELILOT

See *Clover, Sweet*. Usually, from *Melilotus officinalis* and occasionally from *M. alba*. Julius Caesar and his fellow Romans used its extracts for its volatile oil freely in their skin anointments. One commercial extract is known as *Esberiven*.

Untoward effects: A 25-year-old **female** developed abnormal clotting and mild clinical bleeding after consuming a herbal tonic containing it and other over-the-counter products, such as ***tonka beans*** (***Dipteryx odorata*** - q.v.) and ***sweet woodruff*** (q.v.) (*Asperula odorata*).

MELITRACEN HCL

= Dixeran = Melixeran = Trausabun = U-24973A

***Antidepressant.* A tricyclic. Oral and IM.**

Untoward effects: Disturbance of accomodation (10/10), headache and vertigo (7/10), constipation (7/10), trembling (7/10), xerostomia (6/10); pain in legs, chest, and precordial region (3/10), diarrhea (1/10), nausea (1/10), meteorism (1/10), paresthesia of leg (1/10), mild transient extrapyramidal symptoms after 3 weeks, hallucinations (5/10), anxiety (2/10), stupor (1/10), and circulatory problems (2/10). Tachycardia and sweating have also occurred. An 18-year-old **female** overdosed with 20–30, 25 mg tablets, leading to moderate agitation, hallucinations, xerostomia, tachypnea, and slight mydriasis. No cardiotoxicity, **she** recovered after treatment.

LD_{50}, **rat**: 96 mg/kg; **mouse**: 315 mg/kg; IV, LD_{50}, **mice**: 52 mg/kg.

MELITTIN

= Forapin = Mellitin

Use: As an antirheumatic.

Untoward effects: The principle component of **bee** (q.v.) venom. Contains a hemolysin.

MELOCHIA sp.

Untoward effects: ***M. hermannoides***: After drinking the juice of the macerated roots tid/3 days, **women** of Paraguay became infertile for 3 months.

M. pyramidate: **Cattle** in El Salvador eating this weed develop a posterior paralysis referred to as derrengue. On three farms, 40/500 died. Post-mortem revealed SC edema, edema around the sciatic nerve, excess pericardial fluid, disintegration of muscle fibers, and nerve degeneration.

MELOXICAM

= Metacam = Mobic

Anti-inflammatory. Non-steroidal anti-inflammatory drug.

Untoward effects: Gastrointestinal (diarrhea, nausea, ulcers, bleeding) and skin reactions (erythema multiforme, bullous lesions of conjunctiva and genital mucosa; maculopapular lesions on palms and soles) in **man**. Can interfere with platelet function and prolong bleeding. ***Non-steroidal anti-inflammatory drugs*** are not recommended in **women** trying to become pregnant since they interfere with implantation of fertilized eggs in **animals**.

Drug interactions: Slight decrease in ***angiotensin-converting enzyme inhibitor*** antihypertensive effect; significant increase in ***warfarin*** anticoagulant effect in one patient; and ***cholestyramine*** can decrease its serum concentration by ~50%.

Stomach ulceration and hemorrhage in **dogs**.

LD_{50}, **mouse**: 470 mg/kg.

MELPHALAN

= Alanine Nitrogen Mustard = Alkeran = CB-3025 = Melfalan = NSC-8806 = L-PAM = Sarcoclorin = L-Sarcolysine

***Antineoplastic.* A *nitrogen mustard* derivative for oral use.**

Untoward effects: Mild nausea, occasionally vomiting and gastric pain, anorexia, and hypersensitivity reactions, including hemolytic anemia, are early toxic effects. Bone marrow depression, thrombocytopenia, leukopenia, pancytopenia, amenorrhea, sterility, leukemia, alopecia, pleural effusion, pulmonary infiltrates, mucositis, and fibrosis are delayed effects. Potentially teratogenic. Sepsis and death also occur. Can affect the nail bed and plate leading to brown nails or nails with longitudinal brown bands.

Also used in regional perfusion and IA leading to tissue necrosis requiring amputation in **some**, angioneurotic edema, thrombophlebitis and burning at injection site, tingling of finger tips, and acute epithelial edema of the cornea with decreased visual acuity, diplopia, and inappropriate antidiuretic hormone secretion.

TD_{LO}, **man**: 67 mg/kg over 4.3 years.

Drug interactions: General anesthesia apparently exacerbated epithelial edema of the cornea in 60-year-old **female**. ***Cyclosporine*** with it may cause renal failure. It has exacerbated ***cyclophosphamide***-induced cystitis.

Causes papillomas in **mouse** skin; lung tumors and lymphosarcomas in **mice**; and peritoneal sarcomas in **rats**.

Thrombocytopenia in **cats** and **dogs**.

MEMANTINE

= Akatinol = D 145 = DMAA

Untoward effects: Use in parkinsonism has caused pharmacotoxic psychosis.

MENISPERMUM CANADENSE

= Moonseed = Texas Sarsaparilla = Vine Maple = Yellow Parilla

Untoward effects: In eastern North America, the black or purple 3/8" fruit with crescent-shaped grooved seeds and the roots of this perennial twining vine found in in wooded areas are poisonous when ingested by **humans**.

Children have died, mistaking the fruits for wild grapes. Severe gastrointestinal upsets and abdominal pain occur, and the sharp pits can cause gastrointestinal injury. Convulsions have occurred before death. The toxic principles are ***isoquinoline*** alkaloids, including ***dauricine*** with ***curare***-like action.

Birds appear to eat the fruit and seeds without ill-effects.

MENOGARIL

= *Menogarol* = *NSC-269148* = *7-OMEN* = *Tomoscar* = *TUT-7* = *U-52047*

***Antineoplastic.* An anthracycline antibiotic.**

***Untoward effects*:** Transient increase in mean arterial pressure during and shortly after the infusion. Slight decrease in cardiac output and substantial increase in peripheral resistance in 2/4 **patients**. Concentration-dependent phlebitis, mild hair loss, alopecia, granulocytopenia, and transient atrial fibrillation in a **patient** after **his** fifth course of treatment.

Cardiotoxicity in **rabbits** and **mice** is 1/15 that of ***doxorubicin***.

MENOTROPINS

= *8126CNB* = *Humegon* = *Neo-Pergonal* = *Pergonal* = *Pregova* = *Repronex*

Human post-menopausal gonadotropins.

***Untoward effects*:** Rapid ovarian enlargement with ascites and occasionally pain and/or pleural effusion, ovarian hyperstimulation syndrome, ovarian cysts, bloating, abdominal pain, multiple pregnancy, urticaria, fever, swelling at injection site, nausea, dizziness, rash, tachycardia, arterial thromboembolism, hypotension, oliguria, hemoperitoneum, hypercoagulability, flatulence, birth defects, and hyperthyroidism. **Men** may develop breast enlargement.

MENTHOL

= *Peppermint Camphor*

***Antipruritic, Skin "coolant", germicide.* Found naturally in oil of *peppermint*.**

Untoward effects*:** Contact dermatitis from its use in toothpastes; urticaria with hives all over body and swelling of hands and upper lip in 18-year-old **female** smoking 1/2 pack of mentholated ***cigarettes/day with frequent eating of ***mint***-flavored candy, using menthol cough drops, ***mint***-flavored toothpaste, and mentholated topical ointment. Recurrence after use of ***mint***-flavored iced ***tea***. Laryngospasm with labored inhalation, cyanosis, and inability to cry in two **infants** (12 days and 4 1/2 months) after rubbing a mentholated ointment on **their** chests, backs, upper lips, and nostrils. Intranasal application of a 2% mentholated ointment and a 1% solution has caused **infant** fatality, as well as **infant** asphyxia. Application to the nostrils of young **children** has caused reflex apnea, laryngospasm, and instant collapse. Urticaria, flushing, and headache in 31-year-old **female** after exposure to medicated toothpaste, facial cream, and ***cigarettes***. Cheilitis and stomatitis is also reported in **individuals** who become sensitized to it in mouthwashes, candy, chewing gum, liquors, and mixed drinks. A **woman** developed an irritation of **her** thumbs and first two fingers of each hand from handling the filter of the TRUE® brand of menthol ***cigarettes***. Hypersensitive **patients** develop a severe eczema with pruritus. Ingestion leads to abdominal pain, vomiting, staggering, stupor, facial flushing, bradypnea, flaccid or rigid muscles, and motor aphasia. Anuria occasionally occurs.

LD, adult **human**: ~2 g, LD_{LO}, **human**: 50 mg/kg.

In **animals**, absorption can occur from topical use. Overdosing can cause convulsions and, eventually, death.

***dl-Menthol*:** LD_{50}, **rats**: 2.9–3.18 g/kg; **cats**: 1.5–1.6 g/kg.

***l-Menthol*:** LD_{50}, **rats**: 3.3 g/kg; **cats**: 0.8–1 g/kg.

MENTHYL ANTHRANILATE

Sunscreen.

***Untoward effects*:** Has caused photosensitivity reactions and occasional irritation. Avoid contact with the eyes.

LD_{50}, **rats**: > 5g/kg.

MENTHYL SALICYLATE

= *SALA 4*

***Sunscreen.* Primarily, an ultraviolet-absorber, as are all *salicylates*.**

***Untoward effects*:** Does not promote total protection against the erythemogenous part of the ultroviolet spectrum. Contact dermatitis has been reported.

MENZIESIA FERRUGINEA

= *False Azalea* = *Fool's Huckleberry* = *Mock Azalea* = *Western Minniebush*

Untoward effects*:** A ***huckleberry-like shrub 3–15 ft high of the northwestern U.S. and Canada, whose leaves are poisonous to **sheep**, especially when eaten along trails, leading to frothing at the mouth, vomiting, and incoordination.

MEPARFYNOL

= *Allotropal* = *Anti-Stress* = *Apridol* = *Atemorin* = *Atempol* = *Dalgol* = *Dorison* = *Dormalest* = *Dormidin* =

Dormigen = Dormiphen = Dormison = Dormosan = Formison = Hesofen = Hexofen = Imnudorm = Methylparafynol = Methypentynol = Oblivon = Pentadorm = Perlopal = Riposon = Seral = Somnesin

Hypnotic, sedative.

Untoward effects: At 2 g/day/5 days, the upper therapeutic dose, leads to nystagmus, ptosis, loss of facial muscle tone, ataxia, dysarthria, tremors, mood changes, decreased concentration and attention, electroencephalogram abnormalities, and, possibly, hallucinations. Therapeutic doses are stomach irritants and cause belching and unpleasant after-taste. Death has occurred with a 5 g dose, but recovery occurred after a 10 g dose. Can cause hepatitis and dermatitis after repeated usage.

TD_{LO}, **human**: 70 mg/kg; LD_{LO}, **human**: 500 mg/kg.

LD_{50}, **rat**: 300 mg/kg; **mouse**: 760 mg/kg; **guinea pig**: 534 mg/kg; SC, LD_{50}, **mice**: 0.56 g/kg.

MEPAZINE

= III-2318 = Lacumin = Mepasin = MPMP = Nothiazine = P-391 = Pacatal = Pacatol = Paxital = Pecazine

***Antipsychotic, tranquilizer.* A tricyclic.**

Untoward effects: Agranulocytosis, leukopenia, thrombocytopenic purpura, aplastic anemia, pancytopenia, photosensitivity, itching eyes due to drying of conjunctiva, cholestasis, hepatocanalicular jaundice (5%), hypertonia (40%), and delirium.

LD_{LO}, 50 mg/kg.

Use in **animals** requires it NOT be used in conjunction with ***organophosphorus cholinesterase depressors*** or ***procaine***, as they may potentiate each other's toxicity, nor use it with ***epinephrine***, which may cause further hypotension.

IV, LD_{50}, **mouse**: 70 mg/kg; **rabbit**: 20 mg/kg.

MEPENZOLATE BROMIDE

= Cantil = Cantril = Gastropidil = Trancolon

***Anticholinergic.* Contains 19% *bromine*.**

Untoward effects: Noncomedone-type of **adult** facial acne and xerostomia. Erythematous, persistent, papular lesions on cheeks and chin in 53-year-old **female** after taking it for several months. Impotence reported.

LD_{50}, **rat**: 742 mg/kg; **mouse**: 900 mg/kg.

MEPERIDINE

= Algil = Alodan = Centralgin = Demerol = Dispadol = Dolantin = Dolcontral = Dolestine = Dolosal = Dolsin = Isonipecaine = Mefedina = Pethidine

Analgesic, sedative, narcotic.

Untoward effects: Addiction and Abuse: Myopathy in an **addict** who used 1 g/day/10 years. Stress and frustrations of practice has led to dependence on it in many **physicians** and **anesthetists**. Increasing difficulty in flexing elbows fully (18 months) and knees (8 months) in a 60-year-old **male** self-injecting up to 1 g/day/several years. Biopsy of left rectus femoris and right tricep showed gross widespread interstitial fibrosis, almost completely replacing muscle fibers in some areas, and broad bands of mature collagen, with some atrophic muscle fibers distributed at random at injection sites. **Addicts** who have exceeded 200 mg every 2 h have had serious convulsive reactions.

Cardiovascular: Orthostatic hypotension after 1 h; initial increase in mean systemic arterial pressure, vascular resistance and heart rate, followed by a decrease in these after 10–15 min, with decrease in cardiac output and stroke volume. Palpitations, cardiomyopathy, leukopenia, and syncope.

Clinical test interference: Causes increased or false positive sulfobromophthalein and amylase results.

Fetus and neonate: Readily passes through the placenta and detected in **fetal** blood within 2 min of IV to the **mother**. Less than 1% of **maternal** IM dose reaches the **fetus**. If more than 1 h since IM to **mother**, the **newborn** is significantly depressed and respirations are decreased. Obstetrical use may increase **infantile** colic. Multiple congenital anomalies reported when it has been used daily in early pregnancy. Only traces appear in breast milk (< 0.1 mg/100 ml, compared to plasma or serum levels of 0.07–0.1 mg/100 ml). ***Narcotic*** withdrawal syndrome with hyperkinesia, hyperirritability, vomiting, and mucous hypersecretion in **neonate**. Although it is useful in obstetric practice, its effect on the **fetus** is related to placental transfer, **maternal** and **fetal** metabolization, and degree of **fetal** central nervous system penetration.

Hypersensitivity: Disorientation, delusions, perivenous skin erythema, itching, decreased blood pressure, headache, xerostomia, vomiting, dyspnea, and shock.

Miscellaneous: Mild nausea, vomiting, hypotension (especially in the **elderly**), xerostomia, giddiness, sweating, apprehension, syncope, miosis, cold and clammy feeling after 20 min, anorexia, depression (especially in the **elderly**), lens opacities, decreased lower esophageal sphincter pressure, seizures (especially in **children**), reversible parkinsonism, tightness in chest, urinary incontinence, gynecomastia, and increased lactate dehydrogenase. A **gunman** entered a pharmacy and consumed toxic amounts. **He** became comatose for 3 days, avoiding harm to employees. Deaths in 2/10 severe **asthmatics**. A 19-year-old **female** repeatedly within 20–30 s had a ***licorice*** taste in

her mouth after injections of 25–50 mg. Use may aggravate glaucoma. Biliary duct spasms are common. Localized venous thrombosis, transient spasms, skin-blanching, and pain at injection site. Accumulation of its toxic metabolite, ***normeperidine*** (15–20 h half-life), occurs leading to dysphoria, tremors, myoclonus, occasional seizures, and, eventually, decreased renal function. In the **elderly**, it can impair mental performance and induce confusion and delirium.

Nervous system: Can cause or accentuate tremors. Headache, weakness, myoclonic jerks, convulsions, delirium and hallucinations (especially in postoperative period), drowsiness, dizziness (~60%), euphoria, dysphoria, and paradoxical excitement. As pain therapy via controlled pump in 82-year-old **female** led to confusion within 48 h and totally disoriented after 96 h, but no ill-effects afterwards from ***morphine***.

Respiration: In doses that significantly decrease pain, respiratory depression starts within 20 min, peaking after 60 min, and remaining up to 2 h. Apnea (often transitory) during anesthesia reported. Dyspnea, cough, pharyngitis, and cyanosis. Allergic life-threatening reactions in 15-year-old **female**, with respiratory arrest, itching, and eye swelling.

Skin: Eruptions, thrombocytopenic purpura, facial flushing, and allergic rash at injection site.

Therapeutic serum levels are reported as 0.03–0.1 mg%; toxic at 0.5 mg%; and lethal at 3 mg%.

LD_{LO}, **human**: 5 mg/kg.

Drug interactions: ***Anesthetics***: Anoxia in mature **infants** of **mothers** on the combination for delivery. Respiratory acidosis, hypoxia, and hypotension in the **mothers** premedicated with meperidine given ***halothane*** by inhalation; slight tendency toward tachycardia when given with ***nitrous oxide***. The combination of meperidine, ***promethazine***, and ***chlorpromazine*** for sedation/anesthesia caused prolonged sedation (> 7 h) in 63/95 **pediatric patients**.

Antacids and Sodium bicarbonate: Potentiates its oral action by increasing its absorption.

Atropine: Combination causes some increase in apprehension, dizziness, excitement, tachycardia, and hypotension; increased effect of ***atropine***.

Chlordiazepoxide: Significant respiratory depression, lasting 2 h.

Chlorpromazine: Marked lethargy and decrease in systolic and diastolic blood pressure by enhancing meperidine effects.

Cimetidine: Markedly decreases meperidine clearance and may be particularly serious in **patients** with renal failure.

Diazepam: Danger of cardiovascular collapse when meperidine IV is used with it. IV combination causes respiratory depression and transient comatose state (especially in the **elderly**).

Isoniazid: Can cause an additive anticholinergic effect and be dangerous in glaucoma.

Levallorphan: Combination has caused a slight increase in adverse effects.

Mebanazine: Death, following shock and pulmonary hemorrhage in 73-year-old **male** given the drugs as premedication for routine bladder examination.

Miscellaneous: ***Nitrates*** and ***nitrites*** increase its effects. ***Phenothiazines*** increase its effect and have caused shock and **fetal** death, as well as enhancing its sedative and respiratory depression effects. Use with ***tricyclic antidepressants*** is dangerous in glaucoma.

MOAIs (***Iproniazid, Isocarboxid, Nialamide, Pargyline, Phenelzine, Tranylcypromine,*** etc.): Serious reactions occur, resembling meperidine overdosage (respiratory depression, cyanosis, hypotension, coma) or can cause immediate excitement, restlessness, violence, muscle twitching, sweating, Cheyne–Stokes respiration, coma, shock, rigidity, hyperpyrexia, severe hypertension, headache, myocarditis, cerebral hemorrhage, increased tendon reflexes, convulsions, and death. These reactions are likely to occur for up to 2 weeks after discontinuing ***MOAI*** therapy, due to their inhibition of enzymes that break down meperidine. Acidification of urine may help increase elimination of unchanged meperidine. A life-threatening interaction with ***selegiline*** in 56-year-old **male** with Parkinson disease reported. ***Furazolidone***, by its ***MOAI*** properties, can also potentiate its effects after 3 or 4 days of use. Use with ***iproniazid*** has caused malignant hyperpyrexia, a variety of neurological symptoms, excitation, and coma.

Phenazocine: The combination causes respiratory depression postoperatively.

Promazine: Several severe cases of hypotension and collapse following its use during labor. Despite numerous successful trials, reports of decreased Apgar scores and **fetal** deaths exist.

Ritonavir: Decreased meperidine metabolization and can increase its toxicity.

Overdoses (twice recommended dosage in **cats**) may cause excitement in many species. Increased heart rate in **cats** and **horses**. It is metabolized in the liver, therefore, decrease dosage in known or suspected hepatic disease. Has caused sudden death in treated **monkeys**.

MEPHENESIN

= *Atensin* = *Avosyl* = *Avoxyl* = *BDH-312* = *Curythan* = *Daserol* = *Decontractyl* = *Dioloxol* = *Glyotol* = *Glykresin* = *Kinavosyl* = *Lissephen* = *Memphenesin* = *Mepherol* = *Mephesin* = *Mephson* = *Mervaldin* = *Myanesin* = *Myanol* = *Myodetensine* = *Myolysin* = *Myopan* = *Myoserol* = *Myoten* = *Oranixon* = *Prolax* = *Relaxar* = *Relaxil* = *Renarcol* = *Rhex* = *Sansdolor* = *Sinan* = *Spasmolyn* = *Stilalgin* = *Thoxidil* = *Tolansin* = *Tolax* = *Tolcil* = *Tolhart* = *Tolosate* = *Toloxyn* = *Tolserol* = *Tolseron* = *Tolulexin* = *Tolulox* = *Tolyspaz* = *Walconesin*

Skeletal muscle relaxant.* The *carbamate* form = *Tolseran.

Untoward effects: Use discontinued, due to adverse effects, such as eosinophilia, pulmonary infiltrates, and, with the *carbamate* form, **brunettes** turn blonde in ~3 months, reverting to its original color ~3 months after treatment withdrawal. Transient decrease in intraocular pressure, ocular palsies, diplopia, and nystagmus. Possible case of induced thrombocytopenic purpura reported. The *carbamate* form has caused lassitude, fever, leukopenia, anorexia, nausea, vomiting, allergic reactions, vertigo, and syncope. Overdoses lead to blurred vision, numbness of mouth and lips, hypotension, heart-block, and respiration paralysis. May cross-react with ***carbromal***.

Causes paralysis in **cats**, **mice**, and **chicks**.

LD_{50}, **rat**: 945 mg/kg; **mouse**: 890 mg/kg; **hamster**: 821 mg/kg.

MEPHENOXALONE

= *AHR-233* = *Control-Om* = *Dorsiflex* = *Dorsilon* = *Ekilan* = *Lenetran* = *Methoxadone* = *Methoxydon(e)* = *Metoxadone* = *OM-518* = *Placidex* = *Riself* = *Tranpoise* = *Trepidone* = *Xerene*

Anxiolytic, skeletal muscle relaxant.

Untoward effects: Nausea, vomiting, headache, insomnia, somnolence, dizziness, anxiety, tinnitus, drowsiness, dry mouth, dysuria; occasionally skin eruptions, leukopenia, and thrombocytopenia. Analgesic actions of ***aminopyrine***, ***dihydrocodeine***, and ***morphine*** are enhanced by it.

LD_{50}, **rat**: 3.8 g/kg; LD_{LO}, **dog**: 480 mg/kg.

MEPHENTERMINE

= *Mephine* = *Wymine*

Anti-hypertensive, α-adrenergic agonist.

Untoward effects: Abused, with paranoid or paranoid hallucinatory syndromes from ingestion of 2–6 inhalers' contents/day/3–7 days. Inadvertent IA injection of 45–120 mg in a **drug addict** attempting to inject **his** left antecubital vein has also caused dull pain in distal forearm, ice-cold hand and fingers, cyanotic nail beds, weak left radial pulse, and lack of ulnar pulse. Head ache, flushing, nervousness, palpitations, and tremors reported.

Drug interactions: Hypertensive crisis is exaggerated by it when used with ***monoamine oxidase inhibitors***. ***Methyldopa*** potentiates its action and ***reserpine*** antagonizes it. Antagonizes the hypotensive effect of ***guanethidine*** and ***bretylium*** in **man** and **animals**. Arrhythmias occur when used with ***halothane***. Its effectiveness is decreased by ***guanethidine***. ***Phenelzine*** enhances its effect and causes hypertension and tachycardia. ***Reserpine*** decreases its effect.

IP, LD_{50}, **mice**: 100 mg/kg.

MEPHENYTOIN

= *3-Ethylnirvanol* = *Gerot-Epilan* = *Insulton* = *Mesantoin* = *Mesontoin* = *Methoin* = *Phenantoin* = *Sacerno* = *Sedantoinal* = *Triantoin*

Anticonvulsant.

Untoward effects: Lymphadenopathy, aplastic anemia, hemolytic anemia, bone marrow depression, fever, vertigo, muscular incoordination, ataxia, gastric discomfort, hyperirritability, restlessness, weight decrease, dermatitis, hyperpigmentation, drowsiness, agranulocytosis, pancytopenia, eosinophilia, leukocytosis, plasmacytosis, systemic lupus erythematosis, arthritis of ankles, polymyalgia, polyarthralgia, hemolytic jaundice, toxic epidermal necrolysis, weight increase, edema, photophobia, conjunctivitis, gum hyperplasia, lymphoma-like syndrome, nephrotic syndrome (with swollen lymph nodes, splenomegaly, and exanthema in 2½-year-old **male**), and thrombocytopenia reported. A 3-year-old **male**, after 3 months treatment, developed nephrotic syndrome with hemorrhagic skin rash and lymph gland enlargement. Hypersensitivity reaction with coughing, eosinophilia, pleural effusions, sore throat, diffuse macular rash (15%), and increased serum transaminases in 20-year-old **female**. ***Vitamin D*** deficiency and osteomalacia after chronic use. Extreme pancytopenia and aplastic bone marrow in 10-year-old **female** with no prior blood checks during 8 months of treatment. "Fetal hydantoin syndrome" in **infant** of **mother** treated with it during pregnancy. May cause lens opacities.

TD_{LO}, **woman**: 1.1 g/kg.

Drug interactions: ***Barbiturates*** decrease its effect. ***Disulfiram*** increases its effect.

LD_{50}, **rat**: 850 mg/kg, IP, 270 mg/kg; **mouse**: 550 mg/kg; **cat**: 190 mg/kg; **guinea pig**: 380 mg/kg; **rabbit**: 430 mg/kg.

MEPHOBARBITAL

= *Isonal* = *Mebaral* = *Methylphenobarbital* = *Phemiton* = *Prominal*

Untoward effects: ***Cardiovascular***: Bradycardia, syncope, hypotension, and megaloblastic anemia.

Central nervous system: Drowsiness, somnolence (1–3%), agitation, confusion, hyperkinesia, ataxia, nervousness, insomnia, anxiety, dizziness, nightmares, hallucinations, psychiatric disturbances, and headache.

Gastrointestinal: Nausea, vomiting, and constipation.

Miscellaneous: Hypersensitivity (angioedema, rashes, exfoliative dermatitis, fever, liver damage). After 100 mg/day/ several months, a 71-year-old **female** developed submassive hepatic necrosis.

Converted in **man** to ***phenobarbital*** (q.v.). Overdoses lead to respiratory and central nervous system depression, slight miosis (paralytic mydriasis in severe cases), oliguria, tachycardia, hypotension, hypothermia, and coma. In very severe cases, apnea, circulatory collapse, respi-ratory arrest, and death can occur. *In utero* exposure may have caused cleft lip and/or palate in a **fetus**, as well as a heart defect in **another**.

LD_{LO}, **human**: 50 mg/kg; TD_{LO}, pregnant **woman**: 3.1 g/kg.

Drug interactions: ***Alcohol, antihistamines, tranquilizers***, and other ***central nervous system depressants*** lead to increased central nervous system depression. With ***anticoagulants***, decreased prothrombin time; decreased effect of ***corticosteroids, digitalis, digitoxin, doxycycline***, and ***tricyclic antidepressants***. ***Monoamine oxidase inhibitors*** increase its effect. It decreases ***griseofulvin*** absorption and the effectiveness of ***oral contraceptives***.

LD_{50}, **mouse**: 300 mg/kg.

MEPINDOLOL

= *Betagon* = *Corindolan* = *Mepicor* = *SH-E-222*

Antihypertensive, antianginal, β-adrenergic blocker.

Untoward effects: IV use leads to 27–36% drop in portal pressures, 14% decrease in heart rate, 8% decrease in cardiac output. ~1% of **maternal** dose appears in breast milk.

MEPIVACAINE

= *Carbocaine* = *Chlorocain* = *Meaverin* = *Mepicaton* = *Mepident* = *Mepivacine* = *Mepivastesin* = *Optocain* = *Scandicain*

Local anesthetic.

Untoward effects: A few **patients** became nauseous and vomited. Moderate hypotension (4%), tremors, blurred vision, miosis, anxiety, tinnitus, and allergic reactions occur. Bradycardia, ventricular arrhythmias, decreased cardiac output, and heart-block from excessive dosage or accidental intravascular injection. Drowsiness, lassitude, and amnesia systemic reactions are less than with ***lidocaine***. Rapid placental transfer has caused high blood levels in the **fetus**, as well as **fetal** bradycardia. Intoxication at birth from paracervical use leads to severe ataxia, dysarthria, and delayed development. **Fetal** and **neonate** apnea, bradycardia, convulsive movements, flaccidity, areflexia, mydriasis, occasionally tachycardia, cyanosis, hypotension, and depression occur. Sialism, tranquilization, and drowsiness from injection use in oral surgery. A 53-year-old **female** developed excitation and restlessness after 1.3 g topically. Retrobulbar use in cataract extraction leads to rise in intraocular pressure in 30% of **patients**. Therapeutic serum levels are 0.25 mg%, and toxic at 1.0 mg%. Use caution in its use as it is very toxic.

LD_{50}, **rat**: 500 mg/kg; IV, 30 mg/kg; SC, **mouse**: 300 mg/kg; IV, 35 mg/kg; IV, **rabbit**: 53 mg/kg.

MEPROBAMATE

= *Acroban* = *Amosene* = *Andaxin* = *Aneural* = *Artolon* = *Atraxin* = *Ayeramate* = *Bamo* = *Biobamat* = *Calmiren* = *Cap-O-Tran* = *Cirpon* = *Cyrpon* = *Desabam* = *Ecuanil* = *Equanil* = *Fas-Cile* = *Gadexyl* = *Holbamate* = *Kesso-Bamate* = *Klort* = *Mar-Bate* = *Mepavlon* = *Meposed* = *Meprin* = *Meprindon* = *Meproban* = *Meprocompren* = *Meprol* = *Meprospan* = *Meprotabs* = *Meprotan* = *Meproten* = *Meprotil* = *Meptran* = *Mesmar* = *Miltaun* = *Miltown* = *Morbam* = *My-trans* = *Nervonus* = *Oasil* = *Panediol* = *Perequil* = *Pertranquil* = *Placidon* = *Probamyl* = *Procalmdiol* = *Procalmidol* = *Promate* = *Qidbamate* = *Quaname* = *Quanil* = *Reostral* = *Restenil* = *Robamate* = *Setran* = *SK- Bamate* = *Sowell* = *Tamate* = *Trankvilan* = *Tranlisant* = *Tranquilan* = *Tranquiline* = *Urbilat* = *Viobamate*

Sedative, anxiolytic. **Introduced in 1955; 10 years later, 600 tons consumed in the U.S.**

Untoward effects: Sedation is most common especially with doses > 1.2 g/day. Hypotension, edema, anorexia, dizziness, stupor, depression, apathy, personality changes, collapse, urinary frequency, ptyalism, thirst, coma, and respiratory failure occur.

Addiction and withdrawal problems: A 40-year-old **female** with severe psychosomatic pain for 3 years ingested 400 mg qid, leading to visual hallucinations, confusion, disorientation, and memory loss after sudden withdrawal. Epilepsy with grand mal seizures and transient weakness of face and upper extremities after sudden withdrawal of 400 mg/tid/3 weeks. Withdrawal after prolonged use of 25, 400 mg tablets/day caused fatal cardiovascular collapse with hemoglobinuric nephrosis.

Blood: Agranulocytosis, thrombocytopenia, leukopenia, thrombocytopenic purpura, pancytopenia, eosinophilia, leukocytosis, and neutropenia. A fatal case of aplastic anemia in a **female** taking 200 mg/day/1 week. Other cases reported with 800–1200 mg/day/8 days-1 year.

Cardiovascular: Tachycardia, arrhythmias, palpitations, hypotension, and facial flushing.

Clinical tests: Falsely high ***17-ketosteroid*** and ***17-hydroxysteroid*** results, due to high urinary chromogen levels. Can increase chemical urinary ***estriol*** results.

Eye: Concentric constriction of the visual field. Blurred vision, mydriasis, and temporary diplopia.

Fetus and neonate: Conflicting results reported; increase in severe **fetal** malformations when used during the first 42 days of pregnancy, such as cleft lip and palate, Down's syndrome, malformed joints, omphalocele, defective anterior abdominal wall, malformed diaphragm, dislocated hip, retarded development, heart disease, and behavioral disorders. In **human** breast milk at levels comparable to **maternal** plasma levels, and, due to the **neonate's** immature hepatic conjugating mechanisms, it can lead to jaundice.

Hypersensitivity: Anaphylaxis, erythematous rash, chills, fever, nausea, vomiting, leukocytosis, itching, and dyspnea reported. Occasionally joint swelling or edema of lower extremities and rice-water diarrhea.

Miscellaneous: Rhabdomyolysis, gastrointestinal upsets, "feeling drunk", weakness of extremities, profuse sweating, hypotension, cardiovascular collapse and death from suden withdrawal, headaches, impotence, menstrual irregularities, nasal stuffiness. A 400 mg tablet may permit a **person** to lie without alerting a polygraph's breathing pattern, blood pressure, pulse, and skin resistance centers. Continued use stimulates its own metabolism. Has formed gastric concretions in **man** and **rats**.

Poisoning and suicide: A 46-year-old **male** ingested 20 g and 500 ml of a ***cellulose*** solvent, leading to deep coma, cyanosis, hyperpnea, hemoptysis, pulmonary edema, and shock. After 14.4 g ingested by a 19-year-old **female**, she developed 60 h coma, followed by a polyneuritis of **her** legs, lasting several months. Two **females**, 17 and 18, after ingestions of 2 g each, suffered respiratory and cardiac arrest, followed by anuria and cardiac weakness. Hundreds of suicide attempts, some fatal, reported. Serum levels of 180–200 μg/ml lead to deep coma. Serum levels of 0.8–2.4 mg% are considered therapeutic; 5–10 mg% as toxic, and 14–35 mg% (or 1.4–3.5 mg/dl) potentially lethal. Other reports indicate these numbers should be considerably less (up to $^1/_{10}$).

Skin: Bullous lesions, urticaria, itching, acute intermittent porphyria, exanthema, purpura, fixed-drug eruptions, Stevens–Johnson syndrome, and erythema multiforme. Urticarial or erythematous maculopapular eruption may occasionally become purpuric. Ingestion of 600 mg led to sudden onset of purpura with dermatitis, pyrexia, and oliguria, with fatal results.

Drug interactions: Addiction potential and greater depressive effect when used with ***alcohol, ethyl*** and the latter inhibits the rate of the former's metabolization. Minimal interaction with ***coumarins*** and ***warfarin*** until the fourth week, with little clinical significance, due to increased metabolism of ***coumarins***. By inducing liver enzymes, it may cause failure of protective effect of ***oral contraceptives***. ***Barbiturate*** action is potentiated by it. Use with ***benactyzine*** increases drowsiness and vertigo. Fixed-drug reaction in 67-year-old **male**, due to cross-reaction with ***carisoprodol***. Tremors and stupor reported when used with ***promazine***. Muscle relaxation is enhanced when taken with ***monoamine oxidase inhibitors***. Decreased effectiveness of ***warfarin*** and ***oral anticoagulants***.

LD_{LO}, **human**: 500 mg/kg; TD_{LO}, 280 mg/kg.

Withdrawal convulsions can occur in **dogs** on high dosage. Chronic use stimulates its own metabolism in **rats** and **man**, and stimulates the metabolism of other drugs administered to **rats**. Chronic usage decreases prothrombin times in **dogs**. ***Chlorpromazine, glutethimide, phenaglycodol, phenobarbital, phenytoin, triflupromazine***, and ***urethane*** pretreatment increase its metabolism in **rats**. Causes increased fetal weight and increased time to acquire conditioned avoidance behavior in **rat pups**. Causes digital anomalies in the fetus of **mice**. ***Imipramine*** potentiates its sedative effect in **rats**.

LD_{50}, **rat**: 1 g/kg; **mouse**: 750 mg/kg; **hamster**: 1.4 g/kg.

MEPRYLCAINE

= *Oracaine*

Local anesthetic.

Untoward effects: Generalized convulsions, twitching, numbness of extremities, face, and trunk, dysarthria, sweating, euphoria, disorientation, and dyspnea after IV.

MEPTAZINOL

= *Meptid* = *WY 22,811*

Analgesic.

Untoward effects: Nausea and dysphoria.

MERALLURIDE

= *Dilurgen* = *Mercardan* = *Merculıydrin* = *Mercuretin*

***Diuretic*. Contains ~32% *mercury* (q.v).**

Untoward effects*:** Severe, acute fibrinogenopenia in 61-year-old **male** also given ***sodium diatrizoate. Neutropenia and thrombocytopenia also reported. Severe nocturnal leg cramps in 60-year-old **female** after two injections, probably due to ***sodium*** depletion. Significantly decreased serum ***protein-bound iodine*** levels. Can displace ***bilirubin*** from ***albumin***. Can cause SC irritation.

IM, LD_{LO}, **human**: 314 mg/kg total during an 18 days period.

Contraindicated in acute nephritis, severe renal insufficiency, and possible ***mercurial*** hypersensitivity.

IM, LD_{50}, **rat**: 12 mg/kg; SC, 28 mg/kg.

MERBROMIN

= *Asceptichrome* = *Chromargyre* = *D.O.M.F.* = *Flavurol* = *Gallochrome* = *Gynochrome* = *Mercuranine* = *Mercurescein* = *Mercurocol* = *Mercurome* = *Mercurophage* = *Planochrome*

***Antiseptic*. Contains 26.7% *mercury* (q.v.).**

Untoward effects*:** Fatal poisoning in a 5-day **neonate** with giant omphalocele painted with 1% solution every 3 h/three times, then, 2% twice daily. Oliguria on day 2, anuria on day 3 with pink urine and diffuse sclerema. Respiratory arrest on day 5 with blood ***mercury 30 mg/dl. Another similar case with large, infected omphalocele was treated with ***povidone iodine*** dressings qid/2 days, followed by 2% merbromin qid/4 days, then 0.5% until the **neonate** died. Extensive skin peeling with bullous lesions, edema, and fever were noted 3 days after merbromin therapy. On postmortem, ***mercury*** in the blood was 83 µg/100 ml (normal – ≤ 50 µg), brain – 27.5 µg/100 g (normal – 0 µg), kidney – 212 µg/100 g (normal – ≤ 50 µg), liver – 77 µg/100 g (normal – ≤ 6 µg). Eczematous contact dermatitis in 19-year-old **male** after its use in preparation for surgery. Cheilitis from its oral use to identify plaque. See ***mercury*** for additional toxicity.

LD_{LO}, **human**: 50 mg/kg.

2-MERCAPTOBENZOTHIAZOLE

= *Captax* = *Dermacid* = *MBT* = *Mebethisol* = *Mebethysol* = *Mertax* = *Thiotax*

Fungicide, preservative, rubber vulcanizing accelerant.

Untoward effects*:** Contact dermatitis from its use in machine cutting oils and in ***rubber-sensitive **individuals** from ***Spandex***® fibers, ***rubber*** bands, erasers, gloves, boots, condoms, girdles, etc. Also see ***thiram***.

In **dogs**, allergic dermatitis, ataxia, weakness, and death.

LD_{50}, **rat**: 100 mg/kg.

MERCAPTODIMETHUR

= *Bay 5024* = *Bayer 37,344* = *Draza* = *DRC 736* = *ENT 25,726* = *Esurol* = *H-321* = *Mesurol* = *Methiocarb* = *N-Methyl-carbamate* = *Metmercapturon* = *SD9228*

***Insecticide, molluscicide, bird repellant*. A *carbamate*.**

***Untoward effects*:** Potent emetic for **birds**. Tenseness, ataxia, running and falling, disorientation, ptosis, dyspnea, and opisthotonus in **birds**. In **dogs**, symptoms of toxicity were collapse, severe diarrhea, miosis, congestion of mucous membranes, bradycardia, and irregular pulse.

LC_{50}, **rat**: inhalation 535 mg. LD_{50}, **rat**: 60 mg/kg; **mouse**: 34 mg/kg; **guinea pig**: 13 mg/kg; **redwing black-birds**: 4.6 mg/kg; **mallard ducks**: 12.8 mg/kg; **pheasant**: ~270 mg/kg.

MERCAPTOMERIN SODIUM

= *Diucardyn Sodium* = *Thiomerin Sodium*

***Diuretic*. Contains 33.1% *mercury*.**

***Untoward effects*:** Nephrotoxic. Stomatitis, gastric upsets, nausea, vomiting, diarrhea, febrile reactions, facial flushing, chills, headache, and skin eruptions. Rarely, bone marrow depression, neutropenia, and agranulocytosis. Weakness, hypotension, and hypochloremic alkalosis after chronic usage. IV use can cause severe hypotension, electrolyte imbalance, cardiac arrhythmias, and sudden death.

LD_{50}, **rat**: IM, 15 mg/kg; IV, 250 mg/kg.

MERCAPTOPURINE

= *Leukerin* = *Mercaleukin* = *6MP* = *NSC 755* = *Purine-6-thiol* = *Puri-Nethol* = *Purinethol*

***Antineoplastic, antimetabolite, immunosuppressant*. A pro-drug that must be converted intracellularly into its active nucleotide form, *6-mercaptopurine ribose phosphate (6-MPRP)*. *Azathioprine* is metabolized to mercaptopurine.**

Untoward effects*:** Bone marrow depression, anemia, agranulocytosis and pancytopenia, leading to hemorrhages. Control the level of dosage based on degree of leukopenia and thrombocytopenia. Slight anorexia; nausea, vomiting (25%), and diarrhea in **adults**. Cholestatic jaundice in up to 35% of **patients**, which usually disappears on cessation of therapy. Hepatic necrosis has occurred. If serum ***bilirubin rises, discontinue therapy. Rash, alopecia, photosensitivity, hyperpigmentation of skin, oral pellagra, interstitial pneumonitis, intestinal ulceration, microangiopathic hemolytic anemia in a **newborn**, **fetal** malformations (especially if used during the first trimester) and low birth weight **infants** from **maternal** use, hyperuricemia from rapid cell lysis, headache, panmyelopathy, asthenia, mental confusion, severe chills, fever, dysuria, cystitis, oligospermia, weight gain or loss, increased susceptibility to infections, hematuria, and crystalluria. Possible cause of a fatal, acute, hemorrhagic pancreatitis. Interferes with ***uric acid*** assays and gives increased or false positive ***bilirubin*** results.

Drug interactions*: *Allopurinol inhibits xanthine oxidase, which metabolizes ***6MP*** and requires 25–30% dose decrease of ***6MP***. Adrenal insufficiency followed this combined therapy in 2 **patients**. Potentiates ***warfarin's*** anticoagulant effect. Inhibition of ***warfarin's*** anticoagulant effect reported in 62-year-old **male**. Possible additional bone marrow suppression with ***olsalazine***.

Potentiates hypoprothrombinemic effect of ***warfarin*** and ***ethylbiscoumacetate*** in **rats**. Postnatal treatment of **rats** with high dosage (2 mg/kg/day, SC) leads to severe atrophic muscle degeneration. Causes increased incidence of certain tumors in the hematopoietic system in **rats** and **mice** and chromosomal mutations in **mice**. Facial defects in **chicks**; ophthalmic, limb, and tail defects and fetal deaths in **rats** and **mice**; encephalocele, spina bifida, syndactyly, ectromelia, phocomelia, amelia, and fetal death in **rabbits** and **pikas**. Lethargy, depression, and anorexia reported in treated **dogs**. Resistance to infection and parasitism is decreased in treated **animals**.

IP, LD_{50}, **rat**: 250 mg/kg; **mice**: 157 mg/kg; **hamsters**: 364 mg/kg.

MERCURIALIS ANNUA

= *Annual Mercury* = *French Mercury* = *Girl's Mercury* = *Mercury Herb*

In Europe and, occasionally, in North America.

Untoward effects*:** The fresh plant can be very toxic to **people**, probably from its ***methylamine (q.v.), ***mercurialin*** (a volatile oil), ***trimethylamine*** (q.v.), and ***saponin*** content. Its toxicity is lost when dried or heated. In Italy, where it is used as a laxative, diuretic, and analgesic, its aeroallergens have caused rhinitis, bronchial asthma, and/or hypersensitivity pneumonitis in **man**. Use of it colors urine red.

Sheep grazing heavily on it develop hemoglobinuria, hematuria, ataxia, incoordination, constipation, loss of wool, polyuria, coughing, and death. Post-mortem demonstrates hepatic swelling; centrolobular degeneration and hyperemia; swollen and hyperemic kidneys; pulmonary edema and pneumonic foci; and 100–150 ml of yellow pericardial fluid. **Calves** develop inappetence, severe hematuria and jaundice, kyphosis, and tachycardia. Post mortem findings were similar to those in **sheep**. Anemia, dark red urine, yellowish mucous membranes, and decreased milk production reported in affected **cows**. Affected **horses** have shown icterus and hemoglobinuria.

MERCURIALIS PERENNIS

= *Dog's Mercury*

In Europe.

Untoward effects*:** Its leaves, unfortunately, have been consumed as a vegetable pot-herb in Germany, and its volatile oil, ***mercurialin, also known as ***Oil of Euphorbia***, has caused severe gastroenteritis in **man** and **livestock**. Some **ewes** eating it for a few days developed hemoglobinuria and pale conjunctiva. Post-mortem revealed liver and kidney degeneration, reticular inflammation, and abdominal edema.

MERCURIC ACETATE

Contains ~63% *mercury* (q.v.).

***Untoward effects*:** IP to **hamsters** at 2, 3, or 4 mg/kg during day 8 of gestation leads to fetal growth retardation, SC edema, cleft lip and palate, rib fusion, syndactyly, encephalopathy, and resorption.

LD_{50}, male **rats**: 76 mg/kg; **mouse**: 62 mg/kg.

MERCURIC AMMONIUM CHLORIDE

Contains 53% *mercury* (q.v.).

Antifungal.

Untoward effects*:** Use by **female** African **nurses** in a skin-lightening cream led to urine ***mercury levels of 100–220 µg/l without evidence of renal damage.

MERCURIC CHLORIDE

= *Bichloride of Mercury* = *Bichloro de mercurio* = *Chlorure Mercurique* = *Corrosive Sublimate* = *Mercury Bichloride* = *Mercury Perchloride* = *Quecksilberchlorid*

Sporocidal antiseptic, caustic, parasiticide, irritant absorbant, mordant for furs, and wood preservative.

Untoward effects*:** Can be fatal if ingested, due to extensive gastrointestinal pathology and nephrosis after symptoms of severe nausea, metallic taste, tachycardia, weak pulse, burning in throat, vomiting, hematemesis, abdominal pain, hematuria, oliguria, melena, diarrhea, nasal bleeding, bradypnea, shallow breathing, severe acidemia, leukocytosis, shock, and prostration. Blue-black gum line, liver damage, tremors, and partial face and limb paralysis are delayed and chronic symptoms. Rhabdomyolysis and renal failure after ingesting 2 g in a failed suicide attempt. Poisoning in 3-year-old **male** after drinking milk containing it for a butter fat assay. A ***polyvinyl alcohol

preservative for routine collection of specimens for ova and parasite examination contained 4.5% mercuric chloride. UCLA gave a **mother** three 20 ml bottles with instructions in Spanish and English. **She** misinterpreted the instructions and gave the 4½-year-old **male child** the contents of two bottles, with a total of 1.8 g of mercuric chloride. **He** experienced a burning sensation and vomited immediately. His **mother** took **him** back the next day to the clinic because of abdominal pain and bloody stool. Accidental insufflation of it, instead of ***mercurous chloride*** (***Calomel***), in the eye caused loss of vision and widespread necrosis of conjunctiva. Ingestion of 2.5 g has been associated with abortion in a 31-year-old **female**. Dermatitis from skin contact (medical **personnel**, dust of impregnated wood), blue or brownish lunulae of nails from topical use followed by sunlight. Use in a hair wash caused several cases of coloring of hair, due to formation of black ***mercuric sulfide***. Corrodes metal instruments. Precipitates proteins, including living cells in wounds. Irritant to raw wounds. Commercial preparations are made a distinctive blue color, usually with ***methylene blue***, although the chemical itself is white. In 1982, it was found as a contamination in deliberately contaminated ***Excedrin***®. Another 5-year-old **male** was accidentally given 10 ml of a ***PVA*** preservative with 430 mg of mercuric chloride. **He** immediately experienced abdominal pains and vomited. Despite clinical care, **he** developed acute renal failure and was a month in intensive care. Other cases are in the literature. Fatal cases followed its application (80 grains or 5.2 g/1 oz ***alcohol***) to the scalp for ringworm. Gives spuriously high urinary ***albumin*** readings. During the early part of the twentieth Century, it was the most prevalent poison used in suicides. Mean lethal dose in **humans** is 1–4 g, but fatalities have been reported from as little as 500 mg.

Glomerulonephritis in **dogs**, **horses**, and **rats**; thin egg shells in **quail**, immunosuppression in **rabbits**, and depression of muscle contractility in **frogs**. Methylation dramtically increases its toxicity in **man** and **animals**.

It and ***mercurous chloride*** have been used as disinfectants and insecticides on plant roots. Soil organisms decompose them, liberating ***mercury*** vapors. This is accelerated by increased soil temperature and pH. ***Mercury*** has been recovered from the leaves of plants exposed to soil moistened with it.

LD_{50}, **rat**: 20 mg/kg; **mouse**: 10 mg/kg; LD, **cow**: 4–8 g (9–18 mg/kg); **horse**: 8 g (10–20 mg/kg); **sheep**: 4 g (60–80 mg/kg); **dog**: 100–350 mg; LD_{LO}, **pigeon**: 2.2 g/kg; **rabbit**: 40 mg/kg.

MERCURIC CYANIDE

= *Cianuria*

Contains 79.4% *mercury* (q.v.) plus *cyanide* (q.v.).

Use: In electroplating, instrument disinfecting, and as a topical antiseptic or treatment for syphilis in **man**.

Untoward effects: The mercuric and ***cyanide*** ions are separately toxic and must be simultaneously detoxified in poisoning cases. Emesis, tachypnea, restlessness, convulsions, and ataxia are, primarily, from the ***cyanide***, while hematuria and renal damage are from the ***mercury*** component. A **chemist**, 30-year-old **male** accidentally swallowed 10–15 ml of a 20% solution while suctioning it up in a burette, leading to severe poisoning with gastric ulceration, necrosis of oral mucosa, renal tubular necrosis, and acute renal failure. Hospitalized for 51 days during treatment; found well 5 months later. A 14-year-old **male** developed acute renal failure after ingestion of 1.5 g.

TD_{LO}, **man**: 27 mg/kg; **woman**: 10 mg/kg.

LD_{50}, male **rats**: 26 mg/kg; **dogs**: 2.71 mg/kg.

MERCURIC IODIDE, RED

= *Biyoduro de mercurio* = *Mercury Biniodide* = *Hydrargyrum Iod. Rub.* = *Iodine Mecurique* = *Quecksilberjodid*

***Counterirritant, vesicant, topical antiseptic.* Contains 44% *mercury* (q.v.).**

Untoward effects: Can be absorbed through intact or blistered skin.

LD_{LO}, **human**: 357 mg/kg.

Ingestion from licking by treated **horses** has caused stomatitis, gastrointestinal corrosive damage, diarrhea, steatorrhea, anorexia, alopecia, and renal tubular necrosis.

LD_{50}, **rat**: 40 mg/kg; **mouse**: 80 mg/kg. Estimated LD in **horses**: 10 g.

MERCURIC NITRATE

= *Mercuric Pernitrate*

Use: In felt manufacturing; as a fungicide and miticide; and in ophthalmic use (***Citrine Ointment***). Contains ~62% ***mercury*** (q.v.).

Untoward effects: The terms "Mad Hatter" and "Mad as a Hatter" developed from **hatters** using mercuric nitrate in treating the felt for top hats, with its ensuing encephalopathy.

Freshwater **fish** exposed to 3 mg/l survived, but 71% became blind after 4 weeks of exposure.

LD_{50}, **rats**: ~10 mg/kg.

MERCURIC OXIDE, YELLOW

Red on heating, yellow again on cooling. Contains 92.6% *mercury* (q.v.).

Untoward effects: Pulmonary toxicity in **man** from inhaling the vapors, with absorption at the alveolar and bronchial level, leading to capillary damage, pulmonary edema, and desquamation of surface cells. Chronic exposure leads to central nervous system toxicity and renal insufficiency. Corneal opacity and cataracts have been reported.

LD_{50}, male **rats**: 18 mg/kg.

MERCURIC OXYCYANIDE

Contains 85.5% *mercury* (q.v.).

Untoward effects: Can explode by percussion or flame, unless mixed with an excess of ***mercuric cyanide*** (q.v.) for use in the ***leather*** industry.

MERCURIC SULFIDE, RED

= Chinese Red = C.I. 77,766 = Cinnabar = C.I. Pigment Red 106 = Vermillion

Use: Used to color ***cheese*** rinds and anchovy sauce in Britain until outlawed. In 1831, an article entitled "Poisoned Confectionary" appeared in the *Lancet*, describing its use to color candies and the paper in which they were wrapped. Now, used as a pigment in tattoos, colored papers, and plastic. Contains 86.22% ***mercury*** (q.v.).

Untoward effects: A photoallergic reaction with an intense pseudolymphomatous inflammation reaction occurs from tattooage in hypersensitive **people**. Ulceration and extrusion of these pigmented cells and generalized eczematous eruptions (after minor laceration of tattooed area 9 years after tattooage). In Hawaii, 15 **male** military **personnel** with 20 tattoos developed pruritus and nodular or verrucose lesions after sun exposure. Biopsy showed a sarcoid-like response. Commercial material usually contains some ***cadmium sulfide***, a known photosensitizer, to produce a more intense red color. Unfortunately, has been used topically as a folk medicine treatment for eczema in India. Repeated use of a 4% preparation in a number of **patients** to bleach facial freckles led to gray pigmentation of periorbital region, temples, eyelids, lips, chin; stomatitis with gray gingival margins, cardiac arrhythmias, paresthesias of extremities, colic, insomnia, and uncontrolled aggressiveness, alternating with anxious timidity.

MERCUROPHEN

***Antiseptic, disinfectant.* Contains 53.11% *mercury* (q.v.).**

Untoward effects: A 23-year-old **female** treated with 1 : 7500 as wet dressings for pyoderma developed an acute dermatitis venenata.

IV, LD_{LO}, **rat**: 8 mg/kg; **rabbit**: 4 mg/kg; IM, **rat**: 12 mg/kg.

MERCUROUS CHLORIDE

= Calogreen = Calomel = Précipité Blanc = Mercury Monochloride = Mercury Protochloride = Mercury Subchloride = Mild Mercury Chloride

***Cathartic, diarrhetic, and antiseptic.* Contains 85% *mercury* (q.v.).**

Untoward effects: Dr. Benjamin Rush, a **physician** in Philadelphia, Pennsylvania, and a signer of the Declaration of Independence, used it to remove "poisons" from Yellow Fever **patients** in the 1790s, until **they** began to salivate. **Their** hair and teeth often fell out. **Some** survived the "cure". Its use as a laxative has been criticized. Two **women** who had taken two laxative tablets/day for years, with a daily dose of 120 mg, developed personality changes, tremors, failure of cognition and hearing before their deaths, renal failure, and necrosis of the colon. Chronic ingestion as a laxative leads to dementia, erethism, colitis, and renal failure. Acrodynia, Pink Disease, in **infants** and young **children** has dramatically diminished with its decline in usage in teething powders. Autonomic hyperactivity and increased urinary ***mercury*** levels were symptoms. Will pass into breast milk in sufficient quantities to affect the bowels of ½ the **infants**. A 14-year-old **male** developed Feer's disease (sweating, cyanosis of extremities, motor weakness, tremors, tachycardia, and insomnia, after its repeated use in a nasal ointment. Should be used cautiously in the eye, as it may contain undesirable impurities; and, in the presence of light and moisture, it is unstable and can form ***mercuric chloride***. After oral use, will color feces yellowish-green. ***Crema de Belleza-Manning***, a beauty cream from Mexico containing it, has caused ***mercury*** poisoning in **users** in Arizona, California, New Mexico, and Texas. It is absorbed through the skin. Urinary ***mercury*** levels were > 200 μg/l in 27 **users**. Symptoms included fatigue, memory loss, tremors, weakness, and loss of vision or taste. Another dangerous source is ***Tse Koo Choy*** powder, a traditional Chinese **children's** medication. Bacteria can convert it to ***methylmercury***, a more toxic form, that can cross the placenta.

LD_{LO}, **human**: 5 mg/kg.

Application to a **calf's** eye for keratitis can cause mercurialism. **Cattle** and **hogs** have died from consuming ***oats*** and other seed grains treated with various mercurial fungicides.

TD, **cow**: 8–12 g or 18–36 mg/kg; **horse**: 10–20 g; **sheep**: 1–5 g; **goat**: 1–2 g; **dog**: 0.4–2 g; LD_{50}, **rat**: 210 mg/kg.

MERCUROUS IODIDE

= Mercury Protoiodide = Yellow Mercury Iodide

Contains 61.25% *mercury* (q.v.).

Untoward effects: Now rarely used, but cases of nephrotic syndrome (puffy eyes, swollen ankles, weight increase, and fatigue) in **patients** given it as oral treatment for flat warts. Hypertension and tachycardia also reported. Oral use with soluble iodides forms the more dangerous ***mercuric iodide***.

LD_{50}, **mouse**: 110 mg/kg; IP, 50 mg/kg.

MERCURY

= *Hydrargyrum* = *Liquid Silver* = *Quicksilver*

Untoward effects: "Ashes to ashes, dust to dust. If DDT doesn't get you, then mercury must". ANON It is the only metal that is liquid at room temperature. Many of its adverse effects are due to increased levels in the environment, but, fortunately, many of its compounds are no longer in frequency medical usage. So-called normal serum levels are 0.0–8.0 μg%, toxic levels – 100 μg%, and lethal levels are 600 μg% and above. Other reports cite toxic levels as > 20 μg%.

Blood and cardiovascular: Has lytic effect on **human**, **sheep**, **guinea pig**, and **rabbit** erythrocytes. Causes a hypersensitivity vasculitis or periarteritis nodosa with hemorrhagic, purpural, or necrotic dermal lesions. When systemic, it can involve vital organs, which is often fatal, as well as causing a glomerulonephritis. Has direct injurious effect on the heart, and directly damages the heart muscle by damaging the coronary circulation. Hematopoietic marrow depression; thrombocytopenia and leukopenia improved with mercury elimination. Aplastic anemias, neutropenia, and arrhythmias reported. Hypertension in 9-year-old **male** exposed to it from a borrowed broken sphygmomanometer.

Clinical test interference: Decreases serum ***sodium*** results.

Central nervous system: Excitement, tremors, headache, dizziness, convulsions, coma, ataxia, myoclonus; and bizarre movement disorders preventing walking, sitting, or feeding **oneself**. Early neurological symptoms are often mistaken to be neurotic complaints and the damage is cumulative and irreversible. Distal sensory involvement is often an early symptom.

Ear: Can damage the inner ear. Irrigation of the external ear canal with ***thimerosal*** has caused a death and several instances of increased serum mercury levels. A **patient** vomited as a Miller–Abbott tube containing mercury ruptured while being inserted as an in-dwelling drainage tube.

Eye: Nystagmus, constricted visual field, and impaired visual acuity. Mercury deposits in conjunctiva and cornea. Photophobia.

Fetus: Placental transfer occurred in an **infant** from a **mother** who had eaten mercury-contaminated ***pork***. Mercury was found in the urine, serum, and cerebrospinal fluid of **humans** who ate the ***pork***. Epidemics and case reports of **children** exposed as fetuses that later developed brain damage or neurological disorders. See below details of epidemics, starting in Japan in 1953, in the USSR in 1968, and in Iraq in 1971. Teratogenic. Congenital acquired toxicity affects the **infant's** central nervous system, causing tremors, involuntary movements, irritability, and a high-pitched cry. Abnormal electroencephalogram can occur and spontaneous abortions have been reported.

Miscellaneous: Systemic autoimmune disease in 63-year-old **male** after 35 years exposure while manufacturing fluorescent tubes. Skin burns and blisters on buttock after cleaning with ***thimerosal*** and ***aluminum*** foil diathermy electrode under buttocks. The mercury was a catalyst for the oxidation of the ***aluminum*** with production of heat. Can interfere with uptake of ***radioiodine*** by the thyroid for 7 days. Can decrease the effect of ***coumarin anticoagulants***. Hypersensitivity to it can manifest itself as asthma. Mucoid diarrhea and diarrhea after excessive intake.

Muscle: Brief exposures can masquerade as Lou Gehrig's disease (amyotrophic lateral sclerosis or ALS).

Neonates and children: Poisoning and death have followed from its absorption after use as an antiseptic on umbilical stumps and omphaloceles. Acrodynia, and even a fatal case, from its use in teething powders and ointments for diaper rash. Indoor use of mercury-containing paint caused acrodynia in 5-year-old **male** from inhalation its vapors. A **neonate** received high exposure to mercury from inhaling vapors produced by a broken expansion switch in the heating unit of an **infant** incubator. Acrodynia in 11-year-old **male** from repeated treatment of rashes with 10% ***ammoniated mercury*** ointment. At age 14, **he** developed Guillain–Barré syndrome from 3 years chronic exposure to mercury vapor (~0.17 mg/m^3/~8 h per night). Mexican-American **infants** have, unfortunately, been given metallic mercury for suspected gastrointestinal upsets or *empacho* (food boluses) by **curanderas** (folk doctors). Droplets of mercury and radiographs have confirmed this congenital exposure has caused blindness, cerebral palsy, and mental retardation. Excretion in milk (3–6% of **maternal** plasma levels) can make it hazardous to an **infant**. In Argentina, 1600 were poisoned by dermal absorption, after a laundry used a mercurial disinfectant on **their** diapers.

Poisoning: ***From cosmetics***: See ***Mercurous Chloride*** for Mexican facial cream. Has been used in toothpastes and eye cosmetics. U.S. limits were 65 ppm for use in eye area if no other effective or safe products were available for use.

From foods: The best known is the devastating Minimata disease, named from the Minimata Bay on Kyushu Island in Japan, where mercury-laden industrial wastes were dumped, poisoning **fish**, **people** (**adults** and **babies**), **cats**, **birds**, **rats**, and **mice**. This caused fatal neural disorders. Mercury from industrial wastes is changed by water organisms to the more toxic ***methyl mercury***. Of 6530 hospitalized cases, 459 **people** in Iraq died of eating ***flour*** and ***bread*** from imported mercury-treated ***wheat*** seed. Panic-stricken **farmers** dumped the seed into the Tigris River, and **people** sickened after eating **fish** from these waters. Symptoms included sore mouths, metallic taste, blue gum line, vomiting, diarrhea, polyuria, fever, insomnia, hand tremors, dysarthria, ataxia, Babinski reflex, hyperreflexia, hyporeflexia, limb atrophy, blindness, and deafness. **Fish** from Scandinavian waters raised the mercury content of the red blood cells in **people** who ate them. Levels up to three times tolerances have been found in **fish** from Lake St. Clair, the Detroit River, and Lake Erie. In 1975, analysis of **tuna fish**, **sea perch**, and **swordfish** indicated that pregnant **women** should not eat them, and other **people** should limit themselves to 1 such **fish** meal a week. **Sperm whale** meat, sold in Japan in 1978, had levels six times recommended maximum. Illness in a Georgia **family** was traced to **their** having eaten meat from a **pig** fed on waste feed grain treated with a mercurial fungicide. Mercury was detected in **their** urine, serum, and cerebrospinal fluid. Placental transfer was demonstrated. Gross tremulous movements of limbs was noted in an **infant** at birth, **who**, at 6 months, developed myoclonic convulsions; hypotonic, irritable, and nystagmus at 8 months; and was unable to sit or see 1 year later. **Mother's** ingestion of mercury was estimated at 390 mg daily over a 100 days period. In New Mexico, **members** of a family that ate meat from **hogs** fed mercury-treated seed suffered brain and muscular damage. Exposure from eating mercury contaminated eggs also reported.

From injections: A 20-year-old **male** received ***gamma globulin*** injections with a mercurial preservative for ~15 years and developed symptoms of acrodynia. Fatal poisoning in 23-year-old **female** after deliberate forearm injection of 1–2 ml unsterilized mercury, leading to tender, red, and swollen forearm with median nerve damage in 3 weeks. SC leads to granuloma. Suicide attempt by SC injection. A 34-year-old **male** had toxicity for up to 12 years after an IV. **He** had occasional hand tremors and memory disturbances, and mild neuropathy of the legs. Lung biopsy 4 years after the injection indicated fibrosis. **He** also had a history of asthenozoospermia, and his **wife** had a miscarriage. Acute lung embolization and disease in a young **boxer** after deliberate IV injection.

Oral: Small ***batteries*** in watches, hearing aids, cameras, and calculators have been a source or problems after ingestion, especially in **children**.

Sources: Organic mercurials from antiseptics, viscose rayon plants, fungicides, etc., rather than elemental mercury, are now a key public health problem. Some ***batteries*** contain 5 g ***mercuric oxide***, other dry-cell ***batteries*** contain 20 mg (1/5 minimum lethal dose for a **child**). Paints have been a source of mercurial vapors. ***Copper*** smelters, ***coal***-burning, and garbage incinerators have contaminated the environment with mercury dust. Metallic and/or mercury embolism and/or aspiration has occurred from cardiac catheterization, and blood gas and intestinal tube equipment ruptures or leakage. **Children** may swallow droplets or have prolonged skin contact from a broken thermometer (~0.3 ml mercury). Patent medicines and ***dolomite*** dietary supplements have also been a source. Five pints were found by **children** in an abandoned van in Florida. A total of 477 **persons** were potentially exposed; 86 were evacuated from **their** homes. A school was closed. Extensive decontaminated procedures followed. Some feces were tar-colored.

Vapors and dusts: Spilling it on a floor, carpet, or counter; breaking a thermometer in an incubator, and heating mercury-***gold*** amalgams. Acute poisoning and fatality from respiratory failure after high-density vapor exposure in 7-months-old **female** after heating it in a home. Acrodynia in 5-year-old **male** from water-base paint vapors. Indoor use of a marine anti-fouling paint caused acute stomatitis and laryngotracheobronchitis in 3-year-old **male**; 3 days later, **he** was hospitalized with mild tubular necrosis and transient ataxia. A **mother** and 3 **children** were exposed to its vapors for ~16 h from a freshly-painted gas heater in **their** closed bedroom. One **child** was dead on arrival at the hospital, the second died in 30 h, and the third on the fifth day. The **mother** was discharged on the twenty eighth day after treatment. **All** had symptoms of tachypnea, cough, fever, gastrointestinal upsets, and central nervous system problems. Post-mortem revealed erosive bronchitis and bronchiolitis with severe interstitial pneumonitis. Vapor poisoning has simulated mucocutaneous lymph node syndrome with tenderness of palms and soles, generalized rash, desquamation, fever, and cervical and axillary adenopathy. High incidence of mercury poisoning in some dental offices (recirculation of air), **dentists**, and **dental personnel**. Vapors from a contaminated carpet led to hyperthyroidism in a **dentist** after several years exposure. See ***Neonates and children*** for toxicity related to a faulty incubator switch.

Renal: Nephrotoxic syndrome, anuria, and oliguria.

Skin, mouth, and nails: A number of mercurials are red, and that should be a reminder of their potential danger.

Has caused contact sensitization, particularly from ***ammoniated mercury***. Lichenoid eruptions, exfoliative erythroderma, erythema multiforme, exanthematous eruptions, fixed-drug eruptions, anaphylactoid or serum-sickness-like reaction, urticaria, and vesico-bullous reactions reported. Dermatitis reported after using mercurial ointments for lice and around ***cinnabar*** tattoos. Cheilitis from hypersensitivity to dental amalgams, and mouthwashes. Oral pigmentation, stomatitis, sialism, and metallic taste have followed systemic absorption, and imbedded amalgam fragments in gingival tissue ("amalgam tattoo"). Facial pigmentation has followed prolonged use of a mercurial facial cream. Black mottling of teeth and brown-black discoloration of **carroter's** finger and thumb nails from ***mercuric nitrate*** treatment of fur for felt hats. Local application of mercurials can affect the nail bed and turn nails brown. Toxic exposures can cause clubbing of nails.

National Institute for Occupational Safety and Health, vapors, 0.05 mg/m^3; TWA, immediate danger to life or health, 10 mg/m^3.

Mercury has been found in high concentrations in some wild **animals** and **birds**. Accumulates in eggs, feathers, liver, and kidneys of **birds**. **Raptors** that fed on these **birds** developed weakness, incoordination, and weight decrease. Neurological changes, ototoxicity, and death in **cats**. See Minimata discussion above. A **cat** died as the result of a thermometer breaking in its rectum. **Cattle** are highly susceptible - with eczema; petechia in mucous membranes, digestive tract, and uterus; nervous symptoms, tachycardia, and renal failure. **Dogs** have died from eating meat from **pigs** poisoned by it. **Cats**, **dogs**, **cattle**, and **horses** have become toxic from licking mercurial ointments. Two **dogs**, 200 **chickens**, and five **people** died from inhalation of vapors after it was heated on a kitchen stove. Residues in **fish** have helped kill American **bald eagles**. Mercury in rivers and lakes is converted by microorganisms into more toxic ***methyl*** and ***dimethylmercurys***. See ***Poisoning, From Foods*** discussion above. Little excretion in **cow** and **goat** milk. A **horse** has been poisoned by excessive licking of a blistering agent, ***Mercuric Iodide, Red*** (q.v.). A large number of Swiss army **horses** treated topically with mercurial ointments developed papular eruptions, exanthema, itching, desquamation, inappetance, and edema of the prepuce. Vapors cause cardiac muscle damage and renal damage in **rabbits** and **man**. Acute poisoning has caused fatal renal damage. Mercury-treated ***corn*** or seed causes **hog** cholera-like symptoms and postmortem lesions, and blindness in **swine**. **Ferrets** have died after consuming **poultry** meat from asymptomatic **birds** fed mercury-treated ***wheat***. Placenta passage reported in **monkeys**.

MERCURY, AMMONIATED

= Aminomercuric Chloride = Mercuric Chloride, Ammoniated = Mercury Amide Chloride = Mercury Ammonium Chloride = White Precipitate = White Mercuric Precipitate

Untoward effects: Contact sensitization is common. Highly poisonous. Accidental ingestion has caused gastric pain, nausea, and purging. Healthy 9-month-old **female** with diaper rash treated with it (2.5% ointment) topically for 2 weeks leading to renal tubular acidosis, self-limiting with acute anorexia, vomiting, constipation, polyuria and weight decrease, increased plasma chloride, decreased plasma bicarbonate, decreased blood pH, and persistent alkaline urine, with ***mercury*** level of 30 μg/l. Pink disease in 11-month-old **patient** with irritability, photophobia, pyrexia, lassitude, increased systolic blood pressure, sleeping pulse of 160/min, sluggish knee-jerk reflexes, pink swelling of hands and feet, and excessive sweating after use of 1 oz of 2% ointment for > 10 days in treatment of seborrheic dermatitis. A month later, another ounce was used for > 1 week. Resultant toxicity required hospitalization. Slate-gray pigmentation of skin in 79-year-old **female** using it (4%) at night, every night, for 50 years. Motor polyneuropathy, irritability, and eosinophilia in 4-year-old **male** treated for 2 months on scalp and body for insect bites. Nephrotic syndrome in Bantu 48-year-old **female** after possible use of 288 g mercury during 8 years. May cause graying of nail plates. Prolonged topical use in **females** for at least 50 years (***Kremola***, ***Nodinole***, ***Stillman's Bleach Cream***) led to discoloration of skin (hydrargyria) in 3/3, neuropsychiatric changes in 2/3, chronic thrombocytopenic purpura in 1/3, and laryngeal ulceration in 1/3.

LD_{LO}, **human**: 5 mg/kg.

Gradual loss of body condition, ataxia, and paresis in **cattle** ineffectively treated with it (5%) topically for ringworm.

MERSALYL

= Mercuramide = Mercusal = Mersalin = Salyrgan

Diuretic.

Untoward effects: Nephrotic syndrome after 9 months of IV therapy with classic features of cardiac failure. Stomatitis, gastrointestinal upsets with occasionally bloody diarrhea, vertigo, headache, fever, cutaneous reactions, and thrombocytopenia. Anaphylactic reactions reported after injection. Has been used in a fixed 2 : 1 combination with ***theophylline***, as the latter increases its absorption, enhancing immediate and total urinary output and decreasing local irritation. Mercurial periodontitis in 38-year-old **female** with postural edema receiving weekly injections for ~5 years.

IV, LD_{50}, **rat**: 17 mg; IM, 24 mg/kg; IV, **mouse**: 73 mg/kg; IM, 97 mg/kg; IP, 118 mg/kg; IM, **rabbit**: 24 mg/kg.

MERULIUS LACRYMANS

Untoward effects: Positive skin reactions to this mold in 60 **patients**, 14–60 years, with bronchial asthma.

MERYTA sp.

= *Lau Fagufagu*

Untoward effects: A Samoan piscicide.

MESACONITINE

An ester of *aconitine*.

Untoward effects: Slightly more toxic in the **mouse**, **rabbit**, and **frog** than ***aconitine*** (q.v.).

MESCALINE

= *Big Chief* = *Mesc* = *Mezcaline* = *TMPE* = *TMPEA*

***Psychomimetic, hallucinogen.* The active crystalline alkaloid in *peyote* (q.v.). Also found in *Agave cacti* from which a Mexican *beer, Pulque*, is made, and in the San Pedro cactus, *Trichocereus pachanoi*.**

Untoward effects: Hallucinations, nausea, vomiting, tremors, perspiration, deep sleep, vertigo, headache, chest pains, poor coordination, euphoria, time distortion, hyperflexia of limbs, decreased blood pressure, dryness of lips and tongue, flushing, focusing difficulties, scanty urine, cardiac area pain, irregular pulse, dyspnea, indigestion, nasal discharge, numbness, and giddiness. Serious cases may also have severe diarrhea and body odor. Motor paralysis by central nervous system depression and death by respiratory failure. It was made famous by Aldous Huxley, who, in 1954, wrote a book under its influence. On the "street", mescaline is rarely mescaline. It is apt to be ***LSD***, ***PCP***, or ***STP***.

TD_{LO}, **human**: oral and IV, 5 and 7 mg/kg, respectively.

Causes tremors in **cats**; hyper- and hyporeactivity, ataxia, weakness, tremors, bizarre postures, clonic convulsions, salivation, piloerection, dyspnea, and hyperpnea in **mice**; the same, except for salivation, in **rats**; hyporeactivity, bizarre postures, salivation, catalepsy, ataxia, weakness, tremors, convulsions, mydriasis, and piloerection in **dogs** and **monkeys**. The latter did not show mydriasis, but became very vocal. Repeated IV doses to **dogs** led to bradycardia. Use in pregnant **hamsters** led to decreased growth rate of neonates. In **guinea pigs**, increased fetal resorption rate, decreased intrauterine fetal growth rate, and increased central nervous system defects.

LD_{50}, **rat**: 132 mg/kg; **guinea pig**: 328 mg/kg; **mouse**: 212 mg/kg; IV, LD_{50}, **dog**: 54 mg/kg; **mouse**: 110 mg/kg; IP, LD_{50}, **rat**: 270 mg/kg; estimated LD_{50}, **monkey**: 130 mg/kg.

MESEMBRYANTHEMUM CRYSTALLINUM

= *Diamond Fig* = *Diamond Plant* = *Ice Plant* = *Khanna*

Untoward effects: In South Africa, ingestion of this plant leads to torpor, due to its alkaloid content. It is uncertain if this is the same ***khanna*** that South African **Hottentots** chewed hundreds of years ago to induce visions. Now, found in Europe, Australia, and California. Has a saline, somewhat nauseating taste.

M. expansum contains an alkaloid, ***mesembrine***, with ***cocaine***-like activity.

MESITYL OXIDE

***Solvent*. Over 1 million lbs production annually in the U.S.**

Untoward effects: Eye, skin, and mucous membrane irritant. Inhalation of vapors causes headache, dizziness, nausea, dyspnea, narcosis, and coma. Chronic exposure leads to dryness, irritation, and inflammation of the skin. Exposure limits set by OSHA are 25 ppm and 10 ppm (40 mg/m^3) by National Institute for Occupational Safety and Health. ***Honey***-like odor detectable by **humans** at 0.45 ppm.

Inhalation TC_{LO}, **human**: 25 ppm, eye irrition 25 ppm/15 min.

Acute inhalation by **mice** and **rabbits** leads to narcosis, liver necrosis, pulmonary hemorrhage and edema, intestinal distension, and death. Subchronic inhalation by **rats**, **mice**, and **guinea pigs** leads to liver congestion and kidney tubule pathology.

LD_{50}, **rat**: 1.12 g/kg; **rabbit**: 1 g/kg; Inhalation LC_{LO}, **rat**: 1,000 ppm/4 h.

MESNA

= *D-7093* = *Mesnex* = *Mesnum* = *Mistabron* = *Mistabronco* = *Mucofluid* = *UCB-3983* = *Uromitexan*

***Mucolytic*. Also used IV for *ifosfamide* and *cyclophosphamide*-induced hemorrhagic cystitis. Produced naturally by intestinal bacteria.**

Untoward effects: Abdominal discomfort, nausea, vomiting, diarrhea, headache, fatigue, limb pain, and hypotension. IV leads to rare allergic reactions – 0.001% (fever, sinus tachycardia, conjunctivitis, increased aminotransferases). Itching, urticaria, toxic encephalopathy, and a bad taste effect in the mouth. Use may cause false positive ***ketone*** results on ***Labstix***® reagent test.

MESORIDAZINE BESYLATE

= *Lidanar* = *Lidanil* = *NC 123* = *Serentil* = *TPS 23*

Psychotherapeutic.* A metabolite of *thioridazine.

Untoward effects: Extrapyramidal (Parkinson-like) symptoms, drowsiness, weakness, occasionally constipation, transient granulocytopenia, leukopenia, eosinophilia, thrombocytopenia, pancytopenia, anemia, aplastic anemia, headache, dry throat, akathesia, blurred vision, lens and corneal opacities, miosis, dizziness, slight confusion, tremors, allergic rhinitis, nasal mucosal swelling, photophobia, facial edema, fatigue, exanthema, vertigo, tachycardia, hypotension, collapse, palpitations, lymphocytosis, electrocardiogram abnormalities on 75 mg/day, insomnia, galactorrhea, inhibition of ejaculation, impotence, priapism, rash, itching, hypertrophic tongue papillae, angioneurotic edema, jaundice, fever, increased SGOT, and generalized epileptic seizures have occurred. Mild red-green color vision defect in a **schizophrenic**. Crosses the placenta and has been associated with extrapyramidal symptoms in the **neonate**. Hepatotoxicity is condsidered to be a hypersensitivity reaction, which disappears in a few weeks after treatment withdrawal. A 25-year-old **male** ingested 6 g in two suicide attempts, leading to first degree atrioventricular block, prolonged QRS duration, and supraventricular and ventricular arrhythmias. Successful suicides reported with 2.5–8 g ingestion.

Drug interactions: ***Phenobarbital*** induces a significant decrease in its plasma levels.

LD_{LO}, **human**: 42 mg/kg; LD_{50}, **rat**: 664 mg/kg; **mouse**: 560 mg/kg; **rabbit**: 7.8 g/kg.

MESQUITE

= *Algarrobo* = *Prosopsis sp.*

Untoward effects: Its aeroallergens have caused allergic rhinitis, bronchial asthma, and/or a hypersensitivity pneumonitis. A 31-year-old **male** took a tablespoonful of purchased **bee** pollen to relieve **his** chronic rhinitis. Within 20 min, **he** had hives and started to wheeze. **He** was in acute respiratory distress when **he** arrived at the emergency room for treatment. It was later determined that **he** was extremely sensitive to mesquite pollen, which was the principle consituent of the **bee** pollen **he** bought.

Extracts of ***P. fracta*** have a hypoglycemic effect in **rats**. **Cattle** consuming mesquite (***P. glandulosa***) in the southwestern U.S., Mexico, and Central America have become anemic, emaciated, and nervous; they salivate, show protruding and paralyzed tongues, and may die. Many affected **cattle** have difficulty in chewing. **Goats** develop similar symptoms.

In Peru, they have also suffered from "coquera", a form of ***hydrocyanic acid*** poisoning from consuming ***P. juliflora*** pods. An enzyme in the pod, when fed to **cattle** eating ***sugar cane***, releases ***hydrocyanic acid*** from the cane.

MESTEROLONE

= *Androviron* = *Mestoranum* = *Proviron* = *SH 723*

Androgen.

Untoward effects: Erythema, cessation of beard growth, nipple pain, and acute gastritis.

Drug interactions: Use with ***clomiphene*** associated with development of a testicular seminoma in 27-year-old **male**.

Abnormal retention of sulfobromophthalein in treated **pigs**.

MESTRANOL

= *Menophase* = *Norquen* = *Ovastol*

Estrogen.* Used with many progesteroids in *oral contraceptives.* Metabolized by liver to *ethinyl estradiol.

Untoward effects: Intrahepatic cholestasis, increased serum transaminases, increased sulfobromophthalein retention, diabetogenicity, nausea, vomiting, urinary incontinence, uterine hemorrhages, vaginitis, nipple pigmentation, sodium and water retention; slight increase in blood pressure. **Employees** at an Ortho Pharmaceutical Corporation mestranol-manufacturing plant developed adverse effects from it and ***norethindrone***: 20 **women** developed unusual vaginal bleeding and 5/25 **men** who had direct contact with the powdered estrogen developed hyperestrogenism, including three with enlarged breasts. Eliminated in breast milk. See ***Diethylstilbestrol*** for similar effects in a manufacturing plant.

Drug interactions: Decrease in ***ascorbic acid, cyanocobalamin, folic acid, pyridoxine***, and ***riboflavin*** serum levels.

Oral dosing to **mice** leads to increased incidence of pituitary and malignant mammary tumors; increased incidence of mammary tumors in female **rats** and **dogs**. In grain at 0.1–0.5%, it decreases **rodent** reproductive capacity.

MESUPRINE

= *MJ 1987*

Uterine relaxant, vasodilator.

Untoward effects: Marked tachycardia, slight hypotension, palpitations, and restlessness caused discontinuation of its use.

METAHEXAMIDE
= *Melanex*

Hypoglycemic.* A *sulfonylurea.

***Untoward effects*:** Induced severe intrahepatic cholestasis (5%), hypoglycemic shock with extreme diaphoresis, headache, dizziness, bradycardia, convulsions, nausea, and vomiting; occasionally drug fever and papular and macular rashes. Intentional overdosing has caused deaths and loss of sanity. Now banned in the U.S.

METALDEHYDE
= *Metacetaldehyde*

In slug and snail toxic baits, and as a compressed fuel.

Untoward effects*:** Usually within ½–3 h, severe gastritis with salivation, nausea, vomiting, and abdominal pain, followed by central nervous system disturbances and convulsions. Exaggerated reflexes, trismus, tongue-biting, facial flushing, increased temperature, muscle rigidity, opisthotonus, cyanosis, and coma (often with hyperthermia); death within 5–24 h. Impedes aldehyde reductase function, resulting in toxic levels of ***acetaldehyde (q.v.). A 3½-year-old **female** ate unwashed ***strawberries*** treated with it, leading to sudden arm pain, nausea, vomiting, respiratory distress, cyanosis, cold and flexed extremities, flaccid muscles, becoming unconscious and in shock. Recovered in 48 h after treatment.

LD_{LO}, **human**: 43 mg/kg.

Narcotic effects in **cats**, **dogs**, **ducks**, and **geese**. Toxic to **birds** and **wildlife** in treated areas. In **cats** and **dogs**, sudden onset of poisoning (1–2 h after ingestion) with hypersalivation, frothing at the mouth, hind limb stiffness, circling, muscle tremors, mydriasis, nystagmus, tachycardia, tachypnea, opisthotonus, increased temperature (102–107°F), cyanosis, convulsions, and death from respiratory failure. Poisoning in **calves** and **cows** recorded leading to profuse salivation, incoordination, hyperesthesia, partial blindness, and frothy diarrhea. Another group of **cattle** entered a barn and ate 150 lb of the bait. The next morning, ten were found dead. Postmortem revealed acute pulmonary congestion, and extensive endo- and epicardial hemorrhages and enteritis. Ataxia, agitation, hyperesthesia, twitches, convulsive spasms, and exaggerated leg movements in **horses**. **Donkeys** and **mules** are similarly affected. Poisoned **sheep** suffer from salivation, epileptiform convulsions, tremors in legs and neck, ataxia, trismus, nystagmus, dyspnea, and death. **Rats** show transverse lesions in the spinal cord and dose-related posterior paralysis.

LD, **dogs**: ~200 mg/kg; **chickens**: 2 g/kg; **ducks**: 500 mg/kg; **geese**: 800 mg/kg; LD_{50}, **rat**: 630 mg/kg.

METANIL YELLOW
= *C.I. Acid Yellow* = *C.I. 13065* = *Ext. D&C Yellow No. 1* = *Tropaeolin G*

***Untoward effects*:** Use no longer permitted by the FDA. Damages seminiferous tubules and decreases rate of spermatogenesis in **rats**.

METAPROTERENOL
= *Alotec* = *Alupent* = *Metaprel* = *Novasmasol* = *Orciprenaline* = *TH-152*

β_2-Adrenergic agonist bronchodilator.* Synthetic *sympathomimetic.

Untoward effects*:** Tachycardia, palpitations, hypertension, nervousness, tremors, spasmodic cough, nausea, and vomiting, as with other sympathomimetics. Occasionally tolerance to it may force use of increased amounts. Bowel distension when used in premature labor. Oral and IV forms can cause **fetal** tachycardia. In 1989 Boehringer Ingleheim replaced its suspending agent, ***sorbitan trioleate with ***soya lecithin***, that prompted a product recall, possibly due to choking or gagging sensations in some **people**. Mild diarrhea, sleep disturbances, visual hallucinations, sweating, odd taste, migraine, dizziness, and deaths also reported. Unsuccessful suicide attempts have been made with it.

In **cats**, toxicity to an overloaded heart was ~15 times that of a normal heart. "Ingestion" and/or inhalation from a puntured inhaler by a **dog** led to tachycardia, weakness, lethargy, tachypnea, frequent premature ventricular contractions, and vomiting. Arrhythmogenic in **rats**. Overdose in **dogs** may cause paradoxical bronchospasm.

METARAMINOL BITARTRATE
= *Aramine* = *Icoral B* = *Pressorol*

***Sympathetic amine.* For parenteral use.**

***Untoward effects*:** Can cause local ischemia and necrosis after extravasation. Drowsiness, dizziness, rash, jaundice, leukopenia, hemolytic anemia, headache, excitation, facial flushing, arteriolar constriction increased blood pressure, premature ventricular contractions, atrioventricular block, faintness, arrhythmias, and chest pain. In addition, overdosage can also cause nausea, vomiting, chest tightness, euphoria, pulmonary edema, tachycardia or bradycardia, diaphoresis, cerebral hemorrhages, myocardial infarction, convulsions, and cardiac arrest. Mollaret's meningitis associated with recurrent hereditary polyserositis in 14-year-old **male** induced by 10 mg IV infusion.

Drug interactions: Use with inhalation ***anesthetics*** can enhance its dysrhythmic effects. Infrequently, it can cause significant ***monoamine oxidase inhibitor*** interaction and ***tricyclic antidepressants*** increase its effect leading to hypertensive crisis, which can be fatal. ***Reserpine*** depletes stored ***catecholamines*** and, therefore, antagonizes its effects to some degree. Combined with ***digitalis*** preparations can cause ectopic arrhythmias; use with ***meperidine*** can cause abnormal reactions. ***Guanethidine*** decreases its effectiveness and the combination causes nausea and headaches. Mild potentiation of ***methyldopa***. Causes ***polyethylene*** devices to discolor.

Causes cardiac lesions and tubular necrosis in **rats** and ventricular fibrillation in **dogs**.

IP, LD_{50}, **rat**: 41 mg/kg; SC, 117 mg/kg; SC, LD_{50}, **mouse**: 92 mg/kg; IV, 51 mg/kg; IP, 440 mg/kg.

METAXALONE

= *AHR 438* = *Skelaxin* = *Zorane*

Skeletal muscle relaxant.

Untoward effects: Gastrointestinal upsets with nausea, vomiting, and indigestion; drowsiness, dizziness, headache, polyuria, painful micturation, leukopenia, intensified muscle cramps, jaundice, and rashes. Causes spuriously high readings on sulfobromophthalein retention, urinary ***albumin*** and ***glucose*** tests.

TD_{LO}, **human**: 16 mg/kg.

Hemolytic anemia and vomiting in **dogs**.

LD_{50}, **rat**: 775 mg/kg; IP, **mouse**: 490 mg/kg.

METEPA

= *MAPO*

Use: In crease-proofing and flame-proofing textiles; in chemical sterilization of **flies** and **boll worms**.

Untoward effects: Only moderately toxic to **rats** by the oral route, but dermal toxicity is high. Daily use of 5 mg/kg/day (~4% of LD_{50}) decreased fertility of male **rats** in 22 days and caused sterility within 77 days. Carcinogenic and teratogenic in **rats**.

LD_{50}, **rats** male: 136 mg/kg; female: 213 mg/kg; **mouse**: 292 mg/kg; skin, **rat**: 183 mg/kg.

METERGOLINE

= *Contralac* = *FI 6714* = *Liserdol* = *Nicergoline* = *Sermion*

Prolactin inhibitor.

Untoward effects: Nausea (5%), headache (4%), vomiting (1%), paresthesia, precordial pain, decreased blood pressure, and transient tachycardia and weakness. Blocks α_1***-adrenergic agonist*** effects of ***propranolol***.

Undesirable effects (aggressiveness toward **owners**) reported in **dogs**.

METFORMIN

= *Diabetosan* = *Diabex* = *DMGG* = *ER 102* = *Glucophage* = *LA-6023* = *Metiguanide* = *NNDG*

***Antidiabetic.* For non-*insulin*-dependent diabetes mellitus. Acts by stimulating the peripheral metabolic response to endogenous *insulin*, decreasing intestinal absorption of *glucose*.**

Untoward effects: The most serious is lactic acidosis. Subtle warnings can include unexplained hyperventilation, myalgia, malaise, and unusual somnolence. It possesses ~1/10 the risk of ***phenformin*** in leading to lactic acidosis. Adverse gastrointestinal effects (3–4%) include metallic taste, nausea, vomiting, diarrhea, abdominal bloating, flatulence, and anorexia. Long-term use causes a decrease (~10%) in **patients'** absorption of ***vitamin*** B_{12} and ***folic acid***; thrombocytopenia, megaloblastic anemia, hemolysis, and jaundice also occur. Therapeutic serum levels reported as 0.1–1.3 µg/ml; comatose state or potentially fatal at 85 µg/ml.

Drug interactions: ***Cimetidine*** increases its serum concentration. Causes spuriously high urine ***acetone*** test readings.

LD_{50}, **rat**: 1 g/kg; SC, **mouse**: 230 mg/kg.

METHACHOLINE CHLORIDE

= *Amechol* = *Mecholine* = *Mecholyl* = *Provocholine*

Cholinergic, diagnostic aid in respiratory function, parasympathomimetic, bronchoconstrictor.

Untoward effects: Ingestion leads to nausea, vomiting, transient dyspnea, substernal pain, and involuntary micturation and defecation. Although rapidly destroyed in the gastrointestinal tract, large doses can precipitate hypoglycemia, syncope, asthma, and cardiac arrest. Intranasal use requires low dosage to prevent the same ill-effects, which can also include coughing, pharyngeal itching, light-headedness, and throbbing headache. Injection leads to flushing of face, neck, and chest; salivation, sweating, lacrimation, rhinorrhea, intestinal hyperactivity, and asthmatic attacks in some **people**. Hypotension and tachycardia may be followed by bradycardia. Can cause falsely high amylase and lipase serum results. Serious arrhythmias and ventricular fibrillation when it is used to provoke Prinzmetal variant angina for diagnostic purposes. It can lead to irreversible coronary occlusion and death in **some**.

Can produce acute and chronic ulcerative colitis and increased intraocular pressure in **dogs**; chromodacryorrhea in **rats**.

LD_{50}, **rat**: 750 mg/kg; **mouse**: 1.1 g/kg; IV, **rat**: 20 mg/kg; **mouse**: 15 mg/kg.

METHACRYLIC ACID

Untoward effects: Acrid repulsive odor. Inhalation leads to eye, skin, and mucous membrane irrition. Corrosive and contact can cause eye or skin burns. Several **patients**-21 months, 2.5 years, and 27 years, after using three brands of artificial nail primer containing it, developed chemical burns in airway and gastrointestinal tract and on skin. Numerous esters, salts, and derivatives are widely used in industry, surgery, and dentistry.

Odor can be recognized in tissues of affected **animals**.

IP, LD_{50}, **mouse**: 0.048 ml/kg.

METHACYCLINE

= *Adriamicina* = *Bialatan* = *Ciclobiotic* = *Germiciclin* = *Londomycin* = *Metacycline* = *Metadomus* = *Metilenbiotic* = *MIC 100* = *Optimycin* = *Physiomycine* = *Rindex* = *Rondomycin*

Antibiotic. Oxytetracycline **derivative.**

Untoward effects: ***Blood***: Hemolytic anemia, thrombocytopenia, neutropenia, and eosinophilia.

Gastrointestinal: Anorexia, nausea, vomiting, diarrhea, glossitis, stomatitis, dysphagia, enterocolitis, tympanitis, heartburn, and inflammation lesions with monilial overgrowth in the anogenital region.

Hypersensitivity: Angioneurotic edema, prurigo, urticaria, anaphylaxis, purpura, pericarditis, and exacerbation of systemic lupus erythematosis.

Miscellaneous: Bulging fontanels in young **infants** on heavy dosage that disappears after cessation of treatment. May cause brown-black microscopic discoloration of thyroid glands with no abnormalities of function. Staining and structural damage to developing teeth and bones may be expected. Monilial balanitis, lymphedema of hands, paresthesia, fever, and transient increase in SGOT and SGPT. Some passage into breast milk.

Renal: Dose-related increase in blood urea nitrogen.

Skin: Maculopapular and erythematous rashes. Photosensitivity. Exfoliative dermatitis is rare.

Drug interactions: ***Ferrous sulfate*** seriously impairs its absorption, as do ***antacids*** containing ***aluminum, calcium***, and ***magnesium***. ***Dairy products, food***, and ***sodium bicarbonate*** also decrease its absorption.

Can cause teratogenic defects in **mice**. Limited adverse effects on tooth development in **rats**.

IP, LD_{50}, **rats**: 252 mg/kg; **mice**: 288 mg/kg.

METHADONE HCl

= *Adanon* = *AN-148* = *Algidon* = *Algolysin* = *Amidon* = *Butalgin* = *Depridol* = *Dolls* = *Dolly* = *Dolophine* = *Fenadone* = *Heptadon* = *Heptanon* = *K 174* = *Ketalgin* = *Mecodin* = *Mephenon* = *Miadone* = *Moheptan* = *Phenadone* = *Physeptone* = *Tussol*

Narcotic, analgesic. **Oral and parenteral.**

Untoward effects: The most serious are respiratory depression, circulatory depression, shock, and respiratory and cardiac arrest. Also reported are light-headedness, dizziness, nausea, vomiting, sweating, weakness, headache, insomnia, agitation, euphoria, dysphoria, disorientation, visual disturbances, miosis, anorexia, constipation, dry mouth, biliary tract spasm, facial flushing, syncope, palpitations, bradycardia, decreased libido in **male** and **female**, decreased potency; ejaculation difficulty, gynecomastia in **male**, anorgasma in **male** and **female**, urinary retention, antidiuretic effect, skin rashes, urticaria, pruritus, edema, immune hemolytic anemia, eosinophilia, positive direct Coomb's tests, menstrual irregularities, and lens opacities. After overdosage, cyanosis, stupor, coma, severe miosis, pulmonary edema, flaccid skeletal muscles, myoglobinuria, rhabdomyolysis, apnea, circulatory collapse, cardiac arrest, and death may occur. Withdrawal symptoms reported in **adults** and **newborns** of **mothers** taking it. As little as 10–20 mg in **children** and 75 mg in an **adult** has caused death. Accidental ingestion by both groups occurred primarily when it was consumed in ***Tang***®, thinking **they** were drinking ***orange*** juice. **Mothers** receiving 10–40 mg/day had 50–170 ng/ml of breast milk leading to average intake of 57 µg/day by **infant**. Thrombocytosis and increased rate of sudden infant death syndrome in **infants** born to methadone-addicted **mothers**. Breast milk and **infant's** plasma need to be monitored when **mothers** receive > 20 mg/day. Tolerance to it can develop. Therapeutic serum levels range from 30–110 µg% and toxic levels reported as 0.08–0.2 mg%. Some reported toxicity at 1/10 this level. **Addicts** may sell **their** prescribed supply and when given it in prison, it can become an overdose; has caused the death of a **prisoner**.

Drug interactions: ***Rifampin*** decreases its plasma concentration and increases urinary excretion of its major metabolite. A 76-year-old **male** showed major potentiation of it by ***cimetidine***. ***Phenytoin*** has increased its metabolism, causing ***methadone*** withdrawal symptoms. ***Ritonavir*** decreases its metabolism, increases its potential toxicity. The reverse effect was reported in a 51-year-old **male**

with HIV. A combination of ***lopinavir*** and ***ritonavir*** can lower its serum concentration. ***Fluoxetine*** causes a slight increase in its plasma levels. Methadone withdrawal symptoms within a week in 32-year-old **female** after taking ***nevirapine*** 200 mg twice daily, requiring 60% increase in methadone dosage.

Readily crosses the placenta in pregnant **animals**. Intrauterine growth retardation, increased neonatal death rate, and decreased postnatal growth in **rats**. Withdrawal symptoms (aggressiveness, tremors, and weight decrease) can be demonstrated in **rats**. So-called tolerance in **mice** is due to increased metabolization. Congenital malformations in **hamsters** increase as dosage is increased. **Cats**, **horses**, and **mice** have shown excitement, especially with increased dosage. Vomiting, defecation, bradypnea, bradycardia, and salivation occur in **dogs**.

LD_{50}, **rat**: 86 mg/kg; **mouse**: 70 mg/kg; IV, LD_{50}, **rat**: 11 mg/kg; SC, 28 mg/kg; **mouse**: IV, 20 mg/kg; SC and IP, 35 mg/kg; IV, LD_{LO}, **dog**: 26 mg/kg.

METHALLIBURE

= *Aimax* = *Almas* = *Almax* = *AY 61122* = *ICI 33,828* = *Match* = *Metallibure* = *TCH* = *Turisynchron*

Pituitary gonadotrophic inhibitor, oral contraceptive, estrus synchronizer.

Untoward effects: Do NOT feed to **boars**. Teratogenic effects have been reported when given to **gilts** or **sows** during day 20 to day 50 of gestation. **Boars** accidentally ingesting treated feed will develop drowsiness, increased thirst, inappetence, weight decrease, decreased libido, and smaller testicular size, but permanent ill-effects are unlikely. Diaphragmatic hernias in baby **pigs** of treated **sows**.

TD_{LO}, **sows** (29–48 days pregnant): 20 mg/kg.

METHAMIDOPHOS

= *Bayer 71628* = *Chevron Cmpnd 9006* = *ENT-27396* = *Hamidop* = *Monitor* = *Ortho 9006* = *RE 9006* = *SRA-5172* = *Tamaron*

***Insecticide, acaricide.* An *organophosphate*.**

Untoward effects: The FDA once banned importation of Mexican ***bell peppers***, due to contamination with it. It is a metabolite of, and more toxic than, ***acephate***. Neurotoxic effects noted in 12–96 h in exposed **patients**. Symptoms of poisoning by any route include headache, dizziness, extreme weakness, ataxia, miosis, muscle twitches, tremors, nausea, bradycardia, pulmonary edema, sweating, heart-block, respiratory depression, and influenza-like illness with anorexia and malaise. To a lesser extent, blurred or dark vision, confusion, incontinence, convulsions, unconsciousness, vomiting, abdominal cramps, diarrhea, tightness in chest, wheezing, productive cough, rhinorrhea, tearing, and lacrimation. Poisonous if swallowed, inhaled, or absorbed through the skin. Rapid percutaneous absorption. Eye irritant. **Users** should wear rubber gloves, protective clothing, and goggles.

Mallard ducks show poor balance, ataxia, falling, wing-drop, wings spread, tremors, nutation, and wing-beat convulsions within 10 min of exposure; death usually occurs 50–90 min after exposure.

LD_{50}, 7.5 mg/kg; **mouse**: 30 mg/kg; **rabbit**: 10 mg/kg; **chicken**: 25 mg/kg; **mallard ducks**: 8.48 mg/kg; **quail**: 99 ppm in feed. LD_{50} dermal, **rabbit**: 118 mg/kg; dermal, **rat**: 50 mg/kg.

METHAMPHETAMINE

= *Amphedroxyn* = *Benzefit* = *Blue-collar Cocaine* = *Bombita* = *Corvitin* = *Crank* = *Cristy* = *Crystal* = *Daropervamin* = *Desfedrin* = *Desoxyephedrine* = *Desoxyn* = *Destim* = *Dexies* = *Doe* = *Doxephrin* = *Drinalfa* = *Efroxine* = *Gerobit* = *Hiropon* = *Ice* = *Isophen* = *Madrine* = *Metamphetamine* = *Meth* = *Methampex* = *Methedrine* = *Methylbenzedrine* = *Methylisomyn* = *Neopharmedrine* = *Norodin* = *Pervitin* = *Philipon* = *Semoxydrine* = *Soxysympamine* = *Speed* = *Splash* = *Syndrox* = *Tonedron* = *Zip*

***Sympathomimetic, central nervous system stimulant.* Oral or IV (smoked, snorted, swallowed, or injected). Inhalers have also been used. The stench of *hydrogen sulfide* and *ammonia* is often noted near illicit manufacturing sites. *Ice*, a liquid form popular in the Far East, was dropped on the end of a *cigarette* and smoked.**

Untoward effects: ***Addiction***: **Users** are often called "the living dead". Violence, punching holes in walls, weight decrease, sleeplessness for days, and severe dermatoses are many additional ill-effects. Combination with ***barbiturates*** facilitates the abuse. Habituation has led to daily dosing of 2000 mg. Delirium has been noted on withdrawal. Acute psychotic reaction reported after inhalation of the drug. IV use leads to shock symptoms, disseminated intravascular coagulation, and rhabdomyolysis with myoglobinuria and azotemia. Chills, fever, sweating, nausea, and abdominal cramps have occurred shortly after, and vomiting, myalgia, paresthesia, neck pain, photophobia, headache, and orthostasis have developed in a few hours. Cardio-respiratory arrest, accelerated bleeding, fatal intracerebral hemorrhage, and pulmonary edema have also been reported. IA use in **addicts** leads to necrotizing angiitis, tachycardia, pyrexia, severe epigastric pain, polyneuritis, polymyalgia, leukocytosis, eosinophilia, and increased amylasemia. Pott's puffy tumor in

34-year-old **female** inhaling it after 15 years of intranasal use.

Cardiovascular: Tachycardia, palpitations, increased blood pressure, and dilated cardiomyopathy. Intranasal inhalation in a 35-year-old **male** led to acute myocardial infarction.

Central nervous system: Adverse effects on central nervous system are greater than with ***amphetamine*** (q.v.). Dysphoria, euphoria, insomnia, tremors, restlessness, and headache. Exacerbation of tics and Gilles de la Tourette's syndrome. Psychotic episodes are rare, except on overdoses, and this has occurred on as little as 55 mg. Has caused symptoms of paranoid schizophrenia that may last up to 48 h, and suddenly reappear months after use has ceased. Pressor effects may cause permanent neurological deficit. Sustained use causes **addicts** to become suspicious, have mood swings and gaps in **their** memories, become suicidal or kill **others**, and can develop myocardial infarction with overdoses. Use by General Montgomery's army and the Royal Air Force prompted a London newspaper in 1941 to headline "Methedrine Wins the Battle of London".

Gastrointestinal: Mouth dryness, unpleasant taste, diarrhea, and constipation.

Miscellaneous: Urticaria, "bugs under skin", decreased libido, loss of hair, and growth retardation. Hepatitis and other infections and endocarditis reported from illicit IV use. A **female addict** was convicted of killing her 2 months **infant** by administering a lethal quantity of it by breast feeding, despite the fact that **she** often skipped days without feeding the **infant**. At autopsy, the **infant's** blood level was 39 ng/ml. Microencephaly, mental retardation, motor dysfunction, and withdrawal symptoms reported in **newborns**. A male overdosed with 140 mg orally, leading to cardiovascular collapse, acute renal failure, pyrexia, and death. Talkativeness, hyperactive reflexes, sweating, mydriasis; unusually bright, shiny eyes; and anorexia also reported on overdoses or in **addicts**. In Japan, 14% of **addicts** develop a chronic psychosis which, on lobotomy or postmortem, showed evidence of permanent brain damage. Toxic serum concentrations in **adults** are reported as 0.06–0.5 mg%. Psychosis and death reported with doses < 100 mg.

Drug interactions: Reduces effectiveness of ***guanethidine***, ***methyldopa***, and ***reserpine***; cardiac arrhythmias with ***halothane***; ***reserpine*** can decrease its effectiveness as well. Headache, hypertension, and death reported with ***tranylcypromine***. Will discolor ***polyethylene*** devices.

Intoxication in **dogs** leads to excitement, agitation, fever, and convulsions. In **horses**, leads to tachycardia, tachypnea, and excitement. Aroused behavior in **rabbits** and induced-hyperactivity in **rats** and **mice**. **Rats** intoxicated on it have shown abnormal behavior for up to 8 weeks.

LD_{50}, **mouse**: 34 mg/kg; **dog**: 10 mg/kg; IV, **dog**: 5.9 mg/kg; LD_{LO}, **rat**: 70 mg/kg; SC, **cat**: 50 mg/kg.

METHAM SODIUM

= *A 7 Vapam* = *Carbam* = *Carbathion* = *Karbathion* = *Mapasol* = *Metam* = *Nematine* = *Sistan* = *SMDC* = *Sometam* = *Trimaton* = *Vapam* = *VPM*

Soil fumigant. A carbamate. Cholinesterase inhibitor.

Untoward effects: Contact dermatitis. Skin, eye, nose, and throat irritant. Aerosoled material can cause a variety of respiratory problems. Prolonged contact can cause severe skin irritation. Can be fatal after dermal absorption.

Toxic to **fish**.

LD_{50}, **rat**: 820 mg/kg; **mouse**: 50 mg/kg; **rabbit**: 320 mg/kg; **guinea pig**: 815 mg/kg; acute dermal, **rabbit**: 800 mg/kg.

METHANDRIOL

= *Anabol* = *Andriol* = *Androdiol* = *Androteston-M* = *Crestabolic* = *Diolandrone* = *MAD* = *Masdiol* = *Megabion* = *Mestenediol* = *Metandiol* = *Metendiol* = *Methanabol* = *Methandiol* = *Methostan* = *Methylandrostenediol* = *Metidione* = *Metildiolo* = *Metocryst* = *Nabadiol* = *Neosteron* = *Neutrormone* = *Neutrosterone* = *Notandron* = *Protandren* = *Stenediol*

***Anabolic agent, androgen.* IM use.**

Untoward effects: Masculinization (clitoral enlargement, labioscrotal fusion) of **female fetus**, especially when the **mother** is given it during the first trimester of pregnancy. Sublingual use exacerbated chronic interstitial pneumonia and chronic osteomyelitis. Transient glycosuria has also occurred. Prolonged use can cause jaundice.

Hypertension and vascular and cardiac disease in **rats**.

METHANDROSTENOLONE

= *Danabol* = *Dianabol* = *Methandienone* = *Nabolin* = *Nerobol* = *Stenolon*

***Anabolic steroid, androgen.* Topical, parenteral, and oral use.**

Untoward effects: Prolonged use will cause intrahepatic cholestasis, cholangiolitis, and excessive cardiac stimulation. A chronic **abuser** for 13 years probably induced cor bovinum with ventricular wall hypertrophy and dilation of all cardiac chambers, found at necropsy. Topical application of 1% leads to local and diffuse eczema. Mild acne, delayed menses, heartburn, hoarseness, Stevens–Johnson syndrome, hypertrichosis, decrease in fasting blood ***glucose***, increased serum ***cholesterol***, slight increases in serum ***potassium*** and ***sodium*** occur, and even a decrease in libido. A 38-year-old **male** developed Wilm's tumor as a result of high-dose injections over a period of years, as well as other oral ***anabolic steroids***. Abuse has caused pulmonary embolism and psychotic episodes in **male bodybuilders**. Unprescribed oral drug was given by a **mother** to 4½-year-old **male** as **"*vitamins*"**, leading to irreversible sexual precocity. Severe depression can follow withdrawal in an **abuser**.

Drug interactions: By inhibiting metabolic enzymes, it increases the activity of ***oral contraceptives*** (with hypertensive effects, fluid retention, and diabetogenic action), ***coumadin, oxyphenbutazone*** (> 30%), ***phenindione***, and ***warfarin***. It decreases the effectiveness of ***phenylbutazone***.

TD_{LO}, **human**: 580 mg/kg over 7.4 years.

METHANE

= *Marsh Gas* = *Methyl Hydride* = *Natural Gas* = *North Sea Gas*

Untoward effects: Odorless, non-toxic, but, if inhaled in high concentration leads to narcosis and death by asphyxiation. Has a narrow margin of safety; unconsciousness at 87% by volume and death at 90%. Has been involved in numerous homicides and suicides. Combustion in a limited supply of air leads to ***carbon monoxide*** (q.v.); complete combustion leads to ***carbon dioxide*** (q.v.). Concentration of 5.53–15% in air is explosive. An odorant, such as ***methanethiol*** (q.v.), is usually added to methane used for cooking or lighting. ***Nitrous oxide*** and it form an explosive mixture and can be fatal in the abdomen during laparoscopic sterilization by diathermy. Farm **workers** have been asphyxiated by it in manure pits.

Inhalation causes respiratory distress in **hogs**. Lighter than air, it accumulates near ceilings, unventilated corners, and tops of manure pits. Explosions have occurred.

METHANEARSONIC ACID

= *MAA*

Herbicide.

Untoward effects: A spray-rig **operator** accidentally had the front of **his** clothes saturated with it and got some in **his** mouth, which **he** spat out and rinsed **his** mouth with water. That night, **he** developed stomach cramps. On the way to the hospital, paralysis, irregular breathing, numbness, and faintness occurred. Apparently normal 5 days later. Can produce typical ***arsenic*** (q.v.) poisoning with gastrointestinal symptoms, weakness, cardiovascular collapse, and rapid death.

The *sodium salt* = ***Ansar 6.6*** = ***Ansar 529*** = ***Bueno*** = ***MSMA*** = ***Super Arsonate*** = ***Weed Hoe.***

The *disodium salt* = ***Ansar 8100*** = ***Arrhenal*** = ***Arsinyl*** = ***Clout*** = ***Crab*** = ***E-Rad*** = ***Dal-E-Rad*** = ***DSMA*** = ***Sodar.***

LD_{LO}, **human**: 50 mg/kg. LD_{50}, **rat**: 1.8 g/kg.

It is extremely toxic to **honey bees.** Anorexia in **calves** in 3–6 days after being fed diets with 4 g/kg.

METHANETHIOL

= *Mercaptomethane* = *Methyl Mercaptan* = *Methyl Sulfhydrate*

Use: In the manufacture of ***methionine***, jet fuels, pesticides, fungicides, and plastics.

Untoward effects: Fatal poisoning in 53-year-old **male** inhaling its fumes while emptying several tanks of it. Found in a coma 1 h later, **he** had a brief, acute, severe hemolytic anemia, and methemo-globinemia, and was probably glucose-6-phosphate dehydrogenase-deficient. **He** died 28 days later. When a glass gauge on a storage tank broke, (exposing three **workers** and a **fireman**) the **owner** of a chemical company passed out, lost some hair, and recovered, as did two **others** hospitalized overnight after a 500–600 gallon leakage. Glucose-6-phosphate dehydrogenase deficiency may be a factor in its development of hemolytic anemia. Eye, skin, and respiratory system irritant. Narcosis, cyanosis, and convulsions occur. Contact with the liquid can cause frostbite. American Conference of Governmental Industrial Hygienists, TLV, TWA, 0.5 ppm. Immediate danger to life or health, 150 ppm.

Inhalation LC_{50}, **human**: 675 ppm.

Acute inhalation experiments with **rats** and **mice** led to restlessness, tachypnea, muscle weakness, respiratory depression, paralysis, convulsions, and death. Pulmonary edema in **monkeys**, and persistent hepatitis and cellular changes of liver, lungs, and kidneys in **mice**.

METHANTHELINE BROMIDE

= *Asabaine* = *Avagal* = *Banthine Bromide* = *Doladene* = *Gastron* = *Gastrosedan* = *MTB-51* = *Metanyl* = *Metaxan* = *Methanide* = *Methanthine Bromide* = *SC-2910* = *Uldumont* = *Vagamin* = *Vagantin* = *Xanteline*

Anticholinergic, ganglionic-blocking agent.

Untoward effects: Impotence, chills, fever, flushing, inability to urinate, paralytic ileus, decreased lower esophageal sphincter tone, agitation, tachycardia, and tachypnea. High dosage causes most of the adverse effects in **adults**, but moderate doses in **infants** and **children**

can elicit them. Large doses can have a ***curare***-like effect on skeletal muscles.

Blocks ovulation in **rats** and **rabbits**.

LD_{50}, **rat**: 1.66 g/kg; **mouse**: 40 mg/kg; IP, LD_{50}, **mouse**: 76 mg/kg; IV, 8 mg/kg; IV, LD_{LO}, **dog**: 23 mg/kg.

METHAPYRILENE

= *AH-42* = *Lullamin* = *Paradormalene* = *Pyrathyn* = *Pyrinistab* = *Pyrinistol* = *Rest-On* = *Restryl* = *Emikon* = *Sleepwell* = *Thenylene* = *Thenylpyramine* = *Thionylan*

Antihistamine. An ethylenediamine.

Untoward effects: Carcinogenicity in **rats** prompted its withdrawal from several markets for **humans**. This probably occurs when it reacts with ***nitrites***. Lethargy and drowsiness in **child** overdoses. Anorexia and rashes reported. Nearly 1% of **patients** (5/413) evidenced skin hypersensitivity reactions to it. Nausea, vomiting, drowsiness, hypotension, disorientation, and convulsions have occurred. Tolerance to it has developed. A 20-month-**child** accidentally ingested 800 mg leading to cyanosis, unconsciousness, convulsions, and decreased cardiorespiratory function; recovered. A 16-month-old **female** accidentally ingested 100 mg, leading to nausea, drowsiness, and vomiting, followed by excitement, tremors, convulsions, and death. Post-mortem revealed cerebral edema, upper nephron nephrosis, anuria, and great increase in ***non-protein nitrogen***. ***Alcohol*** ingestion may potentiate its toxicity. Considerable misuse, intoxication, hallucinations, and fatalities occurred in over-the-counter combinations with ***scopolamine***. Leukocytosis, confusion, seizures, and hypertension after ingestion of 100 ***Sleep-Eze***® tablets (with ***scopolamine***) in a suicide attempt. Aplastic anemia and thrombocytopenia after ***Nytol***® (with ***salicylamide***).

Therapeutic serum levels reported to be 0.2–0.4 mg%; toxic levels at 3–5 mg%, and lethal levels 5 mg% and above. Deaths have occurred with levels of 1.2–3.0 mg% when given with other drugs.

Carcinogenicity in **rats** occurred when they received doses equivalent to **man** getting 200 sleep-aid tablets every night for many years; no tumors at half that level.

~LD_{50} in mg/kg; **rat**: 450, **mouse**: 182; **guinea pig**, **cat**: 150; IP, **rat** and **guinea pig**: 70; **mouse**: 77; **cat**: 55; SC, **mouse**: 60; **guinea pig** and **dog**: 25; IV, **mouse**: 13, **cat**: 10.

METHAQUALONE

= *Cateudyl* = *Citexal* = *Dormigoa* = *Dormogen* = *Dormutil* = *Dorsedin* = *Fadormir* = *Holodorm* = *Hyminal* = *Hypocol* = *Hyptor* = *Ipnofil* = *Ludes* = *MAOA* =*Melsed* = *Melsedin* = *Melsomin* = *Mequelon* = *Mequin* = *Metadorm* = *Methased* = *Metolquizolone* = *Mollinox* = *Motolon* = *Mozambin* = *MTQ* = *Nobedorm* = *Noctilene* = *Normi-Nox* = *Omnyl* = *Optimal* = *Optimil* = *Opti-Noxan* = *Ortonal* = *Parminal* = *Parest* = *Paxidorm* = *Quaalude* = *Quads* = *Quas* = *QZ-2* = *Revonal* = *RIC-272* = *Riporest* = *Rorer 148* = *Roulone* = *Rouqualone* = *Sedaquin* = *Sindesvel* = *Sleepinal* = *Soapers* = *Somberol* = *Somnafac* = *Somnium* = *Somnomed* = *Sonal* = *Sopes* = *Sopor* = *Soverin* = *Toquilone* = *Toraflon* = *Torinal* = *TR-495* = *Tuazol* = *Tuazolone*

Sedative, hypnotic, central nervous system depressant. **Highly lipophilic.**

Untoward effects: As Shakespeare said, "it increaseth the desire, but taketh away the performance", despite its over-glamorization on the street. Inhibits ejaculation in **males** and decreases libido in **females**. It does not contain mythical sexual powers. Headache, hypotension, dizziness, "hangover", drowsiness, anorexia, nausea, transient tingling and paresthesia, gout, dry mouth, tachycardia, blurred vision, and skin reactions reported and, rarely, aplastic anemia with therapeutic doses. Agranulocytosis, neutropenia, polyneuropathy, and addiction also reported. Overdosage leads to restlessness, delirium, retinal hemorrhages, purpura, hypertonia, and convulsions or coma. Occasionally decreased cardiac function, respiratory depression, hepatitis, pulmonary and cutaneous edema, decreased renal function, bleeding, and spontaneous vomiting. Death from central nervous system depression, shock, respiratory arrest and interaction with ***alcohol*** intake. Diaphoresis and bromhidrosis reported. Withdrawal problems, including delirium tremens. Mid-cycle increased metabolization in premenopausal **women**, not observed when **they** are using ***oral contraceptives***. Crosses the placenta. Involved in suicide attempts and suicides, often with other drugs. Therapeutic serum levels reported as 0.3–0.6 mg%; toxic levels as 1–2 mg%; and lethal levels above that.

Hypotensive effect of IV use in **cats**. Vertebral and rib defects in the offspring of treated **rats**, and, to some extent, in **rabbits**.

LD_{50}, **rat**: 300 mg/kg; **mouse**: 400 mg/kg; **guinea pig**: 360 mg/kg.

METHARBITAL

= *Endiemal* = *Gemonil* = *Gemonit* = *Metharbitone*

Barbiturate, anticonvulsant.

Untoward effects: Drowsiness, irritability, rash, vertigo, and gastrointestinal upsets. Less hypnotic than ***phenobarbital***. Fatal neonatal hemorrhage in **male infant** whose **mother** took 150 mg/day and ***methsuximide*** 900 mg/day throughout **her** pregnancy.

TD_{LO}, **human**: 30 mg/kg; LD_{LO}, **human**: 50 mg/kg.

LD_{50}, **mouse**: 500 mg/kg.

METHAZOLAMIDE

= *Neptazone* = *Theraplix*

***Diuretic.* Carbonic anhydrase inhibitor.**

Untoward effects: A 69-year-old **male** treated with it developed intrahepatic cholestasis on the third week and aplastic anemia on the fifth week. Agranulocytosis, leukopenia, thrombocytopenia, neutropenia, and pancytopenia have occurred. A depressed 74-year-old **male** became drowsy, confused, and delirious, starting 2 days after beginning treatment with it at 50 mg twice daily for glaucoma. Hyperventilation, hypokalemia, hyperuricemia, acidosis, and, rarely, kidney calculi. Decreased libido in **male** and **female**, impotence, melena, photosensitivity, hematuria, glycosuria, convulsions, and transient myopia reported. Acute overdosage leads to anorexia, nausea, vomiting, malaise, fatigue, drowsiness, headache, vertigo, tinnitus, mental confusion, depression, and paresthesias of fingers, toes, hands, and feet. A 62-year-old **female**, after a single 50 mg dose, developed Stevens–Johnson syndrome. Fatalities have occurred.

Teratogenic in **rats** with absence of digits and metacarpals. Thus, its use in pregnant **women** was contraindicated.

METHAZOLE

= *Oxydiazole* = *Paxilon* = *Probe* = *VCS 438*

Herbicide.

Untoward effects: Chloracne in 90% of **workers** manufacturing it, probably from exposure to an extraneous intermediate, ***TCBA*** (***3,4,3´,4´-tetrachloroazoxybenzene***).

LD_{50}, **albino rats**: 250 mg/kg.

METHDILAZINE

= *Dilosyn* = *Disyncran* = *Tacaryl*

Tranquilizer, antihistamine, antipruritic.

Untoward effects: Lens opacity, respiratory depression, and convulsant. Lethal doses have caused hypnosis, dyspnea, and clonic convulsions. Potentiates ***thiopental*** effects.

LD_{50}, **rat**: 162 mg/kg; **mouse**: 225 mg/kg; **guinea pig**: 263 mg/kg; IP, **rat**: 183 mg/kg.

METHENAMINE

= *Aminoform* = *Ammoform* = *Cystamin* = *Cystogen* = *Formin* = *Hexamethylenamine* = *Hexamine* – *HMT* – *HMTA* = *Uritone* = *Urotropin*

Use: As a ***rubber*** accelerator; in shell molding, solid fuel tablets, and manufacturing of ***cyclonite*** (q.v.) explosives, pesticides, and pharmaceuticals. Chemical stabilizer and urinary antiseptic.

Untoward effects: Nausea, rashes, fixed-drug eruptions, and dysuria. **Workers** handling raw, warm ***rubber*** can develop a dermatitis commonly called "rubber itch" that resembles ***poison ivy*** eruptions. It usually begins on wrists, extends up forearms, and even to face and neck. Severe at night. Being water-soluble, it penetrates the skin and is oxidized, in part, to ***formic acid*** (q.v.), which gets into perspiration, adding to the local irritation, unless it is neutralized with ***sodium bicarbonate***. Urinary treatment effectiveness is based on release of ***formaldehyde*** (q.v.), which can have adverse effects. Use can cause a falsely increased urine ***catecholamine*** and serum ***estriol*** assays. Interferes with ***vanillylmandelic acid*** assay procedures to diagnose pheochromocytoma. Excreted in the milk of **humans**, **cows**, **goats**, and **sows**. After elimination by the kidneys, it slowly releases ***formaldehyde*** and ***ammonia*** if the urine is acid. ***Formaldehyde*** is irritant and, unless dosage is reduced, it can cause secondary nephritis and cystitis, especially in **cats**, who often develop anorexia and vomiting from it.

LD_{LO}, **human**: 500 mg/kg.

Drug interactions: Pharmacological incompatability with some soluble ***sulfonamides*** (***sulfamethizole***, ***sulfathiazole***, etc, but not with ***sulfisoxazole***, ***sulfisomidine***, and ***sulfadimethoxine***) causing the formation of insoluble precipitate, decreasing their therapeutic efficacy, and increasing potential hazard of calculus formation and renal blockage. Effectiveness as a urinary antiseptic is antagonized by ***alkalinizing agents***, such as ***sodium bicarbonate*** and increased by acidification (***ammonium chloride***, ***methionine***, ***sodium acid phosphate***, etc). Gives false positive or spuriously high urine ***albumin***, ***propanediol*** urine, urinary ***vanillylmandelic acid***, and urinary blood test results; false negative or spuriously low urinary ***estriol*** and ***5-HIAA*** test readings. ***Sulfathiazole*** or ***sulfamethizole*** with it forms an insoluble precipitate in urine.

Many forms exist.

The *hydrochloride* = ***Dowicil (R) 200***, used as a preservative. Eye and skin irritant; burning on urination reported. Flammable.

The *hippurate* = *Haiprex* = *Hipeksal* = *Hippramine* = *Hippuran* = *Hiprex* = *Urex* = *Urotractan* = *Viapta*

The *mandelate* = *Cedulamin* = *Hexydaline* = *Mandacon* = *Mandamina* = *Mandalay* = *Mandelamine* = *Mandoz* = *Mandurin* – *Mantropine* = *Prov-U-Sep* = *Purerin* = *Reflux* = *Renelate* = *Uro-Cedulamin* = *Uromandelin* = *Uronamin*

Untoward effects: Drastically decreases ***estriol*** content of **maternal** urine to levels indicative of intrauterine death or an ancephalic **fetus**. Burning on urination has occurred. Hemorrhagic cystitis and mild azotemia in 2½-year-old **male** after accidental ingestion of at least 8 g. Lipoid pneumonia in two senile **hemiplegics** after receiving 15 ml of oily suspension qid.

METHENOLONE ACETATE

= *Metenolone* = *Nibal* = *Primobolan* = *Primono-bol* = *SH 567* = *SQ 16,496*

Anabolic steroid.

Untoward effects: Virilization in 50% of **females** treated, weight increase, increased ***sulfobromophthalein*** retention, hepatoxicity, and prostatic and testicular hypertrophy. Use cautiously in growing **animals** and **children** to avoid possible premature epiphyseal closure. Maintain adequate protein intake during therapy. Androgenic effects are possible.

METHETOIN

= *Deltoin* = *N-3*

Anticonvulsant. A phenylhydantoin.

Untoward effects: Drowsiness, ataxia, skin rashes (scarlatinal – 5/44), slurred speech, transient nausea, dizziness, transient mental confusion, and slight leukopenia. Failed suicide attempt with ingestion of 4 g.

LD_{50}, **mouse**: 500 mg/kg; **rabbit**: 1.4 g/kg.

METHIDATHION

= *ENT 27,193* = *GS 13,005* = *Somonil* = *Supracide* = *Ultracide* = *Ustracide*

Acaricide, insecticide. Organophosphate cholinesterase inhibitor.

Untoward effects: Poisoning reported in 60 kg 25-year-old **male**. Recovered in 3 weeks with therapy. No delayed neurotoxicity noted. Symptoms in poisoned **humans**: headache, blurred vision, miosis, weakness, nausea, cramps, chest discomfort, giddiness, nervousness, salivation, sweating, tearing, muscle twitches, pulmonary edema, dyspnea, pappiledema, ataxia, loss of reflexes and sphincter control, convulsions, and coma. Can cause skin irritation and irreversible eye damage. Six **human** volunteers safely ingested up to 0.11 mg/kg/day/6 weeks.

Animals have similar symptoms. Within 10 min, **birds** show goose-stepping ataxia, leg weakness, dyspnea, lacrimation, salivation, ataraxia, seizures with spread wings, and terminal opisthotonus.

LD_{50}, 20 mg/kg; **mouse** and **guinea pig**: 25 mg/kg; **hamster**: 30 mg/kg; **rabbit**: 63 mg/kg; **Canadian goose**: 8.41 mg/kg; **mallard duck**: 23.6 mg/kg; **pheasant**: 33.2 mg/kg; **chukar**: 225 mg/kg. Some **chukars** have died with 77.9 mg/kg; Dermal LD_{50}, **rat**: 25 mg/kg; **rabbit**: 375 mg/kg.

METHIMAZOLE

= *Basolan* = *Danantizol* = *Favistan* = *Frentirox* = *Mercazole* = *Mercazolyl* = *Metazolo* = *Methizole* = *Tapazole* = *Thacapsol* = *Thacapzol* = *Thiamazole* = *Thycapsol* = *Strumazol*

Antihyperthyroid.

Untoward effects: The most serious are bone marrow depression with agranulocytosis, granulocytopenia, leukopenia, pancytopenia, neutropenia, thrombocytopenia, eosinophilia, and aplastic and hemolytic anemia. Also a lupus-like syndrome, ***insulin*** autoimmune syndrome with hypoglycemic coma, hepatitis with jaundice (acute intrahepatic cholestasis), periarteritis, and hypothrombinemia. Rarely, nephritis. More frequently, but less serious, are skin rashes, pemphigus, urticaria, pruritus, nausea, vomiting, sore throat, polyarthritis, arthralgia, myalgia, bone pain, fever, paresthesia, drowsiness, headache, loss of hair, agusia, increased alkaline phosphatase, skin pigmentation, neuritis, vertigo, edema, decreased libido in **female**, sialadenopathy, and lymphadenopathy. Hypothyroidism, congenital goiter, and scalp lesions in the **fetus** of treated **mothers**. May also cause **fetal** encephalopathy and mental retardation. Breast milk contains enough to transfer 7–16% of the **mother's** dose to an **infant**.

Drug interactions: May increase prothrombin time with ***anticoagulants***. Use may require decrease in ***β-blocker*** dosage. Spuriously low ***radio-iodine*** uptake test results.

Depressed thyroid activity decreases the ability of **ruminants** to convert ***carotene*** into ***vitamin A***, and additional supplementation may be required. Hypothyroidism may also depress ***vitamin A*** in other species. ***Phenobarbital*** causes a striking increase in its metabolization in **rats**. Pulmonary edema in **rabbits** given 2–4 g/kg. Occasionally poor appetite, vomiting, thrombocytopenia, hepatotoxicity, scratching at face and neck, leukopenia, eosinophilia, autoimmune hemolytic anemia, and encephalopathy in **cats** and in **dogs**. Methimazole-sensitive **cats** have also become completely anorexic.

TD_{LO}, **rat**: 1100 mg/kg over 2 years period, **mouse**: 33 g/kg over a 1 year period.

METHIODAL SODIUM

= *Abrodil* = *Diagnorenol* = *Radiographol* = *Segosin* = *Skiodan*

Radiopaque.

Untoward effects: Convulsions in **man** from urography have prevented its acceptance in the U.S.

Hyperesthesia, tremors, and opisthotonus in a **horse** after myelography.

IV, LD_{50}, **rat**: 4.8 g/kg; **mouse**: 3.9 g/kg.

METHIONINES

L form = Acimethin = Met = Methionine

DL form = Amurex = Banthionine = Dyprin = Hydan = Liquimeth = Lobamine = Mepron = Metione = Odor-Trol = Pedameth = Racemethionine = Uranap = Urimeth

Hydroxy analog form = Alimet = MHA

Amino acid, urinary acidifier.

Untoward effects: Non-prescription chelating agents, such as sulfhydryl amino acids, ***cysteine*** and ***methionine*** have been promoted as detoxicants by the lay press, and are ultimately self-prescribed by **people**. They leach out essential, as well as toxic, minerals. Fatigue, weakness, drowsiness, lethargy, and nausea reported. Methionine has caused neurological deterioration in 7/9 **patients** with portal cirrhosis and chronic portal systemic encephalopathy. Metabolic breakdown of it causes a fetor hepaticus, due to formation of ***dimethyl sulfide*** and its excretion in the breath of **patients** with liver disease. It can provoke or aggravate megaloblastosis. Hypermethionemia in three **siblings**, all of whom died within a few months after birth. **They** were irritable and restless, followed by increased somnolence, with hypoglycemia and hemorrhagic diathesis. Plasma levels were 30–100 times normal with amino-aciduria, leading to an excess of ***keto-γ-*****methylbutyric acid** in body fluids, including plasma, urine, and sweat, which causes a peculiar odor from the **infants** and their excreta. Post-mortem revealed liver cirrhosis, renal tubular dilation and cellular degeneration, and pancreatic islet hypertrophy. Severe symptoms can be caused by 30 g IV: nausea, vomiting, hypotension, tachycardia, and disorientation followed by azotemia.

Drug interactions: Use with ***tranylcypromine***, a ***monoamine oxidase inhibitor***, leads to hypertensive crisis with headache, myocarditis, hallucinations, delusions, insomnia, drowsiness, cold sweats, teeth chattering, nausea, vomiting, cerebral hemorrhage, respiratory arrest, and coma. Use will give false positive or spuriously high urinary ***acetone*** test results.

In **animals**, avoid excessive oral intake in liver cirrhosis, where it may paradoxically cause a hepatic-type coma and/or renal atrophy, and decrease weight gains. IV injections must be given slowly. **Dogs** are particularly adversely sensitive to the oxidized ***methionine sulfoximine*** form, which was the toxic form that caused "fits" from feeding ***Agene*** (***nitrogen chloride***) bleached ***flour*** during World War II. It may cause nausea in large oral doses on an empty stomach. **Cats** may slowly adapt to anemia produced by continuous feeding of excessive doses. Daily doses over 0.5 g in **cats** may cause an anemia and methemoglobinemia. ***Calcium*** injections may induce coma in **cattle** receiving supplemental methionine.

METHISAZONE

= BW 33T57 = Compound 33T57 = IBT = Marboron = NSC 69,811 = Viruzona

Antiviral. A thiosemicarbazone.

Untoward effects: Weakness, nausea, vomiting, diarrhea, decreased appetite, paresthesia, and increased aminotransferases. Possible teratogenicity.

METHIXENE

= Cholinfall = Methixart = Methyloxan = SJ 1977 = Tremaril = Tremarit = Tremonil = Tremoquil = Trest

Antispasmodic.

Untoward effects: Xerostomia, cycloplegia, urinary retention, rashes, slight increase in intraocular pressure, nausea, constipation, tiredness, dizziness, headache, ataxia, and confusion.

LD_{50}, **mouse**: 370 mg/kg; IP, 144 mg/kg; IV, 18 mg/kg.

METHOCARBAMOL

= AHR-85 = Delaxin = Etroflex = Forbaxin = Neuraxin = Miolaxene = Lumirelax = Relestrid = Robamol = Robaxin = Romethocarb = Spenoxin = Traumacut = Tresortil

Skeletal muscle relaxant, central nervous system depressant. **For oral or parenteral use.**

Untoward effects: Drowsiness, dizziness, light-headedness, vertigo, fainting, rash, urticaria, anorexia, conjunctivitis, hypotension, headache, bradycardia, fever, nasal congestion, diplopia, and nystagmus. Seizures and anaphylaxis following IM or IV. Finger or tongue numbness, paralytic ileus; and green, brown, or black urine reported. Some is excreted in breast milk and it can cross the placenta. Can interfere with ***5-HIAA*** urinary ***vanillylmandelic acid*** assays by false increase in ***vanillylmandelic acid***.

Sialorrhea, weakness, shaking, ataxia, and vomiting, in **cats** and **dogs**.

LD_{50}, **rat**: 1.3 g/kg; **mouse**: 1.5 g/kg, **dog**: 2 g/kg; **hamster**: 1.4 g/kg.

METHOHEXITAL SODIUM

= *Brevital* = *Brevimytal* = *Brietal* = *Methohexitone*

Untoward effects: Thrombophlebitis, circulatory and respiratory depression, arrhythmias, laryngospasm, apnea, shivering, anxiety, bronchospasm, dyspnea, rhinitis, salivation, hiccups, twitching, headache, rashes, urticaria, pruritus, erythema, porphyria, nausea, and vomiting. Hypersensitivity hepatitis is rare.

Parenteral Use: Thrombosis, thrombphlebitis, phlebitis, hypotension, tachycardia, fainting, decreased liver function, and pain at injection sites. Convulsions in susceptible **people**.

IV, TD_{LO}, **human**: 25 μg/kg.

Excessive excitement during recovery when it is used alone for **horses**. Use in **animals** for 1–5 days leads to leukopenia.

METHOMYL

= *DuPont 1179* = *Insecticide 1179* = *Lannate* = *Mesomile* = *Nudrin* = *SD 14,999* = *WL 18,236*

Insecticide, nematocide.* A *carbamate.

Untoward effects: Highly toxic. A 20-year-old **male** became ill (nausea, vomiting, hyperhidrosis, and miosis) 2½ h after accidentally splashing some water containing it on **his** arms and face. After 5 days of aerial spraying with it, three **pilots** became ill; one developed eye problems. In marshy areas of Mexico, a report indicates 58 agricultural **workers** were poisoned by it. **Fishermen** on the South Coast of Jamaica ate ***roti*** (unleavened ***bread*** made of ***flour***, water, and ***salt***) and, within < 3 h, 3/5 became seriously ill, with frothing at the mouth, twitching, and trembling; **they** were DOA at the hospital. One **victim** had vomited, defecated, trembled, and had visual disturbances. A **dog** that ate the vomitus also died. Apparently, a sack of methomyl was mistakenly used as ***salt*** in making the ***bread***. Post-mortem of the dead **men** revealed hazy cornea, mydriasis, and severe congestion of stomach mucosa, lungs, trachea, and bronchi. Toxicity reported in **workers** at a pesticide manufacturing plant.

Cattle have been poisoned by grazing on pastures that had the chemical drifted onto it from nearby vegetable fields and citrus groves. Highly toxic to **honey bees** and **reptiles**. Affected **birds** develop drowsiness, ataxia, tachypnea, dyspnea, salivation, tenseness, diarrhea, tremors, and tetany or wing-beat convulsions. Poisoned **rats** develop a syndrome similar to that in the poisoned Jamaican **fishermen**.

LD_{50}, male **rat**: 17 mg/kg; **mallard duck**: 15.9 mg/kg; **pheasant**, **duck**: 15 mg/kg; **mule deer**, 11–22 mg/kg; LD_{LO}, **chicken**, **quail**: 15 mg/kg; Inhalation LC_{50}, **rat**: 77 ppm; **duck**: 1890 ppm; **quail**: 3680 ppm.

METHOPHENAZINE

= *Frenolon* = *Metofenazate*

Psychotherapeutic.

Untoward effects: Mild orthostatic hypotension and parkinsonism; tachypnea, tachycardia, apathy, rigidity, tremors, and intermittent coma. Intoxicated **children** are often misdiagnosed with encephalopathy, encephalitis, or epilepsy.

TD_{LO}, **rat** (23 days pregnant): 500 mg/kg.

METHOPHOLINE

= *Metofoline* = *RO 4-1778/1*

Analgesic.

Untoward effects: Nausea, vomiting, excitement, abdominal cramps, diarrhea, and constipation.

LD_{50}, **rat**: 400 mg/kg; **mouse**: 180 mg/kg; SC, **rat**: 400 mg/kg.

METHOPRENE

= *Altocid* = *ENT 70,640* = *Entocon* = *Pharorid* = *Precor* = *ZR 515*

Insecticide, insect growth regulator.

Untoward effects: Goose-stepping ataxia, sitting and reluctance to move in **mallard ducks**. Teratogenic in **rats** at 2 g/kg; in **mice** after 1 g/kg IP.

LD_{50}, **rat**: > 34.6 g/kg; **mallard duck**: > 2 g/kg; **dog**: 5–10 g/kg.

METHOTREXATE

= *A-Methopterin* = *Amethopterin* = *CL-14377* = *Emtexate* = *Folex* = *Ledertrexate* = *Methylaminopterin* = *Methyltrexate* = *Mexate* = *MTX* = *NSC 740* = *Rheumatrex*

***Antineoplastic, antirheumatic, antimetabolite. Folic acid* antagonist.**

Untoward effects: They are quite common. Early symptoms are anorexia and nausea (20%), vomiting, abdominal cramps, diarrhea, fever, stomatitis, gingivitis, hematemesis, melena, mucositis, anaphylaxis, and hepatic necrosis. Most toxicity is related to length of treatment, not necessarily the dose. Major delayed effects include oral and gastrointestinal ulceration with occasional perforations; ***calcium***, ***carotene***, ***cholesterol***, ***lactose***, ***vitamin*** B_{12}, and ***d-xylose*** malabsorption; bone marrow depression, pleuritis, pulmonary infiltrates and fibrosis, leukopenia, thrombocytopenia, neutropenia, pancytopenia, eosinophilia;

megaloblastic, aplastic, and immune hemolytic anemia; hypogammaglobulinemia, decreased hematocrit, increased prothrombin time, and vasculitis. Hypersensitivity reactions occur. Within 10 min after starting it by IV, a 56-year-old **female** developed an anaphylactoid reaction (tight, dry throat; chest tightness, dyspnea, agitation, and increased blood pressure). Flaccid paralysis and sensory loss after an intrathecal injection in a **child**. Fatal neurological reaction in 63-year-old **female** after intraventricular injection. Intrathecal use has caused meningeal irritation and paraplegia. Preservatives, such as ***benzyl alcohol***, can be a factor. Others are renal and hepatic toxicity, liver cirrhosis (5%), liver fibrosis (10%), increased lactate dehydrogenase, osteoporosis, bone pain, alopecia, rashes, skin depigmentation or hyperpigmentation; erythematous indurated papules, mostly on proximal areas of extremities; bullous pemphigoid, erythema multiforme, Stevens–Johnson syndrome, hives, conjunctivitis, menstrual disorders, encephalopathy; infections and lymphomas, due to its immunosuppressive action; impotence, oligospermia, gynecomastia in **male**, photosensitivity, renal failure, headache, blurred vision, keratitis, cataracts, confusion, convulsions, brown nail plates and nail bed, white lines in finger nails, acute paronychia, and acute gouty arthritis. It has also been reported to have caused or accentuated parkinsonism. Direct intraventricular injections in treatment of primary brain tumors has caused encephalopathy and leukoencephalopathy as late as 3–15 months afterwards. Intrathecal injections have caused transient and permanent ascending paralysis. A **mother** using it in an attempted abortion had a **female child** with growth retardation and abnormal bony structures. Teratogenicity, including agenesis of frontal bones, cranial synostosis, rib anomalies, unusual facies, cleft palate, digital defects, encephalocele, oxycephaly, wide fontanel, hypertelorism, as well as postnatal growth retardation. Passes into breast milk. Appears in breast milk (1% of **maternal** dose) and can cause immunosuppression. Has caused false increase in ***sulfobromophthalein*** retention and cerebrospinal fluid protein assays. Sunburn dermatitis has been reactivated by it.

Oral, TD_{LO}, **human**: 43 mg/kg, over a 5 year period, **child**: 2 mg/kg, over 12 days, **human** (8–10 weeks pregnant): 625 μg/kg; IV, TD_{LO}, **child**: 100 mg/kg, over 4 h period, **human**: 4.65 mg/kg, over 4 weeks period; IM, TD_{LO}, **human**: 200 mg/kg, over 5 years period.

Drug interactions: The mechanism is usually of interference with its absorption, ***folic acid's*** metabolization, ***pyrimidine*** synthesis, or its renal excretion. Use with ***trimethoprim-sulfamethoxazole, sulfisoxazole*** can cause potentiation of its effect, with symptoms of anorexia, weight decrease, bloody diarrhea, leukopenia, pancytopenia, and coma, due to its displacement from ***albumin*** binding sites. This can be fatal as therapeutic doses are often close to the toxic dose. ***Antibiotics***, such as ***neomycin, nystatin, paromomycin, polymyxin B***, and ***vancomycin*** can decrease its absorption by 30–50%. Interferes with activity of ***5-fluorouracil*** and ***floxuridine***. Its activity is increased by ***chloroquine, hydroxychloroquine, non-steroidal anti-inflammatory drugs (aspirin*** and other ***salicylates, indomethacin, ketoprofen, naproxen, phenylbutazone***), ***PABA***, ***probenecid***, and many ***sulfonamides***. These interactions have been life-threatening and even fatal. Combination can decrease ***phenytoin*** concentration and can increase ***warfarin's*** anticoagulant effect. ***Azapropazone*** use has precipitated methotrexate toxicity. ***Penicillins*** (***amoxicillin, piperacillin***) have decreased its renal clearance. Combined with ***misoprostol*** can cause abortions. Causes increased ***gallium citrate*** uptake by kidney, lung, and bone, but decreased tumor or abscess uptake. Respiratory failure and coma after a ***cocktail*** and methotrexate ingestion. Accumulation of methotrexate is greater in **patients** with fibrosis or cirrhosis, due to heavy ***alcohol*** intake. Blood dyscrasias reported with ***aspirin*** and ***salicylates***. ***Non-steroidal anti-inflammatory drugs*** can increase its toxicity.

Hepatotoxicity reported in **animals**. Large doses to **rats** lead to vacuolar degeneration of hepatocytes. Delays wound-healing in **man** and **animals**. Anorexia, nausea, vomiting, myelosuppression, and renal tubular necrosis in treated **dogs**. Teratogenic in **rats**. ***Aminoglycoside antibiotics*** and some ***sulfonamides*** decrease its absorption and increase its intestinal excretion in **mice**.

TD_{LO}, **cat** (11–14 days pregnant): 2 mg/kg; IV, LD_{50}, **rat**: 14 mg/kg; IP, LD_{50}, **mouse**: 94 mg/kg.

METHOTRIMEPRAZINE

= *Levomepromazine* = *Levomeprazine* = *Levoprom* = *Levoprome 2-Methoxytrimeprazine* = *Milezin* = *Minozinan* = *Neozine* = *Neuractil* = *Neurocil* = *Nirvan* = *Nozinan* = *RP-7044* = *Sinogan-Debil* = *Sofmin* = *Tisercin* = *Viractil*

***Analgesic, tranquilizer.* For IM use.**

Untoward effects: Highly addictive. Orthostatic hypotension, dizziness, weakness, slurring of speech, abdominal discomfort, nausea, vomiting, akathesia, sedation, and corneal opacities. Agranulocytosis, decrease in systolic and diastolic blood pressure, palpitations, tachycardia, tremors, athetosis, and nasal stuffiness. A 45-year-old **male** developed a grand mal seizure and died. After 20 mg, severe agitation, disorientation, and crying within 1 h; recovery in 4 h. A death after 10 mg IM in 72-year-old **male**. Potentially teratogenic and may cause extrapyramidal symptoms in the **neonate**. Passes into breast milk. Therapeutic serum concentrations - 0.03–0.3 mg/l; toxic concentration – 0.5 mg/l.

LD_{LO}, **human**: ~11 mg/kg.

Drug interactions: Adynamic ileus with ***tricyclic antidepressants*** or ***antiparkinson drugs***. Potentiating interactants that increase toxicity are ***alcohol, analgesics, anesthetics, anticholinergics, anticholinesterases, antidiabetics, antihistamines, antihypertensives, aspirin, atropine, β-adrenergic blockers, benzodiazepines, benzodiazepines, central nervous system depressants, dipyrone, epinephrine, furazolidone, griseofulvin, guanethidine, haloperidol, hypnotics, monoanine oxidase inhibitors, meperidine, morphine, muscle relaxants, narcotics, nylidrin, oral contraceptives, phenothiazines, piperazine, quinidine, Rauwolfia alkaloids, scopolamine, sedatives, thiazide diuretics, thiopental***, and minor ***tranquilizers***.

Injections cause local irritation in **steers**, **heifers**, and **pigs**.

LD_{50}, **rat**: 1.1 g/kg; **mouse**: 380 mg/kg; IV, 65 mg/kg.

METHOXAMINE

= *Vasoxine* = *Vasoxyl* = *Vasylox*

***Antihypotensive, sympathomimetic, and α-adrenergic agonist.* Used IV or IM.**

Untoward effects: Hypertension with headache, nausea, and vomiting (often projectile). Occasionally bradycardia and ectopic ventricular beats, followed by tachycardia. Sweating, piloerection, urinary urgency, and **fetal** bradycardia. Large doses also cause mydriasis, blurred vision, irritability, fever, opisthotonus, tingling of extremities, sensation of coldness, convulsions, coma, gasping, and respiratory failure.

Drug interactions: Hypertensive reactions with ***monoamine oxidase inhibitors***.

Cardiomyopathies (myolysis, infarcts), tubular and occasionally glomerular necrosis, dissecting aneurysm, skeletal and visceral muscle degeneration after high dosage (3 mg twice daily) to **rats**. Bradycardia, hypotension, and heart-block in treated **dogs**.

LD_{LO}, **mouse**: 135 mg/kg; IV, 15 mg/kg.

METHOXSALEN

= *Ammidin* = *Ammoidin* = *Geroxalen* = *Houva* = *Meladinine* = *Meloxine* = *8 or 9-Methoxypsoralen* = *8-MOP* = *8-MP* = *Oxsoralen* = *Puvaderm* = *Puvamet* = *Xanthotoxin*

***Skin pigmenter and tanning agent.* Usually from *Ammi majus* (q.v.). It increases the body's production of *melanin*. A *furocoumarin*.**

Use: Promotes protection against sunlight-induced blistering and burning, and in treating vitiligo. Inhibits psoriasis.

Untoward effects: Photosensitivity with erythema, pruritus, blistering, and acute vesicular dermatitis, especially when high concentrations are used topically or with excessive exposure to sunlight and UVA. Nausea is minimized by ingesting it with food or ***milk***. Nervousness, insomnia and depression, edema, headache, dizziness, nephrotic syndrome, urticaria, milaria, folliculitis, contact dermatitis, light allergy, and leg cramps are also reported. Lens opacity, acne, and nail pigmentation when used with ultraviolet light. Carcinogenicity (basal cell and squamous cell carcinomas and malignant melanoma) noted with topical use and UVA treatment. An unsuccessful suicide attempt reported in a **female** after ingesting 850 mg. Serum levels were 606 ng/ml 9 h later, and 21 ng/ml after 32 h.

Drug interactions: Inhibits ***caffeine's*** metabolization and elimination. Can dramatically increase ***nicotine*** blood levels.

In combination with ultraviolet light, it increases skin tumors in **mice**, as well as after oral or IP use. **Ducks** fed ***Ammi majus*** seeds developed photosensitization with beak and web deformities, mydriasis, and eccentric location of pupils. Photosensitization also reported in **geese**, **turkeys**, **chickens**, **cows**, and **swine**.

LD_{50}, IP and SC, **rat**: 470 mg/kg.

p-METHOXY ACETOPHENONE

= *Acetanisole* = *p-Acetylanisole*

Use: As a fragrance in soaps, detergents, cosmetic creams and lotions, and in foods.

Untoward effects: No irritation or sensitization in **human volunteers** at 6% in petrolatum.

LD_{50}, **rat**: 1.72 g/kg; acute dermal, **rabbit**: > 5 g/kg.

4-METHOXYAMPHETAMINE

= *MA* = *Methoxy-α-methylphenethylamine*

Sometimes referred to as PMA, which is also a designation used for *phenylmercuric acetate*.

Untoward effects: As for other ***amphetamines***, and much like ***dextroamphetamine***. Estimated hallucinogenic effect is 5X that of ***mescaline*** and more than ***methylenedioxyamphetamine*** (q.v.). Lethal serum levels for **man**: 0.2–0.4 mg%.

LD_{50}, **mouse**: 284 mg/kg; IV, LD_{50}, **mouse**: 49 mg/kg; **dog**: 7 mg/kg; estimated LD, **monkey**: 10 mg/kg.

METHOXYCHLOR

= *Chemform* = *Dimethoxy-DT* = *DMDT* = *ENT 1716* = *Marlate* = *Methoxo* = *Methoxy-DDT* = *Metox* = *Moxie*

***Ectoparasiticide.* analogue of *DDT* (q.v.).**

Untoward effects: Toxic symptoms in **man** are expected to be similar to those of ***DDT***. Environmental Protection Agency has set a limit of 0.04 ppm for drinking water and 0.05 ppm maximum for 1 day for **children**. **Adults** are asked not to drink water over 0.02 ppm for period > 7 years. FDA has set a limit of 0.1 ppm in bottled drinking water. OSHA sets a maximum of 15 mg/m^3 in workplace air for an 8 h day and 40 h work week. American Conference of Governmental Industrial Hygienists recommends a maximum level of 10 mg/m^3 in workplace air.

LD_{LO}, **human**: 6.43 g/kg; TD_{LO}, skin, **human**: 2414 mg/kg.

It is less toxic to warm-blooded species (**mammals**, **birds**), **bees** and more toxic to **fish**, **amphibians**, and **reptiles**. Trembling, convulsions, and kidney and liver damage reported in poisonings. Teratogenic in **rats** leading to rib deformities and delayed ossification, plus decreased maternal weight gain; wavy ribs or bent ribs in **mice** and **rats**, and fetotoxic in **rat** trials.

LD_{50}, **rat**: 2–6 g/kg; **mouse**: 1.85 g/kg; **quail** and **pheasant**: > 2 g/kg. 96 h LC_{50}, **trout**: 55–69 μg/l; acute dermal, female **rat**: > 66 g/kg.

σ-METHOXYCINNAMIC ALDEHYDE

= *β-(σ-Methoxyphenyl)acrolein*

In fragrances for soaps, detergents, cosmetic creams and lotions, perfumes, and foods.

Untoward effects: No sensitization or irritation at 4% in petrolatum on **human volunteers**.

LD_{50}, **rat** and acute dermal **rabbit**: > 5 g/kg; **mouse**: 400 mg/kg.

5-METHOXYDIMETHYL-TRYPTAMINE

= *5-MeO-DMT = 5-MDMT*

The most abundant hallucinogenic ***tryptamine*** in many New World snuffs or ingested pellets, made from *Virola theiodora* and *Piptadenia peregrina* or *Anadenenthera peregrina*. It is not active orally, unless taken with a ***monoamine oxidase inhibitor***, which is usually supplied by a ***β-carboline*** alkaloid, present in very small amounts in the same plant. Also found along with other ***tyramines*** in *Justica*, *Phalaris*, and other *Virola*.

Untoward effects: Australian **workers** found it to be one of the ***tryptamine*** alkaloids in *Phalaris tuberosa*, causing "*Phalaris* staggers" in **sheep**. Symptoms include hyperexcitability, salivation, head-nodding, mydriasis, incoordination, lateral recumbency, and acute cardiac failure. Injection of 1–2 mg/kg is lethal to **sheep**.

IP, LD_{50}, **mouse**: 115 mg/kg.

2-METHOXYETHANOL

= *Ethylene Glycol Monomethyl Ether = 2-ME = Methyl Cellosolve® = Methyl Oxitol®*

***Solvent*. In varnishes, *paints*, wood stains, *inks*, resins, *leather*dyeing, antistall agent in *gasoline*, anti-icing additive in brake fluids and aviation fuels, and, formerly, in cosmetics.**

Untoward effects: Vapors are eye, nose, and throat irritants. Headache, drowsiness, weakness, ataxia, tremors, somnolence, and anemic pallor also reported. OSHA skin exposure limit – 25 ppm.

LD_{LO}, **human**: 3380 mg/kg; TC_{LO}, **human**: 25 ppm.

Long-term exposure has caused liver, kidney, teratogenicity, and reproductive problems in **rats**, **mice**, and **rabbits**.

LD_{50}, **rat**: 2,460 mg/kg; **rabbit**: 890 mg/kg; **guinea pig**: 950 mg/kg; LC_{50}, inhalation, **mouse**: 1480 ppm; LC_{LO}, inhalation, **rat**: 2000 ppm/4 h.

METHOXYETHYL ACETATE

= *Ethylene Glycol Monomethyl Ether Acetate =2 MEA = Methyl Cellosolve® Acetate*

Solvent.

Untoward effects: An abused solvent by sniffing. Irritant to eyes, respiratory tract, and skin. May cause renal damage.

TC_{LO}, **human**: 1000 mg/m^3.

Inhalation LC_{LO}, **rat**: 7000 ppm/4 h; **cat**: 6000 mg/m^3/7 h.

METHOXYETHYLMERCURY CHLORIDE

= *Ceresan™*

Fungicide.

Untoward effects: In Germany, a 42-year-old **female** ingested 125 g of ***Ceresan™***, containing 3.5% ***mercury*** from it (5682 mg = 4,375 mg ***mercury***), in a suicide attempt. **She** was admitted 2 h later into an intensive care unit; was somnolent, and vomited a blue liquid. **She** recovered after treatment.

LD_{LO}, **human**: 5 mg/kg.

Azotemia, enzymuria, contact dermatitis, stomatitis, and local cauterization after 22–44 mg/kg in **rats**.

LD_{50}, **rat**: 30 mg/kg; **mouse**: 47 mg/kg.

METHOXYETHYL MERCURY SILICATE

= *Cerasan*™ = *MEMS*

Fungicide. Cerasan*™ is a seed dressing containing 1.75%-5.25% *mercury.

Untoward effects: Four daily doses of 20 mg/kg on ***wheat*** has been lethal to **sheep**.

LD_{50}, **rat**: 1.14 g/kg.

METHOXYFLURANE

= *DA-759* = *Methoxane* = *Metofane* = *Pentec* = *Pentrane* = *Pentrano*

***Inhalation anesthetic.* Fruity aroma. Used since the early 1960's.**

Untoward effects: ***Blood***: Eosinophilia (5–28%) associated with hepatotoxicity. Hemolytic anemia in 27-year-old **male** with glucose-6-phosphate dehydrogenase deficiency, after its use as an anesthetic for 4½ h.

Cardiovascular: Cardiac contractility decreased, as is cardiac output and arterial pressure. Only moderately sensitizes the heart to ***catecholamines***. Rare ventricular arrhythmias, occasionally bradycardia (***atropine*** responsive), transient nodal rhythm premature ventricular contractions in 7-year-old **male** after starting anesthesia. Dysrhythmias with ***carbon dioxide*** retention.

Contamination or adulteration: In-series vaporizers have delivered anesthetic levels of ***halothane*** along with methoxyflurane. Foam formation due to ***Dow-Corning #11 Sealant*** in ***copper*** kettles, which was potentially lethal. ***Fetus, Neonate, and Children***: Mild **fetal** tachycardia. Cough, extrasystoles, post-op rigors, emesis, transient excitation, respiratory depression, sialism, vomiting, and hiccups in **neonates** and **children**. Persistent drowsiness in **mother** and narcotic effect on **newborn** due to its slow elimination.

Liver: Jaundice, hepatic function failure, and liver necrosis reported. Moderate increase in SGOT, SGPT, and sulfobromophthalein retention. These may be hypersensitivity reactions.

Miscellaneous: Post-op rigidity, respiratory depression, pallor, nausea, and vomiting are common. Subclinical myasthenia gravis. Malignant hyperpyrexia, acidosis, hyperkalemia, neuromuscular blockade, increased cerebrospinal fluid pressure, eyelid ptosis; increased serum ***creatinine, fluorides***, and ***urates***. Emergence somnolence or excitement. Occasionally nausea and vomiting. Rarely abused. In the operating room environment, 2–10 ppm has been found near the **anesthesiologist** and 1–2 ppm near the **surgeon**. When used for obstetric anesthesia, there is a significant increase in urinary ***fluoride*** in delivery ward **personnel**. Inhalation by a 33-year-old nurse exacerbated a subclinical myasthenia gravis that persisted for 2–3 h. Thrombophlebitis has resulted from its use as an IV emulsion.

Renal: Nephrotoxic (5/17,334 anesthesias lasting > 1 h), increased blood urea nitrogen, renal oxalosis, interstitial fibrosis; and polyuria (especially if surgery is > 2 h) occur.

Respiratory: Laryngospasm and bronchospasm.

Drug interactions: Use with drugs that have ***β-adrenergic*** activity and/or ***sympathomimetics*** can cause dangerous or fatal arrhythmias. These can include ***tricyclic antidepressants, l-dopa***, and ***monoamine oxidase inhibitors*** that increase ***noradrenaline*** levels at postganglionic sympathetic nerve endings. Pretreatment with ***secobarbital*** in 57-year-old **male** probably altered its metabolization after 2½ h of anesthesia, leading to ***vasopressin***-resistant nonoliguric renal insufficiency. Fatal nephrotoxicity increase, probably due to increased ***oxalate*** levels, when ***demeclocycline, minocycline***, and ***tetracycline*** are used with it. Hardening of ***PVC*** plastic endotracheal catheters, often with great numbers of methoxyflurane droplets, which dissolved ***dioctylphthalate*** plasticizer, causing irritation of tracheobronchial mucosa. Explosion danger from re-use of disposable anesthetic circuits after its use.

LD_{LO}, **human**: 140 mg/kg; TC_{LO}, inhalation, **human**: 3,500 ppm/1 h.

Avoid use of ***β-adrenergic drugs*** or ***epinephrine***, as they may induce or aggravate cardiac arrhythmias. The possibility of hepatotoxicity, excessive transplacental passage, and adverse interaction with ***tetracycline*** or renal function should not be ignored. Some **dogs** recover quietly, others with excitation, thrashing, and whinings. **Cats** may also recover with excessive agitation; wrapping the body of **cats** in a "cocoon" formed from newspaper during recovery helps minimize the problem, similarly, wraps on **guinea pigs** during induction. Chronic use in **rats** stimulates liver microsomal activity, shortens ***hexobarbital*** hypnotic effects, etc. **Monkeys**, particularly *M. arctoides*, may vomit during induction, with the inherent danger of aspiration pneumonia, unless premedicated with ***atropine***. Beginners tend to over-anesthetize patients. Reduce vaporizer settings as soon as surgical anesthesia is reached, and close it entirely, when surgery is nearly completed. Elimination by the lungs and liver. Cardiovascular and respiratory depression has occurred in anesthetized **swine**. **Guinea pigs** exhibit squirming motions during anesthetic induction. Chronic administration to **mice** can influence the carcinogenic effect of ***diethylnitrosamine***, altering the tumor type induced. Post-op renal toxicity may be from the ***fluoride*** ion released by the catabolism of the drug. This can be severe enough to make its use contraindicated in 344 female **rats**, where it produces a diabetes insipidus-like syndrome. A prolonged anesthetic procedure (9 h) has caused hepatorenal pathology problems, as a result of hypersensitivity (this normally follows multiple exposures) in a **dog**. Nephrotoxicity has been reported in **dogs**, when given with ***flunixin***. Avoid high concentrations (> 2%) during induction in **dogs**. Produces neonatal depression in **cats**.

Inhalation LC_{50}, **rat**: 123 g/m^3/4 h, **mouse**: 118 g/m^3/2 h.

6-METHOXYHARMALAN

Untoward effects: Synthetic material is hallucinogenic at 1.5 mg/kg orally.

10-METHOXYHARMALAN

Use: Solvent. Made by removing a molecule of water from ***melatonin***, which is derived from ***serotonin***.

Untoward effects: Highest concentration of ***serotonin*** is found in the brain of **mental patients**. Weak ***serotonin*** inhibitor.

Causes abnormal behavior in **rats**; maze errors in well-trained **rats** with low dosage and hour-long tremors after larger doses.

METHOXYMETHYLAMPHETAMINE

Untoward effects: A synthetic hallucinogen. Mydriasis, increased deep-tendon reflexes, agitation, confusion, delusions, and difficulty in naming colors. Toxic psychosis generally disappears in a few days, after ceasing use of the drug, but it can persist for weeks or months.

IP, LD_{50}, **mouse**: 96 mg/kg.

METHOXYPHENAMINE

= *Orthoxine* = *Proasma*

Brochodilator, sympathomimetic.

Untoward effects: Nausea, vomiting, chills, irritability, fever, cyanosis, tachycardia, initial hypertension, mydriasis, blurred vision, opisthotonus, spasmodic jerking, dyspnea-gasping, convulsions, coma, and death from respiratory failure.

TD_{LO}, **human**: 714 μg/kg.

4-METHOXYPHENOL

= *p-Hydroxyanisole* = *Mequinol*

Mixed *caramel* and *phenol* odor. Waxy solid.

Untoward effects: Eye, skin, nose, throat, and upper respiratory tract irritant on exposure or inhalation. Central nervous system depression and eye and skin burns. National Institute for Occupational Safety and Health exposure limit of 5 mg/m^3.

4-(*p*-METHOXYPHENYL)BUTAN-2-ONE

= *Anisyl Acetone*

As a fragrance (sweet, floral odor) in soaps, detergents, cosmetic creams and lotions, and foods. From *aloe* wood (*Aquilaria agallocha*).

Untoward effects: No irritation or sensitization at 5% in petrolatum on **human** volunteers.

Non-irritating when applied full-strength for 24 h under occlusion on intact or abraded **rabbit** skin.

LD_{50}, **rat** and acute dermal, **rabbit**: > 5 g/kg.

N-[2-*m*-METHOXY PHENYL)-2-ETHYL BUTYL-(1)] γ-HYDROXYBUTYRAMIDE

Untoward effects: Synthetic with ***curare***-like action, paralyzing striated skeletal and respiratory muscles, useful parenterally with ***mebezonium iodide*** and ***tetracaine***, as euthansia in **animals**.

METHOXYPROMAZINE

= *Methopromazine* = *Mopazine* = *RP 4632* = *Tentone* = *Vetomazin*

Tranquilizer, antipschotic.

Untoward effects: Use by pregnant **mother** associated with significant increased risk of having a **child** with malformations, such as syndactly, microcephaly, club feet or hands, abdominal muscle aplasia, brachymesophalangy, clinodactyly, hydrocephalus, cardiac malformations, hypospadia, and cleft lip. If taken during the last trimester, extrapyramidal symptoms may develop in the **neonate**.

LD_{LO}, **human**: 50 mg/kg.

LD_{50}, **rat**: 730 mg/kg.

5-METHOXYPSORALEN

= *Bergapten(e)* = *Heraclin* = *Majudon* = *5-MOP* = *Psoraderm*

Use: Similar to ***methoxsalen*** and in sunscreens.

Untoward effects: Delayed erythema, cutaneous eruptions, and phytophoto-dermatitis. A number of case reports indicate probability of it inducing skin cancers. Dermatitis and blisters in **workers** handling ***parsley***, ***parsnips***, and ***rue***.

Causes skin cancers in **mice** also receiving UVA radiation.

METHOXYPYRIDOXINE

Untoward effects: Neurotoxic to the hippocampus. Oral or IV causes generalized seizures in **rabbits**.

METHSUXIMIDE

= *Celontin* = *Petinutin* = *Mesuximide*

Anticonvulsant.

Untoward effects: ***Blood and bone marrow***: Eosinophilia, leukopenia, monocytosis, thrombocytopenia, and pancytopenia (with or without bone marrow depression). Hyperemia and increased prothrombin time. Aplastic anemia, including a fatal case.

Central nervous system: Drowsiness, dizziness, ataxia, irritability, headache, photophobia, blurred vision, hiccups, insomnia, confusion, erratic behavior, and visual hallucinations.

Fetus and neonate: Possibly, along with ***metharbital***, induced fatal **neonatal** hemorrhage in **male-infant** after **mother** took 900 mg and 150 mg, respectively, throughout **her** pregnancy. Nail and distal phalanges hypoplasia reported. Decreased scores on gross motor index, fine motor index, language index, and personal social index.

Gastrointestinal: Anorexia, nausea, vomiting, diarrhea, epigastric or abdominal pain, constipation, and weight decrease.

Genitourinary: Proteinuria and microscopic hematuria.

Miscellaneous: Periorbital edema, increased temperature, cervial adenopathy, hepatitis, and diaphoresis. Severe osteomalacia in 22-year-old **female** after 6 years (900 mg/day/2 years, then, 450 mg/day). Improvement noted 2 months after treatment ceased. Occasionally renal damage reported.

Poisoning: An 18-year-old **female** ingested < 10 g in a suicide attempt. Half-life is < 3 h. **She** was in a rousable semistupor for 9 h before onset of deep coma that lasted about 60 h, possibly due to its metabolite, ***2-methyl-2-phenylsuccinamide*** (***N-desmethyl methsuximide***), suggesting the need for forced diuresis.

Skin: Urticaria, pruritic erythematous rashes, porphyria, and systemic lupus erythematosis.

Therapeutic serum levels reported as 0.25–0.75 mg%; toxic levels at > 1 mg%; and comatose at 2.5 mg%.

Induces ***porphyrin*** accumulation in young **chickens**, but not in **mice**.

LD_{50}, **mouse**: 1.4 g/kg.

METHYCLOTHIAZIDE

= *Diutensin* = *Methylchlorothiazide* = *MK 6096*

Diuretic, antihypertensive.

Untoward effects: ***Blood and Bone Marrow***: Leukopenia, agranulocytosis, thrombocytopenia, and aplastic anemia.

Cardiovascular: Orthostatic hypotension can be exacerbated by ***alcohol, barbiturates***, and ***narcotics***. Vasculitis (necrotizing angiitis).

Central nervous system: Dizziness, vertigo, paresthesias, headache, and xanthopsia.

Fetus and neonate: Crosses the placenta. A report indicates its use in the latter part of pregnancy may have been associated with thrombocytopenia in the **fetus** and **neonatal** death.

Miscellaneous: Hyperglycemia, glycosuria, hypercalcemia, hyperuricemia, hypokalemia, weakness, restlessness, muscle spasms, anorexia, cramping, nausea and vomiting, backache or leg pains, frequent urination, jaundice, and pancreatitis.

Skin: Hypersensitivity, photosensitivity, rash, urticaria, and cutaneous angiitis.

Drug interactions: May increase ***tubocurarine*** effect, decrease response to ***norepinephrine***, and alter ***insulin*** requirements.

LD_{50}, IV, **mouse**: 400 mg/kg; IP, 870 mg/kg; IP, **rat**: 2 g/kg.

METHYL ABIETATE

= *Abalyn*

Thick liquid by methylation of the resinous residue of *turpentine*.

Use: Tonnage is used in fragrances for soaps, detergents, cosmetic creams and lotions, and perfumes.

Untoward effects: No sensitization of 2% in petrolatum on 25 **human** volunteers.

LD_{50}, **rat** and acute dermal on **rabbit**: > 5 g/kg.

N-METHYLACETAMIDE

= *MMAC*

***Solvent.* In solubilizing *phenobarbital*.**

Untoward effects: Embryotoxic after maternal topical application on **rats**.

IV, LD_{LO}, **rabbit**: 17 g/kg; **chicken**: 14 g/kg. TD_{LO}, skin, **rabbit** (12–34 days pregnant): 1.2 g/kg.

METHYL ACETATE
= *Methyl Ethanoate*

Solvent.

Use: In ***lacquers***, ***paint*** removers, and adhesives.

Untoward effects: Inhalation leads to eye, skin, nose, and throat irritation; headache, drowsiness, unconsciousness, vision defects, optic nerve atrophy, and chest tightness. Symptoms from inhalation or ingestion resemble those of ***alcohol, methyl*** (q.v.). **Cementers** using it in adhesives, as in shoes, develop a characteristic eruption on the dorsum of the second and third fingers and the fourth metacarpophalangeal joint. National Institute for Occupational Safety and Health and OSHA exposure limits of 200 ppm (610 mg/m^3). LEL – 3100 ppm.

TC_{LO}, **human**: 15 g/m^3.

Narcotic to **animals**. Lethal (100%) to **cats** at 10,000 ppm/22 h and 22,000 ppm/2½ h.

LC_{LO}, inhalation, **cat**: 6700 mg/m^3/1 h. LD_{50}, **rabbit**: 3.7 g/kg; LD_{LO}, **rat**: 4.8 g/kg.

METHYL ACETONE

Solvent.* Actually, a mixture of *acetone, methyl acetate,* and *methyl alcohol.

Untoward effects: Ingestion can be fatal or cause blindness. 350 ppm irritates eyes. Flammable.

LD_{LO}, **human**: 500 mg/kg; TC_{LO}, inhalation, **human**: 100 ppm/5 min.

LD_{50}, **rats**: 3.4 g/kg; skin, **rabbits**: 13 g/kg.

p-METHYLACETOPHENONE
= *Methyl p-tolyl Ketone*

Use: Tonnage is used in fragrances for soaps, detergents, cosmetic creams and lotions, perfumes, and foods.

Untoward effdcts: A patch-test using it at full-strength for 24 h on 15 **humans** produced only one case of irritation; 6% in petrolatum on 25 **human** volunteers produced no sensitization.

Slightly irritating at full-strength on abraded or intact **rabbit** skin for 24 h under occlusion.

LD_{50}, **rat**: 1.4 g/kg; acute dermal, **rabbit**: > 2 g/kg.

METHYL ACETYLENE
= *Allylene* = *Propine* = *Propyne*

Untoward effects: Inhalation leads to respiratory irritation, tremors, hyperexcitability, and anesthesia. Contact with the liquid causes frostbite. National Institute for Occupational Safety and Health/OSHA limits exposure to 1000 ppm (1650 mg/m^3). Immediate danger to life or health is 1700 ppm. Flammable, sweet-smelling gas.

METHYL ACRYLATE
= *Methyl Propenoate*

Polymer.

Untoward effects: Acrid odor. Inhalation leads to eye, upper respiratory tract, and skin irritation. Contact dermatitis from airborne exposure in 39-year-old **male**. Use in manufacturing illicit ***fentanyls***. National Institute for Occupational Safety and Health/OSHA exposure limits for skin are 10 ppm (35 mg/m^3) TWA. Immediate danger to life or health, 250 ppm.

Inhalation TC_{LO}, **human**: 75 ppm.

Inhalation LC_{LO}, **rat**: 1000 ppm/4 h, **rabbit**: 2522 ppm/1 h. LD_{50}, **mouse**: 9.6 mmol/kg; **rat**: 300 mg/kg.

METHYL ACRYLONITRILE

Odor of bitter *almonds*.

Untoward effects: Eye and skin irritant in **man**. National Institute for Occupational Safety and Health sets exposure limits for skin of **man** at 1 ppm (3 mg/m^3).

Lacrimation, convulsions, and hind limb ataxia in **animals**.

LD_{50}, **rat**: 250 mg/kg; LD_{LO}, **mouse**: 15 mg/kg; LC_{50}, inhalation, **rat**: 328 ppm/4 h, **mouse**: 36 ppm/4 h, **rabbit**: 37 ppm/4 h, **guinea pig**: 88 ppm/4 h.

METHYLAL
= *Dimethoxymethane* = *Formal*

Volatile flammable liquid with *chloroform*-like odor and pungent taste.

Use: In perfumery and in resin manufacturing.

Untoward effects: Inhalation and vapors lead to eye, skin, upper respiratory tract irritation, and anesthesia. National Institute for Occupational Safety and Health/ OSHA exposure limits of 1000 ppm (3100 mg/m^3). Immediate danger to life or health, 2200 ppm.

METHYLAMINE
= *Aminomethane* = *Monomethylamine*

Colorless gas with fish- or ammonia-like odor.

Use: Millions of lbs used in manufacturing explosives, insecticides, and surfactants, and in ***leather*** tanning.

Untoward effects: One of the many adulterants in ***olive oil*** that caused > 1000 **people** to be hospitalized in Spain in 1981, with respiratory distress. Inhalation leads to eye, skin, and respiratory system irritation. Coughing, skin and mucous membrane burns, dermatitis, and conjunctivitis. The "liquid" can cause frostbite. National Institute for Occupational Safety and Health/OSHA exposure limits are 10 ppm (12 mg/m^3). Immediate danger to life or health of 100 ppm. Flammable. Used in the illicit manufacture of ***amphetamines***.

SC, LD_{LO}, **mouse**: 2.5 g/kg; **frog**: 2 g/kg.

β-*N*-METHYLAMINO-*L*-ALANINE
= *BMAA*

Amino acid.

Untoward effects: Ingestion of this uncommon ***amino acid*** in the cytotoxic ***cycad*** (q.v.) seeds may be responsible for the neuronal degeneration noted with Guam amyotrophic lateral sclerosis-parkinsonism-dementia in **humans**.

The seeds induce in **monkeys** a neurodegenerative syndrome closely resembling the Guam ALS-PD of **man**.

3-METHYL-4-AMINO-*N*-DIETHYL-ANILINE HCL
= *CD-2*

Color film developer. **Also used as dye intermediates and in copying processes for blueprints.**

Untoward effects: Lichen planus-like reaction from contact.

2-METHYLAMINOETHANOL
= *2 MAE*

Untoward effects: Corrosive, viscous liquid with a fishy odor. Skin, eye, and mucous membrane irritant and caustic. Believed to penetrate nail plate and cause discoloration.

Intubation of **mice** leads to teratogenicity.

p-METHYLAMINOPHENOL SULFATE
= *Metol*

Black and white photographic developer. **Used since the 1890's.**

Untoward effects: Contact with skin has caused eczema in **photographers**.

LD_{LO}, **human**: 50 mg/kg.

LD_{LO}, **rat**: 200 mg/kg.

METHYL AMYL ALCOHOL
= *Methyl Isobutyl Carbinol* = *MIBC*

Solvent. **In dyes, resins, *lacquers*, oils, and gums.**

Untoward effects: Vapors are irritant to eyes, nose, and throat. Sharp odor. Can cause headache, dizziness, nausea, drowsiness, and unconsciousness. Goggles and protective gloves are recommended for **handlers** and respirators near spills. TWA–TLV, skin – 25 ppm.

Causes narcosis in exposed **animals**.

LD_{50}, **rat**: 2.6 g/kg; LD_{LO}, **mouse**: 1 g/kg; inhalation LC_{LO}, **rat**: 2000 ppm/4 h.

N-METHYLANILINE
= *N-Methylbenzenamine*

Solvent. **Also used in organic synthesis.**

Untoward effects: Vapor inhalation leads to weakness, dizziness, headache, dyspnea, cyanosis, methemoglobinemia, pulmonary edema, and liver and kidney pathology.

Fed to **mice** with ***nitrites***, produces ***nitrosamines***, which are then absorbed and produce lung tumors.

METHYL ANTHRANILATE
= *MA*

Use: In perfumes, as a **bird** and **mouse** repellent, and in cosmetics and foods as a flavorant.

Untoward effects: No irritation or sensitization of 10% in petrolatum on **human** volunteers.

Causes taste aversion in **birds** and **mice**. Full-strength to intact or abraded **rabbit** skin for 24 h under occlusion was moderately irritating.

LD_{50}, **rat**: 2.9 g/kg; **mouse**: 3.9 g/kg; **guinea pig**: 2.78 g/kg; acute dermal, **rabbit**: > 5 g/kg.

METHYL ASPARTATE

Amino acid.

Untoward effects: A very potent neuroexcitant and neurotoxic ***amino acid***, which, based on trials in **mice**, may be responsible for seizures in **man**, especially after head injuries. Antagonists are being investigated.

METHYLATED SPIRITS

Normally, used industrially. A mixture usually containing *alcohol, methyl* and *alcohol, ethyl*, often at a 3–5%: 95% ratio.

Untoward effects: A premature **infant** (985 g – 27th week of gestation) died as a result of its use as a body cleanser. Plum-colored necrotic lesions on the buttocks and lumbar area, probably because **her** poorly keratinized skin readily absorbed the ***alcohol*** and **she** was lying on **her** back, preventing evaporation. Asthma has been reported from occupational exposure to fumes. Spore-forming bacilli have been isolated from hospital samples.

2-METHYLAZIRIDINE

= *Propyleneimine*

Use: Volatile chemical intermediate for ***paper***, textile, ***rubber***, photographic dye, oil refining, rocket fuel, and pharmaceutical industries. Improves binding of adhesives. Produced in tonnage.

Untoward effects: Inhalation, ingestion, and percutaneous absorption are routes of **human** exposure. Flammable. Heating causes emission of toxic ***nitrogen oxides*** (q.v.). American Conference of Governmental Industrial Hygienists lists exposure limit for it at 2 ppm TWA.

Orally in male **rats** leads to leukemia and intestinal adenocarcinomas; mammary adenomas in female; and gliomas and squamous cell carcinomas of the ear canal in male and female.

LD_{50}, **rat**: 19 mg/kg; LC_{LO}, inhalation, **rat**: 500 ppm/4 h, **guinea pig**: 500 ppm/1 h.

METHYLBENZETHONIUM CHLORIDE

= *Bactine* = *Diaparene Chloride* = *Hyanoine 10X*

***Topical antiseptic.* A quaternary, usually used at a 1:1,000 aqueous dilution.**

Untoward effects: The pure material is corrosive and causes severe eye and skin damage. Serious adverse effects from ingestion are decreased by consuming large quantities of water, milk, egg whites, or gelatin solutions. Circulatory shock and convulsions can follow.

LD_{LO}, **human**: 50 mg/kg.

Central nervous system depression and death after gavage in **mice** and **rats**. Eye and skin irritant on **rabbits**.

LD_{50}, **rat**: 828 mg/kg; **mouse**: 750 mg/kg.

METHYL BENZIMIDAZOLE-2-CARBAMATE

= *Bavistin* = *Carbendazim* = *Carbendazole* = *Correx* = *Derosal* = *G 665* = *HOE 17,411* = *Lignasan BLP*

Fungicide, anthelmintic.

Untoward effects: Eye, nose, throat, and skin irritant. Environmental Protection Agency restricted its use, as a treatment of Dutch Elm Disease, to trained **applicators**.

In **cats**, **rabbits**, **guinea pigs**, and **mice** leads to decreased hematopoiesis, gastrointestinal upsets, nervousness, and decreased growth rate.

METHYL BENZOATE

= *Niobe Oil*

Use: In perfumery, as a solvent, and as a chemical intermediate.

Untoward effects: As a ***coccidiostat*** (resistance to it develops quickly) in **poultry**.

LD_{50}, **rat**: 1.35 g/kg; **mouse**: 3.3 g/kg; **rabbit**: 2.17 g/kg; **guinea pig**: 4.1 g/kg.

3-METHYL-2-BENZOTHIAZOLINONE HCL

= *MBTH*

Reagent.

Untoward effects: Sluggishness, tremors, ataxia, and excessive salivation in **rats** and **rabbits**; convulsions also noted in **rabbits**. Moderate primary irritant after topical application on **rabbit** eyes. Percutaneous absorption demonstrated.

LD_{50}, male **rat**: 308 mg/kg; female: 149 mg/kg; male **rabbit**: 177 mg/kg; female: 268 mg/kg.

1-METHYL-2,3-BIS-HYDROXYMETHYL-3-PYRROLINE CARBAMATE

Untoward effects: Highly toxic, leading to acute and chronic liver and lung lesions in **rats**, identical to those seen with ***monocrotaline*** (q.v.) poisoning.

METHYL BROMIDE

= *Bromo-O-Gas* = *Bromomethane* = *Dowfume MC-2* = *Embafume* = *Fumigant 1* = *Kayafume* = *Me Br* = *Meth-O-Gas* = *Pestmaster* = *Profume*

***Fumigant.* Also for degreasing wool and extracting oils from nuts, seeds, and flowers. In fire-extinguishers and used as a refrigerant. A chlorinated hydrocarbon colorless gas with a chloroform-like odor. U.S. production is > 60 million lbs/year.**

Untoward effects: Poisonous! Contact leads to eye and skin irritation. Burning and vesicular skin eruptions. The liquid (compressed gas) can cause frostbite. Inhalation leads to respiratory tract irritation, headache, visual disturbances, dizziness, vertigo, nausea, vomiting, severe weakness, malaise, hand tremors, bizarre myotonic states, behavioral and emotional disturbances, dyspnea, pulmonary edema, slurred speech, confusion, convulsions, coma, and even death from circulatory and respiratory failure. Delayed symptoms, even years later, leading to memory loss, hallucinations, and temporary blindness. By 1976, > 50 deaths from it were reported. In 1988, 1/32 cylinders was punctured when a truck over-turned. **People** were treated for inhalation-type exposure (7) and for dermal exposure (2), leading to nausea, vomiting, dyspnea, headache, dizziness, burning throat, coughing, and chest tightness in six that were hospitalized. Symptoms 2–48 h after exposure to as little as 35 ppm reported. OSHA sets a maximum skin exposure of 20 ppm (80 mg/m^3). American Conference of Governmental Industrial Hygienists sets inhalation limit of 5 ppm TWA and 15 ppm for skin exposure. Immediate danger to life or health is 250 ppm. A tentative phase-out for its use has been set for the year 2001. ***Chloropicrin*** (q.v.) has been intentionally added as an irritant warning agent.

LD_{LO}, **human**: 50 mg/kg; TC_{LO}, inhalation, **human**: 35 ppm.

Target organs in **rats**, **mice**, and **guinea pigs** are liver, kidney, spleen, and brain. In **rats**, gastric gavage leads to squamous cell carcinomas and hyperplasia of the stomach. Has caused accidental deaths in **poultry**. **Cattle**, **goats**, and **horses** eating volunteer ***oat*** hay treated with it as a soil fumigant became lethargic, weak, and ataxic. ***Bromide*** content of the hay was 6800–8400 ppm. Used in treating **dog** food, grains, and citrus pulp for feed use. The FDA limits residues from such treatments at 400, 125, and 90 ppm, respectively. Premises must be well aerated after use to prevent possibility of pulmonary edema or central nervous system disturbances, and suitable gas masks must be worn during application.

Inhalation LC_{LO}, **rat**: 3120 ppm/15 min, **guinea pig**: 300 ppm/9 h, **rabbit**: 500 ppm/11 h.

METHYL-*tert*-BUTYL ETHER

= *MTBE*

Octane booster, cholelitholytic agent. **Billions of gallons produced annually.**

Untoward effects: Neurotoxic, hematologic, and oncogenic effects from chronic inhalation. An acute renal failure reported after extravasation during dissolution of gallstones, due to leakage alongside the catheter. Dizziness, headache, and rash in **many** exposed to it in industry. Will disintegrate ***rubber*** gloves. Exposure to carcinogenic doses, 8000 ppm vapors leads to decreased body, pituitary, and ovary weight; estrus cycle lengthened.

METHYL BUTYL KETONE

= *2-Hexanone* = *MBK* = *MnBK*

Solvent. **In adhesives, *lacquers, paints,* and vinyl and acrylic coatings.**

Untoward effects: Peripheral neuropathy with sensory loss, motor weakness, and degeneration of nerve fibers. Its major neurotoxic metabolite is ***2-hexanedione***. Caused polyneuropathy in 138 **employees** of a factory making plastic-coated and color-printed fabrics. Muscle weakness and electromyographic abnormalities were primarily distal. Sensory deficits were mostly distal and limited to pain, touch, temperature discrimination, and occasionally loss of vibration sense. Several cases occurred in **spray-painters**. Has also caused neuropathies in Italian ***shoe*** factory **workers** and **glue-sniffers**. Irritant to eyes and nose. Can cause weakness, paresthesia, dermatitis, and headache. American Conference of Governmental Industrial Hygienists, TWA, 5 ppm; OSHA, TLV, TWA, 5 ppm. Percutaneous absorption in **man** and **dogs**. **Workers** exposed to it in the air for nearly a year developed weakness, numbness, and skin tingling of hands and feet.

Peripheral neuropathy development after 3 months air exposure of 200–600 ppm in **cats**, **rats**, and **chickens**. Exposure of **monkeys** and **rats** to 98 ppm/6 h/day, 5 days/week led to decreased sciatictibial nerve conduction, after 10 months in **monkeys**, and 7 months in **rats**. Relative sensitivity to its toxicity is **chickens** > **cats** > **dogs** > **monkeys** > **rats**. All **guinea pigs** died after exposure to 5000 ppm for 45 min. Peripheral neuropathy in **rats** inhaling 1300 ppm 6 h/day, 5 days/week for 5 months.

LD_{50}, **rat**: 2.6 g/kg; LD_{LO}, **mouse**: 1 g/kg; **guinea pig**: 914 mg/kg.

5-METHYL-3-BUTYLTETRAHYDROPYRAN-4-YL ACETATE

Use: In fragrances for soaps, detergents, cosmetic creams and lotions, perfumes, and foods.

Untoward effects: No irritation or sensitization of 8% in petrolatum on **human** volunteers.

Mildly irritating on abraded or intact **rabbit** skin for 24 h under occlusion.

LD_{50}, **rat** and acute dermal **rabbit**: > 5 ml/kg.

METHYL CELLULOSE

= *Bagolax* = *Celevac* = *Cellothyl* = *Cellucon* = *Cellulose Methyl Ether* = *Cellumeth* = *Cologel* = *Hydrolose* = *Methocel MC* = *Nicel* = *Syncelose* = *Tylose* = *Vet-Sorb*

Laxative, suspending agent.

Untoward effects: Diarrhea can occur in obese **patients** who think "more will do better" in satiating appetite. Severe diarrhea can occur in **children** who eat large quantities of **their** parents' dietetic "candy". If swallowed dry or chewed, esophageal obstruction has occurred. Will form complexes with ***p-chloro-m-xylenol***, making it unavailable as an antiseptic.

METHYL CHLORIDE

= *Chloromethane*

***Local anesthetic, refrigerant.* U.S. annual production capacity is > 900 million lbs/year. Natural biomass annual production estimated at ~5 million tons annually.**

Use: Solvent, refrigerant, aerosol propellent, and in manufacturing ***silicones, tetramethyl lead***, and ***tetraethyl lead***.

Untoward effects: Even at dangerous concentrations, this colorless gas is not noticable, despite its faint, sweet odor. Inhalation leads to dizziness, nausea, vomiting, blurred vision, severe headache, ataxia, staggering, slurred speech, convulsions, coma; anemia with polymorphonuclear leukocytosis after refrigeration leaks; liver and kidney damage (often leading to mistaken diagnosis of botulism, meningitis, or drunkeness). Poisoning occurred in a **crew** of 15 on a fishing trawler at sea, when a refrigerator in the forecastle leaked; one died; three became mentally depressed, and two of **those** committed suicide; five had personality changes; five had decreased tolerance to ***alcohol***, and nine had delayed central nervous system effects. Rapidly hydrolyzed to ***methyl alcohol*** (q.v) and ***hydrochloric acid***. Therefore, it is not surprising that many symptoms are similar to those of the former. Fumes are often released released by **hobbyists** heating ***polystyrenes***. Percutaneous absorption occurs. Even years after exposure, **patients** have developed memory loss, hallucinations, and temporary blindness. Direct contact with the liquid compressed gas can cause frostbite, skin burns, and irritation. National Institute for Occupational Safety and Health suggests it be considered as a potential carcinogen and teratogen. OSHA, TWA exposure limit of 100 ppm. Immediate danger to life or health, 2000 ppm.

Inhalation adversely affects **guinea pigs**, **rats**, **rabbits**, **mice**, **dogs**, **monkeys**, **cats**, and **chickens**, with **dogs** being the most sensitive and **rats** the least. Hemolytic anemia after IP injections in **rats**.

Inhalation LC_{50}, **mouse**: 3146 ppm/7 h; LC_{LO}, **dog**: 15,000 ppm/7 h; **guinea pig**: 20,000 ppm/2 h.

2-METHYL-4-CHLOROANILINE

= *4-Chloro-orthotoluidine*

A metabolite of *chlordimeform* (q.v.).

Untoward effects: Severe acute hemorrhagic cystitis, hematuria, and abdominal pain in **workers** exposed to the parent compound.

Less severe effects in **cats**.

SC, LD_{LO}, **cat**: 310 mg/kg.

METHYL CHLOROCARBONATE

= *Methyl Chloroformate*

Made from phosgene (q.v.) and alcohol, methyl. May contain 0.5% Phosgene.

Untoward effects: Vapors are strongly irritating to respiratory tract and eyes. Heavy exposure can cause pulmonary edema. It can burn mucous membranes and exposed skin. A 48-year-old **male** cleaned an empty tank and took **his** respirator off, leading to coughing, tachypnea, shallow breathing, and tachycardia. Post-mortem revealed extensive pulmonary edema, left heart hypertrophy, fatty liver, and gastrointestinal petechial hemorrhages. Liver, kidney, and spleen were shrunken.

LD_{50}, **rat**: 110 mg/kg; inhalation, 88 ppm/1 h.

METHYL CHLOROPHENOXY ACETIC ACID

= *Agritox* = *Agroxone* = *Chiptox* = *Cornox* = *Dikotex* = *MCP* = *MCPA* = *Methoxone* = *Weedex* = *Verdone*

Herbicide.

Untoward effects: A depressed 65-year-old **male** died within 24 h after ingesting 57 ml or 250 mg/kg in a suicide. Suicidal poisoning in two **females**, 33-year-old and 51, 1–4 days after ingesting it with ***2,4-D***. Symptoms included vomiting, deep coma, cerebral edema, and shock before death. A 32-year-old **male** who ingested ~350 mg/kg was frothing at the mouth, unrousable, unresponsive to painful stimuli, had limb jerks, facial twitching, grand mal seizures, opisthotonus, and rapid pulse. Despite gastric lavage, myoclonic limb twitchings continued. **He** was anuric and unconscious until **his** death about 20 h after admission. Usually toxic at 100 µg/ml; 180 µg/ml may be fatal.

LD_{LO}, **human**: 500 mg/kg.

Neurological symptoms can be produced in exposed **rats** and dogs. Hatchability and viability of **chicks** and **pheasants** decrease with increasing exposure. Early deaths in experimental **animals** are due to ventricular fibrillation, whereas paralysis, clonic jerking, and coma precede delayed death.

LD_{50}, **rat**: 680 mg/kg; acute dermal, **rabbit**: > 1 g/kg.

4(2-METHYL-4-CHLOROPHENOXY) BUTYRIC ACID

= *Bexone* = *Cantrol* = *4(4-Chloro-2-methylphenoxy)-butanoic Acid* = *MCPB* = *2M-4Kh-M* = *Thitrol* = *Tropotox*

***Herbicide.* Susceptible plants metabolize it to *MCPA*.**

Untoward effects: LD_{50}, **rat**: 680 mg/kg; acute dermal, **rabbit**: 1 g/kg; LC_{50} in feed, **quail**, **pheasant**, **mallard duck**: > 5000 ppm.

3- or 20-METHYLCHOLANTHRENE

= *3-MECA*

Experimental carcinogen. **Research chemical synthesized from bile.**

Untoward effects: SC leads to sarcomas in **rats** and **mice**, papillomas in **mice**, fibrosarcomas at injection sites in **monkeys**, colon cancer in **male hamsters**, and liposarcomas and leukopenia in **guinea pigs**. Oral gavage leads to mammary adenocarcinomas in **rats**; intratracheal use leads to lung cancers in **mice**; IP leads to liver enlargement in **rats**. Intra-amniotic use in **mice** leads to monstrosities, tail and feet deformities, and SC hemorrhages. Topical cervical application in **monkeys** causes cervical dysplasias; skin cancers in **mice**, and cutaneous hemangiomas in **ducks**. Passes into the milk of **rats** (~0.2%), **rabbits** (~0.003%), and **sheep** (0.01%).

TD_{LO}, **rat**: 280 mg/kg; **mouse** (20 days pregnant): 40 mg/kg.

METHYL CINNAMATE

= *Methyl-3-phenylpropenoate*

Fragrance used in soaps, detergents, cosmetic creams and lotions, perfumes, and foods. Fruity, balsamic odor similar to *strawberries*.

Untoward effects: No sensitization or irritation of 10% in petrolatum on **human** volunteers. Higher concentration reported as irritant.

LD_{LO}, **human**: 500 mg/kg.

Non-irritating at full-strength on intact or abraded **rabbit** skin for 24 h under occlusion. No conjunctival irritation when applied to **rabbit** eyes.

LD_{50}, **rat**: 2.61 g/kg; acute dermal, **rabbit**: > 5 g/kg.

METHYLCINNAMIC ALCOHOL and ALDEHYDE

Fragrance in soaps, detergents, cosmetic creams and lotions, and perfumes. The *aldehyde* is also approved for food use.

Untoward effects: ***Alcohol***: No irritation at 2% in petrolatum and no sensitization at 8% on **human** volunteers.

Mildly irritating after full-strength application to intact or abraded **rabbit** skin for 24 h under occlusion.

LD_{50}, **rat**: 2.4 ml/kg; acute dermal, **rabbit**: > 5 g/kg.

Aldehyde: No irritation or sensitization of 8% in petrolatum on **humans**.

Non-irritating when applied full-strength to intact or abraded rabbit skin for 24 h under occlusion.

LD_{50}, **rat**: 2 g/kg; acute dermal, **rabbit**: > 5 g/kg.

METHYL-2-(CINNAMIDOOXY) PROPIONATE

= *MVP*

Untoward effects: Slight, reversible liver toxicity and retinopathy after repeated intragastric intubation in **rats**.

METHYLCLONAZEPAM

A clonazepam derivative, which acts directly to kill schistomas.

Untoward effects: Unfortunately, in **man**, its central nervous system effect has been deep sedation and, often, inability to function.

METHYLCONIINE

Untoward effects: A poisonous alkaloid along with ***coniine*** (q.v. for toxicity) in ***Conium maculatum*** (q.v.).

LD_{50}, **mouse**: 100 mg/kg leading to death within 12 min.

6-METHYLCOUMARIN

= *6-Methyl-1,2-benzopyrone*

Fragrance in soaps, detergents, cosmetic lotions and creams, and perfume. Also approved for food use at 30 ppm in Europe.

Untoward effects: No irritation or sensitization by 4% in petrolatum in **human** volunteers. Photoallergic contact, eczematous dermatitis reported.

Mildly irritating to intact or abraded **rabbit** skin for 24 h under occlusion. No toxicity to **rats** in 2-years feeding study, unless level is at 7500 ppm (growth depression) in **male** and in **female** at 15,000 ppm. A **male dog** fed 150 mg/kg required euthanasia after 39 days due to weakness, emaciation, dehydration, hepatitis, and muscle atrophy.

LD_{50}, **rat**: 1.68 g/kg; acute dermal, **rabbit**: > 5 g /kg.

METHYLCYCLOHEXANE

= *Cyclohexylmethane* = *Hexahydrotoluene* = *MCH*

Solvent. **Used in organic synthesis.**

Untoward effects: Eye, nose, throat, and skin irritation; light-headedness, drowsiness, and unconsciousness from inhalation or contact. OSHA exposure limit of 500 ppm TWA; National Institute for Occupational Safety and Health, 400 ppm; LEL, 1200 ppm.

Narcosis in exposed laboratory **animals**.

LD_{50}, **mouse**: 214 mg/kg.

METHYLCYCLOHEXANOL

= *Hexahydrocresol* = *Hexahydromethylphenol*

***Solvent*. For *cellulose* esters, in *lacquers*, in textile soaps and detergents, and as an antioxidant for lubricants.**

Untoward effects: Vapors are harmful, leading to eye, skin, and upper respiratory tract irritation; headache. National Institute for Occupational Safety and Health, 50 ppm, OSHA, 100 ppm upper exposure limits. Immediate danger to life or health, 500 ppm.

TC_{LO}, inhalation, **human**: 500 ppm.

Narcosis and liver and kidney damage in exposed laboratory **animals**.

LD_{50}, **rat** and **rabbit**: 1.7 g/kg.

METHYLCYCLOHEXANONE

***Solvent*. Used in *lacquers*. *Acetone*-like odor.**

Untoward effects: Eye and mucous membrane irritation, narcosis, and dermatitis. National Institute for Occupational Safety and Health, 50 ppm, OSHA, 100 ppm upper exposure limits. Immediate danger to life or health, 600 ppm.

LD_{LO}, **rabbit**: 1 g/kg; skin, **rabbit**: 4.9 g/kg.

METHYLCYCLOPENTADIENYL-MANGANESE TRICARBONYL

= *MMT*

Use: Anti-knock or octane-booster fuel additive.

Untoward effects: Inhalation, ingestion, or contact leads to eye, skin, and respiratory system irritation. Chest tightness, dyspnea, wheezing, cough, dry throat, pulmonary edema, and skin blisters. National Institute for Occupational Safety and Health upper exposure limit of 0.01 ppm.

Bronchiolar epithelial cell necrosis and alveolar damage in **mice**, **rats**, and **hamsters**.

LD_{50}, **rat**: 58 mg/kg.

S-METHYL CYSTEINE SULFOXIDE

= *SMCO*

Untoward effects: The primary intravascular hemolytic toxin (***Kale Anemia Factor***) in ***kale*** (q.v.). Substantial amounts also found in the roots and tops of ***broccoli***, ***brussel sprouts***, ***cabbage***, ***kohlorabi***, ***radishes***, ***rutabagas***, ***turnips***, and other *Brassicas*.

Its fermentation, as in the rumen of **cattle**, leads to ***dimethyl disulphide***, which is the active secondary hemolysin. Its ingestion has decreased plasma ***cholesterol*** activity in **rats**, and, because of the plant's high level of consumption in **man** in Japan, **their** level of cardiac disease may be lower. In **goats**, 1.5–1.9 g/10 kg causes serious hemolytic effects. **Rats** fed a 3% level in their feed have enlarged spleens, and 4% level caused death and alimentary bleeding in a week.

16α-METHYL DICHLORISONE ACETATE

***Hair-restorer*. Ten times potency of *dichlorisone* (q.v.).**

Untoward effects: Cushingoid features at 40 mg qid, but not at half that dosage. Increased appetite and, possibly, lens opacity.

METHYLDICHLOROPHOSPHONIC ACID

Use: In poisonous artillery shells. Related to ***sarin*** (q.v.). The U.S. Army has not released details of its toxicity, which should approach that of ***sarin***. It is one of the basic compounds in the binary GB shell system. Each 155 mm shell contains 6.4 lbs. In the early 1980s, the U.S. Army proposed buying up to 6 million lbs.

METHYLDIHYDROMORPHINONE

= *Metopon*

***Analgesic*, *narcotic*. Effective orally. Analgesia 3X that of *morphine*.**

Untoward effects: Abuse liability similar to that of ***morphine***.

SC, LD_{50}, **mouse**: 25 mg/kg.

METHYL DIPHENYL ETHER

= *2-Methoxy Biphenyl* = *σ-Phenyl Anisole*

***Fragrance*. In soaps, detergents, cosmetic creams and lotions, and perfume.**

Untoward effects: No irritation or sensitization of 2% in petrolatum on **human** volunteers.

Slightly irritating applied full-strength to intact or abraded **rabbit** skin for 24 h under occlusion.

LD_{50}, **rat**: 3.6 g/kg; acute dermal in **rabbits**: > 5 g/kg.

METHYLDOPA

= *Aldomet* = *Dopamet* = *Dopegyt* = *Presinol*

***Antihypertensive*.**

Untoward effects: Sedation, depression, and inability to perform complex tasks are the most common adverse effects (~50%). Can be especially severe in the **elderly**.

Untoward effects: Blood: Hemolytic anemia, positive direct antiglobulin (Coomb's) test, aplastic anemia, decreased hemoglobin, agranulocytosis, pancytopenia, leukopenia, thrombocytopenia, and occasional increase in prothrombin time and blood urea nitrogen.

Cardiovascular: Hypotension. Can aggravate angina pectoris. Given IV, it causes sudden release of ***catecholamines*** into the circulation, leading to hypertension. Carotid sinus hypersensitivity and hypersensitivity myocarditis reported. Sinus node depression, bradycardia, heart-block, and paradoxical hypertension can occur. A 66-year-old **female** given 250 mg twice daily developed carotid sinus hypersensitivity causing a recurrent syncope.

Clinical test interference: May cause false increase in ***catecholamines*** (up to 10 days after the last dose), ***creatinine***, and ***uric acid***. Dose-related positive Coomb's test in ~20% (negative Coomb's test in 38 Chinese **patients** on 1–3 g/day/1–2 years). Can interfere with ***vanillylmandelic acid***, ***glucose***, and ***SGOT*** assays as well as interfering with cross-matching of blood.

Fetus and neonate: No apparent effect on **fetal** growth, but can disturb immunological responses of the **fetus**. Crosses the placenta in **animals**. **Infant** may get 0.02–1.14 μg/ml of or an average intake of 0.02% of the **maternal** dose in breast milk, and must be watched for possible depression of respiration, blood pressure, and alertness, although these effects are unlikely. A British report states that **mothers** who received it during pregnancy delivered **infants** with smaller head circumference, but had no evidence of mental retardation when checked at 4 years of age.

Galactorrhea and gynecomastia: Due to induced hyperprolactemia.

Gastrointestinal: Acute colitis, diarrhea, flatulence, pancreatitis, and reversible malabsorption syndrome. Nausea, vomiting, xerostomia, metallic taste, sialadenitis, indigestion, anorexia, and sore or "black tongue".

Liver: Hepatocellular jaundice, granulomatous hepatitis; liver necrosis, fibrosis, and cirrhosis.

Miscellaneous: Forgetfulness, fever, mental agitation, chest discomfort, blurred vision, ocular grittiness, choreoathetotic movements, dizziness, edema (weight increase), myalgia, mild arthralgia, parotid pain, headache, weakness, mild ***sodium*** retention, nasal congestion, chills, feeling of warmth, parkinsonism, psychotic reactions (delusions, restlessness, violence), and reactivation of schizophrenia. Retroperitoneal fibrosis in 39-year-old **female** after 2 g/day/3 years. May color urine red, brown, or black, and has caused acute interstitial nephritis and renal calculi. A 70-year-old **male** treated with 250 mg/day/12 days developed septic shock, which disappeared 48 h after discontinuance; accidental ingestion 12 h later led to fever and transitory collapse. Other case reports confirm this adverse effect. An 81-year-old **male** developed severe hyponatremia (due to ***syndrome of inappropriate antidiuretic hormone secretion***) and bone marrow granulomatosis while taking it. Sudden, sharp, excrutiating pain in knees, shoulders, and wrist; ache in arms and swelling of hand and ankle joints. Poisoning in a 19-year-old **male** led to coma, hypothermia, hypotension, bradycardia, and dry mouth. Therapeutic levels are 0.25 mg% and toxic levels 1.0 mg%. A 27-year-old **female** developed asthma from occupational inhalation of the powder in a manufacturing facility. A process in the manufacturing of it for Merck killed three **workers** and hospitalized four **others**. Suspect as a cancer-promoter. Withdrawal reactions (insomnia, agitation, headache, and nausea) occur.

Sexual dysfunction: Inability to maintain an erection, decreased libido (**male** and **female**), and ejaculation difficulty, due to antiadrenergic effects (~89% in black **men**, ~62% in Hispanic **men**, and ~30% in other **men**).

Skin: Lupus erythematous and systemic lupus erythematosis photosensitivity, lip ulcer (plasma cell cheilitis, a fixed-drug eruption), lichenoid and eczematous eruptions, erythematous lesions, epidermal necrolysis, granulomatous skin lesions and arthralgia, purpuric rashes, alopecia, dysesthesia, paresthesia, yellow skin or eyes, and easy bruising or bleeding. Usual therapeutic serum levels are 1–5 μg/ml and toxic at 9 μg/ml.

Drug interactions: Use with ***antipsychotic agents***, including ***trifluoperazine***, may cause confusion; behavioral deterioration and hypertension. ***Chlorpromazine*** increases its hypotensive effect, central nervous system depression, sedation, and dizziness. Use with ***alcohol*** has also caused these same effects, plus epigastric pain, hallucination, jaundice, increased liver enzymes, and dark urine. ***Mephentermine*** and ***metaraminol*** cause mild potentiation. Potentiated by ***norepinephrine***. Efficacy has been decreased when taken with ***oral contraceptives***, ***tricyclic antidepressants***, and ***methamphetamine***. Tachycardia, palpitations, agitation, fine hand tremors, and hypertension with ***amitriptyline***. Use with ***monoamine oxidase inhibitors*** and ***furazolidone*** (if given for over 4–5 days) leads to excitation, hypertension, hallucinosis, and occasionally fibrillation. It can increase effect of ***anesthetics*** and ***sympathomimetic amines***. Use with ***metoprolol*** or ***propranolol*** leads to hypertension. It potentiates ***levarterenol***. Dementia syndromes with tremors, confusion, and extrapyramidal reactions reported in **patients** when ***haloperidol*** was added to

their treatment regimen. Marked postural hypotension when used with ***l-dopa***. Can potentiate minor ***tranquilizers*** and ***lithium carbonate***. Use with ***digoxin*** can cause bradycardia, when neither one alone does; urinary incontinence in 50-year-old **female** when used with ***phenoxybenzamine***, but not with either when used alone. ***Barbiturates*** may enhance its metabolization. Severe hypertensive reaction in a **patient** also given ***phenylpropanolamine***. Blood dyscrasias with ***tolbutamide***. Potentiates ***bishydroxycoumarin's*** anticoagulant effect and ***guanethidine*** effect.

LD_{50}, **rat**: 5 g/kg; **rabbit**: 713 mg/kg.

METHYLENEANILINE

Use: In manufacturing ***rubber*** goods.

Untoward effects: Commercial material contains traces of ***aniline*** (q.v.) and ***formaldehyde*** (q.v.), which can contribute to its induced contact dermatitis.

N, N´-METHYLENE-bis-ACRYLAMIDE

= *MBA*

Use: In grouting and chromatography.

Untoward effects: Causes anemia and porphyria in **rats** and **mice**; testicular atrophy and neurotoxicity in **mice**.

4,4´-METHYLENE-bis[2-CHLOROANILINE]

= *DACPM* = *MBOCA* = *MOCA*

Use: A widely-used synthetic for setting ***glues*** and adhesives, and used as a curing agent for resins, and in cancer research. For ***polyurethane*** manufacturing in a wide variety of industrial, military, and home uses.

Untoward effects: Although the situation has been rare, large amounts may cause toxic effects in **man**. A **man**, accidentally sprayed in the face with hot material, became nauseous and developed nephrotoxicity. Methemoglobinemia and cyanosis reported. Skin absorption appears to be the major source of **human** exposure. May be a potent bladder carcinogen in **man**. National Institute for Occupational Safety and Health limit of 0.01 ppm (11 mg/m^3) exposure.

Dietary intake by male and female **rats** leads to lung adenocarcinomas; liver tumors in male; and malignant mammary tumors in female.

LD_{50}, **mouse**: 880 mg/kg; **rat**: 25 g/kg, over 89 weeks.

METHYLENE BIS(4-CYCLOHEXYLISOCYANATE)

= *HMDI*

Untoward effects: Eye, skin, and respiratory irritation; chest tightness, dyspnea, cough, dry throat, pulmonary edema, and skin blisters. National Institute for Occupational Safety and Health exposure limit of 0.01 ppm.

4,4´-METHYLENEBIS(*N,N*-DIMETHYL)BENZENAMINE

= *Michler's Base*

Use: Intermediate in dye manufacturing, an analytical reagent for ***lead***, and as an antioxidant in greases and oils. Millions of lbs used annually.

Untoward effects: Inhalation and dermal contact are the main routes of exposure for **man**. It emits toxic fumes of ***nitrogen oxides***. Dietary intake leads to hepatocellular adenomas and carcinomas in male and female **rats**.

METHYLENE BLUE

= *CI 52,015* = *CI Basic Blue 9* = *Methylthionine Chloride* = *Solvent Blue 8* = *Swiss Blue* = *Tetramethylthionine Chloride* = *Urolene Blue*

Use: Antidote for ***nitrite***, ***nitrate***, ***sodium chlorate***, and, possibly, ***cyanide*** poisonings to decrease methemoglobinemia; as a weak antiseptic; as an antioxidant, and as a diagnostic agent.

Untoward effects: Orally, can cause gastrointestinal irritation and bladder irritation. Nausea, vomiting, and diarrhea occur infrequently. Use in tubal patency test leads to mild abdominal discomfort, occasionally vaginal spotting, and a "chemical salpingitis". Intra-amniotic use leads to ileal stenosis in **neonates**. Hemolytic Heinz-body anemia in **neonates** after intra-amniotic use or from use in a nasojejunal tube. A pregnant 33-year-old **female** received an infusion of it instead of ***indigo carmine*** for the detection of premature membrane ruptures, and hemolytic anemia developed in the **baby** after delivery. A 12.5 kg 4-year-old **child** survived a 1 g IV dose during fistula surgery. Sudden increase in heart rate to 120/min, blood pressure 80/50, and apparent cyanosis. Even after treatment, the **patient** remained intensely blue-stained for 6 days and had dark blue stools. Discharged 8 days later with faint blue skin. A 920 g **female** born at 26 weeks of gestation had a bluish color, interfering with a diagnosis of hypoxia. The discoloration lasted > 2 weeks. Amniocentesis 18 h before delivery to determine the possibility of membrane ruptures was done with 1 ml of 1% solution. Hemolysis in glucose-6-phosphate dehydrogenase-deficient **persons** (10–30% of **Negroes**, 3–30% of **Sardinians**, 25% of **Iraqi**, 60% of **Kurdish**, and 12% of **Filipinos**) given relatively small doses IV. A photosensitizer. Extensive radiculomyelopathy after intrathecal instillation in 32-year-old **male**. Colors urine green when the body eliminates it. A 62-year-old **male** suffered necrotic SC abscesses after a 1% solution was injected between the toes to outline pelvic lymphatics.

Causes short-term reversible anemia in **dogs**. ***Phenobarbital*** use accelerates its metabolization in **rabbits**. Hemolytic anemia has been reported in **cats** treated with it for cystitis. It is well absorbed after oral use in **man**, but poorly in **dogs** (53–97% versus 2.4–3.8%). Fatal dose for **dogs** is > 1 g/kg.

IV, LD_{LO}, **dog**: 500 mg/kg; **monkey**: 10 mg/kg; LD_{50}, **rat**: 1.2 g/kg; LD_{LO}, **mouse**: 150 mg/kg; **dog**: 500 mg/kg; **rabbit**: 1 g/kg.

METHYLENE CHLORIDE

= *Dichloromethane* = *Methylene Dichloride*

Use: In ***paint*** removers and refrigeration; degreasing, dewaxing; solvent and aerosol propellents, and in manufacturing photographic film. U.S. consumption has varied from ~500 million to 1.5 billion lbs annually.

Untoward effects: Abused by inhalation, leading to eye and skin irritation, fatigue, weakness, somnolence, lightheadedness, vertigo, headache, nausea and vomiting, numbness and tingling of limbs, respiratory depression, anemia, eosinophilia, anorexia, and visual disturbances (from slight blurring to diplopia). ***Formate*** is found in the urine. Cyanosis, coma, arrhythmias, and death by cardiac arrest. Improper ventilation in areas of use has caused fatalities in plant **workers** and in **people** in their workshops. Delusions, hallucinations, and cognitive dysfunction, such as decreased attention span, lack of alertness, memory loss, and personality changes are occasional neurotoxic effects. Metabolized in **man** to ***carbon monoxide*** (q.v. for toxicity), and **humans** exposed to levels below the TLV of 500 ppm had abnormal levels (6–8%) of ***carboxyhemoglobin*** in blood and central nervous system depression. **People** with chronic exposure to 50 ppm have as much ***carboxyhemoglobin*** in **their** blood as **those** who smoke 2 packs of ***cigarettes***/day. A maximum residue of 10 ppm is permitted in decaffeinated ***coffees***. OSHA suggests exposure limit of 500 ppm – TWA – 2000 ppm/5 min maximum peak in any 2 h period. Immediate danger to life or health, 2300 ppm.

LD_{LO}, **human**: 500 mg/kg; TC_{LO}, inhalation, **human**: 500 ppm/1 year and 500 ppm/8 h.

In determining inhalation toxicity in **monkeys**, skin irritation was noted. Induces liver and alveolar/bronchiolar neoplasms in male and female **mice**. Induces benign fibroadenomas of the mammary gland in female **rats**; small increase in malignant salivary gland tumors in male **rats**. It sensitizes **mice** and **dogs** to cardiac arrhythmias and induces cardiac arrhythmias in **mice**. Minimum lethal dose for **guinea pigs** is 75 ppm/72 h.

Inhalation LC_{50}, **mouse**: 26,700 ppm/20 min; 14,500 ppm/2 h. Inhalation LC_{LO}, **dog**: 20,000 ppm/7 h; **guinea pig**: 5000 ppm/2 h; LD_{50}, **rat**: 167 mg/kg; 2.3 ml/kg; LD_{LO}, **dog**: 3 g/kg; **rabbit**: 1.9 g/kg.

4,4´-METHYLENE DIANILINE

= *DDM* = *p,p´-Diaminodiphenylmethane* = *MDA*

Use: In producing ***isocyanates*** used in the manufacture of ***polyurethane*** foams. In ***Spandex***® fibers, in azo dyes, and as a corrosion-inhibitor. ***Epoxy resin*** hardener. Over 200 million lbs production annually in the U.S.A. and ~1/2 billion lbs used annually in the U.S. ***MDA*** is also a designation for ***methylenedioxyamphetamine***.

Untoward effects: When heated, it eliminates toxic fumes of ***aniline*** (q.v.) and ***nitrogen oxides***. It is a potent allergic sensitizer that stains the skin and fingernails a deep yellow. Exposure is limited to the workplace. May be a **human** carcinogen (bladder and intestinal - lymphosarcoma. Reticulosarcoma, jaundice, bile duct inflammation). Causes cholestasis and hepatic necrosis in many **animals**. Epping jaundice (with weakness, severe right upper-quadrant abdominal pain, nausea and/or vomiting, anorexia, high fever, and chills), when, in 1965, 84 **persons** in Epping, England ate ***bread*** made with ***flour*** contaminated with it. Its acute hepatotoxic properties can follow absorption via the oral or percutaneous routes. A 20-year-old **male** had chest and abdominal pain for 5 days. For 2 weeks previously, **he** had handled large quantities of ***MDA***. The week before admisssion, an air filter malfunction spewed yellow dust into the air. While it was being repaired, **he** dropped the front of **his** protective full-sleeve coveralls and lunched. The next morning, **he** awoke with severe supraumbilical pain, a pruritic, macular rash encircling the forearms and ending abruptly at the sleeve level. After a 2 days lull, the sharp pains returned, along with a headache. In the emergency room, **he** had a temperature of 103°F, icterus, tender liver, and a macular rash. After another 3 days, **he** was hospitalized with intense parasternal chest pain, and an enlarged, tender liver. Acute myocardiopathy was eventually diagnosed that took a year to return to normal.

TD_{LO}, **man**: 84 mg/kg.

Thyroid follicular cell carcinomas and neoplastic nodules in the liver of male **rats**; follicular cell and C-cell thyroid adenomas in female **rats**; thyroid follicular cell adenomas, alveolar bronchiolar adenoma, and hepatocellular carcinomas in male and female **mice**; adrenal pheochromocytomas in male **mice**; hepatocellular adenomas and malignant lymphomas in female **mice**.

LD_{50}, **rat**: 335–830 mg/kg; **mouse**: 210 mg/kg; **guinea pig**: 260 mg/kg; **rabbit**: 620 mg/kg.

METHYLENEDIOXYAMPHETAMINE

= *Adam* = *E* = *EA 1299* = *Ectasy* = *Entactogen* = *Love Pill* = *MDA* = *MDMA* = *"Rolls"* = *X* = *XTC*

***Hallucinogen, stimulant.* An *amphetamine* analog. A "designer drug" (with poor quality control). Banned in the U.S. *MDA* is also a designation for *methylenedianiline*. Street pills often have stamped logos, such as *Calvin Klein, Mitsubishi, Motorola, Nike*, and *Tweety Bird*. Sold in 1914 as an anorexiant.**

Untoward effects: Popular in drug-oriented **individuals**. Large doses have ***amphetamine***-like stimulating effects and intoxication with increased acoustic, visual, and tactile sensory perceptions. Hallucinogenic effects similar to ***mescaline***. Chronic paranoid psychosis in 40% of 89 **subjects** (mean age 22.8 years) **who** used it 6 months–3 years, including irritability, depression, and panic attacks. Acute urinary retention in 19-year-old **male who** ingested 15 doses in 36 h. Acute renal failure in 23-year-old **male** after abusing the drug. Hyponatremia in 23-year-old **female** and 17-year-old **female**, and catatonic stupor in a 17-year-old **female**. Cerebral edema and semiconsciousness in two **female teenagers** after ingesting some with ***alcohol***. Severe generalized reactions in 32-year-old **female** after ingestion of 100–150 mg as an oral solution. Depression with suicidal thoughts in 23-year-old **male** after four doses 2–3 weeks apart. Drug-induced psychosis with aggression in 24-year-old **male** after 200 mg/day/4 years. Gross pneumomediastinum, pneumoretroperitoneum, chest pain, vomiting, and leukocytosis in 17-year-old **male** after ingesting two tablets. Hyperpyrexia, with subsequent rhabdomyolyis and disseminated intravascular coagulation in several **teenagers**. Fatal ingestion reported after visual hallucinations, mydriasis, agitation, rigidity, seizure, parkinsonism, hypertension, hyperpyrexia, tachycardia, hematemesis, rhabdomyolysis, disseminated intravascular coagulation and bleeding, refractory shock, acute respiratory distress, and acute renal failure. Risks of hyperthermia often associated with its ingestion and violent exercise, including dancing. Acute toxicity in two pregnant **mothers** and congenital anomalies were also reported among **infants** of 74 **women** taking it alone during pregnant. A survey, in 1987, of a California University's **undergraduates** indicated 143/369 (39%) had tried it at least once. After July 1, 1985, its use in the U.S. constitutes a felony. Abused orally and IV, occasionally smoked. Young **partygoers** in Europe who use it with ***marijuana*** do poorly on intelligence tests weeks later. Severe near-fatal cases in several **drug-abusers** after 500 mg orally. Overdoses with fatalities have been reported. Marked hypertension, diaphoresis, altered mental status, and hypertonicity when ingested with 15 mg ***phenelzine***. A 17-year-old **male** died within a few hours after signs of intoxication, mistakenly believed to be due to ***alcohol***, followed by unconsciousness, spastic movements, and mydriasis. Post-mortem revealed visceral congestion, edema, and petechiae on the heart. Can cause nystagmus in **some**. Called the ***"love drug"*** because it supposedly was found at the site of the Sharon Tate killings. JFK and Newark airports are favorite entry ports for illicit material from sources in Belgium and Holland. Some street material has contained ***antihistamine*** laced with an insecticide. An air **traveller** flying into JFK was arrested having swallowed 70 condoms containing 2800 pills. Deaths in Chicago have been associated with pills laced with deadly ***PMA*** (***paramethoxyamphetamine***). The combination is known as ***Double Mitsubishi***. Serum levels of 0.04–1.0 mg% are considered lethal. A fatal interaction in a **male** taking 600 mg ***ritonavir*** twice daily 4 h after 180 mg ***MDMA***; **his *MDMA*** plasma concentration was 4.56 mg/l. Thrombotic thrombocytopenic purpura after acute liver failure in 29-year-old **male** after ingestion of one tablet.

Hyperactivity, salivation, and pilo-erection in **mice**. IV use in **dogs** leads to tremors, mydriasis, and salivation, followed in 15 min with clonic, then tonic, convulsions with opisthotonus, respiratory paralysis, and death. Has damaged brain neurons in **monkeys** and **rats** at doses/kg used by **humans** for recreational purposes.

METHYLERGONOVINE

= *Basofortina* = *Metenarin* = *Methergin(e)* = *Methylergobasine* = *Methylergobrevin* = *Methylergometrine* = *Ryegonovin* = *Spametrin-M*

***Oxytocic.* The *maleate* is usually used.**

Untoward effects: Oral use has a large inter-**individual** variability in bioavailability, making IV or IM route the reliable way to use it. IM use in a 25-year-old **female**, 2 weeks after normal pregnancy and delivery, led to acute renal failure (probably, due to renal vasospastic reaction) and microangiopathic hemolytic anemia. Severe headache, restlessness, mental clouding, vomiting; blood pressure 80/130 in 28-year-old **female** and severe headache; and blood pressure 210/130, followed by unconsciousness and Jacksonian convulsions lasting 3 min, in 32-year-old **female**. Used IM in both cases to aid expulsion of placenta after spontaneous delivery. Given IM or IV in fourth stage of labor leads to increased systolic blood pressure and increased incidence of emesis. Bronchospasm in 35-year-old **male** after test dose of 25 mg in diagnosing coronary artery spasms; normal again after 15 min. Use during delivery leads to significantly higher blood pressure increase in hypertensive **females** than in **those** with normal blood pressure. Variant angina in 33-year-old **female** after oral use for 4 days. Other case reports confirm this. Umbilical cord pressure > 1/3 higher from it in 23 **females** during delivery. Erroneous administration to **neonates** led to convulsions, respiratory depression, and apnea. Found in breast milk of treated **mothers** at 4.6% of **maternal** dose. Severe hypertension and headaches when used with ***methoxamine***.

LD_{50}, **rat**: 93 mg/kg; **mouse**: 187 mg/kg; IV, **rat**: 23 mg/kg; **mouse**: 85 mg/kg.

N-METHYLETHANOLAMINE

Use: In textile chemicals and pharmaceuticals.

Untoward effects: Considered to be moderately toxic if ingested. Can cause skin irritation; prolonged or repeated contact can result in chemical burns.

LD_{50}, **rat**: 2.34 g/kg.

METHYL EUGENOL

= *4-Allylveratrole* = *Eugenyl Methyl Ether*

Use: ~25 tons used annually in the U.S. as a fragrance in soaps, detergents, cosmetic creams and lotions, perfumes, and foods. Present in many edible plants.

Untoward effects: No sensitization or irritation on **human** volunteers at 8% in petrolatum. An ultraviolet light-absorber. May be one of the hallucinogenic substances in ***nutmeg*** (1 mg/2 g).

Full-strength application to intact or abraded **rabbit** skin for 24 h under occlusion was irritating. Carcinogenic in **rodents** and some metabolites are mutagens. A synthetic pheromone for the **male oriental fruit fly**.

LD_{50}, **rat**: 0.81–1.56 g/kg; acute dermal, **rabbit**: > 5 g/kg.

α-METHYL FENTANYL

= *Mefentanyl*

Untoward effects: A "designer drug". An illicit ***fentanyl*** analog, ~3500 times more potent than ***morphine*** and 1500 times the potency of ***heroin***. Was first reported on the streets of Orange County, California in 1979, and misrepresented as **"*China White*"** or ***heroin***. Street names are ***"Synthetic Heroin"***, ***"Shooting Speed"***, ***"White Dope"***, and **"*Bathtub Gin*"**. Often used to spike ***heroin***. This is often called **"*Tango and Cash*"**. Incredible potency makes it easy for **addicts** to overdose. Many (> 100) fatal overdoses and thousands of emergency room cases occurred. Metabolization within 30 s of injection. Symptoms including muscle spasms, gritting of teeth, nervousness, headaches, gastro-intestinal upsets, seizures, and unconsciousness. Death has occurred within seconds or minutes of injection, with respiratory arrest and muscular rigidity. Acute pulmonary edema noted on postmortem.

METHYL FLUOROACETATE

= *MFA*

***Rodenticide*. Water-soluble.**

Untoward effects: Highly toxic. See ***fluoroacetic acid*** for its toxicity. Secondary toxicity can occur in **cats** and **dogs**.

LD_{50}, **rat**: 4 mg/kg; **rabbit** and **guinea pig**: 500 μg/kg; **cat**: 300 μg/kg; LD_{LO}, **dog**: 100 μg/kg; **monkey**: 10 mg/kg.

METHYL FLUOROSULFONATE

= *Magic Methyl* = *Methyl Fluorosulfate*

Untoward effects: Extremely toxic. A 25-year-old chemist spilled a few milliliters on **his** clothing. Inhalation of vapors proved fatal, due to pulmonary edema.

N-METHYLFORMAMIDE

= *MMF* = *Monomethylformamide* = *NSC 3051*

Solvent.

Untoward effects: **Mice** (11 days pregnant) given 2.5–10 mg orally or 5–10 mg percutaneously developed increased fetal and neonatal mortality, high frequency of malformations (encephalocele, kyphosis, spina bifida, and other vertebral defects), but it was not toxic to the mothers. Often used in the illicit manufacture of ***amphetamines***.

LD_{50}, **mouse**: 2.6 g/kg; IP, 2.3 g/kg; IM, 2.7 g/kg; IV, 1.58 g/kg; TD_{LO}, **rat** (11 days pregnant): 1 g/kg.

METHYL FORMATE

Fumigant and larvicide.

Use: On dried fruits, such as ***raisins*** and ***dates***, and in making military poisonous gas.

Untoward effects: Flammable. Systemic toxicity from its degradation product, ***methanol*** (q.v.). Vapors have irritated eyes and nose and cause dyspnea, vision problems, and central nervous system depression.

LD_{LO}, **human**: 500 mg/kg.

Exposure of **animals** leads to narcosis and pulmonary edema. **Guinea pigs** exposed to 0.15% vapors developed nasal and conjunctival irritation, vomiting, narcosis, and death. Post-mortem revealed congested and edematous lungs, and hyperemia of liver, kidneys, and brain surface.

LD_{50}, **rabbit**: 1.6 g/kg; inhalation LC_{LO}, **guinea pig**: 10,000 ppm.

N-METHYL-*N*-FORMYLHYDRAZINE

Found in the commonly eaten wild *mushroom, false morel* (*Gyromitra esculenta*, q.v.) at a concentration of 50 mg/100 g (500 ppm or 226.8 mg/lb).

***Untoward effects*:** Ingestion causes lung tumors in **mice** at low dietary intake of 20 mg/day and liver tumors in **hamsters**.

LD_{50}, **mouse**: 118 mg/kg.

3-METHYLFURAN

A naturally occurring atmospheric contaminant occasionally found in urban smog.

***Untoward effects*:** Highly reactive; binds irreversibly to tissues in **rat, mouse**, and **hamster** inhalation trials, causing extensive damage to nasal mucosa. It or its metabolite may be a problem after **human** exposure to smog.

METHYL GALLATE

= *Gallicin*

Antioxidant.

Untoward effects*:** A cause of contact dermatitis from handling certain thermofax ***papers.

METHYLGLUCAMINE ANTIMONATE

= *Glucantime* = *Protostib* = *RP 2168*

***Antiprotozoal*. IV, IM, or intralesional. A *pentavalent antimony*.**

***Untoward effects*:** Renal tubular dysfunction.

METHYL GREEN

= *Heptamethyl-p-rosaniline Chloride*

***Untoward effects*:** Rare cases of shock in **man** after its use topically.

METHYL HEPTINE CARBONATE

Synthetic perfume with *violet*-like aroma and *fruit-berry* taste.

Untoward effects*:** Dermatitis from its use in ***lipstick and facial creams.

METHYL HEPTYL KETONE

= *5-Methyl-2-octanone*

***Solvent*. In the coating industry as a solvent for *alkyd resins, cellulose nitrate, epoxy, melamine, polyesters*, and *vinyl copolymers*.**

***Untoward effects*:** Neurotoxic. After 60 days of oral intake, treated **rats** began dragging one or both hind limbs.

LD_{50}, **rat**: 3.2 g/kg.

METHYL HYDRATROPALDEHYDE

Synthetic fragrance used in soap, detergents, cosmetic creams and lotions, perfumes, and foods. Several tons/year used in the U.S.

***Untoward effects*:** Sensitization in ~10% of **human** volunteers by 4% in petrolatum. Other trials showed none. No irritation of 4% in petrolatum after 48 h closed-patch test on **humans**.

LD_{50}, **rat**: 3.5 g/kg; acute dermal, **rabbit**: > 5 g/kg.

METHYL HYDRAZINE

= *MMH* = *Monomethylhydrazine*

***Use*:** In rocket fuel and as a chemical intermediate.

Untoward effects*:** Flammable. Inhalation leads to eye, skin, and respiratory system irritation. Vomiting, diarrhea, tremors, ataxia, anoxia, cyanosis, and convulsions reported. If ingested, as in ***Gyromitra (q.v.), ***Helvella***, and some species of ***Paxina*** (poisonous ***mushrooms***), there is a latent period of 6–12 h followed by bloating, lower abdominal cramps, weakness, lassitude, headache, and, in severe cases, vertigo, jaundice, tachycardia, and fever occur. Usually fatal unless promptly treated. Hematopoietic toxicity, nausea, vomiting, and paresthesias in **patients** with Hodgkin's disease or malignant lymphomas. A ***pyridoxine*** antagonist and it acts as a hemolytic agent or toxin. Can cause a ***disulfiram*** (q.v.)-like reaction with ***alcohol***. National Institute for Occupational Safety and Health/OSHA upper exposure limit of 25 ppm. Immediate danger to life or health, 100 ppm.

Causes lung tumors in **mice**, tumors of liver Kupffer cells and the cecum of **hamsters**. Causes marked renal damage in **dogs**, but not in **monkeys**. Inhalation by **dogs** leads to respiratory irritation, pulmonary hemorrhage and edema, central nervous system stimulation, and convulsions. Subchronic inhalation by **dogs, monkeys, rats**, and **mice** can cause symptoms similar to **man's**.

LD_{50}, **mouse**: 33–57 mg/kg; **rat**: 33 mg/kg; LC_{50}, inhalation, **rat**: 74 ppm/4 h; **mouse**: 56 ppm/4 h; **dog**: 96 ppm/1 h.

4-METHYLIMIDAZOLE

Untoward effects*:** It is suspect as a major neurotoxic principle in ammoniated feeds fed to **cattle**. **Calves** and **cows** appeared to be "going crazy". They ran into fences for about 15 min, then appeared normal for a while following ingestion of green feed treated with 3% ***anhydrous ammonia. Post-mortem revealed (aside from physical injuries) histopathology indicating perivascular and intramyelinic edema. The condition is known as "**bovine** bonkers".

LD_{50}, **mouse**: 370 mg/kg.

METHYL IODIDE

= *Iodomethane*

Use: In converting ***coal*** to gas; in chemical synthesis, ***fire-extinguishers***, and as a light-sensitive etching compound for electronic circuitry.

Untoward effects: Irritant to skin, respiratory system, and eyes. Vapors have caused nausea, vomiting, vertigo, ataxia, drowsiness, and slurred speech. Can burn skin on contact.

LD_{LO}, **human**: 50 mg/kg; skin, 1 g/10 min.

LD_{LO}, **rat**: 150 mg/kg.

METHYL IONONE

Use: As a fragrance in soaps, detergents, cosmetic creams and lotions, perfumes, and foods. U.S. consumption is > 100 tons/year.

Untoward effects: No sensitization of 10% in petrolatum on 25 **human** volunteers. Patch-testing of full-strength on 16 **people** revealed no irritation.

Only moderately irritating when applied full-strength to intact or abraded **rabbit** skin for 24 h under occlusion.

LD_{50}, **rat** and acute dermal on **rabbit**: > 5 g/kg.

METHYL ISOAMYL KETONE

= *MIAK*

Solvent.

Untoward effects: Combustible. Prolonged or repeated inhalation of vapors leads to eye, skin, and mucous membrane irritation; headache, narcosis, and coma. National Institute for Occupational Safety and Health suggests 50 ppm as upper limit of exposure; OSHA, 100 ppm. Hepato-toxicity and renal toxicity in exposed laboratory **animals**.

LD_{50}, **rat**: 4.76 ml/kg; **rabbit**: 10 ml/kg; LD_{LO}, **mouse**: 3.2 g/kg; LC_{LO}, inhalation, **rat**: 2000 ppm/4 h.

2-METHYLISOBORNEOL

Untoward effects: Responsible for the muddy flavor found in **fish** and some lake water. Produced by microorganisms in the water.

METHYL ISOBUTYL KETONE

= *Hexone* = *Isopropylacetone* = *4-Methyl-2-pentanone*

***Solvent.* Faint *camphor*-like odor. Consumption varies – 150 ± million lbs/year in the U.S.**

Untoward effects: Flammable. Vapors lead to skin, eye, respiratory tract, and mucous membrane irritation; nausea, vomiting, headache, dizziness, incoordination, narcosis, and coma. Central nervous system affective syndrome including fatigue, memory impairment, irritability, difficulty in concentrating, and mild mood disturbance. Greater exposure has caused chronic toxic encephalopathy. Long-term abuse cases have developed a severe, chronic encephalopathy. Odor threshold level is < 100 ppm; eye irritation at 200 ppm/15 min, nose and throat irritation at > 200 ppm; initial headache and nausea at ~100 ppm, with tolerance developed during continued work exposure. Undiluted material splashed onto eyes leads to painful irritation. Long-term exposure has caused defatting, dryness, inflammation, and desquamation of the skin. Local legislation has prohibited its use in some areas. Many ***alcohol***-based colognes are denatured with a 1% solution. Upper limits of exposure are 50 ppm – National Institute for Occupational Safety and Health – 75 ppm/15 min; OSHA - 100 ppm. LD_{LO}, **human**: 500 mg/kg.

Causes weakness, ataxia, and tremors in **guinea pigs**. Exposure to 4000 ppm/4 h killed 6/6 **rats**, but 0/6 at 2000 ppm. Saturated vapors at room temperature killed **rats** within 30 min. Some skin and eye irritation by vapors or liquid to **rabbits**. Temporary immobilization of exposed **mice**.

LD_{50}, **rat**: 2 g/kg; **guinea pig**: 1600–3200 mg/kg.

METHYL ISOCYANATE

= *MIC*

Use: In the manufacture of ***carbamate pesticides*** and other organic synthesis.

Untoward effects: Vapors cause severe skin and respiratory tract irritation. Eye, nose, and throat irritation; coughing, chest pain, dyspnea, asthma, pulmonary edema with effusions, and temporary blindness occur. OSHA and National Institute for Occupational Safety and Health set upper exposure limit for skin at 0.02 ppm (0.05 mg/m^3). Immediate danger to life or health, 3 ppm. Exposure to 2 ppm for 1–5 min has caused lacrimation and severe irritation. On December 2, 1984 near midnight, **someone** (apparently, in error or in sabotage) opened a valve allowing > 125 gallons of water into a tank of it, setting up chemical chain reactions that spewed it as a poisonous gas over the city of Bophal, India. Estimates vary, but 2000–4000 **people** died that night and ~3000 **others** died afterwards. Hospitals reported treating 60,000 gas **victims**, and governmental officials estimated up to 200,000 were physically affected. High-level exposures apparently trigger a specific immune response. Pulmonary fibrosis has been noted in **survivors**. Several gas releases occurred in West Virginia, U.S.A. on August 27, 1981 and on December 31, 1983; 19 other releases were < 1 lb–> 10 lbs. Thousands of **people** heard the plant's warning alarm and stayed safely indoors while the fumes dissipated; six exposed **employees** had severe eye irritation and 125 area **residents** sought emergency care at hospitals. It is considered to be the most poisonous ***isocyanate*** used industrially. It is non-volatile and must be heated.

LD_{50}, **rat**: 71 mg/kg; **mouse**: 120 mg/kg; single skin application, **rabbit**: 33 ppm; LD_{50}, inhalation, **rat**: 5 ppm/4 h.

METHYL ISOEUGENOL

= *4-Propenylveratrole*

Fragrance in soap, detergents, cosmetic creams and lotions, perfume, and foods; > 1 ton/year used in the U.S. For carnation-like aroma.

Untoward effects*:** No irritation or sensitization of 8% in petrolatum on **human** volunteers. A minor hallucinogen in ***nutmeg (11 mg/20 g).

Full-strength on intact or abraded **rabbit** skin for 24 h under occlusion was irritating.

LD_{50}, **rat**: 1.5–2.5 g/kg; acute dermal, **rabbit**: > 5 g/kg.

METHYL ISOPROPYL KETONE

= *3-Methyl-2-butanone* = *MIPK*

Use*:** In perfume, pesticide, and ***cephalosporin manufacturing.

***Untoward effects*:** Flammable. Eye, skin, respiratory tract, and mucous membrane irritation from contact or inhalation. Causes coughing. National Institute for Occupational Safety and Health suggests exposure limit of 200 ppm.

Irritant on **rabbit's** belly and causes corneal injury to **rabbits**.

LD_{50}, **rat**: 5.66 ml/kg; acute dermal, **rabbit**: 6.5 ml/kg; inhalation, **rat**: 5700 ppm/4 h.

2-METHYL-4-ISOTHIAZOLINE-3-ONE

***Antimicrobial, fungicide.* In dermatologicals, latex *paint*, and in cooling waters.**

Untoward effects*:** It and a methylchloro form are present in ***Eucerin creams used topically for dermatitis and can cause a vesicular dermatitis.

METHYLISOTHIOCYANATE

= *Methyl Mustard Oil* = *Trapex* = *Vorlex* = *Vortex*

***Use*:** In pesticides.

***Untoward effects*:** Highly irritating contact dermatitis.

LD_{LO}, **human**: 50 mg/kg.

LD_{50}, **rat** and **mouse**: 97 mg/kg; skin, **rabbit**: 33 mg/kg; **mouse**: 1820 mg/kg.

METHYL MERCURY

= *Monomethyl Mercury*

Untoward effects*:** A disease of **man**, now called Minamata disease, first described in Japan in 1953, due to the chemical being discharged into Minamata Bay. Of 121 **people** initially poisoned, 46 died. Neurological damage occurred in **families** who ate ***fish and ***shellfish*** (q.v.) caught in the bay, in the **fish** themselves, and in **birds** and **cats** eating the contaminated **fish**. Many deaths were reported. **Adults** and **children** develop cerebellar ataxia; severe, generalized paralysis; chorea, tremors, seizures, paresthesia, asteriogenesis, myoclonus, irritability, dysarthria, insomnia, impaired hearing and mental powers, and blindness. Readily crosses the placenta. At least 19 **babies** of **mothers** who ate contaminated ***fish*** and ***shellfish*** had congenital defects; 25 had brain damage. From 1955 to 1959, it was estimated that 6% of the **children** born in the Minamata area had cerebral palsy. A small epidemic of cerebral palsy occurred in newborn **infants** whose **mothers** had consumed contaminated ***fish***, but didn't show clinical symptoms. Passes into breast milk. Later, 26 cases of poisoning occurred at Niigata, Japan. Through chemical analysis of umbilical cords stored since 1927 in Japan, it was demonstrated that, 20 years before the Minamata episode, a similar episode had occurred unnoticed. Mercurial compounds and liquid ***mercury*** are, apparently, converted by bacteria in the mud and by **fish** intestinal contents to the lethal methyl mercury. In Iraq, it was reported that 459/6530 **people** died in 1971–1972 from eating ***bread***, baked from ***wheat*** and ***barley*** flour, after fields were treated with methyl mercury compounds. In 1971, Iraqi **farmers** washed the water-soluble red dye off the ***wheat*** that was to identify it as treated with mercurials, not realizing that the ***mercury*** was not equally soluble. **They** then fed it "safely" to **chickens** for a few days, not realizing that a lengthy latency period was involved. Prior to this outbreak, cases of such poisoning were reported in 1956 and 1960 in Iraq from its use as an antifungal seed-dressing. In the 1960 outbreak, it was estimated that 1000 **people** were poisoned and 370 hospitalized. During 1963–1965, 45 **people** in Guatemala were affected and 20 died. It was originally misdiagnosed as viral encephalitis. A similar outbreak occurred in Pakistan from the same cause, ***methylmercury dicyandiamide***. In 1968, a similar epidemic occurred in Russia. In 1964, a 38-year-old **female**, 62" in height and weighing 165 lb, adopted a fad diet for 10 months that consisted of 12–14 oz/day of

swordfish. It worked; **she** was then reduced to 120 lb, but **she** noticed that **she** was becoming lethargic, had frequent headaches, blurred vision, and trembling hands. **She** thought the weakness was from weight loss. During the next 5 years for periods of 4–8 weeks, **she** resumed the ***swordfish*** diet. **Her** symptoms became worse. These included marked dizziness, greatly aggravated by lateral head movement; poor memory, severe hand tremors, "quivering" tongue, light sensitivity, difficulty in focusing, impaired speech and writing, explosive speech, and incomplete sentences. **She** rarely ate any other ***fish*** and used no weight control medicines. An examination at a neurological institute stated **she** was "probably suffering from psychosomatic complaints". **They** did not do a mapping of the visual fields, which might have shown characteristic lesions of ***mercury*** poisoning. After nearly 7 years, **she** read about ***mercury*** poisoning in ***swordfish*** and discontinued eating it. After 5 months, most of **her** adverse symptoms, except mild dizziness, had disappeared. Analysis of **her** hair showed 42 ppm ***mercury***. **Tuna fish** and **fish** in the Great Lakes between the U.S. and Canada have also had increased levels of methyl mercury, probably from industrial use of waterways and lakes as sewers. Chromosome breakage reported from Sweden in **humans** exposed to it via ***fish*** consumption. A report from Scotland indicated increased blood levels in **dentists**. Biological half-life in **man** is ~65–70 days. Reports from Iraq indicate 35–189 days and 240 days in Minamata **victims**.

The **cat** is more susceptible to its toxicity than **man**. After 0.4–1 mg/kg/day, toxicosis developed in **cats**, **dogs**, and **rabbits** in 10–90 days. Minamata disease occurred in **cats** eating **fish** from contaminated waters of the English River in Ontario, Canada. They showed neurological symptoms (ataxia, abnormal movements, convulsions, glassy stare, and hypersalivation). Postmortem lesions were the same as those in Japan. ***Mercury*** blood concentrations > 100 ppb were found in 26% of 110 area **Indians**. In lactating **cows**, little ingested methyl mercury appears in the milk, but muscles had 72% of their total body ***mercury***, which could be hazardous to **human** health. Whole meat has contained ~2 ppm. Japanese regulations allowed up to 0.4 ppm in **fish** and the U.S. and Canada allowed up to 0.5 ppm as ***mercury***. In 1979, the U.S. raised the action level to 1 ppm. **Tuna** usually contains < 1 ppm. Marine **fish** also contain 0.4–0.9 ppm ***selenium***, which apparently decreases methyl mercury toxicity. Methyl mercury interferes with the development of **tadpoles**, and regeneration of **fish** fins and limbs in **crabs**. Chronic low-level exposure of **monkeys** impairs spatial vision. Ingestion by **hens**, **quails**, and **ducks** leads to thin-shelled eggs and impaired embryonic development. Biological half-life in **hens** was 36 days. **Rat** and **mouse** trials indicate decreased immune responses, decreased reproductive capability, embryotoxicity, teratogenicity and encephalopathy, and brain and liver damage in **rat** fetus. Strain differences in excretion in **mice** reported. Nephrotoxicity and weakness reported in **rats**. Lactating and non-lactating **ewes** fed treated grain (8.5 ppm ***Hg***) for 65 days developed exfoliative dermatitis and gingival bleeding with biological half-life of 29 days. In **swine**, it may take days or months for clinical symptoms to appear; postmortem lesions resemble those of **hog** cholera.

METHYL MERCURY CHLORIDE

= *Methyl Mercuric Chloride*

***Untoward effects*:** Ingestion has caused learning and behavior disorders in **children**.

A **horse** received daily doses in the feed (~430 µg/kg) for 5 days/week for 10 weeks. In ~35 days, the **horse** developed a bilaterally symmetrical, exudative dermatitis with crusting, erythema, and depigmentation. On day 50, the **horse** was lethargic, moved with difficulty, had hair loss, had decreased vision, and had lost weight. On day 70, it was ataxic and showed head-nodding. **Cattle**, **pig**, **sheep**, **mink**, **cat**, **monkey**, **dog**, **pigeon**, **quail**, **duck**, and **guinea pig** toxicity has been reported. Reproductive capability decreased in male **mice** and **rats**. Teratogenic in **rats** and **mice** and increase in stillbirths and neonatal deaths in **mice** without evidence of maternal clinical signs. Fed to **hens**, it causes decreased body and egg weight; decreased egg production, fertility, and hatchability; and misshapen eggs.

TD_{LO}, **rat** (9 days pregnant): 2 mg/kg; **mouse** (6 days pregnant): 30 mg/kg; **cat** (10–15 days pregnant): 4 g/kg.

METHYLMERCURYDICYANDIAMIDE

= *Cyano(methyl-mercuri)guanidine* = *EP 227* = *MMD* = *Morsodren* = *Panogen*

***Disinfectant, fungicide*. For use on plant seeds.**

Untoward effects*:** One of the many mercurial seed treatments used on cereal grains, whose flour was, unfortunately, used in making ***bread in Pakistan and other countries (see ***Methyl Mercury*** for symptomology). Accidental feeding of **poultry** with treated grain for 4 weeks created a public health problem for Canada. Tolerance levels were set at 0.1 ppm. Old **hens** and their eggs met this within 6 weeks and **pullets** within 2 months. Ataxia, partial blindness, agitation, and slurring of speech, progressing to semi- or complete coma in 8-year-old **female**, 13-year-old **male**, 20-year-old **female**, and **male infant**. 40-year-old **female** had gross tremulous movement of limbs at birth and, later, myoclonic convulsions; unable to sit and see at 1 year of age in **family** of nine eating ***pork*** daily from **hogs** fed grain contaminated with it. Over a period of 100 days, it was estimated that each **family member** had ingested ~390 mg ***mercury***.

LD_{LO}, **human**: 5 mg/kg.

The meat of **chickens** which showed no signs of poisoning when fed it, poisoned both **ferrets** and **goshawks** who ate their flesh. **Sheep** and **cattle** fed 330 μg/day showed mild toxicity within 1½–2 months. Decreased egg production in **hens** and **mallard ducks** during the breeding season; decreased hatchability and poor **duckling** survival when fed to **black ducks** during the breeding season. IP dosing of **mice** led to growth and developmental retardation, cleft lip and palate, and deaths of fetuses and neonates. Decreased reproductive capability of male **rats** and **mice**. A month after experimental feeding of 33 ppm as ***mercury***, the highest mortality was in **pheasants** (90%), **ducks** (85%), and only 7.5% in **chickens**. At 35 days, the heart contained the greatest concentration, followed by breast muscle, liver, and kidney. Tissue concentration and retention were highest in **ducks**, less in **pheasants**, and lowest in **chickens**. Demyelination, neuron shrinkage, necrosis and meningeal hemorrhages in the brains of dead **ducklings**.

LD_{50}, **rat**: 32 mg/kg; **mouse**: 20 mg/kg.

METHYL MERCURY HYDROXIDE

= *Methyl Mercuric Hydroxide*

***Fungicide*. Used on seed grains.**

Untoward effects: In **man** and **rats**, ingestion damages peripheral sensory nerve fibers, more than motor fibers, and the peripheral neuropathy precedes central nervous system effects (see ***Methyl Mercury***). Has caused nephrotoxicity. OSHA limits air exposure to 10 μg as ***mercury***/m^3 – TWA.

Has caused nephrotoxicity in **rats** after injections. Clinical signs of toxicity in **swine** leading to ataxia, blindness, convulsions, paresis, and death. Post-mortem revealed neuronal necrosis, neuronophagia, cortical vacuolation, axon swelling, gliosis, leptomeningitis, and vascular fibrinoid necrosis. Highest concentration found in the liver, followed by kidney, muscle, spleen, and brain. Similar symptoms and postmortem findings in **cats**. Transplacental passage to **kittens** demonstrated.

METHYL METHACRYLATE

Use: In acrylic plastics and resins, in molding and extrusion compounds, and in surface coatings. ***Lucite***®, ***Plexiglas***®, and ***Perspex***® are polymerized forms. U.S. production is ~1½ billion lbs annually.

Untoward effects: Use as a bone cement in hip arthroplasty has caused severe and occasionally fatal cardiovascular depression, due to systemic absorption of the monomer and the associated peripheral vasodilation and hypotension, due to the build-up of heat and pressure in the femoral canal while the polymer was curing. Transient pulmonary embolism has also occurred. These effects were minimized by allowing some of the curing process to procede before inserting the prosthesis. **Surgeons, dental technicians**, and **dentists** have developed a contact dermatitis from using it, as well as a distressing paresthesia of **their** fingertips for weeks to several months after the dermatitis subsided. Nail-extenders or elongators also used it and have caused permanent scarring about the nails and often with permanent nail loss. It is now banned in the U.S. and replaced with ***butyl*** and ***ethyl methacrylates***. Can penetrate ***latex rubber*** gloves, but, apparently, not ***neoprene rubber***. When used to fill openings between bone fracture fragments, it has interfered with normal healing processes. Fumes can irritate skin, eyes, nose, and throat. A **patient** failed to recover after its use in arthroplasty. Post-mortem indicated cerebral fat embolism and infarction. Bladder carcinoma adjacent to and 3 months after its intrapelvic use. Bacterial contamination and infections prompted ***antibiotic*** admixtures. Bone and tissue necrosis and sciatic nerve damage reported adjacent to areas of use. Inhalation of vapors may have been associated with cystitis, bladder cancer, and lung granulomas in hospital **workers** making contact lenses. Full and partial-thickness burns on **patients'** abdomens from contact with excess cement during arthroplasty. OSHA upper exposure limit is 100 ppm, TWA. Immediate danger to life or health, 1000 ppm.

Inhalation TC_{LO}, **human**: 125 ppm; oral LD_{LO}, **human**: 5 g/kg.

Accidental exposure of a **rhesus monkey** to its vapors caused its death. Post-mortem revealed mottled liver, pulmonary edema, and atelectasis, and yellow fluid in the thorax. As with **man**, use in **dogs** leads to hypotension. Neurological and behavioral changes in experimental **animals**, leading to marked impairment of locomotor activity and learning with an increase in aggressive behavior.

LD_{50}, **mouse**: 52 mmol/kg; **guinea pig**: 6.3 g/kg; LD_{50}, **rat**: 8 g/kg; **dog**: 5 g/kg. IP, LD_{50}, **mouse**: 1.2 ml/kg; **rat**: 1.3 g/kg; **guinea pig**: 2 g/kg.

METHYLMETHANE SULFONATE

= *MMS = NSC 50,256*

Untoward effects: Ingestion of as little as 1.42 mg/kg leads to intractable nausea and vomiting within 1 h. Hepatic damage with increased liver enzymes and ***sulfobromophthalein*** retention.

SC in **rats** leads to local sarcomas, squamous cell carcinomas, and a nephroblastoma.

LD, female **quail**: 75 mg/kg; LD_{50}, **mouse**: 100 mg/kg; IM and SC, **rat**: 125 mg/kg.

METHYLNAPHTHALENE

= *Methyl Naftalen*

Solvent.

Untoward effects: Causes a leukocytosis in **dogs**. IP use in **mice** leads to bronchiolar necrosis.

LD_{LO}, **human**: 500 mg/kg, LD_{50}, **rat**: 4.4 g/kg.

β-METHYL NAPHTHYL KETONE

= *Oranger Crystals*

Fragrance used in soaps, detergents, cosmetic creams and lotions, perfumes, and in foods, adding *jasmin, narcissus,* and *wisteria* notes.

Untoward effects: No sensitization reactions of 2% in petrolatum on 25 **human** volunteers. Full-strength for 24 h produced irritation in 1/24 **subjects**.

No-effect level when fed to **rats** for 12 weeks was 34.2 mg/kg.

LD_{50}, **rat**: 599 mg/kg.

METHYL NICOTINATE

= *Midalgan*

***Rubefacient.* Peripheral vasodilator.**

Untoward effects: The pure compound is an eye and skin irritant. An 18-year-old **female** used it (1%) topically with ***capsaicin*** (0.12%) for muscle pain, leading to internal pain, nausea, and syncope. Skin-testing revealed it was due to hypersensitivity to methyl nicotinate.

N-METHYL-*N*′-NITROSOGUANIDINE

= *MNNG*

Bacterial mutagen, experimental carcinogen.

Untoward effects: Orally in drinking water to **rats** leading to adenomas, adenocarcinomas, leiomyomas, and carcinomas of the stomach. Also carcinogenic to **dogs**, **mice**, **guinea pigs**, **hamsters**, and **rabbits** after oral intake. Found as a toxic pollutant in water.

1-METHYL-1-NITROSOUREA

= *NSC 23,909*

Untoward effects: Carcinogen. Simple erythema up to severe dermatitis after topical use versus mycosis fungoides in **man**.

Weekly IV to boxer **dogs** led to malignant neurinomas. Syrian **hamsters** given 5 mg developed retinal changes within 1 week. Repeated IV to **rats** led to peripheral nervous system, ovary, lung, uterine, colon, jaw, pituitary gland, and mammary gland neoplasms. Teratogenic to **rat** fetuses. Bladder instillation in **rats** led to tumors. Adenomas or carcinomas of midventral sebaceous gland pad of **guinea pigs**. Long-term oral use in **swine** led to stomach tumors. May be the toxic element of ***streptozotocin***.

LD_{50}, **rat**: 180 mg/kg; oral and SC, **mouse**: 200 mg/kg; TD_{LO}, **rat**: 90 mg/kg; **mouse**: 112 mg/kg/10 weeks; **monkey**: 20 g/kg over a 6-years period, **guinea pig**: 800 mg/kg, 67 weeks; **hamster**: 282 mg/kg, 17 weeks.

METHYL NONYL KETONE

= *2-Hendecanone* = *Undecanone-2*

A fragrance for use in soaps, detergents, cosmetic creams and lotions, perfumes, and food. *Oil of Rue* contains 90%. See *Rue*.

Untoward effects: No sensitization by 5% in petrolatum on 25 **human** volunteers. No irritation after 48 h closed-patch test on **humans**. Slight irritation by it in ***rue oil***.

Non-irritating when applied full-strength to the backs of hairless **mice** and **swine** or to intact or abraded **rabbit** skin under occlusion for 24 h.

LD_{50}, **mouse**: 3.9 mg/kg; **rat** and acute dermal in **rabbit**: > 5 g/kg.

METHYL OCTINE CARBONATE

= *Methyl-2-Nonynoate*

A fragrance for use in soaps, detergents, cosmetic creams and lotions, perfumes, and food.

Untoward effects: No sensitization and no irritation by 2% in petrolatum on 50 **human** volunteers.

Non-irritating after full-strength for 24 h to intact or abraded **rabbit** skin.

LD_{50}, **rat**: 2.22 g/kg; skin, **rabbit**: 5 g/kg.

METHYL PARATHION

= *Azophos* = *BAY 11,405* = *BAY E-601* = *Bladan M* = *"Cotton Poison"* = *Dalf* = *Dimethyl Parathion* = *ENT 17,292* = *E601* = *Folidol-M* = *Metacide* = *Metaphos* = *Methaphos* = *Metron* = *MPT* = *Nitrox* = *σ,σ-Dimethyl-σ-(4-nitrophenyl)phos-phorothioate* = *Parathion-methyl* = *Parton M* = *Penncap M* = *Tekwaisa* = *Wofatox*

***Insecticide.* Organophosphate cholinesterase inhibitor.**

Untoward effects: Very toxic. Inappropriate use in a home to kill **spiders** caused hospitalization of seven **children** with abdominal pain, lethargy, increased salivation, miosis, and the death of two of **them**. In addition, vapors cause eye and skin irritation, diarrhea, nausea, headache, giddiness, vertigo, weakness, rhinitis, chest tightness, blurred vision, muscle twitching, tremors, bradycardia, pulmonary edema, sweating, tearing, and anorexia. A Ghent emergency room treated 53 **patients** poisoned by it. One was a case of intended murder. 12 (23%) died; **they** were from the intended suicide group. National Institute for Occupational Safety and Health sets 0.2 mg/m^3, TWA as upper exposure limit for skin.

LD_{LO}, **human**: 5 mg/kg.

Found in **cow's** milk after exposure. Highly toxic to **bees**, **fish**, and **poultry**. Fetotoxic in **rats** and **mice**.

LD_{50}, **rat**: 6 mg/kg; **mouse**: 9.3 mg/kg; **rabbit**: 420 mg/kg; **guinea pig**: 1.27 g/kg; **iguana**: 82.7 mg/kg; **kestrel**: 3 mg/kg; **starling**: 7.5 mg/kg; **red-winged blackbird**: 10 mg/kg; acute dermal, **rabbit**: 67 mg/kg; LC_{50} in feed, **Bobwhite quail**: 90 ppm; **Japanese quail**: 46 ppm; **pheasant**: 116 ppm; **mallard duck**: 682 ppm.

METHYLPHENIDATE

= *Ciba 4311b* = *Contedrin* = *Metadate ER* = *MPH* = *Ritalin*

***Antidepressant, Central nervous system stimulant.* Street names are *R-Ball* and *Vitamin R*. Extended release formulation = *Concerta*.**

Untoward effects: Therapeutic levels are reported as 1–6 μg%; toxic levels as 80 μg%; and lethal levels as 230 μg%.

Abuse: Drug abuse (snorting) has led to foreign-body reactions to ***talc*** or ***cornstarch***. Deaths from pulmonary granulomata, vasculitis, and thrombosis. Endocarditis, eosinophilia, retinopathy, amputation of fingers, bacteremia (*Serratia Staphylococcus*), meningitis, glomerulonephritis, empyema, arthritis, hemiplegia, and headache after IV use of crushed tablets. IV **drug-abusers** have developed circulatory collapse, fever, leukemoid reactions, disseminated intravascular coagulation, azotemia, and rhabdomyolysis. A 28-year-old **female**, after IV use of a crushed tablet, developed flaccid quadriplegia and loss of posterior column sensation; **she** died 8½ months later.

Cardiovascular: Atrial fibrillations, tachycardia, palpitations, and angina. Arrhythmias may develop within minutes after IV. Marked pressor effects may follow IV and, sometimes, accompanied by motor restlessness, akathesia, dry throat, and insomnia. Myocardial abnormalities with widespread membranous cytoplasmic lamellations within myocardiocytes reported. After increase in blood pressure, occasional diaphoresis may occur, followed by hypotension.

Miscellaneous: Anorexia, nausea, dizziness, dyskinesia, encephalopathy, headache, palpitations, euphoria, disorientation, weight decrease, depersonalization, abdomial pain, insomnia, ankle and patellar clonus, chest pain, tremulousness, apprehension, coughing, addiction, nervousness, psychotic behavior, fatigue, mydriasis, increased appetite, leukopenia, bed-wetting, increased deafness, convulsions, decreased growth, ***histamine***-like conjunctivitis, formication, constipation, tics, transient proteinuria, hallucinations, and depression. Cerebral arteritis and infarction after 10 mg twice daily/7 years in 12-year-old **male** as treatment for attention deficit disorder. Hepatotoxicity with right hemiplegia in 6-year-old **female** after 10 mg tid for depression. May precipitate Gilles de la Tourette's syndrome.

Skin: May have been the causative agent of erythema multiforme (bullous-type) and exfoliative dermatitis. An 8-year-old **male** and a 10-year-old **male** with attention deficit disorder treated with 10 mg/day developed fixed-drug eruption of the scrotum. Rechallenge confirmed the cause. Necrotizing vasculitis, thrombocytopenic purpura, and patchy alopecia also occur.

LD_{LO}, **human**: 50 mg/kg.

Drug interactions: Potentiates the anticoagulant effect of ***coumarins*** (***ethylbiscoumacetate, dicoumarol***, and probably ***warfarin***) and the action of ***phenytoin, primidone, phenobarbital, phenylbutazone, serotonin***, and ***tricyclic antidepressants*** such as ***desipramine***, and ***imipramine*** by decreasing their metabolization. Can, theoretically, potentiate ***oral contraceptives*** by decreasing their hepatic metabolization. ***Monoamine oxidase inhibitors*** and ***pressor agents*** can extend its effects leading to acute hypertensive crisis with possible intracranial hemorrhage, hyperthermia, convulsions, coma, and, occasionally, death. Ingestion with ***food*** accelerates its absorption. May antagonize ***bretylium's*** and ***guanethidine's*** hypotensive effect and induce cardiac arrhythmias and ventricular tachycardia.

Long-term (2 years) use of very high oral doses in **mice** led to increased incidence of hepatocellular adenomas. Decreased food intake by **rats**. Requires at least 12–24 h for negative results on **horse** urine.

LD_{50}, **rat**: 450 mg/kg; 70 mg/kg IV, 170 mg/kg SC, **mouse**: 190 mg/kg; 40 mg/kg IV, 150 mg/kg SC, **rabbit**: 900 mg/kg; 30 mg/kg IV, 170 mg/kg SC.

METHYL PHENYLACETATE

= *Methyl α-toluate*

Fragrance in soaps, detergents, cosmetic creams and lotions, perfumes, and foods. U.S. consumption is ~ 2 ton/year.

Untoward effects: No irritation or sensitization on **human** volunteers at 8% in petrolatum.

Slightly irritating when applied full-strength to abraded or intact skin for 24 h under occlusion on **rabbits**.

LD_{50}, **rat**: 2.55 g/kg; acute dermal, **rabbit**: 2.4 g/kg.

1-METHYL-4-PHENYL-1,2,5,6-TETRAHYDROPYRIDINE

= *MPTP*

Neurotoxin.* A derivative of *meperidine.

Untoward effects: Homemade synthetic ***heroin*** containing this contaminant gave many **street-drug users** irreversible symptoms of parkinsonism, due to brain neuron destruction in the substantia nigra, after its conversion to a toxic metabolite, ***MPP***, by monoamine oxidase B. The same symptoms have been shown in non-human **primates** by dopaminergic neurons. Cutaneous absorption and/or inhalation of vapors by a 49-year-old chemist led to Parkinson's disease.

METHYLPREDNISOLONE

= *Medrate* = *Medrol* = *Medrone* = *Metastab* = $\Delta^1 6\alpha$-*Methylhydrocortisone* = *Metrisone* = *Promacortine* = *Suprametil* = *Urbason*

The acetate form = *Depo-Medrate* = *Depo-Medrol* = *Depo-Medrone* = *Mepred* = *Vetacortyl*

The phosphate disodium salt form = *Medrol Stabisol*

The succinate form = *Urbason-Solubile* = *Solu-Medrol*

The aceponate form = *Advantan*

Glucocorticoid.

Untoward effects: ***Blood***: Eosinopenia, increased blood urea nitrogen and transaminases, transient leukocytosis, anemia, and possibly non-thrombocytopenic purpura.

Central nervous system: Headache, vertigo, convulsions, arachnoiditis, neuromuscular weakness, and aggressive psychosis reported.

Endocrine: Decreased growth in **children**. Cushingoid effect (moon facies), menstrual abnormalities, decreased ***carbohydrate*** tolerance can aggravate a latent diabetes, increased blood ***glucose***, decreased serum ***testosterone*** in older **men** due to changes in ***GnRH*** secretion.

Eye: Posterior subcapsular cataracts, and increased intraocular pressure. Partial but permanent amaurosis in a **child** after injection into tonsillar fosa, probably due to large particles in older preparations. Reversible steroid glaucoma 3 weeks after retrobulbar or subconjunctival injection. Exophthalmos. Topically leads to transient stinging, burning, lacrimation, itching, and feeling of dryness.

Fluid and electrolytes: Fluid and ***sodium*** retention, hypokalemic alkalosis, and congestive heart failure.

Hypersensitivity: Has caused several cases of an immediate anaphylactic reaction with urticaria, angioedema, and bronchospasm. A 40-year-old **male** developed anaphylaxis after 250 mg of the *sodium succinate* form, but not the *acetate*.

Miscellaneous: Asymptomatic pleocytosis of spinal fluid after intrathecal use, osteoporosis, fat necrosis at injection site, increased susceptibility to certain bacterial and mycotic infections, nausea, vomiting, backache, gastrointestinal ulceration and perforations, myopathy, increased appetite, weight increase, leg pains, arthralgia, difficulty in micturation, metallic taste, decreased serum ***IgG*** and immunity, pancreatitis, cardiac arrest, spontaneous ruptures of Achilles tendon, hiccups, spuriously high ***sulfobromophthalein*** retention time, spuriously low ***protein-bound iodine*** readings.

Skin: Decreased wound-healing, petechiae, ecchymoses, acne; thin, fragile skin; hyperhidrosis, facial erythema, and depigmentation at injection site.

Drug interactions: ***Oral contraceptives, erythromycin, itraconazole***, and ***troleandomycin***, decrease its body clearance. Use increases ***cyclosporine*** plasma levels. Prolonged neuromuscular blockade when used with ***atracurium***.

Lymphocytopenia in **dogs** persisting up to 2 weeks after use. Immunosuppressive in **poultry** and **animals**. IM or SC is teratogenic in **mice**, but not **rats** or **rabbits**. In **dog** trials, 30 mg/kg produced less toxicity than three doses of 10 mg/kg over 72 h. If used near the end of pregnancy, it can induce premature parturition, retention of fetal membranes, and abortions in **animals** and, possibly, in **women**.

SC, TD_{LO}, **mouse** (11–15 days pregnant): 87 mg/kg.

METHYL PROPYL KETONE

= *Ethyl Acetone* = *MPK* = *2-Pentanone*

Solvent.

Untoward effects: Short-term exposure leads to headache, dizziness, nausea, vomiting, incoordination, and eye and respiratory tract irritation. Dermatitis from chronic exposure to the liquid. Exposure can occur from inhalation of vapors (narcosis and central nervous system depression), ingestion, and dermal or eye contact. OSHA set upper exposure limits at 200 ppm and National Institute for Occupational Safety and Health at 150 ppm. Immediate danger to life or health – 1500 ppm. Extremely flammable.

Narcosis and congestion, hemorrhage and edema of lungs, liver, and kidneys after acute inhalation by **guinea pigs**. 100% lethal to **guinea pigs** at 30,000 ppm/45 min.

LD_{50}, **rat**: 3.73 ml/kg; acute dermal, **rabbit**: 8 ml/kg; inhalation LC_{50}, **rat**: 2000 ppm/4 h.

METHYL SALICYLATE

= *Oil of Betula* = *Oil of Gaultheria* = *Oil of Sweet Birch* = *Oil of Teaberry* = *Oil of Wintergreen*

Use: Flavoring, counterirritant, and in perfumery.

Untoward effects: A teaspoonful contains 2.7 g of ***salicylate***. ***Salicylate*** toxicity in 62-year-old **male** after applying it in a topical ointment to **his** thigh twice daily/3 weeks. Urticaria due to sensitivity to it in toothpaste. Skin and muscle necrosis and interstitial nephritis in 62-year-old **male** after using it as an 18.3% ointment with 16% ***menthol*** (q.v.) and brief heating pad applications. An 81-year-old **male** died after drinking juice with it from an air freshener. A 22-month-old **female** received second and third degree mucosal burns after ingestion of it. Has caused prolonged prothrombin time, hemorrhages, and even the need for transfusions, after topical use on **patients** stabilized on ***warfarin***. Munching on a wintergreen mint, such as ***Certs***® or ***Life- Savers***® produces a spark when **one** bites down firmly. This can produce dire consequences in operating rooms using certain gases for anesthesia. Poisoning from ingestion, especially in **children**, often occurring with vomiting, lethargy, flushing, and hyperpnea. Because of its candy-like odor, it has frequency caused poisoning in **children**. A high percentage (45±%) of poisoning cases from ingestion have been fatal. **Adults** have died after ingesting 4–40 ml; < 4 ml has killed a **child**. Persistent vomiting, abdominal cramps, bloody diarrhea, hyperhidrosis, dehydration, thirst, tachypnea, tachycardia, restlessness, and delirium often followed by stupor, drowsiness, dyspnea, cyanosis, methemoglobinemia, dark brown urine, convulsions, coma, and death reported. Renal and hepatic toxicity in **survivors**.

LD_{LO}, **human**: 170 mg/kg.

Skeletal anomalies, increased fetal resorption, eye defects, craniorrachischisis, vertebral and rib abnormalities, gastroschisis, decreased fetal growth, exencephaly, hydrocephaly, and facial clefts in **rats**. Toxicity in **dogs** increased by ***pentobarbital*** and ***barbital***. Orally to **dogs** at 1.2 g/day led to fatality in 3–4 days; 500–800 mg/kg/day fatal in 1 month; 350 mg/kg/day/2 years leads to decreased weight and liver enlargement.

LD_{50}, **rat**: 887 mg/kg; **guinea pig**: 1 g/kg.

METHYL SILICATE

Untoward effects: Severe eye irritant causing corneal damage, even on short-term exposure to vapors. Has caused pulmonary edema and lung and kidney injury. National Institute for Occupational Safety and Health upper limit for exposure set at 1 ppm.

Severe irritation of **rabbit** eye by 250 µg.

LD_{LO}, **rat**: 700 mg/kg; inhalation, 250 ppm/4 h.

α-METHYL STYRENE

Use: Primarily in ***polyester resins*** and in production of synthetic ***rubber***. Demand is for well over 100 million lbs/year.

Untoward effects: Has caused moderate skin burns. Vapors lead to eye, nose, and throat irritation; drowsiness, and dizziness. Chronic exposure leads to central nervous system disturbances, temporary moderate leukemia, and hematologic disorders. National Institute for Occupational Safety and Health sets upper exposure limit at 50 ppm; OSHA, 100 ppm; immediate danger to life or health, 700 ppm.

METHYL TESTOSTERONE

= *Android* = *Glosso-Stérandryl* = *Metandren* =*Neohombreol M* = *Orchisterone-M* = *Oreton Methyl* = *Perandren* = *Testred*

***Androgen, anabolic agent.* Oral and buccal use.**

Untoward effects: ***Fetus***: Pseudohermaphrodism in **female fetus**, especially if given during the first trimester.

Liver: Cholestasis (hepatocanalicular jaundice), hepatitis, cirrhosis, peliosis, and increased alkaline phosphatases. Probably associated with malignancies, such as angiosarcoma and hepatocellular carcinoma.

Miscellaneous: Adverse activating mood and behavioral effects, decreased high-density lipoprotein, pain in lower abdomen and along inguinal canal, and slight scrotal edema in 45 cryptorchid **patients** (17 days–14 years), headache, hypercalcemia, myocarditis, nausea, amenorrhea, menstrual irregularities, clitoral enlargement, pubic hair, and phallic enlargement in Mongoloid **children**. Seminoma in 36-year-old **male** who illicitly took 10–400 mg/day/8 years.

Skin: Acne, allergic rashes, enlarged sebaceous glands, hirsutism, and **male** pattern baldness.

Drug interactions: Enhanced bleeding with ***phenprocoumon*** and ***warfarin***. May cause a paranoid reaction in **men** with depression receiving ***imipramine*** and virilization with hyperthyroidism when used with ***thyroid*** for stimulation of growth in short **children**. May increase toxicity of ***demeclocycline*** (q.v.).

TD_{LO}, **woman** (12–24 weeks pregnant): 39 mg/kg; **man**: 470 mg/kg over 4.1 years.

Increased ***sulfobromophthalein*** retention and SGOT levels in **rabbits**. Induces hypertension and vascular and cardiac disease in **rats**. In **mice**, increased ***hexobarbital*** sleeping time. Fed to young **fish** leads to monosex which are more desirable in the market place.

TD_{LO}, **dog** (0–63 days pregnant): 9.6 mg/kg.

METHYLTHIOURACIL

= *Alkiron* = *Antibason* = *Basecil* = *Basethyrin* = *Methiacil* = *Methicil* = *Methiocil* = *MTU* = *Muracil* = *Prostrumyl* = *Strumacil* = *Thimecil* = *Thiothyron* = *Thyreostat I*

Thyroid inhibitor.

Untoward effects*:** Systemic lupus erythematosis, itching, urticaria, rashes, edema, fever, leukopenia, neutropenia, granulocytopenia (42/3336), eosinophilia, purpura, lymphatic glandular enlargement, hypothrombinemia, hypersensitivity vasculitis, anosmia, ageusia, acute cholestasis, lymphocytosis, lupus erythematosus, thrombocytopenia, anemia, aplastic anemia, and spuriously low ***radioiodine uptake test results. Can cause **neonatal** goiter and mental retardation.

TD_{LO}, **woman** (20–39 weeks pregnant): 1.12 g/kg.

Drug interactions*:** Potentiates ***coumarin anticoagulants.

Goitrogenic in **animals** with increased water retention and ***cholesterol*** levels. Morphological changes in **rabbit** fetal hearts and parakeratosis in **swine**. Cretinism in **chin-chillas**, neonatal goiter in **rabbits** and thyroid carcinogenesis in **hamsters**.

TD_{LO}, **rat**: 9.1 g/kg over a 2 year period, pregnant **rabbit**: 50 mg/kg; **mouse**: 196 g/kg over a 1 year period. LD_{LO}, **rabbit**: 2.5 g/kg.

METHYL TRITHION

= *ENT 25,886* = *Methyl Carbophenothion* = *R 1492* = *Tri-Me*

***Organophosphorus insecticide.* Cholinesterase inhibitor.**

Untoward effects*:** Toxicity similar to, but less on a weight basis, to ***carbophenothion (q.v.).

LD_{LO}, **human**: 50 mg/kg.

LD_{50}, **rat**: 98–200 mg/kg; acute dermal, **rabbit**: 190–250 mg/kg; **mouse**: 390 mg/kg; male **chicks**: 544 mg/kg; **wild birds**: 18 mg/kg; **quail**: < 4000 ppm in feed.

METHYLTYRAMINE

***Sympathomimetic amine.* Found in *peyote* (q.v.), *Acacia berlandieri* (q.v.), and other plants.**

***Untoward effects*:** Mild hallucinogen.

LD_{LO}, **human**: 300 μg/kg.

METHYLURACIL

Untoward effects*:** Tympany and stilted gait within 4 h in **sheep** given 250 mg/kg. After four daily doses, post-mortem revealed gastroenteritis, liver congestion, and friable adrenals. See ***Bromacil and ***Isocil***.

TD_{LO}, **woman** (20–39 weeks pregnant): 1.12 g/kg.

LD_{LO}, **rabbit**: 2.5 g/kg; TD_{LO} (pregnant): 50 mg/kg.

METHYL VIOLET

A mixture of *rosanilines*.

***Untoward effects*:** Contact with intact or bruised skin can cause local acneiform lesions and eczema. Occasionally epitheliomata. Contact by licking of indelible pencils leads to inflamed lips and gums. Inflammation, swelling, cellulitis, and tissue necrosis can follow imbedding of particles in injured skin. Corneal clouding and ulceration, conjunctivitis, and swollen eyelids occur from contact. Ingestion can cause mucosal irritation, ulceration, headache, vomiting, and weakness.

LD_{LO}, **mouse**: 25 mg/kg; **rabbit**: 75 mg/kg.

METHYLXANTHINES

Untoward effects*:** See discussions under ***aminophylline, ***caffeine***, ***chocolate***, ***cocoa***, ***coffee***, ***coke***, ***dyphylline***, ***oxtriphylline***, ***tea***, ***theobromine***, and ***theophylline***.

METHYPRYLONE

= *Dimerin* = *Methylprylone* = *Noctan* = *Noludar*

Sedative, hypnotic.

Untoward effects*:** Dizziness, some gastrointestinal upsets, confusion, ataxia, slurred speech, hallucinations, nightmares, rashes, porphyrias, headache, and paradoxical excitement. Usual **adult** dose is 200–400 mg at bedtime. Transient bone marrow depression in two sisters, 38 and 42-year-old, after 900 mg/night in one and 600 mg/night in the **other**. Use with ***alcoholic beverages, including ***beer***, can further impair **patient** performance and even cause euphoria. Overdoses have caused coma, respiratory failure, and, possibly, agranulocytosis. These cases are usually in the emergency room within 4–6 h of ingestion. Some of these evidence miosis, epigastric distress, neuromuscular twitching or seizures, and lateral nystagmus. Myocardial infarction, hepatic dysfunction, and increased SGOT also reported. A 57-year-old **female** ingested 22.5 g, leading to deep coma, blood pressure 60/32, pulse 68/min, bradypnea, shallow breaths, and absence of most reflexes. Mean blood concentration in fatal cases-76.7 μg/ml (range 1–771). Appears in breast milk and can be sedative to a **baby** and impair **its** feeding ability. Psychological and physical dependence often occurs after long-term use. Convulsions have followed sudden withdrawal. **Adult** lethal dose is ~5–6 g. Therapeutic serum levels are 0.5–1.5 mg%; toxic levels at 3–6 mg%. and lethal levels at 10 mg%.

METHYRIDINE

= *Dekelmin* = *Metyridine* = *Promintic*

Anthelmintic.

Untoward effects: Percutaneous absorption can occur. Promptly wash off with cold water any spilled material. Avoid contact with paintwork, metals, ***rubber*** goods, etc., as the drug is a strong solvent. Avoid spilling it within 2 weeks of prior medication with organophosphorus anthelmintic-type compounds. NOT to be used with other anthelmintics (high toxicity when used with ***diethylcarbamazine***). Do NOT administer to **horses**, as it is highly irritating to them. Avoid its use in **calves** under 3 month of age or **cows** within 1 month before or after calving. Its use will taint milk. Ataxia is common in **cats**, **monkeys**, and exotic species.

SC, LD_{50}, **mouse**: 1.64 g/kg.

METHYSERGIDE

= *Deseril* = *Desernil* = *Sansert* = *UML-491*

***Serotonin antagonist, antimigraine.* The *maleate* is usually used.**

Untoward effects: Cardiovascular: Endocardial fibrosis associated with development of heart murmurs, valvular heart disease and fibrotic thickening of the aortic root, angina, and myocardial infarction. A 48-year-old **male** developed tricuspid and mitral valve lesions after 2 mg twice daily/4 years. A young **woman** developed severe mitral regurgitation after 4 years therapy, which required mitral valve replacement. Mild congestive heart failure in 43-year-old **female**. Thrombophlebitis, intestinal and myocardial ischemia, decreased femoral pulse, cold and numb extremities, palpitations, immune hemolytic anemia, thrombocytopenia, lupus erythematous, neutropenia, and eosinophilia.

Central nervous system: Speech-slurring, staggering, dizziness, giddiness, euphoria, insomnia, numbness and facial drooping. Hip claudication and impotence in 36-year-old **male** 7 months after 2 mg tid/3 months, then 8 mg/day/4 months, then 10 mg/day/10 weeks. Confusion, nightmares, and mild hallucinations reported.

Eye: Retinal vasospasm lasting 30 min in 34-year-old **female** after one dose of 0.25 mg. Eye ptosis and nystagmus.

Fetus and Neonate: Suspected as a cause of malformation in **newborn** that died within 48 h. It was the only drug the **mother** took during **her** pregnancy.

Fibrosis: Has induced retroperitoneal, pleuropulmonary, renal, and cardiac fibrosis. Retroperitoneal fibrosis associated with ureteral obstruction, hydronephrosis, oliguria, anuria, uremia, ascites, iliac artery obstruction, unilateral leg edema, pelvic venous obstruction, hypertension, nausea, vomiting, and limb pains. Pleuropulmonary fibrosis leading to chest pain, dyspnea, pleural effusions, and pleurisy. Poor inhibition of ***serotonin*** receptors in the vessels and fibrous tissues of the chest, abdomen, and extremities leads to excessive ***serotonin*** in these areas, manifesting by vasoconstriction and fibrotic reactions.

Miscellaneous: Weight increase, edema, tightness of chest, dysuria, psychiatric complications (severe depressive behavior, violence, and suicidal tendencies), light-headedness, leg and muscle cramps, metallic taste, and impotence. Increased risk of arterial spasm and occlusion when taken with ***ergot*** alkaloids, ***β-adrenergic-blockers, erythromycins, dopamine***, or ***troleandomycin***. Use with ***chlorpromazine*** leads to depression. May cause spuriously high ***blood urea nitrogen*** test results.

Skin: Generalized pruritus, urticaria, facial flushing, and reddened, painful shins and feet. Multiple red confluent spots on face, neck, arms, and hands with hyperkeratosis and slight edema of wrists and hands in 71-year-old **female** who ingested 1.65 mg tid/1 year, then twice daily/1 year, then once/day/3 years. Biopsy suggested subacute lupus erythematous. Recurred after rechallenge. Reversible alopecia in 30-year-old **male** .

LD_{LO}, **human**: 500 mg/kg.

METHYSTICODENDRON AMESIANUM

= *Culebra borrachera*

Hallucinogen.

Untoward effects: Widely used by Indian **natives** and **medicine men** high in the Colombian Andes of South America. An infusion of the leaves and stems contains 0.3% total alkaloids, of which up to 80% is ***scopolamine*** (q.v.).

METIAMIDE

= *SKF 92,058*

Antihistamine.

Untoward effects: Neutropenia and fatal agranulocytosis, which led to its withdrawal from the marketplace.

Constipation, headache, nausea, vomiting, diarrhea, acne, sore throat, and increased plasma ***creatinine*** also reported.

Potentiates systemic anaphylaxis in **horses**.

METOCLOPRAMIDE

= *AHR-3070-C* = *Cerucal* = *Clopromate* = *DEL-1267* = *Duraclamid* = *Elieten* = *Emetid* = *Emperal* = *Eucil* = *Gastrese* = *Gastrobid* = *Gastromax* = *Gastronerton* = *Gastrosil* = *Gastro-Tablinen* = *Gastrotem* = *Gastro-Timelets* = *Imperan* = *Maxeran* = *Maxolon* = *Meclopran* = *Metamide* = *Metoclol* = *Metocobil* = *Metramid* = *MK 745* = *Moriperan* = *Mygdalon* = *Parmid* = *Paspertin* = *Peraprin* = *Plasil* = *Pramiel* = *Primperan* = *RD 1267* = *Reglan* = *Relivran*

Antiemetic.

Untoward effects: Blood: Agranulocytosis in 61-year-old **female** after 10 mg qid/2 weeks. A 3-week-old infant received ten times normal accidental overdose (1 mg qid for 36 h) leading to methemoglobinemia, responding to IV ***methylene blue***.

Cardiovascular: Transient hypertension after IV (240/170 returning to normal 110/70 within 5 min; another **patient** had 150/90 for ~1 h, then returned to normal 110/70). Supraventricular tachycardia after 10 mg IV before spinal anesthesia and tubal ligation. Complete heart-block and asystole after 10 mg IV in 54-year-old **male**. Sinus arrest in 51-year-old **female** receiving IV push of 5 mg qid for gastroparesis; 10 mg doses after 96 h; after another 24 h, dosage increased to 20 mg. **She** then developed bradyarrhythmias (15–20 min duration). Rechallenge later confirmed the cause. A 62-year-old **male** received 2.5 mg IV and a 71-year-old **male** received 10 mg IV. **Both** developed total heart-block.

Central nervous system: Restlessness, drowsiness, fatigue, and lassitude are common (10%). Headache, dizziness, confusion, insomnia, dystonia, mania, akathisia, depression, and suicidal ideation. Convulsions in **children** receiving more than the suggested dosage. Use in the **elderly** causes confusion and symptoms mimicking Parkinson's disease and may have caused misdiagnoses. Neuroleptic malignant syndrome occurs.

Endocrine: Amenorrhea, gynecomastia, galactorrhea, nipple tenderness, and impotence in **men**, and decreased libido in **women**, secondary to hyperprolactinemia and transient increase in ***aldosterone***, leading to fluid retention, weight increase, and increased respiratory effort.

Extrapyramidal reactions: Acute dystonia and hyperkinesis (face, epiglottis, limbs), torticollis, protrusion of tongue, trismus, opisthotonus, oculogyric crisis, dyspnea, stridor, respiratory failure, agitation, akathesia, jitteriness, pacing, fever, and parkinsonism. Familial incidence of dystonic reactions reported. A few deaths from acute dystonic reactions have been reported.

Gastrointestinal: Nausea and diarrhea, dry mouth. Constipation in **some**.

Hypersensitivity: Anaphylactoid symptoms and wheezing in **patients** sensitive to ***sodium bisulfite***, a preservative, in an injectable form.

Miscellaneous: Sweating, bronchospasm, mild hypotension, macular rash, urticaria, joint aches, cervical lymphadenopathy, ***salt*** retention with edema, glossal or periorbital edema. Respiratory failure in asthmatic 32-year-old **female** after two 10 mg doses. Reversible nonthrombocytopenic purpura in 72-year-old **male** after 10 mg tid/2 days. Found in breast milk at therapeutic **infant** doses.

Renal: Incontinence and increased urinary frequency. Decreased renal plasma flow.

Drug interactions: Exacerbation of acute intermittent porphyria with ***valproate***. Effects on absorption and gastric passage appear to cause decreased ***cimetidine*** absorption (25–30%), decreased serum ***digoxin***, decreased efficacy of ***l-dopa, ethambutol, quinidine***, and ***sulfonamides***. Increases ***acetaminophen, aspirin, diazepam, l-dopa, ethanol, lithium sustained release, pivampicillin, propranolol***, and ***tetracycline*** absorption and increases bioavailability of ***cyclosporine*** by 20%. When added to ***clobutinol*** and ***bacampicillin*** therapy for a 31-year-old **male** and a 32-year-old **female** led to severe extrapyramidal symptoms. Acute dystonia in 14-year-old **male** with anorexia nervosa when it was added to ***fluvoxamine*** therapy. Reduces plasma concentrations of ***atovaquone*** by 40–50%. Because it accelerates ***food*** absorption, it can increase ***insulin*** dosage needed and its timing in **diabetics**.

A 4-year-old **female macaw** developed ataxia, torticollis, and opisthotonus; died after several days. Can aggravate seizures in **cats** and cause tremors and anxiety or behavioral changes. Increases rate of ***chlorpromazine*** absorption in **rats**. Long-term use can cause constipation in **cats** and **dogs**. Rapid IV in **cats** and **dogs** has induced extrapyramidal symptoms and their frenzy can injure **people**.

LD_{50}, **mouse**: 280 mg/kg.

METOLACHLOR

= *CGA 24,705* = *Dual* = *Pennant*

Herbicide.

Untoward effects: Sialism, diarrhea, exophthalmus, piloerection, and trismus in male and female **rats** given > 2 g/kg. Post-mortem revealed hemorrhagic lungs.

Primary skin irritant on **rabbits**. Acute emetic dose$_{50}$ in **dogs**: 19 mg/kg.

LD$_{50}$, **rat**: 2.78 g/kg; acute dermal, **rabbit**: > 10 g/kg; inhalation LC$_{50}$, **rats**: > 1.75 mg/l of air 4 h.

METOLAZONE

= *Diulo* = *Metenex* = *Microx* = *SR 720-22* = *Zaroxolyn*

Diuretic, antihypertensive.* A *sulfonamide.

Untoward effects*: *Blood: Hemoconcentration, aplastic anemia, agranulocytosis, leukopenia, decreased blood ***potassium***, hyperglycemia, hyponatremia, hypochloremia, hyperuricemia, transient hypoplastic anemia in 74-year-old **female** after 5 mg/day/4 months; increased ***blood urea nitrogen*** and ***creatinine***.

Cardiovascular: Orthostatic hypotension, palpitations, venous thrombosis, chest pain and/or discomfort, decreased blood volume; pulmonary embolus.

Central nervous system and peripheral nervous system: Syncope, vertigo, paresthesias, neuropathy, encephalopathy, drowsiness, light-headedness, fatigue, weakness, headache, seizures, stupor, and coma.

Gastrointestinal: Hepatitis, cholestatic jaundice, pancreatitis, epigastric discomfort, nausea, vomiting, bloating, anorexia, diarrhea, and constipation.

Miscellaneous: Weakness, long-acting (1–3 days after single dose); hyperosmolar, non-ketotic diabetes mellitus, toxic epidermal necrolysis, glycosuria, muscle cramps, myalgia, decreased absorption in **patients** with cardiac failure, and erythema-multiforme-like dermatitis, chills, acute gouty attacks, cutaneous hypersensitivity angiitis, photosensitivity, and transient blurred vision.

Drug interactions: Renal toxicity in renal transplant **patient** after use with ***cyclosporine***. Potentiates the effect of ***furosemide*** and related to a sudden death in one **patient**. May potentiate ***warfarin***. Use with ***captopril*** suspected as a cause of renal failure in 65-year-old **female** with rheumatic heart disease. Possible increase in serum ***lithium*** and toxicity. In theory, cross-reactivity may occur with ***quinethazone***, ***sulfonamides***, and ***thiazides***.

TD$_{LO}$, **woman**: 150 mg/kg.

METOMIDATE

= *Hypnodil* = *Methomidate* = *Methoxymol* = *R 7315*

Hypnotic, narcotic.

Untoward effects: Most commonly used in **swine**, where it has caused slight tremors lasting up to a few hours. Doses > 30 mg/kg in 3–7-week old **chicks** led to spasms of the neck and extremities. Death in 74/253 at 50 mg/kg.

LD$_{50}$, **mallard duck**: 120 mg/kg. IV, LD$_{50}$, **rat**: 50 mg/kg.

METOPROLOL

= *Beloc* = *Betaloc* = *CGP 2175* = *H 93/26* = *Lopresor* = *Lopressor* = *Prelis* = *Seloken* = *Selopral* = *Selo-Zok* = *Toprol-XL*

***Antihypertensive, antianginal, antiarrhythmic.* β_1*-Adrenergic-blocker.* Has very little β_2 activity. The *tartrate* is usually used.**

Untoward effects*: *Cardiovascular: Palpitations, bradycardia (3–9%), peripheral edema, hypotension (4%), congestive heart failure, cold extremities, Raynaud's disease, first degree heart-block (3%), claudication, and peripheral gangrene in hypertensive 58-year-old **female** after 100 mg twice daily/7 months. Has, on occasion, aggravated arrhythmias.

Central nervous system: Dizziness (10%), tiredness (10%), depression (5%), mental confusion, loss of short-term memory, headache, nightmares, insomnia, leg tremors, and catatonia.

Gastrointestinal: Diarrhea (5%), nausea, dry mouth, gastric pain, constipation, flatulence, and heartburn.

Miscellaneous: Blurred vision, musculoskeletal pain, tinnitus, decreased libido, hypoglycemia, thrombocytosis, and sweating. Although excreted in breast milk, it is unlikely to have significant effect on an **infant** because only 1.7–3.3% of the **maternal** dose can be passed to the **infant** in 24 h. Pain and soreness of both eyes in **female** taking 100 mg/day/4 months for tachycardia. Retroperitoneal fibrosis, pseudoscleroderma, and arthritic conditions. Increased blood urea nitrogen, serum transaminases, alkaline phosphatase, and lactate dehydrogenase; thrombocytopenic purpura and nonthrombocytopenic purpura. Rarely, dry eyes, visual disturbances, agranulocytosis, and Peyronie's disease.

Respiratory: Bronchospasm with wheezing and dyspnea (1%).

Skin: Pruritus and rashes (5%), worsening of psoriasis, and, rarely, alopecia. Eczematous and psoriasisiform eruptions in **male** after long-term therapy, which disappears slowly after drug withdrawal.

Massive intoxication after 19-year-old **male** ingested 200, 50 mg tablets (160 mg/kg) leading to peripheral cyanosis, weak heart sounds, unregisterable blood pressure, normal electrocardiogram, and abnormal blood gases. Recovery after 12 h of emergency treatment.

Drug interactions: Use with ***catecholamine***-depletors, such as ***reserpine***, can cause hypotension, bradycardia, vertigo, syncope, and postural hypotension and can potentiate ***insulin***-induced hypoglycemia in **diabetics**. ***Pentobarbital*** significantly decreases its plasma concentration. Absorption is increased by ***food*** intake. ***Cimetidine*** may inhibit its clearance to some degree, but appears to be of no clinical significance. Slight increase in serum concentrations when taken by **those** using ***oral contraceptives***. Can block the benefit from ***bronchodilators***. ***Paroxetine*** increases its bioavailability.

METOSERPATE

= *Avicalm* = *Pacitran* = *SU 9064*

Tranquilizer, sedative, neuroleptic.

Untoward effects: Restlessness is the primary adverse effect in **man**. Do NOT use for laying **birds** or replacement **birds** over 16 weeks of age, or slaughter treated **birds** within 72 h of treatment. The FDA allows a tolerance of up to 0.02 ppm for its negligible residues in uncooked edible tissues of **poultry**.

LD_{50}, **rat**: 182 mg/kg.

METRIBUZIN

= *Lexone* = *Sencor*

Herbicide.

Untoward effects: Eye and skin irritant.

Central nervous system depression, liver and kidney damage in experimental **animals**. Can alter thyroid function.

LD_{50}, **rat**: 2.2 g/kg; **mouse**: 698 mg/kg; **guinea pig**: 250 mg/kg; LC_{50}, **rainbow trout**: > 10 ppm.

METRIZAMIDE

= *Amipaque* = *WIN 39,103*

Radiopaque diagnostic agent. **Non-ionic. Water-soluble.**

Untoward effects: Various types of seizures, asterixis, transient confusion, decreased blood pressure, nephrotoxicity, exacerbation of hypertension, hyperreflexia, cortical blindness, delirium, nausea, vomiting, pleocytosis, cervical myelopathy, headache, hallucinations, subdural hematoma, innappropriate secretion of ***antidiuretic homone***, leg pain, dizziness, tinnitus, diplopia, depression and urinary retention. Limited passage into breast milk.

Drug interactions: Occasional seizures reported when ***chlorpromazine***, ***prochlorperazine***, or ***thioridazine*** was used with it.

Neurotoxicity, including seizures, in **cats**, **dogs**, **rabbits**, and **rhesus monkeys**. Causes fluid to accumulate in the distal air spaces of the lungs in **dogs**, which clears in ~6 h. Seizures have been potentiated by ***phenothiazines***, especially ***chlorpromazine***.

METRONIDAZOLE

= *Arilin* = *Bayer 5630* = *Clont* = *Deflamon* = *Elyzol* = *Flagyl* = *Fossyol* = *Gineflavir* = *Klion* = *Metocream* = *MetroGel* = *Metrolag* = *Metrolyl* = *Metrotop* = *Metryl* = *Orvagil* = *Protostat* = *Rathimed* = *Sanatrichom* = *Trichazol* = *Trichocide* = *Tricho Cordes* = *Tricho-Gynaedron* = *Trichopal* = *Tricocet* = *Trivazol* = *Unipazole* = *Vagilen* = *Vagimid* = *Zadstat*

A *nitroimidazole.*

Antiprotozoal, antibacterial. **Microbiocide/algicide for cooling tower water and to destroy anaerobes in manure pits of hog confinement buildings. An enzyme produced by *Heliobacter pylori* converts it to *hydroxylamine* that causes severe DNA damage to and death of the bacterium.**

Untoward effects: Blood and bone marrow: Neutropenia (1–3%) – usually reversible in 2 weeks; leukopenia, eosinophilia, decreased serum lipids, and hemolytic-uremic syndrome; occasionally leukocytosis reversible thrombocytopenia. A possible case of aplastic anemia reported.

Cardiovascular: Slight T-wave flattening and palpitations; thrombophlebitis after IV.

Eye: Oculogyric crisis and tremors in 33-year-old **patient** after three doses of 200 mg. Transient myopia after therapy for 11 days; confirmed by rechallenge.

Fetus and neonate: Despite its known mutagenic and carcinogenic properties, congenital defects reported could not be definitely attributed to the drug. Crosses the placenta and passes into breast milk with potential for adverse neurological and hematologic effects, anorexia, and vomiting, particularly in an older **child** consuming large quantities of **mother's** milk. Levels in breast milk are equal to **maternal** plasma levels.

Liver: Jaundice in apparently healthy 46-year-old **male** self-medicated 250 mg/day/5 days upon suggestion of **paramour** taking it for trichomoniasis. On the fifth day, led to malaise, chills, and fever, followed 5 days later with jaundice, conjunctival and cutaneous icterus, enlarged liver; abnormal sedimentation rate; SGOT, and serum ***bilirubin*** assay errors. An overdose in a 58-year-old **female** led to anorexia, nausea, and hepatitis.

Miscellaneous: Dizziness, headache, nervousness, insomnia, nausea, vomiting, diarrhea, anorexia, nasal congestion, vertigo, hallucinations, epigastric distress, depression, agitation, crying, disorientation, joint pain, arm pain, moniliasis, hot flashes, weakness, chills, fever, pseudomembranous colitis, pneumonitis, and tinnitus. Bilateral breast tenderness in 36-year-old **male** after 250 mg tid/2 months; breast tenderness recurred on subsequent rechallenge with 250 mg/day. Pancreatitis on two separate occasions in 29-year-old **female** after 250 mg tid; also in 23-year-old **female** after 250 mg tid and later on several occasions after 500 mg twice daily. Vaginal burning, dryness of vulva and vagina, nasal congestion, pruritus, and constipation. Disorientation and depression in 19-year-old **female** after 7 days treatment with 250 mg tid. Decreased libido reported. Grand mal seizures, coma, and prolonged behavior abnormalities in 12-year-old **male** after 250 mg IV qid/4 days. In China, adverse reactions in 47.7% with 600 mg/day, in 80% after 1.2 g/day, and in 90% after 1.8 g/day. Can induce delirium in the **elderly**.

Neoplasms: In one study, an excess of lung cancer has been reported and it may have also been a factor in increased incidence of cervical cancer. Breast malignancy and cholangiocarcinoma in three **patients** (27–32 years) with Crohn's disease after 250–1200 mg/day/ 9 months–3½ years, with total dosage 275, 340, and 720 g, respectively.

Neuropathy: Cerebellar dysfunction with ataxia and sensory neuropathy after 750 mg qid in 45-year-old **female**, which essentially disappeared within 6 days after withdrawal of therapy. An 81-year-old **male** on 500 mg tid/8 weeks developed peripheral neuropathy with paresthesia and numbness of lower legs and both hands. A 26-year-old **female** with Crohn's disease developed debilitating neuropathy after 2 g/day/50 days. The syndrome improved after withdrawal, but still remained 2 years later. In a 43-year-old **male**, 400 mg tid/3½ months led to peripheral neuropathy with total loss of sensation from 6 cm above both knees and elbows. Residual ankle effects remained 7½ months after discontinuance of treatment. Central and peripheral neurotoxicity in 81-year-old **male** after 37.2 g during 29 days leading to axonal degeneration, cerebellar limb incoordination, and gait ataxia. Nearly normal 5 months later. Hyperalgesia and convulsions reported.

Poisoning and suicide: Vomiting in 15-year-old **female** a few hours after ingesting 18 tablets (200 mg each) and in 19-year-old **female** ingesting 60 tablets (200 mg each) 24 h previously.

Renal: Dark brown urine, urethral burning, cystitis, polyuria, pyuria, dysuria, and incontinence.

Skin and mouth: Dermatitis, urticaria, vesicular eruption on chin, eczematous and/or psoriasisiform eruption after long-term therapy in **male**, altered smell and metallic taste, sore furry tongue, black tongue, diaphoresis, photosensivite skin rashes, dry mouth, and flushing.

Therapeutic serum concentrations are 10–30 mg/l, and 200 mg/l is toxic.

Drug interactions: At high dosage, it inhibits metabolization of ***cimetidine***. With ***cimetidine*** 800 mg/day, its plasma clearance is decreased. ***Chloroquine*** induces dystonic reactions in its presence. ***Disulfiram*** (q.v.)-like effect when taken with ***ethanol***, due to its inhibition of alcohol dehydrogenase and other ***alcohol***-oxidizing enzymes. The combination has also had a potential for abuse. Some **abusers** (usually white middle-class **females**) found pleasure in the resulting "rushes", giddiness, excitement, and flushing (rarely with nausea), due to increased circulating ***acetaldehyde*** (q.v.) levels. When IV solutions contained 10% ***alcohol, ethyl*** they could cause ***disulfiram***-like reactions. Use with ***disulfiram*** has caused ataxia and neuropsychiatric symptoms. ***Phenobarbital*** use increases its clearance. ***Lithium*** levels increased by it, causing ***lithium*** toxicity. Increases ***warfarin*** serum levels by decreasing its metabolization, resulting in 40% increase in prothrombin time and prolonged half-life of ***warfarin*** from 35 to 50 h. Dizziness, diplopia, and nausea when 250 mg tid given to 49-year-old **female** receiving 1 g/day ***carbamazepine***. Use increases ***quinidine*** serum levels. ***Cyclosporine*** blood levels are increased by it. ***Tetracycline*** increases the concentration of its metabolite in the saliva and urine ~1.7 times. Causes false decrease in SGOT values. Use with ***cisapride*** can cause torsades de pointes and prolong QT interval. Can decrease ***astemizole*** and ***terfenadine*** metabolization, increasing their serum concentration and potential for toxicity. ***Aluminum-*** and ***magnesium***-containing ***antacids*** can decrease its absorption and serum concentration. Can induce metabolization of and decrease effectiveness of ***oral contraceptives***.

TD_{LO}, **human**: 3.57 mg/kg/day; 1030 mg/kg over 8 weeks.

Hepatic and brain lesions and neuropathies reported in **monkeys**. Carcinogenic (pulmonary tumors and malignant lymphomas) in **mice** and **rats** (mammary tumors), and mutogenic in **bacteria**; chromosomal aberrations in the lymphocytes of treated **human patients**. Pituitary, testicular, and liver tumors in **rats**. Carcinogenic TD_{50}, **rat**: 542 mg/kg; **mouse**: 506 mg/kg; LD_{50}, **mouse**: 3.8 g/kg. When **dogs** do not respond to 5 days treatment for giardiasis and treatment is extended to 8–30 days, it leads to ataxia, tremors, nystagmus, and seizures. Death has occurred during seizures. Therapeutic levels in **hamsters** approach toxic levels. Has caused salivation, vomiting, anorexia, weight decrease, and neurotoxic effects in **cats**. Thrombocytopenia in **cats** and **dogs**. Loss of appetite in ~2% of treated **horses**.

METUREDEPA

= AB 132 = Turloc

Antineoplastic.

Untoward effects: Nausea, diarrhea, "nightmares", hematopoietic toxicity, leukopenia, thrombocytopenia, and apnea with poorly reversible inhibition of anticholinesterase, especially when given with ***succinylcholine***.

METYRAPONE

= *Metopirone* = *Metroprione* = *SU 4885*

Pituitary function diagnostic aid, corticosteroid inhibitor.

Untoward effects: Gastric pain, nausea, perspiration, slight mental disturbance, vomiting, hypotension, drowsiness, sedation, light-headedness, headache, dizziness, vertigo, rash, and, rarely, thrombophlebitis after IV.

The death rate from it in **mice** was highest at 21.00 h, showing that drug toxicity may vary according to the time of day or a subject's biological clock.

IV, LD_{50}, **mouse**: 261 mg/kg.

METYROSINE

= *Demser* = *H44/68* = *α-Methyl-p-tyrosine* = *Methyltyrosine* = *MK 781* = *α-MPT*

Untoward effects: ***Central nervous system***: Sedation, decreased alertness and motor coordination, extrapyramidal signs (tremors, speech difficulty, drooling), occasionally with trismus and parkinsonism, depression, disorientation, hallucinations, and confusion. Supresses ***catecholamine*** synthesis by 50–80%.

Miscellaneous: Diarrhea, decreased salivation, nausea, abdominal pain, vomiting, impotence, ejaculation failures, nasal stuffiness, slight breast swelling, and galactorrhea, occasional crystalluria, transient dysuria, hematuria, and ***dopamine*** depletion, leading to neuroleptic malignant syndrome. Rarely, peripheral edema, eosinophilia, increased SGOT, urticaria, and pharyngeal edema. Potentiates neuroleptic drugs such as ***haloperidol***, increasing its toxicity. Potentiates ***thioridazine***. Decreases euphoric effect of ***amphetamines*** and blocks ***morphine*** and ***dextroamphetamine*** self-administration behavior.

Induces tremors and catatonia in **monkeys**.

MEVINPHOS

= *CMPD* = *Duraphos* = *ENT-22374* = *Gesfid* = *Menite* = *OS-2046* = *PD 5* = *Phosdrin* = *Phosfene*

***Insecticide, acaricide.* An organophosphorus anticholinesterase.**

Untoward effects: Highly toxic. Hypersalivation, abdominal cramping and gastrointestinal hypermotility, diarrhea, vomiting, sweating, dyspnea, cyanosis, miosis, muscle twitching, convulsions, and tetany, followed by weakness and paralysis. Bradycardia in some cases. Hypoxia, due to bronchoconstriction can cause death. Exposure of an 18-year-old **female** to its use as a spray led to, within 30 min, vomiting, erythema, sweating, miosis, salivation, and clonic contractions. **She** recovered after treatment. Death of a 58-year-old **male** within 10 min after ingesting 100 mg. A case of criminal action by injecting some through the cork of a ***wine*** bottle led to the death of a **female**. Occupational poisonings, including fatalities, have occurred in agricultural **workers**. During 1982–1990, 495 poisoning cases due to it were reported in California; 26 cases from its use in Washington ***apple*** orchards during July–December, 1993. A 69 kg 66-year-old **male**, about to be charged with murder, drank (in front of detectives) a solution estimated to contain 28 g (~100 LDs) and was dead in less than 1 min, apparently, from respiratory failure.

Estimated LD, **human**: 4 mg/kg; 690 μg/kg/28 days. Highly toxic to **bees**, **fish**, and **birds**. **Cattle** have been poisoned when spray has drifted onto their pastures. Congenital foot deformities in **ducklings** and **chicks**. Ataxia, curled toes, tachypnea, dyspnea, salivation, diarrhea, tremors, phonation, tetany and violent wing-beat convulsions (within 2 min in **ducks** and 5 min in **pheasants** and **grouse**), and death. Lethal for **dogs** eating 200 ppm in their feed for 14 weeks.

LD_{50}s in mg/kg: **rat**: 0.89–6.1: dermal 4.2–4.7; **mouse**: 3.2; female **mallard duck**: 4.63, dermal: 11; female **chick**: 7.52, **starling**: 3.9, **grouse**: 1.34, **pheasant**: 1.37, male **gerbil**: 0.45, female **gerbil**: 0.54.

MEXACARBATE

= *Dowco 139* = *ENT 25,766* = *NCI C00544* = *OMS 47* = *Zectran*

***Pesticide, molluscicide, acaricide.* A methyl carbamate.**

Untoward effects: In **birds**, and often in **mammals**, leads to ataxia, imbalance, neck tremors, nystagmus, miosis, decreased responses, ataraxia, falling, convulsions, tachypnea, dyspnea, lacrimation, salivation, diarrhea, tachycardia, hyperthermia. Peak in symptoms within 1 h and recovery in ~3 h in **survivors**.

LD_{50}s in mg/kg: **rat**: 14, **mouse**, **dog**, **guinea pig**: 15, **rabbit**: 37, male **bullfrog**: 566, most **bird** species: 2.64–10, **mule deer**: 12.5–25, and **goat**: 15–30. LC_{50}, **fish**: 0.6–23 mg/l.

MEXENONE

= *Uvistat*

Ultraviolet screen.

Untoward effects: In rare instances, has caused light sensitivity.

MEXILETINE

= *KO 1173* = *Katen* = *Mexitil* = *Ritalmex*

Antiarrhythmic.

Untoward effects: Blood: Thrombocytopenia.

Cardiovascular: Hypotension (14%), bradycardia (14%), increased ventricular or atrial arrhythmias, premature ventricular contractions, palpitations, syncope, atrioventricular block, and hot flashes. Rarely, cardiogenic shock, torsades de pointes, and, paradoxically, proarrhythmias.

Central nervous system: Tremor, nystagmus, blurred vision, dizziness, ataxia, confusion, nervousness, incoordination, weakness, fatigue, paresthesias, headache, short-term memory loss (~1%), hallucinations, depression, psychosis, and convulsions.

Gastrointestinal: Nausea, vomiting, and heartburn. Less commonly, dysphagia, gastrointestinal bleeding, esophageal ulceration, hepatitis, anorexia, constipation, diarrhea, and cramps.

Miscellaneous: Pulmonary fibrosis in 75-year-old **male** after treatment with 200 mg tid/3 months. Discontinued 10 months later and **patient** died from intractable respiratory distress. Possibly causative of pseudothrombocytopenia in 62-year-old **male**. Alopecia, fever, tinnitus, speech impairment, dry mouth, dyspnea, rash, edema, impotence, and decreased libido. Increased plasma concentration noted in congestive heart failure. Slow oral absorption after myocardial infarction.

Poisoning: A 22-year-old **male** swallowed 22, 200 mg tablets and died. Blood level was ~35 μg/ml (therapeutic levels are 0.0008–0.002 mg/ml). Toxicity noted at levels > 3 μg/ml.

Drug interactions: Use usually requires 40–50% decrease in ***theophylline*** dosage, apparently due to inhibition of demethylation (metabolization) of the latter. A 54-year-old **female** with cardiomyopathy received 300 mg twice daily, but became toxic (diplopia, blurred vision, nystagmus, weakness, "thick tongue", slow and and garbled speech, leg muscle fasiculations) after 600 mg oral ***lidocaine*** during a 1 h period. ***Lidocaine*** blood concentration was 26.9 mmol/l 50 min after first signs of toxicity. ***Caffeine*** clearance is decreased by its use. ***Magnesium-aluminum silicate antacids*** decreased its absorption. Its rate of absorption, but not bioavailability, is increased by ***metoclopramide***. Absorption rate is decreased by ***heroin, morphine, meperidine***, and ***pentazocine***. Coadministration with ***ritonavir*** has induced adverse cardiac and neurological events.

LD_{LO}, **human**: 63 mg/kg.

LD_{50}, **rat**: 350–400 mg/kg; **mouse**: 140–310 mg/kg; **rabbit**: 160–180 mg/kg.

MIANSERIN

= *Athymil* = *Bolvidon* = *GB-94* = *Lantanon* = *Norval* = *Org-GB-94* = *Tetramide* = *Tolvin* = *Tolvon*

***Antidepressant, α-adrenergic antagonist.* A tetracyclic.**

Untoward effects: Blood: Agranulocytosis, neutropenia, leukopenia, thrombocytopenia, and hyperglycemia.

Cardiovascular: Tachycardia (substantially less than ***nortriptyline***), increased blood pressure, increased PR interval, flattened T-waves, and occasionally orthostatic hypotension. Heart-block report after overdoses viz. in 39-year-old **female** after ingestion of 58, 10 mg tablets. Repeated episodes of complete heart-block during 16 h after ingestion of 900 mg in 62-year-old **female**; 5 h after ingestion, **her** serum level was 0.89 mg/l.

Central nervous system: Epileptogenic convulsions. Delirium in 70-year-old **female** after 30 mg/day/14 days.

Miscellaneous: Toxic epidermal necrolysis, arthritis after 30 mg/day/5 days in 36-year-old **patient**, hepatic dysfunction, facial edema, glossitis, drowsiness, dizziness, finger tremor, dry mouth, constipation, sweating, insomnia, increased salivary flow, miosis, increased weight, hypophosphatemia, decreased production or increased catabolism of ***calcitriol***. An idiosyncratic pulmonary reaction in 42-year-old **male** after 20 days on 40–60 mg/day leading to pleuritic chest pain, anorexia, fever, diarrhea, abdominal pain, and leukocytosis. Found in breast milk at 0.5–1.4% of the **maternal** dose. No untoward effects noted in the **infants**. ***Carbamazepine*** causes a marked decrease in its plasma concentration.

LD_{50}, **mouse**: 365–390 mg/kg.

MIBEFRADIL

= *Cerate 50* = *Posicor* = *RO 40-5967*

Calcium channel-blocker, antihypertensive.

Untoward effects: Bradycardia, dizziness, light-headedness, and fatigue. Occasionally peripheral edema, asymptomatic Wenckebach-type second degree heart-block. Overdoses lead to increased U-wave and T-wave amplitude.

Drug interactions: Could increase plasma concentrations of ***astemizole, cisapride,*** and ***terfenadine***, causing fatal arrhythmias. Dosage of ***cyclophosphamide, cyclosporine, ifosfamide, tamoxifen, vinblastine, vincristine***, and ***tricyclic antidepressants***, such as ***imipramine***, should be decreased, as concommitant use will also increase their serum levels. Syncope reported in **some** when used with ***β-adrenergic-blockers***. It increases ***triazolam*** peak plasma concentration 80% and increases its elimination half-life by about five times. Use with "***statins***" (***atorvastatin,***

simvastatin) caused rhabdomyolysis. As a result, in June 1988, Roche withdrew the production from U.S. marketing, after 24 deaths that might have resulted from its use. Continued use has inhibited its own metabolization.

MICA

= *Biotite* = *Lepidolite* = *Margarite* = *Muscovite* = *Phlogopite* = *Roscoelite* = *Zimmwaldite*

Annual use in the U.S. is ~140,000 tons.

Untoward effects: Eye irritation. Cutaneous eruptions among industrial **workers** handling it; boils and skin infections in tunneling **workers** from particles penetrating **their** skin. Pneumoconiosis, chronic pulmonary fibrosis, coughing, dyspnea, weakness, and weight decrease reported. National Institute for Occupational Safety and Health upper respiratory limit of 3 mg/m^3; OSHA – 20 mppcf.

MICHLER'S KETONE

Use: In manufacturing dyes and pigments. World-wide production and use is < 1,000,000 lbs annually.

Untoward effects: Inhalation and dermal contact are potential sources for ~2000 industrial **workers** in the U.S.

Oral use in their diets leads to hepatocellular carcinomas in male and female **rats** and female **mice**; hemangiosarcomas in male **mice**.

TD_{LO}, **rat**: 5475 mg/kg, during 1 year.

MICONAZOLE NITRATE

= *Aflorix* = *Albistat* = *Andergin* = *Brentan* = *Conoderm* = *Conofite* = *Daktar* = *Daktarin* = *Deralbine* = *Dermonistat* = *Epi-Monistat* = *Florid* = *Fungiderm* = *Fungisdin* = *Gyno-Daktarin* = *Gyno-Monistat* = *Micatin* = *Miconal Ecobi* = *Micotef* = *MJR 1762* = *Monistat* = *Prilagin* = *R 14,889* = *Vodol*

Antifungal. An imidazole.

Untoward effects: ***Systemic Use***: Chills and fever (10%), dizziness, diarrhea, nausea (18%), vomiting (7%), anorexia, rash (9%), itching (21%), thrombophlebitis (~35%), leukopenia, anemia, thrombocytosis, drowsiness, thrombocytopenia, transient hematocrit decrease, hyponatremia, and hyperlipidemia. IV use (600–1,000 mg in 150–200 ml ***saline*** during an 8-h period) in seven **patients** with hematologic malignancies led to eight episodes of major cardiorespiratory and anaphylactic reactions. Has caused a ventriculitis after intraventricular use. Inhibits ***cortisol*** biosynthesis in the adrenal gland. Tachycardia, bradycardia, arrhythmias, and arachnoiditis also reported. The vehicle is capable of inducing hyperlipidemia. Aggregation of erythrocytes reported with doses > 600 mg tid (rouleau formation). Rarely, its use has been associated with euphoria, anxiety, increased libido, acute toxic psychosis, weakness, hyperesthesia, bitter taste, headache, bleeding gums, increased bruising, eosinophilia, jaundice, increased SGOT and SGPT, increased serum ***phosphorus***, decreased serum ***calcium***, pain at injection site, and retinopathy.

Topical use: Irritation, burning, and maceration.

Vaginal use: Burning, itching, irritation, pelvic cramping, and headaches. Less commonly, skin rashes and hives. Small amounts are absorbed systemically.

Drug interactions: Toxic levels of ***phenytoin*** in 51-year-old **male** after concomitant use with 500 mg IV tid. Confirmed by rechallenge. Possible alteration in ***tobramycin*** pharmacokinetics reported. Effects antagonized by ***amphotericin B***. As an oral gel, it caused a dramatic international normalized ratio increase (from 1.5 to 10) during a 2 week period in a 73-year-old **male** taking ***warfarin***. Another case report reported an international normalized ratio increase from 2.5 to 17.9. Tests in healthy **individuals** showed 80% and 40% decrease in clearance of various ***warfarin*** fractions with oral miconazole. It is a potent inhibitor of CYP2C9 metabolizing enzyme and 1% topically can induce bleeding in **patients** taking ***warfarin***. As a topical gel, it caused a hematological reaction when used with ***fluindione***. Several **patients**, 43–89 years, stabilized on ***acenocoumarol***, had increased international normalized ratios after 100 mg 2% oral gel qid for oral candidiasis. In other **patients**, it decreased ***pentobarbital*** clearance by 50–90%. Suspected as a cause of increased ***cisapride*** plasma concentration with resulting torsades de pointes. Increases serum levels of ***astemizole*** and ***terfenadine***.

The FDA has considered all ***imidazoles*** "potentially carcinogenic".

LD_{50}s in mg/kg: **rat**: > 640, **mouse**: 578, **guinea pig**: 276, **dog**: > 160.

MICROCYSTIS AERUGINOSA

A fresh-water blue-green alga.

Untoward effects: Contains ***microcystin L-R***, a potent hepatotoxin, which exists in the water around the **alga**. During hot, dry, warm weather, **people** will not swim in infested areas, but **dogs** don't hesitate getting into the water, leading to, within 15–30 min, discomfort, severe gastroenteritis (occasionally hemorrhagic), vomiting, abdominal pain, muscle trembling, incoordination, paralysis, occasionally convulsions, and death (within an hour in severe cases). **Waterfowl** are also commonly affected. Cause confirmed by IP injections into **mice**. Lyophilized cells requires 3–5 times the amount on a weight basis to kill **calves**, **lambs**, **chickens**, and **ducks** as for **rats**, **mice**, **guinea pigs**, and **rabbits**. Post-mortem reveals liver congestion, enlargement, hemorrhage, and necrosis. The same effect has been reported in **man**.

MICROPOLYSPORA FAENI

Untoward effects: A common cause of "**farmer's** lung", a hypersensitivity type of pneumonitis, due to a thermophilic oligosporic actinomycete. Found in moldy hay, moldy sugar cane, compost, air-conditioning systems, and home humidifiers. A 53-year-old **female** had a sudden onset of chills, fever, cough, and dyspnea, which was traced to a contaminated humidifier. Other cases reported due to use of a home furnace humidifier, a humidifier placed in a sun-lit window, and a number of cases due to contamination of an office air-conditioning system. Can also be a cause of **mushroom-worker's** disease. **People** working around moldy sawdust and **bird**-breeding areas are also potential **victims**.

A cause of extrinsic allergic alveolitis or pneumonitis in **cattle** and **sheep**.

MIDAZOLAM

= *Dormicum* = *Flormidal* = *Hypnovel* = *Ipnovel* =*RO 21-3981* = *Versed*

***Anesthetic*. Usually IV; occasionally IM. Intranasal and oral use explored. A water-soluble *benzodiazepine*.**

Untoward effects: Apnea often occurs ~2 min after IV. Occasionally local pain or thrombophlebitis. Fatal cases of cardiorespiratory arrest after IV or IM use, especially in the **elderly**. By 1988,~40 such deaths were attributed to its use in doctors' or dentists' offices in **patients** > 60 years, in addition to many cases of non-fatal, but life-threatening cardiorespiratory depression. Withdrawal reactions, especially in **infants** and young **children**. Acute oral dystonia after IV in 14-year-old **male**. Agitation during recovery noted in some **pediatric patients** as well as paradoxical excitation in 16-year-old **female**. Long-term decreased cognitive function in 5% of 262 **patients**, median age 69 years. Euphoria, hallucinations, confusion, and hostility have also been reactions to it. Nausea, constipation, dry mouth, tachycardia, hypotension, drowsiness, apathy, headache, depression, and visual disturbances noted in **some**. In pediatric **patients**, intense burning, irritation, and lacrimation after intranasal use.

Drug interactions*: *Cimetidine, clarithromycin, diltiazem, erythromycin, fluoxetine, grapefruit juice, itraconazole, ketoconazole, roxithromycin, and ***verapamil*** significantly increase and prolong its effects. ***Grapefruit's*** effect only occurs with midazolam's oral, not IV use. Can prolong neuromuscular weakness after high dose ***corticosteroids*** and ***non-depolarizing skeletal muscle-relaxants***, such as ***vecuronium***. Hypotension is induced when used with ***alfentanil***. ***Food*** delays its oral absorption. Even small amounts of ***alcohol*** potentiate its effect and large amounts could be lethal. Its clearance is decreased 37% by ***propofol***. ***Saquinavir*** increases its peak plasma concentration more than twice, with sedative effects and impaired skills. A combination of ***lopinavir/ritonavir*** can also increase its serum concentration.

MIDODRINE

= *Alphamine* = *Amatine* = *Gutron* = *Hipertan* = *Metligine* = *Midamine* = *ProAmatine* = *ST-1085*

***Antihypotensive, vasoconstrictor, α-adrenergic, sympathomimetic*. Prodrug of *desglymidodrine*, which does not stimulate heart or central nervous system.**

Untoward effects: Paresthesias, pruritus (usually on scalp), chills, piloerection, and urinary problems (retention, urgency, pollakiuria, and dysuria). Supine hypertension (occasionally sitting hypertension) can occur and is serious. Infrequently, headache, facial flushing, confusion, xerostomia, and rash. Rarely, dizziness, somnolence, insomnia, skin hyperesthesia, dry skin, canker sores, erythema multiforme, asthenia, backache, leg cramps, nausea, flatulence, heartburn, and visual field defect. Bradycardia occurs and can potentiate similar effect of some cardiac medications.

MIFEPRISTONE

= *Lunarette* = *Mifegyne* = *Mifeprex* = *RU 486* = *RU 38,486*

***Abortifacient, antiprogestagen*. Use being explored in oncology, treating Cushing's syndrome, and other medical problems.**

Untoward effects: May interrupt early pregnancy, but does not always induce a complete abortion. Failed as an abortifacient in only 4% of 2000 **women**, and caused severe bleeding in **one**. After 60,000 abortions in France without mortality, a 31-year-old **female** heavy **smoker**, in early 1991, elected to abort her thirteenth pregnancy. **She** received a synthetic ***prostaglandin*** after **her *RU 486*** tablet, and it, apparently, caused **her** death from cardiac failure. The pill's manufacturer recommends it not be used by heavy **smokers**. Transient nausea, vomiting, diarrhea, crampy abdominal pain, vaginal bleeding, and headache reported. Rarely, endometritis and salpingitis. After the addition of a ***prostaglandin***, cardiovascular upsets and abdominal pain can occur. A **female** who took 400 mg with the intention of terminating **her** pregnancy, changed **her** mind. At 17 weeks of gestation, an ultrasound scan revealed lack of amniotic sac, stomach, gall bladder, and urinary tract. The pregnancy was terminated 1 week later.

Several similar decisions were rewarded with normal **infants**.

Drug interactions: Slowly causes a dramatic increase in ***cyclosporine*** serum levels. One can expect ***St. John's Wort*** and ***rifampin*** to decrease its serum levels, and ***erythromycin, grapefruit juice***, and ***ketoconazole*** to increase its blood levels.

LD_{50}, **rat**: > 1 g/kg.

MILBEMYCIN OXIME

= *CGA 179,246* = *Interceptor*

Parasiticide.

Untoward effects: Transient mild anaphylactoid reactions in **dogs** with a large number of circulating *Dirofilaria* microfilaria. Usually, high dosage in **collie dogs** leads to ataxia, pyrexia, and recumbency.

LD_{50}, **mouse**: 1–1.5 g/kg.

MILBOLERONE

= *Cheque* = *Matenon* = *U 10,997*

***Anabolic, androgen, oral contraceptive.* It is a controlled substance in the U.S.**

Untoward effects: Early trials in **women** indicated it had a potential for causing altered liver function. There is a potential for abuse by **body-builders**.

Use of 50 μg/day for female **cats** will prevent estrus but leads to thyroid dysfunction, thickening of neck skin, and clitoral hypertrophy. Hepatic dysfunction noted at 60 μg/day and mortality at 120 μg/day. Treatment of immature female **dogs** led to clitoral enlargement, white viscid vaginal discharge, mounting behavior, and musk body odor. In older **dogs**, clitoral enlargement and ossification of *os clitoridis* (20%), vaginal discharge (10%), riding behavior (1.6%), watering eyes (5.6%), objectionable body odor (4.3%), icterus, and increased liver function tests. The **Bedlington terrier** has a genetic liver defect and use in them should be avoided. **German shepherds** may also suffer from more side effects than expected.

LD_{50}, **rat**: > 1.6 g/kg.

MILK

Untoward effects: Ethnic backgrounds can make a difference. White **Americans** or **descendants** from Northern European stock usually have no ill-effects from drinking it, but **those** from Asia, Africa, **Middle Easterners** or **those** from non-milking regions of the world are often deficient in lactase and develop gastrointestinal upsets (diarrhea, cramps, bloating) from the ***lactose*** (q.v.) content. Due to its high ***saturated fat*** content, some consider it causes an increase in atherosclerosis. Dairy products, including milk, have triggered migraines and stomach acid production with heartburns in some **people**. A 41-year-old **female** ingested 200 ml of 35% ***formaldehyde***, followed by 1 l of milk and died from suffocation 40 min later. Cause of death was the clotting of the milk, which caused the suffocation. A milk-alkali syndrome reported, primarily in **those** with ulcers overusing it. Hypersensitivity reactions with hives reported.

Drug interactions: Use decreases blood levels of ***erythromycin, iron, norfloxacin, oleandomycin, tetracyclines*** (40–50%), and most oral ***penicillins*** by chelation, but not ***ciprofloxacin*** or ***tosufloxacin***. Emesis induction is delayed when given with ***ipecac***. Serum levels of ***etretinate*** increase 260%. Due to its relative alkalinity, it causes the breakdown of enteric coatings, such as on ***bisacodyl***. Can inactivate ***potassium*** supplements and decrease absorption of ***zinc sulfate***. A 19-year-old **male *heroin* abuser** had immediate worsening of **his** deep stupor, when friends gave **him** a milk IV. It caused deep coma, extreme cyanosis, bradypnea, miosis, and severe pulmonary edema. Canned milks have a high levels of amines and need to be avoided by **people** ingesting ***monoamine oxidase inhibitors***.

Drugs and chemicals can adulterate milk. Chemicals from printing ***ink*** on the carton, at only a few ppm, lead to slight, but detectable, ***butter-toffee*** note. ***Lactose*** increases ***lead*** absorption. **Adults** have been poisoned by drinking milkshakes that contained ***apricot*** kernels. Goitrogens can be passed by **cows** into their milk. Acute ***dimethylnitrosamine*** poisoning, resembling Reye's syndrome, from deliberate criminal contamination of milk and ***lemonade***. The carcinogenic and mutagenic toxin from ***bracken ferns*** (q.v.) can pass into **cow's** milk. ***Iodophor*** residues in milk, due to its overuse as a farm disinfectant, may have helped cause increased incidence of thyrotoxicosis in Tasmania in 1964. Reactions (rashes, headaches) from milk and ***ice creams*** containing ***penicillin*** residues. High ***lead*** content, leached from the can, reported from evaporated milk. Thyroid abnormality incidence slightly increased in 2000 **children** (10–18 years) in Utah radioactivity fallout zone from ***iodine***^{131} in milk. *Klebsiella pneumoniae* infections in five **infants** in a Virginia **newborn** intensive care unit, due to failure to sterilize an electric suction pump. *Yersinia enterocolitica*, *Campylobacter jejuni*, salmonellosis, brucellosis, and listeriosis outbreaks occur from drinking contaminated or raw milk. At one time, raw milk was an important source of tuberculosis. **Cows** fed *Aspergillus*-contaminated feeds, including ***peanut meal***, have excreted a "milk toxin" (***aflatoxin***, q.v.) in their milk. ***Nitrosamines*** have been found in unpasteurized milk. Milk can hasten the absorption of ***p-dichlorobenzene*** (q.v.). During 1955, 12,038 **infants** in Japan suffered from typical ***arsenic*** poisoning from use of a dry milk product containing 30 mg ***arsenous acid***/kg; **128** died. To improve the solubility of the dried milk powder, ***sodium phosphate dibasic*** was added. The factory failed to test the purity of the chemical, which contained ***arsenic***. Milk from **cows** fed ***cottonseed*** contains ***sterculic*** and ***malvalic acid***. ***Carbon tetrachloride*** causes a bad taste to milk and little ***mercury*** is excreted by **cows** in their milk.

Infants may be vulnerable to milk reconstituted with fluorinated water. Hyperchromic megaloblastic anemia has been reported in **infants** and young **children** consuming it. Allergic reactions including urticaria, diarrhea, eczema, angioedema, laryngeal edema, bronchitis, rhinorrhea, vomiting, asthma, shock, and even collapse reported in **children** and **adults** from **cow's** milk. Allergic contact urticaria reported in **adults**. **Milker's** warts or nodules are more apt to be from cowpox exposure. **Maternal** consumption of **cow's** milk has caused colic in breast-fed **infants**. A Japanese **infant** that failed to thrive, was unable to suck, vomited, and had nystagmus due to hypervalinemia (five times normal), had clinical improvement on a low ***valine*** diet, but relapsed when given milk. **Human** breast milk has contained harmful amounts of ***alcohol, ethyl; aloe, amantadine, androgens, antineoplastic drugs, anticoagulants; aspirin, atropine, barbiturates, benzodiazepines, bromides, DDT, senna, cascara, chloral hydrate, corticosteroids, danthron, diuretics, ergot, estrogens, heroin, iodides, isoniazid, lithium, meprobamate, nalidixic acid, polycyclic hydrocarbons, phenytoin, progesteroids, pyrimethamine, radiopharmaceuticals, sotalol***, some ***sulfonamides, theobromine*** (from ***chocolate***), and ***thiouracil***. Eczema in a **baby** whenever its **mother** ingested ***chocolate***. Projectile vomiting reported in a breast-fed **infant**, due to **mother's** milk allergy. Many other drugs have little effect on the **infant** from quantities normally present. Even endogenous ***pregnanediol*** in breast milk has caused hyperbilirubinemia in the **nursling**, by inhibiting hepatic glucuronyl transferase activity. An **infant** known to be allergic to milk protein developed an anaphylaxis after application of a ***casein***-based diaper ointment to the perineal area. Improper dilution of powdered milk has caused fatal hypernatremia. Intestinal obstructions (from curd formation) and deaths reported in **infants**. Contamination of dried milk powder by *Bacillus cereus* and *B. subtilis* has caused gastrointestinal upsets and methemoglobinemia. In the U.S., low-fat ***yogurt*** contains added skim milk powder to thicken it, increasing its ***galactose*** (q.v.) (~14% in whole milk ***yogurt***) content, which can have adverse effects.

Animals have occasionally been allergic to milk and have ingested toxic drugs and chemicals present in milk. A strain of **mice**, bred so that 80% develop mammary tumors, have young who develop a high incidence of breast cancer when nursing them, but not when suckling foster **mothers**. This so-called milk factor may be hormone, virus, or chemical. A similar relationship has been suspect in **women** and their **infants**.

MILKVETCH

= *Poison Vetch*

Untoward effects: ***A. bisculatus*** = ***Two-grooved Milkvetch*** of western Canada rangelands also accumulates ***selenium***, leading to staggering, abdominal pain, diarrhea, and prostration after ingestion by **cattle**, **horses**, **sheep**, and **swine**. Plants containing < 200 ppm of ***selenium*** can cause chronic toxicity.

A. canadensis = ***Canadian Milkvetch*** = ***Astragale du Canada***. After ingestion has killed **cattle**. Contains ***nitrites***, particularly ***3-nitropropionic acid***. LD_{50}, week-old **chick**: 2 g.

Timber milkvetch (***Astragalus miser***) and ***Emory milkvetch*** (***A. emoryanus***) contain ***miserotoxin*** (q.v.) which has poisoned **chicks**, **rabbits**, **cattle**, and **sheep** that ingested them. They are ***selenium***-converters. Acute poisoning and death within 1–4 h in a 1000 lb **cow** eating 2 lbs of the fresh plant. In those ingesting less, leads to muscular weakness, irregular gait, inability to stand, tachycardia, nervousness, frequent urination, pale mucous membranes, coma, and convulsions. **Sheep** often have a rasping dyspnea on inspiration and coughing on expiration. Acute cardiac dilation noted in those that die suddenly. It is in the ***pea*** family and usually grows in the western U.S. at elevations above 6000 ft. **Honeybees** that forage on ***A. miser*** have been poisoned.

See ***Astragalus*** for additional toxic symptoms.

MILLET

Many forms (q.v.) exist: *Echinochloa, Panicum, Paspalum, Pennisetum*, and *Setaria*.

Untoward effects: Known to be ***nitrate-*** and ***cyanogenic glycoside-accumulators***. There is little sickle-cell anemia in African **natives**. One theory is that the ***cyanate*** of **their** ethnic diets (up to 1 g/day) prevents the anemia. U.S. diets contain ~25 mg/day. Esophageal cancer is ~14 times more prevalent in Luo **tribesmen** in western Kenya, who make **their *beer*** from dark ***sorghum*** and dark millet, than among their **neighbors** who consume white or light-brown varieties. Anaphylaxis has been reported after ingestion of millet seeds. Contact dermatoses occur on **broom-makers** using moor-millet. Contains a ***niacin*** inhibitor and is often contaminated with ***aflatoxin*** from ***Aspergillus*** growth on it.

MILLIPEDES

Terrestial arthropods, ranging in length from a few millimeters to 30 centimeters.

Untoward effects: Some species have defensive stink glands and large tropical millipedes can squirt their offensive liquid or gas as far as a meter. This can cause severe skin, eye, and mouth irritation. Contains ***quinone*** (q.v.), ***p-benzoquinones***, and ***toluquinone***. Some produce ***hydrogen cyanide*** (q.v.).

MILRINONE

= *Corotrope* = *Primacor* = *WIN 47,203*

Inotropic and vasodilating agent. **Oral and IV use.**

Untoward effects: ***Oral***: Gastrointestinal upsets, palpitations, headache, dizziness, ventricular arrhythmias, fatigue, muscle weakness, tachycardia/fibrillation, hypotension, and increased liver enzymes. An excess mortality with its use was noted.

IV: Ventricular and supraventricular arrhythmia, tachycardia/fibrillation, hypotension, headache, thrombocytopenia, and hypokalemia.

MILTEFOSINE

= *D 18,506* = *HPC* = *MILTEX*

Antineoplastic.

Untoward effects: ***Topical***: Burning sensation after application, skin dryness, atrophy, and pruritus.

Oral: Frequent (62%) gastrointestinal upsets and increased aspartate transaminase and ***creatinine***.

MIMOSA

Untoward effects: ***M. hostilis*** root of a leguminous shrub in eastern Brazil is used to prepare an hallucinogenic drink, known as ***ajuca***, ***yurema***, or ***vinho de jurema***, containing ***N,N-dimethyltryptamine*** (q.v., once called ***nigerine***). Supposedly inactive orally, unless taken with a ***monoamine oxidase inhibitor***. Apparently, the drink contains an unknown ***monoamine oxidase inhibitor***. ***M. verrucosa***, ***jurema branca*** root extract is often used to enhance the drink, by adding a stupefying effect.

M. invisa has poisoned **water buffaloes** in Indonesia leading to sudden collapse, tachypnea, shallow breathing, fever, and extreme pain. Post-mortem revealed hemorrhages on rumen mucosa, spleen, liver, and kidneys; parenchymatous degeneration of these organs and blood-stained gelatinous fluid in the lymph nodes.

M. pudica = ***Sensitive Plant*** = ***Chhuimui*** = ***Punyosisa*** = ***White Flower***. The leaves and seeds of this species were the first discovered to contain high levels of ***mimosine*** (q.v. for the most toxic effects of this plant). Photosensitization in **sheep** reported from the Fiji Islands.

MIMOSA ABSOLUTE

A fragrance from the flowers of a 20+ ft tall tree, *Acacia decurrens* var. *dealbata* used in soaps, detergents, cosmetic creams and lotions, perfumes, and foods.

Untoward effects: No sensitization or irritation at 1% in petrolatum reported in **human** volunteers. One report indicates possible dermatitis in hypersensitive **individuals**.

No irritation when applied undiluted on the backs of **swine** and hairless **mice**. Slightly irritating when applied full-strength to intact or abraded **guinea pig** skin for 24 h under occlusion.

LD_{50}, **mouse** and acute dermal, **guinea pig**: > 5 g/kg.

MIMOSINE

= *Leucenol* = *NSC 69,188*

First isolated from *Mimosa pudica* (q.v.). Found in *Leucaena sp.* (q.v.) and *Ipil-Ipil* (q.v.). A depilatory *amino acid*.

Untoward effects: Has interfered with keratinizing structures in **human** hair. Sudden loss of hair reported in Australian **Aborigines** eating plant seeds containing it. **Women** who ate *L. glauca* seeds lost **their** hair.

Ingestion of high levels by **livestock** leads to alopecia, inanition, decreased weight gain, cataracts, and infertility. Loss of hair from the mane and tail has been reported in **horses**; loss of weight, ring formation on the hooves, and hemorrhagic gastritis with larger intake. Loss of fertility in **rats** and hair in **mice**. Optimum dosage to **sheep** causes chemical defleecing. A metabolite, ***3-hydroxy-4(1H)-pyri done*** (q.v.) is toxic to Australian **cattle** and goitrogenic in **rats** and **mice**. Consumption of *Leucaena* has caused both defects, fetal resorption and loss of hair, in **swine** from its mimosine content.

TD_{LO}, **rat** (1–18 days pregnant): 5.265 g/kg.

MINERAL OIL

= *Adepsine Oil* = *Alboline* = *Baby Oil* = *Bayol* = *Blandol* = *Clearteck* = *Draked* = *Fractol* = *Glymol* = *Kaydol* = *Kremol* = *Liquid Paraffin* = *Liquid Petrolatum* = *Marcol* = *Paroleine* = *Paraffin Oil* = *Petrolatum, Liquid* = *Splicing Oil* = *White Oil*

Use: As a lubricant, solvent, laxative, emollient, and dust-control agent. Also as a sealant on chicken eggs, release agent for bakery products, and coating for fruits and vegetables. In pharmaceuticals, cosmetics, and hair oils. ***Vitamin E*** is used as an antioxidant in medicinal mineral oils.

Untoward effects: Long-term oral use can lead to malabsorption of ***calcium***, ***phosphorus***, oil-soluble ***carotene***, and ***vitamins A***, ***D***, and ***K***, possibly leading to rickets in **children**, osteomalacia in **adults**, and increased prothrombin time. Can decrease absorption of ***oral contraceptives*** and ***anticoagulants***. Increased absorption of ***ascorbic acid*** and ***surfactants***, such as ***docusate sodium*** and ***poloxamers***. Can delay gastric emptying. Ingestion has caused hepatic lipogranulomas. Use as a glove lubricant has caused salpingitis, uterine and parametrium granulomas with lesions similar to those caused by ***starch*** or ***talc***, and possible ureteral obstruction that can be misdiagnosed as a neoplasm. Pulmonary, intestinal mucosa, splenic, and abdominal lymph node granulomas have also followed chronic ingestion. Finely divided particles, as in emulsified oils, can pass through the intestinal mucosa and act as foreign body material throughout the body. Peritoneal granulomas in pelvis and abdomen of 64-year-old **female** discovered at autopsy, presumably from use as a glove lubricant during an emergency appendectomy at age 16. Multiple paraffinomas after intrapleural injections following pneumonectomy. Other similar case reports reported. Nasal paraffinomas with mobile cystic masses 4–6 months after use in rhinoplasty. Lipoid pneumonias from chronic use as a laxative, inhalation of aerosoled material, and use as a lip gloss (***Kissing Potion***®). Some of these may have resulted from use in **people** with faulty swallowing mechanisms, poor cough reflexes from oily substances in **children**, debility, senility, and from oily nose drops or sprays. Potential aspiration of light mineral oil in **children** from cracks in floating toys. Topical application of baby oils has caused lipogranulomatosis of **infant** genitalia. Local nodules, occasionally severe, and sterile abscesses, which can track through muscle and fascia planes, after its use in oil adjuvant vaccines. Calcified paraffinomas, soft tissue infection, and osteomyelitis in 74-year-old **male** after several decades of self-medication with SC injections. A 24-year-old **male addict**, thinking it contained ***cocaine***, self-injected 10–30 ml, leading to breathlessness, agitation, and generalized aching. Cyanotic, temperature 38.3°C (101°F), rapid shallow breathing (despite ***oxygen***) on day 4. Survived. Self-injection of the penis leads to granulomas. Skin rashes (red papules over back of hands, face, back, and chest) often confused with measles, in large numbers of white (rarely, in **others**) U.S. **children** in the 1960s from a toy (***Flubber***®) that contained mineral oil (27%). A rare irritant or contact dermatitis with folliculitis from either the paraffinic or naphthenic base forms, especially when used as cutting or grinding oils. Oil acne (acne oleosa) and contact eczema are common. Compresses soaked in mineral oil applied to arms lead to redness and swelling of follicles in 8–10 days and acne after 16 days. **Workers** using it as a cutting oil for steel, nuts, and bolts often saturate **their** clothing, and this is, apparently, the cause of scrotal and vulval carcinomas, as well as other skin cancers.

Comedogenic. Large oral doses have caused anal leakage, local pruritus, hemorrhoids, cramping, palpitations, risk of anorectal infections, weakness, and dizziness. Oral use can interfere with antibacterial action of ***sulfasuxidine*** and decreased anthelmintic effect of ***hexylresorcinol*** (now, rarely used) and ***warfarin's*** anticoagulant effect.

Aspiration pneumonia has occurred in many **animals**, especially from careless drenching. Weak comedogenic agent in the external ear canal and an ophthalmic irritant in **rabbits**. Retards wound healing in experimental **pig** trials. In **animals** and **man**, vomitus after mineral oil dosing can readily enter the trachea, because it is oily and fails to stimulate the cough reflex.

MINERAL SEAL OIL

A distilled and refined low-viscosity oil. Boils at ~350°F. In furniture *polishes*, *inks*, floor-sweeping compounds, and as a rust preventative.

Untoward effects: Aspiration pneumonia and pneumonitis with respiratory distress, and, occasionally, central nervous system upsets. Spontaneous or induced vomitus after ingestion readily enters the trachea due to its oiliness and failure to stimulate the cough reflex. **Children** are primarily involved with accidental ingestion and toxicity is similar to that of ***gasoline*** and ***kerosene***. **Pets** have also been poisoned by it. Diarrhea often follows ingestion. If the **patient** coughed while ingesting, chest X-rays and white blood cells counts are indicated. A 12-year-old **female** drank 8 oz of furniture ***polish*** (~99% mineral seal oil) in a suicide attempt. **She** survived after treatment.

LD_{LO}, **human**: 500 mg/kg.

MINERAL WATERS

Untoward effects: Best known by the publicity generated when the French-based Perrier bottling operation found a small amount of ***benzene*** (q.v.) in some U.S. samples and ordered the destruction of 72 million bottles of it. Despite the FDA calling ***benzene*** a dangerous carcinogen, the event caused a classic case of "foot-in-mouthitis" when James Benson, acting U.S. FDA Commissioner, said, "If I had a bottle in the refrigerator, I would drink it". Do not do as I would do, do as I say.

MINERAL WOOL

= *Rock Wool* = *Slag Wool*

Untoward effects: Eye, skin, and respiratory system irritant. Inhalation has caused dyspnea and coughing.

National Institute for Occupational Safety and Health suggests limits of 3 fibers/cm^3; OSHA suggests maximum of 15 mg/m^3 and respiratory exposure limit of 5 mg/m^3.

MINOCYCLINE

= *Dynacin* = *Minocin* = *Minomycin* = *Klinomycin* = *Vectrin*

***Antibiotic.* A semi-synthetic tetracycline for oral and parenteral use.**

Untoward effects*: *Blood: Hemolytic anemia, thrombocytopenia, neutropenia, and eosinophilia. Atypical leukocytosis after 100 mg twice daily/4 weeks in 24-year-old **male**.

Central nervous system: Vestibulotoxic. Light-headedness, weakness, headache, muscle pain, dizziness, vertigo, loss of balance, and ataxia due to acute vestibular effects. Transient memory and inattention problems.

Gastrointestinal: Anorexia, nausea, vomiting, (infrequent) diarrhea, cramping, glossitis, dysphagia, enterocolitis, monilial overgrowth in anal region; rarely, esophagitis.

Hypersensitivity: A rare but serious event. Urticaria, angioneurotic edema, anaphylaxis, anaphylactoid purpura, pericarditis, exacerbation of systemic lupus erythematosus, and pneumonitis. A 35-year-old **female** using 100 mg/day/3 weeks for acne developed a severe allergic reaction, with high fever, neutropenia, eosinophilia, and lymphadenopathy. A 39-year-old **female** tested for acne with 50 mg twice daily/1 month developed a reversible hypersensitivity reaction with pericardial effusion, eosinophilia, and liver injury.

Miscellaneous: Transient bulging fontanels in **infants** during treatment and black-brown discoloration of thyroid glands. Use during dental development (last half of pregnancy, infancy, and up to 8 years) can cause permanency yellow-gray-brown discoloration of teeth, especially after 3 days of treatment. After 4 months therapy for arthritis in 68-year-old **female** there was blue-black staining of lower anterior teeth that was eventually removed with difficulty. Occasional reports of enamel hypoplasia. Can decrease plasma prothrombin; causes loss of ***potassium*** from erythrocytes, transient increase in SGOT and SGPT, acute hepatitis, alopecia, pulmonary infiltrates, and eosinophilia. Antianabolic effect by interfering with ***protein*** synthesis. A 14-year-old **female** developed pseudotumor cerebri after treatment for acne. Symptoms included headaches for 2 weeks, pounding in both ears and eyes, and a 2 days history of diplopia. Acute hepatitis (probably autoimmune), cycloplegia, hot flashes, insomnia, tinnitus and subacute ischemia of a leg reported. Passes into breast milk. Use of 100 mg twice daily/< 4 years in 24-year-old **female** with acne led to secretion of black milk and skin pigmentation.

Renal: Acute interstitial nephritis, increased blood urea nitrogen, exacerbation of uremia in **patients** with renal insufficiency.

Skin: Maculopapular and erythematous rashes. Urticaria, allergic eczematous dermatitis, and photosensitivity. Exfoliative dermatitis is uncommon. Longitudinal melanonychia in 15-year-old **female** after 100 mg/day/6 months for acne. Photo-onycholysis reported. Hyperpigmentation (blue or blue-black) of skin and oral mucous membranes, particularly at sites of previous inflammation and after long-term therapy; may be due to ***iron*** particles, and rarely occurs if dosage is < 200 mg/day.

Drug interactions: Possibility of increased ***theophylline*** serum concentration after IV ***minocycline***. ***Milk*** and ***dairy products*** have less of an effect on its absorption than on ***tetracycline***, but ***aluminum hydroxide-containing antacids*** will complex it.

LD_{50}, **mouse**: 3.1 g/kg; IP, 310 mg/kg; IV, 140 mg/kg.

MINONENE

A food spice and flavoring agent.

Untoward effects: Contact dermatitis of hands.

MINOXIDIL

= *Alopexil* = *Alostil* = *Hairgro* = *Loniten* = *Lonolox* = *Minoximen* = *Normoxidil* = *PDP* = *Pierminox* = *Prexidil* = *Regaine* = *Rivixil* = *Rogaine* = *Tricoxidil* = *U-10858*

Antihypertensive, hair growth restorer, vasodilator.

Untoward effects*: *Blood: Thrombocytopenia and leukopenia are rare.

Cardiovascular: Edema, ***sodium*** retention, congestive heart failure, pericardial and pleural effusions, angina, tachycardia, positive T-wave flattening or inverting and negative T-waves on electrocardiogram, postural hypotension, decreased myocardial uptake of ***radiopharmaceuticals***, and palpitations. Reflex tachycardia (160/min) with no hypotension in 2-year-old **male** after ingesting 100 mg (20, 5 mg tablets).

Gastrointestinal: Nausea.

Miscellaneous: Decreased glomerular filtration rate and increased ***creatinine***, systemic lupus erythematosus syndrome, and possible teratogenicity (cyanotic heart disease and hypertrichosis). Between 1985 and 1991, U.S. Poison Control Center reported 285 cases from oral and topical use; 145 were in **children** < 6 years. The oral tablets were involved in 224 cases, with moderate toxicity in 4.2% and major toxicity in 1.8%. Edema in some **patients** after topical use. Rapidly excreted into breast milk.

Skin: Mild hypertrichosis of face and arms on 10 or 20 mg/day and hirsutism in **female patients**. Pruritus and serosanguinous bullae in 69-year-old **male** after four 5 mg doses. Stevens–Johnson syndrome. Changes in hair color and photosensitivity reported.

High dosage in **rats** leads to ischemic cardiac necrosis and aggravates cardiac hypertrophy. Necrosis also noted in **dogs** and **minipigs**, but not **monkeys**. Initially, hemorrhagic and, subsequently, atrophic lesion in right atrium of **dogs** and **minipigs**, but not **rats** or **monkeys**.

LD_{50}, **rat**: 1.3 g/kg; IV, LD_{50}, **rat** and **mouse**: 50 mg/kg.

MINQUARTIA GUIANENSIS

= *Kobakedwe*

Untoward effects: The Waorani **natives** of Ecuador pound the bark in water to make a **fish** poison.

MINTS

= *Mentha sp.*

Untoward effects: Mint oil from ***M. arvensis*** contains 90% ***menthol*** (q.v. for toxic effects). Contains ***cinnamaldehyde*** (q.v.), a common cause of cosmetic-induced allergies. Japanese report of ***aspirin***-induced asthma provoked by mint-flavored toothpaste and gum. See ***Peppermint***.

MIPAFOX

= *Isopestox* = *Pestox XV* = *Pestox 15*

***Organophosphorus insecticide.* Cholinesterase inhibitor.**

Untoward effects: Delayed peripheral neurotoxicity in **man**. Even a single exposure can possibly alter a **person's** brain activity for a year.

LD_{LO}, **human**: 50 mg/kg.

Delayed peripheral neuropathy in **rats** after 3–4 months.

LD_{50}, **rabbit**: 100 mg/kg; **guinea pig**: 80 mg/kg.

MIRABILIS JALAPA

= *Four O'Clock* = *Itana pa osho (Yoruba)*

Untoward effects: The large tuberous roots contain the alkaloid, ***trigonelline*** (q.v.). The seeds and roots are toxic for **man** and **animals**. Gastroenteritis is the chief symptom. **Children** are more commonly affected.

Pigs can be affected from uprooting and eating the roots.

MIREX

=*CG 1283* = *Compound 1283* = *Dechlorane* = *ENT 25,719*

***Insecticide* (fire ants), *fire-retardant* (plastics, *rubber*, *paint*, *paper*, and electrical items). A *chlorinated hydrocarbon*.**

Untoward effects: Now prohibited for general use in the U.S. Residues have been detected in water, soil, food and beverages, and **human** tissues for as long as 12 years after exposure. Environmental half-life is 5–12 years. Symptoms in **humans** after heavy exposure could include apprehension, headache, disorientation, paresthesia, and convulsions. To a lesser extent, excitability, dizziness, muscle twitching, tremors, respiratory depression, and coma.

Carcinogenic and teratogenic in **fish** where, as in **people**, it bioaccumulates. Maximum allowable limit in **fish** was 0.1 ppm. It often contained some ***kepone***. Transplacental passage demonstrated in **goats**. Maternal toxicity and teratogenic in **rats**; increased incidence of hepatic nodules and lung and mouth tumors in **rats**, pheochromocytomas in male **rats** and mononuclear cell leukemias in female **rats**. When fed to **rats** through gestation and lactation, their **pups** had a dose-related incidence of irreversible cataracts. Increased incidence of hepatomas in **mice** and skin tumors when painted on their backs. Intoxicated **ducks** and **pheasants** developed ataxia within 40 min. Accumulates in the fatty tissues, liver, and eggs of exposed **hens**.

LD_{50}, **rat**: 235 mg/kg; **mallard duck** and **pheasant**: > 2.4 g/kg.

MIRTAZEPINE

= *Org 3,770* = *Remeron*

Antidepressant.

Untoward effects: Sedation, increased appetite, weight increase, xerostomia, dizziness, and constipation. Occasionally asthenia, and increased ***cholesterol***, amino-transferases, and ***triglycerides***. Rarely, agranulocytosis, induced-asthma, sexual dysfunction, and mania. Occasionally confusion reported in the **elderly**.

Drug interactions: Use with ***clonidine*** can induce severe hypertension.

MISEROTOXIN

Untoward effects: This glycoside is hydrolyzed to ***glucose*** and ***3-nitro-1-propanol*** in the rumen of **cattle** and **sheep**. The latter is a more toxic substance. It is then metabolized to yield ***nitrite***, which is also toxic. It is found in ***milkvetch*** (q.v.) and has been poisonous to **chicks** and **rabbits**. Lethargy, incoordination, flaccid paralysis, and death in **chickens**. Respiratory distress, methemoglobinemia, and anemia in **rabbits**.

MISO

An all-purpose, fermented, *soybean-protein* seasoning used as a bouillon or meat stock in soups, stews, sauces, dips, and gravy in Oriental cooking.

Untoward effects: Its ***tyramine*** (q.v.) content has caused a hypertensive crisis with ***monoamine oxidase inhibitors***.

MISONIDAZOLE

= *RO 7-0582*

A *nitroimidazole*. Has an *imidazole* ring structure similar to that of *cimetidine*.

Untoward effects: As a sensitizer in chemical or radiation antineoplastic therapy, it has induced a mild neuropathy. Tissue levels are decreased by pretreatment with ***phenytoin*** or ***phenobarbital***. Reduced clearance of ***5-fluorouracil***.

LD_{50}, **dog**: 8.5 mmol/kg.

MISOPROSTOL

= *Arthrotec* = *Cytotec* = *Rioprostil* = *SC 29,333*

***Antiulcerative, abortifacient.* A synthetic *prostaglandin* E_1 analog. Used intravaginally and orally.**

Untoward effects: ***Gastrointestinal***: Dose-related diarrhea (> 10%), flatulence and abdominal pain, especially in **patients** with Crohn's disease. Occasionally severe nausea. Vomiting, diarrhea, headache, and constipation. A 20-year-old **male** with ***flubiprofen***-induced gastropathy suffered progression of gastropathy and bleeding after 100 µg qid/2 weeks.

Genitourinary: > 1% of each: spotting, cramping, hypermenorrhea, and dysmenorrhea. An abortifacient.

Miscellaneous: Fever in a cirrhotic 61-year-old **female** taking 1.6 mg/day. Oral (400–600 µg) or intravaginal unsucessful use in the first trimester of pregnancy as an abortifacient leads to **infants** with congenital malformation of the frontal and/or temporal skull regions. Occasionally headaches. Tumors, convulsions, dyspnea, palpitations, hypotension, and bradycardia in overdosage.

Drug interactions: ***Cyclosporine*** blood concentrations are decreased by it. Attenuates the anticoagulant effects of ***acenocoumarol*** and ***phenprocoumarin***. Displacement from protein-binding sites by ***aspirin*** is not clinically significant, unless the latter is well over 975 mg.

MISSILE SILO

Untoward effects: In 1980, a Titan II missile silo exploded in Arkansas. The fuel that leaked mixed with an oxidizer to form the propellent, which was a mixture of ***hydrazine*** (q.v.) and ***dimethylhydrazine*** (q.v.). **Residents** 5 miles away complained of nausea, burning sensations in **their** noses, throats, and lungs, and dry salty lips after the light snow-like fog drifted over **them**. ***Beryllium*** (q.v.) can also be a problem from missile propellents.

MISTLETOE, AMERICAN

= *Gui de Chêne* = *Phoradendron*

Untoward effects: All parts, especially the berries, are poisonous. They contain toxic amines, particularly ***tyramine*** (q.v.), and ***β-phenethylamine***, as well as toxic lectins. Deaths in **children** and **adults** have been noted from eating the white berries or drinking ***tea*** made from them. Gastrointestinal upsets and effects on bladder and uterus from direct stimulation of smooth muscle and indirect action on sympathetic nervous system. Affects blood vessels. Nausea, vomiting, abdominal pain, tachypnea, dyspnea, hypertension, hepatotoxicity, delirium, occasionally hallucinations, sweating, convulsions, bradycardia, and cardiovascular collapse. Dehydration and a shock-like syndrome can occur from the profuse diarrhea and vomiting. It may be an important factor in spreading yuletide mononucleosis.

After ingestion, **dogs** have become apprehensive, excited, vomited, and had diarrhea and bradycardia.

MISTLETOE, EUROPEAN or JAPANESE

= *Helixor* = *Iscador* = *Missell* =*Niharu* = *Niyaru* = *Viscum album*

Untoward effects: Contain ***viscotoxins*** and ***viscumin*** (similar in action to ***abrin*** (q.v.), ***modeccin***, and ***ricin*** (q.v.)). Its lectins agglutinate **human** erythrocytes. A poisonous glue (***Ixia*** or ***Birdlime***) was made from the ***viscin*** in the berries. A 49-year-old **female** reportedly developed hepatitis from ingesting a number of 90 mg tablets. A 3 year-old child ingesting berries suffered vomiting, prostration, coma, fixed and somewhat miotic pupils, and convulsions. Causes hypotension and bradycardia. SC use of extracts in 47 **patients** led to induration, swelling, and pruritus. Use of mask, goggles, and gloves recommended for handling the herb or powder.

LD, **dogs**: 4 mg/kg.

MITHRAMYCIN

= *A 2371* = *Aureolic Acid* = *Mithracin* = *NSC 24,559* = *PA144* = *Plicamycin*

***Antineoplastic, antiviral.* Used IV and IA.**

Untoward effects: Epistaxis and thrombocytopenia, which can be followed by serious general and gastrointestinal bleeding. Hepatotoxicity with necrosis, venoclusive disease, and increased liver enzymes; anorexia, nausea, vomiting, diarrhea, erythematous rashes, decreased white blood cells, uremia, hypocalcemia, hypokalemia, nephrocalcinosis, malaise, toxic epidermal necrolysis, sudden occlusion of tibial arteries, increased lactate dehydrogenase, stomatitis, oral ulceration, and fever. Severe necrotic reactions after extravasation. Readily chelates divalent ions.

Has caused bone marow decrease, hypocalcemia, and vomiting in **dogs**, and bone marrow depression and vomiting in **monkeys**.

IV, LD_{50}, **rat**: 1.74 mg/kg; **mouse**: 2.14 mg/kg; IP, **mouse**: 2.9 mg/kg.

MITOBRONITOL

= *DBM* = *Dibromomannitol* = *Myebrol* = *Myelobromol* = *NSC 94,100*

***Antineoplastic.* An alkylating agent.**

Untoward effects: Thrombocytopenia, leukopenia, hemorrhagic diathesis, baldness, weakness, anorexia, limb pain, mild nausea, vertigo, renal colic with calculi, allergic skin reactions, and lens changes, including posterior subcapsular lens opacities.

Cataractogenic effect in **rats**. Peritoneal sarcomas and SC tumors in **rats** and lung tumors and lymphomas in **mice**. Bone marrow decrease in **dogs** and **monkeys**.

MITOLACTOL

= *DBD* = *Dibromodulcit* = *Dibromodulcitol* = *Elobromol* = *Mitolac* = *NSC104800*

Antineoplastic.

Untoward effects: Leukopenia, thrombocytopenia, and gastrointestinal toxicity with nausea in **man**.

Induces lung tumors and lymphomas in **mice** and SC tumors in **rats**.

MITOMYCIN C

= *Ametycine* = *Mitomycin* = *Mutamycin* = *Mytomycin C* =*NSC 26,980*

***Antineoplastic, antiviral.* IV, IP, IA, intravesical, and oral.**

Untoward effects: Severe leukopenia and thrombocytopenia. Alopecia, cellulitis at injection site, nausea, vomiting, fever, anemia, hepatic necrosis, dyspnea, stomatitis, and renal failure with azotemia and glomerular sclerosis. Increased intraocular pressure, glaucoma, amenorrhea, sterility, bladder calcification, hemolytic-uremic syndrome, and occasionally acute pneumonitis and fibrosis. Can cause local necrosis and cellulitis after extravasation and, within 48 h, erythema, burning, and hand pain contralateral to injection site. Skin irritant topically (wear gloves).

Bone marrow depression and vomiting in **dogs** and **monkeys**. Peritoneal sarcoma in **rats**.

IV, LD_{50}, **mouse**: 5 mg/kg; **dog**: 1 mg/kg; IV, LD_{LO}, **monkey**: 1 mg/kg; **cat**: 2.5 mg/kg.

MITOTANE

= *CB 313* = *Chlorditane* = *σ, p′-DDD* = *Lysodren* = *NSC 38,721*

A degradation product of, *σ, p′-DDT* that has antisteroid action. *Antineoplastic.* A *chlorinated hydrocarbon.*

Untoward effects: Causes nausea, vomiting, anorexia, and diarrhea. Chronic use leads to central nervous system depression, dizziness, vertigo, visual disturbances (retinopathy, diplopia, corneal opacity), adrenal insufficiency, hematuria, hemorrhagic cystitis, albuminuria, rashes, skin darkening, muscle twitching, muscle aches, fever, flushing, increased prothrombin time, hypertension, and orthostatic hypotension. Occasionally leukopenia. Memory impairment is common in the **elderly**. A 58-year-old **female** on 4 g/day orally required increased levels of ***warfarin*** to maintain **her** levels of anticoagulation.

Dogs have shown central nervous system depression, emesis, albuminuria, bloody diarrhea, and adrenal atrophy and necrosis.

MITOXANTRONE HCL

= *CL 232,315* = *DHAD* = *Mitozantrone* = *Novantrone* = *NSC 301,739*

***Antineoplastic.* A *doxorubicin* analog and an *anthraquinone* derivative. Half-life ~56 min. Used IV and intrathecally.**

Untoward effects: ***Blood***: Myelosuppression, leukopenia, thrombocytopenia, and febrile neutropenia.

Cardiovascular: Tachycardia, congestive heart failure, arrhythmias, decreased left ventricular ejection fraction, and chest pain. Mild phlebitis (10%). Bradycardia also reported in some **patients**. Because of heart injury risk, dosage should be limited to not more than 8–12 doses over a 2–3 years period.

Gastrointestinal: Nausea, vomiting, stomatitis, mucositis, and oral ulceration. A case of *Clostridium difficile*-related diarrhea appeared to be associated with each 5 days treatment cycle. Constipation also reported.

Hypersensitivity: Hypotension, urticaria, rashes, and dyspnea.

Miscellaneous: Greenish discoloration of urine, blue tinge in sclera after IVs, fatigue, increased liver enzymes, increased ***bilirubin***, fever, bladder infections, menstrual cycle changes, mouth sores, and renal failure. A 38-year-old **male** became paraplegic after intrathecal use.

Skin: Selective alopecia of white hair in **patients** after 5 days treatment, bluish skin, onycholysis, and severe extravasation necrosis.

IV use in **cats** leads to vomiting, anorexia, diarrhea, depression, infections secondary to myelodepression, convulsions, and death. Similar adverse effects in **dogs** plus alopecia and myelosuppression.

MIVACURIUM

= *Mivacron*

***Skeletal muscle relaxant.* IV use.**

Untoward effects: Cutaneous flushing of the face, neck, and chest. Occasionally hypotension, arrhythmias, phlebitis, bronchospasm, wheezing, rash, urticaria, erythema, dizziness, muscle spasms, and (rarely) a severe allergic reaction.

Drug interactions: Prolonged effect can occur if used with cholinesterase inhibitors (such as ***organophosphate insecticides***, ***ecothiophate***, and some ***antineoplastics***), ***oral contraceptives***, ***curare***-like antibiotics and drugs, some ***monoamine oxidase inhibitors***, ***phenothiazine tranquilizers***, ***procaine***, and ***succinylcholine***.

MOBAM

= *ENT 27,041 = MCA 600 = OMS 708*

Pesticide.* A *carbamate.

Untoward effects: Asynergy, myasthenia, ataxia, goose-stepping, stumbling, collapse, tremors, lacrimation, foamy salivation, diarrhea, tachypnea, dyspnea, ataraxia, tetany, convulsions, and phonation in affected **birds**.

LD_{50}, **rat**: 70 mg/kg; **guinea pig**: 50 mg/kg; **mallard duck**: 952 mg/kg; **California quail**: 463 mg/kg; **Japanese quail**: 668 mg/kg; 12 days-male **chick**: 85.4 mg/kg; dermal, **rabbit**: > 6 g/kg.

MOCK ORANGE

= *Philadelphus*

Untoward effects: The ***saponin*** content in its small fruits can be fatal.

MOCLOBEMIDE

= *Aurorix = Manerix = Moclamine = RO 11-1163*

Antidepressant, monoamine oxidase inhibitor.

Untoward effects: Arrythmias and palpitations (especially after ***tyramine***-containing foods). Hypertension and blurred vision. Intrahepatic cholestasis and death reported in 85-year-old **female**. A 22-year-old **female** after 450 mg/day/14 days developed fever, vomiting, diarrhea, headache, pruritus, and diffuse erythematous rash on face and trunk. Agitation, insomnia, restlessness, galactorrhea, and aggressiveness reported in some **patients**. Excreted in breast milk.

Drug interactions: ***Serotonin*** syndrome in **patients** due to interaction with ***citalopram***, ***clomipramine***, or ***imipramine***. Oral availability increased by ***cimetidine***.

Use with ***tyramine*** in **dogs** leads to arrythmias.

LD_{50}, **rat**: 700 mg/kg.

MODAFINIL

= *Alertec = CRL 40,476 = Cephalon = Modiodal = Provigil*

Central nervous system stimulant, psychostimulant, α_1-adrenergic agonist.

Untoward effects: Nausea, diarrhea, xerostomia, headache, and anorexia. Ocasionally chest pain, palpitations, nervousness, anxiety, insomnia, increased serum aminotransferase, dyspnea, and transient T-wave changes.

Embryotoxic in **rats**.

MODECCA DIGITATA

Untoward effects: A flowering wild plant in southern Africa with ***turnip***-like roots resembling edible roots. When eaten leads to lethal, accidental poisonings in **people** and **livestock**. Contains a highly toxic lectin, ***modeccin***, capable of destroying neurons.

MODIOLA CAROLINIANA

= *Bristly Mallow = Ground Ivy*

Untoward effects: Ingestion of the stems and leaves has poisoned **cattle**, **sheep**, and **goats** in Alabama, leading to central nervous system disturbances, incoordination, staggers, posterior paralysis, prostration, and death.

MOEXIPRIL

= *RS 10,085 = Univasc*

Antihypertensive. Angiotensin-converting enzyme-inhibitor. Precursor of moexiprilat, the active form.

Untoward effects: Cough, dizziness, diarrhea, fatigue, pharyngitis, flushing, rash, myalgia, and flu-like syndrome are the most common. Neutropenia, agranulocytosis, abdominal pain, rhinitis, nausea, hyperkalemia, hypotension, angina, myocardial infarction, sweating, hemolytic anemia, and nervousness have also occurred.

MOFEBUTAZONE

= *Acromonal = Butazon = Mobutazon = Mofesal =Monazan = Monophenylbutazone*

Anti-inflammatory.* A derivative of *phenylbutazone.

Untoward effects: Side-effects similar to, but less serious than, those of ***phenylbutazone***. Rashes, pruritus, weakness, vertigo, tinnitus, and blurred vision occur.

IV, LD_{50}, **mouse**: 600 mg/kg.

MOLASSES

Untoward effects: On January 15, 1919, in Boston, a cast-iron tank 50 ft high and 90 ft wide that held 2.2 million gallons of gooey molasses for making ***rum*** ruptured with little warning. A wave of molasses 2 stories high moved down Commercial Street at about 35 miles per hour, killing 21 **persons** and injuring 150. It was known as the "Great Boston Molasses Disaster" and joined Boston's claim to fame with baked beans and a Tea Party.

Excessive oral intake leads to polioencephalomalacia in **bulls** and flaccid posterior paralysis and ruminal parakeratosis in **cows** (called Borrachera or molasses drunkeness in Cuba).

MOLINDONE

= *EN 1733A* = *Lidone* = *Moban*

Psychotherapeutic.

Untoward effects: Dry mouth, mydriasis, cycloplegia, urinary retention, tachycardia, decreased gastrointestinal motility, constipation, memory impairment, akathisia, delirium with high dosage, and extrapyramidal effects. Less commonly, drowsiness, dystonia, anorexia, weight decrease, pallor, flushing, edema of face and hands, nausea, weakness, rigidity, sweating, accentuation of Parkinson's disease, slight epileptogenic effect, agranulocytosis (often with sore throat, fever, and lymphadenitis), rash, tardive dyskinesia, tremors, and menstrual irregularities. Rarely, leukopenia, postural hypotension, electrocardiogram abnormalities, tinnitus, increased SGOT, hepatotoxicity, xerostomia, blurred vision, oculogyric crisis, galactorrhea, priapism, and neuroleptic malignant syndrome.

TD_{LO}, **human**: 714 μg/kg/day.

Drug interactions: ***Lithium*** increases its serum levels.

The FDA considers it a ***prolactin***-inducing drug, with a potential of causing mammary tumors in **rats** and **mice**.

LD_{50}, **rat**: 261 mg/kg.

MOLLUSKS

A large group (~45,000 species) of marine invertebrates exists, and most poisonings result from eating *clams* (q.v.), *mussels* (q.v.), *oysters* (q.v.), and *scallops* (q.v.). Some species are capable of envenomations, such as *octopus* (q.v.) and *cone shells* (q.v.).

See *Lolliguncula brevis* (*Brief Squid*).

MOLYBDENUM

Untoward effects: In **humans**, exposures are primarily to **workers** in foundries. Toxic manifestations are non-specific, but exposed **workers** show increased blood ***uric acid*** and gout-like symptoms in parts of the Soviet Union. Blood levels are increased in uremia and hepatic levels are increased in hemochromatosis. Dust exposure to it and its compounds can cause eye, nose, and throat irritation. Autopsies reveal the highest concentrations in **people** in the liver, kidneys, small intestine mucosa, and adrenal glands. OSHA exposure limits have varied between 5 and 15 mg/m^3.

In **cattle**, excesses are associated with retarded growth, scours, weakness, brittle bones, emaciation, breeding difficulties, arched backs, and hair color changes (**Herefords** from cherry-red to dusty orange; **Angus** from blue-black to dirty gray). This syndrome in **ruminants** is referred to as "Teartness" or "Teart disease". Microcytic anemia, foul and gaseous diarrhea, aspermatogenesis, and developmental anomalies also occur. Forages containing in excess of 6 ppm are generally considered poisonous. Nevada and California areas, Great Lakes and eastern seaboard areas of the United States, and areas in Hawaii and England have reported such excesses. About 35 countries have areas with excessive amounts in the soil. Pasture contamination from industrial smoke, where it was used as a catalyst, has been an unexpected source of excessive intake. Similar stunting effects can occur in **poultry**. In England and Wales, it has been suspect as a cause of calculi and rat-tail syndrome in feeder **calves**. In **cattle**, molybdenosis produces the same symptoms as in **copper** deficiencies.

MOMETASONE FUROATE

= *ECURAL* = *ELOCON* = *Nasonex* = *SCH 32,088*

***Corticosteroid.* Topical cutaneous use and intranasal.**

Untoward effects: Burning, pruritus, irritation, furunculosis, skin-thinning, telangiectas, dryness, acneiform eruptions, hypopigmentation, perioral dermatitis, allergic contact dermatitis, striae, headache, pharyngitis, epistaxis, and secondary infection.

MOMORDICA ANGUSTISEPALA

Untoward effects: Abortifacient. Extracts of its roots are oxytocic with ***ergot***-like alkaloid action in Nigerian **women**.

IP is abortifacient within several hours in **mice** and **guinea pigs**.

MOMORDICA CHARANTIA

= *Balsam Pear* = *Bitter Gourd* = *Boga Ia Kibarianai (Swahili)* = *Karela* = *Wild Balsam Apple* = *Wild Cucumber*

Untoward effects: Extracts are used as a cough syrup in the Ozarks, and the fruit in folk-medicines of Jamaica, Bahamas, Mexico, Peru, and Asia. Its alkaloids cause vomiting, diarrhea (due to ***momordicin*** content, a strong cathartic), weakness, bradypnea, tremors, sweating, urinary incontinence, visceral vasodilation, excitement, and coma in severe intoxication; and hypoglycemia in **man**. Contains ***vicine***.

Drug interactions: Increased hypoglycemic effect in 40-year-old **female** with diabetes mellitus taken concurrently with ***chlorpropamide***.

The seeds and other parts of the plant are hypoglycemic in **rats**, **mice**, and **rabbits** and often fatal at effective dosage. A **puppy** in Florida ingested the fruit pods, seeds, and leaf fragments, leading to convulsions alternating with profound depression. 4–7 g is usually lethal to **dogs**.

MONACETIN

= *Acetin* = *Glycerol Monoacetate* = *Glyceryl Monoacetate* = *Monoacetin*

Use: Dye solvent; in ***leather***-tanning, and manufacture of smokeless powder and dynamite. Formerly, as a supposed antidote for ***fluoroacetate*** poisoning.

Untoward effects: As an antidote, 0.1–0.5 mg/kg IV of a 1% solution was used. It has caused sedation, tachypnea, vasodilation, hemolysis, and capillary disease.

MONENSIN

= *Antibiotic 3823 A* = *Coban* = *Elancoban* = *Lilly 67,314* = *Romensin* = *Rumensin*

Coccidiostat, antibacterial, antifungal antibiotic.

Untoward effects: Withdraw feeding 72 h before slaughtering treated **broilers**, to avoid residues > 0.05 ppm (as the acid) in edible tissues. Do NOT feed to laying **birds**. Wear protective clothing and gloves while mixing. Only for **cattle** in confinement feeding for slaughtering purposes. Ingestion by **horses** has been fatal (LD_{50}, 2–3 mg/kg). May cause a dermatitis in treated **broilers**. Do NOT store mixed feeds over 90 days. Monensin may delay feather maturation in **chickens**, especially at high ambient termperatures. Resistance to some strains of the **turkey** coccidia, *E. meleagrimitis*, has been reported. Not approved for milking **cows** or lactating **goats**, to avoid residue in **human** ***milk*** supply. Remove from feed at calving or kidding time. It is eliminated within 48–72 h from **dogs**. Poor agitation in liquid feeds can lead to intoxication (anorexia, pica, depression, diarrhea, some ataxia, and dyspnea). Official withdrawal times in **poultry** may no longer exist. Toxicosis in **swine**, **turkeys**, and **chickens** can be potentiated by ***tiamulin***, and by ***salinomycin*** in **turkeys**; by ***oleandomycin*** in **chickens**; ***chloramphenicol*** in **turkeys**; and ***sulfonamides*** in **calves**, **turkeys**, and **chickens**. **Cattle**, **sheep**, **swine**, and **goats** are easily poisoned by LD_{50}s of 12–40 mg/kg; and **chickens** by LD_{50}s of ~200 mg/kg. Increasing age seems to increase toxicity in **turkeys** and **swine**. Avoid overdosing **cattle**, as it may cause severe cardiac and gastrointestinal lesions, central nervous system and electrolyte imbalance, and even death. Toxic to **quail**. Contraindicated in **swine** receiving ***tiamulin***.

LD_{50}, **mouse**: 43.8 mg/kg; **chick**: 284 mg/kg.

MONEY

Untoward effects: Facetiously, many comments can be made about its adverse effects on **people**, but dermatologists have noted that **bankers** in the U.S. and Germany have developed scaling eczemas on **their** hands. Paper money **makers** have developed dermatitis from occupational exposure to ***chromates*** and ***formaldehyde*** resins. Also see ***cocaine***.

A detailed case report describes a case of ***zinc***-induced hemolytic anemia in a **pup** caused by ingestion of pennies.

MONOACETOXYSCIRPENOL

Untoward effects: A *Fusarium* toxin which, when ingested by **rats**, **turkey poults**, **swine**, and young **chickens**, has caused illness and death. Causes decrease in food intake, weight, and egg production in **hens**.

MONOAMYLAMINE

Untoward effects: A strong irritant causing contact urticaria, and delayed hypersensitivity reported from its use in Schering Corp.'s ***Tinactin***® (***tolnaftate***) cream. The reaction was confirmed by patch-testing on 41-year-old **female**.

MONOBENZONE

= *Agerite* = *Benoquin* = *Benzoquin* = *Benzyl Hydroquinone* = *Depigman* = *Hydroquinone Benzyl Ether* = *Hydroquinone Monobenzyl Ether* = *p-Hydroxyphenyl Benzyl Ether* = *Monobenzyl Hydroquinone* = *Pigmex*

Skin bleach.

Untoward effects: Generally irreversible depigmentation. Leukoderma at 2–4 years intervals after treatment. Leukoderma reported in **worker** using ***rubber*** gloves containing it as an antioxidant. Confirmed by patch-testing. **Negroes** in a tannery found **their** arms and hands turning white from similar use of ***rubber*** gloves. Erythematous rashes sometimes occur during treatment. Has caused depigmentation, not only at the site of application, but also at distant skin sites. One dermatologist claimed it can make white **people's** skin black. Sensitizer in **people** and **guinea pigs**.

TD_{LO}, **mouse**: 16 g/kg over 78 weeks; SC, TD_{LO}, **mouse**: 1 g/kg; IP, LD_{50}, **rat**: 4.5 g/kg.

MONOCHLORODIFLUOROETHANE

= *Dymel 142* = *FC 142b*

Aerosol propellent.

Untoward effects*:** Hypotension, tachycardia, and increased pulmonary resistance by inhalation of 20% concentration in experimental **animals**. Also see ***Freons.

MONOCHLORODIFLUO-ROMETHANE

= *Chlorodifluoromethane* = *F 22* = *FC 22* = *Freon 22*

High-pressure aerosol propellent, refrigerant.

Untoward effects*:** Escaping gas from refrigeration equipment aboard a deep sea fishing boat killed six **fishermen**, apparently from displacement of available ***oxygen. Post-mortem revealed fine fat droplets in hepatocytes, which could also have occurred from ***alcohol***, as the **sailors** had been drinking ***sake***. **Teenagers** have shown paddling movements after sniffing it. A 14-year-old **male** was found dead with **his** mouth around a ***Freon 22*** tank.

Exposed **mice** evidenced reeling, forelimb weakness, falling down, tumbling, and mucous from mouth and nose. Post-mortem revealed fat droplets in hepatocytes. Low incidence of SC fibrosarcomas in salivary gland region of male **rats** inhaling 50,000 ppm/2 years, but not in female **rats** or **mice** of both sexes.

Inhalation LC_{LO}, **rat**: 25 pph/4 h; **dog**: 70 pph.

MONOCROTALINE

= *Crotaline* = *MCT*

Toxic *pyrrolizidine* alkaloid in *Crotalaria sp.* (q.v. for detailed toxicity). Found in all parts of the fresh and dried plant with highest concentration in the ripe pods.

***Untoward effects*:** Hemorrhagic necrosis of liver, ascites, pulmonary edema, and pleural effusions in **rats**. Hepatotoxic in **mice** and most affected **animals**. As a dried contaminant of hay or seeds in their grain, it has caused acute toxicity within a few hours to days in farm **animals** and **poultry**. Chronic poisoning also occurs. Endothelial dihiscence, fragmentation, and hypertrophy of endothelial and muscular component of pulmonary vasculature in **rats**. They die from severe exudative lesions and cor pulmonale. SC injections into infant **stumptail monkeys** (*Macaca arctoides*) caused severe pulmonary dysfunction, endocardial fibrosis, and cor pulmonale. **Guinea pigs** are relatively resistant to its ill effects.

LD_{50}, **rat**: 66 mg/kg.

MONOCROTOPHOS

= *Azodrin* = *C-1414* = *Crisodrin* = *ENT-27129* = *Monocron* = *Nuvacron* = *SD-9129*

***Organophosphorus insecticide*. An anticholinesterase.**

***Untoward effects*:** Muscular weakness and blurred vision in 19-year-old **male**, 28 h after 1 pint of emulsion splashed onto **his** bare chest and arms, in spite of immediate washing with water. Miosis, pallor, sweating, continued weakness, increased salivation, and dry retching 38 h after exposure. Treated; acute symptoms gone by day 3, but blood cholinesterase values were abnormally low for 8 weeks. Neurotoxic effects in ten **patients**, 20–60 years. Ingestion with suicidal intent in nine, and one was from spraying exposure. Symptoms leading to paralysis of proximal limb muscles, neck flexors, motor cranial nerves, and respiratory muscles, after a cholinergic phase with three **patients** dead.

Fatal secondary poisoning noted in **birds** of prey eating poisoned **voles**. Poisoned **birds** have fluffed feathers, eyes closed, ataxia, lacrimation, salivation, polydipsia, dyspnea, tracheal congestion, defecation, mydriasis, tremors, convulsions, tetany, or opisthotonus. Poisoning of **ducks**, **geese**, and other **birds** has been common, due to drifting of the chemical after spraying of ***alfalfa***, ***cotton***, ***rice***, and other crops. **Mammals** show ataxia, miosis, hyporeactivity, quivering, immobility, tracheal congestion, tachypnea, dyspnea, and phonation.

LD_{50}, **rat**: 21 mg/kg; **mouse**: 14.4 mg/kg; young **chick**: 3.54 mg/kg; **golden eagle**: 188 mg/kg; **mallard duck**: 4.76 mg/kg; **Canadian goose**: 1.58 mg/kg; **pheasant**: 2.83 mg/kg; **goat**: 20–50 mg/kg; **duck**: 3.36 mg/kg; **quail**: 4 mg/kg; skin, **rat**: 112 mg/kg; skin, **rabbit**: 354 mg/kg. LC_{50} in feed, **quail**: 2.4 ppm.

MONOCTANOIN

= *Capmul 8210* = *Moctanin* = *Monooctanoin*

Cholelitholytic.

***Untoward effects*:** Noncardiogenic pulmonary edema in 29-year-old **female** during intraductal administration of 10 ml into an obstructed common bile duct. Recovery in 12 h without treatment. Complete dissolution of gallstones is uncommon in diabetic **patients**. Excessive perfusion pressure can cause dangerously ascending cholangitis.

MONOETHYLHEXYL PHTHALATE

= *MEHP*

The principle metabolite of *bis(2-ethylhexyl)phthalate* (q.v.).

Untoward effects: Fed to **rats**, it leads to severe testicular atrophy and decreased ***zinc*** concentration in their accessory sex organs. Similar results in **mice**. Not teratogenic in **rats** and **rabbits**, except at unusually high dosage.

LD_{50}, male **rat**: 1.8 g/kg; female **rat**: 1.34 g/kg.

MONOLINURON
= *Aresin*

Herbicide, pesticide.

Untoward effects: Lilly, with Schering, imported some for formulation. Symptoms of hypoxia reported in three **workers**. Methemoglobinemia was diagnosed in two. A 63-year-old **female** drank a cupful of herbicide containing 140 g/l, and also develop methemoglobinemia.

Has caused methemoglobinemia in **cats** and **dogs**. Waxy rib teratogenicity reported in **rats** and **mice**. Toxic to **fish**.

LD_{50}, **rat**: 1.8 g/kg; **dog**: 500 mg/kg; acute dermal in female **rat**: > 1.5 g/kg; **hen**: 240 mg/kg.

MONOMETHYLNAPHTHALENE SULFONATE

Untoward effects: Widely used in a home carpet cleaning system (***RinseNvac***®). Apparently volatile; when inhaled, has caused bronchospasms in household **members**.

MONOSODIUM METHANEARSONATE
= *Ansar* = *Asazol* = *Bueno* = *Daconate* = *Dal-E-Rad* = *Herban M* = *Merge* = *Mesamate* = *Methanearsonic Acid, Sodium Salt* = *Monosodium Methyl Arsonate* = *MSMA* = *Phyban* = *Silvisar* = *Weed 108* = *Weed-E-Rad* = *Weed Hoe*

***Herbicide.* Highly toxic pentavalant arsenical.**

Untoward effects: See ***Arsenic*** and ***methanesulfonic acid***. Severe peripheral neuropathy with decreased and absent deep-tendon reflexes and decreased position sense, muscle wasting, anemia, leukopenia, and dyserythropoiesis after several days exposure to it from overhead aerial ***cotton*** spraying in a 25-year-old **male**. **His** hands, neck, head, and feet were uncovered. Nausea, vomiting, and swollen feet also occurred. In a suicide attempt, a 20-year-old **male** drank ~500 ml of a 16% solution and vomited more than nine times. **He** recovered with no side-effects from 5 days treatment.

Cattle given 10mg/kg/day developed toxic nephrosis and hemorrhagic gastritis. Symptoms in poisoned **pets** and most **animals** include vomiting, diarrhea, abdominal pain, weakness, ataxia, anorexia, gastrointestinal irritation and hemorrhage, anuria, trembling, salivation, posterior paralysis, decreased temperature, collapse, and death. **Hares** are easily killed by ingestion of it.

LD_{50}, **rat**: 700 mg/kg; **bobwhite quail**: 3.3 g/kg. LC_{50} in feed, **quail**: > 5000 ppm.

MONOSULFIRAM
= *Kutkasin* = *Monosulfide* = *Tetmosol* = *Tetraethylthiuram* = *TTMS*

Ectoparasiticide.

Untoward effects: ***Disulfiram***-like adverse reactions with ***alcohol***-based solutions after topical use. Skin swelling, flushing, sweating, and tachycardia reported. Topical use also associated with toxic epidermal necrolysis. Possible skin sensitizer to the **person** handling the chemical. Cross-sensitivity to other ***thirams*** and ***rubber*** goods is also possible.

MONOTROPA UNIFLORA
= *Corpse Plant* = *Indian Pipe*

Untoward effects: ***Honey*** from it can cause mild to serious poisoning in **man**.

MONSTERA sp.
= *Ceriman* = *Cutleaf Philodendron* = *Fruit Salad Plant* = *Mexican Breadfruit* = *Swiss Cheese Plant*

Cultivated aroid and house plant.

Untoward effects: Their tissues contain irritant juices and crystals of ***calcium oxalate*** (q.v.) that can cause a contact dermatitis and, if ingested, can cause discomfort and swelling in the mouth and throat, hoarseness, blistering, and aphonia.

MONTELUKAST
= *L 706,631* = *MK 0476* = *MK 476* = *Singulair*

Antiasthmatic, antileukotriene.

Untoward effects: ~25% of **patients** show no response. Churg-Strauss vasculitis noted with it may have been induced by ***corticosteroid*** withdrawal.

MONURON
= *Chlorfenidim* = *CMU* = *Karmex Monuron Herbicide* = *Monurex* = *Telvar*

Herbicide.

Untoward effects: **Cattle** are poisoned by a single oral dose of 500 mg/kg, leading to diarrhea, bloat, incoordination, and prostration. **Sheep** receiving 250 mg/kg/day/2 days develop incoordination, excitability, anorexia, torticollis, and prostration. After 100 mg/kg/day/8 days, **chickens** die and show congestion of intestinal mucosa after death. Post-mortem in **sheep** reveals severe congestion of lungs, respiratory mucosa, liver, kidneys, and, occasionally, in the meninges.

LD_{LO}, **human**: 500 mg/kg.

LD_{50}, **rat**: 1.48 g/kg; dermal, **rabbit**: > 2.5 g/kg; LC_{50} in feed, **quail** and **mallard duck**: > 5000 ppm. LD_{LO}, **guinea pig**: 670 mg/kg.

MOPERONE

= *Luvatrene* = *Methylperidol* = *R 1658*

Psychotherapeutic.

Untoward effects: Dyskinetic, hypertonic, extrapyramidal reactions (severe cramps in muscles of extremities, opisthotonus, and trismus) in 27-year-old **male**. Akinesia, severe fatigue and somnolence, hypotension, severe parkinsonism, senile dementia, sudden rigidity, and agitation in **others**.

MOPS

Untoward effects: Massive bacterial contamination throughout a hospital by wetmopping techniques. Mops stored wet supported high bacterial growth, not adequately decontaminated by chemical disinfection alone.

MORAEA sp.

= *Tulp*

Untoward effects: ***M. carsonii*** and ***M. erici-rosenii*** ingestion have killed **sheep**. Arrhythmias, tachycardia, frothing at the mouth, constant movement and picking up of its feet, and dyspnea preceeded death. Post-mortem revealed pulmonary emphysema, prescapular and suprapharyngeal lymph gland hemorrhages, tracheal froth, and petechiae in SC fascia and conjunctiva.

M. zambesiaca (***M. schimperi***) = ***Alubasa kwadil***, a common cause of cattle poisoning in Nigeria and Rhodesia. A small **ox** can be killed by 4–5 oz of green tulp and adult **heifers** within 48 h after eating 4 oz of the dried tulp hay. Gastroenteritis and bloating are early symptoms. Highly toxic to **sheep**.

MORANTEL TARTRATE

= *Banmith II* = *Bovhelm* = *CP 12,009-18* = *Exhelm E* = *Nematel* = *Ovithelm* = *Paratect* = *Rumatel*

Anthelmintic.

Untoward effects: Do NOT use in lactating or dairy **cattle** of breeding age, or within 14 days of slaughter, to meet U.S. residue maximum of 1–2 ppm in muscle, 2–4 ppm in liver, 3.6 ppm in kidney, 48 ppm in fat, and 0.4 ppm in milk.

LD_{50}, **rat**: 926 mg/kg.

MORCHELLA DELICIOSA

A gourmet's *mushroom* delicacy.

Untoward effects: None directly, but history tells of barbarous times in Italy when the Borgias and Medici tried to eliminate each other. Substitution of ***Amanita phalloides*** (q.v.) for this delicacy produced the desired effect. ***4-Amanitin*** has a delayed effect with the first 24 h post-ingestion being symptomless. After another 24 h, hepatic function fails and death ensues. A **monarch**, delighted with the taste of Morchella, issued a royal decree that, henceforth, they were to be the only ***mushrooms*** on **his** plate. That was **his** last decree.

MORFAMQUAT

= *Morfoxone* = *Morphamquat* = *PP 745*

Herbicide.

Untoward effects: As a ***dipyridyl*** compound, it can bind to and injure epithelial tissues, especially if in the concentrated form.

Reversible tubulopathy in **dogs**. Glomerular damage also in the **rat**.

LD_{50}, **rat**: 345 mg/kg; **mouse**: 325 mg/kg; **cat**: 160 mg/kg; **chicken**: 367 mg/kg.

MORICIZINE

= *EN 313* = *Ethmosine* = *Ethmozin* = *Ethmozine* = *Moracizine*

Antiarrhythmic.* A *phenothiazine.

Untoward effects: Proarrhythmic, ventricular, and supraventricular arrhythmias; ventricular fibrillation, palpitations, tachycardia, prolonged PR (27%) and QRS (10%) on electrocardiogram, nausea, dizziness, vertigo, incoordination, headache, congestive heart failure, dysuria, urinary retention or urinary incontinence, rash, blurred vision, nystagmus, diplopia, abdominal pain, vomiting, flatulence, nervousness, asthenia, paresthesia, bitter taste, dry mouth, speech disorders, tinnitus, loss of memory, confusion, seizures, euphoria or depression, urticaria, dry skin, swelling of lips and tongue, periorbital edema, impotence, sweating, cough, apnea, and asthma. Drug fever, renal failure, and hepatic toxicity are rare.

Drug interactions: Varying reports, but **users** should be alerted that increased prothrombin time can occur with ***warfarin***. ***Cimetidine*** increases its plasma concentration. May decrease ***theophylline*** serum concentration. ***Food*** can delay its time to reach peak plasma concentration, but not its bioavailability.

MORINAMIDE

= *B 2310* = *Morphazinamide* = *Nicoprazine* = *Piazofolina* = *Piazolina*

Antitubercular, antibacterial.

Untoward effects: Urticaria, pruritus, nausea, anorexia, headache, slight hematuria, decreased blood pressure, decreased renal clearance of ***urates*** with temporary increased uricemia, and slight transient hypochromic anemia.

MORINDA RETICULATA

Untoward effects: A ***selenium*** (q.v.)-accumulator, even when growing on ***selenium***-deficient soils, leading to chronic ***selenium*** poisoning, with lameness, hoof loss, and loss of mane and tail hair in **horses**.

MORNING GLORY

= *Liseron*

Untoward effects: Hallucinogenic properties in the seeds of several species and varieties of the genera ***Ipomoea*** and ***Rivea*** (q.v.), due to indole alkaloids (~0.04%), ***lysergamide*** (q.v.), ***isoergine***, ***elymoclavine*** (q.v.) and other related alkaloids, which cause actions similar to ***lysergic acid diethylamide***. Usual ingestion has been of 200–300 seeds. **Thrill-seekers** chew or ingest > 50 seeds for the mental effects. Symptoms include depersonalization, memory loss, feelings of transcendence and euphoria, visual and tactile hallucination-like states, panic reactions, nausea, anorexia, abdominal discomfort, explosive diarrhea, decreased deep-tendon reflexes, numbness of extremities, muscle tightness, mydriasis, uterine stimulation, and polyuria. Several species of *Rhynchosia* in Mexico are also known as morning glories.

I. hederacea: Also called ***Heavenly Blue*** and ***Pearly Gates***; popular in the U.S. drug culture community. Suicide by 24-year-old **male** during recrudescence of psychomimetic effects 3 weeks after chewing 300 seeds. Contains ***elymoclavine*** and *d*-***lysergic acid amides***, which vary in amounts in different cultivars. Mydriasis, facial flushing, hypotension, hallucinations, nausea, and diarrhea in **humans**. Excitement, staring, and meowing in a **cat** that had eaten the seeds.

I. purpurea: Also contains alkaloids with ***LSD***-like action, but of lower potency. Tuberous roots contain a strong purgative resin.

I. violacea = ***Badoh Negra*** (Aztec) = ***Flying Saucers*** = ***Heavenly Blues*** = ***Ololiuqui*** (Aztec) = ***Pearly Gates*** = ***Wedding Bell***: Hallucinogenic effects last 8½–16 h, then completely disappear, recurring later (even after a few days). Contains ~60 mg alkaloids/100 g of seeds (~2500 seeds), of which 40 mg is ***lysergamide*** (q.v.). Injection into **mice** leads to loss of motor coordination, tremors, ataxia, piloerection, ptosis, hypersensitivity to touch, and slight cyanosis.

MORPHINE

= *Dolcontin* =*Duromorph* = *Kapanol* = *Morfin* = *Morphia* = *Morphina* = *Morphium* = *Nepenthe* = *Oramorph* = *Reliadol*

Street names include: Christ's Opium = *Cube* = *Dreamer* = *Emsel* = *First Line* = *Hocus* = *M* = *Melter* = *Miss Emma* = *Monkey* = *Morf* = *Morphie* = *Morpho* = *Mud* = *Tab* = *Unkie*

Narcotic, analgesic, hypnotic, antitussive. **Usually used SC, IM, and IV; occasionally, orally or smoked. Isolated in 1803, after extraction from *opium* sap, which was the beginning of the scientific era, although, as the prime constituent of *opium*, it was used as a narcotic for thousands of years, and by ancient Egyptians to calm fretful infants. Commercially ~90% is now converted into *codeine*. Named for Morpheus, the Greek god of dreams and sleep. Acetylation converts it into *heroin* (q.v.). Morphine is the active metabolite of *codeine*.**

Untoward effects: Light-headedness, dizziness, sedation, nausea, vomiting, ileus, urinary retention, abdominal distention, skin rashes, fixed-drug eruptions, pruritus, urticaria, headache, and hypothermia frequency occur with usual or single large doses. Dysphoria, euphoria, dulling of senses, syndrome of inappropriate antidiuretic hormone secretion, constipation, dry mouth, thirst, orthostatic hypotension, small decrease in urinary ***vanillylmandelic acid*** excretion, itching of skin (especially the nose), and (rarely) anaphylaxis. In addition, excessive use or overdosage has also caused pulmonary edema, bradycardia, tachycardia, decreased blood pressure, bronchiolar constriction with subsequent asthmatic attacks, miosis, diaphoresis, myoclonus, respiratory depression, dyspnea, hallucinations, slight increase in prothrombin time, increased urinary ***norepinephrine*** excretion, corneal opacity, fever, increased lactate dehydrogenase, increased SGOT, immune thrombocytopenia, postural rigidity, decreased release of pituitary ***gonadotropin*** with decrease in circulating ***androgens*** and impaired sexual function, hyperprolactinemia, enhanced tone in the sphincter of Oddi, increased or false positive sulfobromophthalein retention, increased serum amylase when no food has been ingested in previous 4+ h, leading to false diagnosis of pancreatitis. Seizures during high-dose IV in 56-year-old **female** due to a ***sodium bisulfite*** preservative. Abuse liability is high and, in addition to the above, some **addicts** show anxiety, panic, slurred speech, tremors, poor coordination, runny eyes and nose, insomnia, diarrhea, convulsions, cyanosis, unconsciousness, and hepatitis. **Some** have died and mydriasis may occur with asphyxia before death. Mean lethal level in a review of cases was 0.09 (range 0.01–0.2) mg/100 ml. In a group of 121 fatal cases, mean blood concentration was 53 (range 0.9–1640) μg/100 ml. Symptoms from oral overdoses start within 15–20 min. **People** not habituated to it may die from 1–2 grains (65–130 mg) and **some** have been poisoned by $<$ 1 grain. Some **addicts** have taken up to 10 fatal doses without apparent injury. The first report of its addiction was noted in 1879 in a German publication. The drug activates bronchial asthma and prostatic ureteral obstruction and is more toxic in liver and kidney disease. Hyperalgesia and allodynia in 21-month-old **female** after increased IV dosage.

It crosses the placenta and can cause addiction *in utero*, severe respiratory distress and miosis in the **newborn**, and withdrawal symptoms (irritability, tremors, high-pitched cry, hyperactivity, wakefullness, diarrhea, poor sucking reflex, yawning, lacrimation, respiratory alkalosis, hiccups, sneezing, twitching, myoclonic jerks, and seizure) after birth. In the pregnant **addict**, sudden withdrawal can lead to premature labor and intrauterine death. With normal doses, ~16 mg, a small amount (0.1 mg) appears in breast milk 4 h later. **Maternal** passage into breast milk leads to mean intake of 2.9 μg/kg for the **infant**. Excretion levels in breast milk can vary (up to 12% of a **maternal** dose to **her infant**), and the **infant** may not show withdrawal symptoms. The SC or IV therapeutic dose for an **infant** is 100–200 μg/kg.

Human minmimum lethal dose, 200–300 mg.

Drug interactions: Prolonged and exaggerated ***morphine*** response with ***cimetidine***, including near-lethal effect. ***Alcohol*** can have an additive central nervous system effect with severe respiratory depression. ***Monoamine oxidase inhibitors*** and ***phenothiazine tranquilizers*** potentiate its analgesia and respiratory decrease. Decreased ***theophylline*** and ***aspirin*** clearance. Enhances effect of ***coumarin anticoagulants***. Potentiates ***propranolol*** and ***phenothiazine tranquilizers***, even causing unexpected deaths in **patients**. Produces false positive or spuriously high serum amylase, sulfobromophthalein retention, urine ***glucose*** (Benedict's), and serum lipase test readings. Slows down the absorption of many drugs, including ***acetaminophen*** and ***mexiletine***.

Addictive qualities have been demonstrated in experimental **animals**. 3 mg/kg SC, qid led to physiological dependence in **rhesus monkeys**. As addicts, they eventually self-administered 50–75 mg/kg daily. Addiction noted in **guinea pigs**, **mice**, **rats**, **rabbits**, and **monkeys**. Chronic use in female **rats** before mating led to retarded growth in their offspring, apparent only after 3–4 weeks of age. In rare cases in **dogs**, where it causes excitement, it still decreases ***pentobarbital*** required for anesthesia. Overdosage may unnecessarily decrease respiration, produce a hypotensive state, or even produce excitement (the **cat** is a classical example of the latter). May be *contraindicated* in nephritis, as the urine is the principle route of excretion in a number of species tested. May cause severe skin irritation (pads and soles) and anorexia in **monkeys**. May produce excitement in **horses** at higher dosage, unless premedicated with ***xylazine***. Avoid use in **animals** with severe liver disease, as it is metabolized in the liver. Offspring of addicted **rabbits** may be weak, cyanotic, and show limb malformations. May be detected for at least a week after injections in **race horses**!!

MORPHOCYCLINE

Antibiotic.

Untoward effects: In **dogs**, IV use leads to transient flushing, tachypnea, and tachycardia. Hypotensive after IV in **cats**.

IV, LD_{50}, **mouse**: 260 mg/kg; SC, 820 mg/kg.

MORPHOLINE

= *Diethylene Oximide*

***Solvent.* Used in *rubber*, cosmetics, soap, wax, corrosion-inhibitors, and optical brighteners, and in the illegal manufacturing of *phencyclidine*.**

Untoward effects: Eye, skin, nose, and respiratory irritant. Has caused visual disturbances and coughing as an occupational irritant and sensitizer, but these have not been observed in cosmetic formulations. High concentration can cause pulmonary edema. Absorbed material causes liver necrosis and degeneration of renal tubules. Local necrosis on skin contact and gastrointestinal mucosal damage after ingestion. National Institute for Occupational Safety and Health and OSHA skin exposure limits of 30 and 20 ppm, respectively.

LD_{LO}, **human**: 50 mg/kg.

Reacts easily with nitrosating agents leading to ***N-nitrosomorpholine***, which can cause liver tumors in **rats** and lung adenomas in **mice**.

LD_{50}, **rat**: 1 g/kg; acute dermal, **rabbit**: 500 mg/kg.

MORTIERELLA sp.

Untoward effects: Deep systemic mycosis in **man**. Mycotic placentitis, abortion, and pneumonia in **cattle**.

MOSS

Untoward effects: Allergic contact dermatitis from some species.

MOTH BALLS or POWDER

Untoward effects*:** Hemolytic anemia in glucose-6-phosphate dehydrogenase-deficient **people**, usually from ***naphthalene (q.v.) or ***p-dichlorobenzene*** (q.v.). Wrapping **infants** in clothes or using diapers stored in moth balls has caused hemolysis and jaundice in susceptible **neonates**. Anorexia and nausea after ingestion, followed by toxic degenerative changes in liver and kidneys.

MOTHS

***Untoward effects*:** An outbreak of acute contact dermatitis from the species *alinds Druce* in **employees** of 17 tourist hotels reported from Mexico. Similar effects reported in **people** from the genus *Hylesia* in the U.S., Venezuela, Peru, and Mexico. Reactions also reported from the stings of adult moths.

MOTOR OIL

It can penetrate into and through human nails.

Untoward effects*:** Used motor oil can cause ***lead poisoning. Repeated use of motor oil on **sows** has caused severe acanthosis, particularly of their ears. Emissions from space heaters burning the used oil can contain high levels of ***arsenic, barium, cadmium, chromium, nitrogen oxides, polycyclic aromatic hydrocarbons, sulfur***, and various solvents, which can also cause respiratory impairment.

MOXALACTAM

= *Festamoxin* = *Lamoxactam* = *Latamoxef* = *LY 127935* = *Moxam* = *S-6059* = *Shiomarin* = *Shionogi 60595*

***Antibiotic*. A third generation semi-synthetic cephalosporum. A *β-lactam*. IV or IM.**

Untoward effects*:** Has a ***methylthiotetrazole side chain which inhibits platelet aggregation and prolongs prothrombin time, with bleeding events that appear to be dose-related. Eosinophilia, thrombocytopenia, hypoprothrombinemia, anemia, and leukopenia also reported. Mild thrombocytosis in three cases. Hypersensitivity reactions (rash, fever, positive Coomb's test, anaphylaxis) in ~3% of **patients**, who usually were allergic to ***penicillin***. Diarrhea and other gastrointestinal upsets, including pseudomembranous colitis and fungal superinfection, in about 2–3%. A **few** have increased liver enzymes and nephrotoxicity in ~0.05%. Severe myoclonus, generalized seizures, and marked stupor in 39-year-old **female** with chronic renal insufficiency after 2 g IV, tid/1 week. A 91-year-old male exhibited tonic-clonic seizures after 3 g/day/7 days. Local pain at injection site in ~3%. Can potentiate ***warfarin's*** anticoagulant effect and precipitate a mild ***disulfiram*** reaction in some **patients** when taken with ***ethanol***. Breast milk levels after 2 g tid in eight **women** were up to 3.24 µg/ml on day 4. An **infant** ingesting 550 ml/day would receive 2 mg/day.

Hypoprothrombinemia in treated **rats**.

MOXIDECTIN

= *CL 301,423* = *Cydectin* = *Proheart* = *Vetdectin*

***Use*:** Semi-synthetic macrolide antibiotic, anthelmintic in **cattle** and **sheep**. Heartworm preventative for **dogs**.

***Untoward effects*:** In **dogs**, nausea, vomiting, ataxia, anorexia, diarrhea, lethargy, nervousness, itching, and increased thirst.

MOXIFLOXACIN

= *Avelox*

An oral *fluoroquinolone*.

***Untoward effects*:** Skin rashes, gastrointestinal upsets, and prolonged QT interval. Confusion, dizziness, hallucinations, insomnia, seizures, and tendinitis reports can be expected as its use increases.

MOXONIDINE

= *BDF 5,895* = *BE 5,895* = *Physiotens*

***Antihypertensive*.**

***Untoward effects*:** Itching, nausea, jaundice, maculopapular rash, and cholestatic hepatitis in 83-year-old **male** after 0.2 mg/day/9 months.

MPPP

A *meperidine* analog and synthetic *opiate*.

Untoward effects*:** In November 1976, a 23-year-old **male** chemistry graduate **student** injected **himself** with a homemade batch over several days. It is a by-product of the synthesis of toxic ***MPTP (q.v.). **He** was admitted to a psychiatric ward of a hospital with an initial diagnosis of catatonic schizophrenia, and successfully treated with anti-Parkinson drugs. **He** continued to abuse drugs and died of an overdose in September, 1978. Post-mortem revealed severe loss of ***dopamine***-producing cells in the substantia nigra and an eosinophilic inclusion, both characteristic of Parkinson's disease.

MPTP

A contaminant of a street-version of a synthetic *heroin*. A by-product of *MPPP* production.

Untoward effects*:** Rapidly and selectively destroys the same group of brain cells that slowly degenerates in natural Parkinson's disease. It does not kill the brain cells, but a metabolic derivative, ***MPP+ or ***MPPP*** (see above) may be the actual cause. Within a week of first parenteral use, visual hallucinations, jerking of limbs, and stiffness occurred, followed in a few days by difficulty in moving. After injecting it for a few days, two **brothers** had extensive brain damage and, for a week, lay helplessly in a bed, able to move only **their** eyes. **They** responded dramatically to treatment with L-***dopa***, an antiparkinsonism drug. This syndrome was not reproduced by injections in **rats**, but it did in **monkeys**, leading to slowed movements, tremor, rigidity, and hunched posture. Post-mortem revealed extensive destruction of substantia nigra cells that produce ***dopamine***, a neurotransmitter, but leaving the adjacent mesolimbic area undamaged. They also responded to L-***dopa***.

MUCOITIN POLYSULFATE ESTER

= *Adequan* = *Arteparon* = *Eleparon* = *Flexequin*™ *IA* = *Hirudoid* = *MSPS* = *Mucopolysaccharide* = *Polysulphated Glycosaminoglycane*

Proteolytic enzyme-inhibitor.

Untoward effects: Intra-articular use leads to irritation and sensation of heat, effusion-resorbed without aspiration (20%), and some tenderness of joint for several hours. Pain is usually associated with periarticular or superiostal injections. Infections have followed some intra-articular injections.

Swelling, tenderness, and warmth noted after intra-articular use in **horses**.

MUCUNA sp.

M. pruriens = *Agbala (Ibo)* = *Cigu (Thai)* = *Cowitch* = *Cowhage* = *Horseye Bean* = *Kiwach* = *Mijeh (Thai)* = *Sijeh (Thai)* = *Werepe (Yoruba)*

The seed pods or fruit (trichomes) have ~5,000 barbed spines or hairs on them and contain a highly irritating proteolytic enzyme, *mucunain*. The dried pods, even those in herbarium collections are stil toxic. The fruit contains a histamine releaser, *serotonin*, that causes wheals, itching, and irritation.

***M. sloanei* = *Ojo de Venado* (eye of the deer in Spanish).**

Untoward effects: A single seed can contain > 350 mg L-***dopa***, leading to choreiform and/or dystonic movements; occasionally cardiac arrhythmias, palpitations, orthostatic hypotension, bradykinesia, paroid ideation, psychotic episodes, dementia, depression, and urinary retention. The seeds of it, ***Sophora secundiflora*** (q.v.), and ***Abrus precatorius*** (q.v.) are in a Mexican good luck charm to ward off *mal de ojo* or evil eye, and *el aire* or bad and ominous air. These are very dangerous to inquisitive **children** and **adults**.

MUCANAIN

Untoward effects: This is the itch chemical, a proteolytic enzyme, found in ***Mucuna pruriens***.

MULBERRY

= *Morus sp.*

Untoward effects: **Humans** that have eaten large quantities of ***M. rubra***, the red mulberry, become mildly intoxicated. Its aeroallergens have caused allergic rhinitis, bronchial asthma and/or a hypersensitivity pneumonitis in some **people**. Some mulberries have a laxative effect.

Stems, leaves, and green fruit have caused severe central nervous system disturbances in **livestock** that have grazed them.

MUMMIES

The word mummy is derived from *mumii* or *Bitumen of Judea*.

Untoward effects: A hypersensitivity pneumonitis from an unknown antigen on the mummy wrappings.

MUNDULEA SERICEA

= *Swahili: Mkwaia* = *Mkwaya* = *Mtupawa-pori* = *Yoruba: Igun* = *Laktu*

Untoward effects: Contains ***rotenone*** (q.v.), rotenoids, and unknown toxic agents in the bark and seeds. Used as a **fish** and **crocodile** poison. These are then unwholesome to eat. A homicidal poison in East Africa, with cardio-respiratory failure preceding death.

MUNITIONS

Untoward effects: In addition to the obvious, **workers** are exposed to ***dinitrotoluene*** (q.v.). Waterborne pollution by ***nitrocellulose, nitroglycerin*** wastes, ***white phosphorus*** manufacturing, and pink water from ***TNT*** production are the more common problems.

MUPIROCIN

***Antibiotic.* For topical use.**

Untoward effects: Burning, stinging or pain (1.5%), itching (1%), rash, erythema, dry skin, tenderness, swelling, contact dermatitis, and occasionally increased exudate. Local reactions are, primarily, from its ***polyethylene glycol*** vehicle.

MUROCTASIN

= *DJ7041* = *MDP-Lys(L18)* = *Nopia* = *Romurtide*

Immunostimulant.

Untoward effects: Passive cutaneous anaphylaxis, sytemic anaphylaxis, and delayed skin reactions in **guinea pigs** and **rabbits**. SC use in **mice** and **dogs** leads to leukocytosis, splenic white pulp, irritation of axillary lymph nodes, and lesions at injection site; anemia and synovitis in **dogs**. With toxic IV and SC doses in **mice**, **rats**, and **dogs**, screaming, loss of motor activity, hair loss, weakness, and necrosis and ulceration at injection site. No teratogenicity noted after SC in **mice** and **rabbits**, but offspring had below normal birth weights.

SC, LD_{50}, male **rat**: 761 mg/kg; female **rat**: 801 mg/kg; male **mouse**: 436 mg/kg; female **mouse**: 625 mg/kg; **dog**: > 200 mg/kg.

MURRAYA PANICULATA

= *Jiu-li-xiang*

Found in India and China.

Untoward effects: Oral or SC use causes anti-implantation and abortifacient effect in **rats**, **mice**, and **rabbits**, due to ***yuehehukene***, an alkaloid in it.

MUSCARINE

***Cholinergic, parasympathomimetic.* Toxic alkaloid in *Amanita muscaria, A. pantherina, A. verna, Russula emetica, Boletus luridus, B. Satanus, Inocybe sp.*, and some *Clitocybe sp. I. patonillardii* contained 0.037% of muscarine chloride or 150–200 times that of *A. muscaria*. Dried *C. dealbata* contains 0.15% and *C. olearia* contains about 0.001%. A *ptomaine*.**

Untoward effects: Can cause behavioral changes, delirium, miosis, dyspnea, abdominal pain, nausea, vomiting, diarrhea, lacrimation, salivation, thirst, vertigo, confusion, bradycardia, diaphoresis, paresis of oculomotor nerve, diplopia, chromatopsia, bronchospasm, headache, excessive urination, oliguria or anuria, hypotension, dehydration, liver and kidney damage, decreased tendon reflexes, decreased temperature, cardiovascular shock, and terminal convulsions. Symptoms usually occur within 15–120 min after ingestion; occasionally, as long as 10–12 h later. Death has occurred in **people** with underlying pulmonary and cardiac disease. Resembles ***pilocarpine*** (q.v.) in its physiological effects.

Antidiuretic effect in **goats**.

LD_{50}, **mouse**: 750 mg/kg; **rabbit**: 200 mg/kg; SC, **cat**: 10 mg/kg; **rabbit**: 27 mg/kg.

MUSCAZONE

Untoward effects: An ***isoxazole*** neurologic toxin from ***Amanita muscaria*** (q.v.), but less potent than ***muscarine*** (above).

MUSHROOMS

They are discussed seperately in this text; ~100 of > 5000 grown in the U.S. are poisonous. Many, such as *Agaricales, Amanita, Psilocybe, Pholiata spectabilis, Conocybe, Paneolus*, and *Stropharia* are hallucinogenic. *Boletus, Calvatia*, and *Flammulina* are carcinostatic. *Lycoperdon* is narcotic and also caused respiratory illness in eight people (16–19 years) at a party who inhaled and chewed them. *Agaricus bisporus, A. campestris*, and *Paxillus involutus* are edible. *Gyromitra sp.*, and *Pleurotus ostreatus* can be edible or poisonous. *Crotonarius* are poisonous. *Amanita phalloides* has caused many deaths when ingested accidentally or by criminal intent (see *Morchella*).

Untoward effects: One of the most common poisons in **man**. During a 3-year period in Poland, 1050 **people** were poisoned by them and 73 died. Drinking of ***Kombucha tea*** made by incubating the ***Kombucha*** mushroom in sweet black ***tea*** was associated with unconsciousness and illness in two **patients**. Allergic alveolitis or "**mushroom-worker's** lung" or "**mushroom-picker's** lung" is, primarily, due to inhalation of fungal spores in moldy compost. **They** also develop a slight dermatitis and conjunctivitis from ***ammonia*** (q.v.) vapors. **Goats** and **snails** eat large quantities of many poisonous mushrooms without any ill-effects. The toxin can be indirectly transmitted by **goat *milk*** and to **those** eating ***snails***. ***Disulfiram***-like reactions when some species are ingested with ***alcohol***. Consumption of one bottle of ***beer*** after a dinner of **"*inky-cap*"** mushrooms led to ***disulfiram***-like reaction. A 40-year-old **male**, who had previously eaten *Catharellus floccosus*, *Clavaria formosa*, and **Boletus** *variipes* without ill-effect, became toxic from them after administration of ***hydroxychloroquine***. Home-canned mushrooms have been a source of *Clostridium botulinum*. In 1989, staphylococcal food poisoning caused many outbreaks in **people** in the U.S. that consumed mushrooms from the PRC. **Patients** with ***chromate*** dermatitis need to avoid mushrooms, to avoid a recurrence of **their** lesions. Acute adverse gastric effects, nausea, salivation, vomiting, abdominal pain, followed by liver and renal pathology from toxic mushrooms.

"There are old mushroom hunters, there are bold mushroom hunters, but there are no old, bold mushroom hunters."

Bufotenine (q.v.), ***psilocybin*** (q.v.), ***ibotenic acid*** (q.v.), ***muscarine*** (q.v.), ***muscazone*** (q.v.), and ***amanitins*** (q.v.) have been found in toxic mushrooms.

MUSK MELON

= *Cucumis melo* = *Magje*

Untoward effects: Oral ingestion by **dogs** of 2 g/kg/3 days of powdered seed led to marked diuretic and chloruretic effects; increasing the dosage to 5 g/kg/3 days leads to toxicity with gradual decrease in urine output, which was dark red or dark brown and turbid. Crystalluria, diarrhea, and progressive anorexia noted. Post-mortem of one that died 13 days after the start of treatment showed chronic nephritis and hepatic cirrhosis.

MUSK XYLOL

= *Musk Xylene*

Use: As a fragrance in soaps, detergents, cosmetic creams and lotions, and perfume. Approved for food use in Europe. ~75 tons used annually in the U.S.

Untoward effects: Mild irritation by 5% in petrolatum after a 48-h closed-patch test on **human** volunteers. No sensitization by the same concentration on 25 **volunteers**.

Minimum lethal dose, skin **human**: 5 mg/48 h.

Not irritating when applied full-strength for 24 h under occlusion on intact or abraded **rabbit** skin.

LD_{50}, **rat**: > 10 g/kg; acute dermal, **rabbit**: > 15 g/kg.

MUSSELS

= *Mytilus sp.*

Untoward effects: ***Saxitotoxin***, a neurotoxin, is produced by ***Gonyaulaux***, the red tide organism, and concentrated by **shellfish** during warm weather (May-October in California waters). The **shellfish** eating it can then poison **man**. Symptoms occur within 30–120 min after ingestion and include paresthesia of mouth and limbs with tingling, itching, and numbness; paralysis, giddiness, difficulty in swallowing, and loss of voice. Severe cases develop respiratory paralysis and death within 6–10 h. Outbreaks have also occurred around the British Isles and Spain, where they have also killed **birds**, **ducks**, and **sand eels**. In December 1987, about 150 **Canadians** became sick after eating mussels that contained high levels of ***domoic acid*** (q.v.). Of these, 4 died and 12 **survivors** had permanent memory loss, similar to that of Alzheimer's disease. Post-mortem revealed neuron damage in the hippocampus, the brain area involved with memory.

A mussel poison has the highest toxicity for **mice** of many known toxins. Zebra mussels, small freshwater mollusks, have spread from Russia to the Great Lakes and U.S. rivers, where they clog the intake systems of power plants, etc.

MUSTARD

Brassica juncea = Chinese Mustard; B. nigra = common Black Mustard; Sinapis alba = B. alba = senvre = White Mustard = Yellow Mustard. Broccoli, cabbage, charlock, kale, kohlrabi, rape, and turnips are also in the Brassica family.

The seeds and oil contain *sinigrin* (q.v.), leading to *allylisothiocyanate* (q.v.).

Untoward effects: Pungent odor and irritant topically (oils and mustard plasters). ***Brassica*** has caused photodermatitis. Both genera have increased pancreatic secretions and peristalsis, and can cause emesis. Irritant effects have been utilized as rubefacients, counterirritants, and as a deterrent to ***glue***-sniffing by its addition (which causes eye, nose, and sinus irritation). ***Isothiocyanates*** are in mustard oil. Among the ***Brassica***, mustard seeds often cause poisonings in **children**. Allergic reactions to mustard in foods. A 13-year-old **female** developed a severe systemic reaction (flushing, angioedema, urticaria, localized pruritus, dyspnea, and severe hypotension), apparently as a result of a ***sodium benzoate*** (q.v.) preservative. Confirmed by testing. Contact dermatitis has been reported. When the seeds are ground or crushed without prior heat treatment, an enzyme in the seed, myrosinase, catalyzes the ***sinigrin*** and ***sinalbin*** in them, to form ***allylisothiocyanate***, commonly known as mustard oil, which is the key irritant and flavorant in mustard. Mustard greens are high in ***oxalates***. Disagreeable taste and unpleasant odor in ***milk***, milk products, and meat of **animals** that grazed it.

Mustard seed cake from ***Chinese Mustard*** poisoned 99 **cattle** from 14 herds in Denmark, 27 of which died or had to be killed. The minimum lethal dose of the cake was 300 g, which contained 2.5 g of the oil. ***Kohlrabi*** fed with it contained myrosinase, which liberated the ***allylisothiocyanate*** from the cake. TD of the latter for **cattle** was 2–3 mg/kg, and the LD was 5–20 mg/kg. Similar reports have come from Canada. **Horses** and **pigs** have also been adversely affected by seeds.

Sinapis arvensis = Brassica sinapis = Charlock = Moutard des Champs = Wild Mustard, a common weed in cornfields, etc. Its aeroallergens can cause rhinitis, asthma, and hypersensitivity pneumonitis in sensitive **people**. Hay containing large amounts of its seeds eaten by **cattle** leads to salivation, incoordination, collapse, and death from their ***glucosinolate*** content.

MUSTARD GAS

= *Blistering Gas = Dichlorodiethyl Sulfide = Dichloroethyl Sulfide = H = HD = HS = King of the Battle Gases = King of War Gases = Lost = Senfgas = Sulfur Mustard = Yellow Cross Shells = Yperite*

Use: In chemical and biological warfare.

Untoward effects: Severe tissue vesicant, irritant, affecting eyes, skin (erythema, edema, vesicles, ulcers, and necrosis - **Negroes** less sensitive than **whites**, **blondes** most sensitive), respiratory tract (including pulmonary edema), and gastrointestinal tract. Some symptoms are delayed. Causes oat-cell lung cancers. During World War II, an explosion of stored mustard gas in Naples, Italy harbor caused the deaths of many exposed **soldiers** from atrophy of **their** immune systems and loss of bone marrow cells. This led to the development of ***nitrogen mustard*** for the treatment of Hodgkin's disease. A **male nurse** attempted suicide by ingesting 5 ml. When the **patient** vomited, the attending **physicians** and **nurses** were injured by the chemical in the vomitus. The military classifies it as a "lethal and incapacitating agent". The density of the vapors is 5.5 times that of air and it can contaminate an area for days or weeks. The vapors can penetrate clothes, ***leather***, and even ***rubber***. A concentration of 0.15 mg/l of air is lethal in 10 min and within 1 h with concentrations of as little as 1 µg/l caused a severe conjunctivitis. Skin irritation is delayed for 2–10 h after exposure, with maximum irritation reached by the 3–5th day. Occasionally death resulted from asphyxia, secondary to severe swelling of the glottis induced by vapors. Iraq dropped it on **Kurds** in Halabja in 1988, killing ~5000 **people**. In the 1970s, the U.S. military destroyed ~6 million lbs by incineration.

Inhalation LC_{50}, **human**: 1.5 g/m^3/min; LC_{LO}, 23 ppm/10 min.

Reactions in **horses**, **mules**, **dogs**, **rabbits**, and **pigs** are similar to those in **man**. The **camel** is relatively tolerant to it. Inhalation or IV leads to increase in lung tumors in **mice**, and SC leads to local fibrosarcomas or sarcomas. **Animals** have been poisoned by eating contaminated feeds leading to profuse salivation, lacrimation, conjunctivitis, rhinitis, anorexia, pain in the mouth and pharynx, tachycardia, tachypnea; necrotic lesions of the mouth, pharynx, esophagus, and stomach; and local skin burns. Death can occur after a few days.

Inhalation LC_{50}, **rat**: 420 mg/m^3/2 min; LD_{50}, skin **rat**: 9 mg/kg; **mouse**: 92 mg/kg; **dog** and **guinea pig**: 20 mg/kg.

MU TONG

A Chinese herbal remedy.

Untoward effects: Nephrotoxicity is due to its ***aristolochic acid*** (q.v.) content. Toxicity varies when it is derived from different plants.

MYCOPHENOLATE MOFETIL

= *CellCept* = *MMF* = *RS 61,443*

***Immunosuppressant*. Hydrolysed *in vivo* to *mycophenolic acid*.**

Untoward effects: Dose-related incidence of diarrhea, leukopenia, and cytomegalovirus tissue-invasive infection. Gastrointestinal hemorrhage has occurred.

Drug interactions: Plasma levels increased by ***tacrolimus***.

MYLABRIS CICHORII

= *Chinese Blistering Flies* = *Chinese Cantharides*

***Vesicant*. Contains 1.0–3.5% *cantharidin* (q.v. for toxic effects).**

MYOPORUM sp.

Untoward effects: ***M. acuminatum*** and ***M. tetrandrum*** = ***Boobialla***, ***M. deserti*** = ***Ellangowan Poison Bush*** = ***Turkey Bush***, ***M. laetum*** = ***Ngaio Tree*** = ***Transparente***, and ***M. serratum*** in South Pacific areas contain hepatotoxic furanoid sesquiterpenes, including ***ngaione*** (q.v.). ***M. laetum*** is a common shrub in Uruguay, where it is used as a fence. It contains emenogogues and is used to induce abortions in **women**. An 11-month-old **child** died after ingestion of it lead to persistent vomiting, convulsions, coma, and death after a few days. In New Zealand, its toxin, ***ngaione*** (q.v.), poisons **horses** and **ruminants**. ***M. deserti***'s sesquiterpene, ***dehydromyodesmone***, produces periportal necrosis in the liver of **calves**. Hereford X Ayrshire (60–100 kg) **calves** given 0.08–0.12 ml/kg became dull and anorexic within 24 h; by 48 h, they had no desire to move and 2/4 had yellow conjunctiva; one died on day 3, and another on day 5. Post-mortem revealed periportal necrosis and periportal sinusoids distended with blood. The liver congestion in **calves** is also noted with ***M. acuminatum***. ***M. deserti*** = ***Turkey Bush*** has caused midzonal necrosis of the liver in **mice**; submandibular edema, jaundice, and marked hypoglycemia in **sheep**. After 48 h, post-mortem revealed extensive hemorrhagic areas and early cytolytic necrosis of the liver. ***M. laetum*** has caused liver damage, icterus, photosensitivity, and facial edema in **cattle**, **sheep**, **horses**, and **pigs**. ***M. tetrandrum*** leaves contain 0.25–0.5% of the sesquiterpenes, most of which are ***dehydrongaion***, leading to extensive hemorrhagic centrilobular liver necrosis in **calves** and centrilobular or periportal liver lesions, occasionally with pulmonary edema, in **sheep**.

MYRCENE

Use: Fragrance in soaps, detergents, cosmetic creams and lotions, perfumes, and foods. An acyclic terpene.

Untoward effects: No irritation or sensitization of 4% in petrolatum on **human** volunteers.

Full-strength to intact or abraded **rabbit** skin for 24 h under occlusion was moderately irritating.

LD_{50}, **rat** and acute dermal, **rabbit**: > 5 g/kg.

MYRISTICIN

An active aromatic principle in *carrots* (q.v.), *dill* (q.v.), *mace*, *nutmeg* (q.v.), and *parsley* (q.v.).

Untoward effects*:** Hallucinogen (210 mg/20 g ***nutmeg). Metabolization in the body to ***elemicin*** and ***amphetamine*** compounds (***MMDA*** and ***TMA***). Central nervous system toxicity after ingestion of 4–5 g, leading to headache, dizziness, facial flushing, edema of face and eyelids, thirst, disorientation, stupor, numbness of limbs, abdominal pains, nausea, vomiting, oliguria, cold sweats, and clammy skin. Can be fatal in **infants** who eat as little as two ***nutmegs***, with symptoms resembling ***alcohol*** intoxication.

TD_{LO}, **human**: 5.7 mg/kg.

LD_{LO}, **cat**: 40 mg/kg.

MYRRH

Untoward effects*:** **People** with eczematous reactions to ***benzoin (q.v.) can cross-react with myrrh gum. Can cause catharsis after ingestion.

N

NABAM

= *AAgrunol* = *Chem Bam* = *D-14* = *Disodium Ethylene Bisdithiocarbamate* = *Dithane* = *DSE* = *ENT 9,106* = *Groningen* = *Nabasan* = *Parzate* = *Spring-Bak*

***Fungicide.* A *carbamate.* Related to *maneb* (q.v.) and *zineb* (q.v.). A cholinesterase inhibitor.**

Untoward effects: Moderately toxic to **vertebrates**. Minor irritation of skin and eyes of agricultural **workers**. Maximum daily intake of 5 µg/kg (WHO). Symptoms usually develop rapidly.

LD_{LO}, **human**: 50 mg/kg.

In **birds**, ataxia, tremor, tachypnea, myasthenia, salivation, miosis, tenesmus, diarrhea, and piloerection. In feed for **quail** at 0.5%, causing death in 11%.

LD_{50}, **rat**: 395 mg/kg; **mouse**: 580 mg/kg; **bullfrog**: 420 mg/kg; **pheasants**: 707 mg/kg; **Japanese quail**: 2,120 mg/kg; **goats**: > 800 mg/kg, 100% lethality **guppies**: 10 mg/l.

NABILONE

= *Cesamet* = *Lilly 109,514*

***Antiemetic.* A *cannabinol* derivative. A synthetic analog of *tetrahydrocannabinol*.**

Untoward effects: Euphoria (16%), somnolence, dry mouth, dizziness, marked hypotension with 4 or 5 mg dose, hallucinations (> 2%), and behavioral disturbances.

Unexpected neurological toxicity in **dogs** at high dosage (≥ 0.5 mg/kg/day/> 2 months) causes seizures associated with sudden death in some. No apparent brain pathology. Half-life in **dogs** may be ~20 h and 6 h in **humans**.

NABUMETONE

= *Arthaxan* = *Balmox* = *BRL-14777* = *Consolan* = *Nabuser* = *Relafen* = *Relifen* = *Relifex*

Analgesic, anti-inflammatory.

Untoward effects: Abdominal pain, nausea, gastrointestinal bleeding, diarrhea, indigestion, occasionally ulcerogenicity, photosensitivity, pseudoporphyria with bullous lesions, dizziness, headache, edema, rash, tinnitus, nightmares, membranous nephropathy, interstitial nephritis, and pulmonary fibrosis reported. Cardiac arrest in 70-year-old **female**, due to 500 mg twice daily/5 weeks, plus renal insufficiency and high intake of ***potassium***. **She** recovered. Significant international normalized ratio increase and hemarthrosis in 72-year-old **maie** after 1500 mg/day added to a long-term stable regimen of ***warfarin*** 34 mg/week.

NADOLOL

= *Anabet* = *Corgard* = *Solgol*

Antihypertensive, antianginal. β-Adrenergic blocker.

Untoward effects: ***Blood***: Agranulocytosis, thrombocytopenia, and non-thrombocytopenic purpura.
Cardiovascular: Bradycardia (< 40/min – 2%), Raynaud disease (2%), cardiac failure, hypotension, and rhythm or conduction disturbances (1%). Rarely, heart-block.
Central nervous system: Depression, fatigue, dizziness, catatonia, visual problems, hallucinations, disorientation, short-term memory loss, and headache.
Gastrointestinal: Abdominal discomfort, nausea, vomiting, flatulence, indigestion, anorexia, constipation, raised liver enzymes, ischemic colitis, and mesenteric arterial thrombosis.
Miscellaneous: Rash, pruritus, decreased libido, facial edema, slurred speech, increased weight, nasal stuffiness, blurred vision, sweating, tinnitus, bronchospasm, and dry mouth, eyes, and skin in 0.1–0.5% of **patients**. A 62-year-old **male** taking 40 mg/day developed infiltrative dermatitis and alopecia. May alter blood ***glucose*** serum levels. Can cause high levels in breast milk.

Drug interactions: May increase ***salicylate*** blood levels with ***aspirin***. May potentiate anaphylactic reactions when taken with oral ***penicillin V***. Can antagonize bronchodilator and hypoglycemic action of other drugs. Exaggerates hypotension of ***general anesthetics***.

Embryotoxic and fetotoxic to **rabbits** at 100 and 300 mg/kg orally.

NAFAMOSTAT MESYLATE

= *FUT 175* = *Futhan*

Protease inhibitor.

Untoward effects: A 77-year-old **male** treated with a continuous IV infusion of 150 mg/day for acute pancreatitis developed hyperkalemia.

NAFARELIN ACETATE

= *Nasanyl* = *RS 94,991-298* = *Synarel*

Synthetic gonadotropin releasing hormone agonist.

Untoward effects: During intranasal therapy, significant increase in bone resorption after 3–6 months of treatment.

A 38-year-old **female** using 400 μg/day intranasally developed severe hyperkalemia. The nasal spray has also decreased short-term memory, decreased libido, and caused impotence, leukopenia, decrease in breast size, hot flashes, ***estrogen*** deficiency, vaginal dryness, depression, headache, weight gain or loss, insomnia, emotional instability, dyspnea, chest pain, body odor, oily skin, rashes, seborrhea, hives, pruritus, muscle pain, excessive growth of pubic hair, ovarian cysts, occasional nasal irritation, acne, and increased serum triglycerides and ***cholesterol***. May fail to totally suppress ovulation and may be teratogenic.

NAFENOPIN

= *SU 13,437*

Hypolipidemic.

Untoward effects: Nausea, vomiting, diarrhea, muscle pain and tenderness in calves, urticaria, and transitory increase in SGOT in **humans**.

Hepatomegaly in treated **mice** and liver nodules in treated **rats** caused cessation of clinical tests in the U.S.

NAFRONYL OXALATE

= *Citoxid* = *Di-Actane* = *Dusodril* = *EU-1806* = *LS-121* = *Praxilene*

Vasodilator.

Untoward effects: Occasional severe thrombophlebitis after IV. Oral use has induced acute esophageal ulceration and hepatotoxicity. IM, IV, and oral use have caused insomnia, nausea, epigastric pain, diarrhea, headache, dizziness, vertigo, decreased concentration, and allergic skin reactions.

NALBUPHINE

= *EN 2234A* = *Nubain*

Narcotic, analgesic. IV use.

Untoward effects: Sedation (36%), sweaty or clammy feeling (9%), nausea (6%), dizziness or vertigo (5%), dry mouth (4%), headache (3%), abuse potential; paradoxical nightmares and pain in 25-year-old **male** receiving 30 mg every 2 h, but not with 15 mg every 2 h. Nervousness, depression, confusion, crying, hallucinations, dysphoria, numbness, pulmonary edema, hyper- or hypotension, tachy- or bradycardia, bitter taste, dyspnea, asthma, itching, urticaria, flushing, blurred vision, urinary urgency, and anaphylaxis have occasionally been reported. Perinatal use during labor associated with cyanosis, hypotonia, bradycardia, and bradypnea in **infants**.

NALIDIXIC ACID

= *Betaxina* = *Dixiben* = *Eucistin* = *Innoxalon* = *Nalidicron* = *Nalidixinic Acid* = *Nalitucsan* = *Nalixan* = *Narigix* = *NegGram* = *Negram* = *Nevigramon* = *Nicelate* = *Nogram* = *Poleon* = *Specifin* = *Uriben* = *Uriclar* = *Uralgin* = *Urodixin* = *Uroman* = *Uroneg* = *Uropan* = *Win-18320* = *Wintomylon*

Antibacterial. A fluoroquinolone.

Untoward effects: Nausea, vomiting, gastrointestinal bleeding, anorexia, rash, pruritus, and urticaria are the most common.

Central nervous system: Muscular weakness, headache, drowsiness, dizziness, vertigo, paresthesia, myalgia, excitement, restlessness, insomnia, giddiness, confusion, hallucinations, and visual disturbances (blurred vision, diplopia, photophobia [1.5%], brighter color perception, blue tint to vision, esotropia, poor accomodation, overbrightness of light, increased intraocular pressure, temporary loss of vision [30 min to 72 h], nystagmus, and papilledema). Overdoses lead to convulsions. Bulging fontanel with widening of skull sutures, irritability, and vomiting from increased intracranial pressure and palsy of the 6th cranial nerve. Peripheral neuropathies with numbness and tingling or burning sensations of the extremities have occurred.

Miscellaneous: Thrombocytopenia, eosinophilia, leukopenia, purpura, hemolytic anemia (with or without glucose-6-phosphate dehydrogenase deficiency) – even in a **newborn child**, due to its excretion into breast milk. False positive or spuriously high SGOT, blood urea nitrogen, ***thymol*** turbidity, blood ***glucose***, and urinary ***glucose*** results with ***copper*** reduction tests (***Clinitest***®). Interferes with urinary ***17-ketosteroid*** determinations; angioneurotic edema, asthma, and interstitial pneumonitis. Its effectiveness is decreased by ***antacids*** and ***urinary alkalinizers***. A hemolytic crisis in a 20-year-old **male** intermittently exposed to its dust in an unventilated area of a pharmaceutical plant. Intermittent therapy in a **patient** led to fatal autoimmune hemolytic anemia. Rarely, cholestatic jaundice and transient central vision disturbances. Overdoses cause hyperglycemia, glycosuria, convulsions, abnormal behavior, lactic acidosis, and coma.

Skin: Photodermatitis (bullous eruptions after sun exposure, even after ceasing treatment for several months [mostly in **female**], petechiae, milia, and skin fragility), toxic epidermal necrolysis, and pseudoporphyria cutanea tarda. Diffuse erythema with burning and pruritus after 500 mg tid for 1 day only in 71-year-old **female**. An lupus erythematous-like reaction of skin in 70-year-old **female** after 1 g/day/28 days.

Therapeutic serum levels are ~10–30 mg/l and toxic levels reported to be 40–50 mg/l. TD_{LO}, **woman**: 160 mg/kg during 2 days.

Drug interactions: Potentiates anticoagulant effects of ***acenocoumarol***, ***dicumarol***, and ***warfarin*** by displacing them from serum ***albumin***-binding sites. ***Anticholinergics, antihistamines, antimalarials, cardiac depressants, narcotic analgesics, phenothiazines, probenecid, sympathomimetics***, and ***tricyclic antidepressants*** increase its effect in urinary infections and increase the risk of systemic toxicity. ***Barbiturates, clofibrate, ethacrynic acid, mefenamic acid, nitrofurantoin, phenylbutazone, salicylates, sulfonamides, sulfonylureas***, and ***thyroxine*** decrease its effect in urinary infections. ***Aluminum, calcium***, and ***magnesium*** ions in ***antacids*** and ***alkalinizing agents***, such as ***sodium bicarbonate***, increase its urinary excretion and may decrease its absorption. ***Acidifying agents***, such as ***ammonium chloride***, potentiate its effect by decreasing urinary excretion. ***Alcohol*** has caused toxic effects ingested in combination with it.

No adverse effects noted on the fetuses of **monkeys**. ***Fenbufen*** increases its serum, brain, and cerebrospinal fluid levels in **rats**.

LD_{50}, **rat**: 1.35 g/kg; IV, 250 mg/kg; **mice**: 572 mg/kg; IV, 176 mg/kg; SC, 500 mg/kg.

NALMEFENE

= *Nalmephene* = *Nalmetrene* = *ORF 11,676* = *Revex*

***Narcotic antagonist*. An analog of *naltrexone*. Veterinarians have used it successfully for behavior modification in dogs and horses.**

Untoward effects: Nausea and vomiting may occur, especially on high dosage. A 2 mg dose in 52-year-old **female** caused hallucinations, but 0.5 mg twice daily did not. Tachycardia, fever, dizziness, headache, chills, flushing, and pain also reported.

NALORPHINE

= *Allorphine* = *N-Allylnormorphine* = *Anarcon* = *Antorphine* = *Lethidrone* = *Nalline* = *NANM* = *Norfin*

***Narcotic antagonist*. For parenteral use.**

Untoward effects: Use has essentially been discontinued, due to a high incidence of undesirable and, occasionally, bizarre psychotic effects. Ventricular fibrillation, drowsiness, dizziness, dysphoria, disorientation (20%), daydreams and visual hallucinations (20%), pallor, respiratory depression, nausea, diaphoresis, anxiety, and panic attacks. Crosses the placenta and becomes a possible hazard to a ***narcotic***-addicted **fetus**.

SC, TD_{LO}, **human**: 71 μg/kg.

Causes emesis, salivation, and hyperirritability in ***morphine***-dependent **monkeys**, but not in controls.

SC, LD_{50}, **rat**: 474 mg/kg; **mouse**: 500 mg/kg; IV, LD_{50}, **rat**: 226 mg/kg; **mouse**: 127 mg/kg; SC, LD_{LO}, **monkey**: 400 mg/kg; IV, 100 mg/kg.

NALOXONE

= *Nalone* = *Narcan* = *Narcanti*

***Narcotic antagonist*. Possesses 10–15 times the potency of *nalorphine* in many species.**

Untoward effects: Rapid IV use leads to nausea, vomiting, tachycardia, and hypertension. Pulmonary edema and arrhythmias are rare. Some sedation in **non-addicts** and can precipitate a withdrawal syndrome, which is self-limited (15–60 min) because of its short half-life. Ventricular fibrillation and cardiac arrest in two young **females** postoperatively after 0.4 mg and two doses of 0.2 mg IV. IM use (0.8 mg) in a 42-year-old **male** ingesting 40–60 capsules/day of ***propoxyphene*** and ***aspirin*** (***Darvon 65***®), and a 26-year-old **male** consuming 20 capsules/day/7 years caused uncontrollable yawning, profuse diaphoresis, lacrimation, rhinorrhea, myalgia, severe abdominal cramps, nausea, hypertension, mydriasis, and hyperthermia. Oral, IV, and SC in **others** led to mild depression, lethargy, irritability, transient muscle twitches, tachycardia, constipation, paresthesias, increased libido, thrombocytopenia, and slight miosis.

SC, LD_{50}, newborn **rats**: 260 mg/kg; **mouse**: 368 mg/kg.

NALTREXONE

= *Antaxone* = *Celupan* = *EN-1639A* = *Nalorex* = *ReVia* = *Trexan* = *UM-792*

Narcotic antagonist.

Untoward effects: Abdominal cramps, nausea, diarrhea, restlessness, agitation, anxiety, bone or joint pain, and myalgia, often as part of a ***narcotic*** withdrawal syndrome. Headache, dizziness, fatigue, heartburn, diarrhea, weight loss, anorexia, chills, increased ***insulin*** requirement, mydriasis, rhabdomyolysis, piloerection, delayed ejaculation and decreased potency, and yawning. Possibly, hepatotoxic.

Drug interactions: Lethargy and somnolence in **patients** also taking ***thioridazine*** (q.v.).

Pruritus reported in a **dog** with tail-chasing behavior. Drowsiness and withdrawal from owner in a **dog** treated for acral lick dermatitis. Low dosage (1 mg/kg) in **rats** delayed spontaneous motor and sensorimotor behavior, but high dosage (50 mg/kg) accelerated them.

NAMOXYRATE

= *Namol Xenyrate* = *W 1760A*

Antiarthritic.

***Untoward effects*:** Gastrointestinal irritation, heartburn, nausea, light-headedness, and drowsiness were the main complaints.

LD_{50}, **mice**: 1,150 mg/kg.

NANDROLONE

= *19-Nortestosterone*

***Anabolic agent*. Many esters are used. Usually IM.**

Untoward effects*:** Intrahepatic cholestasis. Can cause false positive urinary ***calcium tests.

Cyclohexane propionate form = ***Sanabolicum***. Has caused beard growth in three **females** and acne in **another**. High dosage has decreased spermatogenesis in **men**.

Decanoate form = ***Deca Durabol*** = ***Deca Durabolin*** = ***Deca Hybolin*** = ***Hybolin Deconate*** = ***Retabolin***. In **men**, prepubertal phallic enlargement and increased frequency of erections; postpubertal induces testicular atrophy, oligospermia, impotence, priapism, gynecomastia, epididymitis, and bladder irritation. In **females** causes hirsutism, **male** pattern baldness, increased libido, clitoral enlargement, hypertension, and deepening of voice. Both **males** and **females** develop nausea, increased or decreased libido, acne in **females** and prepubertal **males**, habituation, excitement, sleeplessness, chills, and premature closure of epiphyses in **children**, urticaria and induration at injection site, furunculosis, and alterations of many laboratory results.

Phenpropionate form = ***Activin*** = ***Durabol*** = ***Durabolin*** = ***Nandrolin*** = ***Strabolene*** = ***Superanabolin***. Marked virilization with facial hair, voice changes, and coarser skin in ~30% on 100–300 mg/week.

p-Hexyloxyphenylpropionate form = ***Anador*** = ***Anadur***. IM use in five healthy **men** 21–25 years with 100 mg/week/3 weeks, and then, 200 mg/week/10 weeks led to azoospermia in 7–13 weeks after start of treatment, which persisted for 4–14 weeks after last injection.

Drug interactions*:** Potentiates ***oral anticoagulants.

NAN-LIEN CHUIFONG TOUKU-WAN

A Chinese over-the-counter product available in the U.S.

Untoward effects*:** Has contained ***diazepam (q.v.).

NAPHAZOLINE

= *Ak-Con* = *Albalon* = *Clera* = *Coldan* = *Iridina Due* = *Naphcon* = *Niazol* = *Opcon* = *Privine* = *Rhinantin* = *Rhinoperd* = *Sanorin* = *Sanorin-Spofa* = *Strictylon*

α-Adrenergic agonist, topical vasoconstrictor, sympathomimetic.

***Untoward effects*:** Topical intranasal use often causes severe rebound congestion, paralysis of nasal cilia, and anosmia. Occasionally stinging sensations, sneezing, nausea, headache, and allergic reactions. Many accidental poisonings reported in Japan from the ingestion of dermatologic preparations containing it causing somnolence, hypertension, and bradycardia. Arrhythmias, transient hypertension with rebound hypotension, bradycardia, agranulocytosis, diaphoresis, and drowsiness have been systemic effects. Deep sleep, hypothermia, and coma can easily occur in young **children**.

Drug interactions*:** Will react with ***aluminum in an atomizer.

SC, LD_{50}, **rat**: 385 mg/kg; **mouse**: 514 mg/kg; **rabbit**: 950 μg/kg.

NAPHTHA

High-density naphthas = *Heavy Naphtha* = *Ligroin* = *Mineral Spirit* = *Petroleum Spirits* = *Solvent Naphtha* = *Stoddard Solvent* = *Texsolve S* = *Varnish Naphtha* = *Varsol 1*

Low-density naphthas = *Benzin* = *Dry Cleaner's Naphtha* = *Light Naphtha* = *Naphtholite* = *Naphthol Spirits* = *Petroleum Benzin* = *Petroleum Naphtha* = *VM & P Naphtha*

Untoward effects*:** Vapors cause eye, skin, and nose irritation; light-headedness, drowsiness, nausea, vomiting, ataxia, cyanosis, weak pulse, convulsions, and coma. Ingestion causes many of the same symptoms. Hemor-rhagic pneumonitis within 2 h after 3 ml IV in a suicide attempt. Homicide due to IV of 25 ml of ***Energine®, a petroleum distillate spot remover reported. The culprit was apprehended. Aspiration can cause pneumonia. Peripheral neuropathy, secondary to inhalation abuse by **adolescents**. Ingestion has also caused giddiness, mydriasis, flushing, dyspnea, and delirium, and is a serious problem in **children**.

LD_{LO}, **human**: 50 mg/kg. OSHA TWA for ***coal tar naphtha*** is 100 ppm; immediate danger to life or health level is 1,400 ppm.

Continuous exposure of **dogs**, **monkeys**, **rabbits**, **rats**, and **guinea pigs** to the same concentration of vapors leads to toxicity only in **guinea pigs**. Subacute inhalation trials in **rats** and **monkeys** produced hematological abnormalities, decreased body weight gain, and mucous membrane irritation. Eye irritation, salivation, tremors, mydriasis, incoordination, and convulsions at high concentration in **rats**, **cats**, and **dogs**.

LD_{50}, **rat**: 500 mg/kg (***petroleum naphtha***), 100 mg/kg (***coal tar naphtha***).

NAPHTHALENE

= *Naphthalin* = *Tar Camphor* = *White Tar*

Use: In repellent moth balls, insecticides, vermicides, dusting powders, and in chemical manufacturing. U.S. annual production ~350 million lbs.

Untoward effects: Acute hemolytic toxicity and anemia after accidental ingestion by **infants**, especially **those** with glucose-6-phosphate dehydrogenase deficiency. Similar cases reported due to diapers stored with naphthalene-containing mothballs. Methemoglobinemia, hemoglobinuria, and jaundice also reported in glucose-6-phosphate dehydrogenase-deficient **infants**, and has been called "full-moon disease" in Hong Kong. The birth of a Chinese **child** is celebrated at the first full moon after the **baby's** 30th day of life. The **child** is usually dressed in heirloom robes which, in modern Hong Kong, are frequently impregnated with naphthalene. The **child** is often found sick and jaundiced after the party, if it is glucose-6-phosphate dehydrogenase-deficient. It is readily absorbed through the skin and **infant** deaths have occurred from such innocent exposures. Acute hemolytic anemia in a 21-month-old **male** after ingesting only part of one naphthalene mothball, required several transfusions and 6 days of hospital treatment before recovery. Often scattered on lawns and around shrubs to discourage **animals**, which presents a great risk for **children**. Severe hemolytic anemia, methemoglobinemia, Heinz-Ehrlich bodies, severe poikilocytosis, anisocytosis, and coma in 4-year-old **male** after inhalation therapy with ***Vaporin*** (containing naphthalene), as a cough remedy. About 1 million "Komfy Kid®" dolls from Taiwan were sold in the U.S. They contained ***phenol*** and naphthalene. The aroma of mothballs of 100% naphthalene permeated an elderly **woman's** home for years. Recurrence during the winter and resolution during the summer explained **her** clinical history of pain, paresthesias of hands and feet, and tremors so severe **she** couldn't move a limb or stand without losing **her** balance. A **woman** whose apartment contained 20 ppb, as a result of ~300–500 mothballs in **her** apartment, had **friends** who became ill (headache, nausea, and vomiting) while visiting **her** place. **People's** sensitivities differ. A 6-year-old **child** died after ingesting 2 g over 2 days and a **16-year-old** survived 6 g. Asthmatic allergic reactions in a 29-year-old **female** were attributed to **her** use of it in the ***Rinse 'n Vac® Carpet Cleaning System***. Toxic hemolytic anemia in pregnant 21-year-old **female** with pica for toilet air freshener blocks whose main constituent was naphthalene. **She** ingested 1–2 blocks/week throughout **her** pregnancy, without any apparent harm (renal or hepatic) to **herself** – both labor and **infant** were normal. Cataractogenic. A German report indicates a possible relationship between its regular use as a cleaning agent with malignant tumors. ***Naphtha*** and ***limestone*** were used for "Greek Fix" in the siege of Acre. Its key metabolite acts as a quinone, another, as naphthols, which are chemical hemolysins. Can color urine brown to a deep brown, and cause painful urination, eye and skin irritation, abdominal cramps, nausea, vomiting, diarrhea, headache, confusion, tremors, fatigue, kidney and liver damage, jaundice, diaphoresis, convulsions, coma, and death by respiratory failure. Aplastic anemia has also been reported after several years exposure to its crystals. OSHA and National Institute for Occupational Safety and Health set upper limits of 10 ppm TWA. Immediate danger to life or health: 250 ppm.

LD_{LO}, **child**: 100 mg/kg; **adult**: 50 mg/kg.

Causes an offensive taste in **fish** exposed to it in polluted waters. **Cats** and **dogs** have been adversely affected with nausea, vomiting, depression, and hemolysis. Testicular atrophy in **rats** and cataract formation in **mice**.

LD_{50}, **rat**: 1.78 g/kg.

NAPHTHALENE DIISOCYANATE
= *Naphthylene Diisocyanate*

Untoward effects: Eye, nose, throat, and respiratory tract irritant. Chest pain, coughing, dyspnea, and asthma. National Institute for Occupational Safety and Health sets exposure limits of 0.005 ppm TWA and 0.02 ppm/10 min.

LC_{50}, > 10 g/kg.

1-NAPHTHOL
= *α-Naphthol*

Use: In manufacturing dyes, chemical intermediates, and synthetic perfumes.

Untoward effects: Systemic absorption from intact skin occurs in **man**. Used as a ***rubber*** additive. Has contained ***1-naphthylamine*** (q.v.) as an impurity, which is carcinogenic. Destroys erythrocyte stability in **people** sensitive to ***primaquine*** (q.v.).

LD_{50}, **rat**: 2.6 g/kg.

2-NAPHTHOL
= *β-Naphthol*

Use: As an antiseptic; an anthelmintic. Danish ointment contains 5.5%.

Untoward effects: Use as a ***rubber*** additive revealed it has contained a carcinogenic impurity, ***2-naphthylamine*** (q.v.). Cataractogenic and causes hemolytic anemia in ***primaquine*** (q.v.)-sensitive **individuals**. Overexposure may cause abdominal pain, diarrhea, nausea, vomiting, convulsions, nephritis, albuminuria, liver pathology, decreased blood pressure, brownish-red urine (due to

betanaphthoquinone), and occasionally paralysis after ingestion or dermal exposure.

LD_{LO}, **human**: 50 mg/kg.

LD_{50}, **rat**: 2.4 g/kg; **rabbit**: 3.8 g/kg.

NAPHTHOQUINONES

Untoward effects: Hemolytic anemia in ***primaquine*** (q.v.)-sensitive **individuals**. Cataractogenic. A photosensitizer and allergic contact dermatitis cases reported.

Drug interactions: Can potentiate ***anticoagulants***.

1-NAPHTHYLAMINE

= *1-Aminonaphthalene* = *Naphthalenamine* = *Naphthalidine* = *α-Naphthylamine*

Aromatic amine.

Use: In dye and ***1-(1-Naphthyl)-2-thiourea*** (q.v. below) manufacturing and testing for ***nitrites***. Hard to manufacture without traces of the *β*-form, which is a proven carcinogen.

Untoward effects: Hemorrhagic cystitis, and urinary tract and bladder carcinogen. A 52-year-old **female** died from hematuria and a bladder tumor after working with ***rubber*** and exposure to *α*- and *β*-forms as impurities in ***Nonox-S***, an antioxidant. Other cases of bladder tumors in 15 **men** in the same factory. Dermatitis, dyspnea, ataxia, methemoglobinuria, and dysuria have also occurred.

Its *N*-hydroxymetabolite has induced bladder cancer in **dogs** after instillation by catheter.

LD_{50}, **rat**: 779 mg/kg.

2-NAPHTHYLAMINE

= *2-Aminonaphthalene* = *β-Naphthylamine* = *2-Naphthalenamine*

***Aromatic amine.* Was used in dye and *rubber* manufacturing. No longer produced commercially or used in the U.S.**

Untoward effects: Can form in the body from ***azo dyes***, including ***2-nitronaphthalene***. It and its metabolite, ***1-amino-2-naphthol***, are carcinogenic. Four red dyes (#'s 10, 11, 12, and 13) and one yellow dye (***yellow 1***) were banned in the U.S. in the 1970s because they contained *β*-naphthylamine. ***Nonox-S*** cited above (q.v.) had ~2% of the *α*- and *β*-forms as impurities. It is a carcinogen found in ***cigarette*** smoke. **Chemical worker** exposure in the 1980s resulted in > $300 million of lawsuits. In regard to **human** bladder cancers, it is estimated that 6 months exposure to the *β*-form is equivalent to 5 years of the *α*-form. The carcinogenic effect can occur, but delayed, after only 3 months exposure. With enough exposure, ~50% of the **male workers** may develop bladder cancer. The U.S. industry then switched to ***PBNA*** or ***N-phenyl-2-naphthaleneamine*** (q.v.), which, in the **human** body, can also be metabolized to 2-naphthylamine. Has caused bladder tumors in **man**, **dog**, **monkey**, and **hamsters**, but not in the **rat**; hepatomas in **mice**; and hematuria in **cows**.

LD_{50}, **rat**: 727 mg/kg, TD_{LO}, **dog**: 3.9 g/kg during 2 years; **monkey**: 14 g/kg, during 12 years; **rabbit**: 40 g/kg, during 5.2 years; **hamster**: 365 g/kg, during 43 weeks; **rat**: 31 g/kg, during 1 year.

1-NAPHTHYLISOTHIOCYANATE

= *ANIT*

Use: As an insecticide, usually with ***pyrethrum***.

Untoward effects: When sprayed in a room, some **people** have developed a dermatitis. Potentially hepatotoxic for **workers** using or manufacturing it.

Hepatotoxic to **sheep**, **calves**, **mice**, and even the offspring of treated female **rats**. Bile duct proliferation, focal necrosis of hepatocytes, increased liver enzymes, increased ***bilirubin***, and hepatocarcinogenesis in **rats**.

LD_{50}, **rat**: 200 mg/kg; **mouse**: 245 mg/kg.

1-(1-NAPHTHYL)-2-THIOUREA

= *ANTU* = *Anturat* = *Bantu* = *Chemical 109* = *Krysid* = *α-Naphthylthiocarbamide* = *Rattrack*

Rodenticide.

Untoward effects: Ingestion of large doses (4 g/kg) causes vomiting, dyspnea, cyanosis, coarse pulmonary rales, variations in hair growth and skin pigmentation, pulmonary edema, hyperglycemia, hypothermia, and some liver damage. Bladder carcinoma in two pesticide **operators**. Lethal doses have varied between 1 and 40 g. National Institute for Occupational Safety and Health/OSHA exposure limits: 0.3 mg/m^3 TWA. Immediate danger to life or health: 100 mg/m^3.

Selective toxicity. Very toxic with rapid onset (10–30 min) in **rats**, **mice**, **dogs**, but **guinea pigs**, **rabbits**, and **monkeys** are resistant. Marked hydrothorax, pulmonary edema, coughing, weakness, cyanosis, dyspnea, diarrhea, vomiting, and incoordination in most affected species. Larger particle material seems to be more toxic than smaller particles, for reasons unknown.

LD_{50} in mg/kg: **rat**: 6, **mouse**: 35, **guinea pig**: 143, **monkeys**: 4.2, **cat**: 75–100, **pig**: 40, **horse**: 50, **poultry**: > 2,000; LD_{50}, **dog**: 380 μg/kg.

NAPROXEN

= *Aleve* = *Anaprox* = *Apranax* = *Artragen* = *Axer Alfa* = *Bonyl* = *Diocodal* = *Dysmenalgit* = *Equiproxen* = *Flanax* = *Floginax* = *Gynestrel* = *Laraflex* = *Laser* = *Miranax* =

MNPA = Naixan = Nalyxan = Napren = Naprium = Naprius = Naprosyn(e) = Naprotis = Naprux = Naxen = Nycopren = Pranoxen = Prexan = Primeral = Proxen = Proxine = Reuxen = RS-3540 = RS-3650 = Synflex = Veradol = Xenar

***Anti-inflammatory, analgesic, antipyretic.* A propionic acid.**

Untoward effects*: *Blood*:** Agranulocytosis, eosinophilia, thrombocytopenia, aplastic anemia, neutropenia, and decreased platelet aggregation, thereby prolonging bleeding time. A 16-year-old **female** ingested 10 g in an apparent suicide attempt, and developed hypothrombinemia, probably, due to inhibition of synthesis of ***vitamin K-dependent clotting factors. Occasionally slight increase in hemoglobin and hematocrit. Hemolytic anemia has been reported in two **people**.
***Cardiovascular*:** Edema, palpitations, increased blood pressure (by increasing fluid volume in circulation).
***Gastrointestinal*:** (20–30%). Microbleeding 2.3%, gastric mucosal irritation and gastroduodenal ulcers, nausea, heartburn, stomatitis, diarrhea, increased thirst, vomiting, melena, constipation, and reactivation of colitis and proctitis.
***Liver*:** A rare cause of clinical liver disease, but two studies indicate the incidence of induced abnormal liver enzymes to be ~4%. Acute liver injury reported in 3.8/100,000 **users**. Jaundice has been uncommon and has occurred after treatment for several months.
Miscellaneous*:** Headache, dizziness, vertigo, tinnitus, sore throat, drowsiness, depression, cognitive dysfunction, giddiness, personality changes in some **elderly people**, allergic reactions in an ***aspirin-sensitive **patient**, exacerbation of idiopathic Parkinson disease in 74-year-old **female** receiving 250 mg tid, impotence, ejaculatory dysfunction, agoraphobia, menstrual disturbances, aseptic meningitis (with headache, fever, shaking chills, and nuchal rigidity), and severe hyponatremia and fluid retention in a **neonate** after a **maternal** overdose 8 h before delivery. Cross-allergenicity to ***aspirin*** is ~75–100%. Only ~1% of **maternal** plasma concentration is excreted in breast milk. Exacerbation of glaucoma in 65-year-old **female** after 275 mg/tid/3 weeks. Other visual disturbances reported. Rarely, angioedema; and ankle, feet, or leg swellings. Can induce a positive reading for occult blood. Aseptic meningitis reported.
***Renal*:** Oliguric renal failure in 85-year-old **female** after 250 mg twice daily/5 weeks. A 41-year-old marathon **runner** receiving 250 mg tid/1 week, ending 36 h before a race, developed acute renal failure after 17 miles. Numerous case reports of interstitial nephritis and proteinuria with nephrotic syndrome.
Respiratory*:** Severe bronchospasm asthmatic attacks, even after a single 250 mg dose, especially in ***aspirin-sensitive **people**. Eosinophilic pulmonary infiltration and pneumonia occur, which may be a hypersensitivity reaction – confirmed by rechallenge, X-ray, and ***gallium*** scan.
Skin*:** Pseudoporphyria cutaneous tarda with bullae, erythema, and increased skin fragility. Fulminant and necrotizing vasculitis, pustules, subacute cutaneous lupus erythematous, itching, photosensitivity, and alopecia. Use of ***quinoline yellow (q.v.) as a naproxen tablet coloring agent is banned in some countries because of induced urticarial reactions in some **people**.

Therapeutic serum levels are reported as 20–70 μg/ml and toxic at 414 μg/ml.

TD_{LO}, **woman**: 20 g/kg.

Drug interactions*: *Antacids, but not ***sodium bicarbonate*** or a commercial mixture of ***magnesium*** and ***aluminum hydroxides*** (***Maalox***®), delay its absorption. Other reports indicate the last two compounds decrease its absorption. ***Cholestyramine resin*** also delays its absorption. Use with ***aspirin*** causes slight decrease in its serum levels by increasing its rate of excretion. It decreases ***furosemide***-induced urine and ***sodium*** excretion; increases ***glyburide***-induced hypoglycemia. Hypertensive crisis and cerebral thrombosis in an 80-year-old **female** after a single dose of 250 mg added to a 25 mg/day routine of ***hydrochlorothiazide***. Increases circulating ***lithium*** concentration by 20–60%. Serious or fatal interactions reported with ***methotrexate***. ***Ciprofloxacin*** may cause adverse neurological effects when taken with it. ***Probenecid*** decreases its metabolism and renal clearance leading to higher naproxen plasma levels. Minor increase in blood pressure of **patients** taking ***atenolol***. Decreases total plasma level of ***valproic acid***. Severe nausea, vomiting, and abdominal cramps within 20 min in a 46-year-old **male** after a single dose of it and ***zomepirac***. Effect on ***anticoagulants*** has not been agreed upon. It may displace ***coumarins*** from their plasma-binding sites and potentiate ***oral anticoagulants***. Can decrease effectiveness of most ***angiotensin-converting enzyme inhibitors***.

Many case reports of its toxicity in **dogs** (duodenal ulcers, renal and hepatic dysfunction, vomiting, diarrhea, melena, anemia). Neuropathies in **rats**, **mice**, and **rabbits** at high dosage, but not in **monkeys** or **miniature pigs**. Ulcerogenic in **rats** at 7.1 mg/kg. Overdosing by well-intentioned **owners** has led to toxicosis (usually vomiting, diarrhea, and gastrointestinal hemorrhage) in **pets**. Avoid use in **dogs**. Half-life in **dogs** is ~74 h.

LD_{50}, **rats**: 543 mg/kg; **hamsters**: 41 g/kg; **dogs**: > 1 g/kg; **mice**: 1,234 mg/kg.

NAPTALAM

= ACP 332 = Alanap = Nip-A-Thin = NPA = Peach-Thin

Pre-emergence herbicide, fruit thinner.

***Untoward effects*:** Irritation, particularly of the eyes. **Users** are warned not to get it into **their** eyes and on **their** skin or clothing. Harmful if inhaled. Percutaneous absorption occurs.

LD_{LO}, **human**: 500 mg/kg.

LD_{50}, **rat**: 1.8 g/kg.

NARASIN

= *Monteban*

***Coccidiostat*. An *ionophore*.**

Untoward effects*:** Although useful in broiler **chickens** at 54–72 g/ton of feed, it is fatal for adult **turkeys** and **equines**. In laying **hens**, it decreased egg weights. Used with ***tiamulin (q.v.) in **chickens** produced toxicity.

NARCEINE

In *opium* at 0.1–0.5% level.

Untoward effects*:** Hypersensitivity with weeping and scaling eczemas in native French **workers**, but not in **Negro workers** from North Africa. Narcotic effects once reported were probably from impure material contaminated with ***morphine.

NARCISSUS sp.

Consists of *daffodils* (q.v.), *jonquils* (q.v.), and narcissus.

Untoward effects*:** Due to their alkaloids, ***buphanine, ***lycorin***, and ***narcissine*** content. Ingestion usually due to mistaking them for ***onions***, which rapidly causes dry mouth, nausea, repeated and violent vomiting, cramps, abdominal pain, diarrhea, dyspnea, trembling, collapse, convulsions, coma, and occasionally death. Hepatic toxicity reported. Contact dermatitis from repeated handling of the bulbs by ~25% of greenhouse **workers**.

Dogs are often poisoned by eating the bulbs during planting season. In 1939–1945, during World War II, because of the shortage of feed for **cattle**, the Dutch fed the bulbs to them, causing serious toxicity and death.

NARCOBARBITAL

= *Enibomal* = *Eunarcon* = *Narcotal* = *Narkotal* = *Pronarcon*

Sedative, hypnotic, IV anesthetic.

***Untoward effects*:** Muscle twitching, tautness, and hiccups during anesthesia.

NARTHECIUM OSSIFRAGUM

= *Bog Asphodel*

***Untoward effects*:** **Sheep** grazing it developed jaundice, photosensitization, photophobia, eye irritation, drooping and edematous ears, and decreased appetite. The syndrome is called "alveld", "head grit", or "yellows". The plant grows in wet moors in Scotland, Ireland, Norway, and other countries.

NASTURTIUM

= *Tropaeolum majus*

***Untoward effects*:** Rubefacient. Has caused dermatitis of **gardeners'** hands.

NATAMYCIN

= *Antibiotic A 5283* = *CL-12,625* = *Mycophyt* = *Myprozine* = *Natacyn* = *Pimafucin* = *Pimaricin* = *Synogil* = *Tennecetin*

Topical antifungal.

***Untoward effects*:** Mild irritation and burning during initial treatments. Some blood cell damage after IV usage. Ingestion causes nausea, vomiting, anorexia, and diarrhea. Conjunctival chemosis and one case of ophthalmic hyperemia reported.

NAUCLEOPSIS sp.

Untoward effects*:** Its latex is used as poisons on arrows or darts in northwestern Amazonia (Brazil and Columbia). Contains a mixture of cardiotonic glycosides, mainly, ***β-antiarin and some ***α-antiarin*** and ***antioside***. The **natives** call the poison ***kierátchi*** (**Noanama Indians**) or ***pakurú-niaará*** (**Emberá Indians**).

NEALBARBITAL

= *Censedal* = *Nealymal* = *Nevental*

***Sedative, hypnotic*. Long-acting (8–12 h) *barbiturate*.**

Untoward effects*:** Nausea, dizziness, fatigue, unsteadiness, and occasionally drowsiness. See ***phenobarbital for any additional adverse effects.

NEBURON

***Herbicide*. A substituted *urea*. Used in Europe.**

***Untoward effects*:** Low **mammalian** toxicity. Can adversely affect hemoglobin and red blood cell production.

LD_{50}, **rat**: 11 g/kg.

NECTARINE

Untoward effects*:** Contain small natural amounts of physiologically and pharmacologically active ***salicylates. Allergic reactions to them occur in **man**. Contain cyanogenic glycosides, but less than in other *Prunus* sp.

Potentially dangerous for **livestock**, especially during droughts or after frosting.

NEDOCROMIL SODIUM

= *FDL 59,002KP* = *Rapitil* = *Tilade* = *Tilarin*

Antiasthmatic, antiallergic. **Inhalant.**

Untoward effects: Unpleasant taste, headache, and occasionally nausea, vomiting, and dyspepsia. Possibly, rash, arthritis, tremors, and sensation of warmth. A case of anaphylaxis and one of pneumonitis with eosinophilia. Has induced whealing in skin by iontophoresis at 0.002–2% solution strength. Topically for allergic conjunctivitis causes headache, unpleasant taste, and nasal congestion. Occasionally local irritation, stinging, or burning.

NEFAZODONE

= *BMY 13,754* = *Dutonin* = *MJ 13,754-1* = *Nefazone* = *Serzone*

Antidepressant, serotonergic. **A *phenylpiperazine.***

Untoward effects: Headache, blurred vision, agitation, confusion, dizziness, light-headedness, constipation, dry mouth, nausea, drowsiness, tiredness, and weakness. Less common complaints are clumsiness, unsteadiness, postural hypotension, dysuria, memory problems, difficulty in concentration, rash, pruritus, taste alterations, paresthesia, decreased libido, insomnia, anorgasmia, abnormal ejaculation, impotence, painful menstruation, breast pain, heartburn, tinnitus, flushing, increased appetite, constipation, lethargy, liver failure, and strange dreams. Rarely, akathesia, lip-smacking, hand and arm gesturing, and mania. After emergency treatment, a 27-year-old **female** survived ingesting 30 100 mg tablets. Visual disturbances including periods of shimmering or strobe-like brightness undulation in peripheral visual fields in a 51-year-old **male**, and jibby lines in both peripheral fields in a 52-year-old **male**.

Drug interactions: Increases serum ***cyclosporine*** levels. A 44-year-old **male** on ***simvastin*** 40 mg/day/19 weeks developed myositis, rhabdomyolysis, and tea-colored urine a month after 100 mg/twice daily of nefazodone was added to **his** treatment. Asymptomatic after its withdrawal and no problems with resumption of ***simvastin*** treatment. ***Triazolam*** is contraindicated with it, as nefazodone inhibits oxidative metabolism by ***CYP3A***, the mechanism by which ***triazolam*** is metabolized. Neurotoxic, increased ***creatinine***, and increased ***tacrolimus*** serum levels when nefazodone was added to therapy. As a CYP3A4 inhibitor, it can dramatically increase ***lovastatin*** serum levels. Increases creatine kinase levels when used with ***pravastatin***.

NEFOPAM

= *Acupan* = *Ajan* = *Benzoxazocine* = *Fenazoxine* = *Nocipam* = *R 738*

Muscle relaxant, analgesic, antidepressant.

Untoward effects: Mild dizziness, hypothermia, tachycardia, insomnia, decreased appetite, nervousness, and xerostomia. Occasionally sweating, nausea, and light-headedness. Nausea and vomiting is more common with dosage of 120 mg or over. Confusion reported in the **elderly**. Pink discoloration of urine noted in a young 20-year-old **volunteer** given 300 mg in 24 h. Passes into breast milk with **infant** exposure of < 0.05 mg/kg/day.

Therapeutic serum levels range up to 0.1 μg/ml; toxic at 4 mg/ml; and a comatose state at 12 μg/ml.

Drug interactions: Mild increase in diastolic blood pressure when given with ***propoxyphene***. It has inhibited metabolism of ***cisapride***. Serotonin syndrome occurred when taken with ***Hypericum*** (***St. John's Wort***).

Peak serum concentration in **dogs** in ~30 min with half-life of 6.8–14.2 h. **Dogs** have tolerated 100 mg/kg twice daily.

LD_{50}, Carworth **rats**: 123 mg/kg; Charles River **rats**: 80–85 mg/kg; **mice**: 119 mg/kg; **dog**: ~150 mg/kg.

NELFINAVIR

= *Viracept*

Protease inhibitor. **For treating HIV patients.**

Untoward effects: New, or exacerbation of, diabetes mellitus cases, hyperglycemia, increased thirst and hunger, decreased weight, polyuria, diarrhea (67%), headache (23%), increased hepatic transaminases (7%), neutropenia, fatigue and weakness (15%), rash, and dry and itchy skin.

Drug interactions: Decreases metabolism of ***astemizole, benzodiazepines, cisapride, rifabutin, ritonavir***, and ***saquinavir***, increases their toxicity. ***Delavirdine*** increases its toxicity by decreasing its metabolism; the combination decreases ***delavirdine's*** effect, and neutropenia has occurred. Use with ***indinavir*** decreases the metabolism of both and can increase the toxicity of both. ***Nevirapine*** increases its metabolism, decreasing its effect.

NEOARSPHENAMINE

= *Arsevan* = *Collunovar* = *Miarsenol* = *N.A.B.* = *Neo-Arsoluin* = *Neosalvarsan* = *Novarsan* = *Novarsenol* = *Novarsenobenzol* = *Novarsenobillon* = *Vetarsenobillon*

Protozoacide. **Contains 32.14% *arsenic* (q.v.). For IV use.**

Untoward effects: Nausea, vomiting, prurigo, icterus, hemolytic anemia in glucose-6-phosphate dehydrogenase-deficient **patients**, hepatic cirrhosis (18%), fixed-drug eruptions, flushing, burning taste, tongue and eyelid edema, hyperhidrosis, dyspnea, cyanosis, and precordial distress. Unconsciousness in some severe cases. Many symptoms disappear in ~30 min. Some don't appear until 1–4 h after treatment. Dermatitis and fixed-drug eruptions usually appear 1–14 days after treatment. Exfoliative dermatitis and nephritis are rare. Severe and fatal reactions can occur within a few days to 1 month, and include aplastic anemia, granulocytopenia, yellow atrophy of the liver, jaundice, and acute purpura. Severe central nervous system upsets have occasionally occurred ≥ 1 month after treatment. Use freshly made solutions, as the oxidized (discolored) solution is more toxic.

NEOBOUTONIA CANESCENS

Untoward effects: A toxic African plant whose roots can kill within 10 min after ingestion.

NEOCARZINOSTATIN

= *NSC 69,856* = *Zinostatin*

***Antineoplastic*. Experimental. For IV use.**

Untoward effects: Caused pulmonary toxicity (prominent endothelial cells, thrombosis of small veins, and, later, intimal thickening of arterioles and dyspnea).

LD_{50}, **mice**: 1 g/kg; IV, 0.96 mg/kg.

NEODYMIUM

***Thromboplastic*. A rare earth metal.**

Untoward effects: Thrombocytopenia and prolonged clotting time.

Intracerebral TD_{LO}: 17 μg/kg.

NEOMYCIN

= *Fradiomycin* = *Kolimycin* = *Kolmitisin* = *Mycifradin* = *Neolate* = *Neomas* = *Neomin* = *Pimavecort* = *Vonamycin Powder V*

***Antibiotic*. A complex of A, B, and C forms.**

Untoward effects: ***Blood***: Dose and duration of therapy is unknown in rare cases of hemolytic anemia and thrombocytopenia reported.
Ear: Ototoxic with nearly total deafness in a 53-year-old **female** after 9 g during 2 days postoperatively. Even with minimal absorption after oral use, large doses or prolonged exposure can be ototoxic. Has caused severe damage and destruction, primarily of sensory cells of the inner ear and eighth cranial nerve, often dose- and duration of therapy-related. Topically, it has caused contact dermatitis, superimposed on otitis externa in many **patients**. Readily absorbed through the skin of burn **victims** and, after prolonged use over 30% of a 9-year-old **female's** body was covered with skin ulcerations. Chronic oral use in prophylaxis of chronic portal encephalopathy has caused deafness. Similar untoward effects have followed its use in wound irrigation. Its ill-effects can be additive to those of ***colistin***, ***dihydrostreptomycin***, ***gentamicin***, ***kanamycin***, ***polymyxin B***, and ***streptomycin***.
Eye: Allergic reactions after its topical use.
Malabsorption syndrome: Due to mucosal damage, steatorrhea, and decreased absorption of ***fats***, ***lactose***, ***sucrose***, ***vitamin A***, ***vitamin*** B_{12}, and ***xylose*** occurs. When I first reported this in **dog** trials for a basic manufacturer, I was told by their Research Director that I was wrong and the data would **not** be reported. Later, a report came from physicians in Texas confirming this finding.
Miscellaneous: Anaphylactoid reactions. Possible cause of swelling in mouth and necrotic leg ulcerations after oral use of a troche (***Tetrazet***). Vertigo, decreased **maternal *estrogen*** excretion. Low incidence of nausea, vomiting, abdominal pain, weight loss, or diarrhea after preoperative oral use for bowel sterilization. Proliferation of *Staphylococcus aureus*, *Candida*, *E. coli*, and *Clostridium difficile*. Causes increased or false positive blood urea nitrogen and urinary ***albumin*** test results. Contact ocular dermatitis and conjunctivitis reported. Steatorrhea from precipitation of bile salts. Wheezing reported.
Neuromuscular block: Also decreased blood pressure and respiration, paralysis of respiratory muscles, diaphragmatic breathing, muscle flaccidity, and central nervous system depression. These have occurred after retrograde pyelography with 50 ml contrast material containing 2.5% ***neomycin***, after IP or intraluminal use during surgery, and 3 h after accidental spillage of some into the peritoneal cavity. Induces weakness, hypoactive or absent deep tendon reflexes, and mydriasis, especially in myasthenia gravis **patients**. Interferes with release of ***acetylcholine*** at neuromuscular junctions.
Renal: Nephrotoxicity with uremia and acute renal failure is usually related to dose and duration of therapy. Has followed oral and parenteral use, and systemic absorption from irrigating solutions.
Skin: Allergic contact dermatitis (10–50%; others state ~1–6%) with greater risks if under occlusion or on damaged skin and eczemas. Toxic epidermal necrolysis, eczematous eruptions, and diaper dermatitis. Cross-reactions or co-sensitization reported with ***bacitracin***, ***butirosin***, ***framycetin***, ***gentamicin***, ***kanamycin***, ***paromomycin***, ***spectinomycin***, ***streptomycin***, and ***tobramycin***. Reactions have followed use of a neomycin powder on a 1-week-old **infant** after circumcision. Reactions may also be caused by ***ethylenediamine*** (q.v.) in a neomycin cream applied to an **infant's** diaper area.

Drug interactions: Oral use decreases absorption or bacterial production of ***vitamin K***, leading to potentiation of ***coumarin*** and ***warfarin*** anticoagulant effects. Alteration of normal intestinal flora and damage to the intestinal wall can decrease ***methotrexate*** absorption by up to 50%, decreases fat absorption, decreases ***vitamin A*** and ***vitamin*** B_{12} absorption, decreases absorption of ***penicillin V***, and the bioavailability of ***digitalis*** glycosides. Its neuromuscular blockade is enhanced by ***alcuronium*** (q.v.), ***amobarbital*** (q.v.), ***edetic acid*** (q.v.), ***promethazine*** (q.v.), and ***curariform drugs***, such as ***curare*** (q.v.), ***decamethonium*** (q.v.), ***ether*** (q.v.), ***gallamine*** (q.v.), ***pancuronium*** (q.v.), ***succinylcholine*** (q.v.), and ***tubocurarine*** (q.v.). Enhances the effect of ***clofibrate, gallamine***, and ***triiodothyronine***. Decreased ***digoxin*** bioavailability and ***digoxin*** toxicity occurs when prolonged neomycin therapy is withdrawn.

Prolonged or excessive oral use can cause undesirable sterilization of the gastrointestinal tract, malabsorption of nutrients, excessive shedding of hair, destruction of ***vitamin***-producing bacteria, steatorrhea, etc. Associated with the depression of a number of intestinal enzymes. Prolonged topical use may permit overgrowth of nonsusceptible organisms. Systemic use can cause ototoxicity (deafness, vertigo). Excessive systemic use or systemic absorption can cause, and any amount can theoretically potentiate, neuromuscular block and respiratory paralysis, particularly those associated with ***curariform drugs, ether***, or ***barbiturate*** anesthesia. Nephrosis and general toxicity can occur with systemic neomycin in the presence of renal impairment. Its ototoxicity may be additive to other drugs affecting the eighth cranial nerve. Contrary to popular advertising, some bacterial strains have developed resistance to it, but not as often and as rapidly as against ***streptomycins***. U.S. regulations permit a negligible residue (0.25 ppm) in edible tissues of **calves** and 0.15 ppm in milk of treated **animals**. ***Calcium*** has corrected some adverse neuromuscular transmission effects (these can be associated with a dose-dependent decrease of systemic blood pressure, cardiac output, and heart rate) of high dosages in **monkeys**, **baboons**, and **dogs**. (20 mg/kg IV, in **owl, monkeys** and ~25 mg/kg, IV, in **baboons** causes a 50% decrease in muscle response.) ***Heparin*** in large concentration decreases the antibiotic's effectiveness versus *Staphylococcus aureus*. In France, Russia, and some Russian satellite countries, the word "***Colimycin***" frequently refers to the antibiotic neomycin, and *not* ***colistin***. Elsewhere, ***colimycin*** *is* ***colistin*** (q.v.). 30 days withdrawal time prior to slaughter is suggested, after use in **cattle**; 5 days in **broilers**; 14 days for **turkeys** and **laying hens**; and 20 days in **swine** and **sheep**. Caution should be used in parenteral treatment of post-parturient **cows** prone to milk fever, because it decreases total and bound blood ***calcium*** without affecting unbound ***calcium***. Similar precautions should apply to other **animal** species, as well.

LD_{50}, **rat**: 2.75 g/kg; **mouse**: 2.9 g/kg; SC, 120–265 mg/kg; IV, 15–53 mg/kg.

NEOSTIGMINE

= *Proserine* = *Prozerin* = *Synstigmin*

Bromide form = *Juvastigmin* = *Neoesserin* = *Neostigmin* = *Normastigmin* = *Prostigmin*.

Methyl sulfate form = *Intrastigmina* = *Javastigmin* = *Metastigmin* = *Neostigmin* = *Normastigmin* = *Prostigmin* = *Stiglyn*.

***"Reversible" cholinesterase inhibitor, cholinergic, miotic.* Antidote for *atropine, curare,* and *tubocurarine*.**

Untoward effects: By inactivating cholinesterase and preventing ***acetylcholine*** hydrolysis or destruction produces excessive ***acetylcholine*** (q.v.), prolonging depolarization of the postjunctional membrane and potentiating the action of depolarizing drugs, which overstimulates nerves, muscles, and glands. If in further excess, causes convulsions and fibrillations, flaccid paralysis, and even death. Increases gastrointestinal transit time. Has induced pulmonary edema and, after ophthalmic use, pigmentary changes in iris, cataracts, miosis, nystagmus, and conjunctival thickening. Anaphylactoid reactions, decreased or increased blood pressure, bradycardia, dizziness, vertigo, weakness, vomiting, purging, cramps, diarrhea, rhinorrhea, headache, salivation, diaphoresis, lacrimation, pale and clammy skin, muscle twitching, dyspnea, rash, alopecia, hyperhidrosis, flushing, vocal problems or difficulty in speaking, and urinary incontinence reported. Overdose in a myasthenia gravis **patient** led to inability to expand chest fully and choking sensation. Gross overdosing leads to cholinergic crisis, agitation, mental clouding, and coma. The *bromide* form given in low dosage (7.5 mg tid) in treatment of functional megacolon led to severe and prolonged poisoning.

SC, LD, **human**: 60 mg; IV, ~0.43 mg/kg.

Drug interactions: Potentiates polarization block caused by ***succinylcholine***. Use with ***kanamycin*** and ***polymyxin B*** causes neuromuscular block. Mydriasis effect when used with ***sympathomimetics***. Severe bradycardia can occur when used with ***propranolol***. Inability of neostigmine to reverse ***curare***-induced paralysis and it can potentiate curarization when used with ***colistin***. ***Quinidine*** may antagonize neostigmine's cholinergic effect. Cardiac arrhythmias when used with ***atropine***.

In **animals**, overdoses may cause respiratory distress, colic, and bradycardia. Walk treated **horses** for 10–15 min after treatment, to prevent their rolling on the ground.

Atropine can prevent excessive intestinal stimulation. Potentially dangerous in the hands of the untrained.

LD_{50}, **rat**: 51 mg/kg; **mice**: 7.5 mg/kg; IV, 0.16 mg/kg; SC, 0.42 mg/kg.

NEPETA HEDERACEA

= *Cat's Foot* = *Creeping Charlie* = *Gill-Over-The-Ground* = *Glechoma hederacea* = *Ground Ivy* = *Lierre Terrestre* = *Robin-Runs-Away*

Perennial common in England, Europe, and the U.S.

Untoward effects: Contact dermatitis in susceptible **people**.

Pulmonary edema, dyspnea, cyanosis, salivation, mydriasis, hyperhidrosis, and enteritis in **horses** eating large quantities of the fresh plant or when in their hay, due to an unidentified irritant oil. Has been fatal.

NEPETALACTONE

***Psychopharmacological*. The active ingredient in the volatile oil of *catnip* (q.v.).**

Use: Some **cats** become excited or intoxication after smelling or eating it. Although its mode of action is not known, it is believed that its rapid oxidation alters skin sensitivity, similar to ***estrogen***, with it causing face-rubbing, body-rolling, and rubbing against the plant, typical of estrus.

NEREISTOXIN

A *sulfur*-containing amine found in the marine worm (annelid), *Lumbriconereis heteropoda*.

Untoward effects: **Japanese**, noting that **flies** died after coming into contact with the dead **worms**, developed insecticides from it. One, called ***Padan***, had low **mammalian** toxicity, but was highly toxic to **rice stem-borers** and other **insects** by respiratory paralysis. Annual production was ~1,500 tons.

NEROL

Fragrance used in soaps, detergents, cosmetic creams and lotions, perfumes, and foods. An alcohol in *neroli oil*.

Untoward effects: No irritation or sensitization by 4% in petrolatum on **human** volunteers.

Vasodilator on skin of **dogs**, **rabbits**, and **mice**.

LD_{50}, **rat**: 4.5 g/kg and acute dermal, **rabbit**: > 5 g/kg; IM, **mouse**: 3 g/kg.

NETILMYCIN

= *Certomycin* = *N-Ethyl Sisomicin* = *Netillin* = *Netilyn* = *Netromycine* = *Netromycin* = *Nettacin* = *Sch-20,569* = *Vetacin* = *Zetamicin*

***Antibiotic*. A semi-synthetic *aminoglycoside*. IV or IM.**

Untoward effects: Neurotoxic and nephrotoxic (~10–14%), with increased serum blood urea nitrogen and ***creatinine***, decreased ***creatinine*** clearance, proteinuria, cylinduria, and oliguria are the worst. The former is manifested as ototoxicity on auditory and vestibular branches of the eighth cranial nerve. Peripheral neuropathy or encephalopathy with numbness, skin tingling, muscle twitches, convulsions, and myasthenia gravis-like syndrome, acute muscular paralysis, apnea, dizziness, vertigo, vomiting, Meniere's syndrome, cochlear damage, hearing loss, and nystagmus. Unfortunately, severe ototoxicity is usually irreversible. In addition to increased SGOT and SGPT, alkaline phosphatases, and ***bilirubin***; rash, pruritus, fever, eosinophilia, thrombocytosis, and increased prothrombin time can occur. Occassionally anemia, lowered hemoglobin, leukopenia, leukemoid reactions, immature circulating white blood cells, hyperkalemia, vomiting, diarrhea, palpitations, dependent edema, hypotension, headache, disorientation, blurred vision, paresthesias, and severe induration, phlebitis, or hematomas at injection site. Fanconi-like syndrome with aminoaciduria and metabolic acidosis has also been reported. Crosses the placenta and small amounts are present in breast milk, putting **infants** at risk for otoxicity, based on experience with ***streptomycin***.

Has caused waxy rib defects in the offspring of treated **rats**, but not **rabbits**.

NEURINE

Untoward effects: A *ptomaine*. Has caused poisoning in **man** from putrefaction of proteins in edible ***mushrooms***, even if kept in the refrigerator for over 24 h after harvesting. Has a fishy odor.

SC, LD_{LO}, **mouse**: 46 mg/kg.

NEUROLAENA sp.

Untoward effects: The leaves and stems of ***N. lobata*** are used as piscicides and insecticides in the West Indies and Central America. Hypoglycemic in **mice**.

NEUTRAL RED

= *C.I. Basic Red*

Untoward effects: Allergic contact dermatitis with pruritic erythematous dermatitis on genitals of a 31-year-old **male** after accidental spill – **he** had used it successfully to treat herpes genitalis lesions and in a 38-year-old **female** after topical application of 3% solution to skin herpes simplex of the nose.

Has caused decreased plasma ***calcium*** levels after IV in **sheep** and **goats**.

IV, LD_{50}, **rat**: 112 mg/kg; **mouse**: 142 mg/kg; **rabbit**: 97 mg/kg.

NEVIRAPINE
= BI-RG 587 = Viramune

Antiviral, reverse transcriptase inhibitor.

Untoward effects: Early rash and Stevens-Johnson syndrome that can be life-threatening. Fever, blistering, oral lesions, conjunctivitis, swelling, muscle or joint aches, general malaise, abnormal liver function tests, headache, and nausea. Late onset (after 4–6 weeks) acute hepatitis and hepatic failure.

Drug interactions: Induces increased CYPA3A metabolism of ***oral contraceptives, indinavir, nelfinavir, ritonavir***, and ***saquinavir***. ***Methadone*** withdrawal symptoms (abdominal pain, cramps, rhinorrhea, and tremors) in a 34-year-old **male** with HIV on ***methadone*** maintenance after starting nevirapine therapy. Use associated with Stevens–Johnson syndrome or toxic epidermal necrolysis in some HIV **patients**.

NEWSPAPERS

Untoward effects: The odor has triggered migraines in some **people**. When burned, pulmonary irritants, such as ***acetaldehyde, acrolein***, and ***formaldehyde*** (q.v.), are produced as well as the toxic gases ***acetic acid*** (q.v.), ***carbon monoxide*** (q.v.), ***formic acid*** (q.v.), and ***methane*** (q.v.).

Cows that eat newspaper bedding ingest small amounts of mutagens, with insignificant amounts passing into their ***milk***. Colored ***ink*** can include diazo compounds, some of which are derivatives of possible carcinogens, such as ***benzidine yellow***.

NEWT

Untoward effects: The rough-skinned newt (*Taricha granulosa*) contains a non-protein neurotoxin, ***tarichatoxin***, identical to the **puffer-fish** poison ***tetrodotoxin*** (q.v.) in its skin and other organs. Ingestion of the toxin can be fatal. A drunk 20-year-old **male** swallowed one of these 20 cm (8") newts on a dare. Within 10 min, **his** lips tingled; numbness and weakness developed in the next 2 h. **He** then developed cardiopulmonary arrest. **His** pupils became fixed and dilated, and **he** died the following day, despite treatment. Additional cases with similar outcomes are in the literature. A single gram of the newt eggs contains enough ***tarichatoxin*** to kill 23 **mice**. Also see ***Salamanders***.

NGAIONE

A toxin found in *Myoporum laetum* (q.v.), the toxic *Ngaio Tree* of New Zealand and the South Pacific.

Untoward effects: After storms blow branches down, **cattle** eat the leaves leading to abdominal pain, constipation, dullness, anorexia, swollen faces, photosensitization with inflammation of skin on their teats and udder. Liver damage, icterus, and photosensitivity reported in affected **cattle**, **sheep**, **horses**, and **pigs**. Post-mortem demonstrates hemorrhagic abomasum and intestines. Mid-zonal liver necrosis in **mice**.

LD_{50}, **mice**: 300 mg/kg.

NIALAMIDE
= Espril = Niamid = Niamidal = Niaquitil = Nuredal = Nyazin

***Antidepressant.* A *monoamine oxidase inhibitor.* Oral, IV, and IM.**

Untoward effects: Headache, especially when taken with ***tyramine-*** and other ***pressor amine- (dopamine, serotonin)*** containing foods (***bananas***, young ***broad beans*** with pods, aged ***cheeses, chicken livers, chocolate***, pickled ***herring, yeast*** supplements, and ***yogurt***) and drinks (***beer*** and red ***wines***). Insomnia, nausea, drowsiness, dizziness, vertigo, ataxia, occassionally weight gain, hyperhidrosis, dry mouth, tinnitus and reduced hearing, hypertension, extrapyramidal symptoms, and fever. Pain and induration at IM injection sites. A 27-year-old **male** swallowed 200 25 mg tablets (total of 5 g) in a suicide attempt. After 12 h, **he** became nauseous and evidenced severe agitation and motor activity, pyrexia (43 °C – 109.4 °F), flushed skin, disorientation, and hallucinations. **He** finally died after cardiac asystole.

Drug interactions: Flushing, palpitations, light-headedness, and hypertension with ***L-dopa***. Use with sympathetic amines (***amphetamine, ephedrine, metaraminol, methylphenidate, phenylephrine***, and ***phenylpropanolamine***, leads to acute hypertensive crisis with possible hyperthermia, intracranial hemorrhage, convulsions, coma, and even death. It seemed to have potentiated the effects of ***meperidine*** in a mentally retarded **phenylketonuric** 27-year-old **male**. Severe agitation in a 49-year-old **female who** took it with ***phenelzine*** (q.v.), another ***monoamine oxidase inhibitor***. ***Reserpine*** (q.v.) and ***tetrabenazine*** (q.v.) can displace ***norepinephrine*** from its binding or storage sites and create the same effect as with sympathetic amines.

Reserpine antagonizes its effects on **cats** and it antagonizes the ***reserpine***-induced hypothermia in **chickens**. Hyperthermia when given with anticholinergic ***antiparkinson drugs*** (***benztropine*** and ***procyclidine***) in **rabbits**.

High dosage potentiates the hypoprothrombinemic action of ***acenocoumarol, ethyl biscoumacetate***, and ***warfarin*** in **rabbits** and **rats**, but no evidence of such interaction in **man**. Antagonizes the cataleptic action of ***chlorpromazine*** and ***haloperidol*** in **rats** and **mice**.

LD_{50}, **rat**: 1.7 g/kg; **mouse**: 590 mg/kg.

NIAX® CATALYST ESN

A mixture of *dimethylaminopropionitrile* 95% and *bis(2-dimethylamino) ethyl ether*.

Untoward effects: Extremely hazardous. Sales of it and related forms, that were used in the manufacture of flexible ***polyurethane*** foams, were discontinued by Union Carbide Corporation. **Worker** exposure has been by inhalation, percutaneous absorption, and ingestion. It caused neurotoxicity problems (muscle weakness and paresthesias of hands and feet), dysuria, lassitude, nausea, vomiting, retention of urine, and eye irritation from vapors.

NICANDRA PHYSALODES

= *Apple of Peru* = *mu Gumacembere* (Shona) = *Wild Gooseberry*

Found in Africa, Australia, North America, and Europe. In the *nightshade* (*Solanaceae*) family.

Untoward effects: The plant is eaten when forage is scarce. Causes bloating in **sheep**.

NICARBAZIN

= *Nicarb* = *Nicoxin* = *Nicrazin*

Coccidiostat.

Untoward effects: Causes depigmentation of colored egg shells and decreases hatchability in **chickens**. Do NOT use as a therapeutic agent, in flushing mashes, or for egg-laying **birds**. Discontinue feeding 4 days before marketing to prevent undesirable levels in edible tissues. Its use may increase the mortality of heat-stressed **birds**. When ambient temperatures increase significantly above 36 °C (96.8 °F), i.e., at 40 °C (104 °F), **chicks** fed it showed increased mortality above ***amprolium***-treated **birds**. This can be avoided by feeding 50 ppm with ***narasin***.

NICARDIPINE

= *Barizin* = *Bionicard* = *Cardene* = *Dacarel* = *Lecibral* = *Lescodil* = *Loxen* = *Nerdipina* = *Nicant* = *Nicardal* = *Nicarpin* = *Nicapress* = *Nicodel* = *Nimicor* = *Perdipina* = *Perdipine* = *Ranvil* = *Ridene* = *RS-69,216* = *Rycarden* = *Rydene* = *Vasodin* = *Vasonase* = *YC-93*

***Calcium channel blocker, antihypertensive, antianginal.* Oral and IV.**

Untoward effects: Flushing, headache, asthenia, palpitations, reflex tachycardia, pedal edema, and dizziness. Decreased cardiac contractility reported in 12/14 **patients** in one report. Nausea, abdominal discomfort, vomiting, dry mouth, dermatitis, dizziness, peripheral neuropathy, drowsiness, fatigue, insomnia, edema, arrhythmias, congestive heart failure, angina, tachycardia, myocardial infarction, frequent urination, diaphoresis, muscle cramps, body aches, flu-like symptoms, impotence, coughing, and sore throat have also been reported. Hyperkinetic muscular dyskinesia in a 74-year-old **male** after dosage slowly increased to 60 mg twice daily. Urinary retention and erythromelalgia have been reported as well.

Drug interactions: Causes increased ***cyclosporine*** and ***digoxin*** blood levels and its plasma concentration is increases by ***cimetidine***. ***Grapefruit juice*** increases its bioavailability.

LD_{50}, in mg/kg: **rats**: male: 634, female: 557; IV, male: 18.1, female: 25; **mice**: male: 634, female: 650; IV, male: 20.7; female: 19.9.

NICKEL

U.S. annual consumption is > 600 million lbs and > 1 million lbs goes into the atmosphere.

Untoward effects: The most common and aggravating effects are those on the skin. Pompholyx or dishydrotic eczema, a vesicular or vesicopustular eruption on hands and feet has followed a high dietary nickel intake or skin contact. The usual induced dermatitis is the result of ear-piercing, contact with nickel-containing jewelry, IV cannulae, metal instruments, skin clips, pacemakers, orthopedic prostheses or screws, garments, eyeglass frames, wristwatches, hair dressings, detergents, nickel coins, transcutaneous electrical nerve stimulation pads, electroplating, acupuncture needles, fingernail polishes (from nickel leaching from stainless steel mixing balls), shoe eyelets, garter snaps, brassiere cup wires, zippers, metal chairs, arm rests, dentures, white ***gold*** crowns (some yellow ***gold***), needles, and electric wiring. ***Platinum*** normally contains no nickel, except as an impurity or in a soldering agent. Stainless steel contains firmly-bound nickel, so that sensitive **people** do not react to it. An allergic stomatitis and cheilitis has followed **people** holding nickel-plated needles, pins, bobby pins, and metal lipstick-holders in **their** mouths. **Some** have developed the same lesions from nickel-plated mouthpieces of musical instruments. **Patients** have experienced swelling of **their** mouths for hours or days after dentists have given **them** an acid ***local anesthetic*** solution that reacted with metal ions leached from the syringe's metal plunger. Injections of **human** serum ***albumin*** have caused iatrogenic hypernickelemia. Nickel is a catalyst in the hydrogenation of ***fats*** and ***oils*** and some residues remain in the foods. Nickel can produce a dermatitis at the contact site, as well as a generalized dermatitis from local contact, since the blood can carry it. Nickel dermatitis has been due to a foreign body in the stomach. In poisoning by oral or parenteral routes it causes nausea, vomiting, and purging, as well as central nervous system depression. Causes falsely increased or spuriously high serum ***copper*** tests. Mill **workers** in nickel refining and smelters, or in nickel conversions, have an increased incidence of neoplasms of the nasal sinuses, pharynx, and lungs. Inhalation of dusts and soluble salts are implicated in precipitating asthma and impairing the body's cellular immune system. Serum levels in most species are usually 1–5 µg/l, except for **hogs** (5.3), **rabbits** (9.3), and **Maine lobsters** (8.3–20.1) µg/l. OSHA upper exposure limits are 1 mg/m^3 TWA; National Institute for Occupational Safety and Health are 15 mg/m^3 TWA.

Sebertia accuminata of New Caledonia is a hyperaccumulator of nickel (25.74% of the latex on a dry weight basis). **Cattle** downstream of a nickel-plating plant died. ***Cyanide*** may also have been involved. **Calves** fed 1,000 ppm/8 weeks ate little and lost weight; those fed at 250 ppm/8 weeks had 13% decrease in food consumption and 11% decrease in weight. Feeding 62.5 ppm had no ill-effects.

NICKEL ACETATE

***Use*:** Catalyst, textile mordant, and in nickel-plating.

***Untoward effects*:** Levels of 500 or 1,000 ppm in their diet decreased growth, packed-cell volume, hemoglobin, and tissue cytochrome oxidase and alkaline phosphatase activities in weanling **rats**. Decreased growth of **chicks** at 700 g/kg.

NICKEL CARBONYL

= *Nickel Tetracarbonyl*

Formed by combination of *nickel* and *carbon monoxide*.

Untoward effects*:** Exposure of industrial **workers** to it by inhalation led to pulmonary edema and high incidence of respiratory cancer (squamous cell) of the lung and nose. Exposure of 100 **men** repairing a reactor at a Texaco refinery required ***BAL antidote therapy in 32, **one** of whom died. First symptoms were headache and dizziness, eased by moving the **victim** to fresh air. From 10 h to 8 days later, a paroxysm of coughing precedes the more serious symptoms of extreme weakness and dyspnea. Death, due to pulmonary and cerebral edema, within 4 days of exposure to the gas in a 47-year-old **male**. Urine ***nickel*** was 53.5 µg/100 ml of urine 24 h after exposure. Apparently, a major cause of Legionnaire Disease that killed 28 **persons** and made 150 **others** ill in July, 1976, in Philadelphia, Pennsylvania, due to its formation during incomplete combustion of business forms coated with a ***nickel***-containing duplicating material. OSHA and National Institute for Occupational Safety and Health exposure limits are 7 µg/m^3 TWA or 1 ppb. This is the least detectable level.

Inhalation TC_{LO}, **human**: 7 mg/m^3; LC_{LO}, 30 ppm/30 min.

Rats inhaled 4 ppm three times weekly/1–3 years leading to metastasizing lesions (squamous cell carcinoma, adenocarcinoma, and anaplastic carcinoma) and intra-alveolar edema and swelling. Prenatal exposure of pregnant **rats** causes their offspring to have extraocular anomalies, including anophthalmia and microphthalmia. In **dogs**, inhalation led to intra-alveolar edema and swelling.

Inhalation LC_{50}, **rat**: 35 ppm/30 min; **cat**: 1,900 mg/m^3/30 min; LC_{LO}, **dog**: 360 ppm/90 min.

NICKEL CHLORIDE

Use*:** For ***nickel-plating ***zinc*** or as an ***ammonia*** absorbant in gas masks.

***Untoward effects*:** IP to pregnant **mice** caused fetal deaths and decreased fetal weights.

LD_{50}, **rat**: 105 mg/kg; IV, LD_{LO}, **dogs**: 10 mg/kg.

NICKEL SULFATE

Use*:** As a mordant in dyeing and printing fabrics, in ***nickel-plating, and blackening ***brass*** and ***zinc***.

Untoward effects*:** Allergic contact dermatitis in > 10% in cheap costume jewelry-**wearers**. Acne reported in **electroplaters**, baby carriage **worker**, and an acute exacerbation of eczema from it in ***hepatitis B vaccine. Asthma symptoms within 2 weeks in a 24-year-old **male *nickel* electroplater**.

Mallard ducklings fed it at 1,200 ppm developed tremors and paresis within 2 weeks and mortality 71% within 8 weeks. It adversely affected ***calcium*** deposition in bone. Triggers pustule development when applied to scratched abdominal areas of **rabbits**. Orally to **mice** it increased mortality, due to encephalomyocarditis virus. Large oral doses cause severe gastritis, convulsions, cachexia, and conjunctivitis in **rabbits** and **dogs**.

NICLOSAMIDE

= *B 2353* = *Bayer 2353* = *Clonitralid* = *Fenesal* = *Lintex* = *Mansonil* = *Phenasal* = *Preparation 391* = *Radeverm* = *Vermitin* = *Yomesan*

Anthelmintic, taeniacide, lampricide, molluscicide, piscicide.

Untoward effects: Nausea and vomiting (~4%), mild abdominal pain and loss of appetite (~3%), foul-smelling and (~1.5%) mucoid and watery diarrhea, drowsiness, dizziness, or headache. Less frequently, skin rashes, pruritus ani, oral irritation, fever, rectal bleeding, weakness, sweating, palpitations, bad taste in mouth, constipation, alopecia, and backache. Some of the ill-effects are from the passage or dissolution of the worms. I personally supervised its use on a young **teenager** who passed ~30 intact dog tapeworms with scoleces up to 3 ft in length and numerous partially disintegrating ones in **her** foul-smelling, mucoid stool. Of the ingested drug, ~12% is detected in the urine of treated **humans**.

Mammalian toxicity is low. Occassionally emesis in treated **cats**. Masseter tenseness, regurgitation, polydipsia, hyporeactivity, ataraxia, ataxia, tachypnea, dyspnea, tremors, and tetanic convulsions within 15 min and death within 30 min to 3 h later in poisoned **ducks**, **quail**, and **gulls**. At 0.3–1.0 ppm as a molluscicide, it has killed some **frogs**.

LD_{50}, **rat**: 500 mg/kg.

NICORANDIL

= *Adancor* = *Ikorel* = *Perisalol* = *SG-75* = *Sigmart*

Antianginal; potassium channel-opener.

Untoward effects: Major aphthous ulcers.

LD_{50}, **rat**: 1.25 g/kg.

NICOTIANA sp.

Untoward effects: Various species contain varying amounts of the alkaloid ***nicotine*** (q.v.), and are usually smoked, chewed, or inhaled (snuff), which causes central nervous system stimulation, followed by central nervous system depression.

N. attenuata = ***Coyote Tobacco*** = ***Wild Tobacco***, found in the western and southwestern U.S., has poisoned **cattle** and **sheep**.

N. glauca = ***Palán-palán*** = ***Tree Tobacco*** = ***Tumbaku*** (Ethiopia), native to Argentina, but introduced into southern U.S. and Mexico. Also called ***Wild Tobacco*** in Australia and Zimbabwe. Contains the toxic alkaloid ***anabasine*** (q.v). Has caused serious poisonings and fatalities in **man**, especially from eating the boiled greens or leaves in salads. The boiled leaves were given to a **woman** by her **son-in-law** as a poison, in the guise of cooked greens. They tasted good, but **she** died. A 76-year-old **male** ingested some and developed neuromuscular blockade and respiratory failure. **He** recovered after treatment. Symptoms develop rapidly including salivation, nausea, vomiting, sensation of sweating, and dizziness. Sudden hypotension and apnea can occur early. Occassionally convulsive movements.

It is poisonous to most **animals** (primarily, **cattle** and **horses**, occassionally, **sheep**, **giraffes**, and **ostriches**); causes weak pulse, staring, hyperesthesia, unsteadiness, trembling, stumbling, spasms, colic, palpitations, diarrhea, bloating, salivation, frequent urination, dyspnea, convulsions, and death. 10 oz of fresh leaves will kill a **sheep** within 3 h.

N. rustica = ***Aztec Tobacco*** = ***Small Tobacco*** = ***Turkish Tobacco*** = ***Yellow Henbane***. When smoked, it can have hallucinogenic properties in **man** due to its alkaloids, ***harman*** (q.v.) and ***norharman***. It is also used for its insecticidal properties with ***N. tabacum*** in Nigeria. Although not palatable, is has poisoned **cattle**.

N. sauveolens has poisoned **horses** and **sheep** in Australia.

N. tabacum = ***Large Tobacco*** = ***Taba*** (Hausa) = ***Tobacco*** = ***Wai-munoh*** (Amazonia). Its carcinogenic potential with fatal malignant polyps of the nose was first described in 1761 by Sir John Hill, a physician and botanist, from its excessive use as snuff. ***Tobacco*** contains > 4,000 organic chemicals in its leaves and at least 16 have been identified as carcinogens. ***Chewing tobacco*** has been associated with oral carcinomas and leukoplakia in **man**. Other toxicity in **man** is primarily from its ***nicotine*** and ***anabasine*** content. African **natives** overusing it orally develop nausea, vomiting, prostration, convulsions, and respiratory paralysis. Contains 0.6–0.9% ***nicotine*** in its leaves. Despite its poor palatability, **livestock**, particularly **cattle**, develop a taste for it, leading to retching, bloating, salivation, hyperhidrosis, marked dyspnea, clonic spasms, trembling, muscular incoordination, dullness, stupor, palpitations, coma, and death. Causes arthrogryposis in **piglets** of **sows** that consumed it and in **calves** of **cows** that consumed it. In addition to ***nicotine*** (q.v.) and ***anabasine*** (q.v.), it also contains ***anatabine***.

N. trigonella = ***N. trigonophylla*** = ***Desert Tobacco*** found in southwestern U.S., Mexico, and Central America has poisoned **cattle** and **sheep**.

NICOTINE and SALTS

The active constituent in various patches and chewing gums to help control smoking addiction.

Habitrol = *Nicabate* = *Nicoderm* = *Nicolan* = *Nicopatch* = *Nicorette* = *Nicotell TTS* = *Nicotinell* = *Nicotine Polacrilex* = *Tabazur*

***Black Leaf 40* is a 40% solution of *nicotine sulfate*.**

Untoward effects: One of the most deadly poisons known. Rapidly absorbed after ingestion, from the respiratory tract, mucous membranes, and intact skin. The pure alkaloid is very volatile and more rapidly absorbed than the salts. Both are water soluble and act rapidly in blocking ***acetylcholine***, thus decreasing transmission of nerve impulses at the synapses in all sympathetic and parasympathetic ganglia at nerve–muscle junctions. There is usually a transient excitatory phase and, with high enough dosage, it is followed by a secondary and persistent decrease of nerve impulse transmission leading to skeletal muscle paralysis. Constricts arteries.

Burning sensations in the mouth, throat, and stomach are followed by increased salivation, nausea, vomiting, abdominal pain, violent diarrhea and purging, headache, sensations of sweating (but, which does not occur), flushing, decreased knee-jerk reflex, fatigue, auditory and visual disturbances, miosis, bronchoconstriction, deep and rapid respirations, increased blood pressure, increased ***oxygen*** requirement by heart muscle, tachycardia followed by bradycardia, mydriasis, and, then, exhaustion, which precedes collapse, with or without convulsions. This is followed by prostration, extreme collapse, terminal convulsions, and death from respiratory paralysis, which can occur within 2 h after the first symptoms or exposure. Blood is usually bright red and mesenteric blood vessels are severely hyperemic on post-mortem.

Prenatal exposure forces adrenal cells to mature prematurely, and, in the first year of the **child's** life, it is unable to secrete stress hormones on demand. Can cause small and premature **infants**. Use by the **mother** causes **fetal** tachycardia and decreased blood flow to it. **Babies** of nursing **mothers** who smoke 20 ***cigarettes***/day may show signs of restlessness, diarrhea, vomiting, tachycardia, and decreased ***oxygen***-carrying ability of **their** blood. **Some** receive 9.3 µg/kg/day. Also decreases milk secretion. Ingesting one or two ***cigarette*** butts in **children** leads to nausea and vomiting. Some ***cigarettes*** contain 20–30 mg nicotine in each and can be lethal to a small **child**. Suspect as a cause of spontaneous abortion and **fetal** and **neonatal** death. Breast milk of **smokers** contains ~200–500 ppb.

The use of nicotine insecticides and snuff extracts has caused accidental, homicidal, and suicidal deaths, in addition to many cases of poisoning. A 54-year-old **male** who crushed five to ten ***cigarettes*** in a pan of water, heated it, and used the solution as an enema for constipation developed nicotine poisoning. Eliminated from the body in ~16 h. Excretion is increased in acid urine and decreased in alkaline urine.

Reversible widespread segmented cerebral artery narrowing with severe headache, transient neurological deficits, nausea, and visual hallucinations in a 62-year-old **female** after use of a 30 mg nicotine patch. Severe anxiety, agitation, nausea, and vomiting in a schizophrenic 27-year-old **female** after use of a patch while still smoking. Recurrence of a myocardial infraction in a 47-year-old **male** using a 21 mg patch and smoking for ~5 min. Atrial fibrillation in a 55-year-old **female** after use of a single 52.5 mg patch. A 60-year-old **female** on 250 mg ***niacin*** twice daily/3 years developed flushing and shaking 3 days after a 21 mg patch/day. All of these **patients** used the patches for smoking cessation.

Palpitations, premature ventricular contractions, paroxysmal atrial fibrillation, pulmonary mucous membrane irritation, chronic bronchitis, pulmonary emphysema, constriction of coronary and superficial limb blood vessels, oliguria, hyponatremia, contact dermatitis, "skin aging", increased clotting tendency of blood, aggravation of Crohn's disease, increased risk of lung cancer and mouth lesions, stimulation of liver microsomal enzyme activity, tolerance to it, brown discoloration of nails from handling it and increased basal metabolic rate. Secondary smoke problems in **non-smokers**, addiction, and up to a three times increase in CYP1A2 drug metabolizing activity are additional problems associated with its use. Rarely, amblyopia. Hiccups in a 51-year-old **female** after repeated use of nicotine gum; jaw soreness and gastrointestinal upsets in **others**. It constricts arteries helping their clogging by ***fats*** and ***cholesterol***. Use can cause night sweats. Toxicity in a 40-year-old **male** after covering two transdermal patches with ***Saran Wrap***® while showering. The nicotine inhaled by **smokers** will continue to increase blood pressure for at least 30 min afterwards. Addiction can occur within a few days.

Toxic levels are 0.5–1.0 mg% and lethal levels reported as 0.5–5.2 mg%. Minimum lethal dose, **human**: 0.36 mg/kg; **adult**: 40 mg (~0.5–1 mg/kg); topically, **adult**: 65 mg; **child**: 4 mg. National Institute for Occupational Safety and Health/OSHA limits for skin are 0.5 mg/m^3.

In **animals**, although initial nerve cell ganglionic stimulation may occur, depression and progressive cerebrospinal (downward progression) paralysis is the rule on overdoses or deliberate immobilizing dosage. Sudden death can occur. Considerable individual and species variation exists in regard to toxicity, benefits, and rate of elimination. It is also extremely dangerous in **man**. Keep from unauthorized **persons**. Acute and chronic administration to **rats** increases liver microsomal enzyme activity. In immobilizing **animals**, it is important to remember that they remain conscious and it may increase the **animal's** fear. Vapors from the use of the *sulfate* on perches have killed **poultry** within a few minutes. They show dizziness and tremors, and stagger, become unconscious, and die. Poisoned **chicks** also show mydriasis and congestion of their nictitating membranes. Inhibits spinal reflexes in **cats**. **Puppies** have been poisoned by chewing on cigar ends. Use of the *salicylate* form as an immobilizing agent in **goats** and beef **heifers** caused trembling, violent shakes, and increased salivation. Carcinogenicity trials in **mice**, **rats**, and **guinea pigs** were negative. Given to pregnant **rabbits**, it caused fetal resorption, decreased fetal size, and increased the number of stillbirths. SC or IP to pregnant **mice** caused death in their progeny, without total resorption. Teratogenic (adactyly, brachydactyly, cleft palates), in **mice** if given before limb bud development. Tolerance to it or addiction noted after 6 days in **rats**. Limb deformities in **pigs** from treated **sows**. Fetal brain stem cell damage after its use in drinking water for **rats**, due to anorexia, suggesting it may be a cause of sudden infant death in **humans**. Addiction in **rats** when given an opportunity to selfadminister. Medial calcification of blood vessels has followed chronic use in **dogs**.

~Minimum lethal dose, **horse** and **cow**: 200–300 mg; **sheep**: 100–200 mg; **cat** and **dog**: 20–100 mg; 40 g **mouse**: 5 μl or 5/70 drop (5 mg). ~Minimum lethal dose of the *sulfate* - 18-month-old **lambs**: 200–300 mg; adult **sheep**: 0.7–1.0 g, 1–1½-year-old **bulls** and **heifers**: > 1 g.

LD_{50}, **rat**: 53 mg/kg; **mice**: 16 mg/kg; **dog**: 9.2 mg/kg; **pigeon** and **duck**: 75 mg/kg.

NICOTINYL ALCOHOL

= *Nicotinic Alcohol* = *NU 2121* = *2-Pyridine-methanol* = *Roniacol* = *Ronicol*

***Peripheral vasodilator. Riboflavin* solubilizer.**

The *tartrate* is *Niltuvin* = *Radicol* = *Tigacol*

Untoward effects: Transient flushing of face and neck, pruritus, rash, urticaria, paresthesias, slight facial swelling, hypotension, gastric pain, nausea, vomiting, and anorexia. Light-headedness, occassionally severe headache, and hepatocellular toxicity. Suspect as a causative agent in increased intraocular pressure and glaucoma.

LD_{50}, **mice**: 3 g/kg.

NIFEDIPINE

= *Adalat(e)* = *Adapress* = *Aldipin* = *Alfadat* = *Anifed* = *Aprical* = *Bay a 1040* = *Bonacid* = *Camont* = *Chronadalate* = *Citilat* = *Coracten* = *Cordicant* = *Cordilan* = *Corotrend* = *Duranifin* = *Ecodipin* = *Hexadilat* = *Introcar* = *Kordafen* = *Nifedicor* = *Nifedin* = *Nifelan* = *Nifelat* = *Nifensar XL* = *Orix* = *Oxcord* = *PA 20* = *Pidilat* = *Procardia* = *Sepamit* = *Tibricol* = *Zenusin*

Antihypertensive, antianginal, calcium channel blocker. A ***dihydropyridine***.

Untoward effects: ***Cardiovascular***: Palpitations and pulmonary hypertension edema; occassionally arrhythmias, increased angina after first dose or an increase in dosage, hypotension, syncope, cerebral ischemia, congestive heart failure, angina, tachycardia, hypokalemia, myocardial infarction, aplastic anemia, and death.
Central nervous system: Nervousness, anxiety, acute psychosis, headache, paresthesia, insomnia and somnolence, vertigo, paroniria, and stroke. Occasionally ataxia, hypoesthesia, depression, dizziness, headache, migraine, and tremors.
Gastrointestinal: Dry mouth, upset stomach, nausea, vomiting, flatulence, GERD, abdominal pain, diarrhea, constipation, and melena. Near total obstruction of small intestines in a 77-year-old **male** by an extended release from ***Procardia XL***.
Miscellaneous: Gingival hyperplasia, myalgias, erythromelalgia, leg cramps, back pain, decreased libido, impotence, priapism, fatigue, and gout. Facial, periorbital, ankle, and pulmonary edema. Lacrimation, tinnitus, altered taste and smell, vision difficulties, hyperhidrosis, dyspnea, coughing, epistaxis, sinusitis, hyperglycemia, menorrhagia, parotitis, esophageal injury, bezoars from fragments of extended-release tablets, gynecomastia (unilateral or bilateral), breast pain, polyuria, dysuria, nocturia, and hematuria. Diabetogenic in a 60-year-old **male** who had to increase daily ***insulin*** by 30%. Cases of severe chest pain simulating myocardial infarction or acute angina within 13 min after 10 mg. Acute hypertension with pallor, dyspnea, sweating, and pulmonary edema has followed abrupt withdrawal of treatment. Rarely, hepatitis. A case of bilateral occipital lobe infarction with bilateral blindness precipitated by a single 20 mg dose. Within 30 min of a deliberate overdose with ingestion of 900 mg, hypotension, bradycardia, hyperglycemia, and sinus and atrioventricular node dysfunction occurred in a 59-year-old **male**. A 14-month-old **child** ingested ~800 mg and was unresponsive, markedly hypotensive, hyperglycemic, and had a cardiac arrest at the emergency room, with recovery after treatment. Exacerbation of myasthenia gravis reported. A 77-year-old **female** had peptic stricture due to a bezoar of extended-release tablets taken a year earlier.
Skin: Rash, pruritus, bullous fixed-drug reactions (pemphigus), and exfoliative dermatitis; occasionally ankle, feet, or leg swelling, alopecia, purpura, flushing, telangiectasis, and photosensitivity reactions.

Drug interactions: Use with ***β-blockers*** (***acebutalol***, ***alprenolol***, ***atenolol***, ***betaxolol***, ***metoprolol***, ***nadolol***, ***propranolol***, and topical ***timolol***) has caused hypotension, cardiac failure, myocardial infarction, and even thyrotoxicosis. Withdraw ***β-blocker*** therapy slowly over several weeks before starting on nifedipine therapy. Use with ***digoxin*** has caused dramatic increases in the latter's serum concentration by up to 45%. I found no reports of such interaction with ***digitoxin***. Interaction with ***quinidine*** varies from slight increases in serum levels of both, to inability to sustain ***quinidine***

levels. Profound hypotension and neuromuscular blockade in **women** receiving IV ***magnesium sulfate***. ***Cimetidine*** decreases its hepatic metabolism, increasing its antihypertensive effect. Many negative ill-effects reported with ***theophylline***, but a clinical report indicated it decreased ***theophylline*** clearance by 9%. A **patient** with angina had anginal symptoms exacerbated by coadministration or rechallenge with ***rifampin***. ***Tacrolimus*** dosage in liver transplant **recipients** could be decreased by 25–35% in nifedipine-treated **patients**. ***Diltiazem*** prolongs its elimination half-life. Use with ***terfenadine*** precipitated an acute anginal attack in a 43-year-old **female**. The action of ***insulin*** and ***oral hypoglycemics*** is interfered with by nifedipine. ***Grapefruit juice*** augments its bioavailability by 8–69%. The red **man** phenomenon (macular rash, hypotension, and headache) within 20 min of starting a 1 g ***vancomycin*** infusion. Gingival hyperplasia and increased blood urea nitrogen when given with ***cyclosporine***.

TD_{LO}, **human**: 200 μg/kg.

Profound hypotension and reflex tachycardia in **dogs** with overdoses. Massive overdoses cause decreased nodal conduction leading to bradycardia.

LD_{50}, **rat**: 1 g/kg; **mouse**: 494 mg/kg.

NIFLUMIC ACID

= *Actol* = *Forenol* = *Landruma* = *Nifluril* = *UP-83*

Anti-inflammatory.

Untoward effects: Nausea, gastric pain, vomiting, diarrhea, unexplained decrease in hemoglobin, and gastrointestinal hemorrhages.

LD_{50}, **rat**: 370 mg/kg; **mouse**: 455 mg/kg.

NIFURATEL

= *Inimur* = *Macmiror* = *Magmilor* = *Methylmercadone* = *Nitrofuratel* = *Omnes* = *Polmiror* = *Tydantil*

***Antifungal, antibacterial, antitrichomonal.* Oral and intravaginal use.**

Untoward effects: Occassionally abdominal discomfort, nausea, vomiting, anorexia, metallic taste, headache, breathlessness, rash, drowsiness, and memory loss reported. Nausea and indigestion when taken with ***alcohol*** or ***beer***.

NIFUROXIME

= *Micofur*

Antifungal, antitrichomonad.

Untoward effects: Generalized urticaria sensitization, irritation, burning, erythema, pruritus, and cholestatic hepatitis in **women** after intravaginal use.

No acute toxicity in **chicken** and **rabbit** trials, but some weight loss in chronic trials.

IP, LD_{LO}, **mouse**: 100 mg/kg.

NIFURPRAZINE

= *Carofur* = *Furenazin* = *NF 1002* = *HB 115*

***Antibacterial.* For topical use.**

Untoward effects: Initial burning sensation after use reported in many **patients**. Contact allergies occur. Suspect carcinogen for man.

NIFURTIMOX

= *Bayer 2502* = *Lampit*

Use: In treatment of acute American trypanosomiasis (Chagas' disease, *T. cruzi*).

Untoward effects: Reactions are generally mild and more common in **adults** than **children**. Nausea, vomiting, anorexia, weight loss, loss of memory, tremors, excitation, vertigo, headache, rashes, paresthesias, weakness, myalgia, sleep disorders, and peripheral polyneuritis. Rarely, fever, pleural effusions, pulmonary infiltrates, and convulsions.

No evidence of carcinogenicity in **rat** trials. At very high dosage, **rats** evidenced ruffled coat, hind limb paralysis, dyspnea, and convulsions before death; ruffled coats in **mice**, vomiting in **cats** and **dogs**, and myoclonic convulsions in **dogs**.

LD_{50}, **rat**: 4 g/kg; **mouse**: 3.72 g/kg; **rabbit**: ~2 g/kg; **cat**: > 2 g/kg; **dog**: > 4 g/kg.

NIHYDRAZONE

= *Furiton* = *HC 064* = *NF 64* = *Nidrafur* = *Zonifur*

***Coccidiostat.* Fed to poultry at a level of 0.011% (100 g/ton).**

Untoward effects: Some growth decrease in **chicks** at > 256 g/ton of feed. Insignificant decrease of fertility or hatchability of eggs at 180 g level, but marked decrease at 240 g/ton level. Toxic symptoms in **poultry** with overdosing include incoordination and curled toes. **Swine** eating it at 100 g level showed no toxicity. **Dogs** can tolerate a single dose of 400 mg/kg, but was lethal when given on three consecutive days. **Dogs** consuming 25–50 mg/day/28 days were normal. Doses of 100 mg/kg/day were lethal to **goats**, but a 25 mg level/28 days showed no ill-effects. May be tumorigenic in **laboratory animals**.

LD_{50}, **chickens**: 300–400 mg/kg.

NIKETHAMIDE

= *Anacardone* = *Astrocar* = *Carbamidal* = *Cardiamid* = *Cardimon* = *Coracon* = *Coractiv N* = *Coramine* =

Cordiamin = Corediol = Cormed = Cormid = Corvitol = Corvotone = Diethylnicotinamide = Dynacoryl = Eucoran = Inicardio = Niamine = Nicamide = Nicor = Nicorine = Nicotinic Acid Diethylamide = Nikardin = Nikorin = Pyricardyl = Salvacard = SD 1 = Stimulin = Ventramine

***Central nervous system and respiratory stimulant.* A derivative of *nicotinic acid*. Usually IV in man. IV, IM, or SC in animals.**

Untoward effects*:** Convulsive in therapeutic doses. No damage to cells outside the central nervous system. Sweating, nausea, vomiting, flushing, sneezing, coughing, muscle twitching (especially facial), restlessness, fear, tachycardia, increased blood pressure, skin irritation, anxiety, hyperhidrosis, and it has induced acute attacks in porphyrias. A vasoconstrictor, and it is not to be used in treating comatose **children** from ***phenothiazine tranquilizer poisoning. Topical use has caused redness and pustules, with subsequent hyperpigmentation. Large doses have caused epileptiform convulsions, followed by central nervous system depression.

LD, **human**: ~12 g IV or 175 mg/kg.

Visible excitement in albino **rats**, **guinea pigs**, **cats**, **rabbits**, and **pigeons** with 100–250 mg/kg SC. Convulsions in some species at 125 mg/kg.

Oral LD_{50}, **mouse**: 188 mg/kg; IV, LD, **dog**: 150–200 mg/kg; IP, LD_{50}, **rat**: 272 mg/kg.

NILUTAMIDE

= Anandron = Nilandron = RU 23,908

Antiandrogen.

Untoward effects*:** Nausea, vomiting, interstitial pneumonitis, hot flashes, and ***alcohol intolerance, followed by gynecomastia, hepatotoxicity, and delayed visual adaptation to darkness.

NIM OIL

= Neem Oil = Margosa Oil = Veepa Oil

A fixed oil with ***garlic***-like odor from the seed-kernel of the Indian ***neem tree***, *Azadirachta indica*, found in India, Pakistan, Malaysia, and parts of Africa. **Robins** are narcotized by eating the fruit or the bark; **sheep** and **pigs** become paralyzed with respiratory depression; and **frogs** and **rabbits** are unaffected. Used as a tonic, antipyretic, insect repellent, antiseptic, in toothpastes, and topically on skin diseases.

***Untoward effects*:** Reye's syndrome from poisoning by 5–30 ml for minor ailments in 13 **infants** and **children** (aged 21 days to 4 years). Symptoms included drowsiness, tachypnea, and generalized seizures with a duration of a few minutes to several hours. Liver biopsies demonstrated fatty degeneration and mitochondrial damage.

In **rats** and **rabbits**, the lungs and central nervous system are target organs for its toxicity. Hypoglycemic in **rats**. **Dogs** have died after two drops were put on their gums. The vapors from a glass rod moistened with it killed a small **bird** when the rod was placed near its beak. Resistant to putrefaction and can be isolated from dead **animals** 3 months after death.

LD_{50}, **rat**: 14 ml/kg; **rabbit**: 24 ml/kg.

NIMESULIDE

= Aulin = Flogovital = Mesulid = Nisulid = R 805

Anti-inflammatory. Non-steroidal anti-inflammatory drug.

Untoward effects*:** Usually mild, and primarily gastrointestinal. A few superficial gastric erosions and ulcerations reported. Minor renal toxicity noted on 800 mg dosage, but not at 400 or 600 mg. Thrombocytopenia in a 29-year-old **male** with HIV infection. Of 940 **patients** (15–77 years) receiving 100 mg twice daily/mean of 10 days, only 75 (8%) reported adverse effects and only 26 (~2.8%) withdrew from treatment. Fulminant liver failure in a 58-year-old **female** ingesting 100 mg twice daily on two occasions required a liver transplant. Death followed in 12 h from a multiorgan failure. No clinically significant interactions in **patients** taking 200 mg/day/7 days and ***acenocoumarol, digoxin, theophylline, or ***warfarin***. ***Non-steroidal anti-inflammatory drug***s are not recommended in **women** trying to become pregnant, since they interfere with implantation of fertilized eggs in **animals**. Caution should be used if taken with drugs that adversely affect renal function.

NIMODIPINE

= Adalat = Admon = Bay e-9736 = Nimotop = Periplum

***Cerebral vasodilator, calcium channelblocker.* A *dihydropyridine*.**

Untoward effects*:** Bradycardia in a 65-year-old **female** receiving 80 mg tid after an acute ischemic stroke. Increased risk of intraoperative and postoperative hemorrhage in 149 **patients**, due to 30 mg 12 h before open-heart surgery. Inadvertent IV injection of capsule contents intended for nasogastric administration in a 58-year-old **male** caused bradycardia and aberrant conduction on electrocardiogram. Low incidence of adverse effects. Hypotension, thrombocytopenia, acne, tachycardia, anemia, nausea, indigestion, gastrointestinal reflux hemorrhage, palpitations, flushing, wheezing, dizziness, psychiatric disturbances, hematoma, headache, bradycardia, rebound vasospasms, and hypertension. Adverse effects tend to increase with increased dosage. Limited amounts pass into breast milk. No significant clinical interactions with ***diazepam, ***glyburide***, ***indomethacin***, ***nifedipine***, and ***tirilazad***.

Drug interactions: ***Propranolol*** decreases its mean peak plasma level by nearly 25%. ***Grapefruit juice*** increases its bioavailability and serum levels 24–51%.

LD_{50}, **rat**: 6.6 g/kg; **mouse**: 3.56 g/kg.

NIMORAZOLE

= *Acterol* = *Esclama* = *K 1900* = *Naxofem* = *Naxogin* = *Nitrimidazine* = *Nulogyl*

Trichomonacide, antiprotozoal. **For intravaginal use.**

Untoward effects: Nausea and anorexia. Occassionally rashes, vertigo, drowsiness, and hypersensitivity. Use with ***alcohol*** will increase nausea.

LD_{50}, **rat**: 3.2 g/kg; **mouse**: 1.5 g/kg.

NIRIDAZOLE

= *Ambilhar* = *BA 32,644* = *Ciba 32,644-Ba* = *Nitrothiamidazol*

Schistosomicide, trichomonacide, amebocide, antimicrobial. **Also used to treat *Dracunculus* infections.**

Untoward effects: Abdominal cramping and vomiting (20/25), diarrhea and anorexia (7–30%), are common during treatment. Occassionally eosinophilia, headache, dizziness, erythematous rashes, 44% decrease in serum ***uric acid*** concentration, myalgia (10%), arthralgia (3%), reddish or brown-black urine, flushing, sweating, tachycardia, off-taste, unpleasant body odor, electrocardiogram changes (lengthening of QT interval and flattening or inversion of T wave in 43/75 **patients**), and neuropsychiatric upsets (insomnia, confusion, giddiness, hallucinations, anxiety, and convulsions). Hemolytic anemia may occur in glucose-6-phosphate dehydrogenase-deficient **patients**. Allergic reactions (dyspnea and pulmonary infiltrations) and paresthesia have also been reported. Epileptiform convulsions are rare, but can be serious and have caused death. **Children** usually experience fewer reactions than **adults**. Suspect carcinogen in **man**.

Carcinogenic in **mice** and **hamsters** (forestomach and urinary tract). Administration of 30 mg/kg/day orally to **dogs** for 4–5 weeks caused severe toxic bone marrow depression and hemorrhagic disorders in about ½ the test **animals**. Similar levels and chronic trials in **goats**, **sheep**, **monkeys**, and **rats** led to no discernible myelotoxicity. Temporary impairment of spermatogenesis has followed its use in **goats**, **sheep**, and **mice**.

NISOLDIPINE

= *Bay K-5522* = *Baymycard* = *Sular* = *Syscor* = *Zadipina*

Calcium channel blocker, antihypertensive, antianginal, peripheral vasodilator. A ***dihydropyridine***.

Untoward effects: Dramatic increase in peak plasma concentration of both when given with ***propranolol*** and increases peak plasma concentration of ***atenolol*** by ~20%. Its bioavailability decreased 25% in the presence of ***quinidine***. ***Grapefruit juice*** or ***naringin*** increase its maximum serum concentration and decrease its time to reach maximum serum concentration.

NITARSONE

= *Hep-A-Stat* = *Histostat*

Histomonostat. **Contains ~30% *arsenic* (q.v.). An aid in preventing blackhead (histomonads) in chickens and turkeys, and in the prevention and treatment of bloody enteritis in swine.**

Untoward effects: Maintain adequate fresh water availability, especially near feeding areas, to help decrease potential toxicity. Do NOT feed to other classes of **birds**. Use NO other feed or water sources of ***arsenic***. Withdraw treated feed 5 days before slaughtering **birds** for **human** consumption. Do NOT feed to laying **flocks**. Not FDA approved for **swine**. Dangerous to **ducks**, **geese**, and **dogs**. Overdosage or lack of adequate water intake has caused leg weakness and paralysis.

LD_{50}, **rat**: 100 mg/kg.

NITHIAZIDE

= *Hepzide*

Protozoastat. **As an aid in the prevention and treatment of blackhead (histomonads) in chickens and turkeys and hexamitiasis in turkeys.**

Untoward effects: **User** should avoid inhaling its irritant dust, which may result in inflammation of the eyes, nose, throat, and skin. Withdraw medication 24 h before slaughtering **birds** for **human** consumption to avoid tissue residues. U.S. regulations require that, if used for laying **hens**, eggs are to be used for hatching purposes only. Occassionally causes tumors of the skin, zymbal gland, preputial/clitoral gland, and mammary gland of **rats**.

LD_{50}, **mouse**: 2,150 mg/kg.

NITRAMINE

= *Methyltetranitroaniline* = *Tetralite* = *Tetryl*

Use: As an explosive in detonators.

Untoward effects: **Workers** handling it have a distinct yellow staining of **their** hair and skin of **their** hands and neck. Skin rashes, pruritus, and erythema. Edema of nasal folds, neck, and cheeks. Keratitis, sneezing, anemia, fatigue, coughing, headache, lassitude, insomnia, anorexia, nosebleeds, nausea, and vomiting reported. Many of these reactions reported in military facilities during World Wars I and II. May cause liver and kidney damage. National Institute for Occupational Safety and Health/OSHA upper limit for skin exposure is 1.5 mg/m^3. Immediate danger to life or health: 750 mg/m^3.

SC, LD_{LO}, **dog**: 5 g/kg.

NITRATES

Untoward effects: Its presence in drinking water and foods has regularly been a cause of alarm. It is a common colorless and odorless contaminant of rural shallow wells and high level intake has been associated with methemoglobinemia and anemia, especially in **infants** (Blue Baby Syndrome, with bluish tinge on nose, lips, face, fingers, toes, and ear tips; diarrhea, lethargy, and occassionally deaths) whose intestinal bacterial flora don't prevent rapid conversion of nitrates to ***nitrites***, and also due to lower gastric acidity than in **adults**, which permits certain bacteria to convert nitrates to ***nitrites***. Storage of opened ***baby food*** or cooked ***vegetables*** reduces the nitrate content to ***nitrites***. "Normal" intestinal flora develop between 6–12 months of age. ***Fertilizer*** run-off, feedlots, and barnyards can be the source. Intake has been associated with miscarriages in pregnant **women**. Water containing > 10 mg/l as ***nitrogen*** or 45 mg/l as nitrate should not be used for **infants** or pregnant **women**. Cyanosis occurs when the level of ***methemoglobin*** reaches 10% and hypoxia is obvious at levels > 20%. In a series of 278 cases, 39 **infants** died; in 30 other cases, eight died. Toxic methemoglobinemia in 56-year-old **male** on home dialysis using well water with nitrates at 94 mg/l. ***Beets, celery, lettuce, radishes, rhubarb***, and ***spinach*** contain ~0.1–0.2% nitrate. Preserved meats including ***bologna, hot dogs***, and ***salami*** contain high levels and have caused "***hot dog*** headache", a pounding vascular headache starting within a few minutes after ingestion. Use in such meat products is essential to prevent botulism and intake by **humans** is usually less than from ***vegetables***. Conversion to ***nitrites***, then to ***nitrosamines*** can cause gastric cancer in **man**. Methemoglobinemia (9 g/100 ml) in a 2-week-old **Negro male** who consumed 500 ml ***carrot juice*** in the previous 24 h. Remaining ***carrot juice*** was assayed 48 h later and it had 525 ppm nitrate and 775 ppm ***nitrite***. ***Spinach*** stored in the refrigerator has contained 150 mg ***nitrite***/100 g causing methemoglobinemia of 54% with cyanosis, drowsiness, tachypnea, vomiting, pulse 160/min, and blood pressure 100/70 in a 2-year-old **male** after ingesting 300–400 g twice weekly. Bacterial action on the nitrates in the ***spinach*** produced ***nitrites***. Similar cases in a 2-month-old **male** and a 3½-month-old **male** with cyanosis 45 min and 2 h, respectively, after eating mashed ***vegetables*** containing ***spinach***. Unlike **babies**, **adults** usually absorb all nitrate before it is chemically reduced. Methemoglobinemia in three **males** from absorption of nitrate salts (***sodium nitrate*** and ***potassium nitrate***) through burned skin in an industrial accident. Cyanosis in **all** and death in **one** with ***methemoglobin*** concentration of 65%. Dermatitis from skin contact can be due to a true allergy or ulcerations when the chemical, such as ***calcium nitrate***, enters abrasions. Intake of nitrates has been associated with increased intraocular pressure and glaucoma. Organic nitrate poisoning in ~10% of **workers** in a ***dynamite*** factory, possibly, from reusing surgical gloves contaminated with ***ethylene nitrate***. Also see specific nitrates and various discussions under different botanicals.

Cattle are frequently poisoned by it (LD: ~500 mg/kg or 2% of their diet). **Sheep** and **horses** are also poisoned by it. **Animals** often safely ingested it at 0.5% of their diet, but 1% can be toxic. Nitrate ***fertilizers*** have been consumed with toxic effects. As with **humans**, high unsafe levels have been consumed in drinking water and in plants (total content of up to 10–30%). Even crop plants such as ***corn, oats***, or ***sorghum*** can contain high levels.

NITRAZEPAM

= Benzalin = Calsmin = Demethyldiazepam = Eatan = Eunoctin = Imeson = Insomin = Ipersed = LA-1 = Mogadan = Mogadon = Nelbon = Neuchlonic = Nitrados = Nitrenpax = Noctesed = Pelson = Radedorm = Remnos = Ro-4-5360 = Ro 5-3059 = Sameko = Somnased = Somnibel = Somnite = Sonebon = Surem = Unisomnia

Hypnotic, anticonvulsant. A benzodiazepine.

Untoward effects: Drowsiness, light-headedness, and hangover effect. Confusion, disorientation, giddiness, vivid nightmares, and hypothermia can occur in older **patients**. Ataxia, vertigo, asthenia, headache, fatigue, and occassionally emesis and restlessness with high dosage. Impaired work and driving performance for over 24 h after a 10 mg dose. **Children** receiving high dosage may have excessive salivation and mucous secretion, ataxia, aspiration pneumonia, slurred speech, and irritability. May aggravate grand mal seizures and impair renal function. Paresthesias, acute gout attacks, worsening of respiratory problems, and skin rashes occur. Addiction occurs, but is unusual. Central nervous system stimulation is unrelated to age, but it is to dosage. Placental transfer occurs and use during the first trimester of pregnancy was associated with cleft lips, with or without cleft palates. Floppy **infant** syndrome after **maternal** use during pregnancy. Lethargy and weight loss in a nursing **infant** after the **mother** took 10 mg tid/2 days. Self-poisoning from overdoses (lengthy coma with bullae development) and suicides have resulted in some deaths. Plasma levels peak in 1–12 h. Plasma half-life in **man** is reported as 7–48 h. Delirium tremens and opisthotonus have followed withdrawal of treatment. Therapeutic concentrations are 0.05–0.15 mg/l and toxic levels reported as 0.2–0.5 mg/l.

Drug interactions: Use with ***alcohol*** impairs attention. Given to a **male slow-acetylator** stabilized on ***phenelzine***, **he** developed symptoms typical of ***monoamine oxidase inhibitor*** toxicity. Its hepatic clearance is decreased by ***cimetidine***. Use with ***phenytoin*** causes hyperglycemia and increased ***phenytoin*** toxicity. Its metabolism can be decreased by ***oral contraceptives***. Can give spuriously high ***catecholamine*** test readings.

LD_{50}, **rat**: 825 mg/kg; **mouse**: 1.8 g/kg; IP, LD_{50}, **rat**: 733 mg/kg; **mouse**: 275 mg/kg; IV, LD_{50}, **mouse**: 130 mg/kg; **rabbit**: 520 mg/kg.

NITRENDIPINE

= *Bay E 5009* = *Bayotensin* = *Baypress* = *Bylotensin* = *Deiten* = *Nidrel*

Calcium channel blocker, antihypertensive. A dihydropyridine.

Untoward effects: Headache, flushing, tachycardia, edema, fatigue, polyuria, and agranulocytosis requiring discontinuance in 5–10% of **patients**.

Drug interactions: Its metabolism is inhibited by ***grapefruit juice*** and ***cimetidine***. ***Grapefruit juice*** increases its bioavailability by 40–100%.

LD_{50}, **rat**: > 10 g/kg; **mouse**: 2.54 g/kg.

NITRIC ACID

= *Aqua Fortis* = *Azotic Acid* = *Engraver's Acid* = *Hydrogen Nitrate*

In manufacturing *nitrates*, fertilizers, dyes, and explosives. Also used in metal etching and manufacture of felt *hats*.

Untoward effects: Caustic! Violent oxidizing agent. Even when diluted, it first blanches the skin, then turns it an intense yellow. Fumes cause severe eye, respiratory tract, and skin irritation, and dental erosions. After exposure, these irritations usually cease, but fluid continues to build in pulmonary alveoli. Immediate coagulation of burned skin. Ingestion has caused acidosis, nausea, bradypnea, convulsions, shock and death can occur. Corrosive action in the mouth and on gastrointestinal mucous membranes. Tachycardia, pulmonary edema, laryngitis, bronchitis, pneumonitis, and dyspnea from inhalation of fumes. Improperly packaged material in an aircraft cargo hold leaked into sawdust packaging; the fumes (smoke) entered the cockpit and obscured the **crew's** vision, causing a Pan-Am 707 cargo plane to crash, killing the entire **crew**. Has been ingested in suicides; when the intent was homicidal, the **victim** was either a **child** or an **adult** rendered unconscious by sleep or drunkenness. An **adult** is killed by 10–12 ml in 2–24 h. **Children** die from spasmodic closure of the larynx. OSHA/ National Institute for Occupational Safety and Health upper exposure limit is 2 ppm (5 mg/m³) TWA. Immediate danger to life or health level is 25 ppm.

LD_{LO}, **human**: 430 mg/kg.

NITRIC OXIDE

= *Nitrogen Monoxide*

Use: In manufacturing rocket propellents, neoprene ***rubber***, bleaching rayon, and vasodilation in treatment of cyanotic attacks due to pulmonary hypertension.

Untoward effects: The main ingredient in smog and is also found in silo gas. Eye, nose, throat, and skin irritant. Inhalation leads to drowsiness, unconsciousness, pulmonary edema, and methemoglobinemia, decreasing the blood's ***oxygen***-carrying capacity. Pleural bleeding in a 37-year-old **female**; intracranial bleeding and death in a 59-year-old **female** 2 days after inhalation treatment for pneumonia. May be a factor in inducing lower motor neuron disease in an **alcoholic** 58-year-old **female** after lung transplantation. Rebound pulmonary hypertension reported after treatment withdrawal. Neurodegenerative diseases associated with increased body levels and has been linked to impotence. Topically, it is pro-inflammatory and toxic to DNA. Exposure to air produces ***nitrogen dioxide*** (q.v.), which is very toxic. Has caused deaths from its contamination of anesthetic ***nitrous oxide***. National Institute for Occupational Safety and Health/ OSHA upper exposure limit is 25 ppm (30 mg/m³). Immediate danger to life or health level is 100 ppm.

Inhalation LC_{LO}, **mouse**: 320 ppm; LC_{50}, **rabbit**: 315 ppm/15 min.

NITRILOTRIACETIC ACID

= *NTA* = *Triglycine*

Use: Chelating agent in metal plating and cleaning, pulp and ***paper*** processing, ***leather*** tanning, synthetic ***rubber***, and as a substitute for phosphates in synthetic detergents.

Untoward effects: Fed to **rats** causes nephritis and nephromas; carcinomas of the ureter in both sexes, of the kidney in **male**, of the bladder in **female**. Oral use in **dogs** increased urinary excretion of ***chondroitin sulfate***. Readily absorbed from the gastrointestinal tract of **rats** and **dogs**, but not by **rabbit**, **monkey**, or **man**.

LD_{50}, **rat**: 1.46 g/kg; **mouse**: 3.16 g/kg.

NITRITES

Many have therapeutic uses and their own specific toxicity.

Untoward effects: In general, adverse effects concentrate on the fact that they convert (oxidize) hemoglobin to ***methemoglobin***, which is ineffective in transporting ***oxygen*** to tissues. Symptoms include nausea, vomiting, colic, headache, diarrhea, hypotension, flushing, respiratory depression, tachypnea, tachycardia, tremors, staggering, cyanosis, stupor, coma, and death by asphyxiation. Bacterial reduction of ***nitrate*** (q.v.) in **human** saliva can produce large quantities of nitrites, an average intake of 9 mg/day versus 2.4 mg supplied by ***cured meats***. Other sources are from use as a food preservative, in drinking water, and vegetables from industrial exposures and endogenous formation by bacteria in the intestines or the acidic environment of the stomach, varying especially in **babies**. They then combine with various ***amino acids*** and ***amines*** to produce ***nitrosamines***, some of which have been shown to be mutagenic and carcinogenic. ***Methemoglobin*** levels 50–90% are, invariably, fatal and symptoms are noticeable at a 30% level. Suspect as a cause of esophageal cancer in **Chinese** and stomach cancer in **man**. Various forms have become drugs of abuse by inhalation.

Conversion of ***nitrates*** to nitrites is excessive in open ***zinc*** galvanized water tanks. **Cattle**, **sheep**, and **goats** are especially susceptible to poisoning by ***nitrate***-accumulating plants, as ruminal bacteria convert it to nitrite, with rapid death from ingestion of ***oat hay*** and ***cornstalks***.

NITROANILINES

Dyestuff intermediates.

Untoward effects: *Meta* form = ***3-Nitrobenzenamine***. Very toxic. Percutaneous absorption occurs with methemoglobinemia and cyanosis. Liver damage can follow chronic exposure.

LD_{50}, **rat**: 535 mg/kg; **mouse**: 308 mg/kg.

Ortho form =***o-Nitraniline***.

LD_{50}, **rat**: 3.56 g/kg.

Para form = ***p-Nitroniline*** = ***4-Nitrobenzenamine*** = ***PNA***. Used as a ***rubber*** antioxidant, ***gasoline*** additives, in dyes, ***Para Red***, and pharmaceuticals. Strong nose and throat irritant. Causes methemoglobinemia, cyanosis, ataxia, tachycardia, dyspnea, vomiting, diarrhea, anemia, jaundice, respiratory arrest, and convulsions.

LD_{LO}, **human**: 5 mg/kg.

LD_{50}, **rat**: 3.25 g/kg; **mouse**: 812 mg/kg; **birds**: 75 mg/kg.

NITROBENZENE

= *C.I. Solvent Black 5* = *Essence of Mirbane* = *Nitrobenzol* = *Oil of Mirbane*

Use: Almost entirely (~1.5 billion lbs) for ***aniline*** production. Some in shoe ***polishes***, ***inks***, ***paint*** solvent, and glass cleaners, and as an artificial ***almond***-like flavor and odorant. Has been called ***Artificial Oil of Bitter Almonds***.

Untoward effects: Rapidly absorbed through the skin, respiratory tract, and gastrointestinal mucosa. Adverse effects often delayed for several days. Death can occur within a few hours if the **patient** is not treated after heavy exposure. Dizziness, headache, malaise, irregular pulse, cold skin, bluish nails, and cyanosis due to methemoglobinemia; and dermatitis from ***skin contact***.
Inhalation causes headache, tiredness, nausea, vomiting, vertigo, numbness, hyperalgesia, sleepiness, hypotension, anorexia, respiratory tract irritation, dyspnea, hepatosplenomegaly, slight jaundice, methemoglobinemia, muscular incoordination, and respiratory arrest.

Ingestion causes mouth and throat burning, gastric pain, nausea, vomiting, meteorism, hemorrhagic diarrhea, sweating, mydriasis, pallor, cold and blue-black extremities, methemoglobinemia, cyanosis, tachycardia followed by bradycardia, arrhythmias, hypotension, dark urine, severe headache, central nervous system depression, delirium, giddiness, coma, and convulsions.

Ingestion, inhalation, or dermal absorption has also caused anemias. 1 g or 15 drops can be fatal. Mortality ~20%. National Institute for Occupational Safety and Health/OSHA upper exposure limit and TLV is 1 ppm; immediate danger to life or health level is 200 ppm.

TD_{LO}, **human**: 200 mg/kg; LD_{LO}, **human**: 5 mg/kg.

Liver, kidney, and testicular damage reported in **animals**.

LD_{50}, **rat**: 640 mg/kg; LD_{LO}, **dog**: 750 mg/kg; **cat**: 2 g/kg; **rabbit**: 700 mg/kg. Skin LD_{50}, **rat**: 2.1 g/kg; skin LD_{LO}, **cat**: 25 g/kg; **rabbit**: 600 mg/kg.

p-NITROBIPHENYL

= *p-Phenylnitrobenzene* = *PNB*

Untoward effects: **Human** carcinogen on Ames test. Overexposure leads to lethargy, headache, dizziness, dyspnea, weakness, ataxia, methemoglobinemia, urinary burning, and acute hemorrhagic cystitis.

Bladder tumors reported in experimental **animals**.

LD_{50}, **rat**: 2.23 g/kg; **rabbit**: 1.97 g/kg; TD_{LO}, **dog**: 7 g/kg over 2 year period.

NITROCHLOROBENZENES

Dye intermediates.

Untoward effects: *Ortho* form = ***ONCB***. Causes methemoglobinemia, cyanosis, and anemia.

LD_{50}, **rat**: 288 mg/kg; **mouse**: 135 mg/kg.

Para form = ***PCNB*** = ***PNCB***. ~100 million lbs used annually in manufacturing agricultural and ***rubber*** chemicals, dyes, and pharmaceuticals. Very toxic. Rapidly absorbed through the skin, respiratory tract, and after ingestion. Severe methemoglobinemia, anemia, and cyanosis after very small quantities. Anoxia, unpleasant taste, throat dryness, blue-black mucous membranes, slate-gray skin, severe headache, nausea, vomiting, weakness, confusion, ataxia, vertigo, tinnitus, disorientation, heart-blocks, arrhythmias, painful urination, hematuria, oliguria, allergic contact dermatitis, shock, and coma occur. Soluble in oils and water.

In **animals**, spleen, kidney, bone marrow, and reproductive ill-effects.

LD_{50}, **rat**: 420 mg/kg; **mouse**: 650 mg/kg; SC, **rat**: 16 g/kg.

NITROETHANE

= *Nitroetan*

***Solvent*. Used in chemical synthesis.**

Untoward effects: Dermatitis in **man**. National Institute for Occupational Safety and Health/OSHA exposure limits of 100 ppm. Immediate danger to life or health level is 1,000 ppm.

Inhalation by **animals** leads to lacrimation, dyspnea, pulmonary rales, edema, narcosis, and liver and kidney pathology.

LD_{50}, **rat**: 1.1 g/kg; LD_{LO}, **rabbit**: 500 mg/kg.

NITROFEN

= *2,4-Dichlorophenyl-p-nitrophenyl Ether* = *TOK*

No longer manufactured or sold in the U.S. as a herbicide.

Untoward effects: Occupational exposure during manufacture and use was by dermal contact, inhalation, and ingestion.

Carcinogenic in **animals** after incorporation into their diets causing hepatocellular carcinomas and adenomas in **mice**, hemangiosarcomas of liver and spleen in male **mice**, and pancreatic adenocarcinomas in female **rats**.

LD_{50}, **rat**: 740 mg/kg; **mouse**: 450 mg/kg; **rabbit**: 1.62 g/kg; LD_{LO}, **cat**: 300 mg/kg.

NITROFURANTOIN

= *Berkfurin* = *Chemiofuran* = *Cyantin* = *Cystit* = *Dantafur* = *Fua-MEd* = *Furachel* = *Furalan* = *Furadantin* = *Furadantine MC* = *Furadoin* = *Furantoina* = *Furobactina* = *Furophen T-Caps* = *Ituran* = *IVADANTIN* = *J-Dantin* = *Macrobid* = *Macrodantin* = *Orafuran* = *Parfuran* = *Sarodant* = *Trantoin* = *Urantoin* = *Urizept* = *Urodin* = *Urolong* = *Uro-Tablinen* = *Welfurin*

Antibacterial.

Untoward effects: Relatively high incidence (> 10%).
Allergic: Acute asthma, pneumonitis, alveolar infiltrates, pneumonia (2 h to 1 year after treatment), pleural effusions, pulmonary hemorrhage, acute pancreatitis, dyspnea, cough, pulmonary edema and fibrosis, fever, chills, chest pain, eosinophilia, leukocytosis, granulomatous vasculitis, mild cyanosis, circulatory collapse, pruritus, urticaria, erythematous maculopapular rashes, calf tenderness, lupus syndrome, angioneurotic edema, and anaphylaxis have been reported. Rarely, chronic pulmonary fibrosis develops insidiously after prolonged therapy. A case of fatal fulminant hemorrhagic pneumonitis in a chronic **alcoholic** has occurred. These reactions have been more common in **women** than in **men**.
Blood: Megaloblastic anemia in **infants**, **adults**, and hemolytic anemia, especially in ***primaquine***-sensitive, ***fava bean***-sensitive, or glucose-6-phosphate dehydrogenase-deficient **individuals**. Thrombocytopenic purpura, granulocytopenia, leukopenia, pancytopenia, bone marrow depression, neutropenia, aplastic anemia, and agranulocytic pneumonia. Fatal aplastic anemia in a 78-year-old **male** after 300 mg/day/27 days. A very small amount has been measured in breast milk (in 1/5 after 100 mg; in 2/4 after 200 mg). This can be enough to trigger hemolytic reactions in glucose-6-phosphate dehydrogenase-deficient **infants**. Its use increases ***lactate dehydrogenase*** levels. Eosinophilia has occasionally persisted for weeks after discontinuation of therapy. Can inhibit blood coagulation by inhibiting platelet function.
Clinical test interference: Causes increased or false positive serum ***bilirubin***, urinary ***glucose*** and ***creatinine*** test results.
Eye: Nystagmus and amblyopia.
Fetus and neonate: Crosses the placenta causing **fetal** hemolysis, hyperbilirubinemia, and yellow teeth. The concentration in blood is higher in pregnant than non-pregnant **women**, and serum levels in the **embryo** and placenta were higher than in the **mother**. Causes hemolysis in glucose-6-phosphate dehydrogenase-deficient **infants**. Possible cause of icterus and kernicterus in **infants**.
Gastrointestinal: Epigastric pain, nausea, vomiting, diarrhea, erosive gastritis, and anorexia. Taken with ***food*** decreases these adverse effects.

Liver: Hepatitis, cholestatic jaundice, increased alkaline phosphatase and ***bilirubin***, and hepatic necrosis. Hepatic failure required a liver transplant in a 40-year-old **female** after 200 mg twice daily/1 month. Symptoms of hepatic encephalopathy occurred 2 weeks after starting treatment.
Miscellaneous: Bright yellow pigmentation of teeth during their eruption. Burning feet syndrome, porphyrias, transient alopecia, amblyopia, sweating, aching joints, and decreased **male** fertility. Deaths have been reported, secondary to liver failure and respiratory effects. Reabsorption can occur from alkaline urine. Intracranial hypertension reported in a 10-month-old **patient** receiving 25 mg tid/ 7 days. Will color urine rust brown or brownish. Contraindicated in **patients** with marked decrease in renal function, in **infants** < 1 month of age, and in **patients** with glucose-6-phosphate dehydrogenase deficiency.
Neuropathy: Polyneuropathy and hyperesthesias during 40 days of treatment with 400 mg/days. Some similar cases have often been severe and irreversible. Can develop in a few days. Acute psychosis and mental confusion after 250 mg/day/7 days. Paresthesia, headache, dizziness, drowsiness, and seizures reported.
Renal: Crystalluria reported in elderly **patients** receiving 50 mg twice daily. Acute interstitial nephritis in a 68-year-old **male** after 100 mg twice daily/14 days. Urinary tract fibrosis after several months of usage.
Skin: Diffuse, erythematous, maculopapular eruptions; exfoliative dermatitis, and toxic epidermal necrolysis. Also see allergic effects above.

TD_{LO}, **human**: 80 mg/kg.

Drug interactions: Toxic effects with ***alcohol***, as it prevents oxidation of ***acetaldehyde*** (q.v.), an ***alcohol*** metabolite. Can also cause a ***disulfiram***-like reaction when taken with ***alcohol***. Can decrease absorption of ***nalidixic acid***. ***Antacids*** (***aluminum hydroxide, aluminum phosphate, calcium carbonate, magnesium oxide, sodium bicarbonate***, etc.) can antagonize it by complexing and by decreasing its absorption and increasing its urinary excretion. ***Acidifying agents*** potentiate its effect by decreasing its urinary excretion. ***Phenobarbital*** decreases its effectiveness. ***Pyridoxine*** increases its excretion and concentration in urine; ***propantheline*** also increases its excretion. May alter response to oral ***anticoagulants*** by increasing prothrombin time. Break-through bleeding reported in **women** taking it with ***oral contraceptives*** and decreases the effectiveness of the latter. Given with ***antidiabetic agents***, it can enhance uremic metabolic acidosis. Its action is increased when given with ***anticholinergics, antihistamines, antimalarials, cardiac depressants, narcotic analgesics, phenothiazines, pro-benecid, sympathomimetics***, and ***tricyclic antidepressants***. Its action is decreased by ***barbiturates, clofibrate, ethacrynic acid, mefenamic acid, phenylbutazone, salicylates, sulfonamides, sulfonylureas***, and ***thyroxine***.

High oral dosage leads to serious neuropathy in the **rat** and this may produce nausea or central nervous system depression in **dogs**. The possibility of decreased fertility in **males** being treated should not be overlooked. Systemic and urinary alkalinizers decrease its effectiveness by increasing its excretion rate. Systemic effectiveness increases if urine is acid. Vomiting reported in **cats**. Resistant strains have also been reported. Dramatically decreases bacteriostatic effect of ***nalidixic acid***. ***Selenium*** and not ***vitamin E*** can help protect **chicks** from its toxicity.

LD_{50}, **rat**: 604 mg/kg; **mouse**: 360 mg/kg.

NITROFURAZONE

= *Aldomycin* = *Amifur* = *Chemofuran* = *Chixin* = *Coxistat* = *Furacilin* = *Furacinetten* = *Furacoccid* = *Furalone* = *Furaplast* = *Furazin* = *Furazina* = *Furazol W* = *Furesol* = *Ibiofural* = *Mammex* = *Nefco* = *NF-7* = *NFZ* = *Nifucin* = *Nifuzon* = *Nitrofural* = *Nitrofuran* = *Nitrozone* = *Otofural* = *Babrocid* = *Vitrocin* = *Yatrocin*

Untoward effects: Nausea, vomiting, polyneuritis, joint pain, headache, thrombocytopenia, and decreased spermatogenesis. Hemolytic anemia reactions in glucose-6-phosphate dehydrogenase-deficient **patients**. After topical use, contact dermatitis, allergic reactions (usually after several days' application), maculopapular rash, pruritus, local edema, and bulbous and exudative lesions at site of application. Severe, near-fatal exfoliative dermatitis with systemic toxic reaction in a 37-year-old **male** after topical application of soluble dressing to a wound. Other toxicity can occur in addition to the ***NFZ*** component, since soluble dressings contain ***polyethylene glycol*** (q.v.), which can be absorbed through denuded skin, and may not be excreted normally by a compromised kidney, leading to increased blood urea nitrogen, metabolic acidosis, increased serum ***calcium***, and further renal damage or failure. Can cause false positive or spuriously high qualitative or quantitative urine ***sugar*** tests. Sensitization has persisted for more than a year in some cases. Has caused gross hematuria in 19% after bladder irrigation. Graying of eyelashes and depigmentation of eyelid skin after its use as a solution for eyedrops. Peripheral neuropathy occurs.

Its use may interfere with certain saliva tests in **horses**. Oral use in **rats** caused benign mammary tumors in **females**, decreased spermatogenesis in **males**, hepatorenal lesions, hyperirritability, hyperreflexia, tremors, weakness and convulsive seizures, and death from respiratory paralysis. **Pheasant chicks**, **goslings**, and **ducklings** are more susceptible to the drug's toxic effects than are **chickens**. The drug can potentiate ***β-aminopropionitrile*** toxicity for **poults**. It is unstable in galvanized waterers. When fed to **poultry** at 0.022% level it causes hyperexcitability, frequent falling forward, and loud squeaking. Fed at 0.033% level to **turkeys** it causes ascites, cardiac dilation, and death. Residues will appear in **poultry** eggs. A 5 day withdrawal before slaughtering **chickens** and **turkeys** has been suggested. Slight increase in stillbirths when fed to pregnant **rabbits**. Of 28 **cattle** receiving it orally, all developed ataxia, collapse, and seizures, and one died. IV to **dogs** at 20–75 mg/kg led to lacrimation, salivation, emesis, diarrhea, excitement, weakness, ataxia, and weight loss. Convulsions and death after 100 mg/kg IV.

LD_{50}, **rat**: 590 mg/kg; **mouse**: 380 mg/kg.

NITROGEN

Untoward effects: At sea level, air is made up of ~79% nitrogen and 21% ***oxygen***. It acts as an anesthetic for **man** at 20 atmospheres, so, in a sense, most of us are chronically exposed to 1/20th of a dose of nitrogen anesthesia. In **man**, the higher levels in dives or at high altitudes cause the development of the "bends", "raptures of the deep", or so-called "***helium*** shakes" on decompression. This does not occur in **seals**, as they exhale air before diving, then, at a depth of ~30 m, their lungs collapse and they have limited nitrogen uptake. The increased blood hemoglobin dissolves the nitrogen and distributes it to blubber and other tissues. Nitrogen level in the blood is then insufficient to cause narcosis. Gases remaining in the blood are compressed during deep dives, and during ascent, they expand, opening the collapsed lungs. Nitrogen rash with tender, pruritic, erythematous lesions (primarily on **their** elbows and flanks) occurs in scuba **divers** exceeding one atmosphere. Cases of emphysema reported in two elderly **patients** shortly after the use of liquid nitrogen on actinic keratoses lesions on **their** cheeks. Painful, large, hemorrhagic bullae often follow over-zealous freezing of warts on palms and soles with liquid nitrogen. Temporary damage (necrosis) has occurred from treatment of warts near fingernails.

Rats continuously exposed to 2 ppm in air survived for their normal lifespan with persistent tachypnea, but usually died of non-pulmonary diseases. **Dog** trials indicated it effective for euthanasia by hypoxia, when they inhale a maximum of 1.5% ***oxygen*** with nitrogen. They were unconscious in ~40 s and dead in ~204 s. Vocalization and blinking reflex noted during this brief period. Probably not advisable for euthanasia in newborn **puppies** and **kittens**, as they have been accustomed to low ***oxygen*** levels *in utero* and may be fairly resistant to hypoxic conditions. **Turtles** of the *Pseudemys* genus can survive an atmosphere of 100% nitrogen for up to 72 h and can readily change from aerobic to anaerobic respiration.

NITROGEN CHLORIDE

= *Agene* = *Nitrogen Trichloride*

Formerly used in bleaching and aging of *flour*; use is now prohibited in the U.S.

Untoward effects: As early as the 1920s, it caused "fright disease", "hysteria", or "running fits" in **dogs**. Only after World War II, was it identified as the causative agent. Toxic to **dogs**, **ferrets**, and **rabbits**, but not to **guinea pigs**, **cats**, **monkeys**, and **humans**. In those species that are adversely affected, patchy necrosis was noted microscopically in the deeper parts of the cerebellar cortex, due to oxidation of the ***methionine*** in ***wheat*** protein. Causes anoxia, cyanosis, methemoglobinemia, weakness, dizziness, and liver and kidney pathology in affected **animals**. A potential oxidant for spacecraft propulsion led to its testing on **rats**. Lethal to them after exposure to a 1% level in air for 1 h.

NITROGEN DIOXIDE

= *Nitrogen Peroxide*

Untoward effects: The key agent inducing **silo-filler's** disease, but a hazard as well to **welders**, **firemen**, farm and chemical plant **workers**, and **people** using ***acetylene*** torches or gas ranges in confined areas. Also found in ***cigarette*** smoke, aircraft and auto exhaust, smog, and solar flares; it is heavier than air. Toxic amounts are formed when plastic furnishings and interior design fixtures, wood, or wallpaper burn in residential or hotel fires. In a 1929 fire of "nitrocellulose" X-ray films at the Cleveland Clinic, 123 **people** died. Use of electric cautery and X-ray machines during ***nitrous oxide*** anesthesia can increase its concentration in operating room air. When silage ferments, the key end product is poisonous nitrogen dioxide. The ill-effects often occur to farm **workers** entering a partially filled silo, during silo filling, or being near the bottom of a silo chute. Dead **flies** or **birds** near the bottom of the chute are an indication of toxic fumes. Symptoms include severe coughing and burning or choking pains in the throat and chest, hemoptysis, extreme weakness, and nausea. The **victim** is then free of discomfort for 5–12 h, then, a severe recurrence with aveolitis, bronchitis, acute emphysema, tachypnea, tachycardia, pulmonary edema, bronchoconstriction, increased respiratory infections, cyanosis, and shock can occur. Some of the pulmonary damage is permanent, and deaths have been reported. The gas is yellow-brown or reddish-brown in color and it can be recognized at air concentrations of 100 ppm or higher. It smells like laundry bleach. Standard gas masks will not afford great protection against it. Methemoglobinemia has been reported. Ice-resurfacing machines in unventilated indoor ice hockey arenas have been a source of many outbreaks of intoxication in **players** and **spectators**. Submarines have considered 0.5 ppm (0.9 mg/m^3) as a safe upper exposure limit for 60 days. OSHA suggests 5 ppm and National Institute for Occupational Safety and Health lists 1 ppm/15-min exposure. Immediate danger to life or health level is 20 ppm.

LC_{LO}, **human**: 200 ppm/1 min.

Toxicity has been reported in **cats**, **dogs**, **cattle**, **guinea pigs**, **monkeys**, **chickens**, **mice**, **rabbits**, **rats**, **sheep**, and **swine**, with respiratory problems similar to those reported in **man**.

Inhalation LC_{50}, **rat**: 88 ppm/4 h; **rabbit**: 315 ppm/15 min; **guinea pig**: 30 ppm/1 h; LC_{LO}, **mouse**: 250 ppm/30 min; **monkey**: 44 ppm/6 h.

NITROGEN FLUORIDE

= *Nitrogen Trifluoride* = *Trifluoramine* = *Trifluorammonia*

Untoward effects: Inhalation of vapors by **animals** leads to anoxia, methemoglobinemia, weakness, dizziness, cyanosis, and liver and kidney pathology. National Institute for Occupational Safety and Health/OSHA exposure limit of 10 ppm (29 mg/m^3). Immediate danger to life or health level is 1,000 ppm.

Inhalation LC_{50}, **guinea pig**: 900 ppm/1 h.

NITROGEN TETROXIDE

= *Hyponitric Acid* = *Nitrogen Peroxide*

It is a commercial brown, liquid, compressed material of *nitrogen dioxide* (NO_2) (q.v. above) and is $N_2O_4 \rightleftharpoons 2NO_2$, a rocket propellent.

Untoward effects: A deadly poison and an occupational hazard. Very corrosive causing severe skin and eye burns, even after momentary contact. Immediate symptoms of inhalation include irritation of eyes, nose, and upper pharynx with nausea, slight cough, chronic bronchitis, or chest tightness. May be followed in a few hours with severe illness, including pulmonary edema, dysuria, delirium, and death.

Inhalation LC_{50}, **rabbit**: 315 ppm/15 min.

NITROGLYCERIN

= *Adesitrin* = *Angibid* = *Anginine* = *Angiolingual* = *Angorin* = *Aquo-Trinitrosan* = *Blasting Gelatin* = *Blasting Oil* = *Cardamist* = *Cordipatch* = *Corditrine* = *Coro-Nitro* = *Deponit* = *Diafusor* = *Discotrine* = *Gilucor* = *Glonoin* = *Glycerol Nitric Acid GTN* = *Triester* = *Glyceryl Trinitrate* = *Klavikordal* = *Lenitral* = *Lentonitrina* = *Millisrol* = *Minitran* = *Myoglycerin* = *NG* = *Nitradisc* = *Nitran* = *Nitriderm-TTS* = *Nitrobaat* = *Nitro-Bid* = *Nitrocine* = *Nitrocontin* = *Nitroderm TTS* = *Nitrodisc* = *Nitro-Dur* = *Nitrofortin* = *Nitrogard* = *Nitro-Gesanit* = *Nitroglin* = *Nitroglycerol* = *Nitroglyn* = *Nitrol* = *Nitrolan* = *Nitrolande* = *Nitrolar* = *Nitrolent* = *Nitrolingual* = *Nitro Mack* = *Nitromex* = *Nitronal* = *Nitrong* = *Nitro-PRN* = *Nitrorectal* = *Nitroretard* = *Nitrosigma* = *Nitrospan* = *Nitrostat* = *Nitrozell-retard* = *Nysconitrine* = *Percutol* = *Perlinganit* = *1,2,3-Propanetriol Trinitrate* = *Reminitrol* = *S.N.G.* = *Susadrin* = *Suscard* = *Sustac* = *Sustonit* = *Transderm-Nitro* = *Transiderm-Nitro* = *Tridil* = *Trinalgon* = *Trinitrin* = *Trinitroglycerol* = *Trinitrosan* = *Vasoglyn*

Mixed with *diatomaceous earth* 24.5% and *sodium carbonate* 0.5%, it becomes *dynamite*; mixed with *β-lactose*, it becomes the familiar *antianginal* and *coronary vasodilator*.

Untoward effects: When, in 1846–1847, Sobrero made it, **he** discovered that a small amount on the tip of **his** tongue, and those of **his assistants**, caused an immediate intense headache lasting several hours. Later, others discovered that < 1/300th of a drop (< 0.0002 ml) caused severe headache and tachycardia, followed by bradycardia. Interindividual variation is common. Tachycardia in **some** sublingually from 1/1,000th drop; **others** require 1/10th or 1/5th drop to double **their** heart rates. Doses of 1/30–1/300th drop cause headaches within ~1 min, chest pains radiating down the left arm, followed by feelings of warmth or stabbing pains in the head, neck, and shoulders lasting 10–60 min. Therapeutic use of sublingual tablets has been associated with headaches, vertigo, weakness, postural hypotension, syncope, increased intraocular pressure, flushing, tachycardia, rash, and occassionally palpitations, nausea, and vomiting. Cerebral infarction and fatal intracerebral hemorrhage, paresthesia of throat and chest, localized sublingual erythema and burning, exacerbation of gouty arthritis, diplopia, flatus, and asystole have also been reported. IV use can also cause bradycardia and hypoxemia. Long-time exposure leads to cardio-cerebrovascular diseases (even a death within 1 week after a normal electrocardiogram) and giddiness in ***dynamite* workers**. **Workers** at an Army ordinance plant developed a physical dependence on it. Abnormal pulse waves in plethysmography in 143/1,446 ***dynamite*** factory **workers**. Has caused increase in ***epinephrine, norepinephrine***, and ***vanillylmandelic acid*** urinary excretion. Long-term use (> 5 months) has been associated with thrombocytopenia. Overdoses can cause methemoglobinemia. Contact dermatitis with macular erythematous rashes at application site of transdermal form and ointment form. The **patient** may later be unable to tolerate sublingual use. Contact allergy to it first reported in 1978. Many

patients are not allergic to its sublingual use, but ~50% may react to it when a paste form is applied to the chest skin. A 54-year-old **female** was hospitalized after intercourse. A nurse found her **husband's** transdermal patch had stuck to **her** buttocks. Severe occupational exposure can induce delirium and central nervous system depression, numbness, paresthesia, and Raynaud's phenomenon of the fingers. Disturbances of the cardiovascular system, with death as a sequela 36–72 h after the last exposure and sudden withdrawal from exposure. An electric arc between a defibrillator and the ***aluminum*** covering on a transdermal patch caused a loud explosion with a flash and yellow smoke on a **patient** without ill-effects. Similar reports of such arcing to nitroglycerin ointment also reported. A serious medication error occurred when a prescription for ***Transderm-Nitro***, two 5 mg patches twice daily was filled by a **pharmacist** with ***Transderm-V***, a long-acting ***scopolamine*** product. Due to its slow deterioration by exposure to air and inactivation by light, heat, or moisture, 2 months should be considered as a time limit for storage by a **patient**. Except for what must be carried, the reserve should be refrigerated. A second degree burn reported on a **patient** from exposure of his ***Transderm-Nitro*** patch to a leaking microwave oven. 75–80% can be bound to various medical plastics. OSHA limit of 0.2 ppm (2 mg/m^3) exposure to skin, while National Institute for Occupational Safety and Health sets a 15-min exposure maximum of 0.1 mg/m^3. Immediate danger to life or health level is 75 mg/m^3.

LD_{LO}, **human**: 5 mg/kg.

Drug interactions*:** Orthostatic hypotension, vertigo, faintness, flushing, and pallor can be profound when taken with ***alcohol, ***antihypertensives***, ***whiskey***, and other ***vasodilators***. A **patient** became stuporous after IV use of ***Nitro-Bid***, due to its ***alcohol*** content; and a 72-year-old **female** developed lactic acidosis, central nervous system depression, and hemolysis after IV use of ***Nysconitrine***, due to its ***propylene glycol*** content (322 ml/day). ***Sympathomimetics*** decrease its antianginal effect. In a number of reports, it enhances the activated partial thromboplastin time response to ***heparin***. With ***histamine***, it decreases ***histamine's*** effect; with ***norepinephrine***, decreases the pressor effect; and with ***acetylcholine***, decreases cholinergic receptor stimulation. ***Phenobarbital*** causes more rapid excretion of its metabolite, ***glyceryl mononitrate***, and headaches appear less frequently. Dry mouth associated with ***imipramine*** has delayed dissolution of sublingual tablets. Visual hallucinations after addition of ***propranolol*** and ***ranitidine*** to treatment with sublingual nitroglycerin. ***Oxolinic acid*** antibacterial activity is antagonized by it. Use with ***sildenafil*** in a 70-year-old **male** caused severe hypotension and ischemia leading to acute myocardial infarction.

Has been used in ***"speed balls"*** for doping **horses**. Excessive amounts of it after topical ointment has caused weakness and ataxia, secondary to hypotension. **Users** should wear gloves. **Cattle** have died after licking abandoned ***dynamite***.

SC, LD_{LO}, **cat**: 150 mg/kg; **rabbit**: 400 mg/kg.

NITROMERSOL

= *Metaphen*

***Antiseptic*. Contains 57% *mercury*.**

Untoward effects*:** After topical use on gums, ***mercury was found 24 h later in urine. Contact dermatitis reported after topical use.

NITROMETHANE

= *Nitrocarbol* = *Nitrometan*

Use*:** In rocket fuel and as a solvent. A ***nitroparaffin with a disagreeable odor.

***Untoward effects*:** Contact dermatitis. Inhalation of high concentration of vapors causes mild respiratory tract irritation and symptoms of intoxication with headache, nausea, and vomiting. OSHA exposure limit of 100 ppm (250 mg/m^3). Immediate danger to life or health level is 750 ppm.

LD_{LO}, **human**: 500 mg/kg.

In **animals**, eye and respiratory tract irritant, narcosis, convulsions, and hepatic pathology.

LD_{50}, **rat**: 344 mg/kg. Inhalation LC_{LO}, **guinea pig** and **monkey**: 1,000 ppm; inhalation LC_{50}, **rat**: 1,280 ppm/1 h; **mouse**: 5,940 ppm/1 h.

2-NITRONAPHTHALENE

= *β-Nitronaphthalene*

Untoward effects*:** Skin and respiratory system irritant. Can form ***methemoglobin with some cyanosis and anemia. Metabolized to a known carcinogen, ***β-naphthylamine*** (q.v.).

TD_{LO}, **dog**: 2.4 g/kg, during 34 weeks; **monkey**: 340 g/kg, during 4.5 years.

NITROPHENIDE

= *Megasul*

Untoward effects*:** Use as a coccidiostat for **chickens** no longer popular or FDA-approved in the U.S. because usual use levels are toxic to **swine** and **dogs** and may be toxic to **birds**. May break down ***vitamin potency of feeds by interference with antioxidants. Toxicity is enhanced by restricting water intake.

p- or 4-NITROPHENOL

Use: Primarily in manufacturing ***acetaminophen*** and ***phenacetin***. Some in ***leather***-tanning and dyestuffs. U.S. exported it for ***parathion*** manufacture elsewhere in the world.

Untoward effects: As a metabolite of ***parathion***, assaying the urine for it permits evaluation of exposure of **humans**, **fish**, **cattle**, **poultry**, and **rats**, etc. Absorbed from the gastrointestinal and respiratory tracts, and by application to the skin. Causes tachypnea, dyspnea, flushing, collapse, and coma. Redness, burning, and blisters, and can stain skin and hair yellow after skin contact.

LD_{50}, 350 mg/kg; **mouse:** 467 mg/kg.

1-NITROPROPANE

Solvent.

Untoward effects: Eye irritation, headache, nausea, vomiting, and diarrhea usually from inhalation or ingestion. OSHA/National Institute for Occupational Safety and Health maximum exposure level 25 ppm TWA.

LD_{LO}, **human**: 500 mg/kg; inhalation TC_{LO}, **human**: 150 ppm.

Liver and kidney damage reported in experimental **animals**.

LD_{LO}, **rat**: 250 mg/kg.

2-NITROPROPANE
= *Dimethylnitromethane*

Solvent.

Untoward effects: Several deaths from irreversible liver damage reported from use in confined or enclosed spaces. Eye, skin, nose, and respiratory tract irritant. Nausea, headache, vomiting, and diarrhea also reported. OSHA upper exposure limit is 25 ppm TWA; immediate danger to life or health level is 100 ppm.

Inhalation TC_{LO}, **man**: 20 ppm.

Long-term exposure of **rats** to high concentrations (207 ppm/6 months) causes liver cancers. No cancers in lifetime studies with **rats** exposed to 25 ppm.

LD_{LO}, **rat** and **rabbit**: 500 mg/kg; inhalation LC_{LO}, **rat**: 1,513 ppm/5 h; **cat**: 714 ppm/5 h; **rabbit**: 2,381 ppm/5 h; **guinea pig**: 4,622 ppm/5 h.

3-NITRO-1-PROPANOL

A very toxic metabolite of *miserotoxin* (q.v.), found in *milk-vetch* (q.v.), which is lethal to cattle, sheep, chicks, and rabbits.

β-NITROPROPIONIC ACID

Untoward effects: This mycotoxin is produced by several species of fungi and some higher plants, poisoning **animals** that consume them. May also be a cause of liver atrophy, hemorrhage, and necrosis in neonatal **foals** after they evidence listlessness, jaundice, diarrhea, epistaxis, hematuria, hyperexcitablity, blindness, and ataxia.

NITROPYRENES

Formed during grilling and charring *fish* or *meat*. In diesel exhaust, *fly ash*, and *kerosene* heater emissions.

Untoward effects: Some are carcinogens and mutagens.

4-NITROQUINOLINE-1-OXIDE
= *4 NQO*

Untoward effects: Mutagen and carcinogen. Gastric carcinogen in **dogs**, **rats**, and **mice**. Topically on **mouse** skin, it converts benign cutaneous papillomas to squamous cell carcinomas. Reported to induce uterine tumors (probably oral, species not stated).

TD_{LO}, **mouse**: 576 mg/kg.

N-NITROSOBIS (2-OXYPROPYL)AMINE
= *BOP*

Untoward effects: Weekly SC injections of 10 mg/kg/6 weeks to **hamsters** led to high incidence of pancreatic duct cancers.

LD_{50}, **hamster**: 142 mg/kg; SC, 89 mg/kg.

N-NITROSODIBUTYLAMINE
= *DBN*

Untoward effects: Carcinogenic in **rat**, **rabbit**, **hamster**, **guinea pig**, and **mouse** trials, by oral, SC, IP, and IV causing bladder, liver, lung, kidney, esophagus, stomach, and intestinal cancers and neoplasms, as well as leukemia. One of its metabolites is carcinogenic in **dogs**, **rats**, **mice**, and **hamsters**; another in **rats**.

LD_{50}, **rat**: 1.2 g/kg; SC, **rat**: 1.2 g/kg.

N-NITROSOBUTYLUREA

Untoward effects: Leukemia and acoustic duct tumors in **rats**.

LD_{50}, **rat**: 400 mg/kg; TD_{LO}, **rat**: 2.5 g/kg.

N-NITROSODIETHANOLAMINE

= *NDELA*

Has been found in cosmetics (shampoos, bubble baths, creams, and lotions) at levels up to 260 mg/kg, possibly, by a reaction of *nitrite* with *diethanolamine* or *triethanolamine* at up to 48 ppm. Also in metal cutting or grinding synthetic cutting fluids and in *tobacco* leaves and smoke.

***Untoward effects*:** Hepatocellular carcinomas and renal adenomas in **rats**. SC to **hamsters** leads to nasal adenocarcinomas, hepatocellular carcinomas, papillary tumors of the trachea, and fibrosarcomas at injection site.

SC, LD_{50}, **hamsters**: 11 g/kg.

N-NITROSODIETHYLAMINE

= *DEN* = *DENA* = *Diethylnitrosamine* = *NDEA*

Patents exist for its use in *gasoline*, lubricants, and as a stabilizer in plastics.

***Untoward effects*:** Induced liver tumors in **dogs**, **rabbits**, **rats**, **mice**, **guinea pigs**, **monkeys** (hepatocellular carcinoma), and **fish**, as well as kidney, esophagus, and nasal cavity tumors in **rats**; and lung, stomach, and esophagus tumors in **mice**. Inhalation caused liver tumors in **rats**; tracheal, bronchial, and lung tumors in **hamsters**; SC or IM to various species of **hamsters** led to tumors of the trachea, larynx, nasal cavities, stomach, lung, and liver. SC or IM caused liver tumors in **parakeets** and upper respiratory tract and liver tumors in **hamsters**, **gerbils**, **guinea pigs**, and **hedgehogs**.

LD_{50}, **guinea pig**: 250 mg/kg; SC, **hamsters**: 246 mg/kg; TD_{LO}, **rat** (22 days pregnant): 150 mg/kg; **dog**: 560 mg/kg; IP, **monkey**: 18 g/kg; during a 2 year period.

N-NITROSODIMETHYLAMINE

= *Dimethylnitrosamine*= *DMN* = *DMNA*

Rocket fuel intermediate.

Untoward effects*:** Some small amounts are ingested by **humans**, especially in foods stored under acidic conditions or brines used in preserving ***fish or ***meat***. A 2½-year-old **male** was hospitalized on the 3rd day of **his** progressive vomiting and listlessness, followed by hypoglycemia, cerebral edema, seizures, liver failure, thromobocytopenia, and severe coagulopathy, indicative of a typical Reye's syndrome. **He** died on the 4th day. Post-mortem revealed centrilobular necrosis with little fatty degeneration of the liver (in Reye's syndrome there is peripheral zone liver necrosis and intense fatty degeneration). A day later, his 24-year-old uncle died after similar events. Five other family **members** also became acutely ill. On the day of onset, the common denominator was that **they** shared ***milk*** and ***lemonade***. **Those** who had not were unaffected. A year later, a former cancer research **assistant**, an ex-suitor of the **aunt**, was convicted of first degree murder. These were the first cases of acute N-nitrosodimethylamine poisoning. High air concentrations were found in ***leather*** tanneries and a ***rubber*** tire manufacturing plant without adequate ventilation. Found in cooked ***bacon*** and its drippings, ***cigarette*** smoke, ***frankfurters***, some ***hams***, ***moonshine whiskey*** from Africa, Europe, and Canada, broiled dried ***squid***, and some Scotch ***whiskies*** and many ***beers***. There have been Zambian esophageal cancer clusters around stores that sell a local ***corn*** liquor, ***kachasu***, that contains ***DMN***. Failure to have an in-line ***activated carbon*** filter led to levels up to 32 µg/l ***DMN*** in ***dialysate*** from a number of units and increased ***DMN*** in **patients'** blood. ***Nitrous acid*** in saliva and some foods reacts with ***aminopyrine*** to produce ***DMN***, and has led to its ban from Japanese over-the-counter drugs in 1966. A 42-year-old **female** evidenced severe nausea, vomiting, epigastric pain, night sweats, and slightly increased temperature, with death after 2½ years of chronic poisoning by **her** husband, a chemistry **teacher**. Some of **her** canned ***blackberries*** contained 200 mg in ~400 g.

Orally in **mice** caused hemangiosarcomas of the liver, hepatocellular carcinomas, and kidney and lung tumors; kidney and bile duct tumors in **rats**; hepatocellular carcinomas and bile duct tumors in **guinea pigs**, **rabbits**, and **hamsters**; liver hemangiomas in **ducks**; and liver adenomas and adenocarcinomas in **fish**. The **cat** is very susceptible to its oral toxicity.

IM led to hemangiosarcomas of the liver and abdominal tissues, and lung tumors in adult **mice**; parenchymal cell and vascular liver tumors, and lung tumors, in newborn and suckling **mice**; kidney tumors in adult **rats**; and kidney and liver tumors in newborn and suckling **rats**. Acute hepatic necrosis and chronic hepatic insufficiency in **dogs**; hemangiosarcomas in the liver and bile duct, nasal cavity tumors, and cholangiocarcinomas in **hamsters**; bile duct tumors in **mastomys**.

IP led to liver tumors and increased incidence of lung and liver tumors in **mice**; kidney and nasal cavity tumors in **rats**; and liver tumors in **newts**.

Inhalation caused lung, liver, and kidney tumors in **mice**; lung, liver, kidney, and nasal cavity tumors in **rats**. ***Copper*** deficiency increased hepatic neoplasms in **rats**. Transplacental induction of carcinogenesis demonstrated in **rats**, **mice**, and **hamsters**. **Mink** and **foxes** eating ***herring meal*** containing it lost their appetite, became pot-bellied, had liver degeneration and blood vessel tumors, and often died. **Ducklings** have decreased weight gains, abdominal distention, ascites, hydropericardium, liver atrophy and necrosis, bile duct cell proliferation, enlarged spleens, and lymphoid hyperplasia. Fatalities in **sheep** caused by it in ***sodium nitrite***-preserved ***fish*** meal in their diet. Centrolobular liver necrosis and hemorrhages noted. Has also formed in ***formaldehyde*** and ***sodium benzoate***-preserved ***fish*** and when the meal was fed to **pigs**, it caused cirrhosis of the liver and vaso-obstructive diseases. **Pigs** seem slightly more resistant to its ill-effects in their diet than other **animals**.

LD_{50}, **rat**: 26 mg/kg; **rabbit**: ~15 mg/kg; **hamster**: 28 mg/kg; LD_{LO}, **dog**: 20 mg/kg; **guinea pig**: 25 mg/kg. Inhalation LC_{50}, **rat**: 78 ppm/4 h, **mouse**: 57 ppm/4 h.

p-NITROSO-*N,N*-DIMETHYLANILINE

= *Accelerine* = *Oxynone*

Use: Accelerator in vulcanizing ***rubber***, and manufacturing of ***methylene blue***, and organic chemicals.

Untoward effects: Vesicular dermatitis of the penis, slowly spreading, followed by exfoliation.

LD_{50}, **rat**: 65 mg/kg; LD_{LO}, **guinea pig**: 650 mg/kg.

N-NITROSO-2, 6-DIMETHYLMORPHOLINE

Untoward effects: A powerful SC pancreatic carcinogen in Syrian **hamsters**.

p-NITROSODIPHENYLAMINE

Use: Accelerator in vulcanizing ***rubber***; an intermediate in synthesizing dyes and pharmaceuticals.

Untoward effects: Causes liver cancers in male **mice** and **rats**, but not female **mice** and **rats**.

LD_{50}, **rat**: 2.14 g/kg; IV, **mouse**: 178 mg/kg.

N-NITROSODI-*n*-PROPYLAMINE

= *DPNA*

Found in µg quantities in chemical factory waste waters, *cheese*, *brandy*, liquors, and pesticides or herbicides, such as *isopropalin* and *trifluralin*.

Untoward effects: Pesticide and research **workers** might be exposed by inhalation, ingestion, or dermal contact. It undergoes photolytic degradation, and, when heated, it emits toxic ***nitrogen oxide*** fumes.

In **rat** drinking water it caused carcinomas of liver and tongue; papillomas and carcinomas of the esophagus. SC to **rats** caused benign and malignant tumors of the liver, kidney, esophagus (squamous cell papillomas), nasal and/or paranasal cavity, and lungs (adenomas and carcinomas). SC to **hamsters** caused adenomas and one adenocarcinoma in the kidneys; neoplasms of the nasal and paranasal cavities, laryngobronchial tract and lung. Bi-monthly IP injections (40 mg/kg) to **monkeys** caused liver cancer within 22–33 months.

LD_{50}, 480 mg/kg; SC, **mouse**: 689 mg/kg; **hamster**: 600 mg/kg.

NITROSOEPHEDRINE

A nitrosation product of *ephedrine*.

Untoward effects: A potent liver carcinogen in **mice** after three doses of 200 mg/kg. 13/15 had liver cell carcinomas and eight had pulmonary metastases.

N-NITROSO-*N*-ETHYLUREA

= *ENU* = *N-Ethyl-N-nitrosourea*

Use: In chemical synthesis, plant growth promotion, and as an experimental mutagen and antineoplastic.

Untoward effects: Dermal contact is the chief exposure route for research laboratory **workers** and manufacturing **chemists**. Orally in **rats** produced malignant neurogenic tumors in brain, spinal cord, and peripheral nervous system; papillomas and sarcomas of the stomach, sarcomas of the colon, mammary adenocarcinomas, and leukemia in male and female. Transplacental induction of carcinogenesis reported in **rats**, **mice**, and **hamsters**. Prenatal oral exposure of female **rats** caused brain, spinal cord, and trigeminal nerve tumors in their offspring. Prenatal IP exposure of female **mice** caused pulmonary adenomas, endocrine gland tumors, central nervous system and peripheral nervous system tumors in their offspring. Prenatal IV exposure of female **rats** produced malignant neurogenic and neuroectodermal tumors in their **pups** and nervous system tumors in **hamsters**. SC in **rats** caused hepatomas, hepatocellular carcinomas, lung adenomas; brain, spinal cord, and peripheral nervous system tumors. IP in pregnant **sows** produced massive skeletal malformations in offspring; to pregnant **rats** caused necrotic foci in telencephalon and elongated heads in their **pups**. Skin papillomas and sebaceous adenomas after transplacental or IP to **mice**; central nervous system tumors in **rats** after oral or IP; renal adenomas in **rabbits** after IP; papillary adenocarcinomas of the thyroid in **hamsters**; carcinoma of the thyroid and ovary in **dogs**; and lung tumors in **mice** after oral or IP. IP in **rats** caused thymic lymphomas and myeloid leukemia; in **mice**, produced intracranial neurogenic and renal epithelial neoplasms and malignant tumors of the liver, kidney, ovary, lung, harderian gland, stomach, and lymphoreticular system. IV in **rats** caused leukemia, gliomas, malignant uterine and vaginal tumors; in **monkeys** caused malignant tumors of ovary, uterus, bone, bone marrow, vascular endothelium, and skin.

LD_{50}, **rat**: 300 mg/kg; TD_{LO}, **hamster** (11 days pregnant): 60 mg/kg; SC and IV, LD_{50}, **rat**: 240 mg/kg.

N-NITROSOHEPTAMETHYL-ENEIMINE

= *NHMI*

***Untoward effects*:** Chronic SC injections to European **hamsters** produced adenomas and carcinomas.

LD_{50}, **rat**: 283 mg/kg.

4-(N-NITROSOMETHYLAMINO)-1-(PYRIDYL)-1-BUTANONE

= *NNK*

It is an oxidation and nitrosation product of *nicotine*.

***Untoward effects*:** In **rats**, multiple SC injections led to nasal neuroblastomas and rhabdomyosarcomas, adenocarcinoma and adenosquamous cell carcinomas of the lung, hepatocellular carcinomas and liver hemangiosarcomas, esthesioneuroepitheliomas, and assorted carcinomas and sarcomas of the nasal tract.

N-NITROSOMETHYLANILINE

= *NMA*

***Untoward effects*:** Chronic treatment of **rats** led to sessile papillomas and squamous cell carcinomas of the alimentary tract, especially esophagus, pharynx, and tongue. Incidence in male > female.

LD_{50}, male **rat**: 336 mg/kg; female **rat**: 225 mg/kg; male **hamster**: 150 mg/kg.

N-NITROSOMETHYL BENZYLAMINE

***Untoward effects*:** Induces esophageal cancers in **rats**.

LD_{50}, **rat**: 18 mg/kg.

N-NITROSO-*N*-METHYL-*N*-CYCLOHEXYLAMINE

= *N-Methyl-N-nitroso-N-cyclohexylamine* = *NMC*

***Untoward effects*:** Chronic treatment of **rats** produced sessile papillomas and squamous cell carcinomas of the alimentary tract, especially esophagus, pharynx, and tongue. Incidence in male > female.

LD_{50}, male **rat**: 80 mg/kg; female **rat**: 180 mg/kg; **mice**: 57 mg/kg; male **hamster**: 168 mg/kg; IP, **rat**: 28 mg/kg.

N-NITROSO-*N*-METHYLUREA

= *NMU*

Use*:** In research and in cancer treatments with ***cyclophosphamide.

***Untoward effects*:** Orally in **rats** caused squamous cell carcinomas of the stomach, sarcomas and gliomas of the brain, neurosarcomas and a spinal cord neurinoma; in **guinea pigs**, caused stomach carcinomas and sarcomas, pancreatic adenocarcinomas, malignant ear duct tumors, a neurinoma of the lumbar nerve, and leukemia; in **pigs**, caused malignant stomach tumors; and in **monkeys**, caused squamous cell carcinomas of the oropharynx and/or esophagus. Intragastric intubation in **rats** produced malignant kidney, stomach, intestinal, jaw, and skin tumors; in **hamsters**, odontogenic tumors, epidermoid carcinomas in the mouth, and intestinal adenocarcinomas; in **guinea pigs**, adenocarcinomas of the stomach, pancreas, and colon, and mesenteric lymph node lymphomas, and a hepatocellular carcinoma. Topically on **mice** produced leukemia and malignant skin tumors; on **rats**, squamous and basal cell carcinomas of the skin; and on **hamsters**, squamous cell carcinomas of the skin. Has caused cleft palate formation in the offspring of treated pregnant **rats**.

LD_{50}, **rat**: 180 mg/kg.

N-NITROSOMETHYLVINYLAMINE

***Untoward effects*:** Orally in **rats** caused papillomas and squamous cell carcinomas of the esophagus, and carcinomas of the tongue and pharynx. Inhalation by **rats** caused carcinomas and cholesteatomas of the nasal cavity, squamous cell carcinomas and esophageal papillomas.

LD_{50}, **rat**: 24 mg/kg.

N-NITROSOMORPHOLINE

= *NMOR* = *NNM*

Untoward effects*:** Presence in ***margarines was taken from the paper wrappings. Plants manufacturing ***rubber*** tires had air levels of 0.5–27 $\mu g/m^3$ equal to daily **human** exposure of 50–250 μg. Inhalation, ingestion, and dermal contact are potential sources of **human** exposure.

Orally in **rats** led to hepatocellular carcinomas, cystadenomas, cholangiofibromas, cholangiocarcinomas, liver hemangiosarcomas, epithelial kidney tumors, and neoplasms of the tongue and esophagus; in male **mice**, to benign hepatocellular neoplasms and lung adenomas. In tank water, produced liver neoplasms in **fish**. SC in **hamsters** caused tracheal neoplasms, squamous cell papillomas, and carcinomas of the nasal cavity, olfactory neuroepitheliomas, epidermoid carcinomas and assorted neoplasms. IV in **rats** produced hepatocellular and ethnoturbinal carcinomas.

N'-NITROSONORNICOTINE

Untoward effects*:** The first organic carcinogen (1.9–88.6 ppm) isolated from ***tobacco, and it is the most abundant carcinogen in ***tobacco*** smoke.

Orally to **rats** produced esophageal carcinomas, papillomas and carcinomas of the nasal cavity; in **hamsters**, papillomas of the nasal cavity and trachea; in male **rats**, esthesioneuroepitheliomas and squamous cell carcinomas of the nasal cavity, and squamous cell carcinomas of the esophagus. SC to **rats** caused olfactory neuroblastomas, rhabdomyosarcomas, esthesioneuroepitheliomas, squamous cell and anaplastic carcinomas, spindle cell sarcomas, and lung adenomas; in **hamsters** caused tracheal papillomas. IP in **mice** produced multiple pulmonary adenomas; lung adenomas in female **mice**; nasal cavity tumors and tracheal papillomas in male **hamsters**.

N-NITROSOPIPERIDINE

***Use*:** In manufacturing epoxy resins and cardiovascular implants. Some used in research.

***Untoward effects*:** Potential **human** exposure is by inhalation, ingestion, and dermal contact.

Orally in male **mice** produced squamous cell carcinomas of the stomach, esophageal papillomas, hepatocellular adenomas and carcinomas, and liver hemangioendotheliomas; lung adenomas in male and female **mice**; esophageal carcinomas and hepatocellular carcinomas in **rats**; and hepatocellular carcinomas in **monkeys**. SC in **rats** caused squamous cell carcinomas and other tumors of the nasal cavity, esophageal squamous cell carcinomas and papillomas; in **hamsters**, nasal cavity, trachea, lung, tongue, palate, esophagus, stomach, and liver tumors. Transplacental carcinogenesis noted in **hamsters**.

LD_{50}, **rat**: 200 mg/kg.

N-NITROSOPYRROLIDINE

= *NOPYR*

Untoward effects*:** **Human** exposure has been from heating and eating of ***nitrite-treated foodstuffs, such as ***bacon***. Most of it is released in the vapors. Found in ***cigarette*** smoke, premixed dry spice cures, and in air at a tire manufacturing plant.

Orally in **mice** caused lung adenomas; in **rats**, hepatocellular carcinomas, leukemia, cholangiocarcinomas, and olfactory carcinomas; in male **rats**, tunica vaginalis papillary mesotheliomas, interstitial cell tumors, and a cavernous hemangioma of the testis; in **hamsters**, tumors of the trachea and lung.

LD_{50}, **rat**: 900 mg/kg.

N-NITROSOSARCOSINE

Research chemical found in foodstuffs, particularly in *meat* at concentrations of 2–56 µg/kg.

***Untoward effects*:** Orally in **rats** produced papillomas and squamous cell carcinomas of the esophagus; in **mice**, squamous cell carcinomas of the nasal region.

LD_{50}, **rat**: 5 g/kg.

NITROTOLUENE

= *Methylnitrobenzene*

***Use*:** In manufacturing dyes and chemicals.

***Untoward effects*:** *Meta*, *ortho*, and *para* forms have similar adverse effects from inhalation, skin contact, or ingestion, leading to anoxia, cyanosis, headache, weakness, dizziness, ataxia, dyspnea, tachycardia, nausea, and vomiting. National Institute for Occupational Safety and Health upper limits for skin exposure are 2 ppm and OSHA gives it as 5 ppm. Immediate danger to life or health level is 200 ppm.

LD_{50}, **rat**, *meta*: 1 g/kg; *ortho*: 890 mg/kg; *para*: 2.14 g/kg; **mouse**, *meta*: 330 mg/kg; *ortho*: 2.46 g/kg; *para*: 1.23 g/kg; **guinea pig**, *meta*: 3.6 g/kg; **rabbit**, *meta*: 2.4 g/kg.

NITROUS OXIDE

= *Laughing Gas*

Inhalation anesthetic, analgesic.

Untoward effects*:** Chiefly due to lack of ***oxygen and duration of use.
***Abuse*:** So-called recreational or fun use from pharmaceutical channels or consumer products. Malignant hyperthermia is a serious potential threat in **those** who possess the genetic make-up leading to this often fatal result. **Anesthesiologists** and **thieves** have frequently misused it, and the former rarely suffer ill-effects. Inhalation has been unpleasant to 44% of 110 **women** and 18% of 259 **men**. Several young **men** inhaled it for an hour/day (30 min to 6 h), often while napping in **their** offices, over periods ranging from a few months to 6 years. **One**, a **dentist**, who abused it for about 6 years developed posterior paralysis and slowly regained **his** ability to walk. **Others** had impotence, urinary incontinence, numbness in hands and feet, poor coordination, and sensations of "electric shock" along the spine. Said to be habituating, but not physically addicting.

Blood and bone marrow: Use for > 3 days, as in treating tetanus, has caused a delayed reversible megaloblastic bone marrow depression, anemia, and pancytopenia.
Cardiovascular: Systemic vascular resistance tends to rise during its use in **patients** with heart disease. Slight increase in urinary ***catecholamines*** without very significant cardiovascular changes. Transient arrhythmias reported, especially when used with other ***anesthetics***, ***atropine***, and ***muscle relaxants***. Premature ventricular contractions in 7-year-old **male** reported during anesthetic induction.
Fetus and neonate: Higher than normal rates of **fetal** deformity, miscarriages, and infertility among **female anesthesiologists**, but no proof that it was due to nitrous oxide exposure. Although it reaches the **fetus** rapidly, little adverse effects can be expected from 10–15 min exposures. Levels of 330–9,700 ppm found near **anesthesiologists** and 310–550 ppm near **surgeons** in operating rooms. **Female dental assistants** with > 5 h exposure/week had dramatic decrease in conception rate.
Miscellaneous: Unexplained jaundice and fever in 62-year-old **male** after repeated anesthetic procedures for treatment of a gangrenous hallux. Hearing acuity decreased, increased intraocular pressure, malignant hyperpyrexia, ketoacidosis, oxidation and inactivation of ***vitamin*** $\boldsymbol{B_{12}}$ with subsequent decrease of serum ***methionine*** levels, occassionally decreased temperature, psychotic sensations during narcosis (some lasting ~1 year), delusions, or hallucinations, euphoria, myeloneuropathy, and nervous laughter. A mix-up in an emergency room's supply tubes may have caused the deaths of five **patients** receiving it instead of ***oxygen***. Its diffusion into an endotracheal tube increased the cuff size, with pressure causing an obstruction in 58-year-old **male**; therefore, periodic deflation is recommended. Dizziness, nausea, vomiting, hyperventilation, postanesthetic nausea, and increased intracranial pressure reported. Contamination with ***nitric acid*** and ***ammonia*** leads to methemoglobinemia, cyanosis, and intense pulmonary edema. Contact with it as a liquid can cause frostbite. Postanesthetic hypoxia with embolism can occur after high concentrations or prolonged anesthesia, especially if insufficient supplemental ***oxygen*** is given during recovery. Usually, the first danger signs during anesthesia are noisy and irregular respirations. Usually requires addition of other agents to alleviate pain during many surgical procedures. Inhalation has caused ataxia, respiratory depression, coma, and deaths. Death, probably from asphyxia, in 19-year-old **male** found dead with head and neck covered in a large plastic bag, inside which **he** was holding a partially opened cylinder of nitrous oxide. Chronic inhalation can cause bronchitis, emphysema, and wearing down of teeth. Inhalation of pressurized material leads to pneumomediastinum with pleuritic chest pain, and symptoms of upper respiratory infection.
Neuropathy: Myeloneuropathy has, essentially, been in **abusers**. In one group, 14/15 were **dentists**; 13 had abused it for 3 months to several years, and two had been exposed due to poor ventilation. Symptoms included Lhermitte sign, loss of balance, leg weakness, ataxia, impotence, and sphincter disturbances. Spastic paresis in 14-year-old **male** with phenylketonuria after its use as an anesthetic.

National Institute for Occupational Safety and Health sets upper exposure limit of 25 ppm.

TD_{LO}, **human**: 24 mg/kg/2 h.

Embryotoxic in **rats** at usual anesthetic concentrations. Its use has shown teratogenicity in **rabbits** and **rats**, as have other common inhalation anesthetics. This is, apparently, not related to decreased oxygenation. Exposure of preg-nant **rats** on day 9 led to fetal resorption, skeletal anomalies, and gross lesions, including encephalocele, anophthalmia, microphthalmia, and gastroschisis. Prolonged anesthesia with it has decreased white cell formation and white blood cells counts in **rats**, in contrast to the fact that even short exposures to most inhalation anesthetics lead to leukocytosis. Like ***halothane***, it can produce hyperthermia in **swine**. Cortical cell damage, especially in the occipital region in **rats** and testicular functions with chronic exposure in **mice**.

NITROXOLINE

= *Enterocol* = *Nibiol* = *5-Nitro-8-quinolinol* = *Noxi-biol* = *Uritrol* = *Urocoli*

Antibacterial.

Untoward effects: Gastrointestinal upsets rare. One with tachycardia on 800 mg/day in 200 **patients** with urinary infections receiving 150–800 mg/day/2–3 months.

NITROXYNIL

= *Dovenix* = *Trodax*

Fasciolicide.

Untoward effects: May cause irreversible skeletal muscle inhibition in **animals**. Small doses may cause hypertension and overdoses may cause irreversible neuromuscular, cardiac, and respiratory failure without evidence of any anticholinesterase effects, and various forms of these reactions may be noticed during clinical trials. May cause SC irritation and abscesses in **dogs** and **elephants**. There is a narrow margin between therapeutic and toxic dosage.

LD_{LO}, **mammal**: 125 mg/kg; SC, 50 mg/kg.

NIVALENOL

A trichothecene mycotoxin (the final toxin is probably *fusarenone*). Occurs naturally in *Fusarium sp.* on foodstuffs.

***Untoward effects*:** Said to be one of the toxins air-dropped in yellow rain. Symptoms of exposure include nausea, vomiting, diarrhea, fever, leukopenia, sepsis, bleeding, necrotic skin, and mucous membrane lesions.

IP, LD_{50}, **mouse**: 4.1 mg/kg; SC, 5.2 mg/kg.

NIZATIDINE

= *Axid* = *Calmaxid* = *Cronizat* = *Distaxid* = *Gastrax* = *LY-139,037* = *Naxidine* = *Nizax* = *Nizaxid* = *Zanizal* = *ZE-101* = *ZL-101*

***Antiulcerative, H_2-receptor antagonist.* Reduces gastric acid secretion.**

Untoward effects*:** Somnolence, sweating, urticaria, rashes, exfoliative dermatitis, anemia, thrombocytopenia, impotence (3 weeks after 300 mg at night in 40-year-old **male**), central nervous system reactions (reversible mental confusion), increased liver enzymes, eosinophilia, hyperuricemia, fever, nausea, and allergic reactions (also with ***famotidine and ***ranitidine*** that have similar chemical side chains to their ring structure). Rarely, gynecomastia. During a 12 h period, < 0.1% of **maternal** dose passes into breast milk.

Drug interactions*:** It increases blood levels of ***theophylline. Hemorrhage and increased prothrombin time in a single **patient** taking ***warfarin***. Other trials indicated no interaction with ***warfarin***. High dosage may increase serum ***salicylate*** levels.

Cyclosporine blood levels and hepatotoxicity are increased by it in **rats**. Given to lactating **rats**, causes their **pups** to have decreased growth rate. Interactions may occur with ***alcohol*** and ***ketoconazole***.

LD_{50}, **rat** and **mouse**: ~1.68 g/kg; IV, **rat**: > 300mg/kg; **mouse**: 265 mg/kg.

NODULARIA SPUMIGENA

A blue-green algae.

Untoward effects*:** **Ducks** 35 days of age that had free access to an infected pond developed cloacal hemorrhages, inappetance, and paralysis. 400/500 died within 5 days. **Mice** given 0.2–1.0 ml of the filtrate died within 2 days, after showing excitement, fainting, paralysis, spasms, and bleeding from their noses and feet. Toxicity was, apparently, due to an algal pigment, ***plycocyanin. **Sheep** given water containing it developed pyrexia and leukopenia before death. Post-mortem revealed acute hepatic necrosis, extensive hemorrhages, and yellowish fluid in body cavities. **Guinea pigs** died with periacinar liver necrosis. **Dogs** swimming in infected waters had a lymphocyte increase from 3–24% with a slight decrease in leukocytes and red blood cells.

NOLINA TEXANA

= *Bear Grass* = *Sacahuista*

A perennial in the lily family, common in Texas, the Southwest, and northern Mexico.

***Untoward effects*:** Hepatogenic photosensitivity with kidney and liver damage from this and other species, primarily in **sheep** and **goats**, and occasionally in **cattle**. Depression, anorexia, yellow nasal and eye discharge, jaundice, and occasionally swelling of face and ears 16–24 h after ingestion of the buds, flowers, and fruit, but not the leaves. Sloughing of skin, pruritus, dark or purplish-red urine, and, occasionally, a purplish band appears near the top of the hoof above the coronary band. Post-mortem revealed jaundice, yellow-brown liver, and greenish-brown to greenish-black swollen kidneys.

NOMIFENSINE MALEATE

= *Alival* = *HOE-984* = *Hostalival* = *Merital* = *Neurolene* = *Psicronizer*

Antidepressant.

Untoward effects*:** Has caused elation (maniac states), drowsiness, tremors, renal failure (acute tubular necrosis), sudden fever, hepatitis (33%), nausea and vomiting (5%), nervousness and restlessness (10%), dermatitis, headache, extrapyramidal movements, paranoid symptoms, hypersensitivity alveolitis, thrombocytopenia, necrotizing vasculitis, eosinophilia, and, finally, acute hemolytic anemia and fatalities, that forced its withdrawal from world markets. Plasma levels were decreased in **epileptics** treated with ***anticonvulsants. Therapeutic serum levels reported as 0.01–0.1 μg/ml and toxic at 8 μg/ml.

Cardiovascular toxicity reported in **rabbits**. **Rats** have shown abuse of it by IV self-administration.

LD_{50}, **rat**: 430 mg/kg; **mouse**: 400 mg/kg.

γ-NONALACTONE

= *Coconut Aldehyde* = *Prunolide*

Several tons/year used in the U.S. as a fragrance in soaps, detergents, cosmetic creams and lotions, perfumes, and foods.

Untoward effects*:** No irritation or sensitization by 10% in petrolatum on **human** volunteers. No irritation of 32% concentration in ***acetone on **human** skin.

No ill-effects from 2 year feeding trial of 61.4 mg/kg/day to **rats**. Pure material topically on the skin of **rabbits** is severely irritating, but not on **guinea pig** or **miniature swine** skin.

LD_{50}, **rat**: 6.6 g/kg; **guinea pig**: 3.44 g/kg; acute dermal LD_{50}, **rabbit**: > 5 g/kg.

NONANE

= *Nonyl Hydride*

Untoward effects: Inhalation causes eye, skin, nose, and throat irritation; headache, drowsiness, dizziness, confusion, nausea, tremors, incoordination, and dyspnea. Aspiration of the liquid leads to pneumonia. Flammable liquid. National Institute for Occupational Safety and Health upper exposure limit of 200 ppm.

1-NONANETHIOL

= *1-Mercaptonomane* = *n-Nonylmercaptan* = *Nonylthiol*

Untoward effects: Eye, skin, nose, and throat irritation; tachypnea, weakness, cyanosis, nausea, drowsiness, headache, and vomiting after inhalation or absorption. Combustible liquid. National Institute for Occupational Safety and Health upper exposure limit of 0.5 ppm/15 min.

5-NONANONE

A ketone.

Untoward effects: Dosing of six **rats** with 1 ml/kg resulted in three dying instantly, two within 1 min, and one within 2 min. Lungs were severely congested. Metabolized to ***2,5-hexanedione*** (***acetonyl acetone*** [q.v.]), a neurotoxic agent.

NONOXYNOL-9

= *C-Film* = *Conco NI-90* = *Dowfax 9N9* = *Encare* = *Gynol II* = *Igepal CO-630* = *Intercept* = *Neutronyx 611* = *Semicid* = *Staycept* = *Tergitol TP-9*

Spermaticide.

Untoward effects: Associated with increased risk of congenital malformations when used near the time of conception. At one time, Ortho's over-the-counter vaginal ***contraceptive*** contained gritty, irritating material. In addition to some irritation, an increased vaginal discharge was noted after use of certain formulations. The irritation can affect either sex **partner**. Lesions in vaginal and cervical epithelium, leaving a **woman** more vulnerable to HIV infection. A 24-year-old **female** inadvertently placed a contraceptive suppository in **her** urethra, causing severe cystitis.

Rapid absorption from the vaginal wall into the systemic circulation in **rats** and **rabbits**.

n-NONYL ACETATE

= *Acetate C-9* = *Pelargonyl Acetate*

Use: In fragrances (~2 ton/year in the U.S.) in soaps, detergents, cosmetic creams and lotions, perfumes, and foods.

Untoward effects: No sensitization of 2% in petrolatum on 25 **human** volunteers.

LD_{50}, **rat**: > 5 g/kg.

NORBOLETHONE

= *Genabol* = *WY 3475*

Anabolic steroid.

Untoward effects: Break-through spotting, slight acne; occasionally edema and weight gain, hirsutism, canalicular jaundice and some increase in SGOT, and, for about 3 weeks, decreased excretion of sulfobromophthalein. Decreases effectiveness of ***dicumarol***.

LD_{50}, **mouse**: 750 mg/kg.

NORBORMIDE

= *McN-1025* = *Raticate* = *Shoxin*

Rodenticide.

Untoward effects: Very large doses (300 mg) in **humans** may cause hypotension and hypothermia.

LD_{LO}, **human**: 5 mg/kg.

Toxic to the Norway **rat**, which is one of the few species with a smooth muscle receptor for it, leading to irreversible vasoconstriction, tissue ischemia, and death. **Mallard ducks** given toxic doses developed regurgitation, dyspnea, polydipsia, excessive preening, and slight ataxia. **Dogs** fed it at 0.1% of their total diet/5 months showed no ill-effects; at 1% level, none died, but they had a decreased appetite and they were weak. Tests on 36 species, including domestic **livestock**, **monkeys**, **chimpanzees**, **poultry**, and **fish** showed no ill-effects. **Cats**, **dogs**, **poultry**, and **monkeys** given 1 g/kg/day had no ill-effects. **Guinea pigs** and **hamsters** of the above 36 species are most susceptible. **Pigs** and **dogs** eating poisoned **rats** showed no ill-effects.

LD_{50}, male **rats**: 4.4 mg/kg; female: 3.5 mg/kg; **mouse**: 2.25 g/kg; **hamster**: 140 mg/kg; **mallard ducks**: > 3 g/kg.

NORCHLORCYCLIZINE

A metabolite of *chlorcyclizine* (q.v.) and *meclizine* (q.v.).

Untoward effects: Crosses the placental barrier and considered to be the teratogenic agent when the above drugs were used. In **rat** trials, caused cleft palate, micrognathia, microstomia, and glossopalatine fusion. In **man**, ***meclizine*** is not metabolized to norchlorcyclizine.

NORDAZEPAM

= *A-101* = *Calmday* = *Desmethyldiazepam* = *DMDZ* = *Madar* = *Nordaz* = *Nordiazepam* = *Praxadium* = *Ro-5-2180* = *Stilny*

***Anxiolytic.* A long-acting metabolite of *diazepam*.**

Untoward effects*: *Antacids*, *food*,** and ***radiation therapy decrease its absorption, and ***cimetidine*** can dramatically impair its elimination. Medical examiner reports indicate levels of 0.1–7.2 mg/l (mean of 0.9 mg/l) in drug deaths. Found in breast milk at low levels.

LD_{50}, **rat**: > 5 g/kg; **mice**: 1.3 g/kg.

NORDEFRIN

= *Levonordefrin* = *Neo-Cobefrin*

Vasoconstrictor.

Untoward effects*:** Hypertensive crisis with pulmonary edema and myocardial infarction in 57-year-old **male** after its inadvertent intravascular injection with ***mepivacaine in a dental procedure.

IV, LD_{50}, **mouse**: 12.6 mg/kg; IV, LD_{LO}, **rabbit**: 11 mg/kg; SC, LD_{LO}, **rat**: 3 mg/kg.

NOREPINEPHRINE

= *Adrenor* = *Arterenol* = *Levarterenol* = *Levophed* = *Noradrenalin(e)* = *Sympathin*

The *dextro* bitartrate form = Aktamin = Binodrenal is relatively inactive.

Untoward effects*:** Has caused local vasoconstriction, ischemia, and tissue necrosis at injection site. Necrosis of skin and toes after IV treatment for shock. Premature ventricular contractions, increased lower esophageal sphincter pressure, hyperprolactinemia, increased urinary protein test results and hot flashes in **female**, decreased intraocular pressure reported after parenteral use. Vascular damage and intestinal hemorrhage after treatment of shock in some **patients**. Bowel infarction and paroxysmal tachycardia after use in cardiac failure or myocardial infarction **patients**. Administration to pregnant **women** leads to cord levels 2–18% of **maternal** levels, causing **fetal** bradycardia after its use. Decreases mammary blood flow, decreasing ***lactose and ***potassium*** in **her** milk, while increasing the ***sodium chloride*** content.

Drug interactions*:** Induces ventricular arrhythmias when given during ***halothane anesthesia. ***Antidepressants*** have a potentiating effect on it and have produced palpitations, sweating, and persistent headaches. ***Antihistamines*** potentiate its cardiovascular effects. ***Alcohol, guanethidine, methyldopa, oxytocin***, and ***reserpine*** potentiate its effects. Hypertensive crisis and associated problems (headache, myocarditis, cerebral hemorrhage, respiratory arrest, and coma) when used with ***iproniazid*** and other ***monoamine oxidase inhibitors***. In 1958, several **workers** demonstrated that prior IA injection of ***phenoxybenzamine*** or ***tolazoline*** could block its vasoconstrictor effect.

SC or IM use may cause local necrosis and sloughing, and should be avoided, despite textbook suggestions for it. Excessive levels can cause a shock syndrome (tissue anoxia from vasoconstriction) and should be avoided in pregnant **guinea pigs**, **rabbits**, or **bitches**, where it may even jeopardize fetal oxygenation. Underdose rather than overdose. Especially avoid its use during ***halothane*** (q.v.) anesthesia in **dogs** and **man**; and ***chloroform, thiamylal***, and ***thiopental*** anesthesia in **dogs**, where it can induce fatal arrhythmias, cardiac fibrillation, and cardiac arrest.

LD_{50}, **mouse**: 20 mg/kg; SC, 7.6 mg/kg; IV, 3.7 mg/kg; IV, **rat**: 10 mg/kg.

NORETHANDROLONE

= *Nilevar* = *Solevar*

Androgen, anabolic steroid.

Untoward effects*:** Intrahepatic cholestasis, jaundice, increased serum ***bilirubin, increased alkaline phosphatase and SGOT. Decreased libido, impotence, hepatocellular carcinoma, increased skeletal maturation, and menstruation disorders.

Potentiates the anticoagulant effect of ***dicumarol*** or ***warfarin***.

NORETHINDRONE

= *Conludag* = *Menzol* = *Micronor* = *Micronovum* = *Mini-Pe* = *"Mini-pill"* = *Norcolut* = *19-Norethisterone* = *Noriday* = *Norlutin* = *Norpregneninolone* = *Nor-QD* = *Primolut N* = *Utovlan*

The acetate form = *Aygestin* = *Milligynon* = *Norlutate* = *Primolut-Nor.*

Progestogen.

Untoward effects*:** Allergic reactions, headache (~1%); nervousness, dizziness, and mastalgia (~0.4%). Delayed menstruation, virilization, acne, voice hoarseness, **fetal** masculinization (> 20% in first trimester exposure; ~4% after the 12th week – enlarged clitoris, large and scrotal labia), polycythema, tachycardia, dyspnea, restless leg syndrome, anemia, nausea, vomiting, cholestasis, hepatocanalicular jaundice, break-through bleeding, increased blood ***glucose and ***insulin*** levels. False or spuriously high sulfobromophthalein retention test time readings. Occupational exposure may be by ingestion, inhalation, or dermal contact. Found in working environment air at 0.3–59.56 μg/m^3.

Drug interactions: May increase ***cyclosporine*** serum levels and suppresses ***methaqualone*** levels.

Marked **rat**, **monkey**, and **dog** fetal masculinization has been observed, as in the **human**. Excessive dosage during pregnancy may also cause fetal death, teratogenicity, and delayed parturition. In male **mice** and male **rats**, caused benign liver cell tumors; benign and malignant mammary tumors in male **rats**; pituitary tumors in female **mice**; and after SC in **mice**, led to granulosa cell ovarian tumors.

SC, TD_{LO}, **mouse**: 163 mg/kg, over 78 weeks.

NORETHYNODREL

***Progestogen.* Usually used with an *estrogen*, such as *mestranol*, in *oral contraceptives*.**

Untoward effects: Rarely, **fetal** hypospadia and masculinization. Stimulates microsomal drug metabolism. Has estrogenic effects after metabolism. Passes into breast milk. Can cause hepatocellular carcinoma or adenoma, and focal nodular hyperplasia.

NORFLOXACIN

= *AM-715* = *Baccidal* = *Barazan* = *Chibroxin(e)* = *Chibroxol* = *Floxacin* = *Fulgram* = *Gonorcin* = *Kyorin* = *Lexinor* = *MK-366* = *Noflo* = *Nolicin* = *Noracin* = *Noraxin* = *Norocin* = *Noroxin(e)* = *Norxacin* = *Sebercim* = *Uroxacin* = *Utinor* = *Zoroxin*

***Antibacterial.* A *fluoroquinolone*. A more potent antibacterial derivative of *nalidixic acid*.**

Untoward effects: Nausea (~4%), headache (~3%), dizziness (~2%), abdominal cramps (1.6%), asthenia (> 1%), tendonitis, Achille's tendon rupture, exacerbation of myasthenia gravis, anaphylactoid reactions, myalgia, arthritis, arthralgia, dyspnea, acute interstitial nephritis, anorexia, diarrhea, hyperhidrosis, constipation, anal and rectal pain, flatulence, indigestion, vomiting, finger paresthesia, dry mouth, rash, pruritus, toxic epidermal necrolysis, erythema multiforme, exfoliative dermatitis, photosensitivity, hepatitis, jaundice, increased SGOT, increased alkaline phosphatase, stomatitis, bitter taste in mouth, swollen tongue, tinnitus, diplopia, myocardial infarction, peripheral neuropathy, neutropenia, leukopenia, hemolytic anemia, thrombocytopenia, eosinophilia, psychotic reactions, convulsions, confusion, and may interfere with ***glucose*** test results.

Drug interactions: Use with ***fenbufren*** can induce convulsions. Decreases ***caffeine***, ***cyclosporine***, and ***theophylline*** metabolism. Prolongs prothrombin time and a possible fatal result reported from its use with ***warfarin***. ***Sucralfate*** decreases its bioavailability by 91%. Its absorption is decreased by ***antacids***, ***milk***, ***iron***, ***magnesium***, and ***aluminum*** ions. Antagonizes the antifungal effect of ***amphotericin*** and ***mepartricin***. ***Boric acid, EDTA, PEG 4,000***, and ***PEG 6,000*** complex it and decrease its activity.

High dosage (60 mg/kg/day) to 3–5-month-old **dogs** led to lameness with erosion and blister formation in the articular cartilage and crystalluria. Hypertrophy of chondrocytes also noted. Crystalluria in **rats** with 200 mg/kg/day.

NORGESTREL

= *Neogest* = *Norplant* = *Ovrette* = *Wy 3707*

Progestogen.

Untoward effects: Intermenstrual spotting, breast tenderness, slight weight gain, occasionally weight loss, headache (> 20%), nausea (~7%), tiredness, bloated feeling, depression, irritability (14%), abdominal discomfort, vomiting, edema, blurred vision, ovarian cysts, upset menstrual cycles (35%), dizziness (~9%), decreased libido (~3%), acne (~4%), chloasma (~2%), dysmenorrhea (2.6%), strong masculinizing action (hirsutism, alopecia), breast discharge, mastalgia, cervicitis, leukorrhea, vaginitis, mood swings, depression, and muscle pain. In breast milk at ~0.1% of the **maternal** dose. As with ***norethynodrel***, liver problems can develop.

TD_{LO}, **mouse**: 57 mg/kg, over 82 weeks.

NORMEPERIDINE

= *Norpethidine*

Untoward effects: A ***meperidine*** metabolite with less analgesic, but more convulsant, activity. After multiple doses of ***meperidine*** in **patients**, especially **those** with cancer, it accumulated, due to its half-life being five to eight times that of the parent drug causing central nervous system excitation with tremors, twitches, multifocal myoclonus, and grand mal seizures. Its toxicity is not reversed by ***naloxone*** and may be exacerbated by it. ***Ritonavir*** decreases its metabolization.

NORMETHANDRONE

= *Metahitin* = *Methalutin* = *Methylestrenolone* = *Methylnortestosterone* = *NSC 69,948* = *Orgasteron* = *U 11,828*

Androgenic progesteroid.

Untoward effects: Virilization, increased libido, peripheral edema, hepatocanalicular jaundice, and dizziness. Can cause **fetal** masculinization.

NORPSEUDOEPHEDRINE

= *Adiposetten* = *Amorphan* = *Cathine* = *Exponcit N* = *Fasupond* = *Fugoa* = *Katine* = *Minusin* = *ψ Norephedrine*

***Anorexic.* An isomer of *phenylpropanolamine.* Found in the leaves of *Catha edulis* (q.v.).**

Untoward effects: Profuse sweating in 16-year-old **female**. As a metabolite of ***ephedrine***, it can produce a false positive result in urinary screening for ***amphetamine*** intake.

IP LD_{50}, **mouse**: 161 mg/kg; SC, 275 mg/kg; IV, LD_{LO}, **rabbit**: 75 mg/kg.

NORTRIPTYLINE

= *Acetexa* = *Allegron* = *Antilev* = *Ateben* = *Avantyl* = *Aventyl* = *Desitriptilina* = *Lilly 38,489* = *Noritren* = *Nortrilen* = *Norzepine* = *Pamelor* = *Psychostyl* = *Sensaval* = *Sensival* = *Vividyl*

***Tricyclic antidepressant.* An active metabolite of *amitriptyline*.**

Untoward effects: Blood and bone marrow: Bone marrow depression, agranulocytosis, eosinophilia, thrombocytopenia, and purpura. Rarely, leukopenia.
Cardiovascular: Palpitations, arrhythmias, tachycardia, heart-block, myocardial infarction, hypotension, hypertension, stroke, and flushing.
Central nervous system: Ataxia, tremors, peripheral neuropathy, paresthesias of extremities, seizures, electroencephalogram alterations, tinnitus, extrapyramidal symptoms, confusion, hallucinations, disorientation, delusions, anxiety, agitation, restlessness, panic, insomnia, nightmares, and headache.
Endocrine: Gynecomastia, breast enlargement and galactorrhea in **female**, increased or decreased libido, impotence, testicular swelling, increased or decreased blood ***glucose***, and syndrome of inappropriate antidiuretic hormone secretion.
Fetus and neonate: Urinary retention after birth in **neonates** whose **mothers** had ingested 25 mg qid or 25 mg/day. Withdrawal effects include colic, cyanosis, irritability, hyperhidrosis, and dyspnea. Use during first trimester suspected as a cause of central nervous system, craniofacial, and skeletal anomalies. ~0.5% of **maternal** dose in breast milk, produces **infant** daily dose of 8.3 μg.
Gastrointestinal: Epigastric distress, abdominal cramps, nausea, vomiting, anorexia, diarrhea, dark tongue, stomatitis, peculiar "cotton" taste, dry mouth, sublingual adenitis, and paralytic ileus.
Miscellaneous: Dizziness, weakness, vertigo, nocturia, jaundice, blurred vision, mydriasis, urinary retention, acute intermittent porphyria, and altered weight. Ototoxic (eighth cranial nerve damage) in 8-year-old **female** given 20 mg/day/several months for enuresis.
Poisoning and suicide: Hypotension, coma, arrhythmias, and bradycardia in suicide attempts after 600–5,000 mg. Therapeutic serum levels are reported as 12–16 μg%; toxic levels as 0.05 mg%; and lethal levels as 1.3 mg%.
Skin: Rash, petechiae, pruritus, urticaria, photosensitivity, general edema or edema of face and tongue.

Drug interactions: Possible increased sedation with ***alcohol, ethyl***. ***Carbamazepine, oral contraceptives, pentobarbital,*** and ***phenobarbital*** increase its metabolism, decreasing its plasma levels. Cardiac disorders and deaths can follow use of ***local anesthetics*** containing ***norepinephrine*** or ***epinephrine***. Serum concentrations are increased by ***antipyrine, aspirin, chloramphenicol, chlorpromazine, cimetidine, dextropropoxyphene, dicumarol, divalproex sodium, fluconazole, fluoxetine, hydrocortisone, perphenazine*** (by 55%), and ***valproic acid***. May potentiate ***morphine*** and negate ***guanethidine*** action. Allow 2 week interval between its use and ***monoamine oxidase inhibitors***, to prevent possibility of hyperpyrexia, tremors, profuse sweating, delirium, excitation, convulsions, coma, and death from their combined effects. Has caused hypoglycemia in a **patient** stabilized on ***chlorpropamide***. Paradoxically, addition of ***reserpine*** to therapy has stimulated some **patients**. May alter ***anticoagulant*** control (no adverse reports with ***warfarin***, but has increased bioavailability of ***dicumarol***). In the **elderly** with ***anticholinergics***, antiparkinson therapy, and some ***antihistamines*** (***diphenhydramine, phenindamine***, and ***promethazine***), increases risk of ***atropine***-like adverse effects. ***Terbinafine*** can cause a dramatic two to three times increase in its serum levels within 2–3 weeks and can increases its toxicity.

IV infusions of 0.5 mg/kg/min led to death within 60 min in **cats**, due to bradycardia, arrhythmias, hypotension, and cardiac conduction defects. **Dogs** showed no deleterious effects of 20 mg/kg/day for 1 year.

LD_{50}, **rat**: 502 mg/kg; **mouse**: 327 mg/kg.

NOSCAPINE

= *Capval* = *Coscopin* = *Coscotabs* = *Longatin* = *Lyobex* = *Narcompren* = *Narcosine* = *Narcotine* = *Narcotussin* = *Nectadon* = *Nicolane* = *Nipaxon* = *Noscapalin* = *NSC-5366* = *Opian* = *Opianine* = *Terbenol* = *Tusscapine* = *Vadebex* = *Vetinol*

***Antitussive.* A natural, and one of the principle alkaloid, ingredients (6–10%) of *opium*.**

Untoward effects: Cough suppressant effect can lead to fluid retention, atelectasis, and pneumonitis. Half-life in **man** and **mice** is ~9 min, but it can still contribute to excessive drowsiness in young **children**. Additive depressant effect when taken with any ***central nervous system depressant***, including ***alcohol***. Large and repeated dosage can cause some drowsiness, dizziness, headache, and nausea.

Doses of 1.6 and 3.2 g/kg to **rats** caused death in 3/5 and 5/5, respectively, with dyspnea and clonic convulsions before death. Post-mortem revealed lungs filled with mucous. Antitussive doses in **man** and **dogs** are 1 mg/kg.

NOTHOLAENA SINNUATA

= *Cloak Fern* = *Jimmy Fern* = *Rock Fern*

Evergreen erect fern found in Texas, New Mexico, Arizona, and northern Mexico on rocky slopes and crevices.

Untoward effects: Affects **sheep**, **cattle**, and **goats** in decreasing order. **Animals** tremble (ranchers called it the "jimmies") ~48 h after eating ~0.5% of their body weight of the fern and 10–60 min of exercise. Symptoms include arched back, stilted hind leg movement, and tachypnea. **Cattle** often recover after 5–19 days; affected **sheep** and **goats** often die suddenly.

NOVOBIOCIN

= *Albamix* = *Albamycin* = *Biodry* = *Biotexin* = *Cardelmycin* = *Cathocin* = *Cathomycin* = *Crystallinic Acid* = *Drygard* = *Inamycin* = *PA-93* = *Robiocina* = *Spheromycin* = *Streptonivicin* = *Streptonivocin* = *U-6591* = *Vulcamicina* = *Vulcamycin* = *Vulkamycin*

Antibiotic.

Untoward effects: Idiosyncratic agranulocytosis, mild diarrhea, and hyperbilirubinemia. Hepatitis as a hypersensitivity reaction secondary to fever, rash (erythematous, papular, and urticarial), pruritus, and eosinophilia reported in 7–20+%. Pneumonitis and myocarditis also reported. Occasionally leukopenia and Henoch-Schonlein purpura; rarely, aplastic and hemolytic anemia, pancytopenia, neutropenia, and thrombocytopenia. Hemolytic anemia primarily in **those** with glucose-6-phosphate dehydrogenase deficiency. Despite reports of synergism with ***tetracycline***, the latter's activity versus certain strains of *Staphylococcus aureus* is decreased when they are used in combination. Severe jaundice and yellow discoloration of skin and sclera were frequent in premature and newborn **infants**, because it interferes with the metabolization of ***bilirubin***. Levels in breast milk (0.36–0.54 mg%) 6–30 h after **mother** received an initial oral dose of 500 mg, followed by 250 mg every 6 h. **Mother's** serum level was 1.2–5.2 mg% 2 h after dosing. Can cause spuriously high serum ***bilirubin*** or ***phenolsulfonphthalein*** test results.

Use should be limited to where other drugs have failed, or where sensitivity testing indicates its potential usefulness. Jaundice produced by it in newborn **rats** appears to be similar to that reported in **human** infants. Do NOT feed to egg-laying **birds**. The U.S. established a zero tolerance for it in milk, eggs, or uncooked edible tissues of treated **birds**, and later allowed 0.1 ppm in milk from **dairy cows**, and 1 ppm in uncooked edible **animal** and **poultry** tissues. The withdrawal time is 4 days (3 days for **ducks**) prior to slaughter of treated **birds**. Resistant cases of staphylococcal mastitis in **cattle** and staphylococcal synovitis in **turkeys** are becoming more prevalent. **Rat** and **mouse** trials indicate high levels inhibit glucuronyl transferase, the enzyme involved in conjugation of ***bilirubin***, thus, producing hyperbilirubinemia. 30 days preslaughter withdrawal time, when used in an oily vehicle for dry **cow** treatment.

NOXYTHIOLIN

= *Noryflex* = *Noxytiolin*

Antiseptic.

Untoward effects: Allergic reactions to 5–10% concentration in powder. Hematuria after bladder instillation of 1 or 2.5% solution. Fetid vegetable-like odor on breath after its use in nine **patients** on continuous ambulatory peritoneal dialysis.

NUPHAR

Untoward effects: ***N. polysepalum*** = ***Yellow Pond Lily***. Narcotic plastic rigidity, and tendency to maintain abnormal postures.

N. luteum = ***Nymphaea lutea*** = ***Yellow Water Lily*** and ***N. alba*** = ***White Water Lily***. When the flowers or powdered rhizomes are ingested, they induce a deep sleep in **mice**, **dogs**, and **eels**, and appear to be hypnotic and decrease libido in **man**, due to a compound called ***nupharine***, explored as an ***opium*** substitute in the early stages of World War II.

NUTS

Untoward effects: Urticaria, pruritus, angioedema, laryngeal edema, bronchospasm, vomiting, abdominal cramps, diarrhea, shock, and collapse in some **people**. ***Cashew*** (q.v.) nuts, related to *Rhus sp.* (***poison ivy***, ***poison oak***, and ***poison sumac***), are a common cause of dermatitis. Brazil nuts cause an atopic allergic reaction. **Patients** with ***nickel*** or ***cobalt*** dermatitis and **those** on ***lysine*** therapy should avoid nuts. ***Aflatoxin*** (q.v.) has been found (1–400 ppb) on nuts. Many contain ***oxalic acid*** (q.v.) and can lead to renal calculi. The nutty taste of some Swiss ***cheeses*** are due to ***pyrazines***. ***Apricot*** and ***peach*** seeds are related botanically to ***almonds***, and may not be tolerated by ***almond***-sensitive **patients**. Some **people** report skin and eye irritation after handling ***black walnuts***. ***Peanuts***, in the legume family, are usually well-tolerated. **Children** rarely outgrow **their** allergy to nuts. Excessive oral intake of nuts reported to aggravate chronic prostatitis in **men** and urethrotrigonitis in **women**. Causes spuriously increased urinary ***vanillymandelic acid*** test results. The Roman poet, Lucretius, some 2,000 years ago, stated, "What is food to one **man** is bitter poison to **another**".

NUTMEG

= *Muscade*

The dried, ripe seed kernels of a tall Pacific area evergreen tree, *Myristica fragrans*. *Madashuanda* (meaning "narcotic fruit") in Ayurvedic medicine.

***Use*:** As a carminative, in condiments, seasonings, fruitcakes, eggnogs, soaps, and cosmetics.

Untoward effects*:** Narcotic, and a popular hallucinogen and euphoric agent. Has caused a delayed eczematous skin reaction. Contains 210 mg ***myristicin (q.v.)/20 g, which is metabolized to ***elimicin*** and ***amphetamine*** compounds. After ingestion, few serious effects are noted for at least 2–5 h. Next morning, the **person** has a "hangover"-type of headache, dry mouth, tachycardia, dizziness, and general malaise. A group of **male prisoners** in New Jersey ate large quantities. Disorientation, confusion, and auditory and visual hallucinations in **two**, recovering from **their** toxic psychosis in ~6 months. A 17-year-old **female** consumed 25 g and a 22-year-old **female** consumed 15 g. **Both** had impaired visual perception and dream-like feelings. The 17-year-old **female** slept for 40 h and awoke euphoric. **Her** symptoms lasted 10 days. No further reactions in the 22-year-old **female**. Usual untoward effects last only 2–3 days. Has been fatal to **infants** eating as little as two nutmegs, with symptoms resembling ***alcohol*** intoxication; one and a half nutmegs have caused severe symptoms in **adults**. Auditory hallucinations and diplopia are common. Sensations of floating or flying reported, as well as separation of the limbs from the body. Has caused liver damage, severe vomiting, depersonalization, time disorientation, limb vibrations, miosis in **some**, palpitations, vertigo, contact dermatitis, flushing, nasal congestion, constipation, ecbolic effect, and death. ***Monoamine oxidase inhibitor***-like action. Contains ***safrole*** (q.v.), a suspect carcinogen. Wear mask, goggles, and protective gloves when handling it.

Hepatic necrosis in **cats**. Its ***myristicin*** causes ataxia and disorientation in **monkeys** and enhances ***morphine***-induced rage in **cats**.

IP, LD_{50}, **rat**: 500 mg/kg.

NUX VOMICA

= *Bachelor's Buttons* = *Dog Buttons* = *Poison Nuts* = *Quaker Buttons* = *Vomit Nut*

***Stimulant, bitter tonic.* Most sources contain material with 1–4% *strychnine* (q.v.) and other alkaloids, such as *brucine* (q.v.).**

Untoward effects*:** These are primarily from its ***brucine and ***strychnine*** content. Large doses cause paralysis of peripheral nerves, convulsions, violent changes in blood pressure, and respiratory failure. Used by **Ticuna Indians** of Brazil on **their** ***curare***-tipped arrows in killing their **enemies**.

LD_{LO}, **human**: 50 mg/kg.

It has been used to kill **sheep**-marauding **dogs**. It causes tetaniform seizures, mydriasis, and death in them. Lethal doses in most species, other than **ruminants**, are ~20–100 mg/kg.

NYLIDRIN

= *Arlidin* = *Bufedon* = *Buphedrin* = *Dilatal* = *Dilatol* = *Dilatropon* = *Dilydrin* = *Opino* = *Penitardon* = *Perdilatal* = *Rudilin* = *Rydrin* = *SKF-1700-A* = *Tocodilydrin* = *Tocodrin*

Peripheral vasodilator. β_2-Adrenergic.

***Untoward effects*:** Nervousness, trembling, dizziness, flushing, sensation of chilliness, postural hypotension, palpitations, cottony feeling in mouth, nausea, and vomiting.

Drug interactions*:** Because it releases ***phenothiazines from non-reactive sites, it potentiates their antipsychotic effect.

LD_{50}, **mouse**: 250 mg/kg; IV, 40 mg/kg.

NYSTATIN

= *Biofanal* = *Candex* = *Candio-Hermal* = *Diastatin* = *Fungicidin* = *Moronal* = *Mycostatin* = *Nilstat* = *Nystan* = *Nystavescent* = *O-V Statin*

Antibiotic.

Untoward effects*:** Contact dermatitis, fixed-drug eruption, exacerbation of periungular pustular eruptions, exacerbation of chronic bronchial asthma, and eosinophilia after inhalation. Oral intake may possibly cause nausea, diarrhea, vomiting, and stomach ache. Can decrease ***methotrexate absorption by 30–50%. U.S. regulations establish a zero tolerance for its residue in or on the uncooked tissues or by-products (including eggs) of treated **animals** or **poultry** if used for **human** consumption.

Local irritation and necrosis after SC in **mice**.

IP, LD_{50}, **mouse**: 20–26 mg/kg; **rat**: 8.4 mg/kg.

O

OAKS

= *Aranyérre* = *Chêne Gubacs* = *K'ao* = *Kéreg* = *Li* = *Niu* = *Quercus sp.*

Untoward effects: Large amounts of any parts of the oak have caused poisoning in **humans**. Symptoms include constipation, bloody stools, abdominal pain, extreme thirst, tachycardia, weak pulse, frequent urination, and delayed liver and kidney pathology. Symptoms are often delayed for days or weeks after ingestion. Young **children** are often affected from chewing acorns, which contain ***quercitin, quercitrin***, and ***tannin***. A common cause of tree pollen-induced allergies. Disagreeable taste and unpleasant odor in ***milk***, milk products, and meat of **animals** that eat acorns or bark. ***Black oak*** and ***red oak*** contain > three times the level of ***phenolics*** of ***white oak*** in their acorns.

The leaves, shoots, and buds in early spring and acorns in the fall are particularly toxic to **cattle**, **sheep**, and **water buffalo** when eaten in large quantities; occasionally, to **goats** and **horses**, and, rarely, to **hogs** and **deer**. Symptoms include constipation, mucoid and blood-stained fecal pellets, anorexia, emaciation, anemia, bradycardia, hypocalcemia, and edema of neck, pelvis, udder, scrotal and anal areas. Urine is often light-colored. Symptoms are often delayed, as in **humans**. Oak browse > 50% causes toxicity; > 75% is fatal. Major renal damage and petechiae in perirenal fat on post-mortem. Poisoned **sheep** often show ventral edema, polydipsia, and polyuria.

OAKMOSS

= *Ervenia sp.*

Untoward effects: Contact dermatitis from this lichen used in soaps, detergents, cosmetic creams and lotions, and perfumes (> 30 tons/year in U.S.). Contains ***thujones***. ***Thujone***-free material is permitted in U.S. foods and low-***thujone*** material is permitted in foods by the Council of Europe. No sensitization or irritation of 10% in petrolatum on **human** volunteers.

Non-irritating by undiluted material on backs of hairless **mice** and **swine**; nor on intact or abraded **rabbit** skin for 24 h under occlusion.

LD_{50}, **rat**: 2.9 g/kg; acute dermal, **rabbit**: > 5 g/kg.

OATS

= *Avena sativa*

Untoward effects: Contact dermatitis reported in some **people**; sensitivity is low in **children**. A 15-year-old **male** became ill the day **he** threshed moldy oats and died 24 h later. *Aspergillus fumigatus* was cultured from **his** lungs. Asthmatic symptoms have followed inhalation of its dust in ***flour*** mills. Using the wrong pesticide tainted millions of bushels, and required destruction by the manufacturers. Many celiac **patients** can tolerate its ***gluten*** without becoming ill.

Sporadic cases of photosensitization reported in **cattle**, **pigs**, **goats**, and **sheep**. After contamination with *Aspergillus*, it accumulates large quantities of ***oxalates***, that have caused intoxication of **mice**, **rabbits**, and **guinea pigs**. Can concentrate ***nitrates***, and consumption of oat hay has caused ***potassium nitrite*** poisoning and fatalities in **cattle**, **sheep**, and **horses**. Symptoms can include weakness, trembling, cyanosis, and death. *Stachybotrys, Ustilago, Fusarium*, etc. contamination has caused poisoning in **pigs**, **poultry**, **dogs**, and **horses**. Has concentrated ***lead*** when planted in contaminated soil. The terminal spines on oats have caused gastric hemorrhages in **hamsters**. **Dogs** often get the spines imbedded into their eyelids.

Estrone has been isolated from it. ***Ergot*** has been found on oats at Ames, Iowa.

OCGODEIA sp.

Untoward effects: Used by **natives** of northwestern Amazonia on **their** arrow poisons for its curarizing effects from its ***cardenolide*** content.

OCHRATOXINS

Nephrotoxins produced by a number of *Aspergillus* and *Penicillium* molds on stored grains. *Ochratoxin A* is the most serious, particularly outside the U.S. Early reports centered on swine and poultry in Denmark and Sweden, but it is a world-wide problem. *Ochratoxin B* and *C* are less serious.

Untoward effects: It appeared to be a cause of Balkan endemic nephropathy in **people** of Yugoslavia. Uremia and kidney dysfunction in various areas during 1957–1982 was ~2.5–3.0%, with an increased incidence of epithelial tumors of the renal pelvis and ureter. Out of > 600 blood samples from **people** in an endemic region in Yugoslavia, ~7% were positive for ochratoxin, with up to 40 ng/g serum.

Up to 28% of the toxin in the raw materials used was found in the final brewed ***beer***. Residues have been found in meat from **animals** freshly slaughtered after eating contaminated feed. Levels of 10–920 µg/kg have been found in ***bacon***, ***ham***, and ***sausage***.

Toxicosis from it has caused death from uremia in **cattle**. Post-mortem occasionally reveals non-bacterial pneumonia and perirenal edema. Histopathology demonstrates nephrosis with hyaline casts, dilated tubules, and kidney fibrosis; occasionally fatty liver degeneration. In **ducks**, symptoms include staggers, diarrhea, and death. Fatty infiltration of the liver occurs. Weight gains are decreased by as little as 0.5 ppm in the feed for **chicks**; 4–8 ppm leads to high mortality and nephrotoxicity. Transplacental exposure of **mice** led to cerebral necrosis. At low dosage, it was teratogenic and at high dosage it was embryocidal. Causes wavy ribs in offspring of exposed **mice** and **rats**, but not **hamsters**. Induced renal adenomas and carcinomas, and hepatic carcinomas in **mice**. Renal tubular cell adenomas and carcinomas in **rats**, and mammary fibroadenomas in female **rats**. Nephrotoxic and hepatotoxic to **swine**. Symptoms include diarrhea, polyuria, polydipsia, neutrophilia, slow growth, and dehydration. Congenital tumors in **pigs** after giving it to sows at 200 ppm in feed. **Dogs**, **hens**, and **rainbow trout** have had similar nephrotoxic and hepatotoxic effects from its ingestion.

LD_{50}, **rat**: 20 mg/kg; **mouse**: 58.3 mg/kg; **duck**: 0.5 mg/kg; **chicks**: 3.6 mg/kg; **chicken**: 3.3 mg/kg. TD_{LO}, **hamster** (7 days pregnant): 5 mg/kg.

Ochratoxin B, LD_{50}, **chicken**: 54 mg/kg.

OCRYLATE

= *COAPT Tissue Adhesive* = *N-Octyl-2-cyanoacrylate*

Tissue adhesive.

Untoward effects: Wound edges must be in apposition and the adhesive put on top of the wound, because it acts as a foreign body if it gets between the edges. When used on hand or joint areas, it can peel off in a few days. Spontaneous disunion of eye lenses bonded to corneal surfaces.

OCTACHLORONAPHTHALENE

= *Perchloronaphthalene*

Solvent.

Untoward effects: See ***Chlorinated Naphthalenes***. OSHA and National Institute for Occupational Safety and Health upper limits for skin contact are 0.1 mg/m^3 and National Institute for Occupational Safety and Health 15-min exposure is 0.3 mg/m^3.

OCTADECANETHIOL

= *Mercaptooctadecane* = *Octadecyl Mercaptan* = *Stearyl Mercaptan*

Untoward effects: Inhalation or skin absorption causes eye, skin, and respiratory irritation, headache, dizziness, weakness, cyanosis, nausea, and convulsions. National Institute for Occupational Safety and Health sets 0.5 ppm as upper limit of exposure/15 min.

OCTAMOXIN

= *Nimaol* = *Ximaol*

Antidepressant. A monoamine oxidase inhibitor.

Untoward effects: Decreased blood pressure, decreased centroretinal tension, and, at doses > 10 mg/day, causes orthostatic syndrome. Insomnia, tachycardia, palpitations, and dysuria reported. *Warning:* Monitor carefully, if used with other ***monoamine oxidase inhibitors***.

OCTANE

Untoward effects: Inhalation, skin contact, and ingestion lead to eye, nose, and skin irritation, drowsiness, unconsciousness, and aspiration pneumonia. National Institute for Occupational Safety and Health exposure limit of 75 ppm TWA and 385 ppm/15 min. OSHA: 500 ppm.

Severe blister formation and surface necrosis after application to the skin of **miniature swine**.

OCTANETHIOL

= *Mercaptooctane* = *Octyl Mercaptan* = *Octylthiol*

Untoward effects: Inhalation, skin contact, and ingestion lead to eye, skin, nose and throat irritation, weakness, tachypnea, headache, drowsiness, nausea, vomiting, and cyanosis. National Institute for Occupational Safety and Health sets exposure limit of 0.5 ppm/15 min.

OCTOGEN

= *HMX*

***Explosive.* Widely used by the military. Mixed with *aluminum* in a Nonel or shock tube.**

Untoward effects: A spark starts the explosion, its dust is then pulled off the inner walls of the tube, which then ignites. The shock wave moves at ~6,000 ft/s. Only $\frac{1}{10}$ grain of the ingredients are used /ft of tubing.

Hyper- and hypokinesia and clonic convulsions in **rabbits** at 168 mg/kg.

LD_{50}, **rat**: ~6 g/kg; **mouse**: 1.5 g/kg; **guinea pig**: 300 mg/kg.

OCTOPAMINE

= *Epirenor* = *ND 50* = *Norden* = *Norfen* = *Norphen* = *Norsympatol* = *Norsynephrine* = *WV-569*

Adrenergic. **Originally found in the octopus, but has since been found in many invertebrates, vertebrates, and *foods*.**

Untoward effects: Readily transformed into ***epinephrine*** (q.v.) and ***norepinephrine*** (q.v.). Temporary increase in blood pressure, headache, and nausea reported.

Drug interactions: Can decrease effectiveness of ***monoamine oxidase inhibitors***.

IP, LD_{50}, **mouse**: 600 mg/kg; IV, 75 mg/kg.

OCTOPUS

Untoward effects: Octopus bites are infrequent, and consist of two small puncture wounds, caused by their parrot-like chitinous jaws. Profuse bleeding may be due to an anticoagulant in their saliva. Hypotension occurs in some cases from ***eledoisin*** (q.v.). Recovery is usually uneventful. Some species are venomous. One is found in the Gulf of Mexico. The small blue-ringed octopus, *Hapalochlaena maculosa*, off the coast of Australia, has a neurotoxin, ***maculotoxin***, in its salivary glands, similar to ***tetrodotoxin***. Flaccid paralysis often develops within 5–10 min. **Some** experience sting-like burning sensations, numbness of the mouth and tongue, blurred vision, dysphonia, dysphagia, paresthesias, ataxia, paralysis, coagulopathies, nausea, vomiting, urticaria, and respiratory failure. Many **human** fatalities have occurred, especially after handling them outside the ocean waters.

OCTREOTIDE
= *Longastatin* = *Sandostatin* = *SMS 201–995*

A synthetic analog of, with action similar to that of, the hormone, *somatostatin*. Long-acting peptide inhibiting secretion of pancreatic, gastrointestinal, and pituitary hormones. Usually SC or IV.

Untoward effects: ***Cardiovascular***: Sinus bradycardia (~25%) in **acromegalics**, arrhythmias and conduction problems (~10%), QT prolongation, and worsening of congestive heart failure. Occasionally thrombophlebitis, tachycardia, chest pain, palpitations, dyspnea, and hypertension.
Gastrointestinal: Nausea, abdominal discomfort, vomiting, flatulence, and diarrhea in $> \frac{1}{3}$ of **patients**. Delayed effects are steatorrhea and gallstones. Gastrointestinal bleeding, pancreatitis, jaundice, and acute hepatitis also reported.
Miscellaneous: **Maternal–fetal** transfer has occurred; hypoglycemia, hyperglycemia, hyperkalemia, hypothyroidism, headache, dizziness, chest pain, galactorrhea, gynecomastia, amenorrhea, hematuria, anemia, anaphylactoid reactions, and pain at injection site. Biliary colic reported on abrupt withdrawal of therapy.

OCTYL ACETATE
= *Acetate C-8*

A few ton/year used in fragrances for soap, detergents, cosmetic creams and lotions, perfumes, and foods. Also as a wax, oil, nitrocellulose, and resin solvent.

Untoward effects: No irritation or sensitization of 8% in petrolatum on **human** volunteers.

Slightly irritating when applied full-strength to intact or abraded **rabbit** skin for 24 h under occlusion.

LD_{50}, **rat**: 3 g/kg; acute dermal, **rabbit**: > 5 g/kg. Minimum lethal dose, skin, **rabbit**: 500 mg/24 h.

OCTYLDIMETHYL PABA
= *Padimate O* = *Suncare* = *Sundown* = *Urasorb DMO*

Sunscreen, UVA absorber. **Usually used at 3.3–8.0% concentration.**

Untoward effects: Allergic "ocular stinging", even when applied to the face, not necessarily when near the eyes. Has the ability to act as a precursor to ***n-nitrosamine*** formation. May stain some fabrics.

Minimum lethal dose, skin, **human**: 15 mg/3 days.

OCTYL ISOBUTYRATE

Small amounts used as a fragrance in soaps, detergents, cosmetic creams and lotions, perfumes, and foods.

Untoward effects: No sensitization or irritation of 2% in petrolatum on **human** volunteers, and non-irritating when applied full-strength to intact or abraded **rabbit** skin.

LD_{50}, **rat** and acute dermal, **rabbit**: > 5 g/kg.

OCTYL PALMITATE
= *Ceraphyl 368*

Emollient.

Untoward effects: Comedogenic and a contact allergen. Found in **preteen's** cosmetics.

OCTYL STEARATE

Emollient.

Untoward effects: Comedogenic and a contact allergen. Found in **preteen's** cosmetics.

Poorly-tolerated on skin and acute ocular irritant on **rabbits**.

OENANTHE sp.

= *Water Dropworts*

Contain a resin poisonous to man and livestock. Common in Europe.

Untoward effects*: *O. aquatica* = *Fine-leaved Water Dropwort*.** Oral overdoses of its seeds in **man** lead to vertigo and narcotic effects. After ingestion, has poisoned **cattle**. Symptoms include profuse salivation, hyperexcitability, and occasionally paralysis. A **guinea pig** fed 2 g of the roots and stems developed cramps, tachycardia, tachypnea, and died within about 20 min. The plant contains poisonous ***phellandrene and a resin, ***oenanthine*** (found mostly in roots), which is probably the same as ***cicutoxin***, found in ***Cicuta*** (q.v.). Symptoms are the same as those from ***Cow Bane* = *Cicuta virosa***.

O. crocata* = *Dead Men's Fingers* = *Dead Tongue* = *Water Lovage. The Irish government has said it can be fatal if ingested by **humans**, especially **children**. In **people**, death within 2 h of ingestion is common after short periods of mydriasis, salivation, giddiness, convulsions, respiratory failure, and cardiac arrest. Applied externally, the root causes redness and irritations of the skin with eruptions.

The roots are the most poisonous part and are eaten by **cattle**, etc., after ditching or draining operations. Symptoms in **horses** and **cattle** include hypersialosis, mydriasis, and spasmodic convulsions. Some show no symptoms and have a rapid death. Half the affected **sheep** recover slowly after diarrhea for several days. Death is usually sudden with no symptoms in **pigs**, although a few vomit. In mg/kg, toxic doses are **horses**: 1, **cattle**: 1.25, **sheep**: 2, **pig**: 1.5, and **rabbit**: 20.

O. lachenalii* = *O. fistulosa* = *Parsley Water Dropwort, also of England, Ireland, and Scotland cause similar symptoms, but less severe than ***O. crocata's***. In an **adult**, $\frac{1}{2}$ grain or 32 mg leads to long, continual irritation of the fauces and 65 mg causes occasional vomiting.

O. sialaifolia of Greece has caused toxicity in **people** ~8 h after eating the cooked plant.

Extracts tested IV in **dogs** and **rabbits** caused hypotension, salivation, bradycardia, shallow respirations, and miosis. All the **rabbits** died. Toxicity in decreasing order was flowers, roots, fruits, and stems.

OENTHERA BIENNIS

= *Evening Primrose*

Use: An herbal oil remedy in cosmetic preparations and for oral use and topical treatment of various atopic eczemas.

Untoward effects: Seizures, manic and hypomanic reactions reported after oral use of unknown amounts of its seed oil may be related to some factor other than its ***linoleic acid*** and ***γ-linolenic acid*** content.

OFLOXACIN

= *DL-8280* = *Exocin* = *Flobacin* = *Floxil* = *Floxin* = *MW 361* = *Ocuflox* = *Oflocet* = *Oflocin* = *Ofloxacine* = *Oflozet* = *Oxaldin* = *Tabrin* = *Tarivid* = *Visiren*

Antibiotic. A fluorinated quinolone.

Untoward effects: Gastrointestinal upsets, particularly nausea (3–10%), diarrhea and vomiting (4%), flatulence, constipation, abdominal pain and cramps, headache (up to 9%), insomnia (up to 7%), dizziness (up to 5%), pruritus and vaginitis (up to 6%), rash (0.5–2%), anorexia, visual disturbances, dysgeusia, dry mouth, tendonitis, Achilles tendon rupture, hepatitis, hypersensitivity vasculitis (one fatal case), leukocytosis, leukopenia, eosinophilia, lymphocytopenia, agranulocytosis, thrombocytopenia, psychonosema, tremors (including Gilles de la Tourette-like syndrome), seizures, hypo- and hyperglycemia, glucosuria, hematuria, proteinuria, pyuria, insomnia, photosensitivity and joint pain (40% after 3–6 months therapy), increased γ-glutamyl transpeptidase, increased aspartate and alanine aminotransferases, increased ***creatinine***, ulcerative colitis, psychotic reactions, hallucinations, and shock. A 26-year-old **female** accidentally was given 3 g IV leading to (within 9 h) some severe central nervous system symptoms with nausea, drowsiness, dizziness, facial swelling and numbness, hot and cold flushes, slurred speech, and disorientation. Appears in breast milk. Sudden total rupture of the long head tendon of **his** right biceps in a 60-year-old **male** followed 400 mg/day/5 days.

Drug interactions: Use with ***aluminum-magnesium antacids*** decreases its absorption by 73%. Hematuria and dramatic increase in prothrombin time (21–77 s in one case report) in **patients** taking ***warfarin***. ***Probenecid*** prolongs its serum levels.

Causes erosion of joint cartilage in **dogs** and **rats** causing the suggestion that it should not be used in **humans** < 18 years.

LD_{50}, **rat**: 1493 mg/kg; **mouse**: 1842 mg/kg.

OIL ORANGES

Use: To color ***gasoline*** and food products.

Untoward effects: Some (***E*** and ***Tx***) were suspect carcinogens.

OKRA

= *Hibiscus esculentus*

Untoward effects: High ***oxalate*** content to be avoided, especially in **women** with vulvodynia. Contains ***malvalic***

and ***sterculic acids***, which can be toxic. Contact dermatitis in many **people**.

OKT 3

= *Muromonab CD3* = *Orthoclone OKT 3*

Immunosuppressant. **A murine monoclonal antibody. Over $100 million in annual U.S. sales.**

Use: In treatment of renal graft, cardiac, and bone marrow transplant rejection.

Untoward effects: Aseptic meningitis, fever (73%), chills (57%), dyspnea (21%), chest pain and/or tightness (14%), wheezing (11%), urticaria, erythema, and anaphylactic shock. Other adverse effects occasionally reported are generalized seizures, optic neuritis, photophobia, confusion, thromboses, headache, nausea, vomiting, encephalopathy, cerebral and pulmonary edema, and hearing loss. Non-Hodgkin's lymphomas prevalent in **patients** receiving it, especially with high dosage.

OLAMINE

= *Antivariz* = *Esclerosina* = *Ethamolin* = *Ethanolamine Oleate*

Sclerosing agent.

Untoward effects: Decreases ***creatinine*** clearance. IV overdose in treatment of varicose veins leads to acute renal failure,nausea, vomiting, and epigastric pain.

OLANZAPINE

= *Lanzac* = *LY 170,053* = *Zyprexa*

Psychotherapeutic.

Untoward effects: Increased SGPT (2%), postural hypotension, tachycardia, constipation, dry mouth, joint pain, somnolence, dizziness, akathesia, amnesia, articulation impairment, stuttering, euphoria, rhinitis, chills and fever, headache, palpitations, increased salivation, nausea, vomiting, weight loss, tardive dyskinesia, neuroleptic malignant syndrome, extrapyramidal symptoms, seizures, neutropenia, dyspnea, hematuria, and urinary incontinence are the more common adverse effects reported in early trials. A schizophrenic 27-year-old **male** developed severe agranulocytosis after 1 week of therapy. A 2½-year-old **male** accidentally ingested one or two 7.5 mg tablets; he slept and was difficult to arouse. Later, agitation, aggression, miosis, tachycardia, hypersalivation, and ataxia, requiring a further 24 h before recovery. After 2 weeks of treatment, a 36-year-old **male** developed pruritic pustular eruptions and tender erythematous plaques on **his** body. Fatal status epilepticus in 41-year-old **female** after 10 mg/day. Tardive dyskinesia in 25-year-old **male** after 20 mg/day/10 years. Severe exacerbation of type 2 diabetes mellitus in a 54-year-old **female** 12 days after initiating treatment with it.

Drug interactions: ***Carbamazepine, ciprofloxacin, fluvoxamine, omeprazole***, and ***rifampin*** can increase its clearance. Can antagonize ***L-dopa*** and ***dopamine*** agonist effects. ***Activated charcoal*** decreases its absorption by ~60%.

OLEANDER

= *Adelfa* = *Berbéria* = *Laurel Rosa* = *Laurier Rose* = *Laurose* = *Nerium oleander* = *Rosa de Berbéria* = *Rose Bay* = *Rosenlorbeer*

Popular ornamental evergreen shrub 5–20 ft tall, native to the Mediterranean region and eastern Asia. Now common in southern U.S., Canada, West Indies, Hawaii, Guam, and Queensland. Also see *Thevetia* for *Yellow Oleander*.

Untoward effects: Toxicity in **humans** is mainly from using small branches to skewer ***meat*** or ***frankfurters*** at outdoor barbecues, where the toxic cardiac glycosides transfer to the ***meat***, or for stirring cooking foods. Symptoms include local mouth irritation, nausea, depression, abdominal pain with frequent cries, bradycardia (as low as 40/min/5 days), arrhythmias, vertigo, dizziness, mydriasis, bloody diarrhea, convulsive movements, paralysis, coma, and even death. A less common cause of toxicity in **man** is from its use in Northern Africa as "arrow poisons". All parts of the shrub (fresh or dried leaves, branches, and flowers) are toxic. **Children** have been seriously poisoned after sucking the nectar from the flowers or chewing the leaves. Other poisoning reports have followed partially chewing a leaf, flower, or cuttings, or licking the sticky sap. The sweet aroma of the flowers has caused nausea and dizziness in some **people**. A **man** developed an acute headache, hypertension, and arrhythmias after trimming the dead wood from six oleander bushes, possibly, from inhaling airborne particles of sap and swallowing them. Smoke from burning oleander is highly toxic, causing serious illness in **firemen** and **others** in Florida. An uncommon complaint is contact dermatitis. Use mask, goggles, and protective gloves when handling it. ***Honey*** made from its nectar can be toxic. The seeds have been used in suicide attempts in India and a nearly successful suicide attempt in Florida from drinking an infusion of 12 leaves. A 40 kg 90-year-old **female**, apparently in a suicide attempt, died after ingesting ~4 g of the leaves, and evidenced increased serum ***digoxin*** levels. One leaf can kill an **adult**. Its cardiac glycosides, ***neriin*** and ***oleandrin***, are similar to ***digitalis*** and ***Strophanthus*** derivatives in action, and have even been used therapeutically, the latter at 0.5–2 mg/day.

Symptoms of poisoning in **horses**, **cattle**, and **sheep** are similar. Convulsions, tremors, diarrhea, colic, weakness, tachycardia, dyspnea, bloody feces, progressive paralysis, and vomiting in **cattle** and **sheep**. As a house plant, eating the leaves has been particularly toxic to pet **birds**, 120 mg killing **canaries** in ~5 min. Abortions and salivation also reported in **monkeys** from the leaves. **Geese**, **ducks**, and **hens** (after consuming 8–15 g of the fresh leaves) developed dullness or excitation, weakness, ataxia, tachycardia, gastroenteritis, wing paralysis, and death. Of the previous year's leaves, 6–11 g had the same effect. The leaves have also been fatal to **zoo animals**. An ounce of leaves has been the ~lethal dose for **farm animals**. **Cattle** have died from drinking water into which the leaves had fallen. Once used in southern Europe as a rodenticide.

OLEANDOMYCIN

= *Amimycin* = *Ammomycin* = *Landomycin* = *Matromycin* =*PA 105* = *RO2-7638* = *Romicil*

Named for a complex sugar molecule in its structure found elsewhere only in the *oleander* bush. Combined with *tetracycline* it is known as *Oletrin* = *Sigmamycin* = *Tetraolean*; with *morphocycline* produces *Olemor-fotsiklin* = *Olemorphocycline*. Its triacetyl derivative is *troleandomycin* (q.v.).

Untoward effects: Hepatotoxic, nausea, vomiting, diarrhea, abdominal discomfort, and occasionally skin rash; increases prothrombin time and may cause increased response to ***anticoagulants***. ***Foods***, especially ***milk***, interfere with its absorption. Cross-resistance with ***erythromycin*** and ***carbomycin***, but strains resistant to it may be susceptible to ***erythromycin***. U.S. regulations established a zero tolerance for its residues on or in the uncooked tissues or by-products from **poultry** or **animals** that have received the drug. 2–3 days elimination from **poultry** meat and eggs.

Do NOT feed to breeding **swine** or laying **hens**. Strongly comedogenic in topical trials on **rabbit** ears.

LD_{50}, **rat**: 460 mg/kg; **mouse**: 4 g/kg.

OLEIC ACID

= *Red Oil*

Untoward effects: Moderately irritating to **human** skin after 15 mg/3 days. As a dispersant in various metered-dose inhalers, it has triggered a cough reflex and bronchospasm in **man**.

Comedogenic after topical application on **man** and **rabbits**. Undiluted, it is depilatory on **mice** and **rabbits**.

OLESTRA

= *Olean*

Fat substitute.

Untoward effects: Can cause abdominal cramping, flatulence, bloating, nausea, and loose stools, and can inhibit the absorption of ***fat***-soluble ***vitamins***.

OLIBANUM

= *Frankincense* = *Gum Thus*

Untoward effects: Eczematous contact dermatitis from its use in adhesive plasters. Cross-sensitivity with ***benzoin*** preparations.

OLIVES

Color is enhanced with *ferrous gluconate*.

Untoward effects: Contains ***benzo[α]pyrene*** (q.v.), ***tannins***, ***urethane***, and ***sodium***. The olive tree, *Olea europa*, or its wood has caused a contact dermatitis. Common privet, ***Ligustrum*** (q.v.), is a member of the olive family.

Olive Oil = ***Sweet Oil***. In 1981, an explosive outbreak of pneumonitis occurred in Spain in ~22,000 **people** from the use of olive oil contaminated with ***rapeseed oil*** adulterated with ***aniline, acetylamide, methylamine***, and ***quinoline***. Peripheral neuropathy, fever, cough, eosinophilia, exanthema, pruritus, myalgia, arthralgia, headache, pulmonary edema, muscular atrophy, sclerodoma-like changes, xerophthalmia, xerostomia, hepatopathy, and osteopenia occurred. Several **people** died. In Germany in 1988, some for pharmaceutical use was found to be contaminated with ***tetrachloroethylene*** (q.v).

High oral concentrations of olive oil help produce experimental gallstones in **rabbits**. Can exacerbate heart (myocarditis and focal fibrosis) lesions, liver lesions (proliferation of bile ducts), and pancreatic tumors in **rats**. Strong comedogenicity on **rabbit** ears.

OLMEDIOPEREBEA SCLEROPHYLLA

Untoward effects: The fruits of this jungle tree were used as a hallucinogenic snuff by **Indians** of the Pariana region of central Amazonia.

OLOPTADINE

= *Patanol*

Antihistamine. **Selective H_1 antagonist for ophthalmic use.**

Untoward effects: Headache, some burning, stinging eye or foreign body sensations, hyperemia, keratitis, lid edema, pruritus, asthenia, taste perversion, rhinitis, pharyngitis, and sinusitis.

OMEPRAZOLE

= *Antra* = *Gastroloc* = *H-168/68* = *Logastric* = *Losec* = *Mepral* = *Mopral* = *OME* = *Omepral* = *Omeprazen* = *Parizac* = *Pepticum* = *Prilosec*

Stomach acid antisecretory agent.

Untoward effects: Despite its frequent use, it is interesting that, in 1984, Astra of Sweden, parent company of U.S.-based Astra Pharmaceutical Products, Inc., dropped its program to develop the drug because of carcinoid side-effects that research was unable to overcome. In 1986, they resumed testing of it.
Blood: Agranulocytosis (some fatal), thrombocytopenia, neutropenia, leukocytosis, pancytopenia, anemias (including hemolytic), hyponatremia, and hypoglycemia.
Cardiovascular: Bradycardia or tachycardia, palpitations, chest pain/angina, increased blood pressure, and peripheral edema.
Central nervous system: Depression, aggression, insomnia, nervousness, confusion, hallucinations, tremors, vertigo, paresthesia and hemifacial dyesthesia, reversible ataxia, headache and dizziness. Confusion and incoherent speech reported in an elderly **patient**.
Eye: Impaired vision in nine **patients** (25–70 years) after oral or IV doses of 20–120 mg/day/up to 7 months. Visual abnormalities appeared within 3 days and up to 7 months after treatment. Irreversible anterior ischemic optic neuropathy, confirmed by fundoscopy in six **patients**. Optic neuritis (papillitis) in a 48-year-old **male** as a possible immune reaction after a previous course of treatment.
Gastrointestinal: Anorexia, colitis, flatulence, pancreatitis (some fatal), dry mouth, mucosal atrophy of the tongue, pharyngeal pain, diarrhea, nausea, and reversible benign polyps of gastric fundic glands.
Genitourinary: Acute interstitial nephritis, renal failure, pyuria, increased urinary frequency, proteinuria, hematuria, glycosuria. Rarely, painful nocturnal erections, gynecomastia, and impotence.
Liver: Increased liver function tests. Rarely, hepatic encephalopathy and even fatal cases of liver failure and necrosis.
Miscellaneous: Weight increase, epistaxis, fever, lethargy, and edema of feet. A fatal dispensing error, due to misreading and dispensing of ***Lasix***®, instead of ***Losec***®. After 6 days of treatment, the **patient** had cardiac arrest and died on day 12. Possible teratogenic effects reported. Has induced a ***vitamin*** B_{12} deficiency. Rarely, acute interstitial nephritis and acute renal failure. Anaphylaxis (malaise, urticaria, diffuse sweating, and swollen tongue and lips) in a 54-year-old **female** and a 61-year-old **male**.
Muscle and skeletal: Subacute myopathy, arthralgias, acute gout, and back pain.
Skin: Severe rashes, toxic epidermal necrolysis, angioedema, urticaria, exfoliative dermatitis, lichenoid eruptions, erythroderma, fixed-drug eruptions, and yellowing of skin or sclera.

Drug interactions: Interethnic differences in its inhibition of ***diazepam*** metabolism (38% decreased clearance in eight **white males** and 20.7% decreased clearance in seven **Chinese males**). Also decreases metabolism of ***antipyrine*** and ***phenytoin***. ***Ketoconazole, itraconazole, oral vasodilating antianginal drugs, ampicillin esters***, and ***iron salts*** absorption is decreased by it. It increases the absorption of ***digoxin***; and ***bismuth*** from ***tripotassium dicitrato bismuthate***. Potentiation of anticoagulant effects of ***acenocoumarol*** in 78-year-old **female** and also with ***warfarin***. Use with ***clarithromycin*** inhibits omeprazole and has caused taste perversion (15%) and occasionally tongue discoloration, rhinitis, pharyngitis, and flu-like syndrome. Caused ***disulfiram*** toxicity and decreased effectiveness of ***prednisone***. ***Artemisinin*** increases its metabolism. Lack of its metabolic enzyme, CYP2C19, in 3–5% of **African-Americans** and **Caucasians** and 15% of **Asians** increases its plasma concentrations in **them**. It induces the metabolism of and decreases the serum levels of ***caffeine, clomipramine***, and ***theophylline***. It is a ***sulfone***-containing drug that can cause a hypersensitivity reaction in **those** sensitive to ***sulbactam***. ***Maalox***® granules, but not the suspension, caused a marked decrease in its plasma levels. It can decrease ***non-steroidal anti-inflammatory drug*** gastrointestinal toxicity.

In **rabbits**, doses of ~7–70 mg/kg led to embryolethality and fetal resorptions. Similar results in **rats** at 14–138 mg/kg.

LD_{50}, **rat**: > 4 g/kg; **mouse**: 80 mg/kg.

OMPHALOTUS ILLUDENS

= *Jack O'Lantern Mushroom*

Untoward effects: Contains a toxic sesquiterpene, ***illudin S***. Ingestion causes vomiting, diarrhea, weakness, tiredness, and feeling cold. Complete recovery in 14 **subjects** in 18 h.

ONDANSETRON

= *GR-C507/75* = *GR-38032F* = *SN-307* = *Vomitron* = *Zofran* = *Zophren*

Serotonin antagonist, antiemetic. **Oral and IV use.**

Untoward effects: Diarrhea (0.9%), headache (1.9%), constipation (1.2%), rash, light-headedness, dry mouth, bronchospasm, mild sedation, increased alanine and aspartate aminotransferases, dizziness, faintness, weakness, sensations of warmth and cold, fever, anorexia, dysgeusia, gastrointestinal cramps, jaundice, and pain and redness at injection site. Rarely, hypo- or hypertension, paresthesias, coughing, choking sensations, dyspnea, hot flashes, pruritus, hyperhidrosis, increased lacrimation, blurred vision, tachycardia, bradycardia, grand mal seizures, angioedema, shock, and extrapyramidal reactions (mouth distortion, lip twitches, involuntary arm-flexing, jerking, and tremors). In one **patient**, therapy was discontinued due to induced cardiac arrhythmia.

Drug interactions: ***Rifampin*** has reduced elimination half-life of IV ondansetron peak serum levels by ~45% and peak serum concentration by ~50%.

ONIONS

= *Allium cepa* = *Oignons*

Untoward effects: Can trigger migraines in some **people** and cause a contact allergic dermatitis (**housewife** eczema), lacrimation, and dyshidrosis. It inhibits platelet aggregation. Contact dermatitis in **worker** at a ***potato chip*** factory from onion powder. Onions contain disulfide goitrogens. **Patients** with ***nickel*** dermatitis should avoid onions to avoid dyshidrotic episodes. Sautéed onions from a restaurant, left uncovered, were a source of botulism in 28 **people**.

Cull onions, cooked or raw, in large quantities have been a cause of poisoning with hemolytic anemia and Heinz body formation in **cattle**, **sheep**, **horses**, **dogs**, **cats**, **rabbits**, **rats**, and **poultry**. Symptoms in **cattle** include diarrhea, hematuria, dullness, tachycardia, tachypnea, and incoordination. Post-mortem reveals anemia, enlarged spleen, fatty liver, intestinal inflammation, cortical discoloration of enlarged kidneys, pulmonary emphysema, and flabby right cardiac ventricle. The odor, especially that of ***Allium validum*** (q.v.), has passed into **cow's** milk. Onions impart a disagreeable taste and unpleasant odor into ***milk***, milk products, and meat of **animals** that have grazed on them.

ONOCLEA SENSIBILIS

= *Sensitive Fern*

Untoward effects: In Alabama, **horses** grazing on it developed incoordination, icterus, colic, and death. Congestion and edema of brain and liver swelling on post-mortem. Older **horses** are more sensitive to its ill-effects. Found in wet areas of the eastern U.S. and Canada.

OONOPSIS sp.

= *Goldenweed*

Only grows in soils high in *selenium*.

Untoward effects: A ***selenium*** (q.v.)-accumulator that can cause poisoning in **livestock**.

OPERCULINA sp.

In the *morning glory* family.

Untoward effects: ***O. tuberosa*** roots are poisonous to **horses**, **cattle**, and **deer**, but not to **goats**.

OPIPRAMOL

= *Dinsidon* = *Ensidon* = *G 33,040* = *Insidon* = *Nisidana* = *RP 8,307*

***Psychotherapeutic.* Oral dosing with 150–300 mg/day.**

Untoward effects: Primarily, dizziness, sleepiness, and fatigue, followed by depersonalization, headache, xerostomia, postural hypotension, syncope, rash, nausea, vertigo, increased blood pressure, atrial fibrillation, elation, and anxiety. Sudden death in a 9 kg 14-month-old **male**, 18 h after ingesting 110 mg/kg, and following initial recovery from induced coma. Oral doses of 5 g in suicides caused deaths in 12–18 h, with convulsions in one **person**. Deaths in four **patients** attributed to sudden cardiac arrest. Avoid use for 2–3 weeks after ***monoamine oxidase inhibitors***. Effectiveness is decreased when given with ***oral contraceptives***. Therapeutic serum levels are 0.05–0.2 mg/l, and toxic concentrations are 0.5 mg/l and it is potentially lethal at 7–10 mg/l.

LD_{50}, **rat**: 1.9 g/kg; **mouse**: 728 mg/kg.

OPIUM

***Narcotic, analgesic.* Known in India, Egypt, Mesopotamia, and China 3,000–5,000 BC. The Sumerians referred to it as the "*Joy Plant*" about 4,000 BC. Dr Thomas Sydenham and Dr van Helmont, 17th Century English physicians, were called Doctor Opiatus because they prescribed opium so frequently. One said, "Who would be callused enough to practice medicine without opium?" On another occasion, he said, "Medicine would be a one-armed man without opium". For legal medical uses, > 1,000 tons are used annually, and as much is used for illicit purposes. High-quality opium poppies contain ~10–12% opium, which contains 9–16% *morphine* and 4–8% *noscapine;* also *codeine, thebaine*, and others at ~0.5–2.5%. Of the > 25 alkaloids present, *morphine* and *codeine* are used as analgesics and antitussives. *Noscapine*, a cough suppressant, and *papaverine*, an intestinal relaxant, have little central nervous system effect. Legal annual imports into the U.S. are ~270 ton. Opium was the "*aspirin*" of ancient Egypt and Egyptians often wore good luck charms shaped like the opium poppy. In the early 1900s, it was an**

over-the-counter item in the U.S. and England, and was often found in "patent medicines", that led to addiction. England fought two wars (the Opium Wars 1839–1860, and the Arrow War, 1860), forcing China to legalize England's lucrative opium trade. The victors also got Hong Kong.

Untoward effects: As with ***morphine*** (q.v.), but usually milder. Addiction (oral or smoked), drowsiness, euphoria, respiratory depression, miosis, and nausea. Fixed-drug eruptions occur (maculopapular and urticarial) with pruritus and toxic epidermal necrolysis are uncommon. Various eczemas, contact dermatitis, and angioedema due to hypersensitivity. Hippocrates stated it caused "uterine suffocation", i.e. embryotoxicity and fetotoxicity. Chronic opium toxicity in **neonates** causes nervousness, tachypnea, and convulsive movements immediately after birth and death during the first week of **their** life. Overdoses in people lead to drowsiness, bradypnea, clammy skin, sleepiness, ataxia, dizziness, convulsions, coma, and occasionally death. Withdrawal symptoms include increased lacrimation, rhinitis, yawning, anorexia, irritability, tremors, panic, chills, hyperhidrosis, cramps, and nausea. In the 1850–1860s, tens of thousands of Chinese **laborers** immigrated to the U.S., bringing **their** habit of smoking opium with them. By 1865, the habit had spread into **whites**, particularly **gamblers**, **prostitutes**, and **criminals**. Opium dens opened in California and labor **contractors** often gave **workers** an incentive of ½ lb/month as a bonus. Use has caused impotence or inhibition of ejaculation in **men** and delayed orgasms in **women**. ***Arsenical*** neuropathy and ***lead*** poisoning from brewing raw opium in a metal pot or IV use of melted-down ***lead acetate*** and powdered opium in ***cocoa butter*** suppositories reported in chronic **users** of contaminated opium. Granulomatosis and hypersensitivity pneumonitis after parenteral use by an **addict**, due to *Scopulariopsis brumptii* fungal spores. Walter Reed General Hospital reported flashbacks in 7/236 military **personnel**. Fatalities reported in **some** from 4–5 grain (260–324 mg). Additional central nervous system depression can occur when taken concomitantly with ***alcohol***. **Users** of ***Laudanum*** (***Tincture of Opium***) have had withdrawal symptoms with worsening of previous diarrhea. A 58-year-old **male** given a 10 cc oral dose of ***Paregoric*** within 2 h became unresponsive and developed miosis and apnea. Its IV abuse leads to thrombophlebitis, pulmonary thrombi, arteritis, bacterial endocarditis, injection site abscesses, and viral hepatitis. Similar symptoms from IV use of ***Blue Velvet***®, due to the talc in the ***tripelennamine*** tablets mixed with ***Paregoric***.

LD_{LO}, **human**: 5 mg/kg.

Opium addiction reported in an obese 1½-year-old **Chihuahua**, after an illness with sporadic constipation and symptoms of severe abdominal pain after eating, receiving 6–8 drops nightly, then, increasing amounts directly from the bottle for 6–8 months. Increased lacrimation was its only withdrawal symptom. Lethal interaction between it and ***propranolol*** reported in **dogs**, **rats**, and **mice**. Apparently, a **camel's** gastric juice dissolved a cork holding opium in a tube, and customs **agents** found 14 lb in it, when the **camel** was weaving uncharacteristically. Overdoses in **animals** lead to respiratory and circulatory depression.

OPLOPANAX HORRIDUM

= *Devil's Club* = *Echinopanax horridum*

Untoward effects: Painful wounds and swelling from contact with its prickles. Considered poisonous.

OPUNTIA sp.

= *Prickly Pear* = *Scurgeon Needle*

Untoward effects: The fruit has been used for making marmalade.

Israeli **cattle** grazing on young "leaves" developed diarrhea. In the southeastern U.S., its spines become imbedded in the tongues of grazing **cattle** and **sheep**, predisposing the tongues to bacterial infection and suppurative granulomas.

ORANGE

Use: Tremendous quantities are used in the U.S. The New York metropolitan area weekly consumes ~26 million gallons of the Tropicana brand of orange juice.

Untoward effects: Contains the flavone, ***tangeretin***, which is embryotoxic, and ***d-limonene*** (q.v.), which is carcinogenic in **rodents**. Often implicated in food allergies and contact dermatitis in **fruit-handlers**, also in allergic contact cheilitis. May trigger migraine headaches in some **people** from its ***tyramine*** content. Also contains ***synephrine*** (an adrenergic vasopressor), ***salicylates***, and ***citral***. Intake of large quantities of orange juice will increase urinary pH. Dermatitis also reported from ***Yellow OB*** (q.v.) dye, once used on oranges. The carcinogen, ***urethane*** (q.v.) at 0.1–0.2 mg/l can form in **people** from filamentous fungi, such as *Fusarium* and ***Aspergillus*** on orange leaves hitting the eye. Orange oils have caused Berlock dermatitis after skin contact and orange peel oil can cross-react with ***clove oil*** and ***cinnamon oil***, producing a delayed eczematous dermatitis. Valencia oranges contain a cholinesterase inhibitor. Large amounts of orange juice can add a moderate amount of ***oxalate*** to a **person's** diet. In North Carolina, 201 **students** and 12 **adults** at a rural school became ill within minutes after drinking reconstituted orange juice with ***sodium fluoride***-treated water from a malfunctioning ***fluoride*** feeder. A 52-year-old **male** became unconscious after mistaking ***methadone***-treated ***Tang***® for orange juice. **He** consumed ~140 mg ***methadone***. Contact dermatitis to orangewood reported in a **manicurist**. Distinct phototoxic effects from ***Bitter Orange Oil***. ***β-Carotene*** is used to color ***Sunny Delight***® and a thirsty **child** drank > 1.5 l/day leading to yellow skin. Some batches of "Unpasteurized" or "Fresh-Squeezed" orange juice have contained *Salmonella enteritiasis* and caused illness.

ORANGE G

= *CI Food Orange 4* = *Novaurantia*

Untoward effects: Allergic dermatitis in tractor factory **workers** from its use in rust-preventative oil. Metabolized to ***anilene*** (q.v.) and excreted in the urine, primarily as ***p-aminophenol*** (q.v.).

Causes Heinz bodies in the erythrocytes of **mice**.

ORBIFLOXACIN

= *Orbax*

Antibiotic. A fluoroquinolone.

Untoward effects: Schering-Plough only suggests its use for **cats** and **dogs**, and not for **humans**. Irritation may follow ocular or dermal exposure in **man**. Induced arthropathy reported in a young **dog**. Convulsive seizures can occur in **animals**. Vomiting, diarrhea, and decreased food intake reported in **cats**.

ORIXA JAPONICA THUMB

= *Stink Tree*

Untoward effects: Contact dermatitis in agricultural **workers**.

ORLAYA PLATYCARPA

In the same botanical family as *Oenanthe* (q.v.), *water dropworts*.

Untoward effects: Poisoning in flocks of **sheep** in Greece with mortality of ~5%, SC leads to severe lassitude or malaise, tremors, ataxia, congestion of conjunctiva and vaginal mucous membranes, swollen vulva, tachycardia, and tachypnea.

ORLISTAT

= *Tetrahydrolipostatin* = *Xenical*

Pancreatic lipase inhibitor.

Untoward effects: Flatulence, fecal urgency, oily spotting; and decreased absorption and serum levels of ***vitamins A, D, E, K***, and ***β-carotene*** in 20–40% of **patients**. In **females** > 45 years, an increase in breast cancer was noted. The FDA advisory panel voted five to five on approval because of its possible relationship with breast cancer. Hypertension in a 40-year-old **female** after 120 mg tid/1 week.

ORMETOPRIM

= *Ormetropen* = *RO 5-9754*

Sulfonamide potentiator.* A *dihydrofolate reductase inhibitor.

Untoward effects: Use with ***sulfonamides*** in veterinary medicine can impair hemostasis and/or induce thrombocytopenia.

A maximum tolerance of 0.1 ppm has been established by the FDA for its negligible residues in edible parts of treated **birds**. Withdraw treatment 5 days before slaughter. May produce goitrogenic effects in **swine** offspring. Do NOT treat food **fish** within 6 weeks of marketing. 5 days withdrawal in **chickens** and **turkeys**.

ORNIDAZOLE

= *Madelen* = *Ornidal* = *RO 7-0207* = *Tiberal*

Antimicrobial.* Effective versus anaerobes. An *imidazole.

Untoward effects: Dizziness, headache, abdominal pain, vomiting, anorexia, weakness, and metallic taste. Rarely, a reversible peripheral neuropathy. Extended half-life in hepatic disease.

LD_{50}, **rat**: 1.8 g/kg; **mouse**: 1.4 g/kg.

ORNIPRESSIN

= *Ornithine-8-vasopressin* = *Orpressin* = *POR 8*

Vasoconstrictor, hemostatic.

Untoward effects: SC or IV leads to peripheral resistance, and slight decrease in pulse rate and cardiac output. After regional infiltration, general pallor in 34%, increased blood pressure in 18%, palpitations, extrasystoles; occasionally sweating, diarrhea, and oliguria. Rebound bleeding in 5.5%.

ORNITHOGALUM UMBELLATUM

= *Dame d'onze Heures* = *Star of Bethlehem*

Native to the Mediterranean region and South Africa; cultivated and naturalized in the U.S. and Canada and now a pasture weed.

Untoward effects: Bulbs, leaves, and flowers are toxic, due to its ***digitalis***-like cardiac glycosides, including ***convallatoxin*** (q.v.) and ***convalloside***. **Children** have been common poisoning **victims** after ingestion has induced tachycardia, arrythmias, dizziness, abdominal pain, nausea, vomiting, diarrhea, dyspena, excitement, tremors, collapse, coma, and death. Its flowers can persist for weeks without water and sold as cut-flowers. They and its sale as a house plant are particularly dangerous for **children** and **pets**.

Cattle and **sheep** have died after eating the bulbs. In a single year in Maryland, > 1,000 **sheep** died from ingesting the bulbs. Permanent blindness and death after ingestion by **cattle** reported from South Africa.
O. thyrosoides = ***Chinkerinchee*** ingestion has been fatal to **horses** and **sheep** in South Africa and the Netherlands.
O. longibracteum of eastern Africa is also extremely poisonous.

OROTIC ACID

= *Animal Galactose Factor* = *Lactinium* = *Oropur* = *Oroturic* = *Orotyl* = *Uracil-6-carbolyic Acid* = *Whey Factor*

Uricosuric.* In *vitamin B_{13}.

Untoward effects*:** Dietary levels as low as 0.001% can alter liver metabolism; it constitutes 0.125% of powdered milk. In **cow's** milk at 50–100 mg/l, not found in **non-ruminant** milk. Fatty livers in **rats** at 1% dietary level unless supplemented with ***adenine.

ORPHENADRINE

= *Banflex* = *Biorphen* = *Brocasipal* = *BS-5930* = *Disipal* = *Mephenamin* = *Norflex* = *X-Otag*

***Antihistamine, skeletal muscle relaxant, anticholinergic.* Related chemically to *diphenhydramine*.**

***Untoward effects*:** Vertigo, diplopia, increased intraocular pressure and acute glaucoma, blurred vision, xerostomia, nausea, vomiting, tremors, drowsiness; occasionally excitement and constipation, headache, dry mouth, tachycardia, palpitations, and anaphylaxis. Mental confusion, hallucinations, sleepiness, and euphorizing effects first noted in **patients** with Parkinsonism. Arrhythmias, hallucinations, delirium, anxiety, dysarthria, short-term memory loss, agitation, tremors, enuresis, muscle weakness, disorientation, convulsions, pulmonary edema, coma; very rarely, aplastic anemia, with overdoses. Ingestion of 2–3 g can be fatal in an **adult** within 6–12 h. Therapeutic serum levels are 5–20 μg%; toxic levels at 0.2 mg%; and lethal levels at 0.4–0.8 mg%; but **children** have survived with blood levels of 0.9–2.12 mg%. A 2-year-old **male** ingested 60–50 mg tablets leading to deep coma, status epilepticus, decreased blood pressure, tachycardia, cyanosis, and mydriasis, with 18 days slow recovery. Within 30 min after a 2-year-old **female** accidentally ingested 8–50 mg tablets, she developed tachycardia, collapse, cardiac arrest, and death. Dependence reported for its euphoric effect; without it, a schizophrenic 25-year-old **male** became tense and anxious, and unable to work properly.

TD_{LO}, **human**: 14 mg/kg.

Drug interactions*:** Stimulates microsomal drug metabolism. Decreases ***aminopyrine, carisprodol, griseofulvin, hexobarbital, meprobamate, and ***phenylbutazone*** effects. Neuropsychiatric adverse effects (confusion, facial dyskinesia or chorea) in 7/12 **patients** when added to ***L-dopa*** treatment. Usually enhances the effect of ***propoxyphene***. Use with ***chlorpromazine*** can cause hypoglycemia. Anxiety, tremors, and mental confusion when used with ***propoxyphene***. Jitteriness, giddiness, occasionally dizziness, and nausea in **patients** receiving it with ***fluphenazine***. ***Flupentixol*** parenterally in a 41-year-old **schizophrenic** led to possible precipitation of neuroleptic malignant syndrome, with syndrome of inappropriate antidiuretic hormone secretion and hyponatremia by addition of oral orphenadrine.

In **rats**, it inhibits (by enzyme induction) ***estradiol*** and ***estrone*** effects on the uterus.

ORYZALIN

= *EL 119* = *Dirimal* = *Ryzelan* = *Surflan*

Pre-emergence herbicide.

***Untoward effects*:** Can be harmful if swallowed, inhaled, or absorbed through the skin. Eye irritant. Misapplication to plantings prevents normal development of lateral plant roots.

LD_{50}, **rat**: 10 g/kg.

OSAGE ORANGE

= *Bois d'arc* = *Hedge Trees* = *Maclura pomifera* = *Toxylon pomiferum*

Untoward effects*:** Some **individuals** develop a dermatitis from contact with milky sap in its stems, leaves, and fruit. The fruit (hedge apples*** or ***hedge balls***) have been a common cause of choking and bloating in **cattle** and **calves**.

OSELTAMIVIR

= *GS 4104* = *RO 640,796* = *Tamiflu*

***Neuramidase inhibitor.* To decrease symptoms and severity of influenza A and B.**

***Untoward effects*:** Nausea (13%), vomiting (2.6%), and headache.

OSMIUM TETROXIDE

= *Osmic Acid*

***Untoward effects*:** Although used occasionally in Europe for the treatment of arthritis, a case of skin necrosis followed its use for chemical synovectomy in 31-year-old **female**. Acrid, pungent odor, eye and respiratory irritation, lacrimation, conjunctivitis and blepharospasms, vision problems (halos around lights), headache, cough, dyspnea, dermatitis (black discoloration of skin and occasionally squamous eczema), and local gangrene after entry through abrasions. Systemic absorption leads to nephritis and hematuria. OSHA exposure limit of 0.002 mg/m^3 TWA. National Institute for Occupational Safety and Health 15 min limit of 0.006 mg/m^3.

Inhalation TC_{LO}, **human**: 100 μg/m^3.

LD_{50}, **rat**: 14 mg/kg; **mouse**: 162 mg/kg. Inhalation LC_{LO}, **rat** and **mouse**: 40 ppm/4 h.

OSMORHIZA CHILENSIS

= *Sweet Cicely*

Untoward effects: Ingestion can be fatal. Other species have been used by **American Indians** medicinally.

OUABAIN

= *Acocantherin* = *Astrobain* = *Gratibain* = *G-Strophanthin* = *Strodival*

Cardiotonic, diuretic.

Untoward effects: Arrhythmias, increased blood pressure, premature ventricular contractions, and generally adverse effects seen with ***digitalis*** and ***digoxin***. Liver pathology and biliary retention can increase its toxicity. Oral doses have little physiological effects, so it is used parenterally. In its crude form, it is used on many African arrow poisons.

LD_{LO}, **human**: 5 mg/kg.

Sheep are particularly sensitive. Usual IV dose in **dogs** is 0.01–0.02 mg/kg and 0.035 mg/kg is the Emetic $Dose_{50}$. Toxicity in **animals** is particularly influenced by cardiac levels of ***catecholamines*** and ***calcium***. IM use in **dogs** is irritating. Arrhythmias are the most common untoward effects in **animals**. In **mice**, it increases ***barbiturate*** effect.

LD_{LO}, **rabbit**: 7 mg/kg; **pigeon**: 14 mg/kg. IV, LD_{50}, **cat**: 90 μg/kg; **rat**: 14 mg/kg; IV, LD_{LO}, **dog**: 54 μg/kg; **guinea pig**: 185 mg/kg.

OXALIC ACID

= *Ethanedioic Acid*

Found in *ink* and stain removers, marble *polishes*, and metal cleaners, as well as in many plants (e.g. *Caladium*, *celery*, *Dieffenbachia*, *dock*, *greasewood*, *halogeton*, *Jack-In-The-Pulpit*, *nuts*, *Oxalis*, *Philodendron*, *Portulaca*, *pothos*, *rhubarb*, *spinach*, etc.) at 15–20% of the plant's dry weight, and as a result of *glycol* ingestion and *methoxyflurane* anesthesia.

Untoward effects: Corrosive within 30 min of ingestion leading to burning pain, nausea, vomiting, fibrillary twitching, and clonic convulsions. Hypocalcemia, oliguria or anuria, ***calcium oxalates*** in renal tubules, bladder pain, and hematuria. Most **people** metabolize ***vitamin C*** rapidly past the oxalic acid stage and excrete it in **their** urine, but ~6–7% have a congenital defect and can't readily get past the oxalic acid stage leading to increased susceptibility to ***oxalate*** kidney stones. The rare genetic disorder of primary hyperoxaluria causes recurrent nephrolithiasis, chronic renal failure, and death from uremia. Irradiation of foods containing ***vitamin C*** destroys it and changes it to oxalic acid. ***Rhubarb*** stalks are safe, but the leaves contain soluble oxalic acid. Deaths from their ingestion have been reported after nausea, vomiting, abdominal pain, burning sensations in mouth and throat, internal bleeding, "coffee grounds" black-brown appearance of stomach contents, weakness, dyspnea, and coma. Has caused a variable blue discoloration and brittleness of nails. Normal serum levels have been reported as 0.05–0.3 mg/l and toxic levels as 0.5–1 mg/l. Has given decreased or false negative amylase, ***calcium***, and alkaline phosphate serum test results. Avoid heavy consumption of ***oxalate***-containing plants and ***monoamine oxidase inhibitors***. National Institute for Occupational Safety and Health and OSHA set upper exposure limits of 1 mg/m^3 and immediate danger to life or health of 500 mg/m^3. The fatal dose varies with its concentration. Usual fatal dose for **man** is $\frac{1}{2}$–1 oz; 1 dram (3.9 g or 71 mg/kg) was fatal in one case and recovery has occurred after ingesting 2 oz. Often ingested in suicides, with death occurring in ~10 min.

Oxalic acid toxicity is more readily produced in **cats** and **dogs** from ingestion of ***ethylene glycol*** or plants. ***Calcium oxalate*** nephrolithiasis, vomiting, ataxia, polydipsia, polyuria, tachypnea, tachycardia, and, eventually, uremia and death have occurred. ***Fescue*** seed screenings have been a source of ***oxalate*** poisoning and renal calculi in **cattle**. **Horses**, **monkeys**, **rabbits**, and **rats** have also been poisoned. Toxic to **cats** at 200 mg and **dogs** at 1 g.

LD_{50}, **rat**: 375 mg/kg; LD_{LO}, **dog**: 1 g/kg.

OXALIPLATIN

= *Eloxatine*

Untoward effects: Nausea, vomiting, paresthesias, and pharyngolaryngeal dysesthesias. Less nephrotoxic than ***cisplatin***. Anaphylaxis with dyspnea, hypotension, tachycardia, and flushing is rare. Delayed toxicity including diarrhea, bone marrow depression, thrombocytopenia, pulmonary fibrosis, and peripheral neuropathy.

OXALIS sp.

= *Lucky Clover* = *Shamrock* = *Oxala* = *Sorrel*

Untoward effects: **O. *cernua*** has a high ***oxalate*** content, and has caused deaths in **sheep** grazing on it in Australia and **cattle** occasionally in Europe.
O. pes-caprae* = *Bermuda Buttercup* = *Soursob has caused chronic ***oxalate*** poisoning, anemia, interstitial nephritis and deaths in **sheep** in Australia and Israel. Avidly eaten by **children** in an Israeli report. Lethality of other species has also been reported.

OXALYL CHLORIDE

= *Ethanedioyl Dichloride*

Untoward effects: ***Oxalic acid*** interacts with ***phosphorus pentachloride*** to form it, a military poison gas. Severely irritating to skin, eyes, and respiratory tract.

OXAMNIQUINE

= *Mansil* = *Vansil* = *UK 4271*

Schistosomicide.

Untoward effects: Nausea, somnolence, insomnia, headache, dizziness, fever, rash, eosinophilia, orange-red urine, and alterations in hepatic enzymes, electrocardiogram, and electroencephalogram. Rarely, convulsions and neuropsychiatric upsets, including hallucinations.

OXAMYL

= *DPX 1410* = *Thioxamyl* = *Vydate*

Insecticide, nematocide, acaricide. A carbamate.

Untoward effects: Toxic to **bees**, **fish**, and **wildlife**. Commercial forms contain ***methanol*** (q.v.), which, by itself, can be fatal or cause blindness if swallowed. The ***methanol*** can be equally toxic after inhalation or skin contact.

LD_{50}, **rat**: 5 mg/kg; **quail**: 4.2 mg/kg.

OXANDROLONE

= *Anavar* = *Lonavar* = *Oxandrin* = *Provitar* = *Vasorome*

Androgen.

Untoward effects: Pubic hair, and phallic enlargement in very young **children**; acne in older **children**. Increased SGOT, increased sulfobromophthalein retention, marked ***salt*** and water retention (edema), exacerbation of hypertension, masculinization in **female**. Many toxic effects occur in **body-builders**, usually in combination with other ***anabolic agents***. Violent paranoid and suicidal events have been reported with them.

OXAPROZIN

= *Alvo* = *Daypro* = *Durapro* = *Duraprost* = *Duraprox* = *Oxapro* = *Wy 21,743*

Non-steroidal anti-inflammatory drug.

Untoward effects: Nausea and dyspepsia (8%), constipation, diarrhea, and rash in 3%. Depression, sedation, somnolence, abdominal pain, anorexia, flatulence, confusion, tinnitus, vomiting, insomnia, and dysuria occur in about 1% of **patients**. Hematological and liver function adverse effects are among many infrequent adverse effects. A 53-year-old **female**, after a single 600 mg oral dose, developed a severe asthmatic reaction. A 56-year-old **female**, after 6 weeks of 600–1200 mg/day for degenerative joint disease, developed nausea, mild anorexia, and fulminant hepatitis. **She** died in about 2 weeks, while awaiting a liver transplant. Renal toxicity, photosensitization, increased blood pressure, and increased prothrombin time also occur. ***Non-steroidal anti-inflammatory drug***s are not recommended in **women** trying to become pregnant since they interfere with implantation of fertilized eggs in **animals**. Can give false positive reading on test strip for occult blood.

Drug interactions: ***Cimetidine*** and ***ranitidine*** can decrease its clearance by ~20%. Can displace ***aspirin*** and other ***salicylates*** from their binding sites, and can cause ***salicylate*** toxicity. Several case reports on its interference with ***phenytoin*** assays, as well as false positive urine tests for ***benzodiazepines***.

Induced gastric mucosal bleeding and submucosal lesions in **rat** trials.

OXATOMIDE

= *Celtec* = *Cobiona* = *Dasten* = *KW 4,354* = *Tinset* = *R 35,443*

Antiallergic, antiasthmatic.

Untoward effects: A 40-year-old **male** and a 68-year-old **male** developed pruritic erythemas after 30 mg/day for 1–12 months. Unusual neurological reactions and altered consciousness (lethargy and somnolence, and a clinical picture of encephalitis) in six **children** 4–54 months on 1 mg/kg in five and tid in one **child**. The reactions appeared within 30–60 h in five **children** and on the 8th day in **another**. Complete recovery within 24 h after treatment withdrawal.

OXAZEPAM

= *Abboxapam* = *Adumbran* = *Alepam* = *Anxiolit* = *Aplakil* = *Azutranquil* = *Benzotran* = *Bonare* = *Durazepam* = *Enidrel* = *Hilong* = *Isodin* = *Lederpam* = *Limbial* = *Murelax* = *Nesontil* = *Noctazepam* = *Nozepam* = *Oxa-Puren* = *Praxiten* = *Propax* = *Quilibrex* = *Rondar* = *Serax* = *Serenal* = *Serenid* = *Serepax* = *Seresta* = *Serpax* = *Sigacalm* = *Sobril* = *Tazepam* = *Uskan* = *Wy-3498* = *Zaxopam*

Anxiolytic and minor tranquilizer.* A *benzodiazepine.

Untoward effects*:** Drowsiness is common. Occasionally urticaria, rash, porphyrias, skin blisters, photosensitization, nausea, fatigue, weakness, dizziness, syncope, ataxia, hypotension, tachycardia, tachypnea, hyperventilation, edema, lethargy, slurred speech, stomatitis, nasal stuffiness, nightmares, amnesia, excitement, and confusion. Rarely, giddiness, leukopenia, eosinophilia, thrombocytopenia, purpura, aplastic anemia, macrocythemia, jaundice, and hepatic dysfunction and hypotension. Half-life is ~9 h (3–25). Use during first trimester has been associated with oral clefts and possibly facial paralysis (Möbius syndrome) in the **neonate**. Can cause false blood ***glucose test values. Accidental poisoning in a 2-year-old **female** fed six 15 mg capsules (90 mg) by an older **sister**. **She** fell out of bed 3 h later and awoke screaming. Hospitalized 18 h after ingestion, with lethargy, ataxia, decreased reflexes, paradoxical excitation, facial edema, and inability to stand. Some severe symptoms persisted for 3–8 days. A 2-year-old **male** was poisoned after ingesting 80 mg. Excitation, ataxia, and falling backwards or to the left 3 h later. Coma, hypertension, and absence of deep reflexes in lower extremities and of planter responses, lasting 4 h in a 16-year-old **male** after ingesting 900 mg (60 15 mg tablets). **He** was treated and recovered. Poisoning (suicide attempt) in a 45-year-old **male**, with symptoms simulating non-keto-acidotic diabetic coma. Withdrawal symptoms can occur for up to a month, and include convulsions, agitation, shaking, visual hallucinations, and incoherence. Overdoses lead to confusion, weakness, dizziness, diplopia, dysarthria, ataxia, coma, and respiratory and cardiovascular depression. Therapeutic serum concentrations are 0.2–2 mg/l and toxic ones are 3–5 mg/l.

Drug interactions*:** Additive sedation or increased ***atropine-like anticholinergic effects, if given with ***tricyclic antidepressants***. ***Heparin*** causes a rapid 150–250% rise in its free serum fraction. ***Diflunisal*** displaces it from plasma protein binding sites. May stimulate ***coumarin*** metabolizing enzymes.

LD_{50}, **rats** and **mice**: > 5 g/kg.

OXCARBAZEPINE

= *GP 47,680* = *Trileptal*

Anticonvulsant, antiepileptic.

Untoward effects*:** Nausea, vomiting, rash, diplopia, somnolence, dizziness, and ataxia. These may be less severe and less common than with ***carbamazepine, except for hyponatremia.

OXETHAZINE

= *Emoren* = *Oxaine* = *Oxetacaine* = *Oxethazaine* = *Storocain* = *Tepilta* = *Topicain* = *Wy-806*

Untoward effects*:** Oral use with ***antacids leads to tympanites, dry tongue, nausea, and constipation.

LD_{50}, **rat**: 626 mg/kg; **mouse**: 400 mg/kg.

OXFENDAZOLE

= *Autoworm* = *Benzelmin* = *Repidose* = *RS-8858* = *Synanthic* = *Systamex*

Anthelmintic.* A *benzimadole carbamate.

***Untoward effects*:** Embryotoxic in **rat** and **sheep**. Teratogenic. Arthrogryposis, ankylosis, and bone hypoplasia in **calves** and palatoschisis in a **pup**.

LD_{50}, **rats** and **mice**: > 6.4 g/kg; **dogs**: > 1.6 g/kg.

OXIBENDAZOLE

= *Anthelcide EQ* = *Equitac* = *Loditac* = *SKF-30,310*

Anthelmintic.* A *benzimadole carbamate.

Untoward effects*:** Occasionally hepatic dysfunction and fatalities in **dogs** when given with ***diethylcarbamazine. Increased alkaline phosphatase and glutamic pyruvic transaminase activity in **dogs**.

LD_{LO}, **mouse**: 32 g/kg.

OXICONAZOLE NITRATE

= *Gyno-Myfungar* = *Myfungar* = *Oceral* = *Oxistat* = *Ro-13-8996* = *Sgd-301-76*

***Antifungal.* An *imidazole* for topical use.**

***Untoward effects*:** Vaginal burning reported in **patients**. Irritation, pruritus, erythema, fissuring, allergic contact dermatitis, folliculitis, and nodules in ~0.5–2% of **patients**.

OXOLAMINE

= *AF-438* = *Bredon* = *Broncatar* = *Flogobron* = *683-M* = *Oxarmin* = *Perebron* = *Prilon*

Respiratory anti-inflammatory, antitussive.

***Untoward effects*:** Nausea and diarrhea. Chronic administration associated with bladder lesions in **animals**.

LD_{50}, **mouse**: 925 mg/kg; IP, 208 mg/kg; SC, 465 mg/kg; IV, 63 mg/kg.

OXOLINIC ACID

= *Emyrenil* = *Inoxyl* = *Nexolin* = *Nidantin* = *Ossian* = *Oxoboi* = *Pietil* = *Prodoxol* = *Urinox* = *Uritrate* = *Uro-Alvar* = *Urotrate* = *Uroxin* = *Uroxol* = *Utibid* = *W 4,565*

Antibacterial.* A *quinolone.

Untoward effects*:** In 24–70% of **patients**, central nervous system stimulation similar to ***nalidixic acid (q.v.), restlessness, metabolic acidosis, eosinophilia, insomnia, dizziness, vertigo, nausea, headache, rash, vomiting, anxiety, confusion, hallucinations, and suicidal ideas, but no convulsions as with ***nalidixic acid***. ***Nitrofurantoin*** antagonizes its antibacterial effect.

Can cause arthropathic syndrome in treated **dogs**.

LD_{50}, **rat**: 525 mg/kg; **mouse**: 1.9 g/kg.

OXOMEMAZINE

= *Doxergan* = *Dysedon* = *Imakol* = *RP 6847*

Untoward effects: Teratogenic in **man** (cleft lip).

LD_{50}, **mouse**: 220 mg/kg; SC, 260 mg/kg; IV, 35 mg/kg.

OXOPHENARSINE

= *Arseno 39* = *Arsenosan* = *Arsenoxide* = *Ehrlich 5* = *Fontarsan* = *Mapharsen* = *Mapharsal* = *Mapharside* = *Oxiarsolan*

***Antiprotozoal*. Contains ~32% *arsenic*. Used IV.**

Untoward effects: Toxic effects are similar to ***neoarsphenamine*** (q.v.). Jaundice, pain at injection site; occasionally nausea, vomiting, and mild neuropathy are the most common adverse effects.

Perivascular injections may cause tissue necrosis and sloughing. Dilute solutions to 4 mg or less/ml. Some consider it too toxic for treating **cats**.

OXOTREMORINE

Use: As a cholinergic, experimental convulsant in studies of Parkinsonism.

Untoward effects: It is the active metabolite of ***tremorine*** (q.v.), a tremor-inducing muscarinic agonist in the brain. Induces hypothermia, tremors, parasympathetic stimulation, analgesia, and salivation in **mice** after parenteral use. In **cats** and **monkeys**, it induces excitement, rage, and hallucinations; has an analgesic effect hundreds of times as great as ***morphine's***; antagonized by ***atropine***. Tremorogenic in adult **dogs**, but not newborn **puppies**. Does not cause tremors in ***vitamin*** B_1-deficient **pigeons**. ***Propranolol*** increases its toxicity in **mice**.

IV, LD_{50}, **mouse**: 1.8 mg/kg; IP, 3 mg/kg.

OXPRENOLOL HYDROCHLORIDE

= *Ba-39,089* = *Coretal* = *Laracor* = *Oxyprenolol* = *Paritane* = *Slow-Pren* = *Tracosal* = *Trasacor* = *Trasicor*

β-Adrenergic blocker, antihypertensive, antianginal, antiarrhythmic.

Untoward effects: Retroperitoneal fibrosis ~60 months after start of therapy of 48 months. Atrial fibrillation within 2 min and lasting 72 h in 1/9 of thyrotoxic **females**. Use has caused severe circulatory collapse, asystole, deaths, psychotic reactions, and can exacerbate psoriasis. Cross-sensitivity in **patients** with skin reactions to ***propranolol***. Generalized hyperpigmentation of skin in a 62-year-old **female** after 240 mg/day/5 months. Raynaud's phenomenon with cold hands and feet in 50% of **patients**. Rashes in a 77-year-old **female** and a 49-year-old **male** after treatments of 5 months and 2 months, respectively. Several **patients** develop painful peripheral skin necrosis. Use may unmask unrecognized myasthenia gravis. A 61-year-old **male** taking 40 mg tid for angina developed progressive shortness of breath during an 18 month period. An **asthmatic** 39-year-old **male**, after a single 40 mg tablet, developed bronchospasm and cyanosis. Cardiac failure, dizziness, headache, and ocular reactions occur. Drug fever in a 65-year-old **male** receiving 20 mg tid/15 days. Thrombocytopenia in a 57-year-old **male** taking 640 mg slow-release form/day/1½ years. Fatal **fetal** cardiac depression after treatment of a 32-year-old **female** gestational **diabetic**. Possible increased danger of anaphylactic shock in a 57-year-old **female** after a **wasp** sting. A 34-year-old **female** overdosed with 2.5–3 g and recovered after 2 h of cardiac massage. Ingestion of 112 40 mg tablets (4.48 g) was fatal for a 57-year-old **female**. Sudden increase in blood pressure in a **patient** given ***phenylpropanolamine***. Usual therapeutic serum levels are 0.05–1 mg/l; toxic at > 2 mg/l; and fatal at > 10 mg/l.

TD_{LO}, 571 µg/kg.

Readily transferred across the **ovine** placenta.

LD_{50}, **mouse**: 184 mg/kg.

OXYBENZONE

= *Benzophenone* = *Eusolex 3573* = *MOB* = *Spectra-Sorb UV 9* = *Uvinul M-40*

***Sunscreen, ultraviolet stabilizer in cosmetic, pharmaceutical, and plastic products*. Ultraviolet B, A, and even into C range, at 6% concentration.**

Untoward effects: Photoallergic contact dermatitis. Urticaria and delayed onset or erythematous papulovesicular eruption of sunlight exposed skin areas.

Irritant to **mouse** skin.

LD_{50}, **rat**: 12.8 g/kg.

OXYBUTYNIN CHLORIDE

= *Cystrin* = *Ditropan* = *Dridase* = *Pollakisu* = *Tropax*

Anticholinergic, urinary antispasmodic, antimuscarinic.

Untoward effects: Induced reflux esophagitis in 36-year-old **female** on oral 10 mg/day/5 years. Heat stroke in a

76-year-old **male**. Excitation (5/50), insomnia (2/50) with diarrhea, nausea, vomiting, and colic. Many adverse effects including constipation (13/50), urinary retention (3/50), blurred vision (2/50), increased intraocular pressure, xerostomia, tachycardia, palpitations, dizziness, drowsiness, hallucinations, night terror, confusion, hypohidrosis, decreased breast milk production, and rash are directly related to its anticholinergic effects. Disorientation, delirium, and memory impairment can occur in the **elderly**.

In **cats** vomiting, constipation, sialism, and urinary retention.

LD_{50}, **rat**: 1.22 g/kg.

OXYCLOZANIDE

= *Nilzan* = *Zanil*

Flukicide.

***Untoward effects*:** May cause slight temporary decrease in yield and specific gravity of **cow's** milk. Some **animals**, particularly **water buffalo**, develop slight diarrhea after treatment.

LD_{50}, **rat**: 1 g/kg.

OXYCODONE

= *Dihydrocodeine* = *Dihydrone* = *Dinarkon* = *Eubine* = *Eucodal* = *Eukodal* = *Eutagen* = *Oxikon* = *Oxycon* = *Oxycontin* = *Pancodine* = *Proladone* = *Tecodin* = *Tekodin* = *Thecodine* = *Thekodin*

***Narcotic analgesic*. Derived from *thebaine* and *codeine*. Oral or parenteral.**

Untoward effects*:** Similar to ***codeine, but with a higher dependence potential. Life-threatening allergic reactions, including pruritus and eye swelling, reported. Miosis in overdoses. Extreme agitation on withdrawal. Fatal toxic epidermal necrolysis after ***Percocet*** (5 mg oxycodone and 325 mg ***APAP***). Therapeutic serum levels are 1.7–3.6 μg%; toxic levels are 20–500 μg%.

SC, LD_{50}, **mouse**: 426 mg/kg; IV, LD_{LO}, **rabbit**: 45 mg/kg.

4,4´-OXYDIANILINE

***Use*:** In manufacturing heat-resistant enamels.

Untoward effects*:** Dietary intake caused Harderian gland and hepatocellular adenomas in **mice**, follicular cell ademomas in female **mice** and **rats** of both sexes, hepatocellular carcinomas and thyroid carcinomas in **rats**. Emits toxic ***nitrogen oxide fumes when heated.

OXYFEDRINE

= *D 563* = *Ildamen* = *Modacur* = *Myofedrin* = *Oxyphedrine*

***Antianginal*. Oral and IV.**

***Untoward effects*:** Increased systolic blood pressure, little change or slight decrease in diastolic blood pressure, and increased heart rate (maximum 10/min) after IV in 106 **patients** 38–78 years. Reported to alter taste.

IV, LD_{50}, **rat**: 46 mg/kg; **mouse**: 29 mg/kg; **dog**: 50 mg/kg; **cat**: 21 mg/kg; **guinea pig**: 340 mg/kg.

OXYGEN

Untoward effects*:** When an oxygen molecule gains an extra molecule, it becomes a very reactive oxygen free-radical, capable of reacting with fat in a cell, altering cell membrane function with a cell's protein, interfering with protein's catalyst function, or altering DNA and changing genes. These cause anemia, brain damage in strokes, heart damage in myocardial infarctions, and tissue damage during surgery. Used as a propellent for nebulized ***albuterol caused deterioration in blood gas values and clinical conditions in two **patients**. Severe respiratory depression in a 58-year-old **patient** with oxygen-propelled ***albuterol***, but not when air was used. After ~6 months, a 60-year-old **male** with respiratory insufficiency on almost continuous self-administered oxygen developed confusion, dysarthria, dimness of vision, headache, and increased somnolence. Inspiration of 50% oxygen, ordinarily not toxic, can be dangerous to **patients** with bronchitis and emphysema. Prolonged use of high concentrations (80–100%) has resulted in squamous metaplasia of the bronchial lining and fibrosis of parenchyal lung tissue. May cause apnea in **patients** under deep anesthesia. Inhaling pure oxygen at 1 atmosphere can produce coughing and chest pains within 8–24 h. Concentrations of 60% can cause the same symptoms in several days. **Infants** exposed to levels > 35–40% can suffer permanent visual impairments or blindness, due to retrolental fibroplasia. Hemolytic anemia in **patients** on hypobaric oxygen (100% at 7.4 psi/2 days) and hyperbaric oxygen (100% at 15–42.5 psi/20 min to 5 h). Cleaning solvent with ***trichloroethylene*** in newly installed oxygen equipment contributed to the deaths of four very ill hospital **patients**. Coliform organisms were cultured from dust in oxygen therapy mask and tent. A sickle-cell anemia **patient** (24 years), after receiving 8 l/min/7 days, developed decreased erythropoiesis. Blindness from retinal atrophy in 32-year-old **male** after 150 days on 80% oxygen. Occupational dermatitis (itching of wrists; redness, then black discoloration of skin) in **workers** measuring oxygen pressures in containers. Intra-arterial use leads to prolonged skin hyperemia. Excessive oxygen (> 40%) in incubators may cause retrolental fibroplasia in the **neonate**, followed by permanent impairment of vision, with tortuosity and dilation of retinal vessels. High oxygen tension in the **mother** can constrict **fetal** blood vessels of the placenta. Pulmonary dysplasia and hemorrhage in **neonate**. Respiratory acidosis in the **fetus** after oxygen administered to the **mother**. In a Boston hospital, 3–9 ft flames shot out from an oxygen mask, due to a short circuit on a floor lamp. The spark from a friction-run toy touched off a fire in a Chicago hospital oxygen tent and killed a 5-year-old **male**. Fire hazard and explosion risks with oxygen demand careful attention to causes. ***Tocopherol*** or ***vitamin E*** deficiency increase susceptibility to oxygen toxicity.

Hyperbaric oxygen (i.e. pressures > 1 atmosphere). Generally, possible irreversible central nervous system injury at pressures > 2 atmospheres; to lungs and retina at > 1½ atmospheres within 2–3 h. After 1–2 h, convulsions, peripheral vasoconstriction, increased blood pressure, and visual changes can occur at 3 atmospheres. The convulsive seizures have ended in death, due to paralysis of respiratory and cardiovascular centers. Severe vascular endothelial necrosis and fibrosis in 50% of normal **subjects** breathing 100% oxygen at 2 atmospheres. Difficulty in equilibrating middle ear pressures, transient tympanic hematoma, ear pain, deafness, claustrophobia, headache, dyspnea, disorientation, refractive errors, persistent myopia, star-like granular posterior subcapsular lenticular changes, retrolental fibroplasia, retinal vasoconstriction and ischemia, metabolic acidosis, progressive atelectasis and pneumonitis, nausea, and upper lip twitching are among the many adverse effects reported. Although necessary for life, too much can be toxic. **Scuba divers**, submarine **personnel**, and high-altitude aircraft **pilots** are at risk of hyperoxia from high-pressure oxygen or treatment with hyperbaric oxygen. Violent emesis, weakness, and sweating in **divers** with decompression sickness. Restlessness, profuse sweating, and nystagmus in ten **newborns** 5–8 days after hyperbaric oxygen. Use in local management of leg ulcers with plastic bags has caused venous stasis, pain, and cyanotic discoloration of skin.

Hyperventilation with pure oxygen can actually decrease cerebral oxygen tension, decrease cerebral blood flow, and cause retinal detachment and other eye problems in **dogs**. Hyperbaric oxygen may cause fetal damage. Retinal detachment, hemorrhage, and iritis in **dogs**. Deaths of 50% after exposure to 100% oxygen. Pulmonary pathology reported in **dogs** and **monkeys**. Hyperbaric oxygen has caused pulmonary edema, hemorrhage, and fibrosis; and seizures in **rats**; hemolysis and central nervous system damage in **mice**.

OXYLOBIUM sp.

Untoward effects*:** ***O. parvifolium = ***Box Poison*** and ***O. spectabile*** = ***Roe's Poison*** are widely distributed and extremely toxic (especially young leaves, flowers, and seeds) to grazing **livestock** in western Australia, due to their high ***fluoroacetate*** content. Lethal to **sheep** at < 1 oz.

OXYMETAZOLINE

= *Afrazine* = *Afrin* = *H-990* = *Hazol* = *Iliadin* = *Nafrine* = *Nasivin* = *Navisin* = *Nezeril* = *Oxilin* = *Rhinofrenol* = *Rhinolitan* = *Sinerol* = *Sinex*

Adrenergic, vasoconstrictor, nasal decongestant. **For topical use.**

***Untoward effects*:** Burning, stinging, smarting, dry throat, nervousness, insomnia, palpitations, hydrorrhea nasalis, headache, nosebleeds, sneezing, increased systolic blood pressure, itching, and, rarely, nausea and vomiting; bradycardia, hypotension, and dizziness in a 73-year-old **male** after several nasal sprayings/day. Rebound congestion occurs, especially after 5–7 days of use. Overuse in a 61-year-old **male** for > 20 years led to chronic hallucinosis. Young **children** and **infants** may be particularly sensitive to central nervous system depression, coma, hypothermia, and shock. Use may be associated with occasionally increase in Hodgkin's disease.

LD_{50}, **rat**: 800 μg/kg; SC, 1.1 mg/kg.

OXYMETHOLONE

= *Adroyd* = *Anadrol* = *Anapolon* = *Anasterone* = *Nastenon* = *Pardroyd* = *Plenastril* = *Protanabol* = *Synasteron*

Androgen, anabolic agent. **Used for erythropoiesis.**

Untoward effects*:** Mild virilization in young **male** and **female**. Clinical jaundice in 80% of **patients**. Varying degrees of hoarseness and hirsutism, clitoral enlargement, ***sodium and water retention, polycythemia, acne, peliosis hepatitis, hepatoma, hepatic adenoma, increased alkaline phosphatase, hyperlipidemia, confusion, and choreiform movements reported.

Drug interactions*:** Increased prothrombin time with ***acenocoumarol and ***warfarin***.

OXYMORPHONE

= *Numorphan*

Narcotic analgesic. **IM, SC, and oral.**

***Untoward effects*:** Nausea and vomiting after pre-surgical use is more severe than with **morphine**. Transitory stinging at injection site, and respiratory depression. Respiratory depression in **newborn** after **mother** received an injection 12 min before delivery. Dizziness, hypotension, bradycardia, urticaria, and miosis can occur. Dependence-producing potential.

In **dogs** causes panting, hyperesthesia, emesis, flatulence, and defecation.

OXYPENDYL

= *D 706* = *Oxipendyl* = *Pervetral*

***Antiemetic.* IM, IV, orally, or rectally.**

***Untoward effects*:** Vertigo, fatigue, orthostatic hypotension, and slight sedation after IV. Slight increase in standing blood pressure and decrease standing pulse.

LD_{50}, **rat**: 1.6 g/kg; IP, 185 mg/kg; LD_{50}, **mouse**: 735 mg/kg; IP, 139 mg/kg; SC, 352 mg/kg; IV, 75 mg/kg.

OXYPERTINE

= *Equipertine* = *Forit* = *Integrin* = *Win-18501-2*

Antidepressant.

***Untoward effects*:** Drowsiness in 5–60%, hypotension, slurred speech, dizziness, vomiting, akathisia, hand tremors, giddiness, nervousness, persistent leukocytosis, photophobia, increased serum transaminase levels, insomnia, eosinophilia, and extrapyramidal ill-effects.

LD_{50}, **mouse**: 2.8 g/kg; IV, 90 mg/kg.

OXYPHENBUTAZONE

= *Californit* = *Crovaril* = *Flogitolo* = *Flogoril* = *Frabel* = *G 27,202* = *Neo-Farmadol* = *Oxalid* = *Rapostan* = *Tandacote* = *Tandearil* = *Visubutina*

Anti-inflammatory, antiarthritic.

***Untoward effects*:** Cyanosis, hyperhidrosis, nausea, vomiting, epigastric pain, gastrointestinal bleeding, heartburn, cholestatic hepatitis, stupor, respiratory depression, chills, fever, agitation, hallucinations, hypertension, hepatomegaly, tinnitus, hearing difficulty, hematuria, oliguria, ecchymoses, rash, toxic epidermal necrolysis, urticaria, headache, cardiac pain, facial swelling, parotitis, edema, and weight gain. Convulsions followed by coma reported. Rarely, anaphylaxis, jaundice, buccal and gastrointestinal ulceration, and reduced vision. Elimination half-life of 27–64 h.

***Blood and bone marrow*:** Deaths from aplastic anemia. Increased prothrombin time, leukopenia, thrombo cytopenia, agranulocytosis, lupus erythematous cells, anemia, leukocytosis, leukemia, pancytopenia, and purpura.

***Cardiovascular*:** Prolonged PR interval, bradycardia, congestive heart failure, and pulmonary edema.

***Miscellaneous*:** Allergic reactions, including angioneurotic edema of lip, conjunctivitis, hepatocellular jaundice, diffuse pulmonary disorders, and hypothyroidism and goiter (after long-term use).

Drug interactions*:** As an inhibitor of drug metabolism, it potentiates ***dicumarol, indomethacin, penicillin, phenindione, phenprocoumon, warfarin, or ***long-acting sulfas***, as well as the antidiabetic drugs ***acetohexamide, chlorpropamide, glipizide, insulin***, and ***tolbutamide***. Its effect is decreased by ***desipramine***. Its effect is potentiated by ***methandrostenolone***, due to decreased metabolism. May be additive to ***insulin*** or ***oral diabetic agents*** in decreasing blood sugar.

TD_{LO}, **human**: 24 mg/kg/4 days.

In **dogs**, half-life is only 30 min; in **horses**: ~7 h.

LD_{50}, **rat**: 1 g/kg; **mouse**: 480 mg/kg.

OXYPHENCYCLIMINE

= *Daricon* = *Enarex*

Anticholinergic.

***Untoward effects*:** Even at recommended doses, anticholinergic effects (xerostomia, blurred vision, mydriasis, tinnitus, constipation, somnolence, bloating, vertigo, malaise, dysuria, tachycardia, flushed skin, and heartburn) occur.

OXYPHENISATIN

= *Acetphenolisatin* = *Bekunis* = *Bisatin* = *Bydolax* = *Cirotyl* = *Contax* = *Diphesatin* = *Isacen* = *Isaphen* = *Isocrin* = *Lavema* = *Laxo-Isatin* = *Lisagal* = *Promassolax* = *Prulet* = *Purgaceen* = *Sanapert*

Cathartic.

***Untoward effects*:** Acute hepatitis in 56-year-old **female** taking 4–6 mg/day/1 year. Hepatotoxicity occasionally in infrequent **users**. Severe jaundice, maculopapular rash and pruritus reported. Using it as an enema (10 mg in water) led to tachycardia, and vomiting; abdominal cramps in 18, sweating and faintness in eight, and mild transient nausea in six of 615 **patients**. Hepatic cirrhosis reported after prolonged use.

OXYPHENONIUM BROMIDE

= *Antrenyl* = *Ba-5473* = *C 5,473* = *Spasmophen*

Anticholinergic.

***Untoward effects*:** Usual adverse effects are blurred vision, mydriasis, xerostomia, dysuria, drowsiness, constipation, and tachycardia.

LD_{50}, **mouse**: 400 mg/kg.

OXYTERIA ACEROSA

= *Copperweed*

***Untoward effects*: Sheep** have been poisoned by eating the tops of these plants on arid western U.S. ranges. Its toxicity increases with age of the vegetation. Poisoning in **cattle** usually occurs in autumn. Weakness, ataxia, and jaundice is noted. Death follows in ~24 h.

OXYTETRACYCLINE

= *Abbocin* = *Alamycin* = *Aquacycline* = *Berkmycen* = *Bio-Mycin* = *Biostat* = *Clinimycin* = *Dabicycline* = *Duphacycline* = *Engemycin* = *Geomycin* = *Glomycin* = *Gynamousse* = *Hydroxytetracycline* = *Imperacin* = *LA 200* = *Liquamycin* = *Macocyn* = *Medamycin* = *Mycoshield* = *Occrycetin* = *Oxatets* = *Oxlopar* = *Oxybiocycline* = *Oxybiotic* = *Oxycyclin* = *Oxydon* = *Oxy-Dumocyclin* = *Oxyject* = *Oxy-Kesso-tetra* = *Oxylag* = *Oxymycin* = *Oxymykoin* = *Oxypan* = *Oxysol* = *Oxyterracin* = *Oxytetrachel* = *Oxytetracid* = *Oxytetrin* = *Riomitsin* = *Stecsolin* = *Stevacin* = *Syntomycin* = *Terrafungine* = *Terraject* = *Terralon-LA* = *Terramycin* = *Terravenos* = *Tetramel* = *Tetran* = *Tetra-Tablinen* = *Toxinal* = *Unimycin* = *Vendarcin*

Antibiotic.

Untoward effects: Of the ***tetracyclines***, it may cause the least amount of objectionable yellow-brown permanent discoloration of permanent teeth, when given to **youngsters**. Hypoplasia of tooth enamel occurs. Staining and discoloration of teeth with enamel hypoplasia common after years of continuous treatment of cystic fibrosis **patients**. Overgrowth of *Candida albicans* with resulting diarrhea and abdominal distention noted in **patients** after long-term therapy. Fatty metamorphosis of liver with increased SGOT in pregnant and puerperal **women**. High dosages and extended use resulted in the death of a 41-year-old **female** after surgery. Pigmentation of teeth and bones of **fetus**, decrease in linear bone growth, hypoplasia of dental enamel, and staining of crowns of permanent teeth in ~50% after **mothers** were treated with it during the last trimester of pregnancy. Australian reports indicate it may possibly be a cause of congenital cataract, cleft palate, microphthalmia, and persistent ductus arteriosis, if taken by the **mother** during early pregnancy. Photosensitivity and skin eruptions reported. Disturbed renal function and uremia in a 79-year-old **male**. Symptoms include black hairy tongue, severe dysphagia, polydipsia, polyuria, weight decrease, fatigue, dehydration, transient hypotension, and anemia. Dizziness, headache, light-headedness, increased blood urea nitrogen, anorexia, glossitis, nausea, vomiting, diarrhea, urticaria, anaphylaxis, hemolytic anemia, aplastic anemia, thrombocytopenia, neutropenia, eosinophilia, leukopenia, lupus erythematous, and purpura also occasionally reported. Bulging fontanels can occur in **infants**. Injections have caused deafness, pain and induration at injection site, and thrombophlebitis after IV use.

TD_{LO}, **man**: 114 mg/kg/4 days; TD_{LO}, **woman** (24–28 weeks pregnant): 420 mg/kg.

Drug interactions: May antagonize ***metronidazole*** effect on *Trichomonas vaginalis*. ***Antacids, milk***, and ***iron*** compounds, complex it and decrease its absorption. ***Vitamins*** B_1 and B_2, and ***chlorpropamide*** decrease its antibacterial effect. May decrease prothrombin activity, causing the need to reduce dosage of ***anticoagulants***.

Guinea pigs and laboratory **rodents** are susceptible to its toxicity resulting from a rapid kill of Gram-positive organisms and overgrowth of Gram-negative toxin-producing organisms. It will discolor ***polyethylene*** equipment. It loses potency in ***riboflavin***-containing solutions, and is easily complexed by ***calcium, iron***, and ***magnesium***, seriously decreasing its oral availability! **Cats**, **dogs**, and **pigs** have occasionally shown nausea and/or diarrhea on high oral dosage. High oral dosage to **ruminants** may be toxic, due to secondary effects from sudden alteration of the rumen flora. Avoid residues in foods for **man** by obeying specific product instructions. Therapeutic levels produce residues in milk. Do NOT feed to **hens** laying eggs for **human** consumption. Do NOT use ***honey*** from **bees** while being treated for foulbrood. Withdraw treated water from **poultry** 3 days before slaughtering. The FDA established 0.1, 0.1, 3.0, 1.0, and 0.1 ppm, respectively, as maximum allowable residues in the uncooked edible tissues of **cattle**, **swine**, kidneys of **poultry**, muscle, liver, fat, and skin of **poultry**, and **fish**. Do NOT slaughter parenterally treated **animals** for **human** consumption until 20–30 days after injection to avoid unnecessary trimming wastage at injection site or possible tissue residues. Cooking destroys all residues in **poultry** muscle, but only 50% of the liver residues. Withdraw treated feed from **poultry** 24–36 h before slaughtering, to prevent residues in their meat. A 5-day withdrawal period of treated feed is advised for any species when in doubt. **Horses** have developed diarrhea, depression, and anorexia, even after IV use. This has nothing to do with oxytetracycline's weak neuromuscular-blocking effect. Cardiovascular collapse after IV use is probably due to the solvents. Oral use as a preventative of disease in new feedlot **cattle** can decrease feed intake and weight gains, while IM injections do not. 44 mg/kg IM of ***LA-200*** led to nephrotoxicity in 400 **calves**. Although **rabbits** tolerate 500 mg/kg orally, double that dosage has been fatal. IV use (10–30 mg/kg) in **dogs** and **rats** may cause severe hypotension; a collapse syndrome has also been reported with ***tetracyclines*** in **man**, **horses**, and **cattle**, and may often be due to the ***propylene glycol*** vehicle. **Cats** showed an initial increase in blood pressure after IV injection of 30 mg/kg. 3 mg/kg, IV, in **dogs** had no noticeable cardiovascular effects. Slow dilute IVs help minimize anaphylactic reactions in **cattle**. Treated **fish** may have tissue residues for up to 1 month after treatment. 3–4 weeks withdrawal after IM use in **swine**; 3 weeks withdrawal after IM use in **cattle**; 4 weeks withdrawal after ***LA-200*** in **swine** and **cattle**. **Pigs** receiving 200 ppm in feed gave positive microbiological assays for oxytetracycline in urine and bone 3 weeks after treatment. Even when restricting the quantity injected at a given site, its IM use in **cattle** has been associated with severe local reactions with necrosis, hemorrhage, and fibrosis. Sick **cows** appear to have a slower drug clearance time than well ones. Residue in **cattle** meat appears to be destroyed by boiling for 15 mins; roasting was not equally effective. Unlike ***tetracycline*** or ***chlortetracycline***, it has a destructive effect on ***vitamin*** $\boldsymbol{B_6}$ stability. Acid salts are stabilized in solution with excess acid. Water-soluble basic salts are not stable in aqueous solution. ***Lidocaine*** is often mixed with it to decrease the sting of IM injections.

LD_{50}, **rat**: 4.8 g/kg; **mice**: 6.7–7.2 g/kg. IV, LD_{50}, **mice**: 100 mg/kg.

OXYTOCIN

= Ocytocin = α-Hypophamine = Intertocine-S = Orasthin = Oxystin = Partocon = Perlacton = Pitocin = P. O. P. = Synpitan = Uteracon = Vetocin

Oxytocic. **Oral, parenteral, buccal, and intranasal.**

Untoward effects*:** Potentially dangerous unless actions are monitored closely. Overuse during labor can cause uterine tetany with uteroplacental blood flow interference, uterine and placental membrane ruptures, laceration of the cervix, amniotic fluid embolism, **infant** trauma (hypoxia, intracranial or subdural hemorrhage), and anaphylactoid and other reactions. Bradycardia, thrombocytopenia, premature ventricular contractions and other arrhythmias, and deaths occur in the **fetus**. **Neonates** of treated **mothers** have an increased incidence of jaundice. Bolus injections lead to 30% decrease in mean arterial blood pressure and 50% decrease in peripheral resistance 40 s after injection. Cardiac output increases > 50%, because heart rate increases 30% and stroke volume 25%. A 33-year-old **female** had a sustained vasospasm during an IV via catheter in **her** left hand of 10 milliUN/min. Headache, nausea, facial pallor, and Guillain–Barré syndrome also reported. Water retention and intoxication with oliguria, and occasionally deaths. Constricts coronary arteries and can cause angina and myocardial infarction. Do NOT use in dystocia without prior correction of abnormal presentation or prior use of ***estrogens, natural dilation, etc. in failure of the cervix to dilate. It is a complex protein and IV use is more apt to precipitate shock-like reactions.

Drug interactions*:** ***Ethyl alcohol inhibits oxytocin-producing centers in the pituitary gland and inhibits uterine activity during labor. Severe prolonged hypertension can follow its use with ***sympathomimetics***. Can potentiate the action of ***muscle relaxants***, such as ***succinylcholine***, as well as the abortifacient properties of ***prostaglandins*** given during mid-trimester pregnancy. ***Halothane*** in concentrations > 1% interferes with its oxytocic effect.

Although it causes a decrease in blood pressure and weakens cardiac contractions in **man** and many **animals**, it has little such effect in **dogs**.

OXYTROPIS

Several hundred species exist.

***Untoward effects*:** Toxicity varies, depending on the soil type they grow on. Common in the U.S. western plains, western Canada, and southward into Mexico. Rare reports from elsewhere, such as China.

O. Lambertii = Crazy Weed = Lambert Loco = Point Loco = Purple Locoweed, a perennial legume whose ingestion causes weight loss, eratic actions, nervousness, weakness, decreased libido, infertility, abortion, hydrops amnii, cardiovascular disease, and incoordination in various grazing domesticated **animals** and **elk**, including abortion in **cattle** and congenital malformations in their **calves**; arthrogryposis in **foals** and **lambs**. The toxic agent has been identified as ***locoine*** or ***swainsonine*** (q.v.), an indolizidine alkaloid first isolated from Australian ***poison peas*** (***Swainsona sp.***, q.v.). The word loco means madness in Spanish. This disorder was first recognized in the 16th century by **DeSoto** and **conquistadores** whose **horses** became frightened or hyperexcited and violent from slight stimulation.

O. sericea has caused abortions and congenital malformations, incoordination, nervousness, and belligerence in **cattle** and **sheep**, and congestive heart failure and right ventricular hypertrophy in **cattle**. Hypoplastic testicles, enlarged seminal vesicles, and decreased libido and fertility in **rams**. **Horses** show incoordination, nervousness, and rough dry coats.

OYSTERS

***Untoward effects*:** They are commonly eaten uncooked, causing major disease outbreaks. *Vibrio cholerae* from them caused bloody diarrhea in Florida outbreaks. *V. vulnificus* often present in raw oysters found in Gulf Coast waters causes septicemia with high fever, hypotension, cellulitis, and occasionally bullae. Progression of illness can be rapid, with deaths in ~40%. **Some** die within 48 h after onset of the illness. Infectious hepatitis A found in oysters in Louisiana bays contaminated with Mississippi River water. Dyshidrosis and allergic reactions (especially with exercise induction) reported. Excessive *arsenic* and *cadmium* levels have been reported in some.

OZONE

Untoward effects: Accidentally created by welders, generators, motors and various electrical equipment, and photochemical reaction of solar rays on automobile exhausts. Damages some growing plants, cracks ***rubber*** products causing premature failure, weakens some fabrics, and is harmful to **humans** and **animals**. Causes major problems and discomfort in **people** with reduced lung function and respiratory difficulties. Headache, chest pain, dryness of upper respiratory tract, pulmonary edema, fatigue, and exertional dyspnea in **people** inhaling 2–3 ppm. It is a component of smog that irritates eyes, nose, throat, and respiratory tract, leading to cough, frontal headache, anosmia, nausea, pulmonary edema, and chronic respiratory disease. Pneumonitis reported in a 51-year-old **welder** after 4 h exposure (1.8 mg/m^3) in poorly ventilated area. **Asthmatics** have become worse when exposed. It is present in toxic concentrations when its odor is detected by smell. Anorexia, vomiting, hyperirritability, reduced visual acuity, and vestibular upsets (lasting as long as 9 months). Because ozone odor was faint, a 52-year-old **male worker** failed to use **his** air-supplied protective mask where 1% ozone (10,000 ppm) was sprayed into a tank, causing dry cough, lacrimation, frontal headache (rapidly increasing in severity), near-unconsciousness, decreased blood pressure, cold skin, and perspiration after a 2 h exposure. National Institute for Occupational Safety and Health and OSHA TWA exposure limits of 0.1 ppm (0.2 mg/m^3). Immediate danger to life or health: 5 ppm.

A susceptible strain of **mice** exposed to 1 ppm developed chronic bronchitis, emphysema, earlier development of lung adenomas, and markedly decreased immunological defense mechanisms.

P

PACHYCEREUS

= *Cawe*

Untoward effects: A cactus of southwestern U.S. and Mexico. Used as a narcotic and hallucinogenic agent, due to its ***carnegine*** or ***pectenine*** content.

PACHYRRHIZUS EROSUS

Untoward effects: The seeds contain ***pachyrrhizid***, ***rotenone***, and ***saponins*** and are toxic for **man**, causing central nervous system depression of respiratory center and large doses cause bradycardia. Used as a piscicide, arrow poison, and insecticide.

PACHYSTIGMA sp.

Untoward effects: The South African plant species ***P. pygmaeum*** and ***P. thamnus*** cause a toxic condition known as "gousiekte" in grazing **sheep**, **cattle**, and **goats**. After a latent period of weeks or months, symptoms develop including inappetence, tachypnea, dyspnea, systolic murmurs and arrhythmias, and death following chronic interstitial myocarditis.

PACLITAXEL

= *NSC 125,973* = *Taxol*

A plant toxoid alkaloid from the western *yew*, *Taxus brevifolia*, that accelerates polymerization of *tubulin* and stops its depolymerization.

Use: In cell research and in the IV infusion treatment of breast, ovarian, non-small cell lung, melanoma, esophageal, colon, bladder, and other neoplasms.

Untoward effects: Hypersensitivity reactions (anaphylaxis, hypotension, angioedema, tachypnea, dyspnea, pneumonitis, transient pulmonary infiltrates, light-headedness, diaphoresis, nausea, severe abdominal cramping, vomiting, pruritus, and urticaria) may be due to the drug and its usual suspending agent, ***Cremophor EL*** (q.v.), and occur during or shortly after administration. Serious severe delayed toxic effects are bone marrow depression, granulocytopenia, neutropenia, leukopenia, thrombocytopenia, anemia, and peripheral neuropathies. Arthralgias, increased intraocular pressure, myalgias, Parkinsonism, heart-block, myocardial infraction, venous or arterial thrombosis, nausea, diarrhea, cystitis, neutropenia fever, and alopecia also occur. Fatal acute tumor lysis in a 36-year-old **female** after 24 h IV infusion of 190 mg (135 mg/m^2). Onycholysis (shortly after treatment in one and between 10th and 13th week in three) in four **females** after IV infusions for ovarian cancer. Hyperkeratosis of fingernails and toenails also reported. Open-angle glaucoma in 31-year-old **female** after repeated use. Radiation recall dermatitis and leaching of ***bis(2-ethylhexyl) phthalate*** (q.v.) from ***polyvinyl chloride*** containers has occurred. Extravasation has caused severe local necrosis during IV infusions and soft-tissue injury during IP use.

Drug interactions: Limited *in vitro* tests indicate ***ketoconazole*** may decrease its metabolism. Can induce ***alcohol*** intoxication. Several workers have described a potential for interaction with ***ritonavir***.

PADIMATES

= *p-Dimethylaminobenzoate esters*

Sunscreens.

Untoward effects: ***Padimate A*** = ***Amyl Dimethyl Paba*** = ***Escalol 506*** = ***208J-2***. In 3%, stinging or burning sensations on face, especially near the eyes in hot weather or when sweating. See ***Octyldimethyl PABA*** for ***Padimate O***.

PAECILOMYCES sp.

Untoward effects: A toxigenic fungus found on cereal and legume products. ***P. lilacinus*** has caused keratitis and endophthalmitis in **man**. Its aeroallergens have caused allergic rhinitis, bronchial asthma, and/or hypersensitivity pneumonitis. ***P. varioti*** exposure to **workers**, where it is used to make ***paper*** industry waste into single cell protein.

Ingestion of ***P. varioti*** has killed **ducklings** within 2 weeks and has been toxic to other species (**monkeys**, **horses**, **cats**, **dogs**, and **tortoises**), causing systemic effects and granulomatous lesions.

PAGAMEA MACROPHYLLA

A tree-like shrub of tropical northern South America called *ma-na-shu-kema* or *ma-nu-su'-ka-ta* by Amazonian Columbians.

Untoward effects: **Medicine men** use its pulverized leaves as a snuff for its alkaloids, which may be hallucinogenic. Other **natives** use a ***tea*** made from scrapings of young branches of ***P. coriacea*** as a strong stimulant to frequently overcome **their** inability to walk, due to a prevalent disease of unknown cause. The active principles have not been identified.

PAHUTOXIN

Untoward effects: The Hawaiian delicacy, ***boxfish*** = ***Pahu*** (Hawaii) = ***moa-moa*** (Tahiti) = ***hako-fugu*** (Japanese), a slow-moving **fish**, secretes a potent **fish** poison. At sublethal concentrations, it is a **fish** repellent. It is a ***choline*** ester, related to ***acetylcholine*** (q.v.). Boiling the **fish** destroys the toxin.

PAINT

Untoward effects: Fungicides added to latex paint often vaporize. In 1990, a Michigan 4-year-old **male** was diagnosed with acrodynia, a rare manifestation of childhood ***mercury*** (q.v.) poisoning. Symptoms included leg cramps, rash, pruritus, hyperhidrosis, tachycardia, irritability, marked personality change, intermittent low-grade-fevers, insomnia, headache, hypertension, redness and peeling of skin on hands, feet, and nose, and weakness of pectoral and pelvic girdles, as well as nerve dysfunction in **his** legs. A 24 h urine collection showed 65 μg/l of ***mercury***. Both **parents** and **siblings** showed high urine ***mercury*** levels, but were asymptomatic. ***Phenylmercuric acetate*** in the paint was identified as the cause. Windows were not opened, but 17 gallons of paint had been used in an air-conditioned house during July. Other homes painted with such latex paints have had levels up to 0.003 mg/m^3. ***Lead*** (q.v.) salts were originally used as accelerators in oil-based paints. Ingestion of dried flakes from toys, cribs, and windowsills has been toxic when eaten in quantity, especially by **children**. Dr Shallenberger of Cornell discovered that it wasn't boredom and poverty that motivated **them** to peel the paint and eat the flakes – they were SWEET. A sign over his workbench read: "AAAH...SWEET (SOUR, SPICY) MYSTERIES OF LIFE...ALIMENTARY, MY DEAR". Exposure to epoxy paint fumes (heavier than air) containing ***glycidyl ether*** as a diluent caused the deaths of three **men** from asphyxia, **who** were waterproofing an underground tank (~66 × 20 × 10 ft) with only one entry and exit. Painting started at 9 p.m.; one **worker** left at 10 p.m. when **he** became drowsy, nauseous, and began vomiting. **Another** left at 10:30 p.m. The **three** that remained were found dead at 11:30 p.m. **They** failed to use electric exhaust fans provided, for fear of electrocution from standing in a wet tank. Primer and artist's acrylic paints contain ***chromium*** (q.v.) as the oxide or ***zinc*** salts, etc. ***Arsenic*** (q.v.) is used in fresco, tempera, watercolor, and oil paints. Self-glow paints usually contain ***tritium*** or ***promethium-147***. These are mixed with ***lacquer*** and applied to hands and dials of watches. Polyneuritis often occurs from accidental chronic exposure to numerous paint thinners and solvents, and from inhalation by **abusers** of these compounds.

PALICOUREA sp.

Many varieties found in Brazilian and Columbian Amazonia.

Untoward effects: A major cause of plant poisoning to **cattle** and **sheep**. Symptoms include sternal recumbency, frothy salivation, dullness, cold extremities, and death. Post-mortem reveals cyanosis, and hemorrhages in meninges, heart, kidneys, and intestines. Laryngeal, esophageal, abomasal, liver, gall bladder, lung, and kidney congestion. Liver dystrophy and necrosis, and hematuria. Death of **cattle** within 5 h after consuming 110 g of seed and leaf extract. Lethal to **guinea pigs** (0.5–8 g of the same extract, IV or oral). ***P. marcgravii*** is the common poisonous species. A drop of ***myotonic acid*** extracted from ***P. rigida*** of Brazil, given parenterally, causes death in a **pigeon**. ***Fluoroacetic acid*** (q.v.) is its toxic agent. Other species are used as piscicides and rodenticides.

PALLADIUM

Untoward effects: Some of its compounds cause cancer in **rats** and **mice**, and are toxic to **mammals**, **fish**, microflora, and some plants. Repeated topical applications of its hydrochloride on **rabbits** causes an allergic contact dermatitis. An allergic reaction also followed an IV of 620 μg/kg. ***Palladium*** or ***Palladous Chloride*** LD_{50}, **rat**: 200 mg/kg; IV, **rat**: 5 mg/kg.

PALMS

Thousands of species exist, including the *coconut* palm and *date* palm whose fruits are edible, and palmettos. The *Carolina palmetto* or *cabbage palmetto* is depicted on the State seal of South Carolina. It commemorates the defeat of a British fleet on June 28, 1776, at Charleston Harbor, by defending American forces in a crude fortification made of *cabbage palmetto* logs and sand. It is also the state tree of Florida.

Untoward effects: ***Calcium oxalate*** raphides from the ***Fishtail Palm*** (*Caryota mitis*) cause an irritant dermatitis. An unusual amino acid, ***β-methylamino-L-alanine***, in the seeds of the ***False Sago Palm*** is neurotoxic in **monkeys** leading to symptoms similar to amyotrophic lateral schlerosis and Parkinsonism in **humans**. **People** in Sri Lanka who eat the germinating seeds of ***Palmyrah Palm*** (*Borassus flabellifer*) have a high incidence of hepatic disorders. Its flour in the diet of **rats** leads to central nervous system disturbances in 2 days and death within 10 days. Thorns of the ***South American Clump Palm*** (*Bactris minor*) have been found in the lungs of **owl monkeys** imported from Columbia. The fleshy seeds of the ***Fern Palm*** (*Cycas circinalis*) or ***Sago Palm*** and

Macrozamia sp. contain poisonous azoxy glycosides (if washed out, the seeds are edible) that have affected **cattle**, **cats**, and **dogs**. See ***Cycads*** and ***Macrozamia*** in the text. The ***Corozo Palm*** = ***Palmas Reales*** (*Schoela zonensis*) of Panama harbors the **kissing bug** (*Rhodinus pallescens*) a primary vector of Chagas disease (trypanasomiasis), where its infested fronds are used as thatch for roofs of dwellings.

Palm oils contain a high level of saturated fat. Carotenoderma is endemic in West Africa, where red palm oil is used for cooking. Pure palm oil contains 0.05–0.2% carotenoids.

PALTHOA TOXICA

Untoward effects: In Hawaii in the early 19th century, the king ordered that a certain reddish poisonous moss be put on the tips of spears to make them lethal. This deadly "seaweed" coral, a coelenterate, found only in one area of Maui, was ***limu-make-o-Hana*** (q.v.). It has since been rediscovered in a few tide pools on Oahu and Maui. Its active principle is ***palytoxin***.

IV, LD_{50}, **rat**: 89 ng/kg; **mouse**: 150 ng/kg; **dog**: 33 ng/kg; **monkey**: 78 ng/kg; **rabbit**: 25 ng/kg; **guinea pig**: 110 ng/kg.

PALYTOXIN

= *PTX*

Untoward effects: The most poisonous non-proteinaceous substance known. It is synthesized by a marine *Vibrio* species, growing symbiotically with ***Palythoa*** (q.v. above). A tumor-promoter.

PAMAQUIN

= *Aminoquin* = *Béprochine* = *Gamefar* = *Pamaquine* = *Plasmochin* = *Plasmoquine* = *Praequine* = *Quipenyl*

Protozoastat, antimalarial, sulfonamide potentiator. An aminoquinolone.

Untoward effects: Toxic reactions in ~10% of 4,361 **patients**. Hemolytic anemia especially in glucose-6-phosphate dehydrogenase deficient and ***primaquine***-sensitive **individuals**. Methemoglobinemia, cyanosis, leucocytosis, leukopenia, agranulocytosis, and neutropenia also reported. Colors urine a dark brown or black; the naphthoate form colors it a rust yellow or brown. Large doses lead to headache, giddiness, nausea, vomiting, jaundice, backache, abdominal pain, and collapse. Death has occurred with doses of only 400 and 600 mg. Its effects and toxicity are enhanced by ***mepacrine*** and ***quinacrine***.

LD_{LO}, **human**: 50 mg/kg.

Cats are very sensitive to its toxicity. Methemoglobinemia developed slowly in **cats** and **dogs**. After 3–6 days of treatment, **dogs** developed enophthalmos, miosis, and relaxed nictitating membrane. Large doses led to hypotension and dyspnea.

LD_{LO}, **dog**: 20 mg/kg; **cat**: 7.5 mg/kg; **rabbit**: 225 mg/kg; **chicken**: 57 mg/kg.

PAMIDRONATE DISODIUM

= *Aminomux* = *APD* = *Aredia* = *GCP 23,339A*

Use: In oral or IV treatment of hypercalcemia, osteoporosis, and Paget's disease. A ***biphosphonate***.

Untoward effects: Transient fever and leukopenia may occur during the first 3 days. General malaise (often lasting several weeks after treatment), nausea, abdominal pain, hearing loss, decreased renal function, anemia, hypokalemia, hypomagnesemia, increase in bone pain for **some** treated for Paget's disease, and rash in < 2% on prolonged oral therapy. Mild thrombophlebitis reported after IV. Acute pseudogout and symptomatic hypocalcemia have occurred.

Can be nephrotoxic in **dogs** at ≥ 10 mg/kg IV.

PANAEOLUS sp.

Untoward effects: ***P. papilionaceus mushrooms*** contain a toxin, ***psilocin*** (q.v.), which causes a hallucinogenic, hyperkinetic state 30–60 min after ingestion, and lasting for several hours. Called the ***Laughing Mushroom*** in Japan, and the ***Laughing Fungus*** in China. Thrifty **farmers** in Maine and Louisiana have also ingested it to get "drunk for nothing". In California and British Columbia, young **people** have eaten ***P. subbalteatus*** with similar results.

PANCRATIUM TRIANTHUM

Untoward effects: **Bushmen** in Botswana rub this lily-like bulb on an incision made on **their** heads to induce visual hallucinations.

PANCREATIC DORNASE

= *Deanase* = *Dinase* = *DNase I* = *Dornavac* = *Pancreatic Desoxyribonuclease*

Debriding agent, mucolytic.* With *fibrinolysin* = *Elase.

Untoward effects: Aerosoled inhalation produces irritation, allergic reactions, bronchospasm, and cyanosis.

PANCREATIN

= *Creon* = *Pancrex V* = *Panzytrat*

Untoward effects: High concentrations can burn buccal mucosa and skin around mouth and anus. Sensitivity reactions (sneezing, wheezing, lacrimation, dysphagia, pyrexia, and dermatitis) from handling the powder.

Occasionally dyspnea and abdominal cramping. A 6-year-old **male** receiving 13–15 thousand Units/kg/day/14 months developed fibrosing colonopathy.

PANCRELIPASE

= *Cotazyme* = *Ilozyme* = *Pancrease* = *Ultrase MT20* = *Viokase*

Untoward effects: Has caused bleeding from tongue, gums, and rectum in **infants** and **neonates**. Occasionally rash, sneezing, tearing, dyspnea, abdominal cramps, and diarrhea. Use caution in its use if **patients** have a hypersensitivity to ***pork*** or **they** may develop asthmatic attacks. The same can occur from occupational exposure. Large doses have induced fibrosing colonopathy in **children**.

PANCURONIUM BROMIDE

= *Miobloc* = *NA 97* = *Org NA 97* = *Pavulon*

***Skeletal muscle relaxant.* IV use.**

Untoward effects: Tachycardia, arrhythmias, hypoxemia, apnea, anaphylaxis, pain and burning at injection site, anxiety, and bronchospasm have occurred. Hypotensive effect noted in a premature **infant**.

Drug interactions: ***β-Blockers, lithium carbonate***, and ***thiotepa*** prolong its neuromuscular blockade. May potentiate ***antibiotic*** neuromuscular blockade. ***Quinidine*** may potentiate its muscle relaxation. Neuromusclar-blocking activity may be increased in hypokalemia. Although 0.2–0.6 μg/ml of serum has been reported as therapeutic, 0.4 μg/ml has been reported as toxic.

IV, LD_{50}, **rat**: 153 μg/kg; **mouse**: 47 μg/kg; **rabbit**: 16 μg/kg.

PANICUM sp.

= *Panic Grasses*

Untoward effects: Contain cyanogenic glycosides (viz. ***hydrocyanic acid***), whose concentration intensifies under certain environmental conditions. They have caused photosensitivity and hepatotoxicity in grazing **animals** and contain ***phytoestrogens***. The latter may be the cause of sterility in **cattle**.

P. antidotale = ***Blue Panicum*** causes acute illness, dyspnea, and early death in **sheep** and **cattle**, especially shortly after early plant fertilization. Post-mortem reveals pulmonary emphysema, edema, and straw-colored fibrinous fluid in pleural cavity. Blood is almost black.

P. capillare = ***Witch Grass*** can contain dangerous ***nitrate*** levels.

P. coloratum = ***Coolah Grass*** = ***Kleingrass*** from South Africa, is now widely planted in Texas. Causes hepatogenous photosensitization in **sheep**, **goats**, and **horses**. Symptoms include swollen heads, ear necrosis, icterus, and lethargy. Post-mortem reveals toxic saponin-filled bile crystals.

P. effusum = ***Hairy Panic*** causes similar problems in Australian **sheep**. They also show liver and colon mucosal congestion.

P. maximum = ***Guinea Grass*** causes osteodystrophia fibrosa in **horses** in Australia, periodontal disease in Brazilian **zebu calves** pastured on it for ~6 months. Its ***hydrogen cyanide*** content increases in drought, and kills **cattle** in Columbia. A ***nitrate***-accumulator. It and ***P. milliaceum*** = ***Broom Corn*** = ***French Millet*** = ***Proso Millet*** have caused "yellow big head" disease or photosensitization in **sheep**, known as "Dikoor" in South Africa. Symptoms there and in Yugoslavia are similar, and include swelling, drooping of ears, central nervous system depression; swelling of face, nose, lips, and (eventually) the whole head, eye and nose discharge, dyspnea, keratitis, corneal opacity, icterus, skin sloughing, and even death within 1–2 weeks in some.

Numerous other species are used for fodder and may occasionally cause toxicity.

PANTOPRAZOLE

= *Protonix*

***Proton pump inhibitor.* Do NOT crush before ingestion. Formulated for intestinal tract absorption.**

Untoward effects: Diarrhea, nausea, constipation, increased aminotransferases, anaphylaxis, fatal toxic epidermal necrolysis, Stevens-Johnson syndrome, erythema multiforme, and gastric cell hyperplasia.

Drug interactions: Can decrease absorption of ***ketoconazole***.

PAPAIN

= *Arbuz* = *Caroid* = *Nematolyt* = *Papase* = *Papayotin* = *Summetrin* = *Tromasin* = *Vegetable Pepsin* = *Velardon* = *Vermizym*

***Proteolytic enzyme.* In original *Adolph's Meat Tenderizer*®, in clarifying *beer* (up to 17 ppm), in contact lens cleaners, and in digestive aids. Much of the so-called crystalline material of commerce has not been true crystalline material, but rather just decolorized material. "Units" may vary greatly between companies, and some refuse to give details of their assay procedures. The U.S. FDA has shown indifference to these problems.**

Untoward effects: Tingling sensation from buccal use as an anti-inflammatory; pain with topical use in debriding skin ulcers. Hypersensitivity in ~1% of allergic

individuals. This can occur after eating treated foods or use of ***chymopapain*** for chemonucleolysis. Symptoms include itching, headaches, gastrointestinal upsets, and conjunctivitis. Inhalation has caused rhinitis, bronchial asthma, hypersensitivity pneumonitis, and shock. Allergic contact urticaria reported. Use of 4 teaspoonfuls of ***Adolph's Meat Tenderizer***® in 200 ml of water every 2 h/1½ days in treatment for a large phytobezoar led to hypernatremia, confusion, and lethargy in a 65-year-old **male**. Use in a 27-year-old **female** to digest an impacted piece of meat led to mediastinitis from perforation of the ischemic esophagus which led to **her** death. May potentiate effects of ***anticoagulants***, including ***warfarin***. No ill-effects reported from eating or handling meat (***Pro-Ten Beef***®) after IV use 30 min before slaughtering.

Causes an emphysema-like condition in **rats** inhaling it.

TD_{LO}, **rat** (13 days pregnant): 375 mg/kg.

PAPAVER sp.

= *Pavot*

Untoward effects: ***P. bracteatum*** is a source for ***thebaine*** (q.v.) and, ultimately, ***codeine*** (q.v.).

P. nudicaule = ***Iceland Poppy*** has poisoned **sheep** with ***hydrogen cyanide***. Symptoms in **cattle** include increased central nervous system excitement effects and acute severe bloat that caused some to die. In New Zealand, when fed to **horses** while the seed-heads were still green caused ataxia, nervousness, and muscle spasms.

P. rhoeas = ***Coquelicot*** = ***Corn Poppy*** = ***Red Poppy*** is rarely eaten in enough quantities by **animals** to cause poisoning, but **people** who ingest an infusion or decoction of the petals and capsules become sedated and narcotized from its alkaloid, ***rheadine***.

P. somniferum = ***Kiskas*** (Arabic) = ***Opium Poppy*** is the commercial source of ***opium*** (q.v.), and eventually ***heroin*** (q.v.) and ***codeine*** (q.v.). Ingestion of fresh seeds has been toxic to **children**. ***Poppyseed*** has caused allergic reactions in some **people**. Tonic–clonic seizures in a 26-year-old **male** after ingesting 2 l ***tea***/day made from 4 kg of seed. Analysis of the tea revealed 0.14 mg ***morphine***/ml. After decreasing **his** intake, **he** still consumed 280 mg (4.3 grains) ***morphine***/day and entered a rehabilitation program. Other symptoms in **humans** are eczemas, miosis, sweating, headache, and cyanosis. ***Poppyseeds*** now sold in stores have been heat-treated to destroy any of their ***codeine***, ***morphine***, and ***protapine*** content. **Cattle** that ingest any parts of the plant develop tachypnea, diarrhea, decreased milk production, nervousness, constant motion, and ataxia. Post-mortem reveals nephritis, intestinal inflammation, and yellowish livers.

PAPAVERETUM

= *Omnopon* = *Pantopon*

Contains *opium* alkaloids (~50% *morphine*, 20% *noscapine*, 5% *papaverine*, and 3% *codeine*). Used IM and IV, occasionally orally.

Untoward effects: High abuse liability. Anaphylaxis, respiratory failure, pulmonary edema, bradycardia, miosis, nausea, vomiting, diarrhea, and coma have occurred. Use for the **mother** during labor leads to clinical depression.

Drug interactions: Use with ***monoamine oxidase inhibitors*** may precipitate serious adverse effects (see ***Meperidine*** with ***monoamine oxidase inhibitors***).

PAPAVERINE

= *Artegodan* = *Cepaverin* = *Cerebid* = *Cerespan* = *Dipav* = *Dynovas* = *Lempav D-Lay* = *NSC 35,443* = *Optenyl* = *Pameion* = *Panergon* = *Papacontin* = *Papital T.R.* = *Pavabid* = *Pavacap* = *Pavacen* = *Pavadel* = *Pavagen* = *Pavakey* = *Pavased* = *Pavatest* = *Pava-Wol* = *Ro-pav* = *Spasmo-Nit* = *Therapav* = *Vasal* = *Vasospan*

Cerebral vasodilator. **Oral or parenteral. A phosphodiesterase inhibitor. Its derivatives, *gallopamil*, *tiapamil*, and *verapamil* are useful *calcium channel blockers*.**

Untoward effects: Hepatotoxicity in 20% of **patients** (others say "rarely"), with increased alkaline phosphatase, increased SGOT and SGPT, eosinophilia, cirrhosis, and abdominal pain that some consider a hypersensitivity reaction. Nausea, constipation, diarrhea, stupor, ventricular tachycardia, fibrillation and arrhythmias, vertigo, ataxia, dry mouth, hyperhidrosis, headache, rash, priapism. Jaundice, facial flushing, atrioventricular heart block, arterial hypoxemia, dyspnea, brachial phlebothrombosis, and eczema also reported. Ingestion of ~15 g in a suicide attempt by a 61-year-old **female** caused severe lactic acidosis. Slowly antagonistic to ***L-dopa's*** therapeutic response. ***Rutin*** and ***horse chestnut*** extract increase its thrombocyte inhibition. Occasionally fatalities reported after IV use. Occasionally abuse and dependence reported.

LD_{50}, **mouse**: 350 mg/kg; IV, male **rat**: 17 mg/kg; male **mouse**: 21.4 mg/kg; **rabbit**: 25 mg/kg; IP, **mouse**: 64 mg/kg; SC, LD_{50}, **rat**: 370 mg/kg; **mouse**: 150 mg/kg.

PAPAYA

= *Chichihualxochitl* (Aztec) = *Gwanda* (Hausa) = *Melon Tree* = *Mutie papayi* (Kiyanzi) = *Okwulu Oyibo* (Igbo) = *Paupaw* = *Pawpaw*

It is the dried latex of the *Carica papaya* fruit.

Untoward effects: Eating it by a **Japanese person** led to yellowish discoloration of only palms and soles. Its seeds have been used in India as an adulterant in ***black pepper***, causing some adverse effects as when using ***papain*** (q.v.).

An extract has strong central nervous system depression in **rats**.

PAPER

Untoward effects: Skin hazards in pulp and paper manufacturing are, primarily, due to alkali burns. Preservatives used, such as ***phenylmercuric nitrate*** (q.v.) and ***o-phenylphenol*** (q.v.), can cause skin irritation. ***Chlorine*** bleaching has caused ***dioxin*** formation in paper mill pulp and sludge, as well as in some finished paper products. **Workers** making paper and paper ***money*** are often exposed to ***chromates***. Paper pulp manufacturers use scrubbing systems to decrease **worker** exposure to ***sulfur dioxide*** (q.v.). Hypersensitivity pneumonitis (***maple*** bark disease), especially in winter, from *Cryptostroma corticale* spores in infected ***maple*** bark. *Alternaria sp.* is the major cause of hypersensitivity pneumonitis (paper mill **worker's** disease or pulpwood **handler's** disease, from handling moldy wood pulp). Occupational eczema in woodpulp and ***cellulose*** manufacturing plants. Keratitis, photophobia, and pain from ***hydrogen sulfide*** (q.v.) use in paper mills. Colored gift-wrapping papers must be kept away from small **children** and **pets** who eat or chew it, as they often contain high levels of ***chromium, copper, lead***, or ***zinc***. The same is true for many colored ***inks*** in comic strips. ***Lead chromate*** is used to color paper yellow, ***copper*** compounds for blue, and ***chromates*** for green. In Denmark, use of ***diphenyl*** (q.v.) as a fungistat in a fruit paper factory caused serious nerve and liver damage, including one death, when **worker** exposure was > threshold exposure limit of 1 mg/m^3. Allergic reactions with headaches, nasal congestion, nausea, and dizziness reported in a **male** after contact with a particular magazine (*Hoard's Dairyman*), and to a lesser degree with comic books. Phytobezoar near sigmoid colon in a 26-year-old **female** from chewing paper facial tissue over a period of several years. Old newspapers and books may contain molds and particulate material that can trigger allergic reactions. ***Dimethyl*** (q.v.) and ***diethyl phthalates***, used as defoaming agents in paper manufacturing, have caused toxic effects. Dermatitis from typing paper due to a gum resin size consisting of ***rosin***. **People** sensitive to this paper are also sensitive to ***juniper tar*** and ***storax***. All three contain ***abietic acid*** (q.v.) as the sensitizing agent. ***Formaldehyde*** (q.v.)-sensitive **people** can develop dermatitis where it and its derivatives are used to improve wet-strength, and water-, shrink-, and grease-resistance. Also found in paper ***money***, high-quality blotting papers, some paper towels, and typewriter correction paper. Newspaper **workers** have developed contact dermatitis on **their** elbows from leaning on phototypesetting paper containing the antioxidant, ***tert-buty catechol***. Similar dermatitis reported in **some** after handling carbonless copy paper and diazo papers. Parenteral solutions are tested to decrease incidence of pulmonary granulomas, due to filter paper fibers, after IV use.

PAPHIOPEDILUM HAYNALDIANUM

= *Lady Slipper*

In the family *Orchidacea*. The genus is sometimes referred to as *Cypripedium*.

Untoward effects: A retired 68-year-old **male** had an eczematous dermatitis on **his** hands and forearms after severe swelling and pruritus, when **he** finished **his** greenhouse potting work on ~500 plants. **He** had some reactions 2 years previously, when handling ***orchids***, and ~30 years earlier, when handling ***primroses***. Other species cause a contact dermatitis, especially in hot weather or when **people** are perspiring. Violent inflammation appears 8–12 h after early irritation. Blisters, as with ***poison ivy***, can eventually follow. Apparently, triggered by a ***fatty acid*** on the glandular hairs of its stems and leaves, especially during and after its flowering season.

PAPRIKA

Annual world-wide use is ~40,000 metric tons/year.

Untoward effects: If consumed in large quantities, urine becomes a deep orange color. Hypersensitivity pneumonitis (paprika **slicer's** disease) from exposure to *Mucor stolonifer* on moldy paprika pods. Reported to cause marked increase in gastric hypoacidity.

PARABENS

= *Parasepts*

Benzyl, butyl, ethyl, heptyl, isobutyl, isopropyl, methyl, octyl, and propyl esters of *p-hydroxybenzoic acid* (q.v.).

Untoward effects: Allergic contact hypersensitivity has been reported with all esters, even when in ***corticosteroid*** creams. Especially dangerous in some **people** when applied to denuded, ulcerated, or eczematous skin. No contact dermatitis reported when used in ***mascara***. A frequent sensitizer on **infants'** skin. Causes contact urticaria. A 12-year-old **male** had an anaphylactic reaction after receiving it in an expectorant. Use in local ***lidocaine*** injections causes erythema and wheal formation. Cross-reaction with ***tartrazine*** may induce asthma. In cosmetic products, methyl paraben, followed by propyl paraben, are responsible for most of the preservative-induced reactions. These can appear very quickly and be very severe.

Propyl: LD_{LO}, **human**: 500 mg/kg.
LD_{50}, **dog**: 6 g/kg; IP, **mouse**: 200 mg/kg; LD_{LO}, **mouse**: 6 g/kg.

Methyl: LD_{50}, **dog**: 3 g/kg; IP, **mouse**: 960 mg/kg; LD_{LO}, **mouse** and **guinea pig**: 3 g/kg.

PARAFFIN

This is a different thing in many countries. In some countries, it means *kerosene* (q.v.), and, in others, paraffin wax, which is a mixture of high molecular weight *hydrocarbons*. *Vaseline*® or *petrolatum* is a mixture of liquid and solid paraffins. ~600 million lbs used annually in the U.S.

Untoward effects: Paraffin wax fumes can cause eye, skin, and respiratory tract irritation and nausea. National Institute for Occupational Safety and Health upper exposure limit of 2 mg/m³. **Children** have had serious pulmonary problems after accidental aspiration of liquid paraffin. Has caused SC and muscle granulomas. Neuropathy in ***shoe*** factory **workers**. Connective tissue disease (scleroderma, rheumatoid arthritis, systemic lupus erythematosus, and polymyositis) has followed cosmetic surgery for breast augmentation and rhinoplasty. A 13-year-old **male** with congenital Netherton's syndrome had daily topical application of ~400 g of various emollients containing ~200 g paraffin which caused oleogranulomatous response in **his** lymph nodes. Resolved after total daily application of 10 g paraffin – then, no recurrence at 100 g level.

PARALDEHYDE

= *Paral* = *Rektidon*

Sedative, hypnotic.

Untoward effects: Can be habit-forming. Toxic symptoms are similar to ***chloral hydrate's*** (q.v.). Has caused hepatic coma. Disagreeable breath and urine for up to 24 h. Large doses cause dizziness, syncope, hypotension, incoherence, muscle relaxation, tachycardia, right side heart failure, tachypnea, miosis, bradypnea, little or no response to light, collapse, acute pulmonary edema, coughing, irritation of throat, esophagus and stomach, nausea, vomiting, progressive renal failure, deep narcosis, coma and death from respiratory failure. Undesirable cutaneous effects include erythema, urticaria, hemorrhagic and eczematous lesions, and exfoliative dermatitis. May cause false positive urine protein test results. It is unstable and decomposes into several toxic compounds, including ***acetaldehyde*** and ***acetic acid***. **People** develop a severe hyperchloremic metabolic acidosis after excessive ingestion. After a long period of habituation to high intake extreme organic aciduria can occur. Hemorrhagic gastritis, irritability, azotemia, oliguria, albuminuria, and leukocytosis also occur. Thrombophlebitis is a frequent complication of IV use. Sterile abscesses, skin sloughing, and permanent sciatic nerve injury after IM use. Apnea, sedation, and respiratory depression in **infant** after use in the **mother**. Characteristic odor of the an **infant's** breath lasts 2–3 days. Toxic serum level 200–500 mg/l, while therapeutic levels range between 20 and 110 mg/l. Fatalities from it, as well as tolerance to it, have occurred.

Drug interactions: Its effect is enhanced by concomitant use of ***tolbutamide***. With ***alcohol***, additive depressant effect. The combination has caused death in some **alcoholics**.

Oral and IV, TD_{LO}, **human**: 14 mg/kg; IM, 71 mg/kg.

LD_{50}, **rat** and **guinea pig**: 1.15 g/kg; **mouse**: 230 mg/kg; **dog**: 3.5 g/kg; **rabbit**: 3.3 g/kg.

PARAMETHADIONE

= *Paradione*

Anticonvulsant. An oral oxazalodine.

Untoward effects: Hypersensitivity reaction includes neutropenia, agranulocytosis, thrombocytopenia, exfoliative dermatitis, rashes, sore throat, hepatocellular jaundice and necrosis (in fatal cases), albuminuria, and nephrotic syndrome. Occasionally photophobia, spontaneous abortions, hiccups, and anorexia. Interferes with urinary ***albumin*** tests, causing false positive or spuriously high results. Teratogenic (cardiac malformations, clefts of palates or lips, microphthalmia, palate and facial dysmorphia, reduced intrauterine growth, and mental retardation) in **offspring** of treated **mothers**. Bleeding problems in some **newborns**. Half-life of > 200 h. A hepatic enzyme inducer.

Can cause nausea, ataxia, and blood dyscrasias in treated **dogs**.

LD_{50}, **mouse**: 1 g/kg; IP, 750 mg/kg.

PARAMETHASONE ACETATE

= *Cassene* = *Cortidene* = *CS 1,483* = *Dilar* = *Dillar* = *Haldrate* = *Haldrone* = *Metilar* = *Monocortin* = *Paramezone* = *Syntecort* = *Stemex*

Glucocorticoid.

Untoward effects: Lens opacities, cataracts, increased intraocular pressure, glaucoma, adrenal–pituitary suppression, weight gain, abnormal fat deposition (Cushing's syndrome), and tinnitus with some nerve deafness. Hyperemia and edema after subconjunctival injection and eyelid edema after retrobulbar injection. After IM use, can cause gastric pain, fatal gastrointestinal tract hemorrhage, cerebrovascular accidents, coma, diabetes, moon facies, limb fatigue, and bone pain. The sodium phosphate ester, IM or intra-articular, causes joint pain, mild headache, and skin rashes.

IP, LD_{50}, **rat**: 127 mg/kg.

PARAQUAT

= *Dextrone* = *Gramoxone* = *Dexuron* = *Esgram* = *Herboxone* = *Orthoquat* = *Orvar* = *Pathclear* = *PP 148* = *PP 910* = *Tenaklene* = *Weedol*

Contact herbicide.

Untoward effects: Vapors irritate the eyes, skin, nose, throat, and respiratory system. Acutely toxic by ingestion, inhalation, or dermal exposure. Accidental splashing onto an eye during spraying caused keratitis, and, ultimately, keratoplasty, in one **patient** and conjunctival necrosis in **another**. Also superficial ocular burns and corneal scarring reported. **Sprayers** have often reported nose bleeds. Scrotal and perineal burns reported. Absorption via a thigh lesion in a **child**. Topically, it has caused raised, itchy, red-brown lesions with vesiculation over areas scratched by grasses. After absorption, it is widely distributed in body tissues, damages and discolors nails, and causes pulmonary intra-alveolar and subpleural hemorrhages, edema, and fibrosis, pulmonary effusions, as well as vomiting, abdominal tenderness, jaundice, hepatic enlargement and failure, ulcerated buccal, tongue, soft palate, pharynx and esophageal mucosa, SC emphysema of neck and chest wall, dyspnea, anemia, leukocytosis, oliguria, cardiac arrest, methemoglobinemia, hemolysis, tachycardia, mydriasis, headache, muscle weakness, and nephrotoxicity. Since it is excreted by the kidney, it can help induce renal failure. Serious symptoms may be delayed a few days. In < 10 years of use, > 120 deaths reported; 50% from accidental ingestion, and 50% from suicidal intent. A mortality rate of up to 50%, even with treatment. **Fetal** death after **maternal** ingestion. Lung damage has resulted from use of paraquat-sprayed ***cannabis***. When smoked, 60–70% of the paraquat is converted to ***dipyridyl*** (q.v.). Adult **human** oral LD has been as low as 1 ml (4 mg/kg) of 20% solution. National Institute for Occupational Safety and Health restricts **human** exposure to 0.1 mg/m^3, and OSHA to $0.5 mg/m^3$. Toxicity noted with serum levels of 0.005–0.85 mg%, and 3.5 mg% is considered lethal. Toxic effects may be mediated by ***oxygen radicals***.

TD_{LO}, **man**: 32 mg/kg.

Extremely toxic to **bees** at 100 ppm. **Cats** and **dogs**, after eating treated grass, developed anorexia, salivation, and ulcerated tongues by day 3; neutrophilia, raised erythrocyte sedimentation rate, and tachypnea in the **cat**; the **dog** started vomiting on day 3, followed by dyspnea. Post-mortem revealed pulmonary edema, congestion of liver, sloughing of tongue epithelium, and lesions in brain and kidney. Within 6 h after ingestion 100 mg/kg **water buffalo calves** developed dullness, depression, ataxia, stiff neck, arrhythmias, dyspnea, bloat, prostration, convulsions, and death. At 250 ppm in their water, **fish** evidenced mass desquamation of epithelial cells of their gills. Poisoning in **guinea pigs**, **monkeys**, and **rats** caused anorexia, adipsia, diarrhea, dyspnea, hypoxia, and tachycardia. Pulmonary, liver, gastrointestinal, and kidney damage also reported. **Cows** grazing treated pasture passed some into their milk on day 1 (0.001–0.015% of ingested dose); **horses** developed mouth lesions and increased mucous secretions. Very toxic to **chickens**, **ducks**, and **turkeys**. Teratogenic in **hens** and **rats**. Fatal doses in **hamsters**, **rabbits**, and **sheep** don't cause pulmonary pathology. Symptoms in poisoned **pigs** include vomiting, dyspnea, pulmonary edema, depression, and jaundice.

LD_{50}, **rat**: 57 mg/kg; **mouse**: 196 mg/kg; **guinea pig**: 22 mg/kg; **monkey**: 50 mg/kg; **dog**: 35 mg/kg; **cat**: 48 mg/kg; **mallard ducks**: 199 mg/kg; **cattle**: 50 mg/kg; **sheep**: 80 mg/kg; **pig**: 75 mg/kg; **chicken**: 362 mg/kg; dermal, **rabbit**: 236 mg/kg.

PARATHION

= *AAT* = *AATP* = *AC 3,422* = *Alkron* = *Alleron* = *Aphamite* = *Bay E-605* = *Bladan* = *Corothion* = *DNTP* = *Ecatox* = *ENT-15,108* = *Etilon* = *Folidol* = *Fosferno* = *Niran* = *Nitrostigmine* = *Orthophos* = *Panthion* = *Paramar* = *Paraphos* = *Parathene* = *Parawet* = *Phoskil* = *Rhodiatox* = *Slathion* = *SNP* = *Soprathion* = *Thiophos*

Organophosphorus insecticide, acaricide, cholinesterase inhibitor.

Untoward effects: Classified as highly toxic, i.e. LD_{50} < 50 mg/kg, inhibiting ***acetylcholine*** metabolism. Symptoms usually appear within the first few hours of exposure to cholinergic over-stimulation of the parasympathetic nervous system, leading to weakness, muscle twitching, sweating, incoordination, vertigo, nausea, abdominal cramps, vomiting, miosis, salivation, headache, giddiness, loss of depth perception, blurred vision, lacrimation, cyanosis, acrocyanosis, bradycardia, dyspnea, occasionally pulmonary edema and asthma-like symptoms, diarrhea with tenesmus, hypertension, convulsions, delirium, coma, decreased reflexes and sphincter control, and urinary concentration of ***p-aminophenol*** (q.v.) of 1–2 ppm, with death due to hypoxia with bronchoconstriction from within minutes to 4 h in serious exposure cases. Its metabolite, ***paraoxon***, is considered to be an additional cause of its toxicity. ***p-Nitrophenol*** (q.v.) is also a metabolite. Only a few drops of this pure pesticide can kill a **man** inhaling the dust or spray, by dermal contact, or from contaminated clothing. In 1957, > 60 fatalities. Used in suicides and murders. Complete or partial amnesia reported in some cases, with up to a year for recovery in **some**. In a 5 year period, > 6,000 cases of toxicity reported world-wide. In Tijuana, Mexico in September, 1967, 17 deaths and > 250 **people**, mostly **children**, became ill from parathion food (***bread***, sweet roll, and cookie) contamination. In November, 1967, another large-scale poisoning in 165 **people** occurred in Columbia with 63 deaths. Environmental temperature extremes can slightly alter its toxicity. These might have been prevented if a dye and/or obnoxious odorant had been added to the chemical. A 41-year-old **male** developed a delayed polyneuropathy after ingesting a large amount. Toxicity can be potentiated by ***phenothiazines***. Toxic serum concentrations are 0.01 mg/l and **patients** with levels > 0.05 mg/l become comatose or die. National Institute for Occupational Safety and Health upper exposure limit of 0.05 mg/m^3 TWA; OSHA: 0.11 mg/m^3 TWA.

LD_{LO}, **human**: 240 µg/kg; TD_{LO}, **woman**: 5.67 mg/kg.

Highly toxic to honey **bees**. Poisoning in **livestock** was commonly due to drifting from sprayed citrus groves, etc. to pastures or from contamination of water supplies. A **farmer** was fined for killing ~6,000 **songbirds** by illegal crop-spraying with it. Highly toxic to **birds** with dermal toxicity very close to oral toxicity. In Texas, consumption of treated ***wheat***, within a day of treatment, killed ~1,500 migrating **geese**. Cholinesterase inhibition occurs more rapidly in exposed **monkeys** than **man**. In **sheep**, exposure of a **ewe** caused a 25% decrease in fetal plasma cholinesterase. In **fish**, LC_{50} varies between 0.02 and 0.2 mg/l. **Ducklings** eating exposed **tadpoles** have died from secondary toxicity. Run-off irrigation water from parathion-treated citrus grove has poisoned **cattle** and **horses**. In **rats**, increased perinatal death, SC hemorrhages, and decreased postnatal growth. Enzyme-inducers, such as ***aldrin, chlorcyclizine, chlordane, lindane, pentobarbital***, and ***phenobarbital***, pretreatment decreased its toxicity in **rats**. **Animals** show many of the same symptoms of poisoning as does **man**.

Lethal dose in 90 lb **pig**: 122 mg/kg; **cats**: 15 mg/kg; **dog**: 3 mg/kg; **guinea pig**: 8 mg/kg; **rabbit**: 10 mg/kg; **sheep**: 20 mg/kg; **calves**: 1.5 mg/kg; **steers**: 75 mg/kg. LD_{50}, **rat**: 2 mg/kg; **mouse**: 6 mg/kg; **red-winged blackbird**: 2.4 mg/kg; **starling**: 5.6 mg/kg; **male mallard duck**: 2.13 mg/kg; **chukar**: 24 mg/kg; **female quail**: 6 mg/kg; **pigeon**: 2.52 mg/kg; **sparrow**: 1.3 mg/kg; **quelea**: 1.8 mg/kg; **goats**: 20–56 mg/kg; **mule deer**: 22–44 mg/kg.

PARATHYROID HORMONE

= *Parathormone* = *PTH*

Normal function is to increase intestinal absorption of *calcium*, increase its resorption from renal tubules and bone. By decreasing reabsorption of *phosphorus* from the renal tubules, it increases body loss of *phosphorus*, and converts inactive *vitamin D* to an active form.

Untoward effects: Hypercalcemia, osteopenia, and increased deposition of ***calcium salts*** in renal tubules. Anaphylaxis in a 49-year-old **male** with suspected hyperparathyroidism. Excessive quantities can cause destructive changes in bones and joints. Can cause false positives in urinary ***calcium*** tests.

Drug interactions: IV ***mithramycin*** may block its peripheral action on gut and bone, and ***cimetidine*** may decrease its production. Can sensitize heart to toxic effects of ***digitalis***.

Animal trials showed it increases ***aluminum*** absorption and its deposition in the brain.

PARBENDAZOLE

= *Helmatac* = *SKF 29,044*

Anthelmintic.

Untoward effects: Teratogenic in **sheep**, **goats** (bowed forelegs), **rats** and **mice**, but not in **rabbits** or **dogs**. In addition, some affected **lambs** had cardiac septal defects, microphthalmia, atresia ani, spina bifida, and costal fusion.

LD_{LO}, **rat** (8–14 days pregnant): 70 mg/kg; **mouse** (8–14 days pregnant): 210 mg/kg.

PARGYLINE

= *Eutonyl*

Antihypertensive. An monoamine oxidase inhibitor.

Untoward effects: Nausea, vomiting, xerostomia, ankle edema, blurred vision, weakness, headache, insomnia, nightmares, nervousness, muscle twitching, dizziness, constipation, hyperhidrosis, paresthesia, chest pain, tachycardia, palpitations, congestive heart failure, impotence, delayed or failed ejaculation, difficulty in achieving orgasm in **female**, diarrhea, pollakiuria, rashes, transient hearing impairment and jaundice, decreased blood ***glucose***, arthralgia, purpura, and weight increase. Overdoses cause severe hypotension, orthostatic hypotension, confusion, hallucinations, and lethargy.

TD_{LO}, **child**: 8.75 mg/kg; **woman**: 15 mg/kg/day of the *hydrochloride*.

Drug interactions: Hypertensive crisis with some ***alcoholic beverages***. It has an intrinsic ***amphetamine***-like activity and potentiates the effects of other ***amphetamines***, foods containing ***tyramine*** (q.v.), and ***sympathomimetics***. Will enhance central nervous system depression of ***anesthetics, barbiturates, chloral hydrate***, and ***meperidine***. Iatrogenic brain syndrome (acute psychosis) and cardiovascular irritability within an hour after ingestion ***imipramine***. Use with ***methyclothiazide, trifluoperazine***, and other ***monoamine oxidase inhibitors*** potentiates its adverse effects. Delirium, violence, and hallucinations reported with ***methyldopa***. Excitation noted with ***reserpine***. Enhances ***cocaine*** effects.

LD_{50}, **rat**: 300 mg/kg; **mouse**: 680 mg/kg; **dog**: 175 mg/kg.

PARIETARIA sp.

= *Pellitory-of-the-wall*

Untoward effects: Pollen from this perennial weed of the nettle family in Greece, Portugal, Spain, and France causes allergic asthma-like respiratory problems, similar to those of ***goldenrod*** in the U.S. Several species grow in warm coastal regions of the U.S., and have been identified as causing seasonal respiratory allergies.

PARIS sp.

= *Daiwa sp.* = *Herb Paris*

In the lily family.

Untoward effects: ***P. quadrifolia*** contains a toxic ***saponin***-like substance, especially in its berries and seeds, which are poisonous for **children** who ingest them.

P. polyphylla = ***Satuwa***. Its rhizome is used as an anthelmintic in India and China. Its ***saponins*** have demonstrated spermicidal action in **humans**, **rats**, and **mice**.

PARMELIA MOLLIUSCULA

Untoward effects: A ground-covering lichen, poisonous to **cattle** and **sheep**, due to its ***usnic acid*** (q.v.) content.

PAROMOMYCIN

= *Amaroral* = *Aminoxidin* = *Aminosidine* = *Antibiotic 1600* = *Antibiotic 4915* = *Catenulin* = *Estomycin* = *Farmiglucin* = *Farminosidin* = *FI 5,853* = *Gabbromicina* = *Gabbromycin* = *Gabbroral* = *Humagel* = *Humatin* = *Humycin* = *Monomycin* = *Paramicina* = *Pargonyl* = *Paricina* = *Sinosid* = *Zygomycin*

***Antibiotic, amebicide.* Poor oral absorption, except perhaps through ulcerated mucosa.**

Untoward effects: Frequent gastrointestinal upsets (diarrhea, abdominal cramps, and nausea), pancreatitis, vertigo, headache, pruritus ani, and skin rashes. Long-term use may cause malabsorption syndrome. Rarely, eighth cranial nerve (auditory) damage; renal damage after IV and oral use. May cross-react and show cross-resistance with ***kanamycin, neomycin***, and ***streptomycin***. May interfere with neuromuscular transmission.

Anemia, gastrointestinal inflammation, and degeneration and atrophy of liver and kidneys in **sheep**. Nephrotoxic after parenteral use in laboratory **animals**.

LD_{50}, **mice**: ~15 g/kg.

PAROXETINE

= *Aropax* = *BRL 29,060* = *Deroxat* = *FG 7,051* = *Paxil* = *Seroxat*

***Antidepressant.* A *selective serotonin reuptake inhibitor.* Lacks the major anticholinergic, antihistaminic, and *α-adrenergic receptor* blocking actions of *tricyclic antidepressants*.**

Untoward effects: Agitation or sedation, nausea, vomiting, hyperhidrosis, tremors, asthenia, insomnia, delayed ejaculation, cough, headache, anorexia, dizziness, diarrhea, constipation, weight gain, postural hypotension, yawning, blurred vision, altered taste, and xerostomia. Rarely, hyponatremia, mania, hypomania, seizures, aggressive and suicidal impulsivity, bruxism, syndrome of inappropriate antidiuretic hormone secretion, hyperglycemia, acute angle glaucoma, bradycardia, alopecia, bleeding, anisocoria, sexual stimulation in **female**, impotence and anorgasmia in **male**, and night terrors. Withdrawal syndrome with abrupt discontinuation. Half-life of ~21 h. Effectiveness and adverse effects unrelated to plasma levels. Found in breast milk.

Drug interactions: Increases bleeding with ***warfarin***, without change in prothrombin time. Can increase ***procyclidine*** absorption. Availability of ***digoxin*** and ***phenytoin*** decreased. ***Cimetidine*** increases its plasma concentration. Use within 2 weeks of ***monoamine oxidase inhibitor*** can cause life-threatening effects. ***Isoniazid*** may inhibit ***monoamine oxidase*** and has a potential for serious interaction with it. Adding it to ***perhexiline*** therapy (~half-life of 23.5 days) in a 86-year-old **female** led to ataxia, falls, lethargy, nausea, and death in ~2 weeks. By inhibiting CYP2D6 enzyme metabolism, it can increase blood levels of ***amitriptyline, clozapine, codeine, desipramine, dextromethorphan, diazepam, haloperidol, hydrocodone, imipramine, metoprolol, mexilitene, perphenazine*** (two to 13 times), ***propafenone, propranolol, thioridazine***, and ***timolol***. Use with ***Hypericum*** (***St. John's Wort***) causes nausea, lethargy, and incoherence.

LD_{50}, **mice**: 500 mg/kg.

PAROXYPROPIONE

= *B-360* = *Frenantol* = *Frenohypon* = *H-365* = *Hypostat* = *Paroxon* = *Possipione* = *Profenone*

***Pituitary gonadotrophic hormone inhibitor.* Oral or parenteral use.**

Untoward effects: Symmetrical myoedema in lower legs of 36-year-old **female**.

Inhibits estrus and causes pseudopregnancy in **sows**; in **mice**, increases lipids in adrenal cortex and decreases thyroid activity.

PARQUETINA NIGRESCENS

Untoward effects: Its copious latex has been used on poison arrows in Africa, due to the action of its very poisonous ***strophanthidin***-based glycosides.

PARSLEY

= *Perejil* = *Petroselinum crispum*

***Culinary herb and diuretic. P. sativaum* is also used.**

Untoward effects: Contains ***psoralen*** (q.v.), a ***furocoumarin*** photosensitizer, and ***myristicin*** (q.v.), an hallucinogen. Also contains ***methoxsalen*** (q.v.) a natural carcinogen, ***oxalates*** (q.v.), and ***bergapten***. A common contact allergen causing cutaneous eruptions. Its seed oil contains ***apiole, myristicin***, and ***α-pinene***.

The herb is toxic to **canaries** when > 600 mg is fed. **Ducks** fed parsley seed develop a dermatitis when placed in the sun.

Seed Oil LD_{50}, **rat**: 3.96 g/kg; **mouse**: 1.52 g/kg. Drowsiness, anuria, dyspnea, and death of **mice** after 0.01 ml/g in 24–60 h.

PARSNIP

= *Panais sauvage* = *Pastinaca sativa* = *Wild Parsnip*

Vegetable.

Untoward effects: Contains 4–5% ***psoralens*** (q.v.) that are not destroyed by cooking and can cause photodermatitis. **People** who harvest them develop a photosensivity and dermatitis vesicular or bullosa from the roots or juice of the green tops. Contains ***oxalates***. It is a ***nitrate***-accumulating plant.

PARSONSIA sp.

Untoward effects: Contains the hepatotoxic ***pyrrolidizine*** alkaloids, ***heterophylline, lycopsamine, parsonsine, spiracine***, and ***spiraline***. They are also ***nitrate***-accumulators.

PARTHENIUM sp.

Untoward effects: ***P. argentatum*** or ***Guayale argentatum*** = ***Guayule*** = ***Mexican Rubber Plant***, contains up to 20% dry weight of *cis*-isoprene ***rubber***, which contains a sesquiterpene ***cinnamic acid*** ester that causes an allergic contact dermatitis in **people** and experimental **animals**.

P. hysterophorus = ***Buttercup*** has caused allergic hay fever-type reactions. In India, a common cause of lichenoid eczema with hypo- and hyperpigmentation. During droughts in India, **cattle** and **water buffalo** feeding on it develop papular erythematous eruptions over most of their bodies, alopecia, depigmentation of neck and shoulders, and eyelid and facial muscle edema. Post-mortem reveals ulcerations of muzzle, dental pads, tongue, and palate. Necrosis and severe congestion of liver and gastrointestinal tract, and pulmonary edema.

PARTHENOCISSUS QUINQUEFOLIA

= *Vigne Vierge* = *Virginia Creeper*

Untoward effects: Leaves and mature berries contain small amounts of ***calcium oxalate***. Other than local irritation, reports of toxicity have been questioned. Illness (vomiting, purging, and tenesmus) and death reported in **children** after swallowing juice of chewed leaves or ingesting ripe berries.

P. tricuspidata = ***Boston Ivy*** might also be suspect of similar adverse effects.

PASPALUM sp.

Untoward effects: An orange or black smut (***ergot-Claviceps***, q.v.) infesting the seed heads causes "paspalum staggers" in **cattle**, leading to muscle tremors, staggering, falling, and often inability to rise. Violent continuous spasms interfere with breathing and death losses of 10% occur. Also adversely affects **horses** and **sheep**. Stunted, wilted, or frosted plants are also a source of ***hydrogen cyanide*** (q.v.) toxicity.

P. dilatatum = ***Dallis-grass*** and ***P. distichum*** in U.S., Australia, and New Zealand also cause "staggers". ***Chanoclavine, ergonovine***, and ***lysergic acid diethylamide***, have been isolated from the sclerotia of the infecting *Claviceps paspali*.

P. scrobiculatum = ***Koda Millet*** = ***Khodoadhan*** = ***Poor Man's Millet*** = ***Varagu Millet*** cultivated in India, Asia, and Africa, has caused tremors, clonic convulsions, coma, and death of **cattle**. A tranquilizing effect has been reported in **water buffalo calves**, **dogs**, and **rats**.

PASSIFLORA sp.

Untoward effects: ***P. alba*** is poisonous for **cattle** and **sheep** eating it.

P. caerula contains cyanogenic glycosides and central nervous system stimulating alkaloids (q.v.) ***harmaline, harman***, and ***harmol***.

P. foetida = ***Ahuemji*** (Igbo) = ***Corona de Birge*** = ***Ikokosin*** (Benin) = ***Ninge Ninge*** (Efik) = ***Kleistubom*** = ***Wild Water Lemon***. Nausea, vomiting, abdominal distention, dyspnea, and collapse reported in **man**. The fruit is edible when ripe, but leaves and green fruit contain a cyanogenic glycoside and an unknown alkaloid, which has been dangerous to **animals** consuming a lot. **Goats** refuse to eat it.

P. incarnata = ***Apricot Vine*** = ***Grenadile*** = ***May-Pop*** = ***Passion Flower***. Smoking it has produced a ***cannabis***-type of high. Contains various pressor amines (***histamine, tyramine, tryptamine, norepinephrine, octopamine***, and ***serotonin***. A **male** who took it in a herbal preparation developed erythematous rash, blisters, and purpura on **his** chest.

P. quadrangularis roots, fruits, leaves, and immature seeds contain ***hydrogen cyanide*** and ***passiflorine*** which have psychoactive and poisonous effects.

PASTRY

Untoward effects: **Bakers** have experienced a dermatitis caused by a ***sugar*** mite. Of 1,004 claims for compensation for occupational-caused dermatitis, 18 were in confectionary and pastry **cooks** from ***sugar*** mites.

PATCHOULI OIL

Use: Large quantities are used in toiletries, such as **men's** cologne and aftershaves for its woody aroma. **Indian weavers** use it on their exotic fabrics.

Untoward effects: Some evidence of phototoxicity.

PATINOA ICHTHYOTOXICA

= *Tëhara*

Untoward effects: The dried and powdered pulp of this tree's fruit is carried in waterproof pouches and cast onto small pools of water by Colombian Amazonian **Indians**. Within 20–30 min, stunned fish float to the surface and are gathered for food. The roasted seeds have been eaten safely, but others insist they cause painful intestinal cramps and diarrhea, perhaps from different methods of preparation.

PATULIN

= *Clavacin* = *Clavatin* = *Claviformin* = *Expansine* = *Mycoin* C_3 = *Penicidin*

Antibiotic. A fungal toxin produced by various *Penicillium* and *Aspergillus sp.*

Untoward effects: Highly toxic and carcinogenic in **animals**. Traces have been found in ***apple*** and ***grape juices*** and in imported ***cheeses***. Fermentation reduces its concentration. Chromosome abnormalities in **human** leukocyte cultures. In view of its **animal** toxicity, it is probably a health hazard, even if small, for **man**. Associated with ***yellow rice*** toxicity in Japan.

Causes tachycardia, hypotension, and decreased urine production in **mice**; malignant tumors, edema, and nephrotoxic in **rats**; liver lesions and intestinal hemorrhages in **chicks**; has caused hemorrhage in the brain and death in **cattle**; ruminal atony, nasal discharge, anorexia, and painful retrosternal area in **sheep**, with a dose of 20 mg/kg killing one and 50 mg/kg causing minor toxicity to another.

LD_{50}, **mouse**: 29 mg/kg; **rat**: 32 mg/kg; 4 day **chick embryos**: 2.4 µg/egg; **White Leghorn cockerels**: 170 mg/kg.

PAULLINIA sp.

Untoward effects: ***P. cururu*** is used to prepare a ***curare***-type immobilizing agent.

P. pinnata = ***Aza*** (Benin) = ***Hannu Biyar*** (Hausa) = ***Kakashenla*** (Yoruba) roots, seeds, and stems contain a slow-acting poison used by **Amazonians** and **people** in tropical Africa against their **enemies**. Eating the leaves has been poisonous to **pigs** in the Camerooons and **cattle** in Nigeria. Also see ***Guarana*** for ***P. cupana***.

PAVETTA sp.

Untoward effects: Ingestion of the leaves from the shrubs ***P. harborii*** and ***P. schumanniana*** by **sheep** and **goats** in South Africa also causes "gousiekte" with the same symptoms, collapse, and death as described under ***Pachystigma sp.***

PAW-PAW

= *Asimina triloba* = *Asiminier trilobé* = *Indian Banana*

Appalachian Indians called it *asimina* and used it for kidney problems. In Florida, Seminole Indians used *A. reticulata* to make *Seminole Tea*, also used for the same purpose.

Untoward effects: In Guadeloupe, regular consumption of the fruit was associated with an increased incidence of Parkinsonism. Severe gastroenteritis in **some** after ingestion. Occasionally dermatitis from handling the mature fruit.

Extract of the twigs at 0.04 ppm can kill **brine shrimp**. It contains ***acetogenins***, one of which, ***asimicin***, can kill numerous **insects** and nematodes. Another, ***bullatacin***, which is undergoing testing, is ~1 million times more potent than ***cisplatin*** in inhibiting **human** ovarian tumor cells transplanted into **mice**. Still another, ***trilobacin***, is being tested versus other neoplasms.

PAXILLINE

Untoward effects: A severely tremorgenic metabolite of *Panicillium paxilli* causing severe tremors in **mice** at 25 mg/kg; lasting several hours. A cause of ***rye grass*** staggers, a neuromuscular disorder in grazing **sheep** leading to tremors, incoordination, staggering, falling, and muscular spasms.

PAYLOOAH

Untoward effects*:** This traditional orange-red powdered ethnic remedy from southeast Asia for fever and rashes is taken in a ***tea or at 2–3 g doses. In California in 1991–1992, it was associated with increased blood ***lead*** levels in 1–4-year-old **children**. In 1984, in a Thai **refugee** camp, samples contained 0.01% ***mercury*** and 0.03% ***arsenic***.

PEAS

Untoward effects*:** ***Cicer arietinum = ***Bengal Gram*** = ***Channa*** = ***Chick Peas*** = ***Garbanzo*** = ***Gram*** = ***Himmas*** = ***Himmis***. The leguminous seeds or peas are a food staple in India, and contain a ***protease inhibitor***. It contains ***pangamic acid*** and a neurotoxic amino acid, ***β-N-oxalyl-L-α,β-diaminopropionic acid***. Consumption has caused allergic reactions in **people**, and, if inadequately cooked, it can lead to ***cyanide*** toxicity.

Pisum sativum = ***Bisillah*** = ***Bislah*** = ***Bizillah*** = ***Common Peas*** = ***Garden Peas*** = ***Matur***. Found as early as 5500 B.C. in Greece. Has caused asthma and dyshidrosis. Contains free ***glutamate*** that may trigger migraines, and moderate amounts of ***oxalates***. Excessive intake can activate or aggravate an irritable bowel syndrome. Can trigger an attack of gout because of its high ***purine*** content that can increase ***uric acid*** blood levels. Some **people** have developed acute, and sometimes chronic, urticaria from eating them. **Pea-picker's** disease has been caused by infections with *Leptospira grippotyphosa*. **Patients** with ***nickel***-associated eczemas should avoid them to prevent dyshidrosis.

Swine fed at levels of at least 50% in their ration develop a ***selenium*** and ***vitamin E*** deficiency. **Lambs** eating peavine silage become paralyzed.

PEACHES

= *Prunus persica*

Untoward effects*:** The kernels, leaves, and twigs contain ***cyanogenic glycosides, amygdalin (q.v.) is the more common, ***prulaurasin***, and ***prunasin***, which yield ***hydrogen cyanide*** (q.v.) upon hydrolysis intraluminally by β-glucosidases, causing ***cyanide*** (q.v.) poisoning. **Children** have died from eating the seeds. Allergic dermatitis reported in ***almond***-sensitive **patients**. A 24-year-old **female** ate a peach before dancing, leading to anaphylactic shock with itching, hives, angioedema, abdominal cramping, diarrhea, and impending loss of consciousness. A 65-year-old **male** was allergic to all fruits containing large pits. **He** developed severe anaphylaxis after eating an ***apple*** that had been lying next to a peach. **His** would be a sensitivity to all members of the ***plum*** family (***plums, almonds, apricots, cherries, nectarines***, and ***peaches***). The Roman poet, Lucretius, stated over 2,000 years ago, "What is food to one man is bitter poison to another". Peach fuzz has caused erythematous reactions and itching in **fruit-handlers**. Peaches contain ***neochlorogenic acid, oxalates, salicylates*** (of significance in **patients** with chronic ***salicylate***-induced urticaria), and some goitrogens.

Livestock grazing on frosted peach vegetation or fruit can develop ***hydrocyanic acid*** poisoning.

PEANUTS

= *Arachis hypogaea* = *Earth Nuts* = *Goobernuts* = *Groundnuts* = *Gyada* (Hausa) = *Okpa Ekele* (Igbo)

Untoward effects*:** Can contain ***aflatoxin (q.v.) contamination, mostly from *Aspergillus flavus* and other species. In different parts of the world, the toxin source has also been from *Sclerotium rolfsii*, *Pithomyces chartarum*, etc. The U.S. carefully monitors peanuts for ***aflatoxin*** contamination. In Africa, large quantities of ***aflatoxin***-contaminated peanut meal have been eaten, especially by **children**, and it is suspect as a cause of hepatocarcinomas in **Bantus** and **others**. Urticaria, angioedema, laryngeal edema, bronchospasm, diarrhea, vomiting, shock, and collapse reported. **People** allergic to peanuts should not take a ***walnut*** from a dish of mixed nuts, because some peanut oil may have rubbed onto it and will trigger an allergic reaction. Cross-sensitization can occur with ***peas***, string ***beans, soybean*** products, and ***lentils***. At a ball game, a young **boy** savored **his** ***hot dog*** and licked **his** fingers. A **boy** next to **him** was eating peanuts and threw the shells on the floor by the first **boy's** camera. Each time the first **boy** put the camera near **his** face, **he** then licked **his** fingers, triggering an intense allergic reaction. Later, **he** ate a peanut cookie and had an immediate and violent allergic reaction. Contact does not have to be direct, as in the case of a **person** who shared a soft drink with a **sister** who had eaten peanuts, and suffered an immediate life-threatening asthma attack. A 22-year-old **male**, allergic to peanuts, ate a meal with satay sauce, causing cardiac arrest and coma, and was diagnosed as brain-dead. As an organ donor, **his** liver and kidney were transplanted into a 35-year-old **male** (with no previous allergic reactions to peanuts) **who** then developed a rash after eating them. Another **recipient**, a 27-year-old **female**, received **his** pancreas and a kidney, but showed no subsequent allergic reactions to peanuts. Sensitivity may be about one in 100 **preschoolers**. **Children** rarely outgrow **their** allergic reactions to it. It is one of the more commonly aspirated foreign bodies causing bronchial obstruction. Also contains high ***oxalate*** levels. Peanut oil in cosmetics is comedogenic, and, in parenterals, has caused lipogranulomas. An Amerindian **tribe** in British Guyana crushed them and "gave" it to "troublesome **people**" causing insanity and death.

Acute respiratory distress syndrome in **cattle** fed peanut-vine hay, possibly, due to toxin contamination from *Fusarium sp.* Acute hepatic toxicity in **ducklings** fed moldy peanuts. Peanut meal contaminated with *Aspergillus flavus* ***aflatoxin*** killed 100,000 **turkeys** (called "**turkey** X disease") in Britain in 1960, which led to major studies of ***aflatoxin*** in **human** foods and in other **animal** species. Topically, the oil was mildly irritating on **rabbit** and **guinea pig** skin; only mildly irritating on **rat** skin, and non-irritating on **humans** and **miniature swine** skin. Fed to **monkeys** they caused thickening of arterial walls; high-level intake in **rats** led to marked fatty infiltration of livers, intraventricular thrombi with myocardial infraction, pulmonary hemorrhages, and pneumonia. Rarely, jaundice in **rats**.

PEARS
= *Pyrus sp.*

Untoward effects*:** Pear seeds contain ***amygdalin (q.v.), the ***cyanogenic glycoside*** which can be hydrolyzed intraluminally in the gut by β-glucosidases, causing ***cyanide*** poisoning. Allergic reactions after ingestion. May cross-sensitize with ***apples***. **People** with ***nickel*** dermatitis should avoid fresh or cooked pears. Some goitrogenic activity noted in pears.

See ***Momordica*** for ***Balsam Pear***.

PEARLS

***Untoward effects*:** Dermatitis and pseudosyphilitic leukoderma in the summertime from pearl necklaces shielding some skin from the sun and allowing surrounding skin to be tanned.

PEAT

***Untoward effects*:** Of 240 **workers** in Russian peat bogs, 35% had pyodermatitis, 72% had intertrigo, and 100% had dry skin and rhagades.

Calves ingest their peat bedding leading to hypomagnesemia.

PEBULATE
= *PMBC* = *Stauffer 2,061* = *Tillam*

Pre-emergence herbicide. A thiocarbamate.

***Untoward effects*:** Overexposure can lead to depression, anorexia, nausea, vomiting, salivation, bloat, dyspnea, muscle spasms, hepatic and pulmonary congestion, ascites, hydrothorax, and hemorrhage in **animals**.

LD_{LO}, **human**: 500 mg/kg.

LD_{50}, **rat**: 1120 mg/kg; **mouse**: 1652 mg/kg; topical, **rabbit**: > 3 g/kg.

PECANS
= *Carya illinoensis* = *Carya pecan*

Untoward effects*:** High in ***oxalates. Pecans cause allergic reactions. *Aspergillus flavus* contamination has been common.

PECILOCIN
= *Leofungine* = *Supral* = *Variotin*

Antifungal antibiotic.

***Untoward effects*:** Skin irritation and eczematous rash after topical use.

PECTIN

Untoward effects*:** In antidiarrheal preparations with ***kaolin absorbs ***digoxin*** (40%), ***erythromycin, lincomycin, tetracycline*** (50%), and ***theophylline***, but not ***clindamycin***.

PEFLOXACIN
= *AM 725* = *EU 5306* = *Peflacine* = *RB 1589*

***Antibacterial. A fluorinated quinolone.* Oral and IV usage.**

***Untoward effects*:** Achilles tendonitis and rupture, arthralgia, destructive polyarthropathy in 17-year-old **male**, peripheral neuropathy (paresthesia of legs, foot weakness, and difficulty in walking) in 0.5–1.5%, gastro-intestinal upsets in ~6% (abdominal pain, nausea, vomiting, and diarrhea), and skin reactions in ~1%. Hallucinations, dizziness, thrombocytopenia, and severe anaphylactoid reaction in a 32-year-old **female** with AIDS, manifested severe hypotension, dizziness, itching, rash, and fever. Can pass through placenta and into amniotic fluid at low concentrations, and, at higher concentrations, into breast milk. Because of the potential for drug-induced arthropathy in **children**, its use should generally be avoided in pregnant or lactating **women**. A few cases of pseudomembranous colitis have been reported with its use.

Drug interactions*:** ***Antacids, metallic ions, and ***sucralfate*** decrease its absorption. Decreases ***theophylline*** clearance by ~20%, and slight decrease in clearance of ***rifampin***.

PEGADEMASE BOVINE
= *Adagen* = *Bovine Adenosine Deaminase PEG* = *PEG-ADA*

Use*:** In treating severe combined immunodeficiency disease due to a rare inherited deficiency of ***ADA, which is usually fatal in < 2 years. An "orphan drug".

***Untoward effects*:** Some pain at injection site. Half-life varies from ~3–7 days. Long-term effectiveness and adverse effects are yet to be determined.

PEGANUM HARMALA

= *African Rue* = *Armel* = *Harmal* = *Harmel* = *Hurmal* = *Syrian Rue* = *Wild Rue*

Unpalatable perennial herb growing on pastured ranges. Found in southwestern U.S., South America, and from the Mediterranean to Manchuria.

Untoward effects: Central nervous system stimulation with tremors and convulsions, due to ***harmine*** and ***harmaline***, in **man**. Poisonous doses depress central nervous system, leading to motor weakness, hypotension, lowered temperature, and depressed respiration. Progressive paralysis is due to ***harmol***.

Cattle and **sheep** have been poisoned after eating it (leaves, stems, fruit, or seeds) when other feeds are scarce, leading to nervousness, trembling, incoordination, knuckling over in hind legs, anorexia, pollakiuria, paralysis, and death. Post-mortem reveals severe gastroenteritis and congestion of lungs, kidneys, liver, and heart. **Guinea pigs** and possibly **horses** have also been poisoned by it. Contains 2.5–4% total of ***vasicine*** and alkaloidal hallucinogens, ***harmaline*** (q.v.), ***harmalol*** (q.v.), and ***harmine*** (q.v.). Reduces litter size and is probably abortifacient in **rats**.

PEGASPARGASE

= *Oncaspar* = *PEG-L-asparaginase*

Antineoplastic. **Used to treat acute lymphoblastic leukemia.**

Untoward effects: Essentially, the same as ***asparaginase*** (q.v.). In limited trials, nausea, vomiting, peripheral edema, allergic reactions, anaphylaxis, hepatic dysfunction, coagulopathy, bleeding, thrombosis (4%), pancreatitis (1%), hyperglycemia (> 3%), and hypoproteinemia or hypoalbuminemia. Leukopenia, pancytopenia, agranulocytosis, thrombocytopenia, disseminated intravascular coagulation, hemolytic anemia, anemia, hyponatremia, hyperuricemia, uric acid nephropathy, headache, seizures, paresthesias, night sweats, increased blood urea nitrogen and ***creatinine***, hemorrhagic cystitis, tachycardia, colitis, itching, rashes, alopecia, whiteness and ridging of nails, dyspnea on exertion (25%), and bradycardia also reported.

PELAGIA sp.

A cnidarian jellyfish. Common in the Mediterranean and off La Paz, Baja California.

Untoward effects: Even after storms or rough water, many of their tentacles break free and drift as small transparent pieces, still capable of firing their nematocysts causing painful, separated, and non-linear wheals lasting several hours or longer, with complete recovery in ~12 h.

PELECYPHORA ASELLIFORMIS

Untoward effects: This small Mexican ***"peyote"*** cactus contains ***mescaline*** (q.v.) and other ***peyote*** alkaloids with psychomimetic effects.

PEMOLINE

= *Azoksodon* = *Betanamin* = *Centramin* = *Cylert* = *Dantromin* = *Deltamine* = *Hyton Asa* = *Kethamed* = *Nitan* = *Phenylisohydantoin* = *PIO* = *Pioxol* = *Pondex* = *Ronyl* = *Senior* = *Sigmadyn* = *Stimul* = *Tradon* = *Volital*

Magnesium Pemoline = *Abbott 30,400*

Central nervous system stimulant. **Low abuse potential.**

Untoward effects: Addictive use led to paranoia, and visual, tactile, and auditory hallucinations in a 38-year-old **male physician**. Fatigue, restlessness, decreased weight, insomnia, anorexia, diarrhea, malaise, tics, itching, excitability, seizures, hypertension, rashes, headache, abdominal pain, nausea, slight dizziness, indigestion, tachycardia, palpitations, Gilles de la Tourette syndrome, hallucinations, decreased growth in **children**, talkativeness, depression, rash, raised liver enzymes, and fulminant liver failure have also been reported. Oculogyric reaction in a 8-year-old **male** after 56.25 mg/day/2 weeks. Half-life in serum of ~12 h.

Doses up to 2 g/1,000 lb IV in **horses** led to restlessness, searching stable floor for food, and increased running speed. Half-life in plasma of **horses** is 150 h, with 494 days final clearance. **Monkeys** will not self-administer it, although they will ***amphetamines***.

LD_{50}, **rat**: 500 mg/kg.

PEMPIDINE TARTRATE

= *M & B 4486* = *Pempidil* = *Pempiten* = *Perolysen* = *Tenormal* = *Tensinol* = *Tensoral*

Antihypertensive. Ganglionic blocker.

Untoward effects: Constipation, xerostomia, blurred vision, weakness, drowsiness, postural hypotension, tremors, diarrhea, dysuria, collapse (and a **few** experienced paralytic ileus), increased serum ***uric acid*** and decreased clearance as urinary pH increases above 4.5. Can pass through the placenta.

LD_{50}, **mouse**: 275 mg/kg; IV, 40 mg/kg.

PENBUTOLOL SULFATE

= *Betapressin* = *HOE-893d* = *HOE-39-893d* = *Levatol* = *Paginol*

β-Adrenergic blocker, antihypertensive, antianginal, antiarrhythmic.

Untoward effects: Bradycardia, hypotension, fatigue (20–30%), constipation, impotence, cold extremities, depression, dizziness (5–33%), headaches (17–25%), and vivid dreams. Occasionally nausea, anorexia, flushing, giddiness, catatonia, short-term memory loss, rashes, agranulocytosis, purpura, mesenteric arterial thrombosis, and ischemic colitis. Severe bronchospasm reported in **asthmatics**. Sudden treatment withdrawal can exacerbate angina or induce an myocardial infraction.

Drug interactions: Use with ***cimetidine*** causes slight increase in its plasma levels. Excessive negative inotropic effects can occur when used with ***diltiazem*** or ***verapamil***.

PENCICLOVIR

= *BRL 39,123* = *Denavir* = *PCV* = *Vectavir*

Antiviral. **For topical use versus several herpes viruses.**

Untoward effects: Headache. Some local erythema may be due to ***propylene glycol*** in its vehicle.

PENDIMETHALIN

= *AC-92,553* = *Herbadox* = *Penoxalin* = *Pre-M* = *Prowl* = *Stomp*

Herbicide.

Untoward effects: During 1 year in Taiwan, 71 poisonings with it reported to Poison Control Center. Most were due to accidental or intentional ingestion (average of 106.1 ml) and two were from skin and eye contact; 20 were asymptomatic, 38 had mild nausea, vomiting, and sore throat; and seven had severe retching, hematemesis, and seizures. **Others** died from ingesting other herbicides as well.

LD_{50}, **rat** and **dog**: > 5 g/kg.

PENFLURIDOL

= *R 16,341* = *Semap*

Neuroleptic.

Untoward effects: Said to enter the brain slowly and "leaves it with reluctance". Transient insomnia, akathisia, agitation, and tension. Extrapyramidal effects were usually dose-related.

Given to nursing **rat** mothers after delivery it led to impaired learning a month later in their **pups**. **Rats** treated with it for 24 months had increased incidence of pancreatic tumors.

LD_{50}, 1 week **mice**: 86.8 mg/kg.

PENICILLAMINE

= *Sulredox* = *β,β-Dimethylcysteine* = *D-3-Mercaptovaline*

Antirheumatic, copper chelator, amino acid. **Chelates heavy metals and solubulizes *cystine*.**

Untoward effects: Despite benefits, at least two-thirds of treated **patients** have early or delayed adverse effects. Allergic reactions, especially skin eruptions, pruritus, lymphadenopathy, and fever; bone marrow aplasia, agranulocytosis, leukopenia, thrombocytopenia, neutropenia, pancytopenia, leukocytosis, thrombocytosis, eosinophilia, sore throat, coryza, hypogeusia and dysgeusia, hemolytic anemia in glucose-6-phosphate dehydrogenase-deficient **patients**. Aplastic anemia, sideroblastic anemia, cholestatic hepatitis and jaundice, liver necrosis, increased lactate dehydrogenase, increased alkaline phosphatase, myasthenia syndrome (by stimulating the production of ***acetylcholine***-receptor antibodies, rather than by an intrinsic receptor-blocking effect), pruritus, muscle weakness, proteinuria, neuropathies (including Guillain-Barre syndrome), systemic lupus erythematosus, abdominal cramps, nausea, vomiting, diarrhea, hypercholesterolemia, arthritis, arthralgia, malabsorption syndrome, stomatitis, mouth ulcers, anorexia, and purpura also occur. Deters healing by delaying the cross-linking of ***collagen*** fibrils. Less commonly, optic neuritis, cataracts, renal damage and failure, chronic lymphocytic leukemia, acute lymphoblastic leukemia, tinnitus, diffuse alveolitis, acute colitis, bronchiolitis, ptosis, breast enlargement in **females**, gynecomastia in **males**, dyspnea, pleural effusion, yellow nail syndrome, pancreatitis, thyroiditis, and anetoderma. Prolonged use reduces circulating ***pyridoxine*** levels, which may cause peripheral neuropathy. Fatal pulmonary hemorrhage in induced Goodpasture's syndrome. Beneficial response in 50–70% of **patients** only after 2–6 months therapy, leading to current recommendations of "go-low, go-slow" to decrease mortality. Its use is contraindicated in **people** allergic to ***penicillins***. A contact reaction (irritation and rash over penis) in a 55-year-old **male** after intercourse with his **wife** who had been treated with it for arthritis. **He** also had a past history of ***penicillin*** sensitivity, indicating a possible cross-reaction. Intolerance in a 13-year-old **male**. Use may have been associated with rapid progression of scleroderma and inducing pemphigus foliaceus in a 49-year-old **female**.

It is teratogenic in **humans** causing connective tissue defects, cutis laxa, broad nasal bridge, low-set ears, flattened face, inguinal hernias, hyperflexibility of joints, vein fragility, short neck, simian creases, flexion contractures of hips and knees, varicosities, mannosidosis, poor wound-healing, and hydrocephalus.

Drug interactions: As a chelating agent, it should not be taken with ***food***, other medications, and various ions, or during pregnancy or before surgery. ***Antacids*** containing either ***magnesium*** or ***aluminum hydroxides*** and ***iron***

supplements can decrease its absorption by 50–65%. Has reduced serum concentration of ***digoxin*** in a **patient** and its similar effect on ***pyridoxine*** is noted above. Its plasma concentration is increased by ***isoniazid*** up to 34% when used with ***hydroxychloroquine*** or ***indomethacin***, and increased by > 50% with ***L-dopa***. **Patients** who react adversely to ***gold sodium thiomalate*** often react to penicillamine, especially if the interval between the two treatments is less than 6 months. It has exacerbated ***diazepam***-induced phlebitis. Has caused hypoglycemia, by triggering ***insulin*** autoimmunity. Its dosage in treating muscular dystrophy in **boys** and **animals** can be decreased when ***vitamin E*** therapy is added.

TD_{LO}, **human**: 546 g/kg.

Excessive doses in **rats** lead to hyperventilation, convulsive seizures, ***pyridoxine*** deficiency, and occasionally gastric erosions. Decreased immune response in **rabbits** and **mice**. Teratogenic in the **rat** leading to agenesis, cleft palate, and malformations of and incomplete mineralization of bones.

TD_{LO}, **rats** (10–17 days pregnant): 13 g/kg.

PENICILLIN

Benzylpenicillin = Penicillin G = Penicillin II = Benzylpenicillinic Acid = Free Benzylpenicillin = Free Penicillin G = Free Penicillin II

Untoward effects: Acid-labile and destroyed in the stomach after oral use, unless ***calcium*** and ***ammonium salts*** were used. ***Sodium*** and ***potassium salts*** are used IV or SC. Neutropenia, leukopenia, pancytopenia, thrombocytopenia, agranulocytosis, reticulocytosis, aplastic anemia, and fatal hemolytic anemia reported. Various oily vehicles, ***aluminum monostearate***, and ***beeswax***, were used in suspensions for IM or SC use, to extend the duration of antibiotic activity. The additives frequently caused residual nodules, indurations, and sterile abscesses. Encephalopathy after 20 million Un/day. Epileptogenic, and has led to generalized convulsions and myoclonic seizures with massive IV doses. Seizures reported in **infants** with meningitis after IV of 500,000–2,000,000 Un; fatal violent convulsions in 3-year-old **male** after 1 million Un intraspinal, instead of 5,000 Un. Crosses the placenta and suspect as a cause of cleft lip and palate in **infants** whose **mothers** received it during their first trimester. Inadvertent intra-arterial injection of the drug with ***aluminum monostearate*** into the right superior gluteal artery caused immediate pain, lameness, gangrene of the glans penis and scrotum, and blueness and coldness of the right foot. Electrolyte upsets and hyperkalemia from the ***potassium*** form; 15 million Un contains 25 mEq of ***potassium*** ions. Older, so-called less pure forms, often caused eczematization at sites of previous interdigital toe fungal infections, and in the **male** groin after 24–48 h. High levels in breast milk are probably not a significant problem, as it is destroyed in the acid environment of the stomach. Hypersensitivity reactions (~5%) include fever, anaphylaxis, rashes, urticaria, toxic epidermal necrolysis, nausea, vomiting, diarrhea, hepatitis, chest pain, hemophilia-like syndrome, breathing difficulties, asthma, systemic lupus erythematosus, delayed vasculitis, purpura, facial swelling, and contact dermatitis. These reactions have followed ingestion of contaminated dairy products, etc. or use of oral throat lozenges. Fever, pruritus, and a maculopapular rash in a 20-year-old **male** followed a blood transfusion containing 2 μg/ml. Many hypersensitivity reactions, including laryngeal edema and shock have been fatal, even from μg quantities. Acute interstitial nephritis with hematuria, and proteinuria. Hepatocellular damage, jaundice and necrosis, pulmonary infiltrates, and eosinophilia have occurred. Penicillin allergy is an ***IgE***-mediated reaction and should not be seen in **children** < 4 years, but **they** may have generalized reactions to it. A 26-year-old **female** had allergic reactions to it, once promptly, and another time, with a delayed peripheral vascular collapse from mucous membrane contact with it. **She** had penicillin once or twice/year without reactions; then, **she** received three to five injections of ***penicillin G benzathine*** for a postpartum infection. 2 years later, **she** had an allergic reaction after ingesting an oral penicillin cold remedy. Subsequently, **she** reacted to tasting, but not to swallowing a ***penicillin V*** suspension. Another reaction followed vaginal contact with her **husband's** mucous membranes, after **his** treatment with ***procaine penicillin G***, and, possibly, another reaction from hand-shaking. Ingestion of locally produced ***butter*** led to numbness and tingling of **her** tongue with urticaria. Postcoital urticaria reported in a 33-year-old **female** from possible transfer from her **boyfriend's** seminal fluids. Other similar case reports. A wrongful death award followed a **nurse's** inadvertent IV, instead of IM, injection with it. A 27-year-old **male** had a rash and periorbital edema, which was misdiagnosed as a penicillin allergy. Muscle biopsy confirmed it was due to trichinosis. Induced Jarisch-Herxheimer reaction, an inflammatory reaction due to toxic products of destroyed microorganisms. Severe necrotic reactions have followed its extravasation. During the first 20 years of use, deaths averaged > 300/year. *Candida* and other fungal infections have followed large and/or prolonged usage. Topical use in the eye has caused local inflammatory reaction and edema of the eyelids. Serum half-life in **man** is 51 min; 30 min in **dogs**, 53 min in **horses**, and 20 min in **swine**. A **patient** abused the drug by taking 15–20 tablets with a glass of ***beer***, for a sensation of euphoria. May cause falsely raised urinary ***glucose*** and protein tests, and reduced ***protein-bound iodine*** test results. Because of **their** decreased renal function, premature **infants**, young **children**, and aged **patients** tend to accumulate penicillin.

Drug interactions*: *Antacids decrease the rate of absorption of acidic drugs, such as ***aspirin, penicillin, nitrofurantoin, oxyphenbutazone, phenobarbital, phenylbutazone***, and ***sulfonamides***. Oral ***chymotrypsin*** increases the absorption of oral penicillin. *p-**Aminobenzoic acid, aspirin, dicumarol, ethyl biscoumacetate, oxyphenbutazone, phenylbutazone, salicylic acid***, and ***sulfonamides*** increase penicillin activity. It is potentiated by ***cloxacillin, methicillin***, and ***nafcillin*** by β-lactamase inhibition. Although some disagree, it has potentiated ***warfarin*** anticoagulant activity, but antagonizes ***heparin's*** activity. Use was associated with a severe toxic reaction to ***methotrexate*** and failures with oral contraception. ***β-Adrenergic antagonists***, such as ***nadolol*** and ***propanolol*** can potentiate anaphylactic reactions to penicillin. **People** who have had anaphylactic reactions to penicillin should not receive ***cephalosporins***, because the latter comes from a related starting chemical. Only about 10% of **children** evidence this cross-reactivity. It and all oral penicillins can inhibit ***vitamin K*** synthesis by gut flora. Penicillin activity is inhibited by ***bacteriostatic antibiotics*** (***chloramphenicol, erythromycin, kanamycin, neomycin, oleandomycin, paromomycin, streptomycin***, and ***tetracyclines***) because penicillin is most effective versus actively growing organisms. ***Dihydrostreptomycin, kanamycin***, and ***streptomycin*** can enhance its bactericidal activity versus many organisms. Displaced from serum-binding sites by ***probenecid***. Effectiveness is enhanced by ***aspirin, indomethacin, sulfaethidole, sulfamethoxypyridazine, sulfaphenazole, sulfasymazine, sulfinpyrazone***, and ***sulfisoxazole***. Effectiveness is decreased by ***antacids***.

Discontinue use at least 5 days before slaughtering **animals** for **human** use. Milk is normally withheld from **human** use during intramammary treatment and for at least 96 h and/or eight milkings after the last treatment, to assure zero penicillin levels in the milk. **Cattle** have shown severe anaphylactic or hypersensitivity reactions to its IV use. Has caused vasculitis in **dogs**.

LD_{50}, **hamsters**: 24 mg/kg. SC, LD_{50}, **hamsters**: 96 mg/kg.

Benzylpenicillin Ammonium

***Untoward effects*:** In **mice**, causes anorexia, weight loss, oligodipsia, oliguria, anuria, diarrhea, irritability, weakness, pallor, stupor, coma, shock, convulsions, and death due to respiratory failure.

LD_{50}, **rat**: 8.4 g/kg; **mouse**: 7.8 g/kg.

Benzylpenicillin, Benethamine = Benapen = Betapen = Benetolin

***Untoward effects*:** Long-acting parenteral form in aqueous suspension has caused SC indurations.

Benzylpenicillin, Benzathine = Benzetacyl = Benzethacil = Bicillin = Cepacilina = Debecillin = Dibenzathine Penicillin = Longicil = Moldamin = Penduran = Penidural = Peniduran = Permapen = Tardocillin = Tardomyocel

Long-acting parenteral in oily or aqueous suspension – usually IM.

***Untoward effects*:** Fever, headache, shivering, perspiration, muscle and joint pain; swelling, pain, and erythema at injection site; exacerbation of psioriatic lesions, urticaria; transitory severe ischemia of leg, buttock, perineum, and abdominal wall in **children**; serum sickness-type of reaction, pericarditis and myocarditis, tachycardia, weakness, dyspnea, urticaria, tongue edema, anemia, and benign intracranial hypertension. An intra-arterial injection in a 4-year-old **female** caused embolism and secondary thrombosis, leading to peripheral gangrene. Hemorrhagic allergic meningoencephalitis in a 27-year-old **male** caused meningism, vomiting, tonic limb spasms, mental confusion, delirium, and aphasia.

Has caused local indurations at SC injection sites. U.S. regulations state: "NOT for use in **animals** which are raised for food production", although it is in wide clinical use as an effective, practical, and often necessary method of controlling susceptible infections. Milk levels in treated **cattle** may persist for 2 weeks. NOT to be used within 30 days of slaughtering **animals** for **human** consumption.

Benzylpenicillin, DAEEH = Bronchocilline = Ephicillin = Estopen = Iomycin = Leocillin = Mamyzin = Penester = Penethamate Hydriodide = Penicillin G,-Diethylaminoethyl Ester Hydriodide

Based on the natural affinity of its *iodine* fraction, it is ideal for pulmonary and mammary infections. Penicillinase resistant.

***Untoward effects*:** Urticaria in a 25-year-old **female**; severe anaphylactic shock in a 56-year-old **male** with chronic frontal sinusitis. Fatal cases reported in 4–5-month-old **infants** after severe convulsions that followed 500,000 Un injections. **Veterinarians** have developed a contact dermatitis while using it in treating **cows**.

Milk withholding times in **cows** are estimated at 3–5 days.

Benzylpenicillin, l-Ephenamine = Compenamine

***Untoward effects*:** Do NOT save milk for **human** consumption for 96 h and eight milkings after the last treatment, or slaughter treated **animals** for at least 5 days after the last treatment.

Benzylpenicillin, Potassium = Burcillin = Cosmopen = Cristapen = Crystapen = Crystapen G (oral) = Dymocillin = Eightpen = Eskacillin = Falapen = Fivepen = Forpen = Genecillin = G-Recillin = Hipercilina = Hyasorb = Hylenta

= *Ka-Pen* = *K-cillin* = *Kesso-Pen* = *K-Pen* = *Matrin* = *M-cillin* = *Megacillin* = *Monopen* = *Neopens* = *Nodopen-500* = *Notaral* = *Novopen G* = *P-50* = *Palocillin 5* = *Pencitabs* = *Penioral-500* = *Penivet* = *Pentid* = *Pfizerpen G* = *Potassium Benzylpenicillinate* = *Potassium Penicillin G* = *SK-Penicillin G* = *Solupen* = *Sugracillin* = *Tabilin* = *Tenpen* = *Wescopen*

Untoward effects: Rapid high IV dosage causes apnea, cyanosis, hypotension, right bundle-branch block on electrocardiogram and asystole. Allergic reactions, hemolytic anemia, grand mal convulsions, and hyperkalemia. Each million Un contains 1.68 mEq ***potassium***. Neurotoxicity with seizures after IV use of 500,000–2,000,000 Un in **infants**. Cardiac arrhythmias and asystole due to an error in giving a **child** 500,000 Un/kg as an IV bolus, with a total of 8.5 mEq ***potassium***. A 60-year-old **female** was resuscitated after an IV push of 5 million Un (8.4 mEq ***potassium***), and later after 4 million Un (6.72 mEq ***potassium***) by IV drip during 10–15 min. Asystole, ventricular fibrillation, and atrioventricular block followed each occurrence.

Fed to **dairy calves** at 500 mg/100 lb of milk replacer it caused reduced growth and death. Toxic doses in **rabbits** cause anorexia, adipsia, bradypnea, pallor of ears, prostration, violent kicking, respiratory failure, and death. Toxic **rats** evidenced diarrhea, anorexia, weight loss, polyuria, glycosuria, albuminuria, and hematuria. Do NOT add ***procaine*** solution to minimize IM pain, as this will form some insoluble depot ***benzylpenicillin procaine***. If necessary, some ***lidocaine*** can be added. Discontinue use 1 day before slaughtering **birds** for **human** food consumption. 0.01 ppm is the maximum permissible residue in uncooked edible tissues of **turkeys**; no residue permitted in any edible eggs, **chickens**, **pheasants**, **quail**, or **swine**. When used IV, administer very slowly, to avoid toxicity of the ***potassium*** ion.

LD_{50}, **rats**: 6.7 mg/kg; **mouse**: 5.6 g/kg; **rabbit**: 5.25 g/kg; IV, **mouse**: 448 mg/kg; **guinea pig**: 303 mg/kg.

Benzylpenicillin, Procaine = *Abbocillin-DC* = *Afsillin* = *Ampinpenicillin* = *Aquacillin* = *Aquasuspen* = *Avloprocil* = *Burcillin-P* = *Cilicaine* = *Crystapen* = *Crysticillin* = *Despacilina* = *Depocillin* = *Distaquaine* = *Dorsallin "A.R."* = *Duracillin* = *Ecmonovocillin* = *Flo-Cillin* = *Hydracillin* = *Ilcocillin P* = *Kabipenin* = *Ledercillin* = *Lenticillin* = *Lentopen* = *Mammacillin* = *Megapen* = *Micro-Pen* = *Mylipen* = *Neoproc* = *Penaquacaine G* = *Pen-Fifty* = *Pfizerpen-AS* = *Premocillin* = *Procanodia* = *Pro-Pen* = *Tu-Cillin* = *Wycillin*

Untoward effects: A 20-year-old **female** with syphilis after IM 400,000 Un/day/8 days developed transient blindness, tinnitus, vertigo, tachycardia, and anxiety. Acute, dramatic pseudoanaphylactic reactions include life-threatening cardiac arrhythmias and seizures, due to microembolization and direct toxicity of free ***procaine*** (q.v.) in **patients** 20–71 years. Anaphylactoid reaction with rectal temperature of 104.6 °F in a 2-year-old **female**, 15 min after IM injection of 300,000 Un for cellulitis; immediate cardiac massage, intracardiac ***epinephrine***, and mouth-to-mouth resuscitation led to recovery. Jarisch-Herxheimer reaction with temperature of 37.7–40.0 °C in 24 syphilitic **patients** within 3–5 h and in 52 before 11th h after first 600,000 Un dose. In general, allergic reactions may be due to the drug, the suspending agents, or the vehicle. A large venereal disease clinic stated that even in their experienced hands, 3/200 IM injections enter a blood vessel. Usual pullback on syringe with a viscous solution may not show blood. They now insert a clean needle and wait a full minute to see if any blood appears, to avoid the possibility of pulmonary embolism and sudden death from the lipoid vehicle, usually misinterpreted as a simple anaphylactic reaction. Irreversible ischemic gangrene of a hand in a 1-year-old **child**, due to an unintentional intra-arterial injection. Acute psychotic reactions of short duration in six **males** with gonococcal urethritis immediately after 2.4 million Un IM. Acute mania, shouting, and exaggerated physical activity in six within 30 s, lasting 2–3 min in five, and 4–5 h in one. **Patients** were bewildered upon recovery. Acute anxiety, hyperactivity, bradycardia, and mydriasis in a 38-year-old **female** within 2 min and lasting ~1 h after **her** 15th IM injection of 1 million Un for actinomycosis of the mandible. Sweating, tachycardia, dyspnea, and severe respiratory distress after 3 days of 600,000 Un IM in a 65-year-old **male**. Treatment of a 68-year-old **female** with pneumonia with 1.2 million Un IM for 3 days caused hepatotoxicity with jaundice, ***bilirubin*** - 8, ***sulfobromophthalein*** 42%, SGOT 226, and Ceph Floc 2+. A 32-year-old **female**, after one IM injection of 1.2 million Un for treating pharyngitis, developed acute polyarthritis, conjunctivitis, fever, and erythema multiforme 4 days later. Stevens–Johnson syndrome with oral, ocular, and penile lesions in a 17-year-old **male** after two IM injections of 600,000 Un for treatment of tonsillitis. Hemolytic anemia – Coombs positive, granulocytopenia, syncope, hypotension, dyspnea, bronchospasms, tachypnea, seizures, laryngeal edema, angioedema of eyelids and lower lip, rashes, transient left bundle-branch block, ischemic ST-T waves, increased cardiac enzymes, sinus tachycardia, and reversible neurological syndrome have also occurred. Hallucinations and wild running in a 30-year-old **male**, 30 s after an intragluteal injection; returned to normal in 10 min with bizarre subjective experience. Intradermal skin tests are not without risk, reactions (non-lethal anaphylaxis and shock) occurring within 15 s after 10–20 Un, and death in a 53-year-old **female** with a history of penicillin reaction 2 years previously. Within a few seconds after 1,000–2,000 Un, **she** became asthmatic, cyanotic, and died within 5–10 min. In single dose studies, adverse effects were reported in 4.2% of **patients**.

In **cows**, aqueous suspensions IM will necessitate milk withholding times of at least 3 and 6 days for the dosage recommended. Oil-***aluminum monostearate*** suspensions will require at least a 5 day milk withholding period for the lower dosage. Do NOT use on pet **birds**, as the procaine in the complex or added procaine may be fatal. This and added ***streptomycins*** may also be involved in conflicting reports on its and/or the added procaine toxicity for **monkeys**. Overdosage is also a key factor in these mortalities. Discontinue use 1 week before competition, to avoid disqualification of **horses**, due to positive tests for procaine in their urine. 5–7 days withdrawal times are suggested by most agencies for parenterally treated **cattle** and **swine**. Young **calves** (3–35 days of age) receiving 25,000 Un/kg IM still had undesirable levels in their livers, kidneys, and at the injection site after 20 days. In general, its parenteral use will produce residues in milk for at least 24 h after serum levels disappear. Growth-promoting levels of ***copper*** in **turkey** feed appear to interfere with those effects of ***penicillin-streptomycin***. The U.S. FDA now objects to the use of ***streptomycin*** and oil-based mixtures for parenteral use in food-producing **animals**.

Benzylpenicillin, Sodium = BRL 3,000 = Crystapen (injectable) = Purapen G

Untoward effects: Bone marrow hypoplasia, leukopenia, hypokalemia, metabolic acidosis, and hypernatremia. Acute hemolytic anemia in a 45-year-old **female** who received ***ampicillin*** 9 years previously. Neurotoxicity post-operatively after 50 million Un in 3/4 after open-heart surgery and cardiopulmonary by-pass, apparently, due to penetration into the brain, inducing epileptogenic activity. Encephalopathy in three **male** 10–13-year-olds, after 1,000,000 Un doses IM, IV, and intrapleural. Various allergic cutaneous reactions reported and hypersensitivity after 5 million Un IV, caused immediate urticaria, bronchoconstriction, peripheral vasodilation, hypotension, and cardiac arrest. Diarrhea reported in ~6% of **patients**. Contains 2 mEq ***sodium*** ion/million Un. See the ***potassium salt*** for discussion on ***procaine*** or ***lidocaine*** to relieve IM pain.

Penicillin V = Acipen-V = Apocillin = Beromycin = Compocillin V = Crystapen V = Distaquaine V = Distaquine V = Fenospen = Fenoxypen = Macromycin = Meropenin = Oracilline = Oratren = Phenopenicillin = Phenoxymethyl Penicillin = Pen-Vee = V-Cillin = Veesyn

***Potassium, calcium, benzathine,* and *hydrabamine* forms have been used. Acid-stable.**

Untoward effects: Occasionally gastric irritation, nausea, vomiting, esophagitis, black hairy tongue, diarrhea, colitis, anaphylaxis, shock, rash, pruritus, urticaria, bullous pemphigoid, periorbital edema, flushed skin, dyspnea, hypotension, syncope, cardiac disorders, and nasal congestion. A 36-year-old **female** had marked systemic reaction, and severe liver function disturbance with jaundice after only 250 mg qid/2 days. Acute generalized erythematous pustulosis in a 2-year-old **female** ~24 h after treatment. A 23-year-old **female** developed anaphylactic shock within 60 s of ingesting two 250 mg tablets. Decreased absorption if taken with ***food***. A 29-year-old **female** had fever and perineal eruptions 8 h after sexual intercourse. **Her** sexual **partner** had taken the drug for 4 days. Similar reaction 2 years previously with fever, swelling, urticaria, and joint pain. Oral use has induced ***oral contraceptive*** failure and break-through bleeding. Oral use for gum inflammation in two hypertensive **patients** taking ***nadolol*** or ***propranolol*** caused anaphylactic reactions and death. **They** had no history of medication allergy or previous exposure to penicillin. The *potassium salt* has also induced taste disorders. Mammary excretion occurs in **women** and **cows**. In **dogs** and **man**, pericardial fluid and serum concentrations are in equilibrium within 2 h. Effectiveness enhanced when used with ***aspirin, sulfaethidole, sulfamethoxypyridazine, sulfasymazine, sulfinpyrazone***, and ***sulfisoxazole***. ***Neomycin*** interferes with its absorption.

TD_{LO}, **woman**: 10 mg/kg/2 days.

PENICILLINS – "NEWER AND SYNTHETICS"

Amdinocillin = Coactin = FL 1060 = Mecillinam = RO 10-9,070 = Selexidin

Semi-synthetic. Used IM and IV. Poor oral absorption except for its *pivoxil* ester, *Pivmecillinam* (q.v. below).

Untoward effects: Swelling of tongue and lips, pruritus, and urticarial exanthema. Eosinophilia and thrombocytosis in ~5%, increased SGOT, SGPT and alkaline phosphatase, thrombophlebitis, anemia, leukopenia, and neutropenia. Occasionally nausea, vomiting, pain at injection site, drowsiness, hypertension, vaginitis, thrombocytopenia, and increased serum ***bilirubin***.

Amoxicillin = Agram = Alfamox = Almodan = Amocilline = Amoclen = Amodex = Amolin = Amopenixin = Amoram = Amoval = Amoxi = Amoxidal = Amoxidin = Amoxil = Amoxillat = Amoxipen = Amoxi-Wolff = Amoxycillin = Amoxypen = AMPC = Anemolin = Ardine = Aspenil = AX 250 = Betamox = Bristamox = BRL- 2333 = Cabermox = Clamoxyl = Cuxacillin = Delacillin = Dura AX = Efpenix = Flemoxin = Grinsil = Helvamox = Hiconcil = Ibiamox = Imacillin = Larocin = Larotid = Moxal = Moxaline = Neamoxyl = Noxypen = Optium = Ospamox = Pasetocin = Penamox = Penimox = Piramox = Polymox = Raylina =

RO 10-8756 = Robamox = Sawacillin = Sigamopen = Silamox = Simoxil = Sumox = Suprapen = Synsepal = Trimox = Uro-Clamoxyl = Utimox = Van-Mox = Widecillin = Wymox = Zamocillin = Zimox

Acid-stable semi-synthetic penicillin.

Untoward effects: Anaphylaxis, maculopapular erythematous skin rashes, erythema multiforme, toxic epidermal necrolysis, Stevens-Johnson syndrome, nausea, vomiting, indigestion, pruritus vulvae, increased serum transaminases, fatigue, sore tongue, headache, moniliasis, colitis, loose stool, frequent bowel movements, flatulence, bitter taste, dizziness, agitation, confusion, aseptic meningitis, acute interstitial nephritis (especially with overdose), Reiter's syndrome (polyarthritis, urethritis, and iritis) after induced *Clostridium difficile* colitis, serum sickness-like reactions and drug fever, hypoprothrombinemia, neutropenia, eosinophilia, and pancytopenia. A fatal case of pseudomembranous colitis in a 5-week **infant** reported. Pseudomembranous colitis after 250 mg tid in a 10-year-old **male**. After 125–250 mg tid/10 days in **infants** it led to two-fold increase in *Candida albicans* isolation from **their** rectums and inguinal folds, and diaper dermatitis of 15.8%. Rarely, hepatotoxicity and cholangitis.

Drug interactions: Can cause ***oral contraceptive*** failure. Has increased ***methotrexate*** serum levels and toxicity. Its addition to ***allopurinol*** therapy was suspected as a cause of fever, exanthema, and death with severe aggravation of ***allopurinol***-induced cholangitis. Can be adsorbed by ***attapulgite, kaolin, magnesium trisilicate***, and ***veegum***. ***Food*** delays its absorption.

Amoksiklav = Augmentin = Ciblor = Clavamox = Co-Amoxiclav = Klavocin = Neo-Duplamox = Spectramox, an often used combination with potassium clavulanate.

Untoward effects: Hepatotoxicity and jaundice in 1/10,000 **patients**. The *clavulanate* can increase gastrointestinal side-effects above amoxicillin's. Headache, rashes, agranulocytosis, and *Candida* vaginitis, as well as the adverse effects of amoxicillin alone.

Discard five milkings after intramammary treatment of **cows** and avoid slaughtering for **human** consumption for 2 weeks; 25 days after parenteral therapy in **cattle**; and 20 days after oral treatment of **calves**. Allergic reactions are rare. Halve the dosage or double the interval between dosing in **animals** with renal disease.

Ampicillin = Albipen = Alpen = Amblosin = Amcill = Amcillin = Amfipen = Amipenix = Ampilag = Ampilar = Ampilin = Ampipenin = Ampitab = Ampi-Tablinen = Amplital = Austrapen = AY 6,108 = Binotal = Bonapicillin = Britacil = BRL 1,341 = Cetampin = Cymbi = Doktacillin = Domicillin = Dumopen = Grampenil = Nuvapen = Omnipen = P-50 = Pen A = Penbristol = Penbritin = Penbrock = Péniciline = Pensyn = Pentrexyl = Polycillin = Princillin = Principen = Rosampline = Tokiocillin = Totacillin = Totalcidina = Totapen = Ukapen = Ultrabion = Vidopen

***Sodium* and *potassium* forms are occasionally used. Acid-stable semi-synthetic penicillin that is inactivated by penicillinase.**

Untoward effects: Agranulocytosis and peripheral monocytosis with marrow histiocytosis in 66-year-old **male** receiving 250 mg qid/8 days for epididymitis. Contains an ***α-aminobenzene*** structure, common to many drugs, causing agranulocytosis. Severe thrombocytopenia and slight megakariocytopenia also occurred. A 30-year-old **female** developed neutropenia and thrombocytopenia after 500 mg qid/4 days. Neutropenia in **patients** 1–57 years, 8–27 days after treatment with doses as low as 40 mg/day, led to rashes, fever, and eosinophilia. Reversible neutropenia 3–12 days after 100–400 mg/kg/day in pediatric **patients**. Thrombocytopenia in a 69-year-old **female** after 500 mg IM qid/2 weeks. Very high IV doses (3 g every 4 h of *sodium* form) caused hypokalemic metabolic acidosis in a 59-year-old **male**. Has caused false elevations of serum ***fibrinogen***. Muscle irritation from IM use leads to increased SGOT and creatinine phosphokinase levels. Pain, hyperemia, and sterile cellulitis have occurred at injection site. High IV doses have caused ampicillinuria, crystalluria, leukopenia, neutropenia, and, rarely, interstitial nephritis. Drugs with the potential of causing anaphylactic reactions should be avoided in term pregnancies. A 43-year-old **female** given 2 g IV to prevent intra-amniotic infection had an **infant** with severe metabolic acidosis, clonic seizures, and brain edema. Neurological deficits were still present after 6 months. Diarrhea and candidal overgrowth noted in **many** on long-term therapy and in breast-fed **infants** of **mothers** receiving treatment. Pseudomembranous colitis and diarrhea have been common sequelae to long-term treatment. A 54-year-old **male**, 2 h after intital dose each morning (250 mg qid/7 days), developed frontal parietal headache, benign intracranial hypertension, hallucinations, mental confusion, disorientation, and mydriasis. **He** was worse after a switch to ***tetracycline***. Numerous case reports of allergic reactions with nausea, vomiting, gastric pain, shock, peripheral circulatory failure, rashes, and respiratory distress syndrome. Dramatic increase in hypersensitivity reactions in **patients** with impaired renal function. Anaphylactoid reaction (dyspnea, apprehension, and urticarial rashes) in **female** with previous allergy to penicillin immediately after opening an ampicillin bottle. Cross-sensitivity with other penicillins. Rashes are usually maculopapular, erythematous, and urticarial. Over 20% of **patients** have some gastrointestinal irritation. Thrombophlebitis after IV, especially in **those** with small veins. Aplastic anemia in a 24-year-old **male**, 6 months after ingestion of eight tablets. A case report of Henoch–Schoenlein purpura with subsequent nephrosis. Assorted induced blood dyscrasias have included anemia, aplastic anemia, hemolytic anemia in a **patient** with glucose-6-phosphate dehydrogenase deficiency, eosinophilia, leukocytosis, leukopenia, disseminated intravascular coagulation, thrombocytopenia, lupus erythematosus, neutropenia, and pancytopenia. Hepatotoxicity is rare.

Drug interactions: ***Allopurinol*** potentiates its induced skin rashes. Occasionally decreases efficacy of ***oral contraceptives***. Some reduced absorption of ***atenolol*** when given together. Has **bilirubin**-displacing effect in serum. Adsorbed by ***attapulgite, kaolin, magnesium trisilicate***, and ***veegum***. ***Foods*** can seriously interfere with its absorption. Decreases effectiveness of ***carbenicillin***, produces spuriously high plasma ***catecholamine*** test results, and its effectiveness is decreased by ***chloramphenicol***.

In the U.S., it is not recommended for use in food-producing **animals** except **swine**. A tolerance of 0.01 ppm for its negligible residues in the uncooked edible tissues of **swine** is permitted by the FDA. Withhold oral treatment for at least 24 h before slaughtering **swine** for **human** consumption. Contraindicated in penicillin-sensitive **animals** or in infestations caused by penicillinase-producing organisms (some forms of *Staphylococcus, Pseudomonas, Proteus, Aerobacter*, and *E. coli*). A 15 day withdrawal is required, to avoid undesirable tissue levels in orally treated **calves**, or **animals** for **human** consumption. Inactivated by ***gentamicin*** in infusion fluids, but in serum, this is clinically insignificant. Severe local reactions at injection sites have been reported in **horses**. IM dosage in **cats** and **dogs** is often painful. In **dairy cows**, the *trihydrate* form causes more irritation than the *sodium* form.

Azlocillin = Azlin = Bay E 6905 = Secoropen

A broad-spectrum semi-synthetic penicillin. The usually used *sodium* form is given IV.

Untoward effects: Local reaction and minor phlebitis in ~3%. Hypersensitivity with 0.3% drug fever, 1.8% cutaneous reactions, and eosinophilia in 1.1%. Hepatotoxicity in 1.7%, diarrhea in 1.9%, leukopenia in 0.3%, and decreased serum ***uric acid***. Occasionally transient mild nausea or vomiting.

Bacampicillin = Ambacamp = Ambaxin = Bacacil = Bacampicine = Lekobacyn = Penglobe = Spectrobid

An *ampicillin* ester with more rapid absorption and fewer adverse effects than *ampicillin*.

Untoward effects: Transient loose stools in ~2%, especially after high dosage, and epigastric upsets 2%. ***Food*** and reduced gastric acidity decrease its bioavailability. Hypersensitivity reactions can parallel those for ***ampicillin***.

LD_{50}, **mice**: 8.5 g/kg; IV, 184 mg/kg; IP, 176 mg/kg; SC, 9.48 g/kg.

Carbenicillin Disodium = Anabactyl = BRL-2064 = Carbapen = Carbecin = CP- 15639-2 = Geopen = Gripenin = Hyoper = Microcillin = Pyocianil = Pyopen

Semi-synthetic penicillin for parenteral use.

Untoward effects: Hypokalemic alkalosis (which has caused arrhythmias), neutropenia, leukopenia, eosinophilia, hemolytic anemia, and granulocytopenia. Hypoprothrombinemia with impaired platelet functions that caused life-threatening hemorrhage. Neurotoxic with generalized tonic–clonic seizures, acute metabolic acidosis, grand mal seizures, anicteric hepatitis with nausea, vomiting and tender enlarged liver, renal toxicity with acute interstitial nephritis, rash, hemorrhagic cystitis, urticaria, bad taste, transient increases in SGOT and SGPT, drug fever and increased temperature, increased lactate dehydrogenase, episodic reduction of vision with flashes of multi-colored lights, and altered taste reported. Pain after injection, especially IM. Cross-sensitivity with other penicillins. Passes into breast milk. Activity is decreased by ***ampicillin***. Has caused a profound decrease in ***gentamicin*** activity *in vitro*, but, in **man** and **dogs**, this was not significant if the drugs were not permitted to first stand as a mixture.

Carindacillin Sodium = Carbenicillin Indanyl Sodium = Carindapen = CP 15,464 = Geocillin = G.U.-Pen

Semi-synthetic acid-stable penicillin.

Untoward effects: Nausea, vomiting, diarrhea, eosinophilia, bad taste and smell of capsules, increased SGOT and SGPT, and pseudomembranous colitis. Skin rashes, urticaria and pruritus, flatulence, dry mouth, furry tongue, vaginitis, and abdominal cramps also occur.

Cloxacillin/Sodium salt = Ankerbin = Austrastaph = BRL-1621 = Bactopen = Bicloxil = Cloxapen = Cloxypen = Dariclox = Ekvacillin = Gelstaph = Orbenin = Methocillin-S = Penstapho N = Prevencillin P = Prostaphlin-A = Staphobristol-250 = Staphybiotic = Syntarpen = Tegopen = Tepogen

Benzathine salt = Boviclox = Dry-Clox = Noroclox DC = Opticlox = Orbenin DC = Triclox

Acid-stable, penicillinase-resistant, semi-synthetic penicillin. A derivative of *oxacillin*.

Untoward effects: High dosage for ~2 months produces urticarial rash. Eosinophilia 4–25%, leukopenia 7.6%, neutropenia 2.5%, agranulocytosis, hemolytic anemia, thrombocytopenia, purpura, proteinuria, febrile reactions, transient diarrhea, nausea, vomiting, moniliasis, vaginal thrush, epigastric fullness, abdominal discomfort, foul taste, headache, thrombocytopenic purpura, auricular fibrillation, cholestatic jaundice without hepatitis, exfoliative dermatitis, maculopapular rash, and lymphadenopathy. A 15-year-old **male** who ingested 3 g/day developed diarrhea, vomiting, dyspnea, sneezing, wheezing, edema, hypertension, and uremic death as a result of induced acute interstitial nephritis. Local induration reported after IM use; phlebitis after repeated IV. Absorption is somewhat impaired if taken with ***food***. Effectiveness is enhanced by ***aspirin***, ***sulfisoxazole***, ***sulfinpyrazone***, ***sulfasymazine***, ***sulfaethidole***, and ***sulfamethoxypyridazine***.

NOT recommended in the U.S. for use in food-producing **animals**. The *benzathine* form is now approved in the U.S. for dry **cow** treatment and treated **animals** may not be slaughtered for **human** food until 28 days (10 days for *sodium* form) after treatment. A similar withdrawal period applies to the use of milk from treated **animals**.

IP, LD_{50}, **rat**: 1.6 g/kg; **mouse**: 1.3 g/kg; IV, **mouse**: 916 mg/kg.

Cyclacillin = Calthor = Ciclacillin = Citosarin = Cyclapen = Syngacillin = Ultracillin = Vastcillin = Vatracin = Vipicil = Wy 4,508 = Wyvital

Semi-synthetic penicillin related to *ampicillin*.

Untoward effects: Diarrhea in ~5%, nausea and vomiting in ~2%, and rashes in < 2% of **patients**. Occasionally headache, dizziness, abdominal pain, increased SGOT, vaginitis, and urticaria. Nephrotoxic in **male**, but not **female**, **rats** after 25 weeks treatment. After comparable treatment, **dogs** and **rhesus monkeys** showed no nephrotoxicity, nor did **humans** given 2 g/day/30 days.

TD_{LO}, **human**: 210 mg/kg/7 days.

LD_{50}, **rat** and **mouse**: 5 g/kg; **dog**: 2.5 g/kg.

Dicloxacillin = BRL 1702 = Maclicine

The sodium salt = Brispen = Constaphyl = Dichlor-Stapenor = Diclocil = Dicloxin = Dycill = Dynapen = Noxaben = P 1011 = Pathocil = Pen-Sint = Stampen = Syntarpen = Veracillin

A semi-synthetic penicillin, better absorbed than *cloxacillin*. Acid-stable and penicillinase-resistant.

Untoward effects: Gastrointestinal upsets including moderate diarrhea, epigastric or abdominal pain, nausea, unpleasant taste, gastrointestinal bleeding, anorexia, cholestasis, erythematous rash, maculopapular lesions, fever, stomatitis, eosinophilia (3–15%), and hemolytic anemia. Placental transfer at term occurs. Pain reported at IV injection sites in **infants**. In two **patients**, accidental intra-arterial injections caused ischemia and gangrene, requiring amputation of the hand and lower arm.

LD_{50}, **rat**: 3.58 g/kg; **mouse**: 5 g/kg.

Diphenicillin Sodium = Ancillin = BL P413 = SKF 12,141

Semi-synthetic penicillin for oral use.

Untoward effects: Drug fever, eosinophilia, nausea, vomiting, and mild diarrhea.

SC, LD_{50}, **rat**: 2.5 g/kg; IV, **mouse**: 1 g/kg.

Epicillin = Dexacillin = Omnisan = Spectacillin = SQ 11,302

Semi-synthetic penicillin for oral use. The sodium salt for IM and IV. Omnisan is also the name for a Yugoslavian quaternary ammonium compound, aralkonium chloride.

Untoward effects: Reactions in 144/1272 and 37/153 (11–24%). Diarrhea or loose stools are the most common ill-effects (7–8%), skin rashes (3%); slight nausea, vomiting, and dizziness also reported.

Floxacillin = Abboflox = Culpen = Floxapen = Flucloxacillin = Flupen = Heracillin = Ladropen = Penplus = PK 900 = Stafoxil = Staphcil = Staphlipen = Staphylex

Semi-synthetic penicillinase-resistant derivative of *oxacillin*, particularly versus penicillin-resistant staphylococci. Oral, IM, or IV.

Untoward effects: Diarrhea, nausea, headaches, rashes, and pseudomembranous colitis. **Patients** receiving it for at least 2 weeks risk cholestatic jaundice (1/15,000). Nearly 300 cases of hepatotoxicity reported in Australia. In **neonates**, kernicterus can result from it displacing ***bilirubin*** from its binding sites on ***albumin***. Breast milk concentrations reported as < 0.1 μg/ml. Case reports of ***oral contraceptive*** failures from its use. Phlebitis in 13% of IV treated **patients**. Concomitant use of ***probenecid*** increases serum levels.

LD_{50}, **mouse**: 3.8 g/kg; SC, 2.2 mg/kg.

Hetacillin = Amplipen = BL-P804 = BRL 804 = Etaciland = Heciline = Hexacilllin = Penplenum = Phenazacillin = Versapen = Versatrex = Vimocillin

The potassium salt = Natacillin = Uropen

Semi-synthetic acid-stable penicillin converted in the body to *ampicillin*. Oral, IM, or IV.

Untoward effects: Hypersensitivity with maculopapular erythematous rash, severe nausea, vomiting, mouth sores, diarrhea, occult blood in stool, eosinophilia, transient leukopenia, increased SGOT, and increased ***thymol*** turbidity test. Dietary fat increases its absorption, but ingestion with food can decrease its blood level by 50%.

In **dogs**, IV use has potentiated ***epinephrine's*** vasopressor effect. Withdrawal time of 10 days required before slaughtering treated **animals** for **human** consumption.

Methicillin Sodium = Azapen = Belfacillin = BRL 1,241 = Celbenin = Celpillina = Cinopenil = Dimocillin = Flabelline = Leukopenin = Meticillin = Penistaph = Staphcillin = X 1497

Semi-synthetic bacteriocidal penicillinase-resistant penicillin. Not acid-resistant. Parenteral use only.

Untoward effects: Less potent antibacterial than ***benzylpenicillin*** and can stimulate penicillinase production. Bone marrow depression, agranulocytosis, eosinophilia, leukopenia, thrombocytopenia, pancytopenia, hemolytic anemia, hypokalemia, lymphocytosis, leukocytosis, and monocytosis. Fever, rash, eosinophilia, anemia, and abnormal liver enzymes reported in treated **infants** and **neonates**. Nephropathy, glomerulitis, interstitial nephritis, acute tubular necrosis, oliguria, chills, and fever. Hematuria, proteinuria, azotemia, dysuria, ureteral inflammation, flank pain, hemorrhagic cystitis, and acute renal failure. Usually causes elevated or false positive ***creatinine***, serum ***ammonia, non-protein nitrogen***, serum ***potassium, uric acid***, eosinophils, inorganic ***phosphate***, and ***blood urea nitrogen*** test values, and decreased ***calcium*** values. Pain at injection site. Drug-induced skin eruptions, diarrhea, mild jaundice, thrombophlebitis, and vasculitis occur. Half-lives in premature **infants** have ranged from 1.4–3.3 h, in term **infants**, 0.9–3.3 h, and in **adults**, 30 min. Bactericidal antagonist when used with ***erythromycin***. Use with ***salicylates, sulfaethidole***, or ***sulfamethoxypyridazine*** enhances its effectiveness.

Mezlocillin = Baycipen = Bay f 1353 = Baypen = Mezlin

Semi-synthetic penicillinase-resistant penicillin related to *azlocillin*. The *sodium salt* is used IV and IM.

Untoward effects: Hypokalemia, leukopenia, thrombocytopenia, neutropenia, eosinophilia, positive Coomb's test, and reduced hemoglobin and hematocrit. Nausea, vomiting, diarrhea, and pseudomembranous colitis reported. Rashes and renal dysfunction with interstitial nephritis are common. Pruritus, urticaria, drug fever, and anaphylactic reactions occur. Abnormal taste. Increased SGOT, SGPT, alkaline phosphatase, ***bilirubin, creatinine***, and ***blood urea nitrogen*** in serum. Thrombophlebitis after IV and pain after IM. Neuromuscular hyperirritability and convulsive seizures (especially with overdoses) have also occurred. Slight increase in prothrombin time reported.

Nafcillin Sodium = Nacil = Unipen = WY 3,277

Acid-stable semi-synthetic penicillinase-resistant penicillin. Low and erratic oral absorption. Primarily used IV or IM.

Untoward effects: Acute interstitial nephritis, deep tissue necrosis and sloughing (after SC extravasation), thrombophlebitis, pseudoproteinuria, and hypokalemic acidosis. High dosage causes reversible bone marrow toxicity, neutropenia, leukopenia, leukocytosis, agranulocytosis, hemolytic anemia, eosinophilia, and coagulopathy with clinical bleeding episodes. Has caused anaphylaxis in 0.004–0.015% with mortality rate of ~10%. Oral or IV causes urticarial or maculopapular erythematous rashes. Oral or IM caused diarrhea in 36/109 **children**. Oral administration produces vomiting, loose stools in 7% or 14/200 in ages 3 months to 18 years, perianal burning, fever, and rash. Injection site irritation. Falsely increased SGOT. Cross-sensitization with other penicillins. Excreted in breast milk (~10% of **mother's** plasma level), and this can stimulate an antigenic response.

Drug interactions: Causes serum binding of ***aztreonam*** to drop by ~5%. Nafcillin decreases hypoprothrombinemic response to ***warfarin***. Use with ***probenecid, salicylates, sulfaethidole***, or ***sulfamethoxypyridazine*** enhances its effectiveness. Potentiated penicillin effect when used with ***benzylpenicillin***.

IP, LD_{50}, **rat**: 920 mg/kg; **mouse**: 1600 mg/kg; IV, **mouse**: 1 g/kg.

Oxacillin Sodium = Bactocill = Bristopen = BRL 1,400 = Cryptocillin = Micropenin = Oxabel = Oxacyclin = Penicillin P-12 = Penstapho = Penstaphocid = Prostaphlin = Resistopen = Stapenor

Acid-stable semi-synthetic penicillinase-resistant narrow-spectrum, bacteriocidal penicillin for oral, IV, or IM.

Untoward effects: Skin rashes, urticaria, serum sickness, anaphylaxis, acute interstitial nephritis, oliguria, albuminuria, proteinuria, hematuria, pyuria, cylinduria, epigastric distress, diarrhea, nausea, vomiting, hypokalemia, agranulocytosis, leukopenia, leukocytosis, hemolytic anemia, thrombocytopenia, eosinophilia, transient neutropenia, and thrombophlebitis (after IV). Hepatocellular damage with cholestatic–hepatocanalicular jaundice with granulomatous lesions, increased SGOT and SGPT reported 2–21 days after treatment. Occasionally glossitis, stomatitis, and rectal moniliasis. Oral absorption is low and erratic. Half-lives in **prematures** and **neonates** are ~1–1.5 h, and 0.5–0.7 h in **adults**. Causes falsely increased sulfobromophthalein retention and ***bilirubin*** test results.

Drug interactions: Use with ***salicylates, sulfaethidole***, and ***sulfamethoxypyridazine*** increases its unbound moiety in plasma and body tissues. ***Probenecid*** prolongs its effectiveness. Oral absorption is inhibited by ***food*** and its oral use should be 1 h before or 2–3 h after eating.

After intramammary use in **cows**, levels persist for up to 30 days.

Phenethicillin Potassium = Alfacillin = Alpen = Bendralin = Brocsil = Broxil = Chemipen = Darcil = Dramcillin-S = Feneticillin = Maxipen = Optipen = Oralopen = Penicillin-152 = Penicillin MV = Peniplus = Penorale = Penova = Pensig = Rocillin = Semopen = Syncillin = Synthecilline = Synthepen

Acid-stable, bacteriocidal, semi-synthetic penicillin.

Untoward effects: Diarrhea, abdominal cramps, vomiting, urticaria, and skin rash. Cross-sensitivity with other penicillins. Oral ***chymotrypsin*** increases its blood levels after oral use.

IV, LD_{50}, **guinea pig**: 324 mg/kg.

Piperacillin Sodium = Avocin = CL 227,193 = Isipen = Penticillin = Pipracil = Pipril = T 1220

Semi-synthetic *carbenicillin* analog.

Untoward effects: Poor oral absorption. Diarrhea or loose stools in 2%, occasionally nausea and vomiting, increased lactate dehydrogenase, SGOT, and SGPT. Hyperbilirubinemia, cholestatic jaundice, moniliasis, and drug fever (up to 24% in cystic fibrosis **patients**). Neutropenia, leukopenia, thrombocytopenia, eosinophilia, increased ***creatinine*** and blood urea nitrogen, rashes, urticaria, and macular and vesicular eruptions with secondary purpura. Low concentrations appear in breast milk of treated **mothers**. IV use led to thrombophlebitis in 4% and seizures in an 11-year-old **male**; after IM, induration, pain, and erythema at injection site. Often falsely increased urinary ***glucose*** results with ***Clinitest***®.

Drug interactions: Inactivates ***tobramycin*** *in vivo*. Combined with ***tazobactam*** = ***Zosyn*** for the latter's antipenicillinase benefits, causes diarrhea in ~11%, allergic skin reactions (rash, pruritus) in ~4%, headache or constipation in ~8%, nausea or insomnia in ~7%, vomiting in > 3%, fever or agitation in > 2%; moniliasis, abdominal pain, dizziness, chest pain, edema, anxiety, dyspnea, and rhinitis in < 2%. Occasionally or rarely, hypotension, ileus, syncope, tachycardia or bradycardia, arrhythmias, myocardial infraction, ventricular fibrillation, gastritis, hiccups, stomatitis, pseudomembranous colitis, tinnitus, arthralgia, bleeding, confusion, hallucinations, vaginitis, photophobia, pulmonary edema, bronchospasm, coughing, hematuria, oliguria, dysuria, and interstitial nephritis. Pain at injection site. ***Probenecid*** extends the piperacillin half-life by 21%, and that of ***tazobactam*** by 71%. It has delayed elimination of ***methotrexate*** in an 8-year-old **male**.

Pivampicillin = Alphacillin = Berocillin = Centurina = Diancina = Inacillin = Maxifen = MK 191 = Pivatil = Pondex = Pondocil = Pondocillin = Sanguicillin

Semi-synthetic penicillin prodrug of *ampicillin*.

Untoward effects: Case reports indicate the most common side-effects are nausea, vomiting, dyspepsia, diarrhea, rash, and ***carnitine*** deficiency with accompanying weakness and pain. Several capsule formulations caused gastric interstitial hemorrhage and mucosal erosions. Use has been associated with ***oral contraceptive*** failures. For other adverse effects, see ***Ampicillin*** above.

LD_{50}, **rat**: 5 g/kg; **mouse**: 3.34 g/kg; SC, 4.5 and 3.6 g/kg; respectively.

Pivmecillinam = Amdinocillin Pivoxil = FL 1039 = Melysin = Negaxid = Selexid

Semi-synthetic penicillin for oral use. Can be taken with or without *food*.

Untoward effects: Decreased absorption in celiac disease. Mild gastrointestinal discomfort and ***carnitine*** deficiency with muscle weakness and pain. Risk of esophageal irritation from tablets getting stuck in the esophagus, especially if taken after, instead of during, a meal.

LD_{50}, **rat**: 10 g/kg; SC, 2 g/kg; IV, 465 mg/kg.

LD_{50}, **mouse**: 3 g/kg; SC, 1.85 g/kg; IV, 475 mg/kg.

Propicillin Potassium = Baycillin = BRL 284 = Brocillin = Cetacillin = Levopropylcillin Potassium = Oricillin = PA 248 = Synthepen-P = Trescillin = Ultrapen

Semi-synthetic bacteriocidal penicillin.

Untoward effects: Vomiting and allergic exanthema are the most common adverse effects reported.

Quinacillin

Semi-synthetic penicillin.

Untoward effects: Limited trials indicated skin rash and death in one **patient**, possibly from agranulocytosis.

Sultamicillin = Bacimex = Bethacil = CP 49,952 = Unacim = Unasyn

A molecular combination of *ampicillin* and the β-lactamase inhibitor, *sulbactam*, to help overcome resistant infections. A prodrug.

Untoward effects: Diarrhea and rash reported. Also see individual components.

Talampicillin = BRL 8988 = Talat = Talpen = Talpicil = Yamacillin

Semi-synthetic prodrug of *ampicillin*.

Untoward effects: Diarrhea, unpleasant taste, heartburn, rash, pruritus, nausea, vomiting, moniliasis, headache, and anaphylactic shock. ***Food*** in the stomach decreases and delays peak blood levels, but not the total amount absorbed. Also see ***ampicillin***.

Ticarillin Disodium = Aerugipen = BRL 2,288 = Monapen = Ticar = Ticarpen = Ticillin

Broad-spectrum semi-synthetic penicillin with poor oral absorption. For IV and IM use.

Untoward effects: Skin rashes, drug fever, pruritus, urticaria, nausea, vomiting, pseudomembranous colitis, anemia, thrombocytopenia, leukopenia, neutropenia, eosinophilia, platelet dysfunction (leading to hemorrhages), increased SGOT and SGPT, neuromuscular excitability and convulsions with high dosage, phlebitis, pain at injection site, hypokalemia, and falsely increased ***glucose Clinitest***® readings. *In vivo* inactivation of ***aminoglycoside antibiotics***, such as ***amikacin***, ***gentamicin***, and ***tobramycin*** which was confirmed *in vitro*.

Teratogenic in **mice**.

A combination with *potassium clavulanate = Betabactyl = BRL 28,500 = Timentin*.

Untoward effects: Hypokalemia and slight decrease in peak ***digoxin*** serum levels. Also reports of thrombocytosis, hyperkalemia, reduced ***chloride*** and ***sodium***, and platelet dysfunction with occasional clinical bleeding.

PENICILLIUM sp.

Untoward effects: Many species produce mycotoxins and contaminate food supplies (especially stored grains) of **man** and **animals**. Common symptoms are anorexia, icterus, weakness, prostration, bloody stools, photosensitization, reproductive failures, and abortions. Unidentified species have caused Balkan nephropathy, dyspnea, pneumonitis, conjunctivitis, otitis, rhinitis, and deaths in **humans**; dermatomycosis in **cats**, **dogs**, **horses**, and some **rodents** and **birds**; nasal penicilliosis and hepatitis in **dogs**; pneumonias, uterine infections, and abortions in **cows**; gastrointestinal hemorrhages in **chicks**, **ducklings**, and **poults** and pneumonomycosis in **chicks**; tremors in **sheep**; vulvovaginitis, nephropathy, hemorrhagic gastritis, hemorrhagic disorders and deaths in **swine**; nervousness in **water buffaloes**; and hyperestrogenism in **sows** and **rabbit does**.

Over 100 years ago, the way was paved by Lister for the discovery of ***penicillin***. Lister's laboratory diary revealed that on November 21, 1871, he started an experiment that showed that, in a glass that had a heavy growth of *Penicillium*, there was little bacterial growth. Many produce toxic antibiotics and some, such as ***P. chrysogenum*** and ***P. notatum***, have been used for production of ***penicillin***, and ***griseofulvin*** from several other *Penicilliums*. Most *Penicilliums* produce bright green or blue pigments. Some of the more toxic ones of the hundreds that have been identified are listed below.

P. articae, on ***apple*** products, produces a toxin that is nephrotoxic and causes edema in **rats**.

P. atrovenetum produces toxic ***nitropropanoic acid*** (q.v.) found in ***Indigofera*** (q.v.) and ***deoxy-hequeinone***, the monomethyl ether of ***atrovenetin***, which is toxic to experimental **animals**.

P. bruneum, on ***rice***, produces the toxic anthraquinonoid pigment, ***rugulosin*** (q.v.).

P. canescens produces a tremorgenic toxin, ***penitrem A*** (q.v.).

P. chartarum produces ***sporidesmin*** (q.v.), causing facial eczema in **sheep**, **cattle**, and **deer**.

P. chermesinum produces toxic ***ergot*** (q.v.) alkaloids.

P. citreo-viride causes a toxin, ***citreoviridin***, that yellows stored ***rice*** and causes an ascending central nervous system paralysis in higher **vertebrates**. An ingredient in the toxic chemical and biological warfare agent, ***yellow rain***. Formerly called ***P. toxicarum***. Central nervous system paralysis and fatal as well as fetotoxic in **rats** at 10 or 15 mg/kg.

P. citrinum produces the toxin, ***citrinin*** (q.v.), a benzopyran pigment that causes hemorrhagic enteritis, weight loss, and liver and kidney pathology.

P. claviforme produces the toxin, ***patulin*** (q.v.). Found on spoiled ***apples***.

P. commune found on ***cottonseed meal*** is neurotoxic to **chicks** due to its metabolites, ***penitrem A*** (q.v.) a tremorgen, and ***roquefortine*** (q.v.), a paralytic neurotoxin.

P. concavorugulosum grows on ***rice*** and produces toxic ***rugulosin*** (q.v.) or ***rugulovasines*** (q.v.).

P. crustosum produces ***penitrem A***, a tremorgen.

P. cyclopium ingested on moldy ***corn*** has caused hemorrhagic syndrome, toxicity, and deaths in **cattle**, **rabbits**, **mice**, and **guinea pigs**; nephrotoxic in **swine**, and its ***tremorins A, B***, and ***C*** cause staggers in **sheep**.

P. decumbens produces a lactone, ***brefeldin A***, toxic to **rats** and **mice**.

P. expansum, found on ***apples***, ***meat***, and vaginal smears from non-pregnant **women**, produces ***citrinin*** (q.v.) and ***patulin*** (q.v.), both of which cause teratogenicity after injection into **hen** eggs.

P. frequentans causes reduced weight gains, nephro- and hepatotoxicity in **mice**, and produces the antineoplastic antibiotic, ***hadacidin***.

*P. **glaucum*** causes *Capsicum* rots and "**chili-splitter's** lung", a toxomycosis, rather than a mycosis.

*P. **islandicum*** produces several potent liver carcinogens and other toxic chemicals. Naturally occurring on moldy ***rice*** produces yellow ***rice*** and hepatotoxic pigment, ***luteoskyrin*** (q.v.) and ***cyclochlorotine*** (q.v.). Fed to **chicks** causes renal and hepatic degeneration; toxic to **rats**, **mice**, and **ducklings**. Liver cirrhosis, hemorrhages, and primary hepatomas in **rats** and **mice**. Suspect carcinogen in **man**. Found in cereals used for ***beer*** and ***bread*** in Ethiopia.

*P. **janthinellum*** produces tremorgens, ***fumitremorgin A*** (q.v.) and ***verruculogen*** (q.v.), leading to ***ryegrass*** staggers in Australian **cattle** and **sheep**. Bronchiolitis obliterans organizing pneumonia in a 38-year-old **male** after accidentally inhaling its spores from contaminated ***orange juice***.

*P. **marneffi*** in large numbers of **patients** in southeast Asia. Disseminated infections with fever, anemia, decreased weight, and skin lesions.

*P. **martensi*** on high-moisture ***corn*** has killed experimental **mice**. Has caused bald spots on limbs and tails of affected **cats**.

*P. **meleagrinum*** on moldy ***oat*** flakes causes rapid emaciation, progressive weakness, anorexia, vomiting, bloody diarrhea, and death in naturally infected **dogs**.

*P. **obscurum*** produces toxic ***gliotoxin***, which is very similar to ***sporidesmin***.

*P. **ochraseum*** induces mycotoxicosis in **mice** leading to reduced weight gains, some renal lesions, extensive hepatic cell necrosis, periductal edema and fibrosis, and death.

*P. **ochrosalmoneum*** produces toxic ***citreoviridin***.

*P. **oxalicum*** produces ***secalonic acid*** and is lethal to **ducklings**, **rats**, and **mice**.

*P. **palitans*** produces highly toxic tremorgens, ***tremorin A*** and ***B***, causing tremors and bursa regression in **broilers**; hemorrhage, tremors, and death in **cows**.

*P. **patulin*** metabolite, ***patulin*** (q.v.) is toxic.

*P. **paxilii*** produces the tremorgens, ***paxilline*** (q.v.) and ***verruculogen*** (q.v.).

*P. **piceum*** on cereals and grains is toxic to **ducklings**, **rats**, and **mice**.

*P. **piscarium*** produces ***verruculogen*** and, possibly, ***fumitremorgin B***, both tremorgens causing, in **sheep**, symptoms similar to ***ryegrass*** staggers.

*P. **puberculum*** is a frequent contaminant of foods with ***puberulic acid***, ***puberlonic acid***, and ***penitrem A***. Within 15–30 min stomach-tubed **ducklings** and **mice** develop incoordination, exaggerated movement, and stiffness of limbs. Albino **mice** had darkened eye color and cyanotic mouth, feet, and tail. After a brief moment of ataxia, convulsions with apnea preceeded death. Oral doses to **calves** cause tremors, ataxia, muscle rigidity, convulsions, and bile duct proliferation in **ducklings**.

*P. **purpurogenum*** fed to **chicks** causes depression, reduced weight gain, and death with internal hemorrhages, vulvovaginitis, and deaths in **swine**.

*P. **roqueforti*** is used in French ***cheese*** ripening and found in fermenting silage. Hypersensitivity with erythema annulare centrifugum of 8 years duration in a 39-year-old **female** during winter months. An atypical strain isolated on ***cheddar cheese*** in Nebraska produced ***penicillic acid*** and ***patulin***. Overripe trimmings of ***blue cheese*** produced ***roquefortine***, a neurotoxic mycotoxin that caused shivering, tetanic spasms, opisthotonus, and death in a **dog**.

*P. **rubrum*** produces ***rubratoxin***, fatal to **ducklings**, **chicks**, **swine**, **mice**, **guinea pigs**, **rabbits**, and **dogs** with hepatoxic, nephrotoxic, and hemorrhagic syndromes. Toxicity also reported in four **horses** and a **goat**. Toxic ***rugulovasines*** indole alkaloids have also been isolated from it. LD_{50}, **mouse**: 8 mg/kg; IV, 4 mg/kg, IP.

*P. **steckii*** produces toxic ***citrinin*** (q.v.) and ***isochroman***.

*P. **urticae*** produces ***patulin*** (q.v.), on malt feed in Japan, caused marked central nervous system stimulation and deaths in **dairy cattle** and major illness with deaths in young **calves** in Russia from contaminated hay. Lethal to **mice**.

*P. **verruculosum*** yields ***verruculogen*** (q.v.) or ***verruculotoxin***, a severely tremorogenic mycotoxin, adversely affecting **mice**, **rats**, **guinea pigs**, and **chickens**.

*P. **viridicatum***, common on stored grains, produces ***citrinin***, ***oxalic acid***, ***viridicatumtoxin***, and ***ochratoxins*** that are toxic to **pigs**, **mice**, **rats**, and **guinea pigs**.

TD_{LO}, **mouse**: 2910 g/kg over 28 weeks (~15 g/kg/day).

See ***Aflatoxins***, ***Citrinin***, and ***Ochratoxins***.

PENICILLOYL-POLYLYSINE

= *PPL* = *Pre-Pen* = *Testarpen*

Use: As an intradermal test for ***penicillin*** hypersensitivity.

Untoward effects: Anaphylactic reaction within 20 min in 30-year-old **male** with history of sensitivity to ***benzylpenicillin potassium***; marked edema, erythema and pruritus on entire forearm, and both hands and feet on a subsequent test. Itching and/or urticaria, localized or general, immediate or delayed, with wheals in some positive tests.

PENITREM A

A tremorgenic fungal metabolite produced by *Penicillium commune, P. canescens, P. cyclopium,* and *P. crustosum.*

Untoward effects*:** Causes outbreaks resembling ***ryegrass staggers in **cattle** and **sheep**. **Dogs** receiving 250–500 μg/kg IP develop tremors and convulsions within 10 min. Hepatic necrosis and hemorrhages (thoracic and peritoneal areas) in **dogs** after 2.5–5 mg/kg. Intoxication in **mice**, **rats**, **guinea pigs**, **hamsters**, **chickens**, and **swine** leading to tremors, irritability, and, with larger doses, convulsions.

IP, LD_{LO}, **dog**: 500 μg/kg.

PENMESTEROL

= *Pandrocene* = *RP 12,222*

***Androgen.* Oral use.**

***Untoward effects*:** In 53 **males** and seven **females**, aged 12–65 years, caused nervousness and palpitations in one, slight chin hirsutism in one **female** (even after discontinuance), and bilateral gynecomasty with bleeding from nipples in a 15-year-old **male** after 1 month of therapy. Slight jaundice when treatment repeated.

PENNISETUM sp.

***Untoward effects*: *P. clandestinum* = *Kikuyu Grass*.** Acute rumen atony, alkalosis, tremors, depression, abdominal pain, constipation, hemoconcentration, cyanosis, and death of **cattle** in New Zealand and South Africa. Osteodystrophia fibrosa reported in affected **horses**. **Sheep** appear to be less affected than **cattle**.

P. pupureum* = *Elephant Grass*.** Grows in Colombia, where it is a ***nitrate-accumulator. **Cattle** grazing it develop colic, diarrhea, muscle spasms, increased salivation, stupor, opisthotonus, cyanosis, and die.

***P. spicatum* = *Gero* = *Maiwa*.** Cyanogenic. Toxic to **cattle** in Nigeria.

P. typhoides* = *Bulrush Millet* = *Mpondu*. *Ergot develops on its seeds and inhibits normal mammary development of **sows** during pregnancy, with resulting death of their **piglets** from starvation.

PENNYROYAL

= *Mentha pulegium*

Herbal fragrance, mint-like flavoring agent, and insect repellent.

Untoward effects*:** Toxicity is due to its ketone, ***pulegone, which irritates the kidneys and bladder, and reflexly excites uterine contractions. Its oil is used, unfortunately, as an abortifacient and to induce menstruation. It has been ingested directly as the oil and indirectly in herbal ***teas***. As little as 3 ml has induced convulsions; larger doses have caused bleeding, diathesis, shock, disseminated intravascular coagulation, renal toxicity, massive hepatic necrosis and fatty degeneration, cardiopulmonary arrest, and death. **Infants** and young **children** have developed hepatic and neurological damage, as well as death, after ingestion of its ***tea***. ***Methofuran*** is a toxic metabolite. ***Pulegone*** is also found in ***Hedeoma sp.*** (q.v.). Pennyroyal use in now banned in many countries.

LD_{LO}, **human**: 50 mg/kg.

The oil causes pulmonary and hepatic necrosis and neuropathology in **rats** and **mice**. Topical use on a **dog** for flea control led to emesis within 2 h and death within 2 days. Post-mortem revealed evidence of epistaxis, gastrointestinal bleeding, and erythematous areas on the abdomen and medial aspect of the thighs.

LD_{50}, **rat**: 400 mg/kg; acute dermal **rabbit**: 4.2 g/kg.

PENTABORANE-9

= *Pentaboron Nonahydride*

Rocket propellent.

***Untoward effects*:** Ignites spontaneously in air. Inhalation leads to dizziness, drowsiness, light-headedness, incoordination, tremors, eye and skin irritation, and tonic spasms of face, neck, abdominal muscles and limbs. Convulsions, behavioral changes, increased blood pressure, tachycardia, and dyspnea are also caused by inhalation. Convulsions in many **people** and death in a **few** after 1 min exposure to 1,000 ppm. National Institute for Occupational Safety and Health and OSHA set exposure limits of 0.005 ppm TWA. 15-min exposure: 0.015 ppm.

Estimated LD_{50}, **human**: 72 μg/kg.

LC_{50}, **mice**: 14 ppm/2 h; **rats**: 7 ppm/4 h.

PENTACHLOROETHANE

= *Ethane Pentachloride* = *Pentalin*

***Solvent.* For oil and grease removal from metals.**

***Untoward effects*:** Eye and skin irritation, weakness, narcosis, restlessness, irregular respiration, incoordination, and liver, kidney, and lung pathology in experimental **animals**.

LD_{LO}, **human**: 50 mg/kg.

LD_{LO}, **dog**: 500 mg/kg. Inhalation LC_{LO}, **rat**: 4,238 ppm/2 h; **mouse**: 35 mg/m^3/2 h.

PENTACHLORONAPHTHALENE

= *Halowax 1013*

Untoward effects: Inadequate exhaust ventilation and poor personal hygiene put a large number of **workers** at risk while applying it as wax insulation on wire coils in an electric manufacturing plant. Symptoms included headache, fatigue, vertigo, anorexia, pruritus, chloracne, jaundice, and liver necrosis. National Institute for Occupational Safety and Health and OSHA exposure limits of 0.5 mg/m^3.

Emissions from a factory making waxes contaminated a nearby pasture, causing the loss of ~150 **cattle** and **calves** over a 3 year period, as well as abortions and infertility. Symptoms included typical hyperkeratosis or X-disease.

PENTACHLOROPHENOL

= *Chlorophen* = *Dowicide 7* = *Dowicide G* = *EP 20* = *PCP* = *Penchlorol* = *Penta* = *Pentacon* = *Penwar* = *Sanituko* = *Santobrite* = *Santophen* = *Sinituho* = *Weedone*

***Insecticide, wood preservative, molluscicide, herbicide.* The *sodium* form is also used. *PCP* is a term also used for *polychlorpine* and *phencyclidine*.**

Untoward effects: Found in inside air of treated log homes and their cistern water, and **children** < 12 years living in them excrete it in their urine at a higher rate than **those** > 12 years. Numerous deaths reported after overexposure of **people**. Poisoning with pyrexia, acidosis, aminoaciduria, and ketonuria in a 3+-year-old **female** from bath water from contaminated tank. At one time, the grocery store was an excellent **human** exposure source from contaminated ***food*** and ***soft drinks***. In Florida, it was found at 20 ppb in the urine of 60 **students** tested and at 70 ppb in the seminal fluid of seven **students** tested. Severe toxicity reported in **infants** after absorption from **their** diapers at a hospital misusing a laundry product containing it as an antimildew agent in final rinse of diapers and bed linens leading to percutaneous absorption, profuse generalized diaphoresis, fever, tachycardia, tachypnea, hepatomegaly, and acidosis in nine; two died, and seven improved dramatically on exchange transfusion. 11 other **infants** during a 5 month period had similar, but milder, symptoms. Flushed skin, headache, abdominal pain, conjunctivitis, sneezing, collapse, coma, aplastic anemia, red blood cell aplasia, acute leukemia, and Hodgkin's disease also reported. In 1952, a case of monocular retrobulbar neuritis reported in a **male** and two cases in **female**s after using it as an insecticide. Some commercial material also contained ***dioxins*** and ***dibenzofurans*** as contaminants. Sytemic toxicity in **people** can appear in **adults** when **their** blood and urine concentrations reach 1 ppm. Half-life is ~33 h in **humans**. Suicides have been reported and minimum lethal dose is ~29 mg/kg. National Institute for Occupational Safety and Health and OSHA upper exposure limts are 0.5 mg/m^3/skin.

TD_{LO}, **human**: 196 mg/kg.

It can volatilize from the surface of treated wood into the air for many years, and **cows** in confinement have received it this way or from rubbing or licking treated wood. Spontaneous abortions, dyspnea, persistent infections, liver and kidney damage, and deaths have occurred in some exposed **cattle**. **Horses** have been poisoned from chewing on treated wood. Contaminated wood shavings in the nests decrease hatchability of **hen** eggs and contaminated sawdust may have poisoned 11/80 **cats** and can cause violative levels in slaughtered **hogs**. Poisoning causes tachypnea, nervousness, tremors, vomiting, and death in young **pigs** from rubbing on "bleeding" treated lumber or chewing on it. Skin and mucous membrane burning in **pigs** on contact and decreased growth; skin irritation on older **hogs**. **Goats** have been similarly affected. Treated hives have killed **bees** and their ***beeswax*** and ***honey*** have contained high concentrations. After 2 years, ***honey*** from these hives averaged 143 ppb. Rapidly taken up into the bodies of exposed lake **trout**, **goldfish**, etc.; causes chloracne in **mice**, and **chick** edema. Polydipsia, tachypnea, wing twitches, shakiness, ataxia, tremors, spasms, and deaths in **ducks** and **pheasants**, and regurgitation in **mallards**. **Rabbits** develop tachypnea, hyperpyrexia, hyperglycemia, glycosuria, weakness, and cardiac failure. Toxic to many aquatic and estuarine species, especially **crustaceans**, causing inhibition of limb regeneration and variations in shell thickness.

LD_{50}, **rat**: 50 mg/kg; **calves**: 140 mg/kg; **sheep**: 120 mg/kg; **mallard ducks**: 380 mg/kg; **pheasants**: 504 mg/kg. LD_{LO}, **rabbit**: 70 mg/kg.

PENTACYNIUM

= *Presidal*

Antihypertensive. Ganglionic-blocking agent.

Untoward effects: Xerostomia, constipation, postural hypotension, nausea, vomiting, impotence, malaise, and blurred vision.

PENTADECYLCATECHOL

= *Tetrahydrouroshiol*

Untoward effects: An oleoresin in the plant genus *Toxicodendron* that includes ***poison ivy***, ***poison oak***, and ***poison sumac***, that is a frequent cause of contact dermatitis (cell-mediated delayed hypersensitivity). First time reactions usually occur 2–3 weeks after contact; subsequent episodes often occur within 8–72 h. Indirect exposure can occur from pet's fur, clothing, or burned plants. Reactions are rare in **children** < 1 year. The allergens in ***cashew nut*** oil, ***cardol***, and ***anacardic acid*** are immunochemically related to it. Cross-sensitization with ***p-tert-butyl catechol*** (q.v.) in ***paper*** used in heat-process duplicators, such as Thermofax machines.

IP, LD_{LO}, **mouse**: 50 mg/kg.

PENTAERYTHRITOL

Use: In resins, synthetic lubricants, and ***PETN***. Production and demand vary, but, currently, ~150 million lbs annually.

Untoward effects: Exposure to fumes causes eye and respiratory tract irritation.

PENTAERYTHRITOL DICHLOROHYDRIN

= *Dispranol*

Tranquilizer, muscle relaxant.

Untoward effects: Decreased prothrombin time.

PENTAERYTHRITOL TETRANITRATE

= *Angijen* = *Angitet* = *Angitrate* = *Cardiacap* = *Dilanca* = *Dilcoran-80* = *Hasethrol* = *Lentrat* = *Mycardol* = *Naptrate* = *Neo-Corovas* = *Niperyt* = *Nitropenta* = *Nitropenton* = *Pentafin* = *Pentanitrine* = *Penthrit* = *Pentitrate* = *Pentral 80* = *Pentraspan* = *Pentrite* = *Pentritol* = *Pentryate* = *Pentylan* = *Pergital* = *Perihab* = *Peritrate* = *Perityl* = *PETN* = *Prevangor* = *Quintrate* = *Rate* = *Subicard* = *Terpate* = *Vasodiatol*

Coronary vasodilator, military explosive.

Untoward effects: Occasionally transient headache, fixed-drug eruptions, anorexia, gastric pain, nausea, vomiting, diarrhea, palpitations, auricular fibrillation, vertigo or dizziness, skin flushing and sweating, postural hypotension, and occasionally cerebral edema.

TD_{LO}, **human**: 1.67 g/kg/8 years.

PENTAGASTRIN

= *Acignast* = *AY 6608* = *Gastrodiagnost* = *ICI 50,123* = *Peptavlon*

Gastric acid secretion stimulant. **Parenteral use, generally, SC. Has also been used as a snuff.**

Untoward effects: Flushing, tachycardia, thrombocytopenia, dizziness, light-headedness, syncope, hypotension, drowsiness, headache, fatigue, blurred vision, abdominal pain, nausea, vomiting, borborygmi, sensation of needing to defecate, chills, sweating, tingling in fingers, dyspnea, injection site pain, acute interstitial nephritis, bradycardia or tachycardia, regur tion of gastric tube, and increased peristalsis. Adverse effects occur readily with dosage > 6 μg/kg. ***Cimetidine*** decreases intrinsic factor output under basal or pentagastrin stimulation. Atrial fibrillation in 51-year-old **male** given ***calcium gluconate*** 2 mg/kg IV over 1 min and pentagrastin 0.5 μg/kg over 5 s period.

In lactating **cows**, IV use causes marked hypocalcemia and hypophosphatemia.

PENTAMETHYLMELAMINE

= *PMM*

Antineoplastic.

Untoward effects: Severe dose-limiting gastrointestinal and central nervous system toxicity.

PENTAMIDINE ISETHIONATE

= *Aeropent* = *Banambax* = *M&B 800* = *Nebupent* = *Pentacarinat* = *Pentam 300* = *Pneumopent* = *RP 2512*

Antiprotozoal. **Used IV, IM, and by aerosoling. *Lomidine* is the *mesylate* form.**

Untoward effects: Sudden fatal hypoglycemia in 52-year-old **male** with *Pneumocystis carnii* pneumonia ("PCP") after 4 mg/kg/day/5 days IV, then, 200 mg/day/9 days. Pancreatitis and hyperglycemia after IV in 38-year-old **male** with AIDS. Nephrotoxicity and hyperkalemia in 22 AIDS **patients** after 4 mg/kg/day IV or IM. Similar adverse effects in 37 **patients** (25–54 years) with AIDS. Bradycardia and hypotension in 22-year-old **male** after IV of 240 mg/day/2 weeks. Facial numbness in 21-year-old **male** with "PCP" after IV of 300 mg/day. Vein irritation, hematuria, toxic epidermal necrolysis, raised ***creatinine***, hyperuricemia, rashes, megaloblastosis, dizziness, headache, vomiting, dyspnea, formication, syncope, fecal and urinary incontinence, epileptiform twitchings, and facial edema. Leukopenia, cardiac arrhythmias, hypomagnesemia, hypocalcemia, cardiac arrest, thrombocytopenia, and anemia also occur. Rarely, anaphylaxis, acute pancreatitis, and fatal Herxheimer-type reactions. Occasionally confusion in the **elderly**.

After IM, toxic epidermal necrolysis, azotemia, increased blood urea nitrogen, hyperkalemia, hyperuricemia, and decreased renal function; abcess formation, ulceration, pain, and erythema at injection sites. Renal toxicity of 32% after IM and 14% after IV in 167 **patients** with "PCP". Hyperglycemia in 56-year-old **male** with "PCP" after IM of 4 mg/kg/day/2 weeks. Tachycardia and torsades de pointes in two **patients** with AIDS on IM or IV treatment with 4 mg/kg/day/13–20 days. Hypotension is more common after IM than IV injection. Nausea and tachycardia are frequent. Sciatic nerve damage has occurred.

Aerosoled causes nausea, vomiting, abdominal discomfort, anorexia, diarrhea, hypoglycemia, toxic epidermal necrolysis, eosinophilic pneumonia, pulmonary interstitial fibrosis, cough, wheezing, bitter taste, burning sensation in throat, increased blood urea nitrogen, increased serum ***creatinine***, severe pancreatitis, digital necrosis, disseminated and extrapulmonary "PCP", bronchial bleeding, hepatitis, urticarial rashes, and conjunctivitis. Bronchospasm has occurred in **nurses** and **others** in the same room with **patients**, due to their use of aerosoled material.

The *mesylate* form has caused fatal acute pancreatitis in a 49-year-old **male** after 200 mg/day IM; fatal pancreatitis in "PCP" 27-year-old **patient** after 4 mg/kg/day/12 days IM. It has caused megaloblastic bone marrow changes and renal toxicity.

All routes of administration inhibit dihydrofolate reductase, preventing folate's metabolism into an active form. IV pentamidine increases ***didanosine's*** distribution into pancreas and muscle, but use with ***didanosine*** increases renal clearance of pentamidine. Caution must be used in treating of **diabetics** with it.

TD_{LO}, **woman**: 6 mg/kg/day.

PENTANE

Use: In artificial ice manufacturing, as an anesthetic, solvent in shoe factory glues, in aerosol propellents, in foam manufacturing, and in low-temperature thermometers.

Untoward effects: In **humans**, irritating to eyes, nose, and skin; drowsiness, and unconsciousness. **People** appear to show no symptoms from 5,000 ppm for moments. OSHA sets exposure limit at 1,000 ppm TWA and immediate danger to life or health of 1,500 ppm.

LC_{LO}, **human**: 130,000 ppm; TC_{LO}, **human**: 90,000 ppm/5 min.

Narcotizing to **animals** (90,000–120,000 ppm/5–60 min in **mice**), defatting effect on skin; eye, skin, nose, and respiratory tract irritation, and rapidly fatal to **mice**. Lethal to **rats** at > 6,000 ppm and **mice** at 128,200 ppm/37 min. Delayed axonal degeneration in exposed **rats**.

1-PENTANETHIOL

= *n-Amyl Mercaptan*

Use: An odorant in gas lines to detect leaks.

Untoward effects: Inhalation leads to eye, nose, throat, respiratory tract, and skin irritation; nausea, dizziness, vomiting, diarrhea, and dermatitis. National Institute for Occupational Safety and Health upper exposure limit of 0.5 ppm.

Inhalation LC_{LO}, **rat**: 2,000 ppm/4 h.

PENTAPIPERIDE METHYLSULFIDE

= *Crylène* = *Hycholin* = *Quilene*

Anticholinergic, antispasmodic.

Untoward effects: Xerostomia, difficult micturation, constipation, nausea, visual blurring, and tachycardia. A limited number of reports of headache, loss of libido, cutaneous burning sensation, allergic reaction, severe vomiting, and obstructive uropathy. Avoid use in **patients** with glaucoma, gastric retention or obstruction, urinary bladder neck obstruction, prostatic hypertrophy, megaesophagus, stenosing peptic ulcer, coronary insufficiency, tachycardia, and arrhythmias that can be aggravated by vagal blockade.

High dosage (50–150 mg/day) to **rats**, **mice**, and **rabbits** showed no teratogenic effects.

PENTAQUINE

= *SN 13,276*

An *aminoquinoline*, once used as an antimalarial.

Untoward effects: Clinically significant hemolytic anemia in **people** and in a **fetus** with glucose-6-phosphate dehydrogenase deficiency. Anemia, methemoglobinemia, leukocytosis, leukopenia, and neutropenia also reported.

Causes methemoglobinemia in the **dog**, but not in the **monkey**. Use with ***quinine*** markedly increases its toxicity in the **rat** and **monkey**.

IP, LD_{50}, **mouse**: 150 mg/kg.

PENTASTARCH

= *Pentaspan*

Use: As a blood volume-expander and in harvesting leukocytes in leukopheresis.

Untoward effects: Hypersensitivity reactions (wheezing, urticaria, hypotension, and anaphylaxis). In leukapheresis, headache, diarrhea, nausea, weakness, insomnia, fatigue, dizziness, edema, acne, paresthesia, chills, nasal congestion, and tachycardia occasionally reported.

PENTAZOCINE

= *Fortral* = *Fortwin* = *Liticon* = *NSC 107,430* = *Pentagin* = *Sosegon* = *Talwin* = *Win 20,228*

***Narcotic, analgesic.* The *hydrochloride* and *lactate* are also used. Oral, IM, and IV with *morphine*-like adverse effects, but without its strong addicting qualities. A *benzomorphan* derivative.**

Untoward effects: Respiratory depression in 1% (3% in **neonates**), hallucinations and disorientation in 0.1%.

Bronchospasms in an **asthmatic** with ***aspirin*** sensitivity; apnea in **some** after IV. Acute pulmonary edema immediately after 30 mg IV in a **patient** with recent myocardial infarction. Bradycardia, miosis, and coma reported. Therapeutic serum levels are 0.01–0.06 mg%; 0.2–0.5 mg% is toxic, and 1–2 mg% has been lethal. Others report these values as 50% less. Despite statements by some to the contrary, drug dependence reported in **many**, and severe withdrawal symptoms (abdominal cramps, depression, insomnia and headache, nausea, vomiting, yawning, sneezing, lacrimation, diaphoresis, paresthesia, agitation, tremors, itching, rhinorrhea, anorexia, restlessness) within 12–24 h of last dose, peaking at 48–72 h and up to 10 days later in **some**. Some cases of abuse have been reported, including IV injection of crushed tablets, with resultant bacterial endocarditis, pulmonary embolism, and pulmonary edema. Winthrop Laboratories was found liable in a Texas Superior Court in the death of a **person**, for misrepresentation in their advertizing that the drug would not cause physical dependence. That **patient's** long-term therapy was based on that assumption. Tolerance to it has developed. It is a habituating drug with addiction potentials. **Abusers** have accidentally injected **themselves** IA, requiring amputation of fingers. A **neonate** born to a **mother** who took 50 mg tid throughout **her** pregnancy showed withdrawal symptoms, recovered after treatment, but died of sudden infant death syndrome at 3 months of age. Adverse effects of withdrawal in **neonate** include irritability, tremors, high-pitched cry, hyperactivity, diarrhea, insomnia, poor sucking reflexes, lacrimation, respiratory alkalosis, hiccups, sneezing, and seizures. Can cause **neonatal** respiratory depression after repeated administration. Agranulocytosis after 1–3 months ingestion. Eosinophilia in 3/20 after 8–10 months of treatment. A few case reports of neutropenia, thrombocytopenia, leukopenia, and non-thrombocytopenic purpura reported. A case report of aplastic anemia and fatality in 77-year-old **female** after 90 tablets of ***Fortalgesic®*** (***pentazocine*** 15 mg, ***acetaminophen*** 550 mg) over an 18 day period. Tachycardia, urinary retention, dyspnea, and occasionally hypotension reported. Hypotension within 30 min in a 57-year-old **female** after 30 mg IM, and sudden hypertension in a 56-year-old **female** within minutes after 30 mg IM. Increased pulmonary arterial pressure after IV of 30 or 60 mg in 15 **patients**; reduced cardiac output in another group of 13 **females**. Grand mal seizures reported within minutes after IV in several **patients**. Numerous case reports of visual hallucinations and disorientation, feelings of formication, dizziness, vertigo, tachycardia, anxiety and fear, sweating, chills, flushing, auditory hallucinations, delirium, blurred vision, nystagmus, diplopia, miosis, insomnia, syncope, constipation, tremors, tinnitus, paranoia, nausea, vomiting, xerostomia, cramps, leukopenia, taste alterations, and weakness. Muscle stiffness in a 50-year-old **male** after 50 mg qid orally. Fibrous myopathy in IM **abusers**. Fever (up to 106 °F) in a 26-year-old **male** after oral use. Acute urinary retention in a 32-year-old **male** after 120 mg IM during 48 h. Pruritus, toxic epidermal necrolysis, facial edema, porphyria, scleroderma-like reactions, indurations, and depressions at IM sites are among the uncommon skin adverse effects precipitated by it. Can exacerbate any form of porphyria. Possibly significant association with cancers of the mouth floor, pharynx, and lungs in 70/1673 **patients**. Jaundice has been reported rarely. Adverse effects in > 20% of **patients**. Can cause false positive or elevated test results in amylase tests for pancreatitis. Fatal cases usually had blood levels of 3.4–21 mg/100 ml and toxic levels reported at 0.7–3.0 mg/100 ml. Metabolized more rapidly in **smokers** and **city-dwellers** than in **non-smokers**.

Drug interactions: Potentiated by ***alcohol***, miscellaneous ***sedatives, sleep-inducers, tranquilizers***, and, possibly, ***monoamine oxidase inhibitors***. Use with ***methadone*** can precipitate an immediate withdrawal reaction; ***acetaminophen*** and ***mexiletine*** decrease its absorption; and by its strong inhibition of gastric emptying, it decreases the rate of ***acetaminophen*** absorption. May potentiate ***central nervous system depressants***, and is physically incompatible with a number of barbiturates in solution.

LD_{LO}, human: 18 mg/kg.

In **dogs**, **cats**, **goats**, **ponies**, and **swine**, varying degrees of sialosis, mydriasis, pollakiuria, emesis, polypnea, and central nervous system depression after IM use. In **horses**, transient incoordination, ataxia, localized muscle twitching, and nervousness, particularly at IM and IV dosing above recommended levels. Tachycardia, increased arterial blood pressure and cardiac output, and diaphoresis, followed by euphoria in **horses** at recommended doses. Detectable in **horse** urine 5 days after treatment. Potentiated by ***monoamine oxidase inhibitors*** in **animals**. **Monkeys** have demonstrated weak withdrawal symptoms. Prolonged clearance time of nearly 1 week in treated **horses**. Addictive properties have been demonstrated in **dogs**. Use only the lowest recommended dose, IV, for **horses**, to decrease ataxia and/or nervousness or euphoria. Follow with similar dose, IM, in 15–20 min, if needed. Avoid use in food-producing **animals**. Accidental intra-arterial, instead of anterior vena cavae injections, have caused deaths in **swine**.

LD_{50}, **rat**: 340 mg/kg; IV, 21 mg/kg; SC, 174 mg/kg; **mouse**: 335 mg/kg; IV, 22 mg/kg.

Many drug combinations with it (***P***), particularly with ***tripellenamine*** (***T***) = ***Ts and Bs*** or ***Ts and Blues*** cause many other adverse effects. ***Ts and Bs***, an important street drug of abuse causing euphoria, disorientation, hallucinations, paranoia, irritability, aggressiveness, severe anxiety, suicidal states, and **fetal** growth reduction.

PENTETIC ACID

= *DTPA*

***Chelating agent.* Mostly in hemosiderosis. The *calcium trisodium salt* is used in chelating *plutonium*.**

***Untoward effects*:** IV causes mild nausea, diarrhea, and skin rash.

IM causes headache, anorexia, pruritus, conjunctivitis, stomatitis, gingivitis, slight mental confusion, and painful injection sites. Although results were excellent in ***lead*** poisoning, adverse effects were too great.

PENTOBARBITAL

= *Diabutal* = *Ethaminal* = *Halatal* = *Mebubarbital* = *Mebumal* = *Mebumalon* = *Napental* = *Narcoren* = *Nebralin* = *Nembutal* = *Neodorm* = *Pentobarbitone* = *Pentobrocanal* = *Pentol* = *Pentone* = *Pentosol* = *Praecicalm* = *Sagatal* = *Somnopentyl* = *Sopental* = *Sopentyl*

A common street name is *"Yellow Jackets"*.

***Sedative, hypnotic, anesthetic (IV), euthanasic (IV).* Also used orally. Rarely, IM.**

Untoward effects*:** Addiction, bradycardia, syncope, hypotension, slight decreased intraocular pressure, nystagmus, drowsiness, slurred speech, respiratory depression, central nervous system depression, headache, coma, hyperreflexia, Babinski's sign, and Stevens–Johnson syndrome. Statistically significant association with lung cancer in chronic **users**. Acute renal failure with muscle necrosis and calcification in a 35-year-old **male** from oral overdose. Withdrawal psychosis resembling ***alcohol (q.v.) withdrawal delirium tremens. Use is associated with frequent falls and fractures in the **elderly**. Neutropenia, thrombocytopenia, purpura, and isolated cases of anemias and eosinophilia reported. During IV anesthesia, salivation (1%), vomiting (1% during pre-op), pyrexia, and apnea. Passes into breast milk. Poisoning in a comatose 27-year-old **male** with sweat gland necrosis, hyperthermia, and erythematous, indurated, cutaneous lesions and bullae. Commonly used in suicide attempts. Blood concentrations in fatal cases have been extremely variable with a range in levels from 1.9–157 mg/l and means of 13.8, 19.5, and 25.2 in several reports. Has also caused poisoning in **neonates** from accidental ingestion. Lactic acidosis after IV use probably due to the ***propylene glycol*** solvent. Fatalities have occurred after IP, oral, and IV use by **people** of euthanasia solutions, ***Toxital***®, ***Lethal***®, and ***Beuthanasia***® (***pentobarbital*** and ***phenytoin***).

LD, **human**: ~1.5 g; IV, 7.1 mg/kg; LD_{LO}, **human**: 36 mg/kg.

Drug interactions*:** Stimulates metabolism of ***alprenolol, oral anticoagulants, meprobamate, quinidine, and ***vitamin C***. ***Diazepam*** decreases its serum levels. ***Alcohol*** intoxication can double its half-life; ***chloramphenicol, chlorpromazine, cimetidine, ginsenoside, imipramine***, and ***tetracycline*** potentiate its effects.

Extravasation may occasionally cause local irritation and necrosis, as usual solutions have a pH of about 10. In general, ***barbiturates*** should be used with caution in the presence of severe liver and/or renal disease. In these cases, pentobarbital is best used in conjunction with other drugs, which markedly decrease its requirements and are detoxified by other mechanisms. Watch for bloat or salivation in **ruminants**. ***Procaine*** potentiates its effects in **guinea pigs** and ***antihistamines*** may potentiate it in **rats**. IV ***glucose*** during recovery from anesthesia has also potentiated its action in **horses**, **cats**, and **dogs**. A predictable marked leukopenia has been observed in anesthetized **dogs** (20% of controls 90 min after induction) and **rats** by Dr J.G. Graca and this author, invalidating considerable data collected from **animals** under this anesthetic. Antidiuretic effects reported in **humans**, **dogs**, and **roosters**. Analgesic effects in **rabbits** are poor. Dosage 50% above recommendations can be toxic, especially with subclinical respiratory problems. Reduce dosage in injured, stressed, and **wild animals**. Increased sleeping time with it, but not with non-metabolized ***barbital***, has been demonstrated in **mice** with various tumors. Also, blood from **rats** with Walker carcinoma 256 contains factor(s) which decrease its metabolism by the liver. Pentobarbital metabolism decreases and toxicity increases in protein-deficiency **rats** of both sexes, but toxicity was more marked in the males. IV use in **cats** causes transient hypotension and slight tachycardia. **Rabbits** vary widely in "anesthetic" requirements (7–83 mg/kg reported). Preanesthesia dosage of **dogs** with ***epinephrine*** decreases anesthetic requirement by 25%; when ***epinephrine*** was given 30 min after anesthetic induction, anesthesia duration is prolonged by about 40%; yet, administration on recovery fails to re-anesthetize.

LD_{50}, **rat**: 129 mg/kg; LD_{LO}, **rabbit**: 175 mg/kg; IV, LD_{50}, **rabbit**: 33 mg/kg.

PENTOLINIUM TARTRATE

= *Ansolysen* = *M&B 2050A* = *Pentilium*

***Ganglionic blocker, antihypertensive.* Oral, IV, or IM.**

***Untoward effects*:** Urinary retention, paralytic ileus, xerostomia, paralysis of pupillary reflex, diarrhea, constipation, shock, coronary thrombosis, transient hemiplegia, decreased libido, brain ischemia, postural hypotension, chronic pulmonary fibrosis, and malignant hypertension.

LD_{50}, **rat**: 890 mg/kg; **mouse**: 545 mg/kg; IP, **mouse**: 81 mg/kg; IV, **mouse**: 65 mg/kg.

PENTOSAN POLYSULFATE

= *Cartrophen* = *CB 8061* = *Elmiron* = *Fibrase* = *Fibrezym* = *Hemoclar* = *Sodium Pentosan Polysulfate* = *SP 54* = *Thrombocid* = *Xylan Hydrogen Sulfate*

Heparinoid, anti-inflammatory. **Semi-synthetic. Used IM, oral, SC, or IV.**

Untoward effects: Local lipodystrophy 6 months after 6 months self-medication SC. Many case reports of alopecia after IV and IM. Nasal and gingival bleeding and increased prothrombin time. Nausea, diarrhea, and abdominal pain reported. Allergic symptoms include thrombocytopenia, after oral use.

In **cats**, weight gain, poor hair coat, and sedation.

PENTOSTATIN

= *CI 825* = *CL 67,310,465* = *2-Deoxycoformycin* = *Nipent* = *NSC 218,321*

Antineoplastic, adenosine deaminase inhibitor. **IV use.**

Untoward effects: Nausea, vomiting, rash, and fever. Delayed toxicity leading to myelosuppression, renal and liver dysfunction, hemolysis, conjunctivitis, central nervous system depression, respiratory failure, photophobia, arthralgia, and myalgia. A fatal hypersensitivity reaction in a 70-year-old **male** after 4 mg/m^2 followed 3 days later by ***allopurinol*** 300 mg/day. A week after the start of treatment he developed fever, hypotension, oliguria, and death.

PENTOXIFYLLINE

= *Azupentat* = *BL-191* = *Durapental* = *Oxpentifylline* = *Rentylin* = *Torental* = *Trental* = *Vazofirin*

Vasodilator. **Increased pliability of red blood cells with maximum benefit after 120 days.**

Untoward effects: Dyspnea, hypotension, edema, xerostomia or sialism, flatulence, belching, headache, dizziness, chest pain, cholecystitis, anorexia, constipation, nausea, vomiting, conjunctivitis, blurred vision, scotoma, earache, bad taste, leukopenia, sore throat and/or swollen neck glands, anxiety, confusion, dizziness, seizures, depression, rash, urticaria, flushing, pruritus, angioedema, and brittle fingernails. Abused by **athletes**. A 67-year-old **female**, after a single 400 mg dose, developed bleeding duodenal ulcer. Bradycardia, first and second degree heart-block in a 22-year-old **male**. In a 70-year-old **female** and an 80-year-old **female**, 400 mg twice daily or tid/3–10 weeks led to aplastic anemia. Secreted into breast milk. It induced re-emergence of panic attacks in a 70-year-old **male**. The possibility of rare association with it exists for angina, arrhythmias, tachycardia, pancytopenia, thrombocytopenia, purpura, leukemia, and jaundice.

Drug interactions: Related chemically to ***theophylline***, and use may increase the toxicity of the latter. Serum sickness (fever, skin rashes, malaise, arthralgias, lymphadenopathy, and gastrointestinal upsets) in a 62-year-old **male** when taking it with ***cefoxitin***. A strange case of contact dermatitis to a ***leather*** watchband in a 63-year-old **male** only when **he** was ingesting 400 mg tid.

Cimetidine decreases its clearance by 37% in **rats**.

LD_{50}, **mice**: 1.385 g/kg.

PENTYLENETETRAZOL

= *Cardiazol* = *Cenalene-M* = *Cenazol* = *Coranormol* = *Corasol* = *Corazol* = *Corvasol* = *Deumacard* = *Gewazol* = *Korazol* = *Leptazol* = *Leptazolum* = *Metrazol* = *Neo-Gerastan* = *Pentazol* = *Pentetrazol* = *Phrenazol* = *PTZ* = *Ventrazol*

Central nervous system stimulant. **Parenteral or oral use.**

Untoward effects: Preconvulsive anxiety, epileptiform convulsions with increased blood pressure, pallor, and unconsciousness. Senility symptoms exaggerated, nausea, dizziness, drowsiness, and acute attacks of porphyria in **some**. Convulsive doses in **man** can raise arterial pressures 100 mmHg.

LD_{LO}, **human**: 50 mg/kg; TD_{LO}, **human**: 2.86 mg/kg; IV and SC, TD_{LO}, **human**: 1.43 g/kg.

Overdosage may cause convulsions. Use with caution in **animals** that have had epileptiform convulsions. Epileptic **monkeys** have lower thresholds for its convulsive effects. Increased environmental temperatures (> 26 °C or 79 °F) markedly increase its convulsive effects and lethality for **rats**. Induced convulsions in **guinea pigs** are associated with lower brain ***ascorbic acid*** levels. Epileptogenic in **cats**, **guinea pigs**, **mice**, **rabbits**, and **rats**. ***Chloral hydrate*** decreases its metabolism.

LD_{50}, **rat**: 140 mg/kg, SC: 85 mg/kg, IP: 62 mg/kg, IV: 45 mg/kg, **mouse**: 88 mg/kg, SC: 97 mg/kg, IP: 71 mg/kg, IV: 51 mg/kg; IV, **rabbit**: 69 mg/kg; LD_{LO}, 40 mg/kg.

p-tert-PENTYLPHENOL

= *p-tert-Amylphenol*

Germicide, fumigant. **Used in organic synthesis, manufacturing oil-soluble resins, and pesticides.**

Untoward effects: Skin pigmentation reported in hospital **workers**. Cross-sensitivity with other ***phenols***.

LD_{50}, **rat**: 1.8 g/kg; topical, **rabbit**: 2 g/kg.

PEONY

= *Paeonia sp.*

Use: The root as an ecbolic, emmenagogue, and uterine hemostatic, apparently, due to ***peonine***, an alkaloid.

Untoward effects: An 8-month-old **female** vomited 1 h after eating three or four flower petals. The roots contain an acrid juice claimed to cause paralysis.

PEPPERS

Over 700 species have been identified. The term pepper usually refers to *Black Pepper*.

Untoward effects: ***Black Pepper*** = ***Piper nigrum*** = ***hu chiao*** = ***Kalimirch***. ***Piperine***, an alkaloid and close chemical relative, of ***safrole*** (q.v.) constitutes about 10% of it by weight, and also small amounts of ***safrole***. Has some hallucinogenic effects. Also contains ***myristicin*** (q.v.), ***nordihydrocapsaicin, homocapsaicin, N-isobutyl trienamides, N-isobutyl dienamides, piperettine, piperoleine***, and ***piperyline***. Overuse may be a factor in recurring prostatitis. ***P. nigrum*** is the source of commercial pepper and primarily from Java. Black pepper is from the unripe fruit, and the ripe fruit without its outer pericarp is the source of white pepper. The plant itself is toxic and a member of the ***nightshade*** family, as are ***eggplants*** and ***tomatoes***. **Human** consumption is ~2 mg/kg/day. Tumors in **mice** after ingestion of 160 mg/kg/day/3 months. Has caused cystic ovaries in **rats**. Crude and purified extracts are effective insecticides. Commercial material has often been adulterated with similar-looking black seeds of *Carica papaya*.

Cayenne Pepper = ***Red Pepper*** or ***Chili*** contains ***capsaicin*** (q.v.) and is a strong gastrointestinal and skin irritant. Fiery peppers and/or their juice can also cause eye burns on contact. Its ***capsaicin*** is 70 times as "hot" as the ***piperine*** of ***Black Pepper***.

Green Pepper = ***American Pepper*** = ***Bisbâs*** (Yemen) = ***Capsicum frutescens*** = ***Polofifisi*** (Tonga) contains ***capsaicin***. Rotten ones have been ingested for psychedelic effects by the "**hippie**" generation. Allergic cross-sensitivity with ***tomatoes***. Contains large amounts of ***oxalates***.

Jalapeno Pepper: After handling, **one** must be careful not to touch mucous membranes, as the act can cause severe burning. A case of **child** abuse with it in Virginia has been reported. Coughing and gagging in **personnel** grinding **macaw** beaks that had eaten peppers.

White Pepper = ***Piper album***. High levels given to **rats** increased the bioavailability of ***verapamil*** twofold.

Salmonella and molds have been found as contaminants. Also see ***Capsaicin, Capsicum, Chillies***, and ***Kava***.

PEPPERMINT

= *Brandy Mint* = *Hoja de Menta* = *Hortelà-Pimenta* = *Lamb Mint* = *Mentha piperita*

U.S. annual production of the oil is ~4–5 million lbs. Used in dentifrices, chewing gum, confectionaries, and pharmaceuticals. The flowers are used as a fragrance and the leaves for making *tea* and flavoring in other *teas* and drinks. Natural peppermint oil contains ~1,500 identifiable compounds. The commercially cultivated forms contain more oil than the wild form.

Untoward effects: Can stimulate esophageal reflux by decreasing lower esophageal sphincter pressure. Contact dermatitis occurs. **People** taking it in capsules should not take it with meals or if **they** suffer from achlorhydria, to reduce heartburn. An acute allergic reaction in the mouth, throat, and neck of a **person** was traced to the oil in a toothpaste used. Recurrent muscle pain in two **patients** after consuming large quantities of peppermint oil-flavored confectionery. Idiopathic auricular fibrillation in **patients** addicted to sucking on peppermints. Peppermint ***tea*** reported as a cause of fixed-drug eruptions. Peppermint water is a consistent cough-provoker in man.

Oil LD_{50}, **rat**: 4.4 g/kg.

PEPPERONI

Untoward effects: An unusual circular-shaped lesion on the palate, due to "pizza pepperoni burn" has been reported.

PERACETIC ACID

= *Peroxyacetic Acid* = *Persteril*

Disinfectant, sporicide, oxidizing agent.

Untoward effects: Strong skin and eye irritant; corrosive to skin and a strong oxidizing agent. Up to 0.5% appeared to be non-toxic on **patient's** dermatoses. Corrodes ***rubber*** and requires stainless steel equipment.

LD_{LO}, **human**: 50 mg/kg.

Spray of 2–4% on **cattle** ringworm lesions led to slight local irritation. Experimental **rats** and **dogs**, exposed by passing through disinfected lock entries, had increased blood pressure and tachycardia for 25–45 min after exposure. Prolonged contact of 1–3% solution caused skin irritation in **rabbits** and **rats**; inhalation of 45% solution was highly toxic to both. Repeated spraying of 1.5% solution on skin of **sows** led to tachypnea, struggling, lacrimation, coughing, reddening of skin with fissures and scaly crusts, hyperkeratosis, and parakeratosis.

Minimum lethal dose, **rat**: 315 mg/kg. LD_{50}, **rat**: 1540 mg/kg; **guinea pig**: 10 mg/kg; topical LD_{50}, **rabbit**: 1.4 g/kg.

PERAZINE
= *P 725* = *Taxilan*

Psychotherapeutic, neuroleptic. **Usually oral, occasionally, IM. A *phenothiazine*.**

Untoward effects: Dyskinesias in 5%, occasionally corneal opacity, hypotension, tendency to collapse, vertigo, thirst, salivation, increased diuresis, allergic dermatitis, photosensitization, leukocytosis, and leukopenia. A case of lupus erythematous after 5 years of treatment with non-specified dosage. Acute neuropathies, 1–4 days after sun exposure in five 23–68-year-old **patients** that included facial weakness and burning sensations of fingers, toes, arms, and legs after 50–400 mg/day for 1–24 months. Therapeutic serum levels are 0.025–0.1 mg/l and 0.5 mg/l is toxic.

LD_{50}, **mouse**: 5 g/kg; IV, 70 mg/kg.

PERCHLORIC ACID

Untoward effects: Can be corrosive to skin and mucous membranes; fumes are irritating to nose, throat, and skin. Can cause abnormal thyroid function. Explodes without warning on contact with some organic material. ***Perchlorates*** in ground water inhibit ***iodide*** uptake decreasing thyroid hormone production.

Feeding ***perchlorate*** to **rats** blocks ***iodine*** metabolism, and can lead to neoplasia in **rats**.

PERCHLOROMETHYL MERCAPTAN
= *Clairsit* = *PCM*

Use: Military poison gas and fumigant versus insects.

Untoward effects: Fumes cause eye, skin, nose, and throat irritation. Lacrimation, coughing, dyspnea, chest pain, and coarse râles. Contact or ingestion leads to vomiting, pallor, tachycardia, acidosis, anuria, and liver and kidney pathology. National Institute for Occupational Safety and Health and OSHA exposure limits of 0.1 ppm or 0.8 mg/m^3 TWA.

Inhalation TC_{LO}, **human**: 45 ppm.

Inhalation LC_{LO}, **rat**: 10 ppm/6 h; **mouse**: 9 ppm/3 h.

PERCHLORYL FLUORIDE
= *Chlorine Oxyfluoride*

Sweet smelling gas. Shipped as a compressed liquid.

Use: Insulator for high voltage systems, as an oxidizing agent, and in organic synthesis.

Untoward effects: Respiratory tract irritant after inhalation. Frostbite from liquid contact. National Institute for Occupational Safety and Health and OSHA set upper exposure limits at 3 ppm TWA. Immediate danger to life or health: 100 ppm.

Methemoglobinemia, cyanosis, weakness, pulmonary edema, pneumonitis, and anoxia in **animals**.

Inhalation LC_{LO}, **rat**: 2,000 ppm/40 min; Inhalation LC_{50}, **mouse**: 630 ppm/4 h.

PERGOLIDE MESYLATE
= *Celance* = *LY 127,809* = *Parkotil* = *Permax*

Antiparkinsonism; dopamine receptor agonist. **Used in treating prolactinomas. Synthetic *ergoline* derivative.**

Untoward effects: Arrhythmias, orthostatic hypotension, confusion, hallucinations, paranoia, anxiety, depression, sleepiness, hypersexuality, priapism, spontaneous ejaculation, and ocular hypotension. **Patients** have developed erythromelalgia after long-term treatment. Retroperitoneal fibrosis in an 83-year-old **female** with Parkinson's disease after 3 mg/day. Some **patients** can safely use it when **they** cannot tolerate ***bromocriptine***.

Drug interactions: Possible induced hypotension when given with ***lisinopril***. Significant increase in dyskinesias with the addition of ***warfarin*** to therapy. Since it is 90% bound to plasma proteins, interactions may occur with ***erythromycin*** and other drugs that may displace it from binding sites.

PERHEXILENE

Coronary vasodilator, diuretic, antianginal. **A non-selective *calcium channel blocker*.**

Untoward effects: Peripheral neuropathy, papilledema, hypoglycemia, lipidosis, weight loss, proximal myopathy, and numerous case reports of hepatic cirrhosis and necrosis. Epigastric pain, nausea, and vomiting were common. Other significant problems include dizziness, headache, lethargy, incoordination, tremors, palpitations, increase in SGOT and alkaline phosphatase, paresthesias, insomnia, anorexia, blurred vision, diplopia, decreased libido, arthralgia, nervousness, dysuria, and increased intracranial pressure. Occasionally rashes.

Drug interactions: After ***paroxetine*** (q.v.) added for 6 weeks to its use in an 86-year-old **female**, perhexilene toxicity developed and **she** died.

TD_{LO}, **man**: 857 mg/kg, over a 40 week period.

LD_{50}, **rat**: > 7 g/kg; **mouse**: 4.37 g/kg.

PERILLA FRUTESCENS

= *Beef Steak Plant* = *Purple Mint* = *Wild Coleus*

Found near woody pasture areas.

Untoward effects: Ingestion of the plant or its hay, particularly when in the seed stage, causes acute interstitial emphysema and pneumonias in **cattle** with death losses of ~50%, due to its content of ketones. **Calves**, unfortunately, acquire a taste for it, possibly, due to ***perillartine*** in its essential oil. ***Perillartine*** is ~200–2,000 times as sweet as ***sucrose*** and used in Japan to flavor ***honey***, ***maple sugar***, and ***tobacco***. Its ketone is its main toxicant.

IP, LD_{50}, **rat**: 10 mg/kg; **mouse**: 2.5 mg/kg.

PERINDOPRIL *tert*-BUTYLAMINE

= *Acéon* = *Conversum* = *Coversyl* = *S 9490-3*

***ACE inhibitor, antihypertensive.* The *diacid* form is *perindoprilat*.**

Untoward effects: Cough, pancreatitis, pneumonitis, visual disturbances, and rarely, angioneurotic edema.

PERLITE

Untoward effects: Inhalation causes eye, skin, throat, and upper respiratory tract irritation. National Institute for Occupational Safety and Health upper exposure limit of 10 mg/m^3 (5 mg/m^3 for respiratory tract); OSHA: 15 mg/m^3 and 5 mg/m^3, respectively.

PERMANENT PRESS

= *Durable Press* = *Creaseproof Agents*

Untoward effects: Hypersensitivity reactions from skin contact include severe itching, burning sensations, and generalized skin eruptions from contact with treated clothing, bed sheets, etc. ***Formaldehyde*** (q.v.) is used in the manufacture of ***dimethyloldihydroxyethyleneurea***, ***dimethylolmethoxydimethylpropylene urea***, and which are sensitizing agents. Other curing resins agents are ***zinc fluoborat***, ***zinc nitrate***, etc.

PERMETHRIN

= *Atroban* = *Corsair* = *Delice* = *Dragnet* = *Ectiban* = *Elimite* = *Eksmin* = *Expar* = *FMC 33,297* = *FMC 41,655* = *Insectrin* = *Kafil* = *Lyclear* = *NIA 33,297* = *NIX* = *NRDC 143* = *Overtime* = *Permaban* = *Permectin* = *Perthrine* = *Pounce* = *PP 557* = *Pramex* = *Pulvex* = *Pynosect* = *Ridect* = *S 3151* = *Sanbar* = *SBP 1513* = *WL 43,479*

***Ectoparasiticide, pediculocide.* A synthetic *pyrethroid*.**

Untoward effects: Occasionally localized burning, stinging, pruritus, rash, erythema, edema, numbness and pain after topical use. May cause eye irritation and can be absorbed through the skin. An outbreak of urticaria occurred in **children** in one household due to airborne exposure.

Highly toxic to **fish**. Withdrawal period of 5 days required before slaughtering treated **hogs** for **human** consumption. Resistance to it has developed.

LD_{50}, **rat**: 410 mg/kg; skin minimum lethal dose, **rabbit**: 500 mg/24 h.

PERNETTYA FURENS

= *Hierba loca* = *Hush-hued*

In the *Ericaceae* or *Heath* family. Common in Chile.

Untoward effects: The fruit or berries are toxic if ingested in quantity leading to mental confusion and madness with intoxication similar to ***Datura's*** (q.v.). Sensations of cold, symptoms of drunkenness, and paralysis also occur.

Cattle have been poisoned by eating the leaves.

PERPHENAZINE

= *Chlorpiprazine* = *Chlorpiprozine* = *Decentan* = *Etaperazine* = *Ethaperazine* = *Fentazin* = *Perfenazine* = *Perphenan* = *PZC* = *Sch-3940* = *Trilafon* = *Trilifan*

Psychotherapeutic, tranquilizer, antidepressant. A phenothiazine–piperazine compound.

Untoward effects: Extrapyramidal effects are common. Lethargy in **infants** whose **mothers** received it before delivery. **Neonates** and **children** have shown neurological complications (rigidity with neck retraction, almost becoming opisthotonus, but with flexor Babinski responses in a 10-month-old **male**, 6 h after 1.6 mg), jaundice, blood dyscrasias, anemias, eosinophilia, agranulocytosis (rare), painful torticollis, and trismus. Risus sardonicus in a 26-year-old **female** after 6 mg twice daily/3 days. Blurred vision, diplopia, lens opacities, optic atrophy with prolonged high dosage, nasal congestion, xerostomia or sialism, headache, drowsiness, electrolyte imbalance, priapism, libidinous sexual drives, impotence or inhibited ejaculation in **males**, decreased sexual responsiveness in **females**, hepatic necrosis, hepatocellular jaundice, weight gain, pseudoparkinsonism, dystonia, dyskinesia, akathisia, painful spasms of facial muscles and legs, photosensitivity, urticaria, weakness, dizziness, agitation, purpura, nausea, paresthesia, hypotension, chromosome abnormalities, galactorrhea, rashes, systemic lupus erythematosus, anticholinergic effects (xerostomia, mydriasis, cycloplegia, tachycardia, urinary retention, decreased gastrointestinal motility, memory impairment), and neuroleptic malignant syndrome occur. An association with enlargement of pituitary gland has been shown. Increases protein-bound iodine in ~25% of **patients**. Causes false positive urinary ***bilirubin*** plasma ***catecholamine*** tests. Menstruation changes noted during first 6 months of treatment in

16/36 **females** and one **patient** lactated for a brief time. A 2-year-old **child** accidentally ingested 120–160 mg, causing drowsiness, followed by excitement, confusion, tachycardia, and pseudoparkinsonism. Therapeutic blood levels are ~0.5 μg% and toxic levels ~100 μg% or 0.001–0.01 mg/l and 0.05 mg/l, respectively.

LD_{LO}, **human**: 50 mg/kg.

Drug interactions: Cross-sensitivity reactions with other ***phenothiazines*** leads to cutaneous eruptions. Use with ***droperidol*** causes hypotensive reaction. Potentiates the activity of ***pargyline*** and ***reserpine***. Large doses decrease ***heparin*** effect. ***Imipramine, nortriptylline*** (with 55% increase in serum levels), ***paroxetine*** (with two to 13 times increase in serum levels), and probably ***amitryptilline*** metabolism is decreased by it, due to its enzyme inhibition and their serum levels are increased or prolonged by it. ***Haloperidol*** potentiates some of the adverse effects and led to fever, rigidity, convulsions, and death in a 17-year-old **female**. ***Disulfiram*** decreases its serum levels. It should NOT be used with ***organophosphates*** and/or ***procaine hydrochloride***, since, in **man** and **animals**, it can potentiate the toxicity of ***organophosphates*** and the activity of ***procaine hydrochloride***.

Sensitive **animals** may show excessive or prolonged depression or even psychic stimulation and restlessness (occasionally in **dobermans** or **greyhounds**). Detrimental effect on temperature regulation may be particularly dangerous in **marine mammals**. Mammotropic effects have been noted in **rats**. *Contraindicated* in hypotensive states, central nervous system depression, and in **horses**. SC injections should not be given as they cause severe local reactions. Dense granules in the eyes of treated **dogs** similar to those seen with ***chlorpromazine*** therapy have been noted. Although drowsiness is the most common reaction in **chimpanzees**, one developed parkinsonian-like syndrome with > 2.3 mg/kg. Others tolerated up to 4.4 mg/kg without ill-effects. In 1970, Schering Corporation recalled its injectable form from veterinary use because of central nervous system disorders, incoordination, and loss of balance in shipped **cattle** and **swine**.

LD_{50}, **rat**: 318 mg/kg; **mouse**: 120 mg/kg; **dog**: > 100 mg/kg; IV, **rat**: 36 mg/kg; **mouse**: 37 mg/kg; **dog**: 51 mg/kg.

PERSICARY

Untoward effects: In Argentina, ingestion of 400 g of white persicary leaves over an 8 week period by two 18-month-old **bullocks** led to anorexia, weight loss, normocytic and normochromic anemia, increased serum ***calcium*** and ***phosphorus***, cachexia, decreased skin elasticity, and severe mucoid degeneration of SC connective tissue. The syndrome is known as "enteque seco".

PERSIMMON

= *Diospyrus sp.*

Untoward effects: Although most phytobezoars occur in **people** who have had gastric surgery or severance of the vagus nerve, they are sometimes caused by persimmons. The material under the skin of the fruit and the unripe pulp contain a substance that reacts with stomach acid, forming an insoluble mass or bezoar.

D. tokoro is used as a piscicide in Japan.

PERTHANE

= *ENT 17,082 = Ethylan = Q 137*

***Chlorinated hydrocarbon*. Phased out of U.S. production by Rohm & Haas Company.**

Untoward effects: Nausea, vomiting, chloracne, and assorted skin rashes in agricultural **workers** on fruit and vegetable farms.

LD_{LO}, **human**: 5 g/kg.

Moderately toxic to **honeybees**. Toxic doses for **calves** are ~250–500 mg/kg and > 1 g/kg for **sheep**.

LD_{50}, **rat** and **mouse**: 6.6 g/kg.

PERUVOSIDE

= *Encordin*

Cardiac glycoside.

Untoward effects: Anorexia, nausea, vomiting, diarrhea, and cardiac arrhythmias.

PETIVERIA ALLIACEA

= *Anamu = Gully Root = Kojo Root*

A favorite folk-remedy plant in Central America and the West Indies. Decoctions are used externally and internally for most ailments and as an abortifacient. Contains *mustard oil* (q.v.) and *allyl sulfide* (see *garlic*).

Untoward effects: Ingestion causes mild bloating, frequent eructation, intermittent diarrhea, weakness, glomerulonephritis, polyuria, glucosuria, and abortions in **cattle**. Causes a ***garlic*** flavor in the meat and ***milk*** of **cows** that graze on it.

PETROLATUM

= *Petroleum Jelly = Soft Paraffin = Vaseline*

Untoward effects: Scalp soreness, burns, alopecia, destruction of hair follicles, and polytrichea from use with hot hair iron. Continued handling of unpurified material ("wax") causes acne, papillomas, and carcinomas. Allergic reactions are rare. "Vaselinomas" or "paraffinomas" after injection are now rare. Comedogenic and possible photosensitizer when used topically. ***Vaseline***® brand appears to cause less problems. Excessive ingestion significantly decreases absorption of fat-soluble ***vitamins***.

Comedogenic topically on external ear canal of **rabbit**. Slightly irritating to **guinea pig** and **rabbit** skin, but not to **miniature swine** or **humans** in cosmetic testing. **Mice** and **rats** are also susceptible to dermatitis from the "wax".

PEYOTE

= *"Big Chief"* = *Hikori* = *Hikuli* = *Huatari* = *Lophophora* = *Mescal Bean* = *'Mescal Button* = *P* = *Seni* = *Wokowri*

Psychomimetic. **Contains at least 15 active alkaloids, including *mescaline*, a hallucinogen. It is the tops of a narcotic cactus. Aztecs called the dried tops *teonanacatl*.**

Untoward effects*:** Intoxication is from all the alkaloids, and, therefore, the symptoms can be highly complex and variable. Incorrectly, ***mescaline is often used synonymously with the word peyote. Peyote consumed legally in a brew by Southwestern **Indians** in a religious ceremony leads to beautiful colors, hallucinations, anxiety, hyperflexia of limbs, mydriasis, hypertension, panic reactions, euphoria, irritability, and static tremors. It is habituating, but, apparently, does not cause addiction or physical withdrawal symptoms. Has an obnoxious taste. Diarrhea and nauseating in large amounts. Flushing, sweating, tachycardia, perceptual distortions, disturbed time and place relationships, and oliguria. Its ***lysergic acid diethylamide (LSD)***-like state usually lasts 6–12 h.

PFIESTERIA PISCIDA

A dinoflagellate found in coastal waters where fresh water mixes with salty seawater.

***Untoward effects*:** Its toxins can rapidly kill **fish** and cause bloody sores on others. Some exposed **watermen** or **fishermen** have developed itchy, red, dime-sized skin lesions, and complained of memory loss, diarrhea, and dyspnea. Often affects **people** within 2 weeks of exposure to estuarine water or certain aquaculture facilities.

PHACELIA sp.

= *Bluebells*

Common annual in deserts and semi-arid areas.

***Untoward effects*:** Many **people** develop a severe dermatitis with erythema, blistering, and pruritus, from contact with its glandular irritating hairs.

PHACETOPERANE

Antidepressant, anorexic. **The *levo* form = *Lidepran* is more active.**

***Untoward effects*:** Psychic disequilibrium, fluctuations in psychometric output, diminution of scholastic efficiency, and regression to infantile behavior in pediatric use.

PHALARIS sp.

Untoward effects*:** Sudden death, cardiac arrhythmias, acute neurological effects, and chronic neurological symptoms due to ***tryptamine alkaloids [*N,N-**dimethyltryptamine*** (q.v.), ***5-methoxydimethyltryptamine*** (q.v.), and ***5-HDMT***] present in this perennial pasture grass are known as Phalaris staggers in affected **sheep** and **cattle** in Australia, New Zealand, South Africa, South America, and the U.S.A. Head-nodding, ataxia, weakness, and tachycardia occur. Young, rapidly growing grass is, apparently, more toxic. Most reports refer to ***P. tuberosa*** = ***Ronpha Grass***.

PHALLOIDIN

Untoward effects*:** A principle toxin of the most poisonous ***mushrooms of the genera ***Amanita*** (q.v.), causing hematuria, proteinuria, gastroenteritis, dehydration, thirst, perspiration, dry tongue, nephritis, and jaundice in **man** and experimental **animals** within 15 min to 15 h. ***Phalloin*** is another thermostable toxin in ***Amanita*** causing similar symptoms.

IM, LD_{50}, **mouse**: 3.3 mg/kg; IP, 2 mg/kg.

PHANQUINONE

= *Ciba 11,925* = *Entobex* = *Phanquone*

Protozoacide.

Untoward effects*:** Nausea, giddiness, slight gastric burning sensation, and transient dizziness. Other adverse effects in treatment of amebiasis are often due to its combination with ***iodochlorhydroxyquin.

PHASEOLUTIN

= *Linamarin*

Untoward effects*:** A poisonous glycoside in cultivated ***flax, *Linum usitatissimum*, and **lima beans**, releasing ***hydrogen cyanide*** (q.v.) by enzyme action.

PHENACAINE

= *Holocaine*

Local anesthetic.

Untoward effects*:** Topically on the eye, it has caused occasionally keratoconjunctivitis medicamentora. Injected before dental surgery, it has caused vasomotor complications. As toxic as ***cocaine when given IV; twice as toxic when given SC.

PHENACEMIDE

= *Epiclase* = *Phacetur* = *Phenacetylurea* = *Phenurone*

Anticonvulsant.

Untoward effects: Dosage is limited by anorexia and insomnia effects. Idiosyncratic effects are hepatocellular jaundice, hepatitis and hepatic necrosis (2%), dizziness, nephritis, decreased weight, psychosis, rash, exfoliative dermatitis, nausea, vomiting, personality changes (17%), paranoia, blood dyscrasias (2%) [leukopenia, aplastic anemia, eosinophilia, thrombocytopenia, leukocytosis, neutropenia, and bone marrow depression], increased serum ***creatinine***, headache, and fever. High dosage has caused muscle weakness, ataxia, mydriasis, injected conjunctiva, and loss of reflexes. Can increase prothrombin time and give spuriously high urinary ***albumin*** test results.

LD_{50}, **rat**: 4 g/kg; **mouse**: 5 g/kg; **rabbit**: 2.5 g/kg. LD_{LO}, **cat**: 2 g/kg; **dog**: 2.5 g/kg.

PHENACETIN

= *Acetophenetidin* = *Femidine* = *Femina* = *Fenacetine* = *Phenidin* = *Phenin*

***Analgesic, antipyretic, and hydrogen peroxide stabilizer.* Its metabolite, *acetaminophen* (q.v.), is its therapeutic agent. Contains an *aniline* (q.v.) ring in its chemical structure.**

Untoward effects: The most common adverse effects have been renal (pyelonephritis, papillary necrosis, interstitial nephritis, and tubular damage, especially of the lower nephrons), particularly with overdoses or long-term use, resulting in uremia (with or without oliguria or anuria). Also chronic or recurrent urinary infections, as well as headache, dizziness, insomnia, acne, anemia, cyanosis, Stevens–Johnson syndrome, toxic epidermal necrolysis, bullous disorders, fixed-drug eruptions, depression, and enlarged spleen. It has probably been associated with increased incidence of renal and bladder cancer. Blood dyscrasias of leukopenia, agranulocytosis, thrombocytopenia, thrombocytopenic purpura, pancytopenia, neutropenia, aplastic and hemolytic anemias; methemoglobinemia, sulfhemoglobinemia, cyanosis, tachycardia, and tachypnea. Methemoglobinemia has been common in exposed **infants**, due to immaturity of the glucose-6-phosphate dehydrogenase enzyme system. Hemolysis commonly produced in glucose-6-phosphate dehydrogenase-deficient **people**. Severe methemoglobinemia in a **family** was shown to be due to a genetic defect causing failure of the drug to be de-ethylated. Abuse or addiction has caused many of the above symptoms, as well as toxic dementia. Hemolytic jaundice. Use can interfere with ***uric acid*** assays and give false positive urinary ***bilirubin*** or ***glucose*** test results. Methemoglobinemia in a young pregnant **woman** ingesting large doses, followed by the death of her **child** with degenerative encephalopathy a month after birth. Can color urine dark brown to black or red-purple, especially on standing. Cross-sensitization with ***meprobamate*** reported. Suspect as a cause of renal pelvis carcinomas. Therapeutic serum levels reported as 5–10 mg/l and toxic as 50 mg/l.

TD_{LO}, **human**: 1 g/kg and 57 mg/kg/day/47 years.

Drug interactions: ***Cigarette*** smoking or ingestion of ***charcoal-broiled beef*** increase its metabolism and its metabolism is decreased in **diabetics**. May interfere with effectiveness of ***oral contraceptives***. Anaphylaxis reported when it was used with ***propoxyphene***.

Long-term feeding of it to **dogs** failed to produce chronic interstitial nephritis, but caused lipofucin-type pigmentation in their livers and kidneys. Nephrotoxicity and increased incidence or urinary tract and nasal cancers in **rats**. Experiments with **dogs** and **rats** indicate that usual doses over limited periods have NOT caused renal damage. Contraindicated in **cats**, as they convert it to ***acetaminophen*** (q.v.).

LD_{50}, **rat**: 1.65 g/kg; **mouse**: 1.22 g/kg; **guinea pig**: 1.87 g/kg. TD_{50}, **rat**: 1.25 g/kg; **mouse**: 2.1 g/kg.

PHENAGLYCODOL

= *Ultran*

***Anticonvulsant, sedative, tranquilizer. Mephenesin*-like action.**

Untoward effects: Light-headedness, dizziness, sleepiness, nausea, and occasionally rashes. Gynecomastia in a 32-year-old **male** after 12–14 300 mg capsules/day/5 years. Additive depression with ***ethyl alcohol*** intake. Use gives spuriously high urinary ***steroid*** results.

LD_{LO}, **human**: 500 mg/kg.

Slight ataxia and weakness in treated **rats**. In **rats**, decreased metabolism of ***pentobarbital*** and increased metabolism of ***meprobamate***.

LD_{50}, **rat**: 832 mg/kg; **mouse**: 514 mg/kg.

PHENAMIDINE ISETHIONATE

= *Lomadine*

***Protozoacide.* In IM or SC treatment of babesiosis or piroplasmosis in elephants, cattle, horses, and dogs.**

Untoward effects: Avoid SC dosage in **horses** (40% solution), as it may produce extensive swellings with local necrosis and sloughing of skin. May cause a brief period of vomiting, shivering, and even central nervous system hemorrhaging in treated **dogs**.

PHENANTHRENE

***Coal tar* derivative.**

Untoward effects: Phototoxic in wavelengths of 3,500–4,000 Angstroms after skin contact causing "tar smarts", erythema, pricking sensation, and melanosis.

Nephrotoxic to **rats** after 150 mg/kg IP and can cause heart-block.

LD_{50}, **rat** and **mouse**: 700 mg/kg; IV, **mouse**: 56 mg/kg.

PHENARSAZINE CHLORIDE

= *Adamsite* = *Diphenylaminochlorarsine* = *DM*

***War gas*. Generally replaced by more harassing agents, such as σ-chloro-benzylidine-malonitrile.**

Untoward effects: Causes skin irritation and lesions, respiratory tract irritation, profuse watery nasal discharge, sneezing, coughing, severe chest, nose and sinus pain, nausea, vomiting, weakness, and marked depression.

Inhalation LC_{LO}, **human**: 30 g/m^3; TC_{LO}, 19 mg/m^3/3 min. LC_{LO}, **human**: 54 ppm/30 min.

Inhalation LD_{50}, **rat**: 14 mg/kg; **mouse**: 18 mg/kg; **guinea pig**: 2.4 mg/kg.

PHENAZOCINE HYDROBROMIDE

= *Narphen* = *NIH-7519* = *Prinadol* = *SKF-6574* = *Xenagol*

***Analgesic, narcotic*. A *benzomorphan*. IM or IV, occasionally, oral or sublingual.**

Untoward effects: Hypotension, light-headedness, nausea, vomiting, constipation, drowsiness, dizziness, pruritus, bradycardia, xerostomia, euphoria, diaphoresis, hot flashes, and dependency (less in **man** and **monkeys** than with ***morphine***), but greater respiratory depression than with ***morphine*** or ***meperidine***. Rarely, tachycardia and hallucinations. **Fetal** depression after **mothers** received it during labor. In **man** and **dogs**, 0.1 mg depressed respiration about the same as 5 mg ***meperidine***. Estimated minimum lethal dose: 100 mg.

IV, LD_{50}, **mouse**: 11 mg/kg.

PHENAZOPYRIDINE

= *Azo-Standard* = *Azo-Stat* = *Diazo* = *Gastrotest* = *Malophene* = *Phenazodine* = *Phenyl-idium* = *Pirid* = *Pyrazofen* = *Pyridacil* = *Pyridiate* = *Pyridium* = *Sedural* = *Uridinal* = *Urodine*

***Urinary antiseptic and analgesic, diagnostic agent*. An *azo dye*. Use estimated at ~100,000 lbs/year.**

Untoward effects: Gastrointestinal upsets, colic, headache, vertigo, rash, pruritus, hepatotoxicity (periportal), Heinz body-type hemolytic anemia and methemoglobinemia (especially after overdoses and in glucose-6-phosphate dehydrogenase-deficient **patients**). Stains clothing, red or orange-colored urine (red in acid pH), orange-red feces, gingival ulceration, and yellow pigmentation of sclera, nails, and skin. Hypersensitivity reactions of the liver are rare. Renal failure and bladder calculi after high dosage. A 56-year-old **male** developed aseptic meningitis with fever and confusion after an earlier course of treatment. Confirmed by rechallenge. May cause false positive test results for urine ***bilirubin***, phenol-sulfon-phthalein, sulfobromophthalein retention, and interfere with ***vanillylmandelic acid*** assays. Poisonings and attempted suicides with it have been reported.

In **animals**, methemoglobinemia is a potential risk of overdosage and/or prolonged therapy. Although indicated and effective in nephritis, cystitis, pyelitis, urethritis, and associated dysuria, avoid excessive doses in nephritis where the kidneys may be unable to eliminate the drug, causing secondary blood and liver pathology. May interfere with ***phenolsulphonphthalein*** (q.v.) kidney function test interpretation. Use may be associated with development of keratoconjunctivitis sicca in **dogs**. Severe hemolysis and icterus with Heinz bodies in red blood cells and death within 6 days in a **cat**, when recommended dosage was exceeded. Has caused colon and rectal adenomas and adenocarcinomas in male and female **rats**, and hepatocellular adenomas and carcinomas in female **mice**.

IP, LD_{50}, **rat**: 560 mg/kg.

PHENCYCLIDINE

= *Angel Dust* = *Angel Hair* = *Aurora Borealis* = *Busy Bee* = *Cadillac* = *CI 395* = *CJs* = *Crystal* = *Cyclone* = *Death Drug* = *D.O.A.* = *Dust Parsley* = *Elephant Tranquilizer* = *Elysion* = *Embalming Fluid* = *Fuel* = *Goon* = *Hawaiian Wood Rose* = *Hog* = *Horse Tranks* = *K.J. Krystal* = *KW* = *LBJ* = *Magic Mist* = *Monkey Dust* = *PCP* = *Peace Pills* = *Rocket Fuel* = *Scuffle* = *Sernyl* = *Sernylan* = *Sheets* = *Shirm* = *Smoke Dust* = *Snorts* = *Soma* = *Super Joint* = *Super Weed* = *Surfer* = *TAC* = *TIC* = *Trank* = *Wobble Weed*

***Anesthetic, analgesic, psychotropic, cataleptic, neuroleptic, and sympathomimetic*. *PCP* is also a designation for *pentachlorophenol* and *polychlorpinene*; *Peace Pills* also refers to a street mixture of *mescaline*, lysergic acid diethylamide (*LSD*), and *cocaine*; and *Killer Weed* is a street mixture of it with *cannabis* or other plant materials, such as *spinach*, *parsley*, and *broccoli*. *Ketamine* (q.v.) is a derivative. Limited use and heavy rate of abuse**

discouraged its manufacture ~30 years ago. Often snorted or smoked by addicts.

Untoward effects: Hallucinations, delirium, and dissociation in ~15–20%; dizziness, ataxia, syncope, incoordination, tachycardia, increased deep reflexes, nystagmus, miosis, and visual disturbances in ~20–30%; paranoia, depression, euphoria, catatonia, feeling of dying, antisocial attitudes, sleepiness, coma, jitteriness, xerostomia, maculopapular rash, pruritus, sweating, flushing, insomnia, anorexia, nausea, vomiting, and abdominal pain in ~10%. Violent aggressive and combative behavior, renal failure, hypertension, intracranial hemorrhage, malignant hyperthermia, sexual dreams, seizures, and acute rhabdomyolysis occur in **some**. Illicit IM self-administration by **addicts**. Illegally manufactured material may contain ***potassium cyanide*** (q.v.) and other impurities. A **prisoner** in an Illinois prison, apparently, swallowed four balloons, each containing 1 g of it, to avoid detection on a strip search; three burst in **his** stomach leading to death with cardiovascular and respiratory failure. A great deal of ill-egal street material has actually been **tetra-hydro-cannabinol**. Use during breast feeding leads to tremors and generalized spasticity in **infants**. Intoxication reported in at least 16 **children** < 5 years from accidental inhalation, when **individuals** near **them** smoked it. **Neonatal** symptoms after **maternal** abuse include jitteriness, hypertonicity, vomiting, and diarrhea. It is highly lipophilic and breast milk levels of 3.9 ng/ml were found in a 20-year-old **female** with plasma levels of 0.77 ng/ml. Teratogenic effects (abnormal facies, dislocated hip, and cerebral palsy) reported.

TD_{LO}, **human**: 71 mg/kg; LD_{LO}, **human**: 14 mg/kg; IV, TD_{LO}, **human**: 10 μg/kg.

Aggressive behavior in sleep-deprived **rats**. As with **man**, **rats** have shown catatonic stupor with repetitive orofacial and limb movements with overdoses. Hyperthermia at low doses and hypothermia at high doses in **rats**. Prolonged ataxia in treated **deer**. Acute behavioral effects in **dogs** are similar to **man's**. Salivation and muscle tremors are main adverse effects in **pigs**. Catatonia and intermittent clonic convulsions in treated **bears**. Hypothermia, shock, and death in *Macaca irus* **monkeys**. In **animals**, the cataleptic state often includes increased tonicity and occasionally convulsive movement, particularly in the neck and forelimbs, followed by trembling and spasms of the entire body. Fatal pulmonary edema and salivation from its use is prevented by concurrent use of ***scopolamine*** or ***atropine***. When used as a **wild animal** immobilizing agent with other drugs that prolong recovery time, care should be taken to prevent deaths due to adverse position or by being suitable prey for predators. IM use is recommened but, in some trials, SC and IP use has been equally effective. NOT for use in **animals** that may be used for food in any way (cases have been reported of serious mental disturbances in **men** who ate experimentally treated **chickens**). The drug is not degraded by cooking temperatures or in frozen foods. Concentration in newborn **pigs** is approximately ten times that of the **sow's**.

LD_{50}, **rats**: 140 mg/kg; **pigeon**: 75 mg/kg; **duck**: 237 mg/kg. IP, LD_{50}, **mice**: 55 mg/kg.

PHENELZINE

= *Nardelzine* = *Nardil* = *β-Phenylethylhydrazine* = *Stinerval* = *W 1544A*

Monoamine oxidase inhibitor, antidepressant.

Untoward effects: Drowsiness, insomnia, postural hypotension, and restlessness are common. Urinary retention, tremors, paresthesias, impotence, retarded or no ejaculation, priapism, delayed or no orgasm in **males** and **females**, xerostomia, nausea, constipation, weight gain, edema, anorexia or increased appetite, photosensitivity, tachycardia, and peripheral neuropathies occur occasionally. Rarely, muscle spasms. Lupus erythematous-like reactions, rashes, increased temperature, tinnitus, leukopenia, alopecia, motor hemiplegia and severe, and often fatal, hepatic necrosis and heptocellular carcinomas. Abnormalities of ***cephalin*** flocculation test noted. Breast enlargement in a 19-year-old **female** after 15 mg tid/3 months. Fatal disseminated intravascular coagulation in a 31-year-old **female** after an overdosage of 450 mg.

TD_{LO}, **child**: 7.5 mg/kg.

Drug interactions: Use with ***tyramine***-containing ***foods*** or ***wine*** leads to dizziness and hypertension. Causes hypoglycemia in **diabetics**. **Slow-acetylators** of ***isoniazid*** (another ***hydrazine***) or ***nitrazepam*** do the same with phenelzine, and increase its toxicity. **Plumbers** and **workers** who use soldering and welding fluxes can become allergic to ***hydrazine*** and can have a recurrence of **their** dermatitis after taking phenelzine. Adverse effects can be potentiated by ***alcohol***. Use with ***amphetamine, chlorphentermine, L-dopa, ephedrine, epinephrine, imipramine, isoproterenol, meperidine, metaraminol, methylphenidate, phentermine, phenylephrine***, and ***phenylpropanolamine*** with it can cause a hypertensive crisis, intracranial hemorrhage, hyperthermia, headache, myocarditis, rigidity, convulsions, coma, respiratory failure, and even death. Use with ***tricyclic antidepressants*** causes hyperpyrexia, excitement, convulsions, as well as death, so that the combined use is contraindicated. Death in a 26-year-old **female** ~5 h after ingestion of 2 oz ***dextromethorphan*** and ~6 h after two 15 mg tablets instead of 1 qid. Potentiates ***droperidol, propanidid***, and ***succinylcholine***. May change ***glucose*** tolerance and ***insulin*** sensitivity. Transient hypertension possible with ***amantadine***. Risk of ***serotonin*** syndrome with ***sertraline***. Malaise, nausea, and headache in **lab workers** exposed to ***amines*** as reagents, while taking it at 15 mg tid.

High dosage in **dogs** (10 mg/kg/day) causes some thyroid hypofunction, liver pathology, and extensive brain pathology and death, but lower dosage (1 mg/kg) appeared to be safe. **Rats**, **mice**, **monkeys**, and **dogs** appear to tolerate single large doses with only transient ill-effects.

LD_{50}, **mouse**: 130 mg/kg; SC, 150 mg/kg; IP, 135 mg/kg.

PHENETURIDE

= *Benuride* = *Ethylphenacemide* = *Trinuride*

Anticonvulsant.

Untoward effects: Somnolence, intellectual impairment, acute facial edema, ataxia, exfoliative dermatitis, and systemic lupus erythematosus reported. May decrease ***vitamin D*** availability, cause renal insensitivity to ***furosemide***, and be teratogenic. Cause slight increase in ***phenytoin*** serum levels by inhibiting its metabolism.

LD_{50}, **rat**: 1.14 g/kg.

PHENFORMIN

= *Azucaps* = *DBI* = *DBI-TD* = *Debeone-DT* = *Debinyl* = *Dibein* = *Dibotin* = *Dipar* = *Feguanide* = *Fenformin* = *Glucopostin* = *Insoral* = *Lentobetic* = *Meltrol* = *Normoglucina* = *PED G* = *PFU* = *Phenethylbiguanide*

Antidiabetic.

Untoward effects: Early side-effects in ~60% were anorexia, nausea, vomiting, metallic taste, and, rarely, diarrhea. After 4–16 weeks, ~35% reported weight loss, weakness, glycosuria, lethargy and malaise. The latter adverse effects disappeared 1–3 days after discontinuance of therapy. Severe lactic acidosis and secondary hyperuricemia caused its recall from general marketing in the U.S. by the FDA. It is still sold and used in European and Caribbean areas. Exercise exaggerates this abnormality. Incidence of deaths from lactic acidosis in the U.S. was estimated to be approximately one-quarter of a case to four cases/1,000 or ~50–700 deaths annually. Transient myopia in a 53-year-old **male**, probably due to allergic reaction of ciliary body. Hypoglycemia with secondary seizures, rash, urticaria, and hyperalaninemia also occur. Its use has been associated with increased risk of death from cardiovascular disease, including hypertension, edema, and cardiac enlargement. Thrombocytopenia, disseminated intravascular coagulation, and megaloblastic anemia reported. Glucose-6-phosphate dehydrogenase deficiency can also alter a treated **patient's** response. Malabsorption of ***vitamin*** B_{12} and acute pancreatitis occur with its use. Severe multiple **fetal** malformations have occurred in **infants** born to treated **mothers**. Occurs in breast milk. Abdominal pain, explosive vomiting, hypoglycemia, circulatory failure, anorexia, and gastric hemorrhage reported from overdosage in attempted suicides. May give false positive or spuriously high urinary ***acetone*** test readings and false negative or spuriously low serum ***bicarbonate*** test results. Therapeutic serum levels reported as ~30 μg% and toxic levels as ~350 μg%.

Drug interactions: Use with ***alcoholic beverages*** causes dramatic increase in lactic acidosis, because ***alcohol*** inhibits hepatic activity, potentiating phenformin's hypoglycemic effect. Potentiates ***vasopressin*** activity. Its hypoglycemic effect is potentiated by co-administration of ***sulfisoxazole***. Fibrinolysis reported when used with ***ethylestranol*** and may increase anticoagulant effect of ***warfarin***. ***Tetracycline*** may have increased its blood levels and precipitated lactic acidosis.

Hypoglycemic effect varies widely in different **animal** species. **Monkeys** and **guinea pigs** are most responsive; the **rat** is quite resistant. In the **rat** and **guinea pig**, the liver is the major site of its biotransformation to ***p-hyxdroxyphenformin***, and urinary excretion is the major path of elimination for it and its metabolite. In **rats**, enhances fibrinolytic activity within 4 h.

LD_{50}, **mouse**: 450 mg/kg; IV, 19 mg/kg.

PHENICARBAZIDE

= *Cryogenine* = *Febrimin* = *Kryogenin* = *Phenylsemicarbazide* = *Vertine*

***Antipyretic.* A *hydrazine* alkaloid isolated from *Heimia salicicifolia*.**

Untoward effects: Acute hemolytic anemia reported in an **alcoholic woman**. Stevens-Johnson syndrome has followed its use.

Powerful pulmonary tumorigenic in **mice**.

IP, LD_{50}, **rat**: 55 mg/kg.

PHENINDAMINE TARTRATE

= *Pernovin* = *Thephorin*

Antihistamine.

Untoward effects: Can cause contact eczema, xerostomia, dizziness; gastrointestinal upsets are common. Can have a stimulant effect and overdoses can precipitate convulsions.

LD_{50}, **rat**: 280 mg/kg; **rabbit**: 577 mg/kg.

PHENINDIONE

= *Accluton* = *Athrombin* = *Bindan* = *Cronodione* = *Dandilone* = *Danilone* = *Dindevan* = *Dineval* = *Diophindane* = *Emandione* = *Eridone* = *Fenhydren* = *Fenilin* = *Hedulin* = *Hemolidione* = *Indema* = *Indon* = *Phenylindandione* = *PID* = *Pindione* = *Pival* = *Rectadione* = *Thromasal* = *Thrombasal* = *Thrombatin* = *Thrombosan*

Anticoagulant, rodenticide.

Untoward effects: Tachycardia, diarrhea and hematuria, allergic reactions (scarlatiniform, morbilliform, purpuric, erythematous, and maculopapular rashes with itching, desquamation, urticaria, and fever), inflammation and ulceration of mucous membranes of mouth, pharynx, and colon. Nose bleeds, melena, hemoptysis, cerebral and meningeal hemorrhage, agranulocytosis, nephritis with acute tubular necrosis, paralysis of ocular accomodation, conjunctivitis, xerostomia, thirst, polyuria, oliguria, anuria, ***Christmas Factor*** deficiency, hemorrhagic cutaneous necrosis, infarcts, ecchymoses, and petechiae. Steatorrhea, hepatitis (hepatocellular cholestasis and jaundice with hepatic degeneration and necrosis, usually after 3–4 weeks of therapy [range: 10 days to 9 weeks]), and occasionally exfoliative dermatitis. Leukopenia, thrombocytopenia, leukocytosis, pancytopenia, neutropenia, eosinophilia, albuminuria, red blood cell aplasia, hematemesis, decreased serum urates, ulcerative colitis, paralytic ileus, abdominal distension, pancreatitis, alopecia, altered taste, thromboembolism, hematomas, neuropathy, myocarditis, lymphadenopathy, dysphagia, edema, impairment of ***thyroxin*** synthesis, severe fatigue, and occasionally fatalities. Excreted in breast milk in sufficient quantities to anticoagulate an **infant**; especially dangerous if surgery is performed on the **infant**. May be associated with congenital malformations. Massive hematomas reported in nursing **infants**. It, or its metabolites, can color urine red or orange-red. Has caused diffuse brown and yellow nail pigmentation and recurrent circumscribed bright orange patches of skin pigmentation, even from merely handling it. Can cause spuriously high or false positive urinary ***albumin*** test results.

Drug interactions: ***Haloperidol*** and ***trifluperidol*** markedly decrease its anticoagulant effect; ***cimetidine*** and ***clofibrate*** prolong and increase its anticoagulation. Increases toxicity of ***demeclocycline***. After 3 days, ***phenyramidol*** markedly prolongs prothrombin time. ***Chlordiazepoxide*** aggravates its non-urticarial induced skin eruptions.

Causes a severe anemia after 1–8 days of treatment in **rats**. ***Monoamine oxidase inhibitors*** increase its anticoagulant effect in **rats**.

LD_{50}, **rat** and **mouse**: 175 mg/kg.

PHENIPRAZINE

= *Catral* = *Catron* = *Catroniazid* = *Cavodil* = *JB 516* = *β-Phenylisopropylhydrazine* = *PIH*

***Antihypertensive*. A *hydrazine*-type *monoamine oxidase inhibitor*.**

Untoward effects: Despite apparent effectiveness, withdrawn from most markets because its use was associated with fatal liver necrosis, red-green color blindness, toxic amblyopia, bilateral optic atrophy, central visual field defect, disc pallor, and visual hallucinations. Nervousness, insomnia, and eczema.

Drug interactions: Blocks the antihypertensive effect of ***guanethidine***.

In the **dog**, its use caused a consistent and striking focal, bilateral degeneration of the inferior olivae, some cerebellar gliosis (not in **cats**, **rabbits**, and **squirrel monkeys**), ataxia, tremors, and nystagmus. Progressive cerebellar ataxia, cerebral stimulation, convulsions, and hemolytic anemia develop over a period of time in these **dogs**.

PHENIRAMINE MALEATE

= *Avil* = *Daneral* = *Inhiston* = *Trimeton*

***Antihistamine*. The *aminosalicylate* form is also used.**

Untoward effects: Excessive dosage can cause hallucinations and hyperactivity, followed by exhaustion, mydriasis, anorexia, akinetic seizures, dysuria, pollakiuria, ventricular extrasystoles, persistent tachycardia, ataxia, abdominal cramps, and toxic psychosis.

TD_{LO}, **woman**: 14 mg/kg; LD_{LO}, **human**: 5 mg/kg.

Has caused fatal hyperpyrexia in ***monoamine oxidase inhibitor***-pretreated **rabbits**.

PHENMETRAZINE

= *A 66* = *Anorex* = *Marsin* = *Neo-Zine* = *Phentrol* = *Preludin* = *Probese P* = *Willpower*

***Anorexic, central nervous system stimulant. Amphetamine*-like, but with less effect on the cardiovascular system.**

Untoward effects: Personality changes after large repeated doses or overdoses; nervousness, pollakiuria, agitation, restlessness, and insomnia common with overdoses. Also exhilaration, excitation, confusion, hallucinations, optical illusions, delirium, tremors, extrapyramidal symptoms, ataxia, muscle cramps, mydriasis, xerostomia, hypertension, tachycardia, palpitations, nausea, vomiting, constipation, diarrhea, friable fingernails, fever, urinary retention, headache, bitter taste, aplastic anemia, and convulsions can occur. Hemolysis has occurred after IV. Abuse and addiction reported with chills, fever, diaphoresis, prostration, tachycardia, and hypotension after IV abuse. A report indicated it may have caused visceral and skeletal anomalies when taken during the 4–12th week of pregnancy. Therapeutic serum levels are reported as 0.015 mg% and lethal levels at 0.4 mg%.

LD_{LO}, **human**: 5 mg/kg.

Drug interactions: Adverse effects are potentiated by other ***central nervous system stimulants*** and ***monoamine***

oxidase inhibitors; increases ***monoamine oxidase inhibitor*** hypertensive crises; stimulation decreased by ***chlorpromazine*** and other ***phenothiazines***.

Propranolol eliminated its central nervous system stimulative effects in **mice**.

LD_{50}, **rat**: 381.2 mg/kg; IP, **mouse**: 165 mg/kg; IV, **rabbit**: 40.7 mg/kg.

PHENOBARBITAL

= *Agrypnal* = *Barbiphenyl* = *Barbipil* = *Barbita* = *Chinoin* = *Eskabarb* = *Fenical* = *Gardenal* = *Lepinal* = *Luminal* = *Luminaletas* = *Mephobarbital* = *Phenemal* = *Phenobal* = *Phenobarbitone* = *Pheno-Squar* = *Sevenal* = *Solfoton*

Sedative, hypnotic, anticonvulsant, liver microsomal enzyme stimulator. **Oral; occasionally, IV and IM.**

Untoward effects: Drowsiness and confusion are the most common (and often transient) ill-effects in **adults**, as well as behavioral changes in **children**. Treated **children** often evidence paradoxical excitement. Osteomalacia, rickets, ***vitamin D*** and ***K*** deficiency, ***folic acid*** deficiency and megaloblastic anemia, thrombocytopenia, ataxia, decreased thyroid hormone, shoulder–hand syndrome, depression, erectile dysfunction, a multi-system hypersensitivity syndrome (with fever, chills, nephritis, urticaria; and pruritic, morbilliform, erythematous, and exfoliative skin rashes), fixed-drug eruptions, toxic epidermal necrolysis, Stevens-Johnson reactions, pemphigus, periarteritis nodosa or allergic vasculitis, erythema multiforme, and exacerbation of acne. Induced jaundice, increased alkaline phosphatase, and hepatitis may be an allergic reaction, and has also been associated with skin reactions. Although used in **newborns** to prevent physiological jaundice and hyperbilirubinemia, by competing with ***albumin***-binding of ***bilirubin***. In older **children** and **adults** with Gilbert's syndrome, a hemorrhage diathesis is occasionally reported. **Neonatal** hypotonia and respiratory difficulties, in addition to deaths from hemorrhage, have occurred. A **neonatal** phenobarbital withdrawal syndrome reported. Readily passes through the **human** placenta, and can cause hypocoagulation and methemoglobinemia in the **newborn** after **maternal** use. Birth defects are usually associated with combined use with ***phenytoin*** in **mothers**. Most common are dysmorphogenic craniofacial features, and growth and mental retardation. Prenatal exposure of **men** led to decreased verbal intelligence as **adults**. Found in breast milk, therefore, **infants** must be monitored for usual (sedation, poor sucking reflexes, decreased responsiveness, and methemoglobinemia) adverse effects and infantile spasms. Use can interfere with certain ***theophylline*** serum assays and cause false positive ***dexamethasone*** suppression test; chronic use interferes with ***nalorphine*** pupil test for ***narcotic*** usage. Large doses can also cause respiratory depression, cyanosis, stupor, anuria, nystagmus, diplopia, circulatory collapse, periorbital edema, fever, adenitis, arthralgias, thrombocytopenic purpura, megaloblastic anemia, eosinophilia, leukocytosis, leukopenia, agranulocytosis, depression, coma, and death. Has been a drug of abuse. Thrombophlebosis due to high dosage IV. A ketogenic diet has increased its serum levels up to 100%.

Debate still exists as to its carcinogenicity. Evidence is usually inconclusive or conflicting. A statistically significant effect of chronic use in increased development of lung cancer. Possibility of increased incidence of brain tumors by *in utero* exposure.

Abuse or attempted suicides with it have been common. Therapeutic serum levels reported as 1.5–3.9 mg%, toxic as 4–6 mg%, and 7.8–15 mg% as lethal. Severe intoxication after ingesting 4–7 g and 6–9 g has proven lethal.

TD_{LO}, **child**: 10 mg/kg; **human**: 214 μg/kg; **pregnant woman**: 466 mg/kg; LD_{LO}, **human**: 50 mg/kg.

Drug interactions: Both it and ***alcoholic beverages*** stimulate hepatic microsomal enzymes. Its effects are potentiated by ***alcohol*** and ***zonisamide***. Decreases ***warfarin*** half-life in > 60% of **patients**, requiring an average increase of 33% in ***warfarin*** dosage. Also decreases ***dicoumarol*** activity. Its activity is decreased by ***rifampin*** therapy. Serum levels or half-lives decreased, by interfering with absorption or increasing metabolism of ***aspirin, aminopyrine, androstenedione, antipyrine, chloramphenicol, chlorpromazine, cimetidine, clonazepam*** (~20%), ***clozapine, cyclophosphamide, cyclosporine, dichlorodiphenyltrichloroethane (DDT), desipramine, desoxycorticosterone, dexamethasone, digitoxin*** (but not ***digoxin***), ***dipyrone, doxycycline, estradiol, estrone, fluroxene, folic acid, furadantin, furosemide, griseofulvin, haloperidol, hexobarbital, hydrocortisone, imipramine, itraconazole, meperidine, meprobamate, mesoridazine, methyldopa, metronidazole, miconazole, misonidazole, nitrofurantoin, nortriptyline, oral contraceptives, pentobarbital, phenylbutazone, phenacetin, prednisolone, prednisone, procaine, progesterone, propallylonal, quinidine, rifampin, sulfadimethoxine, testosterone, theophylline, tridihexethyls***, and ***zoxazolamine***. When used with ***β-adrenergic blockers***, the latter's plasma levels are decreased and phenobarbital's depressant effects are enhanced. ***Monoamine oxidase inhibitors*** and ***sulfonylureas*** enhance its effects. When a metabolism-inducing drug, such as phenobarbital, is withdrawn from a **patient** treated with these drugs, prescribers must remember that the toxicity can increase. ***DDT*** has caused central nervous system arousal in **patients** with phenobarbital intoxication. Decreases phenobarbital clearance by ***chloramphenicol, chloroquine, felbamate, isoniazid***, and ***methylphenidate***, and, possibly, by ***acetohexamide, chlorpropamide,***

tolbutamide, and ***valproic acid***. Increases ***carbon tetrachloride*** hepatotoxicity and potentiates ***acetaminophen*** toxicity. Possibly decreasing ***thyroxine's*** rate of metabolism. Alkalinization of urine by ***acetazolamide***, ***sodium bicarbonate***, and ***thiazides*** increase its rate of excretion. Use with ***diazepam*** leads to increased ***diazepam*** metabolism with serious hypotension and respiratory depression. Use with ***meperidine*** leads to increased sedation and increased concentration of the toxic metabolite, ***normeperidine***. Methemoglobinemia and hemolytic anemia in a 17-year-old **female** after it was added to ***phenacetin*** therapy. ***Foods*** decrease its absorption. In a **newborn**, it decreases ***theophylline*** serum levels and half-life. ***Acetazolamide*** may accelerate osteomalacia induced by its long-term use.

Occasionally, some **dogs** show excitement or whining during induction or wearing off of hypnotic effects. Regular use in **cats** and **dogs** causes thrombocytopenia in some. The drug has shown strong embryotoxic and teratogenic effects in **rabbits**, and daily IP dosage (50 mg/kg) to **rats** was associated with increased fetal death and SC ecchymoses. Daily oral treatment for 2 weeks to infant (4–5 day old) **rats** (15–30 mg/kg) reduced brain growth. Geriatric **cats** and **dogs** show more toxicosis than younger **animals**. Infrequently, chronic use in **dogs** has been associated with liver disease, respiratory depression, and hypotension. Microsomal enzyme stimulation is NOT always advantageous, as shown in one **sheep** trial, where the accelerated metabolism of ***carbon tetrachloride*** appeared to enhance liver necrosis. In another trial, by the same workers, with a low-protein diet, five daily doses IP (40 mg/kg) helped protect against the same drug. Oral ***iodinated glycerol***, ***sodium bromide***, ***sodium iodide***, and/or IP ***glycerin***, ***inorganic iodides***, ***iodinated glycerol***, ***sorbitol***, and ***sucrose*** intensify phenobarbital narcosis in **rats**. Oral ***DL-amphetamine*** in **mice** delays the absorption of and, in a sense, initially antagonizes the effect of, phenobarbital, yet, during the rise of its anticonvulsant effects, they appear to be synergistic. Phenobarbital potentiates ***α-naphthylisothiocyanate*** hyperbilirubinemia in **mice**. In **mice**, phenobarbital may antagonize the analgesic effects of ***morphine***; possesses hyperalgesic properties against electroshocks to the tail area or head-twitching when exposed to ***cigarette*** smoke; and 100–150 mg/kg SC to **mice** and **hamsters** stimulates locomotor activity along with ataxia. A circadian increase in this activity was noted at night. Its use delays ovulation in **hamsters**. **Sheep** eat 10–30% more feed when it was added at the rate of 0.5–2 g/kg of ration. It increases activity in hyperactive **mice** and **children**. In **guinea pigs** pretreated with it, it increased the metabolism of ***diazepam***. In **rats**, it increased the metabolism of ***aminopyrine***, ***androsterone***, ***benzpyrene***, ***bishydroxycoumarin***, ***diethylstilbestrol***, ***diazepam***, ***desoxycorticosterone***, ***hexobarbital***, ***meprobamate***, ***mestranol***, ***methimazole***, ***p-nitrobenzoic acid***, ***norethindrone***, ***norethynodrel***, ***parathion***, ***pentobarbital***, ***progesterone***, and ***testosterone***. Longer drug effect in **female's** versus **male's**. Phenobarbital use in **mice** increased metabolism of ***clomethiazole***, ***diazepam***, ***hexobarbital***, ***ngaione***, and ***phenytoin***. Behavioral response is greater in pregnant than non-pregnant **mice**. ***Lynestrenol*** decreases anticonvulsant effects of phenobarbital and ***phenytoin*** in **mice**. It markedly decreases hyperbilirubinemia in **rats**. In **dogs**, it increases the metabolism of itself, ***antipyrine***, ***digoxin*** (by 30%, but not ***digitoxin***), ***lindane***, and ***testo-sterone***. Phenobarbital pretreatment increases the half-life of ***phenylbutazone*** and ***phenytoin*** in **dogs**. It may be present in the milk of treated **cows** for about a week after use. ***Polychlorinated biphenyls*** increase the metabolism of phenobarbital in **rats**. In **calves**, 20 mg/day/20 days orally stimulated liver microsomal enzymes. **Sheep** pretreated with phenobarbital had increased metabolism of ***antipyrine***, ***bishydroxycoumarin***, and ***sulfobromophthalein***.

Pretreatment of **animals** with phenobarbital accelerated the metabolism of ***benzene*** in **rabbits**; ***2-allyl-2-isopropylacetamide***, ***androgens***, ***chloral hydrate***, ***chlordane***, ***DDT***, ***dieldrin***, ***EPN***, ***estrogens*** (especially ***estrone***), ***glucocorticoids***, ***griseofulvin***, ***progesteroids***, ***progesterone***, and ***zoxazolamine*** in **rats**; ***hydrocortisone*** in **guinea pigs**; and ***bishydroxycoumarin*** in **dogs**. Pretreatment of **rat** dams with phenobarbital during gestation dramatically decreased the acute oral toxicity of ***meprobamate*** and phenobarbital, subsequently given their young offspring. **Dogs** appear to develop a tolerance to its long-term use, as in epilepsy, due to stimulation of liver microsomal enzyme activity, so that dosage and/or frequency of administration must be increased. Avoid sudden withdrawal in these cases, as it may precipitate convulsions. Polydypsia and polyphagia have been reported in **dogs** on long-term therapy. Strong embryotoxic and teratogenic effects in **rabbits**. Do NOT use meat of treated **poultry** until after a 1 week withdrawal time.

LD_{50}, **rat**: 162 mg/kg; **mouse**: 168 mg/kg; **dog**: 150 mg/kg.

PHENOL

= *Carbolic Acid* = *Hydroxybenzene* = *Oxybenzene* = *Phenolum* = *Phenyl Alcohol* = *Phenyl Hydroxide* = *Phenylic Acid*

***Antiseptic, local anesthetic, escharotic.* Over 4 million lbs used annually in the U.S.**

Use: In phenolic resins for plywood, ***alkylphenols***, ***aniline***, ***bisphenol A***, ***caprolactam***, ***parabens***, ***xylenols***, as a disinfectant, and in manufacturing ***salicylic acid***, then, ***aspirin*** and ***adipic acid***.

Untoward effects: It is a potent protoplasmic poison, precipitating proteins, and causing severe burns, skin blanching, and scarring. Burning is often followed by an anesthetic effect, which does indicate the serious effects are over. Rapidly absorbed after skin contact, and systemic effects (primarily, central nervous system, respiratory depression, and shock) can cause death within 30 min to several hours. Use for facial chemexfoliation or rejuvenation has, in some cases, caused scarring, hyperpigmentation (after early sun exposure), cardiac arrhythmias, tachycardia, and death. Percutaneous and mucous membrane absorption occurs from its use as a topical antipruritic (0.5–1.5%), and has followed its use of ***Castellani's paint*** (***carbol-fuchsin*** solution) on seborrheic eczemas of **infants**. **Laboratory workers**, especially **those** working with *Mycobacteria*, using it in concentrated form, have developed poor-healing burns, hypersensitivity, decreased respiratory tract cilial movement, and contact dermatitis. Severe local reactions in **some** from using phenol-containing ***soaps*** on skin and in enemas. A 1-day-old **baby** died 11 h after application of an umbilical bandage moistened with 2% phenol, instead of ***sodium chloride***. Another **baby** treated with a phenol-***camphor*** (q.v.) preparation for a skin ulcer developed circulatory failure, central nervous system disturbances, and methemoglobinemia, requiring exchange transfusion. Many phenolic plant constituents cause an allergic contact dermatitis. Area covered, rather than concentration, appears to be the critical aspect of topical toxicity. Death reported if 64 sq in of body surface is covered. Found in some hair color sprays. Intrathecal use with ***glycerin*** for the relief of pain has occasionally caused paralysis and paresthesias of various areas. Methemoglobinemia and death reported in a 7-day-old **infant** after instillation of 30 ml of 5% to treat hemorrhagic cystitis. Injury from accidental contact with it to cornea and conjunctiva from a dermatological preparation. Inhalation of smoke from it in burning materials is very poisonous. In 1986, a Ciba-Geigy plant in Switzerland emitted a phenol-laden cloud into the atmosphere. Ingestion causes severe local irritation and necrosis, nausea, vomiting, diarrhea, abdominal pain, cyanosis, excitement, alkalosis (followed by acidosis), green or bluish discoloration of urine, hyperpnea, hypotension, tremors, convulsions, ventricular arrhythmias, raised ***uric acid***, pulmonary edema, pneumonia, coma, and death often follows respiratory failure and uremia. Once popular for suicides, where successful deaths usually occurred within 2–3 min and < 2 h, with decreased respirations, collapse, paralysis, and coma. It was a cause of marasmus and poisoning among **surgeons** in the days of Lister. Its use in parenterals has caused few problems, except in the case of high-dosage ***glucagon*** used as an antidote for ***β-adrenergic blocker*** or ***calcium channel blocker*** overdoses, where its accompanying diluent contains 0.2% phenol as a preservative. It is suggested that **users** substitute normal ***saline*** or 5% ***dextrose*** in water to reconstitute it. OSHA and National Institute for Occupational Safety and Health suggest a general upper exposure limit of 5 ppm for skin and a 250 ppm immediate danger to life or health. An oral **human** LD reported as 2 g and 6 g, respectively, for ***Lysol***® and ***cresol***. Urinary levels of 20 mg/l indicate toxicity.

LD_{LO}, **human**: 140 mg/kg.

Adsorbed by ***polypropylene*** and coated ***polyethylene*** containers. After 1 week of contact, 60.5% adsorbed by ***nylon*** syringes. Occasionally found in drinking waters, imparting a sweet, acrid odor and taste, aggravated by chlorination, due to the formation of ***chlorophenols***. Phenol contamination of potable water found, due to an improperly cured liner of a Georgia hospital's solar water tank.

Although **cats** have received serums, anesthetics, and topicals containing phenol without apparent ill-effects, caution must be exercised in its use for them, as they cannot metabolize it to harmless ***glucuronides***. ***Chlorophenols*** impart a horrible taste to ***milk*** products, and they can be and have been formed by reaction of milk house ***chlorine*** with phenol in udder balms. Undetected contamination of a few cans of ***milk*** has caused the further contamination and loss of tankloads of market ***milk***. The ill taste detectable in 1 : 4,000,000 dilution is much more potent than the odor. Readily causes convulsions in **mice**. Experimentally, produced primary glaucoma by subconjunctival injections in 90% of **rabbits**. **Pigs** are frequently poisoned by exposure to phenolic wood preservatives.

LD_{50}, **rat**: 400 mg/kg; **mouse**: 300 mg/kg; **dog**: 50 mg/kg; **horse**: 60 mg/kg. LC_{50}, **goldfish**: 46 mg/l; **rainbow trout**: 11.3 mg/l. LD, **cat**: 50 mg/kg. LD_{LO}, **dog**: 500 mg/kg; **rabbit**: 420 mg/kg. Dermal LD_{50}, **rabbit**: 1.4 g/kg.

PHENOL-FORMALDEHYDE RESINS

The first was known as *bakelite*; they are used in manufacturing adhesives, *paints*, plywood, binders, laminates, etc.; used by hobbyists as well.

Untoward effects: Contact dermatitis is common, and dermatologists regularly use it in standard tray patch-test allergens.

PHENOLPHTHALEIN

= *Chocolax* = *Darmol* = *Ex-Lax* = *Laxin*

***Laxative, cathartic, pH indicator.* Almost colorless in neutral or acidic solutions, and bright purple-carmine in pH 8.6–10 solutions.**

Untoward effects: Its physiological effects were accidentally discovered in Hungary when it was purposely used to color ***wine***. It is now illegal to do so, since many **people** had diarrhea as a result of its use; it depletes ***potassium***, which can induce paralysis of intestinal musculature. It had caused a vicious cycle of overuse. A 51-year-old **male** abusing ***Ex-Lax***® (90 mg/tablet) on 15–20 tablets/day/20 years developed back and hip pain of 6 months duration due to histiologically proven osteomalacia. Normal bowel habits and improvement of osteomalacia after discontinuance. Nephropathy in a 36-year-old **female** ingesting 10–12 doses/day/2–3 years and previous daily abuse of 15–20 years. Although phenolphthalein was used in ***Ex-Lax***® for many years, current formulations in the U.S. have changed. Severe bullous erythema multiforme-type of fixed-drug reaction in a 22-year-old **female** after ingesting ***Feenamint***®, proven by double-blind challenge. Eruptions appeared on lips, mouth, tongue, and fingers with classic, persistent red, brown, blue-gray, and sometimes purple macules, yet no sensitivity to ***phenolsulfonphthalein***. Phenolphthalein was also present in some toothpastes and food colorings. Diarrhea and abdominal pain occur frequently. Albuminuria, hemoglobinuria, non-thrombocytopenic purpura, and eyelid edema reported. Malabsorption syndrome can follow excessive use. Toxic epidermal necrolysis and skin pigmentation changes with itching and burning sensations also reported. Skin lesions slowly fade, but can last for months to years. Yet, some pigmentation can remain. Blue-gray lunulae occur. Acute pancreatitis in a 34-year-old **male** with 2 year history of chronic use following inadvertent ingestion of 2 g. Can cause reddish-brown urine and feces. Anemia (hematocrit of 26.6% and hemoglobin of 7.9 g/100 ml) and gastrointestinal bleeding in a 24-year-old **female** ingesting 150 tablets/week. A 3-year-old **female** who ingested ~1.8 g (18 tablets) was given gastric lavage within 30 min. **She** died 13 h later. Postmortem revealed cerebral and pulmonary edema. Systemic absorption is ~15% and it is occasionally excreted into breast milk, where it has adversely affected some **infants**. Syncope has been reported. Yellow phenolphthalein contains impurities and is two to three times the potency of the white. Has caused increased or false positive sulfobromophthalein retention test results. Use with ***anticoagulants*** increases hypoprothrombinemia. Suspect as a **human** carcinogen because of **mouse** studies.

Accidental ingestion occurs in **pets** that are attracted by the ***chocolate*** (q.v.) on some products. May be carcinogenic in **mice**.

PHENOLSULFONIC ACID

Use: Chemical intermediate and in ***tin*** plating.

Untoward effects: Skin irritant. Can contain traces of ***phenol*** or ***sulfuric acid***.

PHENOLSULFONPHTHALEIN

= *Phenol Red* = *PSP*

Use: ***Diagnostic agent*** for kidney function and intestinal transit time.

Untoward effects: Allergic reactions have occurred. ***Novobiocin***, ***1,8-dihydroxyanthraquinone***, and other drugs and dyes may give falsely increased readings, while ***penicillin***, ***salicylates***, and ***sulfonamides*** may give falsely decreased readings. Causes spuriously high serum ***creatinine*** test results.

PHENOPERIDINE

= *Lealgin* = *Operidine* = *R 1406*

Analgesic. **IM or IV usually with various anesthetics.**

Untoward effects: IV can cause muscle rigidity, disorientation, struggling, respiratory depression, and persistent (7 h) drowsiness. Metabolized to ***meperidine***, and caution dictates it can be undesirable therapy with ***monoamine oxidase inhibitors***.

PHENOPYRAZONE

Analgesic, antipyretic.

Untoward effects: Has induced pseudolupus, a systemic rheumatic disease.

PHENOTHIAZINE

= *AFI-Tiazin* = *Antiverm* = *Dibenzothiazine* = *Fentiazin* = *Helmetina* = *Hippozin* = *Lethelmin* = *Nemazine* = *Orimon* = *Phénégic* = *Phenoverm* = *Phenovis* = *Phenoxur* = *Reconox* = *Souframine* = *Thiodiphenylamine* = *Vermitin*

Use: As an anthelmintic, insecticide, in manufacturing pharmaceutical tranquilizers, and in axle grease.

Untoward effects: In 1940, agricultural **workers** using it as a spray to destroy **coddling moths** developed an intolerance to sunlight with exanthema as a result. Neurotoxic effects reported. Nausea, vomiting, liver pathology with jaundice, acute hemolytic anemia, hematuria, albuminuria, abdominal pain, and tachycardia. Some treated **children** have died from its use.

LD, **human**: ~10 g.

Photosensitivity reactions are common in treated **calves** and adult **cattle**; occasionally, in **pigs**, **sheep**, and **goats**. Some unthrifty **horses** have developed hemolytic anemia, weakness, colic, hepatitis, oliguria, and hemoglobinuria after treatment with it. Urine and milk of treated **animals** may contain the drug or its metabolites. Enough may be present to impart reddish colors to these products and cause serious staining of **sheep *wool***. It is a strong photosensitizing agent in **mammals** and **fish**. Milk should not be used for **human** food for at least eight milkings after treating lactating **animals**. It has generally been suggested that pregnant **animals** NOT be treated during the last months of their gestation period. Toxicity for **birds** is generally very low. Caution must be used in administering it to anemic, weak, or emaciated **animals** or **poultry**. Do NOT use with ***organophosphates***, as it may potentiate their toxicity. Commercial batches may contain enough ***iodine*** to temporarily decrease thyroid function.

PHENOXYBENZAMINE HYDROCHLORIDE

= *Dibenyline* = *Dibenzyline* = *Dibenzyran*

Antihypertensive, α-adrenergic blocker, sympatholytic.

Untoward effects: Postural hypotension, wheezing, râles, rhonchi, hyponatremia, impotence or delayed ejaculation in **males**, decreased libido in **females**, decreased blood pressure, diarrhea, miosis, morning dizziness and nausea, urinary incontinence (due to smooth muscle relaxation and decreased intraurethral pressure), and inhibition of platelet function. If blood volume is not normal, its use can be hazardous and hasten death. An IV infusion of 500 mg in a 70-year-old **male** caused hypertensive crisis (260/170 mmHg) within 30 min, lost consciousness, severe diaphoresis, piloerection, and intense skin vasoconstriction. Rapid development of dry gangrene in fingers and toes in a 52-year-old **male** with Raynaud's phenomenon after 10 mg twice daily. Nasal congestion, headache, giddiness, and weakness in some **patients** with atopic dermatitis after oral dosing. Priapism after the third 10 mg daily dose in a 12-year-old **male** with Fabry's disease. Total urinary incontinence reported when given with ***methyldopa*** in some **patients**, but not when either drug was given alone. It counteracts the vasoconstrictor properties of ***dopamine***. When heated, it emits toxic ***hydrochloric acid*** fumes and ***nitrogen oxides***.

Harmful in postischemic treatment of strokes in **gerbils**. IP use in **mice** and **rats** causes peritoneal sarcomas in both species and lung tumors in **mice**. Can cause hypotension and gastrointestinal upsets in **cats**.

PHENOXYPROPAZINE MALEATE

= *Drazine* = *HP 1275*

Monoamine oxidase inhibitor.

Untoward effects: Diarrhea, nocturnal knee-jerking, dizziness, and xerostomia. Hepatotoxicity and associated deaths caused temporary market withdrawal in England.

LD_{50}, **mouse**: 460 mg/kg; IV, 350 mg/kg.

PHENPROCOUMON

= *Falithrom* = *Liquamar* = *Marcumar*

Anticoagulant.

Untoward effects: Necrosis of skin and SC tissue, hematomas, hematuria, hepatitis with jaundice, and fatal hemorrhages. Plasma half-life of 4–6 days. Usual therapeutic serum levels are 0.16–4 μg/ml and toxic at > 5 μg/ml. ***Cholestyramine resin*** decreases its absorption. ***Barbiturates*** inhibit; ***acetaminophen, allopurinol, aspirin, bezafibrate, clofibrate, metoprolol, oxyphenbutazone, phenylbutazone***, and ***tramadol*** potentiate its action. ***Oral contraceptives*** and ***rifampin*** increase its clearance. It increases ***phenytoin*** half-life. ***Sulfadimethoxine*** and ***sulfinpyrazone*** displace it from ***albumin***-binding sites.

PHENSUXIMIDE

= *Lifène* = *Milontin* = *Mirontin* = *Succitimal*

Anticonvulsant.

Untoward effects: Has caused lymphoma-like syndromes of lymph nodes in the neck. Occasionally nausea, ataxia, vertigo, vomiting, muscle weakness, persistent hiccups, somnolence, skin eruptions, and nephrotoxicity (glomerulotubular). Rarely, microscopic or gross hematuria, leukopenia, thrombocytopenia, and aplastic and megaloblastic anemia.

LD_{50}, **mice**: 960 mg/kg.

PHENTERMINE

The HCl form = *Adipex-P* = *Fastin* = *Wilpo*

The ion exchange resin form = *Duromine* = *Ionamin* = *Linyl* = *Mirapront* = *Noviresin* = *Omnibex*

The hydrogen tartrate form = *Modatrop*

Anorexic, stimulant, sympathomimetic.

Untoward effects: Xerostomia, diaphoresis, motor agitation, nausea, headache, bitter taste, constipation, mydriasis, tachycardia, thrombocytopenic purpura, acute psychotic disorganization, dyspnea, and chest pain.

Drug interactions: The most serious are the interactions with ***fenfluramine*** (***Phen-Fen***), which, when used in tandem, was associated with valvular heart disease (both left and right-sided valves). This followed months of use. An estimated 6,000,000 **Americans** used the combination in 1996, even though the FDA never approved the combination. Unstable angina induced after 30 mg/day/3 weeks in 51-year-old **male**. Deaths from irreversible pulmonary hypertension after only a few weeks of treatment. Severe gastrointestinal upsets can follow withdrawal.

Ischemic colitis in a 36-year-old **female** after 3 months of use. Hypertensive crisis reported when taken with ***phenelzine***. Near-fatal episode of ventricular fibrillation when phentermine taken with ***thyroid*** and ***trichlormethiazide***. Causes sudden increase in blood pressure in **patients** taking ***adrenergic blockers***.

LD_{50}, **mouse**: 105 mg/kg.

PHENTOLAMINE MESYLATE

= *Regitine* = *Rogitine*

Antihypertensive, α-sympatholytic. **Usually IV or IM; occasionally, orally.**

Use: Diagnostic agent for pheochromocytoma and in the prevention and control of hypertension during pre-op preparation or surgical excision of pheochromocytoma. Also in the prevention or treatment of dermal necrosis from ***norepinephrine*** extravasation.

Untoward effects: Sudden and prolonged hypotension, arrythmias, premature ventricular contractions, tachycardia, angina, flushing, weakness, dizziness, nasal congestion, nausea, vomiting, diarrhea, abdominal discomfort, exacerbation of duodenal ulcers, orthostatic hypotension, transient anorexia, decreased lower esophageal sphincter pressure, urinary incontinence, decreased platelet aggregation, and decreased emission of seminal fluid. An acute inferior myocardial infarction with chest pain and blood pressure decrease to 70/60 from 240/138 in 65-year-old **male** after phentolamine test. After overdoses, excitation, headache, diaphoresis, miosis, visual disturbances, cerebrovascular spasms, and occlusion have also been reported.

LD_{50}, **rat**: 1.25 g/kg; **mouse**: 1 g/kg; **rabbit**: 2 mg/kg.

PHENYLALANINE

Essential amino acid. **Combined with *aspartic acid* to form *aspartame*, a popular sweetener.**

Untoward effects: Normal consumption of ***aspartame*** does not produce high enough levels of phenylalanine to cause mental retardation, similar to that resulting from phenylketonuria. A genetic deficiency of the oxidizing agent phenylalanine hydroxylase in **humans** caused phenylketonuria, the well-known metabolic disorder of **babies**, and has stimulated a search for **animal** models. Use with caution, as it and other ***amino acids*** may contribute to the development of hepatoencephalopathy. ***Aspartame***-containing products should carry a warning alert to **phenylketoneurics** that the product contains phenylalanine. In **man**, it is ineffective in reversing ***chloramphenicol*** hematopoietic toxicity.

1% levels in the diets to mother and young **rats** led to a reversible mental retardation. In **rhesus monkeys**, similar trials led to retarded learning performances and other behavioral problems, persisting for 2 years or longer after returning to normal diets. As with other ***amino acids***, it diffuses across the placental membrane, reaching higher fetal than maternal levels. In the **rhesus monkey**, when serum maternal levels are 1–2 mg/100 ml near full-term, there is approximately a 1.5 : 1 diffusion rate, but when maternal levels are abnormally high (25 mg/100 ml), fetal serum levels reach 45 mg/100 ml, to the detriment of the fetus. **Mouse** work indicates it enhances ***chloramphenicol's*** antibacterial action and deficiencies retard formation of precancerous and cancerous cells, as well as humoral, but not cell-mediated, immunity. Its use may be antidotal to ***chloramphenicol***-induced blood dyscrasias. Within 1 h after injection into **rats**, they evidence hyperalgesia, probably due to the lowering of brain ***tryptophan*** and ***serotonin*** synthesis. Deficiency increases the life expectancy of leukemic **mice** by stimulating their immune systems. Blocks ***enkephalin***-inactivating enzyme in **humans**, easing chronic pain. The same may be true in **horses**, where its SC use (500 mg) produces effective analgesia.

IP, LD_{50}, **rat**: 5.3 g/kg.

PHENYLBUTAZONE

= *Ambene* = *Artrizin* = *Azolid* = *Bizolin* = *Butacote* = *Butadion(a)* = *Butapyrazol* = *Butatron* = *Butazolidin* = *Butazone* = *Bute* = *Butoz* = *Buzon* = *Diphebuzol* = *Ecobutazone* = *Equipalazone* = *Exrheudon N* = *Fenibutazona* = *Fenibutol* = *Flexazone* = *G 13,871* = *Intrabutazone* = *Intrazone* = *Mepha-Butazon* = *Phenyzene* = *Robizone-V* = *R-3-ZON* = *Tevcodyne* = *Uzone*

Anti-inflammatory, analgesic. **Oral or IM.**

Untoward effects: Bone marrow depression with leukopenia, pancytopenia, agranulocytosis, eosinophilia, leukocytosis, non-thrombocytopenic purpura, thrombocytopenia, thrombocytopenic purpura, neutropenia, and aplastic anemia rarely occur during a week-long treatment, but is most serious. Prodromal symptoms of this usually include fever, sore throat, or stomatitis. Other adverse effects occur less in **male** than **female**, and are gastrointestinal and oral irritation and ulceration, hematemesis, rashes, water retention and edema, acute myeloid and monocytic leukemias, vertigo, hearing loss, hepatitis, cirrhosis, hepatic granulomas, hypertension, cardiac arrhythmias, myocarditis, transient psychoses, central nervous system disturbances, lethargy, decreased serum urates, megaloblastic anemia, granulomas, lymphadenopathy, thrombosis, hypersensitivity vasculitis, vertigo, optic neuritis, lens opacity, chronic conjunctivitis, retinal hemorrhages, symblepharon, panus, corneal ulceration and scarring, toxic amblyopia, decreased vision, ototoxicity, sudden gastric pain, nausea, pallor, increased blood pressure, renal damage, interstitial nephritis, glomerulitis, papillary necrosis, melena, esophageal injury, increased alkaline phosphatase in 1–5%, parotid gland enlargement, systemic lupus erythematosus, psoriasis, toxic epidermal necrolysis, maculopapular dermatitis, urticaria, erythema multiforme, erythema nodosum, fixed-drug eruptions, plasma-cell cheilitis, exfoliative erythroderma, exanthematous eruptions, pemphigus, porphyria, photosensitivity reactions, xerostomia, decreased taste acuity, bitter taste, goiter, and occasionally myxedema. Acute bronchospasms in ***aspirin***-sensitive **asthmatics**. ***Radiopharmaceutical*** uptake in thyroid imaging decreased by it. Can precipitate or exacerbate congestive heart failure. Pericarditis with effusion, salivary gland enlargement, and atrial fibrillation in a 49-year-old **female** after 100 mg tid/5 days. One report indicated increased chromosomal-damaged cells (2% versus 0.48% in **controls**) in tests on 50 **patients**. A 65-year-old **female** developed flushing, blurred vision, weakness, dyspnea, pruritus, and urticaria after **her** first 100 mg dose due to ***tartrazine*** (q.v.) coloring of tablets. Confirmed by rechallenge. Chinese herbal medicines have contained unlabeled phenylbutazone (up to 30 mg/pill) and were associated with cases of agranulocytosis after taking recommended 12–18/day. Rare cases of congenital goiter from its use and hyperbilirubinemia in an **infant** from its toxic effects on the liver. Breast milk levels are only about one-tenth **maternal** serum levels. Red blood cell aplasia in **infant** of **mother** who ingested 200 mg tid/7 days in 33rd week of pregnancy. The **baby**, at 3 months of age, had a hemoglobin level of 2 g/100 ml and **her** marrow showed absence of precursors. Has caused increased serum ***sodium***. Despite its clinical effectiveness, its use in **humans** is now restricted, because it has been replaced with other effective drugs not causing the unacceptable incidents of deaths (associated with its induced aplastic anemia and agranulocytosis). Elimination half-life is decreased in the **elderly**. Accidental acute poisoning in a 2½-year-old **child** caused transient hyperglycemia for a few hours; coma, convulsions, diarrhea, and cholestatic jaundice during next 10 days. Recovered in 3 weeks. Massive overdosage in **adults** leads to confusion, convulsions, and coma. Usual therapeutic serum levels are 50–100 μg/ml; toxic at 120–200 μg/ml; and coma or fatalities at > 400 μg/ml.

TD_{LO}, **woman**: 276 mg/kg/3 weeks and 4.2 g/kg/65 weeks.

Drug interactions*:** Nausea, indigestion, diarrhea, and oral ulcers with ***acetaminophen. ***Acidifying agents*** decrease its urinary excretion and potentiate its effects. ***Alkalinizing agents***, such as ***sodium bicarbonate***, increase its urinary excretion and decrease its effects. It can decrease gastric absorption of ***antacids***, increasing gastric pH. With ***aminopyrine***, causes thrombocytopenic purpura, allergic agranulocytosis, paroxysmal tachycardia, arrhythmia, hepatitis, and meningoencephalopathy. It enhances metabolism of ***aminopyrine***, its close chemical relative, in **humans** and **rats**. Early use potentiates ***acenocoumarol, cyclocoumarol, dicumarol, ethyl biscoumacetate, phenprocoumon***, and ***warfarin***. Continued use may decrease their effectiveness. ***Phenobarbital*** increases its metabolism in **man**. It displaces ***thiobarbiturates*** from their protein-binding sites, increasing the latter's effect, and decreasing their need in anesthesia by 10–25%. Can decrease anticonvulsant concentrations of ***phenobarbital*** and ***primidone***. ***Cephalosporins'*** urine levels are decreased by it, due to competition for renal tubular secretion. Serum half-life is decreased by ~19% in **workers** exposed to ***DDT*** or ***lindane***. Use with ***chloroquine*** has caused thrombocytopenia, agranulocytosis, and death. ***Cholestyramine*** binds it and decreases its absorption; ***magnesium*** increases its rate of absorption. Has decreased the efficacy of ***oral contraceptives*** with increased incidence of breakthrough bleeding, ***estradiol, estrogens, progesterone***, and ***testosterone***, by increasing their metabolism. ***Methandrostenelone*** potentiates or decreases phenylbutazone action. Increases the metabolism and binding displacement of ***desoxycorticosterone*** and ***hydrocortisone***. It decreases the half-life of ***digitoxin*** and ***griseofulvin***. Enhances ***penicillin's*** antibacterial activity by displacing it from binding sites. Uricosuric effects antagonized by, and anticoagulation increased by, large doses of ***aspirin*** and other ***salicylates***. Dramatically inhibits metabolism of ***lithium*** and ***phenytoin*** and can increasing their toxicity. Use with long-acting ***albumin***-bound ***sulfonamides*** (viz. ***sulfamethoxypyridazine***) can displace them, enhancing antibacterial activity and, possibly, increasing ***sulfonamide*** toxicity. Potentiates hypoglycemic effect of ***tolbutamide*** (half-life increases from 4 to 17 h and occasionally 21½ h) and other ***sulfonylureas*** (***acetohexamide, carbutamide, chlorpropamide, glipizide***, and, possibly ***glyburide***), and can lead to hypoglycemic shock. Since it can alter platelet function, avoid use with ***urokinase***. By enzyme induction, it can increase its own metabolism and that of ***carisprodol, chlorcyclizine, diphenhydramine, meprobamate***, and ***zoxazolamine***. ***Desipramine*** and ***imipramine*** decrease its absorption by complex formation. Half-life decreased by 43% in **smokers**. Implicated in causing lethal ***methotrexate*** toxicity. ***Indomethacin*** competitively displaces it from **human *albumin*** and it is potentially more toxic than ***indomethacin***. Toxicity of ***demeclocycline*** is increased by it. It decreases activity of ***nalidixic acid*** and ***nitrofurantoin***. Fatal allergic pancytopenia in a 67-year-old **male** also treated with ***propoxyphene*** for cervical vertebral osteochondrosis. Fatal malignant reticulosis in a 4-year-old **male** with rheumatoid arthritis given ***aminopyrine*** 800 mg/day during 12th to 8th month before death, and phenylbutazone 200 mg twice daily for approximately the same time.

In some **laboratory animals**, it is metabolized so quickly that "ordinary" doses appear to have no effect, although plasma levels are usually similar at the effective level (viz. 300 mg/kg or 5–10 mg/kg; respectively, daily for **rabbits** and **man** to obtain similar effective anti-inflammatory plasma levels of 100–150 µg/ml). Its half-life in most **laboratory animals** is 3–6 h, compared to 3 days in **man**. After IV use of 4.4 mg/kg (2 g/1,000 lbs) and 6.5 mg/kg in **horses**, half-lives of 3.5 and 7 h, respectively, have been reported. Pretreatment with ***phenobarbital*** or phenylbutazone stimulates its metabolism in **rats**, and repeated doses of the drug to **dogs** over a 1–4 month period cause a 67±% decrease in plasma levels. Initially, **dogs** metabolize ~90% of their daily intake, compared to ~20% in **man**. Its use causes increased metabolism of ***aminopyrine*** in **rats** and **man** and hydroxylation of ***hydrocortisone*** in **guinea pigs** and **man** by liver microsomes. Many stimulators of this enzyme system can, theoretically, accelerate the metabolism of phenylbutazone. This has been demonstrated by pretreating **dogs** with small oral doses of ***chlordane***. Stimulates its own metabolism in **rats** and **dogs**. It potentiates the hypoglycemic effect of ***tolbutamide*** in **rabbits**. In **dogs**, it increases the metabolism of ***antipyrine*** and ***testosterone***; first potentiates, then antagonizes, ***warfarin***, by displacing it from binding sites, and then stimulating its metabolism. Pretreatment of **dogs** with ***chlorcyclizine*** or ***phenobarbital*** increases its half-life and decreases its metabolism. Pretreatment of **horses** with ***chloramphenicol*** or ***quinidine*** did not alter phenylbutazone metabolism. Pretreatment of **rats** with ***chlordane*** or ***DDT*** increased its metabolism. In **rats**, it increased metabolism of ***androsterone, desoxycorticosterone, estradiol, progesterone***, and ***testosterone***. Half-life in **rats** is 6–13 h; **man**, 1–4 days; **monkeys, dogs, rabbits**, and **guinea pigs**: 3–8 h (10 mg and 50 mg/kg, IV); **horse**, 8 h (10 mg/kg) (dose-dependent in **dogs** and **horses**); **swine**, 2–6 h; **baboons**, 5 h; **rabbits**, 3 h; **cattle**, 30–82 h; **goats**, 14.5 h (males) and 19 h (females). In **mice, rabbits, guinea pigs**, and **horses**, it disappears in a few hours. *Contraindicated* in cases with severe cardiac, renal, or hepatic pathology. Daily oral doses of 2–8 g to **horses** have been associated with the development of necrotizing portal phlebitis. Its use is usually prohibited for 48 h before a **horse** race. Local rules must be respected. Usual **dog** dosage may be toxic for **cats**. Do NOT use in food-producing **animals**, to avoid residues for **man**. ***Antacids*** or increases in gastric pH decrease its absorption after oral use. Its use during reproduction in **rabbits** and **rats** at three to five times the recommended dosage for **man** caused some reduction in litter size, some fetal mortality, and a low incidence of teratogenic effects. **Rat** trials indicate that it, like other anti-inflammatory agents, can retard wound-healing. Gastrointestinal upsets, vomiting, jaundice, and lowered white cell count can occur, as well as a generalized fatal hemorrhagic syndrome, particularly in hypertensive **dogs**. Concentrated solution has caused phlebitis and ear-sloughing in **elephants**. Not for **horses** intended for food. Dosage of 10 mg/kg/day has led to toxicity in some **foals** within 3 days. Long-term use in **dogs** has been associated with fatalities. **Thoroughbreds** in training tend to develop acidic urine with lower excretion of phenylbutazone (higher levels in **horses** with alkaline urine). Thrombocytopenia reported in treated **cats** and **dogs**.

LD_{50}, **rat**: 375 mg/kg; **mouse**: 630 mg/kg. TD_{LO}, **rat** (6–15 days pregnant): 38 mg/kg.

4-PHENYL CATECHOL

Untoward effects: Severe and widespread eczemas in two **men** making a powder mixture containing this sensitizing agent for emulsions used in Verifax® copiers (Eastman-Kodak).

PHENYLCYCLOHEXENE

Untoward effects: *Chemical Marketing Reporter* headlined federal employee complaints versus the Environmental Protection Agency with "Who Watches the Watcher". Environmental Protection Agency announced a program to decrease public exposure to chemical emissions from new carpet and carpet installation materials after their headquarter **employees** suffered from eye and respiratory tract irritation, headaches, and dizziness. 2 days before announcing the program, it denied these emissions were responsible. During smoking of ***phencyclidine***, ~50% is hydrolyzed to phenylcyclohexene.

LD_{LO}, **rat**: 5 g/kg.

PHENYLCYCLO-HEXYLPYRROLIDINE

= *PHP*

A *phencyclidine* analog.

Untoward effects: In the 1980s, it was found with increasing frequency in the urine of Los Angeles, California **probationers**. A 31-year-old **male** at an emergency room, 3–4 h after ingestion, presented with horizontal nystagmus, resting tremor of upper limbs, agitation, hostility, panic, and was suspicious and had thoughts of impending doom. On suspicion that it was a case of ***PCP*** intoxication, **he** was given 2 mg ***physostigmine salicylate*** IM. In 20 min, **he** showed progressive improvement and treatment continued at 20 min intervals. Urine analysis confirmed the problem was due to ***PHP***, and no other drug.

PHENYLDICHLORARSINE

***Military poison gas.* Shipped as a liquid.**

Untoward effects: Very irritating sternutator.

Minimum lethal dose, **guinea pig**: < 400 µg/l for 10–30 min; skin, **rabbit**: 8–10 mg/kg; Inhalation LC_{LO},

mouse: 370 mg/m^3. Intense and fatal vesicant burns on **rabbits** after application of < 0.02 cc.

p-PHENYLENEDIAMINE

= *4-Aminoaniline* = *Diaminobenzene*

Use: Photographic color film developing agent, in dye and resin, and organic synthesis.

Untoward effects: Inhalation irritates larynx and pharnyx, and causes bronchial asthma. Causes contact urticaria and allergic contact dermatitis, as well as sensitization reactions. Has occurred in **beauticians** using hair dyes, in ***rubber*** **workers**, where it has been used as a vulcanizing antioxidant, and in **workers** using stamp pads. National Institute for Occupational Safety and Health and OSHA suggest upper exposure limit of 0.1 mg/m^3. Immediate danger to life or health: 25 mg/m^3. Eczematous and lichenoid skin eruptions in color film **developers** and ***leather*** **workers**, where it is used as a dye. Alopecia in **women** from use of hair sprays, dyes, and home permanent solutions. Allergic **people** may cross-react with *p-**aminobenzoic acid, "caine" drugs***, p-aminosalicylic acid, and ***sulfonamides***. **Doctors** and **dentists** were frequently sensitive to ***procaine***. **Laboratory workers** screening for gonorrhea have reacted to *N-**dimethyl-paraphenylenediamine***, a dye. Chronic renal failure reported in two **patients** who were exposed to it in hair dyes for 2 or more years. A 22-month-old child ate an unknown amount and died 3 h later. A 21-year-old **female** ingested a tablespoonful as therapy for constipation, and the following day, 2 h before admission to an emergency room, developed rhabdomyolysis, muscle pain (lasting 7 days), and engorgement of face and tongue. Slow recovery after treatment. A 50-year-old **male** developed rhabdomyolysis and eventually died after drinking, by mistake, a ***coffee***- like hair dye solution made from a hard lump of *p*-phenylenediamine. Intentional ingestion by 21 **patients** caused vomiting, severe facial, neck and pharyngeal edema, dyspnea, and the death of **eight**. In another group of 39 attempted suicides, vomiting, pains, muscular rigidity, convulsions, edema of larynx and tongue, urticarial rash, albuminuria, and paralysis were reported. Dermal toxicity is decreased by combining it with a reducing agent.

LD_{LO}, **human**: 50 mg/kg.

In **dogs**, 10 mg/kg causes vomiting, diarrhea, and fatal coma.

LD_{50}, **rat**: 80 mg/kg; LD_{LO}, **cat**: 100 mg/kg; **rabbit**: 250 mg/kg.

PHENYLEPHRINE HCL

= *Adrianol* = *Ak-Dilate* = *Ak-Nefrin* = *Isophrin* = *Meta Synephrine* = *Mezaton* = *Neophrin* = *Neo-Synephrine* = *Prefrin* = *Pyracort D* = *Mydfrin* = *m-Sympatol*

Sympathomimetic **or** ***α_1-adrenergic agonist, mydriatic, vasoconstrictor.*** **Synthetic. Structurally similar to *ephedine* and *epinephrine*.**

Untoward effects: Local application to the eye causes transient pain, transitory increase then decrease of intraocular pressure, angle-closure glaucoma, floaters and occasionally eyelid retraction, and cycloplegia. Prolonged pupil dilation has occurred; one such **patient** was taking ***guanethidine***. Use of > 0.25% for nasal decongestion often has a rebound effect within a few days. Systemic reactions include hypertension (especially in **neonates** and the **elderly**), tachycardia, angina, ventricular arrhythmias and extrasystoles, sinus bradycardia, pancytopenia, stroke, myocardial infarction, cardiac arrest, and subarachnoid hemorrhage (particularly with high concentration [10%] or use on a ***cotton*** conjunctival pack). Even 2.5% has caused severe hypertension in premature **infants** and **patients** with orthostatic hypotension. Nailbed purpura, xerostomia, pulmonary embolism, psychotic state, and hallucinations reported. Hyperglycemia, glycosuria, and acetonuria simulating diabetes in very young **neonates** taking it in a syrup for coughs or gastrointestinal infections. Hypertensive crises from chronic intoxication or abuse with its use in nasal decongestants and cough medicines. Overdoses have caused palpitations, headache, and vomiting, in addition to hypertension.

Drug interactions: ***Hydrocortisone*** potentiates its effect on topical eye application by increasing its affinity for ***adrenergic receptors***. ***Debrisoquin*** with it can cause life-threatening hypertension. Pressor response is increased by ***β-adrenergic blockers***, ***monoamine oxidase inhibitors*** (phenylephrine is rapidly metabolized by ***monoamine oxidases***), and ***tricyclic antidepressants***. Use with ***guanethidine*** potentiates its action and prolongs mydriasis. Partially reverses beneficial effects of ***nitroglycerin***. Cardiac arrhythmias with ***halothane***.

Use in pregnant **dogs** has impaired maternal side of fetal circulation, leading to hypoxia. Will narcotize many freshwater **rotifers** at 0.1–0.5% concentration in acidic water; poor effect in alkaline waters.

LD_{50}, **rat**: 350 mg/kg; **mouse**: 120 mg/kg; IV, **mouse**: 21 mg/kg; **rabbit**: 500 µg/kg. IV, LD_{LO}, **rat**: 6.8 mg/kg.

PHENYL ETHER

= *Diphenyl Ether* = *Diphenyl Oxide* = *Oxybisbenzene* = *Phenoxy Benzene* = *Phenyl Oxide*

Use: In organic synthesis, heat-transfer fluids, antimicrobials, and surfactants. Heavy use (~100,000 lbs/year) as a perfume agent (0.05–0.4%) with ***geranium***-like odor for soaps, detergents, and perfumes.

Untoward effects: Its vapors are very irritant to eyes, nose, and throat. Overexposure causes nausea. No sensitization in petrolatum on 25 **human volunteers**.

Eye and nose irritation in **rats** and **rabbits**, but not **dogs**, exposed to 20 and 10 ppm, respectively.

LD_{50}, **rat**: 3.37 g/kg; acute dermal, **rabbit**: > 5 g/kg.

PHENYLETHYL ACETATE

= *β-Phenethyl Acetate*

Use: In fragrances (~50,000 lbs/year) for soaps, detergents, cosmetic creams and lotions, and foods (0.01–1.0%).

Untoward effects: No sensitization of 10% in petrolatum on 25 **human volunteers** and no irritation when applied undiluted to the skin of 20 **volunteers**.

Slightly irritating to intact or abraded **rabbit** skin after 24 h of full-strength material under occlusion.

LD_{50}, **rat**: > 5 g/kg; acute dermal, **rabbit**: 6.21 g/kg.

PHENYLETHYL BENZOATE

Use: Limited (~1,000 lbs/year in U.S.) as a ***rose*** or ***orange*** aroma in soaps, detergents, cosmetic creams and lotions, perfumes, and foods at 0.01–0.8% concentrations.

Untoward effects: No irritation after 48 h closed-patch test of 8% in petrolatum on **humans** and no sensitization reactions of the same on 25 **human volunteers**.

Non-irritating when full-strength material was applied to the backs of **rabbits**, **hairless mice**, and **swine**. No phototoxic effects on **hairless mice** or **swine**.

LD_{50}, **rat**: 5 g/kg; acute dermal, **rabbit**: > 5 g/kg.

PHENYLETHYLHYDANTOIN

= *Nirvanol*

***Anticonvulsant*. A major metabolite of *mephenytoin* (q.v.).**

Untoward effects: Fever, fixed-drug eruptions, serum sickness, eosinophilia, and capillary injury in 90% led to its disuse.

PHENYLETHYL ISOVALERATE

Use: Limited (~1,000 lbs/year in U.S.) as a fragrance (0.001–0.2%) in soaps, detergents, cosmetic creams and lotions, perfumes, and foods (5 ppm).

Untoward effects: No sensitization by 2% in petrolatum on 26 **human volunteers**.

Non-irritating for 24 h of full-strength material under occlusion on intact or abraded **rabbit** skin.

LD_{50}, **rat** and acute dermal, **rabbit**: > 5 g/kg.

PHENYLETHYL PHENYLACETATE

= *Phenethyl Phenylacetate*

Use: Estimated at 10,000 lbs/year as a fragrance in soaps, detergents, cosmetic creams and lotions, and perfumes (up to 0.8%), and in foods (10 ppm).

Untoward effects: No irritation or sensitization of 2% in petrolatum on **human volunteers**.

No adverse effects of feeding it to **rats** at 1,000–10,000 ppm of their diet/17 weeks.

LD_{50}, **rat**: 15.3 g/kg.

PHENYLETHYL PROPIONATE

= *Phenethyl Propionate*

Use: In fragrances (~5,000 lb/year) for soap, detergents, cosmetic creams and lotions, perfumes, and foods (usually at 0.002–0.8%).

Untoward effects: No sensitization of 8% in petrolatum on 25 **human volunteers**.

Full-strength to intact or abraded **rabbit** skin for 24 h under occlusion was not irritating.

LD_{50}, **rat**: 4 g/kg; acute dermal, **rabbit**: > 5 g/kg.

PHENYLETHYL TIGLATE

= *Phenethyl Tiglate*

Use: Limited (~1,000 lbs/year) as a fragrance in soaps, detergents, cosmetic creams and lotions, perfumes, and foods, usually at 0.01–0.6%.

Untoward effects: No sensitization or irritation of 6% in petrolatum on 26 **human volunteers**.

Full-strength to intact or abraded **rabbit** skin for 24 h under occlusion was mildly irritating.

LD_{50}, **rat** and acute dermal, **rabbit**: > 5 g/kg.

PHENYLHYDRAZINE

= *Hydrazinobenzene*

Use: In manufacturing ***antipyrine***, dyes, reagents, and in a stabilizer for explosives.

Untoward effects: Hemolytic anemia (each gram destroys ~6 g of hemoglobin), especially in glucose-6-phosphate dehydrogenase-deficient **people**. Acute symptoms include vomiting, diarrhea, headache, vertigo, anorexia, fatigue, severe dyspnea, asthenia, eye and skin irritations, and pruritus. Chronic exposure causes skin sensitization; eczematous dermatitis with redness, swelling, and rash; splenic pain, hemolytic jaundice and anemia; renal damage; bilirubinemia, ***iron*** increase in liver and spleen, and increased urinary ***urobilin*** and ***urobilinogen***. Urine turns darker on standing.

Causes hemolytic anemias and gastroenteritis in **cats**, **guinea pigs**, **rabbits**, **poultry**, **rats**, **lambs**, and **sheep**. Produces hyperbilirubinemia in **horses** and porphyrias in **rabbits**. Chronic oral use in **mice** associated with increase in lung cancers. Thin blood slides have a distinctive green color. Possibly, carcinogenic in some **animals**.

LD_{50}, **rat**: 188 mg/kg; **rabbit** and **guinea pig**: 80 mg/kg. LD_{LO}, **dog**: 200 mg/kg; **mouse**: 175 mg/kg.

PHENYL ISOCYANATE

= *Carbanil* = *Phenyl Carbimide* = *Phenyl Mustard Oil*

Use: Reagent for identifying ***alcohols*** and ***amines***.

Untoward effects: Severe eye irritant. ***Methyl isocyanate*** (q.v.) in the Bophal disaster is the most toxic of the isocyanates, and the phenyl is about the fourth. ***Phenyl isothiocyanate*** can be extracted from ***turnips*** and is an effective insecticide.

LD_{50}, **rat**: 940 mg/kg; skin, **rabbit**: 7.13 g/kg.

PHENYLMERCURIC ACETATE

= *Acetoxyphenyl mercury* = *Agrosan* = *Agrozan* = *Cekusil* = *Ceresan Slaked Lime* = *Gallotox* = *HL 331* = *Hong Nien* = *Liquifene* = *Lorophyn* = *Mersolite* = *Nylmerate* = *Pamisan* = *Phenmad* = *Phenylmercury Acetate* = *Phix* = *PMA* = *PMAC* = *PMAS* = *Riogen* = *SC 110* = *Scutl* = *Shimmer-ex* = *Tag Fungicide* = *Tag HL-331*

The designation *PMA* is also used for *methoxyamphetamine*.

Use: Fungicide, herbicide, pharmaceutical preservative, in latex ***paint***, eye cosmetics, and treating ***leather***. Contains ~60% ***mercury***.

Untoward effects: Occasionally allergic contact dermatitis. Repeated topical use raises tissue levels and symptoms of ***mercury*** (q.v.) poisoning. Also toxic orally; can cause lens opacity. Can damage contact lenses. Vapors in confined areas can be dangerous.

LD_{LO}, **human**: 5 mg/kg.

In **quail**, dietary concentration of 414 ppm is lethal to 20%, and 1,087 ppm to 90%. Depression, slow movement or inability to move, inappetence, frequent fecal passage, and some lacrimation are symptoms of acute poisoning in **chicks**. It concentrates in their livers and kidneys. Poisoning in large numbers of **veal calves** in the Netherlands, Italy, and France from contamination of a milk substitute. **Mink** were poisoned by eating starch manufactured for wallpaper paste that contained it. It took 3 weeks to deplete it from body tissues. Chromosome disorders dramatically increased in **pigs** fed it. Vaginal application of 0.1 mg to pregnant **mice** produced fetal death, malformations of spinal cord, and abnormal tails. Rapid and extensive gastrointestinal absorption in a **horse**, with bad breath, anorexia, decreased weight, degeneration and necrosis of proximal renal tubules, necrotic foci in facial and masticating muscles, splenic trabeculae, and myocardium after dosing with 400 µg ***mercury***/day/190 days.

LD_{50}, **mouse**: 25 mg/kg; **rat**: 30 mg/kg; **mallard duck**: 878 mg/kg; **pheasant**: 169 mg/kg; **roosters**: 3.2 g/kg; **male chicks**: 100 mg/kg. LD_{100}, **roosters**: 3.6 g/kg.

PHENYLMERCURIC BENZOATE

Use: Preservative on surgical sutures and a fungicide.

Untoward effects: Dermatitis in 12/21 factory **workers** exposed to it. See ***Mercury***.

PHENYLMERCURIC BORATE

= *Famosept* = *Hydromerfen* = *Merfen* = *Merphen*

Disinfectant, antiseptic, antifungal.

Untoward effects: Allergic contact dermatitis in **nurses** using 2% solution for hand disinfection. See ***Mercury***.

PHENYLMERCURIC CHLORIDE

***Agricultural fungicide*. Contains ~64% *mercury* (q.v.).**

Untoward effects: Daily doses of 0.19–4.56 mg ***mercury***/kg/day for up to 90 days in **swine** led to weight loss, diarrhea, necrotic typhlitis and colitis, and nephrosis.

LD_{LO}, **human**: 5 mg/kg.

LD_{50}, **rat**: 60 mg/kg.

PHENYLMERCURIC NITRATE

= *Phenmerzyl* = *PMN*

***Antiseptic, antifungal, bactericide*. Contains ~63% *mercury* (q.v.).**

Use: Widely used preservative is cosmetics, eye preparations, ***paints***, and ***paper*** manufacturing slimicides.

Untoward effects: Lens opacity (mercurialentis). Local irritant. Occasionally cross-sensitivity in eye preparations to ***thimerosal***. Dosage of 40 mg qid/1 day led to diarrhea in a **man** and **his** urine inhibited bacterial growth. Vapors in confined areas can be dangerous.

LD, **rabbit**: 10 mg. IV, LD_{50}, **mouse**: 27 mg/kg. IV, LD_{LO}, **rabbit**: 5 mg/kg.

PHENYLMERCURIC PROPIONATE

Antifungal.

Untoward effects: Contact urticaria. A 5-year-old **male** who helped his **mother** paint part of a kitchen and bedroom was hospitalized with acrodynia and evidenced "pink disease" of the hands and feet, as well as peripheral neuritis. ***Mercury*** (q.v.) was found in the **boy's** urine. It emphasizes the need for ***paints*** containing mercurials to be used outside the home, where ill-effects of mercurial vapors are minimized.

IP, LD_{LO}, **mouse**: 2 mg/kg.

N-PHENYL-*β*-NAPHTHYLAMINE

= *2-Anilinonaphthalene* = *β-Naphthylphenylamine* = *PBNA*

Antioxidant for neoprene rubber. **Commercial material contains 20–30%** ***β-naphthylamine*** **(q.v.).**

Untoward effects: Can cause eye and skin irritation, acne, leukoplakia, and hypersensitivity to sunlight. Metabolized to ***β-naphthylamine*** (q.v.), a known carcinogen. In the 1970s and 1980s, it was estimated 15,000 **workers** were at risk during its manufacture and use.

LD_{50}, **rat**: 1.63 g/kg; **mouse**: 1.23 g/kg; inhalation, **rat**: 9 g/kg.

o-PHENYLPHENOL

= *Dowicide 1* = *2-Hydroxydiphenyl* = *Orthoxenol*

Bacteriostat, fungicide, antiviral germicide.

Use: In cutting fluids, ***glues***, pastes, soaps, detergents, ***paints***, ***paper***, ***leather***, surface disinfectants, dyes, in the ***rubber*** industry, and on ***oranges*** in transit.

Untoward effects: Leukoderma with occlusive patch-testing. Widespread dermatitis over **his** whole body in a young **man** after applying a popular hand cream containing it. **Humans** appear to tolerate up to 5% concentrations in ***sesame oil*** on skin.

The *sodium tetrahydrate* form = ***Dowicide A*** = ***Natriphene*** is irritant to **human** skin at concentrations > 0.5%, and can cause corneal necrosis. Maximum daily oral intake needs to be < 1 mg/kg.

LD_{LO}, **human**: 500 mg/kg.

Dietary levels of 2% for 2 years caused slight retardation, kidney tubule dilation, and residues in the kidneys of **rats**. **Dogs** survived on 1 and 3 g/kg; **cats** died within 15 and 6 h, respectively. Symptoms of intoxication include incoordination, muscle fasiculations, depression, and coma with respiratory and cardiac depression. Pathology was similar in **cats** and **dogs**, leading to hemorrhagic gastroenteritis and ulceration, tubular nephrosis, and glomerular protein leakage. Toxic doses in **rats** cause central nervous system depression.

The *sodium* form has been carcinogenic in **rats**.

LD_{50}, **rat**: 2.5 g/kg.

PHENYLPHOSPHINE

= *PF* = *Phosphaniline*

Untoward effects: **Laboratory animal** trials indicate anemia, testicular degeneration, anorexia, diarrhea, lacrimation, tremors of hind legs, and dermatitis. Foul-smelling liquid, capable of spontaneous combustion of high vapor concentrations in air. National Institute for Occupational Safety and Health sets upper exposure limit of 0.05 ppm.

Inhalation LC_{50}, **rat**: 38 ppm/4 h.

PHENYLPROPANOLAMINE

= *Acutrim* = *Kontexin* = *Monydrin* = *Mydriatin* = *dl-Norephedrine* = *Obestat* = *PPA* = *Proin* = *Propadrine*

Anorexic, decongestant, bronchodilator, vasoconstrictor, sympathomimetic, α-adrenergic. **Found in nearly 100 products.**

Untoward effects: The most serious is hypertension, which can lead to encephalopathy or intracerebral hemorrhage. Severe hypertension and postural hypotension has followed ingestion of a single tablet containing 85 mg. Arrhythmias (premature atrial and ventricular contractions), tachycardias, second degree atrioventricular-block, palpitations, insomnia, psychotic reactions, grand mal seizures, tremors, restlessness, increased motor activity, agitation, hallucinations, rhabdomyolysis, diuresis, nephritis, acute renal failure, non-hemorrhagic stroke, headache, dizziness, nausea, vomiting, abdominal cramps, hot flashes, dyspnea, depression, diaphoresis, blurred vision, exacerbation of schizophrenia, confusion, and Raynaud's phenomenon. Overuse has caused rebound exacerbation of symptoms and addiction. An association has been reported with increased Hodgkin's disease, lymphoma, and leukemia. Adverse symptoms in **some** from doses as low as 17.5 mg/kg (10 mg/kg when used with ***caffeine***). **Some** have attempted suicides with it. Pseudopheochromocytoma and cardiac arrest in a 16-year-old patient. Fatal intracranial hemorrhage after ingestion of 300 mg (as four 75 mg capsule). The small irreversible risk of hemorrhagic stroke or bleeding in the brains of young **women** has caused the FDA to request that **people** discontinue its use. Therapeutic serum levels reported as 0.1–0.5 μg/ml and toxic at 2 μg/ml.

Drug interactions: Use with ***monoamine oxidase inhibitors*** leads to hypertensive crisis with secondary effects, hyperthermia, convulsions, coma, and even death, as

monoamine oxidase is irreversibly inhibited. ***Caffeine*** magnifies the toxic effects of each and has often been called ***"Legal High"***. A 27-year-old **female** safely taking 85 mg/day/several months, developed a severe bifrontal headache within 15 min after taking 25 mg ***indomethacin***; 30 min later, **her** systolic pressure was 210 mmHg and **her** physician was unable to record **her** diastolic pressure. **She** was given 10 mg IV ***morphine*** and hospitalized. Within 15 min, **her** blood pressure was 150/80, rising to 200/100, and later falling to 160/90. **She** was discharged with intermittent headaches and anxiety. Rechallenge produced blood pressure of 200/150. Both drugs can raise blood pressure and inhibition of ***prostaglandin*** synthesis by ***indomethacin*** may have exacerbated the problem. Can antagonize hypotensive effects of ***bethanidine, debrisoquine, guanethidine***, and some of ***diazepam's*** effects. Psychic disturbances reported with ***brompheniramine***; severe hypertension in 31-year-old **male** when it was added to therapy with ***methyldopa*** and ***oxprenolol***; papular, pustular, bullous eruptions, hemorrhagic stomatitis, and shock from various combinations with ***chlorpheniramine, isopropamide, methapyrilene, pheniramine***, or ***sulfadiazine***. Use with ***chlorpheniramine*** has caused excessive dryness of nose, throat, or mouth, headache, rash, weakness, palpitations, angina, increased or decreased blood pressure, thrombocytopenia, leukopenia, agranulocytosis, hemolytic anemia, drowsiness, insomnia, nervousness, dizziness, irritability, ataxia, tremors, convulsions, visual disturbances, nausea, vomiting, abdominal pain, diarrhea or constipation, anorexia, dysuria, and chest tightness. Sudden increases in blood pressure with ***oxprenolol***. Death due to respiratory distress in a 15-year-old **female** after deliberate ingestion of 400 × 450 mg with 1–1.8 mg ***belladonna alkaloids***.

In **animals**, adverse effects and warnings would be similar to ***ephedrine's*** (q.v.).

LD_{50}, **rat**: 1.5 g/kg.

PHENYLQUINONE

= *2-Phenyl-1,4-benzoquinone*

Untoward effects: For experimental induction of writhing by IP use in **rats** and **mice**, for testing ***analgesics*** and analgesic activity of ***narcotic antagonists***.

PHENYL SALICYLATE

= *Musol* = *Salol*

Use: Enteric coating for tablets, urinary acidifier, ultraviolet light absorber in suntan oils and creams, in ***lacquers, polishes***, waxes, and adhesives, to stabilize and plasticize resins, and as an analgesic, antipyretic, and anti-inflammatory.

Untoward effects: ***Aspirin***-sensitive **people** may also show urticaria from it. Dusts should not exceed 10 mg/m^3, to prevent eye, skin, and respiratory irritation. Because of the large amount of ***phenol*** it contains, it is more toxic than a corresponding dose of ***salicylic acid***. Fatalities with its use have been reported. It is especially dangerous in **patients** with renal disease. Breaks down into ***phenol*** and ***salicylic acid***, and will color urine dark green.

LD_{LO}, **human**: 50 mg/kg.

LD_{50}, **rat**: 3 g/kg. LD_{LO}, **rabbit**: 3 g/kg.

PHENYLTHIOUREA

= *NCI-CO 2017* = *Phenylthiocarbamide* = *PTU* = *U 6324* = *USAF EK 1569*

Use: In medical genetic studies on taste perception. ***Herbicide***.

Untoward effects: A bitter substance that ~30% of **Caucasians** (including **people** from Japan and India) cannot taste. These **non-tasters** are more likely to have goiters and dental caries. All the captive **chimpanzees** in England were tested and found to have the same percentage of non-tasters as the **humans**. When researchers were asked, "How did you find out?", the answer was quite simple: "we offered them a sugar solution containing ***PTU***. If they were tasters, they spat it in our faces; it was an all or nothing reaction." African and American **Indian** populations contain few **non-tasters**.

In **quail**, a dietary concentration of 1,500 ppm killed 10% and 3,000 ppm killed 80%. LC_{50} was 2,214 ppm. Pulmonary edema and pleural effusion in **rats** (LD: 3 mg/kg) and **mice** (LD: 37 mg/kg). The dominant gene that confers **insect** resistance to ***benzene hexachloride, DDT***, and ***parathion*** has also conferred abnormal susceptibility to the lethal effects of phenylthiourea.

LD_{50}, **rat**: 3 mg/kg; **mouse**: 10 mg/kg; **rabbit**: 40 mg/kg.

PHENYLTOLOXAMINE

= *Antin* = *Bristamin* = *C 5581 H* = *Phenoxadrine* = *PRN*

Antihistamine.

Untoward effects: Has some anticholinergic and sedative effects. Occasionally nausea at high dosage. Overdose problems have primarily been in association with other drugs, such as ***hydrocodone*** in a sustained-release resin complex antitussive, ***Tussionex***®. Deaths occurred.

Has caused central nervous system depression, ataxia, muscle weakness, emesis, convulsions, and death by respiratory paralysis in **dogs** and **mice**. No obvious toxic manifestations in **dogs** given 10, 20, or 40 mg/kg/day/1 year or in **rats** given 25 or 50 mg/kg/day/> 1 year.

LD_{50}, **rat**: 840 mg/kg. IP, LD_{50}, **mouse**: 246 mg/kg; IV, **dog**: 50 mg/kg.

PHENYRAMIDOL

= *Abbolexin* = *Anabloc* = *Analexin* = *Cabral* = *Fenyramidol* = *IN 511* = *Miodar* = *MJ-505* = *NSC-17777*

***Analgesic, skeletal muscle relaxant.* Oral, IM, and rectal use.**

Untoward effects: Nausea, epigastric distress, drowsiness, rashes, and pruritus. Dizziness and weakness have followed IM. Cross-sensitivity with ***salicylates***.

Drug interactions: Potentiates ***oral anticoagulants*** (***anisindione, dicumerol, phenindione***, and ***warfarin***), ***phenytoin*** and ***tolbutamide***, and, probably other ***sulfonylureas***, by inhibiting enzyme metabolism. Has also potentiated the side-effects of ***oral contraceptives***.

Prolongs the half-life of ***dicumerol*** in **rabbits** and **mice**.

SC, LD_{50}, 405 mg/kg.

PHENYTOIN

= *Dihydantoin* = *Di-Lan* = *Di-Phen* = *Diphenylhydantoin* = *DPH* = *Enkefal* = *Epelin* = *Epinat* = *Fenatoin* = *Fenytoin* = *Hydantoin* = *Novodiphenyl* = *Phentoinum* = *Phenydantoin* = *Phytoin* = *Sodanton*

The name *Dilan* is also used for an insecticidal combination of *Bulan* and *Prolan*. *Diphen* is a disinfectant.

Anticonvulsant, antiepileptic.

Untoward effects: ***Blood***: Erythroid aplasia, anemia, megaloblastic anemia, hemolytic anemia, macrocytic anemia, leukopenia, delayed fatal agranulocytosis. Thrombocytopenia, pancytopenia, leukocytosis, eosinophilia, decreased serum ***folate*** levels in ~20% of **patients**, hypoprothrombinemia, increased serum alkaline phosphatases, hypercalcemia (25–55% have decreased serum ***calcium***), increased serum ***copper***, increased serum ***urates***, increased serum ***cholesterol***, decreased plasma ***glucagon***, increased serum γ-glutamyl transpeptidase; reticuloendothelial reactions mimicking malignant lymphomas, failure of terminal differentiation of B-lymphocytes, myelofibrosis, acute intermittent porphyria, and mononucleosis syndrome reported.

Bone: Osteomalacia, osteoporosis of metaphyseal areas of long bones with hypophosphatemia and hypocalcemia, calvarial thickening, decreased bone mineral mass, rachitic changes in wrists, forearms, or ribs, rickets, bowed legs, decreased ***calcifediol***, fractures, bone marrow granulomata, and abnormalities of dental roots.

Cardiovascular: After oral use sinus bradycardia, arrhythmias, idioventricular rhythm and no discernable atrial activity, anteroseptal infarction, shortened A-H interval, hypotension, and intermittent sensation of heart beat with bigeminus rhythm have occurred; disseminated intravascular coagulation in **female** taking it as an anticonvulsant (also developed exfoliative dermatitis, hepatitis, cutaneous hypersensitivity, vasculitis, and microangiopathic hemolytic anemia).

After pulmonary IA use paroxysmal coughing and substernal burning sensation from 250 mg in 5 ml diluent in 5/12 **patients** was reported.

IV use for treatment of cardiac arrhythmias caused catastrophic adverse effects, including hypotension, shock, ventricular arrhythmias and fibrillation, impaired left ventricular function, asystole, and respiratory and cardiac arrest with fatalities. Occasionally yawning, somnolence, pain at injection site, nausea, light-headedness; nystagmus, dizziness, angiospasm, and disorientation, especially if plasma levels > 20 μg/ml. The intensity of these reactions is often related to its concentration and rate of administration.

Children: Erythema multiforme (Stevens–Johnson syndrome) in a 2½-year-old **male** treated for convulsions. Drug therapy with it should be discontinued in the presence of such skin eruptions. Failure to do so, in this case, was fatal. Intoxication in a 3-year-old **male** with flushing, dazed condition, and ataxia after 2 h; coma, absent tendon and light reflexes, mydriasis, choreiform movements, vomiting, and tonic–clonic convulsions after 4 h, followed by hypothermia, tachypnea, and intestinal atony. Transient gingivitis and bleeding of gums 4 days later. Same course reported in a 2½-year-old **female**. Athetosis in a 3⅔-year-old **male** and a 3½-year-old **male**. An infectious mononucleosis-like syndrome in several young **children** 3–7 weeks after starting therapy. Slight increase in serum lymphocytes reported in many **children**. Prolonged toxicity (13 days) in a 6-year-old **female**, even with low serum levels.

Central nervous system: Often bizarre extrapyramidal symptoms, mental clouding, schizophrenic and toxic psychoses, loss of deep tendon reflexes, peripheral neuropathy, nystagmus, ataxia, astasia, abasia, tremors, asterixis, vertigo, headache, dizziness, cerebellar speech disorders, opisthotonus, dementia, orofacial dyskinesias or choreoathetosis, neuromuscular blockade, cerebellar atrophy and degeneration, hallucinations, and depression. Hallucinations in a 65-year-old **female** with blood levels of 32 μg/ml. Many of these symptoms are associated with serum levels of 30–50 μg/ml for a long period of time. Right-sided hemiparesis and facial palsy in a 31-year-old **male**, 7 days after treatment with it for post-traumatic epilepsy. A **patient** had an impulse for word reversal after starting therapy with it. Has aggravated symptoms in a 55-year-old **male** with stable tardive dyskinesia after taking 300 mg/day.

Diagnostic tests: Maximum decrease in ***protein-bound iodine*** test results after 2 weeks of treatment. Falsely low ***theophylline*** blood levels (up to 10 μg/ml). Can increase plasma ***glucose*** by ~30% and plasma ***insulin*** levels by ~75%. False negative ***glucagon*** stimulation test results in insulinoma **patients**. ***Clove Oil*** and ***Sassafras Oil*** in ***Dr Hand's Teething Lotion***® caused false positive phenytoin serum levels; confirmed in **dog** study. Interferes with ^{14}C ***aminopyrine*** test for hepatic microsomal activity and impaired response to ***dexamethasone*** in ***dexamethasone*** suppression test. ***Oxaprozin*** may cause erroneously high phenytoin serum levels.

Eye: Nystagmus (usually lateral), double vision, ophthalmoplegia, scotoma, corneal scarring (100 mg/day/5 years), glaucoma (in a **neonate**) and mydriasis and blurred vision on overdoses. Conjunctivitis, ptosis, and hallucinations also reported.

Fetus and neonate: Crosses the placenta, causing cleft lip and/or palate. These abnormalities and craniofacial anomalies: facial hemihypertrophy, nail, digital, nasal, limb, nipple and pulmonary hypoplasia, agenesis of radius and thumb, short nose with anteverted nares, flat nasal bridge, simian crease, mongoloid facies, short or webbed neck, hydrocephalus, microcephaly, anencephaly, brachycephaly, triginocephaly, pilonidal sinus, clinodactyly, abnormal **male** genitalia (including hypospadias, cryptorchidism, and micropenis), prenatal-onset growth deficiency and mental deficiency, postnatal growth retardation, mental retardation, esophageal and ileal atresia, tracheoesophageal fistula, epicanthic folds, ocular abnormalities and agenesis, wide fontanels, talipes, equinovarus, ptosis, strabismus, low-set ears, wide mouth with prominent lips, cardiac malformations, aortic stenosis, a solitary kidney, hiatus, inguinal and diaphragmatic hernias, tetralogy of Fallot, ventricular-septal defect, patent ductus arteriosus, omphalocele, myelomeningocele, congenital dislocated hip, bladder extrophy, rectourethral fistula, imperforate arms, wide-set or accessory nipples, neonatal hemorrhage, and decreased scores for gross motor index, fine motor index, language index, and personal-social index are often lumped together as "**fetal** hydantoin syndrome", seen in 11–40% of **offspring** of **mothers** who took the drug during pregnancy (especially in the first trimester). Has interfered with ***folic acid*** metabolism and caused hemorrhagic disease and neuroblastoma of the **newborn**. Excreted in breast milk at 6 μg/ml when **maternal** serum level is 28 μg/ml; the **infant** can ingest significant amounts to induce microsomal enzyme induction. Drug has been in these **neonates** for up to 5 days postpartum. Hematuria in a **neonate** associated with abnormal prothrombin time. A **mother** who took the drug during **her** pregnancy gave birth to a **child** who had no eyes; with organic brain dysfunction, anxiety; speech, language, and learning difficulties; tremors of **her** left hand and head, and was awarded $7 million in Federal Court. Vomiting and melena in a 4-day-old **male** responded to ***vitamin K***. With plasma or serum levels of 0.3–4.5 mg/ml, breast milk showed 0.6–1.8 mg/100 ml and ~1.4% of the **mother's** drug appeared/day in the milk. Other **neonates** showed tremors, drowsiness, rashes, ecchymoses, methemoglobinemia, and decreased ***bilirubin***. Failure of physicians to check existing literature on phenytoin's risk of birth defects led to a verdict against them for developmental and physical defects in two **infants** born to a **mother** who was prescribed the drug during two pregnancies (U.S. District Court, Washington State and Supreme Court of Washington). **Maternal** serum levels usually decrease during pregnancy.

Gastrointestinal: Has been associated with malabsorption syndrome, stomatitis, xerostomia, ageusia, gingivitis, and gingival hyperplasia in interdental spaces.

Glucose: Hyperglycemia, non-ketotic coma, and death. Impaired ***insulin*** release and/or response associated with hyperosmolar state and hypotension.

Hypersensitivity: A multi-system anticonvulsant hypersensitivity syndrome occurs. High fever (105.4 °F/40.8 °C), vomiting, morbilliform and petechial rash, pitting edema, arthritis, lymphadenopathy, hepatomegaly, albuminuria, and thrombocytopenia in a 7-year-old **male** after 100 mg twice daily, then 50 mg twice daily. Fatal toxic epidermal necrolysis reaction in a 51-year-old **male** after 100 mg qid/~1 month (also receiving ***chlorpromazine***). Delayed general lymphadenopathy, hepatosplenomegaly, exfoliative dermatitis, and fever in a 2½-year-old **female** with phenylketonuria on 8 mg/kg/day/13 days. Also exhibited mild hemolytic anemia and plasmacytosis in the blood. Pseudolymphoma syndrome in a 9-year-old **male** on 4 mg/ day/18 days reversed after discontinuation of treatment. Eosinophilia often accompanies this syndrome.

Immunity: Polyendocrine deficiency syndrome due to decreased ***immunoglobulin A*** levels. Agammaglobulinemia in a 48-year-old **male** taking 100 mg tid/3 months. Chronic therapy often associated with general immunosuppression.

Liver: Hepatitis, cholangitis, increased aklaline phosphatase, slight increase in SGPT, hepatic porphyria cutanea tarda, fatal hepatic failure, centrolobular hepatic necrosis, and granulomatous hepatitis. Many of these are considered to be hypersensitivity reactions. Can be fatal.

Miscellaneous: Fibrous hyperplasia, facial deformity, fever, rhabdomyolysis, systemic lupus erythematosus, vertigo, nausea, low-back pain, polyarteritis nodosa, myopathy, decreased lymphocytes, choreoathetosis, and severe necrotic reactions when extravasated into surrounding tissues by IV use. In 1995, a phenytoin manufacturer paid a $10 million fine for concealing quality problems with it during 1990–1992. Exacerbates myasthenia gravis with oculo-bulbar and limb paralysis. There may be absorption difficulties in the **elderly**. Extravasation can cause edema and cyanosis in the affected arm. Impotence, decreased libido, and priapism also reported. Can cause ***carnitine*** deficiency.

Neoplasms: Malignant lymphomas and lymphosarcomas; neuroblastomas in **neonates** born to **mothers** who ingested it during pregnancy. A case report of malignant mesenchyoma in a **patient** with phenytoin-induced malformations.

Poisoning and suicide: Death in a 4½-year-old **female**, 3 days after accidental ingestion of 2,000 mg, attributed to brain stem compression and necrosis, due to severe cerebral edema. The mistaken use of a 100 mg/5 ml suspension, instead of a 30 mg/5 ml pediatric suspension, led to many cases of toxicity in **children**. A 6-year-old **male** developed ataxia and dysarthria after 100 mg tid/3 days, due to **pharmacist's** dispensing error instead of 30 mg capsules. Similar errors in dispensing 125 mg/5 ml instead of 30 mg/5 ml, as well as failure to shake suspension prior to use. Prolonged toxicity in a 5-year-old **male**, due to desorption from unpassed ***activated charcoal*** for ingestion of 20–30 50 mg tablets, emphasizing the need for a laxative as well. A 24-year-old **male**, in an attempted suicide, ingested > 2 g and was successfully treated with plasmapharesis. A group of 11 Army **aviators** were deliberately given contaminated ***coffee*** and developed mild neurological symptoms; two developed transient nystagmus. A **neonate** was overdosed (50 mg/kg) with injectable phenytoin, instead of a ***calcium gluconate*** solution that was ordered, due to similarity of colors of capsule and labels. Therapeutic serum levels are 1–2 mg%; 2–5 mg% is toxic, and 10 mg% is lethal. Ingestion and smoking of phenytoin adulterated ***cannabis*** by two 22-year-old **males** caused simulated botulism with vomiting, diarrhea, muscle stiffness and pain, xerostomia, dysphagia, blurred vision, and dysphonia. IV phenytoin contains 40% ***propylene glycol*** (q.v.); has a pH of 11, and has been implicated in causing acute hypotension.

Renal: Acute interstitial nephritis in a 63-year-old **female** after 300–600 mg/day/~3 weeks. Renal dysfunction in a 37-year-old **male** after 300 mg/day/3 weeks. Nocturnal enuresis after 350 mg/day (blood levels 21.9 μg/ml) in a **patient**. Occasionally detrusor hyperactivity. Polyarteritis and acute renal failure in a 20-year-old **male** after 300 mg/day/30 days. Can color urine pink, red, or red-brown.

Respiratory: Miliary pulmonary infiltrates and hypoxemia in a 47-year-old **male** after 100 mg qid. Cases of acute pulmonary disease after 4–6 weeks of therapy led to dyspnea, severe hypoxemia, fever, rash, and cough. Reduced pulmonary function in 45% of 40 **individuals** studied. Respiratory arrest after pulmonary eosinophilia, wheezing, fever, and diffuse pulmonary fibrosis. Rarely, hypersensitivity pneumonitis.

Skin: Allergic skin rash, Quincke's edema, acne (due to its androgenic effect), thickening of SC tissues of face and scalp, gross enlargement of lips and nose, hypertrophic retroauricular folds, erythroderma, fixed-drug eruptions, hyperpigmentation, erythema multiforme (black **children** are particularly prone to it), Stevens–Johnson syndrome, exfoliative dermatitis, increased thickening of heel pad, toxic epidermal necrolysis, purpura, photosensitization, acute intermittent porphyria, maculopapular and erythematous dermatitis, urticaria, exanthema, vasculitis, serum-sickness-like reaction, hirsutism, ochre-brown discoloration of nail plate, and diffuse "burning feet". Mycosis fungoides in a 14-year-old **male**, 6 weeks after start of treatment with 300 mg/day. A **child** with therapeutic blood levels developed a fatal skin reaction. Malaise, fever, sore throat (severe Stevens–Johnson syndrome in a 26-year-old **female** who previously had similar reactions to ***carbamazepine*** and ***gabapentin***. Local reactions reported after IV.

Usual therapeutic serum levels are 5–15 μg/ml; toxic at > 20 μg/ml. Coma and fatalities can occur with > 50 μg/ml.

LD_{LO}, **child**: 67 mg/kg; TD_{LO}, **child**: 11 mg/kg. Estimated LD, **human**: ~2 g; TD_{LO}, **human**: 1 mg/kg.

Drug interactions: May have accentuated ***acetaminophen***-induced hepatic necrosis in a 63-year-old **female** and **others**. Use with ***acetazolamide*** produces osteomalacia. Increased rate of metabolism in **alcoholics**. Hodgkin's disease in a **child** whose **mother** had taken 100 mg/day during **her** pregnancy, along with large quantities of ***ethyl alcohol***. ***p-Aminosalicylic acid, isoniazid***, and ingestion of large ***aspirin*** doses decrease enzyme metabolism, leading to increased phenytoin toxicity. ***Amiodarone*** 200 mg/day with phenytoin 300 mg/day for 15 days increased phenytoin toxicity. Several **patients** died from liver failure and necrosis, due to its enhancement of the hepatotoxic effects of the anesthetic, ***fluroxene***. Use with ***halothane*** leads to acute phenytoin toxicity. ***Antacids***, such as ***aluminum hydroxide*** with ***magnesium trisilicate***, initially show no interference with its serum concentrations until after the 5th or 6th day of continuous treatment, when there is a significant decrease in its concentration. Has potentiated the activity of ***anticoagulants*** by inhibiting their metabolism, and they (***dicumarol, phenprocoumon***, and ***warfarin***, except ***phenindione***), in turn, have inhibited its metabolism, leading to increased phenytoin serum levels and toxicity.

A biphasic interaction reported in a 70-year-old **female** after the introduction of phenytoin to a ***warfarin*** regimen; first, with increases in prothrombin time for 6 days, then a decrease below previous baseline. Screaming, confusion, nystagmus, and disorientation after ~4 weeks ***warfarin*** therapy in a 59-year-old **male** on phenytoin 100 mg tid/1 year. Prompt recovery after discontinuing phenytoin. A 60-year-old **male** on 300 mg/day was given ***apazone*** 600 mg twice daily for arthralgia and, after 2 weeks, developed increasing confusion, nausea, diplopia, vertigo, and nystagmus on lateral gaze, apparently, due to inhibition of phenytoin metabolism. ***Barbiturates***, particularly ***phenobarbital***, accelerate its metabolism in **man**, **dogs**, **rats**, and **mice**. Pseudohypoparathyroidism with paradoxical exacerbation of hypocalcemic seizures in a 32-year-old **female** treated for many years with 100 mg tid and ***phenobarbital*** 30 mg tid. Teratogenic potential increased with ***phenobarbital***. Florid rickets in a 16-year-old **female** from combination of phenytoin and ***phenobarbital*** 120 mg/day/8 years, even when receiving 600 IU ***vitamin D***/day. The combination increases the rate of ***vitamin D*** metabolism. **She** responded to 1200 IU/day. In **children**, phenytoin stimulates the enzymatic metabolism of ***hexobarbital***, ***pentobarbital***, and ***phenobarbital***. Macrocytosis, decreased serum ***folic acid***, agranulocytosis, **neonatal** hemorrhage at puncture site, hematemesis, and rectal bleeding in a 2-day-old **male** after treatment of the **mother** with the combination during pregnancy. Another **neonate** developed thrombocytopenic purpura and macrocytic anemia. **Children** born to **mothers** who took both during **their** pregnancy had distal extremity hypoplasia and nail hypoplasia. Bundle-branch block and coma in a 20-year-old **male** from overdoses. Increases ***sulfobromophthalein***, ***indocyanine green*** fractional clearance, and ***dibromsulfophthalein*** in a 43-year-old **male**. By enzyme induction, it decreases ***carbamazepine*** and ***valproic acid*** plasma levels and the latter, in turn, dramatically decreases phenytoin half-life from 10.6 to 6.4 h, yet, erythema multiforme has been reported with the combination. Also a case of extrarenal Wilm's tumor in an **infant** whose 23-year-old mother was on long-term therapy with both. ***Antibiotics***, such as ***chloramphenicol*** and ***tetracycline***, inhibit the enzymes, as do ***amphetamine, apazone, chlorpheniramine, chlorpromazine, cimetidine, cycloserine, diazepam, dicumarol, disulfiram, fluconazole, gabapentin, imipramine, isoniazid*** (in **slow-acetylator phenotypes**), ***itraconazole, ketoconazole, miconazole, omeprazole, oxyphenbutazone, pheneturide, phenylbutazone, phenyramidol, primidone, prochlorperazine, progabide, propoxyphene, ranitidine, sulfamethazole, sulfaphenazole, sulfathiame, thioridazine, tizanidine, tolbutamide, trimethoprim*** with ***sulfamethoxazole***, and ***warfarin***, increasing their serum concentration. Occasionally, ***chlordiazepoxide*** and ***diazepam*** have the same effect. Half-life of ***doxycycline*** is decreased 50% by phenytoin. Severe granulocytopenia reported with high IV dosage of it and ***cimetidine***. ***Diazoxide***, ***rifampin***, and high doses of ***aspirin*** and ***salicylates*** decrease its serum levels. It increases metabolism of ***bromfenac***, ***clonazepam*** (increases clearance by ~50%), ***DDT***, ***disopyramide***, ***estrogens***, ***haloperidol***, ***meprobamate***, ***mesoridazine***, ***methadone***, ***quetiapine***, ***quinidine***, ***rifampin***, and ***tolbutamide***, and decreases efficacy of ***oral contraceptives***. ***Loxapine*** and IV ***ciprofloxacin*** decrease phenytoin serum concentrations. Drugs causing hypokalemia will antagonize its antiarrhythmic action. ***Clonazepam*** clearance is decreased ~50% by it. ***Dexamethasone*** has decreased, and occasionally increased, phenytoin serum levels, and free ***dexamethasone*** concentrations are decreased by the combination. Mutual antagonism with oral ***diazoxide***. It and ***digitalis*** preparations may be additive in slowing the heart rate. ***Digitoxin*** half-life is decreased by it and ***digitoxin*** may also increase its metabolism. A Down's syndrome 53-year-old **male** receiving ***digoxin*** 0.25 mg/day and phenytoin 200 mg/day had the latter increased to 300 mg/day. After 2 months, **he** became lethargic, and showed rigidity, heart block, and coma. A month later, **he** was still unable to walk and evidenced some ataxia, nystagmus, and rigidity. Blocks usual response to ***dopamine*** and ***L-dopa***, and bradycardia and hypotension can occur. ***Folic acid*** lowers phenytoin serum levels 15–50% and can precipitate seizures in stabilized **epileptics**. Phenytoin, in turn, can cause a ***folic acid*** deficiency and precipitate a megaloblastic anemia. As for ***foods***, carbohydrates enhance and protein decreases its absorption. ***Vanilla pudding*** decreases its serum levels. Phenytoin potentiates ***griseofulvin***. It increases the metabolism and decreases the activity of ***hydrocortisone***, ***methylprednisolone***, and ***prednisolone***. Additive myocardial depression and complete sino-atrial arrest in a 57-year-old **male** with heart block, when ***lidocaine*** was used IV with phenytoin. Has caused ***lithium*** toxicity; may increase need for raised dosage of ***meperidine*** and may cause withdrawal symptoms 3–4 days after initiation of ***methadone*** use, as it increases the metabolism of the latter. Many ***phenothiazines*** cause a dramatic increase in phenytoin dosage required to control epilepsy. ***Phenobarbital*** is a major metabolite of ***primidone*** and interactions (above) with phenytoin are similar. Ophthalmoplegia from excessive doses in attempted suicides and megaloblastic anemia with this combination. In a preliminary trial, ***pyridoxine*** decreased its serum levels. Severe hemolytic anemia and toxic epidermal necrolysis reported with ***pyritinol***. ***Quinidine*** half-life decreases by ~50% and severe toxicity can occur with sudden phenytoin withdrawal. ***Sulthiame***, in limited trials, may increase phenytoin toxicity. ***Theophylline*** half-life is decreased by phenytoin and it decreases phenytoin's anticonvulsant concentration. Clinical hypothyroidism in a 32-year-old **female** stabilized on ***thyroxine*** after phenytoin administration. Confirmed by rechallenge. ***Valproic acid*** initially decreases phenytoin serum levels, but becomes normalized after 4–5 months. Others report an increase. ***Valproic acid*** 4 g/day and phenytoin 4.5 mg/kg to a 65-year-old **male** for 2 months led to severe pancreatitis. The importance of any "inerts" in a formulation can be best illustrated by events after an Australian company changed capsule diluent from ***calcium sulfate*** dihydrate to ***lactose***, with ordinary therapeutic doses suddenly becoming toxic. A 46-year-old **male** developed a ***tacrolimus***-phenytoin interaction, increased phenytoin levels and decreased ***tacrolimus*** concentration. Phenytoin toxicity in a 48-year-old **female** receiving 200 mg/day and ***tolbutamide*** 1 g tid/ 2 days, led to headache, nausea, ataxia, and nystagmus. **She** improved when ***insulin*** was substituted for ***tolbutamide***. ***Topiramate*** has increased phenytoin serum levels up to 25% and phenytoin decreases its serum concentration by ~50%. In some **children**, ***methyphenidate*** has increased its serum levels, leading to ataxia. ***Chlorpromazine*** potentiates its anticonvulsant action. It increases ***demeclocycline*** toxicity. Will increase hepatic metabolism of ***thyroxine***, as well as azole concentration, with a potential for failure of antifungal treatment. Can decrease ***clozapine, haloperidol, tirilazad***, and ***verapamil*** levels, and can increase some ***penicillins'*** effectiveness. ***Chloral hydrate*** and ***vigabatrin*** can cause dramatic decreases in its serum levels. Dizziness and tremors reported when used with ***diltiazem***. It can decrease ***furosemide*** absorption by up to 50% and ***furosemide*** decreases its absorption. ***Rifampin*** and ***zidovudine*** have decreased its blood levels and anticonvulsant activity. Can increase metabolism by CYP3A4 enzyme. Coadministration with ***vincristine*** increases the latter's systemic clearance by > 60%. Phenytoin's serum concentration is decreased dramatically when given with ***enteral feeding formulations***. ***Ticlodipine*** inhibits its clearance and downward adjustment of phenytoin dosage may be needed.

In **animals**, excessive dosage may lead to ataxia or gastric upsets. Although ***barbiturates*** may potentiate or accelerate its metabolism (**mice**, **dogs**, and **man**), they also potentiate the therapeutic benefits. ***Phenylbutazone*** and certain ***sulfonamides*** may inhibit its metabolism. May be a circulatory depressant in **cats**. Significant increase in collagen or dermal pathology, oral epithelial keratinization, and tensile strength of wounds occur in some species, especially when on high or continued dosage. Avoid sudden withdrawal of other ***anticonvulsants*** or use of ***lactose*** or ***lactose***-containing drugs or foods. ***Lactose*** may cause dramatic increases in serum levels and their duration. **Cats** metabolize the drug very slowly. In **rats**, phenytoin causes a ***folic acid*** deficiency and interferes with ***folic acid*** uptake by many tissues. Phenytoin pretreatment of **rats** increases liver microsomal enzyme activity, accelerates the metabolism of and decreases effect of ***benzphetamine, DDT***, and ***meprobamate***. ***Isoniazid*** causes a dose-dependent increase in phenytoin levels, and inhibits liver microsomal metabolism of it in **rats**. Pretreatment with ***phenobarbital*** or ***chlorcyclizine*** increase its half-life in **dogs**, but decrease its activity in the **mouse**. ***Lynestrenol*** pretreatment in the **mouse** also increases its metabolism, as well as ***phenobarbital's***, thereby decreasing their effects. Giving **mice** 10 mg/kg simultaneously with, or 30 min later, led to increased ***pentobarbital*** brain levels and sleeping time. The analeptics, ***nikethamide*** and ***picrotoxin***, have similar effects in the **mouse**. Teratogenic in **mice** and **rats**. In **mice** its use leads to increased bone resorption, shortened long bones, intrauterine growth retardation, syndactyly, ectrodactyly, cleft lip and palate, hydronephrosis, hydrocephalus, and orofacial anomalies. Fed to **mice**, it produced increased incidence of thymic and generalized lymphomas in females; increased incidence of thymic and mesenteric lymphomas and leukemias in males and females. A case report indicated a possibility of acute hypersensitivity, with hepatitis and death occurring in about 2 weeks, in a **dog**. Increases ***aniline*** and ***hexobarbital*** metabolism by the liver of **rats**; 0.2 mg/kg/day/10 days to **mice** interfered with conception. Ten times the recommended dosage caused fetal resorption in **rats**. ***Phenobarbital*** pretreatment can decrease half-life by two-thirds in **dogs**. Transplacental transfer occurs and, at 4 h after injections, levels are the highest in the **mouse** and are similar in **rats** and **rabbits**. Also crosses the placental barrier in the **goat**. **Rats** show an increase in seizure susceptibility after prenatal exposure to phenytoin. Diabetogenic in the **rabbit**.

TD_{LO}, **rat**: 1.5 g/kg. LD_{50}, **mouse**: 200 mg/kg. IV, LD_{50}, **dog**: 90 mg/kg; **rabbit**: 64 mg/kg; **mouse**: 117 mg/kg.

PHEOMELANIN

Red or yellow pigment that dominates in red hair; ***eumelanin*** predominates in black hair. Pheomelanin resists ***hydrogen peroxide***, preventing redheads from turning blond. ***Eumelanin*** is readily destroyed by ***peroxide***, lightening black, brown, and blond hair.

PHILODENDRON sp.

A very common houseplant and vine.

Untoward effects: Its tissues contain an irritating juice and/or crystals of ***calcium oxalate*** (q.v.) causing, after ingestion, severe throat and mouth burning sensations and swelling that have led to blisters and asphyxiation. This was due to an enzyme and/or ***asparagine***. Infrequently, it causes death. Chewing of the leaves causes an immediate intense pain. **Children** usually shriek with pain as soon as **they** bite the leaves or stems. **Some** become nauseous, vomit, and have abdominal pain and eye irritation. Occasionally, onset of pain is delayed 5–10 min. **Others** may salivate and have diarrhea. Skin contact with the juice can cause a dermatitis and pruritus. Ingestion by an **infant** produced lip, tongue, and esophageal ulceration. The latter caused cardiac arrest and death, secondary to vagotonia.

Pets, particularly **dogs**, may also evidence diarrhea, salivation, nausea, vomiting, and swelling of the mouth and throat. Some **birds** may die after eating its leaves, and others may show no ill-effects.

PHLOROGLUCINOL

= *1,3,5-Benzenetriol* = *Dilospan S* = *Phloroglucin* = *Spasfon-Lyoc* = *1,3,5-Trihydroxybenzene*

Antispasmodic.

Untoward effects: Dermatitis in **humans**; cross-reacting with ***hexylresorcinol, hydroquinone, phenolics, pyrocatechol, pyrogallol***, and ***resorcinols***.

High doses cause myoclonic convulsions in **mice**.

LD_{50}, **mice**: 4.55 g/kg.

PHOLCODINE

= *Codylin* = *Ethnine* = *Galenphol* = *Galphol* = *Homocodeine* = *Memine* = *Pectolin* = *Weifacodeine*

***Antitussive.* Has been used in many countries other than the U.S., for once or twice daily dosing.**

Untoward effects: Longer half-life in **man** and **rat** than ***codeine***. In **man**, only 20% is excreted in the urine in 26 h and 50% within 1 week. Nausea and drowsiness occasionally occur. Large doses may lead to restlessness, ataxia, and excitement.

SC, LD_{50}, **mouse**: 0.54 g/kg; IP 500 mg/kg.

PHOLEDRINE

= *Paredrinol* = *Pulsotyl* = *Veritol*

Circulatory stimulant, antihypotensive.

Untoward effects: Short duration (20–30 min) effect. Topically (1–5%) on the eye causes mydriasis. A 43-year-old **male** committed suicide by ingesting ~70 tablets.

SC, LD_{50}, **rat**: 400 mg/kg; IP, 100 mg/kg.

PHOLIOTA sp.

Untoward effects: ***P. spectabilis*** is generally not eaten because of a bitter taste. In Massachusetts, a **male**, mistaking them for an edible ***mushroom***, picked, fried, and ate them. Within 15 min, the **person** became light-headed and objects took on odd color changes. His **wife** and a **neighbor** ate some. **She** giggled and the **neighbor** felt "as though **she** had a jag on". Chromatography indicated the presence of ***indole*** derivatives (not ***psilocin*** or ***psilocybin***). Similar case reports on other **subjects**. It has been the subject of Japan's ***"Laughing and Dancing Mushrooms"***, which caused **people** to have fits of laughing and dancing after ingestion. Hallucinations also reported from ingestion of the uncooked ***mushroom***.

People in Michigan who ate ***P. sqarrosa*** accompanied by ***alcohol***, ~4 h later, suffered from severe vomiting and diarrhea.

PHOMA sp.

Untoward effects: A soil-borne fungus, yielding aeroallergens and causing allergic rhinitis, bronchial asthma, and hypersensitivity pneumonitis.

P. hibernica was isolated from a lesion on the leg of a 22-year-old **female** who lived on a farm in southwestern Ontario, Canada. Oral intake of ***griseofulvin*** eventually eradicated the lesion. A 18-month-old **male**, in the same area of Canada, had a crusting, erythematous, perioral eruption for over a month; ***P. eupyrena*** was isolated and, after many antibiotic failures, responded to ***clotrimazole*** solution (10 mg/ml) twice daily/2 weeks.

P. sorghina growing on ***millet*** was associated with thrombocytopenic purpura (onyalai) and may have been included in "yellow rain".

PHOMOPSIS sp.

Untoward effects: A number of species are often found on ***lupines*** (q.v.), producing a ***mycotoxin*** causing lathyrism or lupinosis with extensive liver pathology, weight loss, posterior paralysis, and nephrosis in **sheep**, **cattle**, **horses**, **pigs**, and experimentally in **goats**, **dogs**, **rabbits**, **rats**, **mice**, and **chicks**.

PHORATE

= *Amer Cyan 3911* = *CL 3911* = *EI 3911* = *ENT-24042* = *Granatox* = *L 11/6* = *Rampart* = *Thimet* = *Timet* = *Vegfru*

***Organophosphorous insecticide, cholinesterase inhibitor.* Has a *turnip*-like odor.**

Untoward effects: Potent poison for **humans**. A **family** of five, including two **children**, became ill after eating homemade ***peanut brittle*** after one of **them** made it from phorate-treated seed ***peanuts***. Poisoning reported in an insecticide formulation plant. Inhalation caused eye, skin, and respiratory tract irritation. Ingestion and absorption from skin caused rhinitis, miosis, headache, chest tightness, wheezing, laryngeal spasms, salivation, anorexia, nausea, vomiting, abdominal cramps, diarrhea, diaphoresis, muscle fasciculations, weakness, ataxia, confusion, arrhythmias, hypotension, convulsions, cyanosis, and paralysis. While transporting two **cattle** carcasses to a laboratory, a contaminated dirt sample was placed in the cab of the pick-up truck. The farmer's **wife**, 10 h after a 2 h ride in the cab, showed signs of poisoning including one-sided face and body numbness, dyspnea, and salivation. ***Atropine*** was effective treatment.

LD_{LO}, **human**: 5 mg/kg.

Moderately toxic to **honey bees**. **Cats** and **dogs** eating from fatally phorate-poisoned **pigs** have become poisoned. Drift onto pastures after application to vegetables and citrus groves has poisoned **cattle**. In Colorado, 56 yearling dairy **heifers** died after they ate phorate-contaminated feed. **Guppies** were killed by 0.3 g/l. Corn **rootworms** are killed by application of 1 lb/acre. Improperly stored material has contaminated feeding areas and killed **sheep**.

LC_{50} ppm in feed, **Bobwhite quail**: 373; **Japanese quail**: ~200; **pheasant**: 441; **mallard duck**: 248. LD_{50}, **rat**: 2 mg/kg; **mouse**: 11 mg/kg; **guinea pig**: 20 mg/kg; **mallard duck**: 2.55 mg/kg; **pheasant**: 7.12 mg/kg; **chicken**: 12.8 mg/kg; **bullfrog**: 85.2 mg/kg. Acute dermal LD_{50}, **rat**: 6.2 mg/kg; **rabbit**: 70–300 mg/kg; **mallard duck**: 203 mg/kg. TD, **cattle**: 1 mg/kg; **sheep**: 0.75 mg/kg.

PHORBOL-12-MYRISTATE-13-ACETATE

= *Croton Oil Factor* A_1 = *PMA* = *TPA*

Untoward effects: Tumor-promoter. Phorbol esters are present in *Euphorbiacea*, some of which are used as folk remedies or herbal ***teas***, are potent promoters of carcinogenesis, and may be the cause of esophageal cancer in Curacao and nasopharyngeal cancer in China. Induces virus production from HIV-infected cells. Promotes cell division in the skin.

Assayed by tumor-promoting and inflammation reactions on the ears of **mice**.

PHOSALONE

= *Benzophosphate* = *Embacide* = *RP 11,974* = *Rubitox* = *Zolone*

***Organophosphorus pesticide, acaricide, molluscicide.* Used in *apple*, *grape*, *peach*, and *pear* orchards; and *citrus* and *pecan* groves.**

Untoward effects: **Workers** exposed to residues of up to 7 µg/cm² on ***peaches*** showed no blood cholinesterase reduction.

Dietary concentration of 2,000 ppm killed 1/15 **quail**; 5,000 ppm killed 11/15.

LD_{50}, female **rat**: 110 mg/kg; **mouse**: 180 mg/kg. Dermal LD_{50}, female **rats**: 1.5 g/kg.

PHOSFOLAN

= *AC-47031* = *CL-47031* = *EI-47031* = *ENT-25830* = *Cyolan(e)* = *Dithiolane* = *Phospholan*

Organophosphorus insecticide.

Untoward effects: **Dairy cows** eating lucerne hay contaminated from aerial spraying (usually on ***cotton***), developed anorexia, apathy, weakness, diarrhea, decreased milk production, decreased blood cholinesterase, and some deaths. One group of 300 **cows** developed diarrhea, tremors, pollakiuria, bradycardia, and two deaths. Feeding the hay 8–12 months later to three **calves** produced diarrhea and decreased blood cholinesterase. Fed to **mallard ducks** it led to polydipsia, ataxia, wing-drop and tremors, falling, prostration, bradypnea, dyspnea, tonic convulsions, and opisthotonus within 8 min. Mortalities in ~30 min (LD_{50}, **mallard ducks**: 3.18 mg/kg). Remissions in ~1 week.

LD_{50}, **rat**: 9 mg/kg; **mouse**: 12 mg/kg.

PHOSGENE

= *Carbonic Dichloride* = *Carbonyl Chloride* = *Chloroformyl Chloride*

Use: In many ***diisocyanates*** and ***polycarbonate resins***. Formerly as a war gas. Annual production varies. In 1996 it was > 3,000 million lbs. Odor of newly mown hay. Other odorants are added as a warning.

Untoward effects: Inhalation has caused severe pulmonary edema, scarring, and pneumonia. Even with fatal concentrations, it may not be immediately irritating. It and ***chlorine*** gas killed ~100,000 **people** in World War I. It incapacitates more slowly than ***chlorine***, but, ultimately, is more dangerous. Choking, coughing, dyspnea, bloody sputum, lungs flooded with exudates, cyanosis, and eye irritation occur. Used in Yemen in 1967 and afterwards; in Laos in 1978; and in the horrors of **Nazi** concentration camp experiments of 1943 and 1944. Phosgene reacts with moisture in the lungs to form ***hydrochloric acid*** (q.v.) which destroys lung tissue. A 45-year-old **male**, who developed arterial hypoxia and airway obstruction after inhaling it from decomposition of ***trichloroethylene*** (used for degreasing) during ***carbon dioxide*** welding, eventually recovered. Produced during combustion of ***polyvinyl chloride***. **Anesthesiologists** and **physicians** should be aware that ***trichlorethylene*** (q.v.) can yield phosgene and ***trichloroacetylene***. Adequate ventilation is a must. The gas is also generated when plastic furnishings and fixtures burn, as in hotel fires. It has also been incriminated in aircraft cabin and cockpit fires since 1961. ***Fluorocarbon propellants***, upon contact with hot surfaces, can break down into ***hydrofluoric acid*** and phosgene. ***Methylene chloride*** can be converted to phosgene by contact with an open flame or hot surface. ***Chloroform***, in a half-full clear bottle exposed to air and daylight, decomposes rapidly on hot days, forming phosgene. As early as 1900, German **surgeons** complained that **they** were affected by phosgene produced when ***chloroform*** vapors were heated by gas lamps during surgery. A 40-year-old **male** developed burning of **his** eyes and coughing after massive exposure. Within a few minutes after **he** was moved into fresh air and had **his** clothes removed, **he** felt well. **He** started a hacking cough 2 h later, followed by a decrease in plasma volume, glycosuria, and severe dys-pnea, requiring intermittent positive pressure and ***oxygen*** therapy. Improved in 5 days. Military documents showed World War II production of phosgene in two plants at Niagara Falls. Accidental exposure of ten **workers** in a petrochemical plant caused immediate to mild respiratory distress, which, after a latency period of 6–12 h, increased in severity, causing headache, nausea, vomiting, cough, laryngitis, bronchitis, bronchiolitis, asthma, and severe weakness. Only one developed pulmonary edema. **All** survived, but recovery was slow and incomplete. Syncopal episodes and a cerebellar syndrome persisted in two. Contact with the liquid can cause frostbite. An **interior decorator** was exposed to the gas from the action of a hot stove on ***paint***-stripper vapor and was ill for 3 months. After a phosgene gas leak, one **boy** died; several **adults** became ill, and two backyard **hens** and a pet **budgie** died. Post-mortem of the **birds** revealed inflammation and acute pulmonary edema; cloudy swelling and fatty degeneration of the livers in the **birds**, and early necrosis in the **budgie**. National Institute for Occupational Safety and Health and OSHA set permissible exposure limits of 0.1 ppm (0.4 mg/m^3). National Institute for Occupational Safety and Health 15 min exposure limit of 0.2 ppm. Immediate danger to life or health is 2 ppm.

Inhalation LC_{50}, **human**: 3.2 g/m^3. Inhalation TC_{LO}, **human**: 25 ppm/30 min.

Has caused skin lesions on the ears of exposed **rabbits**. In **sheep**, inhalation has caused acute lung injury, pulmonary edema, and leukopenia. **Dogs** can smell it sooner than **man** and run away from it.

LC_{LO}, **rat**: 50 ppm/30 min; **dog**: 80 ppm/30 min; **guinea pig**: 31 mg/m^3/20 min.

PHOSPHAMIDON

= *Ciba 570* = *Dimecron* = *ENT 25,515*

Organophosphorus insecticide. Cholinesterase inhibitor.

Untoward effects: Suicide in an 18-year-old **female** after ingesting 50–75 ml. Within 20 min, developed blood pressure 90/55, pulse 80, miosis, flaccidity, areflexia, copious respiratory secretions, and apnea. **She** died on day 6, despite heroic and proper treatment. Mild poisoning in a 50-year-old **male**, after working for 8 h with shrubs sprayed with it 2 weeks previously, including dizziness, repeated severe vomiting, diaphoresis, lacrimation, mental confusion, weakness, and decreased serum cholinesterase. Recovered several days after treatment.

LD_{LO}, **human**: 5 mg/kg.

Highly toxic to **bees**. Young **calves** (< 2 weeks) severely poisoned by 10 mg/kg; 25 mg/kg caused intoxication within 30 min, followed by death. Adult **cattle** developed mild poisoning at 5 mg/kg. Symptoms include dyspnea, salivation, stiffness, incoordination, and respiratory paralysis. **Sheep** are not clinically poisoned by 5 mg/kg. Affected **birds** develop lacrimation, foamy salivation, miosis, wing-drop, tachypnea, dyspnea, ataxia, immobility, convulsions, tetany, and opisthotonus within 6 min; mortality between 8 and 30 min, and remission took up to 1 day.

LD_{50}, **rat**: 11 mg/kg; **mouse**: 6 mg/kg; **cattle**: 25 mg/kg; **starlings**: 5.6 mg/kg; **pheasants**: 4.24 mg/kg; **mule deer**: 44–88 mg/kg; male **chicks**: 9 mg/kg; female **mallard ducks**: 3.81 mg/kg; **redwing blackbirds**: 1.8 mg/kg; **chukar**: 11.8 mg/kg. LC_{50}, **fish**: 7.8 mg/l.

PHOSPHATES

Untoward effects: After two phosphate enemas for fecal retention, a 23-month-old **female** developed hypocalcemic tetany, hypokalemia, hyperphosphatemia, and dehydration. An 81-year-old **male** with mild renal insufficiency received 135 ml of a ***sodium phosphate*** (q.v.) enema, causing life-threatening hyperphosphatemia and hypocalcemia. A 15-month-old **male** treated with 60 ml of ***Phospho-Soda***® (66% ***sodium phosphate*** and ***sodium biphosphate***) developed phosphate poisoning with immediate vomiting; within 3 h, had severe abdominal distention, dyspnea, dehydration, hypotension, and a fever up to 40.9 °C (105.6 °F). These enemas rarely induce skin eruptions. Presence of phosphates can give decreased or false negative test results for ***potassium, sodium*** and ***alkaline phosphate***. Hyperphosphatemia has been associated with rickets and osteomalacia. Can interfere with release of ***acetylcholine*** at nerve terminals.

Hypertonic phosphate enemas are contraindicated in **cats**. In addition to severe electrolyte disturbances, hyperphosphatemia, hypernatremia, and hypocalcemia, led to weakness, anxiety, tachycardia, hypothermia, and dehydration have occurred.

PHOSPHINE

= *Hydrogen Phosphide* = *Phosphorous Trihydride*

Pesticide, rodenticide, and fumigant. **Shipped as a liquified compressed gas. Pure gas is odorless. Commercial gas has a *fish*- or *garlic*-like odor. The term is occasionally used for *chrysaniline yellow*.**

Untoward effects: Poisonous, explosive gas. Anticholinesterase-like central nervous system effects. ***Inhalation*** causes hypotension, nausea, abdominal pain, vomiting, dyspnea, pulmonary edema, convulsions, and coma. Death can occur within 4 days, and, occasionally, after 1–2 weeks. ***Chronic inhalation*** causes toothache, swollen jaw, and mandibular necrosis. Occasionally anorexia, decreased weight, hemolytic anemia, spontaneous fractures, restlessness, fatigue, thirst, headache, dizziness, coughing, fluorescent sputum, and disturbances of vision, speech, and gait. ***Contact*** with the liquid may cause frostbite. Produced by wet ***phosphide***-contaminated ***lime*** or ***cement***. **Workmen** exposed to this develop gastrointestinal upsets, rashes on **their** foreheads, and epistaxis. Inhalation of toxic fumigant fumes caused acute illness in two **children** and 29/31 freight crew **members**, who developed headache, nausea, vomiting, fatigue, coughing, dyspnea, jaundice, paresthesias, ataxia, intention tremor, and diplopia. A **child** died. Post-mortem revealed focal myocardial infiltration with necrosis, pulmonary edema, and widespread small vessel injury. When brought back home from a grain elevator and placed about the house for **rodent** control, it led to the poisoning of seven **children** and two **teenagers**. Several grain **inspectors** probing ***wheat*** treated with phosphine pellets on a railroad car developed nausea, vomiting, muscle pain, and dizziness; they required hospitalization. A review of 59 poisoning case reports indicate inhalation minimum lethal dose of 5–10 ppm/2–4 h/day/several days. Can ignite spontaneously. National Institute for Occupational Safety and Health and OSHA set upper exposure limits at 0.3 ppm (0.4 mg/m^3) and immediate danger to life or health of 50 ppm.

In **cats**, **dogs**, and **guinea pigs**, behavioral changes, decreased serum cholinesterase, and decreased liver function afer inhalation of 5 ppm. **Cats** and **dogs** are more resistant to its ill-effects than **rats**, **rabbits**, and **poultry**.

PHOSPHORIC ACID

= *Orthophosphoric Acid*

Use: Annual production is > 13,000 tons. As a phosphorous supplement in **animal** feeds; as an acidulant in foods, including ***cottage cheese***; in ***cola***, ***root beer***, and ***fruit drink beverages*** to add tartness; antibacterial and antiviral topically. As a stabilizer in ***hydrogen peroxide*** solutions, in purifying raw ***sugar*** juice by precipitating lime as ***calcium phosphates***, in detergents, rust-proofing metals, in dental cements, in firming jellies for use in baked goods, and to help stiffen whipped toppings.

Untoward effects: Consuming soft drinks containing it increases urinary ***calcium*** and is contraindicated in **people** prone to ***calcium oxalate*** stones. Irritant to eyes, upper respiratory tract, and skin. Can cause eye and skin burns after contact. ***Arsine*** poisoning in a 21-year-old **male** and a 36-year-old **male**, from cleaning an ***aluminum*** trailer tank with phosphoric acid solution. ***Sodium arsenite*** had previously been stored in the tank. National Institute for Occupational Safety and Health and OSHA exposure limit is 1 mg/m^3; immediate danger to life or health: 1 g/m^3. Corrosive.

Inhalation TC_{LO}, **human**: 100 mg/m^3.

Commonly used at 12–15% in ***iodophor*** concentrates for **cow** teat disinfection. Improper dilution has caused teat irritation, cracking, and mastitis.

LD_{50}, **rat**: 1.53 g/kg.

PHOSPHORUS

An essential mineral element found in various rocks, plants, soil, bone, and tissues. In the body, its levels are normally controlled by *calcitonin* and *parathyroid hormone*. Most of it is reabsorbed in the proximal tubules.

Untoward effects: ***White Phosphorus***, sometimes called ***Yellow Phosphorus*** or ***Elemental Phosphorus***, due to impurities in it. Has ***garlic***-like odor. Exposed to air, in the dark, it gives off white fumes and a greenish light. Can ignite spontaneously in moist air at temperatures 10–15 °F above room temperature. ***Red Phosphorus***, used in clandestine laboratories making ***methamphetamine***, has often ignited spontaneously when mishandled. Ingested, it can be converted to the more dangerous white phosphorus. Once popular as a rodenticide paste, making it readily available for attempted suicides. Also used in matches, fireworks, and **roach** poisons. The fumes cause eye and respiratory tract irritation. Other symptoms described below can occur after 1–2 days. Mild dermal contact can cause blisters; greater amounts cause slow healing and second and third degree burns. Ingestion produces nausea, abdominal pain, vomiting, anemia, tender, enlarged liver, jaundice, cachexia, hypotension, skin eruptions, dental and jaw pain, salivation, mandibular necrosis, "Lucifer disease" or "phosphorus necrosis", anorexia, oliguria, hematuria, spontaneous bone fractures, Cheyne-Stokes respiration, delirium, convulsions, coma, and death. These symptoms vary with the amount ingested and if intake is repeated. Death while in a coma can occur within 48 h. Temporary improvement can be followed by relapses. Strongly corrosive; bloody diarrhea and mouth and gastrointestinal burns may occur. A 15-month-old **male** died of cardiac failure 8 h after eating a Chinese firecracker, although **he** vomited 90 min after the incident. In Bogota at a festival, 93 young and healthy **people** were hospitalized after eating small fireworks. After burn injuries, hemolytic anemia reported in three. Vomitus may have a ***garlic***-like odor. **Workers** exposed to it while molding and wrapping a 4–6% paste have died. A **child** died after sucking the heads of two ***matches***; **others** have recovered after sucking up to 10 packages. A **person** died from < 1/5 grains (~13 mg). Deaths from lethal doses have occurred in < 1 h, some within 4 h, and 75% within 1 week. Radioactive phosphorus has precipitated gout attacks and increased the risk of leukemia in **patients** with polycythemia vera.

LD, **human**: 0.05–0.5 g. LD_{LO}, **human**: 1.4 mg/kg. TD_{LO}, **human**: 16 mg/kg; **woman**: 2.6 mg/kg.

Symptomology in **animals** is similar to **man's**. On postmortem, stomach contents, if shaken, may glow in the dark; gastroenteritis and fatty degeneration of the liver occurs. Phosphorus in detergents, manure, and sewage has been blamed for eutrophication and algal growth in waterways and other bodies of water.

White Phosphorus LD_{50}, **rat**: 7 mg/kg; **dog**: 5 mg/kg; **cattle**: 2.5 mg/kg. LD, **horse** and **ox**: 0.5–2 g; **pig**: 0.05–0.3 g; **dog**: 0.05–0.1 g; **poultry**: 0.02 g.

Red Phosphorus is relatively non-toxic.

PHOSPHORUS OXYCHLORIDE

= *Phosphorus Chloride* = *Phosphoryl Chloride*

Over 40,000 tons used annually in the U.S.

Use: Most is used in flame-retardants and plasticizers for plastics and ***urethanes***. Some used in pesticides and lube oil additives. In the manufacture of chemical warfare agents and in the illicit manufacture of ***methaqualone***.

Untoward effects: Corrosive. Exothermic reaction with moisture. An accidental spill of 6,000 gallons at a chemical plant caused seven **employees** and 65 **people** in the neighboring city to be treated for eye and lung irritation from exposure to the vapors; 12 were hospitalized. Ingestion or contact can cause eye irritation, skin burns, dyspnea, pulmonary edema, coughing, headache, dizziness, weakness, nausea, abdominal pain, vomiting, and nephritis. National Institute for Occupational Safety and Health set exposure limit of 0.1 ppm (0.6 mg/m^3) and 15-min exposure: 0.5 ppm.

LD_{50}, **rat**: 380 mg/kg.

PHOSPHORUS PENTACHLORIDE

= *Pentachlorophosphorus* = *Phosphoric Chloride*

Use: In chemical synthesis.

Untoward effects: Inhalation or exposure to fumes causes eye, skin, respiratory tract irritation, and bronchitis.

National Institute for Occupational Safety and Health and OSHA upper exposure limit of 1 mg/m^3. Immediate danger to life or health: 70 mg/m^3.

PHOSPHORUS PENTASULFIDE

= *Phosphorus Sulfide* = *Sulfur Phosphide*

Use: Approximately 90,000 tons annually. In manufacture of lube oil additives, pesticides (such as ***malathion*** and ***chlorpyrifos***), and safety ***matches***.

Untoward effects: Eye, skin, and respiratory tract irritation; keratoconjunctivitis, conjunctival pain, lacrimation, photosensitivity, and corneal vesiculation. Ingestion or excessive inhalation of fumes causes headache, dizziness, irritability, insomnia, and gastrointestinal upsets. Flammable solid which may ignite spontaneously in the presence of water. National Institute for Occupational Safety and Health and OSHA upper exposure limit of 1 mg/m^3; 15-min exposure: 3 mg/m^3; immediate danger to life or health: 250 mg/m^3.

PHOSPHORUS TRICHLORIDE

Use: Pesticide intermediate, in flame retardants, plasticizers, water treatment, and lube oil and ***paint*** additives. > 300,000 tons produced annually in the U.S.

Untoward effects: Reacts violently with moisture. Under export control to prevent use by Middle Eastern countries for chemical weapons manufacturing. Used in making illicit ***methaqualone***. Inhalation leads to eye, skin, nose, and throat irritation. Pulmonary edema, and eye and skin burns have occurred. In 1980, a tank car carrying 13,500 gallons ruptured after being side-swiped by another train in Somerville, Massachusetts, causing evacuation of 7,000 **people** from downtown Boston and Cambridge. 150 sought treatment for eye and respiratory irritation; five were hospitalized. National Institute for Occupational Safety and Health set upper exposure limit at 0.2 ppm and 15-min exposure at 0.5 ppm. OSHA set upper exposure limit at 0.5 ppm. Immediate danger to life or health: 25 ppm.

LD_{50}, **rat**: 550 mg/kg; inhalation LC_{50}, **rat**: 104 ppm/4 h; **guinea pig**: 50 ppm/4 h.

PHOXIM

= *B 9053* = *Bay 5,621* = *Bay 77,488* = *Baythion* = *ENT 27,448* = *Sebacil* = *Valexon* = *Volaton*

Organophosphorus insecticide. Anticholinesterase agent.

Untoward effects: In **mallard ducks**, **grouse**, and **pheasants**, polydipsia, ataxia, hyperexcitability, goose-stepping, running and falling, phonation, asthenia, wing-drop, tremors, tachypnea, dyspnea, wing-beat convulsions, tetany, and opisthotonus. Symptoms have occurred within 5 min, and mortalities, usually 30 min to 4 h after treatment.

LD_{50}, **rat**: 1845 mg/kg; **cat**, **dog**, and **rabbit**: 250 mg/kg; **guinea pig**: 600 mg/kg; **mouse**: 820 mg/kg; **mallard duck**: 546 mg/kg; **grouse**: 35.7 mg/kg; **pheasant**: 46.9 mg/kg. LD, **sheep**: 400–450 mg/kg.

PHTHALIC ANHYDRIDE
= *PAN*

Annual use in the U.S. is ~1,000 million lbs.

Use: In manufacturing plasticizers, ***polyester*** and ***alkyd resins***, and ***benzoic acid***.

Untoward effects: A common inducer of anaphylactic sensitization, leading to asthma and allergic rhinitis after any route of exposure. **Meat-wrappers** inhaled fumes after electrically heated wires sealed ***polyvinyl chloride*** film, causing acute bronchospasm, exertional dyspnea, and productive cough. Has occurred in a **chemist** exposed while working with it in industry. In addition, acrid vapors have caused eye, skin, and upper respiratory tract irritation; conjunctivitis, nasal ulcer bleeding, bronchitis, and sneezing. National Institute for Occupational Safety and Health set upper exposure limit at 1 ppm (6 mg/m^3); OSHA at 2 ppm (12 mg/m^3); immediate danger to life or health: 60 mg/m^3.

Liver and kidney damage reported in **experimental animals**.

LD_{50}, **rat**: 4 g/kg. LD_{LO}, **guinea pig**: 100 mg/kg.

PHTHALIMIDE

Use: As an intermediate in production of dyes, pigments, photocopy chemicals, and fungicides, such as ***captan*** and ***folpet***.

Untoward effects: Eye and respiratory tract irritation.

One study showed teratogenicity in **mice** at 25 mg/kg.

LD_{50}, **rat**: > 10 g/kg; **mouse**: 5 g/kg; acute dermal **rabbit**: > 8 g/kg; LC_{50}, **trout** and **bluegills**: ~50 mg/l.

m-PHTHALODINITRILE
= *mPDN*

Untoward effects: Inhalation causes nausea, headache, and confusion. National Institute for Occupational Safety and Health set upper exposure limit at 5 mg/m^3.

Eye and skin irritant in **experimental animals**.

PHTHALOFYNE
= *NSC 25,614* = *Whipcide*

Anthelmintic.

Untoward effects: Severe ataxia and hypnosis can occur in **dogs**, especially after IV use (in approximately half the cases), and vomiting can occur with either oral or IV methods. The drug's stinking **skunk**-like odor is often noted from the feces of treated **animals**.

PHTHALOPHOS
= *Amdax* = *Bay 9,002* = *Bayer 25,820* = *Bayer S940* = *ENT-25,567* = *Maretin* = *Naftalofos* = *Naphthalophos* = *Rametin* = *S 940*

***Organophosphorus anthelmintic, cholinesterase inhibitor.* In Russia, the same name is used for *phosmet* (q.v.).**

Untoward effects: LD_{LO}, **human**: 50 mg/kg. Keep away from **children**.

Drug interactions: Do NOT use in conjunction with or within a few days of (before and after) any other ***cholinesterase inhibitors*** and avoid its use with ***arsenicals***, ***phenothiazine***, ***phenothiazine tranquilizers***, ***purgatives***, or drugs producing purgation as a side-effect.

Transient depression and inappetence may occur in **sheep** and **goats**. Toxic at effective dosage in **poultry**, **horse**, and **swine** trials. **Brahma cattle** are generally recognized as having increased sensitivity to this class of compound. Do NOT contaminate feeds, except as in product use directions. Avoid use in constipated **animals**, those with cirrhotic livers, or those with acute infectious diseases.

LD_{50}, **rat**: 147 mg/kg; **mouse**: 26 mg/kg; **guinea pig**: 200 mg/kg; **chicken**: 0.7 g/kg.

PHTHALYLSULFATHIAZOLE
= *AFI-Ftalyl* = *Entexidina* = *Ftalazol* = *Intestiazol* = *Phthalazole* = *Phthalylsulfonazole* = *Sulftalyl* = *Sulfathalidine* = *Taleudron* = *Talidine* = *Thalazole* = *Ultratiazol*

Untoward effects: Dermatitis medicamentosa in 64-year-old **female** with diverticulitis receiving 100 × 0.5 g tablets during 2 weeks. Pruritic, erythematous eruption developed on arms and trunk 1 week later; 5 weeks later, facial erythema, many red papules, and macular lesions with some scaling. Severe bullous toxic epidermal necrolysis in a 16-year-old **female** with a history of allergies. Erythema multiforme has been more common. A 33-year-old **female**, after 1 g during 1 day, developed aplastic anemia. A 20-year-old **male**, after 2–3 g/day for ulcerative colitis, had a positive lupus erythematous cell test, with fever, weakness, arthralgia, and pleural effusion. After 10 g/day and 5 g/day rectally as a retention enema, a 27-year-old **male** developed neutropenia. Pancytopenia after ingestion for 1 week by another **patient**. **Patients** receiving 12 g/day often have severe abdominal cramping, diarrhea, and an increase in number of stool evacuations.

PHYLLANTHUS sp.

Untoward effects: ***P. abnormis*** = ***Abnormal Leafflower*** Bark is used as a suicidal agent for **man** in Tanzania. Although relatively unpalatable, they have poisoned **cattle** in the southern U.S. **Sheep** and **goats** are less commonly affected. Symptoms include severe gastric upsets from an acrid irritant, hematuria, depression, inappetence, cachexia, slight icterus, and diarrhea. Toxic hepatitis progresses to cirrhosis and acute tubular nephritis on post-mortem.

P. engleri roots are mixed with ***Albizia petersiana*** (q.v.), and smoked in Tanzania, causing unconsciousness and even death.

P. gasstroemii of Australia contains dangerous levels of ***hydrogen cyanide*** (q.v.).

P. nirusi = ***Herva pombinha*** of South America contains ***phyllanthin***, a piscicide.

P. piscatorum = ***Dzin-zi-a-pa*** is cultivated in Columbia as a **fish** poison.

P. urinaria of West Bengal is also a piscicide. Many species are used in native medicinals.

PHYLLOERYTHRIN

Untoward effects: In **ruminants**, this breakdown ***porphyrin*** product of ***chlorophyll*** in ingested plants has caused secondary or hepatogenous photosensitivity with liver dysfunction and bile duct obstruction. The liver pathology is particularly serious producing depression, icterus, anorexia, and diarrhea. Some **animals** die before photosensitivity with hyperemia and edema occurs. In **cattle**, weight loss, decreased ***milk*** production, and submandibular edema are common; and liver abscesses, parasites, and neoplasia have been associated with photosensitization.

PHYSALEAMIN

= *Physalemin*

Untoward effects: It is the most potent natural hypotensive and vasodilator ***kinin***-like ***polypeptide*** found in the skin of South American **toads**, *Physalaemus fuscumaculatus*, and ten times as potent as ***eledoisin*** (q.v.) in some species (three to four times in **dogs** and **rabbits**). Causes sialism and intestinal contractions. It is 100–700 times more potent than ***bradykinin*** and 200–400 times more potent than ***histamine***.

PHYSALIA

= *Portuguese-Man-of-War*

Untoward effects: ***P. physalis*** or ***P. pelagica***, a hydroid, moves by wind, current, or tides in tropical and subtropical waters. It floats by means of colorful air sacs or balloons 4–12" in length with masses of long (usually 30–40 ft, but up to 100 ft) tentacles reaching downward. They contain stinging cells (nematocysts) which emit a neurotoxin when accidentally touched by **swimmers** causing a linear patch of urticarial wheals, vesicular dermatitis, and ulcerations after an immediate stinging sensation with severe pain, burning, nausea, vomiting, numbness, cramps, "bear hug" sensation in chest, anxiety, and even occasionally paresthesia, loss of voice, muscle pain, dyspnea, delirium, and coryza. Rarely, death, occasionally from drowning. A small **fish**, the **nomeus**, is unaffected by the neurotoxin and the **loggerhead turtle** can eat the ***Portuguese-Man-of-War*** without ill-effects. After being washed up on beaches, the venom in the nematocysts is still active for many days. Permanently pigmented scarring of involved areas occurs.

PHYSALIS sp.

= *Alkekengi* = *Bladder Cherry* = *Chinese Lantern* = *Ground Cherry* = *Husk Tomato* = *Strawberry Tomato* = *Winter Cherry*

Untoward effects: ***P. alkekengi*** = ***Chinese Lantern's*** mature fruits are apparently edible. Immature fruits contain ***solanine*** and, when eaten by **children**, cause a scratchy feeling in the back of the throat, gastroenteritis, diarrhea, and fever.

P. heterophylla and ***P. subglabrata*** have poisoned **animals** during periods of drought, when they eat the tops and unripe berries. Effects are related to anticholinergic activity.

P. minima contains dangerous concentrations of ***nitrates***. ***Cape-gooseberry*** is the edible fruit of ***P. peruviana***. Ingestion by **children** of the immature fruits will cause the same ***solanine***-induced symptoms as with ***Chinese Lanterns***.

PHYSOSTIGMA VENENOSUM

= *Calabar Bean* = *Chopnut* = *Esere Nut* (Efik) = *Faba Calabarica* = *Isho* (Yoruba) = *Mberge Musir* (Kiyanzi) = *Nongu* (Kikongo) = *Ordeal Bean*

Native to West Africa and introduced to Brazil and India.

Untoward effects: **People** in Nigeria use it only as a poison, particularly as an ordeal poison. In Africa, when given by the head **man**, it is usually fatal to the **accused**, unless frequent vomiting occurs. A draught containing 19 seeds, pounded and in water, killed a **man** within an hour. In Liverpool, 70 **children** ate some of the beans thrown on a waste heap. Vomiting occurred or was induced in all but **one**, in whom four beans caused death. Nausea and vomiting occur within 30 min and nervous symptoms in < 1 h. Course muscular tremors (violent enough to suggest convulsions), reduced reflexes, and progressive loss of motor activity occurred. Has parasympathetic activity with sweating, abdominal pain, miosis, and salivation. Also in arrow and dart poisons. The bean contains ~0.1–0.15% of ***physostigmine*** (q.v.).

PHYSOSTIGMINE

= *Cogmine* = *Eserine* = *Esromiotin* = *Isoptoeserine* = *Physostol*

Parasympathomimetic agent, anticholinesterase, miotic.* The *salicylate* form = *Antilirium.

Untoward effects*:** It readily crosses the blood–brain barrier. By inactivating cholinesterase overdoses, it prevents the breakdown of ***acetylcholine, leading to overstimulation of nerves, muscles, and glands, followed by convulsions, fibrillation, flaccid paralysis, and death. Thus, proper dosage makes it an effective antidote of the anticholinergic syndrome. Tremors, hyperreflexia, muscle twitches, hyperperistalsis, occasional miosis, nystagmus, lens opacity, nausea, abdominal pain, vomiting, diarrhea, loss of sphincter control, lacrimation, salivation, palpitations, cold extremities, vertigo, hypotension, fainting, depression, irritability, confusion, hallucinations, bradycardia, pulmonary edema, asphyxia, bronchoconstriction, apnea, cardiac arrest, and death can occur with overdoses. Totally metabolized in ~2 h. Topical use has caused pupillary and accomodative spasms, hyperemia of conjunctiva and iris, and eyelid twitching. Eyelid dermatitis with depigmentation of lid margins in several **Negroes** with open-angle glaucoma. It will prolong ***succinylcholine's*** neuromuscular blockade and paralysis. Causes increased or false positive (and occasionally falsely negative) ***glucose*** test results.

In **rats** treated with ***lithium chloride***, it produces sustained limbic seizures and widespread brain damage. Causes teratogenicity in **chicks** leading to parrot-beaks, micromelia, syndactyly, clubbed down, vertebral defects, brachymelia, and spina bifida. Contraindicated in pregnancy, respiratory diseases, or in weak or debilitated **animals**. Avoid using it simultaneously with other ***anticholinesterase*** agents (***organophosphorus insecticides*** or ***anthelmintics, levamisole, phenothiazine, phenothiazine tranquilizers, succinylcholine***, etc.), and in those with liver disorders (metabolized in the liver). Overdosage is usually characterized by miosis, muscle weakness, colic, bowel evacuations, and vomiting (in species capable of it). Exposure to light or storage in plastic syringes will inactivate it.

LD_{50}, **mouse**: 4.5 mg/kg; IV, **mouse**: 450 µg/kg; IP, **mouse**: 640 µg/kg.

PHYTANNIC ACID

= *Phytansäure*

Untoward effects*:** It, a derivative of ***chlorophyll, accumulates in an autosomal recessive disease (Refsum's disease), characterized by retinitis pigmentosa, demyelinating peripheral neuropathy, deafness, nystagmus, cerebellar ataxia, and skin and bone changes.

PHYTEUMA

= *Devil's Claw Root* = *Harpagophytum procumbens*

Imported from South Africa and found in herbal *teas*.

***Untoward effects*:** Has oxytocic properties and must be avoided during pregnancy.

PHYTIC ACID

Untoward effects*:** Naturally occurring chelating agent, making ***calcium, copper, magnesium, phosphorus, zinc, and ***vitamin D*** less available for normal metabolism. Low protein–high phytate diet increases risk of ***zinc*** deficiency in dysmature **infants**. Potentiates the action of ***warfarin*** by chelating ***vitamin K***.

PHYTOLACCA sp.

= *Garget* = *Inkberry* = *Pocan* = *Pigeonberry* = *Pokeberry* = *Poke Salad Plant* = *Pokeweed* = *Skoke*

Untoward effects*: *P. americana contains triterpene ***saponins*** (one of which is ***phytolaccigenin***, an alkaloid), ***phytolaccine***, and ***phytolascotoxin***. It is found in the eastern U.S., Texas, and Arizona. **People** have been poisoned by eating cooked parts of the roots with attached young shoots (and occasionally berries and leaves). If the first cooking water is discarded, they are used as ***asparagus*** substitutes. The seeds are toxic and **children** eating the berries are often poisoned, while **birds** appear to eat them without ill-effects. Violent nausea, abdominal pain, burning in the mouth, and vomiting occur about 2 h after ingestion, often followed by purging, spasms, amblyopia, visual disturbances, weak pulse, hypotension, peripheral blood plasmacytosis, and occasionally convulsions. When death occurs, it is preceded by coma and respiratory paralysis. Other parts of the plant are less poisonous. Gastroenteritis in a 32-year-old **male** who inadvertently ingested ten leaves, thinking it was ***spinach*** salad. Found in home remedies for arthritis and rheumatism. **Adults** have been poisoned by its ***teas*** purchased in health food stores. Green parts and roots are irritant to skin and conjunctiva. Because of its mitogenic properties, its use as a potherb (even after two changes of water) should be avoided. ***Pokeweed*** salad prepared from young leaves, boiled, drained, and reboiled the next morning, was eaten by

52 **campers** and **counselors** in a "nature group". Nausea and stomach cramps in 18/21 (86%) ill **campers**, vomiting in 17/21 (81%), headache in 11/21 (52%), dizziness in 10/21 (48%), burning in the stomach or mouth 8/21 (38%) and 6/21 (29%) with diarrhea. Illness occurred within 30 min to 5.5 h, with symptoms lasting 1–48 h; and was associated with eating at least 1 teaspoonful. Gloves should be worn to prevent dermal absorption of its toxic principles.

When grazing is poor, **animals** may browse close to the plant and get part of the succulent poisonous roots and young stalks. **Cattle** and **sheep** are more seriously affected, occasionally, **horses**, **goats**, and **swine**. ***P. americana*** is nephrotoxic to **rats** and **mice** (1 seed/30 g), causing tubular damage and necrosis; sustained hypertension in young **mice**. Growth is retarded and mortality occurs in **turkeys** fed the berries. General symptoms in **livestock** may include gastroenteritis, salivation, diarrhea, tachypnea, vomiting, ataxia, convulsions, and abortions.

P. dodecandra = ***Awueli*** (Igbo) = ***Shibti*** (Ethiopia) has been toxic to **livestock** in Kenya, Zimbabwe, the Netherlands, and the United Arab Emirates; intense gastrointestinal irritation, vomiting, dysentery, and hypotension occurred in people after incorporation into soup or eaten with ***yams*** in Nigeria. **Calves** and **sheep** died within 5 days after daily feeding of > 15 g of leaves and/or roots. Symptoms include salivation, muscle spasms, tachypnea, shallow respirations, coughing, and blood-stained diarhhea. Post-mortem reveals froth in the trachea, pulmonary hemorrhage and edema, and hemorrhagic gastroenteritis. Causes violent vomiting and depressed respiration and circulation in **cats**.

P. esculenta has caused poisonings in Japan and as has ***P. thyrsiflora*** in Brazil.

PICKLES

Untoward effects: It is now illegal to color pickles with ***copper salts*** (coppering). Many fatalities occurred. A **female** who enjoyed eating quantities of them, developed stomach pains, and died 9 days later. Some pickles, especially the Japanese ***fukujinzuke***, have caused marked changes in the stomach mucosa, and are suspect as possible causes of gastric cancer. A case report of bromhidrosis in a 40-year-old **male** whose axillae smelled "like sour pickles", despite bathing, but, eventually, responded to topical ***clindamycin***. High in ***sodium***.

PICLORAM

= *Amdon* = *ATCP* = *Borolin* = *Grazon* = *M 3179* = *Tordon*

***Systemic herbicide*. Used in *Agent White* in Vietnam.**

Untoward effects: Eye, skin, and respiratory system irritant. Poorly absorbed through **human** skin; > 90% of oral doses of 0.5 or 5 mg/kg to **human** volunteers excreted within 72 h. Ingestion leads to nausea. OSHA set upper exposure limit at 15 mg/m^3; respiratory tract exposure at 5 mg/m^3.

Honey bees can tolerate usual use levels without ill-effects. Regurgitation soon after ingestion by **mallard ducks**. Mild ataxia and fasciculations in **pheasants** after treatment. Liver and kidney pathology in poisoned **animals**. Anorexia was the main symptom in **sheep** and **cattle**.

LD_{50}, **rat**: 3.75 g/kg; **mouse**: 1.5 g/kg; **guinea pig**: 3 g/kg; **chicken**: 4 g/kg; **rabbit**, **mallard duck**, and **pheasant**: > 2 g/kg; acute dermal, **rabbit**: > 4 g/kg. LC_{50} in feed, **Bobwhite** and **Japanese quail**, **mallard duck**, and **pheasant**: > 5,000 ppm. No ill-effect level of 250 mg/kg in **cattle**.

PICRASMA QUASSOLOIDES

Untoward effects: In Asia, extract of the bark is used as a stomachic, as a parasiticide for **lice** and **fleas**, and used to poison **deer**. The **Ainus** of Japan used it with ***aconite*** on arrowheads.

PICRIC ACID

= *Carbazotic Acid* = *Melinite* = *2,4,6-Trinitrophenol*

Antiseptic, astringent.

Use: Topically in medicine. In explosive, ***match***, ***battery***, and colored ***glass*** manufacturing; in the ***leather*** industry; and as a textile mordant.

Untoward effects: Can destroy the viability of normal cells; picric acid-impregnated sutures can destroy the viability of cancer cells. Contact can cause eye irritation, dermatitis, skin eruptions, and severe pruritus. Percutaneous absorption leads to nausea, abdominal pain, vomiting, diarrhea, and, eventually, hepatitis, anemia, coughing, headache, anorexia, nosebleeds, myalgia, anuria, polyuria, albuminuria, nephritis, bitter taste, and increased metabolic rate with secondary pyrexia. Has caused yellow-red urine, yellow color of gastrointestinal mucous membranes, and even of feces. Nails can develop a brownish color. Xanthopsia or yellow vision also reported. Yellowish discoloration of hairs on head and breast, in addition to bullous dermatitis of hands, forearms, face, and neck in a munitions **worker**, after pouring heated material into large shells. The eczematous dermatitis has been called "picric itch". Can explode if heated rapidly or by percussion. Ingestion of 1–2 g has caused serious symptoms, yet, a **person** recovered after ingesting 6 g. National Institute for Occupational Safety and Health and OSHA set upper exposure limit for skin at 0.1 mg/m^3; 15-min exposure at 0.3 mg/m^3; immediate danger to life or health at 75 mg/m^3.

LD_{LO}, **human**: 5 mg/kg.

LD_{LO}, **rabbit**: 120 mg/kg.

PICROTOXIN

= *Cocculin*

Central nervous system and respiratory stimulant. **Found in the seeds of *Anamirta*, in southeastern Asia.**

Untoward effects: After IV use, diaphoresis, nausea, vomiting, coughing, and skin flushing have occurred. Occasionally anxiety, skin irritation, and muscle twitching. Epileptiform convulsions followed by central nervous system depression, palpitations, shallow respirations, confusion, stupor, and unconsciousness have followed large doses. Then, after 30 min to 3 h, trembling, tonic and clonic convulsions. Some cases proceed into paralysis and death by asphyxia. Toxic dose for **man**: ~20 mg.

If given to **mice** with or within 30 min after ***pentobarbital*** anesthesia, it increases brain levels leading to increased sleeping time. Its antidotal action in **animal** anesthetic is due to its convulsive action.

LD_{LO}, **cat**: 1.75 mg/kg; LD_{50}, 15 mg/kg. IP, LD_{50}, **mouse**: 3 mg/kg; IV, **rabbit**: 1.2 mg/kg; **dog**: 1.5 mg/kg.

PIE(S)

Untoward effects: Custard pies have been frequently involved in bacterial food poisonings. A **baker** developed a dermatitis on **his** hands whenever **he** prepared meringue pies. It was a rare sensitivity reaction to ***karaya gum***, used as a stabilizer and pastry filler.

PIERIS sp.

Untoward effects: Toxic ***grayanotoxins***, including ***asebotoxin*** or ***andromedotoxin***, are found in its leaves, twigs, pollen grain, and ***honey***. It causes cardiac arrhythmias, salivation, vomiting, and ataxia, followed by convulsions, dyspnea, progressive paralysis, and death. ***Honey*** from it, ***Kalmia***, and ***Rhododendron*** has caused one of the earliest poisoning cases recorded in history.

A fatal and a non-fatal case of experimental ***P. japonica*** poisoning in **goats** fed it at 0.1% level of their body weight. Symptoms included colic and nausea. Postmortem revealed inhalation pneumonia.

PIGEONS

Untoward effects: **Pigeon breeder's** disease or **pigeon fancier's** disease (and other similar names) is common in **children** raising **pigeons**, **budgies**, **parrots**, and **hens** as a hobby, causes chronic respiratory problems, due to inhalation of antigens in fungal spores found in **bird** droppings. Onset is slow (3–30 days). Acute symptoms include fever, chest pains, and chills. Chronic symptoms include dyspnea, coughing, crepitant râles, and weight loss.

PILDRALAZINE

= *Atensil* = *ISF 2123* = *Propyldazine*

Antihypertensive, peripheral arterial vasodilator.

Untoward effects: Usually mild, often transient, headache, tachycardia, and dizziness.

PILLOWS

Untoward effects: Papular urticaria and pruritus in a 21-year-old **male** from feathers infested with *Dermatophagoides scheremetewskyi*. In homicides for suffocating **people**.

PILOCARPINE

The hydrochloride is *Akarpine* = *Almocarpine* = *Isopto Carpine* = *Pilopine HS* = *Pilostat* = *Salagen*.

The nitrate is *Chibro Pilocarpine* = *Licarpin* = *Pilo* = *Pilofrin* = *Pilagan*. Also available as *Mi-Pilo* = *Ocusert* = *Pilocar* = *Pilogel*.

Parasympathomimetic, cholinergic, miotic, and antiglaucoma agent.

Untoward effects: ***Eye***: Stinging, local irritation, burning, itching, red eye, difficulty with night vision or in dim light, keratoconjunctivitis, increased intraocular pressure, ciliary spasm, myopia, ocular pain, miosis, blurred vision, and cataractogenic effects have occurred from topical use. Rarely, retinal detachment. Topical ocular use has caused confusion, cognitive problems, short-term memory loss, dyspnea, and agitation.

Systemic toxicity: Has included tachycardia (followed by sinus bradycardia), headache, vasodilation, hypotension, salivation, lacrimation, atrioventricular-block, bronchoconstriction and spasms, sweating, urethral burning, muscle tremors, nausea, and vomiting, particularly in elderly **patients** given repeated doses to treat acute angle closure before surgery. It apparently caused malfunction in the eustachian tube and middle ear upsets in a 76-year-old **female**. As a ***histamine*** liberator, it can induce urticaria. Has caused leukonychia. A survey showed 53/100 samples to be contaminated with bacterial and mycotic isolates. Deliberate use as a ***food*** adulterant in a hospital geriatric ward caused death in two **patients** from cardiac paralysis, pulmonary edema, and bronchopneumonia. ***Corticosteroids*** decrease its effect. Use with ***β-blockers*** increases risk of asthma.

SC, LD_{LO}, **human**: 143 µg/kg. LD **human**: ~60 mg intraocular; oral, 5 mg/kg.

In **animals**, constricts bronchi and is *contraindicated* in respiratory distress, pregnancy, cardiac disease, or disturbances of swallowing mechanisms. Can cause diarrhea, vomiting, and increased salivation in treated **animals**. Will make glaucoma worse if it is due to inflammation, such as uveitis. IV use has caused increases in plasma ***glucose*** levels in **dogs**, probably due to increased production of ***epinephrine***. Skeletal anomalies reported in **chicks**.

IV, LD_{LO}, **rabbit**: 120 mg/kg.

PIMELIA sp.

Untoward effects: The leaves of ***P. altior*** have poisoned **calves** in Australia.

P. continua causes diarrhea, weakness and anemia, SC edema, and hydrothorax in Australian **cattle** ingesting it.

P. decora is highly toxic to **horses** in Australia, causing abdominal pain, diarrhea, and death. Post-mortem reveals intestinal hemorrhage and necrosis, and hemorrhage in adrenal cortex and spleen.

P. prostata = ***Strathmore Weed*** has killed **horses** in New Zealand. Symptoms include abdominal pain, depression, anorexia, and watery diarrhea. An irritant in the plant causes ulceration of the mouth, tongue, and esophagus.

P. simplex = ***Desert Rice Flower*** causes diarrhea, severe anemia, hypoproteinemia, decreased appetite, bloody stool, and edema of jaws, neck, face, and brisket in Australian **cattle** (St. George disease).

Also in Australia, ***P. trichostachya*** causes similar symptoms to that of ***P. simplex***.

PIMENTO
= *Allspice* = *Jamaica Pepper* = *Pimenta*

Grown in East and West Indies, and Central and South America. A Mexican variety yields *tabasco*.

Use: In soaps, detergents, cosmetic creams and lotions, perfumes, and foods.

Untoward effects: No irritation, sensitization, or phototoxic effect of 12% leaf oil in **human volunteers**.

Undiluted to the backs of **hairless mice** was not irritating, but severely irritating to intact or abraded **rabbit** skin for 24 h under occlusion.

LD_{50}, **rat**: 3.6 ml/kg; acute dermal, **rabbit**: 2.8 ml/kg.

PIMINODINE
= *Alvodine* = *NIH 7590* = *Pimadin* = *WIN 14,098*

***Analgesic, narcotic.* A *meperidine* derivative.**

Untoward effects: Similar to ***morphine*** (q.v.). Overdoses lead to tachycardia, tachypnea, nausea, vomiting, depressed respiration, and urticaria reported.

Drug interactions: Potentiated by ***anesthetics, anticholinergics, barbiturates, muscle relaxants, narcotics***, and ***phenothiazines***.

PIMOZIDE
= *Opiran* = *Orap* = *R 6238*

Pschotherapeutic, central dopaminergic blocker.

Untoward effects: Slight akathisia, and rigidity and tremor of extremities at above optimum dosage; at even higher doses, causes lack of facial expression, mental depression, and hypochondria. Motor restlessness, dystonia, hyperreflexia, oculogyric crises, opisthotonus, headache, cogwheel rigidity, drowsiness, insomnia, tardive dyskinesia, glycosuria, dizziness, syncope, blurred vision, cataracts, nocturia, decreased libido, impotence, priapism, menstrual irregularities, galactorrhea, altered taste, xerostomia, thirst, hypo- or hypertension, salivation, belching, handwriting changes, postural hypotension, sweating, rashes, palpitations, tachycardia, seizures, nausea, vomiting, anorexia, gingival hyperplasia, chest pain, periorbital edema, and hemolytic anemia also occur.

Drug interactions: Sudden deaths and cardiac toxicity when ***clarithromycin*** was added to treatment with it, as ***clarithromycin*** probably inhibits its metabolism and pimozide's clearance is also decreased. Use with ***ritonavir*** is contraindicated. ***Ethyl alcohol*** exacerbates its extrapyramidal effects. Pimozide decreases ***somatotropin*** secretion stimulation by ***bromocriptine***.

LD_{50}, **rat**: 11 g/kg; **guinea pig**: 245 mg/kg; **dog**: 40 mg/kg.

PINACIDIL
= *P 1134* = *Pindac*

Antihypertensive, potassium channel opening activator, vasodilator.

Untoward effects: Similar to ***nifedipine***, but with an increased incidence of fluid retention-related effects.

LD_{50}, **rat**: 570 mg/kg; **mouse**: 600 mg/kg.

PINAVERIUM BROMIDE
= *Dicetel* = *LAT 1717*

Antispasmodic.

Untoward effects: Has caused esophageal injury with mucosal damage and ulceration, if not taken with adequate volume of fluid or if lying down within 5 min.

LD_{50}, **mouse**: 1.4 g/kg.

PINDOLOL

= *Calvisken* = *Decreten* = *LB 46* = *Pynastin* = *Visken*

Antiarrhythmic, antianginal, sympathomimetic, β-adrenergic blocker, antiglaucoma agent.

Untoward effects: Psoriasiform reactions, systemic lupus erythematosus-like syndrome, pulmonary fibrosis, tremors, orthostatic hypotension, depression, intrinsic sympathomimetic activity, arrhythmias, insomnia, nervousness, joint pain, muscle cramps and leg muscle fatigue, restless leg, nausea, abdominal discomfort, diarrhea, headache, xerostomia, vivid dreams, hallucinations, hypoglycemia, and exacerbation of premature ventricular contractions. Occasionally bronchospasm (especially in **asthmatics**); paradoxically, large doses may cause hypertension. Severe hypertension in a **patient** who took 500 mg in a suicide attempt. Appears in breast milk. Crosses the placenta. Serum levels reported as therapeutic at 0.02–0.15 μg/ml and toxic at 0.7–1.5 μg/ml. Use in **patients** with chronic ***thioridazine*** therapy increases ***thioridazine*** serum levels by 47 ± 29% and increases expected pindolol serum levels. This combination of drugs requires close monitoring.

LD_{50}, **rat**: 263 mg/kg; **mouse**: 254 mg/kg; IV, **rat**: 51 mg/kg; **mouse**: 36 mg/kg.

PINDONE

= *tert-Butyl Valone* = *Chemrat* = *Pivacin* = *Pival* = *Pivaldione* = *Pivalyl Indandione* = *Pivalyl Valone* = *Pivalyn* = *Tri-Ban*

Rodenticide, insecticide.

Untoward effects: Single doses not likely to cause death from internal hemorrhage. Overexposure in **humans** causes epistaxis, excessive bleeding from minor cuts and bruises, smoky urine; black, tarry stools, and back and abdominal pain. National Institute for Occupational Safety and Health/OSHA set upper exposure limit at 0.1 mg/m^3; immediate danger to life or health: 100 mg/m^3.

LD_{LO}, **human**: 50 mg/kg.

Secondary poisoning in **mink** and **dogs** eating meat from pindone-killed **nutria**.

LD_{50}, **rat**: 50 mg/kg; **dog**: 4 mg/kg/5 days; **rabbit**: 150 mg/kg. LD_{LO}, **dog**: 5 mg/kg.

PINE

= *Pinus sp.*

Untoward effects: **People** have had allergic reactions to pine wood resin. **Indian women** in the western U.S. ingested Ponderosa pine needles to induce abortions. Oil on pine needles triggers allergic reactions in **people**. This is most common at Christmas time. Exposure to pine dust decreases pulmonary flow rates.

Symptoms of poisoning after ingestion of ***Pine Oil*** include gastroenteritis, nausea, vomiting, halitosis, leukocytosis, somnolence, erythema, ataxia, pareses, headache, impaired consciousness, psychomotoric excitation, delirium, tachycardia, nephritis, and renal failure. Exhaled air has a ***violet***-like odor. **Patients** have attempted suicide by drinking large quantities.

Pine Oil, LD_{LO}, **human**: 500 mg/kg.

Causes serious losses in western U.S. and Canadian **cattle** and **deer** that consume Ponderosa pine needles; symptoms include abortions, premature births, weak or dead **calves**, retained placenta, perivulvar swelling, and premature udder development. Birth defects noted in **lambs** after **ewes** have eaten it. ***Morphine***-treated **mice** show more hepatotoxicity after being put on pine bedding. Pine pollen has caused allergic reactions in **dogs**. Eating needles or drinking water at the base of Christmas pine trees has poisoned **cats**. Newborn **poults** develop lethargy, dehydration, inanition, coma, and death after eating their white pine shavings used as litter. In the **horse**, rapid IV administration of 0.1 ml/kg was rapidly fatal, leading to massive pulmonary edema and death within minutes.

PINEAPPLE

= *Ananus comosum*

Untoward effects: Pineapple dermatitis is caused by its juice that contains ***bromelain*** (q.v.), a proteolytic enzyme. Contains pharmacologically active pressor vasoactive amines that can be involved in Drug interactions. Pneumonias reported in **children**, due to aspiration while ingesting the oil. A 2-year-old **male** nibbled on a pineapple leaf and nuzzled it, scratching **his** cheek which led to flushing and an exaggerated wheal reaction lasting several hours.

PINENE

Untoward effects: It is the key irritant and cause of eczemas in **painters**, **typographers**, **machinists**, etc. using ***turpentine*** (q.v.).

Topical use of ***turpentine oil*** and ***l-pinene***, its major constituent, promotes the development of skin tumors on **rabbits**, but not **mice**. **Guinea pigs** can be successfully sensitized to it.

PIOGLITAZONE

= *Actose*

A *thiazolidinedione anti-Type 2 diabetic agent.*

***Untoward effects*:** Occasionally edema, weight gain, mild anemia, and slight increase in low-density lipoprotein.

Drug interactions*:** ***Ketoconazole can increase its serum concentration.

PIPAMAZINE

= *Mornidine* = *Nausidol* = *Nometine* = *SC 9387*

A ***piperidine tranquilizer and emetic.***

Untoward effects*:** Similar to ***chlorpromazine (q.v.). IM use for post-op **patients** has caused hypotension.

Pre-incubation inoculation of **chicken** egg yolk led to only a 7% hatch and teratogenic effects.

LD_{50}, **rat**: 620 mg/kg; **mouse**: 370 mg/kg.

PIPAMPERONE

= *Dipiperon* = *Floropipamide* = *Piperonil* = *Propitan* = *R-3345*

Psychotherapeutic. Serotonin-S_2 antagonist.

***Untoward effects*:** Drowsiness, increased weight, anorexia, pacing, restlessness, rocking, facial rigidity, body spasm, facial rash, tremor, and increased SGOT occurred. Low incidence of extrapyramidal effects.

Serum levels reported as therapeutic at 0.1–0.4 μg/ml and toxic at > 0.5 μg/ml.

LD_{50}, 160 mg/kg; IV, 48 mg/kg; IV LD_{50}, **mouse**: 66 mg/kg.

PIPAZETHATE

= *D-254* = *Lenopect* = *Selvigon* = *Selvjgon* = *Theratuss*

Antitussive.

***Untoward effects*:** Vomiting and diarrhea in 4/47 under age 18. Slight drowsiness and anal erythema reported. Acute **childhood** poisoning has occurred.

LD_{50}, **rat**: 560 mg/kg.

PIPECURONIUM BROMIDE

= *Arduan* = *RGH 1106*

Neuromuscular-blocking agent, skeletal muscle relaxant.

***Untoward effects*:** Bradycardia in ~3% of surgical **patients**. Muscle weakness persists longer than expected in some **patients** postoperatively.

IV, LD_{50}, **rat**: 1726 mg/kg; **mouse**: 29.7 mg/kg.

PIPER sp.

***Untoward effects*:** Some species are narcotic.

P. betle leaves in Taiwan are steamed and commonly used as a facial dressing to induce skin bleaching leading to contact leukomelanosis.

In Panama, leaves of ***P. darienensis*** are used to poison **fish**.

P. methystium = ***Kava*** (q.v.) in large amounts is hallucinogenic.

PIPERACETAZINE

= *PC 1421* = *Psymod* = *Quide*

Psychotherapeutic.

***Untoward effects*:** Drowsiness, dizziness, weakness, orthostatic hypotension, syncope, and extrapyramidal symptoms are the most common. Akathisia, slight increase in blood pressure, transient increase in alkaline phosphatase and SGPT, mild Parkinson rigidity, weight gain, photosensitivity, lens or corneal opacity, frequent euphoria, headache, nausea, vomiting, bradycardia, decreased libido; rarely, xerostomia, nasal stuffiness, galactorrhea, amenorrhea, leukopenia, thrombocytopenia, urinary retention, pedal edema, convulsions, and jaundice. T-wave flattening and prolongation of QT interval reported in 7/10. Fatal aplastic anemia in a 62-year-old patient after 17 months of treatment with 10 mg/day.

Can cause mammary gland cancers in **rats** and **mice**. Tachycardia in unanesthetized **dogs**. Doses of up to 5 mg/kg/day/1 year to **dogs** caused only ataxia, depression, scleral congestion, and lacrimation.

LD_{50}, **rat**: 390 mg/kg; **mouse**: 820 mg/kg. IV, LD_{50}, **dog**: > 25 mg/kg and no lethality at 100 mg/kg.

PIPERAZINE

= *Arthriticin* = *Diethylenediamine* = *Dispermin* = *EM 1960* = *Hexahydropyrazine* = *MK 7617* = *Oxyzin* = *Piperazidine*

The adipate = *Entacyl* = *Eravern* = *Helminthex* = *Mapiprin* = *Nometan* = *Oxurasin* = *Oxypaat* = *Oxypate* = *Oxyzin* = *Pipadox* = *Toxivers* = *Vermicompren.*

The citrate = *Antepar* = *Helmezine* = *Multifuge* = *Oxucide* = *Parazine* = *Pinozan* = *Pipizan Citrate* = *Pipracid* = *Rhomex* = *Tasnon* = *Ta-Verm* = *Tripiperazine Dicitrate* = *Worm Away.*

The phosphate = *Antepar* = *Pincets* = *Pinrou* = *Pinsirup* = *Piperverm* = *Piperazate* = *Tasnon.*

The tartrate = *Noxiurotan* = *Piperate.*

Other less commonly used forms are the *dithiocarbamate, gluconate, fluosilicate, glycolylarsanilate, monochloride,* and *sulfate.*

***Anthelmintic. Ketoconazole* is an *imidazole–piperazine* compound. Also found in the chemical structure of**

some *phenothiazine antineoplastics, tranquilizers,* and an *estrogen*.

Untoward effects*:** Urticaria, headache, vomiting, diarrhea, tremors, incoordination, muscle weakness, and erythema multiforme have occurred in ~5% of **patients**. Allergic reactions also include purpura, fever, and arthralgia. A 2½-year-old **female** became unsteady, fell, and stared vacantly each time **she** was given appropriate doses as an anthelmintic. Gastrointestinal, allergic, and neurological adverse effects reported. The latter, especially after accidental or deliberate overdoses leading to delirium and cerebellar-type ataxia called "worm wobble". Can cause blurred vision, paralytic strabismus, and irritate or accelerate formation of cataracts or lens opacities. Metabolism can cause cholestatic hepatitis and ***nitrosamine formation. It can cause allergic drug eruptions in ***ethylenediamine***-sensitive **individuals**. A case report of convulsions after administration of ***chlorpromazine***. Chemical plant **workers** have had eye, skin, and respiratory tract irritation, including asthma and skin burns from exposure to its dust. Oral intake leads to gastrointestinal upsets, headache, nausea, vomiting, incoordination, hyporeflexia, paresthesia, electroencephalogram abnormalities, and muscle weakness. Hypersensitivity vasculitis in a 56-year-old **male** after piperazine hydrate 1 g/day/7 days. Involuntary choreiform movements, decreased muscle tone, and psychic disturbances are worse in **patients** with anemia, uremia, and severe arteriosclerosis. In **epileptics**, it increases disposition to convulsions. Hemolytic anemia reported in African **children** with glucose-6-phosphate dehydrogenase deficiency. Has caused false negative or spuriously low ***uric acid*** blood test results.

Serum levels reported as therapeutic at 0.01–0.02 μg/ml and toxic at > 0.5 μg/ml.

LD_{50}, **human**: 5 g/kg.

Vapors and skin or eye contact with concentrated material must be avoided. It may exaggerate extrapyramidal effects of ***phenothiazine tranquilizers***. Clinically, this could occur most frequently in **Dobermans** and **German shepherds**. Vomiting and diarrhea is occasionally reported in **cats** and **dogs**, especially if overdosed. Renal insufficiency may delay excretion. A **mountain lion** wormed with the *citrate* form developed hypocalcemic tetany. Avoid the *citrate* form in hypocalcemic **animals**.

LD_{50}, **rat**: 3.8 g/kg.

PIPERIDINE

A poisonous amine alkaloid in *lupines*. Small amounts also found in *black pepper* plants.

Untoward effects*:** The main ingredient in the illicit manufacture of ***phencyclidine. Can form ***nitrosamines*** in the stomach.

LD_{50}, **rat**: 400 mg/kg.

PIPERIDIONE

= *Dihyprylone* = *Sedulon* = *Tusseval*

Antitussive, sedative.

***Untoward effects*:** In a Swedish study of 150 **patients** on 161 occasions, two cases of poisoning occurred from its derivatives, requiring an artificial kidney.

PIPERIDOLATE

= *Crapinon* = *Dactil* = *JB 305*

Anticholinergic.

Untoward effects*:** Similar to ***atropine (q.v.).

IV, LD_{50}, **rat**: 100 mg/kg; **mouse**: 75 mg/kg; **guinea pig**: 16 mg/kg.

PIPERIDYL BENZILATE

= *Ditran* = *JB 329*

***Psychomimetic.* A synthetic *atropine* derivative.**

***Untoward effects*:** Delirium and hallucinations.

IV, LD_{50}, **rat**: 10 mg/kg; **mouse**: 22 mg/kg.

PIPEROCAINE

= *Metycaine*

***Local anesthetic.* A *cocaine* derivative.**

***Untoward effects*:** Grand mal convulsion in 18-year-old **female** receiving 5 cc of a 0.5% solution caudally for obstetrical delivery. Excessive irritation after topical ophthalmic use.

IV, LD_{LO}, **mouse**: 26 mg/kg; SC, 590 mg/kg.

PIPERONYL BUTOXIDE

= *Butacide* = *ENT 14,250* = *Ethanol Butoxide* = *FAC 5273* = *FMC 5273* = *NIA 5273* = *Pyrenone 606*

Pyrethroid and rotenone synergist.

Untoward effects*:** Occasionally allergic reactions may be due to ***pyrethrins or solvents used with it. A 24-year-old **male**, 2 weeks after excessive use of it and ***pyrethrum*** in a tent for 1 week, developed pancytopenia and paroxysmal nocturnal hemoglobinemia, that lingered for 6 months, until reversed by a bone marrow transplant.

LD_{LO}, **human**: 5 g/kg.

Splenomegaly in **rats** after SC injection. Doses up to 500 mg/kg by esophageal intubation in pregnant **rats** caused no teratogenicity or untoward effects. Use in **rats** and **rabbits** inhibits metabolism of many drugs, and, in **rats**, prolongs the half-life of ***p-chloroamphetamine***. Has caused thrombocytopenia and occasional increase in reticulocytes in **dogs**. It enhances the acute toxicity of ***benzo[α]pyrene***, ***freons***, and ***griseofulvin*** in infant **mice**. Anorexia, vomiting, diarrhea, hemorrhagic enteritis, inanition, and mild central nervous system depression in **laboratory animals**.

LD_{50}, **rat**: 11.5 g/kg; **mouse**: 3.8 g/kg; **rabbit**: 7.5 g/kg; skin, **rabbit**: 200 mg/kg.

β-PIPERONYL ISOPROPYL HYDRAZINE

= *Safrazine*

Antidepressant.

***Untoward effects*:** Insomnia, decreased blood pressure, constipation, transient orthostatic hypotension. Some extrapyramidal symptoms lead to dizziness, hydrodipsia, and vomiting.

PIPOBROMAN

= *A-8103* = *Amedel* = *NSC 25,154* = *Vercyte*

Antineoplastic. **An alkylating agent in limited use since 1960.**

Untoward effects*:** Primarily hematopoietic, with dose-related bone marrow depression, leukopenia, reduced red blood cells, and thrombocytopenia. Transient hemolytic anemia; nausea, vomiting, abdominal cramps, diarrhea, dose-related anemia, anorexia, heartburn, proteinuria, and occasionally skin rash, sore throat, and fever. Can cause false positive or spuriously high serum ***bilirubin test results.

PIPOSULFAN

= *A 20,968* = *Ancyte* = *NSC-47774*

Antineoplastic.

***Untoward effects*:** Anemia, leukopenia, thrombocytopenia, and diarrhea. In some cases, liver function tests were abnormal. Gastrointestinal complaints were rare.

PIPOTIAZINE

= *Piportil* = *Pipothiazine* = *RP 19,366*

Psychotherapeutic. **Has antihistaminic and antiserotonin activities.**

Untoward effects*:** IM use of the *depo palmitate* form (RP 19,552***) causes hand tremor rigidity, dystonia or hypokinesia, a depressive syndrome, and local reactions. A 26-year-old **female** on 25 mg orally every 2 weeks developed mania.

Therapeutic serum levels reported as 0.001–0.06 μg/ml and toxic at 0.1 μg/ml.

LD_{50}, **mouse**: 440 mg/kg; SC, 360 mg/kg; IP, 108 mg/kg.

PIPRADOL

= *Gerodyl* = *Meratran* = *MRD 108* = *Pipradrol*

Central nervous system stimulant, antidepressant. **Without action on the autonomic nervous system.**

***Untoward effects*:** Has caused nausea, anorexia, weight loss, hyperexcitability, insomnia, pallor, tachycardia, aggravation of anxiety, epigastric discomfort, dizziness, macular skin rash, delusions, and hallucinations. A 25 lb 2½-year-old child ingested 15 mg (~30 times usual dose). Continuous vocal expression, frenzied activity, excitability, voracious appetite, and continual brushing of hand across forehead, even while asleep. Slow return to normal blood pressure and pulse rate after 2 days. Chronic use of 2.5 mg tid produced anxiety symptoms in 15%.

Drug interactions*:** Antagonizes hypotensive effect of ***guanethidine.

LD_{LO}, **human**: 50 mg.

Lethal doses IV or orally to **rats**, **mice**, **guinea pigs**, and **rabbits** lead to tremors and convulsions; death by respiratory failure during convulsions. Lethal doses orally in **dogs** and SC in **rats** lead to continuous intense activity and sudden death during hyperactivity. In **rats**, decreases food intake.

LD_{50}, **rat**: 180 mg/kg; **mouse**: 168 mg/kg.

PIPROCURARINE

= *Brevicurarine* = *LD 2,480*

Curarimimetic. **Used during anesthesia, for short duration effect.**

***Untoward effects*:** Frequently tachycardia and apnea.

PIRACETAM

= *Avigilen* = *Axonyl* = *Cerebroforte* = *Encetrop* = *Gabacet* = *Genogris* = *Geram* = *Nootron* = *Nootrop* = *Nootropil* = *Nootropyl* = *Normabraïn* = *Norzetam* = *Pirroxil* = *Pyramem* = *Pyracetam*

Nootropic.

***Untoward effects*:** In latent psychotic conditions, it has provoked reactions, aiding diagnosis. Induces headaches.

Drug interactions*:** Hemorrhage and increased prothrombin time reported with ***warfarin.

PIRBUTEROL

Bronchodilator, sympathomimetic, vasodilator. **An analog of *albuterol*. High selectivity of β_2 pulmonary, rather than cardiac adrenergic receptors.**

***Untoward effects*:** Tremors in one-third of **patients**, dizziness, neck pain, palpitations, insomnia, polyuria, and tachycardia in **others**. High dosage can cause hypokalemia.

TD_{LO}, **human**: 1.6 mg/kg over 4 weeks.

PIRENZEPINE HYDROCHLORIDE

= *Abrinac* = *Duogastral* = *Durapirenz* = *Gasteril* = *Gastrozepin* = *Leblon* = *LS 519* = *Maghen* = *Renzepin* = *Tabe* = *Ulcuforton* = *Ulcosan*

Antiulcerative, antimuscarinic.

***Untoward effects*:** Slight xerostomia; paradoxical salivation has been reported.

PIRIBEDIL

= *ET 495* = *EU 4,200* = *Trivastal*

Peripheral vasodilator, dopamine receptor stimulator.

***Untoward effects*:** Dyskinesia, nausea, drowsiness, confusion, hallucinations, and psychosis.

In the **mouse**, IP injection, and in the **rat**, IP, SC, and IV decrease body temperature (hypothermia).

IP, LD_{50}, **mouse**: 690 mg/kg.

PIRIBENZYL METHYL SULFATE

= *Acabel* = *Bevonium Methyl Sulfate* = *CG 201*

***Anticholinergic, antispasmodic, bronchodilator.* Oral, IM, IV, or in suppositories.**

***Untoward effects*:** Pain at injection site, tachycardia, auricular fibrillation; decreased tone and motility in esophagus, stomach, and small intestine. Xerostomia, urinary retention, nausea, vomiting, palpitations, and fatigue.

Weight loss becomes dramatic at 100 mg/kg orally, SC, IP, and IV in **mouse**, **rat**, **guinea pig**, **rabbit**, and **dog**.

SC, LD_{50}, **rat**: 2.4 g/kg.

PIRIMIPHOS METHYL

= *Actellic* = *Actellifag* = *Blex* = *Silosan*

Organophosphorus insecticide, cholinesterase inhibitor.

***Untoward effects*:** Can be absorbed through bronchial mucosa and skin. Usual symptoms include miosis, vomiting, abdominal pain, diarrhea, wheezing, dyspnea, "tightness" in chest, and excessive salivation and bronchial secretions occur. Bradycardia, muscle weakness and fasciculations, and convulsions also noted after greater exposure. Eye irritant.

LD_{50}, **rat**: 2 g/kg; **mouse** and **rabbit**: 1.2 g/kg; **guinea pig**: 1.0 g/kg.

PIROMEN

= *Pyromen*

***Untoward effects*:** Headache, chills, fever, myalgia, acute asthmatic attacks, nausea, and vomiting.

PIROXICAM

= *Brexic* = *CHF 10/21* = *CHF 1,251*

Anti-inflammatory; antiarthritic. Non-steroidal anti-inflammatory drug.

***Untoward effects*:** Most frequent (20–30%) are epigastric distress and nausea. Also abdominal pain, constipation or diarrhea, and flatulence. Peptic ulcers in about 1%, but increase with dosage > 20 mg/day. Peripheral edema (2%), inhibition of platelet aggregation, increased prothrombin time, and irritation of esophageal mucosa. Occasionally gastrointestinal bleeding and subsequent anemia, dizziness, somnolence, headache, tinnitus, anorexia, stomatitis, vertigo, depression, blurred vision and eye irritation, pruritus, rashes, and increased blood urea nitrogen. An arthritic 85-year-old patient on 20 mg/day/8 months developed severe hyperkalemia. A 64-year-old **male** on 20 mg/day/6 weeks developed acute interstitial nephritis. The drug was introduced by Pfizer into the U.S. in 1982. By the end of 1985, the FDA received 2,621 non-fatal and 182 fatal adverse reaction reports. A 74-year-old **female**, after topical use of 180 g of the 3% gel during 3 weeks, and a 57-year-old **female**, after 400 g over a 4 month period, developed renal failure. Liver disease, pancreatitis, photosensitivity, toxic epidermal necrolysis, palpitations, exacerbation of angina and congestive heart failure, angioedema, bronchospasm, vasculitis, erythroderma, urticaria, onycholysis, alopecia, paresthesia, and lichenoid eruptions reported. A 73-year-old **male** who ingested 20 mg/day/~2 weeks developed fatal pemphigus vulgaris. Fatal subacute hepatic necrosis in a 66-year-old patient after 40 mg/day/3 days. Dangerous adverse effects occur in **people** 60 years and older, and **those** with peptic ulcers or renal disease. Reversible agranulocytosis in a 65-year-old **female** after 20 mg/day/3 days. Fatal aplastic anemia in a 29-year-old **female** after 10–20 mg/day/~ 1 year. Azotemia and severe hyperkalemic acidosis also reported. Poisoning with it has occurred. A 2-year-old **male** ingested 100 mg causing vomiting, irritability, dehydration, acidosis, and generalized seizure within 2 h. Serious multisystem toxicity 3 days later, included liver damage, renal toxicity, pancytopenia, leukopenia, eosinophilia, and bone marrow aplasia. Found in breast milk. Usual therapeutic serum levels reported as 2–6 µg/ml and toxic at 14 µg/ml.

Drug interactions: Slight potentiation of ***lithium carbonate, metoprolol*** (not ***atenolol***), and ***warfarin***. Can decrease effectiveness of angiotensin-converting enzyme ***inhibitors***, some ***β-blockers, diuretics, furosemide***, and ***vasodilators***. Its plasma levels are increased by ***cimetidine*** in **humans** and **rats**. Cross-sensitivity can occur with ***aspirin*** and ***prostaglandin inhibitors***, and it is contraindicated in **patients** with angioedema or bronchospasm induced by these drugs. ***Aspirin*** can cause slight decrease in its plasma concentration by decreasing its absorption. ***Ritonavir*** decreases its metabolism and may increase its tox. Its elimination is enhanced when used with ***cholestyramine***.

Gastric ulceration and hematemesis reported in treated **dogs**.

LD_{50}, **mouse**: 360 mg/kg.

PIRPROFEN

= *Rangasil* = *Rengasil* = *Seflenyl* = *SU 21,524*

Anti-inflammatory, antipyretic, analgesic.* A *non-steroidal anti-inflammatory drug.

Untoward effects: Mostly gastrointestinal, peptic ulceration and hemorrhage, nausea, indigestion, constipation or diarrhea, and abdominal pain. Headache, tinnitus, vertigo, and light-headedness are less frequent than with ***aspirin***. Severe urticaria, pruritus, and hepatitis are less common complaints.

PISTACIA sp.

= *Sultân*

The *pistachio* nut *P. vera* is grown commercially in Turkey, Iran, Syria, Afghanistan, and the western U.S.

Untoward effects: ***Aflatoxin*** contamination has been common. Allergic reactions from eating the nuts have been reported.

PITCH

= *Coal Tar Pitch*

By-product of the *coke* industry.

Untoward effects: Pitch volatiles have been a common cause of urticaria and dermatitis, especially of the legs of potroom **workers** in factories using recycled ***aluminum***. It causes acne by blocking pores leading to infection and eventual destruction of sebaceous glands. Has caused photosensitization reactions on sunlight exposed parts of the body in electrical conduit and ***battery*** manufacturing **workers**, and is often referred to as "tar smarts" in roofing and paving **workers**. Occupational melanosis, conjunctivitis, keratitis, skin ulcerations, and epitheliomas reported. May have been a factor in scrotal cancers in **chimney sweeps**. Bronchitis reported from aerosoled particles. National Institute for Occupational Safety and Health and OSHA upper limits of exposure are ~0.1–0.2 mg/m^3.

Coal tar pitch deaths in **hogs** within ~2 weeks after ingestion of ***"clay pigeons"*** or "blue rocks" used in trap shooting; symptoms included mottled necrotic livers, anemia, decreased red blood cells, decreased hemoglobin, jaundice, lymph node edema, and ascites. The targets are an amalgam held together by ***coal tar pitch***.

PITHOMYCES sp.

Untoward effects: ***P. chartarum's*** major ***mycotoxin, sporidesmin*** (q.v.), on moldy grass in pastures, has caused facial eczema, diarrhea, photosensitization, jaundice, weight loss, cystitis, liver necrosis and cirrhosis, and death in **sheep** and **cattle** in Australia and New Zealand. The eczema is caused by failure of the diseased liver to metabolize a ***phytoporphyrin*** that induces photosensitivity. A similar disease noted in **cattle** of the southeastern U.S. feeding on moldy (*Periconia minutissima*) ***Bermuda grass***.

PITUITARY

Untoward effects: Use of the powder as a snuff (***Pitressin snuff***) causes pituitary **snuff-taker's** lung, due to bovine and porcine serum proteins and pituitary antigens, leading to anaphylaxis, hypersensitivity pneumonitis or allergic alveolitis. Can progress to irreversible pulmonary fibrosis. Fever, chills, dyspnea, coughing, sneezing, myocardial ischemia, abdominal cramps, nausea, and anorexia can occur. Overdosage can cause fluid retention and edema.

PITUITARY, POSTERIOR

= *Di-Sipidin* = *Hypophysin* = *Pituidrol* = *Pituitrin*

Contains *vasopressin* (q.v.) and *oxytocin* (q.v.).

Untoward effects: Parenteral use has caused shock and myocardial ischemia. Cerebral edema and death in 4-year-old **male** after receiving a liter of water along with the drug by injection. Water intoxication resulted from the antidiuretic effect of its ***vasopressin*** component. In **patients** treated for diabetes insipidus or psychogenic polyuria, and **lab workers** exposed to it by inhalation or injection, can develop allergic rhinitis, asthma, angioneurotic edema, and urticaria.

In **animals**, unnecessary pressor effects may occasionally be serious; avoid its use in hypertensive states. It should NOT be used to induce labor without careful checks to assure that cervical dilation has occurred and that mechanical blockage of the canal (including faulty position of the **offspring**) does not exist. *May be contraindicated* in nephritis, arteriosclerosis, and coronary disease. Caution is advised in extreme respiratory distress and extreme environmental heat situations.

PITURANTHOS TRIRADIATUS

Untoward effects: This perennial desert plant is common in the Negev Desert in Israel and grazed occasionally by **goats**, **gazelles**, **camels**, and, rarely, by **sheep** with resultant photosensitization. Severe sensitization with erythema of beak and conjunctiva (followed in 2 days by increased severity, blepharitis, and exudate on and closure of eyes) in **ducklings** that eat the seeds. The plant and seeds contain ***furocoumarins***, including ***methoxypsoralens*** (q.v.), which, together with ultraviolet light, is used in treating psoriasis on **humans**. Photosensitization with retinal atrophy reported in **albino rats**.

PIZOTYLINE

= *BC 105* = *Mosegor* = *Pizotifen* = *Sandomigran* = *Sanmigran* = *Sanomigran*

Anabolic, orexigenic, antidepressant, antiserotonin, migraine prophylaxis.

Untoward effects: Somnolence, constipation, impotence, tachycardia, palpitations, listlessness, large weight gain, xerostomia, nausea, vertigo, excitation, general malaise, tingling of extremities, cold extremities, and facial flushing; enuresis (in a **child**).

PIZZA

Untoward effects: An unusual and distinctive circular-shaped lesion on the palate has been called "pizza pepperoni burn".

PLACEBO

Untoward effects: A controlled clinical trial with ***mephenesin*** and placebo in courses of 2 weeks revealed light-headedness, drowsiness, and anorexia as major complaints by **those** taking the placebo. **Some** reported weakness, nausea, palpitations, and rashes. It is said that **patients** taking the placebo are often the lucky ones – **they** get better without risking the side-effects of the new drug. In 1998, the Brazilian government fined Schering millions of dollars for failure to control placebo ***oral contraceptive*** tablets (650,000 packages of 21 tablets each), after ten **women** became pregnant after taking them for birth control.

PLACENTAL EXTRACTS

= *EP 50* = *Fibracel*

Untoward effects: Use for rejuvenation by injection led to bullous dermatitis in 80-year-old **male**.

PLANTAGO SEED

= *Metamucil* = *Plantain Seed* = *Psyllium Seed* = *Testa Ispaghula*

~10 million lbs of the husks imported annually.

Untoward effects: Although the entire seeds have been used as a laxative, the outer seed coat (***Psyllium Husks*** = ***Ispaghula Husks***) contains most of the water-soluble mucilage and is irritant. Esophageal obstruction and asphyxiation reported with water-soluble gums in laxative and weight control over-the-counter products. ***Psyllium hydrophilic mucilloid*** decreases ***digoxin*** and ***carbamazepine*** absorption. A 33-year-old **female** developed urticaria and anaphylactic shock after a single oral dose of a granulated form. Anaphylactic shock in a 60-year-old **female nurse**; four **nurses** in a Veterans Administration hospital acquired allergic reactions (wheezing, chest congestion; swollen, watery, and pruritic eyes; and sneezing) after several years of dispensing it. A 60-year-old **female** developed eosinophilia after 6 weeks of intake. Nasal congestion, sneezing, wheezing, and acute bronchospasm in pharmaceutical plant **workers**, apparently, from handling it.

Rats and **cats** fed it at 4% and 19%, respectively, in their diets developed marked kidney pigmentation, but urea clearance was normal.

PLASMOCID

= *Antimalarine* = *710F* = *Fourneau 710* = *Rhodoquine* = *SN 3,115*

A *quinoline antimalarial*.

Untoward effects: Optic atrophy and central nervous system toxicity in **man**; lesions of oculogyric, vestibular, and cochlear nuclei in **rhesus monkeys**; SC or IP in **rats** led to focal myocardial necrosis and inflammation.

PLATINIC CHLORIDE

= *Chloroplatinic Acid* = *Hydrogen Hexachloroplatinate*

Use: In electroplating, etching, indelible ***inks***, coloring ceramics, and in microscopy.

Untoward effects: Hay fever–asthma-like syndrome, scaly erythema, eczematous dermatitis, and urticaria.

PLATINUM

Use: In automotive catalytic converters, as a catalyst in chemical manufacturing, in electroplating, prostheses, jewelry, and dentistry. In the U.S., ~100 tons used annually.

Untoward effects: Exposure to soluble salts by contact or inhalation causes eye irritation, rhinorrhea, sneezing, dyspnea, tightness of chest, wheezing, coughing, dermatitis, and cyanosis. Can cause sensitization of the body's cellular immune system so that repeated exposure has caused a contact dermatitis in **adults** and **children**. A 37-year-old **male** with past history of asthma, after exposure to it in refining, developed acute facial eruption. National Institute for Occupational Safety and Health upper exposure limit of 1 mg/m^3.

Parenteral use of its salts causes transient central nervous system and muscle stimulation, followed by central nervous system depression and paralysis in **laboratory animals**.

PLEUROTIS sp.

Untoward effects: A case report of 25 cases of poisoning from ingestion of ***P. olearius***, the ***Oyster mushroom***, reported. In Zagreb in 1968, 15 **people** from five separate incidents were poisoned by it. ***P. japonicus*** has caused poisoning by it after ingestion.

PLUM

= *Prunus sp.*

Untoward effects: The seeds or pits contain ***hydrocyanic acid*** (q.v.), and ***amygdalin*** (q.v.), causing ***cyanide*** (q.v.) poisoning.

P. americana = ***American Plum*** = ***Wild Plum*** = ***Wild Yellow Plum***. Its leaves, stems, bark, and seed pits contain cyanogenic glycosides, causing nausea, vomiting, abdominal pain, diarrhea, dyspnea, muscle weakness, dizziness, stupor, and convulsions. A 65-year-old **male** was allergic to all fruit with large pits, including plums. Plums contain ***oxalates***, ***caffeic acid*** (50–200 ppm), and ***neochlorogenic acid*** (50–500 ppm) which is converted to ***caffeic acid***. The latter two are carcinogenic in **rodents**. Red plums contain ***tyramine***, ~6 μg/g. **People** with ***chromate*** dermatitis should avoid prunes and canned plums.

P. augustifolia = ***Wild Plum*** and ***P. umbellata*** = ***Hog Plum*** has caused poisoning in **livestock**, particularly **hogs**. **Birds** eating fermented purple plums lying on the ground develop intoxication and were unable to fly.

See ***Cherry*** for ***P. emarginata***, ***P. laurocerasus***, ***P. serotina***, ***P. sphaerocarpa***, and ***P. virginiana***.

PLUTONIUM

Untoward effects: Difficult to handle safely and readily combines with ***oxygen***, producing ***plutonium oxide***, explosions, and fires; one fire was so extensive, it caused $45 million of damages in Rocky Flats, Colorado. Its natural radioactivity is weak and due to alpha particles. Inhaled, it has caused cancers in **animals** and has been suspect as a cause of lung cancer in **man**. The major source of environmental plutonium is from atmospheric testing of nuclear weapons. Occasionally, small amounts have been accidentally released by power plants or maliciously into water, soil, and plants. From 1944 to 1974, the U.S. government injected 18 **civilians** with it in clear violations of ethical and moral standards. In 1947, informed consent was given by one **man**. During explosive accidents, small particles may imbed in tissues, especially face neck, and hands, causing a foreign-body reaction and, eventually, nodules and scar tissue. Inhalation, ingestion, and plutonium-contaminated wounds are the main direct routes of entry into **man**. **Human** kidneys are 50 times less efficient than **animal** kidneys at removing it. After entering the body, plutonium enters the blood stream and is deposited primarily in the liver and bone, with a half-life of ~200 years. **Survivors** of the government experiment had painful osteoporosis and urinary tract infections. Autopsies revealed bones "that looked like Swiss cheese". Nuclear fuel is usually ***plutonium-238***; it has also been used to power cardiac pacemakers. ***Uranium-235*** was in the Hiroshima bomb.

Plutonium-239 metabolism in **man** is similar to a **dog's**.

In **dog** experiments, lymphosarcoma was associated with nodal concentrated ***plutonium dioxide***-contaminated puncture wounds. Aerosoling **dogs** with ***plutonium-239 nitrate*** caused dyspnea, weight loss, pleural fibrosis, arterial necrosis, and epithelial and mesothelial hyperplasia. After 2 weeks, lymph node scarring and decreased white blood cells. Fertility is decreased in **mice** after IV.

PODOPHYLLOTOXIN

= *Condylox* = *Podofilox*

Untoward effects: Transient inflammation, erosions, pruritus, pain, and burning sensations after topical use in > 50% of **patients**.

LD_{50}, **mouse**: 90 mg/kg.

PODOPHYLLUM

= *American Mandrake* = *Devil's Apple* = *Hog Apple* = *Indian Apple* = *Mandragoro* = *Mandrake Root* = *May Apple* = *Umbrella Plant* = *Vegetable Calomel* = *Vegetable Mercury* = *Wild Lemon*

Untoward effects: The ripe yellow fruit is edible, but ingesting the green fruit, shoots, or roots leads to congested mucous membranes, arrhythmias, severe vomiting, diarrhea and severe purging, and other gastrointestinal symptoms in **man**, **cattle**, **hogs**, and **sheep**. It interferes with cellular mitosis and should not be used as a laxative in pregnant **women**. Mask, goggles, and gloves should be worn by pharmaceutical manufacturing **workers**. Contains ***scopolamine*** (q.v.). Occasionally a contaminant in herbal ***ginseng*** products, causing dehydration, severe diarrhea, weakness, coma, and peripheral neuropathy, usually within 3–7 h after one or two cups of ***tea***. The **Romans** knew it as *Atropa mandragoro*. **Greeks** and **Romans** believed it was endowed with spiritual powers because its root resembled a **human** abdomen with thighs, and, occasionally, the **male** genitals. Its ancient popularity was suggested by the Old Testament, in which Leah uses the root to "lie with Jacob". A **male**-suicide ingested ~10 g of the extract leading to confusion, comatose, and death after 39 h.

In **rats**, fetal death, resorption, and multiple deformities. Severe lacrimation, mydriasis, salivation, moaning, excitement, congested mucous membranes, and agitation in a **cow** after ingestion.

PODOPHYLLUM RESIN

= *Podophyllin* = *Posalfilin*

Use: As a topical caustic for various warts.

Untoward effects: Not only caustic to warts, but also to surrounding skin if applied carelessly. Intense, painful local inflammation. The dust has caused severe eye irritation and conjunctivitis. Dermatitis in factory **workers** grinding it. Systemic absorption after topical use. Polyneuritis, nausea, and vomiting after ingestion. A 15-year-old **female** developed leukopenia and thrombocytopenia after topical treatment with 20% for condyloma acuminata. A 19-year-old **female** developed weakness and polyneuritis of legs 2 days after application of 25% solution. Recovery took 10 weeks. An 18-year-old **female** became comatose 24 h after 25% ointment on vulvar warts; died 8 days later. A 20-year-old **female** with extensive condylomata acuminata of labia majora, labia minora, and vaginal mucosa treated with 10% in ***collodion*** for 6 h on a tampon developed paresthesia, toe and finger weakness, and **she** was febrile. The polyneuritis-induced paresis disappeared in 2 weeks. Delirium with visual hallucinations 2 months after treatment of anogenital warts. A 2-year-old child took an oral overdose of a Turkish cathartic containing it leading to transient visual hallucinations. Extreme nausea and vomiting, persistent sore throat, and floating sensation after topical treatment with 10% in an ***alcohol-acetone*** mixture for black hairy tongue. A 25-year-old **male** with extensive condylomata acuminata of the prepuce and perianal area treated with 16% in ***alcohol*** vomited within 2 h and became unconscious and cyanotic. **He** drank a lot of ***alcohol*** after the treatment; its effect, if any, on the outcome has been a subject for discussion. Other cases of toxicity include nausea, vomiting, hyperpnea, shallow breathing, confusion, peripheral neuropathy, swelling, exudation, pain, cough, fever, tachycardia, oliguria, anuria, adynamic ileus, coma, death, and the birth of a stillborn **child**. A few hours after application on venereal warts, they become blanched; in 24–48 h, they appear necrotic, and they begin to slough on day 3, disappearing without scarring. Fatality in a 60-year-old **female** after ingesting ~5 grains (324 mg) or 5 mg/kg. Symptoms included vomiting and purging, followed by coma, slight increase in temperature, and hemoglobinuria.

Fetal death and teratogenicity in **rats** and probably in **humans**.

LD_{50}, **mouse**: 68 mg/kg.

POINCIANA GILLIESSI

= *Mexican Bird of Paradise*

Untoward effects: The green seed pods or seeds cause gastroenteritis (nausea, vomiting, abdominal pain, and profuse diarrhea). In the U.S. gulf states and westward, **children** have been severely poisoned by eating the pods.

POINSETTIA

= *Christmas Flower* = *Easter Flower* = *Euphorbia pulcherrima* = *Lobster Flower* = *Mexican Flame Leaf* = *Mexican Flame Tree* = *Mexican Flower Plant*

Untoward effects: Hypersensitivity to its sap is not uncommon in Florida and is an occupational hazard to greenhouse **workers** who can develop dermatitis. **Infants** ingesting small amounts of the sap develop oral and anal irritation and burning sensations; **children**, after eating a leaf or sucking nectar from a few flowers, develop vomiting, abdominal cramps, diarrhea, gastrointestinal bleeding, and, possibly, delirium. A case report of a 2-year-old **child** who died after eating the leaves. The leaves contain ***euphorbin*** and ingestion of one leaf can kill a young **child** (reported in 1919). The sap loses toxicity on drying or heating. Temporary blindness if the sap is rubbed in the eyes. An 8-month-old **female** chewed a leaf and developed burns of **her** buccal mucosa.

In **rats**, **cats**, **dogs**, and **humans**, the toxicity after ingestion is usually gastrointestinal in nature. Ingestion by an older **dog** led to nausea, protracted vomiting, renal failure, coma, and death.

POISON IVY

= *Herbe à la Puce* = *Markweed* = *Poison Vine* = *Rhus radicans* = *Rhus toxicodendron* = *Toxicodendron radicans*

Found in most of the U.S., southern Canada, Central America, the Bahamas, and Bermuda.

Untoward effects*:** Severe contact and allergic dermatitis with pruritus, swelling, leukocytic infiltration, blistering, and weeping; occasionally delayed for 2–3 days, rarely, 9 days. Toxic epidermal necrolysis, fever, and headaches develop in some **people**. Black ***lacquer-like deposits may occur on the lesions. ***Urushiol*** is the main constituent in the toxic, oily, resinous sap of the plant. Smoke from the burning plant has had toxic effects. Contaminated ***shoes***, clothing, and tools can be sources of dermatitis months after contact with the plant. Handling **cows** and **pets** that have contacted it can also be a source of **human** exposure. Some **people** develop a severe cheilitis and, occasionally, a mucositis of the tongue and mouth after chewing its leaves inadvertently or to produce a hyposensitization effect. Rarely, **others** develop gastrointestinal irritation, proctitis, vulvitis, nephritis, and narcosis. Raw ***cashew nuts*** sold in health food stores contain the shell oil on their surface, which is antigenically similar to poison ivy resin. When eaten in large quantities by sensitive **people**, it causes generalized eruptions. Sensitive **individuals** may crossreact with ***cashew nut tree*** (and incompletely processed ***cashew nuts***), ***India marking nut tree, Japanese lacquer tree, mango tree, poison oaks, poison sumacs***, and ***Renghas tree***. Parenteral injections of an extract for treatment and prevention have been used. Hypersensitivity reactions including swelling of joints, pruritus, purpuric rashes, and bullous exanthema have been reported.

POISON OAK

= *Toxicodendron diversilobum (western) = Toxicodendron toxicarium (eastern)*

Some consider *Rhus toxicodendron* to be poison oak, rather than *poison ivy*.

Untoward effects*:** Allergic contact symptoms are similar to those of ***poison ivy (q.v.) and also due to its ***urushiol*** content. An extract, ***Angerex***, has been used IM to decrease an **individual's** sensitivity to it. Mild to moderately severe pain at injection site. SC injections can cause longer lasting pain and nodule formation. Diaphoresis and palpitations also reported.

POLDINE METHYLSULFATE

= *IS 499 = McN R726-47 = Nactate = Nacton*

Anticholinergic.

Untoward effects*:** Side-effects, as with ***atropine, include xerostomia, tachycardia, blurred vision, extrasystoles, and slight dysuria. Occasionally headache, decreased libido, and belching.

POLISHES, FURNITURE

Untoward effects*:** High fever and illness in **children** from inhalation and ingestion ***Hawes Lemon Oil, due to ***hydrocarbon*** and ***aniline*** content. Many case reports of ***hydrocarbon*** pneumonitis, lethargy, hyperpnea, drowsiness, vomiting, bronchopneumonia, cyanosis, tonic convulsions, coma, and death. Also references to poisoning by ***Holders Supreme Wax Oil Polish®, O'Cedar All Purpose Red Polish®, Old English Red Furniture Polish®***, and ***Johnson's Car Plate®***. Occupational exposure to ***chromates*** leads to dermatitis in furniture **polishers**.

The main toxicant in most is ***mineral seal oil*** (q.v.).

POLISHES, SHOE

Untoward effects*:** Ingestion by 3-year-old **male** (10–20 ml, containing 3% ***nitrobenzene and 27% ***aniline***) led to cyanosis, and bilateral Babinski signs within 1 h. Elation within 10 min, followed by severe headache in school **children** sniffing it. Migraine headaches and hallucinations in some **users** after sniffing. A case report of eczematous reactions on feet due to sensitivity to shoe polish. I diagnosed a ***sulfonamide*** reaction from pearlescent shoe polish on Italian-made shoes.

POLONIUM

Untoward effects*:** All isotopes from 197 to 218 are radioactive. Radioactive polonium-210 is found in ***cigarette smoke and ***tobacco*** and chronic exposure may be a factor in bronchial or lung cancer. Radioactive polonium in ***phosphate fertilizers*** used on ***tobacco*** plants or airborne fallout on ***tobacco*** leaves is also a source.

Lung cancer has been induced in **hamsters** by intratracheal instillation of ***polonium-210***, an *alpha* emitter.

POLOXALENE

= *Bloat Guard = Exocorpol = Pluronic F-68 = Poloxalkol = Poloxamer 188 = Rheoth R = SKF 18,667 = Therabloat*

***Surfactant*. Non-ionic.**

***Untoward effects*:** Apparently, safe for IV use at 1.28–2.56% in perfluorochemical perfusion emulsions and plasma expanders in **human**, **rat**, and **dog** trials. Safe use of IV fat emulsions with 0.2% in **humans**. Pain, injection site irritation, and nausea in **volunteers** given 10–90 mg/kg/h.

LD_{50}, **mouse**:> 5 g/kg.

POLYANTHES TUBEROSA

= *Tuberose*

***Untoward effects*:** Growing in groups or on still humid days, its strong sweet-smelling aroma often causes headaches and nausea in some **people**.

POLYBROMINATED BIPHENYLS

= *Firemaster* = *PBBs*

Fire retardants.

Untoward effects: In 1973, some 2,000 lbs was mistaken for ***magnesium oxide*** and mixed into **dairy cattle** feed in Michigan, exposing **those** who consumed ***milk***, dairy products, and meat from these **animals**. Breast milk in Michigan **women** showed levels up to 1.2 ppm. Found in **catfish** in the Ohio River, plants, soil, and **human** hair. In **humans** causes muscle weakness, memory loss, incoordination, fatigue, skin dryness, decreased libido, rashes, porphyrias, chloracne, alopecia, itching, brittle bones, headaches, painful joints, blurred vision, nervousness, sweating, increased toenail and fingernail growth, and immunologic deficits. Long-term persistence in **human** tissue. Neurotoxic. Thousands of **cattle**, **hogs**, and **sheep**, 25 million **chickens**, and 5 million ***eggs*** were destroyed to prevent its spread into the food chain. In China, ***PBB*** poisoning reported in 122 **people** who ingested contaminated ***cooking oil***. Suggested safety limits are 20 μg/day or 0.3 μg/kg/day, and less proportionately for **infants** and **children**. No longer produced in the U.S.

Cows suffered from anorexia, decreased milk production, hyperkeratosis, prolonged gestation, spontaneous abortions, birth defects in their **calves**, and death. High levels ingested by **rats** retarded growth, increased liver size, led to porphyria, and interfered with synthesis and secretion of ***thyroxine***. In **monkeys**, decreased growth rate, failure to conceive, abortions, and stillbirths reported. Follicular hyperkeratosis from 60 μg applied to **rabbit** ears. Enlarged livers in some **guinea pigs** fed it. Hepatocellular carcinomas in **rats** and **mice**; cholangiocarcinomas and neoplastic nodules in **rats**; trabecular hepatocellular carcinomas in offspring of treated **rats**.

TD_{LO}, **rat** (pregnant 12 days): 800 mg/kg; **mouse** (pregnant 7–18 days): 2.4 g/kg.

POLYBUTADIENE

Use: 750,000 metric tons used/year for manufacturing tires and treads, ***polystyrene***, conveyor belts, hoses, seals, and gaskets.

Untoward effects: Skin rashes with papules in industrial **workers** and in **children** from a toy, ***Flubber***®, which contains 73% with 27% ***mineral oil***.

POLYCHLORINATED BIPHENYLS

= *Arochlors* = *Clophen* = *Delors* = *Kanechlors*

The mixtures are colorless to dark brown oils, viscous liquids, or sticky semi-solids.

Use: In transformers and capacitators made before the U.S. ban in 1977.

Untoward effects: Primary non-occupational source is food, especially ***fish*** from contaminated waters. Since 1929, 1.25 billion lbs have been produced in the U.S. with about 450 million lbs entering the environment. Readily absorbed, but poorly metabolized. Toxicity affects skin (chloracne – weeks or months after exposure) and liver. The acneform lesions have lasted for up to 30 years on chin, arms, thighs, genitalia, buttocks, and periorbital and malar areas. Asymptomatic hepatomegaly with increased SGOT, aspartate aminotransferase, and γ-glutamyl transpeptidase. Developmental deficits in **offspring** of exposed **mothers** usually disappear by age 2. In exposed **men**, decreased sperm motility and fertility problems. **Offspring** of **women** who ate two or three Lake Michigan **fish** a month for at least 6 years before **their** pregnancies (and **those** of the Yusho incident in Japan) delivered preterm with lower birth weights, smaller skull circumference, abnormal tooth development, dark brown skin and nail pigmentation, hypotonicity, hyporeflexivity, and behavioral deficits. Breast milk has contained up to 0.96 ppm. Exposed industrial **workers** had chloracne and skin pigmentation (40%), headaches (30%), dizziness, nausea, eye discharge, swollen eyelids, and musculoskeletal problems. In June, 1968, a 3-year-old **female** was brought to Kyushu University Hospital with a skin rash. Soon after, her **parents** and a **sister** developed a similar rash and the above symptoms. ~1,300 **people** and their **babies** in western Japan became affected from ingesting contaminated ***rice bran*** cooking oil manufactured in February, 1968. Contamination was from a heat exchanger leak. Yusho is Japanese for cooking oil and now refers to "oil sickness". Incidence of cancer has been higher in the exposed **population**. Induced cognitive defects in exposed **human** and experimental **monkeys** have lasted up to 12 years, and, generally, permanent. Only 82% of contaminated Yusho **victims** showed chloracne. In central Taiwan, 2,000 **residents** became ill with debilitating and disfiguring symptoms, including stomach pain, chloracne, and nerve inflammation – Yu Cheng incident, traced to a machine leaking ***polychlorinated biphenyls*** while clarifying ***rice oil***. In 1992, 17 years after the incident, exposed **women** had blood levels 50 times greater than unexposed **women**. Depending on the amount of ***chlorine*** in the compound, OSHA set upper exposure limits of 0.5–1 mg/m^3 TWA.

LD_{LO}, **human**: 500 mg/kg.

Dietary intake by **rats** leads to trabecular cell carcinomas, neoplastic nodules, and cholangiomas of the liver; hepatocellular carcinomas and liver adenocarcinomas in female **rats**. Hepatomas and hepatocellular carcinomas in male **mice**. Exposure of male **rats** by their **mother's** milk decreases their fertility in later life. Dying **dolphins** and **seals** had unusually high tissue levels. Many **fish**, **poultry** species, and **mammals** have suffered from exposure to ***polychlorinated biphenyls***.

POLYCLADA sp.

Untoward effects: Pupae of these leaf-eating **beetles** are used as a toxic poison on the arrows of Bushman groups of Botswana and northern Namibia of South Africa. The dried pupae are mixed into a paste and their contents are squeezed out and applied to the arrow shaft.

POLYESTER

= *Mersilene*

Untoward effects: A polyester pants or slacks paresthesia syndrome reported. Polyester foam binds ***magnesium***, inhibiting the growth of *Staphylococcus aureus* strains, but causes excessive production of its toxin and possibly associated with toxic shock syndrome in menstruating **women**. Its resins used in clothing and fabrics, when burning, release ***hydrogen chloride*** (q.v. for symptoms).

POLYETHYLENE

= *Biopor* = *Courlene* = *Dermalene* = *Epolene* = *Marketote* = *Marlex* = *Microthene* = *Petrothene* = *Polyfilm* = *Polythene*

Production has increased dramatically every year and, at this writing, is ~14 billion lbs of high- and low-density material.

Untoward effects: Lipid-soluble preservatives or therapeutic drugs can migrate into it from solutions. Danger of explosion from reuse of disposable anesthetic circuits after use of ***methoxyflurane*** and other volatile anesthetic agents. Systemic toxicity reported in **people** from soft plastic food containers. Miliaria rubra, folliculitis, and boils after its use as a covering in various dermatoses. Not suitable for occlusion of lichen planus hypotrophicus, as it exacerbates the pruritus. Hypopigmentation (hypomelanosis) from an antioxidant in the polyethylene occlusive dressings. Contact dermatitis reported from its use in hearing aids.

Elicits adhesions in **rabbit** abdomens after implantation.

Implant, TD_{LO}, **rat**: 2.12 g/kg; **mouse**: 330 mg/kg.

POLYLETHYLENE GLYCOL

= *Carbowax* = *Lutrol E* = *Macrogol* = *PEG* = *Pluracol E* = *Poly G* = *Polyglycol E*

Untoward effects: Allergic contact urticaria reported. Allergic dermatitis in a 54-year-old **female** who developed eczematous lesions below cellophane (***Sellotape***®) tape from ***PEG*** component. Contact urticaria in **people** from topical use of ***PEG 300*** and ***PEG 400***. ***PEG*** is an irritant and sensitizer. Eye contact by ***PEG 400*** can cause slight transitory pain and irritation, as with a mild soap. Acute pulmonary edema and severe respiratory distress in an 8-year-old **female** after 1 l of solution via nasogastric tube the night before a scheduled colonoscopy. Symptoms also included vomiting and tachypnea. **She** required intubation and ventilatory support. Use of large quantities in parenterals can cause local inflammation and hemorrhagic reactions; chronic IV use in a 57-year-old **male** of ***PEG 400*** with ***lorazepam*** led to acute tubular necrosis. A fatal syndrome consisting of renal failure, metabolic acidosis, increased serum ***calcium***, and increased anion and osmolal gaps has been reported, due to absorption from an ***antimicrobial*** cream on burn **patients**. The results were duplicated after 7 days use on denuded **rabbit** skin, even when used without any ***antimicrobials***. After ingestion, absorption of ***PEG 4000*** is ~2.5%. May adversely affect the stability of ***penicillins***.

LD_{LO}, **human**: 5 g/kg.

Skin reactions in **mares** given parenteral ***fenprostalene*** was due to ***PEG 400*** vehicle. In the **dog**, IV ***PEG 600*** enhanced the response to ***acetylcholine*** through its effect on the peripheral vasculature.

LD_{50}, **rat**: 26 g/kg; **mouse**: 31 g/kg; IV, **rat**: 6.4 g/kg; **mouse**: 6.6 g/kg; IP, **mouse**: 4.2 g/kg; **rat**: 13 g/kg.

POLYGALA sp.

Untoward effects: ***P. angulata*** has been an adulterant in misbranded ***ipecac***.

P. klotzchii, a milk wort in Brazil, has poisoned and killed large numbers of **cattle** that have grazed it. Incoordination, tremors, spasms, dyspnea, and diarrhea occur before death. Post-mortem reveals congestion of gastrointestinal mucosa with some petechiation, congestion and hemorrhage of cerebral vessels, pulmonary hemorrhage, and necrosis of lymph nodes, Peyer's patches, and spleen.

Podophyllotoxins (q.v.) have been isolated from ***P. macradenia*** and ***P. polygama***. Some species contain ***methyl salicylate***.

Also see ***Senega snakeroot*** (***Seneca snakeroot***).

POLYGLYCOLIC ACID

= *Dexon*

Untoward effects: Lacks systemic toxicity when used as sutures, causing only mild irritation in **humans**, **horses**, and **dogs** in surgery on skin, fascia, muscle, peritoneum, and gastrointestinal tract. Serious post-operative complications were manifested by capsular wound disruption, joint effusion, and knee instability in 10/928 **people** undergoing arthrotomies of the knee during a 7 year period, apparently, due to inadequate suture tensile strength.

Tissue reactions in 2/9 **dogs** from suturing with it in bladder surgery. Total absorption after 90 days in **dogs** and 110 days in **horses**.

POLYGONATUM BIFLORUM

= *Conquer-John* = *Sealwort* = *Solomon's Seal* = *Whitewort* = *Yuzhu*

Untoward effects: Contains an ***anthroquinone*** in its berries, causing vomiting and purgation after ingestion; can cause hypoglycemia. The flowers are also reported as acrid and poisonous.

POLYGONUM sp.

= *Dooryard Weed* = *Knotweeds* = *Mat Grass* = *Smartweeds* = *Water-pepper*

The rhizome of *P. Bistorta* is known as *Adder's Wort* = *Bistort* = *Snakeweed* = *Snakewort*, and is used in Europe and Asia as an intestinal astringent in diarrhea, due to its 20% *tannin* content.

Untoward effects: It gets its name, ***smartweed***, because its pungent plant juices often cause smarting and irritation of eyes and nostrils; occasionally, skin rashes in some **people** from contact with its leaves. Common in ***corn*** rows and moist areas.

P. hydropiper has antiovulatory effects in female **rabbits**. In the U.S. Southwest, ***P. persicaria* = *Lady's-thumb*** has caused photosensitization in **cattle** and **sheep**.

P. punctatum* = *Water Smartweed. The leaves contain ~7% ***calcium oxalate*** and are poisonous for **man** and can be fatal to **livestock**. Polygonum species, used as a food and in folk medicines, have been suspect as a cause of esophageal cancer in Asia, West Indies, and South Africa. Some species used as an insecticide and anthelmintic. Seeds mixed in grain or ground feed have poisoned **livestock**. The plant (weed) contains high levels of ***nitrates*** and, after grazing, has caused abortions in **cattle**. Ingestion can induce photosensitization in **livestock**.

POLY(HEXAMETHYLENE BIGUANIDE HCL)

= *Baquacil* = *Cosmocil CQ* = *PHMB*

Untoward effects: Methemoglobinemia. Eye, skin, and respiratory tract irritant on inhalation. A skin sensitizer.

In **animals**, mouth, pharyngeal, esophageal, and stomach irritation can follow ingestion. Toxic to **fish**.

LD_{50}, **rat**: 2.5 g/kg.

POLY I : C

= *Polyribinosinic and Polyribocytidylic Acids*

***Interferon inducer.* Parenteral use.**

Untoward effects: Nausea, vomiting, and anemia in ~30%; fever 6–8 h after injection, peaking within 36 h in ~65%. Occasionally bronchospasm and flushing.

Endotoxin-like effects in **mice** and pathology of cerebellum in **chicks**. IV use in **cats** leads to toxicity within a few hours; mucoid diarrhea, weakness, trembling, and occasionally death.

IV, LD_{50}, **mouse**: 8 mg/kg; IP, 30 mg/kg. IV, LD_{LO}, **dog**: 3 mg/kg; **monkey**: 15 mg/kg.

POLYMYXIN B

= *Aerosporin* = *Bacillosporin B* = *Polytrim*

Untoward effects: Decreased serum ***albumin*** and hypocalcemia with occasionally tetany in some **patients** after parenteral use. Urticaria, angioedema, and fever from induced histamine release. Nephrotoxic with renal tubular dysfunction. Myasthenic syndrome, weakness, shock, facial flushing, paresthesias, peripheral neuropathy, visual disturbances, dizziness, ataxia, lethargy, anorexia, increased or false positive blood urea nitrogen, malabsorption syndrome (especially of ***fat*** after oral use), albuminuria, eosinophilic leukemoid reaction, hyponatremia, hypochloremia, hypokalemia, azotemia, pruritus, nausea, dyspepsia, corneal opacities, hearing loss or vestibular disturbances. The powder may cause eye, skin, or respiratory irritation. Avoid breathing the dust. Pain at injection site reported.

Drug interactions: Orally, it can decrease ***methotrexate*** absorption by 30–50% and decrease absorption of ***carotene, hexose sugars, iron***, and ***vitamin*** B_{12}. Can inhibit hepatic enzymes, enhancing the effect of ***general anesthetics***. Significant potentiation of ***curare*** and other ***non-depolarizing agents*** and ***depolarizing agents*** by interfering with ***acetylcholine*** at the neuromuscular junction. Can potentiate nephrotoxic and neurotoxic effects of ***gentamicin, kanamycin, neomycin, streptomycin***, and ***viomycin***.

Parenteral use in pregnant **animals** should be avoided according to **rat** studies. Neuromuscular-blocking effects have been reported in **animals**. The FDA has established a tolerance of 2 Units/ml in milk of treated **animals**. Glomerular and tubular nephrotoxicity in **dogs** after a single IV of 2.5 mg/kg or 5 mg/kg/day IM. Nephrotoxicity is due to depletion of methyl groups and ***dl-methionine*** can reduce this effect.

IV, LD_{50}, **mouse**: 8 mg/kg.

POLYOXYETHYL LAURYL ETHERS

= *Briz J 30 and 35* = *Laureth 4, 9, and 23* = *Lauromacrogol 400* = *Liposec* = *Polidocanol*

Non-ionic emulsifier and drying agent, spermaticide, antipruritic, sclerosing agent, and topical anesthetic.

***Untoward effects*:** Injection for sclerotherapy of varicose veins in a 41-year-old **female** caused reversible ischemic neurological deficit 1 h later. Symptoms included warm sensation from abdomen to head, unpleasant taste, paresthesia in right half of body, loss of right upper quadrant of the visual field, latent right arm paresis, and hyperesthesia of ulnar side of the right hand and forearm lasting 2 days.

Percutaneous absorption reported in **mice**.

POLYPODY FERN

= *Polypodium virginianum* = *Resurrection Fern*

***Untoward effects*:** Ingestion by **livestock** causes a bradycardia from an unknown chemical.

POLYSORBATES

= *Drewpones* = *Polyoxyethylene Sorbitan Esters* = *Sorboxethenes* = *Tweens*

Non-ionic surfactant in medicinals, vaccines, and foods.

Untoward effects*:** Have caused contact urticaria and various hypersensitivity reactions. In **humans**, oral use of 4.5–6 g/day for up to 4 years was without ill-effects. ***Tween 80 can neutralize the bactericidal effects of ***cetylpyridinium chloride***.

IP use in **rats** kills by respiratory paralysis. Reactions in **animals** due to its use in parenterals may be due to its release of foreign proteins eluted from syringes normally considered clean. ***Polysorbate 80*** produces marked corneal edema when injected into the anterior chamber of **rabbit** eyes.

Tween 80: LD_{50}, **mouse**: > 25 g/kg. IP, LD_{50}, **rat**: 6.3 ml/kg; **mouse**: 7.5 ml/kg; IV, **mouse**: 5.8 ml/kg.

POLYSTYRENE

= *Dylene* = *Tricyte*

***Use*:** ~7,000 million lbs annually for packaging, electronics, furniture and building construction, etc.

Untoward effects*: *Lemon-flavored ***tea*** can dissolve a polystyrene cup. The ***tea*** may be safe, but the dissolved polystyrene isn't. It is carcinogenic in **laboratory animals**. Constipation and intestinal bezoars and obstructions from ingestion of ***polystyrene sodium sulfonate*** with ***aluminum hydroxide*** gel. Polystyrene disposable syringes react with many ***iodine-containing contrast media, aromatic hydrocarbons, ketones, esters***, and ***aldehydes***. Polystyrene containers adsorb ***phenylmercurials***. On heating, it can release ***methyl chloride***, which can irritate eyes, nose, and respiratory system; and cause headaches, fatigue, dizziness, narcosis, ataxia, a defatting dermatitis, and, possibly, liver pathology. A fire hazard.

Small pherules released by industry may cause intestinal blockage in **fish**.

Implant: TD_{LO}, **rat**: 19 mg/kg.

POLYTEF

= *Fluon* = *Goretex* = *Polytetrafluoroethylene* = *PTFE* = *Silverstone* = *Teflon* = *Tetran*

Untoward effects*:** In prosthetic material, leads to early fibroplastic invasion of mesh grafts with mild inflammatory reaction, causing organized scar tissue between graft and fascia. Polymer fume fever (sudden coughing, breathlessness, shivering, shaking, chills, fever, pulmonary edema, sore throat, myalgia, and leg weakness) occurs from inhalation of its pyrolysis products. In the workplace when particles or dust settles on lighted ***cigarettes, even from particles on unwashed hands. Inhalation of fumes increases body temperature 0.7–3.0 °C. In fires, it produces ***hydrogen fluoride*** (q.v.) and ***octafluoro-isobutylene***.

Overheating non-stick surfaces produces toxic fumes killing **birds** within min. **Cockatiels** have died within 30 min after exposure to fumes from an overheated ***Teflon***-coated frying pan. Post-mortem revealed pulmonary edema and degenerative changes in liver and myocardium. The **owner** developed polymer fume fever. Death of a **budgie** also reported after 3 h of exposure to fumes. Periurethral injections into female **dogs** and male **monkeys** caused granulomatous reactions at the site and at points of distant migration.

Implant: TD_{LO}, **rat**: 80 mg/kg; **mouse**: 1.14 g/kg.

POLYTHIAZIDE

= *Drenusil* = *Nephril* = *Renese*

***Diuretic, antihypertensive.* Oral or IM.**

Untoward effects*:** Use has been associated with hypercalcemia, hypokalemia, hyperuricemia, hypochloremic alkalosis, hyponatremia, decreased ***glucose tolerance, and slight hyperglycemia. Some of these contribute to the symptoms of xerostomia, thirst, weakness, drowsiness, lethargy, muscle fatigue and/or cramps, oliguria, tachycardia, sweating, nausea, and vomiting. Aplastic anemia, thrombocytopenia, neutropenia, pancytopenia, cholestatic hepatitis and jaundice, photosensitivity, and skin rashes also reported. Contraindicated in **patients** sensitive to other ***thiazides*** or other ***sulfonamide*** derivatives.

LD_{50}, **dog**: 450 mg/kg; IP, **rat**: 400 mg/kg.

POLYURETHANES

= *Duraspan* = *Glospan* = *Lycra* = *Lyomousse* = *MRD 535* = *Ostamer* = *Spandelle* = *Spandex* = *Synthaderm*

Use: In insulation, seat cushions, carpeting, wall coverings, medical implants, clothing, and in fire and explosion prevention in combat jets.

Untoward effects: Allergic contact dermatitis (often with sharply demarcated, red, weeping areas with occasional depigmentation), especially in **females**, from its fibers in brassieres, girdles, swimsuits, waistbands, hosiery, socks, suits, sportswear, and blouses. Occasionally bronchitis and bronchospasms in plant **workers** manufacturing its foam, probably from ***toluene diisocyanate*** (q.v.) used in its manufacture. Small amounts of ***2,4-toluenediamine*** (q.v.), a carcinogen, have been released from polyurethane foam covering ***silicone*** breast implants in **women**. Pyrolysis releases ***hydrogen cyanide*** (q.v.), ***carbon monoxide*** (q.v.), and ***isocyanates*** (pulmonary irritants).

Excess ***ostamer*** in oral fracture repairs in **dogs** led to ischemia, necrosis, and ulceration. IP implants of the foam in **rats** caused adenocarcinomas of the colon, and sarcomata and areas of fibroplastic proliferation in **sponges**. Autoclaving foam test tube plugs released volatile amines, originally incorporated as catalysts, inhibiting the growth of some microorganisms.

TD_{LO}, *implant* (***Y #'s***), **rat**: 6.75 g/kg.

POLYVINYL ACETATE

Untoward effects: Implicated as one of the causes of pulmonary granulomas in **hairspray users**.

POLYVINYL ALCOHOL

= *Akwa Tears* = *Corvol* = *Cryogel* = *Elvanol* = *Ethenol Homopolymer* = *Gelvatol* = *Ivalon* = *Liquifilm* = *Mowiol* = *Optifilm* = *Polyviol* = *PVA* = *Sno Tears* = *Vinarol* = *Vinol*

Over 300 million lbs used annually. Widely used in the plastic industry, cosmetics, printing ***inks***, intestinal parasite preservation, and artificial tears; and to prolong contact of drugs in ophthalmic preparations and to increase viscosity of pharmaceuticals.

Untoward effects: Its routine use (5%) in a preservative solution for protozoa and other intestinal parasites has, unfortunately, led to serious acute ***mercury*** poisoning in **children** from ***mercuric chloride*** 4.5% (q.v.), after ingesting (430–860 mg), due to a misunderstanding by **parents** of the Mexican-American **children** (ages 4½–5 years).

SC use in **rats** caused sarcomas and fibrosarcomas; nephrosis increased with increases in molecular weights.

SC, TD_{LO}, **rat**: 2.5 g/kg.

POLYVINYL CHLORIDE

= *Breon* = *Chloroethene Homopolymer* = *Chloroethylene Polymer* = *Crinovyl* = *Envilon* = *Fibravyl* = *Geon* = *Isovyl* = *Koroseal* = *Marvinol* = *Movyl* = *Nip* = *PVC* = *Retractyl* = *Rhovyl* = *Sicron* = *Tevilon* = *Thermovyl* = *Vestolit* = *Welvic*

Use: Covering for electrical wires and cables; upholstery, tubing, raincoats, ***shoe*** soles, credit cards, automobile and aircraft interiors, transfusion blood bags; tubing in heart, lung, and kidney dialysis machines; and food wrappings in the amount of ~13,000 million lbs annually.

Untoward effects: It can contain organic ***tin*** preparations, phthalate esters (such as ***dioctyl phthalate*** and ***DEHP*** (q.v.), a carcinogen and residual ***vinyl chloride*** monomer [q.v.]). Suddenly in 1986, the FDA (your protectors) decided, in spite of the Delaney Clause, to allow carcinogens in food packaging material under the principle of *de minimis*. **Meat-wrappers** asthma, a bronchospasm and bronchitis, from inhalation of fumes occurs from cutting ***PVC*** wrappings with a hot wire. ***Diphenylthiourea***, a heat stabilizer additive in it, is also a potential allergen. ***PVC*** plant **workers** with bronchospasms, dyspnea, and coughing, have shown alveolar hyperplasia in **their** lungs. Inhalation of fumes has caused dizziness. Contact dermatitis and burns also reported. Eczematous reactions from its film between the filter ***paper*** and adhesive tape used in routine patch-testing. Repeated events of cutaneous necrotizing dermatitis in a 59-year-old **male** starting 8 h after its use in hemodialysis. Ohio studies indicated a slight increase in congenital malformations in **children** born to **mothers** living near a ***PVC*** manufacturing plant. During early production times in Europe, manufacturing plant **workers** reportedly had an increased incidence of liver, spleen, and bone diseases, and a questionable increase in angiosarcoma, a rare liver cancer. ***PVC beer***, soda pop, etc. containers, when burned, produce ***hydrochloric acid*** (q.v.), ***phosgene*** (q.v.), and ***chlorine*** (q.v.), which are pulmonary irritants. ***PVC*** containers vary widely in pH, with leaching of chemicals into solutions and into the ***PVC***. Avoid their use with ***diazepam*** and ***benzalkonium chloride***. Water vapors can pass through containers and increase the concentration of drugs in them.

TD_{LO}, *implant*, **rat**: 100 mg/kg.

POLYVINYLPYRROLIDONE

= *Collocral* = *Enterodex* = *Isoplasma* = *Kollidon* = *Luviscol* = *MK 7879* = *Periston* = *Plasmosan* = *PVP* = *Subtosan* = *Toxobin* = *Viniril*

***Suspending agent, plasma volume-extender.* Different molecular weights are used.**

Untoward effects: Confluent pruritic red-yellow papules on forehead, chin, cheeks, neck, and submammary folds in a diabetic insipidous 43-year-old **female**, 2 months after SC ***posterior pituitary extract***. Osteolysis and thesaurismosis with cutaneous manifestations after similar use or IV and IM in other **patients**. The reticulo-endothelial system accumulates it after massive chronic use. In France and Belgium, "Storage Disease", "Polyvinyl Disease", and "Dupont-Lachapelle Syndrome" are all the same, and occur after years of parenteral use of high mw ***PVP*** with ***pituitrin***. Dermatological symptoms usually occur after > 2,000 g IM, for at least 6–10 years. After inhalation, it may exacerbate asthma, as it can cause the release of ***histamine***. Lung pathology and blood dyscrasias, if any, from its use in hair sprays can be related to its concentration and duration of exposure. A 37-year-old **male**, after intra-articular use, developed anaphylactic shock. In general, it can elicit foreign-body, granulomatous, or delayed hypersensitivity tissue responses.

Drug interactions: Use with ***demeclocycline*** increases the latter's toxicity. Prolongs the blood level of ***tetracyclines***.

LD_{50}, **rat**: > 100 g/kg; IP, **mouse**: 12 g/kg. IV, TD_{LO}, **monkey**: 5.3 g/kg.

POLYVINYLPYRROLIDONE IODINE COMPLEX

= *Betadine* = *Betaisodona* = *Braunol* = *Braunosan H* = *Disadine D.P.* = *Disphex* = *Efodine* = *Havlodine* = *Inadine* = *Isodine* = *Operand* = *Povidone Iodine Complex* = *Proviodine* = *PVP-Iodine* = *Traumasept* = *Ultradine* = *Vagidine* = *Vidine*

***Topical anti-infective.* An iodophor.**

Untoward effects: Excessive absorption after topical use can cause systemic effects, such as iodism. Use on body burns covering 20% or more, or in the presence of renal failure, leads to severe metabolic acidosis. The same can follow intraperitoneal use for treatment of peritonitis. Hypersensitivity reactions can also follow the ***iodine*** release. Allergic contact dermatitis reported. Dermatitis herpetiformis triggered by its use in preoperative skin preparation. Pseudobacteremia reported in 16 **patients** during a 10 week period, due to contamination of a 10% solution by *Pseudomonas cepacia*. Its use as a scrub was more toxic to cochlear function than rinsing the area with a solution. Use as a mediastinal lavage after surgery for pulmonary stenosis and patent ductus arteriosis in a 22-day-old **male** led to hyperthyroidism. Use topically qid on an **infant's** large intact exomphalos led to hypothyroidism. Can also be absorbed through the intact skin of 2–11-year-old **children**. A 7½-month-old breastfed **infant** had an ***iodine*** odor emanating from it and increased serum and urine levels of ***iodine*** after topical use of a ***Betadine*** vaginal gel by the mother for 6 days. Use of 2% solution for prenatal and perinatal perineal cleansing caused transient hypothyroidism in a large number of **infants**. In 1979, the FDA received a large number of reports indicating skin erythema, tissue necrosis, and dermatological problems from a revised 10% formulation. Skin sensitivity, reddening, desquamation, and pustular ioderma. Skin reactions on **surgeons** by 10% solution can often be eliminated by use of 1% solution or diatomic ***iodine***. Thyrotoxicosis has followed twice daily washes over a 6 month period in a 72-year-old **male**. More readily absorbed after vaginal use in pregnant, rather than non-pregnant, **women**. Occasionally pain after application to wounds. Renal failure, increased ***protein-bound iodine***, neurological deterioration, hypernatremia, tachyarrhythmias, pericarditis, and false positive tests for occult blood in feces and urine. Anaphylaxis in a 25-year-old **female** from cleansing prior to an abortion. A seizure reported in a 62-year-old **male** associated with continuous mediastinal irrigation after by-pass surgery. Do NOT mix with ***hydrogen peroxide*** and leave in a stoppered container!! A hospital explosion has occurred from such a procedure. A number of surfactants decrease its bactericidal activity.

Skin TD_{LO}, **human**: 3.4 g/kg/24 h.

POMEGRANATE

= *Apple of Carthage* = *Khiman* (Egyptian) = *Malum punicum* = *Punica granatum* = *Roman* (Ethiopia and Portugal) = *Rimon* (Hebrew) = *Ruman* (Arabic) = *Rumanu* (Aramaic)

A symbol of ancient Hebrews and Canaanites.

Untoward effects: Although roots and leaves were once used as an anthelmintic, they contain the alkaloid ***isopelletierine*** and others, which have caused severe digestive upsets after ingestion, especially by **children**. Extracts also contain ***estrone*** and have been used as a source of hormones in Near Eastern and Oriental cultures. ***Pelletierine Tannate* = *Punicine Tannate*** is a mixture of its alkaloids used as an oral taeniacide. It has caused headache, vertigo, diplopia, drowsiness, weakness, leg cramps, colic, nausea, vomiting, and diarrhea at therapeutic dosage. Overdoses have also caused central nervous system depression, partial blindness, profound prostration, convulsions, respiratory failure, and death.

Bark LD_{LO}, **human**: 500 mg/kg.

PONCEAU 3R

= CI 80 = Ext. D & C Red #15 = FDC Red #1

***Coal tar dye.* Formerly used primarily for coloring maraschino *cherries* and *sausages*.**

Untoward effects*:** Found to be carcinogenic in Wistar **rats**, due to its metabolites, ***trimethyl aniline derivatives, and now deleted by the FDA as safe for foods, drugs, and cosmetics. Still permitting for dyeing ***wool***.

TD_{LO}, **rat**: 730 mg/kg during 2 years; **mouse**: 80 mg/kg during 68 weeks.

PONCEAU 4R

= CI 16,255 = Coccinillerot A

***Untoward effects*:** Urticaria, dermatitis, and benign pigmentation of **user's** stools.

In **animal** trials, decrease in erythrocytes and hemoglobin concentration.

LD_{50}, **rat** and **mouse**: > 8 g/kg.

PONCEAU MX

***Untoward effects*:** Suspect carcinogenic dye. In **mice** and **rats**, increased renal weight; hepatic cellular necrosis and nodular hyperplasia.

TD_{LO}, over 62 weeks period, **rat**: 35 g/kg; **mouse**: 87 g/kg.

PONCEAU SX

= CI 14,700 = CI Food Red 1 = FD&C Red #4

Use in foods is prohibited or restricted to limited uses in many countries.

***Untoward effects*:** Sublingual testing in **humans** indicated it was allergenic.

May induce urinary bladder polyps and adrenal atrophy in **dog** trials (1% dietary level for 7 years) led to chronic follicular cystitis, adrenal atrophy, hemosiderotic foci in the liver, and hematomatous projections into the urinary bladder. After 6 months to 2½ years, 3/5 **dogs** died after consuming it at the 2% level in their feed. In **rats**, 4% in the diet for 18 months was associated with a mesenteric lymphosarcoma in some **rats**. Repeat trials with three other strains for 2 years revealed no malignancies.

LD_{50}, **rat**: > 2 g/kg.

PONGAMIA PINNATA

= Karanj

A cultivated plant in southern U.S., especially California and Florida. Native to India.

Untoward effects*:** It and ***P. glabra extracts are used as piscicides and assumed to be possibly poisonous for **man**.

POPLAR

= Chinese White Poplar = Populus tomentosa

Untoward effects*:** Its aeroallergens have caused allergic rhinitis, bronchial asthma, and hypersensitivity pneumonitis. Leaves contain ***benzoic acid, populin, salicin, and ***tremuloidin***.

POPPY

Poppies on plantations in Greece, southern Spain, northeast Africa, Egypt, and Mesopotamia contain *opium* with three times the *morphine* content than those of the Far East.

See *Argemone* for the *Mexican Prickly Poppy* and *Papaver sp.* for *Corn Poppy*, *Iceland Poppy*, and other poppies. The seeds are devoid of narcotic properties and used as food. On rare occasions, rumor has it that they may cause false positive urine test results for *morphine*. It is doubtful if the seeds are from *Papaver somniferum*, the *opium* poppy, as they are presumed to be. This has led to the so-called "poppy seed defense" in people found positive in drug tests for *opiates* after eating poppy seeds on pastries, rolls, and bagels or in cakes.

PORFIROMYCIN

= N-Methylmitomycin C = NSC 56,410 = U 14,743

Antineoplastic, antibiotic, trypanocide.

***Untoward effects*:** After IV use, causes leukopenia, thrombocytopenia, local tissue necrosis, thrombophlebitis, and occasionally gastrointestinal upsets.

IV, TD_{LO}, **human**: 1.5 mg/kg.

Bone marrow depression and vomiting reported in **dogs**.

LD_{50}, **rat**: 68 mg/kg.

PORK

Untoward effects*:** Can trigger migraine headaches in some **people**, often from ***sodium nitrate (q.v.), used to preserve the color of pork. After extensive curing with ***sodium nitrate*** and ***sodium nitrite***, some cured pork contains ***nitrosamine*** levels of 48 ppb. Immediate hypersensitivity pneumonitis reactions from aerosoled pork collagen from use in pituitary snuffs, etc. Cooking pork and foods in microwave ovens relys on distribution and excitation of water molecules. The presence of bones in pork can prevent even distribution of temperatures over 170 °F (176 °F for 30 min to help prevent botulism and kill trichinae larvae and *Toxoplasma gondii*). Pork fat has been found to contain 0.3–21 mg ***estrogen***/100 g by various investigators. Some pork may contain undesireable amounts of **boar** odor. Normally, this is used in ***salamis***. Some Italian-type dried and aged ***salamis*** have been found to be contaminated with ***aflatoxin***-producing *Aspergillus flavus*. Allergic reactions reported in **people** using pork-derived ***insulin***. **Pig's** blood can be a cause of contact allergic dermatitis in laboratory and slaughterhouse **workers**. **Swine** erysipelas infections and group R streptococci meningitis have occurred in **people** working with **hogs**.

Consumption of raw or cooked pork has caused allergic reactions in **dogs**.

PORTULACA OLERACEA

= *Bara laniya* = *Efere Makara* (Efik) = *Fasa K'aba* (Hausa) = *Papa San* (Yoruba) = *Purslane* = *Pussley*

Untoward effects: Contains high levels of soluble ***oxalates*** (q.v.) and often cyanogenic glycosides in varying amounts. Acute ***oxalate*** poisoning and death in Australian **cattle** and **sheep** after ingestion.

P. filifolia in Australia has caused similar problems.

POTASSIUM

= *Kalium*

Untoward effects: Iatrogenic hyperkalemia from excessive use of potassium supplements, ***angiotensin-converting enzyme inhibitors***, induced ***aldosterone*** deficiency, ***amiloride–hydrochlorothiazide*** combination, ***blood*** and ***plasma tranfusions, captopril, cyclosporin, heparin, nonsteroidal anti-inflammatory drugs*** (***indomethacin***), ***potassium-sparing diuretics, salt substitutes, spironolactone, triamterene***, respiratory or metabolic acidosis; vasoconstriction with muscle necrosis (after injection by ***heroin*** **addicts**); leading to weakness, flaccidity, and electrocardiogram abnormalities. A near-fatal hyperkalemia resulted from an inadvertent 17.2 g dose of ***Morton's Salt Substitute***® (containing 90% ***potassium chloride*** = 208 mEq potassium or 26 mEq/kg). Excess can cause acute renal failure, decreased alkaline phosphatase, and contact dermatitis. An 84-year-old **female** ingested 540 mEq in a suicide attempt.

Cardiovascular: Occasionally prolongation of PR interval and sinoauricular block a dangerous sequelae to ***digitalis*** intoxication. Ventricular tachycardia, fibrillation, asystole, and cardiac arrhythmias reported. Blood transfusions, especially older blood, and concentrated plasmas have caused intoxication, cardiac arrests, and death from potassium content. With serum potassium levels > 5.5 mEq/l, the T-wave is narrowed and peaked; spiked T-waves and widened QRS at 6.5 mEq/l; P-waves wide and flat with increase PR interval at 7.0 mEq/l and the P-wave may be absent when levels are > 8.5 mEq/l, followed by cardiac arrest. High concentrations cause spasms of vascular (venous and arterial) smooth muscle.

Gastrointestinal: Ulceration and stenosis of small intestines. Enteric-coated forms have occasionally failed to release potassium uniformly, and have been associated with gastrointestinal ulceration. Irritation, indigestion, and heartburn from ingestion of effervescent potassium (***KCl***) tablets. Nausea, vomiting, and diarrhea also reported.

Drug interactions: ***Aminosalicylic acid*** and ***tetracyclines*** react with serum potassium to give falsely negative values. ***Isoniazid*** reacts with serum potassium to give falsely positive values. ***Potassium penicillin V***, in high dosage, can add a high potassium (1.7 mEq/million Units) burden to the body. An 8.5 kg 11-month-old **male** was given two oral doses of ***potassium penicillin V*** as prophylaxis for otitis surgery. In the morning, the **anesthesiologist** was assured an IV vial contained ***penicillin G***, but it contained 250,000 Units ***potassium penicillin G***; after an IV bolus ventricular arrhythmia and, within 2 min, asystole requiring cardiopulmonary resuscitation occurred. Excessive levels from such use have been found in cerebrospinal fluid. ***Calcium, methicillin, penicillin G, sodium***, and ***spironolactone*** can increase serum potassium tests. Use of ***thiazides*** and other ***oral diuretics*** can deplete potassium potentiating curarization by ***muscle relaxants***. Elemental potassium can ignite spontaneously and violently when exposed to ***water*** or ***water vapor*** and cause skin burns. Irrigating these burns with ***water*** only intensifies the pain and burn. ***Arginine*** can shift potassium from cells to the extracellular compartment, increasing potassium excretion, but, in **patients** with renal insufficiency, severe hyperkalemia occurs.

Rapid IV of potassium in **dogs** leads to hypertension and tachycardia, due to stimulation of ***epinephrine*** secretion. IV fluids for **horses** should be limited to 20 mEq/l at a rate of 5 l/h. Malicious poisoning of a **horse** with IV potassium. Rapid IV to **dogs** of ***potassium penicillin G*** has caused ventricular fibrillation and cardiac arrest. Feed levels of 1.25% have been toxic to **turkey poults**.

POTASSIUM ACETATE

= *Kalii Acetas*

Untoward effects: Death in an aged **patient** 90 min after ingesting it with other potassium salts (6–8 oz ***Potassium Triplex***, containing at least 540 mEq ***potassium***).

LD_{50}, **rat**: 3.25 g/kg.

POTASSIUM-*p*-AMINOBENZOATE

= *KPABA* = *Potaba*

***Antifibrotic.* Used to treat scleroderma, Peyronie's diease, pemphigus, dermatomyositis, and scars.**

Untoward effects: Briny-bitter taste. Occasionally anorexia, nausea, fever, and rash. Hypoglycemia if food intake is decreased.

POTASSIUM-*p*-AMINOSALICYLATE

= *Fenamisal* = *Pavasal Potassium* = *Paskalium* = *Paskate* = *Tebamin*

Tuberculostat.

Untoward effects: Methemoglobinemia and hemolytic anemia, hypersensitivity reactions, nausea, vomiting, diarrhea, drug fever, skin rashes, jaundice, liver necrosis, albuminuria, hematuria, and anuria have also been reported. Occasionally lymphadenopathies, hypoprothrombinemia, and thrombocytopenia.

POTASSIUM AMMONIUM BIFLUORIDE

Wood preservative.

Untoward effects: **Cows** had decreased ***milk*** production and acute diarrhea from ***fluorine*** (q.v.) poisoning, after eating hay that had lain on the wooden floor of a hay loft that had been treated with it as a wood preservative.

POTASSIUM ARSENITE SOLUTION

= *Fowler's Solution* = *Liquor Arsenicalis* = *Liquor Potassi Arsenitis*

***Antineoplastic, alterative, hematinic.* Contains 1% *arsenic trioxide* (~0.76% elemental *arsenic*).**

Untoward effects: Has induced skin cancers (squamous cell carcinomas, Bowen's cell carcinoma), lung cancers, and ***folic acid*** deficiency in **humans**, usually after many years of oral intake. Many case reports of intrahepatic portal hypertension without cirrhosis from chronic use (4–29 years), yet, liver function tests were normal. Gastrointestinal bleeding from varices in **all** and one **patient** was anemic; **all** had skin pigmentation and hyperkeratosis, and nearly 50% had malignant skin lesions. Hepatic angiosarcoma and lung carcinoma in a 67-year-old **patient**, 40 years after 4 years of intake (~10 g ***arsenic trioxide***). Delayed palmar and plantar keratoses in many **patients** after years of intake. Dermatitis reported in **workers** while loading it into freight cars. Brief sensation of burning, pain, and erosions from topical use on condylomata acuminata.

TD_{LO}, **human**: 74 mg/kg; 30 mg/kg/over 3 week period; 215 mg/kg during 15 years

Metastatic adenocarcinomas in **mice** and also their offspring. Also see ***Arsenic Trioxide*** and ***Arsenic*** for additional **human** and **animal** adverse effects.

LD_{50}, **rat**: 14 mg/kg.

POTASSIUM AZIDE

= *Azide* = *Kazide* = *Kazoe* = *PPG 101* = *Smite*

Herbicide, fungicide, nematocide, insecticide, and bactericide.

Untoward effects: Can be neurotoxic to **people** working with it.

Ingestion by **ducks**, **quail**, and **pheasants** leads to ataxia, falling, sitting, polydipsia, ptosis, ataraxia, myasthenia, nutation, tremors, spasms, spread wings, phonation, opisthotonus, bradypnea, dyspnea, and wing-beat convulsions within 5 min after a toxic dose. Mortalities in **birds** occur between 15 min and 17 h after ingestion.

POTASSIUM BISULFITE

Antioxidant.

Untoward effects: Ingestion of sulfiting agents in processed foods and beverages, ***beer***, ***cider***, ***fruit juices***, raw ***fruit***, salad bars, ***shrimp*** and other seafoods, soft drinks, vinegar, ***wines***, and medicinals, has caused urticaria, angioedema, airway constriction, and death in **some**. These **individuals** may evidence the same symptoms from aerosoled exposure to ***sulfur dioxide*** (q.v.) or air pollution.

POTASSIUM BROMATE

Use: Oxidant facilitating gas retention and softness of ***bread***, usually at 10–40 ppm (0.001–0.004%); in photographic film developers, and as a sedative.

Untoward effects: An outbreak of food poisoning with abdominal pain, vomiting, diarrhea, xerostomia, weakness, and dizziness in 816 **people** after ingesting ***bread*** made with excessive amounts (1.1%). Apathy, irritability, loss of consciousness, central nervous system depression, loss of tendon reflexes, convulsions, tachycardia, decreased blood pressure, decreased body temperature, methemoglobinemia, anuria, albuminuria, tachypnea, shallow respiration, pulmonary edema, hepatitis, and myocarditis can occur after ingestion. Poisoning in 55 **patients** due to **their** ingestion of contaminated ***sugar***. **Bakers** have developed a contact dermatitis from handling it. Use in hair permanent neutralizing preparations caused accidental poisoning, with severe renal failure requiring dialysis, after ingestion by four **children** (18 months to 2 years). Hearing loss in two case reports.

POTASSIUM BROMIDE

***Sedative, anticonvulsant.* Contains ~67% *bromine*.**

Use: In photography, engraving, lithography, and in medicines.

Untoward effects: Severe cases of bromism from oral ingestion of ***bromine***-containing ***Bromo-Seltzer***® leading to lethargy, slurred speech, somnolence, mental deterioration, confusion, paranoia, and skin rash. ***Bromo-Seltzer***® abuse for 5 years by a **mother** caused retarded growth and small heads in two **male children**. Chronic brain syndrome and schizophrenia reported after long-term use. Therapeutic doses yield non-harmful breast milk levels of 1–3 mg/100 ml. Allergic contact dermatitis with pustules reported. In general, accumulation in the body leads to muscle weakness, mental and body sluggishness, sleepiness, loss of memory, apathy, decreased temperature, fetid breath, and acne-like eruptions. Symptoms disappear rapidly after treatment withdrawal, but urine tests remain positive up to 1 month. Any cardiac depressant action is due to the ***potassium*** ion, not the ***bromine***.

LD_{LO}, **human**: 500 mg/kg.

Drug interactions: May decrease therapeutic benefits of ***oxyphenbutazone***.

In **dogs**, 3 g leads to listlessness, ataxia, muscle weakness, and occasionally eczematous eruptions; **horses** require 30 g for the same effect.

POTASSIUM CHLORATE

= *Kalium Chlorium* = *Potassii Chloras*

Untoward effects: Fixed-drug eruptions, fatty degeneration of the liver, jaundice, nausea, vomiting, abdominal pain, and congestion of the spleen and ***iron*** metabolites in the spleen, liver, and kidneys. Occasionally acute hemolytic anemia, hemorrhages, methemoglobinemia, and anuria. A 1-month-old **infant** died after a compounding of errors. A welfare clinic suggested ***sodium citrate*** to "settle" **her** stomach; the **father** misread it as ***sodium chlorate***, and **he** was mistakenly dispensed ***potassium chlorate*** (31.9% ***potassium***). A gastrointestinal and renal irritant. Can cause red blood cell hemolysis and hemoglobinemia. Most **human** poisonings are accidental, a few are suicidal or homicidal. A 61-year-old **female** sucked on 20 five grain tablets daily for 6–10 weeks leading to acute renal failure and death. A 6-week-old **female** ingested 30–35 grains during 2½ days and died 12 h after hospital admission. Ingestion of 13 g by a 54-year-old **male** led to jaundice, hepatomegaly, and acute renal failure; Heinz bodies, fragmented red blood cells, and hemolysis for 20 days with normal blood test results after 2 months. Strong oxidizing agent and may also have a direct toxic effect on red blood cell membranes. A fatal dose in a 3-year-old **child** was 46 grains (2.9 g); 1 g was fatal in another **child**; 11.65 g has been fatal to an **adult**. Effects are more severe if dose is given in divided doses. Death has been reported in ~5 h, but usually occurs after several days from nephritis. Symptoms of poisoning (including overdoses) include violent vomiting, profuse diarrhea, severe dyspnea, poor cardiac function, marked cyanosis; often marked nervous symptoms, including cramps, delirium, and coma; dark reddish-brown colored urine with casts, red blood cell debris, and increased ***albumin***.

Its use as a herbicide caused the death of 15 **cattle** consuming it, they developed fatty degeneration around heart and kidneys, hyperemia, cardiac and gastrointestinal hemorrhages, kidney parenchyma, and heart muscle degeneration.

LD_{LO}, **rat**: 7 g/kg; **dog**: 1.2 g/kg; **rabbit**: 2 g/kg.

POTASSIUM CHLORIDE

= *AHR 3261* = *Chloropotassuril* = *Cloreto de Potássio* = *Diasal* = *Diffu-K* = *Enseal* = *Kaleorid* = *Kalii Chloridum* = *Kalion 39* = *Kalitabs* = *Kalitrans* = *Kalium Chloratum* = *Kalium-Duriles* = *Kaochlor SF* = *Kaon-Cl* = *Kaskay* = *Kato* = *Kayback* = *Kay-Cee-L* = *Kay-Ciel* = *Kay Contin* = *Klor-Con* = *Kloride* = *Klorvess* = *Klotrix* = *K-Lyte/CL* = *K-Norm* = *K-Tab* = *Lento-Kalium* = *Micro K* = *Muriate of Potash* = *Nu-K* = *Pan-Kloride* = *Peter-Kal* = *PfiKlor* = *Potassii Chloridum* = *Potassion* = *Rekawan* = *Repone K* = *Slow-K* = *Span-K*

Use: In photography, as a ***potassium*** supplement and electrolyte-replenisher, in dialysis solutions, salt substitutes, soft drinks, meat and poultry tenderizers, and as a viscosity-enhancer in cosmetics.

Untoward effects: Esophageal mucosa irritation with hyperemia, hemorrhage, strictures, erosions, dysphagia, and spontaneous perforation reported. Can become concentrated on small areas of esophageal mucosa in **patients** with achalasia leading to ulceration and obstruction with stricture formation. Erosions and ulceration of the upper gastrointestinal tract. Nausea, vomiting, and upset stomach when taken without food. Vomiting may lead to cardiac complications. Some **patients** have developed life-threatening ventricular arrhythmias after receiving 15–20 ml orally of a 10% solution. Effects can vary with different dissolution rates. Enteric-coated tablets have caused an excessive amount of intestinal ulceration. Many **patients** have developed ulceration and strictures of the small intestines. Perforation and gastric ulcers also reported. Accidental ingestion by a **child** caused small bowel strictures. Ventricular fibrillation, dyspnea, and unconsciousness in a 62-year-old **male** after 6 g orally. Heartburn and hematemesis also reported. Acidified ileal contents decreased ***vitamin*** $\boldsymbol{B_{12}}$ absorption in 18/60 **patients**. A 27-year-old **female** ingested 40 coated tablets during 2 years of weight reduction and 6–8 tablets in the month before abdominal pains, vomiting, tachycardia, constipation, and diarrhea. Appendectomy gave no relief. A month later, a resection of several centimeters of ulcerated ileum was performed. Life-threatening hyperkalemia in three **patients** after overdoses of 600–630 mg in slow-release tablets. A book, *Let's Have Healthy Children*, advised that colicky **infants** be given large doses of potassium chloride, citing a study of 653 sick **infants** successfully treated for it with 3,000 mg potassium chloride. A **woman** gave her **son** 2,500 mg one day and 500 mg the next day. After the last dose, the **infant** developed cardiac arrhythmias and was rushed to the hospital, where **he** died 4 days later. It was particularly toxic because the **infant** was dehydrated and the book failed to give that as a contraindication. Severe, near-fatal hyperkalemia in **people** and **infants** given ***salt substitutes***. An 8-month-old **infant** was inadvertently given a possible 17.2 g dose of ***Morton's Salt Substitute***® (containing 90% potassium chloride = 208 mEq ***potassium*** or 26 mEq/kg). A hypokalemic 32-year-old **female** died after taking 47 tablets. A 4.8 kg **child** was given 3,000 mg, followed by 1,500 mg; it died 28 h later. Sucking on sustained-release tablets caused multiple, painless, deep ulcerations of oral mucosa in a 69-year-old **female**. IV infusion in **patients** with heart failure has led to decreased cardiac output and increased central venous and mean arterial blood pressure. Many reports of life-threatening arrhythmias and some deaths from use on day 1 of myocardial infarctions. Rarely, asystole reported. Many case reports of fatal iatrogenic errors, due to IV of it instead of prescribed ***sodium chloride*** or other drugs. Rapid injections (bolus effect) can be fatal. Serious hyperkalemia due to its insertion into parenteral fluid containers without adequate mixing, again, can lead to bolus adverse effects. A **physician** prevented **his** own death immediately after a **nurse** added 40 mEq into a 500 ml bag of $\frac{1}{2}$ normal saline without mixing. While still conscious, **he** bit the tubing and pulled the needle out of **his** vein. Gas gangrene in **patient** receiving IM instead of IV. Hyperkalemia has developed in **patients** with normal renal function after ingestion of extended-release preparations. Its use can give decreased or false negative fasting ***glucose*** test results, followed by increased results. An association with lung cancer has been reported. Rarely, it has caused skin eruptions.

LD_{LO}, **human**: 500 mg/kg; **infant**: 938 mg/kg over 2 day period.

Drug interactions: Use to enhance the diuretic effect of ***thiazides*** or ***mercurial compounds*** can be toxic, especially if renal function is impaired. Pancreatitis after use with ***cyclopenthiazide***. Use with ***spironolactone*** can enhance its effects and be lethal.

Excessive use in dry **cows** may be associated with udder edema at calving time. Typical gastrointestinal ulceration, as in **man**, has occurred in **dogs**, **rhesus monkeys**, and **baboons** receiving the drug orally for only a few days. Uncoated material is more apt to produce gastric lesions and enteric-coated material is more apt to produce intestinal ulcers. **Fish** in rivers containing high concentrations from industrial wastes have developed severe gastrointestinal ulcers. Avoid use in renal insufficiency.

LD_{LO}, **rat**: 2.43 g/kg; LD_{50}, **guinea pig**: 2.5 g/kg.

IV, LD_{50}, **mouse**: 117 mg/kg.

POTASSIUM CHROMATE

Untoward effects: Occupational dermatitis with bird's eye-like necroses on fingers of a **worker** using it to help grind and polish ***amber***. Exposure to its dust in chrome-***leather*** tanning has caused ulceration and perforation of nasal septums.

LD_{LO}, **human**: 50 mg/kg.

TD_{LO}, **mouse**: 1.6 g/kg during 62 weeks.

POTASSIUM CITRATE

= *Kajos* = *Kalium Citricum* = *Seltz-K* = *Tripotassium Citrate* = *Urocit K*

Untoward effects: Case reports of hyperkalemic cardiac arrhythmias in the **elderly** after ingestion of large amounts. Impaired ***vitamin*** $\boldsymbol{B_{12}}$ absorption in ~30% of **patients**, but less severe than that caused by the more acidifying ***potassium chloride***.

IV, LD_{50}, **dog**: 167 mg/kg.

POTASSIUM CLAVULANATE

A *β-lactamase inhibitor*.

Untoward effects: For general adverse effects, see ***Clavulanic Acid***. Candidiasis in ~16% of **female patients** using it in ***Augmentin***® tablets.

POTASSIUM CYANIDE

= *KCN*

Untoward effects: Massive poisoning in a 30-year-old **male robber** who hid in a tank that had contained it. A 15-year-old **female** ingested a fatal dose of 5 or 10 g (324 or 648 mg). ***KCN*** became highly publicized after ~930 Reverend Jones' **followers** ingested it and killed themselves in Jonestown, Guayana. A **photographer** drank some containing 2.75 oz that **he** mistook for lemonade. **He** convulsed and detectable breathing and heart action stopped. **He** recovered after 90 min of cardiopulmonary resuscitation and IV antidotes. It is a potent respiratory paralytic agent causing rapid unconsciousness. Chronic poisoning in a 21-year-old **male** using it to clean ***gold*** objects. Vapors cause eye, skin, and upper respiratory tract irritation. Ingestion causes varying degrees of nausea, vomiting, confusion, tachypnea, dyspnea, weakness, headache, confusion, and asphyxiation. Death has occurred – estimated LD of 200–500 mg. Its use in 1986 to contaminate bottles of ***Tylenol***® led to a **woman's** death and a new era of tamper-proofing foods and medicinals. Onset of symptoms is slower than with ***hydrocyanic acid*** (q.v.), allowing more time for treatment. Gastrointestinal symptoms usually precede convulsive and narcotic effects. Post-mortem findings are usually marked congestion of stomach from its strong alkalinity. Penetration of the salt into the blood leading to a bright red hue in the throat, esophagus, and stomach.

LD_{LO}, **human**: 2857 μg/kg.

Used as a fumigant in **armadillo** dens. Repeated SC injections of **rat** with 0.5 mg every week for 22 weeks led to neuronal degeneration, pale myelin, and cell loss in brain areas. A 6-ton **elephant** at Dreamland, Coney Island in New York, survived a 400 grain (< 1 oz avoir) oral dose.

LD_{50}, **rat**: 4 mg/kg; **dog**: 2 mg/kg; **rabbit**: 5 mg/kg; **mouse**: 8.5 mg/kg.

POTASSIUM DICHROMATE

= *Potassium Bichromate*

Untoward effects: Oral use can provoke ***chromate*** eczema. Contact dermatitis, usually severe and persistent. Involved in some cases of ***shoe*** dermatitis from its use in treating ***leather***. Many **immigrants** in the U.S. working with ***cement*** have shown dermatitis with it in patch tests. Otitis externa in many **patients** who have cleaned their ear canals with ***match*** heads. Positive reactions to it in **people** working with concrete or tanned ***leather***. A case report of oral ingestion indicated enlarged tender liver, skin eruptions, and ***chromium*** in the urine. A 20-year-old **male** ingested ~30 g intended as a radiator descaler, causing yellow vomitus, followed by blood-stained vomitus and diarrhea. Tachycardia, decreased blood pressure, and collapse 3 h later. Liver and kidney damage, somnolence, torpor, and anorexia, required 6 months of hospitalization. A 14-year-old **male**, 3 days after ingesting ~1.5 g, was hospitalized with acute gastroenteritis, toxic hepatitis, bleeding diathesis, renal failure, and death. ***Chromium*** has been found in some ***flours*** and caused sensitivity. As a preservative in ***milk*** for testing, it has caused dermatitis in **"milk-testers"**. Respiratory irritation from its acrid vapors, excessive sneezing, nose septal destruction, ozena, and skin eruptions and excoriations have led to chronic ulcers. Overdoses cause violent irritation and corrosive effects, severe vomiting, violent abdominal pains, frequent dark, bloody stools, heart failure, collapse, and coma. Unconsciousness within 5 min after ingestion of < 1 oz, with death 35 min later. **Workers** making it have had painful hand ulcerations.

LD_{LO}, **human**: 50 mg/kg; **child**: 26 mg/kg.

Acute toxicity and deaths of **hogs** chewing on wood treated with it as a preservative. SC injection causes > 90% nephrotoxicity in **mice**. Dietary concentration of 1,000 ppm causes ~30% mortality; 5,000 ppm leads to > 90% mortality in **quail**. In water, 300 ppm is toxic to **goldfish** and lower concentration is toxic to **catfish**. **Cows** licking on telephone poles have become toxic, due to its use as a wood preservative. SC doses of 50 mg are lethal to **poultry**. A **horse** inadvertently given 10 g, instead of ***sodium bicarbonate***, developed anorexia, transient stiffness, tachycardia, arrhythmias, increased temperature, bradypnea (followed by tachypnea), cyanotic mucous membranes, abdominal pain, intense thirst, and death. Post-mortem revealed extensive erosion and sloughs of gastrointestinal mucous membranes.

LD_{LO}, **dog**: 2.8 g/kg; **guinea pig**: 163 mg/kg.

POTASSIUM FERRICYANIDE

= *Racun Besi* = *Red Prussiate of Potash*

Untoward effects: Ingestion by a 50-year-old **male** induced ***cyanide*** (q.v.) poisoning that was successfully treated.

Soon after drinking from a pail of water used for cooling ***iron*** at a blacksmith's shop in Indonesia, two **horses** became seriously ill, developing pain, excitement, dyspnea, collapse, tachycardia, congested mucous membranes, profuse sweating, and dilated pupils. One died from anoxia within 10 min. ***Orange Racun Besi***, when heated, is used to harden ***iron***.

LD_{LO}, **rat**: 1.6 g/kg.

POTASSIUM GLUCONATE

= *Gluconsan K* = *Kalimozan* = *Kaon* = *Katorin* = *K-IAO* = *Potasoral* = *Potassuril* = *Tumil-K*

Untoward effects: ***Potassium*** ion in it has caused jejunal ulceration.

POTASSIUM HYDROXIDE

= *Caustic Potash* = *Lye* = *Potassium Hydrate*

Use: In manufacturing soap, oven and drain cleaners, ***paint*** and varnish strippers, electroplating, photoengraving, printing ***inks***, chemical synthesis, in microscopic work, and as a caustic to dehorn baby **calves**.

Untoward effects: Corrosive. Tissue burns in the presence of moisture. Topically, causes skin damage, including softening and necrosis with pain and erythema. Strong solutions or solid particles cause rapid corneal destruction. 5.5% solution has a pH of 14. Ingestion leads to corrosion and ulceration of mucous membranes of mouth, esophagus, stomach, and duodenum; nausea, abdominal pain, blood-flecked vomitus, systemic acidosis, cardiovascular shock, and occasionally peritonitis. Usually, esophageal stricture occurs in **survivors**. Fatal, complete pulmonary necrosis in a 5-year-old **male** after ingesting an unknown amount of solution. A 12-year-old **male** had extensive esophageal damage after using an empty spoon **he** had dropped into a solution, and used it for taking a dose of cough syrup. Ingestion of miniature hearing aid or camera ***batteries*** by small **children** presents a risk of potassium hydroxide toxicity, when their ***nickel***-plated steel coverings become corroded. ***Cement*** burns have resulted in necrotic ulcers, due to kneeling on wet ***cement***, when its ***potassium monoxide*** content reacts with moisture to form potassium hydroxide. Its use as a degreasing agent in electroplating and watch-making has caused various occupational dermatoses. National Institute for Occupational Safety and Health set upper exposure limit at 2 mg/m^3 TWA.

Ingestion by **dogs** leads to similar gastrointestinal and oral problems.

POTASSIUM HYDROXYQUINOLINE SULFATE

= *Chinosol*

Antiseptic, fungistat, spermicide.

Untoward effects: After topical use, some **people** show sensitivity to it leading to irritation and erythema. Also see ***8-Hydroxyquinoline***.

POTASSIUM IODATE

Iodine source, oxidizing agent in wheat breads.* Contains *59.3% potassium.

Untoward effects: U.S. limits use to 75 μg/g of ***bread*** and it tends to be at highest levels in lower-priced, fluffy, spongy white ***breads*** causing increased incidence of thyrotoxicosis in endemic goiter areas.

LD_{LO}, **human**: 50 mg/kg.

Near-lethal doses to **dogs** cause emesis. Long-term use of 36–108 mg every 2 weeks to pregnant **ewes** leads to hypertrophy and hyperplasia of follicular epithelium of their thyroids.

Minimum lethal dose, fasted **dogs**: 200–250 mg/kg. LD_{LO}, **mouse**: 531 mg/kg; **guinea pig**: 400 mg/kg.

POTASSIUM IODIDE

= *KI* = *Iodure de Potassium* = *Jodid* = *Joptone* = *Kalii Iodidum* = *Thyroblock* = *Thyrojod*

Iodine source, expectorant, prophylaxis versus radioactive iodine, and antifungal.

Untoward effects: A **male** ingested 600 ml of solution (15 g ***iodine***) and within 12 h developed swelling of mouth, neck, and face; short bursts of ventricular tachycardia, ventricular bigeminy, and ectopic auricular beats required 10 days of hospitalization to restore normal beats. Concentrated solutions irritate gastric mucosa, occasionally causing persistent gastric distress with nausea and vomiting severe enough to cause discontinuance of treatment. Mild allergic reactions occur. There is always a small risk of hypersensitivity, goiter, hyperthyroidism (Jod-Basedow phenomenon), hypothyroidism, and ioderma. Transient acneiform eruptions and thyroid enlargement in **children** receiving iodotherapy for asthma. Large doses have been associated with fever, anorexia, and hemoptysis. Others report swollen parotid and submaxillary salivary glands ("iodide mumps"), coryza, acne, headache, diarrhea, chills, rigors, leukocytosis, and eosinophilia. A small thyroid adenoma in a 34-year-old **male** after 2.4 g/day/3 years for asthma. Generalized pustular psoriasis in two **patients** after 500 mg. Hypothyroidism in 10% of **patients** ingesting a saturated solution. It can cause urticarial vasculitits, but nodular vasculitis responds to treatment with it. Most **patients** showing idiosyncratic reactions to it do so on very small doses and are less likely to develop severe reactions. Breast milk concentration varies (~15% of **mother's** intake) and toxicity can be produced in an **infant** if the **mother's** dose is high.

Infants of **women** who have ingested it from the 14th week of pregnancy have goiter, cyanosis, respiratory distress, and mental retardation. A **few** died within 1–2 days after birth. One case report suspects ***lithium*** intake with potassium iodide may accelerate hypothyroidism. Topically, it has induced pustules.

LD_{LO}, **human**: 500 mg/kg.

In **animals**, iodism has occurred on therapeutic doses. In **dairy cows**, 106 mg/day increases milk ***iodine*** levels to ~50 times those of controls. **Dogs** given 120–180 grains (~10 g) in water usually vomited and died in a few days; **rabbits** did the same after 60 grains (4 g).

LD_{LO}, **mouse**: 1862 mg/kg.

POTASSIUM METABISULFITE

= *Potassium Pyrosulfite*

***Antioxidant, antifermentive*, and preservative.**

Untoward effects*: *Fruit, fruit juices, and vegetable preservative; to prevent spoilage of ***beer, wines***, and parenterals; and in bleaching straw.

Untoward effects: Certain **individuals**, particularly **asthmatics**, have had severe reactions to it, leading to generalized flushing, faintness, weakness, severe wheezing and labored breathing, chest tightness, coughing, cyanosis, urticaria, hives, nausea, angioedema, and unconsciousness. As little as 7.5 mg has triggered reactions. Estimated consumption of metabisulfites by **Americans** is 2–3 mg/day; ***wine***- and ***beer*-drinkers** get an additional 5–10 mg/day. Most intake is from restaurant salads; vegetables (particularly ***potatoes***), and ***avocado*** dips providing an additional 25–100 mg. It is used to enhance the foods' crispness and prevents their browning. It is estimated that 400,000 **people** in the U.S. are allergic to it, and twice that number may be sensitive to it.

Oral doses of 1 g/day to each of ten **guinea pigs** led to death in 1–4 days. Post-mortem revealed gastrointestinal and liver congestion and hemorrhage. **Dogs** receiving 200–500 mg/day/2 months developed vomiting, bleeding gums, and congestion of stomach and liver.

POTASSIUM NITRATE

= *Kalii Nitras* = *Niter* = *Saltpeter* = *Sal Prunelle*

Use: In pickling and preserving the color of meats, fireworks ***matches***, ***glass***, gunpowder, blasting powder, fertilizers, tempering steel, treating ***tobacco*** (to make it burn evenly and decrease atmospheric pollution), as a diuretic, and in tooth desensitizers.

Untoward effects: A 2-year-old **male** ingested 1 g, leading to methemoglobinemia and cyanosis after 11 h. Reduction to ***sodium nitrite*** (q.v.) in the digestive tract causes the methemoglobinemia. Other effects of ingestion are gastroenteritis, abdominal pain, vomiting, weakness, irregular pulse, albuminuria, oliguria, hematuria, nephritis, headache, vertigo, cyanosis, convulsions, and collapse. Fatalities often occur after ingestion of 15–30 g. Several cases of toxic methemoglobinemia and one fatality resulted from absorption of molten nitrate salts through burned skin areas in an industrial accident. Poisoning reported in **infants** given contaminated well water (upper safe limit has varied between 10 and 20 ppm of ***nitrates***). The absence of ***hydrochloric acid*** in the stomach of **infants** readily permits reduction of the ***nitrate*** to ***nitrite***. This reduction can also occur in vegetables allowed to remain at room temperature after harvesting. The change does not take place in frozen or canned ***spinach***, but readily occurs in ***nitrate***-rich unprocessed ***spinach***. Methemoglobinemia reported in an **infant** fed soley on ***carrot juice*** containing 525 ppm ***nitrate*** and 775 ppm ***nitrite*** from Florida-raised ***carrots***. California ***carrots*** had 22 ppm ***nitrate*** and 0.2 ppm ***nitrite***. So-called "organically-grown" ***carrots*** had 322 ppm ***nitrate*** and 0.2 ppm ***nitrite***. A 22-year-old **female** developed methemoglobinemia after ingesting 50 g during a 12-day period. Estimated safe daily intake of 5 mg/kg. Can cause leukonychia. 4 g has been fatal, **others** have recovered after 30 g. When death occurs, it is usually within 2 h.

LD_{LO}, **human**: 500 mg/kg.

Nitrates are absorbed from the soil by plants that normally rapidly convert the ***nitrogen*** into proteins. Yet, many plant species concentrate large quantities in the form of potassium nitrate. ***Salt***-hungry **animals** readily eat or lick ***nitrate fertilizers*** or consume forage after heavy application of fertilizer. Sublethal levels cause decreased ***milk*** production in **cattle**. **Cattle** consuming 1.3 g/kg develop tremors, diuresis, cyanosis, and collapse. Deaths in 3/9 **dairy heifers** receiving 140 g/day/2 days; 100 g/day after several days led to abortions. Paralysis, dyspnea, muscle weakness, and convulsions in **dogs**. Large numbers of **pigs** died after consuming feed containing 2,605 ppm. In **goslings**, 8 mg/kg caused weakness, diarrhea, and ataxia; 0.6–1 g/day was lethal.

TD, **horses** and **cattle**: 180 g; **sheep**: 23 g; **dogs**: 5 g. LD_{50}, **cattle**: 50 g (Cornell); **rabbits**: 3 g/kg.

POTASSIUM NITRITE

Untoward effects: See discussion under ***Potassium Nitrate*** above. U.S.D.A. finally, in 1985, required a reduction from the originally approved 120 ppm in ***bacon***. The nitrite decreases reproductive capability in **animals** and causes methemoglobinemia in **infants**, **sheep**, **hogs**, **poultry**, **guinea pigs**, etc.

LD_{50}, **rabbit**: 200 mg/kg.

POTASSIUM OXALATE

Use: In cleaning and bleaching formulas and as a blood coagulant in laboratory work.

Untoward effects: Combines with ***calcium*** to prevent blood coagulation. It also decreases the size of the red blood cells, thus, giving falsely lower packed-cell volume readings. Ingestion by **man** or **animals** can lead to salivation, dysphagia, thirst, edema of the glottis, severe abdominal pain, vomiting, difficulty in swallowing, muscle fibrillation, circulatory collapse, and uremic convulsions. Early fatal doses can kill within minutes. Causes distortion of leukocyte nuclei and may cause falsely increased blood ***ammonia*** test results. Fed at 2–16 g/kg/week to **cattle**, causes partial anorexia, increased central nervous system reactions, and death within a week at highest level. Lower levels lead to albuminuria and mild degenerative changes in the liver. One of the poisonous ***oxalates*** in ***rhubarb*** leaves. *Rumex Acetosa* = ***Sour Dock*** can poison **animals** eating it, if their diet is low in ***calcium***. It is often eaten by **people** as pot-herbs or greens, and appears to be harmless, if **their** diets contain adequate ***calcium***. Some susceptible **people** develop a dermatitis from handling the ***Sour Dock*** leaves.

POTASSIUM PERCHLORATE

= *Perchloracap* = *Peroidin*

Antithyroid drug. **In treatment of thyrotoxicosis.**

Use: In pyrotechnics, explosives, ***matches***, photography, and analytical chemistry.

Untoward effects: Bone marrow depression, agranulocytosis, granulopenia, anemia, thrombocytopenia, aplastic anemia, leukopenia, erythropenia, eosinophilia, pancytopenia, uncomfortable oral lesions, fever, purpura, enlarged lymph nodes, sore throat, jaundice, erythema nodosum, cutaneous reactions, weakness, restlessness, tachycardia, and gastrointestinal upsets. Anemia and neutropenia in a 43-year-old **female** after 1 g/day/3 months.

POTASSIUM PERMANGANATE

= *Cairox* = *Chameleon Mineral* = *Kalium Permanganicum*

Antiseptic, astringent, deodorant, bleach, and oxidant.

Untoward effects: Corrosive burns in the mouth and laryngeal edema in a 3-year-old **male** after ingesting 5–10 g. Extubated after 3 days. On day 4, **his *manganese*** plasma level was 4.1 µg/l; after 1 month, 1.5 µg/l. Death due to poisoning in another 3-year-old **male**, 3 days after ingesting the same solution. At autopsy, amounts in mg/100 ml were 45.2 for intestine, 10.7 for liver, 2.9 for kidneys, 3.0 for brain, < 0.07 for blood, < 0.06 for urine; ***manganese*** concentration was up to 100 times the norms. Pyloric stenosis reported in an 18-year-old **male** after ingesting 3 g. A 2½-year-old child was inadvertently given a teaspoonful, instead of ***activated charcoal***, causing severe pharyngeal corrosion, cardiovascular failure, and death within 3 h. Vaginal bleeding has followed self-medication with it as an abortifacient within 3½–4 h in 23 **women** seen in one hospital during a 2 year period. Anemia and hypotension were striking signs and shock occurred in seven. Only one aborted. A medication error by a **nurse** attempting to give prophylactic eye treatment to an **infant** by putting two drops onto the **baby's** eyes from a ***silver nitrate*** stick (75% ***silver nitrate*** and 25% potassium permanganate), led to red, edematous eyelids with thick, sanguinous, yellowish, purulent secretions within an hour, areas of corneal opacification, and friable, easy-to-bleed conjunctiva. Nearly total improvement 1 year later. Bilateral chemical eye burns in a 22-year-old **male** after accidental spill of crystals. Eventual complete recovery. Variable brown or yellow nail plate discoloration. College fraternity **members** "initiated" a **freshman** by painting obscene phrases on **his** belly. Ashamed of **their** actions, **they** poured ***hydrogen peroxide*** on the words and caused a serious explosion, injuring **those** leaning over **him** and only scaring the **freshman**.

TD_{LO}, **woman**: 2.4 mg/kg/day.

Magenta-colored, but often causes red-brown stains.

Brown solutions are not antiseptic. Dilute ***oxalic acid***, ***vinegar***, or ***hydrogen peroxide*** may help decolorize stains caused by it. Strong oxidizing agent!! Violent reactions can occur when ***formalin*** is added to such a strong oxidizing agent. Follow directions! To avoid risk of fire, use a metal container with sides at least three times the depth of chemicals used and place it several feet away from inflammable material.

LD_{50}, **rat**: 1.1 g/kg.

POTASSIUM PERRHENATE

Untoward effects: IV use in **dogs** caused slight hypertension and tachycardia. Similar results in **cats** at 10–50 mg/kg; at 60–70 mg/kg caused hypotension, bradycardia, and bradypnea. Death from cardiovascular collapse and respiratory failure in 80% of **cats** at 70 mg/kg.

POTASSIUM PERSULFATE

= *Anthion*

Use: In bleaching fabrics, in hair bleaches (to accelerate ***peroxide*** action), in some European countries to whiten ***flour***, in chelating ***thiosulfate***, and in analytical chemistry.

Untoward effects: Irritation, allergic contact dermatitis, generalized urticaria, rhinitis, asthma, and syncope. The scalp dermatitis can cause temporary hair loss. Edema of scalp, face, and neck reported. Hand contact dermatitis in **bakers** who add it to ***flour*** to inhibit proteolytic enzymes from attacking ***gluten*** films which surround ***carbon dioxide*** bubbles formed during fermentation, which make ***bread*** light. Chronic eczema from contact with it in ***flour***, as well.

POTASSIUM SELENIDE

Untoward effects: A **chemist** who worked with ***selenium*** for many years developed an erythematous, edematous, vesicular eruption on **his** face, hands, and genitals from hot vapors. Patch tests with pure ***selenium*** were negative.

POTASSIUM SILICATE

Untoward effects: Highly alkaline. Skin irritation is common in sensitive **people**.

POTASSIUM SORBATE

= *Sorbistat*

Use: As a ***mold*** and ***yeast*** inhibitor and preservative in foods (***cheeses***, ***gelatins***, meringues, jams, and ***nuts***), cosmetics, electrocardiogram conduction jelly, and medicinals.

Untoward effects: A 1% concentration causes an allergic dermatitis in many sensitive **people**. A **baker** has shown a hypersensitivity to it in the ***flour*** **he** used.

POTASSIUM SULFATE

= *Salt of Lemery*

Use: In manufacturing fertilizers, ***glass***, ***alum***, fire-extinguishing compounds, munitions, ***fruit juice*** preservative, as a ***sulfur*** source in **livestock** feeds, and as a laboratory reagent and cathartic. ~1.5 million lbs used annually in the U.S.

Untoward effects: Gastrointestinal irritant, especially at high concentrations. Contains ~45% ***potassium*** (q.v. for other possible adverse effects). Ingestion of ~45 g has been fatal.

LD_{LO}, **woman**: 800 mg/kg.

Cattle wandered into a field treated with it leading to early death in one and another after 2 days. Symptoms included prostration, paresis, hypothermia, loss of rumen atony, and constipation. Post-mortem revealed congested mucous membranes of abomasum and intestines.

POTASSIUM THIOCYANATE

= *Potassium Rhodanide* = *Potassium Sulfocyanate* = *Rhocya*

Use: In textile dyeing and printing, photography, fireworks, manufacturing artificial ***mustard*** oil, and as a hypotensive.

Untoward effects: Diarrhea, trembling, mental confusion, and coma after 12.5 g. Intoxication in a self-medicating 62-year-old **female** led to severe hematological, thyroidal, and neurological abnormalities. Use in alcoholomania causes anorexia, weight loss, and asystole. In a suicide attempt by a 19-year-old **male** resulted in abdominal pain, vomiting, weakness, confusion, and xanthopsia 4½ h after ingesting 30 g in water. Disorientation, hallucinations, and anxiety peaked about the 8th hour. Has caused skin eruptions and alopecia.

LD_{LO}, **human**: 80 mg/kg.

Lambs ingesting it developed thyroid impairment, goiter, and 20% had myopathy. ***Wool*** growth in **sheep** is decreased by its combination with ***flumetasone***.

LD_{50}, **rat**: 854 mg/kg; **mouse**: 594 mg/kg.

POTATO

= *Aalu* = *Irish Potato* = *Solanum tuberosum* = *White Potato*

Untoward effects: The new sprouts on the tubers and the "eyes" contain ***solanine***, a saponin, and ***solanidine***. Tubers growing on the soil surface turn green from sunlight exposure. These green potatoes are unfit for **human** consumption and cases of fatal poisoning after eating them have been reported. Green potatoes have high levels of glycoalkaloids, the most toxic of which is ***solanine*** (q.v.) and has been found at levels of 80–100 mg/100 g potato. Commercial potatoes are treated to inhibit sprouting, and new varieties are tested for alkaloid content. Hundreds of chemicals are found in potatoes, including ***vitamin C***, alkaloids (***α-solanine, α-chaconine, demissidine, solanidine, α-*** and ***β-solamarine, tomatidenol***), terpenes (***lubimin, phytuberin, rishitin, rishitinol, vitispiranes***), sterols (***24-methylene-iophenol, 24-methyllophenol, 31-norlanosterol, sitosterol, stigmasterol***), and miscellaneous compounds, such as ***arsenic, caffeic acid, chlorogenic acid, chlorophyll, nitrates, oxalic acid, quincic acid, scopoletin, scopolin***, and ***tannins***. The maximum acceptable alkaloid content permitted in the U.S. is set at 20–25 mg/100 g; 10–40 times this is required to cause pathology in **animals** and most whole potatoes contain < 10 mg/100 g; most of this is in the peel. Artificial light can also turn potato skin green. Eating green potatoes causes weakness, paralysis, mydriasis, clammy skin, mental confusion, cardiac depression, prostration, and dyspnea. Potato "eyes" and sprouts and the plants themselves can be poisonous to **man** and **beast**. In 1845–1848 in Ireland, *Phytophora infestans*, a fungus, infected their potatoes, causing a potato famine that led to the death of a million **people** from starvation and malnutrition and the emigration of 1.5 million **people**. Mortality was extremely high in these **émigrés** from starvation, and louse-borne typhus and dysentery in so-called "coffin ships". A Mrs John Ford was **one** who died. Her **husband** wandered to the Detroit, Michigan area, where his **grandson**, Henry, became well-known for **his** horseless carriage. An association between consumption of blighted potatoes and spina bifida in **man** has been debated, yet, positive results were reported in experimental **rabbits** and **swine**; occasionally, in **rats** and **marmosets**, but not in **hamsters**. After eating potatoes at lunch during the fall term, 78 **schoolboys** became ill; 17 were hospitalized, three were seriously ill. Symptoms within 7–19 h included vomiting, diarrhea (occasionally severe), abdominal pain (one received an appendectomy). **They** had low plasma cholinesterase levels. The potatoes were left over from the summer term, and high alkaloid levels accumulated during storage, apparently, from exposure to light. *Listeria monocytogenes* has been found on the surface of some potatoes. Allergic contact dermatitis with urticaria has been reported. **People** sensitive to ***sulfites***, after eating french fries or hashed-browns in restaurants, can develop severe asthmatic reactions. One death reported in a

person from ***sulfites*** in canned potatoes. ***Sulfites*** are now banned on potatoes in the U.S. **Patients** with ***chromate*** dermatitis should avoid potatoes. **Diabetics** should be aware that baked potatoes can cause a significant increase in blood ***glucose***. A number of **atopics** told of itching, tingling and/or edema of lips, mouth, and tongue or irritation of **their** throats, including hoarseness, when eating a raw potato. Gastrointestinal upsets, rhinitis, angioedema, and exacerbation of hand dermatitis in **some**. Cooking, apparently, destroyed the potato's allergenic properties. ***Pyrazines*** occur in potato chips. Contact dermatitis of potato chip factory **workers** has been traced to ***cheese*** powder, ***onion*** powder, and ***wheat*** filler. Potato chips prepared in cooking oil containing ***methyl polysiloxane*** as an additive have decreased ***phenindione*** and ***warfarin*** anticoagulant effects.

Half-cooked or raw potatoes eaten by **dogs** may contain similar alkaloids, leading to similar symptoms to those of **man**. Some **dogs** have shown allergic reactions (respiratory and dermatologic) to potatoes. Potato vine ingestion by **cows** has caused serious incoordination and convulsions, due to their anticholinesterase content. **Cattle** that eat too many green potatoes may also develop jaundice. **Pigs** are more commonly poisoned by it, but **horses**, **cattle**, **sheep**, and **poultry** have also been adversely affected. Unripe, old, green, rotten, or sprouted tubers, kept for a long time can be dangerous. ***Solanine*** is water-soluble, and comes out when boiling potatoes, but not by baking. **Pigs** have been poisoned by such water from commercial processing plants. Exanthematous lesions, ulcerative stomatitis, conjunctivitis, and dermatitis occurred in affected **pigs** and **cattle**. **Pigs** have been poisoned after consuming moldy "mashed potato flakes" from a potato processing plant. **Swine** fed sprouted potatoes developed diarrhea, nervousness, dullness, mydriasis, salivation, incoordination, decreased temperature, and anorexia. Potato sprouts have been teratogenic (exencephaly, harelip or cleft palate, cebocephalic nasal defects, spina bifida, and microphthalmia) in some **hamster** strains. Avoid giving **pets** or any **animals** access to potato peelings or sprouts.

Green parts, LD_{LO}, **hamster**: 2.7 g/kg.

PRACTOLOL

= *AY 21,011* = *Dalzic* = *Eraldin* = *ICI 50,172*

β-Adrenergic blocker, class II antiarrhythmic.

Untoward effects: Therapy with it has caused serious eye, skin, oral, cardiac, renal, nasal mucous membrane, ear, and peritoneal adverse effects.

TD_{LO}, **woman**: 4 mg/kg.

Blood: A limited number of reports of hemolytic anemia, antinuclear antibody, leukocytosis, purpura, and thrombocytopenia.

Eye: Conjunctivitis, decreased tearing (xerophthalmia), corneal ulceration, and decreased intraocular pressure in ~20% of **patients**. Exacerbation of keratoconjunctivitis sicca and blindness. This led to its early market withdrawal.

Miscellaneous: Nephrotic syndrome, bronchospasm (18%), bronchoconstriction, hypotension, ototoxicity, pneumatosis coli, liver cirrhosis, constipation, nausea, fatigue, Raynaud's phenomenon, arthralgia, giddiness, myathenic syndrome, muscle cramps, and joint effusions.

Pericarditis, peritonitis, and pleuritis: Fibrinous and sclerosing peritonitis (15%), abdominal pain, vomiting, and meteorism. Pleural effusions and thickening, pulmonary fibrosis, and obstructive lung disease.

Poisoning and suicide: A 39-year-old **male** ingested 39 g in a suicide attempt. **He** survived with no special therapy.

Skin: Scaly exfoliative dermatitis, psoriatic or lichenoid rashes, systemic lupus erythematosus-like syndrome, and subungal blotchy erythema of the nail bed.

Drug interactions: Asystole after injection of 7 mg ***verapamil***. IV practolol (10 and 20 mg), followed by slow IV ***disopyramide*** caused sinus bradycardia in one **patient** and death due to asystole in **another**. Use can increase ***lithium***-induced tremors.

Exacerbates ***potassium chloride***-induced arrhythmias in **dogs**. When added to β-amino-proprionitrile in the feed of **turkeys**, it increased the number of aortic ruptures. When evidence of cancer was found in treated **mice**, the FDA notified the manufacturers on 3/11/1970 that **human** testing should cease. It didn't, and, a month later, they were notified again. It wasn't effective, and in January, 1971 (9 months later) the FDA tried, again, to persuade them to cease and desist **human** testing. The company, apparently, refused to notify the doctors using the drug that it was potentially carcinogenic. Finally, on August 27, 1971, 8 months after the last warning and 17 months after the first notice, the testing was stopped.

IV, LD_{50}, **mouse**: 69 mg/kg.

PRAJMALINE TARTRATE

= *GT 1012* = *Neo-gilurytmal* = *NPAB*

Antiarrhythmic.

Untoward effects: Induced a cholestatic syndrome in some **patients**. Within 24 h after 100 mg in divided doses, a 67-year-old **male** developed cerebral confusion. Usual dosage is 20–40 mg tid–qid. Dosages of 80–140 mg/day have been associated with gastrointestinal upsets and visual disturbances. At these dosages, a **patient** had sino-auricular block and delayed atrioventricular conduction.

LD_{LO}, **woman**: 22 mg/kg; TD_{LO}, **man**: 1.4 mg/kg.

LD_{50}, **mouse**: 43 mg/kg; IV, 1.7 mg/kg.

PRALIDOXIME

= *2-PAM* = *Protopam*

Cholinesterase reactivator.

Use: The *chloride*, *iodide*, and *mesylate* (*methanesulfonate* = ***Contrathion*** = ***2-PAMM*** = ***P_2S*** = ***7676 R.P***) forms have been used as antagonists for drug or toxin cholinesterase inhibitors, such,as nerve gases and certain pesticides. The *chloride* is more commonly used; usually IV, occasionally IM, SC, or IP, and rarely, orally or topically. The U.S. Army has issued it in three automatic IM injectors (600 mg/2 ml) in addition to three ***atropine*** injectors.

Untoward effects: Hypertension and slight tachycardia, followed occasionally by some hypotension 3–4 h later. Occasionally blurred vision, diplopia, dizziness, drowsiness, nausea, increase blood pressure, hyperventilation, and muscle weakness. High dosage or rapid IV cause supraorbital headache, tachycardia, arrhythmias, respiratory depression, alteration in visual accomodation, sluggish light reflex, and epigastric discomfort. Pralidoxime itself is a weak cholinesterase and caution is advised in treating continued weakness. Use in a case of ***mevinphos*** poisoning caused a psychotic reaction (including motor agitation, ataxia, anxiety, confusion, combativeness, and hallucinations), runs of cardiac arrhythmias, and urticarial rash. Subconjunctival use leads to local pain; increase in flare, cells, and pigment in anterior chamber, and maximum mydriasis. Half-life in **man** is 1.7 h and 360 h in **dogs**.

Drug interactions: Infusion of ***thiamin*** decreases its urinary excretion, increasing its half-life and serum concentration.

IV, TD_{LO}, **man**: 14 mg/kg.

IM in **rabbits** led to myonecrosis, inflammation, and hemorrhage. In **rat** studies, it enhanced the toxicity of ***carbaryl***, leading to a general statement cautioning against its use in ***carbamate*** poisonings in **man** and **beast**, although it appeared to be effective against ***dimetilan*** and ***isolan*** poisoning.

LD_{50}, **mouse**: 4.1 g/kg; IV, **rat**: 96 mg/kg; **mouse**: 90 mg/kg; **rabbit**: 95 mg/kg.

IV *chloride* LD_{50}, **rat**: 96 mg/kg; **mouse**: 115 mg/kg; **rabbit**: 95 mg/kg.

IV *iodide* LD_{50}, **mouse**: 140 mg/kg.

IV *mesylate* LD_{50}, **rat**: 109 mg/kg; **mouse**: 118 mg/kg.

PRAMIPEXOLE

= *Mirapex* = *SND 919*

***Antiparkinsonian.* A *dopamine* agonist.**

Untoward effects: Nausea, somnolence, and can worsen dyskinesias. Hallucinations in ~10% in early cases and 20% in **those** with more advanced symptoms. Occasionally orthostatic hypotension. Sudden, irresistible sleep onset in some **patients** while driving leading to accidents; or during business meetings or while on the telephone.

PRASTERONE

= *Dehydroepiandrosterone* = *DHEA* = *Diandrone* = *EL 10* = *GL 701* = *Ketovis 17*

***Androgen, immunomodulatory adrenal hormone.* Some capsules have contained only a small fraction of the labeled amount.**

Untoward effects: Increased sebum production in **women** after topical use. Excessive intake in "diet ***teas***" has been suspect as a cause of illness in some **people**. Cardiac arrhythmias in 55-year-old **male** after 50 mg/day/2 weeks, taken to enhance sexual potency and energy. High dosage 25–50 mg/day caused palpitations and arrhythmias in **others**. In **female**, decreased high-density lipoprotein and in **male**, increased prostrate size and growth of small prostrate tumors. Can induce hirsutism, facial acne, and virilization.

PRAVASTATIN

= *CS-514* = *Elisor* = *Eptastatin* = *Lipostat* = *Mevalotin* = *Oliprevin* = *Pravachol* = *Pravaselect* = *Selectin* = *Selipran* = *SQ 31,000* = *Vasten*

***Antihyperlipoproteinemic.* Inhibits *cholesterol* synthesis.**

Untoward effects: Occasionally rashes, transient gastrointestinal upsets, headache, and sleep disturbances. A 63-year-old **female** had symptoms of diabetes mellitus including polyuria, polydipsia, and increase blood ***glucose***. Reversible impotence in 57-year-old **male** after 3 weeks of therapy with 20 mg/day. A 47-year-old **female** had reversible neuropathy after 13 months of ***lovastatin*** 20 mg/day. Treatment was discontinued for 8 weeks and restarted with pravastatin 20 mg/day. Paresthesias returned in 2 weeks after pravastatin, progressing to include the upper extremities to the elbows and lower ones to the knees. A syndrome resembling dermatomyositis developed over a 5-week period in a 66-year-old **female** after 10 mg/day/5 months. Rarely, a severe myopathy has developed. Has occasionally been implicated in cases of depression. Transient increases in serum aminotransferase at the start of treatment, occasionally continuing, in 1–2% of **patients**. Rarely, symptomatic hepatocellular and cholestatic hepatitis, and increased ***creatine kinase***.

Drug interactions: Use with ***alcohol*** can cause a hepatitis. ***Propranolol*** decreases its serum levels. A 64-year-old **female** developed an increased international normalized ratio (10.2) and hematuria when pravastatin 10 mg/day/5 days was added to **her** stabilized anticoagulation (international normalized ratio, 3.8) with ***fluindione***. Use with ***nefazodone*** further increases elevated plasma ***creatine kinase*** levels.

Dogs treated with high dosage have developed cataracts.

PRAWNS

Untoward effects: Immediate contact urticaria on intact or abraded skin in some **people**. Severe asthma-like reactions in 18/50 **prawn-processors** in Great Britain, 15 min to 6 h after exposure to prawn meat, which was blown out of the tails by compressed air jets. Symptoms included wheezing, coughing, mucoid sputum, breathlessness, sneezing, nasal congestion, and itching eyes and hands. After substitution to water jets, only three **workers** continued to have symptoms.

PRAZEPAM

= *Centrax* = *Demetrin* = *Lysanxia* = *Prazene* = *Sedapran* = *Settima* = *Trepidan* = *Verstran* = *W-40020*

Anxiolytic.

Untoward effects: Fatigue > 10%, dizziness < 10%, weakness 8%, drowsiness and light-headedness 7%, and ataxia 5%. Occasionally headache, confusion, tremor, vivid dreams, slurred speech, palpitations, excitement, xerostomia, diaphoresis, amnesia, pruritus, transient skin rashes, joint pains, edema of feet, mild nausea, blurred vision, transient abnormal liver function tests and slight blood pressure reduction. Rarely, jaundice, impotence in **males**, and blood dyscrasias may be associated with its use. Poisoning has been more common in **children** causing sleepiness, agitation, and ataxia; hypotonia in severe poisoning. Usual therapeutic serum levels reported as 0.2–0.7 μg/ml and toxic levels are > 1 μg/ml.

Drug interactions: Metabolism may be decreased if taken concurrently with ***cimetidine***.

PRAZIQUANTEL

= *Biltricide* = *Cesol* = *Distocide* = *Droncit* = *Pyquiton*

Anthelmintic, cestocide, schistosomicide.

Untoward effects: Malaise, headache, and dizziness occur frequently; occasionally abdominal discomfort, nausea, vomiting, sedation, sweating, fever, decreased renal function, fatigue, and eosinophilia. Rarely, rash, urticaria, and pruritus. Adverse effects affected 67.8% of 6,134 **subjects** treated in the Philippines.

Drug interactions: Coadministration with ***sodium bicarbonate*** or ***dextrose*** tended to decrease the serum concentration of praziquantel and ***cimetidine*** to increase it.

SC or IM injection in **dogs** may cause transient local pain and may be toxic to some pet **bird** species. SC use in **cats** a day after ***fenthion*** (***Spotton***®) treatment may cause generalized fasciculations, vomiting, diarrhea, salivation, and miosis. **Cats** may pull out hair at injection site. Avoid use in **cats** and **dogs** until 1 or 2 months of age, respectively.

PRAZOSIN HYDROCHLORIDE

= *Alpress LP* = *CP-12299-1* = *Duramipress* = *Eurex* = *Hypovase* = *Minipress* = *Peripress* = *Sinetens*

α_1-Adrenergic agent, antihypertensive, sympatholytic.

Untoward effects: Dizziness > 10%, headache and drowsiness ~7.6%, weakness 6.5%, palpitations and nausea ~5%. Vomiting, diarrhea, constipation, edema, orthostatic hypotension, angina, dyspnea, syncope, vertigo, nervousness, depression, rash, pollakiuria, epistaxis, blurred vision, amblyopia, reddened sclera, pigmentary mottling and serious retinopathy, xerostomia, and nasal congestion in 1–4%. Pancreatitis, abdominal discomfort, abnormal liver function tests, tachycardia, paresthesias, hallucinations, lichen planus, alopecia, pruritus, erythema multiforme and Stevens–Johnson syndrome, incontinence, positive antinuclear antibody titer, fecal incontinence, angioedema, coma, Cheyne-Stokes respiration, sinus bradycardia, eosinophilia, ***sodium*** and water retention, impotence, priapism, and arthralgia in < 1%. Accidental ingestion of > 50 mg by a 2-year-old child caused profound drowsiness and decreased reflexes. Usual therapeutic serum levels are 0.001–0.02 μg/ml and toxic at 0.9 μg/ml.

TD_{LO}, **human**: 285 μg/kg.

Drug interactions: Synergism reported with ***thiazide diuretics*** and ***β-adrenergic blockers***. A **patient** treated for 13 weeks had a transient collapse after taking a ***nitroglycerin*** tablet.

PREDNISOLONE

= *Codelcortone* = *Cortalone* = *Decaprednil* = *Decortin H* = *Δ¹-Dehydrocortisol* = *Delta-Cortef* = *Deltacortril* = *Delta F* = *Deltasolone* = *Deltastab* = *Flamasone* = *Hydeltra* = *Hydeltrone* = *Δ¹-Hydrocortisone* = *Hydrodeltalone* = *Hydroretrocortine* = *Klismacort* = *Leocortal* = *Metacortandralone* = *Meticortelone* = *Paracortol* = *Precortancyl* = *Precortilon* = *Precortisyl* = *Prednicen* = *Predniretard* = *Predonine* = *Solone* = *Sterolone* = *Zenarid*

The 21-Acetate = *Ak-Tate* = *Hostacortin H* = *Inflanefran* = *Pred Forte* = *Pred Mild* = *Scherisolon* = *Sterane.*

The 21-tert-Butylacetate = *Cordelcortone* = *Hydeltra-T.B.A.* = *Predalone T.B.A.* = *Prednisolone Tebutate.*

The 17-Ethylcarbonate-21-propionate = *HOE 777* = *Prednicarbate.*

The 21-Hydrogen Succinate = *Fiasone* = *Prednisolone Hemisuccinate.*

The Sodium Phosphate = *Ak-Pred* = *Codelsol* = *Cortisate* = *Hefasolon* = *Hydeltrasol* = *Inflamase* = *Metreton* = *Prednesol* = *Predsol* = *Solucort* = *Solu- Predalone.*

The Sodium Metasulfobenzoate = *Cortico-Sol* = *F 75,980* = *Predenema* = *Predfoam* = *Prednisolut* = *Solupred.*

The 21-Stearoylglycolate = *Erbacort* = *Erbasona* = *Estilsona* = *K 1557* = *Lentosone* = *Prednisolone Steaglate* = *Rolisone* = *Sintisone.*

The 21-Succinate Sodium = *Di-Adreson F* = *Meticortelone* = *Prednisolone Sodium Succinate* = *Solu-Decortin-H* = *Solu-Delta-Cortef.*

The 21-Trimethylacetate = *Prednisolone 21-Pivalate* = *Ultracortenol* = *Vetcortenol.*

The 17-Valerate = *Prenival* = *W 4,869.*

Anti-inflammatory corticosteroid, glucocorticoid, immunosuppressive.

Untoward effects: Blood: Many case reports of leukocytosis, non-thrombocytopenic purpura, eosinopenia, anemia, lymphopenia, and one of neutropenia.

Endocrine: Menstrual irregularities, development of Cushingoid state, suppression of growth in **children**, secondary adrenocortical and pituitary unresponsiveness (particularly in times of stress, as in trauma, surgery, or illness), decreased carbohydrate tolerance, manifestations of latent diabetes mellitus, increased requirement for ***insulin*** or ***oral hypoglycemic agents*** in **diabetics**.

Eye: Posterior subcapsular cataracts, increased intraocular pressure, glaucoma, and exophthalmos.

Fetus and neonate: It crosses the placenta and **fetal** blood levels are about 10% of **maternal** levels. Different studies have varying results, but, in general, increases antenatal death rate and decreases length of gestation. Up to 0.23% of **maternal** dose/l of breast milk over 48 h. Theoretically, could suppress endogenous ***corticosteroid*** production in the **infant**.

Fluid and electrolytes: ***Sodium*** and water retention, congestive heart failure in susceptible **patients**, ***potassium*** loss, hypokalemic alkalosis, and hypertension.

Gastrointestinal: Peptic ulcer with possible subsequent perforation of hemorrhage; pancreatitis, abdominal distension, and ulcerative esophagitis. Intestinal obstruction due to enteric-coated tablets in 58-year-old **male**.

Metabolic: Negative ***nitrogen*** balance due to protein catabolism, rare instances of blindness associated with intralesional therapy around the face and head. Hyperpigmentation or hypopigmentation, SC and cutaneous atrophy, sterile abscesses, postinjection flare (following intra-articular use), and Charcot-like arthropathy. Mild to moderate fat deposition in 50% of **patients**.

Miscellaneous: Maximum increase in polymorphonuclear leukocytes (1,600–7,500 cells/mm^3) 4–6 h after a dose; lymphocyte count will decrease 70% and monocyte count decrease 90%. Leukocytes return to normal 24 h after last dose, increasing susceptibility to infection. Adrenal atrophy reported after long-term use. Myocardial ischemia in an **asthmatic** 29-year-old **male** when dosage increased to 30 mg IM and 30 mg orally on alternate days. Thromboembolic complications led to death in 4/71 **patients** on it and other ***corticosteroids***. Suppression of Kveim test in 27/28 **patients** with active sarcoidosis. Use may be associated with pancreatitis and cystic adrenal glands. Decrease in spermatogenesis and motility after 15 days, requiring 6 months for recovery. Possible tendon ruptures due to immunosuppression and after repeated injections near tendons. **Patients** have unknowingly taken herbal remedies contaminated with it. Allergic reaction in a 29-year-old **female** with fever and butterfly erythema from it and the *sodium succinate* form.

Musculoskeletal: Muscle weakness, ***steroid*** myopathy, loss of muscle mass, osteoporosis, vertebral compression fractures, aseptic necrosis of femoral and humeral heads, and pathological fracture of long bones.

Neurological: Convulsions, increased intracranial pressure with papilledema (pseudotumor cerebri) usually after treatment, and psychosis.

Skin: Impaired wound healing, thin fragile skin, petechiae and ecchymoses, facial erythema, increased sweating, and may suppress reactions in skin tests. Exacerbation of erythematous and bulbous lesions. In England, a 58-year-old **female** taking a new generic brand developed a morbilliform rash on **her** arms, legs, and trunk within 6 days due to its ***dextran*** and ***stearic acid*** content.

Drug interactions: Its metabolism is decreased with a dramatic increase in its plasma concentration by ***oral contraceptives***. By its inhibition of metabolizing enzymes, ***cyclophosphamide*** activity is increased. It markedly decreases ***antipyrine*** half-life and decreases ***isoniazid*** plasma concentrations. ***Sodium metabisulfite*** accelerates its decomposition. ***Phenytoin*** increases its metabolic clearance rate and decreases its half-life by 45%; ***rifampin*** increases its plasma clearance, decreasing its half-life and decreasing its tissue availability by 66%, requiring doubling of its usual dosage. After 2 weeks of therapy with either ***indomethacin*** or ***naproxen***, plasma levels of prednisolone increase. Use with ***ketoconazole*** can be expected to increase the immunosuppressive effect.

In **animals**, side reactions such as polydipsia, polyuria, and fluid retention associated with ***sodium*** retention are the most common. Weight loss or gain and ***potassium*** decrease can occur on long-term therapy. Anorexia and diarrhea occur infrequently. Theoretically, it should not be used in viral infections, but clinicians often find it very useful with other therapy, especially in some chronic viral infections or during recovery periods. Its anti-inflammatory effects may mask latent infections, but, used with ***antibiotics***, it aids survival in acute **hog** "flus" or **cattle** "shipping fever". It often may be *contraindicated* in congestive heart failure, diabetes, or osteoporosis. Except for emergency life-saving use, it should be omitted in tuberculosis, chronic nephritis, Cushingoid syndromes, and peptic ulcer cases. The FDA has established a zero tolerance for its residues in milk of treated **animals**. **Rabbit** trials confirm suspicions in **man** that injections to relieve tendosynovitis leave steroid deposits, decrease tensile strength, and predispose them to rupture. 3 mg/kg on one eye of a **rabbit** delays wound healing by 40% in the control eye by systemic effect. When used to prevent post-operative abdominal adhesions, it also slows down healing of surgical incisions. Avoid use in **animals** before slaughter. Administration during the last trimester of pregnancy may precipitate premature parturition with dystocia, fetal death, metritis, and retained placenta. Its use in pregnancy has caused cleft palate, high-arched palate, increased perinatal death rate, and umbilical hernias in **rats**; cleft palate in **mice**.

SC, LD_{50}, **rat**: 147 mg/kg; **mouse**: 3 g/kg.

PREDNISONE

= *Ancortone* = *Colisone* = *Cortancyl* = *Δ^1-Cortisone* = *Dacortin* = *Decortancyl* = *Decortin* = *Δ^1-Dehydrocortisone* = *Delcortin* = *Deltacortisone* = *Deltacortone* = *Delta E* = *Deltasone* = *Deltison* = *Deltra* = *Di-Adreson* = *Encorton* = *Hostacortin* = *Lisacort* = *Metacortandracin* = *Meticorten* = *NSC-10,023* = *Nurison* = *Orasone* = *Paracort* = *Prednilonga* = *Pronison* = *Rectodelt* = *Retrocortine* = *Servisone* = *Sone* = *Ultracorten* = *Zenadrid*

The 21-Acetate = *Delta-Cortelan* = *Hostacortin*

Anti-inflammatory corticosteroid, glucocorticoid, immunosuppressive.

Untoward effects: Blood: Many case reports of anemia, nonthrombocytopenic purpura, lymphopenia, monocytopenia, leukocytosis, eosinopenia, and a few of thrombocytopenia and eosinophilia.

Endocrine: Menstrual irregularities, development of Cushingoid state, secondary adrenocortical and pituitary unresponsiveness (particulary in times of stress, as in trauma, surgery or illness), suppression of growth in **children**, decreased ***carbohydrate*** tolerance, manifestations of latent diabetes mellitus, and increased requirement for ***insulin*** or ***oral hypoglycemic agents*** in **diabetics**.

Eye: Posterior subcapsular cataracts, increased intraocular pressure, glaucoma, exophthalmos, and bluish scleral discoloration.

Fetus and neonate: Abnormal lymphocytic chromosomes at birth, slowly disappearing by 32 months. Small malpositioned kidneys, reduced intrauterine fetal growth, decreased birth weight, occasionally decreased gestation time, hydrocephaly, and adrenal insufficiency have been reported. Death of a premature 4-hour-old **male** after his **mother** received 60 mg/day/18 days and 40 mg/day/6 days for sarcoidosis. Post-mortem revealed hypoplastic adrenals with cortical necrosis, hemorrhage, and cyst formation. 2 h after 10 mg oral dose, breast milk contained a total of 28.3 μg of it and ***prednisolone***/l.

Fluid and electrolytes: ***Sodium*** and fluid retention, congestive heart failure in susceptible **patients**, ***potassium*** decrease, hypokalemic alkalosis, and hypertension.

Gastrointestinal: Peptic ulcer with possible perforation and hemorrhage, pancreatitis, abdominal distension, and exacerbation of existing gastrointestinal diseases. Rarely, ulcerative esophagitis and other esophageal damage.

Metabolic: Negative ***nitrogen*** balance due to protein catabolism and even severe hyperglycemia and acidosis.

Miscellaneous: Bison obesity, truncal obesity, moon facies, fatty infiltration of the liver (with massive fat embolism and sudden death in a **child** after 11 weeks therapy with it and ***hydrocortisone***), decreased slow wave sleep, increased blood pressure, hyperuricemia, immunosuppression by doses > 40–60 mg/day, hypertension, decreased **male** reproductive ability, euphoria, pheochromocytoma, chondrocyte degeneration, panniculitis, tremors, psychological dependence, adrenocortical failure, mood cycles, increased serum ***cholesterol***, disseminated Kaposi's sarcoma, increased rate of infection, and thromboembolism.

Musculoskeletal: Muscle weakness, steroid myopathy, loss of muscle mass, osteoporosis, vertebral compression fractures, aseptic necrosis of femoral and humeral heads, and pathological fracture of long bones.

Neurological: Increased intracranial pressure with papilledema (pseudo-tumor cerebri) usually after treatment, headache, vertigo, convulsions, psychosis, mania, euphoria, and depression; catatonia in an 11-year-old **male** after 50 mg/day. In elderly **patients** taking 60–100 mg/day it caused confusion, memory impairment, and delirium.

Skin: Impaired wound healing, thin fragile skin, petechiae and ecchymoses, facial erythema, increased sweating, may suppress reactions to skin tests, purpura on the skin, spontaneous tearing of skin, and necrosis. Can temporarily decrease the inflammation of acne vulgaris but, eventually, doses of 20 mg/day can exacerbate existing acne or induce it.

Drug interactions: Antagonizes the anticoagulant effect of ***coumarins*** and ***indandiones***. Use with ***azathioprine*** associated with increased risk in skin cancers (sarcomas). ***Phenobarbital*** and ***phenytoin*** increase its metabolism by enzyme induction. A drug combination of ***trifluoperazine*** and prednisone induces extrapyramidal symptoms. Pulmonary toxicity reported when used with ***amiodarone***. May interfere with contraceptive effect of ***intrauterine devices***. ***Omeprazole*** may decrease its effectiveness. It antagonizes the action of ***vitamin D*** and decreases the absorption of ***calcium***. May interact with ***aspirin***, increasing the likelihood of gastrointestinal bleeding, especially when prednisone is withdrawn. ***Aluminum-magnesium hydroxides*** can significantly decrease its absorption. ***Anesthetics, anticholinergics, tricyclic antidepressants, sympathomimetics***, and ***modified-live virus vaccines*** in **asthmatics** can increase its toxicity. Since it can increase ***potassium*** loss, it may, in turn, increase ***digoxin*** toxicity.

In **animals**, especially **dogs**, side reactions such as polydipsia, polyuria, and fluid retention associated with ***sodium*** retention are the most common. Weight loss or gain and ***potassium*** loss can occur on long-term therapy. Euphoria and increased food and water intake often occur. Anorexia and diarrhea occur infrequently. Theoretically, it should not be used in viral infections, but clinicians often find it very useful with other therapy, especially in some chronic viral infections or during recovery periods. Its anti-inflammatory effects may mask latent infections, but, used with ***antibiotics***, it aids survival in acute **hog** "flus" or **cattle** "shipping fever". It often *may be contraindicated* in congestive heart failure, diabetes, or osteoporosis. Except for emergency life-saving use, it should be omitted in tuberculosis, chronic nephritis, Cushingoid syndromes, and peptic ulcer cases. The FDA has established a zero tolerance for its residue in milk of treated **animals**. **Rabbit** trials confirm suspicions in **man** that injections to relieve tendosynovitis leave ***steroid*** deposits, decrease tendon tensile strength, and predispose them to rupture. 3 mg/kg on one eye of a **rabbit** delays wound healing by 40% in the control eye by systemic effect. When used to prevent post-operative abdominal adhesions, it also slows down healing of surgical incisions. Avoid use in **animals** before slaughter. Administration during the last trimester of pregnancy may precipitate parturition with associated dystocia, fetal death, metritis, and retained placenta. Birth weights and intrauterine growth of **rats** and **mice** are decreased. Transient diabetes mellitus has been associated with its use in **dogs**. Plasma ***thyroxine*** and ***triiodothyroxine*** concentrations decreased significantly in **dogs** after only three IM injections of 2.2 mg/kg every other day. Acute pancreatitis in **dogs** given it with ***azathioprine***. It apparently exacerbates ***flunixin*** gastric mucosal injury in **dogs**. Use associated with hyperglycemia and increased production of angiotensin-converting enzymes in **rabbits**.

PRENYLAMINE LACTATE

= *Angormin* = *Bismetin* = *Carditin-Same* = *Coredamin* = *Corontin* = *Crepasin* = *Daxauten* = *Diaphril* = *Diaphryl* = *Herzcon* = *Hostaginin* = *Incoran* = *Irrorin* = *Lactamin* = *NP 30* = *Nyuple* = *Onlemin* = *Plactamin* = *Rausetin* = *Reocorin* = *Roinin* = *Seccidin* = *Sedolatan* = *Segontin* = *Synadrin*

Coronary vasodilator, calcium channel blocker.

Untoward effects: Sedation, lassitude, epigastric pain and sensation of heat, vomiting, diarrhea, headache, giddiness, flushing of skin, erythematous rashes, bradycardia, atrioventricular and intra-ventricular conduction disorders, prolonged QT interval, ventricular arrhythmias, syncope, intention tremors, dizziness, and extrapyramidal symptoms. Excessive doses or overdoses lead to hypotension, tachycardia, and convulsions. Withdrawn from the British market by Hoechst in 1988.

Drug interactions: A 50–100 mg bolus of ***lidocaine*** after prior treatment with prenylamine caused atrioventricular-block in two elderly **patients**.

In **guinea pigs**, ***propranolol*** increased the chance of atrioventricular-block while receiving it.

LD_{50}, **rat**: 1 g/kg; **mouse**: 580 mg/kg; IP, **rat** and **mouse**: 40 mg/kg.

PRESTONIA AMAZONICA

Untoward effects: Use as a hallucinogenic snuff in the Amazon is attributed to its *N,N-**dimethyltryptamine*** (q.v.) content.

PRETZELS

Untoward effects: Illustrates the fact that nothing in this world may be truly safe. In 1977, the FDA recalled 19,000 cases sold under various brand names. ***Lye*** crystals sprinkled on them caused chemical burns to the mouths and tongues of **consumers**. ***Lye*** and water are used to give the pretzels their glazed look. Properly carried out, the procedure is safe, because the pretzels absorb ***sodium hydroxide***, where it is converted to harmless ***sodium carbonate*** when they are baked. Apparently, the ***lye*** was not mixed properly to dissolve all of it, permitting it to get into packages along with the pretzels.

PRILOCAINE

= *Astra 1512* = *Citanest* = *Ditanest* = *L 67* = *Propitocaine* = *Xylonest*

Local anesthetic. **IV, regional analgesia, peridural.**

Untoward effects: Discomfort, shivering, nausea, **maternal** and **infant** cyanosis, **fetal** arrhythmias and bradycardia, dizziness, anaphylaxis, methemoglobinemia, occasionally decreased (maximum of 15%) blood pressure and heart rate, central nervous system depression in **fetus** and **newborn**, and seizure reported. Usual therapeutic serum levels reported as 0.5–2 μg/ml with toxicity above 5 or 6 μg/ml. Central nervous system depression and potential fatalities at ~20 μg/ml.

Methemoglobinemia also reported after IV in **dog** trials. Central nervous system toxicity with seizures after IV infusions in **rhesus monkeys**. Chronic IP use in **mice** caused focal necrosis of liver, renal tubular damage, and casts in the kidneys.

IV, LD_{50}, **mouse**: 62 mg/kg; IP, 30 mg/kg.

PRIMAQUINE

= *Chinocid* = *Quinocide* = *SN 13,272*

Antimalarial.

Untoward effects: Glucose-6-phosphate dehydrogenase-deficient **patients** develop a serious intravascular hemolysis and hematuria. This can also occur in **patients** with certain hemoglobinopathies. Cases of Mediterranean and Asian glucose-6-phosphate dehydrogenase deficiency are more severe than in **black** Africans. Hemolysis in **neonates** due to poor development of **their** glucose-6-phosphate dehydrogenase enzyme system. Abdominal discomfort, nausea, dizziness, headache, altered visual accomodation, pruritus, neutropenia, depression, hemolytic jaundice, liver failure. Diarrhea, weakness, cyanosis, methemoglobinemia, and leukocytosis also reported. Colors urine brown. Rarely, leukopenia, agranulocytosis, hypertension, and arrhythmias.

LD_{LO}, **human**: 50 mg/kg.

Drug interactions: ***Quinacrine*** enhances its activity and toxicity as well as increasing ***quinacrine*** toxicity.

The *d* isomer is four times as toxic as the *l* isomer in **monkeys**. **Dogs** were rather sensitive to it and developed increased serum transaminases, decreased fasting blood ***glucose*** levels, and inflammation and degenerative changes in the liver and kidneys. **Rats** were less sensitive and **monkeys** were intermediate – all with the same symptoms. Pneumonia and increased serum ***haptoglobin*** in the **dog**; methemoglobinemia, thrombocytopenia, degenerative and inflammatory changes in striated muscle of the **dog** and **rat**. Reticulocytosis, nucleated erythrocytes, and bile duct hyperplasia in **rats**. Edema and gliosis of cerebral cortex and erythrocytopenia in some **rhesus monkeys**. Lymphoid depletion in **dogs** and **monkeys**.

LD_{50}, **mouse**: 100 mg/kg.

PRIMIDONE

= *Liskantin* = *Medi-Pets* = *Mylepsin* = *Mysoline* = *Neurosyn* = *Primaclone* = *Resimatil* = *Ro-Primidone* = *Sertan*

Anticonvulsant.

Untoward effects: Long-term therapy has occasionally caused megaloblastic anemia, decreased serum ***folates***, increased alkaline phosphatase, lymphadenopathy, acute intermittent porphyria, connective tissue diseases, osteomalacia, eosinophilia, lymphocytosis, thyroid enlargement, and decreased serum ***calcium***, ***phosphorus***, and active ***vitamin D***. Common adverse effects are ataxia, drowsiness, vertigo, nausea, vomiting, exacerbation of hyperactivity in **children**, systemic lupus erythematosus-like syndrome, diplopia, decreased cellular immunity, morbilliform rashes, toxic epidermal necrolysis, eyelid edema, difficulty in concentrating, disturbed behavior, decreased libido, giddiness, and acute porphyrias. Progressive headache, dysarthria, ptosis, diplopia, strabismus, nystagmus; exaggerated light reflexes in a 25-year-old **female** after accidentally ingesting three 250 mg tablets. Impotence reported in a **male**. Macrocytosis in 42%. Ophthalmoplegia with overdoses or in suicide attempts. Poisoning causes semicoma, areflexia, crystalluria, and asterixis. Therapeutic serum levels 0.5–1.2 mg%; 5–8 mg% is toxic, and 10 mg% has been fatal. Large amounts excreted in breast milk leading to somnolence and drowsiness in **infants**. Also **neonatal** coagulation factor defects, clinical bleeding, and, rarely, death. Use during pregnancy produces increased congenital heart disease, cleft lip and/or palate, abnormalities of face, chest, abdomen, and extremities; low broad nasal bridge, low-set posteriorly rotated ears, down- or up-slanted palpebral fissures, short forehead, hypoplastic mandible, and long prominent philtrum in ~8% of their **children**. Growth retardation, hirsutism, microcephaly, mental retardation, hypoplastic nails, and lymphadenopathy, including lymphoma-like syndrome also occur. Visual and auditory hallucinations in a severely retarded 19-year-old **female** receiving 250 mg tid/7 years. Delirium has followed overdoses.

Drug interactions*:** ***Carbamazepine, chloramphenicol, chloroquine, methylphenidate, phenytoin, and ***valproic acid*** increase its plasma concentration. Delayed fatal agranulocytosis in an **epileptic** 40-year-old **male** receiving 400 mg ***phenytoin*** and 750 mg primidone/day/17 years. Use with ***phenobarbital*** and ***phenytoin*** has also been associated with macrocytic anemia. It is a microsomal enzyme stimulating the metabolism of ***dexamethasone, meprobamate, phenylbutazone***, and ***vitamin D,*** and decreasing the efficacy of ***oral contraceptives***. It decreases the absorption of ***folic acid*** and/or interferes with its metabolism in the tissues. Decreases ***diazepam*** half-life.

Cats show neurotoxicity (ataxia, paralysis) on usual **dog** doses. Some clinicians have used it successfully in **cats** at reduced dosage. Adjust dosage downward if staggering or drowsiness occurs in treated **cats** and **dogs**. May be associated with hepatic necrosis after long-term therapy in **dogs**.

LD_{50}, **mouse**: 280 mg/kg.

PRIMULA sp.

= *Primrose*

Many related plants, such as *Cyclamens, Lysimachia*, and *Angalis* contain poisonous *saponins*.

Untoward effects*:** ***P. obconica leaves and stems have glandular hairs; their heads contain the skin irritant ***primin***. As little as 20 μg can trigger an allergic reaction in some **people**. Its pollen has also caused a dermatitis which can be associated with erythema, blistering, conjunctivitis, fever, and eczema.

P. Auricula, P. malacoides, P. mollis, P. reticulata, P. Sieboldi, and ***P. sinensis*** also cause a contact dermatitis due to hypersensitivity of the skin and oral mucous membranes in ~6% of exposed **people**. Sensitivity to it can persist for many years. ***Oil of Primrose*** has been touted as a cure (with questionable benefits) for over a dozen medical conditions. It may be a suitable source of essential ***fatty acids***.

PRISTANE

= *Norphytane* = *Robuoy*

***Use*:** Lubricant, anticorrosion agent.

***Untoward effects*:** Topically causes erythema and swelling of **human** skin.

Severely irritating topically to **rat**, **rabbit**, and **guinea pig** skin and moderately irritating to the skin of **miniature swine** causing slight acanthosis and hyperkeratinization in the epidermis and weak vasodilation and edema at the upper dermis. Implantation in **mice** induces a plasmacytoma incidence of 70%. Intrathoracic injections of 0.5 ml in **mice** produced dyspnea and pleuritis in ~3 weeks.

PROBENECID

= *Benemid* = *Benuryl* = *Probalan* = *Probecid* = *Proben* = *Probenimead* = *Robenecid*

Uricosuric.

***Untoward effects*:** Poisoning after suicide attempt in a 49-year-old **male** after ingestion of 95 × 0.5 g tablets led to several episodes of copious vomiting followed by progressive stupor, coma, grand mal seizures, mydriasis, flushed skin, and slight hyperreflexia. Visual hallucinations, mild tremors, fever, and aspiration pneumonia on day 3.

***Blood*:** Hemolytic anemia in glucose-6-phosphate dehydrogenase-deficient **persons**; aplastic anemia, leukopenia, neutropenia, and thrombocytopenia.

***Central nervous system*:** Dizziness, headache, and flushing.

***Fetus*:** Crosses the placenta near term.

***Gastrointestinal*:** Nausea, vomiting, anorexia, and sore gums.

***Genitourinary*:** Nephrotic syndrome, hematuria, pollakiuria, and renal colic.

***Hypersensitivity*:** Anaphylactoid reaction, anaphylaxis, fever, urticaria, and pruritus.

***Liver*:** Hepatocellular jaundice. Fatal hepatic necrosis has occurred.

Miscellaneous*:** Can decrease ***estriol test results; can cause false positive ***Clinitest***® urinary ***glucose*** results. Gouty arthritis in ~30% of **patients** during first 6–9 months of therapy.

***Skin*:** Dermatitis, urticaria, pruritus, and erythema multiforme.

Drug interactions*:** By blocking renal tubular excretion, it increases ***penicillin, salicylates, and many ***cephalosporin*** levels and half-lives. By increasing ***aspirin*** and other ***salicylate*** serum levels by up to 50%, it increases their toxicity. It inhibits the transport of ***p-aminosalicylic acid, cephalothin***, and ***penicillin***. ***Salicylates*** decrease its therapeutic efficacy. Inhibits the excretion of ***acetaminophen, adinazolam, allopurinol, anticoagulants*** (***coumarins*** and ***indandiones***), ***captopril*** (14%), ***carprofen, chlorpropamide, ciprofloxacin, cisplatin*** (60%), ***clofibrate*** (79%), ***dapsone, dyphylline, indomethacin*** (64%), ***ketoprofen*** (400%), ***lorazepam, mandelic acid, meclofenamate, methotrexate*** (340%), ***nalidixic acid*** (288%), ***naproxen*** (50%), ***nitrofurantoin, rifampin, sulfinpyrazone, sulfonamides***, and ***sulfonylureas*** and increases their plasma levels. ***Ethacrynic acid*** and ***xanthines***

antagonize its effects and it antagonizes the hyperuremic effect of ***thiazides***. After initial decrease, it intensifies ***furosemide's*** effect. May decrease effectiveness of ***allopurinol*** in some cases. Occasionally increases the hypoglycemic effect of ***tolbutamide*** and generally has little effect on its metabolism. It stimulates its own metabolism after repeated usage. Its effectiveness is decreased by ***aspirin*** and ***xanthines***. Clearance of ***ofloxacin*** is increased 15% by it. Prolongs half-life of ***piperacillin*** 20% and of ***tazobactam*** by 71%.

LD_{LO}, **human**: 500 mg/kg.

Reduced renal toxicity in **rats**. In **animals**, use caution in administering with other drugs, especially ***salicylates***. In species where it or ***acetylsalicylic acid*** alone may decrease serum urate levels, their use in combination may completely inhibit uricosuric activity. It inhibits the metabolism and renal tubular secretion of ***zidovudine*** in **monkeys** and **rabbits**. It increases ***indomethacin*** serum concentration by 20% in **rats**. Also increases serum levels of ***dyphylline*** in **chickens** and **rats**; and ***acetaminophen, ketoprofen, lorazepam***, and ***rifampin*** in **rats**.

LD_{50}, **rat**: 1.6 g/kg; SC, LD_{LO}, 611 mg/kg; IP, LD_{LO}, 394 mg/kg. IP, LD_{50}, **mouse**: 1 g/kg; oral, LD_{LO}: 1.67 mg/kg; SC, LD_{LO} 11.6 g/kg; IV, LD_{LO} 458 mg/kg. IV, LD_{LO}, **dog**: 230 mg/kg.

PROBUCOL

= *Biphenabid* = *DH-581* = *Lorelco* = *Lurselle* = *Sinlestal*

Antihyperlipoproteinemic.

Untoward effects: Gastrointestinal upsets, diarrhea (10%), flatulence, abdominal pain, nausea, vomiting, and gastrointestinal bleeding. Dizziness, palpitations, syncope, chest pain; headaches, paresthesias, and eosinophilia in ~2%. Occasionally (0.1–0.6%)) rash, pruritus, impotence, insomnia, conjunctivitis, tearing, blurred vision, tinnitus, altered taste and smell, anorexia, QT interval prolongation, ecchymoses, petechiae, thrombocytopenia, hyperhidrosis, fetid sweat, nocturia, increased SGOT, SGPT, ***bilirubin***, alkaline phosphatase, and creatine phoshokinase, and peripheral neuritis, and angioneurotic edema. Decreases serum ***cholesterol*** by only 10–15% and, unfortunately, also high-density lipoprotein, which led to its abandonment as a useful drug.

PROCAINAMIDE

= *Amisalin* = *Novocainamide* = *Novocamid* = *Procaine Amide* = *Procamide* = *Procan-SR* = *Procapan* = *Pronestyl* = *Sub-Quin*

***Antiarrhythmic, Class 1A.* Oral, IV, or IM.**

Untoward effects: Agranulocytosis with soreness or ulceration of mouth, throat, or gums; and symptoms of upper respiratory tract infection occasionally follow repeated use. Fatalities also reported. Hemolytic anemia in glucose-6-phosphate dehydrogenase-deficient **patients**. It can induce or exacerbate a systemic lupus erythematosus-type syndrome with increased SGOT, abdominal pain, chills, fever, nausea, vomiting, pleural and pericardial effusions, splenomegaly, and acute hepatomegaly. Pericarditis in ~15% of the systemic lupus erythematosus-type cases. Occasionally hepatocanalicular jaundice and granulomatous hepatitis reported. Transient pancytopenia in a 73-year-old **female** (250 mg qid). Neutropenia and bone marrow granulomas in a 77-year-old **male** treated with it for 50 days. A 67-year-old **male**, after 500 mg qid/2 months, developed agranulocytosis and thrombocytopenia. A 50-year-old **female**, after 2 g/day, developed autoimmune hemolytic anemia. Antinuclear antibodies in 8/10 **patients** in one study. Pancreatitis, myocarditis, pericarditis, angioneurotic edema, drug fever, eosinophilia, leukopenia, maculopapular edema, torsades de pointes, decreased blood pressure, ventricular tachycardia, premature ventricular contractions, ventricular fibrillation, arrhythmias, cardiac conduction defects, decreased cardiac output, prolonged QT intervals, widening of the QRS interval (after an IV), blurred vision, and slurred speech have also been reported. Paradoxical acceleration of ventricular tachycardia after IV treatment. Vasculitis after IV therapy required amputation of distal digits of both index fingers in 53-year-old **male**. Large doses have also caused anorexia, nausea, photosensitivity, urticaria, pruritus, depression, giddiness, mania, psychosis, hallucinations, and nephrotic syndrome with membranous glomerulonephritis. Hypotension noted after IV use is rare after oral use. It has a postsynaptic ***curare***-like effect and exacerbates myasthenic weakness and apnea, and induces arthralgias, myopathy, tremors, diaphragmatic paralysis, and peripheral neuropathy. A reported induced pseudo-obstruction of the bowel was probably due to its inherent anticholinergic effect. Poisoning cases have been reported. A 67-year-old **female** self-medicated with ~7 g causing severe hypotension, lethargy, renal insufficiency, and life-threatening cardiac problems that eventually responded to hemodialysis. Therapeutic serum concentrations are 4–8 mg/l and toxic concentrations are 10–15 mg/l (1 mg%). Cerebellar ataxia reported after a 63-year-old **male** ingested 2 g sustained-action in 6 h. A symptomatic reactive hypoglycemia, possibly associated with induced lupus erythematous, reported. Urticaria, pruritus, and pain at injection site occur. Angioedema can be life-threatening if airway is obstructed. Skin eruptions occur occasionally. Its use has produced false positive serological tests for syphilis and increased serum alkaline phosphatase test results. Delirium noted in some elderly **patients**.

LD_{LO}, **human**: 50 mg/kg; TD_{LO}, 10 g/kg during 1 year; TD_{LO}, 2.28 g/kg, during a 22 week period.

Drug interactions*: *Cimetidine and ***ranitidine*** inhibit its renal tubular excretion. ***Acetazolamide*** and ***sodium bicarbonate*** increase its renal absorption due to urinary pH and can raise its serum levels. Use with ***aminoglycoside antibiotics*** (***colistin, kanamycin, neomycin, streptomycins, tetracyclines***, etc.) or ***depolarizing muscle relaxants*** enhances its inherent neuromuscular blockade. Enhances ***d-tubocurarine*** effects. It increases ***warfarin's*** anticoagulant effects and potentiates the negative inotropic effect of ***quinidine***. Its antiarrhythmic action is antagonized by drugs causing hypokalemia. Oral use with ***phenytoin*** is associated with agranulocytosis and fatalities may follow its use with IV ***phenytoin***. Adsorbent ***antidiarrheals*** can decrease bioavailability by ~30%. Use concurrently with ***amiodarone*** may decrease procainamide dosage needed by 20–30%.

In **animals**, dosage must be adjusted downward as response is noted. Cross-sensitivity to ***procaine*** and related drugs may occur. It has induced antinuclear antibodies in **beagle dogs**. Thrombocytopenia reported in **cats** and **dogs**.

LD_{50}, **mouse**: 890 mg/kg; LD_{LO}, **dog**: 2.2 g/kg.

IV, LD_{50}, **rat**: 121 mg/kg; **mouse**: 137 mg/kg.

PROCAINE

= *Aethocaine* = *Allocaine* = *Alokain* = *Anestil* = *Aslavital* = *Atoxicocaine* = *Bernocaine* = *Biocaine* = *Chlorocaine* = *Clinocaine* = *Dentoraline* = *Enpro* = *Ethocaine* = *Genocaine* = *Gerioptil* = *Gero* = *Gerovital* = *H3* = *Herocaine* = *Hormo-Gerobion* = *Isocaine* = *Jenacaine* = *Juvocaine* = *Kerocaine* = *Medaject* = *Mynol* = *Naucaine* = *Neocaine* = *Neo-dentoraline* = *Novocain* = *Omnicain* = *Paracaine* = *Plaocaine* = *Prokopin G* = *Rocain* = *Syncaine* = *Syntocain*

***Local anesthetic.* Usually parenteral.**

Untoward effects*:** Oral use can cause epigastric pain, vomiting, vertigo, diarrhea, and occasionally delirium. Thrombocytopenia in 28-year-old **male** after oral use. Most fatalities after parenteral use involve its use with ***epinephrine; inadvertent intravascular injections have caused some serious reactions. Fatal anaphylaxis has occurred in sensitive **patients** after injection into an accidentally punctured vein. Infiltration of peritonsillar tissues with 25 ml of 1% and 2 drops of 1:1000 ***epinephrine*** led to convulsions and apnea. Use in dental procedures has led to edema at injection site. Acute hemolytic anemia, thrombocytopenia, fever, disorientation, drowsiness, irritability, and choreiform movements ~1 week after generalized convulsion 10 min after injection of 15 ml of 2% solution. **Patient** had been on ***sulfathiazole*** 3 years previously, and it was probably a manifestation of a cross-reaction with the antibodies against the ***sulfonamide***, both of which are ***p-aminobenzene*** derivatives. Procaine dermatitis was common in **dentists** from leaky syringes. Use of a 5% ointment in treating rectal pruritus caused marked edema, especially of the face, resembling nephritic edema. Causes transient inactivation of pseudocholinesterase. Paravertebral injection of it has led to severe (sometimes irreversible) spinal cord and cauda equina damage in several cases, apparently, from penetration through the intervertebral foramen into the spinal canal. Fatal respiratory paralysis and shock in two **parturients** after spinals (120 mg and 150 mg). Areflexic paraplegia after spinal in a 57-year-old **female**. It is reported to cross-react with "caine"-type local anesthetics, such as ***benzocaine, butacaine, cyclomethycaine***, and ***tetracaine***, as well a topical sunscreens containing ***p-aminobenzoic acid***. Since it is a vasoconstrictor, IV use may cause up to a 50% increase in blood pressure. Transient amaurosis reported after mandibular nerve block. Transient blindness due to suspected intra-arterial injection in dentistry. Anaphylactic and dermatological reactions (usually eczematous and urticarial) reported. Depressed **infants**, by direct drug effect or indirectly, due to **maternal** hypotension, after spinal or epidural procedures on the **mother**. A considerable amount of it is an adulterant in imported Mexican ***heroin***. When injected to relieve local pain in sports injuries, masking the pain permits the **athlete** to stress injured limbs. Fever, chills, vertigo, contact hypersensitivity, sweating, chest pain, dyspnea, and euphoria occasionally reported.

LD_{LO}, **human**: 500 mg/kg; IM TD_{LO}, **woman**: 1.6 mg/kg.

Drug interactions*:** It antagonizes antibacterial activity of ***sulfonamides because bacteria may utilize its metabolite, ***p-aminobenzoic acid***, instead of the sulfa drug. Use with ***ammonium chloride*** increases its urinary excretion; decreased urinary excretion with ***sodium bicarbonate***. It increases ***succinylcholine***-induced apnea, interferes with formation or release of ***acetylcholine***, and increases effects of ***decamethonium***. ***Phenobarbital*** decreases its effectiveness.

Just as in **humans**, neonatal **animals** have low plasma cholinesterase, which helps metabolize procaine. Avoid unnecessary quantities in small **lab animals** (0.4 ml of 2% in **gerbils**, although not fatal, was toxic). Behavioral excitation, noted in **horses** after IV infusion, occurs at plasma levels ~5% of those associated with central nervous system stimulation in **humans**. Use of 0.5–2.0% causes SC inflammatory reactions and significant atrophy of muscle fibers in **rabbits**. **Race horses** given it for leg injuries have often won their races despite leg injuries. After IV infusion of **horses**, plasma levels of 600 ng/ml lead to behavioral excitement, including pawing at the ground, pacing, and deep-blowing respiratory sounds. At plasma levels of 1,500 ng, produces uncontrollable excitement. **Horses** are 20 times more sensitive to its central nervous system effects than **man**. In **cattle**, epidural injection of extremely large doses (100–200 ml 2% solution) lead to tachycardia, hypotension, and tachypnea. Highly toxic to **birds**, even from traces of it in ***procaine penicillin G***. Some ***penicillins*** had extra procaine added. Injections readily cause convulsions in **rats**, **mice**, and **guinea pigs**. Acute IV toxicity in **dogs** varies with the speed of and amount injected.

LD_{50}, **mouse**: 500 mg/kg; SC, 530 mg/kg; IV, 45 mg/kg. IP, LD_{LO}, **rat**: 280 mg/kg; SC, 1.6 mg/kg; IV, 50 mg/kg. SC, LD_{LO}, **cat**: 450 mg/kg; IV, 45 mg/kg. SC, LD_{LO}, **guinea pig**: 400 mg/kg; IV, 75 mg/kg. IV, LD_{LO}, **rabbit**: 30 mg/kg.

PROCARBAZINE

= Ibenzmethyzin = Matulane = MBH = MIH = Natulan = NSC 77,213 = RO 4-6467

***Antineoplastic, cystostatic.* A *methyl hydrazine* derivative.**

Untoward effects: Initially, a high incidence of nausea and vomiting (**patients** have less of these effects later), anorexia, stomatitis, and diarrhea, followed in **some** by pain, fever, weakness, drowsiness, respiratory symptoms and infections, leukemia, facial flushing, pneumonia, tachycardia, hypotension, vertigo, bleeding, leukopenia, thrombocytopenia, eosinophilia, basophilia, anemia, and skin rashes with erythema, papules, and exfoliation. Papilloedema, alopecia, pruritus, angioedema, urticaria, toxic epidermal necrolysis, skin hyperpigmentation, impaired liver function, paresthesias, nervousness, peripheral neuropathy, decreased reflexes, ataxia, euphoria, and confusion. Occasionally tremors, convulsions, granulomatous hepatitis, intestinal ulceration, acute psychosis, amenorrhea, and coma. Rarely, severe infections and leukopenia lead to death. Use in early pregnancy led to an **infant** normal at birth, but, in one case report, the **baby** was born with a hemangioma. It readily crosses the blood–brain barrier and causes confusion and memory impairment in the **elderly**. **Human** exposure can be by ingestion, inhalation, dermal contact, and in its manufacturing or dispensing. Acute myelofibrosis (with hypoplastic anemia and thrombocytopenia) and chromosomal aberrations in a 66-year-old **male** after 6 months of treatment. Some of its metabolites are both carcinostatic and carcinogenic.

Drug interactions: Small malpositioned kidneys in an **infant** whose **mother** also received ***nitrogen mustard*** and ***vincristine***. As a weak ***monoamine oxidase inhibitor***, it interacts with certain drugs and foodstuffs, leading to hypertension. Potentiates ***alcohol, barbiturates, insulin, meperidine, phenothiazines, tricyclic antidepressants***, and ***sympathomimetics***. ***Disulfiram*** (q.v.)-type reaction and central nervous system depression after ingestion of ***ethyl alcohol***. Use with non-prescription ***sympathomimetics*** must be avoided to prevent hypertension. Interaction with ***bananas, cheese, cigarettes, coffee, cola*** drinks, ***desipramine, imipramine, local anesthetics, nortriptyline, protriptyline, sympathomimetics, tea, tricyclic antidepressants***, and ***wine*** can cause a toxic response.

Myelosuppression, leukopenia, and central nervous system and gastrointestinal toxicity in **dogs** and **rhesus monkeys** after oral use for 4 weeks. Some hepatotoxicity was also noted in both species. Used experimentally to induce acute myelogenous leukemia in **rhesus monkeys**. Leukemia, lymphoma, and hemangiosarcoma in **Cynomolgus monkeys** after oral, IV, and SC use. Oral and IP use in **rats** caused cancers of breast, lungs, and blood vessels, and malignant lymphomas; in **mice**, led to leukemia, benign lung tumors, olfactory neuroblastomas, and renal tumors. After use in female **rats**, their **pups** have had limb and eye defects, anencephaly, microencephaly, and other central nervous system defects.

LD_{50}, **rat**: 785 mg/kg.

PROCHLORPERAZINE

= Bayer A 173 = Buccastem = Compazine = Meterazine = Prochlorpemazine = RP 6,140 = SKF 4,657 = Stemetil = Vertigon

***Tranquilizer, antiemetic, anticholinesterase, psychotherapeutic.* The *maleate* is usually used. A *piperazine–phenothiazine* compound. Oral, rectal, and parenteral use.**

Untoward effects: Sedation and hypotension is less than with ***chlorpromazine***, but extrapyramidal effects are more. Dystonia, especially in **children**, particulary ill **children**. Psychoneurologic symptoms include nuchal rigidity, opisthotonus, akathisia, jaw stiffness, inability to close mouth or swallow, angioneurotic edema of tongue, trismus, spasms of extremities, oculogyric crisis, pigmentary retinal changes, lens opacity, myopia, night blindness, drowsiness, photosensitivity, motor dyskinesia, mental confusion, agitation, slurred speech, headaches, hyperhidrosis, cyanosis, urinary retention, prostatic hypertrophy, cholestatic hepatitis, hepatic necrosis, hives, jaundice, constipation, seizures, protrusion of tongue, spontaneous dislocation of mandible, anxiety, hyperextension and rotation of neck, carpopedal spasms, restless leg syndrome, catatonia, impotence in **males**, and decreased libido in **females**. Aplastic anemia in a 42-year-old **female** after ingesting 50–75 mg/day/64 days. Death 26 days later. Neutropenia, eosinophilia, leukocytosis, and leukopenia in many case reports. Bone marrow aplasia, agranulocytosis, thrombocytopenia, and thrombocytopenic purpura. Congenital limb abnormalities may be related to **maternal** use during pregnancy. Found in breast milk. IV push of 50 mg ***diphenhydramine*** has been recommended to reverse its dystonic effects. Toxic serum levels are reported as 0.1 mg% and therapeutic levels as 0.01–0.04 mg%. Others report levels $\frac{1}{10}$–$\frac{1}{5}$ of these.

LD_{LO}, **human**: 500 mg/kg.

Drug interactions: Has an additive effect to ***alcohol, chloral hydrate, chlordiazepoxide, diazepam, glutethimide, haloperidol, meperidine, morphine***, and ***phenobarbital*** sedation. Decreases effects of ***monoamine oxidase inhibitors***. Additive orthostatic hypotension effect when given with ***antihypertensives***. With ***antacids***, decreases absorption; with ***l-dopa***, decreases antiparkinsonian effect; with ***psilocybin***, unusual visual effects and decreases in mydriasis; and with ***phenytoin***, increases ***phenytoin*** intoxication as it decreases ***phenytoin*** metabolizing enzymes. Use can cause interference with urine ***17-hydroxycorticosteroid*** tests and prevent accurate readings.

Epinephrine may further decrease, rather than increase, blood pressure in **animals** on this ***phenothiazine*** derivative. Teratogenic effects have been produced by its administration to pregnant **rats** and **mice** (**rats**: cleft palate, reduced fertility, litter size, and birth weight; **mice**: cleft palate). Albuminuria in 3/7 **dogs**; significant sedation, hypotension, bradycardia, and bradypnea. Chronic use of high levels (200 mg/kg) in **dogs** caused liver and renal pathology. **Dogs** given 25 mg/kg/day for 5 days/week for 13 weeks developed mild depression, xerostomia, diarrhea, and relaxation of nictitating membrane.

LD_{50}, **mouse**: 1.08 g/kg; IV, 46 mg/kg; SC, 500 mg/kg. LD_{LO}, **dog**: 102 mg/kg; IV, 100 mg/kg. LD_{50}, **rat**: 1.8 g/kg; IV, LD_{LO}, **rat:** 20 mg/kg. IV, LD_{LO}, **rabbit:** 5 mg/kg.

PROCYCLIDINE

= *Apricolin* = *Kemadrin* = *Lergine* = *Osnewan* = *Tricoloid* = *Tricyclamol*

Anticholinergic.

Untoward effects: Urinary retention, fecal impaction from atonic colon, constipation, xerostomia, mydriasis, blurred vision, tachycardia, ataxia, dizziness, giddiness, hallucinations, skin rash, increased intraocular pressure, nausea, vomiting, parotitis, and weakness. Occasionally leading to elation and manic states. Abuse of it by teenage **boys** leads to disorientation, agitation, and violence. Chewing of ***betel nuts*** (***Areca***) with its ***arecoline*** content by two schizophrenic **patients** while taking it caused increases in dyskinesias. Can potentiate parasympathetic block by ***atropine***.

Causes hyperthermia in **rabbits** and prolongs ***ketamine*** catalepsy in them.

IP, LD_{50}, **mouse**: 131 mg/kg; IV, 60 mg/kg.

PROFLAVINE

= *3,6-Acridenediamine*

Topical antiseptic dye. **In treatment of herpes genitalis.**

Untoward effects: Dozens of cases of allergic, eczematous, pigmentous, contact dermatitis when used as a preoperative surgical skin preparation.

SC, LD_{50}, **mouse**: 140 mg/kg; IV, LD_{LO}, **cat**: 11 mg/kg.

PROFLURALIN

= *B 4576* = *CGA 10,832* = *ER 5461* = *Pregard* = *Premasone* = *Tolban*

Herbicide.

Untoward effects: May have ***aromatic solvents*** or ***petroleum distillates*** in various formulations. Chemical pneumonitis can occur if aspirated.

Undiluted, it is extremely irritating to **rabbit** eyes; slightly irritating at use concentration (4% w/v).

LD_{50}, **rat**: 2.2 g/kg; acute dermal, **rabbit**: > 10.3 g/kg.

PROGESTERONE

= *Colprosterone* = *Corlutina* = *Corlutone* = *Corluvite* = *Corpomone* = *Corpus Luteum Hormone* = *Cyclogest* = *Flavolutan* = *Gesterol* = *Gestiron* = *Gestone* = *Glanducorpin* = *Glanestin* = *Gonadyl* = *Gynlutin* = *Gynolutin* = *Lipo-Lutin* = *Lucorteum* = *Luteine* = *Luteogan* = *Luteohormone* = *Luteomersin* = *Luteosan* = *Luteostab* = *Luteosterone* = *Lutin* = *Lutocyclin M* = *Lutoform* = *Lutogyl* = *Lutren* = *Lutromone* = *Macrogestin* = *Midgestone* = *Nalutron* = *Neolutin* = *Progestasert* = *Progesterol* = *Progesteronum* = *Progestin* = *Progestogel* = *Progestol* = *Progestone* = *Progestronaq* = *Prolidon* = *Prolutin* = *Proluton* = *Syngenstrone* = *Syngesteron(e)* = *Utrogestan*

Progestogen. **Usually IM. Also SC, IP, rectally, vaginally, and, rarely, orally.**

Untoward effects: Masculinization of the **fetus** has occurred. In 17/21 cases of **female newborns** with pseudohermaphroditism, their **mothers** had been treated with it during pregnancy. **One** had an enlarged clitoris (2 cm in length), enlarged and fused labia majora and her **mother** had received a total of 30 mg in **her** 3rd month of pregnancy. Similarly, some newborn **male infants** have had mild to moderate hypertrophy of penis and scrotum, exceptional muscularity, increased neuromuscular development, and some hyperkinesis, irritability, and gastrointestinal problems. Several case reports of malformed **children** born to **mothers** who had progesterone pregnancy tests. Acne vulgaris, erythema multiforme, autoimmune progesterone dermatitis of pregnancy, chloasma and melasma after sun exposure, edema, weight gain, gynecomastia, breast regression, hirsutism, acute porphyrias, hyperbilirubinemia, headache, exhaustion, fainting, lethargy, depression, abdominal cramps, decreased sperm motility, decreased libido, urticaria, optic neuritis, pruritus vulvae, vaginal secretions, candidiasis, altered menstrual patterns, and increased blood clotting. Pain at injection site. Jaundice, leg weakness, and low-grade fevers have been rare. Long-term use has been suspect in cases of hypertension and pre-eclampsia.

Drug interactions: Use with ***antihistamines***, particularly ***chlorcyclizine***, increases its hydroxylation and may decrease its effectiveness. Similar decrease in effectiveness when ***phenobarbital*** and ***phenylbutazone*** are given due to enzyme induction. Potentiates ***phenothiazine***-type ***antihistamines***. If given before ***lysergic acid diethlamide*** (LSD), it decreases the latter's behavioral effects.

U.S. regulations prohibit any residues in the uncooked edible tissues or by-products of **lambs** or **steers** given the drug in growth-promoting implants. These forms are NOT to be used within 60 days of slaughtering such **animals** for food. Experimentally, it has been used to produce porphyria in **rats** (300 mg/kg IP) and this can be prevented by pretreatment with ***phenobarbital***. Lethargy, obesity, polyuria, and polydipsia has been reported in treated **birds**. The **dog** and the **ferret** are the only known **animal** species where it can produce a substantial lobuloalveolar growth of mammary tissue. In the **rat**, ***chlorcyclizine, chlordane, DDT***, and ***phenobarbital*** stimulate its metabolism by microsomal enzymes. ***DDT*** and ***dieldrin*** do the same in the **pigeon**. Use can cause slight temperature increase in **cattle**. In **mice**, induces mammary carcinomas. Offspring of treated females develop various genital tract lesions in females. In **dogs**, chronic use leads to endometrial and mammary hyperplasia, fibroadenomatous mammary gland nodules, and inhibition of ovarian development. **Rats** and **mice** had increased incidence of mammary and vaginal-cervical tumors when used after certain triggering chemicals.

IV, LD_{50}, **mouse**: 100 mg/kg.

PROLINE

= *Pro*

***Amino acid.* Essential for chicks, but not for people.**

Untoward effects: Hyperprolinemia occurs as a rare inborn error of metabolism, and increased levels then also appear in the urine. Type I is characterized by hereditary renal disease and congenital anomalies of the genitourinary tract, as well as mild mental retardation. Type II has mild mental retardation. It can cause bile duct hyperplasia in **rats**. It is found in high concentrations in *Fasciola*, *Schistosoma*, and all trematodes tested. Implicated in the development of liver fibrosis.

PROLINTANE

= *Katovit* = *Promotil* = *SP 732*

Antidepressant, central nervous system stimulant.

Untoward effects: Headache (14/20), nervousness (12/20), and blurred vision (4/20) after total dose of 20 and 40 mg. Panic state in a 17-year-old **female** with headache, palpitations, nausea, vomiting, sweating, dizziness, and feeling of total paralysis and imminent death after 20 mg/day/3 days.

In **rats**, placental and fetal concentrations are more than maternal concentrations.

LD_{50}, **rat**: 160 mg/kg; **mouse**: 257 mg/kg.

PROMAZINE

= *Intrazine* = *Liranol* = *Pipolphen* = *Prazine* = *Promwill* = *Protactyl* = *Romtiazin* = *RP-3276* = *Sparine* = *Starazin* = *Talofen* = *Tranquazine* = *Tranzine* = *Verophene* = *Wy-1094*

***Tranquilizer, antiemetic.* A non-halogenated *phenothiazine* of the *propyldimethylamine* series of *tranquilizers*. Oral, IV, and IM.**

Untoward effects: Agranulocytosis. A fatal case in a 62-year-old **male** after 1 g during 33 days. Leukopenia, thrombocytopenic purpura, and aplastic anemia with pancytopenia also reported. Localized thrombophlebitis, vascular spasm, and localized cellulitis after IV. Seizure activity in ~1%, especially if dosage is > 1 g/day. **Epileptics** are more apt to have induced seizures. Cholestatic jaundice, liver dysfunction, and skin rashes are rare. It has induced photosensitivity, neuroleptic malignant syndrome, neuromuscular blockade, hypotension, low-grade fevers, exanthema, and pruritus vulvae. Inadvertent intra-arterial injection led to arterial thrombosis with gangrene. Its use in **women** has been associated with congenital dislocation of the hip in the **neonate**. Can cause jaundice in the **neonate**. Hypothermia, increases in serum ***prolactin***, extrapyramidal effects, decreased uterine tone, chemical burns with ulcers on oral mucosa, slurred speech, dizziness, tachycardia, nausea, vomiting, pain at injection site, lens opacity, and miosis are among many symptoms reported occasionally after oral and/or parenteral use. Skin necrosis at IM injection site. Thrombophlebitis occurs readily with undiluted solutions. Fatal overdose in a mentally defective **female** (> 100 × 100 mg tablets) and in **others**. After treatment with a total of 31.7 g during 48 days, a 48-year-old **female** died. Post-mortem revealed pulmonary edema, atelectasis, fatty degeneration of the liver, agranulocytosis, and ulcerative colitis. Toxic at > 0.1 mg% and can be fatal at five times this level.

LD_{LO}, **human**: 50 mg/kg.

Drug interactions: Iatrogenic brain syndrome in a 30-year-old **female** on 50 mg IM every hour and 25 mg ***chlordiazepoxide*** qid, leading to depressed and apathetic; tremendous improvement 1 week after discontinuance of treatment. ***Oral contraceptives*** may decrease its metabolism. Use IV or IM with ***meperidine*** causes dangerous hypotension, shock, and collapse. It can deplete plasma cholinesterase and caused the death of a **farmer** exposed to ***anticholinesterase organophosphorus insecticides***. It caused marked potentiation of ***piperidyl benzilate, pridinol, succinylcholine***, and ***thiopental*** effects. Use it with caution in **patients** receiving ***depolarizing muscle relaxants***. Sudden apnea followed 25 mg IV in a 50-year-old **female** on ***succinylcholine chloride*** infusion. ***Epinephrine*** intensifies its hypotension. Use can interfere with urinary ***17-hydroxycorticosteroid*** test, preventing accurate readings. Large doses can decrease effectiveness of ***heparin***. On occasions, **horses** have shown an allergic response following IV use or some local swellings after IM use. Do NOT use ***epinephrine*** as an antidote to any promazine effects, as it may potentiate the latter's hypotensive effect; use ***norepinephrine***. Avoid rapid IV or arterial injections. Has increased the toxicity of ***organophosphorus*** or other ***acetylcholinesterase inhibitors*** and ***procaine***. Has produced marked arrhythmias in both anesthetized and active unanesthetized **dogs** (of shorter duration when under anesthesia). FDA regulations establish a zero tolerance for its residues on or in the uncooked tissues of food-producing **animals**. Do NOT use on lactating **animals**, and **birds** and **animals** intended for **human** consumption to avoid undesirable ***milk*** or meat residues in countries permitting such use. IM use of commercial solutions has been associated with myositis and neuritis in **rabbits**. An intracarotid injection has been associated with the death of a **horse**. **Rat** pups have increased postnatal mortality after their mothers received it. Phototoxicity in treated **mice**.

LD_{50}, **rat**: 650 mg/kg; **mouse**: 485 mg/kg; IV, **rat**: 32.5 mg/kg; IV, **mouse**: 45 mg/kg; IV, **rabbit**: 21 mg/kg.

PROMECARB

= *Carbamult* = *Minacide* = *SN 34,615*

Insecticide, acaricide. A methyl carbamate cholinesterase inhibitor.

Untoward effects: In **cattle** sprayed one to eight times with a 0.2% emulsion at intervals of 3 days or fed it at 2 ppm or 20 ppm/day/20 days and slaughtered 30 days afterwards, sprayed **animals** had omental fat residues of 0.9–1.5 ppm and 0.4–1.9 ppm of a metabolite. Perirenal fat had residues of 0.8–1.9 ppm and 0.9–1.6 ppm, respectively. Toxic symptoms in **mice** caused decreased movement, increases in fecal and urine elimination, salivation, bristling of hair, excitement, weight loss, and convulsions, followed by death. **Hens** are easily poisoned by single doses of ≥ 60 mg/kg.

LD_{50}, **rat**: 74 mg/kg; IV, 5 mg/kg; IP, 27 mg/kg; skin, 450 mg/kg; **mouse**: 16 mg/kg.

PROMETHAZINE

= *Atosil* = *Diprazin* = *Dorme* = *Duplamin* = *Fargan* = *Fellozine* = *Fenazil* = *Genphen* = *Hiberna* = *Lergigan* = *Methzine* = *Phencen* = *Phenergan* = *Phensedyl* = *Proazamine* = *Procit* = *Promazinamide* = *Prorex* = *Prothazine* = *Provigan* = *Remsed* = *RP-3277* = *RP-3389* = *Tanidil* = *Thiergan*

***Antihistamine, antiemetic, central nervous system depressant, serotonin agonist.* The *hydrochloride* is usually used.**

Untoward effects: ***Blood***: Use during or prior to labor may inhibit **fetal** platelet function. Jaundice and agranulocytosis in a 42-year-old **female** after earlier therapy with ***chlorpromazine***; was fatal. Occasional cases of anemia, aplastic anemia, neutropenia, and thrombocytopenia.

Cardiovascular: Delirium, dysrhythmia, pyrexia, and tachycardia in 5½-year-old **female** after 200–300 mg. Bradycardia or tachycardia, increased or decreased blood pressure, hypotension after parenteral use. Occasionally venous thrombosis at injection site. IA can induce gangrene of the extremity.

Clinical tests: High concentrations have caused false positive ***cannabinoid*** test results. False negative results with ***Prepuerin***® and ***DAP***® pregnancy tests; false positives in ***Gravindex*** test.

Central nervous system: Drowsiness and pronounced sedation are very common; lassitude, tinnitus, dizziness, incoordination, fatigue, blurred vision, diplopia, acute oculogyric crisis, euphoria, nervousness, excitation, restlessness, tardive dyskinesia, acute dystonia, insomnia, tremors, convulsions, grand mal seizures, vertigo, hysteria, catatonic-like states, hyperreflexia, and impaired hand–eye coordination occur. Radiculitis in 2/19 **patients** given spinal anesthesia with it. Occasionally extrapyramidal symptoms. Uncontrollable limb movements and psychosis in **children** and speech problems in **adults**. Dizziness and drowsiness in ~20–25% after 150–250 mg/day.

Fetus and neonate: Use of high dosage in **mothers** may cause **fetal** immunoincompetence and hemolytic disease in the **neonate**. Use possibly associated with sudden infant death syndrome. Reaches equilibrium in **fetus** within 15 min and remains in **fetal** blood for ~4 h. The incidence may not be significant, but includes endocardial fibroelastosis, brachymesophalangy, clinodactyly, microcephaly, cleft palate, and supernumerary phalanx of the fourth digit.

Miscellaneous: Cholestatic jaundice, renal disturbances with urine retention, oliguria, proteinuria and microscopic hematusion, pain at injection site, nausea, vomiting, abdominal pain, nasal stuffiness, dry throat, xerostomia, hypokalemia, restless leg syndrome, lens opacity, decreased sperm motility, transient myopia, akathisia, and may cross-react with ***chlorpromazine*** and ***tripelennamine***.

Poisoning: A 3-year-old **female** ingested 100 ml (1 mg/ml) of the syrup leading to drowsiness and lethargy. Ingestion of eight 25 mg tablets in a 12-year-old **male** led to hospitalization 7½ h later. Symptoms included visual and tactile hallucinations, profound anxiety, extreme hyperesthesia and hyperalgesia. Dermal absorption of 10–15 g of a 2% cream (200–300 mg of drug) on a 16-month-old **male** and a 44-month-old **female** caused central nervous system depression, acute excitomotor reactions, screaming, ataxia, visual hallucinations, and peripheral anticholinergic effects. Topical application of three tubes (90 g) of a 2% cream on a generalized urticaria on a 14-year-old **male** led to unconsciousness. Later symptoms included being dazed, disoriented, agitated with myoclonus, and xerostomia. Electrocardiogram indicated premature ventricular contractions. Anticholinergic adverse effects. Toxic psychosis in a 40-year-old **female**. After ingestion of 20 tablets, a 20-year-old **female** developed drowsiness, weakness, vertigo, and epigastric pain. Apnea reported in young **children** after use in a cough syrup. Therapeutic serum concentration reported as 0.1–0.4 mg/l, and toxic levels as 1–3 mg/l. Lethal levels are ~12–19 mg/l. A 2-year-old child developed agitation, confusion, convulsions, and stupor after ingestion of 250 mg.

Skin: Allergic contact dermatitis, lupus erythematous, photosensitivity, and photoallergic reactions. Facial flushing, and eczematous and papular lesions.

Drug interactions: It can enhance the neuromuscular blockade of ***colistin***, ***kanamycin***, ***neomycin***, and ***streptomycin*** leading to apnea and muscle weakness. The first oral dose of 25 mg to a **patient** with allergic reactions to ***turpentine*** was well-tolerated, but a second dose caused fever, hematuria, anuria, asthma, coma, and death. Induced failure of ***oral contraceptives*** can be due to its stimulation of liver microsomal enzymes. Use with ***meperidine*** or ***morphine*** increases the sedative, but not necessarily the analgesic or respiratory, effects, and the narcotic dosage can be decreased by 25–50%. Large doses can decrease the effectiveness of ***heparin***.

In **animals**, high dosage is sedative. Antagonizes ***acetylcholine*** and ***histamine***. May enhance neuromuscular blockade by ***antibiotics*** (decreases ***acetylcholine*** release), and produces apnea and/or muscle weakness. High environmental temperatures dramatically increase the drug's toxicity in **rats**. Will react with ***polyethylene*** tubing. Parenteral use in the **dog**, **guinea pig**, and **mouse** increase body temperature. It is a potent ***histamine***-releaser after IP use in **guinea pigs** leading to gastric ulcers (125 mg/kg causing 85% ulceration). Use of 250 mg/kg in **rats** led to 25% open palates in fetuses by the 18th day. **Dogs** have shown pain at injections site, disorientation, lack of awareness, and reduced release of ***gastrin***. Topically, it is extremely irritating to a **rabbit's** eye.

LD_{50}, **rabbit**: 580 mg/kg; **guinea pig**: 640 mg/kg. IV, LD_{50}, **rat**: 45 mg/kg; **mouse**: 40 mg/kg; **rabbit**: 19 mg/kg; **guinea pig**: 27 mg/kg. SC, LD_{50}, **rat**: 700 mg/kg; **mouse**: 400 mg/kg; **guinea pig**: 520 mg/kg.

PROMETONE

= *G-31,435* = *Gesafram* = *Methoxypropazine* = *Pramitol* = *Primatol* = *Prometon*

Non-selective herbicide. **For post-emergence use on non-cropped land. A *triazine*.**

Untoward effects: Excess exposure can cause skin and mucous membrane irritation in and is corrosive on the eye of **humans**. Deliberate ingestion of large amounts can be expected to cause nausea, vomiting, abdominal distress, diarrhea, circulatory shock, respiratory depression, and convulsions. Can be absorbed through the skin.

LD_{LO}, **human**: 500 mg/kg.

Cattle were poisoned by a single oral dose of 25 mg/kg or by 10 mg/day/8 days. **Sheep** are less affected, requiring 25 mg/kg/day/4 days or 50 mg/day for symptoms to show. **Chickens** have decreased weight gain after 25–250 mg/kg. Poisoning symptoms in **cattle** and **sheep** are primarily anorexia and salivation. Usual amounts on sprayed hay fields do not adversely affect them. Field spraying has caused edema and stunted growth in hatched **mallard ducks**.

LD_{50}, **rat**: 1.75 g/kg; **mouse**: 2.2 g/kg.

PROMETRYNE

= *Caparol* = *G-34,161* = *Gesagard* = *Prometrex*

Triazine herbicide. **In *celery* and *cotton* fields.**

Untoward effects: Eye and skin irritant. Harmful if absorbed through the skin or inhalation. Protective clothing should be worn by **applicators**.

LD_{LO}, **human**: 500 mg/kg.

Doses of 50 mg/kg/day/2 days required to poison **cattle**. Fatal to them at 100 mg/kg. Fatal to **sheep** after

100 mg/kg/day/6 days. **Chickens** had decreased weight after 50–100 mg/day/10 days and were poisoned by 250 mg/kg/day/2 days. In poisoned **cattle** and **sheep**, anorexia and depression occurred, and, occasionally, incoordination. **Rats** given 500–5,000 mg/kg developed ataxia, stupor, and tachypnea.

LD_{50}, **rat**: 2.1 g/kg; **mouse:** 3.75 g/kg; dermal, **rabbit**: > 1 g/kg.

PROMOXALONE

= *Dimethylane* = *Meproxol* = *Telvol*

Muscle relaxant, central nervous system depressant.

Untoward effects**:** Can cause gastrointestinal upsets and drowsiness in **humans**.

LD_{LO}, **human**: 500 mg/kg.

Has caused transient paralysis in the **mouse**.

PRONAMIDE

= *Kerb* = *Propyzamide* = *RH 315*

Selective pre-emergence herbicide.

Untoward effects**:** Eye, skin, nose, and throat irritant. **Users** must wear protective clothing.

Causes cancer in **mice**.

LD_{50}, **rat** male: 8.35 g/kg; female: 5.62 g/kg.

PRONETHALOL

= *Alderlin* = *ICI 38,174* = *Nethalide*

β-Adrenergic blocker, antiarrhythmic, antianginal, antihypertensive. **Oral and IV.**

Untoward effects**:** Nausea, vomiting, diarrhea, paresthesias, heart-block, dizziness, fatigue, ataxia, skin rashes, headache, sweating, giddiness, insomnia, and tinnitus reported. Small margin between therapeutic and toxic effect dosages.

Potent carcinogen in the **mouse**, but not the **rat**, and, possibly, not the **dog**. High incidence of malignant tumors of the thymus gland in **mice** receiving dietary intake of 0.05–0.2%.

LD_{50}, **rat**: 900 mg/kg; **mouse**: 337 mg/kg, IV, 46 mg/kg.

PROPACETAMOL

= *Pro-Dafalgan* = *UP-34,101*

Analgesic, antipyretic. **Injectable prodrug of** ***acetaminophen.***

Untoward effects**: Nurses** experience an eczematous contact dermatitis from preparing it for injection. Confirmed by patch tests.

PROPACHLOR

= *Bexton* = *CP-31,393* = *Ramrod*

Selective pre-emergence herbicide.

Untoward effects**:** Causes eye, nose, throat, and skin irritation. Allergic contact dermatitis.

Toxic to **fish** and **wildlife**.

LD_{50}, **rat**: 710 mg/kg; **mouse**: 290 mg/kg; **rabbit**: 710 mg/kg; skin, **rabbit**: 380 mg/kg.

PROPAFENONE

= *Arhythmol* = *Pronon* = *Rhythmol* = *RO 2-5803* = *Rytmonorm*

Antiarrhythmic. Class IC. **Weak *β-adrenergic blocker* and *calcium channel blocker.***

Untoward effects**:** Gastrointestinal, central nervous system, and cardiovascular. Dysgeusia with metallic or bitter taste in 15–20%. Several case reports of mania, disorientation, dizziness, incoordination, tremors, delusions, paranoia, agitation, anorexia, nausea, vomiting, xerostomia, blurred vison, drug fever, constipation, headache, paresthesias, fatigue, sweating, acute pleuritis, myoclonus. Increased alkaline phosphatase, aspartate aminotransferase, alanine aminotransferase, and γ-glutamyl transferase, indicative of liver injury. Cardiac conduction failures (AV-block or bundle branch-block), exacerbation of congestive heart failure, ventricular arrhythmias and tachycardia, premature ventricular contraction increase, palpitations, hypotension, widening of QRS interval, and bradycardia. Rarely, bronchospasm and asthma (risk increased with doses > 400 mg/day). Coughing and wheezing in a 50-year-old **female**, 1 week after starting on 150 mg tid. Lupus erythematous-like syndrome in a 63-year-old **male** after ingestion of 300 mg tid. Fatal ventricular tachycardia in a 63-year-old **patient** after two 150 mg doses at an 8 h interval. Transient global amnesia in a 61-year-old **male** after 150 mg tid/6 days. Amnesia resolved 6–7 h after discontinuing therapy. Severe poisoning in a 24-year-old **female** 1 h after intentionally swallowing 2.7 g caused convulsions and cardiac arrhythmias.

Therapeutic serum levels reported as 0.06–1 μg/ml; toxic in **many** at 2–3 μg/ml, and fatalities can occur at 7.7 μg/ml.

Drug interactions**:** Causes significant increases in ***digoxin*** blood levels, usually up to 37% on 450 mg/day and 85% on 900 mg/day, requiring a reduction in the latter's dosage. Use with ***warfarin*** produces substantially enhanced anticoagulant effect. In some **people**, ***quinidine*** can increase its plasma concentration by > two times. It doesn't have a significant effect on ***lidocaine*** metabolism, but the combination causes serious central nervous system toxicity. Use with ***food*** increases its bioavailablity.

Its plasma concentrations are decreased when used with ***rifampin***. It decreases ***theophylline*** clearance, possibly, from enzyme inhibition. ***Cimetidine***, in limited trials, increased its plasma levels by ~20%. ***Lopinavir*** and ***ritonavir*** decrease its metabolism and may increase its toxicity. ***Fluoxetine*** impaired the metabolism of its enantiomers in healthy **Chinese volunteers**.

Large IV doses cause reversible reduced spermatogenesis in **rabbits**, **dogs**, and **monkeys**, and had a similar effect in some **human** volunteers.

PROPAMIDINE

= *Brolene* = *M&B 782*

Protozoacide, antiseptic, fungicide.

***Untoward effects*:** Contact dermatitis in **humans**.

PROPANE

= *Bottled Gas* = *Dimethylmethane* = *Propyl Hydride*

Untoward effects*:** Although classified as a high-pressure propellent with intermediate toxicity, it has caused serious toxicity and deaths after inhalation in **abusers**, often **teenagers**. Inhalation caused ventricular tachycardia in a 2-year-old **child** from an aerosoled deodorant. Dizziness, confusion, and excitation occur, in addition to asphyxiation. Dizziness within a few minutes after exposure to 10% concentration in air. Chemical pneumonitis from its fumes as a propellent in ***Wilson's Leather Protector® in ~550 cases reported in 1992. Narcosis at higher concentrations. Frostbite from the liquid. It is a strong oxidizing agent and flammable; air concentrations of 2.2–9.5% are explosive. National Institute of Occupational Safety and Health/OSHA upper exposure limits of 1,000 ppm (1,800 mg/m^3). LEL of 2,100 ppm.

Early respiratory depression in the **monkey** after 10% propane by inhalation. Propane sensitizes a **dog's** heart to the adverse effects of ***epinephrine***.

1,3-PROPANE SULTONE

***Untoward effects*:** Potential **human** exposure is by inhalation and accidental ingestion in chemical **workers** making it or using it in chemical synthesis, and making detergents, lathering agents, bacteriostats, and corrosion inhibitors.

It increased the incidence of astrocytomas (malignant gliomas) in **rats**, mammary adenocarcinomas in female **rats**, granulocytic leukemia, small intestinal adenocarcinomas, and squamous cell carcinomas of the ear. Eye and skin irritant in **animals**.

TD_{LO}, **rat**: 1.05 g/kg, over 35 weeks.

PROPANIDID

= *BA 1420* = *Bayer 1420* = *Epontil* = *FBA-1420* = *Propantan* = *Sombrevin* = *2180 TH*

***Anesthetic.* IV use.**

Untoward effects*:** Marked vasodilation, acute hypotension, arrhythmias, placental passage, anaphylactoid reactions, direct skin sensitivity, and acute porphyrias are among the many adverse effects, some of which are due to the solvent ***Cremophor EL (q.v.). Occasional seizures. Prolongs ***succinylcholine*** paralysis by 50%. Anticholinesterases (***ecothiopate*** eye drops, many ***pesticides***), ***cyclophosphamide***, and ***phenelzine*** can prolong the duration of its anesthesia. Many hypersensitivity reactions. After 1983, withdrawn from most of the world's markets.

IV, TD_{LO}, **human**: 5 mg/kg.

Marked excitation in **dogs**.

LD_{50}, **rat**: 700 mg/kg.

PROPANIL

= *Chem Rice* = *DPA* = *FW-734* = *Rogue* = *S 10,165* = *Stam* = *Stampede* = *Surcopur*

Herbicide, nematocide.

***Untoward effects*:** High incidence of chloracne in pesticide plant **workers**. Eye and skin irritant.

Industrial plants manufacturing it have contaminated the environment with ***hexachlorobenzene*** (q.v.), with a potential for entering the food chain.

Spraying of **mallard duck** embryos led to limb and neck abnormalities in survivors. Oral drenching of 1-year-old **water buffalo calves** with 100 or 250 mg/kg/day led to central nervous system excitement. Post-mortem revealed vascular and retrogressive changes in liver, heart, kidney, intestines, and brain. Toxic to **fish**. Hepatotoxicity in exposed **rats**.

LD_{50}, **rat**: 560 mg/kg; **dog**: 1,217 mg/kg. LC_{50} for various **test** fish: 3.8–7.6 mg/l. LC_{50} in the diet of **quail**: 2,294 ppm.

PROPANTHELINE BROMIDE

= *Banlin* = *Corrigast* = *Eacoril* = *Ercotina* = *Neo-Banex* = *Neo-Metantyl* = *Norpanth* = *Pantheline* = *Pro-Banthine* = *Ropanth*

Anticholinergic.

Untoward effects*:** Xerostomia, blurred vision, cycloplegia, mydriasis (unilateral in two cases; one after splashing it on one eye and possibly from hand contact in the other), increased intraocular pressure, impotence (in a 38-year-old **male** after six tablets/day/1 year), decreased libido, tachycardia, palpitations, constipation, flatulence, paralytic ileus, urinary retention, ageusia, headache, nervousness, confusion, drowsiness, decreased lactation, epistaxis, decreased esophageal tone, and exfoliative dermatitis of skin, mouth, and tongue; papular lesions on face, neck, and upper chest of a 32-year-old **female** receiving six tablets/day (probably due to its ***halogen content). Overdosage often causes central nervous system upsets, respiratory depression, paralysis, and coma. A 2-year-old child ingested 375 mg leading to extensive rash, increased blood pressure for 8 h, tachycardia and mydriasis for 20 h, despite having gastric lavage 1 h after symptoms were noted.

Drug interactions*:** Use with slow-dissolving ***digoxin tablets may increase ***digoxin*** serum levels and, in turn, the latter's toxicity. May decrease absorption and/or rate of absorption of ***alcohol, acetaminophen, cimetidine, diazepam, l-dopa, lithium, pivampicillin, propranolol, sulfamethoxazole***, and ***tetracycline***. Enhances mydriasis and bronchial relaxation production by ***sympathomimetics***. Its absorption is enhanced by ***sodium bicarbonate*** and ***foods***. Increases ***nitrofurantoin*** absorption. Delays the absorption of ***riboflavin***, but increases the total amount absorbed.

Dry mucous membranes, increased intraocular pressure, constipation, vomiting, decreased intestinal motility, ileus, tachycardia, sialism, urinary retention, and pupillary dilation occur from its anticholinergic action in **cats** and **dogs**. Slows the absorption of ***carvedilol*** in **rats**. Excessive use can cause bromism.

LD_{50}, **rat**: 370 mg/kg; **mouse**: 445 mg/kg; **rabbit**: 750 mg/kg.

PROPARACAINE HYDROCHLORIDE

= *Ak-Taine* = *Alcaine* = *Ophthaine* = *Ophthetic*

***Topical ophthalmic anesthetic.* A non-para-aminobenzoic ester.**

***Untoward effects*:** Hypersensitivity in ~1/1,000 **patients** causing contact dermatitis, dryness, fissuring, and bleeding of fingertips in one **male** and peeling and scaling in two **who** handled it during intraocular pressure testing of **animals**. Keratitis after its ophthalmic use in 24-year-old **female**. May cause sloughing of corneal epithelium.

PROPARGITE

= *BPPS* = *Comite* = *Cyclosulfyne* = *DO-14* = *ENT-27,226* = *Omite* = *Propargil*

***Acaricide.* It is a non-cholinesterase-inhibiting miticide.**

Untoward effects*:** Second most frequently reported pesticide-induced dermatitis in California agricultural **workers**. Severe dermatitis in **orange-pickers**, ranging from 23% (6/26) to as high as 78% (28/36) in six work crews affected (58% – 114/198). Because of **animal** carcinogenicity, Environmental Protection Agency in 1996 banned its use on dried ***figs and dried ***tea***. Corrosive to eyes. Inhalation can be fatal. **Users** must wear protective clothing, face masks, and goggles. Flammable.

Toxic to **fish** and **shrimp**.

LD_{50}, **rat**: 1.48 g/kg; dermal, 250 mg/kg.

PROPAZINE

= *Gesamil* = *Milogard* = *Primatol P* = *Prozinex*

***Herbicide.* A *triazine*.**

***Untoward effects*:** LD_{LO}, **human**: 500 mg/kg.

Oral intake of 25–45 mg/kg/30 days in **pigs** led to poor weight gain, catarrhal or hemorrhagic enteritis, lymphocytosis, and monocytosis. **Sheep** show depression and anorexia, and slowly recover. **Heifers** show salivation, diarrhea, anorexia, and depression.

LD_{50}, **rat**: > 5 g/kg.

PROPERICIAZINE

= *Aolept* = *Neulactil* = *Neuleptil* = *Periciazine* = *Pericyazine* = *RP 8909* = *SKF 20,716*

***Psychotherapeutic.* Oral and IM.**

***Untoward effects*:** Orthostatic hypotension, sedation, extrapyramidal effects, mild Parkinsonism, dyskinesia, dystonia, fatigue, headache, muscular rigidity, dizziness, nausea, vomiting, heartburn, akathisia, drooling, impotence, agitation, tremors, slight decrease in blood pressure, restlessness, oculogyric crisis, nasal congestion, corneal opacity, tachycardia, leg cramps, xerostomia, and pruritus. Therapeutic serum levels 0.005–0.05 mg/l; toxic at 0.1 mg/l.

Strong mammotropic action in **cows**.

LD_{50}, **rat**: 395 mg/kg; **mouse**: 530 mg/kg.

PROPETAMPHOS

= *Blotic* = *Safrotin* = *SAN 3,221*

***Organophosphorus ectoparasiticide.* Potent cholinesterase inhibitor.**

***Untoward effects*:** Use in the U.S. is limited to licensed professional **applicators** for premise treatment against fleas, ticks, and roaches, because of its strong anticholinesterase activity.

LD_{50}, **rat**: 75 mg/kg.

PROPHAM

= *Chem-Hoe* = *IFK* = *INPC* = *IPC* = *isoPPC* = *Tuberite*

Soil herbicide. A carbamate.

Untoward effects: After large single oral dose, 3-month-old **mallard ducks** developed goose-stepping ataxia, falling when walking, walking with the aid of wing tips, high carriage, and regurgitation persisting for 2–4 weeks. Found in the egg yolks of exposed **hens**.

LD_{50}, **rat**: 1–9 g/kg; **mouse**: 940 mg/kg; **mallard ducks**: > 2 g/kg.

β-PROPIOLACTONE

= *Betaprone* = *BPL* = *2-Oxetanone*

***Sterilant.* For vaccines, tissue grafts, surgical instruments, blood plasma, and as vapor-phase disinfectant in enclosed spaces. Also as a chemical intermediate.**

Untoward effects: High level of exposure leads to skin irritation, blistering, burns, hair loss, scarring, corneal opacity (which can be permanent), pollakiuria, dysuria, and hematuria. May cause significant decrease in tensile strength of **human** pericardium strips.

By gavage in **rats**, caused squamous cell stomach carcinomas. Topically, caused papillomas and squamous cell carcinomas in **mice**; papillomas, melanomas, keratocanthomas, and squamous cell carcinomas in male **hamsters**. After SC in **rats** and **mice**, caused injection-site sarcomas, fibrosarcomas, adenocarcinomas, and squamous cell carcinomas in female **mice**. Keratocanthomas and a melanoma after application to the skin of **guinea pigs**. Oral or IP in **rats** caused muscle spasms, dyspnea, convulsions, and death. IV in **rats** led to kidney tubule and liver damage. IP in suckling **mice** led to lymphatic tumors and liver cancer.

PROPIOMAZINE

= *CB 1678* = *Dorevan* = *Indorm* = *Largon* = *Propavan* = *WY 1,359*

***Tranquilizer, sedative, hypnotic. A phenothiazine.* Usually IM or IV.**

Untoward effects: Mild tachycardia after IV. Severe prolonged hypotension in 2/365. Severe hypotension after sudden tilting to 60° in 3/9. Pain at IM injection site, dizziness, tachycardia, nausea, vomiting, and hypotension as preoperative medication. Tardive dyskinesia and restless leg syndrome also reported. **Mothers** taking it during pregnancy bore **infants** with an insignificant increased incidence of malformations, and, if taken during the last trimester, some of their **neonates** showed extrapyramidal symptoms. Potentiates ***thiopental***.

LD_{LO}, **human**: 50 mg/kg.

PROPIONIC ACID

= *Interprop* = *Luprosil* = *Methylacetic Acid* = *Mono Prop* = *Propanoic Acid*

***Preservative, mold-inhibitor.* In hay, grain, and food preservation, and in making *non-steroidal anti-inflammatory drugs*. > 200 million lbs used annually.**

Untoward effects: Can cause severe eye, skin, nose, and throat burns; and blurred vision. Ingestion can cause abdominal pain, nausea, and vomiting.

Fed to **rats** caused hyperplastic ulcers and papillomas of stomach.

LD_{50}, **rat**: 4.29 mg/kg.

PROPIONITRILE

= *Ethyl Cyanide*

Untoward effects: Poisonous when heated or on contact with acids liberating ***cyanide*** (q.v.). Fumes can cause eye, skin, and respiratory tract irritation. Dermal absorption or ingestion lead to nausea, vomiting, chest pain, weakness, stupor, and convulsions.

Ingestion has caused duodenal ulcers in **rats**. Liver and kidney damage reported in **animals**.

LD_{50}, **rat**: 39 mg/kg; Acute inhalation LC_{50}, **mice**: 153 ppm.

PROPIONYLPROMAZINE

= *1497 CB* = *Combelen* = *Propiopromazine* = *Tranvet*

Tranquilizer.

Untoward effects: Long-lasting priapism in **horses** after injections have lasted, in some cases, for ~2 years.

LD_{50}, **rat**: 1.1 g/kg; IP, 160 mg/kg; **mice**: 650 mg/kg; IP, 170 mg/kg.

PROPIRAM FUMARATE

= *Algeril* = *BAY 4,503* = *Dirame* = *FBA 4,503*

Narcotic analgesic.

Untoward effects: Sleepiness, sweating, nausea, and headache. Occasionally vertigo, flushing, vomiting, grogginess, or allergic skin reactions.

LD_{50}, **rat**: 1657 mg/kg; **mouse**: 874 mg/kg.

PROPOFOL

= *Diprivan* = *Disoprofol* = *ICI 35,868* = *Rapinovet*

***IV anesthetic.* An emulsion containing *glycerol, egg lecithin*, and 10% *soybean oil*.**

Untoward effects: In ~50% of **patients**, pain or discomfort on injection unless premedication is given. Residual respiratory depression can last 30–120 min after surgery. In **children**, cardiac arrest, hypotension, and oliguria have occurred. Pediatric **patients** aged from 4 weeks to 6 years developed metabolic acidosis, bradycardia, heart failure, and death. IV infusion of a 5% emulsion in a 66-year-old **male** caused erythema of chest and then the whole body within 2 min. Also laryngeal and facial edema after the first milliliter. Ventricular tachycardia in a 39-year-old **male** after 10 mg. Sexual disinhibition. A 48-year-old **male** given 30 mg developed recurrent, amorous, disinhibited behavior and hallucinations with sexual thoughts. Green urine containing ***phenols*** 3 days after infusion. Apnea, acute bronchoconstriction, and bradycardia can be severe. In older **patients**, blood pressure has decreased > 40 mmHg. Clusters of **patients** with fever or infections were traced to failures of meticulous sterility before its administration. Anaphylaxis, angioedema, flushing, diaphoresis, and phlebitis reported. Prolonged use may have induced pancreatitis. High dosage increased amylase and lipase values. Adverse cardiovascular effects occur mostly in the **elderly**, **children**, and cardiac **patients**. Acute dystonia in an **epileptic** 16-year-old **male** undergoing surgery. Placental transfer occurs.

Drug interactions: Its use has reversed ***warfarin's*** anticoagulant effect, due to its ***soybean oil*** content. Motor neuron disease reported in a 71-year-old **female** on ***enflurane/ propofol*** who consumed > 70 g ***alcohol***/day. Risk of dystonia increases in **patients** taking ***phenytoin***.

Transient, but profound, decrease of cardiopulmonary function in **cats**. Bradycardia, central nervous system stimulation, opisthotonus, seizures, hypotension, respiratory depression, tachypnea, salivation, cyanosis, emesis, and panting in **dogs**. During recovery, tremors, face-rubbing, paddling movements, and vocalization in **dogs**. Rapid single or repeat bolus IV use in **dogs** can cause apnea, and hypotension. Undesirable incidence of tachypnea and apnea reported as 43% and 26%, respectively, in clinical use on **dogs**. Anaphylactoid response reported in **pigs**.

PROPOLIS

= *Bee Glue*

Production by *Apis mellifera* bees from a gummy resin from tree buds. Their digestive enzymes then convert it to propolis.

Untoward effects: **Bee-masters** develop a dermatitis from collecting it for the cosmetic industry.

LD_{50}, **mouse**: > 7.34 g/kg.

PROPOXUR

= *Aprocarb* = *Bay 9,010* = *Bay 39,007* = *Baygon* = *Bifex* = *Bolfo* = *Blattanex* = *ENT 25,671* = *Invisi-Gard* = *OMS 33* = *Para-Ban* = *Propyon* = *Sendran* = *Suncide* = *Tendex* = *Unden*

***Insecticide*. A *carbamate*. The term *Unden* is also used for *estrone*.**

Untoward effects: Most bone marrow failures and dystonias are due to exposures jointly with other insecticides. A cholinesterase inhibitor. Miosis, blurred vision, diaphoresis, nausea, vomiting, weakness, abdominal cramps, diarrhea, salivation, headache, and muscle twitching. National Institute of Occupational Safety and Health sets upper exposure limit as 0.5 mg/m^3.

LD_{50}, **rat**: 83 mg/kg; **mouse**: 37 mg/kg; **guinea pig**: 40 mg/kg; **pheasant**: 20 mg/kg; **mallard ducks**: 7.38–14.6 mg/kg; **chicks**: 46.5 mg/kg; **pigeons**: 60 mg/kg; **quail**: 28 mg/kg; **bullfrog**: 595 mg/kg; and **goats**: > 800 mg/kg.

PROPOXYPHENE

= *Abalgin* = *Antalvic* = *Darvon* = *Dextropropoxyphene* = *Deprancol* = *Develin* = *Dolene* = *Dolocap* = *Dolotard* = *Depronal* = *Erantin* = *Femadol* = *Harmar* = *Propoxychel* = *Proxagesic* = *SK 65*

Narcotic analgesic.

Untoward effects: Cutaneous reactions, addictions, and abuse often in combination with ***alcohol*** or ***psychotropics***; depression, agranulocytosis, thrombocytopenic purpura, increased lactate dehydrogenase, and hypoglycemia. A 22-year-old **male** swallowed 130 mg every 2 h for 6 days, instead of the prescribed 65 mg qid leading to tinnitus, deafness, and ataxia. Structurally related to ***methadone***. A hospitalized 40-year-old **female** was found by the lab to be ***methadone***-positive, apparently from taking eight 65 mg capsules/day until the time of admission. Withdrawal symptoms included shivering, diarrhea, pain, and fear. **Neonates** whose **mothers** took it during pregnancy developed irritability, high-pitched cry, restlessness, ravenous appetite, vomiting, diarrhea, nasal congestion, tremulousness, hyperreflexia, hypertonicity, and some tachypnea. Breast milk from treated **mothers** can give an **infant** up to 1 mg/day. Treated **mothers** can have an increased incidence of omphalocele, defective anterior abdominal wall, diaphragmatic defect, congenital heart disease, and congenital dislocated hips in their **infants**. Pulmonary edema occurs rarely, even with overdoses. Overdoses cause (within 30 min to 1 h) miosis and intractable seizures with knee jerks. Within minutes nausea, vomiting, drowsiness, gasping, respiratory depression, dizziness, headache, and coma occur. Hepatotoxicity, increased alkaline phosphatase, acute cholestasis, jaundice, brown urine, portal inflammation, periorbital edema, pruritus of hands and feet, erythema multiforme, anemia, hemolytic anemia, disseminated intravascular coagulation, leukopenia, neutropenia, thrombocytopenia, polyuria, and rhinitis have also occurred. A potentially fatal case in a 4-year-old **male** led to severe respiratory depression, convulsions, and coma after inadvertently swallowing 585 mg. Near-fatal poisoning in a 2-year-old **male** after ingesting 260 mg. A low toxic dose of 160 mg reported in a comatose 3¼-year-old **female. Juveniles** represent a large group who deliberately overdose. A 15½-year-old **female** ingested 88 ***Darvon 65*** capsules (also contains ***aspirin*** and ***caffeine***) in an attempted suicide, leading to complete cardiorespiratory arrest, bilateral basilar bronchopneumonia, and tonic–clonic seizures with residual brain damage after treatment. Bilateral optic atrophy and spasticity of all limbs 6 months later. Case reports of IV poisonings include symptoms of pulmonary edema, intravascular hemolysis, disseminated intravascular coagulation, and renal failure. IA injections in **drug-abusers**, while trying to find a vein, causing instant severe pain with blanching, cyanosis, swelling, and occasionally gangrene. A popular dangerous scenario among thrill-seeking **drug-abusers** was to take 8–12 capsules with a fifth of cheap ***wine***. Early signs of poisonings often were bradycardia, respiratory depression, miosis, convulsions, and coma. Deaths from overdoses can occur early, without any of these symptoms. In Scotland, **women** aged 31–50 were more frequently associated with deaths from it. It has been estimated that up to 2,000 deaths/year occur in the U.S. from it alone or with other drugs. Deaths have occurred with amounts as low as 25–30 mg/kg. Various authors cite different figures, but, in general, serum levels of 5–20 μg% are therapeutic; 30–60 μg% are toxic, and 80–200 μg% has been lethal. Slow-release forms can be especially dangerous, because **patients** can take excessive amounts, seeking quicker therapeutic results.

TD_{LO}, **human**: 20 mg/kg.

Drug interactions: Use with ***acetaminophen*** (in ***Distalgesic***®) has caused aplastic anemia, pancytopenia, and hemolytic anemia. A 38-year-old **female** became psychotic after overdosing with 20 ***Distalgesic***® tablets. A 27-year-old **female**, after taking 25–30 ***Distalgesic***® with half a bottle of ***sherry*** collapsed with miosis, and cessation of respiration; treated and recovered. Ingestion with excessive amounts of ***alcoholic beverages*** can cause coma and even death. Similar results can occur if ingested with ***alprazolam, antidepressants, sedatives***, or ***tranquilizers***. It can decrease the metabolism of ***phenobarbital*** and ***carbamazepine***, increasing their serum concentration by 45–77% increasing their toxicity. ***Doxepin*** serum levels increase more than 50% after receiving it. ***Antipyrine*** levels are increased by it and ***antipyrine*** half-life is prolonged by ~20%. ***Amphetamine*** and other ***central nervous system stimulants*** are contraindicated, as they may exacerbate the convulsions. ***Orphenadrine*** may have additive central nervous system (anxiety, tremors, mental confusion) effects. May have increased toxicity in combination with ***propranolol***. Less effective in heavy **smokers**. In some **patients**, use with ***warfarin*** has increased the latter's serum level, as both are metabolized by the same hepatic microsomal enzyme system. ***Warfarin's*** anticoagulant effect was definitely increased by use with ***Distalgesic***®. Absorption is increased by ***food*** intake. Induced asthma may occur in cross-**reactors** with ***sodium benzoate*** or ***tartrazine***.

Hepatic enlargement with fatty liver changes in **rats**. Congenital malformation in **hamster** fetuses increased with increasing dosage. It increased ***hexobarbital*** sleeping time in **rats, dogs**, and **rhesus monkeys**. Its metabolite, ***norpropoxyphene***, can cause heart-block in anesthetized **dogs**. Administration to male **rats** prior to mating led to decreased survival in their offspring.

LD_{50}, **rat**: 18 mg/kg.

PROPRANOLOL

= *Anapryline* = *Angilol* = *Apsolol* = *Avlocardyl* = *AY-64,043* = *Bedranol* = *Beprane* = *Berkolol* = *Beta-Neg* = *Beta-Tablinen* = *Beta-Timelets* = *Cardinol* = *Caridolol* = *Deralin* = *Dociton* = *Duranol* = *Efektolol* = *Elbrol* = *Euprovasin* = *Frevken* = *Inderal* = *Indobloc* = *Intermigran* = *Kemi S* = *Obidan* = *Obsidan* = *Oposim* = *Prano-Puren* = *Prophylux* = *Propranur* = *Pylapron* = *Rapynogen* = *Reducor* = *Sagittol* = *Servanolol* = *Sloprolol* = *Sumial* = *Tesnol*

β-Adrenergic blocker, antihypertensive, class II antiarrhythmic, and antianginal. **Usually oral and IV.**

Untoward effects: Nausea, vomiting, diarrhea, fatigue, headache, fainting, and dizziness have been common complaints.

Blood: Agranulocytosis, especially at high dosage. Decreases high-density lipoprotein by ~13%, increases triglycerides by ~24%, increases serum ***uric acid*** by ~10%, decreases platelet aggregation, increases plasma ***potassium***, decreases serum ***melatonin*** and ***triiodothyronine*** concentrations, decreases affinity of hemoglobin for ***oxygen***, eosinophilia, leukocytosis, neutropenia, and hypoglycemia (especially in **diabetics** and pediatric **patients**).

Cardiovascular: Bradycardia, congestive heart failure, heart block, hypotension, cold extremities, Raynaud's phenomenon, paresthesias of the hands, thrombocytopenic purpura, angina, and atrial fibrillation and insufficiency. Sudden withdrawal of treatment can markedly increase the frequency and severity of angina and, in some cases, cause myocardial infarction, serious arrhythmias, or sudden death. In hypertensive **patients** with borderline cardiac compensation, it can aggravate or precipitate heart failure. The **Chinese** may only require one-tenth to a half the dosage of U.S. **men**. Leukocytoclastic vasculitis in alcoholic 69-year-old **male** given 40 mg twice daily.

Clinical tests: Causes clinical confusion due to giving falsely raised ***bilirubin*** concentration with diazo reaction tests.

Central nervous system: Light-headedness, hallucinations (mostly visual), and illusions in 11/63 on ~200 mg/day, ataxia, may unmask or aggravate myasthenia gravis, memory loss, paresthesias, induces depression, forgetfulness, insomnia, weakness, fatigue, nightmares, acute psychosis with paranoia, combativeness, confusion, disorientation, seizures, and schizophrenoid effects. Occasionally a ***curare***-like effect.

Eye: Gritty feeling, soreness, redness, photophobia, stinging, corneal perforation, and internuclear ophthalmoplegia. Blindness reported in several **patients** after treatment of malignant hypertension.

Fetus and neonate: Intake during pregnancy associated with intrauterine growth retardation, neonatal respiratory depression (apnea), bradycardia, floppiness, postnatal hypoglycemia, impaired drug metabolism, small placenta, polycythemia, poor responses to anoxic stress (usually requiring a few minutes of ***oxygen*** after birth), increases in **fetal** or **neonatal** deaths, vomiting, loose stools, and Fallot's tetralogy. It appears in breast milk of treated **mothers** occasionally at levels comparable or more than in maternal plasma.

Gastrointestinal: Nausea, vomiting, abdominal cramping, ischemic colitis, xerostomia, mesenteric thrombosis, brownish discoloration of tongue, constipation, or diarrhea, and indigestion.

Hypersensitivity: Severe anaphylaxis, angioedema, urticaria, profound shock, and bradycardia. Pharyngitis, fever, erythematous rash, pulmonary edema, dyspnea, bronchospasm, laryngospasm, respiratory distress, and fever with sore throat.

Liver: Hepatic encephalopathy in cirrhotic 69-year-old **female**.

Miscellaneous: Although effective against migraines in some **people**, it can cause headaches in **others**. Peyronie's disease, retroperitoneal fibrosis, Dupuytren's contracture, sclerosing peritonitis, hyperglycemia, ***sodium*** retention, weight gain, muscle cramps, and ototoxicity reported. Hypoglycemia with sweating, tachycardia, and myoclonic jerks in **children** taking 1 mg/kg every 6 h. ***β-Blockers*** deplete liver ***glycogen*** stores. Erectile dysfunction in 22/50 **males** after 3–72 months of therapy and decreased libido in **males** and **females**. Myotonia, myasthenia, thyrotoxicosis, lupus erythematous-like reactions, and, rarely, systemic lupus erythematosus.

Poisoning: Overdoses and suicide attempts with it reported. Therapeutic levels reported as 2.5–20 μg% and lethal at ≥ 0.8–1.2 mg%. Suicide attempts have often been with 1.2–7.2 g. Symptoms include generalized intermittent convulsions, bradycardia, weak pulse, severely decreased blood pressure, shock, cyanosis, and supraventricular dysrhythmia. A 47-year-old **female** was intoxicated by 240 mg/day leading to hypotension, coma, and seizures.

Renal: Can cause acute interstitial nephritis and proteinuria.

Skin: Psoriasiform cutaneous eruptions, erythema multiforme, alopecia, cheilostomatitis, peripheral skin necrosis, thrombocytopenic and non-thrombocytopenic purpura, abnormal pigmentation, and peeling of light-exposed areas, and pruritic maculopapular eruptions on limbs, torso, and scalp. Cross-sensitivity with other ***β-blockers*** reported.

Drug interactions: It has completely inhibited all the cardiac and metabolic responses to ***albuterol***. A large intersubject variation exists, but, in general, its maximum plasma levels are increased and its liver metabolism is increased by ***ethyl alcohol***. It can attenuate or antagonize some responses to ***aminophylline*** therapy. Antagonizes the stimulation and other effects of ***amphetamine***, but potentiates its anorexigenic effect. Decreases renal clearance of ***antipyrine***; increases rate of absorption of ***acetaminophen*** and decreases bioavailability if given with ***aluminum hydroxide*** gel. Its hemodynamic effects can be blocked by IV ***atropine***. ***Neostigmine*** can cause severe bradycardia with it. Use with ***anesthetics***, such as ***chloroform***, ***cyclopropane***, ***ether***, ***halothane***, and ***methoxyflurane*** can increase its adverse effects. Drug interactions or antagonism at the same effector site reported for its use with ***catecholamines*** and ***adrenergics***, especially ***epinephrine*** and ***isoproterenol*** leading to bradycardia, atrioventricular block, hypertension, and stroke. ***Cimetidine*** significantly decreases liver blood flow, decreases liver microsomal enzyme activity and clearance of oral and IV ***propranolol***. The 95% increase in ***propranolol*** serum levels can increase its toxicity. ***Ranitidine*** showed no effect on its blood levels. Renal failure reported after ***antihypertensives***, such as ***propranolol***, added to ***cisplatin*** therapy. May interfere with the hypotensive effect of ***guanethidine***. Use with ***clonidine*** has induced a paradoxical hypertension and death. Prolonged postoperative curarization in two thyrotoxic **patients** receiving ***curare*** and 120 mg/day ***propranolol***. Prolongs the effect of non-depolarizing ***muscle relaxants***, such as ***atracurium***, ***pancuronium***, and ***vecuronium***. Use with ***calcium channel blockers***, such as ***diltiazem***, can cause severe hypotension, myocardial infarction, and death. Use with ***nimodipine*** has decreased the latter's peak plasma level by ~25%, while ***nisoldipine*** has increased ***propranolol's*** maximum concentration by ~50%. Adverse effects increase in **patients** with poor left ventricular function when ***verapamil*** is added to therapy. May increase blood levels of ***chlorpromazine*** and enhance the latter's antipsychotic effect. ***Cigarettes*** and ***tobacco*** smoking, by enzyme stimulation, decrease its half-life and serum concentration. ***Digitalis*** and ***digoxin*** intoxication with enhanced bradycardia, atrioventricular block, shock, and death in **patients** shortly after IV propranolol. Enhances ***l-dopa***-induced plasma growth hormone levels. ***Etintidine*** decreased its elimination and prolonged the elimination of its metabolite, ***4-hydroxypropranolol***. ***Ergotamine tartrate*** in ***Cafergot***® was used rectally twice daily/6 years by a 61-year-old **male** to control severe migraines. When propranolol 30 mg/day was added to **his** treatment regime, **his** feet became progressively more purple and painful after each successive suppository. Lower extremity ischemia in a 48-year-old **female** with propranolol 10 mg qid and 2 mg ***ergotamine tartrate*** sublingually as needed. ***Halofenate*** decreases its plasma concentration. Oral ***hydralazine*** enhances its systemic availability by altering its first-pass hepatic clearance and, given with ***hydralazine***, it prevents the latter's tachycardia. Enhanced hypoglycemic effect when given with ***insulin*** and ***sulfonylureas***, such as ***acetohexamide, chlorpropamide***, and ***tolbutamide***. Its hypotensive effects are partially antagonized by concomitant ***indomethacin*** (100 mg/day/3 weeks) use. ***Isoproterenol's*** and ***levarterenol's*** actions are antagonized by it. Its addition to sublingual ***isosorbide dinitrate*** in treating angina pectoris increases untoward effects, such as left ventricular failure, dyspnea, headaches, nausea, depression, blurred vision, and transient diarrhea. ***Lidocaine*** plasma clearance significantly decreased by it. It decreases the serum levels of ***lovastatin*** and ***pravastatin*** by ~18% and 23%, respectively. ***Heroin*** **addicts** who self-administer ***heroin*** after use of propranolol precipitated withdrawal symptoms, instead of expected euphoric effects. Hypertensive crisis has occurred with ***methyldopa***. Transient diarrhea, increased hypotension, bradycardia, nausea, and vomiting have occurred when combined with ***quinidine*** therapy. Both have a negative inotropic effect on the heart. Use with ***monoamine oxidase inhibitors*** can worsen hypertension. With ***ether, hexobarbital, morphine***, and ***urethan*** its effects are potentiated and death has been reported. The death of two **patients** aged 48 and 51 years, **who** respectively received ***nadolol*** and ***propranolol***, had anaphylactic reactions to oral ***penicillin V*** for treatment of gum inflammation. **Neither** had a history of medication allergy or prior exposure to ***penicillin***. It blocks ***phentolamine's*** stimulation of ***adenylate cyclase*** in **human** leukocytes. ***Phenytoin*** has an additive cardiac depressant effect, especially after IV use. Severe postural hypotension ~1 week after ***prazosin*** addition to propranolol therapy. Markedly decreases ***terbutaline***-induced hyperglycemia. May inhibit ***theophylline*** metabolism. Severe paranoia in a 51-year-old **male** when used with ***tocainide***. After 80 mg twice daily/2 weeks, ***warfarin*** concentrations were increased 15% and bleeding episodes occurred in three **patients** 6 weeks after its addition to ***phenindione*** therapy. Potentially significant interactions reported with ***aspirin***, but not clearly defined. Excessive sedation and gastrointestinal bleeding when given with ***reserpine***. In addition, its effects are also increased by ***cimetidine, oral contraceptives, morphine, phenobarbital, quinidine***, and ***rifampin***; and decreased by ***dobutamine, dopamine, epinephrine, isoproterenol, methimazole, nicotine, norepinephrine, nonsteroidal anti-inflammatory drugs, propylthiouracil, salicylates***, and ***thyroid hormones***. Use with ***haloperidol*** in a schizophrenic 48-year-old **female** led to three episodes of hypotension and two of cardiopulmonary arrest. ***Avitriptan*** and ***zileuton*** increase its serum concentration. Prolongs ***diazepam*** half-life from 49 h to 58 h.

In **animals**, may aggravate asthmatic or heave syndromes by increasing bronchoconstriction. In **mice**, **rabbits**, and **swine**, ***verapamil*** enhances its toxicity, which can be antagonized by IV ***calcium***. Will enhance toxic effects of ***narcotics*** in **mice**, **rats**, and **dogs**!! It is very toxic in hypothyroid **mice**, and hyperthyroid **mice** can safely tolerate dosages unsafe for the normothyroid. Acute central nervous system toxicity of ***hexobarbital*** is increased by it in **mice**. Potentiates ***morphine*** toxicity in **rats**. In **cats**, ***tubocurarine*** neuromuscular blockade is increased by it in magnitude and duration. Contraindicated with sinus bradycardia and first degree heart block. Start with low dosage and increase it to desired effect. Taper dosage downward when discontinuing.

IV, LD_{50}, **mouse**: 22 μg/g; IP, 80 mg/kg.

n-PROPYL ACETATE

Use: Solvent in ***paint*** thinners and in manufacturing flavors and perfumes. Has a mild, fruity flavor.

Untoward effects: An airline **pilot** painted a model glider in a heated garage with an epoxy-based ***paint*** containing it. **His** total exposure time was ~4 min. Within 30 min, **he** was aware of a strong odor of algae that was apparent only to **him**. Within 2 h, **he** had chest pains that resulted in **his** hospitalization in a coronary unit, due to severe pulmonary irritation from the vapors. Vapors have caused eye, nose, throat, and skin irritation, and weakness, drowsiness, and unconsciousness. National Institute of Occupational Safety and Health/OSHA upper exposure limits of 200 ppm or 840 mg/m³. Immediate danger to life or health: 1700 ppm.

LD_{LO}, **human**: 500 mg/kg; inhalation TC_{LO}, **human**: 1000 mg/m³.

Eye, nose, throat, and skin irritation, and narcosis in experimental **animals**. A 30 min exposure of **cats** to 24,500 ppm led to narcosis and death.

LD_{50}, **rat**: 9.4 g/kg; **mouse**: 8.3 g/kg; **rabbit**: 6.64 g/kg. Inhalation LC_{LO}, **rat**: 8,000 ppm/4 h; **cat**: 38 g/m³/~5.5 h.

n-PROPYL ALCOHOL

= *Ethyl Carbinol* = *Optal* = *1-Propanol* = *Propylic Alcohol*

***Solvent.* Approximately 0.47 ml in a gallon of *whiskey*.**

Use*:** In pigments, ***inks, resins, and ***cellulose*** esters.

***Untoward effects*:** Eye, nose, throat, and skin irritant; dry, cracking skin from contact. Ingestion or excessive absorption from contact or fumes causes drowsiness, headache, ataxia, abdominal pain and cramps, nausea, vomiting, and diarrhea. Local cutaneous erythema noted in patch tests on **humans**. National Institute of Occupational Safety and Health/OSHA upper exposure limits of 200 ppm or 500 mg/m³. Immediate danger to life or health: 800 ppm.

LD_{LO}, **human**: 500 mg/kg.

Narcosis reported in exposed **animals**.

LD_{50}, **rat**: 1.87 g/kg. Inhalation LC_{LO}, **rat**: 4,000 ppm/4 h.

n-PROPYLBENZENE

= *Isocumene* = *1-Phenylpropane*

***Use*:** Solvent and in textile dyeing and printing.

***Untoward effects*:** Eye irritation from its vapors.

Pulmonary irritation in **rats** from vapors.

LD_{LO}, **rat**: 4.8 g/kg; inhalation LC_{LO}, **mouse**: 4,100 ppm.

n-PROPYLCARBAMATE

= *Propyl Urethane*

***Untoward effects*:** High incidence of lung tumors after IP in **mice** and teratogenic after single IP in **hamsters**.

TD_{LO}, **mouse**: 1 g/kg; IP, 650 mg/kg during 13 weeks; IP, **hamster** (8 days pregnant): 500 mg/kg. SC, LD_{50}, **mouse**: 1.3 g/kg.

n-PROPYL DISULFIDE

Untoward effects*:** Wild or domestic ***onions contain the oxidant *n*-propyl disulfide and excessive intake has caused Heinz body anemia in **cats**, **dogs**, and **cattle**. The oxidant causes hemolytic anemia by oxidative degradation of hemoglobin with formation of Heinz bodies. Wild ***onion*** tops, when fed experimentally to a **horse**, caused hemolytic anemia. See ***Onions*** for symptoms, postmortem findings, and other species affected.

PROPYLENE

= *Methylethene* = *Methylethylene* = *2-Propene*

Use*:** > 26 billion lbs annually. In chemical synthesis of ***acrylonitrile, ***isopropanol***, ***polypropylene***, ***propylene oxide***, etc.

***Untoward effects*:** Contact with the liquid causes skin burns. Inhalation of the vapors causes a mild anesthetic and asphyxiant effect. Unconsciousness in **people** within 2 min by 50% concentration; 37–40% leads to anesthesia. Explosion of a tanker truck in Spain killed > 100 **people**.

Fumes sensitize a **dog's** heart to ***epinephrine***-induced arrhythmias. Ectopic ventricular beats in exposed **cats** and **dogs**.

PROPYLENE CARBONATE

***Use*:** Solvent and as a gellant for clays.

***Untoward effects*:** 100 mg/day/3 days irritant to **human** skin and 60 mg topically on **rabbit** eye is irritant. Can cause necrosis from IM injections in pectoral muscle of **broilers**.

LD_{50}, **rat**: 29 g/kg.

PROPYLENE DICHLORIDE

= *1,2-Dichloropropane*

***Solvent.* In dry-cleaning fluids, synthetic *rubber*, scouring compounds, and as a fumigant. Flammable, *chloroform* odor.**

***Untoward effects*:** Eye, skin, and particularly a respiratory tract irritant on exposure to vapors. Skin absorption or ingestion causes liver and kidney pathology, drowsiness, and light-headedness. OSHA sets limit of exposure at 75 ppm TWA and immediate danger to life or health of 400 ppm. Potential occupational carcinogen.

LD_{LO}, **human**: 50 mg/kg.

In **mice**, increased SGOT; after exposure to 2,200 ppm, **guinea pigs** showed extensive coagulation and focal hemorrhagic necrosis of the adrenal cortex, congestion and hemorrhagic necrosis of the adrenal medulla; **guinea pigs** and **rats** had fatty degeneration of the liver. In anesthetized **dogs**, inhalation of 0.25% decreased myocardial contractility.

LD_{50}, **rat**: 1.19 ml/kg; **mouse**: 860 mg/kg; inhalation LC_{LO}, **rat**: 2,000 ppm/4 h.

PROPYLENE GLYCOL

= *Methyl Glycol* = *1,2-Propanediol* = *Sirlene*

1 billion lbs used annually.

***Use*:** In food, pharmaceuticals, and cosmetic applications as a solvent, emulsifier, humectant, preservative, and antifreeze; in brake and hydraulic fluids, animal feeds, windshield de-icing fluids, inks, and semi-moist pet foods.

Untoward effects*:** Stinging and burning due to its use as a vehicle in various topicals. Facial and scalp eczemas, pruritus, and allergic contact dermatitis. Allergic vulvitis and proctitis in a 29-year-old **female** due to propylene glycol after examination with gloved hands lubricated with ***K-Y Jelly® after previous sensitivity reactions due to its presence in several topical creams prescribed for chronic vaginitis. Dermatitis of the abdomen, vulva, and rectal areas also occurred after **she** ate ***salad dressing*** containing propylene glycol. A 55-year-old **veterinarian** developed a severe pruritic dermatitis of the hand after using ***K-Y Lubricating Jelly***® before doing rectal palpations on **horses** and **cattle**. A 27-year-old **female** also suffered a pruritic dermatitis after using the same lubricant under a nerve-stimulation device for shoulder pain. Frequency of skin reactions is often concentration-dependent. Perioral dermatitis in **children** from use on **children's** lips of ***Lip Smackers***®. Patch-testing with propylene glycol at 2–10% under occlusion causes mild erythema in most **people**. It is estimated that ~5% of the population is allergic to its presence in lip balms, hair dyes, stick deodorants, and creamy facial cosmetics. The FDA considers it generally regarded as safe for **humans** and **animals**, (perhaps, when compared to ***ethylene glycol***). It may be more toxic to **humans** and more toxic to **dogs** than **rats**. Hyperosmolality with cardiopulmonary arrest in an 8-month-old burn **patient** attributed to it in a topical ***silver sulfadiazine*** preparation. Use as a vehicle for ***vitamin C*** and ***D*** in **infants** has, within 8 days, caused irregular apical heart rate and sinus arrhythmia, followed in 2 days by tachypnea, tachycardia, diaphoresis, and unresponsiveness. Some of these **children** became stuporous and had seizures. Tinnitus reported from its topical use in otic preparations. Use as a solvent (322 ml/day) for a ***nitroglycerin*** infusion in a 72-year-old **female** led to severe hyperosmolality, lactic acidosis, central nervous system depression, and hemolysis. A 15-year-old **male** probably became toxic from 2.25 ml/h/42 days in a ***lorazepam*** infusion to maintain sedation. An adverse reaction with increased anion gap due to high-dose ***pentobarbital*** infusion with propylene glycol as the solvent. Injectable ***diazepam*** has caused allergic reactions due to its 40% propylene glycol content. Phlebitis at injection site in many **patients**, due to it in ***diazepam*** and ***phenytoin*** injections. Its use in local anesthetics prolongs their effect by damaging the nerves. High hemolytic potential. Large IV dosage is narcotic. Central nervous system toxicity also reported from its ingestion with ***vitamin D*** in 12-year-old **male**. Skin irritation in **people** after 104 mg topically/day/3 days.

TD_{LO}, **child**: 91 mg/kg during 56 week period.

Large oral doses have been toxic to **horses**. A **horse** inadvertently given 3.8 l (7.6 ml/kg) instead of ***mineral oil*** developed abdominal pain, diaphoresis, sialism, ataxia, circling, rapid shallow breathing, cyanosis, and death 28 h after ingestion. **Cattle** have developed hypotension. IV use for **sheep** (as in ***pentobarbital*** solutions) causes hemoglobinuria. After oral use, traces appear in the milk. Avoid epidural use, where it may cause nerve degeneration. Tissue damage is often reported from its parenteral usage. Cardiac depression from IV use has been reported in **calves**, **cats**, **horses**, etc. Rapid IV in **large animals** caused sudden collapse, hemolysis, and death. In **sheep**, IV use has caused decreases in ionized ***calcium*** in the blood. IV use in the **rabbit** caused decreased blood clotting time, decreased lymphocytes, and increases in circulating polymorphs. IV use in **dogs** has a tranquilizing and depressing effect (~one-third that of the same quantity of IV ***alcohol***). No longer recognized as a safe additive in soft-moist **cat** foods. Heinz body-induced acceleration of red blood cell destruction, methemoglobinemia, and leukemia in **cats** is dose-dependent. Topically on **mice** leads to skin lesions, slight inflammation and ulceration. Very high dosage in **mice** leads to deep depression and death within 45 min. Slight ocular irritation in **rabbits**. Topically on **guinea pig** middle ear mucosa causes severe inflammation and hair cell loss.

LD_{50}, **rat**: 21 g/kg; **mouse**: 11.4 g/kg; **guinea pig** and **rabbit**: 19 g/kg; **dog**: 22 g/kg. IV, LD_{50}, **mouse**: 8 g/kg; **rabbit**: 4.2 g/kg.

PROPYLENE GLYCOL 1,2-DINITRATE

= PGDN

Potential substitute for *ethylene glycol dinitrate* in manufacturing antifreeze dynamite. A major component in the U.S. Navy's torpedo and other weapons fuels.

***Untoward effects*:** Potent vasodilator. Exposure to vapor concentrations of 0.2 ppm or more causes visual disturbances, eye irritation, conjunctivitis, headache, nasal congestion, methemoglobinemia, nausea, dizziness, dyspnea, impaired balance (after 0.5 ppm), and poor eye–hand coordination. National Institute of Occupational Safety and Health upper skin exposure limit of 0.05 ppm. Potentially explosive.

Inhalation TC_{LO}, **human**: 1500 ppb/5 min and 200 ppb/6 h.

Liver and kidney pathology in experimental laboratory **animals**. Methemoglobinemia in the **rat**.

LD_{50}, **rat**: 250 mg/kg; SC, **rat**: 463 mg/kg; **mouse**: 1.2 g/kg; **cat**: 200 mg/kg. IV, LD_{LO}, **monkey**: 410 mg/kg.

PROPYLENE GLYCOL MONOMETHYL ETHER

= *Arcosolv PM* = *Dowanol PM* = *Dowtherm 209* = *Glycol Ether PM* = *Methoxypropanol* = *PGME*

Solvent.

Untoward effects: Vapors induce eye, skin, nose, and throat irritation; headache, nausea, light-headedness, drowsiness, incoordination, vomiting, and diarrhea. National Institute of Occupational Safety and Health upper exposure limit of 100 ppm (360 mg/m^3). Short-term exposure limit 150 ppm.

LD_{50}, **rat**: 7.5 g/kg; **mouse**: 11.7 g/kg; **rabbit**: 8 g/kg; LD_{LO}, **dog**: 10 g/kg. Inhalation LC_{LO}, **rat**: 7,000 ppm/4 h, **rabbit** and **guinea pig**: 15,000 ppm/7 h.

PROPYLENE GLYCOL MONOSTEARATE

Untoward effects: Allergic eczematous contact dermatitis reported in some **people** after use in topical medicaments.

LD_{LO}, **human**: 5 g/kg.

PROPYLENE OXIDE

= *1,2-Epoxy Propane* = *Propene Oxide*

Solvent, fumigant, soil sterilant, and chemical intermediate. **Annual U.S. production > 4 billion lbs/year.**

Untoward effects: Liquid forms can cause severe skin damage (blisters and burns). Excessive absorption can cause systemic illness due to central nervous system, liver, and kidney toxicity. Often formulated with ***carbon dioxide*** to reduce explosion hazard. Exposure to vapors can irritate eyes, skin, and respiratory system; ataxia and secondary lung infections, corneal burns, and contact dermatitis can occur.

Nasal tumors in **rats** and **mice** after inhalation, and forestomach squamous cell carcinomas in **rats** after gavage.

LD_{50}, **rat**: 1.14 g/kg; **guinea pig**: 690 mg/kg; dermal, **rabbit**: 1.5 ml/kg. Inhalation LC_{50}, **rat**: 9486 mg/m^3, **mouse**: 4126 mg/m^3.

PROPYL ETHER

= *Dipropyl Ether* = *1,1´-Oxybispropane*

Use: In the ***rubber***, plastics, ***paints***, coatings, petroleum, chemicals, perfume, and cosmetic industries.

Untoward effects: Vapors produce anesthesia and pulmonary irritation in **mice**.

PROPYL GALLATE

= *N-Propyltrihydroxybenzoate*

Antioxidant. **In cosmetics, foods, fats, oils, and *vitamin A*. Common in cereals, snack foods, and pastries.**

Untoward effects: Has caused a ***lipstick*** dermatitis and contact dermatitis in **bakers** using it in dough. Daily applications of 20% for 24 days caused pruritus and erythema in 5/10 **patients**.

Increased incidence of lung tumors reported in **mice**. Contact sensitization in **guinea pigs** and egg yolk mottling when fed to **hens**.

LD_{50}, **rat**: 2.6 g/kg; **cat**: 400 mg/kg; IP, **rat**: 380 mg/kg.

PROPYLHEXEDRINE

= *Obesin* = *PPH*

Adrenergic, nasal decongestant, vasoconstrictor.

Untoward effects: The use of ***Benzedrex***® inhalers, with 250 mg of the drug and 12.5 mg of ***menthol*** in each, has been abused by **people** who have chewed their contents, ingested them after mixing it daily in ***coffee*** with ***sugar***, or injected them IV often causing psychosis and death. Its central nervous system stimulation is similar to ***amphetamine's***. A 22-year-old **male**, after swallowing 250 mg, developed violent palpitations, headache, and severe central chest pain for several hours, followed by "shock lung". **Others** have suffered acute paranoid psychosis, auditory and visual hallucinations, fatigue, depression, disturbed sleep, and exacerbation of quiescent schizophrenic state. A 3-year-old **child** developed nausea, insomnia, tremor, tachycardia, decreased reflexes, and muscular hypertonicity. IV use has caused left ventricular dysfunction in some **patients** and sudden death in many **others**, with pulmonary edema, foreign-body granulomas, fibrosis, and pulmonary hypertension noted on post-mortem. Cyanosis, ecchymoses, and petechiae have also been reported. Lethal blood levels are reported as 0.2–0.3 mg%.

IV, LD_{50}, **mouse**: 32 mg/kg.

PROPYLHEXEDRINE ETHYLPHENYLBARBITURATE

= *Maliasin*

Untoward effects: Immediate intolerance in **some**; tiredness, and falling asleep in 30-year-old **male** after 100 mg injection; excitement after 150 mg/day in 40-year-old **male** and 100 mg/day in 48-year-old **female**. Restlessness, tiredness, palpitations, irritability, vertigo, diaphoresis, anorexia, altered personality, and weight loss in 26% of **patients**.

PROPYLIODONE

= *Dionosil*

Radiopaque contrast media. **In bronchography as a 50% aqueous suspension and a 60% *arachis oil* suspension for injection. Contains 56.8% *iodine*.**

Untoward effects: Aqueous suspension causes slight febrile reactions for 24–36 h; occasional coughing, dyspnea, malaise, nausea, vomiting, and, rarely, transient pulmonary lobar collapse. Use can interfere with thyroid function tests. Avoid use in **patients** allergic to ***iodine***, but the frequency of adverse reactions in these **patients** was less than with other ***iodinated contrast media***.

IV, LD_{50}, **mouse**: 300 mg/kg.

n-PROPYL NITRATE

Use: Rocket fuel propellent, ignition-promoter in fuels.

Untoward effects: Sweet, sickly, ***ether***-like odor. Occupational skin irritation; yellow skin discoloration after repeated contact. High vapor concentrations are respiratory irritants and can cause dyspnea, weakness, headache, dizziness, and methemoglobinemia. High volatility lessens probability of toxic sytemic effects. Flammable. Explosive when heated. National Institute of Occupational Safety and Health/OSHA upper exposure limits of 25 ppm. Immediate danger to life or health: 500 ppm.

Also leads to eye irritation in laboratory **animals**.

Inhalation LC_{50}, **rat**: 9,000 ppm/4 h; **mouse**: 4,000 ppm/4 h; **dog**: 2,000 ppm/4 h.

PROPYLTHIOURACIL

= *Propacil* = *Propycil* = *Propyl-Thyracil* = *PTU* = *Threostat II*

Antihyperthyroid. **Inhibits formation of *thyroxine* and *triiodothyronine*. Also blocks conversion of T_4 to T_3.**

Untoward effects: TD_{LO}, pregnant **women**: 1.68 g/kg.

Blood and bone marrow: A 53-year-old **female** developed pancytopenia and severe bone marrow aplasia after treatment with 100 mg qid/40 days. Agranulocytosis (~10%), leukopenia (~10%), disseminated intravascular coagulation, thrombocytopenia, neutropenia, aplastic anemia, hemolytic anemia, intraoperative bleeding, hypoprothrombinemia, granulocytopenia, eosinophilia, and hypocalcemia. Acute myeloblastic leukemia was fatal in a 74-year-old **female** after 50 mg tid/several years.

Fetus and neonate: Neonatal goiter, intrauterine death, abortions, Down's syndrome, cryptorchidism, hypothyroidism, mental retardation, and cretinism. Readily crosses placental barrier after 14th week of pregnancy.

Liver: Severe hepatitis and jaundice, occasionally epigastric pain, nausea, vomiting, weakness, anorexia, and palpitations. Cirrhosis and centrilobular necrosis reported.

Miscellaneous: Altered ^{131}***I*** uptake test results for up to 1 week after treatment, which has caused false negative or spuriously low test results. Use interferes with ***uric acid*** assay on SMA/Autoanalyzer II. A hyperthyroid 2½-year-old **female** receiving 25 mg tid developed hypothyroidism, sexual precocity, menstruation, and galactorrhea. Splenomegaly, unilateral sensorineural hearing loss, arthralgia, headache, ageusia, paresthesias, neuritis, alopecia, edema, vertigo, dizziness, skin pigmentation, sialadenopathy, lymphadenopathy, periarteritis, visual and auditory hallucinations, and sore throats. Suspect carcinogen in some **patients** (malignant thyroid lesions and a case of malignant myeloblastic leukemia). A 12-year-old **female** ingested 100–260 × 50 mg tablets had a benign course, but in a **patient** with normal thyroid gland or impaired renal function, adverse effects might have been serious. Can induce or activate systemic lupus erythematosus.

Skin: Systemic lupus erythematosus-like syndrome with rash, urticaria, fever, cutaneous ulcers, and vasculitis. Sweet's syndrome in a 37-year-old **female** after 100 mg tid/3 months.

Drug interactions: Potentiates activity of ***oral anticoagulants*** (***coumarins***, ***indandiones***, including ***warfarin***) by its induction of decreased prothrombin. May cause decreases in dosage required of some ***β-blockers***.

Goitrogenic in **animals**, crosses the placental barriers, and passes into the milk of treated **animals**. Overdoses to **rats** lead to salivation, lacrimation, tremors, ataxia, decreased body temperature, convulsions, and death. Audiogenic seizures and cochlear damage in **rats** after perinatal treatment. Adenomas of anterior pituitary and thyroid carcinomas in **rats**. Both reported in treated **hamsters** and **guinea pigs**. Thyroid hyperplasia and thrombocytopenia in treated **dogs**.

PROPYPHENAZONE

= *Arthrodestal* = *Budirol* = *Causyth* = *Cibalgina* = *Eufibron* = *Isoprochin P* = *Isopropylphenazone*

Anti-inflammatory, analgesic, and antipyretic. **Over-the-counter in some countries, often with *phenylbutazone*, *prednisone*, etc.**

Untoward effects: Serious self poisonings with fatalities reported in Spain. Anaphylaxis frequently reported from drug combinations containing it. Therapeutic serum levels reported as 0.005–0.2 mg/l and toxic levels as 0.4 mg/l.

LD_{50}, **rat**: 2.4 g/kg; **mouse**: 1 g/kg.

PROSCILLARIDIN

= *Caradrin* = *Cardion* = *Carmazon* = *Compound 386-2* = *Proscillan* = *Prostasin* = *Prostosin* = *Proszine* = *PSC 801* = *Purosin-TC* = *Sandoscill* = *Scillacrist* = *Simeon* = *Solestril* = *Stellarid* = *Talucard* = *Talusin* = *Urgilan* = *Wirnesin*

***Cardiotonic.* A glucoside extracted from (*Squill*) *Scilla bulbus* or *Scilla maritima*.**

***Untoward effects*:** Nausea, vomiting, bigeminal or trigeminal pulse, heart block, bradycardia, supraventricular extrasystoles, diarrhea, dyspnea, and atrial fibrillation. Local reactions or thrombophlebitis after IV use.

LD_{50}, **rat**: 56 mg/kg; **cat**: 732 mg/kg; IV, LD_{50}, **rat**: 9 mg/kg; **cat**: 185 μg/kg.

PROSTACYCLIN

= *Epoprostenol* = *Flolan* = *PGI_2* = *PGX* = *Prostaglandin I_2* = *U 53,217*

Found in the inner lining of blood vessels, where it blocks platelet aggregation.

***Untoward effects*:** Nausea, vomiting, abdominal pain, headache, migraine, hypotension, chest pain, depression, facial flushing, bradycardia or tachycardia, dyspnea, musculoskeletal pain, drowsiness, dizziness, restlessness, and anxiety after or during IV infusions.

PROTAMINE SULFATE

A life-saving drug derived from *protamine* in salmon sperm, inhibiting thromboplastin formation and activity. Despite possessing anticoagulant properties, it has a strong antiheparin effect.

Untoward effects*:** Bradycardia, hypotension, and dyspnea can occur as a reaction to a foreign protein during IV use. Anaphylactic reactions to its IV administration reported in a **patient** that had no identifiable risk factors (prior protamine exposure, ***fish allergy, or vasectomy). Rapid (5–10 min) IV in a 74-year-old **male** after coronary artery bypass caused immediate non-cardiogenic pulmonary edema. Due to ***heparin's*** short half-life, it is easy to overdose if treatment is delayed leading to anticoagulation. Gives false positive or spuriously high plasma ***catecholamine*** test results.

Pronounced decrease in serum ***calcium*** in young or lactating **animals**. Anticoagulation effect in **dogs** given four times the amount necessary to neutralize ***heparin***.

IV, LD_{50}, **mouse**: 44 mg/kg.

PROTEIN HYDROLYSATES

= *Breogamine* = *Crotein* = *Hydropot* = *Progestimil* = *Zymino*

***Nutrient, amino acid source.* Some types are *Amigen, Aminokrovin, Aminonat, Aminosol, Hyprotigen, Lacotein, Neo-Protostan, Parenamine, Protigenyl I, Protolysate, Travamin, Trophysan,* and *Virex*. Oral and IV use. From blood, fibrin, meat, *casein*, cowhide *gelatin*, *yeast*, or *beans*.**

Untoward effects*:** Hypersensitivity reactions can occur. IV use of partially digested protein or peptides can be toxic. An excess of certain ***amino acids over normal values can add to such toxic reactions (viz. ***glutamic*** and ***aspartic acids*** can cause vomiting). ***Glutamic, aspartic***, and ***cysteic acids***, SC, were associated with acute degeneration of hypothalamic nerves, and their oral use was associated with similar hypothalamic lesions. In general, do NOT add other drugs, especially ***antibiotics***, to parenterally or orally administered protein hydrolysate solutions, as protein binding can occur and desired drug effects may not be attained. Most have an unpleasant taste for **people**. High anion gap acidosis in a 28-year-old **female** after 4800 ml/day of 5% ***casein hydrolysate*** and 5% ***dextrose***. Acute hypersensitivity, hyperosmotic coma, rash, shock, pruritus, conjunctival edema, low-grade fever, weakness, tingling, and numbness of extremities, circumoral area, face, and tongue; hypophosphatemia, coma, convulsions, ***folic acid*** deficiency, and death are among the many adverse effects reported with hyperalimentation infusions. In **infants**, many of the above symptoms have occurred, as well as increased liver enzymes, progressive cholestasis, increased serum ***ammonia***, decreased hemoglobin, flapping tremors, alkalosis, and deaths. Deaths have occasionally been attributed to sepsis from the IV catheters. Incompatible with at least 85 drugs for parenteral use.

Liquid protein diets were associated with at least 58 deaths in the U.S. during the latter half of 1977 and the early part of 1978, apparently, due to induced cardiac arrhythmias and ventricular fibrillation. Alopecia and thrombophlebitis apparently followed the use of liquid protein diets in some **people**.

PROTHIPENDYL

= *D 206* = *Dominal* = *Timovan* = *Tolnate* = *W 1962*

Psychotherapeutic.

Untoward effects*:** Drowsiness, nocturnal insomnia, orthostatic hypotension, headache, tachycardia, and sweating. Use with ***thiopental enhances the latter's effect. Therapeutic serum concentrations reported as ~0.05–0.2 μg/ml and toxicity at ~0.5–1 μg/ml.

LD_{50}, **mouse**: 415 mg/kg; IP, 155 mg/kg.

PROTIONAMIDE

= *Ektebin* = *Peteha* = *Prothionamide* = *RP 9778* = *TH 1321* = *Trevintix*

Tuberculostatic.

Untoward effects: Nausea, vomiting, headache, prurigo, increased SGPT and SGOT, hepatotoxicity, altered taste (metallic), confusion, purpura, impotence, gynecomastia, and amenorrhea.

LD_{50}, **rats**: 1.32 mg/kg; **mouse**: 1 mg/kg.

PROTOANEMONIN

A glycoside, *ranunculin*, in *Anemone pulsatilla* and *Ranunculus sp.* (q.v.) is hydrolyzed to protoanemonin, a vesicant oil.

Untoward effects: It is antibacterial, but, when ingested by **livestock** especially **cattle** on pastures, causes salivation, hematuria, gastroenteritis, bradycardia, depression or excitement, convulsions, and death.

PROTOKYLOL

= *Caytine* = *JB 251* = *Ventaire*

***Bronchodilator, sympathomimetic.* Oral doses are effective in 30–90 min and the effects continue for 3–4 h.**

Untoward effects: Palpitations, tachycardia, precordial pain, hypotension, dizziness, nausea, fainting, insomnia, headache, nervousness, tremors, dysuria, flushing, and dermatitis have been reported.

In **cats** and **dogs**, orally, IM, or SC causes occasional palpitations, tachycardia, apprehension, tenseness, and irritability.

LD_{50}, **rat**: 1 g/kg.

PROTOVERATRINES A and B

Antihypertensive.

Untoward effects: Severe nausea and vomiting, epigastric pain, cold sweats, vertigo, shallow respiration with overdoses and, occasionally, with therapeutic doses. Postural hypotension, arrhythmias, and bradycardia occur. Placental passage occurs and it can cause **fetal** bradycardia.

Causes muscle paralysis in **frogs**.

PROTRIPTYLINE

= *Concordin* = *Maximed* = *MK 240* = *Triptil* = *Tryptil* = *Vivactil*

***Antidepressant.* A *tricyclic tranquilizer*.**

Untoward effects: Occur more frequently in **women** and **older patients**.

Anticholinergic: Paralytic ileus, hyperpyrexia, urinary retention, micturation difficulties, constipation, blurred vision, mydriasis, xerostomia, and increased intraocular pressure.

Blood: Agranulocytosis, bone marrow depression, leukopenia, thrombocytopenia, purpura, eosinophilia, and disseminated intravascular coagulation.

Cardiovascular: Myocardial infarction, stroke, heart block, arrhythmias, hypotension or hypertension, tachycardia, and palpitations.

Central nervous system: Seizures, incoordination, ataxia, tremors, peripheral neuropathy, numbness, tingling, paresthesias of extremities, extrapyramidal symptoms, drowsiness, weakness, fatigue, dizziness, headache, inappropriate antidiuretic homone secretion syndrome, tinnitus, and altered electroencephalogram.

Gastrointestinal: Nausea, vomiting, anorexia, diarrhea, altered taste, stomatitis, abdominal cramps, and black tongue.

Hypersensitivity: Drug fever, petechiae, skin rash, urticaria, pruritus, photosensitization, and edema.

Miscellaneous: Jaundice, parotid swelling, flushing, nocturia, alopecia, pollakiuria, sweating, alterations of blood ***glucose*** levels, impotence, failure to ejaculate, altered libido, **male** gynecomastia and testicular swelling, **female** breast enlargement and galactorrhea, and hiatal hernias. A Dutch reference lists therapeutic serum levels as 0.05–0.2 mg/l; toxic levels as 0.4 mg/l; a U.S. reference lists 12 μg% and 50–200 μg%, respectively. Fatalities reported at 1 mg/l.

Drug interactions: Use with ***oral contraceptives*** or ***guanethidine*** decreases its serum levels and effects. Excitation, tremor, hyperhidrosis, delirium, hyperpyrexia, clonic and tonic convulsions, and even coma and death have occurred when used with ***monoamine oxidase inhibitors***. Potentiates ***dextroamphetamine, epinephrine*** (three-fold), and ***noradrenaline*** (nine-fold). Similar effects can occur with ***lidocaine*** and ***mepivacaine***.

Cardiotoxic in **guinea pigs** and **rats**. Increases concentration of ***dextroamphetamine*** in the brain of **rats**. Tachycardia in **dogs**.

LD_{50}, **rat**: 299 mg/kg; IV, **mouse**: 30 mg/kg.

PRUNE

A useful nutritious fruit that stimulates peristalsis; contains soft fibers and an *anthraquinone*, *oxyphenysatin* (q.v.).

Untoward effects: After ingestion in large quantities, they are potent gas-formers (flatulence), and can cause griping and indigestion. Decreases urinary pH. Contraindicated in nursing **mothers** and in **patients** with ***chromate*** dermatitis.

PSEUDOEPHEDRINE

= *Afrinol* = *Disophrol* = *Eltor* = *Galpsend* = *Isoephedrine* = *Novafed* = *Rhinalair* = *Sinufed* = *Sudafed*

Sympathomimetic, nasal decongestant. **Oral and topically intranasal. ~1 Billion dollars worth used annually.**

Untoward effects: Primarily, stimulates α- and some β-receptors; its oral use can easily cause hypertension, vasoconstriction, tachycardia, palpitations, central nervous system stimulation, and urinary retention. Occasionally mania, hallucinations, tremors, convulsions, and even death with overdoses. Headache, dizziness, nausea, anxiety, nervousness, insomnia, and hypotension also reported. Use with caution in hypertensive **patients**. Listlessness, lethargy, confusion, and ataxia in a 4-year-old **male** after oral dose. Recurrent toxic shock syndrome in an 18-year-old **female** from it in a cough medicine. Fixed-drug eruptions on fingers of a 41-year-old **male**, 12 h after ingestion. Recurrent pseudo-scarlatina reactions in a 32-year-old **female**. **Women** taking it during their first trimester of pregnancy have had three times greater risk of having gastroschisis. In breast milk at ~4% of **maternal** dose. Topically, it has caused few hazardous side-effects, but overuse causes rebound exacerbation of symptoms, additional overuse, and virtual addiction to it. Overdoses from various cold and sinus medications, herbal ***teas*** and supplements lead to symptoms of restlessness and agitation within 30 min to 3 h.

Drug interactions: Its effects are increased by use of ***monoamine oxidase inhibitors*** and often accompanied by headache and marked hypertension; in some cases, acute left ventricular failure or fatal intracerebral hemorrhage, which are decreased by ***adrenergic blockers***. Antihypertensive effect of ***bethanidine, debrisoquin, guanethidine, mecamylamine, methyldopa, reserpine***, and ***veratrum alkaloids*** are decreased by it. A 37-year-old **female** presented at an emergency room with stroke-like symptoms and severe hypertension after taking excessive amounts of ***phenylpropanolamine*** (150 mg twice daily) for weight reduction plus a pseudoephedrine product as a nasal decongestant. A 2-year-old **male** was given 1½ teaspoonful every 6 h of an over-the-counter cough and cold preparation that also contained ***dextromethorphan***; he awakened an hour later with hyperexcitability, irritability, babbling, and poor balance control.

Since the dosage for **dogs** is only 15–30 mg twice daily–tid, it is a common cause of poisoning in them when **owners** think it is innocuous to give their **pets** the **human** dosage of 60–120 mg, leading to central nervous system stimulation, tachycardia, vasoconstriction, mydriasis, tremors, vocalization, seizures, and urinary retention.

Used in the illicit manufacture of ***methamphetamine***.

PSILOCIN

= *Psilocyn* = *Psilotsin*

Hallucinogen. **Extracted from Mexican *mushrooms, Psilocybe sp.*, and other *mushrooms*; used by ancient people in religious ceremonies.**

Untoward effects: After repeated use, **one** needs increased doses to get the same hallucinogenic effect, which lasts ~6 h. It and the ***mushrooms*** it comes from have become substances of abuse in the U.S., Mexico, Australia, and Europe.

IV, LD_{50}, **rat** and **mouse**: 75 mg/kg; **rabbit**: 7 mg/kg.

PSILOCYBIN

= *Indocybin* = *Psilotsibin* = *Teonanactyl*

Psychomimetic, hallucinogen. **Also a *tryptamine* from the same *mushrooms* as noted above. Originally used by the Aztecs.**

Untoward effects: Visual hallucinations without clouding of consciousness; mydriasis. The "high" induced is less intense, but lasts longer than that of lysergic acid diethylamide (***LSD***). Adversely affects higher cognitive functioning. Severe persistent anxiety and anticholinergic syndrome in a 21-year-old **male** and a 24-year-old **male** after ingesting 30 and 25 *Psilocybe* ***mushrooms***, respectively. Ingestion of 4–8 mg (in 20–30 ***mushrooms***) leads to hallucinations, distortions and chromatic aberrations, and depersonalization.

Human: TD_{LO}, 60 μg/kg; IM, 75 μg/kg; IP, 37 μg/kg.

IV, LD_{50}, **rat** and **mouse**: 280 mg/kg; **rabbit**: 13 mg/kg.

PSILOSTROPHE sp.

= *Paperflowers*

Untoward effects: **Sheep** are frequently poisoned by grazing it on southern ranges of Texas, New Mexico, and southern Utah. Stumbling, sluggishness, coughing, vomiting, and green nasal discharge are symptoms. **Cattle** and **goats** can also be affected. Dry and young plants are more toxic than mature blooming plants.

PSORALEA sp.

Untoward effects: ***P. argophylla*** = ***Scurf Pea*** toxic principals are probably ***coumarins***, and ingestion of its seeds probably caused severe poisoning in a **child**.

P. corylifolia = ***Bavachi*** = ***Bawchang*** = ***Malay Tree*** has caused a photodermatitis in **man** and contains ***psoralen***.

P. tenuiflora found from Illinois to Texas and westward has poisoned **cattle** and **horses** grazing on the plant.

PSYCHOTRIA sp.

Found in the southwestern Amazon basin.

Untoward effects*:** Enhances hallucinogenic beverages due to its ***N,N-dimethyltryptamine (q.v.) content.

P. emetica of Columbia contains an ***ipecac***-type drug.

PTEROIS VOLITANS

= *Lionfish*

Also known as *featherfish, firefish, scorpionfish, tigerfish, turkeyfish,* and *zebrafish*. In nature, it inhabits temperate and subtropical seas.

Untoward effects*:** Rare in the U.S., but a 15-year-old **male** cleaning algae from a **fish** tank in a Las Vegas, Nevada, pet store was stung on a finger by a 10" long ***lionfish. Immediately, excruciating pain radiated up **his** arm. At the emergency room, erythema and edema of the finger and a red streak up to **his** wrist was noted. ***Lionfish*** respond aggressively to perceived threats by erecting their venomous stinging spines. Severe pain, vasodilation, muscle weakness, nausea, vomiting, hypotension, respiratory arrest, and death have been reported. Occasionally necrosis and sloughing of tissues at the wound site. Tetanus can develop. An antivenin is available. See ***Fish***.

PUMICE

***Untoward effects*:** Self-induced fibrous penile lesions (sclerogranuloma) in a 40-year-old **male** after 25–30 SC injections of a suspension.

PUMILIOTOXINS

Cardiotonic.

Untoward effects*: *Pumiliotoxin A is found on poisonous arrows and darts.

SC, LD_{50}, **mouse**: 2.5 mg/kg.

PUMPKIN

= *Cucurbita pepo*

***Untoward effects*:** Hypersensitivity to it runs in families – usually a dermatitis.

After being fed to male **rats** for 2 weeks caused 60% decrease in spermatazoa motility. Some diarrhea in **sheep** fed 400 g/day/3 days.

PUROMYCIN

= *CL 16,536 = P 638*

Antibiotic.

Untoward effects*:** Causes memory loss. Nephrotoxic, protein synthesis inhibitor, and can cause fatty liver. Enhances ***clofibrate effects and the anticholesterol effect of ***liothyronine***.

Causes long-term memory loss in **fish**. Blocks memory of maze-learning in **mice**. Causes fatty livers in **rats**. In **rats**, 25 mg/kg IP led to weight loss, weakness, and anuria. Renal and bone marrow lesions at 100 mg/kg. **Dogs** tolerate long-term treatment with 10 mg/kg without ill-effects, but a single dose of 50 mg/kg; orally or IV, caused repeated vomiting. In **cats**, 25 mg/kg IV led to hypotension.

LD_{50}, **mouse**: 675 mg/kg; **guinea pig**: 600 mg/kg; IV, LD_{50}, **mouse**: 335 mg/kg; **guinea pig**: 202 mg/kg.

PUTTY

Untoward effects*:** Old sticks of window glazing putty contain a high percentage of ***lead (q.v.). The dried sticks resemble lollipops or candy bars to **children** who suck on them or chew them. If > 2–3 oz are ingested, there is a danger of mechanical intestinal obstruction and chemical toxicity. ***Silly Putty®*** contains ***silicones*** and 1% ***boric acid*** (q.v.). Intestinal obstruction can occur.

PYRACANTHA sp.

***Untoward effects*:** Incidence of toxicity in the U.S. from ingestion of its berries has been 3–8% of Poison Control Center suspected plant poisonings. Its glycosides lead to inflammation of mucous membranes and vomiting.

P. coccinea = ***Gratego*** in Uruguay is said to be nontoxic, but cases of poisoning from eating the fruit have been reported. Symptoms include nausea, ataxia, hypotonia, miosis, and coma. Extrasystoles reported in one case.

Pet **animals** can be easily affected, as it has been a cultivated house plant. **Canaries** and **budgies** have been poisoned by eating its fruits and leaves. Prolonged exposure to the vapors from the flowers can cause severe irritation to **rabbit** eyes.

PYRAN COPOLYMER

= *NSC 46,015*

A synthetic polyanion or plastic that stimulates white blood cells and the reticulo-endothelial system.

Untoward effects*:** Hemolytic-uremic syndrome, macular rash interfering with ***fibrin formation and inducing hemorrhage, and febrile response in **children**.

PYRANTEL

The pamoate = *Banminth P* = *Cobantril* = *Combantrin* = *CP-10423-16* = *Early Bird* = *Helmex* = *Helmintox* = *Lumbriareu* = *Nemex* = *Piranver* = *RFD*

The tartrate = *Ban-A-Worm* = *Banminth* = *CP-10423-18* = *MK* = *Nemosan* = *Strongid*

The embonate = *Trilumbrin*

Nematocide.

Untoward effects: Mild and transient symptoms include nausea, vomiting, anorexia, abdominal pain or discomfort, tenesmus, diarrhea, headache, dizziness, and increased SGOT. Occasionally drowsiness, insomnia, rashes, and fever. Has caused an increase in ***theophylline*** serum levels.

Causes neuromuscular block in *in vitro* and *in vivo* trials. It is of a depolarizing type and is reversible (similar to ***succinylcholine***). It is a strong cholinesterase inhibitor and may cause hypertension (reversed by ***hexamethonium*** [q.v.], not ***atropine***). Use with caution in weak or debilitated **animals**. The U.S. now permits its use in **swine**, but limits its residues in their edible tissues at 10 ppm for liver and kidney, and 1 ppm in muscle. This is accomplished by withholding any treated feeds for at least 24 h before slaughtering. Causes significant decrease of serum alkaline phosphatase activity in **greyhounds**. Mild colic and watery diarrhea may occur in nearly 5% of the **horses** treated.

Pamoate LD_{50}, **rat** and **mouse**: 172 mg/kg.

PYRAZINAMIDE

= *Aldinamide* = *D-50* = *Pezetamid* = *Piraldina* = *Pirilène* = *Py Bracco* = *Pyrafat* = *Pyrazinecarboxamide* = *Pyrazine Carboxylamide* = *Tebrazid* = *Unipyranamide* = *Zinamide*

Tuberculostatic.

Untoward effects: Blood: Thrombocytopenia, sideroblastic anemia (hematocrit ~10%), increased serum ***urates***, increased plasma ***fibrinogen***, increased ***protein-bound iodine***, decreased plasma ***globulin*** and ***albumin***, decreased plasma ***prothrombin***, increased prothrombin time, hyperuricemia, and azotemia.

Clinical tests: False positive serum ***bilirubin***, icteric index, ***protein-bound iodine***, SGPT, and ***uric acid***; false negative or spuriously low ***17-ketosteroids*** urine test results.

Liver: Hepatotoxicity (incidence of 3–5%), jaundice, liver necrosis, increased SGOT, increased SGPT, and deaths.

Miscellaneous: Acute gouty arthritis, acute interstitial nephritis, dysuria, increased ***urobilinogen***, nausea, vomiting, anorexia, headache, diffuse macular rash from hypersensitivity, and a lichenoid photodermatitis.

Drug interactions: Blocks hypouremic action of ***allopurinol***. ***Alcohol*** abuse markedly enhances its toxic effects on the liver.

PYRAZOLE

= *1,2-Diazole*

Use: As a chelating agent and in chemical and dye sythesis.

Untoward effects: An active diuretic in **man**.

Drug interactions: Use with ***analgesics*** can add to gastrointestinal distress, renal cortical necrosis, and ***sodium*** retention.

Inhibits ***alcohol*** dehydrogenase completely, preventing its oxidation *in vitro* and *in vivo* (**rats** and **mice**); induces thyroid necrosis in **rats**. Although effective in decreasing toxicity of ***ethylene glycol*** in **dogs**, **monkeys**, and **rodents**, it causes emesis, anorexia, and liver damage, but less than that of ***ethylene glycol***. Can block ***bromobenzene*** hepatonecrosis in **mice**.

LD_{50}, **rat** and **mouse**: 21 and 22 mmol/kg, respectively.

PYRENE

A *polycyclic aromatic hydrocarbon* found in *coal tar*; a common air and water pollutant.

Untoward effects: Phototoxic, killing half the organisms (**tadpoles**, **brine shrimp**, **water fleas**, and **minnows**), after exposure to a few ppb (i.e. 1 lb in 110 million gallons). It is a cocarcinogen.

TD_{LO}, skin **mouse**: 0.5 g/day/3 weeks.

PYRETHRUM, PYRETHRINS, and PYRETHROIDS

Pyrethrum is usually obtained from the flowers of *Chrysanthemum cinerariifolium* = *Persian Insect Powder*, which contains ~1% pyrethrins, readily destroyed by sunlight. Oily vehicles help protect it from chemical degradation. *Pyrethrins I* and *II*, *cinerins I* and *II*, and *jasmolins I* and *II* are the main insecticidal pyrethrins.

Untoward effects: The flowers have been pulverized and used as an insecticide. Most toxic case reports are due to other added insecticides and carriers. Some **workers** processing it for several years have developed a sensitivity to it. Contact dermatitis and mental confusion in factory **workers** where it is ground or in those **people** who sleep in closed apartments where it is used. A hypersensitivity pneumonitis has developed after repeated exposure to aerosoled pyrethrum. A very few mild bronchospastic reactions in **people** with similar reactions to ***ragweed***. Hypersensitive **people** may also develop numbness of lips and tongue, sneezing, vomiting, diarrhea, tinnitus, headache, restlessness, incoordination, clonic convulsions, stupor, respiratory paralysis, and death after contact or inhalation. These reactions were noted especially when it was used parenterally. Conjunctival application led to chemosis. Use in shampoo against head lice has, within 1 h, been associated with an anaphylactoid reaction (periorbital edema, and later, dyspnea, chest tightness and numbness). Pyrethrum can contain varying amounts of ***pyrethrosin***, a contact allergen without any insecticidal properties. A **patient** who abused pyrethrin by repeated inhalation developed aplastic anemia and leukemia. Pyrethroids include ***allethrin***, ***fenvalerate***, ***permethrin***, ***phenothrin***, ***resmethrin***, and ***tetramethrin***. National Institute of Occupational Safety and Health/OSHA exposure limits are 5 mg/m^3; immediate danger to life or health: 5 g/m^3.

Despite repeated statements and clinical experience affirming its low **mammilian** toxicity (even in fogging trials), hematotoxic effects have been demonstrated within 1 day after fogging trials in **dogs**. Although safely used on a wide variety of zoological and laboratory **animals**, repeated applications to **snakes** may prove toxic. In fogging trials with **mice** and **birds**, it appeared that the high toxicity of the pyrethrins was due to sulfonated residues present. **Roaches** and **caterpillars** react within 1½ min after application with extreme excitement, then, incoordination, paralysis, and death. Pyrethrum is poisonous to **fish** and **frogs**.

Pyrethrum, LD_{50}, **rat**: 200 mg/kg.

Pyrethrin, LD_{LO}, **human**: 500 mg/kg.
LD_{50}, **rat**: 1.2 g/kg.

PYRIDINE

= *Azabenzene* = *Azine*

***Solvent and chemical intermediate. Coal tar* derivative.**

Untoward effects*:** Has a burnt and pungent **fish**-like odor with recognition threshold of 0.021 ppm. A respiratory irritant in ***tobacco smoke. Inhalation, percutaneous absorption, or even ingestion can lead to eye irritation, headache, nervousness, dizziness, insomnia, nausea, anorexia, polyuria, dermatitis, photosensitivity; central nervous system, liver, and kidney pathology. Dyspnea followed by shallow respiration after inhalation. National Institute of Occupational Safety and Health/OSHA upper exposure limits of 5 ppm. Immediate danger to life or health of 1,000 ppm.

LD_{LO}, **human**: 500 mg/kg.

Enlarged livers with high water content in **rats** ingesting it for 28 days at a 0.6% level in their feed. Necrosis and fatty liver changes in other **rat** feeding trials. High doses in the **rat**, **guinea pig**, and **cat** cause hypotension without peripheral vasodilation. In anesthetized **dogs**, IV use of lethal doses led to hypotension, marked tachycardia, and death due to respiratory failure. Chronic inhalation of vapors by **cats** (1 h/day/3–4 months) led to pulmonary emphysema, chronic bronchitis, and slight fatty infiltration of liver and kidneys.

LD_{50}, **rat**: 891 mg/kg; LD_{LO}, **guinea pig**: 4 g/kg. IV, LD_{50}, **dog**: 880 mg/kg. 1 h Inhalation LC_{50}, male **rat**: 9,010 ppm (4,000 ppm/4 h).

PYRIDOSTIGMINE BROMIDE

= *Kalymin* = *PB* = *Regonol* = *RO 1-5130*

***Cholinergic.* A *carbamate* analog of *neostigmine* (q.v.). *Cholinesterase inhibitor* with a short half-life. Contains 30.6% *bromine*.**

Use*:** In treating myasthenia gravis, ***atropine-like adverse effects of ***tricyclic antidepressants***; in antagonizing ***d-tubocurarine*** and ***pancuronium***-induced neuroblockade; and as a pre-exposure antidote to chemical and biological warfare nerve gases.

Untoward effects*:** After high doses, diarrhea and increased glandular secretions. Has induced hypertension, bradycardia, salivation, and widening of the pulse pressure. Toxic alopecia reported from its use in a 69-year-old **female** with myasthenia gravis. Confirmed by rechallenge. During Operation Desert Storm of the Gulf War, > 40,000 U.S. **troops** received (in violation of an agreement between the military and the FDA) 30 mg tid for 1–7 days while under threat of a gas attack. Induced symptoms in ~50% were flatulence, abdominal cramps, soft stools, chronic fatigue, muscle pain, memory loss, sleep disorders, and urinary urgency; ~1% wanted medical attention; and < 0.1% discontinued **their** medication because of ill-effects. ***Bromide (q.v.) intoxication with paranoid delusions, visual hallucinations, and increased ***bromide*** blood levels in a 59-year-old **female** given 150 mg every 3 h plus 180 mg sustained release at night for treatment of myasthenia gravis. Crosses the placenta decreasing **fetal** plasma cholinesterase activity and inducing **neonatal** myasthenia gravis and drug clearance in the **fetus** is slower than in the **mother**.

Drug interactions*:** Antagonizes mydriasis effect of ***sympathomimetics and the muscle relaxation of ***non-depolarizing agents***. ***Methocarbamol*** may have antagonized its action in a myasthenia gravis **patient**. Will potentiate ***depolarizing muscle relaxants***, leading to symptoms of excess amounts of ***acetylcholine***. Induced prolonged paralysis in 82-year-old **male** treated with ***succinylcholine***.

Diarrhea and emesis reported after clinical use in a **dog**.

LD_{50}, **rabbit**: 54 mg/kg; IM, 2.8 mg/kg; IV, 1.9 mg/kg.

PYRIDOXAL-5-PHOSPHATE

= *Codecarboxylase* = *Pysomijin* = *Sechvitan* = *Vitazechs*

Biologically active form of *pyridoxine*.

***Untoward effects*:** Rapid IV injections have caused severe nausea.

PYRILAMINE

= *Mepyramine* = *Pyranisamine* = *RP 2786*

The **male**ate form = *Antamine* = *Anhistol* = *Anthimin* = *Anthisan* = *Dorantamin* = *Enrumay* = *Histalon* = *Histan* = *Histapyran* = *Histatex* = *Neo-Antergan* = *Paraminyl* = *Parmal* = *Pyramal* = *Stamine* = *Stangen* = *Thylogen*

***Antihistamine*. An *ethylenediamine* (q.v.). Has anticholinergic activity.**

Untoward effects*:** A deliberate overdose (10 g) in a 46-year-old **patient** led to cardiogenic shock. A case report of lethality after 30 g. Apparently, coma and cardiorespiratory collapse precede death. Occasionally slight drowsiness. SC causes 6–12 times blood ***histamine levels for 24 h. Transient side-effects reported in 25–30% of **patients**. Only 3.6% discontinued therapy due to them. Can cause allergic contact dermatitis or photosensitivity. Acute toxicity produces dizziness, jitteriness, somnolence, nausea, abdominal cramping, weakness, faintness, fatigue, palpitations, dry nose, numbness, cold extremities, ***chloroform*** taste, mydriasis, narcolepsy, acute hysterical reaction, tendency to hemorrhage, sore tongue, hot flushes, early menses, and dermatitis medicamentosa. In **maternal** breast milk at ~0.6% of administered drug. A case report of aplastic anemia in a 65-year-old **male** after 75 mg/day/~3½ months.

Acute toxicity in **rat** and **mouse** leads to generalized tremors, incoordination, increased activity, restlessness, squeaking, chronic convulsions, and death by asphyxia.

LD_{50}, **rat**: 36 mg/kg; **mouse**: 235 mg/kg; IV, 30 mg/kg; IP, 102 mg/kg; SC, 100 mg/kg.

PYRIMETHAMINE

= *Chloridin* = *Daraprim* = *Malocide* = *RP-4753* = *Tindurin*

Sulfonamide synergist, protozoastat, folic acid antagonist.

Untoward effects*: *Folic acid deficiency, blood dyscrasias (thrombocytopenia, leukopenia, megaloblastic anemia) and decreased hematopoiesis, fever, atrophic glossitis, buccal mucosa ulceration, alopecia, vomiting, anorexia, urticaria, rashes, arthralgia, and shock. Nausea, vomiting, abdominal pain, exfoliative dermatitis, diarrhea, leukopenia, convulsions, coma, and death, especially in **children**. A 27-month-old **male** accidentally ingested 250–300 mg; after ~2 h he developed severe convulsions, apnea, cyanosis, tachycardia, and loss of consciousness. Death occurred ~3 h later, despite extensive therapy. Administered drug appears in **mother's** milk and has been effective in prevention and treatment of malaria in some **infants**. Occasionally teratogenic reports, but usually when combined with other drugs. A case report of a **male infant** with no thumbs, syndactyly of the first and second toes on a foot, bilateral undescended testicles, hypoplastic phallus, and a widely open anterior frontanelle born to a **mother** who received a single dose every week during **her** first 5 months of pregnancy.

Drug interactions*: *Calcium carbonate and ***magnesium trisilicate*** can decrease its absorption. By inhibition of dihydrofolate reductase, a ***folic acid*** deficiency is created. It is roughly 2,000 times as toxic to this enzyme system in malarial parasites than to the **human** host. Use with ***p-aminobenzoic acid*** and ***folic acid*** decreases its effectiveness. It displaces bound ***quinine*** and increases ***quinine*** toxicity.

Teratogenic in **miniature swine**. Used alone as a coccidiostat for **poultry**, it is too toxic, leading to poor growth and feathering, decreased erythropoiesis, lymphocytopenia, perosis, severe hyperchromic anemia, and deaths. A strong ***folic acid*** antagonist, teratogenic effects in pregnant **monkeys**, and bone marrow depression and leukopenia in **dogs** are understandable side-effects.

LD_{50}, **mouse**: 128 mg/kg; LD_{LO}, **hamster**: 250 mg/kg.

PYRIMINIL

= *DLP 787* = *PNU* = *N-3-Pyridylmethyl-N-p-nitrophenylurea* = *RH 787* = *Vacor*

***Rodenticide*. Inhibits *nicotinamide* (*niacinamide*). A yellow-green powder.**

Untoward effects*:** A neurotoxin requiring antidotal therapy within in an hour. Nausea, vomiting, abdominal pain, hypotension, chest pain, dysphagia, cardiac arrhythmias, loss of bladder control, nystagmus, peripheral neuropathy, gastrointestinal perforation, pneumonia, and hyperglycemia with or without ketosis. Most symptoms occur 4–48 h after ingestion. Ketoacidosis and diabetes mellitus (from necrosis of β-cells in the pancreas' islets of Langerhans) have often followed. Many deaths followed its ingestion in attempted and often successful suicides reported from South Korea and California. Deaths in 6/12 aged 19–50 years, after ingesting 0.39–7.02 g. A 25-month-old child was accidentally given some orally, causing lethargy, unresponsiveness, acute vomiting, seizures, hypoglycemia, evidence of autonomic and peripheral neuropathy, and ***glucose intolerance. Several months required for recovery. Lethal doses have ranged from 5–140 mg/kg. ***p-Nitroaniline*** (q.v.) has been found in the livers of **humans** poisoned by it.

High acute toxicity to **rodents** after a single feeding, with relatively low toxicity to non-target **animals**. **Dogs** have vomited after eating it; one became blind, and several developed glycosuria. Temporary glycosuria in poisoned **horses**. **Cats** developed bilateral mydriasis.

LD_{50}, **rat**: 4.5 mg/kg; **mouse**: 80 mg/kg; **cat**: 62–200 mg/kg; **dog** and **rhesus monkey**: 2–4 g/kg; **swine** and **ruminants**: 300 mg/kg; **poultry**: 700 mg/kg.

PYRITHIAMIN
= *Neopyrithiamine*

A *pyridine* antagonist analog of *thiamin*.

Untoward effects*:** Feeding to **mice** leads to symptoms of ***thiamin deficiency. Rapid depletion of brain ***thiamin*** after SC administration of 500 µg/kg/day in **rats** or 50 µg/day in **mice** decreases nerve conduction or transmission.

PYRITHIONE
= *Omadine* = *PTO* = *2-Pyridinethiol-1-oxide* = *SQ 2113*

***Antibacterial, antifungal.* The *sodium* form = *Fonderma* = *SQ 3277* is frequently used.**

***Use*:** In shampoos, cutting fluids, and as a fiber lubricant.

Untoward effects*: *2-Methylsulfonylpyridine is the terminal plasma metabolite in **humans** manufacturing it, as well as in the plasma of **rats**, **rabbits**, **dogs**, and **monkeys**.

Toxicity studies with it and its *sodium salt* as daily doses in **rats**, **mice**, **dogs**, and **monkeys** revealed in **dogs** sporadic emesis, lacrimation, conjunctival erythema, and a specific optic sensitivity with prolonged mydriasis, loss of pupillary response to light, and apparent blindness. These changes became irreversible if dosing was continued, and were probably associated with a ***pyridoxine*** deficiency. **Rats** and **monkeys** had transient signs of this toxicity, without marked ocular effects. Retinal detachment also reported in **dogs**.

Its *dimer* form = ***Dipyrithione*** = ***Omadine Disulfide*** = ***OMDS*** dermally to pregnant **rats** and **rabbits**, caused toxicity in **rabbits** at the 5 mg/kg/day level and 10 mg/kg/day in **rats**. Impaired hind leg function at the 7.5 mg/kg level in **rats**. Oral use in **rats** and dermal use on **swine** of its ***magnesium*** adduct led to fetal anomalies.

Sodium form LD_{50}, 875 mg/kg; **mouse**: 1172 mg/kg.

PYRITINOL DIHYDROCHLORIDE
= *Biocefalin* = *Enbol* = *Encephabol* = *Enerbol* = *Life*

Psychotropic.

***Untoward effects*:** Motor excitation, tension, insomnia, lichen planus-like photosensitive dermatitis, pemphigus, ageusia, proteinuria, and lupus-like syndromes.

PYROCATECHOL
= *1,2-Benzenediol* = *Pyrocatechin*

***Antiseptic.* A phenolic metabolite of *tannic acid*.**

Untoward effects*:** Topical use as an antiseptic has caused hypopigmentation. Causes an eczematous dermatitis and cross-reacts with other ***phenolics on patch-testing. Systemic effects as with ***phenols***.

LD_{LO}, **human**: 50 mg/kg.

Growth depressing for **chicks** at dietary level of 0.1% and nearly 100% mortality at 2% level. As an oxidizing agent in **horse** urine, it can cause red spots where a **horse** urinates on the snow – it is not hematuria.

LD_{50}, **rat**: 3.9 g/kg; **mouse**: 250 mg/kg; **dog**: 130 mg/kg; **cat**: 100 mg/kg; **rabbit**: 1 g/kg. SC, LD_{50}, **mouse**: 108 mg/kg.

PYROGALLOL
= *Pyrogallic Acid*

Untoward effects*:** By action on the nail plate, it colors treated nails brown. Allergic contact dermatitis when used in hair dyes and can cross-react with other ***phenolics and ***p-phenylenediamine***. Colors urine green to deep yellow, brown, or black. Systemic symptoms similar to ***phenol*** (q.v.). Stains skin and hair black. Ingestion has caused methemoglobinemia, hemolysis, gastrointestinal irritation, vomiting, diarrhea, chills, glycosuria, renal and hepatic (fatty degeneration) damage, cyanosis, convulsions, circulatory collapse, and death. Within 5 min after a **patient** covered ~two-thirds of the **patient's** body with it as an ointment in treating psoriasis, the **patient** collapsed and died due to percutaneous absorption of ~10 g of the drug. Fatal poisoning occurs after application to intact skin and has occurred even after earlier use without ill-effects.

LD_{LO}, **human**: 50 mg/kg.

Acorn ***tannins*** are transformed in the rumen of **cattle**, **sheep**, and **goats** by bacteria into the toxic pyrogallol.

LD_{50}, **rat**: 789 mg/kg; **rabbit**: 1.6 g/kg. LD_{LO}, **dog**: 25 mg/kg.

PYROLAN
= *ENT 17,588* = *Geigy 22,008*

Carbamate insecticide. Cholinesterase inhibitor.

***Untoward effects*:** LD_{LO}, **human**: 5 mg/kg.

Calves (< 2 weeks), 6 min after 0.25% water-based spray developed severe intoxication and death 6 min later. **Calves** survived after 0.05% spray. **Angora goats** dipped in a 0.25% solution died within 20 min. Symptoms included profuse salivation, running into fences, dyspnea, cyanosis, and hypoxia.

LD_{50}, **rat**: 50 mg/kg; **mouse**: 90 mg/kg; **chicken**: 11 mg/kg.

PYRROLIZIDINES

Alkaloids (*amsinckine, floridanine, heliotrine, heleurine, indicine, intermedine, jacobine, jacoline, jacozine, lasiocarpine, lycopsamine, macrophylline, retrorsine, ridelline, senecionine, seneciphylline, supinine, symphytine*, etc.) in many plants (*Amsinckia, Arnebia, Borago, Crotalaria, Cynoglossum, Echium, Heliotropium, Senecio, Symphytum, Trichodesma*, etc.).

Untoward effects*:** A naturally occurring liver carcinogen found at 0.01–0.6% concentration in herbal ***teas. At least 260 are present in ~6,000 plants and ~90 in medicinal plants are toxic. Commonly in herbal medicines and ***"bush teas"***, especially in Asia, Africa, India, and the West Indies, where liver disease is common. Toxicity has been reported in > 2,000 **people** in Afghanistan. Has also caused cell damage in the lungs and other organs. The term seneciosis is now reserved for this toxic effect in **horses**, **cattle**, **sheep**, **goats**, **pigs**, **quail**, **chickens**, and **doves**. The term hepatic veno-occlusive disease is used in **man**. The main pathology in all are occlusion of the centrilobular hepatic vein with centrilobular hemorrhagic necrosis, hepatocellular megalocytosis, bile duct hyperplasia, and cirrhosis. **Women** in Tanzania ingest ***Heliotropium supinum*** after childbirth. **Infants** are more susceptible to its ill-effects than **adults**. Lactating **mothers** may show no ill-effects, but their **sucklings** may develop tumors. Use of such herbals has been correlated with the high incidence of liver and pancreatic tumors among **African Bantus**. **Infants** of Mexican Americans in Arizona developed the fatal disease after being given herbal ***teas***. The calculated intake by a 6-month-old **infant** in the 2 weeks before hospitalization was 70–147 mg of the alkaloid and its ***nitrogen oxide*** derivative (equally toxic) from the ***gordoloba yerba*** (***Senecio longilobus***) ***tea***. Extract of the plant revealed 3 mg of free alkaloid and 10.5 mg of the ***nitrogen oxide***/g of plant. Another 2-month-old **infant**, fed the ***tea*** for 5 days as a cough medicine, also became toxic. Severe liver damage documented in a **patient** due to a supplier providing a more toxic plant for the ***tea***. Liver damage and death in a **woman** who drank large quantities of ***maté tea*** for a long period of time. Liver pathology in **children** is likely from drinking ***milk*** from **cows** that have consumed ***Senecio***. Has been found in ***comfrey-pepsin*** capsule preparations found in health food stores. Has also been found in ***honey*** produced by **bees** and mixed with cereal grains.

First cutting hays tend to have more of the plant sources. Liver cirrhosis and death have developed in 2–6-month-old **calves** after consuming such hay, while adult **cattle** were unaffected. It appears to predispose **sheep** to hemolysis associated with chronic ***copper*** poisoning. Grazing plants with high levels of the alkaloids has killed some **cows** within a few days. Lower levels lead to fibrosis, bile duct hyperplasia, and megalocytosis. ***Milk*** transfer was ~1–17 μg/100 ml of ***milk***. In Australia, ingestion has been the major cause of "walkabout" in **horses** with central nervous system excitement and even violence. A single dose in **rats** can induce chronic liver lesions and hepatomas that may not manifest themselves for 1½–2 years. Severe hepatotoxicity and, usually, dullness, tachycardia, tachypnea, and weakness, occur in all **animals** except the **guinea pig**. These plants are bitter and normally only grazed when feed is scarce, or if it is in hay. Occasionally, 6 months may pass before toxic symptoms are noted.

LD_{50}, **rats**: 30–90 mg/kg; occasionally 300 mg/kg.

PYRULARIA PUBERA

= *Buffalo Nut* = *Oilnut*

Decidous 5–15 ft shrub in the mountains of northeastern Alabama northward to the Virginias.

***Untoward effects*:** The fruits, especially its seeds, are very poisonous when ingested by **man** and **livestock**.

PYRVINIUM PAMOATE

= *Molevac* = *Neo-Oxypaat* = *Pamovin* = *Pamoxin* = *Polyquil* = *Poquil* = *Povan* = *Povanyl* = *Pyrcon* = *Pyrvin* = *Pyrvinium Embonate* = *Tru* = *Vanquin* = *Vermitiber* = *Viprynium*

Anthelmintic. A cyanine dye.

***Untoward effects*:** The *chloride* form should no longer be used as it is four times as toxic as the *pamoate*. Can color feces, vomitus, and clothing red. Slight abdominal pain, nausea, vomiting, and diarrhea in some **patients**. Hypersensitivity leads to photosensitization, pruritic polymorphous rash, fever, periorbital swelling, edema, and arthralgia. May cause false or spuriously high readings for blood in feces.

Q

QUARTZ

Untoward effects: **Man** is exposed to inhalation of its aerosoled particles, primarily in mining and quarrying operations, although exposure also comes from some spackling and drywall taping compounds. Symptoms would be the same as for ***silicon dioxide*** (q.v.), and include respiratory fibrosis, and probably esophageal and gastrointestinal cancers.

In **rat** experiments, inhalation causes particles to enter pulmonary interstices and penetrate regional lymph nodes, causing fibrosis.

Inhalation TC_{LO}, **human**: 16 million particles/cu ft. LC_{LO}, **human**: 300 μg/m³.

QUAZEPAM

= *Doral* = *Dormalin* = *Oniria* = *Prosedar* = *Quazium* = *Sch-16,134* = *Selepam*

Hypnotic, sedative.* A *benzodiazepine.

Untoward effects: Drowsiness, lethargy, headache, fatigue, xerostomia, and dizziness. Uncommonly, ataxia, confusion, tremors, depression, nervousness, anorexia, euphoria, decreased libido, nightmares, taste alteration, nausea, pruritus, rash, and urinary incontinence. Small amounts of it and its metabolites have been found in breast milk.

Limited ataxia and central nervous system stimulation in **cat** and **squirrel monkey** trials.

LD_{50}, **mouse**: > 5 g/kg.

QUETIAPINE FUMARATE

= *IGI 204,636* = *Seroquel* = *ZD 5077* = *ZM 204,636*

Psychotherapeutic.

Untoward effects: Postural hypotension, agitation, headache, xerostomia, drowsiness, insomnia, dizziness, constipation, weight increase, and increased serum aminotransferase and alanine transaminase. Occasionally tardive dyskinesia, extrapyramidal reactions, and increased ***T4*** concentration. After 2 months of treatment, a 46-year-old **female** with schizoaffective disorder developed increased thyroid stimulating hormone. Shortness of breath, hyperventilation, tachypnea, and acute respiratory alkalosis after 50 mg at bedtime in 69-year-old **female**.

Drug interactions: ***Phenytoin*** increases its clearance. A similar, but lesser, effect by ***thioridazine***.

QUILLAJA

= *China Bark* = *Soap Bark*

Traces are used to maintain foaming in beverages.

Untoward effects: This ***saponin*** can be irritant in all tissues, killing muscle and nervous tissue by direct action. Orally to **mammals** leads to violent gastrointestinal irritation, tonic and clonic convulsions, disturbances of circulatory and respiratory systems, and post-mortem rigidity. One of its active principles is ***sapotoxin***, which can cause weakness, convulsions, and death.

QUINACRINE

= *Acrinamine* = *Acriquine* = *Atabrine* = *Atebrimum* = *Antimalarina* = *Chinacrina* = *CI 423* = *Mepacrine* = *RP 866* = *SN 390* = *Triquine*

Protozoacide, anthelmintic, antimalarial.

Untoward effects: ***Blood and bone marrow***: Hypersensitivity bone marrow damage with agranulocytosis and, rarely, aplastic anemia. Hemolytic anemia, especially in glucose-6-phosphate dehydrogenase-deficient **subjects**.

Clinical tests: Interference with false positive or spuriously high ***bilirubin***, icterus index, and adrenal function.

Eye: Retinal toxicity reported.

Liver: Hepatic necrosis and jaundice is usually a result of hypersensitivity; increases alkaline phosphatase.

Miscellaneous: Severe nausea, vomiting, epigastric pain, diarrhea, central nervous system stimulation, dizziness, headache, and psychotic reactions, including mania, peripheral neuritis, insomnia, and bizarre dreams. Although used in treating petit mal, it can exacerbate grand mal seizures. Transvaginally led to hyperirritability lasting ~4 h in 1/85 **females**. Resorbed if urine is alkaline. Yellow urine if urine is acid. Intrauterine use has induced tubal fibrosis and occlusion. Teratogenic at high dosage.

Poisoning: A **soldier** swallowed 250 tablets (25 g). Vomiting, collapse, and stupor followed. Recovered after treatment.

Skin: Yellow skin discoloration (especially of the face and dorsum of the hand), yellow–brown pigmentation of the palate and oral mucosa, variable blue–brown nails with diffuse colored tranverse bands, delayed (12–27 days) exfoliative dermatitis, lichenoid skin reactions, contact dermatitis, exanthematous eruptions, urticaria, alopecia, occasional yellowing of conjunctiva, and photosensitivity. Many with severe lichen planus-like lesions develop anhydrosis, scaly red plaques and fungating warty growths, primarily on **their** hands, and, eventually, squamous cell carcinomas.

Drug interactions: Its use enhances the effects of ***pamaquine***. Its effects are enhanced by ***primaquine*** and the latter's toxicity and that of other ***8-aminoquinolines*** is increased by it. ***Disulfiram***-like reaction with ***alcohol*** intake, because it inhibits oxidation of ***acetylaldehyde***, a metabolite of ***alcohol***.

Cholinesterase inhibitor and has caused anorexia, nausea, dermatitis, stomatitis, fever, and lethargy in **cats** and **dogs**. Marked decrease in renal elimination can occur in basic, as compared to acid, urine. Urinary alkalinizers can increase its activity and toxicity. Its activity may be potentiated by ***primaquine*** and other ***8-aminoquinolines***. Increases fetal death rate in treated **rats**.

LD_{50}, **mouse**: 1.3 g/kg; **chicken**: 714 mg/kg.

QUINAGOLIDE
= *CV 205-502* = *Norprolac* = *SDZ 205-502*

***Prolactin inhibitor. Non-ergot dopamine D_2 receptor agonist* in treatment of hyperprolactinemia, but not for suppression of puerperal lactation.**

Untoward effects: Usually transient and occurring during the first few days of treatment or with increased dosage leading to nausea, vomiting, headache, dizziness, and fatigue. Anorexia, abdominal pain, constipation or diarrhea, insomnia, flushing, edema, nasal congestion, and hypotension in 1–10% of **patients**. Rarely, acute psychotic reactions.

QUINAPRIL
= *Accupril* = *Accuprin* = *Accupro* = *Acequin* = *Acuitel* = *CI 906* = *Korec* = *PD 109,452-2* = *Quinazil*

Angiotensin-converting enzyme inhibitor, antihypertensive.

Untoward effects: Headache (2–6%), dizziness (2–7%), dry cough (3.1–5%), nausea and vomiting (0.8–6%), pruritic skin rash (< 1%), taste disturbances (< 0.5%), fatigue, diarrhea, constipation, decreased ***aldosterone*** secretion, and hypotension reported. Rarely, sore throat, chest pain, tingling or swelling of hands and feet, and yellowing of eyes or skin.

Drug interactions: Severe hyperkalemia and renal failure in 74-year-old **male** after addition of ***trimethoprim-sulfamethoxazole*** twice daily for treatment of mild acute pyelonephritis.

QUINAPYRAMINE
= *Antrycide* = *M7555* = *Piraldin* = *Pyraldin*

Trypanocide.

Untoward effects: Young **animals** and **horses** cannot tolerate overdosage. European **cattle** breeds seem more subject to reactions than tropical breeds. The *chloride* form leads to more local reactions than the *methyl sulfate*. The *sulfate* is noticeably irritant to **horses** and very irritant to **dogs**.

QUINCE
= *Cydonium sp.*

Untoward effects: Anaphylaxis can occur. An immediate type of hypersensitivity has been reported from the allergens in quince seeds. The latter is found as a suspending and stabilizing agent in hair and cosmetic preparations.

QUINDOXIN
= *Quinoxaline 1,4-dioxide*

Use: Formerly, as a growth-promoter for **pigs** and **poultry**.

Untoward effects: Withdrawn from the market because **people** handling the feed develop eczematous reactions. Feeding excessive (above recommended) amounts leads to nose and liver tumors in **animals.**

QUINESTROL
= *Estrovis* = *Pentovis* = *Quinestradiol* = *W 3566*

Estrogen.

Untoward effects: Mucorrhea, headache, nausea, vomiting, dizziness, lassitude, depression, abdominal cramps, bloating, weight increase, breast engorgement, edema, nervousness, breakthrough or withdrawal bleeding, rash, hyperpigmentation of areolae, pruritus, urticaria, and jaundice. Rebound lactation after treatment withdrawal. Azospermia in 28-year-old **male**.

QUINETHAZONE
= *Aquamox* = *CL 36,010* = *Hydromox*

Diuretic, antihypertensive.

Untoward effects: Photosensitivity dermatitis, hyperuricemia, nausea, weakness, dizziness, anorexia, headache, xerostomia, cholestasis, and menstrual irregularities with increased flow and shorter cycles. Enhances the effects of ***gallamine*** and ***tubocurarine***. Urine tests show glycosuria.

QUINIDINE

***Antiarrhythmic.* Class Ia. Many forms exist. The more common ones are:**

The bisulfate = *Biquin* = *Chinidin-Duretter* = *Chinidin-Durules* = *Kiditard* = *Kinichron* = *Kinidin Durules* = *Kiniduron* = *Quiniduran* = *Quinora*

The gluconate = *Duraquin* = *Gluquinate* = *Quinaglute* = *Quinate*

The polygalacturonate = *Cardioquin* = *Galactoquin* = *Naticardina*

The sulfate = *Cin-quin* = *Quinidex* = *Quinicardine* = *Quiniduran* = *Systodin*

Untoward effects*: *Blood and bone marrow: Thrombocytopenic purpura, bone marrow depression, hypoplastic anemia, agranulocytosis, hemolytic anemia, immunohemolytic anemia, thrombocytopenia with hemoptysis, melena, subretinal hemorrhage, hematuria, ecchymoses, petechiae, bloody blisters in the buccal mucosa, and epistaxis; non-thrombocytopenic purpura, drug fever, aplastic anemia, pancytopenia, lupus erythematous, and neutropenia. Small doses lead to hemolysis and anemia in glucose-6-phosphate dehydrogenase-deficient **persons**.

Cardiovascular: Ventricular fibrillation and tachycardia, syncope, extrasystoles, hypotension, flushing, prolonged QT interval, widened QRS complex, torsades de pointes, atrial flutter, slowed atrial rate, and conduction defects. Fatalities have been induced. Use in the nursing **mother** can cause arrhythmias in her **infant**.

Ear and eye: Tinnitus and transient loss of hearing in 3/1000. Impaired vision due to optic neuritis (toxic amblyopia), retinal damage, sicca syndrome, diplopia, photophobia, scotoma, mydriasis, nyctalopia, and acute decompensation of cornea.

Fetus: Large doses have caused abortions, intrauterine deaths, teratogenic effects (central nervous system defects, especially hydrocephalus; eye, ear, limb, face, and genitourinary tract malformations), deafness, and thrombocytopenia.

Gastrointestinal: Esophagitis, nausea, vomiting, diarrhea, abdominal pain, and anorexia.

Hypersensitivity: Fever, chills, tremors, gastrointestinal symptoms, shock, asthma, and swelling of face, neck, hands, and feet.

Liver: Rare instances of induced hepatic pathology, probably a type of hypersensitivity reaction. Fever, increased SGOT, increased SGPT, increased lactic dehydrogenase, increased alkaline phosphatase, and increased leucine aminopeptidase. Granulomatous hepatitis also reported.

Miscellaneous: Polyarthritis, unconsciousness, fatalities, headache, confusion, vertigo, syncope, fever, joint or muscle pain, sweating, excitement, incoordination, dementia, and delirium. Severe tremors with toxic levels. The **elderly** are particularly susceptible to central nervous system effects.

Poisoning: Ataxia, lethargy, convulsions, severe hypotension, broad-notched P wave on electrocardiogram; prolonged PR, QRS, and QT intervals; broad-base T wave in 57-year-old **female** who ingested 20, 200 mg tablets. Serum level was 9.7 mg/l 6 h after ingestion and 0.35 mg/l at 80th hour. Unconscious for 9 h; recovery after 4 days. Therapeutic serum levels quoted as 0.03–0.6 mg%, toxic at 1 mg%, and 3–5 mg% as lethal (some have reported half these levels as fatal). Other self-poisonings include more than one drug. Adverse symptoms after doses of 1 or more grams.

Renal: Cases of induced lupus nephritis (49-year-old **female** receiving 200 mg qid) and irreversible glomerulonephritis.

Respiratory: An 80-year-old **male** developed respiratory arrest after 400 mg, and a 42-year-old **female** apnea after 1.2 g IM. **Both** revived. Has induced pulmonary hemorrhage.

Skin: Exacerbation of psoriasis; lichen planus, systemic lupus erythematosus, lupus erythematosus, photosensitivity with delayed erythema, fixed-drug eruptions, purpura, exanthematous eruptions, vasculitis; erythematous papules and nodules that may ulcerate and become gangrenous; acneform eruptions, pruritus, exfoliative dermatitis, toxic epidermal necrolysis, and pigmentation.

Drug interactions: Susceptible to hepatic microsomal enzyme inducers, such as ***barbiturates*** and ***phenytoin***. Decreased cardiac output with ***propranolol***. Bradycardia due to its interaction with ophthalmic ***timolol***. ***Antacids*** increase its absorption and toxicity. Use with ***acetazolamide*** or ***sodium bicarbonate*** increases its reabsorption due to increased urinary pH. Quinidine excretion is decreased and its potential toxicity increased in alkaline versus acid urine. Additive neuromuscular-blocking effects with apnea and muscle weakness with ***clindamycin***, ***colistin***, ***kanamycin***, ***lincomycin***, ***neomycin***, ***streptomycin***, and ***tobramycin***. ***Erythromycin*** may inhibit its hepatic metabolism. Combined with ***verapamil*** may cause significant hypotension and increased quinidine plasma levels. ***Rifampin*** increases oral or IV quinidine rate of metabolism. Use with ***reserpine*** has led to cardiac arrhythmias. Synergistic or additive to ***warfarin's*** anticoagulant effect and it decreases intestinal synthesis of ***vitamin K***. A potentially fatal interaction reported with ***digoxin*** due to increase (occasionally two times) in serum ***digoxin*** levels was first recognized in 1978. Syncope, paroxysmal ventricular tachycardia, extrasystole, increased premature atrial contractions and premature ventricular contractions. Occasionally no interaction with ***digitoxin***; in other trials, increase in ***digitoxin*** elimination half-life by 50%. ***Phenobarbital*** and

phenytoin decrease its half-life by about 50%. Atypical ventricular tachycardia (torsades de pointes) with prolonged QT interval when used with ***amiodarone***. ***Triamterene*** with it leads to spuriously high quinidine serum test levels. ***Cimetidine*** and, to a lesser extent, ***ranitidine***, decrease its clearance, increasing its serum concentration. Chronic administration of low doses may enhance the systemic effect of ***dextromethorphan***. In some **patients**, it decreases the clearance of ***flecainide*** and ***propafenone***. Sinoatrial arrest and atrioventricular junctional escape rhythm in 68-year-old **male** 30 min after ***lidocaine*** infusion. Use with ***nifedipine*** has caused variable results in different **patients**, ranging from ~40% decrease in quinidine levels to slight increase in both. It can attenuate the bioavailability of ***nisoldipine***. ***Amiloride*** exaggerates its effects on QRS duration. Inhibits ***amantadine*** renal clearance in **males** only. ***Ciprofloxacin*** and ***metronidazole*** increased its serum concentration in 51-year-old **female**. ***Ritonavir*** decreases its metabolism and increases its toxicity. ***Ketoconazole*** caused increased quinidine plasma levels in a 68-year-old **patient**. ***Itraconażole*** has the same effect. A rare case of sensitivity to it due to a previous intake of a cocktail containing ***quinine***, with resulting thrombocytopenia reported. It can induce neuromuscular blockade and potentiate both ***non-depolarizing*** (***gallamine*** and ***tubocurarine***) and ***depolarizing*** (***decamethonium*** and ***succinylcholine***) ***muscle relaxants*** and can increase respiratory depression and cause apnea. By inhibiting CYP2D6 metabolizing enzyme, it can increase serum concentrations of ***amitriptyline***, ***codeine***, ***desipramine***, ***dextromethorphan***, ***haloperidol***, ***hydrocodone***, ***imipramine***, ***metoprolol***, ***mexiletine***, ***paroxetine***, ***propafenone***, ***propranolol***, ***thioridazine***, and ***timolol***.

LD_{LO}, **human**: 50 mg/kg.

Inhibits liver microsomal enzyme metabolism of ***pentobarbital*** in tissues of **goats**, **mice**, and **rats**. Has caused laminitis and other allergic reactions in **horses** (use only 20% of initial recommended dose to test for hypersensitivity). Has caused nausea, vomiting, diarrhea, and hypotension in **dogs**. In **dogs**, decreases ***digoxin*** levels by at least 1/3 (but not with ***digitoxin***). May exacerbate congestive heart failure.

LD_{50}, **rat**: 263 mg/kg.

QUININE

= *Chininum*

Antipyretic, protozoacide, skeletal muscle relaxant.

Untoward effects: Cinchonism.

LD_{LO}, **human**: 50 mg/kg; TD_{LO}, **woman** (4–5 weeks pregnant): 20 mg/kg.

Blood and bone marrow: Thrombocytopenic purpura (from as little as 0.3 g in 16-year-old **female**), thrombocytopenia, agranulocytosis, hemolytic anemia, granulocytopenia, eosinophilia, pancytopenia, neutropenia, epistaxis, purpuric spots on skin and mucous membranes, disseminated intravascular coagulation, lymphocytopenia, postpartum hemorrhage, and methemoglobinemia. Hemolytic anemia in many glucose-6-phosphate dehydrogenase-deficient **patients**. A 23-year-old **female** consumed ~5 oz quinine water (8 mg/100 ml) in cocktails leading to thrombocytopenia. ***Tonic water*** and bitter ***lemon*** cocktail mixes both contain quinine and, in sensitized **people**, can induce precipitous thrombocytopenia. Can interfere with blood group compatibility testing.

Cardiac: Bradycardia, widened QRS interval, and arrhythmias. As a common adulterant in ***heroin*** leading to clinical decrease of myocardial excitation.

Ear: Ototoxicity. May worsen tinnitus or deafness and dizziness. Hearing loss is usually transient.

Eye: Retinopathy, retinal vasoconstriction, amblyopia (often from its use as a ***heroin*** adulterant), and toxic amaurosis. Amblyopia in **female** ingesting 5.6 g to induce menses. Total blindness in a **female** ingesting 8 g within 24 h as an abortifacient. Self-poisoning with severe visual loss in **patients** 38 and 60 years ingesting 2 and 4 g, respectively. After abstinence, visual acuity improved, but no full return to normal. Iris abnormalities with abnormal stromal pattern and loss of pupillary margin pigment after 17–20 years. Diplopia, keratopathy, and mydriasis.

Fetus: Congenital malformations (brain, skull, and spine; diaphragmatic hernia, extra liver lobe, hypoplastic lung, intestinal defects, left displacement of heart, patent foramen ovale, ventricular septal defect, meningocele, umbilical hernia, glaucoma, wide distance between eyes, underdeveloped nose, cleft palate, deafness, thrombocytopenia, hypoplastic optic nerve, hydrocephalus, and genitourinary defects), and **fetal** death in pregnant **women** ingesting it at high dosage as an abortifacient.

Liver: Hemolytic jaundice and granulomatous hepatitis.

Miscellaneous: Speech difficulties, confusion, flushing, giddiness, staggering, headache, delirium, weakness, paralysis, and collapse. Its toxicity in late pregnancy is not due to its oxytocic action, but rather to its causing the release of ***insulin***. Dysphagia and respiratory distress in myasthenia gravis **patients**. Quinine-containing drinks can hamper the diagnosis of malaria. Fibrous nodules 1.5–6 cm in buttocks of 58-year-old **female**, 47 years after 18 injections for malaria. Gastrointestinal upsets (nausea, vomiting), fever, chills, hypotension, hypoglycemia, and shock reported. Causes weakness by interfering with ***calcium*** influx at the nerve terminal, which inhibits ***acetylcholine***

release. A 62-year-old **male** nearly died after drinking one ***gin and tonic*** cocktail. **His** only previous allergies were to *Rhus* and an over-the-counter cold remedy. By increasing ***catecholamine*** levels, it can cause interference with fluorimetric urinary assays. Can aggravate myasthenia gravis. As a ***heroin*** adulterant because of its bitter, narcotic-like taste, and **users** get a flash sensation from it by IV use, probably by dilation of facial blood vessels. Convulsions, coma, and death have also followed its IV use when mixed with ***heroin***. Colors urine brown to black.

Neonate: Found in breast milk (up to 1.6 μg/ml). Has induced thrombocytopenia.

Poisoning: Amblyopia with nearly total blindness after ingestion of 4.5 g of the *sulfate* in a suicide attempt. Total blindness in 44-year-old **male** after ingestion of 34, 200 mg (6.8 g). Convulsions and cardiac arrest in 2-year-old **female** after ingestion of nearly 2 g. Resuscitated. **Her** fixed dilated pupils returned to normal after 3 weeks. Many case reports of renal failure and other symptoms in **female**s taking it as an abortifacient. One source lists therapeutic plasma levels as 0.18 mg% and lethal levels at 1.2 mg%. 8 g is reported as fatal, but a **patient** has survived ingestion of 30 g. Hyperthermia (viz. saunas) and ***alcohol*** can potentiate its toxicity. Use with water can increase its plasma concentration. Death within a few hours from acute toxic doses. Large doses in treating malaria lead to "black water fever", characterized by hyperpyrexia and hemoglobinuria. A 17-year-old **male** ingesting 5 g of the *bisulfate* developed fatal respiratory distress syndrome, despite treatment.

Skin: Toxic epidermal necrolysis; lichenoid, eczematoid, acneform, and exanthemous eruptions; contact dermatitis, Stevens–Johnson syndrome, fixed-drug eruptions, urticaria, edema of mucous membranes, cutaneous vasculitis, and photosensitivity. Purpura in 27-year-old **female** ingesting cocktails (***Dubonnet-on-the-Rocks***®) containing 7 mg/100 ml (not stated on the label). It is a ***histamine*** (q.v.) releaser.

Drug interactions: Excreted if urine has acidic pH and reabsorbed if urine is alkaline. ***Acidifying agents***, such as ***ammonium chloride***, increase its urinary excretion and ***sodium bicarbonate*** decreases its urinary excretion. Significantly inhibits ***amantadine*** renal clearance in **males**. Decreases ***digoxin*** clearance by 26% and increases ***digoxin*** serum concentration. Enhances ***antipyrine*** metabolism and decreases its half-life. Has potentiated the anticoagulant effect of ***acencoumarol*** and ***warfarin***. ***Pyrimethamine*** can displace quinine bound to tissue sites, enhancing its activity and toxicity. ***Cigarette*** smoking and ***rifampin*** pretreatment markedly increase its urinary clearance.

In **animals**, discontinue if any eye pathology is noted. Will be present in milk of treated **animals**. Its use increases the sleeping time of ***phenobarbital***-treated **mice** and **rats**, increases the plasma half-life of ***pentobarbital*** in **goats**, and decreases the LD_{50} of ***pentobarbital*** in **mice**.

LD_{LO}, **rabbit**: 4 g/kg; **frog**: 1.5 g/kg.

QUINOLINE YELLOW

= *D&C Yellow #10 = E104 = Food Yellow 13*

Use: Synthetic dye for woollens, nylons, and ***silks***. FDA approved for drugs and cosmetics (except near the eye).

Untoward effects: A 56-year-old **male** developed severe urticaria with ***naproxen*** tablets colored with this dye, but not with the same medication in a white tablet. Banned in some countries. Found in ***Mylanta® Gelcaps*** and ***Ogilvie Hair Repair Lotion***.

Quinoline Yellow Spirit Soluble = D&C Yellow #11, also approved by the FDA for use in externally applied drugs and cosmetics, has caused some degree of allergic dermatitis in ~6% of **people** using a popular brand of soap.

QUINONE

= *1,4-Benzoquinone*

Use: In photography, dye and ***hydroquinone*** manufacturing, tanning hides, and as a reagent.

Untoward effects: Strong oxidizing agent. Irritating vapors lead to eye irritation, conjunctivitis, keratitis, and skin irritation. National Institute for Occupational Safety and Health/OSHA upper exposure limit of 0.1 ppm (0.4 mg/m^3). Immediate danger to life or health, 100 mg/m^3. LD_{LO}, **human**: 40 mg/kg.

LD_{50}, **rat**: 130 mg/kg; IV, 25 mg/kg.

QUINOXALINE DIOXIDE

Untoward effects: Severe contact eczemas in Lurgan, Northern Ireland **patients** handling **pig** feeds containing it as a growth-promoter. Rash maximal in, but not restricted to, light-exposed areas. Lesions were dry and tended to develop fissures. All strongly positive to standard and photopatch tests with 0.5% material.

QUINTOZENE

= *Avicol = Botrilex = Brassicol = Folosan = Kobutol = PCNB = Pentagen = PKhNB = Terraclor = Tilcarex = Tritisan*

Soil and seed fungicide.

Untoward effects: LD_{LO}, **human**: 500 mg/kg.

Hexachlorobenzene (q.v.) is a trace contaminant of commercial material. Purified material shows maternal toxicity in **mice** and cleft palates in **mouse** fetuses; also fetal malformations in **rats**. 180 or more ppm in diet of **dogs** leads to hepatosis and necrosis. Possible carcinogen. Ataxia, tenseness, jerkiness, jitteriness, hyperexcitability, disorientation, goose-stepping ataxia, and sitting within 1 h in **mallards** and 3 days in **pheasant**. Remission in up to 15 and 7 days, respectively.

LD_{50}, **female rat**: 1.65 g/kg; **male**: 1.71 g/kg; **mallard ducks** and **pheasants**: > 2 g/kg.

QUINUCLIDINYL BENZILATE

= *Bravo Zulu* = *BZ* = *QNB* = *Ro 2-3308*

***Chemical and biological warfare nerve gas.* Used as aerosoled particles or dust.**

***Untoward effects*:** Long-lasting, incapacitating agent, hallucinogen, leading to dizziness, erratic eye movements, poor reflexes; burning sensation in nostrils, lips, and throat, in tests with Army **personnel**. The U.S. Army, in 1985, had 50 tons stockpiled. American Citizens for Honesty in Government, part of the Church of Scientology, obtained a vial containing 100 mg from the manufacturers, simply by asking for it. This was enough to incapacitate 400–500 people for several days (~200 µg each). Psychosis and uncontrolled violence also reported. Upsets the body's water balance and temperature regulation.

Disorientation, aimless running, and screaming in exposed **cats**, **dogs**, **mice**, **rats**, **guinea pigs**, and **rabbits**.

IV, LD_{50}, **mouse**: 18 mg/kg; **dog**: 15 mg/kg.

QUINUPRISTIN

Use*:** In a 30 : 70 combination with ***dalfopristin, known as ***Synercid***, for IV use versus life-threatening ***vancomycin***-resistant *Enterococcus faecium* infections. Both are ***streptogramin antibacterials***.

Untoward effects*:** Pain, inflammation, and edema at injection site; thrombophlebitis. Arthralgias, myalgias, and increased conjugated ***bilirubin occur frequently.

Drug interactions*:** May increase serum concentrations of ***cyclosporine, ***midazolam***, and ***nifedipine*** and potentiate ***cisapride*** increase of QTc interval.

QUINURONIUM

= *Acaprin* = *Babesan* = *Bayer 205* = *1,3-Di-6-quinolyl-urea* = *Ludobal* = *Pirevan* = *Zothelone*

***Piroplasmocide*. The *sulfate* is usually used.**

***Use*:** Effective versus *Babesia* infections in **cattle**, **horses**, **sheep**, **swine**, and **dogs**.

Untoward effects*:** Strong ***parasympathomimetic. It potentiates neuromuscular block by drugs such as ***succinylcholine***. Usual dose in **sheep** inhibit about 40% of their cholinesterase levels. ***Atropine*** is protective against some of these side-effects and is also antidotal. Avoid use with ***phenothiazines***, ***procaine***, ***organophosphorus insecticides*** or ***anthelmintics***, or any drugs depressing cholinesterase. Although effective versus many *Babesia* and *Theileria spp.*, it is not effective versus *Theileria parva* (East Coast Fever).

SC, LD_{50}, **rat**: 8 mg/kg.

R

R-11
= *MGK Repellent 11*

Insect repellent.

Untoward effects*:** Widely used for > 30 years in **human** and **animal** products, yet, the Environmental Protection Agency panicked and suggested U.S. suppliers not use it any more, because **rats** and **mice** fed LARGE doses developed adverse reproductive effects, ovarian atrophy, and oncogenicity. To equal these levels, a **dog** would have to drink more than one cup of ***flea and tick spray/day/1 year.

LD_{50}, **rat**: 2.5 g/kg.

RABEPRAZOLE
= *Aciphex*

Benzimidazole proton pump inhibitor.

Untoward effects*:** Occasionally nausea, abdominal pain, flatulence, constipation, diarrhea, and headache. Infrequently, myalgias, rash, and breast enlargement. Gastric hyperplasia in **animals** and **humans**; carcinoid tumors in **animals**, and secretion into ***milk of **animals**. Disagreeable taste and unpleasant odor in ***milk***, milk products, and meat of **animals** that graze on it. Oral tablets are formulated to prevent inactivation by stomach acids.

Drug interactions*:** Increases ***digoxin peak plasma levels ~30% and decreases ***ketoconazole*** bioavailability ~30%.

RADISH
= *Raphanus sp.*

Untoward effects*: *R. sativus, the common edible radish, contains high levels of ***nitrates*** (~200 mg/100 g), which have been implicated in the etiology of stomach and esophageal cancer, as they convert to ***nitrosamines***. Also contains ***sulforaphane*** which is a growth deterrent and anticarcinogen, the toxic alkaloid, ***solanine*** (q.v.), and interferes with the use of ***iodine***. A clinically noticeable effect in ***iodine*** metabolism can be expected from ingestion of ~50 g. Cases of allergic contact dermatitis reported. Can cause flatulence. In attempting to uncover the source of a *Listeria* outbreak, FDA investigators found about 26% of ***potatoes*** and 30% of ***radishes*** in two Minneapolis supermarkets were contaminated with *L. monocytogenes*. ***R. raphanistrum*** or ***Wild Radish***, a common weed in the U.S., Europe, and Australia, contains goitrogenic substances and ***mustard oil*** (***allylisothiocyanate*** – q.v.). All parts, especially the seeds, are poisonous when grazed or eaten in hay by **cattle**, **horses**, **sheep**, and **swine**. Symptoms are severe gastroenteritis, diarrhea, salivation, anorexia, photosensitivity, weight loss, and cardiopulmonary failure. Post-mortem shows jaundice, friable liver, renal congestion, and hemoglobinuria.

RADIUM

Untoward effects*:** Dial **painters**, **chemists**, and **processors** develop malignant bone neoplasms. In the past, iridescent watch-dial **painters**, usually **girls**, who pointed the tips of small black brushes with **their** lips (after dipping the brushes into radium ***paint) slowly accumulated enough radium in **their** bodies to cause bone malignancies 30 years later. Radium **miners** and ***tobacco*-smokers** have an increased risk of bronchial and pulmonary cancers from radium. Blood dyscrasias, burns, rhagades, hyperkeratosis, ulcerations, Bowen's disease, skeletal defects, bone absorption, and cataracts have also followed exposure. Thyroid carcinomas and leukemias have also been reported. An acute leukemia in a 3-year-old **female** whose **mother** was irradiated for an enlarged spleen in **her** fourth month of pregnancy. When it was used in treating carcinoma of the uterus, vesical ulcers, vesicovaginal fistulae, stenosis of the ureters, hematuria, and fever occurred. Has localized in bone leading to chronic osteitis in which osteosarcomas have supervened. Cataracts and bone tumors developed in **dogs** 1–3 years after experimental exposure. **Rats** and **mice** have developed similar lesions.

RADON

A gaseous product of natural decay of *uranium* and *radium*. Found in numerous other unrelated mines.

Untoward effects*:** Occupational exposure of **miners** gives strong evidence of respiratory tract tumors. Possible increased risk in **people** living in homes having higher levels of it or its decay products. Many homes are sources of exposure, with Grand Junction, Colorado being among the worst, where ***uranium mill tailings were used as subfill and backfill around those built between 1952 and 1966. Environmental Protection Agency has estimated 5000–20,000 lung cancer deaths/year are linked to it. Radiation proctitis and enterocolitis in 74-year-old **female**, 30 years after radon implants for tumor of uterine cervix. Cases of leukemia, skin and renal cancers have also been attributed to exposure to it. "Spent" radon seed (once used for cancer therapy) in ***gold*** rings eventually caused squamous cell carcinomas of the fingers, and radiodermatitis with erythema and hyperkeratosis. Old ***gold*** jewelry should be checked for radioactivity. Carcinomas on or near the eyelids also reported from ***gold*** seed implants.

Inhalation by **male hamsters** led to bronchiolar basal cell hyperplasia, malignant lung tumors (adenomas, adenocarcinomas, squamous cell carcinomas), fibrosis, and interstitial pneumonia. In **dogs** and **rats**, inhalation led to nasal carcinomas, bronchio-alveolar carcinomas, and fibrosarcomas. Tumors of the upper lip and urinary tract of **rats** also reported.

RAFFINOSE

Untoward effects*:** A simple sugar found in legumes and ***beans that is poorly digested by **man** leading to offensive intestinal gas.

RAFOXANIDE

= *Bovanide* = *Duofas* = *Flukamide* = *Ranide*

Fasciolicide.

***Untoward effects*:** Blindness has been reported in some treated **sheep** on therapeutic dosage, and experimentally only after 200 mg/kg dosage; in **dogs** given 100 mg/kg/day/3–11 days and blindness reported in **zebu cattle** at 60 mg/kg. Transient amaurosis in a **mare** after 3 mg/kg dose has been reported.

RAGWEED

= *Ambrosia sp.*

Untoward effects*:** Its pollen is one of the most allergenic in North America leading to rhinitis or "hay fever" and a contact dermatitis from an oil it contains. A gram of its pollen contains ~90,000,000 pollen grains. Photoxicity occurs. Disagreeable taste and unpleasant odor in ***milk, milk products, and meat from **animals** that graze on it. ***Echinacea*** can cause allergic reactions, particularly in ragweed-allergy **sufferers**.

Causes asthmatic-type of allergic reaction in some **dogs**. Plants are ***nitrate***-accumulators (> 1.5%) and, if **animals** eat it at > 0.05% of their body weight, it can be fatal. **Cattle**, **sheep**, and **goats** are more commonly affected, **horses**, less often. Symptoms include weakness, ataxia, collapse, tachypnea, shallow breathing, tachycardia, blue-brown discoloration of the white of the eye, tongue, and lips; coma and death. Blood is chocolate brown. Ingestion of western ragweed by **cows** leads to bitter milk. Passive transfer of anaphylaxis to a normal **dog** with serum from ragweed-sensitive **dogs**.

RAISIN

Untoward effects*:** A source of food allergies in some **people**, as are ***grapes. Both are also potent sources of intestinal gas.

RALOXIFENE

= *Evista* = *Keoxifene* = *LY 139,481*

***Antiosteoporotic*. For use in postmenopausal women. *Estrogen* agonist effect on bone and an antagonist effect on breast and uterus. The *hydrochloride* form was originally referred to as *LY 156,758*.**

***Untoward effects*:** Hot flashes (~25%, especially during the first 6 months of therapy), leg cramps (4%), vaginal discharge, hepatitis, fatigue, mood changes, and increased venous thromboembolism manifested as leg swellings, sudden chest pain, shortness of breath, coughing up of blood, and vision changes. Teratogenic.

Drug interactions*: *Cholestyramine decreases its absorption by ~60%. ***Ibuprofen*** increases its serum concentration by 20–30%.

RAMARIA FLAVO-BRUNNESCENS

***Untoward effects*:** Ingestion of this toadstool-like fungus on pastures in Brazil leads to "mal do eucalipto" in **cattle** causing alopecia on tail, salivation, anorexia, corneal opacity, blindness, abortion, rumen atony, atrophy and desquamation of tongue and mouth, papillae, loosening of horns and hooves, and extravasation of blood in the anterior chamber of the eye, and death. Post-mortem reveals esophageal necrosis, petechiae on the myocardium, pulmonary emphysema, and dark-colored blood that fails to clot. Symptoms were induced by 5 g/kg and a single dose of 36 g/kg was lethal in **cattle**.

RAMIPRIL

= *Altace* = *Cardace* = *Delix* = *HOE-498* = *Pramace* = *Quark* = *Ramace* = *Triatec* = *Tritace* = *Unipril* = *Vesdil*

Angiotensin-converting enzyme inhibitor, antihypertensive.

***Untoward effects*:** Hypotension, dizziness; non-productive cough that can last for at least 6 months after discontinuance of treatment; angioedema (eye, face, lips, tongue, and larynx), syncope, nutropenia, and hyperkalemia. Occasionally abdominal pain, constipation, diarrhea, anorexia, xerostomia, dysphagia, increased salivation, taste upsets, photosensitivity, rash, pruritus, anxiety, amnesia, convulsions, neuralgia, paresthesia, somnolence, tinnitus, tremor, vertigo, arthralgia, epistaxis, myalgia, and hyperhidrosis. Rarely, decreased hemoglobin and hematocrit.

LD_{50}, **rat**: > 10 g/kg; **mouse**: 10.5 g/kg.

RANITIDINE

= *AH-19065* = *Antak* = *Azantac* = *Melfax* = *Noctone* = *Raniben* = *Ranidil* = *Raniplex* = *Sostril* = *Taural* = *Terposen* = *Toriol* = *Trigger* = *Ulcex* = *Ultidine* = *Zantac* = *Zantic*

***Antiulcerative, histamine H_2-receptor antagonist.* Inhibits gastric acid secretion.**

Untoward effects*: *Blood: Thrombocytopenia, granulocytopenia, agranulocytosis, neutropenia, leukopenia, aplastic anemia, bone marrow hypoplasia, eosinophilia, and immune hemolytic anemia.

Cardiovascular: Bradycardia, hypotension, atrioventricular block, premature ventricular contractions, chest pain, tachycardia, and cardiac arrest.

Central nervous system: Occasionally depression, delirium, lethargy, confusion, somnolence, disorientation, chorea, throbbing bifrontal headache, agitation, aggressive behavior, non-responsiveness, hallucinations, exacerbation of glaucoma, and blurred vision. Muscle spasms and rigidity after second dose of 150 mg reported in 26-year-old patient. Renal failure and nephrotic syndrome has occurred.

Endocrine: Painful gynecomastia, decreased libido, no ejaculation, and impotence in **men**.

Liver: Hepatitis and jaundice incidence of 0.06%–0.08% with increased SGPT after 2–5 weeks. Death has occurred.

Miscellaneous: Polyarticular pain and swelling, myalgia, anaphylactic shock, hypergastrinemia, fever, salmonellosis, pancreatitis, abdominal discomfort, constipation, diarrhea, nausea, vomiting, bronchospasm, and slight increase in serum ***creatinine***. Has induced a carcinoid tumor. Found in breast milk of treated **mothers**.

Skin: Rashes, Stevens–Johnson syndrome, toxic epidermal necrolysis, and rare instance of alopecia.

Drug interactions: Interferes with ***itraconazole*** and ***ketoconazole*** absorption, since they require gastric acidity for absorption. Impaired absorption decreases ***diazepam*** serum levels 10–25% and ***enoxacin*** by 26–40%. May cause dramatic increase in serum ***phenytoin*** levels; peak ***metoprolol*** plasma levels increase by 33%, and half-life increases from 3.9 to 6 h. Increases blood ***alcohol*** levels by 34%, ***propantheline bromide*** increases its bioavailability by 22%, increases ***glyburide's*** hypoglycemic effect, and causes dramatic increase in ***theophylline*** serum levels and toxicity by decreasing the latter's metabolism. Slows metabolism of ***thiobarbiturates***, and ***benzodiazepines*** requiring decrease of 10–25% in anesthetic dose. Increases ***triazolam's*** absorption from 67% to 87% and increases ***didanosine*** and ***midazolam*** bioavailability. Possible synergism with ***acetaminophen*** leading to cholestatic hepatitis.

Can increase ***caffeine*** serum levels when taken with ***coffee*** or ***colas*** leading to jitters. Has induced neurotoxicity in **mice**.

RANUNCULIN

Untoward effects: A lethal glycoside present in many species of true ***buttercup*** (***Ranunculus***, q.v.) plant, which is hydrolyzed to ***protoanemonin*** (q.v.).

RANUNCULUS

= *Buttercup* = *Crowfoot* = *Jaljari* = *Spearworts*

Untoward effects: The entire plant contains ***ranunculin*** (q.v.), which hydrolyzes to an oil, irritant and blistering to tissue, and causes gastric irritation, diarrhea, vomiting, and convulsions from large amounts. **Beggars** used it topically to ulcerate **their** feet and arouse sympathy. Acrid taste has discouraged **children** from eating it and being seriously injured by it. Irritancy peaks during flowering stages. Disagreeable taste and unpleasant odor for **man** of the ***milk***, milk products, and meat of **animals** that have grazed on it.

Poisoned **cows** give less ***milk***, that is bitter and reddish; when severe leads to colic, diarrhea, black foul-odored feces, blistering of mouth, photosensitivity, nervousness, ear and lip twitching, dyspnea, bradycardia, salivation, hematuria, abortions, and eventually convulsions and occasional blindness. **Horses**, **goats**, and **sheep** show similar symptoms. **Horses** also often show mydriasis. Poisoned **sheep** may suddenly collapse; **swine** show little gastrointestinal upsets, but suddenly become paralyzed and blind. The toxin is volatile and when in dry hay, it causes no problems.

RAPAMYCIN

= *AY 22,989* = *NSC 226,080* = *Rapamune* = *Sirolimus*

Immunosuppressant.

Untoward effects: Dose-dependent hyperlipidemia, leukopenia, and thrombocytopenia. Arthralgias, hypokalemia, hypotension, rashes, thrombocytopenic purpura, and mucosal herpes simplex are common. Undesirable taste and interstitial pneumonitis reported by **some**.

Drug interactions: Use with ***Neoral*** formulation of ***cyclosporine*** causes significant increase in the latter's serum levels. ***Grapefruit juice*** increases its serum concentration and ***rifampin*** decreases its serum concentration by 71%. ***Diltiazem*** and ***ketoconazole*** delay its time to peak concentration.

Testicular adenomas, decreased sperm counts, and decreased testicular weight in **male laboratory animals**.

RAPE

= *Brassica napus* = *Navette* = *Swedes*

Untoward effects: A ***nitrate***-accumulator. Causes photosensitivity with jaundice and hemoglobinuria in **cattle**,

sheep, and **swine**. A goitrogen. Contains toxic ***erucic acid*** (q.v.) that caused pathology in **poultry** (enlarged thyroid, congested liver, and decreased weight gain) and ***glucosinolates*** (q.v.). Intestinal stasis, anorexia, hemoglobinuria, dyspnea, icterus, SC and pulmonary emphysema, blindness, hemolytic anemia, weakness, paralysis, coma, and death have been reported in **cattle**.

RAPISTRUMRUGOSUM

= *Turnip Weed*

Untoward effects: ***Glucocheirolin*** and its aglucone, ***cheirolin***, are in the fruit and leaves of this Australian pasture weed, causing goitrogenic effects in **people** from drinking the ***milk*** of **cows** that ingested the weed.

Cheirolin is the principal component of the thioglycosides in the plant, and is goitrogenic to **rats**. ***Isothiocyanate*** in ***cheirolin*** is converted in the rumen of **cows** to a goitrogenic ***thiourea*** (q.v.).

RASPBERRY

= *Rubus sp.*

Untoward effects: Since it contains ***salicylates***, it can cross-react with ***salicylates*** leading to induced urticaria. Contains ***oxalates*** (q.v.) and ***ellagic acid***. Raspberries from Guatemala and Chile, in 1997, became tainted with *Cyclospora* and sickened ~1000 **people** in the U.S. and Canada.

RAUWOLFIA SERPENTINA

= *Alseroxylon* = *Austrawolf* = *Chandmaruwa* = *Egalin* = *Gendon* = *HBP* = *Hiwolfia* = *Hyper-Rauw* = *Hywolfia* = *Raudixin* = *Rauserpa* = *Rauserpin* = *Rautensin* = *Rauval* = *Rauwolfia* = *Rauwiloid* = *Ra-Valeas* = *Rawfola* = *Rawpentina* = *Rivadescin* = *Ru-Hy-T* = *Serfolia* = *Serpadex* = *Serpina* = *Wolfina*

Contains > 40 alkaloids. Reserpine is one of the best known. Dr Leonard Rauwolf, in the sixteenth century, journeyed to the Orient and introduced the plant and drug to Western civilization, although, at that time, Western scientists refused to accept its therapeutic claims.

Antihypertensive, tranquillizer.

Untoward effects: Bradycardia, mental depression, drowsiness, diarrhea, fluid retention, parkinsonism, salivation, blurred vision, nightmares, ptosis, rhinitis, nasal stuffiness, memory loss, decreased libido, impotence, ejaculation difficulties, decreased ***catecholamine*** and ***vanillylmandelic acid*** body stores, orthostatic hypotension, galactorrhea, and extrapyramidal symptoms. Controls excitement in psychotic **patients**, but, when given to anxiety-ridden **neurotics**, the anxiety is often aggravated. Increases seizures in **epileptics**. **Infants** born to **mothers** who took it during **their** third trimester of pregnancy are frequently lethargic during the first 24 h, and have nasal congestion that lasts about 1 week. May potentiate ***alcohol's*** depressing effect. Use with ***digitalis*** glycosides (***digoxin*** and ***digitoxin***) can induce arrhythmia and undue bradycardia. Potentiated by ***general anesthetics***. Poisoning with psychic depression has resulted in suicides. In Tanzania, extracts of ***R. rosea*** are used on poisonous arrows.

The leaves and roots of ***R. tetraphylla*** (***Bitter Ash***) are lethal to **poultry**.

R. vomitoria = ***Akanta*** (Igbo) = ***Ira Igbo*** (Yoruba) leads to nausea, vomiting, violent purging, drowsiness, and depression.

LD_{LO}, **human**: 5 g/kg.

Rat trials indicate it would be best to avoid it in pregnancy. Ancient **Indian** legend tells that the **mongoose** chews its leaves to mobilize courage, before attacking **cobras**.

RAZOXANE

= *ICI 59,118* = *ICRF 159* = *NSC 129,943* = *Razoxin*

The (+)-form = ADR 529 = Dexrazoxane = ICRF 187 = NSC 169,780 Zinecard

Antipsoriatic, antineoplastic.

Untoward effects: Acute myeloid leukemia, cutaneous squamous cell carcinomata, nausea, lethargy, diarrhea, alopecia, leukopenia, neutropenia, gout, and occasional epistaxis. Use was abandoned due to carcinogenic effects.

REDUVIUS sp.

= *Kissing Bugs*

Untoward effects: Nausea, vomiting, faintness, and swelling at site of the bite, which is extremely painful, comparable to that of a **snake**.

REDWOOD

= *Sequoia sempervirens*

Untoward effects: Occupational asthma from its dusts in **foresters** and **wood-workers**.

REFUIN

= *AME* = *RO 5-9000*

***Antineoplastic.* Usually IV.**

Untoward effects: Pain at injection site, phlebitis of injected veins, nausea, vomiting, diarrhea, chills, fever, irreversible shock, and death. Use abandoned in favor of more effective and less toxic agents.

RENGAS

= *Gluta renghus* = *Rhengas Tree*

Untoward effects: Acute dermatitis in **man**, **rabbits**, and **guinea pigs** from the cut fruit and bark. Dermatitis from handling the wood in furniture manufacturing, and from its resin used in making a ***lacquer*** (***Japan lacquer***) leading to Mah-Jong dermatitis. Can cause skin blisters. Cross-reaction with *Rhus* plants, *Toxicodendron sp.*, ***mango***, and ***cashew nut trees***.

RENOXIPRIDE

= *A 33,547* = *FLA 731* = *Roxiam*

Psychotherapeutic.

Untoward effects: Bone marrow toxicity, cytopenia, red cell aplasia, thrombocytopenia, and extrapyramidal symptoms reported. Interferes with ***L-dopa*** efficacy.

REPAGLINIDE

= *Novonorm* = *Prandin*

***Antidiabetic.* A nonsulfonylurea.**

Untoward effects: Hypoglycemia. Some hematological and cardiovascular problems. Dizziness, visual disturbances, nausea, vomiting, and hypoglycemia in 68-year-old **male** taking 2 mg 1.5–2 h before **his** evening meal. ***Erythromycin*** and ***ketoconazole*** may decrease its metabolism and increase its serum concentration. ***Rifampin*** may decrease its effects.

RESERPINE

= *Crystoserpine* = *Eskaserp* = *Geneserp* = *Hiserpia* = *Lemiserp* = *Raurine* = *Rau-Sed* = *Releserp* = *Reserpoid* = *Rivasin* = *Sandril* = *Sedaraupin* = *Serexal* = *Serfin* = *Serpalan* = *Serpaloid* = *Serpanray* = *Serpasil* = *Serpasol* = *Serpate* = *Serpine* = *Serpiloid* = *Vio-Serpine*

***Antihypertensive, tranquillizer.* An indole alkaloid.**

Untoward effects: LD_{LO}, **human**: 50 mg/kg.

Behavior: Central sedation, sleep disturbances, depression, and personality changes.

Blood: Inhibits platelet function leading to thrombocytopenia, and decreased ***serotonin*** levels. Excessive bleeding during or after surgery. Anemia, lupus erythematosus, and non-thrombocytopenic purpura.

Cardiovascular: Sinus bradycardia, premature ventricular contractions, angina, fluid retention, congestive heart failure, arrhythmias, hypotension, and orthostatic hypotension leading to syncope.

Clinical tests: Can cause spuriously low protein-bound iodine test results, spuriously high ***5-hydroxyindoleacetic acid*** urine test results, and false positive tests with ***phentolamine***.

Central nervous system: Depression, syncope, dullness, nightmares, paradoxical excitement, headache, nervousness, dizziness, catatonic stupor, akathisia, tremors, drowsiness, insomnia, parkinsonism, tardive dyskinesia, and suicidal ideation.

Ear: Acoustic nerve neuropathy and hearing loss.

Eye: IV use leads to decreased intraocular pressure, optic atrophy, retinopathy, secondary glaucoma, blurred vision, lacrimation, conjunctival hyperemia, hyphema, miosis, uveitis, cataracts, and loss of light perception.

Fetus and neonate: Lethargy, nasal congestion, breathing difficulties while nursing, anorexia, substernal muscle retractions, diarrhea; bradycardia, hypothermia, stillbirths, and congenital anomalies. Excreted in breast milk.

Gastrointestinal: Nausea, vomiting, diarrhea, xerostomia or salivation; increased volume and free ***hydrochloric acid*** in gastric juice; erosive gastritis, hemorrhages from duodenal ulcer, anorexia, and intestinal hypermotility.

Miscellaneous: Weight increase, decreased metabolic rate, ***salt*** and ***water*** retention; (in **males**) leads to impotence or delayed ejaculation, muscle aches, gynecomastia, galactorrhea, non-puerperal lactation, dysuria, systemic lupus erythematosus, excessive appetite, leg weakness, and nosebleeds. Conflicting reports on its possible association with increase in breast cancer. Accidental poisonings reported in young **children**. It depletes tissue ***catecholamine*** stores for days after withdrawal. Indian **physicians** have said rauwolfia effects can sometimes last for at least 6 months.

Respiratory: Rhinitis, nasal congestion, dyspnea, epistaxis, and bronchospasm.

Skin: Rash, pruritus, fixed-drug eruptions, lupus erythematosus, photosensitivity, and flushing.

Drug interactions: Caution must be used in **patients** also receiving ***anesthetics*** or ***phenothiazines***, such as ***chlorpromazine***, ***perphenazine***, ***thioridazine***, and ***trifluperazine***, because they enhance its hypotensive and sedative effects. Arrhythmias and undue bradycardia from combination with ***digitalis glycosides*** (***digoxin*** and ***digitoxin***). Excitation, increased blood pressure, tachycardia, and seizures reported with ***monoamine oxidase inhibitors*** and ***furazolidone***. Potentiates the anticoagulant effect of ***warfarin***. ***Thiazide diuretics*** antagonize its ***sodium***- and ***water***-retaining effects. Use with ***alcohol*** can cause rigid extremities, incoordination, and head-drop, even with very low doses of reserpine. Lower ***quinidine*** doses required with ***quinidine***. Excessive sympathetic blockade has occurred with ***propranolol*** or other ***β-blockers***, leading to hypotension and bradycardia. Apparently decreased anticonvulsant activity and increased central depression with ***alcohol*** or ***barbiturates***. With ***monoamine oxidase inhibitors***, can increase effects of ***monoamine oxidase inhibitors***, cause paradoxical excitement, and antagonize reserpine's hypotensive effect. May attenuate cardiovascular effects of ***ephedrine***. Mania and hyperexcitability occurs with ***tricyclic antidepressants***. Central nervous system decrease with ***antihistamines***. Can interfere with ***l-dopa*** antiparkinsonian effects. Use with ***norepinephrine*** leads to uncontrolled blood pressure increase, due to increased sensitivity to ***norepinephrine***.

A zero tolerance has been established by the FDA for its metabolites in or on uncooked edible tissues and eggs of **poultry**. To assure its safe use, regulations exist covering the maximum concentration in premixes. Avoid unnecessary prolonged use on pregnant **animals**. Single injections of 0.8–2.0 mg/kg IM on the ninth or tenth day of gestation in **rats** has been teratogenic. Depletes ***catecholamine*** stores in **man**, **rabbits**, **rats**, **mice**, **monkeys**, **mink**, **cattle**, **horses**, **sheep**, **swine**, **poultry**, **cats**, **dogs**, **fish**, and **snails**, and possesses subtle long-term effects. Young **rats** recover quicker than older ones and ***norepinephrine*** repletion requires at least 2 weeks after simple short-term depletion studies. May take many months for full recovery in some species, including **man**. Relaxed nictitating membrane in **cats** is associated with nearly complete depletion of its ***norepinephrine*** content. Blocks ovulation in **rats**. Potentiates ***parathion*** and ***carbaryl*** toxicity in **rats**. Toxicity dramatically increases with either increase or decrease of normal ambient room temperatures in **rat** trials. Ptosis and diarrhea are common untoward effects in **cats** and **dogs**. Ptosis also occurs in **rodents**. Bradycardia production by it in **dogs** is easily neutralized by ***atropine***. Although used experimentally to block conditional avoidance responses, it can have serious long-term effects on memory. In **horses**, 5 mg/kg caused violent colic. A 900 kg **bull** died after 50 mg IV. Tranquillizing doses in **poultry** (100–200 μg/kg) caused decreased body temperature, bradycardia, hypotension, and diarrhea. A single oral dose of 10 mg/kg is fatal to **dogs**. Symptoms include severe depression and diarrhea. Oral and SC use in **rats** and **mice** leads to assorted carcinomas.

LD_{50}, **mouse**: 390 mg/kg; IV, LD_{50}, **dog**: 500 μg/kg; **rat** and **rabbit**: 15 mg/kg; **guinea pig**: 17 mg/kg.

RESORANTEL
= *HOE 296V* = *Terenol*

Anthelmintic.

Untoward effects: Some treated **cattle** may show transient diarrhea and inappetance for 24 h.

RESORCINOL
= *1,3-Benzenediol* = *Resorcin*

Keratolytic, antipruritic, antiseptic, surfactant. A phenolic.

Untoward effects: Irritant to skin, eyes, and mucous membranes of the nose, throat, and upper respiratory tract. Absorption through the skin or ulcerated areas or after ingestion leads to methemoglobinemia, cyanosis, convulsions, restlessness, tachycardia, dyspnea, dizziness, tinnitus, drowsiness, hypothermia, tremors, anemias, hemoglobinuria, dermatitis, sweating, stinging, scaling, pruritus, myxedema, and pathology of liver, kidney, and spleen. Conjunctivitis and keratitis after local application. Can color urine dark green and nail plates become brown. When used as hair lotion for dandruff (2.5%) on light-colored hair, or on traces of soap, the hair develops greenish discoloration. Resorcinol, a known thyroid inhibitor, is in ***coal***-processing wastes. Areas of Colombia, South America and Kentucky, U.S. have a high goiter incidence, despite adequate ***iodine*** intake, and it has been found in the well water of these areas. Collapse, convulsions, and unconsciousness after a **person** ingested 10 g; eventually recovered.

In laboratory **animals**, it causes tremors, epileptiform convulsions, and unconsciousness. With larger doses, these symptoms are more violent, respiratory distress, and, eventually, death. Urine becomes olive-green to black.

Ingestion by **dogs** leads to nausea, vomiting, abdominal pain, and methemoglobinemia. Topical application has caused proliferative, benign, and malignant ear tumors in **rabbits**, but not **mice**.

LD_{50}, **rat**: 301 mg/kg; skin, **rabbit**: 3.4 g/kg. SC, LD_{50}, **mouse**: 213 mg/kg.

RETEPLASE
= *BM 06,022* = *Rapilysin* = *Retavase*

Thrombolytic.

Untoward effects: Bleeding (~15%; significant in ~5%; 1% may require transfusion) including intracranial hemorrhage. Serum half-life of 13–16 min (***alteplase*** – 5 min, ***anistreplase*** – 90 min, ***streptokinase*** – 23 min).

RETRORSINE

Untoward effects: Hepatotoxic ***pyrrolizidine*** (q.v.) alkaloid found in ***Senecio sp.***, due to its pyrrolic metabolites. Given to lactating **rats** caused death in their **sucklings** with striking liver lesions, even when their **mothers** showed no such lesions. Toxicity decreased with age. Most severe liver lesions in 3–7 week-olds. **Guinea pigs** showed no toxicity at up to 800 mg/kg. Health hazard from its presence in a Paraguayan herbal ***tea*** made from *Senecio grisebachii*.

IP, LD_{50}, **male mouse**: 65 mg/kg; **male quail**: 279 mg/kg; **male hamster**: 81 mg/kg.

RHIZOCTONIA LEGUMINICOLA

Untoward effects: This microscopic fungus on certain forages, especially ***red clover***, an alkaloidal parasympathomimetic "slobber or salivation factor", now called ***slaframine*** (q.v.), which causes **cows** to salivate profusely, plus pollakiuria, diarrhea, stiff joints, bloat, and anorexia; some die. **Sheep**, **swine**, **goats**, **chickens**, **guinea pigs**, **rats**, and **mice** are also affected.

RHIZOPUS sp.

Untoward effects: A number of nosocomial surgical wound infections reported in many hospitals after use of elasticized adhesive tape (***Elastoplast***®) over wounds. ***Ethylene oxide*** sterilization of these rolled bandages was not effective. ***Cobalt*** irradiation was substituted, but sterility is not guaranteed. Fatal cases of cerebral mucormycosis were attributed to it in hemodialysis-induced ***aluminum*** overload therapy with ***deferoxamine***, which, apparently, served as a growth factor for *Rhizopus sp.* Has been a cause of abortions in **cows**.

RHODAMINE B

= *CI Basic Violet 10* = *D&C Red 19*

Untoward effects: Use is not common in the U.S. (occasionally in ***lipsticks*** and mouthwashes, as long as ingestion did not contribute to > 1% in **human** diet). It colors urine red. Allergic contact dermatitis reported in **people**. Aerosoled environmental exposure of 17 **people** for a mean of 26 min, with 16 complaining of burning of **their** eyes, excessive lacrimation, nasal burning or itching, chest pain or tightness, rhinorrhea, cough, dyspnea, burning of throat and skin, pruritus, headache, or nausea lasting ~24 h.

As a visible fluorescent dye qualitative marker of **animal** bait consumption, it stains the pelage of **jackrabbits** for at least 6 weeks and their gastrointestinal tract, feces, and urine for 6–8 days. Claws and hair of **coyotes** are marked for at least 175 days; claws and hairs of **beavers** and **gophers**, and feathers of **chickens** for several weeks. A photosensitizer. Pathology (including carcinogenicity) of kidney, spleen, and liver in experimental **animals**. Up to 20 ppb is used to check water flow and is allowed for use with food **fish**.

LD_{LO}, **rat**: 500 mg/kg.

RHODIUM

Untoward effects: Inhalation of fumes may cause respiratory sensitivity.

The *trichloride* is a commonly available soluble form, in IV **rabbit** and **rat** trials, slight weight decrease, waning respiration, and death reported without any parenchymal disease in vital organs, suggesting death from central nervous system effects.

Chloride: IV, LD_{50}, **rat**: 198 mg/kg; **rabbit**: 215 mg/kg.

RHODODENDRON

Untoward effects: Contains a toxic resinoid cardiac glycoside (***andromedotoxin*** – q.v.). All parts of the plant are toxic. **Children** have been poisoned from sucking on the leaves or ingesting a "***tea***" made from the leaves. **American Indians** on the eastern U.S. seaboard used a ***tea*** from the related ***mountain laurel*** as a suicide agent. **Children** often chew on the rhododendron leaves, leading to transitory mouth burning, vomiting, diarrhea, paresthesias, sialism, headache, muscle weakness, and vision problems. Bradycardia and hypotension in more severe cases. ***Honey*** made from its nectar has caused secondary poisoning in a 27-year-old **female**, leading to cardiac arrhythmias requiring a temporary pacemaker. Such ***honey*** was responsible for poisoning Greek **soldiers** in 400 B.C. and 11 **people** in Turkey in the 1980s. A 4-year-old **male** was poisoned after ingesting an unknown amount of leaves and flowers. **He** vomited and experienced abdominal pain 2½ h later, became somnolent and unresponsive with bradycardia and hypotension. Ingestion can cause nausea, vomiting, anorexia, lacrimation, abdominal pain, diarrhea, depression, dyspnea, prostration, limb paralysis, convulsions, bradycardia, and death.

Sheep have been poisoned after eating a few ounces of its leaves. Symptoms include salivation, vomiting, and weakness. In New Zealand, 7/300 **sheep** died after eating its cuttings. Symptoms included slow, groaning expiration; followed by short, rapid, gasping inspiration after excitement; frothy salivation, ataxia, and collapse. **Others** lose weight. Post-mortem revealed pulmonary congestion, subendocardial hemorrhages, subpleural hematomas, and reddening of tracheal and bronchial mucosa. Deaths may occur over a 2 week period. Pet **birds**, **goats**, and **zoo animals** have also been poisoned by it.

RHODOPHYLLUS LIVIDUS

Untoward effects: This poisonous large toadstool ***mushroom*** caused 40 cases of poisoning in **people** in Zagreb, Yugoslavia in 1962, after being mistaken for edible ***mushrooms***. It has a short incubation time for its gastroenterocolitic syndrome.

RHODOTORULA

Untoward effects: This supposedly non-pathogenic ***yeast*** has caused fungemia in **people**. A 7-year-old **male**

developed an endocarditis from its localization. Fungemia with it in two, 39- and 47-year-old **male patients** after long-term IV ***antibiotic*** treatment for subacute bacterial endocarditis. Enteric isolate in **children** with enteritis, from conjunctiva and urine. Septicemia with hypotension and shock traced to ***R. glutinis*** contaminated arterial catheters during long-term IV therapy. Its aeroallergens cause an allergic rhinitis, bronchial asthma and/or a hypersensitivity pneumonia. Involved in mastitis of **cows** and **mares**.

RHODOTYPOS SCANDENS

= *Jetbead* = *Jetberry*

Untoward effects: Contains cyanogenic glycosides. Its fruits, when eaten, cause dyspnea, loss of voice, twitches, spasms, shaking, and coma from its ***amygdalin*** (q.v.) content.

RHUBARB

= *Rheum rhaponticum* = *Chu-Lo* = *Dolu*

Untoward effects: The stalk is edible and contains traces of ***oxalic acid*** and high levels of ***nitrates***, but the leaves contain large amounts of ***oxalic acid*** (q.v.) and can be very toxic. Ingestion of large amounts of raw or cooked leaves leads to convulsions, corrosion of the gastrointestinal tract, local mouth irritation, vomiting, severe abdominal pain, weakness, muscle cramps, anuria, and coma, followed rapidly by death. The ***anthraquinone*** (q.v.) glycosides in the leaves are also toxic, causing purgation. A **female** ~5 years was admitted to an emergency room in a semiconscious state. First symptoms were drowsiness and nausea. **She** showed positive for ***oxalic acid*** in **her** urine. Having tasted rhubarb leaves and liking them, **her** playmates fed them to **her**. **She** died 2 days after hospitalization. Gross hematuria on two occasions in a **family** eating large amounts of rhubarb stalks. Dyshidrosis also reported from such ingestion. Ingestion can cause red urine if it is alkaline; orange, if acid. "**Dieters**" ingesting excessive amounts of its roots in herbals develop severe diarrhea, abdominal pain, and dehydration. Feces can become yellow or greenish yellow. **Patients** with ***nickel*** dermatitis should avoid eating rhubarb. During World Wars I and II, many **people** in Europe died from eating the leaves. **Children** have become seriously ill from eating 20–100 g of leaves and some stalks. Abortion and death in a **woman** after ingestion of the leaves.

The rhubarb used in medicinals is **not** the common vegetable garden food rhubarb, but from special roots grown in China and Tibet.

RIBAVIRIN

= *ICN-1229* = *Ribetol* = *Ribovirin* = *RTCA* = *Tribavirin* = *Vilona* = *Viramid* = *Virazid* = *Virazole*

***Antiviral*. Synthetic nucleoside used as an aerosol, oral, or IV.**

Untoward effects: Eye irritation, skin rashes, headaches, and dyspnea. ***Water*** intoxication cases in young **infants** after treatment with nebulized drug led to seizures, cyanosis, and apnea. Contact lens damage of soft contact lenses of a pediatric unit **nurse** exposed to the aerosoled drug over a 1 month period caused atopic conjunctivitis. Systemic use has caused hypotension and a dose-related hemolytic anemia with decreased hemoglobin, hematocrit, and red blood cell count in **man** and **rhesus monkeys**. Can induce rash, pruritus, coughing, and depression. Use in ***Rebetron*** (also contains ***interferon*** ***α-2b***) has decreased white blood cells.

Teratogenic, embryotoxic (**rabbit** and **mouse**), mutagenic and carcinogenic (benign mammary, pancreatic, pituitary, and adrenal tumors) in experimental trials. Has induced cardiac lesions in **rats** and **mice**. Vomiting, diarrhea, depression, rough coat, and ataxia in treated **rats**, **mice**, **guinea pigs**, and **monkeys**. Gastroschisis in **rat** offspring and anomalies of limbs, ribs, eyes, and central nervous system in **hamster** offspring.

LD_{50}, **rat**: 3.9 g/kg; **rhesus monkey**: > 10 g/kg; **male guinea pig**: 2.3 g/kg; **mouse**: 2 g/kg.

RICE

= *Oryza sativa* = *Chawal*

Untoward effects: Toxic "yellow rice" is primarily a result of metabolites of *Penicillium islandicum* and other molds growing on rice, causing liver pathology in **humans** and test **animals** in Japan. In India, 43 **people** became ill and 13 died from eating discolored rice. Symptoms were paresthesias, anorexia, vomiting, diarrhea, lethargy, weakness, and black pigmentation of hands and feet. **Some** had glycosuria, jaundice, and impaired liver function tests. These metabolites include ***citrinin***, ***citroveridin***, ***islandotoxin*** (a carcinogen), ***luteoskyrin***, and ***patulin***. ***Talc*** containing ***asbestos*** was added to rice intended for Japanese consumption in the Los Angeles area, and in Japan probably contributed to the high incidence of gastric cancer (seven times the rate of U.S. **men**). A severe intestinal reaction in 4½-month-old **male** after ingestion. Post-operative rice starch granulomatous peritonitis from its use on surgical gloves. ***Cola***-colored **babies** in Japan (rice-oil disease or Yushu) from grain contaminated with ***polychlorinated biphenyls***. Rice diet enhances ***chloroquine*** bio-availability. Incidence of **human** leprosy is higher in rice-growing areas where rice consumption is high, as in Louisiana, where infected **armadillos** are found. *Bacillus cereus* in fried rice has been serious cause of emetic-type food poisoning in the U.S. Excessive consumption of brown rice (macrobiotic diet) has been lethal. Dangerously high levels of ***cadmium*** have been found in some Japanese rice.

Contaminated rice-oil caused the deaths of 100,000 **broilers** in Japan in 1968. Despite advertising to the contrary, **dogs** have shown allergic reactions to rice in their diets.

RICIN

Untoward effects: This ***glucoside phytotoxin***, ricin, is concentrated mostly in the seeds of the ***castor bean*** (q.v.). It is among the most toxic known, and its parenteral LD in **animals** is ~0.01 μg. This toxin is destroyed or inactivated by moderate heat. See ***Castor Bean*** for symptomatology of poisoning in **man** and **animals**. Red blood cell hemolysis only if injected. **Children** are the usual victims of poisoning. Secret **police** have used it for assassinations. Bulgarian secret **police** used it to kill a **broadcaster**.

LD_{LO}, **human**: 2 mg/kg.

LD_{LO}, **rabbit**: 500 μg/kg; **rat, horse**: 100 mg/kg. IV, LD_{50}, **mouse**: 3 ng/kg.

RIFABUTIN

= *Ansamycin* = *Ansatipine* = *LM 427* = *Mycobutin*

Antibiotic, tuberculostatic. **Related to *rifampin*.**

Untoward effects: Neutropenia, gastrointestinal upsets, and rashes. May permanently stain contact lenses and transiently color body fluids and skin brown–orange. Occasionally thrombocytopenia, hepatitis, arthralgia, myositis, ageusia, flu-like syndrome, dyspnea, and chest pain. *Clostridium difficile*-associated diarrhea and uveitis reported.

Drug interactions: Decreases plasma concentration of ***zidovudine*** and its concentration may be increased by ***fluconazole***, ***itraconazole***, and ***ketoconazole***. ***Nelfinavir*** decreases its metabolism and increases its toxicity, suggesting a 50% reduction in dosage. ***Delavirdine***, ***indinavir***, and the combination ***lopinavir/ritonavir*** may also decrease its metabolism and increase its toxicity, and rifabutin may increase ***delavirdine*** and ***indinavir*** metabolism and decrease their concentration and effect. Early reports indicate it may increase the effectiveness of ***clarithromycin*** and ***dapsone***, possibly by preventing the emergence of resistant strains. ***Ritonavir*** induces a large increase in rifabutin serum concentration and ***clarithromycin*** can double rifabutin serum concentration. It decreases plasma levels of ***ethinyl estradiol*** and ***norethindrone*** in ***oral contraceptives***.

RIFAMIDE

= *M14* = *Rifocin M*

Antibiotic. **Semisynthetic.**

Untoward effects: Transient icterus. Yellow discoloration of skin and mucous membranes and dark urine, especially at high dosage. Pain at injection site. Necrosis and peripheral zones of hypertrophied hepatic cells in pregnant **mice**, **rats**, and **rabbits** after high single IP dosage (15–100 mg).

LD_{50}, **rat**: > 4 g/kg; **mouse**: 2.45 g/kg.

RIFAMPIN

= *Abrifan* = *Eremfat* = *Rifa* = *Rifadin* = *Rifaldazine* = *Rifaldin* = *Rifamycin AMP* = *Rifaprodin* = *Rifoldin* = *Rimactane* = *Rimapen* = *Rifampicin*

Antibiotic, tuberculostatic. **Semisynthetic of *Rifamycin SV* inhibiting DNA-dependent bacterial RNA polymerase. Oral and IV use.**

Untoward effects: These are often associated with simultaneous use of other ***antituberculars***.

Blood: Severe immune thrombocytopenia with nosebleeds, tongue spots, throbbing headache, spontaneous bruising, massive hemolysis, erythema, macules, leukopenia, agranulocytopenia, neutropenia, eosinophilia, decreased hemoglobin, and methemoglobinemia reported. Thrombocytopenia is often more frequent in **patients** receiving intermittent therapy (viz. 1200 mg twice per week) than in **those** receiving daily doses. Cerebral hemorrhage and death when continued after purpura noted. Increased blood urea nitrogen and serum ***uric acid***.

Clinical tests: Inhibits standard assays for serum ***vitamin*** B_{12} and ***folate***. Has caused a false positive urinary ***melanin*** test result and falsely increases ***bilirubin*** test results.

Central nervous system: Drowsiness, fatigue, dizziness, ataxia, headache, muscle weakness, pain in extremities, numbness, behavioral changes, inability to concentrate, and mental confusion.

Eye: Exudative conjunctivitis; eyes painful, tender, red, and congested; visual disturbances.

Fetus: Serum concentrations are less than **maternal**. A case report of multiple **fetal** abnormalities after **maternal** use in early pregnancy. Use near the end of pregnancy can cause postnatal hemorrhages in the **mother** or **infant**.

Gastrointestinal: Nausea, vomiting, heartburn, epigastric pain, dysphagia, anorexia, flatulence, cramping, diarrhea, pseudomembranous and eosinophilic colitis, and sore mouth and tongue.

Liver: Hepatotoxicity. Hepatocyte injury with ballooning or necrosis in ~10%. Jaundice, hyperbilirubinemia, increased alkaline phosphatase, increased ***sulfobromphthalein***, and increased SGOT; occasionally bilirubinuria and urobilinogen in urine. Some of these reactions and most liver failures occurred when used with ***isoniazid***.

Miscellaneous: Orange–reddish coloration of contact lenses, urine, and tears. Anaphylactoid reactions to it or ***ciprofloxacin*** reported in five **patients** with symptomatic HIV disease. Drug fever, menstrual disturbances, muscle weakness, generalized numbness, pain in extremities, and edema of face and extremities reported.

Neoplasms: Lung cancers, apparently associated with drug-induced immunosuppression.

Poisoning: Fatal poisoning after 27 h from suicide following ingestion of 15 g in 33-year-old **male**. Other cases from combination with other drugs.

Renal: Interstitial nephritis, often after intermittent therapy; vomiting, diarrhea, lumbar pain, chills, anuria, and proteinuria. Fatal kidney failure due to papillary necrosis, glomerulosclerosis, and granulomatous interstitial nephritis in 32-year-old **female** treated with 450 mg/day/14 months. Only ~13% of an administered dose is eliminated unchanged in the urine.

Skin: Acneiform lesions, pemphigus, urticaria, pruritus, toxic epidermal necrolysis, and fixed-drug eruptions.

Drug interactions: Given with ***p-aminosalicylic acid***, its serum concentration is decreased by ~50% because of decreased absorption, which, in part, may be due to the use of ***bentonite*** as an excipient in *p*-aminosalicylic acid. Fatigue, headache, and tachycardia induced by ***bunazocin***, an α_1-adrenoreceptor antagonist, is eliminated by rifampin. Rifampin increases the hepatic enzyme metabolism of many drugs. During coadministration, decreases serum levels and effectiveness of ***acenocoumarol***, ***acetohexamide***, ***chloramphenicol***, ***cortisone***, ***cyclosporine*** (decreases ~60%), ***digitoxin***, ***disopyrimide***, ***ethinyl estradiol***, ***fluconazole***, ***glibenclamide***, ***hexobarbital*** (half-life of 407 min decreases to 171 min), ***itraconazole***, ***ketoconazole*** (by 80–90% after long-term use), ***methadone*** (leading to ***narcotic*** withdrawal symptoms), ***methylprednisolone***, ***norethisterone***, ***oral contraceptives***, ***oxytriphylline***, ***phenobarbital***, ***phenprocoumon***, ***phenytoin***, ***prednisolone*** (decreases 40–60%), ***prednisone***, ***quinidine***, ***ritonavir***, ***tamoxifen*** (decreases 86%), ***theophylline***, ***tolazemide***, ***tolbutamide*** (half-life of 292 min decreases to 160 min), ***toremifene*** (decreases 87%), ***verapamil***, ***warfarin*** (~50%); non-nucleoside reverse transcriptase inhibitor, ***delavirdine***; and possibly ***clarithromycin***, ***dapsone***, ***diazepam***, ***doxycycline***, ***midazolam***, ***nevirapine***, ***nifedipine***, ***propranolol***, ***metoprolol***, ***tenoxicam***, and ***zidovudine***. A nearly fatal case of shock-producing hepatic reaction when given after anesthetic induction with ***halothane***. ***Ketoconazole*** can also decrease rifampin absorption. Oral ***probenecid*** increases its serum levels by > 85%. It increases activity of CYP2C9, CYP2C19, and CYP3A4 metabolization enzymes. It decreases ***amiodarone*** and ***losartan*** serum concentrations and increases metabolism of ***indinavir***. It decreases elimination half-life and peak plasma levels of IV ***ondansetron*** by ~45% and nearly the same after oral ***ondansetron***. Use with ***tacrolimus*** led to renal allograft dysfunction in 61-year-old **male**. ***Atovaquone*** plasma concentration decreased 40–50% by rifampin.

TD_{LO}, **human**: 180 mg/kg; **man**: 13 mg/kg/2 days.

In **dogs**, altered liver function tests at 25 mg/kg. ***Glyburide*** half-life and absorption decreased by it in **rabbits**. Myopathy and teratogenicity in **rats** and **mice**. Increases hepatomas in **mice**.

LD_{50}, **rat**: 1.7 g/kg; **mouse**: 500 mg/kg; IV, 300 and 260 mg/kg, respectively.

RIFAPENTINE

= *Priftin*

***Antibiotic, tuberculostatic*. Long-acting analog of *rifampin* (q.v.).**

Untoward effects: Hyperuricemia, increased aminotransferase, red–orange discoloration of body fluids. Will stain contact lenses. Drug interactions can be expected.

RILUZOLE

= *PK 26,124* = *RP 54,274* = *Rilutek*

***Neuroprotective*. Oral use for amyotrophic lateral sclerosis.**

Untoward effects: Nausea, vomiting, diarrhea, dizziness, vertigo, somnolence, mouth paresthesia, anorexia, abdominal pain, increased spasticity, asthenia, and icteric toxic hepatitis. Neutropenia in 3/4000 **patients**. Reversible granulocytopenia in 63-year-old **male** after inadvertent increase of dosage from 100 to 200 mg/day. Deliberate overdosage with 2.8 g led to dyspnea, peripheral cyanosis, and methemoglobinemia.

LD_{50}, **mouse**: 67 mg/kg; IP, 46 mg/kg.

RIMANTADINE

= *EXP-126* = *Flumadine* = *Meradan(e)* = *Roflual*

***Antiviral*. An *amantadine* analog.**

Untoward effects: Nausea, vomiting, abdominal pain, anorexia, peripheral edema, xerostomia, dizziness, asthenia, nervousness, anxiety, insomnia, light-headedness, and poor concentration are most common. Occasionally seizures, hallucinations, and confusion. All of these are more common and more severe in the **elderly**, and usually dose-related. A case report of glaucoma in 18-year-old **male** after 2 years treatment for congenital rubella infection. Acute overdoses lead to anticholinergic-type effects.

Teratogenic in **experimental animals**.

RIMEXOLONE

= *ORG 6216* = *Rimexel* = *Trimexolone* = *Vexol*

***Local anti-inflammatory.* A *corticosteroid* originally developed for intra-articular use. Now used topically on the eye.**

***Untoward effects*:** Topically leads to blurred vision; eye discharge, discomfort, pain, increased intraocular pressure, foreign-body sensation, hyperemia, and pruritus.

SC of 0.5 mg/kg/day to **rabbits** (~twice **human** ophthalmic dose) was embryotoxic and teratogenic.

RISEDRONATE

= *Actonel*

Oral treatment of Paget's disease of bone. A *biphosphonate.*

Untoward effects*:** Mild diarrhea, nausea, abdominal pain, esophagitis, heartburn, esophageal ulcers, asthenia, myasthenia, and iritis. Flu-like symptoms in 10%, arthralgia in 30%. ***Calcium supplements taken at the same time decrease its absorption.

RISPERIDONE

= *R 64,766* = *Risperdal*

Psychotherapeutic.

Untoward effects*:** Vertigo, dizziness, sedation, asthenia, drowsiness, decreased concentration, impotence, rhinitis, rash, headache, blurred vision, anorexia, and nausea. Occasionally neuroleptic malignant syndrome, extrapyramidal symptoms, tardive dyskinesia, leg edema, asthma, tachycardia, hypotension, weight increase, galactorrhea, increased serum ***prolactin, agitation, anxiety, QTc prolongation, insomnia, and constipation. A 31-year-old **male schizophrenic**, after initial success in treatment, had a re-emergence of auditory hallucinations. A 34-year-old **female schizophrenic**, 5 days after initiating therapy, had an acute exacerbation of chronic schizophrenia, then a fatal pulseless cardiac electrical activity. Hepatotoxicity with jaundice in 81-year-old **male** after two doses of 0.5 mg. Mania induced in several **females**.

Drug interactions*:** Use with ***donepezil in an elderly Alzheimer **patient** may have induced significant drowsiness and a shuffling gait.

LD_{50}, **rat**: 56–113 mg/kg; **mouse**: 63–82 mg/kg; **dog**: 18.3 mg/kg.

RISTOCETIN

= *Ristomycin* = *Riston* = *Spontin*

***Antibiotic complex.* IV use.**

***Untoward effects*:** Thrombocytopenia, agranulocytosis, thrombocytopenic purpura, allergic hypersensitivity, toxic epidermal necrolysis, ototoxicity, and occasionally phlebitis. IM is painful.

IV, LD_{50}, **mouse**: 500 mg/kg.

RITODRINE

= *DU-21,220* = *Miolene* = *Prempar* = *Pre-Par* = *Utemerin* = *Utopar* = *Yutopar*

***Tocolytic, β-sympathomimetic.* Usually IV.**

***Untoward effects*: *Blood*:** Neutropenia and hypokalemia.

***Cardiovascular*:** Tachycardia (**maternal** and **fetal**), increased pulse pressure, myocardial ischemia, profound hypotension, cerebral vasospasm, asymptomatic cardiac arrhythmias, migraine headache, chest pain, dysrhythmias in **infant**, thickens cardiac septum in **neonates** if **maternal** treatment is > 7 weeks. May unmask occult heart disease.

***Fetus and neonate*:** **Fetal** death associated with severe ketoacidosis in **diabetic** 28-year-old **female** receiving continuous IV infusion. Has also induced acidosis and hyperlactacidemia in the treated **mothers** and jaundice in their **infants**.

Miscellaneous*:** Increased serum transaminase, fever, enlarged salivary gland, ***sodium retention, palpitations, tremors, nervousness, increased ***insulin*** requirement, and myotonic muscular dystrophy.

***Respiratory*:** Pulmonary edema. Has been fatal.

***Skin*:** Acute cutaneous vasculitis in pregnant 30-year-old **female** after IV treatment for > 4 weeks for preterm labor. Case reports of induced erythema multiforme with maculopapular rashes.

Drug interactions*:** Use with ***corticosteroids can induce **maternal** pulmonary edema. Can alter a **patient's** ***insulin*** requirement.

IV, LD_{50}, **rabbit**, non-pregnant: 64 mg/kg; **mouse**, non-pregnant: 540 mg/kg; pregnant **rat**: 85 mg/kg.

RITONAVIR

= *A 84,538* = *Abbott 84,538* = *ABT 538* = *Norvir*

Antiviral. Protease inhibitor.

***Untoward effects*: *Cardiovascular*:** Postural hypotension, tachycardia, palpitations, headaches, syncope, hypotension, peripheral vascular problems, hemorrhage, anemia, thrombocytopenia, and leukopenia.

Central nervous system: Agitation, amnesia, anxiety, ataxia, asthenia, amnesia, confusion, hallucinations, hyperesthesia, circumoral and peripheral paresthesias, euphoria, decreased libido, tremors, vertigo, and seizures.

Ear: Pain, impaired hearing, increased cerumen, and tinnitus.

Endocrine: Non-ketotic hyperglycemia and diabetes along with increased thirst and hunger, weight decrease, increased urination, fatigue, and dry, itchy skin that may occur after ~11 weeks of treatment and as early as 4 days.

Eye: Diplopia, amblyopia, blurred vision, blepharitis, iritis, uveitis, and photophobia.

Gastrointestinal: Xerostomia, altered taste, cheilitis, gingivitis, esophagitis, gastritis, nausea, vomiting, diarrhea, bloody diarrhea, cholangitis, and colitis.

Miscellaneous: Lymphadenopathy, lymphocytosis, cachexia, photosensitivity, chest or back pain, facial edema or pain, neck rigidity or pain, enlarged abdomen, pancreatitis, increased hepatic aminotransferase, increased triglyceride and ***cholesterol*** levels, muscle cramps or weakness, myositis, arthralgia, and hiccups. Latent dysbetalipoproteinemia precipitated by it and other HIV-***protease inhibitors*** reported after 1 month of treatment. Hypermenorrhea, buffalo humps, and severe hepatotoxicity also occur.

Renal: Acute renal failure, dysuria, nocturia, polyuria, urethritis, and pollakiuria.

Respiratory: Dyspnea, asthma, coughing, interstitial pneumonia, rhinitis, parosmia, and epistaxis.

Skin: Contact dermatitis, dry skin, eczema, folliculitis, acne, maculopapular rash, pruritus, psoriasis, molluscum contagiosum, and urticaria.

Drug interactions: An overwhelming number of interactions reported within the first 2 years of use. It decreases the metabolism and may increase toxicity of ***amiodarone***, ***astemizole***, ***benzodiazepines***, ***bepridil***, ***bupropion***, ***cisapride***, ***clozapine***, ***flecainide***, ***lovastatin***, ***meperidine***, ***methadone***, ***midazolam***, ***piroxicam***, ***propafenone***, ***quinidine***, ***rifabutin***, ***simvastatin***, ***terfenadine***, ***triazolam***, ***tricyclic antidepressants***, and ***zolpidem***. Use with ***ergotamine*** may cause ergotism for the same reasons and it is now contraindicated with ***pimozide***. Adverse cardiac and neurological effects when coadministered with ***disopyramide***, ***fluoxetine***, ***mexiletine***, and ***nefazodone***. By decreasing their metabolism, problems can develop when given with ***clarithromycin*** (especially with decreased renal function), ***desipramine***, ***nelfinavir***, ***nifedipine***, ***rifabutin***, and ***sildenafil***. Can decrease protection by ***oral contraceptives*** containing ***ethinyl estradiol***. Plasma concentration of ***saquinavir*** increases by > 20 times when used with ritonavir 400 or 600 mg twice daily. Increases metabolism of ***nevirapine***, ***rifampin***, and ***theophylline***. The oral solution contains ***alcohol*** and can elicit ***disulfiram***-like reactions with certain drugs, such as ***metronidazole***. A **patient** taking 600 mg twice daily took 180 mg ***methylene-dioxyamphetamine*** (q.v.) ~4 h before hospitalization and evidenced profuse sweating, tachypnea, tachycardia, hypertonia, and cyanosis. **She** died shortly thereafter. It inhibits P4503A4 enzyme and **her** ***MDMA*** blood levels were 4.56 mg/l. ***Methadone*** effect decreased by it in 51-year-old **male**. Interactions with ***carbamazepine*** and ***fluvoxamine*** increase their serum levels.

RITUXIMAB

= *Mab Thera* = *Rituxan*

Monoclonal antibody for IV treatment of low-grade B-cell non-Hodgkin's lymphoma.

Untoward effects: Chills, nausea, vomiting, hypotension, fever, headache, bronchospasm, severe dyspnea, hypoxia, rigors, angioedema, urticaria, pruritus, myalgia, and pain at lymphoma site, usually during infusion. Occasionally cardiac arrhythmias. Uncommon, but thrombocytopenia, anemia, and neutropenia can occur later. Breathlessness, hypotension, rigors, and death of a leukemic 71-year-old **female** after 25 mg/h followed by 37.5 mg/h during the next hour.

RIVEA CORYMBOSA

= *Ololiuqui*

Botanical vine. **Used by Aztecs and other Mexican Indians.**

Untoward effects: Early Spanish writings described by Schultes and Hofmann stated "when the Aztec **priests** want to commune with their gods and to receive a message from them, **they** eat this plant to induce delirium, during which time a thousand visions and satanic hallucinations appear to **them**." Its primary active alkaloid agent is ***lysergamide*** (q.v.). The seeds contain 0.0065%. (1 or 2 mg leads to psychomimotic effects with a narcotic component).

A few seeds induce mild narcosis in **frogs**.

ROBENIDINE

= *Cycostat* = *Robenz* = *Robenzidine*

Coccidiostat.

Untoward effects: The FDA permits a maximum of 0.2 ppm for its residues in skin and fat of treated **birds** and a 0.1 ppm tolerance for it in other edible tissues. ***Bentonite*** in feeds may interfere with its assay. To insure adequate potency, finished feeds containing the drug must be used within 50 days of manufacture. NOT for use in laying **hens**. Resistance to it often develops rapidly. 5 days withdrawal in **poultry**.

ROBIN

Untoward effects: A ***phytotoxin*** in *Robinia pseudo-acacia* (***False Acacia, Black Locust***). It and a glycoside, ***robitin***, are in the inner bark, young leaves, and seeds and poison **humans**, especially **children** who suck on the twigs and eat the seeds and soft inner bark. Symptoms include stupor, purging, nausea, vomiting, epigastric pain, gastroenteritis, weak pulse, facial flushing, xerostomia, mydriasis, irregular respirations, and depressed cardiac action. Convulsions and muscle fasciculations have occurred. Poisoning from chewing the inner bark reported in one case report on 32 **boys**.

Horses, **mules**, **cattle**, **sheep**, and **chickens** have been poisoned by eating the bark (1.6% robin) and twigs, but deaths have been rare. Symptoms include colic, diarrhea, excitability, shock, and paralysis. Laminitis, anorexia, and circulatory problems in **horses**. Post-mortem reveals gastroenteritis, liver and kidney degeneration, and phytoagglutination of red blood cells. The seeds may not be toxic to **animals**.

ROCURONIUM BROMIDE

= *Esmeron* = *Org 9426* = *Zemuron*

Skeletal muscle relaxant. Non-depolarizing neuromuscular-blocking agent.

Untoward effects: Fast action (within 1 min); its effects are prolonged. It requires up to 67 min for 25% recovery, while others used require only up to 23–45 min. Should have similar Drug interactions as ***atracurium***, ***mivacurium***, and ***vecuronium***.

ROFECOXIB

= *MK 966* = *Vioxx*

Anti-inflammatory, analgesic. A selective COX-2 inhibitor. An non-steroidal anti-inflammatory drug.

Untoward effects: Gastrointestinal upsets including nausea, diarrhea, abdominal pain, indigestion, and occasionally gastrointestinal bleeding. Lower extremity edema, hypertension, anemia, increased aminotransferase, and renal toxicity. In a ***naproxen*** and rofecoxib trial (> 8000 **patients**) the incidence of myocardial infarctions was 0.1% versus 0.4%, respectively. ***Non-steroidal anti-inflammatory drugs*** are not recommended in **women** trying to become pregnant since they interfere with implantation of fertilized eggs in **animals**.

Drug interactions: ***Antacids*** decrease its maximum serum concentration ~20%. It increases ***methotrexate*** serum concentration ~20% and increases ***warfarin***-induced prothrombin time by ~10%. Contraindicated in **patients** allergic to ***sulfonamides***.

Excreted in ***milk*** of **animals** and increases mortality of nursing **neonates**.

ROLITETRACYCLINE

= *Reverin* = *Syntetrin* = *Tetraverin* = *Transcycline* = *Velacycline*

The nitrate form = *Bristacin* = *Tetrim* = *Tetriv*

Antibiotic.* Semisynthetic derivative of *tetracycline.

Untoward effects: Respiratory and cardiac failure after slow IV in **patients** with severe cardiac decompensation and tachycardial arrhythmia. Bitter or acid taste, abdominal pain, nausea, vomiting, transient giddiness, hot flushes, facial reddening, and occasional peripheral circulatory failure after rapid IV. Initially increases blood pressure. Has a ***curare***-like effect and can exacerbate myasthenia gravis. Appears in breast milk. Found in newly formed bone, calcifying cartilage matrix, and teeth.

Dogs have shown extreme sensitivity at injection site.

IV, LD_{50}, **mouse**: 87 mg/kg.

ROLLER COASTERS

Untoward effects: Aside from the expected injuries and risks, a previously healthy 48 kg 13-year-old **female** suffered a stroke after **she** rode upside down, supported by a padded metal shoulder support made for older **people**. **Her** neck had apparently been forced against the supports leading to left carotid artery trauma with formation of intimal thrombus, that embolized shortly thereafter.

ROMULEA ROSEA

= *Onion Grass*

Untoward effects: **Sheep** grazing it on pasture develop infertility, abortion, and paralysis. Decreases breeding efficiency in **female mice** and decreases fertility in **male guinea pigs**. A fungal pathogen, *Helminthosporium biseptatum*, causes leaf spot and may be the actual cause of the toxicity.

RONNEL

= *Dermafos* = *Ectoral* = *ENT 23,284* = *ET 14* = *ET 57* = *Etrolene* = *Fenchlorphos* = *Fenchlorvos* = *Grenol* = *Korlan* = *Nankor* = *Rid-Ezy* = *Rovan* = *Trolene* = *Vetemac* = *Viozene* = *Z-50*

Insecticide.* Organic *phosphorus* compound. *Cholinesterase inhibitor.

Untoward effects: Topically and orally in treating creeping eruption (cutaneous larva migrans) in **humans** leads to nausea, weakness, and serpiginous ulcers. Eczematoid dermatitis in severe cases. Clinically, topical use was generally ineffective. National Institute for Occupational Safety and Health and OSHA set exposure limits as 10 mg and 15 mg/m^3, respectively. Immediate danger to life or health, 300 mg/m^3.

In **animals**, eye irritation and liver and kidney damage can occur. Poisoned **mallard ducks** and **pheasants** develop ataxia, slowness, running, falling, tenseness, tremors, ptosis, and tetanic seizures within 1 day. When deaths occur, they are usually 1–6 days after treatment. Salivation and dyspnea in adversely affected **livestock**. Do not use in conjunction with or within a few days or weeks of treatment with ***arsenicals***, ***phenothiazines***, ***procaine***, ***purgatives***, or ***succinylcholine*** in **man** and **animals**. The FDA calls for a zero tolerance in meat or milk of treated **animals**. 28 days withdrawal for **sheep**. Topical ronnel may produce hepatic necrosis and pyoderma in **dogs**. ***Lime-sulfur*** and alkaline agents accelerate degradation in dip solutions.

LD$_{50}$, **rat**: 906 mg/kg; **mouse**: 2 g/kg; **rabbit**: 420 mg/kg; **guinea pig**: 1.4 mg/kg; **chicken**: 6.5 g/kg; **mallard duck**: ~2 g/kg; **pheasant**: 611 mg/kg; **turkey**: 500 mg/kg; topical, **rat**: 2 g/kg; **rabbit**: 1 g/kg; **guinea pig**: 1.4 g/kg.

ROOT BEER

Untoward effects: Was often flavored with oil of ***sassafras*** containing ~75% ***safrole*** (q.v.). ***Safrole*** is present in many edible plants. It is carcinogenic in **rodents** and some of its metabolites are mutagenic. Currently prohibited as an additive in the U.S. **People** allergic to root beer are usually allergic to ***eggs***, whose whites are used to make its froth. **People** should be aware that ***salicylates*** are found in regular and dietetic root beers.

ROPINIROLE

= *Requip* = *SKF 101,468*

***Antiparkinsonian.* A *dopamine* agonist.**

Untoward effects: Syncope and bradycardia in ~12%. Nausea, somnolence, postural hypotension, and worsening of dyskinesias. **Patients** can suffer a sudden sleep attack from it while driving, causing accidents.

Drug interactions: ***Ciprofloxacin*** increases its serum concentration.

ROPIVACAINE

= *LEA 103* = *Naropin*

Local anesthetic.

Untoward effects: In clinical trials, > 5% led to nausea, hypotension, bradycardia, **fetal** bradycardia, vomiting, back pain, and paresthesia; 1–5% with fever, headache, pain, dizziness, pruritus, urinary retention, oliguria, tachycardia, anxiety, chest pain, and **fetal** tachycardia. **Neonatal** jaundice, tachypnea, fever, vomiting, and respiratory problems. Convulsions after inadvertent IV. Toxicity reported with serum concentrations of 1–2 µg/ml.

ROQUEFORTINE

Untoward effects: It and ***penitrem A*** are two neurotoxic metabolites isolated from a fungus, *Penicillium commune* growing on ***cottonseed***. It is a paralytic neurotoxin, also found on ***blue cheese***. A **dog**, mistakenly diagnosed as suffering from (shaking, convulsing) and dying from ***strychnine*** poisoning, was found on post-mortem to have eaten overripe ***blue cheese***.

ROSAMICIN

= *4′-Deoxycirramycin A* = *Rosamycin* = *Rosaramicin*

***Antibiotic.* Related to *erythromycin*.**

Untoward effects: Swelling of hands and feet, pruritic rash, erythematous patches on face and dorsal surface of hands and feet, nausea, abdominal pain, and myalgia associated with hepatotoxicity reported in 21-year-old **female**. Secreted in breast milk.

LD$_{50}$, **mouse**: SC, 625 mg/kg; IP, 350 mg/kg; IV, 155 mg/kg.

p-ROSANILINE

= *Fuschine* = *Magenta*

***Dye, antifungal.* Well-known ingredient in *Castellani's Paint* used for ¾ of a century.**

Untoward effects: Considered carcinogenic and manufacturing plant **workers** are asked to have semi-annual medical examinations, including exfoliative cytology of **their** urine. It has been incriminated as a cause of malignant bladder neoplasms in the dye and ***rubber*** industry. See ***Methyl Violet***.

ROSE BENGAL

Dye, biological stain, hepatic and nasolacrimal function and eye injury diagnostic aid.

Untoward effects: Slight stinging sensation when used topically on the eye. Photosensitivity from its use in ***lipsticks*** and dental plaque-staining agents.

ROSEMARY

= *Romero* = *Rosmarinus officinalis*

Untoward effects: Emmenagog and may induce abortion. A panacea of the past. **Bees** like its nectar and it gives their ***honey*** a peculiar balsamic flavor.

LD$_{50}$, *oil*, **rat**: 5 g/kg.

ROSEWOOD

= *Palisander Wood*

Untoward effects: The Brazilian and East Indian wood contains ***dalbergiones***, strong sensitizers, causing allergic contact dermatitis. A typical case of eczematoid dermo-epidermitis of the neck and forearms after wearing a necklace and bracelet made of it.

ROSIGLITAZONE

= *Avandia*

***Antidiabetic.* Resistance to *insulin* is decreased by it.**

Untoward effects: Edema, slight weight increase, mild anemia, 12–19% increase in low-density lipoprotein and high-density lipoprotein. Hepatomegaly, hyperbilirubinemia, hypoalbuminemia, increased transaminases, and slight increase in international normalized ratio should be an alert of probable hepatotoxicity, as with other ***thiazolidinediones***.

ROSIN

= *Colophony* = *Resin* = *Yellow Resin*

***Protective wound dressing.* A residue left after distilling off the *oil of turpentine* from the oleoresin of various *pine* tree species. Called *Wood Resin* when obtained from southern United States *pine* stumps, and *Gum Resin* when obtained from the exudate of living *pine* trees. A similar product, known as *Tall Oil Resin*, is obtained as a by-product of wood pulp or kraft *paper* processing.**

Untoward effects: Potential skin sensitizer in some adhesive tapes or plaster. Cross-reaction with ***clove oil*** or ***cinnamon oil***. A cause of ***shoe***, typing ***paper***, and other ***paper*** sensitivity. A case report of a **person** who consumed it with ***alcohol***. It can be a source of hypersensitivity to certain solders, string instrument rosin, ***lacquers***, varnishes, glues, waxes, adhesives, carpentry nails, sealing wax, and fireworks.

ROSOXACIN

= *Acrosoxacin* = *Eracine* = *Eradacil* = *Eradacin* = *Win-35,213* = *Winuron*

Antibacterial.

Untoward effects: Symptoms of central nervous system toxicity, such as dizziness, drowsiness, and light-headedness reported in 51% of **patients**.

ROTENONE

Pesticide, piscicide.

Untoward effects: Contact and stomach poison. Inhalation leads to eye, skin, and respiratory tract irritation and numbing of mucous membranes. Ingestion by **man** leads to nausea, vomiting, abdominal pain, diarrhea, tremors, tachypnea, incoordination, clonic convulsions, and stupor. Liver pathology. Symptoms occur within a few minutes or hours. After ingestion, death has occurred as early as 5 h or can be delayed 10 days. **Workers** manufacturing it without suitable masks and proper ventilation often develop violent dermatitis of genital region, rhinitis with anosmia, and irritation of **their** tongues and lips. National Institute for Occupational Safety and Health and OSHA set exposure limits of 5 mg/m^3, TWA; immediate danger to life or health 2500 mg/m^3.

LD_{LO}, **human**: 50 mg/kg.

Mammalian toxicity is enhanced by vaporization and heat. May cause eye irritation. Drinking freshly treated waters may be without ill-effects on **wild** or most **domesticated animals**, but can be toxic to **pigs**. **Cats** and **dogs** may vomit after licking quantities of the powder. Very toxic to **fish**, **snakes**, and **insects**. As an arrow poison in Sumatra. Poisoned **ducks** and **pheasants** develop ataxia, nutation, dyspnea, polyuria, fluffed feathers, wing-drop, and immobility. Regurgitation after 1.5 g/kg. Vomiting, central nervous system depression, tremors, seizures, and death in **rats**. **Guinea pigs** are most susceptible to its toxicity.

LD_{50}, **rat**: 13–132 mg/kg; **mouse**: 350 mg/kg; **mallard duck**: ~2.2 g/kg; **pheasant**: 1.7 g/kg.

ROTTBOELLIA EXALTATA

= *Itchgrass*

Untoward effects: This aggressive 8–10 ft annual grass in Florida and Louisiana causes severe contact sensitivity with severe pruritus and urticaria.

ROTTENSTONE

= *Terra Cariosa* = *Tripoli*

Untoward effects: Occupational dermatoses from its use as a polishing agent in the aircraft industry. Inhalation of these porous siliceous rock particles can cause respiratory problems.

ROXARSONE

= *3-Nitro* = *Nitronic* = *NSC 2101* = *Ren-O-Sal*

***Growth-promoter, coccidiostat, antibacterial.* An *arsenical*.**

Untoward effects: Do NOT administer to **ducks**, **geese**, laying **hens**, or **dogs**. Discontinue use 5 days before treated **animals** are slaughtered for **human** consumption. United States regulations limit total ***arsenic*** residues in **poultry** muscle tissues to 0.5 ppm and 1 ppm in edible by-products. In **poultry**, its relative toxicity is about twice that of ***arsanilic acid***. Its safe use demands an adequate water intake and no other sources of ***arsenic***. Incoordination, ricket-like appearance, and paralysis can occur under these conditions.

LD_{50}, **rat**: 155 mg/kg; **chicken**: 110 mg/kg.

ROXATIDINE ACETATE

= *Altat* = *Gastralgin* = *HOE 760* = *Neo H2* = *Pifatidine* = *Roxit* = *TZU 0460*

Antiulcerative.

Untoward effects: Mild diarrhea, constipation, and skin rash. Only 0.22% of **maternal** dose excreted in breast milk.

LD_{50}, **male mouse**: 1 g/kg.

ROXITHROMYCIN

= *Assoral* = *Claramid* = *Forilin* = *Overal* = *Rossitrol* = *Rotramin* = *RU-965* = *RU-28,965* = *Rulid* = *Surlid*

Antibiotic.

Untoward effects: Several cases of acute hepatitis, acute pancreatitis, Churg–Strauss syndrome or allergic granulomatous vasculitis, facial edema, and meteorism.

Drug interactions: Reports vary, but a significant number of **patients** had increased prothrombin time and hemorrhages when given to **patients** stabilized on ***warfarin***. Causes a non-significant or variable increase in ***cyclosporin*** and ***midazolam*** serum levels.

ROYAL JELLY

= *Apilak*

Untoward effects: Allergic reactions include acute anaphylaxis in **patients** allergic to **bee** pollen or **bees**. Has increased the urinary excretion of ***17-hydroxycorticoids***. Severe respiratory distress in 31-year-old **female** with mild asthma. Confirmed by rechallenge.

RUBBER

= *Caoutchouc* = *Elastica* = *Kautchak*

Natural rubber is, supposedly, not allergenic (yet see *Latex*), but over 1000 chemicals in vulcanizers, accelerators, activators, and antioxidants can cause contact dermatitis.

Untoward effects: Erythematous and edematous bands on wrist, dermatitis of eyelid from a rubber electric cord and an eyelash curler; allergic-type burns from toy rubber balloons; ***shoe*** dermatitis from adhesives or linings; and dermatitis from adhesive tapes reported. A bleach, ***monobenzone*** (q.v.), used to prevent oxidation of rubber caused exposed arms and hands of **Negroes** to turn white after **their** use of rubber ***gloves***. The rubber accelerator, ***zinc dibenzyl dithiocarbamate***, in elasticized underwear becomes a powerful skin sensitizer when it reacts with ***chlorine bleach***. Carcinoma of bladder in 58-year-old **male** after ~20 year exposure to industrial antioxidants, such as ***2-naphthylamine*** (q.v.) and ***Nonox S***. **People** working with rubber in the electric-cable industry develop bladder cancer and hematuria. An association noted between rubber solvents and lymphatic leukemia. Bladder cancer was associated with handling and mixing the raw materials and stomach cancer with processing the "green" precured rubber. In British tire and general rubber goods manufacturing **workers**, a small excess of esophageal cancer was observed. A number of carcinogenic impurities exist, such as ***4-aminodiphenyl*** (q.v.), ***2-naphthylamine*** (q.v.), and ***1-naphthylamine*** (q.v.). ***Procaine HCl*** has leached ***thiram*** (q.v.) from rubber closures of vials. ***Thiram***-sensitive **people** must avoid many rubber-containing products, such as bunion pads, eyelash curlers, mattresses, stethoscope ear tips, condoms, pessaries, ***gloves***, goggles, dress shields, ***shoes***, adhesives, dentures, bathing caps, garters, girdles, brassières, elasticized garments, gum erasers, toys, and aprons. A positive allergic reaction to ***thirams*** (q.v.) is an absolute contraindication for systemic ***disulfiram*** (q.v.) administration and if drinking ***alcoholic beverages***. Another rubber accelerator, ***mercaptobenzothiazole*** (q.v.), is another major cause of rubber ***glove*** contact dermatitis. ***Diethylthiourea***, also an accelerator, has caused outbreaks of superficial keratitis. Bacteriostats have been absorbed by them and fragments have been drawn into syringes. Cytotoxicity from some rubber closures. Possible leukoderma from contact with alkylphenols in the rubber industry or from wearing rubber ***gloves***. Contaminants have leached into parenteral solutions from the rubber plunger seals of disposable syringes. Allergic cheilitis can develop from a ***mercaptobenzothiazole*** (q.v.) in rubber-sensitive **individuals** who chew on rubber erasers of pencils. Persistent, severe, diffuse, plantar dermatitis in **people** allergic to ***ethyl butyl thiourea*** used on the insole of a popular athletic ***shoe*** and ***neoprene*** rubber products. ***Calcium***, ***magnesium***, and ***zinc*** have also been leached from closures into acidic solutions. ***Zinc*** had a pyrogenic effect in **rabbit** trials. Soviet **workers** in the production of synthetic rubber from ***α-methylstyrene*** (q.v.) for > 4 years developed central nervous system disturbances, leukemia, and other blood disorders. **Animal** trials showed its toxicity increases with increasing ambient temperatures. Black rubber contains antioxidants, such as ***p-phenylenediamine*** (q.v.). ***Benzene*** (q.v.) and ***polynuclear aromatic hydrocarbon*** (PAH) exposure is also common in tire manufacturing operations. Accelerators and stabilizers are added in many baby bottle nipples and rubber pacifiers which cause unacceptably high levels of potentially carcinogenic ***nitrosamines*** (up to 390 ppb). By 1985, U.S. and Canada set upper exposure limit of 10 ppb. A

45-year-old **male** was hospitalized for uncontrolled epistaxis. **He** noticed a strong solvent odor 3 days earlier while **his** log home was being caulked with a ***toluene*** and ***petroleum distillate*** butyl rubber compound. **He** developed a severe headache, nausea, dizziness, and disorientation until profuse nosebleeds occurred before **his** hospitalization. His **wife** and two **sons** had similar, but milder, symptoms after sleeping upstairs in the uncaulked floor of the home. ***Acetylcysteine*** reacts with rubber in mucolytic misting equipment. ***Dimethyl sulfoxide*** can penetrate rubber ***gloves***. Gum rubber absorbs ***ethylene oxide***, which can burn hands, cause hemolysis of blood, and unaerated endotracheal tubes can cause tracheitis and tissue necrosis. Millions of pounds of ***salicylic acid*** used as a rubber retarder, to reduce the tendency of a rubber compound to vulcanize prematurely. Reclaiming of old rubber is done with strong compounds, such as ***phenylhydrazine*** (q.v.) to break it down for spreading in making rubberized cloth. These **workers** suffer from edema, infiltration and fissuring of fingers, with separation of the free borders of **their** nails. Continued exposure leads to erythema and edema of face, arms, and legs. Rubber band ligation of hemorrhoids has caused pain, bleeding, recurrences, and possibility of tetanus and other infections. Rubber cement, used in many occupations, has been a cause of various contact dermatoses, and **glue-sniffers** have found its solvent a source of ***benzene*** (q.v.), ***hexane*** (q.v.), ***toluene*** (q.v.), and other toxic fumes.

Silicone rubber (***Silastic***) tubing can readily absorb ***halothane*** and reuse for other purposes can be dangerous. Injected, it can disrupt dermal or SC architecture and stimulates connective tissue leading to thickening of dermal collagen. Its extravasation backward through a needle pathway has caused a lump and orbital deformity. **Children** have put rubber bands on the ends of **cattle**, **cat**, and **dog** tails; and **animal** ears and necks, with sometimes disastrous results. **Cattle** have developed hyperkeratosis from contamination of rubber mats with chlorinated ***naphthalenes*** (q.v.) and ***polychlorinated biphenyls*** (q.v.). **Dogs** have had loss of pigment of their noses (planum nasale) after eating from rubber dishes, due to the presence of the antioxidant, ***monobenzone*** (q.v.). Gastrointestinal upsets and obstructions reported in **dogs** eating rubber toys and **horses** eating frayed rubber fencing. ***B-D***® syringe rubber was found to be toxic to **equine** sperm. Scraps of rubber used for ***shoe*** soles were highly fatal after ingestion by **chickens** due to ***σ-tricresyl phosphate***. Symptoms were inappetence, fetid diarrhea, leg paralysis, and death within 3–4 weeks.

RUBIDIUM

Untoward effects: Reacts violently with ***oxygen*** or any liquid, including water, containing ***oxygen***. Small amounts are present in normal food intake. Unlike ***lithium***, another earth metal, it produced hyperactivity, increased aggressiveness, and electroencephalogram activation in **monkeys**.

RUBRATOXIN A and B

From *Penicillium rubrum* and *P. purpurogenum*.

Untoward effects: Occasionally found on moldy ***corn*** and has poisoned **poultry** and **swine**. It is a hepatotoxin and causes hemorrhage and kidney damage. Single IP dosage of ***rubratoxin B*** to **mice** on day 6–12 of gestation led to growth retardation, embryocidal, and teratogenic when examined on day 18. Exencephaly, malformed pinnae and jaws, umbilical hernia, and eye defects were common. **Guinea pigs** given 2–10 mg/day showed considerable weight decrease, increased prothrombin time, decreased ***β-globulin***, and increased ***γ-globulin***. Low toxicity in **broilers** leading to liver hypertrophy, atrophy of bursa of Fabricus, anemia, decreased serum protein, increased serum ***cholesterol***, and capillary fragility after 1 mg/g of diet. Post-mortem on poisoned **animals** revealed congestive, hemorrhagic, and degenerative lesions of liver and spleen. Lethal doses in **dogs** cause extensive hepatic necrosis and mild degenerative changes in renal tubular epithelium. **Ducklings** (1 day old) had hepatic lesions at 1 mg ***rubratoxin B*** dosage.

B: IP, LD_{50}, **rat**: 350 μg/kg; **mouse**: 270 μg/kg; **guinea pig**: 480 μg/kg; **cat**: 200 μg/kg.

RUDBECKIA sp.

Untoward effects: ***R. hirta*** = ***Black-eyed Susan*** and ***R. laciniata*** (***Cone Flower***, ***Dormilón***, ***Golden Glow***), eaten in late spring or early summer in low-lying pastures by **hogs**, **cattle**, and **sheep** has caused gastroenteritis; occasionally anorexia and incoordination.

RUE

= *Herb of Grace* = *Herbygrass* = *Ruda* = *Rueworr* = *Ruta graveolens*

The common herb or garden rue. In the Aztec and Mayan materia medicas.

Untoward effects: Ingestion of it and its oil (***bergapten***) used as an emmenagog and abortifacient. Its ***furocoumarin*** content is a photosensitizer, leading to phytodermatitis, particularly if the skin is wet and exposed to the sun, as in the case of 16 **women** picking the herb at a Czech agricultural farm. **They** developed a dermatitis and blisters. The oil is a potent local irritant to skin and mucous membranes. Although traces were once used as a food flavoring and antispasmodic, large oral doses cause violent gastric pains, vomiting, and prostration, as well as inflammation of the digestive tract mucous membranes and kidneys. Used in criminal abortions, often with fatal results.

Antifertility effect in **rats**, but not **hamsters**. **Animals** are rarely poisoned.

Oil, minimum lethal dose, skin, **rabbit**: 500 mg/kg.

RUELENE

= *Crufomate* = *Dowco 132* = *Kempak* = *Montrel*

Anthelmintic. Organic phosphorus anticholinesterase.

Untoward effects: Considered moderately toxic. Allows accumulation of ***acetylcholine*** at cholinergic neuro-effector junctions (muscarinic effect) and skeletal muscle myoneural junctions and in autonomic ganglia (nicotinic effect) while also impairing central nervous system function. Toxicity occurs from inhalation, ingestion, and percutaneous absorption. Serious toxicity would include headache, dizziness, extreme weakness, ataxia, early miosis, twitching, tremors, nausea, bradycardia, pulmonary edema, hyperhidrosis, etc. Maximum acceptable daily intake for **man** has been 100 μg/kg. Do NOT use in conjunction with or within a few days of (before and/or after) any other cholinesterase ***inhibitors*** and avoid use with ***arsenicals***, ***phenothiazine***, ***phenothiazine tranquillizers***, ***purgatives***, ***succinylcholine***, or drugs producing purgation as a side-effect in **man** and **animals**.

In **cattle** warble control, there is a seasonal variation in toxicity, which appears to be associated with the death and absorption of the larvae, rather than the toxicity of the insecticide *per se*. Use the drug *before* the larvae reach the gullet and epidural areas. This is September for most of the United States, Italy, Bulgaria, and Ireland. The drug will pass into the milk of treated **animals**. Do NOT apply to lactating **cows**. To avoid residues in food for **man**, withdrawal periods of 14 days, 28 days, and 35 days, respectively, have been suggested for recommended usage in **goats** and **sheep**, **cattle**, and **reindeer**. **Dairy calves** < 2 weeks poisoned by 50 mg/kg; adult **cattle** and **angora goats** by 100 mg/kg; **sheep** by 200 mg/kg (killed by 300 mg/kg), and **horses** by 50 mg/kg. Severe poisoning in **pigs** by 15 mg. Water-based topical sprays or dips of 2% or more can be toxic to **cattle**, **sheep**, and **goats**. Many of the above symptoms in **man** also occur in **animals**.

LD_{50}, **male rat**: 635 mg/kg; **female rat**: 460 mg/kg.

RUGS

Untoward effects: Allergic reactions from many sources, such as jute in inexpensive rugs and carpeting, ***dichlorophen*** on rug backing, ***naphthalene*** (q.v.) in the ***Rinsen vac® Carpet Cleaning System***, and the Kawasaki syndrome (mucocutaneous lymph node syndrome) from rug shampoos. Dust or rug mites are also a cause.

RUGULOVASINES

Untoward effects: These toxic indole alkaloids are produced by a number of ***Penicillium sp.*** and are acutely toxic to day-old **chicks**.

RUMEX sp.

= *Docks* = *Sorrels*

Untoward effects: Its roots contain ***anthraquinones*** and are used in herbal mixtures and ***teas*** in many countries as a cathartic, where it sometimes has caused severe diarrhea, dehydration, weakness, coma, and peripheral neuropathy. Disagreeable taste and unpleasant odor for **man** in ***milk***, milk products, and meat of **animals** that have grazed it.

Has caused ***oxalate*** poisoning, mostly in **cattle** and **sheep**, leading to dullness, depression, colic, dyspnea, tremors, ataxia, prostration, coma, and death. Post-mortem reveals echymoses and petechial hemorrhage on abdominal serosa, large amounts of yellowish fluid in the abdominal cavity, abomasitis, enteritis, pale and edematous kidneys, congested lungs, nephrosis, and fatty infiltration of the liver.

R. acetosella = ***Sheep Sorrel*** = ***Sour Dock*** = ***Red Sorrel*** irritates the skin, mucous membranes, and the digestive system of **sheep** and **horses**. Excessive ingestion leads to muscle spasms, ataxia, nasal discharge, dyspnea, coma, and death.

RUSSULA

Untoward effects: Ingestion of this ***mushroom***, within 1–2 min, causes gastroenterocolitic and parasympathetic symptoms and hallucinations.

R. emetica also produces the toxic alkaloid, ***muscarine*** (q.v.). ***Muscarine*** causes ***pilocarpine*** (q.v.)-like effects.

RYANIA sp.

Untoward effects: The root and wood is insecticidal and ***R. speciosa*** is particularly effective versus insecticide-resistant **cockroaches** and **ticks**. It is known as ***Capansa*** or ***Kapahasa***, an effective piscicide, but used mostly as a poison to **humans** and **animals** by **Brazilian Indians**. It is also lethal to **animals**.

R. dentata is used on poison arrows by **Colombian Indians**. The insecticidal principal, an alkaloid, is known commercially as ***ryanodine***.

Bradycardia, dyspnea, weakness, diarrhea, vomiting, tremors, convulsions, coma, and death occur in experimental **animals**. The alkaloid is ~700 times as potent as the stem wood.

LD_{50}, **rat**: 750 mg/kg; **mouse** and **rabbit**: 650 mg/kg; **guinea pig**: 2.5 g/kg.

RYE

= *Secale cereale*

Untoward effects: Reactions can be in the form of allergic contact dermatitis, bronchial asthma, rhinitis, hypersensitivity pneumonia, and upsets from ingestion, especially if it is contaminated with fungi and their ***mycotoxins***. See ***Bread***, ***Claviceps***, and ***Ergot***. Rye is a ***nitrate***-accumulator. Contact with rye straw has caused contact dermatitis in some **people**. Has estrogenic activity. Contains ***gluten*** (q.v.), intolerance to which leads to celiac disease or sprue.

Sprouting rye poisoned ~250 **sheep**. Symptoms were apathy, inappetence, stiffness, and gnashing of teeth 24 h after grazing on it in a stubble field. 24 died.

RYE GRASS

= *Lolium sp.*

Untoward effects: ***L. perenne*** = ***Perennial Rye Grass*** = ***Timothy Grass*** causes facial eczema in **sheep**, a type of hepatogenous photosensitivity after ingestion of rye grass infected with a fungus, *Pithomyces charatum*. The toxic principal is ***sporodesmin*** and the photosensitizing agent ***phylloerythrin*** increases in their blood after liver damage. **Cattle** and **horses** are only occasionally affected. Symptoms occur usually within 5 days, including head-shaking, stamping of feet, rubbing themselves on rough surfaces; swollen eyelids, face, ears, and legs; and watery secretions from skin cracks, which later become encrusted. Some die after tetanic spasms and convulsions. Post-mortem reveals liver pathology and fibrosis; distended gall bladder, and thickened ducts. Rye grass staggers is a syndrome of **sheep**, **cattle**, and **horses** from their grazing new shoots or seedlings leading to slight spasms or stiffness. In severe cases, tetanic spasms, falling with legs stiffly extended, unable to eat, and starvation.

L. rigidium = ***Annual Rye Grass*** ingestion by **cattle** and **sheep** causes symptoms similar to those of the staggers syndrome, and appear to be due to toxic galls on the plants.

L. temulentum = ***Darnel*** = ***Poison Rye Grass*** is parasitized by ***Claviceps sp.***, causing ergotism and toxicity from its alkaloids, ***temuline*** and ***loliine***. Ingestion by **horses** leads to mydriasis, incoordination, collapse, convulsions, and death. Rye grass may also cause hemolytic anemia.

RYNCHOSIA sp.

Untoward effects: **Aztecs** used the seeds as narcotics and hallucinogens.

S

SABADILLA

= *Caustic Barley* = *Cevadilla*

Insecticide.

Untoward effects: All parts of the plant, especially the roots are poisonous. Mucous membrane irritant. May cause sneezing. Contains ***veratrine*** (q.v.) and other ***veratrum*** alkaloids which are potent hypotensive agents.

LD_{LO}, **human**: 50 mg/kg.

Can cause paralysis in **rabbits** and **frogs**.

LD_{50}, **rat**: 4 g/kg.

SACCHARIN

= *Benzosulfimide* = *Garantose* = *Glucid* = *Gluside* = *Hermesetas* = *Saccharimol* = *Saxin* = *Sykose*

Non-caloric synthetic sweetener.

Untoward effects: Although up to 500 times as sweet as ***sucrose***, its bitter, objectionable taste requires neutralization by 9–10 parts ***cyclamates*** to one part saccharin. When ***cyclamates*** were banned, ***glycine*** was substituted. **Human** safety, careful weighing of the benefits versus risks, and the awkward legal battle trying to decide if it was a drug or a food additive, kept it in the marketplace, despite evidence of carcinogenicity in **rats** and **mice**. After nearly 100 years of use, some reports of photosensitization with erythema of the face, urticaria, fixed-drug reactions, and alterations of cardiac rhythm have been reported. Its influence on the incidence of bladder cancer in **man** has been debated, as are the effects of possible co-carcinogen contaminants, ***p-sulfamidobenzoate*** and ***σ-toluene sulfonamide***. **Human** exposure was primarily by ingestion and occasionally by dermal exposure. At equipotent sweetening effects, ***sucrose*** is 375 times more likely to cause cancer. Very small amounts are used to stabilize some ***phenothiazines***.

Ingestion by **rats** leads to transitional cell carcinomas of the bladder and an increase in leukemias and lymphomas. Carcinogenicity carried over into male **offspring**; by **mice**, leads to thyroid adenocarcinomas.

SAFFLOWER OIL

Untoward effects: Allergenicity of water-soluble protein fraction of the seed hull, but not the pure oil. ***Liposyn***, a 10% solution used in total parenteral nutrition caused sinus bradycardia in 23-year-old **female** within 4 h after start of infusion; 9/21 **children** on long-term therapy developed gallstones, and five required cholecystectomy.

SAFFRON

= *Crocus sativus*

Contains a yellow dye, also called *saffron*. Food dye and flavoring agent. 140 stigmata to make 1 g of the dye.

Untoward effects: Toxic for **man** leading to facial flushing, vertigo, bradycardia, epistaxis, vomiting, and stupor. Metrorrhagia and death have followed its use as an abortifacient. Basle, Switzerland was once the center of the saffron trade, and, when ~360 kg from Italy was seized, Italy sent an army to get it back in what became known as the Saffron War of 1374.

LD_{LO}, **human**: 5 g/kg.

Crocin, one of its glycosides, prolongs blood coagulation time in **mice**, inhibits platelet aggregation in **rabbits**, and leads to thrombus formation in **rats**.

SAFRANINES

Untoward effects: Cheilitis from the use of ***tolu-safranine*** or ***dimethylsafranine*** in ***lipstick***.

SAFROLE

Untoward effects: **Human** exposure has been by ingestion in foods, pharmaceuticals, and beverages; also by dermal contact. In the U.S., now prohibited in foods. Found in many edible plants, in black ***pepper***, ***nutmeg***, and ***mace***, and in many essential oils. May cause a contact dermatitis. Estimated carcinogenic dose for 70 kg **man**: 6.6 mg.

Weak hepatotoxic and carcinogenic agent in chronic feeding trials in **rats**, **mice**, and **dogs**.

LD_{50}, **rat**: 1.95 g/kg; **mouse**: 2.35 g/kg.

SAGE

= *Salvia officinalis*

Untoward effects: Allergic dermatitis from its pollen reported in **man** and **dogs**. A report indicates it may have psychoactive properties, but not confirmed by others. Small quantities appear to be safe for cooking, but reports indicate long-term use of low dosages (to relieve hot flashes and to decrease excessive perspiration) or short-term use of high dosages in herbals leads to adverse mental and physical effects, convulsions, and loss of consciousness.

Oil of sage, being bactericidal, when ingested by **deer** decreases rumen microorganisms leading to poor digestion and anorexia. IP to **rats** of the oil (200 μl to 4 ml/kg) leads to convulsions and death.

Oil, LD_{50}, **rat**: 2.6 g/kg.

SALAMANDERS

= *Newts*

Untoward effects: The skin of these common **children's** pets contains toxins including ***tetrodotoxin*** (q.v.), that can be dangerous if ingested. Fortunately, after severe burning sensations in the mouth, one of the first ill-effects is vomiting, which rids the body of most of the toxin. The alkaloids, including ***saladarine***, in the skin toxin can cause violent convulsions.

SALAMI

Untoward effects: Causes migraine headaches, due to the ***sodium nitrate*** (q.v.), added to preserve its color. In November–December, 1994, 20 laboratory-confirmed cases of diarrhea caused by *Escherichia coli* 0157:H7 after consumption of dry-cured salami.

SALICYL ALCOHOL

= *Saligenin*

***Local anesthetic. A benzoic acid* (q.v.) derivative.**

Untoward effects: Allergic reactions include asthma, angioedema, urticaria, facial erythema, pruritus, dermatitis, rhinitis, and dyspnea, especially in ***aspirin***-sensitive **people**.

SALICYLAMIDE

= *Amid-Sal* = *Cidal* = *Dropsal* = *2-Hydroxybenzamide* = *Liquiprin* = *Salamid* = *Salrin* = *Salicym* = *Salizell* = *Salymid* = *Samid* = *Urtosal*

***Analgesic, antipyretic.* It is not a *salicylate*.**

Untoward effects: Dizziness, drowsiness, gastrointestinal upsets, and vomiting. Rarely, thrombocytopenia, leukopenia, tachycardia, and sleepiness. Encephalopathy in 12-year-old **female** receiving 3 g/day/8 days causing hallucinations, coma, and bilateral Babinski sign. Gross urticaria in two **infants** using a teething jelly with 8% level. Hemorrhage, purpura, and death in 62-year-old **female**. Along with other drugs, has been involved in suicides.

Except for some gastrointestinal irritation occasionally in **dogs** and **man**, it is well tolerated by the **mouse**, **rat**, **pig**, **rabbit**, **dog**, and **man**, but not **cats**. Fatal uremic poisoning (oliguric necrosis) in the **cat** at $\sim\frac{1}{10}$ the toxic dose for other **animals**. Has been used to induce experimental papilledema in **dogs**.

SALICYLAZOSULFAPYRIDINE

= *Azulfidine* = *Colo-Pleon* = *Salazopyrin* = *Sulfasalazine* = *Sulcolon* = *Sulphasalazine*

***Antimicrobial.* The active therapeutic moiety in Crohn's disease is probably *5-aminosalicylic acid* and not *sulfapyridine*.**

Untoward effects: ***Blood and bone marrow***: Hemolytic, megaloblastic, and aplastic anemia; leukopenia, agranulocytosis, thrombocytopenia, purpura, hypoprothrombinemia, methemoglobinemia, red cell aplasia, myelodysplastic syndrome, reticulocytosis, macrocythemia, megakaryocytic aplasia, and bone marrow necrosis. Hemolysis in glucose-6-phosphate dehydrogenase-deficient **subjects**.

Clinical tests: Interferes with yellow froth test for bile in urine. Can yield positive results on an antinuclear antibody test.

Central nervous system: Guillain–Barré syndrome, peripheral neuropathy, depression, drowsiness, vertigo, dizziness, tinnitus, hearing loss, headache, insomnia, ataxia, hallucinations, confusion, meningitis, and convulsions.

Gastrointestinal: Gastric upsets, nausea and vomiting (up to 40%), diarrhea, abdominal pain, neutropenic enterocolitis, stomatitis, decreased taste acuity, decreased ***folic acid*** and ***digoxin*** absorption, xerostomia, mouth petechiae, hepatitis, pancreatitis, and occasional hepatic necrosis.

Miscellaneous: Fever, lymphadenopathy, lymphocytosis, splenomegaly, malaise, cyanosis (3%); orange yellow–brown color in alkaline urine; joint aches, conjunctivitis, anorexia, pleuritis, cough, pneumonitis, shivering, night sweats, pulmonary eosinophilia, dyspnea, bronchospasm, tracheolaryngitis, temporary reversible sterility in **males**, irreversible staining of soft contact lenses, bronchiolitis fibrosa obliterans, impotence, and Raynaud's phenomenon. Rarely, goiter is possible. *E. coli* resistance has developed. In some cases, exacerbation of ulcerative colitis. Cases of possible teratogenicity reported. In breast milk at up to 30% of simultaneously drawn **maternal** serum level.

Renal: Calculi, bilirubinuria, nephritis, toxic nephrosis with oliguria and anuria, nephrotic syndrome, hematuria, crystalluria, and proteinuria. Left renal thrombosis in a **newborn** after a **mother** received it and ***prednisone*** during **her** last 5 months of pregnancy.

Skin: Rash, pruritus, urticaria, erythema multiforme, exfoliative dermatitis, paresthesias, epidermal necrolysis, vasculitis, systemic lupus erythematosus syndrome, photosensitization, lichen planus, and erythrodermic psoriasis.

Drug interactions: May potentiate ***coumarins***, ***hypoglycemics***, and increased bone marrow suppression of ***methotrexate***.

Found in the breast milk of treated **mothers** and placental transfer occurs. Therapeutic serum levels at up to 50 mg/l; toxic at 60–100 mg/l.

Long-term use in **dogs** has a high risk of producing permanent keratoconjunctivitis sicca and/or allergic dermatitis, cholestasis, and emesis. **Cats** may show hypersensitivity to it.

SALICYLIC ACID

= *2-Hydroxybenzoic Acid* = *Keralyt* = *Occlusal* = *Verrugon*

Keratoplastic, keratolytic, antiseptic, and antifungal. **> 30 million lb/year manufactured; 60% is used in making *aspirin*. First isolated from *Spirea*, and called *spirin*, which Bayer acetylated and put an "a" in front of, coining the word *aspirin*.**

Untoward effects: ***Miscellaneous***: Low concentration in breast milk, crosses the placental barrier. Hyperuricemia, vomiting, dyspnea, abdominal pain, acidosis, and central nervous system upsets after ingestion. Therapeutic doses occasionally cause tinnitus, mental confusion, headache, hyperhidrosis, and dim vision. Excessive doses lead to hypothermia, weak pulse, dyspnea, and coma.

Poisoning: Percutaneous absorption occurs. Potentially hazardous in **people** with impaired liver and kidney functions or for small **children**. Somnolence, polypnea, alkalosis, agitation, fever, and tachycardia in 7-year-old **female** with ichthyosis 2 days after topical treatment with 15% ointment; recovery in 2 days. Bilateral sensorineural hearing loss of 50–60 decibels for pure tones in case report of **females** with generalized psoriasis after 3% crude ***coal tar*** (***Zetar***®) with 5% ointment tid leading to mild respiratory alkalosis and metabolic acidosis. Its salts are readily absorbed from the gastrointestinal tract causing high serum concentrations in 30 min; ~$\frac{1}{2}$ absorbed in 1 h, and peak serum levels in 2–4 h leading to local gastrointestinal irritation, central nervous system respiratory center stimulation, increased metabolic rate, and interference with normal ***carbohydrate*** metabolism and blood coagulation. Application of 100 ml of 6% under an occlusive dressing overnight resulted in salicylism (36.6 mg/dl) and tinnitus. A 7-year-old **male** treated daily with 150 g 10% ointment developed hyperventilation and hyperpyrexia (43°C) on day 4, and was comatose. Serum ***salicylate*** level was 670 mg/l on day 5 and brain edema, coagulopathy, and decreased blood pressure. Hypertonic, opisthotonic, and extension spasms during following weeks. Chronic toxicity above 300 mg/l; acute toxicity with hyperventilation 350–500 mg/l 6 h after ingestion. For **children**, decrease these numbers by 50 mg. Death of two **patients** after 50% of **their** bodies were painted with 20.7% alcoholic solution for tinea infection. Accidental overdosing and its once widespread use in preserving food and drinks have caused pain and irritation of pharynx and stomach, difficulty in swallowing, vomiting, diarrhea, facial flushing, roaring in ears, feeling of head fullness, dimmed vision, confusion, delirium, and eventually coma. ***Albumin*** in urine, which is discolored by ***hematin***; weak pulse, dyspnea, anorexia, diarrhea or constipation, and eczema. LD, **human**: 50 mg/kg; LD_{LO}, **human**: 50 mg/kg.

Skin: Contact dermatitis, erythema, diaphoresis, and urticaria. Local irritation and bullous pemphigoid lesions after topical cream and sunlight. Occasionally scaling and purpuric or vesicular lesions. Can retard wound-healing. Cross-sensitivity with other ***salicylates***.

Drug interactions: Enhances ***sulfamethoxypyridazine*** antibacterial activity. Excretion is decreased by acidifying agents, such as ***ammonium chloride*** and increased with alkalinizing agents, such as ***sodium bicarbonate***. ***Para-aminobenzoic acid*** decreases its metabolism. Displaces bound ***methotrexate*** and ***phenytoin***, increasing their toxicity. May increase activity of ***penicillins***. Antagonizes uricosuric action of ***probenecid*** and ***sulfinpyrazone***. After 1 week of contact, 80% is absorbed by ***nylon*** syringes. ***Dimethyl sulfoxide*** decreases its topical penetration rates.

Gastric irritant in **animals**. A zero tolerance for it in milk is established by the FDA. It can appear in milk from its use in teat bougies or by passive diffusion from blood. It can also be dietary in origin. **Cats** may be particularly sensitive to its toxic renal effects. **Rabbit** trials indicate percutaneous absorption can occur.

IV, LD_{50}, **mouse**: 500 mg/kg. LD_{50}, **rat**: 891 mg/kg; **mouse**: 480 mg/kg; **cat**: 400 mg/kg.

SALINOMYCIN

= *Bio-Cox* = *Coccistac* = *Coxistac* = *Ovicox* = *Saccox* = *Salgain* = *Usten* = *Yusuchin*

Coccidiostat, antibiotic, ionophore.

Untoward effects: Although approved and effective in **broilers** and **chickens** receiving it continuously at 40–60 g/ton of complete feed, it has been fatal to **turkeys** and **horses**. A feed mill supplied a grower a **broiler** level for his 12-week-old **turkeys**. A total of 2000 died, starting in a few days to 2 weeks. Symptoms included cyanosis, depression, decreased weight, and sternal recumbency with one or both legs extended. Toxicity increased in **swine** also given ***tiamulin***.

LD_{50}, **mouse**: 50 mg/kg.

SALMETEROL

= *Arial* = *GR-33343X or G* = *Salmetedur* = *Serevent*

β_2-Adrenergic, bronchodilator.

Untoward effects: Headache, tremor, nervousness, dizziness, giddiness, tachycardia, cough, and awareness of heartbeat or palpitations. Exacerbation of asthma, as well as two deaths reported. Lower U.S.-approved doses than those of other countries reflects U.S. concern about higher doses. A fatal case of induced respiratory arrest reported in a 15-year-old **patient** with a history of respiratory disease. Loses its lasting effectiveness versus exercise-induced asthma attacks. High dosage leads to hypokalemia. Limited trials indicate a 50% incidence of adverse effects.

SALMON

Contains *aluminum sulfate* as a firming agent in canned salmon.

Untoward effects: Salmon poisoning disease in **dogs** reported along the western seacoast of the U.S. and on Vancouver Island in Canada leading to lethargy, anorexia, diarrhea, pyrexia, lymph node enlargement, and death in up to 90% of untreated **animals**. Caused by *Neorickettsia helminthoeca*, as a result of eating **salmon**, **trout**, or **steelhead** that harbor a vector fluke. **Foxes** are also affected.

SALSALATE

= *Disalcid* = *Disalicylic Acid* = *Disalgesic* = *Mono-Gesic* = *Nobacid* = *NSC 49,171* = *Salflex* = *Salicylsalicylic Acid* = *Salisal* = *Salyphec* = *Salysal* = *Sasapyeine*

Analgesic, anti-inflammatory; non-steroidal anti-inflammatory drug.

Untoward effects: Tinnitus, hearing loss, vertigo, and dizziness, particularly with high dosage. Abnormal thyroid function tests and occult gastrointestinal bleeding (less than with ***aspirin***). Occasional rash, nausea, dyspepsia, and heartburn. Multiple ulcerations of small intestine in 47-year-old **female** after 750 mg twice daily. After sublingual use of 750 mg tid, an edentulous 73-year-old **male** developed ulcerated lingual lesions. Acute renal failure in **patients** with chronic renal insufficiency receiving short-term treatment with dose > 3 g/day. Minimal change nephrotic syndrome in 73-year-old **male** after 1.5–2 g/week/2 years. Alkaline urine will increase its excretion and acid urine decreases it. Reye's syndrome may develop if used in **people** with viral infections. Large amounts in **mother's** breast milk. Anecdotal reports of many other possible drug-induced untoward effects.

Drug interactions: Other ***salicylates*** can increase its toxicity. Can antagonize effects of uricosuric drugs and potentiate ***liothyronine***, ***methotrexate***, ***naproxen***, ***penicillin***, ***phenytoin***, ***sulfinpyrazone***, and ***thyroxine***.

Dosages 4–5 times recommended **human** doses have been embryocidal and teratogenic in **rats**.

SALSOLA sp.

= *Russian Thistle* = *Saltwort* = *Tumbleweeds*

Untoward effects: Its aeroallergens cause allergic rhinitis, bronchial asthma and/or hypersensitivity pneumonia in **people**.

Contains > 1.5% ***nitrate*** (q.v.), which is potentially toxic to **livestock**. ***Oxalates*** (q.v.) in ***S. Kali***. Its inhalant allergens probably cause skin reactions in **dogs**. Pregnant **ewes** that ate 0.9 kg of leaves and twigs of ***S. tuberculata*** var. ***tomentosa***, a shrub in South Africa, for 10–50 days during any stage of their pregnancy, developed Groot-lamsiekte, a specific syndrome of prolonged gestation and retarded udder development. Their **lambs** evidenced hypophysial, adrenal, and thymic atrophy; hypertrophy of **female** genitalia; polyfollicularity of ovaries, Leydig cell hyperplasia; long ***wool*** coat; erupted inciscors, and pigmentation, especially of kidneys and lymph nodes. Causes prolonged gestation and diestrus phase of the estrus cycle in **rats**.

SALVIA

Untoward effects: In Australia, ingestion of ***S. coccinea*** has killed **cattle**.

S. divinorum = ska Maria Pastora is used by the **Mazatec Indians** of Mexico as a hallucinogen.

In **mice** and **rats**, ***S. haematodes*** leads to decreased motor activity, sedation, analgesia, anticonvulsive activity, and hypoxia. Another group found oral use of the root extract increased penile erection and frequencies of mounting and intromission in male **rats**.

S. miltiorrhiza = Danschen. ***Warfarin*** anticoagulation is increased by its coadministration.

S. reflexa has killed Australian **sheep**.

S. rosmarinoides is poisonous to **livestock** in Brazil when other forage is scarce.

S. sclarea = Clary in the early 1800s in England was used as a substitute for ***hops*** for sophisticating ***beer***, giving it bitterness and intoxication properties which gave the ***beer*** an "insane exhilaration of spirits, succeeded by severe headaches".

For ***S. officinalis***, see ***Sage***.

SAMARIUM-153 LEXIDRONAM

= *Quadramet* = *Samarium-153 EDTMP*

A man-made *radionucleotide* and *chelate*.

***Untoward effects*:** White blood cells and platelets decrease ~50% within 3–5 weeks and return to previous levels by the eighth week. Within a few days after injection, 10% of **patients** have a flare-up of mild bone pain.

The EDTMP chelator has caused hypocalcemia and electrocardiogram effects in **dogs**.

SAMBUCUS sp.

Untoward effects*:** ***S. niger* = *Elder and ***S. ebulus* = *Danewort* = *Dwarf Elder*** contain an emetic and purgative oil with a strong and repulsive odor. Eating of raw berries has caused nausea and vomiting in **humans**. Cyanogenic. **Children** have been poisoned by using its hollow stems as pipes. Heating destroys its toxins, ***sambunigrin*** and ***vicianin***.

Ordinarily not eaten by **livestock** unless starved, leading to superpurgation. Poisoning in **swine** causes vomiting, abdominal pain, dyspnea, tachycardia, tachypnea, salivation, trembling, weakness, and occasionally death.

SAND

Untoward effects*:** Dermatitis in **men** making "core", a sand mix used in foundries, and its dust from sandpapering. Foundry sand can be contaminated with ***chromium (q.v.). Pulmonary fibrosis and chronic dyspnea reported in **sandblasters**. An association in these **workers** also noted with mycobacterial disease by *Mycobacterium kanasii*.

Inhalation LC_{LO}, **human**: 300 μg/m^3 for 10 years.

Sand-induced enteropathy with diarrhea and weight decrease in a young **foal**. Colic and death in a **horse** from > 200 lb of sand in its gastrointestinal tract 4–5 years after grazing in such an area.

SANDALWOOD OIL

The heart wood of *Santaluna album*, a small evergreen in mountainous regions of India and the Malay Archipelago is dried and ground.

***Untoward effects*:** Its oil causes photosensitivity reactions and is used in soaps, detergents, cosmetic creams and lotions, and perfume. No irritation to 18 **volunteers** of full-strength material in 48 h closed-patch tests. No sensitization of 10% in petrolatum in 25 **volunteers**.

Slightly irritating to the backs of hairless **mice** and it was irritating when applied full-strength to intact or abraded **rabbit** skin for 24 h under occlusion.

LD_{50}, **rat**: 3.8–5.58 g/kg and acute dermal LD_{50}, **rabbit**: > 5 g/kg.

SANDBUR

= *Cenchrus sp.*

***Untoward effects*:** Its burs have caused mechanical injury, inflammation, and infection of the skin of **man** and **animals**.

Horses grazing it have developed osteodystrophia fibrosa.

SAND FLIES

= *Culicoides sp.* = *No-See-Ums* = *Punkies*

***Untoward effects*:** Blood-sucking, often with burning and painful bites. Hypersensitivity reactions reported.

SANGUINARIA CANADENSIS

= *Bloodroot* = *Puccoon* = *Red Root* = *Tetterwort*

Untoward effects*:** The underground stem, thick horizontal roots, and their red juicy contents are toxic primarily from its anticholinesterase alkaloid, ***sanguinarine, leading to nausea, vomiting, salivation, burning sensations in mucous membranes, extreme thirst, dermatitis, faintness, vertigo, dimness of vision, unconsciousness, coma, depressed cardiac action, weakness, paralysis, and death. Too bitter for most **people** to eat. Once used therapeutically as a nauseant.

Hogs and **cattle** ingesting it in early spring develop violent vomiting, excessive thirst, dyspnea, mydriasis, collapse, and death.

SANGURA

= *Hyptis suaveolens*

Untoward effects*:** Contains ***menthol (q.v.). Although used in one of the commonest ***bush teas*** of Curacao, it is suspected of poisoning **cattle** and causing illness in Queensland **horses**.

SANITARY NAPKINS

Untoward effects*:** Some ***tampons were involved in a toxic shock syndrome in **women**. A vulvar dermatitis reported from perfumes used on the napkins. See **Cellulose**.

SANSEVIERIA sp.

= *Mother-in-Law's Tongue* = *Snake Plant*

Untoward effects*:** The leaves of ***S. thyrsiflora* = *Tulalal* = *Turada* = *Tural of Venezuela and Trinidad are used externally, but reported to be poisonous if they, or their decoctions, are taken internally.

SANTONIN

Anthelmintic.

Untoward effects*:** Fixed-drug eruptions, disturbed color vision, xanthopsia, toxic amblyopia, mydriasis, mental confusion, giddiness, pallor, sweating, disturbed hearing, headache, nausea, vomiting, trembling, hemolytic anemia, variable brown nail pigmentation, convulsions, opisthotonus, emprosthotonus, respiratory failure, and yellow or greenish urine that turns red when made alkaline. Synergistic with ***α-kainic acid; mutual antagonist with ***hexylresorcinol***.

SAPINDUS sp.

Untoward effects: ***S. saponaria*** = ***Soap Berry*** and other species crushed seeds are used in Venezuela and India as a piscicide. It and ***S. Drummondii*** in the U.S. can cause an irritant allergic dermatitis in some **people** after handling the fruit. Contains ***saponins***. **Bees** enjoying the nectar of ***S. emarginatus*** have poisoned their ***honey*** for **man**.

SAPIUM sp.

Untoward effects: When the roots of ***S. fragrans*** = ***Collihuay*** are burned, the pleasant fragrance causes headaches. Its milky latex is so caustic it has caused **woodcutters** to lose **their** sight, and it has been used in the Americas on poison arrows and darts.

S. sebiferum = ***Chinese Tallow Tree*** = ***Tallow Tree***, native to China, but flourishing in southern U.S. states, often escapes cultivation. Its green or blackish-brown berries and occasionally its leaves (but not its white-hulled seeds) cause poisoning, leading to rapid, intense gastrointestinal upsets, nausea, and vomiting.

Domestic **animals** develop severe diarrhea, anorexia, weakness, dehydration, and increased blood urea nitrogen. Post-mortem reveals small lesions in gastrointestinal tract and renal tubular nephritis.

SAPONARIA VACCARIA

= *Cow Cockle* = *Cow Herb*

Untoward effects: Normally, **animals** refuse to eat the seeds or grain or screenings containing them, as they contain a distasteful ***sapotoxin***-like substance causing severe gastrointestinal upsets (nausea, vomiting, diarrhea), vertigo, and dyspnea.

SAPONIN

Causes foaming and emulsifying and is used for those purposes.

Untoward effects: Local irritant to epithelial tissues and mucous membranes; alters the permeability of cell walls. Poorly absorbed after ingestion and causes gastrointestinal irritation, hemolytic jaundice, vomiting, and purging. Bitter taste. Saponins have been added to toxic substances on poison arrow and spear heads. In certain *Agrostemma* seeds, ingestion leads to headache, vertigo, vomiting, hot skin, tachycardia, weak pulse, muscle weakness, and death. The larvae of the Leafcutter **bee**, a wild **bee**, are killed by the saponin in ***alfalfa*** leaves. When injected into **animals**, saponins destroy red blood cells by hemolysis. Excessive intake of ***alfalfa*** saponins by **ruminants** leads to growth retardation, bloat, and dyspnea; and decreased egg production by **hens**.

LD_{LO}, **mouse**: 3 g/kg; SC, 900 mg/kg; IV, 1 g/kg; IV, **rabbit**: 40 mg/kg.

SAQUINAVIR

= *Fortovase* = *Invirase* = *Ro 31-8959*

Antiviral, protease inhibitor.

Untoward effects: More common in **patients** receiving high dosage, and are usually mild and reversible. Nausea, vomiting, diarrhea, abdominal pain, and increased aminotransferases in < 4% of **patients**. Poor oral absorption and extensive first pass metabolism. Occasionally acute pancreatic diabetes, buffalo humps, acute paranoid psychosis, and renal disease.

Drug interactions: Plasma concentration increases 20 times when given with ***ritonavir*** (q.v.). ***Nevirapine*** may increase its metabolism and decrease its effect. Possible synergism with ***zalcitabine***. Hypoprothrombinemia with ***warfarin***, requires a 20% decrease of the latter's dosage. ***Indinavir*** and ***nelfinavir*** decrease its metabolism and may be therapeutically useful, but increase its toxicity. ***Delavirdine*** also decreases its metabolism, increases its toxicity, and may cause hepatotoxicity. ***Rifabutin*** and ***rifampin*** decrease its effectiveness. Saquinavir decreases metabolism of ***astemizole***, ***cisapride***, and ***terfenadine***, which may induce serious ventricular arrhythmias. ***Grapefruit juice*** can increase its effects. ***Food*** increases its bioavailability. Clearance of IV ***midazolam*** is decreased 56% by it.

SARALASIN

= *P 113* = *Sarenin*

Antihypertensive, angiotensin II antagonist.

Untoward effects: Occasionally, a mild sustained increase in blood pressure; it is very dramatic in **patients** with pheochromocytoma. Severe hypotension and risk of stroke with IV infusion.

IP use blocks ovulation in **rats**.

SARAMYCETINE

= *RO2-7758* = *X5079C*

Antibiotic, antifungal. **For deep mycoses.**

Untoward effects: SC leads to acute local swelling and inflammation, prolonged burning sensation, coalescing deep induration, shaking chills with increased temperature, eosinophilia, ***sulfobromophthalein*** retention, increased ***bilirubin***, mild periportal inflammation, and occasional urticaria.

Delayed ***sulfobromophthalein*** clearance in treated **dogs**.

SARCOBATUS VERMICULATUS

= *Chico* = *Greasewood*

A range shrub on alkaline fields in western United States.

***Untoward effects*:** Its aeroallergens can cause allergic rhinitis, bronchial asthma, and hypersensitivity pneumonia in **people**.

Untoward effects due to its ***sodium*** and ***potassium oxalate*** content (equal to 19% ***anhydrous oxalic acid***) in its aerial parts. Toxicity increases as the growing season advances. Small amounts are tolerated without ill-effects. **Cattle** may die after consuming 3–3½ lb of green parts with other forage. Without other forage, 2 lb is lethal. Symptoms usually occur within 4–6 h leading to dullness, colic, depression, dyspnea, prostration, coma, and death from cardiac and renal failure. **Sheep** are commonly affected.

SARCOPHALUS DIDERRICHII

= *Dundaki*

***Untoward effects*:** **Workers** in Africa who process its wood have been poisoned by it, sometimes fatally, due to a cumulative alkaloid causing cardiac disturbances.

SARCOSTEMMA sp.

Untoward effects*:** ***S. australe is a caustic vine on dry hills of limestone rocks of Australia. Toxic to **sheep** who eat 2 oz and to **cattle** and **horses** who consume 2 lb. Symptoms include staggering, recumbency, apparent pain, violent spasms and limb paddling, and bloat before death.

S. viminale = ***Melktou*** in Zimbabwe is poisonous. **Sheep** grazing it stagger, grind their teeth, and have tremors and ventral flexion of their heads. While lying down, they have furious galloping movements before death.

SARDINELLA sp.

= *Tamban*

Untoward effects*:** Poisoning and fatalities of **people**, **cats**, and **dogs** in the Philippines from eating this **fish** (marinated or boiled). Occurs in May and June, at the start of the rainy season, may be due to ***clupeotoxin. **Human** symptoms include nausea, severe stomach pain, headache, weakness, and inability to speak shortly before death. Affected **fish** may taste bitter or metallic.

SARDINES

Untoward effects*:** Contain excessive amounts of ***purine, a ***uric acid*** precursor. Gout **patients** may have to avoid their consumption.

SARIN

= *GB* = *Weteye Bombs*

Chemical and biological warfare nerve gas, organic phosphate, anticholinesterase. **Colorless and odorless.**

Untoward effects*:** Extremely active ***cholinesterase inhibitor, permitting accumulation of ***acetylcholine*** in the tissues and, thus, continuous stimulation of the parasympathetic nervous system. Death can occur within seconds. During World War II, Germany made some but never used it. It is usually released as a spray. The U.S. Army had ~10 lb in each **M-55** rocket. The U.S. also stored it in 1000 lb bomb clusters and each individual device in them contained 2.6 lb and an explosive agent. A thimbleful of vapor can kill six **men**. LD of vapors for **man** is estimated at 10 mg/m^3/10 min. Other **workers** estimate it as 1 mg. **Workers** accidentally exposed to low doses showed increased high-frequency β waves in **their** electroencephalogram lasting > 1 year. Small quantities can kill on contact with skin or by inhalation of its vapors. It was the agent used in Tokyo's 1995 subway terrorist attack. Typical toxic effects can include salivation, tearing, miosis, muscle twitches, and fasciculations. In March 1991, during the Gulf War, U.S. armed forces destroyed Iraqi rockets containing it and ***cyclosarin*** at the Khamisiyah weapons depot. Exposure of nearly 100,000 **troops** may have been associated with the "Gulf War Illness" (memory problems, fatigue, nervous system disorders, rash).

TD_{LO}, **human**: 2 μg/kg. LD_{50}, skin, **human**: 28 mg/kg. Inhalation LC_{50}, **human**: 70 mg/m^3.

Monkeys exposed to a single large dose or a series of asymptomatic low doses showed significant electroencephalogram changes lasting a year. Its test firings on Dugway Proving Grounds may have been one of the causes of 5000 **sheep** suddenly dying in Utah's Skull Valley. Has caused a delayed neuropathy in **hens**.

LD_{50}, **rat**: 550 μg/kg; SC, 113 μg/kg; IV, 45 μg/kg; IM, 200 μg/kg; **mouse**: IP, 450 μg/kg; SC, 100 μg/kg; IM, 222 μg/kg; skin, 1.08 mg/kg; **dog**: IV, 19 μg/kg; **cat**: IV, 22 μg/kg; **rabbit**: IV, 28 μg/kg; SC, 30 μg/kg; skin, 925 μg/kg; **guinea pig**: SC, 38 μg/kg. IV, LD, **calf**: 400 μg/kg. IV, LD_{50}, **pig**: 10 μg/kg.

SARSAPARILLA

The dried roots of *Smilax aristolochiaefolia*, *S. febrifuga*, *S. regelii*, and *S. ornata*. Contains three *saponins*, *starch*,

and small amounts of a volatile oil and resin. Formerly used to treat diseases by stimulating the body's defense mechanism and decreasing *cholesterol*. Later, relegated to use in manufacturing soft drinks.

Untoward effects*:** It contains ***safrole (q.v.), and now in disrepute because of the latter's carcinogenicity in **rats**. Yet, still sold in health food stores to increase stamina, energy, muscle growth, recovery from stress, and (due to its estrogenic content) it, supposedly, stabilizes the menstrual cycle.

SASSAFRAS

= *Ague Tree* = *Saxifrax*

Untoward effects*:** Sassafras oil contains ~80% ***safrole (q.v.). Sassafras herb ***teas*** and other sassafras products are sold in U.S. health food stores and are potential health hazards for **people**. A 47-year-old **female** ingested one teaspoonful of the oil leading to tachycardia, hypertension, vomiting, and tremors, required hospitalization. A 4-month-old **male** used a commercial ***Dr Hand's Teething Lotion***® containing sassafras and ***clove oil***, caused false positive tests and a misdiagnosis of ***phenytoin*** consumption. The volatile oil has narcotic properties.

SAUERKRAUT

Over 2000 years ago, Chinese preserved their *cabbage* in *vinegar*. The Mongols stole the procedure and introduced it to Eastern Europe. There, Austrians found they did not need *vinegar*, could *salt* it, allowed it to ferment, and called it sauerkraut.

Untoward effects*:** It can leach ***cadmium (q.v.) and ***lead*** (q.v.) from various containers. Sauerkraut juice has a high ***histamine*** content and can potentiate ***isoniazid*** and ***monoamine oxidase inhibitor*** hepatotoxic and neurotoxic effects. Commercial material contains 730 mg ***sodium*** and 490 mg ***potassium***/100 g.

SAUNAS

Untoward effects*:** **Sauna-taker's** disease, due to *Pullaria* contaminated sauna bath water occurs. The warmth can enhance the absorption of ***insulin leading to hypoglycemia. Dehydration in a sauna increased serum ***lithium*** levels in a **patient** leading to slurred speech, mild ataxia, and mild impairment of memory.

SAURINE

Untoward effects*:** A toxin which, along with ***histamine, appears in decomposing ***fish*** (q.v.) tissues after *Proteus morganii* breaks down ***histidine***, leading to scombroid fish poisoning.

SAUSAGE

Often contains *sodium nitrite* (q.v.), *sodium nitrate* (q.v.), *sodium acid pyrophosphate*, up to 15% poultry products, and *sorbitol*.

Untoward effects*:** Ingestion can cause a fatal hypertensive reaction with ***monoamine oxidase inhibitor antidepressants, due to its ***tyramine*** content. Acute episodes of methemoglobinemia have followed ingestion of Polish sausage (***kiska***) with high ***nitrite*** levels; these **people** may be at a higher risk for colon cancer. Bacteria (viz. *E. coli*) have been isolated from some uncooked sausages. Botulism and *Proteus vulgaris* toxin were a common cause of sausage poisoning in the early 1900s.

SAVIN OIL

= *Oleum Sabinae*

From *Juniperus sabina*.

***Untoward effects*:** Its use as an emmenagog and abortifacient has caused violent gastrointestinal irritation and occasionally hematuria, bloody vomitus, dyspnea, oliguria, unconsciousness, convulsions, coma, and death. Local irritant. Repeated applications cause severe inflammation.

SAWDUST

Untoward effects*:** Exposure to it has caused Joiner's disease in **man**, a hypersensitivity pneumonitis. Sequoiosis, a granulomatous pulmonary disease, in a 38-year-old **male** sawmill **worker** with disabling cough, diffuse pulmonary infiltrates, and restrictive lung disease with aveolar-capillary block syndrome. Lung biopsy revealed diffuse chronic interstitial pneumonitis and multiple granulomata with giant cells often containing birefractive, incinerable, ***cellulose-digestible foreign material. The **patient** had high levels of antibodies to several antigens in ***redwood*** extracts, one of which was related to an antigen in a ***redwood*** fungus (*Graphium sp.*). Weathered sawdust containing *Thermoactinomyces* has caused a hypersensitivity pneumonitis in **man**. *Klebsiella* infection, apparently from bark sawdust has occurred in **carpenters** and sawmill **operators**.

Klebsiella mastitis and even fatalities in **cows** from lying on fresh sawdust bedding. It also promotes coliform mastitis in **cows**. Extensive mycobacterial infections (especially *M. avium-intracellulare*) reported in **swine** from use of green (not bagged or dried) sawdust, causing MMA problems and financial losses at slaughter-time. In **poultry**, its use was associated with *Salmonella* infections and a leg dermatitis. Particles adhering to the genital mucosa of female **guinea pigs** interfere with copulation. Use of it as bedding for **cats** led to poisoning of 11/80 from ***pentachlorophenol***. Sawdust bedding induced mechanical blockage of suckling **rat** ileo-cecal openings, resulting in

43% mortality. **Rams** on ***redwood*** sawdust bedding developed balanoposthitis with inflammation of preputial mucosa and glans penis.

SAW PALMETTO

= *Fan Palm* = *Sabal serralatum* = *Serenoa serrulate*

Popular herbal and health food store treatment for benign prostatic hyperplasia and failing male sexual prowess.

***Untoward effects*:** A 65-year-old **male** developed acute and protracted cholestic hepatitis with jaundice 2 weeks after taking it for benign prostatic hyperplasia, tender abdomen, and severe pruritus, with slow resolution of the problems after therapy discontinued. Headache, gastrointestinal upsets, hypertension, decreased libido, and impotence have also been reported. Composition of the berry extracts sold have not been standardized. A recent poll of **users** indicated ~50% had a favorable response to its use.

SAXITOXIN

= *Clam Poison* = *Mussel Poison* = *Paralytic Shellfish Poison* = *STX*

Untoward effects*:** A powerful neurotoxin produced by ***Gonyaulax sp. (q.v.) and ingested by **Alaskan butter-clam, California sea mussels**, and **scallops**. Some toxic freshwater blue-green **algae** (*Aphanizomenon flos-aquae*) produce it as well. It is one of the most toxic non-protein substances known. It is very stable, not destroyed by freezing or routine cooking of **shellfish** harboring it. The toxin is ingested and concentrated in these **shellfish** without any apparent ill-effects on them. **Sand** and **horseshoe crabs** also contain a saxitoxin-like poison. Saxitoxin blocks ***sodium*** transport in nerve cells with inhibition of axonal depolarization. Symptoms start within minutes to a few hours of ingestion including skeletal muscle weakness and paralysis of respiratory musculature. Tingling and numbness of lips, tongue, and fingertips; dizziness, incoordination, slurred speech, swallowing difficulty, bradycardia, hypersalivation, and early (within 16 h) death have been reported.

LD_{50}, **rat**: 531 µg/kg; **mouse**: 263 µg/kg.

SCABIOSA SUSSISA

= *Devil's Bit*

Untoward effects*:** A single case report of violent tongue and mouth irritation in **cattle**. **Bees** using the nectar of its flowers yield poisonous ***honey.

SCALLOPS

Untoward effects*:** Have been a cause of ***saxitoxin (q.v.) ingestion and paralytic **shellfish** poisoning. It is suggested that **patients** with ***cobalt*** dermatitis avoid eating them.

SCARLET RED

= *Biebrich Scarlet Red* = *CI 26,105* = *Sudan IV*

***Epithelial stimulant.* An official 5% ointment exists, but it is frequently used at 0.1–1.0% concentration.**

***Untoward effects*:** A 1912 case report of serious poisoning in a **patient** after its application at an unknown concentration over extensive ulcerated areas leading to nausea, vomiting, epigastric pain, fever, and possibly renal irritation.

SCHINUS TEREBINTHIFOLIUS

= *Aroeira nigra* = *Brazilian Pepper* = *Chichita* = *Christmas Berry* = *Copal* = *Corneiba* = *Florida Holly* = *Pink Pepper*

A large, ornamental tree in southern Asia, Argentina, Arizona, Brazil, California, Central America, Cuba, the French Riviera, Guam, Florida, Hawaii, Mediterranean areas, North Africa, and Puerto Rico.

***Untoward effects*:** The pollen is not windborne, but a source of widespread respiratory and skin irritation, especially in **people** who trim it or cut a branch. Eye inflammation and facial swelling occur. Second-degree type of burn lesions from contact by **wood-cutters** or **helpers** who carry away the wood, producing lesions with yellowish fluid exudate, rashes, and severe pruritus.

Birds eating the fruit become intoxicated and cannot fly. An exposed **calf** developed swelling of its head and hemorrhages in its eyes after eating its leaves.

SCHOENOBIBLUS PERUVIANUS

Untoward effects*:** The roots and bright orange fruits of this 4½ ft tall shrub are used by the **Kofán Indians** of Colombia for making ***curare (q.v.). **They** use it on poison arrows and darts, often to kill **birds** and **fish**. The **Tikunas Indians** of Colombia know the stems and roots are toxic, but do not make ***curare*** from them. **Others** use the dried leaves as a poultice on wounds; the plants are rich in anticoagulant ***coumarins***.

SCHOENUS ASPEROCARPUS

= *Poison Sedge*

Untoward effects*:** In Australia, 210/1000 **sheep** died within a few days after grazing on it in newly cleared land. Some had acute respiratory distress before death. Post-mortem revealed large quantities of pale, yellow, fibrinous fluid in the thorax; frothy fluid in the trachea and bronchi; and pulmonary edema. Fed experimentally to **sheep** at 285–500 g caused death in 24–30 h. I suspect it was from ***hydrogen cyanide, as later growth was non-toxic.

SCHRADAN

= *ENT 17,291* = *OMPA* = *Ompacide* = *Ompatox* = *Pestox III* = *Sytam*

***Organophosphate anticholinesterase insecticide.* No longer used in the U.S.**

***Untoward effects*:** After 30 days of 1.5 mg/day, prison **volunteers** had a 23.5% decrease in plasma cholinesterase. Maximum air exposure is ~0.5 mg/m^3 TWA.

LD_{LO}, **human**: 5 mg/kg.

Acute toxicity is decreased in protein-deficient **rats**. Pretreatment of **rats** with ***diethylstilbestrol***, ***estradiol***, ***estrone***, and ***thiopental*** increases OMPA toxicity. In **mallard ducks** and **pheasants**, toxicity symptoms include ataxia, falling, wing-drop, fluffed feathers, loss of righting reflex, tetany, ptosis, apnea, wing-beat convulsions, and opisthotonus within 20–30 min and deaths in ~1 h in **mallards** and up to 4½ days in **pheasants**. **Pigs** are poisoned by 15 mg/kg or 1.5 mg/kg/9 days.

LD_{50}, male **rat**: 5–9 mg/kg; female **rat**: 42 mg/kg (dermal 15 and 44 mg/kg, respectively); **mouse**: 9.6 mg/kg; **guinea pig**: 35 mg/kg; **rabbit**: 25 mg/kg; **mallard duck**: 36.3 mg/kg; **pheasant**: 19 mg/kg.

SCINDAPUS AUREUS

= *Pothos*

Untoward effects*:** All parts of this house plant contain rapid-acting ***oxalates (q.v.) leading to local irritation of mouth, pharynx, and esophagus and contact dermatitis. House **pets** have been similarly affected.

SCIRPUS AMERICANUS

= *Bulrush*

***Untoward effects*:** Suspected of causing pulmonary emphysema in **cattle** in Wyoming and Australia.

SCLEROTINIA SCLEROTIORUM

Untoward effects*:** A pink-rot fungus on ***celery producing a photoreactive ***mycotoxin***, ***methoxsalen***, leading to skin blisters on **man**, **mice**, and **rabbits**. Also found on ***lettuce*** and ***sunflowers***.

SCOPARIA sp.

Untoward effects*: *S. dulcis* = *Ruma Fada (Hausa) is used topically in Central and South America as an insecticide. By rubbing on or other contact with the tube of their blowguns, the branches of ***S. montvidensis*** make the darts more deadly.

SCOPOLAMINE

= *Hyoscine* = *Scop* = *Transcop* = *Transderm*

The hydrobromide = *Joy Rides* = *Kwells* = *Scopolamine Lux* = *Scopos*

The methyl nitrate = *Skopolate*

***Anticholinergic, muscarinic antagonist. Lindane* is *Kwell*®, not to be confused with *Kwells*.**

Untoward effects*:** As for ***atropine (q.v.). Central nervous system stimulation precedes central nervous system depression with drowsiness, dizziness, vertigo, and coma. Blurred vision, cycloplegia, mydriasis, dry mouth (> 60%), amnesia, fatigue, and tachy- or bradycardia. Occasionally urinary retention, constipation, disorientation, excitement (~8%), restlessness, delirium, and hallucinations, even at therapeutic doses. Instilled in eyes with anatomical narrow angles leads to acute angle-closure glaucoma; may increase intraocular pressure in eyes with open-angle glaucoma; and psychosis in **others**. Mydriasis and possibly asthma from handling Australian ***corkwood*** leaves that contain the alkaloid. Anisocoria has been induced by **users** of ***Transderm*** patches, possibly from accidental direct eye contamination. Apparently, associated with exacerbation of Tourette's disorder. Transdermal application of 0.5 mg for vertigo and nausea in 76-year-old **patient** led to toxic delirium.

Fetus, neonate, and children*:** Crosses the placenta and may cause **fetal** mydriasis. **Maternal** doses totaling 1–8 mg during labor and delivery led to fever, tachycardia, lethargy, and barrel chest in newborn **infant**. Decreases Apgar score when used with ***general anesthesia in **parturients**. Minute quantities appear in breast milk. An 8-year-old **child** given given half of an **adult *Kwells*** dose, was followed in 4 h with the other ½ dose to control motion sickness developed after 3 h, facial flushing, visual and auditory hallucinations, mydriasis, choreiform movements, and staccato speech. Acute psychotic reactions (confusion, disorientation, visual and audio hallucinations, aggressive behavior, variable levels of consciousness) in some **children** from instillation in eyes. An acute psychotic reaction in 9-year-old **child** after a 1 mg eye drop dose.

Poisoning*:** New York City **detectives** found various well-to-do **visitors** to their city with symptoms of acute alcoholism and in various stages of undress, suffering from ***anticholinergic poisoning caused by their "**hosts**" or "**hostesses**" spiking **their** drinks with scopolamine eye drops. In New York City, a 33-year-old **female** was found bound and gagged, remembering that a casual **acquaintance** drove **her** home and **she** felt very dizzy after sipping some ***wine***. **Her** apartment was then robbed of furs and jewelry. Liquid in the ***wine*** glass tested positive for scopolamine. Many of the above symptoms reported in attempted suicides from use of over-the-counter items containing it that, perhaps, should be under-the-counter, rather than touted on TV. ***Psycho Drops*** is the street name for it or ***atropine***, and it has been used maliciously in **people's** drinks to induce amnesia, psychosis, etc. Poisoning in two elderly **people** from eating fried green ***tomatoes*** picked from a vine grafted on ***Datura*** stock. Found mixed with many arrow poisons. Overdosage is a common emergency room problem. NOT to be used by **persons** having glaucoma or increased intraocular pressure, **children** under 12 years or many elderly **people**, or **those** with bladder or gastrointestinal obstruction.

LD_{LO}, **human**: 5 mg/kg; SC, TD_{LO}, 2 μg/kg.

Boiling may decompose its solutions and storage in plastic syringes or exposure to light inactivates it.

Cats and **dogs** have a greater sensitivity to its toxic effects than **herbivores**. Adverse effects include xerostomia, mydriasis, cycloplegia, tachycardia, ileus, and urinary retention. Toxicity or overdoses lead to nausea, vomiting, dysphagia, tremors, and hyperexcitability. Depressant rather than excitatory effects are dominant in **animals**, and used in tranquillizing many **zoo animals**.

SC, LD, **dog**: 0.13 mg/kg. IV, LD_{50}, **mouse**: 163 mg/kg.

SCOPULARIOPSIS sp.

Untoward effects: A ***mycotoxin*** causing yellowish peripheral nail pigmentation in **people**. Repeatedly isolated from U.S. Navy **divers** during prolonged submergence studies. These were isolated from three **sea lions**. Isolated from cervico-uterine discharges in **cows**. Apparently, non-pathogenic in **guinea pig**, **rabbit**, **dog**, **sheep**, **pig**, and **hen** trials.

SCORPIONS

650–1500 species identified.

Untoward effects: Deaths from their stings are rare in the U.S., except, perhaps, for young **children** or **adults** sensitive to their venom, but often cause slight–severe local swelling and pain, occasionally with discoloration of tissues. Systemic reactions may occur leading to skeletal muscle choreiform movements, fibrillation or convulsions, pulmonary edema, respiratory arrest, mydriasis, sialism, visceral hyperemia, and gastric hyperdistention. Hemolysis, renal failure, and local necrosis reported in a case report. Can cause hypertension, sweating, priapism, acute pancreatitis, arrhythmias, congestive heart failure, dyspnea, and paralysis. In the U.S., > 9000 scorpion stings reported in 1994. In **children**, symptoms include pain, extreme restlessness; spasticity of abdominal, arm, and leg muscles, tachycardia, fever, cyanosis, and wheezing. Death is common unless treated. In Africa and Malaysia, they have been added as minor ingredients to poison arrows and darts. World-wide, it is estimated they cause 1000–2000 deaths/year. Brazil reports ~5000 stings/year and 48 deaths, 90% of which are in **children** < 14 years. The most dangerous in North Africa and the Middle East are *Leiurus* and *Androctonus*; in South Africa, *Parabuthus*; in Trinidad and Latin America, *Tityus*; in Mexico and the southwestern U.S., *Centruroides*; and in India, *Mesobuthus*. In Greek, ***androctonus*** means "killer of **man**", and these large scorpions, some the size of one's hand, kill several thousand **people** each year. Three cases of envenomation reported in Michigan from stowaways on Florida produce.

SCROPHULARIA sp.

= *Figworts*

Untoward effects: Once used in **human** therapeutics, but the seeds, belonging to the ***digitalis*** group, can be toxic. Young **cows** ingesting ***S. aquatia*** = ***Water Figwort*** develop tachypnea, excitement, mydriasis, slight ulceration and congestion of oral mucous membranes, dysuria, dark fetid diarrhea, thirst, anorexia, and decreased ***milk*** production, probably from the plant's ***saponin*** content.

SCUTELLARIA LATERIFLORA

= *Scullcap* = *Skullcap*

Untoward effects: It and its ***teas*** have been ineffectual for treating many conditions it is recommended for; has caused hepatitis. High doses lead to giddiness, stupor, confusion, and epilepsy-like symptoms.

SEA ANEMONES

Untoward effects: Their nematocysts or stinging cells eject when they are touched. One species, *Actinia equina*, injects ***thalassin***, a toxin that liberates ***histamine*** and slow-reacting substance that have induced bronchoconstriction. *Sargartia* looks like a 2" high flower, often afflicting **sponge-divers**. "**Sponge-divers'** dermatitis" occurs within minutes of contact leading to erythema and painful vesicles. Lesions change to a deep purple and the **diver** may develop nausea, vomiting, headache, and fever. Lesions are slow-healing, often abscessing and sloughing. Several types produce stinging, burning, redness, severe dermatitis around solitary wheals, or papular linear eruptions.

SEA CUCUMBER

Found in the warm shallow waters of the Bahamas, Caribbean, Gulf of Mexico, and Florida and the colder waters near England and Ireland. Vary in size, 6–40" and up to 8" across. A few species are found in deep water.

Untoward effects: Produces a neurotoxin. When handled, causes painful stinging sensation, then mild swelling, itching, and erythema. Their venom in the water causes an inflammation reaction, blindness if it contacts the eyes. A toxic substance for **fish**, but not for **man**, ***holothurin***, has been isolated from some species.

SEA SNAKES

Untoward effects: Venomous and cause **human** fatalities in the Pacific's Indo-Australian region. Related to **cobras**. Usually not dangerous to **man** unless provoked, such as handling them when caught in fish nets or accidentally stepping on them when wading in shallow water. They have short fangs that are readily broken off and bites may not result in envenomation. Their venom is very neurotoxic and contains ***hemotoxins*** and ***myotoxins***. Symptoms of envenomation occur within 5 min or slowly over several hours, including anxiety, pain with movement, malaise, myalgias, ascending paralysis, trismus, dysphagia, dysphonia, euphoria, ptosis, diaphoresis, hemoglobinuria, and myoglobinuria. Death from bulbar paralysis leading to respiratory failure; aspiration pneumonia, hyperkalemia, and acute renal failure. The common sea snake is *Enhydrina schistosa*.

SEA URCHINS

Untoward effects: The ***Sea Egg***, ***White Sea Urchin***, and ***Black Sea Urchin*** are venomous in addition to causing wound injuries. Usually found in shallow waters among rocks and on sandy areas. Envenomation occurs as distal fragments of these calceous spines break off in a **person's** skin leading to pain, erythema, bleeding, and swelling. Some spines are phagocytized and others cause, within weeks, a sarcoid-like granulomatous reaction of pink or flesh-color, later turning brown. Even after surgical removal, it may appear that the spine is still there, due to a purple dye some leave along their course. Sea urchins of the genus *Tripneustes*, found only in the Pacific Ocean, have spines that liberate ***histamine*** and ***bradykinin***-like agents leading to severe pain, dyspnea, and giddiness. Its toxin is neurotrophic for cranial and facial nerves. Local pain lasts ~1 h and facial nerve paralysis may last for 6 h.

SEA WASP

= *Box Jellyfish* = *Charabdea sp.* = *Chironex fleckeri* = *Chiropsalmus quadrigatus*

Untoward effects: Probably the most venomous of **jellyfishes** causing numerous sudden and excrutiating, painful deaths (occasionally within seconds or 30 min) in **swimmers** due to the venom's lethal, dermonecrotic, and hemolytic properties. Mostly in tropical coastal waters around Australia, the Philippines, southern Japan, and Indian Ocean, killing 15–20% of their **victims**. A less dangerous form on the Atlantic coast from North Carolina southward to Brazil. They may weigh over 4 lb and are semitransparent with tentacles that may be up to 10 m in length.

SEBASTIANIA

Untoward effects: Used in poison arrows and darts by **natives** of Central and South America.

SECOBARBITAL SODIUM

= *Barbosec* = *Immenoctal* = *Meballymal Sodium* = *Pramil* = *Quinal-barbitone Sodium* = *Quinalspan* = *Seconal Sodium* = *Sedutain* = *Seotalnatrium*

***Sedative, hypnotic.* Called *Red Birds* and *Red Devils* in street slang.**

Untoward effects: ***Addiction***: Often with drowsiness, nystagmus, ataxia, and slurred speech. A 23-year-old **drug addict** inadvertently injected **his** right radial artery instead of an adjoining vein leading to instant burning pain in **his** right hand, which, within 10 min, became swollen and mottled; within 2 h, the hand was severely swollen, cyanotic, and anesthetized. Eventually, ischemic gangrene of the distal phalanges occurred. Other similar cases reported.

Fetus and neonate: Reaches the **fetus** within 1 min of a **maternal** IV and equilibrium is reached in 2–5 min. **Fetal** levels are ~70% of **maternal** levels. Little or no depression in 90% of **infants**. Withdrawal syndrome with seizures and hyperirritability reported in an **infant's** immediate neonatal period and again at 4 months of age whose **mother** took 1.2–3 g/day during **her** pregnancy.

Miscellaneous: Heterophoria, poor flight performance by **aviators**, hyperreflexia and bilateral extensor plantar responses, bullous skin lesions due to contact on pressure areas during drug stupor, hangovers, cardiorespiratory depression, decreased blood pressure, depression, megaloblastic anemia, and leukocytosis from high dosage. Fixed-drug eruptions from the ***tartrazine dye*** in walls of ***Seconal*** capsules. Some found in breast milk of **mothers** taking the drug.

Neoplasms: Statistically significant associated with lung cancer, but **abusers** are also often ***cigarette*** **smokers**.

Poisoning and suicide: Deliberate overdoses are often with other ***barbiturates*** leading to deep coma, areflexia, and respiratory distress. Therapeutic serum levels are 2–10 mg/l and toxic concentrations are 15–20 mg/l. In **adults**, single doses > 3 g or blood levels > 3.5 mg% can be fatal.

TD_{LO}, 714 μg/kg. LD_{LO}, 50 mg/kg.

Drug interactions: ***Alcohol*** increases its rate of absorption and increases its depth of coma. With ***warfarin*** decreases prothrombin time by increasing ***coumarin*** metabolism. Additive effects with ***cannabis***. Daily use has altered ***methoxyflurane*** metabolism leading to toxic concentrations of its metabolites in the blood of 57-year-old **male**, causing toxic nephropathy. May decrease circulating ***thyroid hormone*** levels leading to hypothyroidism and increase ***corticosteroid*** metabolism. ***Chlorpromazine*** enhances its effects. Use with ***ketamine*** prolongs recovery time 30–40%.

Its solutions discolor ***polyethylene*** equipment, and simultaneous use dramatically accelerates ***salicylate*** toxicity in **dogs**. **Animal** tolerance, disposition, age, and condition frequently alter dosage and duration of effects. Use cautiously in the presence of severe liver and/or renal disease. Refrigerate all solutions to maintain indefinite stability. Additive cerebral depressant effects with ***alcohol*** in **mice**.

LD_{50}, **rat**: 125 mg/kg; **dog**: 85 mg/kg; **cat**: 50 mg/kg; **pigeon**: 133 mg/kg; **duck**: 75 mg/kg.

SELAMECTIN

= *Revolution* = *Selenamectin* = *UK 124,114*

***Parasiticide*. Topically for heartworm prevention in cats and dogs; flea and mite control in cats and dogs; tick control in dogs; nematode treatment in cats; and sarcoptic mange treatment in dogs. The solution contains *alcohol*, *isopropyl*, *butylhydroxytoluene*, and *dipropylene glycol monomethyl ether* (q.v.).**

Untoward effects: Can be irritant to skin, eyes, and respiratory tract of **humans**. Flammable.

Drug interactions: ***Sebum-reducing*** products can decrease its efficacy.

Transient alopecia ~1% in **cats**; < 1% of **cats** and **dogs** develop nausea, vomiting, loose stool (occasionally with blood), anorexia, salivation, tachypnea, muscle tremors, and lethargy.

SELENIUM

An essential element and *lipotrope*.

Use: Its use as an antioxidant, in prevention of cystic fibrosis, in aiding the immune system, in preventing Keshan disease (an endemic cardiomyopathy), in preventing toxicity from ***arsenic***, ***cadmium***, ***lead***, ***mercury***, ***silver***, and ***thallium***, as well as the herbicide, ***paraquat***; in some forms of arthritis, and in cancer prevention must be weighed carefully against excessive doses. "Sola dosis facit venenum" or "Only the dose makes the poison" applies easily to selenium and its compounds.

Untoward effects: Toxicity was first reported in 1857 and chronic toxicity appeared in some **groups** living in high selenium content soil areas, such as South Dakota, Oregon, Montana, Wyoming, and Venezuela, due to excessive content in crops and forage grown there. In the 1940s, the FDA and some **oncologists** considered it a cause of cancer; 30 years later, the FDA reversed its position and permitted limited amounts in **livestock** feeds. In December, 1983, a New York 57-year-old **female** started taking a daily selenium supplement labelled as 150 μg/tablet. Marked loss of scalp hair was noted on day 11, progressing to almost total alopecia over a 2 month period. Horizontal white streaks on the fingernails of the left fifth digit, with tenderness and swelling of the fingertip and purulent discharge from the nail bed occurred 2 weeks later. Other fingernails became involved during the following 3 weeks. After consuming 77 of the 90 tablets, **she** heard on the radio that these tablets were being recalled. Tablet analysis revealed they each contained 27.3 mg (182 times the labelled content). **Her** serum level 4 days later was 528 ng/ml, about four times the normal. Analysis of **her** tablets indicated 31 mg/tablet. **Her** toxicity, fortunately, was limited by **her** ingesting > 1 g ***vitamin C*** daily. Some areas of China have high levels of selenium and **individuals** affected had nail and hair loss, nausea, ***garlic*** breath odor, fatigue, irritability, and hyperreflexia. **Their** daily dietary intake ranged from 3.2 to 6.69 mg. Dusts and fumes are readily absorbed from the lungs. It increases the susceptibility to caries in **man** and **rats**, where prevalence of caries is increased in the states noted above, in which levels in ***eggs*** and ***milk*** are ten times those of control areas. Has been found in **human** breast milk. Extensive industrial exposure occurs due to its wide use in pigment, ***glass***, ceramic, steel, and electronic industries leading to gastrointestinal upsets, giddiness, apathy, lassitude, pallor, nervousness or depression, fever, dyspnea, bronchitis, metallic taste, ***garlic*** breath odor, ***garlic*** odor of sweat, headache, chills, dermatitis, visual disturbances; eye, nose, and throat irritation from fumes; eye and skin burns, loss of feeling and control in arms and legs, brittle hair, alopecia, and brittle nails. Teratogenicity in **man** has been suspected only because it has caused it in some **animals**. Hair loss, weakened nails, and listlessness was reported in a family of **Ute Indians** near Ignacio, Colorado from drinking well water for ~3 months that contained 9 ppm. Their **dog** also lost its hair. In Wyoming, many wells, springs, and seeps carry toxic amounts of selenium, but, fortunately, also have a high level of ***salt***, making it unpalatable. Toxic in 17-year-old **male** and 11-month-old **female**, both cystic fibrosis **patients**, consuming 400 μg selenium yeast tablets/day for 2 weeks, and the **child** 25 μg day/2 months of a selenium yeast complex. Environmental Protection Agency sets upper level in drinking water at 50 ppb and OSHA sets upper workplace exposure limit as 0.2 mg/m^3/TWA. Immediate danger to life or health, 1 mg/m^3. Levels of 5 ppm in food or 0.5 ppm in milk and water can be toxic.

High level intake interferes with normal reproduction in **monkeys**, **rats**, and **sows**, and leads to malformations in **birds**. Many plants are selenium accumulators. Dietary levels of > 5 ppm for several weeks or months can be toxic to grazing **livestock**. It causes "blind staggers" and "alkali disease" in **cattle**, **horses**, **sheep**, and **swine**, causing them to walk into objects, have cracked or sloughing hooves, show lameness, stiff joints, dullness, emaciation, alopecia, and anemia, and stillborn **pigs**. Placental transfer occurs leading to congenital malformations and deformed hooves in **foals** and **lambs**. Nursing **calves** have shown toxicity after drinking **cow's** milk with 3 ppm. Acute poisoning leads to tachycardia, weak pulse, ataxia, colic, bloating, bloody froth on mouth and nose, polyuria, dark diarrhea, cyanosis, coma, respiratory failure, and death. Dietary toxic levels in ppm for **swine** are 5–8, **beef cattle**: 8.5, **dairy cattle**: 5, **horses**: 5–40, **chickens**: 10, **sheep**: 10–20, and **lambs**: 2. Dietary levels of 5 ppm induce liver cirrhosis and liver tumors in **rats**. Inhalation of vapors by **mice** for 30 min caused death. Topically as an ointment on **mice** resulted in dermatitis and slow growth.

Inhalation LC_{LO}, **rat**: 33 mg/kg, over 8 h.

SELENIUM DISULFIDE

= *Dandrex* = *Exsel* = *Seleen* = *Selsun*

Untoward effects: Use as a shampoo has caused eye and skin irritation, generalized itching, and burning sensations. Treated **patients** with a ***Selsun*** ointment had urine selenium levels of 0–0.4 µg/ml. Can induce hair color changes.

May be very irritant to scrotal, eye, and occasionally perianal areas of **dogs**. Excessive use has caused undue hair loss and excessive skin oil secretion. By oral gavage to **rats** and **mice**, caused hepatocellular carcinomas; to female **mice** caused alveolar/bronchiolar carcinomas and adenomas.

LD_{50}, **rat**: 138 mg/kg.

SELENIUM HEXAFLUORIDE

= *Selenium Fluoride*

Untoward effects: Pulmonary irritation and edema in **animals** exposed to 4 ppm of this colorless gas. National Institute for Occupational Safety and Health/OSHA sets exposure limits of 0.05 ppm, TWA; immediate danger to life or health of 2 ppm.

SELENIUM OXIDE

= *Selenious Anhydride* = *Selenium Dioxide*

Untoward effects: Highly irritant to skin and eyes. Formed during burning of ***coal***.

Excessive doses (80 times normal) accidentally given as dietary ***selenium*** supplement to **lambs** led to colic, grinding of teeth, salivation, and death. Post-mortem revealed hyperemia and edema or hemorrhage in various internal organs, including the brain.

SC, LD_{50}, **rabbit**: 4 mg/kg.

SELENOMETHIONINE

= *Sethotope*

Diagnostic radioactive imaging agent and nutritional supplement.

Untoward effects: Cataractogenic after SC in **rats**, apparently due to interference with ***glutathione*** metabolism. In **mallard ducks**, 10 ppm was teratogenic leading to foot and toe defects, ectrodactyly, hydrocephaly, bill defects, microphthalmia, and anophthalmia.

IV, LD_{LO}, **pig**: 7.45 mg/kg.

SEMICARBAZIDE

= *Carbamylhydrazine* = *Hydrazine Carboxamide*

Untoward effects: Mild photic stimulation decreases the IV convulsant dose for **man** from 40 to 25 mg/kg. The approximate parenteral convulsant dose (TD_{LO}) for **man** is 40 mg/kg. Used as a stabilizer in natural ***rubber***.

IV, TD_{LO}, **human**: 40 mg/kg.

Tumorigenic in lungs and blood vessels of **mice** after oral gavage. TD_{LO}, 25 g/kg/over 30 week period. Nearly 100% of cleft palates in a litter of **rat pups** after treatment of their **dams**. Teratogenic for **chick** embryos. White Pekin **ducks** fed it at a 0.1% level developed tremors, ataxia, and paresis after 2 weeks. **Turkey poults** fed it at a 0.03% level showed decreased growth rate, aortic ruptures, and leg deformities. Inhibited egg-laying in **hens**.

LD_{50}, **mouse**: 176 mg/kg; IV, 126 mg/kg.

SEMUSTINE

= *Methyl-CCNU* = *Methyl Lomustine* = *NSC 95,441*

Antineoplastic.

Untoward effects: Nausea, vomiting, renal failure, occasional transient SGOT and SGPT increase, leukemia, leukopenia, and thrombocytopenia. Acute nausea and vomiting occurs in 2–6 h and usually lasts < 24 h. Myelosuppression is delayed. Less frequently, diarrhea, stomatitis, anemia, alopecia, mental confusion, disorientation, delusions, and neuro-ophthalmologic effects occur. A **patient** who received a cumulative dose of 4.1 g over 25 months developed diffuse interstitial pulmonary fibrosis with dyspnea and non-productive cough. Avoid direct contact with capsule contents, as ***nitrosoureas*** are irritating to skin and mucous membranes.

TD_{LO}, **human**: 6 mg/kg.

A single dose to **monkeys** led to delayed nephrotoxicity. Inhibits spermatogenesis in **chicks**.

LD_{LO}, **dog**: 25 mg/kg; **monkey**: 200 mg/kg; IV, **dog**: 14 mg/kg; **monkey**: 45 mg/kg.

SENECIO sp.

= *Ragworts*

Untoward effects: Used in many herbal medicines and ***bush teas*** in Asia, Africa, and West Indies and contain hepatotoxic ***pyrrolizidine*** (q.v.) alkaloids, which may account for the high incidence of liver disease in those areas. These alkaloids are metabolized by the liver to ***pyroles***, very toxic metabolites.

In Colombia, the resinous smoke from burning ***S. abietinus*** causes a long-lasting burning sensation of the nose and mouth.

The rhizomes' ***maturin*** of Mexico's ***S. cervariaefol*** and ***S. grayanus*** causes increased temperature, mydriasis, and violent tetanic spasms. The plants collectively are known as ***matarique*** or ***guerena***.

S. jacobea = ***Stinking Willie*** = ***Tansy Ragwort***, originally a native of Europe, is found in many parts of the world and induces cirrhosis and primary hepatomas in many wild and farm **animals. Japanese quail** showed no signs of illness and their eggs yielded normal **chicks** after consuming large quantities. Fed to young **chicks** leads to megalocytosis, focal necrosis, focal hyperplasia, and portal fibrosis of the liver. Ingestion by **mice** leads to liver, kidney, and lung pathology. Passes into **cows'** and **goats'** ***milk***. Induces hypertensive pulmonary vascular disease in **rats**. The fresh plant is bitter and usually avoided by **animals**. Dangerous in hay, as the bitter taste disappears on drying. **Cattle** and **horses** are usually affected; **sheep**, **goats**, and **rabbits** are partially resistant. Symptoms are delayed for months and include anorexia, emaciation, scabby nose, incoordination, jaundice, and occasionally staggering, extreme nervousness, bile-stained feces, and painful straining. **Animals** may attack nearby moving objects before their deaths. Ingestion caused Molteno disease of **cattle** in Africa; Winton's disease of **horses** and **cattle** in New Zealand; Pictou disease of **cattle** and **horses** in Canada; and Van Es' walking disease of **horses** in Nebraska, U.S. – all eventually discovered to be due to ***Senecio sp***. Post-mortem primarily revealed hepatitis, nephritis, and petechia and ecchymotic hemorrhages on internal organs. **Bees** making ***honey*** from ***S. jacobea*** pose **human** health hazards.

S. longilobus ingestion leads to high percentage of liver tumors. An 6-month-old **infant** given a ***tea*** from ***S. longilobus*** = ***Gordoloba yerba*** = ***Threadleaf Groundsel*** = ***Woolly Groundsel*** developed hepatic veno-occlusive disease after ingesting 70–147 mg of alkaloids during 2 weeks. Also in Arizona, a 2-month-old **male** was treated for cough and nasal congestion with the herb, leading to hematemesis and progressive lethargy 2 weeks later. Hospitalized and 3 days later developed seizures, decerebrate posturing, hypernatremia, hyperkalemia, hyperbilirubinemia, and periods of apnea, followed by death on day 6. Post-mortem revealed liver toxicity. This toxic herb contained 1.5% alkaloids and was mistaken for its look-alike, the harmless plant, *Gnaphalium* (**Gordoloba**) used in folk remedies. The toxicity can mimic the Reye syndrome. The FDA recalled a commercial batch, stating that it was toxic and could cause heart, lung, and liver damage.

Liver disease has been documented in **humans** eating ***bread*** made from grains contaminated with ***Senecio*** seeds. ***Tea*** from these plants may have been associated with increase in esophageal cancer. A **mother** drinking ***tea*** containing ~343 μg ***senecionine*** caused fatal veno-occlusive disease in a newborn **infant**. Acute, subacute, or chronic hepatic disease with prominent ascites in some **people** due to its ***jacobine*** and ***seneciphylline*** alkaloids.

At least ~1000 species exist world-wide that cause delayed liver disease.

SENNA

= *Follicule de Sene* = *Hojas de Sen* = *Hajasen (Mexico)* = *Sennesbälglein*

The dried leaflets of *Cassia acutifolia* (*Alexandria Senna*) and *C. angustifolia* (*Tinnevely Senna*) of 3–5 ft shrubs, respectively, from tropical Africa and southern India.

Laxative, purgative. **Contains *anthraquinones* (q.v.).**

Untoward effects: May color feces yellow or a yellowish-green; alkaline urine is red–violet; and yellow–brown in acidic urine. High dosage to **mothers**, often as ***Senokot***®, may cause purgation in breast-fed **infants**. As a herbal ***tea***, leads to strong catharsis, benign melanosis coli, colic, possible hepatotoxicity, dehydration, coma, and peripheral neuropathy. Mild to moderate cramping, nausea, and an allergic reaction (sudden dyspnea, wheezing, and eyelid edema) in 19-year-old **female** after a single ***Senokot***® suppository. A 21-year-old **female** took three ***Senokot***® tablets/day/3 years to control **her** weight, leading to hypertrophic osteoarthropathy and clubbing of digits, both of which regressed after discontinuing its use. Occasional skin eruptions. Many unofficial varieties exist.

LD_{LO}, **human**: 500 mg/kg.

SERJANIA sp.

Untoward effects: Used as piscicides in Amazonia by beating the stalks, releasing ***saponins***, ***serjanosides***, poisoning the water, as with ***rotenone***.

SEROTONIN

= *Enteramine* = *5-Hydroxytryptamine* = *5-HT* = *Thrombocytin*

Neurotransmitter. **The active ingredient of the hallucinogenic *Mexican Mushroom* and the sacred *Mexican Morning Glory* resembles it. A pressor amine found in many foods. Closely related chemically to *bufotenine* (q.v.). Found in many *Cnidarians* and is a common constituent in many of their venoms.**

Untoward effects: Endogenous production or exogenous parenteral administration may induce stomach ulcers, increase intestinal spasms; induce migraines, hyperprolactinemia, hyperventilation, diaphoresis, shivering, hyperthermia, hyperreflexia, agitation, restlessness, confusion, incoordination, hypomania, coma, myoclonus, tremors, rigidity, tachycardia, fever, bronchospasm, diarrhea, and aid in causing scleroderma. It is a ***histamine***-releaser and causes wheals and irritation. Associated with abortions, **fetal** death, and multiple **fetal** anomalies of skeleton and organs. ***5-Hydroxyindole acetic acid*** is its main metabolite and can give false negative glucose oxidase test results.

Drug interactions: ***Tricyclic antidepressants*** may inhibit its action and ***monoamine oxidase inhibitors*** and ***methylphenidate*** enhance its effects. Brain stores are depleted by ***reserpine***.

In **lab animals** leads to bronchoconstriction and pulmonary vasoconstriction; increased pulmonary arterial pressure is blocked by prior injection of ***cyproheptadine*** or ***methysergide***.

SERTRALINE

= *CP 51,974-1* = *Lustral* = *Zoloft*

Selective serotonin-uptake inhibitor, antidepressant.

Untoward effects: Unlike ***tricyclic antidepressants***, it stimulates the central nervous system leading to agitation, nervousness, tremors, and insomnia. Frequently, nausea, dry mouth, dizziness, somnolence, diarrhea, hyperhidrosis, and ejaculatory delay. Specifically and additionally:

Cardiovascular: Occasional postural hypotension and dizziness, syncope, edema, peripheral ischemia, and tachycardia. Rarely, myocardial infarction, chest pain, and varicose veins.

Ear: Occasional earache. Rarely, otitis media, and tinnitus.

Endocrine: Rarely, gynecomastia, galactorrhea, and exophthalmos.

Eye: Occasional conjunctivitis, diplopia, eye pain, xerophthalmia, and poor accommodation. Rarely, photophobia, abnormal lacrimation, and anicosoria.

Gastrointestinal: Occasionally, eructation and dysphagia. Rarely, fecal incontinence, nausea, gastroenteritis, diverticulitis, glossitis, stomatitis, hiccup, melena, tenesmus, and proctitis.

Genitourinary: Occasionally, dysmenorrhea, intermenstrual bleeding, dysuria, nocturia, polyuria, and urinary incontinence. Rarely, amenorrhea, breast enlargement and pain, leukorrhea, menorrhagia, atrophic vaginitis in **female**; balanoposthitis, anorgasmia, delayed ejaculation, oliguria, urinary retention, and renal pain. Genital anesthesia in 36-year-old **male** 3 days after 50 mg/day.

Hematopoietic and lymphatic: Occasional lymphadenopathy and purpura. Rarely, anemia, blood in aqueous humor, and thrombocytosis.

Miscellaneous: Frequently, asthenia. Occasionally, malaise, generalized edema, weight increase or decrease, and hyponatremia. Rarely, enlarged abdomen, halitosis, yawning, diaphoresis, compulsive water-drinking, panic attacks, and sedation. A 19-year-old **male** ingested ~11.3 g/day/~6 months (~56 times maximum recommended dose) to produce euphoria and excitement. It and its chief metabolite, ***desmethylsertraline***, are passed in very low quantities into breast milk (~1% of **maternal** level). Serotonin syndrome in **children** ingesting 400 mg; acute ingestion of 700–2100 mg in **adults** led to only mild adverse symptoms.

Muscle and joints: Occasionally, arthralgia, arthrosis, muscle cramps, dystonia, and muscle weakness. Rarely, hernias.

Nervous system, autonomic: Infrequently, flushing, mydriasis, salivation, and cold clammy skin. Rarely, pallor.

Nervous system, central and peripheral: Confusion is frequent. Occasionally, incoordination, ataxia, hyperesthesia, hyper- or hypokinesia, migraine, nystagmus, and vertigo. Rarely, coma, convulsions, syncope, delirium, dyskinesia, dysphoria, hypotonia, ptosis, hyporeflexia, local anesthesia, mood changes (from depression to hypomania), stuttering, and akathesia.

Psychiatric: Occasionally, aggression, amnesia, apathy, abnormal dreams, delusions, depression, euphoria, hallucinations, neurosis, sexual dysfunction, teeth grinding, and suicidal ideation. Rarely, hysteria and sleep-walking.

Respiratory: Occasionally, bronchospasm, dyspnea, coughing, and epistaxis. Rarely, bradypnea, hyperventilation, sinusitis, stridor, and rhinitis.

Skin: Occasionally, acne, alopecia, pruritus; erythematous, follicular, or maculopapular rashes, and dry skin. Rarely, bullous eruption, erythema multiforme, hypertrichosis, abnormal hair texture, photosensitivity, skin discoloration, abnormal skin odor, and urticaria.

Drug interactions: Use with ***monoamine oxidase inhibitors*** (***isocarboxide***, ***phenelzine***, and ***tranylcypromine***) leads to ***serotonin*** syndrome (hyperthermia, rigidity, myoclonus, confusion, delirium, and coma). Can occur as well with some ***tricyclic antidepressants***, such as ***amitriptyline*** or the analgesic, ***tramadol***. Priapism in 41-year-old **male** whose sertraline level was raised to 100 mg/day/2 weeks after treatment with ***lithium carbonate*** (600 mg/day) and sertraline (50 mg/day) for ~1 year. Questionable clinical significance of interactions with ***desipramine***, ***diazepam***, ***imipramine***, ***tolbutamide***, and ***warfarin***. A 27-year-old **male** who overdosed with it developed acute liver damage when treated with ***venlafaxine***. ***Serotonin*** syndrome in 12-year-old **male** 4 weeks after adding ***erythromycin*** 400 mg/day to **his** regimen of sertraline 37.5 mg/day. ***Clozapine*** plasma levels are increased by it.

Mydriasis and transient anorexia in **dogs** at 10–20 mg/kg; muscle tremors and increased salivation at 30–50 mg/kg. Half-life was 26 h.

Minimum lethal dose, **dog**: 80 mg/kg.

SESAME
= *Benne* = *Ginglii* = *Teel*

One of the oldest oil crops known to man. Contains ~55% oil.

Untoward effects: Hand contact dermatitis in some **workers**. Anaphylaxis reported after sesame seed ingestion. A 19-year-old **male** ate one piece of ***halva***. Allergic reactions within 5 min included tongue, palate, and pharynx paresthesia, progressing to generalized pruritus, swallowing difficulty, generalized erythema, wheezing, blood pressure 80/40, and pulse 140. Intense reaction after scratch test with seed solution, and none to other ***halva*** ingredients or sesame oil. Similar reaction in 64-year-old **male** 10 min after eating sesame seed crackers and in 40-year-old **female** after a few bites of a roll after the seeds were removed. A toxic polyneuropathy in two **females** in Sri Lanka after ingesting 560–1120 ml of the oil, in accordance with a religious custom. The oil had been stored in containers originally used to store ***tritolyl phosphate*** (q.v.). Use of the oil as a topical emollient associated with the occurrence of tinea versicolor in areas with few sebaceous glands.

Comedogenic on **rabbit** ears. Orally, can interfere with ***zinc*** absorption in **chickens** and cause minimal tissue irritation after IM injection of **broilers**.

SESBANIA sp.
= *Rattlebox*

Untoward effects: In the U.S., ***S. drummondii***, ***S. punicea***, and ***S. vesicaria*** are toxic to **livestock** along the coastal plains from the Carolinas to Texas. Toxicity is similar to that of ***arsenic*** poisoning, and caused by the ingestion of seeds during the fall and winter. Less than 2 oz of seed will kill a **cow**. Many **cows** graze where plants abound and do not consume the seed. Problems occur when new **cattle** get into short pastures with these plants. Some are found dead, others depressed and with diarrhea.

S. drummondii = ***Coffee Bean*** = ***Drummond Sesbane*** = ***Poison Bean*** is a perennial legume found mostly in Florida to Texas and south into Mexico. **Sheep** and **goats** are killed by ~1 oz, and 2 oz kills larger **animals**. Depression, diarrhea, and tachycardia in milder cases; weakness and dyspnea precede death. Symptoms develop about 1 day after ingestion. Death of a **chicken** has been caused by ingestion of nine seeds; **pigeons** by 3–4 seeds. Poison symptoms are weakness, central nervous system depression, anorexia, diarrhea, ruffled feathers, cold feet, and rapid weight loss.

S. exaltata = ***Hemp Sesbania*** in the coastal plains and southern Alabama contain poisonous ***saponins*** in their seeds in summer and fall, adversely affecting **poultry**, **sheep**, **goats**, **cattle**, and **children**.

S. punicea, a common shrub in South Africa and South America causes conjunctival hyperemia, weakness, diarrhea, pyrexia, depression, tachycardia, and terminal renal failure. All parts of the plant are toxic, especially the seed, and affect most domestic **animals** and **poultry**.

Sesbania vesicaria = ***Bagpod Sesbane*** = ***Glottidium vesicarium***, an annual legume, also common in the West Indies. Ingestion of its seeds leads to diarrhea, irregular pulse, depression, and death in **cattle**.

In Australia, feed contaminated with ***S. simpliuscula*** has caused death of **poultry**.

SESSEA BRASILIENSIS

Untoward effects: Ingestion of its fruit by **cattle** causes tremors, inappetence, rumen atony, incoordination, hypothermia, bradycardia, and occasional visual disturbances and convulsions, followed by death in 12–36 h. Post-mortem reveals hemorrhages in lungs, heart, liver, kidneys, gastrointestinal tract, and brain; centrilobular liver necrosis, yellowish pericardial fluid, enlarged and edematous lymph nodes, and distended bladder. Experimentally, poisonous to **guinea pigs**, **goats**, **sheep**, and **pigeons**.

SETARIA sp.
= *Foxtails*

Untoward effects: Mechanical injury to domestic **animals**. Its awns have penetrated the mucous membranes of **horses** and **dogs**. Some species are ***nitrate***-accumulators.

Ingestion of ***S. sphacelata*** of Australia and Costa Rica has caused ***oxalate*** poisoning in **cattle** and osteodystrophia in **horses**. Has contained up to 6.9% ***oxalate***.

S. lutescens = ***Yellow Brittlegrass*** = ***Yellow Foxtail*** and ***S. virides*** have caused stomatitis, gingivitis, awn ulcers, and periodontal disease in **horses** ingesting them.

SEVOFLURANE
= *Sevofrane* = *Ultone*

Inhalation anesthetic.

Untoward effects: Nausea, vomiting, peripheral neuropathy, and hypotension. Rarely, malignant hyperthermia and hepatic toxicity. It will react with ***soda lime*** or ***Baralyme*** yielding a ***vinyl ether*** that is nephrotoxic in **rats**.

SHARK

Untoward effects: Serious traumatic injuries, massive blood loss, and often death from aggressive sharks. Some species are venomous with stings from their dorsal spines leading to local pain, weakness, and numbness lasting 1–2 h. Other species have caused outbreaks of *Ciguatera* poisoning. Poisonous scales on *Scyliorhinus caniculus* shark skin lead to **manta ray** dermatitis. Hypervitaminosis A reported from eating certain Pacific sharks. Shark meat can contain high levels of ***urea*** (q.v.).

SHAVE GRASS

Untoward effects: In herbal diuretic ***teas***. It contains ***nicotine*** (q.v.) and thiaminase.

Horses and other **animals** grazing on it develop excitement, anorexia, muscle weakness, diarrhea, dyspnea, convulsions, coma, and death. Fed to **sheep** gives classic beri-beri symptoms.

SHELLAC
= *Lac* = *Lacca*

Use: In ***lacquers***, varnishes, and manufacturing. Special grades are used as film-formers and in slow-release tablets.

Untoward effects: Occupational dermatoses in **workers** handling it. Use in aerosoled hair sprays leads to thesaurosis eliciting a foreign-body, granulomatous delayed hypersensitivity tissue response. Some crude commercial shellacs contained ***arsenic*** (q.v. for potential toxicity).

SHELLFISH

Untoward effects: Neurological paralytic shellfish poisoning (PSP) after eating unaffected shellfish that ingested ***saxitoxin*** (q.v.) producing planton (red-tide). A diarrheic shellfish toxin has also been isolated. Viral hepatitis, *Vibrio cholerae*, *Vibrio vulnificus*, and *Plesiomonas shigelloides* infections from eating uncooked or inadequately cooked shellfish. Paralytic shellfish poisoning has been lethal to **man**, **sea birds**, and **sand eels**. Can contain large amounts of ***cadmium***, ***copper***, and ***mercury***. Anaphylaxis can occur from the ***sulfites*** or ***sulfur dioxide*** sprayed on them. Exercise-induced anaphylaxis in a long-distance **runner** after eating shellfish. Some **people** react to eating them with urticaria, angioedema, laryngeal edema, bronchospasm, vomiting, diarrhea, shock, and collapse. **Children** rarely outgrow **their** allergies to shellfish. PSP can kill **humans** in 12–14 h and is ~50 times the potency of ***curare***. A cord of blue that the **Lord** commanded **Moses** to put on the corners of **their** garments derives its color from a shellfish, the **banded dye-murex**, on the shores of the Mediterranean. See individual shellfish items: ***Clam***, ***Crab***, ***Lobster***, ***Mussel***, ***Scallop***.

SHERRY

Untoward effects: Contains large amounts of ***histamine*** (q.v.). An extensive history of high intake leading to increased hemoglobin due to decreased plasma volume. Many of its drug interactions are due to its ***ethyl alcohol*** (q.v.) content. A **patient** with an allergic eczematous contact dermatitis from ***alcohol*** also had positive reactions to sherry.

SHIKIMIC ACID

Untoward effects: A carcinogen in ***bracken fern*** (q.v.). *Ilex anisatum* (***bastard anise***) of Japan produces very poisonous fruits (***skikimmi*** or ***skimmi***) containing shikimic acid. When ingested, leads to vomiting, epileptiform convulsions, mydriasis, and cyanotic appearance. Its seeds have been sold as ***star-anise***, with harmful ***picrotoxin*** effects. Excreted (26–56%) as ***hippuric acid*** in **rhesus monkeys**, but not in **rats**.

TD_{LO}, **mouse**: 3.2 g/kg.

SHOES

Untoward effects: Allergic contact dermatitis from ***thiram*** (q.v.) in ***rubber***; ***p-aminoazobenzene*** and ***p-phenylenediamine***

(q.v.) used in dyeing shoe ***leather***; from shoe ***leather*** or polishes and from the ***chromium*** (q.v.) in some ***leather*** treatments. Severe, diffuse, persistent plantar dermatitis due to ***ethyl butyl thiourea*** (q.v.) in ***rubber*** insoles of certain Nike® athletic and street shoes; and contact dermatitis from ***rubber***-based adhesives or shoe linings. Leukoderma attributed to ***alkyl-phenol formaldehyde resins*** or ***p-tert-butylphenol formaldehyde*** used in the manufacture or repair of shoes. Neuropathy in Italian **shoemakers** from contact with ***n-hexane*** (q.v.) and ***methyl butyl ketone*** (q.v.), which have a common metabolite, ***acetonylacetone***. ***Methyl ethyl ketone*** (q.v.) and ***tritolyl phosphate*** (q.v.) have also caused a neuropathy. Serious bone marrow decrease in **shoemakers** from inhalation of ***benzene*** (q.v.) from glues and from inhalation and percutaneous absorption in **Finns** and **Italians** working over open ***benzene*** vats. **Shoemakers** in Italy, Japan, and Spain have developed neuropathies from shoe adhesives. Italian fluorescent-colored shoes, with a pearlescence formerly derived from micronized shells of Mediterranean **shellfish**, had poor quality ***sulfonamides*** (ultraviolet light-absorbers) substituted for the pulverized shells when disease discolored the shells. As a result, I diagnosed a case of shoe dermatitis in a **female** previously allergic to ***sulfonamides***. Severe bullous eruptions on dorsal surface of both feet in 12-year-old **female** from East Indian **buffalo**-hide sandals. Platform shoes are a frequent cause of ankle sprains and fractures in **males** and **females**. "Tennis toe" from sudden stops with good traction court shoes, forcing the foot forward and bruising the toes. Possible increase in carcinomas of the nasal cavity and sinuses in shoe factory **workers**; also increase in bladder cancer. Non-porous synthetic footwear, particularly with **polyvinyl chloride** materials or ***polyurethane***-coated fabrics, are conducive to fungal infections, dermatitis, and hypersensitivity reactions. **Children** have a "soggy sock syndrome" from wearing plastic running shoes leading to wrinkled skin on the soles of **their** feet. ***Nickel*** eyelets on shoes have caused a dermatitis of the dorsum of the foot. Atopic dermatitis on the dorsal aspect of feet due to irritation from ill-fitting shoes.

SHRIMP

Untoward effects: Allergic reactions occur, including immediate contact urticaria. **People** allergic to shrimp are often allergic to ***crabs*** and ***lobsters***. To prevent browning and spoilage of commercially prepared or displayed shrimp, ***sulfites*** are added. Breaded shrimp often contains ***monosodium glutamate***. Shrimp may also contain ***arsenic*** (40–170 ppm), ***copper*** (up to 400 ppm), and ***iodine*** at high levels. Improperly cooked shrimp have contained *Vibrio vulnificus* leading to bullous skin lesions, cellulitis, septicemia, shock, and often death in **people** with impaired immunity, diabetes, or severe liver disease. *V. parahemolyticus* food poisoning in ~1100 13–78-year-old **people** in Louisiana, U.S., eating inadequately cooked shrimp leading to diarrhea, cramps, weakness, nausea, and chills in 55–95%; headache, fever, and vomiting in < 50%. A significant cause of diarrhea in other countries.

SIBRAFIBRIN

Untoward effects: Major bleeding with high dosage.

SIBUTRAMINE

= *BTS 54,524* = *Meridia* = *Reductil*

Anorectic. Monoamine reuptake-inhibitor.

Untoward effects: Withdrawn from the U.S. market in 1997 because it increased blood pressure in some **patients** and it was feared that it might cause heart valve defects, since its mode of action is similar to that of ***Redux***® and ***pondium***, also withdrawn from the U.S. market. Re-introduced in the U.S. market. Headache; back, chest, and neck pain; tachycardia, vasodilation, anorexia, constipation, nausea, dyspepsia, xerostomia, dizziness, hypertension, nervousness, depression, insomnia, paresthesia, rhinitis, pharyngitis, sinusitus, rash, diaphoresis, taste perversion, and dysmenorrhea are the most common adverse effects.

SILDENAFIL

= *UK 92,480* = *Viagra*

Untoward effects: It failed as a drug for hypertension and angina therapy because some cardiac **patients** reported unexpected penile erections. Most common (6–18% in **men**) are headache, flushing, and upset stomach. Mild, temporary visual changes in blue/green color perception or increased sensitivity to light; hypotension, nasal congestion, dizziness, rash, and diarrhea. The Federal Aviation Administration recommends **pilots** not take it within 6 h of flying, as it may make it difficult for **them** to distinguish between blue and green lights. Many deaths have followed its use (one report indicated 522 or 35% of 1473 **users** died – usually 4–5 h after dosing), especially when taken with ***nitroglycerin***. Severe pulmonary hemorrhage can occur in **patients** with chronic interstitial lung disease. Acute cystitis in 15 55–75-year-old **females** after intercourse with **spouses** treated with it.

Drug interactions: Considerable increase in its serum levels when taken with ***erythromycin***, ***grapefruit juice***, and ***verapamil***. Potentiates the hypotensive effect of ***nitroglycerin*** or other organic ***nitrates***. ***Cimetidine***, ***erythromycin***, ***ketoconazole***, ***lopinavir***, and ***ritonavir*** may decrease its metabolism and increase its serum concentration. ***Itraconazole*** and ***mibefradil*** are expected to act the same. ***Rifampin*** is expected to decrease its effect.

In **rats**, use with ***tacrolimus*** may increase hepatotoxicity and decrease renal function.

SILICON

Untoward effects: Inhalation leads to eye, skin, and respiratory tract irritation with coughing, dyspnea, wheezing, and pneumoconiosis or silicosis. OSHA and National Institute for Occupational Safety and Health set upper respiratory limits of 5 mg/m^3 and total exposures, respectively, of 15 and 10 mg/m^3. Silicon nephropathy in **workers** with renal failure.

Inhalation leads to pulmonary reactions in **guinea pigs**.

SILICON DIOXIDE

= *Aerosil* = *Cab-O-Sil* = *Coesite* = *Hi-Sil* = *Nalcoag* = *Quso* = *Silcron* = *Silica* = *Silicium Oxide* = *Syloid*

Untoward effects: Slowly (10–30 years) developed silicosis in exposed **miners**, foundry **workers**, pottery **manufacturers**, sandstone and granite **cutters**, crushing and milling plant **workers**, and **sandblasters** inhaling crystalline material leading to dyspnea, nodular or conglomerate fibrosis, and coughing. Occasionally sarcoid-like pulmonary changes with non-caseating granulomata. National Institute for Occupational Safety and Health suggests maximum of 50 $\mu g/m^3$, TWA.

Amorphous or colloidal silica causes considerably less pulmonary, eye, and topical irritation.

Pulmonary lesions by crystalline material have been reproduced in **rats**, **mice**, and **rabbits**. Spontaneous silica uroliths reported in **dogs**, **steers**, and **sheep**.

SILICONES

Untoward effects: Inflammation at injection sites in breasts and legs leading to cellulitis, phlebitis, pulmonary embolism, and siliconoma. Gravitational migration. Adenocarcinoma of the breast after augmentation mammoplasty. **Children** breast-fed by **mothers** with gel implants may develop symptoms of an autoimmune attack. The ***polyurethane*** foam covering the implant can release ***2,4-toluenediamine*** (q.v.), a carcinogen. Penile and scrotal sclerosing lipogranuloma in 51-year-old **male**, 4 years after suprapubic injection for impotence. Silicone fragments from dialysis pump tubing may have caused liver disorders. Blepharitis from accidentally rubbing eyes with fingers contaminated by it in window and spectacle cleaners, hair sprays, water-repellent ***lipsticks***, and hand lotions.

Injection of gel into **mice** stimulates growth of rare tumors.

SILICON TETRACHLORIDE

= *Silicon Chloride*

Untoward effects: A fuming liquid with a suffocating odor used for smoke screens in warfare with eye and respiratory tract irritation. Very corrosive to metals in the presence of moisture. A major incident involved its accidental release. In another incident, rain contacted it, releasing ***hydrochloric acid*** mist. In 1979, a 15,000 gallon spill occurred near Tuscola, Illinois, leaving a corrosive ***hydrogen chloride*** precipitate on buildings and cars.

SILICON TETRAFLUORIDE

= *Silicon Fluoride*

Untoward effects: A pungent, suffocating, colorless gas forming dense clouds of ***hydrogen fluoride*** with moist air. Has caused ***fluorine*** (q.v.) poisoning in **man** and **cattle**, due to large quantities emitted from a Swiss ***aluminum*** plant. A small drop on an eye can cause its permanent loss of vision.

SILK

Untoward effects: Allergic contact urticaria. Occupational dermatitis in 60% of **silk-reelers** with thinning or thickening of finger skin; edematous, purulent, and itching lesions, primarily of the right hand fingers. Severe corneal inflammatory reaction from 8-0 silk sutures in keratoplasty. ***Ammonia***, ***carbon monoxide***, ***hydrogen cyanide***, and ***hydrogen sulfide*** are toxic products of its combustion. Primary allergic reactions to silk extract, an additive in vaccines (no longer used in the U.S.), can precipitate a reaction in a previously sensitized **individual**.

SILO

Untoward effects: Older silos often were lined with ***tar***, which frequently contained ***polychlorinated biphenyls*** (q.v.). Pieces often broke off and were eaten. Also see ***Coal Tar*** and ***Creosote***. **Silo-filler's** disease or **silo-filler's** lung is caused by silo gases (***nitric oxide*** (q.v.), ***nitrogen dioxide*** (q.v.), and ***nitrogen tetroxide*** (q.v.)) leading to acute pulmonary symptoms, coughing, chest tightness, bronchitis fibrosa obliterans, nausea, vascular collapse, shock, and death. Has even been reported in 6-year-old **male**. Do NOT go into or near the silo if a brown or yellowish, irritating gas is noted by it. Consider a silo dangerous up to 14–20 days after filling. Normal gas masks will not protect **anyone** from these gases. A **silo-unloader's** syndrome from moldy fodder stored in the silo occurs 2–9 h after exposure leading to myalgia, fever, malaise, headache, cough, tightness in chest, and dyspnea.

Coughing **dairy cattle** may be due to inhalation of the heavier-than-air silo gases present in the barns. **Cows** are quite sensitive to re-exposure. **Pigs** inhaling high concentrations for over 10 min died in 21–72 min.

SILVER

Untoward effects: Cumulative effect leads to argyria, a slate gray or bluish discoloration of the skin, gastric mucosa, and liver from its oral or intranasal use; use as a topical adsorbant in 60-year-old **male** who chewed pieces of photographic film for ~6 months. General argyria due to ten silver amalgam dental fillings in 18-year-old **male**. Once widely used for gastritis, peptic ulcers, cystography, sexually transmitted diseases, and nose and throat infections. Argyria often occurs with ophthalmic changes (silver staining of the conjunctiva, especially of the inner canthus and Descemet's membrane) and impaired immune defenses. Argyria 10 years after brief exposure to silver needles during acupuncture in 39-year-old **female**. Allergic contact dermatitis reported. Occupational dermatitis in **watch-makers**, in **those** handling silver with accidental tattooing. Causes blue–gray azure lunulae of the nails. Long-term exposure to ***cyanide*** (q.v.) by silver-reclaiming (photographic film and X-ray film) **workers** has caused acute and chronic poisoning. The dust or powder is flammable. Can irritate eyes and skin. Long-term inhalation or use of silver-containing lozenges has caused blue–gray eyes, nasal septum, and gastrointestinal tract. Skin irritation and ulceration after contact. Certain types of silver solder release toxic ***cadmium*** (q.v.) fumes. So-called silver stools do not contain silver, but are a rare sign of carcinoma of Vater's ampulla. It is nephrotoxic. National Institute for Occupational Safety and Health/OSHA exposure limit is 0.1 mg/m^3; immediate danger to life or health, 10 mg/m^3. Flash fires in aircraft when de-icing fluids (***ethylene glycol*** and water) contact silver-coated electric circuitry, wiring, switches, or circuit-breakers.

Silver foundry's ***arsenic*** emissions have caused the deaths of **animals** grazing downwind. Dietary supplementation of it increases severity of ***selenium–vitamin E*** deficiency and growth rate in **swine**, **rats**, **poults**, **chicks**, or **ducklings**.

96 h LC_{50}, **rainbow trout**: 6.5 μg/l in soft water and 13 μg/l in hard water.

SILVER IODIDE

= *Neo-Silvol*

Use: In medicine, photography, and artificial rain-making.

Untoward effects: See ***Silver*** above. Irreversible argyria in 69-year-old **male** using a 20% solution intranasally six times/day/9 years leading to slate gray discoloration of body, especially on light-exposed skin. Reduces tolerance to sun.

SILVER NITRATE

= *Argol*

Untoward effects: Ingestion of large doses leads to violent gastroenteritis, dyspnea, coma, convulsion, and paralysis. Accidental swallowing when used topically for throat infection leads to instant throat and stomach pain, prompt emesis, and purging. After absorption, causes vertigo, spasms, dyspnea, and coma; one **child** died. Repeated topical use can cause argyria. Caustic and irritant to skin and mucous membranes. Numerous case reports of methemoglobinemia and electrolyte imbalance (leaching of ***sodium chloride***) after its topical use on burn **victims** – 4 months to 55+. Ingestion of 30 grains (2 g) by an **adult** was fatal. Chemical conjunctiviitis 6–8 h after routine prophylactic gonococcal ophthalmic use in **newborns**, usually disappearing in 24–48 h. Nail discoloration has followed local application.

Due to a pharmacy compounding error and its mislabelling as ***sodium sulfate*** 20%, a 15-year-old **male** and a 15-month-old **male** ingested some and, fortunately, promptly vomited. A **nurse** put two drops from a silver nitrate stick into a **newborn's** eyes leading to serious purulent discharge, corneal opacification, red and edematous eyes, and friable and bleeding eyelids. After a year of treatment, some corneal opacity remained. Use of a silver nitrate stick to burn the base of a chalazion caused inflammation of eyelids and nearly complete corneal opacity, requiring surgery 5 years later to restore vision. A pharmacy substituted a 25–35% ammoniacal silver nitrate dental caustic solution for a 1% solution leading to damage to the eyes of two **children**. Strong solutions or "sticks" are no longer considered safe for cautery of corneal ulcers.

Drug interactions: Incompatibility with ***alkalis***, ***ferrous compounds***, ***halogen acids*** or ***salts***, ***phosphates***, ***tannins***, and ***tartrates***.

SILVER PROTEIN, MILD

= *Argyrol* = *Silver Nucleinate* = *Silver Vitellin* = *Stillargol* = *Veraseptyl*

Untoward effects: Local and general argyria, increased intraocular pressure, and conjunctival and lens pigmentation after topical use. The argyria has been mistaken for cyanotic heart disease.

SILVER SULFADIAZINE

= *Flamazine* = *Protosulfil* = *Silvadene*

Untoward effects: Topically (1%) on burn **patients** leads to rash, tachycardia, leukocytosis, and leukopenia. The latter may actually be due to thermal injury. Extensive use of it in a base of ***propylene glycol*** caused cardiorespiratory arrest in a **patient**. Argyria and nephrotic syndrome have resulted from excessive use. Has been implicated in causing toxic epidermal necrolysis, neutropenia, and delayed epithelialization.

SILVEX
= *Fenoprop* = *Kuron* = *2,4,5-TP*

Herbicide.

Untoward effects*:** It and its various esters and salts lead to eye, skin, and mucous membrane irritation. Now banned in the U.S., as commercial batches contain ***dioxin (q.v.), a known **animal** carcinogen.

Sheep and yearling **cattle** were killed by 100 mg/kg/day/ 29 days.

LD_{50}, **rat**: 650 mg/kg; **mallard duck**: > 2 g/kg.

SIMAZINE
= *Algi-Gon* = *Aquazine* = *Gesatop* = *Primatol S* = *Princep*

Herbicide.

***Untoward effects*:** Ingestion of 500 mg/kg killed 4/6 **sheep**. Symptoms (muscle weakness, trembling, ataxia, and posterior paralysis) are delayed 3–20 days after ingestion. Post-mortem reveals lung, kidney, and liver congestion. Fatal to **chickens** after 250 mg/kg/day/8 days. In **rats**, 5 mg/kg caused drowsiness and irregular respiration.

LD_{50}, **rat**, **mouse**, **rabbit**, **chicken**, and **pigeon**: > 5 g/kg; LC_{50} in feed/5 days, **Bobwhite quail**, **pheasant**, and **mallard duck**: > 5000 ppm; **Japanese quail**: > 3720 ppm.

SIMVASTATIN
= *Denan* = *Liponorm* = *Lodalès* = *MK-733* = *Simovil* = *Sinvacor* = *Sivastin* = *Zocor* = *Zocord*

Cholesterol synthesis inhibitor.

***Untoward effects*:** In ~6.5% of **patients**. Headache, lightheadedness, constipation, cramping, nausea, and flatulence are common. Diarrhea, anorexia, jaundice, and inflammation of mouth and pancreas. Rarely, thrombocytopenia, thrombocytopenic purpura, hepatitis, and liver cancer. Muscle weakness, rhabdomyolysis, aches or pains, shoulder and joint pain, hypothyroidism, dizziness, tingling, memory loss, tremors, psychic disturbances, altered taste, blurred vision, progression of cataracts, pruritus, alopecia, rashes, dryness and discoloration of skin, decreased libido, impotence, breast enlargement, and various hypersensitivity reactions reported. Nephrotoxicity potential (glomerular-type proteinuria), transient symptomatic hypotension, and depression. Lupus-like syndrome and Raynaud's phenomenon in 39-year-old **male** after 20 mg/day/4 years.

Drug interactions*:** Addition of ***mibefradil or ***nefazodone*** (q.v.) to regime of simvastatin leads to serious rhabdomyolysis. Clinical significance of any increase in prothrombin time with ***warfarin*** has been questioned. Its body clearance is decreased by ***cyclosporine***, increasing its plasma concentration and increasing the potential for myopathy. ***Lopinavir*** and ***ritonavir*** can decrease its metabolism and increase its serum concentration.

SISOMICIN
= *Antibiotic 6640* = *Baymicin* = *Extramycin* = *Mensiso* = *Pathomycin* = *Rickamicin* = *Sch-13475* = *Siseptin* = *Sisobiotic* = *Sisolline* = *Sisomin*

Antibiotic.

***Untoward effects*:** Has induced renal tubular damage.

LD_{50}, **mice**: IV, 34 mg/kg; IP, 221 mg/kg; SC, 288 mg/kg.

SIUM SUAVE
= *Water Parsnip*

Untoward effects*:** **Cattle** have been poisoned by it in the U.S. and Canada. In Europe, ***S. latifolium causes local irritation, colic, excitement or depression, and tainted ***milk*** in **cows**.

SKATOLE
= *3-Methylindole*

Untoward effects*:** It is one of the most unpleasant stenches known to **man**. Has an odor threshold of 0.0000004 mg/m^3 of air. An excess of the ***amino acid, ***tryptophan***, in lush pastures is broken down to it by ruminal bacteria leading to acute **bovine** pulmonary edema, a typical interstitial pneumonia, and emphysema in **cattle** and **goats**. **Sheep** are more resistant to its effects. Also formed in the digestive tract of **man**, **horses**, and other **mammals** and in ***cigarette*** smoke.

SKUNK CABBAGE
= *Chou Puant* = *Fetid Hellebore* = *Skota* = *Skunk Weed* = *Symplo-carpus foetidus*

Untoward effects*:** The leaves and short erect rootstocks contain the poisonous principle which is destroyed by boiling, heating, or thorough drying, still leaving needle-like ***calcium oxalate (q.v.) crystals, which easily penetrate oral tissues leading to severe irritation, burning sensation; swelling of mouth, lips, tongue, and throat; as well as airway closure, dysphagia, and aphonia. Inhalation of the pungent rhizome oils or ingestion of the powdered rhizome were used as headache remedies; overdoses lead to nausea, vomiting, vertigo, headaches, and blurred vision.

SLAFRAMINE
= *Slobber Factor*

Untoward effects*:** A toxic substance from a microscopic black fungus, *Rhizoctonia leguminocola*, particularly on ***red

clover pasture or hay, whose ingestion by **cattle** leads to excessive slobbering, decreased feed intake, diarrhea, pollakiuria, lacrimation, bloat, stiff joints, and death. The factor is destroyed in hay stored for a year; longer for silage. Except for profuse salivation, symptoms are similar in **rats**; **guinea pigs** may exhibit all the symptoms in addition to pilo-erection.

SMILAX sp.

Untoward effects: Some, ***sarsaparilla*** (q.v.), containing ***safrole*** (q.v.) are still sold. Many species are used worldwide in herbal ***teas*** and native remedies. The aroma of ***S. herbacea*** flowers leads to headaches and even nausea. Other species native to Alabama, U.S. contain ***smilagenin*** steroidal precursors causing abortions and changes in **women's** estrous cycle.

SNAKE(S)

Untoward effects: All snakes can bite and all species have teeth (usually four rows in the upper jaw and two in the lower). Non-venomous snakes often wiggle their tails when provoked or alarmed. **Bullsnakes** and U.S. **hogsnakes** often hiss loudly. The latter may strike out, but do not bite. In the U.S., a snake is venomous if (1) it is marked with red, yellowish-white and black rings with yellow rings on each side of the red ones (A saying goes, "red against yellow will kill any fellow; red against black a poison does lack".), (2) it has a rattle on the end of the tail, (3) it has a deep pit on the side of the head between its nostril and eye, and (4) most of its scales under the tail look like the flat, broad plates on the belly. One species, the non-venomous **longnose snake** has the broad, flat plates under its tail, but none of the other features listed above. **Coral snakes** (*Micrurus sp.*) and **pit vipers** (including **copperheads**, **rattlesnakes**, and **water moccasins**) in the U.S. are venomous and have fangs. The venom of **coral snakes** is primarily neurotoxic and death of the bite **victim** is usually due to respiratory failure. It has been inflicting 6000–8000 bites annually. The **coral snake** has small teeth and its bite marks are small, often looking like scratches. Its venom lacks proteolytic enzymes and fails to induce a local edema. Emergency rooms may discharge the **patient** before toxicity is noted. Central nervous system toxicity and paralysis may be delayed for 12 h. In Central and South America, **pit vipers** are mostly the **Fer-de-Lance** or ***Bothrops*** type. **Rattlesnake**, *Crotalus sp.*, venom contains a myotoxin, and hemolytic agents causing paresthesia, vomiting, severe local swelling, tissue necrosis, decreased neuromuscular transmission conduction, pulmonary emboli, shock, and occasionally death. Has been used on poison arrows and darts. An overwhelming infection by *Arizona hinshawii* in a Hispanic 61-year-old **female** cancer **patient** given **rattlesnake** powder capsules daily along with the prescribed medication. Heavy flooding has forced snakes to higher ground, causing bite epidemics in Pakistan, India, Bangladesh, and the Mississippi delta. Conjunctival edema, facial weakness, ptosis, muscle tremors, incoordination, shock, pulmonary edema, mydriasis, cardiac arrhythmias, and myoglobinuria often follow bites by venomous snakes. Spitting, or, more correctly, spraying a fine jet stream of venom by a snake into a **person's** eye has caused chemosis, blepharitis, corneal opacity, and blindness. This spraying can be extremely accurate, even at a distance of 2–3 m. The supposed success of many treatments may be due to the fact that often only 80% of bites by venomous snakes may involve envenomation. Snake bites are one of the most common causes of defibrination syndrome, which, by itself, is benign and not a cause of spontaneous bleeding. It is estimated that > 20,000 fatalities occur annually in Asia, mostly from **viper** envenomation. In Africa, < 1000 **people** die annually from **mamba** and **pit viper** bites; ~4500 are killed annually in Central and South America from **Fer-de-Lance** and **Brazilian rattlesnake** bites; and usually < 20 die annually in the U.S. from venomous snake bites.

Although many **animals** are injured or die from snakebites, the **woodrat** of the southwestern U.S. has considerable resistance to **rattlesnake** venom.

Snake antivenins: Proteinaceous in nature and from blood of hyperimmunized **animals** can cause immediate hypersensitivity, including anaphylaxis, skin manifestations, fever, cough, or delayed serum sickness reactions in **people**.

SNAKEROOT, WHITE

= *Eupatorium rugosum* = *Fall Poison* = *Richweed*

Untoward effects: Contains ***tremetol***, a mixture of fat-soluble toxic substances, found primarily in the green plant parts and decreased with drying. It can be a cumulative poison in **cattle**, **horses**, **sheep**, **goats**, and **swine** and passes into their ***milk*** leading to "***milk*** sickness", "vomiting disease", or "trembles" in **humans** with constipation, weakness, dizziness, nausea, frequent vomiting, polydipsia, abdominal pain, swollen and coated tongues, dry skin, bradypnea, decreased temperature, weakness, collapse, and delirium. A strong ***acetone*** or fruit-like breath odor often occurs. Death may follow in 10–25%. I helped confirm a case in a **child** by checking the family **cow** and the pasture. President Abraham Lincoln's mother, Nancy Hanks Lincoln, died in 1818 from its ill-effects. **Patients** who recover seem to have lasting debility. In Nigeria, *E. odoratus* may have caused similar symptoms in **people**.

Ruminants tremble (especially nose and legs), lose weight, and are depressed and inactive; they have dyspnea, constipation, nausea, and weakness, and become unable to stand. Partial throat paralysis can occur. Can be cardiotoxic in **horses** and they usually die in 2–3 days.

SNOWBERRY

= *Symphoricarpus albus* = *Waxberry*

Untoward effects: Toxicity from ingestion of the white, waxy, berry-like fruits in **children** reported from U.S., Canada, and Europe. Semicomatose delirium, dizziness, and vomiting following ingestion of the berries with their ***isoquinoline poppy*** alkaloid, ***chelidonine***, as well as other alkaloids.

SNUFF

Untoward effects: In the U.S., India, etc., dipping or sniffing snuff has been associated with development of oral cancer and has become common in **women**. Many South American snuffs contain hallucinatory amines. Pituitary snuffs have caused allergic alveolitis.

SOAP

= *Sapo*

An anionic detergent often mixed with many potentially toxic agents affecting some people.

Untoward effects: Potential photoallergic agents or inducers of contact dermatitis from ***bithional*** (q.v.), ***fenticlor*** (q.v.), ***halogenated carbanilides***, ***halogenated salicylanilides***, and ***hexachlorophene*** (q.v.) in some **people**. Soapsuds enemas have caused acute inflammation reactions of the colon. A 23-year-old **female** received a soapsuds enema while in active labor and developed a colitis. Rectal irritation can last for weeks. A 26-year-old **female** developed erythema and hypotension from a single Castile soap enema on day 3 postpartum. **She** had lower abdominal cramps and pruritus of hands and neck. While expelling the enema, **she** fainted, lacked peripheral pulse and blood pressure, and erythema covered **her** entire body. Other case reports of rectal gangrene, massive fluid loss, anaphylaxis, hypotension, bradycardia, and death. Intrauterine use to induce abortions leads to uterine perforation, peritonitis, ileus, septicemia, acidosis, jaundice, metritis, nephrosis, fatty degeneration of the liver, pulmonary thrombosis and embolism, myocarditis, and even death. Bubble baths have caused irritation of lower urinary tract in many **girls** and occasionally in young **boys**. Hospital soaps and lotions have been contaminated with *Staphylococcus aureus*. The odors of some soaps have induced migraines in some **people**. ***Cresol*** is ineffective when soap mixed with it exceeds $\frac{1}{3}$ the ***cresol*** component. The result of proper mixing is ***Lysol***® or ***Liquor Cresolis Saponatus*** types.

SODIUM

***Essential mineral.* The principle cation of extracellular fluids. A mEq is 23 mg.**

Untoward effects: There is substantial disagreement about the relationship of excessive intake and hypertension. ***Antacids*** may contain substantial amounts and can be a factor in congestive heart failure. Other drugs cause sodium retention and pulmonary edema. A neurological syndrome (hypocalcemic tetany, muscular hypertonicity, convulsions, oliguria, renal tubule necrosis, and expansion of extracellular fluid retention) in ~35% of **those** with severe hypernatremia. The use of high dosage (> 20,000 U) ***sodium penicillin*** has induced increased cerebrospinal fluid levels of sodium. Renal tubule function in **infants** is poor and, thus, excess sodium intake can easily make **them** edematous. Passes rapidly into breast milk. **Mothers** consuming certain herbal ***teas*** had increased breast-milk sodium leading to poor suck reflex, weak cry, lethargy, cachexia, apnea, and bradycardia. Contact with elemental sodium can cause severe burns. Extra intake as ***salt*** can increase urinary ***calcium*** excretion, increase the possibility of ***calcium oxalate*** urinary tract calculi in some **people**. Can potentiate ***warfarin*** action.

Clinical tests: Increased or false positive test results with ***anabolic agents***, ***calcium***, ***corticosteroids***, ***iron***, ***mannitol***, ***phenylbutazone***, ***potassium***, and ***protein***; decreased or false negative test results with ***oral diuretics***, ***ethacrynic acid***, ***heparin***, ***iron***, ***laxatives*** (excessive use), ***mercurial diuretics***, ***phosphates***, and ***sulfate***.

In **animals**, excesses are toxic. ***Corn oil*** may help sensitize **animals** to cardiac necrosis in the presence of excess sodium. Sudden heavy ingestion by **swine** or **poultry** has been very toxic.

SODIUM ACETRIZOATE

= *Acétiodone* = *Bronchoselectan* = *Cystokon* = *Diaginol* = *Iodopaque* = *Pyelokon-R* = *Salpix* = *Thixokon* = *Tri-Abrodil* = *Triiotrast* = *Triopac* = *Triumbren* = *Triurol* = *Urokon Sodium* = *Vesamin* = *Visotrast*

***Radiopaque contrast medium.* Contains ~65% *iodine*.**

Untoward effects: Ventricular fibrillation, decreased renal function, anuria, and paraplegia have been reported after aortograms, etc. Newer agents are now preferred. May alter thyroid function tests. Severe renal insufficiency with anuria ending in death in a **patient** during pyelography.

May cause renal damage in **dogs**, which may be directly due to cytotoxicity or by interaction with the red blood cells. Pulmonary hypertension and cardiac arrhythmias can occur. Hypotensive effect in **cats**.

IV, LD_{50}, **cat**: 5.6 g/kg; **dog**: 6.3 g/kg; **rabbit**: 5.2 g/kg.

SODIUM ALGINATE

= *Algin* = *Kelgin*

Untoward effects: Soreness and swelling of injected areas when used as a vaccine adjuvant.

IV, LD_{50}, **rat**: 1 g/kg; **mouse**: 200 mg/kg; **rabbit**: 100 mg/kg.

SODIUM ARACHIDONATE

Untoward effects: IV of 1.4 mg/kg in **rabbits** led to platelet aggregation and death within 3 min; 90 mg/kg IV in **mice** is lethal, and doses > 0.45 mg/kg into the carotid artery of heparinized **rats** led to cerebrovascular occlusion or stroke within 1 min.

SODIUM ARSANILATE

= *Arsamin* = *Atoxyl* = *Nuarsol* = *Pro-Gen Sodium* = *Protoxyl* = *Soamin* = *Sodium Anilarsonate* = *Sonate* = *Trypoxyl*

***Protozoastat, growth-promoter, alterative.* Contains 31.34% *arsenic*.**

Untoward effects: Members of an Illinois farm **family** had high levels of ***arsenic*** (q.v.) in **their** blood and hair from a chemical feeder in line with the household water line; some sodium arsanilate had siphoned back into the **family's** water supply. In the U.S., blindness from optical neuritis led to its abandonment as a therapeutic agent for **man**.

Poison. In **animals**, withdraw treatment at least 5 days before slaughter. Be certain no other feed or water source of ***arsenic*** is given at the same time. Do NOT feed to egg-laying **birds** or administer to **ducks** or **geese**. Avoid contamination with any other **livestock** or **human** feeds.

SC, LD_{50}, **rat**: 75 mg/kg; **mouse**: 400 mg/kg.

SODIUM ARSENATE

= *Disodium Arsenate*

***Insecticide, alterative.* Contains 40.3% *arsenic*.**

Untoward effects: Has caused poisoning in **man** and after ingestion by **children** from its use as an **ant** poison.

LD_{LO}, **human**: 5 mg/kg.

After 2 years of chronic feeding (as 250–400 ppm ***arsenic***) to **rats** caused enlargement of common bile duct, decreased weight, and decreased survival; similar effects in **dogs** fed > 50 ppm as ***arsenic***. Embryotoxic and teratogenic (central nervous system, eyes, skeleton, and genitourinary tract) in **mice**, **rats**, and **hamsters**. Orally toxic to fasted **sheep** at 0.5–1 g/20–30 lb. **Cats** and **dogs** have been poisoned after licking ant poisons. Symptoms include vomiting, muscle weakness, and trembling.

LD_{50}, **rabbit**: 12.5 mg/kg.

SODIUM ARSENITE

= *Atlas A* = *Chem Pels C* = *Chem-sen* = *Kill-All* = *Penite*

***Insecticide, herbicide, alterative.* Contains 57.7% *arsenic*.**

Untoward effects: Occupational dermatitis from spraying with shoulder pumps to kill **grasshoppers** leads to erythema, papules, and ulcerations of wetted skin. Dermatitis reported in **workers** loading it into freight cars. Headache, malaise, abdominal pain, weakness, chills, nausea, vomiting, and dark urine within ½–3 h after exposure in **male workers** who entered a storage tank by an ***aluminum*** ladder. ***Arsine*** (q.v.) was probably produced. Anuria, oliguria, severe hemolysis, and bronzing of the skin occurred. Exposure was 2–15 min. Skin cancer increase in sheep-dip **workers** involved in its manufacture. In a suicide attempt, a 44-year-old **male** ingested ~400 mg in an ***ant poison*** leading to sharp abdominal cramps, vomiting, and watery brown diarrhea, followed by hypotension, oliguria, and decreased renal function. Hemodialysis was successful. Readily absorbed from all body surfaces.

LD_{LO}, **human**: 5 mg/kg.

Poisoned **ducks**, **quail**, and **pheasants** develop ataxia, goosestepping, asthenia, slowness, jerkiness, falling, loss of righting reflex, ptosis, and tetanic seizures as early as 1 h after ingestion. Weeds were sprayed with it for ~3 ft outside a Missouri establishment fence, killing 24/26 **cattle** grazing in an adjacent pasture. A **cattle** owner, thinking an old drum contained ***molasses***, mixed a 40% solution with feed and killed 18/25 **calves**. A rancher, thinking his **cattle** were eating dirt that poisoned them (gastroenteritis and diarrhea), discovered that they had uncovered a corroded drum with other buried trash. Swelling of the face and muzzle with severe ulceration and dry necrosis of the ears, feet, prepuce, and mucocutaneous junction of the upper lips with listlessness, soft feces, rough hair coat, decreased weight, and innappetance in a male-**hound dog** after exposure to some leaking into the dog house. Mortality can occur 1–6 days later. Do NOT use milk from dipped **animals** for several days after dipping or within a week after worming with it. Do NOT slaughter **animals** within 1 week after treatment. When used as a herbicide for aquatic plants, remember that pond water will kill **livestock** when **fish** are unaffected. Highly toxic. Use a dust mask when handling.

LD_{50}, **rat**: 41 mg/kg; **rabbit**: 7.5 mg/kg; **mallard duck**: 323 mg/kg; **California quail**: 47.6 mg/kg; **pheasant**: 386 mg/kg. LD, **horse**: 1–3 g/day/14 weeks; **cow**: 1–4 g, 25 mg/kg; **sheep** and **goats**: 200–500 mg; **pig** and **dog**: 50–100 mg; **poultry**: 10–100 mg.

SODIUM AZIDE

= *Smite*

Use: Fungicide, air bag propellant, in automatic blood counters, herbicide, and reagent preservative.

Untoward effects: Can cause rapid hypotension, tachycardia, tachypnea, severe headaches, and convulsions. A **technician** pipetting a 10% solution swallowed 1.5 ml, within 5 min, dyspnea and tachycardia. **Another** developed nausea, vomiting, headache, restlessness, and diarrhea within 15 min. The next day, the diarrhea was worse and **her** blood showed a leukocytosis. Residual symptoms lasted 4 weeks. Can be fatal. Violent explosions reported with use of automatic blood cell counters. In 1985, 100,000 l of Austrian ***wine*** were marketed with it as a preservative. A poison information center was sued as a result of a **student** dying after ingesting a buffered saline solution containing it as a preservative. National Institute for Occupational Safety and Health suggests skin exposure limit of 300 $\mu g/m^3$. Can accumulate in plumbing systems as explosive ***lead azide*** or ***copper azide***.

LD_{LO}, **human**: 5 mg/kg.

LD_{50}, **rat**: 45 mg/kg; **mouse**: 27 mg/kg.

SODIUM BENZOATE

***Preservative, solubilizer, color and flavor protectant, urinary antiseptic and acidifier, and in treating hyperammonemic infants.* U.S. annual consumption is ~20 million lb. Usually in foods at 0.05–0.1%. Some states exclude its use in meat, dairy products, or catsup.**

Untoward effects: Occasionally urticarial reactions in **people** with previous urticarial reactions. Can interfere with ***prostaglandin*** synthetase and provoke asthma and urticaria. Anaphylactic reactions with laryngeal edema and general urticaria in **female** immediately after ingesting certain store-bought products. Previously, **she** was considered neurotic and emotionally unstable. Many **patients** sensitive to it are also sensitive to ***aspirin*** and ***tartrazine***. Severe anaphylactic reaction included flushing, angioedema, dyspnea, and severe hypotension in 19-year-old **female** 30 min after eating treated ***chipped beef, mustard sauce***, and ***potatoes***. Urticaria and pruritus after rechallenge with a provocative dose. A case report of death due to hypersensitivity to it.

Production of ***fish meal*** for **livestock** from sodium benzoate and ***formaldehyde***-preserved catches yields ***dimethylnitrosamines***, toxic to **pigs**.

LD_{50}, **rat**: 4.1 g/kg.

SODIUM BICARBONATE

= *Acidoser* = *Baking Soda* = *Sel de Vichy* = *Sodium Acid Carbonate*

Use: U.S. demand is for ~600,000 tons annually, of which 35% goes into foods and baking, 25% into **animal** feeds, 10% in chemical manufacturing, 8% in cleaning products, 7% in pharmaceuticals (***antacids, urinary*** and ***systemic alkalizer***, and as a ***mucolytic agent***) and personal care items, 5% in water treatment, 5% in ***fire-extinguishers***, and 5% for miscellaneous uses, such as ***paint*** blasting, carpet deodorizers, beverages, toothpastes, fishery anesthetic, etc. Contains > 27% ***sodium***.

Untoward effects: **Dog** trials indicate that, without adequate oxygenation, its IV use can increase rather than decrease acidosis. This explains the failure in **man** to correct acidosis after cardiopulmonary resuscitation. Large oral or IV doses can cause systemic alkalosis, increased ***sodium*** (q.v.) load, fluid retention, and death. Orally or rectally, it can cause bloating, ***carbon dioxide*** gas release, and gastrointestinal cramping. A 31-year-old **male** with stomach overload and indigestion drank $\frac{1}{2}$ teaspoonful in $\frac{1}{2}$ glass of water, and had **his** stomach rupture within 1 min. Was the food or medication the cause? A decrease in its excessive use in **newborns** (with or without hypernatremia) caused a decrease in intracranial hemorrhage from 13.4 to 2.6% in a 2 year period. Although often recommended in bath water for pruritic dry skin, it is very drying and has been a cause of eczema craquelé. Long-term contact has been a cause of **baker's** eczema. Track **athletes** often took large quantities a couple of hours before a race to improve **their** performance by buffering the ***lactic acid*** produced in the run, but it often causes severe diarrhea thereafter. Addition to an **infant's** formula may cause proteinuria. Absence seizures in 2½-year-old **female** given 4 mEq/day/1 month for renal tubular necrosis. Halving the dosage resolved the problem. Hypokalemic metabolic acidosis in 4-month-old **infant** secondary to transcutaneous absorption after its topical use for diaper rash. Has caused severe tissue necrosis after extravasation. Unnecessary use during ***halothane*** anesthesia leads to severe hypotension. Although excessive use can cause systemic alkalosis, the body usually splits the bicarbonate radical into water and ***carbon dioxide***, and the latter can easily be eliminated via the lungs.

LD_{LO}, **human**: 500 mg/kg.

LD_{50}, **rat**: 4.22–6.29 g/kg.

Drug interactions: It increases urinary pH and decreases renal excretion of ***amphetamines, antihistamines, antipyrine, chloroquine, ephedrine, fenfluramine, imipramine, levorphanol, mecamylamine, meperidine, procainamide, procaine, pseudoephedrine, quinidine, quinine, rifampin***, and ***theophylline***. It increases excretion of ***chlorpropamide, lithium, nalidixic acid, nitrofurans, phenobarbital, phenylbutazone, salicylates***, and ***sulfonamides***. Increases absorption of ***L-dopa, naproxen***, and ***propantheline***, and increases rate of absorption of ***aspirin***. Decreases absorption of ***barbiturates***, ***chlorpromazine***, ***dicumarol***, ***folic acid***, ***nitrofurantoin***, ***penicillins***, ***phenylbutazone***, ***sulfonamides***, ***tetracycline*** (50% if in a capsule) and ***warfarin***. Increases solubility of ***uric acid*** and ***sulfonamides*** decreasing calculi formation. Decreases serum concentration of ***epinephrine*** and ***praziquantel***. Increases urinary activity of ***chloramphenicol*** and ***streptomycin***.

SODIUM BISULFITE

= *Sodium Acid Sulfite*

***Preservative, antioxidant.* Antibrowning agent on fresh *fruits*, vegetables, and seafood. Found in *beers* and *wines*.**

Untoward effects: Eye, skin, and mucous membrane irritant. Liberates ***sulfuric acid*** when exposed to moisture or steam and emits very toxic fumes of ***sulfur oxide*** when heated. Certain **individuals**, especially **asthmatics**, have had severe reactions from it. Symptoms include flushing, faintness, weakness, wheezing, dyspnea, chest tightness, swelling of the tongue, swallowing difficulty, coughing, dizziness, cyanosis, and loss of consciousness. Allergic reactions to many parenterals, such as ***gentamicin*** and ***morphine***, have, apparently, been due to its use as a preservative. Blood dyscrasias may develop in sensitive **individuals**. In parenterals, can degrade ***vitamin B*** and inactivate ***penicillin***. ***Sulfite***-preserved bronchodilator solutions liberate ***sulfur dioxide*** when nebulized. A 31-year-old **female asthmatic** had a paradoxical response to ***isoetharine*** containing it as a preservative. **Crewmen** of a **shrimp**-fishing vessel collapsed after entering a hold containing treated **shrimp**. Several deaths in other similar cases. Airway constriction, angioedema, and urticaria first described in 1976. Deaths have been reported. Has been used in the illicit manufacture of ***phencyclidine*** and analogs. National Institute for Occupational Safety and Health upper exposure limit of 5 mg/m^3.

LD_{LO}, **human**: 500 mg/kg.

SODIUM BROMIDE

= *Sedoneural*

***Sedative, hypnotic.* Contains 77.65 *bromine*.**

Untoward effects: Fungoid bromodermia with paleness of facial skin; tumor-like formations on face, scalp, and extremities; fever (38°C) in 8-year-old **male** after ~25 g in 2½ months for epilepsy. Depression with serum ***bromide*** (q.v.) level of 365 mg% in 6-day-old infant whose **mother** had ingested unknown amounts of ***Miles Nervine***® prior to delivery. **Infant** rapidly recovered from transplacental bromism, but **mother** experienced prolonged toxicity delirium. Skin rashes have occurred. Therapeutic serum levels are close to toxic levels.

LD_{LO}, **human**: 500 mg/kg.

LD_{50}, **rat**: 3.5 g/kg; **mouse**: 7 g/kg.

SODIUM CACODYLATE

= *Agent Blue* = *Arsecodile* = *Arsicodile* = *Arsycodile* = *Rad-e-cate* = *Silvisar* = *Sodium Dimethylarsenate*

***Herbicide, alterative.* Contains 46.83% *arsenic*. *Cacodylic acid* (q.v.) contains ~15% more *arsenic*.**

Untoward effects: More toxic orally than parenterally because of the rapid release of inorganic ***arsenic***. Produces a strong ***garlic***-like odor to breath, urine, and perspiration, especially after oral use. Toxicity is similar to that of ***arsenic trioxide*** (q.v.) and large dosage has caused nephritis, albuminuria, and hematuria.

LD, **human**: ~1.5 g/kg.

Single IP of 1.2 g/kg in **mice** caused fetal skeletal malformations; 0.9–1 g/kg IP in **hamsters** led to increase in fetal malformations and some maternal deaths; 1–2 mg/egg caused developmental abnormalities in **chicks**; fetal and maternal deaths in **rats**.

LD_{50}, **rat**: 2.6 g/kg; **mouse**: 4 mg/kg; **dog**: 850 mg/kg.

SODIUM CARBONATE

= *Sal Soda* = *Soda Ash* = *Solvay Ash* = *Washing Soda*

***Disinfectant, antipruritic, antacid.* U.S. annual demand is > 12 million tons.**

Untoward effects: Strongly alkaline and caustic. Inhalation has caused laryngeal, tracheal, and upper bronchial irritation with coughing and dyspnea. Massive destruction of cells and tissues can cause swelling of lips and tongue, acute hemorrhage, and unconsciousness. Ingestion in a 22-year-old **female** suicide attempt caused atrioventricular block, atrial rate 75/min, and ventricular rate 43/min. Amounts used to adjust swimming pool water should not cause pulmonary damage in near-drowning **victims**. Chronic skin exposure leads to dermatitis, eczema, irritation, burns, and ulceration. Vesiculation after contact with abraded skin.

LD_{LO}, **rat**: 4 g/kg.

SODIUM CARBOXYMETHYLCELLULOSE

= *Carbose* = *Cellulose Gum* = *CMC Sodium*

Untoward effects: Oral use to decrease hunger intensity often leads to constipation, thirst, and nausea. Use in parenterals may cause local inflammation, hemorrhagic reactions, and sterile abscesses. An enzyme, β-glucosidase, produced by **human** tissues and vaginal bacteria, was able to hydrolyze the carboxymethylcellulose in a particular brand of tampons yielding ***glucose*** that supported the growth of toxin-producing *Staphylococcus aureus* associated with the toxic shock syndrome.

SODIUM CHLORATE

= *Atlacide* = *Defol* = *Dervan*

Herbicide, wood pulp bleach. **U.S. and Canadian annual demand of ~2 million tons.**

Untoward effects: Irritating to skin and mucous membranes. Ingestion can be fatal. Symptoms include nausea, vomiting, abdominal pain, cyanosis, gastritis, methemoglobinemia, thrombocytopenia, leukopenia, hemolysis, anemia, convulsions, collapse, nephritis, and renal failure. In England, when the amount ingested was > 100 g, mortality was 64%. A 28-year-old **male** accidentally ingested 1.5 oz or ~42 g. **He** survived despite extensive hemolysis and a period of anuria. Estimated **human** LD is ~30 g. Accidental ingestion of 20 g by 15-year-old **male** caused diarrhea, abdominal pain, vomiting, fever, cyanosis, clouding of consciousness, anuria, decreased blood pressure, and hemolysis. Fatal poisoning reported in several suicidal attempts by **people**. Murder of a 64-year-old **male alcoholic** by administration of 20–30 g in **his *beer*** over a 5 week period. An oxidizing agent capable of causing fires and explosions.

Many case reports of poisoning and death in **cattle** from eating treated weeds or from accidental admixture to feed leading to dyspnea, apprehension, marked distress, incoordination, diarrhea, dyspnea, hematuria, cyanosis, convulsions, and death. Methemoglobinemia, nephropathy, and death in **cats** and **dogs**.

LD, **horses**: 250 g; **cow**: 500 g; **sheep**: 100 g; **dog**: 60 g; **chicken**: > 5 mg/kg; and **rat**: 12 g/kg.

SODIUM CHLORIDE

= *Salt*

Untoward effects: Potentially dangerous when used to induce emesis, as it can cause ***sodium*** toxicosis in **man** and **animals**. Death reported from an enema of a saturated solution. Clinical signs of ***salt*** poisoning or water deprivation include nausea, vomiting, weakness, tachycardia, cerebral edema, pulmonary edema, and seizures. Excessive intake in **infants** and **children** can cause permanent brain damage. This can occur with poorly compounded ***electrolyte mixtures***, from some ***baby foods***, or use instead of ***sugar***. Hyperemia after saline enemas. Cases of severe hypernatremia in **children** reported as a result of **child** abuse by administration of excessive ***salt*** and water deprivation. A well-defined dermatitis localized on the legs of ***salt*** mine **workers** apparently caused by the necrotic, caustic, and mechanical action of salt. Chronic urticaria and eczema often aggravated by up to 10 g/day. Chronic partial separation of fingernails caused by handling cowhides preserved in brine. Can increase intraocular pressure. Radioactive material appears in 20 min in breast milk, peaking at 2 h, and detected for up to 96 h. ***Salt***-koshered ***meat*** can be soaked again in cold water to decrease the ***salt*** content. **Seamen** are aware of the hazards of drinking ***salt*** water (4% level). **Chinese** have swallowed saturated ***salt*** solutions as a traditional method of committing suicide. Low-***salt*** diets can cause ***lithium*** retention and toxicity. *Pseudomonas pickettii* colonization found in respiratory therapy solution used on hospitalized **infants**. A major recall was made of a seasoned ***salt*** because it contained 6.25% ***sodium nitrite*** (q.v.). Extra intake can increase urinary ***calcium*** excretion, increasing the probability of ***calcium oxalate*** urinary tract calculi in some **people**.

Hypertonic intrauterine solutions to induce abortions may cause convulsions, coma, dyspnea, shock, arrhythmias, fever, drowsiness, tachycardia, nausea, vomiting, defibrination syndrome, disseminated intravascular coagulation, severe bleeding, anemia, intense thirst, intracerebral hemorrhage, endometriosis, and occasionally deaths. Dispensing errors using it instead of ***Ringer's Solution*** during eye surgery have caused partial blindness.

LD_{LO}, **human**: 500 mg/kg.

Excessive intake can cause serious pathology and death. Food grades are not necessarily safe for parenteral use due to anticaking additives. ***Ferric ferrocyanide***, ***sodium chromate***, and ***sodium hexametaphosphate*** are frequently used as corrosion and caking preventatives in street use grades. SC use of concentrated solutions can cause sarcomas in **rats** and **mice**. Avoid excessive IV usage without attention to ***magnesium***, ***potassium***, etc. needs. Topical instillation of isotonic saline on the corneas of albino **rabbits** interfered with their epithelial regeneration by unknown mechanism(s); this deleterious effect can be overcome by addition of ***polyvinyl alcohol*** 1.4%. Swedish **pigs** fed waste containing ***salt*** developed symptoms of acute and chronic meningoencephalitis. Mixing errors and purposeful overtreatment of **hogs** with dysentery using 2.5% in water for 48 h instead of 6 h led to blindness, frothing at the mouth, twitching, sitting, circling, pushing, and high mortality. Physiological ***salt*** solution (0.9%) as sole drinking water for **chicks** is fatal within a few days.

LD_{50}, **rat**: 3 g/kg; **dog** and **mouse**: 4 g/kg.

SODIUM CHROMATE

Corrosion inhibitor.

Untoward effects: Use in air conditioning or recirculating water heating systems has a potential for feedback into public water supplies leading to severe nausea.

LD_{LO}, **human**: 50 mg/kg.

Use in oilfield drilling muds exposes grazing **livestock** to ***chromate*** poisoning and deaths of **cattle**. Embryotoxic to **mice**.

SODIUM CITRATE

= *Citnatin* = *Citrosodine* = *Trisodium Citrate* = *Urisal*

Anticoagulant, electrolyte, systemic and urinary alkalinizer, diuretic, expectorant, laxative, pH buffer, and chelating agent. 26.7% sodium.

Untoward effects: Hyperemia after use in enemas. Indiscriminate use in **infant** formulas increases risk of hypernatremia and catharsis.

IV, LD_{50}, **mouse**: 71 mg/kg; **rabbit**: 418 mg/kg.

SODIUM CYANATE

Untoward effects: Only gastrointestinal distress and drowsiness reported when treating sickle-cell anemia **patients** at 10–35 mg/kg/6–18 months. Dosages of 41 or 44 mg/kg/day/14–20 months caused motor and sensory neuropathies with weight decrease, weakness, and paresthesias of extremities.

TD_{LO}, **human**: 5.4 g/kg, over a 24 week period.

LD_{LO}, **mouse**: 4 mg/kg.

SODIUM CYANIDE

Untoward effects: **Workers** using it in reclaiming ***silver*** from X-ray and photographic film had acute and residual toxic effects, including abnormalities of ***vitamin*** $\boldsymbol{B_{12}}$, ***folic acid***, and thyroid function; headache, dizziness, nausea and vomiting from frequent skin contact, and ingesting food and drink in production areas. ***Almond*** breath odor in 22-year-old **male** after ingesting 1.5 g in a suicide attempt. One can assume it is a cause of poisoning in a small **child** after eating **rat** poison containing it. Symptoms include deep and labored respirations, coma, and corrosive burns about the mouth. Giddiness, staggering, headache, mydriasis, palpitations, unconsciousness, violent convulsions, and death, especially after inhalation of its fumes. Often used in the illicit manufacture of ***phencyclidine*** and its analogs.

Fatal poisoning reported in **cows** drinking from a river containing it as factory waste. Used on **sheep** collars to kill wild **predators**. Marked differences in toxicity among different species of **birds**.

LD_{50}, **sheep**: 3.8 mg/kg; **rat**: 4.6–15 mg/kg.

SODIUM CYCLAMATE

= *Sucaryl*

A *p-aminosulfonamide* sweetener.

Untoward effects: Its metabolite, ***cyclohexylamine*** (q.v.), can induce chromosome breaks *in vitro* in **human** cells. Chronic rhinitis in 53-year-old **male** from using it. An 18-year-old **male** with bulimia and anorexia nervosa after abusing the sweetener (21.5 kg over 4 years) had histologic and radiologic abnormalities of the small intestine and a malabsorption syndrome which were reversible within 3 months after discontinuing sodium cyclamate.

Drug interactions: Use decreases absorption of ***lincomycin*** by ~75%.

SODIUM DEHYDROCHOLATE

= *Decholin*

Choleretic.

Untoward effects: Severe hypersensitivity-type reactions after IV. Hypotension, shock, and deaths have been reported. Some serious clinical reactions may be due to direct myocardial toxicity. Headache, faintness, dyspnea, tachycardia, sweating, chills, fever, nausea, vomiting, diarrhea, abdominal discomfort, erythema, and pruritus also reported. Contraindicated in bronchial asthma and biliary obstruction.

SODIUM DICHROMATE

= *Sodium Bichromate*

Use: ***Antioxidant corrosion inhibitor***, in oil well-drilling muds, ***leather*** tanning, and stump removal. Annual U.S. consumption ~150,000 tons.

Untoward effects: Irritant and caustic to skin. Allergic contact dermatitis in railroad **workers** using it to prevent rusting of radiators and pipelines in diesel locomotives. Fatal ingestion in a **child** reported.

LD_{LO}, **human**: 50 mg/kg.

SODIUM EDETATE

= *Chelon 100* = *Sodium Edethamil* = *Sodium EDTA* = *Sequestrene* = *Tetracemin* = *Tetrine* = *Versene 100*

Chelating agent.

Untoward effects: Nausea, diarrhea, and muscle cramps after IV. Occasionally pain at injection site, fever, headache, and muscle pain. High concentration or rapid IV infusion rate leads to thrombophlebitis, drowsiness, malaise, vertigo, hypocalcemic tetany, and cardiac arrest. Albuminuria, oliguria, and renal failure have occurred.

LD_{LO}, **human**: 500 mg/kg.

SODIUM ETIDRONATE

= *Didrocal* = *Didronel* = *Etidron*

Calcium resorption inhibitor.

Untoward effects: Allergic skin rashes in **patients** with Paget's disease. Occasional bone pain at pagetic sites. Alopecia, arthralgia, arthritis, esophagitis, glossitis, angioedema, assorted rashes, urticaria, amnesia, confusion, depression, bone fractures, hallucinations, paresthesias, headache, gastritis, diarrhea, nausea, and leg cramps have been possible undesired ill-effects. Rarely, agranulocytosis, pancytopenia, and leukopenia. A case report of confusion and hallucinations in an **elderly person**.

Dogs treated with 2 mg/kg/day showed no pathology, but, by the end of week 3, some **dogs** at 4, 6, and 8 mg levels showed slight muscular or skeletal weakness. Radiography revealed irregular bone mineralization with increased width of epiphyseal plates. Levels of 15 mg/kg/day to **rats** decreased weight gain and feed consumption.

SODIUM FLUORIDE

= *Chemifluor* = *Dentafluoro* = *Duraphat* = *Florocid* = *Fluoros* = *Flura* = *Karidium* = *Lemoflur* = *Luride* = *Ossalin* = *Ossin* = *Osteo-F* = *Osteofluor* = *Villiaumite* = *Zymafluor*

Untoward effects: Perioral dermatitis, a common rosacea-like dermatitis, was first reported in the mid 1950s. Although probably multifactorial in origin, fluorides in dentifrices appear to be the precipitators. Malfunction of a relay switch controlling dilution of a school's sodium fluoride water treatment caused acute illness in 34 **people** in a Los Lunas, New Mexico elementary school, shortly after drinking the water with levels of 375 and 93.5 ppm and designed to supply it at 1–5 ppm. The bitter salty taste of the water deterred consumption. Symptoms included nausea, vomiting, and abdominal pain. A similar outbreak of illness followed consumption of beverages made with treated water at a Vermont elementary school and at a North Carolina school. Headaches, dizziness, and diarrhea were also reported. Accidental ingestion of tablets reported in many **children**. Giant cells in bone marrow of elderly **females** on 16–150 mg/day/1–36 months. Poisoning in 25-year-old **male** after ingesting 120 g **rat** poison (97% sodium fluoride) leading to tetany, multiple episodes of ventricular fibrillation, and esophageal stricture. Potential for serious toxicity from swallowing large amounts of mouthwashes containing it. Has caused acneform lesions. Ulceration of skin and fingernail injury and loss caused by a window cleaning fluid containing it. Chronic excesses or levels > 4 ppm may cause mottling of teeth. Weakness, convulsions, coma, hypocalcemia, hypomagnesemia, hypokalemia, leg or facial paralysis, acidosis, albuminuria, and anuria have also followed overdosing. Do NOT drink water that contains > 0.7 ppm. The chief cause of death is inactivation of body ***calcium***. Severe symptoms have followed ingestion of < 1 g and 1–4 g has been fatal.

TD_{LO}, **human**: 4 mg/kg. LD_{LO}, **human**: 75 mg/kg.

Accidental ingestion of excessive quantities by **pigs** leads to vomiting, colic, diarrhea, muscle weakness, shock, and death. Post-mortem reveals hemorrhagic gastroenteritis. Use in slop instead of dry feeds can cause uneven distribution and death of some **pigs**. **Mice** and **rats** fed amounts up to 20 times those that **humans** would consume in treated drinking water. Some male **rats** had osteosarcomas at the highest levels. None in female **rats** or any of the **mice**. Toxic in **rats** leading to ataxia, depression, followed by muscle weakness, loss of righting reflex, tremors, cyanosis, bradypnea, and death. Pustules on scratched **rabbit** skin by 0.5% solution. Fed to **owls** leading to reproductive disorders. Dental fluorosis and lameness in **cattle**. In **chicks** and **turkeys**, it decreases growth. High IV doses to **rabbits** cause retinal degeneration.

LD_{50}, fasted **rabbit**: 180 mg/kg; **dog**: 75 mg/kg; **lambs**: 145 mg/kg.

SODIUM GLUTAMATE

= *Accent* = *Ajinomoto* = *Chinese Seasoning* = *Glutacyl* = *Glutavene* = *Monosodium Glutamate* = *MSG* = *RL-50* = *Vetsin*

***Flavor enhancer, treatment of hepatic coma.* The *levo* form is the seasoning agent. Annual U.S. consumption of ~45,000 metric tons.**

Untoward effects: As a common ingredient in oriental cuisine, it has been associated with the Chinese Restaurant Syndrome. Within 20 min leads to headache, nausea, chest pains, paresthesias, asthma, tremors, and depression. Seldom reported in China. ***Wonton soup*** is loaded with it and usually served first at meals in the U.S., giving the stomach a high initial dose. In the Orient, the soup is brought out later in the meal. An 18-month-old **child** began crying with apparent abdominal pain 10 min after eating ***wonton soup***. A 3-year-old **female**, on three occasions, had 30–45 min episodes of inappropriate behavior, confusion, slight ataxia, headache, nausea, vomiting, or abdominal pain after eating ***wonton soup***. The **family** has a strong history of migraines. Tachycardia, glove-and-stocking paresthesia, and depersonalization in 54-year-old **male**,10 min after ingestion with food. A case report of induced bronchospasm, as seen with ***sulfite***-sensitive **people**. ***Bortsch***; strained, canned whole, and pickled whole ***beets***, ***French dressing***, ***mayonnaise***, ***mushrooms***, ***peas***, breaded ***shrimp***, ***tomatoes***, and canned ***tuna***, have high levels of free glutamate. ***Glutamic acid*** is a natural component of protein foods. Symptoms are more severe in **infants** and **children**. Considerable variation in oral threshold doses in **humans**. Some **people** have a recessive adult-onset olivopontocerebellar degeneration associated with a partial deficiency of glutamate dehydrogenase.

TD_{LO}, **human**: 43 mg/kg.

Injected IP in **rats** leads to seizures at ~50 times normal **human** intake. **Humans** consume it orally, and it is primarily detoxified in the gastrointestinal tract. SC in **mice** at 0.5–4 mg/g leads to acute neuronal necrosis of the brain and retina in newborns and stunted skeletal development, marked obesity, and female sterility. Similar brain lesions after only SC use in an infant **rhesus monkey** and **rabbits**, but not in **chickens**.

LD_{50}, **rat**: 17 g/kg; **mouse**: 13 g/kg.

SODIUM HYALURONATE

= *ARTZ* = *Connettivina* = *Equron* = *Healon* = *Healonid* = *Hyacid* = *Hyalgan* = *Hyalgin* = *Hyalovet* = *Hylartin* = *Hyonate* = *Ial* = *Legend* = *Opegan* = *Provisc* = *Synacid*

A normal constituent of vitreous humor and synovial fluid. Molecular weights vary.

Untoward effects: Pain, swelling, warmth, and effusions within 24 h at injection site which may last up to 3 weeks. Pseudogout immediately after injection due to deposition of ***calcium pyrophosphate dihydrate*** crystals. Non-fatal anaphylactic reactions and septic arthritis reported. May cause increased intraocular pressure after use in the eye.

After intra-articular injections in **horses** caused heat in 15%, transient edema in 12%, and pain in 9% around the treated joint. Reactions were mild and subsided in 1–5 days. No antigenicity or sensitivity noted.

SODIUM HYDROXIDE

= *Caustic Soda* = *Lye* = *Soda Lye* = *Sodium Hydrate* = *White Caustic*

Use: In petroleum refining; making cellophane and viscose rayon, soaps, detergents, and reclaiming ***rubber***; as a disinfectant. U.S. annual demand is > 15 million tons.

Untoward effects: Very corrosive. Causes irritation, erythema, dermatitis; and severe burns to skin, eyes, and mucous membranes. Skin and corneal damage occurs during the first 60 s of exposure. Cosmetics for **children** and cuticle removers have caused eye damage. Fumes have caused pneumonitis. Death due to accidental or suicidal ingestion leading to esophageal burns, perforations, and stricture formation; pyloric stenosis, emphysematous gastritis, gastric necrosis; burns in pharynx, on tongue, lips, chin, and chest; massive necrosis and multiple perforations of small and large intestine; aortic rupture after esophageal perforation, and carcinomas reported at site of esophageal strictures. In Georgia, 30 gallons of a 50% solution were accidentally siphoned into a city water well due to malfunctioning equipment. Mouth burning and sores, abdominal pain, nausea, and vomiting occurred in **users**. Retrosternal burning pain, dysphagia, and esophageal stricture in **children** ingesting it and in an 82-year-old **female** with failing vision **who** ingested a ***Clinitest***® tablet containing 38% sodium hydroxide instead of an intended ***aspirin*** tablet. Small ***mercury batteries*** are a hazard to **children** and **pets** who ingest them because they contain ***sodium*** or ***potassium hydroxide*** in addition to ***mercuric oxide***. ***Lye*** crystals were found on ***pretzels*** (q.v.) and caused a major product recall. A 30-year-old **male robber** hid in a tank containing ***lye***, ***silver cyanide***, and ***potassium cyanide*** (q.v) and suffered from severe ***cyanide*** poisoning. Poisoning reported in 12-year-old **male** who used a teaspoon that fell into a basin of water and ***Liquid Plumber***® to take a dose of cough medicine, leading to esophageal burns and, 2 weeks later, strictures. **Children** at Granite City, Illinois slipped through the ice of a ditch and developed severe leg burns due to leakage of caustic from an oil company. Contamination of a hospital's anesthetic machine led to bronchospasms in a **patient**. Cases of endophthalmitis were associated with implantation of sodium hydroxide-disinfected intraocular lens protheses. Used in the illicit manufacture of ***amphetamines***. OSHA upper exposure limit set at 2 mg/m^3.

Usual disinfectant solutions are ***ineffective*** against tuberculosis and spore-forming organisms. Wear protective clothing and eyeglasses. Can react violently with water under certain conditions and with acids.

SODIUM HYPOCHLORITE

Disinfectant, bleach, oxidizing agent.

Use: In ***Clorox***, ***Daykin's solution***, ***Dazzle***, ***Gipokhlor***, ***Hyposan***, ***Javelle Water***, ***Milton***, ***Parozone***, ***Purex***, and ***Zonite***.

Untoward effects: Esophageal burn and possible stricture after ingestion in **some**. A 7-year-old **female** had intermittent vomiting, abdominal pain, bronchopneumonia, fever, and respiratory distress after sucking on socks washed with it since **she** was 14 months of age. Asymptomatic for 5 months when **she** quit the habit. Recurrence with rechallenge, but not with boiled socks. Ingested strong solutions are very alkaline and caustic and mild gastrointestinal symptoms follow ingestion of weaker solutions. Small amounts of ***sodium hydroxide*** (q.v.) are present with it and add to its caustic potential. In a suicide attempt, a 20-year-old **male** ingested 650 ml ***Purex***® (5.25% ***NaOCl***) leading to loss of consciousness and gasping breathing (2/min) within 2½ h. Complete atrioventricular block (atrial rate 75/min and ventricular rate 43/min) 2 months after ingestion by pregnant 22-year-old **female**. Upon contact with gastric juices or acids, it releases ***hypochlorous acid*** leading to irritation and corrosion of mucous membranes, with pain, vomiting, hypotension, delirium, or coma. A swimming pool **attendant** developed onycholysis of **her** fingernails whenever **she** used a 16% solution in the pool. Topical use on wounds leads to transient pain or irritation. Use on **men's** underwear treated with certain ***rubber accelerators*** leads to dermatitis and contact urticaria. Will darken bleeding edges of a wound without impairing the healing. A 61-year-old **female** undergoing hemodialysis was accidentally exposed after undiluted ***Clorox***® (5.25% ***NaOCl***) soaked the membranes (for < 2 min) **her** blood passed through leading to massive hemolysis, hyperkalemia, cyanosis, and cardiopulmonary arrest. Commercial laundry bleaches, known as ***caustic soda bleach***, are usually 15–16% with a pH of > 11. Accidental splashing on the eye, unlike the weaker solutions, can cause bluish edema and discharge. Accidental IV infusion of 150 ml of a 1% solution caused bradycardia, mild hypotension, and tachypnea. Its use is contraindicated with ***vinegar***, various caustic agents, and acidic ***toilet bowl cleaners***, as it will cause sudden release of ***chlorine*** gas. Inactivated by highly alkaline water and excessive organic debris. ***Chloramines*** are released when mixed with household ***ammonia***.

May cause chapping of **cow's** teats when used as an after milking teat dip. With irrigation, strong solutions on the eyes of **rabbits** leads to slight corneal epithelial haze and conjunctival edema, clearing within 24–48 h. If not irrigated, causes more extensive corneal and conjunctival edema with small hemorrhages.

SODIUM IODIDE

= *Anayodin* = *Ioduril*

Untoward effects: Goiters in most **patients** within 3 months to 12 years (usually 2–3 years). IV or orally, the ^{131}I radioactive drug reportedly leads to hyperparathyroidism, hypothyroidism, and carcinomas. Cretinism in 4⅓-year-old **female** whose **mother** received four courses of ^{131}I during pregnancy. Thyrotoxicosis in **infant** whose **mother** ingested 30 ml of 3.6% w/v preparation/day for many years for treatment of asthma. Occasional angioneurotic reactions immediately after administration or several hours later.

LD_{LO}, **human**: 500 mg/kg.

In **animals**, monitor iodism. Avoid unnecessary use in advanced pregnancy since it may produce goiter in the newborn. Appears rapidly in the ***milk*** of treated **animals** and guidelines are needed for the consumption of such ***milk***. In the absence of such guidelines, I suggest avoiding the use of ***milk*** from treated **animals** for at least 72 h and six milkings after such treatment. May potentiate ***barbiturate*** anesthesia in **rats**. Inadvertent normal rate IV injection can cause nearly instant severe reactions.

LD_{50}, **rat**: 4.34 g/kg; LD_{LO}, **mouse**: 1.65 g/kg.

IV, LD_{LO}, **rat**: 1.3 g/kg.

SODIUM LAURYL SULFATE

= *Duponal QC* = *Irium* = *Maprofix* = *Pendit WA* = *SDS* = *Sodium Dodecyl Sulfate* = *Stepanol WA 100*

Anionic surfactant.

Untoward effects: Intolerance to it in 3/57 **patients** with various dermatoses. Can induce skin damage. Contact allergy reported. Can be comedogenic. A skin defatting agent. Avoid eye contact. When combined with ***erythromycin propionate***, it becomes the hepatotoxic ***erythromycin estolate***. It increases gastrointestinal absorption of ***aspirin***. May complex cationic agents including commonly used ***quaternary ammonium compounds***. This can be beneficial as such a reaction decreases the skin sensitizing properties of ***bacitracin***. 1% solutions are definite irritants to skin and eyes. IV toxicity in **animals** on their lungs, liver, and kidneys.

LD_{50}, **rat**: 1.3 g/kg.

SODIUM METABISULFITE

= *Sodium Pyrosulfite*

***Antioxidant.* A strong reducing agent in parenterals, foods, *beers*, *fruit juices*, soft drinks, *wines*, vinegar, *potato* chips, dried *fruit*, *shrimp*, etc.**

Untoward effects: A sulfiting agent, releasing ***sulfur dioxide*** after contact with moisture. Symptoms include flushing, hypotension, faintness, weakness, wheezing, dyspnea, chest tightness, cough, colic, diarrhea, cyanosis, contact dermatitis, urticaria, pruritus, angioedema, airway constriction, and loss of consciousness. An **asthmatic** 10-year-old **female** had a severe asthmatic attack after eating a ***guacamole*** salad in a Mexican restaurant that contained 950 ppm, a level that was then (1985) approved by the FDA. Bronchospasm in 67-year-old **female** immediately after eating a salad with ***vinegar*** containing sodium metabisulfite. Confirmed by rechallenge. **Others** have had similar reactions from sodium metabisulfite in infusions of ***dexamethasone*** and ***gentamicin***; inhalation of ***albuterol***, ***cromolyn***, ***nedocromil***, and ***terbutaline***; and ***metoclopramide*** injections. Large doses can cause colic, diarrhea, and death. Accelerates decomposition of ***thiamine*** and ***epinephrine*** (it is oxidized before the ***epinephrine***). Inactivates ***vitamin E***. Fumes can irritate eyes, mucous membranes, and skin. National Institute for Occupational Safety and Health upper limit of exposure 5 mg/m^3.

Mink fed ***fish*** treated with 2% had decreased fertility. Minimum lethal dose for **sheep**: 2.25 g/kg; 100% LD, 3 g/kg. Symptoms of acute poisoning are nervousness, dullness, anorexia, ruminal atony, tachycardia, tachypnea, cyanosis, and foamy nasal discharge. Post-mortem reveals hemorrhages and hyperemia of viscera. Long-term feeding of 0.5–2% in feeds to **pigs** caused hyperplasia of gastric and cecal epithelium and black discoloration of cecal mucosa. Levels > 0.5% to **rats** led to gastric lesions. **Cows** fed 20 g/day/15 weeks showed no ill-effects.

IV, LD_{LO}, **rabbit**: 192 mg/kg.

SODIUM METABORATE

Herbicide. **Contains ~16.4% *boron* (q.v.).**

Untoward effects: Highly alkaline. Poisoning symptoms include nausea, vomiting, weakness, diarrhea, rashes, ecchymoses, headache, shock, coma, and death.

LD_{LO}, **infant**: 1 g/kg; **human**: 500 mg/kg; **man**: 709 mg/kg.

LD_{50}, **rat**: 2.7 g/kg.

SODIUM METASILICATE

Untoward effects: Extremely alkaline and caustic; a 1.2% solution has a pH of 12.6 and can rapidly damage corneal epithelium. Use in detergents is often the primary agent in them responsible for esophagitis, gastritis, and occasionally oral burns.

LD_{LO}, **human**: 500 mg/kg.

In the drinking water, it decreases growth of **chicks**, decreases reproduction in female **rats**, and decreases growth of female **sheep**.

LD_{LO}, **dog** and **pig**: 250 mg/kg.

SODIUM METRIZOATE

= *Isopaque* = *Ronpacon* = *Triosil*

Radiopaque ionic diagnostic aid. **Intracardiac, intra-arterial, and intravenous use.**

Untoward effects*: *Iodine (q.v.) hypersensitivity with severe laryngeal edema and asphyxia in 59-year-old **male**. Severe pulmonary hypertensive episodes in 3-year-old **male** and 3-month-old baby. Transient tachycardia, sensation of warmth, and a few seconds T-wave inversion. Proteinuria, often massive, after its use in nephroangiography in **man** and **dogs**.

SODIUM NITRATE

Use: Meat preservative; in manufacturing ***glass***, pottery, enamels, ***fireworks***, explosives, and ***matches***; and as a fertilizer. The crude natural form is called ***Chile saltpeter***.

Untoward effects: Gut flora can reduce the nitrate to the more toxic ***nitrite***. A cause of ***hot dog*** headaches, where it is used to preserve color and help prevent botulism. These usually occur within 30 min after ingestion. Can cause violent combustion or ignition with organic or readily combustible materials. Toxic methemoglobinemia and cyanosis after absorption through burned skin areas in three **males** after an accidental industrial explosion. Its preservative capabilities are evident when archeologists digging in Chile's Atacama coastal desert (which contains large amounts of sodium nitrate in the soil) uncovered bodies thousands of years old with preserved skin, hair, and internal organs.

LD_{LO}, **human**: 500 mg/kg.

Lambs fed a diet containing 3% had liver ***vitamin A*** stores 46% below controls. Often eaten by **cows** off the ground after use as a fertilzer or mistaken for ***salt*** with poisonous results. Levels of ~2% in the feed decrease growth of **swine**; fatal in 1–4 days in **quail** given 5300 ppm in their drinking water. Toxic symptoms in **animals** include weakness, incoordination, tachycardia, dyspnea, cyanosis, convulsions, coma, and death. Occasional abortion in **sheep**.

LD_{50}, **rat**: 200 mg/kg; **cattle**: 700 mg/kg; **sheep**: 1 g/kg.

SODIUM NITRITE

Use: Corrosion and rust preventative in cold sterilization solutions. A preservative and color fixative in meat or fish products (not to exceed 20 ppm or 0.002%).

Untoward effects: Under acidic conditions, it can react with ***amines*** yielding carcinogenic ***nitrosamines***. Can oxidize hemoglobin forming ***methemoglobin***. Nausea, vomiting, dizziness, syncope, urticaria, diarrhea, weakness, headache, methemoglobinemia, cyanosis, hypotension, and coma reported. Recurrent attacks of muscle pain in a **person** drinking an Australian ***beer*** containing it. Mislabelling store repackaged material as ***monosodium glutamate*** sent 19 **people** to the hospital. Adulteration of ***fish*** adversely affected > 40 **people**, including one death. A 36-year-old **male** ingested two tablets leading to cyanosis, perspiration, and distress requiring hospitalization and treatment. Contamination of food by it in cooling fluids led to severe hypotension, methemoglobinemia, cyanosis, vomiting, and two fatalities. A massive recall of a seasoned salt occurred as it contained a 6.25% level. Causes a significant anemia in glucose-6-phosphate dehydrogenase-deficient **people**. Mistakenly added to a ***sugar*** bowl and each of three **children** (2-year-old female and 4-year-old twin boys) ingested ~1 teaspoonful in ***tea*** leading to methemoglobinemia. **They** recovered after IV ***methylene blue*** – 2 mg/kg.

TD_{LO}, **human**: 14 mg/kg.

Slow continuous infusion of 2.2 or 4.4% will induce cardiac arrest in **cattle**. Some will abort. Oral LD in **dogs**: 60–100 mg/kg; SC, 40–60 mg/kg. Ingestion by **guinea pigs** leads to anorexia, decreased weight gain, decreased hemoglobin concentration; fetal death from hypoxia. Increases blood ***glucose*** in **horses** and **rabbits**. Developmental toxicity in **rats**.

LD_{50}, **rat**: 85 mg/kg; **mouse**: 214 mg/kg; **dog**: 330 mg/kg; **cattle**: 160 mg/kg; **cat**: 1.5 g/kg. IV, LD_{50}, **rat**: 65 mg/kg; IV, LD_{LO}, 15 mg/kg; **rabbit**: 80 mg/kg.

SODIUM NITROPRUSSIDE

= *Nipride* = *Nitropress* = *Sodium Nitroferricyanide*

***Antihypertensive.* For IV infusion. Causes rapid direct peripheral vascular relaxation.**

Untoward effects: IV use has caused nausea, vomiting, abdominal pain, retrosternal pain, restlessness, agitation, anorexia, apprehension, headache, flushing, nasal stuffiness, tinnitus, metabolic acidosis, dizziness, disorientation, perspiration, palpitations, muscle weakness, twitching, and methemoglobinemia. Reversible hypothyroidism after prolonged use. Rarely, rashes. Cyanosis of nails and perioral area unrelated to hypoxia in 53-year-old **male**. Severe lactic acidosis in 66-year-old **female** and 52-year-old **female** with severe hypertension. Severe hypotension, confusion, and delirium can occur with accidental overdosage or in the **elderly**. Severe oliguric prerenal azotemia in 65-year-old **male** when infusion dosage increased from 0.5 to 2 µg/kg/min for cardiac afterload reduction. A 6-year-old **male** with blood pressure 200/150 and nephrotic syndrome was treated with it until vomiting and hallucinations started when **his *thiocyanate*** plasma level reached 11.5 µg/dl. It is converted in the body to ***cyanides***, which are metabolized in the liver by rhodanase into ***thiocyanate***, which is slowly excreted in the urine. A potential risk of ***cyanide*** poisoning occurs with its use. A surgical **patient** died after 400 mg IV (10 mg/kg) to induce hypotension. The authors concluded 7 mg/kg was lethal and 3.5 mg/kg was safe. **Human** and **dog** studies have since suggested early toxicity can be recognized at doses > 1.5 mg/kg.

Neurotoxicity reported in **mouse** cerebral cortex.

IV, LD_{LO}, **human**: 10 mg/kg; **dog** and **cat**: 1 mg/kg; LD_{LO}, oral, **rat**: 20 mg/kg; **rabbit**: 40 mg/kg.

SODIUM OXYBATE

= *Gamma OH* = *Sodium γ-Hydroxybutyrate* = *Somatomax PM* = *WY 3478*

***Anesthetic.* Primarily for IV use.**

Untoward effects: Transient apnea. After awakening, nausea, vomiting, dizziness, restlessness, mental disturbances, twitching, flushing, and increased blood pressure. A 30-year-old **female**, after ingestion of 25 g/day/2 years, abruptly decreased **her** dose to 10 g/day and was hospitalized with withdrawal symptoms including tremors, anxiety, and insomnia.

IP, LD_{50}, **rat**: 1.65 g/kg.

SODIUM PERBORATE

= *Dexol*

Untoward effects: Due to its alkalinity, has caused inflammation and chemical burns of the mouth's mucous membranes when used as a dentifrice or mouthwash. Liberates ***hydrogen peroxide*** when wet. Eczematogenic when used in ***soap***.

LD_{LO}, **human**: 50 mg/kg.

SODIUM PEROXIDE

= *Solozone*

Untoward effects: A powerful oxidizer. Mixtures with organic materials can easily explode. Friction or a small amount of water can ignite it.

SODIUM PERSULFATE

Bleaching or oxidizing agent.

Untoward effects: Highly irritating to skin and mucous membranes.

IP, LD_{50}, **mouse**: 226 mg/kg; IV, LD_{LO}, **rabbit**: 178 mg/kg.

SODIUM PHOSPHATE, DIBASIC

= *Disodium Hydrogen Phosphate*

***Laxative, pH buffer, phosphorus source.* Contains ~22% *phosphorus.* It is the same as *Sodium Phosphate, U.S.P.* A 1% solution has a pH of 9.1.**

Untoward effects: Diarrhea, nausea, epigastric pain, phosphaturia, calculi (if urine pH > 7.5), hypocalcemia, and convulsions. Can exacerbate symptoms of hypermagnesemia and then cause respiratory failure due to peripheral neuromuscular blockade. Fatal hypocalcemia and coma due to overdosing with it in a laxative preparation. It has been especially dangerous in **infants** and **children**. In the summer of 1955, 12,038 Asian **infants** suffered from ***arsenic*** (q.v.) poisoning and 128 died. A factory added sodium phosphate, dibasic into a ***milk*** powder to improve its solubility. They failed to test its purity. It contained ***arsenic***.

Use of hypertonic solution as an enema leads to hypernatremia, hyperphosphatemia, hypocalcemia, hyperglycemia, depression, ataxia, vomiting, hemorrhagic diarrhea, tachycardia, hypothermia, pallor, and stupor, usually within 30 min in treated **cats**. Has been fatal. Use of 1% in feed for **sheep** increases urinary calculi.

IP, SC, and IM, LD_{LO}, **rat**: 1 g/kg; IV, LD_{LO}, **dog**: 298 mg/kg; **rabbit**: 1075 mg/kg.

SODIUM PHOSPHATE, MONOBASIC

= *Monosodium Phosphate* = *Sodium Acid Phosphate* = *Sodium Biphosphate* = *Sodium Dihydrogen Phosphate*

Urinary acidifier, laxative, pH buffer. **The pH of a 1% solution is 4.6. Contains ~22% *phosphorus*.**

***Untoward effects*:** Tetany, convulsions, hypernatremia, and dehydration have followed repeat enemas in **infants** and young **children**. Hypertonic solutions orally in **infants** lead to coma, diaphoresis, fever, hypocalcemia, hyperphosphatemia, and tetany.

Adults have also reported epigastric pain, nausea, and diarrhea.

In **animals**, excessive use may cause acidosis and laxative action when not desired.

SODIUM PHOSPHATE, TRIBASIC

= *Oakite* = *Trisodium σ-Phosphate* = *TSP*

Detergent, water softener.

***Untoward effects*:** Very alkaline. Eye and skin irritant.

Moderate irritation to skin of **rabbits** and **guinea pigs** of 500 mg/kg/24 h.

SODIUM POLYSTYRENE SULFONATE

= *Kayexalate* = *Resonium*

Cation exchange resin. **Each gram has an exchange capacity of 3.1 mEq *potassium*.**

Untoward effects*:** Constipation, anorexia, nausea, vomiting, hypokalemia, and hypocalcemia. Excessive intake or prolonged retention of it in an enema leads to hypokalemia, weakness, and even paralysis. Aspiration of particles leads to acute bronchitis and pneumonitis. Colonic necrosis followed 50 g in 200 ml of 20% ***sorbitol solution as an enema. Oral administration in ***orange juice*** to treat hyperkalemia was ineffective, as it picked up the ***potassium*** from the ***orange juice***. Systemic alkalosis reported if given with ***magnesium-*** and ***calcium-containing antacids***.

SODIUM SALICYLATE

= *Alysine* = *Brocacef* = *Enterosalicyl* = *Enterosalil* = *Entrosalyl* = *Idocyl* = *Magsalyl*

Analgesic, antipyretic, antirheumatic, sclerotic.

***Untoward effects*:** Tinnitus, acid–base imbalance, rashes, gastrointestinal bleeding, retinal hemorrhage and temporary blindness, hypoprothrombinemia, and, rarely, eosinophilia, leukopenia, and thrombocytopenia; but mostly gastrointestinal irritation reported. Traces are found in breast milk. Suspect teratogen in 35-year-old **female** with frequent ingestion of unknown quantities.

Drug interactions*:** It can block the uricosuric effect of ***probenecid and may potentiate ***chlorpropamide*** and ***tolbutamide*** action. May suppress the uricosuric effect of ***sulfinpyrazone***. Usually incompatible with ***acids***, ***alkaloids***, free ***iodine***, and ***iron salts***.

Minimum lethal dose ~25 g. LD_{LO}, **human**: 700 mg/kg.

Rat and **mouse** trials indicate teratogenic effects during pregnancy (at levels probably greater than those normally used).

LD_{50}, **rat**: 1.6 g/kg; **mouse**: 900 mg/kg; **rabbit**: 1.7 g/kg; LD_{LO}, **dog**: 450 mg/kg.

SODIUM SELENITE

Untoward effects*:** More toxic than the *selenate* form. Contains ~46% ***selenium. In South Dakota, fed to **cattle** as 15 ppm ***selenium*** leading to in 1/5 failure to gain weight, sore feet, and deformed hooves. Also in South Dakota, **rats** fed 5–15 ppm as ***selenium*** developed anemia due to hemolysis and increased incidence of caries in the offspring.

LD_{50}, **rat** and **mouse**: 7 mg/kg; **rabbit**: 2.25 mg/kg; **guinea pig**: 5 mg/kg; LD_{LO}, **dog**: 4 mg/kg.

SODIUM SILICATE

= *Water Glass*

***Untoward effects*:** Strongly alkaline. Caustic and irritating to skin and mucous membranes. Contact urticaria in 57-year-old **male**. Ingestion can cause nausea, vomiting, diarrhea, and gastrointestinal distention.

LD_{LO}, **human**: 500 mg/kg.

Fed to **lambs**, it increases incidence of urinary calculi. Ingestion by **dogs** leads to emesis.

LD_{LO}, **dog** and **pig**: 250 mg/kg.

SODIUM SILICOFLUORIDE

= *Salufer* = *Sodium Fluosilicate* = *Sodium Hexafluorosilicate*

Insecticide, pediculicide, rodenticide, and ascaricide. ~60% fluorine.

Use*:** Can be fatal if ingested. Toxicity as for ***sodium fluoride (q.v.). Avoid inhalation of dusts or contamination of feedstuffs. LD of ~5 mg/kg. After ingestion or topical exposure, symptoms are delayed for about 2 h and include nausea, vomiting, apprehension, tremors, twitching, arrhythmias, convulsions, coma, and death after cardiac failure. Fatal to a **man** who ingested $\frac{1}{2}$ teaspoonful.

Feed off-loaded from a railway car that had previously contained a shipment of it caused the deaths of 15 **cattle** that had eaten it; 20 tons of contaminated feed remained. Fatal within 7 h when 400 mg/kg was given to a **ewe lamb**. Field cases also reported in **horses** and **pigeons**. Extremely irritant topically on **rabbit** skin.

Acute LD for **rabbits**, **goats**, and **dogs**: 150–200 mg/kg. LD_{50}, **rat**: 125 mg/kg.

SODIUM SULFATE

= *Glauber's Salt* = *Salt Cake*

***Cathartic.* U.S. annual demand is ~1 million tons/year. Up to 6% in some *lards* as an antimicrobial.**

Untoward effects*:** Well water that contained excessive quantities (> 60 mg/100 ml of ***sulfates) led to diarrhea in young **children**, but not in older **children** or **adults**. Hypernatremia in **patients** after IV treatment for hypercalcemia. Adverse effects are usually from the ***sodium*** (q.v.).

High ambient temperatures increased water consumption in a **cattle** feedlot where well waters contained ***sulfate*** levels of 3500 and 5203 ppm leading to deaths due to polioencephalomalacia. **Cattle** drinking water containing 5000 ppm had a 450% increase in ***methemoglobin*** and increased diuresis. **Pigs** drinking water with ~27 g/gallon (7500 ppm) had diarrhea. Higher feed levels in **pigs** caused muscle tremors, convulsions, and deaths.

IV, LD_{LO}, **mouse**: 1.22 g/kg; **rabbit**: 4.47 g/kg.

SODIUM SULFIDE

= *Sodium Sulfuret*

Depilatory, caustic, antioxidant.

***Untoward effects*:** Contact urticaria and caustic burns. Caustic effects from inhalation or swallowing of dusts. Can explode on rapid heating or percussion.

LD_{LO}, **human**: 50 mg/kg.

SODIUM SULFITE

***Antioxidant, antiseptic.* U.S. annual demand > 100,000 tons.**

Use*:** In pulp and ***paper manufacturing; water treatment, photography, oil recovery, textile bleaching, food preservation, and chemical manufacturing.

***Untoward effects*:** A sulfiting agent in foods and drinks leading to flushing, chest tightness, weakness, difficulty in swallowing, dyspnea, wheezing, and unconsciousness. An especially serious problem for **asthmatics**.

LD_{LO}, **rabbit**: 2825 mg/kg; IV, LD_{50}, **rat**: 115 mg/kg; **mouse**: 130 mg/kg; **rabbit**: 65 mg/kg; **hamster**: 95 mg/kg.

SODIUM TELLURATE

***Untoward effects*:** Large (0.5–1 g) oral dose to **dogs** retards gastric digestion and induces violent vomiting, anorexia, and somnolence.

IP, LD_{LO}, **rat**: 37 mg/kg. LD_{50}, **rat**: 385 mg/kg; **mouse**: 165 mg/kg; **rabbit**: 104 mg/kg.

SODIUM TETRADECYL SULFATE

= *H1E1Z9* = *Sotradecol* = *Tergitol 4* = *Trombavar* = *Trombovar*

***Sclerosing agent.* IV use.**

***Untoward effects*:** Extravasation has caused sloughing. High dosage IV leads to pain, ulceration, and hemolysis; occasionally thrombosis, pulmonary embolism, dizziness, nausea, vomiting, hives, headache, slight discoloration of treated area, dyspnea, syncope, and anaphylactic shock. Rarely, death. IV, LD_{50}, **mouse**: 56 mg/kg.

SODIUM THIOSULFATE

= *Ametox* = *Antichlor* = *Hypo* = *Sodium Hyposulfite* = *Sodothiol* = *Sulfothiorine*

Reducing agent, cyanide antidote.* Often added to *table salt.

Untoward effects*:** Large oral doses are cathartic. Contact dermatitis. Used in the illicit manufacture of ***amphetamines.

LD_{50}, **human**: 500 mg/kg.

In **dogs**, 3 g/kg IV led to rapid development of metabolic acidosis, hypoxemia, hypernatremia, electrocardiogram and blood pressure changes, with death in 2/5. Physically incompatible with IV ***calcium salts*** causing precipitation of ***sulfur***.

SODIUM TRIPOLYPHOSPHATE

= *STPP*

***Use*:** In detergents, foods, and moisture binding in meats, fish, and poultry. U.S. annual use is ~350,000 tons.

***Untoward effects*:** pH of 1% solution < 10. Vesiculation of abraded sites on 1/6 **people**, but only slight reactions on **rabbit** and **guinea pig** skin.

LD_{LO}, **human**: 500 mg/kg.

LD_{50}, **rat**: 6.5 g/kg; **mouse**: 3.2 g/kg.

SODIUM URATE

Untoward effects: Intra-articular injections of the crystals in **humans**, **dogs**, and **pigeons** lead to acute arthritis.

SODIUM VALPROATE

= *Convulex* = *Depakene* = *Depakin* = *Depakine* = *Di-n-propylacetate* = *Epilim* = *Ergenyl* = *Eurekene* = *Labazene* = *Leptilan* = *Orfiril* = *Sodium 2-n-Propyl Pentanoate* = *Valcote*

Anticonvulsant, antiepileptic. **Oral and IV.**

Untoward effects: ***Blood***: Thrombocytopenia, red blood cell aplasia, decreased erythrocyte sedimentation rate, decreased ***fibrinogen***, decreased serum ***zinc***, and symptomatic hyperammonemia.

Central nervous system: Drowsiness, sedation, ataxia, headache, nystagmus, diplopia, "spots before the eyes", muscle weakness, tremors, dizziness, coma, depression, psychosis, aggression, hyperactivity, and behavioral changes.

Fetus and neonate: Teratogenic effects, especially spina bifida, when taken during the first trimester. **Infantile** liver failure after 300 or 500 mg/day during pregnancy. Fatal afibrinogenemia in a **neonate** whose **mother** took 200 mg tid during pregnancy. Post-mortem indicated widespread hemorrhages into scalp, neck muscles, and internal organs. Excreted in breast milk of treated **mothers**.

Gastrointestinal: Nausea, vomiting, and indigestion are common and transient during the beginning of therapy. Constipation, cramping, and diarrhea also reported.

Liver and pancreas: Malaise, weakness, lethargy, facial edema, anorexia, weight decrease, vomiting, jaundice, and easy bruising are often prodromal symptoms of impending hepatic failure and resultant death. Non-fatal hepatic dysfunction is estimated in 3–44% of **patients** and, apparently, not dose-related. Acute pancreatitis has also been associated with its use. The drug's chemical structure is similar to ***hypoglycin*** (q.v.) which, in Jamaica, causes fatty degeneration of the liver, encephalopathy, and vomiting.

Miscellaneous: Exacerbation of intermittent acute porphyria, nocturnal enuresis in **children**, weight increase, ankle edema, hyponatremia, increase in urinary ***D-glucaric acid*** excretion, and hypothyroidism. An **infant** given 12.5–25 mg/kg/day for treatment of generalized tonic–clonic seizures developed noticeable gynecomastia and galactorrhea after 3 days.

Poisoning: Accidental in 23-month-old patient after a 5 g oral dose. A 20-month-old **male** who ingestion 15 g died from cardiorespiratory failure 46½ h later. Serum levels of > 130 mg/l or 1000 μmol/l are considered toxic.

Skin: Rashes, cutaneous vasculitis, hair loss and several reports of hair-curling, and ***zinc*** deficiency.

Drug interactions: Its sensitivity to enzyme induction and high affinity for serum proteins induces many interactions. ***Amitriptyline*** causes a slight but significant increase in its half-life. It decreases the clearance and increases the half-life of ***antipyrine***. ***Aspirin***, especially in pediatric **patients**, decreases its metabolism. ***Chlorpromazine*** induces a 15% increase in its trough plasma levels. It has caused absence status with ***clonazepam***. It has decreased ***corticotropin*** secretion in **patients** with Nelson's syndrome. Inhibits ***diazepam*** and ***ethosuximide*** metabolism. It increases the risk of ***isoniazid*** toxicity. Dramatically increases ***lamotrigine*** half-life. Exacerbation of acute intermittent porphyria in 15-year-old **female** given 10 mg ***metoclopramide*** IM. It will displace ***midazolam*** and ***tolbutamide*** from their plasma binding sites. ***Naproxen***, ***phenylbutazone***, and ***salicylic acid*** displace it from its plasma binding sites. Interactions with other ***anticonvulsants*** are significant. It increases ***phenobarbital*** serum levels and also requires decrease in ***primidone*** dosage. Displaces ***phenytoin*** and ***carbamazepine*** from their plasma binding sites and inhibits their hepatic metabolism. In HIV **patients**, clearance of ***zidovudine*** was decreased by it.

High dosage decreases spermatogenesis and causes testicular atrophy in **rats** and **dogs**. In male **rats**, an increase in SC fibrosarcomas and an increase of benign pulmonary adenomas in male **mice** with high dosage in long-term studies.

LD_{50}, **mouse**: 1.7 g/kg; TD_{LO}, **rat** (6–15 days pregnant): 1.5 g/kg; **mouse** (6–15 days pregnant): 6 g/kg; and **rabbit** (6–18 days pregnant): 4 g/kg.

SOLANDRA sp.

Untoward effects: ***S. nitida*** = ***Chalice Vine*** = ***Trumpet Flower***, a native of Mexico with 9" long white to yellow fragrant flowers, yet a common winter-blooming Florida vine from fall to late spring. Its round 2½" wide fruit is non-toxic, but the sap, foliage, and flowers contain the poison, ***salandrine***, with anticholinergic action similar to ***atropine's*** (q.v.). In the evening, the odor of the flowers leads to nausea, headache, and dizziness. Accidental or deliberate ingestion of the flowers (as a ***tea***) leads to violent intoxication, mydriasis, delusions, and numbness of hands and feet.

Can be poisonous to **pets** that ingest it.

SOLANINE

= *Solatunine*

A toxic alkaloid found in *nightshades*, green or sunburned *potatoes*, *wild celery*, *European bittersweet*, *tomato* plants, and the *Jerusalem cherry*.

Untoward effects: Contains a ***cholinesterase inhibitor*** interfering with transmission of nerve impulses. Vomiting, diarrhea, headache, abdominal pains, prostration, weakness, mental confusion, clammy skin, dyspnea; cardiac depression and other anticholinergic symptoms from eating the eyes, sprouts, berries, green discolored skin, or sun-exposed Irish ***potatoes*** (q.v.). Once a common problem when ***potatoes*** were a major part of **one's** diet.

Human toxic dose, ~2.8 mg/kg; LD_{LO}, 5 mg/kg.

Cattle that eat green or sprouted ***potatoes*** or ***potato*** peelings can become jaundiced and die. **Pigs**, **horses**, and **poultry** can also become poisoned. Death to young **rats** whose **dams** were fed ***potato*** sprouts at a 10% level in their diet. **Dogs** eating half-cooked or raw ***potatoes*** have developed a ***strychnine***-like syndrome. Blood dyscrasias in **sheep** ingesting 225 mg/kg.

LD_{50}, **rat**: 590 mg/kg.

SOLANUM sp.

Over 1000 species have been described.

Untoward effects: The ***nightshades*** are an important group. All parts are toxic, especially the unripe berries, causing intense digestive upsets, nervous symptoms, and fatalities. In many species, ***solanine*** (q.v.) is the main intoxicant.

S. americanum and ***S. nigrum*** are called ***Black Nightshade*** and ***Deadly Nightshade*** (a term also used for ***Atropa belladonna***, which causes classic anticholinergic effects) leading to salivation, dyspnea, and paralysis. ***S. americanum*** also contains ***atropine*** and related alkaloids leading to fever, tachycardia, and mydriasis; and hot, flushed, dry skin.

S. nigrum is also called ***Pepper Bush*** in the Bahamas, ***Mako*** in India, ***Yerba Mora*** in Venezuela, ***Woody Nightshade*** in Denmark, and ***Stubble Berries*** in the Dakotas. It is toxic to **cattle**, **sheep**, **calves**, **pigs**, **goats**, **chickens**, **ducks**, and **geese** leading to thirst, diarrhea, bloating, nausea, abdominal pain, constipation, anorexia, weakness, narcosis, incoordination, convulsions, and death from respiratory paralysis. Ripe berries are relatively harmless and have even been made into jam or used in pies. Poison is destroyed by cooking. Garden strains grown for their edible fruits are sold as ***Garden Huckleberry*** or ***Wonder Berry*** (***S. nigrum*** *var.* ***guineense***, now called ***S. melanocerasum***). Young **children** and **dogs** often ingest the unripe green berries and develop classical gastrointestinal and nervous symptoms. It is also a ***nitrate***-accumulator.

Ingestion of ***S. auriculatum*** has killed **cattle** in Australia and is used as an insecticide in Africa.

In Australia and New Zealand, ***S. aviculare*** = ***Australian Kangaroo Apple*** = ***Bullibulli*** = ***Poroporo*** contains ***purapurine*** or ***solasonine***, and ***solamargine*** in the leaves and unripe fruit. The green unripe fruit is poisonous, but the ripe fruit has been used in making jams. In Russia and Hungary, it was grown to convert the alkaloids into ***cortisone*** and other steroid drugs.

S. carolinense in the U.S. and Canada = ***Apple of Sodom*** = ***Bull Thistle*** = ***Horse Nettle*** = ***Purple Horse Nettle*** = ***Sand Brier*** = ***Stickerweed*** contains ***solanine*** (mostly in the roots and berries). **Children** have eaten the poisonous unripe fruit causing stomach pain, vomiting, diarrhea, headache, hypothermia, mydriasis, paresthesia, and cardiorespiratory impairment. Life-threatening shock in severe cases. Has caused stupor and spinal convulsions in **man**. It is generally unpalatable unless browsed with other forages and is poisonous if in hay. **Cattle** often like the berries. Dried berries cling to the plants over winter and are potent enough to kill **cattle** the following spring. Eating large quantities in a short time leads to slobbering, vomiting, bloating, sleepiness, paralysis, and death. Small quantities daily lead to unthriftiness, rough hair coat, weight decrease, marked constipation, and anorexia. Hunting **dogs** are often affected by the prickles on the stems and leaves developing salivation, gagging, pawing at their mouths, occasional vomiting, occasional alkaline urine, and often walking with difficulty.

S. dimidiatum = ***Potato Weed*** = ***Western Horsenettle*** of Texas causes crazy-**cow** syndrome from ingestion of ripe fruits leading to permanent incoordination, especially when excited.

S. dulcamara = ***Climbing Nightshade*** = ***Dulcamara*** = ***European Bittersweet*** = ***Scarlet Berry*** = ***Woody Nightshade*** is also known as the ***Marriage Vine*** or ***Matrimony Vine***. Its unripe berries and leaves contain ***solanine*** (q.v.). Causes vomiting, diarrhea, abdominal pain, burning in the mouth and throat, drowsiness, tremors, weakness, dyspnea, dizziness, mydriasis, salivation, and paralysis, usually in **children** and occasionally in **animals**. Also contains a small quantity of an ***atropine***-like alkaloid and the ***saponin***, ***dulcamarin***.

S. eleagnifolium = ***Silver-Leafed Nightshade*** is found in Missouri to Texas, and California south into Mexico. The leaves and fruit (green or ripe) contain ***solanine*** leading to heavy death losses of **cattle** in Texas. It is less toxic to **sheep** and **goats** tolerate ten times the toxic dose for **cattle** (900 mg/kg).

S. escuriale of Australia is poisonous and a cause of "humpy back" in **sheep** (a form of calcinosis).

Digestion of ***S. glaucophyllum*** by **cattle** in Argentina, Brazil, France, and the U.S. leads to calcinosis or "enteque seco" with arterial and lung calcifications.

S. incanum = ***Bitter Apple*** = ***Guatan kura*** (Hausa) = ***Igba AJA*** (Yoruba) = ***UM DULUKWA*** (Ndebele) =

Afufa OYIBO (Igbo) = ***Engule*** (Ethiopia) = ***Airsum*** (Yemen) accumulates ***nitrosamines*** and its ***nitrate*** levels are nearly toxic. The unripe berries contain ***solanine***, a gastrointestinal irritant, with nausea, colic, and diarrhea and, after absorption, leads to hemolysis, tachycardia, central nervous system stimulation followed by depression, and mental confusion. Symptoms in poisoned **sheep** include salivation, diarrhea, colic, bloating, tachycardia, tachypnea, and paralysis. Occasionally vesicular skin lesions and conjunctivitis.

S. malacoxylon = ***Duraznillo Blanco*** = ***South American Eggplant*** may be identical to ***S. glaucophyllum*** and is common in Argentina and Brazil leading to calcification of soft tissues, especially the heart, aorta, kidneys, and lungs due to a ***1,25-dihydroxycholecalciferol***-glycoside. **Animals** become stiff and thin and have hypercalcemia and hyperphosphatemia. The syndrome is called "enteque seco" in Argentina and "espichamento" in Brazil, and affects 10–40% of grazing **cattle**. Also affects **sheep**, **pigs**, **hens**, and **guinea pigs**.

S. panduriforme = ***Ntutrwa*** (Ndau) of Rhodesia readily kills **sheep** after ingestion of the ripe fruit with symptoms similar to ***S. incanum***.

S. pseudocapsicum = ***Jerusalem Cherry*** = ***Natal Cherry*** = ***Winter Cherry*** has attractive scarlet–orange berries poisonous to **children** due to its ***solanine*** and ***solanocapsine*** content leading to cardiac depression, gastrointestinal irritation, nervousness, anticholinergic effects, shock, and unconsciousness. The leaves and unripe fruit are very toxic.

S. rostratum = ***Beaked Nightshade*** = ***Buffalo Bur***, common in western U.S. and Mexico, causes gastrointestinal irritation from grazing on its prickles and suspected as a cause of perirenal edema in **swine**.

S. sodomaeum is extremely toxic to **chicks** and **Japanese quail** when the dried powdered ripe fruit is fed at the 2% level; decreases growth and feed efficiency at the 0.5% level.

S. stellatum spines produce blisters full of lymph and then pus in **people** of Peru.

S. torvum ingestion by **cattle** in Papua, New Guinea leads to calcinosis.

Eggplant (q.v.), ***potatoes*** (q.v.), and ***tomatoes*** (q.v.) are also ***Solanum*** species. ***Tomato*** and ***potato*** vines contain high levels of ***solanine***. ***Apples*** and ***sugar beets*** also contain ***solanine***.

SOLASULFONE

= *Cimedone* = *Novotrone* = *RP-3668* = *Solapsone* = *Sulphetrone* = *Sulphonazine*

Leprostatic, antibacterial.

Untoward effects: Nephritis, acute parenchymatous hepatitis, hypochromic anemia, erythema nodosum, urticaria, and iridocyclitis reported.

SOLDER

Untoward effects: **Solderers** are frequently exposed to ***hydrazine*** (q.v.) in fluxes. In the U.S., ***lead***-based solder for potable water systems was outlawed in 1988. Its use in putting together illegal stills led to gout and ***lead*** poisoning in heavy **drinkers** of ***moonshine whiskey***. ***Yeast*** contaminants in the sour mash yield ***acetic acid*** when heated that leaches out the ***lead*** to produce ***lead acetate***. Chronic bronchitis and emphysema from heating ***silver*** solder, a ***silver***-brazing alloy yielding ***cadmium oxide*** (q.v.) fumes. Deaths have also been reported from the fumes. Contact dermatitis reported in **solderers**. Inhalation of fumes from ***Kyrol Aluminum Soldering Flux*** have caused instant asthma.

SOMAN

= *GD*

Chemical warfare agent, acetylcholinesterase inhibitor. **A *sarin* variant.**

Untoward effects: By inhibiting the breakdown of ***acetylcholine***, allowing it to accumulate at peripheral and central nervous system cholinergic receptors, leads to muscle fasciculations, twitching, weakness, salivation; increased nose, eye, airway, and intestinal secretions; weakness, apnea, convulsions, unconsciousness, and paralysis. Bradycardia or tachycardia occurs. Symptoms vary with the amount of liquid skin contact or vapor exposure, and can include miosis, ocular pain, conjunctivitis, blurred vision, sweating, bronchoconstriction, and often delayed nausea, vomiting, abdominal cramps, and uncontrolled defecation. Death from asphyxia can occur in minutes. The degree of permanent brain injury, psychiatric sequelae, and incapacitation in **human** survivors is not well known. A 33-year-old **male** was working with small amounts of a 25% solution when some splashed into and around **his** mouth. **He** washed and rinsed **his** mouth with water. **He** was asymptomatic for about 10 min, and then **he** felt "the world was caving in on me", and collapsed. **He** was comatose, mildly cyanotic with miosis, markedly injected conjunctiva, marked oral and nasal secretions, some trismus and nuchal rigidity, muscle fasciculations, hyperactive deep tendon reflexes, tachycardia, and bronchoconstriction. Delayed nausea, vomiting, atrial fibrillation, torticollis, and athetoid movements.

Inhalation LC_{LO}, **human**: 70 mg/m^3; LD_{LO}, skin, 18 mg/kg.

Rat and **mouse** trials indicated that there are four different isomers, each with different toxicity. LD_{50} in **rats** ranged from 38 to 156 μg/kg. Acute neuropathy in poisoned **rats**. Acute SC toxicity in the **dog** is > **guinea pig** > **rat** > **mouse**. The toxicity in **dogs** was 12.6 times that of **mice**. The LD_{50} in eight strains of **mice** varied from 83 to 161 μg/kg. Induced seizures in **rats** led to a more than four times increase in brain ***glucose*** use. Severe delayed neuropathy in **hens**.

LD_{50}, **rat**: 90 μg/kg; **guinea pig**: 38.1 μg/kg; **hen**: 10 μg/kg.

SOMATOSTATIN

= *Aminopan* = *GH-RIF* = *Growth Hormone Release Inhibiting Hormone* = *Modustatin* = *Panhibin* = *Somatofalk* = *Somatotropin Release Inhibiting Factor* = *SRIF* = *Stilamin*

Untoward effects: Secreted by the hypothalamus, inhibiting the pituitary's ***growth hormone***, ***insulin***, and ***glucagon***, as well as ***gastrin*** and ***secretin*** production. Marked water retention and hyponatremia, venospasm, hyperglycemia, abdominal pain, dizziness, diarrhea, and decreased platelet aggregation reported in some **patients** after IV use.

Immunization against it has increased growth rate of **calves** and **lambs**. Thrombocytopenia in **baboons** after IV treatment with it.

SOMATREM

= *Genotropin* = *Methionyl Human Growth Hormone* = *Protropin*

***Growth Hormone*. IM or SC use.**

Untoward effects: As a protein, antibodies have developed against it in some **patients**. Potential reactions include infrequent mild peripheral edema, carpal tunnel syndrome, pancreatitis, gynecomastia, and increased growth of nevi.

SOOT

Untoward effects: In 1775, Sir Percival Potts, a London physician, claimed that chronic exposure to it, as in the case of **chimney sweeps**, caused the development of scrotal cancer. Since then, ***coal*** soot has been incriminated as causing larynx and lung cancers. In cases of arson, soot can be examined to help determine the nature of the accelerant. ***Polychlorinated biphenyls*** remaining after fires and explosions require special handling.

Rat and **rabbit** trials indicate it is a bronchial irritant with normal precipitation on and retention by bronchial epithelium.

SOPHORA sp.

Untoward effects: Often called ***Mountain Laurel***, ***Coral Bean***, or ***Colorines***, which can also be ***Kalmia***, ***Erythrina***, and ***Erythrina***, respectively. Its fruit is a type of "***rosary beads***" and is poisonous. Small **children** are also at risk from choking on the beads.

S. secundiflora = ***Frijolillo*** = ***Frijolito*** = ***Mescal Beans*** = ***Red Beans*** = ***Red-hots*** of southwestern U.S. and adjacent Mexico contain a toxic ***pyridine***, ***cytisine*** (q.v.) leading to after ingestion, a symptomless period followed by excessive salivation, thirst, nausea, burning of the mouth and esophagus, severe abdominal pain, tympanism; violent, persistent, and occasionally hemorrhagic vomiting; tachycardia, followed by bradycardia and arrhythmia, dyspnea, cold and clammy skin, headache, ataxia, vertigo, somnolence, muscle twitches and cramps, aphasia, visual problems, delirium, hallucinations, and unconsciousness with deep sleep sometimes lasting for 2–3 days. Respiratory paralysis, oliguria, anuria, and uremia precede death. The seeds were probably used as psychotropic agents thousands of years B.C. Use of the seeds in Mexican good luck charms (buena suerte) that can be bought in Mexico and many U.S. border states, can be a disaster waiting to happen. A single seed can cause death.

Cattle are very suceptible to poisoning by ingesting its leaves or seeds. **Sheep** and **goats** are more tolerant. The seeds are very poisonous if crushed, but due to their hardness, they are usually swallowed whole and are unaffected by **ruminants'** digestive tract and are passed in the feces unchanged. **Animals** may have a stiffened gait, especially when excited.

S. sericea = ***Silky Sophora*** in the western U.S. has poisoned **cattle**.

S. tomentosa = ***Mbaazi-mwitu*** or ***Mnuka-vundo*** (Swahili) in East Africa is used as a piscicide and also contains ***cytisine***.

Seeds, LD_{50}, **mouse**: 1.4 g/kg.

Seeds, LD, **horse**: 0.5 g/kg; **dog**: 6 g/kg.

Cytisine, LD, **cat** and **dog**: 3–4 mg/kg.

SORBIC ACID

= *Sorbistat*

***Mold and yeast inhibitor*. In foods and *cheeses*. Used to prevent secondary *yeast* fermentation in *wines* after bottling and as a defoaming agent. Metabolizes to *carbon dioxide* and *water*.**

Untoward effects: Contact skin irritant and allergen leading to urticaria and angioneurotic edema. A **baker** had an eczematous contact dermatitis of the hand from hypersensitivity to it. Another **baker** developed an immediate reaction to it, with redness and itching of hands when **he** handled ***flour*** containing it. Weakens ***nylon*** containers and syringes. An allergic reaction to its presence in a contact lens solution reported.

Can cause injection site sarcomas in **rats**. No toxic effects from 2% level in **cat** diets. Significant **rabbit** eye irritation from the powder.

LD_{50}, **rat**: 7.36 g/kg.

SORBINIL

= *CP 45,634*

Aldose reductase inhibitor, anticataract agent.

Untoward effects: Serious skin rashes. Hypersensitivity reactions from its metabolites. Elimination half-life longer in **male** than **female**, and long elimination half-life in the **elderly**, especially with chronic dosing.

SORBITAN TRIOLEATE

= *Arlacel 85* = *Span 85*

Surfactant.

Untoward effects: IM use in **chicken** pectoral muscle leads to necrosis.

SORBITOL

= *Crystosol* = *D-Glucitol* = *L-Gulitol* = *Resulax* = *Sionon* = *Sorbilande* = *Sorbilax* = *Sorbit* = *Sorbitur* = *Sorbo* = *Sorbostyl*

Annual U.S. demand is for > 500 million lb of 70% solution. *Humectant, osmotic, laxative, diuretic, nutrient, sweetener.*

Untoward effects: Its use in oral medications, dietetic foods and candies, meats, shredded ***coconut***, whips, frappés, baked goods, ***soft drinks***, gums, candies, other foods, and even toothpastes can cause diarrhea, abdominal cramps, and defecation urgency. A 4-year-old **female** chewed 1–2 pieces of sugarless gum several times/week leading to paroxysmal, periumbilical cramping abdominal pain several times/week. Use with ***activated charcoal*** during treatment of poisoning in 2-year-old **child** caused large volume of stool fluid loss. After discontinuance, an inadvertent access to a sorbitol ***sugar***-free gum caused a prompt return of abdominal pain. Conventional gum had no such effect. Fatal hyperosmolar coma after 72 h of continuous peritoneal dialysis with hypertonic 4.25% dialysate in uremic, severely edematous 46-year-old **male** with diabetic nephropathy. IV use in **females** with ketosis can produce lactic acidosis. Angiopathy, neuropathy, retinopathy, cataract, and a high concentration in **diabetic** tissues due to high dietary intake. Osmotherapy reported to cause a case of anaphylactic shock.

SORGHASTRUM NUTANS

= *Indiangrass*

Untoward effects: As with ***sorghum***, it contains ***dhurrin***, the source of ***hydrocyanic acid*** (q.v.) poisoning in **livestock**.

SORGHUM

Untoward effects: Contains ***cyanogens***, especially in the sprouts, drought-stunted or frosted plants. The average fatal dose of ***hydrocyanic acid*** for **man** is 50–60 mg and this can easily be obtained from the sprouts from 100 g of seed and is within the realm of possible consumption by some **people**. Genetically, the sickle-cell anemia trait gives protection versus malaria. There may be few problems from it in Africa, possibly due to the ethnic diet (***sorghum***, ***millet***, ***yams***, and ***cassava***) high in ***cyanates*** (usually over 1000 mg/day versus a U.S. diet of ~25 mg/day).

S. almum = ***Columbus Grass***, a grain sorghum's toxicity reported in **cattle** and **horses** in Argentina, South Africa, Australia, New Guinea, Nigeria, Algeria, and the U.S.

S. halpense = ***Johnson Grass*** poisoning is due to its ***dhurrin*** content, which releases ***hydrocyanic acid*** quickly if **animals** drink water shortly after ingesting the plant causing dyspnea, weakened cardiac function, vertigo, and muscle spasm, particularly in **cattle** and **horses**, and perosis in **poultry**. Found in the Mediterranean areas, Rhodesia, and North America.

S. vulgare var. ***sudanense*** = ***Broomcorn*** = ***Durra*** = ***Fatarita*** = ***Jawar*** = ***Jowar*** = ***Kafir Corn*** = ***Milo*** = ***Sorgho*** = ***Stapf*** = ***Sorghum sudanense*** = ***Sudan grass***, when drought-stunted or frosted, can kill grazing **animals** within minutes with ***hydrocyanic acid***. When plants are short, the acid content is high. **Sheep** have developed photosensitivity grazing it. **Horses** have developed cystitis and urinary incontinence grazing it. **Mares** have had dystocia at term; dead **foals** were born with arthrogryposis or ankylosis of joints. A **mare** developed a posterior ataxia and aborted after grazing on it. Poorly stored grain developed a high mold content and caused infertility in **hogs**. Black tongue induced in **dogs** fed on a diet containing 65%/30–80 days. Striking deterioration in mental condition of pellagra **patients** given 10–30 g of grain/day/6–10 days leading to altered electroencephalogram with excess theta activity, and sometimes delta activity, due to excessive ***leucine*** in **their** diet. Also contains high ***oxalate*** content.

In poisoned **cattle** muscle tremors, spasms, apathy, staggering gait, salivation, mydriasis, and bradypnea occur. Post-mortem reveals icterus in SC tissues, muscles, and organs; and epicardial hemorrhage. Irregular estrus and failure to conceive also reported. Most of the non-toxic bound ***dhurrin*** glycoside in the plant is acted on by another chemical, ***emulsin***, in the plant to release ***hydrocyanic acid***. Cystitis, urinary incontinence, and ataxia in **horses** grazing sorghum. Pellagra-like symptoms in **chicks** eating a sorghum-based diet; increased incidence of bone abnormalities and decreased growth rate in **chicks** fed a high ***tannin*** sorghum. Urolithiasis in **sheep** eating a diet containing 45% sorghum grain. Mass poisoning of **swine** reported from Russia. Decreased rate of weight gain and decreased feed efficiency in **pigs** eating a sorghum diet. Poisoning of **water buffaloes** in Palestine and Syria 30 min after ingesting the leaves covered with secretions from **locusts**. In general, there is initial excitement, tremors, tachypnea, dyspnea, salivation, lacrimation, urination, defecation, and clonic convulsions. Blood is bright cherry-red. Death can occur within 15 min to several hours. Sorghums are also ***nitrate***-accumulators.

SORSACA

= *Annona muricata* = *Corossol* = *Kowosol* = *Sorsaka* = *Soursop*

Untoward effects: The leaves are used in native ***bush teas*** in the Bahamas, Curaco, Cuba, Ecuador, Jamaica, and Trinidad, where it is sedative and sudorific. **Children** have been fatally poisoned by it. Contains some ***arecoline*** (q.v.), ***triacontanol***, and a volatile oil.

Bark extract is a cardiac depressant and respiratory stimulant in **rabbits**. Leaf extract SC in **rats** leads to local fibrosarcomas.

A. cherimola* = *A. purpurea* = *A. senegalensis* = *Custard Apple stem cortex extract and seeds used for piscicides and insecticides. Contains ***apomorphine*** alkaloids.

SOTALOL

= *Beta-Cardone* = *Betapace* = *Darob* = *MJ-1999* = *Sotacor* = *Sotalex*

β-Adrenergic-blocker, antihypertensive, antianginal, antiarrhythmic (Class II and III).

Untoward effects: Fatigue, bradycardia, dyspnea, proarrhythmia, dizziness, and asthenia in 2–4%. Incidence of torsades de pointes (~4%) is increased by high dosage, rapid infusion rate, and hypokalemia in **women**. Supraventricular tachyarrhythmia, prolongation of QT interval, episodes of unconsciousness, proximal myopathy, syncope, mental depression, bradycardia, and skin reactions. Distressing hypnagogic and hypnopompic auditory and visual hallucinations after 600 mg/day/3 days. Self-poisoning in 39-year-old **male** with 4.48 g after ingesting a ***glycrrhiza*** preparation leading to severe ventricular tachyarrhythmias and in a 58-year-old **female** who died after ingesting 14.4 g. High concentrations excreted into breast milk can result in a **neonate** ingesting a pharmacologically active dose. Mean **maternal** plasma concentration was 2.3 μg/ml and **her** breast milk 10.5 μg/ml. IV use of 10 mg in **patients** with chronic obstructive pulmonary disease increases airway resistance. Rare anecdotal reports of thrombocytopenia, eosinophilia, leukopenia, photosensitivity, myalgia, incoordination, and vertigo.

Drug interactions: Profound bradycardia after addition of ***diltiazem*** to treatment with it. Its antihypertensive effect is decreased by previous ***alcohol*** intake. Use with 25 mg ***hydrochlorothiazide*** twice daily leads to severe hypokalemia after 10 weeks. Paradoxical hypertension in some **patients** when used in combination with ***clonidine***. Acidification of urine prolongs its half-life. Food decreases its absorption. Therapeutic serum concentration of 0.5–3 μg/ml; toxic at 7.5–16 μg/ml; comatose or fatal at 40 μg/ml.

Incidence of aortic rupture induced by feeding 0.7% ***β-aminopropionitrile*** (q.v.) in young **turkeys** reduced by feeding 0.2% sotalol. Will block the cardiac effects of ***isoproterenol*** in **dogs** and ***epinephrine***-induced hyperglycemia in **rats**. ***Cimetidine*** decreases its renal clearance ~50% in **rats**.

LD_{50}, **rat**: 3.45 g/kg; **mouse**: 2.6 g/kg.

SOTERENOL

= *MJ 1992*

***Bronchodilator, sympathomimetic.* Used as an aerosoled drug.**

Untoward effects: Dry mouth in 6/10, nausea in 3/10, abdominal cramps in 2/10; increased wheezing, headache, and vertigo in 1/10.

SOUTHERN PRICKLY ASH

= *Zathoxylum clavaherculis*

Untoward effects: The bark, when ingested by **cattle**, is lethal.

SOYBEAN

= *Soja* = *Sahuca Bean*

After a member of the Japanese Court presented Commodore Perry with a soybean plant that he brought back to the Western World, it was called the *Japan Pea*.

Untoward effects: Although soy ***milk*** is a staple in Asia and elsewhere, especially in **infants** allergic to **cow's *milk***, in ***lactose*** intolerant **infants**, and **infants** with galactosemia, untoward effects have occurred. Acrodermatitis enteropathica exacerbated by it in a 5-week-old **infant** with bilious vomiting, diarrhea, abdominal distention, and perioral excoriations, became angry, eczematous with deep fissures of chin. Perianal dermatitis in **infants** receiving it in **infant** formulas. Cases of a Bartter-like syndrome in **infants** were traced to **their** ingestion of Syntex's ***NeoNull-Soy***® **infant** formula leading to metabolic alkalosis, failure to gain weight, anorexia, or constipation, probably due to inadequate ***chloride*** content. Two case reports of generalized erythema of short duration with dryness of throat, light-headedness, and intense burning sensation of entire body skin after ingestion of soybean ***bread***. Has decreased availability of ***thyroid hormone***. Triggers migraine in some **people**. Contains goitrogenic substances (destroyed by heating or cooking) that can lead to an ***iodine*** deficiency. A 7-week-old **female**, while on an ***Isomil***® formula developed hypoprothrombinemia and bleeding. Some formulas may have inadequate amounts of ***vitamin K***. A 37-year-old **female** had repeated attacks due to sensitization to dust inhalation at work leading to asthma, allergic rhinitis, and conjunctivitis. Subsequently, **she** had severe hives and asthma induced by eating soybean pancakes. Atopic dermatitis in **adults** from soybeans in **their** diets. High ***bioflavonoid*** levels in soybean-based **infant** formulas in Japan and Hong Kong may be the cause of **infant** acute myelogenous leukemia or acute lymphoblastic leukemia at three times the U.S. rate. Contains ***trypsin*** inhibitors, usually destroyed by heating.

Its ***trypsin*** inhibitors lead to pancreatic hypertrophy in **rats** and **chicks**, but not in **dogs** or **calves**. Hypertrophy and hyperplasia of pancreas and hyperplastic nodules and adenomas, as well as a few cases of pancreatic cancer in **rats** fed continuous diets containing raw ***soya*** flour. Growth inhibition and diarrhea reported in **calves**, **chicks**, **rats**, and **mice**. Pancreatic hypertrophy also noted in **chicks**. Goitrogenic in **pigs**. Uncooked soybeans decrease growth rate of **pigs**. The plant is a ***nitrate*** accumulator. Heavy losses in **cattle** from hemorrhage due to ***trichloroethylene*** (q.v.) extracted soybean meal. This method was used in Scotland in 1916 and toxicity to **cattle** was noted; its use in Europe was discontinued. In the late 1940s, someone in the U.S. started using this extraction method again with extensive **livestock** losses and aplastic anemia in **cattle**, **horses**, **sheep**, and **chickens**. Some **calves** have a hypersensitivity to soybean protein in their diet.

SOYBEAN OIL

Untoward effects: Use in Japan as a machine oil leads to contact dermatitis. Its use as ***Intralipid***®, an IV emulsion for hyperalimentation use, has been associated with anorexia, pyrexia, somnolence, hyperlipidemia, icterus, gastrointestinal hemorrhages, chills, vomiting, thrombopenia, strange tastes, thrombophlebitis, pain in thorax or back, dyspnea and cyanosis, sinus bradycardia, tachycardia, pruritic urticaria, decreased blood pressure, gallstones, fat embolism, Wernicke's encephalopathy, and diarrhea are among the many adverse effects reported. To what extent these reactions are also due to the ***glycerin*** and ***egg lecithin*** are unknown. It decreases ***warfarin's*** anticoagulant effect.

Use in feed for **calves** increased blood ***cholesterol***, decreased weight gain, increased incidence of diarrhea, and caused hair loss. Increases cardiac fibrosis in **rats**.

SPARFLOXACIN

= *AT 4140* = *CI 978* = *PD 131,501* = *RP 64,206* = *Spara* = *Zagan*

Antibiotic. A fluoroquinolone.

Untoward effects: Photosensitivity (8%), tendonitis, tendon rupture, prolonged QT interval, and torsades de pointes. As a ***fluoroquinolone***, one can eventually expect case reports of skin rashes, gastrointestinal upsets, pseudomembranous colitis, insomnia, dizziness, confusion, hallucinations, seizures, rash (4%), vasculitis, fever, toxic epidermal necrolysis, Stevens–Johnson syndrome, arthralgia, nephritis, anemias, thrombocytopenia, leukopenia, agranulocytopenia, pancytopenia, and anaphylaxis. Adverse effects reported in ~50% of 400 U.S. **patients**.

Drug interactions: Coadministration with ***cisapride*** increases its absorption and use with ***sucralfate*** decreases its bioavailability 44%. ***Aluminum compounds***, ***iron***, ***magnesium***, and ***zinc*** given at the same time will decrease absorption. An interaction with ***amiodarone*** is probable.

Has caused arthropathies in young **animals**.

IV, LD_{50}, **rat**: 2.25 mg/kg; **mouse**: 4.32 mg/kg; **dog**: 0.5–1 mg/kg.

SPARTEINE

= *Actospar* = *Depasan* = *Perivar* = *Spartocin* = *Synastrin* = *Tocosamine*

***Oxytocic.* Usually IM; occasionally oral or IV. An alkaloid from *Sarathammus scoparius*.**

Untoward effects: Uterine tetany, nausea and vomiting, uterine rupture, **fetal** distress, cervical lacerations, anaphylaxis, blurred vision, dizziness, and tachycardia. Most of these reactions are usually due to excessive stimulation of uterine musculature. Use with ***vasoconstrictors*** leads to severe, persistent hypertension with rupture of cerebral blood vessels. Therapeutic serum levels: 0.5–1 mg/l.

Rat trials confirm toxicity leading to bradycardia, abrupt decreased blood pressure, and sudden cessation of QRS complexes with ensuing death.

SPARTIUM JUNCEUM

= *Ginestra* = *Retama* = *Spanish Broom*

A popular plant in Uruguay, Ecuador, Italy, and Turkey.

Untoward effects: An M.D. in Uruguay reported ingestion of the seed has caused vomiting and abdominal pain. Elsewhere, others report severe cases with tachycardia, hypotension, purging, renal irritation, and weakness. Contains small amounts of ***cytisine*** (q.v.) and ***sparteine*** (q.v.).

SPATHYPHYLLUM sp.

= *Peace Lilies*

Untoward effects: **Animals** or **children** chewing on its leaves are injured by its high ***oxalate*** (q.v.) content.

SPECTINOMYCIN

= *Actinospectacin* = *CHX 3101* = *M 141* = *Spectam* = *Spectogard* = *Stanilo* = *Togamycin* = *Trantan* = *Trobicin* = *U 18,409*

Antibiotic.

Untoward effects: Cardiac arrest and anaphylaxis shortly after an injection in 33-year-old **male**; cardiopulmonary resuscitation was successful. Weakness and giddiness immediately after IM with mild urticarial, itchy rash and local pain. Cholestatic jaundice from an idiosyncratic reaction. Occasionally nausea, vomiting, headache, dizziness, fever, urticaria, and pruritus. Will cross-react with ***neomycin***. Possible increase in ***lithium*** serum levels.

IP, LD_{50}, **rat**: 7.5 g/kg; **mouse**: 5.7 g/kg.

SPERMACETI

= *Cetaceum* = *Spermwax*

Emollient, emulsifier.

Untoward effects: A known sensitizer causing hand dermatitis.

SPHAGNUM MOSS

= *Bog Moss* = *Peat Moss*

Untoward effects: In Vermont, 14 **persons** who handled it became infected with sporotrichosis (*S. schenkii*). In central Wisconsin, one of the largest areas of its production in the U.S., four **male** high school **students**, working part-time at a garden center making grave sprays, developed ulcerating lesions on hands or wrists and 3/4 had non-tender ascending lymphangitis. *Sporothrix* or *Sporotrichum schenkii* was isolated. In Puerto Rico, a 55-year-old **female**, **who** was a **hobbyist** making sphagnum moss animals, had a chancreform pyoderma of the flexor surface of **her** left arm.

SPIDERS

Untoward effects: Their poison sacs, located in their cephalothorax, are connected to two horny fangs introducing a poison with their fatal bites, while they suck fluids from their victims. Most spiders contain extremely small amounts of venom, useful only against **insects** and other **arachnids**. They live almost everywhere from the Arctic to the tropics, and only a few are dangerous for **man**. In a 10 year period ending in 1959, 65 fatalities were reported in the U.S. from their bites. Death is usually due to direct toxicity to the central nervous system and/or cardiovascular system, rather than from anaphylaxis. In northern Japan, poisonous spiders were added to a mixture with **their** arrow poisons. Ordinary house spider bites may occasionally leave an intense hemorrhagic reaction. Very little mortality reported from North and South America and Australia, but many bites reported. Necrotizing spider bites may be associated with hemolysis, hematuria, poor coagulation, and anuria. Hypertension can be induced by some bites. Thousands of venomous spider species exist in the U.S.; ~50 have fangs capable of penetrating **human** skin. Only ***Widow*** and ***Brown Recluse*** spiders regularly cause significant damage.

Atrax sp.: In ***A. formidabilis***, the tree-dwelling or north-coast funnel-web spider of Northern South Wales and southern Queensland in Australia, both **male** and **female** bites are toxic and have caused deaths in **people**. ***A. robustus, Sidney funnel-web spider***, is slightly less toxic and many **people** have died, especially from the **males** who inject more venom. The principle toxin in their venom is ***atraxotoxin*** to which **monkeys** and **man** are highly susceptible. Initally, nausea, vomiting, abdominal pain, profuse sweating, hypersalivation, lacrimation, and dyspnea. Fibrillary twitching is common and occasionally violent. Hypotension, pulmonary edema, and decreased temperature occur. Urticaria in many non-fatal cases. Eventually, coma, asphyxia, and cardiac arrest. Several **infants** have died within 15–90 min; **adults** may take up to at least 30 h before dying. These aggressive spiders have even attacked **lizards**. Although indigenous to Australia, it can enter the U.S. and other countries from cargo ships.

Black-Widow Spider = Hour-glass Spider = Lactrodectus mactans, a venomous spider with a black glossy-looking body and a red–orange hour glass shape on the underside of its abdomen. Found under stones, logs, and brush piles, and in barns, garages, and privies. Its venom is primarily neurotoxic and contains ***α-latrotoxin***. Needle-prick-type bites are often unnoticed until 30–120 min later, when pain (that can be excruciating) occurs. Within another 2–3 h, painful abdominal cramps and fasciculations of bitten extremity, followed by muscle spasms in upper body in arm bites; abdominal cramps after leg bites.

Weakness, dyspnea, headache, chest pain, and paresthesia, often misdiagnosed as myocardial infarction or acute abdomen. Board-like abdominal rigidity, piloerection, priapism, hypotension, vomiting, bradycardia, and ptosis. **People** using outdoor privies often have bites on **their** buttocks or genitals. The **male** has poorly developed venom glands and symptoms, therefore, come from the bites of **females**. Called a widow, because, after mating, the **female** kills and devours the **male**. **Dogs** are frequently bitten leading to weakness, dyspnea, anorexia, paralysis, and often death occurring in 4–6 h or after several days.

Brown Recluse or ***Hairy Brown Spine*** = ***Fiddleback Spider*** = ***Loxosceles reclusa*** = ***Violin Spider***, once found only in Central America and the southern U.S., is now found in most states, including Hawaii, and even a few species in South Africa and Australia. Although found outdoors, it loves dark corners, as in closets, storerooms, hen houses, barns, attics, and around furniture. It can easily bite **people** reaching for **their** clothing in **their** closets. It has a dark violin-shaped mark on the thorax or "head" and a very potent venom. It is ~$\frac{3}{4}$" in size (1–5 cm leg to leg) and varies in color from light yellow to dark brown. A $\frac{1}{2}$ microliter (~65 μg) of it eventually causes pain and necrosis at the bite site, nausea, fever, vasospasms, malaise, myalgia, and a delayed (up to 10 days) neurotoxicity and circulatory collapse. The venom contains four enzymes (as in **snake** venoms), a protease, an esterase, ***sphingomyelinase***, and ***hyaluronidase*** (q.v.). The latter breaks down cellular glue, permitting the spread of the other enzymes. ***Sphingomyelinase*** can erode the endothelium of blood vessels, causing clotting. Within 30 min of envenomation with pain and a burning sensation, a wheal develops. In 1–12 h, a "bull's-eye"-like vesicle develops with a ring of erythema and necrosis. Itching, swelling, and tenderness occur. Eventually, the damaged skin and tissues die, leaving a scab which sloughs off, leaving a raw, gaping wound that can reach as deep as the bone and take months to heal in ~20% of **patients**, often requiring plastic surgery. Hemolytic anemia has occurred after such bites and even requires blood transfusions. Where outdoor privies are used, bites are frequently on the buttocks.

Chiricanthium sp. are found outdoors and are Europe's second most toxic spider, and are also reported from Hawaii, Fiji, and Australia. Bites have caused intense local pain and edema, erythema, small areas of skin necrosis, enlarged and tender regional lymph nodes, malaise, pyrexia, and sensation of chest compression, pruritus, and weakness. Death can occur.

Dysdera crocata, a European spider, is also found in Australia and Tasmania. The bite is painful and put a **child** in the hospital for several days.

Gray Widow Spider, ***Brown Widow Spider*** (***L. geometrius***), and ***Red Widow Spider*** (***L. bishopi***) cause symptoms similar to, but slightly less toxic than from the ***Black Widow Spider***.

Japanese Spider = ***Jashukka*** and ***Mekuragame***, both of which are poisonous.

Lampona cylindrata, widespread under tree bark, stones, crevices, and under boards or boxes in Australia and New Zealand. Its bites lead to local discoloration, headache, pyrexia, and pruritus.

Missulaena occatoria, a ***Trapdoor Spider*** of Australia and New Zealand, as well as the genera of ***Arbamitis*** and ***Blackistonia***, also trapdoor spiders, are squat with short legs and mostly of a black or velvety blue–black to brown appearance that bite, but only cause minor effects. In Western Australia, another trapdoor spider, ***Agamippe raphiduca*** has caused severe reactions in **man**.

Miturga, a genus in Australia and New Zealand. All its known species are large and live outdoors. A case report of envenomation after a bite on a finger leading to severe, instantaneous pain and, after 20 min, difficulty in walking and stiff knee joints. Within 3 h, walking was even more difficult. The next day, the knee stiffness improved, but wrist and finger joints became painful. Aching back, shoulders, and legs with intermittent muscle cramps occurred on subsequent days. Free of symptoms after 45 days.

Mopsus sp., a "jumping spider" results in painful swellings at the bite site with discoloration lasting ~1 week.

Phoneutria = ***South American Banana Spider*** often arrives on ***banana*** boats. Causes symptoms similar to those of the ***Black Widow Spider***, especially intense local pain at the bite site, sneezing, lacrimation, salivaton, mydriasis, visual disturbances, priapism, muscle spasms, convulsions, drowsiness, and ataxia. Mortality rate is high.

Red-Back Spider = ***Lactrodectus hasselti*** in Australia is related to the widow spiders, and, until an antivenin was developed, its bite was occasionally fatal. Found under logs, stones, gas-meters, and outdoor privies. Pain occurs in ~90%, perspiration in ~38%. In ~40%, stiffness, incoordination, paresthesia, and tremors. Less frequently reported were nausea, vomiting, dizziness, syncope, local erythema and edema, muscle weakness in ~11%, tachycardia, palpitations, rigors, pyrexia, insomnia, muscle spasms, and temporary hypertension (~2%). Localized skin sweating at the bite site is often diagnostic. Some **patients** remained lethargic for weeks.

Selenocosmia stirlingi = ***Barking Spider*** of Australia actually makes a whistling sound; it is a dry soil burrower, and ~5–6" in length, with tufts of reddish-gold hair on its eight legs, with a head and body the size of a **mouse**. Their usual prey is **frogs**, **birds**, and small **reptiles**. **Human** bites are rare, but cause severe local reaction. This is related to the **bird**-catching ones of South America.

Steatoda paykulliana is a European spider with a potent venom. Within 5–10 min after biting, **guinea pigs** developed snout rubbing (regardless of where on their body they were bitten), then neck distortion, motoric restlessness with clonic cramps, followed by curving the whole body to one side, extension of hind limbs, and eventually paralysis; excessive salivation, gradually returning to normal after 4–5 h. After about 15 simultaneous bites, similar symptoms, ataxia, and death.

Tarantula refers to a number of species. In the U.S., ***Eurypelma henizii*** bites are painful. They only bite to defend themselves and are, apparently, non-poisonous. Other hairy spiders of Australia also rarely bite **people**. South American species can cause toxic local reactions to their bites.

SPIGELIA ANTHELMIA

= *Erve Aran Aparan* (Yoruba)

Anthelmintic. **Andoque Indians of Colombia use infusions of the root as a tranquillizer for children.**

Untoward effects*:** Overdoses cause dimness of vision, mydriasis, eye and face muscle spasms, giddiness, tetanic convulsions, cyanosis, and death, especially in **children**. **Cuma Indians** in Panama and Colombia use it to execute **criminals** with its ***strychnine-like effects. **They** also use it as a rodenticide, ***INA NUSU***.

The seeds will kill **rabbits**. The fresh plant is poisonous to Nigerian **livestock** within 2–3 h.

SPINACH

= *Spinacia oleracea*

Untoward effects*:** Carotenemia has occurred after excessive intake. A cup of cooked spinach contains 10 mg ***carotene. A vegetable **dealer** had contact dermatitis from handling it and asthma after eating it. For several years, **workers** employed in packing spinach developed eruptions on **their** hands and forearms similar to those caused by ***poison ivy***. It is a possible food offender causing dyshidrosis, and it is goitrogenic. It has also been used to cut ***cocaine*** or ***PCP***. Contains 0.1–0.2% ***nitrate***, a known carcinogen after formation of ***nitrosamines***. It contains no ***nitrites*** and when eaten fresh, is generally well-tolerated by **babies**, but the water in which it is cooked does, and in Germany it is used in making a spinach purée. A number of **infants** 2–10 months old in Germany became poisoned because of high initial ***nitrate*** content in the spinach or because, when it is kept at room temperature, bacterial reduction of the ***nitrate*** to ***nitrite*** occurs. This reduction does not occur in frozen or canned spinach. In **infants** < 3 months, **their** upper gastrointestinal flora can easily reduce the ***nitrate*** leading to methemoglobinuria. Even a 2-year-old **male**, after ingesting 300–400 g twice a week, which was stored in a refrigerator, to overcome obstipation developed to methemoglobinemia of 54% with cyanosis, drowsiness, tachypnea, vomiting, and tachycardia (160/min). It contains high levels of ***oxalates*** that impair ***calcium*** absorption, and may induce renal calculi. ***Oxalic acid*** greatly decreases the availability of its ***iron*** content, defusing the "Popeye the sailor man" myth. It contains large amounts of ***vitamin K***, interfering with the action of ***anticoagulants***, such as ***warfarin***. Hypertensive reaction in 60-year-old **female** receiving 45 mg/day ***phenelzine*** after ingesting 500 g whole New Zealand spinach. May color urine green. It is suggested that **people** with ***nickel*** dermatitis avoid it. The type and amount of farm fertilizer used on the plants can alter the ***nitrate*** content.

Allergic reactions have been reported in **dogs**. Suppression of prolonged prothrombin time in **rabbits** given spinach extract while on warfarin. Its ***calcium***-binding ***oxalates*** can adversely affect pet **birds**. When used as their sole diet, it killed **rats** in 3 days.

SPIPERONE

= *R 5147* = *Spirodecanone* = *Spiroperidol* = *Spiropitan* = *Suanovil*

Psychotherapeutic, tranquillizer, dopamine receptor blocking agent.

***Untoward effects*:** Extrapyramidal symptoms, poor accommodation, increased lacrimation and salivation, increased thirst and appetite after IM.

LD_{50}, **mouse**: 600 mg/kg; IV, LD_{50}, **rat**: 14 mg/kg; **mouse**: 29 mg/kg.

SPIRAMYCIN

= *Foromacidin* = *Rovamycin* = *RP 5,337* = *Selectomycin*

Antibiotic.

***Untoward effects*:** A 27-year-old **male** received 500 mg qid/5 days leading to severe abdominal cramps and diarrhea. Slight diarrhea in 6/418, oral moniliasis in four. Diarrhea in 4/110 in another trial. Occupational dermatitis in handling it by **veterinarians**. Dermatitis when applied under occlusive bandage. Has induced QT interval prolongation. Diarrhea, stomach ache, nausea, and vomiting in 2–3% of **patients**. Can induce asthma-like symptoms.

Listlessness, coma, and respiratory failure at high dosage in **guinea pigs**, **rats**, **mice**, **rabbits**, **cats**, and **dogs**. Polycythemia and anemia in overdosed **rats**. SC to **cats** leads to excitement and convulsive movements and, within 24 h, weakness, apathy, anorexia, and weight decrease, followed by prostration and bradypnea before death a day later. Oral dosing to ten **dogs** of 500 mg/kg/day/8 weeks led to systemic toxemia and death of eight after progressive anorexia, vomiting, diarrhea, irritability, pallor, dim vision, prostration, green–black stools, and fecal incontinence. Spermatogenesis inhibited.

LD_{50}, **rat**: 9.4 g/kg; **dog**: 5.2 g/kg; **mouse**: > 5 g/kg; SC, LD_{50}, **rat**: 3.5 g/kg; **mouse**: 1.5–2 g/kg; **cat**: 1 g/kg; IV, **mouse**: 152 mg/kg.

Adipate, LD_{50}, **rat**: 4.85 g/kg; **mouse**: 3.13 g/kg; **rabbit**: 4.33 g/kg; **guinea pig**: 3.5 g/kg.

SPIRONOLACTONE

= *Aldace* = *Aldactone* = *Aldopur* = *Almatol* = *Altex* = *Aquareduct* = *Deverol* = *Diatensec* = *Dira* = *Duraspiron* = *Euteberol* = *Lacalmin* = *Lacdene* = *Laractone* = *Nefurofan* = *Osiren* = *Osyrol* = *Sagisal* = *SC-9420* = *Sincomen* = *Spiretic* = *Spiroctan* = *Spiroderm* = *Spirolone* = *Spiro-Tablinen* = *Supra-Puren* = *Suracton* = *Urusonin* = *Verospiron* = *Xenalon*

***Diuretic, aldosterone antagonist, antihypertensive.* Promotes renal excretion of *sodium* and not *potassium*. Structural similarity to *digoxin*, *estrone*, and *progesterone*. Originally, given only by injection, then an oral form was developed that smelled like rotten eggs – a secretary fainted from the odor – then research developed the form that is used today.**

Untoward effects*:** Its major metabolite is ***canrenone (q.v.). Adverse effects have slowly discouraged its use.

Blood*:** Agranulocytosis, hemolytic anemia, pancytopenia, neutropenia, and thrombocytopenia occur. A 57-year-old **male** had leukocytes decrease from 8.2 to 2.3×10^9/l with 6% neutrophils 4 days after the start of treatment; improved when treatment was discontinued. Hyperkalemia and hyponatremia have occurred. It decreases renal clearance of ***urates and increases their serum level.

***Cardiac*:** Clinical congestive heart failure with fatigue, orthopnea, dyspnea on exertion, and edema due to pumping ability of the heart, and increase in left ventricular filling pressure.

Clinical test interference*:** Has led to a false diagnosis of Cushing's syndrome in 57-year-old **male alcoholic**. As a ***steroid, it has caused a false elevation in tests of plasma ***progesterone*** and a false negative effect on ***estrogen***. It may cause elevated ***potassium*** results and it interferes with ***11-deoxycorticosterone*** assays.

Endocrine*:** Often associated in **men** with ***estrogen-like adverse effects, including decreased libido, impotence, delayed ejaculation, gynecomastia and breast tenderness (10%); in **women**, menstrual irregularities, secondary amenorrhea, breast tenderness, decreased libido, and inhibition of vaginal lubrication; in **children**, increased plasma ***gonadotropin*** levels. Gynecomastia in **male** usually occurs after several months of therapy, but it developed in 51-year-old **male** after only 100 mg/day/15 days. Weight loss occurs.

***Gastrointestinal*:** Cramps, diarrhea, gastritis, anorexia, gastric bleeding and ulceration, and vomiting. A few cases reported of cholestatic/hepatocellular toxicity with periportal hepatitis.

Miscellaneous*:** Absorption is increased when taken with ***food. Breast milk contains 0.2–1.2% of the **maternal** dose.

***Neurological*:** Headache, drowsiness, lethargy, asterixis, confusion, drug fever, and ataxia.

***Skin*:** Rash; maculopapular, erythematous, and systemic lupus erythematosus-like eruptions; and urticaria.

Drug interactions*:** Potentiates the action of other ***diuretics, ***antihypertensives***, ***hydralazine***, ***mecamylamine***, ***pentolinium***, ***potassium salts***, and ***veratrum alkaloids***. Variable increase in ***digoxin*** concentration of ~25–30%. A similar effect noted with ***digitoxin***. ***Aspirin*** may antagonize up to 30% of its diuretic effect. It decreases ***antipyrine*** half-life and increases ***lithium*** retention. Severe hyperkalemia noted with ***angiotensin-converting enzyme inhibitors*** and ***indomethacin***. It decreases ***norepinephrine*** effectiveness. Use with other ***potassium-sparing diuretics*** or ***potassium chloride*** can cause dangerous hyperkalemia. Although it does not affect ***warfarin*** serum concentrations, it decreases its hypoprothrombinemic effect, probably by increasing hematocrit and concentration of circulating clotting factors. ***Dicumarol*** anticoagulant effects are decreased by it. Destroys ulcer-healing capability of ***carbenoxolone***.

Use in chronic toxicology studies with high dosage (~350 times **human** therapeutic dosage) in **rats** led to benign thyroid and testicular tumors, but not in **rhesus monkeys**. It decreases plasma ***testosterone*** concentrations in **dogs** and **men**. It blocks ovulation and decreases implantation in **mice**. It has little diuretic effect in **rodents**, as they have low ***aldosterone*** levels. Given in early pregnancy to **rats**, it can have an abortifacient effect; if given after delivery, it decreases lactation. Can induce malignancies in **mice**.

SPONDIANTHUS UGANDENSIS

Untoward effects*:** The bark and seeds are used as a **rat** poison in Nigeria and can kill a **man**. Kills in ~5 h. Called ***Obo ekute or ***Owe*** in Yoruba, and ***Ibok-eku*** in Efik.

The leaves are poisonous to **cattle**, **goats**, and **dogs**. ***S. preussii*** is also toxic.

SPONGES

Untoward effects: Of ~4000 species, only ~13 cause a contact dermatitis with itching, burning, prickly sensation, and erythema, followed in a few hours by local pain, swelling, and stiffening of finger joints, and eventually vesiculation, serous or purulent fluid in the blisters, and desquamation. In addition to a toxin, many contain ***silicon dioxide*** or ***calcium carbonate*** spicules that break off when handled. Most common are the West Indian and Hawaiian fire sponge, ***Tedania ignis***; the poison bun sponge, ***Fibula nolitangere***, and in the northeastern U.S., the red sponge, ***Microciona prolifera***, whose lesions may last for months. Sponge **fishermen** develop a dermatitis from the often attached ***sea anemones*** (q.v.).

Surgical sponges, often gauze, have been left in **patients** for 1–40 years and were often associated with granulomas, abscesses, back pain, a fatal outcome, and many malpractice lawsuits.

SQUALANE

= *Cerumene* = *Cosbiol* = *Hexamethyltetracosane* = *Perhydrosqualene* = *Robane* = *Sebumsol* = *Spinacane*

Ceruminolytic, lubricant, suspending agent.

Untoward effects: When pure material was applied to clipped **rabbit** skin, slight erythema noted after 24 h and only moderately irritating to **guinea pig** skin. No irritation noted on **rat**, **miniature swine**, or **human** skin. A 2% level in hair spray for 90 days on confined **rabbits** led to no histopathology or lesions on X-ray.

SQUALENE

= *Supraene*

Untoward effects: Comedogenecity noted on **human** (slight to moderate) and **rabbit** (slight to strong) skin is enhanced by ultraviolet radiation.

SQUALUS ACANTHIUS

= *Dogfish Shark* = *Spiny Dogfish*

Untoward effects: Extremely dangerous to **man**, as it has sharp stingers at at the forward end of its dorsal fins that inject a venom leading to excruciating pain that lasts for hours and occasionally death. Found in tropical and temperate waters on both sides of the Atlantic and Pacific Oceans.

SQUASH

= *Cucurbita maxima* = *Dubba*

Untoward effects: A ***nitrate***-accumulator. Excessive intake can cause carotenemia. Summer squash is high in ***oxalates***.

SQUILL

= *Bulbus Scillae* = *Meerzwiebel* = *Red Squill* = *Urginea maritima*

Rodenticide, emetic.

Untoward effects: Rarely poisonous for **man**, as it is bitter and induces emesis. Can cause gastrointestinal upsets, such as nausea, vomiting, and violent purging; bradycardia, and cardiac arrhythmias. Given IV 800 µg of its glycosides has the cardiac effect of 900 µg ***digitoxin***. The powdered glycosides are poisonous and irritating to the mucous membranes of the respiratory tract. A bark decoction of ***U. burkei*** in South Africa is used to induce abortions.

Red Squill contains the same cardiac glycosides, but is more lethal to **rats**, due to its ***scilliroside*** content, than the white bulbs. **Rodents** like the red variety; they do not vomit and it is, therefore, very toxic to them, acting on their central nervous system. Generally "non-toxic" to **animals** that can vomit, but poisoning has been reported in **cattle**, **sheep**, **swine**, **chickens**, **cats**, and **dogs**. **Pigs** normally will not ingest it voluntarily because of its bitterness and, after accidentally mixing in their feed, have died within 24–72 h with an LD of 200 mg/kg. Post-mortem reveals congestion of abdominal and thoracic organs, including kidneys, liver, lungs, and myocardium. ***U. altissima*** has proven dangerous to **livestock** in Africa. ***U. brachystachys*** is used in arrow poisons in Tanzania. Symptoms in **dogs** include depression, incoordination, collapse, twitching, and convulsions with death in 2–3 days. Post-mortem shows hyperemia, gastrointestinal inflammation, liver and kidney pathology, and bronchopneumonia.

LD_{50}, **rat**: 430 µg/kg.

Red Squill, LD_{LO}, **human**: 500 mg/kg; **cat**: 100 mg/kg; **rabbit**: 300 mg/kg; **pig**: 300 mg/kg; TD_{LO}, **man**: 1414 mg/kg.

STACHYS ARVENSIS

= *Field Nettle*

Untoward effects: Ingestion by **cattle** causes central nervous system disorders.

STAMPS and STAMP PADS

Untoward effects: Recurrent contact eczema due to use in stamp pad red inks of ***acid red 3*** and *p*-***phenylazoaniline***, which uses *p*-***phenylenediamine*** (q.v.) in its formation. ***Chromium*** (q.v.) dermatitis from its use in postage stamps.

STANLEYA PINNATA

= *Prince's Plume*

Untoward effects: As it only grows in soils high in ***selenium*** (q.v.), it is not only an "indicator plant", but its ***selenium*** content can also be toxic. It is not palatable to **livestock** and only eaten when vegetation is scarce. Can adversely affect **cattle**, **sheep**, **horses**, and **poultry**. Only small amounts lead to weight decrease, rough hair coat, impaired vision, and straying from the herd. This is followed by pronounced blindness, depraved appetite, wandering in circles, and pushing forward into any obstacle encountered, such as buildings or fences.

STANNOUS CHLORIDE

= *Chloride of Tin* = *Stannochlor* = *Tin Dichloride*

Use: As a preservative and to retain color in foods, usually not > 15 ppm as ***tin***. In manufacturing pharmaceuticals, dyes, and sensitized ***paper***. Annual U.S. consumption is < 50,000 lbs.

Untoward effects: A violent irritant and caustic in **man**.

Tin can concentrate in bone and decrease its compressive strength, as demonstrated by feeding **rats** a 13 week diet containing 0.1% equivalent to 530 ppm ***tin***. The same results noted after 4 weeks on 300 ppm. Another 13 weeks feeding trial with 0.1%, with weekly increase of 0.1% until a final level of 0.8% caused slight growth retardation, gastrointestinal mucosal irritation, slight anemia, and severe damage to exocrine pancreatic activity and decreased ***insulin*** secretion.

LD_{50}, **rat**: 700 mg/kg; **mouse**: 1.2 g/kg; LD_{LO}, **dog**: 500 mg/kg; **rabbit**: 40 mg/kg.

STANNOUS FLUORIDE

= *Fluoristan* = *Tin Difluoride*

Use: In dental caries prophylaxis.

Untoward effects: Pigmentation (light brown to black) of 696/2504 teeth in 105 **children** ~9–19 years from aqueous tooth applications and use of a stannous fluoride toothpaste for 3 years. Use of it in dentifrices (usually 0.4%) can exaggerate pre-existing edema or inflammation. Stomatitis, glossitis, and perioral dermatitis are common from ***fluoride*** dentifrices leading to usually superficial ulcerations, whitish exudate, and vesicles.

STANOLONE

= *Anabolex* = *Andractin* = *Androlone* = *DHT* = *4-Dihydrotestosterone* = *Neodrol*

Androgen, anabolic steroid.

Untoward effects: Sebaceous glands convert circulating ***testosterone*** into it; it is the tissue androgen and probably causes acne, baldness, and hirsutism. Other adverse effects are similar to those caused by ***testosterone*** (q.v.).

STANOZOLOL

= *FZ 348* = *NSC 43,193* = *Stanazol* = *Stromba* = *Win 14,833* = *Winstrol*

***Androgen, anabolic steroid.* Synthetic derivative of *testosterone*.**

Untoward effects: In **children** on long-term therapy, premature closing of epiphysis; phallic enlargement, increased erectile frequency, and acne in **males**; in **females** leads to hirsutism and **male** pattern baldness. In **women** leads to clitoral enlargement, voice-deepening, gynecomastia, irregular menstruation, and decreased libido which are often irreversible. In **men** leads to testicular atrophy, oligospermia, chronic priapism, impotence, epididymitis, and activation of plasmin-induced fibrinolysis. The more serious adverse effects are peliosis hepatitis, severe cholestasis, hepatocellular carcinoma, and angiosarcoma. Virilization, edema, nausea, vomiting, diarrhea, ***iron*** deficiency, and intracranial hypertension. Clinically significant increase in prothrombin time when given to **patients** on ***warfarin***.

Androgenic effects in female **monkeys** of 1 mg/kg and 5 mg/kg in **males**. Vomiting and diarrhea in < 1% of treated **dogs**. Parenteral use in a **horse** may have induced nervousness and muscle fasciculations.

STARCHES

= *Amylum*

From *arrowroot*, *barley*, *cassava*, *corn* or *maize*, *potato*, *rice*, *sorghum*, *sweet potato*, *wheat*, etc. Many specialty forms designed for special uses.

Untoward effects: Amylases are usually plentiful in **people** so that starches are converted into ***glucose*** very rapidly, thus, as far as a **diabetic** is concerned, ***potatoes*** are like candy. Contact urticaria, even from spray starches. It colors feces and urine yellow. When starches and ***sucrose*** are added to the diet, usually in ***fruits*** or ***vegetables***, some **infants** and **children** with congenital sucrase–isomaltase deficiency develop a chronic diarrhea. Use as a ***glove*** lubricant has been associated with granulomatous peritonitis, skin sensitization and cutaneous reactions, renal petechiae and focal glomerulitis after handling a ***contrast agent***, and granulomas in middle ear and mastoid after surgery. Starch granules in lungs of 50% of **newborns** after treatment with positive pressure ventilation. Serious pulmonary problems in **infants** can also follow inhalation of particles after its use as a "**baby** dusting powder". Chest pain, coughing, and rhinitis also occur. Use as a drying agent on skin can be a nutrient for fungal and bacterial growth. Raw ***corn starch*** ingestion leads to flatus, borborygmus, and megaloblastic and ***iron***-deficiency anemia. Retinopathy and emboli in the eye, as well as in the lung and brain of **drug-abusers** after IV of crushed tablets or capsules. National Institute for Occupational Safety and Health/OSHA upper limits of respiratory exposure: 5 mg/m^3.

STARFISHES

Untoward effects*:** It is one of > 6000 **echinoderms**, of which 80 are venomous. Its venom contains ***acetylcholine-like substances, ***serotonin***, and ***steroid glycosides***. There are many species. The most important one is the Giant Starfish, ***Acanthaster planci***, known as the "***Crown of Thorns***" among the coral reefs of the South Pacific. It can reach a diameter of > 50 cm (> 20") and may have up to 21 arms whose quills contain a venom that has caused vomiting and paralysis in **swimmers**. Fascinating is the fact that the **pied crabs**, only 5 cm in length, within a few minutes can turn the starfish on its back and eat it within a 24 h period. Also see ***Fishes***.

STAVUDINE

= *BMY 27,857* = *D4T* = *Zerit*

***Antiviral.* An *antiretroviral nucleoside* analog.**

***Untoward effects*:** Painful peripheral neuropathy – dose-related and usually reversible. Occasional nausea, vomiting, and increased serum aminotransferase. Rarely, pancreatitis, gynecomastia, and in a 32-year-old **female** it caused hepatic steatosis and lactic acidosis. Possibility of abnormal fat distribution leading to buffalo hump, supraclavicular fullness, protuberant abdomen, and weight increase.

STEARIC ACID

= *Octadenoic Acid*

Untoward effects*:** Use in facial cosmetics has interfered with normal skin exfoliation and may be a cause of acne. Although it is a saturated fat, it is unusual in that it does not increase concentration of ***cholesterol in the blood.

Comedonergic in **rabbit** ears.

STEEL

Untoward effects*:** **Workers** in steel foundries can be exposed to ***fluoride fumes. Poisoning in **five** cutting steel bolts plated with ***cadmium*** to prevent corrosion led to malaise, shivering, sweating, and chest pain and dyspnea within a few hours, followed by pulmonary edema and pneumonitis. **One** died on day 5. Post-mortem revealed pulmonary edema, thrombi in arteries, cellular proliferation and metaplasia in alveoli, necrosis of renal cortex, tubular degeneration, and glomerular infarction. Hypersensitivity to ***nickel*** in stainless steel screws used in orthopedic surgery requires their removal. Corrosion in stainless steel implants and screws is decreased by increasing ***nickel*** content to 11%. Most ***nickel***-sensitive **people** can tolerate 2% ***nickel*** in **their** stainless steel implants. Sterilization can decrease its hardness. Manufacturing **workers** have burns and intertrigo from the heat, acid (now usually ***HCl***) burns from the pickling solutions, alkali burns from ***lime*** used in pickling and cutting compounds, and steel splinters. Fingernail polishes usually contain ***nickel*** from stainless steel mixing balls and can contribute to irritation or sensitization caused by them.

Osteosarcomas after 6 and 11 years in two **dogs** after stainless steel type 316L tibia repair with plate and screws in the first and a similar intramedullary pin in the second. Corrosion of the pin was noted. Dust emitted from Slovakian steelworks led to diarrhea, apathy, and anorexia in exposed **pigs**; in **sheep**, the same symptoms and intestinal mucous membrane hypertrophy and coniosis of mesenteric lymph nodes; increased hemoglobin, hematocrit, erythrocytes, and leukocytes; and **cows** developed hypocalcemia.

STEGANOTAENIA ARALIACEA

***Untoward effects*:** This plant is one of many used by African **natives** in the treatment of leprosy and has caused severe penile burns.

STENBOLONE

= *Anatrofin* = *RS 2106*

Anabolic.

***Untoward effects*:** Irreversible symptoms of early virilization; increased incidence of erections in **male children** and some clitoral hypertrophy in **female children**. Oligomenorrhea in ten **females** 17–39 years with malnutrition.

STEPHANIA TETRANDA

Untoward effects*:** In Belgium, its use with *Magnolia officinalis* in a Chinese herbal medicine for weight loss was associated with an outbreak of progressive interstitial fibrosis and terminal renal failure that may have been due to contamination with ***aristolochic acid (q.v.), derived from *Aristolochia fungi* or ***fang-chi***, a potentially nephrotoxic Chinese herb.

STERCULIA sp.

Untoward effects*:** Dermatitis venenata from working with ***mansonia wood (***S. altissima***). Instant dermatitis reported from contact with irritant hairs in the fruit of the ***Panama Tree***.

S. foetida bark and leaves are used as an insect repellent and chemosterilant in Africa. The **Creoles** of French

Guiana and **Bushnegroes** of Surinam use the ash of ***S. diversifolia*** = ***kurrajong*** in ***tobacco*** as a psychotomimetic. Fixed-drug eruptions reported from Sterculia gum. ***Karaya*** gum is exuded by the ***S. urens*** tree.

Gum LD_{50}, **rat**: 9.1 g/kg; **mouse**: 8.2 g/kg; **rabbit**: 6.4 g/kg; **hamster**: 6.9 g/kg.

STERIGMATOCYSTIN

Untoward effects: It is a toxic mutagenic and carcinogenic mold metabolite and a precursor of ***aflatoxin*** B_1. It is estimated that 300–500 μg/day will be carcinogenic in 50% of **those** exposed.

A hepatotoxin and liver carcinogen to **rats**; usually found on ***cereal grains*** and ***cheese*** rinds. Teratogenicity reports.

LD_{50}, **rat**: 120 mg/kg; **mouse**: > 800 mg/kg.

STERNO®

Untoward effects: Contains ***alcohol***, ***methyl*** (q.v.). Skid-row **derelicts** have strained it through ***bread*** to remove the ***calcium stearate*** gel, leaving relatively pure ***methanol*** to be mixed with ***wine***. Sterno® eventually reminded **buyers** that it only be used as a fuel and had a skull and crossbones printed on it.

STIBOCAPTATE

= *Astiban* = *Ro 4-1544/6* = *TWSb*

Anthelmintic, schistosomacide. **The commercial product contains 25–26% *trivalent antimony*. IM and occasionally IV.**

Untoward effects: Nausea, vomiting in 35–70%, skin rash and eruptions, pruritus, transient inversion of electrocardiogram T-wave, extrasystoles, abdominal pain in ~40%, malaise in ~24%, giddiness and pruritus in ~13%, headache in ~10%, anorexia in ~8%, diarrhea and facial edema in ~2.6%, increased SGOT, joint and back pain at the injection site, diarrhea, and, in a **few**, a metallic taste. Occasional anaphylaxis.

SC, LD_{50}, **mouse**: 500 mg ***Sb***/kg; IP, 1.4 g ***Sb***/kg.

STIBOPHEN

= *Antimosan* = *Fantorin* = *Fouadin* = *Fuadin* = *Heyden 611* = *Neoantimosan* = *Repodral* = *Sdt-91*

Anthelmintic, schistosomacide. **IM, occasionally IV, oral. Contains about 13.6% *antimony* in commercial material.**

Untoward effects: Occasionally nausea, vomiting, bradycardia, epigastric pain, shock, thrombocytopenic purpura, hemolytic anemia, eosinophilia, leukocytosis, non-thrombocytopenic purpura, optic neuropathy with visual disturbances, and fundus changes (hemorrhage and exudate). Immunohemolytic anemia and sulfhemoglobinuria in **several** after IM and death has followed in **some** after a repeated course of treatment. After oral use, increase in erythrocyte sedimentation rate, SGOT (59%), and SGPT (57%); disturbances of repolarization in electrocardiogram (87.5%), and some toxic necrosis of liver and myocardium. Use with caution in nephritis, hepatitis, and cardiac and thyroid disease. Recommended air exposure limit of 500 μg/m^3, TWA.

IM, LD_{50}, **rabbit**: 91 mg/kg; IV, LD_{50}, **rabbit**: ~90 mg/kg; **mouse**: 1 g/kg.

STILBAMIDINE

= *M&B 744*

Antiprotozoal. **Formerly used in parenteral treatment of leishmaniasis and trypanosomiasis. A *pentavalent antimony* compound.**

Untoward effects: Yellow atrophy of the liver; long-lasting, severe residual sensory and motor neurological defects (often delayed), trigeminal nerve neuropathy, and nonspecific immediate skin responses.

IV, TD_{LO}, **man**: 39 mg/day/29 days.

IP, LD_{50}, **rat**: 43 mg/kg; **mouse**: 317 mg/kg.

STILBAZIUM IODIDE

= *BW 61-32* = *Monopar*

Nematocide.

Untoward effects: Vomiting in 2/52 **children** receiving 10 mg/kg, which had a high cure rate versus *Enterobius vermicularis*. Enteric coating can reduce this.

Negligible vomiting in **dogs** and **cats** with doses under 50 mg base/kg.

LD_{50}, **mouse**: 1.36 g/kg.

STILLINGIA sp.

Untoward effects: The plants contain high levels of ***hydrocyanic acid*** (q.v.).

S. sylvatica was once used as a nauseant and emetic, as well as a purgative in mixtures for **man**.

S. treculiana = ***Queen's Delight*** = ***Trecul Queen's Delight*** causes numerous **sheep** losses in Texas.

S. texana = ***Texas Queen's Delight*** causes similar losses, as does ***S. sylvatica***, which is widely distributed in Texas and in North America, where the dried plants have been scattered as a **flea** repellent.

STINGRAYS

Untoward effects: ***Altobatus narimari*** = ***Spotted Stingray*** has a slender finless tail longer than its disc-shaped body and is armed with a venomous spine or stinger at the base of its tail. Found in tropical waters; some are 11 ft wide and weigh up to 450 lb.

Dasyatis akajei = ***Aikor Chiep*** = ***Aka-ei*** = ***Japanese Stingray***. Its stinger or spine is ~5" (12 cm) in length. Its venom was used alone or with ***aconite*** on spears and harpoon heads. The toxin leads to atrioventricular block and large doses cause marked vasoconstriction and respiratory depression.

Dasyatis americana = ***Southern Stingray*** has a tail ending with a whip-like appendage that has a venomous spine on it. It is found in brackish or fresh water from New Jersey to Brazil.

Dasyatis brevicandata up to 14 ft wide and weighing up to 965 lb in the tropical Pacific, has caused **human** deaths in Australia and the Pacific Islands.

Dasyatis dipterurus = ***Diamond Stingray*** found in shallow water in the Pacific Ocean from British Columbia to Central America. Its venom is injected from the sheath that covers its stinger, causing (within minutes) pain, hypotension, vomiting, sweating, diarrhea, and paralysis.

Dasyatis sabina = ***Atlantic Stingray*** = ***Stingaree***. A venomous barb is about halfway down its short tail. Found in estuaries and freshwater streams partly burrowed into the mud.

Urolophus jamaicensis = ***Yellow-Spotted Stingray***; ***Gymmura sp.*** = ***Butterfly Rays***; and ***Myliobatis sp.*** = ***Bat Ray*** = ***Eagle Ray*** have also caused toxicity in **people**.

In general, their stings can cause intense pain, often of the entire extremity (usually the foot), discoloration, erythema, cyanosis, muscle cramps, tremors, nausea, vomiting, fever, chills, lymphadenitis, paresthesias, arrhythmias, dyspnea, convulsions, hypotension, syncope, and shock; possible foreign body, necrosis, wound infections, and tetanus; and, rarely, death. The toxin is heat labile. Although not aggressive, their venomous nature has been known for > 2000 years. Aristotle, Pliny, and others called them "***devil fish***" and "***demons of the sea***". They are bottom-feeders and **tourists**, **swimmers**, **fishermen**, or **divers** unintentionally step on them and are stung. This is usually prevented by a "shuffle-walk", i.e. with each step, put the heel down first and swish the foot rapidly from side to side, frightening away stingrays in the path. Also see ***Fish***.

Cats given large amounts of the venom develop increased PR interval, atrioventricular block, ischemia, and vasoconstriction. **Mice**, **cats**, and **monkeys** given lethal doses develop hyperkinesis, prostration, marked dyspnea, blanching of ears and retina, and exophthalmus, followed by complete atonia, cyanosis, dyspnea, gasping, coma, and death.

STIPA sp.

Untoward effects: In Alabama, ***Canadian Spear Grass***, ***Buzzard Grass***, and ***Feather Grass*** awns have penetrated the mucous membranes of **dogs** and caused the deaths of many from infections and lung injury.

S. capillata, a Russian grass, kills **sheep**, the same way ***S. spartea*** does.

S. robusta = ***Sleepy Grass*** = ***S. vaseyi*** is found in the western and southwestern U.S. Both the green and dry plants have ***narcotic***-like effects in **horses**, causing some to sleep up to 7 days or more. **Pioneers** travelling through such areas often had to wait for their **horses** to wake up. Several **members** of a survey party died because their **horses** slept through an **Indian** attack. **Cows** that ate it readily became sleepy and catatonic, sometimes staying in one position with a hoof raised and immobile, oblivious of pestering flies, for as long as 45 min. **Cows** refuse to try it again, so **ranchers** have given them and **sheep** small amounts, so that when they go on the range, they refuse to eat it. Initial exposure also acts as a tranquillizer, calming **cattle** for the drive. **Sheep** are affected to a lesser extent than **cattle**. The ***narcotic***-like substance has been identified as ***diacetone alcohol***.

S. spartea = ***Porcupine Grass*** = ***Needle-Grass*** = ***Hay Needle*** = ***Devil's Darning Needle*** should not be grazed during the seeding stage when its awned sharp-pointed seeds can be injurious to **livestock**, especially **sheep**, where the awns catch in their fleece and work downward into their flesh and under the pleura, causing decreased carcass quality and death.

STIZLOBIUM DEERINGIANUM

= *Velvet Bean*

Untoward effects: Contains 25 g ***L-dopa*** (q.v.)/kg of beans.

Fed as ⅓ their diet leads to decreased growth rate in **chicks** and egg production in **hens**. High mortality and decreased weight gain in **rats** fed it at a 5–15% level.

STODDARD SOLVENT

= *Mineral Spirits*

Use: A form of ***naphtha*** (q.v.) used in dry cleaning, metal cleaning, etc.

Untoward effects: Aplastic anemia and death in 31-year-old **male** after 16 year exposure while a heavy equipment

mechanic. Since it did not contain ***benzene***, its myelotoxic properties were not recognized until ~1970. Follicular dermatitis of hands and forearms with redness, roughness, dehydration, and desquamation after 2 weeks in a cleaning establishment with hands wet with it. Later, the **man** developed jaundice and subacute yellow atrophy of **his** liver. Inhalation irritates eyes, nose, and throat; dizziness can occur. Aspiration leads to chemical pneumonia. National Institute for Occupational Safety and Health and OSHA upper exposure limits 350 and 500 mg/m³, respectively. Immediate danger to life or health, 20,000 mg/m³.

STONE FISH

Most are *Synańceja sp.* and some are in the genus, *Ininicus*.

Untoward effects: The ***Deadly Stonefish = S. horrida, S. trachynis***, or ***S. verrucosa*** is extremely venomous (heat labile) and its sting causes instant excruciating pain, syncope, shock, and anaphylaxis; takes months to heal. See ***Fish*** for details. A hyperimmune **horse** serum is available to prevent fatalities.

STORAX

= *Styrax = Sweet Gum*

Untoward effects: Sensitivity to the sizing agent in typing ***paper*** causes dermatitis and sensitivity to storax. May cross-react with ***Balsam of Peru***. ***Styrax argenteus*** bark is well-known as a poison in the San Francisco area Mexican community, where **women** drink a ***tea*** made from it leading to amennorhea, emaciation, and weakness, with resultant sterility that may last up to 7 years.

LD_{LO}, **human**: 500 mg/kg.

STOVES

Untoward effects: Wood-burning stoves to heat homes contributed to persistent coughing and wheezing in young **children** in a report from Michigan State University College of Human Medicine.

STRAMONIUM

Untoward effects: Contains ***hyoscyamine*** with small amounts of ***hyoscine*** and ***atropine***. U.S. annual imports are ~350,000 lb. The plants are native to India and, apparently, carried to Europe by wandering **gypsies** who used the seeds in **their** magic arts. Non-prescription medications, such as ***Asthmador®*** (containing ***scopolamine*** and ***belladonna alkaloids***), when used by irresponsible **individuals**, have caused **many** to be hospitalized. It was intended for inhalation, but **those** who ingest it develop acute psychosis, agitation, confusion, hallucinations (visual and auditory; ~50%), hyperactivity or combativeness, disorientation, ataxia, mydriasis, flushing and occasionally stupor, delirium (~25%), fixed skin eruptions, coma, amnesia, and death. A serious problem of drug abuse in the 1960s and 1970s. Deaths have occurred due to impaired judgement and physical incoordination.

LD_{LO}, **human**: 50 mg/kg.

STRAWBERRY

= *Fragaria vesca*

Untoward effects: Contain natural ***salicylates*** and have caused allergic contact urticaria. Some strawberry-flavored soft drinks contain ***tartrazine*** (q.v.). Rich in ***oxalates***.

Severe angioedema in a 17-month-old **male** lowland **gorilla** after eating two strawberries 12 h earlier. Cause confirmed by a scratch test.

STRELITZIA REGINAE

= *Bird of Paradise*

Untoward effects: Ingestion of its plant parts by **cats** leads to abdominal pain, nausea, salivation, and vomiting.

STREPTOKINASE

= *Kabikinase = Streptase*

***Thrombolytic.* IV use.**

Untoward effects: Bleeding and phlebitis have occurred at injection sites. An incidence of 0.3–0.5% of major internal bleeding (gastrointestinal, liver, genitourinary, retroperitoneal, intracerebral, and meningeal – others have reported up to 16%), and death in **some**. Vitreous hemorrhage, iritis, uveitis, blindness, pulmonary emboli, Guillain–Barré syndrome, spontaneous splenic rupture, backache, and serum sickness-like reactions have also been reported. Allergic reactions (1–4%) with fever, shivering, respiratory difficulties, bronchospasm, nausea, periorbital edema, angioneurotic edema, urticaria, pruritus, flushing, malaise, sweating, hypotension, headache, musculoskeletal pain, vasculitis, interstitial nephritis, anuria, oliguria, glycosuria, proteinuria, hematuria, and rectal hemorrhage. May potentiate ***anticoagulants***, such as ***warfarin*** and ***aspirin***, increasing bleeding. Use with ***streptodornase*** increases allergic reactions and increases blood urea nitrogen.

STREPTOMYCIN

Antibiotic, tuberculostatic.

Untoward effects: Potentially nephrotoxic and neurotoxic, including ototoxicity. Renal immaturity in **premature**

infants, **infants**, and **neonates** increases risk of nephrotoxicity, as well as in **patients** with impaired renal function or prerenal azotemia. Crosses the placenta and is found in the breast milk of treated **mothers**.

TD_{LO}, **woman** (17 weeks pregnant): 680 mg/kg; IP, 143 mg/kg.

Blood: Leukopenia, neutropenia, pancytopenia, eosinophilia, thrombocytopenia, agranulocytosis, hemolytic and aplastic anemia, thrombocytopenic purpura; decreases ***vitamin K*** production by intestinal bacteria; lymphocytosis, leukocytosis; and anemia in glucose-6-phosphate dehydrogenase-deficient **patients**.

Cardiovascular: Myocarditis, acute obstruction of radial and ulnar arteries, vasculitis, and polyarteritis nodosa.

Clinical tests: Gives false positive ***glucose*** readings in tests using ***copper*** and false positive ***phentolamine*** test results.

Central nervous system: Dizziness, fever, peripheral neuritis, circumoral or perioral paresthesias, flushing, pyschosis; confused state with choreiform movements; loss of consciousness, vertigo, ataxia, and induced myasthenia gravis syndrome. Encephalopathies are rare.

Ear and eye: Vertigo, tinnitus, deafness (usually at high dosage), toxic amblyopia, optic neuritis (usually reversible), and lens opacity.

Fetus and neonate: Eighth nerve damage, hearing loss, micromelia, and multiple skeletal anomalies. In **neonate**, stupor, flaccidity, deep respiratory depression, deaf mutism, alteration of gut flora, and occasionally coma. With **mother's** serum level of 2–3 mg/100 ml, **her** breast milk contains 1–3 mg/100 ml.

Gastrointestinal: Nausea and vomiting.

Liver: Necrosis and jaundice.

Miscellaneous: Angioneurotic edema, anaphylactic shock, muscle weakness, and pain with IM injection. Cross-reactivity reported with other ***aminoglycosides***. Alkaline urine increases its activity. Dyspnea and hypersensitivity-type asthma. Asthmas and rhinitis in manufacturing plant **workers**. Endotoxic shock, pulmonary infiltrates and eosinophilia, anosmia, and metabolic alkalosis. When used as an inhalant, contact dermatitis and throat irritation was noted. Use has increased secondary fungal infections. Rarely, reversible damage to proximal renal convoluted tubule.

Skin: Rash, exfoliative dermatitis, exanthematous and lichenoid eruptions, alopecia, facial edema, toxic epidermal necrolysis, systemic lupus erythematosus-like lesions, fixed-drug eruptions, toxic erythema with follicular pustules, and urticaria. Contact dermatitis and eczema was once common among **nurses** and manufacturing plant **workers**.

Drug interactions: Neuromuscular blocking effects with some general ***anesthetics***, parenteral ***magnesium***, and ***skeletal muscle relaxants***, such as ***alcuronium***, ***decamethonium***, ***gallamine***, ***succinylcholine***, and ***tubocurarine***, and causing prolonged respiratory depression and apnea. ***Amobarbital***, ***chlorpromazine***, ***ethylenediaminetetra-acetic acid***, and ***promethazine*** enhance its ***curare***-like effect. Potentiates ***warfarin's*** anticoagulant effect. ***Dimenhydrinate*** masks its symptoms of toxicity. Its ototoxicity is enhanced by ***ethacrynic acid*** and ***furosemide***. Potentiates the antitubercular action of ***aminosalicyclic acid*** and ***isoniazid***.

It should not be given parenterally within 30 days of slaughtering **animals** for **human** consumption, and ***milk*** from treated quarters should not be saved for **human** consumption for at least eight milkings or 96 h after the last treatment. The FDA requires up to 4, 2, 4, and 3 days withdrawal periods, respectively, before slaughtering **chickens**, **calves**, **swine**, and **turkeys** after oral use of the drug to meet the requirements of a zero tolerance for the drug in their uncooked edible tissues and products. Toxicity increases in nephritis and in any interference with renal clearance. May increase the curarizing effects of ***muscle relaxants*** and cause neuromuscular blocks during anesthesia. May discolor ***polyethylene*** devices. Oral solutions have poor water stability and should be mixed fresh daily. Simultaneous feeding with ***copper*** to **poultry** destroys growth response to streptomycin. Cardiovascular decrease by IV use may be a function of decreased plasma ***calcium*** ions. IV use causes varying degrees of hypotension and bradycardia in many species (**cats**, **dogs**, **non-human primates**). Anaphylactic reactions have been noted in **cattle**. Clinical doses can induce dyspnea and apnea in ***pentobarbital*** anesthetized **rabbits** and **cats**. The U.S. FDA is still confused and vacillates about allowing it for **animal** and **human** use. In **rats** and **guinea pigs** leads to scratching at snout, chewing, and apathetic condition; **guinea pigs** develop bronchospasm; hypotension in **dogs** and bradycardia at high dosage. Toxic in **cats** and **dogs** leading to nausea, vomiting, and ataxia and the **cat** is very sensitive to its adverse effects. Fatal enterotoxemia after its use in **rabbits**.

LD_{50}, **mouse**: 9 g/kg; SC, 520 mg/kg; IV, 160 mg/kg; **rat**: 9 g/kg.

STREPTONIGRIN

= *Nigrin* = *NSC 45,383*

Antineoplastic, antibiotic.

Untoward effects: After IV, nausea, occasional vomiting, fever, transient and infrequent diarrhea, severe dermatitis, urticaria, esophagopharyngitis, facial erythema and edema, thrombophlebitis at injection site, anorexia, proteinuria, alopecia, central nervous system toxicity, azotemia, leukopenia, and thrombocytopenia; occasionally severe bone marrow depression; and stomatitis occurring rarely.

Teratogenic in **rats**.

IP, LD_{50}, **mouse**: 800 μg/kg.

STREPTOZOCIN

= NSC 85,998 = Streptozotocin = U 9889 = Zanosar

***Antineoplastic.* For IV use; not active orally. A *nitrosourea* analog.**

Untoward effects*:** High emetogenic potential, abdominal cramps, diarrhea, fever, and pain at injection site are early toxic effects. Extravasation leads to severe necrotic reactions. Delayed nephrotoxicity (azotemia, anuria, hypophosphatemia, glycosuria, proteinuria, Fanconi syndrome, and tubular acidosis) and hepatotoxicity in ~65%. Also hypo- or hyperglycemia, anorexia, and eosinophilia; occasionally bone marrow depression, lens opacity, cardiac arrhythmia, arm pain, flushing, fever, hypotension, syncope, rash, pruritus, excitation, and nephrogenic diabetes insipidus. Occasionally a life-threatening acute hypoglycemia caused by great increase in the level of circulating ***insulin.

TD_{LO}, **human**: 1044 mg/kg/5 days. IV, TD_{LO}, **woman**: 170 mg/kg/3 weeks; IV, LD_{LO}, **woman**: 440 mg/kg/65 weeks.

A potent pancreatic *β*-cell toxin inducing experimental diabetes in **rats**, **cats**, **dogs**, ***Octogon degus***, and **monkeys**. Induces hepatotoxicity in **dogs**, **monkeys**, and **hamsters**; pancreatic tumors in **rats**; liver tumors in **hamsters** and female **rats**; uterine tumors in female **mice**; lung tumors in **mice**; kidney tumors in **mice** and **rats**; and peritoneal sarcomas in male **rats**. Rapid placental transfer in **monkeys**.

LD_{50}, **mouse**: 264 mg/kg; IV, LD_{50}, **rat**: 138 mg/kg; **dog**: 50 mg/kg; **mouse**: 275 mg/kg; IP, LD_{50}, **mouse**: 275 mg/kg; IV, LD_{LO}, **dog**: 25 mg/kg; TD_{LO}, **rat**: 25 mg/kg.

STROBANE®

= Compound 3961 = Terpene Polychlorinate

***Insecticide.* Contains 65% *chlorine*. A *chlorinated hydrocarbon*.**

***Untoward effects*:** Dermatitis. In **man** and **animals**, it can induce fasciculations, muscle spasms, restlessness, convulsions, and fever followed by cyanosis, depression, coma, and death.

LD_{LO}, **human**: 50 mg/kg.

Can be found in **cows'** milk after exposure. Highly toxic. Avoid inhalation or contamination of feed and water supplies. Do NOT allow **poultry** to eat killed **insects**. Do NOT use on buildings housing **dairy cattle**.

LD_{50}, **rat**: ~200 mg/kg; dermal, **rabbit**: > 20 mg/kg; female **chicks**: 139 mg/kg; LC_{50} in feed, **mallard duck**: 451 ppm; **pheasant**: 800 ppm; **quail**: ~750 ppm.

STRONTIUM

Untoward effects*:** Radioactive strontium-90 is from nuclear fallout, with a long half-life of ~28 years; strontium-89 occurs in negligible amounts from atomic plants. **Infants** are protected from large amounts of radioisotopes of strontium by ***calcium in the ***milk***. Highest levels of strontium-90 suddenly occurred in 1963 after extensive U.S. and Russian atomic weapon tests that year. Strontium intake can interfere with the absorption of ***tetracyclines***. Radioactive strontium seeks out bones and can cause cancer. Dietary ***calcium*** is preferred by **pigs** and **sheep** to strontium-90.

STRONTIUM CHLORIDE

Untoward effects*:** **Pigs** on a low ***calcium intake when fed it develop weakness, incoordination, paralysis, and gross bone abnormalities.

LD_{50}, **rat**: 2250 mg/kg; **mouse**: 3.1 g/kg; IV, LD_{50}, **mouse**: 148 mg/kg.

STRONTIUM HYDROXIDE

= Strontium Hydrate

Use*:** In ***sugar refining and as a plastic stabilizer.

***Untoward effects*:** Strongly alkaline solution with a pH of ~13.5. Eye contact can cause serious corneal edema and opacification. Confirmed in **animal** trials.

STRONTIUM SULFIDE

***Untoward effects*:** A 20–25% concentration topically will soften hair in 3–4 min; 15% will work in 5–7 min, enabling hair to be removed easily.

LD_{LO}, **human**: 50 mg/kg.

STROPHANTHIDIN

A cardiac glycoside from *Strophanthus*.

Untoward effects*:** Nausea, vomiting, and cardiac effects similar to ***digitalis (q.v.) and ***oubain*** (q.v.). Poisonous.

IV, LD_{LO}, **cat**: 280 μg/kg; **rabbit**: 110 μg/kg; TD, **dog**: 50 μg/kg.

STROPHANTHIN

= *Kombetin* = *Myokombin* = *Strophanthin K* = *K-Strophanthoside*

A cardiac glycoside from *Strophanthus kombe*. Usually IM or IV.

Untoward effects: Nausea, vomiting, and diarrhea reported on oral, IV, or IM use. Toxic effects and warnings similar to ***digitalis*** (q.v.) and ***oubain*** (q.v.).

LD_{LO}, **human**: 5 mg/kg.

IV, LD_{50}, **rat**: 15 mg/kg; **mouse**: 2.5 mg/kg; **guinea pig**: 270 µg/kg; **frog**: 690 µg/kg; IV, LD_{LO}, **cat**: 160 µg/kg; **rabbit**: 200 µg/kg.

STROPHANTHUS sp.

Untoward effects: Misuse or overdose of this cardiotonic will cause cardiac arrhythmias and reproduce other toxicities of ***digitalis*** (q.v.), ***oubain*** (q.v.), or ***strophanthin*** (q.v.). Species are used in African arrow poisons (***Msugu*** (Dorobo), ***Ota*** (Ibo), ***Uta*** (Igbo), ***Mchoki*** (Swahili), ***Isha*** (Yoruba), ***Kwankwani*** (Hausa), ***Mlibiti*** (Zigua)), and in Philippine, South-East Asian, and Indo-Chinese arrow poisons. The seeds are the most potent part.

STROPHARIA sp.

Untoward effects: A hallucinogenic ***mushroom*** found in the southern U.S., southward into South America, containing ***psilocin*** (q.v.) and ***psilocybin*** (q.v.). Ingestion has also caused a narcotic syndrome in **man**.

STRYCHNINE

Analeptic, rodenticide, stimulant. **An alkaloid in *Strychnos sp.* with direct effect on the central nervous system and spinal cord.**

Untoward effects: Toxicity can have a diurnal variation. Overdoses lead to hyperesthesias, violent convulsions lasting 30–120 s and tetanic spasms culminating in respiratory failure, asphyxia, cyanosis, and death. In rigor mortis, **animal** and **human** limbs are extended and the neck is opisthotonic. Adulteration of "street drugs" with it has caused fatalities. Found in tonics, cathartics, and stimulants. Lethal dose varies in different reports as 5 mg or 15–30 mg and a single reference to 75 mg. Toxic serum concentration reported – 100–300 µg/l. After mistaking it for ***cocaine***, eight young **adults** inhaled it at a party and within 30 min, convulsions began. A depressed 53-year-old **female** drank some ***mole poison*** with a 0.5% w/v level and convulsed twice within 30 min and developed abnormal eye movements. Accidental ingestion of many medicinals containing it poisoned numerous **children**. Mammary excretion occurs in **humans**. Induces fixed-drug eruptions. Rapid absorption from the gastrointestinal tract. Many laxative pills contained it and were often sugar-coated, posing a serious problem for **children**. A constipated **person** can take several and become toxic. Only 10–20 min elapses before symptoms develop with fatal doses and death occurs within 40 min after ingestion. 0.2 mg% in serum is considered toxic and 0.9–1.2 mg% as lethal. A **woman** died after ingesting 5 mg (1/12 grain) and a **male** recovered after 1 g (15 grains). 60–90 mg (1–1.5 grains) is usually fatal. In the 1940s, ~75 **children** were accidentally killed by it annually in the U.S. One of the alkaloids of the poison arrows of the **pygmies** of Central Africa. Urinary acidification increases excretion.

IV, LD_{LO}, **human**: 160–340 µg/kg (11.5–23 mg).

In 1866, a scientist was able to help **whalers** kill **whales** in 10 min with it, but **they** preferred their harpoons. **Cats** and **dogs** are often affected within 10–120 min by ingesting poisoned baits; symptoms include apprehension, nervousness, excitation, salivation, mydriasis, tetanic spasms, sawhorse appearance, and convulsions. Spasms can be spontaneous or due to external stimuli (touch, sound, hand-clapping, hyperacusis, or light). Tachycardia, weak pulse, anoxia and cyanosis, and eventually death can occur. **Dogs** rarely vomit or void urine or feces and do not having running fits or extremes of yelping, but do have intermittent periods of relaxation or flaccidity. Rate of appearance of symptoms depends on dose size and amount of stomach contents, with death in **dogs** within 1–2 h and it is usually eliminated from the body in 1–3 days. Its use was once widespread for the control of **rodents** (in poisoned ***wheat***), **salamanders**, **bats**, **wolves**, **foxes**, **moles**, **gophers**, **squirrels**, **pigeons**, **birds**, **prairie dogs**, **skunks**, and **rabbits**, and marauding **coyotes** and **dogs**. Most uses are now prohibited in the U.S. Secondary poisoning of **coyotes** occurs. Wild **geese** have fallen from the sky, convulsed and died after eating strychnine grain bait for **rodents**. Symptoms in poisoned **birds** include feathers fluffed or held tightly against body, ataxia, asynergy, fasciculations, wing-drop, tail pointed down, salivation, tremors, hyperacusis, muscle tenderness, intermittent tetanic seizures or convulsions, anorexia, tachycardia, and opisthotonus within 10 min of ingestion, and death usually in 1–5 h, except **California quail** (up to 7 days) and **mule deer** (up to 4 days). Has been used to kill **mink** at pelting time. **Animals** have relapsed into fatal convulsions several hours after apparent recovery.

LD_{50}, **mallard duck**: 2.9 mg/kg; **pheasant**: 8.5 mg/kg; **chuckar**: 16 mg/kg; **Cotornix quail** and **Japanese quail:** 22.6 mg/kg; **pigeon**: 21.3 mg/kg; **sparrow**: 4.18 mg/kg; **bullfrog**: 2.21 mg/kg; **mouse**: 500 µg/kg;

LD, **rat** and **horse**: 1 mg/kg; **cat** and **dog**: 0.75 mg/kg; **cattle**: 1.5 mg/kg; **pig**: 0.5 mg/kg; **golden eagle**: 5–10 mg/kg; **California quail**: 112 mg/kg; **pheasant**: 8.5–25 mg/kg; **mule deer**: 17–24 mg/kg; **rabbit**: 600 μg/kg; LD_{100}, **mouse**: 0.5–1.25 mg/kg.

STRYCHNOS

Untoward effects: ***S. nux-vomica***, a source of ***strychnine*** (q.v.), has been used in Asia from Bengal south for arrow poisons. Other species have been used in west and central Africa as an arrow poison and as an ordeal poison; and in South America as an arrow and dart poison. Some African species have a muscle relaxant effect. ***S. brachinata***, ***S. cogens***, ***S. guianensis***, ***S. javariensis***, ***S. Mitscherlichii***, ***S. panurensis***, ***S. Peckii***, and others are sources of ***curare*** (q.v.) found in their roots or bark by the **Indian natives** of Colombia and used on poison arrows or darts.

STRYPHNODENDRON OBOVATUM

= *Barbatimao*

Untoward effects: During June to August, a Portugese report made .of an outbreak of photosensitization in **cattle**, with salivation, constipation, dullness, and erosions of buccal mucosa with high morbidity and mortality. **Cattle** ate the fruit during a dry season. Experimental feeding of 500 g of the triturated fruit/12 days led to photosensitization after a total of 3 weeks. Post-mortem revealed skin lesions, jaundice, ruminal mucosa hyperemia, and renal petechiae.

STYRENE

= *Cinnamene* = *Cinnamol* = *Ethenylbenzene* = *Phenylethylene* = *Styrol* = *Styrolene* = *Vinylbenzene*

U.S. annual demand > 14 billion lb.

Use: In manufacturing plastics, synthetic ***rubber***, ***rubber*** cement, adhesives, insulation, packaging, wall panels, carpet backs, and resins. Tremendous amounts of ***benzene*** are used in its manufacture. Occurs naturally in ***beef***, ***coffee***, ***oats***, ***peaches***, ***peanuts***, ***strawberries***, and ***wheat*** at < 10 ng/g; ***cinnamon*** contains > 150 ng/g and Indonesian samples had > 36,000 ng/g. Oily liquid with a sweet, pungent floral odor.

Untoward effects: Inhalation is irritant to eyes, nose, and especially to the respiratory tract; causes headache, fatigue, dizziness, confusion, nausea, malaise, drowsiness, incoordination, narcosis, and ocular toxicity. A different taste is noted after it enters into the body. Abnormal electroencephalograms in **workers** after exposure to high levels. Defatting and myeloneurotoxicity on skin contact. In smoke it leads to dyspnea, cough, wheezing, and central nervous system symptoms. It can migrate out of ***polystyrene*** food packaging. National Institute for Occupational Safety and Health and OSHA set upper exposure limits, TWA as 50 and 100 ppm, respectively. National Institute for Occupational Safety and Health 15 min exposure, 100 ppm. Immediate danger to life or health, 700 ppm.

LD_{LO}, **human**: 500 mg/kg; inhalation TC_{LO}, 376 ppm; LC_{LO}, 10,000 ppm/10 min.

In male **beagle dogs** fed 400 or 600 mg/day/9 months or in female **beagle dogs** fed 600 mg/day/9 months, red blood cells and hemoglobin levels were decreased. Will decrease hatchability of **chicken** eggs.

LD_{50}, **rat**: 5 g/kg; **mouse**: 316 mg/kg.

SUCCINYLCHOLINE

Muscle relaxant.

The bromide form = *Brevidil M* = *Compound 48/268* = *IS-370* = *LT-1* = *M&B 2207* = *Suxamethonium Bromide.*

The chloride form = *Butanedioyl Chloride* = *Succinyl Dichloride* = *Succylene* = *Suxamethonium Chloride* = *Sux-Cert.*

The iodide form = *Celocurine* = *Ditilin* = *Diacetylcholine Diiodide* = *Diacetylcholine Iodide.*

Relatively short-acting depolarizing drug for IV use.

Untoward effects: Adverse reactions are generally an extension of its pharmacological effects. Plasma pseudocholinesterase catalyzes its hydrolysis and influences the amount that reaches its action site on the myoneural junction. Serious problems occur when it is used in **patients** taking or exposed to anticholinesterase compounds (viz. ***bunamidine***, ***butonate***, ***cythioate***, ***diisopropyl fluorophosphate***, ***echothiophate***, ***edrophonium***, ***neostigmine***, ***physostigmine***, ***procaine***, ***prochlorperazine***, ***promazine***, ***pyridostigmine bromide***, and assorted ***aminoglycoside antibiotics***, ***organophosphate insecticides***, and ***anthelmintics***). Short-acting ***anesthetics***, such as ***eugenols***, ***ketamine***, and ***propanidid***, may potentiate its neuromuscular blocking effects by competing with it for plasma pseudocholinesterase. It is fast-acting (even in 2–3 min) and can disappear in as little as 5 min. ***Thiazide diuretics***, ***furosemide***, and ***Cinchona alkaloids***, such as ***quinidine***, also potentiate it. ***Thiopental*** hydrolyzes it. ***Amikacin***, ***aprotinin***, ***bacitracin***, ***cimetidine***, ***cyclophosphamide***, ***dexpanthenol***, ***dihydrostreptomycin***, ***fluorinated anesthetics***, ***gramicidin***, ***hexafluorinium***, ***kanamycin***, ***lithium***, ***lysergic acid diethylamide***, ***neomycin***, ***phenelzine***, ***polymyxin B***, ***streptomycin***, ***tacrine***, ***trimethaphan***, and some ***antineoplastics*** can also inhibit serum cholinesterase and prolong muscle relaxation. In a nearby hospital, it was administered to a farmer's **wife** during surgery, but **she** failed to destroy it rapidly and had to be kept alive by a pulmotor for many days. Apparently, **their** home was plagued with **flies** and **she** had a number of fly strips containing ***dichlorvos*** hanging all over **her** kitchen and home. The **anesthesiologist** was surprised that I decided that was the cause of the dilemma. Some **people** (1/2500–1/3000) inherit a modified type of plasma esterase that is much less efficient than other **people's**, and **they**, too, can remain paralyzed for a longer period of time than most **patients**. A rare **few** inherit almost no plasma esterase and **they** are particularly sensitive to the drug. Fatal apnea has occurred. **Infants** < 1 year are more sensitive to its effects. Some **newborns** have been resistant to its effects. Severe bronchospasm in 18-year-old **female** with Pott's disease after 100 mg IV. Its injection can induce a sudden release of ***potassium*** from muscles, causing ventricular arrhythmias. Marked hyperkalemia reported in burn **patients**, in **patients** with uremia or renal failure, or **those** with extensive trauma. Myoglobinuria in 7-year-old **male** after two SC doses of 50 mg; myoglobinemia in 40% of **children** and 3% of **adults**; significant increase in creatine phosphokinase, tachycardia or bradycardia, asystole, nodal rhythmn hypertension, and cardiac arrest have occurred. Has increased intraocular pressure by contracting extraocular muscles and size of anterior chamber and decreasing the lens thickness. It is a ***histamine***-releaser leading to hypotension, bronchospasm, and excessive bronchial and salivary secretions. Although it crosses the placenta very poorly, it may cause hypoventilation in the **newborn** and has been suspect as a cause of skeletal deformities. Has caused hyperpyrexia, acidosis, disseminated intravascular coagulation, renal failure, QT interval prolongation, and ventricular irritability in a digitalized **patient**. Since it may stimulate the sympathetic nervous system, it can increase blood pressure, especially after continuous infusion. Hypersensitivity reactions with erythema, hypotension, muscle tenderness, cutis anserina, bronchospasm, and anaphylaxis have occurred. "Absolute foolishness that lethal and non-lethal drugs should have been kept together" led to the death of 1-day-old **male** after a **nurse** mistook it for distilled water.

Avoid use in **animals** with depressed cholinesterase levels, as might occur from recent (as long as 1 month) use of ***organophosphorus insecticides*** or ***anthelmintics***; ***bunamidine***, ***nitroxynil***, ***methyridine***, ***pyrantel***, ***tetramisole***, or ***phenothiazine anthelmintic*** use, ***procaine*** use, ***phenothiazine tranquillizer*** use; cholinesterase inhibitors, such as ***edrophonium***, ***neostigmine***, ***physostigmine***, etc.; or any ***curare***-like acting drugs. Momentary apnea frequently occurs during the period of peak effects. If prolonged apnea occurs, artificial respiration is required. Administer with caution in some breeds or species (**cattle**, **goats**, **sheep**, **bison**, **deer**) with naturally low plasma cholinesterase levels. Experimental **dog** trials confirm its production of hyperkalemia in **humans** with central nervous system injury and subsequent paralysis. Solutions are physically incompatible with alkaline agents, such as ***thiopental sodium***. Use of fresh solutions may eliminate many untoward effects reported. Aqueous solutions are highly acidic. They hydrolyze on standing with loss of activity and increase in acidity. Refrigerate solutions. Avoid use in **horses** with cardiac disease since the drug produces cardiac arrhythmias, hypertension, and respiratory distress. Electrocardiogram and post-mortem reports on **horses** are controversial. Some indicate acute myocardial injury. Cardiac irregularities are less pronounced in anesthetized **horses**. Prevent downed **animals** from aspirating stomach contents. Prevent accidental injection of **man** (100 mg SC or IM may cause serious distress). Hypothermia potentiates its effects. May contribute to malignant hyperthermia in **dogs**, **pigs**, and **man**. **Users** must remember it is NOT an anesthetic. Treated **animals** still perceive sensations of pain, touch, and pressure. Although used by the U.S. government to kill **swine** in controlling disease, it is not really humane as it causes death by respiratory paralysis or suffocation.

LD_{50}, **mouse**: 125 mg/kg; IV, 5 mg/kg; IP, 2.1 mg/kg; IV, **rabbit**: 800 μg/kg; IV, LD_{LO}, **dog**: 300 μg/kg.

SUCCINYLSULFATHIAZOLE

= *Sulfasuxidine*

Antimicrobial, sulfonamide. **"Non-absorbable" for enteric use.**

Untoward effects*:** Allergic reactions in **patients** previously sensitized to ***sulfonamides. Erythema multiforme or Stevens–Johnson syndrome and toxic epidermal necrolysis.

Drug interactions*:** ***Mineral oil, ***laxatives***, and ***purgatives*** can interfere with its action.

SUCKLEYA SUCKLEYANA

= *Poison Suckleya*

***Untoward effects*:** The drier the soil, the more stressed the plant becomes, the more toxic (cyanogenic) the plant becomes. Lethal to **cattle** ($\frac{1}{2}$ lb of green plant/cwt) and other **livestock** in Colorado and New Mexico. Symptoms include salivation, dyspnea, muscle tremors, arrhythmias, ataxia, bloating, convulsions, and death from respiratory failure, yet the heart may still be beating. Post-mortem reveals bright red blood and pulmonary congestion.

SUCRALFATE

= *Antepsin* = *Carafate* = *Citogel* = *Hexagastron* = *Keal* = *Succosa* = *Sucralfin* = *Sucrate* = *Sugast* = *Sulcrate* = *Ulcar* = *Ulcermin* = *Ulcogant*

***Antiulcerative*. Complex salt of *sucrose sulfate* and *aluminum hydroxide*.**

Untoward effects*:** Like all ***aluminum-containing drugs, it can cause hypophosphaturia and constipation (2–3%). Very little is absorbed systemically, but may increase plasma and urinary levels of ***aluminum*** to cause ***aluminum*** toxicity in **patients** with chronic renal failure. A severe case of encephalopathy in 50-year-old **female** taking 2 g/day and eight capsules of ***aluminum hydroxide***/day. Hard, putty-like gastric bezoar and esophageal bezoars in 79-year-old **male** ingesting 4 g/day. Cases of probable anaphylactoid reactions have occurred in 10 min–2 h after ingesting 1 g. Occasional dry mouth, flatulence, cramping, indigestion, nausea, vomiting, diarrhea, pruritus, rash, hives, facial swelling, dyspnea, dizziness, insomnia, sleepiness, vertigo, back pain, and headache reported in < 0.5% of **patients**. Coughing up or vomiting bright red material can occur. Red or black stools.

Drug interactions*: *Aluminum-containing antacids are best given at least 2 h before it because they decrease its absorption. A very significant interaction exists between it and ***fluoroquinolone antibiotics***, decreasing their bioavailability, such as ***ciprofloxacin*** (12.5%), ***enrofloxacin***, ***fleroxacin*** (25%), ***grepafloxacin***, ***lomefloxacin*** (26%), ***norfloxacin*** (30%), and ***sparfloxacin*** (44%). Apparently binds to ***cimetidine*** and ***levothyroxine***, decreasing their absorption. Delays absorption of ***naproxen***. It decreases bioavailability of ***aminophylline***, ***amitriptyline***, ***digoxin***, ***ketoconazole***, ***phenytoin***, ***procainamide***, ***quinidine***, ***tetracycline***, and ***theophylline***. Many case reports of important interactions causing decreased plasma concentration and prothrombin time with ***warfarin***.

Causes constipation in **dogs** and is unpalatable. Similar interactions, as in **man**, can be expected. **Rats** show the same accumulation of ***aluminum*** in bone as does **man**.

SUCROSE

= *Beet Sugar* = *Cane Sugar* = *Misri* = *Rock Candy* = *Saccharose* = *Sugar* = *Sittamisiri*

***Nutrient, sweetener, osmotic agent*. Its consists of one molecule of *glucose* and one molecule of *fructose* with a molecule of water removed.**

Untoward effects*:** "How sweet it is", but long-time **workers** in ***sugar mills develop "sugar boils". Within a few minutes after ingestion, the blood ***dextrose*** is increased, which can be very important to **diabetics**. Excessive intake associated with dental caries where dental plaque *Streptococcus mutans* uses it for growth. Excessive dietary intake has been a predisposing factor in increased incidence of coronary heart disease. Renal toxicity after IV use has caused a dramatic decrease in IV use. Dietary intake of 500 g/day/25 days led to hypoglycemia, epigastric discomfort, shoulder and arm aches, sweating, increased serum lipids, and anorexia $\frac{1}{2}$ h post-prandial in seven, 21–41-year-old **males**. Primary intolerance in 5-year-old **males** causing persistent diarrhea. Congenital sucrose–isomaltase or sucrase deficiency, a common congenital disaccharidase deficiency, causes chronic diarrhea. The undigested disaccharides are split into ***lactic acid*** and volatile lower ***fatty acids*** by intestinal bacteria leading to intestinal irritation, increased peristalsis, increased mucous formation, and watery acid stools. Some **people** are allergic to it. Its intake has been suspect as a cause of hyperactivity or attention deficit disorder in **children**, but not only has that been dispelled, but **they** were found to be significantly less active and had a calming effect while consuming ***sugar***. Possible proneness to diabetes and hyperinsulinism reported. Caramelized ***sugar*** can cause brown nail pigmentation from contact and ingestion indicates it can contain DNA-damaging agents and presumptive carcinogens. **Users** exploring new uses should be aware that so-called pure "***sugars***" on the market may contain any one of the approved additives, such as ***calcium silicate*** for anticaking effects and traces of antimicrobials. Manufacturing methods of sucrose, such as a switch from acid to enzyme hydrolysis, can cause major problems in pharmaceuticals using it. ***Sugar*** can not only absorb radiation but can also transmit its lethal effects. Air exposure limits are 5 mg/m^3. Its dust can irritate **workers'** eyes, skin, and respiratory tract and induce coughing. Repeated IV injection can cause renal tubular damage. ***Sugar cane*** fibers have caused a hypersensitivity pneumonitis or allergic alveolitis from *Thermoactinomyces vulgaris* and other organisms.

In young **animals** unable to digest it, oral administration will cause or accent diarrhea. **Sheep** given 10–20 g/kg developed indigestion; 30 g/kg led to indigestion and severe toxicosis; and 40 g/kg was lethal. **Robins** lack sucrase and develop an aversion to high sucrose-content in fruits.

LD_{50}, **rat**: 29.7 g/kg.

SUCROSE OCTACETATE

Use*:** An anhydrous adhesive used in ***lacquers and as a bitter ***alcohol*** denaturant.

***Untoward effects*:** Contact dermatitis.

SUDAN I
= *1-Phenylazo-2-naphthol*

Use: Once used to color ***margarine***. Derivatives are used as food colors.

Untoward effects: A mild carcinogen when injected into **mice**, but preneoplastic bladder changes when fed to them. Metabolized to ***aniline*** (q.v.) which is excreted in the urine mostly as ***p-aminophenol*** (q.v.). No longer permitted in foods in the U.S. and Great Britain.

SUDAN III
= *D&C Red 17* = *1(p-Phenylazophenylazo)-2-naphthol*

Use: A red external use only dye.

Untoward effects: Ingestion leads to pathological lesions in kidney, spleen, and liver of laboratory **animals**.

LD_{LO}, **rabbit**: IP, 250 mg/kg; SC, 1 g/kg.

SUFENTANIL
= *R 30,730* = *Sulfentanil*

The citrate form is *R 33,800* = *Sufenta*.

***Narcotic analgesic.* A *fentanyl* derivative.**

Untoward effects: Generalized and focal tonic–clonic seizures in **patients** after anesthetic induction with 40 and 60 μg. Similar dosage IV for induction of anesthesia induced tonic–clonic movements in 79-year-old **female** with Parkinson's disease and a 67-year-old **male** on chronic ***metoclopramide*** therapy. Can cause respiratory decrease and muscle rigidity, especially of the neck and extremities and chest wall (3–9%). Cardiovascular adverse effects (3–9%) include bradycardia, hyper- or hypotension, and possibly cardiac arrest. Pruritus in 25% and nausea and vomiting in 3–9%. Apnea, bronchospasm, depression, chills, erythema, arrhythmias, and tachycardia in up to 1%.

SULBACTAM
= *Begalin* = *Betamaze* = *CP 45,899* = *Duocid* = *Penicillanic Acid Sulfone*

Semisynthetic β-lactamase inhibitor.

Untoward effects: Has caused an osmotic type of diarrhea and an important change in bacterial flora. Average concentrations in breast milk are 0.5 μg/ml.

SULCONAZOLE NITRATE
= *Exelderm* = *Myk* = *RS-44872* = *Sulcosyn*

Antifungal.

Untoward effects: Topical use of the 1% cream or solution leads to pruritus, burning, and stinging or redness in 1–3% of **patients**.

SULFABENZAMIDE
= *Sulben* = *Sulfabenzide*

Antimicrobial, sulfonamide.

Untoward effects: Photosensitivity reported in **man**.

Accurate withholding times after treating **animals** have not been established to avoid undesirable tissue residues for **man**. In the intermim, 5 days is suggested. Intake of fluids should be encouraged.

SULFABROMOMETHAZINE
= *Sulfabrom*

Antimicrobial, coccidiostat, long-acting sulfonamide.

Use: Orally, to maintain prolonged blood levels in **cattle**. Oral or IP dosage of 214 mg or 3.3 grains/kg gives blood plasma levels of 5 mg/100 ml or over by about the fourth and first hour, respectively, lasting ~44 h. Individual **animals** may carry much longer levels. IV use is not particularly recommended as it only gives adequate blood levels for 24 h in **cattle** after a dosage of 66 mg or 1 grain/kg. Higher dosage has occasionally produced untoward effects (convulsions). These reactions are less severe in **calves** than in adult **cattle**.

Untoward effects: Intake of adequate fluids should be encouraged. Pregnant **cows** may abort. IV use over 16 mg/lb in **horses** may produce adverse reactions. ***Milk*** of treated **animals** should not be used for **man** until at least 96 h after the last treatment, and **animals** should not be slaughtered for food for an arbitrary 18 days after treatment.

SULFACETAMIDE SODIUM
= *Ak-Sulf* = *Albucid* = *Antébor* = *Beocid- Puroptal* = *Bleph 10* = *Cetamide* = *Isopto Cetamide* = *Locula* = *Op-Sulfa* = *Prontamid* = *Sebizon* = *Sodium Sulamyd* = *Sulf-10* = *Sulten-10* = *Urosulfon*

Antimicrobial, sulfonamide.

Untoward effects: Has induced hemolysis in glucose-6-phosphate dehydrogenase-deficient **subjects** and hemolytic anemia after 5 g/day in ***primaquine***-sensitive **individuals**. In ophthalmic ointments or drops it has caused sensitivity reactions on the skin around the eyes. Local treatment of eye can cause some temporary burning or stinging. Several case reports of Stevens–Johnson syndrome, including eosinophilia, after using ophthalmic drops. Can easily pass the placental barrier if used systemically and increase the risk of kernicterus in the **newborn**. Photosensitivity can occur. Half-life in **man** is 12.8 h.

Drug interactions: In some **individuals** and **animals**, ***neomycin*** potency, by unknown mechanisms, is knocked down by its presence.

Shelf-life of finished formulations is less than formerly assumed and *should* be dated. When used for systemic urinary tract effects, assure adequate fluid intake.

LD_{50}, **dog**: 8 g/kg.

SULFACHLORPYRIDAZINE

= *Ba 10,370* = *Ciba 10,370* = *Cosulid* = *Cosumix* = *Nefrosul* = *Prinzone* = *Vetisulid*

Antimicrobial, sulfonamide.

Untoward effects: Its primary use in **man** was in treating urinary tract infections and an occasional adverse effect report included tinnitus, dry throat, epigastric discomfort, headache, weakness, nausea, and loose stool. A case report of an elderly **man** who reported confusion after treatment with it.

Transient agranulocytosis, hemolytic anemia, leukopenia, jaundice, and peripheral neuritis may have occurred in some treated **cattle**. Prolonged use is *contraindicated* and has altered red blood cell counts in **dogs**. IV use in **calves** requires a 5 day withdrawal period, and oral use in **calves** and **swine**, respectively, requires 7 and 4 day withdrawal periods before slaughter to meet U.S. requirement of not over a 0.1 ppm maximum for its negligible residues in uncooked edible tissues destined for **human** consumption.

SULFACYTINE

= *CL 636* = *Renoquid*

***Antimicrobial.* In treating urinary infections.**

Untoward effects: Severe mottling of skin and pruritus in one **patient**. Photosensitivity reported.

SULFADIAZINE

= *Adiazin* = *Debena* = *Diazyl* = *Pyrimal* = *Sulfapyrimidine* = *Sulfolex*

Antimicrobial.* A *sulfapyrimidine.

Untoward effects: ***Blood***: Agranulocytosis and hemolytic anemia. Thrombocytopenia (10,000/mm^3) in 25-year-old **male** with advanced HIV. Neutropenia, eosinophilia, leukocytosis, leukopenia, and plasmacytosis also reported.

Hypersensitivity: Induced fixed-drug eruptions, erythema multiforme, and dermatitis medicamentosa with profuse generalized, petechial, pruritic eruption on trunk and extremities with bullae and erosions on foreskin and eroded vesicles on hard palate. Non-fatal shock and possible fatalities from slowly progressing types of sensitization. Loeffler's syndrome in 29-year-old **female** from the use of it in a vaginal cream leading to recurring fever, pulmonary infiltrations, chest pain, and coughing, apparently from its absorption through inflammation membranes.

Liver: Hepatocanalicular jaundice, cholestasis, necrosis, and globular fatty changes in < 0.5% of **patients**. Often preceded by nausea, vomiting, increased temperature, and macular skin rash. Has been fatal.

Miscellaneous: Placental passage occurs readily leading to high levels in **fetal** blood. A 50-year-old **male** with bronchitis developed swelling of parotid glands and floor of **his** mouth, in addition to plugging of **his** ears within 2 h of taking **his** first dose of a 350 mg/5 ml syrup. It can readily displace ***bilirubin*** from ***albumin*** and may have contributed to problems in **infants**. Absorption is delayed if taken with ***food***. Periorbital edema, conjunctival and scleral injection, chills, arthralgia, allergic myocarditis and periarteritis nodosum can occur.

Renal: Serious crystalluria, hematuria, dysuria, severe urethral pain, and fatalities associated with failure to drink adequate volume of water and to alkalinize the urine. Oliguria and hypoglycemia in 3-year-old **female** after 450 mg/kg/day/2 days and then 300 mg/kg/day/3 days.

Skin: Stevens–Johnson syndrome, systemic lupus erythematosus, toxic epidermal necrolysis, urticaria, pruritus, exfoliative dermatitis, erythema multiforme erythema with conjunctivitis, pharyngitis with signs of myocarditis, and shock reported.

Drug interactions: Slightly increases half-life of ***tolbutamide***.

Keratoconjunctivitis sicca (an incidence of about 4%), with poor healing, corneal scars, and corneal pigmentation have been reported in **dogs** treated for geriatric syndromes. Apparent allergic reactions include non-septic polyarthritis, glomerulonephropathy, focal retinitis, polymyositis, skin rash, fever, anemia, and leukopenia in **Dobermann pinschers**. ***Milk*** from treated **animals** should NOT be used until 72 h after treatment. Use with ***trimethoprim*** may be associated with keratoconjunctivitis sicca in **dogs**.

LD_{50}, **mouse**: 2 g/kg; SC, LD_{50}, **mouse**: 1.3 g/kg.

SULFADIMETHOXINE

= *Agribon* = *Albon* = *Ancosul* = *Arnosulfan* = *Bactrovet* = *Diasulfa* = *Madribon* = *Maxulvet* = *Neostreptal* = *Sudine* = *Sulduxine* = *Sulfabon* = *Sulxin* = *Symbio* = *Ultrasulfon*

***Antimicrobial, sulfonamide.* "Long-acting".**

Untoward effects: Headache, fever, malaise, nausea, vomiting, rash, urticaria, pruritus, systemic lupus erythematosus, fixed-drug eruptions, hematuria, oliguria, arthralgia, angina, Stevens–Johnson syndrome, toxic epidermal necrolysis, leukopenia, anemias, eosinophilia, neutropenia, leukocytosis, diarrhea, interstitial pneumonitis, and thrombocytopenia. A fatal case of toxic epidermal necrolysis reported. Transient myopia, ocular paralysis, iritis, retinal edema and hemorrhage, toxic amblyopia, exudative pemphigoid conjunctivitis, and injected conjunctiva and sclera have also been reported. Nephritis and increased SGOT and SGPT, especially in those **patients** drinking ***alcoholic beverages***. If given near or at term, it can displace ***bilirubin*** from ***albumin*** and increase kernicterus. Excreted in **human** breast milk and in **cows'** ***milk***. Use with ***food*** decreases its rate of absorption. Has caused anemia due to abnormal hemoglobin molecules. Half-life reported as 20.2–41 h by various labs. Enzyme system in **newborns** is unable to dispose of the drug during the first few week of life.

Drug interactions: Displaces a number of drugs from their ***albumin***-binding sites: ***bishydroxycoumarin***, ***phenprocoumon***, and ***warfarin*** leading to increased bleeding; ***methotrexate*** leading to increased ***methotrexate*** toxicity; ***tolbutamide*** leading to hypoglycemia; and ***digitoxin*** leading to increased toxicity. ***Phenobarbital*** decreases its effectiveness.

Insure adequate intake of fluids. Use cautiously in renal diseases. The U.S. does not permit levels above 0.1 ppm in uncooked edible tissues of **cattle**, **chickens**, and **turkeys**, or 0.01 ppm in ***milk*** for **human** consumption. This can be achieved by not slaughtering **cattle** or **poultry** until 6 days after treatment or saving ***milk*** from **animals** during treatment or during the first 48 h and four milkings after treatment. Water solutions may precipitate if highly acidic. If this occurs, add small amounts of 1–2% ***washing soda*** (***Sodium Carbonate***) or household ***ammonia*** to correct the problem. Use cautiously in the **neonate** (biological half-life is several times longer at birth than at or above 2 months of age in **dog** trials). Toxicity in **lambs** has been noted on dosage of 45 or more mg/kg.

LD_{50}, **rat**: > 20 g/kg; **mouse**: > 10 g/kg.

SULFADOXINE

= *Fanasil* = *Fanzil* = *RO 4-4393* = *Sulformethoxine* = *Sulformetoxine* = *Sulforthodimethoxine* = *Sulforthomidine*

***Antimicrobial, antimalarial, sulfonamide.* Ultra long-acting, permitted every third day or every week or two therapy in man for *Plasmodium falciparum* malaria. Oral, IV, and IM.**

Untoward effects: In Africa, incidence of adverse effects is 0.97%. Rash, facial or polymorphos erythema, morbilliform papular eruptions, and pleuri-orifacial erosive ectodermosis in prophylaxis versus epidemic meningitis in Morocco. These were moderate in 10% and severe in another 10% of 997 cases treated. Leukopenia, anemia, and occasionally neutropenia, purpura, and thrombocytopenia, Lyell's syndrome, Stevens–Johnson syndrome, and 11 deaths. Permanent eye damage in 18 **who** recovered developing ectropion, synechiae, and pigmentation lasting 9 months in **some**. Unilateral blindness in two. Occasional nausea, vomiting, granulocytopenia, leukopenia, pain at IM injection site, occipital headache, and mild transient diarrhea. Half-life in **man** is 93–126 h; ~12 h in **goats**; 10.5 h in suckling **pigs**; and 13 h in **calves**.

LD_{50}, **mouse**: 5.2 g/kg; IP and SC, 2.9 g/kg.

SULFAGUANIDINE

= *Abiguanil* = *Aterian* = *Diacta* = *Ganidan* = *Guamide* = *Guanicil* = *Resulfon* = *RP 2275* = *Shigatox* = *Suganyl* = *Sulfaguine* = *Sulfamidinum* = *Sulfanilylguanidine* = *Sulfoguenil*

***Antimicrobial, coccidiostat, sulfonamide.* Widely used by the armed forces in World War II, where it prevented *any* major epidemics of bacillary dysentery.**

Untoward effects: Has induced fixed-drug eruptions in **man**. Erythema multiforme occurred each time of five occasions it was administered to a **patient** in Malaysia. Neutropenia, agranulocytosis, and anemia after occasional use for 7 months in self-treatment of diarrhea. Nausea, vomiting, rash, and drug fever have been reported. Crystalluria in dehydrated **patients**. Can be absorbed into the blood in dangerous amounts in ulcerative colitis.

Although described in the literature as a non-absorbable ***sulfonamide***, usual dosage may produce effective blood levels in **cats** and **poultry**, and among food-producing **animals** it produces highest levels in **swine**. Although withdrawal periods have not been established, do NOT administer (especially to **swine**) just prior to slaughter. May be toxic to **mink**. Goitrogenic in **rats**.

SULFALENE

= *2155TH* = *Abbott AS 18,908* = *Dalysep* = *Farmitalia 204/122* = *Kelfizin* = *Longum* = *Polycidal* = *Sulfamethoxypyrazine* = *Sulfametopyrazine* = *WR 4629*

Untoward effects: Joint pains, papular rash (1.3%), urticaria, pruritus, nausea (0.8%), vomiting (0.13%), abdominal discomfort, headache, malaise (0.06%), and dizziness (0.87%) in 6521 **patients**. Lyell's syndrome and Stevens-Johnson can occur on rare occasions. The long-lasting effect can be dangerous if necessary to abort a reaction. Half-life in 1–10 day term **infants**: 280 h; 11–30 days, 136.8 h; 2 months, 67.8 h; **children** 1–4 years, 44.2 h; **adults**: 36–71.5 h; **apes**, **baboons**, and **monkeys**: ~19 h.

LD_{50}, **mouse**: 2.16 g/kg; IV, 1.41 g/kg.

SULFAMERAZINE

= *2632 RP* = *Debenal M* = *Mebacid* = *Mesulfa* = *Methyldebenyl* = *Methylpyrimal* = *Peroccide* = *Pyralcid* = *Pyrimal M* = *Sulfamerazinum* = *Sumedine* = *Supronal*

Antimicrobial, sulfonamide. **Rapidly absorbed.**

Untoward effects: Fixed-drug eruptions, crystalluria, hematuria, and albuminuria, especially in **children**; anemia, due to abnormal hemoglobin molecules; neutropenia, thrombocytopenia, anemias, photosensitivity, and other ill-effects of other ***sulfonamides*** above. Half-life in **man** is reported by different labs as 23.5–40.8 h. Easily passes through the placental barrier and is found in **human** breast milk of treated **patients**.

Insure adequate fluid intake. Use cautiously in renal disease and follow ***sulfamethazine*** withdrawal guidelines to avoid tissue residues in slaughtered **animals**, **birds**, or in **fish**. Half-life in **cows** is ~7 h and it can be found in their ***milk***.

LD_{50}, **mouse**: 2.5–3.3 g/kg.

SULFAMETER

= *AHR-857* = *Bayrena* = *Durenat* = *Kinecid* = *Kirocid* = *Kiron* = *Methoxypyrimal* = *Sulfamethoxydiazine* = *Sulfametin* = *Sulfametorine* = *Sulfametoxipirimidine* = *Sulfamonomethoxine* = *Sulla* = *Supramid* = *Ultrax*

Antimicrobial, sulfonamide. **Long-acting.**

Untoward effects: Hemolytic anemia, especially in glucose-6-phosphate dehydrogenase-deficient **patients**; crystalluria, hepatotoxicity and jaundice, peripheral neuritis, gestrointestinal upsets, hypersensitivity reactions; acute myalgia in arms and legs; mental depression, psychoses, hypoglycemia, transient allergic exanthema, transient neutropenia and leukopenia, nasal congestion, edema of the lips with thickening of buccal mucosa, fever, urticaria, photosensitivity, vasculitis, and fatal Stevens–Johnson syndrome. Toxic–allergic exanthema with maculopapular erythematous skin eruptions, high fever, clonic convulsions, and death in pulmonary edema in 35-year-old **male** with pharyngitis after other treatment failed. In pediatrics, 11% had adverse effects from tablets or suspensions. ***Paraben, propyl*** (q.v.) and ***paraben, methyl*** (q.v.) in the latter may also have been a factor. Coma, hypoglycemia, hypothermia, and oliguria, thrombopenia in 7-month-old **female** after 2 g/day/3 days (10–30 times normal dose for this age **child**) to treat a sore throat. Stevens–Johnson syndrome in 7-year-old **male** a few days after ingesting an unknown amount of his **mother's** left over tablets. The **child** survived with residual bronchiectasis and entropion of lower eyelid. Passes into **human** breast milk. Half-life in **man**: 30 h.

LD_{50}, **rat**: 1 g/kg; **mouse**: 3 g/kg; **dog**: 2 g/kg; IV, **rat**: 1.2 g/kg; **mouse**: 1.1 g/kg; IP, **rat**: 1.1 g/kg; **mouse**: 1.5 g/kg.

SULFAMETHAZINE

= *Diazil* = *Dimidin R* = *Intradine* = *Neazina* = *S-dimidine* = *Sulfadimerazine* = *Sulfadimezine* = *Sulfadimidine* = *Sulfadine* = *Sulfamezathine* = *Sulmet* = *S-mez* = *Superseptyl* = *Vertolan* = *Vesadin*

Antimicrobial, coccidiostat, sulfonamide.

Untoward effects: Serious toxic effects have been rare. Fixed-drug eruptions in 19-year-old **female**, 2 h after ingestion for treatment of sore throat leading to irritation, erythema, and urticaria. Risk of accumulation in uremic **patients** or if given for extended periods of time. Chances of crystalluria are even less than with ***sulfamerazine***. An African 12-year-old **male** developed Stevens–Johnson syndrome after 6 g during 2 days leading to 2 days later, fulminating skin lesions over 90% of body, and 28 days later leading to acute cardiomyopathy, congestive heart failure, and myocardial damage noted on electrocardiogram, which was probably a hypersensitivity reaction. Thrombocytopenia in **child** aged 3 after a total of 30 g in 11 days. A 7-year-old **male**, after 28 g over a 5 day period developed acute hemolytic anemia, presumed to be a hypersensitivity rather than a direct reaction. Neutropenia, aplastic anemia, thrombocytopenia, purpura, and occasionally lupus erythematosus have occurred. Half-life of 6–8 h.

Drug interactions: Supresses ***cyclosporine*** blood levels. Hypoglycemic coma reported with ***sulfonylureas***.

Insure adequate fluid intake. Use cautiously in renal disease, although less apt to crystallize in either acid or alkaline urine than many commonly used ***sulfonamides***. Parenteral forms (20% or greater concentration) should be limited to IV usage only, although 3–12% solutions have been safely used SC and IP by many without signs of irritation in a number of species. Do NOT use over 5% concentrations IP in young **animals**. Chronic use especially in coccidiosis may require ***vitamin K*** supplementation. The FDA allows not over 0.1 ppm of the drug in the uncooked edible tissues of **cattle**, **swine**, and **poultry**. This is usually accomplished by a 21 day withdrawal period after use of the sustained-release form in **cattle**; avoiding use of milk from **animals** treated systemically until 96 h and eight milkings after the last treatment; and withdrawing treatment 10 days before slaughtering **animals** or **poultry**. Do NOT give to laying **hens**. The drug appears to be very toxic (hypersensitivity-hemorrhage) in **male** Aleutian **mink**. 15 days withdrawal in **swine**. Feed contaminated with sulfamethazine at a 1 ppm level can produce violative tissue residues. U.S. withdrawal times for sustained-release forms (***Calfspan*** = ***Sulfa-Span*** = ***Sustain*** = ***Spanbolet***) were increased from 12 to 28 days.

IP, LD_{50}, **mouse**: 1 g/kg.

SULFAMETHIZOLE

= *Famet* = *Lucosil* = *Methazol* = *Nicene* = *Proklar* = *Renasul* = *Rufol* = *Salimol* = *Sulfametizol* = *Sulfapyelon* = *Thidicur* = *Thiosulfil* = *Urolucosil*

Antimicrobial, sulfonamide.

Untoward effects: Methemoglobinemia, hemolytic anemia, Heinz bodies, hematuria, crystalluria, photosensitivity, and lobular hepatitis. Signs of meningitis and pancreatitis in 23-year-old **female** treated with 2 g/day for abscess of Skene's glands and dysuria. After 1 week of therapy she developed headache, weakness, malaise, and abdominal distress. On separate occasions, **she** later ingested two tablets and within 30 min developed chills and fever followed by hypotension, meningitis, pancreatitis, abdominal pain, and leukocytosis, probably due to hypersensitivity. Agranulocytosis, aplastic anemia, thrombocytopenia, leukopenia, neutropenia, pancytopenia, purpura, sulfhemoglobinemia, hypoprothrombinemia, Stevens–Johnson syndrome, generalized skin eruptions, epidermal necrolysis, urticaria, pruritus, exfoliative dermatitis, anaphylactoid reactions, serum sickness, arthralgia, periorbital edema, conjunctival and scleral injection, allergic myocarditis, peripheral neuritis, depression, tinnitus, vertigo, insomnia, ataxia, hallucinations, convulsions, oliguria, anuria, periarteritis nodosum, and lupus erythematosus have been reported in some **patients**. Rare reactions might include diuresis, hypoglycemia, and goiter. Use cautiously in renal disease and insure adequate fluid intake.

Drug interactions: Use with ***methenamines*** (q.v.) causes a complexing interaction in the kidneys due to ***methenamine's*** slow release of ***formaldehyde*** in acid urine, which then forms an insoluble precipitate with the sulfamethizole decreasing its therapeutic effectiveness and increasing a potential hazard of renal blockage or calculus formation or both. Marked precipitation of it in 9/32 when treated with ***methenamine mandelate***. May inhibit ***warfarin's*** metabolism increasing its anticoagulant effect. Small increase in ***phenytoin*** toxicity by decreasing its rate of metabolism.

SULFAMETHOXAZOLE

= *Gantanol* = *Sinomin* = *Sulfaisomezol* = *Sulfisomezole*

Antimicrobial, sulfonamide. **Bacteriostatic, but synergistic with *trimethoprim* and becoming bactericidal. Usually with *trimethoprim* in a 5 : 1 ratio = *Abacin* = *Apo-Sulfatrim* = *Bactramin* = *Bactrim* = *Bactromin* = *Baktar* = *Chemotrim* = *Comox* = *Co-trimazole* = *Drylin* = *Eusaprim* = *Fectrim* = *Gantaprim* = *Gantrim* = *Imexim* = *Kepinol* = *Laratrim* = *Linaris* = *Microtrim* = *Nopil* = *Oraprim* = *Septra* = *Septrin* = *Sigaprim* = *Sulfotrim* = *Sulprim* = *Sumetrolim* = *Sumetrolium* = *Supracombin* = *Suprim* = *Teleprim* = *Thiocuran* = *Trigonyl* = *Trimesulf* = *Uroplus* = *Uroseptra***

Untoward effects: Hepatotoxicity with jaundice and transient increase in transaminases and alkaline phosphatases in 10% of **patients** and cholestasis without hepatitis. Acute hepatitis in < 0.5% with centrilobular or mid-zonal hepatic necrosis. Chronic active hepatitis and granulomatous hepatitis is rare except in AIDS **patients**. Has caused death due to hepatic failure in 70-year-old **male** after previous hypersensitivity to ***sulfonamides***. Most hepatitis due to it occurs 5–14 days after therapy. Photosensitivity, hemolytic anemia (especially in glucose-6-phosphate dehydrogenase-deficient **patients** on high dosage), nausea, vomiting, fever, morbilliform pruritic and maculopapular rashes, gum lesions, erythema and erythema nodosum, urticaria, Stevens–Johnson syndrome, headache, drug fever, cyanosis, angioneurotic edema, dizziness, leukopenia, neutropenia, thrombocytopenia, pancytopenia, methemoglobinemia, eosinophilia, leukocytosis, macrocytic anemia, and periorbital and facial edema reported. Renal damage includes tubular necrosis, crystalluria, and proteinuria. Use for urinary tract infections may be associated with increased risk of cervical cancer, lymphosarcoma, and death. Possible bilateral ureteral obstruction with medial deviation suggesting early retroperitoneal fibrosis in 67-year-old **female**. Benign intracranial hypertension in 4½-year-old **male** after 2 weeks of treatment. Hyperbilirubinemia or **neonatal** jaundice in **infant** if given to **mother** at or near term. Hemolytic anemia can also occur in glucose-6-phosphate dehydrogenase-deficient **infants**. Adverse reactions are more prevalent in **women** than **men**. Therapeutic serum levels: 30–60 mg/l; toxic levels: 400 mg/l.

Drug interactions: Use with ***phenazopyridine*** (q.v.) leads to hemolytic anemia with Heinz bodies; thrombocytopenia with gross hematuria, rectal bleeding, and petechiae; hemolytic anemia in glucose-6-phosphate dehydrogenase-deficient 64-year-old **female** after 4 tablets/day/14 days. Methemoglobinemia peaking in ~10 h with severe cyanosis, central nervous system depression, and intermittent periods of apnea after ingesting 30 tablets (0.5 g ***sulfamethoxazole*** and 100 mg ***phenazopyridine*** in each) in a failed suicide attempt in 23-year-old **male**. Occasional antagonism with ***erythromycin*** effectiveness. Enhances hypoprothrombinemic effect of ***warfarin***. May complex ***clotrimazole***, decreasing its release and solubility. Displaces ***glyburide*** from ***albumin***. ***Propantheline*** changes gastric emptying rate and decreases sulfamethoxazole rate of absorption.

Side-effects in **cattle** are generally less than with other ***sulfonamides***.

LD_{50}, **mouse**: 5.5 g/kg.

SULFAMETHOXYPYRIDAZINE

= *CL 13,494* = *Depovernil* = *Kynex* = *Lederkyn* = *Midicel* = *Midikel* = *Myasul* = *Quinoseptyl* = *Sulfalex* = *Sulfametossipiridazina* = *Sulfapyridazine* = *Sulfdurazin* = *Sulitrene* = *Sultirene* = *Unosulf*

***Antimicrobial, sulfonamide.* Long-acting.**

Untoward effects: Skin reactions in 2.2% (four-fold if **patient** has had ***sulfonamides*** previously) leading to pustules, morbilliform and urticarial eruptions within 4–20 h. Fine scarlitiniform, non-pruritic eruptions with predilection for axillary, gluteal, and inguinal folds; mild maculopapular eruptions, fixed-drug eruptions, photosensitivity, fatal Stevens–Johnson syndrome, thrombocytopenia, fever, nausea, vomiting, headache, sore throat, hemolytic anemia, Heinz bodies, thrombocytopenic purpura, aplastic anemia, hematuria, pancytopenia, eosinophilia, leukopenia, leukocytosis, neutropenia, purpura, thrombocytosis, giddiness, wheezing, angioneurotic edema, toxic epidermal necrolysis, agranulocytosis, arthralgia, abdominal pain, jaundice, hepatitis (< 0.5%), diarrhea, anorexia, pancreatitis, stomatitis, ataxia, tinnitus, depression, peripheral neuritis, vertigo, insomnia, hallucinations, convulsions, oliguria, anuria, cyanosis, and systemic lupus erythematosus have all been reported in different **patients**. Hemolysis reported in glucose-6-phosphate dehydrogenase-deficient **Negroes**. A **patient** developed a severe exfoliative dermatitis, followed by eyelid edema and crusting of the mucocutaneous junctions, then, bullous and purpuric cutaneous lesions. Fatal aplastic anemia in 67-year-old **female** after 14 g; hemolytic anemia reported in another **patient**. After 1 g/day/7 days in a **patient** induced erythropenia. Fatal allergic myocarditis in several **patients** treated with it for urinary infections. Half-life in **man**: 34.6–63 h. After a single 2 g dose to the **mother**, significant levels remain in the **fetus** for at least 5 days. Very low levels in the amniotic fluid. May cause hyperbilirubinemia in the **newborn**. Will transfer into **human** breast milk and **cow's *milk***.

Drug interactions: Inhibits binding of ***aspirin***, ***phenylbutazone***, and ***probenecid*** to ***albumin***. It displaces from ***albumin*** binding sites: ***tolbutamide*** leading to hypoglycemia; ***methotrexate*** leading to increased ***methotrexate*** (q.v.) toxicity; and ***bishydroxycoumarin*** and ***warfarin*** leading to bleeding. Large doses decrease binding of ***penicillin G***, ***penicillin V***, ***ampicillin***, ***cloxacillin***, ***dicloxacillin***, ***nafcillin***, and ***oxacillin***. Use with ***oral contraceptives*** decreases protection and increases rate of breakthrough bleeding and spotting.

Insure adequate fluid intake. Use cautiously in renal disease. It may decrease the protein binding and duration of effectiveness of a number of ***penicillins***. Use in food-producing **animals** is NOT approved in the United States.

LD_{50}, **mouse**: 2.5–3.5 g/kg.

SULFAMIDOCHRYSOIDINE

= *Prontosil* = *Rubiazol I* = *Streptozon*

***Antimicrobial, azo dye.* Discovered by Gerhardt Domagk in 1932. Active versus streptococci when its *sulfanilamide* moiety is released in the body. He was nominated for the Nobel Prize, but his country's Nazi regime prevented him from accepting it. It paved the way for chemists to develop thousands of derivatives. Ten years later, *sulfonylureas* were created.**

Untoward effects: Occupational hypersensitivity in a chemical plant **worker** leading to weeping eczema.

SULFANILAMIDE

= *1162-F* = *Albexan* = *Deseptyl* = *Prontalbin* = *Prontosil album* = *Prontylin* = *Septoplix* = *Streptocid album* = *Sulmid*

Antimicrobial, sulfonamide.

Untoward effects: ***Blood***: Hemolytic anemia in glucose-6-phosphate dehydrogenase-deficient or ***primaquine***-sensitive **patients**. A 19-year-old **male Negro** with glucose-6-phosphate dehydrogenase deficiency accidentally swallowed some ***contrast media*** containing it and later developed acute hemolysis. WHO reported 3.6 g ingestion by **adult Negroes** with glucose-6-phosphate dehydrogenase deficiency leading to significant hemolysis. Incidence of hemolytic anemia is highest in **children** and **Negroes** and occurs in 24–72 h; occasionally up to 7 days – often with nausea, dizziness, and fever. Aplastic anemia, neutropenia, leukopenia, agranulocytosis, leukocytosis, sulfhemoglobinemia, thrombocytopenia, purpura, and methemoglobinemia. Cyanosis and acidosis also occur.

Eye: Acute transient myopia after applying it in a vaginal cream. Lens opacity.

Liver: Cholestasis, hepatocanalicular-induced jaundice.

Skin: Toxic epidermal necrolysis, erythema, erythema nodosum, photoallergic contact dermatitis with maculopapular and pruritic rashes, and occasional chills, fever, and prostration.

During its early clinical trials at Cornell, toxicity was common while doctors sought to determine the ideal dosage for **man**. Unfortunately, they never tested it on

dogs whose ideal dosage was later determined to be the same as **man's**.

Patients allergic to it are often allergic to ***p-phenylenediamine*** and other sunscreen agents. Adult **diabetics** taking oral ***sulfonylurea hypoglycemic agents***, such as ***Orinase***® (***tolbutamide***), ***Diabinese***® (***chlorpropamide***), and other related drugs can develop a systemic contact dermatitis. This occurred in a 40-year-old **female** taking ***tolbutamide*** for 6 years without any ill-effects, until **she** used a vaginal cream containing sulfanilamide leading to severe pruritic, edematous, eczematous eruption of the vulva area, eventually covering most of the trunk and extremities, and subsiding a few days after discontinuing ***tolbutamide***. Any other ***sulfonamide*** compound, such as ***p-aminobenzoic acid***, ***benzocaine***, ***carbutamide***, ***dapsone***, ***saccharin***, or ***sulfasalazine*** have caused widespread contact dermatitis in sensitized **patients**. Passes into **human** breast milk and also transfers into the milk of **cows** and **rats** treated with it. The concentration in **human** breast milk often equals the **mother's** plasma level and is present in **her** milk for days after **she** discontinues the medication. Placental passage is rapid; absorption is delayed if taken with food. Half-life: 8.8 h. Also see ***Shoes***.

On September 4, 1937, the S.E. Massengill Co. distributed their first batch of a sulfanilamide elixir containing 10% sulfanilamide, 72% of toxic ***diethylene glycol*** (q.v.), 16% water, and small amounts of ***amaranth***, ***caramel***, ***raspberry*** extract, and ***saccharin***, but no ***alcohol*** (required in official elixirs). U.S. government seizure was based on this misbranding. Before it could be withdrawn from sales, > 100 **people**, mostly **children**, died from ingesting it. These deaths prompted the U.S. Congress to enact the 1938 Federal Food, Drug, and Cosmetic Act.

LD_{LO}, **human**: 500 mg/kg.

Toxic **animal** reactions include occasional cyanosis and anemia in **rat** and **mouse**; anemia, nausea, and vomiting in **dogs**; and anemia in **monkeys**. Insure adequate fluid intake. Use cautiously in renal disease. Avoid using ***milk*** for **humans** from treated **animals** for at least 72 h or six milkings after the last oral, intramammary, or intrauterine use. A similar minimum period of time is advisable after oral use before slaughtering treated **animals** for food.

LD_{50}, **mouse**: 3.3 g/kg; **dog** and **rabbit**: 2 g/kg; SC, LD_{50}, **mouse**: 2.8 g/kg.

SULFANILIC ACID

= p-Aminobenzene Sulfonic Acid

Use: Chemical intermediate in making optical brighteners, food dyes, and other dyes.

Untoward effects: Symptoms in **patients** with recurring urticaria or angioedema could be provoked by it. Its dust can cause eye and skin irritation. Dust can be explosive with air in critical proportions and an ignition source. Thermal decomposition can produce ***carbon monoxide*** and/or ***carbon dioxide***.

SULFAPHENAZOLE

= Ciba 17,922 = Eftolon = Orisol = Orisulf = Sulfabid

***Antimicrobial, sulfonamide.* Long-acting.**

Untoward effects: Stevens–Johnson syndrome, especially in **children**. Hemolytic anemia, Heinz bodies, neutropenia, and fixed-drug eruptions reported. A half-life of 8–12 h. Excreted in breast milk. A successful suicide in a 16-year-old **female** after ingesting a large quantity of tablets.

Drug interactions: Inhibits the metabolism of ***acetohexamide***, ***chlorpropamide***, ***tolbutamide***, and possibly ***glyburide*** leading to severe hypoglycemia and coma and may prolong ***tolbutamide's*** half-life from 4 to about 17 h. In some **patients**, it also inhibits the metabolism in the liver of ***phenytoin***, increasing potential ***phenytoin*** toxicity. It displaces ***bishydroxycoumarin*** and ***warfarin*** from ***albumin*** binding sites increasing anticoagulant effect. Has caused an increased ***benzylpenicillin*** half-life. By inhibition of liver enzymes, it can potentiate the side-effects of many ***oral contraceptives***. In seven **subjects** on ***cyclophosphamide***, it inhibited the activity in two, failed to alter it in three, and enhanced it in two.

Avoid using ***milk*** from treated **cows** for **human** consumption for at least 60 h or five milkings after the last treatment. Insure adequate fluid intake. Use cautiously in renal disease and **neonates**.

LD_{50}, **mouse**: 5.8 g/kg.

SULFAPYRIDINE

= Coccoclase = Dagenan = Eubasin = M&B 693 = Pyriamid = Pyribol = Septipulmon

Antimicrobial, sulfonamide.

Untoward effects: Frequent toxicity in ~$\frac{1}{3}$ of **patients**. Severe nausea and vomiting in 25–40%. Leukopenia, agranulocytosis, methemoglobinemia, aplastic anemia, purpura, hypoprothrombinemia, fixed-drug eruptions, renal insufficiency, crystalluria, oliguria, anuria, and hemolytic anemia especially in glucose-6-phosphate dehydrogenase deficiency. Hypersensitivity reactions occur leading to serum-sickness-like syndrome, rashes, plasmacytosis, lymphocytosis, chills, fever, urticaria, disorientation; angioedema of lips, tongue, and hands; multiclonal gammaglobulinopathy, Stevens–Johnson syndrome, exfoliative dermatitis, periorbital edema;

conjunctival and scleral injection; arthralgia, allergic myocarditis, and photosensitization. In some **patients**, abdominal pain, hepatitis, pancreatitis, diarrhea, anorexia, stomatitis, headache, peripheral neuritis, depresssion, ataxia, convulsions, tinnitus, vertigo, hallucinations, insomnia, periarteritis nodosum, cyanosis, and lupus erythematosus. Precipitates easily in acidic urine. Placental passage occurs and it is found in breast milk. Half life: 9.4 h.

Avoid using ***milk*** from treated **cows** for **human** consumption for at least 60 h or five milkings after last treatment. Insure adequate fluid intake. Use cautiously in renal disease and **neonates**. Anemia, cyanosis, and nausea in **rodents**. Nausea, vomiting, anemia, and renal injury in **dog** and **monkey**. Can displace ***bilirubin*** from ***albumin*** binding sites with possible serious consequences in **newborns** with compromised liver metabolism of ***bilirubin***.

LD_{50}, **mouse**: 1.7 g/kg.

SULFAQUINOXALINE

= *Embazin* = *Sulcon* = *Sulfa Q* = *Sulquin*

Antimicrobial, coccidiostat, sulfonamide. **Popular in treating poultry.**

Untoward effects: Its anti-***vitamin K*** activity was utilized as a potentiating agent for ***warfarin*** (0.025% each) in a Monsanto rodenticide. Potentiated by ***pyrimethamine***. Rapid (within 1–2 h) hyperglycemia in **mice**. Levels of 1% in feed or in the water at 0.0645% for **chicks** lead to toxicity with decreased growth, anorexia, and intestinal hemorrhages.

LD_{LO}, **rat**: 1 g/kg.

SULFATHIAZOLE

= *Avisol* = *Cibazol* = *Duatok* = *Eleudron* = *Enterobiocine* = *M&B 760* = *Norsulfasol* = *Norsulfazole* = *Planomide* = *Polyseptil* = *RP-2090* = *Sulfamul* = *Sulfavitina* = *Sulzol* = *Thiazamide*

Antimicrobial, sulfonamide.

Untoward effects: Fixed-drug eruptions, Stevens–Johnson syndrome, urticaria, erythema nodosum-like lesions, drug fever (10%), crystalluria in acid urine, photosensitivity, acute hepatitis (< 0.5%), thrombocytopenia, antileukocytic activity, acute leukemia, leukopenia, aplastic anemia, leukocytosis, eosinophilia, neutropenia, pancytopenia, purpura, nausea and vomiting (10%), hematuria, periarteritis nodosa, and azotemia; hemolytic anemia and agranulocytosis when used with ***sulfapyridine***. Focal necrosis in liver, spleen, bone marrow, lymph nodes, lungs, and kidneys. Intracranial use leads to convulsions. Adverse reactions in ~18% of **patients**. A **patient** received 65 g during 2 weeks leading to hemolytic anemia, spherocytosis, leukopenia, and white blood cell count of 65,800. Half-life of 3.5–3.8 h. Passes into **human** breast milk at 33–50% of **maternal** plasma level. In one trial, an **infant** received ~4 mg/day after the **mother** received 3 g/day.

To avoid undesirable tissue residues for **man**, **poultry** or **animals** should not be slaughtered for **human** consumption for at least 10 days after the last treatment. Insure adequate fluid intake. Use cautiously in renal disease. Prolonged usage may require ***vitamin K*** supplementation (at least five times usual levels). Nausea and vomiting reported in **dogs** and **monkeys**; anemia in **rodents** and **monkeys**; and renal injury in **rodents**, **rabbits**, and **monkeys**.

SULFINPYRAZONE

= *Anturan(e)* = *Anturano* = *Enturen* = *G-28315*

Uricosuric, antithrombotic. **A *pyrazole* compound related to *phenylbutazone*.**

Untoward effects: Nausea, vomiting, rashes, abdominal pain, urticaria, ***uric acid*** renal stones, bronchospasm, acute interstitial nephritis, increased serum ***creatinine***, acute tubular necrosis, oliguria, dysuria, ataxia, dyspnea, convulsions, and coma. May exacerbate peptic ulcers and, at the start of therapy, may also cause acute attacks of gout. Rarely, anemia, leukopenia, agranulocytosis, and thrombocytopenia due to bone marrow depression and fever. It increases platelet survival in certain thromboembolic disorders and interferes with platelet adhesions to ***collagen***. Ataxia, dyspnea, convulsions, and coma have been associated with overdoses. Half life: 8.8 h.

Drug interactions: Prolongs the half-life of ***cloxacillin***, ***nafcillin***, ***penicillin G***, and ***penicillin V***. Uricosuric effects are antagonized by small doses of ***aspirin*** and possibly other ***salicylates***. ***Salicylates*** increase its binding to plasma proteins and decrease its plasma concentration. Sulfinpyrazone probably competes with both ***salicylates*** and ***urates*** for secretory mechanisms in the renal tubules. It increases the clearance of ***acetaminophen***. ***Probenecid*** decreases its renal clearance, prolonging its action, and may increase its toxicity. It blocks the ability of antihypertensive diuretics, such as ***chlorothiazide*** and ***chlorthalidone***) to inhibit renal tubular secretion of ***uric acid***. Fatal blood dyscrasias in 3/25 with gout **who** ingested it for 2.8–14 years and intermittent ***colchicine*** for short intervals. It displaces ***sulfonamides*** and the long-acting ones from serum ***albumin***, decreases their urine levels and increases their serum levels, and potentiates their antibacterial effect. Many case reports of ***warfarin's*** enhanced hypoprothrombinemic effect and increased half-life, requiring a 50% decrease in ***anticoagulant*** dosage. ***Phenprocoumarin*** is also displaced from ***albumin***-binding sites. Potentiates the ***sulfonylureas*** and

insulin. ***Phenolsulfonphthalein*** tubular secretion is inhibited by it.

Increases ***uric acid*** excretion in **man**, but decreases it in the **chicken**.

LD_{50}, **rat**: 375 mg/kg.

SULFISOXAZOLE

= *Gantrisin* = *NU 445* = *Sosol* = *Soxisol* = *Soxomide* = *Sulfacin* = *Sulfafurazol* = *Sulfalar* = *Sulfasol* = *Sulfazin* = *Sulfoxol* = *Sulphafurazole* = *Sulsoxin*

Antimicrobial, sulfonamide.

Untoward effects: Incidence of adverse reactions is higher in **women** than **men**. Erythema and urticaria (3%), pruritus; hemolytic anemia in glucose-6-phosphate dehydrogenase-deficient **patients** and in **carriers** of hemoglobin Hasharon; hepatitis and jaundice, sialadenitis, toxic hepatic necrosis after severe dermatitis, Stevens–Johnson syndrome, skin rashes, eye and facial edema; angioneurotic edema of hands, arms, and feet; sore throat; fixed-drug eruptions, and reversible pseudo-Pelger anomaly in 27-year-old **female**. Eosinophilia and drug fever (2.8%), pancytopenia, leukocytosis, leukopenia, plasmacytosis, thrombocytopenia, agranulocytosis, purpura, aplastic anemia, neutropenia; petechiae and ecchymoses over entire body; hematuria, oliguria, arthralgia, wheezing, dyspnea, fever, and paresthesia of extremities reported. An unusual hypersensitivity reaction in 25-year-old **female** with myositis, myocarditis, vasculitis, and severe eosinophilia required leukopheresis for recovery. Another **patient** had a ***Factor VIII*** inhibitor in high titers. After SC caused kernicterus in **infants**; kernicterus and deaths in premature **infants** from treatment with it. Administered to pregnant **women** at or near term, it passes through the placenta and can cause hyperbilirubinemia in the **newborn**. Some excreted in breast milk. Hydrops fetalis in **male** glucose-6-phosphate dehydrogenase-deficient stillborn **fetus**, possibly due to its **maternal** ingestion. Therapeutic serum levels are 9–10 mg%. Half-life: 4.7–7 h.

Drug interactions: Potentiates ***acetohexamide, chlorpropamide***, ***cloxacillin***, ***nafcillin***, ***oxacillin***, ***penicillin G***, ***penicillin V***, ***phenytoin***, ***thiopental***, and ***tolbutamide*** leading to increased protein binding and hypoglycemic coma with increased half-lives of the drugs. It has the same effect on ***phenformin***. It displaces ***sulfonamides*** from their ***albumin*** binding, enhancing their antibacterial activity. Excessive hypoprothrombinemia after its use by **patients** stabilized on ***warfarin***. Generally, does not require alkalinization of urine; use with ***methenamine*** may not cause crystalluria, but the urine will appear cloudy. It decreases ***methotrexate*** plasma protein binding and may increase the latter's toxicity. Use with ***Azo-Gantrisin***® (***phenazopyridine–sulfisoxazole*** combination) leads to anemia, arthritis, leukopenia, and temporal arthritis. Delayed absorption if taken with ***food***.

Some **cats** may show pemphigus-like skin reactions, lethargy, and depression. Insure adequate fluid intake. Use cautiously (but very useful) in renal disease. ***Vitamin K*** needs may need monitoring with long-term and/or high dosage.

LD_{50}, **mouse**: 6.8 g/kg.

SULFOBROMOPHTHALEIN SODIUM

= *Bromosulfophthalein* = *Bromosulphthaleine* = *Bromsulfalein* = *Bromsulphalein* = *Bromthalein* = *BSP* = *Hepartest*

***Diagnostic agent for liver function.* IV use.**

Untoward effects: Delayed contact allergy after extravasation. Diaphoresis, paleness, dyspnea, bronchospasm, dermatitis herpetiformis, fixed-drug eruptions, urticaria, pruritus, shock, non-palpable pulse, loss of consciousness, bradycardia or tachycardia, nausea, vomiting, fever, chills, flushing, back pain, headache, disorientation, dizziness, vertigo, hypotension, laryngeal stridor, convulsions, and many reports of anaphylactic shock, including deaths. Irritant to veins. Use dilute solutions via small gauge needles or cannulas. Many drugs give falsely increased readings, viz. ***anabolic agents***, ***androgens***, ***azo dyes***, ***barbiturates***, ***oral contraceptives***, ***estrogens***, ***meperidine***, ***morphine***, ***oxacillin***, ***phenolphthalein***, and ***radiopaques***. Fever, shock, and dehydration may also cause false increase. Icteric states prevent interpretation of the test. The adverse effects discouraged continued use. Half-life: ~5.5 min.

IP, LD_{50}, **rat**: 320 mg/kg; **mouse**: 330 mg/kg.

SULFOTEP

= *ASP 47* = *Bayer E 393* = *Bladafum* = *Dithione* = *Dithiophos* = *ENT 16,273* = *TEDP* = *Thiotepp*

Organophosphorus insecticide, miticide. Anticholinesterase.

Untoward effects: Poisonous vapors lead to eye and skin irritation, eye pain, lacrimation, blurred vision, miosis, rhinorrhea, headache, nausea, vomiting, abdominal cramps, salivation, bradycardia, diarrhea, giddiness, anorexia, diaphoresis, weakness, twitching, paralysis, Cheyne–Stokes respiration, hypotension, arrhythmias, convulsions, and cyanosis.

Contamination of a particular ***coumaphos*** production batch with it killed **cattle**. Oral doses of 3.8 or 4 mg/kg to two **calves** led to tremors, convulsions, salivation, and death. A **calf** survived a 4.5 mg/kg dose.

LD_{50}, **rat**: 1–5 mg/kg; **mouse**: 833 mg/kg; SC, LD_{50}, **mouse**: 8 mg/kg; IM, LD_{50}, **rat**: 110 mg/kg; **dog**: 58 mg/kg; LD_{50}, dermal, **rabbit**: 20 mg/kg.

SULFOXONE

= *Aldesulfone* = *Diasone* = *Diazon* = *Novotrone*

Leprostat.

Untoward effects: Hemolytic anemia and hemolysis in glucose-6-phosphate dehydrogenase-deficient **patients**. A 46-year-old **male** with dermatitis herpetiformis developed peripheral neuropathy after 20 years of ingesting 0.66–0.99 g/2–3 times a week. Headaches, nervousness, paresthesia, reversible psychosis, insomnia, hematuria, nausea, vomiting, anorexia, methemoglobinemia, leukopenia, leukocytosis, agranulocytosis, drug fever, pruritus, infectious mononucleosis-like syndrome, blurred vision, and cyanosis can occur. Partially (~30%) metabolized to ***dapsone*** (q.v.).

SULFUR

= *Brimstone* = *Sulphur*

***Parasiticide, antimicrobial, antifungal, keratolytic, nutrient, and laxative.* Annual use in the U.S. is > 12 million long tons.**

Sulfur is found in the body in the amino acids, *cystine* and *methionine*; *chondroitin*, *heparin*, *taurine*, and many *sulfates*. Important in the metabolism and functions in enzymes and body biochemical reactions.

Use: In manufacturing ***sulfuric acid*** (88%) and 12% in production of ***carbon disulfide***, ***sulfur dioxide***, and ***phosphorus pentasulfide***, and in the ***paper*** industry, ***rubber*** vulcanizing, **animal** nutrition, and pharmaceuticals. As a folk medicine, it is a laxative spring tonic.

Untoward effects: Despite its beneficial effects in dermatology, it is a skin irritant and a cause of contact dermatitis in agricultural **workers** using it on fruit trees. **Workers** are often required to wait 3–4 days before re-entering treated fields. Many mill **workers** complain about handling it. Induces comedones on **human** backs after daily application in treating acne vulgaris. Ingestion, when it is used as a pesticide can cause vomiting, abdominal pain, and metabolic acidosis. A 66-year-old **female** eating ***clay*** on a daily basis ingested 200 g of sulfur as a ***clay*** substitute and as a cathartic leading to metabolic acidosis with a high anion gap and marked hyperkalemia; treated with hemodialysis, decreased sulfur serum levels in 3 weeks from 24 mmol/l to 2.6 mmol/l. Progressive lethargy and confusion in 57-year-old **female** eating 250 g during 6 days leading to metabolic acidosis. ***Sulfarthrol***, IV and IM causes diarrhea, exacerbation of chronic skin disease, insomnia, and a feeling of heat. Drinking water in some areas has a high level and smells like rotten eggs; traces help cause an oxidized flavor in ***milk***. ***Onions*** contain a smelly mixture of ***disulfides*** and ***trisulfides***. It is a fire hazard and its dust is ten times as explosive as ***coal*** dust. Contact with molten sulfur leads to severe burns and release of poisonous gas, which can explode in the presence of sparks, flames, or heat. Protective gear must be worn. A brush fire in Cape Town, South Africa ignited a sulfur stockpile, sending up a huge red cloud of ***sulfur dioxide*** (q.v.). It is used to help explode binary sarin (q.v.) and ***VX*** (q.v.) military shells and to produce violet poisonous smoke in warfare.

Although protectant versus ***copper*** toxicity in **cattle** and **sheep**, its oral use will further decrease ***selenium*** availability. ***Hydrogen sulfide***-like breath or fecal odors may precede toxic symptoms. Rarely, produces dermatitis venenata which mimics the original irritation. This is easily avoided by avoiding excessive usage. Continuous use in **poultry** feed at levels in excess of 0.4–0.5% decreases **chick** growth and causes ricket-like syndromes, especially if they are not exposed to sunlight. Sulfur is able to replace ***phosphorus*** in bone mineral, causing "sulfur rickets" following use as a coccidiostat in **chickens**. High intake has been associated with ***vitamin*** $\boldsymbol{B_1}$ deficiency in **cattle**. Toxicosis in **cattle** and **horses** at 250 mg/kg. Minimum lethal dose of colloidal sulfur IV in most **mammals** is 5–10 mg/kg.

SULFUR DIOXIDE

= *Sulfurous Anhydride* = *Sulfurous Oxide* = *Sulfur Oxide*

***Preservative, disinfectant, bleaching agent.* Non-flammable gas. U.S. annual demand is ~400,000 tons.**

Use: In refining ***sugar*** and ***starch***; preservative for ***beer***, ***fruits*** and ***vegetables***, ***meats***, ***shellfish***, and ***white wine***; bleaching agent, oil refining solvent, agricultural fumigant; in refrigeration and manufacturing ***sulfuric acid*** and ***bisulfites***.

Untoward effects: Forms ***sodium bisulfite*** (q.v.) when it enters the body, damaging ***nucleic acids*** in living cells and when it contacts moisture, it ultimately forms ***sulfuric acid*** (q.v.) and acid rain. ***White wine*** or marinating with it can destroy 50–75% of ***vitamin*** $\boldsymbol{B_1}$. Because of this, the U.S. bans its use on ***meat*** or any food generally recognized as a source of the vitamin. Released into the environment from fuel burning, smelters, automobile exhaust, volcanoes, and industry, most of it reacts with moisture; ***soot*** and trace minerals, such as ***magnesium*** and ***vanadium*** leading to acid rain, ***ammonium sulfate***, other sulfates, and smog. Fermenting ***yeasts*** in all ***wines*** and ***beer*** produce sulfur dioxide from inorganic ***sulfates*** in their ingredients (viz. ***grapes***). The U.S. limit is 250 ppm in finished ***wine***. Levels above 10 ppm must be stated on the label. It is the common oxidation product of ***sulfur***-containing compounds in fires. Opening, handling, or eating ***sulfite***-containing packaged foods can cause aerosoled exposure to sulfur dioxide. Nebulization of bronchodilating drugs containing a ***sulfite*** preservative can release traces of sulfur dioxide that can prove disastrous for an asthmatic **patient**. Sensitive **asthmatics** may experience bronchospasms when exposed to 0.1 ppm in air and non-asthmatic **people** may develop it at 6 ppm. Shrimp boat **crewmen** use ***sulfites*** to preserve **their** fresh **shrimp** catches and, when not adequately kept cold, sulfur dioxide is released leading to ***sulfhemoglobin*** in **their** blood causing coma and death after entering the hold by the **crew**. As a gas, it can cause irritation to the eyes, nose, throat, skin, and respiratory system leading to lacrimation, conjunctivitis, eye burns, rhinorrhea, coughing, choking, chest tightness, dyspnea, bronchitis, bronchoconstriction, laryngeal and pulmonary edema, emphysema, and skin burns. Intolerable to **man** well below its lethal concentration: > 500 ppm/10 min. Contact with the liquefied compressed gas leads to frostbite, vesiculation, and deep skin necrosis. Cleaning **workers** in a St. Louis, Missouri manufacturing plant, in attempting to clean a dip tank that had contained ***chromic acid*** solution, poured in some ***sodium bisulfite***, causing a chemical reaction, liberating sulfur dioxide gas that spread rapidly in the plant causing evacuation of 148 **workers**. Of the 40 working near the tank, 78% had chest discomfort with coughing, 78% had throat irritation, and 50% had eye irritation. Symptoms were still present after 2 weeks including chest discomfort in 30%, throat irritation in 18%, cough in 8%, and eye irritation in 1%. Similar symptoms were experienced in Portland ***cement*** manufacturing plants from the burning of ***coal***. Some **workers** became dizzy and had chest pain for up to 8 months. A brush fire ignited a giant ***sulfur*** stockpile at a chemical plant in Cape Town, South Africa, sending up a huge red cloud of poisonous sulfur dioxide, which sent thousands of **people** fleeing to **their** homes; > 100 people were hospitalized. Except for massive exposure, **people** are automatically warned of its presence by early symptoms of throat and nostril irritation when its atmospheric level reaches or exceeds 3.4 ppm. The FDA estimated 500,000 to 1 million **people** may be sensitive to it and 75% of them are asthmatic. Salad bar exposure was one of the worst offenders, and at least 13 **people** died from the exposure, forcing the proposed ban against its and all ***sulfite*** use on raw, packaged or unpackaged ***fruits*** and ***vegetables***. Common adverse effects are nausea, diarrhea, dyspnea, coughing, hypotension, and fatal shock. Exertion can aggravate the symptoms. Other symptoms include flushing, angioedema, laryngeal edema, hives, wheezing, itching, syncope, anaphylaxis, dermatitis, cyanosis, loss of consciousness, bronchospasm, respiratory arrest, and death. Exanthema in 46-year-old **male** exposed to fumes in open air. Leaves a distinct metallic taste after contacting mouth mucous membranes. Considered to be a co-carcinogen – confirmed in **rats**. OSHA sets 5 ppm as upper exposure limit; National Institute for Occupational Safety and Health sets it at 2 ppm, 15 min exposure of 5 ppm, and immediate danger to life or health as 100 ppm.

Inhalation LC_{LO}, **human**: 400 ppm/1 min; TC_{LO}, 4 ppm/1 min, 3 ppm/5 days.

Used to produce experimental subacute respiratory tract inflammation in **rabbits**, in evaluating mucolytic expectorants. Exposed **cats** can readily develop a ***vitamin B_1*** deficiency with vomiting, anorexia, cardiac irregularities, and convulsions. Therefore, it and ***sulfites*** are not permitted as preservatives for **cat** food or for any ingredient material.

Inhalation LC_{LO}, **rat**: 1000 ppm; **mouse**: 6000 ppm/5 h; **frog**: 1%/15 min.

SULFUR HEXAFLUORIDE

= *Ispan* = *Sulfur Fluoride*

Untoward effects: Inhalation has caused tachypnea, tachycardia, peripheral tingling, mild excitement, fatigue, slight ataxia, nausea, vomiting, and convulsions. National Institute for Occupational Safety and Health and OSHA set exposure limits at 1000 ppm, TWA.

SULFURIC ACID

= *Battery Acid* = *Hydrogen Sulfate* = *Oil of Vitriol*

U.S. annual use is ~47 million tons.

Untoward effects: ***Ingestion*** leads to mouth burning, stomatitis, brown scars on lips and mouth, sour taste, corrosion of teeth, pharyngitis, abdominal pain, tachycardia, bradypnea, nausea, vomiting (may be blood-stained), diarrhea, hypotension, shock, laryngeal and esophageal edema and strictures can cause asphyxia and death. Many of these cases were suicidal attempts. A 1-year-old **child** died after 20 drops (1.3 ml).

Inhalation leads to eye, skin, nose, and throat irritation; stomatitis, bronchitis, pulmonary edema after 6–8 h, emphysema, tracheobronchitis, coughing, choking, weakness, giddiness, cyanosis, and hypertension. Repeated inhalation also causes dental erosions. A North Richmond, California train spill dispersed 20,000 **people** to five hospitals, of which ~4500 were affected and 25 were hospitalized. At a hospital, it was used to clean sludge out of a drain under the cast room's sink, and it liberated poisonous ***hydrogen sulfide*** (q.v.). In 1952, smog containing ***sulfur dioxide*** (q.v.) gas and sulfuric acid mist hung over London for 4½ days, killing ~5000 **people** as well as prize **cattle** at the Smithfield Show kept in clean stalls, while **cattle** kept for slaughter with cow dung contributing to a high concentration of ***ammonia*** were not adversely affected. Has been used in one method of manufacturing illicit and legal ***amphetamines***. Fatal dose is ~5 ml, but 1 ml can cause shock and asphyxia. It and its mists are now considered as potential **human** carcinogens.

Topical: Rapid burning sensation if touched. In many cutting oils causing dermatitis, erythematous squamous eczema, folliculitis with papules, pustules, and an occasional furuncle. Felt **hat-formers** used to immerse the felt in hot water containing it and other chemicals and then press it on an ***aluminum*** form with **their** palms leading to callosity. Use in the center of golf balls has caused accidental eye injury. Massive dousing with it caused extensive bilateral ocular damage including conjunctivitis, edema, and corneal destruction and skin burns on 21-year-old **male** who died. Lacrimation in ***onion*-slicers** caused by release of ***propanethiol-S-oxide*** forming sulfuric acid. Cutting it under running water or chilling the ***onion*** first reduces the chemical's volatility. Burns are painful and of full skin thickness with brown–yellow stains. Eye contact has caused blindness.

LD_{LO}, **human**: 3.8 g. National Institute for Occupational Safety and Health and OSHA upper exposure limits are 1 mg/m^3; immediate danger to life or health, 15 mg/m^3.

A case of malicious mutilation of a **horse** by pouring it on the **horse's** back has been reported. This type of criminal action has also occurred in **man**.

LD_{50}, **rat**: 2.14 g/kg, 1 h LC_{50}, **rat**: 420 ppm.

SULFUR MONOCHLORIDE

= *Sulfur Chloride* = *Thiosulfurous Dichloride*

Fuming, oily, yellow-reddish liquid with a pungent odor used in vulcanizing *rubber* and as a chemical intermediate.

Untoward effects: Vapors are strongly irritating and corrosive to eyes, nose, throat, mucous membranes, and skin, causing pulmonary edema, stomatitis, coughing, and lacrimation. Contact dermatitis reported from dress shields where some remains on the ***rubber*** from vulcanization. OSHA and National Institute for Occupational Safety and Health set upper exposure limits as 1 ppm. Immediate danger to life or health, 5 ppm.

SULFUR TETRAFLUORIDE

Untoward effects: Colorless gas leading to eye and mucous membrane irritation; eye and skin burns as it releases ***hydrofluoric acid*** (q.v. for other symptoms) when it contacts moisture. The compressed liquid can cause frostbite and freezing burns on skin contact.

Dyspnea, weakness, lacrimation, and rhinitis in laboratory **animals**.

Inhalation LC_{LO}, **rat**: 19 ppm/4 h.

SULFUR TRIOXIDE

= *Sulfuric Anhydride* = *Triosul*

Use: In the manufacture of detergents and explosives, in sulfonation of organic compounds, and in chemical warfare.

Untoward effects: Strongly corrosive liquid. Strong oxidizing and dehydrating properties capable of damaging clothing and other organic materials. Can ignite combustibles. Volatilizes when warm. A tractor-trailer carrying it sprung a leak in Shelbyville, Indiana, forcing the evacuation of **students** from a school to protect **them** from dangerous fumes.

TC_{LO}, **human**: 30 mg/m^3; LC_{LO}, **guinea pig**: 30 mg/m^3/6 h.

SULFURYL CHLORIDE

= *Sulfuric Oxychloride*

Use: In organic synthesis, pharmaceuticals, and war gases.

Untoward effects: Colorless, pungent liquid whose vapors are corrosive to **human** skin and mucous membranes.

SULFURYL FLUORIDE

= *Vikane*

***Fumigant, pesticide.* Odorless, colorless gas.**

Untoward effects: Acute exposure has caused muscle twitchings and convulsions. Pulmonary and renal damage on repeated exposures. Rhinitis, pharyngitis, conjunctivitis, and paresthesias occur. The compressed gas causes frostbite on contact. A 30-year-old **male** inhaled it for 4 h due to poor workplace ventilation leading to nausea, vomiting, painful abdominal cramps, pruritus, reddened conjunctivae and pharyngeal and nasal mucosae, diffuse rhonchi and pin-prick paresthesia of lateral border of the right leg. **Some** complain of scratchy throat, flatulence, and difficulty in reading. Cargo fumigant against **snakes**.

LD_{50}, **rat** and **guinea pig**: 100 mg/kg; 1 h LC_{50}, **rat**: 3020 ppm.

SULINDAC

= *Aflodac* = *Algocetil* = *Arthrocine* = *Artribid* = *Citireuma* = *Clinoril* = *Clisundac* = *Imbaral* = *MK-231* = *Reumofil* = *Reumyl* = *Sudac* = *Sulinol* = *Sulreuma*

Anti-inflammatory, analgesic, antipyretic. **Inhibits *prostaglandin* synthesis.**

***Untoward effects*: *Blood and bone marrow*:** Severe agranulocytosis in 34-year-old **female** ingesting 150 mg tid/2 weeks and in a 60-year-old **female** after 200 mg twice daily/17 days. Cases of aplastic anemia, neutropenia, pancytopenia, bone marrow depression, leukopenia, thrombocytopenia, epistaxis, and platelet aggregation inhibition with prolongation of bleeding time. Severe immune hemolytic anemia in **females** taking 150 mg/day.

***Central nervous system*:** Case reports of aseptic meningitis, dizziness, headache, nervousness, vertigo, paresthesias, insomnia, drowsiness, syncope, vertigo, neuritis, and convulsions. Acute deterioration of parkinsonism in 61-year-old **male** after 800 mg/day/7 days. Aseptic meningitis reported in connective tissue disease and systemic lupus erythematosus **patients**.

***Ear*:** Tinnitus (2%) and decreased hearing (0.3%).

***Eye*:** Diplopia, cloudy and blurred vision, eyeball swelling, retinal hemorrhage, non-inflamed corneal ulceration, and spots before the eyes.

***Gestrointestinal*:** Abdominal pain (10%), nausea (6%), occasional vomiting, gastric ulceration and lesions (11%), gastric bleeding, diarrhea (4%) or constipation (3%), flatulence (1.7%), anorexia (1%), and indigestion (9%).

***Hypersensitivity*:** Splenomegaly and lymphadenopathy with fever in 61-year-old **male** after 10–12 days of therapy. Reaction recurred upon rechallenge. A 52-year-old **male** receiving 150 mg/day developed fever (40°C/104°F), chills, hypotension, pruritus, peripheral cyanosis, and vomiting. Confirmed by rechallenge. Severe interstitial pneumonitis in 76-year-old **female** after 150 mg twice daily/6 months; also confirmed upon rechallenge. A 26-year-old **female** taking 150 mg twice daily/5 days developed tingling dysesthesias, anorexia, headache, back and chest pain, and mild nausea. Anaphylaxis, angioneurotic edema, bronchial spasms, dyspnea, and vasculitis also reported.

***Liver and pancreas*:** Hepatitis in 12-year-old **female** on two separate occasions after 200 mg. Severe, reversible hepatotoxicity in 18-year-old **female** after 200 mg twice daily/13 days. Hepatitis in 29-year-old **female** taking 150 mg twice daily/18 days; recurring upon rechallenge. This occurs in ~1.5% of **patients**. Pancreatitis mostly in elderly **patients** after 23 days–4 months.

***Miscellaneous*:** Hyperkalemia, muscle weakness, ageusia, glossitis, occasional palpitations, and hypertension. Congestive heart failure can occur in **patients** with poor cardiac function. Reversible gynecomastia in 63-year-old **patient** after 200 mg twice daily/4 months. Hypothyroidism in 90-year-old **patient** after 300 mg/twice daily/4 weeks.

***Psychotic*:** Within 48 h after ingestion of 150 mg twice daily/2 days, a 53-year-old **male** developed delirium with angry outbursts and homicidal threats. Complete resolution in 48 h after treatment ceased. Depression reported.

***Renal*:** Interstitial nephritis, nephrotic syndrome, azotemia, hematuria, proteinuria, crystalluria, and dysuria.

***Skin*:** Stevens–Johnson syndrome, toxic epidermal necrolysis, exfoliative dermatitis, pruritus, bullous eruptions, photosensitivity, petechiae, bullous erythema multiforme; and morbilliform, maculopapular, and urticarial eruptions.

Drug interactions*:** Induces increase in ***cyclosporine blood levels, increasing the latter's toxicity. ***Methotrexate*** tubular secretion is decreased by it; increases ***methotrexate*** toxicity. Use with ***diflunisal*** decreases its plasma levels 30%. Use with ***dimethylsulfoxide*** leads to peripheral neuropathy and blocks sulindac's effectiveness. Potentiates ***warfarin***. It may decrease ***probenecid's*** uricosuric action. Conflicting reports on possible ***lithium***-sparing effect. ***Aspirin*** and ***phenobarbital*** may decrease its effectiveness.

LD_{50}, **rat**: 264 mg/kg.

SULISOBENZONE

= *Benzophenone-4* = *Cyasorb UV 284* = *NSC-60,584* = *Spectra-Sorb UV 284* = *Sungard* = *Uval* = *Uvinul MS-40*

Ultraviolet light absorber, sunscreen. **Also used in cosmetics, *wool*, pesticides, and coatings for lithographic plates. A *benzophenone*.**

Untoward effects*:** A very blond 45-year-old **female**, after use of it in a lotion, but not in ***isopropyl alcohol, developed burning of eyes and bitter taste when sweat carried it to mouth and eyes, indicating the lotion base was responsible for its irritation. Lotion caused no irritation or sensitization in 1000 other **patients**. A hypersensitivity reaction with burning sensations, redness, and swelling of hand within 15 min in a 54-year-old **patient** with a history of broad spectrum of light sensitivity. Both ***benzophenone*** and sulfonic parts of the molecule were required to elicit a contact reaction, while the ***benzophenone*** part alone elicited the urticarial reaction. Comedogenic if used constantly and plays a role in producing acne.

SULPHYDRILIC ACID

***Untoward effects*:** Abnormal nail dyschromy (blue color) from its systemic use.

SULPIRIDE

= *Abilit* = *Aiglonyl* = *Coolspan* = *Dobren* = *Dogmatil* = *Dogmatyl* = *Dolmatil* = *FK 880* = *Guastil* = *Megotyl* = *Meresa* = *Miradol* = *Mirbanil* = *Misulvan* = *Neogama* = *Omperan* = *Pyrikappl* = *Sernevin* = *Splotin* = *Sulpiril* = *Sulpitil* = *Sursumid* = *Synédil* = *Trilan*

***Psychotherapeutic, neuroleptic, ataractic, antidepressant, antiemetic, antiulcerogenic.* Oral, occasionally IV and IM.**

Untoward effects*:** In about 58% of **patients**, adverse effects were similar to those of ***chlorpromazine, although extrapyramidal symptoms were less and insomnia greater. Induced hyperprolactinemia and galactorrhea. In nine **patients**, after IM and oral led to delusion, agitation, and manic behavior; one had nausea and dizziness, two had blurred vision, and in two **others**, tremors, rigidity, and muscle stiffness. In 156 **patients**, after oral and IV therapy, ~16–25% had marked increase in peristalsis of stomach, duodenum, and jejunum. Increased lower esophageal sphincter pressure after IV. ***Antacids*** given 2 h before it decrease its availability by 25%. It is less if they are given 2 h after sulpiride.

Clinical trials in the U.S. were postponed because of its association with mammary tumors in some **mouse** strains.

LD_{50}, **rat**: 9.8 g/kg; **mouse**: 2.3 g/kg; IP, LD_{50}, **rat**: 230 mg/kg; **mouse**: 180 mg/kg.

SULPROFOS

= *BAY NTN 9306* = *Bolstar* = *Helothion*

Organophosphorus insecticide, anticholinesterase.

***Untoward effects*:** **Cows** that ingest it have residues in their livers. Immersion of **mallard** eggs leads to scoliosis, lordosis, blisters, and edema of **hatchlings**.

LD_{50}, **rat**: 227 mg/kg; LC_{50}, **quail**: 477 ppm in feed.

SULTHIAME

= *Elisal* = *Ospolot* = *Riker 594* = *Trolone*

Anticonvulsant.

***Untoward effects*:** Tachypnea, hyperpnea, anorexia, weight decrease, lethargy, ataxia, metabolic acidosis, headache, giddiness, paresthesia of extremities, confusion, tachycardia, fatal uremia, aggression, incontinence, ptosis and edema of eyelids, abdominal pain, leukopenia, liver damage, and Stevens–Johnson syndrome have been reported. A 17-year-old **male** took 20 tablets (20 g) in a suicide attempt. **He** vomited bloody mucus and had severe headache and vertigo, rash on trunk, some conjunctival injection, photophobia, and ataxia ~6 h after ingestion. May cause false positive serum tests for infectious mononeucleosis. Therapeutic serum levels are generally 0.5–12 mg/l and greater than that are usually toxic.

Drug interactions*:** A 19-year-old **female** effectively treated for many years with ***phenytoin for epilepsy developed permanent cerebellar damage after a few days of sulthiame therapy, due to marked potentiation of ***phenytoin***. The combination has also caused nausea, abdominal pain, persistent severe paresthesias and headache, ataxia, drowsiness, impotence, and confusion.

LD_{50}, **rat**: 65 mg/kg; **mouse**: 4852 mg/kg; **rabbit**: 1 g/kg; IP, **rat**: 1650 mg/kg; **mouse**: 1668 mg/kg.

SUMAC, POISON

= *Poison Dogwood* = *Poison Elder* = *Rhus vernix* = *Swamp Sumac* = *Toxicodendron vernix*

Untoward effects*:** Stupor and coma in 5-year-old **female** after ingestion of its berries. After ingestion, it has narcotic properties leading to mydriasis, vomiting, drowsiness, stupor, convulsive movements, delirium, and fever. Dermatitis can occur without direct plant contact, apparently from pollen and plant hair in the air, on clothing, tools, or **animals**. Causes a violent allergic contact dermatitis similar to that produced by ***poison ivy (q.v.), due to its ***urushiol*** (q.v.) content. It can cross-react with ***cashew nut tree*** (q.v.), ***India marking nut tree***, ***mango*** (q.v.), ***poison oak*** (q.v.), ***rengas*** (q.v.), and ***staghorn sumac***. Other sumacs are used in scenting ***tobacco***.

SUMATRIPTAN SUCCINATE

= *GR 43,175C* = *Imigran* = *Imiject* = *Imitrex*

***Antimigraine, selective vascular serotonin agonist.* Mostly for SC use. Occasionally intranasal or oral.**

***Untoward effects*:** Nausea and vomiting when used orally or parenterally. Incidence of 28% adverse effects of 100 mg oral in 421 **patients** 18–63 years.

***Blood*:** Pancytopenia and thrombocytopenia.

***Cardiovascular*:** Chest tightness and chest pain. Occasional arrhythmia, electrocardiogram changes, palpitations, tachycardia, and hypo- or hypertension. Rarely, angina, cerebral infarction with hemiparesis, bradycardia, heart-block, myocardial infarction, cardiac arrest, ischemic colitis, pulmonary embolism, subarachnoid hemorrhage, and shock. Atrial fibrillation and arrhythmia in 34-year-old **male** due to its use as a nasal spray.

***Miscellaneous*:** Dizziness (16%); in 13%, nausea or vomiting; drowsiness, sedation, and chest pain. Anxiety, fear; ear, nose, and throat discomfort; deafness, thirst, weakness, dysuria, tingling, and sense of burning at injection site. Appears in breast milk. Occasionally, diarrhea, constipation, gastroesophageal reflux disease, myalgia, muscle cramping, breast tenderness, cough, fever, xerostomia, ischemic optic neuropathy, periorbital edema, retinal artery occlusion, and dysuria. Rarely, galactorrhea, lactation suppression, mydriasis, gastrointestinal bleeding, hematemesis, melena, peptic ulcer, anemia, tetany, nipple discharge, edema, lymphadenopathy, renal failure, angioneurotic edema, cyanosis, and death.

Neurological: Photophobia and phonophobia. Occasionally, depression, confusion, alterations of smell or taste, euphoria, facial pain, dysarthria, lacrimation, ataxia, monoplegia, shivering, syncope, sleep disturbances, and tremors. Rarely, apathy, aggressiveness, bradyarthria, headache, dependency, dystonia, decreased or increased appetite, hyperesthesia, hallucinations, facial paralysis, neuralgia, paralysis, twitching seizures, radiculopathy, rigidity, memory failures, and suicidal intent. One report stated a headache recurred in 45% in the 24 h after treatment.

Respiratory: Dyspnea; occasional asthma; rarely, hiccups.

Skin: Diaphoresis and photosensitivity. Occasionally, rash, flushing, erythema, numbness, and pruritus. Rarely, skin wrinkling or tightness.

Drug interactions: Use with ***monoamine oxidase inhibitors*** (e.g. ***deprenyl***, ***furazolidone***, ***iproniazid***, ***isocarboxid***, ***nialamide***, ***pargyline***, ***phenelzine***, and ***toloxatone***) increases its half-life and decreases its excretion. With ***loxapine***, dystonic reaction reported. It delays absorption of ***acetaminophen***. Causes a striking increase in ***somatotropin***. ***Selective serotonin reuptake inhibitors*** (e.g. ***fluoxetine***, ***fluvoxamine***, ***paroxatine***, and ***sertraline***) may rarely with it induce weakness, incoordination, and hyperreflexia.

Oral use in **dogs** leads to opacities of cornea and defects in corneal epithelium at 2 mg/kg/day (about five times **human** oral dosage). No evidence of carcinogenicity in **rats** and **mice** or interference in fertility trials with **rats**. Embryotoxicity in **rabbits** at comparable SC dosage in **people**.

SUMITHION

= *Ag 147* = *Accothion* = *Agrothion* = *Am Cyan 47,300* = *Bayer 41,831* = *Bayer S 5660* = *CP 47,114* = *Cyfen* = *Cytel* = *Cyten* = *ENT 25,715* = *Doff* = *Fenitrothion* = *Folithion* = *MEP* = *Metathion* = *Methinitrophos* = *Methylnitrofos* = *Metilnitrofos* = *Novathion* = *Nuvanol* = *OMS 43* = *S 1102A* = *Sumitomo*

Insecticide, organophosphorus cholinesterase inhibitor, acaricide.

Untoward effects: **Farmers** that sprinkled the chemical developed pains in the backs of **their** eyes, xerostomia, weight decrease, and lassitude. It is especially deadly for **children**. Symptoms include miosis, vomiting, colic, diarrhea, sensation of constriction in chest, wheezing, dyspnea, copious bronchial secretions, bradycardia, muscle weakness, fasciculations, and convulsions. Accidental inhalation and dermal absorption by a 33-year-old **female** laboratory **technician** led to nausea, blurred vision, diarrhea, abdominal cramps, muscle weakness, mental confusion, and weakness within 2 days. The symptoms intensified during the next 3 days with increased sialism, nasal stuffiness, continuous lacrimation, fatigue, headache, mental depression, conjunctivitis, photophobia, and injected buccal mucosae.

In **birds** leads to ataxia, wing-drop, falling, tremors, salivation, loss of righting reflex, dyspnea, tetanic seizures, lacrimation, miosis, and wing-beat convulsions.

LD_{50}, **rat**: 250 mg/kg; **mouse**: 750 mg/kg; **guinea pig**: 1.8 g/kg; **poultry**: 445 mg/kg; **chick**: 280 mg/kg; **red-winged blackbird**: 25 mg/kg; **mallard duck**: 1.2 g/kg; **pheasant**: 55.6 mg/kg; **quail**: 28 mg/kg; Minimum lethal dose, **carp**: 6 μg/l; LC_{50}, **fish**: 2 mg/l.

SUNFLOWER

= *Girasol* = *Helianthus annus*

Untoward effects: Sunflower seed allergy occurs within 5 min–2 h, especially in previously hypersensitive **children** due to its ***chlorogenic acid***, ***quercimeritrin***, ***scototenin***, and ***sesquiterpene lactones***. **Aztec Indians** call it ***Chilamacatyl*** and use its seed in treating fevers, but, when **they** ingest too much, **they** develop headaches. Contact dermatitis occurs with cross-sensitivity with ***chrysanthemums***, ***marigolds***, and ***zinnias***.

Feeding the seeds to **psittacine birds** leads to secondary hyperparathyroidism. ***Lysine*** deficiency develops when the meal is fed to **pigs**. It is a ***nitrate*** accumulator and immature seeds have poisoned **cattle**.

SUPROFEN

= *Masterfen* = *R-25,061* = *Srendam* = *Sulprotin* = *Supranol* = *Suprocil* = *Suprol* = *Sutoprofen* = *Topalgic*

***Non-steroidal anti-inflammatory drug, analgesic.* IV, rectally, and mostly orally. No longer marketed in the U.S. and Europe.**

Untoward effects: Induced transient flank pain, acute renal failure, hematuria, hemolytic anemia, crystallization of ***uric acid*** in renal tubules, abdominal pain, polyuria, polydipsia, oliguria, proteinuria, diaphoresis, and occasional small transient increase in serum ***creatinine***. Small amounts in breast milk. Self-poisoning in 18-year-old **female who** ingested 40 tablets with uncomplicated recovery.

SURAMIN SODIUM

= *309 F* = *Antrypol* = *Bayer 205* = *Fourneau 309* = *Germanin* = *Moranyl* = *Naganin* = *Naganol* = *Naphuride*

Trypanocide. **Used IV for 5 weeks.**

Untoward effects**:** Due to its toxicity, it is no longer used in treating onchocerciasis, since the advent of the safer ***ivermectin***. Toxicity can vary with the nutritional state of the **patient**. Vomiting, edema, urticaria, and pruritus can occur almost immediately. Within a day, **some** report paresthesia, hyperesthesia of hands and feet, photophobia, lacrimation, peripheral neuritis, occasional amblyopia, toxic keratopathy, optic atrophy, nausea, chills, fever, and headache. Renal damage, albuminuria, aminoaciduria, casts, and proteinuria demand caution in further treatment. Agranulocytosis, hemolytic anemia, hemoglobinuria, anaphylaxis; tenderness of soles of the feet and palms of the hands; adrenal cortex degeneration, pyrexia, conjunctivitis, arthropathy, myalgia, shock, dizziness, and exfoliative dermatitis are occasionally reported. Rarely, selective necrosis of the adenal cortex. Therapeutic serum concentrations are > 100 mg/l and toxic levels are said to be ~300 mg/l. May cause spuriously high test readings for urinary ***albumin*** or blood.

Drug interactions**:** Use with ***furosemide*** decreases its body clearance by ~36%.

Toxicity is rare at recommended doses in **camels**, but may be frequent in **donkeys**. Due to development of resistant strains, the drug is often combined with other efficacious drugs.

IV, LD_{LO}, **mouse**: 40 mg/kg.

SURMA

Untoward effects**:** This Indian black cosmetic powder that looks like mascara is applied to the inner lower lid, usually contains 23–26% ***lead*** (q.v.), mostly as ***lead sulfide*** (contains 16% ***lead***) and has led to plumbism, including convulsions, in Indian and Asian **children**.

SUSHI

Untoward effects**:** Parasitic organisms can burrow into **fish** muscle, often giving unsuspecting **sushi-lovers** acute abdominal pain after 12 or more hours; occasionally with nausea. Abdominal surgery after a misdiagnosis of appendicitis because of severe abdominal pain in a college **student** led to a surprise – a 1½ inch bright red worm of the genus *Eustrongylides* crawled onto the drape.

SUTILAINS

= *BAX 1515* = *Ravase* = *Travase*

Proteolytic enzymes from *Bacillus subtilis*, used as a debriding agent.

Untoward effects**:** Topical use leads to mild fleeting pain or burning sensation, paresthesias, and transient dermatitis. Capillary bleeding may occur from below the necrotic tissue.

SWAINSONINE

Untoward effects**:** A toxic alkaloid in ***Astragalus*** (q.v.), ***Oxytropis*** (q.v.), and ***Swainsona*** adversely affecting grazing **cattle**, **sheep**, and **horses**. **Animals** develop symptoms identical to those in **humans** with a genetic deficit in lysomal mannosidase, that can cause axonal degeneration.

SWEET POTATO

= *Ipomoea batatas* = *Mah$_v$-coe^x* (Thai)

Untoward effects**:** In some African countries, it is as much a food staple as were ***potatoes*** for the **Irish**. Moldy (particularly *Fusarium solani*, *Fusarium javanium*, and *Ceratostomella fimbriata*) sweet potatoes produce the toxic ***ipomeamarone*** (q.v.) or ***ipomeanine*** and its toxic metabolite, ***4-ipomeanol***.

Early symptoms in **cattle** include atypical interstitial pneumonia, pulmonary edema, dyspnea, frequent deep coughs, tachypnea, and grunting, with death usually in 2–5 days. Severe dental blackening and erosion seen in **cattle** feeding on fermented sweet potato cannery waste. The toxin is an enantiomer of ***ngaione*** (q.v.). **Mice** and **rats** become dyspneic with aveolar edema and pleural effusion, with liver, spleen, and kidney pathology, dying in 8–24 h. It is high in ***oxalate*** (q.v.) content and all parts contain ***hydrogen cyanide*** (q.v.). When the tops are eaten in large quantity, death has occurred. May have hypoglycemic effect in some species. Use may cause spuriously high icterus index test readings.

IP, LD_{50}, **mouse**: 31 mg/kg.

SWISS CHARD

= *Beta macrocarpa or Beta vulgaris*

Untoward effects**:** A vegetable with a very high ***oxalate*** (q.v.) content (120 μg%) that can cause problems in **man**. Can accumulate ***nitrates*** under certain environmental conditions.

SYMPHYTINE

Untoward effects**:** A carcinogenic pyrrolizidine alkaloid found in ***comfrey*** (q.v.). A cup of ***comfrey*** herb tea contains ~750 μg of pyrrolizidine alkaloid and 38 μg of symphytine. ***Comfrey–pepsin*** tablets contain 200 μg.

TD_{50}, **rat**: 1.9 mg/kg.

SYNADENIUM GRANTII

= *African Milk Bush* = *Tiha* (Giriama) = *Tupa* (Digo) = *Kinyunyu* (Swahili)

Succulent shrub, native to Africa, growing up to 4 m in height; transported to Britain over 100 years ago as a houseplant; and now grown in the U.S.

Untoward effects: Intense skin irritation from contact with its milky latex, a characteristic of many other ***Euphorbias***. The irritant in the latex has been identified as ***12-σ-tigloyl-4-deoxyphorbol-13-isobutyrate***. In the U.S., a **babysitter** found her charge, a 5 month **infant** in a walker, had pulled over a large plant, held a torn dry leaf in **her** hand and none in **her** mouth, without any latex on **her** mouth or hands. Shortly afterward, the **infant** vomited spontaneously. Syrup of ***ipecac*** then failed to produce any evidence of plant parts. The **infant's** face, especially cheeks, under the eyes, and around the right ear became erythematous and the mouth was red and swollen. A physician administered 10 mg ***diphenhydramine*** 2 h after the incident. The **infant** could not swallow any liquid and **her** mouth was still edematous after exposure. This was followed by increased drooling and fine papular rash near the erythematous zones. **Her** rectal temperature was 103°F, lips still swollen and still dysphagic 8 h after the incident. ***Diphenhydramine*** 12 mg was given orally. Fluid-filled blisters on the lobe and pinna of the right ear and in the web of **her** right hand's fourth and fifth digits were noted 19 h after exposure. Slow improvement followed, but scabs developed around the ears and fingers that lasted about 4 days. Delayed (4 h) irritation and burning sensations in a **gardener** increased in severity by the eighth hour, with erythema and blisters on the hands and forearm. In Africa, the latex has been used as a poison, a piscicide, and a molluscicide. Strong digestive tract irritant and suspected of inducing deaths.

In the 1950s during the Mau Mau Kenyan uprising, the latex was suspected as being maliciously introduced into stab wounds on 33 poisoned **steers**; eight died. They developed ventral edema of their upper legs.

SYNEPHRINE

= *Analeptin* = *Ethaphene* = *Oxedrine* = *Parasympatol* = *Simpalon* = *Synephrin* = *Synthenate*

***Adrenergic, vasopressor amine.* IV, IM, or SC.**

Untoward effects: Its presence in foods such as ***oranges*** or ***lemons*** can alter metabolic test results.

IV, LD_{50}, **mouse**: 270 mg/kg.

SYNHEXYL

= *Parahexyl* = *Pyrahexyl*

Various routes of administration.

Untoward effects: A hallucinogenic semisynthetic analog of ***tetrahydrocannabinol***, with slower onset and only $\frac{1}{3}$ of its potency. Greater sedation, increased euphoria and duration, and more pronounced dream-like events than ***lysergic acid diethylamide***. Reddened conjunctiva, decreased blood pressure, increased pulse rate, impaired strength, and "hangover" effect reported. When 15 mg was smoked in a ***cigarette*** caused xerostomia, burning sensation in throat, tingling of extremities, decreased concentration, light-headedness, palpitations, tremulousness, and mental impairment.

IV, LD_{50}, **mouse**: 170 mg/kg; **rabbit**: 143 mg/kg; **dog**: 223 mg/kg.

SYROSINGOPINE

= *Isotense* = *Londomin* = *Raunova* = *Seniramin* = *Singoserp* = *Siringina* = *Su-3118* = *Syringopine*

***Antihypertensive.* A synthetic.**

Untoward effects: Similar to, but milder than ***reserpine*** (q.v.), with little nasal stuffiness and drowsiness.

IV, LD_{50}, **rat**: 50 mg/kg.

T

T-2 TOXIN

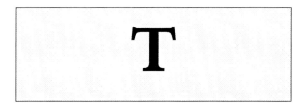

= *4,5-Diacetoxy-8(3-methylbutyryloxy)-3α-hydroxy- 12,13-epoxy-Δ⁹-trichothecan-3-d* = *T-2 Fusariotoxin*

Untoward effects: Present in food and the *Fusaria* that can grow at near-freezing temperatures, which are too low for most fungal growth. It has commonly been found on ***barley, maize***, silage, and ***safflower*** seeds. It is dermonecrotic, hepatotoxic, and a hematopoietic reducer. It destroys rapidly-dividing cells, such as those in bone marrow, gut epithelium, and dermis. Chronic low-level ingestion has led to decreased weight, and hematopoiesis; and caused coagulation defects, as well as nervous system effects. Crude inexpensive extracts of it have been suspect as an ingredient in Russian "Yellow Rain", used in aerial attacks in Vietnam, Laos, Kampuchea, and Afghanistan. Russian **peasants** who ate moldy over-wintered grain in the spring of 1944 died slowly from its effects. It and alimentary toxic aleukia in Russia and "staggering grain toxicosis" in eastern Siberia, known for hundreds of years, were probably due mainly to T-2 toxin. Symptoms are skin lesions, agranulocytosis, necrotic angina, hemorrhagic diathesis, sepsis, bone marrow failure, nausea, and vomiting.

IV half-life in **cattle** and **swine** is ~10 min and undetectable within a few hours. Its metabolite, ***HT2***, is essentially gone within < 1 day. Yet, **Kampuchean's** blood contained it 18 days after exposure, suggesting that, unlike in **animals**, it may be protein-bound or **they** picked it up from **their** hair, lungs, or environment. Beef **calves** given half that level had decreased feed consumption; this was equal to 10 ppm in their total diet. At the higher level, there definitely was decreased ***immunoglobulin A*** and ***M*** production. Has caused a lethal toxicosis in **dairy cows** and, after chronic IM, led to hemorrhaging in a **cow**. Oral dosing to **cats** caused alimentary toxic aleukia with pancytopenia, bone marrow aplasia, hemorrhagic diathesis, decreased hemostasis, nausea, vomiting, bloody feces, weakness, ataxia, and dyspnea. Will kill **fish**. In the **chick**, the liver is the main organ of excretion, via the bile into the intestines, causing severe inflammatory reactions, sloughing of the intestinal epithelium, necrotic lesions in the crop and gizzard, decreased growth, decreased food consumption, decreased spleen size, abnormal feathering, hemolytic anemia, decreased ***vitamin E*** absorption, and nervousness. Levels of 1 μg/g or more of feed caused raised yellow-white caseous lesions on the palate, beak, and tongue within 1 week. Daily dietary intake by **chicks** of 4–16 μg/g of feed led to neural toxicity (impaired righting reflex, hysteroid seizures), similar to that of alimentary toxic aleukia in **man** and **horses**. In **hens**, decreased egg production and thinner egg shells. Hemorrhagic syndrome, immune system suppression, skin necrosis, vomiting, and refusal of feed in **pigs**. **Sows** showed decreased litter size, fetal deaths, abortions, and lethargy. Post-mortem showed petechiae and ecchymoses on serosal surfaces of body cavities, in the muscles, and on the liver surface. Engorgement of cerebral, cardiac, gastric, intestinal, and visceral blood vessels also occurs. Icterus, amber fluid in body cavities, and pale edematous kidneys. Hemorrhagic edematous lymph nodes and gelatinous edema of the colon also occur in chronic cases.

LD_{50}, **swine** and **rats**: 4 mg/kg; **rainbow trout**: 6.5 mg/kg; **mice**: 5.2 mg/kg.

TABACU DI PISCADO

= *Tournefortia gnapholodes*

Untoward effects: In Cuba and Curaçao, it is taken as an abortifacient. **Vendors** normally will not sell it to young **girls**.

TABERNAEMONTANA sp.

Untoward effects: In Africa, the root of ***T. penduliflora*** is lethal within 30 min after ingestion. Contains a cyanogenic alkaloid and can augment hallucinogens used in South America. In the Amazon, contact with the latex of ***T. stenoloba*** leads to dermatitis.

TABERNANTHE IBOGA

Untoward effects: Chewing the roots leads to hallucinogenic effect, mental confusion, excitement, and even paralysis from its ***ibogaine*** (q.v.) content. It is used in the initiation rites of African secret societies, including the well-known Bwiti, and is the center of local religion, which has hampered the **natives'** acceptance of Christianity. Chewing the root by **natives** of Gabon offsets hunger and fatigue. Ingesting its extracts permits **natives** stalking game to endure motionless periods up to 2 days while staying mentally alert. Contains at least 12 alkaloids which, in large doses, lead to mental confusion, excitement, and "drunken madness characterized by prophetic utterances".

TABUM

= *GA* = *Le 100* = *MCE* = *T2104* = *TL 1578*

***Military nerve gas, organophosphorus compound.* First manufactured in Germany in the 1930s.**

Untoward effects: Potent ***anticholinesterase agent*** with accumulation of ***acetylcholine*** and its continuous stimulation of the parasympathetic nervous system. It is a liquid, volatilizes at above ambient temperatures, or when aerosoled by explosion, it can be inhaled. Exposure to the liquid or vapors causes ***acetylcholine***-induced excessive secretions from the eyes, nose, and mouth, and in the intestines; muscle fasciculations, miosis, twitches, and tremors; these symptoms may be delayed for at least 12 h. Heavy exposure can also cause sudden unconsciousness, convulsions, apnea, paralysis, sudden dim or blurred vision, and bronchoconstriction. It was one of the agents used by **terrorists** on 20 March 1995 in the Tokyo, Japan subways. U.N. **observers** found evidence of it in Iraq at the border with Iran.

Inhalation LC_{LO}, **human**: 150 mg/m^3; LD_{LO}, skin, **human**: 23 mg/kg; IV, LD_{LO}, **human**: 14 μg/kg.

LD_{50}, **rat**: 3.7 mg/kg; **rabbit**: 16 mg/kg; LD_{LO}, **dog**: 5 mg/kg; IV, LD_{50}, **rat**: 66 μg/g; **mouse**: 150 μg/kg; **dog**: 84 μg/kg; **rabbit**: 100 μg/kg; SC, LD_{50}, **mouse**: 340 μg/kg; **rabbit**: 500 μg/kg; **guinea pig**: 120 μg/kg; LD_{50}, skin, **mouse**: 1 mg/kg; **monkey**: 9 mg/kg; **rabbit** and **guinea pig**: 35 mg/kg; LD_{LO}, skin, **rat**: 18 mg/kg; **dog**: 30 mg/kg.

TACRINE

= *Cognex* = *Tetrahydroaminoacridine* = *THA*

Nootropic, respiratory stimulant, anticholinesterase agent.

Untoward effects: Serum alanine aminotransferase or SGOT levels increased in ~50% of **patients** and $\frac{1}{2}$ of **these** had levels three times normal. Hepatotoxicity, jaundice, and hepatocellular necrosis within 1–3 months in some **patients**. Nausea, vomiting, diarrhea, constipation, abdominal pain, flatulence, myalgia, dizziness, fatigue, decreased weight, asthenia, purpura, confusion, insomnia, somnolence, tremors, rhinitis, rash, flushing, pollakiuria, urinary incontinence, and occasional vagal over-activity with hypotension, apnea, bradycardia, frequent micturition, shivering, agitation, hostility, and convulsions also reported. ***Food*** decreases its absorption and systemic availability, but decreases incidence of gastrointestinal adverse effects. Readily crosses the blood–brain barrier.

Drug interactions: ***Fluvoxamine*** and ***cimetidine*** decrease its clearance from the body. Will potentiate the toxicity of ***succinylcholine*** (q.v.)-type muscle relaxants, ***phenothiazine***, ***aminoglycoside antibiotics***, and ***organophosphorus compounds***. ***Theophylline*** elimination half-life is decreased, which increases ***theophylline*** serum levels ~2 times. Plasma concentrations increased ~60% 10 days after hormone replacement therapy in ten **females**.

TACROLIMUS

= *FK 906* = *FR 900,506* = *Prograf* = *Protopic*

Immunosuppressant.* Oral, topical, and IV. A *macrolide antibiotic.

Untoward effects: IV use adverse effects > 50% and < 15% for oral use. ***Central nervous system neurotoxicity*** led to headache, tremors, paresthesias, circumoral tingling, dizziness, insomnia, irritability, altered mental state, aphasia, dysarthria, and coma; focal weakness in ~8% of **patients**, mostly after IV use. Symptoms subside quicker after oral therapy. Persistent coma in ~1%. Occasional tinnitus, photophobia, nightmares, blurred vision, and fatigue. Chronic inflammatory demyelinating polyneuropathy in a 62-year-old **male** transplant **patient**. ***Gastrointestinal upsets*** lead to nausea, vomiting, diarrhea, abdominal bloating and pain, anorexia, and gastrointestinal bleeding.

Miscellaneous: Hyperglycemia, new onset of diabetes mellitus, hypomagnesemia, hyper- or hypokalemia, hypophosphatemia, hyperlipemia, chest pain, palpitations, hypertension, dyspnea, coughing, pleural effusion, atelectasis, anemia, leukocytosis, thrombocytopenia, hair loss, increased hair, pruritus, rash, flushing, night sweats, arthralgia, back pain, peripheral edema, and fever are among the many other adverse effects that can occur with oral or IV. Anaphylactic reactions, bradycardia, and hypertrophic cardiomyopathy reported after IV. As levels are increased, so are nephrotoxicity, hepatic rejection, severe cholestasis, coma, and delirium.

Renal problems associated with its use lead to increased ***creatinine*** and blood urea nitrogen, oliguria, and urinary tract infections.

Therapeutic levels reported as 0.005–0.02 mg/l; toxic at 0.02–0.025 μg/l.

Drug interactions: ***Non-steroidal anti-inflammatory drugs*** may increase risk of renal damage. Many drugs reported to increase its serum concentration: ***calcium channel-blockers*** (***diltiazem***, ***nicardipine***, ***nifedipine***, ***verapamil***), ***antifungals*** (***clotrimazole***, ***fluconazole***, ***itraconazole***, ***ketoconazole***), other ***macrolide antibiotics*** (***clarithromycin***, ***erythromycin***, ***rapamycin***, ***troleandomycin***), ***grapefruit juice***, ***gusperimus***, ***gastrointestinal prokinetic agents*** (***cisapride***, ***metoclopramide***), and other drugs such as ***bromocriptine***, ***cimetidine***, ***cyclosporine***, ***danazol***, ***methylprednisolone***, ***nefazodone***, and ***protease inhibitors***. The ***anticonvulsants*** (***carbamazepine***, ***phenobarbital***, ***phenytoin***), ***rifabutin***, and ***rifampin*** can decrease its serum concentration. Unless its dosage is reduced, more serious toxicity can occur when used with ***cyclosporine***. Use with ***mycophenolate mofetil*** causes increased plasma levels of ***mycophenolic acid***.

Oral use in the **dog** caused acute fibroid vasculitis, vomiting, and anorexia; in the **rat** it caused hyperglycemia; thymic medullary atrophy proportionate to dosage. IM use in the **rat** led to thymic medullary atrophy proportionate to dosage; in the **baboon** it led to hyperglycemia, anorexia, and lethargy; emaciation can lead to death. Oral use in **cynomolgus monkeys** caused decreased weight, centrilobular vesicular steatosis of hepatocytes; IM caused similar hepatocyte damage, cardiomyopathy, acute tubular necrosis, and hyalination of pancreatic islets. IM use in the **dog** led to vascular cardiac changes and cell changes in the proximal renal tubules. Toxicity of ***cyclosporine*** and ***simvastatin*** in **rats** increased by its coadministration leading to hepatic toxicity and decreased renal function.

TAGETES MINUTA

= *Mexican Marigold* = *Kuzam* (Yemen)

Common in southeastern African highlands and subtropical America.

Untoward effects: **Farmers** coming in contact with it developed an allergic reaction leading to lichenoid eczema with post-inflammation hyperpigmentation and hypopigmentation.

TAIPOXIN

A phospholipase A_2 (*PLA*) isolated from snake venom.

Untoward effects: This presynaptic basic neurotoxin is extremely poisonous. First, it inhibits, then increases ***acetylcholine*** release and then decreases it leading to irreversible blockade of transmission at the motor nerve terminal. It blocks ***potassium*** ion channels, allowing for a rise in intracellular ***calcium*** ions.

LD_{50}, **mouse**: 300 μg/kg; IV, 2 μg/kg or ~8 μg and 0.05 μg, respectively/**mouse**.

TAMARIND

= *Hamar* (Somali) = *Ma-khaam* (India) = *Mkwaja* (Swahili) = *Raka* (Sanya) = *Ukwaju* (Bajun)

Untoward effects: The fruit pulp is used as a souring spice in Indian ***curry*** and has caused allergic contact dermatitis. Has a laxative effect with ***citric acid*** and other organic acids.

TAMOXIFEN CITRATE

= *ICI 46,474* = *Kessar* = *Noltam* = *Nolvadex* = *Nourytam* = *Tamofen* = *Tamoxasta* = *Zemide*

***Non-steroidal estrogen antagonist*. Oral, mostly for postmenopausal therapy.**

Untoward effects: In **females** leads to hot flashes (16–30%), nausea and vomiting (10–20%), coma, leg cramps, and transient increase in bone or tumor pain. Delayed effects include vaginal bleeding or discharge (3–10%) and menstrual irregularities (10%), depression, dizziness, headache, pruritus, distaste for food (3%), edema, cataracts, optic neuropathy, macular retinopathy, retinal opacity, decreased visual acuity, hyperlipoproteinemia, purpuric vasculitis, thromboembolism (5–7%, including pulmonary), thrombocytopenia (3%), leukopenia (4%), increased SGOT, > 5% weight loss, hair thinning, vaginal dryness, sexual dysfunction, hypercalcemia, small increase in gastrointestinal cancer in Swedish and Danish trials, endometrial cancer, ovarian cysts, and pruritus vulvae. Endometriosis and suspicious PAP smears have been reported. Rarely, Stevens–Johnson syndrome, steatohepatitis, cirrhosis, and bullous pemphigoid. Exposure *in utero* leads to ambiguous genitalia and oculoauriculovertebral dysplasia in **infants**. Can suppress lactation in **mothers** and can induce ovulation in premenopausal **females**. Priapism has occurred in a **male patient** and decreased spermatogenesis and motility in three **males**. Risk of endometrial cancer can increase with increased duration of therapy. Hair color changes may also occur.

TD_{LO}, **woman**: 200 μg/kg/day.

Drug interactions: Life-threatening interaction with ***warfarin***, increasing its anticoagulant effect. Asthma in **patients** with ***non-steroidal anti-inflammatory drug***-induced asthma. ***Rifampin*** decreases its serum concentration ~86%.

In **rats**, 10 times the **human** dose on a weight basis led to precancerous liver changes (hyperplastic nodules) after 1 year of treatment. Anti-implantation action in **mice**. Has estrogenic activity in **ferrets**. IM in **hens** caused molting and decreased egg production.

LD_{50}, **rat**: 600 mg/kg; IP, 62.5 mg/kg IV; **mouse**: 200 mg/kg; IP, 62.5 mg/kg IV.

TAMSULOSIN

= *Alna* = *Amsulosin* = *Flomax* = *Josir* = *Omix* = *Omnic* = *Pradif* = *Urolosin* = *YM 617* = *YM 12,617*

***α-1A adrenoceptor antagonist*. Used in the treatment of benign prostatic hyperplasia.**

Untoward effects: Dizziness in 15–17%, orthostatic hypotension in 0.2–0.4%, and vertigo in 0.6–1%. Rhinitis, retrograde ejaculation or decreased volume, ejaculation failure, skin rashes, pruritus, urticaria, and angioedema of tongue, lips, and face.

Drug interactions: Its serum concentration and potential toxicity is increased by concomitant use with ***cimetidine***.

TAMUS COMMUNIS

= *Black Bryony*

Hedge climber in England and Europe.

Untoward effects: Irritant purgative. Ingestion of the roots and berries has caused rapid death. **Children** are especially at risk from eating the berries. Symptoms are the same as for ***Bryonia*** (q.v.).

TANAECIUM sp.

Untoward effects: ***T. nocturnam*** = ***Koribó*** = ***Pum-ap*** = ***Samedo-ap*** is the main ingredient in a narcotic snuff used by the **Paumarí Indians** of Brazil. In a confined room, the vapors from the vine have caused severe dizziness, headache, and a desire to throw **oneself** immediately into a lake. Toasting it for snuff apparently removes the ***cyanides***, but leaves other intoxicating compounds. **Natives** stay away from the fresh material because of its high ***hydrogen cyanide*** (q.v.) content and fumes.

T. exitiosium causes weakness, pollakiuria, tachycardia, weak pulse, staggering, collapse with muscle spasms, and high mortality in Brazilian **cattle** that have grazed it for 1–2 weeks. It is also toxic to **goats**, **horses**, and **sheep**.

TANNIC ACID

= *Gallotannic Acid* = *Gallotannin*

Use: In ***inks***, pharmaceuticals, ***rubber*** manufacturing, tanning, electroplating, clarifying agent in ***wines*** and brewing, deodorizing crude oils, photography, and in medicine externally as an astringent; internally in treating diarrhea; and as an antidote for metallic and alkaloidal poisoning. Also in cosmetics, antiperspirants, hair tonics, foot treatments, and sunscreens.

Untoward effects: Use with ***barium*** enemas to improve X-ray images has been banned in the U.S. because it was associated with decreased liver function, severe liver zonal necrosis, and deaths of **patients** when absorbed in sufficient quantities. Nausea and weakness also occurred in some of those **patients**. Topical application to bruised, burned, or sensitive areas can cause burning and stinging sensations and it can be absorbed from burned areas leading to hepatic zonal necrosis in ~3 days after application to large burned areas. Severe bullous eruption on dorsal surface of both feet in area of contact with sandals made of East Indian buffalo hide in a 12-year-old **female**. Several months later she developed a contact dermatitis to ***leather*** lining of an American-made ***shoe***; the lining also originated in India, where they used vegetable tannins for tanning. Tannins in ***tea*** cause insoluble complexes with 12 marketed ***antidepressants*** and it causes an irreversible binding of ***vitamin*** B_1. It binds ***activated charcoal*** and makes the "universal antidote" virtually worthless. Incompatible with salts of ***antimony***, ***copper***, ***iron***, ***lead***, ***mercury***, and ***silver***; ***albumin***, ***alkaloids***, ***chlorates***, ***gelatin***, ***iodine***, ***iodoform***, ***lime water***, ***permanganates***, ***spirit nitrous ether***, and ***starch***.

LD_{LO}, **human**: 500 mg/kg.

Excessive dietary intake by **animals** can decrease growth rate, decrease ***thiamin*** and ***iron*** absorption, and cause mottling of **chicken** eggs. Doses of 50–300 g given by stomach tube lead to colic, jaundice, gastric mucosa necrosis, myocardial degeneration, nephritis, and hemolytic anemia. Source of toxicity in many **animals** eating various plants and ***oak*** leaves.

LD_{50}, **rat**: 5 g/kg; **mouse**: 6 g/kg.

TANSY

= *Bitter Buttons* = *Golden Buttons* = *Ponso* = *Tanacetum vulgare*

Culinary herb containing volatile oil.

Untoward effects: Contains ***thujone*** (q.v.) and historically has been used to repel flies and rubbed on meat to help prevent decay, and as an emmenagogue and abortifacient. Ingestion of leaves or flowers also causes tachycardia, weak pulse, convulsions, upset stomach, and even death. Ingestion of 15–30 ml has caused death in 2–4 h. Symptoms of poisoning also include abdominal pain, vomiting, violent epileptiform convulsions followed by deep coma, mydriasis, dyspnea, and paralytic asphyxia.

Animals rarely eat it due to its very bitter taste.

TANTALUM

Untoward effects: Use of tantalum thread on fractures inhibits normal development of osseous calluses. The dust can irritate eyes and skin. Chronic urticaria 10 months after surgical implantation of tantalum staples. OSHA and National Institute for Occupational Safety & Health set upper exposure limits to 5 mg/m^3, TWA; 15-min exposure 10 mg/m^3, and immediate danger to life or health as 2.5 g/m^3.

In some **animal** species, it induced tumors when introduced into bone, muscle, or SC. Pulmonary irritant in **animal** inhalation studies.

LD_{50}, **rat**: of the *oxide*, 8 g/kg.

TAPIOCA

It is heated *cassava* (q.v.) starch.

Untoward effects: The raw roots and leaves have high concentrations of ***hydrocyanic acid*** (q.v.) in its cyanogenic glycoside, ***linamarin*** (q.v.), and can cause respiratory difficulties, loss of voice, twitching, shaking, convulsions, coma, and death. **Amazonian Indians** devised a way to make the poisonous tubers, now a staple in **their** diet, safe to eat. The roots are peeled and after several changes of immersing in cooking water, they are safe to eat.

TARAXACUM OFFICINALE

= *Blowball* = *Cankerwort* = *Dandelion* = *Fortune Teller* = *Horse Gowan* = *Soffione* = *Tuki Phool*

Untoward effects: Despite questionable lay claims for its benefits in menstrual problems, it does have medicinal use as a mild laxative and diuretic, and was reportedly used as such by **American Indians** in New England. The greens are often used in salads or ***tea*** and contain high ***oxalate*** (q.v.) content, which causes adverse effects in **females** with vulvodynia. Contact dermatitis is common, as with other members of the *Compositae* family (***chrysanthemums, ragweed, tansy***, etc.).

Stimulates bile flow in **rats**.

TARAXEIN

Untoward effects: The only hallucinogen extracted from **humans** and may cause schizophrenia (2/3 **schizophrenics** have it in **their** blood). It is a complex protein and, apparently, an antibody.

TARRAGON

= *Estragon*

Untoward effects: The dried leaves and flowers of *Artemisia dracunculus* grown in Europe and Siberia as such or its oil as a condiment for foods and liqueurs. Also in soaps, detergents, cosmetics, creams, lotions, and perfumes at 0.01–0.4%.

Untoward effects: No irritation of the oil after 4% in petrolatum used on 25 **humans** and no sensitization of the same material on 25 **human** volunteers. No phototoxic effects from the oil.

Full-strength oil topically on intact or abraded **rabbit** skin for 24 h under occlusion was irritating. The oil undiluted on the backs of hairless **mice** was irritating. See ***Estragole***.

Oil, LD_{50}, **rat**: 1.9 ml/kg; acute dermal, **rabbit**: > 5 ml/kg.

TARTRAZINE

= *CI 640* = *CI 19,140* = *CI Acid Yellow 23* = *CI Food Yellow 4* = *DFG 64* = *E102* = *FD & C Yellow No. 5* = *Hydrazine Yellow* = *Poly T-128* = *Schultz 737*

Use: Orange-yellow ***azo dye*** in foods, soft drinks, pharmaceuticals, cosmetics, and ***wool*** and ***silk*** fabrics.

Untoward effects: A known sensitizing agent, and some **patients** sensitive to ***aspirin, aminopyrine, indomethacin, p-hydroxybenzoic acid, mefenamic acid***, and ***sodium benzoate*** can cross-react to it. Hundreds of drug products, soft drinks, ***ice creams***, sherberts, desserts, salad dressings, bakery products, confections, candies, assorted foods, and cosmetics contain it. Some **people** are allergic to it and reactions to it are occasionally life-threatening. Although very infrequent, reactions have included severe generalized urticaria and pruritus, itching of tongue and vulva, edema of lips, nausea, vomiting, nasal congestion, severe headache, tickling of the throat, wheezing, coughing, bronchospasm (in up to 15–60% of ***aspirin***-sensitive **patients**), severe asthmatic attacks, angioedema, and vascular purpura and these usually occur 6–14 h after exposure, often lasting up to 1 day. A 23-year-old **female** within 15 min after taking a 0.5 mg tablet of ***dexamethasone*** containing tartrazine as a coloring agent developed severe generalized pruritus, itching of **her** tongue and vulva, wheezing, and urticaria. A hospitalized pregnant 15-year-old **female** was given a liquid castile soap enema upon admittance. In 5 min **she** became dizzy, sweating profusely, with blood pressure 96/60. **She** collapsed after being taken to **her** bed and became unconscious with no recordable blood pressure, tachycardia, and weak pulse, followed by bright red erythema over **her** whole body. After expelling the enema (tartrazine in the soap) **she** became nauseous, conscious with dull perception, and within minutes developed generalized urticaria and respiratory difficulties. There was **fetal** distress during this time. It may induce lupus erythematosus. Nail staining reported by a **patient** from its use in an adhesive plaster applied to **her** thumbnail. A **physician** called me in panic after giving a **patient** in his office two ***tetracycline*** capsules, where **he** collapsed "even before they could have reached **his** stomach". The ***gelatin*** capsule used contained tartrazine. The **patient** recovered from this unusual case of rare hypersensitivity.

LD_{LO}, **human**: 14 µg/kg.

LD_{50}, **mouse**: 12.8 g/kg.

TATRADYMIA sp.

Untoward effects: Common in arid areas of the U.S. where ***T. glabrata*** = ***Coal-Oil Brush*** = ***Horsebrush*** = ***Spring Rabbit Brush***, generally unpalatable to **cattle** and **horses**, but **sheep** graze on the buds and leaves in early spring leading to depression, asthenia, and occasionally death within a few hours. In others, head irritation, followed by swelling of the head, neck, ears, eyelids, lips, and throat, commonly called "big-head" or "swellhead" by ranchers. Generalized edema can cover the entire body if **animals** are exposed to sunshine. Death toll is very high and due to liver failure.

T. canescens = ***Spineless Horsebrush*** causes similar problems in grazing **sheep** and serious photosensitization problems, but without liver pathology.

TAURINE

Untoward effects: Intolerance in **patients** with psoriasis leading to moderate to intense pruritus with increased redness and scaling of lesions and development of new lesions lasting 3–8 days after 2 g in testing on 21 **patients**.

TAXINE

Untoward effects: A mixture of cardiotoxic alkaloids from the berries of a poisonous tree, *Taxus baccata* (***Yew*** - q.v.). Also found in the leaves and shoots. The red pulp is not considered to be especially poisonous, but the seeds contain taxine and ***formic acid*** (q.v.). The taxine is a depressant leading to bradycardia, cardiac or respiratory arrest, and death. **Children's** habits make **them** especially prone to poisoning by the berries leading to nervousness, gastrointestinal upsets, diarrhea, muscle tremors, mydriasis, dyspnea, asthenia, and collapse. Death can be sudden and often preceded by convulsions. Also found in other ***yew*** species.

Except for a few **birds**, ingestion can be lethal to various **animal** species, with the **horse** and **pig** being most sensitive to it.

TAXOTERE

= *Docetaxel* = *NSC 628,503* = *RP 56,976*

Antineoplastic. **For IV infusion.**

Untoward effects: Nausea, vomiting, fever, peripheral edema, alopecia, leukopenia, decreased liver function, granulocytopenia, paresthesia, dysesthesia, pain, mouth sores, phlebitis, myalgia, pleural effusions, neutropenia, thrombocytopenia, febrile neutropenia, atrial fibrillation, thrombophlebitis, hypotension, and tachycardia are among those more commonly reported. Open-angle glaucoma in a 31-year-old **female** after reported use. Severe hypersensitivity noted despite pretreatment with ***corticosteroids*** and ***antihistamines***.

Abnormal mitosis and necrosis of various organs in **rats** given it at a comparable **human** mg/m^2 dosage.

TAZAROTENE

= *AGN 190,168* = *Tazorac* = *Zorac*

Topical retinoid, antipsoriatic, antiacne. **For topical use.**

Untoward effects: Incidence of adverse effects leading to mild to moderate burning, pruritus, stinging, and erythema at the treatment site, which is related to the concentration and dosing frequency (13% with 0.05% gel once daily versus 30% treated with 0.1% gel twice daily). After 4 months, some **patients** had exacerbation of the disease and increased erythema after sun exposure. Some **patients** develop rashes, contact dermatitis, desquamation, fissuring, dry skin, and bleeding. Teratogenic after systemic use.

TEA

= *Thé* = *Thea* = *Thee*

This three letter word means many different things to many people.

Untoward effects: **Herbalists** expound the benefits of ***Herbal Teas***, ***Bush Teas***, or ***Medicinal Teas*** because they supposedly contain no ***caffeine*** or ***tannins***, but teas brewed from ***wax myrtle*** leaves, cherrybark ***oak***, ***sweet gum*** tree leaves, and the roots of ***marsh rosemary*** also contain ***tannins; sassafras*** and ***comfrey*** teas contain a carcinogen. ***Herbal teas*** might contain only a single ingredient and sometimes contain up to ~20 different kinds of leaves, seeds, and flowers. ***Diuretic Teas*** usually contain ***buchu***, ***dandelion***, ***quackgrass***, or ***horsetail*** (*Equisetum*). A Crohn's disease 13-year-old patient drank ***comfrey*** tea for several months leading to thrombosis of the hepatic veins. A 68-year-old **female** ingested a wild ***germander*** preparation leading to fatal hepatitis. Severe diarrhea in six **people** in New York and Pennsylvania **who** consumed an ***herbal tea*** containing ***buckthorn*** bark and ***senna*** leaves, flowers, and bark. ***Chamomile*** tea has caused contact dermatitis, anaphylaxis, and severe hypersensitivity reactions, especially in **people** allergic to *Compositae* plants, such as ***asters***, ***chrysanthemums***, ***goldenrod***, ***marigold***, ***ragweed***, and ***yarrow***. Photodermatitis has developed after drinking ***St. John's Wort*** leaf tea. ***Lobelia*** tea causes bradypnea and tachycardia, and, if drunk in excess, leads to coma or death. ***Bush teas*** or ***Honig Thee*** are popular in Asia, Africa, the West Indies, and other ethnic groups elsewhere from these regions. These are often made from ***Crotalaria***, ***Heliotropium***, and ***Senecio***, and contain up to 0.6% ***pyrrolizidine*** alkaloids; responsible for the high incidence of liver disease in those areas. It mimicked Reye's syndrome in an Arizona **child**. ***Maté*** tea also contains these alkaloids and has caused liver damage and death in a **woman** who consumed this tea for many years. ***Juniper*** berry tea can cause gastrointestinal irritation. Ingestion of ½ cup ***burdock*** tea led to severe anticholinergic effects with mydriasis, xerostomia, blurred vision, dysuria, and hallucinations. Ingestion of teas from > 100 plant species in Curaçao led to esophageal cancer in 20.9/100,000. ***Croton*** leaf tea is common there. U.S. Health & Human Services claims a close relationship exists between esophageal cancer and ***tannins*** in Chinese tea. Similar adverse effects reported in Japan due to high consumption of tea–***rice*** gruel. In the mid-eighteenth century, tea was the national drink in Holland and, at that time, **they** had an increased level of esophageal obstructions and cancers. It has been suspect as a cause of recurring prostatitis in **males** and urethrotrigonitis in **females**. It has also been associated with nasopharyngeal cancer in China. Tea contains a high level of ***oxalates*** (650–700 mg%) and can cause vulvodynia in **females** and ***calcium oxalate*** renal calculi. The danger from ***fluorides*** in tea has been overemphasized. Its contents in decreasing order are oolong, Russian, China blacks, Indian blacks, black blends, Ceylonese blacks, instant or canned teas. Its ***xanthine*** alkaloid content may cause insomnia, anorexia, irritability, and low-grade irregular fever. Liver fibrosis, splenomegaly, and ascites from nibbling on tea leaves. Excessive intake may be associated with benign fibrocystic breast disease in **females**. A hypersensitivity pneumonitis (**tea-graver's** disease) from an unknown antigen in handling tea plants. So-called ***sun tea*** brewing requires refrigeration an hour after completion to prevent a syrupy substance developing, which is actually mold growing on its nutrients. A mailed sampling of a dish-washing detergent, ***Sunlight***®, in a yellow container with a picture of a ***lemon*** on it, led to a major misunderstanding that had **people** adding it to **their** iced tea. Extracellular dehydration, probably from its ***theophylline*** content and secondary increase in intestinal fluid absorption led to constipation in 12 **volunteers** drinking 2 l/day. Ingestion has caused decreased lower esophageal sphincter tone, stomach upsets, and dyshidrosis. A 70-year-old **male** drinking 14 l/day developed myositis due to hypokalemia. The U.S. Burea of Foods has found ***strontium***-90, ***potassium***-40 and ***cesium***-137 in tea samples. The Russian Chernobyl nuclear accident contaminated Turkey's prime tea-growing region with ***cesium***-137. After 6 years, levels dropped to 10% of the original. In England, 18 **adults** collapsed within 5–10 min after drinking tea made with water from a ***copper***-lined boiler. Elsewhere in London, 20 **workmen** brewed tea with similar water leading to acute vomiting and diarrhea. Occasionally, tea leaves have undesirable levels of ***insecticides*** on them. The U.S. has now banned ***propargite*** on dried tea.

Polystyrene cups begin to dissolve when filled with hot ***lemon*** tea. ***Limonene oil*** in the ***lemon*** does it. ***Tannins*** in teas can cause the development of an insoluble and unavailable complex with a number of ***antidepressants***. *In vitro*, it precipitated ***antihistamines***, ***propranolol***, and ***hydrazines***. A 48-year-old **male** developed a severe reaction to ***phenobarbital*** with massive edema of legs and blue-black erythematous rash. The same reaction followed ingestion of tea, possibly because they both contain phenyl rings. Spuriously increased vanillylmandelic acid levels if the test is based on ***phenolic acids*** and diazotized *p*-***nitroaniline***. Tea contains anti-***thiamin*** activity, decreases ***iron*** absorption, and contains ***caffeine***, ***theobromine***, and ***theophylline***. Insignificant quantities of its ***caffeine*** content can be found in breast milk. Associated with migraine headaches in some **people**. **Patients** with ***chromate***, ***cobalt***, or ***nickel*** dermatitis should avoid tea, as it has caused flares within 4 days in at least 40% of these **patients**. Use is associated with falsely increased plasma ***catecholamine*** levels and falsely increased urine vanillylmandelic acid levels. A 24-year-old **male** ingesting $\frac{1}{2}$–1 gallon of green tea decreased **his** international normalized ratio 50% within several weeks while taking ***warfarin***. ***Tannic acid*** in tea and ***tannins*** decrease ***iron*** absorption.

In **rats** and **mice**, anti-***thiamin*** activity and decreased ***iron*** absorption; effects on ***psychotherapeutics*** are the same as in **people**.

TEA TREE

= *Melaleuca alternifolia*

An Australian tree.

Untoward effects*:** Its volatile oil is highly germicidal and used in dentistry, surgery, and soaps, but ingestion of ~$\frac{1}{2}$ teaspoonful by a 60-year-old **male** caused petechial rash of hands, elbows, upper torso, and inside of arms; marked neutrophil leukocytosis; and swelling of face, hands, and feet. **He** had ingested it several times before and occasionally had itchy palms. The oil contains ***eucalyptol and ***terpinen-4-ol*** as its main active ingredients. Useful topically as an insecticide for fleas on **cats** leading to hypothermia, incoordination, coma, and dehydration.

TEAK

= *Tectonia grandis*

Untoward effects*:** Allergic contact dermatitis from ***desoxylapachol, a strong sensitizer in teakwood. Exposure can cause both immediate and delayed hypersensitivity. The former was a wheal-and-flare reaction. Contact eczema and/or severe itching in furniture **workers** and **carpenters** from it.

TECHNETIUM-99m

A short-lived *radionucleotide* for diagnostic purposes.

***Untoward effects*:** Half-life of ~6 h requires a delay of 12 or preferably 48 h to decrease significant **infant** levels after breast feeding.

TEGAFUR

= *Citofur* = *Coparogin* = *Exonal* = *Fental* = *Franrose* = *FT 207* = *Ftorafur* = *Fulaid* = *Fulfeel* = *Furafluor* = *Furofutran* = *Futraful* = *Lamar* = *Lifril* = *MJF-12,264* = *Neberk* = *Nitobanil* = *NSC-148,958* = *Riol* = *Sinoflurol* = *Sunfural* = *Tefsiel C* = *UFT*

***Antineoplastic*. IV use.**

***Untoward effects*:** Nausea, vomiting, diarrhea, stomatitis, leukopenia, and thrombocytopenia.

TEICOPLANIN

= *L 12,507* = *MDL-507* = *Targocid* = *Targosid* = *Teichomycin A_2*

***Antibiotic.* IM and IV.**

Untoward effects: Leukopenia and neutropenia in a 73-year-old **male** after 9 mg/kg or 600 mg/day/20 days. Hypersensitivity reactions reported. Occasional cross-reactions in **patients** sensitive to ***vancomycin*** or ***trimethoprim-sulfamethoxazole*** combination. Ototoxicity in a 39-year-old **male** with Down's syndrome after IM use. Other case reports of ototoxicity. Pain at IM site for up to 12 h. Transient increase in serum aminotransferases, drug fever, skin rash, bronchospasm, eosinophilia, and thrombophlebitis have been reported infrequently.

TELEOCIDIN

Untoward effects: A, B, and other isomers are poisonous substances isolated from several strains of *Streptomyces*. They are used experimentally as potent tumor-promoters. Teleocidin B is lethal to Japanese **killifish**, *Oryzias lapites*, within 1 h at 0.01 μg/ml. Topically, intensely irritating to **rabbit** skin and causes severe irritation and eruptive vesications on **human** skin.

IV, LD_{50}, **mice**: 220 μg/kg.

TELFAIRIA OCCIDENTALIS

= *Fluted Pumpkin*

Untoward effects: The young tender leaves are eaten in southern Nigeria. The seeds are cooked and can be eaten as such or in soups, but the roots are a potent **human** and **fish** poison.

TELIOSTACHYA LANCEOLATA

= *Toe Negra*

Untoward effects: Cultivated by **Indians** in Peru as a narcotic and hallucinogen, when it can cause loss of sight for 3 days.

TELITOXICUM PERUVIANUM

= *Bo-de´-mee-see*

Untoward effects: The bark is the main ingredient of an Amazonian ***curare*** mixture with ***Strychnos*** on poison arrows and darts.

TELLURIUM

= *Aurum paradoxum* = *Metallum problematum*

Untoward effects: Inhalation of tellurium *dioxide* fumes or tellurium dust leads to xerostomia, garlicy breath (due to ***dimethyl telluride***), diaphoresis, nausea, headache, somnolence, ***garlic*** odor of sweat, giddiness, and metallic taste. Ingestion also causes drowsiness, nausea, constipation, and anhydrosis. During retrograde pyelography, it was injected into the ureters of three **patients**; two of three died and had renal pain, stupor, cyanosis, vomiting, garlicy breath, and unconsciousness. National Institute for Occupational Safety & Health/OSHA upper exposure limit of 0.1 mg/m^3; immediate danger to life or health of 25 mg/m^3 as tellurium.

Its administration in their feed at 500–3000 ppm to pregnant **rats** led to neuropathy and hydrocephaly in their offspring. Use in **pig** and **duckling** diets caused ***selenium–vitamin E*** deficiency. Large oral doses of tellurium *oxide* or ***sodium tellurate*** to **dogs** leads to violent vomiting, anorexia, somnolence, and decreased gastric digestion. Tellurium *dioxide* in **rats** caused decreased growth rate, hair loss, redness and edema of digits, temporary paralysis of hind legs, hepatocyte necrosis, and necrosis of renal tubule epithelial cells. Tellurium *tetrafluoride*, 500 ppm, fed to **male ducklings** for 3 weeks led to anorexia, weakness, and death with lesions in the gizzard, intestine, skeletal muscle, and heart.

Tellurites are more toxic than ***tellurates***.

TELLURIUM HEXAFLUORIDE

= *Tellurium Fluoride*

Untoward effects: A colorless gas with a pungent repulsive odor for **man**, which after inhalation leads to headache, dyspnea, and garlicy breath. National Institute for Occupational Safety & Health/OSHA limits of exposure are 0.02 ppm; immediate danger to life or health, 1 ppm.

Inhalation TD_{LO}, **human**: 714 mg/kg.

Inhalation by laboratory **animals** leads to pulmonary edema.

TELMISARTAN

= *BIBR 277* = *Micardis*

***Angiotensin receptor-blocker.* Unlike *angiotensin-converting enzyme inhibitors*, it does not induce coughing.**

Untoward effects: Angioedema.

Drug interactions: It increases plasma concentration of ***digoxin*** ~50%.

TEMAFLOXACIN

= *A-63004* = *Omniflox* = *T-1258* = *Teflox* = *Temac*

Untoward effects: This ***fluoroquinolone*** was voluntarily recalled from the U.S. marketplace in less than 4 months after its introduction, as its use was associated with ~30 cases of hemolytic anemia, renal failure, thrombocytopenia, liver dysfunction, allergic reactions, shock, anaphylaxis, and three deaths.

TEMAZEPAM

= *ER-115* = *Euhypnos* = *Euipnos* = *Gelthix* = *K-3917* = *Levanxene* = *Levanxol* = *Normison* = *Oxydiazepam* = *Perdorm* = *Planum* = *Remestan* = *Restoril* = *Ro-5-5345* = *Wy-3917*

***Sedative, hypnotic, tranquilizer.* A *benzodiazepine* and *diazepam* (q.v.) metabolite.**

***Untoward effects*:** Drowsiness, dizziness, confusion, and ataxia are the most common, especially with overdoses and in the **elderly** or **debilitated**. Amnesia, lethargy, hangover, anxiety, euphoria, excitement, and weakness occur occasionally. Rarely, hypotension, palpitations, dyspnea, arrhythmias, horizontal nystagmus, restlessness, and hallucinations. Generalized lichenoid dermatitis in a 76-year-old patient after 10 mg/day; problem resolved 10 days after medication discontinued. Excreted in breast milk and can be detected in **infant's** plasma. A statistically significant preference for **men** for its hypnotic efficacy if given in a yellow capsule, while **women** felt either the yellow or green capsules were of equal efficacy. Rate of IV administration can influence saccadic eye movements and electroencephalogram effects. IV drug abuse with it from capsule contents leading to leg ischemia, rhabdomyolysis, edema, macular rash, and fatal pulmonary microembolism has occurred. Accidental IA injections in **drug-abusers** have caused compartment syndromes, rhabdomyolysis, deep vein thrombosis, and pulmonary embolism, sometimes requiring amputation of digits and a leg, and fasciotomies. Half-life is generally 6–14 h, but occasionally as long as 25 h. Therapeutic serum levels have been reported as 0.02–0.5 mg/l; toxic at 1 mg/l, and at 8.2–14 mg/l in a comatose state or dead.

Drug interactions*:** Concurrent use with ***oral contraceptives can cause a marked (40%) decrease in its efficacy. Concurrent use with ***disulfiram*** increased its toxicity in a 34-year-old **male**. May potentiate other ***hypnotics*** and central nervous system suppression. Overdoses with ***alcohol*** can be fatal. Peak serum concentrations are reduced by ***oral contraceptives***.

An unexpected stillbirth at term in < 8 h after a **female** ingested 30 mg of it and ***diphenhydramine*** 50 mg. In 13 pregnant **rabbits**, this effect was reproduced; 81% of fetuses were stillborn or died shortly after birth. Both drugs can cross the placenta. High dosage in **rats** and **rabbits** caused decreased fertility and teratogenesis.

LD_{50}, **rat**: 1.8 g/kg; **mouse**: 1.2 g/kg; **rabbit**: > 2 g/kg; IP, **mouse**: 85 mg/kg.

TEMEPHOS

= *Abaphos* = *Abat* = *Abate* = *Abathion* = *Amer. Cyan 52,160* = *Biothion* = *Difenthos* = *Difos* = *ENT 27,165* = *Lypor 20* = *Nimitex* = *Temefos* = *Tetrafenphos*

***Organophosphorus insecticide, mosquito larvicide, anticholinesterase.* In treating drinking water for *Dracunculus* eradication in Ghana and Nigeria.**

***Untoward effects*:** Inhalation of spray leading to eye irritation and blurred vision. Absorption through the skin causes dizziness, confusion, miosis, salivation, and dyspnea; ingestion also causes abdominal cramps, nausea, vomiting, and diarrhea. National Institute for Occupational Safety & Health and OSHA upper exposure limits at 10 and 15 mg/m^3, respectively; 5 mg/m^3 as inhalation limits.

LD_{LO}, **human**: 500 mg/kg.

When widely used in coastal mosquito control, its application coincides with **waterfowl** nesting and rearing. **Duckling** survival is decreased. Symptoms in most **bird** species are asthenia, ataxia, fluffed feathers, fasciculations, tremors, salivation, lacrimation, miosis, tracheal congestion, weakness, tachycardia, tachypnea, tetany, and immobility. Moderately toxic to **honeybees** and **fish**. Pregnant **heifers** given it at 10 or 20 ppm in their drinking water during early gestation showed abnormal development and early deaths of their **calves**. In others fed 1–1.5 mg/kg/day/12 months it caused swelling and ulceration of extremities, decreased weight, incoordination, and convulsions. **Sheep** and **lambs** showed no ill-effects from 80 mg/day/14 months.

LD_{50}, male **rat**: 1–2 g/kg; female **rat**: 2.3 g/kg; **mouse**: 223 mg/kg; **mallard duck**: 79.4 mg/kg; **California quail**: 18.9 mg/kg; **Japanese quail**: 84.1 mg/kg; **pheasant** and **house sparrow**: 35.4 mg/kg; **chukar**: 240 mg/kg; **starling**: 100 mg/kg; **pigeon**: 50 mg/kg.

TEMOZOLOMIDE

= *CCRG-81045* = *M&B 39,831* = *Methazolastone* = *NSC 362,856* = *Temodal* = *Temodar* = *Temodol* = *TMZ*

***Antineoplastic.* For treating anaplastic astrocytomas.**

***Untoward effects*:** Myelosuppression (neutropenia, thrombocytopenia), nausea, vomiting, fatigue, weakness, constipation, headache, and mild immunosuppression frequently occur.

TENECTEPLASE

= *TNKase*

Recombinant variant of *human plasminogen activator*, as IV bolus.

Untoward effects: Intracranial hemorrhage and other bleeding, especially with 5 s bolus and increases with increased dosage. Rarely, allergic reactions.

TENIPOSIDE
= *ETP* = *NSC 22,819* = *PTG* = *Vehem-Sandoz* = *VM 26* = *Vumon*

***Antineoplastic.* Semi-synthetic derivative of *podophyllotoxin* (q.v.). IV, IM, and oral.**

Untoward effects: Nausea, vomiting, diarrhea, phlebitis, severe hypersensitivity reactions, anaphylaxis, mucositis, hypotension, hemolysis or hemolytic anemia, and renal failure. Delayed reactions include bone marrow depression, leukopenia, thrombocytopenia, alopecia, peripheral neuropathy, and leukemia.

TENOXICAM
= *Alganex* = *Dolmen* = *Liman* = *Mobiflex* = *Rexalgan* = *Ro-12-0068* = *Tilatil* = *Tilcotil*

Analgesic, anti-inflammatory.

Untoward effects: In various trials, 12–28% and closer to 15% in an 8567 **patient** trial. Incidence was higher in **females** and 5.2% dropped out of the study because of adverse effects. Acute hepatitis in a 77-year-old **female**. Epigastric pain, gastrointestinal upsets, and decreased renal function can occur. Excreted in breast milk with **infant's** dose on a μg/kg basis ~3% of the **mother's** dose. Therapeutic serum levels are ~5–10 mg/l.

Drug interactions: ***Aspirin*** can significantly decrease its plasma concentration. It causes a 20–30% increase in ***rifampin*** plasma concentration within 1 h. ***Food*** and ***antacids*** decrease its rate of, but not its total, absorption.

TEONANACATL
= *Flesh of the Gods* = *Sacred Mushroom* = *Teon*

Untoward effects: Psychostimulant used as a sacrament in Aztec religious rites. Contains 0.2–0.4% of the hallucinogenic alkaloid ***psilocybin*** (q.v.). The latter has ~1/200th the hallucinogenic effect of ***LSD***.

TEPHROSIA sp.

Untoward effects: In the southeastern U.S., ***T. virginiana***, called ***Goat's Rue*** (not *Galega*) = ***Devil's Shoestring*** = ***Hoary Pea*** = ***Turkey Pea*** contains ***rotenone***, is poisonous to **fish**, and, if ingested, can be lethal. It is also associated with the devil in the world's folklore.

T. vogelii = ***Igun*** (Yoruba) = ***Iwele*** (Igbo) = ***Kibaazi*** (Swahili) = ***Maginfa*** (Hausa) = ***Mtupa-wa-mrima*** (Swahili) = ***Poison Fish Bean*** contains ***flavanones***, ***tephrosin***, and ***rotenoids***, making it an excellent piscicide when the mashed plant is poured into the water at the head of the estuary as the tide is ebbing. **People** who wade into the streams to pick up the **fish** show anesthetic effect on limbs and skin roughness. Other species are used as **fish** poisons and ***insecticides*** in India, Samoa, Africa, Brazil, and Venezuela. Some species are called ***avasa***, meaning forbidden.

TEQUILA

Mexican Indians use the *Agave* cacti to make tequila, *mescal* or *pulque*. The latter is a *beer* of low *alcohol* content and a considerable amount of *mescaline*.

Untoward effects: A 26-year-old **male** drank a pint of it over a 4 h period leading to temporary unilateral sensorineural hearing loss and tinnitus. Slow recovery.

TERAZOSIN
= *Abbott 45975* = *Heitrin* = *Hitrin* = *Hytracin* = *Hytrin* = *Hytrinex* = *Itrin* = *Urodie* = *Vasocard* = *Vasomet* = *Vicard*

***α_1-Adrenergic antagonist, antihypertensive.* Useful in treating benign prostatic hypertrophy.**

Untoward effects: Dizziness, weakness, syncope, palpitations, nausea, headache, peripheral edema, blurred vision, and nasal congestion are common adverse effects and influenced by dosage and **patient's** age. Orthostatic hypotension can be serious, especially with the first few doses. Impotence has occurred and relaxation of the bladder neck can lead to retrograde ejaculation. Use with ***food*** can delay its absorption and decrease peak levels of the drug.

TERBINAFINE
= *Lamisil* = *SF 86-327*

***Antifungal.* Topical use.**

Untoward effects: Burning, irritation, urticaria, pruritus, erythematous rash, exfoliation, taste and smell disorders ("everything tasted mildewy"), diarrhea, abdominal pain, fatigue, parotid swelling, and hepatitis. Occasional reports of nausea, headache, chest pain, dizziness, and insomnia. Rarely, leukopenia, thrombocytopenia, systemic lupus erythematosus and cutaneous lupus erythematosus, hypersensitivity reactions, and anaphylaxis.

Drug interactions: It inhibits the metabolism of ***ethoxycoumarin*** and ***tolbutamide*** by < 5%, ***caffeine*** by 19%, ***cyclosporin*** by 12–15%, and ***ethinylestradiol*** by 35%. It leads to ***nortriptyline*** intoxication. Conflicting reports on its interaction with ***warfarin*** from nil to > 20% decrease in the latter's dosage after several weeks.

TERBUFOS
= *AC 92,100* = *Counter*

Oragnophosphorus insecticide. Cholinesterase inhibitor.

***Untoward effects*:** Symptoms in exposed **workers** are headache, weakness, miosis, blurred vision, salivation, nausea, vomiting, abdominal cramps, and diarrhea.

Accidental exposures have caused poisoning of **cattle** from use of the rootworm chemical in a mineral mix, and as a feed contaminant led to miosis, diarrhea, colic, drooling, central nervous system depression, dyspnea, muscle fasciculations, and ataxia. **Quail** are often killed by concentrations > 280 ppm in their feed, usually within 12 h to 1 week after ingestion.

LD_{50}, **cattle**: ~1.6 mg/kg; **rat**: 1.6 mg/kg; **quail**: 15 mg/kg; **mouse**: 3.5 mg/kg; **dog**: 4.5 mg/kg.

TERBUTALINE SULFATE

= *Brethaire* = *Brethine* = *Bricanyl* = *Butaliret* = *KWD 2019* = *Monovent* = *Terbasmin* = *Terbul*

***Bronchodialtor, tocolytic, β_2-adrenergic receptor agonist. A synthetic sympathomimetic amine.* Oral and SC.**

Untoward effects*:** Nervousness, tremors, tachycardia, palpitations, arrhythmias, muscle twitches and cramping, headache, drowsiness, sweating, nausea, vomiting, decreased T-wave amplitude, lactic acidosis, hyperglycemia, hypokalemia, hypophosphatemia, and insomnia. Pharyngeal blistering in a 66-year-old **patient** after oral inhalation with it and ***ipratropium. Paradoxical bronchospasm due to use of ***sodium metabisulfite*** (q.v.) as a preservative (1984) in a nebulizer. A 7-year-old **patient** maintained on 7.5 mg tid (1 mg/kg/day)/6 months stopped treatment for 4 days. Agitation, tremors, and a generalized left-sided focal seizure started 2 h after restarting therapy. Yet, too frequent use may induce tolerance. Apparent slow-release tablets passed in the stool are only matrices containing no active drug. A **physician** error in a 67-year-old **female** who received 2.5 mg instead of 0.25 mg SC led to chest pain and electrocardiogram changes of myocardial ischemia. A 15-year-old **male** aerosoled ~2 mg of inhaler material on a groin patch of tinea cruris after 0.5 mg by inhalation to relieve asthma symptoms leading to transcutaneous absorption with rapid, regular palpitations; chest discomfort, hypokalemia, hyperglycemia, and sinus tachycardia. Hypersensitivity vasculitis and increased high-density lipoprotein reported. Bacterial contamination common in nebulizers without preservatives.

As a ***tocolytic*** in preterm labor, adverse effects of ~5% occur leading to arrhythmia, myocardial ischemia, atrial fibrillation, cardiomyopathy, profound hypotension, pulmonary edema, cerebral vasospasms, hyperglycemia, hypokalemia, chest pain, tachycardia, acidosis, and ***carbohydrate*** intolerance. Dyspnea, cardiac arrhythmias, and pulmonary edema occurred as part of sudden **maternal** deaths. A paradoxical case of marked uterine hypertonus after 0.25 mg IV in a 19-year-old **female**. **Maternal** use has been associated with dysrhythmias and hypoglycemia in **newborns**. Cardiomyopathy and myocardial necrosis in a **neonate** after **maternal** infusion of 0.5 mg/h beginning at 25th week of gestation. Treated and recovered after 1 month. Diabetic ketoacidosis and transient resistance to its effect in a **diabetic** 25-year-old **female** after infusion of 2–3 mg/24 h. Small amounts are secreted in breast milk leading to 0.6–0.7 µg/kg/day in an **infant**.

Drug interactions*: *Oxprenolol antagonist its hypokalemic effect and decrease its terbutaline clearance by 65%. Oral ***theophylline***, but not ***aminophylline***, potentiates its adverse effects of hypokalemia, hyperglycemia, tachycardia, and increased systolic blood pressure. In **children**, it decreases ***theophylline*** serum levels. ***Nifedipine*** potentiates its bronchodilation. ***Pafenolol*** increases its tachycardia effects. ***β-Blockers*** antagonize its effects.

In **dogs**, toxicosis leads to tremors, tachyarrhythmias, and prolapse of the nictitating membrane.

LD_{50}, **mouse**: 205 mg/kg; SC, 240 mg/kg; IV, 47 mg/kg.

TERCONAZOLE

= *Fungistat* = *Gyno-Terazol* = *R-42,470* = *Terazol* = *Tercospor* = *Triconazole*

***Topical antifungal.* For treatment of vulvovaginal candidiasis.**

***Untoward effects*:** 5–16% absorbed after topical use led to headache, dysmenorrhea, burning, contact dermatitis, urticaria, and pruritus. Vertigo, chills, collapse, fever, and nausea in a 56-year-old **female**.

Large oral doses are embryotoxic in **rodents**.

LD_{50}, male **rat**: 1.7 g/kg; female **rat**: 849 mg/kg; male **dog**: ~1.3 g/kg; female **dog**: ~640 mg/kg.

TERFENADINE

= *Allerplus* = *Cyater* = *MDL-9918* = *Nebralin* = *RMI 9,918* = *Sandane* = *Seldane* = *Teldane* = *Teldanex* = *Terdin* = *Terfex* = *Ternadin* = *Triludan*

H_1-Antihistamine.

***Untoward effects*:** Most common were headache and xerostomia. Supposedly, non-sedating because it did not readily cross the blood–brain barrier, but occasionally sedation was noted. It has prolonged the QTc interval, and caused ventricular tachycardia and torsades de pointes. Severe exacerbation of erythrodermic psoriasis in an 80-year-old **male** after 60 mg twice daily/80 days. Itchy erythematous eruptions on neck, the body, and extremities in a 57-year-old **male** after 60 mg twice daily/30 days.

Urticaria, dermatoses, and photosensitivity also reported. Stevens–Johnson syndrome in a 20-year-old **female** after 240 mg/day. Alopecia reported in a **patient** after 60 mg twice daily/3 months. Acute hepatitis and severe cholestatic hepatitis have occurred. Has induced neutropenia and tongue ulceration in a 42-year-old **male**. Generalized tonic-clonic seizures in a 27-year-old **male** after 60 mg twice daily/6 weeks. An overdose of 240 mg/day/6 days in a 21-year-old **female** led to convulsions and cardiac arrhythmias. Many case reports of adverse central nervous system reactions. Accidental ingestion by 1–5-year-old **children** reported. Possibility of transposition of great blood vessels if ingested in the first trimester. Therapeutic serum levels reported as < 0.01 mg/l and toxic levels at 0.06 mg/l. Its manufacturer, Hoechst, because of increased risk or serious cardiac problems and drug interactions, suggested **users** of its product switch to the active metabolite, ***fexofenadine***, developed by Supracor and now marketed by Hoechst as ***Allegra***®, that lacks the serious adverse effects of terfenadine.

Drug interactions: Serious cardiovascular events and deaths have followed concommitant use with ***ketoconazole*** and probably ***erythromycin***. Other ***imidazoles***, such as ***itraconazole*** and possibly ***fluconazole***, can also increase terfenadine serum levels and its toxicity. Similar adverse effects reported with other ***macrolide antibiotics***, such as ***clarithromycin***, ***troleandomycin***, and probably ***dirithromycin***. ***Fluoxetine*** and ***sertraline***, as with the above, inhibit the cytochrome P-450 enzyme system to increase terfenadine toxicity. The ***protease inhibitors delaviridine***, ***nelfinavir***, ***ritonavir***, and ***saquinavir*** decrease its metabolism and can increase its toxicity. Use with ***paroxetine*** leads to electrocardiogram changes. An 18-year-old **female** taking ***carbamazepine*** showed confusion, disorientation, visual hallucinations, nausea, and ataxia after taking 60 mg terfenadine. ***Grapefruit juice*** has increased its plasma concentration and prolonged QTc prolongation, but with considerable interindividual variation. ***Troglitazone*** has decreased its serum concentration. May potentiate the ***calcium channel-blocker nifedipine***.

TERLIPRESSIN

= *Glypressin* = *GVP* = *Triglycyllypressin*

Vasopressor. **For treating uterine and esophageal bleeding varices.**

Untoward effects: Induced mucocutaneous skin necrosis at the injection site and even elsewhere, such as the tongue and scrotum, less than 48 h after treatment. Case reports of hypokalemia in a 39-year-old **female** and a 57-year-old **male**.

Drug interactions: Has decreased blood levels of ***mitomycin***.

TERMINALIA sp.

Untoward effects: ***T. oblongata*** = ***Yellow Wood*** leaf ingestion led to "MacKenzie river disease" of **cattle** in Australia. Symptoms are abdominal pain, photophobia, photosensitization, and mild jaundice in acute cases; in chronic cases excessive eye blinking and lacrimation, dry and cracking muzzle, dehydration, weight loss, urinary incontinence, and occasionally edema. Post-mortem showed chronic interstitial nephritis and fibrosis; swollen, congested, and orange-colored liver; and green pigmentation and ulceration in the abomasum. A nervous syndrome with tetanic spasms and collapse in affected **sheep**.

A ***T. sericea*** root decoction has caused **human** deaths in East Africa.

TERODILINE

= *Bicor* = *Mictrol* = *Micturin* = *Micturol* = *Terolin*

Calcium channel-blocker, anticholinergic.

Untoward effects: Serious cardiac arrhythmias and torsades de pointes ventricular tachycardia (particularly in elderly **patients**) led to its withdrawal from many world markets. It has also caused xerostomia, blurred vision, tremors, and general tachycardia.

TESTOLACTONE

= *Fludestrin* = *NSC 23,759* = *SQ 9,538* = *Teslac*

Antineoplastic. **Oral and IM.**

Untoward effects: Mild alopecia and trophic changes in nail growth on high oral dosage: 2 g/day/10 months. Edema of extremities, nausea and vomiting, hypertension, paresthesia, glossitis, maculopapular erythema, and pruritus have also occurred. Like ***androgens***, it produces anabolic effects.

TESTOSTERONE

= *Andro* = *Androderm* = *Androgel* = *Mertestate* = *Oreton* = *Testoderm* = *Testolin* = *Testro AQ* = *Tostrex* = *Virosterone*

Androgen, anabolic agent. **Many esters exist. Usually IM, transdermal, or sublingual.**

Untoward effects: Increased ***nitrogen*** retention, skeletal weight, ***sodium*** and water retention, hypercalcemia, increased libido in elderly **men**, and priapism. Large doses decrease spermatogenesis and lead to seminiferous tubule degenerative changes. Large and/or repeated doses in early puberty have caused epiphyseal closure and decrease of linear growth. Polycythemia, cholestatic icterus, prostatic cancers, gynecomastia, hypercholesterolemia, increased high-density lipoprotein, increased low-density lipoprotein, and urticaria have been associated with its use. Can precipitate congestive heart failure in the presence of cardiovascular disease. **Male athletes** had precordial pain, generalized puffy edema, and exaggeration of **their** Gilles de la Tourette's syndrome tics while using it as an anabolic agent. Inhibits anterior pituitary function in **women**, suppressing ovulation and menstruation, increasing virilism, causing breast and endometrial tissue atrophy, deepening of voice, hirsutism, acne, clitoral hypertrophy, edema, nausea, vomiting, vaginal bleeding, and increased blood ***cholesterol***. A **woman** complained of excessive **male** pattern facial hair growth due to transdermal exposure from her **husband's** use of topical testosterone cream for hypogonadism. Given during the first trimester of pregnancy, it has caused **fetal** masculinization of female **offspring**. Suspect as a cause of kernicterus in **infants** and **rhesus monkeys** after treatment of their **mothers**. Inhibits the growth of the brain's left hemisphere and possibly associated with development of left-handedness. Unjustified administration to **boys** < 10 years old led to precocious puberty. High plasma levels of unconjugated ***etiocholanolone***, a metabolite of testosterone, and ***androstenedione*** can cause urticaria during menses, fever, myalgia, arthralgia, abdominal pain, diarrhea, and vomiting. An HIV-infected 39-year-old **male** self-injected 500 mg of the *enanthinate* every 2 weeks leading to acute myocardial infarction. Topical use on **male** can increase levels in **female contacts** with potential androgenic effects, including those on a **fetus**.

TD_{LO}, **woman** (pregnant 7–14 weeks): 35 mg/kg.

Drug interactions: ***Barbiturates, chlorcyclizine***, and ***phenylbutazone*** enhance its metabolism by enzyme induction. ***Cimetidine*** blocks testosterone synthesis.

May suppress production of sperm in **animals**. Do NOT use any testosterone-containing implants within 60 days of slaughtering **animals**, or in any lactating **animals** with milk being used for **human** consumption. U.S. regulations establish a zero tolerance for residues of the drug in edible tissue and by-products from **heifers**. Legal clarification is needed on residues in other **animals**. Do NOT salvage injection site for food. Avoid use in pregnant **animals** as it can have androgenic and teratogenic effects on the fetus. Under some conditions, "riding" or "bulling" effects noted may be undesirable. Numerous reports of excessive bulling, riding, and rectal prolapse after use of ***testosterone–estradiol*** for fattening **heifers**. Decreased liver microsomal metabolism of ***hexobarbital*** and increased sleeping time in **mice**; ***phenobarbital*** and ***chlorcyclizine*** increase its metabolism in **rats**; ***DDT*** and ***dieldrin*** increase its metabolism in **pigeons**; and ***phenobarbital*** and ***phenylbutazone*** increase its metabolism in **dogs**.

TETRABENAZINE

= *Nitoman* = *RO-1-9569*

Psychotherapeutic, antidyskinetic, presynaptic dopamine antagonist.

Untoward effects: Drowsiness and akathisia. Occasional depression, postural hypotension, and extrapyramidal effects. Dysphagia and choking associated with its use in a **patient** with Huntington's chorea. Fatigue, confusion, agitation, nausea, photophobia, salivation, amenorrhea, and ptosis have also occurred. Overdosage leads to sedation, sweating, hypotension, and hypothermia. A **patient** ingested 1 g (five times recommended daily dose) leading to drowsiness, marked sweating, and urinary incontinence; recovered in ~3 days.

Drug interactions: Use with ***L-dopa*** can decrease its dopaminergic effect.

Induces catalepsy and ptosis in **rats** and **mice**.

LD_{50}, **mouse**: 550 mg/kg; IV, 150 mg/kg.

sym-TETRABROMOETHANE

= *Acetylene Tetrabromide* = *TBE*

***Solvent.* Used in microscopy and on aviation instruments.**

Untoward effects: One of the most toxic ***halides***. A ***halogenated hydrocarbon*** whose vapors can have anesthetic and narcotic actions. Symptoms are eye, nose, and throat irritation; dizziness, severe headaches, nausea, abdominal pain, anorexia, jaundice, mental confusion, fatigue, stupor, and monocytosis. Skin contact can cause dermatitis. OSHA sets upper exposure limit at 1 ppm; immediate danger to life or health, 8 ppm.

Inhalation of 0.01 and 0.1 mg/l vapors by **rats** led to biochemical changes and pathology in liver, kidneys, and brain.

LD_{50}, **rabbit** and **guinea pig**: 400 mg/kg; IP, **mouse**: 443 mg/kg.

TETRABROMOSALICYLAMILIDE

***Antibacterial.* Used in soaps.**

Untoward effects: Photodermatitis and, rarely, photohemolysis.

TETRACAINE HYDROCHLORIDE

= *Amethocaine* = *Ametop* = *Anethaine* = *Decicaine* = *Dicain(e)* = *Pantocaine* = *Pontocaine* = *Tonexol*

***Local and topical anesthetic.* A derivative of *p-aminobenzoic acid* (q.v.).**

Untoward effects: Hypersensitivity reactions have been common leading to asthma, eczema, contact dermatitis, hypotension, excitement, anaphylactoid reactions (bronchospasms, coughing, dyspnea, apnea, tachycardia, dysphagia), and even death. Cross-sensitivity with ***benzocaine, butacaine, cyclomethycaine***, and ***procaine***. It is hydrolyzed in plasma by pseudocholinesterase leading to p-aminobenzoic acid, which cross-reacts with ***sulfonamides, sulfonylureas***, and ***thiazides***. A 62-year-old **male** sucked on a tetracaine hydrochloride lozenge without any adverse effects ½ h before **his** throat was sprayed with 0.5–1 ml of a tetracaine hydrochloride solution leading to immediate convulsions and death within 3 min. Sudden death in a 62-year-old **male** 1 min after gargling with a 2% solution in preparation for gastroscopy. Generalized seizure, cardiac arrest, and death after topical application for urethral dilation in a 53-year-old **male**. Photophobia, pain, tearing, miosis, keratitis, or corneal opacity from ophthalmic use. Intranasal use of 2% drops reduces ability to smell for 30–70 min. Intraspinally, it causes hypotension, gagging, headache, and aseptic meningitis. Overdoses cause restlessness, excitement, twitches, nausea, salivation, tachycardia, weak pulse, paleness or cyanosis, convulsions, coma, and even death. Intraspinal injection of 5 mg in a pregnant 25-year-old **female** led to cardiac arrest within 6 min. Resuscitated and **she** remained comatose with hyperpyrexia, oliguria, and pulmonary edema. Died 83 h postpartum. Post-mortem showed extensive nervous system damage including brain stem necrosis. Several case reports of paraplegia after epidural use. May cause depression of **infant** after **maternal** spinal or epidural use.

Drug interactions: ***Organophosphates*** or low serum cholinesterase activity prolong its action. ***p-Aminobenzoic acid esters***, such as tetracaine, can interfere with the therapeutic efficacy of ***sulfonamides*** and, since they may inhibit the bacterial growth, should not be used prior to collecting samples for bacteriology. ***Xanthines*** increase its half-life.

May cause transient corneal edema in **dogs**. Should not be used with ***lidocaine*** (increased toxicity in **mouse** trials). Unnecessary ophthalmic use should be avoided. Even 0.25% solutions are extremely potent for epidural anesthesia. 0.5% inhibits growth of almost all test bacteria. Epidural use in **cattle** has caused hind limb paralysis.

SC, LD, **rabbit**: 20 mg/kg; IV, 6 mg/kg. LD_{50}, **mouse**: SC, 35 mg/kg; IP, 70 mg/kg; IV, 13 mg/kg.

1,1,2,2-TETRACHLORO-DIFLUORETHANE

= Arcton 112 = Freon 112 = Freon MU = Halocarbon 112 = Ledon 112 = Refrigerant 112

Untoward effects: In **animals**, vapors lead to eye and skin irritation, conjunctivitis, narcosis, and pulmonary edema. Toxic to lactating **cows** at 0.1 g/kg and to **sheep** at 1–2 g/kg. Prolonged clotting time after 48–72 h. Post-mortem shows widespread hemorrhages, liver and kidney damage, myocardial degeneration, venous congestion, and anasarca. Repeated exposures of **rats** cause death from asphyxia due to pulmonary hemorrhage.

Inhalation LC_{LO}, 15,000 ppm/4 h.

1,1,2,2-TETRACHLOROETHANE

= Acetylene Tetrachloride = Bonoform = Cellon

Solvent, soil sterilant, grain fumigant. **Non-flammable heavy liquid. Current U.S. production only as a chemical intermediate.**

Untoward effects: Sweet, suffocating, ***chloroform***-like odor. Ingestion leads to cardiac arrhythmias, nausea, vomiting, abdominal pain, headache, psychosis, convulsions, nervousness, jaundice, hepatitis, monocytosis, and tremors of fingers. Breathing vapors leads to narcotic effects, fatigue, vomiting, dizziness, headache, stomach aches, anorexia, coma, and liver damage. ***Alcohol*** increases severity. Toxic amounts can be absorbed from skin contact. Eye contact can cause irritation and more severe damage if not thoroughly washed with water. Poisoning, suicide, and deaths in several British **workers** after ingesting 3–4 ml of this, one of the most poisonous ***chlorinated hydrocarbons*** with pulmonary and gastrointestinal hemorrhage. Many accidental exposures due to mistaking it for ***tetrachloroethylene***. National Institute for Occupational Safety & Health and OSHA skin upper exposure limits of 1 and 5 ppm, respectively.

Liver tumors in **mice**, but not **rats**. Used to produce experimental liver, intestinal tract, and kidney pathology.

TETRACHLOROETHYLENE

= Ankilostin = Didakene = Ethylene Tetrachloride = K_2R = Nema = Perc = Perchloroethylene = Perclene = Perk = Tetracap = Tetracholoethene = Tetrachlorethylene = Tetropil

Anthelmintic, solvent. **U.S. annual needs are ~350 million lbs.**

Use: As a chemical intermediate, in metal cleaning and degreasing, in dry cleaning, and textile processing.

Untoward effects: Vapors can irritate eyes, nose, and throat. Overexposure to vapors or dermal absorption causes nausea, vomiting, abdominal cramps, flushing of face and neck, headache, dizziness, vertigo, frontal sinus congestion, ataxia, sleepiness, erythema, abnormal liver function tests, liver enlargement, fatty liver, fatigue, impaired memory, confusion, giddiness, irritability, anorexia, tremors, numbness, scleroderma, and possible increased menstrual problems and abortions in **women**. Can pass into breast milk. A breast-fed **infant** had obstructive jaundice at 6 weeks of age as a result of the nursing **mother** exposed to it by daily visits to a dry cleaning plant. Excessive systemic absorption leads to central nervous system depression with coma and, ultimately, death from respiratory paralysis or circulatory failure. Direct skin contact causes skin dryness, cracking, erythema, burning, and blistering. **Students** have abused it by inhaling it from ***shoe*** cleaners, etc., to get a "high". Anesthesia and chemical burns in a 68-year-old **male** launderette **worker** who spilled some on **his** clothing. Pulmonary edema and coma in a 21-year-old **male** laundry **worker** exposed to its fumes for 7 h when the cleaning system overheated. A 29-year-old **male** was overcome by fumes and became unconscious from a faulty dry cleaning machine. **He** fell on some of the chemical leading to severe corrosive burns, erythema, and blistering over 25–30% of **his** body. Other case reports of pulmonary edema and coma after accidental exposure to fumes. Mild hepatitis from such heavy vapor exposure in a 25-year-old workman wearing a respirator while cleaning a tankcar with it. Moderate hypertension with T-wave changes, increased SGOT, leukopenia, hematuria, and albuminuria also reported in Brooklyn, New York **firefighters**. Acute brain syndrome in a 21-year-old **male**, 1 h after ingestion of 5 ml in treatment of parasitosis leading to dizziness and vertigo, followed by agitation, irrational behavior, and amnesia over a 2 h period, followed by progressive improvement. A 16-year-old **male** died after intoxication from a freshly dry cleaned, inadequately aired sleeping bag. Increased risk of cancer noted in dry cleaning **workers**. OSHA upper exposure limit of 100 ppm; 300 ppm/5 min in any 3 h period; and immediate danger to life or health of 150 ppm.

LD_{LO}, **human**: 500 mg/kg; inhalation TC_{LO}, **human**: 200 ppm.

Restrict dietary fat within 2 days before and after use to avoid enhanced absorption of this fat soluble liver toxicant. *Contraindicated* in febrile diseases or in debilitated **animals**. Strong mucosal irritant. Breaking capsules in the mouth has produced ataxia, convulsions, and anesthesia. *Will detonate* on contact with particles of granular ***barium***. Central nervous system depressant in **greyhounds**. Inhalation led to hepatocellular carcinomas and adenomas in **mice**, mononuclear cell leukemia in **rats** of both sexes and renal tubular cell neoplasms in male **rats**.

LD_{50}, **mouse**: 8.9 g/kg; LD_{LO}, **dog** and **cat**: 4 g/kg; **rabbit**: 5 g/kg; inhalation LC_{LO}, 4000 ppm/4 h.

TETRACHLOROSALICYLANILIDE

= *Impregnon* = *Irgasan* = *TCSA*

***Bacteriostat.* Used in soaps, shampoos, deodorants, and petroleum products.**

Untoward effects: Contact dermatitis and photodermatitis with erythema and swelling. May cross-react with other halogenated germicidal drugs, such as ***bithional, dichlorophene***, and ***hexachlorophene***. Banned for cosmetic use in the U.S.

LD_{50}, **rat**: 243 mg/kg.

TETRACOSACTIDE

= *Actholain* = *Cortrosyn* = *Cosyntropin* = *Synacthen* = *Tetracosactrin*

***Synthetic adrenocorticotropic hormone.* Parenteral use.**

Untoward effects: Hypersensitivity reactions in ~1/30,000 doses leading to dermal and systemic allergic reactions, acute polyarthritis, pruritic rash, wheezing, apnea, muscle cramps, shivering, leukocytosis, adrenal gland hypertrophy, and edema of face, hands, and feet. Pain, swelling, local erythema, and pruritus at zinc depot from injection site; occasionally Cushingoid facies, weight increase, melanoderma, and hot flashes.

IN, LD_{LO}, **woman**: 50 µg/kg.

TETRACYCLINE

= *Abricycline* = *Agromicina* = *Ambramycin* = *Bio-Tetra* = *Cyclomycin* = *Deschlorobiomycin* = *Dumocyclin* = *Neocycline* = *Oletetrin* = *Omegamycin* = *Orlycycline* = *Sigmamycin* = *Tetradecin* = *Tetraverine* = *Tsiklomitsin*

The hydrochloride = *Achro* = *Achromycin* = *Achromycin V* = *Ala Tet* = *Ambracyn* = *Ambramicina* = *Bristaciclina* = *Cancycline* = *Cefracycline* = *Criseociclina* = *Cyclopar* = *Diocyclin* = *Economycin* = *Helvecyclin* = *Hostacyclin* = *Imex* = *Mediletten* = *Mephacyclin* = *NCI-C55561* = *Oppacyn* = *Paltet* = *Panmycin* = *Partrex* = *Polycycline* = *Polyotic* = *Purocyclina* = *Quadracyclin* = *Remicyclin* = *Riocyclin* = *Ro-Cycline* = *Sanclomycine* = *Solvocin* = *Steclin* = *Sumycin* = *Supramycin* = *Sustamycin* = *Tefilin* = *Tetrabakat* = *Tetrabid* = *Tetrablet* = *Tetrabon* = *Tetrachel* = *Tetracompren* = *Tetracyn* = *Tetrakap* = *Tetralution* = *Tetramavan* = *Tetramycin* = *Tetrosol* = *Topicycline* = *Totomycin* = *Triphacyclin* = *U-5965* = *Unicin* = *Vetquamycin*

The phosphate = *Bristacyclina A* = *Hexacycline* = *Panmycin P* = *Sumycin* = *Telotrex* = *Tetradecin Novum* = *Tetrex* = *Upcyclin*

Antibiotic.

Untoward effects: ***Blood***: A ***folate*** deficiency with megaloblastic anemia with ingestion of 250 mg twice daily/3 years in an 18-year-old **female**. A 13-year-old **female** ingested 1 g/day/3 weeks leading to Coombs' positive hemolytic anemia; 15 months later, **she** ingested 1 g and developed the same type of anemia. **Her** bone marrow showed erythroid hyperplasia. Cases of aplastic anemia have also been reported. Eosinophilia with fever, rash, and proteinuria after ingestion of 1 g/day/6 days and **she** collapsed 3 days later with peripheral circulatory failure. Leukopenia, leukocytosis, neutropenia, pancytopenia, agranulocytosis, thrombocytopenia, thrombopenic purpura, and lupus erythematosus also reported after oral treatment. Prothrombin plasma levels showed some decrease after IV use leading to delayed blood coagulation. Significant increase in blood urea nitrogen is not important under ordinary conditions, but can be serious in the presence of renal failure. Azotemia in elderly **patients** being treated with it for peptic ulcer disease. Hypoglycemia in hepatorenal failure. As a potent chelator, it can cause hypocalcemic tetany if infused too rapidly.

***Brain*:** Benign intracranial hypertension and bulging fontanel with severe headache, fever, irritability, anorexia, nausea, vomiting, and photophobia, especially in **infants** and young **children**, although it has occurred in **teenagers** and a 63-year-old **female**.

Clinical tests*:** False positive or spuriously high urine ***catecholamines (Hingerty method) and serum inorganic ***phosphate***. False decrease in serum ***potsassium***. Can interfere with *Lactobacillus casei* assay for serum ***folate***. Can cause false laboratory test results for urinary ***albumin, amino acids, bilirubin, glucose***, and ***estrogen***.

Eye*:** Transient myopia and blurred vision in a 35-year-old **female** after three separate treatments and considered to be of allergic origin. Permanent yellow discoloration of **fetal** corneas and lenses after total dose of 6–8 g to **mother** during **her** first 8–12 weeks of pregnancy. Congenital cataracts, photophobia, and retrobulbar pain reported. After 10 weeks' therapy for acne, a **patient** had great discomfort from contact lenses; clearing on withdrawal and returning on readministration. Irritation, redness, and blepharitis from ***thimerosal-preserved contact lenses while taking tetracycline in nine **patients**. A 63-year-old **female** treated for acne rosacea with 250 mg developed benign intracranial hypertension, headache, and rings and lines over **her** whole visual field. Partial obscuring of vision several times/day, a couple of episodes of complete amaurosis for ~5 min, papilledema, and increased intraocular pressure.

Fanconi syndrome*:** Has occurred with degraded material either from its age, storage in hot places, or in the tropics. Usually consists of nausea, vomiting, acidosis, proteinuria, glycosuria, renal tubular defects, phosphatemia, lupus erythematosus, hypokalemia, amino-aciduria, dehydration, lethargy, and disorientation. Degradation has caused > 65% change to nephrotoxic ***anhydrotetracycline and ***epianhydrotetracycline***. Outdated material containing ***ascorbic acid*** has been particularly prone to inducing this syndrome.

Fluorescence – bone, organs, and teeth*:** Nearly ½ of **children** exposed to it *in utero* from 29th week of pregnancy to term showed enamel hypoplasia and yellow–brown staining of **their** teeth with dental mottling and increased incidence of caries. Bone is also discolored and can have defective mineralization. Widespread systemic (skull, abdominal and thoracic lymph nodes, and thyroid) pigmentation found at autopsy (after a traffic accident) in a 19-year-old **male** who received 732 g over 33 months and 68 g ***minocycline for 7 months for treatment of acne. If the **mother** receives it close to term, the crowns of permanent teeth can be stained. The degree of placental passage varies and **fetal** levels are 25–75% of **maternal** plasma levels. Causes a 40% decrease of skeletal growth in premature **infants** during the second and third trimester. Multiple courses of therapy or long-term therapy intensifies these effects. Only ~0.3% of administered dose appears in breast milk and hasn't caused problems in **infants**. Most suggest avoiding its use in **children** under 8 years; others suggest 12 years. Its fluorescence in organs has been utilized in tests for cancer. Examination of bone samples of **Sudanese Nubians** from ~14 centuries ago revealed high levels of tetracycline fluorescence. The source may have been from *Streptomycetes*, a natural producer of tetracycline on the ***barley, millet***, and ***wheat*** from the flood plains of the Nile, and stored in mud bins, from which they made ***beer*** and ***bread***.

***Fungal infections*:** Overgrowth of *Candida albicans* after long-term therapy led to diarrhea, vulvovaginitis, and bronchopulmonary and mouth infections.

Gastrointestinal*:** Nausea, vomiting, diarrhea, and colitis, especially with high and continuous dosage. May cause esophageal irritation and gastro-esophageal reflex disease. Alters intestinal flora in terms of numbers, resistance (particularly staphylococci), increases population of *Candida*, streptococci, *Proteus*, *Klebsiella*, and *Pseudomonas*, causing oral candidiasis, pruritus ani, dehydration, fulminant enteritis, and rarely death. Pancreatitis after prolonged therapy in **man** and **pigeons**. In the latter, impaired synthesis and secretion of pancreatic enzymes have been demonstrated. Pseudomembranous colitis and brown–black discoloration of tongue, and hairy tongue reported. It can increase ***vitamin C excretion and inhibit bacterial synthesis of ***vitamin K***. Cases of esophageal ulceration and fistula, including a fatality in a 73-year-old **female**. Metallic taste reported by 32-year-old **male** ingesting 500 mg qid/10 days.

***Hypersensitivity*:** Phototoxicity, anaphylactoid reactions and anaphylaxis, lupus erythematosus, burning sensations, asthma, and pruritus and erythema of fingers and penis. Hyperpigmentation of fingerwebs. Balanoposthitis and hyperthermia reported. Has caused angioedema with occasional deaths.

Liver: Hepatotoxicity with fatty liver changes, hepatocellular jaundice, hyperbilirubinemia, increased serum transaminases, increased alkaline phosphatase, nausea, vomiting, tachypnea, hypoglycemia, and coma. Serious cases occur frequently after high dosage, oral or IV, and particularly in pregnancy or postpartum, or in **patients** with nephrotoxicity. Fatalitites reported.

Malabsorption: ***Iron*** absorption is limited due to formation of ***iron***–tertacycline chelates. Induced changes in intestinal flora leading to malabsorption of ***glucose*** and ***vitamin*** B_{12}, and proteins; and suppression of ***vitamin*** K_1 production.

Miscellaneous: Peritoneal adhesions in a 72-year-old **female** with chronic Q fever after ingesting > 10 kg intermittently as 500 mg qid. It can aggravate or unmask myasthenia gravis. It is an anabolic. Rarely, dizziness, myesthenia, and tinnitus.

Neoplasms: Little evidence. A 16-year-old **female** ingesting 250 mg twice daily/~2½ year for acne developed malignant lymphoepithelioma (not common at this age) of the palate. Pseudotumor cerebri reported.

Renal: Nephrotoxicity (with or without oliguria), blood urea nitrogen increase and retention can lead to hallucinations and delirium, and increased ***creatinine***. Renal failure after normal dosage in dehydrated **patients** or in renal impairment.

Skin and nails: Triad of photosensitivity, nail discoloration, and onycholysis. Fixed-drug eruptions, urticaria, hair loss, toxic epidermal necrolysis, acneiform eruptions, bullous dermatitis; vasculitis with erythematous papules that may ulcerate and become gangrenous; exanthematous eruptions progressing to exfoliative dermatitis; perioral or nasolabial dermatitis, contact dermatitis, penile pigmentation, lupus erythematosus, lichenoid eruptions, erythema multiforme, acneiform eruptions with red papules and pustules, anal or vulvar pruritus, splinter hemorrhages, and brown–red discoloration of separating fingernails reported. Blue nodules in the skin from induced osteosis cutis. Oral mucosal ulcerations after 1 day in a 38-year-old **male** after gargling with it and ingesting 250 mg qid. Allergic reaction to coloring agents (***amaranth, cochineal red***, and/or ***tartrazine***) in tablets led to pigmented plaques on shoulders and back in a 31-year-old **female** taking 2 years to clear. If the face is washed before topical application, tingling occurs in 30% of **patients**. Extravasation of IV solution has caused skin necrosis.

Teratogenicity: Micromelia, syndactyly, cleft lip, and cleft palate. Bone mineralization begins in the fourth month of pregnancy.

TD_{LO}, **woman** (32–33 weeks pregnant): 240 mg/kg.

Drug interactions: Use with the anesthetic ***methoxyflurane*** has increase nephrotoxicity, renal failure, and fatalities; decreases serum ***calcium*** and the combination is contraindicated. Use with ***antacids*** of ***aluminum, calcium, magnesium***, or ***sodium bicarbonate*** forms a non-absorbable chelate, decreasing absorption by 50–90%, some of which is due to decreased absorption from an alkaline pH. The same chelation occurs with ***dairy products, iron***, and ***milk***. 70% of the **maternal** serum level appeared in breast milk where it is chelated, preventing use of most of it by the **infant** leading to only 70 μg/ml of serum. Use with ***zinc sulfate*** reduces its absorption ~30%; its absorption was decreased 34% by ***bismuth subsalicylate***, and 50% by ***Kaopectate***®. ***Cimetidine*** by decreasing gastric acidity has, in some cases, decreased its absorption by up to 40%. ***Sucralfate*** decreases its bioavailability. Concommitant use with ***diuretics*** leads to significant increase in blood urea nitrogen. Dramatic increase of ***lithium*** blood levels (2–3 times) when tetracycline was ingested. One case report indicated a slight decrease. Serum concentration and ***theophylline*** toxicity increased by it. A report to the contrary exists. Potentiates ***dicumarol*** and ***warfarin's*** anticoagulant effects. By inhibition of liver microsomal enzymes, it also potentiates the actions of ***acetanilide, aminopyrine, codeine, pentobarbital, phenytoin***, and ***tolbutamide***. Differing results reported on absorption effects with different proteolytic enzymes, such as ***bromelain, chymotrypsin***, and ***trypsin***. May increase ***phenformin*** levels. Decreases effectiveness of ***penicillins***. Because ***digoxin*** is inactivated by gut bacteria in some **people**, use with tetracycline increases its serum levels. ***Metoclopramide*** increases its rate of absorption. ***Ascorbic acid*** increases its serum concentration. It apparently accelerates the metabolism of ***estrogen*** by unknown means and decreases the effectiveness of many ***oral contraceptives***, causing breakthrough bleeding and, rarely, pregnancy. ***Calcium phosphate***, once used as a so-called inert filler in tetracycline capsules, produced a poorly absorbed complex. ***Atovaquone*** plasma concentrations decrease 40–50% by coadministration of tetracycline.

Therapeutic levels for **man** are given as 4–8 mg/l and toxic levels as 30 mg/l.

The FDA does not permit parenteral use in food-producing **animals**. Systemic and oral use can produce ***milk*** levels of the drug. Large oral doses to **ruminants** may be toxic due to secondary effects from sudden alteration of the rumen flora. **Cats** and **dogs** may show gastric upsets and occasionally diarrhea from high and/or chronic oral dosage. After 1 or 2 days of treatment, **cats** may develop pyrexia, depression, and inappetance. If so, discontinue treatment. Do NOT inject into joint cavities, as it may produce an acute inflammatory response. It will cause fluorescence in growing bone and teeth. Increased curarizing effects of ***d-tubocurarine*** in **rabbit** trials. Do NOT slaughter treated **calves**, **sheep**, **poultry**, and **swine** until at least 14, 5, 4, and 4 days, respectively, after the last oral treatment. Poor aqueous stability. Make fresh solution daily. NOT for **birds** producing eggs for **human** consumption. In **pigeons**, it impairs synthesis, transport, or secretion of both basal and stimulated amylase. This association with pancreatitis has been suspect for some time, and the probable cause of diarrhea observed in some **patients**. It may be associated with tetracycline chelation of ***calcium***. May have a direct toxic effect on **dog** nephrons, and high dosage should be used with caution in renal disease. Metabolic acidosis can be produced in **dogs** by toxic IV dosage. May cause a dose-dependent depression of cardiovascular function in **cats**, **dogs**, and **monkeys** during anesthetic procedures. Perfusion of isolated **cat** heart caused a profound drop in contractile force. IM use in **cattle** has caused severe local reactions with necrosis, hemorrhage, and fibrosis. Avoid use in **animals** still with deciduous teeth. Its fluorescence is used as a laboratory marker, and has been used in **poultry** heads to analyze **foxes'** eating habits. Use with ***sodium bicarbonate*** (in ***antacids***) may decrease its solubility and availability. IM use may cause local irritation in **birds**. Use with ***methoxyflurane*** has been associated with liver damage. Rarely, its use in **dogs** has been associated with hepatotoxicity. ***Anhydro-4-epianhydrotetracycline***, but not ***anhydrotetracycline***, IV (10 mg/kg/day 1 and 20 mg/kg/day 2) had frequent episodes of vomiting 80 h after the second dose. The urine was very dark and positive for occult blood up to 6 h after each dose with increased blood urea nitrogen. Treatment of live-bearing tropical **fish** (**black mollies**) with 50 mg/gal led to offspring with dorsal spine and tail deformities, hampering their ability to swim. High levels in the feed decreased immune response in **guinea pigs**. Oral dosage of 500 mg/kg to **hamsters** can be fatal. A single IV dose of 70 mg/kg to a 72 kg **foal** led to acute renal failure. Inhibits growth with hypomineralization of skeleton in **chicks** and **rats**. Leg malformations induced in **chicks** by pre-incubation inoculation of **chicken** egg yolks. Induces fatty livers in **rats** and micromelia and syndactyly in their offspring.

LD_{50}, **rat**: 807 mg/kg; SC, 320 mg/kg; IV, 160 mg/kg; **mouse**: 808 mg/kg; IV, 162 mg/kg.

n-TETRADECANE

Use: In organic synthesis and as a liquid ***hydrocarbon*** solvent.

Untoward effects: Very potent inducer of skin papillomas in **mice**, 25% of which eventually became squamous cell carcinomas. Strong inducer of **rabbit** ear comedones. Closed-patch testing on the skin of **miniature swine** led to severe reddening, induration, and crust formation.

TETRADECYLTRIMETHYL-AMMONIUM BROMIDE

= *Morpan T* = *Myristyltrimethylammonium Bromide* = *Tetradonium Bromide*

***Surfactant, deodorant, disinfectant, laboratory reagent.* Cationic detergent in industry and hospitals.**

Untoward effects: Injurious to **animal** eyes at 0.1–1.0% concentration. It is corrosive.

IV, LD_{50}, **rat**: 15 mg/kg; **mouse**: 12 mg/kg.

TETRADIFON

= *Akaritox* = *Duphar* = *ENT 23,737* = *FMC 5,488* = *Mition* = *NIA 5,488* = *Polacaritox* = *TCDS* = *Tedion* = *V 18*

Insecticide, acaricide.

Untoward effects: LD_{LO}, **human**: 5 g/kg.

Ataxia, wing twitching, and imbalance in **birds** within 1 h. **Ducks**, **quail**, and **pheasants**: only 10% mortality at 5000 ppm in feed.

LD_{50}, **rat**: 556 mg/kg; **mallard duck**: > 2 g/kg; **dog**: 2 g/kg; LD_{50}, dermal, **rabbit**: 10 g/kg.

TETRADYMA sp.

= *Horsebrush*

Untoward effects: Woody shrubs of the western U.S., where eating of ***T. glabrata* = *Coal-oil Brush* = *Dog Brush* = *Littleleaf Horsebrush* = *Lizard Shade* = *Rat Brush* = *Spring Rabbit Brush*** (especially the new buds) by **sheep** causes centrilobular liver necrosis and pruritus; eyelids, lips, ears, and throat swellings and edema after sun exposure; and skin sloughing due to its ***tetradymol*** content. The edematous ears become so heavy they hang down on the side of their heads. The condition is commonly called "bighead" by ranchers and death rate is very high. Sterility often occurs in the survivors. **Goats** are commonly affected and rarely **cattle**. Head swellings usually occur ~18–24 h after ingesting a toxic dose. Ingestion of *Artemesia nova*, ***Black Sagebrush***, is a precondition to the photosensitization reactions. ~$\frac{1}{2}$ lb of leaves causes "bighead".

T. canescens* = *Gary Horsebrush* = *Silvery Horsebrush* = *Spineless Horsebrush is only $\frac{1}{2}$ as toxic and extensive photosensitization and "bighead" are its worst effects. Most survive. Toxicity of both species decreases rapidly after flowering or dormancy.

TETRAETHOXYSILANE

***Untoward effects*:** Highly toxic irritating vapors causing severe damage to the lungs, liver, and kidneys, followed by death of exposed **animals**.

Inhalation LC_{LO}, **rat**: 1000 ppm/4 h; **guinea pig**: 700 ppm/6 h.

TETRAETHYLAMMONIUM CHLORIDE

= *Etamon Chloride* = *T.E.A. Chloride*

***Ganglion-blocking agent*. A *quaternary ammonium compound* for IM use.**

***Untoward effects*:** Vasodilation, shock, and collapse, especially if a **person** suddenly stands. Deaths have occurred. May cause temporary inability to void feces and urine.

In **rabbits** causes dyspnea, tremors, convulsions, and apnea. Irregular muscle contractions and tremors; first fibrillary, then tonic in **frogs**. Larger doses lead to paralysis. **Mammals** show increased excitability and decrease of nerve endings. ***Curare***-like effect in **cats** with blockade of autonomic ganglia and resulting vasodepression.

TETRAETHYL LEAD

= *Lead Tetraethyl* = *TEL* = *Tetraethylplumbane*

***Gasoline additive*. Contains 64% *lead*. ~200 million lbs used annually.**

Untoward effects*:** Inhalation and dermal exposure primarily lead to adversely increased central nervous system activity with tremors, hyperreflexia, spasticity, insomnia, weariness, disorientation, hallucinations, psychosis, mania, restlessness, agitation, confusion, convulsions, and coma. Bradycardia, hypotension, hypothermia, pallor, nausea, anorexia, diarrhea, eye irritation, and contact dermatitis also reported. ***Lead encephalopathy is a great risk to **workers** handling leaded ***gasolines***. Readily absorbed from the lungs. Maximum permissible atmospheric concentration was set at 150 $\mu g/m^3$; National Institute for Occupational Safety & Health and OSHA maximum exposures to skin now at 75 $\mu g/m^3$ as ***lead***. Minimum lethal dose, 100 mg.

Polydipsia, regurgitation, hypoactivity, wing-drop, wing spread, ataxia, geotaxia, running, falling, sitting, fluffed feathers, ptosis, ataraxia, asthenia, mydriasis, tremors, and anorexia in poisoned **mallards** and **Japanese quail**. Vegetation near highways may contain substantial amounts (80–500 ppm). Single injections in **rabbits** led to microscopic brain lesions similar to those of **men** with Alzheimer's disease. Has decreased pregnancy rate in **mice**. Has caused cancers in laboratory **animals**.

LD_{50}, **rat**: 12.3 mg/kg; inhalation LC_{50}, **rat**: 850 mg/m³/1 h.

TETRAETHYL PYROPHOSPHATE

= *Bladan* = *ENT 18,771* = *Fosvex* = *Killax* = *Kilmite* = *Mifos T* = *Mortopal* = *TEPP* = *Tetron* = *Vapotone*

***Organophosphorus insecticide*. *Anticholinesterase*. Explored as a potential war "nerve gas".**

Untoward effects*:** Poisonous vapors lead to eye and skin irritation, eye pain, lacrimation, blurred vision, miosis, rhinorrhea, headache, nausea, vomiting, abdominal cramps, salivation, bradycardia, diarrhea, giddiness, anorexia, diaphoresis, weakness, twitching, paralysis, Cheyne–Stokes respiration, hypotension, arrhythmias, convulsions, and cyanosis. Death can occur within 10 min to 4 h. Aerial dusting of 46 acres of a Washington State vineyard with 1% ***TEPP caused death of two **cattle** as some of the dust drifted at least 700 ft over pasture, crops, and houses for at least 2 h. Affected **cattle** and **humans** suffered from coughing seizures. The two **heifers** convulsed and died within 2 h; others salivated and gasped for air. Breathlessness and some tightness of chest in 14 **people** exposed for at least 30 min.

LD_{LO}, **human**: 432 μg/kg.

A **farmer** sprayed it on **cows** (for fly control) and killed 29 in 15 min. Led to violent convulsions after 0.2–0.5 mg/kg IV in **cats**.

LD_{50}, **rat**: 500 mg/kg; **mouse**: 7 mg/kg; **mallard duck**: 3.56 mg/kg. Dermal LD_{50}, **rat**: 2.4 mg/kg; **rabbit**: 5 mg/kg; **mallard duck**: 64 mg/kg.

TETRAGONIA TETRAGONOIDES

= *New Zealand Prickly Spinach*

Untoward effects*:** Although a 60-year-old **female** had frequently and uneventfully eaten small quantities before, **she** had a typical hypersensitive reaction after eating a large quantity while taking ***phenelzine 45 mg/day leading to blood pressure 240/110 and then falling steadily at the rate of 10 mm every 15 min until it was normal. The plant is high in ***oxalates*** and it may be contraindicated in **patients** on ***monoamine oxidase inhibitors***.

Its ***oxalate*** content has made it poisonous to **livestock** ingesting it.

Δ-9 TETRAHYDROCANNABINOL

= *Dronabinol* = *Marinol* = *THC*

The main active principle (1–10%) in *Cannabis*.

Untoward effects: Although it has accepted medical uses, it has a potential for abuse. Its concentration varies in different parts of the plant and from different geographical regions. Those plants from latitudes south of 30° North have large amounts of it and little ***cannabidiol***; those north have the reverse. It decomposes to the less psychotropic ***cannabinol***. **Human** placental and milk transfer occurs. After ingestion, its effects are noted in 30–60 min, peaking at ~3 h. It is three times as potent when inhaled. Hallucinations, sedation, euphoria, "highs", red eyes, "hangover", fluid retention, discomfort, dizziness, dream-like floating state, sleepiness, decreased temperature, decreased blood pressure, distorted perception of time, disorientation, anxiety, and decreased sperm motility. It disrupts pituitary production of ***follicle stimulating hormone***, ***luteinizing hormone***, and ***prolactin*** leading to alterations of menstrual cycle, including anovulatory cycles in **women** and **monkeys**. Smoking ***cigarettes*** containing ~6 mg leads to deterioration of driving skills and simulated instrument flying ability. Adverse effects are more common as the oral dose exceeds 7.5 mg/m^3. Small doses, but not large doses, impair immediate memory. Memory impairment and confusion reported in the **elderly**. It is abused mostly by inhalation. Inhalation leads to premature ventricular contractions; IV leads to increased sympathetic reflexes, decreased parasympathetic tone, supine tachycardia, increased blood pressure, and upright hypotension. Most ***joints*** (cannabis ***cigarettes***) contain ~20 mg. Oral use of 0.03 cc/lb in 18 **people** with above average IQ induced a marked inconsistency in **their** performance. Tolerance has been developed in **man**. Illicit street material may often be contaminated with ***PCP***. Usually decreases half-life of ***antipyrine***.

Cats vomit after ingestion of 0.5–4 mg/kg. **Monkeys** given 16–64 mg/kg died. Fetal toxic effects and neonatal deaths in **rodents** and **monkeys**. Suckling male **rodents** receiving it in maternal ***milk*** developed permanent endocrine and behavioral changes. IV use in male **rats** led to mydriasis.

LD_{50}, **rat**: 666 mg/kg; **mouse**: 482 mg/kg; LD_{LO}, **dog**: 525 mg/kg; **monkey**: 128 mg/kg; IV, LD_{50}, **rat**: 29 mg/kg; **mouse**: 42 mg/kg.

TETRAHYDROFURAN

= *Diethylene Oxide* = *1,4-Epoxybutane* = *Tetramethylene Oxide* = *THF*

Use: Polymer solvent. U.S. FDA limits residual on food packaging materials to 1.5%.

Untoward effects: Fumes can irritate eyes, nose, and throat; cause dizziness, nausea, and headaches. National Institute for Occupational Safety & Health and OSHA set upper exposure limits at 200 ppm. Immediate danger to life or health at 2000 ppm.

LD_{LO}, **human**: 50 mg/kg; inhalation TC_{LO}, **human**: 25,000 ppm.

LD_{LO}, **rat**: 3 g/kg; inhalation LC_{LO}, **rat**: 28 g/m^3/2 h; **mouse**: 2.4 g/kg/m^3/2 h.

TETRAHYDROFURFURYL ALCOHOL

***Solvent*. Enhances the percutaneous absorption of drugs and used in ophthalmic drops, usually at 0.2–0.3%.**

Untoward effects: Moderately irritating to skin and mucous membranes.

LD_{LO}, **human**: 500 mg/kg.

LD_{50}, **rat**: 2.5 g/kg; **mouse**: 2.3 g/kg; **guinea pig**: 3 g/kg.

TETRAHYDROHARMINE

Untoward effects: A psychotomimetic hallucinogen in the *Banisteriopsis* sp. of Amazonia, often used in combination with ***bufotenine*** (q.v.).

IP, LD_{LO}, **mouse**: 240 mg/kg; SC, 540 mg/kg.

TETRAHYDROZOLINE

= *Rhinopront* = *Tetryzoline* = *Tinarhinin* = *Tyzanol* = *Tyzine* = *Visine* = *Yxin*

Vasoconstrictor, sedative, sympathomimetic, α-adrenergic.

Untoward effects: **Children** receiving small amounts show apathy, bradycardia, bradypnea, drowsiness, cold extremities, coma, miosis and later mydriasis, apnea, and hypotension. Accidental ingestion of 0.05% ocular decongestant of $\frac{1}{2}$–1 teaspoonful by a 1-year-old **female** and 15–30 ml in other **children** led to poisoning. Intranasal administration in an 8-month-old infant led to bradycardia. Acute allergic dermatitis with redness and burning of eyes after topical ophthalmic use reported. Can interfere with urine test for ***tetrahydrocannabinol***.

Minimum lethal dose, 2-year-old **children**: 5 mg intransal; TD, **children**: 0.05 mg/kg.

In **cats**, IV use leads to hypertension.

IV, LD_{50}, **mouse**: 39 mg/kg.

TETRALIN

= *Tetranap*

Solvent.

Untoward effects: Vapors lead to eye, skin, and mucous membrane irritation. Narcotic in high concentration.

LD_{LO}, **human**: 500 mg/kg.

Cataracts in laboratory **animals**.

LD_{50}, **rat**: 2.9 g/kg.

TETRAM®

= *Amiton* = *Inferno* = *Metramac*

Contact insecticide, acaricide. Anticholinesterase.

Untoward effects: Expected symptoms are the same as for ***sulfotep*** (q.v.) in **man**.

LD_{LO}, **human**: 5 mg/kg.

LD_{50}, **rat**: 5 mg/kg; topical, 5 mg/kg; **mouse**: 860 mg/kg.

TETRAMETHYL LEAD

= *TML*

Gasoline antiknock additive.

Untoward effects: Inhalation, ingestion, or skin contact lead to restlessness, anxiety, insomnia, bad dreams, nausea, hypotension, anorexia, delirium, mania, convulsions, and coma. National Institute for Occupational Safety & Health and OSHA upper exposure limits for skin, 75 µg/m^3; immediate danger to life or health, 40 mg/m^3 as ***lead***.

LD_{50}, **rat**: 109 mg/kg; **rabbit**: 24 mg/kg.

TETRAMETHYL SUCCINONITRILE

= *TMSN* = *TSN*

Untoward effects: Inhalation, ingestion, or dermal and eye contact leading to headache, sensation of pressure in **one's** head, dizziness, nausea, vomiting, peculiar taste, dyspnea, fatigue, convulsions, unconsciousness with liver and kidney pathology. Forms ***cyanide*** (q.v.) in the body. National Institute for Occupational Safety & Health and OSHA upper exposure limits for skin, 3 mg/m^3 or 0.5 ppm. Immediate danger to life or health, 5 ppm.

LD_{LO}, **rat**: 25 mg/kg. Inhalation LC_{LO}, **rat**: 60 ppm/2 h; **mouse**: 125 mg/m^3/3.5 h.

TETRAMISOLE

= *Anthelvet* = *Bayer 9051* = *Citarin* = *Concurat* = *Decaris* = *Galinid* = *ICI 50,627* = *McN-JR 8299* = *Nilverm* = *R8299* = *Ripercol*

Anthelmintic.* A racemic mixture with only ½ the activity of *levamisole* (q.v.), as its *dextro* form has virtually no anthelmintic activity. *Cholinesterase inhibitor.

Untoward effects: High dosage in **children** has caused dizziness, which may have been related to **their** malaria. Other **children** have vomited within 1 h after ingestion.

TD_{LO}, **human**: 2 mg/kg.

In some species, greater effectiveness is achieved by SC or IM dosage and at lower dosage than orally, but adverse reactions are much greater. Adverse effects in **cattle** and **sheep** are salivation, restlessness, yawning, frequent defecation and urination, and increased respiratory rate; in **cats**, **mice**, and **rats** causes salivation, dyspnea, hypotension, restlessness, muscular twitchings, convulsions, and death; in **dogs**, hypertension is noted. Ordinarily, even strong adverse symptoms are transient and disappear within 1–2 h.

The drug is a strong ***cholinesterase inhibitor***, causes neuromuscular block, and can potentiate similarly acting drugs. *Contraindicated* with all ***organophosphorus cholinesterase inhibitors, phenothiazines, methyridine, procaine***, and any specific ***neuromuscular blocking agents***. Avoid IV use – may cause excitement, convulsions, opisthotonus, and death, depending on dosage. Local reactions are occasionally severe in **cattle**.

LD_{50}, **rat**: 480 mg/kg; **mouse**: 210 mg/kg.

TETRANITROMETHANE

= *Tetan* = *TNM*

Use: As an oxidizer in rocket propellents and explosives, as an additive to increase the cetane number of ***diesel fuel***, and in chemical synthesis.

Untoward effects: During World War I, **human** inhalation exposure occurred during its use in the manufacture of ***trinitrotoluene*** (q.v.), causing a high incidence of ***TNT*** intoxication in British and U.S. manufacturing plants. Today, it is an atmospheric pollutant as a by-product of explosive manufacturing in U.S. Government-owned factories, where levels have reached up to 2.5 ppm. It has a pungent odor and its fumes irritate skin, eyes, nose, and throat; and cause dizziness, headache, chest pain, dyspnea, methemoglobinuria, cyanosis, and skin burns. Highly explosive. National Institute for Occupational Safety & Health and OSHA upper TWA exposure limits are 1 ppm. Immediate danger to life or health, 4 ppm.

Exposure to vapors led to pulmonary edema in **rats** and pulmonary inflammation in **mice**. **Rats** and **mice** have a dose-related incidence of alveolar/bronchiolar neoplasms at 2 and 5 ppm, respectively, and often these were carcinomas.

Inhalation LC_{50}, **rat**: 18 ppm/4 h; **mouse**: 54 ppm/4 h.

TETRAPTERIS sp.

Untoward effects: ***T. methystica*** is the source of a potent narcotic and hallucinogenic drink in the Rio Tikié in

Brazil due to its ***N-N-dimethyltryptamine*** (q.v.) and ***harmaline*** (q.v.) content. Other species are used in similar drinks and as a mild ***curare*** source.

TETRODOTOXIN

= *Fugu Poison* = *Maculotoxin* = *Spheroidine* = *Tarichatoxin* = *Tetrodontoxin* = *Tetrodoxin* = *TTX*

Untoward effects: Said to be one of the most poisonous substances known, 50 times that of ***strychnine*** and 1000 times that of ***cyanide***, occurring naturally in the ovary, liver, skin, and intestines of **puffer fish**. See ***Fish***, ***Newts***, and ***Salamanders***. Poisonous types are found in tropical waters. **Puffers** caught off the coast of Maryland have been eaten without any evident toxicity. The toxin has also been isolated in the venom from the posterior salivary glands of the blueringed **octopus**, *Hapalochlaena maculosa*. It has strong emetic and vasodilator action. Over 4000 years ago, the **Chinese** used the toxin as an anticonvulsive drug. It was first commercially marketed as an analgesic–anesthetic in 1913, but it wasn't until 1950 that the first purified material was obtained. In modern times, its action has been used to relieve muscle spasms in some terminal cancer **patients**. It is said that the reputations of licensed Japanese ***fugu*** **cooks** hinge on the number of **guests** still alive an hour after eating the delectable ***fish***. It inhibits ***sodium*** ion transport without affecting outward ***potassium*** or ***chloride*** flow. The amount of toxin is greatest during spawning season. A recent Taiwan report of poisoning in 18 **people** after ingestion of ***fish*** during the spawning season. Symptoms are hypertension, miosis, bronchorrhea, facial flushing, weakness, dizziness, pallor, circumoral paresthesia spreading to the extremities, ataxia, and flaccid lower motor neuron paralysis. Death is due to respiratory paralysis. In ancient times, **people** of Asia and Africa used its anesthetic properties in ritualistic voodoo ceremonies. In Haiti, it has caused zombification (a **person** without willpower, character, and personality), a ritual to punish crimes against the community. It decreases ***oxygen*** consumption by the brain, allowing the **victim** to be buried alive and then exhumed in a ritualistic ceremony, if the **victim** managed to survive. Some brain damage usually occurs in addition to great emotional stress; afterwards, the **victim** is occasionally subjected to slave labor. The British secret agent, James Bond, was poisoned by it and then resuscitated by cardiopulmonary resuscitation in an Ian Fleming novel.

The California **newt** *Taricha torosa* is immune to its toxicity. Intraventricular and IV use in **cats** leads to vomiting, hypothermia, and respiratory paralysis. Emesis, dyspnea, and tachycardia noted in **dogs**.

LD, **mouse**: SC and IV, 8 μg/kg; IP, 10 μg/kg. LD_{80}, IV, **dog**: 4.8 μg/kg.

THALIDOMIDE

= *Algosediv* = *Contergan* = *Distaval* = *Grippex* = *K-17* = *Kevadon* = *Lulamin* = *Neurosedyn* = *Pantosediv* = *Peracon Expectorans* = *Polygripan* = *Sedalis* = *Softenon* = *Talimol* = *Thalomid*

Immunosuppressant, sedative, hypnotic, leprostatic.

Untoward effects: The worst has been its teratogenic effect, especially when pregnant **women** ingested it ~28–56 days after **their** last menstrual period. During 1959–1962, > 10,000 malformed (phocomelia, hearing defects, ectromelia, anotia, microtia, facial palsy, external ophthalmoplegia, anophthalmus, microphthalmus, exotropia, ptosis, coloboma, renal and gastrointestinal anomalies, congenital heart disease, inguinal hernias, and hypoplasia and/or absence of bones) **infants** in Germany, 1000 in Japan, 400 in Britain, 280 in Scandanavia, and in a total of 46 countries, due to it were reported in 20–60% of **those** born to **mothers** who reportedly took it during pregnancy. Neuropathies may be delayed for at least a year. In the U.S., 1267 **physicians** received it for investigational use. Of these, 410 made no attempt to locate the **patients** they had given it to. The FDA was unable to locate 99 of the **physicians**. Peripheral neuropathy has occurred with continuous use, but may be reversible. Such neuropathy can occur without the pre-existing conditions of AIDS or leprosy, that it is used to treat. Rash, urticaria, pruritus, edema, toxic epidermal necrolysis, neutropenia, leukopenia, constipation, dizziness, euphoria, asthenia, thrombophlebitis, headache, decreased libido, increased appetite, hypotension, tachycardia or bradycardia, hypothyroidism, and death have been reported. Commonly, sedation, orthostatic hypotension, xerostomia, and dry skin. It is antagonistic to ***folic acid***, ***glutamic acid***, and ***riboflavin***. A 14-year-old **female** whose **mother** ingested it during the first trimester had right eye lacrimation while eating and a lateral gaze palsy. In a **man**, there is the possibility of drug transfer while having sexual intercourse. Life-threatening toxic epidermal necrolysis reported in several **patients** receiving it for treatment of myeloma.

TD_{LO}, **woman** (44 days pregnant): 2 mg/kg.

Limb reduction deformities of extremities and occasionally defects of mandible, ears, and internal organs in **baboons** and six species of **monkeys** given single oral doses in the first few weeks of their pregnancy. Phocomelia, hearing defects, malformations, increased neonatal death rates, and decreased birth weights in offspring of treated **rabbits**. It is less rapidly absorbed and more rapidly eliminated in **rats** than in **rabbits**. It causes increased incidence of fetal deaths and placental damage in **rats**. Limited teratogenic effects in a number of **hamster** strains. Sensitivity of **mice** is close to **rabbits'** in its teratogenic effects. Anomalies in treated **chick** embryos. Its use interfered with the estrus cycle of **sows**, but caused no gross malformations in their **piglets**. At low dosage, comparable to **man's**, **chinchillas** develop similar teratogenic effects.

LD_{50}, **rat**: 113 mg/kg; **mouse**: 2 g/kg; TD_{LO}, **rabbit** (6–16 days pregnant): 825 mg/kg; **monkey** (25 days pregnant): 10 mg/kg; **hamster** (7 days pregnant): 600 mg/kg.

THALLIUM

Untoward effects: A widely dispersed element once widely used in **human** cosmetics and medicines, depilatory scalp or cancer treatment, and to kill **rats**, field **rodents**, and **ants**, that caused severe poisoning in **children**, domestic and wild **animals**, and **birds**. **Children** and **others** often mistook it for ***sugar***. Environmental contamination occurs from ***coal***-burning power plant emissions; ***copper***, ***lead***, and ***zinc*** smelters; ***cement*** plants, and water run-off around ***sulfide*** ore and ***coal*** mines. Use of ***potash*** fertilizers can transmit it into the **human** food chain. It has been estimated that **people** in the U.S. are regularly exposed to ~10 μg/day. Accidental and suicidal ingestion has occurred with thallotoxicosis as peracute, acute, and chronic forms, depending on the quantity and frequency of exposure. Thallium compounds are absorbed from the digestive tract or through the skin. Placental and milk transfer occurs. Symptoms of the acute form last mostly for several days or weeks leading to gastric irritation, severe cramping, nausea, vomiting, diarrhea, conjunctivitis, stomatitis, mouth and nasal ulcerations, intellectual impairment, excessive salivation, fever, muscle tremors, anorexia, hyperesthesias, pain in extremities, incoordination, convulsions, paralysis, blurred vision, corneal opacity and occasionally blindness, ptosis, extraocular muscle palsies, mydriasis, toxic amblyopia, anemia, aplastic anemia, hemolytic anemia, eosinophilia, lymphocytosis, leukopenia, thrombocytopenia, dementia, encephalitis, bronchitis and dyspnea, ecchymoses, and petechial hemorrhages, with full recovery in **many** after several months. In the chronic form, similar, but often milder symptoms with diffuse alopecia, superficial skin necrosis; thickening, dryness, and scaling of skin; decreased perspiration, tachycardia, albuminuria, hematuria, hemolysis, increased urinary cells and casts, leukocytosis, neutrophilia, anemia, excessive thirst, insomnia, dark blue gingival lines, nail dyschromy with transverse white lines (Mee's lines), and parkinsonian-like signs with resting tremors. Increase of dental caries also occurs after several months. The peracute form from ingestion of large amounts leads to hemorrhagic gastroenteritis, colic, vomiting, trembling, motor paralysis, dyspnea, and death within 24 h. **People** poisoned by it often have a ***garlic***-like breath odor. Only 1 oz of a pesticide containing it at a 1% level can kill a 51 lb **child**. In 1998, court testimony revealed that a plan existed to poison **Nelson Mandela** with it to affect **his** nervous system "to reduce his level of intellectuality and effectiveness by inducing brain damage". A 42-year-old **male** and a 48-year-old **male** were poisoned by it as an adulterant in IV ***heroin*** leading to abdominal pain, alopecia, and encephalopathy. Several **patients** were adversely affected after ingesting a Chinese herbal product, ***Nutrien***, with > 3% thallium content. Cellular degeneration is the main effect noted on post-mortem. Lethal doses are given as 5–8 mg/kg as the *sulfate* or 0.5–1 g/**adult**. National Institute for Occupational Safety & Health and OSHA set skin upper exposure limit as 0.1 mg/m^3 and immediate danger to life or health of 15 mg/m^3.

In **man** and **animals**, symptoms from acute poisoning usually peak in the second and third post-exposure weeks. In **dogs**, this can occur within 1–4 days. Alopecia usually occurs after the second week of exposure so that it is uncommon in **dogs**, although they often show brick-red coloration of their mucous membranes. Neurological changes in **man** and **animals** occur between the first and second post-exposure weeks.

Various salts have been reported in **horse**, **cattle**, **sheep**, **cat**, and **dog** poisonings. Use in Denmark as a rodenticide has caused poisoning of **badgers** and wild **foxes**. Has caused both primary and secondary poisoning hazards to **raptors**, **eagles**, and predatory **wildlife**. Was often sold as ***Certox, Gizmo Mouse Killer, GTA Rat Bait, Martin's Rat Stop Liquid, Rat & Mouse Controller Paste, Senco Corn Mix, TAT Ant Trap, TAT Home Guard Roach Poison***, and ***Zelio Paste***. Chronic ingestion by **rats** has caused adverse reproductive effects. Poisoned **rats** are often scavenging risks to **cats**, **dogs**, and **pigs**. Symptoms in **cats** are anorexia, diarrhea, vomiting, abdominal pain, dyspnea, salivation; reddening, crusting, and peeling of skin and alopecia of the face, feet, limbs, head, and torso. Hypersensitivity, tremors, body contortions, paresis, leukocytosis, and anemia. Post-mortem shows hemorrhagic gastroenteritis, fatty degeneration of liver, and nephrotoxicity. Placental transfer occurred. Has caused cataracts in **rabbits** and **rats**. Epilation is common in surviving **dogs** and slightly poisoned **sheep** lose large patches of ***wool***. Despite its earlier ban in the U.S. as a rodenticide, hairless **racoons** in Kentucky were found in the 1990s. Polydipsia, regurgitation, ataxia, high carriage, slowness, imbalance, goose-stepping ataxia, myasthenia, asthenia, hypoactivity, hopping and falling, wing-drop, and loss of righting reflex within 15 min in poisoned **ducks**, **eagles**, and **pheasants**. Death occurred within 1–7 days. Contrary to many poisons, it is more toxic to older **animals** than younger ones. In some countries, it is mandatory that muscles and carcasses of poisoned **animals** be withheld from **human** consumption.

LD_{50}, **rat**: 16 mg/kg; **mouse**: 7 mg/kg; **dog**: 18 mg/kg; **cattle**: 16 mg/kg; **horse**: 27 mg/kg; **sheep**: 9 mg/kg; **lamb**: 20 mg/kg; **pig**: 15–25 mg/kg; **mallard duck**: 37 mg/kg; **golden eagle**: 60–120 mg/kg; **pheasant**: 24 mg/kg.

THEBAINE

Untoward effects: A natural minor alkaloid of ***opium***, chemically close to ***codeine*** and ***morphine***, but has stimulatory, rather than depressant, effects. It is not a drug of abuse, but the U.S. has made it a regulated controlled substance because it can be converted into narcotic drugs, such as ***codeine*** and ***oxycodone***. It has caused ***strychnine***-like spasms or convulsions in **man**.

Non-addictive in **rat** trials, but increases heart rate, heart preload, afterload, and contractility by vagal action in **animals**. Maternal SC dosage > 110 mg/kg on the eighth day of pregnancy was teratogenic in **hamsters**. Convulsant and peripheral nerve paralysis in lower **animals**.

SC, LD_{50}, **mouse**: 117 mg/kg.

THENALDINE TARTRATE

= *AS 716* = *Sandostene*

Antihistamine, antipruritic.

Untoward effects: Anemia and neutropenia. Fatal agranulocytosis has been reported, as well as non-fatal cases – probably hypersensitivity reactions. No longer sold in the U.S.

LD_{50}, **rat**: 1 g/kg; **mouse**: 165 mg/kg.

THENIUM CLOSYLATE

= *Bancaris* = *Canopar*

Anthelmintic.

Untoward effects: May be toxic in suckling, weak, or debilitated **animals**. Rarely used now, as it had a too narrow safety margin. Occasionally vomiting, cramping, pale or cyanotic mucous membranes, intestinal hemorrhage, and deaths after administration to **dogs**.

THEOBROMINE

= *3,7-Dimethylxanthine* = *Santheose* = *Théosalvose*

Contains no *bromine*. It is the principle alkaloid of *chocolate* or *cocoa*.

Untoward effects: Ingestion of large amounts lead to nausea and vomiting. Passes freely into breast milk of **mothers** who eat ***chocolate***. A **mother** who ate a ***chocolate*** candy bar every 6 h could expose **her** nursing **infant** to 10 mg/day, enough to cause adverse effects, such as stimulation of central nervous system and heart muscle, increased urine volume, and relaxation of smooth muscle (such as the intestines) in some sensitive **infants**. **Infants** receive ~20% of **maternal** dose. Unsweetened baking ***chocolate*** has ~10 times the theobromine content of milk ***chocolate***. Therapeutic serum levels are 10–15 mg/l and toxic at 20 mg/l.

TD_{LO}, **human**: 26 mg/kg.

It is the toxic principle in ***chocolate*** for **dogs** that leads to vomiting, thirst, cardiac stimulation, diarrhea, panting, diuresis, urinary incontinence, mydriasis, salivation, depression, central nervous system stimulation (muscle tremors, restlessness, excitement, seizures, and coma), and death, often occurring within 4–6 h. Fatalities have occurred with $2\frac{1}{4}$ oz of unsweetened ***chocolate*** and ~$1\frac{1}{3}$ lb of milk ***chocolate*** in a 10 kg **dog**. Poisoning in a **dog** after consuming ***cacao*** shells used in landscaping. **Calves** have developed excitement, opisthotonus, and convulsions after eating waste plain ***chocolate***. Excreted in **cow's *milk***. Feeding of similar wastes to **horses** can cause disqualification from racing for at least 10 days due to its presence in their urine and they have been killed by 5 g/day. It is genotoxic, potentiating DNA damage and causes testicular atrophy, and abnormalities of spermatogenic cells in **rats**. Feeding 0.15–0.2% levels in feed depressed food intake and weight gain, but lower levels enhanced growth of **lambs** by stimulating food intake.

LD_{50}, **cat**: 200 mg/kg.

THEOPHYLLINE

= *Accurbron* = *Aerobin* = *Aerolate* = *Afonilum* = *Armophylline* = *Austyn* = *Bilordyl* = *Bronchoretard* = *Bronkodyl* = *BY 912* = *Cétraphylline* = *Duraphyllin* = *Diffumal* = *Elan* = *Elixicon* = *Elixifilin* = *Elixophyllin* = *Etheophyl* = *Euphyllin* = *Euphylong* = *LaBID* = *Lanophyllin* = *Lasma* = *Nuelin* = *Physpan* = *Piridasmin* = *Protheo* = *Pro-Vent* = *PulmiDur* = *Pulmo-Timelets* = *Respbid* = *Slo-Bid* = *Slo-Phyllin* = *Solosin* = *Somophyllin T* = *Spophyllin* = *Sustaire* = *Talotren* = *Teofilina* = *Teonova* = *Teosona* = *Theobid* = *Theocap* = *Theoclear* = *Theochron* = *Theo-Dur* = *Theodyl* = *Theolair* = *Theolin* = *Theolix* = *Theolixir* = *Theon* = *Theophyl* = *Theograd* = *Theostat* = *Theovent* = *Unicontin* = *Unifyl* = *Uniphyl* = *Uniphyllin* = *Xanthium* = *Youteshu*

***Bronchodilator, diuretic, cardiac stimulant, and smooth muscle relaxant.* The alkaloid theophylline name was derived from Greek (thea = *tea* and phyllon = leaf) as it was first isolated from *tea*.**

Untoward effects: Essentially, the same as for ***aminophylline*** (q.v.). Therapeutic levels are close to toxic levels and there is a huge interindividual variation. Dosages should be in terms of lean or ideal weight, since it is not distributed into fatty tissue. Nausea, vomiting, gastric irritation and distress, diarrhea, headache, insomnia, arrhythmias, hypotension, restlessness, tremors, and convulsions are directly related to serum levels. Most of these adverse effects are noted when serum levels are > 15 µg/ml. 90% is metabolized by normal livers and clearance is decreased in liver disorders and congestive heart failure. Because elderly **people** often have increased incidence of congestive heart failure and hepatic cirrhosis, it is often said that **they** are more sensitive to the drug. The serious adverse effects, cardiac arrhythmias and convulsions, can often occur before the usual nausea and headache warning signs occur. Tachycardia, atrial fibrillation, faintness, dizziness, hyperthermia, hyperventilation, agitation, diaphoresis, tremors, hypokalemia, hypercalcemia, hyperglycemia, depression, psychiatric disorders, increased urinary ***catecholamines***, metabolic acidosis, porphyrias, mutism, confusion, lethargy, cyanosis, urinary retention, decreased lower esophageal sphincter pressure, increased gastroesophageal reflux, esophageal ulceration, and coma also reported. Rhabdomyolysis has been associated with large overdoses, as in attempted suicide. A 26-year-old **female** attempted suicide after ingesting 70–100 slow-release 200 mg tablets leading to peak of 260 µg/ml serum concentration; **she** survived after treatment. U.S. Air Force **doctors** prescribed an overdose, 200 mg, to a 6-week-old infant leading to seizures, brain hemorrhage, deafness, near blindness, and severe cerebral palsy. **He** died 2 years later. Rarely, paradoxical exacerbation of asthma has been reported. Spurious increase in blood or urinary ***uric acid*** assays occur occasionally with the ***phosphotungstate*** method.

The Association of Trial Lawyers of America (ATLA), recognizing that **patients** unknowingly ingest overdoses of theophylline over-the-counter products that can lead to seizures and irreversible brain damage, urged the FDA to ban them. Drug manufacturers are legally required to inform FDA of dangers regarding their products. ATLA's action was supported by MDs associated with a Chicago Institute of Allergy and Clinical Immunology and a pharmacology professor at the University of Virginia Medical School.

Many fatalities often followed rapid IV infusion in **adults** and suppositories in **children**. It halts the action of phosphodiesterase, an enzyme that degrades cyclic ***adenosine monophosphate*** to ***AMP***. This accumulation stimulates cell growth and formation of mammary fibrous tissue and cystic fluid. Its use has been associated with a statistically significant increase in lung cancers.

Placental and milk transfer occurs. **Maternal** and cord theophylline levels are approximately the same, but heel-stick levels are higher than **maternal**. **Babies** with serum levels > 10 µg/ml from either placental or milk transfer leading to tachycardia, transient jitteriness (up to 6 h after **mother** receives medication), vomiting, irritability, and fretful sleep. Learning and behavior problems reported in **children** 6–13 years. Excessive doses in **infants** and **children** have also caused hematemesis, albuminuria, hematuria, proteinuria, stupor, cardiac arrhythmias, hyperreflexia, convulsions, and death. Unfortunately, in most of these cases, the usual warning signs of toxicity were ignored. Some toxified premature **infants** have survived 40–70 µg/ml. Half-life in preterm **infants** is 30.2 h; in **neonates**, 20–40 h; and 2–6 h in **children**. Therapeutic serum levels are generally 8–18 mg/l; toxic at > 20 mg/l; and lethal at 50 mg/l.

Drug interactions: ***Macrolide antibiotics***, such as ***erythromycin base***, ***erythromycin ethylsuccinate***, ***erythromycin estolate***, ***erythromycin stearate***, and ***troleandomycin*** decreases its clearance with a potential for serious toxicity. The quinolone antibiotics, ***cefaclor***, ***ciprofloxacin***, ***clinafloxacin***, ***enoxacin***, ***fleroxacin***, ***grepafloxacin***, ***norfloxacin***, and ***prulifloxacin*** are also capable of decreasing theophylline clearance and increasing its toxicity. ***Tetracycline*** weakly inhibits theophylline clearance and ***minocycline*** apparently does also. Theophylline 400 mg once daily decreases ***citirizine*** clearance by 16%. ***Cimetidine***, ***ranitidine***, and high doses of ***allopurinol*** have inhibited the cytochrome P-450 liver enzymes, inhibiting its metabolism and increasing its toxicity. ***Oral contraceptives*** and ***pentoxifylline*** may decrease its dosage requirement by $\frac{1}{3}$. ***Amiodarone***, ***isoniazid***, and ***pyrantel pamoate*** may rapidly increase its serum concentration. ***Mexiletine*** increases its serum concentration ~2 times. ***Propranolol*** decreases its clearance by 37%. ***Tocainide*** has decreased its clearance by 10% and increases its elimination half-life. Theophylline half-life in young **smokers** was ~4 h versus 7 h in **non-smokers** who must decrease **their** theophylline dosage (after ~3 months due to prolonged enzyme stimulation effects) if **they** quit smoking. ***Albuterol***, ***carbamazepine***, ***cyclosporine***, ***furosemide***, ***isoproterenol***, ***moricizine***, ***phenobarbital***, ***phenytoin***, and ***rifampin*** also markedly stimulate its metabolism. It increases ***lithium*** excretion and may have accentuated ***lomustine***-induced thrombocytopenia. ***Antacids***, such as ***sodium bicarbonate***, increase its absorption and decrease its urinary excretion. **Human *interferon*** *α-2a* significantly decreases its clearance and increases its elimination half-life. Viral infections or use of ***influenza vaccines*** have also decreased its metabolism and increased its toxicity. ***Diltiazem***, ***fluoxamine***, ***maté*** drinking, ***morphine***, ***nifedipine***, ***nizatidine***, ***propafenone***, ***ticlodipine***, ***verapamil***, ***viloxazine***, ***zafirlukast***, and ***zileuton*** increase its toxic levels by decreasing its clearance. Acidifying agents, such as ***ammonium chloride***, increases its urinary excretion. ***Metaproterenol*** increases its effectiveness with tachycardia occasionally occurring. ***Albuterol*** and ***terbutaline*** have also increased its effectiveness. ***Ethylenediamine*** (q.v.) improves its solubility and is a potent topical sensitizer, but rarely systemically. It is an antagonist of ***dipyridamole***, sharply reversing the coronary vasodilation and decreasing ***dipyridamole's*** diagnostic accuracy in cardiac imaging. Toxicity may be increase when used with ***ephedrine***. In New Zealand in the mid-1970s, use with ***β-agonists*** increased the death rate in treating acute asthma three times. ***Activated charcoal*** enhances its clearance from the body. ***Sulfinpyrazone*** increases its plasma clearance by 22%. Interactions with ***anticoagulants*** are usually clinically insignificant. May antagonize ***digoxin*** cardiac conduction. Theophylline may decrease ***alprazolam*** serum levels. Mean ***erythropoietin*** levels are higher in theophylline-treated premature **infants** than in the nontreated. Use with ***St. John's Wort*** (***Hypericum***) 300 mg/day caused a need for 2.6 times increase in theophylline to maintain adequate blood levels. ***Tacrine*** decreases its metabolism and increases its serum levels ~2 times.

In **animals**, it is *contraindicated* in acute nephritis. Gastric irritant. Administer with or after feeding or as enteric-coated tablets. *Contraindicated* in **dogs** with ***adenosine***. IV infusion of 1–2 mg/kg/min in **sheep** causes increased plasma levels on non-esterified ***fatty acids*** and a hypocalcemia. The latter is probably due to its causing an accumulation of ***adenosine*** in bone tissue, inhibiting bone resorption. Its use was associated with deaths in **calves** treated for respiratory disease complex. It is teratogenic in the **mouse**.

IV, LD_{50}, **mouse**: ~200 mg/kg.

THERMOACTINOMYCES VULGARIS

Untoward effects: A thermophilic antigen found on moldy hay, moldy ***sugar*** cane, and compost, and in air-conditioning systems and home humidifiers leading to **farmers'** lung, bagassosis, and ***mushroom*-workers'** disease. Hypersensitivity pneumonitis in a 65-year-old **male**, due to its contamination of a home humidifier leading to severe chills, fever, cough, dyspnea, weight decrease, and pneumonitis. A factor in some cases of **bovine** respiratory disease and interstitial pneumonitis, thickening of interlobular septa, and peribronchial lymphoid hyperplasia in **sheep**.

THERMOPOLYSPORA POLYSPORA

Untoward effects: Isolated cases and occasional epidemics of **farmers'** lung, bagassosis, and ***mushroom*-workers'** lung due to this thermophilic oligosporic actino-mycete, also commonly found on moldy hay. A cause of "fog fever" (pulmonary emphysema) and abortions in **cattle**.

THERMOPSIS sp.

Untoward effects: ***T. divaricarpa*** = ***Golden Pea*** is a ***selenium*** (q.v.) accumulator.

T. montana, after ingestion, has induced myopathy in **calves**. ***Sparteine*** (q.v.), an abortifacient alkaloid, has been extracted from this genus. Anorexia, swollen eyelids, depression, humped-up appearance, and death can occur in **cattle**.

T. rhombifolia = ***Golden Bean*** of southwestern Canada probably contains ***anagyrine***, ***cytisine***, ***N-methylcystisine***, and ***thermopsine*** in the leaves and seeds. Suspect as a cause of poisoning in a **child**.

THEVETIA sp.

Untoward effects: ***T. nereifolia*** = ***Be-Still-Tree*** = ***Lucky Nut*** = ***Olomiojo*** (Yoruba) = ***Pella Kaner*** = ***Tiger Tree*** = ***Yellow Oleander*** = ***Zuhur*** (Yemen). Its roots and milky sap are very poisonous. The bark, stem, and especially the kernels or nuts are also poisonous, containing various cardiac glycosides, primarily ***peruvoside*** (q.v.) and ***thevetin***. It is a pantropic, beautiful ornamental plant, commonly used as a hedge plant in Florida and Hawaii. One or two nuts can be fatal after cardiac stimulation, dizziness, and vomiting. Used therapeutically in India for 3000 years.

T. Peruviana, also called ***Be-Still-Tree*** and ***Yellow Oleander***, has also caused poisoning due to its cardiac glycoside content leading to bradycardia, arrhythmias, diarrhea, vomiting, and death. Their glucosides have also been used as arrow poisons.

Livestock have been poisoned by these plants.

THIABENDAZOLE

= *Arbotect* = *Bovizole* = *Eprofil* = *Equizole* = *Foldon* = *Lombristop* = *Mertect* = *Mintezol* = *Minzolum* = *MK-360* = *Nemapan* = *Omnizole* = *Polival* = *TBZ* = *Tecto* = *Thiaben* = *Thiabenzole* = *Thibenzole* = *Tiabenzole* = *Top Form Wormer* = *Triasox*

Anthelmintic, fungicide, chelator. **An *imidazole*.**

Untoward effects: ***Blood***: Occasional transient leukopenia and increase or decrease of blood ***glucose*** levels.

Cardiovascular: Hypotension, decreased blood pressure, and bradycardia.

Central nervous system: Vertigo, dizziness, and incoordination. Occasional drowsiness, giddiness, headache, irritability, restlessness, confusion, euphoria, depression, and weakness. Rarely, hallucinations, collapse, and convulsions. Psychosis, paranoia, delusions, agitation, and violence in a 57-year-old **female** after ingesting 1.5 g twice daily/2 days.

Gastrointestinal: Nausea and vomiting. Occasional epigastric or abdominal pain, anorexia, and diarrhea. Rarely, jaundice, intrahepatic cholestasis, and liver failure.

Genitourinary: Crystalluria and urine malodor. Occasionally hematuria or enuresis.

Hypersensitivity: Skin rashes, pruritus, toxic epidermal necrolysis, photosensitivity, erythema multiforme, chills, fever, facial flushing, lymphadenopathy, Stevens–Johnson syndrome, shock, anaphylaxis, and angioneurotic edema.

Miscellaneous: Eye problems (xanthopsia, decreased vision, and blurred vision, especially with overdoses), tinnitus, xerostomia, dry eyes, and olfactory disturbances. Topical use with ***dimethyl sulfoxide*** leads to local dryness, burning, and erythema in most **patients**. Oral use in **patients** on ***theophylline*** will decrease the latter's requirement by about 50%.

The FDA established a 0.1 ppm and a 0.05 ppm tolerance, respectively, for its negligible residues in uncooked edible tissue or ***milk*** of **animals**. This is readily accomplished by avoiding its use within 30 days of slaughtering such **animals** for **human** consumption (only 3 days for medicated mineral for **cattle**), or by avoiding use of their ***milk*** for at least eight milkings of a 96 h post-treatment period. Resistant *Hemonchus* strains may require a dosage of at least 150 mg/kg. The FDA has allowed residues of up to 8 and 3.5 ppm, respectively, in or on ***sugar beet*** products for **livestock** feeding or raw citrus fruits incident to their permitted uses noted above. NOT for use on **horses** intended for **human** consumption. Resistance to it by small strongyles in **horses** has been well demonstrated. Cross-resistance or "tolerance to" is noted with many other ***benzimidazoles*** (***albendazole***, ***cambendazole***, and ***mebendazole***). Symptoms of toxicity in **sheep** are muscle weakness, depression, hypersialosis, and anorexia. Incoordination and dyspnea in those more severely affected. Symptoms in **horses** are depression; moderate abdominal distress, leukocytosis, and hemoconcentration after 600 mg/kg. Post-mortem in **animals** showed fatty degeneration, swelling of and yellow-colored livers; swollen and pale kidneys; hemorrhagic gastroenteritis after excessive dosage. Pulmonary hemorrhages and emphysema.

LD_{50}, **rat**: 3.1 g/kg; **mouse**: 3.6 mg/kg; **rabbit**: 3.8 g/kg; LD, **sheep**: 1.2 g/kg.

THIACETAZONE

= *Amithiozone* = *Berculon A* = *Conteben* = *Livazone* = *Myrizone* = *Neustab* = *Panrone* = *Seroden* = *Tb I-698* = *Tebethion* = *Thioacetazone* = *Thibone* = *Thiocarbazil* = *Thioparamizone* = *Tibione* = *Tibon* = *Tiobicina* = *Tubercazon*

Tuberculostatic and leprostatic.

Untoward effects: Minor adverse effects are nausea, vomiting, abdominal pain, anorexia, headache, blurred vision, conjunctivitis, vertigo, urticaria, and allergic dermatitis. Serious effects are toxic epidermal necrolysis (fatal in 19.1% of **those** affected). Deaths from induced cutaneous reactions were 3.1/1000 in Tanzania. Other serious side effects are marked bile stasis and often liver cell necrosis, Stevens–Johnson syndrome, agranulocytosis, thrombocytopenia, leukopenia, neutropenia, decreased renal function, and reversible hemolytic anemia and cerebral edema. Massive doses have been convulsant. Ototoxic. Rarely, gynecomastia.

Drug interactions: Enhances ***streptomycin*** ototoxicity. It retards the hepatic acetylation of ***isoniazide*** and increases its elimination half-life, increasing the risk of hepatitis.

LD_{50}, **mouse**: 950 mg/kg; SC, 1 g.

THIALBARBITAL

= *Intranarcon* = *Kemithal* = *Thiobarbitone*

***Short-acting anesthetic, barbiturate.* The soluble *sodium* form is used.**

Untoward effects: Concomitant use of ***atropine*** decreases salivation and helps avoid drug-induced cardiac arrhythmias. Avoid perivascular or intra-arterial injections.

Horses and **dogs** may show excitement during recovery. Solutions should be refrigerated and are stable for ~1 week.

LD_{50}, **mouse**: 370 mg/kg; IP, 384 mg/kg; IV, 390 mg/kg.

THIAMBUTENE

= *NIH 4185* = *Thermalon*

Narcotic, hypnotic, analgesic.

Untoward effects: Ordinary doses excite **cats**. **Dogs** in tropical climates require lower dosages than those in temperate climates. Can cause tetany in **dogs** and they often defecate during recovery.

IP, LD_{50}, **mouse**: 90 mg/kg.

THIAMPHENICOL

= *8053CB* = *Dexawin* = *Hyrazin* = *Igralin* = *Neomyson* = *Rigelon* = *Thiamcol* = *Thiocymetin* = *Thionicol* = *Thiophenicol* = *Urfamicina* = *Urfamycine* = *Urophenil* = *Win 5063-2*

***Antibiotic.* Oral and IM, and, rarely, IV or inhalation.**

Untoward effects: Hypovitaminosis, poikilocytosis, transient anemia, urticaria, diarrhea, agranulocytosis, thrombocytopenia, reversible bone marrow depression, and three cases of aplastic anemia reported. Breast milk levels have exceeded **maternal** plasma levels. Chills in one **patient** after IM. Placental transfer occurs in **humans** and **rats**.

THIAMYLAL SODIUM
= *Bio-tal* = *Surital* = *Thiosecomal*

Anesthetic.* An analog of *secobarbital. IV use.

Untoward effects: In **man**, essentially the same as for ***pentobarbital*** (q.v.). The more common adverse effects are circulatory depression, thrombophlebitis, pain at injection site, respiratory depression, apnea, laryngospasm, bronchospasm, salivation, hiccups, emergence delirium, headache, rashes, urticaria, nausea, emesis, and nerve injury adjacent to injection site.

Drug interactions: Use with ***phenothiazines***, such as ***chlorpromazine***, may induce orthostatic hypotension.

May produce erratic anesthesia in **Siamese cats**. Avoid shaking or foaming mixed solutions (stable for 2–3 days). Use clear solutions only. Induces more cardiac arrhythmias in **dogs** than ***thialbarbital*** or ***thiopental*** (85% of one test group), apparently by unknown mechanisms that stimulate ***epineprine*** release. This can be blocked by pretreatment with ***phenoxybenzamine*** or ***hexamethonium*** and ***acepromazine*** (0.4 mg/kg). Avoid use in **whippets**, **greyhounds**, or very thin **dogs** (slower recovery – see ***Thiopental***). IM ***chloramphenicol***, 50–70 mg/kg prolongs (62–178%) its anesthesia in **dogs**, only when given after the ***barbiturate***. Ventricular dysrhythmias in **dogs** are closely related to dose, concentration, and rate of administration, and were noted in ~4%. Older, larger, brachiocephalic **dogs**, or those in poor condition, usually require slightly lower dosage than suggested, whereas young and smaller ones may require more. Liver pathology will delay detoxification. Emergence excitement is common in **dogs** and **horses**, unless a suitable preanesthetic is used. High concentrations, accidental intra-arterial injections, or extravasation may cause serious phlebitis, arteritis, or tissue sloughing. Leukopenia in treated **animals** and abnormalities in **mouse** fetuses reported.

LD_{50}, **mouse**: 180 mg/kg; IP, 109 mg/kg; **dog**: 134 mg/kg; IV, 32 mg/kg; IV and SC, **rat**: 51 mg/kg; IV, **rabbit**: 24 mg/kg.

THIAZESIN
= *SQ 10,496* = *Thiazenone*

Antidepresssant.

Untoward effects: Adverse effects (psychomotor restlessness, generalized erythema with mild pruritus, abdominal distension, hand tremors, grand mal seizure, hypertension, xerostomia, and drowsiness) have limited the usefulness of this mildly effective ***antidepressant***.

LD_{50}, **mouse**: 210 mg/kg; IV, 29 mg/kg.

THIAZOLSULFONE
= *Thiazosulfone*

Leprostat, antibacterial.

Untoward effects: Hemolytic anemia in glucose-6-phosphate dehydrogenase-deficient **patients**. Prolonged treatment has caused thyroid enlargement and, in young **people**, the development of secondary sex characteristics before puberty.

THIETHYLPERAZINE MALEATE
= *NSC 130,044* = *Torecan* = *Toresten* = *Tresten*

***Tranquilizer, antiemetic.* A *piperazinyl phenothiazine*. Oral, IM, IV, and suppositories.**

Untoward effects: Orthostatic hypotension, tachycardia, urinary retention, xerostomia, blurred vision, dystonic reactions, hysteria, cerebral stimulation, dizziness, drowsiness, bizarre dreams, pseudoparkinsonism, torticollis, cataracts, tremors, trismus, opisthotonus, and bizarre postures. A case report of retinal lesions in a 7-month-old infant may have been related to **maternal** ingestion tid/14 days during 8–10th week of pregnancy. Transient dystonic spasms by a 12-year-old **female** ingesting 20 10 mg tablets.

TD_{LO}, **human**: 140 µg/kg.

LD_{50}, **rat**: 1260 mg/kg; **mouse**: 680 mg/kg; **rabbit**: 1 g/kg; IV, **rat** and **mouse**: ~91 mg/kg; **rabbit**: 27 mg/kg.

THIMEROSAL
= *Merseptyl* = *Mersilon* = *Merthiolate* = *Merzonin* = *Sodium Ethylmercurithiosalicylate* = *Thiomersal* = *Thiomersalate* = *Vitaseptol*

***Antibacterial, antifungal.* Contains ~50% *mercury* (q.v.).**

Untoward effects: Ocular reactions (red eyes, irritation, blepharitis) from this preservative on contact lenses, particularly when **patient** is taking ***tetracycline***. It may be due to a hypersensitivity reaction. Hypersensitivity reactions reported from its use in vaccines and ***γ-globulin***. Delayed hypersensitivity reported when used as a throat spray – edema within a few minutes that required a tracheotomy. Irrigation of ears of an 18-month-old **female** with 1 oz solution twice daily/4 weeks with indwelling drainage tubes leading to metabolic acidosis, hyperkalemia, hypotension, renal and hepatic failure, and death. Numerous case reports of allergic contact dermatitis with erythema, pruritus, and eruptions. Many **patients** allergic to it are not allergic to its ***mercury***, but rather to the ***thiosalicylic acid*** component and, therefore, do not necessarily cross-react to other ***mercury*** compounds. Acrodynia in a 20-year-old **male** after receiving ***γ-globulin*** preserved with it for ~15 years. ***Mercury*** in it is a catalyst for oxidation of ***aluminum*** with production of heat leading to burns on buttock of **patient** 1 day after surgery and washing of vaginal and abdominal skin with 1:1000 solution and **her** buttocks lying on paper-backed ***aluminum*** foil diathermy electrode. IM ***chloramphenicol*** injections were fatal in 5/6, due to the solution inadvertently containing 1000 times normal thimerosal quantity. False positive reactions reported from use of old ***tuberculin*** containing it. ***Tincture of merthiolate*** is colored with two dyes, ***fluorescein*** (q.v.) and ***eosin*** (q.v.), which can cause adverse effects. Migrates into ***polyethylene*** and ***polypropylene*** containers.

Drug interactions*: *Sodium thiosulfate neutralizes its bacteriostatic action.

LD_{LO}, **human**: 50 mg/kg.

Use possibly increases fetal death rate in **rabbits** and **rats**. LD_{50}, **rat**: 75 mg/kg.

THIOACETAMIDE

= *Ethanethiomide* = *TAA*

Untoward effects: Induces hepatotoxicity with necrosis in the centrilobular zone of **rats** and **mice** and bile duct hyperplasia in **ducklings**. Parenteral use also causes adrenal gland hypertrophy in the **rat**. Dietary intake leads to hepatocellular carcinomas in **mice** of both sexes, hepatocellular neoplasms in male **rats**, and bile duct or cholangiocellular neoplasms in **rats** of both sexes.

LD_{LO}, **rat**: 200 mg/kg.

THIOCARLIDE

= *CP 919* = *Datanil* = *DATC* = *Disocarban* = *Isoxyl* = *Tiocarlide*

Tuberculostat.

Untoward effects: Nausea, vomiting, anorexia, and occasionally rash, erythema, pruritus, arthralgia, hypoglycemia, leukopenia, monocytosis, purpura, and jaundice.

THIOCOLCHICOSIDE

= *Coltrax* = *Coltromyl* = *Miorel* = *Musco-Ril*

***Skeletal muscle relaxant, antispasmodic.* Semi-synthetic from *colchicoside*, extracted from *Colchicum autumnale*. IM.**

Untoward effects: Mild somnolence and/or vomiting in **patients** who ingest food prior to IM.

THIOCTIC ACID

= *Biletan* = *α-Lipoic Acid* = *Thioctacid* = *Thioctan* = *Thioctidase* = *Tioctan*

IV antidote for amanitin-containing mushrooms, biocatalyst.

Untoward effects: Phlebitis, transient dyspnea, and hypoglycèmia in small number of **patients** with chronic hepatitis or cirrhosis.

THIOCYANATES

Some use this term synonymously with *potassium thiocyanate* (q.v.).

Untoward effects*: *Hydrogen cyanide and alkali ***cyanides*** are detoxified by conversion to thiocyanates that are much less toxic. ***Cauliflower*** and other members of the *Brassica* family contain substantial amounts, which, in excessive intake, lead to fatigue, drowsiness, weakness, depression, apprehension, nausea, vomiting, adominal pain, delirium, nightmares, tinnitus, mild hypothyroidism, and laryngeal edema. Very high intake leads to fever, leukopenia, thrombocytopenia, agranulocytosis, nausea, anemia, stupor, goiter, psychosis, confusion, disorientation, anorexia, rashes, vertigo, slurring of speech, hypotension, albuminuria, anuria, renal failure, convulsions, and even death. Urine levels after medical IV usage of ***sodium thiosulfate*** or ***sodium nitroprusside*** may need to be monitored. Urinary levels > 1 mg/l indicate intoxication. Some state 1–12 mg/l is a normal serum level and severe toxicity at 35–50 mg/l.

THIOGLYCOLIC ACID

= *Mercaptoacetic Acid*

Use: In making cold wave solutions, heat-activated acid home permanent wave solutions, chemical reagents, depilatories, and bacteriological media.

Untoward effects: Skin contact from long-time exposure or dripping off one's head has caused severe irritation, edema, papular rashes, itching, severe burns and blisters. Leukopenia and anemia may have been induced by it in a few cases.

Weakness, gasping, and convulsions in some laboratory **animals**.

LD_{50}, 0.15 ml/kg; **mouse**: 250 mg/kg; **rabbit** and **guinea pig**: 126 mg/kg.

THIOGUANINE

= *Lanvis* = *NSC 752* = *Tabloid*

***Antineoplastic, antimetabolite.* A *purine* analog. Parenteral and topical have been experimental.**

Untoward effects: Infection associated with leukopenia and hemorrhage with thrombocytopenia, bone marrow aplasia, and anemia. These may often be delayed reactions. Occasionally nausea, vomiting, anorexia, and hyperuricemia. Skin reactions, diarrhea, stomatitis, pharyngitis, gastrointestinal ulceration, crystalluria, and hematuria are rare. Centrilobular hepatic vascular dilation and cholestasis with portal inflammation, necrosis, and veno-occlusive disease. Renal excretion requires dosage decrease of 33% in renal failure.

Drug interactions: When used with ***cytarabine*** at 20–26 weeks of gestation, a **fetus** had chromosomal abnormalities.

Use in pregnant **rats** associated with syndactyly and tibial aplasia in their **pups**.

THIOGUANOSINE

= *NSC 29,422*

Antineoplastic.

Untoward effects: Pancytopenia, mild leukopenia, diarrhea, sore throat, nausea, vomiting, and epigastric pain.

THIOMESTERONE

= *Emdabol* = *Emdabolin* = *Protabol* = *STA307* = *Tiomesterone*

Androgen.

Untoward effects: Acne, deepening of voice, and facial hirsutism in treated **females**.

THIONYL CHLORIDE

= *Sulfinyl Chloride* = *Sulfurous Oxychloride*

Use: In the manufacture of synthetic ***pyrethrum***, synthetic ***vitamin A***, chemical weapons, herbicides, and dyestuffs.

Untoward effects: Vapors or liquid are severe irritants and corrosive to eyes, mucous membranes, and skin. Inhalation exposure leads to effects ranging from coughing to pulmonary edema.

THIOPENTAL

= *Dipentol* = *Hypostan* = *Intraval* = *Nesdonal* = *Penthiobarbital* = *Pentothal* = *Thiomembumal* = *Thionembutal* = *Thiopentalum* = *Thiopentone* = *Trapanal*

***Anesthetic.* Primarily IV use and occasionally suppository or IM use. Occasionally IP in some laboratory animals. "Short-acting".**

Untoward effects: ***Cardiovascular***: Frequent premature ventricular contractions; arrhythmias associated with hypercarbia or anoxia, phlebitis, thrombosis, thrombophlebitis, hypotension, and bradycardia. Incomplete cardiac emptying can precipitate iatrogenic pulmonary edema and/or apnea. Overdose can cause circulatory failure. Phlebitis and other adverse effects decreased by using a 2.5% or less, rather than a 5% solution. Extravasation or IM has caused pain and tissue necrosis. An immune hemolytic syndrome has been reported.

Fetus and neonate: Placental transfer occurs. **Fetal** distress and overdosage, weak contractions, protracted labor, and detected within 45 s in cord blood after IV use in the **mother**. Respiratory decrease noted in **newborns**. Excreted into breast milk (0.75–2 mg/100 ml).

Hypersensitivity: Anaphylaxis, cyanosis, tachycardia, unconsciousness, hypotension, coughing, sneezing, laryngeal spasm, hiccup, bronchospasm, severe generalized erythema, urticaria, and severe bloody diarrhea. Later, persistent drowsiness, confusion, amnesia, and, infrequently, headache and postoperative nausea and vomiting.

Liver: Hepatotoxic after large dosage (900–1500 mg over 60–110 min), but, apparently, not clinically important. Exacerbation or precipitation of liver encephalopathy.

Miscellaneous: IA (~1/3500) can cause gangrene, due to small blood vessels being blocked by precipitate from a reaction with blood. Has precipitated fulminating nervous disturbances leading to complete paralysis, and death in porphyria variegata, but cutaneous porphyrias are hardly influenced. Involuntary muscle movements, tremors, Stevens–Johnson syndrome, and 5% decrease of cognitive function. Half-life ~16 h. **Patients** suffering from either porphyria or dystrophia myotonia can develop severe respiratory depression or exacerbation of **their** disease on ordinary dosage. Activity is increased in **patients** with liver disease. Increases intracranial pressure.

Poisoning and suicide: Death by suicide in a 43-year-old **female** by rectal use of ~6.25 g or 3.6 g/75 lb (therapeutic rectal dose 1 g/75 lb). IV suicides also reported. Vomiting, aspiration, coma, and death in a 30-year-old **female** after 300 mg IV for D&C. Therapeutic levels are usually 1–5 ml/l, although one death reported at 3 mg/l. Levels > 10 mg/l are usually fatal.

TD_{LO}, **woman**: 60 mg/kg; **human**: 430 μg–6.4 mg/kg; LD_{LO}, **human**: 5 mg/kg.

Drug interactions: ***Reserpine*** with it causes hypotension and bradycardia. Use of 5% ***dextrose*** IV with it decreases its anticonvulsant effect. **Patients** pretreated with ***β-adrenergic-blockers*** have thiopental-induced bronchospasm.

Emergence excitement can occur, especially in **horses**. Although concentrations > 2.5% (4–10%) are often used in large **animals**, this increases the probability of serious venospasms, phlebitis, and/or arteritis and thrombosis, especially if accidentally given intra-arterially or due to extravasation, which can also lead to serious tissue sloughing. In **dogs**, anesthetic effects are reversed by 1 mg ***dextroamphetamine*** and 0.125 mg ***yohimbine***; also ***nikethimide***. ***Chloramphenicol*** 50 mg/kg; IV, 15 min before IV thiopental increased greatly the duration of anesthesia. ***Atropine*** 0.5 mg/kg causes dramatic (two times) prolongation of sleeping time in the **dog**, or decreases the anesthetic requirement by ~25%. Diethylstilbestrol injection 30 min before anesthesia enhances its duration in male **dogs**. ***Atropine sulfate***, but not the *methylnitrate*, decreases the anesthetic dosage in **dogs**. ***Propranolol*** decreases the anesthetic dosage and sleeping time in **dogs**. Anesthetic duration in **dogs** may be dangerously prolonged by ***epinephrine***, ***glucose***, or ***sodium lactate***, only if these are given in high concentration when the **dogs** are awakening. Causes bronchoconstriction in **dogs**, **rats**, and **man**. Avoid its use in **greyhounds** and **whippets**, where it is initially distributed rapidly into body fat and **animals** may sleep for 24 h. Hypotension and tachycardia reported in **cats**. ***Epinephrine*** or ***norepinephrine*** causes ventricular arrhythmias in **dogs**. In **rabbits**, raised blood ***glucose*** for up to 3 h after anesthetic induction. Tachycardia and leukopenia in anesthetized **dogs**, and tachycardia, increased blood ***glucose***, and leukopenia in **horses**.

LD_{50}, **mouse** and **rabbit**: 600 mg/kg; IV, LD_{50}, **mouse**: 70 mg/kg; IP, LD_{50}, **guinea pig**: 50–55 mg/kg; **rats**: 115–130 mg/kg; IP, minimum lethal dose, **guinea pig**: 40 mg/kg.

THIOPERAZINE

= *Majeptil* = *SKF 5,883* = *Thioproperazine* = *Vontil*

Psychotherapeutic, neuroleptic, antiemetic, tranquilizer, anticholinesterase.* A *propylpiperazine.

Untoward effects: Can stimulate hebephrenia, extrapyramidal symptoms after low dosage (5–25%), neuroleptic malignant syndrome, lactation, and cataracts.

Cataleptic state in **rats** for ~7 h after 16 mg/kg.

LD_{50}, **rat**: 750 mg/kg; **mouse**: 440 mg/kg.

THIOPROPAZATE

= *Dartal* = *Dartalan*

Psychotherapeutic, antiemetic, tranquilizer.* A *halogenated propylpiperazine.

Untoward effects: Has induced cutaneous reactions and shows cross-sensitivity with other ***phenothiazines***. Has caused cataract development. In a suicide attempt, a **man** ingested 15–10 mg tablets leading to excitement and irritability. Pseudoparkinsonism at doses > 40 mg/day.

LD_{LO}, **human**: 50 mg/kg.

Depression and tremors in **dogs** on 15–25 mg/kg/day/15 days; 5/6 **dogs** given 10 mg/kg/day/3 weeks showed parkinsonian tremors. Growth retardation in **rats** fed 10 mg/kg/day/9 months.

LD_{50}, **mouse**: 279 mg/kg.

THIORIDAZINE

= *Aldazine* = *Mallorol* = *Mellaril* = *Melleretten* = *Melleril* = *Novoridazine* = *Orsanil* = *Ridazin* = *Stalleril* = *TP-21*

Psychotherapeutic, tranquilizer, adrenergic-blocker.* A *piperidine.

Untoward effects: Anticholinergic effects of xerostomia, mydriasis, cycloplegia, urinary retention, tachycardia, ventricular fibrillation, premature ventricular contractions, memory failure, and delirium; drowsiness, increased weight, ejaculation problems (65% of **males**), priapism, decreased libido, anorgasmia, urinary incontinence, and postural hypotension. Occasional photosensitivity, even after 400 mg/day/several months; cutaneous rashes; some patients develop a metallic blue-gray discoloration of skin and conjunctiva after exposure to sunlight. Initially, the skin appears inflamed, tan-colored; extrapyramidal symptoms, tardive dyskinesia, neuroleptic malignant syndrome, menstrual changes, and galactorrhea. Although rare, reversible pigmentary chorioretinopathy has occurred within 1–2 months after 1200 mg or more/day, and has happened after only 200 mg/day/several months. Large doses accumulate in the uveal tract and have caused blindness, night blindness, and decreased central vision. Cataracts, annula scotoma, acute agitation with delusions, and visual and auditory hallucinations have also been induced. Also rare are blood dyscrasias, agranulocytosis, leukopenia, pancytopenia, thrombocytopenia, and increased platelet aggregation; dystonia, cholestatic hepatitis and jaundice, rash, syndrome of inappropriate antidiuretic hormone secretion, convulsions, torsades de pointes, prolonged QTc interval; flattening or broadening and rounding of T-waves; atrioventricular block, premature ventricular contractions, ventricular arrhythmia (often fatal), and disturbed temperature regulation. Painless bilateral parotid enlargement in a 21-year-old **female**. Pigmentation on face, mouth mucosa, breasts, and buttocks of a 40-year-old **female** ingesting 300 mg/day/2½ years. Withdrawal of high-dose therapy has caused gastritis, nausea, vomiting, dizziness, and tremors. Congenital cardiovascular malformations (transposition of great vessels and ventricular septal defect) in an **infant** after the **mother** received 25 mg/day/3 months before conception and during the first 5½ months of **her** pregnancy. **Fetal** deaths also reported after **maternal** ingestion of 100–300 mg/day/18 months. Has caused extrapyramidal syndrome in some **neonates** after **maternal** use. Holoprosencephaly in a **child** whose **father** took it and hemolytic anemia in a **neonate** after prior **maternal** use. ***Prolactin*** secretion increase and galactorrhea reported in both **men** and **women**. Only a few successful suicides with it. Deliberate ingestion of 5 g was fatal. Rhabdomyolysis due to self-poisoning in a 22-year-old **male** ingesting 9.4 g; survived after treatment. Therapeutic serum concentration reported as 0.25–1.5 mg/l; toxic at 2–5 mg/l; and fatal at 5–10 mg/l.

Drug interactions: Interacts with ***desipramine*** leading to hypertension; with ***amitriptyline*** leads to sedation, lethargy, and rigidity due to enzyme inhibition. Use with ***tricyclic antidepressants*** and/or an ***antiparkinson drug*** can cause fatal adynamic ileus. Low doses caused interference with ***imipramine*** in a 12-year-old **male**. ***Phenylpropanolamine*** administered as a nasal decongestant in ***Contac C*** has caused ventricular arrhythmias and death in a 17-year-old **female**. Can exacerbate anticholinergic activity. ***Pindolol*** and ***propranolol*** increase its serum levels. ***Antacids*** decrease its absorption; ***hydroxyzine*** decreases its effectiveness; and it decreases antiparkinson effect of ***L-dopa***. Toxic neurological syndrome in a 45-year-old **female** stabilized on ***lithium*** when thioridazine was added. Use may induce ***phenytoin*** intoxication. Preoperative ***meperidine*** in a 12-year-old **male** maintained on thioridazine led to respiratory failure and cyanosis. Use with ***hydrochlorothiazide*** caused compulsive water-drinking (~96 glassfulls/day), cardiac failure, and acute water intoxication leading to eventual death in a 37-year-old **male**. Causes significant increase in plasma concentration of ***trazodone*** and its metabolites. Can interfere with ***bromocriptine's*** and ***monoamine oxidase inhibitor's*** effectiveness.

Also see ***Mesoridazine***, its active metabolite.

TD_{LO}, **children** and **adults**: ~24 mg/kg; LD_{LO}, **human**: 43 mg/kg.

A hypersensitivity reaction manifested by local ulceration and photosensitivity has occurred after SC use in **rats**. May cause accumulation of some granules in corneal stroma and aberrant motor behavior in treated **dogs**. IV overdoses in **rats** caused cardiorespiratory suppression and hemolysis.

LD_{50}, **rat**: 995 mg/kg.

THIOSEMICARBAZIDE

Reagent, rodenticide.

Untoward effects: Severe seizures in **dogs** induced by 20 mg/kg; in **chickens** injected with 9.2 mg/kg; and in **mice** with 8.8 mg/kg IP.

LD_{50}, **rat**: 9.2 mg/kg; **cat**: 20 mg/kg; **dog**: 10 mg/kg.

THIOTEPA

= NSC 6,396 = Tespamin = Thioplex = TIOFOSFAMID = Tiofosyl = TPPA = Triethylenephosphoramide = TSPA

Antineoplastic. **Usually IV, intravesicular, intracavitory, and topical use.**

Untoward effects: ***Blood and cardiovascular***: Severe bone marrow suppression, thrombocytopenia, anemia, decreased plasma pseudocholinesterase, pancytopenia, and slight leukopenia. Fatal acute myeloblastic leukemia and deaths reported from adverse effects on bone marrow. Some of these reactions may be delayed for at least 1 month. Thrombophlebitis after accidental extravasation.

Genitourinary: Dysuria, urinary retention, hemorrhagic cystitis, menstrual dysfunction (including amenorrhea), and decreased spermatogenesis.

Miscellaneous: Stomatitis, metallic taste in mouth, nausea, vomiting, anorexia, abdominal pain, laryngeal edema, wheezing, asthma, anaphylactic shock, dizziness, headache, blurred vision, conjunctivitis, corneal opacity, fatigue, weakness, fever, local pain at injection site, transient paresthesia after intralumbar use, myelopathy and decreased reflexes in a few **patients** after intrathecal use.

Skin: Poliosis and periorbital skin and eyelash depigmentation after topical use. Contact dermatitis, delayed wound healing, alopecia, rash, urticaria, and pain at injection site.

Drug interactions: May cause or potentiate neuromuscular blockade caused by ***pancuronium***, ***succinylcholine***, and other ***muscle relaxants***.

Use in **pregnant rats** at ~3 mg/kg/IP led to anomalies of extremities in their **pups**. The same dosage was lethal to **rabbit** fetuses. Chemosterilant of male **blackbirds** and **starlings**.

LD_{50}, male **quail**: 237 mg/kg; IV, LD_{50}, **mouse**: 1 µg/g; **rat**: 15 mg/kg; **dog**: 1–5 mg/kg.

THIOTHIXENE

= Navane = Orbinamon = P 4,657B = Tiotixene

Psychotherapeutic, neuroleptic.

Untoward effects: ***Blood and cardiovascular***: Decreased prothrombin time, leukopenia, leukocytosis, tachycardia, postural hypotension, and syncope. Occasionally anemia and neutropenia.

Central nervous system: Restlessness, agitation, and mild drowsiness. Extrapyramidal symptoms in 5–15%, including pseudo-parkinsonism, akathesia, and dystonia; sedation, tardive dyskinesia with torticollis, akathesia, muscle rigidity, and rhythmic involuntary movements of tongue, face, mouth, and jaw has occurred and may not be reversible if advanced. Neuroleptic malignant syndrome with hyperthermia, muscle rigidity, autonomic dysfunction, altered consciousness, hypertension, and coma. Convulsions, ataxia, and delirium at high dosage.

Endocrine: Occasionally gynecomastia, lactation, and amenorrhea. In **males**, sexual dysfunction, spontaneous ejaculations, impotence, and priapism.

Hypersensitivity: Rash, pruritus, and urticaria. Rarely, anaphylaxis.

Liver: Occasionally transient increase in serum transaminases and alkaline phosphatase.

Miscellaneous: Photosensitivity reactions; occasionally lens-corneal opacity; blurred vision, oculogyria, mydriasis, xerostomia, nasal congestion, sialism, constipation, hyperhidrosis, hyponatremia, incontinence, rash, hives, pruritus, nausea, vomiting, weakness, and fatigue. IM in a catatonic schizophrenic 34-year-old **male** led to fever, tachycardia, and hypertension.

Drug interactions: ***Guanethidine's*** antihypertensive effects are decreased by it. **Rat** trials indicate that ***cyclosporine*** inhibits its metabolism by the liver.

IP, LD_{50}, **rat**: 55 mg/kg; **mouse**: 100 mg/kg.

THIOURACIL

= *Deracil* = *Tiouracilo*

Thyroid depressant.

Untoward effects: Occur in > 10% of **patients**. Decreased ***nucleic acid*** metabolism, agranulocytosis, leukopenia, thrombocytopenic purpura, pancytopenia; aplastic and hemolytic anemia (especially in **neonates** with glucose-6-phosphate dehydrogenase deficiency); Stevens–Johnson syndrome, acneiform eruptions, urticaria, lupus erythematosus-like lesions, hepatocanalicular jaundice, anaphylactoid or asthmatic and serum-sickness-like reaction, vasculitis, polyarteritis, alopecia, decreased ***protein-bound iodine*** serum values, chills, fever, nausea, abdominal cramps, dizziness, headache, joint pains, parotitis, diarrhea, thyroiditis, and edema. Deaths have occurred from its use.

Placental transfer occurs. The fetal thyroid develops early in **humans** and **maternal** use during early gestation can be especially undesirable (confirmed in **rat** studies). Milk transfer occurs with at least three times higher concentration in breast milk (9–12 mg/100 ml), than in **maternal** blood (3.4 mg/100 ml), and can cause hypothyroidism, goiter, agranulocytosis, and mental retardation in the nursing **infant**.

Drug interactions: Has potentiated ***oral anticoagulant*** effects.

Atherosclerosis in **dogs** can be induced by feeding it for at least 1 year. In **animals**, overdosage will decrease growth and induce severe hypothyroid states, enlarged thyroid, and agranulocytosis. In lactating **animals**, milk levels can be many times higher than blood levels. Use has caused testicular atrophy in **rats**, and higher than normal levels have caused hepatomas and thyroid carcinomas in certain strains of **mice**. Potentiates teratogenicity of hypervitaminosis A in **mice**. Perosis associated with its feeding to **turkeys**. Parakeratosis when 1 g/day fed to **piglets**. Causes neutropenia in **rhesus monkeys**.

LD_{50}, **rat**: 1 g/kg; IP, LD_{100}, **rat**: 1.5 g/kg; LD_{LO}, **rabbit**: 3.7 g/kg.

THIOUREA

= *Thiocarbamide*

Use: ***Rubber*** vulcanizing accelerator (especially for wetsuits), in manufacturing resins, in photography to remove stains from negatives, as a metal chelator, boiler scale remover, corrosion inhibitor, in hair dyes, as an antioxidant to prevent yellowing in ***paper***, in ***silver*** tarnish removers, and antithyroid drug. Millions of lbs used annually in the U.S.

Untoward effects: Acneiform eruptions, rash, drug fever, eosinophilia, neutropenia, agranulocytosis, cholestatic jaundice, photosensitive dermatitis, swelling of hair, nausea, headache, arthralgia, sore throat, periarteritis nodosa, and allergic vasculitis.

TD_{LO}, **woman**: 1.7 g/kg during 5 weeks.

Causes thyroid enlargement in **rats**. Induces thyroid adenomas and carcinomas in both male and female **rats**, squamous cell carcinomas of the Zymbal gland in male and female **rats**. IP or in drinking water causes mixed sarcomas of Zymbal gland in **rats**. Gross reproductive lesions in young **goats**, 3–4 months old, given 100–250 mg/day/90 days. Many **kids** died after 3–6 weeks of treatment; **females** had poorly developed ovarian germinal layers and uterine edema; **males** showed no spermatogenesis and small seminiferous tubules. Degenerative changes were most pronounced at the lower dosage, emphasizing the effect of subclinical hypothyroidism on sterility problems.

LD_{50}, **rat**: 125 mg/kg; **mouse**: 8.5 mg/kg; LD_{LO}, **rabbit**: > 7 g/kg.

THIRAM

= *Arasan* = *Delsan* = *ENT-987* = *Fermide* = *Fernasan* = *Mecuram* = *Nomersam* = *Nomersan* = *Pomarsol* = *Puralin* = *Resifilm* = *Rezifilm* = *SQ-1489* = *Tersan* = *Tetrapon* = *Thimer* = *Thiosan* = *Thiurad* = *Thiuram* = *Thiuramyl* =

Thiotex = Thirame = Thirasan = Thylate = Tirampa = TMTD = Trametan = Tuads = Tulisan = Vanicide TM-95

Fungicide, bacteriostat, rubber vulcanizing accelerator.

Use: **Deer**, **rabbit**, **bird**, and **rodent** repellent, as a fungicide and bacteriostat in dipping **sheep**, intrauterine in **cattle**, **sheep**, and **swine**; as a seed, crop, and turf fungicide; and in vulcanizing ***rubber***.

Untoward effects: Contact allergen (nose, throat) and skin sensitizer in **man** (***rubber*** **workers**, **golfers**; **users** of gum erasers, bathing capsules, garters, mammary prostheses, condoms, ***gloves***, surgical ***gloves***, goggles, stethoscope eartips, eyelash curlers, bunion pads, and dentures). Will irritate eyes, skin, and mucous membranes. Bullous and eczematous eruptions in some ***rubber*** industry **workers** first reported in 1920. A case of Henoch–Schönlein purpura in a Mexican agricultural **worker** handling and planting tree seedlings treated with a 42% solution. The dermatitis can flare up when ***disulfiram*** (q.v.) is used. Ingestion causes nausea, vomiting, diarrhea, anorexia, ataxia, hyperexcitability, hypothermia, flaccid paralysis, and respiratory failure. Toxicity is increased by liposolvent vehicles. Small amounts cause marked intolerance to ***alcohol*** with an adverse ***disulfiram*** (q.v.)-like reaction. Limited exposure can still trigger itching, redness, hives, and fixed-drug reactions. ***Rubber*** **workers** exposed to it develop flushing of face and wrists and hypotension when drinking any form of ***alcohol***. Adverse effects have also been reported from its leeching out of ***rubber*** stoppers, especially into ***procaine*** solution. Avoid inhalation or ingestion. National Institute for Occupational Safety & Health and OSHA upper exposure limits of 5 mg/m^3; immediate danger to life or health 100 mg/m^3.

LD_{LO}, **human**: 50 mg/kg; inhalation TC_{LO}, 30 μg/m^3/day/5 years. Toxic, TWA, 5 mg/m^3.

Although it is illegal to treat grains for food use with the chemical, such grains occasionally find their way into the food supply. Treated ***corn*** contains ~630 ppm. In some **poultry** areas when such toxic feeds were fed, visible symptoms of toxicity ranged from 10 to 15% in growing **poults**; 25–35% in **broilers**; and 85–100% in **hens**. Decreased growth rate was also observed, leading to extreme lack of uniformity of growth in affected flocks. Feeding 37.5 ppm level to **chicks** caused mild growth depression but no leg or hock disorders in one trial. In another trial, 40 ppm led to some leg disorders in **chicks**. A 150 ppm level caused severe growth depression and leg disorders in **chicks** and **goslings**. **Turkeys** tolerated a 200 ppm level without any evidence of toxicity, even when fed continuously from 1 day of age. Strain differences have been noted. A level of 10–20 ppm in the feed stimulated the growth rate of broad-breasted bronze **poults**, but had no noticeable effect on broad-breasted white **poults**. **Poults** can develop a tolerance to fairly high levels (300–400 ppm) if continuously fed from 1 day of age, but intermittent feeding, as might occur under field conditions, can cause toxicity. The distinctive leg deformity of **chicks** led to a syndrome of sitting on their enlarged hocks; crooked and curled toes; with some slipped tendons and spraddle-legs. Forty to 100 ppm drastically decreased egg production in **hens** and produced a high incidence of soft-shelled eggs within 48 h of exposure. Some may die from egg peritonitis. Growth of hatched **chicks** was retarded. In LD_{50} trials, diarrhea and photophobia was noted in poisoned **poultry**. Liver ***vitamin A*** levels were subnormal in affected **hens**. Feeding trials in **cattle** and **lambs** indicated that 78% of the chemical was degraded within 6 h to ***carbon disulfide*** (q.v.), plus possibly some ***hydrogen sulfide*** and ***dimethylamine***; 4% and 1.5% unchanged chemical appeared respectively in their feces and urine. Poisoned **cattle** developed ataxia, salivation, spasms, lacrimation, and somnolence. Cardiac irregularities may occur in **cattle** even 1 month after apparent recovery. After the 45th day, decreased appetite and growth rate has been noted in fattening **swine** given 1 g/day. One month later, necropsy revealed kidney and liver lesions. **Pigs** fed treated ***corn*** developed a "goose-stepping" gait typical of ***vitamin*** B_5 deficiency and **cattle** eating treated ***corn*** developed typical ***vitamin A*** deficiency symptoms. This prompted trials with vitamins in poisoned **poultry** with beneficial results. At 250 mg/kg it was teratogenic to **hamsters** and **mice**. One g/day/120 days in the food of eight pregnant **ewes** was associated with abortion in four of them between the 116th and 146th day of pregnancy. The others had weak **lambs**. Symptoms in **sheep** are anorexia, dyspnea, weight loss, and lethargy. Thyroid enlargement and pathological changes were noted in the abomasum, liver, and kidney of the **ewes**. Deaths began occurring in pregnant **rats** after the 120th day. Proliferative lesions were noted in the stomach. Classified as slight–moderately toxic to **rats**. Injection of doses as low as 10 μg/**chick** embryo led to paralysis, shortened extremities, muscular atrophy, dwarfing, and death, with microscopic evidence of peripheral neuropathy. Commercial pesticides are often colored blue.

Oral, LD_{50}, **rats**: 375–875 mg/kg; **mice**: 1.35 g/kg; **rabbit**: 210 mg/kg; **guinea pigs**: 1.53 g/kg; **poultry**: **cocks**: 1 g/kg; **chicks**: 800 mg/kg. Oral, LD_{100}, **pigs**: ~1 g/kg; LC_{100}, **fish** (**guppies**): 1 mg/liter of water. Oral LD_{LO}, **cats**: 230 mg/kg; IP, LD_{LO}, **mice**: 200 mg/kg. LD, **sheep**: 225 mg/kg.

THLASPI ARVENSE

= *Fanweed = Field Pennycress = Mithridate Mustard = Stinkweed = Tabouret des Champs*

Found in the U.S., Canada, and England. A member of the *Brassiacea* or *Crucifera* family.

Untoward effects*:** Due to its "mustard oil" or ***allyl isothiocyanate (q.v.) and other ***isothiocyanates***. Photosensitization reactions on white areas of **cattle**. The seeds are especially toxic. Hematuria reported in **cattle** after ingestion. Fatal in **cattle** at ~65 mg/kg. Gastroenteritis, salivation, diarrhea, abortions, and goiter can also occur after ingestion by **cattle**, **horses**, and **pigs**.

THONZYLAMINE HYDROCHLORIDE

= *Anahist* = *Neohetramine* = *Tonamil*

Antihistamine.

Untoward effects*:** Has caused fixed-drug eruptions. Immunochemically related to ***ethylenediamine (q.v.). Accidental ingestion of an unknown quantity by a 10 kg 18-month-old **male** led to grand mal seizures. A case of aplastic anemia has been reported.

LD_{50}, **mouse**: 245 mg Base/kg; **guinea pig**: 493 mg Base/kg.

THORIUM OXIDE

= *Thoria* = *Thorium Dioxide*

Thorotrast = *Tordiol* = *Umbrathor are 20–25% suspensions.*

Radiopaque diagnostic aid.

Untoward effects*:** Long latency period of 20 years or more of induced hemangioendothelioma of the liver; average of 36 years for angiosarcomas of the liver. Induced carcinomas of the bladder and leukemia, as well as other neoplasms in **humans**, **rabbits**, **hamsters**, **rats**, **mice**, and **guinea pigs**, in various organs, locations, and bone. **Human** exposure also occurs from inhalation, ingestion, and dermal contact, as it is found as a by-product of various mining operations, ***coal-burning, and in the manufacture of ceramics, incandescent lamps, metal refining, and vacuum tubes, and by nuclear reactor **workers**. Paranoid psychotic reaction in a **patient** after 11 injections. Has caused chronic osteitis and osteosarcomas.

Progressive myelopathy 3½–7 years after use in myelograms in **dogs**, where it slowly becomes permanently associated with the meninges. Use dramatically increased sensitivity in **monkeys** to staphylococcal ***enterotoxin*** and **rabbits** and **mice** to *Brucella* ***endotoxin***.

THRELKELDIA PROCIFLORA

Untoward effects*:** The plants concentrate ***nitrates (q.v.) and ***oxalates*** (q.v.) and have poisoned **sheep** in Australia.

THUJA OCCIDENTALIS

= *American Arborvitae* = *Arborvitae* = *Tree of Life* = *White Cedar* = *Yellow Cedar*

U.S. annual consumption varies up to 10,000 lbs/year, mostly as a fragrance in soaps, detergents, cosmetic creams and lotions, and foods at a 0.1–0.4% level.

Untoward effects*:** Its essential oil, called ***Cedar Leaf Oil or ***Thuja Oil***, is an emmenagogue and has caused poisoning of **humans** with tonic-clonic or solely clonic convulsions and photosensitivity. Its insidious cumulative effect noted in a case report of a 50-year-old **female** with acne refractory to usual treatment, **who** went to a **"naturologist"** who advised **her** to take 40 drops of a 1 : 100 dilution of a ***thuja oil***/day. **She** forgot to have it diluted and took 20 drops twice daily without any ill-effects until 30 min after the tenth dose and, while standing, had a tonic seizure leading to an occipitoparietal fracture. Contains ***thujone*** (q.v.). No sensitization reported in 25 **volunteers** using 4% in petrolatum.

Moderately irritating when applied full-strength to intact or abraded **rabbit** skin under occlusion for 24 h.

LD_{50}, **rat**: 830 mg/kg; acute dermal, **rabbit**: 4.1 g/kg.

THUJONE

Found in many essential oils, *Absinthium* (q.v.), *Artemesia sp.* (q.v.), and *Thuja* (q.v.). Also in *Cannabis*, *Sage*, and *Tansy* (q.v.).

Untoward effects*:** Psychoactivity causing central nervous system damage, mental deterioration, mania, hallucinations, convulsions, etc. described in detail in the ***Absinthium, ***Artemesia***, and ***Tansy*** discussions.

SC, LD_{50}, **mouse**: α, 87.5 mg/kg; β, 442.2 mg/kg; IP, LD_{LO}, **rat**: 120 mg/kg.

THURFYL NICOTINATE

= *Nicotofuryl* = *Trafuril*

Untoward effects*:** Its use as a ***rubefacient has caused contact dermatitis with the release of ***histamine***, leukocytoclastic vascular damage, inflammation, local pruritus, and burning with dermatographism.

THYLACHIUM AFRICANUM

= *Mdudu* (Swahili) = *Mtongi* (Boni)

A shrub or tree 20 ft high, often found growing on termite hills in Nigeria.

***Untoward effects*:** Although its tuberous roots are an important source of **human** food in times of famine, ingestion has been fatal if improperly cooked. Adequate boiling destroys its toxic principle.

THYMIDINE

***Untoward effects*:** Mild symptoms after IV infusions leading to myelosuppression, nausea, vomiting, headache, drowsiness, and occasional phlebitis at injection site, decrease in serum concentration, and increase in leukocytes and platelets.

THYMINE

= *5-Methyluracil*

***Untoward effects*:** Injections in **rats** caused neuromuscular block similar to that of **human** myasthenia gravis. IP use in **guinea pigs** led to myositis and neuromuscular block.

THYMOL

= *Thyme Camphor*

Untoward effects*:** False positive increase in values of the turbidity test will occur with heavy fat intake or administration of ***erythromycin, ***nalidixic acid***, or ***triacetyloleandomycin***. Gastric and renal irritant and may cause central nervous system depression in high dosage. High oral intake can cause liver damage. Ingestion causes spuriously high test readings of urine for bile and ***albumin***; spuriously low test readings for blood urea nitrogen. Thymol poisoning symptoms are nausea, vomiting, headache, ringing in the ears, dizziness, bradypnea, muscle weakness, thready pulse, and decreased body temperature.

THYMOXAMINE

= *Arlitene* = *Moxisylyte* = *Moxyl* = *Opilon* = *Uroalpha* = *Vasoklin*

***α-Adrenergic-blocker*. Oral, IV, and topical.**

***Untoward effects*:** Facial flushing, vertigo, headache, palpitations, shivering, nausea, diarrhea, tingling of extremities, and miosis reported. Postural hypotension and palpitations may occur after IV use.

LD_{50}, **mouse**: 208 mg/kg.

THYROID

= *Getrocknete Schilddruse* = *Proloid* = *Thyradin* = *Thyrocrine* = *Thyroglobulin(e)* = *Thyroidine* = *Tiroide* = *Tiroidina*

***Untoward effects*:** In adjusting metabolic rate too rapidly, anginal pain, palpitations, skeletal muscle cramps and weakness, tachycardia, auricular fibrillation, arrhythmias, restlessness, insomnia, induced or accentuated tremors, headache, facial flushing, sweating, dyspnea, nocturia, anxiety, heat intolerance, fever, gastrointestinal hypermotility and diarrhea, vomiting, and weight loss. Collapse and coma have been reported. Use as an agent for weight loss has caused occasional periodic paralysis. Additionally:

***Blood and cardiovascular*:** Tendency to hypocoagulability. Myocardial infarction with severe cardiac ischemia and shock in hypothyroid 68-year-old **female** after 260 mg/day. High dosage can precipitate angina and myocardial infarction in **patients** with ischemic cardiac problems.

***Bone*:** Dose-related decrease in bone mineral density in **women** 50–98 years old on long-term therapy. Stimulates epiphyseal maturation in **children**.

Diagnostic tests*:** Will decrease ***cholesterol test levels and excess can increase alkaline phosphate test results.

***Fetus*:** Although it does not cross the placenta in early pregnancy, it does minimally in late pregnancy. Yet, it has been suspect in causing congenital defects, including cardiac effects.

***Miscellaneous*:** Inflammatory acneiform eruption, urticaria, skin pigmentation, and exudative diathesis. Can induce neuromuscular blockade and aggravate or unmask myasthenia gravis. Mania in the **elderly** from high dosage. Thyrotoxicosis in a 44-year-old **male** was associated with abdominal pain, constant spitting, and difficulty in managing **his** oral secretions. Xerostomia has been reported in others. In 1985, an outbreak of fatigue, tachycardia, anxiety, and weight loss was reported in 140 **residents** of Minnesota, South Dakota, and Iowa after eating "gullet trimmed" meat that contained thyroid glands. The United States Department of Agriculture now bans it in the U.S. Thyrotoxic myopathy with decreased hand strength. Significant amounts are excreted in breast milk and can cause severe hypothyroidism in an athyrotic **infant**. Splenomegaly and gymecomastia also reported.

***Neoplasms*:** Osteosarcoma and two times increase in breast cancer incidence.

***Poisoning*:** Agitation, sinus tachycardia, occasional extrasystoles, and slight fever in a 2-year-old **female** and a 3-year-old **female** after ingestion of a total of 1.9 g. A 51-year-old **female** ingested 48 g in 3–4 days leading to thyroid storm syndrome with hyperpyrexia, semicoma, tachypnea, and rapid atrial fibrillation.

Drug interactions*:** Can potentiate ***desipramine, other ***tricyclic antidepressants***, ***catecholamines***, and ***coumarin anticoagulants*** including ***warfarin***. ***Cholestyramine*** impairs its absorption and it decreases ***digoxin*** serum levels. ***Soybean*** derivatives, such as ***Mull-Soy***, have increased its fecal excretion and decreased its effectiveness. Extreme hypertension and supraventricular tachycardia reported when used with IV ***ketamine***.

May deteriorate with age. Avoid excessive doses in cases with weak hearts; decreased cardiac and respiratory rates, and weight loss may occur with overdosage in **dogs**.

Contraindicated in adrenal insufficiency unless ***cortico-steroid*** therapy is initiated first. Thyroid preparations enhance ***catecholamine's*** (***epinephrine***, etc.) adverse cardiovascular effects. Hormone content of generic preparations may vary widely in their T_3 and T_4 content, therefore use only the same brand.

THYROTROPIN

= *Actyron* = *Amibinon* = *Dermathycin* = *Thyratrop* = *Thyrogen* = *Thyroid Stimulating Hormone* = *Thyroliberin* = *Thyrothrophin* = *Thyroton* = *Thyrotropic Hormone* = *Thytropar* = *TSH*

***Hormone.* SC or IM. Reconstituted solutions are stable under refrigeration for ~2 weeks.**

***Untoward effects*:** Occasional nausea, vomiting, urticaria, transitory hypotension, headache, cardiac arrhythmias, and, after repeated uses, allergic reactions.

THYROTROPIN RELEASING HORMONE

= *Antepan* = *Lopremone* = *Protirelin* = *Relfact TRH* = *Stimu-TSH* = *Thypinone* = *Thyrefact* = *Thyroliberin* = *Thyrotropin Releasing Factor* = *TRF* = *TRH*

Untoward effects*:** IV use led to vague sensations of abdominal discomfort, occasional nausea and vomiting, desire to urinate; pleasurable genitourinary sensations lasting 1–3 min in 16, 18–64-year-old women, facial flushing, dizziness, chest tightness, and altered taste. Rarely, hypoglycemia. Increased serum ***cholesterol levels in ~50% of **patients**, mild euphoria, hyperprolactinemia, and relaxation in some. Tachycardia, increased blood pressure, tremors, transient amaurosis, and headache reported. Dyspnea and audible wheezing in **patient** with bronchial asthma. Several case reports of severe hypotension, unconsciousness, and convulsions. Transient nausea after oral use.

Although it causes a transient increase of body temperature in **man**, it causes hypothermia in **cats**, **rats**, and non-hibernating **ground squirrels**.

THYROXINE

= T_4 = *Panthroid* = *Panthyroid* = *3,5,3′,5′-Tetraiodothyroinine* = *Tiroxina*

Usually refers to the biologically active thyroid-like *levo* form. Similar relative activity of the *dextro* form is estimated at 5–15% in some species; its chief indication appears to be in aiding the liver in decreasing circulating *cholesterol*. The *sodium levo* form is marketed as *Eltroxin* = *Euthyrox* = *Laevoxin* = *Letter* = *Levaxin* = *Levothroid* = *Levothyrox* = *Levothyroxine* = *Oroxine* = *Panthroid* = *Soloxine* = *Synthroid* = *Thyroxevan* = *Tirossina* = *Tiroxina*. Also see *Dextrothyroxine*.

***Untoward effects*:** Osteopenia and increased risk of osteoporosis in postmenopausal **women**. Restlessness, anxiety, irritability, sweating, headache, fever, tremors, irregular menses, nausea, sleeplessness, tachycardia, arrhythmias, dizziness, weepiness, hunger, thyroid enlargement, fever, diarrhea, palpitations, agitated depression, poor ability to concentrate, increased incidence of breast cancer; significant amounts secreted in breast milk that can cause severe hypothyroidism in athyrotic **infant**; subclinical liver pathology, ophthalmopathy, and pseudotumor cerebri. Limited crossing of the placenta can lead to kernicterus in **infants** and dystonic movements in a **cretin**. A 53-year-old **female** with Turner's syndrome developed partial complex status epilepticus after 0.15 mg/day/2 weeks. Acute Addison's disease after initiation of therapy in two hypothyroid **patients**. Fatal cardiac arrest in a 63-year-old **female** after 300 μg IV. Medication errors of 0.5 mg instead of 0.05 mg led to tachycardia, tremors, and chest pains. **Patients** with pre-existing psychopathology can show thyroxine-induced mania. Cardiac arrest has occurred in a 63-year-old **female** with myxedema coma after it was given IV. Fatal doses are usually preceeded by intractable vomiting, coma, and collapse and have occurred in **patients** who previously recovered from myocardial infarctions.

Poisoning reported after its use in weight reduction programs with schizophreniform psychosis, auditory hallucinations, and confusion. Thyrotoxicosis in a 50-year-old **female** after mistaking her **dog's** prescription for **her *enalapril***. Nervousness and dizziness followed after 0.6 mg qid. Severe anxiety with low toxicity in a 12-month-old **infant** after ingesting ~15, 0.8 mg tablets. Thyrotoxicosis in a 31-year-old **female** after a single ingestion of 100, 0.1 mg tablets. A 3-year-old **male** ingested 44, 300 μg tablets (13.2 g) leading to vomiting, diarrhea, mild hyperpyrexia, and tachycardia. Intentional ingestion of 900, 0.8 mg tablets (720 mg) of *L*-thyroxine led to severe thyrotoxicosis in a 34-year-old **male**.

Drug interactions*: *Cholestyramine decreases its availability for ~4–5 h. By anion competition leads to protein displacement of ***clofibrate***, decreasing the latter's metabolism and increasing its activity. ***Phenobarbital*** also decreases its rate of metabolism. Increases hypothrombinemic action of ***coumarins***, such as ***dicoumarol*** and ***warfarin***, and increases toxicity of ***heparin***. Heavy ***amphetamine*** abuse increases its serum concentration. ***Fenclofenac, lovastatin, rifampin***, and ***sertraline*** decrease its serum levels. ***Aluminum hydroxide, calcium carbonate***, and ***sucralate*** decrease its absorption. ***Catecholamines*** and ***ferrous sulfate*** decrease its efficacy. ***Chloroquine*** and oral ***antidiabetics*** increase its serum levels. ***Estrogens*** and ***oral contraceptives*** increase protein-bound thyroxine. It decreases the effectiveness of ***nalidixic acid*** and ***nitrofurantoin***. IV use with ***ketamine*** leads to severe hypertension and supraventricular tachycardia.

It decreased metabolism of ***hexobarbital*** and ***morphine*** in **rats** and decreased metabolism of ***progesterone*** and ***testosterone*** only in male **rats**. It increased susceptibility of **mice** to audiogenic seizures. It increases ***zoxazolamine*** metabolism (not by liver enzymes) and decreased induced sleeping time in **rats**. Maternal use has caused cataracts in **rat** pups. Severe tachycardia in **rats**. Use in **cows** can impart an oxidized flavor to their ***milk***. Fatal to a **dog** that ate 40 grains (2.6 g). Symptoms are tachycardia (250/min), respirations 120/min, temperature 106.4°F, mydriasis, circling, running into objects, and death within 12 h. Post-mortem showed congested and enlarged thyroid, congested kidneys, pale liver, and inflamed stomach.

TIAGABINE

= *A-70,569* = *Gabitril* = *NNC-05-0328* = *NO-328* = *NO-05-0328* = *TGB*

Anticonvulsant, GABA uptake inhibitor.

Untoward effects: Decreased by administration with food. Headache, dizziness, asthenia, nervousness, tremors, somnolence, and mood depression reported. Occasionally confusion, irrational thinking, and concentration difficulties.

TIAMULIN

= *Denagard* = *Dynalin* = *Dynamutilin* = *81,723 hfu* = *SQ 22,947* = *Tiamutin* = *Tiotilin*

Antibiotic, animal growth promoter.

Untoward effects: Its toxicosis in **swine**, **turkeys**, and **chickens** can be potentiated by ***ionophores, lasalocid, monensin, narasin***, and ***salinomycin***. May cause a dermatitis, vomiting, and salivation in sick **pigs**; central nervous system disturbances in **calves**. FDA requires a 3 day withdrawal before slaughtering **pigs** for **human** consumption, and 1 day in other species. May irritate skin and mucous membranes of **husbandrymen**.

TIAPAMIL

= *RO 11-1781*

***Calcium channel-blocker, antihypertensive.* A *papaverine* derivative.**

Untoward effects: Asthenia, vertigo, palpitations, headache, and dyspnea. Significant increase in ***digoxin*** plasma levels reported and it attenuated ***digoxin's*** vasoconstricting effect on the splanchnic vascular bed.

TIAPROFENIC ACID

= *FC-3001* = *RU-15,060* = *Suralgan* = *Surgam* = *Surgamic* = *Surgamyl*

***Anti-inflammatory, analgesic, antipyretic.* Oral, IV, and IM.**

Untoward effects: Some side-effects, usually mild and transient, in ~25% of **patients** including dizziness, nausea, vomiting, heartburn, and dermatitis. A 52-year-old **female** died with Stevens–Johnson syndrome after ingesting 300 mg twice daily/14 days in treating chronic back pain. Has increased systolic blood pressure. Gastrointestinal blood loss has been less with sustained-action capsules than with tablets. Urinary reactions have included chronic interstitial cystitis and ureteric obstruction. Liver dysfunction reported in a 56-year-old **female** given 300 mg twice daily. An 89-year-old **female** on 200 mg twice daily developed increased joint pain and reversible delirium. A 14-year-old **female**, overdosed with 2 g orally, had plasma levels of 55 μg/ml with no adverse clinical effects. Dramatic increase of serum ***lithium*** concentration from 0.36 mEq/l to 0.65 mEq/l after 3 days' oral treatment.

TICLOPIDINE HCL

= *Anagregal* = *Aplaquette* = *Caudaline* = *Panaldine* = *Ticlid* = *Ticlodix* = *Ticlodone* = *Ticlosin* = *Tiklid*

***Antithrombotic and antiplatelet actions.* Action is delayed; 1 to several days for onset and continuing 4–8 days after treatment withdrawal.**

Untoward effects: Abdominal cramps, delayed mild–severe chronic diarrhea and weight loss, rashes, slowly reversible hepatotoxicity with cholestatic jaundice and increased liver enzymes, delayed severe neutropenia, agranulocytosis and secondary *Pseudomonas* bacteremia, thrombotic thrombocytopenic purpura that has been fatal, aplastic anemia that has also been fatal, bronchiolitis obliterans organizing pneumonia with dyspnea, 8–10% increase in serum ***cholesterol***, and decreased renal function. It is a postoperative hemorrha gic risk factor. Vomiting, pruritus, flatulence, dizziness, anorexia, headache, epistaxis, tinnitus, and asthenia may occur in ~1% of **patients**.

Drug interactions: In some **individuals**, it altered R-***warfarin*** concentrations with little clinical significance, but ***aspirin*** potentiates its antiplatelet activity. It inhibits the metabolism of ***caffeine, clomipramine, phenytoin***, and ***theophylline***, increasing their serum concentration. ***Antacids*** can decrease its absorption by 18% and ***cimetidine***

decrease its clearance by 50%. It decreases the effectiveness of ***cyclosporine*** and decreases ***digoxin*** serum levels by ~15%. It increases ***antipyrine*** plasma half-life by ~30%. Taken with ***food*** decreases induced gastrointestinal upsets and increases its absorption by ~20%.

IV use in **dogs** increases coronary sinus outflow, cardiac output, carotid arterial flow, stroke volume, and heart rate, and decreases blood pressure. Has prolonged ***hexobarbital*** sleeping time in **mice**. Acute toxicity in **rats** and **mice** leading to gastrointestinal hemorrhage, convulsions, incoordination, ataxia, decreased temperature, and dyspnea.

LD, **rat**: 1.6 g/kg; **mouse**: 500 mg/kg.

TICRYNAFEN

= *ANP-3624* = *CE-3624* = *Diflurex* = *Selacryn* = *SKF-62,698* = *Thienylic Acid* = *Tienilic Acid*

Diuretic, hypouricemic, antihypertensive.

Untoward effects: Anorexia, nausea, vomiting, diarrhea, constipation, dizziness, headache, vertigo, chills, fever, somnolence, insomnia, hypotension, palpitations, gout, hyponatremia, hypokalemia, acute interstitial nephritis, and acute renal failure (due to tubular obstruction by ***urates***). U.S. pharmaceutical manufacturers were required to report unexpected reports of toxicity to the FDA within 15 days of receipt, but some reports of liver damage were delayed by ~3 months, despite their previously having similar reports from France and England. Numbers vary, but at least 510 **people** were adversely affected during its limited sales period and 34 died from **their** liver damage at the time the Federal Court sentenced guilty parties. It was voluntarily withdrawn from marketing.

TIEMONIUM IODIDE

= *Visceralgine*

***Anticholinergic, antispasmodic.* Contains 28.49% *iodine*. Used orally, IM, IV, or rectally.**

Untoward effects: Paralytic ileus in one of 32 clinical **patients**. Oral use in 30 **infants** led to meteorism in one and hypochromic anemia in three.

TILETAMINE

= *CI 634*

***Dissociative anesthetic.* Once used in a synergistic 1 : 1 mixture with *zolazepam*, a *diazepinone tranquilizer*. The mixture was marketed as *Telazol* = *CI 744*. Replaced by *ketamine*.**

Untoward effects: Respiratory decrease, transient apnea, sialism, tachycardia, central nervous system stimulation, convulsions, and clonic movements in **cats**, **dogs**, and other **animal** species.

TILIA sp.

= *Fleurs de Tilleul* = *Flor de Tilo* = *Hepamig* = *Lindenbluten* = *Lime Flowers*

Untoward effects*: *Linden or ***Basswood*** trees = ***T. americana*** and ***T. cordata***. Their flowers are used in making herbal ***teas*** with claims for its sedative, sudorific, spasmolytic, digestive, and circulatory benefits. Oils from the fruits of ***T. cordata***, ***argentea***, and ***platyphyllos*** given orally to **rats** at 1/10 their LD_{50}/day/40 days leading to 10% increase in weight and increased reticulo-endothelial system activity in their lives.

The saccharine exudation on the leaves of the ***European Linden***, ***Tilia europa***, has the same chemical composition of ***manna*** found on Mount Sinai.

Oil, LD_{50}, **rat**: ***T. cordata***, 6.45 g/kg; ***T. argentea***, 12.62 g/kg; ***T. platyphyllos***, 13.3 g/kg.

TILLITIA sp.

Untoward effects: Its spores are aeroallergens causing allergic rhinitis, bronchial asthma, and/or hypersensitivity pneumonitis in **man**.

Spores of this smut fungus, ***T. caries***, have been found in the alimentary tract, feces, urine, and brain of **dogs** with convulsions.

TILMICOSIN

= *EL-870* = *LY-177,370* = *Micotil* = *Pulmotil*

***Antibiotic.* Oral and SC.**

Untoward effects: Although used SC safely in **cows**, Elanco (a division of Eli Lilly), the manufacturer, in an April 1992 brochure, stated it was not to be used parenterally in **humans** where it might be fatal.

On December 30, 1996, Elanco revised their literature, stating that it induces negative inotropic cardiac effects and tachycardia.

On June 1, 1998, an 18-year-old **female** died after apparently ingesting a dose **she** might have thought appropriate, based on mg/kg for **cows**.

The Rocky Mountain Poison Control Center had said ***Micotil***® is highly toxic to the **human** heart and there have been documented cases of suicide with it.

Keep it out of reach of **children**; avoid accidental self-injection; avoid contact with eyes. Additionally, **they** warned that **animals** intended for **human** consumption NOT be slaughterred within 28 days of last treatment; do not use in any dairy **cows** > 20 months of age. IV use in **cattle** can be fatal at dosages as low as 5 mg/kg; not for use in **calves** to be processed for ***veal***. Edema and soreness at injection site in **cows**. In **monkeys**, 20 mg/kg IM led to vomiting and 30 mg/kg was fatal. In **swine**, IM use of 10 mg/kg increased respiration, emesis, and a convulsion in **pigs**; 20 mg/kg caused death in 3/4, and 30 mg/kg killed all **pigs** tested. Other reports indicate 2 cc (600 mg) IV euthanized a 1000 lb **horse**. In June 1997, Elanco warned **people** to avoid inhalation, oral exposure, and direct contact with the feed grade for growing **pigs**. **Those** mixing it or handling it should use protective clothing, impervious gloves, goggles, and a National Institute for Occupational Safety & Health-approved dust mask. Within a few days after the 18-year-old **female** died, **Dr. David Dykstra** of Pecatonica, Illinois, started a one-**man** crusade to inform professionals, the FDA, TV, and Elanco about preventing a repeat of this tragedy.

LD_{50}, **rat**: 2.25 g/kg; **mouse**: 800 mg/kg. SC, LD_{50}, **rat**: 185 mg/kg; **mouse**: 97 mg/kg.

TILUDRONATE

= *Skelid*

Bone calcium regulator. A biophosphonate.

Untoward effects: Occasional abdominal pain, diarrhea, nausea, vomiting, flatulence, dysphagia, esophagitis, and ulceration, but more frequently than with ***etidronate***. Causal relationship not confirmed for possibly induced rhinitis, asthenia, fatigue, hypertension, rashes, cataracts, conjunctivitis, and glaucoma.

Drug interactions: Bioavailability decreased 80% with ***calcium*** and 60% by some ***aluminum–magnesium antacids***, and by ***food*** if taken within 2 h of these.

TIMOLOL

= *Aquanil* = *Betim* = *Blocadren* = *MK-950* = *Proflax* = *Temserin* = *Tenopt* = *Timacor* = *Timoptic* = *Timoptol*

Antihypertensive, β-adrenergic blocker, class II antiarrhythmic, sympatholytic antianginal, antiglaucoma agent.

Untoward effects: Nausea, vomiting, fatigue, anorexia, dizziness, carotid sinus hypersensitivity, and diarrhea are most commonly reported. During 1979–1986, > 3000 reports of adverse reactions to its topical ophthalmic use; 450 serious respiratory and cardiovascular reactions with 32 deaths; 136 had cardiac arrhythmias and 131 had bronchospasm.

Cardiovascular: Bradycardia, heart-block, palpitations, cold extremities, hypotension, Raynaud's phenomenon, and paresthesias.

Central nervous system: Depression, disturbed sleep, headache, confusion, dizziness, dry eyes, poor vision, and hallucinations.

Miscellaneous: Topical ophthalmic use in a 95-year-old **female** of 1 drop 0.5% twice daily led to syncope, shortness of breath, and sinus bradycardia. Arthropathy in a 61-year-old **female** and **others** after prolonged ophthalmic use of 0.25–0.5% twice daily in both eyes. Ophthalmic use in a 72-year-old **male** led to hyperkalemia on two occasions. Faintness, occasional loss of consciousness, and impending sinus arrest in a 75-year-old **male**. A 78-year-old **female** had syncope and visual hallucinations after ophthalmic use; worsening of myasthenia gravis, double vision, and ptosis in a 71-year-old **male** within 24 h of ophthalmic use. Congestive heart failure in a 94-year-old **female** after ~2 weeks of 2 drops 0.5% solution twice daily in each eye. Impotence reported in many **patients** after ophthalmic use. Mental changes, seizures, and altered hypoglycemic response from ophthalmic use in a diabetic 65-year-old **male**. After twice daily ophthalmic use of 0.5% solution, breast milk had ~6 times **maternal** serum levels. This is dangerous, as an **infant** developed apnea after small doses. IV and oral use has induced retroperitoneal fibrosis in a 60-year-old **male**. Oral use of 10 mg tid/18 months led to sclerosing peritonitis in a 50-year-old **male** requireing major surgical intervention. IV use has induced decreased tremors in essential tremor. Acute suicidal depression reported in a 65-year-old **female** after ophthalmic 0.25% twice daily. Systemic lupus erythematosus reported. Most of the **patients** described recover within 2 days after treatment ceases. Case report of decreased libido in **males** and **females**, and erectile disorder.

Respiratory: Topical ophthalmic use has induced bronchoconstriction and dyspnea in **asthmatics**. Respiratory arrest in a 62-year-old patient after single topical ophthalmic application.

Skin: Erythema after topical transdermal patch. Ophthalmic use has induced or potentiated psoriasis. Alopecia, maculopapular rashes, urticaria, and nail pigmentation changes after topical ophthalmic use; local irritation, hypersensitivity; swelling of eyes, eyelids, and surrounding areas; rashes and hives.

Drug interactions: Ophthalmic use causes severe bradycardia in **patients** taking ***verapamil***. Use will exacerbate ***propranolol***-induced psychosis.

TIMONACIC

= *ATC* = *Celepat* = *Detoxepa* = *Hepalidine* = *Heparegen* = *Norgamen* = *NSC-25,855* = *4-Thiazolidinecarboxylic Acid* = *Thioproline* = *Tiazolidin*

Hepato-protective and regenerative agent.

Untoward effects: Diarrhea in 3/15, vertigo and nausea in 1/15, and transient erythema in 1/15 **patients** 35–61 years. Oral use of 20 mg/kg/day/21–42 days in 22 cancer **patients** and IM use of 40 mg/kg/day/14–42 days in 26 **patients** led to neurological symptoms and abnormal renal function in 18/48. Central nervous system upsets in 10. Ingestion by 40 **patients** of 30 mg/kg caused status epilepticus, coma, and hyperthermia. A $2\frac{1}{2}$-year-old child ingested an unknown quantity leading to cyanosis, seizures, and acidosis.

LD_{50}, **mouse**: 400 mg/kg.

TIMOTHY

= *Phleum pratense*

Untoward effects: Contains allergens A, B, and D that cause allergic contact dermatitis in some **people**. Respiratory symptoms are usually from ***aflatoxin***-producing fungi growing on it.

TIN

Untoward effects: Overexposure to inorganic compounds or tin flakes or powder leads to skin, eye, and respiratory irritation; to its organic compounds may cause eye irritation, headache, vertigo, sore throat, cough, abdominal pain, nausea, vomiting, diarrhea, oliguria, anemia, psychoneurological upsets, paralysis, skin burns, and pruritus. Organotin plasticizers have eluted into injectable solutions. It is found in water (1–2 μg/l), **fish**, plants, and limitedly in air. Most tin in tissues comes from ingesting it, especially in **people** primarily using canned foods with daily intake estimated at 1–38 mg, with average American intake of 3–4 mg/day, of which only 3% is absorbed. Tin cans are 99% steel and 1% tin. Stannosis reported in **hearth-tinners** with **their** X-rays showing dense opacities without fibrosis associated with hilar shadows. Lung carcinomas also reported in **them**. The suspected mechanism is that fumes caused by ***ammonium chloride*** powder contacting the molten tin lead to high atmospheric concentration of tin. Lung cancer in tin **miners** is related to **their** ***radon*** gas exposure. Use of ***putty powder***, a higher oxide of tin, caused the death of a **chemist** who, in error, used it as ***pepper***. Ingestion of large amounts decreases ***zinc*** and ***copper*** absorption, cellular immune reactions, and ***immunoglobulin*** in blood. "Normal" serum levels are ~13 μg%.

TD_{LO}, implant, **rat**: 7.6 g/kg; **mouse**: 18 g/kg.

TIN ARSENATE and TIN ARSENITE

***Anthelmintic.* Orally for tapeworming sheep.**

Untoward effects: Do NOT slaughter treated **animals** until at least 1 week after treatment to avoid unnecessary tissue ***arsenic*** levels.

TINIDAZOLE

= *CP-12,574* = *Fasigin* = *Fasigyn* = *Pletil* = *Simplotan* = *Sorquetan* = *Tricanix* = *Tricolam* = *Trimonase*

***Protozoacide, trichomonocide, antibacterial.* A *metronidazole* and *nitroimidazole* derivative. Oral, intravaginal, and occasional rectal use.**

Untoward effects: Mild occasional nausea, vomiting, anorexia, dizziness, headache, metallic taste, xerostomia, diarrhea or constipation, and rash. Esophagitis, hepatotoxicity, and hypersensitivity reactions with generalized urticaria; facial, laryngeal, and periorbital edema; hypotension, bronchospasms, and dyspnea. Placental transfer occurs. Levels in breast milk vary (0.62–1.39 times **maternal** serum levels) and breast-feeding should be avoided for 72 h after the last **maternal** dose. Colors urine dark brown.

TINZAPARIN SODIUM

= *Innohep*

Use: Low molecular weight ***heparin*** for treating deep vein thrombosis; in prevention of thromboembolic events, and for use in hemodialysis.

Untoward effects: Major bleeding in ~1%, severe thrombocytopenia in 0.13%, asymptomatic aminotransferases increase in ~10%, and priapism reported in eight **patients**.

TIOCONAZOLE

= *Fungibacid* = *Gyno-Trosyd* = *Trosyd* = *Trosyl* = *UK-20,349* = *Vagistat* = *Zoniden*

Topical antifungal.

Untoward effects: Mild maculopapular rash and burning sensation at application site; local irritation of vulva and perineum.

TIOPRONIN

= *Acadione* = *Capen* = *Epatiol* = *Mucolysin* = *Thiola* = *Thiosol*

Mucolytic, liver protectant, antidote for heavy metals. **Oral, IM, and IV.**

Untoward effects: Pemphigus, nausea, vomiting, abdominal pain, diarrhea, paresthesias. Injections lead to chills, hypotension, erythema, pyrexia, and sensation of pressure on the chest. Blood dyscrasias can occur.

LD_{50}, **mouse**: 3.1 g/kg.

TIOXOLONE

= *Camyna* = *Gelacnine* = *Thioxolone*

Antiseborrheic, antiacne.

Untoward effects: Topical use leads to allergic reactions, slight initial burning sensation, irritation, folliculitis, hairy skin sensitivity, pruritus, and facial irritation.

TIROFIBAN

= *Aggrastat* = *L 700,462* = *MK 383*

Antithrombotic.

Untoward effects: Reversible thrombocytopenia and increases rate of major bleeding.

SC, LD_{50}, **mouse**: 80 mg/kg.

TISSUE PLASMINOGEN ACTIVATOR

= *Actilase* = *Alteplase* = *ATZ* = *Duteplase* = *Fibrino-kinase* = *t-PA* = *TPA*

Thrombolytic.

Untoward effects: Small but significant increase in stroke and cerebral hemorrhage, compared to treatment with ***streptokinase***. Intracranial hemorrhage, anaphylactoid reactions, ***cholesterol*** embolization; cerebral infarction due to partial lysis of a thrombus; iridocyclitis, and transient renal impairment also reported.

Drug interactions: Potentiation of interactions with ***anticoagulants***, such as ***heparin*** and ***warfarin***.

TITANIUM DIOXIDE

= *C.I. 77,891* = *C.I. White Pigment 6* = *Rutile* = *Titanium Oxide* = *Titanium Peroxide* = *Unitane*

Sunscreen, opacifier. **U.S. annual demand is > 1,000,000 metric tons for *paints*, varnishes, *lacquers, inks, rubber*, ceramics, floor coverings, foods, and pharmaceuticals. Isolated soy protein contains 0.1% as an approved food additive to distinguish it from poultry or animal proteins.**

Untoward effects: **Factory workers** handling it had deposits of it in pulmonary interstitium, and it was associated with cell destruction and slight fibrosis. Found in nearby lymphatics, indicating that clearance may be by the lymphatic system.

Minimum lethal dose, **human**: skin, 300 µg during 3 days.

Intratracheal instillation has caused a few benign papillomas in **hamsters**.

LD_{50}, **mouse**: > 10 g/kg.

TITANIUM TETRACHLORIDE

U.S. production has been > 2 million tons annually and 25 metric tons/year has been discharged into the air from U.S. manufacturing and processing plants.

Untoward effects: It is a liquid with odorous fumes, reacting with water and moisture in factory air producing heat, ***hydrochloric acid*** (q.v.), ***titanium hydroxide***, and ***titanium oxychlorides***. Very irritating to skin, eyes, mucous membranes, and lungs. Inhalation of large quantities can cause severe lung damage and death. Symptoms are coughing, chest tightness, mild eosinophilia, bronchitis, congestion of mucous membranes in upper respiratory tract, pneumonia with long-term effects of tracheal and laryngeal narrowing, depending on the degree and duration of exposure. Accidental spraying of it on a 50-year-old **male** chemical engineer caused an erroneous, but normal, reaction, i.e. removing **his** protective mask to clean **himself**. This instantly exposed **him** to the vapors leading to erythema of the conjunctiva, tongue, pharynx, and respiratory tract. **His** eyes and body symptoms were severe because **he** mistakenly used water to wash them, resulting in second and third degree burns. Death reported after severe dermal exposure in a plant **worker**.

Similar symptoms noted in exposed **rats**, including a tumorigenic response after chronic inhalation.

Inhalation LC_{LO}, **mouse**: 10 mg/m^3/2 h.

TIXOCORTOL PIVALATE

= *JO 1016* = *Pivalone* = *Rectovalone* = *Tiovalon*

Anti-inflammatory, topical corticosteroid.

Untoward effects: Hypersensitivity eczematous reactions. It contains a 21-thiol instead of a 21-hydroxyl group and used in Europe to detect allergic contact dermatitis to ***hydrocortisone***.

TIZANIDINE

= *DS 103-282* = *Sirdalud* = *Zanaflex*

Skeletal muscle relaxant.

Untoward effects: Depression in multiple sclerosis in a 37-year-old **male** after 4 mg/day/9 days. Confirmed by rechallenge. Visual hallucinations or delusions in ~3%, xerostomia, asthenia, dizziness, hypotension, sedation, increased aminotransferases, and bradycardia. Occasional deaths from liver failure reported. Overdosage can cause atrial and ventricular dysfunction.

Drug interactions: ***Oral contraceptives*** have decreased its clearance by ~50%. ***Alcohol*** and ***diazepam*** may increase its central nervous system depressant effects. Has increased serum concentration of ***phenytoin***.

TOADS

Untoward effects: Their skin secretes various poisonous substances, particularly ***bufogenins, bufotenine*** (q.v.), ***bufotoxins, epinephrine*** (q.v.), ***norepinephrine*** (q.v.), and ***serotonin*** (q.v.), etc. in varying quantities.

Bufo alvarius, the ***grapefruit***-sized Colorado River toad, secretes a potent ***strychnine***-like substance from its skin glands, which causes severe pain and death in predatory **dogs**.

Bufo marinus, a grotesque-looking toad known as the Marine Toad, Cane Toad, or *Bufo aqua*, ~5–7 inches in size and often weighing > 3 lbs, was introduced into the U.S. from the Caribbean and South America. As late as 1995, the toxin was sold in U.S. grocery stores and ***tobacco*** shops as an aphrodisiac and hallucinogen by rubbing it on **men's** genitals. Some **men** died after taking it orally. **People** have licked this toad for supposed benefits. **Mayan Indians** use it as an hallucinogen in religious ceremonies. It is now found all over Florida. When the toads are attacked by **dogs**, the **dogs** can absorb the toxin through their buccal mucosa with immediate profuse salivation, trembling, convulsions, and even death within 15 min. A **woman** trying to clean her **dog's** mouth after it had bitten one scratched **her** thumb on one of her **dog's** teeth. "I began to feel ill almost instantly ... in minutes, my right arm was swollen (and it) felt paralyzed. Then my head began to ache and I was extremely nauseated". Her **dog** was dead within 90 min and **she** required a month to recover from the neurotoxin. Another **human** case involved a **man** whose mower cut the toad to pieces, whipping some of the poison onto **his** face. Within minutes, **his** lips and cheeks became painfully swollen. They failed to eat agricultural pests after being imported into Australia from Hawaii, but their toxic skin killed countless native **animals**.

Bufo regularis in Ethiopia causes violent vomiting, salivation, depression, incoordination, paralysis, and death within 3–6 days in **cats** and **dogs**.

Bufo vulgaris in Britain and the tropics is not as lethal for **dogs** or **cats**, and usually induces vomiting, decreased temperature, shallow respirations, and decreased cardiac function, due to its ***digitalis***-like ***bufotenines*** (q.v).

The skin of the South American toads contains the potent ***physaleamin*** (q.v.). Numerous other toxic toad venoms reported from Africa, Vietnam, China, and Japan. Toads tolerate much larger doses of cardiac glycosides than do **frogs**.

TOBACCO

= *Nicotinia tabacum* = *Tabaco* (Espanola) = *Yetl, Picietyl, or Piciete* (Aztec) = *Kutz* (Mayan) = *Sayri* (Peru) *and hierba sagrada and nicociana* (Spaniards)

Untoward effects: Tobacco smoke has > 4000 compounds in it; 16 are known carcinogens. Addiction from smoking or chewing it. Toxic tobacco amblyopia and other neuropathies from prolonged exposure to ***cyanides*** and depletion of ***vitamin*** B_{12} by heavy **smokers**. *N′*-***Nitrosonornicotine*** (q.v.) is an abundant carcinogen in tobacco smoke. Smoking has been associated with an increased incidence of lung and laryngeal cancer, acute cardiovascular events, gastric ulcers, decreased **infant** birth weights, perinatal mortality, low Apgar scores, congenital anomalies, neurological and developmental problems, sudden infant death syndrome, coughing, emphysema, and respiratory infections. **Infants** born to smoking **mothers** have higher than normal concentrations of ***cadmium, lead***, and ***thiocyanate***. An 8-month-old **female** ingested ***cigarette*** butts leading to depression. Toxicity reported from the ingestion of wild tobacco leaves in salads. Its ***nicotine*** (q.v.) content leads to peripheral vasodilation, increased arterial tension and accelerated pulse; transient decreased skin temperature of fingers and toes; and increased gastric secretion. Thus, smoking is contraindicated in Buerger's disease (thrombo-angitis obliterans), Raynaud's disease, hypertension, angina, postcoronary problems, and peptic ulcers. Tar in the smoke contains many carcinogens. Degree of inhalation can also affect toxicity. Smoking can interfere with normal development of **human** sperm and cause polycythemia and carboxyhemoglobin. Chewing smokeless tobacco has increased the risk of oral cancers, peridontal disease, dental decay, staining of tongue papillae, ***nicotine*** (q.v.) addiction, and hypokalemia due to its ***licorice*** content. A hypersensitivity pneumonitis, called Tobacco Grower's Disease, has been from an unknown antigen on tobacco plants. Green Tobacco Sickness occurs from dermal exposure of tobacco **harvesters** to dissolved ***nicotine*** from the wet leaves leading to nausea, vomiting, weakness, dizziness, headache, giddiness, arrhythmias, and fluctuating blood pressure. Rare cases of allergic contact dermatitis from its leaves reported. **East Indians** smoking "backfat" tobacco show diffuse pulmonary fibrosis.

Drug interactions: Probably of clinical importance are increased clearance of ***chlorpromazine, fluphenazine decanoate*** (133%), ***haloperidol*** (~50%), ***insulin, propoxyphene, propranolol,*** and ***theophylline***, and 18 others are similarly described, but not clinically significant. Chronic eczematous contact dermatitis has occurred on fingers of **smokers** and also reported when ***cigarettes*** contained ***diethylene glycol*** as a humectant. Free ***hydrazine*** (q.v.) has also been present on tobacco from its use with ***maleic hydrazide*** used on it as a growth retardant. Stimulates CYP1A2 metabolic enzymes to decrease serum levels of ***caffeine, clomipramine***, and ***theophylline***. ***Cigarette*** smoking can increase CYP1A2 activity up to three times due to ***nicotine*** and ***polycyclic aromatic hydrocarbon*** products of combustion.

Causes teratogenicity, such as cleft palate, when grazed by **ewes** and **cows**. Symptoms also include those of ***nicotine*** (q.v.). **Sows** that ate burly tobacco stalks in early pregnancy had malformed **pigs** with skeletal deformities, including arthrogryposis. An 8 lb 3-month-old **dog** ate a package of ***cigarettes*** leading to severe depression, salivation, increased peristalsis, pale mucous membranes, dyspnea, and vomiting. The **dog** survived a three times minimum lethal dose, after being treated. ***Nicotine*** poisoning in **calves** bedded on wet tobacco scrap led to paralysis, increased temperature, and necrotic liver areas. Tobacco poisoning in **guinea pigs** led to loss of hearing due to cochlear lesions. In the middle of the 19th century, French and German scientists reported hearing deficiencies, buzzing in the ears, and vertigo in **human** tobacco smokers. Tobacco smoke is carcinogenic in **hamsters**. Topical use of tobacco tar on **rabbits** led to warts and carcinomas. Tobacco extracts on the skin of **mice** led to tumors. Fungal growth on ***cigarette*** tobaccos led to emphysema in **mice**.

TOBRAMYCIN

= *A 12,253A* = *Gernebcin* = *Nebramycin Factor 6* = *NF 6* = *Tenebrimycin* = *Tenemycin* = *Tobracin* = *Tobradistin* = *Tobralex* = *Tobramaxin* = *Tobrex*

The sulfate = *Nebcin* = *Nebicina* = *Obracin* = *Tobra*

***Antibiotic.* Has been used IV, IM, orally, topically in ophthalmic solutions, and by inhalation.**

Untoward effects: Nephrotoxicity (increased blood urea nitrogen, increased non-protein nitrogen, increased ***creatinine***, oliguria, cylinduria, and proteinuria), and ototoxicity (eighth nerve vestibular and auditory branches with dizziness, vertigo, tinnitus, and roaring in ears). Hearing loss is usually irreversible. Readily aggravated by other ***aminoglycosides***; azotemia (in ~20%), thrombocytopenia, granulocytopenia, anemia, headache, fever, rash, pruritus, urticaria, nausea, vomiting, pain at injection site, confusion, increased SGOT, increased SGPT, increased serum ***bilirubin***, and delirium have also been induced by it. Many studies indicate increased ***creatinine*** and nephrotoxicity in 10–28% of **patients** and ototoxicity in ~10%. Hypomagnesemic tetany after 3 days' IV treatment in a 25-year-old **female**. Hypersensitivity reactions reported in an 18-year-old **male** with cystic fibrosis after IV leading to (within seconds) shaking, urticaria, pruritus, and **his** arm turning white after a third injection. Can cause neuromuscular blockade and postoperative respiratory depression. Breast milk concentration reported as negligible to 0.85 μg/l. Therapeutic levels reported as 4–10 mg/l and toxic at 12 or above.

Drug interactions: Caution is indicated in **patients** on ***muscle relaxants***. May cross-react with ***neomycin*** after ophthalmic use. Transient renal failure may occur in **children** with cystic fibrosis on maintenance treatment with ***ibuprofen***. *In vivo* inactivation by ***piperacillin*** reported; and by ***ticarcillin*** reported in a 65-year-old **female** with impaired renal function; during concomitant therapy, **her** tobramycin half-life was 7 h and 35 h without ***ticarcillin***. Use with ***miconazole*** may decrease its serum levels. ***Furosemide*** has increased its ototoxicity and nephrotoxicity. Use with ***cephaloridine*** or ***cephalothin*** increases its nephrotoxicity.

Nephrotoxicity occurs readily in **cats**, **salmon**, and **trout**. LD_{50}, **mouse**: IV, 118 mg/kg.

TOCAINIDE

= *Taquidil* = *Tonocard* = *W-36,095* = *Xylotocan*

***Class 1B antiarrhythmic.* Analog of *lidocaine*.**

Untoward effects: Gastrointestinal and central nervous system problems occur in 10–20% of treated **patients**; nausea, dizziness, and vertigo in ~15%; paresthesias and tremors in ~9%. Confusion, short-term memory loss; bleeding from lips and gingiva, nightmares, paranoia, psychosis, coma, seizures, arthralgia, fever, maculopapular lupoid reactions, and rash in 12%; dermal necrosis, alopecia, dystonia, myopathy, interstitial pneumonitis, granulomatous hepatitis, agranulocytosis, leukopenia, eosinophilia, anemia, aplastic anemia, thrombocytopenia, arrhythmias, fatal cardiac arrhythmias (with overdoses), bradycardia, congestive heart failure, blurred vision, tinnitus, diaphoresis, lethargy, ataxia, anorexia, loss of taste, abdominal pain, and diarrhea. Rarely, lupus erythematosus-like syndrome with neutropenia. Therapeutic serum concentrations are generally reported as 6–10 mg/l or 1–5 mg/l; toxic at 13–20 mg/l and 6 mg/l in another report; and fatal at 140 mg/l. Fatalities were 21% while the U.S. permitted it for "compassionate use".

Drug interactions: ***Rifampin*** may increase its metabolism. It causes slight inhibition of ***theophylline*** metabolism. Use with ***propranolol*** may increase incidence of seizures. The same may occur when used with ***lidocaine***.

Central nervous system excitement, tremors, seizures, nausea, vomiting, hypotension, muscle weakness, and arrhythmias have occurred in treated **dogs**.

SC, LD_{50}, **rat**: 969 mg/kg; **mouse**: 367 mg/kg.

TOFENACIN HCL

= *BS 7331* = *Elamol* = *Tofacine*

Antidepressant, antiparkinsonism.

Untoward effects: Anticholinergic effects, xerostomia, dysuria, tremors, and accommodation difficulties. Occasional gastrointestinal upsets, drowsiness, vertigo, agitation, and rashes. Therapeutic serum concentrations reported as 0.025–0.2 mg/l and toxic concentrations as 0.5–1 mg/l.

LD_{50}, **rat**: 400 mg/kg; **mouse**: 182 mg/kg.

TOFISOPAM

= *EGYT 341* = *GRANDAXIN* = *Seriel* = *Tofizopam* = *Tofosam*

Anxiolytic.* A *benzodiazepine.

Untoward effects: Sleep disturbances are common. After repeated doses, it exhibits some stimulant action.

TOFU

Untoward effects: This pasty ***soybean*** curd, a 2000 year old protein staple in Asia, unfortunately is very high in ***oxalates*** (q.v.) and should be avoided by **women** with vulvodynia.

TOLAZAMIDE

= *Diabewas* = *Norglycin* = *Tolanase* = *Tolazolamide* = *Tolinase* = *U-17,835*

Antidiabetic.* A *sulfonylurea.

Untoward effects: Nausea, vomiting, epigastric pain, weakness, headache, paresthesia, and hypoglycemia. Hypersensitivity reactions include photosensitivity, fever, jaundice, transient increase of SGPT, skin rashes, systemic lupus erythematosus, pruritus, agranulocytosis, leukopenia, eosinophilia, thrombocytopenia, aplastic anemia; increased cardiovascular complications and even death, apparently, from them; nausea, dizziness, headache, anorexia, diarrhea, constipation, insomnia, lethargy, drowsiness, palpitations, diaphoresis, and increased intraocular pressure. Severe lactic acidosis and semicoma; Wernicke's encephalopathy also reported. May also cause an intolerance to ***alcohol***, manifested by a ***disulfiram*** (q.v.)-like reaction with facial flushing and spuriously high serum alkaline phosphatase test results. Original batches contained an impurity, ***n-nitrosohexamethyleneimine***, which was claimed by the manufacturer to be "non-toxic", but it was shown to be carcinogenic.

TOLAZOLINE HCL

= *Lambral* = *Priscol* = *Priscoline* = *Vaso-Dilatan*

α or α_1*-Adrenergic-blocker, peripheral vasodilator.* Usually IV, occasionally oral, SC, or IM.

Untoward effects: Gastrointestinal bleeding, gastric hypersecretion, aggravation of dormant peptic ulcers, severe hyperchloremic alkalosis, headache, acute renal failure, abdominal discomfort, piloerection, flushing, miosis, tingling, chills, shivering, diaphoresis, orthostatic hypotension, nausea, vomiting, diarrhea, and exacerbation of peptic ulcer reported. Marked hypertension, arrhythmias, tachycardia, palpitations, anginal pain, and it has precipitated a cardiac infarction. IA has induced limb burning sensations. The U.S. has withdrawn permission for its oral use.

Drug interactions: It inhibits ***alcohol's*** metabolism leading to ***disulfiram*** (q.v.)-like effect.

IV, TD_{LO}, **human**: 150 µg/kg.

Salivation is often noted 30 min after injection in **cats** and **dogs**, and a high percentage of **dogs** developed a mucoid diarrhea.

LD_{50}, **mouse**: 60 mg/kg.

TOLBUTAMIDE

= *Arcosal* = *Artosin* = *Butamide* = *Diaben* = *Diabetol* = *Diasulfon* = *Dolipol* = *Glycemex* = *Glyconon* = *Glycotron* = *Hypoglymol* = *Ipoglicone* = *Melitol* = *Mobenol* = *Neo-Dibetic* = *Novobutamide* = *Orabet* = *Oramide* = *Orinase* = *Oterben* = *Pramidex* = *Rastinon* = *Tolbugen* = *Tolbusal* = *Tolbutol* = *Tolbutone* = *Toluina*

Hypoglycemic, antidiabetic.

Untoward effects: ***Blood***: ~0.2% hemolytic anemia in glucose-6-phosphate dehydrogenase-deficient **patients**, pancytopenia, thrombocytopenia, agranulocytosis, cytopenia, leukopenia, immune hemolytic anemia, aplastic anemia, macrocytic anemia, polyuria, leukemia, purpura and non-thrombocytopenic purpura. Severe thrombocytopenic purpura in diabetic 48-year-old **female** after 1 week of treatment; also increases bone marrow megakaryocytes. Occasional ketonemia and reticulocytosis.

Cardiovascular: Increased risk of congestive heart failure and deaths from cardiovascular disease ~2 times placebo-treated **patients**.

Clinical tests: Elevated false positive ***sulfobromophthalein*** retention, alkaline phosphatase, and ***glucagon*** stimulation test results.

Eye: Lens opacity, optic neuritis, and transient myopia reported, but are rare occurrences.

Fetus and neonate: Placental passage occurs with high concentration in the **newborn**. Can potentiate hypoglycemia in the **newborn** and cause thrombocytopenia. Displaces ***bilirubin*** from its ***albumin*** binding sites and can cause kernicterus in **newborns**. Levels in breast milk reported as 9–40% that of **maternal** serum levels. Teratogenicity and dysmorphia (testicular torsion and hydrocele, tetralogy of Fallot, hydrocephalus, agenesis of the hemidiaphragm, atrial and ventricular septal defects, syndactyly, malformed ears and atretic external auditory canals) can also be due to the diabetes per se.

Gastrointestinal: Hemorrhage and melena reported. Nausea, vomiting, diarrhea, and heartburn are usually dose-related. Occasional griping abdominal pain after meals with increase in watery diarrhea, metallic taste, and anorexia.

Hypoglycemia: Hypoglycemic coma, stomach pains, nausea, dizziness and confusion, diaphoresis, and headache.

Liver: Cholestatic jaundice with incidence of 0.1% and increased alkaline phosphatase. Hepatocyte necrosis and biliary cirrhosis are rare.

Miscellaneous: Drug fever, weight decrease, tremor of forearms and hands, cogwheel rigidity of upper limbs, dysarthria, and mask-like facial expression. Diabetic acidosis has been precipitated by it in juveniles.

Poisoning and suicide: Cramps, vomiting, restlessness, and pallor in a non-diabetic 20-year-old **female** after ingesting 100 tablets (50 g). Died after ~18 h. Post-mortem revealed severe pulmonary edema, congestion of liver and spleen, lower nephron nephrosis, and severe atrophy of islets of Langerhans. Death of a 60-year-old **male** 12 days after ingesting 5 g and a large amount of ***beer*** and ***alcohol*** in intended suicide. Hypoglycemic coma in a 15-year-old **male** after ingesting 3 tablets, thinking they were stimulants. Fatal hypoglycemic coma with permanent brain damage in successful suicide. Hypoglycemic reaction after a **pharmacist** dispensed ***Orinase***® instead of ***Annanase***® (***bromelains***). Inadvertent substitution of it 0.5 g instead of 5 mg ***terbutaline*** also reported.

Renal: Nephrotic syndrome with edema, proteinuria, and, rarely, hyponatremia reported. Has an antidiuretic effect. Hypersensitivity reaction may have induced asymptomatic renal microgranulomas, and a case report of a fatal case of proliferative glomerulonephritis.

Respiratory: Pulmonary infiltrates or densities with eosinophilia.

Skin: ~1% photoallergic eczematous contact dermatitis with transient erythema, pruritus, urticaria, cutaneous porphyria, morbilliform or maculopapular eruptions, and toxic epidermal necrolysis. Porphyria cutanea also reported. Peripheral neuropathy with numbness, tingling, and burning sensations of the extremities also occur.

Thyroid: Rarely, it has induced hypothyroidism, with lethargy, waxy complexion, dry and coarse skin, sluggish reflexes, hoarseness, and undue intolerance to cold.

Therapeutic serum levels are given by various authors as 5.3–9.6 mg% and 50–100 ml/l; toxic as 64 mg% and 120 mg/l and above; and 640 mg/l leads to comatose state or lethality.

TD_{LO}, **woman** (1–37 weeks pregnant): 130 g/kg.

Drug interactions: Vasomotor crises after ***alcohol*** ingestion due to competition for oxidative metabolic sites increased ***tolbutamide*** serum levels by 50% leading to hypoglycemic convulsions. It inhibits alcohol dehydrogenase which ***alcohol*** metabolizes, which increases ***ethanol*** intolerance leading to facial flushing, headache, tachycardia, weakness, nausea, and dizziness in a ***disulfiram***-like reaction. Urine ***albumin***, serum alkaline phosphatase, and ***sulfobromophthalein*** retention tests give false positive or spuriously high readings after its use. ***Clofibrate*** increases its half-life. ***Apazone*** can inhibit its metabolism leading to hypoglycemic coma. ***Chloramphenicol*** increases its serum level and half-life leading to hypoglycemia. ***Dicumarol*** displaces tolbutamide from its protein-binding sites leading to 2–4 times increase in tolbutamide serum levels with sudden and dangerous decrease in blood ***sugar***. ***Warfarin*** and ***phenindione*** do not inhibit its metabolism. ***Cortisone*** has a diabetogenic effect and may require an increase in tolbutamide dosage in diabetic **patients**. Possible additive effect on aggravation of gastrointestinal ulcers. ***Digitoxin*** is displaced off serum ***albumin*** by it, but, apparently, without clinical significance. ***Fluconazole*** increases its serum concentration. It causes a rise in active circulating ***insulin*** and induces ***insulin*** release from the granules in the β-cells of the islets of Langerhans. ***Ketoconazole*** increases its serum concentration and half-life. The clinical significance is uncertain of its altering the plasma binding of ***methotrexate*** and can compete with it for renal tubular secretion. Use with ***monoamine oxidase inhibitors*** can increase or prolong hypoglycemia and may require altered dosage. Blood dyscrasias reported with ***methyldopa***. Enhances the effect of ***paraldehyde***. ***Oxyphenbutazone*** and ***phenylbutazone*** can potentiate its hypoglycemic effect. Hemolytic anemia in a 58-year-old **female**, a regular **user** of ***phenacetin*** after adding tolbutamide (0.5 g tid/ 14 months). Potentiates ***phenobarbital*** and ***phenprocoumon*** effects. ***Phenyramidol*** inhibits its metabolism, increasing tolbutamide half-life from 7 to 18 h. Its interaction with ***phenytoin*** increases the latter's unbound portion by 25% in **adults** and 40% in umbilical serum. Headache, nausea, ataxia, and nystagmus in a 48-year-old **female** on 200 mg ***phenytoin***/day and tolbutamide 1 g tid/2 days. ***Probenecid*** increases its hypoglycemic effect by interfering with urinary tubular excretion of its metabolite ***carboxytolbutamide***. ***Propranolol*** may block some of its hypoglycemic effects and at most other times causes a hypoglycemic reaction, but without usual warning signs, such as sweating and tachycardia. ***Rifampin*** decreases its hypoglycemic effect and its half-life from 292 to 160 min. ***Salicylates***, including ***aspirin*** and ***sodium salicylate***, potentiate its activity. Small decrease in its clearance (16%; from elimination half-life of 6.9 to 8.6 h) by ***sertraline*** 200 mg/day in **volunteers**. ***Sulfonamides***, particularly long-acting ones, such as ***sulfaphenazole***, can potentiate its hypoglycemic effect and dramatically increase its half-life by ~4 times. ***Sulfadimethoxine***, ***sulfamethizole***, and ***sulfisoxazole***, when used with it, have caused severe hypoglycemia. Tolbutamide-treated **patients** can become sensitized to a wide variety of topical ***p-amino compounds***, including ***p-phenylenediamine*** (q.v.) found in hair dyes, sunscreens with ***p-aminobenzoic acid*** (q.v.) or its esters, ***local anesthetics***, such as ***ethyl aminobenzoate*** (q.v.), various vaginal creams, ***saccharin***, ***calcium cyclamate***, and ***sodium cyclamate***, causing a contact dermatitis and/or systemic effects. ***Tetracycline*** can inhibit its metabolizing enzymes. A diabetic 69-year-old **male** successfully treated with it for 4 years developed a hypersensitivity induced jaundice when given with ***thioridazine***, ***meperidine***, and ***dextroamphetamine***. It potentiates ***warfarin***.

Stimulates liver microsomal enzymes in **rabbit**, **rat**, and **dog** trials and markedly decreased ***antipyrine*** half-life in **monkey** trials. Inhibits hepatic ketogenesis and intrahepatic lipolysis in **rats**. The drug stimulates its own metabolism in **dog** trials and first potentiates and then may decrease the hypoprothrombinemic effect of ***warfarin***. Chronic administration of 30 mg/kg or over to normal **dogs** is hepatotoxic and may preclude its long-term use in diabetes. Some teratogenic effects and increased **infant** mortality has been noted in **rabbit** and **rat** trials using high dosage.

LD_{50}, **mouse**: 2.6 g/kg; SC, 980 mg/kg; TD_{LO}, **rabbit** (8–16 days pregnant): 1.8 g/kg.

TOLCAPONE

= *RO 40-7592* = *Tasmar*

Catechol-σ-methyltransferase inhibitor, antiparkinsonism agent.

Untoward effects: Confusion, diarrhea (~17%), hallucinations (~10%), hematuria (3%), fulminant hepatitis, and increased liver enzymes reported.

Drug interactions: Use with ***L-dopa*** increases many of the latter's adverse effects and it increases the duration of ***benserazide***-induced motor effects. Contraindicated with many ***monoamine oxidase inhibitors***, such as ***phenelzine*** and ***tranylcypromine***. Vitiligo in a 50-year-old **male** after adding 300 mg/day to therapy with 375 mg/day ***L-dopa***.

σ-TOLIDINE

= *3,3´-Dimethylbenzidine*

Use: In laboratories as an extremely sensitive reagent to test for the presence of ***gold*** and ***chlorine***; and in manufacturing > 400 dyes. Millions of lbs used annually.

Untoward effects: Eye and nose irritant. When heated, it emits toxic fumes of ***nitrogen oxides***. Suspect as a cause of bladder cancer in **man**, as ***benzidine*** is the end product of its metabolism. Can be absorbed by inhalation of dust or vapors, percutaneously, ingestion from contaminated hands, and contact with soiled clothing or surfaces.

SC use in **rats** led to carcinomas of ear canal and Zymbal gland. Also, hepatic and renal toxicity reported in experimental **animals**.

LD_{50}, **rat**: 404 mg/kg; LD_{LO}, **dog**: 600 mg/kg.

TOLMETIN

= *McN-2559-21-98* = *Reutol* = *Tolectin* = *Tolmene*

***Anti-inflammatory, non-steroidal anti-inflammatory drug.* Chemically similar to *indomethacin*.**

Untoward effects: Gastric irritation with hemorrhage and peptic ulcers in ~11%, nausea, vomiting, diarrhea, abdominal pain, constipation, retrosternal pain, and esophageal ulceration after lying down or swallowing without water. Headache, dizziness, tinnitus, weakness, chest pain, hypertension, thrombocytopenia, granulocytopenia, skin rashes, pruritus; ***sodium*** and water retention with edema in 2%, drug fever, toxic epidermal necrolysis, urticaria, aseptic meningitis, proteinuria; reversible renal failure and acute interstitial nephritis; increased blood urea nitrogen; many case reports of anaphylaxis and anaphylactoid reactions; hepatic toxicity, immune hemolytic anemia, and changes in vision. Causes a slight transient decrease in hemoglobin and hematocrit, and a small, but significant, decrease in platelet adhesiveness. A case report of a fatal fulminant autoimmune hemolytic anemia reported in an 80-year-old **male**. Aplastic anemia has also been reported. A 50-year-old **female** with a history of rheumatoid arthritis ingested 400 mg tid/5 weeks leading to fatal agranulocytosis. Found in breast milk.

TD_{LO}, **human**: 8 mg/kg.

Drug interactions: Has caused a marked increase in prothrombin time in a 64-year-old **male** who was stabilized on ***warfarin***. May enhance toxicity of ***methotrexate***.

Causes gastric erosions in **rats**.

TOLNAFTATE

= *Aftate* = *Chinofungin* = *Focusan* = *Fungistop* = *Hi-Alarzin* = *Sch-10144* = *Sporiline* = *Timoped* = *Tinactin* = *Tinaderm* = *Tinavet* = *Tonoftal*

Topical antifungal.

Untoward effects: Allergic contact dermatitis; sensitivity reactions can consist of either acute dermatitis or spreading of existing lesions. Delayed hypersensitivity with urticaria in a 41-year-old **female** using a Schering Corporation cream containing ***monoamylamine***. Transient local irritation and intense burning with cream or lotion. The cream is usually less irritating than the solutions or powder.

LD_{50}, **rat**: > 6 g/kg; **mouse**: > 10 g/kg.

TOLONIUM CHLORIDE

= *Blutene Chloride* = *Toluidine Blue O* = *C.I. 52,040* = *C.I. Basic Blue 17* = *Klot* = *Tolazul*

Hemostatic, wool and silk dye, biological stain.

Untoward effects: IV led to mild transient decrease of S-T segment of electrocardiogram, acute methemoglobinemia, and self-limiting hemolytic anemia in glucose-6-phosphate dehydrogenase-deficient **patient**. Has caused blue-green urine. A pregnant 39-year-old **female**, in an attempt to stop profuse vaginal bleeding, probably ingested 88, 100 mg tablets leading to chest pain, dyspnea, and vomiting. Uneventful recovery after D & C.

Drug interactions: Can interact with various ***anticoagulants***.

IV, LD_{50}, **rat**: 28.93 mg/kg; **mouse**: 27.56 mg/kg; **rabbit**: 13.44 mg/kg.

TOLOXATONE

= *Humoryl* = *MD-69,276* = *Perenum*

Antidepressant.

Untoward effects: Can potentiate the hypertension of high doses of ***tyramine***. Addition of ***phenylephrine*** to treatment with it has caused life-threatening episodes of hypertension, vasoconstriction, status epilepticus, and other symptoms suggestive of pheochromocytoma in a 72-year-old **female**.

LD_{50}, **mouse**: 1.5 g/kg.

TOLPERISONE HCL

= *Abbsa* = *Arantoik* = *Atmosgen* = *Besnoline* = *Isocalm* = *Kineorl* = *Menopatol* = *Metosomin* = *Minacalm* = *Muscalm* = *Mydeton* = *Mydocalm* = *N-553* = *Naismeritin* = *Tolisartine*

Skeletal muscle relaxant.

Untoward effects: Faintness, transient insomnia, asthenia, dizziness, trembling, tachycardia, and depression.

SC, LD_{50}, **mouse**: 620 mg/kg.

TOLRESTAT

= *Alredase* = *AY-27,773* = *Lorestat* = *Tolrestatin*

Aldose reductase inhibitor.

Untoward effects: Hepatocellular damage with increased liver enzymes in ~2%. A 41-year-old **female** died from hepatic necrosis 80 days after treatment with 200 mg/day/50 days. After this and another fatal case of liver necrosis and poor clinical efficacy, the manufacturer withdrew it from world-wide sales.

TOLTERODINE TARTRATE

= *Detrol* = *Detrusitol* = *PNU 200,583*

Muscarinic receptor antagonist.

Untoward effects: Xerostomia (40%), dry eyes, headache, and dyspepsia. Disorientation, delirium, memory impairment, and confusion can occur in the **elderly**. Avoid use in **patients** with narrow-angle glaucoma or urinary retention.

Drug interactions: A 72-year-old **male** and an 83-year-old **male** stabilized on ***warfarin*** developed prolonged international normalized ratios ~2 weeks after starting treatment with tolterodine 2 mg/days.

TOLUENE

= *Methacide* = *Methylbenzene* = *Phenylmethane* = *Toluol*

Solvent, gasoline octane enhancer, aromatic hydrocarbon. **Millions of metric tons are used annually in *paints*, spray *paint*, *paint* thinners, *lacquers*, *glues*, fingernail polishes, *gasoline*, tire industry, etc. Sweet, pungent smell noted at 8 ppm in the air and tasted in water at 0.04–1 ppm. In the U.S., > 10 million lbs used annually.**

Untoward effects: Vaporizes easily and inhaled in the workplace, in automobile exhausts, and in ***cigarette*** smoke. Handling it or drinking contaminated water are also sources. Vapors are irritant to eyes and nose and have caused weakness, fatigue, coughing, lacrimation, confusion, dizziness, light-headedness, palpitations, nausea, anorexia, headache, euphoria, mydriasis, nervousness, insomnia, paresthesia, dermatitis, serious liver and kidney damage, occasional visual hallucinations, narcosis, coma, leukocytosis, anemia, and even death. Inhalation of high concentrations led to vision and hearing problems, permanent brain and speech damage, ataxia, and **babies** born to heavily exposed **mothers** developed neurological problems and decreased growth rate and development. Acute poisoning reported in 26 **men** on a merchant ship transporting drums that leaked leading to narcosis and anethesia. Lymphocytosis, macrocytosis, anemia, eosinophilia, leukocytosis, and infrequently, aplastic anemia, basophilia, leukopenia, neutropenia, and thrombocytopenia in industrial exposures. It has been a substance of abuse by **sniffers** with adverse effects including periorbital edema, large subconjunctival hemorrhage, vomiting, hepatomegaly, jaundice, increased serum ***bilirubin***, anuria, increased prothrombin time, hematemesis, abdominal pain, pneumonitis, pneumonia, pulmonary edema; hypokalemic periodic paralysis, secondary to renal tubular acidosis; hypophosphatemia; neuropsychiatric disorders, including schizophreniform psychosis; loss of consciousness, peripheral nerve damage and paresthesias, rhabdolysis, leg pain, ataxia, adiadochokinesis, Fanconi's syndrome, sinus bradycardia, cerebral cortical atrophy, and irreversible brain damage. Accidental inhalation as a solvent in insecticidal spray led to ventricular tachycardia in a 38-year-old **male**. Persistent cerebellar ataxia reported after exposure. Most **abusers** use it for short terms before severe neurological changes occur. Repeated exposures for 1–20 years caused neurological changes. Often contains ***benzene*** (q.v.) as a process impurity at > 0.1%. Severe progressive multifocal central nervous system damage in four **abusers** revealed unusual ocular motor impairment, such as opsoclonus, flutter, and dysmetria. Metabolites in **man** include ***cresols***, ***hippuric acid***, and ***phenols*** and are excreted by the kidneys. Probable toxicity from it in a **family** from inhalation as a result of using a butyl caulk that contained it in **their** new log home. Topical use as a solvent-extender is dehydrating and occasionally irritating, especially on nail applications. Skin contact suspected as inducing scleroderma in some **patients**. Lipid-soluble and can appear in breast milk, unnecessarily exposing an **infant** to it for at least 1 h after **maternal** exposure. **Maternal** abuse for > 4 years led to **infant** growth retardation, prematurity, low birth weight, and microcephaly. Although not a mutagen, it is photo-oxidized to produce mutagens. Serum levels of 1.0 mg% reported as lethal. National Institute for Occupational Safety & Health suggests limit of exposure 100 ppm (375 mg/m^3); 15-min exposure 150 ppm (560 mg/m^3); immediate danger to life or health 500 ppm. OSHA suggests limit of exposure 200 ppm or 10 min of 500 ppm.

LD_{LO}, **human**: 50 mg/kg; TC_{LO}, **human**: 100 ppm.

Moderate inhalation exposure of **gerbils** and **rats** to its vapors led to irreversible astroglial hypertrophy and/or proliferation. Eye irritation in **rabbit** trials. Exposure of **rats** has caused irreversible hearing loss. Fumes and overdosage of **cats, dogs, calves**, and **horses** caused weakness and ataxia. Inhalation by **mice** decreases sinoatrial rate, prolongs P-R interval, and sensitizes the heart to asphyxia-induced atrioventricular block. Capable of reactivating antibody neutralized virus.

LC_{50}, **fish**: 16.5–58 mg/l; LD_{50}, 14 days; **rats**: 3 ml/kg; adults, 6.4–7.4 ml/kg; inhalation LC_{LO}, 4000 ppm/4 h.

2,4-TOLUENEDIAMINE

= 2,4-Diaminotoluene = 2,4-Diaminotoluol = 2,4 TDA

Use: Millions of lbs used annually. In manufacturing ***toluene diisocyanate*** used in manufacturing ***polyurethane***, and in manufacturing dyes for fibers, ***leather***, fur, wood, and hair.

Untoward effects: Exposure is primarily by absorption after dermal contact and occasionally after inhalation. Has been absorbed during hair dyeing. It may have been a factor in inducing aplastic anemia in a 54-year-old **female** using hair dye containing it. Has caused incurable eczema in a **hairdresser** from contact with the hair dye in ***Kleenol***®. It leaches out of ***polyurethane*** foam-covered breast implants and the ***Today***® contraceptive sponge. In 1995, the FDA decided the lifetime risk from this **animal** carcinogen in **women** from the implants was 1/1,000,000. Inhalation has caused eye, skin, nose, and throat irritation. Skin contact or ingestion lead to dermatitis, sensitization, ataxia, tachycardia, nausea, vomiting, convulsions, respiratory depression, methemoglobinemia, cyanosis, headache, fatigue, dizziness, and possible hepatotoxicity.

LD_{LO}, **human**: 50 mg/kg.

Dietary use caused hepatocellular carcinomas or neoplastic nodules in **rats** of both sexes; SC fibromas in male and increased mammary gland carcinomas or adenomas in female **rats**, and increased lymphomas in female **mice**.

LD_{LO}, **rat**: 500 mg/kg.

TOLUENE-2,4-DIISOCYANATE

= TDI = 2,4-TDI

Use: 1 billion lbs annual U.S. use; 85% in flexible ***urethane*** foams for furniture, carpet underlay, bedding, etc.; also in ***polyurethane*** coatings, elastomers, ***Spandex***® and rigid foams, flexible packaging, and printing.

Untoward effects: Ingestion, contact, or inhalation lead to conjunctivitis, lacrimation, nose and throat irritation, coughing, paroxysmal choking, chest pain, retrosternal soreness, nausea, vomiting, abdominal pain, bronchitis, bronchospasm, dyspnea, hemoptysis, acute asthma attacks, pulmonary edema, dermatitis, and occasional skin sensitization. 50% decrease of respiratory rate at intolerable exposure level of 0.24 ppm. Hypersensitivity reactions with a death have been reported. Neurological complications followed a single severe exposure in a **worker**. Thrombocytopenic purpura reported. Its decompostion products (***carbon dioxide, cyanides***, and ***nitrogen oxides***) are released during fires. Odor threshold, 0.17 ppm; eye irritation, 0.05–0.1 ppm.

LD_{LO}, **human**: 5 g/kg; inhalation TC_{LO}, **human**: 0.5 ppm; inhalation TC_{LO}, **human**: 0.2 ppm over a 2 year period.

LD_{50}, **rat**: 6.2 g/kg; inhalation LC_{50}, **rat**: 600 ppm/6 h, **mouse**: 10 ppm/4 h; **guinea pig**: 13 ppm.

σ-TOLUENE SULFONAMIDE

= *σ-Toluenesulfamine* = *σ-Toluenesulfonamide*

Use: In manufacturing ***saccharin***, plasticizers, resins, and in organic synthesis.

Untoward effects: A toluene sulfonamide formaldehyde resin as an enamel nail polish and hardener is a common sensitizer and has mimicked onychomycosis. It improves gloss, adhesion, and flow, and yet can cause severe irritation on contacting skin during application. It does not cross-react with ***formaldehyde*** or ***sulfonamides***. It is an impurity in ***saccharin*** and suspect in causing bladder cancer in **man**.

LD_{50}, **rat**: 4.9 g/kg.

p-TOLUENE SULFONIC ACID

Use: In manufacturing dyes and antidiabetic drugs.

Untoward effects: Highly irritating to skin and mucous membranes on contact.

LD_{50}, **rat**: 2.5 g/kg; **mouse**: 400 mg/kg.

TOLUIDINE

= *Aminotoluene* = *Methylaniline* = *Tolylamine*

Use: *meta*: In manufacturing dyes and organic chemicals.

ortho: In manufacturing dyes, ***saccharin***, organic chemicals, pesticides, acid-fast colors for textiles, and as a vulcanizing accelerator.

para: In manufacturing dyes, reagents, and organic chemicals.

Untoward effects: *meta*: Methemoglobinemia, cyanosis, and mild anemia. Eye and skin irritation, nausea, vomiting, hypotension, weakness, and convulsions also reported. LD_{50}, **rat**: 974 mg/kg.

ortho: Linked to an increase in bladder cancer in industrial **workers**. In **man**, it can be absorbed after inhalation of dust or vapors, percutaneously, by ingestion from contamination of hands, or contact with contaminated surfaces or clothing. In the 1960s, bladder tumors occurred in Russian **workers** manufacturing it or the *para* form. **Their** working environment contained 0.5–28.6 mg/m^3. Induces a low level of methemoglobinemia, cyanosis, anemia, eye irritation and burns, headache, anoxia, weakness, dizziness, drowsiness, microhematuria, and dermatitis in exposed industrial **workers**. A major ingredient in the illicit manufacture of ***methaqualone***.

Dietary intake by **mice** led to increase in hepatocellular carcinomas and adenomas in females and hemangiomas. Its hydrochloride caused SC fibromas and mesotheliomas in male **rats**; splenic sarcomas, transitional cell papillomas, carcinomas of the bladder, mammary gland fibroadenomas and adenomas in female **rats**. A cause of hepatomas, cholangiomas, and bladder cancers in **hamsters**. LD_{50}, **rat**: 900 mg/kg; LD_{LO}, **cat**: 300 mg/kg.

para: Strong inducer of methemoglobinemia and cyanosis in **man**. Hematuria, eye and skin irritation, dermatitis, nausea, vomiting, hypotension, anemia, weakness, and convulsions also reported. LD_{LO}, **human**: 50 mg/kg.

LD_{50}, **rat**: 656 mg/kg.

TOMATO

= *Love Apple* = *Lycopersicon esculentum* = *Solanum lycopersicum*

The edible berry is a fruit, but the U.S. Supreme Court in 1893 labeled it a vegetable for the purpose of trade. Originated in Peru, Ecuador, and Bolivia, then, ~2000 years ago, they were transported to Central America and Mexico, and, eventually to Spain, Italy, and Europe by early explorers. In America, it was considered poisonous until 1821, when a wealthy farmer, Robert Gibbons Johnson, ate a tomato in front of the Salem, New Jersey Courthouse and lived.

Untoward effects: Botanically, it is a member of the ***deadly nightshade*** family, *Solanacea*, that consists of ***belladonna***, ***mandragora***, etc. The vines contain poisonous ***solanine*** (q.v.), but the fruit does not. Often implicated in food allergies, esophageal reflux, dyshidrosis, fissured contact dermatitis, and contact urticaria. **People** allergic to tomatoes are often allergic to green ***peppers*** and ***latex***. A high percentage of **atopics** have experienced pruritus; tingling and/or edema of lips, mouth, and tongue; or irritation of throat. Although they contain toxic (if injected) vasoactive amino compounds, such as ***histamine, 5-hydroxytryptamine*** (***serotonin*** - q.v.), ***norepinephrine***, and ***tyramine*** (4 μg/g), these are detoxified in the intestines by the naturally occurring **human** ***monoamine oxidase***, unless a **person** is taking ***monoamine oxidase inhibitors***. Green tomatoes contain potent cardiac depressing alkaloids, but these are destroyed when fried. The juice of tomato vines can cause a mild contact dermatitis in susceptible **individuals**. Tomatoes, tomato juices, and tomato sauces contain moderate amounts of ***oxalates*** and may need to be avoided by **women** with vulvodynia. Ingestion by **women** can increase the pain of interstitial cystitis. Being acidic, they and their juices can leach out ***cadmium*** and ***lead*** from certain storage containers. **Patients** with ***nickel*** dermatitis can develop vesicular hand eczemas after consuming tomatoes. Undigested material may color feces red.

They are common ingredients in **dog** foods, but often caused a dermatitis. The vines have killed **livestock** eating them, due to their ***solanine*** content and **children** have been poisoned after taking a decoction made from the leaves.

TONKA BEANS

= *Tonco Beans* = *Tonquin Beans*

These are the ripe seeds of large, leguminous trees. *Coumarouna odorata* (*Dipteryxodorata*) of Guiana and Brazil yields *Dutch Tonka* and *C. oppositifolia* of the Amazon yields *English Tonka*.

Untoward effects: They contain 3–20% ***coumarin*** (q.v.) and were once used in ***chocolate*** bars and foods for its ***vanilla***-like flavor in the U.S. In the 1940s, I called the FDA's attention to the fact that it could cause abnormal clotting and bleeding in **people**. They said it was harmless, and nearly 10 years later, on March 5, 1954, they banned its use in foods. Causes bradypnea and bradycardia. So-called ***Mexican Vanilla*** may contain it.

Rat and **dog** trials showed it caused extensive liver damage, retarded growth, and testicular atrophy.

LD_{50}, **rat**: 1.4 g/kg.

TOOTHPICKS

Untoward effects: Although apparently innocuous, ~8176 injuries from them were reported in the U.S. during 1972–1982. The greatest incidence occurred in **children** up through 14 years, and eight deaths were reported. Ear and eyeball injuries were common; less common were gastrointestinal punctures.

TOPIRAMATE

= *McN-4853* = *RWJ-17021-000* = *Topamax*

Anticonvulsant.

Untoward effects: Most common are central nervous system effects (~10%) including dizziness, mental slowing (poor word-finding and impaired concentration), somnolence, fatigue, ataxia, confusion, paresthesia, transient hemiparesis, diplopia, nervousness, irritability, and depression, which often increase with increasing dosage. Paresthesia and nephrolithiasis (1.5%) are associated with its inhibition of carbonic anhydrase. Weight decrease of 2–8%, peaking after 12–15 months of therapy and often due to induced anorexia.

Drug interactions: Has decreased ***digoxin*** serum levels. A 39-year-old **female** taking ***carbamazepine*** for 2 years without any liver problems received 50–300 mg/day topiramate/~4 months to control complex seizures leading to oliguria, severe encephalopathy and coagulopathy with fulminant liver failure, requiring a liver allograft.

Bladder tumors reported in **mice**. Malformations of limbs and decreased weight in **rodent** offspring, without adverse effects on fertility.

TOPOTECAN

= *Hycamptamine* = *Hycamptin* = *SKF 104,864*

Antineoplastic. **Inhibits a tumor growth enzyme. IV use.**

Untoward effects: Early onset of nausea (< 20%), vomiting (10%), diarrhea, headache, and flu-like symptoms. Dose-related and dose-limiting delayed bone marrow depression with leukopenia, neutropenia, anemia, and thrombocytopenia. Hematuria, stomatitis, asthenia, abdominal pain, and alopecia can also be delayed sequelae.

TOREMIFENE CITRATE

= *Fareston* = *FC 1157a*

Antiestrogen, antineoplastic. **Structurally similar to *tamoxifen*.**

Untoward effects: Nausea, sweating, hot flashes, and vaginal discharge are the most common. Occasionally peripheral edema, vomiting, dizziness, vaginal bleeding, and hypercalcemia. Less common are dry eyes, cataracts, and increased aminotransferase levels. Rarely, thromboembolic problems. A case report noted congestive heart failure after 1 month of therapy.

Drug interactions: ***Carbamazepine*** and ***phenytoin*** decrease its serum concentration and ***erythromycin*** and ***ketoconazole*** can increase it. May increase prothrombin time in **people** stabilized on ***warfarin***. ***Rifampin*** can decrease its serum levels ~87%. Embryotoxic and fetotoxic in experimental **animals**.

TORSEMIDE

= *AC-4464* = *BM-02015* = *Demadex* = *JDL-464* = *Toradiur* = *Torasemide* = *Torem* = *Unat*

***Diuretic, antihypertensive.* A *pyridine-sulfonylurea*. Oral and IV.**

Untoward effects*:** Headache, dizziness, and rhinitis occur frequently. Less common are hypokalemia, hypovolemia, hypotension, hypochloremic acidosis, hyperuricemia, nausea, vasculitis, and muscle cramps. Rarely, hearing loss. Allergic reactions can occur in **people** allergic to ***sulfonamides.

Drug interactions*:** Decreased renal clearance of ***spironolactone has little clinical significance. ***Indomethacin*** partially inhibits its natriuretic effect only in the presence of low dietary ***sodium*** intake. ***Probenecid*** decreases its secretion into the proximal tubule, decreasing its diuretic effect. May increase toxicity of high ***salicylate*** dosage, ototoxicity of ***aminoglycoside antibiotics*** and probably ***lithium*** compounds.

Cholestyramine decreases its oral absorption in **animals**. Transient hearing loss in 50% of **cats** receiving 20.8 mg/kg.

TOXALBUMIN

Untoward effects*:** In many parts of the world, various plants (q.v. ***Abrus precatorius, Cassia, castor bean, Croton, Jatropha's Locust, and ***Tung Oil Tree***) contain many toxic types. In general, these plants have hard seed coats which, fortunately, must be broken to release their severe toxin. Symptoms vary with the specific toxin. Chewing some causes corrosion at various contact points in the gastrointestinal tract. Severe symptoms, such as convulsions, shock, hemolysis, nephrotoxicity, and hepatotoxicity occur several days after exposure.

TOXAPHENE

= *Alltox* = *Chlorten* = *Chlorinated Camphene* = *Camphechlor* = *Geniphene* = *Hercules 3956* = *Motox* = *Phenacide* = *Phenatox* = *Polychlorocamphene* = *Strobane-T* = *Toxakil*

***Chlorinated hydrocarbon insecticide.* Contains ~68% *chlorine*. By 1974, world-wide cumulative use since its introduction in the 1940s was ~450,000 metric tons and U.S. annual usage nearly reached 100 million lbs. It contains nearly 200 components, with no single one being the only active ingredient. Now banned in many countries.**

Untoward effects*:** Stored in fatty tissues of the body. Very slow breakdown in soils and water. Widespread distribution in the environment can cause reproductive and endocrine-disrupting effects. **Human** exposure was by inhalation, percutaneous absorption, and accidental or deliberate ingestion. Within 1 h, nausea, confusion, agitation, tremors, pyrexia, convulsions, unconsciousness, respiratory failure, and dry red skin occurred. Rapid central nervous system stimulation is enhanced by external stimulation. Increased frequency of chromosomal aberrations of lymphocytes in exposed **workers**. Case reports of aplastic anemia after dermal exposure to a toxaphene-chlordane*** and a toxaphene-***lindane*** mixture. Resistance to ***warfarin's*** hypothrombinemic effect in a 53-year-old **male** on two occasions after heavy exposure to a 5% toxaphene and 1% ***lindane*** mixture. Foodstuffs with levels > 7 ppm have been dangerous.

Estimated LD, 2–3 g. LD_{LO}, **human**: 40 mg/kg.

Exposed **mice** showed increased incidence of hepatocellular carcinomas; in exposed **rats** led to increased incidence of thyroid follicular cell adenomas. Large numbers of **animals** were poisoned by its indiscriminate use by untrained **people** in Africa. Used successfully, but illegally, to kill **leopards** in Africa. Aerial spraying of ***cotton*** fields was often toxic to **birds**. Extremely toxic to **fish** and **marine biota**. Concentrations as low as 0.0005 ppm caused broken back syndrome, a collagen deformity, in **fish**.

LD_{50}, **rat**: 40 mg/kg; **mouse**: 112 mg/kg; **dog**: 15 mg/kg; **guinea pig**: 250 mg/kg; **mallard duck**: 71 mg/kg; **pheasant**: 40 mg/kg; LC_{50}, **trout** and **bluegills**: 0.0030.006 ppm.

TOXIFERINE

Untoward effects*:** The most potent physiologically active alkaloid from ***curare is used in South American arrow and dart poisons.

IV, LD_{50}, **mouse**: 23 μg/kg; **monkey**: 8900 ng/kg; IM, **monkey**: 18 μg/kg; **guinea pig**: 14 μg/kg.

TOXOGONIN

= *BH 6* = *Lü H6* = *Obidoxime* = *Toksobidin* = *Toxonin*

Cholinesterase reactivator.

Untoward effects*:** Weakness of facial muscle with warmth 5–10 min after IM. ***Menthol-like sensation of cold in nasopharyngeal area on inspiration in 12, 20–28-year-old volunteers after self-injection of 1 ml 25% solution. Facial symptoms disappeared within 2–3 h. Cholestatic jaundice after injection of 250 mg in a 34-year-old **female** after suicide attempt with 15 ml ***parathion*** orally.

IM, LD_{50}, **rat**: 205 mg/kg; **mouse**: 172 mg/kg; **cat**: 135 mg/kg; **guinea pig**: 79 mg/kg; IV, LD_{50}, **rat**: 133 mg/kg; **mouse**: 70 mg/kg; **cat** and **rabbit**: 100 mg/kg.

TRACHEMENE GLAUCIFOLIA

= *Wild Parsnip*

Untoward effects: Its ***furocoumarins*** or ***psoralens*** act as a photosensitizer on **people**, especially if **their** skin is moist from water or sweat. Usually misdiagnosed as ***poison ivy*** dermatitis. Lesions with a burning erythema develop a day after exposure. Edematous areas studded with small vesicles develop which coalesce into bullae the following day, enlarging over the next few days. Residual hyperpigmentation may be intense and is diagnostic of severe cases.

Ingestion of it by **sheep** in Australia led to "Bent Leg Syndrome". Oral intake of ***T. othracea*** decreases fertility in **rats** and **mice**.

TRACHYANDRA SALTII

= *Anthericum saltii*

A lily with grass-like leaves in Zimbabwe.

Untoward effects: Experimental feeding to **sheep** has caused deaths within 36 h. Anorexia, depression, dyspnea and frothing at the mouth were noted.

TRAGACANTH GUM

Use: As a ***hydrocolloid*** from Middle Eastern and Turkish ***Astragulus***, it has a long shelf-life as a suspending agent in medical emulsions, pharmaceutical jellies, toilet creams, lotions, dental creams, ***ice cream***, candy, syrups, jellies, salad dressings, sauces, chutney, ***paper*** and dye sizing, binding agent in cigars, stiffening felt, printing ***inks***, and ***polishes***.

Untoward effects: Often contains up to 0.1% ***sodium benzoate*** (q.v.). Despite its relatively safe use for > 2000 years and its generally recognized as safe status, the FDA wanted it banned from over-the-counter laxative and weight control drug products, because, taken with insufficient water, it and/or other gums caused some reports of esophageal obstruction and asphyxiation. It has induced immediate hypersensitivity reactions in some **people**. A Canadian case report described a life-threatening reaction in a 35-year-old **female** after consuming it as an additive in a sauce on a well-known brand of beefburger.

LD_{50}, **mouse**: 10 g/kg; **rabbit**: 7.2 g/kg; **hamster**: 8.8 g/kg; LD_{LO}, **rat**: 20 g/kg.

TRAGIA sp.

= *Noseburn*

Untoward effects: Has stinging irritating hairs that cause a dermatitis in **man** and **dogs**. Often causes salivation and incapacitation of hunting **dogs**.

TRAMADOL

= *CG 315E* = *Crispin* = *E 265* = *Topalgic* = *Tramal* = *U 26,225A* = *Ultram*

***Analgesic.* Oral, IM, and SC use.**

Untoward effects: ***Cardiovascular***: Orthostatic hypotension, syncope, tachycardia, vasodilation, flushing, palpitations, hypertension, and myocardial ischemia.

Central nervous system: Anxiety, confusion, dizziness, vertigo, incoordination, euphoria, nervousness, sedation, headache, tremors, depression, cognitive dysfunction, concentration difficulties, hallucinations, paresthesia, and seizures.

Gastrointestinal: Nausea, abdominal pain, anorexia, flatulence, constipation and occasional vomiting, gastrointestinal bleeding, diarrhea, xerostomia, stomatitis, and hepatitis.

Genitourinary: Menopausal symptoms, pollakiuria, and urinary retention.

Miscellaneous: Hypertonia, visual disturbances, cataracts, dyspnea, anaphylaxis, decreased weight, suicidal tendency, potential for addiction, and tinnitus. After a single 100 mg oral dose, a 32-year-old **male** showed a debilitating reaction within 3 h with ataxia, mydriasis, tremulousness, numbness in the legs, and dysphoria lasting ~4 h. Coma with doses > 500 mg.

Skin: Rash, diaphoresis, Stevens–Johnson syndrome, toxic epidermal necrolysis, urticaria, vesicles, and pruritus.

Drug interactions: ***Carbamazepine*** induces marked increase in its metabolism. It may increase ***digoxin*** toxicity. Conflicting reports on possible interaction with ***phenprocoumon***. Possible ***serotonin*** syndrome with ***sertraline***. Probable interaction reported increasing ***warfarin's*** anticoagulation. Avoid use with ***monoamine oxidase inhibitor***s. Therapeutic serum concentration 0.1–1.0 mg/l and fatal at 2 mg/l.

LD_{50}, **rat**: 228 mg/kg; **mouse**: 350 mg/kg; **dog**: 450 mg/kg; **rabbit**: 500 mg/kg; **guinea pig**: 850 mg/kg. SC, LD_{50}, **rat**: 286 mg/kg; **mouse**: 200 mg/kg; **guinea pig**: 245 mg/kg.

TRANEXAMIC ACID

= *AMCA* = *AMCHA* = *Anvitoff* = *Cyklokapron* = *Emorhalt* = *Exacyl* = *Frenolyse* = *Hexapromin* = *Hexatron* = *Rikavarin* = *RP-18,429* = *Spiramin* = *Tranex* = *Tranexan* = *Transamin* = *Trasamlon* = *Ugurol*

***Antifibrinolytic, hemostatic, thrombin and plasminogen inhibitor.* Usually oral; occasionally IV, and, experimentally, rectally. U.S. orphan drug.**

Untoward effects: Gastrointestinal upsets with diarrhea, abdominal pain, nausea, and occasional vomiting; dizziness, hypotension, exanthema, and meteorism.

Massive pulmonary embolism in a 62-year-old **female** with subarachnoid hemorrhage **who** received 25 g/day/3 days, an extremely high dose. An 83-year-old **male** with idiopathic thrombocytopenic purpura developed thrombophlebitis after 1.5 g/day/16 months. Fatal thrombosis of left common carotid artery after several years of intermittent therapy in a 39-year-old **female**. Case reports of intracranial thrombosis in two **females** taking it for menorrhagia during **their** days of bleeding. A single 500 mg IV caused firm, bullous eruptions in a 36-year-old **female**. Very low amounts secreted in breast milk.

Oral or IV treatment for up to 2 weeks of laboratory **animals** at doses 3–40 times those recommended for humans led to focal areas of retinal degeneration.

LD_{50}, **rat**: 3 g/kg; IV, LD_{50}, **rat**: 1.2 g/kg; **mouse**: 1.5 g/kg.

TRANSMISSION FLUIDS

Untoward effects: Contact irritation from some. **Drug-abusers** like certain ones, such as ***Transgo***®, for its euphoric effect upon inhalation.

TRANYLCYPROMINE

= *Parnate* = *SKF-trans-385* = *Tylciprine*

Antidepressant, monoamine oxidase inhibitor.

Untoward effects: May cause the release or concentration of high levels of vasoactive amines (***dopamine, epinephrine, histamine, norepinephrine, tryptamine, tyramine***) in the blood leading to palpitations, severe occipital headaches, hypertension, photophobia, nausea, vomiting, and prostration. These are aggravated by high amine-containing foods, such as ***ale, avocadoes, bananas, bean curd, beer, broad beans, canned milk, canned raisins, caviar,*** blue ***cheese*** and other aged ***cheeses***, ***chicken livers***, ***chocolate***, ***figs***, dried ***fish***, ***Marmite***®, tenderized ***meats***, ***miso soup***, ***pickled herring***, ***pineapple***, fermented ***sausages*** and ***wines*** (such as chianti), ***shrimp paste, soy sauce, tomatoes***, and ***yogurt***, causing the release of excessive amounts of ***norepinephrine*** and ***serotonin*** (q.v.). Occasionally, the hypertensive reactions are very serious and even fatal. Blindness reported from its use with ***cheeses***. This led to its temporary withdrawal from marketing in the U.S. Later, it was reintroduced for hospital use only. Hepatocellular jaundice, orthostatic hypotension, dizziness, xerostomia, hallucinations, constipation, impaired ejaculation, painful ejaculation, impotence, and paroxysmal hypertension. By stimulating ***antidiuretic homone*** secretion, it may cause hyponatremia. Rarely, marked weight increase is reported. Addiction, angina, and hypotension reported. Many fatal cases of self-poisoning reported. Symptoms are malignant hyperpyrexia, acute renal failure, muscle fasciculations, thrashing, sweating, flushing, hypotension, bizarre grimaces, and hyperactive deep-tendon reflexes.

Drug interactions: Use with the sympathomimetic agents (***amphetamines, ephedrine, isoproterenol, metaraminol, methylphenidate, phenylephrine***, and ***phenylpropanolamine***) prolongs and potentiates ***dopamine*** and ***tyramine*** effects, and releases excessive amounts of stored ***norepinephrine*** (q.v.), causing a hypertensive crisis, which also occurs when used with tricyclic ***antidepressants***, such as ***amitryptiline, desipramine, imipramine, nortriptyline***, and ***protriptyline***. Excitation, hyperpyrexia, weakness, convulsions, and fatalities also occur when tricyclic ***antidepressants*** have been given with it. Use with ***L-dopa*** also causes excessive release of ***norepinephrine*** and use with ***insulin*** can increase the latter's hypoglycemic response. Severe idiosyncratic reactions with ***meperidine***. Will block effectiveness of ***guanethidine***. Adverse neurological symptoms reported with ***amobarbital***. Fatigue, weakness, and drowsiness when given with ***methionine*** or ***L-tryptophan***. Potentiates ***chloral hydrate*** and ***cocaine***. Use with ***sertraline*** can induce chills, incoordination, confusion, restlessness, and memory disturbances. Toxic levels reported as 0.5 mg/l; comatose and fatal at 0.7–5 mg/l.

Moderate hyperthermia, up to 41.4°C, in some **rabbits** given ***procyclidine*** after pretreatment with tranylcypromine. In **animals**, it may cause serious hypertensive effects as with other ***monoamine oxidase inhibitors***. Effects may continue for weeks after the drug is discontinued. Combination with tricyclic ***antidepressants*** (***imipramine***, etc.) is *contraindicated*, as it will produce excitement, convulsions, and hyperpyrexia. Its use will increase cardiac arrhythmias with many ***anesthetics***.

LD_{50}, **mouse**: 38 mg/kg; IP, 41 mg/kg; IV, 37 mg/kg.

TRASTUZUMAB

= *Herceptin*

***Antineoplastic.* IV recombinant human monoclonal antibody for treating metastatic breast cancer.**

Untoward effects: Cardiotoxic, especially with ***anthracyclines***. Chills, nausea, vomiting, pain, asthenia, and headache. Diarrhea, anemia, leukopenia, and respiratory infections, especially with other treatments. Dyspnea, cough, peripheral edema, decreased ejection fraction, and paroxysmal nocturnal dyspnea in ~7%; 28% when used with ***anthracyclines***. Can increase bleeding problems in **patients** receiving ***warfarin***.

TRAZODONE HCL

= *AF-1161* = *Bimaran* = *Desyrel* = *Molipaxin* = *Pragmazone* = *Thombran* = *Tombran* = *Trazolan* = *Trittico*

***Antidepressant. α-adrenergic blocker.* Usually oral.**

Untoward effects: Cardiovascular: Tachycardia, palpitations, occasional sinus bradycardia during long-term use, syncope, shortness of breath, congestive heart failure, hemolytic anemia, atrial fibrillation, methemoglobinemia, ventricular arrhythmias, peripheral edema, angina, premature ventricular contrations, and orthostatic hypotension. Agranulocytosis in a 40-year-old **male** after 100 mg/day/ 1 month.

Central nervous system: Anger, hostility, confusion, decreased concentration, disorientation, delirium, dizziness, excitement, nervousness, fatigue, drowsiness, headache, involuntary jerking, insomnia, ataxia, nightmares, hallucinations, paranoia, mania, and muscle weakness. Parkinsonism in a 74-year-old **female** after 150 mg twice daily/2 months. Prolonged drowsiness in the **elderly** or the **obese**. Multiple tonic-clonic seizures in a 50-year-old **female** receiving 50 mg/day/18 days. Epilepsy in a 76-year-old **male** after 100 mg/day.

Gastrointestinal: Nausea, vomiting, flatulence, diarrhea, constipation, and anorexia.

Miscellaneous: Peripheral edema, xerostomia, and weight increase. Rarely, priapism in **males**, clitoral priapism in **females**, and increased libido in **males** and **females**. Retrograde ejaculation; ejaculatory inhibition in a 51-year-old **male** receiving 50–100 mg h/day/7 days. Blepharospasm after 100 mg/day. Dyspnea, hepatotoxicity, tongue and buccal lesions, malaise, fever, sialisis, pollakiuria, and hematuria. Insignificant levels in breast milk. After 6 weeks of treatment, leads to decreased hematocrit, hemoglobin, serum ***albumin***, ***calcium***, and ***cholesterol***. Ingestion by **adults** of 2–3 g led to respiratory depression/ apnea; occasionally, only ataxia and sedation.

Skin: Pustular psoriasis, leukonychia, erythema multiforme, bullae on distal extremities, and photosensitivity reactions.

Usual therapeutic serum levels are 0.5–1.6 mg/l; toxic at 4 mg/l; comatose or fatalities at 15 mg/l or above.

Drug interactions: Has induced toxic ***digoxin*** levels. May interact with ***monoamine oxidase inhibitor***s, as do tricyclics. ***Thioridazine*** increases its plasma concentration. ***Carbamazepine*** concentration increased by it in a 53-year-old **patient**. May increase or decrease ***warfarin*** effect. Pretreatment with it has decreased adverse ***cocaine*** effects. Inhibits hypotensive effect of ***clonidine***.

Overdoses in **dogs** caused hypotension, sedation, ataxia, tremors, and seizures. At low dosage (< 1 mg/kg), it is a ***serotonin antagonist***, and at doses of 6–8 mg/kg; it is a ***serotonin agonist***.

LD_{50}, **rat**: 486 mg/kg; **mouse**: 610 mg/kg; **rabbit**: 560 mg/kg; **dog**: 500 mg/kg.

TREMA ASPERA

= *Peach-leaf Poison Bush* = *Poison Peach*

Untoward effects: This Australian shrub or small tree varies in toxicity and, when grazed, its ***trematoxin*** has killed **cattle**, **goats**, and **sheep**. Some **cattle** develop a contact dermatitis and photosensitivity effects. Painful death occurs ~2–3 days after ingestion. Post-mortem shows myocardial, subepicardial, and subendocardial hemorrhages. Straw-colored fluid often found in body cavities, along with subserous hemorrhages.

TREMETOL

Untoward effects: A fat-soluble toxin found in ***Eupatorium sp.*** (q.v.), ***Goldenrod*** (q.v.), and ***Snakeroot*** (q.v.), along with another toxin, ***trematone***. Aside from poisoning and killing thousands of **livestock** grazing on it, it is best known as a cause of "***milk*** sickness" with a 10–25% mortality in **humans**. It is eliminated in the ***milk*** and the **nursling** is seriously affected before its **mother**. See above plants for symptoms. It also induces ketosis.

TREMORINE

Use: As a research tool in developing antiparkinson agents.

Untoward effects: Although pharmacologically inactive, it is metabolized by liver microsomes to its active metabolite, ***oxotremorine***, some of which is metabolized to inactive metabolites, and some is transported to the brain, where it has a Parkinson-like action. Oral, IV, IP, or SC of 20 mg/kg to **rats** and **mice** led to tremors within 10–30 min lasting 3–4 h. In the **dog** and **monkey**, tremors develop within 1 h and last for 2–3 days. Tremorine does not deplete brain ***dopamine***, as occurs in Parkinsonism in **man**; brain levels of ***serotonin*** decrease in Parkinsonism, but are augmented by tremorine. ***Atropine*** blocks all experimental tremorine effects, but in **man**, it only relieves the rigidity of Parkinsonism and voluntary movements by **man** decrease Parkinsonian tremor; the reverse occurs in **animals** receiving tremorine. In **cats** and **monkeys**, ***oxotremorine*** induces excitement, rage, and hallucinations with analgesic effect hundreds of times as great as ***morphine***. ***Desipramine*** inhibitis metabolism of tremorine and ***oxotremorine*** increases the latter's concentration in the brain. It causes hypothermia in **mice**.

TRIACETOXYANTHRACENE

= *Anthralin Triacetate* = *Exolan*

Antipsoriatic, antifungal.

Untoward effects: Pain, blurred vision, conjunctivitis, keratitis, and epithelial edema after accidental rubbing onto eyes.

TRIAMCINOLONE

= *Aristocort* = *Cinolone* = *CL-19,823* = *Kenacort* = *Ledercort* = *Omcilon* = *Tricortale* = *Volon*

The acetonide = *Adcortyl* = *Azmacort* = *Delphicort* = *Extracort* = *Ftorocort* = *Kenacort-A* = *Kenalog* = *Ledercort Cream* = *Nasacort* = *Respicort* = *Rineton* = *Solodelf* = *Tramacin* = *Triam* = *Tricinolon* = *Vetalog* = *Volon A* = *Volonimat*

The hexacetonide = *Aristospan* = *CL-34,433* = *Hexatrione* = *Lederlon* = *Lederspan* = *TATBA*

Glucocorticoid, anti-inflammatory. **Numerous forms exist. Topical, intralesional, intranasal, subconjunctival, oral, intra-articular, IM, SC, and IV.**

Untoward effects*: *Inhalation leads to local irritation, headache, dry mucous membranes, throat discomfort, sneezing, and nasosinus congestion. Rarely, dysphonia, and periodic aphonia, ulceration of nasal septum (from excessive use), thrush, and oral candidiasis.

Intralesional leads to panniculitis, skin atrophy, telangiectasis, hypopigmentation or hyperpigmentation, atrophic striae, sharply demarcated blanching line, and hirsutism; skin ulcers and stellate pseudoscars after long-term therapy. Rarely, blindness associated with intralesional injections around the face and head.

Oral leads to Cushing's syndrome, vertigo, insomnia, headache, drowsiness, adrenal suppression, ***sodium*** and fluid retention, hypokalemic alkalosis, hypertension, muscle weakness and myopathy, muscle cramps, loss of muscle mass, osteoporosis, vertebral compression fractures, aseptic necrosis of humeral and femoral heads, impaired wound-healing, thin fragile skin, petechiae, ecchymoses, facial erythema, anemia, eosinophilia, nonthrombocytopenic purpura, leukocytosis, thrombocytopenia, diaphoresis, suppressed reactions to skin tests, pseudotumor cerebri, gastroduodenal erosions, growth suppression in **children**, menstrual irregularities, decreased ***carbohydrate*** tolerance, increased requirement for ***insulin*** or ***oral hypoglycemic agents*** in **diabetics**, adrenal insufficiency, increased intraocular pressure, glaucoma, exophthalmus, posterior subcapsular cataracts, papilledema, and retinal hemorrhage.

Topical is a common sensitizer. Being a ***fluorinated steroid***, it can cause permanent vasodilation on the face, especially in **females**. Permanent stretch marks on the inner thighs and axillae. Hypertrichosis, senile-type purpura, exacerbation of acne, and transient increase in intraocular pressure, herpes or fungal infections in 30% of **patients** applying it for 2–3 weeks. Adrenocortical suppression due to large quantities or some under occlusive dressings. Sensitivity to other ***acetonide corticosteroids***, such as ***desonide***, ***fluocinonide***, and ***halcinonide***.

Drug interactions*:** Toxicity potentiation by ***anesthetics, ***anticholinergics***, ***tricyclic antidepressants***, ***sympathomimetics*** (in **asthmatics**), and ***live attenuated vaccines***.

Orally to **guinea pigs**, to cause experimental dermatophytosis lesions and to **mice** to cause muscle fiber atrophy. Cleft palate in treated **rats** and **mice**. Craniofacial and brain defects in **monkeys**. Teratogenicity during early pregnancy of **mice** has been demonstrated at doses proportionate to those used for **man**. Avoid use during active viral infections, in tuberculosis, Cushingoid syndrome, osteoporosis, diabetes, congestive heart failure, gastrointestinal bleeding, or the last trimester of pregnancy. Diarrhea, weight loss, anorexia, polydipsia, and polyuria are known to occur in some **animals** on high dosage and/or chronic administration. Reduce dosage if Cushingoid syndrome or comedone formations are noted in **dogs**. Osseous metaplasia can occur in **horses** following intra-articular injections. May inhibit estrous behavior and/or shorten its duration. Given during the last trimester of pregnancy, it may precipitate early parturition, and is associated with problems such as dystocia, fetal death, metritis, and retained placenta. Urticaria, dyspnea, collapse, shock, seizures, anaphylaxis, and adrenocortical suppression in **dogs** receiving high dosages.

TRIAMTERENE

= *Ademin(e)* = *Dyren* = *Dyrenium* = *Dytac* = *Jatropur* = *NSC-77,625* = *Pterofen* = *Pterophene* = *SKF-8542* = *Teriam* = *Triteren* = *Uretren* = *Urocaudal*

***Diuretic. Potassium*-sparing.**

Untoward effects*:** Occasional weakness, fatigue, xerostomia, dizziness, headache, jaundice, increased SGOT, increased SGPT, nausea, vomiting, diarrhea, hyper- or hypokalemia, increased blood urea nitrogen, increased ***creatinine, hypercalciuria, renal calculi, acute interstitial nephritis, decreased glomerular filtration rate, erythematous macular rash, slight bluish fluorescent coloration of urine, leg cramps, hyperuricemia, mild hypotension, azotemia, anorexia, hyponatremia, megaloblastosis, immune hemolytic anemia, fever, thrombocytopenia, decreased ***folic acid***, anaphylaxis, and photosensitivity reactions.

Drug interactions*:** Blocks glomerular filtration and tubular secretion of ***digoxin, thereby increasing its serum concentration. Can increase ***lithium*** toxicity. ***Indomethacin*** may unmask its nephrotoxic potential. Enhances the effects of ***hydralazine***, ***mecamylamine***, and ***veratrum*** alkaloids. Dangerously high ***potassium*** levels can occur with ***spironlactone***.

Found in the ***milk*** of treated **animals**.

TRIANTHEMA sp.

= *Gadon machiji* (Hausa)

As a pot herb during periods of scarcity and drought.

Untoward effects: Decoctions of the roots lead to mild uterine contractions and, in large doses, abortifacient in **women**. Poisonous in part due to its ***nitrate*** and ***oxalate*** content for **man** and **beast** leading to hematuria, vomiting, diarrhea, paralysis, and death due to acute nephritis.

TRIATOMA sp.

Confusion exists in parasitology nomenclature. They are often called *Asian bugs, assassin bugs, big bed bugs, kissing bugs, Mexican bed bugs*, and *sucking cone-nose bugs*, and are ~$\frac{3}{4}$–1$\frac{1}{4}$ inches in length. They are common in the southern and south-western U.S.

Untoward effects: Its initial bite is often not painful, but it causes an agonizing bite if pulled, rather than flicked off. Its saliva causes an allergic reaction – massive swelling at the bite site, and even of the face, lips, and eyelids. The bite of one South American species spreads trypanosomiasis or Chagas' disease. *Platymerus*, a large assassin bug, can accurately spit its saliva up to 12 inches, causing blindness in **man.**

TRIAZIQUONE

= *Bayer 3231* = *Trenimon*

***Antineoplastic*. IV. Occasionally topical, intracutaneous, and intralesional.**

Untoward effects: Leukopenia, thrombocytopenia, thrombophlebitis, pancytopenia, rash, nausea, vomiting, fatigue, edema, vertigo, and lens opacity.

IV, LD_{50}, **rat**: 470 μg/kg.

TRIAZOLAM

= *Clorazolam* = *Halcion* = *Novodorm* = *Someze* = *Songar* = *U 33,030*

***Sedative, hypnotic*. A *benzodiazepine*.**

Untoward effects: Most common in decreasing order of frequency have been drowsiness, headache, dizziness, nervousness, light-headedness, ataxia and balance coordination defects, nausea, and vomiting. Cognitive impairment, confusion, memory loss, amnesia, bizarre behavior, agitation; auditory, visual, and tactile hallucinations; suicidal and violent behavior, delirium, oversensitivity to light and sound, tachycardia, rebound insomnia, anxiety, and ***steroid*** psychosis from sudden treatment withdrawal. It has been an abused drug. Fatal intrahepatic cholestasis in a 44-year-old **male** after 0.25 mg/h/twice weekly or three times weekly/~6 months. Fatal self-poisoning with overdoses reported. Elderly **patients** seem more sensitive to its effects. A 62-year-old **male** with an existing hyponatremic state and given 0.25 mg was discovered on the floor 4–5 h later in an acute confusional state with several fractures. Murder charges against a **woman** for killing her **mother** were dropped by a Utah judge who said **her** addiction to the prescribed drug contributed to the killing. Severe and protracted delirium in a 65-year-old **male** after abrupt withdrawal.

TD_{LO}, **human**: 7 μg/kg.

Drug interactions: Coadministration with ***grapefruit juice*** increases its bioavailability by 50%. ***Cimetidine, diltiazem, erythromycin, fluoxetine, itraconazole, ketoconazole, mibefradil, nefazodone, ranitidine***, and ***troleandomycin*** increase the intensity and duration of its effects. ***Ethanol*** with it leads to greater psychomotor recall. ***Rifampin*** dramatically decreases its serum concentration and half-life from 2.8 to 1.3 h. ***Lopinavir*** and ***ritonavir*** decrease its metabolism, increasing its serum concentration. ***Oral contraceptives*** can cause a clinically insignificant increase in its peak plasma levels.

TRIBENOSIDE

= *Alven* = *Ba 21,401* = *Ciba 21,401Ba* = *Flebosan* = *Glyvenol* = *Hemocuron* = *Venex*

***Sclerotic*. Oral. Occasionally local.**

Untoward effects: Heartburn, anorexia, nausea, mild diarrhea, vomiting, pruritus, burning sensations in varicosities, generalized dermatitis and eczema, urticaria, angioneurotic edema, erythema multiforme, headache, deafness, anxiety, tachycardia, vertigo, thirst, polyuria, and leg edema reported in some cases.

TRIBROMOETHANOL

= *Avertin* = *Bromethol*

***Anesthetic, narcotic*. Usually rectal administration.**

Untoward effects: Initial hypotension, often rapid and severe, lasting ~10–15 min, and bradypnea. High dosage can cause respiratory paralysis, decrease cardiac function, and decrease liver function. Decomposed solutions form highly irritating ***dibromoacetaldehyde***, and ***hydrogen bromide*** (q.v.). Can decrease **fetal** respiration, with **infants** remaining inactive for up to 4 days after birth. Lethal serum levels reported as 9 mg% or above. Most recorded deaths occurred with doses of 75–100 mg/kg.

Abdominal adhesions, peritonitis, diarrhea, and intestinal stasis reported in treated **rats** and **mice**. Cyanosis and respiratory and circulatory depression in treated **birds**.

TRIBROMSALAN

= *TBS* = *Temasept* = *Trisanil* = *Tuasol*

***Bacteriostat*.**

Untoward effects: Severe contact photodermatitis resulted in the FDA banning its use in cosmetics.

TRIBULUS TERRESTRIS

= *Caltrop* = *Da Ogun Duro* (Yoruba) = *Devil's Thorn* = *Goathead* = *Puncture Vine* = *Qutiba* (Arabic) = *Tisido* (Yoruba)

In India, it is known as *Burra Gukeroo* or *Burra Gokhroo*.

Untoward effects: Contains hepatotoxic ***saponins*** throughout the plant leading to photosensitization with epithelial hyperplasia of bile ducts, jaundice, erythematous swollen head and ear skin 16–24 h after ingestion with serous seepage and occasional sloughing with a high mortality rate. Common in South Africa, Texas, New Mexico, Mediterranean areas, western Asia, and Hawaii in **sheep** and **goats** and occasionally in **cattle**. Called geeldikkop, geelsiekte, "thick yellow head", "thick ear", or tribulosis in South African **sheep**. The flattened fruit breaks into five nutlets, each with two strong spines, and these have become a mechanical danger on foot paths and waste places for both **man** and **animals**. In the U.S., its effect on **sheep** is called "big head" or "swell head". Preflowering, flowering, and fruiting plants are toxic. The udders of **cows** are easily affected. The wilting and immature plants are particularly toxic. Toxicity in **dogs** reported in India. Hypoglycemic effects in normal and diabetic **mice**.

TRIBUTYL PHOSPHATE

= *TBP* = *TNP*

Use: Plasticizer and solvent for cellulose acetate, ***lacquers***, chlorinated ***rubbers***, resins, and as an additive in aircraft ***hydraulic fluids***, and in vaccine production.

Untoward effects: In **man**, irritant to eyes, skin, and respiratory system; headache from inhalation. Nausea from excessive inhalation or ingestion. Persists in anaerobic soils and sediments. National Institute for Occupational Safety & Health sets upper exposure limit at 0.2 ppm and OSHA sets it at 0.4 ppm; immediate danger to life or health, 30 ppm.

Acute oral toxicity in experimental **animals** with general weakness, dyspnea, pulmonary edema, coma, and bladder hyperplasia.

LC_{50}, **rainbow trout**: 5–9 mg/l; LD_{50}, **rat**: 3 g/kg.

S,S,S-TRI-*N*-BUTYL PHOSPHOROTRITHIOATE

= *B 1776* = *Chemagro B1776* = *DEF* = *De-Green*

Defoliant.

Untoward effects: ***Organophosphate*** with potential for delayed neurotoxicity in **man**. Causes this in **hens**. Toxic symptoms in **mallard ducks** and **pheasants** are jerkiness, neck stretching, lacrimation, slowness, wing-drop, fluffed feathers, and prolonged ataxia.

LD_{50}, **rat**: 150 mg/kg; **mallard duck**: 2.9 g/kg; **pheasant**: 273 mg/kg; male **guinea pig**: 260 mg/kg.

TRICHILIA EMETICA

= *Anona* (Boran) = *Jan Saiwa* (Hausa) = *mnwamaji or mitamaji* (Swahili) = *Mudimdi or Nyamadze* (Digo)

Untoward effects: Very tall evergreen tree in Nigeria, where its fruit is used as food or chewed as ***cola***. Toxic effects from excessive intake leading to vomiting, diaphoresis, and diarrhea. **Natives** prepare an emetic from the bark and is poisonous if the mixture is too strong.

T. gilgiana in Zaire and other species in South America are used in treating respiratory diseases. In Amazonia, bark decoctions of various species are used as febrifuges, and in Cuba, ***T. havanensis*** ingestion has killed **sheep**.

TRICHLORFON

= *Agroforotox* = *Anthon* = *Arpalit* = *Bayer 2349* = *Bayer L 1359* = *Cekufon* = *Chlorofos* = *Chlorphos* = *Ciclosom* = *Combot* = *Danex* = *Delicia* = *Dipterex* = *Diptetes* = *Ditrifon* = *Dylox* = *Dyrex* = *Dyron* = *ENT 19,763* = *Forotox* = *Foschlorem* = *Masoten* = *Metrifonate* = *Metriphonate* = *Neguvon* = *Proxol* = *Soldep* = *Trichlorphene* = *Trinex* = *Tugon* = *Wohlfahrtol* = *Wotexit*

Cholinesterase inhibiting organophosphorus insecticide and anthelmintic.

Untoward effects: Nausea, colicky pain, tremors, vomiting, eye pain, blurring of vision, fainting sensation, vertigo, headache, and diarrhea for ~24 h after treatment. Some **patients** also had mild pruritus, erythema, and lymph gland swelling. Has caused poisoning from accidental or suicidal attempts leading to neuronal degeneration and polyneuropathy. The most important event is a reversible plasma cholinesterase inhibition.

LD_{LO}, **human**: 50 mg/kg.

Do NOT use in conjunction with or within a few days of (before or after) any other ***cholinesterase inhibitors***; avoid use with ***arsenicals, phenothiazine, phenothiazine tranquilizers, purgatives, succinylcholine***, or drugs producing purgation as a side-effect (2 weeks may be advisable in **horses**). ***Atropine*** is antidotal. Do NOT excite toxic **animals**. Do NOT use ***morphine***. Avoid use in **colts** under 4 months of age. **Brahma cattle**, **greyhound** and **whippet dogs** have been generally recognized as having increased sensitivity to this class of compound. Possibility of adverse reaction in heartworm-infested **dogs** has been suggested. Do NOT contaminate feed or water except as in product use directions. Apply to **cattle** as directed only *after* milking. Skin sensitivity has been reported in **humans**, and on **pets** wearing flea collars. Avoid use in very young, debilitated, and constipated **animals**, or those with cirrhotic livers or acute infectious diseases. Keep away from **children**. Since use for **fish** often kills many organisms on which **fish** feed, supplemental feeding must be used for 10 days after water treatment, to permit normal fauna to recover. Blindness can follow overdosing in **fish**. Aerosoling **sheep** 4 g/cm^3 for 1–6 h leads to 7 day tissue and 5 day ***milk*** residues. 1% level on **swine** led to tissue residues for nearly 2 weeks and fat residues for 3–4 weeks. Recommended 8% pour-on for **cattle** leads to ***milk*** residues of 0.03 ppm after 1 day; 0.01 ppm after 2 days; and < 0.01 ppm after third day. 0.06% solution as a spray leads to 0.04 ppm residues between third and sixth hours, and < 0.01 ppm by eighteenth hour. **Horses** often show increased peristalsis and diarrhea as dosage is increased. Use in **sows** has been associated with trembling in newborn **pigs**. Emesis, recumbency, and death reported in **dogs**. In **horses**, cholinergic effects lead to sialasis, nervousness, dyspnea, and recumbency. Deaths have been reported.

LD_{50}, **rat**: 450 mg/kg; **mouse**: 300 mg/kg; **rabbit**: 1450 mg/kg; **mallard duck**: 36.8 mg/kg; **pheasant**: 96 mg/kg; **bobwhite quail**: 22.4 mg/kg.

TRICHLORMETHIAZIDE

= *Achletin* = *Anatran* = *Anistadin* = *Aponorin* = *Carvacron* = *Diurese* = *Esmarin* = *Fluitran* = *Flutra* = *Hydrotrichlorothiazide* = *Intromene* = *Kubacron* = *Metahydrin* = *Naqua* = *Salurin* = *Tachionin* = *Tolcasone* = *Trichloromethiazide* = *Triflumen*

Diuretic, antihypertensive.

Untoward effects: Similar to ***chlorothiazide*** (q.v.). Thrombocytopenia, non-thrombocytopenic purpura, leukopenia, photosensitivity, mild hyperglycemia, hypercalcemia, and increased serum urates by inhibiting tubular secretion. Nausea, flushing, and muscle cramps occasionally reported.

Do NOT use ***milk*** of treated **cows** for **human** consumption until six milkings and 72 h after the last treatment.

LD_{50}, **rat**: > 20 g/kg; **mouse**: 3 g/kg.

TRICHLOROACETIC ACID

= *TCA* = *Trichloroethanoic Acid*

Use: Herbicide, decalcifier in microscopy, chemical peelant, medical caustic for warts, etc., and as a vesicant.

Untoward effects: Highly irritant and corrosive. Inhalation leads to eye, skin, nose, throat, and respiratory system irritation; coughing, dyspnea, salivation, and delayed pulmonary edema. Ingestion causes severe pain, violent vomiting, hematemesis, diarrhea, gastrointestinal corrosion, and acidosis. Topical leads to dermatitis and skin burns. National Institute for Occupational Safety & Health sets upper exposure limit of 1 ppm. It is the major metabolite of ***chloral hydrate*** and potentiates ***warfarin's*** anticoagulant effect.

LD_{50}, **rat**: 5 g/kg; **mouse**: 5.6 g/kg.

TRICHLOROBENZENES

Use: In flame retardants, insecticides, herbicides, dyes, and pharmaceuticals.

Untoward effects: Eye, skin, and mucous membrane irritant. Nephrotoxic and hepatotoxic in **animals** and possibly teratogenic.

LD_{LO}, **human**: 500 mg/kg.

LD_{50}, **rat**: 650 mg/kg; **mouse**: 615 mg/kg; **rabbit**: 812 mg/kg; **guinea pig**: 1.2 g/kg.

1,1,1-TRICHLOROETHANE

= *Aerothene TT* = *Chlorothene* = *Methyl Chloroform* = *Tri-Ethane*

Use: Solvent, in **inks** and **shoe polishes**, cutting oil coolant and degreaser. Has contained **dioxane**, a carcinogen, as a stabilizer. U.S. annual demand has decreased to 200 million lbs.

Untoward effects: Can deplete the ***ozone*** layer in the atmosphere and stratosphere for ~10 years. Excessive exposure causes eye and skin irritation, headache, central nervous system depression, lassitude, ataxia, dermatitis, cardiac arrhythmias, hepatotoxicity, unconsciousness, labored stridulous respirations, generalized body tremors, nephrotoxicity, cerebral edema, dyspnea, constricting chest pressure, chest pain, cough, myalgia, atalectasis, hypoxemia, leukocytosis, lymphocytosis, purpura, eosinophilia, anemia, hematuria, ventricular fibrillation, tachycardia, and cardiac arrest. Use in aerosol sprays, usually for purposeful abuse, has caused many deaths and caused its removal from the marketplace for this purpose. In 18 deaths from its use in a first aid spray, it was not listed as an ingredient because it was considered "inactive". Also absorbed from the gastrointestinal tract. It causes heart muscle to be more sensitive to normal levels of ***epinephrine***. OSHA and National Institute for Occupational Safety & Health set upper exposure limits at 350 ppm TWA and immediate danger to life or health at 700 ppm. Use without adequate ventilation has caused toxicity. Lethal blood levels have been 10–100 mg%.

Inhalation TC_{LO}, **man**: 350 ppm; 920 ppm/70 min.

In **dogs**, exposure to 0.5% air concentration sensitizes the heart to ***epinephrine***-induced arrhythmias; 2.5% causes tachycardia and 10% reduces myocardial contractility. In **monkeys**, 5% led to decreased myocardial contractility with tachycardia.

LD_{50}, **rat**: 14 g/kg; **dog**: 750 mg/kg; **rabbit**: 5.7 g/kg; **guinea pig**: 9.5 g/kg. Inhalation LC_{LO}, **rat**: 1000 ppm; **mouse**: 11,000 ppm/2 h.

1,1,2-TRICHLOROETHANE

= *Vinyl Trichloride*

Untoward effects: Overexposure of this solvent form leads to eye and nose irritation, central nervous system depression, dermatitis, nephrotoxicity, and hepatotoxicity. OSHA and National Institute for Occupational Safety & Health set upper exposure limits at 10 ppm and immediate danger to life or health at 100 ppm.

Carcinogenic in laboratory **animals**.

TRICHLOROETHANOL

***Sedative, hynotic.* It is a metabolite of *chloral hydrate* (q.v.) and *trichloroethylene* (q.v.).**

Untoward effects: In **horses**, alcohol dehydrogenase rapidly metabolizes ***chloral hydrate*** to trichloroethanol, its active form. In **human** volunteers, injection of ***ethanol*** 30 min after ***chloral hydrate*** led to higher and more prolonged concentration of plasma trichloroethanol. Therapeutic serum levels are listed as 5–15 and occasionally 20 mg/l and toxic levels as 40–70 mg/l.

LD_{50}, **rat**: 600 mg/kg.

TRICHLOROETHYLENE

= *Algylen* = *Chlorylen* = *Ethinyl Trichloride* = *Gemalgene* = *Germalgene* = *Narcogen* = *TCE* = *Trethylene* = *TRI* = *Trichlor* = *Trichloren* = *Trichloroethene* = *Tri-Clene* = *Trilene* = *Triline* = *Trimar* = *Westrosol*

Use: ***Inhalation anesthetic, solvent*** for vegetable and animal oils, fats, and wax extraction, metal degreasing, textile scouring in dry cleaning, and as an organic chemical intermediate. U.S. annual demand is for 240 million lbs.

Untoward effects: Eye and skin irritant. Inhalation and percutaneous absorption cause most of the symptoms and occasional ingestion reported leading to headache, vertigo, tremors, visual disturbances, optic nerve damage and atrophy, retrobulbar neuritis, oculomotor nerve paralysis, fatigue, giddiness, somnolence, nausea, vomiting, dermatitis; implicated in scleroderma, excitement, anorexia, paresthesia, cardiac arrhythmias, bradypnea, tachypnea, bradycardia, epigastric pain, conjunctival irritation, reversible dementia, ageusia, and trigeminal (fifth cranial) neuropathy. Hepatorenal toxicity in chronic inhalant **abusers**. Chronic inhalation causes fatty liver, liver necrosis, chills, cramps, anuria, hematuria, oliguria, proteinuria, coma, anemias, eosinophilia, leukocytosis, leukopenia, and death. Use as an inhalation anesthetic also causes increased vagal activity during induction, atrial or ventricular ectopic beat, and pulsus bigeminus. Has triggered a malignant hyperpyrexia syndrome. High anesthetic concentration leads to multifocal ventricular tachycardia and possible fibrillation, tachypnea, especially in **children**. Placental passage occurs and can cause sluggishness in **newborns**, and some adverse effects can last a year. Combines with ***soda lime*** producing ***phosgene*** (q.v.). Delayed absorption and toxicity can occur after oral use. A 62-year-old **male** accidentally ingested 100 ml leading to coma within 10 min and then pulmonary edema lasting 4 and 5 days, respectively. Its presence on surface to be welded caused ***phosgene*** poisoning from inhalation in a 45-year-old **male**. A 36-year-old **male degreaser** lowered a basket of metal into warm trichloroethylene, leading to ageusia, vertigo, analgesia in all divisions of right trigeminal nerve. Prolonged exposure in a **man** led to impotence, gynecomastia, pigmentation lymphadenopathy, peripheral neuropathy, scleroderma, malabsorption syndrome, and Raynaud's phenomenon. Exposure to 100 ppm has caused signifcant decrease in performance ability. A 16-year-old **female** ingested 400–500 ml leading to deep coma, bradypnea, psychomotor agitation, muscle spasms, hyperreflexia, vomiting; caused second degree skin burns, hypernatremia, and hyperchloremia. Tracheostomy and IV treatments saved **her**. Suspected as a cause of secondary deaths in **people** drinking ***milk*** from **cows** fed trichloroethylene-extracted ***soybean meals***. Topically, as a result of splashing onto the eye, caused smarting pain, chemical burns of lids, conjunctiva, and cornea with hyperemia and edema. Contaminated drinking water has caused serious **human** exposure and is the key subject in the book and movie, *A Civil Action*, and is considered a mutagen, teratogen, and cause of increased cardiac malformations, that concerned lawsuits filed against several companies, charging them with causing a cluster of leukemia cases in **children** in Woburn, Massachusetts. It has been an "on again" "off again" agent to decaffeinate ***coffee***. In 1979, **Nyburg** quotes **Paracelsus** (1493–1541), "Sola dosis facit venenum", translated as "Only the dose makes the poison." Minute traces have been found in decaffeinated ***coffee***. The FDA called it a food additive and it has caused cancer in **mice**. Converting those numbers to **humans** indicates that a **human** can reach the same mg/kg intake only by drinking 50 million cups of decaffeinated ***coffee***/day/

70 years. The U.S. manufacturer quietly discontinued its use, in fear of adverse publicity. The FDA permitted them to switch to ***methylene chloride*** (q.v.), which, at that time, had never been tested for carcinogenicity. OSHA upper exposure limit of 100 ppm. Immediate danger to life or health, 1000 ppm.

Inhalation LC_{LO}, **human**: 2900 ppm; inhalation TC_{LO}, **human**: 160 ppm/83 min; LD_{LO}, **human**: 50 mg/kg.

Drug interactions*: Workers** using it for degreasing after a few ***alcoholic drinks showed "**degreaser's** flush"; has also caused symptoms similar to inhalation adverse effects. A 41-year-old **male**, using it for 2 h in a poorly ventilated room, drank a ***beer***, and developed "rusty urine", flank pain, headache, oliguria, myocarditis, increased blood urea nitrogen, and pulmonary congestion. ***Sympathomimetics*** can induce fatal ventricular arrhythmias with it. Can detonate if it contacts finely divided ***barium***.

Inhalation exposure of **Mongolian gerbils** to 60 ppm/3 months led to evidence of brain damage. Cause of anemias, aplastic anemia, weakness, ataxia, jaundice, thrombocytopenia, and neutropenia in **cows** fed fat-extracted ***soybean meal***. ***Epinephrine*** induces cardiac arrhythmias in exposed **rabbits**. Causes hepatocellular carcinomas in **mice**.

LD_{50}, **rat**: 4.9 g/kg; **dog** and **cat**: 5.9 g/kg; **rabbit**: 7.3 g/kg. LC_{50}, **minnows**: 41–67 mg/l.

TRICHLOROFLUOROMETHANE

= *CFC 11* = *Fluorotrichloromethane* = *Refrigerant 11*

Refrigerant, aerosol propellent.

Untoward effects*:** Incoordination, tremors, cardiac arrhythmias, cardiac arrest, dermatitis, frostbite, asphyxiation, and hallucinations from inhalation and/or liquid content. Hundreds of fatalities reported from inhalation. Contraindicated topically in **people** with pigmented skins leading to hypopigmentation, followed by hyperpigmentation. Granular ***barium in contact with it can detonate. National Institute for Occupational Safety & Health and OSHA set upper exposure limits at 1000 ppm. Immediate danger to life or health 2000 ppm.

In **dogs**, 0.3% aerosoled sensitizes the heart to ***epinephrine***-induced arrhythmias; 1% causes tachycardia; and 5% causes hypotension and decreased cardiac contractility. Tachycardia and decreased myocardial contractility in **monkeys** at 5% level, and 10% level led to cardiac arrhythmias in **mice**. Can cause bronchodilation and severe bradycardia in **dogs**.

Inhalation LC_{LO}, **rat**: 10%/20 min.

TRICHLORONAPHTHALENE

***Use*:** In manufacturing condensers and in machine oils.

***Untoward effects*:** Fumes from open vats led to comedones and acne over face, back, chest, extremities, and behind the ears with pruritus in **workers**. The higher the chlorination, the greater the irritation.

2,4,5-TRICHLOROPHENOL

= *Collunosol* = *Dowicide 2* = *2,4,5-TCP*

Fungicide, bacteriocide.

***Untoward effects*:** Chloracne with or without porphyria cutanea tarda in chemical plant **workers**. Serious explosions have occurred in plants manufacturing it.

LD_{LO}, **human**: 500 mg/kg.

LD_{50}, **rat**: 820 mg/kg; **guinea pig**: 1 g/kg.

2,4,6-TRICHLOROPHENOL

= *Dowicide 2S* = *Omal*

Use*:** Wood and ***glue preservative; in antimildew treatment of textiles, and in manufacturing fungicides. Also as a bactericide, insecticide, and defoliant.

Untoward effects*:** Usual exposure is by dermal contact or inhalation in wood preservation, ***leather tanning, textile finishing, and hospital **workers**. Dietary administration led to increased leukemia or lymphomas in male **rats**; hepatocellular carcinomas and adenomas in **mice**.

2,4,5-TRICHLOROPHENOXYACETIC ACID

= *2,4,5-T*

Formerly sold under the following trade names: *Brush Killer, Brush-Rhap, Dacamine, Ded-Weed Brush Killer, Esteron 245, Fence Rider, Forron, Fruitone A, Inverton 245, LineRider, Reddon, Tormona, Tributon, Trinoxol, Trioxone, Weedar,* and *Weedone.*

Herbicide.* No longer sold in the U.S. Used in Vietnam in *Agents Orange, Pink,* and *Purple.

Untoward effects*:** Chloracne and possible increase in gastric cancers in manufacturing plant **workers**. Contamination with ***dioxin (q.v.) up to 30 ppm caused its market withdrawal. **Soldiers** complained of rashes, joint stiffness and pain, extreme fatigue, hypersomnolence, bradycardia, premature ventricular contractions, anorexia, nausea, vomiting, hematemesis, gastrointestinal ulcers, abdominal pain, diarrhea, constipation, hepatitis, jaundice, dizziness, headache, tingling, numbness, autonomic dyscontrol, depression, violent rages, sudden lapses of memory, suicidal attempts, miscarriages, decreased sperm counts and abnormal forms, brown urine, hematuria, blurred vision, dyspnea, gynecomastia, galactorrhea, and decreased libido. Teratogenicity has included myelomeningocele. In both **man** and **animals**, **fetal** ill-effects may be associated with the degree of ***dioxin*** contamination. Toxic serum levels are ~100 mg/l.

LD_{LO}, **human**: 500 mg/kg.

Teratogenic in **mice** and one strain of **rats**. Cleft palates in newborn **mice** and hemorrhagic gastrointestinal tract in **rat** fetuses. Half-life in **rats** and **chicks** is 4.7 h and 23.1 h in **humans**. Increase in cystic kidneys in a strain of **mice** and **rats**. Feeding 1000 or 2000 ppm/28 days to **sheep** and **cattle** led to anorexia and decreased weight gain.

LD_{50}, **rat**: 300–800 mg/kg; **mouse**: 389 mg/kg; **dog**: 100 mg/kg; **guinea pig**: 381 mg/kg; **chicken**: 310 mg/kg.

1,2,3-TRICHLOROPROPANE

= *Allyl Trichloride* = *Glycerol Trichlorohydrin* = *Trichlorohydrin*

Use: Synthetic chemical used as an industrial solvent, ***paint*** and varnish remover, degreasing agent, and in chemical synthesis.

Untoward effects: Short-time exposure to 100 ppm in air or to 50 ppm for an entire workday caused eye, nose, and throat irritation; and central nervous system depression. Chronic exposure to lower levels in **rats** and **mice** led to eye, nose, and lung irritation; gastric irritation, hepatotoxicity and nephrotoxicity. Higher levels led to increased liver, stomach, and mammary gland tumors. National Institute for Occupational Safety & Health sets upper exposure limit at 10 ppm and OSHA at 50 ppm. Immediate danger to life or health, 100 ppm. Persists in ground water.

Inhalation LC_{LO}, **rat**: 1000 ppm/4 h; LD_{50}, skin, **rabbit**: 1.8 g/kg; LD_{50}, 320 mg/kg; LD_{LO}, **dog**: 200 mg/kg.

TRICHLOROTRIFLUOROETHANE

= *Arcton 113* = *CFC 113* = *Freon 113* = *Genetron 113* = *Halocarbon 113* = *Refrigerant 113* = *TTE*

Aerosol propellent, refrigerant; degreasing solvent.

Untoward effects: Skin and throat irritant. Central nervous system depression, drowsiness, and dermatitis. During the period 1983–1990, 12 **workers** died from exposure to it in poorly ventilated workplaces. Narcosis, cardiac arrhythmias, and asphyxiation caused **their** deaths. National Institute for Occupational Safety & Health and OSHA set upper exposure limit at 1000 ppm. Immediate danger to life or health, 2000 ppm.

Inhalation TC_{LO}, **human**: 4500 ppm.

Cardiac arrhythmias and narcosis reported in experimental **animals**.

Inhalation LC_{LO}, **rat**: 87,000 ppm/6 h; **mouse**: 25%/1.5 min; **guinea pig**: 50,000 ppm/1 h.

TRICHOCEREUS PACHANOI

= *San Pedro*

Untoward effects: This ***mescaline*** (q.v. for toxicity) and ***hordenine***-containing cactus is used by Peruvian and Ecuadorian **faith-healers**. It is used in making ***cimora***, an hallucinogenic drink.

TRICHODESMA sp.

Untoward effects: Contain hepatocarcinogenic ***pyrrolizidine*** (q.v.) alkaloids. ***T. incanum*** seeds contain the toxic alkaloids ***incanin*** and ***trichodesmin***, and its contamination of forage and fodder causes mass poisoning of **dogs**, **horses**, **pigs**, and **poultry**. Contamination of cereal grains is potentially dangerous for **man**. In Russia, a nonpurulent encephalitis in **men** has been attributed to it. The rumen of **cattle** and **sheep** apparently destroys the alkaloids and they are resistant to its ill-effects. Young affected **calves** without developed rumens showed hyperemia and petechiae in heart and lungs, swelling and desquamation of bronchial epithelium, cloudy swelling in the myocardium, and catarrhal enteritis. Numerous publications indicate the alkaloids first affect the nervous and vascular systems leading to severe irreversible pathological changes and metabolic disorders. It is one of the commonest food-origin poisons in central Asia. It is the cause of "suilyuk disease" of **horses** in Russian Asia. Symptoms are bronchopneumonia; proliferative changes in the alveolar and bronchial epithelium; slight degeneration and atrophy of liver, kidneys, and heart; and catarrhal enteritis. Depression, fever, pollakiuria, dysuria, tachycardia, occasionally cough, and anorexia in affected **pigs**. Post-mortem showed bloody exudate in body cavities, hemorrhagic necrotizing nephritis, and gastrointestinal hemorrhages. **Chickens**, **ducks**, **geese**, and **turkeys** in decreasing order of susceptibility are also affected.

TRICHOLOMA sp.

Untoward effects: ***Muscarinic*** (q.v.) ***mushrooms*** of Asia, Europe, and U.S. **Human** toxicity reported for ***T. aggregatum***, ***T. muscaria***, ***T. pardinum***, ***T. terreum***, and ***T. venenatum*** leading to mild to severe nausea, abdominal cramps, vomiting, and diarrhea usually ½–2 h after ingestion and often persisting for several hours. Symptoms are greater with raw rather than cooked ones. ***Atropine*** is said to be contraindicated in treatment.

TRICHOMYCIN

= *Cabamicina* = *Hachimycin* = *Tricomicina* = *Trichonat*

Antifungal antibiotic. **Usually topical.**

Untoward effects: Oral use was tested and it was highly toxic. Acute IV intoxication leads to tachypnea. Should only be used topically.

After IV, hemolytic in **rabbits**.

LD_{LO}, **mouse**: 1 g/kg; IP, LD_{50}, **mouse**: 4–5 mg/kg or 0.05 mg/10–12 g. IV, LD_{50}, **mouse**: 0.28 mg/kg; SC, LD_{50}, **mouse**: ~3 mg/kg.

TRICHOSANTHIN

= *Compound Q* = *GLQ223*

Use: In the experimental treatment of the wasting effects of AIDS by inhibiting the virus replication; an herbal extract used as an abortifacient IV or IM.

Untoward effects: Fever (25%) after IM. It is the main active modality in an ancient traditional Chinese herbal medicine, ***Tien Hua Fen***, from the root of *Trichosanthes kirilowii* (***Chinese cucumber*** or ***bitter cucumber***).

TRICHOTHECENE MYCOTOXINS

Untoward effects: These metabolites are toxic for **man**, **cattle**, **donkeys**, **goats**, **horses**, **sheep**, **swine**, **fish**, **poultry**, **ducks**, **pigeons**, **geese**, **cats**, **dogs**, **guinea pigs**, **mice**, **rabbits**, **rats**, and **insects**. *Fusaria* and other fungi produce > 40 of them, with ***T-2*** (q.v.), ***vomitoxin*** (q.v.), ***nivalenol*** (q.v.), and ***diacetoxyscirpenol*** (q.v.) found naturally in foodstuffs and being the most common causes of clinical symptoms. Implicated in a wide variety of **human** and **animal** health problems responsible for the deaths of hundreds of thousands of **people** and used in chemical warfare. Not naturally found in Southeast Asia. General symptoms of the three lethal ones above are initial vomiting, hemorrhagic mucous membranes, bloody diarrhea, anorexia; severe skin paresthesia and pruritus with multiple blisters; often followed by death.

TRICLOBISONIUM CHLORIDE

= *Triburon*

Antiseptic.

Untoward effects: Low incidence of sensitivity or allergic reactions in **humans**. Oral doses of 10 mg/kg/day/14 days caused vomiting and diarrhea in **dogs**. **Rats** tolerate a single oral dose of 400 mg/kg.

LD_{50}, **mouse**: 375 mg/kg; SC, 154 mg/kg; IV, 12.5 mg/kg.

TRICLOCARBAN

= *Cutisan* = *Nobacter* = *Solubacter* = *TCC* = *3,4,4´-Trichlorocarbanilide*

Germicide, antiseptic, disinfectant.

Untoward effects: At high temperature, it can decompose into toxic ***chloroanilines*** that can be absorbed through skin. A cluster of methemoglobinemia and hemolytic anemia occurred in 18 **infants** (12 were premature) in a nursey within a 5-week period when the laundry used it as a 2% chemical rinse. Methemoglobinemia in eight **patients** after an enema with a soap that contained 2%. Prior heating of this solution under alkaline conditions can also form primary amines capable of producing methemoglobinemia. Low-grade skin sensitizer.

TRICLOFENOL

Untoward effects: Chloracne in > 70 **workers** within a few months after a factory explosion during its manufacture; cases still severe after 1½ years; **many** still showed signs 3 years later. Within 4 weeks after cleaning damaged plant, a **man's** 4-year-old son and the **wife** of a 41-year-old **male** had facial comedones 4 months and 11 months later. A ***dioxin*** was found in residue after the explosion.

A 0.001% solution caused acneiform lesions on **rabbit** ears. Was hepatotoxic in **animal** trials.

TRICLOFOS

= *Sch 10,159* = *Trichloroethyl Phosphate* = *Trichloryl* = *Triclos*

Sedative, hypnotic. **Related to *chloral hydrate*.**

Untoward effects: Drowsiness, gastric irritation, hypotension, headache, hypothermia, and respiratory depression. Has caused acute toxicity in a preterm **infant**. Rapidly hydrolyzed to ***trichlorethanol***.

Drug interactions: Potentiates ***warfarin's*** hypoprothrombinemic effect.

LD_{50}, **rat** and **mouse**: 1.4 g/kg.

TRICLOSAN

= *Aquasept* = *CH 3635* = *Gamophen* = *Irgasan DP 300* = *Sapoderm* = *Sterzac* = *Zalclense*

Antiseptic, disinfectant, bacteriostat, preservative.

Untoward effects: Contact dermatitis. Vaginal sprays with it may have been associated with hepatotoxicity and increased vulnerability to fungus infections. Oral use in **rats** indicated it might adversely affect renal function.

LD_{50}, **rat**: 4.4 g/kg.

TRICROMYL
= *Cromonalgina* = *Methyl-3-chromone*

Antispasmodic, vasodilator.

Untoward effects: Acute reversible Fanconi syndrome with renal failure in a $3\frac{1}{2}$-year-old **male** who ingested 3.5 g (normal **adult** dose is up to 600 mg/day). Similarities in chemical structure to ***tetracycline***.

TRIDIHEXETHYL CHLORIDE
= *Pathilon*

Anticholinergic.

Untoward effects: Xerostomia, glaucoma, allergic reactions, blurred vision, cycloplegia, increased intraocular pressure, ageusia, urinary hesitancy or retention, tachycardia, palpitations, headache, weakness, drowsiness, dizziness, insomnia, nausea, vomiting, impotence, urticaria, ataxia, rashes, leukopenia, petechiae, eosinophilia, and non-thrombocytopenic purpura. Although rare, more severe reactions include hyperpyrexia, chills, angioneurotic edema, bronchospasm, anuria, oliguria, delirium, confusion, erythema multiforme, bullous dermatitis, Stevens–Johnson syndrome, stomatitis, proctitis, agranulocytosis, aplastic anemia, and thrombocytopenic purpura.

Tridihexethyl Iodide = ***Claviton*** has similar untoward effects.

TRIENTINE
= *Cuprid* = *Syprine* = *TETA* = *TRIEN* = *Triethylenetetramine*

Chelating agent.

Use: Orphan drug in treating Wilson's disease, epoxy curing, additive for lubricating oil, and analytical reagent for ***copper*** and ***nickel***.

Untoward effects: Highly alkaline and irritating; gastrointestinal upsets, skin rash, and a case of acute rhabdomyolysis reported. Toxic ***ethylenimine*** (q.v.) has been detected in it at very low concentrations.

Drug interactions: Can cross-react with ***ethylenediamine*** (q.v.).

TRIETHANOLAMINE
= *TEA* = *Trolamine*

Solvent, emulsifier.

Untoward effects: Inhaled vapors are irritating. Skin contact irritant and allergen. Percutaneous absorption occurs. Can cause severe eye injury on contact (strong base). Nausea, abdominal discomfort, and collapse after ingestion. The FDA has been concerned about their nitrosation in cosmetics. Commercial synthetic cutting fluids have contained up to 45% and 18% ***sodium nitrite*** leading to ***nitrosamine*** exposure of **workers** and a potential for cancer formation.

LD_{LO}, **human**: 5 g/kg.

A product, ***Flee***®, sold in Florida to control **cat** and **dog** fleas, contained ***alkanolamines***, according to the FDA, which may be ***TEA*** or ***diethanolamine*** (q.v.), that caused hypersensitivity reactions, ataxia, posterior weakness, muscle twitching, and convulsions in **cats** and **dogs**. Some **cats** died. The FDA states it can cause adverse effects on the liver, kidney, pancreas, and nervous system.

LD_{50}, **rat**: 8.7 g/kg; **guinea pig**: 8 g/kg.

TRIETHYLAMINE
= *N,N-Diethylethanamine* = *TEA* = *U 11,555A*

Use: In synthesis of ***quaternary ammonium compounds***; in manufacture of accelerator activators for ***rubber***, corrosion inhibitors, herbicides, and as a solvent.

Untoward effects: Inhalation of vapors led to eye, respiratory system, and skin irritation. **Human** exposures of 18 mg/m³/8 h led to unpleasant visual hazes, halos, and corneal edema that can cause work and traffic accidents, but wears off in a few hours. It is a ***monoamine oxidase inhibitor*** inhibiting liver and brain ***monoamine oxidase*** in **mouse** studies. It is a ***non-steroid anti-estrus agent*** and said to decrease libido. Current published OSHA upper exposure limits at 25 ppm (100 mg/m³); immediate danger to life or health 200 ppm. In 1990, they were 10 ppm, 15-min exposure of 15 ppm, and immediate danger to life or health of 1000 ppm.

Single oral dose of 2.5 mg/kg within the first 4 days after insemination in **rats** previously pregnant by preventing the cleavage of fertilized eggs. Myocardial, renal, and liver damage reported in experimental **animals**.

LD_{50}, **rat**: 460 mg/kg; **mouse**: 546 mg/kg; inhalation LC_{LO}, **rat**: 1000 ppm/4 h.

TRIETHYLENEDIAMINE
= *Dabco*

Use: A ***polyamine*** used in amine curing agents or hardeners in epoxy resin systems, and as a catalyst in making ***urethane*** foams.

Drug interactions: Can cross-react with ***ethylenediamines*** (q.v.).

TRIETHYLENE GLYCOL
= *TEG*

Use: U.S. annual demand is for > 130 million lbs. Many esters have specialty uses. In natural gas dehydration, as a vinyl plasticizer, solvent, humectant, and aerial bactericide.

Untoward effects: Slight skin irritation after prolonged contact requires protective wear and goggles. A 15-year-old **female**, in a suicide attempt, drank 200 mg brake fluid (55% triethylene glycol and 10% ***diethylene glycol***). Full recovery within 2 days after emergency treatment.

LD_{50}, **rat** and **mouse**: ~18 g/kg; **rabbit**: 8.4 g/kg; **guinea pig**: 7.9 g/kg.

TRIETHYLENE GLYCOL MONOMETHYL ETHER

Untoward effects: Used primarily in brake fluid, and has caused severe fulminant blistering allergic dermatitis in sensitive **individuals**. A single drop on the antecubital fossa of such an **individual** led to a tremendous adverse dermatological event covering the entire medial and volar aspect of ⅔ of **his** arm and forearm, that spread to the side of **his** chest and abdomen. Other components of the brake fluid were negative in closed-patch tests.

LD_{50}, **rat**: 11 g/kg; skin, **rabbit**: 8 g/kg.

TRIETHYLENE MELAMINE

= *NSC 9,706* = *Peristol* = *TEM* = *Tretamine*

***Antineoplastic, alkylating agent, antimiotic, antiviral, and chemosterilant.* For parenteral use. Rarely used orally, due to poor absorption.**

Untoward effects: Very variable. Irreversible medullar aplasia has occurred after minimal oral doses. Corneal opacities, leukopenia, hemolytic anemia, and thrombocytopenia after various parenteral routes. Thrombophlebitis after accidental extravasation. IV of 0.25 mg/kg has been fatal. Other rare reports of death.

Mammalian, **avian**, and **insect** chemosterilant. Teratogenic in **rats** (abnormalities of extremities). Mutagenic in **rats** and **mice**. Sarcomas at injection sites in **rats**.

LD_{50}, **rat**: 14 mg/kg; IP, 1 mg/kg; **mouse**: 15 mg/kg; IP, 2.8 mg/kg; LD_{LO}, **dog**: 1 mg/kg; IV, LD_{LO}, **dog**: 400 µg/kg; **monkey**: 100 µg/kg; **cat**: 1 mg/kg.

TRIETHYLENE OLEATE POLYPEPTIDE CONDENSATE

= *Cerumінex* = *Kerumenex* = *Xerumenex*

Untoward effects: Contact dermatitis in ~1% with erythema, pruritus, burning sensations, swelling, and eczematoid reactions of external auditory canal after topical use to loosen cerumen. In one, 40-year-old **female**, swelling of the lateral aspects of **her** face also occurred. In a 43-year-old **male** who left it in **his** ears overnight without flushing it out in 15–30 min as recommended, caused bilateral acute otitis externa, swollen eyelids, and dermatitis of face, neck, and upper chest areas.

TRIETHYLENE PHOSPHORAMIDE

= *Aphoxide* = *APO* = *ENT 24,915* = *TEPA*

***Antineoplastic, chemosterilant, insecticide.* Also used in crease-proofing and flame-proofing textiles.**

Untoward effects: Induces acoustic duct tumors in **rats**. In **birds** leads to ataxia, wing-drop, tachypnea, dyspnea, nutation, tremors, ataraxia, and wing-beat convulsions. Oral doses of 50 or 100 mg to **sheep** led to depression, inappetance, and diarrhea followed in 2–3 days by dyspnea, incoordination, epistaxis, salivation, prostration, and cyanosis before dying with assorted pathology of the liver.

LD_{50}, **rat**: 37 mg/kg; **mouse**: 420 mg/kg; **goose**: 13 mg/kg; **mallard duck**: 8.54 mg/kg; **pheasant**: 30 mg/kg; **chukar**: 64 mg/kg; **bullfrog**: 500 mg/kg.

TRIETHYL LEAD

Untoward effects: Highly cytotoxic and selective neurotoxic derivative of the ***gasoline*** antiknock additive, ***tetraethyl lead*** (q.v.). It alters sensitivity and/or behavioral responses to auditory, visual, and somatosensory stimuli documented in **rat** trials. Experimentally, the *chloride* form is usually used.

LD_{50}, **rat**: mg ***Pb***/kg.

TRIETHYL TIN

In experimental trials, the *bromide* is the most commonly used; occasionally, the *sulfate*.

Untoward effects: In 1954 in France, intoxication in 210 **workers** exposed to triethyl tin salts with > 100 deaths. Non-lethal triethyl tin poisoning in 1972 in five **workers** in a German chemical laboratory. Symptoms were persistent headache, nausea, visual abnormalities, and electroencephalogram abnormalities that slowly resolved. Other **humans** exposed also had rapid weight loss, vertigo, painful sensations, and motor deficits, ranging from muscle weakness to paralysis.

Well-recognized as neurotoxic and verified by **rat** and **mouse** experiments causing cerebral edema, swelling of astrocytes and axons, and intramyelinic transient swelling. After several days of it in drinking water exposure, **rats** showed electroencephalogram abnormalities, hind limb weakness, and decreased motor activity.

LD_{50}, **rat**: 5 mg/kg/day/3 days.

TRIFLUOMEPRAZINE

= *Nortran* = *RP 7746* = *TFM*

***Phenothiazine tranquilizer, antiemetic.* Oral and IM.**

Untoward effects*:** In **man**, central nervous system depression, transient tachycardia in 85%, weakness, dizziness, anxiety, and restlessness. Occasionally hypotension, galactorrhea, extrapyramidal symptoms, and edema. A fatal case of agranulocytosis reported. Use with ***monoamine oxidase inhibitors led to hypotension and extrapyramidal symptoms.

Occasional unfriendly or viscious demeanor, shivering, lacrimation, hyperexcitability, and catalepsy in **dogs** and **cats** after IM or IV; nausea in **cats** after oral use. Motor restlessness has occurred at dosage too low to tranquilize.

LD_{50}, **mouse**: 1.35 g/kg; IV, LD_{50}, **mouse**: 29 mg/kg; **dog**: 60 mg/kg.

TRIFLUOPERAZINE DIHYDROCHLORIDE

= *Eskazine* = *Eskazinyl* = *Jatroneural* = *Modalina* = *Stelazine* = *Terfluzine* = *Triftazin* = *Triphthasine* = *Triphthazine*

Phenothiazine tranquilizer, psychotherapeutic.

Untoward effects*:** Extrapyramidal effects are common (~60%). Xerostomia, mydriasis, anterior cortical lens opacities, blurred vision, cycloplegia, oculogyria, pigmentary changes of eyelids; brown pigmentation of conjunctiva and whitish or yellowish spots on lens and cornea; decreased peristalsis, paralytic ileus, ischemic colitis, tachycardia, urinary retention, ventricular arrhythmias, torsades de pointes, memory impairment, photosensitivity, increased serum ***prolactin, galactorrhea, menstrual changes, breast tenderness, postural hypotension, rash, fixed-drug eruptions, anorexia, tardive dyskinesia, dyspnea, nasal stuffiness, drowsiness, insomnia, poor temperature regulation (heatstroke, etc.), dystonias, akathesia, and altered electrocardiogram results. Cholestatic jaundice, hepatitis, increased SGOT and SGPT, leukocytosis, neutropenia, anemia, leukopenia, gastritis, neuroleptic malignant syndrome, nausea, vomiting, painful ejaculation, decreased libido, spontaneous ejaculation, case report of fatal agranulocytosis, salivation, trismus, localized muscle pain, stiff neck, torticollis, facial spasms, mandibular dislocation, seizures, swallowing difficulties, protrusion of tongue, dysarthria, and altered taste and smell. Withdrawal from high dosage caused dizziness and tremulousness. High dosage can lead to delirium. Teratogenicity appears to be rare, if truly associated with its use (case report of phocomelia of upper extremities, transposition of great vessels, and ventricular septal defect). Case report of **neonatal** parkinson-like syndrome. Found in **human** and **animal** breast milk. In 1971, the manufacturer and the prescribing **physician** were ordered to pay $180,000 to a **woman** after having grimacing, blinking, and tremors; the manufacturer stated side-effects were rare and readily reversible. Toxic serum levels vary. Some reports state toxic levels as 0.1 mg/l, 0.12–0.3 mg%, and 0.1–0.2 mg/l. Fatal serum levels are stated as 0.3–0.8 mg%.

LD_{LO}, **human**: 500 mg/kg.

Drug interactions*:** ***Antacids decrease its absorption. Decreases ***L-dopa's*** antiparkinson effect. Use with ***tranylcypromine*** and other ***monoamine oxidase inhibitors*** increase extrapyramidal reactions and hypotension. When given to a **female** taking ***methyldopa***, **she** developed hypertension. Tumors and rigidity increased when ***physostigmine*** was given with it. It can potentiate the hypnotic effect of ***barbiturates, chloral hydrate***, and ***glutethimide***. It can also potentiate the analgesia of ***haloperidol*** and ***morphine***, and caused respiratory depression. It decreases mydriasis and central nervous system stimulation, and causes unusual visual experiences. It potenitates ***pargyline*** and ***reserpine*** action. Use with ***trazodone*** in a 72-year-old **female** may have induced fatal hepatic necrosis. Use with ***ephedrine*** has caused death of a **patient**.

In **animals**, its untoward effects are similar to ***triflupromazine*** below. Mammotropic in **rats**. NOT approved for use in food-producing **animals** in the U.S. Apt to cause motor restlessness and extrapyramidal symptoms in wild or zoo **animals**.

LD_{50}, **rat**: 543 mg/kg; **mouse**: 424 mg/kg.

TRIFLUPERIDOL

= *Flumoperone* = *McN-JR 2498* = *Psicoperidol* = *Psychoperidol* = *Trifluoperidol* = *Triperidol*

Psychotherapeutic.

***Untoward effects*:** Extrapyramidal symptoms in 30–50%, including tremors, rigidity, akathesia, muscular rigidity, catatonic-like state, dystonia, and parkinsonian-like syndrome. Occasional nausea, vomiting, weight decrease, xerostomia, drowsiness, insomnia, palpitations, fainting, constipation, confusional states, seborrheic dermatitis, and cutaneous allergic symptoms. A questionable role for it in causing cataracts or lens opacities has been reported.

Drug interactions*:** Has antagonized ***coumarin and ***phenindione*** anticoagulation effect.

LD_{50}, **rat**: 140 mg/kg; IV, 14 mg/kg; SC, 70 mg/kg.

TRIFLUPROMAZINE

= *Adazine* = *Fluopromazine* = *Fluorofen* = *Jatrosom* = *Psyquil* = *Siquil* = *Trifluopromazine* = *Vespral* = *Vesprin* = *Vetame*

***Psychotherapeutic, tranquilizer. A phenothiazine.* Oral and occasionally IM.**

Untoward effects*:** Extrapyramidal and sedative effects are greater than with ***chlorpromazine, but less hypotensive. Parkinsonism, paralysis agitans, akathisia, nuchal spasms, vertigo, anticholinergic effects, photosensitivity, skin pigmentation, decreased libido in **female**; impotence and inhibited or painful ejaculation, priapism, and spontaneous ejaculation in **males**; menstrual changes, galactorrhea, rashes, weight increase occasionally. Rarely, altered electrocardiogram, convulsions, cholestatic jaundice, agranulocytosis, eosinophilia, leukocytosis, leukopenia, neutropenia, lupus erythematosus, neuroleptic malignant syndrome, nausea, vomiting, dizziness, poor body temperature regulation, and lens deposits or opacity. After IM use, pain at injection site, tachycardia, hypotension, nausea, dizziness, excitement or restlessness, and vomiting. Therapeutic serum concentrations in **man** are reported as 0.03 and 0.1, and occasionally up to 0.2 mg/l; toxic concentrations given as 0.3–0.5 mg/l.

Drug interactions*:** Has stimulated metabolism of ***meprobamate by enzyme induction. See warning next paragraph.

In **animals**, avoid rapid IV (may cause severe hypotension and cardiovascular collapse), and intra-arterial injections (disorientation, convulsions, and deaths may occur), especially in **horses**. May increase the toxicity of ***organophosphorus*** or other ***acetylcholinesterase inhibitors*** and ***procaine***. Do NOT use ***epinephrine*** to combat its hypotensive or depressant effects, as it potentiates them. Do NOT consume ***milk*** taken respectively within 10 or 24 h after, or from the first milking after an IV or IM injection. To meet a United States zero tolerance for its residues on or in the uncooked tissues of food-producing **animals** requires no slaughtering within 10 days of the last treatment for **cattle** and **sheep** and 14 days for **swine**.

LD_{50}, **mouse**: 245 mg/kg; IV, 44 mg/kg.

TRIFLURALIN

= *Elancolan* = *ENT 28,203* = *L 36,352* = *Lilly 36,352* = *Olitref* = *Su Seguro Carpidor* = *Trefanocide* = *Treficon* = *Treflan* = *Triflurex* = *Trim*

Herbicide.

Untoward effects*:** Original production was highly contaminated with carcinogenic ***nitrosamines, and by 1980, Lilly eliminated the nitrosating agent used in its manufacture. It is now widely distributed in the environment and has caused some reproductive and endocrine upsets, particularly in **birds**. Mild ataxia has followed heavy exposures. High oral dosage experimentally has caused some depression, anorexia, diarrhea, and prostration.

LD_{50}, **rat** and **mouse**: 5 g/kg.

TRIHEXYPHENIDYL

= *Aparkane* = *Artane* = *Broflex* = *Cyclodol* = *Hexyphen* = *Pacitane* = *Panpal* = *Paralest* = *Pargitan* = *Parkinane* = *Parkopan* = *Peragit* = *Pipane* = *Sedrena* = *Tremin* = *Trihexane* = *Triphedinon* = *Triphenidyl* = *Tsiklodol*

Anticholinergic, antiparkinsonism.

Untoward effects*:** Minor undesirable anticholinergic effects, such as xerostomia, blurred vision, mydriasis, tachycardia, increased intraocular pressure, increased glaucoma in the **elderly**, and dizziness in up to 50%; occasionally skin rashes, colonic dilation, paralytic ileus, volvulus, constipation, drowsiness, headache, weakness, urinary retention, excitement, penile irritation, erectile disorders, confusion, memory impairment, delirium, disorientation, visual and auditory hallucinations (especially with overdoses and in the **elderly**, often triggered by photic stimulation). A case report of a chronic schizophrenic 36-year-old **male** Jewish Oriental who, paradoxically, developed bradycardia after treatment. Has been a popular drug of abuse along with ***ethanol and ***methaqualone*** for euphoria or hallucinations. Toxic psychosis in a 34-year-old **female** after ingesting 300 mg in a failed suicide attempt (usual **adult** dosage is 1–20 mg/day). Symptoms before treatment were severe mydriasis, dry skin, visual hallucinations, and depression. Withdrawal symptoms are anxiety, tachycardia, headache, and photophobia. Toxic serum levels reported as 0.5 mg/l.

TD_{LO}, **human**: 14 μg/kg.

Drug interactions*:** Use with ***diphenhydramine can increase xerostomia, with secondary dental caries and loss of teeth. Use with ***monoamine oxidase inhibitor***s can cause tremors and severe sweating by unexplained mechanisms. Lowers plasma level of ***chlorpromazine***.

Markedly increased ***mevinphos*** anticholinesterase activity and decreased ***oxime HI-6*** effect in the **rat**.

LD_{50}, **mouse**: 365 mg/kg; IV, 39 mg/kg; SC, 152 mg/kg; IP, 162 mg/kg.

TRIIODOTHYROACETIC ACID

= *Tiratricol* = *Triac* = *Triacana*

Antihypothyroid, antiobesity.

Untoward effects*:** Angina pectoris and aggravation of stenocardiac syndrome. Abuse by a 29-year-old **female body-builder** who ingested three tablets/day/2 months while on a diet developed 7 kg weight loss. Within 48–72 h, a 38-year-old **female**, after 0.75 mg tid and ***dexfenfluramine 25 mg tid developed progressive diplopia.

TRIMAZOSIN
= *Cardovar* = *CP 19,106* = *Supres*

***Antihypertensive, vasodilator, α_1-receptor antagonist.* Dilates both veins and arteries.**

Untoward effects: Palpitations and transient dizziness; occasional tachycardia.

TRIMEBUTINE MALEATE
= *Cerekinon* = *Debridat* = *Digerent* = *Foldox* = *Polibutin* = *Spabucol* = *TM-906* = *TMB* = *Trimedat*

Antispasmodic.

Untoward effects: Its addictive potential stressed in a case report of a 26-year-old **female** who self-injected it IV in the groin, legs, toes, and into **her** jugular.

TRIMELLITIC ANHYDRIDE
= *TMA (also used for trimethylamine)* = *TMAN*

Use: As a curing agent for epoxy and other resins; in vinyl plasticizers, ***paints***, coatings, polymers, polyesters, dyes, pigments, surfactants, agricultural and pharmaceutical chemicals, it exposes at least 20,000 U.S. **workers** annually.

Untoward effects: 'Extremely toxic and has caused non-cardiac pulmonary edema, immunologic sensitization with immediate onset rhinitis or asthma, a respiratory systemic syndrome with cough, dyspnea, malaise, muscle and joint aches, chills and fever after 4–8 h and worse at night (often called "***TMA*** flu"), and/or a pulmonary disease with hemolytic anemia, delayed onset cough, dyspnea, hemoptysis, pulmonary infiltrates, and hypoxemia. Eye and skin irritation also reported. Occasionally, a latency period of years can exist before symptoms and sensitization occur. **Workers** should wear protective clothing, goggles, and respirators. It is flammable. National Institute for Occupational Safety & Health sets upper exposure limit at 0.005 ppm (0.4 mg/m^3).

Causes eye burns in **albino rats**. In **rats**, inhalation of concentrations as low as 0.054 mg/m^3 caused intra-alveolar hemorrhages.

LD_{50}, **rat**: 5.6 g/kg; **mouse**: 2.2 g/kg.

TRIMEPRANOL
= *Betamann* = *Disorat* = *Glauline* = *Metipranolol* = *Methypranol* = *Optipranolol* = *Salpadyn* = *Torrat* = *VVPB 6453*

Antihypertensive, β-adrenergic blocker, antiglaucoma.

Untoward effects: Topical ophthalmic use leads to bronchoconstriction, uveitis, acute pulmonary edema, diaphoresis; transient stinging, particularly with the 0.6% concentration; conjunctivitis, eyelid dermatitis, blepharitis, blurred vision, tearing, conjunctival leukoplakia, photophobia, edema, and brow ache have occasionally been reported. Rarely, systemic effects, such as hypotension, bradycardia, and fatigue.

IV, LD_{50}, **mouse**: 31 mg/kg.

TRIMEPRAZINE
= *Alimemazine* = *Bayer 1219* = *Panectyl* = *Repeltin* = *RP 6549* = *Temaril* = *Theralene* = *Vallergan* = *Vanectyl*

Tranquilizer, antipruritic.* A *phenothiazine.

Untoward effects: Photosensitizer with delayed erythema, eczematous reactions, and slate gray pigmentation. Stevens–Johnson syndrome, neuroleptic malignant syndrome, corneal deposits or opacity, agranulocytosis, leukopenia, neutropenia, cytopenia, drowsiness (13%), headache, tinnitus, tremors, excitement, euphoria, depression, giddiness, hypotension, erectile disorder, decreased libido, and inhibition of ejaculation reported occasionally. Rarely, xerostomia, bad taste, gastrointestinal upsets, cholestatic jaundice, blurred vision, and skin rashes. Spontaneous hypoglycemia, coma, convulsions, mydriasis, and absent tendon reflexes in a 5-year-old **male** after 20 mg before sleep 3 days in a row. A case report of hypospadias in an **infant** whose **mother** used it during pregnancy. Other malformations are possible and, if taken during the last trimester, extrapyramidal symptoms may occur in the **neonate**. See warning below. Toxic levels reported as 0.5 mg/l.

In **animals**, avoid rapid IV and intra-arterial injections. May increase the toxicity of ***organophosphorus*** or other acetylcholinesterase inhibitors and ***procaine***. Do NOT use ***epinephrine*** to combat its hypotensive or depressant effects, as it potentiates them. In the U.S., it it NOT permitted for use in food-production **animals**. Zoo **workers** considered it very suitable for *Equinae*, as it was not too depressant. Overdosage in **animals** leads to depression, and incoordination. Occasionally excitement, aggression, vasodilation, hypotension, and shock.

LD_{50}, **mouse**: 230 mg/kg; IV, LD_{50}, 33 mg/kg.

TRIMETHADIONE
= *Absentol* = *Epidione* = *Petidon* = *Ptimal* = *Tridione*

Anticonvulsant.

Untoward effects: No longer popular for petit mal and psychomotor seizures because of serious, even if uncommon, adverse effects. Rash with acneiform eruptions, exfoliative dermatitis, hemeralopea, photophobia, glare in bright light, visual fading of colors, leukopenia, agranulocytosis, aplastic anemia, pancytopenia, nervousness, dizziness, epistaxis, malaise, stupor, drowsiness, nephrosis, albuminuria, hepatitis, nausea, megaloblastic anemia, eosinophilia, neutropenia, lymphocytosis, thrombocytopenia, sore throat, alopecia, systemic lupus erythematosus, and myasthenia gravis. Embryotoxic with high incidence of **fetal** malformations (microcephaly, conductive hearing loss, atrial and ventricular septal defects, patent ductus arteriosus, epicanthal folds, "V"-shaped eyebrows, low-set malformed ears, broad nasal bridges, cleft palate, cleft lip, irregular teeth, ocular anomalies, simian crease, lumbosacral meningomyelocele, clinodactyly, and inguinal hernias), growth delay, mental retardation, speech difficulty, and developmental delay in **children** whose **mothers** took the drug during pregnancy. Will cause spuriously increased urinary ***albumin*** test results.

Drug interactions: Aplastic anemia has occurred when used with a ***hydantoin***.

TD_{LO}, **woman** (pregnant): 6.2 g/kg.

LD_{50}, **rat**: 2.3 g/kg; **mouse**: 2 g/kg.

TRIMETHAPHAN CAMSYLATE

= *Arfonad* = *NU 2222* = *Trimetaphan*

***Antihypertensive, ganglionic blocker, cholinesterase inhibitor.* For IV use.**

Untoward effects: Can cause orthostatic hypotension and unpredictable as to degree of decreased cardiac output. Tachyphylaxis can develop with prolonged infusions during treatment for shock. Respiratory arrest has occurred due to a ***curare***-like effect at the neuromuscular junction. Occasional paralytic ileus, constipation, xerostomia, impaired visual accomodation; decreased renal blood flow, glomerular filtration, and protein synthesis; erectile disorder, decreased libido, and ejaculatory failure. Induced-asthma in known **asthmatics** due to ***histamine*** release. Risk increases with > 5 mg/min IV. Cerebral accidents with paresis reported in two **patients**, cardiac arrest in one after 2 cc of 0.1% solution given simultaneously with surgical manipulation of the carotid sinus.

Drug interactions: Unpredictable results can occur when given with other ***antihypertensives*** and ***anesthetics***, especially spinal ***anesthetics***. Concommitant use with ***diuretics*** can cause marked increase in its effects, both good and undesirable.

LD_{50}, **mouse**: 1.25 g/kg; **dog**: 400 mg/kg; IV, LD_{50}, **rat** and **mouse**: 21 mg/kg; **dog**: 750 µg/kg.

TRIMETHOBENZAMIDE

= *Anaus* = *Ro-2-9578* = *Tigan* = *Xametina*

***Antiemetic.* IM or suppository. Occasionally, oral. Bioavailability of the latter is ~32%.**

Untoward effects: Drowsiness and dizziness. Has induced extrapyramidal dysfunction, listlessness, hypotonia, and cyanosis after three 100 mg doses/24 h rectally in a 2-week-old **male neonate**. His **mother** had experienced extrapyramidal reactions to ***prochlorperazine*** 3 years earlier. Therapeutic serum levels in **adults** are stated as 0.1–0.2 mg%. Dysarthria and athetoid movements in a 15-year-old **female** after 200 mg IM. Neuroleptic malignant syndrome in a 48-year-old **male** after 250 mg orally/day/5 days. A 70-year-old **female**, after 200 mg IM, showed extreme nervousness, shaking, trembling, crying, and perspiring. Mixed hepatocellular jaundice and spotty necrotic acute hepatitis after a second course of 250 mg qid orally/2 days. A 4-year-old **male** who received 75 mg showed abdominal pain, painful swellings of wrists and knees, and puffiness of face and eyes. Headache, hypotension, blurred vision, diarrhea, opisthotonus, and skin reactions also reported. Pain, burning, stinging, swelling, and redness after IM. Studies indicate its use during pregnancy can lead to a significantly higher incidence of several congenital anomalies and increased **neonatal** death rate secondary to the anomalies.

LD_{50}, **mouse**: 1.6 g/kg; IP, 400 mg/kg; SC, 464 mg/kg; IV, 122 mg/kg.

TRIMETHOPRIM

= *Alprim* = *Instalac* = *Methoprim* = *Monotrim* = *Primsol* = *Proloprim* = *Syraprim* = *Tiempe* = *Trimanyl* = *Trimogal* = *Trimopan* = *Trimpex* = *Triprim* = *Uretrim* = *Wellcoprim*

***Antibacterial, sulfonamide potentiator.* A basic drug potentiated in acid urine.**

Untoward effects: Can impair nutrient metabolism and utilization, prolonging a ***folate*** deficiency and the accumulation of ***phenylalanine***. By inhibiting the enzyme dihydrofolate reductase leads to megaloblastic anemia. Thrombocytopenia, mostly in the **elderly** and **those** taking ***diuretics***. Eosinophilia, neutropenia, leukocytosis, leukopenia, and anaphylaxis also occur. Aseptic meningitis, photosensitivity, and hepatotoxicity reported. Excreted into breast milk at higher concentrations than in **maternal** serum, with **newborns** receiving ~750 µg–1 mg/day. Most reports of adverse events relate to its use with various ***sulfonamides***, particularly ***sulfamethoxazole***. Its therapeutic serum levels have generally been reported as 1.5–2.5 mg/l, and occasionally up to 5 mg/l. Toxic concentrations have been 10–20 mg/l.

Drug interactions: Absorption is decreased when given with ***cholestyramine*** or ***colestipol***. Hyponatremia in a 48-year-old **male** and a 65-year-old **female** also taking ***hydrochlorothiazide***. It is also synergistic with ***colistin***.

A case of drug-induced hepatitis has been reported in a **dog** at recommended dosage. Long-term therapy in **dogs** for 8 weeks causes a reversible drop in ***folate*** levels, red cell hemoglobin and volume; increase in blood urea nitrogen and serum ***creatinine***. Withdrawal time in **cattle** is 36 h; readily enters the ***milk*** of **cattle**. Some suggest 60 days if with ***sulfadoxime***. Withdrawal time in **calves** is 72 h. Trimethoprim levels persist in treated **fish** for at least 11 weeks. Resistance to trimethoprim is being reported and is transferable.

LD_{50}, **mouse**: 7 g/kg.

Use: Mostly in combination with ***sulfonamides***. Popular ratios are 1 : 5–1 : 9. Oral, IV, IM, and SC.

1 : 5 with ***Sulfachlorpyridazine*** and an ***antihistamine*** = ***Sulfotrim***

1 : 5 with ***Sulfamethoxazole*** = ***Bactrim*** = ***Biseptol*** = ***Cotrimazole*** = ***Eusaprim*** = ***RO 6-2580/9*** = ***Septra*** = ***Sumetrolim*** = ***Sumetrolium***

1 : 5 with ***Sulfadiazine*** = ***Cotrimazine*** = ***Di-Trim*** = ***Tribrissin*** = ***Tri-Globe*** = ***Trizine***

1 : 5 with ***Sulfadoxine*** = ***Animar*** = ***Borgal*** = ***Duoprim*** = ***Gorban*** = ***Trimethoprim S*** = ***Trivetrin***

5 : 4 with ***Sulfalene*** = ***Kelfiprim***

1 : 2.5 : 2.5 with ***Sulfamethazine*** and ***Sulfathiazole*** = ***F/Tnl*** = ***Vetoprim***

Many other ***sulfonamide*** combinations are used (viz. with ***sulfadimethoxine*** = ***Aescotrim*** = ***Trafigal***; with ***sulfamoxazole*** = ***Supristol***; ***sulfasoxizole***, ***sulfatroxazole***, etc.).

Trimethoprim with Sulfamethoxazole

Untoward effects: ***Blood and bone marrow***: Agranulocytosis, thrombocytopenia, decreased serum ***folate*** in 30%, neutropenia, purpura, megaloblastic anemia, abnormal red blood cell and white blood cell morphology, thrombocytopenic purpura with hematuria, acute severe hemolysis, hemolytic anemia, leukopenia, eosinophilia, leukocytosis, anemia, methemoglobinemia, cyanosis, pancytopenia, increased ***creatinine***, and aplastic anemia.

Clinical tests: Can cause misleading test results for ***methotrexate*** serum levels that are based on inhibition of bacterial dihydrofolate reductase.

Fetus and neonate: Hemolytic anemia and **neonatal** jaundice, especially during **their** first weeks of life in glucose-6-phosphate dehydrogenase-deficient **infants** from the ***sulfonamide*** component that can appear in breast milk. Use of ***Kelfiprim Syrup***® in **infants** causes nausea and vomiting.

Gastrointestinal: Pseudomembranous colitis, nausea, vomiting, diarrhea, epigastric pain, and anorexia. Disproportionate absorption rate of the components occurs in Crohn's disease, altering the ratio to 1 : 20.

Hypersensitivity: Reactions in **children** have been common. Anaphylactic shock, severe thrombopenia, renal failure, and hepatitis within 2 h. The **patient** had received it 2 months previously without any reactions. A 84-year-old **female** on the second day of treatment showed acute allergic reaction involving **her** skin, kidney, liver, and central nervous system. **Patients** have developed hypersensitivity rashes and acute renal failure after high dosage. **Some** died. Conjunctival photosensitivity, pneumonitis, vasculitis, pruritus, fever, weakness, and syncope have also been reported. Use of the azo dye ***FD & C Yellow No.6*** (q.v.) as a tablet coloring agent has induced nausea, leukopenia, and rashes.

Immunosuppression: Significant decrease in ***tetanus antitoxin*** titer in 40 **volunteers** treated twice daily/ 4 days after vaccination.

Liver: Jaundice and hepatomegaly 10 days after treatment of an 80-year-old **male** for orchitis. Fatal massive hepatic centrilobular necrosis on 25th day of jaundice. In a number of **patients** treated for 10–30 days, led to intrahepatic cholestasis.

Miscellaneous: Systemic lupus erythematosus, cough, fever, mental changes, decreased sperm counts, sore tongue, glossitis, xerostomia, convulsions, delirium, psychosis, paresthesias, joint pain, vaginal irritation and discharge, headache, decreased thyroid hormone concentration, aseptic meningitis, weakness, hyponatremia, acute pancreatitis, hiccups, esophageal ulcerations; disorientation and catatonia-like state in a 72-year-old **female**; tremors, apathy, hallucinations, ataxia, ankle clonus, periorbital edema, hypoglycemia, hyperkalemia, rhabdomyolysis; local pain or irritation after IV; tinnitus, vertigo, and pulmonary infiltrates reported. Tremors after 2 days of IV treatment with 18.5 mg/kg/day and 92.4 mg/kg/day of ***sulfamethoxazole***.

Renal: Nephrotoxicity with interstitial nephritis and decreased renal function (occasionally reversible), uremia, and increased serum ***creatinine***. An obstructing stone (5 × 2 × 2 mm) blocking the ureter contained a metabolite of the drug combination; transient oliguria. High dosage associated with glomerular and tubular crystallization in a 39-year-old **female** renal transplant **patient**.

Skin: Photosensitivity, toxic epidermal necrolysis, Stevens–Johnson syndrome (has, on occasions, been fatal), peeling skin, ulceration, pruritus, exanthema, maculopapular eruptions, and mild alopecia.

Drug interactions: A ***disulfiram***-type reaction reported after consumption of ***beer*** or taken with ***metronidazole***. Use with ***methotrexate*** led to megaloblastic pancytopenia; with ***phenytoin*** led to ***phenytoin*** toxicity. Potentiates bone marrow effect of ***azathioprine*** and the anticoagulant effect of ***warfarin***. Can induce the metabolism of and decrease the efficacy of ***oral contratraceptives***.

TD_{LO}, **woman**: 346 mg/kg; LD_{LO}, **human**: 274 mg/kg.

Reversible keratitis sicca reported and a polysystemic induced immune hypersensitivity with non-septic polyarthritis and fever in the **Doberman**. A non-reversible retinitis has also occurred. Hepatitis reported in a **St. Bernard** after 1 week of 480 mg/day of trimethoprim–sulfadiazine (***Tribrissin***). IV use can cause ataxia, tremors, respiratory depression, and convulsions in **animals**.

Tribrissin (with ***sulfadiazine***): LD_{50}, **rat** and **mouse**: > 5 g/kg; IV ***Bactrim*** (with ***sulfamethoxazole***): LD_{50}, **rat**: 700 mg/kg.

TRIMETHOXYAMPHETAMINE

= *TMA*

Untoward effects: An hallucinogenic ***amphetamine*** derivative. The psychedelic and hallucinogenic effects of ***calamus*** are due to its ***asarone*** content becoming aminated in the body to produce ***TMA***.

TD_{LO}, **human**: 800 µg/kg.

TRIMETHYLAMINE

Use: A petrochemical derivative used in making ***choline chloride*** and ***quaternary ammonium*** compounds; as an insect attractant and as a warning odor in natural gas.

Untoward effects: ***Choline***-containing dietary foods (***eggs, liver***, and ***soybeans***) are degraded by intestinal bacteria to the foul-smelling trimethylamine, which is then transported and metabolized to its odorless oxide. Some **people** poorly metabolize it and the rotten ***fish***-like smelling chemical is eliminated via the urine, perspiration, and breath. Eye, skin, nose, throat, and respiratory tract irritant leads to coughing, dyspnea, delayed pulmonary edema, blurred vision, corneal necrosis, and skin burns. Contact with the liquid leads to frostbite. It is odorous at 0.6 ppb. Overloading a **person** with 20 g ***choline*** orally can exceed a normal **person's** ability to metabolize the gut's bacteria-produced trimethylamine. National Institute for Occupational Safety & Health sets upper exposure limit at 10 ppm; 15-min exposure 15 ppm.

It has caused a foul fishy taint to **hen's** eggs, especially brown eggs, when the **hen's** liver metabolizing ability has been overwhelmed. Its content in ***fish*** meal fed to **mink** has caused anemia. Putrefaction of ***molasses*** transformed its ***betaine*** content to trimethylamine and poisoned **cattle**. It has tainted **cow's *milk***. Acts as a pheromone in **frogs** and stimulates the growth of plant tumors.

IV, LD_{50}, **mouse**: 325 mg/kg.

1,3,5-TRIMETHYLBENZENE

= *Mesitylene*

A liquid found in *coal tar* and *petroleum* crudes.

Untoward effects: Vapors cause eye, skin, nose, throat, and respiratory irritation, bronchitis, hypochromic anemia, headache, drowsiness, fatigue, dizziness, nausea, ataxia, vomiting, and confusion. Aspiration pneumonia has occurred. Other ***trimethylbenzenes*** cause similar toxicity. National Institute for Occupational Safety & Health sets upper exposure limit at 25 ppm.

Inhalation TC_{LO}, **human**: 10 ppm.

Inhalation LC_{LO}, **rat**: 2240 ppm/24 h.

TRIMETHYL PHOSPHITE

= *Methyl Phosphite* = *Trimethoxyphosphine*

Use: In making chemical warfare weapons.

Untoward effects: Inhalation causes eye, upper respiratory tract, and skin irritation. Dermatitis after skin contact. Neurotoxic. National Institute for Occupational Safety & Health sets upper exposure limit at 2 ppm (10 mg/m^3).

Teratogenic in laboratory **animals**.

LD_{50}, **rat**: 840 mg/kg; **mouse**: 1.47 g/kg; **rabbit**: 1 g/kg.

TRIMETOZINE

= *Opalène* = *Trimethoxazine* = *Trioxazine*

Anxiolytic.

Untoward effects: Evening fatigue, mental instability after high dosage, insomnia, excitement, nausea, anorexia, dizziness, headache, facial edema, and a case report of hyperglycemia.

LD_{50}, **rat**: 2.9 g/kg; **mouse**: 2.45 g/kg.

TRIMETREXATE

= *CI-898* = *JB-11* = *NSC-249,008* = *TMQ*

The glucuronate = *Neutrexin* = *NSC 352,122* = *Oncotrex*

***Antineoplastic, anti-Pneumocystis carinii, folate antagonist.* IV and oral use.**

Untoward effects: Occasional rash, nausea, vomiting, anorexia, increased aminotransferases, fever, chills, and diarrhea. Peripheral neuropathy, bone marrow depression, mucositis, neutropenia, pruritus, hyperpigmentation, nephrotoxicity, and hepatitis are often delayed symptoms. Potentially lethal in the treatment of *Pneumocystis carinii* pneumonia and usually averted by concurrent treatment with ***leucovorin***.

Drug interactions: Its metabolism is inhibited and its toxicity increased when used with ***clotrimazole, ketoconazole***, and ***miconazole***.

TRIMIPRAMINE

= *RP-7162* = *Sapilent* = *Trimeprimine* = *Trimeproprimine*

The maleate and methanesulfonate = *IL 6001* = *Stangyl* = *Surmontil*

***Tricyclic antidepressant*. Oral and IV.**

Untoward effects: Hypotension and tachycardia in ~5%. Xerostomia, sweating, sedation, weight increase, decreased libido, euphoria, pollakiuria, mild tremor of upper extremities, nausea, vomiting, constipation, rashes, confusion, and visual distortions (blurring) in 5–30%. Jaundice, leukopenia, leukocytosis, agranulocytosis, and eosinophilia in 0.5–1.0%. Salivary adenitis, convulsions, photosensitivity, decreased gastric acid secretion, extrapyramidal symptoms, myocardial infarction, arrhythmias, heart block, stroke, paresthesias, breast enlargement and galactorrhea; impotence, inhibited ejaculation, testicular swelling, and gynecomastia in **males**; and fluctuations in blood ***glucose*** reported. Therapeutic serum levels reported as 0.05–0.2 mg/l and toxic at or above 0.4 mg/l.

Drug interactions: Its serum levels and therapeutic effectiveness decreased by ***oral contraceptives***. Can block effectiveness of ***guanethidine***. Use with ***alcoholic*** beverages can increase its toxicity. ***Cimetidine*** and ***quinidine*** may inhibit its elimination, increasing its toxicity.

LD_{50}, **mouse**: 500 mg/kg; IV, 42 mg/kg; IP, 145 mg/kg; SC, 285 mg/kg.

1,3,5-TRINITROBENZENE

Untoward effects: **Human** exposure comes from contaminated food, air, soil, and water near Army munitions plants or chemical manufacturers. Symptoms after sufficient exposure are expected to be the same as for ***dinitrobenzene*** (q.v.) leading to anemia, methemoglobinemia, cyanosis, headache, nausea, and dizziness. Explosive.

LD_{LO}, **human**: 5 mg/kg.

LD_{50}, **rat**: 505 mg/kg; **mouse**: 572 mg/kg.

2,4,6-TRINITROTOLUENE

= *TNT* = *Tolit* = *Trilit* = *Trinitrotoluol* = *Trotyl*

Untoward effects: Explosive. Flammable. Contact can cause skin and mucous membrane irritation. Percutaneous absorption occurs and systemic ingestion can cause nail bed purpura, cataracts, methemoglobinemia, cyanosis, and anemia; even small doses cause hemolysis in glucose-6-phosphate dehydrogenase-deficient populations. A ***TNT*** explosion caused anemia in nearby Iraqi glucose-6-phosphate dehydrogenase-deficient **people** (~25% are deficient). Occupational toxicodermia with yellowish discoloration of hands, hairs on the head and chest, and a bullous dermatitis of hands, forearms, face, and neck in a munitions **worker** 2 months after **he** started handling it. Coughing, sneezing, jaundice, sore throats, hepatic necrosis, nausea, vomiting, diarrhea, anoxia, peripheral neuropathy, muscle pain, nephrotoxicity, leukocytosis, leukopenia, monocytosis, eosinophilia, aplastic anemia, sulfhemoglobinemia, neutropenia, pancytopenia, thrombocytopenia, cardiac arrhythmias, and hypersensitivity dermatitis have also occurred in exposed **workers**. Fatalities of 25–30% in sick exposed munitions **workers**. National Institute for Occupational Safety & Health and OSHA upper exposure limits on the skin are 0.5 mg and 1.5 mg/m^3, respectively. Immediate danger to life or health 500 mg/m^3.

Hematuria in exposed experimental **animals**; decreased liver function in **rats** and **rabbits**. Accidental ingestion by **sheep** led to fever and methemoglobinemia, clinically resembling anthrax or clostridial disease. **Dogs** and **cats** are more susceptible, showing gastrointestinal irritation, bladder irritation, ataxia, nystagmus, moderate anemia, hemosiderosis, methemoglobinemia, and fatty degeneration of the liver.

LD_{LO}, **rat**: 700 mg/kg; **cat**: 1.85 g/kg; **rabbit**: 500 mg/kg.

TRIOXSALEN

= *NSC 71,047* = *TMP* = *4,5′,8-Trimethylpsoralen* = *Trioxysalen* = *Trisoralen*

***Pigmentation-inducer, photosensitizer of skin to UVA*. An isomer of *furocoumarin*, a naturally occurring ubiquitous compound.**

Untoward effects: Essentially, the same as with ***methoxsalen*** (q.v.), but it is a more potent photosensitizer.

TRIPARANOL

= *Acosterina* = *Clotrox* = *Diticil* = *Drenaren* = *Hipocolestina* = *MER-29* = *Metasclene* = *Metasqualene* = *Sclane* = *Trianel* = *Trikosterol* = *Triparin* = *Tropalin* = *Valip* = *Verdiana*

Antilipemic.

Untoward effects: Irreversible lenticular changes and cataracts caused numerous lawsuits and forced the manufacturer to withdraw it from the marketplace in 1962 after 12 years of use. Skin dryness, conjunctivitis, hair turning blond, alopecia, leukopenia, neutropenia, lymphocytosis, ichthyosis, cutaneous xanthomata, decreased libido, malabsorption syndrome, occasional nausea and vomiting, and decreased liver microsome activity.

***Drug interactions*: Increases side-effects of *oral contraceptives*.**

Caused cataracts in **dogs** and **rats**; congenital defects in **rats** included short tail and anophthalmia.

LD_{50}, **rat**: 2 g/kg.

TRIPELENNAMINE

= *Aescobenzamine* = *"Blues"* = *Ciba 63* = *Pyrizine*

The hydrochloride = *Azaron* = *Dehistin* = *PBZ* = *Pyribenzamine* = *Vetibenzamine*.

With *barbiturates*, *paregoric*, or *pentazocine*, street names are *B´s & T´s*, *Blue Velvet*, *T´s & Blues*, *Tops & Bottoms*, *Teddies & Betties*.

***Antihistamine*. The tablets are a light blue color. *T´s* are *Talwin* or *pentazocine*.**

Untoward effects*:** Some drowsiness, xerostomia, gastrointestinal upsets, irritability, nervousness, insomnia, and disorientation reported. Prolonged hallucinations followed a modest overdose in a 5-year-old **female** with intense pruritus. Excitement, agitation, confusion, visual hallucinations, mydriasis, blurring of discs, and myoclonic jerks in an 8-year-old **male** 3 h after topical aerosol (1.75% concentration) treatment of ***poison ivy lesions, indicating significant percutaneous absorption. Urethral instillation of 15 cc of 2% solution in a 70-year-old **male**, within 15 min, led to tachycardia (148/min) and severe hypertension (254/150). Restlessness and disorientation. Within 2½ min, the same treatment in another 70-year-old **male** led to irrational state, twitching, coughing, retching, pulse 140/min, and blood pressure 240/140. Tingling, blurred vision, and transitory paralysis during recovery. Contact dermatitis, fixed-drug eruptions, neutropenia, pancytopenia, aplastic anemia, hemolytic anemia, and anaphylaxis reported. A case of liver hypersensitivity leading to cholestatic jaundice noted. Fever, urticaria, and arthralgia in a 49-year-old **female** after 13 days' treatment. Has anticholinergic activity and triggers an anticholinergic syndrome. Leukopenia, agranulocytosis, thrombocytopenic purpura, pancytopenia, and aplastic anemia have occurred after excessive use. When abused with other drugs, usually IV, pulmonary complications, such as angiothrombosis from inerts in the drugs; hypertension, nephrotic syndrome, convulsions, cramps, restlessness, and euphoria or depression have sent **many** to the emergency room. Estimated LD, 25–50 mg/kg.

Drug interactions*:** It can cross-react with ***ethylenediamine, ***hydroxyzine***, ***methapyrilene***, ***phenergan***, and ***pyrilamine***. Can interact with ***guanethidine*** to antagonize the latter's adrenergic-blocking action.

Overdosage in **animals** leads to ataxia, excitement, and convulsions. Causes a fatal hyperpyrexia in **rabbits** pretreated with ***monoamine oxidase inhibitors***.

LD_{50}, **rat**: 515 mg/kg; **mouse**: 235 mg/kg; IV, LD_{50}, **rat** and **mouse**: 12 mg/kg; **dog**: 3 mg/kg; **rabbit**: 9 mg/kg.

TRIPHENYLPHOSPHATE

= *TPP*

Use*:** Fire-retarding agent and in plasticizing ***cellulose, nitrocellulose, ***lacquers***, and varnishes.

Untoward effects*:** Some changes in blood cholinesterase in **man** and **animals**. Several reports of contact dermatitis in **people** from its use in ***cellulose acetate eyeglass frames. National Institute for Occupational Safety & Health and OSHA set upper exposure limit of 3 mg/m^3. Immediate danger to life or health 1000 mg/m^3.

LD_{LO}, **human**: 50 mg/kg.

Muscle weakness and paralysis in **cattle**, **goats**, **rabbits**, and **chickens**, as well as dyspnea, tympanites, incoordination, and some deaths.

LD_{LO}, **rat**: 3 g/kg (4.7 g/kg of a commercial mixture).

TRIPHOLIUM ALEXANDRINUM

= *Barseem* = *Trifolium alexandrinum*

***Untoward effects*:** Excessive feeding of **water buffaloes** in India led to severe illness, gross anemia, hemoglobinuria, jaundice, and leukocytosis.

TRIPROLIDINE

= *Actidil* = *Actidilon* = *295C51* = *Pro-Actidil* = *Pro-Entra* = *Venen*

***Antihistamine*.**

Untoward effects*:** Subjective central nervous system effects after 1½ h, morning fatigue in 50%. Excitement noted, mostly in **children**. Occasional lichenoid skin eruptions after 4–10 days of therapy, xerostomia. Skin eruptions with poikiloderma in two **females**. Confirmed by rechallenge and more marked in exposed skin areas or after exposure to sunshine. Passes into breast milk – 0.9% of **maternal** dose, leading to **infant** dose of ~0.36 μg/kg/day. Therapeutic **infant** dose is 1.25 mg/kg. Near-fatal case from ***Actifed Compound Linctus® in a 3-month-old **baby** (6.6 mg/kg ***codeine*** in two doses; 1–2 mg/kg usually prescribed). Visual hallucinations lasting 1 week–4 months in three **children** 2½–3½-years-old after recommended dosage, probably due to the ***pseudoephedrine*** content. A case report of aplastic anemia.

Potentiates ***serotonin*** leading to hyperpyrexia in **rabbits** receiving ***monoamine oxidase inhibitors***.

TRIPTERYGIUM WILFORDII

Untoward effects: This Chinese herbal's ***polyglycosides*** are slowly effective in a high percentage of **human** rheumatoid arthritis and psoriasis cases, but cause skin reactions in many **patients** and amenorrhea in 31.3% of 70 treated **patients**. Its active compounds are ***tripdiolide*** and ***triptolide***.

It suppresses spermatogenesis, which is reversible, in **rat**, **mouse**, and **dog** trials, and has caused decreased fertility in female **mice**. The crude root extract also causes lesions in the heart, liver, and testes.

TRIPTORELIN

= *Decapeptyl*

***Luteinizing hormone-releasing hormone, gonadotropin-releasing hormone agonist, antineoplastic.* IM or SC.**

Untoward effects: Hot flushes and night sweats in **females** and erectile failure, hot flushes, and decreased bone density in **men**. Ovarian hyperstimulation has induced a myocardial infarction in a 35-year-old **female**.

TRIS(2-3-DIBROMOPROPYL) PHOSPHATE

= *Tris BP*

***Flame retardant.* Formerly sold in the U.S. under various *Apex, Flammex, Firemaster,* and *Fyrol* names. Now banned.**

Untoward effects: Having caused genetic changes in bacteria and renal tumors in **mice** and **rats**, testicular atrophy and spermatogenesis in **rats** and **rabbits**, and hepatoxicity in **rats**, it was eventually banned for use in **children's** pajamas. Contained several carcinogenic contaminants. **Children** wearing treated and washed pajamas absorbed it and had ***dibromopropanol*** in **their** urine. After its ban in the U.S., millions of treated garments were shipped to countries in Latin America, Asia, and Africa.

LD_{50}, **rat**: 1 g/kg.

TRISETUM FLAVESCENS

= *Green Golden Oat Grass* = *Yellow False Oat* = *Yellow Oat Grass*

Untoward effects: Severe calcification of the left endocardium, the aorta, and other arteries of **sheep** grazing pastures in Germany and Austria, containing more than 25%. Emaciation, ataxia, stiff forelimbs, kyphosis, protracted kneeling posture, tachycardia; calcifications of the diaphragmatic lobe of the lungs, kidneys, and foreleg digital flexor tendon. Morbidity can reach 50%. **Cattle**, **chicks**, and **rabbits** have also been affected.

TRIS(HYDROXYMETHYL) NITROMETHANE

= *HNP* = *Hydroxymethyl-2-nitro-1,3-propanediol* = *Tris Nitro*

Bactericide.

Untoward effects: Slowly releases ***formaldehyde*** (q.v.). Skin and mucous membrane irritant. Cross-reacts with ***bronopol*** (q.v.). **Machinists** often develop a contact allergy to it from its presence in machine ***cutting oils***. Premises must be well ventilated until sprayed surfaces are dry.

LD_{LO}, **human**: 500 mg/kg.

LD_{50}, **mouse** and **rat**: 1.9 g/kg.

TRISODIUM EDETATE

= *Trisodium Versenate*

Untoward effects: Use as a chelating agent can cause the same adverse effects as ***sodium edetate*** (q.v.).

LD_{50}, **rat** and **mouse**: 2.15 g/kg; IP, **mouse**: 300 mg/kg.

TRITOLYL PHOSPHATE

= *Celluflex 179* = *Kronitex* = *Lindol* = *PX917* = *TCP* = *Tricresyl Phosphate*

Use: Plasticizer, instrument sterilant, lube additive, fire retardant, and additive in ***gasolines***. Usually a mixture of isomers, particularly *meta* and *para*.

Untoward effects: Contact dermatitis from its use in ***carbon paper***. Early mixtures often contained some *ortho* material. Wastes from a defunct ***TCP*** manufacturer and discarded service station oils were sprayed on the soil of a riding arena, causing illness and death of many **horses**. Hemorrhagic cystitis in a 6-year-old **female**, chloracne in two **boys**, and recurrent headaches, skin lesions, and polyarthralgia reported in three **individuals**.

Dyspnea, tympanites, incoordination, paralysis, and neurological disturbances reported in **cattle** exposed to it in a mixture of ***triaryl phosphates***.

LD_{LO}, **rat**: 4.7 g/kg; **dog**: 500 mg/kg; **rabbit**: 100 mg/kg.

TRI-σ-TOLYL PHOSPHATE

= *Tri-σ-cresyl Phosphate*

Use: Plasticizer in varnishes and ***lacquers***; fireproofing agent.

Untoward effects: In 1930, this was the first ***organophosphorus compound*** proven experimentally to be neurotoxic and the cause of ***Ginger Jake*** (q.v.). ***Paralysis*** in ~50,000 **people**. Adulteration of cooking oils by aircraft engine oils (with 3% mixed isomers) caused a similar neurotoxic syndrome in > 10,000 **Moroccans**. In 1969, peripheral neuropathy in 56 in Fiji, with 35% still incapacitated nearly 2 years after sacking ***flour*** in contaminated sacks. Romanian literature in 1972 describes the delayed neuropathy in 16 **patients** after accidently drinking ***alcohol*** polluted with it and again in 1980, 12 additional cases. In 1980 in Sri Lanka, 20 young **females** ingested 560–1120 ml of contaminated ***sesame oil*** as part of a religious ceremony caused polyneuropathy. Polyneuritis in the 1930s after use of ***apiole*** (q.v.) as an abortifacient, was apparently due to contamination with it. In Hungary, a 56-year-old **male** was poisoned after consuming 700 ml homemade ***brandy*** that had been passed in boiling condition through ***polyvinyl chloride*** tubing made with 50% ***TOCP*** leading to toxic polyradiculitis with ataxia and absent triceps, radial, ulnar, and Achilles tendon reflexes; atrophy of hand muscles and, to a lesser extent, of foot and peroneal muscle groups. Peripheral neuritis in a 20-year-old **male** motor mechanic, due to 3 years' addiction to oral and inhalation abuse of ***gasoline*** with it as an additive. A 24-year-old **male** using it for > 2 years in fireproof applications slowly developed weakness of distal muscles, wrist and foot drop, loss of plantar reflex, and, eventually, muscle atrophy and flaccid paralysis. Paraplegia in **man**, **dogs**, **pigs**, **goats**, **fowls**, and **mynah birds** in West Bengal villages after access to ***TOCP***-contaminated raw or fried ***flour*** preparations. **Humans** complained of cramps in **their** calves, stomach pain, diarrhea, and paralysis. Ataxia is delayed and irreversible. Contact dermatitis in a 21-year-old **female**, due to its use as a plasticizer in ***polyvinyl chloride*** surgical tape. Use as a plasticizer in ***cellulose acetate*** spectacle frames reported as a cause of dermatitis. Symptoms are due to inhibition of cholinesterase and demyelination. Death has followed respiratory paralysis, and degenerative changes of muscles and spinal cord are found on post-mortem. Nausea and vomiting follow ingestion and 0.5 g or more can also cause the above serious consequences. National Institute for Occupational Safety & Health and OSHA set upper exposure limits at 0.1 mg/m^3. Immediate danger to life or health, 40 mg/m^3.

Drug interactions: Potentiates toxicity of ***dimethoate*** and ***malathion***.

Feeding of ***molasses*** that had been stored in barrels that contained it in ***hydraulic fluids*** led to progressive posterior flaccid paralysis in 42 Louisiana beef **cows**. Urine dribbling was common. Post-mortem showed axonal degeneration of ventral horn gray matter and demyelination in the lumbar spinal cord. Similar cause and case reports from Italy. Severe toxicity from it has followed topical use of 1–2 l of old motor oil on **cattle**. **Hens** are particularly susceptible and used experimentally in studying its delayed neurological effects. In India, sacks of contaminated ***flour*** declared unfit for **human** use caused paralysis when fed to **pigs**, **goats**, **dogs**, and **fowl**. **Cats** are also adversely affected.

LD_{50}, **rat**: 1.15 g/kg; **rabbit**: 3.7 g/kg; **chicken**: 100 mg/kg.

TROGLITAZONE

= *CI 991* = *CS 045* = *Rezulin*

Antidiabetic. **For type II diabetics, especially in those resistant to *insulin*. A *thiazolidinedione*.**

Untoward effects: Although rare, potentially serious liver problems (35/650,000 **patients**) in 1997 caused its withdrawal from marketing in the U.K. The FDA and Parke-Davis/Warner Lambert agreed in March 2000 to withdraw it from the U.S market after liver failure was associated with death in 61 **patients** and seven liver transplants in **those** who took the drug. Nausea, vomiting, fatigue, edema, slight weight gain, mild anemia, anorexia, dark urine, polycythemia in 1–2%, increase in plasma volume, and in 2% increase in liver enzymes. Pleuritis and possible interstitial pneumonitis reported in a hypertensive **patient**. Use may also be associated with heart failure.

Drug interactions: Given at the same time, ***cholestyramine*** decreases its absorption by ~70%. It has caused a dramatic increase in ***warfarin's*** anticoagulation and decreased serum concentrations of ***oral contraceptives*** and ***terfenadine*** due to its stimulation of CYP3A4 metabolic enzymes.

TROLEANDOMYCIN

= *Cyclamycin* = *Evramycin* = *TAO* = *Triacetyloleandomycin* = *Triocetin* = *Wytrion*

Macrolide antibiotic.

Untoward effects: Cholestatic jaundice and hepatic dysfunction in ~50% of **patients**, especially in young **people** or after 10–14 days of therapy. Anorexia, nausea, vomiting, abdominal cramps, eosinophilia, leukocytosis, fever, skin eruptions, urticaria, diarrhea, and heartburn have occurred. Rarely, anaphylactic shock and Stevens–Johnson syndrome. Occasionally causes falsely increased urinary ***catecholamines***, SGOT, SGPT, thymol turbidity, and serum ***bilirubin*** values.

Drug interactions: It inhibits ***ergotamine*** enzymatic metabolism, leading to toxicity with acute classical ergotism and it caused ischemia in all four limbs of an **adolescent**. May increase the risk of arterial spasm and occlusion when given with ***methysergide***. Use with ***terfenadine*** and ***astemizole*** can prolong QT interval, cause ventricular arrhythmias, and cardiac arrest. It decreases body clearance of ***methylprednisolone*** and doubles serum ***theophylline*** concentrations. Hepatotoxicity with severe pruritus and jaundice when given with ***oral contraceptives***. Causes a rapid increase in serum ***carbamazepine*** and its toxicity. Inhibits CYP3A4 metabolic enzymes increasing serum levels and toxicity of many drugs, including ***benzodiazepines***. Inhibits metabolism of ***cisapride***.

Very high doses to **dogs** cause nausea. Continuous feeding to **rats** and **monkeys** leads to reversible fatty metamorphosis of livers and growth decrease. Use with ***dihydroergotamine*** in **minipigs** leads to ergotism.

TROLNITRATE PHOSPHATE

= *Angitrit* = *Bentonyl* = *Duronitrin* = *Metamine* = *Nitretamin* = *Ortin* = *Praenitron* = *Vasomed*

Antianginal.

Untoward effects: Cutaneous vasodilation, flushing, severe headaches; transient dizziness, weakness, and cerebral ischemia. Occasional marked hypotension, pallor, collapse, nausea, vomiting, and increased intraocular pressure.

Drug interactions: Adverse effects may be enhanced by ***alcoholic*** beverages. May antagonize some of the effects of ***acetylcholine, antihistamines***, and ***norepinephrine***.

LD_{50}, **rat**: 21 mg/kg; **mouse**: 318 mg/kg.

TROMETHAMINE

= *Addex-Tham* = *Talatrol* = *THAM* = *TRIS* = *Trisamine* = *Tris-Amino* = *Trisaminol* = *Tris Buffer* = *Trizma* = *Trometamol* = *Tromethane*

***Alkalizer.* IV infusion.**

Use: Biological and medical buffer, as an emulsifying agent, and in the chemical synthesis of pharmaceuticals, surfactants, and vulcanizing agents.

Untoward effects: Extravasation can cause severe tissue damage. Venospasm, phlebitis, thrombosis, and local inflammation reported. Occasional hypoglycemia, respiratory depression, vomiting, diarrhea, hyperkalemia, and hypercalcemia. Anemia and alopecia after prolonged treatment. Hypertonic solutions can increase red blood cell fragility, renal tubular damage, and alteration of tissue fluids. Unintentional IA in two **female newborns** led to severe hemorrhagic necrosis; one died. Bladder necrosis in an **infant** given it in 10% ***dextrose*** solution via umbilical artery catheter. Many cases of apnea within 2 min, intraventricular hemorrhage and hemorrhagic liver necrosis in **newborns** with acidosis. Undiluted material can cause skin irritation on contact and severe irritation and burns of the eyes if not immediately washed out with cold water. Has low vapor pressures unless heated.

LD_{LO}, **human**: 500 mg/kg.

Causes severe irritation on contact with intact or abraded **rabbit** skin.

LD_{50}, **mouse**: 5.5 g/kg; IP, 3.3 g/kg; IV, 1.2 g/kg; LD_{LO}, **rabbit**: 1 g/kg.

TROPICAMIDE

= *Mydriacyl* = *Mydriaticum*

***Mydriatic, cycloplegic, anticholinergic.* Topical use only.**

Untoward effects: Grand mal seizure in a 24-year-old **male**, with no history of previous seizures, within 20 s after instilling a drop into each conjunctival sac. Anaphylactic shock immediately after instillation in each eye of 1 drop of 0.5% in a 10-year-old **male** with prior history of bronchial asthma **who** previously had been treated with ***atropine*** and/or ***cyclopentalate***. Can increase intraocular pressure and cause blurred vision.

TROPISETRON

= *ICS 205-930* = *Navoban* = *Novaban*

Serotonin receptor antagonist, antiemetic.

Untoward effects: A 42-year-old **female**, after an IV dose of 20 mg, within 36 h developed fever, mild hypotension, macular rash, joint aches, and cervical lymphadenopathy. Confirmed by rechallenge 2 months later.

TROVAFLOXACIN

= *CP 99,219* = *Trovan*

***Antibiotic.* A *fluoroquinolone* for oral use. IV *alatrofloxacin* is rapidly converted to it.**

Untoward effects: Dizziness (3% with 100 mg and 11% with 200 mg dose). Occasional nausea, vomiting, diarrhea, exfoliative dermatitis, and headache. Rarely, rash and vaginitis. Acute eosinophilic hepatitis in a 66-year-old **male** after 100 mg/day/4 weeks for chronic sinusitis. After IV pain, pruritus, thrombophlebitis, flushing, hyperesthesia, and skin irritation around injection site.

Drug interactions: ***Aluminum hydroxide, citrates, citric acid, iron compounds, magnesium hydroxide***, IV ***morphine***, and ***sucralfate*** decrease its intestinal absorption. Use of IV form in **patients** on ***anticoagulants*** leads to dramatic increase in international normalized ratio.

TROXERUTIN

= *HR* = *Paroven* = *Posorutin* = *Relvene* = *Ruven* = *Trioxyethylrutin* = *Varemoid* = *Vastribil* = *Veinamitol* = *Veniten* = *Venoruton*

Untoward effects: Mild nausea (6%), diarrhea, and dermatitis after oral use; mild dermatitis after topical use; and increased appetite, weight increase, and hot flushes after IV use.

TRYPAN BLUE

= *Congo Blue* = *Naphthylamine Blue* = *Niagra Blue*

Protozoacide, biological stain.

Untoward effects: Teratogenic in **hamsters**, **mice**, **rabbits**, and **rats**, also carcinogenic in **rats** after IV use. Extravasation has caused tissue sloughing. Induces vomiting in **cats** and **dogs**. Given to experimental pregnant **animals**, caused congenital anomalies.

IV, LD_{LO}, **rat**: 300 mg/kg; **rabbit** and **mouse**: 100 mg/kg.

TRYPARSAMIDE

= *CPA* = *Glyphenarsine* = *Tryparsone* = *Tryponarsyl* = *Trypothane*

Protozoacide.

Untoward effects: Nausea, vomiting, headache, tinnitus, and dizziness. Most serious is damage to the optic nerve with narrowing field of vision, optic neuritis, atrophy, and, rarely, sudden blindness. Occasional fever, allergic reactions, exfoliative dermatitis, and bradycardia.

LD_{LO}, **cat** and **rabbit**: 200 mg/kg; **guinea pig**: 150 mg/kg; IV, **rat**: 2 g/kg; IV, LD_{50}, **rabbit**: 700 mg/kg.

TRYPSIN

= *Parenzyme* = *Parenzymol* = *Tripsina* = *Tryptar* = *Trypure*

***Proteolytic enzyme*. The crystalline trypsin in commerce is usually of bovine or porcine origin and contains substantial amounts of *chymotrypsin* (q.v.). Topical, IV, and oral, but usually IM.**

Untoward effects: Anaphylactoid reactions, shock, urticaria, skin rashes, angioneurotic edema, fever, and local reactions to a foreign protein. Rarely, death. Pain and induration at injection sites and severe burning sensation from topical use. Reactions can be severe (***histamine***-like reaction), if used in closed body cavities. Aerosoled use in treating silicosis or from occupational exposure led to urticaria, fever, asthma-like dyspnea, and transient hoarseness. Allergic and other reactions have occurred in treatment of some **females** for tubal obturation. Its use in buccal tablets led to soreness and ulcerations of the mouth in a high percentage of cases. Can give spuriously high serum ***cholesterol*** test results. Serum contains trypsin inhibitors, and these decrease the drug's effectiveness in some sanguine conditions. Yet, this is often inadequate to protect viable healthy tissue from the enzyme effects, despite commercials to the contrary. Further evidence of this is seen in the cases of pancreatitis due to its overzealous autodigestion. Solutions deteriorate rapidly (75% loss in 3 h). This can be decreased by refrigerating between use and by using material stabilized with ***gelatin***.

Drug interactions: Use of ***chymotrypsin*** increases serum levels of orally administered ***tetracycline***. Use enhances the effects of ***coumarin anticoagulants***.

Mammitis noted with its intramammary use in normal **cow** udders, disproving the statement that normal tissue is not adversely affected by it. Can increase the toxicity of ***Clostridium Type D protoxin***. Anesthetized **dogs** have shown profound blood pressure decrease, congestion of viscera, particularly liver, and increase of ***histamine*** in the blood, and their blood fails to coagulate. Unanesthetized **dogs** may vomit, urinate, defecate, and collapse. Toxic circulatory effects in **cats**, **dogs**, and **guinea pigs** (acute pulmonary embolism with lung collapse) after IV use, are not due to digestion of blood or tissues, as lethal effects are noticed within seconds. Activated trypsin plays a role in pathogenesis of acute pancreatitis. Trypsin inhibitors in raw ***soybeans*** interfere with a **pig's** ability to digest protein. Therefore, **pigs** are fed them or ***soybean*** meal after heat processing, which destroys the inhibitors.

IP, LD_{50}, **mouse**: 650 mg/kg.

TRYPTAMINE

Untoward effects: South American **Indians** discovered it to be an effective hallucinogen only when it is used as a snuff from some *Acacia, Mimosa, Piptadenia*, and *Virula sp.*, and then it is easily converted into ***bufotenine*** (q.v.), ***dimethyltryptamine*** (q.v.), ***ibogaine*** (q.v.), and ***psilocin***, all of which can be hallucinogenic, apparently by prolonging ***serotonin*** effect in the body.

L-TRYPTOPHAN

= *Ardeytropin* = *Kalma* = *Optimax WV* = *Pacitron* = *Sedanoct* = *Trofan* = *Trp* = *Tryptacin* = *Tryptan* = *Tryptocalm* = *W*

***Amino acid, nutritional supplement*. Essential for humans and rats. A precursor of *serotonin*.**

Untoward effects: Metabolized to ***indoleacetic acid*** (q.v.), ***quinolinic acid***, and ***skatole*** (q.v.). The latter apparently helps to induce Huntington's disease, a degenerative neurological disorder (jerky movements, irritability, and profound mental deterioration). After being sold in health food stores to treat insomnia, depression, premenstrual symptoms, stress, and obesity, many **users** developed eosinophilia-myalgia syndrome (EMS) with severe muscle and joint pain, muscle cramps, swelling of arms and legs, paresthesias, eosinophilic leukocytosis, fatigue, low-grade fever, weakness, skin rashes and indurations, and connective tissue inflammation. Some **patients** died and **many** suffered the debilitating symptoms for up to

2½ years. The FDA, on November 17, 1989, recalled all **human** products in which it was the sole or main compound supplying a daily intake of 100 mg or more. In March, 1990, the FDA extended the recall to all **human** products with less than 100 mg. Average daily **human** dietary intake is 1.13 g of tryptophan. Most illnesses were traced to material produced by a Japanese company, who altered their fermentation procedure, although that did not appear to be the cause. The cause of EMS was apparently a contaminant, ***1,1′-ethylidenebistryptophan***. U.S. nationwide illness cases in ~1 year were estimated as at least 5000 with 1536 cases of EMS and 27 deaths. Other **patients** reported alternating drowsiness, euphoria, mania, sustained nystagmus, and pneumonitis.

Drug interactions: Nausea, vomiting, headache, xerostomia, and hypertension in **patients** taking ***isocarboxazid, mebanazine, pargyline***, and ***tranylcypromine***.

Caused bladder cancer when implanted in the bladders of **mice**. Large doses intraruminally caused interstitial pulmonary edema and emphysema in **cattle** due to its metabolite, ***3-methylindole***. Excessive intake can lead to hepatoencephalopathy. After 6 weeks of feeding Showa Denko's contaminated product, **rats** developed fascitis and periomyositis; two types of inflammation in EMS of **man**. Experimental IV use in **ponies** led to tachypnea, acute hemolytic anemia and hemoglobinuria.

TRYPTOQUIVALINE and TRYPTOQUIVALONE

Untoward effects: Both are tremorgenic toxins produced by *Aspergillus clavatus* and have been found on moldy ***rice*** in a Thai household where a young **male** died from toxicosis. Tremorgens, ***nortryptoquivaline*** and ***nortryptoquivalone***, are also produce by *A. clavatus*.

TSE KOO CHOY

Untoward effects: This traditional Chinese **children's** medicinal powder can be very toxic and contains ***mercurous chloride*** (q.v. for toxicity).

TUBERCIDIN

= *7-Deazaadenosine* = *NSC 56,408* = *Sparsamycin A*

***Antifungal, antineoplastic, tuberculostatic.* For IV use.**

Untoward effects: Nephrotoxic in ~19%, local irritation and thrombosis of veins in ~13%, anorexia, nausea, vomiting, and mild leukopenia.

d-TUBOCURARINE CHLORIDE

= *Curarin-HAF* = *Delacurarine* = *Intocostrin* = *Jexin* = *Tubadil* = *Tubarine*

***Non-depolarizing muscle relaxant, parasympatholytic. Tubadil's* name is derived from the South American name for the bamboo tubes in which the natives store *curare*. It is now synthesized. Given IM or IV.**

Untoward effects: Usually, decrease in blood pressure and slight increase in heart rate. Face and neck flushing, fatigue, weakness, voluntary muscle paralysis, usually beginning with eyes, then face, neck, extremities, abdomen, and, eventually, intercostal muscles and diaphragm within 3–5 min after IV and usually lasting ~20 min. Full muscle strength returns in ~40 min. Unfortunately, apnea can last much longer in some **people**. Respiratory paralysis in 7-10 min and regurgitation of stomach contents can occur, due to relaxation of esophageal muscle and sphincter. ***Histamine*** release occurs. Hypersensitivity reactions (rash, hives), asthma, anaphylaxis, cyanosis, malignant hyperthermia, and hypokalemia have occurred. Myasthenic **patients** are extremely sensitive to 1/20–1/10th of its usual curarizing dose. Half-life is ~13 min. Crosses the placenta within 6–10 min. High death rate in tracheostomized **infants**.

Drug interactions: Use with various ***antibiotics***, such as ***bacitracin, colistin, gentamicin, kanamycin, neomycin, polymixin B, streptomycin***, and ***tetracycline; furosemide, lidocaine, magnesium***, other ***muscle relaxants, phenytoin, procainamide, propranolol, quinidine, thiazide diuretics***, and ***halothane*** anesthesia potentiate its paralytic effects.

In **animals**, very slow IV use is essential. It is given as much as necessary in very dilute solution over a 2 min period. Avoid use in **animals** recently exposed to ***organic phosphorus anticholinesterase compounds*** and ***phenothiazines***. ***Streptomycin*** and ***tetracycline***, but neither ***chloramphenicol*** nor ***benzyl penicillin sodium***, potentiate its neuromuscular blockade in **rabbits**. ***Neomycin*** synergizes the neuromuscular block in **rats** and **cats**. Salivation, retching, defecation, bronchospasm, and intestinal hemorrhage reported in **dogs** are probably, in part, associated with the drug's ***histamine*** release. Avoid use with ***ether*** and ***atropine***, and vigorous manipulation or stimulation of the pharynx. Its use will interfere with corneal, conjunctival, and pedal reflexes usually used as guidelines in anesthetic induction. It increases salivation and bronchial secretions, hypotension, and bronchial spasm; occasional deaths in **cattle**. **Cats** are more resistant to its effects than other **mammals**. In **monkeys**, ***dihydrostreptomycin, gentamicin, kanamycin, neomycin***, and ***streptomycin*** potentiate its neuromuscular blocking effect.

IV, LD_{50}, **mouse**: 180 μg/kg; **dog**: 500 μg/kg; **rabbit**: 20 μg/kg.

TULIP

Untoward effects: An **adult** can usually consume up to five bulbs with only mild gastrointestinal upsets; 10 or more cause severe gastrointestinal symptoms. A 3-year-old **male** accidentally consumed one, leading to nausea, vomiting, leg cramps, and slight cyanosis. Sterility in **women** in Holland near the end of World War II attributed to the consumption of bulbs with estrogenic activity. During World War II, the Dutch supplemented **their** diet with the bulbs to help prevent starvation. Acute poisoning did not occur, but chronic poisoning did. Being short of ***onions***, a **woman** in Yugoslavia substituted five tulip bulbs to make a goulash for six in **her** family. Within 10 min after ingestion they experienced nausea, feeling of warmth, sweating, salivation, dyspnea, and palpitations. **Some** were weak for up to 5 days. A **dog** that ate some of it vomited profusely and refused to eat for several days. An irritant in tulips has caused "tulip fingers" with hyperkeratosis and onycholysis in **florists** and bulb-**handlers**. This can occur within 12 h after handling the bulbs and, in serious cases, lead to intolerable tingling and tenderness of the fingertips beneath the nails. Serum exudes, then granulation occurs between the nail and its bed. Later, the nail separates. Contact hypersensitivity dermatitis on the hands, face, and genitals of a **gardener** from handling the bulbs. The causative agent may be ***α-methylene-γ-butyrolactone***. In sensitive **individuals**, cross-sensitivity with ***garlic*** and ***onions*** can occur. Tulip bulbs contain the alkaloid ***tulipine***, closely related to ***colchicine*** (q.v.) and ***solanine*** (q.v.).

TUNG TREE

= *Aleurites fordii* = *Tungoil Tree* = *Tung Nut*

Untoward effects: The foliage, sap, fruit, and untreated tung meal all contain toxic substances, water-soluble ***saponins***, and an alcohol, acetone, or ether-extracted fraction. It is grown in China, Japan, Florida, and other coastal plains. The seeds are especially toxic, leading to severe stomach pain, vomiting, diarrhea, dyspnea, weakness, and poor reflexes; occasionally, convulsions and tachycardia. Other species are also toxic. *A. trisperma* = ***Lumbang Nut*** of the Philippines and Florida cause burning sensations from the mouth to the stomach, severe vomiting, and diarrhea; occasionally, tenesmus and extreme thirst within 30–60 min after ingestion and continuing for at least 6 or 7 h. Glucosuria, increased specific gravity of urine, epithelial cells, and casts are evidence of renal complications. In severe poisonings tachypnea, pyrexia, irregular respirations, mydriasis, paresthesia, and decreases reflexes. Severe headaches may develop after 24 h.

Poisoned **cattle** have symptoms similar to **man's** and after 3–7 days are primarily anorexia, emaciation, listlessness, and watery diarrhea, often with blood. Most die in ~4 weeks if they have eaten at least 2.5 oz of foliage. Chronic cases develop dyspnea, nasal discharge, cracking of muzzle skin, and progressive emaciation. **Horses** and **chickens** have also been affected. In China, the seeds are used in killing **rats**.

TUNG SHUEH

Untoward effects: The US Drug Enforcement Administration seized 44,000 packages of this Asian "herbal" (Cow's Head brand) in 1993 and found it to contain the non-herbals ***diazepam, diclofenac, indomethacin***, and ***mefenamic acid***, at levels leading to central nervous system depression and gastrointestinal and renal toxicity.

TUNGSTEN

= *Wolfram*

Untoward effects: Exposure to fine dust causes eye, skin, and respiratory system irritation, pulmonary fibrosis, anorexia, nausea, coughing, and anemia. National Institute for Occupational Safety & Health sets upper exposure limit of 5 mg/m^3; 15-min exposure of 10 mg/m^3.

Maximum safe level in complete **animal** feeds is 20 ppm. Very low levels secreted in **cow's *milk*** (0.4% of a single oral dose by 84 h after dosing).

TUNGSTEN CARBIDE

Untoward effects: Exposure to finely divided material leads to progressive diffuse interstitial pneumonia with unproductive cough, dyspnea on exertion, and deaths. Also eye, skin, and respiratory tract irritation; anorexia, nausea, and anemia. **People** hypersensitive to ***cobalt*** and ***nickel*** may react to it because of its contaminants.

TURKEY

= *Meleagris gallapavo*

Spaniards, thinking the New World was India, called the turkey ***"toka"***, meaning **peacock** in the Tamil language of India. Jewish **merchants** called it by the Hebrew word, ***"tukki"***, which was eventually pronounced by the **laity** as "turkey".

Untoward effects: An immediate contact urticarial reaction in some **people** from turkey meat or skin. Turkey-**handler's** disease is a hypersensitivity pneumonitis with turkey proteins as the antigen. A large number of **people** at two restaurants developed salmonellosis from eating turkey giblets that had been stored under refrigeration for several days and, because they had been oxidized, they appeared cooked and misled **employees** into using them without prior cooking. The delectable aroma from cooking turkey is due to trace amounts of a toxic chemical, ***acrolein*** (q.v.), and, when grilled, it contains ***benzopyrene*** (q.v.), a carcinogen. Ingestion by **people** can induce a gout attack in **some**.

TURMERIC

= *Curcuma longa* = *Haldi* = *Haledo* = *Massala*

***Condiment and coloring agent.* A common substitute for *tartrazine*.**

***Untoward effects*:** Allergic contact dermatitis in **humans**. Extracts of the rhizomes provide 100% antifertility activity in **rats** at 200 mg/kg.

TURNERA APHRODISIACA

= *Damiana* = *Mexican Damiana* = *Old Woman's Broom* = *T. diffusa*

Untoward effects*:** It contains ***arbutin, a diuretic, and is considered an herbal therapeutic hoax. Has contained ***caffeine***, a central nervous system stimulant. Some **users** have claimed hallucinogenic effects.

TURNIP

= *Brassica campestris*

Untoward effects*:** It is a ***nitrate-accumulator and contains moderate amounts of ***oxalates***. A goitrogenic substance, ***goitrin***, was isolated from it ~50 years ago. Raw turnips are rich in peroxidase and **people** consuming them can have false positive Hemoccult® tests. The investigator said that, "it has been known for years that you can't get blood from a turnip". The roots of ***water hemlock*** (***Cicuta*** [q.v.]) have caused toxicity when mistaken for turnips or ***parsnips***. Myocardial infarction in a 35-year-old **female** receiving ***warfarin*** for a prosthetic aortic valve 5 weeks after a weight reducing diet of ***broccoli***, ***lettuce***, and turnip greens, all of which are high in ***vitamin K***. It actually contains a ***progoitrin***, which is converted by enzymatic action to the active ***goitrin***, a growth-inhibiting and goitrogenic compound. Seed contains up to 8 g/kg. Disagreeable taste and unpleasant odor for **man** in ***milk***, milk products, and meat of **animals** fed it or grazing on it.

Cows eating Ethiopian ***linseed*** cake contaminated with turnip seed showed acute fatal hemorrhagic gastroenteritis and increased ***thiocyanate*** content of their ***milk***.

TURPENTINE

= *Terebinthinae*

Contains at least 35 *terpenes*.

***Untoward effects*:** Absorbed from skin, lungs, and intestines.

Ingestion*:** Mostly in **children** and in suicide attempts by **adults** leading to severe mouth and abdominal burning sensations, nausea, vomiting, dysuria, diarrhea, colic, bradypnea, bronchopneumonia, mydriasis, extreme excitement, ataxia, rashes, convulsions, coma, albuminuria, hematuria, violet-colored urine, hemoglobinuria, glycosuria, and occasionally aspiration, pulmonary pneumatoceles, and death. Its ***thujone (q.v.) content leads to psychic and motoric upsets by damaging the central nervous system. Thrombocytopenia induced in two cases after exposure to fumes and ingestion of an unknown amount.

Inhalation*:** Thrombocytopenic purpura, migraine, anorexia, mental confusion, gastritis, giddiness, tinnitus, anxiety, excitement, bronchitis, and heavy exposure leading to conjunctivitis, eyelid edema, and blepharospasm. ***Topically it has caused a weeping eczema, erythema, pruritus, and occasional vesication and has been common in **painters** and **artists**. **Machinists** washing items in it developed painful inflammation of **their** fingertips, loss of tactile and heat sensations; dryness and fissuring of the skin; and paresthesia of the fingers lasting for several weeks. Severe irritation after contact with mucous membranes. Much of this is due to its oxidation of ***Δ-3-carene*** (q.v.), one of its *terpenes*, and dehydration of the skin. Accidental splashing onto the eye led to immediate severe pain and blepharospasm, followed by conjunctival hyperemia, and slight transient injury to the corneal epithelium.

Intrauterine use in attempted abortions was successful, but peritonitis and pulmonary edema followed. Characteristic violet-like breath odor and similar aroma from the **patient's** vomitus. A **diabetic** 15-year-old **female** abusing it by self ***injection*** showed nodular nonsuppurative panniculitis in extremities and abdominal wall. National Institute for Occupational Safety & Health/OSHA set upper exposure limit at 100 ppm; immediate danger to life or health, 800 ppm.

Estimated LD, **child**: ~15 ml, **adults**: 100–180 ml. Inhalation TC_{LO}, **human**: 175 ppm.

Do NOT administer to **animals** that may be slaughtered shortly for food, as the meat may be tainted, even if aspiration has not occurred. Avoid use in gastritis, enteritis, or nephritis, due to its irritant action. Aspiration pneumonia has been a common sequela to careless administration. In the **rabbit**, turpentine oil and its major component, ***l-pinene***, after topical application, promoted skin tumor development, but not in the **mouse**.

TURTLES

Untoward effects*:** The ***Green Turtle = ***Chelonia mydas***, found in the salt waters of the Indo-Pacific and Texas Gulf of Mexico, is poisonous upon ingestion.

The ***Hawksbill Turtle*** = ***Eretmochelys imbricata*** is about 4 ft long, found in the same areas, and can also be poisonous upon ingestion.

The ***Leathery Turtle*** = ***Leatherback Turtle*** = ***Democheelys coriacea***, found in the same areas, grows to > 6 ft and then weighs > 1000 lbs, and has similar toxicity.

The Softshell Turtle = ***Pelochelys bibroni*** in sea waters from southeast Asia to New Guinea, has similar toxicity. Clinical symptoms are nausea, vomiting, epigastric pain, sweating, tachycardia, diarrhea, vertigo, cold extremities; burning of the lips, mouth, and throat with hypersalivation and dysphagia reported in **some**. Stomatitis, mouth ulceration, and fetid breath may last for weeks to months. Hepatomegaly, icterus, and skin desqua- mation also reported. Symptoms resemble those of ***ciguatoxin*** (see ***Ciguatera***). Fatal cases have hepatic necrosis with centrilobular congestion and fatty degeneration, plus ulcerations in the upper gastrointestinal tract. The livers are especially toxic for **man**. Occasionally, they can be eaten with impunity. Death occurs in ~44% of **those** poisoned. Their bites are traumatic, but non-venomous. Pet turtles have infected **people** with salmonellosis.

Snapping Turtles: ***Chelydra serpentina*** and ***Macroclemys temmincki*** are fresh-water turtles common in the U.S. On land, when disturbed, they are very aggressive, lunging and biting savagely.

TUTIN

Untoward effects: This alkaloid is chemically related to ***picrotoxin*** (q.v.) and found in ***Coriaria* sp.** (q.v. for symptoms), which include central nervous system stimulation, delirium, convulsions, giddiness, mania-like symptoms, loss of memory, stupor, and coma after ingestion. **Cattle** are poisoned by eating the plant's young shoots and **children** by eating its seeds.

IP, LD_{50}, female **mouse**: 3 mg/kg.

TYBAMATE

= *Benvil* = *Nospan* = *Solacen* = *Tybatran*

Tranquilizer, anxiolytic.* A *propanediol carbamate.

Untoward effects: Frequent, drowsiness, dizziness, nausea, insomnia, and euphoria. Occasionally, skin rash, urticaria, and pruritus. Rarely, flushing, hyperactivity, fidgeting, tachycardia, ataxia, muscle hypotonia, tremors, confusion, headache, vertigo, paresthesias, glossitis, and xerostomia.

Drug interactions: Grand mal or petit mal seizures have occurred when given with ***phenothiazines*** or other ***central nervous system depressants***. Decreases half-life of ***coumarin anticoagulants***.

TD_{LO}, **human**: 18 mg/kg/day.

LD_{50}, **rat**: 1 g/kg; **mouse**: 850 mg/kg.

TYLOPHORA sp.

Untoward effects: ***T. indica*** of India ingestion confers long-lasting relief of nasobronchial allergic symptoms, but has blistering and vesicant properties. Sore mouth, loss of taste for ***salt*** and/or morning nausea and vomiting reported in 53% of 110 asthmatic **patients** who chewed one leaf.

T. fasciculata is used as a **rat** bait and poison in India.

Tylocrebrine from Australian ***T. crebriflora*** is an ***antineoplastic***.

TYLOSIN

= *Tylan* = *Tylocine* = *Tylon*

***Antibiotic.* Oral and IM.**

Untoward effects: **Animal** studies indicate that accidental injection in **man** could cause pain and swelling at injection site, and that it could be irritating to the eyes on contact. Hypersensitivity to it can occur in some **individuals**. Cases of contact dermatitis reported. One **patient** had prior sensitivity to ***nitrofurazone***. Injectable solutions contain ***benzyl alcohol*** (q.v.) and ***propylene glycol*** (q.v.).

In the U.S., neglible residue tolerances for the drug in edible **animal** products is limited to 0.2 ppm in the uncooked fat, muscle, liver, and kidneys of **cattle**, **swine**, **chickens**, **turkeys**, and in **poultry** eggs. A 0.05 ppm tolerance has been established for ***milk*** residues. ***Milk*** taken from **cows** during treatment and for 96 h and eight milkings after the last injection should not be used for **human** consumption. Do NOT slaughter **cattle**, **swine**, **chickens**, and **turkeys** for 8, 4, 3, and 5 days, respectively, after the last injection to avoid excessive tissue residues. In **hogs**, it may be advisable to wait 3 weeks to avoid excessive trim waste due to irritation at the injection site. Rectal "rosettes", including edema, erythema, prolapse, diarrhea, and pruritus can occur, particularly on oral use in **swine**. In the U.S., it is NOT permitted for **pullets** or **hens** producing eggs for **human** consumption. Pretreatment of experimental **dogs** with tylosin increased their incidence of ventricular tachycardia and fibrillation during acute myocardial ischemia. In water solutions, should be mixed fresh every 3 days to assure adequate potency. The parenterally available base form assures longer blood levels than aqueous tartrate solutions. Limited topical (2%) use on wounds and eyes. Occupational dermatitis caused by tylosin reported in **veterinarians**. Contact dermatitis reported in **people** handling tylosin-medicated feed should serve as an alert against ***milk*** and tissue levels. 21 days' withdrawal in **swine**. Injections may cause local irritation, ataxia, ptyalism, mydriasis, nystagmus, muscle tremors, and tachycardia in **cattle**. Oral feeding at 5–20 ppm (2.25–9 mg/lb) in the diet has been associated with gray-green feces, ruminal paralysis, and decreased ***milk*** production.

TYLOXAPOL

= *Alevaire* = *Superinone* = *Triton A-20* = *Triton WR-1339* = *Tyloxypal*

Surfactant, mucolytic. **Aerosoled.**

Untoward effects: Pharyngitis, bronchospasm, maculopapular rash, increased congestion in chest, and pharyngeal burning. Rarely, bronchospasms.

Induces hyperlipidemia, hypercholesterolemia, and arteriosclerosis after IV or IP into **rats** and hyperlipidemia in **rabbits**, **guinea pigs**, **mice**, and **dogs**, causing their deaths in ~4 months. It shows selective toxicity to cell cultures of malignant origin.

TD_{LO}, **mouse** (6–8 days pregnant): 600 mg/kg; IP, TD_{LO}, **rat** (7–9 days pregnant): 400 mg/kg; **mouse** (6–8 days pregnant): 600 mg/kg.

TYLUS

Untoward effects: This botanical, used in Argentinean folklore treatments, has caused poisoning in **children**.

TYPEWRITER RIBBONS

Untoward effects: Authentic cases of dermatitis from handling them are rare. Their odor, in some cases, has triggered migraines.

TYRAMINE

= *Tyrosamine*

Sympathomimetic amine, indirect adrenergic, vasopressor.

Untoward effects: A highly potent pressor amine found in many foods, such as ***avocados***, ***bananas***, ***beer***; ***broad beans***, aged cheddar, Camembert, brie, and boursin ***cheeses***; ***chicken liver***, ***chocolate***, ***coffee***, canned ***figs***, pickled ***herring***, ***Marmite***®, ***onions***, ***oranges***, ***pork***, ***seafood***, ***tea***, ***wines*** (especially Chianti), ***yeasts***, and ***yogurt*** can precipitate headaches, photophobia, gastrointestinal distress, and severe acute hypertension when taken with ***monoamine oxidase inhibitors***. Many ***cactus*** species contain it and have been used in folklore treatments.

Drug interactions: Use with ***monoamine oxidase inhibitors*** or drugs such as ***furazolidone*** with that activity increase tyramine body levels and increase the release of ***norepinephrine*** from peripheral body stores leading to tremendous arterial contraction and a hypertensive crisis. IV ***amphetamine*** leads to 2–3 times increase in sensitivity to it.

IV, LD_{50}, **mouse**: 229 mg/kg; **rabbit**: 300 mg/kg.

TYROTHRICIN

Antibiotic. **Contains two *antibiotics*, *gramicidin* (q.v.) and *tyrocidine*, in approximately a 4 : 1 ratio in current production batches. Some technical aspects in the older literature may no longer be significant due to a significant change in manufacturing methods. Does not require refrigeration or dating. May be synergistic with *bacitracin*, *neomycin*, *nitrofurazone*, and *polymyxin B*.**

Untoward effects: In **humans**, anosmia and/or parosmia have been reported. Parenteral use may be toxic. Intramammary use of concentrated material often leads to some inflammatory response and, rarely, may cause hemolysis and hemorrhage. Avoid excessive use on nose areas of search and rescue, guard, or hunting **dogs**. May be inactivated by ***bisulfites***, ***cephalin***, ***lecithin***, ***phospholipids***, ***polysorbates***, ***spans***, and ***sulfites***. ***Tyrocidine*** has been explored as an antimold agent in jet fuel; it hemolyzes red blood cells.

LD_{LO}, **human**: 500 mg/kg.

LD_{LO}, **mouse**: IP, 10 mg/kg; IV, 1.2 mg/kg.

U

UBIDECARENONE

= *Coenzyme Q10* = *E-0216* = *Neuquinone* = *Ubiquinone 10*

Cardiotonic.

***Untoward effects*:** Nausea, anorexia, palpitation, and cold hands and feet.

Drug interactions*:** Has caused a decreased effect of ***warfarin in **patients**.

γ-UNDECALACTONE

= *Aldehyde C-14* = *Peach Aldehyde* = *Peach Lactone* = *Persicol*

***Use*:** Fragrance permitted in soaps, detergents, cosmetic creams and lotions, perfumes, and as a flavoring in foods in the U.S. and Europe.

***Untoward effects*:** No sensitization or irritation of 2% in petrolatum in **human** volunteers.

Severe irritation when used topically on **rabbits** and moderately irritating on **guinea pigs**. Fatty infiltration of liver when fed at 13–115 mg/day/5–9 days to **rats**.

LD_{50}, **rat**: 18.5 g/kg.

1-UNDECANEDITHIOL

= *Undecylmercaptan*

***Untoward effects*:** Eye, skin, and respiratory system irritation from inhalation. Percutaneous absorption or ingestion causes eye, skin, and respiratory tract irritation; confusion, dizziness, headache, drowsiness, nausea, vomiting, weakness, and convulsions. Combustible liquid. National Institute for Occupational Safety & Health sets upper exposure limit at 0.5 ppm (3.9 mg/m^3)/15 min.

UNDECYLENIC ACID

= *Declid* = *Renselin* = *Sevinon*

***Antifungal.* Also used in perfumery, flavoring materials, and plastics, and as a lubricant additive and plasticizer.**

***Untoward effects*:** Occasional irritation with burning and maceration at the epidermophytosis application site, requiring discontinuance of treatment. May cause burning if eyes or mucous membranes are contacted.

LD_{LO}, **human**: 500 mg/kg.

LD_{50}, **rat**: 2.5 g/kg; **mouse**: 8.5 g/kg.

URACIL MUSTARD

= *Demethyldopan* = *Desmethyldopan* = *NSC-34,462* = *U-8,344* = *Uramustine*

Antineoplastic.

***Untoward effects*:** Nausea, vomiting, diarrhea, pruritus, hyperpigmentation, and alopecia. Bone marrow suppression with leukopenia, thrombocytopenia, and anemia often occurring 2–4 weeks after treatment discontinued. This damage may be irreversible if total dosage reaches nearly 1 mg/kg. Occasional nervousness, irritability, amenorrhea, decreased spermatogenesis, and depression reported. Rarely, hepatotoxicity. A case of reversible nerve deafness reported.

Calves are more sensitive to its ill-effects than **rats** or **mice**.

LD_{50}, **rat**: 7.5 mg/kg; IP, LD_{50}, **rat**: ~1.25–2.5 mg/kg.

URANIUM

Untoward effects*:** **Miners** working with the dusty raw material are, apparently, 20 times more susceptible to lung cancer than the general population. **Miners** who smoke have even greater risk. Coagulation upsets with increased clotting time, platelet decrease, decreased hemoglobin and tendency to macrocytosis, renal damage, and dermatitis reported. Up until 1971, it was used in glazes for ceramic dishes and produced extremely low levels of radioactivity during their use. Many plants (viz. *Astragalus thompsonae, A. confertiflorus, A. preussi, A. pattersoni* [all milk vetches]; *Aster venustus* [Woody Aster***], *Stanleya pinnata, Oryzopsis hymenoides* [***Rice Grass***], *Allium acuminatum* [***Wild Onion***], and *Zygadenus gramineus*) are indicators of soil uranium, but their uranium content seldom exceeds 100 ppm. Particles of uranium and ***plutonium*** metals can enter the body through skin wounds following explosive accidents. A worse danger affects the **worker** when hot particles or sparks penetrate the apparently undamaged skin. In time, these particles underlying the skin form painless nodules, usually on the face, neck, and hands, which gradually enlarge. Radiation detectors can locate them even before nodules form.

Pulmonary hyalinosis in **dogs** exposed to ***radon*** daughters with uranium dust. After a 3-month-old pup ate particles of a concentrated uranium ore, which was part of a geological collection, it showed dullness, orange-colored injected mucous membranes, poor appetite, and intermittent vomiting for ~4 days. Alopecia followed and entire skin areas could be denuded painlessly. Bones are the chief storage areas in **cattle**, **pig**, and **fowl** tests.

URANYL NITRATE

= *Uranium Nitrate*

Use: Photographic intensifier, porcelain glazing, and reagent.

Untoward effects: Fire hazard. Explosive on contact with organic material or on exposure to sunlight. Lacrimation, conjunctivitis, dyspnea, coughing, râles, nausea, vomiting, skin burns, albuminuria, increased urinary alkaline phosphatase, increased blood urea nitrogen, and acute renal failure. Estimated LD of 1–5 g; years ago, 3 g/day was given therapeutically in early treatment of diabetes mellitus.

Powerful nephrotoxic agent in **rat** trials, damaging the distal half of the proximal tubules. Urinary excretion of alkaline phosphatase increases dramatically (11 times) and increased lactate dehydrogenase (90 times) in **rats** receiving lethal doses.

URAPIDIL

= *B-66,256* = *Ebrantil* = *Eupressyl* = *Mediatensyl* = *Uraprene*

Antihypertensive.

Untoward effects: Usually mild effects of drowsiness, tachycardia, nausea, vomiting, and enuresis.

LD_{50}, **rat**: 140 mg/kg; **mouse**: 565 mg/kg.

UREA

= *Aquacare* = *Aquadrate* = *Basodexan* = *Carbamide* = *Carbonyldiamide* = *Hyanit* = *Keratinamin* = *Nutraplus* = *Onychomal* = *Pastaron* = *Ureaphil* = *Ureophil* = *Urepearl*

***Diuretic, wound debriding, lymphagogue, nutrient, and humectant.* In 1928, it was the first organic substance to be synthesized. Millions of tons used annually in the U.S. in fertilizers, as a source of non-protein *nitrogen* for ruminants, and in resins and manufacturing *barbiturates*.**

Untoward effects: Orally, it has caused nausea, vomiting, polyuria, and polydipsia in some **patients**. Oral use as an ocular hypotensive caused pulmonary edema and anginal pain, due to sudden increase in blood volume. IV use has been associated with headaches, nausea, vomiting, acute pulmonary edema (with rapid IV), intravascular hemolysis, parotiditis, syncope, confusion, and hypotension. Repeated IV use caused hyponatremia and hypokalemia. Hyperosmotic solutions lead to venous thrombosis, disseminated intravascular coagulation, or phlebitis at injection site, especially if small veins are used. Rebound cranial swelling after temporary profound and rapid decrease in intracranial fluids and cerebrospinal fluid. Extravasation has caused necrosis or sloughing. Topical use can be irritating and its use in hair preparations can cause swelling of hair. It can cause elevated or false positive ***ammonia*** test results and decreased blood urea nitrogen and ***bilirubin*** serum test results.

Drug interactions: IV use increases ***lithium*** excretion, decreasing the latter's serum levels.

In **animals**, avoid systemic use in nephritis or hepatitis. Extravasation may cause serious local necrosis. Do NOT feed to young **ruminants** until rumen development has occurred. Follow feeding directions carefully. **Cattle** unaccustomed to eating it are more easily poisoned due to its breakdown into ***ammonia*** (q.v.). Even after adapting to its dietary intake, ~0.2 g/lb or 0.45 g/kg is toxic to **cattle** and **sheep**. It is particularly toxic if given with ***soybean*** meal, as the urease in the latter accelerates the release of ***ammonia***. IV in **dogs** led to immediate marked hypotension and bradycardia, followed by tachycardia and polyuria. Monogastric **animals** are not readily susceptible to urea poisoning, but may have increased blood ***ammonia*** concentrations. After ingestion, **dogs** may evidence hypersalivation, gastroenteritis, and abdominal pain. Occasionally, weakness, tremors, and methemoglobinemia have occurred.

IV, LD_{LO}, **dog**: 3 g/kg; **rabbit**: 4.8 g/kg.

URECHITES sp.

Untoward effects: Common pasture poisonous plants for Cuban **ruminants** leading to "Muerte subita" (sudden death). ***Loroquin***, a ***pyrrolizide*** alkaloid, has been isolated from some species.

U. suberecta* = *Savannah Flower* = *Yellow-flowered Nightshade, abundant in West Indian Islands; some grow in Florida, and used in Jamaica by the **natives** as a poison leading to violent nausea, vomiting, and purging with diaphoresis, and convulsions; first, increases arterial pressure, then, decreases it. Eventually, death from cardiac arrest with large doses. Contains two ***digitalis***-like glycosides, ***urechetin*** and ***urechetoxin***, and a toxic alkaloid, ***urechitine***.

UREDOFOS

= *Diuredosan* = *Sansalid*

Anthelmintic, cholinesterase inhibitor. A phosphoramidate.

Untoward effects: In **animals**, one must avoid use within a few days of exposure to any ***cholinesterase inhibitors***. Despite early successful trials, American **veterinarians** reported adverse reactions and deaths related to its use.

The U.S. FDA finally withdrew its approval of the product, based on its unacceptable reaction risk.

URETHANE

= *Ethyl Carbamate* = *Ethyl Urethane* = *Urethan*

Antineoplastic, anesthetic, hypnotic, sedative, diuretic, solubilizer.

Untoward effects: Potential **human** exposure is by inhalation, ingestion, and dermal contact. Annual production in the U.S. was once > 6 million lbs. Anorexia, drowsiness, nausea, vomiting, and dizziness. After continuous use, bone marrow depression, anemia, neutropenia, leukopenia, pancytopenia, thrombocytopenia, hepatocellular liver damage, jaundice, increased alkaline phosphatase, and occasionally, a fulminant necrosis of the liver. The leukopenia decreases the body's defense against infections. After IV nystagmus, slurred speech, euphoria, and some nausea or vomiting (60%). ***Diethylpyrocarbonate*** (q.v.), once a popular preservative (0.1–0.2 mg/l of ***orange juice*** and 1–3 mg/l of ***white wine*** and ***beer***; also once used in ***ale, bread, olives, soft drinks, soy sauce***, and ***yogurt***) reacts with ***ammonia*** producing ***uretham***, a suspect carcinogen for **man**.

Drug interactions: May increase prothrombin time when used with ***anticoagulants***. It enhances ***propranolol*** effects and the combination can be lethal.

Given in the drinking water to **mice** led to lung adenomas, lymphomas, liver angiomas and hemangiomas, papillomas, skin sebaceous carcinomas, harderian gland tumors, squamous cell tumors, mammary carcinomas, and malignant mesenchymal tumors of their fat pads. Many similar neoplasms induced in **rats** and **hamsters**. Teratogenic in **hamster** trials.

LD_{50}, **mouse**: 2.7 g/kg; TD_{LO}, **mouse**: 160 mg/kg; IP, LD_{50}, **rat**: 1.5 g/kg; IP, LD_{LO}, **mouse**: 1.2 g/kg.

URGINEA sp.

Untoward effects: ***U. altissima*** = ***Maerman*** = ***Slangkop*** of Africa is poisonous (severe diarrhea and fatalities) for **rabbits**, **sheep**, and other **livestock**.

U. burkei = ***Transvaal Slangkop***, a common poisonous plant in Zimbabwe, has killed **cattle** and **sheep**.

U. indica has had the same rodenticidal effect as ***White Squill*** and is used in India to induce alternating convulsions and paralysis, and, eventually, death of **rats**. For ***U. maritima***, see ***Squill***.

U. sanguinea is also highly poisonous from its ***digitalis***-like glycoside. The young, succulent stalks are commonly eaten and cause large **livestock** losses in Zimbabwe. With all these species, post-mortem shows pulmonary edema and hyperemias, acute catarrhal enteritis, and myocardial and endocardial petechiae. **Bushmen** in Africa use the juice of the bulbs on **their** poison arrows or darts. When the young flower head pushes up from the ground, it resembles a snake's head (***slangkop***).

URIC ACID

Untoward effects: A normal metabolic product of cellular destruction that often accumulates in the body leading to gout and needle-like crystals, causing severe hot, red, swollen, and painful joints. Crystals can accumulate in mounds (tophi) under the skin and, in rare occasions, on the heart valves. Nephrotoxicity can also occur. These **patients** must avoid food items with high levels of ***purine*** (a uric acid precursor), such as ***anchovies, bacon, bouillon, brains, gravies, herring, kidneys, liver, meat*** extracts, ***sardines, scallops, thymus sweetbreads, turkey***, and ***veal***. ***Alcohol*-abusers** (especially ***beer***) have increased serum uric acid levels. Acidic drugs (***chlorthiazide, penicillin, salicylic acid***, and acetylated ***sulfonamides***) can increase uric acid tubular reabsorption. Hyperuricemia can potentiate ***ampicillin*** rashes in some **patients** and excessive urinary excretion can turn urine a reddish-yellow or milky hue. Inhibits enzyme tests for ***glucose***. Use with ***ascorbic acid***, blood transfusions, ***nitrogen mustards, pyrazinamide, salicylates, theophylline***, and ***thiazides*** gives false or spuriously high blood uric acid test results; ***ascorbic acid, dicumarol, corticosteroids, mercaptopurine, probenecid, salicylates, sulfinpyrazone, theophylline***, and ***triamterene*** gives false or spuriously high urinary uric acid test results. ***Coumarin anticoagulants, dicumarol, oxalates***, and ***piperazine*** give false or spuriously low blood uric acid test results; ***acetazolamide, allopurinol, ethacrynic acid***, and ***thiazides*** give false or spuriously low urinary uric acid test results.

Uric acid calculi contain ***ammonium*** from ***urea*** decomposition, ***sodium urates***, and uric acid, and form in acid urine. They are very common in Dalmation **dogs**, which excrete very large amounts of uric acid in their urine instead of metabolizing it to ***allantoin***, as done by most **mammals**, except **primates**.

UROKINASE

= *Abbokinase* = *Actosolv* = *Breokinase* = *Persolv* = *Purochin* = *Ukidan* = *Uronase* = *Win-22,005* = *Win-Kinase*

Plasminogen activator, thrombolytic agent.

Untoward effects: Anaphylactoid and hypotensive reactions with dyspnea after IV use. Has also increased intraocular pressure. Superficial and internal bleeding in **some** and fatalities due to intracranial or retroperitoneal hemorrhage have occurred during IV therapy.

UROSTACHYS SAURURUS

Untoward effects: A botanic, folkloric treatment in Argentina that Buenos Aires Poison Control Center reports as poisoning **children**.

URTICA DIOICA

= *Ardica* = *Bichhubooti* = *Great Nettle* = *Nettle* = *Ortica* = *Stinging Nettle*

Untoward effects: Its poisonous stinging hairs, on contact, cause immediate intense pain and urticaria in **people**. The hollow hair-like projections from the leaf contain ~10 μg of fluid containing ***histamine*** 1 : 500–1 : 2000, dilating capillaries, increasing permeability and causing a wheal, enhancing rapid absorption of its second component, ***acetylcholine*** 1 : 100, to produce the stinging sensation on the body nerve endings and throat, dyspnea, bradycardia, and muscle weakness.

It and ***U. chamaedryoides*** and other species are ***nitrate***-accumulators. The combination of chemicals released has caused trembling, slobbering, dyspnea, vomiting, sneezing, local pain, pawing at the mouth, bradycardia, arrhythmia, muscle weakness, and deaths in affected **dogs**.

USNIC ACID

= *Usnein*

Untoward effects: Contact dermatitis from contact with ***lichen*** on bark. Common in "**wood-cutter's** disease" and "***cedar*** poisoning" in **forest workers**. Has caused a delayed hypersensitivity.

IV, LD_{50}, **mouse**: 25 mg/kg.

USTILAGO

= *Smut*

Untoward effects: Releases aeroallergens leading to allergic rhinitis, bronchial asthma, and hypersensitivity pneumonitis. Common illnes in North America.

U. avenae = ***Loose Smut*** often parasitizes ***oats*** leading to poisoning of **cattle** and other **livestock**.

U. hordei on ***barley*** has caused toxicity in **calves**, **colts**, **pigs**, **sheep**, **mice**, **hamsters**, and **rabbits** in many Russian reports. Pulmonary edema, vomiting (in **swine**), erythropenia, leukocytosis, and encephalitis have occurred.

U. maydis = ***U. zeae*** = ***Corn Smut*** leads to bronchopneumonia in **cattle** fed high levels; convulsions, conjunctivitis, and coughing in **shoates**; and foul-tasting and unpleasant odorous eggs in **poultry**. A German report describes a condition resembling ergotism in **children** after eating infected ***corn***, but it does not contain any ***ergot***-type compounds.

V

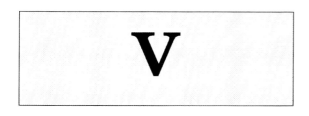

VALACYCLOVIR

= *BW-256U87* = *Valaciclovir* = *ValACV* = *Valtrex* = *Zelitrex*

Antiviral. **Almost completely metabolized to *acyclovir* (q.v.).**

Untoward effects: Nausea, vomiting, headache, insomnia, dizziness, and fatigue are common. Facial edema, tachycardia, hypertension, confusion, agitation, auditory and visual hallucinations, aggressive behavior, mania, increased ***creatinine***, renal failure, erythema multiforme, thrombocytopenia, and aplastic anemia have also been reported. Immunocompromised **patients** receiving long-term high dosage treatment developed thrombotic thrombocytopenic purpura/hemolytic uremic syndrome, which, in some cases, has been fatal.

VALERIANA sp.

Used medicinally in China, India, Europe, South America, Mexico, and the U.S.

Untoward effects: Of the ~200 species, ***V. officinalis*** is the one commonly used medicinally and is widely naturalized in northeastern North America and also known as ***All-Heal, Amantilla, Capon's Tail, Garden Heliotrope, Neurelax, Setwall, Setewale***, and ***Vandal Root***, and the dried root, ***Valerian*** (***Kesso*** in Japanese), has been used for over 1000 years for its sedative, hypnotic, and tranquilizing properties, as well as mood elevation. France buys ~50 ton annually. Despite **marketers'** claims that it is safe, cases of severe liver damage from its use have been reported. Ataxia, hypothermia, extensive muscle relaxation, fatigue, abdominal cramps, mydriasis, and tremors reported after an attempted suicide in a **patient** who had prior ***marijuana*** use. Controlled clinical trials have also reported headaches, excitability, uneasiness, and cardiac changes (bradycardia, increased blood pressure, and increased cardiac force). Withdrawal complications led to cardiac failure and delirium in a 58-year-old **male** who ingested 530 mg–2 g of the root extract five times/day. IV use of an alcoholic extract of the roots by a **drug abuser** caused fever, chills; abdominal, back, and flank pain; neck stiffness, headache, leukocytosis, and abnormal liver function. In Hong Kong, 24 **patients** overdosed on over-the-counter ***valerian*** products leading to central nervous system depression, vomiting, and some anticholinergic effects. Has potentiated ***cyclobenzaprine*** effects in a **patient**.

V. hardwickii = ***Nakali jatamansi*** of India and ***V. wallichii*** = ***muskbala*** (Kashmir) = ***Sugandhwal*** (India) = ***Samyo*** (Kumaon) = ***Sumyon*** (Garwahl) roots have similar properties and are used as incense in magico-religious rites.

VALINOMYCIN

Antibiotic.

Untoward effects: Not for "drug" use; laboratory research only. Dangerous. Work with it carefully under a hood. It weakens the integrity of epithelial cells, as on the cornea, leading to hazy vision and "halo" effect near lights within a few hours, becoming very painful punctate keratitis with edema.

VALNOCTAMIDE

= *Axiquel* = *Ethyl Methyl Valeramide* = *McN X-181* = *Nirvanil*

Antihypertensive, tranquilizer, anticonvulsant.

Untoward effects: Tingling and numbness in legs, dizziness, blurred vision, drowsiness, mild edema of lower extremities, mild syncope and malaise. Toxic serum concentration is 40 mg/l.

VALPROIC ACID

= *Acide Valproique* = *Mylproin* = *Valproïnezuur*

The sodium salts = *Abbott 50,711* = *Convulex* = *Depakene* = *Depakin* = *Dépakine* = *Depakote* = *Divalproex Sodium* = *DPA Sodium* = *Epilim* = *Ergenyl* = *Eurekene* = *Labazene* = *Leptilan* = *Orfiril* = *Sodium Hydrogen Bis(2-propylpentanoate)* = *Sodium Valproate* = *Valcote*

Anticonvulsant, antiepileptic. **Originally used as a solvent and serendipitously, when used in testing a drug's antiepileptic properties, it was discovered that the so-called inert solvent protected rabbits and mice against *pentylenetetrazole*-induced seizures.**

Untoward effects: Most common are nausea, vomiting, indigestion, and abdominal cramps and/or constipation, which are usually transient or decreased by slowly titrating the drug upwards or giving it with food. Central nervous system effects are uncommon except for sedation and include "spots in front of the eyes", nystagmus, diplopia, ataxia, headache, dizziness, coma, tremors, dysarthria, and asterixis. Depression, psychosis, aggression, hyperactivity, muscle weakness, weight increase, hair loss, rash, inhibition of secondary platelet aggregation, petechiae, epistaxis, prolonged bleeding after surgery, thrombocytopenia, leukopenia, pancytopenia, eosinophilia, methemoglobinemia, hemolytic anemia, ketonuria, IgA deficiency, cholestatic hepatitis, increased SGOT, increased SGPT, liver necrosis; fatal acute fulminant hepatic failure; ***carnitine*** depletion, systemic lupus erythematosus, menstrual disturbances, polycystic ovaries, and hyperandrogenism in **females** also reported. Rarely, pancreatitis, Fanconi syndrome, acute intermittent porphyrias, and fatal hepatotoxicity occur, which more frequently is in **children** < 2 years. Valproic acid syrup, because of its ***sorbitol, sucrose***, and ***glycerin*** content, can cause diarrhea in **children**. Antenatal exposure within the first trimester has caused neural defects, especially spina bifida and other teratogenic effects. Can increase viral replication in HIV **patients**. An 18-year-old **patient** after 800 mg/day/10 years developed leg edema. Changes in hair color reported. After 500 mg tid, hyperammonemia and coma in a 69-year-old **female**. Ingestion of 200 mg/kg led to coma in some **adults** and **children**. Severe toxicity in **adults** after cumulative dose of 19 g or more. Low levels found in breast milk of treated **mothers**.

Drug interactions: ***Aspirin*** decreases its oxidation and clearance by ~30%. Use with ***phenytoin*** occasionally may increase the latter's toxicity, and at other times, decrease its serum levels. Its serum levels are increased by coadministration of ***chlorpromazine*** and decreased with ***carbamazepine*** and ***phenobarbital***. It decreases ***topiramate*** serum levels by ~15% and has decreased the response to ***erythropoietin***. Anxiety and confusion reported in a 38-year-old **female** when given with ***erythromycin***. It accentuates the depressant effect of ***ethanol*** and ***phenobarbital*** and inhibits the metabolism of ***diazepam, lamotrigine***, and ***lorazepam***, increasing their toxicity. It increases ***amitriptylline*** and ***nortriptyline*** serum levels. It increases the oral bioavailability of ***zidovudine***. ***Felbamate*** increases its blood levels. ***Cholestyramine*** significantly decreases its absorption and ***antacids*** may increase its bioavailability. Metabolism increased by ***clonazepam***. Osteomalacia in a 22-year-old **female** after ~10 years of treatment with ***carbamazepine***.

In **rats**, it increases ***clonazepam*** serum concentration.

LD_{50}, **rat**: 670 mg/kg; IP, LD_{50}, **mouse**: 470 mg/kg; *Sodium Salt*: LD_{50}, **rat**: 1.53 g/kg; **mouse**: 1.7 g/kg; IP, LD_{50}, **rat**: 790 mg/kg; **mouse**: 1 g/kg; **rabbit**: 1.2 g/kg; **cat**: 565 mg/kg.

VALRUBICIN
= AD 32 = Valstar

Antracycline antibiotic, antineoplastic.

Untoward effects: Intravesicular use leads to bladder irritation with dysuria and pollakiuria.

VALSARTAN
= CGP 48,933 = Diovan = Nisis = Tareg

Antihypertensive, angiotensin II receptor antagonist.

Untoward effects: Angioedema and photosensitive pruritic rash induced in a 71-year-old **female** given 80 mg/day/3 months. May be less effective in black **patients**. Occasionally hepatic dysfunction, increased aminotransferases, neutropenia, and hyperkalemia. Not recommended for use during pregnancy because it interferes with the renin–angiotensin system, which can cause **fetal** death. Rarely, causes coughing and intestinal gas.

Food and drug interactions: ***Grapefruit*** can aggravate the induced flatulence. ***Bayberry*** has antagonized its effect.

VALSPODAR
= PSC 833

Drug interactions: Induces dramatic decrease in ***digoxin*** body clearance and increases serum levels.

VANADIUM

Use: In manufacturing rust-resistant hardened steel. An essential trace element for **chicks**, **rats**, and probably **humans**.

Untoward effects: Inhalation of or exposure to dusts causes skin irritation, eczema, eye and throat irritation, green-black discoloration of tongue, metallic taste, rhinorrhea, sneezing, paroxymal coughing, sore chest, wheezing, bronchitis, pneumonitis, fine râles, dyspnea, emaciation, cachexia, hyposmia, mild hepatic dysfunction, ulcerative pharyngitis and neurasthenia, hypertension, nervousness, giddiness, anemia, anorexia, hemorrhagic nephritis, albuminuria, and tremors of fingers. Inhalation of fumes duplicates many of the above symptoms. Ingestion of its compounds can cause epistaxis, dyspnea, ptyalism, convulsions, and diarrhea. Some **people** have shown a hypersensitivity to its use in joint prostheses. National Institute for Occupational Safety & Health suggests upper exposure limit of 0.05 mg/m^3.

Has helped induce liver necrosis in **rats**. Acute and chronic symptoms induced in **rats** by its salts leading to nose and intestinal hemorrhages, severe diarrhea, dyspnea, immediate intense distress, hind limb paralysis, and convulsions. Acute desquamative enteritis noted on postmortem. Synergism of its toxicity in **chicks** has been demonstrated with ***ochratoxin A*** (q.v.). Accidental ingestion of ***vanadium tetrachloride*** by grazing **cattle** in an industrial-polluted area led to 19 with digestive disorders and impaired locomotion after 3–5 days and lasted for 8–10 days. Watery diarrhea with blood clots, flatulence, oliguria, difficulty in standing upright, hesitant and uncoordinated gait, and limping. After 3 months, all but one recovered.

VANCOMYCIN

= *Vancocin* = *Vancoled*

***Antibiotic*. Usually IV use. Occasionally IP, IM, oral, and intrathecal. It is poorly absorbed after oral administration, and such use is reserved for treatment of colitis. Estimated U.S. annual sales are 100 ± million dollars and worldwide sales ~200 million dollars.**

***Untoward effects*: *Blood*:** Neutropenia (2%), pancytopenia, thrombocytopenia, agranulocytosis, leukopenia, eosinophilia, and granulocytopenia.

***Cardiovascular*:** Hypotension most common with bolus infusions and has occasionally occurred with slow infusions as well. Cardiac arrest reported in a 57-year-old **patient** after 1 g IV as a bolus in ~2 min. **Fetal** distress in pregnant 21-year-old **female** secondary to vancomycin 1 g inducing **maternal** hypotension. Tachycardia and thrombophlebitis (13%) after IV. Has caused cardiac arrest in a 2-year-old **child**.

***Ear*:** Ototoxicity (damages to and destruction primarily of inner ear) even with normal renal function and 2 g/day, and occasionally tinnitus, vertigo, and dizziness. Loss of hearing, due to vestibular and cochlear damage, has occurred in azotemic **patients**, due to its rising blood concentration from faulty urinary excretion. Since it readily crosses the placenta, it can cause deafness in the **newborn**. Ototoxicity is more common when serum concentrations are > 80 μg/ml.

Miscellaneous*:** Caused increased or false blood urea nitrogen test results. Temporary paresthesia and low back pain when given at 16.7 mg/min, but not 2 weeks later in the same 36-year-old **female** when given at a slower infusion rate, 5.6 mg/min. Shock and rash within 5 min in two **newborns** given 15 mg/kg over a 20 min period. Recently, a case of priapism reported in a diabetic 37-year-old **male** after two IV doses of 1 g/day for bursitis after a cadaveric kidney–pancreas transplant. Adsorption by IV ***Teflon tubing led to a 13-year-old **male** receiving a four times normal concentrations causing severe headache, high fever, hypotension, and seizures within 15 min. **Patients** allergic to it may develop allergic cross-reactivity to ***teicoplanin***. Pain at local injection site. Anaphylaxis reported.

Red Man or Neck Syndrome*:** So-called because of an initial red rash (erythematous maculopapular) on the posterior part of the neck, followed by lethargy, facial flushing, itching, hypotension, and cardiac depression, especially when an infusion is given in < 1 h, exceeds 500 mg at a concentration > 5 mg/ml, and is apparently associated with release of ***histamine and is an allergic reaction.

Renal*:** Interstitial nephritis in 5–15%. Prior to 1970, many impurities contributed to a higher level of toxicity. Nephrotoxicity potentially increased when given with other ***aminoglycosides. Has shown nephrotic synergism with ***tobramycin***.

***Skin*:** Maculopapular rashes with pruritus in an 82-year-old **female** with renal dysfunction after 125 mg orally qid for *Clostridium difficile* colitis leading to rashes on legs, torso, abdomen, and arms after 8 days of treatment. Cases of severe maculopapular, bullous, and exfoliative dermatitis, occasionally with intermittent fevers, lymphadenopathy, peripheral eosinophilia, oral and vaginal mucosal ulcers of Stevens–Johnson-like reaction, and toxic epidermal necrolysis.

Therapeutic serum levels are reported as 5–12 mg/l and toxic levels at > 40 mg/l.

Drug interactions*:** It may decrease ***methotrexate absorption by 30–50% when both are given orally. Use with ***vasodilator*** therapy, such as ***nifedipine***, can add to its hypotensive effects. Synergism has been demonstrated with ***streptomycin*** and ***gentamicin*** versus certain organisms. Neurotoxicity and nephrotoxicity potential with ***cephaloridine, colistin, gentamicin, kanamycin, neomycin, paromomycin, polymyxin B, streptomycin, tobramycin***, and ***viomycin***. Consistent failure of vancomycin to reach therapeutic serum concentrations when added to therapy with ***aminophylline, furosemide***, or ***theophylline***.

IV, LD_{50}, **rat**: 319 mg/kg; **mouse**: 400 mg/kg.

VANILLA

Use*:** As a flavoring agent and odorant in foods, pharmaceuticals, and beverages. The genuine vanilla bean = ***Vanilla fragrans or ***planifolia*** and is a native of Mexico and surrounding areas, Madagascar, and Indonesia. It is the unripe fruit of a perennial orchid.

Untoward effects*:** **Bakers** have developed vanillism, a contact dermatitis from it. It causes erythema and a marked edema; occasionally accompanied by rhinitis, asthma, and vertigo. Some **workers** have developed extensive eczemas only when working with synthetic vanilla. In 1985, 50 brands of Mexican vanilla extracts contained ***coumarin, ethyl vanillin, umbelliferone, and ***vanillin***. In 1981, the University of Texas found 89.5% of 850 samples adulterated with ***coumarin*** (q.v.) at up to a 0.61% level. It or ***tonka beans*** (q.v.) may also be an illegal adulterant in imitation vanilla extracts in the U.S. Spuriously elevated urinary vanillylmandelic acid levels after ingesting it if test is a phenolic acid/diazotized *p*-nitroaniline one. It can cross-react with ***Balsam Peru, cinnamon oil***, and ***clove oil***. Customary mixing of ***phenytoin*** with vanilla pudding decreases the former's serum levels. Reports of serious poisoning from vanilla pudding turned out to be ptomaine poisoning.

LD_{LO}, **human**: 500 mg/kg.

LD_{50}, **rat**: 1.6 g/kg; **guinea pig**: 1.4 g/kg; LD_{LO}, **rabbit**: 3 g/kg.

VARNISH

Although many forms exist, they all usually contain the same ingredients with slight variations. Usually made by heating and drying oils, such as gas-proofed *tung oil* or bodied *linseed oil*, with a resin (usually a *phenolic*, but occasionally a wood or gum form), blending this with *mineral spirits* and catalysts (*cobalt, manganese*, or *zinc naphthenate*) to accelerate oxidation. Solvents may include *glycol ethers* and *alcohol* and *cobalt* driers.

Untoward effects: Contact dermatitis is usually due to one of its oils. Some have contained ***polychlorinated biphenyls*** with their associated toxicity. Illegal adulteration with ***methanol*** and its subsequent ingestion led to nausea, vomiting, sudden hypotension, respiratory failure, blindness, cardiac arrest, and the deaths of hundreds of **people** in Madras, in southern India. Inhalation of varnish solvents (***acetone, methanol, methylene chloride, toluene, turpentine***, and ***xylene***) has caused intoxication and other adverse central nervous system effects in **abusers**.

VASOPRESSIN

= *ADH* = *β-Hypophamine* = *Leiormone* = *Pitressin* = *Tonephin* = *Vasotocin*

***Antidiuretic hormone, vasopressor hormone, hemostatic.* Produced by the posterior lobe of the pituitary gland. It is responsible for the antidiuretic and blood pressure increase effects of posterior pituitary extract. IV, IM, SC, intracervical, and intra-arterial.**

Untoward effects: Allergic reactions have been reported due to the nature of the **animal** source.

IA use caused vasoconstriction of gastroduodenal, left gastric, splenic, and superior mesenteric arteries in a **patient**. Gangrene of the feet after infusion in the mesenteric arteries of another **patient**. Gastric necrosis in a 59-year-old **female** after infusion for gastrointestinal hemorrhage.

IM use caused diarrhea, cramps, and diaphoresis in ~27%. IM use of the tannate in oil form has induced extensive skin eruptions, tenesmus, and constipation in some **patients**.

Intranasal use as a spray or snuff caused local vasoconstriction, possible nasal congestion, irritation and occasional ulceration of nasal mucosa. The solution preservative, ***chlorobutanol*** (q.v.), may have caused pruritus in a **patient**.

IV use caused increased systolic blood pressure, ventricular dysrhythmia, prolonged QT interval, asystole, constriction of coronary arteries, myocardial infarction, angina, reversible ischemic colitis, hyponatremia, nipple necrosis, cutaneous necrosis, and gangrene. Bronchospasm, abdominal cramps, skin rash, and uterine contractions have been reported.

After *SC* caused ischemic gangrene in treatment of esophageal varices.

Drug interactions: Its effects are potentiated by ***acetaminophen, chlorpropamide***, and ***phenformin***, and its excretion is increased by ***cyclophosphamide***.

Crosses placental barrier in **rats**. The hyperosmotic state of the renal medulla, due to increased concentration of ***sodium***, is the probable cause for pituitary stimulation and pituitary release of vasopressin. This permits water reabsorption in the distal tubules and collecting ducts of the kidneys. This hemodilution causes a decrease in plasma protein concentration and plasma oncotic pressure, with a loss of fluid into interstitial spaces, resulting in edema and ascites, stressing the need for ***sodium*** intake control in cardiac problems.

VECURONIUM BROMIDE

= *Musculax* = *NC 45* = *Norcuron* = *ORG-NC 45*

***Skeletal muscle relaxant.* Non-depolarizing. A *pancuronium* derivative.**

Untoward effects: Caution, contains 12.5% ***bromide*** (q.v). Neuromuscular blockade in ~2.5–3 min and recovery after 25–40 min if liver metabolism is normal. Hypersensitivity reactions (bronchospasm, hypotension, tachycardia, urticaria, and erythema) have occurred. Prolonged muscle weakness and paralysis, especially after long-term infusions. Asthmatic **patients** on medical wards developed respiratory distress requiring resuscitation and mechanical ventilation after injections of ***hydrocortisone sodium succinate*** that was, apparently, contaminated with the ***muscle relaxant*** manufactured in the same plant.

Drug interactions: Use with ***dantrolene*** or the ***β-adrenergic blockers, atenolol, esmolol***, and ***propranolol***, prolongs recovery time.

VENLAFAXINE

= *Effexor* = *Wy 45,030*

Antidepressant. A selective serotonin reuptake inhibitor.

Untoward effects: Nausea, sedation, headache, diaphoresis, dizziness, nervousness, anxiety, anorexia, mydriasis, weakness, insomnia, sexual dysfunction, abnormal ejaculation, and impotence are the most common. Occasionally xerostomia, constipation, weight loss, dose-dependent increase in blood pressure, blurred vision, bad taste in mouth, ear pain, tinnitus, bronchitis, pollakiuria, pruritus, alopecia, menstrual upsets, fluid retention, and weight increase. Rarely, increased QT interval, hypotension, increased intraocular pressure, inappropriate antidiuretic homone secretion, hepatitis, and seizures (especially after overdoses). Hyponatremia, urinary incontinence, neuroleptic malignant syndrome, and severe suddden withdrawal symptoms (nausea, incapacitating positional vertigo and light-headedness) within 36 h also occur. Neuroleptic malignant syndrome in a 44-year-old **male** after a single 75 mg dose.

Drug interactions: It should not be used within 14 days after discontinuing treatment with a ***monoamine oxidase inhibitor*** and allow at least 7 days after cessation of its use before using a ***monoamine oxidase inhibitor***. Burning fever, stiff muscles, and a near-fatal seizure in a 34-year-old **female** taking it after discontinuing ***tranylcypromine***. A case of ***serotonin*** syndrome with confusion, twitching, tremors, diaphoresis, ankles in plantar flexion, difficulty in speaking, hallucinating after taking a 37.5 mg tablet within 24 h of **his** last 30 mg ***phenelzine*** dose. Similar symptoms in a 43-year-old **male** who twice suffered toxicity after taking ***isocarboxid***. Drugs inhibiting cytochrome P450, such as ***quinidine***, can increase their own serum concentration and toxicity.

Mild depression in **dogs** at 1 mg/kg and tremors occur at 10 mg/kg. Half-life in **dogs** is 2–4 h.

VERAPAMIL

The hydrochloride = *Arpamyl* = *Berkatens* = *Calamth* = *Calan* = *Cardiagutt* = *Cardibeltin* = *Cordilox* = *Covera-HS* = *D365* = *Dignover* = *Drosteakard* = *Finoptin* = *Geangin* = *Ikacor* = *Iproveratril* = *Isoptin* = *Quasar* = *Securon* = *Univer* = *Vasolan* = *Veracim* = *Veramex* = *Veraptin* = *Verelan* = *Verexamil*

***Calcium channel-blocker, class IV antiarrhythmic, antianginal, antihypertensive.* Oral and occasionally IV.**

Untoward effects: ***Cardiovascular***: Hypotension and bradycardia can occasionally occur since it slows conduction across the atrioventricular node. Most of these significant events, congestive heart failure, acute cardiac failure, arrhythmias; ventricular tachycardia and cardiac arrest in Wolf–Parkinson–White syndrome; worsening of congestive heart failure, and torsades de pointes have followed IV use. It has prolonged P-R interval. Acute congestive heart failure reported in two **female** 71- and 81-year-old hypertensives after 240 mg sustained-action tablets/4–5 days. Peripheral edema in ~2%. Pulmonary edema is part of the congestive heart failure complex reported in < 2% or after overdoses. Third degree atrioventricular block in < 1%. Myocardial infarction, palpitations, purpura, and vasculitis have also been associated with its use.

Central nervous system: Transient headache and vertigo in ~2%, dizziness in 3.3–3.6%, light-headedness, weakness, anxiety, nervousness, mental confusion, paresthesias, acute dystonia, and psychotic disorders.

Gastrointestinal: Slight to severe constipation in 7–10%; nausea in 2–3%, rarely, vomiting. Gastrointestinal ulceration and hemorrhage in a 22-year-old patient after 3200 mg. Decreased lower esophageal sphincter action. Possible gastroesophageal reflux and decreased gastric motility in an 84-year-old **female** ingesting 240 mg/day/6 weeks.

Liver: After an extremely high rate of oral absorption, its bioavailability is only 20–35% after its first pass extensive metabolism in the portal circulation. **Patients** with liver dysfunction often have elimination half-lives prolonged by 14–16 h above a norm of ~3–7 h; this may be related to the occasional reports of hepatotoxicity. Striking increase of serum transaminases and some increase in serum ***bilirubin*** and alkaline phosphatase.

Miscellaneous: Fatigue in 1–2%, dyspnea or asthma in 1.4%, hyperprolactinemia, ankle edema (not associated with congestive heart failure), maculopapular skin rash, pruritus, flushing, burning sensation in the limbs; hyperprolactinemia and galactorrhea in a 22-year-old **female** after 5 days of oral treatment with 80 mg qid; impotence, depression, gingival hyperplasia, hyperglycemic metabolic acidosis, nightmares; decreased ***testosterone*** serum concentration and impotence in a 74-year-old **male**; Parkinsonism, xerostomia, ecchymoses, spotty menstruation, polyuria, blurred vision, hypertrichosis, and possible hair color change. Stevens–Johnson syndrome reported ~10 days after 240 mg/day in a 59-year-old **female**. Exfoliative dermatitis also reported in a few **people**. IV infusion during second stage of labor caused **fetal** umbilical serum concentration 50% of **maternal**. In a 70-year-old **male**, irregular, repetitive myoclonic movements in **his** shoulder girdle, biceps, and triceps after 80 mg qid orally. Progressive respiratory failure required mechanical ventilation in a 63-year-old **female**. Cerebral infarction reported in a 39-year-old **female** attempting suicide by ingesting 2280 mg. Other suicide attempts with > 3 times that amount. Mydriasis reported in suicide cases. Topical application of 8% in ***propylene glycol*** on tuberculin-induced hypersensitivity skin reactions inhibits the response. Fatalities have occurred after ingestion of

4 g or 7 g of sustained release form. Yet, some **patients** have survived the ingestion of 16 g and 9.5 g of sustained release form.

Therapeutic serum levels are usually near 0.2 mg/l with toxic levels at 1 mg/l, and fatal levels have occurred at 3 mg/l.

Drug interactions: It can increase blood ***alcohol*** concentrations, prolonging ***alcohol's*** effects. The antibiotics ***ceftriaxone*** and ***clindamycin*** given IV to a 59-year-old **male** stabilized for 2 years on verapamil led to acute verapamil toxicity and complete heart block requiring CPR and a temporary pacemaker. On two separate occasions, a 77-year-old **female** stabilized on 80 mg twice daily received ***clarithromycin*** and ***erythromycin*** leading to bradycardia. Severe hypotension and bradycardia, as well as periods of dizziness and syncope events in a 53-year-old **male** hemodialysis **patient** while receiving ***clarithromycin*** 250 mg/twice daily and verapamil 120 mg twice daily. Two reports of atrioventricular block followed addition of ***clonidine*** 0.15 mg twice daily to verapamil therapy. Fatigue; spontaneous increase in SC bleeding after several weeks has occurred in **patients** regularly taking ***aspirin***. A number of case reports reflect the need for caution in using it with ***β-adrenergic-blockers***, as both induce bradycardia. ***Atenolol*** 100 mg/day with verapamil 360 mg/day led to complete heart block in a 44-year-old patient. Sinus arrest with ventricular escape rhythm and shock in a 42-year-old **male** after ***atenolol*** 50 mg/day and verapamil 40 mg tid and an extra 100 mg ***atenolol*** the previous night. Adding verapamil 120 mg tid to ***metoprolol*** 100 mg tid doubled the latter's plasma levels in a 59-year-old **female**. Asystole within 1 min after verapamil 7 mg IV over 7 min in a 57-year-old **male** receiving ***practolol***. In a 70-year-old **male**, 5 mg IV over 5 min added to ***practolol*** treatment led to bradycardia and complete asystole, and in a 6-month-old child, cardiac massage, etc. was required after verapamil; **both** had prior treatment with ***digitalis*** preparations. History of ***propranolol*** use caused an increased incidence of adverse effects in **patients** given verapamil. In **mice** given ***propranolol*** after pretreatment with verapamil 3 mg/kg, it increased the former's LD_{50} eight times. Severe bradycardia and atrial arrhythmia reported with the addition of topical ophthalmic ***timolol***. Has increased ***carbamazepine*** (q.v.) concentrations by a mean of 50%, causing many adverse effects including ataxia, eye problems, headache, hallucinations, dizziness, and skin and respiratory problems. Single doses of ***cimetidine*** have minor variable effects, but multiple doses (300 mg qid/5 days) decrease IV verapamil clearance by 21% and increase its half-life by 50%. ***Cyclosporine*** (q.v.) serum concentrations are increased by it, as verapamil decreases its metabolism. It gradually increases serum ***digoxin*** levels by 53–77% in **patients** receiving verapamil 240 mg/day. A ***digitoxin*** report indicates its serum level is increased ~30% by verapamil. ***Disopyramide*** should not be given 48 h before or 24 h after verapamil, as the combination causes marked negative inotropic and cardiac depressant effects. ***Grapefruit juice*** has increased its absorption and bioavailability by ~40%. Hazardous neurotoxic interaction with ***lithium*** treatment. Verapamil can double ***midazolam*** serum levels and prolong the latter's elimination. Its serum levels may be decreased by ***phenytoin***. ***Prazocin*** bioavailability is increased by verapamil treatment. Verapamil can cause dramatic and toxic increase in ***quinidine*** serum levels. ***Rifampin*** has caused a dramatic decrease in verapamil oral bioavailability. It increases ***simvastatin*** serum concentrations. It inhibits ***sulfonylurea***-induced ***insulin*** release. ***Terfenedine*** may potentiate its effects. ***Theophylline*** serum levels have been increased up to 50% by its oral use. ***Vincristine*** toxicity is increased by it. Potentiates adverse effects of ***muscle relaxants*** and possibly ***opioids***. Limited trials indicate it decreases ***talinolol*** serum levels. Visual disturbances after a drug interaction with ***metildigoxin*** in a 65-year-old **female**. Subjective effects of ***buspirone*** are reportedly increased by it.

TD_{LO}, **woman**: 64 mg/kg.

Lenticular changes at 30 mg/kg/day and frank cataracts at 62.5 mg/kg in **beagle dogs**, but not in **rats**. Embryocidal and decreased fetal growth rate in **rats** given ~6 times normal **human** dosage.

IV, LD_{50}, **rat**: 16 mg/kg; **mouse**: 1.5–8 mg/kg; IP, **mouse**: 94 mg/kg.

VERATRALDEHYDE

Use: Generally recognized as safe in soaps, detergents, cosmetic creams and lotions, perfumes, and foods in the U.S.

Untoward effects: No sensitization reactions after a 48-h closed-patch test by 15% in petrolatum on 20 **human** volunteers.

Moderately irritating applied full-strength to intact or abraded **rabbit** skin for 24 h under occlusion.

LD_{50}, **rat**: 2 g/kg; acute dermal LD_{50}, **rabbit**: > 5 g/kg.

VERATRAMINE

Untoward effects: This alkaloid in some ***Veratrum*** (q.v.) sp., after plant ingestion by pregnant **ewes**, causes deformed legs in their newborn **lambs**. The deformities include slightly bowed legs, marked flexure of knee joints, loose leg joints, and lack of muscle control.

VERATRINE

A mixture of the alkaloids *cevadilline, cevadine, sabadine*, and *veratridine* obtained from the seeds of *Schoencaulon officinale* (*cevadilla* seeds).

Use: In **human** medicine as a topical counterirritant and as a ruminatoric in veterinary medicine.

Untoward effects: Strong irritant to skin and mucous membranes. Inhalation causes violent sneezing. Poisonous.

Symptoms in **animals** are flatulence, colic, restlessness, arrhythmias, hypotension, and dyspnea.

TD, **horses** and **cattle**: 1–3 g; **dog**: 0.07–0.33 g; **rat**: 4 g/kg.

VERATRUM sp.

= *Varaire*

The term *False Hellebore* is used for *V. album*, *V. californium*, and *V. viride*.

Untoward effects: ***V. album* = *Cöpleme* = *European Hellebore* = *White Hellebore* = *White Helleborine*** contains ***protoveratrines*** and five other alkaloid hypotensive agents; once used in southwestern Europe on arrows and darts. ***Veratrine*** (above) causes most of its toxicity. Its pulverized roots in German sneezing powders (or ***Niespulver***) caused severe intoxication in young Scandinavian (Denmark and Sweden) **children**. Inhalation led to nausea, repeated vomiting, hypotension, syncope, cold sweats, mydriasis, bradycardia or tachycardia, transient blindness, fatigue, ptyalism, dyspnea, and confusion that usually lasted for 6 h. Ingestion led to nausea, repeated vomiting, dizziness, somnolence, bradycardia, mydriasis, and ptyalism lasting from a few to 15 h. Eye contact led to severe irritation for several days, corneal corrosion, and pain and swelling around the eye. Ancient **Romans** used it as a rodenticide. In **animals**, symptoms are the same as those caused by ***veratrine*** (above).

V. californium* = *Indian Corn* = *Skunk Cabbage* = *Western Hellebore* = *Wild Corn contains the teratogenic alkaloids ***cyclopamine*** (q.v.), ***cycloposine***, and ***jervine***. If **sheep** or **goats** eat 6–12 oz of green stems or leaves, they often show symptoms of poisoning within 2–3 h (ptyalism, frothing, ataxia, weakness, vomiting, dyspnea, convulsions, and prostration) that last 3–4 h. Larger amounts will kill. It induces teratogenic effects in **lambs** leading to skull deformity with cyclopian or monkey-face effects and defects described under ***veratramine*** (above). In Alaska, cyclopian **foals** are found where malformed **calves** are born to **cows** that have grazed *E. eschcholtizii*. Veratrum has also had teratogenic effects in **hamsters**, **rabbits**, **rats**, and **chick embryos**.

V. viride* = *American Hellebore* = *Green Hellebore* = *Indian Poke* = *Poke Root* = *Swamp Hellebore. Often induced transient (a few minutes) arrythmias 10–15 min after injection. Once a common antihypertensive in many oral and parenteral preparations leading to sinus bradycardia, nodal rhythms, ventricular extrasystoles, bigeminal rhythms, heart block, and dizziness. Nausea and epigastric burning was a warning of overdosage, followed by vomiting, ptyalism, singultus, decreased body temperature, prostration, pallor, severe stomach pain, diarrhea, seizures, and blurred vision, which led to its decreased clinical use. Skin and mucous membrane irritant. Ingestion of ***chocolate***-coated pills has caused the deaths of 1- and 2-year-old **females** within 2 h. Used by **Indians** in British Columbia to commit suicide.

Drug interactions: Its effects are potentiated by ***general anesthetics***, ***ethacrynic acid***, ***spironolactone***, ***thiazides***, and ***triamterene***. ***Tricyclic antidepressants*** decrease its effects.

Livestock usually do not eat it, but some **cattle**, **sheep**, and **poultry** have been poisoned by eating its young shoots, roots, and rhizomes containing the steroidal alkaloids, ***germidine*** and ***jervine***. Bradycardia, hypotension, abdominal pain, diarrhea, temporary blindness, and occasionally death in **cattle**. Nausea and salivation reported in **sheep** and oligodactylism in **rodents**.

VERBASCUM THAPSUS

= *Aaron's-Rod* = *Bullock's Lungwort* = *Common Mullein* = *Flannel Leaf* = *Moth Mullein* = *Tasso barbasso* = *Recchie d'asine* = *Velvet Plant*

Untoward effects: Its aeroallergens have caused allergic rhinitis, bronchial asthma, and/or hypersensitivity pneumonitis. Its wooly hairs, after ingestion, have caused "hairballs" in **animals**.

VERBENA

= *Vervain*

Untoward effects: Ingestion by **livestock** of ***V. officinalis*** causes severe gastrointestinal irritation, and its use in herbal ***teas*** for **man** must be limited. Photosensitivity reported in South Africa from it. ***V. rigida*** of western Australia, after ingestion, probably induces constipation, fever, photosensitization, and swelling under the throat and neck of **cattle**.

VERBESINA ENCELIOIDES

Untoward effects: This wild plant in Australia and India contains dangerously high levels of ***nitrate*** (q.v.) and a toxic principle, ***galegine***, that has caused the deaths of **equines** that have grazed it.

VERMOUTH

Untoward effects: Its formulas vary, but its ***absinthium*** (q.v.) (***wormwood***) content can cause serious psychoactive effects. A **patient's** dermatitis flared whenever **he** drank vermouth containing ***cinnamon***.

VERNONIA sp.

Untoward effects: Although some species appear to be edible, ***V. ampla*** = ***Zimbabwe Pride*** of Zimbabwe is highly poisonous and lethal for **sheep** and occasionally for **cattle**. Symptoms are muscle tremors with limbs rigidly extended, severe opisthotonus, eyelid twitching, and frothing at the mouth. Post-mortem showed hemorrhagic small intestine mucosa and in the thymus gland. Also reported: hemorrhage in parotid salivary glands, hyperemia of lungs, and cyanosis. Marked central nervous system involvement, but no deaths reported in **pigs**.

Ingestion of ***V. mollissima*** by **cattle** in southern Brazil has also caused central nervous system disturbances and deaths. The roots of ***V. nigritiana*** = ***Batiator Root*** of West Africa contain ***vernonin***, a poisonous cardiac glycoside with ***digitalis***-like action.

V. hildebrandtil is also used as an arrow poison in West Africa.

V. fasciculata in North America is an ***Ironweed***.

VERRUCULOGEN

Untoward effects: A tremorgenic mycotoxin metabolite isolated from *Penicillium verrucolosum* and *Aspergillus sp.* growing on foodstuffs. Neurological effects in **sheep**, **mice**, and **chickens** include tremors, limb weakness, ataxia, and convulsions. A toxin, ***verruculotoxin***, is also produced by the same fungi and causes similar symptoms.

LD_{50}, **mouse**: 126.7 mg/kg; **chicken**: 365.5 mg/kg.

VERTEPORFIN
= *Visudyne*

Injectable for photodynamic therapy of macular degeneration.

Untoward effects: Visual acuity disturbances (1–4%), headache, and injection site reactions. Occasionally back pain and photosensitivity reactions.

VESNARINONE
= *Arkin* = *OPC 8212*

Inotropic cardiac agent. **In U.S. Phase III trials.**

Untoward effects: Agranulocytosis in 1.2% and 0.2% of 3833 **patients** on 60 or 30 mg/day. Reversible neutropenia in 2.5% of 239 **patients** given 60 mg/day. A significant increase in early mortality reported at 120 mg/day. Sudden deaths may be due to arrhythmias. Despite improved quality of life at 60 mg/day, researchers found fewer deaths in the placebo group (18.9%) versus 22.9% in the 60 mg **group**.

VETIVER
= *Vetivert* = *Vetiveryl*

Use: The *acetate* is used as a fragrance in soap, detergents, cosmetic creams and lotions, and perfume. < 100,000 lbs/year U.S. annual use.

Untoward effects: A maximization sensitization test of 20% in petrolatum led to reactions on 3/25 **human** volunteers. A repeat test on additional **volunteers** led to reactions on 2/25. No irritation by 8% in petrolatum in 48 h closed-patch test on **humans**.

Application of full-strength material on intact or abraded **rabbit** skin for 24 h caused moderate irritation. Undiluted on hairless **mice** skin was not irritating.

LD_{50}, **rat** and acute dermal in **rabbits**: > 5 g/kg.

The *oil* is used in fragrances, as above, and in foods in the same volume.

Untoward effects: No irritation or sensitization of 8% in petrolatum on **human** volunteers. **Animal** LDs are the same.

VIBURNUM PRUNIFOLIUM
= *Black Haw* = *Hobblebush* = *Nannyberry* = *Sheep Berry* = *Stag Bush* = *Wild Raisin*

Untoward effects: The dried bark has been used as an antispasmodic, but ingestion of the raw berries leads to gastrointestinal irritation, vomiting, diarrhea, weakness, bradypnea, syncope, tremor, urinary incontinence, diaphoresis, visceral vasodilation, excitement, and coma in severe cases.

V. opulus = ***Guelder Rose*** = ***Obier***, a cultivated ornamental shrub, has caused mild diarrhea and vomiting after ingestion of unripe berries or large quantities of berries.

In **animals** causes muscle weakness, hypotension, loss of reflexes, and paralysis.

VICIA
= *Vetch*

Untoward effects: In general, the various species induce photosensitivity and hepatotoxicity.

V. augustifolia contains a neurotoxic lathyrogen, ***β-cyano-L-alanine*** and cyanogenic glycosides.

V. sativa = ***Common Vetch***. Ingestion of the seeds has been very toxic to **horses**, due to the same cyanogenic lathyrogen; they become hypersensitive to sunlight, adversely affecting their feet and muzzles; it causes tongue and mucous membrane ulcerations. Ingestion by **horses** also leads to abdominal pain, weakness, skin lesions, alopecia, posterior ataxia, and, on post-mortem, enlarged liver. **Cattle** and **sheep** are also affected. See ***Hydrogen Cyanide*** for general toxic symptoms. It is lethal to **chicks** and **poults**. **Chicks** that ate its seeds showed pronounced chirping, blindness, convulsions, and death.

V. villosa = ***Hairy Vetch***. Ingestion by **cattle** caused dermatitis, conjunctivitis, pruritus, and diarrhea. Eventually, thickend pustular skin, alopecia, ulceration, self-mutilation, coughing, and death. In field cases, morbidity was 6–8% and mortality nearly 50%. The seeds fed at 10–40% of their ration were not lethal to many **chicks** and **poults**, only decreasing their growth rate. Others developed severe convulsions and died. Common in the southeastern quarter of the U.S. Salivation, excitement, and pain have been reported in farm **animals**. **Horses** grazing it showed systemic granulomatous inflammation. Conjunctivitis, lip and eye edema, and corneal ulcerations developed.

See ***Beans*** for ***V. faba*** discussion.

VIDARABINE

= *Adenine Arabinoside* = *ara-A* = *Arabinosyladenine* = *Arasena-A* = *CI-673* = *NSC 404,241* = *Spongoadenosine* = *Vira-A*

***Antiviral.* Usually IV; occasionally, IM and topical ophthalmic solution and ointment. Largely replaced by *acyclovir*.**

Untoward effects: Parenteral use leads to neurotoxicity (disorientation, confusion, myoclonic jerks, generalized rigidity, tremor, dizziness, paresthesia, hallucinations, psychosis, and ataxia); has been fatal, especially at high dosage. Neurological impairment reported in 12% of treated **newborns**. Nausea, vomiting, anorexia, diarrhea, decreased hemoglobin, decreased hematocrit, decreased white blood cells, thrombocytopenia, skin rashes, megaloblastic anemia, hyponatremia, and syndrome of inappropriate antidiuretic hormone secretion also occur. Pneumonitis developed in some **patients**. Pain at injection site, central nervous system abnormalities, decreased hematocrit, increased SGOT, and increased creatinine phosphokinase after IM. Topical ointments have produced transient burning, stinging, and hyperemia.

IV, TD_{LO}, **human**: 300 μg-3 mg/kg.

Drug interactions: It increases ***theophylline*** serum levels ~60% and ***theophylline*** half-life > 2 times. ***Allopurinol*** apparently decreases its metabolism and increases its toxicity. It has induced birth defects in experimental **animals** and liver tumors in **rats**.

LD_{50}, **mouse**: > 8 g/kg.

VIGABATRIN

= *4-Amino-5-hexenoic Acid* = *Gamma-vinyl GABA* = *GVG* = *MDL-71,754* = *RMI-71,754* = *Sabril* = *γ-Vinyl-γ-aminobutyric Acid*

Anticonvulsant.

Untoward effects: Severe persistent visual field constriction, often occurring with > 2 years therapy and often receiving other ***anticonvulsants***. Aggressive behavior (7.5%), agitation, psychosis, confusion, dysphoria, irritability, involuntary movement, hyperkinesia, and inappropriate laughing have occurred. Aminoaciduria and pyroglutamicaciduria and parkinsonism occasionally reported. Gradual withdrawal is necessary to avoid renewed seizures. **Infants** receive < 1% of a therapeutic **infant** dose after breast feeding on a treated **mother**.

Drug interactions: Coadministration of ***carbamazepine*** increases vigabatrin clearance. Vigabatrin decreases plasma ***phenytoin*** levels 23%.

Intramyelinic edema noted in **rats**, **mice**, **dogs**, and **monkeys**, but in limited tests, not in **humans**. In **rats**, coadministration with ***phenytoin*** or ***carbamazepine*** decreased the latter's plasma levels.

VIGUIERA ANNUA

= *Annual Goldeneye*

Untoward effects: The flowers and upper branches of this 1–2 ft southwestern U.S. foothill or mountain range plant causes deaths of **cattle** in the fall, due to its ***nitrate*** and/or its ***hydrocyanic acid*** content.

VILOXAZINE

= *Catatrol* = *ICI 58,834* = *Vicilan* = *Vivalan* = *Vivarint*

Antidepressant.

Untoward effects: Nausea, vomiting, xerostomia, and migraine-type headaches are most commonly reported. Diarrhea, constipation, fatigue, sweating, tremors, dizziness, insomnia, palpitations, sinus tachycardia, agitation, confusion, bad dreams, abnormal electrocardiogram and electroencephalogram, urticaria, pruritus, orthostatic hypotension, and convulsive seizures occasionally reported.

TD_{LO}, **human**: 25 mg/kg/6 days.

Drug interactions: Use has induced a 2.5 times sharp increase of ***carbamazepine*** serum levels and a dramatic decrease in viloxazine serum concentration. ***Theophylline***

clearance is reduced and serum levels increased when viloxazine is also given.

LD_{50}, **mouse**: 1 g/kg.

VINBARBITAL

= *Delvinal* = *Sonuctane*

Sedative, hypnotic.

Untoward effects: Addiction potential. The key adverse effect is in **neonates**, where **their** cord blood levels quickly reached **maternal** levels with severe depression in > 50% of 29 **infants**.

LD_{50}, **rat**: 130 mg/kg; **mouse**: 190 mg/kg.

VINBLASTINE

= *29,060-LE* = *Exal* = *Lilly 29,060* = *NSC 49,842* = *Velban* = *Velbe* = *Vincaleukoblastine* = *VLB*

***Antineoplastic, antimitotic.* An alkaloid from the *Madagascar periwinkle* (*Catharanthus roseus* or *Vinca rosea*). IV use.**

Untoward effects: Primarily due to myelosuppression, usually 3–6 days after start of therapy, leading to leukopenia, neutropenia, and occasionally thrombocytopenia and hemolytic anemia. Nausea, vomiting, diarrhea, stomatitis, strange taste in the mouth, anorexia, alopecia, thrombophlebitis and cellulitis at injection site, and myocardial ischemia and infarction. Raynaud's phenomenon (especially in combination with chemotherapy), photosensitization, palmar–plantar erythrodyseathesia syndrome, diuresis, decreased serum ***urates***, osteonecrosis, joint aches, weakness; edema on face, neck, and soft palate; acute pneumonitis, bronchospasm, pulmonary edema, pruritic rash, thrombotic microangiopathy, ototoxicity, syndrome of inappropriate antidiuretic hormone secretion, and decreased fertility also occur. Local SC infiltration causes severe pain, occasionally not manifested for 24 h. Neurological effects often occur at higher dosage leading to peripheral sensory and motor neuropathy (occasionally severe), jaw pain, myopathy, paresthesias of fingers and toes, loss of deep tendon reflexes, muscle atrophy, muscular paralysis, headache, abdominal cramps, constipation, paralytic ileus, sore throat, hoarseness, dysphagia, vocal cord paralysis, depression, anxiety, parotid gland pain, convulsions, oculomotor dysfunction, and retinal degeneration. Tinnitus and hearing loss in a 29-year-old **male** after 13 mg every 2 weeks/12 times.

IV, TD_{LO}, **human**: 140 μg/kg; ocular: 14 μg/kg.

Fetal resorption and gross abnormalities of face, liver, limbs, skeleton, brain, and eyes in **hamsters**, **mice**, and **rats**. Hard to evaluate acute toxicity in **mice** because it first provokes a leukocytopenia that leads to bacteremia and death. Embryotoxic in **mice**.

LD, **mouse**: 330 μg/10 g; IP, 30 μg/10 g; IV, 9.5 mg/kg.

VINCA ROSEA

= *Catharanthus roseus* = *Consumption Bush* = *Madagascar Periwinkle* = *Periwinkle* = *Pink Flower* = *Red Rose* = *Sailor's Flower*

Untoward effects: Testing it for its alleged antidiabetic effect in Surinam and the West Indies led to the discovery of ***vinblastine*** (q.v.) and ***vincristine*** (q.v.), two of its 50+ alkaloids useful against various neoplastic conditions, yet its use in herbals has led to neurological, liver, and kidney damage in **man**, in addition to myelosuppression, granulocytopenia, vomiting, sedation, skin flushing, and hallucinations.

Ingestion has poisoned **cattle** in India and Australia. Its key alkaloids have been teratogenic in **hamsters**, **rats**, **mice**, **rabbits**, and **monkeys**.

Vinca minor* = *Lesser Periwinkle, an evergreen common in forests, thickets, and gardens, contains many alkaloids and the entire plant is poisonous when ingested.

VINCLOZOLIN

= *BAS 352F* = *Ronilan* = *Vorlan*

Fungicide.

Untoward effects: Contact causes eye and skin irritation in **man**. Wear protective clothing when using it.

Causes male **rat** pups to develop female genitalia characteristics, vaginal pouches, and penile abnormalities.

LD_{50}, **rat**: 10 g/kg.

VINCOFOS

= *SD 15,803* = *Vingard*

Organophosphorus anticholinesterase anthelmintic.

Untoward effects: Despite rigorous satisfactory premarketing testing, within 4 months of original approval, Shell withdrew the drug from the marketplace after a high incidence of adverse effects, including vomiting, diarrhea, ptyalism, muscle tremors, and fatalities in 72/131 **dogs** with adverse reactions. **Pony foals** on high dosage showed fluid diarrhea, muscle tremors, circling, and caudal ataxia.

VINCRISTINE

= *Kyrocristine* = *LCR* = *Leurocristine* = *Lilly 37,231* = *NSC 67,574* = *Oncovin* = *VCR* = *Vincosid* = *Vincrex*

***Antineoplastic.* IV use only.**

Untoward effects: It shares those of ***vinblastine*** above; it has less adverse hematological effects, but the neurological and neuromuscular adverse effects are more severe and dose-related. Oculomotor paralysis has occurred. Gynecomastia has been reported. Intrathecal injections have been fatal. Pain at injection site. Severe perivascular tissue reaction after extravasation. Overdoses often associated with syndrome of inappropriate antidiuretic hormone secretion. Amenorrhea in **women** and inhibition of spermatogenesis in **men**. Occasionally constipation, paralytic ileus, and convulsions. Small malpositioned kidney in the **neonate** of a treated **mother**. Unintentional intrathecal use instead of ***methotrexate*** led to death in a 12-year-old **child**.

IV, TD_{LO}, **child**: 500 μg/kg.

Drug interactions: Severe neurotoxicity when ***isoniazid*** was added in treating an 85-year-old **female**. ***Corticosteroids, cyclosporine, itraconazole, nifedipine***, and ***verapamil*** decrease its metabolism and increase its toxicity.

Teratogenic in **hamsters** (skeletal and eye defects, exencephaly, spina bifida), **rats** (eye defects, microcephaly, neural tube closure defects), and **monkeys** (syndactyly, encephalocele). Respiratory paralysis reported in treated **rabbits**. Can induce anorexia, constipation, and peripheral neuropathy in **cats** and vomiting in **dogs**. ***Verapamil*** can increase its toxicity in **animals**.

IP, LD_{50}, **mouse**: 5.2 mg/kg; IV, 2.1 mg/kg.

VINDESINE SULFATE

= *Eldisine* = *Fildesin* = *Lilly 99,094* = *LY 099094* = *NSC 245,467*

***Antineoplastic.* A synthetic *vinca alkaloid* from *vinblastine*. IV use.**

Untoward effects: Bronchospasm and acute pneumonitis occur frequently. Myelosuppression and neuropathy occur with leukopenia, peaking about day 7, and neurotoxicity is primarily paresthesias and constipation. Infrequent are paralytic ileus, jaw pain, and decreased deep tendon reflexes. Alopecia, nausea, vomiting, fever, lethargy, phlebitis, and cellulitis also reported. Rarely, stomatitis and skin rashes. Potent vesicant if extravasated. Contact with eyes can cause severe irritation and corneal ulceration.

In **rats**, high dosage leads to anorexia, atrophic intestinal mucosa, pancytopenia, and decreased spermatogenesis. Chronic administration in **dogs** causes leukopenia, decreased spermatogenesis, skeletal muscle degeneration, and increased lactic dehydrogenase.

IV, LD_{50}, **mouse**: 6.3 mg/kg; **rat**: 2 mg/kg.

VINORELBINE

= *Eunades* = *Navelbine*

***Antineoplastic.* Semi-synthetic *vinca alkaloid* for IV use.**

Untoward effects: Although thrombocytopenia is rare, ~50% of **patients** develop neutropenia. Phlebitis, pain at injection site, some peripheral neuropathy, alopecia, constipation, anorexia, stomatitis, asthenia, diarrhea, dyspnea, pulmonary edema, bronchospasms, syndrome of inappropriate antidiuretic hormone secretion, anemia, skin discoloration, rash, flushing; jaw, joint, or muscle pain; ataxia, nausea, and vomiting have occurred. Dosage may have to be halved in cases with granulocytopenia.

VINYL ACETATE

= *Ethenyl Acetate* = *Ethenyl Ethanoate* = *Vinyl Ethanoate*

Use: In manufacturing ***polyvinyl*** compounds, resins, ***paints***, and alcohols. U.S. annual demand is > 3 billion lbs.

Untoward effects: Inhalation of high concentrations causes eye, nose, throat, and skin irritation. Hoarseness, coughing, anosmia, skin blisters, and eye burns often follow. The latter two events are common after accidental spills or splashings. National Institute for Occupational Safety & Health sets maximum exposure limit of 4 ppm/15 min.

High dosage is carcinogenic in **rats**.

LD_{50}, **rat**: 2.9 g/kg; Inhalation LC_{LO}, 4000 ppm/4 h.

VINYL BROMIDE

= *Bromoethene* = *Bromoethylene*

Use: In fire-retardant thermoplastics.

Untoward effects: Supplied as a compressed gas, frost-bite-type injury on skin and eyes can occur with excessive exposure. Inhalation can induce dizziness, confusion, incoordination, narcosis, nausea, and vomiting.

Rats exposed to concentrated vapors developed angiosarcoma of the liver; squamous cell carcinoma of the zymbal gland; metastatic angiosarcoma, bronchioalveolar carcinomas, and bronchioalveolar adenomas of the lungs; adenocarcinoma of the mammary glands, angiosarcoma of the mesenteric lymph node, and lymphosarcoma of the lymphatic system.

LD_{50}, **rat**: 500 mg/kg; inhalation TC_{LO}, 250 ppm/1 year.

VINYL CHLORIDE

= *Chlorethylene*

Use: U.S. annual demand of this colorless, flammable, sweet-smelling gas is > 15 billion lbs with > 90% used in manufacturing ***polyvinyl chloride*** (q.v.). Once used as an anesthetic for **man** (often for induction in **children** before ***ether*** anesthesia), but found to be a myocardial irritant.

Untoward effects: Exposure in **factory workers** has commonly been due to its release during its polymerization to ***polyvinyl chloride***. Tumor development usually follows 15–20 years of chronic exposure. **People** living in the area of such factories have also suffered from aerosoled exposure. Attention to its toxic effects came as a result of induced liver angiosarcomas (it must be metabolized to become carcinogenic), hepatic hyperplasia, hypertrophy of hepatocytes and sinusoidal cells, sinusoidal dilation, increased ***sulfobromophthalein*** retention, increased alkaline phosphatase, increased γ-glutamyl transpeptidase, and hepatic portal fibrosis. Typical acro-osteolysis cases reported in **workers** engaged in its polymerization. Inhalation has caused dizziness, eye irritation, light-headedness, drowsiness, memory loss, dulled vision and hearing, nervousness, paresthesia of fingers or toes, and euphoria. Inhalation also induces gastrointestinal upsets, leading to nausea, anorexia, abdominal cramps, esophageal or stomach varices, bloody vomitus, and black stools. Higher rates of spontaneous abortion in **women** who have had direct or indirect contact with it. Canadian studies reported a significantly higher level of birth defects (central nervous system malformations, anencephaly, spina bifida, and hydrocephalus) in **residents** of a community near a vinyl chloride polymerization plant than in three other communities. In the U.S., **wives** of vinyl chloride polymerization **workers** had a 2–3.8 times increase in stillbirths and miscarriages. **Adults** in towns where the ***polyvinyl chloride*** was manufactured had a three times normal level of brain tumors (glioblastoma multiforme or astrocytomas). It has also induced hypertension, Raynaud's phenomenon, coughing, sneezing, râles, emphysema, pulmonary fibrosis, dyspnea, anemia, reticulocytosis, leukopenia, thrombocytopenia, splenomegaly, calf and joint pain, scleroderma-like skin changes, contact dermatitis, diaphoresis, fatigue, weight decrease, weakness, impotence, cold sensations in fingers and hands, clubbing of nails, and increased chromosomal aberrations. The compressed gas can cause frostbite on contact. Use of it as a propellent and solvent in aerosols has virtually been discontinued world-wide, where its carcinogenic effect became suspect in automobile tire manufacturing in the early 1970s. ***Hair sprays*** were reported by Dr. Wolfe of Nader's Health Research Group (HRG) to cause air concentrations of > 250 ppm, enough to be carcinogenic in **rats**. The FDA played down its possible danger, and the manufacturers, in 1974, said it would not yield to Health Research Group demands to remove it from their productions. Within a month after a trade media press release, they capitulated and recalled the affected products from the marketplace. OSHA now sets a maximum permitted exposure of 1 ppm (5 ppm/15 min).

Inhalation TC_{LO}, **human**: 500 ppm/1 year.

Angiosarcomas in **mice**, **rats**, and **hamsters** have been induced by exposure to its vapors. **Rats** pretreated with ***phenobarbital*** are susceptible to liver damage at $\frac{1}{2}$ the exposure level of untreated **rats**. A single **rat** suffered damage at $\frac{1}{20}$th the exposure level of untreated **rats**. Because of the enzyme induction involved, if one can extrapolate between species, vinyl chloride **workers** should not drink ***alcoholic beverages*** or take ***phenobarbital*** before working. Inhalation has also induced mammary gland carcinomas, squamous cell skin carcinomas, nephroblastomas, zymbal gland carcinomas; adenocarcinomas and angiosarcomas of the lungs in **rats**. In the **mouse**, anaplastic and squamous metaplasia of the mammary gland, bronchioalveolar adenoma of the lung, hepatic cell carcinoma, renal adenoma, and keratoacanthoma of the skin. In **hamsters**, lymphomas, trichoepitheliomas, and basaliomas reported from vapor exposure.

Inhalation TC_{LO}, **rat**: 250 ppm/4 h/130 weeks; 6000 ppm/4 h (12–18 days pregnant); **mouse**: 250 ppm/35 weeks; **hamster**: 500 ppm/4 h/30 weeks.

VINYLCYCLOHEXENE DIOXIDE

Use: In manufacturing epoxy resins and adhesives.

Untoward effects: Reported as a mild to moderate skin irritant in **humans**. National Institute for Occupational Safety & Health sets upper exposure limit of 10 ppm.

Topically on the skin of **rats** and **mice** causes primarily squamous cell carcinomas. Orally to **mice** causes benign and malignant ovarian tumors with the latter metastisizing to the lungs. Vapors have caused eye and skin irritation in **rabbits**. Acute respiratory irritation, testicular atrophy, leukopenia, and necrosis of the thymus also reported in **animals**.

LD_{50}, **rat**: 2.1 g/kg; inhalation LC_{50}, **rat**: 800 ppm/4 h.

VINYLIDENE CHLORIDE

= *1,1-Dichloroethane* = *1,1-Dichloroethylene*

Use: In the production of various plastics, such as ***Saran***®. Liquid vaporizes easily above 89°F.

Untoward effects: Vapors or liquid contact causes eye, skin, and throat irritation, followed by dizziness, headache, nausea, dyspnea, and pneumonitis.

Inhalation TC_{LO}, **human**: 25 ppm.

In **animals**, inhalation or ingestion of high concentrations caused liver, kidney, and lung pathology; some of their **offspring** had an increased incidence of birth defects after inhalation, but not after ingestion. In **mice**, led to angiosarcoma of liver, bronchioalveolar adenomas of the lungs, and kidney adenocarcinomas. In **rats**, led to angiosarcomas of the mesenteric lymph nodes, mammary tumors, and carcinomas of the zymbal gland.

LD_{50}, **rat**: 200 mg/kg; LD_{LO}, **dog**: 5.75 g/kg; inhalation LC_{LO}, **rat**: 10,000 ppm/24 h; inhalation LC_{50}, **mouse**: 98 ppm/22 h.

VINYL TOLUENE

= *Methyl Styrene*

Use: In fast-drying oils, ***paints***, and ***varnishes***.

Untoward effects: Vapors are painful eye, skin, and upper respiratory tract irritants. Can cause drowsiness. Safety glasses are advised for **workers**. Strong, disagreeable odor discourages willing inhalation. Erythema, swelling, and severe rashes or blisters after extensive skin contact. National Institute for Occupational Safety & Health/OSHA sets upper exposure limit at 100 ppm. Immediate danger to life or health, 400 ppm.

Inhalation TC_{LO}, **human**: 400 ppm.

Acute vapor toxicity in **rats** and **guinea pigs** leading to narcosis, eye and nasal irritation, salivation, and nervousness.

LD_{50}, **rat**: 4 g/kg.

VIOMYCIN

= *Celiomycin* = *Florimycin* = *Tuberactinomycin B* = *Vinactane* = *Viocin*

Aminoglycoside antibiotic. **Bacteriostatic, tuberculostatic. Usually IV.**

Untoward effects: Similar to ***streptomycin***, but often more severe. Dosage > 2 million Un every 3 days usually causes toxicity. Anaphylactoid reactions with dyspnea, facial and laryngeal edema, flushing, urticaria, hypotension, and shock. Fatal cases have shown renal tubular necrosis, fever, and rashes. Albuminuria, eighth nerve damage, eosinophilia, electrolyte imbalance (due to renal loss of ***potassium***, ***chloride***, and ***calcium***), dizziness, vestibular damage, and vertigo in 5–10%; tinnitus, hearing loss, ***nitrogen*** retention, acidosis, electrocardiogram changes, decreased ***creatinine*** clearance, hematuria, and cylindruria occur. Has a curariform-like action decreasing neuromuscular transmission. Rarely, tetany.

Systemic and topical use in **monkeys** led to vestibular damage.

SC, LD_{50}, **mouse**: 1.4 g/kg; IV, LD_{50}, **mouse**: 174 mg/kg.

VIRGINIAMYCIN

= *Antibiotic 899* = *Eskalin* = *SKF 7988* = *Stafac* = *Staphylomycin* = *Trepomycin* = *Virgimicin*

Antibiotic. **Feed additive for swine, poultry, and calves.**

Untoward effects: Topical use of a 2% powder or a 0.5% ointment led to local pruritus, transient urticarial rash, and fatigue in a few **patients**.

IP, LD_{50}, **mouse**: 450 mg/kg.

VIROLA sp.

Untoward effects: South American **Indians** use its resin made from the bark as an hallucinogen, due to its various ***tyramines***, including ***dimethyltryptamine***, ***methylbufotenine***, and ***methylmethoxytyramine***. Symptoms are initial excitement, numbness of limbs, twitching of facial muscles, nausea, various hallucinations, and, eventually, a deep disturbed sleep. **Scientists** are amazed that native South American **Indians** found it hallucinogenic only when used as a snuff (***epéna***, ***nyakwana***, ***parića***, and ***yakee*** are names for the tree and the snuff) and inactive orally.

V. theiodora (***V. elongata***) bark resin is a source of the above hallucinogenic snuff and is also used in arrow poisoning.

VITAMIN A

= *Acon* = *Afaxin* = *Agiolan* = *all-trans-Retinol* = *Alphalin* = *Anatola* = *Anti-infective Vitamin* = *Antixerophthalmic Vitamin* = *Aoral* = *Apexol* = *Apostavit* = *Arovit* = *Atav* = *Avibon* = *Avita* = *Avitol* = *Axerol* = *Axerophthol* = *Biosterol* = *"Dohyfral" A* = *Epiteliol* = *Lard-factor* = *Nio-A-Let* = *Optovit* = *Prepalin* = *Retinol* = *Retinyl* = *Testavol* = *Vaflol* = *Vesanoid* = *Vi-Alpha* = *Vitpex* = *Vogan* = *Vogan-Neu*

Retinoic Acid = *Aberel* = *Airol* = *Aknoten* = *Cordes Vas* = *Demairol* = *Epi-Aberel* = *Eudyna* = *Retin-A* = *Tretinoin* = *Vesanoid* = *Vesnaroid* = *Vitamin A Acid*

Fat-soluble vitamin. Isotretinoin, a retinoid, **is discussed above.**

Untoward effects: Hypercalcemia (14.4 mg%) in a 29-year-old **female** after ingestion of 50,000 Un/day/3 months then 100,000 Un/day/1 month. Normochromic macrocytic anemia in a 44-year-old **male** with severe pityriasis rosea after 150,000 IU/day. **He** also developed a perioral exfoliating dermatitis with local sclerosing keratitis, glossitis, and a smooth swollen tongue. High intake (six times that of southern **Europeans**) led to increased risk of osteoporosis and hip fractures in **Scandinavians**. Liver cirrhosis after 25,000 Un/day/6 years. Neural crest defects in **infants** of **mothers** who have ingested > 10,000 Un/day. Hypervitaminosis A can increase prothrombin time and can result in bone-growth retardation in **children**. The overenthusiastic use of vitamin drops is the usual cause. Benign increased intracranial hypertension in a 26-month-old child from chronic overdosage with 5000 IU qid in ***cod liver oil*** and 1 tsp/day of a multivitamin syrup for 6 months. Symptoms were cranial swelling, anorexia, dry skin, irritability, listlessness, and papuloerythematous rash. Twin **infants** fed 120 g ***chicken liver*** containing 36,000 IU/day/4 months showed irritability, vomiting, and anterior bulging fontanels. Once weekly feeding of 60 g/week with fortified milk and supplements is probably a safe maximum. In an 8-year-old **male**, 200,000 Un qid led to hydrocephalus. Excessive dosage can cause premature epiphyseal closure. Even in older **children** excessive dosage led to SC swelling of forearms, shanks, and feet; anorexia, pruritus; dry scaly lips; bleeding and tenderness of gingiva and at the corners of the mouth; coarse hair, and infantile hyperostosis corticalis. Pseudotumor cerebri can also be caused by a deficiency, but generally 80,00–500,000 IU/day has caused the symptom in **children**; 18,000–60,000 IU/day has induced it in **infants**; and 90,000–200,000 IU/day has caused the syndrome in **teenagers**. Toxicity reported in a 42-year-old **female** ingesting 3–4 pints ***carrot juice***/day (90,000 IU/day). By inhibiting intestinal synthesis of ***vitamin*** $\boldsymbol{K_2}$ excess intake induces hypoprothrombinemia. A 65-year-old **male** ingested 200,000–300,000 IU/day/4 years as a cancer prophylaxis, and it induced chronic liver disease, followed by death. Acute symptoms can follow single doses of 300,000 IU in **infants** and **children**, or with several million IUs in an **adult**. Chronic toxicity in three 6-month-old **infants** after 18,500 IU water-dispersed/day for 1–3 months; in **adults**, 1,000,000 IU/day/3 days or 500,000 IU/day/2 months, or 50,000 IU/day/> 18 months. In **adults**, pain and swelling over the long bones and cortical thickening. May give spuriously high ***cholesterol*** and ***bilirubin*** serum test results. Large doses have been implicated in increased intraocular pressure, retinal hemorrhages, papilledema, yellow discoloration of eyelids, strabismus, altered color vision, anemia, hypothrombinemia, thrombocytopenia, leukopenia, ascites, hypercalcemia, psychotic reactions, hepatosplenomegaly, asthma, osteoporosis, and headaches. In Alaska, ingestion of **polar bear *liver*** is acutely toxic for **man** and **cats**, due to its high level of vitamin A (8300 µg/g). Some Pacific **sharks**, southern Australian **seals**, **whales**, and Antarctic **huskies** also contain large amounts of vitamin A and acute intoxication has followed eating their livers. **Eskimos** have known that the livers of their large **dogs** (**huskies**), **Arctic foxes**, and **wolves**, as well as certain **seals**, can be poisonous if eaten. A 15-year-old **female** taking 200,000 IU/day/2 years for facial acne developed poisoning with headache, vomiting, diplopia, strabismus, papilledema, and great increase in cerebrospinal fluid pressure. Some **babies** do not excrete it well and have non-pathological yellow skin. Hepatomegaly is associated with fatty infiltration of the liver. Unfortunately, a **woman**, took 25,000 IU/day during **her** first trimester of pregnancy, then 50,000 IU/day. Her **infant** had congenital kidney malformations. In **others** receiving high dosage during pregnancy, it caused cleft palate, eye damage, and syndactly. Excessive intake leads to diffuse hair loss and dry hair. Liver cirrhosis in a 68-year-old **male** after ingestion of 400,000 IU/day/8 years. Severe psychotic reaction in an 18-year-old **female**, followed almost immediately by pseudotumor cerebri reaction after 50,000 IU twice daily or tid. After topical use, the acid form leads to erythema, pruritus, scaling and dryness of skin, allergic contact dermatitis, premenstrual symptoms, and vaginal bleeding. A **mother** used ***tretinoin*** cream 0.05% during a month before **her** last menstruation and during **her** first 11 weeks of pregnancy leading to **female infant** born with a crumpled, hypoplastic ear and atresia of the external auditory meatus on the right side. A 39-year-old **female** self-administered large amounts of ***tretinoin*** 0.025% cream topically leading to neurotoxicity. Later, resumption of selftreatment reproduced previous symptoms. Orally, it has caused headache, diarrhea, myositis, hypercalcemia, fever, pulmonary infiltrates, pleural effusions, fatal pneumonitis, peripheral edema, hypotension, arthralgia, cheilitis, pseudotumor cerebri, hypertriglycerides, and thrombophlebitis with increased doses.

Drug interactions: Use with ***ethinyl estradiol*** or ***mestranol oral contraceptives*** increases serum vitamin A. Large doses may be a potentiating factor for pseudotumor cerebri in **people** taking ***tetracycline*** for acne. Exposure to highly ***chlorinated naphthalenes*** and ***neomycin*** can decrease its serum levels. Consumption of ***mineral oil*** decreases its systemic availability. It has been adsorbed onto plastic IV bags and tubing. It has inhibited the anticoagulant effect of ***anticoagulants***. Potential hepatotoxicity increases after chronic ***ethanol*** consumption. ***Cholestyramine*** may decrease its bioavailability.

Excessive intake decreases growth in **trout**, induces cartilaginous hyperplasia with exostosis and a deforming cervical spondylitis in **cats**. The latter problems are rare and reported only in **cats** and **humans**. Excessive use in **calves** leads to tachycardia, decreased cerebrospinal fluid pressure, decreased skeletal bone growth, hyperemia, lameness, and retarded horn growth. In dairy **cattle**, massive intake accumulates in the liver causing hepatic dysfunction and decreased vitamin A plasma clearance. Bone deformities in **chickens** and **turkeys** with excessive intake; also causes increase in **chick** encephalomalacia, decreased egg pigmentation and production in **hens**, increased feather abnormalities in **chicks** and **Japanese quail**; reddened and crusty eyelids and mouth of **chicks**, and hydropericardium in **chicks**. In **rats**, decreased fertility, increased stillbirths and increased incidence in their **pups** of gross deformity of their skull and brain with high incidence of cleft palates, exencephaly, exophthalmia, and ear deformities; ***vitamin K*** deficiency decreases prothrombin synthesis leading to massive hemorrhages and death; hypertrophy of adrenal glands, and decreased size of thymus. Embryo malformations from excessive intake by pregnant female also demonstrated in **mice**, **rabbits**, and **guinea pigs**. Excessive parenteral injection into **rams** led to increased dead and abnormal sperm. Hypervitaminosis A in **pigs** caused decreased growth rate, gait changes, lameness, osteodystrophy, deformed limb bones, malaise, rough hair coat, scaly irritated skin, lacrimation, petechial hemorrhages on legs and abdomen, and hematuria. Excessive doses to experimental **animals** have also caused decalcification, multiple bone fractures, mucous membrane inflammation, trophic skin changes, exophthalmos, and renal and visceral pathology.

LD_{50}, **rat**: 7 g/kg; LD_{LO}, **mouse**: 3.5 g/kg; TD_{LO}, **monkey** (pregnant 20–44 days): 250 mg/kg.

VITAMIN B COMPLEX

See individual B vitamins.

***Untoward effects*:** Due to their intrinsic bad taste, they often change the taste sensation of food. Severe or moderate hypersensitivity with urticaria, pain, and pruritus; angioedema of the skin and mucous membranes – painful and even life-threatening if it involves the larynx. A 50-year-old **female**, after ingesting one or two capsules showed severe pruritic dermatitis of body and limbs, but not **her** face. Photosensitivity reported.

Drug interactions*:** It increases the inactivation of ***estrogen and decreases ***L-dopa*** effects. It potentiates ***oral anticoagulants*** and retards ***ampicillin*** absorption.

IP and SC, LD_{LO}, **mouse**: 3 mg/kg.

VITAMIN B_1

= *Aneurex* = *Aneurine* = *Anti-beriberi Vitamin* = *Antineuritic Vitamin* = *Benavit* = *Benerva* = *Betabion* = *Betalin* = *Betaxin* = *Bewon* = *Metabolin* = *Pharmaneurine* = *Thiaminium* = *Thiamol* = *Vitaneurin*

Thiamin Propyl Disulfide = *Alinamine* = *Dithiopropylthiamine* = *Nevrinton* = *Nevriton* = *Prosultiamine* = *TPD* is unaffected by thiaminase. Given orally and also effective parenterally. Has a garlic-like odor.

Benfotiamine = *s-Benzoylthiamine σ-Monophosphate* = *BTMP* = *8088 CD*, a synthetic stable thiamine derivative, it is also readily absorbed after oral administration and leads to higher and longer lasting thiamine and carboxylase levels in various tissues. Less toxic than the conventional hydrochloride form for parenteral use.

***Thiamine Tetrahydrofurfuryl Disulfide* = *Judolor* is another long-acting form in clinical use.**

Water-soluble vitamin.

Untoward effects*:** In manufacturing, its dust is 10 times more explosive than ***coal dust. Its reported ***curare***-like action is due to its inhibiting the formation of cholinesterase, thus increasing ***acetylcholine's*** activity at the nerve endings and increasing the transmission of impulses by some nerves. Excessive intake increases urinary excretion of ***vitamin B_2*** and even a deficiency of the latter. Intolerance to it with fatalities has been reported after parenteral use. Toxic reactions have followed 50 mg parenterally and death has occurred after 100 mg IV. These anaphylactoid reactions may be due to excessive formation of ***acetylcholine*** leading to occasional headache, hyperirritability, insomnia, tachycardia, ataxia, dyspnea, muscle tremor and weakness, tightness of throat, nausea, edema, and feeling of warmth, especially after oral doses of 17–40 mg/day. It can act directly on mast cells releasing ***histamine***, leading to urticaria. Severe anaphylactic shock in a 46-year-old **female** 5 min after 100 mg IM. Later, **she** developed anaphylactic shock after 100 μg/1 day. Anaphylactic shock in a 65-year-old **male** with history of allergy to ***fish***, 1 min after 10 mg IV. Has caused allergic eye reactions with irritation and dermatitis of the eyelids and conjunctivitis, IM of low dosage leading to hypertension and large doses leading to bradycardia, shock, vasodilation, hypotension, and slight bronchoconstriction. IV use has caused flushing and tingling in the extremities. Metabolic acidosis in an **infant** after 20 mg IV. Collapse and sudden death reported in a **patient** after IV of 100 mg. Presence in fortified ***bread*** has provoked an atopic dermatitis. Large doses IV give false increase in ***theophylline*** serum levels. Oral and parenteral use of thiamin propyl disulfide lead to tinnitus and dizziness (may have been due to arteriosclerosis), and nausea in ~2%. Use cautiously in deep anesthetic states. Avoid unnecessary IV use, as it can cause collapse, cardiac irregularities, vasodilation, respiratory paralysis, and death.

Drug interactions*:** ***Pralidoxime IV after vitamin B_1 infusion increases the oxime's half-life and plasma concentration. ***Probenecid*** delays its excretion. ***Antacids*** decrease its intestinal absorption. Rapidly and completely destroyed by alkaline drugs and ***sulfites*** (common antioxidant in pharmaceuticals), and easily inactivated in aqueous solutions above pH 5.5. Do NOT store in ***polyethylene*** devices.

In **dogs**, IV is hypotensive and an ***antihistamine*** with it causes additional hypotension. Death in experimental **animals** is from respiratory failure after shock, muscle twitching, clonic spasms, and dyspnea.

IV, LD, **rat**: 250 mg/kg; **mouse**: 125 mg/kg; **rabbit**: 300 mg/kg; **dog**: 350 mg/kg.

VITAMIN B_2

= *Belflavin* = *Flavaxin* = *Lactoflavin* = *Riboflavin* = *Vitamin G*

Water-soluble vitamin, food and drug coloring agent.

Untoward effects: In 17 atherosclerotic and hypertensive **patients** 40–72 years, oral intake led to increased ***proconvertin*** activity, increased prothrombin index, and increased fibrinogen. Can interfere with ***catecholamine*** assays in diagnosing pheochromocytoma and, in very high concentration, can yield false positive results for ***cannabinoids***. Large doses turn urine a bright yellow and can cause serious errors in evaluating jaundice. High urinary excretion occurs with excessive intake of ***vitamin*** B_1.

Drug interactions: ***Aspirin***, ***acetaminophen***, and ***antipyrine*** can complex it and decrease its availability. ***Probenecid*** delays its excretion. Physiological levels are decreased by ***oral contraceptives***. Can decrease availability of ***oxytetracycline*** and ***tetracycline***.

IP, LD_{50}, **rat**: 560 mg/kg; SC, 790 mg/kg.

VITAMIN B_3

The acid form = *Akotin* = *Daskil* = *Neopeviton* = *Niacin* = *Niacor* = *Niaspan* = *Nicacid* = *Nicangin* = *Nicobid* = *Nicolar* = *Niconacid* = *Nicospan* = *Nicotinic Acid* = *Vitamin PP* = *Wampocap*

The amide form = *Aminicotin* = *Benicot* = *Dipegyl* = *Nicamindon* = *Nicobion* = *Niacinamide* = *Nicotamide* = *Nicotinamide* = *Pelmin* = *Vitamin PP*

***Vitamin.* The term *niacin* in literature often refers to either form.**

Untoward effects: Use of ***nicotinic acid*** orally or parenterally decreases ***triglycerides***, decreases low-density lipoprotein ***cholesterol*** and total ***cholesterol***, and increases high-density lipoprotein ***cholesterol***. Has caused severe facial flushing (flushing reactions are often minimized by ***aspirin*** taken 30 min before); sensation of heat, head pounding, tingling, urticaria, pruritus, gastrointestinal upsets (cramps, diarrhea, nausea, vomiting), anorexia, hypotension, dizziness, amblyopia, blurred vision, malaise, decreased ***glucose*** tolerance, mild diabetes or aggravation of existing diabetes, increased glycemia, glycosuria, increased serum ***ketones***, jaundice, decreased liver function, hyperuricemia, hyperpigmentation (brown), dry eyes or hyperlacrimation, increased intraocular pressure, palpitations, tachycardia, arrhythmias, atrial fibrillation, headache, paresthesias, acanthosis nigricans, ichthyosis, reactivation of peptic ulcers, nervousness, myopathy (leg cramps and painful muscles), and mydriasis. Ingestion of 1.5 and 3 g in two **patients** who previously had pleasant effects from ***cannabis*** or ***lysergide*** had frightening psychotic experiences. Cystoid macular edema after massive oral doses. Clotting factor synthesis deficiency and coagulopathy without significant hepatotoxicity reported. Can interfere with ***catecholamine*** assays used in diagnosing pheochromocytoma. Excessive amounts in ***pumpernickel bagels*** caused toxic events in New York **consumers**. In 1963, it was banned in Great Britain for meat or meat product coloration because excessive use had caused some toxic reactions. Transient contact dermatitis reported in industrial **users**. An outbreak of food "poisoning" reported from the possible excess of ***sodium nicotinate***, commonly added to commercial ***cornmeal***.

The *amide* may rarely induce some facial flushing. A **patient** apparently had hepatic toxicity from taking 3 g/day/18 months.

Drug interactions: Can potentiate oral ***anticoagulants***. ***Isoniazid*** may inhibit its benefits leading to pellagra.

Amide, LD_{50}, **rat**: 3.5 g/kg; IV, 2.2 g/kg.

VITAMIN B_6

= *Adermine* = *Bécilan* = *Benadon* = *Bonasanit* = *Campoviton 6* = *Hexabetalin* = *Hexabione* = *Hexavibex* = *Hexermin* = *Hexobion* = *Postadoxine* = *Pyridipca* = *Pyridox* = *Pyridoxine*

Vitamin.

Untoward effects: Treatment of premenstrual symptoms with 300 mg/day in 27 **women** led to depression in 92%, headache in 70%, bloatedness in 70%, irritability in 44%, neuropathy in 37%, and puffy eyes in 26%. Significant decrease in symptoms after 2 months. Some **people** taking daily doses as low as 200 mg for extended periods of time (> 5 months) have developed a sensory peripheral neuropathy with numbness, loss of sensation, severe crippling effects, ataxia; and tingling sensations in neck, legs, and soles of feet. Megadoses have induced liver disease. Blackening of growing scalp hair in an 18-year-old **female** with homocystinuria given 500 mg/day; after withdrawal of treatment, prompt return to yellow-brown color. Other such **patients** treated similarly had decreased serum and red blood cell ***folate*** levels after 2–5 months. A **mother** self-medicated with 80 mg/day during pregnancy and intermittently while breast feeding the **baby** at 1–10 weeks of age led to seizures and irritability in the **infant**. A **mother** who was given large parenteral doses during **her** pregnancy gave birth to an **infant** with ***pyridoxine***-dependent convulsive seizures. Peak levels of it in breast milk 3–5 h after supplementation; 0.1–0.4 mg/l common in breast milk, and daily recommended intake for **infants** < 12 months is 0.3–0.4 mg.

Drug interactions: Somnolent reaction in 2/3 after vitamin B_6 IV, following ***isoniazid*** treatment for tuberculosis. ***Isoniazid*** inhibits ***pyridoxal-5-phosphate***, the active form of vitamin B_6. It contains both estrogenic and ***progesterone***-like effects and acts synergistically with them. Use with ***oral contraceptives*** can cause a vitamin B_6 deficiency in 75% of **users**. It increases ***nitrofurantoin*** urinary concentration by 62%. ***Cycloserine*** increases vitamin B_6 urinary excretion. It antagonizes ***L-dopa*** beneficial effects in parkinsonism. Up to 50% decrease in ***phenytoin*** serum levels with 200 mg/day of vitamin B_6.

Megadoses will cause liver disease in **rats**. Doses > 1 g/kg (~1000 times therapeutic doses) in **rats**, **rabbits**, and **dogs** lead to tonic convulsions, poor coordination, weakness, and ataxia. Lethal dose for **pigeons** is ~200 mg/kg. Local pain after IM use in **mice**. Massive doses inhibit the activity of ***quinine*** and ***atabrine*** in **avian** malaria.

LD_{50}, **rat**: 4 g/kg; SC, 3.1 g/kg; IV, 660 mg/kg. LD_{50}, **mouse**: IV, 543 mg/kg; IP, 966 mg/kg.

VITAMIN B_{12}

= *Anacobin* = *Antipernicin* = *Antipernicious Anemia Principle* = *Arphos* = *Bedodeka* = *Bedoz* = *Behepan* = *Berubi* = *Berubigen* = *Betalin-12* = *Betolvex* = *Cobalin* = *Cobamide* = *Crystamine* = *Cyanocobalamin* = *Cykobemin* = *Cytacon* = *Cytamen* = *Cytobion* = *Depivar* = *Docémine* = *Docibin* = *Docigram* = *Docivit* = *Dodex* = *Ducobee* = *Extrinsic Factor* = *Fresmin* = *Lactobacillus lactis Dorner Factor* = *LLD Factor* = *Macrabin* = *Millevit* = *Nascobal* = *Redisol* = *Rubesol* = *Rubramin PC* = *Sytobex* = *Vibalt* = *Vitarubin*

Hydroxocobalamin = *alphaCobione* = *Alpha-Ruvite* = *Axlon* = *Cobalin-H* = *Codroxomin* = *Droxomin* = *Duradoce* = *Duralta-12* = *Hydroxocobemine* = *Neobetalin 12* = *Neo-Cytamen* = *OHB*$_{12}$ = *alphaRedisol* = *Redisol H* = *Sytobex-H* = *Vitamin* B_{12a}

Vitamin co-enzyme, hematopoietic. **Before its chemical structure was known, it was called *Cow Manure Factor* because of its abundance in cow manure, and as *Animal Protein Factor* because of its anabolic effects in various mammals, fish, and poultry. In man, oral and IV use.**

Untoward effects: The FDA delayed approving it as a drug because huge doses failed to kill any experimental **animals** it was tested on. Occasional pain, hirsutism, easy bruisability, epigastric discomfort, acneiform rash, folliculitis, headache, allergic reactions and anaphylactic shock, hot flushes, dizziness, bronchospasm, urticaria, pulmonary edema, thrombosis, congestive heart failure, transient diarrhea, polycythemia vera, pruritus, and transitory exanthema after IM. Many ***nickel*** (q.v.)-sensitive **patients** are sensitive to ***cobalt*** (q.v.), an essential part of the vitamin B_{12} molecule. A **patient** with ***cobalt*** sensitivity developed a pruritic eruption at the vitamin B_{12} injection site. That area flared after ingestion of the vitamin. Rapid development of severe optic atrophy has followed use of the *cyano* form in **patients** with Leber's disease, rather than the recommended ***hydroxocobalamin***. Chronic ***cyanide*** poisoning can follow continued use of the *cyano* form in **people** with pre-existing ***cyanide*** toxicity. USA **mothers** on normal diets showed breast milk levels of 0.03–0.5 ng/ml.

Drug interactions: ***p-Aminosalicylic acid, antihistamines, bacitracin, cimetidine, colchicine, cholestyramine, neomycin, phenformin, polymixin***, and ***estrogen***-containing ***oral contraceptives*** decrease its intestinal absorption. Megadoses of ***vitamin C*** can destroy 50–95% of vitamin B_{12} in foods ingested.

IP, LD_{50}, **mouse**: 1.5 mg/kg; IP or SC, LD_{100}, **mouse**: 3 mg/kg.

VITAMIN B_{15}

= *Aangamik 15* = *Dimethyl Diacetyl Gluconic Acid* = *Pangametin* = *Pangamic Acid* = *Sopangaminé* = *Spur 15*

Untoward effects: Its designation as a vitamin has been seriously questioned, partly because it is a variable mixture of several compounds. IM and IV use has caused circulatory collapse in **humans**. Has neuromuscular-blocking activity in the **rabbit** and **chicken**, and is hypotensive in the **dog**. Its therapeutic value in **man** and **animals** is also seriously questioned, and some have considered its promotion as fraudulent.

IV, LD_{50}, 225 mg/kg; IP 1349 mg/kg.

VITAMIN C

= *Adenex* = *Allercorb* = *Antiscorbutic Vitamin* = *Ascorbef* = *Ascorin* = *Ascorteal* = *Ascorvit* = *Cantan* = *Cantaxin* = *Catavin C* = *Cebicure* = *Cebion* = *Cecon* = *Cegiolan* = *Celaskon* = *Celin* = *Cenetone* = *Cereon* = *Cergona* = *Cescorbat* = *Cetamid* = *Cetebe* = *Cetemican* = *Cevalin* = *Cevatine* = *Cevex* = *Cevimin* = *Ce-Vi-Sol* = *Cevitamic Acid* = *Cevitan* = *Cevitex* = *Cewin* = *Ciamin* = *Cipca* = *Concemin* = *C-Vimin* = *Davitamon C* = *Duoscorb* = *Hybrin* = *Laroscorbine* = *Lemascorb* = *Planavit C* = *Proscorbin* = *Redoxon* = *Ribena* = *Scorbacid* = *Scorbu-C* = *Testa-Scorbic* = *Vicelat* = *Vitacee* = *Vitacimin* = *Vitacin* = *Vitamist* = *Vitascorbol* = *Xitix*

Vitamin, antioxidant, chelator. **U.S. annual demand is for > 40 million lbs with > 50% used in pharmaceuticals, 35% in foods and beverages, and 10% in animal feed.**

Untoward effects: Dose of 1 g/day can induce a diarrhea, some diuresis, and the urinary acidification of 4–12 g/day can cause precipitation of ***urate***, ***oxalate***, or ***cystine*** stones in the urinary tract. Large doses decrease serum ***bilirubin*** levels and could mask early signs of liver disease; causes glycosuria, and occasionally skin allergies. Infertility reported in **women** taking > 2 g/day/6–17 months and, in Russia, 16/20 **women** aborted after taking 6 g in 3 days. Cerebrovascular and coronary episodes in five geriatric **patients** of 538 in a double blind study indicated a greater incidence in **those** receiving extra vitamin C. A 40-year-old **female** ingesting high dosage to ward off colds frequently precipitated sickle-cell thalassemia. Intake of 2–3 g/day/> 3 years led to ***vitamin*** B_{12} deficiency in three **patients**. Long-term dosage of 1 g/day can induce hemochromatosis in **patients** with pyruvate kinase deficiency. Fatal cardiomyopathy in a 29-year-old **male** with hemochromatosis ingesting at least 1 g/day/12 months. Allowing tablets to dissolve on the tongue's surface has produced ulceration and can cause esophageal mucosal injury if entrapped. Intake of 3 g/week has aggravated hyperoxalemia in **patients** on chronic hemodialysis. After 1 g, it can cause false positive results with ***Benedict's solution*** or false negatives with enzyme dip test for glycosuria, increased ***uric acid*** results with colorimetric procedures, false negative results in hematuria dip tests and occult fecal blood, as well as increased ***creatinine*** results. **Diabetics** can get into serious errors trying to establish **their** ***insulin*** needs when taking megadoses of the vitamins. Small doses can cause hemolysis in glucose-6-phosphate dehydrogenase-deficient **people**. Many **Orientals**, **Greeks**, **Arabs**, **Sephardic** and **Kurdish Jews**, **Sicilians** and black **people** have this deficiency and can develop hemolytic anemia by megadoses. Hemolytic anemia with acute renal failure in a 68-year-old **male** with glucose-6-phosphate dehydrogenase deficiency after 80 g/day/2 days. A pregnant **woman** who, prior to delivery, ingested 250–500 mg/day/2 weeks had hydrops fetalis in newborn **male infant** with glucose-6-phosphate dehydrogenase deficiency. Megadoses have induced ***iron***-deficiency anemias. Newborn **infants** can develop scurvy when birth removes **them** from a high maternal exposure, where **they** develop mechanisms to destroy the excesses **they** were getting, and, after birth, continue to do so. Rebound scurvy can also occur in **adults** on withdrawal from megadoses, unless intake is decreased 10–20%. A 28-year-old **female** psychiatric **patient** received megadoses during **her** first 10 weeks of pregnancy, and at 17 weeks, was discovered to be pregnant with a 17 week anencephalic conceptus. Interruption of daily intake of 6 g in a 32-year-old **male** led to acute onset of headache. Megadoses are often ingested by ***cannabis*** **addicts** who believe it interferes with urine assays for ***cannabis*** metabolites. A 32-year-old **male** with HIV treated IV with 40 g/three times a week and 20–40 g orally/1 month had **his** IV dosage increased to 80 g, leading to black urine, breathlessness, and fever. After being entered into the hospital for the oxidate hemolysis, it was determined **he** was glucose-6-phosphate dehydrogenase-deficient. A 70-year-old **male** with renal insufficiency received 2.5 g IV over 5 h and developed permanent renal failure. Warmth and some degree of flushing is associated with large IV doses. IV use can be hypotensive in glaucoma **patients**. A 58-year-old **female** received a single 45 g IV dose leading to acute renal failure due to ***calcium oxalate*** tubular obstruction. Rapid IV has caused transient faintness and dizziness. Topically, its use in ***flour*** has caused a dermatitis in some **bakers**. Use of ***sodium ascorbate*** (11.6% ***sodium***) as a source of the vitamin can be undesirable for **patients** with hypertension or renal disease, or in pregnancy. As a so-called "natural" preservative in cosmetics, it causes allergic reactions in 0.3%. A single report indicates excessive intake may cause thickening of carotid artery walls.

TD_{LO}, **man**: 2.3 g/kg/2 days.

Drug interactions: Rapid decrease in excretion of ***acetaminophen*** after 3 g of vitamin C. Decreased half-life of ***antipyrine*** after 500 mg. ***Aspirin*** decreases its plasma levels. ***Oral contraceptives*** stimulate the rate of vitamin C breakdown and megadoses of the vitamin can increase the risk of cardiovascular disease in **women** on the pill. Supplementation decreases plasma levels of ***fluphenazine*** 25% and inhibits ***tricyclic antidepressants***. Can react with ***chromium picolinate*** leading to a reduced form of ***chromium***, capable of producing DNA mutations. Use with ***pentobarbital*** can increase its metabolism and excretion. ***Atropine, salicylates, sulfonamides***, and ***tetracycline*** can also increase its excretion. Megadoses can increase ***cyanide*** production from ***laetrile***. It rapidly increases serum levels of ***chlortetracycline, oxytetracycline***, and ***tetracycline***. Several grams with ***warfarin*** decrease prothrombin time, and increased ***warfarin*** dosage failed to alter it. It increases bioavailablity of dietary ***iron***; this may be due to its ability to reduce it to the ferrous state. Severity of ***alcohol–disulfiram*** reaction is decreased by it. When used to improve smell and palatability of oral ***iodine*** solutions, it prevented ***radioiodine*** uptake by the thyroid gland of a neuroblastoma **patient**. Can increase intestinal absorption of ***iron*** and lead to false negative Hemoccult® test results.

IV, LD_{50}, **mouse**: 518 mg/kg.

VITAMIN D

= *Sunshine Vitamin*

***Vitamin*. Fat-soluble. Its status as a true vitamin (must be part of an individual's diet) may be questioned, but it will continue as such in the literature, even though it is usually formed in the skin and**

converted to an active or hormonal form in the kidney and liver.

VITAMIN D_2

= *Calciferol* = *Condol* = *D Activated Plant Sterol* = *Decaps* = *Dee-Ron* = *Deltalin* = *De-Rat* = *Deratol* = *Detalup* = *Diactol* = *Divit Urto* = *Drisdol* = *D-Tracetten* = *Ergocalciferol* = *Ergorone* = *Ertron* = *Fortodyl* = *Hi-Deratol* = *Infron* = *Irradiated Ergosterol* = *Metadee* = *Mina* D_2 = *Mulsiferol* = *Mykostin* = *Ostelin* = *Radiostol* = *Radsterin* = *Shock-Ferol* = *Sterogyl* = *Uvesterol-D* = *Vio-D* = *Viosterol*

A synthetic form prepared from *ergosterol* by ultraviolet irradiation.

VITAMIN D_3

= *Activated 7-Dehydrocholesterol* = *CC* = *Cholecalciferol* = *Colecalciferol* = *D-Activated Animal Sterol* = *Delsterol* = *Duphafral* D_3 = *Ebivit* = *Micro-Dee* = *Neo Dohyfral* D_3 = *Oleovitamin* D_3 = *Provitina* = *Ricketon* = *Trivitan* = *Ultranol* = D_3*-Vicotrat* = *Vi-De-3-hydrosol* = *Vigantol* = *Vigorsan*

Humans, cattle, horses, sheep, dogs, and swine utilize vitamins D_2 or D_3 with equal efficiency; poultry and New World monkeys utilize D_3 very efficiently and D_2 poorly.

Untoward effects: Usually due to oversupplementation in **children** (daily vitamins, sunshine, baby food; fortified cereal, milk, and dairy productions; ***liver, cod liver oil***) or in **adults** or **children** receiving massive doses in disease states. Symptoms are lassitude, anorexia, nausea, vomiting, weight decrease, polyuria, nocturia, albuminuria, diarrhea, diaphoresis, polydipsia, vertigo; loss of vision (due to band-shaped corneal degeneration) with ***calcium*** deposits in the cornea and conjunctiva; increased ***calcium*** and inorganic ***phosphorus*** serum levels; decreased alkaline phosphatase values; increased ***cholesterol***, pyuria, fever, headache, acute cerebellar ataxia, anemia, eosinophilia, leukocytosis, dysarthria, dizziness, facial paralysis, metallic taste, glycosuria, decreased growth in **children**, ***calcium phosphate*** and/or ***calcium carbonate*** renal and bladder deposits, acute pancreatitis, constipation, pruritus, pallor, exacerbation of depression in **psychotics**, and even potentially fatal calcification of the heart, lungs, and kidneys. The metastatic calcification has caused a diffuse yellow coloration of the nail bed. Excessive intake during pregnancy induces severe idiopathic hypercalcemia in **infants** and a mild form of infantile hypercalcemia or a severe form with supravalvular aortic stenosis, generalized arteriosclerosis, left ventricular hypertrophy, mental retardation, craniofacial and dental abnormalities (widely spaced eyes, epicanthic folds, turned down mouth, malocclusion, caries, constricted blood vessels, prominent low-set ears), increased blood pressure, impaired kidney function, systolic murmurs, tachycardia, dyspnea, cyanosis, enlarged liver and spleen, and diffuse myocarditis. Has caused false positive or spuriously high urine test dipstick results. In **neonates** and **children**, electrocardiogram abnormalities, increased calcification of metaphyses, osteoporosis, and hypotonicity. Pregnant **women** receiving 50,000 IU/day showed high levels of the metabolite ***25-hydroxyvitamin D*** in **their** breast milk leading to possibility of **infant** hypercalcemia. Oral use in ***peanut oil*** in infancy can cause sensitization to ***peanut oil***. Large doses can create a type of dependence and, when the excess is stopped, withdrawal-type symptoms can occur. A 41-year-old **female** ingesting ***calciferol*** (***vitamin*** D_2)/day/8 years may have induced calcification of the tympanic membrane and deafness. It has been speculated that the high incidence of cardiovascular disease in Japan is due to the very high daily intake of vitamin D ***fish oils*** in frying ***fats*** and ***margarine***. Vitamin D intoxication in **adults** is commonly associated with long-term daily ingestion of 100,000 IU or more (4000 IU or more IV in **children**). A 2-year-old child receiving excess ***cholecalciferol*** (6,800,000 Un total) developed saddlenose and micrognathia, hypercalcemia, uremia, growth retardation, and cachexia.

Drug interactions: Use with ***anticonvulsants***, such as ***pheneturide, phenobarbital, phenytoin***, and ***primidone*** can interfere with its conversion into active forms and cause deficiency symptoms. ***Cholestyramine*** with it can induce osteomalacia, as will ***glutethimide***, by enzyme induction and irritant ***cathartics*** by increased peristalsis and damaging the intestinal wall. Chronic use of ***mineral oil*** as a laxative can induce a deficiency of this fat-soluble vitamin and subsequent symptoms. ***Norethindrone*** and ***mestranol oral contraceptives*** increased ***cholecalciferol*** (vitamin D_3) half-life in 3/4 **patients**. After a single 300 mg dose of ***isoniazid***, the concentration of ***1α-25-dihydroxyvitamin*** D_3, the most active metabolite, decreased 47%. ***Bendroflumethiazine*** or ***methyclothiazide*** given with it has been a common cause of hypercalcemia if there is a sustained decrease in ***calcium*** excretion. A 70-year-old **female** with controlled atrial fibrillation by ***verapamil*** had a recurrence after ingestion of ***calcium adipinate*** 1.2 g and ***calciferol*** (***vitamin*** D_2) 3000 IU/day prescribed for **her** osteoporosis. ***Prednisone*** antagonizes its benefits by decreasing production of its active metabolite and decreasing the intestinal absorption of ***calcium***. Its use decreases alkaline phosphatase serum test results.

A generalized calcinous or metastatic calcification (withdrawal of ***calcium*** from bones and deposition in soft tissues) of organs and blood vessels occurs in **goats**, **sheep**, **swine**, **green iguanas**, **leopards**, **lizards**, **monkeys**, **rabbits**, **rats**, **tigers**, **cats**, and **dogs** receiving excess quantities of vitamin D. These quantities have also produced increased incidence of urinary calculi in **sheep** and are teratogenic in **rabbits**. Excessive quantities or longer than recommended periods of prophylaxis against milk fever in **cows** can lead to a severe osteoporosis. Avoid overdosing **cats**, a species with very low requirement, unless suffering from diseases that interfere with emulsification and absorption of fat. Hypervitaminosis D in pregnant **rats** leads to increased incidence of low birth weight and abnormal bone development. Multiple early bone fractures and impaired healing often occur in the pups. As in **humans**, excessive intake during pregnancy by **rabbits** leads to craniofacial abnormalities including shortening of skull and premature fontanel closure, irregular dental development, low birth weights, supravalvular aortic stenosis, and hypercalcemia. Excessive intake by pregnant **rats** leads to increased fetal death rate, retardation of intrauterine growth by placental calcification, decreased bone mineralization, kyphoscoliosis, and shortened long bones. Young **dogs**, **leopards**, and **tigers** have died after excessive intake by their mothers or from **human**-type baby milk formulas. Post-mortem showed massive deposits of ***calcium*** in stomach, kidneys, lungs, media of arteries, cornea, thyroid, and occasionally heart and liver. ***Cholecalciferol*** (vitamin D_3) ability to induce hypercalcemia in **rats** made it useful as a rodenticide, as it causes delayed liver and kidney calcification and pathology, hemorrhage, and occasionally seizures. Death after 6 days–3 months in **mice**, **rats**, **guinea pigs**, **rabbits**, **cats**, and **dogs** after 0.2–10 mg/day. Post-mortem showed massive ***calcium*** deposits in arteries, kidneys, heart, and other organs. Massive doses (314,000–530,000 IU/kg) D_2 to young **dogs** – nearly half were dead within 2 weeks. Symptoms were anorexia, polyuria, polydipsia, bloody diarrhea, and prostration. Post-mortem showed lung calcification, as well as some in the heart and kidneys, with marked abnormalities of teeth and jaw.

D_2, LD_{LO}, **dog**: 4 mg/kg; IM and IV, 5 mg/kg.

VITAMIN E

= *α-Tocopherol*

Vitamin, antioxidant.* Oil-soluble. *"Tocopherol"* from the Greek, *Tokos*, means childbirth; *phero* means to bring forth or bear; *ol* designates it as oil-soluble, thus, tocopherol – the oil of fertility. In 1923, this substance was referred to by the letter *X* or *Antisterility Factor.

Untoward effects*:** Despite food and cosmetic faddists touting it as a universal panacea, others have stated that overdoses or megadoses can throw chronic rheumatic hearts into failure, cause fatigue, headaches, nausea, dizziness, stomatitis, diarrhea, chapped lips, muscle weakness, low blood sugar, rash, thrombophlebitis, pulmonary embolishment, gynecomastia, breast tumor, vaginal bleeding, increased bleeding tendencies (by prolonging prothrombin time and interference with ***vitamin K in producing clotting factors), blurred vision, necrotizing enterocolitis in **infants**, and inhibition of multiple immune functions. Appears in breast milk. Unfortunately, a high-potency IV product, ***E-Ferol***, for IV use in **infants** (without FDA approval) was linked to the deaths of 38 premature **infants**. Topically, it has caused delayed allergic contact dermatitis, contact urticaria, and delayed wound-healing. Cosmetic use in a Mennen deodorant was withdrawn from marketing 10 months after its introduction because of numerous reports of skin inflammation, pruritus, and erythematous and edematous rash of axillae, with rapid spread to large areas of trunk and arms, far-removed from the areas of use. Many **surgeons** ask **patients** not to take vitamin E for 2 weeks prior to surgery.

Drug interactions*:** Malabsorption can occur if taken with ***cholestyramine. With inadequate intake, **women** taking ***oral contraceptives*** may become anemic. Synergism between it and ***digitalis*** may require 50% decrease in ***digitalis*** maintenance dose. ***Mineral oil*** can carry it out of the body and both stomach acidity and ***iron*** preparations decrease its availability. It, in turn, can impair the hematological response to ***iron*** therapy. Can enhance ***cyclosporine*** absorption in liver transplant **recipients**. The antioxidant, ***sodium metabisulfite***, can inactivate it. **Diabetics** taking ***insulin*** can be adversely affected by large doses of vitamin E. Use (400 IU qid) can dramatically decrease effectiveness of ***nitroglycerin*** in angina. Its use may have eliminated the need for ***nitroglycerin*** in 22 angina **patients**. It increases the hypoprothrombinemic effect of ***warfarin***. Plasma levels decrease in the presence of ***chlorinated naphthalenes***.

VITAMIN K

= *Koagulationsvitamin*

***Vitamin.* Oil-soluble. Two natural products have Vitamin K activity and are now referred to as:**

Vitamin K_1 = *Aquamephyton* = *Konakion* = *Phylloquinone* = *Phytomenadione* = *Phytonadione*

Plant-formed, but now synthesized. Oil-soluble.

Vitamin K_2 = *Menaquinone* = *Menatetrenone*

Bacterially synthesized from *fish* meals. Found in bacteria. Oil-soluble. Commercially unimportant.

***Menadiols* (water-soluble) and *Menadiones* (fat-soluble) are the most important fully synthetic compounds with Vitamin K activity for clinical use.**

Vitamin K_3 = *Kappaxin* = *Menadione* = *Mennaphthone* = *Panosine* = *Synkay*

Vitamin K_3 Dimethyl Pyrimidinol Bisulfite = *Hetrazeen;* contains about 45% K_3.

Vitamin K_3 Sodium Bisulfite = *Clotin* = *Hykinone* = *Klotogen* = *Vicasol*

Each gram of pure material contains ~33% Vitamin K_3 activity.

Vitamin K_4 = *Acetomenaphthone* = *Kappaxan* = *Prokayvit* = *Synkavit(e)*

Vitamin K_4 is water-soluble and has ~$\frac{1}{2}$ the activity of K_3.

Untoward effects: Hyperbilirubinemia after 1 mg K_1 IM in **newborns**, peaking at about the fourth day. Crosses the placenta with increased risk of severe jaundice, particularly in premature **babies**. Single IV or IM doses of 10–50 mg in many **adults** led to chills, fever, flushing, chest pain, apnea, dyspnea, immediate (type 1) hypersensitivity with asthma, hypotension cyanosis, pruritus, dizziness, tachycardia, facial edema, and anaphylactic shock (within seconds or a few minutes). Circulatory collapse, and one death reported from its IV use. Rapid infusion rate often precipitates the worst of these symptoms and possible seizures. Itching papules have occurred 8–10 days post-injection. Some reactions have been precipitated by preparations containing ***Cremophor EL*** (q.v.). Some cases of "***warfarin*** resistance" have been due to its unrecognized ingestion in products such as ***Ensure***®. A similar problem has been reported in a 64-year-old **female** after ingesting ***Isocal***®, an enteral food preparation. It has caused hemolysis in some glucose-6-phosphate dehydrogenase-deficient **patients**. A small, but possibly positive association, may exist between its IM use and childhood leukemias and other cancers. K_3 **maternal** doses of 70 mg caused hyperbilirubinemia and kernicterus in **neonate**; given to **newborns** led to similar adverse effects and hemolytic anemia. Large doses of K_4 IM of only 1 mg/min given to the **mother** at term or to the **newborn** can cause hyperbilirubinemia in the **neonate**. ***K_3 sodium bisulfite*** IM to the **mother** has also caused brain damage in her **infant**. Each time 10 mg K_4 was given IM to a 45-year-old **male** with alcoholic cirrhosis of liver, **his** prothrombin decreased to as low as 13%, illustrating a paradoxical aggravation of hypoprothrombinemia. Cerebral arterial thrombosis in two **patients** with malabsorption syndrome, due to celiac disease after IM and IV K 10 mg. It may increase the risk of thrombosis in **infants** whose hematocrit is high, as in Down's syndrome. Some passes into breast milk. In **newborns**, use of its analogues at 10 mg frequently causes kernicterus. A current fad recommends various K creams, particularly K_1, for topical use on senile purpura, spider-veins, and bruises, with potential adverse effects in **patients** stabilized on ***anticoagulants***. After 18 months, a **patient** had almost infiltrative plaque-like reaction at the injection site, which mimics a T-cell lymphoma.

Drug interactions: Its intestinal absorption can be decreased by ***p-aminosalicylic acid, cholestyramine***, and ***phytates***. ***Antibiotics***, such as ***chloramphenicol, chlortetracycline, neomycin, tetracycline***, and various ***sulfonamides*** can decrease K production by intestinal bacteria and can dangerously potentiate hypoprothrombinemic effect of ***anticoagulants***, such as ***warfarin*** or other ***coumarins***. ***Cefoperazine*** has, on rare occasions, induced a K deficiency. Serum levels are increased by ***oral contraceptives***. Regular use of oral ***mineral oil*** decreases its absorption and, in pregnancy, can cause serious hypoprothrombinemia.

VITAMIN O

No such *vitamin* recognized by scientists. The U.S. Federal Trade Commission sued the manufacturers, alleging false and unsubstantiated health claims. The companies agreed to a permanent injunction until they are able to provide reliable scientific evidence of their product health claims and to pay $375,000.00 in customer redress. Other companies have made similar claims, calling theirs "stabilized" or "aerobic" *oxygen*.

VITAMIN U

= *Ardesyl* = *Cabagin U* = *Epadyn U* = *MMSC* = *Vitas U*

Use: In treatment of gastric and small intestinal mucosal disease and esophagitis.

Untoward effects: Intolerance in ~8% is the most common complaint.

VITEX sp.

Untoward effects: ***V. agnus-castus*** = ***Chasteberry*** = ***Chaste Tree***, primarily of the Mediterranean area, since antiquity has been a symbol of chastity and one of the few folk medicines (extract of dried fruit and leaves) that is reported to be an aphrodisiac. Occasionally induces rashes and gastrointestinal discomfort. A 45-year-old **female** may have had seizures related to its ingestion.

V. mediensis bark extract in Africa has caused exophthalmus.

V. negundo = ***Nirgundi*** seed extract inhibited spermatogenesis in Indian **dogs**.

V. trifolia in southeastern U.S. has caused respiratory irritation, dizziness, and nausea in **man** and **animals**.

VOCHYSIA sp.

Trees in Colombia, South America.

Untoward effects: *V.* ***columbiensis*** = ***ka-ho-gaw*** bark is used by nomadic **Indians** as an arrow poison.

The pulverized leaves and bark of *V.* ***lomatophylla*** = ***ka-kweé-gaw-ya*** are used by the **Barasana Indians** as an ***abortifacient***. In Peru, it is used by the **Campa tribe** as a ***contraceptive***.

VORICONAZOLE

= *UK 109,496*

Use: An ***azole antifungal*** derivative of ***fluconazole*** for oral or topical use. In phase III trials.

Untoward effects: Transient dose-related visual problems, i.e. blurred vision, increased brightness in 8–10%, increased liver enzymes, photosensitivity, and discoid lupus vulgaris-like skin lesions.

Drug interactions: Are expected to be similar to those of other ***azoles***.

VX

Untoward effects: A lethal chemical warfare ***anti-cholinesterase agent***, used as an oily liquid, aerosoled, or sprayed with its contamination still active for days or weeks. An air attack with it would affect ~40 km^2. Theoretically, a few oz could kill millions of **people**. It has ~10 times the lethality of ***sarin*** (q.v.). Percutaneous absorption can occur with an estimated LD of 2–10 mg. Death is by paralysis. It can be used in shells; 75,000 tons was confiscated in Nazi Germany. IV in **soldiers** of 1.5 µg/kg caused mild transient light-headedness or dizziness within 1–2 min, lasting 10–15 min. Occasional nausea ~1 h later. The U.S. Army never admitted liability, but paid **ranchers** for 6400 **sheep** made acutely ill and 4500 killed in Skull Valey, Utah due to a VX airplane release from the Army's Dugway Proving Ground the previous day. Symptoms were marked neck weakness, torticollis, thoracolumbar scoliosis, profuse nasal discharge, myasthenia when driven, incoordination, inability to respond to noise or physical stimuli, pollakiuria, and decreased red blood cell cholinesterase. **Cattle** and **horses** in the area had decreased cholinesterase, but no toxicity symptoms or deaths. Supposedly, due to a wind shift, the vapors had drifted 29 miles, rather than 8 miles that had been calculated.

TD_{LO}, **man**: 4 µg/kg. Skin LD_{LO}, **human**: 86 µg/kg.

LD_{50}, **rat**: 121.9 µg/kg; IV, 17 µg/kg; IM, 23.6 µg/kg; IP, 54.5 µg/kg. SC, LD_{50}, **guinea pig**: 8400 ng/kg. LD_{50}, **dog**: 10 µg/kg. SC, LD_{50}, **rabbit**: 15.4 µg/kg.

W

WAI-LING-SIN

Untoward effects*:** Neuropathies reported due to ***podophyllotoxin contamination of this Chinese herbal by *Podophyllum hexandrum* = ***Gwai-Kou*** = ***P. emodi*** in 12 **patients**. This contaminant found in 22/234 samples caused its retail sales in Hong Kong to be prohibited.

WALKING STICK

Untoward effects*:** This larger striped herbivorous **insect**, *Anisomorpha buprestoides* of the southern U.S. has defensive glands in its thorax and can spray its active principle, ***anisomorphal, 30–40 cm (12–16 in) into the face and eyes of predators. The spray is lachrymogenic and irritating if inhaled. A **person** was sprayed in the eye leading to light-sensitivity for 48 h and 5 days of impaired vision, apparently due to corneal lesions with secondary uveitis. Keratitis reported in a **dog** so sprayed from the spray and/or self-inflicted trauma.

WALLPAPER

Untoward effects*:** Old papered areas have been well-known as a source of ***arsenical pigments and contact dermatitis from ***nickle*** and ***chromium*** in printings.

WALNUT

= *Juglans sp.*

Untoward effects*:** The leaves and hulls contain ***tannins and ***naphthoquinones*** (***juglone***, etc.) and is a skin and nail plate surface brownish colorant. ***J. nigra*** = ***American Walnut, Black Walnut*** = ***Noyer Noir***. Ingestion has caused allergic reactions in some **people**. It is a photosensitizer. Contact dermatitis of the fingers and eye irritation from handling the nuts still in their green outer shell. Juices in the shells and leaves have triggered a dermatitis in some **people**.

Ingestion of its fresh wood shavings and sawdust when used as bedding for **horses** has caused laminitis, lameness, pitting edema of distal limb areas, abdominal distress, increased respiratory rate, anorexia, and lethargy. This may be due to the ***juglone*** and/or an aqueous extract. Its pollen may also be a cause of laminitis in **horses**.

WANDERING JEW

= *Zebrina pendula*

***Untoward effects*:** Native to Mexico and Central America, but grown anywhere as a houseplant nibbled by **children** and **pets**. Many asymptomatic ingestions by very young **children**; occasionally gagging and vomiting in **others**.

WARFARIN

= *Athrombin* = *Compound 42* = *Co-rax* = *Coumadin* = *Coumafene* = *Deathmor* = *End-O´-Rat* = *Liqua-Tox* = *Marevan* = *Panwarfarin* = *Prothromadin* = *Rodex* = *Tintorane* = *Warfilone* = *Waran*

***Anticoagulant*. It and *dicoumarol* are closely related chemically to *vitamin K*.**

Untoward effects*:** The greatest risk is hemorrhage in the skin (petechiae, ecchymoses, hemorrhagic infarcts, maculopapular purpuric or vesicular lesions, purple toe syndrome, and rarely necrotizing vasculitis with gangrene; rashes at injection site), and hematomas (intracranial, epidural, intramuscular, intramural, bleeding in pulmonary cysts, renal pseudotumors, intrathoracic, below tongue, retropharyngeal, and iliopsoas with femoral neuropathy after osteopathic spinal manipulations). Hematuria, orange urine, acute gastrointestinal hemorrhagic infarcts, gastrointestinal bleeding, IP bleeding, hemarthrosis, salivary gland hemorrhage; airway obstruction due to retropharyngeal and submandibular hemorrhage; increased menstrual bleeding, hemoptysis, retinal hemorrhages, blindness and hyphemas, and delayed wound-healing are some of the many adverse effects of induced increased prothrombin time. Use of 52.5 mg during 5 days after amputation of lower left leg for ischemic ulceration and acute ascending cellulitis in a 61-year-old **female**; mammary necrosis developed and a mastectomy was required. Oral or IV use reported by some **workers** to cause dizziness and light-headedness 1 h after administration and lasting 1 h. Livedo reticularis, muscle pain, necroses, and ulcerations of the lower extremities 3 weeks–3 months after starting warfarin therapy may be due to ***cholesterol crystal embolization from insufficient thrombus formation over eroded arteriosclerotic plaques. Many case reports of femoral or sciatic nerve paralysis from local hemorrhage. Anticoagulant malingerers reported. Discoloration of glans penis and foreskin on third day of therapy advancing to mid-shaft on fourth day, after 50 mg/day/2 days then 25 mg/day/2 days. Penis was amputated on day 32. Has caused underestimation of ***theophylline*** blood levels. Peripheral eosinophilia in a 70-year-old **male** after 7 mg/day/18 days. Acquired ichthyosis, part of Conradi's syndrome, reported. **Maternal** use has caused hemorrhages, abortions, stillbirths, and **fetal** death due to placental passage and interferes with prothrombin synthesis by the **fetal** liver. Placental hemorrhages have occurred. A **fetal** warfarin syndrome with symptoms of chondroplasia punctata (Conradi's syndrome), dysmorphic facial features, hypoplastic nasal structures, broadened nose, nail hypoplasia, short limbs, growth and mental retardation, microphthalmia, optic atrophy, blindness, X-ray findings of stippled epiphysis, hydrocephalus, absent nasal septum, short humerus and femur, poor external ear development, brachydactyly, renal malformations, large protuberant eyes, subdural hemorrhages, intraventricular hemorrhage, and asplenia reported. Intrahepatic cholestatic jaundice has been reported in a few **patients**. Dramatic liver failure in a 61-year-old **female** after 2–4 mg/day/12 days. Attempts at suicide with ***rodenticides*** containing it have failed because of its low concentration, requiring many repeated ingestions. Percutaneous absorption occurs. A 24-year-old **farmer** using a 0.5% solution on dry ***bread*** contaminated **his** hands and failed to wash it off for several hours. Hematuria, hematomas; hemorrhage from nose, lip, palate, mouth, and pharynx 2 days later. Ingestion of so-called superwarfarin ***rodenticides*** has been associated with transient abdominal pain, vomiting, and heme positive stools in **children**. A 32-year-old **male** was murdered. **He** suffered from fatal hemorrhagic diathesis, patchy liver necrosis, and cerebral edema after ingestion of 1 g over a 15 day period. A 49-year-old **female** ingesting 2.5–5 mg/day/2 months developed alopecia. Necrosis of cutaneous and subcutaneous tissues is a relatively rare occurence from its use, and is more common in **women** than **men**. Vasculitis, purpura, purplish nail bed, urticaria, pruritus, asthma, and rash have also been reported. After **maternal** use, < 4.4% is found in breast milk. Therapeutic serum concentrations have been reported as 0.6–7 (generally 1–3) mg/l; toxic at 10 mg/l; and lethal at 100 mg/l.

TD_{LO}, **women** (1–37 weeks pregnant): 26 mg/kg. LD_{LO}, **human**: 50 mg/kg.

Drug interactions: It is said to interact with > 250 compounds, only some of which are clinically significant. ***Acarbose*** increases warfarin's international normalized ratio, probably by increasing the latter's absorption. **Patients** ingesting at least 4–325 mg ***acetaminophen*** tablets/day/1 week or more can expect > 30% increase in **their** international normalized ratio. ***Allopurinol*** inhibits warfarin metabolism with serious increase in prothrombin time in some **patients**, while not in **others**. Acute ***ethanol*** ingestion can increase international normalized ratio by 30%; chronic alcoholism can decrease warfarin half-life by an average of 33% and occasionally up to 70%. Use with ***amiodarone*** within 2 weeks has increased prothrombin time 22–108%. ***Antacids*** decrease the absorption rate of acidic drugs, such as warfarin. ***Antipyrine***, a now rarely used ***analgesic***, enhances its metabolism. ***Angelica sinensis, apazone, azithromycin, clarithromycin***, and ***levofloxacin*** markedly potentiate it. ***Barbiturates***, including ***amobarbital, butabarbital, heptabarbital, pentobarbital***, and ***secobarbital***, in normal sedative doses increase its metabolism by enzyme induction, decrease its prothrombin time and half-life. ***Carbamazepine*** increases its metabolism by enzyme induction, decreasing its effects. ***Cefamandole*** caused extremely high prothrombin time in 3/15 heart valve surgical **patients**. Its hypoprothrombinemic effect is increased by 40–80% when ***chloral hydrate*** is given. ***Chloral hydrate*** given concurrently can induce a transient potentiation, usually noticed after 12 h. ***Chloramphenicol*** has potentiated its anticoagulant effect by decreasing its metabolism. Ocular ***chloramphenicol*** has had the same effect. Warfarin activity is potentiated by ***chlorpromazine***, as the latter impairs platelet formation. ***Cholestyramine***, an ***ion exchange resin***, has a strong affinity for acidic drugs and impairs the absorption of warfarin. If use is required, it is best to administer the warfarin at least 6 h after it. ***Cimetidine*** potentiates its hypoprothrombinemic effect by decreasing its hepatic metabolism and plasma clearance, resulting in a 20% increase in prothrombin time. ***Cisapride*** 40 mg/day for gastroesophageal disease for 3 weeks after heart valve surgery increased international normalized ratio to 10.7. An 83-year-old **female** had dangerous increase in **her** international normalized ratio after ***clarithromycin*** 500 mg/twice daily/14 days, requiring ***vitamin K*** 20 mg IM. ***Clofibrate*** inhibits its metabolism and prothrombin synthesis in the liver, generally requiring a 33% decrease in warfarin dosage to maintain a stabilized international normalized ratio. The onset may be immediate or delayed. Use with ***oral contraceptives*** can decrease its anticoagulant effect by stimulating production of ***procoagulants***. ***Corticotropin*** and ***corticosteroids*** (Abbott literature) have potentiated its anticoagulation. Other literature reports inhibition by ***corticosteroids***. ***Cyclosporine*** decreases its anticoagulation effect. Delayed bleeding complications after use of ***danazol*** for endometriosis in **patients** stabilized on warfarin. ***Diazepam*** and ***lorazepam*** slightly increase prothrombin time by 1–2 s in **people** receiving warfarin. ***Dextrothyroxine***, ***disulfiram***, and ***d-propoxyphene*** inhibit its metabolism, potentiating its effect. It decreases the availability of ***digitoxin***. ***Diazoxide*** displaces warfarin from ***albumin*** binding sites, promoting increased hypoprothrombinemia and bleeding. ***Dichloralphenazone*** accelerates warfarin metabolism. Ingestion of ***erythromycin*** (base, lactobionate, and stearate) has caused dramatic increase (up to six times normal) in warfarin anticoagulation in several 1980 reports about 18 **people**. In 1976, a **patient** stabilized on warfarin was given a prescription for 10 days' treatment with ***erythromycin***. A knot developed in **his** stomach, enlarging; it made **his** stomach black. The symptoms disappeared during the next 7 months. **He** died within 3 years from internal bleeding and cardiac problems. The **widow** sued and lost because, in 1976, there had been no other evidence of such interaction. Related ***antibiotics***, such as ***azithromycin***, have induced hypoprothrombinemia in a 41-year-old **female**; dramatic increase in international normalized ratio by ***clarithromycin*** or ***roxithromycin*** with it in several **patients**. ***Dirithromycin*** is expected to act similarly. An *in vitro* interaction with ***ethacrynic acid*** has been demonstrated, and, in a 38-year-old **female**, several episodes of dramatic increase in prothrombin time when given warfarin. Significant decrease in prothrombin time when given to **patients** receiving ***ethchlorvynol***. ***Ethylestrenol*** increases warfarin's hypoprothrombinemic effect. Use with ***famotidine*** (***Pepcid***) increased prothrombin time and hemorrhage. Oral ***felbamate*** has caused increased anticoagulation in several **patients**, requiring ~50% decrease in warfarin dosage. ***Fenbufen*** 400 mg twice daily led to 14% decrease in warfarin serum concentration. Dramatic increase in international normalized ratio with ***fenofibrate*** 200 mg/day. Nausea, epigastric discomfort, and ***tea***-colored urine. ***Fluconazole*** has potentiated its effects with > 3 times increase in prothrombin time. Interactions with ***fluoroquinolones*** reported. Several case reports of ***ciprofloxacin*** increasing its prothrombin time by 25–60%. Serious nature of this interaction confirmed by **rat** studies. ***Enoxacin*** only inhibits warfarin's less active enantiomer. ***Ofloxacin*** 600–800 mg/day has caused increased and prolonged anticoagulation with it. Use with ***5-fluorouracil*** in a 59-year-old **male** with adenocarcinomas led to multiple mucous membrane bleedings, epistaxis, hematuria, hematochezia, and hematemesis. Use with ***fluoxetine*** in an 83-year-old **male** associated with increased international normalized ratio. ***Diazepam*** interaction with ***fluoxetine*** may have also occurred. ***Foods***, especially ***vegetables*** with a high ***vitamin K*** content, can easily destabilize an international normalized ratio. ***Grapefruit juice*** has had no effect in some cases, and definitely increased the international normalized ratio in a 65-year-old **male**. Possible inhibition of hypoprothrombinemic response in a 49-year-old **female** after ingesting ***ice cream***. IV fat emulsions have interfered with its anticoagulant effect. In England, **patients** who ate ***french fries*** had decreased warfarin absorption due to ***dimethicone*** in the cooking oil. It gives the fries a crisp, dry appearance. ***Gemcitabine*** use led to dramatic and rapid increase in international normalized ratio in a **patient** on warfarin. A 47-year-old **male** whose international normalized ratio ranged between 3 and 4 took three ***ginseng*** (***Ginsana***®) capsules/day/2 weeks and **his** international normalized ratio decreased to 1.5. After discontinuing ***ginseng***, **his** international normalized ratio returned to 3.5 in 2 weeks. **Patients** taking ***Hypericum perforatum*** (***St. John's Wort***) have also had a decrease in **their** international normalized ratio. In most **patients** taking IV ***glucagon***, 25 mg/day for only 2 days, it caused marked enhancement of warfarin's hypoprothrombinemia. ***Glutethimide*** and, in some **patients**, ***griseofulvin***, decrease its half-life due to induction of enzyme metabolism, requiring increased warfarin dosage to maintain anticoagulation. ***Griseofulvin*** also decreases its absorption, inhibiting its anticoagulant effect. ***Haloperidol*** reported to antagonize its anticoagulation. Adverse reactions with ***influenza vaccine*** are rare; they suggest a decreased prothrombin time and have not been serious. There is evidence that resistance to its hypoprothrombinemic effects occurs from ***chlorinated hydrocarbon insecticides***, which induce its metabolism. Prothrombin time increased in 52-year-old **female** also receiving ***α-interferon*** for 14 days. Increased ***isoniazid*** dosage precipitated hemorrhage in a 35-year-old **male**. ***Ketoconazole*** 200 mg twice daily prolonged coagulation in a 75-year-old **female**. Use with ***ketoprofen*** prolongs prothrombin time and has caused gastrointestinal hemorrhage and necrotic skin lesions, especially in older **patients**. Use with ***fluorouracil*** and ***levamisole*** in a 60-year-old **female** caused a two times increase in prothrombin time. ***Mefenamic acid*** decreases its ***albumin*** binding and can double its free serum levels, prolonging its prothrombin time and potentiating warfarin anticoagulation. ***Meprobamate*** shortens its half-life by inducing hepatic microsomal activity; the clinical significance of the interaction has been questioned. ***Mercaptopurine*** inhibited its anticoagulant effect in a 62-year-old **male**. ***Mesalamine*** 800 mg tid/4 weeks strongly decreased warfarin's anticoagulant effect. An exaggerated anticoagulant response noted in **patients** receiving ***methandrostenolone***, an ***androgen***, when given warfarin. Similar potentiation reported with ***norethandrolone***. ***Metolazone*** caused a sharp rise in the international normalized ratio of an 86-year-old **male**. Several case reports of significant increase in prothrombin time by ***metronidazole*** with bleeding episodes, SC hemorrhages, and bruising. Use with ***miconazole*** 125 mg gel qid over a 2 week period caused > 6 times increase in international normalized ratio in a 73-year-old **male**. ***Mitotane*** 4 g/day use demands increased doses of warfarin to maintain international normalized ratio. **Patients** receiving ***moricizine*** had an increased warfarin half-life from 34.2 to 37.6 h. Insufficient *in vivo* data on ***nalidixic acid*** interaction with it, but *in vitro* trials indicate it can increase free serum warfarin by 66–400%. Oral ***neomycin*** potentiates its action, probably by decreasing production by intestinal bacteria of ***vitamin K***, thus decreasing the amount absorbed. A **patient** taking ***nizatidine*** with it developed increased prothrombin time and hemorrhage. The ***non-steroidal anti-inflammatory drugs antipyrine***, ***aspirin***, ***ibuprofen***, and ***indomethacin*** often inhibit warfarin's enzymatic metabolism, increasing prothrombin time and hemorrhage, especially if taken continously. ***Omeprazole*** inhibits the metabolism of the less active r isomer of warfarin. ***Oxymetacine*** 100 mg tid potentiated warfarin anticoagulation in 4/12 **patients**. ***Oxymetholone*** has increased its prothrombin time within 7 days. ***Oxyphenbutazone*** and ***phenylbutazone*** strongly inhibit prothrombin synthesis in the liver leading to serious hemorrhages in **people** stabilized on warfarin. Interactions reported with some ***penicillins***. In some **patients**, IV use of ***cloxacillin*** requires a dramatic increase in warfarin to maintain **their** international normalized ratio; prothrombin time decreased 17% within 4 days in a 41-year-old **male** after ***dicloxacillin*** 500 mg qid; and resistance to warfarin anticoagulation noted with IV ***nafcillin***. In a 49-year-old **male**, large parenteral doses of ***penicillin G*** caused a marked potentiation of warfarin's anticoagulation. Severe hematuria 3 months after starting ***phenformin*** slow-release capsules 50 mg/day in a 64-year-old **female** stabilized for 9 months on warfarin. ***Phenformin*** increased fibrinolytic activity and the increased fibrinolysis may also have been a cause of the hemorrhage. ***Phenyramidol*** inhibits its metabolism and causes a marked potentiation of anticoagulation and prolongation of prothrombin time after 3 days. Although there is considerable inter-**patient** variability, it decreases ***phenytoin*** metabolism, causing ***phenytoin*** intoxication and an increase in its half-life. Metabolism of warfarin has also been inhibited leading to hemorrhage. ***Piroxicam*** decreased prothrombin time in a 60-year-old **male** when it was added to warfarin drug regimen. Adsorption on plastic increases in the presence of ***dextrose*** in the solution. Enhanced anticoagulant effect when given with ***propafenone***. Increased hypoprothrombinemia and hemorrhage with warfarin from decrease of ***vitamin K*** clotting factors by ***quinidine*** in some **patients**. Other ***cinchona*** alkaloids, ***quinine*** and ***cinchophen***, have the same effect on ***vitamin-K***-dependent factors. ***Ranitidine*** effect with it has been variable, inducing substantial increase or decrease in anticoagulation. ***Rifampin*** enhances its elimination from the body and has caused a 2–3 times need for increase in warfarin dosage. ***Ritinovir*** 800 mg/day has also caused a need for doubling warfarin dosage to maintain a prior international normalized ratio. The ***salicylates aspirin***, ***salsalate***, and ***sodium salicylate*** cause varying degrees of increased prothrombin time; ***aminosalicylic acid*** has caused up to a 10 times increase and topical ***methyl salicylate*** and ***trolamine salicylate*** also prolong and increase prothrombin time. Increased hypothrombinemia when ***saquinavir*** 600 mg tid added to a warfarin-stabilized international normalized ratio requiring a 20% decrease in warfarin. Use with ***sertraline*** needs to be monitored as it has caused an increase in unbound warfarin, the clinical significance of which has not yet been determined. Smoking has increased its clearance by ~10%, but did not alter prothrombin time. Case reports of increased sensitivity to warfarin effect by ***stanozolol*** requires decrease in warfarin dosage. The ***statins***, ***fluvastatin*** and ***lovastatin***, increase warfarin's prothrombin time and have induced bleeding episodes. Conflicting reports exist on ***sucralfate*** interaction with it. Given simultaneously, evidence indicates warfarin absorption is interfered with. ***Sulfadimethoxine***, ***sulfamethizole***, ***sulfamethoxazole***, ***sulfamethoxypyridazine***, ***sulfaphenazole***, ***sulfinpyrazone***, and ***sulfisoxazole*** cause clinically significant increase in warfarin-induced hypoprothrombinemia. Excessive hypoprothrombinemia after ***sulindac*** 100–200 mg twice daily in **patients** stabilized on warfarin. Life-threatening interaction with excessive bleeding in **patients** taking ***tamoxifen***. A serious interaction with ***terbinafine*** in a 68-year-old **female**, decreasing international normalized ratio. The **manufacturer**, Novartis-Pharma AG, refutes the report. Use of ***testosterone propionate*** 2% topical vaginal ointment increased prothrombin time 78% in a 68-year-old **female** stabilized on warfarin. Prothrombin time returned to normal after discontinuing the ointment. Oral ***tetracycline*** and ***doxycycline*** induced potentiation of warfarin anticoagulation. ***Thyroid*** hormones increase catabolism of ***vitamin K***-dependent clotting factors, and can be dangerous in starting therapy in a previously anticoagulated **patient**, but there is little danger in giving warfarin to a **patient** stabilized on a ***thyroid*** replacement. Use with ***ticlopidine*** increased mean R-warfarin concentrations with no changes in S-warfarin concentrations. Its clinical significance has yet to be determined. ***Ticrynafen*** augments prothrombin time and concentration of S-warfarin. Warfarin activity is potentiated by ***tolbutamide***. ***Tolmetin*** has caused an increased prothrombin time of 70.2 s in a 64-year-old **male** taking warfarin. Case reports of ***tramadol*** causing dramatic increase in prothrombin time and international normalized ratio in **patients** taking warfarin. In a 40-year-old **female**, ***trazodone*** with warfarin decreased prothrombin time 30% and decreased thromboplastin time 18%. ***Triclofos***, as is ***chloral hydrate***, is metabolized primarily to ***trichloroacetic acid***, which displaces warfarin from inactive binding sites on ***albumin*** molecules, significantly potentiating warfarin's anticoagulant effect and prolonging prothrombin time by 49%. Use of ***troglitazone*** in a 51-year-old **male** taking warfarin increased **his** international normalized ratio by 65%. Several case reports of sudden decrease in international normalized ratio in **patients** taking warfarin about 2 weeks after starting therapy with ***ubidecarenone***. ***Omeprazole*** has potentiated warfarin's anticoagulatory effect. After drinking $\frac{1}{2}$–1 gal/day of green ***tea***, a 44-year-old **male** had ~65% decrease in **his** international normalized ratio. ***Rofecoxib*** taken with it led to 10% increase in prothrombin time. ***Celecoxib*** with it also increases international normalized ratio. ***Vitamin C*** in large doses (> 1 g/day) decreased prothrombin time in warfarin-treated **patients**. ***Vitamin E*** can increase prothrombin time, and when used with warfarin, slightly potentiates the latter's anticoagulation effect. Interaction with ***vitamin K*** is well-documented. Warfarin decreases synthesis of clotting ***Factors II, VII, IX***, and ***X***, dependent on ***vitamin K***. The latter antagonizes warfarin's anticoagulation effect. Use with ***zafirlukast***, a ***leukotriene-receptor-antagonist***, increases warfarin's anticoagulation effect and can induce bleeding. ***Zileuton***, a ***leukotriene synthesis-inhibitor***, has caused a clinically significant increase in warfarin serum concentration and prothrombin time. ***Trastuzumab*** use associated with increased bleeding in **patients** stabilized on warfarin. ***Azathioprine, broccoli, Brussel sprouts, cabbage, lettuce***, and ***liver*** decrease its anticoagulation effect; ***garlic, ginger, Ginkgo biloba, nabumetone***, and ***papain*** increase its anticoagulation effect. Warfarin inhibits ***protein C*** and **people** already low on it (~4% of the population) may be thrombotic-prone for a few days. Topical ***miconazole***, after absorption, can decrease its metabolism, increasing bleeding.

Simultaneous or prior use of ***chloral hydrate*** and ***barbiturates*** may decrease its anticoagulant effectiveness by stimulating hepatic microsomal enzymes. Percutaneous absorption can occur (**guinea pigs, humans, rabbits**). Secondary toxicity can occur from **animals** ingesting the meat of poisoned **animals**. This has been noted in **nutria, mink**, and **dogs**. ***Cholestyramine*** use decreases its absorption in **rats** and **man**. ***Mercaptopurine*** potentiates its hypoprothrombinemic effects in **rats**. This may be through inhibition of ***coumarin***-metabolizing enzymes or ***vitamin K***-dependent clotting elements. Large doses of ***hydrochlorothiazide***-enhanced warfarin effects in **rats**, probably by decreasing ***coumarin*** binding to plasma protein. ***Meprobamate*** has decreased its plasma half-life in **dogs**, and a similar effect may have been noticed in **man**. (**Dog** ***albumin*** binds warfarin to a lesser degree than does **human** ***albumin***). ***Clofibrate*** increases the plasma concentration and prolongs warfarin half-life 50% in **dogs**. Similar potentiation has been noted in **man**. ***Phenylbutazone*** and ***tolbutamide*** enhance warfarin's effect in **man**, but decrease it in **dogs** (after short initial potentiation by release from binding sites, followed by stimulation of metabolism). ***Amobarbital, glutethimide, secobarbital***, and some other ***barbiturates*** can decrease warfarin half-life in **dogs** and **man**. In subacute cases, causes anemia, pale mucous membranes, dyspnea, moist râles, bloody feces, ataxia, and brain and spinal cord hemorrhage.

LD_{50}, **mouse**: 374 mg/kg; **rabbit**: 800 mg/kg; **guinea pig**: 182 mg/kg; **cat**: 5–30 mg/kg; **dog**: 20–300 mg/kg. When fed daily for 5 days, **rat**: 1 mg/kg; **cat**: 3 mg/kg; **dog**: 5 mg/kg.

WASABI

Untoward effects*:** This Asian green plant is chopped and mixed with small amounts of ***soy sauce, making a ***horseradish***-like condiment, and placed "on the side" of ***sushi***. A 63-year-old **male**, enjoying the taste, ate the entire serving; **he** became pale, confused, sweated profusely, staggered from the restaurant and collapsed on the sidewalk. Hospitalized and recovered.

WASPS

Untoward effects*:** The ***Paper Wasp = Polistes* sp.** and the ***Yellow Jacket often sting without provocation and have caused anaphylactic reactions, cerebral infarction, hemolytic anemia, myocardial infarction, and deaths in **some**. Egyptian **King Menes**, about 5000 years ago, died from a wasp sting, according to the markings on **his** tomb. A **person** stung in the eye by a mud dauber wasp developed mydriasis, loss of light perception, corneal haze, and eventual recovery. General symptoms in most **people** are local swelling, fear, urticaria, headache, malaise, rhinitis, asthma, and abdominal cramps. Repeat stings often cause anaphylaxis. The common wasp, *Vespula vulgaris*, and the ***hornet*** (q.v.) venoms cause nausea, dizziness, vomiting, diarrhea, chest pain, dyspnea, cyanosis, and occasionally epileptiform convulsions, urticaria, or syncope, usually lasting 20–30 min, followed by fatigue.

WATER

Untoward effects*:** Drinking "soft" water has been associated with increased cardiovascular deaths in **adults** and in **infant** mortality. This may be due to increased release of ***lead from ***lead*** pipes and ***cadmium*** from galvanized pipes. Many ***insecticides, chloroform, benzene***, other chemicals and bacteria are monitored in drinking water. Bacteria or chemicals in bottled drinking water have exceeded recommended limits in ~ $\frac{1}{3}$ of 103 brands tested. ***Nitrates*** (q.v.) and ***nitrites*** (q.v.) have often caused serious problems in **infants**. An aquagenic pruritus in some **people** shortly after brief contacts with water. Enthusiastic **swimmers** and **others** with prolonged contact in water show loss of skin lipids and other agents from the skin; near-drowning exposure has led to hypoxia, and metabolic acidosis; and in **children**, after immersion accidents, several had pulmonary edema, bronchopneumonia, and respiratory distress syndrome. Photosensitizing compounds have been found in some drinking water. Drinking glacier water has caused gastric ulceration in **people**, probably from its content of ***silica*** crystals, a condition analagous to that of **pigs** ingesting ***silica*** crystals from licking food off concrete floors and developing esophagitis. **Babies** will suck on anything when hungry and a number of **babies** were abused and intoxicated by being forced to drink large quantities of water or overdiluted formulas, which precipitated seizures. Prolonged and severe hypotonicity with irreversible brain damage in postoperative **patients** given large quantities of fluids with ***dextrose*** only.

Excessive water consumption (psychogenic polydipsia or compulsive water-drinking), for a short while, has produced feelings of euphoria and light-headedness, convulsive seizures, coma, and, unless treated, in many cases, death. It can follow certain drug therapy, such as high-dosage ***oxytocin***. An overdose with 4 g ***carbamazepine*** for a seizure disorder in a 23-year-old **female** led to water intoxication, possibly from its antidiuretic effect. A 29-year-old **female schizophrenic** drank as much as 4 gal of water/day to "cleanse **her** body" of suspected cancer leading to pulmonary edema, cardiac arrhythmia, severe dyspnea, and death. Chills, fever, hypotension, nausea, vomiting, and myalgia caused 49 reactions in 23/70 dialyzed **patients** in a 3-week period, and was traced to ***endotoxin*** contamination of tap water, possibly due to excess algal levels in the water. **Workers** in fisheries or ***fish***-processing plants are frequently exposed to sea water and often develop skin "disease" from something in the water. Drinking sea water, due to its hypertonicity, and especially its ***sodium*** content can cause death. Seawater begins to freeze at 28.4°F, and the ice crystals are, essentially, ***salt***-free, and the safest when melted to produce fresh water. **Aristotle**, ~350 years before Christ, demonstrated that when ***salt*** water turns to vapors and recondenses, it does not contain ***salt***. Store-bought bottled water has often contained bacteria, microscopic debris, and assorted chemicals.

Experimentally, forcing large quantities of water by stomach tube into **dogs** decreased hemoglobin and serum protein by ~15% and ***chlorides*** by 30% leading to cerebral edema, restlessness, asthenia, diarrhea, salivation, nausea, vomiting, tremors, ataxia, violent convulsions, collapse, and death. Similarly in **rats**, cerebral edema, hypothermia, stupor, coma, and death within 4 h to days.

WATER BOATMAN

= *Notonecta triguttata*

Untoward effects: It is poisonous and found in Japan and Korea. The **Ainu race** of northern Japan used it on **their** poison arrows for hunting.

WATERCRESS

Untoward effects: Like ***radishes, broccoli, kale***, and ***rape***, it contains a ***goitrogen***. On a weight basis, it contains twice the ***vitamin A*** as ***butter*** and an ***anti-vitamin*** that ties up ***vitamin*** B_1. It is also high in ***oxalates*** and should be avoided by **women** with vulvodynia. ***Chlorzoxazone*** elimination half-life was prolonged 53% by inhibiting its hydroxylation after ingesting 50 g watercress. **Patients** with ***chromate*** dermatitis should avoid consuming it.

WEDELIA sp.

Untoward effects: ***W. asperimma = Sunflower Daisy = Yellow Daisy***, abundant in Australia, is normally eaten by **sheep** and **goats** only when hungry. After 2 days traveling by train, 4800 **sheep** were turned into a pasture with an abundance of it. The next day, 1000 were dead. Post-mortem showed straw-colored fluid in body cavities, hemorrhagic inflammation of lower bowel, and blood-stained feces. It contains high levels of ***nitrate*** and an unidentified toxic substance.

W. glauca in Argentina was rapidly poisonous to **pigs**. Symptoms were ataxia, hyperthermia, prostration, and death. Post-mortem showed severe hemorrhages in gastroduodenal mucosa, peritoneum, lymph nodes, endocardium, and skin; necrotic hepatic lesions, and fatty liver.

WELDING

Untoward effects: ***Ozone*** generated by arc welding in confined areas led to pneumonitis. Acid electrodes containing ***manganese*** (q.v.) caused an increased body burden of ***manganese*** in 20 **welders** and IV chelation was required. ***Lead*** tank **welders** have increased blood ***lead*** levels. Arc **welders** and ***acetylene*** **welders** have shown ***chromate*** exposure and increased sensitivity to it. **Welders** in a manufacturing plant breathed ***zinc oxide*** fumes from welding galvanized metal leading to "metal fume fever", "***zinc*** shakes", or ***brass*-founders** ague, with chills, fever, nausea, vomiting, dry throat, coughing, fatigue, yawning, weakness, and arching of head and body. ***Cadmium*** (q.v.) ***oxide*** aerosoled exposure follows welding of galvanized metal and ***silver*** alloys. Welding fluxes have made **workers** allergic to ***hydrazine***. A 45-year-old **male** developed ***phosgene*** (q.v.) poisoning from decomposition of ***trichloroethylene*** (used for degreasing studs) during ***carbon dioxide*** welding. A **welder's** tank of ***argon*** gas was mistakenly connected to an Army hospital's ***oxygen*** system at Fort McClellan, Alabama. Killed three **patients**. **Welders** have shown an excess of lung cancers and mortality from other lung-related diseases, chronic bronchitis with emphysema, chronic interstitial pneumonia and pneumoconiosis, and an increase in bladder cancer. Eye pain, photophthalmia, keratoconjunctivitis, and recurrent pterygium from failure to wear protective eye wear when welding or watching welding.

Blindness affected dairy **cows** watching **welders** working on natural gas pipelines.

WET SUITS

Untoward effects: Severe contact dermatitis from the ***diethylthiourea, ethyl butyl thiourea*** (q.v.), or ***thiourea*** (q.v.), used in vulcanizing neoprene ***rubber***.

WHALES

Untoward effects: Their livers can contain dangerously high levels of ***vitamin A*** (q.v.) and whale meal supplementation of **pig** and laying **hen** feeds was a source of excessive ***mercury*** intake. The meat of some Pacific whales also contains an unidentified poisonous substance. Whale oil supplementation of **calves** led to baldness.

WHEAT

= *Triticum aestivum* = *Triticum sativum* = *Triticum vulgare*

Most U.S. (35%) is Hard Red Winter Wheat for *bread*. Hard Red Spring Wheat (21%) is high in protein and has strong *gluten*, improving the above's baking quality. Soft Red Winter Wheat (27%) is used in baking cakes, pastries, quick *breads*, crackers, cookies, and snack foods. White Wheat (12%) is mostly for export to the Far East, especially for noodles. Durum Wheat is the hardest U.S. wheat and used in making spaghëtti, macaroni, and other pastas (5%). Spaniards brought it to the New World in the 1500s and it was planted as early as 1602 near Martha's Vineyard.

Untoward effects: Consumption of ***methylmercury*** (q.v.)-dressed wheat caused several epidemics of **human** poisoning, sickening thousands and killing hundreds. Isolated cases were also reported in Israel. Many **infant** deaths occurred in 1955–1956 in Turkey after **mothers** ate wheat treated with ***hexachlorobenzene*** (q.v.), which also adversely affected thousands of **people** and many died. Clinical symptoms of porphyria were reported in 132 **males** and 72 **females** 20–30 years after eating such treated wheat. It is a common cause of allergic contact dermatitis, with eczematous and urticarial lesions on **bakers** and **millers**. Similar results with wheat straw. Reactions have occurred with its ingestion. Contact dermatitis reported in a ***potato*** chip factory due to a wheat filler. Wheat hairs and dust are common cause of asthma of **workers** in wheat flour mills. Wheat ***gluten*** is toxic to **celiacs** (1/2000 in the U.S. and 1/300 in western Ireland) where it accents the basic defect of inability to cope with dietary fat. Wheat-sensitive **individuals** with symptoms under control had them reactivated by eating commercial ***bread*** sold as "wheat-free". ***Bread*** and other food products labeled as such may contain ***gluten*** and other wheat proteins. A characteristic rectal inflammatory response to wheat enemas noted in **patients** with celiac sprue. Wheat ***gluten*** is a pathogenic factor in schizophrenia. Wheat contains physiologically and pharmacologically active ***estrogens***. ***Nitrosamines*** and nitroso compounds have been found in wheat plants and the wheat grain. A **wheat-thresher's** lung or **grain-measurer's** lung is a hypersensitivity pneumonitis from exposure to wheat flour containing the weevil *Sitophilus granarius*. Wheat germ contains a high level of ***oxalates*** (q.v.) and can be especially detrimental to **women** with vulvodynia. ***Aflatoxin*** and ***ochratoxin*** have been found on wheat.

Dairy cows readily eat wheat pasture, resulting in fishy-tasting ***milk*** due to its ***trimethylamine*** (q.v.) content. A wheat pasture syndrome in 2 years or older pregnant **cows**, often with **calf** at side. Symptoms are undue excitement, anorexia, ataxia, staggering, viciousness, and 80% occur after 60 days on pasture and before 150 days, due to apparent disturbance in ***calcium, phosphorus***, and ***magnesium*** metabolism. Wheat is also a ***nitrate***-accumulator. **Steers** and **sheep** are also affected. A wheat-sensitive enteropathy has been reported in **dogs**. Wheat, once bleached with ***nitrogen chloride*** (q.v.) induced running seizures in **dogs**. This was replaced by small amounts of ***calcium peroxide, potassium bromate***, and ***potassium iodate*** to accelerate the aging and whitening process. Raw wheat contains a toxin or growth-inhibitor for **chicks**, but is destroyed by heat, pressure cooking, or pelleting.

WHISKEY

= *Spiritus Frumenti*

Untoward effects: Ingestion of moonshine or illegal homemade whiskey has induced many problems, such as ***lead*** (q.v.) poisoning with secondary hyperuricemia and gout, early sterility, spontaneous abortion; small, weak, and neurologically damaged **infants**; anuria, and azotemia from ***arsenic*** (q.v.) poisoning and possibly cancer from its ***dimethyl-N-nitrosamine*** (q.v.) content. The latter has been found in samples from Africa, Europe, and Canada. Spasm of calf muscle and cold feet, followed by gradual paralysis of all limbs in a 56-year-old **male** within a few days after ingesting 700 ml homemade ***brandy*** passed through ***polyvinyl chloride*** tubing containing 50% ***tritolyl phosphate***. Hospitalized with ataxia and absent triceps, radial, ulnar, and Achilles tendon reflexes, and atrophy of hand muscles. ***Ferric chloride*** absorption, but not ***ferrous ascorbate*** or hemoglobin ***iron***, is increased by whiskey or ***brandy***. Consumption of 50 cc ***alcohol*** decreases intraocular pressure and can mask glaucoma. Splashes onto the eye can cause transitory superficial injury with hyperemia and discomfort.

Drug interactions: **Patients** on ***phenformin*** (q.v.) have developed acidosis with increased lactate and pyruvate values.

See ***Alcohol, Ethyl*** for other untoward effects and Drug interactions.

WHITE SPIRIT

Untoward effects: A form of ***naphtha*** (q.v.). An ***aromatic hydrocarbon solvent*** with high volatility has been used in dry-cleaning and has induced scleroderma.

Eye irritant in **rabbits**.

WILD CARROT

Untoward effects: One type, ***Cymopterus watsonii*** = ***Spring Parsley*** = ***Wild Parsley*** in the U.S. contains ***furocoumarins*** in the leaves, but not in the stems or roots, serious photosensitizing agents for **sheep**, affecting skin areas unprotected by ***wool*** or sheared **sheep**. Rarely fatal for **ewes**, but often for their **lambs**, who die of starvation or dehydration, as the **ewe's** teats and udders are very tender and they won't let the **lambs** nurse. **Chickens** and **turkeys** are also affected by eating it or ***C. longipes*** leading to photosensitivity, photophobia; reddening of beak, comb, and feet; loss of periorbital feathers, keratoconjunctivitis, tremors, and gangrene of the toes. Two ***furocoumarins*** were identified, ***isoimperitorin*** and ***oxypencedanin***.

Queen Anne's Lace = ***Daucus carota*** = ***Devil's Plague*** = ***Gajar*** is also known as wild carrot. Some **people** develop a photocontact dermatitis from it after exposure to sunlight. The dried tops have been smoked and reported to effect "tripping" in the drug culture.

WILD CELERY and WILD PARSNIPS

(Trachemene glaucifolia)

Untoward effects: The same photosensitivity agents as found in the paragraph above on ***Queen Anne's Lace*** can cause a contact dermatitis after exposure to sunlight.

WINE

Untoward effects: It contains ***alcohol, fusel oils, tannins, pressor amines***, and ***tyramine*** (q.v.). The latter two can cause a hypertensive crisis and headaches in **patients** taking ***monoamine oxidase inhibitor***s or drugs like ***furazolidone*** with ***monoamine oxidase inhibitor*** action. Red wines and sherries contain large amounts of ***histamine***. The addition of ***phenolphthalein*** (q.v.) into poor quality Hungarian white wine turned the wine red at pH > 8.6 and it then demanded a higher price. The effect of drinking it was diarrhea and this discovery was of a new physiological effect for the chemical. ***Iron*** content of commercial and illicit wines has been as high as 6.9 mg/l and may have contributed to hemochromatosis. **Romans** used ***lead*** (q.v.) in **their** wine vessels and it was a cause of decreased birth rate. ***Lead*** levels in some French wines were 10–100 times higher than in **their** drinking water. **Human** exposures were apparently due to using ***grapes*** grown along highways, ***lead*** foil bottle wrappers, and drinking out of ***lead*** crystal glasses. English upper classes in the seventeenth and eighteenth centuries drank Portuguese port wine with high levels of ***lead*** and developed gout, while lower classes drank ***lead***-free ***gin*** and suffered no gout. A 28-year-old **male** developed ***lead*** poisoning from **his** homemade wine made in a flaking enameled cast ***iron*** bathtub; 70 gal contained 4.3 mg ***lead***/100 ml. Some U.S. levels > 300 ppb, which can pose serious hazards for pregnant and nursing **women**. ***Arsenic*** (q.v.) intoxication has been reported from fruit spray or wine ***yeast*** contamination of wine. Accidental contamination of ***beer*** with ***arsenic*** in England resulted in illness of 6000 **people** and 70 deaths in 1900. It was traced to ***arsenic***-containing ***sulfuric acid*** in preparation of ***sugar***. In 1932, **wine-drinkers** on a French merchant ship were poisoned by ***arsenic*** in it. As a result, a number of European countries and California forbid its use. In 1980, the Medical College of Georgia found ***arsenic*** concentrations of up to 420 ppb in domestic table wine. A Ghanaian 38-year-old **male who** developed a severe sickle-cell crisis from drinking ***gin*** (q.v.) had numerous previous attacks from palm wine, but not after ***whiskey***. Deaths probably related to the use of ***sulfites***, ***sulfur dioxide*** and various ***metabisulfites*** in wine or wine coolers in susceptible **people**. Migration of ***polyvinyl chloride*** (q.v.) chemicals into wine from ***polyvinyl chloride*** containers. ***Quercetin***, found in large quantities in ***tea*** leaves and ***grape*** skins, is carcinogenic in **rats**, but poorly absorbed from the **human** gastrointestinal tract. **Patients** with ***nickel*** or ***chromium*** dermatitis should avoid wines. Chianti red wines have contained as much as 24.5 μg/ml of ***tyramine***. Red wine has caused variable black discoloration of the nail plate surface. **People** with allergic eczematous dermatitis from ***alcohol*** can also react similarly to wine. Flushing has occurred from drinking red wine. Severe encephalopathy in two 3- and 3½-year-old **children** after ingesting 750 and 500 ml, respectively, of red wine on empty stomachs. After a 26 h coma, **one** by the age of 6 regressed to an I.Q. of 50, the **other**, after a 26 day coma, developed cerebral atrophy and vegetative existence. A 19-year-old **male** soldier had seizures after drinking 250 ml of French wine that had flowed through the barrel of a 155 mm gun. It picked up ***tungsten*** from previous firings. Sweet ***sherry*** has caused decalcification of teeth. Fluorosis of the bones reported in eight **Spaniards** induced by continuous ingestion of ***sodium fluoride*** fraudulently added to wine (7.6–72.6 mg ***fluoride***/l) to prevent fermentation. In 1985, Austrian wines and some Italian wines were sweetened with ***diethylene glycol*** (q.v.) leading to nausea, renal failure, and death if excessive quantities were drunk. In 1985, 100,000 l of wine treated with ***sodium azide*** (q.v.), used to make ***lead azide*** detonators, made it to the marketplace with a potential for serious health risks. ***Diethylpyrocarbonate*** (q.v.), added as a preservative in some countries' white wine, reacts with ***ammonia*** to form ***urethane*** (q.v.), a carcinogen. ***Procymidone***, a systemic fungicide on ***grapes***, is found in trace amounts in some French and Italian wines approved for sale in Europe, but French and Italian **vinters** have to certify that wines shipped to the U.S. are free of it. Aging of many white wines causes a musky browning color during maturation. U.S. has permitted polyvinyl-pyrrolidone up to 6 lbs/1000 gal to clarify and stabilize it, and approved various defoaming agents and enzymes to chill-proof wine. ***Sorbitol*** (q.v.) in some wines to prevent clouding caused by ***copper*** and ***iron*** ions. In Japan, a ***sake rice*** wine is fortified with venom from the **mamushi**, or **Japanese viper**. Occasional crystals in white wine are usually ***tartaric acid***; too many indicate improper fermentation. Filtering produces ~2–12 million ***asbestos*** fibers/l. Various enzymes are used in some wines to aid in clarification. A winery in California was closed down by state health officials after excessive ***dibromochloropropane*** (q.v.) was found in their wine.

For drug interactions, see ***Alcohol, Ethyl***.

WISTERIA

Untoward effects: Ingestion of the seeds and pods of various varieties has poisoned many **children**. Symptoms are mild to severe gastroenteritis, abdominal pain, nausea, vomiting, dyspnea, and collapse. Rarely, diarrhea. Serious intoxication from **children** chewing the bark, with serious loss of body fluid from excessive vomiting. Symptoms usually last ~24 h. Ingestion of two seeds has caused serious symptoms from a toxic resin and ***wisterin***, a glycoside. Vomitus is occasionally blood- or bile-tinged. The flowers appear to be harmless if ingested, but their aroma has caused headaches and nausea in some **people**.

Seeds are toxic if ingested by **canaries** and other pet **birds**.

WITHANIA SOMNIFERA

= *Agol* = *Ashwagandha* = *Kalabansa* = *Ubab*

***Herbal.* The leaves and roots are popular in Iraq, Yemen, India, and Ethiopia for analgesic and immunomodulatory effects and treatment of osteoarthritis, especially by Ayurvedic physicians, but it requires caution in its use, as it contains *scopolamine* (q.v.). Overuse has put animals to sleep for days.**

WOOL

Untoward effects: Allergic contact urticaria. Occasionally, certain woolens scratch the skin. Dermatoses reported in **sheep-shearers**, including a case of fatal acute pemphigus. In the 1930s, anthrax was occasionally reported in **sheep-handlers**. An acute vesicular eczema from orthopedic wool was due to its ***docusate sodium*** (q.v.) content. Combustion of wool releases the toxic pulmonary irritants ***ammonia, carbon monoxide, hydrogen cyanide***, and ***hydrogen sulfide***.

WORCESTERSHIRE SAUCE

Untoward effects: Long-time excessive intake (glassful) has led to renal failure. A 10-year-old child habitually consuming 2 qts every month developed gross and microscopic hematuria, massive proteinuria, granular casts, increased blood urea nitrogen, hypertension, bilateral renal calculi, and generalized aminoaciduria.

X

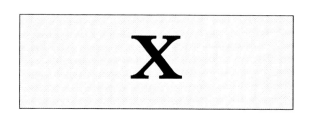

XANTHINOL NIACINATE
= *Angiomin* = *Complamin* = *Sadamin* = *SK-331-A* = *Xanthinol Nicotinate* = *Xavin*

Peripheral vasodilator. **Usually oral.**

Untoward effects: Flushing, burning sensation in epigastrium, rash, urticaria, feeling of prurigo, hypotension, and a **patient** developed rectal bleeding. An elderly **female** developed a stroke 12 h after 1500 mg in 10 ml. Pathological electrocardiogram in 18 **patients** improved in 13 after treatment, but electrocardiogram changes became worse in five **patients**.

XANTHOCILLIN
= *Brevicid*

Antibiotic. **For topical use.**

Untoward effects: Allergic reactions and irritation reported in **man**.

LD_{50}, **guinea pig**: 50 mg/kg; IP, LD_{50}, **mouse**: 40 mg/kg.

XANTHORHIZA SIMPLICISSIMA
= *Yellow Root*

Herbal. **Common in Appalachia, southwestern New York, and down to Kentucky and Florida.**

Untoward effects: A 54-year-old construction **foreman** drank 32–48 oz/day/2 years of a ***tea*** made from it and for 8 months experienced numbness and burning of **his** hands and soles of **his** feet. Progressive weakness forced **him** to use crutches. Samples of **his *tea*** made from the root purchased at a municipal market revealed that it contained ***arsenic*** (q.v.), not a natural constituent of the root.

XANTHORRHEA sp.
= *Grasstree* = *Yacca*

Untoward effects: Ingestion of the flower spike by **cattle** in Australia when feed is scarce leads to encephalopathies, lateral gait; spinning around in one spot; and urine dribbling. Occasionally, toxicity symptoms are delayed for 2–3 weeks.

XANTHOSOMA sp.
= *Blue Ape* = *Coco-Yam* = *Elephant's Ear* = *Indian Kale* = *Malanga* = *Spoon Flower* = *to-vo* = *Yautia*

Untoward effects: As with all ***aroids***, it contains irritant watery sap and the tuberous roots are acrid. Native to tropical America, but now a common ornamental, where, in Florida, it is a major cause of oral injury in **children** or **pets** from biting or chewing a leaf or stem high in ***oxalates***. **Indians** in Rio say "it is poisonous to touch". An edible rooted ***X. caracu*** is difficult to distinguish from the ornamental varieties. A **man** mistakenly cut off the root of the latter and it irritated **his** hands when **he** peeled it for his **wife**. **She** then decided to boil it twice, discarding the first water, then tasting it. **She** swallowed a $\frac{1}{2}$ spoonful which burned **her** mouth, throat, and esophagus. After 2 weeks, **she** had not fully recovered.

XIPAMIDE
= *Aquaphor* = *Aquaphoril* = *Bei-1293* = *Chronexan* = *Diurexan* = *Lumitens*

Diuretic, antihypertensive.

Untoward effects: Nausea, abdominal pain, xerostomia, light-headedness, vertigo, unsteadiness, nervousness, drowsiness, sweating, mild constipation, anorexia, headache, increased blood urea nitrogen, hypokalemia, hyponatremia, hyperglycemia, and secondary ventricular fibrillation.

XYLAZINE
= *Bay 1470* = *Bay Va 1470* = *Narcoxyl* = *Rompun* = *Wh-7286* = *Xylapan* = *Xylasol*

Sedative, analgesic, muscle relaxant, sympathomimetic.

Untoward effects: Severe fatigue, sleepiness, bradycardia, hypotension, ptsosis, loss of muscle tone, pallor, and dry mouth in two **human** volunteers (0.54 mg/kg orally or 0.27–0.68 mg/kg IV). Self-treatment by **insomniac** 34-year-old male with ~10 mg IM led to coma, apnea, and areflexia. Apparent suicide attempt with "10 times **animal** dose" led to minor vertigo a day later. An IM of 13 mg/kg self-administered by a **male** – he became apneic; hospitalized, and comatose for 60 h.

Although promising for **cats**, it produces vomiting before sedation occurs. Repeated SC and IM injections (2% concentration) cause local tissue necrosis in **horses**. A tolerance is shown to repeated injections. A temperature increase (0.5–1.5°C) is usually noted after its use. **Goats** have died after receiving 1 mg/4–5 lbs. Clinical rumors of early parturition in **cows** receiving it in late pregnancy – no such reports in **mares**. Will cause hyperglycemia and may cause bloating in **cattle**. In **horses** and

dogs under ***halothane*** anesthesia, it sensitizes the heart to ***epinephrine***, with resulting fibrillation. Drug possesses an α-sympathomimetic effect and decreases intraocular pressure in **horses**. Xylazine may have a brief pressor effect before hypotensive effects are noted in **horses**, **rabbits**, **rats**, **cats**, and **dogs**. These effects are noticed mostly after IV use and are dose-dependent. Increases contraction, rhythm, and tone of uterine muscle in **cattle**. Respiratory failure has occurred in overheated and overexerted **cattle**. **Brahma cattle** appear sensitive to higher dosage. Combination of xylazine and ***ketamine*** IM in **hamsters** causes extensive tissue necrosis. Decreased heart rate ~30% in **monkeys**, and premedication with ***atropine*** will help prevent falls in cardiac output in many species. Avoid use in **animals** with urinary tract obstruction. Xylazine is a poor anesthetic for **cats** (severe respiratory depression, including fits). Vomiting occurs in most **cats** and **dogs**, particularly after SC or IM dosage. Does NOT produce good skin anesthesia. Infiltration with local anesthetics is often required for surgery. If used as an emetic in **cats**, maximum effectiveness has been reported at 1.1 mg/kg (0.5 mg/lb). Avoid SC use in **horses**. Micturition is occasionally noted after use in **horses** and **cats**.

LD_{50}, **rat**: 130 mg/kg; **mouse**: 240 mg/kg. SC, LD_{50}, **mouse**: 121 mg/kg; IV, LD_{50}, **mouse**: 43 mg/kg.

XYLENE

= *Dimethylbenzene* = *Xylol*

Use: Solvent, octane-enhancer, disinfectant, chemical intermediate, and reagent. Billions of lbs used annually in the U.S.

Untoward effects: Ingestion, inhalation, or percutaneous absorption causes nausea, vomiting, gastrointestinal irritation, abdominal cramps, headache, dizziness, incoordination, disturbed vision, malaise, muscle fatigue, chills, respiratory distress, anorexia, anemia, aplastic anemia, leukocytosis, leukopenia, thrombocytopenia, neutropenia, narcosis, collapse, and coma. Vapors can also cause throat dryness, conjunctivitis, and epistaxis. In three **workers**, inhalation of ***paint*** fumes whose solvent was 90% xylene led to transient liver damage, one died, and two had > 15 h of unconsciousness. Of these one had hypothermia, shivering, flushing, and peripheral cyanosis, the **other** was confused, had slurred speech, amnesia, and ataxia. Temporary renal impairment occurred in one. Estimated vapor exposure was 10,000 ppm. Inhalation characteristically causes decrements in sensorimotor and psychomotor performance and visual acuity deficits with corneal vacuolation. A chemical of abuse in the street culture. Contact with skin causes mild irritation and may induce scleroderma. Flammable liquid. Estimated fatal dosage between 9 and 15 g. Some xylenes contain ***benzene*** (q.v.) and, thus, may adversely affect the cardiovascular system. May be teratogenic in **man** leading to caudal regression syndrome.

Inhalation TC_{LO}, **human**: 200 ppm; LD_{LO}, **human**: 50 mg/kg; LC_{LO}, **human**: 10,000 ppm/6 h.

Residues in water can cause undesirable taste in ***fish***. Teratogenic in **poultry** leading to rumpleness in **chicks** (equivalent to caudal regression syndrome defect in **infants**).

LC_{50}, **trout**: 13.5 mg/l; LD_{50}, **rat**: 4.3 g/kg.

XYLITOL

= *Eutrit* = *Kannit* = *Kylit* = *Newtol* = *Torch Oil* = *Xylite* = *Xyliton* = *Xyranit*

Untoward effects: IV leads to hyperuricemia, lactic acidosis, altered liver function, diuresis, diguria, azotemia, abdominal pain, vertigo, headache, nausea, vomiting, oliguria, and ***calcium oxalate*** crystal deposition. Oral causes diarrhea, meteorism, and flatulence. Dermatitis from skin contact.

Reports of its carcinogenicity in experimental **animals** (bladder cancer in **mice**, renal tumors in **rats**, and liver tumors in **dogs**) may have been due to impurities.

LD_{50}, **mouse**: 25.7 g/kg.

XYLOMETAZOLINE

= *Neo-Rinoleina* = *Novorin* = *Olynth* = *Otriven* = *Otrivin* = *Otrix* = *Therapin* = *Xymelin*

***Adrenergic nasal decongestant.* Topical use.**

Untoward effects: Slight nausea (2%), transient nasal irritation, burning, stinging, sneezing, dry nose, palpitations, disagreeable taste, and headache. Overdoses cause drowsiness, decreased temperature, and coma, particularly in **infants** and **children**. Rebound congestion can occur and often leads to chronic use. A 1-month-old **male** given three times **adult** dose in 1 day showed systolic blood pressure 140 mmHg, gasping respirations, and poor responsiveness.

XYLOSE

= *Wood Sugar* = *Xylomed* = *Xylopfan* = *d-Xylose*

Use: Sweetener, in diabetic foods, tanning, and dyeing; gastrointestinal function diagnostic aid.

Untoward effects: Diarrhea, epigastric discomfort, and it is cataractogenic.

The **goat's** placenta is freely permeable to it. Cataracts in **rats**, if given at high dosage for prolonged periods of time.

Y

YEAST

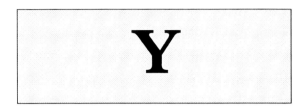

Untoward effects*:** Oral, cutaneous, vaginal, and intestinal infections with various yeasts have been common. This text concentrates primarily on yeasts for ingestion. Many yeast extracts, including the well-known British ***Marmite®, have been consumed by **children** and **adults**. They contain ***tyramine*** (q.v.) and ***histamine*** and if taken with ***monoamine oxidase inhibitors*** lead to hypertension, coma, increased temperature, headache, potentiation of ***hypoglycemic agents*** and ***antihypertensive agents***. It has been a cause of food allergy and diarrhea. Excessive drinking of home-brew ***beer*** with *Acetobacter melangenus* contamination of yeast led to erosive gastritis, nausea, vomiting, and melena. Possible ***selenium*** (q.v.) toxicity in a 17-year-old **male** and an 11-month-old **female**, both cystic fibrosis **patients**, who received 400 μg/day/2 weeks and 25 μg/day/2 months, respectively, of ***selenium*** as a yeast complex were hospitalized and had worsening of **their** pulmonary status. The **female** became unresponsive, hypotensive, and in shock, followed by death. Yeast growth consumes ***oxygen*** and two **men** lost consciousness while trying to repair a malfunctioning agitator in a 28 ft-deep vat in a ***bread*** manufacturing plant. A 36-year-old **female** had a hypersensitivity reaction, apparently from small quantities of *Saccharomyces cerevisae*, in baker's yeast-derived ***hepatitis B vaccine***. A case report of fever of unknown origin secondary to brewer's yeast ingestion reported.

YELLOW AB

= *C.I. 11,380 = C.I. Solvent Yellow 5 = Ext. D & C Yellow no.9 = FD & C Yellow no. 3*

Untoward effects*:** In the U.S., no longer considered safe for internal use, as it contains a carcinogenic impurity, ***2-naphthylamine (q.v.).

YELLOW OB

= *C.I. 11,390 = C.I. Solvent Yellow 6 = Ext. D & C Yellow no. 10 = FD & C Yellow no. 4*

Untoward effects*:** In the U.S., no longer considered safe for internal use, as it contains a carcinogenic impurity, ***2-naphthylamine (q.v.), and it can cause catharsis taken internally. Has caused a contact dermatitis when it was formerly used to color Florida ***oranges***.

YELLOW RAIN

Untoward effects*:** Evidence indicated it was a mixture of ***trichothecenes (q.v.), including ***deoxynivalenol*** (q.v.), ***diacetoxyscirpenol*** (q.v.), ***nivalenol*** (q.v.), and, in some cases, ***T-2*** (q.v.) toxin, that may have been used by the **Soviets** or their **allies** in Southeast Asia and Afghanistan. **Others** have stated it was due to massive defecation by **bees** in flight. Yet, some was supposedly found in Russian cannisters. Symptoms occurred rapidly leading to nausea, dizziness; coughing of blood-tinged material, choking; vomiting massive amounts of blood; shock, and death within minutes–hours. Iraq may have used it against **Iranians** during the Gulf War.

YEW

= *If = Taxus sp.*

Plumyews = Cephalotaxus sp.

Untoward effects*:** **Dioscorides**, a Greek military surgeon and naturalist who traveled with **Emperor Nero** during the first century AD, wrote that "the yew tree is very venemous to be taken inwardly". All parts of the plant, including the green-black seeds, but not the fleshy red cup (aril) contain toxic alkaloids [e.g. ***taxine (q.v.)]. ***T. brevifolia***, the ***Western Yew***, and ***T. baccata***, the ***English Yew*** or ***European Yew***, are two of the most popular types in the U.S. After ingestion, there is a short delay for toxic symptoms of dizziness, xerostomia, and mydriasis, followed by abdominal cramping, sialosis, vomiting, diarrhea, paleness, weakness, and cyanosis; **patient** can become convulsive and comatose. Death has occurred from cardiac and respiratory failure. A 25-year-old **male** ingested ***T. baccata*** leading to coma and arrhythmia (30–250 beats/min) twice requiring defibrillation. ***Taxine*** is rapidly absorbed after ingestion and has caused sudden death. Both fresh or dried plant parts are toxic. Occasionally induced rash reported. ***T. baccata*** leaf, bark, and seed ingestion has poisoned and killed **cattle** and **horses**. Symptoms include incoordination, staggering, trembling, mydriasis, abdominal pain, agitation, collapse, and convulsions.

In British Columbia, several **cattle** became ill and died after eating leaves and twigs of ***T. canadensis*** = ***Canada Yew***.

Taxus cuspidata = ***Japanese Yew*** is very hardy and is, therefore, found in northern U.S. and Canada, where it can also cause similar symptoms. Ingestion has killed many **cattle**, **horses**, other **animals**, and **canaries**. Some yews also contain ***ephedrine*** and ***formic acid***. Most **people** and **animals** with serious symptoms have died. Also see ***Paclitaxel*** and ***Taxotere***.

LD_{50}, *leaves*, **horses**: 2 g/kg; **cattle**: 10 g/kg; **goats**: 12 g/kg; **pigs**: 3 g/kg; or **horses** and **sheep**: 100–200 g; **cattle**: 500 g; **dogs** and **chickens**: 30 g.

YLANG YLANG OIL

= *Cananga Oil*

Fragrance. **In soaps, detergents, cosmetic creams and lotions, perfumes, and foods. In use since the 1880s; variable amounts (76,000 lbs ±) used annually in the U.S.**

Untoward effects: No irritation or sensitization noted on **human** volunteers at 10% in petrolatum.

Acute oral LD_{50} in the **rat** and acute dermal LD_{50} in the **rabbit**: > 5 g/kg.

YOGURT

= *Leben* = *Madzun* = *Yahourt* = *Yaourt* = *Yoghurt*

Untoward effects: Many commercial yogurt products are thickened with dry skim ***milk*** powder leading to excess ***lactose*** (q.v.) and ***galactose*** (q.v.) for some **people**. It is a large part of the diet in India and may be the cause of high incidence of cataracts in that country. Ingestion can trigger migraine attacks and intensify the pain of interstitial cystitis. ***Urethane*** (q.v.) is a natural product of fermentation, is carcinogenic, and present in yogurt.

Food and drug interactions: The absorption of ***ciprofloxacin*** can be decreased by its concomitant use. Contains some ***tyramine*** (q.v.) and if large quantities are consumed while taking ***monoamine oxidase inhibitors***, hypertensive crisis can develop.

An exclusive diet of yogurt for **rats** caused cataracts in all of them, probably due to the high ***galactose*** intake.

YOHIMBINE

= *Aphrodine* = *Corynine* = *Quebrachine*

***α-Adrenergic-blocker, mydriatic, serotonergic stimulant, and vasodilator.* A *cationic indole alkaloid* from *yohimbe*, the bark of *Corynanthe johimbe* (*Pausinystalia yohimba*), indigenous to the Cameroons, is chemically related to *reserpine*.**

Untoward effects: It is hallucinogenic in large doses. Despite its vasodilation, it has caused hypertension. Overdosage with 350 mg in a 38-year-old **male** with impotence led to atrial fibrillation and amnesia starting 17 h after ingestion. It can increase ***norepinephrine*** activity and cause anxiety, irritability, tremors, and nervousness. Insomnia, dizziness, nausea, incoordination, headaches, muscle cramps, palpitations, tachycardia, flushing, rhinitis, "goose-bumps", and antidiuresis have been reported. Used experimentally in psychiatric studies to induce panic attacks and anxiety states. Infusion in hot-flash-prone **women** caused a series of hot flashes; in **those *saline*-**infused, there were no hot flashes. Significant increase in salivation noted in 14 **males** and six **females** from 60 min to 180 min after administration. Toxic doses can cause central nervous system paralysis and death by respiratory paralysis.

LD_{LO}, **human**: 5 mg/kg.

LD_{LO}, **mouse**: 25 mg/kg.

YUANHAUCINE

Untoward effects: In China, it is extracted from the root and/or flower of *Daphne genkwa*, (***Yuan-Hua***), where in clinical trials on 769 **women** by intra-amniotic administration and 176 cases by intrauterine administration of 70–80 μg, it was 98% successful in inducing abortion; 97.8% with ***yuanhaudine*** at 60–70 μg/**woman** in 333 cases. ***Yuanhaufine*** at 200 μg and ***yuanhautine*** at 50 μg were also active in inducing abortion in **monkeys**. *Daphne mezereum*, or **"*Daphne*"**, a deciduous 4 ft shrub naturalized in the U.S. from Europe, has ¼ in red (and, rarely, yellow) fruit (drupe); only a few will kill a **child**, due to their ***coumarin*** glycoside and a diterpene, ***mezerein***, that cause burning of the throat and stomach, internal hemorrhage, weakness, and death. ***Mezerein*** is carcinogenic in **animals**. The powdered bark of *Daphne gnidium*, native to Mediterranean areas, is an ***abortifacient***.

YUCCA

Untoward effects: Although the flowers of ***Y. elephatipes*** are used as food in Central America, and ***Y. glauca*** roots with its foamy ***saponin*** is used in the western U.S. and Mexico as a soap, **people** or **animals** eating the roots of *Yucca* sp. in Alabama develop a severe anemia which may be fatal.

YUREMA

Untoward effects: This root extract of *Mimosa hostilis* is used by various eastern **Brazilian Indian** cults as an hallucinogen, producing wonderful visions of the spirit world, due to its *N,N*-***dimethyltryptamine*** (q.v.) content.

Z

ZAFIRLUKAST

= Accolate = ICI 204,219

Leukotriene D_4 receptor antagonist, antiasthmatic.

Untoward effects: Occasional mild headache, gastrointestinal upsets, increased aminotransferases, and many case reports of typical and atypical Churg–Strauss syndrome with vasculitis, pulmonary infiltrates, eosinophilia, and cardiomyopathy.

Drug interactions: It increases serum concentration of ***warfarin*** and causes bleeding. ***Theophylline*** serum levels can be increased high enough by it to cause toxicity and ***theophylline*** can decrease its effectiveness. By inhibiting CYP2C9 metabolic enzyme, it increases the concentration and potential toxicity of coadministered ***diclofenac, ibuprofen, losartan, piroxicam***, and ***tolbutamide***.

ZALCITABINE

Nucleoside analog.

Untoward effects: Dose-related severe and persistent peripheral neuropathy. Esophageal ulceration, fever, pancreatitis, rash, and stomatitis have also been reported.

ZALEPLON

= CL 284,846 = L 846 = LJC 10,846

***Hypnotic.* Metabolized in the liver.**

Untoward effects: Adverse effects have been mild and infrequent. These have been transient visual illusions, next-day somnolence, and anterograde amnesia. High dosage (20 mg) has caused rebound insomnia. Can be found in breast milk.

Drug interactions: ***Cimetidine*** increases its serum levels 85% and ***rifampin*** decreases them 80%.

Sudden withdrawal after high dosage in **animals** decreases fertility and pre- or postnatal survival.

ZAMIA PALM

Untoward effects: The fleshy seeds are poisonous, due to their ***azoxy glycosides***, which, when soaked and washed out, have been eaten safely.

Acute gastrointestinal toxicity often causes **cattle** deaths within a few hours. A more chronic syndrome in **cattle** leads to progressive and irreversible paralysis, especially of hind limbs (Zamia staggers) in reports from Australia.

ZANAMIVIR

= GG 167 = Relenza

***Antiviral. Neuramidase inhibitor* for inhalation to prevent influenza symptoms.**

Untoward effects: Dizziness is significantly more than with placebo. Occasionally nose and throat irritation, cough, and headache. Bronchospasm and hypoxia in **asthmatics**. Resistance to it can occur and its efficacy has been questioned.

ZANTHOXYLUM SP.

= Angelica Tree = Pellitory Bark = Prickly Ash = Sea Ash = Tsiao

Untoward effects: Many species are found in West Africa, Italy, North America, and India. At least five different ***coumarins*** have been isolated from these plants. Parts are used as food in North China from at least two species.

Z. americanum = Suterberry = Toothache Tree = Yellow Wood, a popular herbal for treating rheumatism, arthritis, and inflammation contains toxic alkaloids. It is an irritant when ingested leading to a sense of gastric heat and a tendency to diaphoresis.

ZEOLITES

Naturally occurring as *clinoptilolite* and *erionite*.

Untoward effects: Chest X-rays of Montana **workers** exposed to ***vermiculite*** ore, crystalline ***alumino-silicates***, revealed pleural changes. Turkish **researchers** showed that mesotheliomas of the chest and abdominal cavity due to it occurred in an ***asbestos***-free village of 604 **people** and caused at least 11 deaths. Experimental insertion into the lungs of **animals** revealed it to be "more potent than ***asbestos*** in producing mesothelioma".

ZEPHYRANTHES ATAMASCO

= Atamasco Lily

Native from eastern Pennsylvania to Florida in wet grasslands and woods of the Coastal Plain.

Untoward effects: The bulbs and, to a lesser extent, the leaves contain toxic alkaloids leading to staggering, bloody feces, and collapse, and is often fatal in < 48 h after ingestion by **livestock**, especially **horses**.

ZERANOL

= *MK-188* = *P-1496* = *Ralabol* = *Ralgro* = *Ralone* = *Zearalanol* = *Zerano*

Anabolic. Growth-promoter **in cattle and sheep. Used as SC implants.**

Untoward effects: Do NOT salvage implanted site for **human** or **animal** food. Implanted tissues may be harmful to **children**.

Will decrease testicular weights and libido in **bull calves**. Failure to conceive noted in 24/25 treated **cows**; vaginal or rectal prolapse and increased udder development in 255/3051, and signs of estrus noted in 11 **cows**. Withdrawal time of 2 months recommended. Interferes with sperm production in **roosters** and **rats**.

ZIDOVUDINE

= *Azidothymidine* = *AZT* = *Aztec* = *BW-A509U* = *Retrovir*

Antiviral. **The abbreviation AZT has led to some serious confusion when it was used for *azathioprine*.**

Untoward effects: ***Blood and cardiovascular***: Anemia, neutropenia, granulocytopenia, pancytopenia, and cytopenia are usually the dose-limiting toxicities; vasodilation, syncope, erythroid aplasia or hypoplasia, and cardiomyopathy. **Children** born to treated **mothers** have decreased red blood cell counts.

Central nervous system: Headache, confusion, anxiety, nervousness, depression, malaise, fatigue, vertigo, dizziness, decreased mental acuity, paresthesia, somnolence, meningo-encephalitis, myelopathy, Wernicke's encephalopathy, mood swings, mania, and seizures.

Ear and eye: Hearing loss, amblyopia, photophobia, excessive growth of eyelashes, and macular edema.

Fetus: Agenesis of right kidney in 1/46 **infants** of **mothers** given it during **their** first trimester. Pectus excavatum, atrial septal defect, and **fetal *alcohol*** syndrome in 3/47 **infants** of **mothers** given it during **their** second trimester. None noted in **infants** born to 20 **mothers** given it during **their** third trimester.

Gastrointestinal: Nausea, vomiting, gastrointestinal pain, indigestion, mouth ulcers, bleeding gums, tongue edema, sore throat, flatulence, eructation, rectal hemorrhage, constipation and diarrhea, taste perversion, and esophageal ulceration.

Miscellaneous: Lymphadenopathy, fever, myopathy, hepatitis, arthropathy, postural hypotension, non-Hodgkin's lymphoma, and increased reactions to **mosquito** bites attributed to increased T-cell function. It can induce a ***carnitine*** deficiency.

Renal: Dysuria, polyuria, pollakiuria, and urinary hesitancy.

Respiratory: Dyspnea, hoarseness, cough, pharyngitis, rhinitis, epistaxis, and sinusitis.

Skin: Acne; fingernail, skin, toenail, and oral mucosa pigmentation; rash, pruritus, urticaria, hyperhidrosis, alopecia, swollen hands, and leukocytoclastic vasculitis. Rarely, *Mycobacterium* skin infections.

Drug interactions: Risk of hematologic toxicity can increase with ***dapsone, doxorubicin, flucytosine, ganciclovir, α-interferon, vinblastine***, and ***vincristine***. Use with ***acetaminophen*** involves competition for the same metabolizing enzymes and can increase zidovudine toxicity. ***Acyclovir*** with it caused extreme fatigue and sleepiness in a **patient** and in many **others** it was associated with prolonged survival. ***Atovaquone*** increases its plasma concentration by inhibition of its glucuronidation. ***Clarithromycin*** decreases its serum concentration by ~25%. Its serum concentration is increased by ***fluconazole***. Synergistic prolongation of prothrombin time when taken with ***ibuprofen***. High-dose zidovudine is poorly tolerated in combination with ***α-interferon***, despite effective antiviral action. Severe neutropenia from zidovudine in treatment of AIDS was decreased by administration of ***lithium carbonate***. ***Probenecid*** and ***cimetidine*** decrease its renal clearance. ***Rifampin*** decreases its effectiveness by increasing its metabolism. Use with ***sulfamethazine-trimethoprim*** increases the incidence of granulocytopenia. ***Valproic acid*** inhibits its glucuronidation and increases its oral bioavailability. ***Ritonavir*** decreases zidovudine activity by 26%. May increase ***amiodarone*** serum concentration. Many other drugs have had variable responses.

Therapeutic serum levels are 0.1–0.3 mg/l and toxic levels are 2–3 mg/l.

In **monkeys**, ***probenecid*** inhibits its metabolism and also decreases its renal tubular secretion. ***Probenecid*** effects in **rabbits** are similar to those in **man**. High dosage has induced vaginal cancer in **rats** and **mice**.

ZIERIA ARBORESCENS

= *Stinkwood*

Untoward effects: Causes "panting disease" in **cattle** heavily browsing this small shrub, due to deranged permeability of the pulmonary capillary bed, causing massive pulmonary edema and, eventually, death. Enteritis also occurs. **Rabbits** are similarly affected. Other species are cyanogenetic.

ZIGADENUS sp.

= *Death Camas* = *Zygadenus*

Untoward effects: The entire plant, green or dried, and its ***onion***-like bulbs are toxic. It was a chief cause of poisoning deaths among early **settlers** in the western U.S. **Children** have been poisoned by eating the bulbs and flowers. **Adults** have also been poisoned after eating $\frac{1}{2}$–2 bulbs. Symptoms are nausea, excessive salivation, vomiting, dyspnea, prickling skin sensations, bradycardia, headache, dimness of vision, hypothermia, complete prostration, diarrhea, coma, and death due to its alkaloid, particularly ***zygacine***, content. Various species are found in different areas of the U.S. and Canada. In the western U.S., **sheep** and occasionally **cattle** and **horses** grazing it show tachypnea, salivation, weakness, violent gastrointestinal upsets, nausea, vomiting (in species that can), dyspnea, staggering, arrhythmias, weak pulse, convulsions, coma, and death after eating $\frac{1}{2}$–2 lbs.

Z. elegans = ***White Camas*** is ~$\frac{1}{7}$ as toxic as ***Death Camas***.

In Texas, ***Z. nuttallii*** is particularly toxic. Few postmortem lesions; mostly pulmonary and renal congestion.

ZILEUTON

= *A 64,077* = *Abbott 64,077* = *Leutrol* = *Zyflo*

***Antiasthmatic, anti-inflammatory.* Inhibitor of *leukotriene* synthesis.**

Untoward effects: Dyspepsia and occasional nausea, abdominal pain, and asthenia. Liver toxicity with increased alanine aminotransferase and hepatitis with jaundice requiring monitoring.

Drug interactions: Markedly increased serum concentration and decreased clearance of ***acetaminophen, propranolol, terfenadine, theophylline***, and ***warfarin***.

ZIMELIDINE

= *Normud* = *Zelmid* = *Zimeldine*

Antidepressant, selective serotonin reuptake-inhibitor.

Untoward effects: Insomnia, decreased heart rate, prolonged Q-T interval, hypersensitivity and Guillain–Barré syndrome, hepatotoxicity, "drug fever", severe erythema multiforme, and convulsions after overdose led to its withdrawal from the marketplace.

Acute toxicity in **cats**, **dogs**, **rats**, and **mice** leading to tremors, muscular rigidity, and convulsions.

LD_{50}, **rat**: 900 mg/kg; **mouse**: 600 mg/kg.

ZINC

Untoward effects: Excessive oral intake can induce ***copper*** deficiency leading to anemias; also decreases high-density lipoprotein. To hasten healing of a minor wound, a 16-year-old **male** ingested 12 g metallic zinc. **He** became lethargic and light-headed, had difficulty in writing legibly, and had a slightly staggering gait 3 days afterward with increased zinc blood levels and increased amylase and lipase values. Mass food poisoning in California occurred with nausea, vomiting, and diarrhea after storage of food and drink in zinc-lined or galvanized vessels. Acute nausea, vomiting, fever, severe anemia, and increased plasma and erythrocyte zinc concentration in a 32-year-old **female** using water stored in a galvanized tank for home hemodialysis. Nausea, abdominal cramps, vomiting, chills, headache, dizziness, and metallic taste in a group of **students** who consumed an acidic mixture of commercial fruit punch, lemonade, and ginger ale stored overnight in galvanized water containers occurred within a few minutes–24 h. Within 2 years, mass food poisoning with similar symptoms from zinc salts in 300–350 of 400 **people** in one case and 51/100 in another, stressing the need for better understanding concerning the use of galvanized containers for food. Traces of ***cadmium*** (q.v.) in the zinc of galvanized pipes can, through corrosion, enter the water supply. This is accelerated by direct attachment to ***copper*** pipes and may be decreased in areas of hard water supply. Rarely, allergic reactions from contact with it, as in U.S. pennies, can occur. Zinc allergy has occurred from injections of ***Zinc Insulin***, zinc that leached from ***rubber*** stoppers on parenteral solutions, and zinc that leached out of the metal plunger of a **dentist's** syringe into the acidic ***local anesthetic***. Zinc smelters have been a source of air pollution, adversely affecting some **animal** species. ***Zinc oxide*** fumes are produced in welding, cutting, and smelting of zinc alloys or galvanized metal, and inhalation of the fumes leads to "metal fume fever" or "zinc shakes". Symptoms usually start 3–10 h after exposure and last ~24–48 h leading to chills, fever, nausea, vomiting, throat dryness, cough, fatigue, increased salivation, myalgia, aching of head and body, blurred vision, dyspnea, râles, decreased pulmonary function, yawning, back pain, and weakness. A 25-year-old **male** burned zinc wire with an ***oxyacetylene*** torch leading to ***zinc oxide*** fumes and was misdiagnosed with viral respiratory infection. Zinc dust (fumed zinc) or powder (atomized zinc) can also cause dyspnea. Conversion of ***nitrates*** to ***nitrites*** is excessive in open zinc-galvanized water tanks. Pulmonary granulomatosis from use of ***deodorant*** aerosol spray inhalation. Young (2–5 years) **children** chewing or playing with zinc alloy toy cars became anemic (hemoglobin 7–9.5 mg/100 ml) and had urinary zinc levels 4–7 times normal until **their** metal toys were taken from them.

Toxicity often occurs in confined **animals** from chewing on cage wiring, galvanized or zinc nuts. Fatalities reported in **dogs** ingesting as few as two pennies. Pennies manufactured after 1983 contain 96% zinc. Monopoly® game pieces ingested by **dogs** contain 98% zinc. Heinz body hemolytic anemia reported in **dogs** receiving excessive zinc. Zinc water pollution has been associated with bacterial infections in **fish**. Excessive zinc is teratogenic to **frog** and **fish** embryos. In **poultry** and **avian wildlife**, diets containing > 2 g/kg diet decrease growth and > 3 g/kg diet decrease survival. **Cows** fed it at 2 g/kg of feed had decreased ***milk*** yield and feed intake after several weeks. **Foals** fed 90 mg/kg/day showed anemia, decreased weight gains, enlarged epiphysis, stiffness, and lameness. Very high doses given to female **rats** led to increased fetal deaths, fetal resorptions, and growth retardation in their **pups**. Young **pigs** are more susceptible to zinc poisoning than adults and have decreased performance, arthritis, hemorrhaging, and gastritis. Readily crosses the placental barrier in **rabbits**.

Zinc Acetate (~35% zinc) ingested as an emetic can cause toxicity if vomiting does not occur.

LD_{LO}, **human**: 50 mg/kg.

LD_{50}, **rat**: 2.5 g/kg; LD_{LO}, **rat**: 976 mg/kg.

Zinc Carbonate = ***Smithsonite*** = ***Zincspar***. **Cats** have licked off toxic quantities of it from use in skin lotions or ointments leading to typical gastroenteritis and even death. In **chickens**, at 1500 ppm in their feed, the *carbonate* was more toxic than the *oxide* or *sulfate* forms. Experimental feeding of 3000–12,000 ppm to mallard **ducklings** led to partial limb paralysis, diarrhea, and weight decrease, progressing to severe paralysis, extreme anemia, and a high mortality rate.

Zinc Chloride = ***Butter of Zinc***. Inhalation causes many of the same symptoms as with the *oxide* plus eye, skin, nose, throat, and conjunctival irritation. National Institute for Occupational Safety & Health and OSHA set upper exposure limits at 1 mg/m^3. A young **child** died after accidentally ingesting zinc chloride soldering flux. It is corrosive and used to remove squamous cell carcinomas on **humans**; has caused burns and skin lesions on contact. Inhalation has caused many symptoms (see above), including cyanosis, pulmonary fibrosis, necrosis, edema, subglottic stenosis, and bronchopneumonia. Degree and duration of exposure have been factors in death by it. Corrosive gastritis in an 18-month-old **female** after ingesting less than a cup of zinc chloride fluid; acute stage treated successfully, but subsequently led to narrowing of stomach to complete pyloric obstruction 7 weeks after ingestion, requiring partial gastrectomy with gastroduodenostomy and successful repair of gastro-colic fistula. Accidental infiltration of tonsillar area by 40 cc of 10% solution led to bilateral carotid perforation and rupture with esophageal fistula and tracheal necrotic sloughing, followed by death after 8 days. Corneal edema and some permanent scarring in two **males** after accidentally splashing onto **their** eyes. Death has followed ingestion of 1.5 oz. A **patient** has died from secondary effects of its caustic action from as little as 6 grains. Accidental, as well as suicidal, poisonings reported.

LD_{LO}, **human**: 50 mg/kg.

In **rats**, it caused erosions in the stomach and intestines and produced peritonitis. In **mouse** drinking water at 10–20 mg/l/at least 5 months led to pulmonary adenomas; mammary, uterine, bone marrow, and other tumors. Their **offspring** developed tumors more often than their **parents**.

LD_{50}, **rat** and **mouse**: 350 mg/kg; **guinea pig**: 200 mg/kg.

Zinc Chromate has caused many cases of allergic contact dermatitis in sensitized **workers** using primer ***paints***.

Chronic poisoning in young **calves** from 30–40 mg/kg/day/1 month.

Zinc Dibenzyldithiocarbamate = ***ZDC***, a ***rubber*** accelerator, has been used in manufacturing elastic products, which, after exposure to ***chlorine bleaches***, is a strong sensitizer and causes a contact dermatitis.

Zinc Diethyldithiocarbamate is also a ***rubber*** accelerator and has caused contact dermatitis from its use in ***rubber*** products.

LD_{50}, **rat**: 3.3 g/kg; **rabbit**: 570 mg/kg.

Zinc Oxide = ***Flowers of Zinc*** = ***Zincite*** = ***Zinc White***. See above for some inhalation problems. Can leach into aqueous solutions from ***rubber*** stoppers. Has been a common ingredient in ***baby powders***, with associated aspiration problems. National Institute for Occupational Safety & Health and OSHA set upper exposure limit for the fumes at 5 mg/m^3; for the dust at 15 mg/m^3, and National Institute for Occupational Safety & Health sets 15-min exposure for fumes at 10 mg/m^3.

LD_{LO}, **human**: 500 mg/kg; inhalation TC_{LO}, **human**: 600 mg/m^3.

Severe respiratory distress and acute deaths in dairy **cattle** exposed to its fumes from ***oxy-acetylene*** cutting and arc welding of galvanized pipes in a barn being remodeled. Feed levels > 500 ppm to **ferrets** caused decreased food intake and weight decrease. Levels > 3000 ppm are lethal to **ferrets** within 2 weeks. Post-mortem showed nephrosis, hypochromic anemia, and fatty infiltration of the liver.

Zinc Phosphide = ***Kilrat*** = ***Mous-con*** = ***Phosvin*** = ***Ratol*** = ***Ridall-Zinc*** = ***Rumetan*** = ***Zinc-Tox*** = ***ZP*** is used as a ***rodenticide*** and is lethal to **birds** and **mammals**. Symptoms are abdominal pain, lethargy, wheezing, anorexia, bloody vomitus, dyspnea, weakness, ataxia, diarrhea, tremors, hypoxia, pulmonary edema, convulsions, and death by 24–48 h after ingestion. During 1993–1998, 20 cases of poisoning by it of **people** in India; suicide attempts in 13/20, and accidental ingestion in 7/20. Symptoms were abdominal pain, profuse vomiting, palpitations, dyspnea, tachypnea, sweating, metabolic acidosis, hypotension, shock, and death of five. Liberates ***phosphine*** (q.v.) in contact with stomach acid. Rotten ***fish***-like or ***garlic***-like odor. Pulmonary edema and liver, kidney, and venous congestion on post-mortem. Little potential for secondary poisoning in **mammalian predators**. Symptoms in **birds** often appear within 15 min and death in 2–21 h.

LD_{50} for most species 20–40 mg/kg.

Zinc Pyrithione = ***Desquaman*** = ***Zinc Omadine*** = ***Zinc Pyridinethione*** = ***Zinc Pyrion*** = ***ZPT*** has many industrial uses and topically in hair shampoos. Very little percutaneous absorption. A 59-year-old **male** used ***Head & Shoulders® Shampoo*** containing a 2% level leading to mononeuritis multiplex of **his** left leg and right arm. Slow, incomplete recovery after discontinuing its use. Eye and rectal and vaginal mucosa irritant.

LD_{LO}, **human**: 50 mg/kg.

Muscle atrophy and hind limb weakness by 8–15th day in **rats** fed it at 166 ppm of their diet.

LD_{50}, **rat**: 92 mg/kg; **dog**: ~600 mg/kg.

Zinc Stearate inhalation has caused bronchopneumonia and pulmonary edema. **Infants** are very susceptible to inhalation of it. Can irritate eyes, skin, and upper respiratory tract with resulting coughing. National Institute for Occupational Safety & Health sets upper exposure limit at 10 mg/m^3; OSHA, 15 mg/m^3

Zinc Sulfate = ***Keratol*** = ***Optraex*** = ***Solvazinc*** = ***Solvezink*** = ***White Vitriol*** = ***Zincaps*** = ***Zincate*** = ***Zincomed*** = ***Zinc Vitriol*** = ***Z Span***; ~50,000 tons used annually in the U.S., primarily for agricultural ***fertilizers***, and ~20% for water treatment, chemical manufacturing, and electroplating. It contains ~40% zinc. **Human** U.S. RDA is 15 mg zinc. Megadoses can be toxic, interfering with ***copper*** metabolism and inducing anemia; causes nausea, vomiting, mild diarrhea, liver toxicity, and prostration. A 15-year-old **female**, ingesting 220 mg twice daily/1 week for acne developed bleeding erosive gastritis. A single oral dose of 45 g zinc as the *sulfate* led to dehydration, electrolyte imbalance, abdominal pain, nausea, vomiting, dizziness, muscular incoordination, acute renal failure, and death. In 25 cases, poisoning in eight was due to mistaking it for ***Epsom Salts***, five were suicidal, and four were homicidal. Once used frequently as an emetic; when it failed, serious poisoning ensued. Occupational "bird-eye" ulcers in zinc-plating **workers** with zinc sulfate crystals in the ulcers. In four pregnant **women** given 100 mg tid during **their** third trimester, three delivered prematurely and one gave birth to a stillborn **infant**. Inadvertent IV overdose in a 72-year-old **female** given 7.4 g over 60 h was fatal. Considerable discomfort for 6–12 h when used topically to treat herpetic keratitis. Causes marked decrease in ***tetracycline***, but not ***doxycycline***, serum levels when taken together.

TD_{LO}, **human**: 106 mg/kg.

Excessive intake by way of supplementation in a ***milk***-replacer caused toxicosis and deaths in 95 veal **calves**. Excessive intake induced hypocalcemia and bone resorption in **rats**.

LD_{50}, **rat**: 2.2 g/kg; **calves**: 3 g/kg.

ZINEB

= *Aspor* = *Dipher* = *Dithane Z-78* = *ENT-14,874* = *Hexathane* = *Kypzin* = *Lonacol* = *Parzate C* = *Polyram z* = *Tiesene* = *Tiezene Tritoftorol* = *Z-18* = *Zebtox* = *Zidan* = *Zinosan*

Fungicide.

Untoward effects: A potent allergic sensitizer for **man**. Has caused eye irritation in agricultural **workers**. Acute and chronic respiratory irritation can occur in **workers** re-entering sprayed or storage areas.

LD_{LO}, **human**: 5 g/kg.

Long-term feeding to **dogs** and **rats** caused enlarged thyroid glands; in **poultry** led to colloid thyroid goiter and various malignant tumors; and teratogenic effects in **rats** from ***ethylenethiourea***, a degradation product. Dairy **cattle** eating grass from a sprayed orchard developed a disagreeable taste to their ***milk***. **Cows** and **calves** fed zineb had decreased weight gains and decreased thyroid function. **Mallard ducks** and **pheasants** developed a goose-stepping ataxia, and general myasthenia with wings crossed over their backs.

LD_{50}, **rat**: 1–8 g/kg; **ducks** and **pheasants**: > 2 g/kg.

ZINNIA

Untoward effects: A member of the second largest plant family, *Compositae*, with ***sesquiterpene lactones*** as the major sensitizing agents. Its contact dermatitis has a characteristic clinical picture leading to a lichenoid dermatitis affecting skin not ordinarily covered by clothing almost exclusively in **males** > 40 years, often seasonal, and then may become permanent. Cross-sensitivity often exists with ***marigolds*** and ***sunflowers***.

ZINOPHOS

= *American Cyanamid 18,133* = *Cynem* = *EN-18,133* = *ENT-25,580* = *NCI-CO 2971* = *Nemafos* = *Thionazin*

Organophosphorus insecticide, nematocide, fungicide.

Untoward effects*: *Cholinesterase inhibitor in **man**, **animals**, and **birds**. Toxic in **mallard ducks** and **pheasants** leading to regurgitation, ataxia, running and falling, miosis, masseter tenseness, lacrimation, dyspnea, tremors, tetanic seizures, wing-beat convulsions, and opisthotonus within 5 min and deaths within 80 min.

LD_{50}, **male rat**: 12 mg/kg; **mallard duck**: 1.68 mg/kg; **pheasant**: 2.11 mg/kg; topical, **guinea pig**: 10 mg/kg.

ZIPEROL

= *Antituxil-Z* = *Citizeta* = *Mirsol* = *Respilene* = *Respirase* = *Zitoxil*

Antitussive.

***Untoward effects*:** Non-opioid synthetic, becoming a drug of abuse in Europe; also involved in accidental ingestion and poisoning in **children**. Convulsions and cerebral edema associated with large doses. A 26-year-old **female** with withdrawal symptoms after up to 2 g/day/6 months developed diaphoresis, diarrhea, muscle pains, anxiety, insomnia, and some claustrophobia and mild behavioral upsets.

LD_{50}, **rat**: 471 mg/kg; **mouse**: 300 mg/kg.

ZIRAM

= *Carbazine* = *Corozate* = *Cuman* = *Fuclasin* = *Fuklasin* = *Fungostop* = *Karbam White* = *Methasan* = *Methyl Cymate* = *Methyl Zimate* = *Mezene* = *Parmarsol* = *Tricarbamix Z* = *Vancide* = *Zerlate* = *Zimate* = *Zincmate* = *Zirberk* = *Ziride* = *Zitox* = *ZnDMD*

Agricultural fungicide, rubber vulcanization accelerator.

***Untoward effects*:** Has caused poisoning and skin and mucous membrane irritation in **man**.

LD_{LO}, **human**: 50 mg/kg.

Fed at 0.1% level to broiler **chickens** decreased weight gains and caused their deaths. Post-mortem showed enteritis, renal hypertrophy, bone fragility, and decreased bone calcification. Laying **hens** fed it at 0.001% level had delayed onset of egg-laying and 28% decrease in egg production.

LD_{50}, **rat**: 1.75 g/kg; inhalation, 1.23 g/kg; LD_{50}, **mouse**: 480 mg/kg.

ZIRCONIUM

Untoward effects*:** Its use in topical deodorants elicited foreign-body, granulomatous, or Type IV delayed hypersensitivity tissue reactions, especially in the axillary areas of many **people**. Its use in aerosols has a serious potential for causing an inflammatory cell reaction in the respiratory tract and possibly leads to growth of neoplasms. Extensive sarcoid-like granulomas of glabrous skin in a 15-year-old **female** after topical use of insoluble ***zirconium oxide in an ointment a few months after topical use for severe ***poison ivy*** dermatitis. Another case report of granulomatous hypersensitivity in a 16-year-old **female** using ***zirconium oxide*** in treating ***poison oak*** dermatitis. Underarm use of ***zirconium oxychloride*** by a 27-year-old **female** led to burning lesions with flat, hyperpigmented papules. Some forms of zirconium used in deodorant/antiperspirant aerosols have, after inhalation, induced lung cancer in experimental **animals**.

ZOLADEX

= *Gosarelin* = *ICI 118,630* = *NSC 606,864*

Antineoplastic, luteinizing hormone-releasing hormone agonist.

***Untoward effects*:** Transient increase in urethral obstruction and bone pain in cases of metastatic prostate cancer, hot flashes, and, eventually, impotence, testicular atrophy, and gynecomastia. After treatment with 3.6 mg SC, an 83-year-old **male** showed headache, vomiting, gradual deterioration of consciousness during the following 3 days, and diplopia on day 4. Apparently, it induced pituitary apoplexy in symptomless pituitary tumor. Alopecia may have occurred after injections in treated **females**.

ZOLMITRIPTAN

= *311C90* = *BW 311C9* = *Zomig*

Antimigraine, selective serotonin receptor agonist.

***Untoward effects*:** Narrows cardiac arteries. Can increase blood pressure and should not be used by **people** who have or had cardiac problems. Dizziness, drowsiness, nausea, headache, lack of energy, tingling of skin, and sensations of tightness and pressure in various parts of the body.

Drug interactions*:** Vasospastic effects increased or prolonged with ***monoamine oxidase A inhibitors or ***ergot***-containing drugs. ***Oral contraceptives*** and ***cimetidine*** increase its serum concentration.

ZOLPIDEM TARTRATE

= *Ambien* = *Ivadal* = *Niotal* = *SL-80 0750–23N* = *Stilnoct* = *Stilnox*

Anxiolytic, hypnotic.

***Untoward effects*:** Worldwide, ~4% of nearly 4000 **patients** discontinued treatment because of adverse effects. Its half-life is ~2 h and blood levels peak at ~30 min after ingestion. Adverse effects are dose-related and manifested primarily with central nervous system or gastrointestinal symptoms. Dizziness, light-headedness, lethargy, drowsiness, and incoordination are most common. Others include visual hallucinations, amnesia, macropsia, respiratory depression, and falls in the **elderly**. A 33-year-old **male** started with 30 mg/day and steadily increased **his** dosage to 150–280 mg/day without severe adverse effects until 60–80 mg caused **him** to be hospitalized with convulsions. Tolerance to it has been reported in **others**. Sudden withdrawal after 100 mg/day led to anxiety, nausea, sweating, tremors, and rebound insomnia. Sleepwalking in a 20-year-old **male** reported after 10 mg. Overdosage may cause nausea and vomiting. **Women** may be more susceptible than **men** to its adverse effects. Excreted in breast milk.

Drug interactions*: *Cimetidine may prolong its hypnotic effect. ***Rifampin*** decreases its elimination half-life and peak plasma concentration, decreasing its beneficial effects. Its metabolism is decreased and its toxicity potential increased when ***ritonavir*** is given with it. Use with ***central nervous system depressants*** can accent the latter's effects.

Therapeutic serum concentrations are quoted as 0.08–0.15 mg/l and toxic levels as 0.5 mg/l.

Adverse maternal effects of lethargy and ataxia in **rats** given 20–100 mg base/kg. Incomplete ossification of fetal skull bones noted. Similar findings in **rabbits**.

ZOMEPIRAC

= *Zomax* = *Zomaxin* = *Zopirac*

Analgesic. Non-narcotic non-steroidal anti-inflammatory drug.

Untoward effects*:** Despite good results in comparison with ***narcotics, many adverse effects eventually forced it out of the marketplace (with a voluntary withdrawal by McNeil Labs), viz., nausea, interstitial nephritis, proteinuria, renal failure and necrosis, drowsiness, dry mouth, toxic epidermal necrolysis, erythema multiforme, fever, lymphadenopathy, arthralgias, eosinophilia, agranulocytosis, angioedema; urinary problems (dysuria, pollakiuria, hematuria, cystitis, pyuria), especially after long-term use; decreased platelet adhesiveness and aggregation, prolonged bleeding time, disseminated intravascular coagulation, increased fecal blood loss, stomatitis, and tongue pain. A 29-year-old **female** had a sudden onset of a reaction to it that mimicked an ectopic pregnancy. Ultimately, its fate as a useful drug was destroyed when the FDA received 2200 reports of associated allergic reactions, of which 503 were classified as life-threatening, and 14 or 15 **people** may have died as a result.

Long-term feeding trials in **rats** led to increase in adrenal tumors.

ZONISAMIDE

= *AD-810* = *Aleviatin* = *CI-912* = *Exceglan* = *Exegram*

Anti-epileptic.

***Untoward effects*:** A 26-year-old **male** developed high fever, fatigue, and rash 6 weeks after ingestion. Reaction confirmed by rechallenge leading to erythema, bilateral inguinal lymphadenopathy, and increased eosinophilia. Some side-effects are common. Weight loss may persist.

Drug interactions*:** It increases ***phenobarbital serum levels.

LD_{50}, **rat**: 2 g/kg; **mouse**: 1.9 g/kg.

ZOXAZOLAMINE

= *Deflexol* = *Flexilon* = *Flexin* = *McN-485* = *Zoxamin* = *Zoxine*

Skeletal muscle relaxant, uricosuric.

***Untoward effects*:** Hypersensitivity in < 1% of **patients** led to hepatocellular jaundice and liver necrosis resembling a viral hepatitis, increased alkaline phosphatase, and fatalities. In **patients** given 500 mg tid, it caused chills, fever, burning and tearing of eyes, rash, gastric burning, nausea, dizziness, drowsiness, light-headedness, and central nervous system stimulation. **Children** given 30–140 mg/day/10–210 days showed hypertonia and anorexia. Has been nephrotoxic. Its hepatotoxicity and nephrotoxicity made it unpopular and finally discouraged its use.

TD_{LO}, **human**: 14 mg/kg/day.

Drug interactions*:** Its actions were decreased due to metabolic enzyme induction by ***phenobarbital and ***DDT***.

In **rats**, ***chlorcyclizine***, ***phenobarbital***, and ***thyroxine*** decrease its effects. ***Chlordane*** stimulates its metabolism in **rabbits** and **squirrel monkeys**. Although it increases ***uric acid*** excretion in **man**, it decreases it in **chickens**.

LD_{50}, **rat**: 375 mg/kg; **mouse**: 825 mg/kg; **hamster**: 670 mg/kg.

ZYGOPHYLLUM sp.

Untoward effects*: *Z. ammophilum ingestion associated with deaths of **cattle** in Australia, due to high ***nitrate*** (q.v.) content.

Appendix: Alternative nomenclature

7 → CARBARYL
9 → CYCLOHEXAMINE
13 → CANNABIS
106-7 → CYCLOSERINE
606 → ARSPHENAMINE
612 → ETHOHEXADIOL
1080 → FLUOROACETIC ACID
1081 → FLUOROACETAMIDE
6063 → ACETAZOLAMIDE
7744 → CARBARYL
10040 → CHLORHEXIDINE
71,285 → GEFARNATE
17/147 → AZINPHOS-METHYL
204/122 → SULFALENE
4A65 → IMIDOCARB
2601 A → CHLORPROMAZINE
295C51 → TRIPROLIDINE
311C90 → ZOLMITRIPTAN
378C48 → DIPIPANONE
566C80 → ATOVAQUONE
1497 CB → PROPIONYLPROMAZINE
1522 CB → ACEPROMAZINE
8053 CB → THIAMPHENICOL
8088 CD → VITAMIN B_1
8126 CNB → MENOTROPINS
7E3 → ABCIXIMAB
309 F → SURAMIN SODIUM
1162 F → SULFANILAMIDE
29,060 LE → VINBLASTINE
2632 RP → SULFAMERAZINE
7432 S → CEFTIBUTIN
2155 TH → SULFALENE
141W94 → AMPRENAVIR

A

A 21 → ISOPROTERENOL
A 33 → BENZALKONIUM CHLORIDE
A 66 → PHENMETRAZINE
A 101 → NORDAZEPAM
A 254 → DIPHENHYDRAMINE
A 363 → AMINOCARB
A 649
A 5610 → AZELASTINE
A 655Z → ACYCLOVIR
A 2371 → MITHRAMYCIN
A 4942 → IFOSFAMIDE
A 8103 → PIPOBROMAN
A 12,253A → TOBRAMYCIN
A 20,968 → PIPOSULFAN
A 27,053 → CHROMONAR
A 29,622 → FOSCARNET SODIUM
A 33,547 → RENOXIPRIDE
A 41,070 → LH-RH
A 41-304 → DESOXIMETASONE
A 43,818 → LEUPROLIDE
A 56,268 → CLARITHROMYCIN
A 56,619 → DIFLOXACIN
A 63,004 → TEMAFLOXACIN
A 64,077 → ZILEUTON
A 65006 → LANSOPRAZOLE
A 70,569 → TIAGABINE
A 84,538 → RITONAVIR
A77 1726 → LEFLUNOMIDE
Aacifemine → ESTRIOL
2-AAF → *N*-2-FLUORENYLACETAMIDE
AAgrunol → NABAM
Aak → CALOTROPIN
Aalu → POTATO
Aangamik 15 → VITAMIN B_{15}
Aarane → CROMOLYN SODIUM
Aaron's-Rod → VERBASCUM THAPSUS
AAT → PARATHION
AATP → PARATHION
AB 132 → METUREDEPA
ABA → ABSCISIC ACID
Aba → ANONA
ABACAVIR
Abacin → SULFAMETHOXAZOLE
Abadol → 2-AMINOTHIAZOLE
Abalgin → PROPOXYPHENE
ABALONE
Abalyn → METHYL ABIETATE
ABAMECTIN
Abaphos → TEMEPHOS
Abapresin → GUANETHIDINE
Abar → LEPTOPHOS
Abasin → ACECARBROMAL
Abat(e) → TEMEPHOS
Abathion → TEMEPHOS
Abbocillin-DC → PENICILLIN > *Benzylpenicillin, Procaine*
Abbocin → OXYTETRACYCLINE
Abboflox → PENICILLINS – "NEWER and SYNTHETICS" > *Floxacillin*
Abbokinase → UROKINASE
Abbolexin → PHENYRAMIDOL
Abboticine → ERYTHROMYCIN STEARATE
Abbott 30,400 → PEMOLINE

Abbott 35,616 → CLORAZEPATE DIPOTASSIUM
Abbott 43326 → CARTEOLOL HYDROCHLORIDE
Abbott 43,818 → LEUPROLIDE
Abbott 45,975 → TERAZOSIN
Abbott 46,811 → CEFSULODIN SODIUM
Abbott 50,711 → VALPROIC ACID
Abbott 56,619 → DIFLOXACIN
Abbott 64,077 → ZILEUTON
Abbott 84,538 → RITONAVIR
Abbott AS 18,908 → SULFALENE
Abboxapam → OXAZEPAM
Abbsa → TOLPERISONE HCL
ABCIXIMAB
Abduga → COTTON
ABECARNIL
Abelcet → AMPHOTERICIN B
Abenox → ENOXACIN
Abensanil → ACETAMINOPHEN
Aberel → VITAMIN A
ABIETIC ACID also → PAPER
Abiguanil → SULFAGUANIDINE
Abilit → SULPIRIDE
Abiocine → DIHYDROSTREPTOMYCIN
A Biol → IMIDAZOLIDINYL UREA
Abitol → DIHYDROABEITYL ALCOHOL also → MASCARA
Abnormal Leafflower → PHYLLANTHUS sp.
Abomacetin → ERYTHROMYCIN
Abricycline → TETRACYCLINE
Abrifan → RIFAMPIN
ABRIN also → ABRUS PRECATORIUS, MISTLETOE, EUROPEAN OR JAPANESE
Abrinac → PIRENZEPINE HYDROCHLORIDE
Abrodil → METHIODAL SODIUM
Abroxol → CEFIXIME
ABRUS PRECATORIUS also → ABRIN, CYANIDES, JEWELRY
Abrus à chapelet → ABRUS PRECATORIUS
Abscisin → ABSCISIC ACID
ABSCISIC ACID
Absentol → TRIMETHADIONE
ABSIDIA sp.
Absin → CASSIA
Absinthe → ABSINTHIUM also → CALAMUS, HYSSOP OIL
Absinthin → ABSINTHIUM
ABSINTHIUM also → VERMOUTH
Absolute Alcohol → ALCOHOL, ETHYL
Abstem → CALCIUM CARBIMIDE
ABT 378 → LOPINAVIR
ABT 538 → RITONAVIR
Aburamycin B → CHROMOMYCIN A
ABUTA sp.
AC 1198 → DIMETHADIONE
AC 3422 → PARATHION
AC 3810 → BAMIFYLLINE
AC 4049 → MALATHION
AC 4464 → TORSEMIDE
AC 18,682 → DIMETHOATE
AC 47,031 → PHOSFOLAN
AC 92,100 → TERBUFOS
AC 92,553 → PENDIMETHALIN
ACA → AMMONIACAL COPPER ARSENATE
7-ACA → CEFACLOR
Acabel → PIRIBENZYL METHYL SULFATE
ACACIA
Acacia berlandieri → GUAIJILLO
Acacia georginae → GIDYEA
Acadione → TIOPRONIN
Acamol → ACETAMINOPHEN
ACAMPROSATE
Acamprosate 6473 → ACAMPROSATE
Acanthaster planci → STARFISHES
Acaprin → QUINURONIUM
Acapulco Gold → CANNABIS
Acaraben → CHLOROBENZILATE
Acaracide → ARAMITE
Acaralate → CHLOROPROPYLATE
ACARBOSE also → DIGOXIN, WARFARIN
Acarexx → IVERMECTIN
Acarin → DICOFOL
Acarithion → CARBOPHENOTHION
ACC 4124 → DICAPTHON
Accelerate → ENDOTHALL
Accent → SODIUM GLUTAMATE
Accluton → PHENINDIONE
Accolate → ZAFIRLUKAST
Accothion → SUMITHION
Accupril → QUINAPRIL
Accuprin → QUINAPRIL
Accupro → QUINAPRIL
Accurbron → THEOPHYLLINE
Accutane → ISOTRETINOIN
ACEBUTOLOL also → ACARBOSE, CAPTOPRIL, NIFEDIPINE
ACECAINIDE
ACECARBROMAL
ACECLIDINE
ACECLOFENAC
Acecoline → ACETYLCHOLINE
Acecor → FOSFINOPRIL SODIUM
Acediur → CAPTOPRIL
Acedoxin → ACETYL DIGITOXIN

Acef → CEFAZOLIN SODIUM
ACEFYLLINE PIPERAZINE
Aceite de Algodon → COTTONSEED OIL
ACEMETACIN
Acemethadone → ACETYLMETHADOL
Acemix → ACEMETACIN
Acenalin → CISAPRIDE
Acenocoumarin → ACENOCOUMAROL
ACENOCOUMAROL also → ACETAMINOPHEN, AMINOGLUTETHIMIDE, AMIODARONE, CETIRIZINE, CISAPRIDE, CLOFIBRATE, CLOZAPINE, DIFLUNISAL, DOXYCYCLINE, GLUCAGON, GLUTETHIMIDE, HEPTABARBITAL, ISOCARBOXAZID, MICONAZOLE, MISOPROSTOL, NALIDIXIC ACID, NIALAMIDE, NIMESULIDE, OMEPRAZOLE, OXYMETHOLONE, PHENYLBUTAZONE, RIFAMPIN
Acenterine → ACETYLSALICYLIC ACID
Aceon → PERINDOPRIL *tert*-BUTYLAMINE
ACEPHATE also → METHAMIDOPHOS
Acephen → MECLOFENOXATE
Aceplus → CAPTOPRIL
Acepress → CAPTOPRIL
Acepril → CAPTOPRIL
ACEPROMAZINE also → ETORPHINE
Acequin → QUINAPRIL
Acer sp. → MAPLE
Acerbon → LISINOPRIL
ACESULFAME K
Acesulfame Potassium → ACESULFAME K
ACETAL
ACETALDEHYDE also → ALCOHOL, ETHYL; CHLORPROMAZINE, COPRINE, DISULFIRAM, FAMOTIDINE, METALDEHYDE, METRONIDAZOLE, NEWSPAPERS, NITROFURANTOIN, PARALDEHYDE, QUINACRINE
Acetaldehyde Diethyl Acetal → ACETAL
ACETALDEHYDE ETHYL *trans*-3-HEXENYL ACETAL
Acetalgin → ACETAMINOPHEN
ACETAMIDE
Acetamidoeugenol → ESTIL
3-Acetamido-4-hydroxybenzenearsonic Acid → ACETARSONE
4-Acetamido-4-hydroxy-2-butenoic Acid → BUTENOLIDE
3-Acetamido-4-hydroxyphenylarsonic Acid → ACETARSONE
L-α-Acetamido-β-mercaptoproprionic Acid → ACETYLCYSTEINE
5-ACETAMIDO-3-(5-NITRO-2-FURYL)-6h-1, 2,4-OXADIAZINE
2-ACETAMIDO-5-NITROTHIAZOLE
p-Acetamido-phenol → ACETAMINOPHEN
ACETAMINOPHEN also → ACENOCOUMAROL, ACETANILID, ACETYLCYSTEINE, ACETYLSALICYLIC ACID, ALBUTEROL, ANISINDIONE, BISHYDROXYCOUMARIN, BRUSSEL SPROUTS, CABBAGE, CAFFEINE, CHLORAMPHENICOL, CHLORDIAZEPOXIDE, CHLORMEZANONE, CHLOROZOXAZONE, CHOLESTYRAMINE RESIN, CIMETIDINE, DESIPRAMINE, DIETHYLENE GLYCOL, DIFLUNISAL, DIPHENADIONE, DOXORUBICIN, FENOLDOPAM MESYLATE, FUROSEMIDE, GLUCOSE, HALOTHANE, HEROIN, IMIPRAMINE, ISONIAZID, LAMOTRIGINE, METOCLOPRAMIDE, MORPHILINE, PARAQUAT, PENTAZOCINE, PHENINDIONE, PHENOBARBITAL, PHENPROCOUMON, PHENYLBUTAZONE, PHENYTOIN, PROBENECID, PROPACETAMOL, PROPANTHELINE, PROPOXYPHENE, PROPRANOLOL, PROPANTHELINE, RANITIDINE, SULFINPYRAZONE, SUMATRIPTAN SUCCINATE, THEOPHYLLINE, URIC ACID, VASOPRESSIN, VITAMIN B_2, VITAMIN C, WARFARIN, ZIDOVUDINE, ZILEUTON
p-Acetaminophenol → ACETAMINOPHEN
Acetamox → ACETAZOLAMIDE
ACETANILID also → CHLORAMPHENICOL, TETRACYCLINE
Acetanisole → *p*-METHOXY ACETOPHENONE
Acetarsol → ACETARSONE
ACETARSONE
Acetate C-8 → OCTYL ACETATE
Acetate C-9 → *n*-NONYL ACETATE
Acetate C-10 → DECYL ACETATE
Acetazine → ACEPROMAZINE
ACETAZOLAMIDE also → AMPHETAMINE, DEXTROAMPHETAMINE, FENFLURAMINE, FUROSEMIDE, GALLAMINE TRIETHIODIDE, LITHIUM, LITHIUM CARBONATE, MECAMYLAMINE, PHENOBARBITAL, PHENYTOIN, PROCAINAMIDE, QUINIDINE
Acetazoleamide → ACETAZOLAMIDE

Acetexa → NORTRIPTYLINE
Acethropan → CORTICOTROPIN
ACETIC ACID also → CHLORINE, INKS, LEAD, NEWSPAPERS, PARALDEHYDE, POTASSIUM PERMANGANATE, SODIUM HYPOCHLORITE, SOLDER
Acetic Acid Amide → ACETAMIDE
Acetic Acid Oxime → ACETOHYDROXAMIC ACID
Acetic Aldehyde → ACETALDEHYDE
ACETIC ANHYDRIDE
Acetic-4-chloranilide → ACETAMINOPHEN
Acetic Ether → ETHYL ACETATE
Acetic Oxide → ACETIC ANHYDRIDE
Aceticy → ACETYLSALICYLIC ACID
Acetilum Acidulatum → ACETYLSALICYLIC ACID
Acetin → MONACETIN
Acétiodone → SODIUM ACETRIZOATE
ACETOACETIC ACID
Acetobacter melangenus → YEAST
ACETOCHLOR
Acetogenins → PAW-PAW
ACETOHEXAMIDE also → ALLOPURINOL, FUROSEMIDE, KAOLIN, OXYPHENBUTAZONE, PHENOBARBITAL, PHENYLBUTAZONE, PROPANOLOL, RIFAMPIN, SULFAPHENAZOLE, SULFISOXAZOLE
ACETOHYDROXAMIC ACID
ACETOIN
Acetomethoxane → DIMETHOXANE
ACETONE also → ALCOHOL, ISOPROPYL; CARBON TETRACHLORIDE, GLUE, LINDANE, METFORMIN, METHIONINES, SHOES
Acetone Anil → 1,2-DIHYDRO-2,2,4-TRIMETHYLQUINOLINE
Acetonecarboxylic Acid → ACETOACETIC ACID
Acetone Chloroform → CHLOROBUTANOL
Acetone Peroxide → FLOUR
ACETONITRILE
ACETONYL ACETONE also → 5-NONANONE
Acetophen → ACETYLSALICYLIC ACID
ACETOPHENAZINE
Acetophenetidin → PHENACETIN
ACETOPHENONE
Acetopromazine → ACEPROMAZINE
Acetosal → ACETYLSALICYLIC ACID
Acetosalic Acid → ACETYLSALICYLIC ACID
Acetosalin → ACETYLSALICYLIC ACID
Acetosulfam → ACESULFAME K
Acetosulfone → ACETOSULPHONE SODIUM
ACETOSULPHONE SODIUM
2-Acetoxybenzoic Acid → ACETYLSALICYLIC ACID
3-Acetoxy-6-dimethylamino-4,4-diphenylheptane → ACETYLMETHADOL
Acetoxyl → BENZOYL PEROXIDE
Acetoxyphenyl Mercury → PHENYLMERCURIC ACETATE
17α-Acetoxyprogesterone → 17-α-HYDROXYPROGESTERONE
Acetphenarsine → ACETARSONE
Acetphenolisatin → OXYPHENISATIN
Acetylacetic Acid → ACETOACETIC ACID
ACETYL ACETONE
Acetyl Adalin → ACECARBROMAL
Acetylamide → OLIVES
2-Acetylaminofluorene → N-2-FLUORENYLACETAMIDE
[3-(Acetylamino)-4-hydroxyphenyl]-arsonic Acid → ACETARSONE
2-Acetylamino-5-nitrothiazole → 2-ACETAMIDO-5-NITROTHIAZOLE
N-Acetyl-p-aminophenol → ACETAMINOPHEN
p-Acetylaminophenol → ACETAMINOPHEN
2-Acetylamino-1,3,4-thiadiazole-5-sulfonamide → ACETAZOLAMIDE
p-Acetylanisole → *p*-METHOXY ACETOPHENONE
Acetylated Lanolin → LANOLIN
Acetylbenzene → ACETOPHENONE
Acetylcarbromal → ACECARBROMAL
ACETYL CARENE
ACETYLCHOLINE also → GF, HISTAMINE, NEOMYCIN, NEOSTIGMINE, NITROGLYCERIN, PHOSPHATES, PHYSOSTIGMINE, TROLNITRATE PHOSPHATE, POLYETHYLENE GLYCOL, POLYMIXIN B, PROCAINE, PROMETHAZINE, QUININE, RUELENE, SARIN, TABUM, TAIPOXIN, VITAMIN B_1
4-Acetyl-N[(cyclohexyl amino)-carbonyl]benzenesulfonamide → ACETOHEXAMIDE
ACETYLCYSTEINE also → ACETAMINOPHEN, RUBBER
N-Acetyl-L-cysteine → ACETYLCYSTEINE
2-Acetyl-10-(3-dimethylaminopropyl)phenothiazine → ACEPROMAZINE
ACETYL DIGITOXIN
ACETYLENE
ACETYLENE DICHLORIDE
Acetyleneogen → CALCIUM CARBIDE

Acetylene Tetrabromide → *sym*-TETRABROMOETHANE
Acetylene Tetrachloride → 1,1,2,2-TETRACHLOROETHANE
ACETYL ETHYL TETRAMETHYLTETRALYN
5-ACETYL-1,1,2,3,3,6-HEXAMETHYLINDAN
Acetylhydrazine → ISONIAZID
N-Acetyl-4-hydroxy-m-arsanilic Acid → ACETARSONE
2-Acetyl-10[3-[(β-hydroxyethyl)piperazinyl]propyl] phenothiazine Dimaleate → ACETOPHENAZINE
N-Acetylhydroxylamine → ACETOHYDROXAMIC ACID
N-[3-Acetyl-4-[2-hydroxy-3-[(1-methylethyl)amino] propoxy]-phenylbutanamide → ACEBUTOLOL
ACETYLIMIDAZOLE
Acetylin → ACETYLSALICYLIC ACID
Acetylisoniazid → ISONIAZID
N-Acetyl-3-mercaptoalanine → ACETYLCYSTEINE
ACETYLMETHADOL
Acetylmethyl Carbinol → ACETOIN
N-Acetylnovocainamide → ACECAINIDE
Acetyl Oxide → ACETIC ANHYDRIDE
2-(Acetyloxy)benzoic Acid → ACETYLSALICYLIC ACID
2-(Acetyloxy)-N,N,N-trimethylethanaminium Bromide → ACETYLCHOLINE
2-Acetyloxy-N,N,N-trimethylethanaminium Chloride → ACETYLCHOLINE
ACETYLPHENYLHYDRAZINE
N-(p-Acetylphenylsulfonyl-N´-cyclohexylurea → ACETOHEXAMIDE
1-[(p-Acetylphenyl)sulfonyl]-3-cyclohexylurea → ACETOHEXAMIDE
Acetylphosphoramidothioic Acid σ, S-Dimethyl Ester → ACEPHATE
N-Acetylprocainamide → ACECAINIDE
Acetylpromazine → ACEPROMAZINE
ACETYLPYRIDINE
Acetyl-SAL → ACETYLSALICYLIC ACID
ACETYLSALICYLIC ACID also → ACENCOUMAROL, ACETAMINOPHEN; ALCOHOL, ETHYL; *p*-AMINOSALICYLIC ACID, APSAC, ATENOLOL, BENZOIC ACID, BIRCH, BISHYDROXYCOUMARIN, CAFFEINE, CAPTOPRIL, CHERRY, CHLORPROPAMIDE, CLOFIBRATE, COAL, COFFEE, DEOXYCHOLIC ACID, DICHLOROMETHYLENE DIPHOSPHONATE, DICLOFENAC, DIPYRIDAMOLE; ESTROGENS, CONJUGATED; FENOPROFEN CALCIUM, FEVERFEW, FOLIC ACID, FUROSEMIDE, GINKO BILOBA, GLYBURIDE, GRISEOFULVIN, IBUFENAC, IBUPROFEN, IMIPRAMINE, INDOMETHACIN, INSULIN, ISOSORBIDE DINITRATE, MAGNESIUM STEARATE, MECLOFENAMIC ACID, MEFENAMIC ACID, METHOTREXATE, METHOTRIMEPRAZINE, METOCLOPRAMIDE, MILK, MINTS, MISOPROSTOL, MORPHINE, NADOLOL, NALOXONE, NAPROXEN, NORTRIPTYLINE, OXAPROZIN; PENICILLIN > *Penicillin V*; PENICILLINS – "NEWER and SYNTHETICS" > *Cloxacillin Sodium*, PHENOBARBITAL, PHENPROCOUMON, PHENYLBUTAZONE, PHENYTOIN, PIROXICAM, PREDNISONE, PROBENECID, PROPRANOLOL, SALICYL ALCOHOL, SALICYLIC ACID, SODIUM BENZOATE, SODIUM BICARBONATE, SODIUM LAURYL SULFATE, SODIUM VALPROATE, SPIRONOLACTONE, STREPTOKINASE, SULFAMETHOXYPYRIDAZINE, SULFINPYRAZONE, SULINDAC, TARTRAZINE, TENOXICAM, TICLODIPINE, TOLBUTAMIDE, VERAPAMIL, VITAMIN B_2, VITAMIN C, WARFARIN
Acetylsalicylicum → ACETYLSALICYLIC ACID
(N_1-Acetyl-6-sulfanilymetanilamido) Sodium → ACETOSULPHONE SODIUM
Acezide → CAPTOPRIL
ACHILLEA sp. also → TEA
Achless → FLUFENAMIC ACID
Achletin → TRICHLORMETHIAZIDE
Achro → TETRACYCLINE
Achromycin → TETRACYCLINE
Achromycin V → TETRACYCLINE
Acibilin → CIMETIDINE
Aciclovir → ACYCLOVIR
Acid → LYSERGIC ACID DIETHYLAMIDE
Acid Blue 1 → BLUE VRS
Acid Blue 9 → FDC BLUE #1
Acid Blue 74 → INDIGOCARMINE
Acide Valpoique → VALPROIC ACID
"Acid Heads" → LYSERGIC ACID DIETHYLAMIDE
Acid Oil → COTTONSEED OIL
Acidoser → SODIUM BICARBONATE
Acid Red 3 → STAMPS and STAMP PADS

Acid Red 87 → EOSIN(E)
Acidum Acetylsalicylicum → ACETYLSALICYLIC ACID
Acid Violet 49 → FDC VIOLET #1
Acignast → PENTAGASTRIN
Aciloc → CIMETIDINE
Acimethin → METHIONINES
Acimetten → ACETYLSALICYLIC ACID
Acinil → CIMETIDINE
Acinitrazole → 2-ACETAMIDO-5-NITROTHIAZOLE
Acino → FENFLURAMINE
Acipen-V → PENICILLIN > *Penicillin V*
ACITRETIN
Ack-Ack → HEROIN
Ackee → BLIGHIA SAPIDA
Aclovir → ACYCLOVIR
Acnegel → BENZOYL PEROXIDE
ACOCANTHERA sp.
Acocantherin → OUABAIN
Acokanthera sp. → ACOCANTHERA sp.
Acolen → DEHYDROCHOLIC ACID
Acon → VITAMIN A
ACONITE also → PICRASMA, STINGRAYS
Aconitic Acid → EQUISETUM
ACONITINE
Acortan → CORTICOTROPIN
Acorto → CORTICOTROPIN
Acorus calamus → CALAMUS
Acorus spurius → CALAMUS
Acosterina → TRIPARANOL
ACP 332 → NAPTALAM
ACPM-629 → CHLORAMBEN
ACPM-728 → CHLORAMBEN
Acquinite → CHLOROPICRIN
Acraldehyde → ACROLEIN
ACREMONIUM sp.
Acrex → DINOBUTON
Acricid → BINAPACRYL
3,6-Acridenediamine → PROFLAVINE
9-Acridinamine → AMINACRINE
ACRIDINE
ACRIFLAVINE
Acrilon → HYDROGEN CYANIDE
Acrilonitrile Butadiene Styrene → HYDROGEN CYANIDE
Acrilonitrile Styrene → HYDROGEN CYANIDE
Acrinamine → QUINACRINE
Acrinol → ETHACRIDINE LACTATE
Acriquine → QUINACRINE
Acritet → ACRYLONITRILE
Acroban → MEPROBAMATE
Acrolactine → ETHACRIDINE LACTATE
ACROLEIN also → ACETYLCYSTEINE, ALLYLAMINE, NEWSPAPERS, TURKEY
Acromonal → MOFEBUTAZONE
Acronize → CHLORTETRACYCLINE
Acrosoxacin → ROSOXACIN
Acrylaldehyde → ACROLEIN
ACRYLAMIDE
ACRYLATES and ACRYLIC ACID also → INKS
Acrylic Aldehyde → ACROLEIN
ACRYLONITRILE
Actaea sp. → BANEBERRY
Actamer → BITHIONOL
Actasal → CHOLINE SALICYLATE
ACTEA sp.
Actedron → AMPHETAMINE
Actellic → PIRIMIPHOS METHYL
Actellifag → PIRIMIPHOS METHYL
Actemin → AMPHETAMINE
Acterol → NIMORAZOLE
ACTH → CORTICOTROPIN
Acthar → CORTICOTROPIN
Actholain → TETRACOSACTIDE
Actidil → TRIPROLIDINE
Actidilon → TRIPROLIDINE
Actidione → CYCLOHEXIMIDE
Actifed-C → GLYCERYL GUAIACOLATE
Actifed Compound Linctus® → TRIPROLIDINE
Actiflo → LECITHIN
Actilase → TISSUE PLASMINOGEN ACTIVATOR
Actilin → FRAMYCETIN
Actimmune → INTERFERONS
ACTINEA
ACTINEA ODORATA
Actinea richardsoni → HYMENOXYS sp.
Actinex → MASOPROCOL
ACTINIA
ACTINIA EQUINA also → SEA ANEMONES
Actinomycin A_{IV} → DACTINOMYCIN
Actinomycin C → CACODYLIC ACID
Actinomycin C_1 → DACTINOMYCIN
Actinomycin D → DACTINOMYCIN
Actinomycin I_1 → DACTINOMYCIN
Actinomycin IV → DACTINOMYCIN
Actinomycin X → DACTINOMYCIN
Actinospectacin → SPECTINOMYCIN
Actiq → FENTANYL
Activated Carbon → CARBONS, ICE CREAM
Activated 7-Dehydrocholesterol → VITAMIN D_3
Activin → NANDROLONE
Activol → *p*-AMINOPHENOL
Actogenins → PAW-PAW
Actol → NIFLUMIC ACID
Acton → CORTICOTROPIN

Actonar → CORTICOTROPIN
Actonel → RISEDRONATE
Actosolv → UROKINASE
Actospar → SPARTEINE
Actozine → BENACTYZINE
Actybaryte → BARIUM SULFATE
Actyron → THYROTROPIN
Acuitel → QUINAPRIL
Acular → KETOROLAC
Acupan → NEFOPAM
Acutrim → PHENYLPROPANOLAMINE
Acuvel → ATOVAQUONE
ACV → ACYCLOVIR
Acycloguanosine → ACYCLOVIR
ACYCLOVIR also → CYCLOSPORINE, LITHIUM, ZIDOVUDINE
Acygoxin → DIGOXIN
Acylanid → ACETYL DIGITOXIN
Acylpyrin → ACETYLSALICYLIC ACID
AD 32 → VALRUBICIN
AD 810 → ZONISAMIDE
Adagen → PEGADEMASE BOVINE
Adalat(e) → NIFEDIPINE
Adalin → CARBROMAL
Adam → METHYLDIAMINE, METHAMPHETAMINE
1-Adamantanamine → AMANTADINE
Adamsite → PHENARSAZINE CHLORIDE
Adancor → NICORANDIL
ADAPALENE
Adapin → DOXEPIN
Adapopo → CLEMATIS sp.
Adapress → NIFEDIPINE
Adas → LENTILS
Adazine → TRIFLUPROMAZINE
Adcortyl → TRIAMCINOLONE
ADD-03055 → FELBAMATE
Adder's Wort → POLYGONUM sp.
Addex-Tham → TROMETHAMINE
Addisomnol → CARBROMAL
ADEFOVIR
Adelfa → OLEANDER
Ademetionine → S-ADENOSYLMETHIONINE
Ademin(e) → TRIAMTERENE
Adenazolam → PROBENECID
Adenex → VITAMIN C
ADENIA VOLKENSII
Adeniine → ADENIUM
Adenine Arabinoside → VIDARABINE
Adenine Riboside → ADENOSINE
ADENIUM sp.
Adenocard → ADENOSINE
Adenock → ALLOPURINOL
Adenogen → CARBAZOCHROME SALICYLATE
ADENOSINE also → CARBAMAZEPINE, THEOPHYLLINE
S-ADENOSYLMETHIONINE
Adepril → AMITRIPTYLINE
Adepsine Oil → MINERAL OIL
Adeps Lanae → LANOLIN
Adequan → MUCOITIN POLYSULFATE ESTER
Adergon → GENTIAN VIOLET
Adermine → VITAMIN B_6
Adesitrin → NITROGLYCERIN
Adfeed → FLURBIPROFEN
ADH → VASOPRESSIN
Adiab → GLYBURIDE
Adiaben → CHLORPROPAMIDE
Adiazin → SULFADIAZINE
Adifax → DEXFENFLURAMINE
ADINAZOLAM
Adipan → AMPHETAMINE
Adipex-P → PHENTERMINE
Adipic Ketone → CYCLOPENTANONE
Adipiodone → IODIPAMIDE
Adipomin → FENFLURAMINE
ADIPONITRILE
Adiposetten → NORPSEUDOEPHEDRINE
Adiuretin SD → DESMOPRESSIN
Adizem → DILTIAZEM
ADMA 6 → HEXADECYLDIMETHYLAMINE
Admire → IMIDACLOPRID
Admon → NIMODIPINE
Ad-Nil → LEVAMPHETAMINE
Adofen → FLUOXETINE
Adolph's Meat Tenderizer® → PAPAIN
Adomal → DIFLUNISAL
Ado-met → S-ADENOSYLMETHIONINE
Adonidin → ADONIS sp.
ADONIS sp.
4-ADP → 4-AMINODIPHENYL
4-ADPA → 4-AMINODIPHENYLAMINE
ADR 529 → RAZOXANE
Adran → IBUPROFEN
Adrenalex → CORTISONE
Adrenalin → EPINEPHRINE
ADRENOCHROME
Adrenocorticotropic Hormone → CORTICOTROPIN
Adrenocorticotrop(h)in → CORTICOTROPIN
Adrenomone → CORTICOTROPIN
Adrenor → NOREPINEPHRINE
Adrenosem → CARBAZOCHROME SALICYLATE
Adreson → CORTISONE
Adrestat-F → CARBAZOCHROME SALICYLATE

Adria → DOXORUBICIN
Adriacin → DOXORUBICIN
Adriamicina → METHACYCLINE
Adriamycin → DOXORUBICIN
Adrianol → PHENYLEPHRINE HCL
Adriblastina → DOXORUBICIN
Adroyd → OXYMETHOLONE
Adrucil → FLUOROURACIL
Adumbran → OXAZEPAM
Adurix → CLOPAMIDE
Aduwa → BALANITES sp.
Advacide TMP → FOLPET
Advantage → IMIDACLOPRID
Advantan → METHYLPREDNISOLONE
Advil → IBUPROFEN
Aequamen → BETAHISTINE
Aerobid → FLUNISOLIDE
Aerobin → THEOPHYLLINE
Aerohaler® → LACTOSE
Aerolate → THEOPHYLLINE
Aerolin → ALBUTEROL
Aerolone → ISOPROTERENOL
Aeromatt → CALCIUM CARBONATE
Aeropent → PENTAMIDINE ISETHIONATE
Aeroseb-Dex → DEXAMETHASONE
Aeroseb-HC → HYDROCORTISONE
Aerosil → SILICON DIOXIDE
Aerosol OT → DOCUSATE SODIUM
Aerosporin → POLYMYXIN B
Aerothene TT → 1,1,1-TRICHLOROETHANE
Aerotrol → ISOPROTERENOL
Aerrane → ISOFLURANE
Aerugipen → PENICILLINS – "NEWER and SYNTHETICS" > *Ticarcillin Disodium*
Aescin → ESCIN
Aescotrim → TRIMETHOPRIM
Aescobenzamine → TRIPELLENNAMINE
AESCULUS sp. also → HONEY, PAPAVERINE
Aescusan → ESCIN
Aethocaine → PROCAINE
Aethoform → ETHYL-*p*-AMINOBENZOATE
Aethylis Chloridum → ETHYL CHLORIDE
Aetina → ETHIONAMIDE
AETT → ACETYL ETHYL TETRAMETHYLTETRALYN
AF-438 → OXOLAMINE
AF 983 → BENDAZAC
AF-1161 → TRAZODONE
Afalon → LINURON
Afatin → DEXTROAMPHETAMINE SULFATE
Afaxin → VITAMIN A
Affirm → ABAMECTIN
Afibrin → Σ-AMINOCAPROIC ACID
AFI-Ftalyl → PHTHALYLSULFATHIAZOLE
AFI-phyllin → DYPHYLLINE
AFI-Tiazin → PHENOTHIAZINE
AFLATOXIN also → ASPERGILLUS sp., BREAD, CACAO, CHEESES, CORN, COTTONSEED MEAL, ETHYLENE DIBROMIDE, MILK, MILLET, PEANUTS, PISTACIA sp., PORK, STERIGMATOCYSTIN, TIMOTHY, WHEAT
Afloben → BENZDAMINE
Aflodac → SULINDAC
Aflorix → MICONAZOLE NITRATE
Afnor → CHLOROPHACINONE
Afoforo oyibo → MELIA sp.
Afonilum → THEOPHYLLINE
AFRAMOMUM MELEGUETA
Afrazine → OXYMETAZOLINE
African Climbing Lily → GLORIOSA
African Coffee → FLAIRA
African Milk Bush → EUPHORBIA also → SYNADENIUM GRANTII
African Redwood → MANSONIA ALTISSIMA
African Rue → PEGANUM HARMALA
Afrin → OXYMETAZOLINE
Afrinol → PSEUDOEPHEDRINE
Afsillin → PENICILLIN > *Benzylpenicillin, Procaine*
Aftate → TOLNAFTATE
Afufa OYIBO → SOLANUM sp.
Afungil → CHLORQUINALDOL
AG-3 → CHROMONAR
Ag 147 → SUMITHION
AG-1749 → LANSOPRAZOLE
Agamippe raphiduca → SPIDERS
Agaricales → MUSHROOMS
Agaricus → MUSHROOMS
AGARIN
AGARITINE
AGAVE
Agbala → MUCUNA sp.
AGE → ALLYL GLYCIDYL ETHER
Agene → METHIONINES also → NITROGEN CHLORIDE
Agenerase → AMPRENAVIR
Agent Blue → SODIUM CACODYLATE
AGENT ORANGE also → DIOXIN
Agent Orange II → DIOXIN
Agent Pink → DIOXIN
Agent Purple → DIOXIN
Agent White → PICLORAM
Agerite → MONOBENZONE
Agglutinin → ABRIN
Aggrastat → TIROFIBAN
Agiolan → VITAMIN A

Agit → DIHYDROERGOTAMINE METHANESULFONATE (MESYLATE)
Aglumin → ETHAMSYLATE
Agmatrin → FENVALERATE
AGN 190,168 → TAZAROTENE
AGN 190,342-LF → BRIMONIDINE TARTRATE
Agnin → LANOLIN
Agnolin → LANOLIN
Agol → WITHANIA SOMNIFERA
Agon → FELODIPINE
Agopton → LANSOPRAZOLE
Agotan → CINCHOPHEN
Agram → PENICILLINS – "NEWER and SYNTHETICS" > *Amoxicillin*
Agribon → SULFADIMETHOXINE
Agriferon → INTERFERONS
Agrimek → ABAMECTIN
AGRIMONIA EUPATORIA
Agrimony → AGRIMONIA EUPATORIA
Agripaume cardique → LEONURUS CARDIACA
Agritan → DDT
Agritox → METHYL CHLOROPHENOXY ACETIC ACID
AGROCLAVINE
Agroforotox → TRICHLORFON
Agromicina → TETRACYCLINE
Agrosan → PHENYLMERCURIC ACETATE
AGROSTEMMA GITHAGO also → COCKLES
Agrostemmic Acid → COCKLES
Agrothin → CYPERMETHRIN
Agrothion → SUMITHION
Agroxone → METHYL CHLOROPHENOXY ACETIC
Agrozan → PHENYLMERCURIC ACETATE
Agrylin → ANAGRELIDE
Agrypnal → PHENOBARBITAL
Aguate → AVOCADO
Ague Tree → SASSAFRAS
AH → ALUMINUM HYDROXIDE
AH 42 → METHANTHELINE BROMIDE
AH 2250 → BUPIVACAINE
AH 3365 → ALBUTEROL
AH 19,065 → RANITIDINE
AH 5158A → LABETALOL
AHA → ACETOHYDROXAMIC ACID
AHR 85 → METHOCARBAMOL
AHR 233 → MEPHENOXALONE
AHR 438 → METAXALONE
AHR 504 → GLYCOPYRROLATE
AHR 619 → DOXAPRAM
AHR 712 → BUTAPERAZINE
AHR 857 → SULFAMETER
AHR 965 → FENFLURAMINE
AHR 3018 → APAZONE
AHR 3261 → POTASSIUM CHLORIDE
AHR 10,282 → BROMFENAC
AHR 3070-C → METOCLOPRAMIDE
A-hydroCort → HYDROCORTISONE
AIA → 2-ALLYL-2-ISOPROPYLACETAMIDE
Aida → HYDROQUINONE
Aiglonyl → SULPIRIDE
Aikor Chiep → STINGRAYS
Ail → GARLIC
AILANTHUS ALTISSIMA
Aim® → CINNAMIC ALDEHYDE
AImas → METHALLIBURE
AImax → METHALLIBURE
Aimax → METHALLIBURE
Aimolin → AJMALINE
Airbron → ACETYLCYSTEINE
Airol → VITAMIN A
Airsum → SOLANUM sp.
Airum → FENOTEROL HYDROBROMIDE
Airwick® → CHLOROPHYLL
Aisemide → FUROSEMIDE
Aizumycin → BICOZAMYCIN
Ajacine → ADONIS sp.
Ajan → NEFOPAM
Ajinomoto → SODIUM GLUTAMATE
Ajmalan-17,21-diol → AJMALINE
AJMALICINE
AJMALINE
Ajo → GARLIC
Ajuca → MIMOSA
Aka-ei → STINGRAYS
Akanta → RAUWOLFIA SERPENTINA
Akar → CHLOROBENZILATE
Akaritox → TETRADIFON
Akarpine → PILOCARPINE
Akar Root → DERRIS ROOT
Akashilota → CUSCUTA
Akatinol → MEMANTINE
Ak-Chlor → CHLORAMPHENICOL
Ak-Con → NAPHAZOLINE
Ak-Dilate → PHENYLEPHRINE HCL
Akee → BLIGHIA SAPIDA
Ak-Fluor → FLUORESCEIN
Akidinmo → CROTALARIA
Akineton → BIPERIDEN
Akinobogho → FLAIRIA
Akinophyl → BIPERIDEN
Ak-Mycin → ERYTHROMYCIN
Ak-Nefrin → PHENYLEPHRINE HCL
Aknin → ERYTHROMYCIN
Aknoten → VITAMIN A
Akotin → VITAMIN B_3
Akpa → ERYTHROPHLEUM

Akpaka → BEANS, Lima
Ak-Pentolate → CYCLOPENTOLATE
Ak-Pred → PREDNISOLONE
Akpu → CASSAVA
Akrodeks → CIODRIN
Ak-Sulf → SULFACETAMIDE SODIUM
Ak-Taine → PROPARACAINE HYDROCHLORIDE
Aktamin → NOREPINEPHRINE
Ak-Tate → PREDNISOLONE
Aktedron → AMPHETAMINE
Aktil → AURANOFIN
Akwa Tears → POLYVINYL ALCOHOL
A1-100 → ALLOPURINOL
AL 0361 → HYDROXYPHENAMATE
AL1021 → CARPERONE
ALACHLOR
Ala-Cort → HYDROCORTISONE
Alamon → HYDROXYZINE
Alamycin → OXYTETRACYCLINE
Alanap → NAPTALAM
Alanex → ALACHLOR
Alanine Nitrogen Mustard → MELPHALAN
Alantroot Oil → INULA sp.
Alapril → LISINOPRIL
Alaprin → ALOXIPRIN
Alapurin → LANOLIN
Alar → DAMINOZIDE
Alarcon → GRETA also → LEAD, LEAD OXIDES
Ala Tet → TETRACYCLINE
ALATROFLOXACIN also → TROVAFLOXACIN
Albacar → CALCIUM CARBONATE
Albalon → NAPHAZOLINE
Albamix → NOVOBIOCIN
Albamycin → NOVOBIOCIN
Albego → CAMAZEPAM
Albemap → DEXTROAMPHETAMINE
ALBENDAZOLE also → THIABENDAZOLE
Albetal → LABETALOL
Albexan → SULFANILAMIDE
Albiotic → LINCOMYCIN
Albipen → PENICILLINS – "NEWER and SYNTHETICS" > *Ampicillin*
Albistat → MICONAZOLE NITRATE
ALBITOCIN
ALBIZIA sp. also → PHYLLANTHUS sp.
Alboline → MINERAL OIL
Albon → SULFADIMETHOXINE
Albucid → SULFACETAMIDE SODIUM
ALBUMIN also → PHENACEMIDE, PHENINDIONE, PHENPROCOUMON
Albustix → BROMPHENOL BLUE
ALBUTEROL also → ACETAMINOPHEN, OXYGEN, PROPRANOLOL, SODIUM METABISULFITE, THEOPHYLLINE
Albutest → BROMPHENOL BLUE
Alcaine → PROPARACAINE HYDROCHLORIDE
Alcanfor → CAMPHOR
Alchloquin → IODOCHLORHYDROXYQUINOLINE
ALCHORNEA CORDIFOLIA
ALCLOFENAC
Alcohol C-7 → HEPTANETHIOL
ALCOHOL, ALLYL
ALCOHOL, *n*-AMYL
ALCOHOL, BENZYL also → ARSENAMIDE, CHYMOPAPAIN, METHOTREXATE
ALCOHOL, *n*-BUTYL
ALCOHOL, CETYL
ALCOHOL, ETHYL also → ACENOCOUMAROL, ACETAMINOPHEN, ACETANILD, ACETYLSALICYLIC ACID, ACITRETIN, AMANTIDINE, AMITRIPTYLINE, ANILINE, ANTIPYRINE, ASPERGILLUS sp., BARBITAL, BEER, BENZTHIAZIDE, BENZYL BENZOATE, BREAD, BROMOCRIPTINE, BRUCINE, CALAMUS, CALCIUM CARBIMIDE, CANNABIS, CARBAMAZEPINE, CARBON DIOXIDE, CARBON TETRACHLORIDE, CARBUTAMIDE, CEFAMANDOLE, CEFMETAZOLE, CEFOPERAZONE, CEFOTETAN, CHLORAL HYDRATE, CHLORAMPHENICOL, CHLOROTHIAZIDE, CHLORPHENAMINE, CHLORPROMAZINE, CHLORPROPAMIDE, CHLORTHALIDONE, CLOMETHIAZOLE, CIMETIDINE, CISAPRIDE, CLONAZEPAM, CLONIDINE, CODEINE, COFFEE, COLOGNES, COPRINE, CYANAMIDE, DENATONIUM BROMIDE, DESIPRAMINE, DIAZEPAM, DIETHYLSTILBESTEROL, DIGOXIN, DIMETHYLFORMAMIDE, DIPHENHYDRAMINE, DISOPYRAMIDE, DISULFIRAM, DOXEPIN, EPINEPHRINE, ETHCHLORVYNOL, ETHINAMATE, ETHYLENE GLYCOL, ETHYLENE GLYCOL DINITRATE, FAMOTIDINE, FATS, FOLIC ACID, FURALTADONE, FURAZOLIDONE; FUSEL OIL, GIN; GINGER, JAMAICAN; GLIPIZIDE, GLUTETHIMIDE, GLYBURIDE,

GUANETHIDINE, HALOPERIDOL, HAWTHORN, HEROIN, HEXYLRESORCINOL, HYDROFLUMETHIAZIDE, HYDROXYZINE, IBUPROFEN, IMIPRAMINE, INSULIN, IRON, ISONIAZID, ISOPROTERENOL, KETOCONAZOLE, KOLA, LEVAMISOLE, LIPSTICK, LITHIUM, LITHIUM CARBONATE, LORAZEPAM, LYSERGIC ACID DIETHYLAMIDE, MECLIZINE, MECLOFENOXATE, MELATONIN, MEPHOBARBITAL, MEPROBAMATE, METHAMPHETAMINE, METHAPYRILINE, METHAQUALONE, METHOTREXATE, METHYCLOTHIAZIDE, METHYL CHLORIDE, METHYL DIAMINE, METHYL HYDRAZINE, METHYL ISOBUTYL KETONE, METHYPRYLONE, METOCLOPRAMIDE, METRONIDAZOLE, MIDAZOLAM, MILK, MONOCHLORODIFLUOROMETHANE, MONOSULFIRAM, MORPHINE, MOXALACTAM, MUSHROOMS, NALIDIXIC ACID, NIFURATEL, NIMORAZOLE, NITRAZEPAM, NITROFURANTOIN, NITROGLYCERIN, NIZATIDINE, NOREPINEPHRINE, NORTRIPTYLINE, NOSCAPINE, OXYTETRACYCLINE, PACLITAXEL, PARALDEHYDE, PARGYLINE, PENTAZOCINE, PENTOBARBITAL, PHENELZINE, PHENFORMIN, PHENICARBAZIDE, PHENOBARBITAL, PHENYTOIN, PIMOZIDE, PODOPHYLLUM RESIN, PRAVASTATIN, PROCARBAZINE, PROCHLORPERAZINE, PROPANTHELINE, PROPOFOL, PROPOXYPHENE, PROPRANOLOL, PYRAZINAMIDE, QUINACRINE, QUININE, RANITIDINE, RAUWOLFIA SERPENTINA, RESERPINE, RITONAVIR, ROSIN, RUBBER, SECOBARBITAL SODIUM, SHERRY, SOTALOL, SUCROSE OCTACETATE, SULFADIMETHOXINE, THIRAM, TIZANIDINE, TOLAZAMIDE, TOLAZOLINE HCL, TOLBUTAMIDE, VALPROIC ACID, VERAPAMIL, VINYL CHLORIDE, WARFARIN

ALCOHOL, *n*-HEXYL

ALCOHOL, HYDRATROPIC

ALCOHOL, ISOBUTYL

ALCOHOL, ISOPENTYL

ALCOHOL, ISOPROPYL also → ACETONE, CARBON TETRACHLORIDE, CHLOROXYLENOL, COLCHICUM AUTOMNALE, SELAMECTIN

ALCOHOL, METHYL also → ASPARTAME, FORMALDEHYDE; GINGER, JAMAICAN; METHYL CHLORIDE, OXAMYL CHLORIDE, STERNO®

ALCOHOL, MYRISTYL

ALCOHOL, OCTYL

ALCOHOL, PHENYLETHYL

ALCOHOL, *n*-PROPYL

ALCOHOL, STYRALLYL

Alcolec → LECITHIN

Alcopan-250 → DEXPANTHENOL

Alcopar → BEPHENIUM

Alcopon → CARAMIPHEN

ALCURONIUM also → NEOMYCIN, STREPTOMYCIN

Aldace → SPIRONOLACTONE

Aldactone → SPIRONOLACTONE

Aldara → IMIQUIMOD

Aldazine → THIORIDAZINE

Aldecin → BECLOMETHASONE

"Aldehyde" → ACETALDEHYDE

Aldehyde C-14 → γUNDECALACTONE

Aldehyde C-16 → ETHYL METHYLPHENYLGLYCIDATE

Alder Buckthorn → FRANGULA

Alderlin → PRONETHALOL

Aldesulfone → SULFOXONE

ALDESLEUKIN also → INTERLEUKINS

ALDICARB also → BANANA

Aldifen → 2,4-DINITROPHENOL

Aldinamide → PYRAZINAMIDE

Aldipin → NIFEDIPINE

Aldocorten → ALDOSTERONE

Aldocortin → ALDOSTERONE

Aldomet → METHYLDOPA

Aldomycin → NITROFURAZONE

Aldopur → SPIRONOLACTONE

ALDOSTERONE also → POTASSIUM, SPIRONOLACTONE

ALDRIN also → PARATHION

Ale → MALT

Alecrim → HOLOCALYX sp.

Ale Hoof → IVY

ALENDRONATE

Alentin → CARBUTAMIDE

Alentol → AMPHETAMINE

Alepam → OXAZEPAM

Alercrom → CROMOLYN SODIUM

Alergicide → CHLORCYCLIZINE

Alerion → CROMOLYN SODIUM

Alerlisin → CETIRIZINE
Alertec → MODAFINIL
Aleudrin → ISOPROTERENOL
Aleurites fordii → TUNG TREE
Aleurites trisperma → LUMBANG NUT
Alevaire → TYLOXAPOL
Aleve → NAPROXEN
Aleviatin → ZONISAMIDE
Alexan → CYTARABINE
Alexandria Senna → SENNA
Alexoprine Forte → ALUMINUM ACETYLSALICYLATE
Alfacillin → PENICILLINS – "NEWER and SYNTHETICS" > *Phenethicillin Potassium*
Alfacron → IODOFENPHOS
Alfadat → NIFEDIPINE
Alfadil → DOXAZOCIN MESYLATE
ALFADOLONE ACETATE
ALFALFA also → SAPONIN
Alfamox → PENICILLINS – "NEWER and SYNTHETICS" > *Amoxicillin*
Alfason → HYDROCORTISONE
Alfaspoven → CEPHALEXIN
Alfatil → CEFACLOR
Alfa-Tox → DIAZINON
Alfatrofin → CORTICOTROPIN
ALFAXALONE
Alfenta → ALFENTANIL
ALFENTANIL also → CIMETIDINE, ERYTHROMYCIN, MIDAZOLAM
Alferon → INTERFERONS
Alficetin → COLISTIN
Alficetyn → CHLORAMPHENICOL
Alflorone → FLUDROCORTISONE
Alfombrilla → DRYMARIA sp.
Alfred Hitchcock → DOMOIC ACID
Alganex → TENOXICAM
Algarroba → CAROB
Algeril → PROPIRAM FUMARATE
Algidon → METHADONE HCL
Algi-Gon → SIMAZINE
Algil → MEPERIDINE
Algimycin → COPPER SULFATE
Algin → CREAM also → SODIUM ALGINATE
Alginodia → DIPYRONE
Algistat → DICHLONE
ALGLUCERASE
Algocalmin → DIPYRONE
Algocetil → SULINDAC
Algocor → GALLOPAMIL also → BENZIODARONE
Algolysin → METHADONE HCL
Algosediv → THALIDOMIDE
Algromix → CHLORTETRACYCLINE
Algylen → TRICHLOROETHYLENE
Alhohl → LEAD
Alidase → HYALURONIDASE
Alidine → ANILERIDINE
Alimemazine → TRIMEPRAZINE
Alimet → METHIONINES
Alinam → CHLORMEZANONE
Alinamine → VITAMIN B_1
Aliporina → CEPHALORIDINE
Alisobumal → BUTALBITAL
ALITRETINOIN
Alival → NOMIFENSINE MALEATE
Alkali Disease → SELENIUM
Alkaliweed → GOLDENROD
Alkekengi → PHYSALIS sp.
Alkeran → MELPHALAN
Alkiron → METHYLTHIOURACIL
Alkron → PARATHION
ALKYL-ω-HYDROXYPOLY (OXYETHYLENE)
Alkyl-phenol Formaldehyde Resins → SHOES
Alledryl → DIPHENHYDRAMINE
Allegra → FEXOFENADINE also → TERFENADINE
Allegron → NORTRIPTYLINE
Allercorb → VITAMIN C
Allerclor → CHLORPHENIRAMINE
Allercur → CLEMIZOLE
Allergefon → CARBINOXAMINE MALEATE
Allergina → DIPHENHYDRAMINE
Allergisan → CHLORPHENIRAMINE
Allergocrom → CROMOLYN SODIUM
Allergodil → AZELASTINE
Allerkif → KETOTIFEN FUMARATE
Alleron → PARATHION
Allerplus → TERFENADINE
ALLETHRIN also → BARTHRIN; PYRETHRUM, PYRETHRINS, and PYRETHROIDS
All-Heal → VALERIANA sp.
ALLICIN
Allidochlor → *N,N*-DIALLYL-2-CHLOROACETAMIDE
Alligator Gar → LEPISOSTEUS SPATULA
Alligator Pepper → AFRAMOMUM MELEGUETA
ALLIGATOR TONGUE OIL
Allisan → 2,6-DICHLORO-4-NITROANILINE
Allium cepa → ONIONS
Allium sativum → GARLIC
Allium schoenoprasum → CHIVES
ALLIUM SCORODOPRASUM
ALLIUM TRICOCCUM

ALLIUM VALIDUM, A. CANADENSE, A. CERNEUM also → ONIONS
ALLOBARBITAL
Allobarbitone → ALLOBARBITAL
Allocaine → PROCAINE
Allocor → LANATOSIDE C
Allodene → AMPHETAMINE
Alloferin → ALCURONIUM
Allomaleic Acid → FUMARIC ACID
Allo-Puren → ALLOPURINOL
ALLOPURINOL also → ACETOHEXAMIDE, ANTIPYRINE, ATENOLOL, AZATHIOPRINE, BISHYDROXYCOUMARIN, CAPTOPRIL, CYCLOPHOSPHAMIDE, CYCLOSPORINE, ETHACRYNIC ACID, FERROUS SULFATE, IRON, MERCAPTOPURINE; PENICILLINS – "NEWER and SYNTHETICS" > *Amoxicillin, Ampicillin;* PENTOSTATIN, PHENPROCOUMON, PROBENECID, PYRAZINAMIDE, THEOPHYLLINE, VIDARABINE, WARFARIN
Allopydin → ALCLOFENAC
Allopytine → ALCLOFENAC
Allorphine → NALORPHINE
Allotropal → MEPARFYNOL
ALLOXAN
Allozym → ALLOPURINOL
all-trans-Retinol → VITAMIN A
Allspice → CALYCANTHUS FLORIDUS also → PIMENTO
Alltox → TOXAPHENE
Allural → ALLOPURINOL
Allura Red AC → FDC RED #40
Alluval → BROMISOVALUM
Allvoran → DICLOFENAC
All Weather Wood → AMMONIACAL COPPER ARSENATE
Allybarbital → BUTALBITAL
ALLYLAMINE
p-Allylanisole → ESTRAGOLE
ALLYL BROMIDE
ALLYL CAPROATE
ALLYL CHLORIDE
ALLYL CYANIDE
ALLYL CYCLOHEXYLPROPRIONATE
Allyl 3-Cycloproprionate → ALLYL CYCLOHEXYLPROPRIONATE
Allylene → METHYL ACETYLENE
ALLYLESTRANOL
ALLYL ETHER
ALLYL GLYCIDYL ETHER
Allylguaiacol → EUGENOL
Allyl Hexahydrophenylproprionate → ALLYL HEXAHYDROPHENYLPROPRIONATE
Allyl Hexanoate → ALLYL CAPROATE
dl-2-Allyl-4-hydroxy-3-methyl-2-cyclopenten-1-one → ALLETHRIN
2-ALLYL-2-ISOPROPYLACETAMIDE also → PHENOBARBITAL
ALLYL ISOTHIOCYANTE also → BRASSICA sp., BROCCOLI, BRUSSEL SPROUTS, CABBAGE, ERYSIMUM CHEIRANTHOIDES, GARLIC, HORSERADISH, MUSTARD, RADISH, THLAPSI ARVENSE
Allyl Isosulfocyanate → ALLYL ISOTHIOCYANATE
3-Allyl-2-methyl-4-oxocyclopent-2-enyl Chrysanthemate → ALLETHRIN
Allyl Mustard Oil → ALLYL ISOTHIOCYANATE
N-Allylnormorphine → NALORPHINE
[4-(Allyloxy)-3-chlorophenyl]acetic Acid → ALCLOFENAC
ALLYL PHENOXYACETATE
S-Allyl-2-propenethiosulfinate → ALLICIN
Allyl Sulfide → PETIVERA ALLIACEA
Allyl Trichloride → 1,2,3-TRICHLOROPROPANE
4-Allylveratrole → METHYL EUGENOL
Allypropymal → APROBARBITAL
Almarytm → FLECAINIDE
Almatol → SPIRONOLACTONE
ALMITRINE BISMESYLATE
Almocarpine → PILOCARPINE
ALMONDS also → APRICOT, CHERRY, CYANIDES, FRUITS and JUICES, PEACHES
ALMONDS
Alna → TAMSULOSIN
Alnide → CYCLOPENTOLATE
ALOCASIA
Alodan → MEPERIDINE
ALOE also → MILK
Aloe arborescens → ALOE
Aloe barbadensis → ALOE VERA
ALOE VERA
Aloe Wood → 4-(*p*-METHOXYPHENYL)BUTAN-2-ONE
Aloginan → CLEMASTINE
ALOIN
Aloinum
Alokain → PROCAINE
Alomide → LODOXAMIDE TROMETHAMINE
Aloperidin → HALOPERIDOL

Alopexil → MINOXIDIL
Alopresin → CAPTOPRIL
Aloral → ALLOPURINOL
ALOSETRON
Alositol → ALLOPURINOL
Alostil → MINOXIDIL
Alotec → METAPROTERENOL
ALOXIPRIN
ALOYSIA LYCIOIDES
Alpen → PENICILLINS – "NEWER and SYNTHETICS" > *Ampicillin* also > *Phenethicillin Potassium*
alphaCobione → VITAMIN B_{12}
Alphacillin → PENICILLINS – "NEWER and SYNTHETICS" > *Pivampicillin*
Alphadolone Acetate → ALFADOLONE ACETATE
Alphadrol → FLUPREDNISOLONE
Alphagan → BRIMONIDINE TARTRATE
Alphakil → CHLORALOSE
Alphalin → VITAMIN A
Alphamin → CLEMASTINE
Alphamine → MIDODRINE
ALPHAPRODINE also → CHLORPROMAZINE, DIAZEPAM
alphaRedisol → VITAMIN B_{12}
Alpha-Ruvite → VITAMIN B_{12}
Alphaxalone → ALFAXALONE
Alpine Violet → CYCLAMEN
Alpiny → ACETAMINOPHEN
Alplax → ALPRAZOLAM
Alplucine → JOSAMYCIN
ALPRAZOLAM also → CARBRAMAZEPINE, DIGOXIN, ERYTHROMYCIN, FLUOXETINE, FLUVOXAMINE, HALOPERIDOL, ITRACONAZOLE, KAVA, KETOCONAZOLE, LITHIUM, MELTONIN, PROPOXYPHENE, THEOPHYLLINE
ALPRENOLOL also → NIFEDIPINE, PENTOBARBITAL
Alpress LP → PRAZOSIN
Alprim → TRIMETHOPRIM
ALPROSTADIL
Alredase → TOLRESTAT
Alrheumat → KETOPROFEN
Alrheumun → KETOPROFEN
Alsadorm → DOXYLAMINE SUCCINATE
Alsanate → GEFARNATE
Alseroxylon → RAUWOLFIA SERPENTINA
ALSTROEMERIA LIGTU
Altabactina → FURALTADONE
Altace → RAMIPRIL
Altafur → FURALTADONE
Altan → DANTHRON
Altat → ROXATIDINE ACETATE
Altepin → DOTHIEPIN
Alteplase → PLASMINOGEN also → RETEPLASE, TISSUE PLASMINOGEN ACTIVATOR
Altex → SPIRONOLACTONE
Altiazem → DILTIAZEM
Alticarb → ALDICARB
Altobatus narimari → STINGRAYS
Altocid → METHOPRENE
Altodor → ETHAMSYLATE
ALTRETAMINE
Alubasa kwadil → MORAEA sp.
Aludrin(e) → ISOPROTERENOL
Aluline → ALLOPURINOL
Alumina → ALUMINUM OXIDE
Aluminium → ALUMINUM
ALUMINUM → CALCIUM, DAUNORUBICIN, ERYTHROPOIETIN, ETHYLENE CHLOROHYDRIN, FLUORIDES and FLOURINE, GREPAFLOXACIN, HYDROFLUORIC ACID, LIPASE, MERCURY, METHACYCLINE, METRONIDAZOLE, NAPHAZOLINE, NITROGLYCERIN, NORFLOXACIN, OFLOXACIN, PARATHYROID HORMONE, PHOSPHORIC ACID, PITCH, RHIZOPUS sp., SILICON TETRAFLUORIDE, SUCRALFATE, SULFURIC ACID, TETRACYCLINE, THIMEROSAL
ALUMINUM ACETATE
ALUMINUM ACETYLSALICYLATE
ALUMINUM AMMONIUM SULFATE
Aluminum Aspirin → ALUMINUM ACETYLSALICYLATE
Aluminum Hydrate → ALUMINUM HYDROXIDE
ALUMINUM HYDROXIDE also → ATENOLOL, CHLORDIAZEPOXIDE, CHLORPROMAZINE, CYCLOSPORINE, DEMECLOCYCLINE, ETHAMBUTOL, INDOMETHACIN, LOMEFLOXACIN, MAGNESIUM, MINOCYCLINE, NITROFURANTOIN, PENICILLAMINE, PHENYTOIN, POLYSTYRENE, PREDNISONE, PROPRANOLOL, SUCRALFATE, THYROXINE, TROVAFLOXACIN
ALUMINUM MONOSTEARATE also → PENICILLIN and PENICILLIN > *Benzylpenicillin, Procaine*
ALUMINUM NICOTINATE
ALUMINUM NITRATE

ALUMINUM OXIDE
ALUMINUM PHOSPHATE also → DOXYCYCLINE, NITROFURANTOIN
ALUMINUM POTASSIUM SULFATE
ALUMINUM SILICATE also → MEXILETINE
ALUMINUM SULFATE also → SALMON
Aluminum Trihydrate → ALUMINUM HYDROXIDE
ALUMINUM TRISILICATE also → BETAMETHASONE
Alunex → CHLORPHENIRAMINE
Alunitine → ALUMINUM NICOTINATE
Alupent → METAPROTERNEOL also → LECITHIN
Alurate → APROBARBITAL
Alurene → CHLOROTHIAZIDE
Alutyl → CINCHOPHEN
Alvedon → ACETAMINOPHEN
Alveld → NARTHECIUM
Alven → TRIBENOSIDE
ALVERINE
Alvo → OXAPROZIN
Alvodine → PIMINODINE
Alvonal MR → CYMARIN
Alvora → DIBROM
Alyrane → ENFLURANE
Alysine → SODIUM SALICYLATE
Alzogur → CYANAMIDE
AM 715 → NORFLOXACIN
AM 725 → PEFLOXACIN
AM 833 → FLEROXACIN
Amabevan → CARBARSONE
Amadil → ACETAMINOPHEN
Amal → AMOBARBITAL
Amamelide → HAMAMELIS VIRGINIANA
Amanin → AMANITA sp.
AMANITA sp. also → GALERINA sp., IBOTENIC ACID, MORCHELLA DELICIOSA, MUSCAZONE, PHALLOIDIN
α and β-AMANITINS also → AMANITA sp., GALERINA sp., GYROMITRA, MORCHELLA DELICIOSA, MUSHROOMS, THIOCTIC ACID
AMANTADINE also → BUPROPION, L-DOPA, MILK, PHENELZINE, QUINIDINE, QUININE
Amantilla → VALERIANA sp.
AMARNATH also → GRAPE, TETRACYCLINE
AMARANTHUS
Amarbel → CUSCUTA
AMARETTO
Amarga → ACHILLEA sp.
Amaroral → PAROMOMYCIN
Amarsan → ACETARSONE
Amaryl → GLIMEPIRIDE
AMARYLLIS
Amasust → AMOBARBITAL
Amatine → MIDODRINE
Amaze → ISOFENPHOS
Ambacamp → PENICILLINS – "NEWER and SYNTHETICS" > *Bacampicillin*
Ambamide → MAFENIDE
Ambaxin → PENICILLINS – "NEWER and SYNTHETICS" > *Bacampicillin*
AMBELANIA
Amben → *p*-AMINOBENZOIC ACID
Ambene → PHENYLBUTAZONE
AMBENONIUM
Amber → HYPERICUM
Ambien → ZOLPIDEM TARTRATE
Ambilhar → NIRIDAZOLE
Ambinon → GONADOTROPINS, CHORIONIC
Amblosin → PENICILLINS – "NEWER and SYNTHETICS" > *Ampicillin*
AMBLYGONOCARPUS ANDOGENSIS
Amboclorin → CHLORAMBUCIL
Ambodryl → BROMODIPHENHYDRAMINE
Ambosol → MAGNESIUM SILICATE
Ambox → BINAPACRYL
Ambracyn → TETRACYCLINE
Ambramicina → TETRACYCLINE
Ambramycin → TETRACYCLINE
AMBRETTOLIDE
Ambrocef → CEPHAPIRIN
Ambrosia sp. → RAGWEED
Ambroxol → CEFIXIME
AMBUPHYLLINE
Ambush → ALDICARB
Ambutyrosin → BUTIROSIN
AMCA → TRANEXAMIC ACID
AMCHA → TRANEXAMIC ACID
Amchem 66-206 → CHLORAMBEN
Amchlor → AMMONIUM CHLORIDE
Amcide → AMMONIUM SULFAMATE
Amcidern → AMCINONIDE
Amcill → PENICILLINS – "NEWER and SYNTHETICS" > *Ampicillin*
Amcillin → PENICILLINS – "NEWER and SYNTHETICS" > *Ampicillin*
AMCINONIDE
Am Cyan 47,300 → SUMITHION
Amdax → PHTHALOPHOS
Amdinocillin → PENICILLINS – "NEWER and SYNTHETICS"
Amdinocillin Pivoxil → PENICILLINS – "NEWER and SYNTHETICS"

Amdon → PICLORAM
AMDRO
AME → REFUIN
Amebacilin → FUMAGILLIN
Ameban → CARBARSONE
Amebarsone → CARBARSONE
Amebil → IODOCHLORHYDROXYQUINOLINE
Amechol → METHACHOLINE CHLORIDE
Amedel → PIPOBROMAN
Amekrin → AMSACRINE
AMELANCHIER ALNIFOLIA
Amen → MEDROXYPROGESTERONE ACETATE
Amer Cyan 3911 → PHORATE
Amer Cyan 52,160 → TEMEPHOS
Americaine → ETHYL-*p*-AMINOBENZOATE
American Arborvitae → THRELKELDIA PROCIFLORA
American Citizens for Honesty In Gov't → QUINUCLIOINYL BENZILATE
American Cyanamid 12,880 → DIMETHOATE
American Cyanamid 18,133 → ZINOPHOS
American Cyanamid 38,023 → FAMPHUR
American Elder → ELDERBERRY
American Hellebore → VERATRUM sp.
American Hemp → APOCYNUM CANNABINUM
American Ivy → IVY
American Mandrake → PODOPHYLLUM
American Pepper → PEPPERS
American Plum → PLUM
American Wormseed → CHENOPODIUM
Amethocaine → TETRACAINE HYDROCHLORIDE
A-Methopterin → METHOTREXATE
Amethopterin → METHOTREXATE
Ametop → TETRACAINE HYDROCHLORIDE
Ametox → SODIUN THIOSULFATE
Ametrex → AMETRYN
AMETRYN
Ametycine → NITOMYCIN
AMEZINIUM METHYL SULFATE
Amezinium Metilsulfate → AMEZINIUM METHYL SULFATE
Amfamox → FAMOTIDINE
AMFEPENTOREX
Amfepramone → DIETHYLPROPION
d-Amfetasul → DEXTROAMPHETAMINE SULFATE
Amfipen → PENICILLINS – "NEWER and SYNTHETICS" > *Ampicillin*
Amfomycin → AMPHOMYCIN
Amforal → KANAMYCIN
Amiben → CHLORAMBEN
Amibiarson → CARBARSONE
Amibinon → THYROTROPIN
Amibufen → IBUPROFEN
Amicar → Σ-AMINOCAPROIC ACID
AMICARBALIDE
Amicardine → KHELLIN
Amicazole → DIAMTHAZOLE
Amicos → CLEBOPRIDE
Amidate → ETOMIDATE
Amidazine → ETHIONAMIDE
Amidazophen → AMINOPYRINE
AMIDEPHRINE MESYLATE
Amidionophydrazone → AMDRO
AMIDOAZOTOLUENE
Amidocyanogen → CYANAMIDE
Amidofebrin → AMINOPYRINE
Amidolacetate → ACETYLMETHADOL
Amidon → METHADONE HCL
Amidonal → APRINDINE
Amidone → DIPIPANONE
Amidopyrazoline → AMINOPYRINE
Amidopyrine → AMINOPYRINE
Amid-Sal → SALICYLAMIDE
Amidryl → DIPHENHYDRAMINE
AMIFOSTINE
Amifur → NITROFURAZONE
Amigen → PROTEIN HYDROLYSATES
Amiglycin → AMIKACIN
Amiglyde → AMIKACIN
Amigol → AMINOTRIAZOLE
AMIKACIN also → DIMETHYL TUBOCURARINE IODIDE, INDOMETHACIN; PENICILLINS – "NEWER and SYNTHETICS" > *Ticarcillin Disodium*, SUCCINYLCHOLINE
Amikal → AMILORIDE
Amikin → AMIKACIN
AMILORIDE also → FUROSEMIDE, HYDROCHLOROTHIAZIDE, LITHIUM, POTASSIUM, QUINIDINE
Amimycin → OLEANDOMYCIN
AMINACRINE
Aminarson(e) → CARBARSONE
Aminazine → CHLORPROMAZINE
AMINEPTINE
Aminicotin → VITAMIN B_3
Aminitrozole → 2-ACETAMIDO-5-NITROTHIAZOLE
AMINOACETONITRILE
5-Aminoacridine → AMINACRINE
9-Aminoacridine → AMINACRINE
Aminoacrine → AMINACRINE
1-Aminoadamantane → AMANTADINE

N-[[5-Amino-2-[(4-aminophenyl)-sulfonyl]-phenyl]sulfonyl]acetamide Monosodium Salt → ACETOSULPHONE SODIUM
4-Aminoaniline → *p*-PHENYLENEDIAMINE
2-AMINOANTHRACINE also → 2-ANTHRAMINE
2-AMINOANTHRAQUINONE
4-Aminoantipyrine → AMPYRONE
σ-AMINOAZOTOLUENE
Aminobenzene → ANILINE
α-Aminobenzene → PENICILLINS – "NEWER and SYNTHETICS" > *Ampicillin*
p-Aminobenzenearsonic Acid → ARSANILIC ACID
p-Aminobenzene Sulfonic Acid → SULFANILIC ACID
p-AMINOBENZOIC ACID also → BUCLOSAMIDE, CHLORPROPAMIDE, ETHYL-*p*-AMINOBENZOATE, HYDROFLUMETHIAZIDE, METHOTREXATE, PENICILLIN, *p*-PHENYLENEDIAMINE, PROCAINE, PYRIMETHAMINE, RIFAMPIN, SALICYLIC ACID, SULFANILAMIDE, TETRACAINE HYDROCHLORIDE, TOLBUTAMIDE
4-Aminobenzoic Acid → *p*-AMINOBENZOIC ACID
p-AMINOBENZYL CAFFEINE HYDROCHLORIDE
p-AMINOBIPHENYL also → *p*-BIPHENYLAMINE
4-Aminobiphenyl → *p*-AMINOBIPHENYL
5-Amino-3,4´-bipyridine-6(1H)-one → AMRINONE
2-Aminobutane → *sec*-BUTYLAMINE
4-Aminobutanoic Acid → γ-AMINOBUTYRIC ACID
γ-AMINOBUTYRIC ACID
Σ-AMINOCAPROIC ACID
Aminocaproic Lactam → CAPROLACTAM
AMINOCARB
Aminocardol → AMINOPHYLLINE
4-AMINOCATECHOL
7-Aminocephalosporanic Acid → CEFACLOR
σ-Aminochlorobenzene → σ-CHLOROANILINE
σ-3-Amino-3-deoxy-α-D-glucopy-ranosyl-(1→6) – σ-[6-amino-6-deoxy-α-D-glucopyranosyl-(1→4)] N[1]-(4-amino-2-hydroxy-1-oxobutyl)-2-deoxy-D-streptamine → AMIKACIN
1-Aminodiamantane → AMANTADINE
2-AMINO-5-DIETHYLAMINOTOLUENE
2-Amino-1,9-dihydro-9-[(2-hydroethoxy)methyl]-6H-purin-6-one → ACYCLOVIR
4-Amino-2´,3´-dimethylazobenzene → σ-AMINOAZOTOLUENE
2-AMINO-3,4-DIMETHYLIMIDAZO(4,5-F)QUINOLINE
4-AMINODIPHENYL also → DIPHENYLAMINE, RUBBER
4-AMINODIPHENYLAMINE
2-AMINODIPYRIDOL(1,2-α-3´,2´-d) IMIDAZOLE
Aminodur → AMINOPHYLLINE
Aminoethane → ETHYLAMINE
2-Aminoethanol → ETHANOLAMINE
β-Aminoethyl Alcohol → ETHANOLAMINE
4-(-2-AMINOETHYL)-β-DIAZO-2,4-CYCLOHEXADIENONE
Aminoethylene → ETHYLENEIMINE
2-AMINO-2-ETHYL-1,3-PROPANEDIOL
4-Aminofolic Acid → AMINOPTERIN
Aminoform → METHENAMINE
AMINOGLUTETHIMIDE also → ACENOCOUMAROL
1-2-Amino-5-guanidinovaleric Acid → ARGININE
4-Amino-5-hexenoic Acid → VIGABATRIN
6-Aminohexanoic Acid → Σ-AMINOCAPROIC ACID
p-AMINOHIPPURIC ACID also → DIAZEPAM
2-Amino-1-hydroxybenzene → σ-AMINOPHENOL
3-Amino-1-hydroxybenzene → *m*-AMINOPHENOL
4-Amino-1-hydroxybenzene → *p*-AMINOPHENOL
4-Amino-2-hydroxybenzoic Acid → *p*-AMINOSALICYLIC ACID
1-N-[L(—)-4-Amino-2-hydroxy-butyryl]kanamycin A → AMIKACIN
Aminokrovin → PROTEIN HYDROLYSATES
Aminomercuric Chloride → MERCURY, AMMONIATED
Aminomethane → METHYLAMINE
2-AMINO-6-METHYLDIPYRIDOL(1,2-α:3´,2´-d)IMIDAZOLE
2-AMINO-3-METHYLIMIDAZ(4,5-f) QUINOLINE
2-AMINO-2-METHYL-1-PROPANOL
2-AMINO-3-METHYL-9H-PYRIDO(2,3-b) INDOLE
Aminomux → PAMIDRONATE DISODIUM
1-Aminonaphthalene → 1-NAPHTHYLAMINE
2-Aminonaphthalene → 2-NAPHTHYLAMINE
1-Amino-2-naphthol → 2-NAPHTHYLAMINE
Aminonat → PROTEIN HYDROLYSATES
AMINONITROTHIAZOLE
α-AMINO-β-OXALYLAMINOPROPIONIC ACID
AMINOOXYACETIC ACID

Aminopan → SOMATOSTATIN
6-AMINOPENICILLANIC ACID
AMINOPENTAMIDE HYDROGEN SULFATE
4-Amino-PGA → AMINOPTERIN
Aminophen → ANILINE
Aminophenazone → AMINOPYRINE
m-AMINOPHENOL
σ-AMINOPHENOL
p-AMINOPHENOL also → ORANGE G, PARATHION, SUDAN I
2-Aminophenol → σ-AMINOPHENOL
2-(p-Aminophenyl)-2-ethylglutarimide → AMINOGLUTETHIMIDE
3-(4-Aminophenyl)-3-ethyl-2,6-piperidinedione → AMINOGLUTETHIMIDE
2-Amino-1-phenylpropane → AMPHETAMINE
AMINOPHYLLINE also → ACETALDEHYDE, DILTIAZEM, ETHYLENEDIAMINE, FUROSEMIDE, KETAMINE, LITHIUM, PROPRANOLOL, SUCRALFATE, TERBUTALINE, VANCOMYCIN
AMINOPROMAZINE
2-Aminopropane → ISOPROPYLAMINE
Aminopropazine → AMINOPROMAZINE
3-Aminopropene → ALLYLAMINE
β-AMINOPROPIONITRILE also → LATHYRUS sp., NITROFURAZONE, PRACTOLOL, SOTALOL
β-Aminopropylbenzene → AMPHETAMINE
3-Aminopropylene → ALLYLAMINE
AMINOPTERIN also → LACTOSE
4-Aminopteroylglutamic Acid → AMINOPTERIN
AMINOPURINE also → CHLORAMPHENICOL
AMINOPYRIDINE
AMINOPYRINE also → CHLORDANE, MEPHENOXALONE, *N*-NITROSODIMETHYLAMINE, ORPHENADRINE, PHENOBARBITAL, PHENYLBUTAZONE, TARTRAZINE, TETRACYCLINE
Aminoquin → PAMAQUIN
8-Aminoquinolones → QUINACRINE
6-Amino-9-β-D-ribofuranosyl-9H-purine → ADENOSINE
p-AMINOSALICYLIC ACID also → ACETYLSALICYLIC ACID, *p*-AMINOBENZOIC ACID, BENTONITE, DIPHENHYDRAMINE, FATS, FOLIC ACID, HEXOBARBITAL, IODINE, ISONIAZID, PHENYLENEDIAMINE, PHENYTOIN, POTASSIUM, PROBENECID, STREPTOMYCIN, VITAMIN B_{12}, VITAMIN K, WARFARIN
5-AMINOSALICYLIC ACID also → SALICYLAZOSULFAPYRIDINE
Aminosidine → PAROMOMYCIN
Aminosol → PROTEIN HYDROLYSATES
Aminosuccinic Acid → l-ASPARTIC ACID
N-[5-(Aminosulfonyl)-1, 3,4-thiadiazol-2-yl]acetamide → ACETAZOLAMIDE
2-AMINO-1,3,4-THIADIAZOLE
2-AMINOTHIAZOLE
Aminotoluene → TOLUIDINE
3-Amino-1,2,4-triazole → AMINOTRIAZOLE
Aminoxidin → PAROMOMYCIN
Amiodar → AMIODARONE
AMIODARONE also → ACENOCOUMAROL, CLONAZEPAM, CYCLOSPORINE, DIGOXIN, FLECAINIDE, INDINAVIR, LAMIVUDINE, PHENYTOIN, PREDNISOLONE, PROCAINAMIDE, QUINIDINE, RIFAMPIN, RITONAVIR, SPARFLOXACIN, THEOPHYLLINE, WARFARIN, ZIDOVUDINE
Amipaque → METRIZAMIDE
Amipenix → PENICILLINS – "NEWER and SYNTHETICS" > *Ampicillin*
Amiperm → AMITRIPTYLINE
AMIPHENAZOLE
Amipramidin(e) → AMILORIDE
Amipramizide → AMILORIDE
Amipress → LABETALOL
Amisalin → PROCAINAMIDE
Amitacon → BENACTYZINE
Amithiozone → THIACETAZONE
Amitid → AMITRIPTYLINE
Amiton → TETRAM®
AMITRAZ
Amitrid → AMILORIDE
Amitril → AMITRIPTYLINE
AMITRIPTYLINE also → AMOBARBITAL, BETHANIDINE, BISHYDROXYCOUMARIN, DIAZEPAM, FLUOXETINE, GUANETHIDINE, METHYLDOPA, PAROXETINE, PERPHENAZINE, QUINIDINE, SERTRALINE, SODIUM VALPROATE, SUCRALFATE, THIORIDAZINE, TRANYLCYPROMINE, VALPROIC ACID
Amitrole → AMINOTRIAZOLE
Amizil → BENACTYZINE
Amizol → AMINOTRIAZOLE
Amizyl → BENACTYZINE
AMLEXANOX
AMLODIPINE also → DILTIAZEM, GRAPEFRUIT

Ammate → AMMONIUM SULFAMATE
AMMELINE
Ammicardine → KHELLIN
Ammidin → METHOXSALEN
AMMI MAJUS also → METHOXSALEN
Ammipuran → KHELLIN
Ammivin → KHELLIN
AMMI VISNAGA
Ammivisnagen → KHELLIN
Ammo → CYPERMETHRIN
Ammoform → METHENAMINE
Ammoidin → METHOXSALEN
Ammomycin → OLENADOMYCIN
AMMONIA also → BEER, CALCIUM CARBIMIDE, CHLORINE, CUPFERON, 4-METHYLIMIDAZOLE, MUSHROOMS, NITROUS OXIDE, POTASSIUM OXALATE, PROTEIN HYDROLYSATES, SILK, SODIUM HYPOCHLORITE, SULFURIC ACID, WOOL
AMMONIACAL COPPER ARSENATE
AMMONIUM CHLORIDE also → ANTIPYRINE, CHLOROQUINE, DEXTROAMPHETAMINE, FENFLURAMINE, METHENAMINE, NALIDIXIC ACID, PROCAINE, QUININE, SALICYLIC ACID, THEOPHYLLINE, TIN
Ammonium Chloroplatinate → AMMONIUM PLATINOUS CHLORIDE
AMMONIUM FLUORIDE
AMMONIUM HEXACHLOROPLATINATE
Ammonium Monofluoride → AMMONIUM FLUORIDE
Ammonium Muriate → AMMONIUM CHLORIDE
AMMONIUM NITRATE
AMMONIUM OXALATE
Ammonium Peroxydisulfate → AMMONIUM PERSULFATE
AMMONIUM PERSULFATE also → FLOUR, HAIR
Ammonium Platinic Chloride → AMMONIUM HEXACHLOROPLATINATE
AMMONIUM PLATINOUS CHLORIDE
AMMONIUM SULFAMATE
AMMONIUM SULFATE also → SULFUR DIOXIDE
Ammonium Tetrachloroplatinate → AMMONIUM PLATINOUS CHLORIDE
Ammophyllin → AMINOPHYLLINE
Amnizol(e) → AMINONITROTHIAZOLE
AMOBARBITAL also → IMIPRAMINE, NEOMYCIN, STREPTOMYCIN, WARFARIN
Amoben → CHLORAMPBEN
Amocilline → PENICILLINS – "NEWER and SYNTHETICS" > *Amoxicillin*
Amoclen → PENICILLINS – "NEWER and SYNTHETICS" > *Amoxicillin*
Amodex → PENICILLINS – "NEWER and SYNTHETICS" > *Amoxicillin*
AMODIAQUIN
Amoenol → IDOCHLORHYDROXYQUINOLINE
Amokin → CHLOROQUINE
Amoksiklav → PENICILLINS – "NEWER and SYNTHETICS" > *Amoxicilll*
Amolin → PENICILLINS – "NEWER and SYNTHETICS" > *Amoxicillin*
Amopenixin → PENICILLINS – "NEWER and SYNTHETICS" > *Amoxicillin*
AMOPHENONE
AMOPYROQUIN
Amoram → PENICILLINS – "NEWER and SYNTHETICS" > *Amoxicillin*
Amorphan → NORPSEUDOEPHEDRINE
AMORPHOPEALLIS
Amosene → MEPROBAMATE
Amosyt → DIMENHYDRINATE
Amotril → CLOFIBRATE
AMOTRIPHENE
Amoval → PENICILLINS – "NEWER and SYNTHETICS" > *Amoxicillin*
AMOXAPINE also → LOXAPINE
Amoxi → PENICILLINS – "NEWER and SYNTHETICS" > *Amoxicillin*
Amoxicillin → PENICILLINS – "NEWER and SYNTHETICS" also → CLAVULANIC ACID, METHOTREXATE
Amoxidal → PENICILLINS – "NEWER and SYNTHETICS" > *Amoxicillin*
Amoxidin → PENICILLINS – "NEWER and SYNTHETICS" > *Amoxicillin*
Amoxil → PENICILLINS – "NEWER and SYNTHETICS" > *Amoxicillin*
Amoxillat → PENICILLINS – "NEWER and SYNTHETICS" > *Amoxicillin*
Amoxipen → PENICILLINS – "NEWER and SYNTHETICS" > *Amoxicillin*
Amoxi-Wolff → PENICILLINS – "NEWER and SYNTHETICS" > *Amoxicillin*
Amoxycillin → PENICILLINS – "NEWER and SYNTHETICS" → *Amoxicillin*
Amoxypen → PENICILLINS – "NEWER and SYNTHETICS" > *Amoxicillin*
AMPC → PENICILLINS – "NEWER and SYNTHETICS" > *Amoxicillin*
Amphaethamine → AMPHETAMINE
Amphamite → PARATHION

Amphedrine-M → LEVAMPHETAMINE
Amphedroxyn → METHAMPHETAMINE
AMPHETAMINE also → ACETAZOLAMIDE, AMMONIUM CHLORIDE, ARGININE, BETHANIDINE, CHLORPROMAZINE, COCAINE, DEBRISOQUIN, DESIPRAMINE, DISULFIRAM, DOXAPRAM; ETHER, ETHYL; FURAZOLIDONE, GUANETHIDINE, HYDRALAZINE, ISOCARBOXAZID, MAGNESIUM CARBONATE, MAGNESIUM OXIDE, METHOSUXIMIDE, METHYLAMINE, *N*-METHYLFORMAMIDE, METUREDEPA, NIALAMIDE, NORPSUEDOEPHEDRINE, NUTMEG, PARGYLINE, PEMOLINE, PHENELZINE, PHENOBARBITAL, PHENYTOIN, PROPOXYPHENE, PROPRANOLOL, SODIUM BICARBONATE, SODIUM HYDROXIDE, SODIUM THIOSULFATE, SULFURIC ACID, THYROXINE, TYRAMINE
AMPHETAMINIL
AMPHIACHYRIS DRAUMCULOIDES
Amphicol → CHLORAMPHENICOL
Ampho-Moronal → AMPHOTERICIN B
AMPHOMYCIN
Amphotec → AMPHOTERICIN B
AMPHOTERICIN B also → CYCLOSPORINE, DIGITALIS, FOSCARNET SODIUM, GENTAMICIN, MICONAZOLE NITRATE, NORFLOXACIN
AMPHOTHALIDE
Amphozone → AMPHOTERICIN B
Ampicillin → PENICILLINS – "NEWER and SYNTHETICS" (including *Carbenicillin Disodium*) also → ALLOPURINOL, ALUMINUM HYDROXIDE, ATENOLOL, DESMOPRESSIN, *N,N*-DIMETHYLAMINE, ETHINYL ESTRADIOL, KAOLIN, LANSOPRAZOLE, OMEPRAZOLE; PENICILLIN, *Benzylpenicillin, Procaine*; SULFAMETHOXYPYRIDAZINE
Ampilag → PENICILLINS – "NEWER and SYNTHETICS" > *Ampicillin*
Ampilar → PENICILLINS – "NEWER and SYNTHETICS" > *Ampicillin*
Ampilin → PENICILLINS – "NEWER and SYNTHETICS" > *Ampicillin*
Ampinpenicillin → PENICILLIN > *Benzylpenicillin, Procaine*
Ampipenin → PENICILLINS – "NEWER and SYNTHETICS" > *Ampicillin*
Ampitab → PENICILLINS – "NEWER and SYNTHETICS" > *Ampicillin*
Ampi-Tablinen → PENICILLINS – "NEWER and SYNTHETICS" > *Ampicillin*
Ampliactil → CHLORPROMAZINE
Amplictil → CHLORPROMAZINE
Ampligram → CEPHALORIDINE
Amplipen → PENICILLINS – "NEWER and SYNTHETICS" > *Hetacillin*
Amplit → LOFEPRAMINE
Amplital → PENICILLINS – "NEWER and SYNTHETICS" > *Ampicillin*
Amplivix → BENZIODARONE
Amprace → ENALAPRIL
AMPRENAVIR also → KETOCONAZOLE
Amprol → AMPROLIUM
AMPROLIUM
Amprovine → AMPROLIUM
AMS → AMMONIUM SULFAMATE
Amsebarb → AMOBARBITAL
Amsinckine → PYRROLIZIDINES
Amsulosin → TAMSULOSIN
Amsustain → DEXTROAMPHETAMINE SULFATE
Amudane → GRISEOFULVIN
Amurex → METHIONINES
Amycor → BIFONAZOLE
AMYGDALIN also → AMARETTO, APPLE, APRICOT, CHERRY, CYANIDES, DIPYRONE, LAETRILE, PEACHES, PEARS, PLEUROTIS sp., RHODOTYPOS SCANDENS
Amyl Acetate → ISOAMYL ACETATE
Amyl Dimethyl Paba → PADIMATES
AMYLENE HYDRATE
Amylene Hydroxide → AMYLENE HYDRATE
Amylenol → AMYL SALICYLATE
AMYLIN
n-Amyl Mercaptan → 1-PENTANETHIOL
Amyl Methyl Ketone → 2-HEPTANONE
AMYL NITRITE
Amylobarbitone → AMOBARBITAL
Amylocarbinol → ALCOHOL, *n*-HEXYL
p-tert-Amylphenol → *p-tert*-PENTYLPHENOL
AMYL SALICYLATE
Amylum → STARCHES
AMYRIS OIL ACETYLATED
Amytal → AMOBARBITONE
AN-1 → AMPHETAMINIL
AN-148 → METHADONE HCL
Anabactyl → PENICILLINS – "NEWER and SYNTHETICS" > *Carbenicillin Disodium*
ANABASINE also → NICOTIANA sp.
ANABENA FLOS-AQUAE
Anabet → NADOLOL
Anabloc → PHENYRAMIDOL
Anabol → METHANDRIOL

Anabolex → STANOLONE
ANACARDIACEA
Anacardic Acid → ANACARDIACEA, PENTADECYLCATECHOL
Anacardium occidentale → CASHEW
Anacardone → NIKETHAMIDE
Anacetin → CHLORAMPHENICOL
Anacobin → VITAMIN B_{12}
ANADENANTHERA PEREGRINA
Anadonis Green → CHROMIC OXIDE
Anador → NANDROLONE
Anadrol → OXYMETHOLONE
Anadur → NANDROLONE
Anafebrina → AMINOPYRINE
Anaflon → ACETAMINOPHEN
Anafranil → CLOMIPRAMINE
ANAGALLIS ARVENSIS
ANAGESTONE ACETATE
Anagregal → TICLOPIDINE HCL
ANAGRELIDE
ANAGYRINE also → LUPINES
Anahist → THONZYLAMINE HYDROCHLORIDE
Analate → DIPYRONE
Analeptin → SYNEPHRINE
Analexin → PHENYRAMIDOL
Analgesine → ANTIPYRINE
Analgin → DIPYRONE
Analud → FEPRAZONE
Analux → MECLOFENOXATE
Anamu → PETIVERIA ALLIACEA
Anamycin → ERYTHROMYCIN ETHYL SUCCINATE
Ananase → BROMELAIN
Anandron → NILUTAMIDE
Ananus comosum → PINEAPPLE
Anapap → ACETAMINOPHEN
Anaphylatoxin → BRADYKININ
Anapolon → OXYMETHOLONE
Anaprime → FLUMETHASONE
Anaprox → NAPROXEN
Anapryline → PROPRANOLOL
Anaptivan → CEFUROXIME
Anarcon → NALORPHINE
Anarexol → CYPROHEPTADINE
Anasterone → OXYMETHOLONE
Anastil → GUAIACOL
ANASTROZOLE
Anatabine → NICOTIANA sp.
Anatola → VITAMIN A
Anatran → TRICHLORMETHIAZIDE
Anatrofin → STENBOLONE
Anaus → TRIMETHOBENZAMIDE
Anautine → DIMENHYDRINATE
Anavar → OXANDROLONE
Anayodin → SODIUM IODIDE
Ancef → CEFAZOLIN SODIUM
Anceron → BECLOMETHASONE
Anchoic Acid → AZELAIC ACID
Anchovies → FISH
ANCHUSA ARVENSIS
Ancillin → PENICILLINS – "NEWER and SYNTHETICS" > *Diphenicillin Sodium*
Anco → IBUPROFEN
Ancolan → MECLIZINE
Ancoron → AMIODARONE
Ancortone → PREDNISONE
Ancosul → SULFADIMETHOXINE
ANCROD
Ancylol → DISOPHENOL
Ancyte → PIPOSULFAN
Andantol → ISOTHIPENDYL
Andanton → ISOTHIPENDYL
Andaxin → MEPROBAMATE
Andere → BUFORMIN
Andergin → MICONAZOLE NITRATE
Andiamine → HEXOBENDINE
Andion → BECLOMETHASONE
Andolex → BENZYDAMINE
Andozac → FINASTERIDE
Andractin → STANOLONE
Andramine → DIMENHYDRINATE
Andriol → METHANDRIOL
Andro → TESTOSTERONE
Androcur → CYPROTERONE
ANDROCTONUS AUSTRALIS also → SCORPIONS
Androderm → TESTOSTERONE
Androdiol → METHANDRIOL
Androfluorene → FLUOXYMESTERONE
Androfluorone → FLUOXYMESTERONE
Androgel → TESTOSTERONE
Androlone → STANOLONE
Android → METHYL TESTOSTERONE
ANDROMEDA sp.
ANDROMEDOTOXINS also → AZALEA, HONEY, KALMIA, LEDUM sp., LEUCOTHOE, LYONIA sp ., PIERIS sp., RHODODENDRON
ANDROSTENEDIONE also → PHENOBARBITAL, TESTOSTERONE
Androsterolo → FLUOXYMESTERONE
ANDROSTERONE also → CLOFIBRATE, DDT, PHENOBARBITAL, PHENYLBUTAZONE
Androteston-M → METHANDRIOL
Androtex → ANDROSTENEDIONE
Androviron → MESTEROLONE

Anekain → BUPIVACAINE
Anemolin → PENICILLINS – "NEWER and SYNTHETICS" > *Amoxicillin*
ANEMONE sp.
Anemone pulsatilla → PROTOANEMONIN
Anemonin → CLEMATIS
Anemonine → ANEMONE sp.
Anesthesin → ETHYL-*p*-AMINOBENZOATE
Anesthetic Ether → ETHER, ETHYL
Anesthone → ETHYL-*p*-AMINOBENZOATE
Anestil → PROCAINE
Anethaine → TETRACAINE HYDROCHLORIDE
ANETHOLE also → ANISE
Anethum graveolens . → DILL
Aneto → DILL
Aneural → MEPROBAMATE
Aneurex → VITAMIN B_1
Aneurine → VITAMIN B_1
Anexate → FLUMAZENIL
Anfamon → DIETHYLPROPION
Anflagen → IBUPROFEN
Anflam → HYDROCORTISONE
Angel Dust → CLENBUTEROL also → PHENCYCLIDINE
ANGEL HAIR also → PHENCYCLIDINE
ANGELICA sp.
Angelica Tree → ZANTHOXYLUM sp.
Angeli's Sulfone → GLUCOSULFONE SODIUM
Angel's Trumpet → DATURA also → BRUGMANSIA
Angerex → POISON OAK
Angex → LIDOFLAZINE
Angibid → NITROGLYCERIN
Angijen → PENTAERYTHRITOL TETRANITRATE
Angilol → PROPRANOLOL
Anginine → NITROGLYCERIN
Anginal → DIPYRIDAMOLE
Anginyl → DILTIAZEM
Angio-Conray → IOTHALMATE SODIUM
Angio-Contrix "48" → IOTHALMATE SODIUM
Angiodarona → AMIODARONE
Angiografin → MEGLUMINE DIATRIZOATE
Angiomin → XANTHINOL NIACINATE
Angiolingual → NITROGLYCERIN
Angionorm → DIHYDROERGOTAMINE METHANESULFONATE (MESLATE)
ANGIOTENSIN
Angitet → PENTAERYTHRITOL TETRANITRATE
Angitrate → PENTAERYTHRITOL TETRANITRATE
Angitrit → TROLNITRATE PHOSPHATE
Angizem → DILTIAZEM
Angolon → IMOLAMINE
Angopril → BEPRIDIL
Angorin → NITROGLYCERIN
Angormin → PRENYLAMINE LACTATE
Anguidine → DIACETOXYSCIPENOL
ANGYLOCALYX OLIGOPHYLLUS
Anhiba → ACETAMINOPHEN
Anhistan → CLEMASTINE
Anhistol → PYRILAMINE
Anhydrohydroxyprogesterone → ETHISTERONE
ANHYDROTETRACYCLINE and ANHYDRO-4-EPI-TETRACYCLINE also TETRACYCLINE
Anhydron → CYCLOTHIAZIDE
Anhydrous Hydrobromic Acid → HYDROGEN BROMIDE
Anhydrous Hydrochloric Acid → HYDROGEN CHLORIDE
Anifed → NIFEDIPINE
ANILAZINE also → 2,4-DICHLORO-6-σ-CHLOROANILINO-s-TRIAZINE
ANILERIDINE
ANILINE also → CORN, CRAYONS, INKS, METHYLENEANILINE, 4,4′-METHYLENE DIANILINE, OLIVES, ORANGE G, PHENYTOIN; POLISHES, FURNITURE
Aniline Oil → ANILINE
Aniline Violet → GENTIAN VIOLET
Anilinobenzene → *p*-BIPHENYLAMINE
2-*Anilinonaphthalene* → *N*-PHENYL-β-NAPHTHYLAMINE
Animal Galactose Factor → OROTIC ACID
Animal Protein Factor → VITAMIN B_{12}
Animar → TRIMETHOPRIM
Aniprime → FLUMETHASONE
Anipryl → DEPRENYL
ANISALDEHYDE
Anisatin → ILLICIUM ANISATUM
ANISE also → ANETHOLE
ANISE ALCOHOL
Anise Camphor → ANETHOLE
Anisic Alcohol → ANISE ALCOHOL
Anisic Aldehyde → ANISALDEHYDE
σ-ANISIDINE
ANISINDIONE also → ACETAMINOPHEN, CLOFIBRATE, PHENYRAMIDOL
ANISOLE
Anisomorpha buprestoides → WALKING STICK
Anisomorphal → WALKING STICK
ANISOMYCIN

ANISOTROPINE METHYLBROMIDE
Anisoylated Plasminogen Streptokinase Activator Complex → APSAC
Anistadin → TRICHLORMETHIAZIDE
Anistreplase → APSAC also → RETEPLASE
Anisyl Acetone → 4-(*p*-METHOXYPHENYL)BUTAN-2-ONE
Anisyl Alcohol → ANISE ALCOHOL
ANISYLIDENE ACETONE
ANISYL PROPRIONATE
ANIT → 1-NAPHTHYLISOTHIOCYANATE
Ankebin → FENOFIBRATE
Ankerbin → PENICILLINS – "NEWER and SYNTHETICS" > *Cloxacillin/Sodium salt*
Ankilostin → TETRACHLOROETHYLENE
Annona muricata → SORSACA
Annual Broomweed → AMPHIACHYRIS DRAUMCULOIDES
Annual Goldeneye → VIGUIERA ANNUA
Annual Mercury → MERCURIALIS ANNUA
Anny → ANISE
Anodynine → ANTIPYRINE
Anodynon → ETHYL CHLORIDE
ANONA sp.
Anona → TRICHILIA EMETICA
Anone → CYCLOHEXANONE
ANONIDIUM MANNI
Anoprolin → ALLOPURINOL
Anorel → GUANADREL SULFATE
Anorex → DIETHYLPROPION also → PHENMETRAZINE
ANP 235 → MECLOFENOXATE
ANP 297 → MEFEXAMIDE
ANP 3624 → TICRYNAFEN
Anparton → CLOFIBRATE
Anprolene → ETHYLENE OXIDE
Anquil → BENPERIDOL
Ansaid → FLURBIPROFEN
Ansamycin → RIFABUTIN
Ansar → MONOSODIUM METHANEARSONATE also → CACODYLIC ACID
Ansar 6.6 → METHANEARSONIC ACID
Ansar 529 → METHANEARSONIC ACID
Ansar 8100 → METHANEARSONIC ACID also → DISODIUM METHANEARSONATE
Ansatin → FLUFENAMIC ACID
Ansatipine → RIFABUTIN
Anseren → KETAZOLAM
Ansiacal → CHLORDIAZEPOXIDE
Ansieten → KETAZOLAM
Ansilan → MEDAZEPAM
Ansiolin → DIAZEPAM
Ansmin → DIPHENIDOL
Ansolysen → PENTOLINIUM TARTRATE
Anspor → CEPHADRINE
Antacids → ACETYLSALICYLIC ACID, AMITRIPTYLINE, ATORVISTATIN, BROMODIPHENHYDRAMINE, CIPROFLOXACIN, GREPAFLOXACIN, LOMEFLOXACIN, MAGNESIUM, METRONIDAZOLE, NALIDIXIC ACID, NAPROXEN, NORFLOXACIN, OXYTETRACYCLINE, PEFLOXACIN, PENICILLAMINE, PENICILLIN, PHENYLBUTAZONE, PHENYTOIN, PROCHLORPERAZINE, QUINIDINE, SODIUM, ROFECOXIB, SULPIRIDE, TENOXICAM, TETRACYCLINE, TICLODIPINE, TILUDRONATE
Antadys → FLURBIPROFEN
Antagonate → CHLORPHENIRAMINE
Antagosan → APROTININ
Antak → RANITIDINE
Antallin → CALCIUM DISODIUM EDETATE
Antalvic → PROPOXYPHENE
Antalzyme → LYSOZYME
Antamine → PYRILAMINE
Antastan → ANTAZOLINE
Antasten → ANTAZOLINE
Antaxone → NALTREXONE
ANTAZOLINE also → ETHYLENEDIAMINE
Anteben → NORTRIPTYLINE
Antébor → SULFACETAMIDE SODIUM
Antees → CARBETAPENTANE CITRATE
Antegan → CYPROHEPTADINE
Antemin → DIMENHYDRINATE
Antepan → THYROTROPIN RELEASING HORMONE
Antepar → PIPERAZINE
Antepsin → SUCRALFATE
Antesite → GLYCYRRHIZA GLABRA
Anthelcide EQ → OXIBENDAZOLE
Anthelvet → TETRAMISOLE
ANTHEMIS COTULA also → ASTERS
Anthericum saltii → TRACHYANDRA SALTII
Anthimin → PYRILAMINE
ANTHIOLIMINE
Anthion → POTASSIUM PERSULFATE
Anthisan → PYRILAMINE
Anthium Dioxide → CHLORINE DIOXIDE
Anthocyanin → BEETS
ANTHODISCUS sp.
Antholamine → ANTHIOLAMINE
Anthon → TRICHLORFON
2-Anthracenamine → 2-ANTHRAMINE
ANTHRACENE
ANTHRACITE also → COAL

Anthradione → ANTHRAQUINONE
Anthralin Triacetate → TRIACETOXYANTHRACENE
2-ANTHRAMINE
ANTHRAMYCIN
Anthranilic Acid → ACETIC ANHYDRIDE
ANTHRAQUINONE also → BEETS, CASCARA SAGRADA, CASSIA, INKS, RHUBARB, RUMEX sp., SENNA
Antiangor → CHROMONAR
Antiarins → NAUCLEOPSIS sp.
ANTIARIS TOXICARIA
Antibason → METHYLTHIOURACIL
Anti-beriberi Vitamin → VITAMIN B_1
Antibiotic 899 → VIRGINIAMYCIN
Antibiotic 1600 → PAROMOMYCIN
Antibiotic 4915 → PAROMOMYCIN
Antibiotic 5879 → BICOZAMYCIN
Antibiotic 6640 → SISOMICIN
Antibiotic 3823 A → MONENSIN
Antibiotic A 5283 → NATAMYCIN
Antibiotic M 4209 → CARBOMYCIN
Antibiotic PA 106 → ANISOMYCIN
Anticarie → HEXACHLOROBENZENE
Antichlor → SODIUM THIOSULFATE
Antidrasi → DICHLORPHENAMIDE
Antifebrin → ACETANILID
Antifoam A → DIMETHICONE (*with silicon dioxide*)
Antigal → DIAZINON
Antigestil → DIETHYLSTILBESTROL
Antihemophilic Factor B → CHRISTMAS FACTOR
Antihistal → ANTAZOLINE
Anti-infective Vitamin → VITAMIN A
Antikrein → APROTININ
Antilev → NORTRIPTYLINE
Antimalarina → QUINACRINE
Antimalarine → PLASMOCID
ANTIMONY also → CORN
Antimony Lithium Thiomalate → ANTHIOLIMINE
ANTIMONY OXIDE also → ANTIMONY TRIOXIDE, COFFEE
ANTIMONY PENTACHLORIDE
ANTIMONY POTASSIUM TARTRATE
ANTIMONY SODIUM GLUCONATE
ANTIMONY SODIUM TARTRATE
Antimony Sulfide → ANTIMONY TRISULFIDE
ANTIMONY THIOANTIMONATE
ANTIMONY TRICHLORIDE
ANTIMONY TRIOXIDE
ANTIMONY TRISULFIDE also → MATCHES
Antimosan → STIBOPHEN
Antimycin → CITRININ
ANTIMYCIN A
Antin → PHENYLTOLOXAMINE
Antineuritic Vitamin → VITAMIN B_1
Antinonnin → DINITRO-σ-CRESOL
Antioside → NAUCLEOPSIS sp.
Antipar → DIETHAZINE
Antipernicin → VITAMIN B_{12}
Antipernicious Anemia Principle → VITAMIN B_{12}
Antiphen → DICHLOROPHEN(E)
Antipress → IMIPRAMINE
ANTIPYRINE also → ALCOHOL, ETHYL; ALLOPURINOL, AMINOGLUTETHIMIDE, CAFFEINE, CHLORDANE, CIMETIDINE, DDT, DISULFIRAM, FLURBIPROFEN, GLUTETHIMIDE, INSULIN, KETOCONAZOLE, LABETOLOL, LINDANE, NORTRIPTYLINE, OMEPRAZOLE, PHENOBARBITAL, PREDNISOLONE, PROPOXYPHENE, PROPRANOLOL, QUININE, SODIUM BICARBONATE, SODIUM VALPROATE, SPIRONOLACTONE, TICLODIPINE, TOLBUTAMIDE, VITAMIN B_2, VITAMIN C, WARFARIN
Antirex → EDROPHONIUM
Antirobe → CLINDAMYCIN
Antirrinoside → LINARIA VULGARIS
Antiscorbutic Vitamin → VITAMIN C
Antisedan → ATIPAMEZOLE
Antispasmin → ALVERINE
Antisterility Factor → VITAMIN E
Antistin → ANTAZOLINE
Antistine → ANTAZOLINE
Anti-Stress → MEPARFYNOL
Antitanil → DIHYDROTACHYSTEROL
Anti-tetany substance 10 → DIHYDROTACHYSTEROL
ANTITHYMOCYTE GLOBULIN or SERUM
Antituxil-Z → ZIPEROL
Antivariz → OLAMINE
Antivert → MECLIZINE
Antixerophthalmic Vitamin → VITAMIN A
Antococcid → AMPROLIUM
Antony, Mark → DATURA (*strammonium*)
Antorphine → NALORPHINE
Antra → OMEPRAZOLE
Antrancine 8 → BUTYLATED HYDROXYTOLUENE
Antrancine 12 → BUTYLATED HYDROXYANISOLE
Antrapurol → DANTHRON
Antrenyl → OXYPHENONIUM BROMIDE
Antrycide → QUINAPYRAMINE
Antrypol → SURAMIN SODIUM

ANTS
ANTU → 1-(1-NAPHTHYL)-2-THIOUREA also → 1-NAPHTHYLAMINE
Antuitrin S → GONADOTROPINS, CHORIONIC
Anturan(e) → SULFINPYRAZONE
Anturano → SULFINPYRAZONE
Anturat → 1-(1-NAPHTHYL)-2-THIOUREA
Antiverm → PHENOTHIAZINE
Anvitoff → TRANEXAMIC ACID
Anxiolit → OXAZEPAM
Anxon → KETAZOLAM
Anya Nwono → ABRUS PRECATORIUS
Anzemet → DOLASETRON
Anzief → ALLOPURINOL
Aolept → PROPERICIAZINE
Aoral → VITAMIN A
Aotal → ACAMPROSATE
6-APA → 6-AMINOPENICILLANIC ACID
Apacil → *p*-AMINOSALICYLIC ACID
Apamide → ACETAMINOPHEN
APAMIN
APAP → ACETAMINOPHEN
Aparasin → LINDANE
Aparkane → TRIHEXYPHENIDYL
Aparkazin → DIETHAZINE
Apas → *p*-AMINOSALICYLIC ACID
Apatef → CEFOTETAN
Apaurin → DIAZEPAM
APAZONE also → PHENYTOIN, TOLBUTAMIDE, WARFARIN
APD → PAMIDRONATE DISODIUM
Apertase → HYALURONIDASE
Apesan → CARISPRODOL
Apex → TRIS(2-3-DIBROMOPROPYL)PHOSPHATE
Apexol → VITAMIN A
APH → ACETYLPHENYLHYDRAZINE
Aphanizomenon flos-aquae → SAXITOXIN
Aphilan R → BUCLIZINE
APHOLATE
Aphoxide → TRIETHYLENE PHOSPHORAMIDE
Aphrodine → YOHIMBINE
Aphthasol → AMLEXANOX
Aphtiria → LINDANE
Apikan → DATURA, *metel*
Apilak → ROYAL JELLY
APIOLE also → PARSLEY
Apior → DIEFFENBACHIA sp.
Apisate → DIETHYLPROPION
Apium graveolens → CELERY
A.P.L. → GONADOTROPINS, CHORIONIC
Aplactan → CINNARIZINE
Aplakil → OXAZEPAM
Aplaquette → TICLOPIDINE
Aplexal → CINNARIZINE
Apllobal → ALPRENOLOL
Aplopappus heterophylus → GOLDENROD
APME → ASPARTAME
APO → TRIETHYLENE PHOSPHORAMIDE
APOATROPINE
Apocard → FLECAINIDE
Apocillin → PENICILLIN > *Penicillin V*
APOCYNUM CANNABINUM
Apodol → ANILERIDINE
Apoidine → GONADOTROPINS, CHORIONIC
Apolan → CLOFIBRATE
Apolar → DESONIDE
APOMORPHINE also → CLONIDINE, CORYDALIS, DICENTRA CUCULLARIA, SORSACA
Aponal → DOXEPIN
Aponeuron → AMPHETAMINIL
Aponorin → TRICHLORMETHIAZIDE
Aposepam → DIAZEPAM
Apostavit → VITAMIN A
Apo-Sulfatrim → SULFAMETHOXAZOLE
Apotel → ACETAMINOPHEN
Apotomin → CINNARIZINE
Appex → GARDONA
APPLE also → CYANIDES, DIAMINOZIDE, FRUITS and JUICES, HAZELNUTS, HYDROGEN CYANIDE, PATULIN, PEARS, PENICILLIUM sp., SOLANUM sp.
Apple of Carthage → POMEGRANATE
Apple of Peru → DATURA also → NICANDRA PHYSALODES
Apple of Sodom → SOLANUM sp.
Apranax → NAPROXEN
Apresoline → HYDRALAZINE
Aprical → NIFEDIPINE
APRICOT also → ALMONDS, AMARETTO, CHERRY, CYANIDES, FRUITS and JUICES, HYDROGEN CYANIDE, MILK, PEACHES
Apricot Vine → PASSIFLORA sp.
Apridol → MEPARFYNOL
Aprikern® → APRICOT, HYDROGEN CYANIDE
APRINDINE also → AMIODARONE
Aprinox → BENDROFLUMETHIAZIDE
Apriscolin → PROCYCLIDINE
Aprobal → ALPRENOLOL
APROBARBITAL
Aprocarb → PROPOXUR
APRONALIDE
Apronitine → APROTININ

APROTININ also → SUCCINYLCHOLINE
APSAC
Apsifen → IBUPROFEN
Apsolol → PROPRANOLOL
Aptine → ALPRENOLOL
Aptol Duriles → ALPRENOLOL
Apulonga → ALLOPURINOL
Apurin → ALLOPURINOL
Apurol → ALLOPURINOL
AQ1 → AMODIAQUIN
Aqua ardens → ALCOHOL, ETHYL
Aquacare → UREA
Aquacide → DIQUAT DIBROMIDE
Aquacillin → PENICILLIN > *Benzylpenicillin, Procaine*
Aquacycline → OXYTETRACYCLINE
Aquadiol → ESTRADIOL
Aquadrate → UREA
Aquafen → FLORFENICOL
Aqua Fortis → NITRIC ACID
Aqualin → ACROLEIN
Aquamox → QUINETHAZONE
Aquamycetin → CHLORAMPHENICOL
Aquanil → TIMOLOL
Aquaphor → XIPAMIDE
Aquaphoril → XIPAMIDE
Aquareduct → SPIRONOLACTONE
Aquarius → HYDROCHLOROTHIAZIDE
Aquasept → TRICLOSAN
Aquasuspen → PENICILLIN > *Benzylpenicillin, Procaine*
Aquatag → BENZTHIAZIDE
Aquathol → ENDOTHALL
Aquazine → SIMAZINE
Aquex → CLOPAMIDE
Aquilaria agallocha → 4-(*p*-METHOXYPHENYL)BUTAN-2-ONE
Aquilegia → COLUMBINE
AQUILIDE A also → BRACKEN FERN
Aquirel → CYCLOTHIAZIDE
Aguo-Trinitroson → NITROGLYCERIN
ara-A → VIDARABINE
Arabinosyladenine → VIDARABINE
Arabinosyl Cytosine → CYTARABINE
Arabitin → CYTARABINE
Ara-C → CYTARABINE
Arachchis hypogaea → PEANUTS
Aracide → ARAMITE
Aracytidine → CYTARABINE
Aracytine → CYTARABINE
Aralen → CHLOROQUINE
ARALIA sp.
Aralkonium Chloride → PENICILLINS – "NEWER and SYNTHETICS" > *Epicillin*
Aramine → METARAMINOL BITARTRATE
ARAMITE
Arantoik → TOLPERISONE HCL
Aranyérre → OAKS
Arasan → THIRAM
Arasena-A → VIDARABINE
Arathane → DINOCAP
Aratron → ARAMITE
Arava → LEFLUNOMIDE
Arbamitis sp. → SPIDERS
Arborvitae → THUJA OCCIDENTALIS
Arbotect → THIABENDAZOLE
Arbuz → PAPAIN
Arcadine → BENACTYZINE
Archin → EMODIN
Arcosal → TOLBUTAMIDE
Arcosolv PM → PROPYLENE GLYCOL MONOMETHYL ETHER
Arctium sp. → BURDOCK
Arctons → FREONS
Arcton 12 → DICHLORODIFLUOROMETHANE
Arcton 112 → 1,1,2,2-TETRACHLORODIFLUORETHANE
Arcton 113 → TRICHLOROTRIFLUOROETHANE
Arcton 114 → 1,2-DICHLORO-1,1,2,2-TETRACHLOROETHANE
ARDEPARIN
Ardex → DEXTROAMPHETAMINE SULFATE
Ardeytropin → *L*-TRYPTOPHAN
Ardica → URTICA DIOICA
Ardine → PENICILLINS – "NEWER and SYNTHETICS" > *Amoxicillin*
Ardinex → IBUPROFEN
Arduan → PIPECURONIUM BROMIDE
ARECA also → PROCYCLIDINE
Arechin → CHLOROQUINE
ARECOLINE also → ARECA, PROCYCLIDINE, SORSACA
Aredia → PAMIDRONATE DISODIUM
Arequine → CHLOROQUINE
Aresin → MONOLINURON
Aretit → DINOSEB
Arfonad → TRIMETHAPHAN CAMSYLATE
ARGEMONE MEXICANA
Argilla → KAOLIN
ARGININE also → GLUCOSE, POTASSIUM ARGININE
Argivene → ARGININE
Argol → SILVER NITRATE
ARGON also → WELDING
Argun → ALCLOFENAC

ARGYREIA NERVOSA also → LYSERGAMIDE
Argyrol → SILVER PROTEIN, MILD
Arhythmol → PROPAFENONE
Arial → SALMETEROL
Aribine → HARMAN
Aricept → DONEPEZIL
Arilin → METRONIDAZOLE
Arimidex → ANASTROZOLE
ARIOCARPUS RETUSUS
ARISAEMA sp. also → CALCIUM OXALATE
ARISTIDA sp.
Aristocort → TRIAMCINOLONE
ARISTOLOCHIA sp.
Aristolochia fungi → STEPHANIA TETRANDA
ARISTOLOCHIC ACID also → STEPHANIA TETRANDA
Aristolochine → ARISTOLOCHIC ACID
Aristospan → TRIAMCINOLONE
Aristotle → WATER
Arizona hinshawii → SNAKE(S)
Arkin → VESNARINONE
Arkitropin → HOMATROPINE
Ark of the Covenant → ACACIA
Arlacel 85 → SORBITAN TRIOLEATE
Arlef → FLUFENAMIC ACID
Arlidin → NYLIDRIN
Arlitene → THYMOXAMINE
Armel → PEGANUM HARMALA
Armidexan → IRON DEXTRANS
Armoise → ABSINTHIUM
Armophylline → THEOPHYLLINE
Armoracia rusticana → HORSERADISH
Armyl → LYMECYCLINE
ARNICA also → DORONICUM sp.
Arnicin → ARNICA also → DORONICUM
Arnosulfan → SULFADIMETHOXINE
Arobon → CAROB
Arochlors → POLYCHLORINATED BIPHENYLS
Aroeira nigra → SCHINUS TEREBINTHIFOLIUS
Aroine → ARUM sp.
Aromasin → EXEMESTANE
Aropax → PAROXETINE
Arovit → VITAMIN A
Arozene 5 → 1,1-DIMETHYLHYDRAZINE
Arpalit → TRICHLORFON
Arpamyl → VERAPAMIL
Arphos → VITAMIN B_{12}
Arpimycin → ERYTHROMYCIN ETHYL SUCCINATE
Arquel → MECLOFENAMIC ACID
Arrach → ATRIPLEX sp.
Arret → LOPERAMIDE also → CHOLINE SALICYLATE
Arrhenal → DISODIUM METHANEARSONATE
Arrivo → CYPERMETHRIN
Arrow Crotalaria → CROTALARIA
Arrow, Dart, & Spear Poisons → ACOCANTHERA, ACONITE, ADENIUM sp., ALBIZZIA, ALOYSIA, AMBELANIA, ANTIARIS TOXICARIA, BALANITES sp., BARBASCO, BARRINGTONIA sp., BATRACHOTOXIN, BOÖPHONE, BUPHANE DISTICHA, CALOTROPIN, CARISSA, CASSAINE, CHELIDONIUM MAJUS, CICUTA sp., CLEMATIS sp., CURARE, CYNANCHUM, DAPHNE, DERRIS ROOT, DIAMPHIDIA, DIEFFENBACHIA sp., ERYTHROPHLEUM, EUPHORBIA sp., GNIDIA KRAUSSIANA, HISTRIONICOTOXIN, HURA CREPITANS, HYOSCYAMINE, HYOSCYAMUS, KETUHAS, LIMU-MAKE-O-HANA, LOPHOPETALUM, LYUTIK, MALOUETIA sp., MANICHEEL TREE, MANSONIA ALTISSIMA, NAUCLEOPSIS, NUX VOMICA, OCGODEIA, OLEANDER, OUABAIN, PACHYRRHIZUS EROSUS, PALYTHOA TOXICA, PARQUETINA, PHYSOSTIGMA NENENOSUM, PICRASMA QUASSOLOIDES, PODOPHYLLUM, POLYCLADA sp., PUMILIOTOXINS, RAUWOLFIA SERPENTINA, ROTENONE, RYANIA sp., SAPIUM sp., SAPONIN, SCHOENBIBLUS PERUVIANUS, SCOPOLAMINE, SCORPIONS, SEBASTIANIA, SNAKES, SPIDERS, STINGRAYS, STROPHANTHUS sp., STRYCHNINE, STRYCHNOS, TELITOXICUM PERUVIANUM, THEVETIA sp., TOXIFERINE, URGINEA sp., VERATRUM sp., VERNONIA sp., VIROLA sp., VOCHYSIA sp., WATER BOATMAN
ARROWGRASS also → CYANIDES, HYDROGEN CYANIDE
Arrow Wood → FRANGULA sp.
Arsambiden → CARBARSONE
Arsamin → SODIUM ARSANILATE
Arsaminol → ARSPHENAMINE
ARSANILIC ACID
Arsanyl → GEFARNATE
Arsaphen → ACETARSONE
Arsecodile → SODIUM CACODYLATE

Asceptichrome → MERBROMIN
ASCLEPIAS sp.
Ascorbef → VITAMIN C
ASCORBIC ACID also → ACETYLSALICYLIC ACID
Ascorin → VITAMIN C
Ascorteal → VITAMIN C
Ascorvit → VITAMIN C
AS-E 136 → DIRITHROMYCIN
Asebo-toxin → PIERIS sp.
Asellacrin → GROWTH HORMONES
Asendin → AMOXAPINE
Ash → FRAXINUS sp.
Ashwagandha → WITHANIA SOMNIFERA
Asian Bugs → TRIATOMA sp.
Asimicin → PAW-PAW
Asimina triloba → PAW-PAW
Asiminer trilobé → PAW-PAW
Asiprenol → ISOPROTERENOL
Asistar → FLUMETHASONE
ASL 279 → DOPAMINE
ASL 8052 → ESMOLOL
Aslavital → PROCAINE
Aslos → CARBETAPENTANE CITRATE
Asmalar → ISOPROTERENOL
Asmaven → ALBUTEROL
ASP 47 → SULFOTEP
l-ASPARAGINASE also → CYTARABINE
Asparagine → PHILODENDRON sp.
l-Asparagine Amidohydrolase → l-ASPARAGINASE
Asparaginic Acid → l-ASPARTIC ACID
Asparago → ASPARAGUS
ASPARAGUS also → PHYTOLACCA sp.
Asparagus Bean → BEANS, Winged
ASPARTAME also → BENZALDEHYDE, COLAS, PHENYLALANINE
l-ASPARTIC ACID also → ASPARTAME, PROTEIN HYDROLYSATES
Aspenil → PENICILLINS – "NEWER and SYNTHETICS" > *Amoxicillin*
Aspenon → APRINDINE
Asperage → ASPARAGUS
Asperge → ASPARAGUS
Aspergesic → LYSINE ACETYLSALICYLATE
ASPERGILLIC ACID
ASPERGILLUS sp.
Asperjo → ASPARAGUS
ASPHALT also → HYDROGEN CYANIDE
ASPIDIUM
Aspidol → LYSINE ACETYLSALICYLATE
Aspirin → ACETYLSALICYLIC ACID
Aspirin Lysine Salt → LYSINE ACETYLSALICYLATE
Aspisol → LYSINE ACETYLSALICYLATE
Aspor → ZINEB
Aspro → ACETYLSALICYLIC ACID
Assaren → DICLOFENAC
Assassin Bug → TRIATOMA sp.
Assassins → CANNABIS
Assault → BRODIFACOUM also → BROMETHALIN
Asses' Parsley → CHAEROPHYLLUM sp.
Assimina → PAW-PAW
Assiprenol → ISOPROTERENOL
Assoral → ROXITHROMYCIN
Assûs → GLYCYRRHIZA GLABRA
Asta Z-4942 → IFOSFAMIDE
ASTATINE
Astelin → AZELASTINE
Astemisan → ASTEMIZOLE
ASTEMIZOLE also → CLOZAPINE, DELAVIRDINE, ERYTHROMYCIN, FLUCONAZOLE, FLUVOXAMINE, INDINAVIR, ITRACONAZOLE, KETOCONAZOLE, METRONIDAZOLE, MIBEFRADIL, MICONAZOLE NITRATE, NELFINAVIR MESYLATE, RITONAVIR, SAQUINAVIR, TROLEANDOMYCIN
Asteric → ACETYLSALICYLIC ACID
Asterol → DIAMTHAZOLE
ASTERS also → CHAMOMILE, GOLDENROD, TEA
Asthmador® → BELLADONNA, STRAMONIUM
Asthmanefrin → EPINEPHRINE
Asthma Plant → EUPHORBIA, *hirta*
Asthma Weed → LOBELIA
Asthmalitan → ISOETHARINE
Asthmolysin → DYPHYLLINE
Astiban → STIBOCAPTATE
Astix CMPP → MECOPROP
Astmamasit → DYPHYLLINE
Astra → OMEPRAZOLE
Astra 1512 → PRILOCAINE
Astragale du Canada → MILKVETCH
ASTRAGALUS also → SWAINSONINE
Astridine → ISOSORBIDE DINITRATE
Astrobain → OUABAIN
Astrobot → DICHLORVOS
Astrocar → NIKETHAMIDE
Astroderm → DICHLORISONE
Astrophyllin → DYPHYLLINE
Asturidon → BUTABARBITAL
Astyn → ECTYLUREA
Asucrol → CHLORPROPAMIDE
Asuntol → COUMAPHOS
AT 7 → HEXACHLOROPHENE
AT 10 → DIHYDROTACHYSTEROL
AT 101 → ISOSORBIDE

AT 2266 → ENOXACIN
AT 4140 → SPARFLOXACIN
ATA → AMINOTRIAZOLE
Atabrine → QUINACRINE
Atacand → CANDESARTAN CILEXETIL
Atale → GINGER, JAMAICAN
Atamasco Lily → ZEPHYRANTHES ATAMASCO
Ataractan → AZACYCLONAL
Atarax → HYDROXYZINE
Atasol → ACETAMINOPHEN
Atav → VITAMIN A
ATC → TIMONACIC
ATCP → PICLORAM
Atebrimum → QUINACRINE
Ateculon → CLOFIBRATE
AteHexal → ATENOLOL
Atelor → DIAMTHAZOLE
Atem → IPRATROPIUM BROMIDE
Atemorin → MEPARFYNOL
Atempol → MEPARFYNOL
Atenezol → ACETAZOLAMIDE
Atenol → ATENOLOL
ATENOLOL also → CHLORTHALIDONE, INDOMETHACIN, NAPROXEN, NIFEDIPINE, NISOLDIPINE; PENICILLINS – "NEWER and SYNTHETICS" > *Ampicillin*; VECURONIUM BROMIDE, VERAPAMIL
Atensin → MEPHENESIN
Atensine → DIAZEPAM
Aterax → HYDROXYZINE
Atgan → ANTITHYMOCYTE GLOBULIN or SERUM
Atgard → DICHLORVOS
Atheropront → CLOFIBRATE
Athrochin → CHLOROQUINE
Athrombin → WARFARIN
Atheromide → CLOFIBRATE
Athymil → MIANSERIN
Athyromazole → CARBIMAZOLE
Atilen → DIAZEPAM
ATIPAMEZOLE
Atirin → CEFAZOLIN SODIUM
Ativan → LORAZEPAM
ATLA → THEOPHYLLINE
Atlacide → SODIUM CHLORATE
Atlantsil → AMIODARONE
Atlas A → SODIUM ARSENITE
Atmosgen → TOLPERISONE HCL
Atocin → CINCHOPHEN
Atock → FORMOTEROL
Atophan → CINCHOPHEN
Atorel → INOSINE
ATORVASTATIN also → MIBEFRADIL
Atosil → PROMETHAZINE
ATOVAQUONE also → ZIDOVUDINE
Atoxicocaine → PROCAINE
Atoxyl → SODIUM ARSANILATE
Atoxylic Acid → ARSANILIC ACID
ATRACTYLOSIDE
ATRACURIUM also → ATENOLOL, ESMOLOL, METHYL PREDNISOLONE, PROPRANOLOL
Atranorin → LICHENS
Atravet → ACEPROMAZINE
Atrax sp. → SPIDERS
Atraxin → MEPROBAMATE
Atraxotoxin → SPIDERS
ATRAZINE also → ALACHLOR
ATRIPLEX sp.
Atrichin → CHLOROQUINE
Atroban → PERMETHRIN
Atrobione → CHOLINE SALICYLATE
Atromidin → CLOFIBRATE
Atromid-S → CLOFIBRATE
ATROPA BELLADONNA
Atropa mandragoro → PODOPHYLLUM
Atropamine → APOATROPINE
ATROPINE also → AMBENONIUM, CHLORDANE, CURARE, DATURA, DIPHENOXYLATE, DOBUTAMINE, HALOTHANE, IMIPRAMINE, IRON, ISONIAZID, ISOPROPAMIDE IODIDE, MAGNESIUM TRISILICATE, MANDRAGORA OFFICINARUM, MEPERIDINE, METHOTRIMEPRAZINE, MILK, NEOSTIGMINE, NITROUS OXIDE, NORTRIPTYLINE, PROCYCLIDINE, PROPRANOLOL, PYRANTEL, SCOPOLAMINE, SOLANUM sp., STRAMONIUM, THIALBARBITAL, THIOPENTAL, TROPICAMIDE, VITAMIN C
Atropyltropeine → APOATROPINE
Atrovent → IPRATROPIUM BROMIDE
Atrovenetin → PENICILLIUM sp.
Atroxicam → PIROXICAM
ATS → ANTITHYMOCYTE GLOBULIN or SERUM
ATTAPULGITE also → PENICILLINS – "NEWER and SYNTHETICS" > *Amoxicillin, Ampicillin*
Atysmal → ETHOSUXIMIDE
ATZ → TISSUE PLASMINOGEN ACTIVATOR
Aûd → GLYCYRRHIZA GLABRA
Audax → CHOLINE SALICYLATE

Augmentin → PENICILLINS – "NEWER and SYNTHETICS" > *Amoxicillin*; POTASSIUM CLAVULANATE
Auligen → BISETHYLXANTHOGEN
Aulin → NIMESULIDE
Aulinogen → BISETHYLXANTHOGEN
AURAMINE
AURANOFIN
Aureocina → CHLORTETRACYCLINE
Aureocycline → CHLORTETRACYCLINE
Aureolic Acid → MITHRAMYCIN
Aureomycin → CHLORTETRACYCLINE
Aureotan → AUROTHIOGLUCOSE
Auricidine → GOLD SODIUM THIOSULFATE
Aurocidin → GOLD SODIUM THIOSULFATE
Aurofac → CHLORTETRACYCLINE
Aurolin → GOLD SODIUM THIOSULFATE
Auropex → GOLD SODIUM THIOSULFATE
Aurora Borealis → PHENCYCLIDINE
Aurorix → MOCLOBEMIDE
Aurosan → GOLD SODIUM THIOSULFATE
AUROTHIOGLUCOSE
Aurothion → GOLD SODIUM THIOSULFATE
Auacatyl → AVOCADO
Aubepine → ANISALDEHYDE
Aurum paradoxum → TELLURIUM
Aurumine → AUROTHIOGLUCOSE
Ausocef → CEPHALEXIN
Austracol → CHLORAMPHENICOL
Australian Fever Tree → EUCALYPTUS GLOBULUS
Australian Kangaroo Apple → SOLANUM sp.
Australian Tea Tree → CAJEPUT
Australian Umbrella Tree → BRASSAIA ACTINOPHYLLA
Austrapen → PENICILLINS – "NEWER and SYNTHETICS" > *Ampicillin*
Austrastaph → PENICILLINS – "NEWER and SYNTHETICS" > *Cloxacillin/Sodium salt*
Austrawolf → RAUWOLFIA SERPENTINA
Austyn → THEOPHYLLINE
Autan → *N,N*-DIETHYL-*m*-TOLUAMIDE
Autoprothrombin II → CHRISTMAS FACTOR
Autoworm → OXFENDAZOLE
Autumin → DICYCLOMINE
Autumn Crocus → COLCHICUM AUTUMNALE
Autumn Sneezeweed → HELENIUM sp.
Auxit → BROMHEXINE
Ava-ava → KAVA
Avadex → DIALLATE
Avagal → METHANTHELINE BROMIDE
Avandia → ROSIGLITAZONE
Avantyl → NORTRIPTYLINE
A 7 Vapam → METHAM SODIUM
Avasa → TEPHROSIA sp.
Avatec → LASALOCID
Avazyme → CHYMOTRYPSIN
Avena sativa → OATS
Aveolar-Capillary Block Syndrome → SAWDUST
Averon-1 → CEPHALOTHIN
Aventyl → NORTRIPTYLINE
Avermectin B_1 → ABAMECTIN
Avermectin B_1s → IVERMECTIN
Avertin → TRIBROMOETHANOL
Avibon → VITAMIN A
Avicalm → METOSERPATE
Avicel → CELLULOSE
Avicol → CHLORPHENTERMINE
Avicol → QUINTOZENE
Avid → ABAMECTIN
Aviderm → COLLAGEN
Avigilen → PIRACETAM
Avil → PHENIRAMINE MALEATE
Avipron → CHLORPHENTERMINE
Aviral → ISOPRINOSINE
Avisol → SULFATHIAZOLE
Avita → VITAMIN A
Avitene → COLLAGEN
Avitol → VITAMIN A
Avitriptan → PROPRANOLOL
Avitrol → AMINOPYRIDINE
Avlocardyl → PROPRANOLOL
Avlochlor → CHLOROQUINE
Avloclor → CHLOROQUINE
Avloprocil → PENICILLIN > *Benzylpenicillin, Procaine*
Avlosulfon → DAPSONE
Avlothane → HEXACHLOROETHANE
AVOBENZONE
AVOCADOS also → FRUITS and JUICES, LATEX, POTASSIUM METABISULFITE
Avocin → PENICILLINS – "NEWER and SYNTHETICS" > *Piperacillin Sodium*
Avolin → DIMETHYL PHTHALATE
Avomec → ABAMECTIN
Avonex → INTERFERONS
Avosyl → MEPHENESIN
Avoxyl → MEPHENESIN
Awassida → CUMIN, BLACK
AWD-08-250 → AMRINONE
Awueli → PHYTOLACCA sp.
Awuje → BEANS, Lima
AX 250 → PENICILLINS – "NEWER and SYNTHETICS" > *Amoxicillin*
Axer Alfa → NAPROXEN
Axerol → VITAMIN A
Axerophthol → VITAMIN A

Axid → NIZATIDINE
Axiquel → VALNOCTAMIDE
Axlon → VITAMIN B_{12}
Axonyl → PIRACETAM
Axoril → CEFUROXIME
Axsain → CAPSAICIN
Axuris → GENTIAN VIOLET
AY 5406-1 → BENACTYZINE
AY 6,108 → PENICILLINS – "NEWER and SYNTHETICS" > *Ampicillin*
AY 6608 → PENTAGASTRIN
AY 11,440 → CLOGESTONE ACETATE
AY 21,011 → PRACTOLOL
AY 22,989 → RAPAMYCIN
AY 24,031 → LH-RH
AY 24,236 → ETODOLAC
AY 27,773 → TOLRESTAT
AY 61,122 → METHALLIBURE
AY 62,013 → ETHOGLUCID
AY 62,014 → BUTRIPTYLINE
AY 62,021 → CLOPENTHIXOL
AY 62,022 → MEDROGESTONE
AY 64,043 → PROPRANOLOL
AYAHAUSCA also → CALATHEA, DIPLOPTERYS CABRERANA, JUANOLA OCHRACEA
Ayeramate → MEPROBAMATE
Ayerst → PRACTOLOL
Aygestin → NORETHINDRONE
Aza → PAULLINIA sp.
10-Azaanthracene → ACRIDINE
Azabenzene → PYRIDINE
Azacitidine → 5-AZACYTIDINE
Azacon → LEAD OXIDES
Azactam → AZTREONAM
AZACOSTEROL
AZACYCLONAL
5-AZACYTIDINE
Azacyclopropane → ETHYLENEIMINE
Azaform → AMITRAZ
AZALEA
Azalone → ANTAZOLINE
Azamune → AZATHIOPRINE
Azanin → AZATHIOPRINE
Azantac → RANITIDINE
Azapen → PENICILLINS – "NEWER and SYNTHETICS" > *Methicillin Sodium*
AZAPERONE also → ETORPHINE
Azapren → APAZONE
AZAPROPAZONE also → METHOTREXATE
Azarcon → GRETA also → LEAD
Azaron → TRIPELENENNAMINE
AZARIBINE
AZASERINE
AZATADINE DIMALEATE
AZATHIOPRINE also → ALLOPURINOL, CURARE, PREDNISONE, TRIMETHOPRIM
6-Azauracil Riboside → AZAURIDINE
AZAURIDINE
Azauridine Triacetate → AZARIBINE
AZELAIC ACID
AZELASTINE
Azeptin → AZELASTINE
Azide → POTASSIUM AZIDE
Azidin → DIMINAZENE ACETURATE
Azidothymidine → ZIDOVUDINE
Azimethylene → DIAZOMETHANE
Azine → PYRIDINE
AZINPHOS-ETHYL
AZINPHOS-METHYL
Aziridine → ETHYLENEIMINE
AZITHROMYCIN also → WARFARIN
Azium → DEXAMETHASONE
Azlin → PENICILLINS – "NEWER and SYNTHETICS" > *Azlocillin*
Azlocillin → PENICILLINS – "NEWER and SYNTHETICS" also → AMIKACIN
Azmacort → TRIAMCINOLONE
Azoangin → 2,4-DIAMINOAZOBENZENE
p-Azoaniline → 2,4-DIAMINOAZOBENZENE
AZOBENZENE
Azobenzide → AZOBENZENE
Azobenzol → AZOBENZENE
Azodicarbonamide → FLOUR
Azodrin → MONOCROTOPHOS
Azo-Gantrisine® → SULFISOXAZOLE
Azohel → 2,4-DIAMINOAZOBENZENE
Azoksodon → PEMOLINE
Azol → *p*-AMINOPHENOL
Azolid → PHENYLBUTAZONE
Azolmen → BIFONAZOLE
Azonam → AZTREONAM
Azophenylenebenzene → AZOBENZENE
Azophos → METHYL PARATHION
Azopt → BRINZOLAMIDE
Azoran → AZATHIOPRINE
Azo-Standard → PHENAZOPYRIDINE
Azo-Stat → PHENAZOPYRIDINE
AZOSULFAMIDE
Azothioprine → AZATHIOPRINE
Azotic Acid → NITRIC ACID
Azovan Blue → EVANS BLUE
Azramycin → CHLORAMPHENICOL
AZT → ZIDOVUDINE
Aztec → ZIDOVUDINE
Aztec Tobacco → NICOTIANA sp.
Az-threonam → AZTREONAM

Aztreon → AZTREONAM
AZTREONAM also → CEPHADRINE; PENICILLINS – "NEWER and SYNTHETICS" > *Nafcillin Sodium*
Azubromaron → BENZBROMARONE
Azucaps → PHENFORMIN
Azuglucon → GLYBURIDE
AZULENE
Azulfidine → SALICYLAZOSULFAPYRIDINE
Azupentat → PENTOXIFYLLINE
AzUR → AZARIBINE also → AZAURIDINE
Azutranquil → OXAZEPAM

B

B 9 → DIAMINOZIDE
B 360 → PAROXYPROPIONE
B 518 → CYCLOPHOSPHAMIDE
B 577 → ETOFENAMATE
B 622 → ANILAZINE
B 663 → CLOFAZIMINE
B 995 → DAMINOZIDE
B 1500 → IOPAMIDOL
B 1776 → *S,S,S*-TRI-*N*-BUTYL PHOSPHOROTRITHIOATE
B 2310 → MORINAMIDE
B 2353 → NICLOSAMIDE
B 4576 → PROFLURALIN
B 9053 → PHOXIM
B 66,256 → URAPIDIL
Ba 168 → LOFEXIDINE
BA 1420 → PROPANIDID
Ba 2758 → 2-DEOXY-D-GLUCOSE
Ba 5473 → OXYPHENONIUM BROMIDE
Ba 10,370 → SULFACHLORPYRIDAZINE
Ba 21,401 → TRIBENOSIDE
Ba 29,038 → BOLDENONE
Ba 29,837 → DEFEROXAMINES
BA 30,843 → BENZOTAMINE
BA 32,644 → NIRIDAZOLE
Ba 33,112 → DEFEROXAMINES
Ba 34,276 → MAPROTILINE
Ba 34,647 → BACLOFEN
Ba 39,089 → OXPRENOLOL HYDROCHLORIDE
BAAM → AMITRAZ
Baba → INDIGOFERA sp.
Babesan → QUINURONIUM
BABN → β-AMINOPROPIONITRILE
Babrocid → NITROFURAZONE
Baby Fae → CYCLOSPORINE
Baby Oil → MINERAL OIL
Baby Powder → KAOLIN
Baby Wood Rose → ARGYREIA NERVOSA
Bacacil → PENICILLINS – "NEWER and SYNTHETICS" > *Bacampicillin*
Bacampicillin → PENICILLINS – "NEWER and SYNTHETICS" also → METOCLOPRAMIDE
Bacampicine → PENICILLINS – "NEWER and SYNTHETICS" > *Bacampicillin*
Bacharrine → BACCHARIS
BACCHARIS sp.
Baccidal → NORFLOXACIN
Bacfeed → BICOZAMYCIN
Bach → CALAMUS
Bachelor's Buttons → NUX VOMICA
Bacillosporin B → POLYMYXIN B
Bacillus Calmette-Guérin → BCG
Bacimex → PENICILLINS – "NEWER and SYNTHETICS" > *Sultamicillin*
BACITRACIN also → CURARE, DIMETHYL TUBOCURARINE IODIDE, GENTAMICIN, MAGNESIUM, NEOMYCIN, SODIUM LAURYL SULFATE, SUCCINYLCHOLINE, *d*-TUBOCURARINE CHLORIDE, TYROTHRICIN, VITAMIN B_{12}
BACLOFEN also → IMIPRAMINE, LITHIUM
Baclon → BACLOFEN
Bacon → *N*-NITROSODIMETHYLAMINE, OCHRATOXINS, POTASSIUM NITRITE
Bacteron → BICOZAMYCIN
Bacticlens → CHLORHEXIDINE
Bactidan → ENOXACIN
Bactine → METHYLBENZETHONIUM CHLORIDE
Bactocill → PENICILLINS – "NEWER and SYNTHETICS" > *Oxacillin Sodium*
Bactol → IODOCHLORHYDROXYQUINOLINE
Bactopen → PENICILLINS – "NEWER and SYNTHETICS" > *Cloxacillin/Sodium salt*
Bactrafen → CICLOPIROX
Bactramin → SULFAMETHOXAZOLE
Bactrim → SULFAMETHOXAZOLE also → TRIMETHOPRIM
Bactris minor → PALMS
Bactromin → SULFAMETHOXAZOLE
Bactrovet → SULFADIMETHOXINE
Badil → GENTIAN VIOLET
Badoh Negra → MORNING GLORY
Bafhameritin-M → MEFENAMIC ACID
Bag → HEROIN
BAGASSE
Bagodryl → DIPHENHYDRAMINE
Bagolax → METHYL CELLULOSE
Bagpod → GLOTTIDIUM VESICARIUM

Bagpod Sesbane → SESBANIA sp.
Baifen → HEROIN
BAILEYA MULTIRADIATA
Baizhi → ANGELICA
Bajaten → INDAPAMIDE
Bakelite → PHENOL-FORMALDEHYDE RESINS
Bakenu → MELIA sp.
Bakers → BARLEY, CLOVES, KARAYA, POTASSIUM BROMATE, POTASSIUM PERSULFATE, POTASSIUM SORBATE, PROPYL GALLATE, SORBIC ACID
Bakers' Eczema → SODIUM BICARBONATE
Baking Soda → SODIUM BICARBONATE
Bakontal → BARIUM SULFATE
Baktar → SULFAMETHOXAZOLE
BAL → DIMERCAPROL
Balan → BENFLURALIN
BALANITES sp.
Baldiran → CONIUM MACULATUM
Balfin → BENFLURALIN
Baligoli → LEAD
Balmox → NABUMETONE
BALSAM, CANADIAN
Balsam Capivi → COPAIBA
Balsam Copaiba → COPAIBA
Balsam of Fir → BALSAM, CANADIAN
Balsam Pear → MOMORDICA CHARANTIA
BALSAM PERU also → BENZOIN GUM, BENZYL SALICYLATE, CINNAMIC ALCOHOL, CINNAMIC ALDEHYDE, CINNAMON, COPAIBA, GERANIOL, KRAMERIA, STORAX, VANILLA
BALSAM TOLU
Balsam Weed → IMPATIENS sp.
Bambec → BAMBUTEROL
BAMBUTEROL
BAMETHAN
BAMIFYLLINE
BAMIPINE
Bamo → MEPROBAMATE
Banabin-Sintyal → CHLORMEZANONE
Banafine → BENFLURALIN
Banagerm → BENZETHONIUM CHLORIDE
Banambax → PENTAMIDINE ISETHIONATE
BANANA also → L-DOPA, LATEX, NIALAMIDE, PROCARBAZINE
Banana Oil → ISOAMYL ACETATE
Ban Apple Gas → ISOBUTYL NITRITE
Ban-A-Worm → PYRANTEL
Bancaris → THENIUM CLOSYLATE
Bancolon → FAMOTIDINE
Banded Dye-murex → SHELLFISH
Bandol → CARBIPHENE
BANEBERRY
Banewort → BELLADONNA
Banex → DICAMBA
Banflex → ORPHENADRINE
BANGALA
Banisterine → HARMINE
Banistyl → FONAZINE MESYLATE
Banlin → PROPANTHELINE BROMIDE
Banminth → PYRANTEL
Banminth P → PYRANTEL
Banocide → DIETHYLCARBAMAZINE
Ban-okra → COCKLEBURS
Banol → CARBANOLATE
Banotu → HYOSCYAMUS
Bantenol → MEBENDAZOLE
Banthine Bromide → METHANTHELINE BROMIDE
Banthionine → METHIONINES
Bantu → 1-(1-NAPHTHYL)-2-THIOUREA
Banvel D → DICAMBA
BAPTISIA sp.
Baptitoxine → CYTISINE
Baqs → BUXUS sp.
BAQD-10 → DEQALINIUMS
Baquacil → POLY (HEXAMETHYLENE BIGUANIDE HCL)
Bara laniya → PORTULACA OLERACEA
Baralyme → SEVOFLURANE
Bara sheal-kanta → ARGEMONE MEXICANA
Baratol → INDORAMIN
Barazan → NORFLOXACIN
Barbados Nut → JATROPHA sp.
Barbamate → BARBAN
Barbamil → AMOBARBITAL
Barbamyl → AMOBARBITAL
BARBAN
Barbane → BARBAN
BABAREA VULGARIS
BARBASCO
Barbatimao → STRYPHNODENDRON OBOVATUM
BARBERRY
Barbeton Daisy → GERBERA
Barbiphenyl → PHENOBARBITAL
Barbipil → PHENOBARBITAL
BARBITAL also → ACENOCOUMAROL, BISHYDROXYCOUMARIN, HEMATOPORPHYRIN, METHYL SALICYLATE, PHENOBARBITAL
Barbitone → BARBITAL
Barbiturates → BEMEGRIDE, BENZTHIAZIDE, CALAMUS, CANNABIS, CHLOROTHIAZIDE, DESMOPRESSIN, DIPYRONE, DISULFIRAM,

ETHCHLORVYNOL, ETHYLBISCOUMACETATE, ETHYL ESTRANOL, FURAZOLIDONE, GLUCOSE, HYDROFLUMETHIAZIDE, HYDROXYZINE, MAGNESIUM OXIDE, MEPHENYTOIN, MEPROBAMATE, METHAMPHETAMINE, METHYCLOTHIAZIDE, METHYLDOPA, MILK, NALIDIXIC ACID, NEOMYCIN, NITROFURANTOIN, OUABAIN, PARGYLINE, PHENPROCOUMOM, PHENYTOIN, PIMINODINE, PROCARBAZINE, QUINIDINE, RESERPINE, SECOBARBITAL, SODIUM BICARBONATE, SODIUM IODIDE, SULFOBROMOPHTHALEIN SODIUM, TESTOSTERONE
Barbosec → SECOBARBITAL SODIUM
Barc → ISOBORNYL THIOCYANOACETATE
Bardane → BURDOCK
Bardock → BURDOCK
Bareon → LOMEFLOXACIN
Baridol → BARIUM SULFATE
Baritop → BARIUM SULFATE
BARIUM also → LATEX, MOTOR OIL, TANNIC ACID, TETRACHLOROETHYLENE, TRICHLOROFLUOROMETHANE
BARIUM CARBONATE
BARIUM CHLORIDE
BARIUM HYDROXIDE
BARIUM SULFATE also → CELLULOSE, COCHINEAL, GOLF BALLS
BARIUM SULFIDE
Barizin → NICARDIPINE
"Bark-stripper's Disease" → MAPLE
Bark-stripper's Lung → CRYPTOSTROMA CORTICALE
BARLEY also → MALT, T-2 TOXIN, TETRACYCLINE
Baron → ERBON
Baros → DIMETHICONE (*with silicon dioxide*)
Baros Camphor → BORNEOL
Barosperse → BARIUM SULFATE
Barotrast → BARIUM SULFATE
Barquat → BENZALKONIUM CHLORIDE
Barquinol → IODOCHLORHYDROXYQUINOLINE
Barricade → CYPERMETHRIN
BARRINGTONIA sp.
Barseem → TRIPHOLIUM ALEXANDRINUM
BARTERIA FISTULOSA
BARTHRIN
BAS 352F → VINCLOZOLIN
Basanite → DINOSEB
Basecil → METHYLTHIOURACIL
Basedol → 2-AMINOTHIAZOLE
Basethyrin → METHYLTHIOURACIL
Basic Lead Carbonate → LEAD CARBONATE
Basic Orange 2 → 2,4-DIAMINOAZOBENZENE
Basic Yellow 2 → AURAMINE
BASILIXIMAB also → DACLIZUMAB
BASIL OIL, SWEET
Basodexan → UREA
Basofortina → METHYLERGONOVINE
Basolan → METHIMAZOLE
Bassa → BUTACARB
BASSIA sp.
Basswood Tree → TILIA sp.
Bastard Anise → SHIKIMIC ACID
Bastard Lentil → LENTILS
Bastel → EBASTINE
Bastiverit → GLYBURIDE
Basuco → COCAINE
Basudin → DIAZINON
"Bathtub Gin" → α-METHYL FENTANYL
Batiator Root → VERNONIA sp.
BATRACHOTOXIN also → FROGS
Bat Ray → STINGRAYS
BATTERIES also → MERCURY, POTASSIUM HYDROXIDE, SODIUM HYDROXIDE
Battery Acid → SULFURIC ACID
Baurere → GLORIOSA
Bauxite → ALUMINUM OXIDE
Bavachi → PSORALEA sp.
Bavistin → METHYL BENZIMIDAZOLE-2-CARBAMATE
Bawchang → PSORALEA sp.
Bax → DIPHENHYDRAMINE
BAX 1515 → SUTILAINS
BAX 1526 → CHYMOPAPAIN
Baxan → CEFADROXIL
Baxo → PIROXICAM
BAX 1400Z → DIMETHADIONE
BAX 2739Z → BAMIFYLLINE
Bay 1470 → XYLAZINE
Bay 1500 → MEFRUSIDE
Bay 4503 → PROPIRAM FUMARATE
Bay 5024 → MERCAPTODIMETHUR
Bay 5621 → PHOXIM
Bay 8173 → DEMETON
Bay 9002 → PHTHALOPHOS
Bay 9010 → PROPOXUR
Bay 9027 → AZINPHOS-METHYL
Bay σ 9867 → CIPROFLOXACIN
Bay 10,756 → DEMETON

Bay 11,405 → METHYL PARATHION
Bay 17,147 → AZINPHOS-METHYL
Bay 25,141 → FENSULFOTHION
Bay 25,634 → COUMATETRALYL
Bay 38,819 → GOPHACIDE
Bay 39,007 → PROPOXUR
Bay 44,646 → AMINOCARB
Bay 68,138 → FENAMIPHOS
Bay 70,143 → CARBOFURAN
Bay 77,488 → PHOXIM
Bay 78,418 → EDIFENPHOS
Bay 92,114 → ISOFENPHOS
Bay a 1040 → NIFEDIPINE
Bay b 5097 → CLOTRIMAZOLE
Baycaron → MEFRUSIDE
Baycid → FENTHION
Baycol → CERIVASTATIN
Baycovin → DIETHYL PYROCARBONATE
Baycillin → PENICILLINS – "NEWER and SYNTHETICS" > *Propicillin Potassium*
Baycipen → PENICILLINS – "NEWER and SYNTHETICS" > *Mezlocillin*
Bay E 601 → METHYL PARATHION
Bay E 605 → PARATHION
Bay E 5009 → NITRENDIPINE
Bay E 6905 → PENICILLINS – "NEWER and SYNTHETICS" > *Azlocillin*
Bay e-9736 → NIMODIPINE
Bayer 128 → APROTININ
Bayer 205 → SURAMIN SODIUM also → QUINURONIUM
Bayer 1219 → TRIMEPRAZINE
Bayer 1362 → BUTAPERAZINE
Bayer 1420 → PROPANIDID
Bayer 2349 → TRICHLORFON
Bayer 2353 → NICLOSAMIDE
Bayer 2502 → NIFURTIMOX
Bayer 3231 → TRIAZIQUONE
Bayer 5312 → ETHIONAMIDE
Bayer 5630 → METRONIDAZOLE
Bayer 8169 → DEMETON
Bayer 9051 → TETRAMISOLE
Bayer 16259 → AZINPHOS-ETHYL
Bayer 19,639 → DEMETON
Bayer 21/116 → DEMETON-METHYL
Bayer 21/199 → COUMAPHOS
Bayer 25,820 → PHTHALOPHOS
Bayer 29,493 → FENTHION
Bayer 37,344 → MERCAPTODIMETHUR
Bayer 41,831 → SUMITHION
Bayer 71,628 → METHAMIDOPHOS
Bayer A-128 → APROTININ
Bayer A 173 → PROCHLORPERAZINE
Bayer E-39 → INPROQUONE
Bayer E 393 → SULFOTEP
Bayer L 1359 → TRICHLORFON
Bayer S 940 → PHTHALOPHOS
Bayer S 5660 → SUMITHION
Bay f 1353 → PENICILLINS – "NEWER and SYNTHETICS" > *Mezlocillin*
Bay g-5421 → ACARBOSE
Baygon → PROPOXUR
Bay h 4502 → BIFONAZOLE
Bay K-4200 → ESTAZOLAM
Bay K-5522 → NISOLDIPINE
BAY LEAF also → FRUIT and FRUIT JUICES
Bay m 1099 → MEGLITOL
Baymal® → ALUMINUM OXIDE
Baymicin → SISOMICIN
Baymix → COUMAPHOS
Baymycard → NISOLDIPINE
BAY NTN 9306 → SULPROFOS
BAY NTN 33,893 → IMIDACLOPRID
Bayol → MINERAL OIL
Bayotensin → NITRENDIPINE
Baypen → PENICILLINS – "NEWER and SYNTHETICS" > *Mezlocillin*
Baypress → NITRENDIPINE
Bayrena → SULFAMETER
Bayrogel → ETOFENAMATE
Baytex → FENTHION
Baythion → PHOXIM
Baytril → ENROFLOXACIN
Bay Va 1470 → XYLAZINE
BAY VP-2674 → ENROFLOXACIN
Bay-W 6228 → CERIVASTATIN
Bazooka → COCAINE
BBC 12 → DIBROMOCHLOROPROPANE
BB-K8 → AMIKACIN
BC 51 → HEXAMARIUM BROMIDE
BC 105 → PIZOTYLINE
BCG
BCM → MANNOMUSTINE
BCME → *sym*-DICHLOROMETHYL ETHER also → BIS (CHLOROMETHYL) ETHER
BCNU → CARMUSTINE
BD-4OA → FORMOTEROL
BDF 5,895 → MOXONIDINE
BDH-312 → MEPHENESIN
B-D® Syringes → RUBBER
BE 5,895 → MOXONIDINE
Beaked Nightshade → SOLANUM sp.
BEANS also → AMPHETAMINE, RAFFINOSE
Beans, Fava → BEANS also → ISOCARBOXAZID, L-DOPA, NIALAMIDE
Beans, Kidney → BEANS also → PEANUTS

Beans, Lima → BEANS also → PHASEOLUTIN, Preface
Bean Trefoil → BUCKBEAN
Bean-Vine → DOLICHOS LALAB
Beanweed → LUPINES
Beaprine → CARSALAM
Bearberry Bark → CASCARA SAGRADA
Bear Grass → NOLINA TEXANA
Bearwood → CASCARA SAGRADA
Beatrice Foods → TRICHLOROETHYLENE
Beast → LYSERGIC ACID DIETHYLAMIDE
Bebate → BETAMETHASONE
Bébé → HEXACHLOROPHENE
Beben → BETAMETHASONE
Be-Calcipas → BENZOYLPAS
BECANTHONE
Becaptan → CYSTEAMINE
Becenun → CARMUSTINE
Bécilan → VITAMIN B_6
BECLAMIDE
Beclacin → BECLAMETHASONE
Beclipur → BECLOBRATE
BECLOBRATE
Becloforte → BECLOMETHASONE
Beclomet → BECLOMETHASONE
BECLOMETHASONE
Beclorhinol → BECLOMETHASONE
Becloslerin → BECLOBRATE
Becloval → BECLOMETHASONE
Beclovent → BECLOMETHASONE
Becodisks → BECLOMETHASONE
Beconase → BECLOMETHASONE
Beconasol → BECLOMETHASONE
Becort → BETAMETHASONE
Becotide → BECLOMETHASONE
Bedermin → BETAMETHASONE
Bedodeka → VITAMIN B_{12}
Bedoz → VITAMIN B_{12}
Bedranol → PROPRANOLOL
Bedriol → BIFONAZOLE
Beebrush → ALOYSIA LYCIOIDES
Beech → FAGUS SYLVATICA
Beef → PHENACEMIDE
Beef Patties → HAMBURGERS
Beef Steak Plant → PERILLA FRUTESCENS
Beef Tea → BOUILLON
Bee Glue → PROPOLIS
BEER also → ALMONDS, COBALT, COBALT SULFATE, HAIR, HAZEL NUTS, HISTAMINE, IRON, MALT, MESCALINE, METHYPRYLONE, MILLET, MUSHROOMS, NIALAMIDE, NIFURATEL, *N*-NITROSODIMETHYLAMINE, OCHRATOXINS, PENICILLIN, POTASSIUM METABISULFITE, SALVIA, SODIUM BISULFITE, SODIUM NITRITE, SULFUR DIOXIDE, TEQUILA, TETRACYCLINE
BEES also → DICLOFENAC, HISTAMINE
BEESWAX also → PENICILLIN
Bee-Seventeen® → APRICOT, HYDROGEN CYANIDE
Beethoven → LEAD
BEETS also → ANTHRAQUINONE, NITRATES and NITRITES, SODIUM GLUTAMATE, SOLANUM sp.
Beet Sugar → SUCROSE
Befeniol → BEPHENIUM
Befizal → BEZAFIBRATE
Begalin → SULBACTAM
Beggars → RANUNCULUS
Beggar's Buttons → BURDOCK
BEHENIC ACID
Behepan → VITAMIN B_{12}
Bei-1293 → XIPAMIDE
Bekunis → OXYPHENISATIN
Beleño → HYOSCYAMUS
Belfacillin → PENICILLINS – "NEWER and SYNTHETICS" > *Methicillin Sodium*
Belflavin → VITAMIN B_2
BELLADONNA also → STRAMONIUM
Bellafolina → *l*-HYOSCYAMINE
Bellasthman → ISOPROTERENOL
Bellergal® → ERGOTAMINE
Bellyache Bush → FLAIRA
Belmark → FENVALERATE
Beloc → METOPROLOL
Belseren → CLORAZEPATE DIPOTASSIUM
Belt → CHLORDANE
Belustine → LOMUSTINE
Belvedere → KOCHIA
Bemachol → CHLORAMPHENICOL
Bemaco → CHLOROQUINE
Bemaphate → CHLOROQUINE
Bemasulph → CHLOROQUINE
BEMEGRIDE
Benaalgin → BENZYDAMINE
Benacol → DICYCLOMINE
BENACTYZINE also → MEPROBAMATE
Benadon → VITAMIN B_6
Benadryl → DIPHENHYDRAMINE
Ben-a-hist → ANTAZOLINE
Benapen → PENICILLIN > *Benzylpenicillin, Benethamine*
Benaquin → CHLOROQUINE
Benason → LINDANE
Benavit → VITAMIN B_1
BENAZEPRIL

BENDAZAC
Bendazolic Acid → BENDAZAC
Bendectin → DICYCLOMINE, DOXYLAMINE SUCCINATE
BENDIOCARB
Bendogen → BETHANIDINE
Bendopa → L-DOPA
Bendralin → PENICILLINS – "NEWER and SYNTHETICS" > *Phenethicillin Potassium*
Bendrofluazide → BENDROFLUMETHIAZIDE
BENDROFLUMETHIAZIDE also → LITHIUM, VITAMIN D_3
Bendylaten → DIPHENHYDRAMINE
Benecardin → KHELLIN
Benefex → BENFLURALIN
Benefin → BENFLURALIN
Benemid → PROBENECID
Benerva → VITAMIN B_1
Benetolin → PENICILLIN > *Benzylpenicillin, Benethamine*
BENFLURALIN
Benfofen → DICLOFENAC
Benfotiamine → VITAMIN B_1
Bengal Gram → PEAS
Ben-Hex → LINDANE
Benicot → VITAMIN B_3
Benirol → BENZALKONIUM CHLORIDE
Benisone → BETAMETHASONE
Benlate → BENOMYL
Benne → SESAME
Benocten → DIPHENHYDRAMINE
Benodine → DIPHENHYDRAMINE
BENOMYL
Benoquin → MONOBENZONE
Benoral → BENORYLATE
Benortan → BENORYLATE
BENORYLATE also → ACETYLSALICYLIC ACID
BENOXAPROFEN
Benoxyl → BENZOYL PEROXIDE
BENPERIDOL
BENSERAZIDE also → TOLCAPONE
Benson, James → MINERAL WATERS
BENSULIDE
Bent → CHLORDIAZEPOXIDE
Bentalan → BETAMETHASONE
"Bent Leg Syndrome" → TRACHEMENE GLAUCIFOLIA
Bentominen → DICYCLOMINE
BENTONITE also → ALUMINUM SILICATE, RIFAMPIN, ROBENIDINE
Bentonyl → TROLNITRATE PHOSPHATE
Bentyl → DICYCLOMINE
Bentylol → DICYCLOMINE
Benuride → PHENETURIDE
Benuron → BENDROFLUMETHIAZIDE
Ben-u-ron → ACETAMINOPHEN
Ben-u-ron Baby → ACETAMINOPHEN
Benuryl → PROBENECID
Benvil → TYBAMATE
Benylin DM → DEXTROMETHORPHAN HYDROBROMIDE
Benzacyl → BENZOYLPAS
Benzagel 10 → BENZOYL PEROXIDE
Benzaidin → BETHANIDINE
Benz-aken → BENZOYL PEROXIDE
BENZALDEHYDE
Benzalin → NITRAZEPAM
BENZALKONIUM CHLORIDE also → KAOLIN, MAGNESIUM SILICATE, POLYVINYL CHLORIDE
Benz-all → BENZALKONIUM CHLORIDE
Benz [α] anthracene → 1,2-BENZANTHRACENE
1,2-BENZANTHRACENE
Benzanthrene → 1,2-BENZANTHRACENE
Benzantin → DIPHENHYDRAMINE
Benzapas → BENZOYLPAS
BENZBROMARONE
Benzchlorpropamide → BECLAMIDE
Benzcurine Iodide → GALLAMINE TRIETHIODIDE
Benzedrex® → PROPYLHEXEDRINE
Benzedrine → AMPHETAMINE
Benzefit → METHAMPHETAMINE
Benzelmin → OXFENDAZOLE
Benzenamine → ANILINE
BENZENE also → CYCLOHEXANE, ETHYLBENZENE, GASOLINE, GLUE, MINERAL WATERS, PHENOBARBITAL, RUBBER, SHOES, TOLUENE, XYLENE
Benzeneazobenzene → AZOBENZENE
Benzenecarbonyl Chloride → BENZOYL CHLORIDE
Benzenecarboxylic Acid → BENZOIC ACID
Benzene Chloride → CHLOROBENZENE
1,2-Benzenediol → PYROCATECHOL
1,3-Benzenediol → RESORANTEL
1,4-Benzenediol → HYDROQUINONE
Benzene Ethanol → ALCOHOL, PHENYLETHYL
BENZENE HEXACHLORIDE also → CYCLOHEXANE, DIAZOXIDE, PHENYLTHIOUREA
γ-Benzene Hexachloride → LINDANE
Benzenemethanol → ALCOHOL, BENZYL
BENZENESULFONYL CHLORIDE
1,3,5-Benzenetriol → PHLOROGLUCINOL

Benzenyl Trichloride → BENZOTRICHLORIDE
Benzetacyl → PENICILLIN > *Benzylpenicillin, Benzathine*
Benzethacil → PENICILLIN > *Benzylpenicillin, Benzathine*
BENZETHONIUM CHLORIDE
Benzhydramine → DIPHENHYDRAMINE
BENZIDINE also → 3,3´-DICHLOROBENZIDINE, σ-TOLIDINE
Benzidine Yellow → INKS, NEWSPAPERS
Benzin → GASOLINE also → NAPHTHA
Benzindamine → BENZYDAMINE
Benzinoform → CARBON TETRACHLORIDE
BENZIODARONE also → DIPHENADIONE
Benzitramide → BEZITRAMIDE
BENZNIDAZOLE
BENZOCAINE also → BENZOIC ACID, BUTACAINE, ETHYL-*p*-AMINOBENZOATE, HYDROFLUMETHIAZIDE
BENZOIC ACID also → BENZALDEHYDE, BENZYL BENZOATE, CHLORPROPAMIDE, CRANBERRY, DIAZEPAM, POPLAR, SALICYL ALCOHOL
Benzodiapin → CHLORDIAZEPOXIDE
Benzoepin → ENDOSULFAN
BENZOIC ACID
BENZOIN GUM also → CAROB, GALBANUM GUM, GAMBOGE, MYRRH, OLIBANUM
Benzol → BENZENE
Benzolin → DIBUCAINE
BENZONATATE
BENZONITRILE
BENZO (g, h, i) PERYLENE
BENZOPHENONES also → OXYBENZONE
Benzophenone-4 → SULISOBENZONE
Benzophenone 8 → DIOXYBENZONE
Benzophosphate → PHOSALONE
BENZO [α] PYRENE also → CARBONS, CATECHOL, CROTON, HAM, HYDROGEN PEROXIDE, OLIVES, PHENOBARBITAL, PIPERONYL BUTOXIDE, TURKEY
3,4-Benzopyrene → BENZO [α] PYRENE
Benzopyrone → COUMARIN
Benzoguin → MONOBENZONE
1,4-Benzoguinone → QUINONE
p-Benzoguinones → MILLIPEDES
Benzosulfimide → SACCHARIN
BENZOTAMINE
Benzothiazide → BENZTHIAZIDE
Benzotran → OXAZEPAM
BENZOTRICHLORIDE
Benzoxazocine → NEFOPAM
Benzoxine → BETHANIDINE
BENZOYL CHLORIDE
Benzoylmethylecgonine → COCAINE
BENZOYLPAS
BENZOYL PEROXIDE also → FLOUR
s-Benzoylthiamine → VITAMIN B_1
α-Benzoyltriethylamine → DIETHYLPROPION
Benzperidol → BENPERIDOL
1,2-Benzphenanthrene → CHRYSENE
2,3-Benzphenanthrene → 1,2-BENZANTHRACENE
BENZPHETAMINE also → PHENYTOIN
BENZQUINAMIDE
BENZTHIAZIDE
BENZTROPINE MESYLATE also → HALOPERIDOL, NIALAMIDE
BENZYDAMINE
Benzydroflumethiazide → BENDROFLUMETHIAZIDE
BENZYL ACETATE
Benzylamine → FENTANYL
Benzylbenzene → DIPHENYLMETHANE
BENZYL BENZOATE
Benzyl Carbinol → ALCOHOL, PHENYLETHYL
BENZYL CHLORIDE
Benzylcyanide → CORONOPUS
BENZYL FORMATE
Benzylhydroflumethiazide → BENDROFLUMETHIAZIDE
Benzyl Hydroguinone → MONOBENZONE
BENZYL ISOBUTYRATE
BENZYL ISOEUGENOL
Benzylisothiocyanate → CORONOPUS, CRESS
BENZYL ISOVALERATE
Benzylmercaptan → CORONOPUS
Benzyl-2-methyl Propionate → BENZYL ISOBUTYRATE
Benzylmethylsulfide → CORONOPUS
BENZYL NICOTINATE
Benzylpenicillin → PENICILLIN
Benzylpenicillin Ammonium → PENICILLIN
Benzylpenicillin, Benethamine → PENICILLIN
Benzylpenicillin, Benzathine → PENICILLIN
Benzylpenicillin, DAEEH → PENICILLIN
Benzylpenicillin, l-Ephenamine → PENICILLIN
Benzylpenicillinic Acid → PENICILLIN
Benzylpenicillin, Potassium → PENICILLIN
Benzylpenicillin, Procaine → PENICILLIN also → PENICILLIN, *Potassium Benzylpenicillinate*

Benzylpenicillin, Sodium → PENICILLIN
BENZYL PHENYLACETATE
Benzyl Propanoate → BENZYL PROPIONATE
BENZYL PROPIONATE
Benzyl-Rodiuran → BENDROFLUMETHIAZIDE
BENZYL SALICYLATE
Benzyl Thiocyanate → CRESS
Benzyltol → CHLOROXYLENOL
Benzyl α-toluate → BENZYL PHENYLACETATE
Benzyrin → BENZYDAMINE
Beocid-Puroptal → SULFACETAMIDE SODIUM
Bepadin → BEPRIDIL
Bepanthen → DEXPANTHENOL
BEPHENIUM
Beprane → PROPRANOLOL
BEPRIDIL also → RITONAVIR
Béprochine → PAMAQUIN
BEQUAERTIODENDRON MAGALIESMONTANUM
Berbamine → BARBERRY
Berbéria → OLEANDER
BERBERINE also → BARBERRY, CHELIDONIUM MAJUS, MAHONIA AQUIFOLIUM
Berberis sp. → BARBERRY
Berculon A → THIACETAZONE
Berenil → DIMINAZENE ACETURATE
Bergacaf → CEFAMANDOL
BERGAMOT OIL also → COLOGNES
Bergapten(e) → BERGAMOT OIL, HERACLEUM sp., METHOXSALEN, PARSLEY, RUE
Beristypt → COLLAGEN
Berkatens → VERAPAMIL
Berkfurin → NITROFURANTOIN
Berkmycen → OXYTETRACYCLINE
Berkolol → PROPRANOLOL
Berkomine → IMIPRAMINE
Berkozide → BENDROFLUMETHIAZIDE
Berlandier Nettlespurge → JATROPHA sp.
Berlin Blue → FERRIC FERROCYANIDE
Bermuda Buttercup → OXALIS sp.
Bermuda Grass → CYNODON
Bernice → COCAINE
Bernies → COCAINE
Bernocaine → PROCAINE
Berocillin → PENICILLINS – "NEWER and SYNTHETICS" > *Pivampicillin*
Berolase → COCARBOXYLASE
Beromycin → PENICILLIN > *Penicillin V*
Beronald → FUROSEMIDE
Berotec → FENOTEROL HYDROBROMIDE
Berry Alder → FRANGULA
BERSAMIA sp.
Berubi → VITAMIN B_{12}
Berubigen → VITAMIN B_{12}
BERYLLIUM also → LANTERNS, MISSILE SILO
Besnoline → TOLPERISONE HCL
Bespar → BUSPIRONE
Be-Still-Tree → THEVETIA sp.
Besuric → BENZBROMARONE
Betabactyl → PENICILLINS – "NEWER and SYNTHETICS" > *Ticarcillin Disodium*
Betabion → VITAMIN B_1
Beta-Cardone → SOTALOL
Betacef → CEFOXITIN SODIUM
Beta-Chlor → CHLORAL BETAINE
Betadexamethasone → BETAMETHASONE
Betadine® → IODINE also → HYDROGEN PEROXIDE, POLYVINYLPYRROLIDONE IODINE COMPLEX, POVIDONE
Betadival → BETAMETHASONE
Betafedrina → DEXTROAMPHETAMINE
d-Betafedrine → DEXTROAMPHETAMINE SULFATE
Betafluorene → BETAMETHASONE
Betagon → MEPINDOLOL also → LEVOBUNOLOL
BETAHISTINE
BETAINE
Betaisodona → POLYVINYLPYRROLIDONE IODINE COMPLEX
Betalin → VITAMIN B_1
Betalin-12 → VITAMIN B_{12}
Betaling → BETHANIDINE
Betaloc → METOPROLOL
Beta macrocarpa → SWISS CHARD
Betamann → TRIMEPRANOL
Betamaze → SULBACTAM
BETAMETHASONE
Betamox → PENICILLINS – "NEWER and SYNTHETICS" > *Amoxicillin*
Betanamin → PEMOLINE
Betanaphthoquinone → β-NAPHTHOL
Beta-Neg → PROPRANOLOL
Betanidine → BETHANIDINE
Betanidol → BETHANIDINE
Betapace → SOTALOL
Betapen → PENICILLIN > *Benzylpenicillin, Benethamine*
Betapressin → PENBUTOLOL
Betaprone → β-PROPIOLACTONE
Betaptin → ALPRENOLOL
Betasan → BENSULIDE
Betaserc → BETAHISTINE

Betaseron → INTERFERONS
Betasolon → BETAMETHASONE
Beta-Tablinen → PROPRANOLOL
Beta-Timelets → PROPRANOLOL
Beta vulgaris → BEETS also → SWISS CHARD
Betaxin → VITAMIN B_1
Betaxina → NALIDIXIC ACID
BETAXOLOL also → NIFEDIPINE
BETAXOLOL
BETAZOLE
BETEL also → ARECA
Bethacil → PENICILLINS – "NEWER and SYNTHETICS" > *Sultamicillin*
Bethamidine → PHENYLPROPANOLAMINE
BETHANECOL also → AMITRIPTYLINE,
BETHANIDINE also → AMITRIPTYLINE, AMPHETAMINE, DESIPRAMINE, IMIPRAMINE, PSEUDOEPHEDRINE
Bethrodine → BENFLURALIN
Betim → TIMOLOL
Betnelan → BETAMETHASONE
Betnesol → BETAMETHASONE
Betnesol-V → BETAMETHASONE
Betneval → BETAMETHASONE
Betnovate → BETAMETHASONE
Betolvex → VITAMIN B_{12}
Betoptic → BETAXOLOL
Betoptima → BETAXOLOL
Betsovet → BETAMETHASONE
Betula → BIRCH
Beuthanasia® → PENTOBARBITAL
Bevonium Methyl Sulfate → PIRIBENZYL METHYL SULFATE
Bewon → VITAMIN B_1
BEXAROTENE
Bexide → BISETHYLXANTHOGEN
Bexone → 4(2-METHYL-4-CHLOROPHENOXY) BUTYRIC ACID
Bextasol → BETAMETHASONE
Bexton → PROPACHLOR
BEZAFIBRATE also → PHENPROCOUMON
Bezalip → BEZAFIBRATE
Bezatol → BEZAFIBRATE
BEZITRAMIDE
BGE → *n*-BUTYL GLYCIDYL ETHER
BGH → GROWTH HORMONES
BH 6 → TOXOGONIN
BHA → BUTYLATED HYDROXYANISOLE
Bhang → CANNABIS
Bhangra → COCKLEBURS
BHC → BENZENE HEXACHLORIDE
γ-BHC → LINDANE
BHT → BUTYLATED HYDROXYTOLUENE
BI 58 → DIMETHOATE
BIALAMICOL
Bialatan → METHACYCLINE
Biallyamicol → BIALAMICOL
Bialzepam → DIAZEPAM
Biaxin → CLARITHROMYCIN
Biazolina → CEFAZOLIN SODIUM
Bibenzene → DIPHENYL
BIBR 277 → TELMISARTAN
BICALUTAMIDE
Bichhubooti → URTICA DIOICA
Bichloracetic Acid → DICHLOROACETIC ACID
Bichloride of Mercury → MERCURIC CHLORIDE
Bichloro de mercurio → MERCURIC CHLORIDE
Bichy Nuts → KOLA
Bicillin → PENICILLIN > *Benzylpenicillin, Benzathine*
Bickie-mol → ACETAMINOPHEN
Biclin → AMIKACIN
Bicloxil → PENICILLINS – "NEWER and SYNTHETICS" > *Cloxacillin/Sodium salt*
BiCNU → CARMUSTINE
Bicol → BISACODYL
Bicor → TERODILINE
Bicosal → FLUOCINONIDE
Bicotussin → HYDROCODONE BITARTRATE
BICOZAMYCIN
Bicuculla → CORYDALIS
BICUCULLINE also → CORYDALIS, DICENTRA CUCULLARIA
Bicyclomycin → BICOZAMYCIN
Bidocef → CEFADROXIL
Bidrin → DICROTOPHOS
Biebrich Scarlet Red → SCARLET RED
BIETASERPINE
Biethylamicol → BIALAMICOL
Biethylene → 1,3-BUTADIENE
Bifex → PROPOXUR
Bifiteral → LACTULOSE
BIFONAZOLE
Biformyl → GLYOXAL
Biforon → BUFORMIN
Big Bed Bug → TRIATOMA sp.
Big C → COCAINE
Big Chief → MESCALINE also → PEYOTE
Big D → LYSERGIC ACID DIETHYLAMIDE
"Big Head" → TRIBULUS TERRESTRIS
Big Leaf Sage → LANTANA sp.
Big Sage → LANTANA sp.
Bigunal → BUFORMIN
Bihypnal → DICHLORALANTIPYRINE
Biklin → AMIKACIN
Biletan → THIOCTIC ACID

Bilevon → HEXACHLOROPHENE
Bilibyk → IOBENZAMIC ACID
Bilidren → DEHYDROCHOLIC ACID
Biligrafin → MEGLUMINE IODIPAMIDE
Biligram → IOGLYCAMIC ACID
BILIRUBIN also → CAFFEINE, CHLORPROMAZINE, ETHOXAZINE, MERCAPTOPURINE, NOVOBIOCIN, PHENACEMIDE, PHENOBARBITAL, PIPOBROMAN, SULFADIAZINE, SULFAPYRIDINE, TETRACYCLINE
Bilivistan → IOGLYCAMIC ACID
Biloptin → IPODATE
Bilordyl → THEOPHYLLINE
Bilostat → DEHYDROCHOLIC ACID
Biltricide → PRAZIQUANTEL
Bimaran → TRAZODONE HCL
Bimethyl → ETHANE
BINAPACRYL
Bindan → PHENINDIONE
Bindazac → BENDAZAC
Bini da zugu → JATROPHA sp.
Binnell → BENFLURALIN
Binodrenal → NOREPINEPHRINE
Binotal → PENICILLINS – "NEWER and SYNTHETICS" > *Ampicillin*
BIO 5462 → ENDOSULFAN
Biobamat → MEPROBAMATE
Biocaine → PROCAINE
Biocefalin → PYRITINOL DIHYDROCHLORIDE
Biociclin → CEFUROXIME
Biocide N-521 → DAZOMET
Bio-Cox → SALINOMYCIN
Bio-Dopa → L-DOPA
Biodry → NOVOBIOCIN
Biofanal → NYSTATIN
Biofurex → CEFUROXIME
Biogastrone → CARBENOLOXONE
Bioheprin → HEPARIN
Biohulin → INSULIN
BIOLF-62 → GANCICLOVIR
Biomag → CIMETIDINE
Biomioran → CHLORZOXAZONE
Biomitsin → CHLORTETRACYCLINE
Biomycin → CHLORTETRACYCLINE
Bio-Mycin → OXYTETRACYLINE
Bionicard → NICARDIPINE
Bioperidolo → HALOPERIDOL
Bioplex → CARBENOLOXONE
Biopor → POLYETHYLENE
Bioral → CARBENOLOXONE
Biorphen → ORPHENADRINE
Bioscleran → CLOFIBRATE
Biosinon → ETHAMSYLATE
Biostat → OXYTETRACYCLINE
Biosterol → VITAMIN A
Bio-tal → THIAMYLAL SODIUM
Bioterciclin → DEMECLOCYCLINE
Bio-Tetra → TETRACYCLINE
Biotexin → NOVOBIOCIN
Biotirmone → DEXTROTHYROXINE SODIUM
Biothion → TEMEPHOS
Biotite → MICA
Biovetin → CHLORTETRACYCLINE
Biovit → CHLORTETRACYCLINE
Bioxima → CEFUROXIME
Bioxiran → ERYTHRITOL ANHYDRIDE
Biozolene → FLUCONAZOLE
BIPERIDEN
Biphenabid → PROBUCOL
BIPHENYL
1,1′-Biphenyl → DIPHENYL
p-BIPHENYLAMINE also → *p*-AMINOBIPHENYL
Bipiquin → CHLOROQUINE
BIPP → IODOFORM
2,2′-Bipyridine → α,α'-DIPYRIDYL
Biquin → QUINIDINE
Birana → CROTALARIA, *retusa*
BIRCH
Birdlime → MISTLETOE, EUROPEAN or JAPANESE
Bird Rape → BRASSICA sp.
Bird of Paradise → STRELITZIA REGINA
Bird Repellents → AMINOPYRIDINE, DIMETHYL ANTHRANILATE
Birdsfoot Trefoil → LOTUS CORNICULATUS
Birdsville Indigo → INDIGOFERA sp.
BI-RG 587 → NEVIRAPINE
Birlane → CHLORFENVINPHOS
BISABOLENE
Bisacetylhomotaurine → ACAMPROSATE
BISACODYL also → MILK
Bisatin → OXYPHENISATIN
Bisbâs → PEPPERS
Bis(2-chloroethyl)ether → *sym*-DICHLOROETHYL ETHER
BIS(CHLOROMETHYL) ETHER also → *sym*-DICHLOROMETHYL ETHER
Bis(2-dimethylamino)ethyl Ether → NIAX® CATALYST ESN
[Bis(dimethylamino)-2′-3′-propyl]-10 Phenothiazine → AMINOPROMAZINE
Biseptol → TRIMETHOPRIM
BIS(2-ETHYLHEXYL)PHTHALATE also → METHOXYFLURANE PACLITAXEL, POLYVINYL CHLORIDE

BISETHYLXANTHOGEN
Bishop's Weed → AMMI MAJUS
BISHYDROXYCOUMARIN also → ACETAMINOPHEN, BARBITAL, CHLORAL HYDRATE, CHLORDANE, CHLORPROPAMIDE, CHLORTETRACYCLINE, CLOFIBRATE, CLOVERS, DEXTROTHYROXINE, DIAZEPAM, ETHCHLORVYNOL, ETHYL ESTRANOL, GLUTETHIMIDE, GLYBURIDE, HEPTABARBITAL, MAGNESIUM OXIDE, MAGNESIUM SILICATE, METHYLDOPA METHYLPHENIDATE, NALIDIXIC ACID, NORBOLETHONE, NORTRIPTYLINE, OXYPHENBUTAZONE, PENICILLIN, PHENOBARBITAL, PHENYLBUTAZONE, PHENYRAMIDOL, PHENYTOIN, SODIUM BICARBONATE, SPIRONOLACTONE, SULFADIMETHOXINE, SULFAMETHOXYPYRIDAZINE, SULFAPHENAZOLE, TETRACYCLINE, THYROXINE, TOLBUTAMIDE
2,2-Bis(4-hydroxyphenyl)propane → BISPHENOL
Bislah → PEAS
Bisillah → PEAS
Bismetin → PRENYLAMINE LACTATE
Bismogenol → BISMUTH SUBSALICYLATE
BISMUTH also → OMEPRAZOLE
BISMUTH AMMONIUM CITRATE
BISMUTH GLYCOL ARSANILATE
BISMUTH SODIUM THIOGLYCOLLATE
BISMUTH SODIUM TRIGLYCOLLAMATE
BISMUTH SUBGALLATE
BISMUTH SUBNITRATE also → BISMUTH, IODOFORM
BISMUTH SUBSALICYLATE also → DOXYCYCLINE, TETRACYCLINE
BISMUTH SULFIDE also → BISMUTH
Bisolvon → BROMHEXINE
BISOPROLOL
BISOXATIN ACETATE
BISPHENOL
Bis(pom)PMEA → ADEFOVIR
Bissy Nuts → KOLA
Bis(thiosemicarbazone) → KETHOXAL
Biston → CARBAMAZEPINE
Bistort → POLYGONUM sp.
BIS(TRI-*n*-BUTYLTIN)OXIDE
Bistrimate → BISMUTH SODIUM TRIGLYCOLLAMATE
Bisulfites → GRAMICIDIN
Bitensil → ENALAPRIL
BITHIONOL also → DICHLOROPHEN, HEXACHLOROPHENE, SOAP, TETRACHLOROSALICYLANILIDE
Bitin → BITHIONOL
BITOLTEROL
Bitrex → DENATONIUM BROMIDE
Bitrop → IPRATROPIUM BROMIDE
Bitter Almond → HYDROGEN CYANIDE
Bitter Apple → COLOCYNTH also → SOLANUM sp.
Bitterash → RAUWOLFIA SERPENTINA
Bitter Buttons → TANSY
Bitter Cherry → CHERRY
Bitter Cress → CORONOPUS sp., *didymus*
Bitter Cucumber → COLOCYNTH also → TRICHOSANTHIN
Bitter Gourd → COLOCYNTH also → MOMORDICA CHARANTIA
Bitter Orange Oil → ORANGE
Bitter Rubberweed → HYMENOXYS sp.
Bitter Sneeze Weed → DUGALDIN also → HELENIVM sp.
Bittersweet, American → CELASTRUS SCANDENS
Bitter Vetch → LENTILS
Bitterweed → DUGALDIN also → HELENIUM sp. also → HYMENOXYS sp.
Bitterweed Actinea → ACTINEA ODORATA
Bitterworm → BUCKBEAN
Bitumen → ASPHALT
Bitumen of Judea → MUMMIES
Bivinyl → 1,3-BUTADIENE
Bivitasi → COCARBOXYLASE
Biyoduro de mercurio → MERCURIC IODIDE, RED
Bizillah → PEAS
Bizolin → PHENYLBUTAZONE
BL 5 → CYCLOCOUMAROL
BL 139 → AMINOPENTAMIDE HYDROGEN SULFATE
BL 191 → PENTOXIFYLLINE
BL 41,624 → ANAGRELIDE
Black Balsam → BALSAM PERU
BLACKBERRY also → *N*-NITROSODIMETHYLAMINE
Blackbrush → FLOURENSIA CERNUA
Black Bryony → TAMUS COMMUNIS
Black Cohosh → CIMICIFUGA RACEMOSA
Black Dogwood → FRANGULA
Black Drink → HOLLY
Black Elder → ELDERBERRY
Black-Eyed Susan → RUDBECKIA sp. also → MARIGOLD
Black Haw → VIBURNUM PRUNIFOLIUM

Black Henbane → HYOSCYAMUS
Blackistonia sp. → SPIDERS
Black Jack → AMYL NITRITE also → BUTYL NITRITE
Black Laurel → LEDUM sp. also → LEUCOTHOE
Black Lead → GRAPHITE
Black Leaf 40 → NICOTINE and SALTS
Black Locust → ROBIN
Black Mustard → MUSTARD
Black Nightshade → SOLANUM sp.
Black Pepper → PEPPERS
Black Russian → CANNABIS
Black Sally Wattle → ACACIA
Black Sea Urchin → SEA URCHINS
Black Snakeroot → CIMICIFUGA RACEMOSA
Black Sore → HEXACHLOROBENZENE
Black Tar → HEROIN
Black and White Bleaching Cream → HYDROQUINONE
Blackwort → COMFREY
Bladafum → SULFOTEP
Bladan → TETRAETHYL PYROPHOSPHATE
Bladan M → METHYL PARATHION
Bladder Cherry → PHYSALIS sp.
Bladderpod → GLOTTIDIUM VESICARIUM
Bladderon → FLAVOXATE HCL
Bladex → CYANAZINE
Blandol → MINERAL OIL
Blane, Sir Gilbert → LIME
Blanket Flower → GAILLARDIA
Blastmycin → ANTIMYCIN A
BLASTICIDIN S
Blasting Gelatin → NITROGLYCERIN
Blasting Oil → NITROGLYCERIN
Blattanex → PROPOXUR
Blaud's Mass → FERROUS CARBONATE MASS
Bleaching Powder → CALCIUM HYPOCHLORITE also → CHLORINATED LIME
Bleeding Heart → DICENTRA CUCULLARIA
Bleminol → ALLOPURINOL
Blenoxane → BLEOMYCIN
BLEOMYCIN also → GALLIUM
Bleph 10 → SULFACETAMIDE SODIUM
Blex → PIRIMIPHOS METHYL
BLIGHIA SAPIDA
"Blind Staggers" → SELENIUM
Blistering Beetle → CANTHARIDES
Blistering Fly → CANTHARIDES
Blistering Gas → MUSTARD GAS
Bloat Guard → POLOXALENE
Blocadren → TIMOLOL
Blood Coagulation Factor IX → CHRISTMAS FACTOR
Bloodroot → SANGUINARIA CANADENSIS
Blood Stone → FERRIC OXIDE
Blood Sugar → GLUCOSE
Blotic → PROPETAMPHOS
Blow → COCAINE
Blowball → TARAXACUM OFFICINALE
Blox → LOPERAMIDE
Bloxanth → ALLOPURINOL
BL P413 → PENICILLINS – "NEWER and SYNTHETICS" > *Dipenicillin Sodium*
BL P804 → PENICILLINS – "NEWER and SYNTHETICS" > *Hetacillin*
BL P 1322 → CEPHAPIRIN
BL S 578 → CEFADROXIL
BL S786 → CEFORANIDE
Blubonnets → LUPINES
Blue Ape → XANTHOSOMA sp.
BLUEBELL also → PHACELIA sp.
BLUEBERRIES also → DATURA
Bluebottle Jellyfish → JELLYFISH
Bluebush → KOCHIA
Blue Cardinal Flower → LOBELIA
Blue Cheer → LYSERGIC ACID DIETHYLAMIDE
Blue Cohosh → CAULOPHYLLINE
Blue collar cocaine → METHAMPHETAMINE
Blue-gum Tree → EUCALYPTUS GLOBULUS
Blue Heaven → LYSERGIC ACID DIETHYLAMIDE
Blue Mist → LYSERGIC ACID DIETHYLAMIDE
Blue Nitro → BUTYROLACTONE
Blue Nitro Vitality → BUTYROLACTONE
"Blue Rocks" → PITCH
"Blues" → TRIPELENNAMINE
Blue Sailor's Succory → CHICORY
Bluestone → COPPER SULFATE
Blue Velvet® → OPIUM
"Blue Velvet" → MAGNESIUM SILICATE, TRIPELENNAMINE
Blue Vitriol → COPPER SULFATE
BLUE VRS
Blulan → BENFLURALIN
Blutene Chloride → TOLONIUM CHLORIDE
Bluton → IBUPROFEN
BM 611 → DIMETHYLAMINOPROPYLMETHACRYLAMIDE
BM 02015 → TORSEMIDE
BM 06,022 → RETEPLASE
BM 14,190 → CARVEDILOL
BM 15,075

BM 51,052 → CARAZOLOL
BMAA → β-*N*-METHYLAMINO-L-ALANINE
BMY 26,538 → ANAGRELIDE
BMY 13,754 → NEFAZODONE
BMY 27,857 → STAVUDINE
BMY 28,142 → CEFEPIME
BMY 281-03-800 → CEFPROZIL
B-Nine → DAMINOZIDE
Bo-Ana → FAMPHUR
Bo-de'-mee-see → TELITOXICUM PERUVIANUM
B.O.E.A. → ETHYL BISCOÜMACETATE
Boehringer Ingleheim → METAPROTERENOL
Bog Asphodel → NARTHECIUM OSSIFRAGUM
Bog Bean → BUCKBEAN
Bog Hop → BUCKBEAN
Boga Ia Kibariani → MOMORDICA CHARANTIA
Bog Laurel → KALMIA
Bog Moss → SPHAGNUM MOSS
Bog Myrtle → BUCKBEAN
Bog Nut → BUCKBEAN
Bog Onion → ARISAEMA
Bohr → CALOTROPIN
Bois d'arc → OSAGE ORANGE
BOIS DE ROSE, ACETYLATED
Boja → CALAMUS
Bojho → CALAMUS
BOL-148 → BROMOLYSERGIDE
BOLDENONE
Bolenol → ETHYLESTRENOL
Boletic Acid → FUMARIC ACID
BOLETUS sp.
Boletus → MUSHROOMS
Bolfo → PROPOXUR
Boliden CCA → CHROMATED COPPER ARSENATE
Boliden Salt K-33 → CHROMATED COPPER ARSENATE
BOLOGNA also → NITRATES and NITRITES
Bolstar → SULPROFOS
Bolus Alba → KAOLIN
Bolvidon → MIANSERIN
Bombita → METHAMPHETAMINE
Bomubomu → CALOTROPIN
BOMYL
Bonabol → MEFENAMIC ACID
Bonacid → NIFEDIPINE
Bonadette → MECLIZINE
Bonadorm → DICHLORALANTIPYRINE
Bonaid → BUQUINOLATE
Bonalan → BENFLURALIN
Bonamid → AZATADINE DIMALEATE
Bonamine → MECLIZINE
Bonapicillin → PENICILLINS – "NEWER and SYNTHETICS" > *Ampicillin*
Bonare → OXAZEPAM
Bonasanit → VITAMIN B_6
Bonefos → DICHLOROMETHYLENE DIPHOSPHONATE
Bonine → MECLIZINE
Bonlam → CAMBENDAZOLE
Bonoform → 1,1,2,2-TETRACHLOROETHANE
Bonpac → FONAZINE MESYLATE
Bonpyrin → DIPYRONE
Bonyl → NAPROXEN
Bonzol → DANAZOL
Boobialla → MYOPORUM sp.
BOÖPHONE
BOP → *N*-NITROSOBIS(2-OXYPROPYL)AMINE
Bophal Disaster → PHENYL ISOCYANATE also → METHYL ISOCYANATE
Boracic Acid → BORIC ACID
BORAGO OFFICINALIS
Borassus flabellifer → PALMS
BORAX
Bor-Cefazol → CEFAZOLIN SODIUM
Borea → BROMACIL
Borgal → TRIMETHOPRIM
Borgias → MORCHELLA DELICIOSA
BORIC ACID also → COTTON, NORFLOXACIN, PUTTY
Bor-Ind → INDOPROFEN
2-Bornanone → CAMPHOR
Borneo Camphor → BORNEOL
BORNEOL
Borneol Acetate → BORNYL ACETATE
BORNYL ACETATE
Bornyl Alcohol → BORNEOL
BORNYL ISOVALERATE
Bornyl 3-methyl Butyrate → BORNYL ISOVALERATE
Bornyval → BORNYL ISOVALERATE
Boroethane → DIBORANE
Borofax → BORIC ACID
Borolin → PICLORAM
BORON also → BORAX
BORON TRIFLUORIDE
Borsht → SODIUM GLUTAMATE
Boston Ivy → PARTHENOCISSUS QUINQUEFOLIA
Bothrops → SNAKE(S)
Botor → BEANS, Winged
Botran → DICHLOROBENZALKONIUM CHLORIDE also → 2,6-DICHLORO-4-NITROANILINE

Botrilex → QUINTOZENE
Bottled Gas → LPG
Bottled Gas → PROPANE
Botu je → JATROPHA sp.
Botuje Pupa → FLARIA
BOUGAINVILLEA
BOUILLON
BOUNCING BET
Bourbonal → ETHYL VANILLIN
Bourrache → BORAGO OFFICINALIS
Bourragi → BORAGO OFFICINALIS
Bovanide → RAFOXANIDE
Bovatec → LASALOCID
Bovicam → CAMBENDAZOLE
Boviclox → PENICILLINS – "NEWER and SYNTHETICS" > *Cloxacillin/Sodium salt*
Bovine Adenosine Deaminase PEG → PEGADEMASE BOVINE
"Bovine Bonkers" → 4-METHYLIMIDAZOLE
Bovizole → THIABENDAZOLE
Box → BUXUS sp.
Boxfish → PAHUTOXIN
Box Jellyfish → SEA WASP
Boxol → DALTEPARIN
Box Poison → OXYLOBIUM sp.
Box Thorn → LYCIUM HALIMIFOLIUM
Boxwood → BUXUS sp.
BP → BENZO [α] PYRENE
B-Paracipan → BENZOYLPAS
BPL → β-PROPIOLACTONE
BPMC → BUTACARB
BRACHIARIA
Brachyachne convergens → HYDROGEN CYANIDE
BRACKEN FERN also → MILK, SHIKIMIC ACID
Bradosol Bromide → DOMIPHEN BROMIDE
BRADYKININ also → SEA URCHINS
Brain Tabasco → COCA
Brake Fern → BRAKEN FERN
Brake Fiddlehead → BRACKEN FERN
BRALLOBARBITAL
Bramble Berry → BLACKBERRY
Brandy → DIGOXIN, FERRIC CHLORIDE, FUSEL OIL
Brandy Mint → PEPPERMINT
BRASSAIA ACTINOPHYLLA
Brass-founder's Ague → WELDING
BRASSICA sp.
Brassica alba → MUSTARD
Brassica campestris → TURNIP
Brassica juncea → MUSTARD
Brassica napus → RAPE
Brassica nigra → MUSTARD
Brassica oleracea → BROCCOLI also → BRUSSEL SPROUTS, CABBAGE, CAULIFLOWER
Brassica sinapsis → MUSTARD
Brassicol → QUINTOZENE
Braunol → POLYVINYLPYRROLIDONE IODINE COMPLEX
Braunosan H → POLYVINYLPYRROLIDONE IODINE COMPLEX
Bravo → CHLOROTHALONIL
Bravo Zulu → QUINUCLIDINYL BENZILATE
Brazilian Pepper → SCHINUS TEREBINTHIFOLIUS
BRAZIL NUTS
Brazil Root → IPECAC
Brazil Wax → CARNAUBA WAX
BRC 968 → BUPHANE DISTICHA
BREAD also → CLAVICEPS, ENDRIN, ETHYLMERCURIC CHLORIDE, FLOUR, HEXACHLOROBENZENE, MERCURY, 4,4′-METHYLENE DIANILINE, PARATHION, POTASSIUM BROMATE, STERNO®, TETRACYCLINE, VITAMIN B_1, WHEAT
Bredon → OXOLAMINE
Breelya Poison → GASTROLOBIUM sp.
Brefeldin A → PENICILLIUM sp.
Brek → LOPERAMIDE
Bremen Blue → COPPER CARBONATE
Bremen Green → COPPER CARBONATE
Bremil → HYDROCHLOROTHIAZIDE
Brenal → MECLOFENOXATE
Brentan → MICONAZOLE NITRATE
Breogamine → PROTEIN HYDROLYSATES
Breokinase → UROKINASE
Breon → POLYVINYL CHLORIDE
Brethaire → TERBUTALINE SULFATE
Brethine → TERBUTALINE SULFATE
Bretylan → BRETYLIUM TOSYLATE
Bretylate → BRETYLIUM TOSYLATE
BRETYLLIUM TOSYLATE also → DIGITALIS, MEPHENTERMINE, METHYLPHENIDATE
Bretylol → BRETYLIUM TOSYLATE
BREVETOXINS also → GONYAULAX
Brevibloc → ESMOLOL
Brevicid → XANTHOCILLIN
Brevicurarine → PIPROCURARINE
Brevidil M → SUCCINYLCHOLINE
Brevimytal → METHOHEXITAL SODIUM
Brevital → METHOHEXITAL SODIUM
Brewers → BARLEY
Brexic → PIROXICAM
Bricanyl → TERBUTALINE SULFATE

BROMELAIN also → PINEAPPLE, TETRACYCLINE
Bromelia → 2-ETHOXYNAPHTHALENE
Bromelin → BROMELAIN
BROMETHALIN
Bromethol → TRIBROMOETHANOL
Bromex → DIBROM
BROMFENAC also → LITHIUM, PHENYTOIN
Bromfenacoum → BRODIFACOUM
BROMHEXINE
Bromic Ether → ETHYL BROMIDE
BROMINDIONE
BROMINE also → BROMODYPHENHYDRAMINE, MILK, PYRIDOSTIGMINE BROMIDE, SODIUM BROMIDE
Bromisoval → BROMISOVALUM
BROMISOVALUM
BROMIUM also → ENAMEL
Bromoacetic Acid → BEER
BROMOACETONE
Bromoacil → BROMACIL
Bromo-Benadryl → BROMODIPHENHYDRAMINE
BROMOBENZENE also → PYRAZOLE
BROMOCRIPTINE also → ALCOHOL, ETHYL; CAFFEINE, ERYTHROMMYCIN, PERGOLIDE MESYLATE, PIMOZIDE, TACROLIMUS, THIORIDAZINE
Bromodialone → BROMADIOLONE
2-Bromo-N,N-diethyl-D-lysergamide → BROMOLYSERGIDE
4-BROMO-2,5-DIMETHOXYAMPHETAMINE
4-Bromo-2,5-dimethoxy-m-methylphenethylamine → 4-BROMO-2,5-DIMETHOXYAMPHETAMINE
BROMODIPHENHYDRAMINE
4-Bromo-2,5-DMA → 4-BROMO-2,5-DIMETHOXYAMPHETAMINE
Bromoethane → ETHYL BROMIDE
Bromoethene → VINYL BROMIDE
Bromoethylene → VINYL BROMIDE
9-BROMOFLUORENE
BROMOFORM
Bromo-O-Gas → METHYL BROMIDE
(α-Bromoisovaleryl) urea → BROMISOVALUM
Bromo-LSD → BROMOLYSERGIDE
D-2-Bromolysergic Acid Diethylamide → BROMOLYSERGIDE
BROMOLYSERGIDE
Bromomethane → METHYL BROMIDE
Bromone → BROMADIOLONE
2-Bromo-2-nitropropane-1,3-diol → BRONOPOL
α-Bromo-β-phenylethylene → BROMSTYROL
BROMOPHOS
BROMOPHOS-ETHYL
1-Bromo-2-propanone → BROMOACETONE
3-Bromo-1-propene → ALLYL BROMIDE
3-Bromo-propylene → ALLYL BROMIDE
Bromo-Seltzer® → POTASSIUM BROMIDE
Bromosulfophthalein → SULFOBROMOPHTHALEIN SODIUM
Bromosulphthaleine → SULFOBROMOPHTHALEIN SODIUM
BROMOXYNIL
BROMPHENIRAMINE also → PHENYLPROPRANOLAMINE
BROMPHENOL BLUE also → CIBENZOLINE
BROMSTYROL
Bromsulfalein → SULFOBROMOPHTHALEIN SODIUM
Bromsulphalein → SULFOBROMOPHTHALEIN SODIUM
Bromthalein → SULFOBROMOPHTHALEIN SODIUM also → BUFORMIN
BROMTHYMOL BLUE
Bromural → BROMISOVALUM
BROMUS sp.
Bromuvan → BROMISOVALUM
Bromvaletone → BROMISOVALUM
Bronalide → FLUNISOLIDE
Bronalin → HEXOPRENALINE
Bronate → BROMOXYNIL
Broncatar → OXOLAMINE
Bronchocilline → PENICILLIN > *Benzylpenicillin*, DAEEH
Broncholysin → ACETYLCYSTEINE
Bronchoretard → THEOPHYLLINE
Bronchoselectan → SODIUM ACETRIZOATE
Bronco → ALACHLOR
Broncon → CLORPRENALINE
Broncovaleas → ALBUTEROL
Bronkaid → EPINEPHRINE
Bronkodyl → THEOPHYLLINE
BRONOPOL also → TRIS-(HYDROXYMETHYL)NITROMETHANE
Bronosol → BRONOPOL
Bronsecur → CARBUTEROL
Brontine → DEPTROPINE
Brook Bean → BUCKBEAN
Broom Corn → KAFFIRCORN also → PANICUM sp., SORGHUM
BROOMWEED also → GUTIERREZIA MICROCEPHALA
Brophene → BROMOPHOS

BROTIZOLAN
Brotopon → HALOPERIDOL
Brovalurea → BROMISOVALUM
Brown Dots → LYSERGIC ACID DIETHYLAMIDE
Brown Dragon → ARISAEMA
Browse Tree → LEUCAENA sp.
Broxalax → BISACODYL
BROXALDINE
Broxil → PENICILLINS – "NEWER and SYNTHETICS" > *Phenethicillin Potassium*
Broxolin → BISMUTH GLYCOL ARSANILATE
Broxykinolin → 5,7-DIBROMO-8-QUINOLINOL
BRS 640 → BRADYKININ
BRUCINE
Brufaneuxol → AMINOPYRINE
Brufen → IBUPROFEN
Brufort → IBUPROFEN
BRUGMANSIA
Brumetidina → CIMETIDINE
Brunac → ACETYLCYSTEINE
BRUNFELSIA
Brush Killer → 2,4,5-TRICHLOROPHENOXYACETIC ACID
Brush-Rhap → 2,4,5-TRICHLOROPHENOXYACETIC ACID
BRUSSEL SPROUTS also → ANTIPYRINE, CABBAGE
Bruxicam → PIROXICAM
BRYONIA
Bryony → BRYONIA
Bruzem → DILTIAZEM
BS 572 → CYCLANDELATE
BS 5930 → ORPHENADRINE
BS 6987 → DEPTROPINE
BS 7331 → TOFENACIN HCL
BS 100-141 → GUANFACINE
BSP → SULFOBROMOPHTHALEIN SODIUM
BST → GROWTH HORMONES
"B's & T's" → TRIPELENNAMINE
BT 436 → DEHYDROEMETINE
BTMP → VITAMIN B_1
BTS 18,322 → FLURBIPROFEN
BTS 27,419 → AMITRAZ
BTS 49,465 → FLOSEQUINAN
BTS 54,524 → SIBUTRAMINE
Bubarbital → BUTABARBITAL
Buburone → IBUPROFEN
Bucarban → CARBUTAMIDE
Buccalsone → HYDROCORTISONE
Buccastem → PROCHLORPERAZINE
Bu-Chlorin → *n*-BUTYL CHLORIDE
BUCHU also → TEA
BUCINDOLOL
BUCKBEAN
Buckeye → AESCULUS sp.
Buckthorn Bark → FRANGULA
BUCKWHEAT also → HONEY
Buclifen → BUCLIZINE
Buclina → BUCLIZINE
BUCLIZINE
BUCLOSAMIDE
Bucrol → CARBUTAMIDE
Buctril → BROMOXYNIL
Budeson → BUDESONIDE
BUDESONIDE also → GRAPEFRUIT
Budirol → PROPYPHENAZONE
Budoform → IODOCHLORHYDROXYQUINOLINE
Bueno → METHANEARSONIC ACID also → MONOSODIUM METHANEARSONATE
Bufadienolide Daigremontianin → KALANCHOE
Bufedon → NYLIDRIN
Bufemid → FENBUFEN
BUFENCARB
BUFEXAMAC
Buffalo Bur → SOLANUM sp.
Buffalo Nut → PYRULARIA PUBERA
Bufo aqua → TOADS
Bufogenins → TOADS
Bufon → DIETHYLSTILBESTROL
Bufonamin → BUFORMIN
BUFORMIN
BUFOTENINE also → AMANITA sp., COHOBA, MUSHROOMS, TOADS, TRYPTAMINE
Bufotoxins → TOADS
Bufylline → AMBUPHYLLINE
Bugbane → CIMICIFUGA RACEMOSA
Bugwort → CIMICIFUGA RACEMOSA
Buis → BUXUS sp.
Bukarban → CARBUTAMIDE
Bulbocapnine → CORYDALIS, DICENTRA CUCULLARIA
Bulbold → GLYCERIN
Bulbonin → BUFORMIN
Bulbus Scillae → SQUILL
Bulgarian Rose → CINNAMIC ALDEHYDE
Bullatacin → PAW-PAW
Bullibulli → SOLANUM sp.
Bullock's Lungwort → VERBASCUM THAPSUS
Bull Thistle → SOLANUM sp.
Bulrush → SCIRPUS AMERICANUS
Bulrush Millet → PENNISETUM sp.
Bumblebees → AMPHETAMINE
BUMETANIDE also → INDOMETHACIN

Bumex → BUMETANIDE
Bunaiod → BUNAMIODYL
Bunamidine → BUTAMISOLE, SUCCINYLCHOLINE
BUNAMIODYL
BUNAZOCIN also → RIFAMPIN
Bunchberry → LANTANA sp.
Bunchflowers → MELATHIUM
Bungarotoxin → LOPHOGORGIA sp.
Buniodyl → BUNAMIODYL
1-Bunolol → LEVOBUNOLOL
Bunt-cure → HEXACHLOROBENZENE
Bunt-no-more → HEXACHLOROBENZENE
Bupatol → BAMETHAN
BUPHANE DISTICHA
Buphanine → NARCISSUS sp.
Buphedrin → NYLIDRIN
BUPIVACAINE
Buprenex → BUPRENORPHINE
BUPRENORPHINE also → COCAINE
BUPROPION also → RITONAVIR
BUQUINOLATE
BUR BUTTERCUP
Burcillin → PENICILLIN > *Benzylpenicillin, Potassium*
Burcillin-P → PENICILLIN > *Benzylpenicillin, Procaine*
BURDOCK also → TEA
Burger, Judge Warren → Preface
Burgodin → BEZITRAMIDE
Burinex → BUMETANIDE
Burning Bush → DICTAMNUS ALBUS also → EUONYMUS, KOCHIA
Burnt Ammonium Alum → ALUMINUM AMMONIUM SULFATE
Burnt Lime → CALCIUM OXIDE
Burnt Sugar → CARAMEL
Burow's Solution → ALUMINUM ACETATE
Burr Burr → BURDOCK
Burweed → COCKLEBURS
Buserelin → LH-RH
Bush → CANNABIS
Bushewe → CARISSA
BUSH TEAS also → PYRROLIDIZINES, SANGURA, TEA
Businessman's Special → *N,N*-DIMETHYLTRYPTAMINE
Busodium → BUTABARBITAL
Busotran → BUTABARBITAL
Buspar → BUSPIRONE
Buspinol → BUSPIRONE
BUSPIRONE also → FLUOXETINE, GRAPEFRUIT, VERAPAMIL
BUSULFAN
Busulphan → BUSULFAN
Busy Bee → PHENCYCLIDINE
Butabar → BUTABARBITAL
BUTABARBITAL also → CYCLOCOUMAROL, HYDROCORTISONE, WARFARIN
Butabarbitone → BUTABARBITAL
Butabarpal → BUTABARBITAL
Butabon → BUTABARBITAL
BUTACAINE also → ETHYL-*p*-AMINOBENZOATE, PROCAINE, TETRACAINE HYDROCHLORIDE
BUTACARB
BUTACHLOR
Butacide → PIPERONYL BUTOXIDE
BUTACLAMOL
Butacote → PHENYLBUTAZONE
1,3-BUTADIENE
α,γ-Butadiene → 1,3-BUTADIENE
Butadion(a) → PHENYLBUTAZONE
Butak → BUTABARBITAL
Buta-Kay → BUTABARBITAL
Butalan → BUTABARBITAL
BUTALBITAL also → IMIPRAMINE
Butalgin → METHADONE HCL
Butaliret → TERBUTALINE SULFATE
Butalix → BUTABARBITAL
BUTAMBEN
Butamide → TOLBUTAMIDE
BUTAMISOLE
2-Butanamine → *sec*-BUTYLAMINE
BUTANE
Butanedione → ERYTHRITOL ANHYDRIDE
1,3-Butanediol → 1,3-BUTYLENE GLYCOL
Butanedioyl Chloride → SUCCINYLCHOLINE
1-Butanethiol → *n*-BUTYL MERCAPTAN
Butanex → BUTACHLOR
BUTANILICAINE
Butanol → BUTYRALDEHYDE
1-Butanol → ALCOHOL, *n*-BUTYL
n-Butanol → ALCOHOL, *n*-BUTYL
1,2-Butanolide → BUTYROLACTONE
1,4-Butanolide → BUTYROLACTONE
2-Butanolone-3 → ACETOIN
2,3-Butanolone → ACETOIN
BUTANONE
2-Butanone → METHYL ETHYL KETONE
Butanotic → BUTABARBITAL
BUTAPERAZINE also → ESTROGENS, CONJUGATED
Butaphyllamine → AMBUPHYLLINE
Butapyrazol → PHENYLBUTAZONE
Butased → BUTABARBITAL
Butatran → BUTABARBITAL
Butatron → PHENYLBUTAZONE

Butazem → BUTABARBITAL
Butazolidin → PHENYLBUTAZONE
Butazon → MOFEBUTAZONE
Butazone → PHENYLBUTAZONE
Bute → PHENYLBUTAZONE
Butedrin → BAMETHAN
Butelline → BUTACAINE
BUTENAFINE
3-Butenenitrile → ALLYL CYANIDE
BUTENOLIDE also → FESCUE, TALL
β-Butenonitrile → ALLYL CYANIDE
Butesin → BUTAMBEN
BUTETHAL
BUTETHAMINE
Butex → BUTABARBITAL
Butformin → BUFORMIN
Buticaps → BUTABARBITAL
BUTIROSIN also → NEOMYCIN
Butisol → BUTABARBITAL
Butobarbital → BUTETHAL
Butobarbitone → BUTETHAL
BUTOCONAZOLE
Butoform → BUTAMBEN
BUTONATE also → SUCCINYLCHOLINE
BUTOPYRANOXYL
BUTORPHANOL also → DIPHENHYDRAMINE
2-BUTOXYETHANOL also → ETHYLENE GLYCOL MONOBUTYL ETHER
Butoz → PHENYLBUTAZONE
Butrate → BUTABARBITAL
BUTRIPTYLINE
Butte → BUTABARBITAL
BUTTER
Butter and Eggs → LINARIA VULGARIS
Butter of Antimony → ANTIMONY TRICHLORIDE
Buttercup → PARTHENIUM sp. also → RANUNCULUS
Butterfly Ray → STINGRAYS
Butterfly Weed → ASCLEPIAS
Butter Yellow → *p*-DIMETHYLAMINOBENZENE
Butter of Zinc → ZINC
Buttonbush → CEPHALANTHUS OCCIDENTALIS
n-BUTYL ACETATE
n-BUTYL ACRYLATE
sec-BUTYLAMINE
Butyl Aminobenzoate → BUTAMBEN
tert-BUTYLAMINOETHYL METHACRYLATE
BUTYLATED HYDROXYANISOLE also → MARGARINE
BUTYLATED HYDROXYTOLUENE also → ETHOXYQUINE, HYDROGEN CYANIDE, MARGARINE
BUTYL BENZYL PHTHALATE
p-tert-BUTYL CATECHOL also → PENTADECYLCATECHOL
tert-Butyl Catechol → PAPER
Butyl Cellosolve → 2-BUTOXYETHANOL also → ETHYLENE GLYCOL, MONOBUTYL ETHER
n-BUTYL CHLORIDE
N-Butyl-4-chlorosalicylanide → BUCLOSAMIDE
BUTYL CYANOACRYLATE
4-*tert*-BUTYLCYCLOHEXANOL
4-*tert*-BUTYL-2,6-DIISOPROPYLPHENOL
tert-BUTYLDIMETHYLCHLOROSILANE
1,3-BUTYLENE DIMETHACRYLATE
2-sec-Butyl-4,6-dinitrophenol → DINOSEB
1,3-BUTYLENE GLYCOL
n-BUTYL ETHER
Butyl Ethyl Ketone → ETHYL BUTYL KETONE
n-BUTYL GLYCIDYL ETHER
BUTYLGLYCOL ACETATE
tert-BUTYLHYDROQUINONE
N-BUTYL-*N*-(4-HYDROXYBUTYL) NITROSAMINE
Butylhydroxytoluene → SELAMECTIN
BUTYL ISOBUTYRATE
BUTYL ISOCYANATE
n-BUTYL MERCAPTAN
n-Butyl-2-methylpropanoate → BUTYL ISOBUTYRATE
BUTYL NITRITE
p-tert-BUTYLPHENOL
p-tert-Butylphenol Formaldehyde → SHOES
n-Butyl Phthalate → DIBUTYL PHTHALATE
3-n-Butylphthalide → CELERY
BUTYL STEARATE
tert-Butyl Valone → PINDONE
BUTYRALDEHYDE
Butyric Ether → ETHYL BUTYRATE
Butynorate → DIBUTYL TIN DILAURATE
BUTYROLACTONE
γ-Butyrolactone → BUTYROLACTONE
n-BUTYRONITRILE
Butyrylperazine → BUTAPERAZINE
Bux → BUFENCARB
Buxine → BUXUS
BUXUS sp.
Buylate → 5-ETHYL DIISOBUTYLTHIOCARBAMATE
Buzon → PHENYLBUTAZONE
Buzzard Grass → STIPA sp.
B.V.U. → BROMISOVALUM

BW 33A → ATRACURIUM
BW 33T57 → METHISAZONE
BW 57-322 → AZATHIOPRINE
BW 61-32 → STILBAZIUM IODIDE
BW 256U87 → VALACYCLOVIR
BW 248U → ACYCLOVIR
BW 311C90 → ZOLMITRIPTAN
BW 356-C61 → KETHOXAL
BW 430C → LAMOTRIGINE
BW 467C60 → BETHANIDINE
BW 759U → GANCICLOVIR
BW A509U → ZIDOVUDINE
BW A938U → DOXACURIUM CHLORIDE
BW B759U → GANCICLOVIR
BW 56158 → ALLOPURINOL
BY 912 → THEOPHYLLINE
Bydolax → OXYPHENISATIN
BYK 1512 → HEXOPRENALINE
Bylotensin → NITRENDIPINE
BYSSOCHLAMYS NIVEA
BZ 55 → CARBUTAMIDE
BZ → QUINUCLIDINYL BENZILATE
BZQ → BENZQUINAMIDE
BZW → BENZOYL PEROXIDE

C

C → COCAINE
C_4 → CYCLONITE
C 10 → DECAMETHONIUM BROMIDE
C 709 → DICROTOPHOS
C 1414 → MONOCROTOPHOS
C 1523 → BENZYDAMINE
C 1983 → CHLOROXURON
C 5473 → OXYPHENONIUM BROMIDE
C 5581 H → PHENYLTOLOXAMINE
C 5720 → CARPROFEN
C 5968 → HYDRALAZINE
C 8514 → CHLORDIMEFORM
C 9491 → IODOFENPHOS
CA2 → INFLIXIMAB
CA-7 → ASPERGILLUS
Caapi → AYAHUASCA
Cabamicina → TRICHOMYCIN
CABBAGE also → ANTIPYRINE, BRUSSEL SPROUTS
Cabbage Palmetto → PALMS
CABERGOLINE
Cabermox → PENICILLINS – "NEWER and SYNTHETICS" > *Amoxicillin*
"Cable Rash" → CHLORINATED NAPHTHALENES
Cab-O-Sil → SILICON DIOXIDE
Cabral → PHENYRAMIDOL
CAC → CHROMATED COPPER ARSENATE
CACALIA DECOMPOSITA
CACAO also → CHOCOLATE, TONKA BEANS
CACODYLIC ACID
CACP → CISPLATIN
CACTINOMYCIN
Cadillac → PHENCYCLIDINE
CADMIUM also → ACETYLENE, CHUIFONG TOURUWAN, COTTON, CYSTEINE, GALVANIZED METALS, GLASS, LEAD, LEMON, MOTOR OIL, OYSTERS, RICE, SAUERKRAUT, SELENIUM, SHELLFISH, SILVER, STEEL, TOBACCO, TOMATO, WATER, WELDING, ZINC
CADMIUM ACETATE
CADMIUM CARBONATE
CADMIUM CHLORIDE
Cadmium Dichloride → CADMIUM CHLORIDE
Cadmium Difluoride → CADMIUM FLUORIDE
Cadmium Dinitrate → CADMIUM NITRATE
CADMIUM FLUORIDE
Cadmium Fluorure → CADMIUM FLUORIDE
CADMIUM NITRATE
CADMIUM OXIDE also → CADMIUM, CADMIUM ACETATE, SOLDER
CADMIUM SULFATE
CADMIUM SULFIDE also → MERCURIC SULFIDE, RED
Cadmium Yellow → CADMIUM SULFIDE
Caerulein → CERULETIDE
Caerulin → CERULETIDE
Caeruloplasmin → CERULOPLASMIN
C. A. F. → CHLORAMPHENICOL
Cafergot® → ERGOTAMINE also → PROPRANOLOL
Caffeic Acid → BROCCOLI, BRUSSEL SPROUTS, CABBAGE, CELERY, CHERRY, EGGPLANT, PLEUROTIS sp., POTATO
CAFFEINE also → ACETANALID, ACETYLSALICYLIC ACID, ANTIPYRINE, BROMOCRIPTINE, CHOCOLATE, CIMETIDINE, CIPROFLOXACIN, CLOZAPINE, COCOA, COFFEE, COLAS, DISULFIRAM, ENOXACIN, FAMOTIDINE, FLUVOXAMINE, GRAPEFRUIT, HOLLY, ILEX sp., KOLA, MELATONIN, METHOXSALEN, MEXILETINE, NORFLOXACIN, OMEPRAZOLE, PHENYLPROPANOLAMINE, RANITIDINE,

TEA, TERBINAFINE, TICLODIPINE, TOBACCO
Cafron → BENACTYZINE
Cafta → CATHA EDULIS
Caid → CHLOROPHACINONE
Cairox → POTASSIUM PERMANGANATE
CAJEPUT
Cajeputol → EUCALYPTOL
Cajuda → ANONA sp.
CALADIUM also → CALCIUM OXALATE
Caladryl® → CALAMINE
CALAMINE also → DIPHENHYDRAMINE
Calamth → VERAPAMIL
CALAMUS
Calan → VERAPAMIL
Cal-Aspirin → CALCIUM ACETYLSALICYLATE
CALATHEA VEITCHIANA
Calcamine → DIHYDROTACHYSTEROL
Calcicard → DILTIAZEM
Calcichew → CALCIUM CARBONATE
Calcidia → CALCIUM CARBONATE
Calcifediol → PHENYTOIN
Calciferol → VITAMIN D_2
Calcimar → CALCITONIN
Calcimux → SODIUM ETIDRONATE
Calcined Magnesite → MAGNESIUM OXIDE
Calciparine → HEPARIN
CALCIPOTRIENE also → CALCIPOTRIOL
CALCIPOTRIOL also → ALUMINUM HYDROXIDE
Calcitare → CALCITONIN
Calcitetracemate Disodium → CALCIUM DISODIUM EDETATE
CALCITONIN also → CALCIUM
CALCIUM also → BREAD, DEMECLOCYCLINE, DIGITALIS, DOXYCYCLINE, ETHINYL ESTRADIOL, FLOUR, GREPAFLOXACIN, KANAMYCIN, MAGNESIUM, METHACYCLINE, NICKEL SULFATE, OUABAIN, OXALIC ACID, OXYTETRACYCLINE, PARATHYROID HORMONE, POTASSIUM, POTASSIUM OXALATE, PREDNISONE, PROPRANOLOL, QUININE, RUBBER, SODIUM, SODIUM THIOSULFATE, STRONTIUM, TETRACYCLINE, TILUDRONATE, WHEAT
CALCIUM ACETATE
CALCIUM ACETYLSALICYLATE
CALCIUM ARSENATE
Calcium Aspirin → CALCIUM ACETYLSALICYLATE
CALCIUM BENZOYLPAS
CALCIUM CARBIDE
CALCIUM CARBIMIDE also → ALCOHOL, ETHYL
CALCIUM CARBONATE also → CHOLESTYRAMINE RESIN, NITROFURANTOIN, PYRIMETHAMINE, SPONGES, THYROXINE
CALCIUM CHLORIDE also → FIRE COLORING SALTS
CALCIUM CHROMATE
Calcium Chrome Yellow → CALCIUM CHROMATE
Calcium Cyanamide → CALCIUM CARBIMIDE also → CYANAMIDE
Calcium Cyanide → CALCIUM CARBIMIDE
CALCIUM CYCLAMATE also → LINCOMYCIN, TOLBUTAMIDE
Calcium N-Cyclohexylsulfamate → CALCIUM CYCLAMATE
CALCIUM DISODIUM EDETATE
Calcium Disodium Versenate → CALCIUM DISODIUM EDETATE
CALCIUM EDETATE
Calcium Ethylenediaminetetraacetate → CALCIUM EDETATE
CALCIUM FLUORIDE
CALCIUM GLUCEPTATE
Calcium Glucoheptonate → CALCIUM GLUCEPTATE
CALCIUM GLUCONATE also → DISODIUM EDETATE, PENTAGASTRIN
Calcium Glyconate → CALCIUM GLUCONATE
"Calcium Gout" → DIHYDROTACHYSTEROL
CALCIUM HYDROXIDE also → CALCIUM OXIDE
CALCIUM HYPOCHLORITE
CALCIUM OXALATE also → ARISAEMA, CALADIUM, CALLA LILY, CHOCOLATE, COLOCASIA, COPPER OXALATE, ETHYLENE GLYCOL, GRAPEFRUIT, MONSTERA sp., OXALIC ACID, PHILODENDRON sp., POLYGONUM sp., SKUNK CABBAGE, SODIUM, TEA, XYLITOL
CALCIUM OXIDE also → COCA
Calcium Oxychloride → CALCIUM HYPOCHLORITE
CALCIUM PEROXIDE also → WHEAT
Calcium Phosphate → TETRACYCLINE
Calcium Pyrophosphate → SODIUM HYALURONATE
Calcium Silicate → SUCROSE
Calcium Stearate → STERNO®

Calcium Sulfate → GELATIN, PHENYTOIN
Calcium Versenate → CALCIUM EDETATE
Caldine → LACIDIPINE
Caldon → DINOSEB
Calendula officinalis → MARIGOLD
Calepsin → CARBAMAZEPINE
Caley Pea → LATHYRUS sp.
Calfkill → KALMIA
Calfspan → SULFAMETHAZINE
Calicheamicin → GEMTUZIMAB OZOGAMICIN
Calico Bush → KALMIA
California Barberry → MAHONIA AQUIFOLIUM
California Fern → CONIUM MACULATUM
California Poppy → ESCHOSCHOLTZIA CALIFORNICA
California Sunshine → LYSERGIC ACID DIETHYLAMIDE
Californit → OXYPHENBUTAZINE
Caliment → PIROXICAM
Calisaya Bark → CINCHONA
CALLA LILY also → CALCIUM OXALATE, LILIES
Callidin I → BRADYKININ
CALLILEPIS LAUREOLA
CALLIONYMUS CALAUROPOMUS
Calmabel → ALVERINE
Calmaxid → NIZATIDINE
Calmday → NORDAZEPAM
Calmeran → AZACYCLONAL
Calmipan → GLYCERYL GUAIACOLATE
Calmiran → MEPROBAMATE
Calmoden → CHLORDIAZEPOXIDE
Calmodid → HYDROCODONE BITARTRATE
Calmonal → MECLIZINE
Calmpose → DIAZEPAM
Calnathal → CARBETAPENTANE CITRATE
Calocan → BEANS, *Winged*
Calogreen → MERCUROUS CHLORIDE
Calomel → MERCUROUS CHLORIDE
Calorose → INVERT SUGAR
CALOTROPIN
Calotropis procera → CALOTROPIN
Calpol → ACETAMINOPHEN
Calsekin → FONAZINE MESYLATE
Calslot → MANIDIPINE
Calsmin → NITRAZEPAM
Calsyn → CALCITONIN
Calsynar → CALCITONIN
CALTHA PALUSTRIS
Calthor → PENICILLINS – "NEW and SYNTHETICS"
Caltidren → CARTEOLOL HYDROCHLORIDE
CALUSTERONE
Calvatia → MUSHROOMS
Calvin Klein → METHYLENEDIOXYAMPHETAMINE
Calvisken → PINDOLOL
Calx → CALCIUM OXIDE
Calycanthine → CALYCANTHUS FLORIDUS
CALYCANTHUS FLORIDUS
CAMAZEPAM
CAMBENDAZOLE also → THIABENDAZOLE
Cambenzole → CAMBENDAZOLE
Cambet → CAMBENDAZOLE
Camboatá Tree → GUAREA
Camcolit → LITHIUM CARBONATE
Camoform → BIALAMICOL
Camolar → CYCLOGUANYL PAMOATE
Camont → NIFEDIPINE
Camoquin → AMODIAQUIN
Campari-Orange® → CARMINE
2-Camphanyl Acetate → ISOBORNYL ACETATE
Camphechlor → TOXAPHENE
CAMPHENE
Camphol → BORNEOL
CAMPHOR also → PHENOL
Camposan → ETHEPHON
Campoviton 6 → VITAMIN B_6
Campral → ACAMPROSATE
CAMPSIS RADICANS
Campto → IRINOTECAN
Camptosar → IRINOTECAN
CAMPTOTHECIN
Camyna → TIOXOLONE
Canada Yew → YEW
Canadian Hemp → APOCYNUM CANNABINUM
Canadian Milkvetch → MILKVETCH
Canadian Spear Grass → STIPA sp.
Canada Thistle → CIRSIUM ARVENSE
Canadian Turpentine → BALSAM, CANADIAN
Canamo Indiano → CANNABIS
Cananga Oil → YLANG YLANG OIL
CANARY GRASS
Canavalia ensiformis → JACK BEAN
CANAVANINE
Cancycline → TETRACYCLINE
Candelabra Cactus → EUPHORBIA, *lactea*
Candelabra Aloe → ALOE
Canderel → ASPARTAME
CANDESARTAN CILEXETIL
Candex → NYSTATIN
CANDICIDIN

Candio-Hermal → NYSTATIN
Candle Plant → EUPHORBIA, *poissonii*
Canescine → DESERPIDINE
Canesten → CLOTRIMAZOLE
Cane Sugar → SUCROSE
Canferon → INTERFERONS
Canfodion → DEXTROMETHORPHAN HYDROBROMIDE
Canifug → CLOTRIMAZOLE
Cankerwort → TARAXACUM OFFICINALE
CANNABIDIOL also → Δ^{-9} TETRAHYDROCANNABINOL
Cannabinol → Δ^{-9} TETRAHYDROCANNABINOL
CANNABIS also → ALCOHOL, ETHYL; CATNIP; a,a´-DYPYRIDYL, EPINEPHRINE, PARAQUAT, PHENCYCLIDINE, PHENYTOIN, SECOBARBITAL SODIUM, VITAMIN B_3
Cannabis sativa → CANNABIS
Cannoc → ESTAZOLAM
Canocenta → LOTRIFEN
Canogard → DICHLORVOS
Canola → RAPE also → CARBOXIN, ERUCIC ACID
Canon → ALACHLOR
Canopar → THENIUM CLOSYLATE
CANRENONE also → SPIRONOLACTONE
Cantan → VITAMIN C
Cantavin C → VITAMIN C
Cantaxin → VITAMIN C
Cantelope → ALDICARB
CANTHARIDES
Cantharides Camphor → CANTHARIDIN
CANTHARIDIN
CANTHARIDIN also → MYLABRIS CICHORII
CANTHAXANTHIN
Cantil → MEPENZOLATE BROMIDE
Cantralax → CASANTHRANOL
Cantrex → KANAMYCIN
Cantril → MEPENZOLATE BROMIDE
Cantrol → 4(2-METHYL-4-CHLOROPHENOXY) BUTRIC ACID
Canudo → IPOMEA sp.
Caocobra → COPPER OXIDE, RED
Caoutchouc → RUBBER
CAP → CHLORMADINONE
Capansa → RYANIA sp.
Caparcide → ARSENAMIDE
Caparol → PROMETRYNE
Caparside → ARSENAMIDE
Caparsolate → ARSENAMIDE
Capastat → CAPREOMYCIN
CAPECITABINE
Cape-gooseberry → PHYSALIS sp.
Cape Lilac → MELIA sp.
Capen → TIOPRONIN
CAPERS also → EUPHORBIA
Caper Spurge → EUPHORBIA, *lathyris*
Cape Tulip → HOMERIA sp.
Capeweed → CRYPTOSTEMMA sp.
Capisten → KETOPROFEN
Capitol → BENZALKONIUM CHLORIDE
Capitus → ETHOSUXIMIDE
Caplenal → ALLOPURINOL
Capmul 8210 → MONOCTANOIN
Capon's Tail → VALERIANA sp.
Capoten → CAPTOPRIL
Cap-O-Tran → MEPROBAMATE
CAPPARIS
Capquin → CHLOROQUINE
Capralense → Σ-AMINOCAPROIC ACID
Capramol → Σ-AMINOCAPROIC ACID
CAPREOMYCIN
Caprin → ACETYLSALICYLIC ACID
Caprocid → Σ-AMINOCAPROIC ACID
Caprodat → CARISPRODOL
Caprofen → PROBENECID
Caprokol → HEXYLRESORCINOL
CAPROLACTAM
Caprolin → CAPREOMYCIN
Capromycin → CAPREOMYCIN
CAPRYLIC ACID
Caprylic Alcohol → ALCOHOL, OCTYL
CAPSAICIN also → CHILLIES, METHYL NICOTINATE, PEPPERS
CAPSELLA BURSA-PASTORIS
CAPSICUM
Capsicum annum var. onoides → CHRISTMAS PEPPER
Capsicum frutescens → PEPPERS
CAPTAFOL
Captain Cook → FISH, KAVA
CAPTAN
Captax → 2-MERCAPTOBENZOTHIAZOLE
Captin → ACETAMINOPHEN
Captodiam → CAPTODIAMINE
CAPTODIAMINE
Captodramin → CAPTODIAMINE
Captolane → CAPTOPRIL
CAPTOPRIL also → CIMETIDINE, INDOMETHACIN, METOLAZONE, POTASSIUM, PROBENECID
Captoril → CAPTOPRIL
CAPURIDE

Capval → NOSCAPINE
Carace → LISINOPRIL
Carachol → DEHYDROCHOLIC ACID, *sodium salt*
Caradrin → PROSCILLARIDIN
Carafate → SUCRALFATE
CARAMEL
CARAMIPHEN
CARAPA PROCERA
CARAWAY also → *d*-LIMONENE
CARAZOLOL
CARBACHOL
CARBADOX
Carbam → DIETHYLCARBAMAZINE also → METHAM SODIUM
Carbamaldehyde → FORMAMIDE
Carbamates → CUTTING OILS
CARBAMAZEPINE also → ADENOSINE, ALPRAZOLAM, BUPROPION, CIMETIDINE, CLARITHROMYCIN, CLONAZEPAM, CLOZAPINE, DESMOPRESSIN, DIAMINOZIDE, DILTIAZEM, DOXYCYCLINE, ERYTHROMYCIN, FELBAMATE, FLUOXETINE, FLUVOXAMINE, GRAPEFRUIT, HALOPERIDOL, METHOSUXIMIDE, ISONIAZID, ISOTRETINOIN, ITRACONAZOLE, LAMOTRIGINE, LITHIUM, MEBENDAZOLE, METRONIDAZOLE, MIANSERIN, NORTRIPTYLINE OLANZEPINE, PHENYTOIN, PLANTAGO SEED, PRIMIDONE, PROPOXYPHENE, SODIUM VALPROATE, TACROLIMUS, TERFENADINE, THEOPHYLLINE, TOREMIFENE CITRATE, TRAZODONE, TROLEANDOMYCIN, VALPROIC ACID, VERAPAMIL, VIGABATRIN, VILOXAZINE, WARFARIN, WATER
Carbamazine → DIETHYLCARBAMAZINE
Carbamidal → NIKETHAMIDE
Carbamide → UREA
Carbamidine → GUANIDINE
Carbamult → PROMECARB
Carbamylcholine Chloride → CARBACHOL
Carbamylhydrazine → SEMICARBAZIDE
Carbanil → PHENYL ISOCYANATE
Carbanilides → SOAP
CARBANOLATE
Carbapen → PENICILLINS – "NEWER and SYNTHETICS" > *Carbenicillin Disodium*
CARBARSONE
CARBARYL also → DIAZOXIDE, PRALIDOXIME, RESERPINE
Carbased → ACECARBROMAL
Carbathion → METHAM SODIUM
Carbazine → ZIRAM
CARBAZOCHROME SALICYLATE
Carbazon → CARBARSONE
Carbazotic Acid → PICRIC ACID
Carbecin → PENICILLINS – "NEWER and SYNTHETICS" > *Carbenicillin Disodium*
Carbelan → CARBAMAZEPINE
Carbendazim → METHYL BENZIMIDAZOLE-2-CARBAMATE
Carbendazole → METHYL BENZIMIDAZOLE-2-CARBAMATE
Carbenicillin Disodium → PENICILLINS – "NEWER and SYNTHETICS" also → AMIKACIN, GENTAMICIN; PENICILLINS – "NEWER and SYNTHETICS" > *Ampicillin*
Carbenicillin Indanyl Sodium → PENICILLINS – "NEWER and SYNTHETICS" > *Carindacillin Sodium*
CARBENOXOLONE also → SPIRONOLACTONE
Carbetane → CARBETAPENTANE CITRATE
CARBETAPENTANE CITRATE
Carbicron → DICROTOPHOS
CARBIDOPA also → L-DOPA, FERROUS SULFATE
Carbilizine → DIETHYLCARBAMAZINE
CARBIMAZOLE also → DIGOXIN
Carbimide → CYANAMIDE
Carbinol → ALCOHOL, METHYL
CARBINOXAMINE MALEATE
CARBIPHENE
Carbitol → 2-(2-ETHOXYETHOXY) ETHANOL
Carbocaine → MEPIVACAINE
Carbocholine → CARBACHOL
Carbochromen → CHROMONAR
Carbocromen(e) → CHROMONAR
Carbodiimide → CYANAMIDE
Carbofos → MALATHION
CARBOFURAN
Carbol-Fuchsin → PHENOL
Carbolic Acid → PHENOL
Carbolith → LITHIUM CARBONATE
Carbolithium → LITHIUM CARBONATE
Carbomer → CARBOXYPOLYMETHYLENE
Carbomethane → KETENE
Carbomycin → OLEANDOMYCIN
CARBON also → 1,2-BENZANTHRACENE
Carbon Bisulfide → CARBON DISULFIDE
Carbon Black → CARBONS
CARBON DIOXIDE also → FIRE EXTINGUISHERS, METHANE,

PHOSGENE, POTASSIUM PERSULFATE, PROPYLENE OXIDE, SODIUM BICARBONATE, SULFANILIC ACID, TOLUENE-2,4-DIISOCYANATE, WELDING
CARBON DISULFIDE also → DISULFIRAM, THIRAM
Carbon Hexachloride → HEXACHLOROETHANE
Carbonic Acid Gas → CARBON DIOXIDE
Carbonic Anhydrase Inhibitor No. 6063 → ACETAZOLAMIDE
Carbonic Anhydride → CARBON DIOXIDE
Carbonic Dichloride → PHOSGENE
CARBON MONOXIDE also → 4-AMINODIPHENYL, *n*-BUTYL MERCAPTAN, CARBONS, CARBON DIOXIDE; FUELS, OIL; KEROSENE, METHANE, METHYL CHLORIDE, NEWSPAPERS, POLYURETHANE, SILK, SULFANILIC ACID, WOOL
CARBON PAPER
CARBON TETRABROMIDE
CARBON TETRACHLORIDE also → ACETYLCYSTEINE; ALCOHOL, ETHYL; ALCOHOL, ISOPROPYL; BARIUM, BENZO[α]PYRENE, BISHYDROXYCOUMARIN; 1,3-BUTYLENE GLYCOL; FIRE EXTINGUISHERS, KEROSENE, MALONALDEHYDE, MILK, PHENOBARBITAL
Carbonyl Chloride → PHOSGENE
Carbonyldiamide → UREA
CARBOPHENOTHION also → METHYL TRITHION
CARBOPLATIN also → ETOPOSIDE
Carbopol → CARBOXYPOLYMETHYLENE
CARBOPROST
Carboprost Methyl → CARBOPROST
Carboprost Trometamol → CARBOPROST, *Tromethamine*
Carbose → SODIUM CARBOXYMETHYLCELLULOSE
Carbose D → CELLULOSE, *Sodium Carboxymethylcellulose*
Carb-O-Sep → CARBARSONE
Carboslane → ALCOHOL, ETHYL
Carbostesin → BUPIVACAINE
Carbosylane → ALCOHOL, ETHYL
Carbothion → CARBOPHENOTHION
Carbowax → POLYETHYLENE GLYCOL
Carboxide → CARBON DIOXIDE
CARBOXIN
Carboxyhemoglobin → METHYLENE CHLORIDE
CARBOXYPOLYMETHYLENE
Carboxytolbutamide → TOLBUTAMIDE
Carboxyvinyl Polymer → CARBOXYPOLYMETHYLENE
CARBROMAL also → MEPHENESIN
CARBUTAMIDE also → BUCLOSAMIDE, PHENYLBUTAZONE, SULFANILAMIDE
CARBUTEROL
Carbyne → BARBAN
Carcholin → CARBACHOL
Carcinil → LEUPROLIDE
Cardace → RAMIPRIL
Cardamist → NITROGLYCERIN
CARDAMON
Cardelmycin → NOVOBIOCIN
Carden → CELIPROLOL
Cardenalin → DOXAZOCIN MESYLATE
Cardene → NICARDIPINE
Cardenolide → OCGODEIA sp.
Cardiacap → PENTAERYTHRITOL TETRANITRATE
Cardiamid → NIKETHAMIDE
Cardiagutt → VERAPAMIL
Cardiazol → PENTYLENETETRAZOL
Cardibeltin → VERAPAMIL
Cardigin → DIGITOXIN
Cardilate → ERYTHRITYL TETRANITRATE
Cardiloid → ERYTHRITYL TETRANITRATE
Cardimon → NIKETHAMIDE
Cardinal Flower → LOBELIA
Cardinalspear → ERYTHRINA, *herbacea*
Cardiaol → PROPRANOLOL
Cardio 10 → ISOSORBIDE DINITRATE
Cardiofilina → AMINOPHYLLINE
Cardiogen → LEVOCARNITINE
Cardiografin → MEGLUMINE DIATRIZOATE
Cardio-Green → INDOCYANINE GREEN
Cardio-Khellin → KHELLIN
Cardiomin → AMINOPHYLLINE
Cardion → PROSCILLARIDIN
Cardioquin → QUINIDINE
Cardiorhythmine → AJMALINE
Cardiosteril → DOPAMINE
Cardiotrast → IODOPYRACET
Cardiovanil → ETHAMIVAN
Cardiovet → ENALAPRIL
Cardiox → DIGOXIN
Cardis → ISOSORBIDE DINITRATE
Cardisan → ITRAMIN TOSYLATE
Carditin-Same → PRENYLAMINE LACTATE
Carditoxin → DIGITOXIN
Cardium → BEPRIDIL
Cardivix → BENZIODARONE
Cardizem → DILTIAZEM

Cardo Amarillo → ARGEMONE MEXICANA
Cardol → CASHEW also → PENTADECYLCATECHOL
Cardomec → IVERMECTIN
Cardophylin → AMINOPHYLLINE
Cardophyllin → AMINOPHYLLINE
Cardo Santa → ARGEMONE MEXICANA
Cardotek-30 → IVERMECTIN
Cardovar → TRIMAZOSIN
Cardoxin → DIPYRIDAMOLE
Cardrase → ETHOXZOLAMIDE
Cardular → DOXAZOCIN MESYLATE
Cardura → DOXAZOCIN MESYLATE
Carduran → DOXAZOCIN MESYLATE
Carecin → CINNARIZINE
Careless Weed → AMARANTHUS
Carena → AMINOPHYLLINE
Δ^3 -CARENE
CAREX sp.
Carfene → AZINPHOS-METHYL
CARFENTANIL
Caricide → DIETHYLCARBAMAZINE
Caridolol → PROPRANOLOL
Carindacillin Sodium → PENICILLINS – "NEWER and SYNTHETICS"
Carindapen → PENICILLINS – "NEWER and SYNTHETICS" > *Carindacillin Sodium*
Carisoma → CARISPRODOL
Carisoprodate → CARISPRODOL
CARISPRODOL also → AMINOPYRINE, GLUTETHIMIDE, IMIPRAMINE, MEPROBAMATE, ORPHENADRINE, PHENYLBUTAZONE
CARISSA
Caritrol → DIETHYLCARBAMAZINE
Carmazon → PROSCILLARIDIN
Carmethose → CELLULOSE, *Sodium Carboxymethylcellulose*
CARMINE also → LIPSTICK
Carmubris → CARMUSTINE
Carmurit → ETHOXAZENE
CARMUSTINE also → CIMETIDINE
CARNAUBA WAX
Carnegine → PACHYCEREUS
CARNE SECA
Carnicor → LEVOCARNITINE
Carnitine → LEVOCARNITINE also → PENICILLINS – "NEWER and SYNTHETICS"> *Pivampicillin, Pivmecillanam*; PHENYTOIN
Carnitor → LEVOCARNITINE
Carnum → LEVOCARNITINE
CAROB
Carobronze → CANTHAXANTHIN
Carofur → NIFURPRAZINE
Carogna → COLTSFOOT
Caroid → PAPAIN
Carolina Jassaminen → GELSEMIUM SEMPERVIRENS
Carolina palmetto → PALMS
Carolina Wild Woodbine → GELSEMIUM SEMPERVIRENS
Carolysine → MECHLORETHAMINE
Caryolysine → MECHLORETHAMINE
Carophyll Red → CANTHAXANTHIN
Carophyllus → CLOVES
Carotaben plus → CANTHAXANTHIN
β -CAROTENE also → CARROTS, METHIMAZOLE, SPIGELIA ANTHELMIA
β,β- Carotene-4,4´-dione → CANTHAXANTHIN
CAROTOTOXIN also → CARROTS
CARPERONE
Carpet Weed → KALLSTROEMIA
CARPHENAZINE
Carpidor → BENFLURALIN
CARPOBROTUS CHILENSIS
CARPROFEN
CARRAGEENAN
Carrbutabarb → BUTABARBITAL
Carrel-Dakin's Solution → CHLORINATED LIME
Carrier → LEVOCARNITINE
CARROTS also → NITRATES and NITRITES, POTASSIUM NITRATE
CARSALAM
Carteol → CARTEOLOL HYDROCHLORIDE
CARTEOLOL HYDROCHLORIDE
Carter-Glogau → VITAMIN E
Cartrol → CARTEOLOL HYDROCHLORIDE
Cartrophen → PENTOSAN POLYSULFATE
Cartwheels → AMPHETAMINE
Carum nigrum → CUMIN, BLACK
CARVACROL also → MARJORAM
Carvacron → TRICHLORMETHIAZIDE
Carvanil → ISOSORBIDE DINITRATE
Carvasin → ISOSORBIDE DINITRATE
CARVEDILOL also → PROPANTHELINE
l-CARVEOL
Carya illinoensis → PECANS
Carya orata → HICKORY
Carya pecan → PECANS
Carylderm → CARBARYL
CARYOCAR sp.
Caryophyllic Acid → EUGENOL
Caryota mitis → PALMS
Casabe → CASSAVA
Casanol → CASANTHRANOL

CASANTHRANOL
Casantin → DIETHAZINE
Casave → CASSAVA
CASCARA SAGRADA also → MILK
CASEIN also → MILK
Casein Hydrolysate → PROTEIN HYDROLYSATES
Case Weed → CAPSELLA BURSA-PASTORIS
"Cash" → α-METHYL FENTANYL
CASHEW also → ANACARDIACEA, PENTADECYLCATECHOL, POISON IVY, RENGAS; SUMAC, POISON
Casodex → BICALUTAMIDE
Cassaidine → ERYTHROPHLEUM
CASSAINE also → ERYTHROPHLEUM
CASSAVA also → CYANATES, CYANIDES, HYDROGEN CYANIDE, LINAMARIN, SORGHUM, TAPIOCA
Cassella 4489 → CHROMONAR
Cassene → PARAMETHASONE ACETATE
CASSIA
Cassia acutifolia → SENNA
CASTANEA sp.
Castellani's Paint → *p*-ROSANILINE also → PHENOL
Castle Bean → GLOTTIDIUM VESICARIUM
CASTOR BEAN also → HYDROGEN CYANIDE, JEWELRY, RICIN
CASTOREUM
CASTOR OIL also → CAMPHOR
Castrix → CRIMIDINE
Cataflam → DICLOFENAC
Catanil → CHLORPROPAMIDE
Catapres → CLONIDINE
Catapresan → CLONIDINE
Catarase → CHYMOTRYPSIN
Catatrol → VILOXAZINE
Catclaw → HYDROGEN CYANIDE
CATECHOL
Catenulin → PAROMOMYCIN
Cateudyl → METHAQUALONE
CATHA EDULIS
Catharanthus roseus → VINCA ROSEA
Cathine → NORPSEUDOEPHEDRINE
Cathinone → CATHA EDULIS
Cathocin → NOVOBIOCIN
Cathomycin → NOVOBIOCIN
Catmint → CATNIP
Catnep → CATNIP
CATNIP
Catral → PHENIPRAZINE
Catron → PHENIPRAZINE
Catroniazid → PHENIPRAZINE
Catrup → CATNIP
Cat's Foot → NEPETA HEDERACEA
Catwort → CATNIP
Caucasian Snowdrops → GALANTHAMINE
Caudaline → TICLOPIDINE HCL
CAULIFLOWER also → THIOCYANATES
CAULOPHYLLINE
CAULOPHYLLUM sp. also → ANAGYRINE
Caustic Barley → SABADILLA
Caustic Potash → POTASSIUM HYDROXIDE
Caustic Soda → SODIUM HYDROXIDE
Causyth → PROPYPHENAZONE
CAVIAR
Cavodil → PHENIPRAZINE
Cavonyl → CYCLOBARBITAL
Cawe → PACHYCEREUS
Cayenne Pepper → CAPSICUM also → PEPPERS
Caytine → PROTOKYLOL
Cazabe → CASSAVA
1522 CB → ACEPROMAZINE
1802 CB → BAMIFYLLINE
CB 154 → BROMOCRIPTINE
CB 304 → AZAURIDINE
CB 311 → GROWTH HORMONES
CB 313 → MITOTANE
CB 1048 → CHLORMAPHAZINE
CB 1348 → CHLORAMBUCIL
CB 1678 → PROPIOMAZINE
CB 2041 → BUSULFAN
CB 2201 → AMFEPENTOREX
CB 3025 → MELPHALAN
CB 4306 → CLORAZEPATE DIPOTASSIUM
CB 8061 → PENTOSAN POLYSULFATE
CBDCA → CARBOPLATIN
CC → VITAMIN D_3
CCA → CHROMATED COPPER ARSENATE
CCC → CHLORMEQUAT CHLORIDE
CCI 15641 → CEFUROXIME
CCI 18781 → FLUTICASONE PROPIONATE
CCK 179 → ERGOLOID MESYLATES
CCl_4 → CARBON TETRACHLORIDE
CCNU → LOMUSTINE
CCRG 81,045 → TEMOZOLOMIDE
CD 2 → 3-METHYL-4-AMINO-*N*-DIETHYLANILINE HCL
CD 68 → CHLORDANE
CD 271 → ADAPALENE
CD → CHENODEOXYCHOLIC ACID
C. D. → GLUTETHIMIDE
Cd → CADMIUM
2-CdA → CLADRIBINE
CDAA → *N,N*-DIALLYL-2-CHLOROACETAMIDE
CDC → CHENODEOXYCHOLIC ACID

CDCA → CHENODEOXYCHOLIC ACID
CDM → CHLORDIMEFORM
CDP 777 → GEMTUZIMAB OZOGAMICIN
CE-3624 → TICRYNAFEN
Cebicure → VITAMIN C
Cebion → VITAMIN C
Cebrum → CHLORDIAZEPOXIDE
Cebutid → FLURBIPROFEN
Cecenu → LOMUSTINE
Cecil → COCAINE
Ceclor → CEFACLOR
Cecon → VITAMIN C
Cedad → BENACTYZINE
Cedar Leaf Oil →, THUJA OCCIDENTALIS
"Cedar Poisoning" → USNIC ACID
CEDAR, WESTERN RED
Cedarwood Oil Alcohols → CEDROL
Cedax → CEFTIBUTIN
Cedilanid → LANATOSIDE
Cedilanid-D → DESLANOSIDE
Cedocard → ISOSORBIDE DINITRATE
Cedol → CEFAMANDOLE
CEDRENOL
CEDRENYL ACETATE
CEDROL
CEDRYL ACETATE
Cedulamin → METHENAMINE
Cedur → BEZAFIBRATE
CeeNU → LOMUSTINE
c7E3Fab → ABCIXIMAB
Cefacidal → CEFAZOLIN SODIUM
CEFACLOR also → THEOPHYLLINE
Cefadol → DIPHENIDOL
Cefa-Drops → CEFADROXIL
Cefadros → CEPHALEXIN
CEFADROXIL
Cefadyl → CEPHAPIRIN
Cefa-Iskia → CEPHALEXIN
Cefa-Lak → CEPHAPIRIN
Cefaloridin → CEPHALORIDINE
Cefalotin → CEPHALOTHIN
Cefaloto → CEPHALEXIN
Cefam → CEFAMANDOLE
CEFAMANDOLE also → ALCOHOL, ETHYL; WARFARIN
Cefamar → CEFUROXIME
Cefamedin → CEFAZOLIN SODIUM
Cefamezin → CEFAZOLIN SODIUM
Cefamox → CEFADROXIL
Cefa-Tabs → CEFADROXIL
Cefatrexyl → CEPHAPIRIN
Cefatriaxone → CEFTRIAXONE
Cefazil → CEFAZOLIN SODIUM
Cefazina → CEFAZOLIN SODIUM
CEFAZOLIN SODIUM
CEFDINIR
CEFEPIME
Cefibacter → CEPHALEXIN
Cefiran → CEFAMANDOLE
CEFIXIME
Cefizox → CEFTIZOXIME
Ceflorin → CEPHALORIDINE
CEFMETAZOLE
Cefobid → CEFPERAZONE
Cefobine → CEFPERAZONE
Cefobis → CEFPERAZONE
Cefomonil → CEFSULODIN SODIUM
CEFONICID
CEFOPERAZONE also → ALCOHOL, ETHYL; BEER, DISULFIRAM, VITAMIN K
Cefoprim → CEFUROXIME
Ceforal → CEFADROXIL
CEFORANIDE
Cefosint → CEFOPERAZONE
Cefossim → CEFUROXIME
Cefotan → CEFOTETAN
Cefotax → CEFOTAXIME
CEFOTAXIME
CEFOTETAN also → DISULFIRAM
CEFOXITIN SODIUM also → PENTOXIFYLLINE
CEFPROZIL
Cefracycline → TETRACYCLINE
Cefradrex → CEPHADRINE
Cefrag → CEPHADRINE
Cefro → CEPHADRINE
Cefroxil → CEFADROXIL
Cefspan → CEFIXIME
CEFSULODIN SODIUM
CEFTAZIMIDES
CEFTIBUTIN
Ceftim → CEFTAZIMIDES
Ceftin → CEFUROXIME
CEFTIOFUR
Ceftix → CEFTIZOXIME
CEFTIZOXIME
CEFTRIAXONE also → VERAPAMIL
Cefumax → CEFUROXIME
Cefurex → CEFUROXIME
Cefurin → CEFUROXIME
CEFUROXIME
Cefzil → CEFPROZIL
Cegadera → HETEROPHYLLEA PUSTULATA
Cegiolan → VITAMIN C
Ceglunat → LANATOSIDE
Ceglution → LITHIUM CARBONATE
Cekufon → TRICHLORFON

Cekusil → PHENYLMERCURIC ACETATE
Celadigal → LANATOSIDE C
Celadine → CHELIDONIUM MAJUS
Celance → PERGOLIDE MESYLATE
Celandine → CHELIDONIUM MAJUS
Celandine Poppy → CHELIDONIUM MAJUS
Cela S-1942 → BROMOLYSERGIDE
Celaskon → VITAMIN C
CELASTRUS SCANDENS
Celbenin → PENICILLINS – "NEWER and SYNTHETICS" > *Methicillin Sodium*
Celebra → CELECOXIB
Celebrex → CELECOXIB
CELECOXIB
Celectol → CELIPROLOL
Celepat → TIMONACIC
CELERY also → CYANIDES, LATEX, *d*-LIMONENE, NITRATES and NITRITES, SCLEROTINIA SCLEROTIORUM
Celestan → BETAMETHASONE
Celestan-V → BETAMETHASONE
Celestene → BETAMETHASONE
Celestoderm-V → BETAMETHASONE
Celestone → BETAMETHASONE
Celevac → METHYL CELLULOSE
Celex → CEPHADRINE
Celexa → CITALOPRAM
Celidonia → CHELIDONIUM MAJUS
Celin → VITAMIN C
Celiomycin → VIOMYCIN
CELIPROLOL
CELLASENE
Cellative → MECLOFENOXATE
CellCept → MYCOPHENOLATE MOFETIL
Cellidrin → ALLOPURINOL
Cellolax → CELLULOSE, *Sodium Carboxymethylcellulose*
Cellon → 1,1,2,2-TETRACHLOROETHANE
Cellosize → CELLULOSE, *Ethyl Hydroxyethyl Ether*
Cellosolve → 2-ETHOXYETHANOL
Cellosolve Acetate → 2-ETHOXYETHYL ACETATE
Cellothyl → METHYL CELLULOSE
Cellucon → METHYL CELLULOSE
Celluflex 179 → TRITOLYL PHOSPHATE
CELLULOSE also → COCAINE, PAPER, SAWDUST
Cellulose Gum → SODIUM CARBOXYMETHYLCELLULOSE
Cellulose Methyl Ether → METHYL CELLULOSE
Cellumeth → METHYL CELLULOSE
Celmidol → DIPHENIDOL
Cel-O-Brandt → CELLULOSE
Celocurine → SUCCINYLCHOLINE
Celotin → METHSUXIMIDE
Celpillina → PENICILLINS – "NEWER and SYNTHETICS" > *Methicillin Sodium*
Celtec → OXATOMIDE
Celtics → ACETALDEHYDE
Celtium → HAFNIUM
Celupan → NALTREXONE
Cemado → CEFAMANDOLE
Cemandil → CEFAMANDOLE
CEMENT also → POTASSIUM HYDROXIDE
Cenalene-M → PENTYLENETETRAZOL
Cenazol → PENTYLENETETRAZOL
Cenchrus sp. → SANDBUR
Cenetone → VITAMIN C
Cenomycin → CEFOXITIN SODIUM
Censedal → NEALBARBITAL
Censpar → BUSPIRONE
CENTALLA ASIATICA
Centasium → CENTALLA ASIATICA
CENTAUREA sp.
Centaureé de Russie → CENTAUREA sp.
CENTAURIUM sp.
Centaury → CENTAURIUM sp.
CENTCHROMAN
Centelase → CENTALLA ASIATICA
Cento ℞ → ABCIXIMAB
CENTOXIN
Centralgin → MEPERIDINE
Centramin → PEMOLINE
Centrax → PRAZEPAM
Centrine → AMINOPENTAMIDE HYDROGEN SULFATE
Centrophenoxine → MECLOFENOXATE
Centruroides → SCORPIONS
Centurina → PENICILLINS – "NEWER and SYNTHETICS" > *Pivampicillin*
Century Plant → AGAVE
Centyl → BENDROFLUMETHIAZIDE
Ceosunin → CERULETIDE
CEPA → ETHEPHON
Cepacilina → PENICILLIN > *Benzylpenicillin, Benzathine*
Cepaloridin → CEPHALORIDINE
Cepalorin → CEPHALORIDINE
Cepan → CEFOTETAN
Cepaverin → PAPAVERINE
Cepazine → CEFUROXIME
CEPHA → ETHEPHON
CEPHADRINE
Cephadrol → DIPHENIDOL
Cephalanthin → CEPHALANTHUS OCCIDENTALIS
CEPHALANTHUS OCCIDENTALIS

CEPHALEXIN also → GENTAMICIN
Cephalin → CEPHALNATHUS OCCIDENTALIS, GRAMICIDIN
CEPHALOGLYCIN
Cephalon → MODAFINIL
CEPHALORIDINE also → FUROSEMIDE, GENTAMICIN, KANAMYCIN, TOBRAMYCIN, VANCOMYCIN
Cephalosporinase → β-LACTAMASE
Cephalosporins → ACETAMIDE, PENICILLIN, PHENYLBUTAZONE, PROBENECID
Cephalotaxus sp. → YEW
CEPHALOTHIN also → FUROSEMIDE, GENTAMICIN, KANAMYCIN, PROBENECID, TOBRAMYCIN
CEPHAPIRIN
Cephation → CEPHALOTHIN
Cephos → CEFADROXIL
Cephrol → CITRONELLOL
Cephulac → LACTULOSE
Ceporacin → CEPHALOTHIN
Ceporan → CEPHALORIDINE
Ceporex → CEPHALEXIN
Ceporexine → CEPHALEXIN
Ceporin → CEPHALORIDINE
Cepovenin → CEPHALOTHIN
Ceptaz → CEFTAZIMIDES
Cequartyl → BENZALKONIUM CHLORIDE
Cer → CEPHALORIDINE
Cera Flava → BEESWAX
Ceraphyl 368 → OCTYL PALMITATE
Ceraphyl 1 PI → ISOPROPYL LINEOLATE
Cerasan™ → METHOXYETHYL MERCURY SILICATE
Cerate 50 → MIBEFRADIL
Ceratocephalus testiculatus → BUR BUTTERCUP
Ceratostomella fimbriata → SWEET POTATO
CERCOCARPUS MONTANUS
Cerebid → PAPAVERINE
Cerebolan → CINNARIZINE
Cerebroforte → PIRACETAM
Cerebrose → GALACTOSE
Ceredase → ALGLUCERASE
Ceregulart → DIAZEPAM
Cerekinon → TRIMEBUTINE MALEATE
Cerelose → GLUCOSE
Cereon → VITAMIN C
Cerepar → CINNARIZINE
Ceresan™ → METHOXYETHYLMERCURY CHLORIDE
Ceresan M → *N*-(ETHYLMERCURI)-*p*-TOLUENESULFONANILIDE
Ceresan Slaked Lime → PHENYLMERCURIC ACETATE
Cerespan → PAPAVERINE
Cerevon → FERROUS SUCCINATE
Cerezyme → ALGLUCERASE
Cergem → GEMEPROST
Cergona → VITAMIN C
Ceriman → MONSTERA sp.
CERIUM
CERIVASTATIN
Cero → CHLORMADINONE
Cerone → ETHEPHON
Certinol → *p*-AMINOPHENOL
Certomycin → NETILMYCIN
Certox → THALLIUM
Certs® → METHYL SALICYLATE also → ETHER, ETHYL
Cerubidin → DAUNORUBICIN
Cérubidine → DAUNORUBICIN
Cerucal → METOCLOPRAMIDE
CERULETIDE
CERULOPLASMIN
Cerumene → SQUALANE
Cerumtinex → TRIETHYLENE OLEATE POLYPEPTIDE CONDENSATE
Cervagem(e) → GEMEPROST
Cervidil → DINOPROSTONE
Cerviprost → DINOPROSTONE
Cesamet → NABILONE
Cescorbat → VITAMIN C
CESIUM also → CHEESES, COFFEE, TEA
Cesol → PRAZIQUANTEL
Cesplon → CAPTOPRIL
Cesporan → CEPHADRINE
CESTRUM
Cetab → CETRIMONIUM BROMIDE
Cetaceum → SPERMACETI
Cetacillin → PENICILLINS – "NEWER and SYNTHETICS" > *Propicillin Potassium*
Cetacort → HYDROCORTISONE
Cetadol → ACETAMINOPHEN
Cetamid → VITAMIN C
Cetamide → SULFACETAMIDE SODIUM
Cetampin → PENICILLINS – "NEWER and SYNTHETICS" > *Ampicillin*
Cetaphyl 140 → DECYL OLEATE
Cetavlon → CETRIMONIUM BROMIDE
Cetebe → VITAMIN C
Cetemican → VITAMIN C
Cethylose → CELLULOSE, *Sodium Carboxymethylcellulose*
Cetiprin → EMEPRONIUM
CETIRIZINE also → ACENOCOUMAROL
Cetosanol → LANATOSIDE

Cétraphylline → THEOPHYLLINE
Cetrexin → MECLOFENOXATE
CETRIMONIUM BROMIDE
Cetsim → ALBUTEROL
Cetyl Alcohol → ALCOHOL, CETYL
Cetylamine → CETRIMONIUM BROMIDE
Cetyl Mercaptan → 1-HEXADECANETHIOL
CETYLPYRIDINIUM CHLORIDE also → KAOLIN, MAGNESIUM SILICATE, POLYSORBATES
Cetyl Stearyl Alcohol → ALCOHOL, CETYL
Cevadilla → SABADILLA also → VERATRINE
Cevadilline → VERATRINE
Cevadine → VERATRINE
Cevalin → VITAMIN C
Cevanol → BENACTYZINE
Cevatine → VITAMIN C
Cevex → VITAMIN C
Ceviche → FISH
CEVIMELINE
Cevimin → VITAMIN C
Ce-Vi-Sol → VITAMIN C
Cevitamic Acid → VITAMIN C
Cevitan → VITAMIN C
Cevitex → VITAMIN C
Cewin → VITAMIN C
Cex → CEPHALEXIN
CF → FOLINIC ACID
CFC 11 → TRICHLOROFLUOROMETHANE
CFC 113 → TRICHLOROTRIFLUOROETHANE
C-Film → NONOXYNOL-9
CFPQ → ENROFLOXACIN
CG 1 → CARBIMAZOLE
CG 201 → PIRIBENZYL METHYL SULFATE
CG 315E → TRAMADOL
CG 1283 → MIREX
CGA 10,832 → PROFLURALIN
CGA 24,705 → METOLACHLOR
CGA 72662 → CYROMAZINE
CGA 179,246 → MILBEMYCIN OXIME
CGA 184,699 → LUFENURON
CGP 2175 → METOPROLOL
CGP 3543/E → BICOZAMYCIN
CGP 7174/E → CEFSULODIN SODIUM
CGP 39,393 → HIRUDINS
CGP 48,933 → VALSARTAN
CGS 14,824A → BENAZEPRIL
CGS 20,267 → LETROZOLE
CH 3635 → TRICLOSAN
α-Chaconine → POTATO
CHAEROPHYLLUM sp.
CHAETOMIUM sp.
Chagas Disease → TRIATOMA
Chalice Vine → SOLANDRA sp.
Chalk → CALCIUM CARBONATE also → AMPHETAMINE
CHAMAESYA sp.
Chameleon Mineral → POTASSIUM PERMANGANATE
Chamico → DATURA, *strammonium*
CHAMOMILE also → ASTERS, CHRYSANTHEMUM, GOLDENROD, TEA
Chandmaruwa → RAUWOLFIA SERPENTINA
Channa → PEAS
Channel Black → CARBONS
Chanoclavine → PARTHENIUM
Chanvre Indien → CANNABIS
Chaparral → LARREA sp.
Charabdea sp. → SEA WASP
Charas → CANNABIS
Charcoal, Activated → CARBONS also → ACETYLCYSTEINE, LEFLUNOMIDE, MAGNESIUM CITRATE, MAGNESIUM OXIDE, OLANZEPINE, SORBITOL,TANNIC ACID, THEOPHYLLINE
Chares → CANNABIS
Charlie → COCAINE
Charlock → MUSTARD
Chat → CATHA EDULIS
Chaux Vive → CALCIUM OXIDE
Chawal → RICE
$C_8H_{11}Cl_3O_6$ → CHLORALOSE
Chealamide → DISODIUM EDETATE
Chebutan → KEBUZONE
CHEESES also → CYCLOPIAZONIC ACID, DEBRISOQUIN, HISTAMINE, IMIPRAMINE, ISOCARBOXAZID, ISONIAZID, NIALAMIDE, PATULIN, PHOSPHORIC ACID, POTATO, ROQUEFORTINE, STERIGMATOCYSTIN
CHEILANTHEA SIEBERI
Cheirolin → RAPISTRUM RUGOSUM
Chelaplex III → DISODIUM EDETATE
Chelaton III → DISODIUM EDETATE
Chelen → ETHYL CHLORIDE
Chelidonine → CHELIDONIUM MAJUS also → SNOWBERRY
CHELIDONIUM MAJUS
Chellah → AMMI VISNAGA
Chelon 100 → SODIUM EDETATE
Chelonia mydas → TURTLES
Chelydra serpentina → TURTLES
Chemagro B1776 → *S,S,S*-TRI-*N*-BUTYL PHOSPHOROTRITHIOATE
Chemathion → MALATHION
Chem Bam → NABAM

Chemcef → CEFOTAXIME
Chemform → METHOXYCHLOR
Chem-Hoe → PROPHAM
Chemical 109 → 1-(1-NAPHTHYL)-2-THIOUREA
Chemicetina → CHLORAMPHENICOL
Chemifluor → SODIUM FLUORIDE
Chemiofuran → NITROFURANTOIN
Chemipen → PENICILLINS – "NEWER and SYNTHETICS" > *Phenethicillin Potassium*
Chemochin → CHLOROQUINE
Chemofuran → NITROFURAZONE
Chemonite → AMMONIACAL COPPER ARSENATE
Chemotrim → SULFAMETHOXAZOLE
Chemox DN → DINOSEB
Chemox PE → DINOSEB
Chemox Selective → DINOSEB
Chem Pels C → SODIUM ARSENITE
Chewrat → PINDONE
Chem Rice → PROPANIL
Chem-sen → SODIUM ARSENITE
Chemshield → α-CHLOROBENZYLIDINEMALONITRILE
Chendol → CHENODEOXYCHOLIC ACID
Chêne Gubacs → OAKS
Chenix → CHENODEOXYCHOLIC ACID
Chenocedon → CHENODEOXYCHOLIC ACID
Chenocol → CHENODEOXYCHOLIC ACID
CHENODEOXYCHOLIC ACID
Chenodex → CHENODEOXYCHOLIC ACID
Chenodiol → CHENODEOXYCHOLIC ACID
Chenofalk → CHENODEOXYCHOLIC ACID
CHENOPODIUM also → DATURA
Chenosäure → CHENODEOXYCHOLIC ACID
Chenossil → CHENODEOXYCHOLIC ACID
Chephalotin → CEPHALOTHIN
Chepirol → KEBUZONE
Cheque → MILBOLERONE
Chequered Daffodil → FRITILLARIA MELEAGRIS
CHERRY also → FRUITS and FRUIT JUICES, HYDROGEN CYANIDE
Cherry Laurel → CHERRY also → HYDROGEN CYANIDE
Cherrypie → LANTANA sp.
Chervils → CHAEROPHYLLUM
Chestnuts → CASTANEA sp. also → LATEX
Chetazolidin → KEBUZONE
Chetil → KEBUZONE
Chevrefeuille → LONICERA CAPRIFOLEUM
Chevron Cmpnd 9006 → METHAMIDOPHOS
CHF 10/21 → PIROXICAM
CHF 1,251 → PIROXICAM
CHF 1511 → BECLOBRATE
Chhuimui → MIMOSA
Chibro Pilocarpine → PILOCARPINE
Chibro-Proscar → FINASTERIDE
Chibroxin(e) → NORFLOXACIN
Chibroxol → NORFLOXACIN
Chichihualxochitl → PAPAYA
Chichita → SCHINUS TEREBINTHIFOLIUS
"Chick Edema Factor" → FAT, TOXIC also → DIOXIN
CHICKEN
Chicken Powder → AMPHETAMINE
Chickling Pea → LATHYRUS sp.
Chickling Vetch → LATHYRUS sp.
Chico → SARCOBATUS VERMICULATUS
CHICORY
Chick Pea → LATHYRUS sp. also → PEAS
Chika Saura → CROTOLARIA
Chilamacatyl → SUNFLOWER
Chilbe → FENUGREEK
Chile Saltpeter → SODIUM NITRATE
"Chillagoe Horse Disease" → CROTALARIA
CHILLIES also → CAPSICUM, LEAD CHROMATE
"Chilli-splitter's Lung" → CHILLIES also → PENICILLIUM sp.
"Chimney Sweeps Cancer" → CARBONS
Chimono → LOMEFLOXACIN
Chinaball Tree → MELIA sp.
China Bark → QUILLAJA
Chinaberry → MELIA sp.
China Clay → KAOLIN
Chinacrina → QUINACRINE
China Oil → BALSAM PERU
"China White" → HEROIN also → α-METHYL FENTANYL
Chinese Blistering Flies → MYLABRIS CICHORII
Chinese Blue → FERRIC FERROCYANIDE
Chinese Cabbage → BRASSICA sp.
Chinese Cantharides → MYLABRIS CICHORII
Chinese Cucumber → TRICHOSANTHIN
Chinese Gooseberry → KIWI
Chinese Inkberry → CESTRUM
Chinese Jellyfish → JELLYFISH
Chinese Lantern → PHYSALIS sp.
Chinese Mustard → MUSTARD
Chinese Peanut Tree → JATROPHA sp.
Chinese Red → MERCURIC SULFIDE, RED
Chinese Restaurant Syndrome → SODIUM GLUTAMATE
Chinese Seasoning → SODIUM GLUTAMATE
CHINESE TALLOW TREE also → SAPIUM sp.
Chinese White Poplar → POPLAR
Chingamin → CHLOROQUINE

Chinidin-Duretter → QUINIDINE
Chinidin-Durules → QUINIDINE
Chininum → QUININE
CHINIOFON
Chinkerinchee → ORNITHOGALLUM UMBELLATUM
Chinocid → PRIMAQUINE
Chinoform → IODOCHLORHYDROXYQUINOLINE
Chinofungin → TOLNAFTATE
Chinoin → PHENOBARBITAL
Chinosol → POTASSIUM HYDROXYQUINOLINE SULFATE
Chiptox → METHYL CHLOROPHENOXY ACETIC ACID
CHIRACANTHIUM PUNCTORIUM
Chircanthium sp. → SPIDERS
CHIRONEX FLEKERI also → JELLYFISH, SEA WASP
CHIRONIA PALUSTRIS subsp. TRANSVAALENSIS
Chiropsalmus → JELLYFISH also → SEA WASP
Chittar → GINGER, JAMAICAN
Chittem Bark → CASCARA SAGRADA
Chittim Bark → CASCARA SAGRADA
Chixin → NITROFURAZONE
Chlo-amine → CHLORPHENIRAMINE, *maleate*
Chlomycol → CHLORAMPHENICOL
CHLORACETYL CHLORIDE
Chlorachel → CHLORTETRACYCLINE
CHLORACIZINE
Chloracon → BECLAMIDE
Chloractil → CHLORPROMAZINE
Chloracysin → CHLORACIZINE
CHLORAL BETAINE
Chloraldurat → CHLORAL HYDRATE
CHLORAL HYDRATE also → ACENOCOUMAROL; ALCOHOL, ETHYL; BISHYDROXYCOUMARIN, CYCLOCOUMAROL, ETHYL BISCOUMACETATE, FURAZOLIDONE, FUROSEMIDE, GRISEOFULVIN, MILK, PARALDEHYDE, PARGYLINE, PENTYLENETETRAZOL, PHENOBARBITAL, PHENYTOIN, PROCHLORPERAZINE, TRANYCYPROMINE, TRIFLUPERAZINE DIHYDROCHLORIDE, WARFARIN
Chlorallylene → ALLYL CHLORIDE
Chloralosane → CHLORALOSE
CHLORALOSE also → BROTIZOLAN
CHLORAMBEN
CHLORAMBUCIL
Chloramex → CHLORAMPHENICOL
Chloramfen → CHLORAMPHENICOL
Chloramficin → CHLORAMPHENICOL
Chloramfilin → CHLORAMPHENICOL
Chloraminophene → CHLORAMBUCIL
Chloramiphene → CLOMIPHENE CITRATE
CHLORAMPHENICOL also → ACENOCOUMAROL, ACETAMINOPHEN, ACETANALID; ALCOHOL, ETHYL; ALUMINUM HYDROXIDE, ANTIPYRINE, BISHYDROXYCOUMARIN, CHLORPROPAMIDE, CODEINE, CYCLOCOUMAROL, CYCLOSPORINE, DISULFIRAM, ETHACRYNIC ACID, *N*-2-FLUORENYLACETAMIDE, GLIPIZIDE, HEXOBARBITAL, INSULIN, MAGNESIUM SILICATE, MAGNESIUM STEARATE, MONENSIN, NORTRIPTYLINE, PENICILLIN; PENICILLINS – "NEWER and SYNTHETICS" > *Ampicillin*; PENTOBARBITAL, PHENOBARBITAL, PHENYLALANINE, PHENYLBUTAZONE, PHENYTOIN, PRIMIDONE, RIFAMPIN, SODIUM BICARBONATE, THIAMYLAL SODIUM, THIMEROSAL, THIOPENTAL, TOLBUTAMIDE, VITAMIN K, WARFARIN
Chloramsaar → CHLORAMPHENICOL
CHLORANIL
Chloraquine → CHLOROQUINE
Chlorasept 2000 → CHLORHEXIDINE
Chlorasol → CHLORAMPHENICOL
Chloratex → CHLORAL HYDRATE
Chlorathrombon → CLORINDIONE
CHLORAZANIL
CHLORAZEPATE also → ERYTHROMYCIN, KETOCONAZOLE, MAGNESIUM
Chlorazin → CHLORPROMAZINE
Chlorazinil → CHLORAZANIL
CHLORBENSIDE
Chlorbenzilat → CHLOROBENZILATE
Chlorbutol → CHLOROBUTANOL
CHLORCYCLIZINE also → DEOXYCORTICOSTERONE, GRISEOFULVIN, HEXOBARBITAL, PARATHION, PHENYLBUTAZONE, PHENYTOIN, PROGESTERONE, TESTOSTERONE, ZOXAZOLAMINE
Chlordan → CHLORDANE
CHLORDANE also → DIGITOXIN, PARATHION, PHENOBARBITAL, PHENYLBUTAZONE, PROGESTERONE, ZOXAZOLAMINE
CHLORDANTOIN
Chlordecone → KEPONE
Chlorderazin → CHLORPROMAZINE

Chlordiazachel → CHLORDIAZEPOXIDE
CHLORDIAZEPOXIDE also → ACETAMINOPHEN; ALCOHOL, ETHYL; AMITRIPTYLINE, CIMETIDINE, CODEINE, DDT, DISULFIRAM, L-DOPA, KETOCONAZOLE, MAGNESIUM HYDROXIDE, MEPERIDINE, PHENINDIONE, PHENYTOIN, PROCHLORPERAZINE, PROMAZINE
CHLORDIMEFORM
Chlorditane → MITOTANE
Chlorethane → HEXACHLOROETHANE
Chlorethiazol → CLOMETHIAZOLES
Chlorethyl → ETHYL CHLORIDE
CHLORETHYLBENZENE
Chlorethylene → VINYL CHLORIDE
Chloretone → CHLOROBUTANOL
Chlorex → *sym*-DICHLOROETHYL ETHER
Chlorfenidim → MONURON
CHLORFENVINPHOS
Chlorguanide → CHLOROGUANIDE
CHLORGYLINE also → CLOMIPRAMINE
CHLORHEXIDINE
CHLORHYDROXYQUINOLINE also → Preface
Chloricol → CHLORAMPHENICOL
Chloride of Lime → CALCIUM HYPOCHLORITE
Chloride of Tin → STANNOUS CHLORIDE
Chloridin → PYRIMETHAMINE
Chlor-IFC → CHLORPROPHAM
Chlorimipramine → CLOMIPRAMINE
Chlorimpiphenine → IMICLOPAZINE
Chlorinat → BARBAN
Chlorinated Camphene → TOXAPHENE
CHLORINATED DIPHENYL also → CHLORETHYLBENZENE
CHLORINATED LIME also → CALCIUM HYPOCHLORITE
CHLORINATED NAPHTHALENES
CHLORINE also → ACETIC ACID, CHLOROFORM, CUTTING OILS, FREONS, PAPER, PHOSGENE, POLYVINYL CHLORIDE, RUBBER, SODIUM HYPOCHLORITE, SOYBEANS
CHLORINE DIOXIDE also → FLOUR
Chlorine Oxyfluoride → PERCHLORYL FLUORIDE
Chlorine Peroxide
CHLORIS sp.
CHLORISONDAMINE
CHLORMADINONE
CHLORMAPHAZINE
Chlormene → CHLORPHENIRAMINE, *maleate*
CHLORMEQUAT CHLORIDE
CHLORMERODRIN
Chlormethazone → CHLORMEZANONE
Chlormethiazole → CLOMETHIAZOLES
Chlormethine → MECHLORETHAMINE
Chlormethiuron → DIAZINON
Chlormethylenecycline → CLOMOCYCLINE
CHLORMEZANONE
Chlormite → CHLOROPROPYLATE
CHLOROACETALDEHYDE
CHLOROACETAMIDE
CHLOROACETIC ACID
CHLOROACETONE
CHLOROACETOPHENONE
N-(3-CHLOROALLYL)HEXAMINIUM CHLORIDE
Chloroambucil → CHLORAMBUCIL
p-Chloroamphetamine → PIPERONYL BUTOXIDE
σ-CHLOROANILINE
CHLOROBENZENE
CHLOROBENZILATE
2-(5-CHLORO-2H-BENZOTRIAZOL-2-YL)-4,6-BIS(1,1-DIMETHYLETHYL)-PHENOL
σ-CHLOROBENZYLIDINEMALONITRILE
CHLOROBROMOMETHANE
1-Chlorobutane → *n*-BUTYL CHLORIDE
CHLOROBUTANOL also → CHLORAL HYDRATE
Chlorocain → MEPIVACAINE
Chlorocaine → PROCAINE
Chlorocaps → CHLORAMPHENICOL
Chlorochin → CHLOROQUINE
Chlorocholine → CHLORMEQUAT CHLORIDE
Chlorocid → CHLORAMPHENICOL
Chlorocide → CHLORBENSIDE
Chlorocizin → CHLORACIZINE
4-CHLORO-*m*-CRESOL
p-CHLORO-*m*-CRESOL also → CUTTING OILS
Chlorocyclizine → CHLORCYCLIZINE
Chlorocyzin → CHLORACIZINE
2-Chlorodeoxyadenosine → CLADRIBINE
3-Chloro-1,2-dibromopropane → DIBROMOCHLOROPROPANE
2-Chloro-N-(2,6-diethylphenyl)-N-(methoxymethyl) acetamide → ALACHLOR
Chlorodifluoromethane → MONOCHLORODIFLUOROMETHANE
Chlorodiphenyl → CHLORINATED DIPHENYL
2-Chloro-1-ethanal → CHLOROACETALDEHYDE
Chloroethane → ETHYL CHLORIDE
Chloroethanes → 1,1-DICHLOROETHANE

Chloroethanoic Acid → CHLOROACETIC ACID
2-Chloroethanol → ETHYLENE CHLOROHYDRIN
Chloroethene Homopolymer → POLYVINYL CHLORIDE
2-Chloroethyl Alcohol → ETHYLENE CHLOROHYDRIN
N-(2-CHLOROETHYL)DIBENZYLAMINE
Chloroethylene Polymer → POLYVINYL CHLORIDE
Chloroethylphenamide → BECLAMIDE
CHLOROFORM also → ALCOHOL, ETHYL; EPINEPHRINE, NOREPINEPHRINE, PHOSGENE, PROPRANOLOL
Chloroformyl Chloride → PHOSGENE
Chlorofos → TRICHLORFON
Chlorogenic Acid → COFFEE also → POTATO, SUNFLOWER
CHLOROGUANIDE also → MAGNESIUM TRISILICATE
α-CHLOROHYDRIN
4-Chloro-2-hydroxybenzoic acid-N-n-butylamide → BUCLOSAMIDE
Chloro-IPC → CHLOROPROPHAM
Chloromethane → METHYL CHLORIDE
(Chloromethyl)benzene → BENZYL CHLORIDE
CHLOROMETHYL METHYL ETHER
4(4-Chloro-2-methylphenoxy)-butanoic Acid → 4(2-METHYL-4-CHLORO PHENOXY) BUTYRIC ACID
Chloromycetin → CHLORAMPHENICOL
Chloronase → CHLORPROPAMIDE
Chloronitrin → CHLORAMPHENICOL
CHLOROPHACINONE
Chlorophen → PENTACHLOROPHENOL
Chlorophenamidine → CHLORDIMEFORM
CHLOROPHENOL
Chlorophenothane → DDT
4-(4-CHLOROPHENOXY)-BUTYRIC ACID
p-CHLOROPHENYLALANINE
4-CHLORO-σ-PHENYLENEDIAMINE
CHLORO-2-PHENYLPHENOL
CHLOROPHYLL also → PHYLLOERYTHRIN, PHYTANNIC ACID, POTATO
CHLOROPHYLLUM MOLYBDITES
2-CHLORO-3-PHYTYL-1,4-NAPHTHOQUINONE
Chloro-pic → CHLOROPICRIN
CHLOROPICRIN
Chloroplatinic Acid → PLATINIC CHLORIDE
Chloropotassuril → POTASSIUM CHLORIDE
CHLOROPRENE
CHLOROPROCAINE
3-Chloro-1,2-propanediol → α-CHLOROHYDRIN
3-Chloro-1-propene → ALLYL CHLORIDE
Chloropropham → CHLORPROPHAM
CHLOROPROPYLATE
3-Chloro-propylene → ALLYL CHLORIDE
Chloroptic → CHLORAMPHENICOL
6-CHLOROPURINE
CHLOROPYRAMINE
Chloroquina → CHLOROQUINE
Chloroquinaldol → CHLORQUINALDOL
CHLOROQUINE also → AMMONIUM CHLORIDE, CIMETIDINE, MEFLOQUINE HCL, METHOTREXATE, METRONIDAZOLE, PHENOBARBITAL, PHENYLBUTAZONE, PRIMIDONE, RICE, SODIUM BICARBONATE, THYROXINE
Chloroquinium → CHLOROQUINE
5-CHLORO-8-QUINOLINOL
Chlorosal → CHLOROTHIAZIDE
CHLOROSULFONIC ACID
Chlorosulthiadil → HYDROCHLOROTHIAZIDE
CHLOROTHALONIL
CHLOROTHEN also → ETHYLENEDIAMINE
Chlorothene → 1,1,1-TRICHLOROETHANE
8-CHLOROTHEOPHYLLINE
CHLOROTHIAZIDE also → BUCLOSAMIDE, CHLORPROPAMIDE, GALLAMINE TRIETHIODIDE, LITHIUM, SULFINPYRAZONE
Chlorothion → CHLORTHION
α-Chlorotoluene → BENZYL CHLORIDE
3-CHLORO-*p*-TOLUIDINE
CHLOROTRIANISENE
CHLOROTRIFLUOROETHYLENE
Chloroxifenidim → CHLOROXURON
CHLOROXURON
CHLOROXYLENOL also → METHYL CELLULOSE
CHLOROZOTOCIN
Chlorozoxazone → CHLORZOXAZONE
Chlorparacide → CHLORBENSIDE
Chlorphenamidine → CHLORDIMEFORM
CHLORPHENESIN
CHLORPHENESIN CARBAMATE
CHLORPHENIRAMINE also → ALCOHOL, ETHYL; HEPARIN, MAGNESIUM TRISILICATE, PHENYLPROPANOLAMINE, PHENYTOIN
d-Chlorpheniramine → CHLORPHENIRAMINE
CHLORPHENOXAMINE

CHLORPHENTERMINE also → PHENELZINE
Chlorphos → TRICHLORFON
Chlorphthalidolone → CHLORTHALIDONE
Chlorpiprazine → PERPHENAZINE
Chlorpiprozine → PERPHENAZINE
Chlorpropham → CHLOROPROPHAM
Chlorpromados → CHLORPROMAZINE
Chlor-Promanyl → CHLORPROMAZINE
CHLORPROMAZINE also → ACENOCOUMAROL, ACETAMINOPHEN; ALCOHOL, ETHYL; BETHANIDINE, CAPTOPRIL, COCAINE, DEBRISOQUIN, DEXTROAMPHETAMINE, DIAZOXIDE, DIPYRONE, GALLAMINE TRIETHIODIDE, GLUCOSE, HEMIN, HEXOBARBITAL, IMIPRAMINE, INSULIN, LIDOCAINE, LITHIUM, LYSERGIC ACID DIETHYLAMIDE, MAGNESIUM HYDROXIDE, MAGNESIUM TRISILICATE, MEPERIDINE, MEPROBAMATE, METHYLDOPA, METHYSERGIDE, METOCLOPRAMIDE, METRIZAMIDE, NIALAMIDE, NORTRIPTYLINE, ORPHENADRINE, PENTOBARBITAL, PERPHENAZINE, PHENMETRAZINE, PHENOBARBITAL, PHENYTOIN, PIPERAZINE, PROCHLORPERAZINE, PROMETHAZINE, PROPRANOLOL, RESERPINE, SECOBARBITAL SODIUM, SODIUM BICARBONATE, SODIUM VALPROATE, STREPTOMYCIN, SULPIRIDE, TOBACCO, TRIHEXYLPHENIDYL, VALPROIC ACID, WARFARIN
CHLORPROPAMIDE also → ACETALDEHYDE, ACETYLSALICYLIC ACID, ALLOPURINOL, *p*-AMINOBENZOIC ACID, AMMONIUM CHLORIDE, BISHYDROXYCOUMARIN, BUCLOSAMIDE, CHLORAMPHENICOL, CHLOROTHIAZIDE, CLOFIBRATE, DISULFIRAM, ETHYL-*p*-AMINOBENZOATE, FUROSEMIDE, MOMOMORDICA CHARANTIA, NORTRIPTYLINE, OXYPHENBUTAZONE, OXYTETRACYCLINE, PHENOBARBITAL, PHENYLBUTAZONE, PROBENECID, PROPRANOLOL, SODIUM BICARBONATE, SODIUM SALICYLATE, SULFANILAMIDE, SULFAPHENAZOLE, SULFISOXAZOLE, VASOPRESSIN
CHLORPROPHAM
CHLORPROTHIXENE
CHLORPYRIFOS
CHLORPYRIFOS-METHYL
Chlor-PZ → CHLORPROMAZINE
Chlorquin → CHLOROQUINE
CHLORQUINALDOL also → HALQUINOL, IODOCHLORHYDROXYQUINOLINE
Chlorquinol → HALQUINOL
Chlorsulphacide → CHLORBENSIDE
Chlorten → TOXAPHENE
Chlortermine → CLORTERMINE
CHLORTETRACYCLINE also → VITAMIN K
CHLORTHALIDONE also → CHOLESTEROL, GALLAMINE TRIETHIODIDE, INSULIN, SULFINPYRAZONE
Chlorthiepin → ENDOSULFAN
CHLORTHION
CHLORTHIOPHOS
Chlor-Trimeton → CHLORPHENIRAMINE
Chlor-Tripolon → CHLORPHENIRAMINE
Chlorure Mercurique → MERCURIC CHLORIDE
Chlorurit → CHLOROTHIAZIDE
Chloryl Anesthetic → ETHYL CHLORIDE
Chlorylen → TRICHLOROETHYLENE
CHLORZOXAZONE also → ACETAMINOPHEN, CLENBUTEROL, CLOMETHIAZOLE, ISONIAZID, WATERCRESS
Chlosumdimeprimyl → CLOPAMIDE
Chlotride → CHLOROTHIAZIDE
CHOCOLATE also → CACAO, COCOA, COPPER, ISOCARBOXAZID, MILK, NIALAMIDE, PHENOLPHTHALEIN, THEOBROMINE
Chocolate Chips → LYSERGIC ACID DIETHYLAMIDE
Chocolax → PHENOLPHTHALEIN
Choke Cherry → CHERRY also → CYANIDES, HYDROGEN CYANIDE
Cholagon → DEHYDROCHOLIC ACID
Cholan-DH → DEHYDROCHOLIC ACID
Cholanorm → CHENODEOXYCHOLIC ACID
Cholebrine → IOCETAMIC ACID
Cholecalciferol → VITAMIN D_3
Cholecystokinin → IMIPRAMINE
Choledyl → CHOLINE THEOPHYLLINATE
Cholegrafin → MEGLUMINE IODIPAMIDE
Cholepatin → DEHYDROCHOLIC ACID
Cholestabyl → COLESTIPOL
CHOLESTEROL also → CHOLESTIN, COFFEE, EGGS, FATS, MACKEREL, TISSUE PLASMINOGEN ACTIVATOR
CHOLESTIN™

CHOLESTYRAMINE RESIN also → ACETAMINOPHEN, APPLE, CEPHALEXIN, CERIVASTATIN, CHLOROTHIAZIDE, DIGITOXIN, DIGOXIN, FATS, FOLIC ACID, FUSIDIC ACID, GLIPIZIDE, IMIPRAMINE, LEFLUNOMIDE, LIOTHYRONINE, LOPERAMIDE, MEXOLICAM, NAPROXEN, PHENPROCOUMON, PHENYLBUTAZONE, PIROXICAM, RALOXIFENE, THYROID, THYROXINE, TRIMETHOPRIM, TROGLITAZONE, VALPROIC ACID, VITAMIN A, VITAMIN B_{12}, VITAMIN D_3, VITAMIN K, WARFARIN
Cholimil → IOCETAMIC ACID
CHOLINE
Choline Chloride Carbamate → CARBACHOL
CHOLINE MAGNESIUM TRISALICYLATE
CHOLINE SALICYLATE
CHOLINE THEOPHYLLINATE also → RIFAMPIN
Cholinfall → METHIXENE
Cholinophylline → CHOLINE THEOPHYLLINATE
Chologon → DEHYDROCHOLIC ACID
Chlolografin → IODIPAMIDE
Cholospect → IODIPAMIDE
Choloxin → DEXTROTHYROXINE SODIUM
Cholybar → CHOLESTYRAMINE RESIN
CHONDROITIN
Chondroitin Sulfate → DANAPAROID
Chondrus Extract → CARRAGEENAN
Chopnut → PHYSOSTIGMA VENENOSUM
Choragon → GONADOTROPINS, CHORIONIC
Chorex → GONADOTROPINS, CHORIONIC
Chorgon → GONADOTROPINS, CHORIONIC
Chorigon → GONADOTROPINS, CHORIONIC
Choriogonin → GONADOTROPINS, CHORIONIC
Chorionic Gonadotropins → GONADOTROPINS, CHORIONIC
Choron → GONADOTROPINS, CHORIONIC
Chou Puant → SKUNK CABBAGE
CHQ → HALQUINOL
CHRISTI
Christmas Berry → HETEROMELES ARBUTIFOLIA also → SCHINUS TEREBINTHIFOLIUS
Christmas Bush → ALCHORNEA CORDIFOLIA
CHRISTMAS FACTOR
Christmas Flower → POINSETTIA
CHRISTMAS PEPPER
Christmas Rose → HELLEBORUS NIGER
Christ's Opium → MORPHINE
Chrom-Ar-Cu → CHROMATED COPPER ARSENATE
Chromargyre → MERBROMIN
CHROMATED COPPER ARSENATE
Chrome Green → CHROMIC OXIDE
Chrome Ocher → CHROMIC OXIDE
Chrome Oxide Green → CHROMIC OXIDE
Chrome Yellow → LEAD CHROMATE
Chromia → CHROMIC OXIDE
Chromic Acid → CHROMIUM TRIOXIDE also → SULFUR DIOXIDE
Chromic Anhydride → CHROMIUM TRIOXIDE
CHROMIC CHLORIDE
CHROMIC NITRATE
CHROMIC OXIDE
CHROMIC POTASSIUM SULFATE
CHROMIC SULFATE
CHROMIUM also → ACETYLENE, ALUMINUM HYDROXIDE, APPLE, CEMENT, CHOCOLATE, COCOA, CONCRETE, CRAYONS, ENAMEL, FIRE EXTINGUISHERS, GALVANIZED METALS, GLOVES, HATS, JEWELRY, LEATHER, MARBLE, MATCHES, MONEY, MOTOR OIL, PAINT, PAPER, PLEUROTIS sp.; POLISHES, FURNITURE; POTASSIUM DICHROMATE, POTATO, SAND, SHOES, SODIUM CHROMATE, STAMPS and STAMP PADS, TARAXACUM OFFICINALE, TEA, WINE
CHROMIUM PICOLINATE also → VITAMIN C
Chromium Sesquioxide → CHROMIC OXIDE
CHROMIUM TRIOXIDE
CHROMOMYCIN A
CHROMONAR
Chronadalate → NIFEDIPINE
Chronexan → XIPAMIDE
Chronicin → CHLORAMPHENICOL
Chronogyn → DANAZOL
Chronulac → LACTULOSE
Chrysaniline Yellow → PHOSPHINE
CHRYSANTHEMUM also → CHAMOMILE, DATURA, GOLDENROD, SUNFLOWER, TEA
CHRYSANTHEMUMIC ACID also → BARTHRIN
Chrysanthemum parthenium → FEVERFEW
CHRYSAORA
Chrysaora quiquecinha → JELLYFISH

CHRYSAROBIN
Chrysazin → DANTHRON
CHRYSENE
Chrysoidine Bydrochloride Citrate → 2,4-DIAMINOAZOBENZENE
Chrysomykine → CHLORTETRACYCLINE
CHRYSOPHAMIC ACID
Chrysophanol → CHRYSOPHAMIC ACID
Chrytemin → IMIPRAMINE
CHUIFONG TOUKUWAN
Chu-Lo → RHUBARB
Church of Scientology → QUINUCLIOINYL BENZILATE
Church Steeples → AGRIMONIA EUPATORIA
CHX 100 → MASOPROCOL
CHX 3101 → SPECTINOMYCIN
CHX 3311 → CYTARABINE
Chymar → CHYMOTRYPSIN
Chymetin → CHYMOTRYPSIN
Chymodiactin → CHYMOPAPAIN
Chymolase → CHYMOTRYPSIN
CHYMOPAPAIN also → PAPAIN
CHYMOTRYPSIN also → PENICILLIN; PENICILLINS – "NEWER and SYNTHETICS" > *Phenethicillin Potassium*; TETRACYCLINE, TRYPSIN
CI 80 → PONCEAU 3R
CI 337 → AZASERINE
CI 395 → PHENCYCLIDINE
CI 400 → CYCLOHEXAMINE
CI 423 → QUINACRINE
CI 440 → FLUFENAMIC ACID
CI 456 → DIAPAMIDE
CI 473 → MEFENAMIC ACID
CI 501 → CYCLOGUANYL PAMOATE
CI 581 → KETAMINE
CI 583 → MECLOFENAMIC ACID
CI 583α → MECLOFENAMIC ACID
CI 634 → TILETAMINE
CI 640 → TARTRAZINE
CI 671 → FDC BLUE #1
CI 673 → VIDARABINE
CI 719 → GEMFIBROZIL
CI 744 → TILETAMINE
CI 825 → PENTOSTATIN
CI 874 → INDELOXAZINE
CI 880 → AMSACRINE
CI 898 → TRIMETREXATE
CI 906 → QUINAPRIL
CI 912 → ZONISAMIDE
CI 945 → GABAPENTIN
CI 978 → SPARFLOXACIN
CI 981 → ATORVASTATIN
CI 991 → TROGLITAZONE
CI 1180 → INDIGOCARMINE
CI 11,020 → *p*-DIMETHYLAMINOBENZENE
CI 11,160 → σ-AMINOAZOTOLUENE
CI 11,380 → YELLOW AB
CI 11,390 → YELLOW OB
CI 12,156 → CITRUS RED 2
CI 13,065 → METANIL YELLOW
CI 14,700 → PONCEAU SX
CI 15,985 → FDC YELLOW #6
CI 16,035 → FDC RED #40
CI 16,185 → AMARANTH
CI 16,255 → PONCEAU 4R
CI 19,140 → TARTRAZINE
CI 23,860 → EVANS BLUE
CI 26,105 → SCARLET RED
CI 40,850 → CANTHAXANTHIN
CI 42,040 → BRILLIANT GREEN
CI 42,053 → FDC GREEN #3
CI 42,090 → FDC BLUE #1
CI 42,555 → GENTIAN VIOLET
CI 45,380 → EOSIN(E)
CI 45,430 → ERYTHROSINE
CI 52,015 → METHYLENE BLUE
CI 52,040 → TOLONIUM CHLORIDE
CI 73,015 → INDIGOCARMINE
CI 77,120 → BARIUM SULFATE
CI 77,199 → CADMIUM SULFIDE
CI 77,223 → CALCIUM CHROMATE
CI 77,288 → CHROMIC OXIDE
CI 77,410 → CUPRIC ACETOARSENITE
CI 77,510 → FERRIC FERROCYANIDE
CI 77,600 → LEAD CHROMATE
CI 77,766 → MERCURIC SULFIDE, RED
CI 77,891 → TITANIUM DIOXIDE
CI Acid Blue 74 → INDIGOCARMINE
CI Acid Red 27 → AMARANTH
CI Acid Red 51 → ERYTHROSINE
CI Acid Red 87 → EOSIN(E)
CI Acid Yellow → METANIL YELLOW
CI Acid Yellow 23 → TARTRAZINE
Ciamin → VITAMIN C
Cianuria → MERCURIC CYANIDE
Ciat → CATHA EDULIS
Ciatyl → CLOPENTHIXOL
Ciba 63 → TRIPELENNAMINE
Ciba 570 → PHOSPHAMIDON
Ciba 2059 → FLUOMETURON
Ciba 4311b → METHYLPHENIDATE
Ciba 5968 → HYDRALAZINE
Ciba 8514 → CHLORDIMEFORM
Ciba 9491 → IODOFENPHOS
Ciba 10,370 → SULFACHLORPYRIDAZINE
Ciba 11,511 → GLUTETHIMIDE

Ciba 11,925 → PHANQUINONE
Ciba 17,922 → SULFAPHENAZOLE
Ciba 21,401-Ba → TRIBENOSIDE
Ciba 32,644-Ba → NIRIDAZOLE
Cibacalcin → CALCITONIN
Cibacthen → CORTICOTROPIN
Cibalgina → PROPYPHENAZONE
Cibas → GLUTETHIMIDE
CI Basic Blue 9 → METHYLENE BLUE
CI Basic Blue 17 → TOLONIUM CHLORIDE
CI Basic Green → BRILLIANT GREEN
CI Basic Red → NEUTRAL RED
CI Basic Violet 3 → GENTIAN VIOLET
CI Basic Violet 10 → RHODAMINE B
Cibazol → SULFATHIAZOLE
CIBENZOLINE
Ciberon → CARBINOXAMINE MALEATE
Cibian → INTERFERONS
Ciblor → PENICILLINS – "NEWER and SYNTHETICS" > *Amoxicillin*
Ciboulette → CHIVES
Cicer arietinum → PEAS
Cichorium intybus → CHICORY
Ciclacillin → PENICILLINS – "NEWER and SYNTHETICS" > *Cyclacillin*
Ciclobiotic → METHACYCLINE
CICLOPIROX
Ciclopiroxolamin → CICLOPIROX
Cicloral → CARBUTAMIDE
Ciclosom → TRICHLORFON
Ciclosporin → CYCLOSPORINE
CICUTA sp.
Cicutaire pourpe → CICUTA sp.
Cicutal → CICUTA sp.
Cicutine → CONIINE
Cicutoxin → CICUTA also → OENANTHE sp.
Cidal → SALICYLAMIDE
Cidamex → ACETAZOLAMIDE
Cidanchin → CHLOROQUINE
Cidandopa → L-DOPA
CIDER
Cidocetine → CHLORAMPHENICOL
CIDOFOVIR
Cidomycin → GENTAMICIN
Cidrex → HYDROCHLOROTHIAZIDE
Cifenline → CIBENZOLINE also → BROMPHENOL BLUE
Ciflox → CIPROFLOXACIN
CI Food Blue 1 → INDIGOCARMINE
CI Food Blue 2 → FDC BLUE #1
CI Food Green → FDC GREEN #3
CI Food Orange 4 → ORANGE G
CI Food Red 1 → PONCEAU SX
CI Food Red 9 → AMARANTH
CI Food Red 14 → ERYTHROSINE
CI Food Yellow 3 → FDC YELLOW #6
CI Food Yellow 4 → TARTRAZINE
Cigarettes → ARSENIC, 1,2-BENZANTHRACENE, CLOVES, CYANIDES, IMIPRAMINE, INSULIN, MENTHOL, METHYLENE CHLORIDE, *N*-NITROSODIMETHYLAMINE, PHENACEMIDE, PHENOBARBITAL, POLONIUM, POLYTEF, PROCARBAZINE, PROPRANOLOL, QUININE, SECOBARBITAL SODIUM, SKATOLE, THEOPHYLLINE, TOBACCO, TOLUENE
Cigu → MUCUNA sp.
CIGUATERA also → EEL, MORAY; FISH, GAMBIERDISCUS TOXICUS, MACKEREL
Ciguatoxin → FISH
CILASTATIN
Cilicaine → PENICILLIN > *Benzylpenicillin, Procaine*
Cilifor → CEPHALORIDINE
Cillimycin → LINCOMYCIN
Ciloprost → ILOPROST
CILOSTAZOL
Cimal → CIMETIDINE
Cimedone → SOLASULFONE
Cimet → CIMETIDINE
Cimetag → CIMETIDINE
CIMETIDINE also → ACENOCOUMAROL, ACETAMINOPHEN; ALCOHOL, ETHYL; ALPRAZOLAM, ALUMINUM HYDROXIDE, AMINOPHYLLINE, ANTIPYRINE, BROMFENAC, CAFFEINE, CAPTOPRIL, CARBAMAZEPINE, CARMUSTINE, CARVEDILOL, CEFAZOLIN SODIUM, CHLORAMPHENICOL, CHLORAZEPATE, CHLORDIAZEPOXIDE, CHLOROQUINE, CIBENZOLINE, CITALOPRAM, CLOMETHIAZOLE, CLONAZEPAM, COFFEE, CYCLOSPORINE, DEXAMETHASONE, DIAZEPAM, DIGITOXIN, DIGOXIN, DOFETILIDE, DOXEPIN, ESTAZOLAM, FAMOTIDINE, FLECAINIDE, FLOSEQUINAN, 5-FLUOROURACIL, FLURAZEPAM, FLUVASTATIN, GLIPIZIDE, GLYBURIDE, HALAZEPAM, HYDROXYZINE, IMIPRAMINE, INDOMETHACIN, IRON, ISONIAZID, KETOCONAZOLE, LABETALOL, LIDOCAINE, LORATADINE, MAGNESIUM HYDROXIDE, MECLOBEMIDE, MEPERIDINE, METFORMIN, METHADONE, METOCLOPRAMIDE, METOPROLOL,

METRONIDAZOLE, MIDAZOLAM, MORICIZINE, MORPHINE, NICARDIPINE, NIFEDIPINE, NITRAZEPAM, NITRENDIPINE, NORDAZEPAM, NORTRIPTYLINE, OXAPROZIN, PARATHYROID HORMONE, PAROXETINE, PENBUTOLOL SULFATE, PENTAGASTRIN, PHENINDIONE, PENTOBARBITAL, PENTOXIFYLLINE, PHENOBARBITAL, PHENYTOIN, PIROXICAM, PRAZEPAM, PROCAINAMIDE, PROPAFENONE, PROPANTHELINE, PROPRANOLOL, QUINIDINE, SILDENAFIL, SOTALOL, SUCCINYLCHOLINE, SUCRALFATE, TACRINE, TACROLIMUS, TAMSULOSIN, TESTOSTERONE, TETRACYCLINE, THEOPHYLLINE, TICLODIPINE, TRIAZOLAM, TRIMIPRAMINE, VERAPAMIL, VITAMIN B_{12}, WARFARIN, ZALEPLON, ZIDOVUDINE, ZOLMITRIPTAN, ZOLPIDEM TARTRATE
Cimetum → CIMETIDINE
CIMICIFUGA RACEMOSA also → AMLODIPINE
Cimora → TRICHOCEREUS PACHANOI
Cinaperazine → CINNARIZINE
CINASERIN
Cinazyn → CINNARIZINE
Cinchocaine → DIBUCAINE
CINCHONA also → GALLAMINE TRIETHIODIDE, SUCCINYLCHOLINE
CINCHOPHEN also → BISHYDROXYCOUMARIN, ETYHL BISCOUMACETATE, WARFARIN
Cinconal → CINCHOPHEN
Cinecol → ETHYLENE CHLOROHYDRIN
Cineole → EUCALYPTOL
Cinerins → PYRETHRUM, PYRETHRINS, and PYRETHROIDS
Cinnabar → LEAD; MERCURIC SULFIDE, RED; MERCURY
Cinnacet → CINNARIZINE
Cinnageron → CINNARIZINE
Cinnamal → CINNAMALDEHYDE
CINNAMALDEHYDE also → APPLE, MINTS
CINNAMEDRINE
Cinnamene → STYRENE
CINNAMIC ACID also → PARTHENIUM
CINNAMIC ALCOHOL
CINNAMIC ALDEHYDE also → BALSAM PERU
Cinnamin → APAZONE
Cinnamol → STYRENE
CINNAMON also → BALSAM PERU, FRUITS and FRUIT JUICES, LIPSTICK, ORANGE, ROSIN, VERMOUTH
CINNAMYL ACETATE
Cinnamyl Alcohol → CINNAMIC ALCOHOL
CINNAMYL ANTHRANILATE also → GRAPE
CINNAMYL CINNAMATE
N-Cinnamylephedrine → CINNAMEDRINE
CINNAMYL ISOVALERATE
CINNARIZINE
Cinnipirine → CINNARIZINE
Cinobac → CINOXACIN
CINOBUFAGIN
Cinobufaginol → CINOBUFAGIN
Cinobufotalin → CINOBUFAGIN
Cinolone → TRIAMCINOLONE
Cinopal → FENBUFEN
Cinopenil → PENICILLINS – "NEWER and SYNTHETICS" > *Methicillin Sodium*
Cinopol → FENBUFEN
CINOXACIN
CINOXATE
Cin-quin → QUINIDINE
CiNU → LOMUSTINE
Cioccolata → CHOCOLATE
CIODRIN
CI Oxidation Base 12 → 2,4-DIAMINOANISOLE
CIPC → CHLORPROPHAM
Cipca → VITAMIN C
CI Pigment Blue 27 → FERRIC FERROCYANIDE
CI Pigment Green 17 → CHROMIC OXIDE
CI Pigment Green 21 → CUPRIC ACETOARSENITE
CI Pigment Red 106 → MERCURIC SULFIDE, RED
CI Pigment White 21–22 → BARIUM SULFATE
CI Pigment Yellow → CALCIUM CHROMATE
CI Pigment Yellow 37 → CADMIUM SULFIDE
Ciplamycetin → CHLORAMPHENICOL
Cipó de prata → MASCAGNIA sp.
Cippolle, Dr. Robert J. → Preface
Cipractin → CYPROHEPTADINE
Cipralan → CIBENZOLINE
Cipramil → CITALOPRAM
Ciprobay → CIPROFLOXACIN
Ciprofibrate → IBUPROFEN
CIPROFLOXACIN also → ALUMINUM, ALUMINUM HYDROXIDE, CAFFEINE, CALCIUM, CALCIUM CARBIMIDE, COFFEE, CYCLOSPORINE, DIDANOSINE,

FAMOTIDINE, FOSCARNET SODIUM, IRON, MAGNESIUM CITRATE, MILK, NAPROXEN, OLANZEPINE, PHENYTOIN, PROBENECID, QUINIDINE, RIFAMPIN, SUCRALFATE, THEOPHYLLINE, WARFARIN, YOGURT
Ciproxan → CIPROFLOXACIN
Ciproxin → CIPROFLOXACIN
Circair → DYPHYLLINE
Circanol → ERGOLOID MESYLATES
Circolene → AJMALICINE
Cirotyl → OXYPHENISATIN
Cirpon → MEPROBAMATE
CIRSIUM ARVENSE
CISAPRIDE also → CLOZAPINE, DALFOPRISTIN, DELAVIRDINE, DIGOXIN, DILTIAZEM, ERYTHROMYCIN, FLUCONAZOLE, ITRACONAZOLE, KETOCONAZOLE, METRONIDAZOLE, MIBEFRADIL, MICONAZOLE NITRATE, NELFINAVIR MESYLATE, RITONAVIR, ROPINIROLE, SAQUINAVIR, SPARFLOXACIN, TACROLIMUS, WARFARIN
CISAPRIDE
cis-DDP → CISPLATIN
Cismaplat → CISPLATIN
CISMETHRIN
CI Solvent Black 5 → NITROBENZENE
CI Solvent Red 80 → CITRUS RED 2
CI Solvent Yellow 2 → *p*-DIMETHYLAMINOBENZENE
CI Solvent Yellow 5 → YELLOW AB
CI Solvent Yellow 6 → YELLOW OB
Cisordinol → CLOPENTHIXOL
Cis-Ortho-Hydroxycinnamic Acid Lactone → COUMARIN
CISPLATIN also → ALCOHOL, ETHYL; ALUMINUM, FUROSEMIDE, GENTAMICIN, IFOSFAMIDE, LITHIUM, PAW-PAW, PROBENECID, PROPRANOLOL
cis-Platinum II → CISPLATIN
Cisplatyl → CISPLATIN
Ciste Absolute → CYSTE ABSOLUTE
Cistobil → IOPANOIC ACID
CITALOPRAM also → CIMETIDINE, MECLOBEMIDE
Citanest → PRILOCAINE
Citarin → TETRAMISOLE
Citarin L → LEVAMISOLE
Citation → CYROMAZINE
Citexal → METHAQUALONE
Citilat → NIFEDIPINE
Citireuma → SULINDAC
Citirizine → THEOPHYLLINE
Citizem → DILTIAZEM
Citizeta → ZIPEROL
Citnatin → SODIUM CITRATE
Citobaryum → BARIUM SULFATE
Citodon → HEXOBARBITAL
CITOFOVIR also → FORMIVERSEN SODIUM
Citofur → TEGAFUR
Citogel → SUCRALFATE
Citol → *p*-AMINOPHENOL
Citopan → HEXOBARBITAL
Citoplatino → CISPLATIN
Citosarin → PENICILLINS – "NEWER and SYNTHETICS" > *Cyclacillin*
Citox → DDT
Citoxid → NAFRONYL OXALATE
CITRAL also → ORANGE
Citreoviridin → PENICILLIUM sp.
CITRIC ACID also → TAMARIND, TROVAFLOXACIN
Citrical → CALCIUM CARBONATE
Citrine Ointment → MERCURIC NITRATE
CITRININ also → ASPERGILLUS, PENICILLIUM sp., RICE
CITROBACTER
CITRON OIL
CITRONELLA OIL
CITRONELLAL
CITRONELLOL
CITRONELLYLS
Citrosodine → SODIUM CITRATE
Citroveridin → RICE
Citrovorum Factor → FOLINIC ACID
Citrullus colycynthus → COLOCYNTH
CITRUS also → COLOGNES
Citrus aurantifolia → LIME
Citrus limon → LEMON
Citrus limonus → LEMON
Citrus paradisi → GRAPEFRUIT
CITRUS RED 2
CIVET
CI White Pigment 6 → TITANIUM DIOXIDE
CJs → PHENCYCLIDINE
CL 68 → CLOCORTOLONE PIVALATE
CL 636 → SULFACYTINE
CD 642 → BUTIROSIN
CL 1,848C → ETOXADROL HCL
CL 3911 → PHORATE
CL 12,625 → NATAMYCIN
CL 12,880 → DIMETHOATE
CL 13,494 → SULFAMETHOXYPYRIDAZINE
CL 14,377 → METHOTREXATE

CL 16,536 → PUROMYCIN
CL 19,823 → TRIAMCINOLONE
CL 26,691 → CYTHIOATE
CL 34,433 → TRIAMCINOLONE
CL 34,699 → AMCINONIDE
CL 36,010 → QUINETHAZONE
CL 40,881 → ETHAMBUTOL
CL 47,031 → PHOSFOLAN
CL 62,362 → LOXAPINE
CL 67,772 → AMOXAPINE
CL 71,563 → LOXAPINE
CL 82,204 → FENBUFEN
CL 112,302 → BUPRENORPHINE
CL 206,214 → BUTAMISOLE
CL 227,193 → PENICILLINS – "NEWER and SYNTHETICS" > *Piperacillin Sodium*
CL 232,315 → MITOXANTRONE HCL
CL 284,635 → CEFIXIME
CL 284,846 → ZALEPLON
CL 291,894 → GROWTH HORMONES
CL 297,939 → BISOPROLOL
CL 301,423 → MOXIDECTIN
CL 67,310,465 → PENTOSTATIN
CLADRIBINE
Claforan → CEFOTAXIME
Clairol → VINYL CHLORIDE
Clairsit → PERCHLOROMETHYL MERCAPTAN
Clamoxyl → PENICILLINS – "NEWER and SYNTHETICS" > *Amoxicillin*
CLAMS also → GONYAULAX, LEAD, MOLLUSKS, SAXITOXIN
Clam Poison → SAXITOXIN
Clanzol → CLEBOPRIDE
Claradin → ACETYLSALICYLIC ACID
Claramid → ROXITHROMYCIN
Claripex → CLOFIBRATE
CLARITHROMYCIN also → ACENOCOUMAROL, ASTEMIZOLE, CARBAMAZEPINE, CISAPRIDE, CYCLOSPORINE, DALFOPRISTIN, DELAVIRDINE, DIGOXIN, DILTIAZEM, DISOPYRAMIDE, FLUOXETINE, GRAPEFRUIT, ITRACONAZOLE, KETOCONAZOLE, LANSOPRAZOLE, LOVASTATIN, MIDAZOLAM, OMEPRAZOLE, PIMOZIDE, RIFABUTIN, RIFAMPIN, RITONAVIR, TACROLIMUS, TERFENADINE, VERAPAMIL, WARFARIN, ZIDOVUDINE
Claritin → LORATADINE
Clarityn → LORATADINE
Clary → SALVIA
Clast → CLEBOPRIDE
Clasteon → DICHLOROMETHYLENE DIPHOSPHONATE
Clavacin → PATULIN
Clavamox → PENICILLINS – "NEWER and SYNTHETICS" > *Amoxicillin*
Clavatin → PATULIN
Claventin → CLAVULANIC ACID
Claversal → 5-AMINOSALICYLIC ACID
CLAVICEPS also → PARTHENIUM
Claviformin → PATULIN
Claviton → TRIDIHEXETHYL CHLORIDE
CLAVULANIC ACID
CLAY PIGEON TARGETS also → PITCH
CLEBOPRIDE
Cleboril → CLEBOPRIDE
Cleiton → HYDROCORTISONE
Clearate → LECITHIN
Clearteck → MINERAL OIL
Clemanil → CLEMASTINE
CLEMASTINE
Clematine → CLEMATIS
CLEMATIS sp.
CLEMIZOLE
CLENBUTEROL
Clenil-A → BECLOMETHASONE
Cleocin → CLINDAMYCIN
Cleprid → CLEBOPRIDE
Clera → NAPHAZOLINE
Clérégil → DEANOL
Cleridium → DIPYRIDAMOLE
CLIBADIUM SYLVESTRE
CLIDANAC
CLIDINIUM BROMIDE
Climbing Nightshade → SOLANUM sp.
Clinafloxacin → THEOPHYLLINE
CLINDAMYCIN also → DIMETHYL TUBOCURARINE IODIDE, QUINIDINE, VERAPAMIL
Clinestrol → DIETHYLSTILBESTROL
Clinicide → CARBARYL
Clinimycin → OXYTETRACYCLINE
Clinistix® → LACTOSE
Clinitar → COAL TAR
Clinitest® → CEFAMANDOLE, CEFAZOLIN SODIUM, CEFONICID, CEFOTAXIME, CEFOTETAN, CEFOXITIN SODIUM, CEFTZOXIME, CEPHALEXIN, CEPHALOTHIN, CEPHAPIRIN, CEPHADRINE, IMPENEM, NALIDIXIC ACID; PENICILLINS – "NEWER and SYNTHETICS" > *Piperacillin Sodium, Ticarcillin Disodium*; PROBENECID, SODIUM HYDROXIDE
Clinium → LIDOFLAZINE

Clinocaine → PROCAINE
Clinoril → SULINDAC
Clinoptilolite → ZEOLITES
Clinovir → MEDROXYPROGESTERONE
Clioquinol → IODOCHLORHYDROXYQUINOLINE
Clistin → CARBINOXAMINE MALEATE
Clisundac → SULINDAC
CLITOCYBE sp.
CLIVIA MINIATA
Clivoten → ISRADIPINE
Cl_2MDP → DICHLOROMETHYLENE DIPHOSPHONATE
Cloak Fern → NOTHOLAENA SINNUATA
CLOBAZAM also → ALCOHOL, ETHYL
Cloberat → CLOFIBRATE
Clobesol → CLOBETASOL, *propionate*
CLOBETASOL
Clobren-SF → CLOFIBRATE
CLOBUTINOL also → METOCLOPRAMIDE
Clocete → MECLOFENOXATE
CLOCORTOLONE PIVALATE
Cloderm → CLOCORTOLONE PIVALATE
Clodronate → DICHLOROMETHYLENE DIPHOSPHONATE
CLOFAZIMINE
Clofelin → CLONIDINE
Clofenoxate → MECLOFENOXATE
Clofenotane → DDT
CLOFIBRATE also → ACENOCOUMAROL, BISHYDROXYCOUMARIN, ETHYL BISCOUMACETATE, NALIDIXIC ACID, NEOMYCIN, NITROFURANTOIN, PHENINDIONE, PHENPROCOUMON, PROBENECID, PUROMYCIN, THYROXINE, TOLBUTAMIDE, WARFARIN
Clofinit → CLOFIBRATE
CLOGESTONE ACETATE
Clogestone → CHLORMADINONE
CLOMACRON
CLOMETHIAZOLES also → ALCOHOL, ETHYL; DIAZOXIDE, PHENOBARBITAL
Clomicalm → CLOMIPRAMINE
Clomid → CLOMIPHENE CITRATE
Clomifene → CLOMIPHENE CITRATE
CLOMIPHENE CITRATE also → MESTEROLONE
CLOMIPRAMINE also → FLUVOXAMINE, LITHIUM, MECLOBEMIDE, OMEPRAZOLE, TICLODIPINE, TOBACCO
CLOMIPRAMINE
Clomivid → CLOMIPHENE CITRATE
CLOMOCYCLINE
Clomphid → CLOMIPHENE CITRATE
CLONAZEPAM also → AMIODARONE, CARBAMAZEPINE, ERYTHROMYCIN, KETOCONAZOLE, PHENOBARBITAL, PHENYTOIN, SODIUM VALPROATE, SOTALOL, VALPROIC ACID
CLONIDINE also → DESIPRAMINE, L-DOPA, FLUPHENAZINE, GROWTH HORMONES, IMIPRAMINE, PROPRANOLOL, TRAZODONE HCL, VERAPAMIL
Clonistada → CLONIDINE
Clonitralid → NICLOSAMIDE
CLONITRATE
Clonopin → CLONAZEPAM
Clont → METRONIDAZOLE
CLOPAMIDE
Clopane → CYCLOPENTAMINE
CLOPENTHIXOL
Clophen → POLYCHLORINATED BIPHENYLS
Clophenoxate → MECLOFENOXATE
CLOPIDOGREL BISULFATE
CLOPIDOL
Clopindol → CLOPIDOL
Clopinerin → CLORPRENALINE
Clopixol → CLOPENTHIXOL
Clopoxide → CHLORDIAZEPOXIDE
Clopromate → METOCLOPRAMIDE
CLOPROSTENOL
Cloramicol → CHLORAMPHENICOL
Cloramin → MECHLORETHAMINE
CLORAZEPATE DIPOTASSIUM also → ALCOHOL, ETHYL; CIMETIDINE, DISULFIRAM
Clorazolam → TRIAZOLAM
Cloreto de Potássio → POTASSIUM CHLORIDE
Clorevan → CHLORPHENOXAMINE
CLOREXOLONE
Clorfentermina → CHLORPHENTERMINE
Clorgiline → CLORGYLINE
CLORGYLINE
CLORINDIONE
Clorochina → CHLOROQUINE
Clorocyn → CHLORAMPHENICOL
Cloromisan → CHLORAMPHENICOL
Cloropiril → CHLORPEHNIRAMINE, *maleate*
Clorox → SODIUM HYPOCHLORITE
CLORPRENALINE
CLORTERMINE
Clortetrin → DEMECLOCYCLINE
Close-Up® → CINNAMIC ALDEHYDE
Closina → CYCLOSERINE
Clostilbegyt Dyneric → CLOMIPHENE CITRATE
Clotbur → BURDOCK

Clotbur → COCKLEBURS
CLOTHIAPINE
CLOTIAZEPAM
Clotol → ALCOHOL, *n*-BUTYL
Clotride → CHLOROTHIAZIDE
CLOTRIMAZOLE also → SULFAMETHOXAZOLE, TACROLIMUS, TRIMETREXATE
Clotrox → TRIPARANOL
Cloudberry → BLACKBERRY
Clout → DISODIUM METHANEARSONATE also → METHANEARSONIC ACID
Clover, Red → CLOVERS also → RHIZOCTONIA LEGUMINICOLA, SLAFRAMINE
CLOVES also → BALSAM PERU, ORANGE, PHENYTOIN, ROSIN, SASSAFRAS
Cloxacillin → PENICILLINS – "NEWER and SYNTHETICS" also → ALUMINUM HYDROXIDE, GENTAMICIN, LOVASTATIN, PENICILLIN, SULFAMETHOXYPYRIDAZINE, SULFINPYRAZONE, SULFISOXAZOLE
Cloxiquine → 5-CHLORO-8-QUINOLINOL
Cloxypen → PENICILLINS – "NEWER and SYNTHETICS" > *Cloxacillin/Sodium salt*
Cloxyquin → 5-CHLORO-8-QUINOLINOL
Cloxyquine → Preface
Clozan → CLOTIAZEPAM
CLOZAPINE also → CAFFEINE, CARBAMAZEPINE, ERYTHROMYCIN, FLUVOXAMINE, PAROXETINE, PHENOBARBITAL, PHENYTOIN, RITONAVIR, SERTRALINE
Clozaril → CLOZAPINE
CLTC → CHLORTETRACYCLINE
Clupeotoxin → FISH also → SARDINELLA sp.
CM 6912 → ETHYL LOFLAZEPATE
CMA 676 → GEMTUZIMAB OZOGAMICIN
CMC → CELLULOSE, *Sodium Carboxymethylcellulose*
CMC Sodium → SODIUM CARBOXYMETHYLCELLULOSE
C-Meton → CHLORPHENIRAMINE, *maleate*
CMME → CHLOROMETHYL METHYL ETHER
CMPD → MEVINPHOS
CMPP → MECOPROP
CMU → MONURON
CN 15757 → AZASERINE
2-CNB → CHLOROBENZENE
Cnicus → CIRSIUM ARVENSE
CNIDARIANS
CO → CARBON MONOXIDE
COAL also → ARSENIC, MERCURY, SULFUR, THALLIUM
Coal Oil → KEROSENE
Coal-Oil Brush → TATRADYMIA sp. also → TETRADYMA sp.
COAL TAR also → ANTHRACENE, 1,2-BENZANTHRACENE, BENZO [α] PYRENE, CHRYSENE, CLAY PIGEON TARGETS, CONCRETE, SALICYLIC ACID
Coal Tar Naphtha → BENZENE
Coal Tar Pitch → PITCH
Coactin → PENICILLINS – "NEWER and SYNTHETICS" > *Amdinocillin*
Co-amilofruse → FUROSEMIDE
Co-Amoxiclav → PENICILLINS – "NEWER and SYNTHETICS" > *Amoxicillin*
COAPT Tissue Adhesive → OCRYLATE
Coaxin → CEPHALOTHIN
Cobalin → VITAMIN B_{12}
Cobalin-H → VITAMIN B_{12}
COBALT also → APRICOT, BEER, BEETS, CHOCOLATE, CLOVES, COCOA, COFFEE, RHIZOPUS sp., SCALLOPS, TEA
Cobalt 60 → COBALT, *radioactive*
COBALT ARSENATE
COBALT CHLORIDE
COBALT NAPHTHENATE
COBALT OLEATE
COBALTOUS ALUMINATE
Cobaltous Chloride → COBALT CHLORIDE
Cobaltous Sulfate → COBALT SULFATE
COBALT SULFATE
Cobalt Violet → COBALT ARSENATE
Cobamide → VITAMIN B_{12}
Coban → MONENSIN
Cobantril → PYRANTEL
Cobrentin Methanesulfonate → BENZTROPINE MESYLATE
Cobiona → OXATOMIDE
Cobutolin → ALBUTEROL
COCA
Coca Cola® → CAFFEINE, COCA, COLAS
COCAINE also → AMPHETAMINE, BENZENE, BROCCOLI, CAFFEINE, CALCIUM OXIDE, COCA, EPINEPHRINE; ETHER, ETHYL; FURAZOLIDONE, GUANETHIDINE, MONEY, PARGYLINE, SPIHELIA ANTHELMIA, TRANYLCYPROMINE, TRAZODONE HCL
Cocalose → COCARBOXYLASE
COCARBOXYLASE
Cocartrit → CHLOROQUINE
Cocarvit → COCARBOXYLASE

Coccinillerot A → PONCEAU 4R
Coccistac → SALINOMYCIN
Coccoclase → SULFAPYRIDINE
Cocculin → PICROTOXIN
COCCULUS sp.
COCHINEAL also → TETRACYCLINE
Cockfoot → CHELIDONIUM MAJUS
COCKLES
COCKLEBURS also → HYDROGEN CYANIDE, HYDROQUINONE
Cockle Button → BURDOCK
Cocksfoot → DACTYLIS GLOMERATA
COCOA
COCOA BEAN SHELLS or HUSKS
Cocoa Butter → OPIUM
COCOBOLO
Coco de Mono → LECYTHIS
COCONUT also → SORBITOL
Coconut Aldehyde → γ-NONALACTONE
Coco-Yam → XANTHOSOMA sp.
Codarone → AMIODARONE
Codarone X → AMIODARONE
Codecarboxylase → PYRIDOXAL-5-PHOSPHATE
CODEINE also → AMIODARONE, CHLORAMPHENICOL, KAOLIN, OPIUM, PAPAVER sp., PAROXETINE, QUINIDINE, TETRACYCLINE, THEBAINE
Codelcortone → PREDNISOLONE
Codelsol → PREDNISOLONE
Codethyline → ETHYLMORPHINE HYDROCHLORIDE
Codiaeum variegatum → CROTON
Codinovo → HYDROCODONE BITARTRATE
COD LIVER OIL also → FISH
Codroxomin → VITAMIN B_{12}
Codylin → PHOLCODINE
COELENTERATES
Coenzyme Q10 → UBIDECARENONE
Coesite → SILICON DIOXIDE
Co-Estro → ESTROGENS, CONJUGATED
Coffea → COFFEE
COFFEE also → FAMOTIDINE, HALOPERIDOL, HOMOCYSTEINE, IRON, KENTUCKY COFFEE TREE, PROCARBAZINE, RANITIDINE
"Coffee" → LYSERGIC ACID DIETHYLAMIDE
Coffee Bean → COFFEE also → SESBANIA sp.
Coffee Bean Weed → GLOTTIDIUM VESICARIUM
Coffee Senna → CASSIA
Coffee Weed → CHICORY also → GLOTTIDIUM VESICARIUM
"Coffee-worker's Lung → DUSTS
Cogentin → BENZTROPINE MESYLATE
Cogentinol → BENZTROPINE MESYLATE
COGNAC
COHOBA also → ANADENANTHERA
Cogmine → PHYSOSTIGMINE
Cognex → TACRINE
Cohydrine → DIHYDROCODEINE
COKE also → BENZO[α]PYRENE, COAL
"Coke" → COCAINE
Coke Paste → COCAINE
COLAS also → COFFEE, FAMOTIDINE, KOLA, PHOSPHORIC ACID, PROCARBAZINE, RANITIDINE
Colace → DOCUSATE SODIUM
Cola de caballo → EQUISETUM, *giganteum*
Colamine → ETHANOLAMINE
Colaspase → *l*-ASPARAGINASE
Colcemid → DEMECOLCINE
Colchamine → DEMECOLCINE
Colchicin → COLCHICINE
COLCHICINE also → COLCHICUM AUTUMNALE, GLORIOSA, GRISEOFULVIN, LACTOSE, VITAMIN B_{12}
Colchicinum → COLCHICINE
COLCHICUM AUTUMNALE also → CALCIUM OXALATE
Colchique d'automne → COLCHICUM AUTUMNALE
Coldan → NAPHAZOLINE
Colecalciferol → VITAMIN D_3
Colectril® → AMILORIDE
Colepax → IOPANOIC ACID
Colepur → 5,7-DIBROMO-8-QUINOLINOL also → IODOCHLORHYDROXYQUINOLINE
COLESEVELAM
Colestid → COLESTIPOL
COLESTIPOL also → ATORVASTATIN, CHLOROTHIAZIDE, DIGITOXIN, DIGOXIN, HYDROCHLOROTHIAZIDE, TRIMETHOPRIM
COLESTIPOL
Colestran → CHOLESTYRAMINE RESIN
Colestyramin → CHOLESTYRAMINE RESIN
Coletyl → CARBACHOL
COLEUS
Colfarit → ACETYLSALICYLIC ACID
Colforsin → COLEUS
Colicon → DIMETHICONE (*with silicon dioxide*)
Colifoam → HYDROCORTISONE
Colifuran → FURAZOLIDONE
Colimune → CROMOLYN SODIUM
Colimycin → COLISTIN also → NEOMYCIN
Coliquifilm → CHLOROBUTANOL
Colisone → PREDNISONE

Colisticina → COLISTIN
Colistimethate Sodium → COLISTIN
COLISTIN also → AMOBARBITAL, CURARE, DIMETHYL TUBOCURARINE IODIDE, EDROPHONIUM; ETHER, ETHYL; GENTAMICIN, NEOMYCIN, NEOSTIGMINE, PROCAINAMIDE, PROMETHAZINE, QUINIDINE, *d*-TUBOCURARINE CHLORIDE, VANCOMYCIN
COLLAGEN also → BLEOMYCIN, HYDRALAZINE, PENICILLAMINE
Collasol → COLLAGEN
Collihuay → SAPIUM sp.
Collocral → POLYVINYLPYRROLIDONE
Collunosol → 2,4,5-TRICHLOROPHENOL
Collunovar → NEOARSPHENAMINE
Colme → CALCIUM CARBIMIDE
COLOCASIA sp.
COLOCYNTH
Colocythis → COLOCYNTH
Colofoam → HYDROCORTISONE
Cologel → METHYL CELLULOSE
COLOGNES
Colony Stimulating Factor 2 → GRANULOCYTE MACROPHAGE COLONY STIMULATING FACTOR
Colophony → ROSIN
Colo-Pleon → SALICYCLAZOSULFAPYRIDINE
Coloquinte → COLOCYNTH
Colorines → ERYTHRINA sp. also → SOPHORA sp.
Colorado Rubberweed → HYMENOXYS sp.
Colpogyn → ESTRIOL
Colpro → MEDOGESTONE
Colprone → MEDROGESTONE
Colprosterone → PROGESTERONE
Colrex → GLYCERYL GUAIACOLATE
Coltrax → THIOCOLCHICOSIDE
Coltromyl → THIOCOLCHICOSIDE
COLTSFOOT
COLUBRINA
COLUMBINE
Columbus Grass → SORGHUM
Colyar → CHOLESTYRAMINE RESIN
Coly-Mycin → COLISTIN
Combantrin → PYRANTEL
Combelen → PROPIONYLPROMAZINE
Combot → TRICHLORFON
COMBRETUM sp.
Comfolax → DOCUSATE SODIUM
COMFREY also → DIGITALIS, PYRROLIZIDINES, SYMPHYTINE, TEA
Comite → PROPARGITE
Commodore Perry → SOYBEANS
Common Bean → BEANS, Kidney
Common Hogweed → HERACLEUM sp.
Common Lantana → LANTANA sp.
Common Mullein → VERBASCUM THAPSUS
Common Stinkfish → CALLIONYMUS CALAUROPOMUS
Common Vetch → VICIA
Comox → SULFAMETHOXAZOLE
Compass Plant → LETTUCE
Compazine → PROCHLORPERAZINE
Compenamine → PENICILLIN > *Benzylpenicillin, l-Ephenamine*
Compendium → BROMAZEPAM
Compitox → MECOPROP
Compitox Plus → MECOPROP
Complamin → XANTHINOL NIACINATE
Compocillin V → PENICILLIN > *Penicillin V*
Compound 22/190 → CHLORTHION
Compound 33T57 → METHISAZONE
Compound 42 → WARFARIN
Compound 47–83 → CYCLIZINE
Compound 47–282 → CHLORCYCLIZINE
Compound 48/268 → SUCCINYLCHOLINE
Compound 88R → ARAMITE
Compound 118 → ALDRIN
Compound 269 → ENDRIN
Compound 338 → CHLOROBENZILATE
Compound 347 → ENFLURANE
Compound 386–2 → PROSCILLARIDIN
Compound 469 → ISOFLURANE
Compound 497 → DIELDRIN
Compound 545 → IOHEXOL
Compound 604 → DICHLONE
Compound 711 → ISODRIN
Compound 1080 → FLUOROACETIC ACID
Compound 1189 → KEPONE
Compound 1283 → MIREX
Compound 1861 → AMINOPYRIDINE
Compound 3961 → STROBANE®
Compound 4072 → CHLORFENVINPHOS
Compound 64,716 → CINOXACIN
Compound 83,846 → APRINDINE
Compound 90,459 → BENOXAPROFEN
Compound 99,638 → CEFACLOR
Compound E → CORTISONE
Compound Q → TRICHOSANTHIN
Comprecin → ENOXACIN
Compudose → ESTRADIOL
Comtan → ENTACAPONE
Co-Nav → DIOXATHION
Concanavalin A → JACK BEAN
Concemin → VITAMIN C

Concerta → METHYLPHENIDATE
Conco NI-90 → NONOXYNOL-9
Concor → BISOPROLOL
Concordin → PROTRIPTYLINE
CONCRETE
Concurat → TETRAMISOLE
Condol → VITAMIN D_2
Conducton → CARAZOLOL
Condylox → PODOPHYLLOTOXIN
Cone Fish → CONE SHELL FISH
Cone Flower → RUDBECKIA sp.
Cone Shell → CONE SHELL FISH
CONE SHELL FISH also → MOLLUSKS
Conest → ESTROGENS, CONJUGATED
Confidor → IMIDACLOPRID
Congo Blue → TRYPAN BLUE
CONGO RED
Conhydrine → CONIUM MACULATUM
Conicine → CONIINE
Coniic Acid → CONIUM MACULATUM
CONIINE also → METHYLCONIINE
CONIUM MACULATUM also → ANISE, METHYLCONIINE
Conludag → NORETHINDRONE
Conmel → DIPYRONE
Connetivina → SODIUM HYALURONATE
Conocybe → MUSHROOMS
Conoderm → MICONAZOLE NITRATE
Conofite → MICONAZOLE NITRATE
Conotrane → HYDRARGAPHEN
Conqer-John → POLYGONATUM BIFLORUM
Conradi's Syndrome → WARFARIN
Conray → MEGLUMINE IOTHALAMATE
Conray-400 → IOTHALMATE SODIUM
Conselt → CLORPRENALINE
Consolan → NABUMETONE
Constaphyl → PENICILLINS – "NEWER and SYNTHETICS" > *Dicloxacillin*
Consoude âpre → COMFREY
Consumption Bush → VINCA ROSEA
Consumption Weed → BACCHARIS sp.
"Contact Lens" → LYSERGIC ACID DIETHYLAMIDE
Contalax → BISACODYL
Contamex → KETAZOLAM
Contax → OXYPHENISATIN
Conteben → THIACETAZONE
Contedrin → METHYLPHENIDATE
Contergan → THALIDOMIDE
Contomin → CHLORPROMAZINE
Contrac → BROMADIOLONE
Contraceptives, Oral → ACENOCOUMAROL, ACETAMINOPHEN; ALCOHOL, ETHYL; Σ-AMINOCAPROIC ACID, AMINOPYRINE, ANTIPYRINE, ATORVASTATIN, BCG, BISHYDROXYCOUMARIN, CAFFEINE, CARBAMAZEPINE, CEPHALEXIN, CERULOPLASMIN, CHLORAL HYDRATE, CHLORCYCLIZINE, CHLORDIAZEPOXIDE, CHLORPROMAZINE, CHOLESTEROL, CLINDAMYCIN, CLOFIBRATE, COPPER, CYCLOPENTHIAZIDE, CYCLOPHOSPHAMIDE, DAPSONE, DESIPRAMINE, DESMOPRESSIN, DIAZEPAM, DIHYDROERGOTAMINE METHANESULFONATE, DISULFIRAM, DOXEPIN, ERYTHROMYCIN, ETHCHLORVYNOL, ETHINYL ESTRADIOL, FOLIC ACID, GLIPIZIDE, GLUCOSE, GRISEOFULVIN, GROWTH HORMONES, GUANETHIDINE, HYDROCORTISONE, IMIPRAMINE, INSULIN, IRON, ISONIAZID, ITRACONAZOLE, LORAZEPAM, MAGNESIUM TRISILICATE, MEPHOBARBITAL, MEPROBAMATE, METHANDROSTENOLONE, METHOSUXIMIDE, METHOTRIMEPRAZINE, METHYLDOPA, METHYLPHENIDATE, METHYLPREDNISOLONE, METOPROLOL, MINERAL OIL, MIVACURIUM, NEVIRAPINE, NITROFURANTOIN, NORTRIPTYLINE, OPIPRAMOL; PENICILLIN also → PENICILLIN > *Penicillin V*; PENICILLINS – "NEWER and SYNTHETICS" > *Amoxicillin, Ampicillin, Floxacillin, Pivampicillin*; PHENACEMIDE, PHENOBARBITAL, PHENPROCOUMON, PHENYLBUTAZONE, PHENYRAMIDOL, PHENYTOIN, PLACEBO, PREDNISOLONE, PRIMIDONE, PROMAZINE, PROMETHAZINE, PROPRANOLOL, PROTRIPTYLINE, RIFAMPIN, RITONAVIR, SULFAPHENAZOLE, SULFOBROMOPHTHALEIN SODIUM, TEMAZEPAM, TETRACYCLINE, THEOPHYLLINE, THYROXINE, TIZANIDINE, VITAMIN C, VITAMIN D_3, VITAMIN E, VITAMIN K, WARFARIN, ZOLMITRIPTAN
Contralac → METERGOLINE
Contrapar → KETHOXAL
Contrast Media → POLYSTYRENE
Contrathion → PRALIDOXIME

Contrheuma Retard → ACETYLSALICYLIC ACID
Contristamine → CHLORPHENOXAMINE
Contrix "28" → MEGLUMINE IOTHALAMATE
Control-Om → MEPHENOXALONE
Contrykal® → APROTININ
Conus textile → CONE SHELL FISH
Convallamarin → CONVALLARIA MAJALIS
Convallanin → CONVALLARIA MAJALIS
CONVALLARIA MAJALIS
CONVALLATOXIN also → CONVALLARIA MAJALIS, ORNITHOGALUM UMBELLATUM
Convalloside → ORNITHOGALUM UMBELLATUM
Conversum → PERINDOPRIL *tert*-BUTYLAMINE
Conversyl → PERINDOPRIL *tert*-BUTYLAMINE
CONVICINE also → BEANS
Convulex → SODIUM VALPROATE also → VALPROIC ACID
CONYZA COULTERI
Cooby Wattle → ACACIA
Cooking Oil → POLYBROMINATED BIPHENYLS also → POLYCHLORINATED BIPHENYLS
Coolah Grass → PANICUM sp.
Coolspan → SULPIRIDE
COPAIBA
Copal → SCHINUS TEREBINTHIFOLIUS
Coparogin → TEGAFUR
Co-Pilots → AMPHETAMINE
Copirene → KEBUZONE
COP 1 → GLATIRAMER
Copaxone → GLATIRAMER
Cöpleme → VERATRUM sp.
Copolymer 1 → GLATIRAMER
COPPER also → *n*-BUTYL MERCAPTAN, CHOCOLATE, COBALT, FRUCTOSE, HELIOTROPIN, MERCURY, MOLYBDENUM, NICKEL, *N*-NITROSODIMETHYLAMINE, PAPER, PICKLES, PYRROLIZIDINES, SHELLFISH, SHRIMP, SULFUR, TEA, THALLIUM, TIN, ZINC
Copper Acetate Arsenite → CUPRIC ACETOARSENITE
COPPER ARSENITE
Copper Azide → SODIUM AZIDE
COPPER CARBONATE
COPPER CHLORIDE also → FIRE COLORING SALTS
COPPER CYANIDE
COPPER NAPHTHENATE
COPPER NITRATE
COPPER OXALATE
COPPER OXIDE, RED
COPPER OXYCHLORIDE
Copper Pentoxide → COPPER OXIDE, RED
COPPER SULFATE also → FIRE COLORING SALTS
COPPER SULFIDE
COPPER THIOCYANATE
COPPER TRIETHANOLAMINE
COPPERWEED also → OXYTERIA ACEROSA
Copra → COCONUT
COPRINE also → CLITOCYBE, DISULFIRAM
Coprola → DOCUSATE SODIUM
Coptisine → ECHOSCHOLTZIA CALIFORNICA
COPY PAPER and COPYING MACHINES
Coquelicot → PAPAVER
Coracon → NIKETHAMIDE
Coracten → NIFEDIPINE
Coractiv N → NIKETHAMIDE
Corafurone → KHELLIN
Coragoxine → DIGOXIN
CORALS also → GRETA, LEAD OXIDES
Co-Ral → COUMAPHOS
Coralbeads → COCCULUS sp.
Coral Bean → ERYTHRINA sp. also → SOPHORA sp.
Coral Deshi Deuva → LEAD
Coral Plant → JATROPHA sp.
Coramedan → DIGITOXIN
Coramine → NIKETHAMIDE
Corangin → ISOSORBIDE MONONITRATE
Coranormol → PENTYLENETETRAZOL
Corasol → PENTYLENETETRAZOL
Corathiem → CINNARIZINE
Corax → CHLORDIAZEPOXIDE
Co-rax → WARFARIN
Corazol → PENTYLENETETRAZOL
Corbit → ANTHRAQUINONE
Cordarex → AMIODARONE
Cordemcura → AMRINONE
Cordes → HYDROCORTISONE
Cordes Vas → VITAMIN A
Cordelcortone → PREDNISOLONE
Cordilox → VERAPAPMIL
Cordioxil → DIGOXIN
Cordran → FLURANDRENOLIDE
Cordycepic Acid → MANNITOL
Coredamin → PRENYLAMINE LACTATE
Coreg → CARVEDILOL
Coretal → OXPRENOLOL

Corflazine → LIDOFLAZINE
CORIAL YELLOW
CORIANDER
Coriantin → GONADOTROPINS, CHORIONIC
CORIARIA also → HONEY
Coric → LISINOPRIL
Corid → AMPROLIUM
Corindolan → MEPINDOLOL
Coriovis → GONADOTROPINS, CHORIONIC
Coriphate → FLUOCINOLONE ACETONIDE
Coristin → ERGOLOID MESYLATES
Corkwood → SCOPOLAMINE
"Cork-worker's Lung" → DUSTS
Corlan → HYDROCORTISONE
beta-Corlan → BETAMETHASONE
Corlin → CORTISONE
Corliprol → CELIPROLOL
Corlopam → FENOLDOPAM MESYLATE
Corlutina → PROGESTERONE
Corlutone → PROGESTERONE
Corluvite → PROGESTERONE
Cormax → DILTIAZEM
CORN also → CYANIDES, HYDROGEN CYANIDE, NITRATES and NITRITES, T-2 TOXIN
Corn Campion → COCKLES
Corn Cockle → AGROSTEMMA GITHAGO
Corne de Cerf → ALOE
Corneiba → SCHINUS TEREBINTHIFOLIUS
CORNMINT OIL
Corn Oil → SODIUM
Cornox → METHYL CHLOROPHENOXY ACETIC ACID
"Corn-picker's Pupils" → CORN also → DATURA
Corn Poppy → PAPAVER sp.
Corn Rose → COCKLES
Cornstalks → NITRATES and NITRITES
Corn Sugar → GLUCOSE
CORN SYRUP
Cornus sp. → DOGWOOD
Corodane → CHLORDANE
Corodil → DIPYRIDAMOLE
Corona → MASCAGNIA sp.
Coronarin → DYPHYLLINE
Coronarine → DIPYRIDAMOLE
Corona de Birge → PASSIFLORA sp.
CORONILLA sp.
Coronin → KHELLIN
CORONOPUS sp. also → CRESS
Corontin → PRENYLAMINE LACTATE
Corossol → SORSACA
Corothion → PARATHION
Corotrope → MILRINONE
Corovliss → ISOSORBIDE DINITRATE
Corozate → ZIRAM
Corozo Palm → PALMS
Corpomone → PROGESTERONE
Corpse Plant → MONOTROPA UNIFLORA
Corpus Luteum Hormone → PROGESTERONE
Correx → METHYL BENZIMIDAZOLE-2-CARBAMATE
Corrigast → PROPANTHELINE BROMIDE
Corrosive Sublimate → MERCURIC CHLORIDE
Corsair → PERMETHRIN
Corsodyl → CHLORHEXIDINE
Corson → DEXAMETHASONE
Corson-P → DEXAMETHASONE
Corstiline → CORTICOTROPIN
Cortadren → CORTISONE
Cortaid → HYDROCORTISONE
Cortalone → PREDNISOLONE
Cortancyl → PREDNISONE
Cortate → DEOXYCORTICOSTERONE, *acetate*
Cort-Dome → HYDROCORTISONE
Cortef → HYDROCORTISONE
Cortelan → CORTISONE
Cortenema → HYDROCORTISONE
Cortexilar → FLUMETHASONE
Cor-Theophylline → DYPHYLLINE
Cortico-Sol → PREDNISOLONE
CORTICOTROPIN also → BISHYDROXYCOUMARIN, ETHYL BISCOUMACETATE, WARFARIN
Cortidene → PARAMETHASONE ACETATE
Cortifoam → HYDROCORTISONE
Cortilet → FLUOROMETHOLONE
CORTINARIUS sp.
Cortiphyson → CORTICOTROPIN
Cortiplastol → FLUOCINOLONE ACETONIDE
Cortiron → DEOXYCORTICOSTERONE, *acetate*
Cortisal → CORTISONE
Cortisate → CORTISONE also → PREDNISOLONE
Cortisol → HYDROCORTISONE
CORTISONE also → BISHYDROXYCOUMARIN, ETHYL BISCOUMACETATE, RIFAMPIN, SOLANUM sp., TOLBUTAMIDE
Δ^1-Cortisone → PREDNISONE
Cortistab → CORTISONE
Cortisumman → DEXAMETHASONE
Cortisyl → CORTISONE
Cortogen → CORTISONE
Cortone → CORTISONE
Cortril → HYDROCORTISONE

Cortrophin → CORTICOTROPIN
Cortrosyn → CORTICOTROPIN also → TETRACOSACTIDE
Corulon → GONADOTROPINS, CHORIONIC
Corundum → ALUMINUM OXIDE
Corvasol → PENTYLENETETRAZOL
Corvert → IBUTILIDE FUMARATE
Corvitin → METHAMPHETAMINE
Corvol → POLYVINYL ALCOHOL
Corycavine → CORYDALIS
Corydaline → DICENTRA CUCULLARIA
CORYDALIS
Corydine → CORYDALIS, DICENTRA CUCULLARIA
Corylus americana → HAZEL NUTS
Corynathe johimbe → YOHIMBINE
Corynine → YOHIMBINE
Cosan S → DAZOMET
Cosbiol → SQUALANE
Coscotabs → NOSCAPINE
Coslan → MEFENAMIC ACID
Cosmegen → DACTINOMYCIN
Cosmetol → CASTOR OIL
Cosmocil CQ → POLY (HEXAMETHYLENE BIGUANIDE HCL)
Cosmopen → PENICILLIN > *Benzylpenicillin, Potassium*
Cosprin → ACETYLSALICYLIC ACID
Cossym → CARBETAPENTANE CITRATE
Costimulator → ALDESLEUKIN
COSTUS ROOT OIL also → INULA sp.
Cosulid → SULFACHLORPYRIDAZINE
Cosumix → SULFACHLORPYRIDAZINE
Cosuric → ALLOPURINOL
Cosylan → DEXTROMETHORPHAN HYDROBROMIDE
Cosyntropin → TETRACOSACTIDE
Cotazym → LIPASE
Cotazyme → PANCRELIPASE
Cotinazin → ISONIAZID
Cotnion-methyl → AZINPHOS-METHYL
COTONEASTER
Cotoran → FLUOMETURON
Cotrimazine → TRIMETHOPRIM
Cotrimazole → TRIMETHOPRIM
Co-trimazole → SULFAMETHOXAZOLE
COTTON
Cottonex → FLUOMETURON
COTTONSEED MEAL also → MILK, PENICILLIUM sp., ROQUEFORTINE
COTTONSEED OIL also → GOSSYPOL
"Cotton Poison" → METHYL PARATHION
Cottonweed → ASCLEPIAS
COTYLEDON ORBICULATA
Coughwort → COLTSFOOT
Coulter Conyza → CONYZA COULTERI
COUMACHLOR
Coumadin → WARFARIN
Coumafene → WARFARIN
COUMAFURYL
COUMAPHOS also → SULFOTEPP
COUMARIN also → ACETYLSALICYLIC ACID, Σ-AMINOCAPROIC ACID, *p*-AMINOSALICYLIC ACID, BUTABARBITAL, CHLORAL HYDRATE, CHLORCYCLIZINE, CHLORTETRACYCLINE, CHOCOLATE, CLOFIBRATE, CLOVERS, FATS, GALIUM sp., GLIPIZIDE, LIOTHYRONINE, MEPROBAMATE, MERCURY, MORPHINE, NAPROXEN, NEOMYCIN, OXAZEPAM, PREDNISONE, PROBENECID, PROPYLTHIOURACIL, SALICYLAZOSULFAPYRIDINE, SCHOENBIBLUS PERUVIANUS, SECOBARBITAL, THYROID, THYROXINE, TONKA BEANS
Coumarouna odorata → TONKA BEANS
COUMATETRALYL
Coumopyran → CYCLOCOUMAROL
Coumopyrin → CYCLOCOUMAROL
Counter → TERBUFOS
Courlene → POLYETHYLENE
COUSSAPOA CINNAMOMEA
COUTOUBEA RAMOSA
Covatin(e) → CAPTODIAMINE
Covera-HS → VERAPAMIL
Covicone → DIMETHICONE (*with silicon dioxide*)
Cowage → COWHAGE
Cowbane → CICUTA sp.
Cow Cabbage → HERACLEUM sp.
Cow Cockle → SAPONARIA VACCARIA also → COCKLES
COWHAGE also → MUCUNA sp.
Cow Herb → SAPONARIA VACCARIA
Cowitch → MUCUNA sp.
Cow Manure Factor → VITAMIN B_{12}
Cow Parsnip → HERACLEUM sp.
Cowslip → CALTHA PALUSTRIS
COXIELLA BURNETII
Coxigon → BENOXAPROFEN
Coxistac → SALINOMYCIN
Coxistat → NITROFURAZONE
Coyden → CLOPIDOL
Coyote Tobacco → NICOTIANA sp.
COYOTILLO
Cozaar → LOSARTAN POTASSIUM

Cozyme → DEXPANTHENOL
CP 919 → THIOCARLIDE
CP 1044 J3 → BUFEXAMAC
CP 1215 → BENACTYZINE
CP 6343 → *N,N*-DIALLYL-2-CHLOROACETAMIDE
CP 10423-16 → PYRANTEL
CP 10423-18 → PYRANTEL
CP 12299-1 → PRAZOSIN HYDROCHLORIDE
CP 12,574 → TINIDAZOLE
CP 14,957 → ISOBENZAN
CP 15,336 → DIALLATE
CP 15,464 → PENICILLINS – "NEWER and SYNTHETICS" > *Carindacillin Sodium*
CP 15,467-61 → LITHIUM CARBONATE
CP 15,639-2 → PENICILLINS – "NEWER and SYNTHETICS" > *Carbenicillin Disodium*
CP 31,393 → PROPACHLOR
CP 19,106 → TRIMAZOSIN
CP 45,634 → SORBINIL
CP 45,899 → SULBACTAM
CP 47,114 → SUMITHION
CP 49,952 → PENICILLINS – "NEWER and SYNTHETICS" > *Sultamicillin*
CP 50,144 → ALACHLOR
CP 51,974-1 → SERTRALINE
CP 53,619 → BUTACHLOR
CP 62,993 → AZITHROMYCIN
CP 67,573 → GLYPHOSPHATE
CP 70,139 → GLYPHOSPHATE
CP 99,219 → TROPISETRON
CPA → CYCLOPIAZONIC ACID also → TRYPARSAMIDE
4-CPB → 4-(4-CHLOROPHENOXY)-BUTYRIC ACID
CPDC → CISPLATIN
CPIB → CLOFIBRATE
Cpiron → FERROUS FUMARATE
CPT 11 → IRINOTECAN
CR-1639 → DINOCAP
CRABS also → METHANEARSONIC ACID, SAXITOXIN, SHRIMP
Crab-E-Rad → DISODIUM METHANEARSONATE
Crabs' Eyes → ABRUS PRECATORIUS also → ABRIN
Crack → COCAINE
Crag 974 → DAZOMET
CRANBERRY
Crane's Bill → GERANIUM
Crank → AMPHETAMINE also → METHAMPHETAMINE
Crap → HEROIN
Crapinon → PIPERIDOLATE
Crasnitin → *l*-ASPARAGINASE
Crataegus sp. → HAWTHORN
CRATAEVA BENTHMII
Cravit → LEVOFLOXACIN
Crawford W. Long, M. D. → ETHER, ETHYL
CRAYONS
Crazy Cow Syndrome → SOLANUM sp.
Crazyweed → ASTRAGALUS also → OXYTROPIS
CRC 1820 → CIMETIDINE
CREAM
Cream of Magnesia → MAGNESIUM HYDROXIDE
Cream Nuts → BRAZIL NUTS
Creaseproof Agents → PERMANENT PRESS
CREATINE
Creatinine → CEFOTETAN, CEFOXITIN SODIUM, CIMETIDINE, FURAZOLIUM CHLORIDE, GENTAMICIN, GLUCOSE, GUANIDINE, HYDROCORTISONE, METHYLDOPA, METIAMIDE, NETILMYCIN, PHENOLSULFONPHTHALEIN
Creeping Charlie → IVY also → NEPETA HEDERACEA
Creeping Indigo → INDIGOFERA sp.
Creeping Jenny → IVY
Creeping Lantana → LANTANA sp.
Creeping Spurge → EUPHORBIA, *myrsinites*
Crema de Belleza-Manning → MERCUROUS CHLORIDE
CREMOPHOREL → ALFADOLONE, CYCLOSPORINE, PACLITAXEL, PROPANIDID
CRENOTHRIX POLYSPORA
CREOLIN
Creon → PANCREATIN
Creosedin → BROMAZEPAM
CREOSOTE
Creosote Bush → LARREA sp.
Crepasin → PRENYLAMINE LACTATE
Crepitan → HURA CREPITANS
Cresatin → *m*-CRESYL ACETATE
Crescormon → GROWTH HORMONES
m-CRESIDINE
Creslan → HYDROGEN CYANIDE
CRESOL also → CUTTING OILS, PHENOL, SOAP, TOLUENE
CRESOL, SAPONATED SOLUTION of
CRESS
Crest® → CINNAMIC ALDEHYDE
Crestabolic → METHANDRIOL
m-CRESYL ACETATE

p-CRESYL ACETATE
Cresylic Acid → CRESOL
Cresylol → CRESOL
CRIMIDINE
Crinovaryl → ESTRONE
Crinovyl → POLYVINYL CHLORIDE
Crinuryl → ETHACRYNIC ACID
Crisalbine → GOLD SODIUM THIOSULFATE
Criseociclina → TETRACYCLINE
Crisinar → AURANOFIN
Crisodrin → MONOCROTOPHOS
Crisofin → AURANOFIN
Crispin → TRAMADOL
Cristal → GLYCERIN
Cristallovar → ESTRONE
Cristalomicina → KANAMYCIN
Cristapen → PENICILLIN > *Benzylpenicillin, Potassium*
Cristapurat → DIGITOXIN
Cristy → METHAMPHETAMINE
Crixivan → INDINAVIR
CRL 40,476 → MODAFINNIL
CR 39 monomer → DIALLYL DIGLYCOL CARBONATE
Croalbidine → CROTALARIA
Crocin → ACETAMINOPHEN
CROCUS SATIVUS also → SAFFRON
Crofton Weed → EUPATORIUM, *adenophorum*
Cromax 2 → CHROMIUM PICOLINATE
Cromedazine → CHLORPROMAZINE
CROMOLYN SODIUM also → SODIUM METABISULFITE
Cromonalgina → TICROMYL
Cromovet → CROMOLYN SODIUM
Cronil → ECTYLUREA
Cronizat → NIZATIDINE
Cronodione → PHENINDIONE
CROPROPAMIDE
Crosemperine → CROTALARIA
Crossroads → AMPHETAMINE
CROTACTIN
CROTALARIA
Crotaline → MONOCROTALINE
Crotalus sp. → SNAKE(S)
Crotamide → CROTETHAMIDE
Crotamitex → CROTAMITON
CROTAMITON
Crotein → PROTEIN HYDROLYSATES
CROTETHAMIDE
CROTON also → BENZO[α]PYRENE, HOGWORT, TEA
Croton capitatus → HOGWORT
Croton Oil Factor A → PHORBOL
Crotonyl Isosulphocyanate → BRASSICA sp.
Crotothane → DINOCAP
Crotoxyphos → CIODRIN
Crovaril → OXYPHENBUTAZINE
Crow Chex → COPPER OXALATE
Crow Foot → GERANIUM also → LARKSPUR, RANUNCULUS
Crown Flower → CALOTROPIN
Crown of the Field → COCKLES
Crown of Thorns → EUPHORBIA also → STARFISHES
Crownvetch → CORONILLA sp.
Croysulfone → DAPSONE
Crufomate → RUELENE
CRUSTACEANS also → CRABS, LOBSTERS, SHRIMP
Cruz de Malta → JUSSIAEA
Crylène → PENTAPIPERIDE METHYLSULFIDE
Cryogel → POLYVINYL ALCOHOL
Cryogenine → PHENICARBAZINE
Cryolite → FLUORIDES and FLUORINE
CRYPTENAMINE TANNATE
Cryptocillin → PENICILLINS – "NEWER and SYNTHETICS" > *Sodium Oxacillin*
Cryptocur → LH-RH
Crytopin → DICENTRA CUCULLARIA
CRYPTOSTEGIA GRANDIFLORA
CRYPTOSTEMMA sp.
CRYPTOSTROMA CORTICALE also → MAPLE
Crystal → METHAMPHETAMINE also → PHENCYCLIDINE
Crystal Violet → GENTIAN VIOLET
Crystamine → VITAMIN B_{12}
Crystapen → PENICILLIN > *Benzylpenicillin, Potassium* also *Benzylpenicillin, Procaine* also *Benzylpenicillin, Sodium*
Crystapen V → PENICILLIN > *Penicillin V*
Crystodigin → DIGITOXIN
Crystoids → HEXYLRESORCINOL
Crystoserpine → RESERPINE
Crystosol → SORBITOL
Crytallinic Acid → NOVOBIOCIN
Crytion → GOLD SODIUM THIOSULFATE
CSA → CARSALAM
CS → α-CHLOROBENZYLIDINEMALONITRILE
CS 045 → TROGLITAZONE
CS 514 → PRAVASTATIN
CS 847 → BARBAN
CS 1170 → CEFMETAZOLE
CS 1483 → PARAMETHASONE ACETATE
CSF-2 → GRANULOCYTE MACROPHAGE COLONY STIMULATING FACTOR

CT2336 → ACYCLOVIR
C. T. A. B. → CETRIMONIUM BROMIDE
CTN → CEFOTETAN
Cu → COPPER
Cube → MORPHINE
CUBE ROOT
Cubes → LYSERGIC ACID DIETHYLAMIDE
Cuca → COCA
Cuckold Dock → BURDOCK
Cuckoo Button → BURDOCK
Curcuma longa → TURMERIC
Cucullarine → DICENTRA CUCULLARIA
CUCUMBERS also → ALDICARB
Cucumis melo → MUSK MELON
Cucurbitacins → CUCUMBERS, FLAX
Cucurbita maxima → SQUASH
Cucurbita pepo → PUMPKIN
Cuemid → CHOLESTYRAMINE RESIN
Cuivasil → LIDOCAINE
Culebra borrachera → METHYSTICODENDRON AMESIANUM
Culicoides sp. → SAND FLIES
Culpen → PENICILLINS – "NEWER and SYNTHETICS" > *Floxacillin*
Cuman → ZIRAM
CUMENE
CUMIN
CUMINALDEHYDE
Cuminic Aldehyde → CUMINALDEHYDE
Cuminum cyminum → CUMIN, BLACK
Cumol → CUMENE
Cupcakes → LYSERGIC ACID DIETHYLAMIDE
CUPFERON
CUPRIC ACETOARSENITE
Cupric Arsenite → COPPER ARSENITE
Cupric Carbonate → COPPER CARBONATE
Cupric Chloride → COPPER CHLORIDE
Cupricin → COPPER CYANIDE
Cupric Sulfate → COPPER SULFATE
Cuprid → TRIENTINE
Cuprinol → COPPER NAPHTHENATE
Cupri-Nox → COPPER NAPHTHENATE
CUPRIZONE
Cuprous Cyanide → COPPER CYANIDE
Cuprous Oxide → COPPER OXIDE, RED
Cuprous Thiocyanate → COPPER THIOCYANATE
Curantyl → DIPYRIDAMOLE
CURARE also → AZATHIOPRINE, BACITRACIN, BRUCINE, CHLOROTHIAZIDE, COLISTIN; ETHER, ETHYL; ISOFLURANE, KANAMYCIN, NEOMYCIN, NUX VOMICA, PROPRANOLOL, SCHOENBIBLUS PERUVIANUS, STRYCHNOS, TELITOXICUM PERVIANUM, TETRAPTERIS sp.
Curarin-HAF → *d*-TUBOCURARINE CHLORIDE
Curaterr → CARBOFURAN
Curatin → DOXEPIN
Curcin → FLAIRA, JATROPHA sp.
Curling Factor → GRISEOFULVIN
Curocef → CEFUROXIME
Curosajin → HOMOCHLORCYCLIZINE
Curoxim → CEFUROXIME
Curré → HEDERA HELIX
Curry Powder → LEAD CHROMATE also → TAMARIND
CURVULARIA GENICULATA
Curythan → MEPHENESIN
CUSCUTA sp.
CUSSONIA CORBISIEN
Custard Apple → SORSACA
Cutheparine → HEPARIN
Cutisan → TRICLOCARBAN
Cutivate → FLUTICASONE PROPIONATE
Cutleaf Philodendron → MONSTERA sp. also → CALCIUM OXALATE
Cutrine → COPPER TRIETHANOLAMINE
CUTTING OILS
Cuxacillin → PENICILLINS – "NEWER and SYNTHETICS" > *Amoxicillin*
Cuxanorm → ATENOLOL
CV 205-502 → QUINAGOLIDE
CV 4093 → MANIDIPINE
C-Vimin → VITAMIN C
CVP → CHLORFENVINPHOS
CY 116 → Σ-AMINOCAPROIC ACID
CY 216 → HEPARIN
Cyamopsis tetragonolobus → GUAR GUM
CYANAMIDE
CYANATES
CYANAZINE
Cyanea capillata → JELLYFISH
CYANIDES also → ACRYLONITRILE, ADENIA VOLKENSII, ALMONDS, AMYGDALIN, APPLE, APRICOT, ARSINE, BEANS, CASSAVA, CELERY, CHERRY, ELDERBERRY, FLAX, FRUITS and FRUIT JUICES, GLYCERIA, GRASSES, HYDRANGEA, LINAMARIN, MALONONITRILE, MERCURIC CYANIDE, MILLET, NICKEL, PEAS, PEACHES, PEARS, PLEUROTIS sp., POTASSIUM FERRICYANIDE, PROPIONITRILE, SILVER, SODIUM HYDROXIDE, SODIUM

NITROPRUSSIDE, TETRAMETHYL SUCCINONITRILE, TOBACCO, TOLUENE-2,4-DIISOCYANATE, VITAMIN C
Cyanoacetonitrile → MALONONITRILE
CYANOACRYLATES also → GLUE
Cyanobenzene → BENZONITRILE
Cyanocobalamin → VITAMIN B_{12}
Cyanoethylene → ACRYLONITRILE
CYANOGEN
CYANOGEN CHLORIDE
Cyanohydroxybutene → CABBAGE
Cyanomethane → ACETONITRILE
Cyano(methyl-mercuri)guanidine → METHYLMERCURY-DICYANDIAMIDE
2-Cyanopropane → ISOBUTYRONITRILE
Cyantin → NITROFURANTOIN
Cyasorb UV 284 → SULISOBENZONE
Cyater → TERFENADINE
CYCADALES
CYCADS also → HYDROGEN CYANIDE
Cycas sp. → CYCADS
Cycas circinalis → PALMS
CYCASIN
Cyclacillin → PENICILLINS – "NEWER and SYNTHETICS"
Cycladiene → DIENESTROL
Cyclaine → HEXYLCAINE
CYCLAMATES also → ANILINE, CYCLOHEXYLAMINE, SACCHARIN
CYCLAMEN
CYCLAMEN ALDEHYDE
Cyclamin → ANAGALLIS, CYCLAMEN
Cyclamycin → TROLEANDOMYCIN
Cyclan → CALCIUM CYCLAMATE
Cyanomethylamine → AMINOACETONITRILE
Cyclapen → PENICILLINS – "NEWER and SYNTHETICS" > *Cyclacillin*
Cyclazenine → GUANACLINE SULFATE
CYCLAZOCINE
Cyclergine → CYCLANDELATE
CYCLIZINE also → DIPIPANONE, HEPARIN
CYCLOBARBITAL
Cyclobarbitone → CYCLOBARBITAL
CYCLOBENZAPRINE
Cycloblastin → CYCLOPHOSPHAMIDE
Cyclobral → CYCLANDELATE
Cyclobuxine → BUXUS
CYCLOCHLOROTINE also → PENICILLIUM sp.
Cyclocort → AMCINONIDE
CYCLOCOUMAROL also → GLUTETHIMIDE, PHENYLBUTAZONE
Cyclodan → ENDOSULFAN
Cyclodol → TRIHEXYPHENIDYL
Cyclodorm → CYCLOBARBITAL
Cycloestrol → HEXESTROL
Cyclogest → PROGESTERONE
Cycloguanil Embonate → CYCLOGUANYL PAMOATE
CYCLOGUANYL PAMOATE
Cyclogyl → CYCLOPENTOLATE
CYCLOHEXAMINE
CYCLOHEXANE
CYCLOHEXANOL
CYCLOHEXANONE
Cyclohexatriene → BENZENE
3-cyclohexyl-1-(p-acetylphenylsulfonyl)-urea → ACETOHEXAMIDE
CYCLOHEXYLAMINE also → CALCIUM CYCLAMATE, SODIUM CYCLAMATE
2-CYCLOHEXYL CYCLOHEXANONE
CYCLOHEXYL ETHYLS
Cyclohexylmethane → METHYLCYCLOHEXANE
Cyclolyt → CYCLANDELATE
Cyclomandol → CYCLANDELATE
Cyclomen → DANAZOL
Cyclomethiazide → CYCLOPENTHIAZIDE
CYCLOMETHICONE
CYCLOMETHYCAINE also → PROCAINE, TETRACAINE HYDROCHLORIDE
Cyclomycin → TETRACYCLINE
Cyclonal → HEXOBARBITAL
Cyclonarol → CYCLOPENTAMINE
Cyclonamine → ETHAMSYLATE
Cyclone → PHENCYCLIDINE
CYCLONITE
CYCLOOCTYLAMINE
CYCLOPAMINE
Cyclopane → CYCLOPROPANE
Cyclopar → TETRACYCLINE
Cyclopentacycloheptene → AZULENE
CYCLOPENTADECANOLIDE
Cyclopentadrine → CYCLOPENTAMINE
CYCLOPENTALATE also → TROPICAMIDE
CYCLOPENTAMINE
CYCLOPENTANONE
CYCLOPENTHIAZIDE also → BUCLOSAMIDE, POTASSIUM CHLORIDE
CYCLOPENTOLATE
CYCLOPENTOPHENANTHRENE
CYCLOPHOSPHAMIDE also → ACETYLCYSTEINE, ALLOPURINOL, CHLORAMPHENICOL, CHLORDANE, DOXORUBICIN, FILGRASTIM, MELPHALAN, MIBEFRADIL,

PHENOBARBITAL, PREDNISOLONE, PROPANIDID, SUCCINYLCHOLINE, SULFAPHENAZOLE, VASOPRESSIN
Cyclophosphane → CYCLOPHOSPHAMIDE
CYCLOPIAZONIC ACID also → ASPERGILLUS
CYCLOPROPANE also → BUTYROLACTONE, EPINEPHRINE, FELYPRESSIN, GALLAMINE TRIETHIODIDE, PROPRANOLOL
Cyclosarin → SARIN
CYCLOSERINE also → ISONIAZID, PHENYTOIN, VITAMIN B_6
Cyclospasmol → CYCLANDELATE
Cyclosporin A → CYCLOSPORINE
CYCLOSPORINE also → ALLOPURINOL, AMIODARONE, CHLORAMPHENICOL, CIMETIDINE, CIPROFLOXACIN, CLINDAMYCIN, CLOZAPINE, DIAMINOZIDE, DICLOFENAC, DILTIAZEM, ERYTHROMYCIN, ERYTHROMYCIN SUCCINATE, FLUCONAZOLE, FLUOXETINE, GANCICLOVIR, GRAPEFRUIT, HYPERICUM, JOSAMYCIN, ITRACONAZOLE, KETOCONAZOLE, LOMEFLOXACIN, LOVASTATIN, MELPHALAN, METOCLOPRAMIDE, METOLAZINE, METRONIDAZOLE, MIBEFRADIL, MIFEPRISTONE, MISOPROSTOL, NEFAZODONE, NICARDIPINE, NIFEDIPINE, NIZATIDINE, NORETHINDRONE, NORFLOXACIN, PHENOBARBITAL, POTASSIUM, RAPAMYCIN, RIFAMPIN, ROXITHROMYCIN, SULFAMETHAZINE, SULINDAC, TACROLIMUS, TERBINAFINE, THEOPHYLLINE, THIOTHIXENE, TICLODIPINE, VINCRISTINE, VITAMIN E, WARFARIN
Cyclostin → CYCLOPHOSPHAMIDE
Cyclosulfyne → PROPARGITE
CYCLOTHIAZIDE
Cycloysal → LYMECYCLINE
Cycocel → CHLORMEQUAT CHLORIDE
Cycostat → ROBENIDINE
Cycrin → MEDROXYPROGESTERONE ACETATE
Cydectin → MOXIDECTIN
Cydonium sp. → QUINCE
Cyfen → SUMITHION
Cyflee → CYTHIOATE
Cygon → DIMETHOATE
Cyklodorm → CYCLOBARBITAL
Cyklokapron → TRANEXAMIC ACID
Cyklosal → CYCLOPENTAMINE
Cykobemin → VITAMIN B_{12}
Cylert → PEMOLINE
Cylphenicol → CHLORAMPHENICOL
CYMARIN
Cymbi → PENICILLINS – "NEWER and SYNTHETICS" > *Ampicillin*
Cymbopogon martini → GERANIUM
Cymbush → CYPERMETHRIN
p-CYMENE
Cymevan → GANCICLOVIR
Cymevene → GANCICLOVIR
Cyminii → CUMIN, BLACK
CYMOPTERUS sp.
Cymovar → GANCICLOVIR
CYNANCHUM
Cynara scolymus → ARTICHOKE
Cynem → ZINOPHOS
CYNODON also → HYDROGEN CYANIDE, PITHOMYCES
Cynoff → CYPERMETHRIN
Cynoglossine → CYNOGLOSSUM OFFICIANALE
CYNOGLOSSUM OFFICIANALE
Cynomel → LIOTHYRONINE
Cyolan(e) → PHOSFOLAN
Cyomel → LIOTHYRONINE
Cypercare → CYPERMETHRIN
Cyperkill → CYPERMETHRIN
CYPERMETHRIN also → CALCIUM CARBONATE
Cypersect → CYPERMETHRIN
Cypip → DIETHYLCARBAMAZINE
Cypress Spurge → EUPHORBIA, *cyparrissias*
Cypripedin → CYPRIPEDIUM sp.
CYPRIPEDIUM sp.
CYPROHEPTADINE also → SEROTONIN
CYPROLIDOL
Cyprostat → CYPROTERONE
CYPROTERONE
Cyren A → DIETHYLSTILBESTROL
Cyren B → DIETHYLSTILBESTROL
CYROMAZINE
Cyrpon → MEPROBAMATE
Cystadene → BETAINE
Cystamin → METHENAMINE
CYSTE ABSOLUTE
CYSTEAMINE also → CYSTINE
Cysteic Acid → PROTEIN HYDROLYSATES
CYSTEINE
CYSTINE also → KENTUCKY COFFEE TREE
Cystit → NITROFURANTOIN

Cysto-Conray → MEGLUMINE IOTHALAMATE
Cystografin → MEGLUMINE DIATRIZOATE
Cystogen → METHENAMINE
Cystokon → SODIUM ACETRIZOATE
Cystorelin → LH-RH
Cystospaz → l-HYOSCYAMINE
Cystrin → OXYBUTYNIN CHLORIDE
Cystural → ETHOXAZENE
Cytacon → VITAMIN B_{12}
Cytamen → VITAMIN B_{12}
Cytadren → AMINOGLUTETHIMIDE
Cytel → SUMITHION
Cyten → SUMITHION
Cytar → CYTARABINE
CYTARABINE also → THIOGUANINE
Cytarbel → CYTARABINE
CYTHIOATE also → SUCCINYLCHOLINE
Cythion → MALATHION
CYTISINE also → SOPHORA sp., SPARTIUM JUNCEUM
CYTISUS (GENISTA) CANARIENSIS
CYTISUS LABURNUM
Cytiton → CYTISINE
Cytobin → LIOTHYRONINE
Cytobion → VITAMIN B_{12}
Cytomel → LIOTHYRONINE
Cytomine → LIOTHYRONINE
Cytophosphane → CYCLOPHOSPHAMIDE
Cytosar → CYTARABINE
Cytosine Arabinoside → CYTARABINE
Cytotec → MISOPROSTOL
Cytovene → GANCICLOVIR
Cytoxan → CYCLOPHOSPHAMIDE
Cytrol → AMINOTRIAZOLE

D

D 4T → STAVUDINE
D 14 → NABAM
D 25 → FENTICLOR
D 40TA → ESTAZOLAM
D 50 → PYRAZINAMIDE
D 65MT → ALPRAZOLAM
D 100 → DOXYCYCLINE
D 145 → MEMANTINE
D 201 → ISOTHIPENDYL
D 206 → PROTHIPENDYL
D 254 → PIPAZETHATE
D 365 → VERAPAMIL
D 563 → OXYFEDRINE
D 600 → GALLOPAMIL
D 706 → OXYPENDYL
D 735 → CARBOXIN
D 1221 → CARBOFURAN
D 2083 → DESONIDE
D 4720 → ALLYL ISOTHIOCYANATE
D 7093 → MESNA
D 18,506 → MILTEFOSINE
2,4-D → 2,4-DICHLOROPHENOXYACETIC ACID
DA 688 → GEFARNATE
DA 759 → METHOXYFLURANE
DA 2370 → FEPRAZONE
2,4-DAA → 2,4-DIAMINOANISOLE
DAAM → ACETYLMETHADOL
DAB → *p*-DIMETHYLAMINOBENZENE
Dabco → TRIETHYLENEDIAMINE
Dabicycline → OXYTETRACYCLINE
Dabroson → ALLOPURINOL
Dabylen → DIPHENHYDRAMINE
DAC-2787 → CHLOROTHALONIL
Dacamine → 2,4,5-TRICHLOROPHENOXYACETIC ACID
DACARBAZINE
Dacarel → NICARDIPINE
Dacatic → DACARBAZINE
Daccha → CANNABIS
Dacliximab → DACLIZUMAB
DACLIZUMAB
Daconate → MONOSODIUM METHANEARSENATE
Daconil 2787 → CHLOROTHALONIL
Dacoren → ERGOLOID MESYLATES
Dacortin → PREDNISONE
DACPM → 4,4´-METHYLENE-bis [2-CHLOROANILINE]
Dactil → PIPERIDOLATE
DACTINOMYCIN
D-Activated Animal Sterol → VITAMIN D_3
D-Activated Plant Sterol → VITAMIN D_2
DACTYLIS GLOMERATA
DACTYLOCTENIUM AEGYPTIUM
Dadex → DEXTROAMPHETAMINE SULFATE
Dadibutol → ETHAMBUTOL
DADPS → DAPSONE
Dafalgan → ACETAMINOPHEN
DAFFODILS also → NARCISSUS sp.
Dagadip → CARBOPHENOTHION
Dagenan → SULFAPYRIDINE
Dagga → CANNABIS
Daisies → CHYSANTHEMUM, MARIGOLD
Daitop → ISOTHIPENDYL
Daiwa sp. → PARIS sp.
Dakin's Solution → CHLORINATED LIME

Daktar → MICONAZOLE NITRATE
Daktarin → MICONAZOLE NITRATE
Dalacin C → CLINDAMYCIN
Dalacin T → CLINDAMYCIN
Dalactine → CLINDAMYCIN
Dalbergia → EBONY
Dalbergiones → ROSEWOOD
Dal-E-Rad → DISODIUM METHANEARSONATE also → METHANEARSONIC ACID, MONOSODIUM METHANEARSONATE
Dalf → METHYL PARATHION
DALFOPRISTIN also → CYCLOSPORINE
Dalgen → DEZOCINE
Dalgol → MEPARFYNOL
Dallis-grass → PASPALUM sp.
Dalpac → BUTYLATED HYDROXYTOLUENE
DALTEPARIN
Dalysep → SULFALENE
Dalzic → PRACTOLOL
DAMAR
Dame d'onze Heures → ORNITHOGALUM UMBELLATUM
Dametin → DEHYDROEMETINE
Damiana → TURNERA APHRODISIACA
Damide → INDAPAMIDE
DAMINOZIDE also → APPLE
Dammar → DAMAR
DAMP → BISACODYL
Danabol → METHANDROSTENOLONE
Danantizol → METHIMAZOLE
DANAPAROID
DANAZOL also → CYCLOSPORINE, TACROLIMUS, WARFARIN
Dandelion → TARAXACUM OFFICINALE also → CHRYSANTHEMUM
Dandilone → PHENINDIONE
Dandrex → SELENIUM
Dandruff of the Gods → COCA
Daneral → PHENIRAMINE MALEATE
Danewort → SAMBUCUS sp.
Danex → TRICHLORFON
Danilone → PHENINDIONE
DANISLEUKIN DIFTITOX
Danocrine → DANAZOL
Danol → DANAZOL
Danoval → DANAZOL
Dansida → IBUPROFFEN
DANSYL CHLORIDE
Dantafur → NITROFURANTOIN
Dantamacrin → DANTROLENE
DANTHRON also → DOCUSATE SODIUM, MILK
Dantrium → DANTROLENE
DANTROLENE also → VECURONIUM BROMIDE
Dantromin → PEMOLINE
Dantron → DANTHRON
Danylen → DIETHYLPROPION
Daonil → GLYBURIDE
DAP → DIGITOXIN
Dap® → PROMETHAZINE
Dapa → ACETAMINOPHEN
Daphene → DIMETHOATE
DAPHNE
"Daphne" → YUANHAUCINE
Daphne genkwa → YUANHAUCINE
Daphne gnidium → YUANHAUCINE
Daphné jolibois → DAPHNE
Daphne mezereum → YUANHAUCINE
Daphnetoxin → DAPHNE
Daphnin → DAPHNE
Dapotum D → FLUPHENAZINE DECANOATE
Dap® Preg Test → PROMETHAZINE
DAPSONE also → DIDANOSINE, PROBENECID, RIFABUTIN, RIFAMPIN, SULFANILAMIDE, SULFOXONE, ZIDOVUDINE
DAPT → AMIPHENAZOLE
Daptazile → AMIPHENAZOLE
Daptazole → AMIPHENAZOLE
Daquin → CHLORAZANIL
Darammon → AMMONIUM CHLORIDE
Daranide → DICHLORPHENAMIDE
Daraprim → PYRIMETHAMINE
Darbid → ISOPROPAMIDE IODIDE
Darcil → PENICILLINS – "NEWER and SYNTHETICS" > *Phenethicillin Potassium*
Darenthin → BRETYLIUM TOSYLATE
Dariclox → PENICILLINS – "NEWER and SYNTHETICS" > *Cloxacillin/Sodium salt*
Daricon → OXYPHENCYCLIMINE
Darmol → PHENOLPHTHALEIN
Darmous → FLUORIDES and FLUORINE
Darnel → LOLIUM TEMULENTUM also → RYE GRASS
Darob → SOTALOL
Daropervamin → METHAMPHETAMINE
Dartal → THIOPROPAZATE
Dartalan → THIOPROPAZATE
Darvilen → CEFOTETAN
Darvon 65® → PROPOXYPHENE also → NALOXONE
Darwin, Erasmus → ANTIARIS TOXICARIA
Dasanit → FENSULFOTHION
Daserol → MEPHENESIN

Dasheen → COLOCASIA sp.
Dashin → COLOCASIA sp.
Daskil → VITAMIN B_3
Dasten → OXATOMIDE
Datanil → THIOCARLIDE
DATC → DIALLATE also → THIOCARLIDE
Date Rape Drugs → ALCOHOL, ETHYL; CHLORAL HYDRATE, FLUNITRAZEPAM, γ-HYDROXYBUTYRATE, KETAMINE
Datril → ACETAMINOPHEN
DATURA sp. also → ATROPINE, HONEY, SCOPOLAMINE
Datura metel → DATURA sp. also → Preface
DAUBENTONIA PUMICEA
DAUCUS CAROTA also → CARROTS
Daunoblastina → DAUNORUBICIN
Daunomycin → DAUNORUBICIN
DAUNORUBICIN also → ALUMINUM
DAURICINE also → MENISPERMUM CANADENSE
Davainex → DIBUTYL TIN DILAURATE
Davitamon C → VITAMIN C
Davoxin → DIGOXIN
DAV Ritter → DESMOPRESSIN
Daxauten → PRENYLAMINE LACTATE
Day-blooming Jessamine → CESTRUM
Day Cestrum → CESTRUM
Day Jessamine → CESTRUM
Daykin's Solution → SODIUM HYPOCHLORITE
Daypro → OXAPROZIN
Dazzle → SODIUM HYPOCHLORITE
DAZOMET
Dazzel → DIAZINON
DBA → 1,2:5,6-DIBENZANTHRACENE
DBCP → DIBROMOCHLOROPROPANE
DBD → AZINPHOS-METHYL also → MITOLACTOL
DBI → PHENFORMIN
DBI-TD → PHENFORMIN
DBM → MITOBRONITOL
DBN → *N*-NITROSODIBUTYLAMINE
DBP → DIBUTYL PHTHALATE
DBTD → DIBUTYL TIN DILAURATE
DBV → BUFORMIN
DCA → DEOXYCORTICOSTERONE also → DICHLOROACETIC ACID
DCB → σ-DICHLOROBENZENE
2,4-DCBA → 2,4-DICHLOROBENZYL ALCOHOL
DCCK → ERGOLOID MESYLATES
DCEE → *sym*-DICHLOROETHYL ETHER
DCI → DICHLORISOPROTERENOL
DCMO → CARBOXIN
DCNA → 2,6-DICHLORO-4-NITROANILINE
DCNU → CHLOROZOTOCIN
DCP → DIPHENCYPRONE
D&C Blue 6 → INDIGO
D&C Red 7 also → LIPSTICK
D&C Red 17 → SUDAN III
D&C Red 19 → RHODAMINE B
D&C Red #22 → EOSIN(E)
D&C Yellow #7 → FLUORESCEIN
D&C Yellow #8 → FLUORESCEIN
D&C Yellow #10 → QUINOLINE YELLOW also → DANAZOLE
D&C Yellow #11 → QUINOLINE YELLOW
DDAVP → DESMOPRESSIN
ddCyd → DIDEOXYCYTIDINE
DDD → 1,1-DICHLORO-2,2-BIS(*p*-CHLOROPHENYL)ETHANE
σ,p´DDD → MITOTANE
p,p´-DDD → 1,1-DICHLORO-2,2-BIS (*p*-CHLOROPHENYL)ETHANE
DDE → FISH
ddI → DIDANOSINE
ddIno → DIDANOSINE
DDM → 4,4´-METHYLENE DIANILINE
DDP → CISPLATIN
DDS → DAPSONE
DDT also → BUTTER, FISH, MILK, PHENOBARBITAL, PHENYLBUTAZONE, PHENYLTHIOUREA, PHENYTOIN, PROGESTERONE, TESTOSTERONE, ZOXAZOLAMINE
p,p´-DDT → DDT
DDVP → DICHLORVOS
Deadly Agaric → AMANITA sp.
Deadly Hemlock → CONIUM MACULATUM
Deadly Nightshade → ATROPA BELLADONNA also → ATROPINE, DATURA sp., → SOLANUM sp.
Dead Men's Fingers → OENANTHE sp.
Deadopa → L-DOPA
Dead Tongue → OENANTHE sp.
Deaminooxytocin → DESAMINOOXYTOCIN
Deanase → PANCREATIC DORNASE
Deaner → DEANOL
DEANOL
Deapasil → *p*-AMINOSALICYLIC ACID
Deapril-ST → ERGOLOID MESYLATES
Death Angel → AMANITA sp.
Death Camas → ZIGADENUS sp.
Death Cap → AMANITA sp.
Death Drug → PHENCYCLIDINE
Deathmor → WARFARIN

Death's Herb → BELLADONNA
Death Wine → MANDRAGORA OFFICINARUM
7-Deazaadenosine → TUBERCIDIN
Deba → BARBITAL
Debecacin → DIBEKACIN
Debecillin → PENICILLIN > *Benzylpenicillin, Benzathine*
Débékacyl → DIBEKACIN
Debena → SULFADIAZINE
Debenal M → SULFAMERAZINE
Debeone-DT → PHENFORMIN
Debinyl → PHENFORMIN
Debridat → TRIMEBUTINE MALEATE
DEBRISOQUIN also → AMPHETAMINE, CHEESES, PHENYLEPHRINE HCL, PHENYLPROPANOLAMINE, PSEUDOEPHEDRINE
Debroxide → BENZOYL PEROXIDE
DEC → DECAHYDRONAPHTHALENE also → DIETHYLCARBAMAZINE
DECABORANE
Decaboron Tetradechydride → DECABORANE
Decacil → CHLORDIAZEPOXIDE
Decacort → DEXAMETHASONE
Decaderm → DEXAMETHASONE
DECADIBROMODIPHENYL OXIDE
Decadron → DEXAMETHASONE
Deca Durabol → NANDROLONE
Deca Durabolin → NANDROLONE
Deca Hybolin → NANDROLONE
DECAHYDRONAPHTHALENE
DECAHYDRO-β-NAPHTHOL
Decalin → DECAHYDRONAPHTHALENE
Decalix → DEXAMETHASONE
2-Decalol → DECAHYDRO-β-NAPHTHOL
DECAMETHONIUM BROMIDE also → BACITRACIN, COLISTIN, CURARE, DEXPANTHENOL, DIHYDROSTREPTOMYCIN; ETHER, ETHYL; GRAMICIDIN, HEXAFLUORENIUM BROMIDE, KANAMYCIN, LIDOCAINE, NEOMYCIN, PROCAINE, QUINIDINE, STREPTOMYCIN
DECAMETHRIN
Decamine → DEQUALINIUMS
Decapeptyl → TRIPTORELIN
Decaprednil → PREDNISOLONE
Decapryn Succinate → DOXYLAMINE SUCCINATE
Decaps → VITAMIN D_2
Decaris → LEVAMISOLE also → TETRAMISOLE
Decaserpyl → DESERPIDINE
Decasone → DEXAMETHASONE
Decatylen → DEQUALINIUMS
Deccotane → *sec*-BUTYLAMINE
Deccox → DECOQUINATE
Decemthion → IMIDAN
9-Decenol-1 → DECYLENIC ALCOHOL
Decentan → PERPHENAZINE
Dechlorane → MIREX
Decholin → DEHYDROCHOLIC ACID also → SODIUM DEHYDROCHOLATE
Decicaine → TETRACAINE HYDROCHLORIDE
Decis → DECAMETHRIN
Declid → UNDECYLENIC ACID
Declinax → DEBRISOQUIN
Declomycin → DEMECLOCYCLINE
Decontractyl → MEPHENESIN
DECOQUINATE
Decortancyl → PREDNISONE
Decortin → PREDNISONE
Decortin H → PREDNISOLONE
Decosteron → DEOXYCORTICOSTERONE
Decrelip → GEMFIBROZIL
Decreten → PINDOLOL
Decril → ERGOLOID MESYLATES
Decrysil → DINITRO-σ-CRESOL
DECYL ACETATE
DECYLENIC ALCOHOL
DECYL OLEATE
DECYLTRIMETHYLAMMONIUM BROMIDE
Dedevap → DICHLORVOS
Ded-Weed Brush Killer → 2,4,5-TRICHLOROPHENOXYACETIC ACID
Deep Tan → CINOXATE
DEER LEATHER
Dee-Ron → VITAMIN D_2
Deet → *N,N*-DIETHYL-*m*-TOLUAMIDE
DEF → *S,S,S*-TRI-*N*-BUTYL PHOSPHOROTRITHIOATE
De-Fend → DIMETHOATE
Defenidol → DIPHENIDOL
DEFEROXAMINES also → RHIZOPUS sp.
Deficol → BISACODYL
Défiltran → ACETAZOLAMIDE
Deflamon → METRONIDAZOLE
Deflexol → ZOXAZOLAMINE
Deflorin → CEPHALORIDINE
Defol → SODIUM CHLORATE
Degalol → DEOXYCHOLIC ACID
Deganol → DEMECLOCYCLINE
Degranol → MANNOMUSTINE
"Degreasers' Flush" → TRICHLOROETHYLENE

De-Green → *S,S,S*-TRI-*N*-BUTYL PHOSPHOROTRITHIOATE
Deguelia Root → DERRIS ROOT
Dehacodin → DIHYDROCODEINE
Dehestin → TRIPELENNAMINE
DEHP → BIS(2-ETHYLHEXYL)PHTHALATE
Dehydrobenzperidol → DROPERIDOL
Dehychol → DEHYDROCHOLIC ACID
DEHYDROACETIC ACID
DEHYDROCHOLIC ACID
Δ^1-Dehydrocortisol → PREDNISOLONE
DEHYDROEMETINE
Dehydroepiandrosterone → PRASTERONE
Dehydromyodesmone → MYOPORUM
Dehydrotestosterone → BOLDENONE
Deidrocolico Vita → DEHYDROCHOLIC ACID
Deiten → NITRENDIPINE
Dekacort → DEXAMETHASONE
Dekadin → DEQUALINIUMS
Dekalin → DECAHYDRONAPHTHALENE
Dekamin → DEQUALINIUMS
Dekelmin → METHYRIDINE
Dekoks → DECOQUINATE
Dekrysil → DINITRO-σ-CRESOL
DEL-1267 → METOCLOPRAMIDE
Delacillin → PENICILLINS – "NEWER and SYNTHETICS" > *Amoxicillin*
Delacurarine → *d*-TUBOCURARINE CHLORIDE
Delagil → CHLOROQUINE
Delalutin → 17-α-HYDROXYPROGESTERONE
Delaney Clause → POLYVINYL CHLORIDE
Delaprem → HEXOPRENALINE
DELAVIRDINE also → ASTEMIZOLE, CISAPRIDE, CLOZAPINE, DIDANOSINE INDINAVIR, NELFINAVIR MESYLATE, RIFABUTIN, RIFAMPIN, SAQUINAVIR, TERFENADINE
Delaxin → METHOCARBAMOL
Delepine → ANTHIOLIMINE
Delgesic → LYSINE ACETYLSALICYLATE
Delice → PERMETHRIN
Delicia → TRICHLORFON
Delimmun → ISOPRINOSINE
Delix → RAMIPRIL
Delmadinone → CHLORMADINONE
Demerol → MEPERIDINE
Delmeson → FLUOROMETHOLONE
Delnatex → DIOXATHION
Delnav → DIOXATHION
Delors → POLYCHLORINATED BIPHENYLS
m-Delphene → *N,N*-DIETHYL-*m*-TOLUAMIDE
Delphicort → TRIAMCINOLONE
Delphimix → DICLOFENAC
DELPHININE also → ADONIS sp., LARKSPUR
Delphinium sp. → LARKSPUR
Delsan → THIRAM
Delsterol → VITAMIN D_3
Delsym → DEXTROMETHORPHAN HYDROBROMIDE
Delta-Cortef → PREDNISOLONE
Delta-Cortelan → PREDNISONE
Deltacortisone → PREDNISONE
Deltacortone → PREDNISONE
Deltacortril → PREDNISOLONE
Delta E → PREDNISONE
Delta F → PREDNISOLONE
Deltafluorene → DEXAMETHASONE
Deltalin → VITAMIN D_2
Deltamethrin → CALCIUM CARBONATE
Deltamine → PEMOLINE
Deltan → DIMETHYL SULFOXIDE
Deltasolone → PREDNISOLONE
Deltasone → PREDNISONE
Deltastab → PREDNISOLONE
Deltazen → DILTIAZEM
Deltison → PREDNISONE
Deltoin → METHETOIN
Deltox → DIOXATHION
Deltra → PREDNISONE
Delvinal → VINBARBITAL
Delysid → LYSERGIC ACID DIETHYLAMIDE
Demadex → TORSEMIDE
Demairol → VITAMIN A
DEMECARIUM BROMIDE
DEMECLOCYCLINE also → ALCOHOL, ETHYL; *p*-AMINOSALICYLIC ACID, CHLORAMPHENICOL, CHLOROTHIAZIDE, CHLORPROMAZINE, CINCHOPHEN, FUROSEMIDE, ISONIAZID, METHOXYFLURANE, METHYL TESTOSTERONE, PHENINDIONE, PHENYLBUTAZONE, PHENYTOIN, POLYVINYLPYRROLIDONE
DEMECOLCINE
Demecolin → DEMECOLCINE
Demeso → DIMETHYL SULFOXIDE
Demethyldiazepam → NITRAZEPAM
Demethyldopan → URACIL MUSTARD
DEMETON
DEMETON-METHYL
Demetraciclina → DEMECLOCYCLINE
Demetrin → PRAZEPAM
Demissidine → POTATO
Demochelys coriacea → TURTLES
Demolox → AMOXAPINE

Demon → CYPERMETHRIN
"Demons of the Sea" → STINGRAYS
Demorphan Hydrobromide → DEXTROMETHORPHAN HYDROBROMIDE
Demos-L40 → DIMETHOATE
Demotil → DIPHEMANIL METHYLSULFATE
Demox → DEMETON
Demoxytocin → DESAMINOOXYTOCIN
Demser → METYROSINE
DEN → *N*-NITROSODIETHYLAMINE
DENA → *N*-NITROSODIETHYLAMINE
Denagard → TIAMULIN
Denapol → CINNARIZINE
Denan → SIMVASTATIN
DENATONIUM BROMIDE
Denavir → PENCICLOVIR
Dendrid → IDOXURIDINE
Dendrodochin → DENDROCHIUM TOXICUM
DENDROCHIUM TOXICUM
Denosine → GANCICLOVIR
Denosyl SD4 → *S*-ADENOSYLMETHIONINE
Dentafluoro → SODIUM FLUORIDE
Dentigoa → IBUPROFEN
Dentoraline → PROCAINE
DENZIMOL
DEOXYCHOLIC ACID
4´-Deoxycirramycin A → ROSAMICIN
2-Deoxycoformycin → PENTOSTATIN
DEOXYCORTICOSTERONE also → CANRENONE, CHLORCYCLIZINE, CHLORDANE, DDT, SPIRONOLACTONE
Deoxycortone → DEOXYCORTICOSTERONE
2´-Deoxy-5-fluorouridine → FLOXURIDINE
2-DEOXY-D-GLUCOSE
Deoxy-hequeinone → PENICILLIUM sp.
5-Deoxyingenol → CROTON
DEOXYNIVALENOL also → YELLOW RAIN
Deoxyphorbols → EUPHORBIA sp.
4-DEOXYPYRIDOXINE
DEOXYRIBONUCLEIC ACID
4-DEOXYSCIRPENOL
DEOXYTHIOGUANOSINES
DEP → DIETHYL PYROCARBONATE also → DIETHYL PHOSPHATE, DIETHYL PHTHALATE
Depakene → SODIUM VALPROATE also → VALPROIC ACID
Depakin(e) → SODIUM VALPROATE → also VALPROIC ACID
Dépakine → VALPROIC ACID
Depakote → VALPROIC ACID also → SODIUM VALPROATE
Deparkin → DIETHAZINE
Depasan → SPARTEINE
DEPC → DIETHYL PYROCARBONATE
Depigam → MONOBENZONE
Depivar → VITAMIN B_{12}
Depixol → FLUPENTIXOL
Deploy → GLYPHOSPHATE
Depocillin → PENICILLIN > *Benzylpenicillin, Procaine*
Depo-Clinovir → MEDROXYPROGESTERONE ACETATE
Depo-Medrate → METHYLPREDNISOLONE
Depo-Medrol → METHYLPREDNISOLONE
Depo-Medrone → METHYLPREDNISOLONE
Deponit → NITROGLYCERIN
Depo-Provera → MEDROXYPROGESTERONE ACETATE
Depostat → GESTONORONE CAPROATE
Depovernil → SULFAMETHOXYPYRIDAZINE
Deprancol → PROPOXYPHENE
Deprenil → DEPRENYL
DEPRENYL also → L-DOPA, FLUOXETINE, MEPERIDINE, SUMATRIPTAN SUCCINATE
Déprényl → DEPRENYL
Depressan → DIHYDRALAZINE
Depresyn → DOTHIEPIN
Depridol → METHADONE
Deprilept → MAPROTILINE
Deprinol → IMIPRAMINE
DEPTROPINE
Dequadin Chloride → DEQUALINIUMS
Dequafungan → DEQUALINIUMS
DEQUALINIUMS
Dequavagyn → DEQUALINIUMS
Dequavet → DEQUALINIUMS
Deracil → THIOURACIL
Deracyn → ADINAZOLAM
Deralin → PROPRANOLOL
Derantel → CEPHALEXIN
De-Rat → VITAMIN D_2
Deratol → VITAMIN D_2
Derbac → CARBA
Derbac-M → MALATHION
Dergotamine → DIHYDROERGOTAMINE METHANESULFONATE (MESYLATE)
Dermabet → BETAMETHASONE
Derma-Blanch → HYDROQUINONE
Dermacid → 2-MERCAPTOBENZOTHIAZOLE
Dermacort → HYDROCORTISONE
Dermadex → HEXACHLOROPHENE
Dermaflor → DIFLORASONE
Dermafos → RONNEL

Dermafur → FURAZOLIUM CHLORIDE
Dermagan → DIACETAZOTOL
Dermalar → FLUOCINOLONE ACETONIDE
Deralbine → MICONAZOLE NITRATE
Dermalene → POLYETHYLENE
Dermaplus → FLUOCINONIDE
Dermasorb → DIMETHYL SULFOXIDE
Dermatan Sulfate → DANAPAROID
Dermathycin → THYROTROPIN
Dermaton → CHLORFENVINPHOS
Dermatophagoides scheremetewskyi → PILLOWS
Dermocortal → HYDROCORTISONE
Dermofungin A → 5-CHLORO-8-QUINOLINOL
Dermolate → HYDROCORTISONE
Dermonistat → MICONAZOLE NITRATE
Dermosol → BETAMETHASONE
Dermoval → CLOBETASOL, *propionate*
Dermovaleas → BETAMETHASONE
Dermovate → CLOBETASOL, *propionate*
Dermoxin → CLOBETASOL, *propionate*
Dermoxinale → CLOBETASOL, *propionate*
Deronil → DEXAMETHASONE
Derosal → METHYL BENZIMIDAZOLE-2-CARBAMATE
Deroxat → PAROXETINE
DERRIS ROOT
Dervan → SODIUM CHLORATE
Desabam → MEPROBAMATE
Desace → DESLANOSIDE
Desacetyldigilanide C → DESLANOSIDE
Desaci → DESLANOSIDE
Desalkylflurazepam → FLURAZEPAM
Desametasone → DEXAMETHASONE
DESAMINOOXYTOCIN
Desanden → BENZOYL PEROXIDE
Descarboethoxyloratadine → LORATADINE
Deschlorobiomycin → TETRACYCLINE
DESCURAINIA PINNATA
Desdemin → FUROSEMIDE
Deseptyl → SULFANILAMIDE
Deseril → METHYSERGIDE
Desernil → METHYSERGIDE
Deseronil → DEXAMETHASONE
DESERPIDINE
Deserol → BROMODIPHENHYDRAMINE
Desert Baileya → BAILEYA MULTIRADIATA
Desert Date → BALANITES sp.
Desert Rice Flower → PIMELIA sp.
Desert Rose → ADENIUM sp.
Desert Tobacco → NICOTIANA sp.
Desfedrin → METHAMPHETAMINE
Desferal → DEFEROXAMINES
Desferrioxamines → DEFEROXAMINES
DESFLURANE
Des-I-Cate → ENDOTHALL
DESIPRAMINE also → ACETAMINOPHEN, ATROPINE, BETHANIDINE, BUTAPERAZINE, CLONIDINE, DISULFIRAM, EPINEPHRINE, FLUOXETINE, HYDROCORTISONE, METHYLPHENIDATE, OXPHENBUTAZONE, PAROXETINE, PHENOBARBITAL, PHENYLBUTAZONE, PROCARBAZINE, QUINIDINE, RITONAVIR, SERTRALINE, THIORIDAZINE, THYROID, TRANYLCYPROMINE
Desitriptilina → NORTRIPTYLINE
DESLANOSIDE
DESMANTHUS VIRGATUS
11-Desmethoxyreserpine → DESERPIDINE
Desmethyldiazepam → NORDAZEPAM
Desmethylimipramine → DESIPRAMINE
N-Desmethyl Methsuximide → METHSUXIMIDE
Desmethylsertraline → SERTRALINE
DESMOPRESSIN also → CHLOROBUTANOL, IMIPRAMINE
Desmospray → DESMOPRESSIN
DESOGESTREL
DESONIDE
DeSoto → OXYTROPIS
Desowen → DESONIDE
DESOXIMETASONE
Desoxiribon → DEOXYRIBONUCLEIC ACID
Desoxycholic Acid → DEOXYCHOLIC ACID
Desoxycorticosterone → DEOXYCORTICOSTERONE also → PHENOBARBITAL, PHENYLBUTAZONE
Desoxyephedrine → METHAMPHETAMINE
DESOXYLAPACHOL
Desoxylapachol → TEAK
Desoxymethasone → DESOXIMETASONE
Desoxyn → METHAMPHETAMINE
4-Desoxypyridoxine → 4-DEOXYPYRIDOXINE
Desoxyribonucleic Acid → DEOXYRIBONUCLEIC ACID
Despacilina → PENICILLIN > *Benzylpenicillin, Procaine*
Desquaman → ZINC
Dessin → DINOBUTON
Destim → METHAMPHETAMINE
Destriol → ESTRIOL
Destrone → ESTRONE
Destroying Angel → AMANITA sp.
Desuric → BENZBROMARONE
Desyrel → TRAZODONE HCL
DET → DIETHYLTRYPTAMINE

m-DETA → *N,N*-DIETHYL-*m*-TOLUAMIDE
Detal → DINITRO-σ-CRESOL
Detalup → VITAMIN D_2
Detamide → *N,N*-DIETHYL-*m*-TOLUAMIDE
Detanol → BUNAZOSIN
Detantol → BUNAZOSIN
Detensiel → BISOPROLOL
Dethyrona → DEXTROTHYROXINE SODIUM
Deticene → DACARBAZINE
DET MS → DIHYDROERGOTAMINE METHANESULFONATE (MESYLATE)
DETOMIDINE
Detoxepa → TIMONACIC
Detravis → DEMECLOCYCLINE
Detreomycin → CHLORAMPHENICOL
Detrol → TOLTERODINE TARTRATE
Detrovel → ESTIL
Detrusitol → TOLTERODINE TARTRATE
Dettol → CHLOROXYLENOL
Detyroxin → DEXTROTHYROXINE SODIUM
Deumacard → PENTYLENETETRAZOL
Devegan → ACETARSONE
Develin → PROPOXYPHENE
Deverol → SPIRONOLACTONE
Devil's Apple → DATURA, *strammonium* also → PODOPHYLLUM
Devil's Bit → SCABIOSA SUSSISA
Devil's Claw → HARPAGOPHYTUM
Devil's Claw Root → PHYTEUMA
Devil's Club → OPLOPANAX HORRIDUM
Devil's Darning Needle → STIPA sp.
Devil's Dung → ASAFETIDA
Devil's Ear → ARISAEMA sp.
Devil's Eye → HYOSCYAMUS
"Devil Fish" → STINGRAYS
Devil's Milk → CHELIDONIUM MAJUS
Devil's Shoestring → TEPHROSIA
Devil's Trumpet → DATURA, *strammonium*
DEW GRASS
Dewtry → DATURA, *strammonium*
Dexacillin → PENICILLINS – "NEWER and SYNTHETICS" > *Epicillin*
Dexacortal → DEXAMETHASONE
Dexacortin → DEXAMETHASONE
Dexafarma → DEXAMETHASONE
Dexal → KETOPROFEN
Dexalone → DEXTROAMPHETAMINE SULFATE
Dexa-Mamallet → DEXAMETHASONE
Dexambutol → ETHAMBUTOL
Dexameth → DEXAMETHASONE
DEXAMETHASONE also → ANTIPYRINE, CARBAMAZEPINE, EPHEDRINE, PHENOBARBITAL, PHENYTOIN, PRIMIDONE, SODIUM METABISULFITE, TARTRAZINE
Dexameth-A-Vet → DEXAMETHASONE
Dexampex → DEXTROAMPHETAMINE SULFATE
Dexamphetamine → DEXTROAMPHETAMINE SULFATE
Dexamyl → DEXTROAMPHETAMINE
Dexawin → THIAMPHENICOL
DEXBROMPHENIRAMINE
DEXCHLORPHENIRAMINE
Dexchlorpheniramine → CHLORPHENIRAMINE
Dexedrine Sulfate → DEXTROAMPHETAMINE SULFATE
DEXFENFLURAMINE also → FENFLUAMINE, TRIIODOTHYROACETIC ACID
Dexies → METHAMPHETAMINE
Dexin → IRON POLYSACCHARIDE COMPLEX
Dexinoral → DEXAMETHASONE
Dexol → SODIUM PERBORATE
Dexon → POLYGLYCOLIC ACID
DEXPANTHENOL also → DECAMETHONIUM BROMIDE, SUCCINYLCHOLINE
DEXRAZOXANE also → RAZOXANE
Dexten → DEXTROAMPHETAMINE SULFATE
Dextone → DIQUAT DIBROMIDE
DEXTRAN also → HEPARIN
Dextraven → DEXTRAN
DEXTROAMPHETAMINE SULFATE also → CANNABIS, GUANETHIDINE, PROTRIPTYLINE, THIOPENTAL, TOLBUTAMIDE
Dextrofenfluramine → DEXFENFLURAMINE
Dextroid → DEXTROTHYROXINE SODIUM
DEXTROMETHORPHAN HYDROBROMIDE also → AMIODARONE, ISOCARBOXAZID, PAROXETINE, PHENELZINE, QUINIDINE
DEXTROMORAMIDE
Dextrone → PARAQUAT
Dextropropoxyphene → PROPOXYPHENE also → NORTRIPTYLINE
Dextropur → GLUCOSE
Dextrose → GLUCOSE
Dextrosol → GLUCOSE
DEXTROTHYROXINE SODIUM also → BISHYDROXYCOUMARIN, CLOFIBRATE, GLUCOSE, INSULIN, WARFARIN
Dexuron → PARAQUAT
DEZOCINE
DF 118 → DIHYDROCODEINE

dFdC → GEMCITABINE
dFdCyd → GEMCITABINE
DFF → DEXFENFLURAMINE
DFG 29 → FDC YELLOW #6
DFG 64 → TARTRAZINE
DFG 105 → INDIGOCARMINE
DFMO → EFLORNITHINE
DFOM → DEFEROXAMINES
DFP → DIISOPROPYL FLUOROPHOSPHATE
2-DG → 2-DEOXY-D-GLUCOSE
DGE → DIGLYCIDYL ETHER
DGI → DIAZOACETYL GLYCINE HYDRAZIDE
DH 581 → PROBUCOL
DHAD → MITOXANTRONE HCL
DH-codeine → DIHYDROCODEINE
D. H. E. 45 → DIHYDROERGOTAMINE METHANESULFONATE (MESYLATE)
DHEA → PRASTERONE
DHP → 3-HYDROXY-4(1H)PYRIDONE also → LEUCAENA sp.
DHPG → GANCICLOVIR
DHSM → DIHYDROSTREPTOMYCIN
DHT → STANOLONE
5-α-DHT → 5-α-DIHYDROTESTOSTERONE
Dhub → CYNODON, *dactylon*
DHURRIN also → SORGHASTRUM NUTANS, SORGHUM
Dia-basan → GLYBURIDE
Diabechlor → CHLORPROPAMIDE
Diaben → TOLBUTAMIDE
Diabenal → CHLORPROPAMIDE
Dibenamine → *N*-(2-CHLOROETHYL)DIBENZYLAMINE
Diabeta → GLYBURIDE
Diabetol → TOLBUTAMIDE
Diabetoral → CHLORPROPAMIDE
Diabetosan → METFORMIN
Diabewas → TOLAZAMIDE
Diabex → METFORMIN
Diabinese → CHLORPROPAMIDE
Diabitex → CHLORPROPAMIDE
Diabutal → PENTOBARBITAL
Diabrin → BUFORMIN
Diacarb → ACETAZOLAMIDE
Diacepin → DIAZEPAM
DIACETAZOTOL
Diacetic Acid → ACETOACETIC ACID
N, N´-DI(ACETOACETYL)-σ-TOLIDINE
Diacetone Alcohol → STIPA sp.
Diacetotoluide → DIACETAZOTOL
4,5-Diacetoxy-8(methylbutyryloxy)3α-hydroxy-12,13-epoxy-Δ⁹-trichothecan-3-d → T-2 TOXIN
DIACETOXYSCIRPENOL also → TRICHOTHECENE MYCOTOXINS, YELLOW RAIN
Diacetylcholine Diiodide → SUCCINYLCHOLINE
Diacetylcholine Iodide → SUCCINYLCHOLINE
α,β-Diacetylethane → ACETONYL ACETONE
Diacetylmethane → ACETYL ACETONE
Diacetylmorphine → HEROIN
Diacid → CARBROMAL
Diacort → DIFLORASONE
Diacta → SULFAGUANIDINE
Di-Actane → NAFRONYL OXALATE
Diactol → VITAMIN D_2
Diactyl Sulfosuccinate → ACETYLSALICYLIC ACID
Di-Ademil → HYDROFLUMETHIAZIDE
Diadol → ALLOBARBITAL
Di-Adreson → PREDNISONE
Di-Adreson F → PREDNISOLONE
Diadril → MECLIZINE
Diafusor → NITROGLYCERIN
Diaginol → SODIUM ACETRIZOATE
Diagnorenol → METHIODAL SODIUM
Dial → ALLOBARBITAL
Dial-a-gesic → ACETAMINOPHEN
Dialar → DIAZEPAM
DIALIFOR
Dialifos → DIALIFOR
DIALLATE
5,5-Diallylbarbituric Acid → ALLOBARBITAL
Diallybis(nortoxiferine) → ALCURONIUM
N,N-DIALLYL-2-CHLOROACETAMIDE
DIALLYL DIGLYCOL CARBONATE
Diallyldisulfide → GARLIC
Diallyl Ether → ALLYL ETHER
N,N´-Diallylnortoxiferinium → ALCURONIUM
Diallyltoxiferine → ALCURONIUM
Diamba → CANNABIS
Diamarin → DIMENHYDRINATE
Diamicron → GLICLAZIDE
Diamide → HYDRAZINE
Diamine → HYDRAZINE
2,4-DIAMINOANISOLE
2,4-DIAMINOAZOBENZENE
Diaminobenzene → *p*-PHENYLENEDIAMINE
2,4-DIAMINOBUTYRIC ACID
4,4´-DIAMINODIPHENYLAMINE
p,p´-Diaminodiphenylmethane → 4,4´-METHYLENE DIANILINE
1,2-Diaminoethane → ETHYLENEDIAMINE
3,6-Diamino-10-methylacridinium Chloride mixture with 3,6-Acridinediamine → ACRIFLAVINE
2,4-Diamino-5-phenylthiazole → AMIPHENAZOLE

Dibenzheptropine → DEPTROPINE
7H-DIBENZO(c,g)CARBAZOLE
DIBENZODIOXINS
Dibenzofuran → PENTACHLOROPHENOL
DIBENZO(a,h)PYRENE
Dibenzo[b,e]pyridene → ACRIDINE
Dibenzothiazepine → CLOTHIAPINE
Dibenzothiazine → PHENOTHIAZINE
Dibenzyline → PHENOXYBENZAMINE HYDROCHLORIDE
Dibenzyran → PHENOXYBENZAMINE HYDROCHLORIDE
Dibetos → BUFORMIN
DIBK → DIISOBUTYL KETONE
Diblocin → DOXAZOCIN MESYLATE
Dibondrin → DIPHENHYDRAMINE
DIBORANE
Dibotin → PHENFORMIN
DIBROM
Dibromfos → DIBROM
Dibromoacetaldehyde → TRIBROMOETHANOL
DIBROMOCHLOROPROPANE also → WINE
Dibromodulcit → MITOLACTOL
Dibromodulcitol → MITOLACTOL
1,2-Dibromoethane → ETHYLENE DIBROMIDE
sym-Dibromoethane → ETHYLENE DIBROMIDE
Dibromomannitol → MITOBRONITOL
Dibromopropanol → TRIS(2-3-DIBROMOPROPYL)PHOSPHATE
5,7-DIBROMO-8-QUINOLINOL
DIBROMSALAN also → BITHIONOL, DICHLOROPHEN
4′,5-Dibromsalicylanilide → DIBROMSALAN
Dibromsulfoplthalein → PHENYTOIN
DIBUCAINE
Dibutil → ETHOPROPAZINE
N,N-Dibutyl-4-(hexyloxy)-1-naphthamidine → BUNAMIDINE
3,5-Di-t-butylphenyl-N-methylcarbamate
DIBUTYL PHTHALATE
DIBUTYL SQUARATE
DIBUTYL TIN DILAURATE
DIC → DACARBAZINE
Dicain(e) → TETRACAINE HYDROCHLORIDE
Dicalite → DIATOMACEOUS EARTHS
DICAMBA
Dicaptan → DICAPTHON
DICAPTHON
Dicaptol → DIMERCAPROL
Dicarbam → CARBARYL
Dicarboxylic Acid → BUGHIA SAPIDA
DICENTRA CANADENSIS
DICENTRA CUCULLARIA
Dicestal → DICHLOROPHEN(E)
Dicetel → PINAVERIUM BROMIDE
DICHAPETALUM sp.
Dichinalex → CHLOROQUINE
DICHLOFENTHION
DICHLONE
DICHLORALAN
Dichloralphenazone → DICHLORALANTIPYRINE also → WARFARIN
Dichloran → DICHLOROBENZALKONIUM CHLORIDE also → 2,6-DICHLORO-4-NITROANILINE
Dichloren → MECHLORETHAMINE
Dichlorethanoic Acid → DICHLOROACETIC ACID
Di-chloricide → *p*-DICHLOROBENZENE
DICHLORISONE
Dichlorisoprenaline → DICHLORISOPROTERENOL
DICHLORISOPROTERENOL
Dichlorman → DICHLORVOS
DICHLOROACETIC ACID
1,3-DICHLOROACETONE
3,4-DICHLOROANILINE
DICHLOROBENZALKONIUM CHLORIDE
σ-DICHLOROBENZENE
p-DICHLOROBENZENE also → MILK, MOTH BALLS or POWDER
1,2-Dichlorobenzene → σ-DICHLOROBENZENE
3,4-Dichlorobenzeneamine → 3,4-DICHLOROANILINE
3,3′-DICHLOROBENZIDINE
2,4-DICHLOROBENZYL ALCOHOL
1,1-DICHLORO-2,2-BIS(*p*-CHLOROPHENYL) ETHANE
Dichlorocadmium → CADMIUM CHLORIDE
2,4-DICHLORO-6-σ-CHLOROANILINO-s-TRIAZINE
Dichlorodiethyl Sulfide → MUSTARD GAS
DICHLORODIFLUOROMETHANE
3,3′-DICHLORO-5,5′-DINITRO-σ,σ′-DIPHENOL
1,1-DICHLOROETHANE also → ETHYLIDENE CHLORIDE
1,2-Dichloroethane → ETHYLENE DICHLORIDE
sym-Dichloroethane → ETHYLENE DICHLORIDE
1,2-Dichloroethene → ACETYLENE DICHLORIDE

1,2-Dichloroethylene → ACETYLENE DICHLORIDE
sym-DICHLOROETHYL ETHER
Dichloroethyl Sulfide → MUSTARD GAS
DICHLOROFLUOROETHANE also → FREONS
Dichloromethane → METHYLENE CHLORIDE
DICHLOROMETHYLENE DIPHOSPHONATE
sym-Dichloromethyl Ether → BIS(chloromethyl) ETHER
2,6-DICHLORO-4-NITROANILINE
DICHLOROPHEN(E) also → BITHIONOL, COTTON, CUTTING OILS, RUGS, TETRACHLOROSALICYLANILIDE
2,4-DICHLOROPHENOXYACETIC ACID also → AGENT ORANGE, DIOXIN
2,4-DICHLOROPHENOXYACETIC ACID
2,4-Dichlorophenyl-p-nitrophenyl Ether → NITROFEN
1,2-Dichloropropane → PROPYLENE DICHLORIDE
1,3-DICHLOROPROPENE
Dichlorosal → HYDROCHLOROTHIAZIDE
1,2-DICHLORO-1,1,2,2-TETRACHLOROETHANE
Dichlorovos → DICHLORVOS
DICHLORPHENAMIDE
Dichlor-Stapenor → PENICILLINS – "NEWER and SYNTHETICS" > *Dicloxacillin*
DICHLORVOS also → SUCCINYLCHOLINE
Dichlotride → HYDROCHLOROTHIAZIDE
Dichronic → DICLOFENAC
Dichystrolum → DIHYDROTACHYSTEROL
Diclobenin → DICLOFENAC
Diclocil → PENICILLINS – "NEWER and SYNTHETICS" > *Dicloxacillin*
DICLOFENAC also → CHOLESTYRAMINE RESINS, CYCLOSPORINE, FAMOTIDINE, FLUCONAZOLE, LEFLUNOMIDE, LITHIUM, ZAFIRLUKAST
DICLOFOP-METHYL
Diclo-Phlogont → DICLOFENAC
Diclo-Puren → DICLOFENAC
Diclord → DICLOFENAC
Dicloreum → DICLOFENAC
Diclotride → HYDROCHLOROTHIAZIDE
Dicloxacillin → PENICILLINS – "NEWER and SYNTHETICS" also → SULFAMETHOXYPYRIDAZINE
Dicloxin → PENICILLINS – "NEWER and SYNTHETICS" > *Dicloxacillin*
DICOFOL
Diconal → DIPIPANONE
Dicophane → DDT
Dicorantil → DISOPYRAMIDE
Dicoumarol → BISHYDROXYCOUMARIN also → ALLOPURINOL, ALUMINUM HYDROXIDE, MAGNESIUM HYDROXIDE
DICROTOPHOS
DICTAMNUS ALBUS
Dicumacyl → ETHYL BISCOUMACETATE
Dicumarol → BISHYDROXYCOUMARIN
Dicumol → BISHYDROXYCOUMARIN
Dicural → DIFLOXACIN
DICYCLOHEXYL PHTHALATE
DICYCLOMINE
Dicycloverin → DICYCLOMINE
Dicynene → ETHAMSYLATE
Dicynone → ETHAMSYLATE
Didakene → TETRACHLOROETHYLENE
Didandin → DIPHENADIONE
DIDANOSINE also → CIPROFLOXACIN, DAPSONE, DELAVIRDINE, GANCICLOVIR, INDINAVIR, ITRACONAZOLE, KETOCONAZOLE, PENTAMIDINE ISETHIONATE, RANITIDINE
8,9-Didehydro-6,8-dimethylergoline → AGROCLAVINE
4,4´-Didemethyl-4,4´-di-2-propenyltoxiferine → ALCURONIUM
DIDEOXYCYTIDINE
2´,3´-Dideoxycytidine → DIDEOXYCYTIDINE
Dideoxyinosine → DIDANOSINE
2´,3´-Dideoxyinosine → DIDANOSINE
3´,4´-Dideoxykanamycin B → DIBEKACIN
Didoc → ACETAZOLAMIDE
Didrate → DIHYDROCODEINE
Didrex → BENZPHETAMINE
Didrocal → SODIUM ETIDRONATE
Didrocolo → DEHYDROCHOLIC ACID
Didronel → SODIUM ETIDRONATE
Didroxane → DICHLOROPHEN(E)
DIEFFENBACHIA sp. also → CALCIUM OXALATE
DIELDRIN also → DDT, HEPTABARBITAL, PHENOBARBITAL, PROGESTERONE, TESTOSTERONE
Dieltamide → *N,N*-DIETHYL-*m*-TOLUAMIDE
Diemal → BARBITAL
Diemalum → BARBITAL
DIENESTROL
Dienoestrol → DIENESTROL
Dienol → DIENESTROL
Diepin → MEDAZEPAM
Diepoxybutane → ERYTHRITOL ANHYDRIDE
Diergo → DIHYDROERGOTAMINE METHANESULFONATE (MESYLATE)

DIESEL FUEL also → COCAINE
DIETHANOLAMINE
DIETHAZINE
Diethion → ETHION
Diethyl Acetal → ACETAL
2-DIETHYLAMINOETHANOL
2-Diethylaminoethylbenzylate → BENACTYZINE
DIETHYLCARBAMAZINE also → METHYRIDINE, OXIBENDAZOLE
Diethylenediamine → PIPERAZINE
1,4-Diethylene Dioxide → DIOXANE
DIETHYLENE GLYCOL also → HYDRAULIC FLUIDS, SULFANILAMIDE, TOBACCO, WINE
Diethylene Glycol Monoethyl Ether → 2-(2-ETHOXY ETHOXY)ETHANOL
Diethylene Oxide → TETRAHYDROFURAN
Diethylene Oximide → MORPHOLINE
N,N-Diethylethanamine → TRIETHYLAMINE
Diethyl Ether → ETHER, ETHYL
Di(2-ethylhexyl)Phthalate → BIS(2-ETHYLHEXYL)PHTHALATE also → CYCLOSPORINE
1,2-DIETHYLHYDRAZINE
Diethyl Maleate → ETHYL MALEATE
Diethylmalonylurea → BARBITAL
DIETHYLMERCURIC SULFIDE
Diethylnicotinamide → NIKETHAMIDE
DIETHYL-*p*-NITROPHENYLPHOSPHATE
Diethylnitrosamine → *N*-NITROSODIETHYLAMINE
Diethyl Oxide → ETHER, ETHYL
DIETHYLPHTHALATE also → PAPER
DIETHYLPROPION
DIETHYLPYROCARBONATE also → BEER, WINE
DIETHYLSTILBESTEROL also → ALCOHOL, ETHYL; PHENOBARBITAL, SCHRADAN, THIOPENTAL
DIETHYL SULFATE
Diethylthiourea → WET SUITS also → RUBBER
N,N-DIETHYL-*m*-TOLUAMIDE
DIETHYLTRYPTAMINE
Diethyl Xanthogenate → BISETHYLXANTHOGEN
Dietreen → GARDONA
Difaterol → BEZAFIBRATE
DIFENACOUM
Difenidolin → DIPHENIDOL
Difenthos → TEMEPHOS
Differin → ADAPALENE
Difflam → BENZYDAMINE
Diffu-K → POTASSIUM CHLORIDE
Diffumal → THEOPHYLLINE
Diffusin → HYALURONIDASE
Diffusing Factor → HYALURONIDASE
Difil → DIETHYLCARBAMAZINE
DIFLORASONE
DIFLOXACIN
Diflucan → FLUCONAZOLE
Difludol → DIFLUNISAL
DIFLUNISAL also → HYDROCHLOROTHIAZIDE, INDOMETHACIN, OXAZEPAM, SULINDAC
Difluorodichloromethane → DICHLORODIFLUOROMETHANE
Difluoromethylornithine → EFLORNITHINE
DIFLUOROTETRACHLOROETHANE
Diflupyl → DIISOPROPYL FLUOROPHOSPHATE
Diflurex → TICRYNAFEN
Diformyl → GLYOXAL
Difos → TEMEPHOS
Difosfonal → DICHLOROMETHYLENE DIPHOSPHONATE
Difulal → DIFLORASONE
Digacin → DIGOXIN
DIGALLOYL TRIOLEATE
Digenea simplex → KAINIC ACID
Digerent → TRIMEBUTINE MALEATE
Digicor → DIGITOXIN
Digifortis → DIGITALIS
Digilanide C → LANATOSIDE C
Digilong → DIGITOXIN
Digimed → DIGITOXIN
Digimerck → DIGITOXIN
Digipural → DIGITOXIN
Digisidin → DIGITOXIN
DIGITALIN
Digitalin, crystalline → DIGITOXIN
Digitaline → DIGITOXIN
Digitaline Nativelle → DIGITOXIN
Digitalinium German → DIGITALIN
DIGITALIS also → ACETAMINOPHEN, ACETYLCYSTEINE, ALUMINUM, AMPHOTERICIN B, BENZTHIAZIDE, BRETYLIUM TOSYLATE, BUMETANIDE, CALCIUM, CALCIUM GLUCEPTATE, CARBENOXOLONE, CARBONS, CHLOROTHIAZIDE, CHLORTHALIDONE, COMFREY, CYCLOTHIAZIDE, EDROPHOMIUM, EPHEDRINE, ETHACRYNIC ACID, FUROSEMIDE, GUANETHIDINE, HYDROCHLOROTHIAZIDE, KANAMYCIN, LIOTHYRONINE,

LITHIUM, MAGNESIUM, MEPHOBARBITAL, METARAMINOL BITARTRATE, NEOMYCIN, PARATHYROID HORMONE, PHENYTOIN, POTASSIUM, PROPRANOLOL, SCROPHULARIA sp., VERAPAMIL, VITAMIN E
Digitophyllin → DIGITOXIN
Digitora → DIGITALIS
DIGITOXIN also → AMINOGLUTETHIMIDE, CHOLESTYRAMINE RESINS, FLUOXETINE, HYDROCHLOROTHIAZIDE, MEPHOBARBITAL, PHENOBARBITAL, PHENYLBUTAZONE, PHENYTOIN, QUINIDINE, RAUWOLFIA SERPENTINA, RESERPINE, RIFAMPIN, SPIRONOLACTONE, TOLBUTAMIDE, VERAPAMIL
DIGLYCIDYL ETHER
Diglycol → DIETHYLENE GLYCOL
Dignonitrat → ISOSORBIDE DINITRATE
Dignover → VERAPAMIL
DIGOXIN also → ALCOHOL, ETHYL; ALPRAZOLAM, ALUMINUM HYDROXIDE, *p*-AMINOSALICYLIC ACID, AMIODARONE, ATORVASTATIN, CAPTOPRIL, CARBIMAZOLE, CARVEDILOL, CEFTAZIDIME, CHLORTHALIDONE, CHOLESTYRAMINE RESIN, CLARITHROMYCIN, CLAVULANIC ACID, CLINDOMYCIN, COLESTIPOL, CYCLOSPORINE, DIAZEPAM, DICLOFENAC, DILTIAZEM, ERYTHROMYCIN, ESMOLOL, FLECAINIDE, FUROSEMIDE, GALLOPAMIL, GRAPEFRUIT, HYDROCHLOROTHIAZIDE, HYDROXYCHLOROQUINE, HYPERICUM, IBUPROFEN, IMIPRAMINE, INDOMETHACIN, ITRACONAZOLE, KAOLIN, LANSOPRAZOLE, LEVALBUTEROL, MAGNESIUM CARBONATE, MAGNESIUM PEROXIDE, MAGNESIUM TRISILICATE, METHYLDOPA, METOCLOPRAMIDE, NEOMYCIN, NICARDIPINE, NIFEDIPINE, NIMULSIDE, OLEANDER, OMEPRAZOLE, PAROXETINE, PECTIN, PENICILLAMINE; PENICILLINS – "NEWER and SYNTHETICS" > *Ticarcillin Disodium*; PHENOBARBITAL, PHENYTOIN, PLANTAGO SEEDS, PREDNISONE, PROPAFENONE, PROPANTHELINE, PROPRANOLOL, QUINIDINE, QUININE, RAUWOLFIA SERPENTINA, RESERPINE, SALICYLAZOSULFAPYRIDINE, SPIRONOLACTONE, SUCRALFATE, TELMISARTAN,TETRACYCLINE, THEOPHYLLINE, THYROID, TIAPAMIL, TICLODIPINE, TRAMADOL, TRAZODONE HCL, TRIAMTERENE, VERAPAMIL
Diguanyl → CHLOROGUANIDE
Dihydantoin → PHENYTOIN
Dihydergot → DIHYDROERGOTAMINE METHANESULFONATE (MESYLATE)
Dihydral → DIPHENHYDRAMINE
DIHYDRALAZINE
Dihydrex → BENZTHIAZIDE
Dihydrin → DIHYDROCODEINE
DIHYDROABEITYL ALCOHOL also → MASCARA
DIHYDROANETHOLE
Dihydrobenzocycloheptadiene → AMITRIPTYLINE
DIHYDROCODEINE also → MEPHENOXALONE
Dihydrocodeineone Bitartartrate → HYDROCODONE BITARTRATE
3,7- -Dihydro-1,3-dimethyl-1H-purine-2,6-dione compound with 1,2-ethanediamine (2 : 1) → AMINOPHYLLINE
Dihydrodiethylstilbestrol → HEXESTROL
Dihydroergocornine → ERGOLOID MESYLATES
Dihydroergocristine → ERGOLOID MESYLATES
Dihydroergocryptine → ERGOILOID MESYLATES
DIHYDROERGOTAMINE METHANESULFONATE (MESYLATE) also → HEPARIN
Dihydroethaverine → DROTAVERINE
Dihydroflumethiazide → HYDROFLUMETHIAIZIDE
Dihydrofollicular Hormone → ESTRADIOL
Dihydrofolliculin → ESTRADIOL
Dihydro-2(3H)-furanone → BUTYROLACTONE
Dihydromorphinone → HYDROMORPHONE
Dihydrone → OXYCODONE
Dihydroneopine → DIHYDROCODEINE
1,5-Dihydro-4H-pyrazolo[3,4-d]-pyrimidin-4-one → ALLOPURINOL
Dihydroquinidine → HYDROQUINIDINE
DIHYDROSAFROLE
DIHYDROSTREPTOMYCIN also → KANAMYCIN, NEOMYCIN; PENICILLIN also → PENICILLIN, *Benzylpenicillin, Procaine*; SUCCINYLCHOLINE
DIHYDROTACHYSTEROL
DIHYDROTELEOCIDIN B

DIHYDRO-α-TERPINEOL
DIHYDROTERPINYL ACETATE
4-Dihydrotestosterone → STANOLONE
5-α-DIHYDROTESTOSTERONE
Dihydrotheelin → ESTRADIOL
1,2-DIHYDRO-2,2,4-TRIMETHYLQUINOLINE
DIHYDROXYACETONE
1,8-Dihydroxyanthraquinone → PHENOLSULFONPHTHALEIN
1,2-Dihydroxybenzene → CATECHOL
p-Dihydroxybenzene → HYDROQUINONE
1,25-Dihydroxycholecalciferol → SOLANUM sp.
(11β)-11,21-Dihydroxy-3,20-dioxopregn-4-en-18-al → ALDOSTERONE
1,3-Dihydroxydimethyl Ketone → DIHYDROXYACETONE
Dihydroxyestrin → ESTRADIOL
1,3-Dihydroxy-2-propanone → DIHYDROXYACETONE
l-α-Dihydroxyvitamin D_3 → VITAMIN D_3
Dihyprylone → PIPERIDIONE
Dihyzin → DIHYDRALAZINE
Diiodohydroxyquin → IODOQUINOL
Diiodohydroxyquinoline → IODOQUINOL
DIISOBUTYL KETONE
DIISOBUTYL PHTHALATE
Diisocarb → 5-ETHYL DIISOBUTYLTHIOCARBAMATE
2,4-DIISOCYANATE
1,6-Diisocyanatohexane → HEXAMETHYLENE DIISOCYANATE
DIISOPROPANOLNITROSAMINE
DIISOPROPYLAMINE
Diisopropyl Ether → ISOPROPYL ETHER
DIISOPROPYL FLUOROPHOSPHATE also → SUCCINYLCHOLINE
Diisopropyl Oxide → ISOPROPYL ETHER
Dikantol → DIMERCAPROL
3,20-Diketo-11β,18-oxido-4-pregnene-18,21-diol → ALDOSTERONE
Dikoor → PANICUM sp.
Dikotex → METHYL CHLOROPHENOXY ACETIC ACID
Dilabar → CAPTOPRIL
Dilabron → ISOETHARINE
Dilafurane → BENZIODARONE
Dilacor → DILTIAZEM
Diladel → DILTIAZEM
Di-Lan → PHENYTOIN
Dilanacin → DIGOXIN
Dilanca → PENTAERYTHRITOL TETRANITRATE
Dilar → PARAMETHASONE ACETATE
Dilatal → NYLIDRIN
Dilatol → NYLIDRIN
Dilatrate → ISOSORBIDE DINITRATE
Dilatrend → CARVEDILOL
Dilatroppon → NYLIDRIN
Dilaudid → HYDROMORPHONE
DILAURYL SUCCINATE
Dila-Vasal → BENZIODARONE
Dilavase → ISOXSUPRINE
Dilcit → INOSITOL NIACINATE
Dilcoran-80 → PENTAERYTHRITOL TETRANITRATE
Dilexpal → INOSITOL NIACINATE
DILL also → *d*-LIMONENE
Dillar → PARAMETHASONE ACETATE
Diloderm → DICHLORISONE
Dilor → DYPHYLLINE
Dilospan S → PHLOROGLUCINOL
Dilosyn → METHDILAZINE
DILOXANIDE FUROATE also → BUTETHAL
Diloxanide 2-Furoic Acid Ester → DILOXANIDE FUROATE
Dilpral → DILTIAZEM
Dilrene → DILTIAZEM
Dilta → DILL
DILTIAZEM also → CARBAMAZEPINE, CILOSTAZOL, CISAPRIDE, CYCLOSPORINE, DIGOXIN, DIPYRIDAMOLE, GRAPEFRUIT, MIDAZOLAM, NIFEDIPINE, PENBUTOLOL SULFATE, PHENYTOIN, PROPRANOLOL, SOTALOL, TACROLIMUS, THEOPHYLLINE, TRIAZOLAM
Diluran → ACETAZOLAMIDE
Dilurgen → MERALLURIDE
Dilydrin → NYLIDRIN
Dilzem → DILTIAZEM
Dilzene → DILTIAZEM
Dimacef → CEPHADRINE
Dimantine → DYMANTHINE
Dimapyrin → AMINOPYRINE
Dimate → DIMENHYDRINATE
Dimaval → DIMERCAPTOPROPANE SULPHONATE
Dimazine → 1,1-DIMETHYLHYDRAZINE
Dimazon → DIACETAZOTOL
Dimecron → PHOSPHAMIDON
Dimedrol → DIPHENHYDRAMINE
DIMEFLINE
DIMEFOX
Dimelin → ACETOHEXAMIDE
Dimelor → ACETOHEXAMIDE

Dimenformon → ESTRADIOL
DIMENHYDRINATE also → 8-CHLOROTHEOPHYLLINE, STREPTOMYCIN
Dimeray → IOCARMATE MEGLUMINE
DIMERCAPROL
DIMERCAPTOPROPANE SULPHONATE
Dimerin → METHYPRYLONE
Dimer X → IOCARMATE MEGLUMINE
Dimetate → DIMETHOATE
DIMETHADIONE
DIMETHICONE also → DICYCLOMINE, MEGESTROL ACETATE, WARFARIN
DIMETHINDENE
DIMETHISTERONE
DIMETHOATE also → TRI-σ-TOLYL PHOSPHATE
Dimethogen → DIMETHOATE
Dimethone → DIPYRONE
Dimethothiazide Mesylate → FONAZINE MESYLATE
DIMETHOXANE
3,3´-Dimethoxybenzidine → σ-DIANISIDINE
Dimethoxy-DT → METHOXYCHLOR
Dimethoxymethane → METHYLAL
2,5-DIMETHOXY-4-ETHYLAMPHETAMINE
DIMETHOXYETHYLPHTHALATE
DIMETHOXYMETHYLAMPHETAMINE also → MESCALINE
2,3-Dimethoxystrychnidin-10-one → BRUCINE
10,11-Dimethoxystrychnine → BRUCINE
Dimethyl → ETHANE
N,N-DIMETHYLACETAMIDE
N,N-DIMETHYLAMINE also → THIRAM
Dimethylamino-analgesine → AMINOPYRINE
4-(Dimethylamino)-antipyrine → AMINOPYRINE
p-DIMETHYLAMINOBENZENE
p-Dimethylaminobenzoate esters → PADIMATES
2´,3-DIMETHYL-4-AMINOBIPHENYL
2-DIMETHYLAMINO-3´,4-DIHYDROXYACETOPHENONE HCL
6-(Dimethylamino)4,4-diphenyl-3-heptanol Acetate → ACETYLMETHADOL
Dimethylaminoethanol → DEANOL
N,N-DIMETHYLAMINOETHOXYETHANOL
Dimethylaminophenazone → AMINOPYRINE
4-DIMETHYLAMINOPHENOL
DIMETHYLAMINOPROPIONITRILE
Dimethylaminopropionitrile → NIAX® CATALYST ESN
3-DIMETHYLAMINOPROPYLAMINE
β-[2-(Dimethylamino)propyl]-α-ethyl-β-phenylbenzene-ethanol Acetate → ACETYLMETHADOL
DIMETHYLAMINOPROPYLMETHA-CRYLAMIDE
1-[10-[3-(Dimethylamino)propyl]-10H-phenothiazin-2yl]ethanone → ACEPROMAZINE
10-(3-Dimethylaminopropyl)phenothiazine-3-ethylone → ACEPROMAZINE
10-[3-(Dimethylamino)-propyl]phenothiazin-2yl Methyl Ketone → ACEPROMAZINE
4-DIMETHYLAMINOSTILBENE
4-(*p*-DIMETHYLAMINOSTYRYL) QUINOLINE
4-Dimethylamino-3-toyl-N-methylcarbamate → AMINOCARB
Dimethylane → PROMOXALONE
N,N-DIMETHYLANILINE
DIMETHYL ANTHRANILATE
Dimethylarsinic Acid → CACODYLIC ACID
9,10-DIMETHYL-1,2-BENZANTHRACENE
7,12-dimethylbenz[a]anthracene → 9,10-DIMETHYL-1,2-BENZANTHRACENE
Dimethylbenzene → XYLENE
3,3´-Dimethylbenzidine → σ-TOLIDINE
DIMETHYLBENZYL CARBINOL
DIMETHYLBENZYL CARBINYL ACETATE
DIMETHYLCARBAMOYL CHLORIDE
Dimethyl Carbinol → ALCOHOL, ISOPROPYL
β,β-Dimethylcysteine → PENICILLAMINE
Dimethyl Diacetyl Gluconic Acid → VITAMIN B_{15}
Dimethyl Dichlorobenzyl Ammonium Chloride → DICHLOROBENZALKONIUM CHLORIDE
trans-4,4´-DIMETHYL-α-α´-DIETHYLSTILBENE
DIMETHYL DISULFIDE
DIMETHYLDODECLAMINE OXIDE
7,10-DIMETHYLELLIPTICINE
Dimethylene Glycol → Preface
Dimethylene Oxide → ETHYLENE OXIDE
Dimethylenimine → ETHYLENEIMINE
Dimethyl Ethyl Carbinol → AMYLENE HYDRATE
DIMETHYLFORMAMIDE
Dimethyl Fumarate → MALATHION
DIMETHYLHEPTENAL
1,1-DIMETHYLHYDRAZINE also → HYDRAZINE, MISSILE SILO
1,2-DIMETHYLHYDRAZINE
Dimethyl Ketol → ACETOIN
Dimethyl Ketone → ACETONE
DIMETHYL MERCURY also → MERCURY
Dimethylmethane → PROPANE
DIMETHYL METHYL PHOSPHONATE
2,2-Dimethyl-3-(2-methyl-1-propenyl) cyclopropanecarboxylic acid 2-methyl-4-

oxo-3- (2-propenyl)-2-cyclopenten -1-yl Ester → ALLETHRIN
Dimethylnitromethane → 2-NITROPROPANE
σ,σ-Dimethyl-σ-(4-nitrophenyl)phosphorothioate → METHYL PARATHION
Dimethylnitrosamine → *N*-NITROSODIMETHYLAMINE also → WHISKEY
Dimethyloldihydroxyethyleneurea → PERMANENT PRESS
DIMETHYLOL DIMETHYL HYDANTOIN
Dimethylolmethoxydimethylpropylene Urea → PERMANENT PRESS
Dimethyloxychinizin → ANTIPYRINE
Dimethyloxyquinazine → ANTIPYRINE
N-Dimethyl-paraphenylenediamine → *p*-PHENYLENEDIAMINE
Dimethyl Parathion → METHYL PARATHION
N-Dimethyl-p-phenylenediamine → *p*-PHENYLENEDIAMINE
DIMETHYL PHOSPHATE
DIMETHYL PHTHALATE also → PAPER
Dimethylpolysiloxane → DIMETHICONE
DIMETHYLSAFRANINE also → LIPSTICK, SAFRANINES
DIMETHYL-*p*-STYRYLANILINE
DIMETHYL SULFATE
Dimethyl Sulfide → DIMETHYL SULFOXIDE, METHIONINES
DIMETHYL SULFOXIDE also → KALE, RUBBER, SALICYLIC ACID, SULINDAC, THIABENDAZOLE
Dimethyl Telluride → TELLURIUM
DIMETHYL TEREPHTHALATE
7β,17α-Dimethyltestosterone → CALUSTERONE
N,N-DIMETHYLTRYPTAMINE also → ANADENATHERA, CANNABIS, COHOBA, JUSTICA PECTORALIS var STENOPHYLLIA, MIMOSA, PHALARIS sp., PRESTONIA AMAZONICA, TETRAPTERIS sp., TRYPTAMINE, YUREMA
DIMETHYL TUBOCURARINE IODIDE
DIMETHYLVINYL CHLORIDE
3,7-Dimethylxanthine → THEOBROMINE
Dimeticone → DIMETHICONE *(with silicon dioxide)*
DIMETILAN also → PRALIDOXIME
Dimetindene → DIMETHINDENE
Dimitone → CARVEDILOL
Dimitron → CINNARIZINE
DIMETRIDAZOLE
Dimidin R → SULFAMETHAZINE
DIMINAZENE ACETURATE
Dimocillin → PENICILLINS – "NEWER and SYNTHETICS" > *Methicillin Sodium*
Dimorlin → DEXTROMORAMIDE
Dimorphone → HYDROMORPHONE
Dimpylate → DIAZINON
Dinacrin → ISONIAZID
Dinaplex → FLUNARIZINE HCL
Dinarkon → OXYCODONE
Dinase → PANCREATIC DORNASE
Dindevan → PHENINDIONE
Dineval → PHENINDIONE
Dinezin → DIETHAZINE
Diniket → ISOSORBIDE DINITRATE
m-DINITROBENZENE
p-DINITROBENZENE
2,4-DINITROCHLORBENZENE also → CHLORAMPHENICOL
4,6-Dinitro-2-sec-butylphenol → DINOSEB
DINITRO-σ-CRESOL
Dinitroisopropylphenol → BINAPACRYL
Dinitrol → DINITRO-σ-CRESOL
2,4-DINITRO-3-METHYL-*tert*-BUTYLANISOLE
2,6-DINITRONAPHTHOL
2,4-DINITROPHENOL
α-Dinitrophenol → 2,4-DINITROPHENOL
DINITROTOLUENE also → KEROSENE, MUNITIONS
Dinitrotoluol → DINITROTOLUENE
DINOBUTON
DINOCAP
Dinoleine → IODOQUINOL
Dinolytic → DINOPROST
DINOPROST
Dinoprost Methyl → CARBOPROST
DINOPROSTONE also → GRACILLARIA
Dinormon → DEXAMETHASONE
DINOSEB
Dinsidon → OPIPRAMOL
Diocodal → NAPROXEN
Dioctyl → DOCUSATE SODIUM
Dioctylal → DOCUSATE SODIUM
Dioctyl Phthalate → BIS(2-ETHYLHEXYL)PHTHALATE also → POLYVINYL CHLORIDE
Dioctyl Sodium Sulfosuccinate → DOCUSATE SODIUM
Diocurb → DEXTROAMPHETAMINE SULFATE
Diocyclin → TETRACYCLINE
Diocyl → DICYCLOMINE
Dioderm → HYDROCORTISONE
Diodone → IODOPYRACET
Diodoquin → IODOQUINOLINE

Diodoxylin → IODOQUINOL
Diodrast → IODOPYRACET
Dioform → DIETHYLCARBAMAZINE
Dioform → ACETYLENE DICHLORIDE
Diogyn → ESTRADIOL
Diogyn E → ETHINYL ESTRADIOL
Diolamine → DIETHANOLAMINE
Diolandrone → METHANDRIOL
Diolene → IODOQUINOL
Dioloxol → MEPHENESIN
Dionin → ETHYLMORPHINE HYDROCHLORIDE
Dionosil → PROPYLIODONE
Diophindane → PHENINDIONE
Diophyllin → AMINOPHYLLINE
Diopine → DIPIVEFRIN
Dioprin → DIPYRONE
DIOSCOREA also → SORGHUM
Diosmol → MANNITOL
DIOSPYRUS sp. also → PERSIMMON
Diotilan → DOCUSATE SODIUM
Diovan → VALSARTAN
DIOXANE also → *sym*-DICHLOROETHYL ETHER
DIOXATHION
DIOXIN also → AGENT ORANGE, CHLOROPHENOL, FISH, PAPER, PENTACHLOROPHENOL, SILVEX
DIOXYBENZONE
4,4´-Dioxo-β-carotene → CANTHAXANTHIN
DIPA → DIISOPROPYLAMINE
Dipam → DIAZEPAM
Dipar → PHENFORMIN
Di-Paralene → CHLORCYCLIZINE
Diparcol → DIETHAZINE
Dipav → PAPAVERINE
Dipaxin → DIPHENADIONE
DIPE → ISOPROPYL ETHER
Dipegyl → VITAMIN B_3
DIPENTAMETHYLENE THIURAM DISULFIDE
Dipentol → THIOPENTAL
Diphacin → DIPHENADIONE
Diphacinone → DIPHENADIONE
Diphantine → DIPHENHYDRAMINE
Diphebuzol → PHENYLBUTAZONE
DIPHEMANIL METHYLSULFATE
Diphemin → DIPIVEFRIN
Di-Phen → PHENYTOIN
DIPHENADIONE also → ACETAMINOPHEN, BENZIODARONE, CLOFIBRATE
DIPHENAMID
Diphenasone → DAPSONE
Diphenatil → DIPHEMANIL METHYLSULFATE
DIPHENAZINE
DIPHENCYPRONE
DIPHENHYDRAMINE also → ALCOHOL, ETHYL; *p*-AMINOSALICYLIC ACID, CALAMINE, DDT, GRISEOFULVIN, HEPARIN, HYDROCORTISONE, IMIPRAMINE, NORTRIPTYLINE, PHENYLBUTAZONE, TEMAZEPAM, TRIHEXYPHENIDYL
Diphenicillin Sodium → PENICILLINS – "NEWER and SYNTHETICS"
DIPHENIDOL
DIPHENOXYLATE also → ATROPINE
Di-phenthane-70 → DICHLOROPHEN(E)
DIPHENYL also → PAPER
2-Diphenylacetyl-1,3-indandione → DIPHENADIONE
DIPHENYLAMINE
p-Diphenylamine → 4-AMINODIOHENYL
Diphenylaminochlorarsine → PHENARSAZINE CHLORIDE
Diphenyldiazene → AZOBENZENE
Diphenyldiimide → AZOBENZENE
Diphenylenemethyl Bromide → 9-BROMOFLUORENE
Diphenyl Ether → PHENYL ETHER
Diphenylhydantoin → PHENYTOIN
DIPHENYLMETHANE
p,p-DIPHENYLMETHANE DIISOCYANATE
Diphenyl Oxide → PHENYL ETHER
N,N´-DIPHENYL-*p*-PHENYLENEDIAMINE
DIPHENYLPYRALINE also → ETHYLENEDIAMINE
Diphenylthiourea → POLYVINYL CHLORIDE
Dipher → ZINEB
Diphesatin → OXYPHENISATIN
Diphos → SODIUM ETIDRONATE
DIPIPANONE
Dipiperon → PIPAMPERONE
Dipirin → AMINOPYRINE
DIPIVEFRIN
Dipivoxil → ADEFOVIR
DIPLODIA sp.
DIPLOPTERYS CABRERANA
Dipofene → DIAZINON
Dipophene → DIAZINON
Diprazin → PROMETHAZINE
DIPRENORPHINE
Diprivan → PROPOFOL
Diproderm → BETAMETHASONE
Diprolene → BETAMETHASONE
Diprophos → BETAMETHASONE

Diprophylline → DYPHYLLINE
Di-n-propylacetate → SODIUM VALPROATE
DIPROPYLENE GLYCOL MONOMETHYL ETHER also → SELAMECTIN
Dipropyl Ether → PROPYL ETHER
Diprosis → BETAMETHASONE
Diprosone → BETAMETHASONE
Dipsan → CALCIUM CARBIMIDE
Dipterex → TRICHLORFON
DIPTERYX
Diptetes → TRICHLORFON
Dipteryxodorata → TONKA BEANS
DIPYRIDAMOLE also → ADENOSINE, CAFFEINE, THEOPHYLLINE
Dipyridan → DIPYRIDAMOLE
α,α'-DIPYRIDYL also → PARAQUAT
Dipyrine → AMINOPYRINE
Dipyrithione → PYRITHIONE
DIPYRONE also → AMINOPYRINE, DILTIAZEM, ETHYL BISCOUMACETATE, GLUTETHIMIDE, METHOTRIMEPRAZINE, PHENOBARBITAL
DIQUAT DIBROMIDE
Diquel → ETHYL ISOBUTRAZINE
1,3-Di-6-quinolylurea → QUINURONIUM
Dira → SPIRONOLACTONE
Diralgan → FLOCTAFENINE
Dirame → PROPIRAM FUMARATE
DIRCA PALUSTRIS
DIRECT BLACK 38
DIRECT BLUE 6
DIRECT BROWN 95
Direma → HYDROCHLOROTHIAZIDE
Direx → ANILAZINE
Direxiode → IODOQUINOL
Dirgotarl → DIHYDROERGOTAMINE METHANESULFONATE (MESYLATE)
Dirimal → ORYZALIN
DIRITHROMYCIN also → TERFENADINE
Dirnach → HYPERICUM
Dirocide → DIETHYLCARBAMAZINE
Dirox → ACETAMINOPHEN
Dirythmin SA → DISOPYRAMIDE
Disadine DP → POLYVINYLPYRROLIDONE IODINE COMPLEX
Disal → FUROSEMIDE
Disalcid → SALSALATE
Disalgesic → SALSALATE
Disalicylic Acid → SALSALATE
Disalunil → HYDROCHLOROTHIAZIDE
Disarim → CHLORDIAZEPOXIDE
Discase → CHYMOPAPAIN
Discoid → FUROSEMIDE
Discorides → YEW
Discorine → DIOSCOREA
Discotrine → NITROGLYCERIN
Disipal → ORPHENADRINE
Disocarban → THIOCARLIDE
Disoderm → DICHLORISONE
Disodium Arsenate → SODIUM ARSENATE
Disodium Cromoglycate → CROMOLYN SODIUM
DISODIUM EDETATE also → CALCIUM GLUCONATE
Disodium Ethylene Bisdithiocarbamate → NABAM
Disodium Hydrogen Phosphate → SODIUM PHOSPHATE, DIBASIC
DISODIUM METHANEARSONATE
DISODIUM METHYLARSONATE
Disodium Monomethanearsonate → DISODIUM METHANEARSONATE
Diso-Duriles → DISOPYRAMIDE
Disogenin → BARBASCO
Disonate → DOCUSATE SODIUM
DISOPHENOL
Disophrol → PSEUDOEPHEDRINE
Disoprofol → PROPOFOL
DISOPYRAMIDE also → AMIODARONE, ATENOLOL, CLOZAPINE, ERYTHROMYCIN, ISOSORBIDE DINITRATE, PHENYTOIN, PRACTOLOL, RIFAMPIN, RITONAVIR, VERAPAMIL
Disoquin → IODOQUINOL
Disorat → TRIMEPRANOL
Disorlon → ISOSORBIDE DINITRATE
Dispadol → MEPERIDINE
Dispermin → PIPERAZINE
Disphex → POLYVINYLPYRROLIDONE IODINE COMPLEX
Di-Spidin → PITUITARY, POSTERIOR
Disposaject → DOCUSATE SODIUM
Dispranol → PENTAERYTHRITOL DICHLOROHYDRIN
Dispril → CALCIUM ACETYLSALICYLATE
Disprin → CALCIUM ACETYLSALICYLATE
Disprol → ACETAMINOPHEN
Dispromil → FAMOTIDINE
Dissenten → LOPERAMIDE
Distaclor → CEFACLOR
Distalgesic® → PROPOXYPHENE
Distaquaine → PENICILLIN > *Benzylpenicillin, Procaine*
Distaquaine V → PENICILLIN > *Penicillin V*
Distaquine V → PENICILLIN > *Penicillin V*
Distaval → THALIDOMIDE
Distaxid → NIZATIDINE
Distigmine Bromide → HEXAMARIUM BROMIDE

Distillate → RUBBER
Distocide → PRAZIQUANTEL
Distolon → 3,3´-DICHLORO-5,5´-DINITRO-σ,σ´-DIPHENOL
Distraneurin → CLOMETHIAZOLES
DISULFIRAM also → ACETALDEHYDE; ALCOHOL, ETHYL; ANTIPYRINE, BARBITAL, BEER, BENZYL BENZOATE, CAFFEINE, CALCIUM CARBIMIDE, CARBON DIOXIDE, CEFAMANDOLE, CEFMETAZOLE, CEFOPERAZONE, CEFOTETAN, CHLORAL HYDRATE, CHLORAMPHENICOL, CHLORDIAZEPOXIDE, COLOGNES, DIAZEPAM, DOPAMINE, ETHOTOIN, ETHYLENE DIBROMIDE, ETHYLENE GLYCOL, HAIR, HEXOBARBITAL, ISONIAZID, MEPHENYTOIN, METRONIDAZOLE, OMEPRAZOLE, PERPHENAZINE, PHENYTOIN, QUINACRINE, RUBBER, TEMAZEPAM, THIRAM, WARFARIN
DISULFOTON also → COTTONSEED MEAL
Disulone → DAPSONE
Disyncran → METHDILAZINE
Disynformon → ESTRONE
Di-Syntramine → DICYCLOMINE
Ditanest → PRILOCAINE
Ditch-bank Fescue → FESCUE, TALL
Dithane → NABAM
Dithane Z-78 → ZINEB
Dithiocarbonic Anhydride → CARBON DISULFIDE
1,2-Dithioglycerol → DIMERCAPROL
Dithiolane → PHOSFOLAN
Dithione → SULFOTEP
Dithiophos → SULFOTEP
Dithiopropylthiamine → VITAMIN B_1
Ditical → TRIPARANOL
Ditilin → SUCCINYLCHOLINE
Ditran → PIPERIDYL BENZILATE
Ditranil → 2,6-DICHLORO-4-NITROANILINE
Ditrazin → DIETHYLCARBAMAZINE
Ditrifon → TRICHLORFON
Di-Trim → TRIMETHOPRIM
Ditropan → OXYBUTYNIN CHLORIDE
Ditrosol → DINITRO-σ-CRESOL
Dittany → DICTAMNUS ALBUS
Ditubin → ISONIAZID
Diucardin → HYDROFLUMETHIAZIDE
Diucardyn Sodium → MERCAPTOMERIN SODIUM
Diucen → BENZTHIAZIDE
Diulo → METOLAZONE
Diu-melusin → HYDROCHLOROTHIAZIDE
Diural → FUROSEMIDE
Diurazine → CHLORAZANIL
Diuredosan → UREDOFOS
Diuresal → CHLOROTHIAZIDE
Diurese → TRICHLORMETHIAZIDE
Diureticum-Holzinger → ACETAZOLAMIDE
Diurexan → XIPAMIDE
Diuril → CHLOROTHIAZIDE
Diurilix → CHLOROTHIAZIDE
Diurite → CHLORTHIAZIDE
Diuriwas → ACETAZOLAMIDE
Diutazol → ACETAZOLAMIDE
Diutensin → METHYCLOTHIAZIDE
Divalproex Sodium → VALPROIC ACID also → SODIUM VALPROATE, NORTRIPTYLINE
Divarine → DIPYRONE
Divicine → BEANS
Divinyl → 1,3-BUTADIENE
Divinyl Ether → ETHER, VINYL
Divinyl Oxide → ETHER, VINYL
Divipan → DICHLORVOS
Divit Urto → VITAMIN D_2
Dixanthogen → BISETHYLXANTHOGEN
Dixarit → CLONIDINE
Dixeran → MELITRACEN HCL
Dixiben → NALIDIXIC ACID
Dixina → DIGOXIN
Dixnalate → GEFARNATE
Dizac → DIAZEPAM
Dizol → AMIPHENAZOLE
DJ7041 → MOROCTASIN
DKB → DIBEKACIN
DL 152 → BIETASERPINE
DL 717 IT → LOTRIFEN
DL 8280 → OFLOXACIN
DLP 787 → PYRIMINIL
DM → PHENARSAZINE CHLORIDE
DMA → *N,N*-DIMETHYLACETAMIDE
DMAA → MEMANTINE
DMAC → *N,N*-DIMETHYLACETAMIDE
DMAE → DEANOL
DMAP → 4-DIMETHYLAMINOPHENOL
3-DMAPA → 3-DIMETHYLAMINOPROPYLAMINE
DMAPMA → DIMETHYLAMINOPROPYLMETHACRYLAMIDE
DMAPN → DIMETHYLAMINOPROPIONITRILE
DMBA → 9,10-DIMETHYL-1,2-BENZANTHRACENE

DMCC → DIMETHYLCARBAMOYL CHLORIDE
DMDM Hydantoin → DIMETHYLOL DIMETHYL HYDANTOIN
DMDP → DICHLOROMETHYLENE DIPHOSPHONATE
DMDR → IDARUBICIN
DMDT → METHOXYCHLOR
DMDZ → NORDAZEPAM
DMEP → DIMETHOXYETHYLPHTHALATE
DMES → *trans-4,4′-DIMETHYL-α-α*-DIETHYLSTILBENE
DMF → DIMETHYLFORMAMIDE
DMFO → DIMETHYLFORMAMIDE
DMGG → METFORMIN
DMH → 1,1-DIMETHYLHYDRAZINE
DMMP → DIMETHYL METHYL PHOSPHONATE
DMN → *N*-NITROSODIMETHYLAMINE
DMNA → *N*-NITROSODIMETHYLAMINE
DMO → DIMETHADIONE
DMOB → σ-DIANISIDINE
DMP → DIMETHYL PHTHALATE
DMP 266 → EFAVIRENZ
DMPA → MEDROXYPROGESTERONE ACETATE
DMPS → DIMERCAPTOPROPANE SULPHONATE
DMS → DIMETHYL SULFATE
DMSO → DIMETHYL SULFOXIDE
DMT → DIMETHYL TEREPHTHALATE also → *N,N*-DIMETHYLTRYPTAMINE
DMTT → DAZOMET
DMZ → DIMETRIDAZOLE
DN → DINITRO-σ-CRESOL
DN-289 → DINOSEB
DNA → DEOXYRIBONUCLEIC ACID
DNase I → PANCREATIC DORNASE
DNBP → DINOSEB
DNC → DINITRO-σ-CRESOL
DNCB → 2,4-DINITROCHLOROBENZENE
DNOC → DINITRO-σ-CRESOL
DNOCP → DINOCAP
DNOP → BIS(CHLOROMETHYL)ETHER also → BIS(2-ETHYLHEXYL) PHTHALATE
DNOSBP → DINOSEB
DNP → *2,4*-DINITROPHENOL
DNT → DINITROTOLUENE
DNTP → PARATHION
DO-14 → PROPARGITE
D.O.A. → PHENCYCLIDINE
Dobesin → DIETHYLPROPION
Dobren → SULPIRIDE
Doburil → CYCLOTHIAZIDE
DOBUTAMINE also → DOPAMINE, PROPRANOLOL
DOCA → DEOXYCORTICOSTERONE
DOCETAXEL also → TAXOTERE
Docémine → VITAMIN B_{12}
Docibin → VITAMIN B_{12}
Docigram → VITAMIN B_{12}
Dociton → PROPRANOLOL
Docivit → VITAMIN B_{12}
Docks → RUMEX sp.
Doclizid-T → CHLORAZANIL
DOCP → DEOXYCORTICOSTERONE
Doctor Opiatus → OPIUM
Doctor Opius → OPIUM
Docusate Calcium → DANTHRON
DOCUSATE SODIUM also → DANTHRON, MINERAL OIL, WOOL
Dodder → CUSCUTA sp.
n-DODECANE
1-DODECANETHIOL
Dodecanoic Acid → LAURIC ACID
1-DODECANOL
Dodecyl Alcohol → 1-DODECANOL
Dodecyl Mercaptan → 1-DODECANETHIOL
Dodex → VITAMIN B_{12}
Doe → METHAMPHETAMINE
DOFETILIDE
Doff → SUMITHION
Dog Brush → TETRADYMA sp.
Dog Buttons → NUX VOMICA
Dog Chamomile → ANTHEMIS COTULA
Dog Fennel → ANTHEMIS COTULA
Dogfish Shark → SQUALUS ACANTHIUS
Dogmatil → SULPIRIDE
Dogmatyl → SULPIRIDE
Dogtooth Violet → ERYTHRONIUM
DOGWOOD
"Dohyfral" A → VITAMIN A
Dokim → DIGOXIN
Doktacillin → PENICILLINS – "NEWER and SYNTHETICS" > *Ampicillin*
Dolac → KETOROLAC
Doladene → METHANTHELINE BROMIDE
Dolantin → MEPERIDINE
DOLASETRON
Dolazon → DIPYRONE
Dolcontin → MORPHINE
Dolcymene → *p*-CYMENE
Dolene → PROPOXYPHENE
Dolestan → DIPHENHYDRAMINE
Dolestine → MEPERIDINE
Dolgin → IBUPROFEN
Dolgirid → IBUPROFEN
Dolgit → IBUPROFEN

DOLICHOS LALAB
Dolicur → DIMETHYL SULFOXIDE
Dolipol → TOLBUTAMIDE
Doliprane → ACETAMINOPHEN
Dolisal → DIFLUNISAL
Dolisina → DIETHAZINE
Dolls → METHADONE HCL
Dolly → METHADONE HCL
Dolmatil → SULPIRIDE
Dolmen → TENOXICAM
Dolobasan → DICLOFENAC
Dolobid → DIFLUNISAL
Dolobis → DIFLUNISAL
Dolocap → PROPOXYPHENE
Dolochlor → CHLOROPICRIN
Dolocyl → IBUPROFEN
Dolo-Dolgit → IBUPROFEN
Dolo-Med-Much → ANTIPYRINE
Dolomite → ALUMINUM, LEAD, LIMESTONE, MERCURY
Dolophine → METHADONE HCL
Dolosal → MEPERIDINE
Dolprone → ACETAMINOPHEN
Dolu → RHUBARB
DOM → DIMETHOXYMETHYLAMPHETAMINE
Domafate → DEXTROAMPHETAMINE SULFATE
Domarax → CARISPRODOL
Domestrol → DIETHYLSTILBESTROL
D.O.M.F. → MERBROMIN
Domicillin → PENICILLINS – "NEWER and SYNTHETICS" > *Ampicillin*
Dominal → PROTHIPENDYL
DOMIPHEN BROMIDE
Domitor → MEDETOMIDINE
DOMOIC ACID also → MUSSELS
Domosedan → DETOMIDINE
Domoso → DIMETHYL SULFOXIDE
DOMPERIDONE
DON → 6-DIAZO-5-OXO-L-NORLEUCINE
DONEPEZIL also → RISPERIDONE
Dong Quai → ANGELICA sp.
Donkey Tail → EUPHORBIA
Donmox → ACETAZOLAMIDE
Donnagel PG → KAOLIN
Doolan Wattle → ACACIA
Dooryard Weed → POLYGONUM sp.
DOOTC → DOXYCYCLINE
DOP → BIS(2-ETHYLHEXYL) PHTHALATE
4-DOP → 4-DEOXYPYRIDOXINE
DOPA
L-DOPA also → AMANTADINE, BANANA, BEANS, CARBIDOPA, CLONIDINE, DOMPERIDONE, ENTACAPONE, FERROUS SULFATE, GRAPE, IMIPRAMINE, ISONIAZID, MANGANESE, METHOXYFLURANE, METOCLOPRAMIDE, METHYLDOPA, MUCUNA sp., NIALAMIDE, OLANZEPINE, ORPHENADRINE, PAPAVERINE, PENICILLAMINE, PHENELZINE, PHENYTOIN, PROCHLORPERAZINE, PROPANTHELINE, PROPRANOLOL, RESERPINE, SODIUM BICARBONATE, TETRABENAZINE, THIORIDAZINE, TOLCAPONE, TRIFLUOPERAZINE DIHYDROCHLORIDE, VITAMIN B COMPLEX, VITAMIN B_6
Dopaflex → L-DOPA
Dopaidan → L-DOPA
Dopal → L-DOPA
Dopalina → L-DOPA
Dopamet → METHYLDOPA
DOPAMINE also → CARBAMAZEPINE, CARBON DIOXIDE, CYCLOPROPANE, DEPRENYL, DOBUTAMINE, L-DOPA, HALOPERIDOL, METHYSERGIDE, METUREDEPA, NIALAMIDE, OLANZEPINE, PHENOXYBENZAMINEHYDROCHLORIDE, PHENYTOIN, PROPRANOLOL, RENOXIPRIDE, TRANYLCYPROMINE
DOPAN
Dopaquinone → BEANS
Dopar → L-DOPA
Doparkine → L-DOPA
Doparl → L-DOPA
Dopasol → L-DOPA
Dopastat → DOPAMINE
Dopaston → L-DOPA
Dopastral → L-DOPA
Dopegyt → METHYLDOPA
Dopom → GUANETHIDINE
Dopram → DOXAPRAM
Doprin → L-DOPA
Doral → QUAZEPAM
Doralese → INDORAMIN
Dorantamin → PYRILAMINE
Dorbane → DANTHRON
Dorcalm → CHLORALOSE
Dorcol → HEXAMETAPOL
Dorcostrin → DEOXYCORTICOSTERONE
Dorevan → PROPIOMAZINE
Dorico → HEXOBARBITAL
Doriden → GLUTETHIMIDE
Doriden-Sed → GLUTETHIMIDE
Dorinamin → BENZYDAMINE

Dorison → MEPARFYNOL
Dormalest → MEPARFYNOL
Dormalin → QUAZEPAM
Dorme → PROMETHAZINE
Dormicum → MIDAZOLAM
Dormidin → MEPARFYNOL
Dormigen → MEPARFYNOL
Dormigene → BROMISOVALUM
Dormigoa → METHAQUALONE
Dormilón → RUDBECKIA sp.
Dormin → ABSCISIC ACID
Dorminal → AMOBARBITAL
Dormiphen → MEPARFYNOL
Dormison → MEPARFYNOL
Dormogen → METHAQUALONE
Dormonal → BARBITAL
Dormosan → MEPARFYNOL
Dormutil → METHAQUALONE
Dormwell → DICHLORALANTIPYRINE
Dormytal → AMOBARBITAL
Dornavac → PANCREATIC DORNASE
DORONICUM sp.
Dorsallin "A.R." → PENICILLIN > *Benzylpenicillin, Procaine*
Dorsedin → METHAQUALONE
Dorsiflex → MEPHENOXALONE
Dorsilon → MEPHENOXALONE
Doryl → CARBACHOL
DORZOLAMIDE
Dosberotec → FENOTEROL HYDROBROMIDE
Dostinex → CABERGOLINE
Dosulepine → DOTHIEPIN
DOTHIEPIN also → FLUOXETINE
Double Cross → AMPHETAMINE
Double Mitsubishi → METHYLENEDIOXYAMPHETAMINE
Dovenix → NITROXYNIL
Dove's Foot → GERANIUM
Dove Weed → EREMOCARPUS SETIGERUS
Dovip → FAMPHUR
Dovonex → CALCIPOTRIENE also → CALCIPOTRIOL
Dowanol DE → 2-(2-ETHOXYETHOXY) ETHANOL
Dowanol EE → 2-ETHOXYETHANOL
Dowanol PM → PROPYLENE GLYCOL MONOMETHYL ETHER
Dowco 132 → RUELENE
Dowco 139 → MEXACARBATE
Dowco 179 → CHLORPYRIFOS
Dowco 214 → CHLORPYRIFOS-METHYL
Dowex 1-X2-C1 → CHOLESTYRAMINE RESIN
Dowfax 9N9 → NONOXYNOL-9
Dowfume MC 2 → METHYL BROMIDE
Dowfume W 85 → ETHYLENE DIBROMIDE
Dow General → DINOSEB
Dowicide 1 → σ-PHENYLPHENOL
Dowicide 2 → *2,4,5*-TRICHLOROPHENOL
Dowicide 2S → *2,4,6*-TRICHLOROPHENOL
Dowicide 7 → PENTACHLOROPHENOL
Dowicide A → σ-PHENYLPHENOL
Dowicide G → PENTACHLOROPHENOL
Dowicide Q → *N*-(3-CHLOROALLYL) HEXAMINIUM CHLORIDE
DOWICIDES also → CHLORO-2-PHENYPHENOL
Dowicil 75 → *N*-(3-CHLOROALLYL) HEXAMINIUM CHLORIDE
Dowicil 200 → *N*-(3-CHLOROALLYL) HEXAMINIUM CHLORIDE
Dowicil (R) 200 → METHENAMINE
Dowmycin E → ERYTHROMYCIN STEARATE
Downs, Mrs. Connie → Preface
Dow Selective → DINOSEB
Dowtherm 209 → PROPYLENE GLYCOL MONOMETHYL ETHER
DOXAPRAM
Doxapril → DOXAPRAM
DOXAZOCIN MESYLATE
Doxephrin → METHAMPHETAMINE
DOXEPIN also → ALCOHOL, ETHYL; CHOLESTYRAMINE RESINS, CIMETIDINE, PROPOXYPHENE
Doxergan → OXOMEMAZINE
Doxil → DOXORUBICIN
Doxinate → DOCUSATE SODIUM
Doxol → DOCUSATE SODIUM
DOXORUBICIN also → ACETAMINOPHEN, ACETYLCYSTEINE, ALUMINUM, INTERFERONS, MALONALDEHYDE, ZIDOVUDINE
Doxychel → DOXYCYCLINE
DOXYCYCLINE also → ACENOCOUMAROL, ALUMINUM HYDROXIDE, BISMUTH SUBSALICYLATE, DIDANOSINE, FERROUS SULFATE, FOLIC ACID, MEPHOBARBITAL, PHENOBARBITAL, PHENYTOIN, RIFAMPIN, WARFARIN
DOXYLAMINE SUCCINATE also → DICYCLOMINE
Dozic → HALOPERIDOL
DPA → DIPHENYLAMINE also → PROPANIL
DPA Sodium → VALPROIC ACID
DPC → DIPHENCYPRONE

DPGME → DIPROPYLENE GLYCOL MONOMETHYL ETHER
DPH → J
DPM → DIPROPYLENE GLYCOL MONOMETHYL ETHER
DPNA → *N*-NITROSODI-*n*-PROPYLAMINE
DPPD → *N,N′*-DIPHENYL-*p*-PHENYLENEDIAMINE
D-Pron → DIPYRONE
DPX 1410 → OXAMYL
DQ 2466 → CARVEDILOL
DQ 2805 → IRINOTECAN
DR 3355 → LEVOFLOXACIN
Dracylic Acid → BENZOIC ACID
Dragnet → PERMETHRIN
Dragon Chasing → HEROIN
Dragon Turnip → ARISAEMA sp.
Draked → MINERAL OIL
Dramamine → DIMENHYDRINATE
Dramarin → DIMENHYDRINATE
Dramcillin-S → PENICILLINS – "NEWER and SYNTHETICS" > *Phenethicillin Potassium*
Dramocen → DIMENHYDRINATE
Dramyl → DIMENHYDRINATE
Drapolene → BENZALKONIUM CHLORIDE
Drapolex → BENZALKONIUM CHLORIDE
Drat → CHLOROPHACINONE
Draza → MERCAPTODIMETHUR
Drazine → PHENOXYPROPAZINE MALEATE
Dr. Benjamin Rush → MERCUROUS CHLORIDE
DRC 714 → GOPHACIDE
DRC 736 → MERCAPTODIMETHUR
DRC 1201 → IOCETAMIC ACID
DRC 1327 → AMINOPYRIDINE
DRC 1339 → 3-CHLORO-*p*-TOLUIDINE
Dr. David Livingston → COFFEE
Dream → COCAINE
Dreamer → MORPHINE
Drenaren → TRIPARANOL
Drenison → FLURANDRENOLIDE
Drenusil → POLYTHIAZIDE
Drewpones → POLYSORBATES
Dr. Hand's Teething Lotion® → PHENYTOIN also → SASSAFRAS
Dr. Harvey W. Wiley → BORAX, COPPER SULFATE
Dr. Henry Morton Stanley → COFFEE
Dricol → AMIDEPHRINE MESYLATE
Dridase → OXYBUTYNIN CHLORIDE
Dridol → DROPERIDOL
Drinalfa → METHAMPHETAMINE
Drinox → HEPTACHLOR
Drinupal → CHLOROGUANIDE
Drisdol → VITAMIN D_2
Drocode → DIHYDROCODEINE
Drocort → FLURANDRENOLIDE
Drogenil → FLUTAMIDE
Drolban → DROMOSTANOLONE PROPIONATE
Droleptan → DROPERIDOL
Drometil → AZOSULFAMIDE
Dromisol → DIMETHYL SULFOXIDE
Dromoran → LEVORPHANOL TARTRATE
DROMOSTANOLONE PROPIONATE
Dronabinol → Δ^{-9} TETRAHYDROCANNABINOL
Droncit → PRAZIQUANTEL
Drooping Tulip → FRITILLARIA MELEAGRIS
DROPERIDOL also → L-DOPA, DOPAMINE, PERPHENAZINE, PHENELZINE
Dr. Opiatus → OPIUM
Dropsal → SALICYLAMIDE
Drostanolone Propionate → DROMOSTANOLONE PROPIONATE
Drosteakard → VERAPAMIL
DROTAVERINE
Drouya → COLTSFOOT
Droxarol → BUFEXAMAC
Droxaryl → BUFEXAMAC
Droxol → CHLORDIAZEPOXIDE
Droxomin → VITAMIN B_{12}
Dr. Stanley Livingston → COFFEE
Dr. Sydenham → OPIUM
Drummond Sesbane → SESBANIA sp.
Dr. Withering → DIGITALIS
Dry Cleaner's Naphtha → NAPHTHA
Dry-Clox → PENICILLINS – "NEWER and SYNTHETICS" > *Cloxacillin/Sodium salt*
Drygard → NOVOBIOCIN
Dry High → CANNABIS
Drylin → SULFAMETHOXAZOLE
DRYMARIA sp.
Dryptal → FUROSEMIDE
DS 103-282 → TIZANIDINE
DSCG → CROMOLYN SODIUM
DSE → NABAM
DSMA → METHANEARSONIC ACID also → DISODIUM METHANEARSONATE
Dsmogit/Amuno Gits → INDOMETHACIN
DSS → DOCUSATE SODIUM
DST → DIHYDROSTREPTOMYCIN
DT_4 → DEXTROTHYROXINE SODIUM
DTIC → DACARBAZINE
DTIC-Dome → DACARBAZINE
DTMC → DICOFOL
DTPA → PENTETIC ACID
D-Tracetten → VITAMIN D_2

DU 21,220 → RITODRINE
DU 23,000 → FLUVOXAMINE MALEATE
Dual → METOLACHLOR
Duatok → SULFATHIAZOLE
Dubba → SQUASH
Dubo → CYNODON
DUBOISIA MYOPOROIDES
Duboisine → *l*-HYOSCYAMINE
Dubonnet-on-the-Rocks® → QUININE
Ducobee → VITAMIN B_{12}
Dufalone → BISHYDROXYCOUMARIN
Dufaston → DYDROGESTERONE
DUGALDIN also → HELENIUM sp.
Dugway Proving Ground → VX
Dulcamara → SOLANUM sp.
DULCAMARIN also → SOLANUM sp.
DULCIN
Dulcion → ERGOLOID MESYLATES
Dulcivac → DOCUSATE SODIUM
Dulcolan → BISACODYL
Dulcolax → BISACODYL
Dumb Cane → DIEFFENBACHIA sp.
Dumirox → FLUVOXAMINE MALEATE
Dumitone → DAPSONE
Dumocyclin → TETRACYCLINE
Dumopen → PENICILLINS – "NEWER and SYNTHETICS" > *Ampicillin*
Duncaine → LIDOCAINE
Dundaki → SARCOPHALUS DIDERRICHII
Dunlop, Sir Derrick → Preface
Duocid → SULBACTAM
Duodin → HYDROCODONE BITARTRATE
Duofas → RAFOXANIDE
Duogastral → PIRENZEPINE HYDROCHLORIDE
Duogastrone → CARBENOLOXONE
Duomycin → CHLORTETRACYCLINE
Duoprim → TRIMETHOPRIM
Duoscorb → VITAMIN C
DUP 753 → LOSARTAN POTASSIUM
Duphacycline → OXYTETRACYCLINE
Duphafral D_3 → VITAMIN D_3
Duphar → TETRADIFON
Duphaspamin → ISOXSUPRINE
Duphaston → DYDROGESTERONE
Duphenicol → CHLORAMPHENICOL
Duplamin → PROMETHAZINE
Duponal QC → SODIUM LAURYL SULFATE
Du Pont 753 → LOSARTAN POTASSIUM
DuPont 1179 → METHOMYL
DuPont 1991 → BENOMYL
Du Pont Herbicide 326 → LINURON
DuPont Herbicide 976 → BROMACIL
Dupont-Lachapelle Syndrome → POLYVINYLPYRROLIDONE
dura AL → ALLOPURINOL
Dura AX → PENICILLINS – "NEWER and SYNTHETICS" > *Amoxicillin*
Duraband → CHLORPHENIRAMINE
Durabetason → BETAMETHASONE
Durable Press → PERMANENT PRESS
Durabol → NANDROLONE
Durabolin → NANDROLONE
Durabolin-O → ETHYLESTRENOL
Duracef → CEFADROXIL
Duracillin → PENICILLIN > *Benzylpenicillin, Procaine*
Duraclamid → METOCLOPRAMIDE
Duraclon → CLONIDINE
Duract → BROMFENAC
Duractin → CIMETIDINE
Duradoce → VITAMIN B_{12}
Durafurid → FUROSEMIDE
Duraglucon → GLYBURIDE
Dural Poisoning → FISH OILS
Duralta-12 → VITAMIN B_{12}
Duraluminum → FISH OILS
Duramax → ACETYLSALICYLIC ACID
Duramipress → PRAZOSIN HYDROCHLORIDE
Duranest → ETIDOCAINE HCL
Duranifin → NIFEDIPINE
Duranitrat → ISOSORBIDE DINITRATE
Duranol → PROPRANOLOL
Durapaediat → ERYTHROMYCIN ETHYL SUCCINATE
Durapental → PENTOXIFYLLINE
Duraphat → SODIUM FLUORIDE
Duraphos → MEVINPHOS
Duraphyllin → THEOPHYLLINE
Durapirenz → PIRENZEPINE HYDROCHLORIDE
Durapro → OXAPROZIN
Duraprost → OXAPROZIN
Duraprox → OXAPROZIN
Duraquin → QUINIDINE
Duraspan → POLYURETHANES
Duraspiron → SPIRONOLACTONE
Duravolten → DICLOFENAC
Durazanil → BROMAZEPAM
Durazepam → OXAZEPAM
Duraznillo Blanco → SOLANUM sp.
Durba → CYNODON
Durbid → DISOPYRAMIDE
Durbis → DISOPYRAMIDE
Durenat → SULFAMETER
Duricef → CEFADROXIL

Durogesic → FENTANYL
DUROIA
Durolax → BISACODYL
Duromine → PHENTERMINE
Duromorph → MORPHINE
Duronitrin → TROLNITRATE PHOSPHATE
Durra → SORGHUM
Durrax → HYDROXYZINE
Dursban → CHLORPYRIFOS
Dusodril → NAFRONYL OXALATE
Dust Parsley → PHENCYCLIDINE
DUSTS
Dutch Liquid → ETHYLENE DICHLORIDE
Dutchman's Breeches → DICENTRA CUCULLARIA
Duteplase → TISSUE PLASMINOGEN ACTIVATOR
Dutonin → NEFAZODONE
Duvadilan → ISOXSUPRINE
Duviculine → ISOXSUPRINE
Duvoid → BETHANECHOL
Duxil → ALMITRINE BISMESYLATE
Duxima → CEFUROXIME
D_3-Vicotrat → VITAMIN D_3
DW 61 → FLAVOXATE HCL
DW 3418 → CYANAZINE
Dwale → BELLADONNA
Dwarf Bay → DAPHNE
Dwarf Elder → SAMBUCUS sp.
Dya-Tron → DIPYRONE
Dyazide → AMANTADINE
Dybar → FENURON
Dybenal → 2,4-DICHLOROBENZYL ALCOHOL
Dycholium → DEHYDROCHOLIC ACID
Dycill → PENICILLINS – "NEWER and SYNTHETICS" > *Dicloxacillin*
Dyclone → DYCLONINE
DYCLONINE
DYDROGESTERONE
Dye-Gen → GENTIAN VIOLET
Dyfonate → FONOPHOS
Dygratyl → DIHYDROTACHYSTEROL
Dykstra, Dr. David → TILMICOSIN
Dylate → CLONITRATE
Dylene → POLYSTYRENE
Dylox → TRICHLORFON
Dymadon → ACETAMINOPHEN
DYMANTHINE
Dymel 142 → MONOCHLORODIFLUOROETHANE
Dymelor → ACETOHEXAMIDE
Dymid → DIPHENAMID
Dymocillin → PENICILLIN > *Benzylpenicillin, Potassium*
Dynabac → DIRITHROMYCIN
Dynabiotic → AZTREONAM
Dynacin → MINOCYCLINE
DynaCirc → ISRADIPINE
Dynacoryl → NIKETHAMIDE
Dynacrine → ISRADIPINE
Dynalin → TIAMULIN
Dynamite → NITROGLYCERIN also → AMMONIUM NITRATE
Dynamos → DIGOXIN
Dynamutilin → TIAMULIN
Dynapen → PENICILLINS – "NEWER and SYNTHETICS" > *Dicloxacillin*
Dynarsan → ACETARSONE
Dynatra → DOPAMINE
Dynothel → DEXTROTHYROXINE SODIUM
Dynovas → PAPAVERINE
Dyodin → IODOQUINOL
Dyphlos → DIISOPROPYL FLUOROPHOSPHATE
DYPHYLLINE also → PROBENECID
Dyprin → METHIONINES
Dyren → TRIAMTERENE
Dyrenium → TRIAMTERENE
Dyrex → TRICHLORFON
Dyrene → ANILAZINE
Dyron → TRICHLORFON
Dysdera crocata → SPIDERS
Dysect → CYPERMETHRIN
Dysedon → OXOMEMAZINE
Dysentulin → BISMUTH GLYCOL ARSANILATE
Dysmenalgit → NAPROXEN
Dyspamet → CIMETIDINE
Dyspas → DICYCLOMINE
Dytac → TRIAMTERENE
Dytransin → IBUFENAC
Dzin-zi-a-pa → PHYLLANTHUS sp.
D-z-n → DIAZINON

E

E → METHYLDIAMINE METHAMPHETAMINE also → OIL ORANGES
E 39 → INPROQUONE
E 102 → TARTRAZINE
E 104 → QUINOLINE YELLOW
E 106E → FURFENDREX CYCLAMATE
E 110 → FDC YELLOW #6
E 132 → INDIGOCARMINE

E 141 → ETHAMSYLATE
E 250 → DEPRENYL
E 265 → TRAMADOL
E 600 → DIETHYL-*p*-NITROPHENYLPHOSPHATE
E 601 → METHYL PARATHION
E 643 → BUNAZOSIN
E 0216 → UBIDECARENONE
E 1015 → BUNAZOSIN
E 1059 → DEMETON
E 2020 → DONEPEZIL
E 3314 → HEPTACHLOR
EA 1299 → METHYLENEDIOXYAMPHETAMINE
EACA Kabi → Σ-AMINOCAPROIC ACID
Eacoril → PROPANTHELINE BROMIDE
EACS → Σ-AMINOCAPROIC ACID
Eagle Ray → STINGRAYS
EAK → ETHYL *sec*-AMYL KETONE
Early Bird → PYRANTEL
Earth Nuts → PEANUTS
East Coast Fever → QUINURONIUM
Easter Flower → ANEMONE also → POINSETTIA
Easter Lily → LILIES also → ERYTHRONIUM
Eastern Coral Bean → ERYTHRINA
Eastman 910 → CYANOACRYLATES also → MECRYLATE
Eatan → NITRAZEPAM
Eazaminum → DIETHAZINE
EBA → ETHYL BROMOACETATE
E-Base → ERYTHROMYCIN
Ebastel → EBASTINE
EBASTINE
EBDCs → ETHYLENE BISDITHIOCARBAMATES
Ebivit → VITAMIN D_3
EBONY also → DIOSPYRUS sp.
Ebrantil → URAPIDIL
Ebufac → IBUPROFEN
Ebutol → ETHAMBUTOL
Ecasolv → HEPARIN
Ecatox → PARATHION
ECH → EPICHLORHYDRIN
Echinatine → CYNOGLOSSUM OFFICIANALE
ECHINOCHLOA CRUS-GALLI
ECHINOMYCIN
Echinopanax horridum → OPLOPANAX HORRIDUM
Echinmine → AMSACRINE
ECHIUM PLANTAGINEUM
ECHOTHIOPHATE IODIDE also → MIVACURIUM, PROPANIDID, SUCCINYLCHOLINE
Eciphin → EPHEDRINE
ECM → ACETYLSALICYLIC ACID
Ecmonovocillin → PENICILLIN > *Benzylpenicillin, Procaine*
Ecobutazone → PHENYLBUTAZONE
Ecodipin → NIFEDIPINE
Ecofenac → DICLOFENAC
Ecogramostatin → GRANULOCYTE MACROPHAGE COLONY STIMULATING FACTOR
Ecolid → CHLORISONDAMINE
Ecolid Chloride → CHLORISONDAMINE
ECONAZOLE NITRATE
Economycin → TETRACYCLINE
Ecosporina → CEPHADRINE
Ecostatin → ECONAZOLE NITRATE
Ecotrin → ACETYLSALICYLIC ACID
Ecoval 70 → BETAMETHASONE
Ecovent → ALBUTEROL
Ecrinal → FLECAINIDE
Ecstasy → METHYLENEDIOXYAMPHETAMINE
Ectiban → PERMETHRIN
Ectomin → CYPERMETHRIN
Ectopor → CYPERMETHRIN
Ectoral → RONNEL
Ectrin → FENVALERATE
ECTYLUREA
Ecuanil → MEPROBAMATE
ECURAL → MOMETASONE FUROATE
Eczecidin → IODOCHLORHYDROXYQUINOLINE
Edalene → CIMETIDINE
Edathamil Disodium → DISODIUM EDETATE
EDB → ETHYLENE DIBROMIDE
EDC → ETHYLENE DICHLORIDE
Edecril → ETHACRYNIC ACID
Edecrin → ETHACRYNIC ACID
Edemex → BENZTHIAZIDE
Edemox → ACETAZOLAMIDE
Eden-psich → CHLORDIAZEPOXIDE
Edetate Disodium → DISODIUM EDETATE
Edethamil Calcium Disodium → CALCIUM DISODIUM EDETATE
EDETIC ACID also → NEOMYCIN, NORFLOXACIN, STREPTOMYCIN
Edex → ALPROSTADIL
EDIFENPHOS
Ediphenphos → EDIFENPHOS
Edolan → ETODOLAC

EDROPHONIUM also → COLISTIN, CURARE, KANAMYCIN, MALATHION, SUCCINYLCHOLINE
EDTA → EDETIC ACID
EDTA Calcium → CALCIUM DISODIUM EDETATE
EDTA Disodium → DISODIUM EDETATE
EEAC → 2-ETHOXYETHYL ACETATE
2-EEE → 2-(2-ETHOXYETHOXY) ETHANOL
EEL, MORAY
EEMC → CINOXATE
E.E.S. → ERYTHROMYCIN ETHYL SUCCINATE
Efalexin → CEPHALEXIN
EFAVIRENZ
Efcorlin → HYDROCORTISONE
Efcortelan → HYDROCORTISONE
Efcortesol → HYDROCORTISONE
Efektolol → PROPRANOLOL
Efere Makara → PORTULACA OLERACEA
E-Ferol → VITAMIN E
Effectin → BITOLTEROL
Effekton → DICLOFENAC
Efferalgan → ACETAMINOPHEN
Effexor → VENLAFAXINE
Efflumidex → FLUOROMETHOLONE
Effortil → ETHYL PHENYLEPHRINE
Effusan → DINITRO-σ-CRESOL
EFLORNITHINE
Efodine → POLYVINYLPYRROLIDONE IODINE COMPLEX
Efpenix → PENICILLINS – "NEWER and SYNTHETICS" > *Amoxicillin*
Efrane → ENFLURANE
Efroxine → METHAMPHETAMINE
Eftolon → SULFAPHENAZOLE
Efudex → FLUOROURACIL
Efudix → FLUOROURACIL
Efuranol → IMIPRAMINE
EFV → EFAVIRENZ
Egacin → *l*-HYOSCYAMINE
Egalin → RAUWOLFIA SERPENTINA
Egazil → *l*-HYOSCYAMINE
EGBE → ETHYLENE GLYCOL MONOBUTYL ETHER
EGBEA → ETHYLENE GLYCOL MONOBUTYL ETHER ACETATE
EGDN → ETHYLENE GLYCOL DINITRATE
EGEEA → 2-ETHOXYETHYL ACETATE
EGGPLANT also → PEPPERS
EGGS also → IRON, SOYBEAN OIL
Eglen → CINNARIZINE
EGYT 341 → TOFISOPAM
2EH → 2-ETHYL-1-HEXANOL
Ehrlich 5 → OXOPHENARSINE
Ehrlich 594 → ACETARSONE
Ehrlich 606 → ARSPHENAMINE
EI → ETHYLENEIMINE
EI 3911 → PHORATE
EI 12880 → DIMETHOATE
EI 47031 → PHOSFOLAN
Eicosapentaenoic Acid → DOCOSAHEXAENOIC ACID
Eightpen → PENICILLIN > *Benzylpenicillin, Potassium*
Einalon S → HALOPERIDOL
Eisenhutknollen → ACONITE
Ekan-ekun → ARGEMONE MEXICANA
Ekilan → MEPHENOXALONE
Ekluton → GONADOTROPINS, CHORIONIC
Eksmin → PERMETHRIN
Ektafos → DICROTOPHOS
Èktebin → PROTIONAMIDE
Ektyl → ECTYLUREA
Ekvacillin → PENICILLINS – "NEWER and SYNTHETICS" > *Cloxacillin/Sodium salt*
EL 10 → PRASTERONE
EL 110 → BENFLURALIN
EL 119 → ORYZALIN
EL 349 → GROWTH HORMONES
EL 614 → BROMETHALIN
EL 870 → TILMICOSIN
Elamol → TOFENACIN
Elan → ISOSORBIDE MONONITRATE also → THEOPHYLLINE
Elancoban → MONENSIN
Elanco/Eli Lily → TILMICOSIN
Elancolan → TRIFLURALIN
Elantan → ISOSORBIDE MONONITRATE
Elase → PANCREATIC DORNASE
Elasterin → FENOFIBRATE
Elastica → RUBBER
Elastonon → AMPHETAMINE
Elastoplast® → RHIZOPUS sp.
Elavil® → AMITRIPTYLINE
Elayl → ETHYLENE
Elazor → FLUCONAZOLE
Elbrol → PROPRANOLOL
Elcarmycin → CARBOMYCIN
Elcema → CELLULOSE
ELD 950 → ELEDOISIN
Eldéprine → DEPRENYL
Eldepryl → DEPRENYL
Elder → SAMBUCUS sp.
ELDERBERRY also → CYANIDES
Elderfield Pyrimidine Mustard → DOPAN

Eldisine → VINDESINE SULFATE
Eldopal → L-DOPA
Eldopaque → HYDROQUINONE
Eldopar → L-DOPA
Eldopatec → L-DOPA
Eldoquin → HYDROQUINONE
Elecampane → INULA sp.
Elecampane Oil → INULA sp.
"Electrician's Rash" → CHLORINATED NAPHTHALENES
Electrocortin → ALDOSTERONE
ELEDOISIN also → OCTOPUS, PHYSALEAMIN
Elemicin → MYRISTICIN
Elen → INDELOXAZINE
Elenium → CHLORDIAZEPOXIDE
Eleparon → MUCOITIN POLYSULFATE ESTER
Elephant Ear → CALADIUM also → COLOCASIA sp., XANTHOSOMA sp.
Elephant Grass → PENNISETUM sp.
Elephant Tranquilizer → PHENCYCLIDINE
Elestol → CHLOROQUINE
Eleudron → SULFATHIAZOLE
Elgetol → DINITRO-σ-CRESOL
Elgetol 318 → DINOSEB
Elieten → METOCLOPRAMIDE
Elimite → PERMETHRIN
Elipten → AMINOGLUTETHIMIDE
Elisal → SULTHIAME
Elisor → PRAVASTATIN
Elixicon → THEOPHYLLINE
Elixifilin → THEOPHYLLINE
Elixophyllin → THEOPHYLLINE
Ellagic acid → RASPBERRY
Ellangowan Poison Bush → MYOPORUM sp.
Ellence → EPIRUBICIN
ELM
Elmarin → CHLORPROMAZINE
Elmiron → PENTOSAN POLYSULFATE
Elobromol → MITOLACTOL
ELOCON → MOMETASONE FUROATE
Elodrine → HYDROFLUMETHIAZIDE
Eloxatine → OXALIPLATIN
Elrodorm → GLUTETHIMIDE
Elspar → *l*-ASPARAGINASE
Eltor → PSEUDOEPHEDRINE
Eltroxin → THYROXINE
Elvanol → POLYVINYL ALCOHOL
ELYMOCLAVINE also → ELYMUS sp., MORNING GLORY
ELYMUS sp.
Elysion → PHENCYCLIDINE
Elyzol → METRONIDAZOLE
Elzogram → CEFAZOLIN SODIUM
EM 1960 → PIPERAZINE
Emandione → PHENINDIONE
Emanil → IDOXURIDINE
EMB → ETHAMBUTOL
Embacetin → CHLORAMPHENICOL
Embacide → PHOSALONE
Embafume → METHYL BROMIDE
"Embalming Fluid" → PHENCYCLIDINE
Embanox → BUTYLATED HYDROXYANISOLE
Embarin → ALLOPURINOL
Embazin → SULFAQUINOXALINE
Embequin → IODOQUINOL
Embichen → MECHLORETHAMINE
Embikhine → MECHLORETHAMINE
Embinal → BARBITAL
Embucrilate → BUTYL CYANOACRYLATE
Emconor → BISOPROLOL
Emcor → BISOPROLOL
Emcyt → ESTRAMUSTINE
EMD 33 512 → BISOPROLOL
Emdecassol → CENTALLA ASIATICA
Emdisterone → DROMOSTANOLONE PROPIONATE
Emdabol → THIOMESTERONE
Emdabolin → THIOMESTERONE
Emedan → CARBUTAMIDE
Emedyl → DIMENHYDRINATE
Emeldopa → L-DOPA
EMEPRONIUM
Emerald Green → BRILLIANT GREEN also → COPPER ARSENITE, CUPRIC ACETOARSENITE
Emergil → FLUPENTIXOL
Emes → DIMENHYDRINATE
Emeside → ETHOSUXIMIDE
Emete-Con → BENZQUINAMIDE
Emeticon → BENZQUINAMIDE
Emetic Weed → LOBELIA
Emetid → METOCLOPRAMIDE
EMETINE
Emikon → METHAPYRILENE
Emilene → MAFENIDE
Eminase → APSAC
Emivan → ETHAMIVAN
Emmatos → MALATHION
Emocel → CELLULOSE
EMODIN also → IBUPROFEN
Emoren → OXETHAZINE
Emorhalt → TRANEXAMIC ACID
Emory Milkvetch → MILKVETCH
Emotival → LORAZEPAM
Empecid → CLOTRIMAZOLE

Emperal → METOCLOPRAMIDE
Empirin → ACETYLSALICYLIC ACID
EMPT → FENFLURAMINE
EMQ → ETHOXYQUINE
EMS → ETHYL METHANESULFONATE
Emsel → MORPHINE
Emtexate → METHOTREXATE
Emtryl → DIMETRIDAZOLE
EMU → ERYTHROMYCIN
Emulpan → LINDANE
Emulsin → SORGHUM
E-mycin → ERYTHROMYCIN
E-Mycin E → ERYTHROMYCIN ETHYL SUCCINATE
EMYLCAMATE
Emyrenil → OXOLINIC ACID
EN 141 → JOSAMYCIN
EN 313 → MORICIZINE
EN 1627 → MECLOFENOXATE
EN 1639A → NALTREXONE
EN 1733A → MOLINDONE
EN 2234A → NALBUPHINE
EN 18,133 → ZINOPHOS
Enacard → ENALAPRIL
ENALAPRIL also → ALCOHOL, ETHYL; CLOZAPINE, LOSARTAN POTASSIUM
Enalaprilat → LISINOPRIL
Enaloc → ENALAPRIL
ENAMEL
Enantone → LEUPROLIDE
Enapren → ENALAPRIL
Enarex → OXYPHENCYCLIMINE
ENB → ETHYLIDINE NORBORNENE
Enbol → PYRITINOL DIHYDROCHLORIDE
Enbrel → ETANERCEPT
Encaid → ENCAINIDE
ENCAINIDE also → DILTIAZEM, Preface
Encaprin → ACETYLSALICYLIC ACID
Encare → NONOXYNOL-9
Encephabol → PYRITINOL DIHYDROCHLORIDE
Encetrop → PIRACETAM
Enclomiphene → CLOMIPHENE CITRATE
Encordin → PERUVOSIDE
Encorton → PREDNISONE
Endak → CARTEOLOL HYDROCHLORIDE
Endane → CHLORDANE
Endecril → ETHACRYNIC ACID
Endep → AMITRIPTYLINE
Endercin → ETHACRYNIC ACID
Endiemal → METHARBITAL
ENDIVE
Endocistobil → MEGLUMINE IODIPAMIDE
Endocorion → GONADOTROPINS, CHORIONIC
Endofolliculina → ESTRONE
Endografin → MEGLUMINE IODIPAMIDE
Endokolat → BISACODYL
Endo-Paractol → DIMETHICONE (*with silicon dioxide*)
Endophleban → DIHYDROERGOTAMINE METHANESULFONATE (MESYLATE)
End-O'-Rat → WARFARIN
Endosan → BINAPACRYL
ENDOSULFAN
ENDOTHALL
Endox → COUMATETRALYL also → ENDRIN
Endoxan → CYCLOPHOSPHAMIDE
Endoxana → CYCLOPHOSPHAMIDE
Endrate Disodium → DISODIUM EDETATE
Endrex → ENDRIN
ENDRIN
Endrol → GARDONA
Endyol → ACETYLSALICYLIC ACID
Eneldo → DILL
Enelfa → ACETAMINOPHEN
Enerbol → PYRITINOL DIHYDROCHLORIDE
Energine® → NAPHTHA
Eneril → ACETAMINOPHEN
ENFLURANE also → DIMETHYLTUBOCURARINE IODIDE, EPINEPHRINE, KETAMINE, MIVACURIUM, PROPOFOL
Engemycin → OXYTETRACYCLINE
English Holly → HOLLY
English Ivy → HEDERA HELIX
Engraver's Acid → NITRIC ACID
Engul → SOLANUM sp.
Enheptin-A → 2-ACETAMIDO-5-NITROTHIAZOLE
Enheptin-T → AMINONITROTHIAZOLE
Enhexymal → HEXOBARBITAL
Enhydrina schistosa → SEA SNAKES
Enibomal → NARCOBARBITAL
Enicol → CHLORAMPHENICOL
Enide → DIPHENAMID
Enidrel → OXAZEPAM
Enkade → ENCAINIDE
Enkefal → PHENYTOIN
Enkephalin → PHENYLALANINE
Enlon → EDROPHONIUM
ENOXACIN also → RANITIDINE, THEOPHYLLINE, WARFARIN
ENOXAPARIN
Enoxen → ENOXACIN
ENOXIMONE

Enoxolone → GLYCYRRHETINIC ACID
Enoxor → ENOXACIN
Enpro → PROCAINE
ENROFLOXACIN also → SUCRALFATE
Enrumay → PYRILAMINE
Enseal → POTASSIUM CHLORIDE
Ensidon → OPIPRAMOL
ENT 92 → ISOBORNYL THIOCYANOACETATE
ENT 154 → DINITRO-σ-CRESOL
ENT 987 → THIRAM
ENT 1122 → DINOSEB
ENT 1716 → METHOXYCHLOR
ENT 3776 → DICHLONE
ENT 4225 → 1,1-DICHLORO-2,2-BIS (*p*-CHLOROPHENYL) ETHANE
ENT 7796 → LINDANE
ENT 9106 → NABAM
ENT 14,250 → PIPERONYL BUTOXIDE
ENT 14,874 → ZINEB
ENT 15,108 → PARATHION
ENT 15,152 → HEPTACHLOR
ENT 16,225 → DIELDRIN
ENT 16,273 → SULFOTEP
ENT 16,391 → KEPONE
ENT 16,519 → ARAMITE
ENT 17,034 → MALATHION
ENT 17,082 → PERTHANE
ENT 17,251 → ENDRIN
ENT 17,291 → SCHRADAN
ENT 17,292 → METHYL PARATHION
ENT 17,510 → ALLETHRIN
ENT 17,588 → PYROLAN
ENT 18,771 → TETRAETHYL PYROPHOSPHATE
ENT 19,507 → DIAZINON
ENT 19,763 → TRICHLORFON
ENT 20,218 → *N,N*-DIETHYL-*m*-TOLUAMIDE
ENT 21,557 → BARTHRIN
ENT 22,014 → AZINPHOS-ETHYL
ENT 22,374 → MEVINPHOS
ENT 22,879 → DIOXATHION
ENT 23,233 → AZINPHOS-METHYL
ENT 23,284 → RONNEL
ENT 23,648 → DICOFOL
ENT 23,708 → CARBOPHENOTHION
ENT 23,737 → TETRADIFON
ENT 23,969 → CARBARYL
ENT 23,979 → ENDOSULFAN
ENT 24,042 → PHORATE
ENT 24,105 → ETHION
ENT 24,482 → DICROTOPHOS
ENT 24,650 → DIMETHOATE
ENT 24,717 → CIODRIN
ENT 24,727 → DINOCAP
ENT 24,833 → BOMYL
ENT 24,915 → TRIETHYLENE PHOSPHORAMIDE
ENT 24,945 → FENSULFOTHION
ENT 24,969 → CHLORFENVINPHOS
ENT 24,988 → DIBROM
ENT 25,445 → AMINOTRIAZOLE
ENT 25,515 → PHOSPHAMIDON
ENT 25,540 → FENTHION
ENT 25,545X → ISOBENZAN
ENT 25,567 → PHTHALOPHOS
ENT 25,580 → ZINOPHOS
ENT 25,640 → CYTHIOATE
ENT 25,644 → FAMPHUR
ENT 25,671 → PROPOXUR
ENT 25,705 → IMIDAN
ENT 25,715 → SUMITHION
ENT 25,719 → MIREX
ENT 25,726 → MERCAPTODIMETHUR
ENT 25,766 → MEXACARBATE
ENT 25,784 → AMINOCARB
ENT 25,793 → BINAPACRYL
ENT 25,796 → FONOPHOS
ENT 25,830 → PHOSFOLAN
ENT 25,841 → GARDONA
ENT 25,843 → LANDRIN
ENT 25,886 → METHYL TRITHION
ENT 26,538 → CAPTAN
ENT 26,539 → FOLPET
ENT 27,041 → MOBAM
ENT 27,093 → ALDICARB
ENT 27,127 → BUFENCARB
ENT 27,129 → MONOCROTOPHOS
ENT 27,162 → BROMOPHOS
ENT 27,164 → CARBOFURAN
ENT 27,165 → TEMEPHOS
ENT 27,193 → METHIDATHION
ENT 27,226 → PROPARGITE
ENT 27,311 → CHLORPYRIFOS
ENT 27,318 → ETHOPROP
ENT-27,396 → METHALLIBURE
ENT 27,448 → PHOXIM
ENT 27,520 → CHLORPYRIFOS-METHYL
ENT 27,572 → FENAMIPHOS
ENT 27,967 → AMITRAZ
ENT 28,203 → TRIFLURALIN
ENT 50,852 → ALTRETAMINE
ENT 50,882 → HEXAMETAPOL
ENT 70,640 → METHOPRENE
ENTACAPONE also → EPINEPHRINE
Entactogen → METHYLDIAMINE METHAMPHETAMINE

Entacyl → PIPERAZINE
"Enteque seco" → SOLANUM sp.
Enteramine → SEROTONIN
Enterfram → FRAMYCETIN
Enterobiocine → SULFATHIAZOLE
Enterocol → NITROXOLINE
Enterodex → POLYVINYLPYRROLIDONE
Enterokanacin → KANAMYCIN
ENTEROLOBIUM
Enteromycetin → CHLORAMPHENICOL
Enteroquinol → IODOCHLORHYDROXYQUINOLINE
Enterosalicyl → SODIUM SALICYLATE
Enterosalil → SODIUM SALICYLATE
Enterosalyl → SODIUM SALICYLATE
Enterosarine → ACETYLSALICYCLIC ACID
Enterosept → IODOQUINOL
Entero-Septol → IODOCHLORHYDROXYQUINOLINE
Enterotonin → CARBACHOL
Entero-Vioform → IODOCHLORHYDROXYQUINOLINE
Enterozol → IODOCHLORHYDROXYQUINOLINE
Entex → FENTHION
Entexidina → PHTHALYLSULFATHIAZOLE
Entobex → PHANQUINONE
Entocon → METHOPRENE
ENTOLOMA
ENTOMOPHTHORA CORONATA
Entomoxan → LINDANE
Entramin → AMINONITROTHIAZOLE
Entrokin → IODOCHLORHYDROXYQUINOLINE
Entrokinol → IODOCHLORHYDROXYQUINOLINE
Entrophen → ACETYLSALICYLIC ACID
Entulic → GUANFACINE
Entumine → CLOTHIAPINE
Enturen → SULFINPYRAZONE
Entyderma → BECLOMETHASONE
ENU → *N*-NITROSO-*N*-ETHYLUREA
Enuclen → BENZALKONIUM CHLORIDE
Enup → CYNANCHUM, *caudatum*
Envilon → POLYVINYL CHLORIDE
Enzamin → BENZYDAMINE
Enzaprost F → DINOPROST
Enzeon → CHYMOTRYPSIN
Enzodase → HYALURONIDASE
EOSIN(E) also → LIPSTICK, THIMEROSAL
Eosine Y → EOSIN(E)
Eosine Yellowish - (YS) → EOSIN(E)
Ep → ERYTHROPOIETIN
EP 20 → PENTACHLOROPHENOL
EP 50 → PLACENTAL EXTRACTS
EP 227 → METHYLMERCURYDICYANDIAMIDE
EPA → PHENYLCYCLOHEXENE, R-11
Epasote → CHENOPODIUM
Epatec → KETOPROFEN
Epatiol → TIOPRONIN
Epazote de Comer → CHENOPODIUM
EPE → ETOPOSIDE
EPEG → ETOPOSIDE
Epelin → PHENYTOIN
Epéna → VIROLA sp.
Epha → DIMENHYDRINATE
Ephedra → EPHEDRINE
EPHEDRINE also → AMINOPHYLLINE, AMMONIUM CHLORIDE, CYCLOPROPANE, GUANETHIDINE, HALOTHANE, NIALAMIDE, NORPSEUDOEPHEDRINE, PHENELZINE, RESERPINE, SODIUM BICARBONATE, THEOPHYLLINE, TRANYLCYPROMINE, TRIFLUOPERAZINE DIHYDROCHLORIDE
Ephicillin → PENICILLIN > *Benzylpenicillin, DAEEH*
Epi → EPICHLORHYDRIN
Epi-Aberel → VITAMIN A
Epianhydrotetracycline → TETRACYCLINE
Epibloc → α-CHLOROHYDRIN
Epichloe sp. → ACREMONIUM
EPICHLORHYDRIN
Epicillin → PENICILLINS – "NEWER and SYNTHETICS"
Epiclase → PHENACEMIDE
3,17-Epidihydroxyestratriene → ESTRADIOL
Epidione → TRIMETHADIONE
Epidophyllin → ETOPOSIDE
Epidophyllotoxin → ETOPOSIDE
4´-Epidoxorubicin → EPIRUBICIN
Epidropal → ALLOPURINOL
Epidx → EPIRUBICIN
Epifrin → EPINEPHRINE
d Epifrin → DIPIVEFRIN
Epiglaufrin → EPINEPHRINE
Epileo Petitmal → ETHOSUXIMIDE
Epilim → VALPROIC ACID also → SODIUM VALPROATE
Epi-Monistat → MICONAZOLE NITRATE
Epinal → ALCLOFENAC also → EPINEPHRINE
Epinat → PHENYTOIN
EPINEPHRINE also → ADRENOCHROME; ALUMINUM, BUTACAINE; CHEESES, CHLORPROMAZINE, CHLORTHALIDONE, COCAINE,

CYCLOPROPANE, DEBRISOQUIN, ETHANE, ETHYL CHLORIDE, GLUCOSE, HALOPERIDOL, HALOTHANE, HARMOL, *n*-HEXANE, HYDROCORTISONE, IMIPRAMINE, INSULIN, IODOFORM, ISOBUTANE, ISOPROTERENOL, LABETALOL, LIDOCAINE, LIOTHYRONINE, METHOTRIMEPRAZINE, NORTRIPTYLINE, OCTOPAMINE; PENICILLINS – "NEWER and SYNTHETICS" > *Hetacillin*; PENTOBARBITAL, PHENELZINE, PILOCARPINE, POTASSIUM, PROCAINE, PROCHLORPERAZINE, PROMAZINE, PROPANE, PROPRANOLOL, PROPYLENE, PROTRIPTYLINE, RESERPINE, SODIUM BICARBONATE, SODIUM METABISULFITE, SOTALOL, THIAMYLAL SODIUM, THIOPENTAL, THYROID, TOADS, 1,1,1-TRICHLOROETHANE, TRIFLUPROMAZINE
Epi-Pen → EPINEPHRINE
Epi-Pevaryl → ECONAZOLE NITRATE
Epirenor → OCTOPAMINE
Epirotin → BENZYDAMINE
EPIRUBICIN also → INTERFERONS
Epiteliol → VITAMIN A
Epitol → CARBAMAZEPINE
Epitrate → EPINEPHRINE
Epivir → LAMIVUDINE
EPN
Epnial → ALCLOFENAC
Epo → ERYTHROPOIETIN
Epoade → ERYTHROPIOETIN
Epobron → IBUPROFEN
Epocelin → CEFTIZOXIME
Epodyl → ETHOGLUCID
Epoetin alfa → ERYTHROPOIETIN
Epogen → ERYTHROPOIETIN
Epogin → ERYTHROPOIETIN, *EPOETIN beta*
Epolene → POLYETHYLENE
Epontil → PROPANIDID
Epoprostenol → PROSTACYCLIN
Eporal → DAPSONE
Eposerin → CEFTIZOXIME
1,4-Epoxybutane → TETRAHYDROFURAN
1,2-Epoxyethane → ETHYLENE OXIDE
1,2-Epoxy Propane → PROPYLENE OXIDE
2,3-Epoxy-1-propanol → GLYCIDOL
Epoxypropyl Alcohol → GLYCIDOL
EPOXY RESINS
Epoxy Resin Contact Dermatitis → TRIETHYLENEDIAMINE
Epoxyscillirosidin → HOMERIA
Eppy → EPINEPHRINE
Eprex → ERYTHROPOIETIN
Eprofil → THIABENDAZOLE
Epsamon → Σ-AMINOCAPROIC ACID
Epsikapron → Σ-AMINOCAPROIC ACID
Epsilcapramin → Σ-AMINOCAPROIC ACID
Epsilon-aminocaproic Acid → Σ-AMINOCAPROIC ACID
EPT → ETOPOSIDE
Eptam → EPTC
Eptastatin → PRAVASTATIN
EPTC
EPTIFIBATIDE
Equa® → ASPARTAME
Equanil → MEPROBAMATE
Equiben → CAMBENDAZOLE
Equibral → CHLORDIAZEPOXIDE
Equicol → GLYCERYL GUAIACOLATE
Equigard → DICHLORVOS also → DIMETHOATE
Equi-gel → DICHLORVOS
Equigyne → ESTROGENS, CONJUGATED
Equimate → FLUPROSTENOL
Equipalazone → PHENYLBUTAZONE
Equipertine → OXYPERTINE
Equipoise → BOLDENONE
Equipose → HYDROXYZINE
Equiproxen → NAPROXEN
Equisetin → EQUISETUM
EQUISETUM
Equitac → OXIBENDAZOLE
Equiverm Plus → MEBENDAZOLE
Equizole → THIABENDAZOLE
Equron → SODIUM HYALURONATE
Eqvalan → IVERMECTIN
ER 102 → METFORMIN
ER 115 → TEMAZEPAM
ER 5461 → PROFLURALIN
Eracine → ROSOXACINE
E-Rad → METHANEARSONIC ACID
Eradacil → ROSOXACIN
Eradacin → ROSOXACIN
Eraldin → PRACTOLOL
Erantin → PROPOXYPHENE
Erantone → LEUPROLIDE
Erasmus Darwin → ANTIARIS TOXICARIA
Erasol → MECHLORETHAMINE
Eratrex → ERYTHROMYCIN STEARATE
Eravern → PIPERAZINE
Erazon → PIROXICAM
Erbacort → PREDNISOLONE
Erbasona → PREDNISOLONE
Erbe da fossi → CHELIDONIUM MAJUS
Erbo de Sant Gui → AGRIMONIA EUPATORIA

ERBON
Erco-Fer → FERROUS FUMARATE
Ercoquin → HYDROXYCHLOROQUINE
Ercotina → PROPANTHELINE BROMIDE
Erebile → DEHYDROCHOLIC ACID
Eremfat → RIFAMPIN
EREMOCARPUS SETIGERUS
EREMOPHILA MACULATA
Eretmochelys imbricata → TURTLES
Ergamisole → LEVAMISOLE
Ergate → ERGOTAMINE TARTRATE
Ergenyl → SODIUM VALPROATE also → VALPROIC ACID
Ergobasine → ERGONOVINE
Ergocalciferol → VITAMIN D_2
Ergodesit → ERGOLOID MESYLATES
Ergohydrin → ERGOLOID MESYLATES
Ergoklinine → ERGONOVINE
Ergoline → ARGYREIA
ERGOLOID MESYLATES
Ergomar → ERGOTAMINE TARTRATE
Ergometrine → ERGONOVINE
Ergomimet → DIHYDROERGOTAMINE METHANESULFONATE (MESYLATE)
Ergonil → ERGONOVINE
ERGONOVINE also → PARTHENIUM
Ergont → DIHYDROERGOTAMINE METHANESULFONATE (MESYLATE)
Ergoplus → ERGOLOID MESYLATES
Ergorone → VITAMIN D_2
Ergostat → ERGOTAMINE TARTRATE
Ergostetrine → ERGONOVINE
ERGOT also → DACTYUS GLOMERATA; FESCUE, TALL; HOLCUS LANATUS, MILK, OATS, PARTHENIUM, PENNISETUM sp., ZOLMITRIPTAN
ERGOTAMINE TARTRATE also → CLAVICEPS, INDINAVIR, PROPRANOLOL, RITONAVIR, TROLEANDOMYCIN
Ergotartrat → ERGOTAMINE TARTRATE
Ergotocine → ERGONOVINE
Ergotonin → DIHYDROERGOTAMINE METHANESULFONATE (MESYLATE)
Ergotrate → ERGONOVINE
Eridan → DIAZEPAM
Eridone → PHENINDIONE
Erioglancine → FDC BLUE #1
Erionite → ZEOLITES
Eriosept → DEQUALINIUMS, *chloride*
Eriscel → ERYTHROMYCIN ESTOLATE
Eritrocina → ERYTHROMYCIN
Eritroger → ERYTHROMYCIN ESTOLATE
Eromycin → ERYTHROMYCIN ESTOLATE
Erpalfa → CYTARABINE
Errolon → FUROSEMIDE
Ertron → VITAMIN D_2
ERUCIC ACID also → GLUCOSINOLATES, RAPE
Erun → ERYTHROPHLEUM
Erusticator® → HYDROFLUORIC ACID
Erve Aran Aparan → SPIGELIA ANTHELMIA
Ervenia sp. → OAKMOSS
Ervum lens → LENTILS
Erwinase → *l*-ASPARAGINASE
ERYC → ERYTHROMYCIN
Erycen → ERYTHROMYCIN
Erycin → ERYTHROMYCIN
Erycinum → ERYTHROMYCIN
Ery Derm → ERYTHROMYCIN
Eryliquid → ERYTHROMYCIN ETHYL SUCCINATE
Erymax → ERYTHROMYCIN
Erymysin → ERYTHROMYCIN
Erypar → ERYTHROMYCIN STEARATE
Eryped → ERYTHROMYCIN ETHYL SUCCINATE
Erypo → ERYTHROPOIETIN
Eryprim → ERYTHROMYCIN STEARATE
Erysan → CHLORMAPHAZINE
ERYSIMUM CHEIRANTHOIDES
Ery-Tab → ERYTHROMYCIN
Erythrene → 1,3-BUTADIENE
ERYTHRINA sp.
ERYTHRITOL ANHYDRIDE
ERYTHRITYL TETRANITRATE
Erythrocin → ERYTHROMYCIN also → ERYTHROMYCIN ETHYL SUCCINATE also → ERYTHROMYCIN STEARATE
Erythro ES → ERYTHROMYCIN ETHYL SUCCINATE
Erythro-Holz → ERYTHROMYCIN ETHYL SUCCINATE
Erythrol Tetranitrate → ERYTHRITYL TETRANITRATE
Erythromast 36 → ERYTHROMYCIN
Erythromid → ERYTHROMYCIN
ERYTHROMYCIN also → ALPRAZOLAM, ATORVASTATIN, ASTEMIZOLE, BROMOCRIPTINE, CARBAMAZEPINE, CERIVASTATIN, CHLORAZEPATE, CHOLINE THEOPHYLLINATE, CILOSTAZOL, CIMETIDINE, CISAPRIDE, CLONAZEPAM, CLOZAPINE, CYCLOSPORINE, DIAZEPAM, DIGOXIN, DIHYDROERGOTAMINE METHANESULFONATE, DISOPYRAMIDE, ESTRIOL, FELBAMATE, FEXOFENADINE, FRUITS and FRUIT JUICES,

ITRACONAZOLE, KAOLIN, KETOCONAZOLE, LINCOMYCIN, LORATADINE, LOVASTATIN, METHYLPREDNISOLONE, METHYSERGIDE, MIDAZOLAM, MILK, OLEANDOMYCIN, PECTIN, PENICILLIN; PENICILLINS – "NEWER and SYNTHETICS" > *Methicillin Sodium*; PERGOLIDE MESYLATE, QUINIDINE, REPAGLINIDE, SERTRALINE, SILDENAFIL, SIMVASTATIN, SULFAMETHOXAZOLE, TACROLIMUS, TERFENADINE, THEOPHYLINE, THYMOL, TEREMIFENE CITRATE, TRIAZOLAM, VALPROIC ACID, VERAPAMIL, WARFARIN
Erythromycin A → ERYTHROMYCIN
ERYTHROMYCIN ESTOLATE also → SODIUM LAURYL SULFATE
ERYTHROMYCIN ETHYL SUCCINATE also → ACENOCOUMAROL
ERYTHROMYCIN GLUCEPTATE
Erythromycin Glucoheptonate → ERYTHROMYCIN GLUCEPTATE
ERYTHROMYCIN LACTOBIONATE
ERYTHROMYCIN PROPIONATE also → SODIUM LAURYL SULFATE
ERYTHROMYCIN PROPIONATE
ERYTHROMYCIN STEARATE
ERYTHRONIUM
Erythroped → ERYTHROMYCIN ETHYL SUCCINATE
Erythrophlamine → ERYTHROPHLEUM
Erythrophleine → ERYTHROPHLEUM, IRONWOOD
ERYTHROPHLEUM
Erythrophloeum chlorostachys → IRONWOOD
ERYTHROPOIETIN also → CAPTOPRIL, GROWTH HORMONES, THEOPHYLLINE, VALPROIC ACID
Erythro S → ERYTHROMYCIN STEARATE
ERYTHROSINE also → DANAZOL, L-DOPA, GELATIN, IODINE
Erythrosine B → ERYTHROSINE
Erythrosine BS → ERYTHROSINE
Erythroxylon → COCA
Erytromycin → ERYTHROMYCIN
Esantene → INOSITOL NIACINATE
Esbatal → BETHANIDINE
Esberiven → MELILOT
Escalol 506 → PADIMATES
ESCHOSCHOLTZIA CALIFORNICA
ESCIN
Esclama → NIMORAZOLE
Esclerosina → OLAMINE
Escort → DIAZINON
Escre → CHLORAL HYDRATE
Esculin → AESCULUS sp.
Esdragole → ESTRAGOLE
Esere Nut → PHYSOSTIGMA VENENOSUM
Eserine → PHYSOSTIGMINE
ESF → ERYTHROPOIETIN
Esgram → PARAQUAT
Esidrex → HYDROCHLOROTHIAZIDE
Esidrix → HYDROCHLOROTHIAZIDE
Esilgan → ESTAZOLAM
Esimil → GUANETHIDINE
Esinol → ERYTHROMYCIN ETHYL SUCCINATE
Eskabarb → PHENOBARBITAL
Eskacef → CEPHADRINE
Eskacillin → PENICILLIN > *Benzylpenicillin, Potassium*
Eskalin → VIRGINIAMYCIN
Eskalith → LITHIUM CARBONATE
Eskaserp → RESERPINE
Eskazine → TRIFLUOPERAZINE DIHYDROCHLORIDE
Eskazinyl → TRIFLUOPERAZINE DIHYDROCHLORIDE
Eskel → KHELLIN
Eskusan → AESCULUS sp.
Esmarin → TRICHLORMETHIAZIDE
Esmeron → ROCKURONIUM BROMIDE
Esmind → CHLORPROMAZINE
ESMOLOL also → VECURONIUM BROMIDE
Esoderm → LINDANE
Esophotrast → BARIUM SULFATE
Esperson → DESOXIMETASONE
"Espichamento" → SOLANUM sp.
Espo → ERYTHROPOIETIN
Espril → NIALAMIDE
Espyre → DIPYRONE
Esradin → ISRADIPINE
Esrar → CANNABIS
Esromiotin → PHYSOSTIGMINE
Esrovis → QUINESTROL
Essence of Ginger → GINGER, JAMAICAN
Essence of Mirbane → NITROBENZENE
Estafiate → LARKSPUR
ESTAZOLAM
Ester 25 → DIETHYL-*p*-NITROPHENYLPHOSPHATE
Esteron 44 → 2,4-DICHLOROPHENOXYACETIC ACID
Esteron 245 → 2,4,5-TRICHLOROPHENOXYACETIC ACID

Estigyn → ETHINYL ESTRADIOL
ESTIL
Estilben → DIETHYLSTILBESTROL
Estilsona → PREDNISOLONE
Estinyl → ETHINYL ESTRADIOL
Estomicina → ERYTHROMYCIN ESTOLATE
Estomycin → PAROMOMYCIN
Estopen → PENICILLIN > *Benzylpenicillin, DAEEH*
Estrace → ESTRADIOL
Estracyst → ESTRAMUSTINE
Estraderm → ESTRADIOL
ESTRADIOL also → BEANS, CHLORCYÇLIZINE, CHLORDANE, DDT, KETOCONAZOLE, ORPHENADRINE, PHENOBARBITAL, PHENYLBUTAZONE, SCHRADAN, TESTOSTERONE
β-Estradiol → ESTRADIOL also → GRAPEFRUIT, MELATONIN
cis-Estradiol → ESTRADIOL
ESTRADIOL BENZOATE also → KALE
ESTRAGOLE
Esitragon → TARRAGON
ESTRAMUSTINE
Estratrienediol → ESTRADIOL
Estrin → ESTRONE
ESTRIOL also → BEANS, CORTICOTROPIN, CORTISONE, METHENAMINE, MEPROBAMATE, PROBENECID
Estroate → ESTROGENS, CONJUGATED
Estrobene → DIETHYLSTILBESTROL
Estrobene DP → DIETHYLSTILBESTROL
Estroclim → ESTRADIOL
Estro-Dequavagn → ESTRIOL
Estrodienol → DIENESTROL
ESTROGENS, CONJUGATED also → APPLE, BUTTER, CERULOPLASMIN, CHOLESTEROL, COMFREY, DIAZEPAM, DIGITALIS, DISULFIRAM, ESTRAMUSTINE, FOLIC ACID, GINSENG, GLIPIZIDE, IMIPRAMINE, MILK, OXYTETRACYCLINE, PHENOBARBITAL, PHENYLBUTAZONE, PHENYTOIN, PORK, SPIRONOLACTONE, SULFOBROMOPHTHALEIN SODIUM, TETRACYCLINE, THYROXINE
Estrol → ESTRONE
ESTRONE also → BEANS, CHLORDANE, OATS, ORPHENADRINE, PHENO-BARBITAL, POMEGRANATE, SCHRADAN
Estropipate → ESTRONE, *Piperazine Sulfate*
Estroral → DIENESTROL
Estrosol → DICHLORVOS
Estrosyn → DIETHYLSTILBESTROL
Estrugenone → ESTRONE
Estrusol → ESTRONE
Estulic → GUANFACINE
Estumate → CLOPROSTENOL
Esurol → MERCAPTODIMETHUR
Esuru gudugudu → DIOSCOREA
ET 14 → RONNEL
ET 57 → RONNEL
ET 495 → PIRIBEDIL
Etaciland → PENICILLINS – "NEWER and SYNTHETICS" > *Hetacillin*
Etadrol → FLUPREDNISOLONE
Etamon Chloride → TETRAETHYLAMMONIUM CHLORIDE
Etamsylate → ETHAMSYLATE
ETANERCEPT
Etaperasine → PERPHENAZINE
Etaphylline → ACEFYLLINE PIPERAZINE
Etapiam → ETHAMBUTOL
ETHACRIDINE LACTATE
ETHACRYNIC ACID also → ALLOPURINOL, DIGITALIS, ETHYL BISCOUMACETATE, GALLAMINE TRIETHIODIDE, GENTAMICIN, GLUCOSE, HEPARIN, INSULIN, KANAMYCIN, LITHIUM, MECAMYLAMINE, NALIDIXIC ACID, NITROFURANTOIN, PROBENECID, SODIUM, STREPTOMYCIN, VERATRUM sp., WARFARIN
Ethal → ALCOHOL, CETYL
ETHAMBUTOL also → ALUMINUM HYDROXIDE, BROMELAIN, METOCLOPRAMIDE
Ethamide → ETHOXZOLAMIDE
Ethaminal → PENTOBARBITAL
ETHAMIVAN
Ethamolin → ETHANOLAMINE also → OLAMINE
ETHAMOXYTRIPHETOL
ETHAMSYLATE
Ethanal → ACETALDEHYDE
Ethanamine → ETHYLAMINE
ETHANE
Ethanedial → GLYOXAL
1,2-Ethanediamine → ETHYLENEDIAMINE
Ethanedioic Acid → OXALIC ACID
1,2-Ethanediol → ETHYLENE GLYCOL
1,2-Ethanediol Dinitrate → ETHYLENE GLYCOL DINITRATE
Ethanedioyl Dichloride → OXALYL CHLORIDE
Ethanenitrile → ACETONITRILE

Ethane Pentachloride → PENTACHLOROETHANE
ETHANETHIOL
Ethanethiomide → THIOACETAMIDE
Ethanoic Acid → ACETIC ACID
Ethanol → ALCOHOL, ETHYL
Ethanolamine Oleate → OLAMINE
Ethanol Butoxide → PIPERONYL BUTOXIDE
Ethaperazine → PERPHENAZINE
Ethaphene → SYNEPHRINE
Ethavan → ETHYL VANILLIN
ETHCHLORVYNOL also → ALCOHOL, ETHYL; BISHYDROXYCOUMARIN, WARFARIN
Ethene → ETHYLENE
Ethenol Homopolymer → POLYVINYL ALCOHOL
Ethenone → KETENE
Ethenyl Acetate → VINYL ACETATE
Ethenylbenzene → STYRENE
Ethenyl Ethanoate → VINYL ACETATE
Etheophyl → THEOPHYLLINE
Ether → ETHER, ETHYL
Ether Chloratus → ETHYL CHLORIDE
ETHER, ETHYL also → ATROPINE, CHLORINE, CURARE, NEOMYCIN, PROPRANOLOL
Ether Hydrochloric → ETHYL CHLORIDE
Ether Muriatic → ETHYL CHLORIDE
ETHER, VINYL
Ethimide → ETHIONAMIDE
ETHINAMATE also → ALCOHOL, ETHYL
Ethine → ACETYLENE
ETHINYL ESTRADIOL also → ACETAMINOPHEN, ATORVASTATIN, CYCLOSPORINE, FLUCONAZOLE, MESTRANOL, RIFABUTIN, RIFAMPIN, RITONAVIR, TERBINAFINE, VITAMIN A
17-Ethinylestradiol → ETHINYL ESTRADIOL
Ethinylestrenol → LYNESTRENOL
3-Ethinylnirvanol → MEPHENYTOIN
Ethinyltestosterone → ETHISTERONE
Ethinyl Trichloride → TRICHLOROETHYLENE
Ethiodan → IOPHENDYLATE
ETHIODIZED OIL
Ethiodol → ETHIODIZED OIL
Ethiofos → AMIFOSTINE
ETHION
ETHIONAMIDE also → ALCOHOL, ETHYL; *p*-AMINOSALICYLIC ACID
Ethioniamide → ETHIONAMIDE
ETHIONINE
ETHISTERONE
Ethmosine → MORICIZINE
Ethmozin(e) → MORICIZINE
Ethnine → PHOLCODINE
Ethnor → BUTABARBITAL
Ethocaine → PROCAINE
Ethodin → ETHACRIDINE LACTATE
Ethodryl → DIETHYLCARBAMAZINE
Ethoforme → ETHYL-*p*-AMINOBENZOATE
ETHOGLUCID
ETHOHEPTAZINE
ETHOHEXADIOL
Ethol → ALCOHOL, CETYL
ETHOPROP
Ethoprophos → ETHOPROP
ETHOPROPAZINE
ETHOSUXIMIDE also → DEXTROAMPHETAMINE, ISONIAZID, SODIUM VALPROATE
ETHOTOIN also → DISULFIRAM
Ethotrimeprazine → ETHYL ISOBUTRAZINE
Ethovan → ETHYL VANILLIN
ETHOXAZENE
p-Ethoxychrysoidine → ETHOXAZENE
ETHOXYCOUMARIN also → TERBINAFINE
Ethoxyethane → ETHER, ETHYL
2-ETHOXYETHANOL *also* → ERYTHROMYCIN
2-(2-ETHOXYETHOXY)ETHANOL
2-ETHOXYETHYL ACETATE
1-ETHOXYETHYLIDENE PROPANEDINITRILE
2-Ethoxyethyl-p-methoxycinnamate → CINOXATE
Ethoxylated Lanolin → LANOLIN
2-ETHOXYNAPHTHALENE
ETHOXYQUINE also → HYDROGEN CYANIDE
Ethoxyzolamide → ETHOXZOLAMIDE
ETHOXZOLAMIDE also → LITHIUM CARBONATE
Ethrane → ENFLURANE
Ethrel → ETHEPHON
Ethril → ERYTHROMYCIN STEARATE
Ethryn → ERYTHROMYCIN STEARATE
Ethy 11 → ETHINYL ESTRADIOL
ETHYBENZTROPINE
Ethydan → ETHER, VINYL
ETHYL ACETATE also → COFFEE
Ethyl Acetone → METHYL PROPYL KETONE
ETHYL ACRYLATE
Ethyl Adrianol → ETHYL PHENYLEPHRINE
Ethyl Alcohol → ALCOHOL, ETHYL
Ethylaldehyde → ACETALDEHYDE
ETHYLAMINE
ETHYL-*p*-AMINOBENZOATE also → CHLORPROPAMIDE, CITRIC ACID,

8-HYDROXYQUINOLINE, LIDOCAINE, PROCAINE, SULFANILAMIDE, TETRACAINE HYDROCHLORIDE, TOLBUTAMIDE
ETHYL *sec*-AMYL KETONE
Ethylan → PERTHANE
ETHYLBENZENE
ETHYL BENZOATE
Ethyl Benzol → ETHYLBENZENE
ETHYLBISCOUMACETATE also → BENZIODARONE, CLOFIBRATE, DIPYRONE, ETHACRYNIC ACID, GLUTETHIMIDE, HEPTABARBITAL, ISOCARBOXAZID, MERCAPTOPURINE, METHYLPHENIDATE, NIALAMIDE, PENICILLIN, PHENYLBUTAZONE
ETHYL BROMIDE
ETHYL BROMOACETATE
Ethyl-n-butanoate → ETHYL BUTYRATE
ETHYL BUTYL KETONE
ETHYL BUTYL THIOUREA also → COPY PAPER and COPY MACHINES, RUBBER, SHOES, WET SUITS
ETHYL BUTYRATE
Ethyl Carbamate → URETHANE
Ethyl Carbinol → *n*-PROPYL ALCOHOL
ETHYL CHLORIDE
ETHYL CINNAMATE
Ethyl Cyanide → PROPIONITRILE
α-ETHYL CYANOACRYLATE
Ethyldicoumarol Acetate → ETHYL BISCOUMACETATE
Ethyl Diethylene Glycol → 2-(2-ETHOXYETHOXY)ETHANOL
Ethyl Digol → 2-(2-ETHOXYETHOXY)ETHANOL
N-[1-[2-(4-Ethyl-4,5-dihydro-5-oxo-1H-tetrazol-1-y1 ethyl)]-4-(methoxymethyl)-4-piperidinyl]-N-phenylpropanamide → ALFENTANIL
5-ETHYL DIISOBUTYLTHIOCARBAMATE
Ethyl Dodecanoate → ETHYL LAURATE
Ethyl Dodecylate → ETHYL LAURATE
Ethylemin → DIETHAZINE
ETHYLENE
ETHYLENE BISDITHIOCARBAMATES
ETHYLENEBISTETRABROMOPHTHALIC IMIDE
ETHYLENE BRASSYLATE
Ethylene Bromide → ETHYLENE DIBROMIDE
Ethylene Chloride → ETHYLENE DICHLORIDE
ETHYLENE CHLOROHYDRIN
ETHYLENEDIAMINE also → AMINOPHYLLINE, GRAMICIDIN, HYDROXYZINE, NEOMYCIN, PIPERAZINE, THEOPHYLLINE, THONZYLAMINE HYDROCHLORIDE, TRIENTINE, TRIPELENNAMINE
ETHYLENE DIBROMIDE also → DISULFIRAM
ETHYLENE DICHLORIDE also → 1,1-DICHLOROETHANE
Ethylene Dinitrate → ETHYLENE GLYCOL DINITRATE
ETHYLENE GLYCOL also → CALCIUM OXALATE, FORMIC ACID, GLYOXYLIC ACID, HYDRAULIC FLUIDS, INKS, OXALIC ACID, PYRAZOLE, SILVER
ETHYLENE GLYCOL DINITRATE
Ethylene Glycol Monobutyl Ether → 2-BUTOXYETHANOL
ETHYLENE GLYCOL MONOBUTYL ETHER
ETHYLENE GLYCOL MONOBUTYL ETHER ACETATE
Ethylene Glycol Monoethyl Ether → 2-ETHOXYETHANOL
Ethylene Glycol Monoethyl Ether Acetate → 2-ETHOXYETHYL ACETATE
Ethylene Glycol Monomethyl Ether → 2-METHOXYETHANOL
Ethylene Glycol Monomethyl Ether Acetate → METHOXYETHYL ACETATE
Ethylene Hexachloride → HEXACHLOROETHANE
ETHYLENEIMINE
1,8-Ethylenenaphthalene → ACENAPHTHENE
Ethylene Nitrate → NITRATES and NITRITES
ETHYLENE OXIDE also → ETHYLENE CHLOROHYDRIN, RHIZOPUS sp., RUBBER
Ethylene Tetrachloride → TETRACHLOROETHYLENE
ETHYLENE THIOUREA *also* → ETHYLENE BISDITHIOCARBAMATES, MANEB
Ethylene Undecane Dicarboxylate → ETHYLENE BRASSYLATE
Ethylenimine → ETHYLENEIMINE
ETHYLESTRENOL also → BISHYDROXYCOUMARIN, PHENFORMIN, WARFARIN
ETHYL FORMATE
Ethyl Green → BRILLIANT GREEN
Ethyl Guthion → AZINPHOS-ETHYL
Ethyl Heptazine → ETHOHEPTAZINE
Ethylhexanediol → ETHOHEXADIOL
2-ETHYL-1-HEXANOL
2-Ethylhexyl Alcohol → 2-ETHYL-1-HEXANOL
Ethyl Hydrate → ALCOHOL, ETHYL
Ethyl Hydride → ETHANE

Eucil → METOCLOPRAMIDE
Eucistin → NALIDIXIC ACID
Eucitin → DOMPERIDONE
Eucodal → OXYCODONE
Eucoran → NIKETHAMIDE
Eucupine → EUPROCIN
Eucytol → DEOXYRIBONUCLEIC ACID
Eudemine → DIAZOXIDE
Eudextran → DEXTRAN, *40*
Eudigox → DIGOXIN
Eudyna → VITAMIN A
Eufibron → PROPYPHENAZONE
Euflavine → ACRIFLAVINE
Euflex → FLUTAMIDE
Eugenic Acid → EUGENOL
EUGENOL also → CLOVES
Eugenyl Methyl Ether → METHYL EUGENOL
EUGLENA SANGUINEA
Euglucon → GLYBURIDE
Euhypnos → TEMAZEPAM
Euipnos → TEMAZEPAM
Eukodal → OXYCODONE
Eukraton → BEMEGRIDE
Eukystol → HALOPERIDOL
Eulaxan → BISACODYL
Eulexin → FLUTAMIDE
Eulipos → DEXTROTHYROXINE SODIUM
Eu-Med → ACETAMINOPHEN
Eumelanin → PHEOMELANIN
Eunades → VINORELBINE
Eunarcon → NARCOBARBITAL
Eunoctal → AMOBARBITAL
Eunoctin → NITRAZEPAM
Euonymin → CELASTRUS SCANDENS
EUONYMUS
EUPATORIUM sp. also → SNAKEROOT, WHITE
Euphorbe des jardins → EUPHORBIA sp.
EUPHORBIA sp. also → HONEY
Euphorbia pulcherrima → POINSETTIA
EUPHORBIN also → EUPHORBIA, POINSETTIA
Euphyllin → THEOPHYLLINE
Euphyllin CR → AMINOPHYLLINE
Euphylong → THEOPHYLLINE
Eupractone → DIMETHADIONE
Eupragin → ERYTHROMYCIN ESTOLATE
Eupressyl → URAPIDIL
EUPROCIN
Euprovasin → PROPRANOLOL
Euradal → BISOPROLOL
Eurax → CROTAMITON
Euraxil → CROTAMITON
Eureceptor → CIMETIDINE
EureCor → ISOSORBIDE DINITRATE
Eurekene → SODIUM VALPROATE also → VALPROIC ACID
Eurex → PRAZOSIN HYDROCHLORIDE
Eurodin → ESTAZOLAM
Eurodopa → L-DOPA
European Bittersweet → SOLANUM sp.
European Buckthorn → FRANGULA
European Goldenrod → GOLDENROD
European Hellebore → VERATRUM sp.
European Hemlock → CONIUM MACULATUM
European Mandrake → MANDRAGORA OFFICINARUM
Eurosan → DIAZEPAM
Eurypelma Henizii → SPIDERS
Eusaprim → SULFAMETHOXAZOLE
Eusaprim → TRIMETHOPRIM
Eusmanid → BETHANIDINE
Eusolex 3573 → OXYBENZONE
Euspiran → ISOPROTERENOL
Euspirin → ISOPROTERENOL
Eustidil → HALOXON
Eustrongylides → SUSHI
Eutagen → OXYCODONE
Euteberol → SPIRONOLACTONE
Eutensin → FUROSEMIDE
Eutensol → GUANETHIDINE
Euthyrox → THYROXINE
Eutonyl → PARGYLINE
Eutrit → XYLITOL
Euvaderm → BETAMETHASONE
Euvestin → DIETHYLSTILBESTROL
Euvitol → FENCAMFAMINE
Evacalm → DIAZEPAM
Evacort → HYDROCORTISONE
EVANS BLUE
Evazol → DEQUALINIUMS, *chloride*
Evening Primrose → OENTHERA BIENNIS
Evening Trumpetflower → GELSEMIUM SEMPERVIRENS
Evik → AMETRYN
Evista → RALOXIFENE
EVONOSIDE also → EUONYMUS
Evorel → ESTRADIOL
Evoxac → CEVIMELINE
Evoxin → DOMPERIDONE
Evramycin → TROLEANDOMYCIN
Ewon Agogo → LANTANA sp.
Exacin → ISEPAMICIN
Exacyl → TRANEXAMIC ACID
Exagamer → LINDANE
Exal → VINBLASTINE
Exangit → BENZONATATE
Exceglan → ZONISAMIDE

Excenel → CEFTIOFUR
EXD → BISETHYLXANTHOGEN
Exdol → ACETAMINOPHEN
Exegram → ZONISAMIDE
Exelderm → SULCONAZOLE NITRATE
Exelgyn → BUTOCONAZOLE
EXEMESTANE
Exhirud → HIRUDINS
Exhirudine → HIRUDINS
Exitelite → ANTIMONY TRIOXIDE
Ex-Lax® → PHENOLPHTHALEIN
Exlutena → LYNESTRENOL
Exluton(a) → LYNESTRENOL
Exmigra → ERGOTAMINE TARTRATE
ExNa → BENZTHIAZIDE
Exocin → OFLOXACIN
Exocorpol → POLOXALENE
Exodol → KETOROLAC
Exofene → HEXACHLOROPHENE
Exolan → TRIACETOXYANTHRACENE
Exonal → TEGAFUR
Exosalt → BENZTHIAZIDE
Exotherm Termil → CHLOROTHALONIL
EXP 105-1 → AMANTADINE
EXP 126 → RIMANTADINE
Expansine → PATULIN
Expar → PERMETHRIN
Exponcit N → NORPSEUDOEPHEDRINE
Exposition → CALADIUM
Exrheudon N → PHENYLBUTAZONE
Exsel → SELENIUM
Ext. D&C Red #15 → PONCEAU 3R
Ext. D&C Yellow no. 1 → METANIL YELLOW
Ext. D&C Yellow no. 9 → YELLOW AB
Ext. D&C Yellow no. 10 → YELLOW OB
Extracort → TRIAMCINOLONE
Extramycin → SISOMICIN
Extranase → BROMELAIN
Extraneal → ICODEXTRIN
Extrinsic Factor → VITAMIN B_{12}
Exuril → CHLOROTHIAZIDE
Exypaque → IOHEXOL
Eyebane → EUPHORBIA
Eye Openers → AMPHETAMINES
E-Z-HD → BARIUM SULFATE
Ezonegi → CHIVES
E-Z-Paque → BARIUM SULFATE

F

190 F → ACETARSONE
F-2 Toxin → GIBERELLA ZEAE
F 190 → ACETARSONE
710F → PLASMOCID
1358F → DAPSONE
F 12 → DICHLORODIFLUOROMETHANE
F 22 → MONOCHLORODIFLUOROMETHANE
F 139 → BUTONATE
F 440 → DANTROLENE
F 1991 → BENOMYL
F 2559 → GALLAMINE TRIETHIODIDE
F 75,980 → PREDNISOLONE
FAA → FLUOROACETAMIDE
2-FAA → *N-2*-FLUORENYLACETAMIDE
Faba Bean → BEANS, Faba
Faba Calabarica → PHYSOSTIGMA VENENOSUM
Fabianol → AMIKACIN
Fabrol → ACETYLCYSTEINE
FAC → FLUOROACETIC ACID
FAC 5273 → PIPERONYL BUTOXIDE
FACE TISSUES
Factor II → WARFARIN
Factor VII → DESMOPRESSIN, WARFARIN
Factor VIII → DESMOPRESSIN
Factor IX → WARFARIN
Factor X → WARFARIN
Factor A-377 → CHLORTETRACYCLINE
Factor U → FOLIC ACID
Fado → CEFAMANDOLE
FADOGIA sp.
Fadormir → METHAQUALONE
Fagopyrin → BUCKWHEAT
Fagopyrum esculentum → BUCKWHEAT
FAGUS SYLVATICA
Fairy Gloves → DIGITALIS
Falapen → PENICILLIN > *Benzylpenicillin, Potassium*
Falithrom → PHENPROCOUMON
Fall Crocus → COLCHICUM AUTUMNALE
Fall Poison → SNAKEROOT, WHITE
False Acacia → LOCUST, BLACK also → ROBIN
False Azalea → MENZIESIA FERRUGINEA
False Blushes → AMANITA sp.
False Cactus → EUPHORBIA
False Hellebore → VERATRUM sp.
False Indigo → BAPTISIA sp.
False Iroko → ANTIARIS TOXICARIA
False Morel → GYROMITRA ESCULENTA
False Peyote → ARIOCARPUS RETUSUS
False Poinciana → DAUBENTONIA PUMICEA
False Ragweed → IVA sp. also → MARSH ELDER
False Sago Palm → PALMS
Famet → SULFAMETHIZOLE
FAMCICLOVIR

Famodil → FAMOTIDINE
Famodine → FAMOTIDINE
Famophos → FAMPHUR
Famosan → FAMOTIDINE
Famosept → PHENYLMERCURIC BORATE
FAMOTIDINE also → ALCOHOL, ETHYL; COFFEE, KETOCONAZOLE, NIZATIDINE, WARFARIN
Famoxal → FAMOTIDINE
2-F-ara-AMP → FLUDARABINE PHOSPHATE
FAMPHUR
Famvir → FAMCICLOVIR
Fanasil → SULFADOXINE
Fang-chi → STEPHANIA TETRANDA
Fang Fang → GINSENG
Fanodormo → CYCLOBARBITAL
Fanosin → FAMOTIDINE
Fan Palm → SAW PALMETTO
Fantorin → STIBOPHEN
Fanweed → THLASPI ARVENSE
Fanzil → SULFADOXINE
Farecef → CEFOPERAZONE
Faredina → CEPHALORIDINE
Fareston → TOREMIFENE CITRATE
Farexin → CEPHALEXIN
Farfara → COLTSFOOT
Fargan → PROMETHAZINE
Farlutal → MEDROXYPROGESTERONE ACETATE
"Farmer's Lung" → ASPERGILLUS also → DUSTS, MICROPOLYSPORA FAENI, THERMOPOLYSPORA POLYSPORA
Farmicetina → CHLORAMPHENICOL
Farmiglucin → PAROMOMYCIN
Farminosidin → PAROMOMYCIN
Farmiserina → CYCLOSERINE
Farmitalia → SULFALENE
Farmolisina → DIPYRONE
Farmorubicin → EPIRUBICIN
Farmoxin → CEFOXITIN SODIUM
Farnwurzel → ASPIDIUM
Fasa K'aba → PORTULACA OLERACEA
Fas-Cile → MEPROBAMATE
Fasciolin → HEXACHLOROETHANE
Fasigin → TINIDAZOLE
Fasigyn → TINIDAZOLE
Fastac → CYPERMETHRIN
Fast Green FCF → FDC GREEN #3
Fast Green J → BRILLIANT GREEN
Fastin → PHENTERMINE
Faston → DIMENHYDRINATE
Fastum → KETOPROFEN
Fasupond → NORPSEUDOEPHEDRINE
FATS also → *p*-AMINOSALICYLIC ACID
FAT, TOXIC
Fatarita → SORGHUM
Fat Hen → CHENOPODIUM
Fausse Herb à Poux → IVA sp.
Faustan → DIAZEPAM
Fava Bean → BEANS, Faba
FAVEIRA TREE
Faverin → FLUOXAMINE MALEATE
Favistan → METHIMAZOLE
Fawn Lily → ERYTHRONIUM
FB/2 → DIQUAT DIBROMIDE
FBA 1420 → PROPANIDID
FBA 4503 → PROPIRAM FUMARATE
FC 3 → ETHYLENE DIBROMIDE
FC 22 → MONOCHLORODIFLUOROMETHANE
FC 142b → MONOCHLORODIFLUOROETHANE
FC 1157a → TOREMIFENE CITRATE
FC 3001 → TIAPROFENIC ACID
FCE 21,336 → CABERGOLINE
F-Cortef → FLUDROCORTISONE
FCV → FAMCICLOVIR
FDA → ANETHOLE, APPLE, BENZENE, BREAD (with POTASSIUM BROMATE), CARBADOX, CHLORPYRIFOS METHYL, CHOCOLATE, CHROMIUM PICOLINATE, CIMETIDINE, COFFEE, COUMARIN, CORN, DANTHRON, DEHYDROCHOLIC ACID, DIMETRIDAZOLE, DIOXIN, ENCAINIDE, ETHYLENE DIBROMIDE, GARDONA, JIN BU HUAN, LASERS, LATEX, METHYL TESTOSTERONE, MINERAL WATERS, PAPAIN, POLYVINYL CHLORIDE, PRACTOLOL, PRETZELS, SELENIUM, STREPTOMYCIN, THEOPHYLLINE, TOCAINIDE, TONKA BEANS, TRICHLOROETHYLENE, TROGLITAZONE, UREDOFOS, VINYL CHLORIDE, VITAMIN B_{12}, Preface
FDA-1541 → EPTC
FD & C Blue no. 2 → INDIGOCARMINE
FD & C Red #2 → AMARANTH
FD & C Red no. 3 → ERYTHROSINE also → CHERRY
FD & C Yellow no. 3 → YELLOW AB
FD & C Yellow no. 4 → YELLOW OB
FD & C Yellow no. 5 → TARTRAZINE
FDC BLUE #1
FDC GREEN #3
FDC Red #1 → PONCEAU 3R
FDC Red #4 → PONCEAU SX
FDC RED 40 also → CHERRY

FDC VIOLET #1
FDC YELLOW #6 also → DANAZOLE
FDL 59,002KP → NEDOCROMIL SODIUM
FDS → ISOSORBIDE DINITRATE
5 FDU → FLOXURIDINE
Featherfew → FEVERFEW
Featherfish → PTEROIS VOLITANS
Feather Grass → STIPA sp.
Febrilix → ACETAMINOPHEN
Febrimin → PHENICARBAZIDE
Febrinina → AMINOPYRINE
February Daphne → DAPHNE
Fectrim → SULFAMETHOXAZOLE
Fedex → IRON DEXTRANS
Fedrin → EPHEDRINE
Feenamint® → PHENOLPHTHALEIN
Feguanide → PHENFORMIN
Feinalmin → IMIPRAMINE
Felac → IRON DEXTRINS
Felacrinos → DEHYDROCHOLIC ACID
FELBAMATE also → PHENOBARBITAL, VALPROIC ACID, WARFARIN
Felbamyl → FELBAMATE
Felbatol → FELBAMATE
Feldene → PIROXICAM
FELDSPAR
Fellozine → PROMETHAZINE
Felmane → FLURAZEPAM
Feloday → FELODIPINE
FELODIPINE also → ERYTHROMYCIN, GRAPEFRUIT, ITRACONAZOLE
Felonwort → CHELIDONIUM MAJUS
FELYPRESSIN
Femadol → PROPOXYPHENE
Femadon → IBUPROFEN
Femara → LETROZOLE
Femergin → ERGOTAMINE TARTRATE
Femest → ESTROGENS, CONJUGATED
Femestrone → ESTRONE
Fem-H → ESTROGENS, CONJUGATED
Femidine → PHENACETIN
Femidyn → ESTRONE
Femina → PHENACETIN
Feminone → ETHINYL ESTRADIOL
Femstat → BUTOCONAZOLE
Femulen → ETHYNODIOL DIACETATE
Fenacetine → PHENACETIN
Fenactil → CHLORPROMAZINE
Fenadone → METHADONE HCL
FENAMIPHOS
Fenamisal → POTASSIUM-*p*-AMINOSALICYLATE
Fenamizol → AMIPHENAZOLE
Fenarol → CHLORMEZANONE
Fenarsone → CARBARSONE
Fenasprate → BENORYLATE
Fenate → IRON DEXTRANS
Fenatoin → PHENYTOIN
Fenazil → PROMETHAZINE
Fenazolina → ANTAZOLINE
Fenazoxine → NEFOPAM
Fenbid → IBUPROFEN
FENBUFEN also → NALIDIXIC ACID, NORFLOXACIN, WARFARIN
FENCAMFAMINE
Fence Rider → 2,4,5-TRICHLOROPHENOXYACETIC ACID
Fenchlorphos → RONNEL
Fenchlorvos → RONNEL
FENCLOFENAC also → THYROXINE
Fenclonine → *p*-CHLOROPHENYLALANINE
FENCLOZIC ACID
Fenesal → NICLOSAMIDE
Feneticillin → PENICILLINS – "NEWER and SYNTHETICS" > *Phenethicillin Potassium*
FENFLURAMINE also → ACETAZOLAMIDE, AMMONIUM CHLORIDE, INSULIN, MAGNESIUM CARBONATE, PHENTERMINE, SODIUM BICARBONATE
Fenformin → PHENFORMIN
Fenhydren → PHENINDIONE
Fenibutazona → PHENYLBUTAZONE
Fenibutol → PHENYLBUTAZONE
Fenical → PHENOBARBITAL
Fenicol → CHLORAMPHENICOL
Fenidrim → FENURON
Fenilin → PHENINDIONE
Fenilor → 5,7-DIBROMO-8-QUINOLINOL
Fenistil → DIMETHINDENE
Fenitrothion → SUMITHION
Fenkill → FENVALERATE
FENNEL also → ANETHOLE
Fenobrate → FENOFIBRATE
FENOFIBRATE also → WARFARIN
Fenoform → LINDANE
FENOLDOPAM MESYLATE
FENOPROFEN CALCIUM
Fenopron → FENOPROFEN CALCIUM
Fenoprop → SILVEX
Fenospen → PENICILLIN > *Penicillin V*
Fenostil → DIMETHINDENE
Fenotard → FENOFIBRATE
FENOTEROL HYDROBROMIDE
FENOVERINE
FENOXEDIL
Fenoxypen → PENICILLIN > *Penicillin V*
Fenpidon → DIPIPANONE
Fenprostalene → POLYETHYLENE GLYCOL

Fibrinokinase → TISSUE PLASMINOGEN ACTIVATOR
FIBRINOLYSIN
Fibrotan → HYDRARGAPHEN
Fibrous Glass → FIBERGLASS
Fibula nolitangere → SPONGE
FICIN
Ficortril → HYDROCORTISONE
Ficus aurea → FIG
Ficus carica → FIG
Fiddleneck → AMSINCKIA INTERMEDIA
Field Balm → CATNIP
Field Bean → BEANS, Faba
Field Horsetail → EQUISETUM
Field Nettle → STACHYS ARVENSIS
Field Pennycress → THLASPI ARVENSE
FIG also → ISOCARBOXAZID
Figworts → SCROPHULARIA sp.
Filaribits → DIETHYLCARBAMAZINE
Filaramide → ARSENAMIDE
Filariol → BROMOPHOS-ETHYL
Filazine → DIETHYLCARBAMAZINE
Filban → DIETHYLCARBAMAZINE
Filcide → ARSENAMIDE
Fildesin → VINDESINE SULFATE
FILGRASTIM also → GRANULOCYTE MACROPHAGE COLONY STIMULATING FACTOR, GRANULOCYTE COLONY STIMULATING FACTOR
Filicis Malis → ASPIDIUM
Filix Mas → ASPIDIUM
Filoral → CHOLINE THEOPHYLLINATE
Finalgon → CAPSAICIN
Finaline → BENACTYZINE
FINASTERIDE
Finastid → FINASTERIDE
Fine-leaved Water Dropwort → OENANTHE sp.
Fingerberry → BLACKBERRY
Finger Euphorbia → EUPHORBIA
Finimal → ACETAMINOPHEN
Finlepsin → CARBAMAZEPINE
Finoptin → VERAPAMIL
Fintrol → ANTIMYCIN A
Finuret → HYDROFLUMETHIAZIDE
FIPRONIL
Fireball → KOCHIA
FIRE COLORING
FIRE CORAL also → FORMIC ACID
Firecracker Lily → LILIES
FIRECRACKERS
FIRE EXTINGUISHERS
Firefish → PTEROIS VOLITANS
Firemaster → POLYBROMINATED BIPHENYLS also → TRIS(2-3-DIBROMOPROPYL)PHOSPHATE
Fire-On-The-Mountain → EUPHORBIA
Firewater → BUTYROLACTONE
Fireweed → AMSINCKIA INTERMEDIA also → DATURA, KOCHIA
Firmacef → CEFAZOLIN SODIUM
Firon → FERROUS FUMARATE
First Line → MORPHINE
Fisa → BLIGHIA SAPIDA
Fisalamine → 5-AMINOSALICYLIC ACID
FISH also → GAMBIERDISCUS TOXICUS, HISTAMINE, MERCURY, METHYL MERCURY, *N*-NITROSODIMETHYLAMINE, POLYCHLORINATED BIPHENYLS, SAURINE, SODIUM BENZOATE
Fish Berry → COCCULUS sp.
FISH OILS
Fishtail Palm → PALMS
Fitweed → CORYDALIS
Fivent → CROMOLYN SODIUM
Fivepen → PENICILLIN > *Benzylpenicillin, Potassium*
FK 027 → CEFIXIME
FK 749 → CEFTIZOXIME
FK 880 → SULPIRIDE
FK 906 → TACROLIMUS
FL 1039 → PENICILLINS – "NEWER and SYNTHETICS" > *Pivmecillinam*
FL 1060 → PENICILLINS – "NEWER and SYNTHETICS" > *Amdinocillin*
FLA 731 → RENOXIPRIDE
Flabelline → PENICILLINS – "NEWER and SYNTHETICS" > *Methicillin Sodium*
Flagecidin → ANISOMYCIN
Flagyl → METRONIDAZOLE
FLAIRA
Flake → COCAINE
Flamasone → PREDNISOLONE
Flamazine → SILVER SULFADIAZINE
Flammex → TRIS(2-3-DIBROMOPROPYL)PHOSPHATE
Flammulina → MUSHROOMS
Flamula jovis → CLEMATIS
Flanax → NAPROXEN
Flannel Leaf → VERBASCUM THAPSUS
Flat-Topped Thorn → ACACIA
Flavaxin → VITAMIN B_2
Flavin → FENTICONAZOLE
Flavolutan → PROGESTERONE
Flavoquine → AMODIAQUIN
FLAVOXATE HCL
Flavurol → MERBROMIN

FLAX also → HYDROGEN CYANIDE, PHASEOLUTIN
Flaxedil → GALLAMINE TRIETHIODIDE
Flaxseed → LINSEED
Flebosan → TRIBENOSIDE
FLECAINIDE also → AMIODARONE, QUINIDINE, RITONAVIR, Preface
Flectadol → LYSINE ACETYSALICYLATE
Flectol → 1,2-DIHYDRO-2,2,4-TRIMETHYLQUINOLINE
Flectron → CYPERMETHRIN
Fleequin™ *IA* → MUCOITIN POLYSULFATE ESTER
Flemoxin → PENICILLINS – "NEWER and SYNTHETICS" > *Amoxicillin*
Flenac → FENCLOFENAC
FLEROXACIN also → SUCRALFATE, THEOPHYLLINE
Flesh of the Gods → TEONANACATL
Fleurs de Tilleul → TILIA sp.
Flexal → CARISPRODOL
Flexartal → CARISPRODOL
Flexazone → PHENYLBUTAZONE
Flexeril → CYCLOBENZAPRINE
Flexiban → CYCLOBENZAPRINE
Flexilon → ZOXAZOLAMINE
Flexin → ZOXAZOLAMINE
Flexuosins A & B → HELENIUM sp.
Flindix → ISOSORBIDE DINITRATE
Flixonase → FLUTICASONE PROPIONATE
Flixotide → FLUTICASONE PROPIONATE
Flobacin → OFLOXACIN
Flo-cillin → PENICILLIN > *Benzylpenicillin, Procaine*
FLOCTAFENINE
Flodil → FELODIPINE
Floginax → NAPROXEN
Flogitolo → OXYPHENBUTAZINE
Flogobene → PIROXICAM
Flogobron → OXOLAMINE
Flogoril → OXYPHENBUTAZINE
Flogovital → NIMESULIDE
Flolan → PROSTACYCLIN
Flomax → TAMSULOSIN
Flonatril → CLOREXOLONE
Floraquin → IODOQUINOL
Flor de Tilo → TILIA sp.
Florel → ETHEPHON
Flores martis → FERRIC CHLORIDE
FLORFENICOL
Florid → MICONAZOLE NITRATE
Florida Holly → SCHINUS TEREBINTHIFOLIUS
Floridanine → PYRROLIZIDINES
Floridin → CEPHALORIDINE
Florimycin → VIOMYCIN
Florinef → FLUDROCORTISONE
Florisil → MAGNESIUM SILICATE
Flormidal → MIDAZOLAM
Florocid → SODIUM FLUORIDE
Florone → DIFLORASONE
Floropipamide → PIPAMPERONE
Floropryl → DIISOPROPYL FLUOROPHOSPHATE
Florphenicol → FLORFENICOL
FLOSEQUINAN
Flosequinon → FLOSEQUINAN
Flosin → INDOPROFEN
Flosint → INDOPROFEN
FLOUR also → CARBOPHENOTHION, ENDRIN, MERCURY, METHIONINES, 4,4´-METHYLENE DIANILINE, OATS, POTASSIUM PERSULFATE, POTASSIUM SORBATE, SORBIC ACID
FLOURENSIA CERNUA
Flovacil → DIFLUNISAL
Flovent → FLUTICASONE PROPIONATE
Flowering Spurge → EUPHORBIA
Flower of Paradise → CATHA EDULIS
Flowers of Antimony → ANTIMONY TRIOXIDE
Flowers of Zinc → ZINC
Floxacillin → PENICILLINS – "NEWER and SYNTHETICS" also → FUSIDIC ACID
Floxacin → NORFLOXACIN
Floxapen → PENICILLINS – "NEWER and SYNTHETICS" > *Floxacillin*
Floxicam → ISOXICAM
Floxil → OFLOXACIN
Floxin → OFLOXACIN
FLOXURIDINE also → METHOTREXATE
Floxyfral → FLUVOXAMINE MALEATE
FLUANISONE
Fluanxol → FLUPENTIXOL
Fluanxol Dèpot → FLUPENTIXOL
Fluaton → FLUOROMETHOLONE
Fluatox → ACETYLCYSTEINE
Flubber® → POLYBUTADIENE also → MINERAL OIL
FLUBENDAZOLE
Flubenisolone → BETAMETHASONE
Flubenol → FLUBENDAZOLE
FLUBIPROFEN also → MISOPROSTOL
Flucinom → FLUTAMIDE
Flucloxacillin → PENICILLINS – "NEWER and SYNTHETICS" > *Floxacillin*
Fluconal → FLUCONAZOLE
FLUCONAZOLE also → ACENOCOUMAROL, CISAPRIDE,

CYCLOSPORINE, GLIPIZIDE, GLYBURIDE, INDINAVIR, NORTRIPTYLINE, PHENYTOIN, RIFABUTIN, RIFAMPIN, TACROLIMUS, TERFENADINE, TOLBUTAMIDE, WARFARIN, ZIDOVUDINE
Flucort → FLUMETHASONE
Fluctin → FLUOXETINE
FLUCYTOSINE also → ZIDOVUDINE
Fludara → FLUDARABINE PHOSPHATE
FLUDARABINE PHOSPHATE
Fludestrin → TESTOLACTONE
Fludex → INDAPAMIDE
FLUDROCORTISONE
Fludroxycortide → ,FLURANDRENOLIDE
FLUFENAMIC ACID also → ETOFENAMATE
Flugeral → FLUNARIZINE HCL
Fluibil → CHENODEOXYCHOLIC ACID
Fluidane → BROMINDIONE
Fluidil → CYCLOTHIAZIDE
Fluimucetin → ACETYLCYSTEINE
Fluimucil → ACETYLCYSTEINE
Fluindione → PRAVASTATIN
Fluindostatin → FLUVASTATIN
Fluitran → TRICHLORMETHIAZIDE
Flukamide → RAFOXANIDE
Flumadine → RIMANTADINE
Flumark → ENOXACIN
FLUMAZENIL
Flumazepil → FLUMAZENIL
Flumen → CHLOROTHIAZIDE
Flumesil → BENDROFLUMETHIAZIDE
Flumetasone → FLUMETHASONE
FLUMETHASONE
Flumetholon → FLUOROMETHOLONE
Flumoperone → TRIFLUPERIDOL
Flumoxal → FLUBENDAZOLE
Flumoxane → FLUBENDAZOLE
Flunagen → FLUNARIZINE HCL
FLUNARIZINE HCL
Flunarl → FLUNARIZINE HCL
Flunase → FLUTICASONE PROPIONATE
Fluniget → DIFLUNISAL
FLUNISOLIDE
FLUNITRAZEPAM
FLUNIXIN MEGLUMINE also → METHOXYFLURANE, PREDNISONE
FLUOCINOLONE ACETONIDE
FLUOCINONIDE
Fluodonil → DIFLUNISAL
Fluohydric Acid → HYDROFLUORIC ACID
Fluolon → FLUOROMETHOLONE
FLUOMETURON
Fluon → POLYTEF
Fluonid → FLUOCINOLONE ACETONIDE
Fluopromazine → TRIFLUPROMAZINE
Fluorakil 100 → FLUOROACETAMIDE
Fluorandrenolone → FLURANDRENOLIDE
2,7-FLUORENEDIAMINE
N-2-FLUORENYLACETAMIDE
FLUORESCEIN also → LIPSTICK, THIMEROSAL
Fluorescite → FLUORESCEIN
FLUORIDES and FLUORINE also → ALUMINUM, FISH, FLUOSILICIC ACID, METHOXYFLURANE, POTASSIUM AMMONIUM BIFLUORIDE, SILICON TETRAFLUORIDE, STEEL, TEA, WINE
FLUORINE MONOXIDE
FLUORINE NITRATE
Fluorine Oxide → FLUORINE MONOXIDE
Fluoristan → STANNOUS FLUORIDE
Fluormetholon → FLUOROMETHOLONE
Fluormone → DEXAMETHASONE
FLUOROACETAMIDE also → ACACIA, GASTROLOBIUM sp.
Fluoroacetate → OXYLOBIUM sp.
FLUOROACETIC ACID also → DICEPHAPETALUM, GASTROLOBIUM, GIDYEA, PALICOUREA
Fluoroacetic Acid Amide → FLUOROACETAMIDE
Fluorocitrate → CISPLATIN, FLUOROACETIC ACID, 5-FLUOROURACIL
Fluorodexamethasone → FLUMETHASONE
Fluoroethanoic Acid → FLUROACETIC ACID
Fluorofen → TRIFLUPROMAZINE
Fluoromar → FLUROXENE
FLUOROMETHOLONE
9α-Fluoro-16α-methylprednisolone → DEXAMETHASONE
Fluoroplex → FLUOROURACIL
9-Fluoroprednisolone → ISOFLUPREDONE
Fluoros → SODIUM FLUORIDE
FLUOROSULFONIC ACID
Fluorotrichloromethane → TRICHLOROFLUOROMETHANE
FLUOROURACIL also → CAPACITABINE, FLUCYTOSINE, LATEX, LEVAMISOLE, METHOTREXATE, MISONIAZOLE, WARFARIN
FLUOSILICIC ACID
Fluostigmine → DIISOPROPYL FLUOROPHOSPHATE
Fluothane → HALOTHANE
Fluotic → HALOTHANE
Fluovitef → FLUOCINOLONE ACETONIDE

FLUOXAMINE also → CLOZAPINE, THEOPHYLLINE
Fluoxeren → FLUOXETINE
FLUOXETINE also → ALPRAZOLAM, AMITRIPTYLINE, BUPROPION, CARBAMAZEPINE, CARVEDILOL, CLARITHROMYCIN, CYCLOSPORINE, DEPRENYL, DESIPRAMINE, ENALAPRIL, HALOPERIDOL, IMIPRAMINE, ITRACONAZOLE, LITHIUM, MELATONIN, METHADONE, MIDAZOLAM, NORTRIPTYLINE, RITONAVIR, SUMATRIPTAN SUCCINATE, TERFENADINE, TRIAZOLAM, WARFARIN
FLUOXYMESTERONE
Flupen → PENICILLINS – "NEWER and SYNTHETICS" > *Floxacillin*
Flupenthixol → FLUPENTIXOL
FLUPENTIXOL also → ALCOHOL, ETHYL; ORPHENADRINE
Fluphenacur → LUFENURON
FLUPHENAZINE DECANOATE also → ALCOHOL, ETHYL; ALPRAZOLAM, ALUMINUM HYDROXIDE, AMITRIPTYLINE, CLONIDINE, LITHIUM, ORPHENADRINE, TOBACCO, VITAMIN C
FLUPREDNISOLONE
FLUPROSTENOL
Fluprowit → ACETYLCYSTEINE
Flura → SODIUM FLUORIDE
Fluracil → FLUOROURACIL
FLURANDRENOLIDE
Flurandrenolone → FLURANDRENOLIDE
FLURAZEPAM also → CIMETIDINE, DISULFIRAM
FLURBIPROFEN also → ANTIPYRINE, FUROSEMIDE
Fluril → FLUOROURACIL
Flurobate → BETAMETHASONE
Fluroblastin → FLUOROURACIL
Flurofen → FLURBIPROFEN
Fluropryl → DIISOPROPYL FLUOROPHOSPHATE
FLUROTHYL
Fluro Uracil → FLUOROURACIL
FLUROXENE also → EPINEPHRINE, PHENOBARBITAL, PHENYTOIN
FLUSPIRILINE
Flustar → DIFLUNISAL
Flusulfonic Acid → FLUOROSULFONIC ACID
FLUTAMIDE
Fluted Pumpkin → TELFAIRIA OCCIDENTALIS
FLUTICASONE PROPIONATE
Flutra → TRICHLORMETHIAZIDE
FLUVASTATIN also → WARFARIN
Fluvean → FLUOCINOLONE ACETONIDE
Fluvermal → FLUBENDAZOLE
Fluvert → FLUMETHASONE
Fluvin → HYDROCHLOROTHIAZIDE
FLUVOXAMINE MALEATE also → ASTEMIZOLE, CLOMIPRAMINE, DIAZEPAM, HALOPERIDOL, IMIPRAMINE, METOCLOPRAMIDE, OLANZEPINE, SUMATRIPTAN SUCCINATE, TACRINE
Fluxarten → FLUNARIZINE HCL
Fluxum → HEPARIN
Fluzon → FLUOCINOLONE ACETONIDE
Fly Agaric → AMANITA sp.
FLY ASH
Fly Honeysuckle → HONEYSUCKLE also → LONICERA CAPRIFOLEUM
Flying Saucers → MORNING GLORY
Flypel → *N,N*-DIETHYL-*m*-TOLUAMIDE
FMC 1240 → ETHION
FMC 5273 → PIPERONYL BUTOXIDE
FMC 5462 → ENDOSULFAN
FMC 5488 → TETRADIFON
FMC 10,242 → CARBOFURAN
FMC 30,980 → CYPERMETHRIN
FMC 33,297 → PERMETHRIN
FMC 41,655 → PERMETHRIN
FML → FLUOROMETHOLONE
Fobex → BENACTZINE
Focusan → TOLNAFTATE
Foeniculum vulgare → FENNEL
"Fog Fever" → THERMOPOLYSPORA POLYSPORA
Folacin → FOLIC ACID
Folbex → CHLOROBENZILATE
Folcodal → CINNARIZINE
Foldon → THIABENDAZOLE
Foldox → TRIMEBUTINE MALEATE
Folex → METHOTREXATE
FOLIC ACID also → *p*-AMINOSALICYLIC ACID, CHLORAMBUCIL, CHLOROQUINE, CHOLESTYRAMINE RESINS, CYCLOGUANYL PAMOATE, ERYTHROMYCIN, LAMOTRIGINE, MAGNESIUM TRISILICATE, MESTRANOL, METFORMIN, METHOTREXATE, PHENOBARBITAL, PHENYTOIN, POTASSIUM ARSENITE SOLUTION, PROTEIN HYDROLYSATES, PYRIMETHAMINE, RIFAMPIN, SALICYLAZOSULFAPYRIDINE, SODIUM BICARBONATE, SODIUM CYANIDE, TETRACYCLINE, THALIDOMIDE

Folidol → PARATHION
Folidol-M → METHYL PARATHION
Foligan → ALLOPURINOL
Folikrin → ESTRONE
FOLINIC ACID
Folipex → ESTRONE
Folisan → ESTRONE
Folithion → SUMITHION
Folleststrine → ESTRONE
Follicular Hormone → ESTRONE
Follicular Hormone Hydrate → ESTRIOL
Follicule de Sene → SENNA
Folliculin → ESTRONE
Follicunodis → ESTRONE
Follidrin → ESTRONE
Follutein → GONADOTROPINS, CHORIONIC
Folosan → QUINTOZENE
Folpan → FOLPET
FOLPET
Folvite → FOLIC ACID
FOMIVERSEN SODIUM
Fonatol → DIETHYLSTILBESTROL
FONAZINE MESYLATE
Fonderma → PYRITHIONE
FONOPHOS
Fontarsan → OXOPHENARSINE
Fontego → BUMETANIDE
Fontex → FLUOXETINE
Fonurit → ACETAZOLAMIDE
Food Orange 8 → CANTHAXANTHIN
Food Red 17 → FDC RED #40
Food Yellow 13 → QUINOLINE YELLOW
FOODS
Fool's Huckleberry → MENZIESIA FERRUGINEA
Fool's Mushroom → AMANITA sp.
Fool's Parsley → CONIUM MACULATUM
Foradil → FORMOTEROL
Forane → ISOFLURANE
Forbaxin → METHOCARBAMOL
Fordiuran → BUMETANIDE
Forene → ISOFLURANE
Forenol → NIFLUMIC ACID
Forhistal → DIMETHINDENE
Forilin → ROXITHROMYCIN
Forit → OXYPERTINE
Forlin → LINDANE
For-Mal → MALATHION
Formal → METHYLAL
FORMALDEHYDE also → ACETALDEHYDE; ALCOHOL, METHYL; BIS(CHLOROMETHYL)ETHER, CARBON MONOXIDE, DIMETHYLOL DIMETHYL HYDANTOIN, FACE TISSUES, GELATIN, GLUTARALDEHYDE, LEATHER, METHENAMINE, METHYLENEANILINE, MILK, MONEY, NEWSPAPERS, *N*-NITROSODIMETHYLAMINE, PAPER, PERMANENT PRESS, POTASSIUM PERMANGATE, SODIUM BENZOATE, SULFAMETHIZOLE, TRIS (HYDROXYMETHYL) NITROMETHANE
Formalin → FORMALDEHYDE
FORMAMIDE
FORMIC ACID also → ALCOHOL, METHYL; ANTS, ETHYL FORMATE, FORMALDEHYDE, LAPORTEA sp., METHENAMINE, NEWSPAPERS, TAXINE
Formic Aldehyde → FORMALDEHYDE
N-Formimidoyl Thienamycin → IMIPENEM
Formin → METHENAMINE
Formison → MEPARYFNOL
Formistin → CETIRIZINE
Formol → FORMALDEHYDE
Formonitrile → HYDROGEN CYANIDE
FORMOTEROL
Formula 144 → BENZETHONIUM CHLORIDE
Foromacidin → SPIRAMYCIN
Forotox → TRICHLORFON
Forpen → PENICILLIN > *Benzylpenicillin, Potassium*
Forron → 2,4,5-TRICHLOROPHENOXYACETIC ACID
Forskolein → COLEUS
Fortalgesic® → PENTAZOCINE
Fortamine → CHLORPHENIRAMINE also → DEXCHLORPHENIRAMINE
Fortasec → LOPERAMIDE
Fortaz → CEFTAZIMIDES
Forticef → CEPHADRINE
Fortigro → CARBADOX
Fortodyl → VITAMIN D_2
Fortovase → SAQUINAVIR
Fortral → PENTAZOCINE
Fortrol → CYANAZINE
Fortum → CEFTAZIMIDES
Fortune Teller → TARAXACUM OFFICINALE
Forturf → CHLOROTHALONIL
Fortwin → PENTAZOCINE
Fosamax → ALENDRONATE
FOSCARNET SODIUM also → CIPROFLOXACIN
Foscavir → FOSCARNET SODIUM
Foschlorem → TRICHLORFON
Fosfakol → DIETHYL-*p*-NITROPHENYLPHOSPHATE
Fosfamid → DIMETHOATE

Fosferno → PARATHION
FOSFINOPRIL SODIUM
FOSFOMYCIN TROMETHAMINE
Fossyol → METRONIDAZOLE
Fosten → APROTININ
Fostion MM → DIMETHOATE
Fosvex → TETRAETHYL PYROPHOSPHATE
Fouadin → STIBOPHEN
Fougère Mâle → ASPIDIUM
Fourneau 309 → SURAMIN SODIUM
Fourneau 710 → PLASMOCID
Four O'Clock → MIRABILIS JALAPA
Fovane → BENZTHIAZIDE
Fowler's Solution → POTASSIUM ARSENITE SOLUTION
Fowl Mannagrass → GLYCERIA sp.
Foxalin → DIGITOXIN
Foxetin → FLUOXETINE
Foxglove → DIGITALIS also → COMFREY
Fox Green → INDOCYANINE GREEN
Foxtail → HORDEUM JUBATUM also → SETARIA sp.
FOY → GABEXATE MESYLATE
FPL 670 → CROMOLYN SODIUM
FR 300 → DECADIBROMODIPHENYL OXIDE
FR 860 → DALTEPARIN
FR 13,479 → CEFTIZOXIME
FR 17,027 → CEFIXIME
FR 900,506 → TACROLIMUS
Frabel → OXYPHENBUTAZINE
Fractol → MINERAL OIL
Frademicina → LINCOMYCIN
Fradiomycin → NEOMYCIN
Fragaria vesca → STRAWBERRY
Fragmin → DALTEPARIN also → HEPARIN
Framomycin → FRAMYCETIN
FRAMYCETIN also → NEOMYCIN
Frandol → ISOSORBIDE DINITRATE
FRANGULA
Frangulic Acid → EMODIN
Franidipine → MANIDIPINE
Franke & Fuchs → CARBUTAMIDE
FRANKFURTERS also → NITRATES and NITRITES, *N*-NITROSODIMETHYLAMINE, OLEANDER, PEANUTS, SODIUM NITRATE
Frankincense → OLIBANUM
Franoside → DIETHYLCARBAMAZINE
Franrose → TEGAFUR
Frantin → BEPHENIUM
Franulin → FRANGULA
Fratol → FLUOROACETIC ACID
Fraxin → AESCULUS sp.
Fraxinella → DICTAMNUS ALBUS
FRAXINUS sp.
Fraxiparine → HEPARIN
Free Benzylpenicillin → PENICILLIN
Free Penicillin G → PENICILLIN
Free Penicillin II → PENICILLIN
FREE RADICALS
Freeuril → BENZTHIAZIDE
FREEZE BALLS
Frekentine → DIETHYLPROPION
Frenactyl → BENPERIDOL
Frenantol → PAROXYPROPIONE
French Bean → BEANS, Kidney
French Green → CUPRIC ACETOARSENITE
French Honeysuckle → GALEGA OFFICINALIS
French Lactucarium → LETTUCE
French Mercury → MERCURIALIS ANNUA
French Millet → PANICUM sp.
Frenohypon → PAROXYPROPIONE
Frenolon → METHOPHENAZINE
Frêne Puant → AILANTHUS ALTISSIMA
Frenolyse → TRANEXAMIC ACID
Frenoton → AZACYCLONAL
Frenquel → AZACYCLONAL
Frentirox → METHIMAZOLE
FREONS also → PIPERONYL BUTOXIDE
Freon 12 → DICHLORODIFLUOROMETHANE
Freon 22 → MONOCHLORODIFLUOROMETHANE
Freon 112 → DIFLUOROTETRACHLOROETHANE also → 1,1,2,2-TETRACHLORODIFLUORETHANE
Freon 113 → TRICHLOROTRIFLUOROETHANE
Freon MU → 1,1,2,2-TETRACHLORODIFLUORETHANE
Freoxide → ETHYLENE OXIDE
Fresmin → VITAMIN B_{12}
Frevken → PROPRANOLOL
Frigen 114 → 1,2-DICHLORO-1,1,2,2-TETRACHLORO-ETHANE
Frigens → FREONS
"Fright Disease" → NITROGEN CHLORIDE
Frigiderm → 1,2-DICHLORO-1,1,2,2-TETRACHLOROETHANE
Frijolillo → SOPHORA sp.
Frijolito → SOPHORA sp.
Frisium → CLOBAZAM
FRITILLARIA MELEAGRIS
Fritilllary → FRITILLARIA MELEAGRIS

Froben → FLURBIPROFEN
FROGS
Frone → INTERFERONS
Frontline → FIPRONIL
Frucote → *sec*-BUTYLAMINE
FRUCTOSE also → HONEY
β-D-Fructose → FRUCTOSE
D-Fructose → FRUCTOSE
Fructosteril → FRUCTOSE
Frugalan → FURFENDREX CYCLAMATE
Fruhjahrslorchel → GYROMITRA ESCULENTA
FRUITS and FRUIT JUICES
Fruitone A → 2,4,5-TRICHLOROPHENOXYACETIC ACID
Fruit Salad Plant → MONSTERA sp.
Fruit Sugar → FRUCTOSE
Frusemide → FUROSEMIDE
Frusetic → FUROSEMIDE
Frusid → FUROSEMIDE
FSH → Δ^9-TETRAHYDROCANNABINOL
FSR-3 → ISONIAZID
FT 207 → TEGAFUR
Ftalazol → PHTHALYLSULFATHIAZOLE
Ftalofos → IMIDAN
F/Tnl → TRIMETHOPRIM
FTorafur → TEGAFUR
Ftorocort → TRIAMCINOLONE
5-FU → FLUOROURACIL
Fuadin → STIBOPHEN
Fua-MEd → NITROFURANTOIN
Fuchsia → EREMOPHILA MACULATA also → HYDROGEN CYANIDE
Fucidin → FUSIDIC ACID
Fucidina → FUSIDIC ACID
Fucidine → FUSIDIC ACID
Fucidin Intertulle → FUSIDIC ACID
Fucithalmic → FUSIDIC ACID
Fuclasin → ZIRAM
FUDR → FLOXURIDINE
"Fuel" → PHENCYCLIDINE
Fuel Oil No. 1 → KEROSENE
FUELS, OIL
Fugerel → FLUTAMIDE
Fugillin → FUMAGILLIN
Fugoa → NORPSEUDOEPHEDRINE
Fugu → FISH
Fugu Poison → TETRODOTOXIN
Fuklasin → ZIRAM
Fukujinzuke → PICKLES
Fulaid → TEGAFUR
Fulcin → GRISEOFULVIN
Fulfeel → TEGAFUR
Ful-Glo → FLUORESCEIN
Fulgram → NORFLOXACIN
FULLER'S EARTH
Fuller's Herb → BOUNCING BET
Full Moon Disease → NAPHTHALENE
Fullsafe → FLUFENAMIC ACID
Fulsix → FUROSEMIDE
Fuluminol → CLEMASTINE
Fuluvamide → FUROSEMIDE
Fulvicin → GRISEOFULVIN
Fumadil B → FUMAGILLIN
Fumafer → FERROUS FUMARATE
FUMAGILLIN
Fumar F → FERROUS FUMARATE
FUMARIA sp.
FUMARIC ACID also → FUMARIA
Fumarin → COUMAFURYL
Fumazone → DIBROMOCHLOROPROPANE
Fumidil → FUMAGILLIN
Fumigant 1 → METHYL BROMIDE
Fumigrain → ACRYLONITRILE
Fuminoisin → CORN
Fumiron → FERROUS FUMARATE
Fumisol → COUMAFURYL
FUMITREMORGIN A also → PENICILLIUM sp.
FUMITREMORGIN B also → PENICILLIUM sp.
FUMONISIN
Fundal → CHLORDIMEFORM
Fundescein → FLORESCEIN
Fungarest → KETOCONAZOLE
Fungibacid → TIOCONAZOLE
Fungicidin → NYSTATIN
Fungiderm → MICONAZOLE NITRATE
Fungilin → AMPHOTERICIN B
Fungisdin → MICONAZOLE NITRATE
Fungistat → TERCONAZOLE
Fungistop → TOLNAFTATE
Fungitrol 11 → FOLPET
Fungizone → AMPHOTERICIN B
Fungoral → KETOCONAZOLE
Fungostop → ZIRAM
FUR
Furachel → NITROFURANTOIN
Furacilin → NITROFURAZONE
Furacinetten → NITROFURAZONE
Furacoccid → NITROFURAZONE
Furadan → CARBOFURAN
Furadantin → NITROFURANTOIN
Furadantine MC → NITROFURANTOIN
Furadoin → NITROFURANTOIN
Furafluor → TEGAFUR
Fural → FURFURAL
Furalan → NITROFURANTOIN
Furalone → NITROFURAZONE

FURALTADONE
Furamide → DILOXANIDE FUROATE
2,5-Furanedione → MALEIC ANHYDRIDE
Furantoina → NITROFURANTOIN
Furaplast → NITROFURAZONE
Furasol → FURALTADONE
Furazin(a) → NITROFURAZONE
Furazol W → NITROFURAZONE
FURAZOLIDONE also → ALCOHOL, ETHYL; AMITRIPTYLINE, AMPHETAMINE, BEANS, CARBADOX, CHEESES, CHLORAL HYDRATE, COCAINE, DEXTROAMPHETAMINE, HERRING, INSULIN, LIVER, MEPERIDINE, METHOTRIMEPRAZIME, METHYLDOPA, RESERPINE, SUMATRIPTAN SUCCINATE, WINE
Furazolin → FURALTADONE
FURAZOLIUM CHLORIDE
Furenazin → NIFURPRAZINE
Furesis → FUROSEMIDE
Furesol → NITROFURAZONE
FURFENDREX CYCLAMATE
Furfenorex → FURFENDREX CYCLAMATE
FURFURAL
Furfuraldehyde → FURFURAL
Furfurol → FURFURAL
FURFURYL ALCOHOL
Furiton → NIHYDRAZONE
Furloe → CHLORPROPHAM
Furmethonol → FURALTADONE
Furnace Black → CARBONS
Furobactina → NITROFURANTOIN
Furocoumarins → ACHILLEA sp., CARROTS, CELERY, COLOGNES, DICENTRA CUCULLARIA, LIME, RUE
Furofutran → TEGAFUR
Furophen T Caps → NITROFURANTOIN
Furo-Puren → FUROSEMIDE
Furosedon → FUROSEMIDE
FUROSEMIDE also → ACETALDEHYDE, BUCLOSAMIDE, CAPTOPRIL, CEPHALORIDINE, CEPHAPIRIN, CHLORAL HYDRATE, COLISTIN, DIFLUNISAL, DILTIAZEM, FLURBIPROFEN, GALLAMINE TRIETHIODIDE, GENTAMICN, GLUCOSE, HYDRALAZINE, INDOMETHACIN, INSULIN, KANAMYCIN, LITHIUM, METHOSUXIMIDE, METOLAZONE, NAPROXEN, OMEPRAZOLE, PHENETURIDE, PHENOBARBITAL, PHENYTOIN, PROBENECID, STREPTOMYCIN, SUCCINYLCHOLINE, SURAMIN SODIUM, THEOPHYLLINE, TOBRAMYCIN, *d*-TUBOCURARINE CHLORIDE, VANCOMYCIN
Furovag → FURAZOLIDONE
Furoxane → FURAZOLIDONE
Furoxone → FURAZOLIDONE
"Furrier's Lung" → FUR
Fursemide → FUROSEMIDE
2-Furylmethanol → FURFURYL ALCOHOL
Fusaben → FONAZINE MESYLATE
Fusca → CLORPRENALINE
Fuschine → *p*-ROSANILINE
FUSEL OIL
FUSIDIC ACID
Fussol → FLUOROACETAMIDE
Fustpentane → CARBETAPENTANE CITRATE
FUT 175 → NAFAMOSTAT MESYLATE
Futhan → NAFAMOSTAT MESYLATE
Futraful → TEGAFUR
FW 293 → DICOFOL
FW 734 → PROPANIL
Fyfanon → MALATHION
Fyrol → TRIS(2-3-DIBROMOPROPYL) PHOSPHATE
Fyrol FR2 → 1,3-DICHLOROACETONE
FZ 348 → STANOZOLOL

G

G 11 → HEXACHLOROPHENE
G 25 → CHLOROPICRIN
G 020 → CHLORACIZINE
G 115 → GINSENG
G 665 → METHYL BENZIMIDAZOLE-2-CARBAMATE
G 13,871 → PHENYLBUTAZONE
G 22,355 → IMIPRAMINE
G 22,870 → DIMETILAN
G 23,133 → COUMACHLOR
G 23,611 → ISOLAN
G 23,922 → CHLOROBENZILATE
G 24,480 → DIAZINON
G 25,766 → CLORINDIONE
G 27,202 → OXYPHENBUTAZINE
G 28,315 → SULFINPYRAZONE
G 29,505 → ESTIL
G 30,320 → CLOFAZIMINE
G 31,435 → PROMETONE
G 32,883 → CARBAMAZEPINE
G 33,040 → OPIPRAMOL
G 33,182 → CHLORTHALIDONE
G 34,161 → PROMETRYNE
G 34,586 → CLOMIPRAMINE

G 35,020 → DESIPRAMINE
GA → TABUM
Ga → GALLIUM
GABA → γ-AMINOBUTYRIC ACID also → AGARIN
Gabacet → PIRACETAM
GABAPENTIN also → PHENYTOIN
Gabbromicina → PAROMOMYCIN
Gabbromycin → PAROMOMYCIN
Gabbroral → PAROMOMYCIN
GABEXATE MESYLATE
Gabitril → TIAGABINE
Gadad → ALBIZIA sp.
Gadalin Kura → KALANCHOE
Gadexyl → MEPROBAMATE
GADODIAMIDE
Gadon machiji → TRIANTHEMA sp.
GADOTERIDOL
Gaduol → COD LIVER OIL
Gage → CANNABIS
GAILLARDIA
Gaillot → ANTHIOLIMINE
Gajar → WILD CARROT
Galactin → LUTEOTROPIN
GALACTOFLAVIN
Galactoquin → QUINIDINE
GALACTOSAMINE
GALACTOSE also → CEFOTAXIME, LACTOSE, MILK, YOGURT
β-D-Galactose Pyranosyl → AMARANTHUS
Galagacha → MAERUA ANGOLENSIS
GALALITH
Galantamine → GALANTHAMINE
GALANTHAMINE
GALANTHUS NIVALIS
Galanthus woronowii → GALANTHAMINE
Galantone → GALANTHAMINE
Galatur → IPRINDOLE
GALBANUM GUM
Galecron → CHLORDIMEFORM
GALEGA OFFICINALIS
Galegine → VERBESINA ENCELIOIDES
Galen → Preface
GALENA
Galenphol → PHOLCODINE
GALEOPSIS
GALERINA sp.
Galfer → FERROUS FUMARATE
Galinid → TETRAMISOLE
GALITOXIN
GALIUM sp.
GALLAMINE TRIETHIODIDE also → CHLOROTHIAZIDE, CHLORTHALIDONE, DIAZEPAM
GALLIC ACID also → BISMUTH SUBGALLATE
Gallicin → METHYL GALLATE
Gallimycin → ERYTHROMYCIN also → ERYTHROMYCIN STEARATE
GALLIUM also → NAPROXEN
GALLIUM CITRATE also → METHOTREXATE
Gallochrome → MERBROMIN
GALLOPAMIL
Gallotannic Acid → TANNIC ACID
Gallotannin → TANNIC ACID
Gallotox → PHENYLMERCURIC ACETATE
Galloxon → HALOXON
Galoxone → HALOXON
Galphol → PHOLCODINE
Galpsend → PSEUDOEPHEDRINE
GALVANIZED METALS
Gamanil → LOFEPRAMINE
Gamaphos → AMIFOSTINE
Gamarex → γ-AMINOBUTYRIC ACID
Gamasol 90 → DIMETHYL SULFOXIDE
Gamatran → ALVERINE
GAMBIERDISCUS TOXICUS
GAMBOGE
Gambogia → GAMBOGE
Gamefar → PAMAQUIN
Gamene → LINDANE
Gamibasin → ANABASINE
Gammacorten → DEXAMETHASONE
GammaG → BUTYROLACTONE
Gamma Globulins → MERCURY
Gammalin → LINDANE
Gammalon → γ-AMINOBUTYRIC ACID
Gamma OH → SODIUM OXYBATE
Gammasol → γ-AMINOBUTYRIC ACID
Gamma-vinyl GABA → VIGABATRIN
Gammatin → LINDANE
Gammexane → LINDANE
Gammopaz → LINDANE
Gamonil → LOFEPRAMINE
Gamophen → TRICLOSAN
Ganal → FENFLURAMINE
Ganasag → DIMINAZENE ACETURATE
GANCICLOVIR also → CYCLOSPORINE, DIDANOSINE, ZIDOVUDINE
Gancoa → GLYCYRRHIZA GLABRA
Ganháwla → AMELANCHIER ALNIFOLIA
Ganidan → SULFAGUANIDINE
Ganite → GALLIUM
Ga 67 Nitrate → GALLIUM
Ganja → CANNABIS
Ganor → FAMOTIDINE
Gantanol → SULFAMETHOXAZOLE

Gantaprim → SULFAMETHOXAZOLE
Gantrim → SULFAMETHOXAZOLE
Gantrisin → SULFISOXAZOLE
Garamycin → GENTAMICIN
Garantose → SACCHARIN
Garasin → CEPHALEXIN
Garasol → GENTAMICIN
Garbanzo → PEAS
GARCINIA sp.
Garcinia hanburyi → GAMBOGE also → GARCINIA sp.
Gardenal → PHENOBARBITAL
Garden Bean → BEANS, Kidney
Garden Heliotrope → HELIOTROPIUM sp. also → VALERIANA sp.
Garden Huckleberry → SOLANUM sp.
Garden Lettuce → LETTUCE
Garden Rosemary → HOLOCALYX sp.
Garden Spurge → EUPHORBIA, *lathyris*
Garden Tox → DIAZINON
GARDONA
Garget → PHYTOLACCA sp.
Garland → MASCAGNIA sp.
GARLIC also → ARSENIC, ARSINE, ASFETIDA, CUPFERON, PETIVERA ALLIACEA
Garmian → BAMETHAN
Garra de Léon → CARBOBROTUS CHILENSIS
Garranil → CAPTOPRIL
Garrathion → CARBOPHENOTHION
Gary Horsebrush → TETRADYMA sp.
Gas Black → CARBONS
GASOLINE also → CARBON MONOXIDE, ISOBUTANE
Gas Plant → DICTAMNUS ALBUS
Gaster → FAMOTIDINE
Gasteril → PIRENZEPINE HYDROCHLORIDE
Gastralgin → ROXATIDINE ACETATE
Gastrausil → CARBENOLOXONE
Gastrax → NIZATIDINE
Gastrese → METOCLOPRAMIDE
Gastridin → FAMOTIDINE
Gastrin → PROMETHAZINE
Gastrion → FAMOTIDINE
Gastrobid → METOCLOPRAMIDE
Gastrocrom → CROMOLYN SODIUM
Gastrodiagnost → PENTAGASTRIN
Gastrofrenal → CROMOLYN SODIUM
GASTROLOBIUM sp.
Gastroloc → OMEPRAZOLE
Gastromax → METOCLOPRAMIDE
Gastromet → CIMETIDINE
Gastron → METHANTHELINE BROMIDE
Gastronerton → METOCLOPRAMIDE
Gastronorm → DOMPERIDONE
Gastropen → FAMOTIDINE
Gastropidil → MEPENZOLATE BROMIDE
Gastrosedan → METHANTHELINE BROMIDE
Gastrosil → METOCLOPRAMIDE
Gastro-Tablinen → METOCLOPRAMIDE
Gastrotem → METOCLOPRAMIDE
Gastrotest → PHENAZOPYRIDINE
Gastro-Timelets → METOCLOPRAMIDE
Gastrozepin → PIRENZEPINE HYDROCHLORIDE
Gas-X → DIMETHICONE (*with silicon dioxide*)
Gat → CATHA EDULIS
Gaucho → IMIDACLOPRID
GAULTHERIA ANASTOMOSANA
GB → SARIN also → SULFUR
GB 94 → MIANSERIN
Gbaguda → CASSAVA
GBL → BUTYROLACTONE
GC 1189 → KEPONE
GC 3707 → BOMYL
GCP 23,339A → PAMIDRONATE DISODIUM
G-CSF → GRANULOCYTE COLONY STIMULATING FACTOR
GD → SOMAN
Geangin → VERAPAMIL
Geapur → ALLOPURINOL
Gebraunter Kalk → CALCIUM OXIDE
Gebutox → DINOSEB
Gecolate → GLYCERYL GUAIACOLATE
Geeldikkop → TRIBULUS TERRESTRIS
Geelsiekte → TRIBULUS TERRESTRIS
Gefanil → GEFARNATE
GEFARNATE
Gefarnil → GEFARNATE
Gefarnyl → GEFARNATE
Gefulcer → GEFARNATE
GEIGERIA sp.
Geigy 22,008 → PYROLAN
Gelacnine → TIOXOLONE
GELATIN also → DIGITALIS, GOLF BALLS
Gelbin → CALCIUM CHROMATE
Geldene → PIROXICAM
Gelfoam → GELATIN
Gelfundol® → GELATIN
Gelifundol → GELATIN
Gelocatil → ACETAMINOPHEN
Gelofusine® → GELATIN
Gelovermin → HEXYLRESORCINOL
Gelsemine → GELSEMIUM SEMPERVIRENS
Gelseminine → GELSEMIUM SEMPERVIRENS

Gelsemoidine → GELSEMIUM SEMPERVIRENS
GELSIUM SEPERVIRENS also → HONEY
Gelstaph → PENICILLINS – "NEWER and SYNTHETICS" > *Cloxacillin/Sodium salt*
Gelthix → TEMAZEPAM
Gelvatol → POLYVINYL ALCOHOL
Gemalgene → TRICHLOROETHYLENE
GEMCITABINE also → WARFARIN
GEMEPROST
GEMFIBROZIL also → ATORVASTATIN, BEXAROTENE, LOVASTATIN
Gemonil → METHARBITAL
Gemonit → METHARBITAL
GEMTUZIMAB OZOGAMICIN
Gemtuzimab Zogamicin → GEMTUZIMAB OZOGAMICIN
Gemzar → GEMCITABINE
Genabol → NORBOLETHONE
Genate → 5-ETHYL DIISOBUTYLTHIOCARBAMATE
Gen. Chem 4072 → CHLORFENVINPHOS
Gendon → RAUWOLFIA SERPENTINA
Genecillin → PENICILLIN > *Benzylpenicillin, Potassium*
General Mills → CHLORPYRIFOS-METHYL, OATS
Generlac → LACTULOSE
Geneserp → RESERPINE
Genetron 12 → DICHLORODIFLUOROMETHANE
Genetron 113 → TRICHLOROTRIFLUOROETHANE
Genievre → GIN
Geniphene → TOXAPHENE
Genisis → ESTROGENS, CONJUGATED
Genlip → GEMFIBROZIL
Genocaine → PROCAINE
Genogris → PIRACETAM
Genophyllin → AMINOPHYLLINE
Genotonorm → GROWTH HORMONES
Genotropin → GROWTH HORMONES also → SOMATREM
Genoxal → CYCLOPHOSPHOSPHAMIDE
Genphen → PROMETHAZINE
Gentabac → GENTAMICIN
Genta-Gobens → GENTAMICIN
Gentalline → GENTAMICIN
Gentalyn → GENTAMICIN
GENTAMICIN also → ATRACURIUM, BROMHEXINE, BUTIROSIN, CISPLATIN, DIGITALIS, DIGOXIN, DIMETHYL TUBOCURARINE IODIDE, ETHACRYNIC ACID, FRAMYCETIN, HALOTHANE, HEPARIN, IBUPROFEN, INDOMETHACIN, MAGNESIUM SULFATE, NEOMYCIN; PENICILLINS – "NEWER and SYNTHETICS" > *Ampicillin, Carbenicillin Disodium, Ticarcillin Disodium*; POLYMIXIN B, SODIUM BISULFITE, SODIUM METABISULFITE, SUCCINYLCHOLINE, *d*-TUBOCURARINE CHLORIDE, VANCOMYCIN
Gentamina → GENTAMICIN
Gentamycin → GENTAMICIN
Gentavet → GENTAMICIN
Gentavetina → GENTAMICIN
GENTIAN VIOLET
Gentiaverm → GENTIAN VIOLET
Genticin → GENTAMICIN
GENTISIC ACID
Gentocin → GENTAMICIN
Gentran → DEXTRAN
Gentran 40 → DEXTRAN, *40*
Genurin → FLAVOXATE HCL
Geocillin → PENICILLINS – "NEWER and SYNTHETICS" > *Carindacillin Sodium*
Geomycin → OXYTETRACYCLINE
Geon → POLYVINYL CHLORIDE
Geopen → PENICILLINS – "NEWER and SYNTHETICS" > *Carbenicillin Disodium*
GEORGINA GIDYEA
Georgina River Poisoning → ACACIA
GEOSMIN
Gephyrotoxin → HISTRIONICOTOXIN
Geram → PIRACETAM
Geranial → CITRAL
Geranio → GERANIUM
GERANIOL
GERANIUM
Geranium Oil Bourbon → GERANIUM
GERBERA
Gerioptil → PROCAINE
Germalgene → TRICHLOROETHYLENE
Germall II → DIAZOLIDINYL UREA
Germall 115 → IMIDAZOLIDINYL UREA
German Chamomile → CHAMOMILE
GERMANDER, WILD also → TEA
Germane → GERMANIUM
Germanin → SURAMIN SODIUM
GERMANIUM
German Lactucarium → LETTUCE
Germapect → CARBETAPENTANE CITRATE
Germiciclin → METHACYCLINE
Germidine → HELLEBORUS NIGER, VERATRUM sp.
GERMINE DIACETATE
Germinol → BENZALKONIUM CHLORIDE

Germitol → BENZALKONIUM CHLORIDE
Germitrine → HELLEBORUS NIGER
Gernebcin → TOBRAMYCIN
Gero → PENNISETUM sp. also → PROCAINE
Gerobit → METHAMPHETAMINE
Gerodyl → PIPRADOL
Gerot-Epilan → MEPHENYTOIN
Gerovital → PROCAINE
Geroxalen → METHOXSALEN
Gesafram → PROMETONE
Gesagard → PROMETRYNE
Gesakur → CHLORPROPYLATE
Gesamil → PROPAZINE
Gesapax → AMETRYN
Gesapon → DDT
Gesarex → DDT
Gesarol → DDT
Gesatop → SIMAZINE
Gesfid → MEVINPHOS
Gestageno → 17-α-HYDROXYPROGESTERONE
Gestanin → ALLYLESTRANOL
Gestanol → ALLYLESTRANOL
Gestanon → ALLYLESTRANOL
Gestanyn → ALLYLESTRANOL
Gestapuran → MEDROXYPROGESTERONE ACETATE
Gestatron → DYDROGESTERONE
Gesterol → PROGESTERONE
Gestfortin → CHLORMADINONE
Gestiron → PROGESTERONE
Gestone → PROGESTERONE
GESTONORONE CAPROATE
GETIFLOXACIN
Getrocknete Schilddruse → THYROID
Gevilon → GEMFIBROZIL
Gevramycin → GENTAMICIN
Gewacalm → DIAZEPAM
Gewazol → PENTYLENETETRAZOL
Gexane → LINDANE
GF
G-Farlutal → MEDROXYPROGESTERONE ACETATE
GG 167 → ZANAMIVIR
GGE → GLYCERYL GUAIACOLATE
GH → GROWTH HORMONES
GEB → γ-HYDROXYBUTYRATE
Ghiu Kumari → ALOE
GH Revitalizer → BUTYROLACTONE
GH-RIF → SOMATOSTATIN
Giant Fennel → FERULA COMMUNIS
Giant Fescue → FESCUE, TALL
Giant Hogweed → HERACLEUM sp.
Giant Milkweed → CALOTROPIN
Giant Star Grass → CYNODON, *plectostachyus*
Giardil → FURAZOLIDONE
Giarlam → FURAZOLIDONE
GIBBERELLA ZEAE
Gibbsite → ALUMINUM OXIDE
Gibicef → CEFUROXIME
Gichtex → ALLOPURINOL
GIDYEA
Gifblaar → FLUOROACETIC ACID
Gift of the Sun God → COCAINE
Giganten → CINNARIZINE
Gigromitsina B → HYGROMYCIN B
Gigrovetina → HYGROMYCIN B
GILA MONSTER
Gilemal → GLYBURIDE
Gill-Over-The-Ground → NEPETA HEDERACEA also → IVY
Gilucor → NITROGLYCERIN
Giluritmal → AJMALINE
Gilurytmal → AJMALINE
GIN also → WINE, JUNIPER
Ginarsol → ACETARSONE
Gineflavir → METRONIDAZOLE
Ginestra → SPARTIUM JUNCEUM
GINGER ALE
Ginger Jake Paralysis → TRI-σ-TOLYL PHOSPHATE
GINGER, JAMAICAN
Gingerol → GINGER, JAMAICAN
Ginglii → SESAME
GINKGO BILOBA
Ginsana → GINSENG
GINSENG also → PODOPHYLLUM, WARFARIN
Ginsenoside → PENTOBARBITAL
Gipokhlor → SODIUM HYPOCHLORITE
Girasol → SUNFLOWER
Girl → COCAINE
Girl's Mercury → MERCURIALIS ANNUA
Githagenin → AGROSTEMA, BOUNCING BET
Gittalun → DOXYLAMINE SUCCINATE
Give Tan F → CINOXATE
Giv-Gard DXN → DIMETHOXANE
Gizmo Mouse Killer → THALLIUM
GL 701 → PRASTERONE
GLADIOLA
GLAFENINE
Glandin → DINOPROST
Glandubolin → ESTRONE
Glanducorpin → PROGESTERONE
Glanestin → PROGESTERONE
Glanil → CINNARIZINE

Glaphenine → GLAFENINE
Glasel → ETOFENAMATE
GLASS
GLATIRAMER
Glauber's Salt → SODIUM SULFATE
GLAUCARUBIN
Glaucon → EPINEPHRINE
Glaucostat → ACECLIDINE
Glaucotensil → ETHOXZOLAMIDE
Glauline → TRIMEPRANOL
Glaumeba → GLAUCARUBIN
Glaupax → ACETAZOLAMIDE
Glauposine → EPINEPHRINE
Glazidin → CEFTAZIMIDES
Glechoma hederacea → NEPETA HEDERACEA also → IVY
Gleem® → CINNAMIC ALDEHYDE
Glentonin → ISOSORBIDE DINITRATE
Glevomicina → GENTAMICIN
GLIADIN also → EGGS
Glianimon → BENPERIDOL
Glibenclamide → GLYBURIDE
Glibenese → GLIPIZIDE
Gliben-Puren N → GLYBURIDE
GLICLAZIDE
Gliddophil → LECITHIN
Glidiabet → GLYBURIDE
Glifanan → GLAFENINE
GLIMEPIRIDE
Glimicron → GLICLAZIDE
Glimid → GLUTETHIMIDE
Glimidstada → GLYBURIDE
GLINUS sp.
GLIOCLADIUM sp.
Glioten → ENALAPRIL
Gliotoxin → PENICILLIUM sp.
GLIPIZIDE also → DISULFIRAM, HEPARIN, OXYPHENBUTAZONE, PHENYLBUTAZONE
Gliporal → BUFORMIN
Glissard → LEAD
Globenicol → CHLORAMPHENICOL
γ-Globulin → THIMEROSAL
Globulin G_1 → LYSOZYME
GLOCOTRICHIA ECHINALATA
Glomycin → OXYTETRACYCLINE
Glonoin → NITROGLYCERIN
GLORIOSA
Gloriosine → GLORIOSA
Glory Lily → GLORIOSA
Glospan → POLYURETHANES
Glosso-Stérandryl → METHYL TESTOSTERONE
GLOTTIDIUM VESICARIUM also → SESBANIA sp.
GLOVES
GLQ 223 → TRICHOSANTHIN
Glubate → GLYBURIDE
GLUCAGON also → ACENOCOUMAROL, DIAZOXIDE, INSULIN, LIDOCAINE, PHENOL, PHENYTOIN, WARFARIN
Glucantine → METHYLGLUCAMINE ANTIMONATE
GLUCASAMINE
Glucid → SACCHARIN
Glucidoral → CARBUTAMIDE
Glucinium → BERYLLIUM
D-Glucitol → SORBITOL
Glucobay → ACARBOSE
β-Glucocerebrosidase → ALGLUCERASE
Glucocheirolin → RAPISTRUM RUGOSUM
Glucochloral → CHLORALOSE
α-D-Glucochoralose → CHLORALOSE
Glucofren → CARBUTAMIDE
Glucolin → GLUCOSE
Gluconsan K → POTASSIUM GLUCONATE
Glucophage → METFORMIN
Glucopostin → PHENFORMIN
Glucoremed → GLYBURIDE
GLUCOSE also → ACETAMINOPHEN, CAPRYLIC ACID, CIMETIDINE, CINCHOPHEN, LACTOSE, PHENACEMIDE, PHENELZINE, PHENYTOIN, PHYSOSTIGMINE, PILOCARPINE, POTASSIUM CHLORIDE, POTATO, PRAZEPAM, PROBENECID, SODIUM CARBOXYMETHYLCELLULOSE, STREPTOMYCIN, TETRACYCLINE, THIOPENTAL
GLUCOSE
D-Glucose → GLUCOSE
GLUCOSINOLATES also → BRASSICA sp., RAPE
GLUCOSULFONE SODIUM
Glucosum → CORN SYRUP
(1-D-Glucosylthio)gold → AUROTHIOGLUCOSE
Gluco-Tablinen → GLYBURIDE
Glucotrol → GLIPIZIDE
GLUE also → RUBBER
Glukor → GONADOTROPINS, CHORIONIC
Glumamycin → AMPHOMYCIN
GLU-P-1 → 2-AMINO-6-METHYLDIPYRIDOL (1,2-α:3´,2´-d) IMIDAZOLE
GLU-P-2 → 2-AMINODIPYRIDOL (1, 2-α-3´,2´-d) IMIDAZOLE

Glupax → ACETAZOLAMIDE
Gluquinate → QUINIDINE
Gluside → SACCHARIN
Glutacyl → SODIUM GLUTAMATE
Glutamic Acid → DEXTROAMPHETAMINE, PROTEIN HYDROLYSATES, SODIUM GLUTAMATE, THALIDOMIDE
L-Glutamic Acid 5-[2-[4-(hydroxymethyl)phenyl]-hydrazide → AGARITINE
β-N-[γ-L(+)-glutamyl]-4-hydroxymethylphenyl-hydrazine → AGARITINE
GLUTARALDEHYDE
Glutaral → GLUTARALDEHYDE
Gluta renghus → RENGAS
Glutathione → ACETAMINOPHEN also → SELENOMETHIONINE
Glutavene → SODIUM GLUTAMATE
GLUTEN also → EGGS, FLOUR, OATS, POTASSIUM PERSULFATE, RYE
GLUTETHIMIDE also → ALCOHOL, ETHYL; BISHYDROXYCOUMARIN, CODEINE, CYCLOCOUMAROL, DIPYRONE, EPHEDRINE, ETHYL BISCOUMACETATE, GRISEOFULVIN, MEPROBAMATE, PROCHLORPERAZINE, TRIFLUOPERAZINE DIHYDROCHLORIDE, VITAMIN D_3, WARFARIN
Glybenzcyclamide → GLYBURIDE
GLYBURIDE also → CIMETIDINE, CYCLOSPORINE, ERYTHROMYCIN, FLUCONAZOLE, MAPROTILENE, MEGLITOL, NAPROXEN, NIMODIPINE, PHENYLBUTAZONE, RANITIDINE, RIFAMPIN, SULFAMETHOXAZOLE, SULFAPHENAZOLE
Glycemex → TOLBUTAMIDE
Glyceol → GLYCERIN
GLYCERIA sp.
GLYCERIN also → HYDROGEN PEROXIDE, PHENOBARBITAL, PHENOL, SOYBEAN OIL
Glycerine → GLYCERIN
Glycerol → GLYCERIN
Glycerol Monoacetate → MONACETIN
Glycerol Nitric Acid → NITROGLYCERIN
Glycerol Trichlorohydrin → 1,2,3-TRICHLOROPROPANE
GLYCERYL GUAIACOLATE
Glyceryl Monoacetate → MONACETIN
Glyceryl Mononitrate → NITROGLYCERIN
Glyceryl Monothioglycolate → HAIR
Glyceryl Trinitrate → NITROGLYCERIN
Glycet → MEGLITOL
Glycide → GLYCIDOL
GLYCIDOL
Glycidyl Ether → EPOXY RESINS, PAINT
GLYCIDYL METHACRYLATE
Glycine → SACCHARIN
Glycine Betaine → BETAINE
Glycine Nitrile → AMINOACETONITRILE
Glycinonitrile → AMINOACETONITRILE
Glycobiarsol → BISMUTH GLYCOL ARSANILATE
Glycodex → GLYCERYL GUAIACOLATE
Glycodiathine → GLYMIDINE
Glycodiazine → GLYMIDINE
Glycolande → GLYBURIDE
Glycol Chlorohydrin → ETHYLENE CHLOROHYDRIN
Glycol Dibromide → ETHYLENE DIBROMIDE
Glycoi Dinitrate → ETHYLENE GLYCOL DINITRATE
GLYCOL ETHER also → ENAMEL, INKS
Glycol Ether EE → 2-ETHOXYETHANOL
Glycol Ether PM → PROPYLENE GLYCOL MONOMETHYL ETHER
Glycolic Acid → AZELAIC ACID
GLYCOLONITRILE
Glyconon → TOLBUTAMIDE
Glyconormal → GLYMIDINE
GLYCOPYRROLATE also → ETHANE
Glycopyrronium Bromide → GLYCOPYRROLATE
Glycotron → TOLBUTAMIDE
Glycyrrhetic Acid → GLYCYRRHETINIC ACID
GLYCYRRHETINIC ACID
GLYCYRRHIZA GLABRA also → SOTALOL, TOBACCO
Glycyrrhizic Acid → GLYCYRRHIZA GLABRA
Glycyrrhizin → ABRUS PRECATORIUS
Glydant → DIMETHYLOL DIMETHYL HYDANTOIN
Glydiazinamide → GLIPIZIDE
Glyfyllin → DYPHYLLINE
GLYHEXAMIDE
Glykocellon → CELLULOSE
Glykresin → MEPHENESIN
GLYMIDINE
Glymol → MINERAL OIL
Glyodex 37-22 → CAPTAN
GLYODIN
Glyotol → MEPHENESIN
GLYOXAL
GLYOXYLIC ACID also → FORMIC ACID
Glyphenarsine → TRYPARSAMIDE
GLYPHOSPHATE also → ALACHLOR
Glyphylline → DYPHYLLINE
Glypolix → DEXFENFLURAMINE

Glypressin → TERLIPRESSIN
Glyvenol → TRIBENOSIDE
GM-CSF → GRANULOCYTE MACROPHAGE COLONY STIMULATING FACTOR
Gnaphalium → SENECIO sp.
GNIDIA KRAUSSIANA
GnRN → LH-RH
Gö-3450 → GABAPENTIN
Goa Bean → BEANS, Winged
GOAT FISH
Goat Nut → JOJOBA BEAN
Goat's Rue → GALEGA also → TEPHROSIA sp.
Goat Weed → HYPERICUM
Godalax → BISACODYL
GOE 3450 → GABAPENTIN
Gofman, Dr. John → PLUTONIUM
Goitrin → TURNIP also → GLUCOSINOLATES
GOLD also → ACETYLCYSTEINE, JEWELRY, MERCURY, NICKEL, RADON
Gold Dust → COCAINE
Golden Bean → THERMOPSIS sp.
Golden Buttons → TANSY
Golden Chain Tree → CYSTISUS LABURNUM also → CYTISINE
Golden Glow → RUDBECKIA sp.
Golden Pea → THERMOPSIS sp.
GOLDENROD also → ASTERS, CHAMOMILE, TEA
Golden Seal → HYDRASTIS CANADENSIS
Goldenweed → OONOPSIS sp.
GOLD SODIUM THIOMALATE also → PENICILLAMINE
GOLD SODIUM THIOSULFATE
Gold Thioglucose → AUROTHIOGLUCOSE
GOLF BALLS
"Golf Course Dermatitis" → GOLF BALLS
GOMPHRENA CELOSIOIDES
Gonacrine → ACRIFLAVINE
Gonodactylus bredini → MANTIS SHRIMP
GONADOTROPINS, CHORIONIC
Gonadotraphon L.H. → GONADOTROPINS, CHORIONIC
Gonadyl → PROGESTERONE
Gonamone → GONADOTROPINS, CHORIONIC
Gondafon → GLYMIDINE
Gonic → GONADOTROPINS, CHORIONIC
Gonorcin → NORFLOXACIN
Gontochin → CHLOROQUINE
GONYAULAX also → SAXITOXIN
Goobernuts → PEANUTS
Goof Butts → CANNABIS
Goon → PHENCYCLIDINE
Gooroo Nuts → KOLA
Goose Grass → ARROWGRASS
Goose Tongue → ARROWGRASS
GOPHACIDE
Gopher Plant → EUPHORBIA, *lathyris*
Gorban → TRIMETHOPRIM
Gordoloba yerba → SENECIO sp. also → PYRROLIZIDINES
Gordox → APROTININ
Goretex → POLYTEF
Gosarelin → ZOLADEX
Gossypium sp. → COTTON
GOSSYPOL also → COTTONSEED MEAL
Gotensin → LEVOBUNOLOL
Gotu → KOLA also → CENTALLA ASIATICA
Gousiekte → PACHYSTIGMA sp. also → PAVETTA sp.
Goyl → ACETARSONE
GP 45,840 → DICLOFENAC
GR 109,714X → LAMIVUDINE
GR 2063 → CEFTAZIMIDES
GR 2/1574 → ALFADOLONE ACETATE
GR C507/75 → ONDANSETRON
GR 33,343X or *G* → SALMETEROL
GR 38,032F → ONDANSETRON
GR 43,175C → SUMATRIPTAN SUCCINATE
GR 43,659X → LACIDIPINE
Grabolin → ETHYLESTRENOL
GRACILLARIA
Gradient → FLUNARIZINE HCL
Grafestrol → DIETHYLSTILBESTROL
"Grain-measurer's Lung" → WHEAT
Graisse de Suint Purifée → LANOLIN
Gram → PEAS
Gramaxin → CEFAZOLIN SODIUM
GRAMICIDIN also → CURARE, SUCCINYLCHOLINE
GRAMICIDIN
Gramoxone → PARAQUAT
Grampenil → PENICILLINS – "NEWER and SYNTHETICS" > *Ampicillin*
Granadilla → EBONY
Granatox → PHORATE
GRANDAXIN → TOFISOPAM
GRANISETRON
Granosan → ETHYLMERCURIC CHLORIDE
Granulestin → LECITHIN

GRANULOCYTE COLONY STIMULATING FACTOR
GRANULOCYTE MACROPHAGE COLONY STIMULATING FACTOR
GRAPE also → L-DOPA, PATULIN, RAISIN, SULFUR DIOXIDE
GRAPEFRUIT also → AMLODIPINE, ASTEMIZOLE, CAFFEINE, CARBAMAZEPINE, CILOSTAZOL, CYCLOPHOSPHAMIDE, CYCLOSPORINE, DIAMINOZIDE, DIGOXIN, DILTIAZEM, DOFETILIDE, ESTRADIOL, FELBAMATE, FEXOFENADINE, FRUITS and FRUIT JUICES, LOSARTAN, LOVASTATIN, MIDAZOLAM, NICARDIPINE, NIFEDIPINE, NIMODIPINE, NISOLDIPINE, NITRENDIPINE, RAPAMYCIN, SAQUINAVIR, SIMVASTIN, TACROLIMUS, TERFENADINE, TRIAZOLAM, VERAPAMIL
Grape Sugar → GLUCOSE
GRAPHITE
Graphium sp. → SAWDUST
Grapple Plant → HARPAGOPHYTUM
GRASSES
Grass → CANNABIS
Grass Pea → LATHYRUS sp.
Grasstree → XANTHORRHEA sp.
Grass Vetchling → LATHYRUS sp.
Gratego → PYRACANTHA sp.
Gratibain → OUABAIN
Graveyard Weed → EUPHORBIA
Gravindex → PROMETHAZINE
Gravol → DIMENHYDRINATE
Gray Baby Syndrome → CHLORAMPHENICOL
Grayanotoxins → ANDROMEDOTOXINS, PIERIS sp.
Gray Millet → LITHOSPERMUM RUDERALE
Grazon → PICLORAM
Greasewood → SARCOBATUS VERMICULATUS
Greater Ammi → AMMI MAJUS
Great Fear of 1789 → ERGOT
Great Lobelia → LOBELIA
Great Nettle → URTICA DIOICA
GREAT WEEVER
Great White Shark → CANNABIS
G-Recillin → PENICILLIN > *Benzylpenicillin, Potassium*
"Greek Fix" → NAPHTHA, NAPHTHALENE
Green → KETAMINE
Green Cinnabar → CHROMIC OXIDE
Green Dragon → ARISAEMA also → DATURA
Green Golden Oat Grass → TRISETUM FLAVESCENS
Greenhartin → LAPACHOL
Green Hellebore → VERATRUM sp.
Green Orach → ATRIPLEX sp.
Green Oxide of Chromium → CHROMIC OXIDE
Green Rouge → CHROMIC OXIDE
Greensalt → CHROMATED COPPER ARSENATE
Green Vetch → LATHYRUS sp.
Grenadile → PASSIFLORA sp.
Grenol → RONNEL
GREPAFLOXACIN also → SUCRALFATE, THEOPHYLLINE
GRETA also → LEAD, LEAD OXIDES
Grievous Bodily Harm → γ-HYDROXYBUTYRATE
Griffo → CANNABIS
Grifomin → AMINOPHYLLINE
Grifulvin → GRISEOFULVIN
Grill-cooks → CARBONS
Grinsil → PENICILLINS – "NEWER and SYNTHETICS" > *Amoxicillin*
Gripenin → PENICILLINS – "NEWER and SYNTHETICS" > *Carbenicillin Disodium*
Grippex → THALIDOMIDE
Grisactin → GRISEOFULVIN
Griséfuline → GRISEOFULVIN
GRISEOFULVIN also → ACENOCOUMAROL; ALCOHOL, ETHYL; ANTIPYRINE, BISHYDROXYCOUMARIN, BUTABARBITAL, CYCLOCOUMAROL, CYCLOSPORINE, DESMOPRESSIN, DISULFIRAM, ETHYL BISCOUMACETATE, FATS, MEPHOBARBITAL, METHOTRIMEPRAZINE, ORPHENADRINE, PHENOBARBITAL, PHENYLBUTAZONE, PHENYTOIN, PIPERONYL BUTOXIDE, WARFARIN
Grisovin → GRISEOFULVIN
Grisowen → GRISEOFULVIN
Gris-PEG → GRISEOFULVIN
Groceme → DEQUALINIUMS, *chloride*
Grolean → GROWTH HORMONES
Gromwell → LITHOSPERMUM RUDERALE
Groningen → NABAM
Grootlamsiekte Syndrome → SALSOLA sp.
Grorm → GROWTH HORMONES
Ground Cherry → PHYSALIS sp.
Ground Gourd → COLOCYNTH
Ground IVY → IVY also → MODIOLA CAROLINIANA also → NEPETA HEDERACEA
Groundnuts → PEANUTS

Groundseltree → BACCHARIS sp.
GROWTH HORMONES also → SUMATRIPTAN SUCCINATE
Growth Hormone Release Inhibiting Hormone → SOMATOSTATIN
Grysio → GRISEOFULVIN
GS 504 → CIDOFOVIR
GS 840 → ADEFOVIR
GS 0504 → CIDOFOVIR
GS 1339 → DYMANTHINE
GS 2989 → MECLOCYCLINES
GS 3065 → DOXYCYCLINE
GS 4104 → OSELTAMIVIR
GS 6244 → CARBADOX
GS 13,005 → METHIDATHION
GS 13,332 → DIMETILAN
G-Strophanthin → OUABAIN
GT 41 → BUSULFAN
GT 1012 → PRAJMALINE TARTRATE
GTA Rat Bait → THALLIUM
GTN → NITROGLYCERIN
Guacamole → SODIUM METABISULFITE
GUAIAC
GUAIACOL
Guaiacol Glyceryl Ether → GLYCERYL GUAIACOLATE
Guaiacuran → GLYCERYL GUAIACOLATE
Guaiamar → GLYCERYL GUAIACOLATE
GUAIJILLO also → ACACIA
Guaiphenesin → GLYCERYL GUAIACOLATE
Guamide → SULFAGUANIDINE
GUANABENZ ACETATE
GUANACLINE SULFATE
GUANADREL SULFATE
GUANAFACINE
Guanamprazine → AMILORIDE
Guanatol → CHLOROGUANIDE
GUANETHIDINE also → AMITRIPTYLINE, AMPHETAMINE, BETHANIDINE, CHLORPROMAZINE, COCAINE, CURARE, CYCLOBENZAPRINE, DESIPRAMINE, DEXTROAMPHETAMINE, DIGITALIS, EPHEDRINE, EPINEPHRINE, HYDROCHLORTHIAZIDE, HYDROXYAMPHETAMINE, IMIPRAMINE, INSULIN, MAZINDOL, MEPHENTERMINE, METARAMINOL BITARTRATE, METHAMPHETAMINE, METHOTRIMEPRAZINE, METHYLDOPA, METHYLPHENIDATE, NOREPINEPHRINE, NORTRIPTYLINE, PHENIPRAZINE, PHENYLEPHRINE HCL, PHENYLPROPANOLAMINE, PIPRADOL, PROPRANOLOL, PROTRIPTYLINE, PSEUDOEPHEDRINE, THIOTHIXENE, TRANYLCYPROMINE, TRIMIPRAMINE, TRIPELENNAMINE
GUANFACINE
Guanicil → SULFAGUANIDINE
GUANIDINE
GUANOXAN
Guaramine → CAFFEINE
GUARANGA
GUAREA
GUAR GUM
Guastil → SULPIRIDE
Guatan kura → SOLANUM sp.
Guayule → PARTHENIUM sp.
Guayale argentatum → PARTHENIUM sp.
Guayanesin → GLYCERYL GUAIACOLATE
Gubernal → ALPRENOLOL
Guelder Rose → VIBURNUM PRUNIFOLIUM
Guerena → SENECIO sp.
Guesapon → DDT
Guethine → GUANETHIDINE
Gui de Chêne → MISTLETOE, AMERICAN
Guinea Corn → KAFFIRCORN
Guinea Grass → PANICUM sp.
Gutierrezia sp. → BROOMWEED
Gulf War → PYRIDOSTIGMINE BROMIDE, SARIN
L-Gulitol → SORBITOL
"Gullet-trimmed Meat" → HAMBURGERS
Gulliostin → DIPYRIDAMOLE
Gully Root → PETIVERIA ALLIACEA
Gum Acacia → ACACIA
Gum Arabic → ACACIA
Gumbaral → S-ADENOSYLMETHIONINE
Gum Benjamin → BENZOIN GUM
Gum Camphor → CAMPHOR
Gum Damar → DAMAR
Gum Dragon → ACACIA
Gum Guaiac → GUAIAC
Gum Resin → ROSIN
Gum Thus → OLIBANUM
Gum Wood → EUCALYPTUS GLOBULUS
Gun Blue → COPPER NITRATE
Gunchi → ACENOCOUMAROL
G.U.-Pen → PENICILLINS – "NEWER and SYNTHETICS" > *Carindacillin Sodium*
GURANIA
Guru Nuts → KOLA
Gusathion → AZINPHOSOS-ETHYL also → AZINPHOS-METHYL
Gusperimus → TACROLIMUS
Guthion® → AZINPHOS-METHYL
GUTIERREZIA MICROCEPHALA

Gutron → MIDODRINE
GUTTA SIAC
GVG → VIGABATRIN
GVIA also → CONE SHELL FISH
GVP → TERLIPRESSIN
GV-Eleven → GENTIAN VIOLET
Gwai-Kou → WAI-LING-SIN
Gwanda → PAPAYA
Gwandar daii → ANONA
Gwanja kusa → BLIGHIA SAPIDA
Gwaska → ERYTHROPHLEUM
GX 118 → IMIDAN
GX 1048 → LACIDIPINE
Gyada → PEANUTS
Gymmura sp. → STINGRAYS
Gymnocladus dioica → KENTUCKY COFFEE TREE
Gynamousse → OXYTETRACYCLINE
Gynäsan → ESTRIOL
Gynefollin → DIENESTROL
Gyne-Lotrimin → CLOTRIMAZOLE
Gynergen → ERGOTAMINE TARTRATE
Gynestrel → NAPROXEN
Gynipral → HEXOPRENALINE
Gynlutin → PROGESTERONE
Gynochrome → MERBROMIN
Gyno-Daktarin → MICONAZOLE NITRATE
Gynoestryl → ESTRADIOL
Gynofon → 2-ACETAMIDO-5-NITROTHIAZOLE
Gynofug → IBUPROFEN
Gynokhellan → KHELLIN
Gynol II → NONOXYNOL-9
Gynolett → ETHINYL ESTRADIOL
Gyno-Monistat → MICONAZOLE NITRATE
Gyno-Myfungar → OXICONAZOLE NITRATE
Gynomyk → BUTOCONAZOLE
Gyno-Pevaryl → ECONAZOLE NITRATE
Gynoplix → ACETARSONE
Gynorest → DIDROGESTERONE
Gyno-Sterosan → CHLORQUINALDOL
Gyno-Terazol → TERCONAZOLE
Gynotherax → CHLORQUINALDOL
Gyno-Trosyd → TIOCONAZOLE
GYPSY MOTH
Gyramid → ENOXACIN
GYROMITRA ESCULENTA also → METHYL HYDRAZINE, MUSHROOMS

H

H → HEROIN also → MUSTARD GAS
H3 → PROCAINE
H 44/68 → METYROSINE
H 56/28 → ALPRENOLOL
H 93/26 → METOPROLOL
H 154/82 → FELODIPINE
H 168/68 → OMEPRAZOLE
H 321 → MERCAPTODIMETHUR
H 365 → PAROXYPROPIONE
H 610 → FENCAMFAMINE
H 990 → OXYMETAZOLINE
H 3292 → DISOPYRAMIDE
H 4723 → CLOBAZAM
96-H-60 → HALOXON
H1E1Z9 → SODIUM TETRADECYL SULFATE
Habba Sûdâ → CUMIN, BLACK
Habbat al-barakah → CUMIN, BLACK
Hachimycin → TRICHOMYCIN
Hadacidin → PENICILLIUM sp.
Haelan → FLURANDRENOLIDE
Haemaccel® → GELATIN
Haemalift → IRON DEXTRANS
Haemanthine → BUPHANE DISTICHA
Haemanthus → BOÖPHONE
Haemanthus toxicarius → BUPHANE DISTICHA
Haemiton → CLONIDINE
HAFNIUM
Haglodase → HYALURONIDASE
Hai gen fen → LEAD
Haiprex → METHENAMINE
HAIR
Hairgro → MINOXIDIL
Hairy Caltrop → KALLSTROEMIA
Hairy Panic → PANICUM sp.
Hairy Stinger → JELLYFISH
Hairy Thornapple → DATURA, *metel*
Hairy Vetch → VICIA
Hajasen → SENNA
hako-fugu → PAHUTOXIN
Halatal → PENTOBARBITAL
HALAZEPAM also → CIMETIDINE, DISULFIRAM
Halbmond → DIPHENHYDRAMINE
Halcion → TRIAZOLAM
Haldi → TURMERIC
Haldol → HALOPERIDOL
Haldrate → PARAMETHASONE ACETATE
Haldrone → PARAMETHASONE ACETATE
Haledo → TURMERIC
Halfan → HALOFANTRINE
Halinone → BROMINDIONE
Haloanisone → FLUANISONE
Halocarbon 112 → 1,1,2,2-TETRACHLORODIFLUORETHANE

Halocarbon 113 → TRICHLOROTRIFLUOROETHANE
HALOFANTRINE
HALOFENATE also → CHLORPROPAMIDE, PROPRANOLOL
HALOGETON
Halon → DICHLORODIFLUOROMETHANE
HALOPERIDOL also → ALCOHOL, ETHYL; ALPRAZOLAM, CARBAMAZEPINE, DEXTROAMPHETAMINE, FLUOXETINE, GLUCOSE, GUANETHIDINE, IMIPRAMINE, INDOMETHACIN, ISOPROTERENOL, LITHIUM, METHOTRIMEPRAZINE, METHYLDOPA, METUREDEPA, NIALAMIDE, PAROXETINE, PERPHENAZINE, PHENINDIONE, PHENOBARBITAL, PHENYTOIN, PROCHLORPERAZINE, PROPRANOLOL, TOBACCO, TRIFLUOPERAZINE DIHYDROCHLORIDE
HALOPROGIN
HALOPROPANE
Halopyramine → CHLOROPYRAMINE
Halosten → HALOPERIDOL
Halotestin → FLUOXYMESTERONE
Halotex → HALOPROGIN
HALOTHANE also → ACETYLSALICYLIC ACID, ANTIPYRINE, DIGITALIS, DIMETHYL TUBOCURARINE IODIDE, EPHEDRINE, EPINEPHRINE, HEXAFLUORENIUM BROMIDE, ISOPROTERENOL, KETAMINE, MEPERIDEINE, MEPHENTERMINE, METHAMPHETAMINE, NOREPINEPHRINE, OXYTETRACYCLINE, PHENYLEPHRINE HCL, PHENYTOIN, PROPRANOLOL, RIFAMPIN, RUBBER, *d*-TUBOCURARINE CHLORIDE
Halowax → CHLORINATED NAPHTHALENES
Halowax 1013 → PENTACHLORONAPHTHALENE
Halox → HALOXON
HALOXON
Halprin → IBUPROFEN
HALQUINOL also → IODOCHLORHYDROXYQUINOLINE
Halquivet → HALQUINOL
Halt® → CAPSAICIN
Haltran → IBUPROFEN
HALVA(H) also → SESAME
HAM also → CYCLOPIAZONIC ACID, *N*-NITROSODIMETHYLAMINE, OCHRATOXINS
Hamar → TAMARIND
HAMAMELIS VIRGINIANA
Hamarin → ALLOPURINOL
HAMBURGERS
Hamidop → METHAMIDOPHOS
Hamlet → HYOSCYAMUS
Hamoloshi → MAERUA ANGOLENSIS
Hämovannid → INOSITOL NIACINATE
HAMYCIN
Hanane → DIMEFOX
Handal → COLOCYNTH
Handy Easter Egg Color® → EGGS
Hanfkraut → CANNABIS
Hannu Biyar → PAULLINIA sp.
Hapalochlaena maculosa → OCTOPUS also → TETRODOTOXIN
Happy Dust → COCAINE
Happy Trails → COCAINE
Haricut Bean → BEANS, Kidney
Harmal → PEGANUM HARMALA
HARMALINE also → AYAHUASCA, PASSIFLORA sp., PEGANUM HARMALA, TETRAPTERIS sp.
HARMALOL also → PEGANUM HARMALA
HARMAN also → NICOTIANA sp., PASSIFLORA sp.
Harmar → PROPOXYPHENE
Harmel → PEGANUM HARMALA
Harmidine → HARMALINE
HARMINE also → AYAHUASCA, PEGANUM HARMALA
Harmogen → ESTRONE
HARMOL also → PASSIFLORA sp., PEGANUM HARMALA
Harmonyl → DESERPIDINE
Harodase → HYALURONIDASE
HARPAGOPHYTUM
Harpagophytum procumbens → PHYTEUMA
HARUNGANA MADAGASCARIENSIS
Hasethrol → PENTAERYTHRITOL TETRANITRATE
Hash → CANNABIS
Hashish → CANNABIS
Hasis → CANNABIS
Hassall, Dr. → COPPER ARSENITE
Hatisura → HELIOTROPIUM sp.
HATS
Haukata → DATURA, *metel*
Havlodine → POLYVINYLPYRROLIDONE IODINE COMPLEX
Havoc → BRODIFACOUM
Hawaiian Baby Wood Rose → ARGYREIA NERVOSA

Hawaiian Wood Rose → ARGYREIA NERVOSA also → PHENCYCLIDINE
Hawe's Lemon Oil® → POLISHES, FURNITURE
Haws → HAWTHORN
HAWTHORN
"Hay" → CANNABIS
Hay Needle → STIPA sp.
Haynon → CHLORPHENIRAMINE
Hayp → COCA
Haze → LYSERGIC ACID DIETHYLAMIDE
HAZEL NUTS
Hazol → OXYMETAZOLINE
HB 115 → NIFURPRAZINE
HB 419 → GLYBURIDE
HBF 386 → CACTINOMYCIN
HBP → RAUWOLFIA SERPENTINA
HBW 023 → HIRUDINS
Hc45 → HYDROCORTISONE
HC 064 → NIHYDRAZONE
HC 20,511 → KETOTIFEN FUMARATE
HCB → HEXACHLOROBENZENE
HCBD → HEXACHLOROBUTADIENE
HCCP → HEXACHLOROCYCLOPENTADIENE
HCCPD → HEXACHLOROCYCLOPENTADIENE
HCDD → HEXACHLORODIBENZO-*p*-DIOXINS
HCG → GONADOTROPINS, CHORIONIC
HCH → BENZENE HEXACHLORIDE
γHCH → LINDANE
HCN → HYDROGEN CYANIDE
HCP → HEXACHLOROPHENE
HD → MUSTARD GAS
HDI → HEXAMETHYLENE DIISOCYANATE
5-HDMT → PHALARIS sp.
Head Grit → NARTHECIUM
Healon → SODIUM HYALURONATE
Healonid → SODIUM HYALURONATE
Health Research Group → VINYL CHLORIDE
Heartguard → IVERMECTIN
Heart Leaf → GASTROLOBIUM sp.
Heart On → ISOBUTYL NITRITE
Hearts → AMPHETAMINES
Heavenly Blues → MORNING GLORY
Heaven Dust → COCAINE
Heaven Leaf → COCAINE
Hebanil → CHLORPROMAZINE
Heciline → PENICILLINS – "NEWER and SYNTHETICS" > *Hetacillin*
Hecogenin → AGAVE
Hedacidin → PENICILLIUM sp.
HEDEOMA sp.
HEDERA HELIX
Hederin → HEDERA HELIX
Hederogenin → HEDERA HELIX, IVY
Hedex → ACETAMINOPHEN
Hedge → BUXUS *sp.*
Hedge Apples → OSAGE ORANGE
Hedge Balls → OSAGE ORANGE
Hedge Trees → OSAGE ORANGE
Hedulin → PHENINDIONE
Hefasolon → PREDNISOLONE
Heferol → FERROUS FUMARATE
Heifex → CLOPROSTENOL
Heitrin → TERAZOSIN
Hekbilin → CHENODEOXYCHOLIC ACID
Helecho Maco → ASPIDIUM
Helenalin → HELENIUM sp.
Helénie nudiflore → HELENIUM sp.
HELENIUM sp.
Heleurine → PYRROLIZIDINES
HelfaCat → DIAZINON
HelfaDog → DIAZINON
Helfergin → MECLOFENOXATE
Helianthus annus → SUNFLOWER
Helicon → ACETYLSALICYLIC ACID
Heliopar → CHLOROQUINE
Heliophan → HOMOMENTHYL SALICYLATE
Heliosupine → CYNOGLOSSUM OFFICIANALE
HELIOTRINE also → HELIOTROPIN, PYRROLIZIDINES
Heliotrope → HELIOTROPIUM sp.
HELIOTROPIN
HELIOTROPIUM sp. also → PYRROLIZIDINES, TEA
Helistat → COLLAGEN
Helixor → MISTLETOE, EUROPEAN or JAPANESE
Helleborin → CALTHA PALUSTRIS, HELLEBORUS NIGER
Helleboxein → HELLEBORUS NIGER
Hellebrin → HELLEBORUS NIGER
Hellipidyl → *p*-AMINOSALICYLIC ACID
Helmatac → PARBENDAZOLE
Helmetina → PHENOTHIAZINE
Helmex → PYRANTEL
Helmezine → PIPERAZINE
Helminthex → PIPERAZINE
Helminthosporum biseptatum → ROMULEA ROSEA
Helmintox → PYRANTEL
Helmirone → HALOXON
Heloderma suspectum → GILA MONSTER
Helogaphen → CHLORDIAZEPOXIDE
Helothion → SULPROFOS

Helvamox → PENICILLINS – "NEWER and SYNTHETICS" > *Amoxicillin*
Helvecyclin → TETRACYCLINE
Helvela suspecta → GYROMITRA ESCULENTA
Helvella → METHYL HYDRAZINE
Hemabate → CARBOPROST
HEMATIN
Hematite → FERRIC OXIDE
HEMATOPORPHYRIN
Hemax → ERYTHROPOIETIN
HEMIN
Hemineurin → CLOMETHIAZOLES
Heminevrin → CLOMETHIAZOLES
Hemlock → CONIUM MACULATUM
Hemocaprol → Σ-AMINOCAPROIC ACID
Hemoclar → PENTOSAN POLYSULFATE
Hemocuron → TRIBENOSIDE
Hemodex → DEXTRAN
Hemolidione → PHENINDIONE
Hemometina → EMETINE
Hemopek → CELLULOSE
Hemostop → CARBAZOCHROME SALICYLATE
Hemp → CANNABIS
HEMPA → HEXAMETAPOL
Hemp Dogbane → APOCYNUM CANNABINUM
Hempnettle → GALEOPSIS
Henbit → LAMIUM AMPLEXICAULE
Hen & Chicks → LANTANA sp.
2-Hendecanone → METHYL NONYL KETONE
HENNA
Hens Dulse → CARRAGEENAN
HEOD → DIELDRIN
Heolite → ANTHRAQUINONE
Hepalidine → TIMONACIC
Hepamig → TILIA sp.
Heparan Sulfate → DANAPAROID
Heparegen → TIMONACIC
HEPARIN also → ACETYLSALICYLIC ACID; ALCOHOL, ETHYL; CHLORACIZINE; 4-CHLORO-*m*-CRESOL, CHLORDIAZEPOXIDE, CHLOROBUTANOL, DIHYDROERGOTAMINE, DIPHENHYDRAMINE, GENTAMICIN, GLIPIZIDE, NEOMYCIN, NITROGLYCERIN, OXAZEPAM, PENICILLIN, PERPHENAZINE, POTASSIUM, PROMAZINE, PROMETHAZINE, SODIUM, THYROXINE, TISSUE PLASMINOGEN ACTIVATOR
Hepartest → SULFOBROMOPHTHALEIN
Hep-A-Stat → NITARSONE
HEPATICA sp.
Hepatic Veno-occlusive Disease → PYRROLIZIDINES
Hepatitis B Vaccine → NICKEL SULFATE
Hepin → Σ-AMINOCAPROIC ACID
Heprinar → HEPARIN
Hepsal → HEPARIN
Heptabarb → HEPTABARBITAL
HEPTABARBITAL also → ACENOCOUMAROL, BISHYDROXYCOUMARIN
HEPTACHLOR
Heptadon → METHADONE HCL
Heptadorm → HEPTABARBITAL
Heptamethyl-p-rosaniline Chloride → METHYL GREEN
Heptamul → HEPTACHLOR
n-HEPTANE
1,7-Heptanedicarboxylic Acid → AZELAIC ACID
HEPTANETHIOL
1-HEPTANOL
2-Heptanol → *n*-HEXANE
Heptanon → METHADONE HCL
2-HEPTANONE
3-Heptanone → ETHYL BUTYL KETONE
n-Heptyl Alcohol → 1-HEPTANOL
HEPTYLCYANOACRYLATE
2-*n*-HEPTYL CYCLOPENTANONE
HEPTYL HYDRAZINE
Heptylmercaptan → HEPTANETHIOL
Hepzide → NITHIAZIDE
Her → COCAINE
Heracillin → PENICILLINS – "NEWER and SYNTHETICS" > *Floxacillin*
HERACLEUM sp.
Heraclin → METHOXSALEN
Herb → CANNABIS
Herba → HYPERICUM
Herbadox → PENDIMETHALIN
Herbo à la Puce → POISON IVY
Herban M → MONOSODIUM METHANEARSONATE
Herbesser → DILTIAZEM
Herbicide 282 → ENDOTHALL
Herbicide 283 → ENDOTHALL
Herbisan → BISETHYLXANTHOGEN
Herb of Grace → RUE
Herbizole → AMINOTRIAZOLE
Herboxone → PARAQUAT
Herb Paris → PARIS sp.
Herbygrass → RUE
Herceptin → TRASTUZUMAB
Hercules 528 → DIOXATHION
Hercules 3956 → TOXAPHENE

Hercules 14,503 → DIALIFOR
Herkol → DICHLORVOS
Hermesetas → SACCHARIN
Hernando Cortés → CHOCOLATE
Herocaine → PROCAINE
HEROIN also → ALCOHOL, ETHYL; DEXTROMORAMIDE, LEMON, MEXILETINE, MILK, PAPAVER sp., POPPY, POTASSIUM, PROCAINE, PROPRANOLOL, QUININE, THALLIUM
Heroisch → HEROIN
Herpes-Gel → IDOXURIDINE
Herplex → IDOXURIDINE
HERRING also → FISH, ISOCARBOXAZID, NIALAMIDE, *N*-NITROSODIMETHYLAMINE
Herva pombinha → PHYLLANTHUS sp.
Herzon → PRENYLAMINE LACTATE
Hesofen → MEPARFYNOL
Hetacillin → PENICILLINS – "NEWER and SYNTHETICS"
Heteroauxin → INDOLEACETIC ACID
HETEROMELES ARBUTIFOLIA also → HYDROGEN CYANIDE
HETERONIUM BROMIDE
HETEROPHYLLEA PUSTULATA
Heterophylline → PARSONSIA sp.
Heterosan → DIETHYLCARBAMAZINE
HETP → HEXAETHYLTETRAPHOSPHATE
Hetrazan → DIETHYLCARBAMAZINE
Hetrum Bromide → HETERONIUM BROMIDE
Hevea brasiliensis → LATEX
Hexabendin → HEXOBENDINE
Hexabetalin → VITAMIN B_6
Hexabione → VITAMIN B_6
gamma Hexachlor → LINDANE
Hexachloran → LINDANE
Hexachlorethane → HEXACHLOROETHANE
HEXACHLOROBENZENE also → PROPANIL, QUINTOZENE
Hexachlorobenzol → HEXACHLOROBENZENE
HEXACHLOROBUTADIENE
Hexachlorocyclohexane → BENZENE HEXACHLORIDE
HEXACHLOROCYCLOPENTADIENE
HEXACHLORODIBENZO-*p*-DIOXINS
HEXACHLOROETHANE
HEXACHLORONAPHTHALENE
HEXACHLORONORBORNADIENE
Hexachlorophane → HEXACHLOROPHENE
HEXACHLOROPHENE also → BITHIONOL, DICHLOROPHEN, FENTICLOR, MAGNESIUM SILICATE, SOAP, TETRACHLOROSALICYLANILIDE
Hexacillin → PENICILLINS – "NEWER and SYNTHETICS" > *Hetacillin*
Hexacycline → TETRACYCLINE
Hexydaline → METHENAMINE
Hexadecadrol → DEXAMETHASONE
1-HEXADECANETHIOL
1-Hexadecanol → ALCOHOL, CETYL
6-Hexadecenlactone → AMBRETTOLIDE
Hexadecyl Alcohol → ALCOHOL, CETYL
HEXADECYLDIMETHYLAMINE
Hexadecylmercaptan → 1-HEXADECANETHIOL
Hexadilat → NIFEDIPINE
HEXADIMETHRINE BROMIDE
Hexadrin → ENDRIN
HEXAETHYLTETRAPHOSPHATE
HEXAFLUORENIUM BROMIDE also → HALOTHANE, SUCCINYLCHOLINE
HEXAFLUOROACETONE
Hexafluorodiethyl Ether → FLUROTHYL
HEXAFLURATE
Hexafluronium Bromide → HEXAFLUORENIUM BROMIDE
Hexagastron → SUCRALFATE
Hexahydrobenzene → CYCLOHEXANE
Hexahydrocresol → METHYLCYCLOHEXANOL
Hexahydromethylphenol → METHYLCYCLOHEXANOL
Hexahydrophenol → CYCLOHEXANOL
Hexahydropyrazine → PIPERAZINE
Hexahydrotoluene → METHYLCYCLOHEXANE
Hexalen → ALTRETAMINE
Hexalin → CYCLOHEXANOL
Hexalon → BENZENE HEXACHLORIDE
HEXAMARIUM BROMIDE
HEXAMETAPOL
Hexamethone → HEXAMETHONIUM BISTRIUM
HEXAMETHONIUM BISTRIUM also → PYANTEL
Hexamethylenamine → METHENAMINE
Hexamethylene → CYCLOHEXANE
HEXAMETHYLENE DIISOCYANATE
HEXAMETHYLENE GLYCOL
Hexamethylmelamine → ALTRETAMINE
Hexamethylphosphoramide → HEXAMETAPOL
Hexamethylphophoric Triamide → HEXAMETAPOL
Hexamethyl-p-rosaniline → GENTIAN VIOLET
Hexamethyltetracosane → SQUALANE
Hexamine → METHENAMINE
Hexanaphthene → CYCLOHEXANE
Hexanastab → HEXOBARBITAL
n-HEXANE also → ACETONYL ACETONE, GLUE, JET FUELS, RUBBER, SHOES

2,5-Hexanediol → HEXAMETHYLENE GLYCOL
2,5-Hexanedione → ACETONYL ACETONE also → HEXAMETHYLENE GLYCOL, *n*-HEXANE, METHYL BUTYL KETONE
HEXANETHIOL
Hexanicit → INOSITOL NIACINATE
Hexanicotol → INOSITOL NIACINATE
Hexanoestrol → HEXESTROL
1-Hexanol → ALCOHOL, *n*-HEXYL
2-Hexanone → METHYL BUTYL KETONE also → ACETONYL ACETONE
Hexanurat → ALLOPURINOL
Hexapromin → TRANEXAMIC ACID
HEXAPROPYMATE
Hexastat → ALTRETAMINE
Hexathane → ZINEB
Hexathion → CARBOPHENOTHION
Hexathonide → HEXAMETHONIUM BISTRIUM
Hexatrione → TRIAMCINOLONE
Hexatron → TRANEXAMIC ACID
Hexavibex → VITAMIN B_6
HEX-BCH → HEXACHLORONORBORNADIENE
Hexemal → CYCLOBARBITAL
Hexermin → VITAMIN B_6
HEXESTROL
Hexevan → HEXESTROL
HEXOBARBITAL also → AMINOPYRINE, *p*-AMINOSALICYLIC ACID, CARBIPHENE; CEDAR, WESTERN RED; CEDROL, CHLORDANE, DIETHYL-*p*-PHOSPHATE, DISULFIRAM, GLUCAGON, HEPTACHLOR, IMIPRAMINE, LYNESTRENOL, MAGNOLIAS, METHOXYFLURANE, METHYL TESTOSTERONE, ORPHENADRINE, PHENOBARBITAL, PHENYTOIN, PROPOXYPHENE, PROPRANOLOL, RIFAMPIN, TESTOSTERONE, THYROXINE, TICLODIPINE
Hexobarbitone → HEXOBARBITAL
HEXOBENDINE
Hexobion → VITAMIN B_6
HEXOCYCLIUM
Hexoestrol → HEXESTROL
Hexofen → MEPARFYNOL
Hexogen → CYCLONITE
HEXOL
Hexone → METHYL ISOBUTYL KETONE
Hexonium → HEXAMETHONIUM BISTRIUM
Hexopal → INOSITOL NIACINATE
HEXOPRENALINE
Hexopropymate → HEXAPROPYMATE
Hexosan → HEXACHLOROPHENE
Hexose → FRUCTOSE
HEXYL ACETATES
HEXYLCAINE
HEXYLENE GLYCOL
n-Hexyl-σ-hydroxybenzoate
Hexylmercaptan → HEXANETHIOL
HEXYL METHYL KETONE
HEXYLRESORCINOL also → ALCOHOL, ETHYL; HYDROQUINONE, MINERAL OIL, PHLOROGLUCINOL, SANTONIN
HEXYL SALICYLATE
n-Hexylthiol → HEXANETHIOL
Hexyphen → TRIHEXYPHENIDYL
Heyden 611 → STIBOPHEN
HF → HYDROFLUORIC ACID
HF 1854 → CLOZAPINE
HF 1927 → DIBENZEPIN
HF 2159 → CLOTHIAPINE
HFA → HEXAFLUOROACETONE
81,723 hfu → TIAMULIN
HHDN → ALDRIN
Hi-Alarzin → TOLNAFTATE
Hibanil → CHLORPROMAZINE
Hiberna → PROMETHAZINE
Hibernal → CHLORPROMAZINE
Hibiclens → CHLORHEXIDINE
Hibicon → BECLAMIDE
Hibidil → CHLORHEXIDINE
Hibiscrub → CHLORHEXIDINE
Hibiscus esculentus → OKRA
HIBISCUS ROSA-SINENSIS
Hibitane → CHLORHEXIDINE
Hiconcil → PENICILLINS – "NEWER and SYNTHETICS" > *Amoxicillin*
HICKORY
Hi-Deratol → VITAMIN D_2
Hi-Enterol → IODOCHLORHYDROXYQUINOLINE
Hierba sagrada → TOBACCO
Hiestrone → ESTRONE
High John Conqueror → JALAP
High John Root → JALAP
High John the Conqueror → JALAP
"High Mountain Disease" → ASTRAGALUS
Hihustan M → DEXTROMETHORPHAN HYDROBROMIDE
Hikori → PEYOTE
Hikuli → PEYOTE
Hilactan → CINNARIZINE
Hill, Sir John → NICOTIANA sp.
Hilong → OXAZEPAM
HIMATANTHUS sp.

Himmas → PEAS
Himmis → PEAS
Hindu Datura → DATURA
Hinosan → EDIFENPHOS
Hioxyl → HYDROGEN PEROXIDE
Hipazote → CHENOPODIUM
Hipeksal → METHENAMINE
Hipercilina → PENICILLIN > *Benzylpenicillin, Potassium*
Hipertan → MIDODRINE
Hiphyllin → DYPHYLLINE
Hipoartel → ENALAPRIL
Hipocolestina → TRIPARANOL
Hipoftalin → HYDRALAZINE
Hippeastrum equestre → AMARYLLIS
Hippo → IPECAC
Hippocrates → ACACIA, OPIUM
Hippomane mancinella → MANCHINEEL TREE
Hippozin → PHENOTHIAZINE
Hippramine → METHENAMINE
Hippuran → METHENAMINE
Hippuric Acid → ALLYL CYCLOHEXYLPROPIONATE, BENZOIC ACID, SHIKIMIC ACID, TOLUENE
Hiprex → METHENAMINE
Hiropon → METHAMPHETAMINE
Hirudex → HIRUDINS
HIRUDINS also → LEPIRUDIN
Hirudoid → MUCOITIN POLYSULFATE ESTER
Hiserpia → RESERPINE
Hi-Sil → SILICON DIOXIDE
Hismanal → ASTEMIZOLE
Hislosine → CARBINOXAMINE MALEATE
Histabromamine → BROMODIPHENHYDRAMINE
Histabutizine → BUCLIZINE
Histabutyzine → BUCLIZINE
Histacryl → CYANOACRYLATES
Histacuran → CLEMIZOLE
Histadur → CHLORPHENIRAMINE
Histalen → CHLORPHENIRAMINE
Histalog → BETAZOLE
Histalon → PYRILAMINE
Histamen → ASTEMIZOLE
Histametizine → MECLIZINE
HISTAMINE also → BEER, CHEESES, CIMETIDINE, CODEINE, CURARE, DIMETHYL TUBOCURARINE IODIDE, FISH, GUANIDINE, HISTIDINE, HYDRALAZINE, ICE CREAM, JONQUIL, LAPORTEA sp., NITROGLYCERIN, PASSIFLORA sp., PILOCARPINE, POLYVINYLPYRROLIDONE, PROMETHAZINE, PYRILAMINE, QUININE, SAUERKRAUT, SAURINE, SEA URCHINS, SEROTONIN, SHERRY, SUCCINYLCHOLINE, THURFYL NICOTINATE, TOMATO
Histaminos → ASTEMIZOLE
Histan → PYRILAMINE
Histapyran → PYRILAMINE
Histaspan → CHLORPHENIRAMINE
Histatex → PYRILAMINE
Histazine → ANTAZOLINE
Histazol → ASTEMIZOLE
HISTIDINE also → SAURINE
Histimin → BETAZOLE
Histoacryl Blue → BUTYL CYANOACRYLATE
Histocarb → CARBARSONE
Histomibal → DILOXANIDE FUROATE
Histosep-S → AMINONITROTHIAZOLE
Histostab → ANTAZOLINE
Histostat → NITARSONE
HISTRELIN
HISTRIONICOTOXIN
Hitrin → TERAZOSIN
Hivid → DIDEOXYCYTIDINE
Hiwolfia → RAUWOLFIA SERPENTINA
HL 255 → BROMINDIONE
HL 331 → PHENYLMERCURIC ACETATE
HL 5746 → CHLORPROMAZINE
HMD → CARBIDOPA
HMDI → METHYLENE BIS (4-CYCLOHEXYLISOCYANATE)
HMG-CoA reductase → CHOLESTIN
HMM → ALTRETAMINE
HMPA → HEXAMETAPOL
HMPT → HEXAMETAPOL
HMS → MEDRYSONE
HMT → METHENAMINE
HMTA → METHENAMINE
HMX → OCTOGEN
HN2 → MECHLORETHAMINE
HNP → TRIS(HYDROXYMETHYL) NITROMETHANE
H_2O_2 → HYDROGEN PEROXIDE
Hoary Pea → TEPHROSIA sp.
Hobblebush → VIBURNUM PRUNIFOLIUM
Hocus → MORPHINE
HOE 296 → CICLOPIROX
HOB 296V → RESORANTEL
HOE 304 → DESOXIMETASONE
Hoe 471 → LH-RH
HOE 490 → GLIMEPIRIDE
HOE 498 → RAMIPRIL
HOE 760 → ROXATIDINE ACETATE
HOE 766 → LH-RH

HOE 777 → PREDNISOLONE
HOE 893d → PENBUTOLOL SULFATE
HOE 984 → NOMIFENSINE MALEATE
HOE 2671 → ENDOSULFAN
HOE 2784 → BINAPACRYL
HOE 2810 → LINURON
HOE 2904 → DINOSEB
HOE 17,411 → METHYL BENZIMIDAZOLE-2-CARBAMATE
HOE 23,408 → DICLOFOP-METHYL
Hoe-grass → DICLOFOP-METHYL
Hoelon → DICLOFOP-METHYL
Hoechst → PRENYLAMINE LACTATE, TERFENADINE
Hoeschst 10805 → DIPIPANONE
Hoffman-LaRoche → QUINUCLIDINYL BENZILATE
Hog → PHENCYCLIDINE
Hog Apple → PODOPHYLLUM
Hog Brake → BRACKEN FERN
Hoggar N → DOXYLAMINE SUCCINATE
Hogplum → COLUBRINA also → PLUM
Hog's Bean → HYOSCYAMUS
Hogweeds → HERACLEUM sp.
Ho-Ho-Ba → JOJOBA BEAN
HOI → KAVA
Hoja de Menta → PEPPERMINT
Hojas de sen → SENNA
Holbamate → MEPROBAMATE
HOLCUS LANATUS
Holders Supreme Wax Oil Polish® → POLISHES, FURNITURE
HO LEAF OIL
Hollandaise Sauce → ISOPIMPINELLIN
Hollowstalk → LEONOTIS NEPETIFOLIA
HOLLY also → ILEX sp.
Holmes, Oliver Wendell → ETHER, ETHYL; Preface
Holocaine → PHENACAINE
HOLOCALYX sp.
Holodorm → METHAQUALONE
HOLOTHURIN also → SEA CUCUMBER
Holoxan → IFOSFAMIDE
Homa → GLORIOSA
Homapin → HOMATROPINE
HOMATROPINE
HOMERIA sp.
HOMOBATRACHOTOXIN
Homocapsaicin → PEPPERS
HOMOCHLORCYCLIZINE
Homoclomin →
Homocodeine → PHOLCODINE
HOMOCYSTEINE also → COFFEE
HOMOCYSTINE
Homoginin → HOMOCHLORCYCLIZINE
HOMOMENTHYL SALICYLATE
Homonal → MAFENIDE
Homoolan → ACETAMINOPHEN
Homosalate → HOMOMENTHYL SALICYLATE
Homosulfamine → MAFENIDE
Homovanillic Acid → HALOPERIDOL
Honcho → GLYPHOSPHATE
Honduras Balsam → BALSAM PERU
HONEY also → ACACIA, AESCULUS sp., ALOYSIA, ANDROMEDA, ANDROMEDOTOXIN, ATROPINE, AZALEA, BORAX, BUCKWHEAT, BUXUS sp., CENTAUREA, CROTON, CYTISUS LABURNUM, DAPHNE, DATURA, ECHIUM PLANTAGINEUM, EUPHORBIA, GELSEMIUM SEMPERVIRENS, JACOBINE, JUNIPER, KALMIA, MONOTROPA UNIFLORA, OLEANDER, OXYTETRACYCLINE, PERILLA FRUTESCENS, PIERIS sp., PYRROLIZIDINES, RETRORSINE, RHODODENDRON, ROSEMARY, SAPINDUS sp., SCABIOSA SUSSISA, SENECIO sp.
HONEYSUCKLE also → LONICERA CAPRIFOLEUM
Hong Nien → PHENYLMERCURIC ACETATE
Honig Thee → TEA
HOPS also → SALVIA
"Hop-rash" → HOPS
Hordenine → TRICHOCEREUS PACHANOI
Hornedine → ACACIA, ARIOCARPUS
HORDEUM JUBATUM
Hordeum sp. → BARLEY
Horizon → DIAZEPAM
Hormezon → BETAMETHASONE
Hormoestrol → HEXESTROL
Hormofemin → DIENESTROL
Hormofollin → ESTRONE
Hormo-Gerobion → PROCAINE
Hormomed → ESTRIOL
Hormonisene → CHLOROTRIANISENE
Hormovarine → ESTRONE
HORNETS
Horse → HEROIN
Horse Bean → BEANS, Faba
Horsebrush → TATRADYMIA sp. also → TETRADYMA sp.
Horse Gowan → CHAMOMILE also → TARAXACUM OFFICINALE
Horse Nettle → SOLANUM sp.
HORSERADISH also → ALLYL ISOTHIOCYANATE

Horsetail → EQUISETUM
Horsetail Fern → EQUISETUM
Horse Tranks → PHENCYCLIDINE
Horseye Bean → MUCUNA sp.
Hortelá-Pimenta → PEPPERMINT
Hortensia → HYDRANGEA
Hospa → LEDUM sp.
Hostacain → BUTANILICAINE
Hostacortin → PREDNISONE
Hostacortin H → PREDNISOLONE
Hostacyclin → TETRACYCLINE
Hostadex → DEXAMETHASONE
Hostaginin → PRENYLAMINE LACTATE
Hostalival → NOMIFENSINE MALEATE
"Hot Dog Headache" → NITRATES and NITRITES
Hot Dogs → FRANKFURTERS
Houblon → HOPS
Hound's Tongue → CYNOGLOSSUM OFFICIANALE
Houva → METHOXSALEN
HOYA sp.
HP 1275 → PHENOXYPROPAZINE MALEATE
HPC → MILTEFOSINE
HPP → ALLOPURINOL
HR → TROXERUTIN
HR 376 → CLOBAZAM
HR 756 → CEFOTAXIME
HRG → VINYL CHLORIDE
HS → MUSTARD GAS
H_2S → HYDROGEN SULFIDE
HS 592 → CLEMASTINE
5 HT → SEROTONIN
5 HTP → 5-HYDROXYTRYPTOPHAN
Huatari → PEYOTE
Hu chiao → PEPPERS
Hughes, Howard → CODEINE
Humagel → PAROMOMYCIN
Human Chorionic Gonadotropins → GONADOTROPINS, CHORIONIC
Humatin → PAROMOMYCIN
Humatrope → GROWTH HORMONES
Humegon → MENOTROPINS
Humin → HURA CREPITANS
Huminsulin → INSULIN
Humoryl → TOLOXATONE
"Humpy Back" → SOLANUM sp.
Humulin → INSULIN
Humulina → INSULIN
Humulus lupulus → HOPS
Humycin → PAROMOMYCIN
Hunan Hand → CAPSAICIN
HURA CREPITANS
Huratoxin → HURA CREPITANS, MACHINEEL TREE
Hurmal → PEGANUM HARMALA
Husk-tomato → GROUND CHERRY also → PHYSALIS sp.
Hutton, Mrs. → DIGITALIS
Huxley, Aldous → MESCALINE
HWA 486 → LEFLUNOMIDE
HXM → ALTRETAMINE
Hyacid → SODIUM HYALURONATE
HYACINTH also → CALCIUM OXALATE, CINNAMIC ALDEHYDE
Hyacinth Bean → DOLICHOS LALAB
Hyacinthus orientalis → HYACINTH
Hyadur → DIMETHYL SULFOXIDE
Hyalase → HYALURONIDASE
Hyalgan → SODIUM HYALURONATE
Hyalgin → SODIUM HYALURONATE
Hyalidase → HYALURONIDASE
Hyalovet → SODIUM HYALURONATE
Hyalozima → HYALURONIDASE
HYALURONIDASE also → SPIDERS
Hyamine 1622 → BENZETHONIUM CHLORIDE
Hyamine 3500 → BENZALKONIUM CHLORIDE
Hyanit → UREA
Hyanoine 10X → METHYLBENZETHONIUM CHLORIDE
Hyanthel → HYGROMYCIN B
Hyasmonta → HYALURONIDASE
Hyason → HYALURONIDASE
Hyasorb → PENICILLIN > *Benzylpenicillin, Potassium*
Hyazyme → HYALURONIDASE
Hybolin Deconate → NANDROLONE
Hybrin → VITAMIN C
Hycamptamine → TOPOTECAN
Hycamptin → TOPOTECAN
HYCANTHONE
Hycholin → PENTAPIPERIDE METHYLSULFIDE
Hyclorate → CLOFIBRATE
Hycodan → HYDROCODONE BITARTRATE
Hycozid → ISONIAZID
Hydac → FELODIPINE
Hydan → METHIONINES
Hydantoin → PHENYTOIN
Hydeltra → PREDNISOLONE
Hydeltrasol → PREDNISOLONE
Hydeltra-T.B.A. → PREDNISOLONE
Hydeltrone → PREDNISOLONE
Hydergine → ERGOLOID MESYLATES
Hydol → HYDROFLUMETHIAZIDE

Hydout → ENDOTHALL
Hydracetin → ACETYLPHENYLHYDRAZINE
Hydracillin → PENICILLIN > *Benzylpenicillin, Procaine*
Hydracort → HYDROCORTISONE
HYDRALAZINE also → BENZTHIAZIDE, CHLOROTHIAZIDE, DIAZOXIDE, ETHACRYNIC ACID, FUROSEMIDE, HYDRAZINE, INDOMETHACIN, PROPRANOLOL, SPIRONOLACTONE, TRIAMTERENE
Hydralin → CYCLOHEXANOL
Hydramethylnon → AMDRO
HYDRANGEA also → CYANIDES
Hydrangin → HYDRALAZINE
Hydraphen → HYDRARGAPHEN
HYDRARGAPHEN
Hydrargyrum → MERCURY
Hydrargyrum Iod. Rub. → MERCURIC IODIDE, RED
HYDRASTIS CANADENSIS
Hydrated Alumina → ALUMINUM HYDROXIDE
Hydrated Lime → CALCIUM HYDROXIDE
Hydrated Oxide of Iron → FERRIC HYDROXIDE
"Hydra Tree of Death" → ANTIARIS TOXICARIA
Hydratropyl Alcohol → ALCOHOL, HYDRATROPIC
HYDRAULIC FLUIDS
Hydrazid → ISONIAZID
HYDRAZINE also → HYDRALAZINE, MALEIC HYDRAZIDE, MISSILE SILO, PHENELZINE, SOLDER, TEA, TOBACCO, WELDING
Hydrazine Carboxamide → SEMICARBAZIDE
Hydrazine Yellow → TARTRAZINE
Hydrazinobenzene → PHENYLHYDRAZINE
1-Hydrazinophthalazine → HYDRALAZINE
HYDRAZOBENZENE
HYDRAZOIC ACID
Hydrea → HYDROXYUREA
Hydrenox → HYDROFLUMETHIAZIDE
Hydresis → HYDROCHLOROTHIAZIDE
Hydrid → HYDROCHLOROTHIAZIDE
Hydril → HYDROCHLOROTHIAZIDE
Hy-Drine → BENZTHIAZIDE
HYDRIODIC ACID
HYDROABIETYL ALCOHOL
Hydro-Adreson → HYDROCORTISONE
Hydro-Aquil → HYDROCHLOROTHIAZIDE
Hydroazide → HYDROCHLOROTHIAZIDE
Hydrobromic Ether → ETHYL BROMIDE
Hydrocal → HYDROCORTISONE
Hydrocarbon → POLISHES, FURNITURE
Hydrochinone → HYDROQUINONE
HYDROCHLORIC ACID also → σ-ANISIDINE, CHLOROFORM, CHLOROSULFONIC ACID, DIMETHYLVINYL CHLORIDE, EPICHLOHYDRIN, METHYL CHLORIDE, PHENOXYBENZAMINE HYDROCHLORIDE, PHOSGENE, POLYVINYL CHLORIDE, POTASSIUM NITRATE, RESERPINE, SILICON TETRACHLORIDE, STEEL, TITANIUM TETRACHLORIDE
HYDROCHLOROTHIAZIDE also → BUCLOSAMIDE, CALCIUM CARBIMIDE, CHOLESTEROL, CHOLESTYRAMINE RESINS, CHUIFONG TOUKUWAN, DICHLORPHENAMIDE, DIFLUNISAL, ETHYL BISCOUMACETATE, FLUCONAZOLE, LITHIUM, NAPROXEN, POTASSIUM, SOTALOL, THIORIDAZINE, TRIMETHOPRIM
Hydrocodin → DIHYDROCODEINE
HYDROCODONE BITARTRATE also → AMIODARONE, PAROXETINE, PHENYLTOLOXAMINE, QUINIDINE
Hydroconchinine → HYDROQUINIDINE
Hydrocort → HYDROCORTISONE
HYDROCORTISONE also → ACETAMINOPHEN, AMINOGLUTETHIMIDE, COBALT SULFATE, COLESTIPOL, EPINEPHRINE, FUSIDIC ACID, NORTRIPTYLINE, PHENOBARBITAL, PHENYLBUTAZONE, PHENYLEPHRINE HCL, PHENYTOIN, TIXOCORTOL PIVALATE
Δ^1-Hydrocortisone → PREDNISOLONE
Hydrocortistab → HYDROCORTISONE
Hydrocortisyl → HYDROCORTISONE
Hydrocortone → HYDROCORTISONE
Hydrocortone Acetate → HYDROCORTISONE
Hydrocortone Phosphate → HYDROCORTISONE
Hydrocotyle → CENTALLA ASIATICA
Hydrocyanic Acid → HYDROGEN CYANIDE also → ACACIA, CAREX, HETEROMELES ARBUTIFOLIA, MESQUITE, PLEUROTIS sp.
Hydrodeltalone → PREDNISOLONE
Hydrodiuretic → HYDROCHLOROTHIAZIDE
Hydro-Diuril → HYDROCHLOROTHIAZIDE
HYDROFLUMETHIAZIDE
HYDROFLUORIC ACID also → PHOSGENE, POLYTEF, SILICON TETRAFLUORIDE, SULFUR TETRAFLUORIDE
Hydrofluosilic Acid → FLUOSILICIC ACID

HYDROGEN also → ARSINE, CHLOROSULFONIC ACID

HYDROGEN BROMIDE

HYDROGEN CHLORIDE also → BIS(CHLOROMETHYL)ETHER, POLYESTER, SILICON TETRACHLORIDE

Hydrogen Cyanamide → CYANAMIDE

HYDROGEN CYANIDE also → ACACIA, ALMONDS, AMELANCHIA ALNIFOLIA, ARROWGRASS, BEANS, CASSAVA, CHERRY, CHLORIS, CYANIDES, CYNODON, DACTYLOCTENIUM AEGYPTIUM, DHURRIN, DIOSCOREA, EREMOPHILA MACULATA, ETHOXAZINE, EUCALYPTUS GLOBULUS, FLAX, HOLCUS LANATUS, LATHYRUS sp., LETTUCE, MELICA sp., MILLIPEDES, PANICUM sp., PAPAVER sp., PARTHENIUM, PASSIFLORA sp., PEACHES, PHASEOLUTIN, POLYURETHANE, POTASSIUM CYANIDE, SCHOENUS ASPEROCARPUS, SILK, SORGHASTRUM NUTANS, SORGHUM, STILLINGIA sp., SWEET POTATO, TANAECIUM sp., TAPIOCA, THIOCYANATES, WOOL, Preface

Hydrogen Dioxide → HYDROGEN PEROXIDE

Hydrogen Fluoride → HYDROFLUORIC ACID

Hydrogen Hexachloroplatinate → PLATINIC CHLORIDE

Hydrogen Iodide → HYDRIODIC ACID

Hydrogen Nitrate → NITRIC ACID

HYDROGEN PEROXIDE also → FLOUR, GLYCERIN, PHEOMELANIN, PHOSPHORIC ACID, POLYVINYLPYRROLIDONE IODINE COMPLEX, POTASSIUM PERMANGATE, SODIUM PERBORATE

Hydrogen Phosphide → PHOSPHINE

HYDROGEN SELENIDE

Hydrogen Sulfate → SULFURIC ACID

HYDROGEN SULFIDE also → BISMUTH, EGGS, FISH, PAPER, SILK, SULFURIC ACID, THIRAM, WOOL

HYDROIDS

Hydrokon → HYDROCODONE BITARTRATE

Hydro-long → CHLORTHALIDONE

Hydrolose → METHYL CELLULOSE

Hydromedin → ETHACRYNIC ACID

Hydromerfen → PHENYLMERCURIC BORATE

HYDROMORPHONE

Hydromox → QUINETHAZONE

Hydronol → ISOSORBIDE

Hydroperoxide → HYDROGEN PEROXIDE

Hydropot → PROTEIN HYDROLYSATES

HYDROQUINIDINE

HYDROQUINONE also → AZELAIC ACID; CRESOL, SAPONATED SOLUTION of; HEXYLRESORCINOL, PHLOROGLUCINOL

Hydroquinone Benzyl Ether → MONOBENZONE

Hydroquinone Carboxyatracytloside → COCKLEBURS

Hydroquinone Monobenzyl Ether → MONOBENZONE

Hydro-rapid → FUROSEMIDE

Hydroretrocortine → PREDNISOLONE

Hydroronol → HYDROCHLOROTHIAZIDE

Hydrosaluric → HYDROCHLOROTHIAZIDE

Hydrosarpan → AJMALICINE

Hydrosulfuric Acid → HYDROGEN SULFIDE

Hydrothide → HYDROCHLOROTHIAZIDE

Hydrothol 47 → ENDOTHALL

Hydrothol 191 → ENDOTHALL

Hydroton → CHLORTHALIDONE

Hydrotrichlorothiazide → TRICHLORMETHIAZIDE

Hydroxocobalamin → VITAMIN B_{12}

Hydroxocobemine → VITAMIN B_{12}

N-Hydroxyacetamide → ACETOHYDROXAMIC ACID

HYDROXYAMPHETAMINE

p-Hydroxyaniline → *p*-AMINOPHENOL

2-Hydroxyaniline → σ-AMINOPHENOL

3-Hydroxyaniline → *m*-AMINOPHENOL

p-Hydroxyanisole → 4-METHOXYPHENOL

2-Hydroxybenzamide → SALICYLAMIDE

Hydroxybenzene → PHENOL

p-HYDROXYBENZOIC ACID also → TARTRAZINE

2-Hydroxybenzoic Acid → SALICYLIC ACID

3-Hydroxy-2-butanone → ACETOIN

1-Hydroxy-4-tert-butylbenzene → *p-tert*-BUTYLPHENOL

γ-HYDROXYBUTYRATE

4-Hydroxybutyrate → γ-HYDROXYBUTYRATE

Hydroxycarbamide → HYDROXYUREA

HYDROXYCHLOROQUINE also → BOLETUS, DIGOXIN, METHOTREXATE, MUSHROOMS, PENICILLAMINE

HYDROXYCITRONELLAL

17-Hydroxycorticosterone → HYDROCORTISONE

7-Hydroxycoumarin → COUMARIN

1,4-Hydroxydaunomycin → DOXORUBICIN

5-Hydroxy-N,N-dimethyltryptamine → BUFOTENINE

HYDROXYDIONE SODIUM
2-Hydroxydiphenyl → σ-PHENYLPHENOL
16α-Hydroxyestradiol → ESTRIOL
9-[(2-Hydroxyethoxy)methyl]guanine → ACYCLOVIR
1-(2-Hydroxyethyl)-4-[3-(2-acetyl-10-phenothiazyl)-propyl]-piperazine → ACETOPHENAZINE
2-Hydroxyethylamine → ETHANOLAMINE
β-Hydroxyethylamine → ETHANOLAMINE
β-Hydroxyethylbenzene → ALCOHOL, PHENYLETHYL
1-[10-3-[4-(2-Hydroxyethyl)-1-piperazinyl]propyl-10H-phenothiazin-2yl]ethanone → ACETOPHENAZINE
10-[3-[4-(Hydroxyethyl)-1-piperazinyl]propyl]-phenothiazin-2-yl Methyl Ketonedimaleate → ACETOPHENAZINE
3-Hydroxy-2-ethyl-4-pyrone → ETHYL MALTOL
HYDROXYETHYL STARCH
N-HYDROXY-*N*-2-FLUORENYLACETAMIDE
Hydroxyhemin → HEMATIN
5-Hydroxyindoleacetic Acid → BANANA, GLYCERYL GUAIACOLATE, RESERPINE, SEROTONIN
HYDROXYLAMINE also → METRONIDAZOLE
Hydroxymesterone → MEDRYSONE
Hydroxymethyl-2-nitro-1,3-propanediol → TRIS(HYDROXY METHYL)NITRO METHANE
γ-Hydroxy-β-oxobutane → ACETOIN
HYDROXYPHENAMATE
p-Hydroxyphenformin → PHENFORMIN
p-Hydroxyphenylacetic Acid → CHEESES
p-Hydroxyphenyl Benzyl Ether → MONOBENZONE
17-α-HYDROXYPROGESTERONE also → MELATONIN
4-Hydroxypropranolol → PROPRANOLOL
4-Hydroxypyrazolo-(3,4-d)pyrimidine → ALLOPURINOL
3-HYDROXY-4-(IH) PYRIDONE also → LEUCAENA, MIMOSINE
8-HYDROXYQUINOLINE
HYDROXYSTILBAMIDE
Hydroxytetracycline → OXYTETRACYCLINE
α-Hydroxytoluene → ALCOHOL, BENZYL
5-Hydroxytryptamine → SEROTONIN also → ANADENANTHERA, DIPHENHYDRAMINE
5-HYDROXYTRYPTOPHAN
3-Hydroxytyramine → DOPAMINE
3-Hydroxytyrosine → DOPA
3-Hydroxy-L-tyrosine → L-DOPA
HYDROXYUREA
HYDROXYZINE also → CHLORAL HYDRATE, CIMETIDINE, ETHYLENEDIAMINE, HEPARIN, HEXOBARBITAL, KETAMINE, THIORIDAZINE, TRIPELENNAMINE
Hydrozide → HYDROCHLOROTHIAZIDE
Hyferdex → IRON DEXTRANS
Hygramix → HYGROMYCIN B
Hygromix → HYGROMYCIN B
HYGROMYCIN B
Hygroton → CHLORTHALIDONE
Hygrovetin → HYGROMYCIN B
Hylartin → SODIUM HYALURONATE
Hylase → HYALURONIDASE
Hylenta → PENICILLIN > *Benzylpenicillin, Potassium*
Hylorel → GUANADREL SULFATE
HYMENOCALLIS OCCIDENTALIS
HYMENOPTERA
Hymenoxin → HYMENOXYS sp.
Hymenoxon → HYMENOXYS sp.
HYMENOXYS sp.
Hyminal → METHAQUALONE
Hymorphan → HYDROMORPHONE
Hyonate → SODIUM HYALURONATE
Hyoper → PENICILLINS – "NEWER and SYNTHETICS" > *Carbenicillin Disodium*
Hyoscine → SCOPOLAMINE also → GINSENG, STRAMONIUM
l-HYOSCYAMINE also → ATROPA BELLADONNA, DATURA, DUBOISA MYOPORIDES, GROUND CHERRY, LETTUCE, LYCIUM HALIMIFOLIUM, MAGNESIUM TRISILICATE, MANDRAGORA OFFICIANARIUM, STRAMONIUM
HYOSCYAMUS
Hypaque → DIATRIZOATE SODIUM
Hypague Cysto → MEGLUMINE DIATRIZOATE
Hypaque Meglumine → MEGLUMINE DIATRIZOATE
HYPERICIN
HYPERICUM also → CYCLOSPORINE, INDINAVIR, TEA, THEOPHYLLINE, WARFARIN
Hypericum red → HYPERICIN
Hypersin → BETHANIDINE
Hyperstat → DIAZOXIDE
Hypertensin → ANGIOTENSIN
Hypertil → CAPTOPRIL
Hypertonalum → DIAZOXIDE
Hyper-Rauw → RAUWOLFIA SERPENTINA

Hyphylline → DYPHYLLINE
Hypnodil → METOMIDATE
Hypnogène → BARBITAL
Hypnomidate → ETOMIDATE
Hypnorex → LITHIUM CARBONATE
Hypnovel → MIDAZOLAM
Hypo → SODIUM THIOSULFATE
HYPOCHLOROUS ACID also → SODIUM HYPOCHLORITE
Hypocol → METHAQUALONE
HYPOGLYCIN A also → SODIUM VALPROATE
Hypoglymol → TOLBUTAMIDE
Hyponitric Acid → NITROGEN TETROXIDE
Hypophyseal Growth Hormone → GROWTH HORMONES
Hypophysin → PITUITARY, POSTERIOR
Hypone → ACETOPHENONE
α-Hypophamine → OXYTOCIN
β-Hypophamine → VASOPRESSIN
Hyposan → SODIUM HYPOCHLORITE
Hypostan → THIOPENTAL
Hypostat → PAROXYPROPIONE
Hypothiazide → HYDROCHLOROTHIAZIDE
Hypovase → PRAZOSIN HYDROCHLORIDE
Hypoxanthosine → INOSINE
Hyprotigen → PROTEIN HYDROLYSATES
Hyproval P.A. → 17-α-HYDROXYPROGESTERONE
Hyptis suaveolens → SANGURA
Hyptor → METHAQUALONE
Hypurin → INSULIN
Hyrazin → THIAMPHENICOL
Hyskon → DEXTRAN, *70*
Hysron → MEDROXYPROGESTERONE ACETATE
HYSSOP OIL
Hytakerol → DIHYDROTACHYSTEROL
Hytinic → IRON POLYSACCHARIDE COMPLEX
Hyton Asa → PEMOLINE
Hytone → HYDROCORTISONE
Hytracin → TERAZOSIN
Hytrin → TERAZOSIN
Hytrinex → TERAZOSIN
Hytrol O → CYCLOHEXANONE
Hyvar → BROMACIL
Hywolfia → RAUWOLFIA SERPENTINA

I

I653 → DESFLURANE
I.A. 307 → ACETOSULPHONE SODIUM
Ial → SODIUM HYALURONATE
IAA → INDOLEACETIC ACID
Ibaril → DESOXIMETASONE
IBD → ISOSORBIDE DINITRATE
Ibenzmethyzin → PROCARBAZINE
Ibiala → ICACINA SENEGALENSIS
Ibiamox → PENICILLINS – "NEWER and SYNTHETICS" > *Amoxicillin*
Ibidomide → LABETALOL
Ibifur → FURALTADONE
Ibilex → CEPHALEXIN
Ibinolo → ATENOLOL
Ibiofural → NITROFURAZONE
Ibiotyzil → BENACTYZINE
Ibli → ALOE
IBOGAINE also → TABERNANTHE IBOGA, TRYPTAMINE
Ibok-eku → SPONDIANTHUS UGANDENSIS
IBOPAMINE also → DILTIAZEM
IBOTENIC ACID also → AMANITA sp., MUSHROOMS
IBT → METHISAZONE
Ibu-Attritin → IBUPROFEN
IBUFENAC
Ibumetin → IBUPROFEN
Ibunac → IBUFENAC
Ibuprocin → IBUPROFEN
IBUPROFEN also → ALCOHOL, ETHYL; BACLOFEN, BUTOCONAZOLE, DIGOXIN, FLUCONAZOLE, FUROSEMIDE, HYDROCHLOROTHIAZIDE, LEFLUNOMIDE, LITHIUM, RALOXIFENE, TOBRAMYCIN, WARFARIN, ZAFIRLUKAST, ZIDOVUDINE
Ibustrin → INDOPROFEN
Ibutad → IBUPROFEN
Ibutid → IBUPROFEN
IBUTILIDE FUMARATE
Ibutop → IBUPROFEN
Ibylcaine → BUTETHAMINE
ICACINA SENEGALENSIS
Icacine → DIBEKACIN
ICE also → METHAMPHETAMINE
Ice Balls → FREEZE BALLS
ICE CREAM also → CARBONS, MILK
ICELANDITOXIN
Iceland Poppy → PAPAVER sp.
Iceland Spar → CALCIUM CARBONATE
Ice Plant → MESEMBRYANTHEMUM CRYSTALLINUM
ICI 32,865 → ETHOGLUCID
ICI 33,828 → METHALLIBURE
ICI 35,868 → PROPOFOL

ICI 38,174 → PRONETHALOL
ICI 46,474 → TAMOXIFEN CITRATE
ICI 50,123 → PENTAGASTRIN
ICI 50,172 → PRACTOLOL
ICI 50,627 → TETRAMISOLE
ICI 54,540 → FENCLOZIC ACID
ICI 58,834 → VILOXAZINE
ICI 59,118 → RAZOXANE
ICI 66,082 → ATENOLOL
ICI 79,939 → DINOPROST
ICI 80,996 → CLOPROSTENOL
ICI 81,008 → FLUPROSTENOL
ICI 118,630 → ZOLADEX
ICI 156,834 → CEFOTETAN
ICI 204,219 → ZAFIRLUKAST
ICI-D 1033 → ANASTROZOLE
ICN 1229 → RIBAVIRIN
ICODEXTRIN
Icodial → ICODEXTRIN
Icoral B → METARAMINOL BITARTRATE
ICRF 159 → RAZOXANE
ICRF 187 → RAZOXANE
ICS 205-930 → TROPISETRON
Icterogenin → LANTANA sp., LIPPIA REHMANNI
Idalon → FLOCTAFENINE
Idarac → FLOCTAFENINE
IDARUBICIN
Idexur → IDOXURIDINE
Idocyl → SODIUM SALICYLATE
Idon Zakara → ABRUS PRECATORIUS
Idoxene → IDOXURIDINE
IDOXURIDINE
Idrochinone → HYDROQUINONE
Idroestril → DIETHYLSTILBESTROL
IDU → IDOXURIDINE
Idulea → IDOXURIDINE
Idulian → AZATADINE DIMALEATE
IDUR → IDOXURIDINE
Iduridin → IDOXURIDINE
Iduviran → IDOXURIDINE
IDZA → IMIDAZOLIDINYL UREA
If → YEW
Ifada → FAMOTIDINE
Ifenec → ECONAZOLE NITRATE
Ifex → IFOSFAMIDE
IFK → PROPHAM
IFNs → INTERFERONS
IFNα → INTERFERONS
IFNα2 → INTERFERONS
IFNβ → INTERFERONS
Ifomide → IFOSFAMIDE
IFOSFAMIDE also → ACETYLCYSTEINE, BLEOMYCIN, CISPLATIN, MIBEFRADIL
IFP → GLYCERIN
Ifrasarl → CYPROHEPTADINE
Ig 9659A → BISMUTH GLYCOL ARSANILATE
Igba AJA → SOLANUM sp.
IGE → ISOPROPYL GLYCIDYL ETHER
Igepal CO-630 → NONOXYNOL-9
Ighu → DIOSCOREA
IGI 204, 636 → QUETIAPINE FUMARATE
Igralin → THIAMPHENICOL
Igun → TEPHROSIA sp.
III 2318 → MEPAZINE
IK → POTASSIUM IODIDE
Ikaclamin → CLOMIPHENE CITRATE
Ikaclomine → CLOMIPHENE CITRATE
Ikacor → VERAPAMIL
Ikaran → DIHYDROERGOTAMINE METHANESULFONATE (MESYLATE)
Ikema → CYNANCHUM
Ikokosin → PASSIFLORA sp.
Ikorel → NICORANDIL
Iktorivil → CLONAZEPAM
IL-1 → INTERLEUKINS
IL-2 → ALDESLEUKIN also → INTERLEUKINS
IL-12 → INTERLEUKINS
Ilcocillin-P → PENICILLIN > *Benzylpenicillin, Procaine*
Ildamen → OXYFEDRINE
Iletin II → INSULIN
ILEX sp.
Iliadin → OXYMETAZOLINE
Ilicin → HOLLY
ILLICIUM ANISATUM
Ilomedin → ILOPROST
ILOPROST
Illoxan → DICLOFOP-METHYL
Ilopan → DEXPANTHENOL
Ilosone → ERYTHROMYCIN ESTOLATE
Ilotycin → ERYTHROMYCIN
Ilozyme → PANCRELIPASE
Imacillin → PENICILLINS – "NEWER and SYNTHETICS" > *Amoxicillin*
Imadyl → CARPROFEN
Imagon → CHLOROQUINE
Imakol → OXOMEMAZINE
Imap → FLUSPIRILINE
Imavate → IMIPRAMINE
Imbaral → SULINDAC
Imdur → ISOSORBIDE MONONITRATE
Imeson → NITRAZEPAM
Imex → TETRACYCLINE
Imexin → SULFAMETHOXAZOLE
Imferon → IRON DEXTRANS

IMI 28 → EPIRUBICIN
IMI 30 → IDARUBICIN
IMICLOPAZINE
IMIDACLOPRID
Imidamine → ANTAZOLINE
IMIDAN
IMIDAZOLIDINYL UREA
Imidazoline Urea → IMIDAZOLIDINYL UREA
IMIDOCARB
Imidol → IMIPRAMINE
Imidurea → IMIDAZOLIDINYL UREA
Imiglucerase → ALĞLUCERASE
Imigran → SUMATRIPTAN SUCCINATE
Imiject → SUMATRIPTAN SUCCINATE
Imilanyle → IMIPRAMINE
Iminobenzoquinone → INPROQUONE
Iminourea → GUANIDINE
Imipem → IMIPENEM
Imipemide → IMIPENEM
IMIPENEM also → CEFEPIME
IMIPRAMINE also → ACETAZOLAMIDE; ALCOHOL, ETHYL; AMMONIUM CHLORIDE, AMOBARBITAL, ATROPINE, BETHANIDINE, BUTALBITAL, CHLORPROMAZINE, CHOLESTYRAMINE RESINS, CIMETIDINE, CITALOPRAM, CLONIDINE, COCAINE, CYPROHEPTADINE, DESMOPRESSIN, DIPHENHYDRAMINE, DISULFIRAM, ENALAPRIL; ESTROGENS, CONJUGATED; FLUOXETINE, FLUVOXAMINE, GUANETHIDINE, HALOPERIDOL, HEXOBARBITAL, HYDROCORTISONE, LAMOTRIGINE, MECLOBEMIDE, MEPROBAMATE, METHYLPHENIDATE, METHYL TESTOSTERONE, MIBEFRADIL, NITROGLYCERIN, PARGYLINE, PAROXETINE, PENTOBARBITAL, PERPHENAZINE, PHENELZINE, PHENOBARBITAL, PHENYLBUTAZONE, PHENYTOIN, PROCARBAZINE, QUINIDINE, SERTRALINE, SODIUM BICARBONATE, THIORIDAZINE, TRANYLCYPROMINE
Imiprin → IMIPRAMINE
IMIQUIMOD
Imisine → IMIPRAMINE
Imitrex → SUMATRIPTAN SUCCINATE
Imizin → IMIPRAMINE
Imizol → IMIDOCARB
Immenoctal → SECOBARBITAL SODIUM
ImmuCyst → BCG
Imnudorm → MEPARFYNOL
Imodium → LOPERAMIDE
IMOLAMINE
Imosec → LOPERAMIDE
Imossel → LOPERAMIDE
Imotryl → BENZYDAMINE
IMPATIENTS sp.
Imperacin → OXYTETRACYCLINE
Imperan → METOCLOPRAMIDE
Imperatorin → HERACLEUM sp.
Imperial Green → CUPRIC ACETOARSENITE
Imperialine → FRITILLARIA MELEAGRIS
Imposil → IRON DEXTRANS
Impregnon → TETRACHLOROSALICYLANILIDE also → DIBROMSALAN
Imprenim → CILASTATIN
Improntal → PIROXICAM
Impruvol → BUTYLATED HYDROXYTOLUENE
Imtack → ISOSORBIDE DINITRATE
Impugan → FUROSEMIDE
Imukin → INTERFERONS
Imunoviral → ISOPRINOSINE
Imuran → AZATHIOPRINE
Imurek → AZATHIOPRINE
Imurel → AZATHIOPRINE
IN 511 → PHENYRAMIDOL
IN 1060 → CYPROLIDOL
Inabrin → IBUPROFEN
Inacillin → PENICILLINS – "NEWER and SYNTHETICS" > *Pivampicillin*
Inactisol → BENZETHONIUM CHLORIDE
Inadine → POLYVINYLPYRROLIDONE IODINE COMPLEX
Inalone O → BECLOMETHASONE
Inalone R → BECLOMETHASONE
Inamycin → NOVOBIOCIN
Ina Nusu → SPIGELIA ANTHELMIA
Inapetyl → BENZPHETAMINE
Inapsine → DROPERIDOL
Inbestan → CLEMASTINE
Incanin → TRICHODESMIN sp.
Incital → MEFENOREX
Incoran → PRENYLAMINE LACTATE
Incorporation Factor → GLYCERIN
Incortin → CORTISONE
Indaflex → INDAPAMIDE
Indalone → BUTOPYRANOXYL
INDALPINE
Indamol → INDAPAMIDE
Indanal → CLIDANAC
INDANEDIONE also → *p*-AMINOSALICYLIC ACID, AMOBARBITAL, BUTABARBITAL,

PREDNISONE, PROBENECID, PROPYLTHIOURACIL
INDAPAMIDE
INDELOXAZINE
Indema → PHENINDIONE
INDENE
INDENO(1,2,3-cd)PYRENE
Inderal → PROPRANOLOL
Inderm → ERYTHROMYCIN
India(n) Marking Nut Tree → POISON IVY also → SUMAC, POISON
Indian Apple → PODOPHYLLUM
Indian Balsam → BALSAM PERU
Indian Banana → PAW-PAW
Indian Berry → COCCULUS sp., *indicus*
Indian Celery → HERACLEUM sp.
Indian Corn → VERATRUM sp.
Indian Dogbane → APOCYNUM CANNABINUM
Indian Dye → HYDRASTIS CANADENSIS
Indiangrass → SORGHASTRUM NUTANS
Indian Grass Oil → GERANIUM
Indian Hay → CANNABIS
Indian Hemp → APOCYNUM CANNABINUM
Indian Kale → XANTHOSOMA sp.
Indian Licorice → ABRUS PRECATORIUS also → ABRIN
Indian Lilac Tree → MELIA sp.
Indian Marking Nut Tree → POISON IVY
Indian Mustard → BRASSICA sp.
Indian Pea → LATHYRUS sp.
Indian Physic → APOCYNUM CANNABINUM
Indian Pipe → MONOTROPA UNIFLORA
Indian Poke → VERATRUM sp.
Indian Tobacco → LOBELIA
Indian Tumeric → HYDRASTIS CANADENSIS
Indian Turnip → ARISAEMA
Indicine → PYRROLIZIDINES also → HELIOTROPIN
INDIGO also → CASSIA
Indigo Blue → INDIGO
INDIGOCARMINE
INDIGOFERA sp. also → PENICILLIUM sp.
Indigo Weed → BAPTISIA sp.
INDINAVIR also → DELAVIRDINE, DIDANOSINE, ERGOTAMINE, FLUCONAZOLE, HYPERICUM, NELFINAVIR MESYLATE, NEVIRAPINE, RIFABUTIN, RIFAMPIN, SAQUINAVIR
Indobloc → PROPRANOLOL
INDOBUFEN also → GLIPIZIDE
INDOCYANINE GREEN also → ASTEMIZOLE, PHENYTOIN
Indocybin → PSILOCYBIN
Indoklon → FLUROTHYL
INDOLEACETIC ACID
Indole-3-carbinol → CABBAGE
Indolin → BENZYDAMINE
INDOMETHACIN also → ACEMETACIN, ACENOCOUMAROL, ACETYLSALICYLIC ACID, ALUMINUM HYDROXIDE, AMIKACIN, ATENOLOL, BISHYDROXYCOUMARIN, BUMETANIDE, CAPTOPRIL, CHUIFONG TOUKUWAN, CIMETIDINE, CORTISONE, CYCLOCOUMAROL, DIFLUNISAL, FUROSEMIDE, GENTAMICIN, HALOPERIDOL, LABETALOL, LITHIUM, MAGNESIUM, MELATONIN, METHOTREXATE, NIMODIPINE, OXYPHENBUTAZONE, PENICILLAMINE, PENICILLIN, PHENYLBUTAZONE, PHENYLPROPANOLAMINE, PREDNISOLONE, PROBENECID, PROPRANOLOL, SPIRONOLACTONE, TARTRAZINE, TORSEMIDE, TRIAMTERENE, WARFARIN
Indometacin → INDOMETHACIN
Indometin → INDOMETHACIN
Indon → PHENINDIONE
Indonaphthene → INDENE
INDOPROFEN
INDORAMIN
Indorm → PROPIOMAZINE
Indos → INDOMETHACIN
Indosmos → INDOMETHACIN
Indospicine → INDIGOFERA sp.
Indotard → INDOMETHACIN
Indulines → INKS
INF 1837 → FLUFENAMIC ACID
INF 3355 → MEFENAMIC ACID
INF 4668 → MECLOFENAMIC ACID
Infacol → DIMETHICONE (*with silicon dioxide*)
Infed → IRON DEXTRANS
Inferno → TETRAM®
Infiltrase → HYALURONIDASE
Infiltrine → DIMETHYL SULFOXIDE
Inflamase → PREDNISOLONE
Inflamen → BROMELAIN
Inflanefran → PREDNISOLONE
INFLIXIMAB
Infonutrol → FATS
Infron → VITAMIN D_2
Infusorial Earths → DIATOMACEOUS EARTHS
INGENOL
Inhiston → PHENIRAMINE MALEATE
INH → ISONIAZID
Inicardio → NIKETHAMIDE

Ininicus → STONE FISH
Inimur → NIFURATEL
Iniprol → APROTININ
Inkberry → DRYMARIA sp. also → PHYTOLACCA sp.
INKS
Inkweed → DRYMARIA sp.
Inmetal → AMOBARBITAL
Inmetsin → INDOMETHACIN
Innohep → TINZAPARIN SODIUM
Innovace → ENALAPRIL
Innoxalon → NALIDIXIC ACID
Inocor → AMRINONE
INOCYBE sp.
Inopamil → IBOPAMINE
Inophylline → AMINOPHYLLINE
Inosie → INOSINE
INOSINE
Inosine Pranobex → ISOPRINOSINE
Inosiplex → ISOPRINOSINE
INOSITOL NIACINATE
Inostral → CROMOLYN SODIUM
Inotropin → DOPAMINE
Inovan → DOPAMINE
Inoven → IBUPROFEN
Inoxyl → OXOLINIC ACID
INPC → PROPHAM
Inponutrol → COTTONSEED OIL
INPROQUONE
Insane Root → HYOSCYAMUS
Insecticide no. 497 → DIELDRIN
Insecticide 1179 → METHOMYL
Insecticide no. 4049 → MALATHION
Insectophene → ENDOSULFAN
Insectrin → PERMETHRIN
Insidon → OPIPRAMOL
Insomin → NITRAZEPAM
Insoral → PHENFORMIN
Inspir → ACETYLCYSTEINE
Instalac → TRIMETHOPRIM
Instant Magic → CYANOACRYLATES
Instat → COLLAGEN
Insubeta → INVERT SUGAR
Insulamin → BUFORMIN
Insulase → CHLORPROPAMIDE
INSULIN also → ALCOHOL, ETHYL; ANTIPYRINE, CAPRYLIC ACID, CHLOROTHIAZIDE, CHLORPROMAZINE, CHLORTHALIDONE, CIMETIDINE, DEXTROTHYROXINE, DIAZEPAM, DIMETHICONE, DOBUTAMINE, ETHINYL ESTRADIOL, ETHYL ESTRANOL, FURAZOLIDONE, FUROSEMIDE, ISONIAZID, METHYCLOTHIAZIDE, METOPROLOL, NALTREXONE, NIFEDIPINE OXYPHENBUTAZONE, PENICILLAMINE, PHENELZINE, PRENYTOIN, PORK, PREDNISOLONE, PREDNISONE, PROCARBAZINE, PROPRANOLOL, QUININE, RITODRINE, SAUNAS, STANNOUS CHLORIDE, SULFINPYRAZONE, TOBACCO, TOLBUTAMIDE, TRANYLCYPROMINE, VITAMIN E
Insulton → MEPHENYTOIN
Intal → CROMOLYN SODIUM
Integrilin → EPTIFIBATIDE
Integrin → OXYPERTINE
Intenkordin → CHROMONAR
Intensain® → CHROMONAR
Intensopan → 5,7-DIBROMO-8-QUINOLINOL
Intercept → NONXYNOL-9
Interceptor → MILBEMYCIN OXIME
INTERFERONS also → COBALT SULFATE, 5-FLUOROURACIL, FLY ASH, INSULIN, THEOPHYLLINE, WARFARIN, ZIDOVUDINE
Inter Gen-16 → GENTIAN VIOLET
Interleukin 2 → ALDESLEUKIN
INTERLEUKIN 11
INTERLEUKINS
INTERMEDINE also → AMSACRINE, PYRROLIZIDINES
Intermigran → PROPRANOLOL
Internal Antiseptic no. 307 → ACETOSULPHONE SODIUM
Interomycetine → CHLORAMPHENICOL
Interprop → PROPIONIC ACID
Intertocine-S → OXYTOCIN
Intestiazol → PHTHALYLSULFATHIAZOLE
Intestibar → BARIUM SULFATE
Intocostrin → *d*-TUBOCURARINE CHLORIDE
Intrabilix → MEGLUMINE IODIPAMIDE
Intrabutazone → PHENYLBUTAZONE
Intradex → DEXTRAN
Intradine → SULFAMETHAZINE
Intralipid® → FATS also → SOYBEAN OIL
Intramycetin → CHLORAMPHENICOL
Intranarcon → THIALBARBITAL
Intrapan → DEXPANTHENOL
Intrasporin → CEPHALORIDINE
Intraval → THIOPENTAL
Intrazine → PROMAZINE
Intrazone → PHENYLBUTAZONE
Introcar → NIFEDIPINE
Introl → CROMOLYN SODIUM
Intromene → TRICHLORMETHIAZIDE

Intron A → INTERFERONS
Introna → INTERFERONS
Inula Oil → INULA sp.
INULA sp.
Invasin → HYALURONIDASE
Invenol → CARBUTAMIDE
Inversine → MECAMYLAMINE
INVERT SUGAR
Inverton 245 → 2,4,5-TRICHLOROPHENOXYACETIC ACID
Invesol → INVERT SUGAR
Invirase → SAQUINAVIR
Invisi-Gard → PROPOXUR
Inyesprin → LYSINE ACETYSALICYLATE
Inyi → ERYTHROPHLEUM
IOBENZAIC ACID
IOCARMATE MEGLUMINE
Io Caterpillar → CATERPILLARS
IOCETAMIC ACID
IOCHROMA sp.
Iodalbin → IODINATED I-131 SERUM ALBUMIN, AGGREGATED
IODAMIDE
"Iodide Mumps" → POTASSIUM IODIDE
IODINATED I-131 SERUM ALBUMIN, AGGREGATED
IODINATED GLYCEROL also → PHENOBARBITAL
IODINE also → CABBAGE, CELLASENE, DEXTROTHYROXINE, L-DOPA, DYCLONINE, ERYTHROSINE, INDOCYANINE GREEN, IODINATED GLYCEROL, ISOPROPAMIDE IODIDE, KELP, LAMINARIA, LITHIUM, MERCURY, MILK, PERCHLORIC ACID, PHENOTHIAZINE, POLYSTYRENE, POLYVINYLPYRROLIDONE IODINE COMPLEX, POTASSIUM IODIDE, PROPYLIODONE, RADISH, RESORCINOL, SHRIMP, SOYBEANS, VITAMIN C
Iodine Mercurique → MERCURIC IODIDE, RED
IODIPAMIDE
IODIZED OILS
IODOCHLORHYDROXYQUINOLINE also → HALQUINOL, PHANQUINONE
Iodochlorohydroxyquinoline → IODOCHLORHYDROXYQUINOLINE
Iodochloroxyquinoline → IODOCHLORHYDROXYQUINOLINE
5-Iodo-2′-deoxyuridine → IDOXURIDINE
Iodoenterol → IODOCHLORHYDROXYQUINOLINE
Iodofen → DISOPHENOL
IODOFENPHOS
IODOFORM
Iodohippuric Acid → CYCLOSPORINE
Iodomethane → METHYL IODIDE
Iodopaque → SODIUM ACETRIZOATE
IODOPHORS also → MILK
IODOPYRACET
IODOQUINOL
Iodure de Potassium → POTASSIUM IODIDE
Ioduril → SODIUM IODIDE
IOGLYCAMIC ACID
IOHEXOL
Iomapidol → IOPAMIDOL
Iomycin → PENICILLIN > *Benzylpenicillin, DAEEH*
Ionamin → PHENTERMINE
Ionol CP → BUTYLATED HYDROXYTOLUENE
IONONE
IOPAMIDOL
Iopamiro → IOPAMIDOL
Iopamiron → IOPAMIDOL
IOPANOIC ACID also → ACETYLSALICYLIC ACID
Iophen → IODINATED GLYCEROL
IOPHENDYLATE
IOPHENOIC ACID
Iopyracil → IODOPYRACET
Ioguin → IODOQUINOL
IOTHALMATE SODIUM
Ipacef → CEFUROXIME
Ipado → COCA
Ipamix → INDAPAMIDE
IPC → PROPHAM
IPE → ISOPROPYL ETHER
IPECAC also → MILK, POLYGALA sp.
Ipecacuanha → IPECAC
Ipersed → NITRAZEPAM
Ipertensina → ANGIOTENSIN
Iphosphamid(e) → IFOSFAMIDE
IPIL-IPIL
Ipnofil → METHAQUALONE
Ipnovel → MIDAZOLAM
IPODATE
Ipoglicone → TOLBUTAMIDE
Ipoh → DERRIS ROOT
Ipolab → LABETALOL
IPOMEA sp.
IPOMEAMARONE also → SWEET POTATO
Ipomeanine → SWEET POTATO
4-Ipomeanol → SWEET POTATO
Ipomoea batatas → SWEET POTATO
IPOMOEA VIOLACEA also → LYSERGAMIDE, MORNING GLORY
Iporal → GUANETHIDINE

Ipotensium → CLONIDINE
IPP → INDOPROFEN
IPPD → *N*-ISOPROPYL-*N*-PHENYLENEDIAMINE
Ipradol → HEXOPRENALINE
IPRATROPIUM BROMIDE also → TERBUTALINE
Iprazid → IPRONIAZID
IPRINDOLE
IPRONIAZID also → ACENOCOUMAROL, AMPHETAMINE, CHEESES, ETHYL BISCOUMACETATE, HEXOBARBITAL, MEPERIDINE, NOREPINEPHRINE, SUMATRIPTAN SUCCINATE
Ipronid → IPRONIAZID
Iproveratril → VERAPAMIL
Ipsilon → Σ-AMINOCAPROIC ACID
IQ → 2-AMINO-3-METHYLIMIDAZ-(4,5-f)QUINOLINE
Ira Igbo → RAUWOLFIA SERPENTINA
Iramil → IMIPRAMINE
Ircon → FERROUS FUMARATE
Irescein → FLUORESCEIN
Iretin → CYTARABINE
Irgasan → TETRACHLOROSALICYLANILIDE
Irgasan DP 300 → TRICLOSAN
Iridin → IRIS
Iridina Due → NAPHAZOLINE
IRIDIUM
Iridocin → ETHIONAMIDE
Irifan → CYCLOBARBITAL
IRIGENIN
IRINOTECAN
IRIS
Irish Moss → CARRAGEENAN
Irish Potato → POTATO
Irisone → CORTISONE also → IONONE
Irium → SODIUM LAURYL SULFATE
Iroject → IRON DEXTRINS
Iromon → FERROUS GLUCONATE
IRON also → ACETYLSALICYLIC ACID; ALCOHOL, ETHYL; CIMETIDINE, CIPROFLOXACIN, COBALT, DIGITALIS, EGGS, FISH, FLOUR, GREPAFLOXACIN, LANSOPRAZOLE, MAGNESIUM SILICATE, MAGNESIUM TRISILICATE, MILK, NORFLOXACIN, OMEPRAZOLE, OXYTETRACYCLINE, PENICILLAMINE, POTASSIUM FERRICYANIDE, SODIUM, STANAZOLOL, STARCHES, TANNIC ACID, TEA, TETRACYCLINE
Iron Ammonium Citrate → FERRIC AMMONIUM CITRATE
Iron Amylose Complex → IRON POLYSACCHARIDE COMPLEX
"Iron Bacterium" → CRENOTHRIX POLYSPORA
Irondex → IRON DEXTRANS
IRON DEXTRANS
IRON DEXTRINS
Iron Gallate → INKS
Iron Hexanehexol → IRON SORBITEX
Iron Hy-Dex → IRON DEXTRANS
IRON POLYSACCHARIDE COMPLEX
IRON SORBITEX
Iron Sorbitol → IRON SORBITEX
Iron Trichloride → FERRIC CHLORIDE
IRONWOOD also → ERYTHROPHLEUM
Iron Wort → ANTHEMIS COTULA
Iroquine → CHLOROQUINE
Irox → FERROUS GLUCONATE
Irradiated Ergosterol → VITAMIN D_2
Irrigor → IMOLAMINE
Irrorin → PRENYLAMINE LACTATE
Irtan → CROMOLYN SODIUM
IS 370 → SUCCINYLCHOLINE
IS 499 → POLDINE METHYLSULFATE
Isacen → OXYPHENISATIN
Isactid → CORTICOTROPIN
Isadrine → ISOPROTERENOL
Isalide → JOSAMYCIN
Isaphen → OXYPHENISATIN
Iscador → MISTLETOE, EUROPEAN or JAPANESE
Isdin → ISOSORBIDE DINITRATE
ISDN → ISOSORBIDE DINITRATE
Isepacin → ISEPAMICIN
ISEPAMICIN
Isha → STROPHANTHUS sp.
Ishin → BLIGHIA SAPIDA
Isho → PHYSOSTIGMA VENENOSUM
Isindone → INDOPROFEN
Isipen → PENICILLINS – "NEWER and SYNTHETICS" > *Piperacillin Sodium*
Islandotoxin → RICE
Ismelin → GUANETHIDINE
Ismo → ISOSORBIDE MONONITRATE
Ismotic → ISOSORBIDE
Isoacetophorone → ISOPHORONE
ISOAMYL ACETATE
ISOAMYL ALCOHOL
Isoamyl Cinnamate → AMYL CINNAMATE
Isoamyl-σ-hydroxybenzoate → AMYL SALICYLATE
ISOAMYL PROPIONATE
Isoamyne → AMPHETAMINE
Isoarteril → AJMALICINE

Isobamate → CARISPRODOL
Isobarin → GUANETHIDINE
ISOBENZAN
Isobicina → ISONIAZID
Iso-Bid → ISOSORBIDE DINITRATE
Isobide → ISOSORBIDE
ISOBORNYL ACETATE
ISOBORNYL PROPRIONATE
ISOBORNYL THIOCYANOACETATE
Isobromyl → BROMISOVALUM
ISOBUTANE
Isobutanol → ALCOHOL, ISOBUTYL
ISOBUTYL ACETATE
ISOBUTYL ALCOHOL
ISOBUTYL BENZOATE
Isobutyl Carbinol → ISOAMYL ALCOHOL
N-Isobutyl Dienamides → PEPPERS
ISOBUTYL NITRITE
ISOBUTYL PHENYLACETATE
ISOBUTYL QUINOLINE
ISOBUTYL SALICYLATE
Isobutyl α-toluate → ISOBUTYL PHENYLACETATE
N-Isobutyl Trienamides → PEPPERS
ISOBUTYRONITRILE
Isocaine → PROCAINE
Isocalm → TOLPERISONE HCL
Isocaramidine → DEBRISOQUIN
ISOCARBOXAZID also → ALCOHOL, ETHYL; AMPHETAMINE, BEANS, CHEESES, DESIPRAMINE, DEXTROMETHORPHAN HYDROBROMIDE, DIPHENHYDRAMINE, ETHYL BISCOUMACETATE, FLUOXETINE, IMIPRAMINE, INSULIN, MEPERIDINE, SUMATRIPTAN SUCCINATE, L-TRYPTOPHAN, VENLAFAXINE
ISOCARBOXIDE also → HERRING, SERTRALINE
Isocard → ISOSORBIDE DINITRATE
Isocef → CEFTIBUTIN
ISOCETYL STEARATE
Isochlorothion → DICAPTHON
Isocholin → AMANITIN
Isochroman → PENICILLIUM sp.
Isocid → ISONIAZID
Isocoma heterophylus → GOLDENROD
Isocoma wrightii → GOLDENROD
Iso-Cornox → MECOPROP
Isocorydine → DICENTRA CUCULLARIA
Isocothane → DINOCAP
Iscotin → ISONIAZID
Isocrin → OXYPHENISATIN
Isocumene → *n*-PROPYLBENZENE
ISOCYCLOCITRAL
Isodin → OXAZEPAM
Isodine → POLYVINYLPYRROLIDONE IODINE COMPLEX
Isodiprene → Δ^3-CARENE
Isodormid → APRONALIDE
ISODRIN
Isoendoxan → IFOSFAMIDE
Isoephedrine → PSEUDOEPHEDRINE
Isoergine → ARGYREIA, MORNING GLORY
ISOETHARINE also → SODIUM BISULFITE
ISOETHARINE
ISOEUGENOL
Isoeugenol Ether → BENZYL ISOEUGENOL
Isofedrol → EPHEDRINE
ISOFENPHOS
Isofluorphate → DIISOPROPYL FLUOROPHOSPHATE
ISOFLUPREDONE
ISOFLURANE also → DIMETHYL TUBOCURARINE IODIDE, EPINEPHRINE
Isoflurophate → DIISOPROPYL FLUOROPHOSPHATE
Isoglaucon → CLONIDINE
Isoimperitorin → CYMOPTERUS sp. also → WILD CARROT
Iso-K → KETOPROFEN
Isoket → ISOSORBIDE DINITRATE
Isolait → ISOXSUPRINE
ISOLAN also → PRALIDOXIME
Isolanid → LANATOSIDE
Isolyn → ISONIAZID
Iso-Mack → ISOSORBIDE DINITRATE
Isomeride → DEXFENFLURAMINE
Isomerine → CHLORPHENIRAMINE also → DEXCHLORPHENIRAMINE
ISOMETAMIDIUM
ISOMENTHONE
Isomenyl → ISOPROTERENOL
Isomil® → SOYBEANS
Isomist → ISOPROTERENOL
Isomonat → ISOSORBIDE MONONITRATE
Isomyn → AMPHETAMINE
Isomytal → AMOBARBITAL
Isonal → MEPHOBARBITAL
Isonex → ISONIAZID
ISONIAZID also → ACENOCOUMAROL, ACETAMINOPHEN; ALCOHOL, ETHYL; ALUMINUM, AMINOSALICYLIC ACID, ATROPINE, CARBAMAZEPINE, CHEESES, CHLORAZEPATE, CLOFAZIMINE, CLONAZEPAM, DIAZEPAM, DISULFIRAM, L-DOPA, FISH, GLIPIZIDE, HISTAMINE,

HYDRAZINE, INSULIN, IRON, ITRACONAZOLE, MEPERIDINE, MILK, PAROXETINE, PENICILLAMINE, PHENELZINE, PHENOBARBITAL, PHENYTOIN, POTASSIUM, PREDNISOLONE, RIFAMPIN, SAUERKRAUT, SODIUM VALPROATE, STREPTOMYCIN, THEOPHYLLINE, THIACETAZONE, VINCRISTINE, VITAMIN B_3, VITAMIN B_6, WARFARIN
Isonipecaine → MEPERIDINE
Isonizida → ISONIAZID
Isonorin → ISOPROTERENOL
ISOOCTANE
Isooctanol → ISOOCTYL ALCOHOL
ISOOCTYL ALCOHOL
Isopaque → SODIUM METRIZOATE
Isopelletierine → POMEGRANATE
Isopentyl Alcohol → ISOAMYL ALCOHOL
Isopentyl Proprionate → ISOAMYL PROPRIONATE
Isopestox → MIPAFOX
Isophen → METHAMPHETAMINE
ISOPHORONE
Isophosphamide → IFOSFAMIDE
Isophrin → PHENYLEPHRINE HCL
ISOPIMPINELLIN
Isoplasma → POLYVINYLPYRROLIDONE
isoPPC → PROPHAM
Isoprenaline → ISOPROTERENOL
ISOPRINOSINE
Isoprochin P → PROPYPHENAZONE
Isoprofenamine → CLORPRENALINE
Isopropalin → *N*-NITROSODI-*n*-PROPYLAMINE
ISOPROPAMIDE IODIDE also → PHENYLPROPANOLAMINE
Isopropanol → ALCOHOL, ISOPROPYL
Isoprophenamine → CLORPRENALINE
2-ISOPROPOXYETHANOL
2-Isopropoxypropane → ISOPROPYL ETHER
Isopropydrin → ISOPROTERENOL
ISOPROPYL ACETATE
Isopropylacetone → METHYL ISOBUTYL KETONE
N-ISOPROPYL ACRYLAMIDE
Isopropyl Alcohol → ALCOHOL, ISOPROPYL
ISOPROPYLAMINE
ISOPROPYLAMINODIPHENYLAMINE
ISOPROPYLANILINE
Isopropyl Benzene → CUMENE
Isopropylcarbinol → ALCOHOL, ISOBUTYL
Isopropyl-σ-cresol → CARVACROL
Isopropyl Cyanide → ISOBUTYRONITRILE
ISOPROPYLDIBENZOYLMETHANE
Isopropyl-4,4´-dichlorobenzilate → CHLOROPROPYLATE
ISOPROPYL ETHER
ISOPROPYL GLYCIDYL ETHER
Isopropyl Glycol → 2-ISOPROPOXYETHANOL
4,4´-Isopropylidenediphenol → BISPHENOL
ISOPROPYL LANOLATE
ISOPROPYL LINEOLATE
ISOPROPYL METHYLHEXENES
ISOPROPYL MYRISTATE
ISOPROPYL NEOPENTANATE
Isopropylnorepinephrine → ISOPROTERENOL
ISOPROPYL PALMITATE
2-Isopropyl-4-pentenoyl Urea → APRONALIDE
Isopropylphenazone → PROPYPHENAZONE
N-ISOPROPYL-*N*-PHENYL-*p*-PHENYLENEDIAMINE
ISOPROPYL QUINOLINE
ISOPROTERENOL also → ALCOHOL, ETHYL; CHLORPROMAZINE, DIGITALIS, EPINEPHRINE, GLUCOSE, LACTOSE, LIDOCAINE, PHENELZINE, PROPRANOLOL, SOTALOL, THEOPHYLLINE, TRANYLCYPROMINE
Isoptin → VERAPAMIL
Isopto Carpine → PILOCARPINE
Isopto Cetamide → SULFACETAMIDE SODIUM
Isopto-Dex → DEXAMETHASONE
Isoptoeserine → PHYSOSTIGMINE
Isopto Tears → CELLULOSE, *Oxidized*
ISOPULEGOL
Iso-Puren → ISOSORBIDE DINITRATE
Isorbid → ISOSORBIDE DINITRATE
Isordil → ISOSORBIDE DINITRATE
Isorenin → ISOPROTERENOL
Isorythm → DISOPYRAMIDE
ISOSAFROLE
ISOSORBIDE
ISOSORBIDE DINITRATE also → PROPRANOLOL
ISOSORBIDE MONONITRATE
ISOSPARTEINE
Isostenase → ISOSORBIDE DINITRATE
ISOSULFAN BLUE
Isosystox → DEMETON
Isoten → BISOPROLOL
Isotense → SYROSINGOPINE
Isothan Q15 → LAURYL ISOQUINOLINIUM BROMIDE
Isothazine → ETHOPROPAZINE
Isothiazine → ETHOPROPAZINE
3-Isothiocyanato-1-propene → ALLYL ISOTHIOCYANATE

Isothiocynate → DESCURAINA PINNATA
Isothiocyanic Acid Allyl Ester → ALLYL ISOTHIOCYANATE
ISOTHIPENDYL
Isothymol → CARVACROL
ISOTOMA LONGIFLORA
Isotomin → ISOTOMA LONGIFLORA
Isotox → LINDANE
Isotrate → ISOSORBIDE DINITRATE
ISOTRETINOIN
Isotrex → ISOTRETINOIN
Isotron 2 → DICHLORODIFLUOROMETHANE
ISOTROPIS
Isovalerone → DIHYDROXYACETONE
2-ISOVALERYL-1,3-INDANEDIONE
Isovon → ISOPROTERENOL
Isovue → IOPAMIDOL
Isovyl → POLYVINYL CHLORIDE
ISOXICAM
ISOXSUPRINE
Isoxyl → THIOCARLIDE
Isozid → ISONIAZID
Ispaghula Husks → PLANTAGO SEED
Ispan → SULFUR HEXAFLUORIDE
ISRADIPINE
Issium → FLUNARIZINE HCL
Isteropac E.R. → IODAMIDE
Istizin → DANTHRON
Isuprel → ISOPROTERENOL
itai-itai Disease → CADMIUM
Italian Green Bean → BEANS, Faba
Itamidone → AMINOPYRINE
Itana pa osho → MIRABILIS JALAPA
Itchgrass → ROTTBOELLIA EXALTATA
Ithang → KRATOM
Itobarbital → BUTALBITAL
ITRACONAZOLE also → ASTEMIZOLE, ATORVASTATIN, CILOSTAZOL, CISAPRIDE, CYCLOSPORINE, DIDANOSINE, DIGOXIN, FAMOTIDINE, FELBAMATE, ISONIAZID, LANSOPRAZOLE, LOVASTATIN, METHYL-PREDNISOLONE, MIDAZOLAM, OMEPRAZOLE, PHENOBARBITAL, PHENYTOIN, QUINIDINE, RANITIDINE, RIFABUTIN, RIFAMPIN, SILDENAFIL, TACROLIMUS, TERFENADINE, TRIAZOLAM, VINCRISTINE
ITRAMIN TOSYLATE
Itridal → CYCLOBARBITAL
Itrin → TERAZOCIN
Itrizole → ITRACONAZOLE
Itrop → IPRATROPIUM BROMIDE
Ituran → NITROFURANTOIN
IUDR → IDOXURIDINE
IVA *sp.*
Ivadal → ZOLPIDEM TARTRATE
IVADANTIN → NITROFURANTOIN
Ivalon → POLYVINYL ALCOHOL
Ivaugan → HYDROCHLOROTHIAZIDE
Iva xanthifolia → MARSH ELDER
IVERMECTIN
Ivomec → IVERMECTIN
Ivosit → DINOSEB
Ivy Bush → KALMIA
IVY
Iwalexin → CEPHALEXIN
Iwele → TEPHROSIA sp.
Ixertol → CINNARIZINE
Ixia → MISTLETOE, EUROPEAN or JAPANESE
Izaberizin → CINNARIZINE
Izadrin → ISOPROTERENOL

J

J → CANNABIS
208J-2 → PADIMATES
Jaagsiekte → CROTALARIA
JACARANDA ENVACO
Jack Aroma → ISOBUTYL NITRITE
JACK BEAN also → CYANIDES
Jack-In-The-Pulpit → ARISAEMA
Jack O'Lantern Mushroom → OMPHALOTUS ILLUDENS
JACOBINE also → PYRROLIZIDINES
Jacob's Ladder → CHELIDONIUM MAJUS
Jacoline → PYRROLIZIDINES
Jacozine → PYRROLIZIDINES
Jacutin → LINDANE
Jaguar → GUAR GUM
Jaikin → BORAX
JALAP
Jaljari → RANUNCULUS
Jalovis → HYALURONIDASE
Jam → COCAINE
Jamaica Pepper → PIMENTO
James Benson → MINERAL WATERS
James Bond → TETRODOTOXIN
Jamestown Lily → DATURA
Jamestown Weed → DATURA
Jamylène → DOCUSATE SODIUM
Jane → CANNABIS
Janimine → IMIPRAMINE
JANITHREMS
Jan Saiwa → TRICHILIA EMETICA

Japanese Edelweiss → EUPHORBIA
Japanese Honeysuckle → LONICERA
Japanese Lacquer Tree → POISON IVY
Japanese Millet → ECHINOCHLOA CRUSGALLI
Japanese Mint Oil → CORNMINT OIL
Japanese Star Anise → ILLICIUM ANISATUM also → SHIKIMIC ACID
Japanese Viper → WINE
Japan Lacquer → RENGAS
Japan Pea → SOYBEANS
Jarosse → LATHYRUS sp.
JASMINE ABSOLUTE
Jasmolins → PYRETHRUM, PYRETHRINS, and PYRETHROIDS
Jatroneural → TRIFLUOPERAZINE DIHYDROCHLORIDE
JATROPHA sp.
Jatropha gossypifolia → FLAIRA
Jatrophin → JATROPHA sp.
Jatropur → TRIAMTERENE
Jatrosom → TRIFLUPROMAZINE
Jaune Brilliant → CADMIUM SULFIDE
Jaunet → ENTOLOMA
Java Bean → BEANS, Lima
Javelle Water → SODIUM HYPOCHLORITE
Jawar → SORGHUM
Jay → CANNABIS
JB 11 → TRIMETREXATE
JB 251 → PROTOKYLOL
JB 305 → PIPERIDOLATE
JB 329 → PIPERIDYL BENZILATE
JB 516 → PHENIPRAZINE
JB 8181 → DESIPRAMINE
JD 91 → EMYLCAMATE
J-Dantin → NITROFURANTOIN
JDL 464 → TORSEMIDE
Jectofer → IRON SORBITEX
Jellin → FLUOCINOLONE ACETONIDE
Jelly Beans → AMPHETAMINES
JELLYFISH also → CHRYSAORA, SEA WASP
Jelly Leeks → ALOE
Jenacaine → PROCAINE
Jenacycline → DOXYCYCLINE
Jenotone → AMINOPROMAZINE
Jequirity Bean → ABRUS PRECATORIUS also → ABRIN
Jerusalem Cherry → SOLANUM sp.
Jerusalem Oak → CHENOPODIUM
Jervine → CALTHA PALUSTRIS, VERATRUM sp.
Jesuit's Balsam → COPAIBA
Jesuit's Bark → CINCHONA
Jesuit's Tea → ILEX sp.
Jet → KETAMINE
Jetbead → RHODOTYPOS SCANDENS
Jetberry → RHODOTYPOS SCANDENS
JET FUELS
Jetrium → DEXTROMORAMIDE
Jewel Flower or *Weed* → IMPATIENTS sp.
JEWELRY
Jexin → *d*-TUBOCURARINE CHLORIDE
Jilkon → GALANTHAMINE
Jim Barawo → LEONOTIS NEPETIFOLIA
"Jimmies" → NOTHOLAENA SINNUATA
Jimmy Fern → NOTHOLAENA SINNUATA
Jimmy Weed → GOLDENROD
Jimson Weed → DATURA
JIN BU HUAN
Ji ona → DIOSCOREA
Jiu-li-xiang → MURRAYA PANICULATA
J-Liberty → CHLORDIAZEPOXIDE
JM 8 → CARBOPLATIN
JO 1016 → TIXOCORTOL PIVALATE
Joc Aroma → ISOBUTYL NITRITE
Jodamid → IODAMIDE
Jod-Basedow Phenomenon → POTASSIUM IODIDE
Jodenphos → IODOFENPHOS
Jodfenphos → IODOFENPHOS
Jodid → POTASSIUM IODIDE
Jodomiron → IODAMIDE
Johannisbrotmehl → CAROB
John Conqueror → JALAP
Johnson & Johnson → CISAPRIDE
Johnson Grass → SORGHUM also → HYDROGEN CYANIDE, CYANIDES
Johnson's Car Plate® → POLISHES, FURNITURE
Joiner's Disease → SAWDUST
Joints → CANNABIS
JOJOBA BEAN
Jolt → ETHOPROP
Jomybel → JOSAMYCIN
Jonquille → DAFFODIL
JONQUILS also → NARCISSUS sp.
Jopagnost → IOPANOIC ACID
Jopamiro → IOPAMIDOL
Joptone → POTASSIUM IODIDE
Josacine → JOSAMYCIN
Josamina → JOSAMYCIN
Josamy → JOSAMYCIN
JOSAMYCIN also → CYCLOSPORINE
Josaxin → JOSAMYCIN
Josir → TAMSULOSIN
Jowar → SORGHUM
"Joy Plant" → OPIUM
"Joy Rides" → SCOPOLAMINE

K

KAMALA
Kamaver → CHLORAMPHENICOL
Kaminax → AMIKACIN
Kamóstim → DATURA
Kamycin → KANAMYCIN
Kamynex → KANAMYCIN
Kanabristol → KANAMYCIN
Kanacedin → KANAMYCIN
KANAMYCIN also → CALCIUM, AMOBARBITAL, CURARE, DIGITALIS, DIGOXIN, DIMETHYL TUBOCURARINE IODIDE, EDROPHONIUM, ETHACRYNIC ACID; ETHER, ETHYL; FRAMYCETIN, FUROSEMIDE, GALLAMINE TRIETHIODIDE, GENTAMICIN, NEOMYCIN, NEOSTIGMINE, PAROMOMYCIN, PENICILLIN, POLYMIXIN B, PROCAINAMIDE, PROMETHAZINE, QUINIDINE, SUCCINYLCHOLINE, *d*-TUBOCURARINE CHLORIDE, VANCOMYCIN
Kanamytrex → KANAMYCIN
Kanaqua → KANAMYCIN
Kanasig → KANAMYCIN
Kanatrol → KANAMYCIN
Kanbine → AMIKACIN
Kanechlors → POLYCHLORINATED BIPHENYLS
Kanescin → KANAMYCIN
Kanicin → KANAMYCIN
Kanju → CASHEW
Kannabis → CANNABIS
Kannasyn → KANAMYCIN
Kannit → XYLITOL
Kano → KANAMYCIN
Kantrex → KANAMYCIN
Kantrexil → KANAMYCIN
Kantrim → KANAMYCIN
Kantrox → KANAMYCIN
K-ao → OAKS
Kaochlor SF → POTASSIUM CHLORIDE
KAOLIN also → CODEINE, DUSTS, LINCOMYCIN; PENICILLINS – "NEWER and SYNTHETICS" > *Amoxicillin, Ampicillin;* PECTIN
Kaon → POTASSIUM GLUCONATE
Kaon-Cl → POTASSIUM CHLORIDE
Kaopectate® → LINCOMYCIN, TETRACYCLINE
Kapahasa → RYANIA sp.
Kapanol → MORPHINE
Ka-Pen → PENICILLIN > *Benzylpenicillin, Potassium*
KAPOK
Kappabi → DIBEKACIN
Karanj → PONGAMIA PINNATA
Karathane → DINOCAP
KARAYA also → PIE(S), STERCULIA sp.
"Kara yara" → HEXACHLOROBENZENE
Karbam White → ZIRAM
Karbathion → METHAM SODIUM
Karbofos → MALATHION
Karela → MOMORDICA CHARANTIA also → CHLORPROPAMIDE
Karen Silkwood → PLUTONIUM
Karidium → SODIUM FLUORIDE
Karmex Monuron Herbicide → MONURON
Karóokkoot → DATURA
Karwinskia humboldtiana → COYOTILLO
K'arya → ADENIUM sp.
KASARAKOFFER YELLOW
Kasari → LATHYRUS sp.
Kaskay → POTASSIUM CHLORIDE
Kat → CATHA EDULIS
Kata-Khuturia → AMARANTHUS
Katen → MEXILETINE
Katine → NORPSEUDOEPHEDRINE
Katlex → FUROSEMIDE
Kato → POTASSIUM CHLORIDE
Katorin → POTASSIUM GLUCONATE
Katoseran → CINNARIZINE
Katovit → PROLINTANE
Kautchak → RUBBER
KAVA
Kavadel → DIOXATHION
Kavaine → KAVA
Kava-kava → KAVA
Kawa → KAVA
Kayafume → METHYL BROMIDE
Kayazinon → DIAZINON
Kayback → POTASSIUM CHLORIDE
Kay-Cee-L → POTASSIUM CHLORIDE
Kay-Ciel → POTASSIUM CHLORIDE
Kay Contin → POTASSIUM CHLORIDE
Kaydol → MINERAL OIL
Kayexalate → SODIUM POLYSTYRENE SULFONATE
Kayju → CASHEW
Kazide → POTASSIUM AZIDE
Kazoe → POTASSIUM AZIDE
K-cillin → PENICILLIN > *Benzylpenicillin, Potassium*
KCN → POTASSIUM CYANIDE
Keal → SUCRALFATE
Kebilis → CHENODEOXYCHOLIC ACID
KEBUZONE
Kechipir → BEANS, Winged
Keck → HERACLEUM sp.

Keenimowe → ENTEROLOBIUM
Kefadol → CEFAMANDOLE
Kefandol → CEFAMANDOLE
Kefazim → CEFTAZIMIDES
Kefenid → KETOPROFEN
Keflet → CEPHALEXIN
Keflex → CEPHALEXIN
Keflin → CEPHALOTHIN
Keflodin → CEPHALORIDINE
Keforal → CEPHALEXIN
Kefroxil → CEFADROXIL
Kefspor → CEPHALORIDINE
Keftab → CEPHALEXIN
Kefurox → CEFUROXIME
Kefzol → CEFAZOLIN SODIUM
Keimax → CEFTIBUTIN
Kelamin → KHELLIN
Kelecin → LECITHIN
Kelene → ETHYL CHLORIDE
Kelfiprim → TRIMETHOPRIM
Kelfizin → SULFALENE
Kelgin → SODIUM ALGINATE
Kelicor → KHELLIN
Kelicorin → KHELLIN
Kellin → KHELLIN
Keloid → KHELLIN
KELP also → ASTEMIZOLE
Kelthane → DICOFOL
Kemadrin → PROCYCLIDINE
Kemate → ANILAZINE
Kemdrin → CIODRIN
Kemicetine → CHLORAMPHENICOL
Kemi S → PROPRANOLOL
Kemithal → THIALBARBITAL
Kempak → RUELENE
Kemsol → DIMETHYL SULFOXIDE
Kenacort → TRIAMCINOLONE
Kenalog → TRIAMCINOLONE
Kendall's Compound E → CORTISONE
Kendall's compound F → HYDROCORTISONE
Kenevir → CANNABIS
Kenmycin → ERYTHROMYCIN
KENTUCKY COFFEE TREE also → CYTISINE
Keoxifene → RALOXIFENE
Kepinol → SULFAMETHOXAZOLE
KEPONE also → MIREX
Keralyt → SALICYLIC ACID
Keramik → DIETHYLPROPION
Keramin → DIETHYLPROPION
K'erana → EUPHORBIA, *kamerunica*
Keratinamin → UREA
Keratol → ZINC
Kerb → PRONAMIDE
Kerecid → IDOXURIDINE
Kéreg → OAKS
Kerlone → BETAXOLOL
Kerocaine → PROCAINE
KEROSENE
Kerr-McGee → PLUTONIUM
Kerumenex → TRIETHYLENE OLEATE POLYPEPTIDE CONDENSATE
Keselan → HALOPERIDOL
Keshan Disease → SELENIUM
Kesint → CEFUROXIME
Kessar → TAMOXIFEN CITRATE
Kesso → VALERIANA sp.
Kesso-Bamate → MEPROBAMATE
Kesso-Pen → PENICILLIN > *Benzylpenicillin, Potassium*
Kestine → EBASTINE
Kestrin → ESTROGENS, CONJUGATED
Ketaject → KETAMINE
Ketalar → KETAMINE
Ketalgin → METHADONE
KETAMINE also → DIAZEPAM, HYDROXYZINE, LIOTHYRONINE, MECAMYLAMINE, PHENCYCLIDINE, PROCYCLIDINE, SECOBARBITAL SODIUM, SUCCINYLCHOLINE, THYROID, THYROXINE
Ketanarkon → KETAMINE
Ketanest → KETAMINE
Ketanrift → ALLOPURINOL
KETANSERIN also → ERYTHROPOIETIN, HYDROCHLOROTHIAZIDE
Ketaset → KETAMINE
Ketason → KEBUZONE
Ketavet → KETAMINE
KETAZOLAM
Ketazone → KEBUZONE
KETENE
Ketensin → KETANSERIN
Kethamed → PEMOLINE
KETHOXAL
Ketobun-A → ALLOPURINOL
2-Ketobutyric Acid → ACETOACETIC ACID
3-Ketobutyric Acid → ACETOACETIC ACID
Ketochromin → DIHYDROXYACETONE
KETOCONAZOLE also → ALCOHOL, ETHYL; ANTIPYRINE, ALPRAZOLAM, ASTEMIZOLE, ATORVASTATIN, CHLORAZEPATE, CILOSTAZOL, CIMETIDINE, CISAPRIDE, CLONAZEPAM, CYCLOSPORINE, DIAZEPAM, DIDANOSINE, DOFETILIDE, DONEPEZIL, FAMOTIDINE, FEXOFENADINE, HALOFANTRINE, INDINAVIR,

Klavocin → PENICILLINS – "NEWER and SYNTHETICS" > *Amoxicillin*
Klebcil → KANAMYCIN
Kleenex® → CITRIC ACID
Kleenol → 2,4-TOLUENEDIAMINE
Kleingrass → PANICUM sp.
Kleistubom → PASSIFLORA sp.
Klemidox → CLEMIZOLE
Klexane → ENOXAPARIN
Klimax E → ESTRIOL
Klimicin → CLINDAMYCIN
Klimoral → ESTRIOL
Klinium → LIDOFLAZINE
Klinomycin → MINOCYCLINE
Klion → METRONIDAZOLE
Klismacort → PREDNISOLONE
Klonopin → CLONAZEPAM
Klor-Con → POTASSIUM CHLORIDE
Klorhexidin → CHLORHEXIDINE
Kloride → POTASSIUM CHLORIDE
Klorita → CHLORAMPHENICOL
Klorokin → CHLOROQUINE
Klorpromex → CHLORPROMAZINE
Klort → MEPROBAMATE
Klorvess → POTASSIUM CHLORIDE
Klot → TOLONIUM CHLORIDE
Klotrix → POTASSIUM CHLORIDE
Klot, stainless → ALCOHOL, *n*-BUTYL
K-Lyte/CL → POTASSIUM CHLORIDE
KM 56 → BENZONATATE
Knaeulgras → DEW GRASS
Knapweed → CENTAUREA sp.
Knight's Spur → LARKSPUR
"Knock-out Drops" → CHLORAL HYDRATE, COCCULUS
K-Norm → POTASSIUM CHLORIDE
Knotweeds → POLYGONUM sp.
Knoxweed → 2,4-DICHLOROPHENOXYACETIC ACID
KO 1173 → MEXILETINE
Koa Haole → LEUCAENA sp.
Kobakedwe → MINQUARTIA GUIANENSIS
Kobutol → QUINTOZENE
KOCHIA
Koda Millet → PASPALUM
Kohl → LEAD
Kohlrabi → BRASSICA sp. also → MUSTARD
Kojo Root → PETIVERIA ALLIACEA
KOLA also → COCA, COLAS, EUPHORBIA
Koli → CASTOR OIL
Kolikodal → HYDROCODONE BITARTRATE
Kolimitsin → COLISTIN
Kolimycin → NEOMYCIN
Kollerdormfix → CYCLOBARBITAL
Kollidon → POLYVINYLPYRROLIDONE
Kolmitisin → NEOMYCIN
Kolpolyn → ETHINYL ESTRADIOL
Kolpon → ESTRONE
Kombetin → STROPHANTHIN
KOMBUCHA also → MUSHROOMS
"Komfy Kid® dolls" → NAPHTHA
Kontexin → PHENYLPROPANOLAMINE
Konyne → CHRISTMAS FACTOR
KOONG YICK HUNG FAR OIL
Koo Sar pills → LEAD
Koo So pills → LEAD
Kopertox → COPPER NAPHTHENATE
Kop-Thiodan → ENDOSULFAN
Kopthion → MALATHION
Korama → HELIOTROPIUM sp.
Korazol → PENTYLENETETRAZOL
Korbutone → BECLOMETHASONE
Kordafen → NIFEDIPINE
Korean Ginseng → GINSENG
Korec → QUINAPRIL
Koribó → TANAECIUM sp.
Korlan → RONNEL
Koropo → CROTALARIA
Koroseal → POLYVINYL CHLORIDE
Korum → ACETAMINOPHEN
K'osain rogo → DIOSCOREA
Kowosol → SORSACA
Koyne → CHRISTMAS FACTOR
KP 363 → BUTENAFINE
KPABA → POTASSIUM-*p*-AMINOBENZOATE
KPB → KEBUZONE
K-Pen → PENICILLIN > *Benzylpenicillin, Potassium*
K₂R → TETRACHLOROETHYLENE
KRAMERIA
KRATOM
Krazy Glue → CYANOACRYLATES
Krebiozen → CREATINE
Krebon → BUFORMIN
Kredex → CARVEDILOL
Kremol → MINERAL OIL
Kremola → MERCURY, AMMONIATED
Kresatin → *m*-CRESYL ACETATE
Kriplex → DICLOFENAC
KRN 8601 → GRANULOCYTE COLONY STIMULATING FACTOR
Kronitex → TRITOLYL PHOSPHATE
Krovar II → BROMACIL
Krumkil → COUMAFURYL
Kryocristine → VINCRISTINE
Kryogenin → PHENICARBAZIDE
Krysid → 1-(1-NAPHTHYL)-2-THIOUREA

K-strophanthin-α → CYMARIN
K-Strophanthoside → STROPHANTHIN
K-Tab → POTASSIUM CHLORIDE
KTS → KETHOXAL
Kubacron → TRICHLORMETHIAZIDE
Kunde Nyika → CASSIA
Kurnan masar → MELIA sp.
Kuron → SILVEX
Kurrajong → STERCULIA sp.
Kurū'a → CASTOR BEAN
KUSHTAS
Kutan → CARBARSONE
Kutkasin → MONOSULFIRAM
Kutz → TOBACCO
Kuzam → TAGETES MINUTA
KW → PHENCYCLIDINE
KW 4354 → OXATOMIDE
Kwai → GARLIC
Kwankwani → STROPHANTHUS sp.
Kwarko → ARGEMONE MEXICANA
KWD 2019 → TERBUTALINE SULFATE
KWD 2183 → BAMBUTEROL
Kwell → LINDANE
Kwells → SCOPOLAMINE
Kyanol → ANILINE
Kylar → DAMINOZIDE
Kylit → XYLITOL
K-Y LUBRICATING JELLY® also → PROPYLENE GLYCOL
Kymo-trypure → CHYMOTRYPSIN
Kynex → SULFAMETHOXYPYRIDAZINE
Kynja → CANNABIS
Kyorin → NORFLOXACIN
Kypfos → MALATHION
Kypzin → ZINEB
Kytril → GRANISETRON

L

L 2 → DEFERIPRONE
L 11/6 → PHORATE
L 67 → PRILOCAINE
84 L → DIETHYLCARBAMAZINE
L 395 → DIMETHOATE
L 846 → ZALEPLON
L 1573 → CYSTEAMINE
L 2103 → HEXAPROPYMATE
L 2214 → BENZBROMARONE
L 2329 → BENZIODARONE
L 12,507 → TEICOPLANIN
L 12,717 → LOTRIFEN
L 34,314 → DIPHENAMID
L 36,352 → TRIFLURALIN
L 642,957 → CILASTATIN
L 700,462 → TIROFIBAN
L 706,631 → MONTELUKAST
L 735,524 → INDINAVIR
LA 1 → NITRAZEPAM
LA 200 → OXYTETRACYCLINE
LA 1211 → IMOLAMINE
LA 6023 → METFORMIN
2329 Labaz → BENZIODARONE
Labazene → SODIUM VALPROATE also → VALPROIC ACID
LABDANUM OIL
Labelol → LABETALOL
LABETALOL also → CIMETIDINE, INDOMETHACIN
Labican → CHLORDIAZEPOXIDE
LaBID → THEOPHYLLINE
Lablab Beans → CYANIDES
Labosept → DEQUALINIUMS, *chloride*
Labrocol → LABETALOL
Labstix® → MESNA
Laburnine → CYTISINE
Laburnum → CYTISUS LABURNUM also → CAULOPHYLLINE
Laburnum anagyroides → CYTISUS LABURNUM
Labyrin → CINNARIZINE
Lac → SHELLAC
LAC-43 → BUPIVACAINE
Lacalmin → SPIRONOLACTONE
Lacca → SHELLAC
Lacdene → SPIRONOLACTONE
LACIDIPINE
Lacipil → LACIDIPINE
Lacirex → LACIDIPINE
Laco → BISACODYL
Lacotein → PROTEIN HYDROLYSATES
LACQUER also → PAINT
Lacretin → CLEMASTINE
β-LACTAMASE
Lactamin → PRENYLAMINE LACTATE
LACTARIUS sp.
LACTIC ACID also → AMSACRINE, CABBAGE, SODIUM BICARBONATE, SUCROSE
Lacticare-HC → HYDROCORTISONE
Lactinium → OROTIC ACID
Lactobacillus casei growth factor → FOLIC ACID
Lactobacillus lactis Dorner Factor → VITAMIN B_{12}
Lactoflavin → VITAMIN B_2
Lactogen → LUTEOTROPIN
LACTOSE also → CHEESES, FLURAZEPAM, GALACTOSE, MILK, NEOMYCIN, PHENYTOIN, YOGURT

LACTUCA sp.
Lactuca sativa → LETTUCE
Lactucarium-German → LACTUCA sp.
Lactuflor → LACTULOSE
LACTULOSE
Lacumin → MEPAZINE
Ladakamycin → 5-AZACYTIDINE
La Dama Blanca → COCAINE
Ladogal → DANAZOL
Ladropen → PENICILLINS – "NEWER and SYNTHETICS" > *Floxacillin*
Lady → COCAINE
Lady Slipper → PAPHIOPEDILUM HAYNALDIANUM
Lady's Slippers → CYPRIPEDIUM sp.
Lady's-thumb → POLYGONUM sp.
LAETRILE also → CELERY, LETTUCE
Laevilac → LACTULOSE
Laevoral → FRUCTOSE
Laevosan → FRUCTOSE
Laevoxin → THYROXINE
LAF → INTERLEUKINS
LAGOCEPHALUS sp.
LAGOCHILLUS INEBRIANS
La Grande Peure → ERGOT
Laitue Scariole → LETTUCE
Laktu → MUDULEA SERICEA
Lamar → TEGAFUR
Lamasine → AMSACRINE
LAMB
Lambda-coniceine → CONIUM MACULATUM
Lambert Loco → OXYTROPIS
Lambkill → KALMIA
Lamb Mint → PEPPERMINT
Lambral → TOLAZOLINE HCL
Lambratene → CYSTEAMINE
Lamb's quarter → CHENOPODIUM
Lamictal → LAMOTRIGINE
Lamidon → IBUPROFEN
LAMINARIA
Lamisil → TERBINAFINE
LAMIUM AMPLEXICAULE
LAMIVUDINE
Lamoparin → DANAPAROID
Lamoryl → GRISEOFULVIN
LAMOTRIGINE also → IMIPRAMINE, SODIUM VALPROATE, VALPROIC ACID
Lampit → NIFURTIMOX
Lampocef → CEFAZOLIN SODIUM
Lamp Oil → KEROSENE
Lampomandol → CEFAMANDOLE
Lampona cylindrata → SPIDERS
Lampren(e) → CLOFAZIMINE
LAMPS
Lampterol → LAMPTEROMYCES JAPONICUS
LAMPTEROMYCES JAPONICUS
Lamra → DIAZEPAM
Lamsporin → CEFUROXIME
Lamuran → AJMALICINE
Lanacorden → DIGOXIN
Lanacart → HYDROCORTISONE
Lanain → LANOLIN
Lanalin → LANOLIN
LANATOSIDE C
Lanatoxin → DIGITOXIN
Land Cress → CRESS
Landomycin → OLEANDOMYCIN
LANDRIN
Landruma → NIFLUMIC ACID
Landsen → CLONAZEPAM
Lanesin → LANOLIN
Lanethyl → LANOLIN
Lanex → FLUOMETURON
Lanexate → FLUMAZENIL
Langoran → ISOSORBIDE DINITRATE
Langwood → CHROMATED COPPER ARSENATE
Laniazid → ISONIAZID
Lanichol → LANOLIN
Lanicor → DIGOXIN
Lanimersk → DESLANOSIDE also → LANATOSIDE C
Laniol → LANOLIN
Lannate → METHOMYL
LANOLIN also → CYPERMETHRIN, LIPSTICK
Lanophyllin → THEOPHYLLINE
Lanoxin → DIGOXIN
LANSOPRAZOLE also → CLOZAPINE, FLUVOXAMINE, ITRACONAZOLE, KETOCONAZOLE
Lantanadene A → LANTANA sp.
LANTANA sp.
Lantanon → MIANSERIN
LANTERNS
LANTHANUM
Lanum → LANOLIN
Lanvis → THIOGUANINE
Lanzac → OLANZEPINE
Lanzor → LANSOPRAZOLE
Lapachic Acid → LAPACHOL
LAPACHOL
Lapaquin → CHLOROQUINE
LAPORTEA sp.
Lappa → BURDOCK
Lara → CASTOR BEAN
Laracor → OXPRENOLOL HYDROCHLORIDE

Laractone → SPIRONOLACTONE
Laradopa → L-DOPA
Laraflex → NAPROXEN
Larapam → PIROXICAM
Laratrim → SULFAMETHOXAZOLE
Lard Factor → VITAMIN A
Largactil → CHLORPROMAZINE
Largaktyl → CHLORPROMAZINE
Largeleaf Lantana → LANTANA sp.
Large Tobacco → NICOTIANA sp.
Largon → PROPIOMAZINE
Lariam → MEFLOQUINE HCL
Lariat → ALACHLOR
Laridal → ASTEMIZOLE
Larixin → CEPHALEXIN
LARKSPUR
Larocin → PENICILLINS – "NEWER and SYNTHETICS" > *Amoxicillin*
Laroscorbine → VITAMIN C
Larotid → PENICILLINS – "NEWER and SYNTHETICS" > *Amoxicillin*
Laroxyl → AMITRIPTYLINE
LARREA sp.
Larvacide 100 → CHLOROPICRIN
Larvadex → CYROMAZINE
LAS → LYSINE ACETYLSALICYLATE
LAS 90 → EBASTINE
LASALOCID also → TIAMULIN
Laser → NAPROXEN
Laserdil → ISOSORBIDE DINITRATE
LASERS
Lasilix → FUROSEMIDE
LASIOCARPINE also → HELIOTROPIN, PYRROLIZIDINES
LASIOSIPHON sp.
LASIOSPERMUM BIPINNATUM
Lasix → FUROSEMIDE
Lasma → THEOPHYLLINE
Lasso → ALACHLOR
Lastet → ETOPOSIDE
LAS W90 → EBASTINE
LAS W-090 → EBASTINE
LAT 1717 → PINAVERIUM BROMIDE
Latamoxef → MOXALACTAM
LATANOPROST
LATEX also → BARIUM SULFATE, CELERY, TOMATO
LATHYRUS sp. also → CYANIDES, HYDROGEN CYANIDE
Latibon → DIETHAZINE
Laticort → HYDROCORTISONE
α-Latrotoxin → SPIDERS
LATUA PUBIFLORA
Laudanum → OPIUM
Laudicon → HYDROMORPHONE
Lau Fagufagu → MERYTA sp.
"Laughing and Dancing Mushrooms" → PHOLIOTA sp.
Laughing Fungus → PANAEOLUS sp.
Laughing Gas → NITROUS OXIDE
Laughing Mushroom → PANAEOLUS sp.
LAUREL
Laurel Rosa → OLEANDER
Laurels → KALMIA
Laureth 4,9, and 23 → POLYOXYETHYL LAURYL ETHERS
LAURIC ACID
Laurier Rose → OLEANDER
Lauromacrogol 400 → POLYOXYETHYL LAURYL ETHERS
Lauromicina → ERYTHROMYCIN ESTOLATE
Laurose → OLEANDER
Laurus nobilis → BAY LEAF also → LAUREL
Lauryl Alcohol → 1-DODECANOL
LAURYL GALLATE also → MARGARINE
LAURYL ISOQUINOLINIUM BROMIDE
LAURYL OXYPROPYLAMINOBUTYRIC ACID
LAVANDIN OIL
Lavema → OXYPHENISATIN
LAVENDER
Lavendula sp. → LAVENDER
Laxadin → BISACODYL
Laxagetten → BISACODYL
Laxanin N → BISACODYL
Laxin → PHENOLPHTHALEIN
Laxo-Isatin → OXYPHENISATIN
Laxonalin → BISOXATIN ACETATE
Laxorex → BISACODYL
Lazo → ALACHLOR
LB 46 → PINDOLOL
LB 502 → FUROSEMIDE
LBJ → PHENCYCLIDINE
LC 44 → FLUPENTIXOL
LCR → VINCRISTINE
LD 2480 → PIPROCURARINE
Le 100 → TABUM
LEA 103 → ROPIVACAINE
LEAD also → ALCOHOL, ETHYL; ALUMINUM ACETATE, CHILLIES, CHUIFONG TOUKUWAN, COAL, ENAMEL, FISH, FLOUR, FRUIT and FRUIT JUICES; FUELS, OIL; GALENA, GASOLINE, GLASS, HEMATOPORPHYRIN, INKS, KASARAKOFFER YELLOW, KUSHTAS, MASCARA, MILK, OATS, OPIUM, PAINT, PAPER, PENTETIC ACID, PUTTY, SAUERKRAUT, SELENIUM, SOLDER,

SURMA, TETRAETHYL LEAD, TETRAMETHYL LEAD, THALLIUM, TOBACCO, TOMATO, WHISKEY, WINE
LEAD ACETATE also → OPIUM, SOLDER
LEAD ARSENATE
Lead Azide → SODIUM AZIDE
LEAD CARBONATE also → COCHINEAL
LEAD CHLORIDE
LEAD CHROMATE also → CHILLIES, PAPER
Lead Monoxide → LEAD OXIDES
LEAD OXIDES also → GRETA, LINSEED
Lead Protoxide → LEAD OXIDES
Lead Sulfide → LEAD ACETATE, MASCARA, SURMA
Lead Tetraethyl → TETRAETHYL LEAD
Lead Tetroxide → LEAD OXIDES
Leaf Green → CHROMIC OXIDE
Leafy Spurge → EUPHORBIA, *esula*
Lealgin → PHENOPERIDINE
Leanol → HEXOPRENALINE
Leanstar → GROWTH HORMONES
LEATHER
Leatherstem → JATROPHA sp.
Leatherwood → DIRCA PALUSTRIS
Lebake → CANNABIS
Lebaycid → FENTHION
Leben → YOGURT
Leblon → PIRENZEPINE HYDROCHLORIDE
Lebrufen → IBUPROFEN
Lecasol → CLEMASTINE
Lecedil → FAMOTIDINE
Lecibral → NICARDIPINE
LECITHIN also → COTTONSEED OIL, FATS, GRAMICIDIN, METAPROTERENOL, SOYBEAN OIL
Lecithol → LECITHIN
Lectopam → BROMAZEPAM
LECYTHIS
Ledclair → CALCIUM DISODIUM EDETATE
Ledemol → FLUOCINONIDE
Ledercillin → PENICILLIN > *Benzylpenicillin, Procaine*
Ledercort → TRIAMCINOLONE
Lederfen → FENBUFEN
Lederglib → GLYBURIDE
Lederkyn → SULFAMETHOXYPYRIDAZINE
Lederlon → TRIAMCINOLONE
Ledermycin → DEMECLOCYCLINE
Lederpam → OXAZEPAM
Lederplatin → CISPLATIN
Lederspan → TRIAMCINOLONE
Ledertrexate → METHOTREXATE
Ledon 112 → 1,1,2,2-TETRACHLORODIFLUORETHANE
Ledons → FREONS
Ledopa → L-DOPA
Ledopur → ALLOPURINOL
LEDUM sp.
Lefax → DIMETHICONE (*with silicon dioxide*)
LEFLUNOMIDE
"Legal High" → PHENYLPROPANOLAMINE
"Legal Speed" → CAFFEINE
Legend → SODIUM HYALURONATE
Legumes → RAFFINOSE
Leiormone → VASOPRESSIN
Leipzig Yellow → LEAD CHROMATE
Leiurus → SCORPIONS
Lekobacyn → PENICILLINS – "NEWER and SYNTHETICS" > *Bacampicillin*
Le-Kuo → DIMETHOATE
Lemarone → CITRAL
Lemascorb → VITAMIN C
Lemazide → BENZTHIAZIDE
Lembrol → DIAZEPAM
Lemiserp → RESERPINE
Lemoflur → SODIUM FLUORIDE
LEMON also → FRUIT and FRUIT JUICES, LIME, ISOPIMPINELLIN, SYNEPHRINE, TEA
Lemonade → LEMON also → MILK, *N*-NITROSODIMETHYLAMINE
Lemonol → GERANIOL
Lempav D-Lay → PAPAVERINE
Lemu → LIME
Lendorm → BROTIZOLAN
Lendormin → BROTIZOLAN
Lenidolor → MECLOFENAMIC ACID
Lenirit → HYDROCORTISONE
Lenisarin → BISETHYLXANTHOGEN
Lenitral → NITROGLYCERIN
Lenopect → PIPAZETHATE
LenoxiCaps → DIGOXIN
Lenoxin → DIGOXIN
Lensan A → HYDROGEN PEROXIDE
Lens culinaris → LENTILS
Lens esculenta → LENTILS
Lente → INSULIN
Lenticillin → PENICILLIN > *Benzylpenicillin, Procaine*
Lentil d'Espagne → LATHYRUS sp.
LENTILS also → PEANUTS
Lentin → CARBACHOL
Lentizol → AMITRIPTYLINE
Lentobetic → PHENFORMIN
Lentogest → 17-α-HYDROXYPROGESTERONE
Lento-Kalium → POTASSIUM CHLORIDE
Lentonitrina → NITROGLYCERIN

Lentopen → PENICILLIN > *Benzylpenicillin, Procaine*
Lentosone → PREDNISOLONE
Lentotran → CHLORDIAZEPOXIDE
Lentrat → PENTAERYTHRITOL TETRANITRATE
Lenzacef → CEPHADRINE
Leo 640 → LOFEPRAMINE
Leocillin → PENICILLIN > *Benzylpenicillin, DAEEH*
Leocortal → PREDNISOLONE
Leodrine → HYDROFLUMETHIAZIDE
Leofungine → PECILOCIN
LEONOTIS NEPETIFOLIA
Leopard's Bane → ARNICA also → DORONICUM sp.
Leostesin → LIDOCAINE
Leparan → HEPARIN
Lepargylic Acid → AZELAIC ACID
LEPIDIUM sp.
Lepidolite → MICA
Lepinal → PHENOBARBITAL
LEPIOTA sp.
Lepiota morgani → CHLOROPHYLLUM MOLYBDITES
LEPIRUDIN
LEPISOSTEUS SPATULA
Leponex → CLOZAPINE
Leptanal → FENTANYL
Leptazol → PENTYLENETETRAZOL
Leptazolum → PENTYLENETETRAZOL
Leptilan → SODIUM VALPROATE also → VALPROIC ACID
LEPTOMEDUSAE
Lepton → LEPTOPHOS
LEPTOPHOS
Lergefin → CARBINOXAMINE MALEATE
Lergigan → PROMETHAZINE
Lergine → PROCYCLIDINE
Leritine → ANILERIDINE
Leron → GUANACLINE SULFATE
Lertus → KETOPROFEN
Lescodil → NICARDIPINE
Lescol → FLUVASTATIN
Lesser Periwinkle → VINCA ROSEA
Lestid → COLESTIPOL
Lethal® → PENTOBARBITAL
LETHANES
Lethelmin → PHENOTHIAZINE
Lethidrone → NALORPHINE
LETROZOLE
Let's Have Healthy Children → POTASSIUM CHLORIDE
Letter → THYROXINE
LETTUCE also → NITRATES and NITRITES, SCLEROTINIA SCLEROTIORUM
Lettuce Opium → LETTUCE
Leu → LEUCINE
LEUCAENA sp.
Leucarsone → CARBARSONE
Leucenol → MIMOSINE
LEUCINE also → INSULIN, SORGHUM
Leucomax → GRANULOCYTE MACROPHAGE COLONY STIMULATING FACTOR
Leucomycin A_3 → JOSAMYCIN
LEUCOTHOE
Leucovorin → FOLINIC ACID
Leukaemomycin C → DAUNORUBICIN
Leukeran → CHLORAMBUCIL
Leukerin → MERCAPTOPURINE
Leukine → GRANULOCYTE MACROPHAGE COLONY STIMULATING FACTOR
Leukomycin → CHLORAMPHENICOL
Leukopenin → PENICILLINS – "NEWER and SYNTHETICS" > *Methicillin Sodium*
Leuna → ALLYL ISOTHIOCYANATE
Leuplin → LEUPROLIDE
LEUPROLIDE
Leuprorelin → LEUPROLIDE
Leurocristine → VINCRISTINE
Leustatin → CLADRIBINE
Leuteotropic Hormone → LUTEOTROPIN
Leutrol → ZILEUTON
Levabo → LEVAMPHETAMINE
Levacide → LEVAMISOLE
Levadin → LEVAMISOLE
LEVALBUTEROL
LEVALLORPHAN TARTRATE also → MEPERIDINE
LEVAMISOLE also → PHYSOSTIGMINE, WARFARIN
LEVAMPHETAMINE
Levanil → ECTYLUREA
Levanxene → TEMAZEPAM
Levanxol → TEMAZEPAM
Levaquin → LEVOFLOXACIN
Levarterenol → NOREPINEPHRINE
Levasole → LEVAMISOLE
Levatol → PENBUTOLOL SULFATE
Levaxin → THYROXINE
Levium → DIAZEPAM
LEVOCARNITINE
Levius → ACETYLSALICYLIC ACID
Levodopa → L-DOPA
Levo-Dromoran → LEVORPHANOL TARTRATE
LEVOFLOXACIN

Levomeprazine → METHOTRIMEPRAZINE
Levomepromazine → METHOTRIMEPRAZINE
Levomycetin → CHLORAMPHENICOL
Levomycetina → CHLORAMPHENICOL
Levomycin → ECHINOMYCIN
LEVONANTRADOL
Levonor → LEVAMPHETAMINE
Levonordefrin → NORDEFRIN
LEVONORGESTREL also → CYCLOSPORINE, FLUCONAZOLE
Levopa → L-DOPA
Levophed → NOREPINEPHRINE
Levoprom → METHOTRIMEPRAZINE
Levoprome → METHOTRIMEPRAZINE
Levopropylcillin Potassium → PENICILLINS – "NEWER and SYNTHETICS" > *Propicillin Potassium*
Levorenin → EPINEPHRINE
Levorphan → LEVORPHANOL TARTRATE
LEVORPHANOL TARTRATE also → AMMONIUM CHLORIDE, SODIUM BICARBONATE
Levothroid → THYROXINE
Levothyrox → THYROXINE
Levothyroxine → THYROXINE
Levsin → *l*-HYOSCYAMINE
Levugen → FRUCTOSE
Levulose → FRUCTOSE
LEWISITE
Lexibiotico → CEPHALEXIN
Lexinor → NORFLOXACIN
Lexomil → BROMAZEPAM
Lexone → METRIBUZIN
Lexotan → BROMAZEPAM
Lexotanil → BROMAZEPAM
LF-178 → FENOFIBRATE
LH-RH
Li → OAKS
Libanil → GLYBURIDE
Libigen → GONADOTROPINS, CHORIONIC
Libritabs → CHLORDIAZEPOXIDE
Librium → CHLORDIAZEPOXIDE
Licarpin → PILOCARPINE
LICHENS
Licorice → GLYCYRRHIZA GLABRA
Lidanar → MESORIDAZINE BESYLATE
Lidanil → MESORIDAZINE BESYLATE
Lidepran → PHACETOPERANE
Lidex → FLUOCINONIDE
LIDOCAINE also → CIMETIDINE, DEBRISOQUIN, DECAMETHONIUM BROMIDE, EPINEPHRINE, MEXILETINE, OXYTETRACYCLINE, PARABENS; PENICILLIN > *Benzylpenicillin, Procaine;* PHENYTOIN, PRENYLAMINE LACTATE, PROPAFENONE, PROPRANOLOL, PROTRIPTYLINE, QUINIDINE, TERTACAINE HYDROCHLORIDE, TOCAINIDE, *d*-TUBOCURARINE CHLORIDE
LIDOFLAZINE
Lidone → MOLINDONE
Lidothesin → LIDOCAINE
Liefcort → HYDROCORTISONE
Lierre → HEDERA HELIX
Lierre Terrestre → NEPETA HEDERACEA
Life → PYRITINOL DIHYDROCHLORIDE
Life Iron → IRON DEXTRANS
Lifene → PHENSUXIMIDE
Life-Savers® → METHYL SALICYLATE also → ETHER, ETHYL
Lifril → TEGAFUR
Liga → GRETA also → LEAD, LEAD OXIDES
Lightning → AMPHETAMINES also → BROMADIOLONE
Lignasan BLP → METHYL BENZIMIDAZOLE-2-CARBAMATE
Lignocaine → LIDOCAINE
Lignocaine Benzyl Benzoate → DENATONIUM BROMIDE
Ligroin → NAPHTHA
LIGUSTRUM VULGARE also → OLIVES
Likuden → GRISEOFULVIN
Lillium sp. → LILIES
LILIES
Lilly → MONOLINURON, TRIFLURALIN
Lilly 20,025 → CLORPRENALINE
Lilly 29,060 → VINBLASTINE
Lilly 31,814 → HETERONIUM BROMIDE
Lilly 33,379 → FLURANDRENOLIDE
Lilly 35,483 → CYCLOTHIAZIDE
Lilly 36,352 → TRIFLURALIN
Lilly 37,231 → VINCRISTINE
Lilly 38,489 → NORTRIPTYLINE
Lilly 67,314 → MONENSIN
Lilly 69,323 → FENOPROFEN CALCIUM
Lilly 99,094 → VINDESINE SULFATE
Lilly 99,638 → CEFACLOR
Lilly 109,514 → NABILONE
Lily of the Valley → CONVALLARIA MAJALIS also → CALCIUM OXALATE
Lima Beans → BEANS also → CYANIDES
Liman → TENOXICA
Limarsol → ACETARSONE
Limas → LITHIUM CARBONATE
Limbatril → AMITRIPTYLINE
Limbatrol → AMITRIPTYLINE
Limberleg → GUAIJILLO

Limbial → OXAZEPAM
LIME also → CALCIUM OXIDE, FRUIT and FRUIT JUICES
Lime Flowers → TILIA sp.
Limene → BISABOLENE
LIMESTONE also → NAPHTHA
d-LIMONENE also → LEMON, LIME, ORANGE, TEA
Limonite → FERRIC HYDROXIDE
LIMU-MAKE-O-HANA also → PALYTHOA TOXICA
LINALOOL also → LAVANDIN OIL
LINALYLS
LINAMARIN also → CASSAVA, LINSEED, PHASEOLUTIN, TAPIOCA
LINARIA VULGARIA
Linaris → SULFAMETHOXAZOLE
Lincocin → LINCOMYCIN
Lincolcina → LINCOMYCIN
Lincolnesin → LINCOMYCIN
Lincomix → LINCOMYCIN
LINCOMYCIN also → AZATHIOPRINE, BISMUTH, CALCIUM CYCLAMATE, CYCLAMATES, KAOLIN, ERYTHROMYCIN, PECTIN, QUINIDINE
Lindagan → LINDANE
LINDANE also → CARBOXIN, BENZENE HEXACHLORIDE, PARATHION, PHENYLBUTAZONE
Lindatox → LINDANE
Linden → TILIA sp.
Lindenbluten → TILIA sp.
Lindflor → LINDANE
Linodoxine → LINOLEIC ACID
Lindol → TRITOLYL PHOSPATE
LineRider → 2,4,5-TRICHLOROPHENOXYACETIC ACID
LINEZOLID
Lingraine → ERGOTAMINE TARTRATE
Lingran → ERGOTAMINE TARTRATE
Link, Dr. Paul → CLOVERS
Linodil → INOSITOL NIACINATE
LINOLEIC ACID
Linolic Acid → LINOLEIC ACID
Linoral → ETHINYL ESTRADIOL
LINSEED also → CYANIDES, FLAX, HYDROGEN CYANIDE, LINAMARIN
Lintex → NICLOSAMIDE
Linton → HALOPERIDOL
Lintox → LINDANE
Linum sp. → LINSEED
Linum usitatissimum → FLAX
Linurex → LINURON
LINURON
Linyl → PHENTERMINE
Liomycin → DOXYCYCLINE
Lionfish → PTEROIS VOLITANS
Lion's Ear → LEONURUS CARDIACA
Lion's Mane → JELLYFISH
Lion's Tail → LEONURUS CARDIACA
Lioresal → BACLOFEN
LIOTHYRONINE also → CARBAMAZEPINE, PUROMYCIN, SALSALATE
Lipamone → DIENESTROL
Lipan → DINITRO-σ-CRESOL
Lipanthyl → FENOFIBRATE
Lipantil → FENOFIBRATE
Liparon → DEANOL, *bitartrate*
LIPASE
Lipavlon → CLOFIBRATE
Lip-Dox → DOXORUBICIN
Liphadione → CHLOROPHACINONE
Lipidil → FENOFIBRATE
Lipiodol → ETHIODIZED OIL
Lipiphysan → COTTONSEED OIL also → FATS
Lipitor → ATORVASTATIN
Liple → ALPROSTADIL
Lipoclar → FENOFIBRATE
Lipofene → FENOFIBRATE
Lipofundin → FATS
Lipohepin → HEPARIN
Lipo-Hepinette → HEPARIN
α-Lipoic Acid → THIOCTIC ACID
Lipolase → LIPASE
Lipo-Lutin → PROGESTERONE
Lipomul → FATS
Liponorm → SIMVASTATIN
Liposec → POLYOXYETHYL LAURYL ETHERS
Liposit → FENOFIBRATE
Lipostat → PRAVASTATIN
Liposyn → SAFFLOWER OIL
Lipozid → GEMFIBROZIL
LIPPIA REHMANNI
Liprinal → CLOFIBRATE
Lipsin → FENOFIBRATE
Lip Smackers® → LIPSTICK also → PROPYLENE GLYCOL
LIPSTICK also → METHYL HEPTINE CARBONATE
Liptan → IBUPROFEN
Lipur → GEMFIBROZIL
Liquaemin → HEPARIN
Liquamar → PHENPROCOUMON
Liquamycin → OXYTETRACYCLINE
Liqua-Tox → WARFARIN
Liquémin → HEPARIN
Liquidambar → TEA

Liquid Ectasy → HYDROXYBUTYRATE
Liquid G → HYDROXYBUTYRATE
Liquid Glucose → CORN SYRUP
Liquid Paraffin → MINERAL OIL
Liquid Petrolatum → MINERAL OIL
Liquid Plumber® → SODIUM HYDROXIDE
Liquid Silver → MERCURY
Liquifene → PHENYLMERCURIC ACETATE .
Liquified Petroleum Gas → LPG
Liquifilm → POLYVINYL ALCOHOL
Liquimeth → METHIONINES
Liquiprin → SALICYLAMIDE
Liquiprim → ACETAMINOPHEN
Liquiviton → CARBENOLOXONE
Liquor Arsenicalis → POTASSIUM ARSENITE SOLUTION
Liquor epispasticus → CANTHARIDIN
Liquor Potassi Arsenitis → POTASSIUM ARSENITE SOLUTION
Liranol → PROMAZINE
Lironox → 2,4-DICHLOROPHENOXYACETIC ACID
Lisacef → CEPHADRINE
Lisacortin → PREDNISONE
Lisagal → OXYPHENISATIN
Lisaglucon → GLYBURIDE
Lis des Incas → ALSTROEMERIA LIGTU
Liserdol → METERGOLINE
Liseron → MORNING GLORY
Lisino → LORATADINE
LISINOPRIL also → LITHIUM, LOVASTATIN, PERGOLIDE MESYLATE
Lisium → CHLORHEXIDINE
Liskantin → PRIMIDONE
Liskonum → LITHIUM CARBONATE
Lispamol → AMINOPROMAZINE
Lispine → DISOPYRAMIDE
Lissephen → MEPHENESIN
Lissolamine V → CETRIMONIUM BROMIDE
Lister → PENICILLIUM
Listica → HYDROXYPHENAMATE
LISURIDE also → ERYTHROMYCIN
Litalir → HYDROXYUREA
Litarex → LITHIUM CITRATE
LITCHI
Litchu chinensis → LITCHI
Lithane → LITHIUM CARBONATE
Litharge → LEAD OXIDES
Lithicarb → LITHIUM CARBONATE
Lithionit → LITHIUM SULFATE
Lithiophor → LITHIUM SULFATE
LITHIUM also → ACETAZOLAMIDE, ACYCLOVIR; ALCOHOL, ETHYL; BACLOFEN, BROMFENAC, CARBAMAZEPINE, CHLOROTHIAZIDE, CHLORPROMAZINE, CHLORTETRACYCLINE, CLOMIPRAMINE, DICLOFENAC, DIGOXIN, ETHACRYNIC ACID, FLUOXETINE, FLUPHENAZINE, FLUVOXAMINE, FUROSEMIDE, GELATIN, HALOPERIDOL, HYDROCHLOROTHIAZIDE, IBUPROFEN, INDOMETHACIN, KETOPROFEN, KETOROLAC, LISINOPRIL, MAZINDOL, MEFENAMIC ACID, METOCLOPRAMIDE, METOLAZONE, METRONIDAZOLE, MILK, PHENYLBUTAZONE, PHENYTOIN, POTASSIUM IODIDE, PROPANTHELINE, SAUNAS, SODIUM BICARBONATE, SODIUM CHLORIDE, SPECTINOMYCIN, SPIRONOLACTONE, SUCCINYLCHOLINE, SULINDAC, TETRACYCLINE, THEOPHYLLINE, THIORIDAZINE, TIAPROFENIC ACID, TRIAMTERENE, VERAPAMIL
LITHIUM ACETATE
Lithium Antimony Thiomalate → ANTHIOLIMINE
LITHIUM CARBONATE also → AMPHETAMINE, BENDROFLUMETHIAZIDE, CLOZAPINE, DEXTROAMPHETAMINE, DIAZEPAM, DOPAMINE, IODINE, LISINOPRIL, METHYLDOPA, NAPROXEN, PANCURONIUM BROMIDE, PIROXICAM, SERTRALINE, ZIDOVUDINE
LITHIUM CHLORIDE also → FIRE COLORING SALTS, PHYSOSTIGMINE
LITHIUM CITRATE
Lithium Hydrate → LITHIUM HYDROXIDE
LITHIUM HYDRIDE
LITHIUM HYDROXIDE
LITHIUM SULFATE
Lithnate S → LITHIUM CITRATE
Lithobid → LITHIUM CARBONATE
LITHOCHOLIC ACID
Lithol Rubine BCA → LIPSTICK
Lithonate → LITHIUM CARBONATE
LITHOSPERMUM RUDERALE
Lithostat → ACETOHYDROXAMIC ACID
Lithotabs → LITHIUM CARBONATE
Liticon → PENTAZOCINE
Little Barley → HORDEUM JUBATUM
Little Blue Stagger → DICENTRA CUCULLARIA
Little Devil → COCAINE
Littleleaf Horsebrush → TETRADYMA sp.

Livazone → THIACETAZONE
LIVER also → ISOCARBOXAZID, NIALAMIDE
Liverleaf → HEPATICA sp.
Liverwort → HEPATICA sp. also → AGRIMONIA EUPATORIA
Liviclina → CEFAZOLIN SODIUM
Livid Agaric → ENTOLOMA
Lixidol → KETOROLAC
Lixil → BUMETANIDE
Lizard Shade → TETRADYMA sp.
LJC 10,846 → ZALEPLON
LLD Factor → VITAMIN B_{12}
Lloncefal → CEPHALORIDINE
Llonexina → ÇEPHALEXIN
LM 91 → CHLOROPHACINONE
LM 280 → IOCARMATE MEGLUMINE
LM 427 → RIFABUTIN
LM 637 → BROMADIOLONE
LM 2717 → CLOBAZAM
LM 5008 → INDALPINE
LMD → DEXTRAN, *40*
LMWD → DEXTRAN, *40*
Loads → CODEINE
Lobak → CHLORMEZANONE
Lobamine → METHIONINES
LOBELIA also → TEA
LOBELINE
LOBSTER also → SHRIMP
Lobster Flower → POINSETTIA
Lobufen → IBUPROFEN
Locacorten → FLUMETHASONE
Localyn → FLUOCINOLONE ACETONIDE
Locapred → DESONIDE
Locibis → BEPHENIUM
LOCHNERA PUSILLA
Lockaroma → AMYL NITRITE also → BUTYL NITRITE
Locker Room → AMYL NITRITE also → BUTYL NITRITE, ISOBUTYL NITRITE
Locoid → HYDROCORTISONE
Locoine → OXYTROPIS
Locorten → FLUMETHASONE
Locoweed → ASTRAGALUS also → DATURA
Locula → SULFACETAMIDE SODIUM
Locust Bean Gum → CAROB
LOCUST, BLACK also → ROBIN
Lodalès → SIMVASTATIN
Lodema → COUMARIN
Lodine → ETODOLAC
Loditac → OXIBENDAZOLE
Lodosin → CARBIDOPA
Lodosyn → CARBIDOPA
LODOXAMIDE TROMETHAMINE
LOFENTANIL
LOFEPRAMINE
Lofetensin → LOFEXIDINE
LOFEXIDINE
Loftran → KETAZOLAM
Logastric → OMEPRAZOLE
Logiparin → HEPARIN
Lokalison F → DEXAMETHASONE
Loliine → RYE GRASS
LOLIUM TEMULENTUM also → FLOUR
Lolium sp. → RYE GRASS
LOLLIGUNCULA BREVIS
Lomadine → PHENAMIDINE ISETHIONATE
Lombristop → THIABENDAZOLE
Lomebact → LOMEFLOXACIN
LOMEFLOXACIN also → SUCRALFATE
Lomexin → FENTICONAZOLE
Lomidine → PENTAMIDINE ISETHIONATE
Lomir → ISRADIPINE
Lomoparan → DANAPAROID
Lomotil → DIPHENOXYLATE (*with atropine sulfate*)
Lomper → MEBENDAZOLE
Lomudal → CROMOLYN SODIUM
Lomupren → CROMOLYN SODIUM
Lomusol → CROMOLYN SODIUM
Lomuspray → CROMOLYN SODIUM
LOMUSTINE also → THEOPHYLLINE
LON 798 → GUANFACINE
Lonacol → ZINEB
Lonavar → OXANDROLONE
LONCHOCARPUS sp.
Londomin → SYROSINGOPINE
Londomycin → METHACYCLINE
Long, Dr. Crawford W. → ETHER, ETHYL
Longasa → ACETYLSALICYLIC ACID
Longastatin → OCTREOTIDE
Longatin → NOSCAPINE
Longdigox → DIGOXIN
Longheparin → HEPARIN
Longicid → DIETHYLCARBAMAZINE
Longicil → PENICILLIN > *Benzylpenicillin, Benzathine*
Longifene → BUCLIZINE
Longum → SULFALENE
LONICERA CAPRIFOLEUM
Lonicera sp. → HONEYSUCKLE
Loniten → MINOXIDIL
Lonolox → MINOXIDIL
"Looney Gas" → LYSERGIC ACID DIETHYLAMIDE
Lopantrol → LORCAINIDE
Lopemid → LOPERAMIDE
Lopemin → LOPERAMIDE

LOPERAMIDE
Loperyl → LOPERAMIDE
LOPHOGORGIA sp.
LOPHOPETALUM
Lophophora → PEYOTE
Lophotoxin → LOPHOGORGIA sp.
Lopid → GEMFIBROZIL
LOPINAVIR
Lopirin → CAPTOPRIL
Lopramine → LOFEPRAMINE
Lopremone → THYROTROPIN RELEASING HORMONE
Lopresor → METOPROLOL
Lopress → HYDRALAZINE
Lopressor → METOPROLOL
Lopril → CAPTOPRIL
Loprox → CICLOPIROX
Lopurin → ALLOPURINOL
LORATADINE also → ERYTHROMYCIN, KETOCONAZOLE
Lorax → LORAZEPAM
LORCAINIDE
LORAZEPAM also → ALCOHOL, ETHYL; POLYETHYLENE GLYCOL, PROBENECID, PROPYLENE GLYCOL, VALPROIC ACID
Lorchel → GYROMITRA ESCULENTA
Lord Adrian → FLUOROACETIC ACID
Lords → HYDROMORPHONE
Lords and Ladies → ARISAEMA
Lorelco → PROBUCOL
Lorestat → TOLRESTAT
Lorexane → LINDANE
Lorfan → LEVALLORPHAN TARTRATE
Loridine → CEPHALORIDINE
Lorinal → CHLORAL HYDRATE
Lorinden → FLUMETHASONE
Lorivox → LORCAINIDE
Lormin → CHLORMADINONE
Loromisin → CHLORAMPHENICOL
Loron → DICHLOROMETHYLENE DIPHOSPHONATE
Lorophyn → PHENYLMERCURIC ACETATE
Loroquin → URECHITES sp.
Lorothidol → BITHIONOL
Lorox → LINURON
Loroxide → BENZOYL PEROXIDE
Lorphen → CHLORPHENIRAMINE, *maleate*
Lorsban → CHLORPYRIFOS
Lorsilan → LORAZEPAM
Losalen → FLUMETHASONE
Losec → OMEPRAZOLE
LOSARTAN POTASSIUM also → FLUCONAZOLE, RIFAMPIN, ZAFIRLUKAST
Lospoven → CEPHALOTHIN
Lost → MUSTARD GAS
Lotanstralin → CASSAVA
Lotensin → BENAZEPRIL
Loticort → FLUOROMETHOLONE
Lotrial → ENALAPRIL
LOTRIFEN
Lotrimin → CLOTRIMAZOLE
Lotronex → ALOSETRON
Loturine → HARMAN
Lotusate → BUTALBITAL
LOTUS CORNICULATUS
Loubarb → BUTABARBITAL
Lovalip → LOVASTATIN
LOVASTATIN also → AZITHROMYCIN, CHOLESTIN, CLOZAPINE, ERYTHROMYCIN, GEMFIBROZIL, GRAPEFRUIT, ITRACONAZOLE, LISINOPRIL, NEFAZODONE, PRAVASTATIN, PROPRANOLOL, THYROXINE, WARFARIN
Love Apple → TOMATO
"Love Drug" → METHYLENEDIOXYAMPHETAMINE
Lovenox → HEPARIN also → ENOXAPARIN
Love Pill → METHYLENEDIOXYAMPHETAMINE
Loverine → DEXAMETHASONE
Low Larkspur → LARKSPUR
Lowpston → FUROSEMIDE
Lowquel → DIPHENOXYLATE (*with atropine sulfate)*
Loxacor → LOFEXIDINE
Loxapac → LOXAPINE
LOXAPINE also → PHENYTOIN, SUMATRIPTAN SUCCINATE
Loxen → NICARDIPINE
Loxitane → LOXAPINE
Loxon → HALOXON
Lozol → INDAPAMIDE
LPG
LRF → LH-RH
LRH → LH-RH
LS 121 → NAFRONYL OXALATE
LS 519 → PIRENZEPINE HYDROCHLORIDE
LT 1 → SUCCINYLCHOLINE
L-tetramisole → LEVAMISOLE
LTG → LAMOTRIGINE
LTH → LUTEOTROPIN
LTX → LOPHOGORGIA sp.
LU 5-110 → FLUPENTIXOL
LU 10-171 → CITALOPRAM
LU 1631 → AMEZINIUM METHYL SULFATE

Lubimin → POTATO
LUCANTHOE HYDROCHLORIDE
Lucelan → BUSPIRONE
Lucerne → ALFALFA
Lucidil → BENACTYZINE
Lucidol → BENZOYL PEROXIDE
Lucidril → MECLOFENOXATE
Lucifer Disease → PHOSPHORUS
Lucite® → METHYL METHACRYLATE
Lucky Bean → ABRUS PRECATORIUS also → ABRIN
Lucky Clover → OXALIS sp.
Lucky Nut → THEVETIA sp.
Lucofen → CHLORPHENTERMINE
Lucorteum → PROGESTERONE
Lucosil → SULFAMETHIZOLE
Lucozade → GLUCOSE
Lucretius → NUTS, PEACHES
Lucrin → LEUPROLIDE
Lucro → HOPS
Ludes → METHAQUALONE
Ludiomil → MAPROTILINE
Ludobal → QUINURONIUM
LUFENURON
Lufyllin → DYPHYLLINE
Lü H6 → TOXOGONIN
Lu Hui → ALOE
Lukadin → AMIKACIN
Lulamin → THALIDOMIDE
Luliberin → LH-RH
Lullamin → METHAPYRILENE
LUMBANG NUT also → TUNG TREE
Lumbriareu → PYRANTEL
Lumbriconereis heteropoda → NEREISTOXIN
Lumecycline → LYMECYCLINE
Luminal → PHENOBARBITAL
Luminaletas → PHENOBARBITAL
Lumirelax → METHOCARBAMOL
Lumitens → XIPAMIDE
Lunamin → HEXAPROPYMATE
LUNANIA PARVIFLORA
Lunarette → MIFEPRISTONE
Lunetoron → BUMETANIDE
Lunis → FLUNISOLIDE
Lunkulwe → ARISTOLOCHIA
Lupersol → DIMETHYL PHTHALATE
LUPINES also → PHOMOPSIS sp.
Lupinine → LUPINES
Lupins → LUPINES
Lupinus sp. → LUPINES
Lupron → LEUPROLIDE
Luprosil → PROPIONIC ACID
Lupulo → HOPS
Lurat → COUMAFURYL
Luride → SODIUM FLUORIDE
Luronase → HYALURONIDASE
Lurselle → PROBUCOL
LU-SHEN-WAN also → ARSENIC TRIOXIDE
Lustral → SERTRALINE
Lutal → LH-RH
Luteine → PROGESTERONE
Luteinizing Hormone-Relasing Hormone → LH-RH
Luteogan → PROGESTERONE
Luteogonin B → GONADOTROPINS, CHORIONIC
Luteohormone → PROGESTERONE
Luteomersin → PROGESTERONE
Luteonorm → ETHYNODIOL DIACETATE
Luteosan → PROGESTERONE
LUTEOSKYRIN also → PENICILLIUM sp., RICE
Luteostab → PROGESTERONE
Luteosterone → PROGESTERONE
LUTEOTROPIN also → MELATONIN
Luteran → CHLORMADINONE
Lutin → PROGESTERONE
Lutocyclin → ETHISTERONE
Lutocyclin M → PROGESTERONE
Lutocylol → ETHISTERONE
Lutoform → PROGESTERONE
Lutogyl → PROGESTERONE
Luto-Metrodiol
Lutoral → CHLORMADINONE also → MEDROXYPROGESTERONE ACETATE
Lutormone → GONADOTROPINS, CHORIONIC
Lutren → PROGESTERONE
Lutrol E → POLYETHYLENE GLYCOL
Lutromone → PROGESTERONE
Luvatrene → MOPERONE
Luviscol → POLYVINYLPYRROLIDONE
Luvox → FLUVOXAMINE MALEATE
Luxazone → DEXAMETHASONE
Luxon → HALOXON
Luzerne → ALFALFA
LVD → DEXTRAN, *40*
LW 2159 → CLOTHIAPINE
LY 061188 → CEPHALEXIN
LY 099094 → VINDESINE SULFATE
LY 110,140 → FLUOXETINE
LY 127,809 → PERGOLIDE MESYLATE
LY 127,935 → MOXALACTAM
LY 139,037 → NIZATIDINE
LY 139,481 → RALOXIFENE
LY 156,758 → RALOXIFENE
LY 170,053 → OLANZEPINE
LY 177,370 → TILMICOSIN

LY 177,837 → GROWTH HORMONES
LY 188,011 → GEMCITABINE
LY 237,216 → DIRITHROMYCIN
Lycanol → GLYMIDINE
Lychee → LITCHI
Lychnis githago → AGROSTEMMA GITHAGO
LYCIUM HALIMIFOLIUM
Lyclear → PERMETHRIN
LYCOPERDON also → MUSHROOMS
Lycopersicon esculentum → TOMATO
Lycopsamine → PYRROLOIZIDINES also → ANSACRINE, PARSONSIA sp.
Lycoremine → GALANTHAMINE
Lycorin → NARCISSUS sp.
Lycorine → CLIVIAS MINIATA, GALANTHUS NIVALUS, HYMENOCALLIS OCCIDENTALIS, NARCISSUS sp.
Lycosine → AMARYLLIS
Lycra → POLYURETHANES
Lye → SODIUM HYDROXIDE also → POTASSIUM HYDROXIDE
Lyman → ALOXIPRIN
Lymantria dispar → GYPSY MOTH
LYMECYCLINE
Lymphazurin → ISOSULFAN BLUE
Lymphocyte Mitogenic Factor → ALDESLEUKIN
Lynamine → KHELLIN
LYNESTRENOL also → PHENYTOIN
LYNGBYA sp.
Lyngbyatoxin A → LYNGBYA sp.
Lynoral → ETHINYL ESTRADIOL
Lyobex → NOSCAPINE
Lyogen → FLUPHENAZINE DECANOATE
Lyogen Depot → FLUPHENAZINE DECANOATE
Lyomousse → POLYURETHANES
LYONIA sp.
LYOPHYLLUM CONGOBATUM
Lypor 20 → TEMEPHOS
LYPRESSIN
Lysal → LYSINE ACETYLSALICYLATE
Lysalgo → MEFENAMIC ACID
Lysanxia → PRAZEPAM
Lysedem → COUMARIN
LYSERGAMIDE also → ARGYREIA, MORNING GLORY, RIVEA CORYMBOSA
LYSERGIC ACID DIETHYLAMIDE also → ALCOHOL, ETHYL; CANNABIS, CLAVICEPS, MESCALINE, PARTHENIUM, PROGESTERONE, SUCCINYLCHOLINE
d-Lysergic Acid Amides → MORNING GLORY
Lysergide → LYSERGIC ACID DIETHYLAMIDE also → VITAMIN B_3
Lysergin → ERGOLOID MESYLATES
Lysinate → LISINOPRIL
LYSINE also → SUNFLOWER
LYSINE ACETYLSALICYLATE
LYSINOALANINE
Lysivane → ETHOPROPAZINE
Lysodem → COUMARIN
Lysodren → MITOTANE
Lysol® → CRESOL, SAPONATED SOLUTION of also → SOAP
LYSOZYME
Lysuride → LISURIDE
Lysuron → ALLOPURINOL
Lyteca → ACETAMINOPHEN
LYTECHINUS VARIEGATUS
Lytispasm → ANISOTROPINE METHYLBROMIDE
LYUTIK

M

2-M-4-A → ESTIL
4M20 → 4-(*p*-DIMETHYLAMINOSTYRYL) QUINOLINE
M 14 → RIFAMIDE
M 50-50 → DIPRENORPHINE
M 141 → SPECTINOMYCIN
683-M → OXOLAMINE
M 1028 → HALOPROGIN
M 3179 → PICLORAM
M 4888 → CHLOROGUANIDE
5512 M → ANTAZOLINE
6029 M → BUPRENORPHINE
M 7555 → QUINAPYRAMINE
M → MORPHINE
MA → 4-METHOXYAMPHETAMINE also → METHYL ANTHRANILATE
Ma → CANNABIS
MAA → METHANEARSONIC ACID
Maalox® → GABAPENTIN
Mab Thera → RITUXIMAB
MACADAMIA NUTS
Macasirool → FUROSEMIDE
Mace → SAFROLE
Machete → BUTACHLOR
MACHINE OILS
Machony → CANNABIS
Macho Ya → ABRUS PRECATORIUS
"MacKenzie river disease" → TERMINALIA sp.
MACKEREL
Maclura pomifera → OSAGE ORANGE
Macmiror → NIFURATEL
Macocyn → OXYTETRACYCLINE

Maconha → CANNABIS
Macrabin → VITAMIN B_{12}
Macrobid → NITROFURANTOIN
Macroclemys temmincki → TURTLES
Macrodantin → NITROFURANTOIN
Macrodex → DEXTRAN, *70*
Macrodiol → ESTRADIOL
Macrogestin → PROGESTERONE
Macrogol → POLYETHYLENE GLYCOL
Macromycin → PENICILLIN > *Penicillin V*
Macrophylline → PYRROLIZIDINES
Macrose → DEXTRAN
MACROZAMIA sp.
Macrozamine → MACROZAMIA sp.
Maculotoxin → TETRODOTOXIN also → OCTOPUS
Madagascar Periwinkle → VINCA ROSEA
Madagascar Rubber Vine → CRYPTOSTEGIA GRANDIFLORA
MAD → METHANDRIOL
Mad Apple → DATURA
Madar → NORDAZEPAM also → CALOTROPIN
"Mad as a Hatter" → MERCURIC NITRATE
Madashuanda → NUTMEG
Madecassol → CENTALLA ASIATICA
Madelen → ORNIDAZOLE
Madlexin → CEPHALEXIN
Madribon → SULFADIMETHOXINE
Madrine → METHAMPHETAMINE
Madzun → YOGURT
2 MAE → 2-METHYLAMINOETHANOL
Maerman → URGINEA sp.
MAERU ANGOLENSIS
MAESA sp.
MAFENIDE
Magcal → MAGNESIUM OXIDE
Magenta → *p*-ROSANILINE
Maghen → PIRENZEPINE HYDROCHLORIDE
Magic Methyl → METHYL FLUOROSULFONATE
Magic Mist → PHENCYCLIDINE
Maginfa → TEPHROSIA sp.
Magje → MUSK MELON
Maglite → MAGNESIUM OXIDE
Magma Magnesii → MAGNESIUM HYDROXIDE
Magmilor → NIFURATEL
Magnamycin → CARBOMYCIN
Magnesia → MAGNESIUM OXIDE
Magnesia Alba → MAGNESIUM CARBONATE
Magnesite → MAGNESIUM CARBONATE
MAGNESIUM also → AMMONIUM CHLORIDE, CHLOROTHIAZIDE, *d*-TUBOCURARINE CHLORIDE, CLORAZEPATE DIPOTASSIUM, GREPAFLOXACIN, METRONIDAZOLE, NORFLOXACIN, OFLOXACIN, OXYTETRACYCLINE, RUBBER, STREPTOMYCIN, SULFUR DIOXIDE, TETRACYCLINE, WHEAT
MAGNESIUM BROMOGLUTAMATE
MAGNESIUM CARBONATE also → HALOFANTRINE
MAGNESIUM CHLORIDE also → MAGNESIUM HYDROXIDE
MAGNESIUM CITRATE also → CIPROFLOXACIN
MAGNESIUM HYDROXIDE also → BISHYDROXYCOUMARIN, CHLORDIAZEPOXIDE, CHLORPROMAZINE, DEMECLOCYCLINE, INDOMETHACIN, LOMEFLOXACIN, PENICILLAMINE, PREDNISONE, TROVAFLOXACIN
MAGNESIUM OXIDE also → NITROFURANTOIN, POLYBROMINATED BIPHENYLS
Magnesium Pemoline → PEMOLINE
Magnesium Perhydrol → MAGNESIUM PEROXIDE
MAGNESIUM PEROXIDE
MAGNESIUM SILICATE also → COCHINEAL, DUSTS, GLOVES, MEGESTROL ACETATE, METHYLPHENIDATE, RICE
MAGNESIUM STEARATE
MAGNESIUM SULFATE also → BETAMETHASONE, DIMETHYL TUBOCURARINE IODIDE, NIFEDIPINE
MAGNESIUM TRISILICATE also → FOLIC ACID; PENICILLINS – "NEWER and SYNTHETICS" > *Amoxicillin, Ampicillin;* PHENYTOIN, PYRIMETHAMINE
MAGNOLIAS
Magox → MAGNESIUM OXIDE
Magrene → DIETHYLPROPION
Magrilon → MAZINDOL
Magsalyl → SODIUM SALICYLATE
Mah$_v$-coe^v → SWEET POTATO
MAHONIA AQUIFOLIUM
Ma-Huang → EPHEDRINE
Mah-Jong Dermatitis → LACQUER, RENGAS
Maidenhair → GINKGO BILOBA
Maikohi → CLEMASTINE
Maimonides → ALCOHOL, ETHYL

Maintate → BISOPROLOL
Maipedopa → L-DOPA
Maitotoxin → GAMBIERDISCUS TOXICUS
Maiwa → PENNISETUM sp.
Maize → CORN
Majudon → METHOXSALEN
Makarol → DIETHYLSTILBESTROL
Makaya ma ngu → ALCHORNEA CORDIFOLIA
Ma-khaam → TAMARIND
Maki → BROMADIOLONE
Mako → SOLANUM sp.
Makrocef → CEFOTAXIME
Malabar Tree → EUPHORBIA
Malachite Green G → BRILLIANT GREEN
Malagride → ACETARSONE
Malamar → MALATHION
Malamute → ANONA
Malanga → XANTHOSOMA sp.
Malaphos → MALATHION
Malaquin → CHLOROQUINE
Malaren → CHLOROQUINE
Malarex → CHLOROQUINE
Malaspray → MALATHION
MALATHION also → CARBOPHENOTHION, TRI-σ-TOLYL PHOSPHATE
Malatrol → MALATHION
Malayan Camphor → BORNEOL
Malay Tree → PSORALEA sp.
Malcotran → HOMATROPINE
"Mal do eucalipto" → RAMARIA FLAVO-BRUNNESCENS
Male Blueberry → LYONIA sp.
Malee → EUCALYPTUS GLOBULUS
Male Fern → ASPIDIUM
Male Shield Fern → ASPIDIUM
MALEIC ANHYDRIDE
MALEIC HYDRAZIDE also → TOBACCO
Malformin C → ASPERGILLUS
Maliasin → PROPYLHEXEDRINE ETHYLPHENYLBARBITURATE
Malilum → ALLOBARBITAL
Malimyxin → COLISTIN
Malipuran → BUFEXAMAC
Malix → ENDOSULFAN also → GLYBURIDE
Mallermin-F → CLEMASTINE
Mallorol → THIORIDAZINE
Malocide → PYRIMETHAMINE
Malonal → BARBITAL
MALONALDEHYDE
Malonic Aldehyde → MALONONITRILE
MALONONITRILE
Malophene → PHENAZOPYRIDINE
MALOUETIA sp.
MALT
Maltoryzine → ASPERGILLUS
Malt Workers → BARLEY
"Malt Worker's Lung" → MALT
Malum punicum → POMEGRANATE
Malvalic Acid → MILK also → OKRA
Malysol → BEMEGRIDE
Mamalexin → CEPHALEXIN
Mamallet-A → AMINOPYRINE
Mamba → SNAKE(S)
Mamiesan → DICYCLOMINE
Mammacillin → PENICILLIN > *Benzylpenicillin, Procaine*
Mammex → NITROFURAZONE
Mammotropin → LUTEOTROPIN
Mamushi → WINE
Mamyzin → PENICILLIN > *Benzylpenicillin, DAEEH*
Manaca → BRUNFELSIA
ma-na-shu-kema → PAGAMEA MACROPHYLLA
MANCHINEEL TREE
Mandacon → METHENAMINE
Mandalay → METHENAMINE
Mandamina → METHENAMINE
Mandelamine → METHENAMINE
MANDELIC ACID also → PROBENECID
Mandelonitrile → AMELANCHIA ALNIFOLIA
Mandelonitrile-β-D-gentiobioside → AMYGDALIN
Mandelonitrile Glucuronide → LAETRILE
MANDEVILLA sp.
Mandioca → CASSAVA
Mandokef → CEFAMANDOLE
Mandol → CEFAMANDOLE
Mandolsan → CEFAMANDOLE
Mandoz → METHENAMINE
MANDRAGORA OFFICINARUM
Mandragoro → PODOPHYLLUM
Mandrake → MANDRAGORA OFFICINARUM also → GINSENG
Mandrake Root → PODOPHYLLUM
Mandrin → EPHEDRINE
Mandrogorine → MANDRAGORA OFFICINARUM
Mandrozep → DIAZEPAM
Mandurin → METHENAMINE
MANEB
Manerix → MOCLOBEMIDE
MANGANESE also → COBALT, POTASSIUM PERMANGATE, WELDING
MANGANESE CYCLOPENTADIENYLTRICARBONYL

MANGANESE DIOXIDE
Manganese EBDC → MANEB
Manganese Ethylenebisdithiocarbamate → MANEB
MANGANESE OXIDE
Manganese Peroxide → MANGANESE DIOXIDE
Mangifera indica → MANGO
MANGO also → ANACARDIACEA, POISON IVY, RENGAS; SUMAC, POISON
Mangold → MANGEL
Manicol → MANNITOL
MANIDIPINE
Manihot esculenta → CASSAVA
Maninil → GLYBURIDE
Manioc → CASSAVA
Maniol → DIPHENIDOL
Mann, Dr. George → LOVASTATIN
Manna → TILIA sp.
Manna Sugar → MANNITOL
Mannidex → MANNITOL
Mannite → MANNITOL
MANNITOL also → SODIUM
Mannitol Nitrogen Mustard → MANNOMUSTINE
Mannitol of Mustard → MANNOMUSTINE
MANNOMUSTINE
Manoplax → FLOSEQUINAN
Manroot → MARAH sp.
Mansil → OXAMNIQUINE
MANSONIA ALTISSIMA also → STERCULIA sp.
Mansonil → NICLOSAMIDE
Mantadix → AMANTADINE
Manta Ray Dermatitis → SHARK
MANTIS SHRIMP
Man T'o Lo → DATURA
Mantropine → METHENAMINE
Manuril → HYDROCHLOROTHIAZIDE
ma-nu-su'-ka-ta → PAGAMEA MACROPHYLLA
Manzanila → CHAMOMILE
MANZANILLO
Manzidan® → MANEB
MAOA → METHAQUALONE
Maolate → CHLORPHENESIN CARBAMATE
Mapasol → METHAM SODIUM
Mapharsal → OXOPHENARSINE
Mapharsen → OXOPHENARSINE
Mapharside → OXOPHENARSINE
Mapiprin → PIPERAZINE
MAPLE
Maple Bark Disease → PAPER
Maple Syrup → MAPLE also → PERILLA FRUTESCENS
MAPO → METEPA
Mappine → BUFOTENINE
Maprofix → SODIUM LAURYL SULFATE
MAPROTILINE also → GLYBURIDE
MAQUIRA sp.
MARAH sp.
MARASMUS OREADES
Maratan → BISOXATIN ACTATE
Marazine → CHLORPROMAZINE
Marbate → MEPROBAMATE
MARBLE
MARBOFLOXACINE
Marboron → METHISAZONE
Marcaina → BUPIVACAINE
Marcaine → BUPIVACAINE
Marc Antony → DATURA
Marcoeritrex → ERYTHROMYCIN ESTOLATE
Marcol → MINERAL OIL
Marcumar → PHENPROCOUMON
Mares Tail → EQUISETUM
Maretin → PHTHALOPHOS
Marevan → WARFARIN
Marezine → CYCLIZINE
Marfanil → MAFENIDE
MARGARINE also → LAURYL GALLATE, *N*-NITROSODIMETHYLAMINE
Margarite → MICA
Marguerite → CHRYSANTHEMUM
Margosa Oil → NIM OIL
Maria Luisa → GRETA also → LEAD OXIDES
Marie Antoinette → BREAD
MARIGOLD also → ASTERS, DATURA, SUNFLOWER, TEA
Marihuana → CANNABIS
Marijuana → CANNABIS
Marimara → MELIA sp.
Marinol → Δ^{-9} TETRAHYDROCANNABINOL
MARJORAM
Marketote → POLYETHYLENE
Markweed → POISON IVY
Marlate → METHOXYCHLOR
Marlex → POLYETHYLENE
Marmite → YEAST
Marogen → ERYTHROPOIETIN, *EPOETIN*
Marplan → ISOCARBOXAZID
Marriage Vine → SOLANUM sp.
Marronier → AESCULUS sp.
Marsh Clover → BUCKBEAN
MARSCH ELDER also → ELDERBERRY, IVA sp.
Marsh Gas → METHANE
Marsh Horsetail → EQUISETUM
Marsh Marigold → CALTHA PALUSTRIS
Marsh Penny → CENTALLA ASIATICA
Marsh Trefoil → BUCKBEAN

Marsh Turnip → ARISAEMA
MARSILEA DRUMMONDII
Marsilid → IPRONIAZID
Marsin → PHENMETRAZINE
Marsthine → CLEMASTINE
Martin's Rat Stop Liquid → THALLIUM
Martius Yellow → 2,6-DINITRONAPHTHOL
Marucotol → MECLOFENOXATE
Marvate → DIETHYLPROPION
Marvinol → POLYVINYL CHLORIDE
Mary Jane → CANNABIS
Mary Walkers → CANNABIS
Marzine → CYCLIZINE
MASCAGNIA sp.
Mascagnite → AMMONIUM SULFATE
MASCARA
Maschitt → HYDROCHLOROTHIAZIDE
Masdil → DILTIAZEM
Masdiol → METHANDRIOL
Mashihiri → JUSTICA PECTORALIS var STENOPHYLLA
Masho-hara → JUSTICA PECTORALIS var STENOPHYLLA
Masletine → CLEMASTINE
Masmoran → HYDROXYZINE
Masoor dhal → LENTILS
MASOPROCOL
Masoten → TRICHLORFON
Massala → TURMERIC
Massicot → LEAD OXIDES
Masterfen → SUPROFEN
Masterid → DROMOSTANOLONE PROPIONATE
Masteril → DROMOSTANOLONE PROPIONATE
Masterone → DROMOSTANOLONE PROPIONATE
MASTIC
Mastiphen → CHLORAMPHENICOL
Mastisol → MASTIC
Mastranto → LEONOTIS NEPETIFOLIA
Mastuergo Hembra → CORONOPUS sp.
Matarique → SENECIO sp.
Match → METHALLIBURE
MATCHES
Maté → ILEX sp. also → TEA, THEOPHYLLINE
Mat Grass → POLYGONUM sp.
Matacil → AMINOCARB
Matarique → CACALIA DECOMPOSITA
Matenon → MILBOLERONE
Matricaria chamomilla → CHAMOMILE
Matrimony Vine → LYCIUM HALIMIFOLIUM also → SOLANUM sp.
Matrin → PENICILLIN > *Benzylpenicillin, Potassium*
Matrol → CHLORMADINONE
Matromycin → OLEANDOMYCIN
Matulane → PROCARBAZINE
Matur → PEAS
Maturin → SENECIO sp.
Maui Waui → CANNABIS
Mauve → KETAMINE
Maveral → FLUVOXAMINE MALEATE
Maxaquin → LOMEFLOXACIN
Maxeran → METOCLOPRAMIDE
Maxforce → AMDRO
Maxibolic → ETHYLESTRENOL
Maxicam → ISOXICAM
Maxicholine → LECITHIN
Maxidex → DEXAMETHASONE
Maxifen → PENICILLINS – "NEWER and SYNTHETICS" > *Pivampicillin*
Maxiflor → DIFLORASONE
Maximed → PROTRIPTYLINE
Maxipen → PENICILLINS – "NEWER and SYNTHETICS" > *Phenethicillin Potassium*
Maxipime → CEFEPIME
Maxiton → DEXTROAMPHETAMINE SULFATE
Maxivate → BETAMETHASONE
Maxolon → METOCLOPRAMIDE
Maxulvet → SULFADIMETHOXINE
Max-Uric → BENZBROMARONE
May Apple → PODOPHYLLUM
Maycor → ISOSORBIDE DINITRATE
Maygace → MEGESTROL ACETATE
May Lily → CONVALLARIA MAJALIS
May-Pop → PASSIFLORA sp.
MAYTANSINE
Maytenus ovatus → MAYTANSINE
Mazanillo → MANCHINEEL TREE
Mazanor → MAZINDOL
Mazicon → FLUMAZENIL
Mazildene → MAZINDOL
MAZINDOL also → BETHANIDINE, LITHIUM
M&B 693 → SULFAPYRIDINE
M&B 744 → STILBAMIDINE
M&B 760 → SULFATHIAZOLE
M&B 782 → PROPAMIDINE
M&B 800 → PENTAMIDINE ISETHIONATE
M&B 2050A → PENTOLINIUM TARTRATE
M&B 2207 → SUCCINYLCHOLINE
M&B 4180A → ISOMETAMIDIUM
M&B 4486 → PEMPIDINE TARTRATE
M&B 5062A → AMICARBALIDE

M&B 8430 → CLOREXOLONE
M&B 9302 → CLORGYLINE
M&B 15,497 → DECOQUINATE
M&B 17,803A → ACEBUTOLOL
M&B 39,831 → TEMOZOLOMIDE
MB 46,030 → FIPRONIL
MBA → *N,N´*-METHYLENE-bis-ACRYLAMIDE also → MECHLORETHAMINE
Mbaazi -mwitu → SOPHORA sp.
MBCP → LEPTOPHOS
Mberge Musir → PHYSOSTIGMA VENENOSUM
MBH → PROCARBAZINE
MBK → METHYL BUTYL KETONE
MBOCA → 4,4´-METHYLENE-bis [2-CHLOROANILINE]
Mbokwe → ANONA
MBT → 2-MERCAPTOBENZOTHIAZOLE
MBTH → *3*-METHYL-2-BENZOTHIAZOLINONE HCL
MC 903 → CALCIPOTRIENE
MCA → CHLOROACETIC ACID
MCA 600 → MOBAM
McAac → 2-AMINO-3-METHYL-9H-PYRIDO(2,3,-b) INDOLE
MCE → TABUM
MCH → METHYLCYCLOHEXANE
Mchicha → AMARANTHUS
Mchoki → STROPHANTHUS sp.
M-cillin → PENICILLIN > *Benzylpenicillin, Potassium*
McN 485 → ZOXAZOLAMINE
McN 1025 → NORBORMIDE
McN 2559-21-98 → TOLMETIN
McN 4853 → TOPIRAMATE
McNeil Labs → ZOMEPIRAC
McN JR 1625 → HALOPERIDOL
McN JR 2498 → TRIFLUPERIDOL
McN JR 4263 → FENTANYL
McN JR 4584 → BENPERIDOL
McN JR 4729 → DROPERIDOL
McN JR 7094 → LIDOFLAZINE
McN JR 8299 → TETRAMISOLE
McN R 726-47 → POLDINE METHYLSULFATE
McN X 94 → CAPURIDE
McN X 181 → VALNOCTAMIDE
MCP → METHYL CHLOROPHENOXY ACETIC ACID
MCPA → METHYL CHLOROPHENOXY ACETIC ACID
MCPB → *4(*2-METHYL-4-CHLOROPHENOXY) BUTYRIC ACID
MCPP → MECOPROP
MCT → MONOCROTALINE
MD 141 → ETHAMSYLATE
MD 516 → CINNARIZINE
MD 69,276 → TOLOXATONE
MDA → *4,4´*-METHYLENE DIANILINE also → METHYLENEDIOXYAMPHETAMINE also → 4,4´DIAMINODIPHENYLMETHANE
Mdagu → ADENIUM sp.
M-Det → *N,N*-DIETHYL-*m*-TOLUAMIDE
MDI → *p,p*-DIPHENYLMETHANE DIISOCYANATE
MDL 507 → TEICOPLANIN
MDL 9918 → TERFENADINE
MDL 14,042A → LOFEXIDINE
MDL 16,455A → FEXOFENADINE
MDL 71,754 → VIGABATRIN
MDMA → METHYLDIAMINE METHAMPHETAMINE also → METHYLENEDIOXYAMPHETAMINE
5-MDMT → 5-METHOXYDIMETHYLTRYPTAMINE
MDP-Lys(L18) → MUROCTASIN
Mdudu → THYLACHIUM AFRICANUM
2-ME → 2-METHOXYETHANOL
ME 1700 → 1,1-DICHLORO-2,2-BIS(*p*-CHLOROPHENYL) ETHANE
MEA → CYSTEAMINE
2 MEA → METHOXYETHYL ACETATE
Meadow-cress → ANTHEMIS COTULA
Meadow Pine → EQUISETUM, *arvense*
Meadow Saffron → COLCHICUM AUTUMNALE
Measurin → ACETYLSALICYLIC ACID
Meat → *N*-NITROSODIMETHYLAMINE, OLEANDER
"Meat-wrapper's Asthma" → BIS(2-ETHYLHEXYL)PHTHALATE; 2,5-DI-*tert*-AMYLQUINONE; DICYCLOHEXYL PHTHALATE, POLYVINYL CHLORIDE
Meaverin → MEPIVACAINE
Mebacid → SULFAMERAZINE
Mebadin → DEHYDROEMETINE
Meballymal Sodium → SECOBARBITAL SODIUM
Mebanazine → CHEESES, MEPERIDINE, L-TRYPTOPHAN
Mebaral → MEPHOBARBITAL
Mebendacin → MEBENDAZOLE
MEBENDAZOLE also → CIMETIDINE, THIABENDAZOLE

Mebenvet → MEBENDAZOLE
Meberyt → ERYTHROMYCIN STEARATE
Mebethisol → 2-MERCAPTOBENZOTHIAZOLE
Mebethysol → 2-MERCAPTOBENZOTHIAZOLE
Mebonat → DICHLOROMETHYLENE DIPHOSPHONATE
Me Br → METHYL BROMIDE
Mebubarbital → PENTOBARBITAL
Mebumal → PENTOBARBITAL
Mebumalon → PENTOBARBITAL
Mebutar → MEBENDAZOLE
3-MECA → 3 or 20-METHYLCHOLANTHRENE
Mecadox → CARBADOX
Mecalmin → DIPHENIDOL
MECAMYLAMINE also → ACETAZOLAMIDE, AMMONIUM CHLORIDE, CHLOROTHIAZIDE, ETHACRYNIC ACID, PSEUDOEPHEDRINE, SODIUM BICARBONATE, SPIRONOLACTONE, TRIAMTERENE
MECHLORETHAMINE
Mecholine → METHACHOLINE
Mecholyl → METHACHOLINE CHLORIDE
Mechothane → BETHANECHOL
Meciclin → DEMECLOCYCLINE
Mecilex → CEPHALEXIN
Mecillinam → PENICILLINS – "NEWER and SYNTHETICS" > *Amdinocillin*
Meclan → MECLOCYCLINES
Meclastine → CLEMASTINE
MECLIZINE also → ALCOHOL, ETHYL; NORCHLORCYCLIZINE
MECLOCYCLINES
Mecloderm → MECLOCYCLINES
Meclodol → MECLOFENAMIC ACID
Meclofenamate Sodium → MECLOFENAMIC ACID also → PROBENECID
MECLOFENAMIC ACID
Meclofenoxane → MECLOFENOXATE
MECLOFENOXATE
Meclomen → MECLOFENAMIC ACID
Meclopran → METOCLOPRAMIDE
Mecloprodin → CLEMASTINE
MECLOQUALONE
Meclosorb → MECLOCYCLINES
Meclozine → MECLIZINE
Meclutin → MECLOCYCLINES
Mecodin → METHADONE HCL
Mecodrin → AMPHETAMINE
MECOPROP
MECRYLATE
Mectizan → IVERMECTIN
Mecuram → THIRAM
Mecuroxyl → CHLORMERODRIN
Medaject → PROCAINE
Medamycin → OXYTETRACYCLINE
Medan → BETAHISTINE
Medaron → FURAZOLIDONE
MEDAZEPAM
Medazepol → MEDAZEPAM
MEDETOMIDINE
Mediatensyl → URAPIDIL
MEDIBAZINE
Mediben → DICAMBA
Medicago sativa → ALFALFA
Medicef → CEPHADRINE
Medichol → CHLORAMPHENICOL
Medici → MORCHELLA DELICIOSA
Medicort → HYDROCORTISONE
Medifenac → ALCLOFENAC
Medifuran → FURALTADONE
Mediletten → TETRACYCLINE
Medinal → BARBITAL
Medio-Contrix "38" → IOTHALMATE SODIUM
Medi-Pets → PRIMIDONE
Medipren → IBUPROFEN
Medomin → HEPTABARBITAL
Medoxim → CEFUROXIME
Medrate → METHYLPREDNISOLONE
Medrocort → MEDRYSONE
MEDROGESTONE
Medrol → METHYLPREDNISOLONE
Medrol Stabisol → METHYLPREDNISOLONE
Medrone → METHYLPREDNISOLONE
MEDROXYPROGESTERONE ACETATE
MEDRYSONE
Medusa-head Rye → ELYMUS sp.
Meerzwiebel → SQUILL
Mefedina → MEPERIDINE
Mefenacid → MEFENAMIC ACID
MEFENAMIC ACID also → CHUIFONG TOUKUWAN, LITHIUM, NALIDIXIC ACID, NITROFURANTOIN, TARTRAZINE, WARFARIN
MEFENOREX
Mefentanyl → α-METHYL FENTANYL
Mefexadyne → MEFEXAMIDE
MEFEXAMIDE
MEFLOQUINE HCL also → HALOFANTRINE
Meflosyl → FLUNIXIN MEGLUMINE
Mefoxin → CEFOXITIN SODIUM
MEFRUSIDE

Megabion → METHANDRIOL
Megace → MEGESROL ACETATE
Megacef → CEPHADRINE
Megacert → GABEXATE MESYLATE
Megacillin → PENICILLIN > *Benzylpenicillin, Potassium*
Megaclor → CLOMOCYCLINE
Megalectil → BUTAPERAZINE
Megalocin → FLEROXACIN
Megalone → FLEROXACIN
Megalopyge opercularis → CATERPILLARS
Megapen → PENICILLIN > *Benzylpenicillin, Procaine*
Megaphen → CHLORPROMAZINE
Megasedan → MEDAZEPAM
Megasul → NITROPHENIDE
Megestat → MEGESTROL ACETATE
MEGESTROL ACETATE also → DOFETILIDE, MAGNESIUM SILICATE
Megimide → BEMEGRIDE
MEGLITOL
Meglum → LEVAMISOLE
Meglumine Amidotrizoate → MEGLUMINE DIATRIZOATE
MEGLUMINE DIATRIZOATE also → ETHYLENEDIAMINE
Meglumine Iocarmate → IOCARMATE MEGLUMINE
MEGLUMINE IODIPAMIDE
MEGLUMINE IOTHALAMATE
Megotyl → SULPIRIDE
MEHP → MONOETHYLHEXYL PHTHALATE
Meilax → ETHYL LOFLAZEPATE
ME IQ → 2-AMINO-3,4-DIMETHYLIMIDAZO(4,5-F)QUINOLINE
MEK → BUTANONE
Meladinine → METHOXSALEN
Melaleuca alternifolia → TEA TREE
Melaleuca sp. → CAJEPUT
Melanex → METAHEXAMIDE
MELANIN also → METHOXSALEN
β-Melanocyte-stimulating Hormone → INTERMEDINE
Melanophore-stimulating Hormone → INTERMEDINE
MELANTHIUM
Melapure → MELATONIN
MELARSONYL POTASSIUM
MELARSOPROL
MELATONIN also → PROPRANOLOL
Mel B → MELARSOPROL
Meldane → COUMAPHOS
Meleagris gallapavo → TURKEY
Melfalan → MELPHALAN
Melfax → RANITIDINE
MELIA sp.
MELICA sp.
MELILOT
Melinite → PICRIC ACID
Meliotus sp. → CLOVERS
Melipramine → IMIPRAMINE
Melitase → CHLORPROPAMIDE
Melitol → TOLBUTAMIDE
Melitoxin → BISHYDROXYCOUMARIN also → HONEY
MELITRACEN HCL
MELITTIN
Melixeran → MELITRACEN HCL
Melktou → SARCOSTEMMA sp.
Mellaril → THIORIDAZINE
Melleretten → THIORIDAZINE
Melleril → THIORIDAZINE
Mellow Yellows → LYSERGIC ACID DIETHYLAMIDE also → BANANA
MELOCHIA sp.
Melons → LATEX
Melon Tree → PAPAYA
Melopat → BETAHISTINE
MELOXICAM
Meloxine → METHOXSALEN
MELPHALAN also → INTERFERONS
Melsed → METHAQUALONE
Melsedin → METHAQUALONE
Melsomin → METHAQUALONE
Melter → MORPHINE
Meltrol → PHENFORMIN
Mel W → MELARSONYL POTASSIUM
Melysin → PENICILLINS – "NEWER and SYNTHETICS" > *Piovmecillinam*
MEMANTINE
Memine → PHOLCODINE
Memphenesin → MEPHENESIN
MEMS → METHOXYETHYL MERCURY SILICATE
Menadiol → ACENOCOUMAROL
Menamin → KETOPROFEN
Mendon → CLORAZEPATE DIPOTASSIUM
Mendrin → ENDRIN
Menformon → ESTRONE
Menhydrinate → DIMENHYDRINATE
Menichlopholan → 3,3´-DICHLORO-5,5´-DINITRO-σ,σ-DIPHENOL
MENISPERMUM CANADENSE
Menitazine → BETAHISTINE
Menite → MEVINPHOS
Mennen → VITAMIN E

MENOGARIL
Menogarol → MENOGARIL
Menogen → ESTROGENS, CONJUGATED
Menopatol → TOLPERISONE HCL
Menophase → MESTRANOL
Menotab → ESTROGENS, CONJUGATED
Menrium → ESTROGENS, CONJUGATED also → CHLORDIAZEPOXIDE
MENOTROPINS
Mensiso → SISOMICIN
Menstridyl → CHLORMADINONE
Mentax → BUTENAFINE
Mentha sp. → MINTS
Mentha piperita → PEPPERMINT
Mentha pulegium → PENNYROYAL
p-Menth-8-en-3-ol → ISOPULEGOL
MENTHOL also → MINTS, PROPYLHEXIDRINE, SANGURA
MENTHYL ANTHRANILATE
MENTHYL SALICYLATE
Menyanthes trifoliate → BUCKBEAN
MENZIESIA FERRUGINEA
Menzol → NORETHINDRONE
5-MeO-DMT → 5-METHOXYDIMETHYLTRYPTAMINE
MEP → SUMITHION
Mepacaine → PAMAQUIN
Mepacrine → QUINACRINE
MEPARFYNOL
Mepartricin → NORFLOXACIN
Mepasin → MEPAZINE
Mepavlon → MEPROBAMATE
MEPAZINE
MEPENZOLATE BROMIDE
MEPERIDINE also → ACETAZOLAMIDE, AMMONIUM CHLORIDE, AMPHETAMINE, ATROPINE, CIMETIDINE, CLOMIPRAMINE, DEXTROAMPHETAMINE, DIAZEPAM, DISULFIRAM, FURAZOLIDONE, IPRONIAZID, ISONIAZID, MAGNESIUM OXIDE, MAGNESIUM SILICATE, METARAMINOL BITARTRATE, METHOTRIMEPRAZINE, MEXILETINE, NIALAMIDE, NORMEPERIDINE, PARGYLINE, PHENELZINE, PHENOBARBITAL, PHENOPERIDINE, PHENYTOIN, PROCARBAZINE, PROCHLORPERAZINE, PROMAZINE, PROMETHAZINE, RITONAVIR, SODIUM BICARBONATE, SULFOBROMOPHTHALEIN SODIUM, THIORIDAZINE, TOLBUTAMIDE
15 ME-PGF$_{2a}$ → CARBOPROST
Mepha-Butazon → PHENYLBUTAZONE
Mephacyclin → TETRACYCLINE
Mephenamin → ORPHENADRINE
MEPHENESIN also → PLACEBO
Mephenon → METHADONE HCL
MEPHENOXALONE
MEPHENTERMINE also → METHYLDOPA
MEPHENYTOIN also → FLUVOXAMINE, PHENYLETHYLHYDANTOIN
Mepherol → NEPHENESIN
Mephesin → MEPHENESIN
Mephine → MEPHENTERMINE
MEPHOBARBITAL also → PHENOBARBITAL
Mephruside → MEFRUSIDE
Mephson → MEPHENESIN
Mepicaton → MEPIVACAINE
Mepicor → MEPINDOLOL
Mepident → MEPIVACAINE
MEPINDOLOL
MEPIVACAINE also → NORDEFRIN, PROTRIPTYLINE
Mepivacine → MEPIVACAINE
Mepivastesin → MEPIVACAINE
Meposed → MEPROBAMATE
Mepral → OMEPRAZOLE
Mepred → METHYLPREDNISOLONE
Meprin → MEPROBAMATE
Meprindon → MEPROBAMATE
MEPROBAMATE also → ACENOCOUMAROL, ACETYLSALICYLIC ACID: ALCOHOL, ETHYL; AMINOPYRINE, CARBROMAL, CARISPRODOL, DDT, GLUTETHIMIDE, IMIPRAMINE, MILK, ORPHENADRINE, PENTOBARBITAL, PHENACEMIDE, PHENAGLYCODOL, PHENOBARBITAL, PHENYLBUTAZONE, PHENYTOIN, PRIMIDONE, TRIFLUPROMAZINE, WARFARIN
Meproban → MEPROBAMATE
Meprocompren → MEPROBAMATE
Meprofen → KETOPROFEN
Meprol → MEPROBAMATE
Mepron → ATOVAQUONE also → METHIONINES
Meprospan → MEPROBAMATE
Meprotabs → MEPROBAMATE
Meprotan → MEPROBAMATE
Meproten → MEPROBAMATE
Meprotil → MEPROBAMATE
Meproxol → PROMOXALONE
MEPRYLCAINE
MEPTAZINOL
Meptid → MEPTAZINOL
Meptran → MEPROBAMATE

Mepyramine → PYRILAMINE
Mepyrium → AMPROLIUM
Mequelon → METHAQUALONE
Mequin → METHAQUALONE
Mequinol → 4-METHOXYPHENOL
MER 17 → AZACYCLONAL
MER 25 → ETHAMOXYTRIPHETOL
MER 29 → TRIPARANOL
Meractinomycin → DACTINOMYCIN
Meradan(e) → RIMANTADINE
Meralen → FLUFENAMIC ACID
MERALLURIDE
Meratran → PIPRADOL
Merbentul → CHLOROTRIANISENE
Merbentyl → DICYCLOMINE
MERBROMIN
Mercaleukin → MECAPTOPURINE
Mercamine → CYSTEAMINE
Mercaptamine → CYSTEAMINE
Mercaptoacetic Acid → THIOGLYCOLIC ACID
2-MERCAPTOBENZOTHIAZOLE also → CUTTING OILS, GLOVES, RUBBER
MERCAPTODIMETHUR
Mercaptoethane → ETHANETHIOL
2-Mercaptoethylamine → CYSTEAMINE
MERCAPTOMERIN SODIUM
Mercaptomethane → METHANETHIOL
1-Mercaptonomane → 1-NONETHIOL
Mercaptooctadecane → OCTADECANETHIOL
Mercaptooctane → OCTANETHIOL
Mercaptophos → DEMETON also → FENTHION
1,2,3-Mercaptopropane → DIMERCAPROL
MERCAPTOPURINE also → ALLOPURINOL, AZATHIOPRINE, WARFARIN
6-Mercaptopurine Ribose Phosphate → MERCAPTOPURINE
Mercaptothion → MALATHION
D-*3-Mercaptovaline* → PENICILLAMINE
Mercardan → MERALLURIDE
Mercazole → METHIMAZOLE
Mercazolyl → METHIMAZOLE
Merck → METHYLDOPA
Merck's Digitalin → DIGITALIN
Mercodinone → HYDROCODONE BITARTRATE
Mercuhydrin → MERALLURIDE
Mercuramide → MERSALYL
Mercuranine → MERBROMIN
Mercurescein → MERBROMIN
Mercuretin → MERALLURIDE
Mercurialin → MERCURIALIS ANNUA also → MERCURIALIS PERENNIS
MERCURIALIS ANNUA
MERCURIALIS PERENNIS
MERCURIC ACETATE
MERCURIC AMMONIUM CHLORIDE
MERCURIC CHLORIDE also → POLYVINYL ALCOHOL
Mercuric Chloride, Ammoniated → MERCURY, AMMONIATED
MERCURIC CYANIDE
MERCURIC IODIDE, RED also → MERCUROUS IODIDE, MERCURY
MERCURIC NITRATE also → HATS
MERCURIC OXIDE, YELLOW also → MERCURY, SODIUM HYDROXIDE
MERCURIC OXYCYANIDE
Mercuric Pernitrate → MERCURIC NITRATE
MERCURIC SULFIDE, RED also → MERCURIC CHLORIDE
Mercurocol → MERBROMIN
Mercurome → MERBROMIN
Mercurophage → MERBROMIN
MERCUROPHEN
MERCUROUS CHLORIDE also → TSE KOO CHOY
MERCUROUS IODIDE
MERCURY also → COD LIVER OIL, COTTONSEED MEAL, CORN, EGGS, FISH, LAMPS, MERBROMIN, MERCURIC AMMONIUM CHLORIDE, MERCURIC CYANIDE, METHOXYMETHYL MERCURY CHLORIDE, MILK, PAINT, PAYLOOAH, PHENYLMERCURIC ACETATE, POLYVINYL ALCOHOL, SELENIUM, SHELLFISH, THIMEROSAL, WHALES
Mercury Amide Chloride → MERCURY, AMMONIATED
MERCURY, AMMONIATED also → MERCURY
Mercury Ammonium Chloride → MERCURY, AMMONIATED
Mercury Bichloride → MERCURIC CHLORIDE
Mercury Biniodide → MERCURIC IODIDE, RED
Mercury Herb → MERCURIALIS ANNUA
Mercury Monochloride → MERCUROUS CHLORIDE
Mercury Perchloride → MERCURIC CHLORIDE
Mercury Protochloride → MERCUROUS CHLORIDE
Mercury Protoiodide → MERCUROUS IODIDE
Mercury Subchloride → MERCUROUS CHLORIDE
Mercusal → MERSALYL
Mereprine → DOXYLAMINE SUCCINATE

Meresa → SULPIRIDE
Merfen → PHENYLMERCURIC BORATE
Meridia → SIBUTRAMINE
Merge → MONOSODIUM METHANEARSONATE
Merinax → HEXAPROPYMATE
Merislon → BETAHISTINE
Merit → IMIDACLOPRID
Merital → NOMIFENSINE MALEATE
Meropenin → PENICILLIN > *Penicillin V*
Meroxyl → GENTIAN VIOLET
Meroxylan → GENTIAN VIOLET
Merpan → CAPTAN
Merphen → PHENYLMERCURIC BORATE
Merseptyl → THIMEROSAL
Mersilene → POLYESTER
Mersilon → THIMEROSAL
Mersolite → PHENYLMERCURIC ACETATE
Mertax → 2-MERCAPTOBENZOTHIAZOLE
Mertect → THIABENDAZOLE
Mertestate → TESTOSTERONE
Merthiolate → THIMEROSAL
MERULIUS LACRYMANS
Mervaldin → MEPHENESIN
Mervan → ALCLOFENAC
Mervasine → MECAMYLAMINE
Merxin → CEFOXITIN SODIUM
MERYTA sp.
Merzonin → THIMEROSAL
MESACONITINE
Mesalamine → 5-AMINOSALICYLIC ACID also → WARFARIN
Mesalazine → 5-AMINOSALICYLIC ACID
Mesamate → MONOSODIUM METHANEARSONATE
Mesantoin → MEPHENYTOIN also → DISULFIRAM
Mesasal → 5-AMINOSALICYLIC ACID
Mesc → MESCALINE
Mescal → PEYOTE also → SOPHORA sp., TEQUILA
MESCALINE also → LYSERGIC ACID DIMETHYLAMIDE, PELECYPHORA ASSELLIFORMIS, PEYOTE, TEQUILA
Mesembrine → MESEMBRYANTHEMUM CRYSTALLINUM
MESEMBRYANTHEMUM CRYSTALLINUM
Mesentol → ETHOSUXIMIDE
Mesitylene → 1,3,5-TRIMETHYLBENZENE
MESITYL OXIDE also → ACETONE
Mesmar → MEPROBAMATE
MESNA
Mesnex → MESNA
Mesnum → MESNA
Mesobuthus → SCORPIONS
Mesomile → METHOMYL
Mesonex → INOSITOL NIACINATE
Mesontoin → MEPHENYTOIN
Mesopin → HOMATROPINE
MESORIDAZINE BESYLATE also → BENZTROPINE MESYLATE, PHENOBARBITAL, PHENYTOIN, THIORIDAZINE
Mesotal → INOSITOL NIACINATE
Mesoxalylcarbamide → ALLOXAN
Mesoxalylurea → ALLOXAN
MESQUITE also → CYANIDES
Mesguitegrass → HOLCUS LANATUS
Mestenediol → METHANDRIOL
MESTEROLONE
Mestoranum → MESTEROLONE
MESTRANOL also → PHENOBARBITAL
Mesudin → MAFENIDE
Mesudrin → MAFENIDE
Mesulfa → SULFAMERAZINE
Mesulid → NIMESULIDE
MESUPRINE
Mesural → CHLORDIAZEPOXIDE
Mesurol → MERCAPTODIMETHUR
Mesuximide → METHSUXIMIDE
Mesylith → CHLOROQUINE
Met → METHIONINES
Metabolin → VITAMIN B_1
Metacam → MELOXICAM
Metacetaldehyde → METALDEHYDE
Metacide → METHYL PARATHION
Metacortandracin → PREDNISONE
Metacortandrolone → PREDNISOLONE
Metacresol Acetate → *m*-CRESYL ACETATE
Metacresylacetate → *m*-CRESYL ACETATE
Metacycline → METHACYCLINE
Metadate ER → METHYLPHENIDATE
Metadee → VITAMIN D_2
Metadelphene → *N,N*-DIETHYL-*m*-TOLUAMIDE
Metadomus → METHACYCLINE
Metadorm → METHAQUALONE
METAHEXAMIDE
Metahitin → NORMETHANDRONE
Metahydrin → TRICHLORMETHIAZIDE
Meta-isosystox → DEMETON-METHYL
METALDEHYDE
Metal Fume Fever → WELDING, ZINC
Metalid → ACETAMINOPHEN
Metalkamate → BUFENCARB
Metallibure → METHALLIBURE
Metallum problematum → TELLURIUM
Metam → METHAM SODIUM

Metamide → METOCLOPRAMIDE
Metamin → FLUPENTIXOL
Metamine → TROLNITRATE PHOSPHATE
Metamisole → DIPYRONE
Metamizol → DIPYRONE
Metamphetamine → METHAMPHETAMINE
Metamucil → PLANTAGO SEED
Metandiol → METHANDRIOL
Metandren → METHYL TESTOSTERONE
Metanephrine → BUSPIRONE
METANIL YELLOW
Metanyl → METHANTHELINE BROMIDE
Metaphen → NITROMERSOL
Metaphos → METHYL PARATHION
Metaphyllin → AMINOPHYLLINE
Metaprel → METAPROTERENOL
METAPROTERENOL also → THEOPHYLLINE
METARAMINOL BITARTRATE also → CYCLOPROPANE, HYDROCORTISONE, METHYLDOPA, NIALAMIDE, PHENELZINE, TRANYLCYPROMINE
Metasclene → TRIPARANOL
Metasqualene → TRIPARANOL
Metastab → METHYLPREDNISOLONE
Metastigmin → NEOSTIGMINE
Meta Synephrine → PHENYLEPHRINE HCL
Meta-Systox → METHYL DEMETON
Metathion → SUMITHION
METAXALONE
Metaxan → METHANTHELINE BROMIDE
Metazolo → METHIMAZOLE
Metenarin → METHYLERGONOVINE
Metendiol → METHANDRIOL
Metenex → METOLAZONE
Metenolone → METHENOLONE ACETATE
METEPA
Meterazine → PROCHLORPERAZINE
Meterfer → FERROUS FUMARATE
METERGOLINE
METFORMIN also → CIMETIDINE, FOLIC ACID, MEGLITOL
Meth → METHAMPHETAMINE
METHACHOLINE CHLORIDE
Methacide → TOLUENE
Methacolimycin → COLISTIN
METHACRYLIC ACID
METHACYCLINE also → FERROUS SULFATE
METHADONE also → AMITRIPTYLINE, AMPHETAMINE, CIMETIDINE, FLUOXETINE, FLUVOXAMINE, MAGNESIUM SILICATE, ORANGE, PHENYTOIN, PROPOXYPHENE, RIFAMPIN, RITONAVIR
Methadyl Acetate → ACETYLMETHADOL
Methaform → CHLOROBUTANOL
Methafrone → KHELLIN
Methagon → FLUMETHASONE
Methalutin → NORMETHANDRONE
Methamidophos → ACEPHATE
Methampex → METHAMPHETAMINE
METHAMPHETAMINE also → EPHEDRINE, GUANETHIDINE, HALOTHANE, ISOCARBOXAZID, LEAD, MECAMYLAMINE, METHYLDOPA, PHOSPHORUS, PSEUDOEPHEDRINE
Methampyrone → DIPYRONE
METHAM SODIUM
Methanabol → METHANDRIOL
Methanal → FORMALDEHYDE
Methanamide → FORMAMIDE
Methandienone → METHANDROSTENOLONE
Methandiol → METHANDRIOL
METHANDRIOL
METHANDROSTENOLONE also → BISHYDROXYCOUMARIN, OXYPHENBUTAZONE, PHENYLBUTAZONE, WARFARIN
METHANE also → NEWSPAPERS
METHANEARSONIC ACID
Methanearsonic Acid, Sodium Salt → MONOSODIUM METHANEARSONATE
Methane Carboxylic Acid → ACETIC ACID
METHANETHIOL also → METHANE
Methanide → MENTHANTHELINE BROMIDE
Methanoic Acid → FORMIC ACID
Methanol → ALCOHOL, METHYL
p-Methan-8-ol → DIHYDRO-α-TERPINEOL
METHANTHELINE BROMIDE also → MAGNESIUM TRISILICATE
Methanthine Bromide → METHANTHELINE BROMIDE
p-Methan-8-yl Acetate → DIHYDROTERPINYL ACETATE
Methaphos → METHYL PARATHION
METHAPYRILINE also → CYCLOMETHYCAINE, ETHYLENEDIAMINE, PHENYLPROPANOLAMINE
METHAQUALONE also → ACETIC ANHYDRIDE; ALCOHOL, ETHYL; DIPHENHYDRAMINE, FLUPHENAZINE, NORETHINDRONE, PHOSPHORUS OXYCHLORIDE, PHOSPHORUS TRICHLORIDE, TOLUIDINE

METHARBITAL also → METHSUXIMIDE
Metharbitone → METHARBITAL
Methasan → ZIRAM
Methased → METHAQUALONE
β-Methasone → BETAMETHASONE
Methazol → SULFAMETHIZOLE
METHAZOLAMIDE
Methazolastone → TEMOZOLOMIDE
METHAZOLE
METHDILAZINE
Methedrine → METHAMPHETAMINE
Methetharimide → BEMEGRIDE
METHENAMINE also → ESTRIOL, SULFAMETHIZOLE, SULFISOXAZOLE
METHENOLONE ACETATE
Methergin(e) → METHYLERGONOVINE
METHETOIN
met-HGH → GROWTH HORMONES
Methi → ALFALFA
Methiacil → METHYLTHIOURACIL
Methicil → METHYLTHIOURACIL
Methicillin → PENICILLINS – "NEWER and SYNTHETICS" also → PENICILLIN, POTASSIUM
METHIDATHION
METHIMAZOLE also → PHENOBARBITAL, PROPRANOLOL
Methinitrophos → SUMITHION
Methiocarb → MERCAPTODIMETHUR
Methiocil → METHYLTHIOURACIL
METHIODAL SODIUM
Methionine → METHIONINES
METHIONINES also → DEXTROAMPHETAMINE, ETHIONINE, METHENAMINE, NITROGEN CHLORIDE, NITROUS OXIDE, POLYMIXIN B
Methionyl Human Growth Hormone → SOMATREM
METHISAZONE
Methisoprinol → ISOPRINOSINE
Methium → HEXAMETHONIUM BISTRIUM
Methixart → METHIXENE
METHIXENE
Methizole → METHIMAZOLE
METHOCARBAMOL also → ACETYLSALICYLIC ACID, PYRIDOSTIGMINE BROMIDE
Methocel → CELLULOSE
Methocel MC → METHYL CELLULOSE
Methocillin-S → PENICILLINS – "NEWER and SYNTHETICS" > *Cloxacillin/Sodium salt*
Methofuran → PENNYROYAL
Meth-O-Gas → METHYL BROMIDE
METHOHEXITAL SODIUM also → EPINEPHRINE, HALOTHANE
Methohexitone → METHOHEXITAL SODIUM
Methoin → MEPHENYTOIN
Metholone → DROMOSTANOLONE PROPIONATE
Methomidate → METOMIDATE
METHOMYL
METHOPHENAZINE
METHOPHOLINE
METHOPRENE
Methoprim → TRIMETHOPRIM
D-Methorphan → DEXTROMETHORPHAN HYDROBROMIDE
Methosarb → CALUSTERONE
Methoserpidine → DESERPIDINE
Methostan → METHANDRIOL
METHOTREXATE also → ACETYLSALICYLIC ACID; ALCOHOL, ETHYL; AMIODARONE, APAZONE, BUTOCONAZOLE, CHLOROQUINE, CISPLATIN, CYCLOPHOSPHAMIDE, DICLOFENAC, FOLIC ACID, FOLINIC ACID, FUROSEMIDE, HYDROXYUREA, INDOMETHACIN, KETOPROFEN, LACTOSE, LATEX, LEFLUNOMIDE, NAPROXEN, NEOMYCIN, NYSTATIN, PENICILLIN; PENICILLINS – "NEWER and SYNTHETICS" > *Amoxicillin, Piperacillin;* PHENYLBUTAZONE, POLYMIXIN B, PROBENECID, ROFECOXIB, SALICYLIC ACID, SALSALATE, SULFADIMETHOXINE, SULFAMETHOXYPYRIDAZINE, SULFISOXAZOLE, SULINDAC, TOLBUTAMIDE, TOLMETIN, TRIMETHOPRIM, VANCOMYCIN
METHOTRIMEPRAZINE also → ACEPROMAZINE
Methoxadone → MEPHENOXALONE
METHOXAMINE also → METHYLERGONOVINE
Methoxane → METHOXYFLURANE
Methoxo → METHOXYCHLOR
Methoxone → METHYL CHLOROPHENOXY ACETIC ACID
METHOXSALEN also → AMMI MAJUS, BENZYL SALICYLATE, PARSLEY, SCLEROTINIA SCLEROTORIUM
p-METHOXY ACETOPHENONE
4-METHOXYAMPHETAMINE
4-Methoxybenzaldehyde → ANISALDEHYDE
σ-Methoxybenzamine → σ-ANISIDINE
Methoxybenzene → ANISOLE
4-Methoxybenzenemethanol → ANISE ALCOHOL

p-Methoxybenzyl Alcohol → ANISE ALCOHOL
p-Methoxybenzyl Proprionate → ANISYL PROPRIONATE
2-Methoxy Biphenyl → METHYL DIPHENYL ETHER
METHOXYCHLOR
σ-METHOXYCINNAMIC ALDEHYDE
Methoxy-DDT → METHOXYCHLOR
10-Methoxydeserpidine → DESERPIDINE
5-METHOXYDIMETHYLTRYPTAMINE also → PHALARIS sp.
Methoxydiuron → LINURON
Methoxydon(e) → MEPHENOXALONE
6-Methoxy-erythromycin → CLARITHROMYCIN
2-METHOXYETHANOL
METHOXYETHYL ACETATE
METHOXYETHYLMERCURY CHLORIDE
METHOXYETHYL MERCURY SILICATE
METHOXYFLURANE also → DIMETHYL TUBOCURARINE IODIDE, EPINEPHRINE, FLUORIDES and FLUORINE, HALOTHANE, KANAMYCIN, POLYETHYLENE, PROPRANOLOL, SECOBARBITAL SODIUM, TETRACYCLINE
6-METHOXYHARMALAN
10-METHOXYHARMALAN
METHOXYMETHYLAMPHETAMINE
Methoxy-α-methylphenethylamine → 4-METHOXYAMPHETAMINE
Methoxymol → METOMIDATE
Methoxynal → MECLOFENOXATE
METHOXYPHENAMINE
2-Methoxyphenol → GUAIACOL
4-METHOXYPHENOL
4-(*p*-METHOXYPHENYL)BUTAN-2-ONE
β(σ-Methoxyphenyl)acrolein → σ-METHOXYCINNAMIC ALDEHYDE
N-[2-*m*-METHOXY PHENYL)-2-ETHYL BUTYL-(1)]γ-HYDROXYBUTYRAMIDE
METHOXYPROMAZINE
Methoxypropanol → PROPYLENE GLYCOL MONOMETHYL ETHER
Methoxypropazine → PROMETONE
β-Methoxypsoralen → METHOXSALEN
8 or 9-Methoxypsoralen → METHOXSALEN
Methoxypsoralens → BERGAMOT OIL, CELERY, PITURANTHOS TRIRADIATUS
METHOXYPYRIDOXINE
Methoxypyrimal → SULFAMETER
2-Methoxytrimeprazine → METHOTRIMEPRAZINE
Methoxyverapamil → GALLOPAMIL
Methrazone → FEPRAZONE
METHSUXIMIDE also → METHARBITAL
METHYCLOTHIAZIDE also → PARGYLINE, VITAMIN D_3
Methyglyoxal → COFFEE
METHYL ABIETATE
N-METHYLACETAMIDE
METHYL ACETATE
Methylacetic Acid → PROPIONIC ACID
METHYL ACETONE also → METHYL ETHYL KETONE
p-METHYLACETOPHENONE
METHYL ACETYLENE
METHYL ACRYLATE also → FENTANYL
METHYL ACRYLONITRILE
METHYLAL
Methyl Alcohol → ALCOHOL, METHYL
Methyl Aldehyde → FORMALDEHYDE
METHYLAMINE also → MERCURIALIS ANNUA, OLIVES
β-N-METHYLAMINO-*L*-ALANINE also → CYCADS, PALMS
3-METHYL-4-AMINO-*N*-DIETHYLANILINE HCL
2-METHYLAMINOETHANOL
p-METHYLAMINOPHENOL SULFATE
Methylaminopterin → METHOTREXATE
METHYL AMYL ALCOHOL
Methyl n-Amyl Ketone → 2-HEPTANONE
Methylandrostenediol → METHANDRIOL
Methylaniline → TOLUIDINE
N-METHYLANILINE
METHYL ANTHRANILATE also → GRAPE
METHYL ASPARTATE
METHYLATED SPIRITS
2-METHYLAZIRIDINE
Methylazoxymethanol → CYCADELES, CYCASIN
Methylbenzedrine → METHAMPHETAMINE
Methylbenzene → TOLUENE
METHYLBENZETHONIUM CHLORIDE
N-Methylbenzenamine → *N*-METHYLANILINE
METHYL BENZIMIDAZOLE-2-CARBAMATE
METHYL BENZOATE
6-Methyl-1,2-benzopyrone → 6-METHYLCOUMARIN
3-METHYL-2-BENZOTHIAZOLINONE HCL
α-Methylbenzyl Alcohol → ALCOHOL, STYRALLYL
1-METHYL-2,3-BIS-HYDROXYMETHYL-3-PYRROLINE CARBAMATE
METHYL BROMIDE also → FIRE EXTINGUISHERS
2-Methyl-1-butanol → ALCOHOL, ISOPENTYL
2-Methyl-2-butanol → AMYLENE HYDRATE
3-Methyl-1-butanol → ISOAMYL ALCOHOL

3-Methyl-2-butanone → METHYL ISOPROPYL KETONE
β-Methyl Butyl Acetate → ISOAMYL ACETATE
METHYL-*tert*-BUTYL ETHER
METHYL BUTYL KETONE also → SHOES
5-METHYL-3-BUTYLTETRAHYDROPYRAN-4-YL ACETATE
N-Methylcarbamate → MERCAPTODIMETHUR
Methyl Carbophenothion → METHYL TRITHION
Methyl-CCNU → SEMUSTINE
Methyl Cellosolve® → 2-METHOXYETHANOL
Methyl Cellosolve® Acetate → METHOXYETHYL ACETATE
Methylcellulose → CELLULOSE
METHYL CELLULOSE
Methyl Chavicol → ESTRAGOLE
METHYL CHLORIDE also → POLYSTYRENE
2-METHYL-4-CHLOROANILINE also → CHLORDIMEFORM
METHYL CHLOROCARBONATE
Methyl Chloroform → 1,1,1-TRICHLOROETHANE
Methyl Chloroformate → METHYL CHLOROCARBONATE
METHYL CHLOROPHENOXY ACETIC ACID
4(2-METHYL-4-CHLOROPHENOXY) BUTYRIC ACID
Methylchlorothiazide → METHYCLOTHIAZIDE
3 or 20-METHYLCHOLANTHRENE
Methyl-3-chromone → TICROMYL
METHYL CINNAMATE
METHYLCINNAMIC ALCOHOL and ALDEHYDE
METHYL-2-(CINNAMIDOOXY) PRIOPIONATE
METHYLCLONAZEPAM
6α-Methylcompactin → LOVASTATIN
METHYLCONIINE also → CONIUM MACULATUM
6-METHYLCOUMARIN
Methyl Cyanide → ACETONITRILE
Methyl-2-cyanoacrylate → MECRYLATE
METHYLCYCLOHEXANE
METHYLCYCLOHEXANOL
METHYLCYCLOHEXANONE
METHYLCYCLOPENTADIENYLMANGANESE TRICARBONYL
Methyl Cymate → ZIRAM
S-METHYL CYSTEINE SULFOXIDE also → BRASSICA sp., KALE
S-Methyl-L-cysteine → BRASSICA sp.
N-Methylcytosine → CAULOPHYLLINE
Methyldebenyl → SULFAMERAZINE
Methyl Demeton → DEMETON-METHYL
METHYLDIAMINE METHAMPHETAMINE
16α-METHYL DICHLORISONE ACETATE
METHYLDICHLOROPHOSPHONIC ACID
METHYLDIHYDROMORPHINONE
Methyldinitrobenzene → DINITROTOLUENE
Methyl Dinoprost → CARBOPROST
METHYL DIPHENYL ETHER
METHYLDOPA also → ALCOHOL, ETHYL; AMITRIPTYLINE, ANTIPYRINE, BETHANIDINE, BISHYDROXYCOUMARIN, COCAINE, DIGOXIN, L-DOPA, FENFLURAMINE, FERROUS SULFATE, FURAZOLIDONE, HALOPERIDOL, HYDROCHLOROTHIAZIDE, LITHIUM, MEPHENTERMINE, METARAMINOL BITARTRATE, METHAMPHETAMINE, NOREPINEPHRINE, PARGYLINE, PHENOBARBITAL, PHENOXYBENZAMINE HYDROCHLORIDE, PHENYLPROPANOLAMINE, PROPRANOLOL, PSEUDOEPHEDRINE, TOLBUTAMIDE, TRIFLUOPERAZINE DIHYDROCHLORIDE
N-Methyldopamine Diisobutyrate → IBOPAMINE
METHYLENEANILINE
N,N´-METHYLENE-bis-ACRYLAMIDE
4,4´-METHYLENE-bis-[2-CHLOROANILINE]
METHYLENE BIS(4-CYCLOHEXYLISOCYANATE)
4,4´-METHYLENEBIS(*N*, *N*-DIMETHYL)BENZENAMINE
METHYLENE BLUE also → AZATHIOPRINE, MERCURIC CHLORIDE, SODIUM NITRITE
α-Methylene-γ-butyrolactone → ERYTHRONIUM also → TULIP
METHYLENE CHLORIDE also → COFFEE, PHOSGENE
α-Methylene Cyclopropylglycine → LITCHI
4,4´-METHYLENE DIANILINE
Methylene Dichloride
METHYLENEDIOXYAMPHETAMINE also → RITONAVIR
24-Methylene-iophenol → POTATO
Methylene Oxide → FORMALDEHYDE
Methylergobasine → METHYLERGONOVINE
Methylergobrevin → METHYLERGONOVINE
Methylergol Carbamide → LISURIDE
Methylergometrine → METHYLERGONOVINE
METHYLERGONOVINE
Methylestrenolone → NORMETHANDRONE
Methyl Ethanoate → METHYL ACETATE

N-METHYLETHANOLAMINE
Methylethene → PROPYLENE
Methylethylene → PROPYLENE
4,4´-(1-Methylethylidene)bisphenol → BISPHENOL
Methyl Ethyl Ketone → BUTANONE also → GLUE, SHOES
METHYL EUGENOL
α-METHYL FENTANYL also → FENTANYL
METHYL FLUOROACETATE
Methyl Fluorosulfate → METHYL FLUOROSULFONATE
METHYL FLUOROSULFONATE
Methylflurether → ENFLURANE
N-METHYLFORMAMIDE
METHYL FORMATE
N-METHYL-*N*-FORMYLHYDRAZINE
3-METHYLFURAN
METHYL GALLATE
METHYLGLUCAMINE → ANTIMONATE
Methyl Glycol → PROPYLENE GLYCOL
METHYL GREEN
Methyl-Guthion → AZINPHOS-METHYL
5-Methyl-3-heptanone → ETHYL *sec*-AMYL KETONE
METHYL HEPTINE CARBONATE also → LIPSTICK
METHYL HEPTYL KETONE
Methyl Hexyl Ketone → HEXYL METHYL KETONE
METHYL HYDRATROPALDEHYDE
METHYL HYDRAZINE also → GYROMITRA ESCULENTA
Methyl Hydride → METHANE
Δ¹6α-Methylhydrocortisone → METHYLPREDNISOLONE
Methyl-1-hydroxyethyl Ketone → ACETOIN
4-METHYLIMIDAZOLE
3-Methylindole → SKATOLE
METHYL IODIDE
METHYL IONONE
METHYL ISOAMYL KETONE
2-METHYLISOBORNEOL
Methyl Isobutyl Carbinol → METHYL AMYL ALCOHOL
METHYL ISOBUTYL KETONE
METHYL ISOCYANATE also → PHENYL ISOCYANATE
METHYL ISOEUGENOL
Methylisomyn → METHAMPHETAMINE
METHYL ISOPROPYL KETONE
2-METHYL-ISOTHIAZOLINE-3-ONE
METHYLISOTHIOCYANATE
Methyl Lomustine → SEMUSTINE
24-Methyllophenol → POTATO
Methyllycaconitine → LARKSPUR
Methylmercadone → NIFURATEL
Methyl Mercaptan → METHANETHIOL
Methylmercaptophos → DEMETON-METHYL
Methyl Mercuric Chloride → METHYL MERCURY CHLORIDE
Methyl Mercuric Hydroxide → METHYL MERCURY HYDROXIDE
METHYL MERCURY also → BARLEY, BREAD, FISH, MERCUROUS CHLORIDE, MERCURY, Preface
METHYL MERCURY CHLORIDE
METHYLMERCURYDICYANDIAMIDE
METHYL MERCURY HYDROXIDE
METHYL METHACRYLATE also → GLOVES, HYDROQUINONE
Methylmethane → ETHANE
METHYLMETHANE SULFONATE
Methyl Methylaminobenzoate → DIMETHYL ANTHRANILATE
Methyl-n-methylanthranilate → DIMETHYL ANTHRANILATE
2-Methyl-2-(methylthio)propanal-o[(methylamino) carbonyl]oxime → ALDICARB
2-Methyl-(methylthio)propionaldehyde-o-(methylcar bamoyl)oxime → ALDICARB
N-Methylmitomycin C → PORFIROMYCIN
Methyl Morphine → CODEINE
Methyl Mustard Oil → METHYLISOTHIOCYANATE
Methyl Naftalen → METHYLNAPHTHALENE
METHYLNAPHTHALENE
β-METHYL NAPHTHYL KETONE
METHYL NICOTINATE
Methylnitrobenzene → NITROTOLUENE
Methylnitrofos → SUMITHION
N-Methyl-N-nitroso-N-cyclohexylamine → *N*-NITROSO-*N*-METHYL-*N*-CYCLOHEXYLAMINE
N-METHYL-*N´*-NITROSOGUANIDINE
1-METHYL-1-NITROSOUREA
METHYL NONYL KETONE
Methyl-2-Nonynoate → METHYL OCTINE CARBONATE
Methylnortestosterone → NORMETHANDRONE
5-Methyl-2-octanone → METHYL HEPTYL KETONE
METHYL OCTINE CARBONATE
Methyloxan → METHIXENE
6-Methyl-1,2,3-oxathiazin-4(3H)-one 2,2-Dioxide → ACESULFAME K
Methyl Oxitol® → 2-METHOXYETHANOL

Methylparafynol → MEPARFYNOL
METHYL PARATHION
4-Methyl-2-pentanone → METHYL ISOBUTYL KETONE
Methylperidol → MOPERONE
(15S)-15-Methyl PGF$_{2\alpha}$ → CARBOPROST
dl-2-Methylphenethylamine → AMPHETAMINE
METHYLPHENIDATE also → AMITRIPTYLINE, BISHYDROXYCOUMARIN, DESIPRAMINE, ETHYL BISCOUMACETATE, GUANETHIDINE, IMIPRAMINE, MAGNESIUM SILICATE, NIALAMIDE, PHENELZINE, PHENOBARBITAL, PHENYTOIN, PRIMIDONE, SEROTONIN, TRANYLCYPROMINE
Methylphenobarbital → MEPHOBARBITAL
METHYL PHENYLACETATE
Methylphenyl Carbinol → ALCOHOL, STYRALLYL
α-Methylphenylethyl Alcohol → ALCOHOL, HYDRATROPIC
N-Methyl-β-phenylethylamine → ACACIA, GROWTH HORMONES
Methyl-3-phenylpropenoate → METHYL CINNAMATE
2-Methyl-2-phenyl-succinamide → METHSUXIMIDE
1-METHYL-4-PHENYL-1,2,5,6-TETRAHYDROPYRIDINE
Methyl Phosphite → TRIMETHYL PHOSPHITE
Methyl Phthalate → DIMETHYL PHTHALATE
METHYLPREDNISOLONE also → CYCLOSPORINE, ERYTHROMYCIN, ITRACONAZOLE, KETOCONAZOLE, PHENYTOIN, RIFAMPIN, TACROLIMUS, TROLEANDOMYCIN
2-Methylpropanenitrile → ISOBUTYRONITRILE
2-Methyl-1-propanol → ALCOHOL, ISOBUTYL
Methyl Propenoate → METHYL ACRYLATE
METHYL PROPYL KETONE
Methylprylone → METHYPRYLONE
Methylpyridyl Ketone → ACETYLPYRIDINE
Methylpyrimal → SULFAMERAZINE
Methylquinoline → 1,2-DIHYDRO-2,2,4-TRIMETHYLQUINOLINE
Methylrosaniline → GENTIAN VIOLET
METHYL SALICYLATE also → ABRUS PRECATORIUS, BARBITAL, BIRCH, ISOSAFROLE, POLYGALA sp.
α-METHYL STYRENE also → RUBBER, VINYL TOLUENE
Methyl Sulfate → DIMETHYL SULFATE
Methyl Sulfhydrate → METHANETHIOL
2-Methylsulfonylpyridine → PYRITHIONE
Methyl Sulfoxide → DIMETHYL SULFOXIDE
Methyl Systox → METHYL DEMETON
METHYL TESTOSTERONE also → CYCLOSPORINE
Methyltetranitroaniline → NITRAMINE
Methyl Theobromine → CAFFEINE
S-Methylthioesters → ASPARAGUS
Methylthionine Chloride → METHYLENE BLUE
Methylthiotetrazole → MOXALACTAM
METHYLTHIOURACIL
Methyl α-toluate → METHYL PHENYLACETATE
Methyl-p-tolyl Ketone → *p*-METHYLACETOPHENONE
Methyltrexate → METHOTREXATE
Methyl Tribromide → BROMOFORM
METHYL TRITHION
Methyltryptamine → PSYCOTRIA
METHYLTYRAMINE also → ACACIA, ARIOCARPUS
α-Methyl-p-tyrosine → HALOPERIDOL also → METYROSINE
METHYLURACIL
5-Methyluracil → THYMINE
METHYL VIOLET
METHYLXANTHINES
Methyl Yellow → *p*-DIMETHYLAMINOBENZENE
Methyl Zimate → ZIRAM
Methypentynol → MEPARFYNOL
Methypranol → TRIMEPRANOL
METHYPRYLONE also → ALCOHOL, ETHYL; BEER
METHYRIDINE also → SUCCINYLCHOLINE, TETRAMISOLE
METHYSERGIDE also → DOPAMINE, ERYTHROMYCIN, SEROTONIN, TROLEANDOMYCIN
METHYSTICODENDRON AMESIANUM
Methysticum → KAVA
Methzine → PROMETHAZINE
METIAMIDE
Meticillin → PENICILLINS – "NEWER and SYNTHETICS" > *Methicillin Sodium*
Meticlopindol → CLOPIDOL
Meticortelone → PREDNISOLONE
Meticorten → PREDNISONE
Metidione → METHANDRIOL
Metifex → ETHACRIDINE LACTATE
Metiguanide → METFORMIN
Metilar → PARAMETHASONE ACETATE
Metildigoxin → DIGOXIN also → VERAPAMIL

Metildiolo → METHANDRIOL
Metilenbiotic → METHACYCLINE
Metilnitrofos → SUMITHION
Metilon → DIPYRONE
Metiltriazotion → AZINPHOS-METHYL
Metindol → INDOMETHACIN
Metione → METHIONINES
Metipranolol → TRIMEPRANOL
Metligine → MIDODRINE
Metmercapturon → MERCAPTODIMETHUR
Metoclol → METOCLOPRAMIDE
METOCLOPRAMIDE also → ACEPROMAZINE, CIMETIDINE, CYCLOSPORINE, DIGOXIN, L-DOPA, DOPAMINE, ETHAMBUTOL, FLUVOXAMINE, FOSFOMYCIN, HYDROXYZINE, KETOPROFEN, MAGNESIUM TRISILICATE, MEXILETINE, SODIUM METABISULFITE, SODIUM VALPROATE, SUFENTANIL, TACROLIMUS, TETRACYCLINE
Metocobil → METOCLOPRAMIDE
Metocream → METRONIDAZOLE
Metocryst → METHANDRIOL
Metocurine Iodide → DIMETHYL TUBOCURARINE IODIDE
Metodik → ASTEMIZOLE
Metofane → METHOXYFLURANE
Metofenazate → METHOPHENAZINE
Metofoline → METHOPHOLINE
Metol → *p*-METHYLAMINOPHENOL SULFATE
METOLACHLOR
METOLAZONE also → CAPTOPRIL, WARFARIN
Metolquizolone → METHAQUALONE
METOMIDATE
Metopirone → MTYRAPONE
Metopon → METHYLDIHYDROMORPHINONE
METOPROLOL also → CIMETIDINE, CITALOPRAM, CLONIDINE, L-DOPA, EPINEPHRINE, FLUOXETINE, INDOMETHACIN, METHYLDOPA, NIFEDIPINE, PAROXETINE, PHENPROCOUMON, PIROXICAM, QUINIDINE, RANITIDINE, RIFAMPIN, VERAPAMIL
Metormon → DROMOSTANOLONE PROPIONATE
METOSERPATE
Metosomin → TOLPERISONE HCL
Metosyn → FLUOCINONIDE
Metox → METHOXYCHLOR
Metoxadone → MEPHENOXALONE
Metracin → CIMETIDINE
Metramac → TETRAM®
Metramid → METOCLOPRAMIDE
Metrazol → PENTYLENETETRAZOL
Metreton → PREDNISOLONE
METRIBUZIN
Metrifonate → TRICHLORFON
Metriphonate → TRICHLORFON
Metrisone → METHYLPREDNISOLONE
METRIZAMIDE
MetroGel → METRONIDAZOLE
Metrogestone → MEDROGESTONE
Metrolag → METRONIDAZOLE
Metrolyl → METRONIDAZOLE
Metron → METHYL PARATHION
METRONIDAZOLE also → ALCOHOL, ETHYL; BUTETHAL, CHLOROQUINE, CIMETIDINE, CISAPRIDE, CYCLOSPORINE, DISULFIRAM, LITHIUM, OXYTETRACYCLINE, PHENOBARBITAL, QUINIDINE, RITONAVIR, WARFARIN
Metroprione → METYRAPONE
Metrotop → METRONIDAZOLE
Metryl → METRONIDAZOLE
Metubine → DIMETHYL TUBOCURARINE IODIDE
METUREDEPA
Metycaine → PIPEROCAINE
METYRAPONE also → CHLORPROMAZINE
Metyridine → METHYRIDINE
METYROSINE
Mevacor → LOVASTATIN
Mevalotin → PRAVASTATIN
Mevinolin → LOVASTATIN
Mevinolinic Acid → LOVASTATIN
MEVINPHOS also → PRALIDOXIME
Mevlor → LOVASTATIN
MEXACARBATE
Mexate → METHOTREXATE
MEXENONE
Mexephenamide → MEFEXAMIDE
Mexican Bed Bug → TRIATOMA sp.
Mexican Bird of Paradise → POINCIANA GILLIESSI
Mexican Breadfruit → MONSTERA sp.
Mexican Damiana → TURNERA APHRODISIACA
Mexican Flame Leaf → POINSETTIA
Mexican Flame Tree → POINSETTIA
Mexican Flower Plant → POINSETTIA
Mexican Marigold → TAGETES MINUTA
Mexican Morning Glory → SEROTONIN
Mexican Mushroom → SEROTONIN

Mexican Poppy → ARGEMONE MEXICANA
Mexican Rubber Plant → PARTHENIUM sp.
Mexican Tea → EPHEDRINE
Mexican Vanilla → TONKA BEANS
MEXILETINE also → AMIODARONE, CAFFEINE, HEROIN, MORPHINE, PAROXETINE, PENTAZOCINE, QUINIDINE, RITONAVIR, THEOPHYLLINE
Mexitil → MEXILETINE
Mexocine → DEMECLOCYCLINE
Mezaton → PHENYLEPHRINE HCL
Mezcaline → MESCALINE
Mezene → ZIRAM
Mezerbae → DATURA
Mezerein → DAPHNE, YUANHAUCINE
Mezereon → DAPHNE
Mezlin → PENICILLINS – "NEWER and SYNTHETICS" > *Mezlocillin*
Mezlocillin → PENICILLINS – "NEWER and SYNTHETICS"
MF 10 → DOXEPIN
MF 110 → DOXEPIN
MFA → METHYL FLUOROACETATE
MG 2555 → BROMINDIONE
MGK Repellent 11 → R-11
MH → MALEIC ANHYDRIDE
MHA → METHIONINES
MI 36 → ETHYL PHENYLEPHRINE
Mi 85 → APAZONE
217 MI → ECHOTHIOPHATE IODIDE
Miacalcic → CALCITONIN
Miadone → METHADONE HCL
MIAK → METHYL ISOAMYL KETONE
MIANSERIN also → ALCOHOL, ETHYL
Miaquin → AMODIAQUIN
Miarsenol → NEOARSPHENAMINE
MIBC → METHYL AMYL ALCOHOL
MIBEFRADIL also → ATORVASTATIN, SILDENAFIL, SIMVASTIN, TRIAZOLAM
MIBOLERONE
MIC → METHYL ISOCYANATE
MIC 100 → METHACYCLINE
MICA
Micardis → TELMISARTAN
Micatin → MICONAZOLE NITRATE
Michler's Base → 4,4′-METHYLENEBIS(*N*,*N*-DIMETHYL)-BENZENAMINE
MICHLER'S KETONE
"Mickey-Finns" → CHLORAL HYDRATE
Micloretin → CHLORAMPHENICOL
Micoclorina → CHLORAMPHENICOL
Micofugal → ECONAZOLE NITRATE
Micofur → NIFUROXIME
Micogin → ECONAZOLE NITRATE
Micol → CETRIMONIUM BROMIDE
Miconal Ecobi → MICONAZOLE NITRATE
MICONAZOLE NITRATE also → ACENOCOUMAROL, AMPHOTERICIN B, CISAPRIDE, GLIPIZIDE, KETOCONAZOLE, PHENOBARBITAL, PHENYTOIN, TOBRAMYCIN, TRIMETREXATE, WARFARIN
Micoserina → CYCLOSERINE
Micotef → MICONAZOLE NITRATE
Micotil → TILMICOSIN
Micrest → DIETHYLSTILBESTROL
Microbar → BARIUM SULFATE
Microcetina → CHLORAMPHENICOL
Microcillin → PENICILLINS – "NEWER and SYNTHETICS" > *Carbenicillin Disodium*
Microciona prolifera → SPONGE
MICROCYSTIS AERUGINOSA
Micro-Dee → VITAMIN D_3
Microdots → LYSERGIC ACID DIETHYLAMIDE
Micro K → POTASSIUM CHLORIDE
Microlat → LEVONORGESTREL
Microlysin → CHLOROPICRIN
Micronase → GLYBURIDE
Micronor → NORETHINDRONE
Micronovum → NORETHINDRONE
Micropaque → BARIUM SULFATE
Micro-Pen → PENICILLIN > *Benzylpenicillin, Procaine*
Micropenin → PENICILLINS – "NEWER and SYNTHETICS" > *Oxacillin Sodium*
MICROPOLYSPORA FAENI
Microsystin L-R → MICROCYSTIS AERUGINOSA
Microthene → POLYETHYLENE
Microtin → CEPHALOTHIN
Microtrast → BARIUM SULFATE
Microtrim → SULFAMETHOXAZOLE
Microval → LEVONORGESTREL
Microx → METOLAZONE
Microzul → CHLOROPHACINONE
Micrurus sp. → SNAKE(S)
Mictone → BETHANECHOL
Mictrol → BETHANECHOL also → TERODILINE
Micturin → TERODILINE
Micturol → TERODILINE
Midalgan → METHYL NICOTINATE
Midamine → MIDODRINE
Midamor → AMILORIDE

MIDAZOLAM also → CLOZAPINE, DALFOPRISTIN, ERYTHROMYCIN, GRAPEFRUIT, ITRACONAZOLE, KETOCONAZOLE, RANITIDINE, RIFAMPIN, ROXITHROMYCIN, SODIUM VALPROATE, VERAPAMIL
Midgestone → PROGESTERONE
Midicel → SULFAMETHOXYPYRIDAZINE
Midikel → SULFAMETHOXYPYRIDAZINE
MIDODRINE
Midronal → CINNARIZINE
Midsummer Daisy → FEVERFEW
Mieas → COLTSFOOT
Mielucin → BUSULFAN
Mifegyne → MIFEPRISTONE
Mifeprex → MIFEPRISTONE
MIFEPRISTONE
Miforon → DILOXANIDE FUROATE
Mifos T → TETRAETHYL PYROPHOSPHATE
Migranal → DIHYDROERGOTAMINE METHANESULFONATE (MESYLATE)
Migristènen → FONAZINE MESYLATE
MIH → PROCARBAZINE
Mijeh → MUCUNA sp.
Mikavir → AMIKACIN
Mikedimide → BEMEGRIDE
Mikelan → CARTEOLOL HYDROCHLORIDE
MIL-127,686E → 2-ETHOXYETHANOL
MILBEMYCIN OXIME
Mildex → DINOCAP
Mildison → HYDROCORTISONE
Mildmen → CHLORDIAZEPOXIDE
Mild Mercury Chloride → MERCUROUS CHLORIDE
Miles Nervine® → SODIUM BROMIDE
Milestrol → DIETHYLSTILBESTROL
Milezin → METHOTRIMEPRAZINE
Milfoil → ACHILLEA sp.
Milibis → BISMUTH GLYCOL ARSANILATE
MILK also → ANTHEMIS, CHRYSANTHEMUM, CIPROFLOXACIN, CONIUM, CORONOPUS sp., DEMECLOCYCLINE, EQUISETUM, EUPHORBIA, HYPERICUM PERFORATUM, LACTOSE, LONICERA, LUPINES, *N*-NITROSODIMETHYLAMINE, NORFLOXACIN, OAKS, OLEANDOMYCIN, ONIONS, OXYTETRACYCLINE, PHENOL, RAGWEED, RANUNCULUS, RUMEX, TETRACYCLINE, TURNIP
Milk Bush → EUPHORBIA
Milk of Magnesia → MAGNESIUM HYDROXIDE also → DEMECLOCYCLINE
Milk Purslane → EUPHORBIA
"Milk Sickness" → EUPATORIUM sp., GOLDENROD; SNAKEROOT, WHITE; TREMATOL
Milk Sugar → LACTOSE
MILKVETCH also → MISEROTOXIN, 3-NITRO-1-PROPANOL
Milkweeds → ASCLEPIAS sp.
Millefoglio → ACHILLEA sp.
MILLET also → CYANATES, CYANIDES, PHOMA sp., SORGHUM, TETRACYCLINE
Millevit → VITAMIN B_{12}
Millicorten → DEXAMETHASONE
Milligynon → NORETHINDRONE
Millinese → CHLORPROPAMIDE
MILLIPEDES
Millipora complanata → FIRE CORAL
Millisrol → NITROGLYCERIN
Milo → SORGHUM also → CYANIDES
Milogard → PROPAZINE
Milontin → PHENSUXIMIDE
MILRINONE
Miltaun → MEPROBAMATE
MILTEFOSINE
MILTEX → MILTEFOSINE
Milton → SODIUM HYPOCHLORITE
Miltown → MEPROBAMATE
MIMOSA
MIMOSA ABSOLUTE
Mimosa hostilis → YUREMA
MIMOSINE also → MIMOSA sp., 3-HYDROXY-4(1H)PYRIDONE, IPIL-IPIL, LEUCAENA
Minacalm → TOLPERISONE HCL
Minacide → PROMECARB
Mina D_2 → VITAMIN D_2
Minadex → FERRIC AMMONIUM CITRATE
Minalgin → DIPYRONE
Minaphil → AMINOPHYLLINE
Mindiab → GLIPIZIDE
Mineral Carbon → GRAPHITE
Mineral Green → CUPRIC ACETOARSENITE
MINERAL OIL also → ANTIPYRINE, COTTON, DOCUSATE SODIUM, POLYBUTADIENE, SUCCINYLSULFATHIAZOLE
Mineral Pitch → ASPHALT
MINERAL SEAL OIL also → POLISHES, FURNITURE
Mineral Spirit → NAPHTHA also → STODDARD SOLVENT
MINERAL WATERS

MINERAL WOOL
Miner's Carbide → CALCIUM CARBIDE
Mingoral → ETHOXZOLAMIDE
Mini-bennies → AMPHETAMINES
Minidiab → GLIPIZIDE
Minihep → HEPARIN
Mini-Pe → NORETHINDRONE
"Mini-pill" → NORETHINDRONE
Miniplanor → ALLOPURINOL
Minipress → PRAZOSIN HYDROCHLORIDE
Minirin → DESMOPRESSIN
Minitran → NITROGLYCERIN
Minocin → MINOCYCLINE
MINOCYCLINE also → METHOXYFLURANE, TETRACYCLINE, THEOPHYLLINE
Minomycin → MINOCYCLINE
MINONENE
MINOXIDIL
Minoximen → MINOXIDIL
Minozinan → METHOTRIMEPRAZINE
Minprog → ALPROSTADIL
Minprostin E_2 → DINOPROSTONE
MINQUARTIA GUIANENSIS
Mintacol → DIETHYL-*p*-NITROPHENYLPHOSPHATE
Mintezol → THIABENDAZOLE
MINTS also → MENTHOL
Minugel → ATTAPULGITE
Minuric → BENZBROMARONE
Minusin → NORPSEUDOEPHEDRINE
Minzil → CHLOROTHIAZIDE
Miobloc → PANCURONIUM BROMIDE
Miocaina → GLYCERYL GUAIACOLATE
Miocard → AMIODARONE
Miochol → ACETYLCHOLINE
Miocor → LEVOCARNITINE
Miodar → PHENYRAMIDOL
Miodaron → AMIODARONE
Miolaxene → METHOCARBAMOL
Miolene → RITODRINE
Miolisodal → CARISPRODOL
Miorel → THIOCOLCHICOSIDE
Mioril → CARISPRODOL
Mio-Sed → CHLORMEZANONE
Miostat → CARBACHOL
Miotisal A → DIETHYL-*p*-NITROPHENYLPHOSPAHTE
Mioton → CAPSAICIN
Miotonal → LEVOCARNITINE
MIPAFOX
Mipax → DIMETHYL PHTHALATE
Mi-Pilo → PILOCARPINE
MIPK → METHYL ISOPROPYL KETONE
Miraa → CATHA EDULIS
MIRABILIS JALAPA
Miracil D → LUCANTHOE HYDROCHLORIDE
Miracle Glue → CYANOACRYLATES
Miracol → LUCANTHOE HYDROCHLORIDE
Miradol → SULPIRIDE
Miradon → ANISINDIONE
Mirapex → PRAMIPEXOLE
Miranax → NAPROXEN
Mirapront → PHENTERMINE
Mirasept → HYDROGEN PEROXIDE
Mirbanil → SULPIRIDE
MIREX
Mirfat → FUROSEMIDE
Mirontin → PHENSUXIMIDE
Mirsol → ZIPEROL
MIRTAZEPINE
MISEROTOXIN also → MILKVETCH, 3-NITRO-1-PROPANOL
MISONIDAZOLE also → PHENOBARBITAL
MISOPROSTOL also → METHOTREXATE
Misri → SUCROSE
Missell → MISTLETOE, EUROPEAN or JAPANESE
Miss Emma → MORPHINE
MISSILE SILO
Missulaena occatoria → SPIDERS
Mistabron → MESNA
Mistabronco → MESNA
Mistarel → ISOPROTERENOL
MISTLETOE, AMERICAN
MISTLETOE, EUROPEAN or JAPANESE
Mistura Magnesii Hydroxidi → MAGNESIUM HYDROXIDE
Misulban → BUSULFAN
Misulvan → SULPIRIDE
Mitaban → AMITRAZ
Mitac → AMITRAZ
mitamaji → TRICHILIA EMETICA
Mithracin → MITHRAMYCIN
MITHRAMYCIN also → CALCITONIN, PARATHYROID HORMONE
Mithridate Mustard → THLASPI ARVENSE
Mitigan → DICOFOL
Mition → TETRADIFON
Mitis Green → CUPRIC ACETOARSENITE
MITOBRONITOL
Mitolac → MITOLACTOL
MITOLACTOL
Mitomycin → MITOMYCIN C
MITOMYCIN C also → TERLIPRESSIN
Mitosan → BUSULFAN

MITOTANE also → WARFARIN
Mitox → CHLORBENSIDE
Mitoxana → IFOSFAMIDE
MITOXANTRONE HCL
Mitozantrone → MITOXANTRONE HCL
Mitragnine → KRATOM
Mitragyna speciosa → KRATOM
Mitronal → CINNARIZINE
Mitsubishi → METHYLENEDIOXYAMPHETAMINE
Miturga sp. → SPIDERS
Mivacron → MIVACURIUM
MIVACURIUM
Mixture of Magnesium Hydroxide → MAGNESIUM HYDROXIDE
MJ-505 → PHENYRAMIDOL
MJ 1986 → CLINDAMYCIN
MJ 1987 → MESUPRINE
MJ 1992 → SOTERENOL
MJ 1999 → SOTALOL
MJ 5107 → CHLORAL BETAINE
MJ 5190 → AMIDEPHRINE MESYLATE
MJ 9022-1 → BUSPIRONE
MJ 9067 → ENCAINIDE
MJ 10061 → BENZBROMARONE
MJ 13,754-1 → NEFAZODONE
MJD 30 → ASTEMIZOLE
MJF 9325 → IFOSFAMIDE
MJF 11567-3 → CEFADROXIL
MJF 12,264 → TEGAFUR
MJR 1762 → MICONAZOLE NITRATE
2M-4kh-M → 4(2-METHYL-4-CHLOROPHENOXY) BUTYRIC ACID
MK12-3214 → DROCODE
MK 89 → ANILERIDINE
MK 130 → CYCLOBENZAPRINE
MK 135 → CHOLESTYRAMINE RESIN
MK 185 → HALOFENATE
MK 188 → ZERANOL
MK 191 → PENICILLINS - "NEWER & SYNTHETICS" > *Pivampicillin*
MK 208 → FAMOTIDINE
MK 231 → SULINDAC
MK 240 → PROTRIPTYLINE
MK 264 → FLUVOXAMINE MALEATE
MK 306 → CEFOXITIN SODIUM
MK 360 → THIABENDAZOLE
MK 366 → NORFLOXACIN
MK 383 → TIROFIBAN
MK 421 → ENALAPRIL
MK 0476 → MONTELUKAST
MK 476 → MONTELUKAST
MK 486 → CARBIDOPA
MK 507 → DORZOLAMIDE
MK 521 → LISINOPRIL
MK 595 → ETHACRYNIC ACID
MK 639 → INDINAVIR
MK 647 → DIFLUNISAL
MK 733 → SIMVASTATIN
MK 745 → METOCLOPRAMIDE
MK 781 → METYROSINE
MK 787 → IMIPENEM
MK 803 → LOVASTATIN
MK 870 → AMILORIDE
MK 905 → CAMBENDAZOLE
MK 906 → FINASTERIDE
MK 936 → ABAMECTIN
MK 950 → TIMOLOL
MK 954 → LOSARTAN POTASSIUM
MK 966 → ROFECOXIB
MK 0476 → MONTELUKAST
MK 0791 → CILASTATIN
MK 2018 → CENTCHROMAN
MK 6096 → METHYCLOTHIAZIDE
MK 7617 → PIPERAZINE
MK 7879 → POLYVINYLPYRROLIDONE
MK → PYRANTEL
Mkalamu → GLORIOSA
Mkandume → ABRUS PRECATORIUS
Mkangauchawi → ABRUS PRECATORIUS
Mkonokono mwitu → ANONA
Mkunde → ANTIARIS TOXICARIA
Mkwaia → MUNDULEA SERICEA
Mkwaja → TAMARIND
Mkwaya → MUNDULEA SERICEA
Mlazalaza → ABRUS PRECATORIUS
Mlibiti → STROPHANTHUS sp.
MLT → MALATHION
MM 14, 151 → CLAVULANIC ACID
MMAC → *N*-METHYLACETAMIDE
MMD → METHYLMERCURYDICYANDIAMIDE
MMDA → MYRISTICIN
MMF → *N*-METHYLFORMAMIDE also → MYCOPHENOLATE MOFETIL
MMH → METHYL HYDRAZINE
4-MMPD → 2,4-DIAMINOANISOLE
MMS → METHYLMETHANE SULFONATE
MMT → METHYLCYCLOPENTADIENYLMANGAN-ESE TRICARBONYL
Mnanaha → DATURA
Mnazate → MANEB
MnBX → METHYL BUTYL KETONE
Mnguonguo → ANTIARIS TOXICARIA
MNNG → *N*′-METHYL-*N*′-NITROSOGUANIDINE

MNPA → NAPROXEN
Mnuka Uvundo → CASSIA
Mnuka-vundon → SOPHORA sp.
mnwamaji → TRICHILIA EMETICA
moa-moa → PAHUTOXIN
MOB → OXYBENZONE
MOBAM
Moban → MOLINDONE
Mobenol → TOLBUTAMIDE
Mobic → MELOXICAM
Mobiflex → TENOXICAM
Mobilawn → DICHLOFENTHION
Mobutazon → MOFEBUTAZONE
MOCA → 4,4′-METHYLENE-bis[2-CHLOROANILINE]
Mocap → ETHOPROP
Mock Azalea → MENZIESIA FERRUGINEA
MOCK ORANGE
Moclamine → MOCLOBEMIDE
MOCLOBEMIDE also → CLOMIPRAMINE
Moctanin → MONOCTANOIN
Mod → DOMPERIDONE
Modacur → OXYFEDRINE
MODAFINIL
Modalina → TRIFLUOPERAZINE DIHYDROCHLORIDE
Modamide → AMILORIDE
Modane → DANTHRON
Modatrop → PHENTERMINE
Modecate → FLUPHENAZINE DECANOATE
MODECCA DIGITATA
Modeccin → MODECCA DIGITATA also → ADENIA VOLKENSII; MISTLETOE, EUROPEAN or JAPANESE
Moderatan → DIETHYLPROPION
Modibromide → DOMIPHEN BROMIDE
Modicare → DOMIPHEN BROMIDE
Modimmunal → ISOPRINOSINE
Modiodal → MODAFINIL
MODIOLA CAROLINIANA
Modirax → HEXAPROPYMATE
Moditen → FLUPHENAZINE DECANOATE
Moditen Enanthate → FLUPHENAZINE DECANOATE
Modulor → DIETHYLPROPION
Moduretic → AMILORIDE
Modustatin → SOMATOSTATIN
MOEXIPRIL
Moexiprilat → MOEXIPRIL
MOFEBUTAZONE
Mofenar → BUFEXAMAC
Mofesal → MOFEBUTAZONE
Mogadan → NITRAZEPAM
Mogadon → NITRAZEPAM
Moheptan → METHADONE HCL
MOLASSES
Molatoc → DOCUSATE SODIUM
Molcer → DOCUSATE SODIUM
Moldamin → PENICILLIN > *Benzylpenicillin, Benzathine*
Mole Plant → EUPHORBIA
Molevac → PYRVINIUM PAMOATE
Molgramostatin → GRANULOCYTE MACROPHAGE COLONY STIMULATING FACTOR
MOLINDONE
Molinollo → LEONOTIS NEPETIFOLIA
Molipaxin → TRAZODONE
Mollinox → METHAQUALONE
Molofac → DOCUSATE SODIUM
Molteno Disease → SENECIO sp.
MOLLUSKS
MOLYBDENUM
Momentum → ACETAMINOPHEN
MOMETASONE FUROATE
Momophos → AMPHETAMINE
MOMORDICA ANGUSTISEPALA
Momordicin → MOMORDICA CHARANTIA
MOMORDICA CHARANTIA also → CHLORPROPAMIDE
Momordicin → MOMORDICA CHARANTIA
Momozol → DIETHYLSTILBESTROL
MON 39 → GLYPHOSPHATE
MON 2,139 → GLYPHOSPHATE
MONACETIN
Monacolin A → LOVASTATIN
Monapen → PENICILLINS – “NEWER and SYNTHETICS” > *Ticarcillin Disodium*
Monarch → ALLOPURINOL
Monargan → ACETARSONE
Monasirup → ANETHOLE
Monaspor → CEFSULODIN SODIUM
Monasus purpureus Went → CHOLESTIN™
Monazan → MOFEBUTAZONE
Mondak → DICAMBA
Mondus → FLUNARIZINE HCL
Mondzo → DATURA
MONENSIN also → CHLORAMPHENICOL, TIAMULIN
MONEY also → COCAINE
Monicor → ISOSORBIDE MONONITRATE
Monistat → MICONAZOLE NITRATE
Monit → ISOSORBIDE MONONITRATE
Monitor → METHAMIDOPHOS

Monkey → MORPHINE
"Monkey Children" → HEXACHLOROBENZENE
Monkey Dust → PHENCYCLIDINE
Monkey Fiddle → EUPHORBIA
Monkshood → ACONITE
Monoacetin → MONACETIN
MONOACETOXYSCIRPENOL
Monoacetyldapsone → DAPSONE
MONOAMYLAMINE also → TOLNAFTATE
Mono-Baycuten → CLOTRIMAZOLE
MONOBENZONE also → RUBBER
Monobenzyl Ether → MONOBENZONE
Monobenzyl Hydroquinone → MONOBENZONE
Monobromoethane → ETHYL BROMIDE
Monocaine → BUTETHAMINE
MonoCedocard → ISOSORBIDE MONONITRATE
Monochlorethane → ETHYL CHLORIDE
Monochloroacetaldehyde → CHLOROACETALDEHYDE
Monochloroacetic Acid → CHLOROACETIC ACID
Monochlorobenzene → CHLOROBENZENE
MONOCHLORODIFLUOROETHANE
MONOCHLORODIFLUOROMETHANE
Monoclair → ISOSORBIDE MONONITRATE
Monocor → BISOPROLOL
Monocortin → PARAMETHASONE ACETATE
Monocron → MONOCROTOPHOS
MONOCROTALINE also → CROTALARIA
MONOCROTOPHOS
MONOCTANOIN
Monoethanolamine → ETHANOLAMINE
MONOETHYLHEXYL PHTHALATE
Monofluoroacetamide → FLUOROACETAMIDE
Monofluoroacetic Acid → FLUOROACETIC ACID
Monofluorotrichloromethane → BARIUM
Mono-Gesic → SALSALATE
Monoheparin → HEPARIN
Monoket → ISOSORBIDE MONONITRATE
MONOLINURON
Monolupine → ANAGYRINE
Mono Mack → ISOSORBIDE MONONITRATE
Monomethylamine → METHYLAMINE
Monomethylformamide → *N*-METHYLFORMAMIDE
Monomethyl Mercury → METHYL MERCURY
MONOMETHYLNAPHTHALENE SULFONATE
Monomycin → ERYTHROMYCIN ETHYL SUCCINATE also → PAROMOMYCIN
Monooctanoin → MONOCTANOIN
Monopar → STILBAZIUM IODIDE
Monopen → PENICILLIN > *Benzylpenicillin, Potassium*
Monophenylbutazone → MOFEBUTAZONE
σ-Monophosphate → VITAMIN B_1
Monopril → FOSFINOPRIL SODIUM
Mono Prop → PROPIONIC ACID
Monores → CLENBUTEROL
Monorest → ESTRADIOL
Monosodium Glutamate → SODIUM GLUTAMATE
MONOSODIUM METHANEARSENATE
Monosodium Methyl Arsonate → MONOSODIUM METHANEARSONATE
Monosodium Phosphate → SODIUM PHOSPHATE, MONOBASIC
Monosorb → ISOSORBIDE MONONITRATE
Monosulfide → MONOSULFIRAM
MONOSULFIRAM
Monotrim → TRIMETHOPRIM
MONOTROPA UNIFLORA
Monovent → TERBUTALINE SULFATE
Monsanto → ACRYLONITRILE
MONSTERA sp. also → CALCIUM OXALATE
Monteban → NARASIN
MONTELUKAST
Montgomery, General → METHAMPHETAMINE
Montrel → RUELENE
Monurex → MONURON
Monurol → FOSFOMYCIN TROMETHAMINE
MONURON
Monydrin → PHENYLPROPANOLAMINE
Moon Daisy → CHRYSANTHEMUM
Moonflower → DATURA also → BUCKBEAN
Moonseed → MENISPERMUM CANADENSE also → DAURICINE
Moosewood → DIRCA PALUSTRIS
5-MOP → METHOXSALEN
8-MOP → MEHOXSALEN
MOPERONE
Mopral → OMEPRAZOLE
MOPS
Mopsus sp. → SPIDERS
Mora → GLORIOSA

d-Moramid(e) → DEXTROMORAMIDE
Moranyl → SURAMIN SODIUM
Morbam → MEPROBAMATE
Morbicid → FORMALDEHYDE
Morena → DIHYDROERGOTAMINE METHANESULFONATE (MESYLATE)
Morf → MORPHINE
MORFAMQUAT
Morfin → MORPHINE
Morfoxone → MORFAMQUAT
MORICIZINE also → THEOPHYLLINE, WARFARIN
MORINAMIDE
MORINDA RETICULATA
Morion Wine → MANDRAGORA OFFICINARUM
Moriperan → METOCLOPRAMIDE
Morkit → ANTHRAQUINONE
Mormon Tea → EPHEDRINE, *Ephedra*
Mornidine → PIPAMAZINE
MORNING GLORY
Morocide → BINAPACRYL
Moronal → NYSTATIN
Morpan T → TETRADECYLTRIMETHYLAMMONIUM BROMIDE
Morphamquat → MORFAMQUAT
Morphazinamide → MORINAMIDE
Morpheus → MORPHINE
Morphia → MORPHINE
Morphie → MORPHINE
Morphina → MORPHINE
MORPHINE also → BERBERINE, BUTACAINE, CHLOROBUTANOL, CIMETIDINE, COCA, CYCLOPROPANE, DISULFIRAM, GUANETHIDINE, KAOLIN, MEPERIDINE, MEPHENOXALONE, METHOTRIMEPRAZINE, METUREDEPA, MEXILETINE, NARCEINE, NORTRIPTYLINE, NUTMEG, OPIUM, PAPAVER, PHENOBARBITAL, PINE, POPPY, PROCHLORPERAZINE, PROMETHAZINE, PROPRANOLOL, SODIUM BISULFITE, SULFOBROMOPHTHALEIN SODIUM, THEOPHYLLINE, THYROXINE, TROVAFLOXACIN, Preface
Morphine-3-methyl-ether → CODEINE
Morphium → MORPHINE
Morpho → MORPHINE
MORPXOCYCLINE
MORPHOLINE
Morsodren → METHYLMERCURYDICYANDIAMIDE
MORTIERELLA sp.
Morton, Dr William T. → ETHER, ETHYL
Morton's Salt Substitute® → POTASSIUM also → POTASSIUM CHLORIDE
Mortopal → TETRAETHYL PYROPHOSPHATE
Morus sp. → MULBERRY
Moryl → CARBACHOL
Mosatil → CALCIUM DISODIUM EDETATE
Mosegor → PIZOTYLINE
Moses → INDIGOCARMINE, SHELLFISH
Mota → CANNABIS
Mota Shool → LEONOTIS NEPETIFOLIA
Motens → LACIDIPINE
MOTH BALLS or POWDER
Mother-in-law Plant → CALADIUM, DIEFFENBACHIA sp.
Mother-in-law's Tongue → SANSEVIERA sp.
Mother's Heart → CAPSELLA BURSA-PASTORIS
Motherwort → LEONURUS CARDIACA
Moth Mullein → VERBASCUM THAPSUS
MOTHS
Motiax → FAMOTIDINE
Motilex → CLEBOPRIDE
Motilium → DOMPERIDONE
Motilyn → DEXPANTHENOL
Motolon → METHAQUALONE
MOTOR OIL
Motorola → METHYLENEDIOXYAMPHETAMINE
Motox → TOXAPHENE
Motricit → IBUPROFEN
Motrin → IBUPROFEN
Mountain Ivy → KALMIA
Mountain Laurel → KALMIA also → RHODODENDRON, SOPHORA sp.
Mountain Mahogany → CEROCARPUS MONTANUS
Mountain Pink → CENTAURIUM sp.
Mountain Tobacco → ARNICA
Mouse-con → ZINC
Mouse Pruf II → BRODIFACOUM
Moutard des Champs → MUSTARD
Moutarde de linde → BRASSICA sp.
Movens → MECLOFENAMIC ACID
Movergan → DEPRENYL
Movyl → POLYVINYL CHLORIDE
Mowiol → POLYVINYL ALCOHOL
Moxacef → CEFADROXIL
Moxadil → AMOXAPINE
Moxal → PENICILLINS – "NEWER and SYNTHETICS" > *Amoxicillin*
MOXALACTAM also → ALCOHOL, ETHYL; BEER, DISULFIRAM

Moxaline → PENICILLINS – "NEWER and SYNTHETICS" > *Amoxicillin*
Moxam → MOXALACTAM
MOXIDECTIN
Moxie → METHOXYCHLOR
Moxisylyte → THYMOXAMINE
MOXONIDINE
Moxyl → THYMOXAMINE
Mozambin → METHAQUALONE
6-MP → MERCAPTOPURINE
8-MP → METHOXSALEN
MP-620 → IOCETAMIC ACID
MPA → MEDROXYPROGESTERONE ACETATE
MPC → σ–DICHLOROBENZENE
M. P. Chlorcaps T. D. → CHLORPHENIRAMINE
MPH → METHYLPHENIDATE
MPK → METHYL PROPYL KETONE
MPMP → MEPAZINE
Mpondu → PENNISETUM sp.
MPP+ → MPTP
MPPP
MPRP → MERCAPTOPURINE
α-MPT → ,METYROSINE
MPT → METHYL PARATHION
MPTP → 1-METHYL-4-PHENYL-1,2,5, 6-TETRAHYDROPYRIDINE also → HEROIN
MPV 785 → MEDETOMIDINE
MPV 1248 → ATIPAMEZOLE
Mrambazi → CASSIA
MRD 108 → PIPRADOL
MRD 535 → POLYURETHANES
MRL-41 → CLOMIPHENE CITRATE
Mr. Rat Guard II → CHLOROPHACINONE
Mrs. Hutton → DIGITALIS
MS-752 → LUCANTHOE HYDROCHLORIDE
Msalafu → CASSIA
MSG → SODIUM GLUTAMATE
MSH → INTERMEDINE
MSMA → METHANEARSONIC ACID also → MONOSODIUM METHANEARSONATE
MSPS → MUCOITIN POLYSULFATE ESTER
Msubili → ALOE
Msubiri mwitu → ALOE
Msufari → GLORIOSA
Msugu → STROPHANTHUS sp.
Mtangwa mwiyi → ARISTOLOCHIA
MTB-51 → METHANTHELINE BROMIDE
MTBE → METHYL-*tert*-BUTYL ETHER
Mtongi → THYLACHIUM AFRICANUM
MTQ → METHAQUALONE
MTU → METHYLTHIOURACIL
Mtunguru → MAERUA ANGOLENSIS
Mtupa-wa-mrima → TEPHROSIA sp.
Mtupawa-pori → MUNDULEA SERICEA
Mturituri → ABRUS PRECATORIUS
MTX → METHOTREXATE also → GAMBIERDISCUS TOXICUS
muBupu → COMBRETUM sp.
MUCANAIN also → COWHAGE, MUCUNA sp.
Muclox → FAMOTIDINE
Mucocedyl → ACETYLCYSTEINE
Mucofluid → MESNA
MUCOITIN POLYSULFATE ESTER
Mucolator → ACETYLCYSTEINE
Mucolysin → TIOPRONIN
Mucolyticum → ACETYLCYSTEINE
Mucomycin → LYMECYCLINE
Mucomyst → ACETYLCYSTEINE
Mucopolysaccharide → MUCOITIN POLYSULFATE ESTER
Mucor stolonifer → PAPRIKA
Muco Sanigen → ACETYLCYSTEINE
Mucosolvin → ACETYLCYSTEINE
Mucret → ACETYLCYSTEINE
Mucuma → COWHAGE
MUCUNA sp.
Mud → MORPHINE
Mudimdi → TRICHILIA EMETICA
Muguet → CONVALLARIA MAJALIS
mu Gumacembere → NICANDRA PHYSALODES
MULBERRY
Mulfa Fern → CHEILANTHEA SIEBERI
Mulga → ACACIA
Mull-Soy → THYROID
Mulsiferol → VITAMIN D_2
Mulsopaque → IOPHENDYL
Multifuge → PIPERAZINE
Multimycine → COLISTIN
Multum → CHLORDIAZEPOXIDE
Mumii → MUMMIES
MUMMIES
MUNDULEA SERICEA
MUNITIONS
Mundisal → CHOLINE SALICYLATE
Munobal → FELODIPINE
mu Pfuta → CASTOR BEAN
MUPIROCIN
Muracil → METHYLTHIOURACIL
Muraena helena → EEL, MORAY
Muramidase → LYSOZYME
Murder/Executions/Criminal Use/Ordeal → ABRIN, AMANITA (*phalloides*), ANTIARIS

TOXICARA, ARISTOLOCHIA GRANDIFLORA, CACALIA DECOMPOSITA, CALOTROPIN, CASTOR BEAN, CURARE CYANIDES, NAPHTHA, NICOTINE, PARATHION, PEANUTS, PHYSOSTIGMA VENENOSUM, PILLOWS, POTASSIUM, POTASSIUM CHLORATE, RICIN, RYANIA, SPIGELIA ANTHALMIA, SYNADENIUM GRANTII, TRIAZOLAM, URECHITES sp., WARFARIN, ZINC SULFATE
Murelax → OXAZEPAM
mu Rhka → COMBRETUM sp.
Muriate of Potash → POTASSIUM CHLORIDE
Muriatic Acid → HYDROCHLORIC ACID
Muriol → CHLOROPHACINONE
MUROCTASIN
Muromomab CD3 → OKT 3
Muronab-CD3 → OKT 3
MURRAYA PANICULATA
Murungu → CATHA EDULIS
Musabber → ALOE
Muscade → NUTMEG
Muscalm → TOLPERISONE HCL
MUSCARINE also → BOLETUS, CLITOCYBE sp., INOCYBE sp., MUSHROOMS, RUSSULA
Muscatox → COUMAPHOS
MUSCAZONE also → MUSHROOMS
Muscimol → AGARIN also → AMANITA sp.
Musco-Ril → THIOCOLCHICOSIDE
Muscovite → MICA
Musculax → VECURONIUM BROMIDE
Muse → ALPROSTADIL
MUSHROOMS also → ALCOHOL, ETHYL; AMANITA sp., CHLOROPHYLLUM MOLYBDITES, GALERINA sp., HYDROXYCHLOROQUINE, LYOPHYLLUM CONGOBATUM, METHYL HYDRAZINE, MORCHELLA DELICIOSA, NEURINE, PHOLIOTA sp., PLEUROTIS sp., PSILOCYBIN, RHODOPHYLLUS LIVIDUS, RUSSULA, SODIUM GLUTAMATE, STROPHARIA sp., THIOCTIC ACID
"Mushroom-picker's Lung or Disease" → MUSHROOMS also → MICRO-POLYSPORA FAENI, THEOPHYLLINE, THERMOACTINOMYCES VULGARIS, THERMOPOLYSPORA POLYSPORA
Musk Ambrette → 2,4-DINITRO-3-METHYL-*tert*-BUTYLANISOLE
muskbala → VALERIANA sp.
MUSK MELON
Muskrat Root → CALAMUS
Musk Xylene → MUSK XYLOL
MUSK XYLOL also → COLOGNES
Musol → PHENYL SALICYLATE
Musquash Root → CICUTA sp.
Mussel Poison → SAXITOXIN
MUSSELS also → DOMOIC ACID, GONYAULAX, MOLLUSKS, SAXITOXIN
MUSTARD also → ALLYL ISOTHIOCYANATE, CAPERS, CHEESES, PETIVERA ALLIACEA, SODIUM BENZOATE
MUSTARD GAS
Mustargen → MECHLORETHAMINE
Mustine → MECHLORETHAMINE
Mutabase → DIAZOXIDE
Mutabons → AMITRIPTYLINE
Mutah → CANNABIS
Mutamycin → MITOMYCIN C
Mutie papayi → PAPAYA
MU TONG also → ARISTOLOCHIA sp.
Muzekezeke → ABRUS PRECATORIUS
MVP → METHYL-2-(CINNAMIDOOXY) PROPIONATE
MW-361 → OFLOXACIN
Mwadiga → ADENIUM sp.
MY-301 → GLYCERYL GUAIACOLATE
Myacide → 2,4-DICHLOROBENZYL ALCOHOL
Myalex → FENCLOZIC ACID
Myambutol → ETHAMBUTOL
Myanesin → MEPHENESIN
Myanol → MEPHENESIN
Myasul → SULFAMETHOXYPYRIDAZINE
Mybasan → ISONIAZID
Mycanden → HALOPROGIN
Mycantine → 8-HYDROXYQUINOLINE
Mycardol → PENTAERYTHRITOL TETRANITRATE
Mycelex-G → CLOTRIMAZOLE
Mychel → CHLORAMPHENICOL
Mycifradin → NEOMYCIN
Mycil → CHLORPHENESIN
Mycilan → HALOPROGIN
Mycinol → CHLORAMPHENICOL
Mycivin → LINCOMYCIN
Mycobutin → RIFABUTIN
Mycobutol → ETHAMBUTOL
Mycofug → CLOTRIMAZOLE
Mycoin C_3 → PATULIN
Mycolog® → ETHYLENEDIAMINE
MYCOPHENOLATE MOFETIL
Mycophenolic acid → MYCOPHENOLATE MOFETIL

Mycophyt → NATAMYCIN
Mycoshield → OXYTETRACYCLINE
Mycospor → BIFONAZOLE
Mycosporan → BIFONAZOLE
Mycosporin → CLOTRIMAZOLE
Mycostatin → NYSTATIN
Mydeton → TOLPERISONE HCL
Mydfrin → PHENYLEPHRINE HCL
Mydocalm → TOLPERISONE HCL
Mydplegic → CYCLOPENTOLATE
Mydriacyl → TROPICAMIDE
Mydriaticum → TROPICAMIDE
Mydriatin → PHENYLPROPANOLAMINE
Mydrilate → CYCLOPENTOLATE
Myebrol → MITOBRONITOL
Myeleukon → BUSULFAN
Myelobromol → MITOBRONITOL
Myelographin → IPODATE
Myeloleukon → BUSULFAN
Myelosan → BUSULFAN
Myfungar → OXICONAZOLE NITRATE
Mygdalon → METOCLOPRAMIDE
Myk → SULCONAZOLE NITRATE
Mykostin → VITAMIN D_2
MYLABRIS CICHORII
Mylanta® Gelcaps → QUINOLINE YELLOW
Mylaxen → HEXAFLURONIUM BROMIDE
Mylepsin → PRIMIDONE
Myleran → BUSULFAN
Mylicon → DIMETHICONE (*with silicon dioxide*)
Myliobatis sp. → STINGRAYS
Mylipen → PENICILLIN > *Benzylpenicillin, Procaine*
Mylodorm → AMOBARBITAL
Mylofanol → CINCHOPHEN
Mylone → DAZOMET
Mylosar → 5-AZACYTIDINE
Mylotarg → GEMTUZIMAB OZOGAMICIN
Mylproin → VALPROIC ACID
Mynol → PROCAINE
Mynosedin → IBUPROFEN
Myocaine → GLYCERYL GUAIACOLATE
Myocholine → BETHANECHOL
Myochrysine → GOLD SODIUM THIOMALATE
Myocord → ATENOLOL
Myocrisin → GOLD SODIUM THIOMALATE
Myodetensine → MEPHENESIN
Myodigin → DIGITOXIN
Myodil → IOPHENDYLATE
Myofedrin → OXYFEDRINE
Myofer → IRON DEXTRANS
Myoglycerin → NITROGLYCERIN
Myokombin → STROPHANTHIN
Myolysin → MEPHENESIN
Myopan → MEPHENESIN
MYOPORUM sp.
Myordil → AMOTRIPHENE
Myorexon → ISOSORBIDE DINITRATE
Myoscain → GLYCERYL GUAIACOLATE
Myoserol → MEPHENESIN
Myospan → BACLOFEN
Myostibin → ANTIMONY SODIUM GLUCONATE
Myoten → MEPHENSIN
Myotonic Acid → PALICOUREA
Myotonine → BETHANECHOL
Myotonochol → BETHANECHOL
Mypron → DIPYRONE
Myprozine → NATAMYCIN
MYRCENE
12-Myristate 13-Acetate → PHORBOL
Myristica fragrans → NUTMEG
MYRISTICIN also → DILL, ILLICIUM ANISATUM, NUTMEG, PARSLEY, PEPPERS
Myristyl Alcohol → ALCOHOL, MYRISTYL
Myristyltrimethylammonium Bromide → TETRADECYLTRIMETHYL AMMONIUM BROMIDE
Myrizone → THIACETAZONE
Myroxylon balsamum → BALSAM TOLU
Myroxylon pereirae → BALSAM PERU
MYRRH also → CINNAMIC ALDEHYDE
Mysoline → PRIMIDONE
Mysuran → AMBENONIUM
Mytelase → AMBENONIUM
Mytilus sp. → MUSSELS
Mytomycin C → MITOMYCIN C
My-trans → MEPROBAMATE

N

N-3 → METHETOIN
N-521 → DAZOMET
N-553 → TOLPERISONE HCL
N-714 → CHLORPROTHIXENE
N-746 → CLOPENTHIXOL
N-2790 → FONOPHOS
N-7009 → FLUPENTIXOL
NA 97 → PANCURONIUM BROMIDE
NAB 365 → CLENBUTEROL
N.A.B. → NEOARSPHENAMINE
Nabadiol → METHANDRIOL
NABAM
Nabasan → NABAM
NABILONE
Nabolin → METHANDROSTENOLONE

Naphthol Spirits → NAPHTHA
NAPHTHOQUINONES also → 2-NAPHTHOL
1-NAPHTHYLAMINE also → 1-NAPHTHOL, RUBBER
α-Naphthylamine → 1-NAPHTHYLAMINE
2-NAPHTHYLAMINE also → 2-NAPHTHOL, 2-NITRONAPHTHALENE, *N*-PHENYL-*β*-NAPHTHYLAMINE, RUBBER, YELLOW AB, YELLOW OB
β-Naphthylamine → 2-NAPHTHYLAMINE
Naphthylamine Blue → TRYPAN BLUE
Naphthylene Diisocyanate → NAPHTHALENE DIISOCYANATE
β-Naphthyl Ethyl Ether → 2-ETHOXYNAPHTHALENE
1-NAPHTHYLISOTHIOCYANATE also → PHENOBARBITAL
1-Naphthyl-N-methylcarbamate → CARBARYL
β-Naphthylphenylamine → *N*-PHENYL-*β*-NAPHTHYLAMINE
α-Naphthylthiocarbamide → 1-(1-NAPHTHYL)-2-THIOUREA
1-(1-NAPHTHYL)-2-THIOUREA
Naphuride → SURAMIN SODIUM
Napoton → CHLORDIAZEPOXIDE
Napren → NAPROXEN
Naprinol → ACETAMINOPHEN
Naprium → NAPROXEN
Naprius → NAPROXEN
Naprosyn(e) → NAPROXEN
Naprotis → NAPROXEN
NAPROXEN also → ALUMINUM HYDROXIDE, CLOPIDOGREL BISULFATE, DIFLUNISAL, FUROSEMIDE, HYDROCHLOROTHIAZIDE, LITHIUM, MAGNESIUM OXIDE, METHOTREXATE, PREDNIDOLONE, PROBENECID, SALSALATE, SODIUM BICARBONATE, SODIUM VALPROATE, SUCRALFATE
Naprux → NAPROXEN
Naqua → TRICHLORMETHIAZIDE
NAPTALAM
Naptrate → PENTAERYTHRITOL TETRANITRATE
Naramycin A → CYCLOHEXIMIDE
NARASIN also → NICARBAZIN, TIAMULIN
Narcan → NALOXONE
Narcanti → NALOXONE
Narcaricin → BENZBROMARONE
NARCEINE
Narcissine → DAFFODIL, GALANTHUS NIVALUS, NARCISSUS sp.
NARCISSUS sp. also → CALCIUM OXALATE
Narcissus jonquili → JONQUIL also → DAFFODILS
Narcissus pseudonarcissus → DAFFODIL
NARCOBARBITAL
Narcogen → TRICHLOROETHYLENE
Narcompren → NOSCAPINE
Narcoren → PENTOBARBITAL
Narcosine → NOSCAPINE
Narcotal → NARCOBARBITAL
Narcotic Pepper → KAVA
Narcotile → ETHYL CHLORIDE
Narcotine → NOSCAPINE
Narcotussin → NOSCAPINE
Narcoxyl → XYLAZINE
Narcozep → FLUNITRAZEPAM
Nardelzine → PHENELZINE
Nardil → PHENELZINE
Nardoo Fern → MARSILEA DRUMMONDII
Narigix → NALIDIXIC ACID
Narilet → IPRATROPIUM BROMIDE
Naringin → NISOLDIPINE
Narkotal → NARCOBARBITAL
Narone → DIPYRONE
Naropin → ROPIVACAINE
Narphen → PHENAZOCINE HYDROBROMIDE
Narrow-Leafed Laurel → KALMIA
Narrowleaf Sumpweed → IVA sp.
Narsis → MEDAZEPAM
Nartate → DIPYRONE
NARTHECIUM OSSIFRAGUM
Nartograstin → GRANULOCYTE MACROPHAGE COLONY STIMULATING FACTOR
Naryal → COCONUT
Nasacort → TRIAMCINOLONE
Nasalcrom → CROMOLYN SODIUM
Nasalide → FLUNISOLIDE
Nasanyl → NAFARELIN ACETATE
Nascobal → VITAMIN B_{12}
Nasivin → OXYMETAZOLINE
Nasmil → CROMOLYN SODIUM
Nasonex → MOMETASONE FUROATE
Nastenon → OXYMETHOLONE
NASTURTIUM
Natacillin → PENICILLINS – "NEWER and SYNTHETICS" > *Hetacillin*
Natacyn → NATAMYCIN
Natal Cherry → SOLANUM sp.
Natal Yellow Tulip → HOMERIA sp.
NATAMYCIN
Naticardina → QUINIDINE
Natil → CYCLANDELATE
Natrilix → INDAPAMIDE

Natrionex → ACETAZOLAMIDE
Natriphene → σ-PHENYLPHENOL
Natulan → PROCARBAZINE
Natural Gas → METHANE
Natural Reds → GALIUM sp.
Naturetin → BENDROFLUMETHIAZIDE
Naturine → BENDROFLUMETHIAZIDE
Natyl → DIPYRIDAMOLE
Naucaine → PROCAINE
NAUCLEOPSIS sp.
Nausidol → PIPAMAZINE
Nautazine → CYCLIZINE
Nauzelin → DOMPERIDONE
Navadel → DIOXATHION
Navane → THIOTHIXENE
Navelbine → VINORELBINE
Navette → RAPE
Navicalm → MECLIZINE
Navidrex → CYCLOPENTHIAZIDE
Navidrix → CYCLOPENTHIAZIDE
Navilox → ISOXSUPRINE
Navisin → OXYMETAZOLINE
Navoban → TROPISETRON
Navy Bean → BEANS, Kidney
Naxcel → CEFTIOFUR
Naxen → NAPROXEN
Naxidine → NIZATIDINE
Naxofem → NIMORAZOLE
Naxogin → NIMORAZOLE
Nazi → COCAINE also → PHOSGENE
NBC → *n*-BUTYL CHLORIDE
NC 45 → VECURONIUM BROMIDE
NC 123 → MESORIDAZINE BESYLATE
NC 262 → DIMETHOATE
NCI-CO-0544 → MEXACARBATE
NCI-CO-2017 → PHENYLTHIOUREA
NCI-CO-2971 → ZINOPHOS
NCI-C55561 → TETRACYCLINE
ND 50 → OCTOPAMINE
NDEA → *N*-NITROSODIETHYLAMINE
NDELA → *N*-NITROSODIETHANOLAMINE
2′NDG → GANCICLOVIR
NDGA → MASOPROCOL
NEALBARBITAL
Nealymal → NEALBARBITAL
Neamoxyl → PENICILLINS – "NEWER and SYNTHETICS" > *Amoxicillin*
Neat Oil → EUCALYPTUS GLOBULUS
Neazina → SULFAMETHAZINE
Nebactam → AZTREONAM
Nebcin → TOBRAMYCIN
Neberk → TEGAFUR
Nebicina → TOBRAMYCIN
Nebralin → PENTOBARBITAL also → TERFENADINE
Nebramycin Factor 6 → TOBRAMYCIN
Nebraska Fern → CONIUM MACULATUM
Nebupent → PENTAMIDINE ISETHIONATE
NEBURON
Necatorina → CARBON TETRACHLORIDE
Nectadon → NOSCAPINE
NECTARINE also → FRUITS and FRUIT JUICES, PEACHES
Ne-Dioxanin → DIGOXIN
NEDOCROMIL SODIUM also → SODIUM METABISULFITE
Needle Antimony → ANTIMONY TRISULFIDE
Needle-grass → STIPA sp.
Neem Oil → NIM OIL
NEFAZODONE also → ASTEMIZOLE, LOVASTATIN, PRAVASTATIN, RITONAVIR, SIMVASTIN, TACROLIMUS, TRIAZOLAM
Nefazone → NEFAZODONE
Nefco → NITROFURAZONE
NEFOPAM
Nefrix → HYDROCHLOROTHIAZIDE
Nefrolan → CLOREXOLONE
Nefrosul → SULFACHLORPYRIDAZINE
Neftin → FURAZOLIDONE
Nefurofan → SPIRONOLACTONE
Negaxid → PENICILLINS – "NEWER and SYNTHETICS" > *Pivmecillinam*
Neguvon → TRICHLORFON
NegGram → NALIDIXIC ACID
Negram → NALIDIXIC ACID
Nektrohan → ALLOPURINOL
Nelbon → NITRAZEPAM
NELFINAVIR MESYLATE also → ASTEMIZOLE, CISAPRIDE, DELAVIRDINE, FLUVOXAMINE, INDINAVIR, NEVIRAPINE, RIFABUTIN, RITONAVIR, SAQUINAVIR, TERFENADINE
Nelson Mandela → THALLIUM
Nema → TETRACHLOROETHYLENE
Nemacide → DICHLOFENTHION
Nemacur → FENAMIPHOS
Nemafos → ZINOPHOS
Nemafume → DIBROMOCHLOROPROPANE
Nemagon → DIBROMOCHLOROPROPANE
Nemapan → THIABENDAZOLE
Nemasole → MEBENDAZOLE
Nematine → METHAM SODIUM
Nematolyt → PAPAIN
Nemazine → PHENOTHIAZINE
Nembutal → PENTOBARBITAL
Nemex → BEPHENIUM also → PYRANTEL
Nemicide → LEVAMISOLE

Nemosan → PYRANTEL
Nemurel → ESTAZOLAM
Nendrin → ENDRIN
Neo-Antergan → PYRILAMINE
Neoantimosan → STIBOPHEN
Neo-Arsoluin → NEOARSPHENAMINE
NEOARSPHENAMINE
Neo-Atromid → CLOFIBRATE
Neo-Banex → PROPANTHELINE BROMIDE
Neobar → BARIUM SULFATE
Neobetalin-12 → VITAMIN B_{12}
NEOBOUTONIA CANESCENS
Neo-Bradoral → DOMIPHEN BROMIDE
Neocaine → PROCAINE
NEOCARZINOSTATIN
Neocefal → CEFAMANDOLE
Neochin → CHLOROQUINE
Neochlorogenic Acid → BROCCOLI, BRUSSEL SPROUTS, CABBAGE, KALE, PEACHES, PLEUROTIS sp.
Neocid → DDT
Neocidol → DIAZINON
Neo-Cobefrin → NORDEFRIN
Neo-Codema → HYDROCHLOROTHIAZIDE
Neo-Corovas → PENTAERYTHRITOL TETRANITRATE
Neocycline → TETRACYCLINE
Neo-Cytamen → VITAMIN B_{12}
Neodalit → DIBENZEPIN
Neo-Dema → CHLOROTHIAZIDE
Neo-dentoraline → PROCAINE
Neo-Devomit → CYCLIZINE
Neo-Dibetic → TOLBUTAMIDE
Neodicoumarol → ETHYL BISCOUMACETATE
Neodigitalis → DIGITALIS
Neo Dohyfral D_3 → VITAMIN D_3
Neodorm → PENTOBARBITAL
Neodrol → STANOLONE
Neodrenal → ISOPROTERENOL
Neo-Duplamox → PENICILLINS – "NEWER and SYNTHETICS" > *Amoxicillin*
NEODYMIUM
Neo-Epinine → ISOPROTERENOL
Neo-Erycinum → ERYTHROMYCIN ESTOLATE
Neoesserin → NEOSTIGMINE
Neofamid → MAFENIDE
Neo-Farmadol → OXYPHENBUTAZINE
Neo-Fluimucil → ACETYLCYSTEINE
Neoflumen → HYDROCHLOROTHIAZIDE
Neo-Fulcin → GRISEOFULVIN
Neogama → SULPIRIDE
Neogel → CARBENOLOXONE
Neo-Gerastan → PENTYLENETETRAZOL
Neogest → NORGESTREL
Neo-gilurytmal → PRAJMALINE TARTRATE
Neo H2 → ROXATIDINE ACETATE
Neohetramine → THONZYLAMINE HYDROCHLORIDE
Neohombreol M → METHYL TESTOSTERONE
Neohydrin → CHLORMERODRIN
Neohydriol → ETHIODIZED OIL
Neo-Istafene → MECLIZINE
Neoisuprel → ISOETHARINE
Neolate → NEOMYCIN
Neolexina → CEPHALEXIN
Neoloid → CASTOR OIL
Neolutin → 17-α-HYDROXYPROGESTERONE also → PROGESTERONE
Neomas → NEOMYCIN
Neomercazole → CARBIMAZOLE
Neomestine → FONAZINE MESYLATE
Neo-Metantyl → PROPANTHELINE BROMIDE
Neo-Methiodol → IODOPYRACET
Neomin → NEOMYCIN
NEOMYCIN also → AMOBARBITAL, AZATHIOPRINE, BACITRACIN, BENTONITE, BUTIROSIN, CLOFIBRATE, COLISTIN, CURARE, DIMETHYL TUBOCURARINE IODIDE; ETHER, ETHYL; ETHINYL ESTRADIOL, FATS, FRAMYCETIN, GALLAMINE TRIETHIODIDE, GENTAMICIN, HALOTHANE, KANAMYCIN, LACTOSE, LIPASE, MALATHION, METHOTREXATE, PAROMOMYCIN; PENICILLIN also → PENICILLIN > *Penicillin V*; POLYMIXIN B, PROCAINAMIDE, PROMETHAZINE, QUINIDINE, SPECTINOMYCIN, SUCCINYLCHOLINE, SULFACETAMIDE SODIUM, TOBRAMYCIN, *d*-TUBOCURARINE CHLORIDE, TYROTHRICIN, VANCOMYCIN, VITAMIN A, VITAMIN B_{12}, VITAMIN K, WARFARIN
Neomycin B → FRAMYCETIN
Neomyson → THIAMPHENICOL
Neo-Naclex → BENDROFLUMETHIAZIDE
Neonal → BUTETHAL
Neonicotine → ANABASINE
Neo-Null-Soy® → SOYBEANS
Neo-Oestranol I and II → DIETHYLSTILBESTROL
Neo-Oxypaat → PYRVINIUM PAMOATE
Neopens → PENICILLIN > *Benzylpenicillin, Potassium*

Neo-Pergonal → MENOTROPINS
Neopeviton → VITAMIN B_3
Neopharmedrine → METHAMPHETAMINE
Neophrin → PHENYLEPHRINE HCL
Neoplanta → CYCLOSPORINE
Neoplatin → CISPLATIN
Neopogen → FILGRASTIM
Neoprene → CHLOROPRENE also → ETHYL BUTYL THIOUREA, RUBBER
Neoproc → PENICILLIN > *Benzylpenicillin, Procaine*
Neoprontosil → AZOSULFAMIDE
Neo-Protostan → PROTEIN HYDROLYSATES
Neopyrithiamine → PYRITHIAMIN
Neoral → CYCLOPHOSPHAMIDE also → CYCLOSPORINE, RAPAMYCIN
Neorickettsia helminthoeca → SALMON
Neo-Rinoleina → XYLOMETAZOLINE
Neosalvarsan → NEOARSPHENAMINE
Neosar → CYCLOPHOSPHAMIDE
Neo-Silvol → SILVER IODIDE
Neo-Skiodan → IODOPYRACET
Neosorex → DIFENACOUM
Neostenovasan → DYPHYLLINE
Neosteron → METHANDRIOL
Neostigmin → NEOSTIGMINE
NEOSTIGMINE also → ATROPINE, COLISTIN, CURARE, GALLAMINE TRIETHIODIDE, KANAMYCIN, MALATHION, PROPRANOLOL, SUCCINYLCHOLINE
Neoston → ALCLOFENAC
Neostreptal → SULFADIMETHOXINE
Neo-Synephrine → PHENYLEPHRINE HCL
Neoteben → ISONIAZID
Neo-Tenebryl → IODOPYRACET
Neothylline → DYPHYLLINE
Neothyreostat → CARBIMAZOLE
Neotigason → ACITRETIN
Neotilina → DYPHYLLINE
Neo-Urofort → CHLORAZANIL
Neo-Vasophylline → DYPHYLLINE
Neo-Zine → PHENMETRAZINE
Neozine → METHOTRIMEPRAZINE
Nep → CATNIP
Nepenthe → MORPHINE
Nepeta cataria → CATNIP
NEPETA HEDERACEA
NEPETALACTONE also → CATNIP
Nephramid → ACETAZOLAMIDE
Nephril → POLYTHIAZIDE
Nepresol → DIHYDRALAZINE
Neptall → ACEBUTOLOL
Neptazone → METHAZOLAMIDE
Neravan → BUTABARBITAL
Nerdipina → NICARDIPINE
NEREISTOXIN
Nericur → BENZOYL PEROXIDE
Neriin → OLEANDER
Neriodin → DICLOFENAC
Nerium oleander → OLEANDER
Nero, Emperor → YEW
Nerobol → METHANDROSTENOLONE
Nerofen → IBUPROFEN
NEROL
Nerolin → 2-ETHOXYNAPHTHALENE
Neroli Oil → NEROL
Nerprun Bourdain → FRANGULA
Nervacton → BENACTYZINE
Nervonus → MEPROBAMATE
Nesacaine → CHLOROPROCAINE
Nesdonal → THIOPENTAL
Nesontil → OXAZEPAM
Nethalide → PRONETHALOL
Netilyn → NETILMYCIN
Netillin → NETILMYCIN
NETILMYCIN
Netromycin(e) → NETILMYCIN
Netsusarin → AMINOPYRINE
Nettacin → NETILMYCIN
Nettle → URTICA DIOICA
Neuchlonic → NITRAZEPAM
Neulactil → PROPERICIAZINE
Neuleptil → PROPERICIAZINE
Neumega → INTERLEUKIN 11
Neupogen → GRANULOCYTE COLONY STIMULATING FACTOR
Neuquinone → UBIDECARENONE
Neuracen → BECLAMIDE
Neuractil → METHOTRIMEPRAZINE
Neuraxin → METHOCARBAMOL
Neurelax → VALERIANA sp.
NEURINE
Neurocil → METHOTRIMEPRAZINE
Neurofort → CHLORAZANIL
NEUROLAENA sp.
Neurolene → NOMIFENSINE MALEATE
Neuroleptone → BENACTYZINE
Neurolytril → DIAZEPAM
Neuronika → ACETYLSALICYLIC ACID
Neurontin → GABAPENTIN
Neuroprocin → ECTYLUREA
Neurosedyn → THALIDOMIDE
Neurosyn → PRIMIDONE
Neurotrast → IOPHENDYLATE
Neustab → THIACETAZONE
Neutrafil → DYPHYLLINE
Neutral Acriflavine → ACRIFLAVINE

Neutral Protein Hagedorn Insulin → INSULIN
NEUTRAL RED
Neutrapen → β-LACTAMASE
Neutraphylline → DYPHYLLINE
Neutroflavine → ACRIFLAVINE
Neutronyx 611 → NONOXYNOL-9
Neutrormone → METHANDRIOL
Neutrosterone → METHANDRIOL
Nevax → DOCUSATE SODIUM
Nevental → NEALBARBITAL
Nevigramon → NALIDIXIC ACID
NEVIRAPINE also → INDINAVIR, METHADONE, NELFINAVIR MESYLATE, RIFAMPIN, RITONAVIR
Neviraptine → SAQUINAVIR
Nevralgina → DIPYRONE
Nevrinton → VITAMIN B_1
Nevriton → VITAMIN B_1
New Guinea Lung → GRASSES
NEWSPAPERS
NEWT also → SALAMANDERS
Newtol → XYLITOL
New Zealand Prickly Spinach → TETRAGONIA TETRAGONOIDES
Nexaband → CYANOACRYLATES
Nexabond → BUTYL CYANOACRYLATE
Nexion → BROMOPHOS
Nexolin → OXOLINIC ACID
Nezeril → OXYMETAZOLINE
NF 6 → TOBRAMYCIN
NF 7 → NITROFURAZONE
NF 64 → NIHYDRAZONE
NF 180 → FURAZOLIDONE
NF 260 → FURALTADONE
NF 963 → FURAZOLIUM CHLORIDE
NF 1002 → NIFURPRAZINE
NFZ → NITROFURAZONE
NG → NITROGLYCERIN
Nga → DIAMPHIDIA
NGAIONE also → MYOPORUM sp., PHENOBARBITAL, SWEET POTATO
Ngaio Tree → MYOPORUM sp.
Ngiengeie → ABRUS PRECATORIUS
NHMI → *N*-NITROSOHEPTAMETHYLENEIMINE
Nia → MEGESTROL ACETATE
NIA 5273 → PIPERONYL BUTOXIDE
NIA 5462 → ENDOSULFAN
NIA 5,488 → TETRADIFON
NIA 10242 → CARBOFURAN
NIA 33,297 → PERMETHRIN
Nicacid → VITAMIN B_3
Nicamindon → VITAMIN B_3
Nicangin → VITAMIN B_3
Niacin → VITAMIN B_3
Niacinamide → VITAMIN B_3
Niacor → VITAMIN B_3
Niagara 1240 → ETHION
Niagara 9044 → BINAPACRYL
Niagaril → BENDROFLUMETHIAZIDE
Niagaramite → ARAMITE
Niagestin → MEGESTROL ACETATE
Niagra Blue → TRYPAN BLUE
Niagrathal → ENDOTHALL
NIALAMIDE also → ACENOCOUMAROL, ALCOHOL, ETHYL; AMPHETAMINE, BEANS, CHEESES, DESIPRAMINE, DIPHENHYDRAMINE, ETHYL BISCOUMACETATE, GUANETHIDINE, HERRING, INSULIN, MEPERIDINE, SUMATRIPTAN SUCCINATE
Nialate → ETHION
Niamid → NIALAMIDE
Niamidal → NIALAMIDE
Niamine → NIKETHAMIDE
Niaquitil → NIALAMIDE
Niaspan → VITAMIN B_3
NIAX® CATALYST ESN
Niazol → NAPHAZOLINE
Nibal → METHENOLONE ACETATE
Nibiol → NITROXOLINE
Nicabate → NICOTINE and SALTS
Nicalex → ALUMINUM NICOTINATE
Nicamide → NIKETHAMIDE
NICANDRA PHYSALODES
Nicant → NICARDIPINE
Nicapress → NICARDIPINE
Nicarb → NICARBAZIN
NICARBAZIN
Nicardal → NICARDIPINE
NICARDIPINE also → CYCLOSPORINE, TACROLIMUS
Nicarpin → NICARDIPINE
Nicel → METHYL CELLULOSE
Nicelate → NALIDIXIC ACID
Nicene → SULFAMETHIZOLE
Nicergoline → METERGOLINE
Nicizina → ISONIAZID
NICKEL also → ACETYLENE, ASPARAGUS, CARROTS, CHOCOLATE, COBALT, CORN, HERRING, JEWELRY, LIPSTICK, PEARS, PEAS, POTASSIUM HYDROXIDE, RHUBARB, SHOES, SPIGELIA ANTHELMIA, STEEL, TEA, TOMATO, WINE
NICKEL ACETATE
NICKEL CARBONYL
NICKEL CHLORIDE

NICKEL SULFATE
Nickel Tetracarbonyl → NICKEL CARBONYL
Niclofolan → 3,3´-DICHLORO-5,5´-DINITRO-σ,σ-DIPHENOL
NICLOSAMIDE
Nicobid → VITAMIN B_3
Nicobion → VITAMIN B_3
Nicociana → TOBACCO
Nicodel → NICARDIPINE
Nicoderm → NICOTINE and SALTS
Nicolan → NICOTINE and SALTS
Nicolane → NOSCAPINE
Nicolar → VITAMIN B_3
Nicolen → FURAZOLIDONE
Niconacid → VITAMIN B_3
Niconyl → ISONIAZID
Nicopatch → NICOTINE and SALTS
Nicoprazine → MORINAMIDE
Nicor → NIKETHAMIDE
NICORANDIL
Nicorette → NICOTINE and SALTS
Nicorine → NIKETHAMIDE
Nicorol → FUROSEMIDE
Nicospan → VITAMIN B_3
Nicotamide → VITAMIN B_3
Nicotell TTS → NICOTINE and SALTS
NICOTIANA sp.
Nicotibina → ISONIAZID
Nicotinamide → VITAMIN B_3
NICOTINE and SALTS also → CYCLANDELATE, EQUISETUM, METHANESULFONATE, NICOTIANA sp., PROPRANOLOL, SHAVE GRASS, TOBACCO
Nicotinell → NICOTINE and SALTS
Nicotine Polacrilex → NICOTINE and SALTS
Nicotinia tabacum → TOBACCO
Nicotinic Acid → VITAMIN B_3
Nicotinic Acid Benzyl Ester → BENZYL NICOTINATE
Nicotinic Acid Diethylamide → NIKETHAMIDE
Nicotinic Alcohol → NICOTINYL ALCOHOL
NICOTINYL ALCOHOL
Nicotofuryl → THURFYL NICOTINATE
Nicoumalone → ACENOCOUMAROL also → BENZIODARONE
Nicoxin → NICARBAZIN
Nicrazin → NICARBAZIN
Nidantin → OXOLINIC ACID
Nidaxin → MEDROXYPROGESTERONE
Nidrafur → NIHYDRAZONE
Nidrane → BECLAMIDE
Nidrel → NITRENDIPINE
Nielle → AGROSTEMMA GITHAGO
Niespulver → VERATRUM sp.
Nifedicor → NIFEDIPINE
Nifedin → NIFEDIPINE
NIFEDIPINE also → ALPRENOLOL, CYCLOSPORINE, DALFOPRISTIN, DIGOXIN, DILTIAZEM, DISOPYRAMIDE, GRAPEFRUIT, MAGNESIUM, SULFATE, NIMODIPINE, QUINIDINE, RIFAMPIN, TACROLIMUS, TERBUTALINE, TERFENADINE, THEOPHYLLINE, VANCOMYCIN, VINCRISTINE
Nifelan → NIFEDIPINE
Nifelat → NIFEDIPINE
Nifensar XL → NIFEDIPINE
Niferex → IRON POLYSACCHARIDE COMPLEX
NIFLUMIC ACID
Nifluril → NIFLUMIC ACID
Niftolid → FLUTAMIDE
Nifucin → NITROFURAZONE
Nifulidone → FURAZOLIDONE
NIFURATEL also → BEER
NIFUROXIME
NIFURPRAZINE
NIFURTIMOX
Nifuzon → NITROFURAZONE
Nigalax → BISACODYL
Nigerine → MIMOSA
Night-blooming Jessamine → CESTRUM also → GELSEMIUM SEMPERVIRENS
Nigrin → STREPTONIGRIN
Nigrosines → INKS
NIH 4185 → THIAMBUTENE
NIH 7519 → PHENAZOCINE HYDROBROMIDE
NIH 7590 → PIMINODINE
NIH 8805 → BUPRENORPHINE
Niharu → MISTLETOE, EUROPEAN or JAPANESE
NIHYDRAZONE also → BENTONITE
Nikardin → NIKETHAMIDE
Nike® → METHYLENEDIOXYAMPHETAMINE, RUBBER
NIKETHAMIDE also → PHENYTOIN, THIOPENTAL
Nikion → BENDROFLUMETHIAZIDE
Nikorin → NIKETHAMIDE
Nilandron → NILUTAMIDE
Nilatil → ITRAMIN TOSYLATE
Nilergex → ISOTHIPENDYL
Nilevar → NORETHANDROLONE
Nilodin → LUCANTHOE HYDROCHLORIDE

Nilstat → NYSTATIN
Niltuvin → NICOTINYL ALCOHOL
NILUTAMIDE
Nilverm → LEVAMISOLE
Nilverm → TETRAMISOLE
Nilzan → OXYCLOZANIDE
Nimaol → OCTAMOXIN
NIMESULIDE
Nimicor → NICARDIPINE
Nimitex → TEMEPHOS
NIMODIPINE also → GRAPEFRUIT, PROPRANOLOL
NIM OIL
NIMORAZOLE
Nimotop → NIMODIPINE
Ninge Ninge → PASSIFLORA sp.
Ninjin → GINSENG
Nio-A-Let → VITAMIN A
Nioform → IODOCHLORHYDROXYQUINOLINE
Niolo → ANONA
Nionate → FERROUS GLUCONATE
Niopam → IOPAMIDOL
Niotal → ZOLPIDEM TARTRATE
Nip → CATNIP also → POLYVINYL CHLORIDE
Nipantiox 1-F → BUTYLATED HYDROXYANISOLE
Nip-A-Thin → NAPTALAM
Nipaxon → NOSCAPINE
Nipecotan → ANILERIDINE
Nipent → PENTOSTATIN
Niperyt → PENTAERYTHRITOL TETRANITRATE
Niphtholide → FLUTAMIDE
Niran → CHLORDANE also → PARATHION
NIRIDAZOLE also → DIAZEPAM
Nirvan → METHOTRIMEPRAZINE
Nirvanil → VALNOCTAMIDE
Nisentil → ALPHAPRODINE
Nisidana → OPIPRAMOL
Nisis → VALSARTAN
NISOLDIPINE also → ATENOLOL, GRAPEFRUIT, PROPRANOLOL, QUINIDINE
Nisotin → ETHIONAMIDE
Nisulid → NIMESULIDE
Nitan → PEMOLINE
NITARSONE
Niter → POTASSIUM NITRATE
NITHIAZIDE
Nitobanil → TEGAFUR
Nitoman → TETRABENAZINE
Nitorol → ISOSORBIDE DINITRATE
Nitradisc → NITROGLYCERIN
Nitrador → DINITRO-σ-CRESOL
Nitrados → NITRAZEPAM
Nitraldone → FURALTADONE
NITRAMINE
Nitran → NITROGLYCERIN
ortho- or para-Nitraniline → NITROANILINES
NITRATES also → ACHILLEA sp., AMARANTHUS, AMMI MAJUS, AMMONIUM NITRATE, BARLEY, BASSIA, BEETS, BISMUTH SUBNITRATE, BOLOGNA, BURDOCK, CARROTS, CELERY, COCKLES, CORN, DAUCUS CAROTA; 2, 4-DI-CHLOROPHENOXYACETIC ACID; ECHINOCHOLA CRUS-GALLI, ELDERBERRY, FISH, FLAX, GROUND CHERRY, HELIOTROPIUM, KALE, LIDOCAINE, MANGEL, MEPERIDINE, MILLET, OATS, PANICUM sp., PARSNIP, PARSONSIA sp., PENNISETUM sp., PHYSALIS sp., POLYGONUM sp., POTASSIUM NITRATE, POTATO, RADISH, RAGWEED, RAPE, RHUBARB, RYE, SALSOLA sp., SETARIA sp., SILDENAFIL, SOLANUM sp., SORGHUM, SOYBEANS, SQUASH, SUNFLOWER, SWISS CHARD, ZYGOPHYLLUM sp.
NITRAZEPAM also → PHENELZINE
NITRENDIPINE also → GRAPEFRUIT
Nitrenpax → NITRAZEPAM
Nitretamin → TROLNITRATE PHOSPHATE
NITRIC ACID also → *N*-NITROSODIMETHYLAMINE, NITROUS OXIDE
NITRIC OXIDE also → SILO
Nitriderm-TTS → NITROGLYCERIN
Nitrile → CYANIDES
NITRILOTRIACETIC ACID
Nitrimidazine → NIMORAZOLE
NITRITES also → AMINOPYRINE, ATRAZINE, BEANS, CARROTS, DESCURAINIA PINNATA; *N*, *N*-DIMETHYLAMINE, FISH, GLUCOSINOLATES, HAM, HERRING, MANGEL, MEPERIDINE, METHAPYRILENE, *N*-METHYLANILINE, MISEROTOXIN, POTASSIUM NITRATE, SAUSAGE, SODIUM NITRATE
3-*Nitro* → ROXARSONE
NITROANILINES also → CRAYONS, PYRIMINIL, TEA
Nitrobaat → NITROGLYCERIN
3- or 4-*Nitrobenzenamine* → NITROANILINES

NITROBENZENE also → POLISHES, SHOE
p-Nitrobenzoic Acid → PHENOBARBITAL
Nitrobenzol → NITROBENZENE
Nitro-Bid → NITROGLYCERIN
p-NITROBIPHENYL
Nitrocarbol → NITROMETHANE
Nitrocellulose → MUNITIONS
NITROCHLOROBENZENES also → 2,4-DINITROCHLOROBENZENE
Nitrochloroform → CHLOROPICRIN
Nitrocine → NITROGLYCERIN
Nitrocontin → NITROGLYCERIN
Nitroderm TTS → NITROGLYCERIN
Nitrodisc → NITROGLYCERIN
Nitro-Dur → NITROGLYCERIN
Nitroetan → NITROETHANE
NITROETHANE
NITROFEN
Nitrofortin → NITROGLYCERIN
Nitrofural → NITROFURAZONE
Nitrofuran → NITROFURAZONE
NITROFURANTOIN also → ACETALDEHYDE; ALCOHOL, ETHYL; ALUMINUM HYDROXIDE, CALCIUM CARBIMIDE, CLOFIBRATE, ETHACRYNIC ACID, MAGNESIUM OXIDE, MEFENAMIC ACID, NALIDIXIC ACID, OXOLINIC ACID, PENICILLIN, PHENOBARBITAL, PHENYLBUTAZONE, PROBENECID, PROPANTHELINE, SODIUM BICARBONATE, THYROXINE
Nitrofuratel → NIFURATEL
NITROFURAZONE also → BARIUM, TYROTHRICIN
Nitrofurmethone → FURALTADONE
Nitrogard → NITROGLYCERIN
NITROGEN also → NITRATES, NITRITES
Nitrogen Benzide → AZOBENZENE
NITROGEN CHLORIDE also → METHIONINES, WHEAT
NITROGEN DIOXIDE also → CARBON DIOXIDE, CUPFERON, ICE, NITRIC OXIDE, SILO
NITROGEN FLUORIDE
Nitrogen Monoxide → NITRIC OXIDE
Nitrogen Mustard → MECHLORETHAMINE also → MUSTARD GAS, PROCARBAZINE
Nitrogen Oxides → σ-TOLIDINE, TOLUENE 2,4-DIISOCYANATE
Nitrogen Peroxide → NITROGEN DIOXIDE also → NITROGEN TETROXIDE
NITROGEN TETROXIDE also → SILO
Nitrogen Trichloride → NITROGEN CHLORIDE also → FLOUR, METHIONINES
Nitrogen Trifluoride → NITROGEN FLUORIDE
Nitrogen Trioxyfluoride → FLUORINE NITRATE
Nitro-Gesanit → NITROGLYCERIN
Nitroglin → NITROGLYCERIN
NITROGLYCERIN also → ALCOHOL, ETHYL; HEPARIN, ISOSORBIDE DINITRATE, MUNITIONS, NITRATES, NITRITES, PHENYLEPHRINE HYDROCHLORIDE, PRAZOCIN HYDROCHLORIDE, PROPYLENE GLYCOL, SILDENAFIL, VITAMIN E
Nitroglycerol → NITROGLYCERIN
Nitroglycol → ETHYLENE GLYCOL DINITRATE
Nitroglyn → NITROGLYCERIN
Nitrogranulogen → MECHLORETHAMINE
n-Nitrohexamethyleneimine → TOLAZAMIDE
Nitrol → ISOSORBIDE DINITRATE also → NITROGLYCERIN
Nitrolan → NITROGLYCERIN
Nitrolande → NITROGLYCERIN
Nitrolar → NITROGLYCERIN
Nitrolent → NITROGLYCERIN
Nitrolime → CALCIUM CARBIMIDE
Nitrolingual → NITROGLYCERIN
Nitro Mack → NITROGLYCERIN
NITROMERSOL
Nitrometan → NITROMETHANE
NITROMETHANE
Nitromex → NITROGLYCERIN
Nitronal → NITROGLYCERIN
2-NITRONAPHTHALENE also → 2-NAPHTHYLAMINE
β-Nitronaphthalene → 2-NITRONAPHTHALENE
Nitrong → NITROGLYCERIN
Nitronic → ROXARSONE
Nitropanoic Acid → PENICILLIUM sp.
Nitropenta → PENTAERYTHRITOL TETRANITRATE
Nitropenton → PENTAERYTHRITOL TETRANITRATE
NITROPHENIDE
p- or 4-NITROPHENOL also → PARATHION
Nitropress → SODIUM NITROPRUSSIDE
Nitro-PRN → NITROGLYCERIN
1-NITROPROPANE
2-NITROPROPANE
3-Nitropropanoic Acid → INDIGOFERA sp.
3-NITRO-1-PROPANOL also → MISEROTOXIN
3-NITRO-1-PROPANOL

β-NITROPROPIONIC ACID also → ASPERGILLUS
NITROPYRENES
4-NITROQUINOLINE-1-OXIDE
5-Nitro-8-quinolinol → NITROXOLINE
Nitrorectal → NITROGLYCERIN
Nitroretard → NITROGLYCERIN
Nitrosamines → BEER, BRONOPOL, CHEESES, CIMETIDINE, FISH, MALT, *N*-METHYLANILINE, MILK, NITRATES, NITRITES, OCTYLDIMETHYL PABA, PORK, RADISH, RUBBER, SOLANUM sp., SPIGELIA ANTHELMIA
Nitrosigma → NITROGLYCERIN
N-Nitrosoatrazine → ATRAZINE
N-NITROSOBIS(2-OXYPROPYL)AMINE
N-NITROSOSDIBUTYLAMINE
N-NITROSOBUTYLUREA
N-NITROSODIETHANOLAMINE
N-NITROSODIETHYLAMINE also → ACETYLCYSTEINE, DIETHYLPROPION; *N,N*-DIMETHYLAMINE, KACHASU, METHOXYFLURANE
N-NITROSODIMETHYLAMINE
p-NITROSO-*N,N*-DIMETHYLANILINE
N-NITROSO-2,6-DIMETHYLMORPHOLINE
Nitrosodimethylnitrosamine → HERRING, MILK, SODIUM BENZOATE
p-NITROSODIPHENYLAMINE
N-NITROSODI-*n*-PROPYLAMINE
NITROSOEPHEDRINE
N-NITROSO-*N*-ETHYLUREA
N-NITROSOHEPTAMETHYLENEIMINE
N-Nitrosohexamethyleneimine → TOLAZAMIDE
4-(*N*-NITROSOMETHYLAMINO)-l-(PYRIDYL)-l-BUTANONE
N-NITROSOMETHYLANILINE
N-NITROSOMETHYL BENZYLAMINE
N-NITROSO-*N*-METHYL-*N*-CYCLOHEXYLAMINE
N-NITROSO-*N*-METHYLUREA
N-NITROSOMETHYLVINYLAMINE
N-NITROSOMORPHOLINE also → MARGARINE, MORPHOLINE
N′-NITROSONORNICOTINE also → TOBACCO
N-NITROSOPIPERIDINE
N-NITROSOPYRROLIDINE
Nitrosorbon → ISOSORBIDE DINITRATE
N-NITROSOSARCOSINE
Nitrosoureas → SEMUSTINE
N-(5-Nitro-2-thiazolyl)acetamide → 2-ACETAMIDO-5-NITROTHIAZOLE
Nitrospan → NITROGLYCERIN
Nitrostat → NITROGLYCERIN
Nitrostigmine → PARATHION
Nitrothiamidazol → NIRIDAZOLE
5-Nitro-2-thiazolamine → AMINONITROTHIAZOLE
NITROTOLUENE
NITROUS OXIDE also → 4-AMINODIPHENYL, AMINOPYRINE, σ-ANISIDINE, BEER, FENTANYL; FUELS, OIL; HALOTHANE, HYDROGEN, MEPERIDINE; METHANE, NITRIC OXIDE, NITROGEN DIOXIDE; 4,4-OXYDIANILINE PHENOXYBENZAMINE HYDROCHLORIDE, PYRROLIZIDINES
Nitrox → METHYL PARATHION
NITROXOLINE
Nitroxy Fluoride → FLUORINE NITRATE
NITROXYNIL also → SUCCINYLCHOLINE
Nitrozell-retard → NITROGLYCERIN
Nitrozone → NITROFURAZONE
Nitrumon → CARMUSTINE
Nitryl Hypofluorite → FLUORINE NITRATE
Niu → OAKS
Nivachine → CHLOROQUINE
NIVALENOL also → TRICHOTHECENE MYCOTOXINS, YELLOW RAIN
Nivalin → GALANTHAMINE
Nivano 1-N → IODOFENPHOS
Nivaquine → CHLOROQUINE
Nivaquine B → CHLOROQUINE
Nivelona → DIPHENAMIL METHYLSULFATE
NIX → PERMETHRIN
Nixatidine → KETOCONAZOLE
Niyaru → MISTLETOE, EUROPEAN or JAPANESE
NIZATIDINE also → THEOPHYLLINE
Nizaxid → NIZATIDINE
Nizoral → KETOCONAZOLE
Njemu → CANNABIS
NK 711 → LEPTOPHOS
Nkaya diamba → CANNABIS
Nkoke-Io → ANONA
NLT → ITRAMIN TOSYLATE
NNC-05-0328 → TIAGABINE
NMA → *N*-NITROSOMETHYLANILINE
NMC → *N*-NITROSO-*N*-METHYL-*N*-CYCLOHEXYLAMINE
NMOR → *N*-NITROSOMORPHOLINE
NMU → *N*-NITROSO-*N*-METHYLUREA
NNDG → METFORMIN
NNK → 4-(*N*-NITROSOMETHYLAMINO)-1-(PYRIDYL)-1-BUTANONE

NNM → *N*-NITROSOMORPHOLINE
NO-328 → TIAGABINE
NO-05-0328 → TIAGABINE
Noah → ALCOHOL, ETHYL
Noan → DIAZEPAM
Nobacid → SALSALATE
Nobacter → TRICLOCARBAN
Nobedon → ACETAMINOPHEN
Nobedorm → METHAQUALONE
Nobfen → IBUPROFEN
Nobgen → IBUPROFEN
Nobrium → MEDAZEPAM
Nocipam → NEFOPAM
Noctan → METHYPRYLONE
Noctazepam → OXAZEPAM
Noctec → CHLORAL HYDRATE
Noctesed → NITRAZEPAM
Noctilene → METHAQUALONE
Noctivane → HEXOBARBITAL
Noctone → RANITIDINE
Noctosom → FLURAZEPAM
Nocu → DIAZEPAM
Nodapton → GLYCOPYRROLATE
Nodinol → MERCURY, AMMONIATED also → MERCURY
Nodopen-500 → PENICILLIN > *Benzylpenicillin, Potassium*
NODULARIA SPUMIGENA
Noflo → NORFLOXACIN
Nogos → DICHLORVOS
Nogram → NALIDIXIC ACID
Noin → INDELOXAZINE
Nolicin → NORFLOXACIN
NOLINA TEXANA
Nolipax → FENOFIBRATE
Nolotil → DIPYRONE
Noltam → TAMOXIFEN CITRATE
Noludar → METHYPRYLONE
Nolvadex → TAMOXIFEN CITRATE
Nolvasan → CHLORHEXIDINE
Nomersam → THIRAM
Nomersan → THIRAM
Nometan → PIPERAZINE
Nometine → PIPAMAZINE
NOMIFENSINE MALEATE
γ-NONALACTONE
NONANE
Nonanedioic Acid → AZELAIC ACID
1-NONANETHIOL
5-NONANONE
NoNemic → IRON DEXTRANS
Nongu → PHYSOSTIGMA VENENOSUM
Nonox-S → 1-NAPHTHYLAMINE
NONOXYNOL-9
n-NONYL ACETATE
Nonyl Hydride → NONANE
n-Nonylmercaptan → 1-NONANETHIOL
Nonythiol → 1-NONANETHIOL
Noogoora Bur → COCKLEBURS
Nootron → PIRACETAM
Nootrop → PIRACETAM
Nootropil → PIRACETAM
Nootropyl → PIRACETAM
Nopalmate → HEXAFLURATE
Nopalumate → HEXAFLURATE
No-Pest Strips® → SUCCINYLCHOLINE
Nopia → MUROCTASIN
Nopil → SULFAMETHOXAZOLE
NOPYR → *N*-NITROSOPYRROLIDINE
Noracin → NORFLOXACIN
Noradrenalin(e) → NOREPINEPHRINE
Noranat → INDAPAMIDE
Noraxin → NORFLOXACIN
NORBOLETHONE
Norboral → CARBUTAMIDE
NORBORMIDE
NORCHLORCYCLIZINE
Norcocaine → COCAINE
Norcolut → NORETHINDRONE
Norcuron → VECURONIUM BROMIDE
31-Nordanosterol → POTATO
Nordaz → NORDAZEPAM
NORDAZEPAM
NORDEFRIN
Norden → OCTOPAMINE
Nordialex → GLICLAZIDE
Nordiazepam → NORDAZEPAM
Nordihydrocapsaicin → PEPPERS
Norditropin → GROWTH HORMONES
Norephedrane → AMPHETAMINE
dl-Norephedrine → PHENYLPROPANOLAMINE
ψ Norephedrine → NORPSEUDOEPHEDRINE also → EPHEDRINE
NOREPINEPHRINE also → ANTAZOLINE, BANANA, CARBON DIOXIDE, CHLORPHENAMINE, CHLORPHENTERMINE, CHOCOLATE, CYCLOPROPANE, DEXTROAMPHETAMINE, DIPHENHYDRAMINE, L-DOPA, GUANETHIDINE, HEROIN, HYDROCORTISONE, IMIPRAMINE, INDOMETHACIN, KETAMINE, METHOXYFLURANE, METHYCLOTHIAZIDE, METHYLDOPA, MORPHINE, NIALAMIDE, NITROGLYCERIN, NORTRIPTYLINE,

Noryflex → NOXYTHIOLIN
Norzepine → NORTRIPTYLINE
Norzetam → PIRACETAM
Noscapalin → NOSCAPINE
NOSCAPINE also → ALCOHOL, ETHYL; OPIUM
Nose Bleed → FEVERFEW
Noseburn → TRAGIA sp.
Nose Candy → COCAINE
No-See-Ums → SAND FLIES
Nosim → ISOSORBIDE DINITRATE
No-Spa → DROTAVERINE
Nospan → TYBAMATE
Nospanum → DROTAVERINE
Nostal → ECTYLUREA
Nostradamus → Preface
Nostyn → ECTYLUREA
Nosydrast → IODOPYRACET
Notandron → METHANDRIOL
Notaral → PENICILLIN > *Benzylpenicillin, Potassium*
Notensil → ACEPROMAZINE
Notézine → DIETHYLCARBAMAZINE
Nothiazine → MEPAZINE
NOTHOLAENA SINNUATA
Notonecta triguttata → WATER BOATMAN
Notropone C → DINOSEB
Notul → CIMETIDINE
Nourytam → TAMOXIFEN CITRATE
Novaban → TROPISETRON
Novacid → DIPYRONE
Novacrysin → GOLD SODIUM THIOSULFATE
Novafed → PSEUDOEPHEDRINE
Novalcina → DIPYRONE
Novaldin → DIPYRONE
Novalgin → DIPYRONE
Novamidon → AMINOPYRINE
Novamin → AMIKACIN
Novantrone → MITOXANTRONE HCL
Novapirina → DICLOFENAC
Novarsan → NEOARSPHENAMINE
Novarsenobenzol → NEOARSPHENAMINE
Novarsenobillon → NEOARSPHENAMINE
Novarsenol → NEOARSPHENAMINE
Novasmasol → METAPROTERENOL
Novatec → LISINOPRIL
Novathion → SUMITHION
Novatrin → HOMATROPINE
Novatropine → HOMATROPINE
Novaurantia → ORANGE G
Novazam → DIAZEPAM
Novazole → CAMBENDAZOLE
Novemina → DIPYRONE
Noveril → DIBENZEPIN
Noverme → MEBENDAZOLE
Novex → FENTICLOR
Noviben → CAMBENDAZOLE
Novicodin → DIHYDROCODEINE
Novigam → LINDANE
Novil → DIPYRONE
Novin → DIPYRONE
Noviresin → PHENTERMINE
Novoban → METOCLOPRAMIDE
NOVOBIOCIN also → PHENOLSULFONPHTHALEIN
Novobutamide → TOLBUTAMIDE
Novocain → PROCAINE
Novocainamide → PROCAINAMIDE
Novocamid → PROCAINAMIDE
Novodigal → ACETYL DIGITOXIN
Novodil → CYCLANDELATE
Novodiphenyl → PHENYTOIN
Novodolan → FLOCTAFENINE
Novodorm → TRIAZOLAM
Novodrin → ISOPROTERENOL
Novofluen → ERGOLOID MESYLATES
Novofur → FURAZOLIUM CHLORIDE
Novogent N → IBUPROFEN
Novohydrazide → HYDROCHLOROTHIAZIDE
Novolaudon → HYDROMORPHONE
Novolin → INSULIN
Novomazina → CHLORPROMAZINE
Novomycetin → CHLORAMPHENICOL
Novon → ERBON
Novo-Nastizol A → ASTEMIZOLE
Novonorm → REPAGLINIDE
Novopen G → PENICILLIN > *Benzylpenicillin, Potassium*
Novophone → DAPSONE
Novoridazine → THIORIDAZINE
Novorin → XYLOMETAZOLINE
Novosed → CHLORDIAZEPOXIDE
Novotrone → SULFOXONE also → SOLASULFONE
Novydrine → AMPHETAMINE
Noxaben → PENICILLINS – "NEWER and SYNTHETICS" > *Dicloxacillin*
Noxibiol → NITROXOLINE
Noxigram → CINOXACIN
Noxiurotan → PIPERAZINE
Noxypen → PENICILLINS – "NEWER and SYNTHETICS" > *Amoxicillin*
Noxyron → GLUTETHIMIDE
NOXYTHIOLIN
Noxytiolin → NOXYTHIOLIN
Noyer Noir → WALNUT
Nozepam → OXAZEPAM

Nozinan → METHOTRIMEPRAZINE
NP 13 → BEMEGRIDE
NP 30 → PRENYLAMINE LACTATE
NP 113 → ISOPRINOSINE
NP 297 → MEFEXAMIDE
NPA → NAPTALAM
NPAB → PRAJMALINE
NPH Insulin → INSULIN
NPT 10381 → ISOPRINOSINE
4-NQO → 4-NITROQUINOLINE-1-OXIDE
NRDC 143 → PERMETHRIN
NRDC 149 → CYPERMETHRIN
NRDC 161 → DECAMETHRIN
NSC 185 → CYCLOHEXIMIDE
NSC 740 → METHOTREXATE
NSC 752 → THIOGUANINE
NSC 755 → MERCAPTOPURINE
NSC 762 → MECHLORETHAMINE
NSC 25-855 → TIMONACIC
NSC 1390 → ALLOPURINOL
NSC 2101 → ROXARSONE
NSC 3051 → *N*-METHYLFORMAMIDE
NSC 3053 → DACTINOMYCIN
NSC 3096 → CHLORAMPHENICOL
NSC 5366 → NOSCAPINE
NSC 6,396 → THIOTEPA
NSC 7,635 → 6-DIAZO-5-OXO-L-NORLEUCINE
NSC 8806 → MELPHALAN
NSC 09247 → HYDROQUINONE
NSC 9,706 → TRIETHYLENE MELAMINE
NSC 10,023 → PREDNISONE
NSC 11905 → LAPACHOL
NSC 12198 → DROMOSTANOLONE PROPIONATE
NSC 13,875 → ALTRETAMINE
NSC 15,200 → GALLIUM
NSC 17777 → PHENYRAMIDOL
NSC 19893 → FLUOROURACIL
NSC 23,436 → DOPAN
NSC 23,759 → TESTOLACTONE
NSC 23,909 → 1-METHYL-1-NITROSOUREA
NSC 24,559 → MITHRAMYCIN
NSC 25,154 → PIPOBROMAN
NSC 25,614 → PHTHALOFYNE
NSC 26,271 → CYCLOPHOSPHAMIDE
NSC 26,980 → MITOMYCIN C
NSC 27640 → FLOXURIDINE
NSC 29,422 → THIOGUANOSINE
NSC 29,691 → AMIFOSTINE
NSC 30152 → DIMETHADIONE
NSC 32,065 → HYDROXYUREA
NSC 33,001 → FLUOROMETHOLONE
NSC 33,669 → EMETINE
NSC 34,462 → URACIL MUSTARD
NSC 35,443 → PAPAVERINE
NSC 38,270 → A-649
NSC 38,721 → MITOTANE
NSC 39084 → AZATHIOPRINE
NSC 39415 → DOMIPHEN BROMIDE
NSC 39470 → BETAMETHASONE
NSC 39,661 → IDOXURIDINE
NSC 40725 → ISOSORBIDE
NSC 43,193 → STANOZOLOL
NSC 45,383 → STREPTONIGRIN
NSC 45388 → DACARBAZINE
NSC 46,015 → PYRAN COPOLYMER
NSC 47439 → FLUPREDNISOLONE
NSC 47774 → PIPOSULFAN
NSC 49,171 → SALSALATE
NSC 49,842 → VINBLASTINE
NSC 50,256 → METHYLMETHANE SULFONATE
NSC 56,408 → TUBERCIDIN
NSC 56,410 → PORFIROMYCIN
NSC 58,404 → DIAZOACETYL GLYCINE HYDRAZIDE
NSC 58,514 → CHROMOMYCIN A
NSC 60,584 → SULISOBENZONE
NSC 62,512 → 2-DIMETHYLAMINO-3′, 4-DIHYDROXYACETOPHENONE HCL
NSC 63,878 → CYTARABINE
NSC 64375 → BENZQUINAMIDE
NSC 67,574 → VINCRISITNE
NSC 68,626 → 1-ACETYL-2-PICOLINOYLHYDRAZINE
NSC 69,188 → MIMOSINE
NSC 69,811 → METHISAZONE
NSC 69,856 → NEOCARZINOSTATIN
NSC 69,948 → NORMETHANDRONE
NSC 71,047 → TRIOXSALEN
NSC 77,213 → PROCARBAZINE
NSC 77,625 → TRIAMTERENE
NSC 78502 → MECLOCYCLINES
NSC 82,116 → KETHOXAL
NSC 83,142 → DAUNORUBICIN
NSC 83,653 → AMANTADINE
NSC 85,998 → STREPTOZOCIN
NSC 88,536 → CALUSTERONE
NSC 89,199 → ESTRAMUSTINE
NSC 94,100 → MITOBRONITOL
NSC 95,441 → SEMUSTINE
NSC 102816 → 5-AZACYTIDINE
NSC 104800 → MITOLACTOL
NSC 107,430 → PENTAZOCINE
NSC 109,229 → 1-ASPARAGINASE
NSC 109724 → IFOSFAMIDE
NSC 114901 → DESIPRAMINE

NSC 115944 → ENFLURANE
NSC 119875 → CISPLATIN
NSC 122,819 → TENIPOSIDE
NSC 123,127 → DOXORUBICIN
NSC 125,066 → BLEOMYCIN
NSC 125,973 → PACLITAXEL
NSC 129,943 → RAZOXANE
NSC 130,044 → THIETHYLPERAZINE MALEATE
NSC 141540 → ETOPOSIDE
NSC 148,958 → TEGAFUR
NSC 169,780 → RAZOXANE
NSC 178,248 → CHLOROZOTOCIN
NSC 218,321 → PENTOSTATIN
NSC 226,080 → RAPAMYCIN
NSC 241240 → CARBOPLATIN
NSC 245,467 → VINDESINE SULFATE
NSC 249,008 → TRIMETREXATE
NSC 256439 → IDARUBICIN
NSC 269148 → MENOGARIL
NSC 301,739 → MITOXANTRONE HCL
NSC 328002 → FLUDARABINE PHOSPHATE
NSC 352,122 → TRIMETREXATE
NSC 362,856 → TEMOZOLOMIDE
NSC 404,241 → VIDARABINE
NSC 409962 → CARMUSTINE
NSC 515776 → DIPYRIDAMOLE
NSC 606,864 → ZOLADEX
NSC 612049 → DIDANOSINE
NSC 628,503 → TAXOTERE
NSC 70731 → LINCOMYCIN
NSC 79037 → LOMUSTINE
NTA → NITRILOTRIACETIC ACID
Ntutrwa → SOLANUM sp.
NU 445 → SULFISOXAZOLE
Nu 2121 → NICOTINYL ALCOHOL
NU 2222 → TRIMETHAPHAN CAMSYLATE
Nuarsol → SODIUM ARSANILATE
Nubain → NALBUPHINE
Nubarene → MECLOQUALONE
Nucidol → DIAZINON
Nuctalon → ESTAZOLAM
Nudrin → METHOMYL
Nuelin → THEOPHYLLINE
Nuflor → FLORFENICOL
Nuggets → AMPHETAMINES
Nuital → ETHYL ISOBUTRAZINE
Nu-K → POTASSIUM CHLORIDE
Nulcerin → FAMOTIDINE
Nulogyl → NIMORAZOLE
Nulomoline → INVERT SUGAR
Numal → APROBARBITAL
Numorphan → OXYMORPHONE
Numotac → ISOETHARINE
Nuncital → EMYLCAMATE
Nuophene → DICHLOROPHEN(E)
Nupercaine → DIBUCAINE
NUPHAR
Nupharine → NUPHAR
Nuprin → IBUPROFEN
Nuran → CYPROHEPTADINE
Nuredal → NIALAMIDE
Nurelle → CYPERMETHRIN
NuRexform → LEAD ARSENATE
Nurison → PREDNISONE
Nurofen → IBUPROFEN
Nuromax → DOXACURIUM CHLORIDE
Nutalls salt-bush → ATRIPLEX sp.
Nutinal → BENACTYZINE
NUTMEG also → ISOEUGENOL, METHYL EUGENOL, METHYL ISOEUGENOL, MYRISTICIN, SAFROLE
"Nutmeg Tree" → JATROPHA sp.
Nutracort → HYDROCORTISONE
Nutraplus → UREA
Nutrase → COCARBOXYLASE
Nutrasweet® → ASPARTAME
Nutrien → THALLIUM
Nutropin → GROWTH HORMONES
NUTS
Nuvacron → MONOCROTOPHOS
Nuvan → DICHLORVOS
Nuvanol → SUMITHION
Nuvanol N → IODOFENPHOS
Nuvapen → PENICILLINS – "NEWER and SYNTHETICS" > *Ampicillin*
NUX VOMICA also → BRUCINE, STRYCHNOS
NY 198 → LOMEFLOXACIN
Nyakwana → VIROLA sp.
Nyamadze → TRICHILIA EMETICA
Nyazin → NIALAMIDE
Nyburg, Dr. → Preface
Nycopren → NAPROXEN
Nyctal → CARBROMAL
Nycton → CHLORAL HYDRATE
Nydrane → BECLAMIDE
Nydrazid → ISONIAZID
Nylmerate → PHENYLMERCURIC ACETATE
NYLIDRIN also → METHOTRIMEPRAZINE
Nylon → 4-CHLORO-*m*-CRESOL, PHENOL, SALICYLIC ACID, SORBIC ACID
Nymphaea lutea → NUPHAR
Nysconitrine → NITROGLYCERIN
Nystan → NYSTATIN
Nystavescent → NYSTATIN
NYSTATIN also → METHOTREXATE

Nytol® → METHAPYRILENE
Nyuple → PRENYLAMINE LACTATE

O

Oakite → SODIUM PHOSPHATE, TRIBASIC
OAKS also → TEA
OAKMOSS
Oasil → MEPROBAMATE
Oat Hay → NITRATES, NITRITES
OATS also → NITRATES, NITRITES
o.b.c.t. → DIETHYLPROPION
Obedrex → FENFLURAMINE
Obesedrin → DEXTROAMPHETAMINE SULFATE
Obesin → PROPYLHEXEDRINE
Obestat → PHENYLPROPANOLAMINE
Obier → VIBURNUM PRUNIFOLIUM
Obidan → PROPRANOLOL
Obidoxime → TOXOGONIN
Oblivon → MEPARFYNOL
Obo ekute → SPONDIANTHUS UGANDENSIS
Obracin → TOBRAMYCIN
Obsidan → PROPRANOLOL
Occlusal → SALICYLIC ACID
Occrycetin → OXYTETRACYCLINE
O'Cedar All Purpose Red Polish® → POLISHES, FURNITURE
Oceral → OXICONAZOLE
OCGODEIA sp.
OCHRATOXINS also → ASPERGILLUS, COFFEE, PENICILLIUM sp., VANADIUM, WHEAT
O.C.M. → CHLORDIAZEPOXIDE
OCRYLATE
Octachlor → CHLORDANE
OCTACHLORONAPHTHALENE also → CHLORINATED NAPHTHALENES
OCTADECANETHIOL
Octadecyl Mercaptan → OCTADECANETHIOL
Octadenoic Acid → STEARIC ACID
Octadine → GUANETHIDINE
Octafluoro-isobutylene → POLYTEF
Octaklor → CHLORDANE
Octalene → ALDRIN
Octalox → DIELDRIN
OCTAMOXIN
OCTANE
OCTANETHIOL
Octanoic Acid → CAPRYLIC ACID
1-Octanol → ALCOHOL, OCTYL
2-Octanone → HEXYL METHYL KETONE
Octapressin → FELYPRESSIN
OCTOGEN
Octoil → BIS(2-ETHYLHEXYL)PHTHALATE
OCTOPAMINE also → PASSIFLORA sp.
OCTOPUS also → ELEDOISIN, MOLLUSKS
Octostim → DESMOPRESSIN
Octrane → HYDRARGAPHEN
OCTREOTIDE
OCTYL ACETATE
N-Octylamine → COTTONSEED MEAL
Octylatropine Bromide → ANISOTROPINE METHYLBROMIDE
N-Octyl-2-cyanoacrylate → OCRYLATE
OCTYLDIMETHYL PABA
Octylene Glycol → ETHOHEXADIOL
OCTYL ISOBUTYRATE
Octyl Mercaptan → OCTANETHIOL
OCTYL PALMITATE
OCTYL STEARATE
Octylthiol → OCTANETHIOL
Ocufen → FLURBIPROFEN
Ocuflox → OFLOXACIN
Ocusert → PILOCARPINE
Ocytocin → OXYTOCIN
ODA 914 → DESAMINOOXYTOCIN
Oddibil → FUMARIA
Odemase → FUROSEMIDE
Odorit → CRESOL, SAPONATED SOLUTION of
Odor-Trol → METHIONINES
Oedemex → FUROSEMIDE
Oekolp → ESTRIOL
OENANTHE sp.
Oenanthine → OENANTHE sp.
OENTHERA BIENNIS
Oesipos → LANOLIN
Oestrasid → DIENESTROL
Oestrin → ESTRONE
Oestriol → ESTRIOL
Oestrogel → ESTRADIOL
Oestrogine → DIETHYLSTILBESTROL
Oestromenin → DIETHYLSTILBESTROL
Oestromensyl → DIETHYLSTILBESTROL
Oestromon → DIETHYLSTILBESTROL
Oestrone → ESTRONE
Oestroral → DIENESTROL
Off® → *N,N*-DIETHYL-*m*-TOLUAMIDE
Oflocet → OFLOXACIN
Oflocin → OFLOXACIN
OFLOXACIN also → PROBENECID
S-(-)-Ofloxacin → LEVOFLOXACIN
Ofloxacine → OFLOXACIN
Oflozet → OFLOXACIN
Oftanol → ISOFENPHOS

Ogast → LANSOPRAZOLE
Ogbunmo → CASSIA
Ogen → ESTRONE
Ogilvie Hair Repair Lotion → QUINOLINE YELLOW
Ogostal → CAPREOMYCIN
OHB_{12} → VITAMIN B_{12}
Ohi → ALBIZIA sp.
Ohlexin → CEPHALEXIN
Oignons → ONIONS
Oil of Betula → METHYL SALICYLATE
Oil of Cade → JUNIPER
Oil of Euphorbia → MERCURIALIS PERENNIS
Oil of Gaultheria → METHYL SALICYLATE
Oil Geranium Reunion → GERANIUM
Oil Green → CHROMIC OXIDE
Oil of Mirbane → NITROBENZENE
Oil of Mustard → ALLYL ISOTHIOCYANATE
Oilnut → PYRULARIA PUBERA
OIL ORANGES
Oil of Palma Cristi → CASTOR OIL
Oil of Primrose → PRIMULA sp.
Oil Rose-geranium → GERANIUM
Oil Sickness → POLYCHLORINATED BIPHENYLS
Oil of Sweet Birch → METHYL SALICYLATE
Oil of Teaberry → METHYL SALICYLATE
Oil of Wintegreen → METHYL SALICYLATE
Oil of Vitriol → SULFURIC ACID
Ojo de Venado → MUCUNA sp.
Oju Ologbo → ABRUS PRECATORIUS
Okpa Ekele → PEANUTS
Okpu Ocha → BLIGHIA SAPIDA
OKRA
Oksid → FENTICLOR
OKT 3
Oktadin → GUANETHIDINE
Oktatensin → GUANETHIDINE
Oktatensine → GUANETHIDINE
Oktatenzin → GUANETHIDINE
Okwulu Oyibo → PAPAYA
Olamin → CINNARIZINE
OLAMINE
OLANZEPINE
Old English Red Furniture Polish® → POLISHES, FURNITURE
Old Man's Beard → CLEMATIS sp., *vitalla*
Old Woman's Broom → TURNERA APHRODISIACA
Olean → OLESTRA
OLEANDER
OLEANDOMYCIN also → CARBOMYCIN, MAGNESIUM, MILK, MONENSIN, PENICILLIN
Oleandrin → OLEANDER
Olefiant Gas → ETHYLENE
OLEIC ACID
Olei Mentha arvensis → CORNMINT OIL
OLMEDIOPEREBEA SCLEROPHYLLA
Olemorfotsiklin → OLEANDOMYCIN
Olemorphocycline → OLEANDOMYCIN
Oleo de Algodoeiro → COTTONSEED OIL
Oleomargarine → MARGARINE
Oleophosvel → LEPTOPHOS
Oleovitamin D_3 → VITAMIN D_3
OLESTRA
Oletetrin → TETRACYCLINE
Oletrin → OLEANDOMYCIN
Oleum Gossypi Seminis → COTTONSEED OIL
Oleum Iodatum → ETHIODIZED OIL
Oleum Jecoris → COD LIVER OIL
Oleum Morrhuae → COD LIVER OIL
Oleum Sabinae → SAVIN OIL
Oli → ALBIZIA
OLIBANUM
Olicard → ISOSORBIDE MONONITRATE
Oliprevin → PRAVASTATIN
Olitref → TRIFLURALIN
OLIVES
OLIVE OIL also → FATS, METHYLAMINE
Oliver Wendell Holmes, M.D. → ETHER, ETHYL
Olivin → ENALAPRIL
Olmagran → HYDROFLUMETHIAZIDE
Ololiuhqui → LYSERGIC ACID DIETHYLAMIDE
Ololiuqui → RIVEA CORYMBOSA also → MORNING GLORY
Ololuidu → JATROPHA sp.
Olomiojo → THEVETIA sp.
OLOPTADINE
Olsalazine → MERCAPTOPURINE
Olynth → XYLOMETAZOLINE
OM 518 → MEPHENOXALONE
Omaine → DEMECOLCINE
Omadine → PYRITHIONE
Omadine Disulfide → PYRITHIONE
Omal → 2,4,6-TRICHLOROPHENOL
Omca → FLUPHENAZINE DECANOATE
Omcilon → TRIAMCINOLONE
OMDS → PYRITHIONE
OME → OMEPRAZOLE
Omegamycin → TETRACYCLINE
7-*OMEN* → MENOGARIL
Omepral → OMEPRAZOLE
Omeprazen → OMEPRAZOLE
OMEPRAZOLE also → ALUMINUM HYDROXIDE, FLUVASTATIN, FLUVOXAMINE, ITRACONAZOLE,

KETOCONAZOLE, OLANZEPINE, PHENYTOIN, PREDNISONE, WARFARIN
Omite → PROPARGITE
Omix → TAMSULOSIN
Omnes → NIFURATEL
Omnibex → PHENTERMINE
Omnic → TAMSULOSIN
Omnicain → PROCAINE
Omnicef → CEFDINIR
Omniflox → TEMAFLOXACIN
Omnipaque → IOHEXOL
Omnipen → PENICILLINS – "NEWER and SYNTHETICS" > *Ampicillin*
Omnisan → PENICILLINS – "NEWER and SYNTHETICS" > *Epicillin*
Omniscan → GADODIAMIDE
Omnizole → THIABENDAZOLE
Omnopon → PAPAVERETUM
Omnyl → METHAQUALONE
OMPA → SCHRADAN
Ompacide → SCHRADAN
Ompatox → SCHRADAN
Omperan → SULPIRIDE
OMPHALOTUS ILLUDENS
OMS 29 → CARBARYL
OMS 33 → PROPOXUR
OMS 43 → SUMITHION
OMS 47 → MEXACARBATE
OMS 75 → DIBROM
OMS 597 → LANDRIN
OMS 658 → BROMOPHOS
OMS 708 → MOBAM
OMS 771 → ALDICARB
OMS 864 → CARBOFURAN
OMS 1155 → CHLORPYRIFOS-METHYL
Omtan → ISOBENZAN
Oncaspar → PEGASPARGASE
ONCB → NITROCHLOROBENZENES
Oncotrex → TRIMETREXATE
Oncovin → VINCRISTINE
ONDANSETRON also → RIFAMPIN
Ondena → DAUNORUBICIN
Oneida Instant Weld → CYANOACRYLATES
One-Iron → FERROUS FUMARATE
One Shot → BROMOXYNIL
Onion Grass → ROMULEA ROSEA
ONIONS also → MALEIC HYDRAZIDE, POTATO, *n*-PROPYL DISULFIDE, SULFUR, SULFURIC ACID
Oniria → QUAZEPAM
Onkotin → DEXTRAN
Onlemin → PRENYLAMINE LACTATE
ONOCLEA SENSIBILIS
OONOPSIS sp.
Onquinin → APROTININ
Ontak → DANISLEUKIN DIFTITOX
Onychomal → UREA
Onyxcide → BRONOPOL
Opacist E.R. → IODAMIDE
Opalène → TRIMETOZINE
Oparenol → IODOPYRACET
OPC 1085 → CARTEOLOL HYDROCHLORIDE
OPC 8212 → VESNARINONE
OPC 13,013 → CILOSTAZOL
OPC 17,116 → GREPAFLOXACIN
Opclor → CHLORAMPHENICOL
Opcon → NAPHAZOLINE
Opegan → SODIUM HYALURONATE
Operand → POLYVINYLPYRROLIDINE IODINE COMPLEX
OPERCULINA sp.
Operidine → PHENOPERIDINE
Ophtalmokalixan → KANAMYCIN
Ophthaine → PROPARACAINE HYDROCHLORIDE
Ophthakote → CELLULOSE, *Oxidized*
Ophthalgan → GLYCERIN
Ophthalmadine → IDOXURIDINE
Ophthetic → PROPARACAINE HYDROCHLORIDE
Ophthochlor → CHLORAMPHENICOL
Ophtocortin → MEDRYSONE
Ophtosol → BROMHEXINE
Opian → NOSCAPINE
Opianine → NOSCAPINE
Opiates → OPIUM also → DIGOXIN
Opilon → THYMOXAMINE
Opino → NYLIDRIN
OPIPRAMOL
Opiran → PIMOZIDE
OPIUM also → CANNABIS, POPPY
Opium Poppy → PAPAVER sp.
OPLOPANAX HORRIDUM
Opobalsam → BALSAM TOLU
Oposim → PROPRANOLOL
Oppacyn → TETRACYCLINE
Oprelrekin → INTERLEUKIN 11
Opren → BENOXAPROFEN
Op-Sulfa → SULFACETAMIDE SODIUM
Optal → *n*-PROPYL ALCOHOL
Optenyl → PAPAVERINE
Opticlox → PENICILLINS – "NEWER and SYNTHETICS" > *Cloxacillin/Sodium salt*
Opticrom → CROMOLYN SODIUM
Opticron → CROMOLYN SODIUM
Optifilm → POLYVINYL ALCOHOL

Optiflex → GROWTH HORMONES
Optimal → METHAQUALONE
Optimax WV → *L*-TRYPTOPHAN
Optimil → METHAQUALONE
Optimine → AZATADINE DIMALEATE
Optimycin → METHACYCLINE
Opti-Noxan → METHAQUALONE
Optipect → DEQUALINIUMS, *chloride*
Optipen → PENICILLINS – "NEWER and SYNTHETICS" > *Phenethicillin Potassium*
Optipranolol → TRIMEPRANOL
Optipress → CARTEOLOL HYDROCHLORIDE
Optium → PENICILLINS – "NEWER and SYNTHETICS", > *Amoxicillin*
Optocain → MEPIVACAINE
Optovit → VITAMIN A
Optraex → ZINC
Optrane → HYDRARGAPHEN
Opturem → IBUPROFEN
OPUNTIA sp.
Orabet → TOLBUTAMIDE
Orabilex → BUNAMIODYL
Orabilix → BUNAMIODYL
Orabolin → ETHYLESTRENOL
Oracaine → MEPRYLCAINE
Oracef → CEPHALEXIN
Oracéfal → CEFADROXIL
Oracilline → PENICILLIN > *Penicillin V*
Oracocin → CEPHALEXIN
Oradexon → DEXAMETHASONE
Oradian → CHLORPROPAMIDE
Oradiol → ETHINYL ESTRADIOL
Oradol → DOMIPHEN BROMIDE
Orafen → IBUPROFEN
Oraflex → BENOXAPROFEN
Orafuran → NITROFURANTOIN
Oragest → MEDROXYPROGESTERONE ACETATE
Orageston → ALLYLESTRANOL
Oragrafin → IPODATE
Oragulant → DIPHENADIONE
Oralcid → ACETARSONE
Oramide → TOLBUTAMIDE
Oramorph → MORPHINE
ORANGE also → CARAWAY, CELERY, FRUITS and FRUIT JUICES, PENICILLIUM sp., SODIUM POLYSTYRENE SULFONATE, SYNEPHRINE
ORANGE G
"Orange Mushrooms" → LYSERGIC ACID DIETHYLAMIDE
Orange Racun Besi → POTASSIUM FERRICYANIDE
Oranger Crystals → β-METHYL NAPHTHYL KETONE
Orange RGL → FDC YELLOW #6
Orange Root → HYDRASTIS CANADENSIS
"Oranges" → AMPHETAMINES
Orange Sneezeweed → HELENIUM sp.
Orange Wedges → LYSERGIC ACID DIETHYLAMIDE
Orahexal → CHLORHEXIDINE
Ora-Lutin → ETHISTERONE
Oranil → CARBUTAMIDE
Oranixon → MEPHENESIN
Orap → PIMOZIDE
Oraprim → SULFAMETHOXAZOLE
Orarsan → ACETARSONE
Orasone → PREDNISONE
Orasthin → OXYTOCIN
Orasulin → CARBUTAMIDE
Oratestin → FLUOXYMESTERONE
Ora-Testryl → FLUOXYMESTERONE
Oratrast → BARIUM SULFATE
Oratren → PENICILLIN > *Penicillin V*
Oratrol → DICHLORPHENAMIDE
Orbax → ORBIFLOXACIN
Orbenin → PENICILLINS – "NEWER and SYNTHETICS" > *Cloxacillin/Sodium salt*
Orbicin → DIBEKACIN
ORBIFLOXACIN
Orbinamon → THIOTHIXENE
Orchard Grass → DACTYLIS GLOMERATA
Orchidacea → PAPHIOPEDILUM HAYNALDIANUM
Orchids → PAPHIOPEDILUM HAYNALDIANUM
Orchisterone-M → METHYL TESTOSTERONE
Orciprenaline → METAPROTERENOL
Ordeal Bean → PHYSOSTIGMA VENENOSUM
Ordeal Poison → PHYSOSTIGMA VENENOSUM
Ordeal Tree → ERYTHROPHLEUM
Ordiflazine → LIDOFLAZINE
Ordimel → ACETOHEXAMIDE
Oregano → CANNABIS
Oregon Grape → MAHONIA AQUIFOLIUM
Oreson → GLYCERYL GUAIACOLATE
Orestol → DIETHYLSTILBESTROL
Orestralyn → ETHINYL ESTRADIOL
Oretic → HYDROCHLOROTHIAZIDE
Oreton → TESTOSTERONE
Oreton Methyl → METHYL TESTOSTERONE
ORF 11,676 → NALMEFENE
ORF 17,070 → HISTRELIN

Orfiril → SODIUM VALPROATE also → VALPROIC ACID
Org 2969 → DESOGESTREL
Org 3,770 → MIRTAZEPINE
ORG 6216 → RIMEXOLONE
Org 9426 → ROCKURONIUM BROMIDE
Org 10,172 → DANAPAROID
Orgabolin → ETHYLESTRENOL
Orgaboral → ETHYLESTRENOL
Orgametil → LYNESTRENOL
Orgametril → LYNESTRENOL
Organidin → IODINATED GLYCEROL
Organoderm → MALATHION
Organon → ALLYLESTRANOL, VECURONIUM BROMIDE
Orgaron → DANAPAROID
Orgasteron → NORMETHANDRONE
Orgasuline → INSULIN
Orgatrax → HYDROXYZINE
Org GB 94 → MIANSERIN
Org NA 97 → PANCURONIUM BROMIDE
ORG NC 45 → VECURONIUM BROMIDE
Oribis → BISMUTH SODIUM TRIGLYCOLLAMATE
Oriconazole → ITRACONAZOLE
Orientomycin → CYCLOSERINE
Orifungal M → KETOCONAZOLE
Origanum → MARJORAM
Orimeten → AMINOGLUTETHIMIDE
Orimon → PHENOTHIAZINE
Orinase → TOLBUTAMIDE
Orisol → SULFAPHENAZOLE
Orisulf → SULFAPHENAZOLE
Orix → NIFEDIPINE
ORIXA JAPONICA THUMB
ORLAYA PLATYCARPA
ORLISTAT
Orlon → HYDROGEN CYANIDE
Orlycycline → TETRACYCLINE
Ormeloxifene → CENTCHROMAN
ORMETOPRIM
Ornid → BRETYLIUM TOSYLATE
Ornidal → ORNIDAZOLE
ORNIDAZOLE
Ornidyl → EFLORNITHINE
ORNIPRESSIN
Ornithine-8-vasopressin → ORNIPRESSIN
ORNITHOGALUM UMBELLATUM
Ornitrol → AZACOSTEROL
Oro → EUPHORBIA, *leonensis*
Orobronze → CANTHAXANTHIN
Oroken → CEFIXIME
Oronol → AUROTHIOGLUCOSE
Oro Onigum Meta → EUPHORBIA, *kamerunica*
Oropur → OROTIC ACID
OROTIC ACID
Oroturic → OROTIC ACID
Orotyl → OROTIC ACID
Oroxine → THYROXINE
ORPHENADRINE also → CHLORPROMAZINE, CHLORPROPAMIDE, L-DOPA, GRISEOFULVIN, PROPOXYPHENE
ORPHENADRINE
Orphol → ERGOLOID MESYLATES
Orpidan → CHLORAZANIL
Orpizin → CHLORAZANIL
Orpressin → ORNIPRESSIN
Orsanil → THIORIDAZINE
Orsile → BENDROFLUMETHIAZIDE
Orstanorm → DIHYDROERGOTAMINE METHANESULFONATE (MESYLATE)
Ortacrone → AMIODARONE
Ortazol → FURAZOLIDONE
Ortédrine → AMPHETAMINE
Ortensan → ACETAMINOPHEN
Orthene → ACEPHATE
Orthesin → ETHYL-*p*-AMINOBENZOATE
Ortho 9006 → METHAMIDOPHOS
Ortho 5353 → BUFENCARB
Ortho 12420 → ACEPHATE
Orthoboric Acid → BORIC ACID
Orthocide 406 → CAPTAN
Orthoclone → OKT 3
Orthoclone OKT 3 → OKT 3
Ortho-Dibrom → DIBROM
Ortho-Gynest → ESTRIOL
Ortho-Klor → CHLORDANE
Ortho-Mite → ARAMITE
Orthophosphoric Acid → PHOSPHORIC ACID
Orthoxenol → σ-PHENYLPHENOL
Orthoxycol → HYDROCODONE BITARTRATE
Ortho Pharmaceutical Corp. → MESTRANOL
Orthophos → PARATHION
Orthoquat → PARAQUAT
Ortica → URTICA DIOICA
Ortin → TROLNITRATE PHOSPHATE
Ortisporina → CEPHALEXIN
Ortonal → METHAQUALONE
Ortran → ACEPHATE
Orudis → KETOPROFEN
Orugesic → KETOPROFEN
Oruvail → KETOPROFEN
Orvagil → METRONIDAZOLE
Orvar → PARAQUAT
ORYZALIN
Oryza sativa → RICE
Oryzias lapites → TELEOCIDIN

OS 1897 → DIBROMOCHLOROPROPANE
OS 2046 → MEVINPHOS
OSAGE ORANGE
Osarsal → ACETARSONE
Osarsol → ACETARSONE
Osbil → IOBENZAMIC ACID
Os-Cal → CALCIUM CARBONATE
Oscorel → KETOPROFEN
OSELTAMIVIR
Osiren → SPIRONOLACTONE
Osmitrol → MANNITOL
OSMIUM TETROXIDE
Osmofundin → MANNITOL
OSMORHIZA CHILENSIS
Osmosal → MANNITOL
Osmose K-33 → CHROMATED COPPER ARSENATE
Osmosin → INDOMETHACIN
Osnewan → PROCYCLIDINE
Ospamox → PENICILLINS – "NEWER and SYNTHETICS" > *Amoxicillin*
Ospolot → SULTHIAME
Ossalin → SODIUM FLUORIDE
Ossein → COLLAGEN
Ossian → OXOLINIC ACID
Ossidrochinone → 8-HYDROXYQUINOLINE
Ossin → SODIUM FLUORIDE
Ossiten → DICHLOROMETHYLENE DIPHOSPHONATE
Ostac → DICHLOROMETHYLENE DIPHOSPHONATE
Ostamer → POLYURETHANES
Ostelin → VITAMIN D_2
Osteo-F → SODIUM FLUORIDE
Osteofluor → SODIUM FLUORIDE
Osteol → GEFARNATE
Osvan → BENZALKONIUM CHLORIDE
Osvarsan → ACETARSONE
Osvil → IOBENZAMIC ACID
Osyrol → SPIRONOLACTONE
Ota → STROPHANTHUS sp.
Otavite → CADMIUM CARBONATE
Oterben → TOLBUTAMIDE
OTFC → FENTANYL
Othesin → ETHYL-*p*-AMINOBENZOATE
Othoxine → METHOXYPHENAMINE
Otifuril → FURALTADONE
Otofural → NITROFURAZONE
Otokalixin → KANAMYCIN
Otompy-Kina → CHELIDONIUM MAJUS
Otriven → XYLOMETAZOLINE
Otrivin → XYLOMETAZOLINE
Otrix → XYLOMETAZOLINE
Otrun → DEANOL, *aceglumate*
Ottasept → CHLOROXYLENOL
OUABAIN also → ACOCANTHERA sp., GARCINIA sp.
Ouch-ouch Disease → CADMIUM
Ourari → CURARE
Ovaban → MEGESTROL ACETATE
Ovarid → MEGESTROL ACETATE
Ovastol → MESTRANOL
Overal → ROXITHROMYCIN
Overtime → PERMETHRIN
Ovest → ESTROGENS, CONJUGATED
Ovesterin → ESTRIOL
Ovestin → ESTRIOL
Ovicox → SALINOMYCIN
Ovifollin → ESTRONE
Ovisot → ACETYLCHOLINE
Ovitelmin → MEBENDAZOLE
Ovitrol → FENTICLOR
Ovocyclin → ESTRADIOL
Ovocylin → ESTRADIOL
Ovrette → NORGESTREL
O-V Statin → NYSTATIN
Ovo-Vinces → ESTRIOL
Owe → SPONDIANTHUS UGANDENSIS
Owl's Claws → DUGALDIN also → HELENIUM sp.
Owsley → LYSERGIC ACID DIETHYLAMIDE
Owu → COTTON
Owulu → COTTON
Oxacillin → PENICILLINS – "NEWER and SYNTHETIC" also → SULFAMETHOXYPYRIDAZINE, SULFISOXAZOLE, SULFOBROMOPHTHALEIN SODIUM
Oxaine → OXETHAZINE
Oxala → OXALIS sp.
Oxalaldehyde → GLYOXAL
Oxalates → ALLIUM TRICOCCUM, AMARANTHUS, BEANS, BEER, BEETS, BROCCOLI, CALCIUM CARBIMIDE, CARPOBROTUS, CHERRY, CHOCOLATE, COLAS, COPPER OXALATE, CRANBERRY, CUCUMBERS, DIEFFENBACHIA sp., GERANIUM, GRAPE, GRASSES, KALE, LETTUCE, LILLIES, METHOXYFLURANE, OKRA, PARSLEY, PARSNIP, PEACHES, PECANS, PORTULACA OLERACEA, RASPBERRY, SALSOLA sp., SCINDAPUS SCLEROTIORUM, SETARIA sp., SPATHYPHYLLUM sp., SPIGELIA ANTHELMIA, SQUASH, STRAWBERRY, SWEET POTATO, SWISS CHARD, TARAXACUM OFFICINALE, TEA,

OXYGEN also → BLEOMYCIN, MAGNESIUM CHLORIDE, RUBIDIUM
Oxygen Fluoride → FLUORINE MONOXIDE
Oxyject → OXYTETRACYCLINE
Oxy-Kess-tetra → OXYTETRACYCLINE
Oxy-L → BENZOYL PEROXIDE
Oxylag → OXYTETRACYCLINE
OXYLOBIUM sp.
Oxylone → FLUOROMETHOLONE
Oxymetacine → WARFARIN
OXYMETAZOLINE
OXYMETHOLONE also → WARFARIN
Oxymethylene → FORMALDEHYDE
OXYMORPHONE
Oxymycin → OXYTETRACYCLINE
Oxymykoin → OXYTETRACYCLINE
Oxynone → *p*-NITROSO-*N*, *N*-DIMETHYLANILINE
Oxypaat → PIPERAZINE
Oxypan → OXYTETRACYCLINE
Oxypate → PIPERAZINE
Oxypencedanin → CYMOPTERUS also → WILD CARROT
OXYPENDYL
OXYPERTINE
Oxyphedrine → OXYFEDRINE
OXYPHENBUTAZONE also → ACENOCOUMAROL, BISHYDROXYCOUMARIN, CHLORPROPAMIDE, CYCLOCOUMAROL, ETHYL BISCOUMACETATE, FLUOXYMESTERONE, GLYMIDINE, INSULIN, LITHIUM, METHANDROSTENOLONE, PENICILLIN, PHENPROCOUMON, PHENYTOIN POTASSIUM BROMIDE, TOLBUTAMIDE, WARFARIN
OXYPHENCYCLIMINE
OXYPHENISATIN also → DOCUSATE SODIUM, PRUNE
OXYPHENONIUM BROMIDE also → MAGNESIUM TRISILICATE
Oxypolygelatin → GELATIN
Oxyprenolol → OXPRENOLOL HYDROCHLORIDE
Oxyquinoline → 8-HYDROXYQUINOLINE
8-Oxyquinolinol → 8-HYDROXYQUINOLINE
Oxysept → HYDROGEN PEROXIDE
Oxysol → OXYTETRACYCLINE
Oxystin → OXYTOCIN
Oxytenia acerosa → COPPERWEED
OXYTERIA ACEROSA
Oxyterracin → OXYTETRACYCLINE
Oxytetrachel → OXYTETRACYCLINE
Oxytetracid → OXYTETRACYCLINE
OXYTETRACYCLINE also → ALUMINUM HYDROXIDE, ANTIPYRINE, BROMHEXINE, FERROUS SULFATE, GLUCOSAMINE, VITAMIN B_2, VITAMIN C
Oxytetrin → OXYTETRACYCLINE
OXYTOCIN also → ERGONOVINE, LITHOSPERMUM RUDERALE, NOREPINEPHRINE, WATER
Oxytr:imethylline → CHOLINE THEOPHYLLINATE
Oxytriphylline → RIFAMPIN
OXYTROPIS also → SWAINSONINE
Oyster Mushroom → PLEUROTUS sp.
OYSTERS also → GONYAULAX, MOLLUSKS
Oxyzin → PIPERAZINE
Ozichinolini → 8-HYDROXYQUINOLINE
OZONE also → DUSTS
Ozovit → MAGNESIUM PEROXIDE

P

P 50 → PENICILLIN > *Benzylpenicillin, Potassium* also *Ampicillin*
P 071 → CETIRIZINE
P 113 → SARALASIN
P 165 → AZASERINE
P 391 → MEPAZINE
501-P → GLUCOSULFONE SODIUM
P 607 → CHLORPROPAMIDE
P 638 → PUROMYCIN
P 725 → PERAZINE
P 1011 → PENICILLINS – "NEWER and SYNTHETICS" > *Dicloxacillin*
P 1134 → PINACIDIL
P 1888 → ISOSULFAN BLUE
P 3693A → DOXEPIN
P 4125 → ISOSULFAN BLUE
P 4241 → IMICLOPAZINE
P 4,657B → THIOTHIXENE
P 5048 → DIMETHISTERONE
P → PEYOTE
PA 14 → ALKYL-ϖ-HYDROXYPOLY(OXYETHYLENE)
PA 20 → NIFEDIPINE
PA 93 → NOVOBIOCIN
PA 94 → CYCLOSERINE
PA 105 → OLEANDOMYCIN
PA 144 → MITHRAMYCIN
PA 248 → PENICILLINS – "NEWER and SYNTHETICS" > *Propicillin Potassium*
PAA 701 → BIALAMICOL
PABA → *p*-AMINOBENZOIC ACID

Pabacidum → *p*-AMINOBENZOIC ACID
Pabenol → DEANOL
Pabestrol D → DIETHYLSTILBESTROL
Pacatal → MEPAZINE
Pacatol → MEPAZINE
Pacemo → ACETAMINOPHEN
Pacetyn → ECTYLUREA
Paceum → DIAZEPAM
PACHYCEREUS
Pachyrrhizid → PACHYRRHIZUS EROSUS
PACHYRRHIZUS EROSUS
PACHYSTIGMA sp.
Pacific Labrador Tea → LEDUM sp.
Pacinol → FLUPHENAZINE DECANOATE
Pacinox → CAPURIDE
Pacitane → TRIHEXYPHENIDYL
Pacitran → DIAZEPAM
Pacitran → METOSERPATE
Pacitron → L-TRYPTOPHAN
PACLITAXEL also → ALCOHOL, ETHYL; CASTOR OIL
Pacyl → ISOXICAM
Padan → NEREISTOXIN
Padicide → BENZALKONIUM CHLORIDE
Padimate O → OCTYLDIMETHYL PABA
PADIMATES
PAECILOMYCES sp.
Paediathrocin → ERYTHROMYCIN ETHYL SUCCINATE
Paeonia sp. → PEONY
Pafenolol → TERBUTALINE
PAGAMEA MACROPHYLLA
Paginol → PENBUTOLOL SULFATE
PAH → *p*-AMINOHIPPURIC ACID
Pahu → PAHUTOXIN
PAHUTOXIN
Painful-painful Disease → CADMIUM
PAINT
Pakurú-niaará → NACULEOPSIS
Palán-palán → NICOTIANA sp.
Palaprin → ALOXIPRIN
Palatinol BB → BUTYL BENZYL PHTHALATE
Palatinol M → DIMETHYL PHTHALATE
Palavale → ECONAZOLE NITRATE
Paldesic → ACETAMINOPHEN
Palestrol → DIETHYLSTILBESTROL
Palfium → DEXTROMORAMIDE
PALICOUREA sp.
Palinum → CYCLOBARBITAL
Palisander Wood → ROSEWOOD
PALLADIUM
Palladous Chloride → PALLADIUM
Pallethrine → ALLETHRIN
Palma Christi → CASTOR BEAN
Palmarosa Oil → GERANIUM
Palmas Reales → PALMS
Palmatine → BARBERRY
Palmityl Alcohol → ALCOHOL, CETYL
PALMS also → CALCIUM OXALATE
Palmyrah Palm → PALMS
Palocillin 5 → PENICILLIN > *Benzylpenicillin, Potassium*
Palohex → INOSITOL
Palopause → ESTROGENS, CONJUGATED
Palphium → DEXTROMORAMIDE
Paltet → TETRACYCLINE
Paludrine → CHLOROGUANIDE
Palusil → CHLOROGUANIDE
Palustrine → EQUISETUM
Paluther → ARTEMETHER
Palux → ALPROSTADIL
PALYTHOA TOXICA
PALYTOXIN also → LIMU-MAKE-HANA, PALYTHOA TOXICA
L-PAM → MELPHALAN
2-PAM → PRALIDOXIME
Pamacyl → *p*-AMINOSALICYLIC ACID
PAMAQUIN
Pamaquine → PAMAQUIN also → QUINACRINE
Pamedon(e) → DIPIPANONE
Pameion → PAPAVERINE
Pamelor → NORTRIPTYLINE
PAMIDRONATE DISODIUM also → DOXORUBICIN
Pamisan → PHENYLMERCURIC ACETATE
Pamisyl → *p*-AMINOSALICYLIC ACID
2-PAMM → PRALIDOXIME
Pamovin → PYRVINIUM PAMOATE
Pamoxin → PYRVINIUM PAMOATE
PAN → PHTHALIC ANHYDRIDE
Panacef → CEFACLOR
Panadol → ACETAMINOPHEN
PANAEOLUS sp.
Panais Sauvage → PARSNIP
Panaldine → TICLOPIDINE HCL
Panaleve → ACETAMINOPHEN
Panama Red → CANNABIS
Panama Tree → STERCULIA sp.
Panamicin → DIBEKACIN
Panasorb → ACETAMINOPHEN
Panax → GINSENG
Panchelidon → CHELIDONIUM MAJUS
Pancodine → OXYCODONE
PANCRATIUM TRIANTHUM
Pancrease → PANCRELIPASE
Pancreatic Desoxyribonuclease → PANCREATIC DORNASE

PANCREATIC DORNASE
PANCREATIN
PANCRELIPASE
Pancrex V → PANCREATIN
PANCURONIUM BROMIDE also → ATENOLOL, ESMOLOL, HALOTHANE, KETAMINE, LINCOMYCIN, LITHIUM, NEOMYCIN, PROPRANOLOL, PYRIDOSTIGMINE BROMIDE, THIOTEPA
Pandrocene → PENMESTEROL
Panectyl → TRIMEPRAZINE
Panediol → MEPROBAMATE
Paneolus → MUSHROOMS
Panergon → PAPAVERINE
Panets → ACETAMINOPHEN
Panex → ACETAMINOPHEN
Panformin → BUFORMIN
Panfungol → KETOCONAZOLE
Pangametin → VITAMIN B_{15}
Pangamic Acid → VITAMIN B_{15} also → PEAS
Panhemin → HEMIN
Panheprin → HEPARIN
Panhibin → SOMATOSTATIN
Panicillium paxilli → PAXILLINE
Panimycin → DIBEKACIN
Pan-Kloride → POTASSIUM CHLORIDE
Panmycin → TETRACYCLINE
Panmycin P → TETRACYCLINE
Panofen → ACETAMINOPHEN
Panogen → METHYLMERCURYDICYANDIAMIDE
Panomec → IVERMECTIN
Panoral → CEFACLOR
PanOxyl → BENZOYL PEROXIDE
Panpal → TRIHEXYPHENIDYL
Panpanit → CARAMIPHEN
Panparnit → CARAMIPHEN
Panretin → ALITRETINOIN
Panrone → THIACETAZONE
Pantelmin → MEBENDAZOLE
Pantenyl → DEXPANTHENOL
Pantheline → PROPANTHELINE BROMIDE
Panther Cap → AMANITA sp.
Pantherine → AGARIN also → AMANITA sp.
Panthion → PARATHION
Panthoderm → DEXPANTHENOL
Panthroid → THYROXINE
Panthyroid → THYROXINE
"Panting Disease" → ZIERIA ARBORESCENS
Pantocaine → TETRACAINE HYDROCHLORIDE
Pantomicina → ERYTHROMYCIN STEARATE
Pantopaque → IOPHENDYLATE
Pantopon → PAPAVERETUM
PANTOPRAZOLE
Pantosediv → THALIDOMIDE
Pantothenol → DEXPANTHENOL
Pantothenyl Alcohol → DEXPANTHENOL
Pantovernil → CHLORAMPHENICOL
Panurin → HYDROCHLOROTHIAZIDE
Panwarfarin → WARFARIN
Panzid → CEFTAZIMIDES
Panzytrat → PANCREATIN
Papacontin → PAPAVERINE
PAPAIN also → BEER
Papa San → PORTULACA OLERACEA
Papase → PAPAIN
PAPAVER sp.
PAPAVERETUM
PAPAVERINE also → L-DOPA, OPIUM
PAPAYA
Papayotin → PAPAIN
PAPER
Paper Acid → LYSERGIC ACID DIETHYLAMIDE
Paper Bark → CAJEPUT
Paper Bark Thorn → ACACIA
Paperflowers → PSILOSTROPHE sp.
PAPHIOPEDILUM HAYNALDIANUM
Papital T.R. → PAPAVERINE
PAPRIKA
Papuan Lung → GRASSES
Para-Aminobenzoic Acid → *p*-AMINOBENZOIC ACID
Para-Ban → PROPOXUR
PARABENS also → HYDROCORTISONE, MASCARA, SULFAMETER
Parabis → DICHLOROPHEN(E)
Parabolin → DIETHYLPROPION
Parabuthus → SCORPIONS
Paracaine → PROCAINE
Paracetamol → ACETAMINOPHEN
Parachloramine → MECLIZINE
Paracide → *p*-DICHLOROBENZENE
Paracodin → DIHYDROCODEINE
Paracort → PREDNISONE
Paracortol → PREDNISOLONE
Paradichlorobenzene → *p*-DICHLOROBENZENE
Paradione → PARAMETHADIONE
Paradise → COCAINE
Paradormalene → METHAPYRILENE
Paradow → *p*-DICHLOROBENZENE
PARAFFIN
Paraffin Oil → MINERAL OIL
Paraffinomas → PETROLATUM
Paraflex → CHLORZOXAZONE
Paraflu → FLUFENAMIC ACID
Paraguay Tea → ILEX sp.

Parahexyl → SYNHEXYL
Paraiso → MELIA sp.
Paral → PARALDEHYDE
Paralan → ACETAMINOPHEN
PARALDEHYDE also → ALCOHOL, ETHYL; TOLBUTAMIDE
Paralergin → ASTEMIZOLE
Paralest → TRIHEXYPHENIDYL
Paralgin → DIPYRONE
Paralkan → BENZALKONIUM CHLORIDE
Paralytic Shellfish Poison → SAXITOXIN
Paralyzed Tongue → DESCURAINIA PINNATA
Paramar → PARATHION
Paramenyl → MAFENIDE
PARAMETHADIONE
PARAMETHASONE ACETATE
Paramezone → PARAMETHASONE ACETATE
Paramicina → PAROMOMYCIN
Paraminol → *p*-AMINOBENZOIC ACID
Paraminyl → PYRILAMINE
Paramite → IMIDAN
Paramoth → *p*-DICHLOROBENZENE
Paranol → *p*-AMINOPHENOL
Paraoxon → DIETHYL-*p*-NITROPHENYLPHOSPHATE also → PARATHION
Para-Pas → *p*-AMINOSALICYLIC ACID
Parapel → CHLOROPHACINONE
Paraphos → PARATHENE
Paraplatin → CARBOPLATIN
PARAQUAT also → ACETAMINOPHEN, α, α´-DIPYRIDYL, FULLER'S EARTH, FUROSEMIDE, SELENIUM
Para Red → NITROANILINES
Parasal → *p*-AMINOSALICYLIC ACID
Parasalicil → *p*-AMINOSALICYLIC ACID
Parasalindon → *p*-AMINOSALICYLIC ACID
Parasan → BENACTYZINE
Parasepts → PARABENS
Parason → BENACTYZINE
Parasitex → DIAZINON
Paraspen → ACETAMINOPHEN
Parasympatol → SYNEPHRINE
Parathesin → ETHYL-*p*-AMINOBENZOATE
PARATHION also → BREAD, DIETHYL-*p*-NITROPHENYL PHOSPHATE, *p*-NITRO PHENOL, PHENOBARBITAL, PHENYLTHIOUREA, RESERPINE
Parathion-methyl → METHYL PARATHION
Parathormone → PARATHYROID HORMONE
PARATHYROID HORMONE
Parawet → PARATHION
Paraxin → CHLORAMPHENICOL
Para-zene → *p*-DICHLOROBENZENE
Parazine → PIPERAZINE
Parbetan → BETAMETHASONE
PARBENDAZOLE
Parda → L-DOPA
Pardisol → ETHOPROPAZINE
Pardroyd → OXYMETHOLONE
Paredrine → HYDROXYAMPHETAMINE
Paredrinex → HYDROXYAMPHETAMINE
Paredrinol → PHOLEDRINE
Paregoric → OPIUM also → KAOLIN
Parenabol → BOLDENONE
Parenamine → PROTEIN HYDROLYSATES
Parenterin → CASANTHRANOL
Parentracin → BACITRACIN
Parenzyme → TRYPSIN
Parenzymol → TRYPSIN
Parest → METHAQUALONE
Parfenac → BUFEXAMAC
Parfenal → BUFEXAMAC
Parfuran → NITROFURANTOIN
Pargin → ECONAZOLE NITRATE
Pargitan → TRIHEXYPHENIDYL
Pargonyl → PAROMOMYCIN
PARGYLINE also → ALCOHOL, ETHYL; AMPHETAMINE, BEANS, CHEESES, CHLOROTHIAZIDE, CHLORPROMAZINE, DESIPRAMINE, L-DOPA, IMIPRAMINE, INSULIN, MEPERIDINE, PERPHENAZINE, SUMATRIPTAN SUCCINATE, TRIFLUPERAZINE DIHYDROCHLORIDE, L-TRYPTOPHAN
Parica → VIROLA sp. also → ANADENANTHERA
Paricina → PAROMOMYCIN
PARIETARIA sp.
PARIS sp.
Paris Blue → FERRIC FERROCYANIDE
Paris Green → COPPER ARSENITE also → CUPRIC ACETOARSENITE
Paris Yellow → LEAD CHROMATE
Paritane → OXPRENOLOL HYDROCHLORIDE
Parizac → OMEPRAZOLE
Parkazin → DIETHAZINE
Parke-Davis/Warner Lambert → TROGLITAZONE
Parkemed → MEFENAMIC ACID
Parkin → ETHOPROPAZINE
Parkinane → TRIHEXYPHENIDYL
Parkopan → TRIHEXYPHENIDYL

Parkotil → PERGOLIDE MESYLATE
Parlef → FLUFENAMIC ACID
Parlodel → BROMOCRIPTINE
Parmal → PYRILAMINE
Parmarsol → ZIRAM
PARMELIA MOLLIUSCULA
Parmid → METOCLOPRAMIDE
Parminal → METHAQUALONE
Parminyl → PYRILAMINE
Parmol → ACETAMINOPHEN
Parnate → TRANYLCYPROMINE
Parodyne → ANTIPYRINE
Paroleine → MINERAL OIL
PAROMOMYCIN also → METHOTREXATE, NEOMYCIN, PENICILLIN, VANCOMYCIN
Paroven → TROXERUTIN
PAROXETINE also → CARVEDILOL, METOPROLOL, PERPHENAZINE, QUINIDINE, SUMATRIPTAN SUCCINATE, TERFENADINE
Paroxon → PAROXYPROPIONE
Paroxyl → ACETARSONE
PAROXYPROPIONE
Parozone → SODIUM HYPOCHLORITE
Parphezein → ETHOPROPAZINE
Parphezin → ETHOPROPAZINE
Parpon → BENACTYZINE
PARQUETINA NIGRESCENS
Parquine → JUANULLOA OCHRACEA also → CESTRUM
Parrot Green → CUPRIC ACETOARSENITE
Parsidol → ETHOPROPAZINE
Parsitan → ETHOPROPAZINE
PARSLEY also → CANNABIS, CHELIDONIUM MAJUS, CONIUM MACULATUM, 5-METHOXYPSORALEN, PHENCYCLIDINE
Parsley Water Dropwort → OENANTHE sp.
PARSNIP also → 5-METHOXYPSORALEN
Parsol 1789 → AVOBENZONE
PARSONSIA
Parsonsine → PARSONSIA
Parsotil → ETHOPROPAZINE
Parterol → DIHYDROTACHYSTEROL
PARTHENIUM
Parthenocissus quinquefolia → IVY
Partocon → OXYTOCIN
Parton M → METHYL PARATHION
Partrex → TETRACYCLINE
Partridge Pea → CASSIA
Partusisten → FENOTEROL HYDROBROMIDE
Parvolex → ACETYLCYSTEINE
Parzate → NABAM
Parzate C → ZINEB
Parzone → DIHYDROCODEINE
PAS → *p*-AMINOSALICYLIC ACID
PAS-C → *p*-AMINOSALICYLIC ACID
Pascorbic → *p*-AMINOSALICYLIC ACID
Pasetocin → PENICILLINS – "NEWER and SYNTHETICS" > *Amoxicillin*
Paskalium → POTASSIUM-*p*-AMINOSALICYLATE
Paskate → POTASSIUM-*p*-AMINOSALICYLATE
Pasnodia → *p*-AMINOSALICYLIC ACID
Pasolac → *p*-AMINOSALICYLIC ACID
Pasolind → ACETAMINOPHEN
Pasolind N → ACETAMINOPHEN
Pasote → CHENOPODIUM
PASPALUM sp.
Paspertin → METOCLOPRAMIDE
Pasque Flower → ANEMONE
PASSIFLORA sp.
Passiflorin → HARMAN
Passiflorine → PASSIFLORA sp.
Passion Flower → PASSIFLORA sp.
Pastaron → UREA
Pastinaca sativa → PARSNIP
PASTRY
Patanol → OLOPTADINE
PATCHOULI OIL also → CINNAMIC ALDEHYDE
Patent Blue AE → FDC BLUE #1
Patent Blue V → BLUE VRS
Paterson's Curse → ECHIUM PLANTAGINEUM
Pathclear → PARAQUAT
Pathilon → TRIDIHEXETHYL CHLORIDE
Pathocil → PENICILLINS – "NEWER and SYNTHETICS" > *Dicloxacillin*
Pathomycin → SISOMICIN
PATINOA ICHTHYOTOXICA
Patricin → FLAVOXATE HCL
PATULIN also → ASPERGILLUS, PENICILLIUM sp., RICE
PAULLINA sp.
Paupaw → PAPAYA
Pausinystalia yohimba → YOHIMBINE
Pavabid → PAPAVERINE
Pavacap → PAPAVERINE
Pavacen → PAPAVERINE
Pavadel → PAPAVERINE
Pavagen → PAPAVERINE
Pavakey → PAPAVERINE
Pavasal Potassium → POTASSIUM-*p*-AMINOSALICYLATE
Pavased → PAPAVERINE
Pavatest → PAPAVERINE

Pava-Wol → PAPAVERINE
Pavecef → CEFAMANDOLE
PAVETTA sp.
Pavot → PAPAVER sp.
Pavulon → PANCURONIUM BROMIDE
PAW-PAW also → PAPAYA
Paxanol → DEANOL
Paxate → DIAZEPAM
Paxidorm → METHAQUALONE
Paxil → PAROXETINE
PAXILLINE also → PENICILLIUM sp.
Paxillus involutus → MUSHROOMS
Paxilon → METHAZOLE
Paxina → METHYL HYDRAZINE
Paxipam → HALAZEPAM
Paxistil → HYDROXYZINE
Paxital → MEPAZINE
PAYLOOAH also → LEAD
Payze → CYANAZINE
Pazote → CHENOPODIUM
PB → PYRIDOSTIGMINE BROMIDE
PBBs → POLYBROMINATED BIPHENYLS
PBNA → *N*-PHENYL-β-NAPHTHYLAMINE
PBZ → TRIPELENNAMINE
PC 1421 → PIPERACETAZINE
PCE → CYCLOHEXAMINE
PCM → PERCHLOROMETHYL MERCAPTAN
PCMX → CHLOROXYLENOL
PCNB → QUINTOZENE also → NITROCHLOROBENZENES
PCP → PHENCYCLIDINE also → PENTACHLOROPHENOL also → POLYCHLORPINE
PCPA → *p*-CHLOROPHENYLALANINE
PCV → PENCICLOVIR
PD 5 → MEVINPHOS
PD 109,452-2 → QUINAPRIL
PD 131,501 → SPARFLOXACIN
PDB → *p*-DICHLOROBENZENE
PDCB → *p*-DICHLOROBENZENE
PDDB → DOMIPHEN BROMIDE
mPDN → *m*-PHTHALODINITRILE
PDP → MINOXIDIL
PDU → FENURON
Peace Lillies → SPATHIPHYLLUM sp. also → LILLIES
Peace Maker → CANNABIS
Peace Pills → PHENCYCLIDINE
Peach Aldehyde → γ-UNDECALACTONE
Peach Lactone → γ-UNDECALACTONE
Peach-leaf Poison Bush → TREMA ASPERA
Peach-Thin → NAPTALAM
Pearly Gates → MORNING GLORY
Pearl Moss → CARRAGEENAN
PEAS also → PEANUTS, SODIUM GLUTAMATE
PEACHES also → ALMONDS, CHERRY, CYANIDES, FRUITS and FRUIT JUICES, HYDROGEN CYANIDE, PHOSALONE
Peanut Brittle → PHORATE
PEANUTS also → ALMONDS, ASPERGILLUS, BASSIA, CYANIDES, CYCLOPIAZONIC ACID, DIAMINOZIDE, MILK, PHORATE
PEARLS also → JEWELRY
PEARS also → FRUITS and FRUIT JUICES, HYDROGEN CYANIDE
PEAT
Peat Moss → SPHAGNUM MOSS
Peavine → ASTRAGALUS
PEBULATE
PECANS
Pecazine → MEPAZINE
PECILOCIN
Pecnon → KEBUZONE
Pecram → AMINOPHYLLINE
Pectenine → PACHYCEREUS
PECTIN also → ACETAMINOPHEN
Pectolin → PHOLCODINE
Pedameth → METHIONINES
PED G → PHENFORMIN
Pediamycin → ERYTHROMYCIN ETHYL SUCCINATE
Pediaprofen → IBUPROFEN
Pedisafe → CLOTRIMAZOLE
Peflacine → PEFLOXACIN
PEFLOXACIN
PEG → POLYETHYLENE GLYCOL
PEG-ADA → PEGADEMASE BOVINE
PEGADEMASE BOVINE
Peganone → ETHOTOIN
PEGANUM HARMALA
PEG-L-asparaginase → PEGASPARGASE
PEGASPARGASE
Pegasyl → HYDROGEN PEROXIDE
PELAGIA sp.
Pelargonyl Acetate → *n*-NONYL ACETATE
PELECYPHORA ASELLIFORMIS
Pelentan → ETHYL BISCOUMACETATE
Pella Kaner → THEVETIA sp.
Pelletierine Tannate → POMEGRANATE
Pellidol → DIACETAZOTOL
Pellitory Bark → ZANTHOXYLUM sp.
Pellitory-of-the-wall → PARIETARIA sp.
Pelmin → VITAMIN B_3
Pelochelys bibroni → TURTLES

PELS → ERYTHROMYCIN ESTOLATE
Pelson → NITRAZEPAM
Peluces → HALOPERIDOL
Pemal → ETHOSUXIMIDE
"Pembe Yara" → HEXACHLOROBENZENE
PEMOLINE
Pempidil → PEMPIDINE TARTRATE
PEMPIDINE TARTRATE
Pempiten → PEMPIDINE TARTRATE
Pen-A → PENICILLINS – "NEWER and SYNTHETICS" > *Ampicillin*
Penamox → PENICILLINS – "NEWER and SYNTHETICS" > *Amoxicillin*
Penaquacaine G → PENICILLIN > *Benzylpenicillin, Procaine*
Penbristol → PENICILLINS – "NEWER and SYNTHETICS" > *Ampicillin*
Penbritin → PENICILLINS – "NEWER and SYNTHETICS" > *Ampicillin*
Penbrock → PENICILLINS – "NEWER and SYNTHETICS" > *Ampicillin*
PENBUTOLOL SULFATE
Pencal → CALCIUM ARSENATE also → CARBETAPENTANE CITRATE
Penchlorol → PENTACHLOROPHENOL
PENCICLOVIR
Pencil Tree → EUPHORBIA
Pencitabs → PENICILLIN > *Benzylpenicillin, Potassium*
Pendase → HYALURONIDASE
PENDIMETHALIN
Pendit WA → SODIUM LAURYL SULFATE
Penduran → PENICILLIN > *Benzylpenicillin, Benzathine*
Penecort → HYDROCORTISONE
Penester → PENICILLIN > *Benzylpenicillin, DAEEH*
Penethamate Hydriodide → PENICILLIN > *Benzylpenicillin, DAEEH*
Penetrex → ENOXACIN
Pen-Fifty → PENICILLIN > *Benzylpenicillin, Procaine*
PENFLURIDOL
Penglobe → PENICILLINS – "NEWER and SYNTHETICS" > *Bacampicillin*
Penicidin → PATULIN
PENICILLAMINE also → BENORYLATE, DIGOXIN, INDOMETHACIN, MAGNESIUM HYDROXIDE
Penicillanic Acid Sulfone → SULBACTAM
Penicillic Acid → PENICILLIUM sp.
PENICILLIN also → ACENOCOUMAROL, ACETAMIDE, ACETYLSALICYLIC ACID, ALUMINUM HYDROXIDE, AMIKACIN, *p*-AMINOHIPPURIC ACID, BISHYDROXYCOUMARIN, CARISPRODOL, CITRUS, CLOFIBRATE, CYCLOCOUMAROL, EPINEPHRINE, ERYTHROMYCIN, FATS, HEPARIN, ICE CREAM, MAGNESIUM, MAGNESIUM OXIDE, MEFENAMIC ACID, METHOTREXATE, MILK, MOXALACTAM, OXYPHENBUTAZONE, PENICILLAMINE, PENICILLIN, PHENYLBUTAZONE, PHENYTOIN, POLYETHYLENE GLYCOL, PRIMIDONE, SALICYLIC ACID, SALSALATE, SODIUM BICARBONATE, SODIUM BISULFITE, TETRACYCLINE, WARFARIN
Penicillin II → PENICILLIN
Penicillin-152 → PENICILLINS – "NEWER and SYNTHETICS" → *Phenethicillin Potassium*
Penicillin, Benzyl → PENICILLIN also → SULFAPHENAZOLE
Penicillin G → PENICILLIN also → CALCIUM CARBIMIDE, CHLORTETRACYCLINE, CIMETIDINE, MAGNESIUM OXIDE, POTASSIUM, SULFAMETHOXYPYRIDAZINE, SULFINPYRAZONE, SULFISOXAZOLE
Penicillin G,-Diethylaminoethyl Ester Hydriodide → PENICILLIN > *Benzylpenicillin, DAEEH*
Penicillin G Potassium → PENICILLIN > *Penicillin G* also → POTASSIUM
Penicillin G Procaine → PENICILLIN > *Penicillin G* also → PROCAINE
Penicillin MV → PENICILLINS – "NEWER and SYNTHETICS" > *Phenethicillin Potassium*
Penicillin P-12 → PENICILLINS – "NEWER and SYNTHETICS" > *Oxacillin Sodium*
Penicillin Sodium → PENICILLIN also → SODIUM
Penicillin V → PENICILLIN also → ESTRIOL, ETHINYL ESTRADIOL, FRAMYCETIN, GLUTEN, NADOLOL, NEOMYCIN, PROPRANOLOL, SULFAMETHOXYPYRIDAZINE, SULFINPYRAZONE, SULFISOXAZOLE
Penicillin V Potassium → PENICILLIN > *Penicillin V* also → POTASSIUM
PENICILLINS – "NEWER and SYNTHETICS" also → ACETAMIDE, ACETYLSALICYLIC ACID, ACETYLCYSTEINE, CLOFIBRATE, ERYTHROMYCIN, MAGNESIUM OXIDE, MEFENAMIC ACID, METHOTREXATE, PENICILLAMINE, TETRACYCLINE, WARFARIN

Penicillinase → β-LACTAMASE
PENICILLIUM sp.
Penicin → 6-AMINOPENICILLANIC ACID
Pénicline → PENICILLINS – "NEWER and SYNTHETICS" > *Ampicillin*
Penidural → PENICILLIN > *Benzylpenicillin, Benzathine*
Peniduran → PENICILLIN > *Benzylpenicillin, Benzathine*
Penimox → PENICILLINS – "NEWER and SYNTHETICS" > *Amoxicillin*
Penin → 6-AMINOPENICILLANIC ACID
Penioral-500 → PENICILLIN > *Benzylpenicillin, Potassium*
Peniplus → PENICILLINS – "NEWER and SYNTHETICS" > *Phenethicillin*
Penticillin → PENICILLINS – "NEWER and SYNTHETICS" > *Piperacillin Sodium*
Penistaph → PENICILLINS – "NEWER and SYNTHETICS" > *Methicillin Sodium*
Penitardon → NYLIDRIN
Penite → SODIUM ARSENITE
Penitracin → BACITRACIN
Penitrem A → CHEESES, PENICILLIUM sp., ROQUEFORTINE
Penivet → PENICILLIN > *Benzylpenicillin, Potassium*
PENMESTEROL
Pennant → METOLACHLOR
Penncap M → METHYL PARATHION
Penniclavine → ARGYREIA
PENNISETUM sp.
PENNYROYAL
Pennywort → LINARIA VULGARIS
Penorale → PENICILLINS – "NEWER and SYNTHETICS" > *Phenethicillin Potassium*
Penotrane → HYDRARGAPREN
Penova → PENICILLINS – "NEWER and SYNTHETICS" > *Phenethicillin Potassium*
Penplenum → PENICILLINS – "NEWER and SYNTHETICS" > *Hetacillin*
Penplus → PENICILLINS – "NEWER and SYNTHETICS" > *Floxacillin*
Pensig → PENICILLINS – "NEWER and SYNTHETICS" > *Phenethicillin Potassium*
Pen-Sint → PENICILLINS – "NEWER and SYNTHETICS" > *Dicloxacillin*
Penstapho → PENICILLINS – "NEWER and SYNTHETICS" > *Oxacillin Sodium*
Penstaphocid → PENICILLINS – "NEWER and SYNTHETICS" > *Oxacillin Sodium*
Penstapho N → PENICILLINS – "NEWER and SYNTHETICS" > *Cloxacillin/Sodium salt*
Pensyn → PENICILLINS – "NEWER and SYNTHETICS" > *Ampicillin*
Penta → PENTACHLOROPHENOL
PENTABORANE-9
Pentaboron Nonahydride → PENTABORANE-9
Pentacard → ISOSORBIDE MONONITRATE
Pentacarinat → PENTAMIDINE ISETHIONATE
Pentacef → CEFTAZIDIME
Pentachlorin → DDT
PENTACHLOROETHANE
PENTACHLORONAPHTHALENE
PENTACHLOROPHENOL also → SAWDUST
Pentachlorophosphorus → PHOSPHORUS PENTACHLORIDE
Pentacon → PENTACHLOROPHENOL
PENTACYNIUM
Pentadecanolide → CYCLOPENTADECANOLIDE
PENTADECYLCATECHOL also → *p-tert*-BUTYL CATECHOL
Pentadoll → CLORPRENALINE
Pentadorm → MEPARFYNOL
PENTAERYTHRITOL
PENTAERYTHRITOL DICHLOROHYDRIN
PENTAERYTHRITOL TETRANITRATE
Pentafin → PENTAERYTHRITOL TETRANITRATE
PENTAGASTRIN also → HALOFENATE
Pentagen → QUINTOZENE
Pentagin → PENTAZOCINE
Pentalin → PENTACHLOROETHANE
Pentam 300 → PENTAMIDINE ISETHIONATE
PENTAMETHYLMELAMINE
PENTAMIDINE ISETHIONATE also → DIMINAZENE, FOSCARNET SODIUM
PENTANE
1,5-Pentanedial → GLUTARALDEHYDE
2,4-Pentanedione → ACETYL ACETONE
1-PENTANETHIOL
Pentanitrine → PENTAERYTHRITOL TETRANITRATE
2-Pentanone → METHYL PROPYL KETONE
Pentanyl → FENTANYL
Pentaphen → CARAMIPHEN also → *p-tert*-PENTYLPHENOL
PENTAPIPERIDE METHYLSULFIDE
PENTAQUINE
Pentasa → 5-AMINOSALICYLIC ACID
Pentaspan → PENTASTARCH
PENTASTARCH
PENTAZOCINE also → FLUOXETINE, HYDROXYZINE, MAGNESIUM SILICATE, MEXILETINE
Pentazol → PENTYLENETETRAZOL

Pentec → METHOXYFLURANE
PENTETIC ACID
Pentetrazol → PENTYLENETETRAZOL
Penthiobarbital → THIOPENTAL
Penthrit → PENTAERYTHRITOL TETRANITRATE
Penticort → AMCINONIDE
Pentid → PENICILLIN > *Benzylpenicillin, Potassium*
Pentilium → PENTOLINIUM TARTRATE
Pentitrate → PENTAERYTHRITOL TETRANITRATE
PENTOBARBITAL also → ALCOHOL, ETHYL; ALPRENOLOL, BROTIZOLAN, CALCIUM CARBIMIDE, CANNABIDIOL; CEDAR, WESTERN RED; CEDROL, DIAZEPAM; 1,1-DICHLORO-2, 2-BIS(*p*-CHLOROPHENYL)ETHANE; DIHYDROSTREPTOMYCIN, GLUTETHIMIDE, IMIPRAMINE, METHYL SALICYLATE, METOPROLOL, MICONAZOLE NITRATE, MORPHINE, NORTRIPTYLINE, PARATHION, PHENAGLYCODOL, PHENOBARBITAL, PHENYTOIN, PICROTOXIN, PROPYLENE GLYCOL, QUINIDINE, QUININE, STREPTOMYCIN, TETRACYCLINE
Pentobarbitone → PENTOBARBITAL
Pentobrocanal → PENTOBARBITAL
Pentol → PENTOBARBITAL
PENTOLINIUM TARTRATE also → ETHACRYNIC ACID, SPIRONOLACTONE
Pentone → PENTOBARBITAL
PENTOSAN POLYSULFATE
Pentosol → PENTOBARBITAL
Pentostam → ANTIMONY SODIUM GLUCONATE
PENTOSTATIN
Pentothal → THIOPENTAL
Pentovis → QUINESTROL
PENTOXIFYLLINE also → THEOPHYLLINE
Pentral 80 → PENTAERYTHRITOL TETRANITRATE
Pentrane → METHOXYFLURANE
Pentrano → METHOXYFLURANE
Pentraspan → PENTAERYTHRITOL TETRANITRATE
Pentrexyl → PENICILLINS – "NEWER and SYNTHETICS" > *Ampicillin*
Pentrite → PENTAERYTHRITOL TETRANITRATE
Pentritol → PENTAERETHRITOL TETRANITRATE
Pentryate → PENTAERYTHRITOL TETRANITRATE
tert-Pentyl Alcohol → AMYLENE HYDRATE
Pentylan → PENTAERYTHRITOL TETRANITRATE
PENTYLENETETRAZOL also → CEDAR, WESTERN RED
Pentyl Nitrite → AMYL NITRITE
p-tert-PENTYLPHENOL
Pentymal → AMOBARBITAL
Pen-Vee → PENICILLIN > *Penicillin V*
Penwar → PENTACHLOROPHENOL
Peonine → PEONY
PEONY
Pepcid → FAMOTIDINE
Pepcidina → FAMOTIDINE
Pepcidine → FAMOTIDINE
Pepdine → FAMOTIDINE
Pepdul → FAMOTIDINE
Pepper Bush → SOLANUM sp.
Pepper Grass → LEPIDIUM sp.
Pepperidge Farm → PRENYLAMINE LACTATE, PRETZELS
PEPPERMINT
Peppermint Camphor → MENTHOL
PEPPERONI
PEPPERS also → PAPAYA, SAFROLE, TOMATO
Pepper Turnip → ARISAEMA
Pepperweeds → LEPIDIUM sp.
Pep Pills → AMPHETAMINE
Pepsi-Cola® → COLAS
Peptan → FAMOTIDINE
Peptard → l-HYOSCYAMINE
Peptavlon → PENTAGASTRIN
Pepticum → OMEPRAZOLE
Peptinimid → ETHOSUXIMIDE
Pepto-Bismol® → BISMUTH SUBSALICYLATE
Peptol → CIMETIDINE
Per-Abrodil → IODOPYRACET
PERACETIC ACID
Peracon Expectorans → THALIDOMIDE
Peragit → TRIHEXYPHENIDYL
Perandren → METHYL TESTOSTERONE
Peraprin → METOCLOPRAMIDE
PERAZINE
Perazyl → CHLORCYCLIZINE
Perc → TETRACHLOROETHYLENE
Percaine → DIBUCAINE
Perce-neige → GALANTHUS NIVALUS
Perchloracap → POTASSIUM PERCHLORATE
PERCHLORIC ACID
Perchlorobenzene → HEXACHLOROBENZENE

Perchlorobutadiene → HEXACHLOROBUTADIENE
Perchloroethane → HEXACHLOROETHANE
Perchloroethylene → TETRACHLOROETHYLENE
Perchloromethane → CARBON TETRACHLORIDE
PERCHLOROMETHYL MERCAPTAN
Perchloronaphthalene → OCTACHLORNAPHTHALENE
PERCHLORYL FLUORIDE
Perchnut → JATROPHA sp.
Perclene → TETRACHLOROETHYLENE
Percorten → DEOXYCORTICOSTERONE
Percorten Pivalate → DEOXYCORTICOSTERONE
Percutol → NITROGLYCERIN
Perdilatal → NYLIDRIN
Perdipina → NICARDIPINE
Perdipine → NICARDIPINE
Perdorm → TEMAZEPAM
Perebral → CYCLANDELATE
Perebron → OXOLAMINE
Perejil → PARSLEY
Peremesin → MECLIZINE
Pérénan → ERGOLOID MESYLATES
Perenum → TOLOXATONE
Perequil → MEPROBAMATE
Perfekthion → DIMETHOATE
Perfenazine → PERPHENAZINE
Perfide → ENTOLOMA
Perfluoroacetone → HEXAFLUOROACETONE
Pergital → PENTAERYTHRITOL TETRANITRATE
PERGOLIDE MESYLATE also → ERYTHROMYCIN
Pergonal → MENOTROPINS
Pergotime → CLOMIPHENE CITRATE
PERHEXILENE also → PAROXETINE
Perhydrol → HYDROGEN PEROXIDE
Perhydrosqualene → SQUALANE
Periactin → CYPROHEPTADINE
Periciazine → PROPERICIAZINE
Periciclina → DEMECLOCYCLINE
Pericyazine → PROPERICIAZINE
Peridamol → DIPYRIDAMOLE
Peridex → CHLORHEXIDINE
Peridon → DOMPERIDONE
Peridys → DOMPERIDONE
Perihab → PENTAERYTHRITOL TETRANITRATE
Perilax → BISACODYL
Perilaxin → CASANTHRANOL
PERILLA FRUTESCENS
Perillartine → PERILLA FRUTESCENS
Perindoprilat → PERINDOPRIL *tert*-BUTYLAMINE
PERINDOPRIL *tert*-BUTYLAMINE
Periplum → NIMODIPINE
Peripress → PRAZOSIN HYDROCHLORIDE
Perisalol → NICORANDIL
Peristaltin → CASANTHRANOL
Peristin → CASANTHRANOL
Peristol → TRIETHYLENE MELAMINE
Periston → POLYVINYLPYRROLIDONE
Peritol → CYPROHEPTADINE
Peritrate → PENTAERYTHRITOL TETRANITRATE
Perityl → PENTAERYTHRITOL TETRANITRATE
Perivar → SPARTEINE
Periwinkle → VINCA ROSEA
Perk → TETRACHLOROETHYLENE
Perlacton → OXYTOCIN
Perlatan → ESTRONE
Perlinganit → NITROGLYCERIN
PERLITE
Perlopal → MEPARFYNOL
Perlutex → MEDROXYPROGESTERONE ACETATE
Permaban → PERMETHRIN
Permabond 102 → CYANOACRYLATES
PERMANENT PRESS
Permapen → PENICILLIN > *Benzylpenicillin, Benzathine*
Permastril → DROMOSTANOLONE PROPIONATE
Permax → PERGOLIDE MESYLATE
Permease → HYALURONIDASE
Permectin → PERMETHRIN
PERMETHRIN also → PYRETHRUM, PYRETHRINS, and PYRETHROIDS
Permitil → FLUPHENAZINE DECANOATE
PERNETTYA FURENS
Perneuron → MEFEXAMIDE
Pernovin → PHENINDAMINE TARTRATE
Peroccide → SULFAMERAZINE
Peroidin → POTASSIUM PERCHLORATE
Perolysen → PEMPIDINE TARTRATE
Peroxyacetic Acid → PERACETIC ACID
Peroxydex → BENZOYL PEROXIDE
Perphenan → PERPHENAZINE
PERPHENAZINE also → ALUMINUM HYDROXIDE, AMITRIPTYLINE, DISULFIRAM, DROPERIDOL, HEPARIN, IMIPRAMINE, LITHIUM, NORTRIPTYLINE, PAROXETINE
Per-Radiographol → IODOPYRACET

Perrier® Water → BENZENE also → MINERAL WATERS
Persadox → BENZOYL PEROXIDE
Persa-gel → BENZOYL PEROXIDE
Persantine → DIPYRIDAMOLE
Persea americana → AVOCADO
Persian Bark → CASCARA SAGRADA
Persian Berries → FRANGULA
Persian Insect Powder → PYRETHRUM, PYRETHRINS, and PYRETHROIDS
Persian Violet → CYCLAMEN
PERSICARY
Persicol → γ-UNDECALACTONE
PERSIMMON
Persolv → UROKINASE
Perspex® → METHYL METHACRYLATE
Persteril → PERACETIC ACID
PERTHANE
Perthrine → PERMETHRIN
Pertofran → DESIPRAMINE
Pertofrane → DESIPRAMINE
Pertranquil → MEPROBAMATE
Peruvian Balsam → BALSAM PERU
Peruvian Bark → CINCHONA
Peruvian Lily → ALOSTROEMERIA LIGTU
PERUVOSIDE also → THEVETIA sp.
Pervetral → OXYPENDYL
Pervitin → METHAMPHETAMINE
Pesos → FENFLURAMINE
Pestmaster → METHYL BROMIDE
Pestox III → SCHRADAN
Pestox XV → MIPAFOX
Pestox 15 → MIPAFOX
Petasitenene → COLTSFOOT
Pet-Derm III → DEXAMETHASONE
Peteha → PROTIONAMIDE
Peter-Kal → POTASSIUM CHLORIDE
Peterphyllin → AMINOPHYLLINE
Pethidine → MEPERIDINE
Petidon → TRIMETHADIONE
Petinimid → ETHOSUXIMIDE
Petinutin → METHSUXIMIDE
PETIVERIA ALLIACEA
PETN → PENTAERYTHRITOL TETRANITRATE
Petnidan → ETHOSUXIMIDE
Petrohol → ALCOHOL, ISOPROPYL
Petrol → GASOLINE
PETROLATUM also → PARAFFIN
Petrolatum, Liquid → MINERAL OIL
Petroleum Benzin → NAPHTHA
Petroleum Ether → GASOLINE
Petroleum Jelly → PETROLATUM
Petroleum Naphtha → NAPHTHA
Petroleum Spirits → NAPHTHA
Petroselinum crispum → PARSLEY
Petrothene → POLYETHYLENE
Petty Spurge → EUPHORBIA sp.
Peucedanum graveolens → DILL
Pevaryl → ECONAZOLE NITRATE
PEYOTE also → CYTISINE
Pezetamid → PYRAZINAMIDE
PF 1593 → BUMETANIDE
PF → PHENYLPHOSPHINE
Pfeffer → CAPSICUM
Pfichlor → CHLORTETRACYCLINE
PFIESTERIA PISCIDA
Pfizer-E → ERYTHROMYCIN STEARATE
Pfizerpen-AS → PENICILLIN > *Benzylpenicillin, Procaine*
Pfizerpen G → PENICILLIN > *Benzylpenicillin, Potassium*
PfiKlor → POTASSIUM CHLORIDE
Pfizer → PIROXICAM
PFU → PHENFORMIN
PGA → FOLIC ACID
PGDN → PROPYLENE GLYCOL 1, 2-DINITRATE
PGE_1 → ALPROSTADIL
PGE_2 → DINOPROSTONE
$PGF_{2\alpha}$ → DINOPROST
pGH → GROWTH HORMONES
PGI_2 → PROSTACYCLIN
PGME → PROPYLENE GLYCOL MONOMETHYL ETHER
PGX → PROSTACYCLIN
PH 105 → CELLULOSE
PHACELIA sp.
PHACETOPERANE
Phacetur → PHENACEMIDE
PHALARIS sp. also → CANARY GRASS
PHALLOIDIN also → AMANITA sp., GALERINA sp.
Phalloin → PHALLOIDIN
Phaltan → FOLPET
Phanodorm → CYCLOBARBITAL
Phanodorn → CYCLOBARBITAL
PHANQUINONE
Phanquone → PHANQUINONE
Phanurane → CANRENONE
Pharlon → 17-α-HYDROXYPROGESTERONE
Pharmacoat 606 → CELLULOSE
Pharmaneurine → VITAMIN B_1
Pharmamedrine → AMPHETAMINE
Pharmasorb → ATTAPULGITE
Pharmatinic → IRON DEXTRANS
Pharmolin → KAOLIN
Pharorid → METHOPRENE

Phasal → LITHIUM CARBONATE
Phaseolamin → BEANS
Phaseolunatin → LINAMARIN
Phaseolus vulgaris → BEANS, Kidney
Phaseoulus limensis → BEANS, Lima
Phaseolus lunatus → BEANS, Lima
PHASEOLUTIN
Phasil → DIMETHICONE (*with silicon dioxide*)
Phasin → LOCUST, BLACK
Phazyme → DIMETHICONE (*with silicon dioxide*)
Phellandrene → OĔNANTHE sp.
Phemeride → BENZETHONIUM CHLORIDE
Phemithyn → BENZETHONIUM CHLORIDE
Phemiton → MEPHOBARBITAL
PHENACAINE
PHENACEMIDE
PHENACETIN also → ACETAMINOPHEN, 4-AMINOCATECHOL, BRUSSEL SPROUTS, CABBAGE, PHENOBARBITAL, TOLBUTAMIDE
Phenacetylurea → PHENACEMIDE
Phenacide → TOXAPHENE
Phenadone → METHADONE
PHENAGLYCODOL also → ALCOHOL, ETHYL; MEPROBAMATE
PHENAMIDINE ISETHIONATE
Phenamin → CHLORPHENIRAMINE also → DEXCHLORPHENIRAMINE
Phenamine → AMPHETAMINE
Phenamiphos → FENAMIPHOS
Phenamizole → AMIPHENAZOLE
PHENANTHRENE
Phenantoin → MEPHENYTOIN
Phenaphen → ACETAMINOPHEN
PHENARSAZINE CHLORIDE
Phenasal → NICLOSAMIDE
Phenatox → TOXAPHENE
Phenazacillin → PENICILLINS – "NEWER and SYNTHETICS" > *Hetacillin*
PHENAZOCINE HYDROBROMIDE also → MEPERIDINE
Phenazodine → PHENAZOPYRIDINE
Phenazone → ANTIPYRINE
PHENAZOPYRIDINE also → SULFAMETHOXAZOLE
Phencen → PROMETHAZINE
PHENCYCLIDINE also → BROCCOLI, CANNABIS, CHLORPROMAZINE, CYANIDES, DIAZEPAM, MESCALINE, MORPHOLINE, PHENYLCYCLOHEXENE, PHENYLCYCLOHEXYLPYRROLIDINE, PIPERIDINE, SODIUM BISULFITE, SODIUM CYANIDE, SPIGELIA ANTHELMIA, Δ^9-TETRAHYDROCANNABINOL
Phendextro → CHLORPHENIRAMINE also → DEXCHLORPHENIRAMINE
Phenedrine → AMPHETAMINE
Phénégic → PHENOTHIAZINE
PHENELZINE also → ALCOHOL, ETHYL; AMITRIPTYLINE, AMPHETAMINE, BEANS, CHEESES, CHLORPHENTERMINE, CLOMIPRAMINE, DESIPRAMINE, DEXTROAMPHETAMINE, DEXTROMETHORPHAN HYDROBROMIDE, DIAZEPAM, DIPHENHYDRAMINE, L-DOPA, DROPERIDOL, FLUOXETINE, HERRING, HYDRAZINE, IMIPRAMINE, INSULIN, KOLA, MEPERIDINE, METHYL-ENEDIOXYAMPHETAMINE, NIALAMIDE, NITRAZEPAM, PHENTERMINE, PROPANIDID, SERTRALINE, SPIGELIA ANTHELMIA, SUCCINYLCHOLINE, SUMATRIPTAN SUCCINATE, TETRAGONIA TETRAGONOIDES, TOLCAPONE, VENLAFAXINE
Phenemal → PHENOBARBITAL
Phenergan → PROMETHAZINE also → CHLORPROMAZINE
Phenethicillin → PENICILLINS – "NEWER and SYNTHETICS" also → CHYMOTRYPSIN
β-Phenethyl Acetate → PHENETHYL ACETATE
Phenethyl Alcohol → ALCOHOL, PHENYLETHYL
Phenethylamine → FENTANYL
Phenethylbiguanide → PHENFORMIN
Phenethyl Phenylacetate → PHENYLETHYL PHENYLACETATE
Phenethyl Propionate → PHENYLETHYL PROPIONATE
Phenethyl Tiglate → PHENYLETHYL TIGLATE
PHENETURIDE also → PHENYTOIN, VITAMIN D_3
PHENFORMIN also → ALCOHOL, ETHYL; TETRACYCLINE, VASOPRESSIN, VITAMIN B_{12}, WARFARIN
PHENICARBAZIDE
Phenidin → PHENACETIN
Phenin → PHENACETIN
PHENINDAMINE TARTRATE also → NORTRIPTYLINE
PHENINDIONE also → CLOFIBRATE, DIMETHICONE, ERYTHROMYCIN, HALOPERIDOL, METHANDROSTENOLONE, OXYPHENBUTAZONE, PHENYRAMIDOL,

POTATO, PROPRANOLOL, TOLBUTAMIDE, TRIFLUPERIDOL
PHENIPRAZINE
PHENIRAMINE MALEATE
Phenistix® → CHLORPROMAZINE
Phenmad → PHENYLMERCURIC ACETATE
Phenmerzyl → PHENYLMERCURIC NITRATE
PHENMETRAZINE also → CHLORPROMAZINE
Phenobal → PHENOBARBITAL
PHENOBARBITAL also → ACETAMINOPHEN, ACETAZOLAMIDE, ACETOHEXAMIDE, ACETYLSALICYLIC ACID; ALCOHOL, ETHYL; AMARANTH, AMINOPYRINE, ANTIPYRINE, BISHYDROXYCOUMARIN, CARBAMAZEPINE, CHLORAMPHENICOL, CHLORPROMAZINE, CIMETIDINE, CLOMETHIAZOLES, CLONAZEPAM, CLOZAPINE, CYCLOCOUMAROL, CYCLOPHOSPHAMIDE, CYCLOSPORINE, DDT, DEOXYCORTICOSTERONE, DESIPRAMINE, DEXAMETHASONE, DEXTROAMPHETAMINE, DIAZEPAM, DIGITALIS, DIGITOXIN, DOCUSATE SODIUM, DOXYCYCLINE, ESTRADIOL, ESTRIOL, ESTRONE, FELBAMATE, FOLIC ACID, GRISEOFULVIN, HALOPERIDOL, HEXOBARBITAL, HYDROCORTISONE, IMIPRAMINE, ITRACONAZOLE, LAMOTRIGINE, LEVONORGESTREL, LINDANE, LYNESTRENOL, MEPHOBARBITAL, MEPROBAMATE, MESORIDAZINE BESYLATE, METHIMAZOLE, METHYLENE BLUE, METHYLPHENIDATE, METRONIDAZOLE, MISONIDAZOLE, NITROFURANTOIN, NITROGLYCERIN, NORTRIPTYLINE, PARATHION, PENICILLIN, PENTOBARBITAL, PHENYLBUTAZONE, PHENYTOIN, PREDNISONE, PRIMIDONE, PROCAINE, PROCHLORPERAZINE, PROGESTERONE, PROPOXYHENE, PROPRANOLOL, QUINIDINE, QUININE, RIFAMPIN, SODIUM BICARBONATE, SODIUM VALPROATE, SULFADIMETHOXINE, SULINDAC, TACROLIMUS, TEA, TESTOSTERONE, THEOPHYLLINE, THYROXINE, TOLBUTAMIDE, VALPROIC ACID, VINYL CHLORIDE, VITAMIN D_3, ZOXAZOLAMINE
Phenobarbitone → PHENOBARBITAL
Phenodin → HEMATIN
Phenododecinium Bromide → DOMIPHEN BROMIDE
PHENOL also → NAPHTHA, PHENYL SALICYLATE, PHLOROGLUCINOL, PYROGALLOL, TOLUENE
PHENOL-FORMALDEHYDE RESINS
PHENOLPHTHALEIN also → WINE
Phenol Red → PHENOLSULFONPHTHALEIN
PHENOLSULFONIC ACID
PHENOLSULFONPHTHALEIN also → DANTHRON, ETHOXAZINE, NOVOBIOCIN, PHENAZOPYRIDINE, SULFINPYRAZONE
Phenolum → PHENOL
Phenopenicillin → PENICILLIN > *Penicillin V*
PHENOPERIDINE
Phenoprocoumarin → PHENPROCOUMON
Phenopropazine → ETHOPROPAZINE
PHENOPYRAZONE
Phenoquin → CINCHOPHEN
Pheno-Squar → PHENOBARBITAL
PHENOTHIAZINE and PHENOTHIAZINES also → BENZQUINAMIDE, CARBANOLATE, CHLORPROMAZINE, COUMAPHOS, CURARE, DICHLOFENTHION, DICHLORVOS, L-DOPA, EPINEPHRINE, MIVACURIUM, MORPHINE, NALIDIXIC ACID, NITROFURANTOIN, PARATHION, PHYSOSTIGMINE, PIMINODINE, PIPERAZINE, QUINURONIUM, RONNEL, RUELENE, TETRAMISOLE, TRICHLORFON, SUCCINYLCHOLINE
"Phenothiazine Diabetes" → CHLORPROMAZINE
Phenothrin → PYRETHRUM, PYRETHRINS, and PYRETHROIDS
Phenoverm → PHENOTHIAZINE
Phenovis → PHENOTHIAZINE
Phenoxadrine → PHENYLTOLOXAMINE
Phenoxene → CHLORPHENOXAMINE
Phenoxine → CHLORPHENOXAMINE
Phenoxur → PHENOTHIAZINE
PHENOXYBENZAMINE HYDROCHLORIDE also → METHYLDOPA, NOREPINEPHRINE
Phenoxy Benzene → PHENYL ETHER
Phenoxymethyl Penicillin → PENICILLIN > *Penicillin V*
PHENOXYPROPAZINE MALEATE
Phen-Fen → FENFLURAMINE also → PHENTERMINE
PHENPROCOUMON also → ACETAMINOPHEN, CLOFIBRATE,

METHYL TESTOSTERONE, MISOPROSTOL, OXYPHENBUTAZONE, PHENYLBUTAZONE, PHENYTOIN, RIFAMPIN, TRAMADOL, SULFADIMETHOXINE, SULFINPYRAZONE, TOLBUTAMIDE
Phensedyl → PROMETHAZINE
PHENSUXIMIDE also → METHSUXIMIDE
Phentanyl → FENTANYL
PHENTERMINE also → FENFLURAMINE, PHENELZINE
Phentoinum → PHENYTOIN
PHENTOLAMINE MESYLATE also → COCAINE, ETHAMBUTOL, PROPRANOLOL, RESERPINE, STREPTOMYCIN
Phentrol → PHENMETRAZINE
Phenurone → PHENACEMIDE
Phenvalerate → FENVALERATE
Phenylacrolein → CINNAMALDEHYDE
β-Phenylacrylic Acid → CINNAMIC ACID
PHENYLALANINE also → ASPARTAME, LINEZOLID
p-Phenylalanine → *p*-AMINOBIPHENYL
Phenyl Alcohol → PHENOL
Phenylamine → ANILINE
σ-Phenyl Anisole → METHYL DIPHENYL ETHER
Phenyl Arsenoxide → ARSENOSOBENZENE
p-Phenylazoaniline → STAMPS and STAMP PADS
1-Phenylazo-2-naphthol → SUDAN I
1(p-Phenylazophenylazo)-2-naphthol → SUDAN III
Phenyl Benzene → DIPHENYL
2-Phenyl-1,4-benzoquinone → PHENYLQUINONE
PHENYLBUTAZONE also → ACENOCOUMAROL, ACETAZOLAMIDE, ACETOHEXAMIDE, ALUMINUM HYDROXIDE, AMINOPYRINE, BISHYDROXYCOUMARIN, CALCIUM CARBIMIDE, CARBUTAMIDE, CHLORDANE, CHLORPROPAMIDE, CHOLESTYRAMINE RESIN, CYCLOCOUMAROL, DDT, DEOXYCORTICOSTERONE, DESIPRAMINE, DESMOPRESSIN, DIGITOXIN, ETHYL BISCOUMACETATE, GLYBURIDE, GRISEOFULVIN, GUANETHIDINE, HYDROCORTISONE, IMIPRAMINE, INSULIN, LINDANE, LITHIUM, MAGNESIUM OXIDE, METHANDROSTENOLONE, METHOTREXATE, METHYLPHENIDATE, NALIDIXIC ACID, NITROFURANTOIN, ORPHENADRINE, PENICILLIN, PHENOBARBITAL, PHENPROCOUMON, PHENYTOIN, PRIMIDONE, PROGESTERONE, PYROGALLOL, SODIUM, SODIUM BICARBONATE, SODIUM VALPROATE, SULFAMETHOXYPYRIDAZINE, TESTOSTERONE, TOLBUTAMIDE
Phenyl Carbimide → PHENYL ISOCYANATE
Phenylcarbinol → ALCOHOL, BENZYL
4-PHENYL CATECHOL
Phenyl Chloride → CHLOROBENZENE
Phenylchloroform → BENZOTRICHLORIDE
Phenylcinchoninic Acid → CINCHOPHEN
Phenyl Cyanide → BENZONITRILE
PHENYLCYCLOHEXENE
PHENYLCYCLOHEXYLPYRROLIDINE
Phenydantoin → PHENYTOIN
PHENYLDICHLORARSINE
p-PHENYLENEDIAMINE also → *p*-AMINOBENZOIC ACID, CHLORPROPAMIDE, HATS, HYDROFLUMETHIAZIDE, INKS, RUBBER, SHOES, STAMPS and STAMP PADS, SULFANILAMIDE, TOLBUTAMIDE
PHENYLEPHRINE HCL also → DEBRISOQUIN, HYDROCORTISONE, IMIPRAMINE, ISOCARBOXAZID, NIALAMIDE, PHENELZINE, TOLOXATONE, TRANYLCYPROMINE
Phenylethane → ETHYLBENZENE
1-Phenylethanone → ACETOPHENONE
PHENYL ETHER
PHENYLETHYL ACETATE
β-Phenylethylamine → MISTLETOE, AMERICAN
PHENYLETHYL BENZOATE
Phenylethylene → STYRENE
PHENYLETHYLHYDANTOIN
β-Phenylethylhydrazine → PHENELZINE
PHENYLETHYL ISOVALERATE
PHENYLETHYL PHENYLACETATE
PHENYLETHYL PROPIONATE
PHENYLETHYL TIGLATE
Phenylformic Acid → BENZOIC ACID
PHENYLHYDRAZINE also → RUBBER
Phenyl Hydride → BENZENE
Phenyl Hydroxide → PHENOL
Phenylic Acid → PHENOL
Phenyl-idium → PHENAZOPYRIDINE
Phenylindandione → PHENINDIONE
PHENYL ISOCYANATE
Phenylisohydantoin → PEMOLINE

SULFISOXAZOLE, SULTHIAME, TACROLIMUS, TETRACYCLINE, THEOPHYLLINE, TICLOPIDINE, TIZANIDINE, TOLBUTAMIDE, TOREMIFENE CITRATE, TRIMETHOPRIM, *d*-TUBOCURARINE CHLORIDE, VALPROIC ACID, VANILLA, VERAPAMIL, VIGABATRIN, VITAMIN B_6, VITAMIN D_3, WARFARIN
PHEOMELANIN
Phenyzene → PHENYLBUTAZONE
Phermerol → BENZETHONIUM CHLORIDE
Phetanol → ETHYL PHENYLEPHRINE
Philadelphus → MOCK ORANGE
Philipon → METHAMPHETAMINE
PHILODENDRON sp.
Philodorm → CYCLOBARBITAL
pHisohex → HEXACHLOROPHENE
pHiso-Med → CHLORHEXIDINE
Phix → PHENYLMERCURIC ACETATE
Phleum pratense → TIMOTHY
Phlogopite → MICA
Phloguron → KEBUZONE
Phloridzin → APPLE
Phloroglucin → PHLOROGLUCINOL
PHLOROGLUCINOL also → HEXYLRESORCINOL, HYDROQUINONE
PHMB → POLY (HEXAMETHYLENE BIGUANIDE HCL)
Phobex → BENACTYZINE
Phoenectin → IVERMECTIN
PHOLCODINE
PHOLEDRINE
PHOLIOTA sp.
Pholiata spectabilis → MUSHROOMS
PHOMA sp.
PHOMOPSIS sp.
Phoneutria sp. → SPIDERS
Phoradendron → MISTLETOE, AMERICAN
PHORATE
PHORBOL
Phorbyol → CASTOR OIL
Phorone → GLUE
Phosacetin → GOPHACIDE
PHOSALONE
Phosdrin → MEVINPHOS
Phosethoprop → ETHOPROP
Phosfene → MEVINPHOS
PHOSFOLAN
PHOSGENE also → CARBON TETRACHLORIDE, CHLOROFORM, ETHYLENE DICHLORIDE, FREONS, HYDROFLUORIC ACID, METHYL CHLOROCARBONATE, POLYVINYL CHLORIDE, WELDING
Phoskil → PARATHION
Phosmet → IMIDAN also → CHLORFENVINPHOS
Phosphacol → DIETHYL-*p*-NITROPHENYLPHOSPHATE
PHOSPHAMIDON
Phosphaniline → PHENYLPHOSPHINE
PHOSPHATES also → FLUORIDES and FLUORINE, POLONIUM
Phosphatidyl Choline → LECITHIN
PHOSPHINE also → CALCIUM CARBIMIDE, ZINC
Phospholan → PHOSFOLAN
Phospholine Iodide → ECHOTHIOPHATE IODIDE
PHOSPHORIC ACID
Phosphoric Chloride → PHOSPHORUS PENTACHLORIDE
PHOSPHORUS also → MATCHES, MUNITIONS
Phosphorus Chloride → PHOSPHORUS OXYCHLORIDE
PHOSPHORUS OXYCHLORIDE
PHOSPHORUS PENTACHLORIDE also → OXALYL CHLORIDE
PHOSPHORUS PENTASULFIDE
Phosphorus Sulfide → PHOSPHORUS PENTASULFIDE
PHOSPHORUS TRICHLORIDE
Phosphorus Trihydride → PHOSPHINE
Phosphoryl Chloride → PHOSPHORUS OXYCHLORIDE
Phospho-Soda® → PHOSPHATES
Phosphothion → MALATHION
Phosvel → LEPTOPHOS
Phosvin → ZINC
Phosvitin → EGGS
Photinia arbutifolia → HETEROMELES ARBUTIFOLIA
Photodyn → HEMATOPORPHYRIN
PHOXIM
Phrenazol → PENTYLENETETRAZOL
Phthalamodine → CHLORTHALIDONE
Phthalazole → PHTHALYLSULFATHIAZOLE
PHTHALIC ANHYDRIDE
PHTHALIMIDE
m-PHTHALODINITRILE
PHTHALOFYNE
PHTHALOPHOS
PHTHALYLSULFATHIAZOLE
Phthalylsulfonazole → PHTHALYLSULFATHIAZOLE

Phyban → MONOSODIUM METHANEARSONATE
Phyberythin → CHLORIS sp.
Phygon → DICHLONE
Phylcardin → AMINOPHYLLINE
Phyllanthin → PHYLLANTHUS sp.
PHYLLANTHUS sp.
Phylletten → DEQUALINIUMS
Phyllindon → AMINOPHYLLINE
Phyllocontin → AMINOPHYLLINE
PHYLLOERYTHRIN also → RYE GRASS
Phyone → GROWTH HORMONES
Physalaemus fuscumaculatus → PHYSALEAMIN
PHYSALEAMIN also → TOADS
Physalemin → PHYSALEAMIN
PHYSALIA
Physalia utriculus → JELLYFISH
PHYSALIS sp. also → GROUND CHERRY
Physeptone → METHADONE HCL
Physex → GONADOTROPINS, CHORIONIC
Physic Nut → JATROPHA sp.
Physiomycine → METHACYCLINE
Physiotens → MOXONIDINE
PHYSOSTIGMA VENENOSUM
PHYSOSTIGMINE also → KETAMINE, MALATHION, PHYSOSTIGMA VENENOSUM, SUCCINYLCHOLINE, TRIFLUOPERAZINE DIHYDROCHLORIDE
Physostol → PHYSOSTIGMINE
Physpan → THEOPHYLLINE
PHYTANNIC ACID also → CHLOROPHYLL
Phytansäure → PHYTANNIC ACID
Phytar → CACODYLIC ACID
PHYTEUMA
PHYTIC ACID
Phytoalexins → CELERY
Phytoin → PHENYTOIN
PHYTOLACCA sp. also → DATURA sp.
Phytolaccigenin → PHYTOLACCA sp.
Phytolaccine → PHYTOLACCA sp.
Phytolascotoxin → PHYTOLACCA sp.
Phytuberin → POTATO
Piazofolina → MORINAMIDE
Piazolina → MORINAMIDE
Pic-clor → CHLOROPICRIN
Picfume → CHLOROPICRIN
Piciete → TOBACCO
Picietyl → TOBACCO
Pick Purse → CAPSELLA BURSA-PASTORIS
PICLORAM
PICRASMA QUASSOLOIDES
PICRIC ACID
"Picric Itch" → PICRIC ACID
Picride → CHLOROPICRIN
PICROTOXIN also → COCCULUS sp., PHENYTOIN
Pictou Disease → SENECIO sp.
PID → PHENINDIONE
Pidgex → IRON DEXTRANS
Pidilat → NIFEDIPINE
Pidorubicin → EPIRUBICIN
Pidyrone → DIPYRONE
PIE(S)
Pied d'alouette → LARKSPUR
Pierami → AMIKACIN
PIERIS sp.
Pierminox → MINOXIDIL
Pietil → OXOLINIC ACID
Pifatidine → ROXATIDINE ACETATE
Pigeonberry → PHYTOLACCA sp.
"Pigeon-breeder's Lung" → DUSTS
PIGEONS
Pig Iron → IRON POLYSACCHARIDE COMPLEX
Pigmex → MONOBENZONE
Pignon d'Inde → JATROPHA sp.
Pigweed → AMARANTHUS
Pigwrack → CARRAGEENAN
PIH → PHENIPRAZINE
Pilagan → PILOCARPINE
Pil-Digis → DIGITALIS
PILLOWS
PILOCARPINE also → ACECLIDINE
Pilo → PILOCARPINE
Pilocar → PILOCARPINE
Pilofrin → PILOCARPINE
Pilogel → PILOCARPINE
Pilopine HS → PILOCARPINE
Piloral → CLEMASTINE
Pilostat → PILOCARPINE
Pimadin → PIMINODINE
Pimafucin → NATAMYCIN
Pimaricin → NATAMYCIN
Pimavecort → NEOMYCIN
PIMELIA sp.
Pimelic Ketone → CYCLOHEXANONE
Pimenta → PIMENTO
PIMENTO
PIMINODINE
PIMOZIDE also → CLARITHROMYCIN, CLOZAPINE, RITONAVIR
Pimpinella anisum → ANISE
PINACIDIL
Pinangi → ARECA
PINAVERIUM BROMIDE
Pincets → PIPERAZINE
Pindac → PINACIDIL
Pindione → PHENINDIONE

PINDOLOL also → DILTIAZEM, INDOMETHACIN, THIORIDAZINE
PINDONE
PINE
PINEAPPLE
PINENE also → PARSLEY
Pineroro → DIPHENIDOL
Ping Lang → ARECA
Pini-pini → JATROPHA sp.
Pink Cloud → CALADIUM
Pink Disease → COPPER, PHENYLMERCURIC PROPIONATE
Pink Flower → VINCA ROSEA
Pink Pepper → SCHINUS TEREBINTHIFOLIUS
Pink Rot → CELERY
"Pink Sore" → HEXACHLOROBENZENE
Pinozan → PIPERAZINE
Pinrou → PIPERAZINE
Pinsirup → PIPERAZINE
Pinto Bean → BEANS, Pinto
Pinus sp. → PINE
PIO → PEMOLINE
Pioxol → PEMOLINE
Pipadone → DIPIPANONE
Pipadox → PIPERAZINE
PIPAMAZINE
PIPAMPERONE also → FLUVOXAMINE
Pipane → TRIHEXYPHENIDYL
PIPAZETHATE
PIPECURONIUM BROMIDE
PIPERACETAZINE
Piperacillin → PENICILLINS – "NEWER and SYNTHETIC" also → AMIKACIN, GENTAMICIN, METHOTREXATE, PROBENECID, TOBRAMYCIN
Piperate → PIPERAZINE
Piperazate → PIPERAZINE
Piperazidine → PIPERAZINE
PIPERAZINE also → CHLORPROMAZINE, ETHYLENEDIAMINE, METHOTRIMEPRAZINE
Piperazine Estrone Sulfate → ESTRONE
Piper betel → BETEL
Piperettine → PEPPERS
PIPERIDINE also → LUPINES
PIPERIDIONE
PIPERIDOLATE
Piperidyl → DIPIPANONE
PIPERIDYL BENZILATE also → PROMAZINE
Piperine → PEPPERS
Piper nigrum → PEPPERS
PIPEROCAINE
Piperoline → PEPPERS
Piperonal → HELIOTROPIN
Piperonil → PIPAMPERONE
PIPERONYL BUTOXIDE
β-PIPERONYL ISOPROPYL HYDRAZINE
Piperverm → PIPERAZINE
Piperyline → PEPPERS
Pipizan Citrate → PIPERAZINE
PIPOBROMAN
Pipolphen → PROMAZINE
Piportil → PIPOTIAZINE
PIPOSULFAN
Pipothiazine → PIPOTIAZINE
PIPOTIAZINE
Pipracid → PIPERAZINE
Pipracil → PENICILLINS – "NEWER and SYNTHETICS" > *Piperacillin Sodium*
PIPRADOL also → GUANETHIDINE
γ-Pipradol → AZACYCLONAL
Pipradrol → PIPRADOL
Pipril → PENICLLINS – "NEWER and SYNTHETICS" > *Piperacillin Sodium*
PIPROCURARINE
Piptadenia peregrina → ANADENANTHERA PEREGRINA
PIRACETAM
Piraldin → QUINAPYRAMINE
Piraldina → PYRAZINAMIDE
Piramox → PENICILLINS – "NEWER and SYNTHETICS" > *Amoxicillin*
Piranver → PYRANTEL
PIRBUTEROL
Pirem → CARBUTEROL
PIRENZEPINE HYDROCHLORIDE
Pirevan → QUINURONIUM
PIRIBEDIL
PIRIBENZYL METHYL SULFATE
Piricef → CEPHAPIRIN
Pirid → PHENAZOPYRIDINE
Piridasmin → THEOPHYLLINE
Piridol → AMINOPYRINE
Pirilène → PYRAZINAMIDE
PIRIMIPHOS METHYL
Piriton → CHLORPHENIRAMINE
Pirkam → PIROXICAM
Piroan → DIPYRIDAMOLE
Piroflex → PIROXICAM
PIROMEN
PIROXICAM also → FLUCONAZOLE, FUROSEMIDE, LITHIUM, RITONAVIR, WARFARIN, ZAFIRLUKAST
PIRPROFEN
Pirroxil → PIRACETAM
Piscidia sp. → DOGWOOD

Piss-goed → EUPHORBIA
Piss-grass → EUPHORBIA
Pistachio → PISTACIA sp. also → JATROPHA sp.
PISTACIA sp.
Pisum sativum → PEAS
PITCH
PITHOMYCES sp.
Pitocin → OXYTOCIN
Pitressin → VASOPRESSIN
Pitressin Snuff → PITUITARY
Pituidrol → PITUITARY, POSTERIOR
PITUITARY
Pituitary Growth Hormone → GROWTH HORMONES
Pituitary Lactogenic Hormone → LUTEOTROPIN
PITUITARY, POSTERIOR also → POLYVINYLPYRROLIDONE
Pituitrin → POLYVINYLPYRROLIDINE
PITURANTHOS TRIRADIATUS
Piturine → DUBOISA MYOPOROIDES
Pivacin → PINDONE
Pival → PHENINDIONE also → PINDONE
Pivaldione → PINDONE
Pivalephrine → DIPIVEFRIN
Pivalone → TIXOCORTOL PIVALATE
Pivalyl Indandione → PINDONE
Pivalyl Valone → PINDONE
Pivalyn → PINDONE
Pivampicillin → PENICILLINS – "NEWER and SYNTHETICS" also → METOCLOPRAMIDE, PROPANTHELINE
Pivatil → PENICILLINS – "NEWER and SYNTHETICS" > *Pivampicillin*
Pivmecillinam → PENICILLINS – "NEWER and SYNTHETICS"
Piv2PMEA → ADEFOVIR
Pix Lantharis → COAL TAR
Pizotifen → PIZOTYLINE
PIZOTYLINE
PIZZA
PK 10,169 → ENOXAPARIN also → HEPARIN
PK 26,124 → RILUZOLE
PKhNB → QUINTOZENE
PLACEBO
PLACENTAL EXTRACTS
Placidex → MEPHENOXALONE
Placidon → MEPROBAMATE
Placidyl → ETHCHLORVYNOL
Plac Out → CHLORHEXIDINE
Plactamin → PRENYLAMINE LACTATE
Planadalin → CARBROMAL
Planate → CLOPROSTENOL
Planavit C → VITAMIN C
Plancol → HYDROCORTISONE
Planipart → CLENBUTEROL
Planochrome → MERBROMIN
Planoform → BUTAMBEN
Planomide → SULFATHIAZOLE
PLANTAGO SEED also → LITHIUM
Plantain Seed → PLANTAGO SEED
Plant Make-Crazy → DATURA
Plant Make-Squint → DATURA
Planum → TEMAZEPAM
Plaocaine → PROCAINE
Plaquenil → HYDROXYCHLOROQUINE
Plasil → METOCLOPRAMIDE
Plasma Thromboplastin Component → CHRISTMAS FACTOR
Plasmin → FIBRINOLYSIN
Plasmochin → PAMAQUIN
PLASMOCID
Plasmoquine → PAMAQUIN
Plasmosan → POLYVINYLPYRROLIDONE
Plaster of Paris → GELATIN, HYDROGEN SULFIDE
Platamine → CISPLATIN
Platet → ACETYLSALICYLIC ACID
Plath-Lyse → DICHLOROPHEN(E)
Platiblastin → CISPLATIN
Platinex → CISPLATIN
PLATINIC CHLORIDE
Platinol → CISPLATIN
Platinoxin → CISPLATIN
PLATINUM also → NICKEL
Platistin → CISPLATIN
Platosin → CISPLATIN
Platymerus → TRIATOMA sp.
Plavix → CLOPIDOGREL BISULFATE
Plavolex → DEXTRAN
Plaxin → DIETHYLENE GLYCOL
Plegomazin → CHLORPROMAZINE
Plenastril → OXYMETHOLONE
Plendil → FELODIPINE
Plenur → LITHIUM CARBONATE
Pleocide → 2-ACETAMIDO-5-NITROTHIAZOLE
Plesiomonas shigelloides → SHELLFISH
Pletal → CILOSTAZOL
Pletil → TINIDAZOLE
Pleurisy Root → ASCLEPIAS
PLEUROTIS sp. also → MUSHROOMS
Plexiglas® → METHYL METHACRYLATE
Plicamycin → MITHRAMYCIN
Pliny the Elder → ALOE also → LEAD
PLUM also → CHERRY, CYANIDES, L-DOPA, FRUITS and FRUIT JUICES, HYDROGEN CYANIDE, PEACHES

Plumbago → GRAPHITE
Plumbum → LEAD
Plumyews → YEW
Pluracol E → POLYETHYLENE GLYCOL
Plurexid → CHLORHEXIDINE
Plurimen → DEPRENYL
Pluronic F-68 → POLOXALENE
Pluryle → BENDROFLUMETHIAZIDE
Plusuril → BENDROFLUMETHIAZIDE
PLUTONIUM
PLV 2 → FELYPRESSIN
Plycocyanin → NODULARIA SPUMIGENA
PMA → METHOXYAMPHETAMINE also → PHENYLMERCURIC ACETATE also → PHORBOL
PMAC → PHENYLMERCURIC ACETATE
PMAS → PHENYLMERCURIC ACETATE
Pmatiniya → LASIOSIPHON sp.
PMBC → PEBULATE
P. M. F. → HYDRARGAPHEN
PMM → PENTAMETHYLMELAMINE
PMN → PHENYLMERCURIC NITRATE
PMP → 2-ISOVALERYL-1,3-INDANEDIONE
PN-200-110 → ISRADIPINE
PNA → NITROANILINES
PNCB → NITROCHLOROBENZENES
Pneumopent → PENTAMIDINE ISETHIONATE
PNU → PYRIMINAL
PNU 100,766 → LINEZOLID
PNU 200,583 → TOLTERODINE TARTRATE
Poa aquatica → GLYCERIA sp.
Pocan → PHYTOLACCA
Pod → CANNABIS
Podofilox → PODOPHYLLOTOXIN
Podophyllin → PODOPHYLLUM RESIN
PODOPHYLLOTOXIN also → POLYGALA sp.
PODOPHYLLUM also → ALCOHOL, ETHYL
PODOPHYLLUM RESIN
POINCIANA GILLIESSI
POINSETTIA also → EUPHORBIN
Point Loco → OXYTROPIS
Pois de Senteur → LATHYRUS sp.
Poison Bean → SESBANIA sp. also → DAUBENTONIA PUMICE
Poison Black Cherry → BELLADONNA
Poison Chickweed → ANAGALLIS ARVENSIS
Poison Dogwood → SUMAC, POISON
Poison Elder → SUMAC, POISON
Poison Fish Bean → TEPHROSIA sp.
Poison Hemlock → CONIUM MACULATUM
Poisonings/Death → ADENIA VOLKENSII, AMANITA (*phalloides*), AMPHETAMINE, ARISTOLOCHIA GRANDIFLORA, ARSENIC, BRUCINE, CAFFEINE, CANTHARIDAN, CASTOR BEAN, 4-CHLORO-*m*-CRESOL, CONIUM MACULATUM (*Hemlock*), CYANIDE, DATURA, ENDRIN, ETHYL PARATHION, EUPHORBIA, GLORIOSA, GNIDIA KRAUSSINA, HYDROFLUORIC ACID, HYDROGEN CYANIDE, INSULIN, LASIOSIPHON sp., LUNARIA, LYE, LYSERGIC ACID DIETHYLAMIDE, METHANE, MEVINPHOS, MUNDULEA SERICEA, NICOTIANA sp., NICOTINE, NITRIC ACID, *N*-NITROSODIMETHYLAMINE, PAULLINIA sp., RICIN, ZINC SULFATE
POISON IVY also → CASHEW, PENTADECYLCATECHOL; SUMAC, POISON
Poison Laurel → KALMIA
Poison Lettuce → LACTUCA sp.
Poison Nuts → NUX VOMICA
POISON OAK also → CASHEW, PENTADECYLCATECHOL, POISON IVY; SUMAC, POISON
Poison Moss → LIMU-MAKE-O-HANA
Poisonous Hudson's Bay Tea → KALMIA
Poison Parsley → CONIUM MACULATUM
Poison Parsnip → CICUTA sp.
Poison Peach → TREMA ASPERA
Poison Rye Grass → RYE GRASS
Poison Sage → ISOTROPIS
Poison Sedge → SCHOENUS ASPEROCARPUS
"Poison Squad" → BORAX, COPPER SULFATE
Poison Suckleya → SUCKLEYA SUCKLEYANA
Poison Sumac → SUMAC, POISON also → CASHEW, POISON IVY
Poison Tobacco → HYOSCYAMUS
Poison Vine → POISON IVY
Pokeberry → PHYTOLACCA sp.
Poke Root → VERATRUM sp.
Poke Salad Plant → PHYTOLACCA sp.
Pokeweed → PHYTOLACCA sp.
Polacaritox → TETRADIFON
Polamin → DEXCHLORPHENIRAMINE
Polaramine → CHLORPHENIRAMINE also → DEXCHLORPHENIRAMINE
Polarfen → ACETAMINOPHEN
Polaronil → CHLORPHENIRAMINE also → DEXCHLORPHENIRAMINE
POLDINE METHYLSULFATE
Poleo I → ALOYSIA
Poleon → NALIDIXIC ACID

Polibutin → TRIMEBUTINE MALEATE
Polidocanol → POLYOXYETHYL LAURYL ETHERS
Polik → HALOPROGIN
Polinalin → AMINOPYRINE
POLISHES, FURNITURE
POLISHES, SHOE
Polistin → CARBINOXAMINE MALEATE
Polistes sp. → WASPS
Poliuron → BENDROFLUMETHIAZIDE
Polival → THIABENDAZOLE
Polixima → CEFUROXIME
Pollakisu → OXYBUTYNIN CHLORIDE
Polmiror → NIFURATEL
Polofifisi → PEPPERS
POLONIUM
POLOXALENE
Poloxalkol → POLOXALENE
Poloxamer 188 → POLOXALENE
Polyacronitrile → HYDROGEN CYANIDE
POLYANTHES TUBEROSA
Polybar → BARIUM SULFATE
Polybrene → HEXADIMETHRINE BROMIDE
POLYBROMINATED BIPHENYLS
POLYBUTADIENE
Polycell → CELLULOSE
POLYCHLORINATED BIPHENYLS also → CONCRETE, FISH, HYDRAULIC FLUIDS, MILK, PHENOBARBITAL, RICE, RUBBER, SILO, SOOT
Polychlorocamphene → TOXAPHENE
Polychloroprene → ETHYL BUTYL THIOUREA
Polychols → LANOLIN
Polycidal → SULFALENE
Polycidine → DEQUALINIUMS
Polycillin → PENICILLINS – "NEWER and SYNTHETICS" > *Ampicillin*
POLYCLADA sp.
Polycyclic Aromatic Hydrocarbons → CARBONS, COKE, DIESEL FUEL, GRAPHITE, HAM, MOTOR OIL
Polycycline → TETRACYCLINE
POLYESTER also → ETHYL MALEATE, HYDROGEN CHLORIDE
Polyethoxylated Castor Oil → CREMOPHOR EL
POLYETHYLENE also → BUTYLATED HYDROXYTOLUENE, 4-CHLORO-*m*-CRESOL, CHLOROBUTANOL, METARAMINOL BITARTRATE, METHAMPHETAMINE, OXYTETRACYCLINE, PHENOL, PROMETHAZINE, SECOBARBITAL SODIUM, STREPTOMYCIN, THIMEROSAL, VITAMIN B_1
POLYETHYLENE GLYCOL also → LORAZEPAM, MUPIROCIN, NITROFURAZONE, NORFLOXACIN
Polyfilm → POLYETHYLENE
Poly G → POLYETHYLENE GLYCOL
POLYGALA sp.
Polyglucin → DEXTRAN
Polyglycol → DIETHYLENE GLYCOL
Polyglycol E → POLYETHYLENE GLYCOL
POLYGLYCOLIC ACID
POLYGONATUM BIFLORUM
POLYGONUM sp.
Polygraph → MEPROBAMATE
Polygripan → THALIDOMIDE
Polygris → GRISEOFULVIN
POLY (HEXAMETHYLENE BIGUANIDE HCL)
POLY I : C
Polyject → IRON POLYSACCHARIDE COMPLEX
Polymer Fume Fever → POLYTEF
Polymox → PENICILLINS – " NEWER and SYNTHETICS" > *Amoxicillin*
POLYMIXIN B also → CEFEPIME, DIMETHYL TUBOCURARINE IODIDE, EPINEPHRINE, GALLAMINE TRIETHIODIDE, GENTAMICIN, METHOTREXATE, NEOMYCIN, NEOSTIGMINE, SUCCINYLCHOLINE, *d*-TUBOCURARINE CHLORIDE, VANCOMYCIN, VITAMIN B_{12}
Polymyxin E → COLISTIN
Polynuclear Aromatic Hydrocarbons → RUBBER
Polyotic → TETRACYCLINE
Polyoxyaluminum Acetylsalicylate → ALOXIPRIN
Polyoxyethylated Castor Oil → CREMOPHOR EL
Polyoxyethylene Sorbitan Esters → POLYSORBATES
POLYOXYETHYL LAURYL ETHERS
Polyphlogin → CINCHOPHEN
Polypodium virginianum → POLYPODY FERN
POLYPODY FERN
Polypropylene → CHLOROBUTANOL, PHENOL, THIMEROSAL
Polyquil → PYRVINIUM PAMOATE
Polyram z → ZINEB
Polyribinosinic Acid → POLY I : C
Polyribocytidylic Acid → POLY I : C
Polysan → MAGNESIUM HYDROXIDE
Polyseptil → SULFATHIAZOLE
POLYSORBATES also → CREAM
POLYSTYRENE also → METHYL CHLORIDE, STYRENE, TEA
Polystyrene Sodium Sulfonate → POLYSTYRENE
Polysulphated Glycosaminoglycane → MUCOITIN POLYSULFATE ESTER

Poly T-128 → TARTRAZINE
POLYTEF
Polytetrafluoroethylene → POLYTEF
Polythene → POLYETHYLENE
POLYTHIAZIDE
Polythion → DIOXATHION
Polytrim → POLYMYXIN B
Polytrin → CYPERMETHRIN
POLYURETHANES also → HYDROGEN CYANIDE, SHOES, SILICONES, 2,4-TOLUENEDIAMINE
POLYVINYL ACETATE
POLYVINYL ALCOHOL also → MERCURIC CHLORIDE, SODIUM CHLORIDE
POLYVINYL CHLORIDE also → BIS(2-ETHYLHEXYL)PHTHALATE, CARBAMAZEPINE, CYCLOSPORINE, DAUNORUBICIN, GLOVES, HYDROGEN CHLORIDE, METHOXYFLURANE, PACLITAXEL, PHOSGENE, PHTHALIC ANHYDRIDE, PROPOFOL, SHOES
Polyvinyl Disease → POLYVINYLPYRROLIDONE
POLYVINYLPYRROLIDONE also → IODOPHORS
POLYVINYLPYRROLIDONE IODINE COMPLEX
Polyviol → POLYVINYL ALCOHOL
Pomarsol → THIRAM
POMEGRANATE
Ponalar → MEFENAMIC ACID
Ponalid → ETHYBENZTROPINE
PONCEAU 3R
PONCEAU 4R
PONCEAU MX
PONCEAU SX also → CHERRY
Poncyl-FP → GRISEOFULVIN
Ponderal → FENFLURAMINE
Ponderax → FENFLURAMINE
Ponderex → FENFLURAMINE
Pondex → PEMOLINE also → PENICILLINS – "NEWER and SYNTHETICS" > *Pivampicillin*
Pondimin → FENFLURAMINE
Pondinil → MEFENOREX
Pondinol → MEFENOREX
Pondium → SIBUTRAMINE
Pondocil → PENICILLINS – "NEWER and SYNTHETICS" > *Pivampicillin*
Pondocillin → PENICILLINS – "NEWER and SYNTHETICS" > *Pivampicillin*
PONGAMIA PINNATA
Ponsital → IMICLOPAZINE
Ponso → TANSY
Ponstan → MEFENAMIC ACID
Ponstel → MEFENAMIC ACID
Ponstil → MEFENAMIC ACID
Ponstyl → MEFENAMIC ACID
Pontal → MEFENAMIC ACID
Pontocaine → TETRACAINE HYDROCHLORIDE
Poor Man's Millet → PASPALUM sp.
Poor Man's Weather Glass → ANAGALLIS ARVENSIS
P. O. P. → OXYTOCIN
POPLAR
Poppers → AMYL NITRITE also → BUTYL NITRITE
POPPY also → CHELIDONIUM MAJUS, PAPAVER sp.
Poppyseed Oil, Iodized → ETHIODIZED OIL
Populin → POPLAR
Populus tomentosa → POPLAR
Poquil → PYRVINIUM PAMOATE
POR 8 → ORNIPRESSIN
Porcelain Clay → KAOLIN
Porcupine Grass → STIPA sp.
PORFIROMYCIN
PORK
Poroporo → SOLANUM sp.
Porphyrin → ETHOXAZINE also → PHYLLOERYTHRIN
Portuguese-Man-of-War → PHYSALIA
PORTULACA OLERACEA
Posalfilin → PODOPHYLLUM RESIN
Posdel → BUCLIZINE
Posedrine → BECLAMIDE
Posicor → MIBEFRADIL
Posorutin → TROXERUTIN
Possipione → PAROXYPROPIONE
Postadoxine → VITAMIN B_6
Postafen Tabl → BUCLIZINE
Postafene → MECLIZINE
Pot → CANNABIS
Potaba → POTASSIUM-*p*-AMINOBENZOATE
Potash → THALLIUM
Potasoral → POTASSIUM GLUCONATE
Potassii Chloras → POTASSIUM CHLORATE
Potassii Chloridum → POTASSIUM CHLORIDE
Potassion → POTASSIUM CHLORIDE
POTASSIUM also → CAPTOPRIL, CHEESES, COFFEE, FRUITS and FRUIT JUICES, MILK; PENICILLIN > *Potassium Benzylpenicillinate*; PHENOLPHTHALEIN, PREDNISONE, SAUERKRAUT, SODIUM, SUCCINYLCHOLINE, TEA

POTASSIUM ACETATE
POTASSIUM-*p*-AMINOBENZOATE
POTASSIUM-*p*-AMINOSALICYLATE
POTASSIUM AMMONIUM BIFLUORIDE
Potassium Antimonyl Tartrate → ANTIMONY POTASSIUM TARTRATE
POTASSIUM ARSENITE SOLUTION
POTASSIUM AZIDE
Potassium Benzoate → MARGARINE
Potassium Benzylpenicillinate → PENICILLIN
Potassium Bichromate → POTASSIUM DICHROMATE
POTASSIUM BISULFITE
POTASSIUM BROMATE also → BREAD, FLOUR, WHEAT
POTASSIUM BROMIDE
POTASSIUM CHLORATE also → MATCHES
POTASSIUM CHLORIDE also → FIRE COLORING SALTS, HYDROCLOROTHIAZIDE, POTASSIUM CITRATE, PRACTOLOL, SPIRONLACTONE
POTASSIUM CHROMATE
Potassium Chrome → CHROMIC POTASSIUM SULFATE
POTASSIUM CITRATE
POTASSIUM CLAUVLANATE also → PENICILLINS – "NEWER and SYNTHETIC" > *Amoxicillin, Ticarcillin Disodium*
Potassium Cyanate → CYANATES
POTASSIUM CYANIDE also → PHENCYCLIDINE, SODIUM HYDROXIDE
POTASSIUM DICHROMATE also → CONCRETE
POTASSIUM FERRICYANIDE
POTASSIUM GLUCONATE
Potassium Hexafluoroarsenate → HEXAFLURATE
Potassium Hydrate → POTASSIUM HYDROXIDE
POTASSIUM HYDROXIDE also → BROMTHYMOL BLUE, SODIUM HYDROXIDE
POTASSIUM HYDROXYQUINOLINE SULFATE
POTASSIUM IODATE also → WHEAT
POTASSIUM IODIDE also → IODINE, LITHIUM
POTASSIUM METABISULFITE
Potassium Monoxide → POTASSIUM HYDROXIDE
POTASSIUM NITRATE also → BORAGO OFFICINALIS
POTASSIUM NITRITE also → OATS
POTASSIUM OXALATE also → HALOGETON, SARCODATUS VERMICULATUS
Potassium Penicillin G → PENICILLIN > *Potassium Benzylpenicillinate*
POTASSIUM PERCHLORATE
POTASSIUM PERMANGANATE
POTASSIUM PERRHENATE
POTASSIUM PERSULFATE
Potassium Pyrosulfite → POTASSIUM METABISULFITE
Potassium Rhodanide → POTASSIUM THIOCYANATE
POTASSIUM SELENIDE
POTASSIUM SILICATE
POTASSIUM SORBATE
POTASSIUM SULFATE
Potassium Sulfocyanate → POTASSIUM THIOCYANATE
POTASSIUM THIOCYANATE
Potassium Triplex → POTASSIUM ACETATE
Potassuril → POTASSIUM GLUCONATE
POTATO also → DIMETHICONE, HAZEL NUTS, MALEIC HYDRAZIDE, POTASSIUM METABISULFITE, RADISH, SODIUM BENZOATE, SOLANINE, Preface
Potato Chips → ONIONS
Potato Weed → SOLANUM sp.
Pothos → SCINDAPUS AUREUS
Potts, Sir Percival → SOOT
Pounce → PERMETHRIN
Povan → PYRVINIUM PAMOATE
Povanyl → PYRVINIUM PAMOATE
Poverty Grasses → ARISTIDA sp.
Povidone Iodine Complex → POLYVINYLPYRROLIDONE IODINE COMPLEX
PP 148 → PARAQUAT
PP 383 → CYPERMETHRIN
PP 557 → PERMETHRIN
PP 580 → DIFENACOUM
PP 581 → BRODIFACOUM
PP 745 → MORFAMQUAT
PP 910 → PARAQUAT
PPA → PHENYLPROPANOLAMINE
PPG 101 → POTASSIUM AZIDE
PPH → PROPYLHEXEDRINE
PRACTOLOL also → DISOPYRAMIDE, VERAPAMIL
Pradif → TAMSULOSIN
Praecicalm → PENTOBARBITAL
Praeciglucon → GLYBURIDE
Praenitron → TROLNITRATE PHOSPHATE
Praequine → PAMAQUIN

Pragmazone → TRAZODONE HCL
Pragmoline → ACETYLCHOLINE
Prairie Crocus → ANEMONE
PRAJMALINE TARTRATE
PRALIDOXIME also → LANDRIN, VITAMIN B_1
Prälumin → CYCLOBARBITAL
Pramace → RAMIPRIL
Pramex → PERMETHRIN
Pramidex → TOLBUTAMIDE
Pramiel → METOCLOPRAMIDE
Pramil → SECOBARBITAL SODIUM
Pramine → IMIPRAMINE
PRAMIPEXOLE
Pramitol → PROMETONE
Prandin → REPAGLINIDE
Prandiol → DIPYRIDAMOLE
Pranone → ETHISTERONE
Prano-Puren → PROPRANOLOL
Pranosine → ISOPRINOSINE
Pranoxen → NAPROXEN
Prantal → DIPHEMANIL METHYLSULFATE
Präparat → HYDRALAZINE
PRASTERONE also → MELATONIN
Pravachol → PRAVASTATIN
Pravaselect → PRAVASTATIN
PRAVASTATIN also → NEFAZODONE, PROPRANOLOL
Pravidel → BROMOCRIPTINE
PRAWNS
Praxadium → NORDAZEPAM
Praxilene → NAFRONYL OXALATE
Praxis → INDOPROFEN
Praxiten → OXAZEPAM
Prazene → PRAZEPAM
PRAZEPAM also → CIMETIDINE, DISULFIRAM
Prazine → PROMAZINE
PRAZIQUANTEL also → CHLOROQUINE, CIMETIDINE, SODIUM BICARBONATE
PRAZOSIN HYDROCHLORIDE also → PROPRANOLOL, VERAPAMIL
Precatory Bean → ABRUS PRECATORIUS also ABRIN
Precef → CEFORANIDE
Precipitated Chalk → CALCIUM CARBONATE
Precipité Blanc → MERCUROUS CHLORIDE
Precor → METHOPRENE
Precortancyl → PREDNISOLONE
Precortilon → PREDNISOLONE
Precortisyl → PREDNISOLONE
Precose → ACARBOSE
Predalon → GONADOTROPINS, CHORIONIC
Predalone T. B. A. → PREDNISOLONE
Predef → ISOFLUPREDONE
Predenema → PREDNISOLONE
Predfoam → PREDNISOLONE
Pred Forte → PREDNISOLONE
Pred Mild → PREDNISOLONE
Prednacinolone → DESONIDE
Prednicarbate → PREDNISOLONE
Prednicen → PREDNISOLONE
Prednilonga → PREDNISONE
Predniretard → PREDNISOLONE
PREDNISOLONE also → AMINOGLUTETHIMIDE, CYCLOPHOSPHAMIDE, CYCLOSPORINE, MAGNESIUM OXIDE, PHENOBARBITAL, PHENYTOIN, PREDNISONE, RIFAMPIN
Prednisolone Hemisuccinate → PREDNISOLONE
Prednisolone 21-Pivalate → PREDNISOLONE
Prednisolone Sodium Succinate → PREDNISOLONE
Prednisolone Steaglate → PREDNISOLONE
Prednisolone Tebutate → PREDNISOLONE
Prednisolut → PREDNISOLONE
PREDNISONE also → AZATHIOPRINE, BISHYDROXYCOUMARIN, ETHYL BISCOUMACETATE, OMEPRAZOLE, PHENOBARBITAL, RIFAMPIN, SALICYLAZOSULFAPYRIDINE
Predonine → PREDNISOLONE
Predsol → PREDNISOLONE
Preeglone → DIQUAT DIBROMIDE
Prefamon → DIETHYLPROPION
Preferid → BUDESONIDE
Prefrin → PHENYLEPHRINE HCL
Pregard → PROFLURALIN
Preglandin → GEMEPROST
Pregnancy Urine Extract → GONADOTROPINS, CHORIONIC
Pregnanediol → MILK
Pregneninolone → ETHISTERONE
Pregnesin → GONADOTROPINS, CHORIONIC
Pregnyl → GONADOTROPINS, CHORIONIC
Pregova → MENOTROPINS
Prêle des Champs → EQUISETUM
Prêle des Marais → EQUISETUM
Prelis → METOPROLOL
Prelital → BUTABARBITAL
Preludin → PHENMETRAZINE
Pre-M → PENDIMETHALIN
Premarin → ESTROGENS, CONJUGATED
Premasone → PROFLURALIN
Premerge → DINOSEB

Premier → IMIDACLOPRID
Premocillin → PENICILLIN > *Benzylpenicillin, Procaine*
Prempar → RITODRINE
Prenazone → FEPRAZONE
Prenival → PREDNISOLONE
Prenormine → ATENOLOL
PRENYLAMINE LACTATE also → LIDOCAINE
Prepalin → VITAMIN A
Pre-Par → RITODRINE
Preparation 391 → NICLOSAMIDE
Preparation K → BISETHYLXANTHOGEN
Prepared Chalk → CALCIUM CARBONATE
Prepidil → DINOPROSTONE
Prepouli → COLTSFOOT
Prepuerin® Preg Test → PROMETHAZINE
Prepulsid → CISAPRIDE
Pres → ENALAPRIL
Presamine → IMIPRAMINE
Prescal → ISRADIPINE
Presdate → LABETALOL
Presidal → PENTACYNIUM
Presinol → METHYLDOPA
Presomin → ESTROGENS, CONJUGATED
Pressalin → DIHYDRALAZINE
Pressalolo → LABETALOL
Pressimmune → ANTITHYMOCYTE GLOBULIN or SERUM
Pressorol → METARAMINOL BITARTRATE
Pressunic → DIHYDRALAZINE
Pressural → INDAPAMIDE
PRESTONIA AMAZONICA
Presuren → HYDROXYDIONE SODIUM
Pretor → CEFOTAXIME
Pretty Baby Chalk® → CALCIUM CARBONATE
PRETZELS also → SODIUM HYDROXIDE
Prevacid → LANSOPRAZOLE
Prevangor → PENTAERYTHRITOL TETRANITRATE
Prevencillin P → PENICILLINS – "NEWER and SYNTHETICS" > *Cloxacillin/Sodium salt*
Preventol G-D → DICHLOROPHEN(E)
Preveon → ADEFOVIR
Prevex → FELODIPINE
Prexan → NAPROXEN
Prexidil → MINOXIDIL
Priadel → LITHIUM CARBONATE
Priamide → ISOPROPAMIDE IODIDE
Prickly Ash → ZANTHOXYLUM sp.
Prickly Comfrey → COMFREY
Prickly Lettuce → LACTUCA sp.
Prickly Pear → OPUNTIA sp.
Prickly Poppy → ARGEMONE MEXICANA
Pride of China → MELIA sp.
Pride of India → MELIA sp.
Pridinol → PROMAZINE
Priest's Pintle → ARISAEMA
Priftin → RIFAPENTINE
Prilagin → MICONAZOLE NITRATE
PRILOCAINE
Prilon → OXOLAMINE
Prilosec → OMEPRAZOLE
Primaclone → PRIMIDONE
Primacor → MILRINONE
Primafen → CEFOTAXIME
Primamycin → HAMYCIN
PRIMAQUINE also → ACETANALID, ANTIPYRINE, 1-NAPHTHOL, 2-NAPHTHOL, NITROFURANTOIN, PAMAQUIN, QUINACRINE, SULFACETAMIDE SODIUM, SULFANILAMIDE
Primatene Mist → EPINEPHRINE
Primatol → PROMETONE
Primatol P → PROPAZINE
Primatol S → SIMAZINE
Primaxin → IMIPENEM
Primbactam → AZTREONAM
Primeral → NAPROXEN
PRIMIDONE also → FOLIC ACID, GRISEOFULVIN, LAMOTRIGINE, METHYLPHENIDATE, PHENYLBUTAZONE, PHENYTOIN, SODIUM VALPROATE, VITAMIN D_3
Primin → ANAGALLIS ARVENSIS
Primobolan → METHENOLONE ACETATE
Primofenac → DICLOFENAC
Primogonyl → GONADOTROPINS, CHORIONIC
Primogyn C → ETHINYL ESTRADIOL
Primolut → 17-α-HYDROXYPROGESTERONE
Primolut N → NORETHINDRONE
Primolut-Nor → NORETHINDRONE
Primonobol → METHENOLONE ACETATE
Primperan → METOCLOPRAMIDE
Primrose → PRIMULA sp. also → PAPHIOPEDILIUM HYANALDIANUM
Primsol → TRIMETHOPRIM
PRIMULA sp.
Prinadol → PHENAZOCINE HYDROBROMIDE
Prinalgin → ALCLOFENAC
Princep → SIMAZINE
Prince's Plume → STANLEYA PINNATA
Princillin → PENICILLINS – "NEWER and SYNTHETICS" > *Ampicillin*

Principen → PENICILLINS – "NEWER and SYNTHETICS" > *Ampicillin*
Prinil → LISINOPRIL
Prinivil → LISINOPRIL
Prinzone → SULFACHLORPYRIDAZINE
Prioderm → MALATHION
Priscol → TOLAZOLINE HCL
Priscoline → TOLAZOLINE HCL
Prisilidine → ALPHAPRODINE
PRISTANE
Pristinamycin → CYCLOSPORINE
Privadol → GLAFENINE
Privaprol → LOTRIFEN
Privet → LIGUSTRUM VULGARE
Privine → NAPHAZOLINE
PRN → PHENYLTOLOXAMINE
Pro → PROLINE
Pro-Actidil → TRIPROLIDINE
ProAmatine → MIDODRINE
Proaqua → BENZTHIAZIDE
Proasma → METHOXYPHENAMINE
Proazamine → PROMETHAZINE
Probalan → PROBENECID
Probamyl → MEPROBAMATE
Proban → CYTHIOATE
Pro-Banthine → PROPANTHELINE BROMIDE
Probe → METHAZOLE
Probecid → PROBENECID
Proben → PROBENECID
PROBENECID also → ACETAMINOPHEN, ACETYLSALICYLIC ACID, ADRENOCHROME, ALLOPURINOL, *p*-AMINOHIPPURIC ACID, *p*-AMINOSALICYLIC ACID, BUMETANIDE, CAPTOPRIL, CARPROFEN, CEFAMANDOLE, CEFMETAZOLE, CEFOTAXIME, CEFUROXIME, CEPHALOTHIN, CEPHADRINE, CHLOROTHIAZIDE, CHLORTHALIDONE, CIPROFLOXACIN, CLOFIBRATE, COLCHICINE, DAPSONE, DYPHYLLINE, ETHACRYNIC ACID, FAMOTIDINE, FUROSEMIDE, GLIPIZIDE, HYDROCHLOROTHIAZIDE, INDOMETHACIN, KETOPROFEN, LOMEFLOXACIN, METHOTREXATE, NALIDIXIC ACID, NAPROXEN, NITROFURANTOIN, PENICILLIN; PENICILLINS – "NEWER and SYNTHETICS" > *Floxacillin, Nafcillin Sodium, Oxacillin Sodium, Piperacillin Sodium*; RIFAMPIN, SALICYLIC ACID, SODIUM SALICYLATE, SULFAMETHOXYPYRIDAZINE, SULFINPYRAZONE, SULINDAC, TOLBUTAMIDE, TORSEMIDE, VITAMIN B_1, VITAMIN B_2, ZIDOVUDINE
Probenimead → PROBENECID
Probese P → PHENMETRAZINE
PROBUCOL also → CYCLOSPORINE
PROCAINAMIDE also → ACETAZOLAMIDE, CIMETIDINE, KANAMYCIN, *d*-TUBOCURARINE CHLORIDE, SODIUM BICARBONATE, SUCRALFATE
PROCAINE also → AMMONIUM CHLORIDE, BENZOIC ACID, BUTACAINE, DECAMETHONIUM BROMIDE, ETHYL-*p*-AMINOBENZOATE, ETHYL ISOBUTRAZINE, MALATHION, MIVACURIUM; PENICILLIN > *Benzylpenicillin, Procaine; Benzylpenicillin, Sodium*; PENICILLINS – "NEWER and SYNTHETICS" > *Potassium Benzylpenicillinate;* PENTOBARBITAL, PERPHENAZINE, PHENOBARBITAL, PROCAINAMIDE, PROMAZINE, QUINURONIUM, RONNEL, RUBBER, SODIUM BICARBONATE, SUCCINYLCHOLINE, TETRACAINE HYDROCHLORIDE, TETRAMISOLE, THIRAM
Procaine Amide → PROCAINAMIDE
PROCAINIDE also → AMIODARONE
Procalm → BENACTYZINE
Procalmdiol → MEPROBAMATE
Procalmidol → MEPROBAMATE
Procamide → PROCAINAMIDE
Procanodia → PENICILLIN > *Benzylpenicillin, Procaine*
Procan-SR → PROCAINAMIDE
Procapan → PROCAINAMIDE
PROCARBAZINE also → ALCOHOL, ETHYL; CHEESES, INSULIN
Procardia → NIFEDIPINE
Processine → CINNARIZINE
Procetofen(e) → FENOFIBRATE
Procetoken → FENOFIBRATE
Prochlorpemazine → PROCHLORPERAZINE
PROCHLORPERAZINE also → CHEESES, DOFETILIDE, METRIZAMIDE, PHENYTOIN, SUCCINYLCHOLINE
Procholon → DEHYDROCHOLIC ACID
Procit → PROMETHAZINE
Procorum → GALLOPAMIL
Procrit → ERYTHROPOIETIN
Proctocort → HYDROCORTISONE
Proctor & Gamble → β-CAROTENE
PROCYCLIDINE also → NIALAMIDE, PAROXETINE, TRANYLCYPROMINE
Procyclomin → DICYCLOMINE

Procymidone → WINE
Procytox → CYCLOPHOSPHAMIDE
Pro-Dafalgan → PROPACETAMOL
Prodasone → MEDROXYPROGESTERONE ACETATE
Prodel → DIDROGESTERONE
Prodopa → L-DOPA
Pro Dorm → LORAZEPAM
Prodox → 17-α-HYDROXYPROGESTERONE
Prodoxol → OXOLINIC ACID
Pro-Entra → TRIPROLIDINE
Profamina → AMPHETAMINE
Profasi → GONADOTROPINS, CHORIONIC
Profasi HP → GONADOTROPINS, CHORIONIC
Profemin → FUROSEMIDE
Profenamine → ETHOPROPAZINE
Profenid → KETOPROFEN
Profenil → ALVERINE
Profenine → ALVERINE
Profenone → PAROXYPROPIONE
Profetamine → AMPHETAMINE
PROFLAVINE
Proflax → TIMOLOL
Proflex → IBUPROFEN
PROFLURALIN
Profoliol B → ESTRADIOL
Pro-Forma → GALEGA OFFICINALIS
Profume → METHYL BROMIDE
Progabide → PHENYTOIN
Proge → 17-α-HYDROXYPROGESTERONE
Pro-Gen → ARSANILIC ACID
Pro-Gen Sodium → SODIUM ARSANILATE
Progeril → ERGOLOID MESYLATES
Progesic → FENOPROFEN CALCIUM
Progestasert → PROGESTERONE
Progesterol → PROGESTERONE
PROGESTERONE also → CHLORCYCLIZINE, CHLORDANE, DDT, DISULFIRAM, FLUCONAZOLE, GRAPEFRUIT, PHENYLBUTAZONE, SPIRONOLACTONE, THYROXINE
Progesteronum → PROGESTERONE
Progestimil → PROTEIN HYDROLYSATES
Progestin → PROGESTERONE
Progestogel → PROGESTERONE
Progestol → PROGESTERONE
Progestone → PROGESTERONE
Progestoral → ETHISTERONE
Progestronaq → PROGESTERONE
Proglicem → DIAZOXIDE
Proglycem → DIAZOXIDE
Progon → GONADOTROPINS, CHORIONIC
Prograf → TACROLIMUS
Program → LUFENURON
Proguanil → CHLOROGUANIDE
Progynon → ESTRADIOL
Progynon C → ETHINYL ESTRADIOL
Progynon M → ETHINYL ESTRADIOL
Proheart → MOXIDECTIN
Proin → PHENYLPROPANOLAMINE
PROLACTIN also → BROMOCRIPTINE, CANNABIS, CHLORPROMAZINE, CLOMIPRAMINE, CYSTEAMINE, PHENOBARBITAL, Δ^9-TETRAHYDROCANNABINOL, THIORIDAZINE
Proloid → THYROID
Prolongal → IRON DEXTRANS
Proloprim → TRIMETHOPRIM
Prolutin → PROGESTERONE also → 17-α-HYDROXYPROGESTERONE
Proluton → PROGESTERONE
Proma → CHLORPROMAZINE
Promacetin → ACETOSULPHONE SODIUM
Promacid → CHLORPROMAZINE
Promacortine → METHYLPREDNISOLONE
Promactil → CHLORPROMAZINE
Promanide → GLUCOSULFONE SODIUM
Promaquid → FONAZINE MESYLATE
Promassolax → OXYPHENISATIN
Promate → MEPROBAMATE
Promazil → CHLORPROMAZINE
Promazinamide → PROMETHAZINE
PROMAZINE also → ATROPINE, HEPARIN, MEPERIDINE, MEPROBAMATE, SUCCINYLCHOLINE
PROMECARB
Promecon → BENZQUINAMIDE
PROMETHAZINE also → ETHYLENEDIAMINE, HEPARIN, MEPERIDINE, NEOMYCIN, NORTRIPTYLINE, STREPTOMYCIN
Promethium-147 → PAINT
Prometon → PROMETONE
PROMETONE
Prometrex → PROMETRYNE
PROMETRYNE
Pro-Hance → GADOTERIDOL
Proheart → MOXIDECTIN
Proheptatriene → CYCLOBENZAPRINE
Prokarbol → DINITRO-σ-CRESOL
Proketazine → CARPHENAZINE
Proklar → SULFAMETHIZOLE
Prokopin G → PROCAINE
Prolactin → LUTEOTROPIN
Proladone → OXYCODONE
Prolate → IMIDAN

Prolax → MEPHENESIN
Proleukin → ALDESLEUKIN also → INTERLEUKINS
Prolidon → PROGESTERONE
Proligne → AMFEPENTOREX
Prolin → COUMAFURYL
PROLINE
PROLINTANE
Prolixan → APAZONE
Prolixin → FLUPHENAZINE DECANOATE
Prolixin Decanoate → FLUPHENAZINE DECANOATE
Prolixin Enanthate → FLUPHENAZINE DECANOATE
Promin → GLUCOSULFONE SODIUM
Prominal → MEPHOBARBITAL
Promintic → METHYRIDINE
Promit → DEXTRAN
Promotil → PROLINTANE
PROMOXALONE
Promwill → PROMAZINE
PRONAMIDE
Pronap → DIETHYLENE GLYCOL
Pronarcon → NARCOBARBITAL
Prondol → IPRINDOLE
Pronestyl → PROCAINAMIDE
PRONETHALOL
Pronison → PREDNISONE
Pronon → PROPAFENONE
Pronox → CYCLOBARBITAL
Prontalbin → SULFANILAMIDE
Prontalgin → IBUPROFEN
Prontamid → SULFACETAMIDE SODIUM
Prontobario → BARIUM SULFATE
Prontocalcin → CALCITONIN
Prontosil → SULFAMIDOCHRYSOIDINE
Prontosil album → SULFANILAMIDE
Prontosil S → AZOSULFAMIDE
Prontosil Soluble → AZOSULFAMIDE
Prontylin → SULFANILAMIDE
PROPACETAMOL
PROPACHLOR
Propacil → PROPYLTHIOURACIL
Propaderm → BECLOMETHASONE
Propadrine → PHENYLPROPANOLAMINE
PROPAFENONE also → AMIODARONE, CARVEDILOL, DIGOXIN, QUINIDINE, PAROXETINE, QUINIDINE, RITONAVIR, THEOPHYLLINE, WARFARIN
Propal → ISOPROTERENOL
Pro-Pam → DIAZEPAM
PROPAMIDINE
PROPANE also → CARBON MONOXIDE, ETHANE
Propanedial → MALONALDEHYDE
1,2-Propanediol → PROPYLENE GLYCOL
1,3-PROPANE SULTONE
Propanethiol-S-oxide → SULFURIC ACID
1,2,3-Propanetriol → GLYCERIN
1,2,3-Propanetriol Trinitrate → NITROGLYCERIN
PROPANIDID also → PHENELZINE, SUCCINYLCHOLINE
PROPANIL
Propanoic Acid → PROPIONIC ACID
1-Propanol → *n*-PROPYL ALCOHOL
2-Propanol → ALCOHOL, ISOPROPYL
Propantan → PROPANIDID
PROPANTHELINE BROMIDE also → ACETAMINOPHEN, DIGOXIN, MAGNESIUM TRISILICATE, NITROFURANTOIN, RANITIDINE, SODIUM BICARBONATE, SULFAMETHOXAZOLE
Propaphenin → CHLORPROMAZINE
PROPARACAINE HYDROCHLORIDE
Propargil → PROPARGITE
PROPARGITE also → TEA
Propasa → *p*-AMINOSALICYLIC ACID
Propavan → PROPIOMAZINE
Propax → OXAZEPAM
PROPAZINE
Propecia → FINASTERIDE
Pro-Pen → PENICILLIN > *Benzylpenicillin, Procaine*
2-Propenal → ACROLEIN
2-Propenamide → ACRYLAMIDE
2-Propen-1-amine → ALLYLAMINE
2-Propene → PROPYLENE
2-Propenenitrile → ACRYLONITRILE
1-Propenol-3 → ALCOHOL, ALLYL
2-Propen-1-ol → ALCOHOL, ALLYL
Propene Oxide → PROPYLENE OXIDE
2-Propenyl Hexanoate → ALLYL CAPROATE
4-Propenylveratrole → METHYL ISOEUGENOL
PROPERICIAZINE
Propess → DINOPROSTONE
PROPETAMPHOS
PROPHAM
Prophelan → ALVERINE
Prophenatin → DICLOFENAC
Prophos → ETHOPROP
Prophyllen → DYPHYLLINE
Prophylux → PROPRANOLOL
Propicillin Potassium → PENICILLINS – "NEWER and SYNTHETICS"
Propine → DIPIVEFRIN also → METHYL ACETYLENE
Propiocine → ERYTHROMYCIN PROPIONATE

Propiocine Enfant → ERYTHROMYCIN ESTOLATE
β-PROPIOLACTONE
PROPIOMAZINE
PROPIONIC ACID
Propionic Anhydride → FENTANYL
PROPIONITRILE
Propionyl Erythromycin Lauryl Sulfate → ERYTHROMYCIN ESTOLATE
PROPIONYLPROMAZINE
Propiopromazine → PROPIONYLPROMAZINE
PROPIRAM FUMARATE
Propisamine → AMPHETAMINE
Propitan → PIPAMPERONE
Propitocaine → PRILOCAINE
Proplex → CHRISTMAS FACTOR
PROPOFOL
PROPOLIS also → BEES
Proponex Plus → MECOPROP
Propoquin → AMOPYROQUIN
PROPOXUR
Propoxychel → PROPOXYPHENE
PROPOXYPHENE also → ALCOHOL, ETHYL; AMPHETAMINE, ANTIPYRINE, CAFFEINE, DEXTROAMPHETAMINE, DIAZEPAM, DOXEPIN, NALOXONE, NEFOPAM, ORPHENADRINE, PHENYLBUTAZONE, PHENYTOIN, TOBACCO
PROPRANIDID also → ATROPINE
PROPRANOLOL also → ACETALDEHYDE, ACETAMINOPHEN; ALCOHOL, ETHYL; ALUMINUM HYDROXIDE, ANTIPYRINE, ATROPINE, CHLORAZEPATE, CHLOROFORM, CHLORPROMAZINE, CHLORPROPAMIDE, CIMETIDINE, CLONAZEPAM, CLONIDINE, COCAINE, CURARE, CYCLOPROPANE, DIAZEPAM, DIGITALIS, DILTIAZEM, L-DOPA, EPINEPHRINE; ETHER, ETHYL; FLECAINIDE, FLUOXETINE, FLURBIPROFEN, FURAZOLIDONE, FUROSEMIDE, GLUCOSE, GROWTH HORMONES, HALOFENATE, HALOPERIDOL, HALOTHANE, HEROIN, HEXOBARBITAL, HYDRALAZINE, HYDROCHLOROTHIAZIDE, INDOMETHACIN, INSULIN, ISRADIPINE, KETANSERIN, LIDOCAINE, LIOTHYRONINE, LOVASTATIN, MEGLITOL, METERGOLINE, METHYLDOPA, METOCLOPRAMIDE, MORPHINE, NEOSTIGMINE, NIFEDIPINE, NISOLDIPINE, NITROGLYCERIN, OPIUM, OXOTREMORINE, OXPRENOLOL HYDROCHLORIDE, PAROXETINE, PENICILLIN also PENICILLIN > *Penicillin V*, PHENMETRAZINE, PRAVASTATIN, PRENYLAMINE LACTATE, PROPOXYPHENE, QUINIDINE, RESERPINE, RIFAMPIN, TEA, THEOPHYLLINE, THIOPENTAL, TIMOLOL, TOBACCO, TOCAINIDE, TOLBUTAMIDE, *d*-TUBOCURARINE CHLORIDE, VECURONIUM BROMIDE, VERAPAMIL, ZILEUTON
Propranur → PROPRANOLOL
Propulsid → CISAPRIDE
Propycil → PROPYLTHIOURACIL
2-Propyl Acetate → ISOPROPYL ACETATE
n-PROPYL ACETATE
n-PROPYL ALCOHOL
2-Propyl Alcohol → ALCOHOL, ISOPROPYL
Propylallylonal → PHENOBARBITAL
2-Propylamine → ISOPROPYLAMINE
p-Propyl Anisole → DIHYDROANETHOLE
n-PROPYLBENZENE
n-PROPYLCARBAMATE
n-PROPYL DISULFIDE
PROPYLENE
PROPYLENE CARBONATE
PROPYLENE DICHLORIDE
PROPYLENE GLYCOL also → DIAZEPAM, DIPHENYDRAMINE, ETOMIDATE, K-Y LUBRICATING JELLY®, LIPSTICK, LORAZEPAM, NITROGLYCERIN, OXYTETRACYCLINE, PENCICLOVIR, PENTOBARBITAL, PHENYTOIN, SILVER SULFADIAZINE
PROPYLENE GLYCOL 1,2-DINITRATE
PROPYLENE GLYCOL MONOMETHYL ETHER
PROPYLENE GLYCOL MONOSTEARATE
Propyleneimine → 2-METHYLAZIRIDINE
PROPYLENE OXIDE
PROPYL ETHER
PROPYL GALLATE also → LIPSTICK, MARGARINE
PROPYLHEXEDRINE
PROPYLHEXEDRINE ETHYLPHENYLBARBITURATE
Propyl Hydride → PROPANE
Propylic Alcohol → *n*-PROPYL ALCOHOL
PROPYLIODONE
n-PROPYL NITRATE
PROPYLTHIOURACIL also → PROPRANOLOL

Propyl-Thyracil → PROPYLTHIOURACIL
N-Propyltrihydroxybenzoate → PROPYL GALLATE
Propyl Urethane → *n*-PROPYLCARBAMATE
Propyne → METHYL ACETYLENE
Propyon → PROPOXUR
PROPYPHENAZONE also → PHENACETIN
Propyzamide → PRONAMIDE
Proquamezine → AMINOPROMAZINE
Prorex → PROMETHAZINE
Proscar → FINASTERIDE
Proscillan → PROSCILLARIDIN
PROSCILLARIDIN
Proscorbin → VITAMIN C
Prosedar → QUAZEPAM
Proserine → NEOSTIGMINE
Proserout → MECLOFENOXATE
Proseryl → MECLOFENOXATE
ProSom → ESTAZOLAM
Proso Millet → PANICUM sp.
Pro-Sonil → CYCLOBARBITAL
Prosopsis sp. → MESQUITE
Pro-Spot → FENTHION
PROSTACYCLIN
Prostaglandin E_1 → ALPROSTADIL
Prostaglandin E_2 → DINOPROSTONE
Prostaglandin $F_{2\alpha}$ → DINOPROST
Prostaglandin I_2 → PROSTACYCLIN
Prostal → CHLORMADINONE
Prostap → LEUPROLIDE
Prostaphlin → PENICILLINS – "NEWER and SYNTHETICS" > *Oxacillin Sodium*
Prostaphlin-A → PENICILLINS – "NEWER and SYNTHETICS" > *Cloxacillin/Sodium salt*
Prostarmon-E → DINOPROSTONE
Prostarmin F → DINOPROST
Prostasin → PROSCILLARIDIN
Prostavasin → ALPROSTADIL
Prostide → FINASTERIDE
Prostigmin → NEOSTIGMINE
Prostin/15M → CARBOPROST
Prostin E_2 → DINOPROSTONE
Prostin $F_{2\alpha}$ → DINOPROST
Prostin VR → ALPROSTADIL
Prostosin → PROSCILLARIDIN
Prostrumyl → METHYLTHIOURACIL
Prosultiamine → VITAMIN B_1
Proszine → PROSCILLARIDIN
Protabol → THIOMESTERONE
Protactyl → PROMAZINE
PROTAMINE SULFATE also → CANNABIS, HEPARIN, INSULIN
Protanabol → OXYMETHOLONE
Protandren → METHANDRIOL
Protangix → DIPYRIDAMOLE
Protapine → PAPAVER sp.
Protease 1 → ASPERGILLUS sp.
PROTEIN HYDROLYSATES
Pro-Ten Beef® → PAPAIN
Proternol → ISOPROTERENOL
Proterytrin → ERYTHROMYCIN
Prothazine → PROMETHAZINE
Protheo → THEOPHYLLINE
Prothiaden(e) → DOTHIEPIN
Prothionamide → PROTIONAMIDE
PROTHIPENDYL
Prothromadin → WARFARIN
Protigenyl I → PROTEIN HYDROLYSATES
PROTIONAMIDE
Protirelin → THYROTROPIN RELEASING HORMONE
PROTOANEMONIN
Protoberine → CORYDALIS
PROTOKYLOL
Protolipan → FENOFIBRATE
Protolysate → PROTEIN HYDROLYSATE
Protomin → GLUCOSULFONE SODIUM
Protonix → PANTOPRAZOLE
Protopam → PRALIDOXIME
Protopic → TACROLIMUS
Protopine → CORYDALIS, DICENTRA CUCULLARIA
Protosol → DIHYDROXYACETONE
Protostat → METRONIDAZOLE
Protostib → METHYLGLUCAMINE ANTIMONATE
Protosulfil → SILVER SULFADIAZINE
PROTOVERATRINES A & B
Protoxyl → SODIUM ARSANILATE
PROTRIPTYLINE also → AMITRIPTYLINE, EPINEPHRINE, FLUOEXTINE, TRANYLCYPROMINE, GUANETHIDINE, IMIPRAMINE, PROCARBAZINE
Protropin → SOMATREM also → GROWTH HORMONES
Protropine → CHELIDONIUM MAJUS, ESCHOSCHOLTZIA CALIFORNICA
Pro-Vent → THEOPHYLLINE
Proventil → ALBUTEROL
Provera → MEDROXYPROGESTERONE ACETATE
Proverine → ALVERINE
Provigan → PROMETHAZINE
Provigil → MODAFINIL
Proviodine → POLYVINYLPYRROLIDONE IODINE COMPLEX
Proviron → MESTEROLONE
Provisc → SODIUM HYALURONATE
Provitar → OXANDROLONE

Provitina → VITAMIN D_3
Provocholine → METHACHOLINE CHLORIDE
Prov-U-Sep → METHENAMINE
Prowl → PENDIMETHALIN
Proxagesic → PROPOXYPHENE
Proxen → NAPROXEN
Proxine → NAPROXEN
Proxol → TRICHLORFON
Prozac → FLUOXETINE
Prozerin → NEOSTIGMINE
Prozil → CHLORPROMAZINE
Prozinex → PROPAZINE
Prulaurasin → PEACHES
Prulet → OXYPHENISATIN
Prulifloxacin → THEOPHYLLINE
Prunasin → AMELACHIA ALNIFOLIA also → BRACKEN FERN, PEACHES
PRUNE
Prunolide → γ-NONALACTONE
Prunus sp. → PLUM
Prunus armeniaca → APRICOT
Prunus emarginata → CHERRY
Prunus persica → PEACHES
Prussian Blue → FERRIC FERROCYANIDE
Prussic Acid → HYDROGEN CYANIDE
Pryleugan → IMIPRAMINE
P_2S → PRALIDOXIME
PSC 801 → PROSCILLARIDIN
PSC 833 → VALSPODAR
Psedera quinquefolia → IVY
Pseudocef → CEFSULODIN SODIUM
Pseudoconhydrine → CONIUM MACULATUM
PSEUDOEPHEDRINE also → ACETAZOLAMIDE, ALUMINUM HYDROXIDE, CARBINOXAMINE, SODIUM BICARBONATE
Pseudomonil → CEFSULODIN SODIUM
Pseudopurpurin → GALIUM sp.
Psichial → CHLORDIAZEPOXIDE
Psicopax → LORAZEPAM
Psicoperidol → TRIFLUPERIDOL
Psicosan → CHLORDIAZEPOXIDE
Psicoterina → CHLORDIAZEPOXIDE
Psicronizer → NOMIFENSINE MALEATE
PSILOCIN also → PANAEOLUS sp., PHOLIOTA sp., STROPHARIA sp.
Psilocybe sp. → PSILOCIN also → PSILOCYBIN, MUSHROOMS
PSILOCYBIN also → CHLORPROMAZINE, LYSERGIC ACID DIETHYLAMIDE, MUSHROOMS, PHOLIOTA sp., PROCHLORPERAZINE, STROPHARIA sp.
Psilocyn → PSILOCIN
Psilomelane → MANGANESE DIOXIDE
PSILOSTROPHE sp.
Psilotsibin → PSILOCYBIN
Psilotsin → PSILOCIN
Psiquium → MEDAZEPAM
Psophocarpus tetragonolobus → BEANS, Winged
Psoraderm → METHOXSALEN
PSORALEA sp.
Psoralens → ANGELICA sp., CELERY, PARSLEY, PARSNIP, PSORALEA
Psorcon → DIFLORASONE
Psorcutan → CALCIPOTRIOL
PSP → PHENOLSULFONPHTHALEIN
PST → GROWTH HORMONES
Psychedrine → AMPHETAMINE
"Psycho Drops" → ATROPINE also → SCOPOLAMINE
Psychoperidol → TRIFLUPERIDOL
Psychosan → AZACYCLONAL
Psycho-Soma → MAGNESIUM BROMOGLUTAMATE
Psychostyl → NORTRIPTYLINE
Psychoton → AMPHETAMINE
PSYCHOTRIA sp.
Psychoverlan → MAGNESIUM BROMOGLUTAMATE
Psyllium Seed → PLANTAGO SEED
Psymion → MAPROTILINE
Psymod → PIPERACETAZINE
Psyquil → TRIFLUPROMAZINE
PT-9 → BETAHISTINE
Ptaquiloside → AQUILIDE A
PTC → CHRISTMAS FACTOR
Pteridium aquilinum → BRACKEN FERN
Pteris aquilina → BRACKEN FERN
Pterofen → TRIAMTERENE
PTEROIS VOLITANS also → FISH
Pterophene → TRIAMTERENE
Pteroylglutamic Acid → FOLIC ACID
PTFE → POLYTEF
PTG → ETOPOSIDE also → TENIPOSIDE
PTH → PARATHYROID HORMONE
Ptimal → TRIMETHADIONE
PTO → PYRITHIONE
Ptotoberberine → DICENTRA CUCULLARIA
PTU → PHENYLTHIOUREA also → PROPYLTHIOURACIL
PTX → PALYTOXIN
PTZ → PENTYLENETETRAZOL
Puberlic Acid → PENICILLIUM sp.
Puberlonic Acid → PENICILLIUM sp.
Puccoon → SANGUINARIA CANADENSIS
Puffball → LYCOPERDON
Puffer → FISH

Pularin → HEPARIN
Pulegone → HEDEOMA sp. also → PENNYROYAL
Pulma → CALATHEA VEITCHIANA
Pulmicort → BUDESONIDE
PulmiDur → THEOPHYLLINE
Pulmotil → TILMICOSIN
Pulmo-Timelets → THEOPHYLLINE
Pulpwood Handler's Disease → PAPER
Pulque → MESCALINE also → TEQUILA
Pulsoton → HYDROXYAMPHETAMINE
Pulsotyl → PHOLEDRINE
Pulvex → PERMETHRIN
Pum-ap → TANAECIUM sp.
PUMICE
PUMILIOTOXINS
PUMPKIN
Punica granatum → POMEGRANATE
Punicine Tannate → POMEGRANATE
Punkies → SAND FLIES
Punk Tree → CAJEPUT
Punktyl → LORAZEPAM
Punyosisa → MIMOSA
Puragel → GELATIN
Puralin → THIRAM
Purantix → CLOCORTOLONE PIVALATE
Purapen G → PENICILLIN > *Benzylpenicillin, Sodium*
Purapurine → SOLANUM sp.
Purecal → CALCIUM CARBONATE
Purerin → METHENAMINE
Purex → SODIUM HYPOCHLORITE
Purgaceen → OXYPHENISATIN
Purging Nut → JATROPHA sp .
Purified Araroba → CHRYSAROBIN
Purified Goa Powder → CHRYSAROBIN
Purine → SARDINES
Purine-6-thiol → MERCAPTOPURINE
Purinethol → MERCAPTOPURINE
Puri-Nethol → MERCAPTOPURINE
Puritrid → AMILORIDE
Purochin → UROKINASE
Purocyclina → TETRACYCLINE
Purodigin → DIGITOXIN
PUROMYCIN
Purosin-TC → PROSCILLARIDIN
Purple → KETAMINE
Purple Cockle → AGROSTEMMA GITHAGO also → COCKLES
Purple Foxglove → DIGITALIS
Purple-headed Sneezeweed → HELENIUM sp.
Purple Hearts → AMPHETAMINE
Purple Horse Nettle → SOLANUM sp.
Purple Locoweed → OXYTROPIS
"Purple Microdots" → LYSERGIC ACID DIETHYLAMIDE
Purple Mint → PERILLA FRUTESCENS
Purple Rattlebox → DAUBENTONIA PUMICEA
Purple Sesbania → DAUBENTONIA PUMICEA
"Purple Toe Syndrome" → WARFARIN
Purpurea Glycoside C → DESLANOSIDE
Purshiana Bark → CASCARA SAGRADA
Purslane → PORTULACA OLERACEA
Puss Caterpillar → CATERPILLARS
Pussley → PORTULACA OLERACEA
PUTTY
Putty Powder → TIN
Puvaderm → METHOXSALEN
Puvamet → METHOXSALEN
PVA → POLYVINYL ALCOHOL
PVC → POLYVINYL CHLORIDE
PVL → IRON DEXTRANS
PVP → POLYVINYLPYRROLIDONE
PVP-Iodine → POLYVINYLPYRROLIDONE IODINE COMPLEX
PX917 → TRITOLYL PHOSPHATE
Py Bracco → PYRAZINAMIDE
Pycazide → ISONIAZID
Pydirone → DIPYRONE
Pydrin → FENVALERATE
Pyelokon-R → SODIUM ACETRIZOATE
Pyelosil → IODOPYRACET
Pyknolepsinum → ETHOSUXIMIDE
Pylupron → PROPRANOLOL
Pylumbrin → IODOPYRACET
Pynamin → ALLETHRIN
Pynastin → PINDOLOL
Pynosect → PERMETHRIN
Pyocefal → CEFSULODIN SODIUM
Pyocianil → PENICILLINS – "NEWER and SYNTHETICS" > *Carbenicillin Disodium*
Pyoktan(n)in → GENTIAN VIOLET
Pyopen → PENICILLINS – "NEWER and SYNTHETICS" > *Carbenicillin Disodium*
Pyquiton → PRAZIQUANTEL
Pyrabital → BARBITAL
PYRACANTHA sp.
Pyracetam → PIRACETAM
Pyracort D → PHENYLEPHRINE HCL
Pyradone → AMINOPYRINE
Pyrafat → PYRAZINAMIDE
Pyrahexyl → SYNHEXYL
Pyralcid → SULFAMERAZINE
Pyraldin → QUINAPYRAMINE
Pyralgin → DIPYRONE
Pyramal → PYRILAMINE
Pyramem → PIRACETAM
Pyramidon → AMINOPYRINE

PYRAN COPOLYMER
Pyranisamine → PYRILAMINE
PYRANTEL also → SUCCINYLCHOLINE, THEOPHYLLINE
Pyrathyn → METHAPYRILENE
PYRAZINAMIDE also → ALCOHOL, ETHYL; ISONIAZID
Pyrazinecarboxamide → PYRAZINAMIDE
Pyrazine Carboxylamide → PYRAZINAMIDE
Pyrazines → POTATO
Pyrazofen → PHENAZOPYRIDINE
PYRAZOLE
1H-Pyrazolo[3,4-d]pyrimidin-4-ol → ALLOPURINOL
Pyrcon → PYRVINIUM PAMOATE
PYRENE
Pyrenone 606 → PIPERONYL BUTOXIDE
Pyrethrosin → PYRETHRUM, PYRETHRINS, and PYRETHROIDS
PYRETHRUM, PYRETHRINS and PYRETHROIDS also → BARTHRIN, CHRYSANTHEMUM, PIPERONYL BUTOXIDE
Pyriamid → SULFAPYRIDINE
Pyribenzamine → TRIPELENNAMINE also → PROMETHAZINE
Pyribol → SULFAPYRIDINE
Pyricardyl → NIKETHAMIDE
Pyricidin → ISONIAZID
Pyridacil → PHENAZOPYRIDINE
Pyridamal-100 → CHLORPHENIRAMINE
Pyridiate → PHENAZOPYRIDINE
Pyridin → FENVALERATE
PYRIDINE
4-Pyridineamine → AMINOPYRIDINE
2-Pyridinethiol-1-oxide → PYRITHIONE
2-Pyridinemethanol → NICOTINYL ALCOHOL
1-(3-Pyridinyl)ethanone → ACETYLPYRIDINE
Pyridipca → VITAMIN B_6
Pyridium → PHENAZOPYRIDINE
PYRIDOSTIGMINE BROMIDE also → SUCCINYLCHOLINE
Pyridox → VITAMIN B_6
PYRIDOXAL-5-PHOSPHATE also → VITAMIN B_6
Pyridoxine → VITAMIN B_6 also → AMINOOXYACETIC ACID
N-3-Pyridylmethyl-N-p-nitrophenylurea → PYRIMINAL
Pyrikappl → SULPIRIDE
PYRILAMINE also → ETHYLENEDIAMINE, GUANETHIDINE, TRIPELENNAMINE
Pyrilax → BISACODYL
Pyrilgin → DIPYRONE
Pyrimal → SULFADIAZINE
Pyrimal M → SULFAMERAZINE
PYRIMETHAMINE also → *p*-AMINOBENZOIC ACID, FOLIC ACID, FOLINIC ACID, MILK, QUININE, SULFAQUINOXALINE
2,4,5,6(1H, 3H)-Pyrimidinetetrone → ALLOXAN
PYRIMINAL
Pyrinex → CHLORPYRIFOS
Pyrinistab → METHAPYRILENE
Pyrinistol → METHAPYRILENE
PYRITHIAMIN
PYRITHIONE
PYRITINOL DIHYDROCHLORIDE also → PHENYTOIN
Pyrizine → TRIPELENNAMINE
Pyroacetic Ether → ACETONE
Pyrocatechin → PYROCATECHOL
PYROCATECHOL also → HEXYLRESORCINOL, HYDROQUINONE, PHLOROGLUCINOL
Pyrodine → ACETYLPHENYLHYDRAZINE
Pyroforane → DICHLORODIFLUOROMETHANE
Pyrogallic Acid → PYROGALLOL
PYROGALLOL also → HEXYLRESORCINOL, PHLOROGLUCINOL
Pyrogastrone → CARBENOLOXONE
Pyrojec → DIPYRONE
PYROLAN
Pyrolase → COCARBOXYLASE
Pyroles → SENECIO sp.
Pyrolusite → MANGANESE DIOXIDE
Pyromen → PIROMEN
Pyrrolamidol → DEXTROMORAMIDE
PYRROLIZIDINES also → BORAGO OFFICINALIS, CACALIA, CROTALARIA, CYNOGLOSSUM OFFICIANALE, ECHIUM PLANTAGINEUM, EUPATORIUM, HELIOTROPIN, HONEY, ILEX sp., PARSONSIA sp., RETRORSINE, SENECIO sp., TEA
Pyrrolylene → *1*,3-BUTADIENE
PYRULARIA PUBERA
Pyrus sp. → PEARS
Pyrvin → PYRVINIUM PAMOATE
Pyrvinium Embonate → PYRVINIUM PAMOATE
PYRVINIUM PAMOATE
Pyxos → BUXUS sp.
Pysomijin → PYRIDOXAL-5-PHOSPHATE
PZC → PERPHENAZINE
PZI Insulin → INSULIN

Q

Q 137 → PERTHANE
Qât → CATHA EDULIS
QD-10733 → FLUPHENAZINE DECANOATE
QNB → QUINUCLIDINYL BENZILATE
Q-Pam → DIAZEPAM
Quaalude → METHAQUALONE
Quack Grass → TEA
QIDbamate → MEPROBAMATE
Quadracyclin → TETRACYCLINE
Quadramet → SAMARIUM-153 LEXIDRONAM
Quads → METHAQUALONE
Quait → LORAZEPAM
Quaker Buttons → NUX VOMICA
Quamonium → CETRIMONIUM BROMIDE
Quaname → MEPROBAMATE
Quanil → MEPROBAMATE
Quantalan → CHOLESTYRAMINE RESIN
Quark → RAMIPRIL
Quarter Moon → CANNABIS
QUARTZ
Quarzan → CLIDINIUM BROMIDE
Quas → METHAQUALONE
Quasar → VERAPAMIL
Quat → CATHA EDULIS
Quaternium 15 → *N*-(3-CHLOROALLYL)HEXAMINIUM CHLORIDE
QUAZEPAM
Quazium → QUAZEPAM
Quebec beer-drinkers' myopathy → BEER
Quebrachine → YOHIMBINE
Queckssilberchlorid → MERCURIC CHLORIDE
Quecksilberjodid → MERCURIC IODIDE, RED
Queen Anne's Lace → DAUCUS CAROTA
Queen's Delight → STILLINGIA sp.
Queletox → FENTHION
Quellada → LINDANE
Quemicetina → CHLORAMPHENICOL
Quenopodium → CHENOPODIUM
Quensyl → HYDROXYCHLOROQUINE
Quercimeritrin → SUNFLOWER
QUERCITIN also → OAKS, WINE
Quercitrin → QUERCITIN also → AESCULUS sp., OAKS
Quercus sp. → OAKS
Querto → CARVEDILOL
Quesil → CHLORQUINALDOL
Quest → GROWTH HORMONES
Questran → CHOLESTYRAMINE RESIN
QUETIAPINE FUMARATE also → PHENYTOIN
Quick → CHLOROPHACINONE
Quick Lime → CALCIUM OXIDE
Quicksilver → MERCURY
Quide → PIPERACETAZINE
Quiess → HYDROXYZINE
Quietidin → DIPHENAZINE
Quilene → PENTAPIPERIDE METHYLSULFIDE
Quilibrex → OXAZEPAM
QUILLAJA
Quilonorm → LITHIUM ACETATE
Quilonorm-retard → LITHIUM CARBONATE
Quilonum → LITHIUM ACETATE
Quilonum-retard → LITHIUM CARBONATE
Quimar → CHYMOTRYPSIN
Quimoral → CHYMOTRYPSIN
Quimotrase → CHYMOTRYPSIN
Quimotripsina → CHYMOTRYPSIN
Quimpe → CORONOPUS sp.
Quinachlor → CHLOROQUINE
Quinacillin → PENICILLINS – "NEWER and SYNTHETICS"
QUINACRINE also → ALCOHOL, ETHYL; AMODIAQUIN, PAMAQUIN, PRIMAQUINE
Quinagamin → CHLOROQUINE
Quinagamine → CHLOROQUINE
Quinaglute → QUINIDINE
QUINAGOLIDE
Quinal-barbitone Sodium → SECOBARBITAL SODIUM
QUINALONE also → ALUMINUM
Quinalspan → SECOBARBITAL SODIUM
Quinambicide → IODOCHLORHYDROXYQUINOLINE
QUINAPRIL
QUINAPYRAMINE
Quinate → QUINIDINE
Quinazil → QUINAPRIL
QUINCE also → FRUITS and FRUIT JUICES
Quincic Acid → POTATO
QUINDOXIN
Quinercyl → CHLOROQUINE
Quinestradiol > QUINESTROL
QUINESTROL
QUINETHAZONE also → GALLAMINE TRIETHIODIDE, METOLAZONE
Quingamine → CHLOROQUINE
Quinicardine → QUINIDINE
Quinidex → QUINIDINE
QUINIDINE also → ACENOCOUMAROL, ACETAZOLAMIDE, AMANTADINE,

AMIODARONE, AMMONIUM CHLORIDE, BISHYDROXYCOUMARIN, BRETYLIUM TOSYLATE, CARVEDILOL, CIMETIDINE, CURARE, DECAMETHONIUM BROMIDE, DIGITALIS, DIGITOXIN, DIGOXIN, DIMETHYL TUBOCURARINE IODIDE, DOCUSATE SODIUM, DONEPEZIL, EDROPHONIUM, ERYTHROMYCIN, GALLAMINE TRIETHIODIDE, GRAPEFRUIT, ITRACONAZOLE, KANAMYCIN, KAOLIN, KETOCONAZOLE, LIDOCAINE, MAGNESIUM CARBONATE, MAGESIUM OXIDE, MEFLOQUINE HCL, METHOTRIMEPRAZINE, METOCLOPRAMIDE, METRONIDAZOLE, NEOSTIGMINE, NIFEDIPINE, NISULDIPINE, PANCURONIUM BROMIDE, PENTOBARBITAL, PHENOBARBITAL, PHENYLBUTAZONE, PHENYTOIN, PROCAINAMIDE, PROPAFENONE, PROPRANOLOL, RESERPINE, RIFAMPIN, RITONAVIR, SODIUM BICARBONATE, SUCCINYLCHOLINE, SUCRALFATE, TRIMIPRAMINE, *d*-TUBOCURARINE CHLORIDE, VENLAFAXINE, VERAPAMIL, WARFARIN
Quiniduran → QUINIDINE
Quinilon → CHLOROQUINE
QUININE also → ACENOCOUMAROL, AMANTADINE, AMMONIUM CHLORIDE, ANTIPYRINE, AZATHIOPRINE, CIMETIDINE, DIGOXIN, MEFLOQUINE HCL, PENTAQUINE, QUINIDINE, SODIUM BICARBONATE, VITAMIN B_6, WARFARIN
Quiniodochlor → IODOCHLORHYDROXYQUINOLINE
Quiniofon → CHINIOFON
Quinocide → PRIMAQUINE
Quinodis → FLEROXACIN
Quinofen → CINCHOPHEN
Quinoform → IODOCHLORHYDROXYQUINOLINE
QUINOLINE YELLOW also → OLIVES
8-Quinolinol → 8-HYDROXYQUINOLINE
Quinolor → HALQUINOL
Quinomycin A → ECHINOMYCIN
QUINONE also → MILLIPEDES
Quinophan → CINCHOPHEN
Quinora → QUINIDINE
Quinoscan → CHLOROQUINE
Quinoseptyl → SULFAMETHOXYPYRIDAZINE
QUINOXALINE DIOXIDE
Quinoxaline 1,4-dioxide → QUINDOXIN
Quinoxyl → CHINIOFON
QUINTOZENE also → COTTONSEED MEAL
Quintrate → PENTAERYTHRITOL TETRANITRATE
QUINUCLIDINYL BENZILATE
QUINUPRISTIN also → CYCLOSPORINE, DALFOPRISTIN
QUINURONIUM
Quinvet → LYSINE ACETYLSALICYALTE
Quipenyl → PAMAQUIN
Quitaxon → DOXEPIN
Quixalin → HALQUINOL
Quixalud → HALQUINOL
Quso → SILICON DIOXIDE
QZ-2 → METHAQUALONE

R

R 11
R 48 → CHLORMAPHAZINE
R 79 → ISOPROPAMIDE IODIDE
R 516 → CINNARIZINE
R 738 → NEFOPAM
R 805 → NIMESULIDE
R 818 → FLECAINIDE
R 837 → IMIQUIMOD
R 875 → DEXTROMORAMIDE
R 1132 → DIPHENOXYLATE
R 1303 → CARBOPHENOTHION
R 1406 → PHENOPERIDINE
R 1492 → METHYL TRITHION
R 1504 → IMIDAN
R 1513 → AZINPHOS-ETHYL
R 1582 → AZINPHOS-METHYL
R 1608 → EPTC
R 1625 → HALOPERIDOL
R 1658 → MOPERONE
R 1707 → GLAFENINE
R 1910 → 5-ETHYL DIISOBUTYLTHIOCARBAMATE
R 2028 → FLUANISONE
R 2113 → DESOXIMETASONE
R 3345 → PIPAMPERONE
R 4263 → FENTANYL
R 4318 → FLOCTAFENINE
R 4584 → BENPERIDOL
R 4749 → DROPERIDOL
R 4845 → BEZITRAMIDE
R 5147 → SPIPERONE
R 6238 → PIMOZIDE
R 6700 → ISOBENZAN

R 7094 → LIDOFLAZINE
R 7315 → METOMIDATE
R 8299 → TETRAMISOLE
R 14,827 → ECONAZOLE NITRATE
R 14,889 → MICONAZOLE NITRATE
R 14,950 → FLUNARIZINE HCL
R 15,889 → LORCAINIDE
R 16,341 → PENFLURIDOL
R 16,659 → ETOMIDATE
R 17,635 → MEBENDAZOLE
R 17,889 → FLUBENDAZOLE
R 25,061 → SUPROFEN
R 30,730 → SUFENTANIL
R 33,799 → CARFENTANIL
R 33,800 → SUFENTANIL
R 33,812 → DOMPERIDONE
R 35,443 → OXATOMIDE
R 39,209 → ALFENTANIL
R 41,400 → KETOCONAZOLE
R 41,468 → KETANSERIN
R 42,470 → TERCONAZOLE
R 43,512 → ASTEMIZOLE
R 51,211 → ITRACONAZOLE
R 51,619 → CISAPRIDE
R 64,766 → RISPERIDONE
RA-8 → DIPYRIDAMOLE
Rabon → GARDONA
Rabond → GARDONA
Race-acetylmethadol → ACETYLMETHADOL
Racemethionine → METHIONINES
Racephen → AMPHETAMINE
Racumin → COUMATETRALYL
Racun Besi → POTASSIUM FERRICYANIDE
Radanil → BENZNIDAZOLE
Rad-e-cate → SODIUM CACODYLATE
Radedorm → NITRAZEPAM
Radethazin → AZELASTINE
Radeverm → NICLOSAMIDE
Radiant Yellow → CADMIUM SULFIDE
Radicol → NICOTINYL ALCOHOL
Radioactive Cobalt → COBALT, *Radioactive*
Radiographol → METHIODAL SODIUM
Radiopaque → BARIUM SULFATE
Radiostol → VITAMIN D_2
Radiothur → Preface
RADISH also → CABBAGE, NITRATES, NITRITES
RADIUM also → Preface
RADON also → JEWELRY, TIN
Radsterin → VITAMIN D_2
Rafamebin → IODOQUINOL
RAFFINOSE also → BEANS
RAFOXANIDE
RAGWEED also → CHAMOMILE, CHRYSANTHEMUM, GOLDENROD; PYRETHRUM, PYRETHRINS, and PYRETHROIDS; TARAXACUM OFFICINALE, TEA
Ragworts → SENECIO sp.
Rai 'dore → CASSIA
Raifort → HORSERADISH
Rairai → CASSIA
RAISIN
Raka → TAMARIND
Ralabol → ZERANOL
Ralgro → ZERANOL
Ralone → ZERANOL
RALOXIFENE also → IBUPROFEN
Ramace → RAMIPRIL
RAMARIA FLAVO-BRUNNESCENS
Rametin → PHTHALOPHOS
Ramik → DIPHENADIONE
RAMIPRIL
Ramodar → ETODOLAC
Rampart → PHORATE
Ramps → ALLIUM TRICOCCUM
Ramrod → PROPACHLOR
Ramucide → CHLOROPHACINONE
Ranch Iron Complex → IRON POLYSACCHARIDE COMPLEX
Randa → CISPLATIN
Randolectil → BUTAPERAZINE
Randox → *N*, *N*-DIALLYL-2-CHLOROACETAMIDE
Rangasil → PIRPROFEN
Range Oil → KEROSENE
Ranger → GLYPHOSPHATE
Rangoon Bean → BEANS, Rangoon
Raniben → RANITIDINE
Ranide → RAFOXANIDE
Ranidil → RANITIDINE
Raniplex → RANITIDINE
RANITIDINE also → ALCOHOL, ETHYL; COFFEE, DIDANOSINE, ENOXACIN, FAMOTIDINE, FLUVASTATIN, GLIPIZIDE, GLYBURIDE, KETOCONAZOLE, MEGLITOL, NITROGLYCERIN, NIZATIDINE, OXAPROZIN, PHENYTOIN, PROCAINAMIDE, PROPRANOLOL, QUINIDINE, THEOPHYLLINE, TRIAZOLAM, WARFARIN
Rantudil → ACEMETACIN
RANUNCULIN also → BUR BUTTERCUP, PROTOANEMONIN, RANUNCULUS
RANUNCULUS also → PROTOANEMONIN
Ranvil → NICARDIPINE
Rapacodin → DIHYDROCODEINE

Rapamune → RAPAMYCIN
RAPAMYCIN also → TACROLIMUS
RAPE also → CORN, DIQUAT DIBROMIDE, EGGS, ERUCIC ACID, GLUCOGLUCOSINOLATES, KALE, OLIVES
Raphanus sp. → RADISH
Raphetamine → AMPHETAMINE
Rapid Set → CYANOACRYLATES
Rapifen → ALFENTANIL
Rapilysin → RETEPLASE
Rapinovet → PROPOFOL
RAPISTRUM RUGOSUM
Rapostan → OXYPHENBUTAZINE
Rapynogen → PROPRANOLOL
RASPBERRY
Rastinon → TOLBUTAMIDE
Ratak → BRODIFACOUM also → DIFENACOUM
Ratan Jot → JATROPHA sp.
Rat Brush → TETRADYMA sp.
Rate → PENTAERYTHRITOL TETRANITRATE
Rathimed → METRONIDAZOLE
Raticate → NORBORMIDE
Ratilan → COUMACHLOR
Ratimus → BROMADIOLONE
Ratindon → CHLOROPHACINONE
Rapitil → NEDOCROMIL SODIUM
Rat-A-Way → COUMAFURYL
Rat & Mouse Controller Paste → THALLIUM
Rat Murder → 2-ISOVALERYL-1,3-INDANEDIONE
Ratol → ZINC
Rat Root → CALAMUS
Rattlebox → CROTALARIA also → DAUBENTONIA PUMICEA, SESBANIA sp.
Rattlesnake Violet → ERYTHRONIUM
Rattleweed → CIMICIFUGA RACEMOSA
Rattrack → 1- (1-NAPHTHYL)-2-THIOUREA
Raudixin → RAUWOLFIA SERPENTINA
Raunormine → DESERPIDINE
Raunova → SYROSINGOPINE
Raurine → RESERPINE
Rau-Sed → RESERPINE
Rauserpin → RAUWOLFIA SERPENTINA
Rausetin → PRENYLAMINE LACTATE
Rautensin → RAUWOLFIA SERPENTINA
Rauval → RAUWOLFIA SERPENTINA
Rauvolfia → RAUWOLFIA SERPENTINA
Rauwolf, Dr. Leonard → RAUWOLFIA SERPENTINA
Rauwiloid → RAUWOLFIA SERPENTINA
RAUWOLFIA SERPENTINA also → ALCOHOL, ETHYL; FENFLURAMINE, GINSENG, METHOTRIMEPRAZINE
Rauwolfine → AJMALINE
Ra-Valeas → RAUWOLFIA SERPENTINA
Ravase → SUTILAINS
Raviac → CHLOROPHACINONE
Ravyon → CARBARYL
Rawfola → RAUWOLFIA SERPENTINA
Rawpentina → RAUWOLFIA SERPENTINA
Raxar → GREPAFLOXACIN
Raybar → BARIUM SULFATE
Rayless Goldenrod → GOLDENROD
Raylina → PENICILLINS – "NEWER and SYNTHETICS" > *Amoxicillin*
RAZOXANE
Razoxin → RAZOXANE
RB 1509 → LOMUSTINE
RB 1589 → PEFLOXACIN
R-Ball → METHYLPHENIDATE
RD 1267 → METOCLOPRAMIDE
RD 4593 → MECOPROP
RD 13,621 → IBUPROFEN
RDX → CYCLONITE
RE 4355 → DIBROM
RE 5353 → BUFENCARB
RE 9006 → METHAMIDOPHOS
Reacthin → CORTICOTROPIN
Reactine → CETIRIZINE
Reactrol → CLEMIZOLE
Reasec → DIPHENOXYLATE (*with atropine sulfate*)
Rebelate → DIMETHOATE
Rebetol → RIBAVIRIN
Rebif → INTERFERONS
Rebriden → ISRADIPINE
Rec 7-0040 → FLAVOXATE HCL
Rec 15-1476 → FENTICONAZOLE
Rec 15-1533 → DENZIMOL
Recanescine → DESERPIDINE
Recchie d'asine → VERBASCUM THAPSUS
Receptal → LH-RH
Recheton → KEBUZONE
Rec-hirudin → HIRUDINS
Recidol → IBUPROFEN
Recolip → CLOFIBRATE
Reconin → CLEMASTINE
Reconox → PHENOTHIAZINE
Recormon → ERYTHROPOIETIN, *EPOETIN beta*
Rectadione → PHENINDIONE
Rectodelt → PREDNISONE
Rectovalone → TIXOCORTOL PIVALATE

Red Beans → SOPHORA sp.
"Red Birds" → SECOBARBITAL SODIUM
Red Chickweed → ANAGALLIS ARVENSIS
"Red Devils" → SECOBARBITAL SODIUM
Reddon → 2,4,5-TRICHLOROPHENOXYACETIC ACID
Red Elderberry → ELDERBERRY
Redentin → CHLOROPHACINONE
Redeptin → FLUSPIRILINE
Redergin → ERGOLOID MESYLATES
Red-hots → SOPHORA sp.
Redipaque → BARIUM SULFATE
Red Iron Oxide → FERRIC OXIDE
Redisol (H) → VITAMIN B_{12}
Red Lead → LEAD OXIDES
Red Man or Neck Syndrome → VANCOMYCIN
Red Moonseed → COCCULUS sp.
Red Oil → OLEIC ACID also → HYPERICUM
Redomex → AMITRIPTYLINE
Redoxon → VITAMIN C
Red Pepper → CAPSICUM also → PEPPERS
Red Poppy → PAPAVER sp.
Red Prussiate of Potash → POTASSIUM FERRICYANIDE
Red Root → SANGUINARIA CANADENSIS also → AMARANTHUS
Red Rose → VINCA ROSEA
Red Sage → LANTANA sp.
Redskin → ALLYL ISOTHIOCYANATE
Red Sorrel → RUMEX sp.
Red Spurge → EUPHORBIA
Red Tide → BREVETOXINS, CLAMS, EUGLENA SANGUINEA, GONYAULAX
Reducor → PROPRANOLOL
Reductil → SIBUTRAMINE
Redul → GLYMIDINE
Redupresin → ETHOXZOLAMIDE
REDUVIUS sp.
Redux® → DEXFENFLURAMINE also → SIBUTRAMINE
Red Water Tree → ERYTHROPHLEUM
REDWOOD also → SAWDUST
Red Wings → COMBRETUM sp.
Reed → FESCUE, TALL
Reed Canarygrass → HYDROGEN CYANIDE
Reed Sweet-grass → GLYCERIA sp.
Reefers → CANNABIS
Refkas → ERYTHROMYCIN ETHYL SUCCINATE
Refludon → LEPIRUDIN
Reflux → METHENAMINE
Refobacin → GENTAMICIN
Refrigerant 11 → TRICHLOROFLUOROMETHANE
Refrigerant 112 → 1,1,2,2-TETRACHLORODIFLUORETHANE
Refrigerant 113 → TRICHLOROTRIFLUOROETHANE
Refsum's Disease → CHLOROPHYLL
REFUIN also → ANTHRAMYCIN
Regaine → MINOXIDIL
Regelan → CLOFIBRATE
Regenon → DIETHYLPROPION
Regitine → PHENTOLAMINE MESYLATE
Regletin → ALPRENOLOL
Reglan → METOCLOPRAMIDE
Reglone → DIQUAT DIBROMIDE
Regonol → PYRIDOSTIGMINE BROMIDE
Regulin → MELATONIN
Regulton → AMEZINIUM METHYL SULFATE
Regutol → DOCUSATE SODIUM
Rehmannic → LIPPIA REHMANNI
Reichstein's Substance Fa → CORTISONE
Reichstein's Substance M → HYDROCORTISONE
Reidamine → DIMENHYDRINATE
Rekawan → POTASSIUM CHLORIDE
Rektidon → PARALDEHYDE
Rela → CARISPRODOL
Relafen → NABUMETONE
Relan Beta → BENDROFLUMETHIAZIDE
Relanium → DIAZEPAM
Relasom → CARISPRODOL
Relaxan → GALLAMINE TRIETHIODIDE
Relaxar → MEPHENESIN
Relaxil → MEPHENESIN
Relaxil G → GLYCERYL GUAIACOLATE
Reldan → CHLORPYRIFOS-METHYL
Relenza → ZANAMIVIR
Releserp → RESERPINE
Relestrid → METHOCARBAMOL
Relfact TRH → THYROTROPIN RELEASING HORMONE
Reliadol → MORPHINE
Reliberan → CHLORDIAZEPOXIDE
Relifen → NABUMETONE
Relifex → NABUMETONE
Relisorm L → LH-RH
Relivran → METOCLOPRAMIDE
Relvene → TROXERUTIN
Rely® → CELLULOSE, SODIUM CARBOXYMETHYLCELLULOSE
Remark → BETAHISTINE
Remeflin → DIMEFLINE
Remeron → MIRTAZEPINE
Remestan → TEMAZEPAM
Remicade → INFLIXIMAB

Remicyclin → TETRACYCLINE
Remid → ALLOPURINOL
Remifemin → CIMICIFUGA RACEMOSA
Reminitrol → NITROGLYCERIN
Remivox → LORCAINIDE
Removine → DIMENHYDRINATE
Remnos → NITRAZEPAM
Remsed → PROMETHAZINE
Renarcol → MEPHENESIN
Renasul → SULFAMETHIZOLE
Renazide → CYCLOTHIAZIDE
Renelate → METHENAMINE
Renese → POLYTHIAZIDE
Reneuron → FLUOXETINE
Renewtrient → BUTYROLACTONE
RENGAS also → ANACADIACEA, POISON IVY; SUMAC, POISON
Rengasil → PIRPROFEN
Renitec → ENALAPRIL
Reniten → ENALAPRIL
Renivace → ENALAPRIL
Renografin → MEGLUMINE DIATRIZOATE
Reno M → MEGLUMINE DIATRIZOATE
Renoguid → SULFACYTINE
Ren-O-Sal → ROXARSONE
Renoux, Dr. Gerald → LEVAMISOLE
Renovue-65 → IODAMIDE
Renovue-DIP → IODAMIDE
RENOXIPRIDE
Renselin → UNDECYLENIC ACID
Renshenlutou → GINSENG
Rentylin → PENTOXIFYLLINE
Renzepin → PIRENZEPINE HYDROCHLORIDE
Reocorin → PRENYLAMINE LACTATE
Reomax → ETHACRYNIC ACID
ReoPro → ABCIXIMAB
Reostral → MEPROBAMATE
Reoxyl → HEXOBENDINE
REPAGLINIDE
Reparil → ESCIN
Repel → *N,N*-DIETHYL-*m*-TOLUAMIDE
Repeltin → TRIMEPRAZINE
Repidose → OXFENDAZOLE
Repodral → STIBOPHEN
Repoise → BUTAPERAZINE
Repone K → POTASSIUM CHLORIDE
Repromix → MEDROXYPROGESTERONE ACETATE
Repronex → MENOTROPINS
Repulson → APROTININ
Requip → ROPINIROLE
Rere → CASSIA
Rescriptor → DELAVIRDINE MESYLATE
Resectisol → MANNITOL
RESERPINE also → ALCOHOL, ETHYL; AMITRIPTYLINE, CHLORPROMAZINE, COCAINE, DEOXYCHOLIC ACID, DEXTROAMPHETAMINE, DIGITALIS, DIGITOXIN, DIHYDRALAZINE, L-DOPA, EPHEDRINE, FENFLURAMINE, FURAZOLIDONE, HEXOBARBITAL, IMIPRAMINE, MEPHENTERMINE, METARAMINOL BITARTRATE, METHAMPHETAMINE, METOPROLOL, NIALAMIDE, NOREPINEPHRINE, NORTRIPTYLINE, PARGYLINE, PERPHENAZINE, PROPRANOLOL, PSEUDOEPHEDRINE, QUINIDINE, SEROTONIN, SYROSINGOPINE, THIOPENTAL, TRIFLUOPERAZINE DIHYDROCHLORIDE
Reserpoid → RESERPINE
Resifilm → THIRAM
Resimatil → PRIMIDONE
Resin → ROSIN
Resin Benzoin → BENZOIN GUM
Resin Damar → DAMAR
Resin Tolu → BALSAM TOLU
Resistomycin → KANAMYCIN
Resistopen → PENICILLINS – "NEWER and SYNTHETICS" > *Oxacillin*
Resistox → COUMAPHOS
Resmethrin → PYRETHRUM, PYRETHRINS, and PYRETHROIDS
Resmit → MEDAZEPAM
Resochen → CHLOROQUINE
Resochin → CHLOROQUINE
Resonium → SODIUM POLYSTYRENE SULFONATE
Resoquina → CHLOROQUINE
Résoquine → CHLOROQUINE
RESORANTEL
Resorcin → RESORCINOL
RESORCINOL also → HEXYLRESORCINOL, PHLOROGLUCINOL
Respaire → ACETYLCYSTEINE
Respicort → TRIAMCINOLONE
Respid → THEOPHYLLINE
Respifral → ISOPROTERENOL
Respilene → ZIPEROL
Respirase → ZIPEROL
Restanolon → CLORPRENALINE
Restenacht → EMEPRONIUM
Restenil → MEPROBAMATE
Restetal → EMYLCAMATE
Rest-On → METHAPYRILENE
Restoril → TEMAZEPAM

Restrol → DIENESTROL
Restryl → METHAPYRILENE
Resulax → SORBITOL
Resulfon → SULFAGUANIDINE
Resurrection Fern → POLYPODY FERN
Resyl → GLYCERYL GUAIACOLATE
Retabolin → NANDROLONE
Retalon-Oral → DIENESTROL
Retama → SPARTIUM JUNCEUM
Retavase → RETEPLASE
Retcin → ERYTHROMYCIN
Retensin → GALLAMINE TRIETHIODIDE
RETEPLASE
Retin-A → VITAMIN A
Retinoic Acid → VITAMIN A
13-cis-Retinoic Acid → ISOTRETINOIN
Retinol → VITAMIN A
Retinyl → VITMAIN A
Retolen → ASTEMIZOLE
Retractyl → POLYVINYL CHLORIDE
Retrangor → BENZIODARONE
Retrocortine → PREDNISONE
Retrone → DIDROGESTERONE
RETRORSINE also → PYRROLIZIDINES
Retrovir → ZIDOVUDINE
Reudene → PIROXICAM
Reumachlor → CHLOROQUINE
Reumaquin → CHLOROQUINE
Reumofene → INDOPROFEN
Reumofil → SULINDAC
Reumyl → SULINDAC
Reuxen → NAPROXEN
Reutol → TOLMETIN
REV 532OA → CELIPROLOL
Revasc → HIRUDINS
Revco → VITAMIN E
Reverin → ROLITETRACYCLINE
Reversol → EDROPHONIUM
Revex → NALMEFENE
ReVia → NALTREXONE
Revivarant → BUTYROLACTONE
Revivarant G → BUTYROLACTONE
Revivon → DIPRENORPHINE
Revolution → SELAMECTIN
Revonal → METHAQUALONE
Rexalgan → TENOXICAM
Rexan → CHLORMEZANONE
Rexitene → GUANABENZ ACETATE
Reye's Syndrome → ACETYLSALICYLIC ACID
Rezifilm → THIRAM
Rezipas → *p*-AMINOSALICYLIC ACID
Rezulin → TROGLITAZONE
RFD → PYRANTEL
R-gene → ARGININE
RGH 1106 → PIPECURONIUM BROMIDE
RH 315 → PRONAMIDE
RH 787 → PYRIMINAL
Rhamnus cathartica → FRANGULA
Rhamnus frangula → FRANGULA
Rhatany → KRAMERIA
RNC 2555 → INDAPAMIDE
Rheadine → PAPAVER sp.
Rheaform → IODOCHLORHYDROXYQUINOLINE
Rhematan → CINCHOPHEN
Rhengas Tree → RENGAS
Rheomacrodex → DEXTRAN, *40*
Rheoth ℞ → POLOXALENE
Rheotran → DEXTRAN, *40*
Rheumatism Weed → APOCYNUM CANNABINUM
Rheumatrex → METHOTREXATE
Rheumibis → ACEMETACIN
Rheumin → CINCHOPHEN
Rheumon Gel → ETOFENAMATE
Rheumox → APAZONE
Rheum rhaponticum → RHUBARB
Rhex → MEPHENESIN
rhIL 11 → INTERLEUKIN 11
Rhinalair → PSEUDOEPHEDRINE
Rhinalar → FLUNISOLIDE
Rhinantin → NAPHAZOLINE
Rhinocort → BUDESONIDE
Rhinofrenol → OXYMETAZOLINE
Rhinolitan → OXYMETAZOLINE
Rhinoperd → NAPHAZOLINE
Rhinopront → TETRAHYDROZOLINE
RHIZOCTONIA LEGUMINICOLA also → SLAFRAMINE
RHIZOPUS sp.
Rhocya → POTASSIUM THIOCYANATE
RHODAMINE B
Rhodiatox → PARATHION
Rhodine → ACETYLSALICYLIC ACID
Rhodinus pallescens → PALMS
RHODIUM
RHODODENDRON also → HONEY, PIERIS sp.
Rhodophyllus → ENTOLOMA
RHODOPHYLLUS LIVIDUS
Rhodoquine → PLASMOCID
RHODOTORULA
RHODOTYPOS SCANDENS
Rhombinin → ANAGYRINE
Rhomex → PIPERAZINE
Rhothane → 1,1-DICHLORO-2,2-BIS (*p*-CHLOROPHENYL)ETHANE

Rhovyl → POLYVINYL CHLORIDE
RHUBARB also → ANTHRAQUINONE, CALCIUM OXALATE, NITRATES, NITRITES, OXALIC ACID, POTASSIUM OXALATE
Rhumalgan → DICLOFENAC
Rhus radicans → POISON IVY
Rhus toxicodendron → POISON IVY
Rhus vernix → SUMAC, POISON
Rhythmarone → AMIODARONE
Rhythmodan → DISOPYRAMIDE
Rhythmodul → DISOPYRAMIDE
Rhythmol → PROPAFENONE
Riacen → PIROXICAM
Riball → ALLOPURINOL
RIBAVIRIN
Ribena → VITAMIN C
Ribetol → RIBAVIRIN
Ribo Azauracil → AZARIBINE also → AZAURIDINE
Riboflavin → VITAMIN B_2
9-β-Ribofuranosidoadenine → ADENOSINE
9-β-D-Ribofuranosyl-9H-purin-6-amine → ADENOSINE
Ribomicin → GENTAMICIN
Ribonosine → INOSINE
Riboxin → INOSINE
Ribrain → BETAHISTINE
RIC-272 → METHAQUALONE
RICE also → CHOLESTIN, MAGNESIUM SILICATE, PENICILLIUM sp., POLYCHLORINATED BIPHENYLS, TEA
Richardson-Merrell → TRIPARANOL
Richweed → SNAKEROOT, WHITE also → CIMICIFUGA RACEMOSA, EUPATORIUM
RICIN also → CASTOR BEAN; MISTLETOE, EUROPEAN or JAPANESE
Ricinus Oil → CASTOR OIL
Rickamicin → SISOMICIN
Ricketon → VITAMIN D_3
Rickettsia burnetii → COXIELLA BURNETII
Ricortex → CORTISONE
Ridall-Zinc → ZINC
Ridaura → AURANOFIN
Ridauran → AURANOFIN
Ridazin → THIORIDAZINE
Ridect → PERMETHRIN
Ridelline → PYRROLIZIDINES
Ridene → NICARDIPINE
Rid-Ezy → RONNEL
Rifa → RIFAMPIN
RIFABUTIN also → AZITHROMYCIN, CLOZAPINE, DELAVIRIDINE, FLUCONAZOLE, INDINAVIR, ITRACONAZOLE, NELFINAVIR MESYLATE, RITONAVIR, SAQUINAVIR, TACROLIMUS
Rifadin → RIFAMPIN
Rifaldazine → RIFAMPIN
Rifaldin → RIFAMPIN
RIFAMIDE
Rifampicin → RIFAMPIN
RIFAMPIN also → ACENOCOUMAROL; ALCOHOL, ETHYL; ALUMINUM HYDROXIDE, *p*-AMINOSALICYLIC ACID, AMIODARONE, BENTONITE, CARVEDILOL, CHLORAMPHENICOL, CHOLINE THEOPHYLLINATE, CLOFIBRATE, CORTISONE, CYCLOSPORINE, DAPSONE, DELAVIRDINE, DESMOPRESSIN, DEXAMETHASONE, DIGITOXIN, DIGOXIN, DISOPYRAMIDE, DISULFIRAM, ENALAPRIL, ETHAMBUTOL, ETHINYL ESTRADIOL, EXEMESTANE, FLUCONAZOLE, FLUVASTATIN, GLYMIDINE, HALOTHANE, HEXOBARBITAL, INDINAVIR, ISONIAZID, ITRACONAZOLE, KETOCONAZOLE, LEFLUNOMIDE, LOSARTAN, MAGNESIUM TRISILICATE, METHADONE, NIFEDIPINE, OLANZEPINE, PEFLOXACIN, PHENOBARBITAL, PHENPROCOUMON, PHENYTOIN, PREDNISOLONE, PROBENECID, PROPAFENONE, PROPRANOLOL, QUININE, RAPAMYCIN, REPAGLINIDE, RIFAPENTINE, RITONAVIR, SAQUINAVIR, SILDENAFIL, SODIUM BICARBONATE, TACROLIMUS, TAMOXIFENE CITRATE, TENOXICAM, THEOPHYLLINE, THYROXINE, TOCAINIDE, TOLBUTAMIDE, TOREMIFENE CITRATE, TRIAZOLAM, VERAPAMIL, WARFARIN, ZALEPLON, ZIDOVUDINE, ZOLPIDEM TARTRATE
Rifamycin AMP → RIFAMPIN
RIFAPENTINE
Rifaprodin → RIFAMPIN
Rifloc Retard → ISOSORBIDE DINITRATE
Rifocin M → RIFAMIDE
Rifoldin → RIFAMPIN
Rigecoccin → CLOPIDOL
Rigedal → ISOSORBIDE DINITRATE
Rigelon → THIAMPHENICOL
Rikavarin → TRANEXAMIC ACID
Riker 52G → APROTININ

Riker 546 → DEANOL
Riker 594 → SULTHIAME
Riker Laboratories → FLECAINIDE
Rilansyl → CHLORMEZANONE
Rilaquil → CHLORMEZANONE
Rilassol → CHLORMEZANONE
Rilutek → RILUZOLE
RILUZOLE
Rimactane → RIFAMPIN
RIMANTADINE also → ACETAMINOPHEN
Rimaon → ETHACRIDINE LACTATE
Rimadyl → CARPROFEN
Rimapen → RIFAMPIN
Rimazole → CLOTRIMAZOLE
Rimexel → RIMEXOLONE
RIMEXOLONE
Rimifon → ISONIAZID
Rimitsid → ISONIAZID
Rimon → POMEGRANATE
Rimso-50 → DIMETHYL SULFOXIDE
Rinatec → IPRATROPIUM BROMIDE
Rinderon-DP → BETAMETHASONE
Rindex → METHACYCLINE
Rinesal → CEPHALEXIN
Rineton → TRIAMCINOLONE
Rinlaxer → CHLORPHENESIN CARBAMATE
Rino-Clenil → BECLOMETHASONE
Rinomar → CINNARIZINE
Rinse 'n Vac® → MONOMETHYLNAPHTHALENE SULFONATE, NAPHTHA, RUGS
Riocyclin → TETRACYCLINE
Riogen → PHENYLMERCURIC ACETATE
Riol → TEGAFUR
Riomitsin → OXYTETRACYCLINE
Rioprostil → MISOPROSTOL
Ripcord → CYPERMETHRIN
Ripercol → TETRAMISOLE also → LEVAMISOLE
Ripirin → EMEPRONIUM
Riporest → METHAQUALONE
Riposon → MEPARFYNOL
Ririlim → BENZYDAMINE
Riripen → BENZYDAMINE
Risamal → CISAPRIDE
Rischiaril → DEANOL
Rise → CLOTIAZEPAM
RISEDRONATE
Riself → MEPHENOXALONE
Rishitin → POTATO
Rishitinol → POTATO
Risolid → CHLORDIAZEPOXIDE
Risordan → ISOSORBIDE DINITRATE
Risperdal → RISPERIDONE
RISPERIDONE also → DONEPEZIL
Ristarun → DEANOL
Ristogen → FLUFENAMIC ACID
RISTOCETIN
Ristomycin → RISTOCETIN
Riston → RISTOCETIN
Risumic → AMEZINIUM METHYL SULFATE
Ritalin → METHYLPHENIDATE
Ritalmex → MEXILETINE
Ritmalan → CIBENZOLINE
Ritmocardyl → AMIODARONE
Ritmodan → DISOPYRAMIDE
Ritmoforine → DISOPYRAMIDE
Ritmos → AJMALINE
Ritmusin → APRINDINE
Ritnodar → DISOPYRAMIDE
RITODRINE,
RITONAVIR also → AMIODARONE, ASTEMIZOLE, BEPRIDIL, BUPROPION, CISAPRIDE, CLOZAPINE, DISOPYRAMIDE, ERGOTAMINE, FLECAINIDE, FLUOEXTINE, INDINAVIR, ITRACONAZOLE, MEPERIDINE, METHADONE, METHYLENEDIOXYAMPHETAMINE, NELFINAVIR MESYLATE, NEVIRAPINE, PACLITAXEL, PIMOZIDE, PIROXICAM, PROPAFENONE, QUINIDINE, RIFABUTIN, RIFAMPIN, SAQUINAVIR, TERFENADINE, WARFARIN, ZIDOVUDINE, ZOLPIDEM TARTRATE
Ritordine → BETAMETHASONE
Rituxan → RITUXIMAB
RITUXIMAB
Rivadescin → RAUWOLFIA SERPENTINA
Rivanol → ETHACRIDINE LACTATE
Rivasin → RESERPINE
Rivastatin → CERIVASTATIN
RIVEA CORYMBOSA also → LYSERGAMIDE, MORNING GLORY
Rivixil → MINOXIDIL
Rivotril → CLONAZEPAM
Rize → CLOTIAZEPAM
Rizen → CLOTIAZEPAM
RL-50 → SODIUM GLUTAMATE
RMI 9,918 → TERFENADINE
RMI 71,754 → VIGABATRIN
RMI 71,782 → EFLORNITHINE
RO 1-5130 → PYRIDOSTIGMINE BROMIDE
RO 1-5431/7 → LEVORPHANOL TARTRATE
Ro 1-5470/5 → DEXTROMETHORPHAN HYDROBROMIDE
Ro 1-9334 → DEHYDROEMETINE
RO 1-9569 → TETRABENAZINE

Ro 2-2453 → DIAMTHAZOLE
RO 2-2985 → LASALOCID
Ro 2-3308 → QUINUCLIDINYL BENZILATE
Ro 2-3773 → CLIDINIUM BROMIDE
RO 2-5803 → PROPAFENONE
RO 2-7638 → OLEANDOMYCIN
RO 2-7758 → SARAMYCETINE
Ro 2-9578 → TRIMETHOBENZAMIDE
Ro 2-9757 → FLUOROURACIL
Ro 4-1544/6 → STIBOCAPTATE
Ro 4-1575 → AMITRIPTYLINE
Ro 4-1577 → CYCLOBENZAPRINE
RO 4-1778/1 → METHOPHOLINE
Ro 4-3816 → ALCURONIUM
RO 4-4393 → SULFADOXINE
RO 4-4602 → BENSERAZIDE
Ro 4-5282 → MEFENOREX
Ro 4-5360 → NITRAZEPAM
RO 4 - 6467 → PROCARBAZINE
RO 5-0831 → ISOCARBOXAZID
Ro 5-2180 → NORDAZEPAM
Ro 5-3059 → NITRAZEPAM
Ro 5-3307/1 → DEBRISOQUIN
Ro 5-4023 → CLONAZEPAM
Ro 5-4200 → FLUNITRAZEPAM
RO 5-4556 → MEDAZEPAM
Ro 5-5345 → TEMAZEPAM
Ro 5-6901 → FLURAZEPAM
RO 5-9000 → REFUIN
RO 5-9754 → ORMETOPRIM
RO 6-2580/9 → TRIMETHOPRIM
RO 6-4563 → GLYBURIDE
RO 640,796 → OSELTAMIVIR
RO 7-0207 → ORNIDAZOLE
RO 7-0582 → MISONIDAZOLE
RO 7-1051 → BENZNIDAZOLE
Ro 09-1978 → CAPECITABINE
Ro 10-6070 → ACITRETIN
Ro 10-6338 → BUMETANIDE
RO 10-8756 → PENICILLINS – "NEWER and SYNTHETICS" > *Amoxicillin*
RO 10-9070 → PENICILLINS – "NEWER and SYNTHETICS" > *Amdinocillin*
RO 10-9359 → ETRETINATE
RO 11-1163 → MOCLOBEMIDE
RO 11-1781 → TIAPAMIL
Ro 12-0068 → TENOXICAM
Ro 13-1042 → LORCAINIDE
Ro 13-8996 → OXICONAZOLE NITRATE
Ro 13-9904/001 → CEFTRIAXONE
Ro 15-1788 → FLUMAZENIL
Ro 20-5720/000 → CARPROFEN
RO 21-3981 → MIDAZOLAM
Ro 21-5998 → MEFLOQUINE HCL
Ro 21-8837 → ESTRAMUSTINE
RO 22-7796 → CIBENZOLINE
Ro 23-6240 → FLEROXACIN
Ro 31-8959 → SAQUINAVIR
RO 40-5967 → MIBEFRADIL
RO 40-7592 → TOLCAPONE
Roaccutane → ISOTRETINOIN
Robamate → MEPROBAMATE
Robamol → METHOCARBAMOL
Robamox → PENICILLINS – "NEWER and SYNTHETICS" > *Amoxicillin*
Robane → SQUALANE
Robanul → GLYCOPYRROLATE
Robaxin → METHOCARBAMOL
Robenecid → PROBENECID
ROBENIDINE
Robenz → ROBENIDINE
Robenzidine → ROBENIDINE
Robexitine → KETOCONAZOLE
Robimycin → ERYTHROMYCIN
ROBIN also → LOCUST, BLACK
Robinia pseudoacacia → LOCUST, BLACK also → ROBIN
Robinine → LOCUST, BLACK
Robin-Runs-Away → NEPETA HEDERACEA
Robinul → GLYCOPYRROLATE
Robiocina → NOVOBIOCIN
Robitin → LOCUST, BLACK also → ROBIN
Robitinin → LOCUST, BLACK
Robitussin → GLYCERYL GUAIACOLATE
Robizone-V → PHENYLBUTAZONE
Robuoy → PRISTANE
Roc → COCAINE
Rocain → PROCAINE
Roccal → BENZALKONIUM CHLORIDE
Rocefin → CEFTRIAXONE
Rocephin(e) → CEFTRIAXONE
Rochagan → BENZNIDAZOLE
Roche → MIBEFRADIL
Ro-Chlorozide → CHLOROTHIAZIDE
Rocillin → PENICILLINS – "NEWER and SYNTHETICS" > *Phenethicillin Potassium*
Rock Candy → SUCROSE
"Rocket Fuel" → PHENCYCLIDINE
Rock Fern → CHEILANTHEA SIEBERI also → NOTHOLAENA SINNUATA
Rock Poison → GASTROLOBIUM sp.
Rocksalt Moss → CARRAGEENAN
ROCKURONIUM BROMIDE
Rock Wool → MINERAL WOOL
Rocky Mountain Grape → MAHONIA AQUIFOLIUM
Ro-Cycline → TETRACYCLINE
Rodalon → BENZALKONIUM CHLORIDE

Rouge → FERRIC OXIDE
Rough Chervil → CHAEROPHYLLUM sp.
Rougoxin → DIGOXIN
Roulone → METHAQUALONE
Roundup → GLYPHOSPHATE
Rouqualone → METHAQUALONE
Roussel Laboratories → TIAPROFENIC ACID
Rovamycin → SPIRAMYCIN
Rovan → RONNEL
Rowasa → 5-AMINOSALICYLIC ACID
Roxanthin Red 10 → CANTHAXANTHIN
ROXARSONE
ROXATIDINE ACETATE
Roxiam → RENOXIPRIDE
Roxicam → PIROXICAM
Roxiden → PIROXICAM
Roxion → DIMETHOATE
Roxit → ROXATIDINE ACETATE
ROXITHROMYCIN also → MIDAZOLAM
Roxomicina → ERYTHROMYCIN ESTOLATE
Royal Blue → LYSERGIC ACID DIETHYLAMIDE
Royal Demolition Explosive → CYCLONITE
ROYAL JELLY
Rozol → CHLOROPHACINONE
RP 866 → QUINACRINE
RP 2090 → SULFATHIAZOLE
RP 2168 → METHYLGLUCAMINE ANTIMONATE
RP 2275 → SULFAGUANIDINE
RP 2512 → PENTAMIDINE ISETHIONATE
RP 2786 → PYRILAMINE
RP 2987 → DIETHAZINE
RP 3203 → IODOPYRACET
RP 3276 → PROMAZINE
RP 3277 → PROMETHAZINE
RP 3356 → ETHOPROPAZINE
RP 3359 → CHLOROGUANIDE
RP 3377 → CHLOROQUINE
RP 3389 → PROMETHAZINE
RP 3668 → SOLASULFONE
RP 3697 → GALLAMINE TRIETHIODIDE
RP 3735 → LUCANTHOE HYDROCHLORIDE
RP 3799 → DIETHYLCARBAMAZINE
RP 3828 → AMINOPROMAZINE
RP 4560 → CHLORPROMAZINE
RP 4753 → PYRIMETHAMINE
RP 5015 → ISONIAZID
RP 5337 → SPIRAMYCIN
RP 6140 → PROCHLORPERAZINE
RP 6171 → AMPHOTHALIDE
RP 6484 → ETHYL ISOBUTRAZINE
RP 6549 → TRIMEPRAZINE
RP 6847 → OXOMEMAZINE
RP 6870 → INPROQUONE
RP 7044 → METHOTRIMEPRAZINE
RP 7162 → TRIMIMPRAMINE
RP 7676 → PRALIDOXIME
RP 7746 → TRIFLUOMEPRAZINE
RP 8167 → ETHION
RP 8307 → OPIPRAMOL
RP 8595 → DIMETRIDAZOLE
RP 8909 → PROPERICIAZINE
RP 9715 → CYCLOBENZAPRINE
RP 9778 → PROTIONAMIDE
RP 9921 → APROTININ
RP 9955 → MELARSONYL POTASSIUM
RP 10,192 → DEMECLOCYCLINE
RP 11,641 → CANRENONE
RP 11,974 → PHOSALONE
RP 12,222 → PENMESTEROL
RP 12,833 → CLOREXOLONE
RP 13,057 → DAUNORUBICIN
RP 18,429 → TRANEXAMIC ACID
RP 19,366 → PIPOTIAZINE
RP 19,552 → PIPOTIAZINE
RP 19,583 → KETOPROFEN
RP 54,274 → RILUZOLE
RP 54,563 → ENOXAPRIN
RP 56,976 → TAXOTERE
RP 64,206 → SPARFLOXACIN
RS 2106 → STENBOLONE
RS 3650 → NAPROXEN
RS 3999 → FLUNISOLIDE
RS 8858 → OXFENDAZOLE
RS 10,085 → MOEXIPRIL
RS 21,592 → GANCICLOVIR
RS 35,887 → BUTOCONAZOLE
RS 37,619 → KETOROLAC
RS 44,872 → SULCONAZOLE NITRATE
RS 61,443 → MYCOPHENOLATE MOFETIL
RS 69,216 → NICARDIPINE
RS 94,991-298 → NAFARELIN ACETATE
R-Sonic → ARSANILIC ACID
RTCA → RIBAVIRIN
RU 486 → MIFEPRISTONE
RU 15,060 → TIAPROFENIC ACID
RU 15,750 → FLOCTAFENINE
RU 19,847 → LH-RH
RU 23,908 → NILUTAMIDE
RU 24,756 → CEFOTAXIME
RU 28,965 → ROXITHROMYCIN
RU 38,486 → MIFEPRISTONE
Ruabasine → AJMALICINE
RUBBER also → CHLOROBUTANOL, GLOVES, *N*-ISOPROPYL-*N*-PHENYLENEDIAMINE, LATEX,

MECHLORETHAMINE, 2-MERCAPTOBENZOTHIAZOLE, METHENAMINE, MONOBENZONE, MONOSULFIRAM, *N*-NITROSOMORPHOLINE, *p*-ROSANILINE, SHOES, SODIUM HYPOCHLORITE
Rubber Vine → CRYPTOSTEGIA GRANDIFLORA
Rubberweed → ACTINEA ODORATA
Rubesol → VITAMIN B_{12}
Rubex → DOXORUBICIN
Rubiazol I → SULFAMIDOCHRYSOIDINE
RUBIDIUM
Rubidomycin → DAUNORUBICIN
Rubitox → PHOSALONE
Rubrafer → IRON DEXTRANS
Rubramin PC → VITAMIN B_{12}
RUBRATOXIN A and B also → PENICILLIUM sp.
Rubus sp. → RASPBERRY
Rucaina → LIDOCAINE
Ruda → RUE
RUDBECKIA sp.
Rudilin → NYLIDRIN
Rudotel → MEDAZEPAM
RUE also → 5-METHOXYPSORALEN
Rueda → GRETA also → LEAD, LEAD OXIDES
Rufen → IBUPROFEN
Ruefenac → ALCLOFENAC
RUELENE
Ruewort → RUE
Rufol → SULFAMETHIZOLE
RUGS
RUGULOSIN also → PENICILLIUM sp.
RUGULOVASINES also → PENICILLIUM sp.
Ru-Hy-T → RAUWOLFIA SERPENTINA
Rulid → ROXITHROMYCIN
RUM also → MOLASSES
Ruma Fada → SCOPARIA sp.
Ruman → POMEGRANATE
Rumanu → POMEGRANATE
Rumatral → ALOXIPRIN
Rum Cherry → CHERRY
Rumensin → MONENSIN
Rumetan → ZINC
RUMEX sp. also → POTASSIUM OXALATE
"Running Fits" → NITROGEN CHLORIDE
Rusa Oil → GERANIUM
Rush → AMYL NITRITE also → BUTYL NITRITE, ISOBUTYL NITRITE
Rush, Dr. Benjamin → MERCUROUS CHLORIDE
Russian Comfrey → COMFREY
Russian Fly → CANTHARIDES
Russian Knapweed → CENTAUREA sp.
Russian Thistle → SALSOLA sp.
RUSSULA
Rusyde → FUROSEMIDE
Ruta graveolens → RUE
Rutgers 612 → ETHOHEXADIOL
Ruthenium → CHEESES
Rutile → TITANIUM DIOXIDE
Rutin → PAPAVERINE
Ruven → TROXERUTIN
Ru-Vert-M → MECLIZINE
RVPaque → CINOXATE
RWJ-17021-000 → TOPIRAMATE
RWJ 17,070 → HISTRELIN
RX 67,408 → FENCLOFENAC
RX 6029M → BUPRENORPHINE
RYANIA sp.
Ryanodine → RYANIA sp.
Rycarden → NICARDIPINE
Rycopel → CYPERMETHRIN
Rydene → NICARDIPINE
Rydrin → NYLIDRIN
RYE also → FLOUR, ERGOT
Ryegonovin → METHYLERGONOVINE
RYE GRASS also → JANITHREMS, LOLIUM TEMULTENUM, PAXILLINE, PENICILLIUM sp.
Rynacrom → CROMOLYN SODIUM
RYNCHOSIA sp.
Rytmonorm → PROPAFENONE
Ryzelan → ORYZALIN
R-3-ZON → PHENYLBUTAZONE

S

S 14 → DIMEFOX
S 767 → FENSULFOTHION
S 768 → FENFLURAMINE
S 805 → LOXAPINE
S 940 → PHTHALOPHOS
S 1102A → SUMITHION
S 1320 → BUDESONIDE
S 1520 → INDAPAMIDE
S 1702 → GLICLAZIDE
S 1752 → FENTHION
S 1942 → BROMOPHOS
S 2620 → ALMITRINE BISMESYLATE
S 3151 → PERMETHRIN
S 4105 → MEDIBAZINE
S 5602 → FENVALERATE
S 5614 → DEXFENFLURAMINE

S 6059 → MOXALACTAM
S 9490-3 → PERINDOPRIL *tert*-BUTYLAMINE
S 10,165 → PROPANIL
S 26,308 → IMIQUIMOD
SABADILLA
Sabadine → VERATRINE
Sabal serralatum → SAW PALMETTO
Sabari → MECLIZINE
Saber → ALOE
Sabidal → CHOLINE THEOPHYLLINATE
Sabril → VIGABATRIN
sab simplex → DIMETHICONE (*with silicon dioxide*)
Sacahuista → NOLINA TEXANA
Saccharimol → SACCHARIN
SACCHARIN also → ANILINE, COLAS, DIGOXIN, SULFANILAMIDE, TOLBUTAMIDE, σ-TOLUENE SULFONAMIDE
Saccharomyces cerevisiae → YEAST
Saccharose → SUCROSE
Saccharum Lactis → LACTOSE
Saccox → SALINOMYCIN
Sacerno → MEPHENYTOIN
Sacred Bark → CASCARA SAGRADA
Sacred Datura → DATURA, *metaloides*
Sacred Mushroom → TEONANACATL
Sadamin → XANTHINOL NIACINATE
Saddleback Caterpillar → CATERPILLARS
SAFFLOWER OIL also → T-2 TOXIN
SAFFRON also → DINITRO-σ-CRESOL
Saffron Crocus → CROCUS SATIVUS
SAFRANINES
Safrazine → β-PIPERONYL ISOPROPYL HYDRAZINE
SAFROLE also → CASEIN, NUTMEG, PEPPERS, ROOT BEER, SARSAPARILLA, SASSAFRAS, SMILAX sp.
Safrotin → PROPETAMPHOS
Sagaang Kawhlaa → KALMIA
Sagatal → PENTOBARBITAL
SAGE
Sages → LANTANA sp.
Sagisal → SPIRONOLACTONE
Sagittol → PROPRANOLOL
Sago Palm → PALMS
Sahuca Bean → SOYBEANS
Saila → GLYCYRRHIZA GLABRA
Sailor's Flower → VINCA ROSEA
Saizen → GROWTH HORMONES
Sakaraan → HYOSCYAMUS
Sake → MONOCHLORODIFLUOROMETHANE
SALA 4 → MENTHYL SALICYALTE
Salacetin → ACETYLSALICYLIC ACID
Saladarine → SALAMANDERS
Salad Dressing → PROPYLENE GLYCOL
Salagen → PILOCARPINE
SALAMANDERS
SALAMI also → NITRATES, NITRITES
Salamid → SALICYLAMIDE
Sal Ammoniac → AMMONIUM CHLORIDE
Salandrine → SOLANDRA sp.
Salanil → GEFARNATE
Salazopyrin → SALICYLAZOSULFAPYRIDINE
Salbulin → ALBUTEROL
Salbumol → ALBUTEROL
Salbutamol → ALBUTEROL
Salbutine → ALBUTEROL
Salbuvent → ALBUTEROL
Salcatonin → CALCITONIN
Salcetogen → ACETYLSALICYLIC ACID
Saletin → ACETYLSALICYLIC ACID
Salflex → SALSALATE
Salgain → SALINOMYCIN
Salicin → POPLAR
SALICYL ALCOHOL
SALICYLAMIDE
Salicylanilides → HEXOCHLOROPHENE, SOAP
Salicylates → ACENOCOUMAROL, ACETAZOLAMIDE, AMMONIUM CHLORIDE, APPLE, APRICOT, BLACKBERRY, BISMUTH SUBSALICYLATE, CHERRY, CHLORPROMAZINE, CORTICOTROPIN, CYCLOCOUMAROL, DICHLORPHENAMIDE, DOCUSATE SODIUM, GLIPIZIDE, GLUCOSE, GRAPE, IMIPRAMINE, LIOTHYRONINE, MAGNESIUM HYDROXIDE, MENTHYL SALICYLATE, METHOTREXATE, NALIDIXIC ACID, NECTARINE, NITROFURANTOIN, NIZATIDINE, ORANGE, OXAPROZIN, PEACHES; PENICILLINS – "NEWER and SYNTHETICS" > *Methicillin Sodium, Nafcillin Sodium, Oxacillin Sodium*; PHENOLSULFONPHTHALEIN, PHENYLBUTAZONE, PHENYRAMIDOL, PHENYTOIN, PROBENECID, PROPRANOLOL, RASPBERRY, ROOT BEER, SALSALATE, SECOBARBITAL SODIUM, SODIUM BICARBONATE, STRAWBERRY, SULFINPYRAZONE
SALICYLAZOSULFAPYRIDINE
SALICYLIC ACID also → *p*-AMINOBENZOIC ACID, DITHRANOL, PENICILLIN PHENYL

SALICYLATE, RUBBER, SODIUM VALPROATE
Salicylic Acid Acetate → ACETYLSALICYLIC ACID
Salicylsalicylic Acid → SALSALATE
Salicyluric Acid → *p*-AMINOBENZOIC ACID
Salicym → SALICYLAMIDE
Saligenin → SALICYL ALCOHOL
Salimid → CYCLOPENTHIAZIDE
Salimol → SULFAMETHIZOLE
SALINOMYCIN also → MONENSIN, TIAMULIN
Salipran → BENORYLATE
Salisal → SALSALATE
Salisan → CHLOROTHIAZIDE
Salivation Factor → RHIZOCTONIA LEGUMINICOLA
Salizell → SALICYLAMIDE
Salmetedur → SALMETEROL
SALMETEROL
Salmiac → AMMONIUM CHLORIDE
SALMON
Salmotonin → CALCITONIN
Salofalk → 5-AMINOSALICYLIC ACID
Salol → PHENYL SALICYLATE
Salpadyn → TRIMEPRANOL
Salpix → SODIUM ACETRIZOATE
Sal Prunelle → POTASSIUM NITRATE
Salrin → SALICYLAMIDE
Salsa → IPOMEA sp.
SALSALATE also → WARFARIN
Sal Soda → SODIUM CARBONATE
SALSOLA sp.
Salt → SODIUM CHLORIDE
Saltbush → ATRIPLEX sp.
Salt Cake → SODIUM SULFATE
Salt of Lemery → POTASSIUM SULFATE
Saltpeter → POTASSIUM NITRATE
Salt Rock Moss → CARRAGEENAN
Salufer → SODIUM SILICOFLUORIDE
Salunil → CHLOROTHIAZIDE
Salural → BENDROFLUMETHIAZIDE
Salures → BENDROFLUMETHIAZIDE
Saluretil → CHLOROTHIAZIDE
Saluric → CHLOROTHIAZIDE
Salurin → TRICHLORMETHIAZIDE
Saluron → HYDROFLUMETHIAZIDE
Salutrid → CHLOROTHIAZIDE
Salvacard → NIKETHAMIDE
Salvarsan → ARSPHENAMINE
Salvation Jane → ECHIUM PLANTAGINEUM
SALVIA
Salvia officinalis → SAGE
Salymid → SALICYLAMIDE
Salyphec → SALSALATE
Salyrgan → MERSALYL
Salysal → SALSALATE
Salyzoron → BENZYDAMINE
Salzburg Vitriol → COPPER SULFATE
Salzone → ACETAMINOPHEN
Samarium-153 EDTMP → SAMARIUM-153 LEXIDRONAM
SAMARIUM-153 LEXIDRONAM
Samberu → ERYTHROPLEUM
Sambucine → ELDERBERRY
Sambuco → ELDERBERRY
Sambucus canadensis → ELDERBERRY
Sambucus mexicana → ELDERBERRY
Sambucus nigra → ELDERBERRY
Sambucus racemosa → ELDERBERRY
Sambunigrin → SAMBUCUS sp.
SAMe → S-ADENOSYLMETHIONINE
Sameko → NITRAZEPAM
Samedo-ap → TANAECIUM sp.
Samedrin → CEPHADRINE
Samid → SALICYLAMIDE
Sammuco → ELDERBERRY
Samorin → ISOMETAMIDIUM
Samyo → VALERIANA sp.
SAN 3,221 → PROPETAMPHOS
Sanabolicum → NANDROLONE
Sanamycin → CACTINOMYCIN
Sanapert → OXYPHENISATIN
Sanasthmax → BECLOMETHASONE
Sanasthmyl → BECLOMETHASONE
Sanatrichom → METRONIDAZOLE
Sanbar → PERMETHRIN
Sanclomycine → TETRACYCLINE
SAND
SANDALWOOD OIL
Sandane → TERFENADINE
Sandbox → HURA CREPITANS
Sand Brier → SOLANUM sp.
SANDBUR
SAND FLIES
Sandimmune → CYCLOSPORINE
Sandomigran → PIZOTYLINE
Sandopart → DESAMINOOXYTOCIN
Sandoptal → BUTALBITAL
Sandoscill → PROSCILLARIDIN
Sandostatin → OCTREOTIDE
Sandostene → THENALDINE TARTRATE
Sandril → RESERPINE
Sanedrine → EPHEDRINE
Sang Cya → CYCLOSPORINE
Sang Stat → CYCLOSPORINE
Sanguicillin → PENICILLINS – "NEWER and SYNTHETICS" > *Pivampicillin*

SANGUINARIA CANADENSIS
Sanguinarine → SANGUINARIA CANADENSIS also → CHELIDONIUM MAJUS, ECHOSCHOLTZIA CALIFORNICA
SANGURA
Sanipirol-4 → *p*-AMINOSALICYLIC ACID
SANITARY NAPKINS
Sanituko → PENTACHLOROPHENOL
Sanizol → BENZETHONIUM CHLORIDE
Sanluol → ARSPHENAMINE
Sanmarton → FENVALERATE
Sanmigran → PIZOTYLINE
Sanochrysine → GOLD SODIUM THIOSULFATE
Sanodin → CARBENOLOXONE
Sanogyl → ACETARSONE
Sanoma → CARISPRODOL
Sanomigran → PIZOTYLINE
Sanopron → CHLORPROMAZINE
Sanoquin → CHLOROQUINE
Sanorex → MAZINDOL
Sanorin → NAPHAZOLINE
Sanorin-Spofa → NAPHAZOLINE
Sanotensin → GUANETHIDINE
Sanoxit → BENZOYL PEROXIDE
San Pedro → TRICHOCEREUS PACHANOI
Sanquinon → DICHLONE
Sansalid → UREDOFOS
Sansdolor → MEPHENESIN
Sansert → METHYSERGIDE
SANSEVIERA sp.
Santaluna album → SANDALWOOD OIL
Santheose → THEOBROMINE
Santavy's substance F → DEMECOLCINE
Santicizer 160 → BUTYL BENZYL PHTHALATE
Santobrite → PENTACHLOROPHENOL
Santoflex → ETHOXYQUINE
SANTONIN
Santophen → PENTACHLORPHENOL
Santoquin → ETHOXYQUINE
Sapec → GARLIC
Sapecron → CHLORFENVINPHOS
Sapilent → TRIMIMPRAMINE
SAPINDUS sp.
SAPIUM sp.
Sapium sebiferum → CHINESE TALLOW TREE
Sapo → SOAP
Sapoderm → TRICLOSAN
Saponaria officinalis → BOUNCING BET
Saponaria sp. → BOUNCING BET also → FLOUR
SAPONARIA VACCARIA
Saponarin → BOUNCING BET
SAPONIN also → COCKLES, CONVALLARIA MAJUS, FLOUR, MERCURIALIS ANNUA, MOCK ORANGE, PACHYRRHIZUS EROSOS, QUILLAJA, SERJANIA sp., SCROPHULARIA sp., SESBANIA sp.
Saporubin → BOUNCING BET
Sapotoxin → COLOCASIA also → QUILLAJA
Saprosan → CHLORQUINALDOL
Sapucaia Nuts → LECYTHIS
SAQUINAVIR also → DELAVIRDINE, GRAPEFRUIT, INDINAVIR, NELFINAVIR MESYLATE, NEVIRAPINE, RITONAVIR, TERFENADINE, WARFARIN
Sarafem → FLUOXETINE
SARALASIN
SARAMYCETINE
Saran® → VINYLIDENE CHLORIDE
Sarathammus scoparius → SPARTEINE
Sarba Arba → ADENIUM sp.
SARCOBATUS VERMICULATUS
Sarcoclorin → MELPHALAN
L-Sarcolysine → MELPHALAN
SARCOPHALUS DIDERRICHII
SARCOSTEMMA sp.
SARDINELLA sp.
SARDINES
Sarenin → SARALASIN
Sargramostin → GRANULOCYTE MACROPHAGE COLONY STIMULATING FACTOR
SARIN
Sarisol → BUTABARBITAL
Sarodant → NITROFURANTOIN
Sarolex → DIAZINON
Saroten → AMITRIPTYLINE
Sarotex → AMITRIPTYLINE
Sarrasin Commun → BUCKWHEAT
SARSAPARILLA also → ISOSAFROLE, SMILAX sp.
Sartosona → CEPHALEXIN
Sasapyeine → SALSALATE
Sashimi → FISH
SASSAFRAS also → PHENYTOIN, ROOT BEER, TEA
Sastridex → FLUFENAMIC ACID
Sasulen → PIROXICAM
Satanolon → DIPHENIDOL
Sativa → CANNABIS
Saturn → LEAD
Satuwa → PARIS sp.
Sauco Blanco → CESTRUM
SAUERKRAUT also → HISTAMINE
SAUNAS
SAURINE

SAUSAGE also → CYCLOPIAZONIC ACID, OCHRATOXINS
Savac → IODOPYRACET
Savannah Flower → URECHITES sp.
Saventrine → ISOPROTERENOL
SAVIN OIL
Sawacillin → PENICILLINS – "NEWER and SYNTHETICS" > *Amoxicillin*
SAWDUST
Sawi → CANNABIS
SAW PALMETTO
Saxifrax → SASSÅFRAS
Saxin → SACCHARIN
SAXITOXIN also → CLAMS, CRABS, GONYAULAX, MUSSELS, SCALLOPS
Saxizon → HYDROCORTISONE
Sayri → TOBACCO
Sb → ANTIMONY
SB 5833 → CAMAZEPAM
SB 7505 → IBOPAMINE
SBP 1513 → PERMETHRIN
SC 110 → PHENYLMERCURIC ACID
SC 2910 → METHANTHELINE BROMIDE
SC 7031 → DISOPYRAMIDE
SC 9376 → CANRENONE
SC 9387 → PIPAMAZINE
SC 9420 → SPIRONOLACTONE
SC 10,363 → MEGESTROL ACETATE
SC 11,800 → ETHYNODIOL DIACETATE
SC 13,957 → DISOPYRAMIDE
SC 18,862 → ASPARTAME
SC 29,333 → MISOPROSTOL
SC 47,111 → LOMEFLOXACIN
SC 58,635 → CELECOXIB
Scabene → LINDANE
SCABIOSA SUSSISA
SCALLOPS also → GONYAULAX, MOLLUSKS, SAXITOXIN
Scandicain → MEPIVACAINE
Scandine → IBOPAMINE
Scarlet Berry → SOLANUM sp.
Scarlet Elderberry → ELDERBERRY
Scarlet Pimpernel → ANAGALLIS ARVENSIS
SCARLET RED
SCE 129 → CEFSULODIN SODIUM
Sch 1000 → IPRATROPIUM BROMIDE
Sch 3940 → PERPHENAZINE
Sch 4831 → BETAMETHASONE
Sch 10,144 → TOLNAFTATE
Sch 10,159 → TRICLOFOS
Sch 10,649 → AZATADINE DIMALEATE
Sch 11,460 → BETAMETHASONE
SCH 12,041 → HALAZEPAM
SC 12,937 → AZACOSTEROL
Sch 13,475 → SISOMICIN
Sch 13,521 → FLUTAMIDE
Sch 14,714 → FLUNIXIN MEGLUMINE
Sch 15,719W → LABETALOL
Sch 16,134 → QUAZEPAM
Sch 18,020W → BECLOMETHASONE
Sch 20,569 → NETILMYCIN
Sch 21,420 → ISEPAMICIN
Sch 25,298 → FLORFENICOL
Sch 29,851 → LORATADINE
Sch 30,500 → INTERFERONS
Sch 32,088 → MOMETASONE FUROATE
Sch 39,300 → GRANULOCYTE MACROPHAGE COLONY STIMULATING FACTOR
Sch 39,720 → CEFTIBUTIN
Scheele's Green → COPPER ARSENITE
Schering → LISURIDE, LORATADINE, MASCARA, MONOAMYLAMINE, MONOLINURON, PERPHENAZINE, PLACEBO, THEOPHYLLINE, TOLNAFTATE
Schering 36,268 → CHLORDIMEFORM
Scherisolon → PREDNISOLONE
Scherosone → CORTISONE
Scheroson F → HYDROCORTISONE
SCHINUS TEREBINTHIFOLIUS
Schistomide → AMPHOTHALIDE
Schizophrenics → TARAXEIN
Schoela zonensis → PALMS
Schoencaulon officinale → VERATRINE
SCHOENOBIBLUS PERUVIANUS
SCHOENUS ASPEROCARPUS
Schokolade → CHOCOLATE
Schrad → KOCHIA
SCHRADAN
Schultes & Hofmann → RIVEA CORYMBOSA
Schultz 737 → TARTRAZINE
Schultz 770 → FDC BLUE #1
Schultz 1309 → INDIGOCARMINE
Schwan's Ice Cream® → EGGS
Schweinfurt Green → CUPRIC ACETOARSENITE
Scillacrist → PROSCILLARIDIN
Scilla nonscripta → BLUEBELL
Scilliroside → SQUILL
SCINDAPUS AUREUS
SCIRPUS AMERICANUS
Sclane → TRIPARANOL
Sclerosol → DIMETHYL SULFOXIDE
SCLEROTINIA SCLEROTIORUM
Scolaban → BUNAMIDINE
Scombroid → FISH
Scop → SCOPOLAMINE

SCOPARIA sp.
SCOPOLAMINE also → ALCOHOL, ETHYL; ATROPA BELLADONNA, CANNABIS, CYCLOPENTOLATE, DATURA, DIAZEPAM, DUBOISA MYOPOROIDES, GROUND CHERRY, MANDRAGORA OFFICINARUM, METHAPYRILENE, METHOTRIMEPRAZINE, METHYSTICODENDRON AMESIANUM, PODOPHYLLUM, WITHANIA SOMNIFERA, STRAMONIUM
Scopolamine Lux → SCOPOLAMINE
Scopolin → POTATO
Scopoletin → POTATO
Scopos → SCOPOLAMINE
SCOPULARIOPSIS sp.
Scorbacid → VITAMIN C
Scorbu-C → VITAMIN C
Scorpionfish → PTEROIS VOLITANS
SCORPIONS
Scototenin → SUNFLOWER
Scouring-rush → EQUISETUM
Scrofula Root → ERYTHRONIUM
SCROPHULARIA sp.
SCTZ → CLOMETHIAZOLES
Scuffle → PHENCYCLIDINE
Scullcap → SCUTELLARIA LATERIFLORA
Scurf Pea → PSORALEA sp.
Scurgeon Needle → OPUNTIA sp.
Scuroforme → BUTAMBEN
SCUTELLARIA LATERIFLORA
Scutl → PHENYLMERCURIC ACID
Scyliorhinus caniculus → SHARK
SD 1 → NIKETHAMIDE
SD 51 → AMINACRINE
SD 1750 → DICHLORVOS
SD 3419 → ENDRIN
SD 3562 → DICROTOPHOS
SD 4294 → CIODRIN
SD 4402 → ISOBENZAN
SD 7859 → CHLORFENVINPHOS
SD 8447 → GARDONA
SD 8530 → LANDRIN
SD 8786 → LANDRIN
SD 9129 → MONOCROTOPHOS
SD 9228 → MERCAPTODIMETHUR
SD 14,999 → METHOMYL
SD 15,418 → CYANAZINE
SD 15,803 → VINCOFOS
SD 43,775 → FENVALERATE
S-dimidine → SULFAMETHAZINE
SDMH → 1,2-DIMETHYLHYDRAZINE
SDS → SODIUM LAURYL SULFATE
Sdt 91 → STIBOPHEN
SDZ 205-502 → QUINAGOLIDE
SE 1520 → INDAPAMIDE
SE 5023 → ALMITRINE BISMESYLATE
SEA ANEMONES also → SPONGES
Sea Ash → ZANTHOXYLUM sp.
SEA CUCUMBER
Sea Egg → SEA URCHINS
Sea Fans → LOPHOGORGIA sp.
Seagull → CALADIUM
Sea-Legs → MECLIZINE
Sealwort → POLYGONATUM BIFLORUM
Sea Nettle → JELLYFISH
Searlequin → IODOQUINOL
SEA SNAKES
SEA URCHINS
SEA WASP also → JELLYFISH
Seaweed Dermatitis → LYNGBA sp.
Sea Whips → LOPHOGORGIA sp.
Sebacil → PHOXIM
SEBASTIANIA
Sebercim → NORFLOXACIN
Sebertia accuminata → NICKEL
Sebizon → SULFACETAMIDE SODIUM
Sebumsol → SQUALANE
Secale cereale → RYE
Secale cornutum → ERGOT
Secalip → FENOFIBRATE
Secalonic Acid → PENICILLIUM sp.
Secbutobarbitone → BUTABARBITAL
Seccidin → PRENYLAMINE LACTATE
Sechvitan → PYRIDOXAL-5-PHOSPHATE
Seclar → BECLAMIDE
Seclodin → IBUPROFEN
SECOBARBITAL SODIUM also → BISHYDROXYCOUMARIN, KETAMINE, METHOXYFLURANE
Seconal Sodium → SECOBARBITAL SODIUM
Secoropen → PENICILLINS – "NEWER and SYNTHETICS" > *Azlocillin*
Secorvas → FOSFINOPRIL SODIUM
Secrosteron → DIMETHISTERONE
Sect-A-Chlor → CHLORPYRIFOS
Sectral → ACEBUTOLOL
Securit → LORAZEPAM
Securon → VERAPAMIL
Sedaform → CHLOROBUTANOL
Sedalande → FLUANISONE
Sedalis → THALIDOMIDE
Sedamyl → ACECARBROMAL
Sedanoct → L-TRYPTOPHAN
Sedantoinal → MEPHENYTOIN
Sedapam → DIAZEPAM
Sedaplus → DOXYLAMINE SUCCINATE

Sedapran → PRAZEPAM
Sedaquin → METHAQUALONE
Sedaraupin → RESERPINE
Sedatine → ANTIPYRINE
Sedatival → LORAZEPAM
Sedatromin → CINNARIZINE
Sedazin → LORAZEPAM
Sédeval → BARBITAL
Sednotic → AMOBARBITAL
Sedolatan → PRENYLAMINE LACTATE
Sedoneural → SODIUM BROMIDE
Sedor → DICHLORALANTIPYRINE
Sedormid → APRONALIDE
Sedotussin → CARBETAPENTANE CITRATE
Sedral → CEFADROXIL
Sedrena → TRIHEXYPHENIDYL
Sedulon → PIPERIDIONE
Sedural → PHENAZOPYRIDINE
Sedutain → SECOBARBITAL SODIUM
Seduxen → DIAZEPAM
Sefacin → CEPHALORIDINE
Seffein → CARBARYL
Seflenyl → PIRPROFEN
Sefona → CHINIOFON
Sefril → CEPHADRINE
Séglor → DIHYDROERGOTAMINE METHANESULFONATE (MESYLATE)
Segontin → PRENYLAMINE LACTATE
Segosin → METHIODAL SODIUM
Segurex → BUMETANIDE
Selacryn → TICRYNAFEN
SELAMECTIN
Seldane → TERFENADINE
Sel de Vichy → SODIUM BICARBONATE
Selecor → CELIPROLOL
Selectin → PRAVASTATIN
Selectol → CELIPROLOL
Selectomycin → SPIRAMYCIN
Seleen → SELENIUM
Selegiline → DEPRENYL
Selenamectin → SELAMECTIN
Selenious Acid → COPPER NITRATE
Selenious Anhydride → SELENIUM OXIDE
SELENIUM also → ASTERS, COD LIVER OIL, CYSTINE, DIOXATHION, FISH, IVA sp., MILKVETCH, MORINDA RETICULATA, OONOPSIS, PEAS, SILVER, STANLEYA PINNATA, SULFUR, TELLURIUM, YEAST
Selenium Dihydride → HYDROGEN SELENIDE
Selenium Dioxide → SELENIUM OXIDE also → COAL
SELENIUM DISULFIDE
Selenium Fluoride → SELENIUM HEXAFLUORIDE
SELENIUM HEXAFLUORIDE
Selenium Hydride → HYDROGEN SELENIDE
SELENIUM OXIDE
Selenocosmia stirlingi → SPIDERS
SELENOMETHIONINE
Selepam → QUAZEPAM
Seles Beta → ATENOLOL
Selexid → PENICILLINS – "NEWER and SYNTHETICS" > *Pivmecillinam*
Selinon → DINITRO-σ-CRESOL
Selipran → PRAVASTATIN
Sellotape® → PROPYLENE GLYCOL
Selobloc → ATENOLOL
Seloken → METOPROLOL
Selopral → METOPROLOL
Selo-Zok → METOPROLOL
Selsun → SELENIUM DISULFIDE
Seltz-K → POTASSIUM CITRATE
Selvigon → PIPAZETHATE
Selvjgon → PIPAZETHATE
Selye, Hans → DIHYDROTACHYSTEROL
Semap → PENFLURIDOL
SEMICARBAZIDE
Semicid → NONOXYNOL-9
Semi-Lente → INSULIN
Seminole Tea → PAW-PAW
Semolin → HALVA (H)
Semopen → PENICILLINS – "NEWER and SYNTHETICS" > *Phenethicillin Potassium*
Semoxydrine → METHAMPHETAMINE
Sempera → ITRACONAZOLE
SEMUSTINE
Senarmontite → ANTIMONY TRIOXIDE
Sencephalin → CEPHALEXIN
Senco Corn Mix → THALLIUM
Sencor → METRIBUZIN
Sendoxan → CYCLOPHOSPHAMIDE
Sendran → PROPOXUR
SENECIO sp. also → JACOBINE, PYRROLIZIDINES, RETRORSINE, TEA
Senecio longilobus → PYRROLIZIDINES
Senecionine → PYRROLIZIDINES, SENECIO sp.
Seneciosis → PYRROLIZIDINES
Seneciphylline → PYRROLIZIDINES, SENECIO sp.
Senfgas → MUSTARD GAS
Seni → PEYOTE
Senior → PEMOLINE
Seniramin → SYROSINGOPINE
SENNA also → CASSIA, MILK
Sennesbälglein → SENNA
Senokot® → SENNA
Sensaval → NORTRIPTYLINE

Sensidyn → CHLORPHENIRAMINE *also* → DEXCHLORPHENIRAMINE
Sensitive Fern → ONOCLEA SENSIBILIS
Sensitive Plant → MIMOSA
Sensival → NORTRIPTYLINE
Senskikene → COLTSFOOT
Senvre → MUSTARD
Seotalnatrium → SECOBARBITAL SODIUM
Sepamit → NIFEDIPINE
Sepan → CINNARIZINE
Sepiolite → MAGNESIUM SILICATE
Sepsinol → FURALTADONE
Septicid → MAFENIDE
Septicol → CHLORAMPHENICOL
Septigen → GENTAMICIN
Septipulmon → SULFAPYRIDINE
Septoplix → SULFANILAMIDE
Septotan → HYDRARGAPHEN
Septox → DEMETON
Septra → SULFAMETHOXAZOLE also → TRIMETHOPRIM
Septrin → SULFAMETHOXAZOLE
Sequestrene → SODIUM EDETATE
Sequoia sempervirens → REDWOOD
Sequoiosis → SAWDUST
Seral → MEPARFYNOL
Serax → OXAZEPAM
Serc → BETAHISTINE
Serefrex → KETANSERIN
Serenace → HALOPERIDOL
Serenal → OXAZEPAM
Serenase → HALOPERIDOL
Serenesil → ETHCHLORVYNOL
Serenid → OXAZEPAM
Serenium → ETHOXAZENE
Serenoa serrulate → SAW PALMETTO
Serentil → MESORIDAZINE
Seren Vita → CHLORDIAZEPOXIDE
Serepax → OXAZEPAM
Seresta → OXAZEPAM
Serevent → SALMETEROL
Serexal → RESERPINE
Serfin → RESERPINE
Serfolia → RAUWOLFIA SERPENTINA
Sergetyl → ETHYL ISOBUTRAZINE
Seriel → TOFISOPAM
SERJANIA sp.
Serjanosides → SERJANIA sp.
Sermaka → FLURANDRENOLIDE
Sermion → METERGOLINE
Sernevin → SULPIRIDE
Sernyl → PHENCYCLIDINE
Sernylan → PHENCYCLIDINE
Seroden → THIACETAZONE
Seromycin → CYCLOSERINE
Seronon-Bagren → BROMOCRIPTINE
Serophene → CLOMIPHENE CITRATE
Seropram → CITALOPRAM
Seroquel → QUETIAPINE FUMARATE
Seroten → AMITRIPTYLINE
Serotinex → CLOFIBRATE
SEROTONIN also → BANANA, CHEESES, CHLORPHENTERMINE, CHLORPROMAZINE, CHOCOLATE, DIPLOPTERYS CABRERANA, FENFLURAMINE, FLUOXETINE, GARLIC, 10-METHOXYPHARMALAN, METHYLPHENIDATE, METHYSERGIDE, MUCUNA sp., NIALAMIDE, PASSIFLORA sp., PHENYLALANINE, TOADS, TOMATO, TRIPROLIDINE
Seroxat → PAROXETINE
Serpadex → RAUWOLFIA SERPENTINA
Serpalan → RESERPINE
Serpaloid → RESERPINE
Serpanray → RESERPINE
Serpasil → RESERPINE
Serpasol → RESERPINE
Serpate → RESERPINE
Serpax → OXAZEPAM
Serpina → RAUWOLFIA SERPENTINA
Serpine → RESERPINE
Serpiloid → RESERPINE
Serral → DIETHYLSTILBESTROL
Sertan → PRIMIDONE
Serten → BENZALKONIUM CHLORIDE
SERTRALINE also → DEPRENYL, DESIPRAMINE, DIAZEPAM, ERYTHROMYCIN, IMIPRAMINE, PHENELZINE, SUMATRIPTAN SUCCINATE, TERFENADINE, THYROXINE, TOLBUTAMIDE, TRAMADOL, WARFARIN
Servanolol → PROPRANOLOL
Serviquin → CHLOROQUINE
Servisone → PREDNISONE
Servizepam → DIAZEPAM
Serzone → NEFAZODONE
SESAME
Sesamin → HALVA (H)
Sesbane → DAUBENTONIA PUMICEA
SESBANIA sp.
Sesquilactones → FEVERFEW
SESSEA BRASILIENSIS
Sestron → ALVERINE
SETARIA sp.
Setewale → VALERIANA sp.
Sethotope → SELENOMETHIONINE

Sethyl → HOMATROPINE
Setonil → DIAZEPAM
Setran → MEPROBAMATE
Settima → PRAZEPAM
Setwall → VALERIANA sp.
Sevenal → PHENOBARBITAL
Serviceberry → AMELANCHIER ALNIFOLIA
Sevin → CARBARYL
Sevinon → UNDECYLENIC ACID
SEVOFLURANE
Sevofrane → SEVOFLURANE
Sexocretin → DIETHYLSTILBESTROL
SF 86-327 → TERBINAFINE
SG 75 → NICORANDIL
Sgd 301-76 → OXICONAZOLE NITRATE
SH 419 → IOGLYCAMIC ACID
SH 567 → METHENOLONE ACETATE
SH 582 → GESTONORONE CAPROATE
SH 617L → IPODATE
SH 717 → GLYMIDINE
SH 723 → MESTEROLONE
SH 714 → CYPROTERONE
SH 863 → CLOCORTOLONE PIVALATE
SH 881 → CYPROTERONE
SH 80881 → CYPROTERONE
Shad-scale → ATRIPLEX sp.
Shakespeare → BURDOCK, CICUTA sp., METHAQUALONE
Shallenberger, Dr. → PAINT
Shamrock → OXALIS sp.
SHARK
Sharon Tate → METHYLENEDIOXYAMPHETAMINE
SHAVE GRASS
SH-E-222 → MEPINDOLOL
Sheep Berry → VIBURNUM PRUNIFOLIUM
Sheep Bur → COCKLEBURS
Sheep Laurel → KALMIA
Sheep Sorrel → RUMEX sp.
Sheets → PHENCYCLIDINE
Shelanski → IODOPHORS
Shell → SUCCINYLCHOLINE, VINCOFOS
Shell 4294 → CIODRIN
SHELLAC
Shell Compound 4072 → CHLORFENVINPHOS
SHELLFISH also → COPPER, GONYAULAX, MUSSELS, SHOES
Shengjiang → GINGER, JAMAICAN
Shepherd's Clock → ANAGALLIS ARVENSIS
Shepherd's Purse → CAPSELLA BURSA-PASTORIS
Shepherd's Weather-glass → ANAGALLIS ARVENSIS
Sherifa → ANONA
Sherpa → CYPERMETHRIN
SHERRY also → HISTAMINE, ISOCARBOXAZID, PROPOXYPHENE
Shesha → CANNABIS
Shg 1942 → BROMOPHOS
Shibti → PHYTOLACCA sp.
Shigatox → SULFAGUANIDINE
Shikimi → ILLICIUM ANISATUM
SHIKIMIC ACID also → BRACKEN FERN
Shimmer-ex → PHENYLMERCURIC ACETATE
Shimshar → BUXUS sp.
Shinseng → GINSENG
Shiomarin → MOXALACTAM
Shionogi 60595 → MOXALACTAM
Shiosol → GOLD SODIUM THIOMALATE
Shirm → PHENCYCLIDINE
Shittim Wood → ACACIA
Shock-Ferol → VITAMIN D_2
"Shock Lung" → PROPYLHEXEDRINE
SHOES also → *p-tert*-BUTYLPHENOL, METHYL BUTYL KETONE
"Shooting Speed" → α-METHYL FENTANYL
Showa Denko KK → L-TRYPTOPHAN
Showy Crotalaria → CROTALARIA
Shoxin → NORBORMIDE
SHRIMP also → SODIUM GLUTAMATE
Sibazon → DIAZEPAM
Sibelium → FLUNARIZINE HCL
Sibol → DIETHYLSTILBESTROL
SIBUTRAMINE
Sicklepod → CASSIA
Sicklepod Senna → CASSIA
Sicol → DIMETHICONE (*with silicon dioxide*)
Sicol 160 → BUTYL BENZYL PHTHALATE
Sicron → POLYVINYL CHLORIDE
Sierra Laurel → LEUCOTHOE
Sigacalm → OXAZEPAM
Sigapedil → ERYTHROMYCIN ETHYL SUCCINATE
Sigaperidol → HALOPERIDOL
Sigaprim → SULFAMETHOXAZOLE
Sigmacort → HYDROCORTISONE
Sigmadyn → PEMOLINE
Sigmamycin → TETRACYCLINE also → OLEANDOMYCIN
Sigmaopen → PENICILLINS – "NEWER and SYNTHETICS" > *Amoxicillin*
Sigmart → NICORANDIL
Signal Grass → BRACHIARIA
Sijeh → MUCUNA sp.
Silain → DIMETHICONE (*with silicon dioxide*)
Silamox → PENICILLINS – "NEWER and SYNTHETICS" > *Amoxicillin*
Silastic → RUBBER

Silastic 200 → DIMETHICONE (*with silicon dioxide*)
Silbephylline → DYPHYLLINE
Silcron → SILICON DIOXIDE
SILDENAFIL also → NITROGLYCERIN
Silentium → DEXTROMETHORPHAN HYDROBROMIDE
Silbesan → CHLOROQUINE
Silibrin → CHLORDIAZEPOXIDE
Silica → SILICON DIOXIDE also → CONCRETE, SILICON
Siliceous Earth → DIATOMACEOUS EARTHS
Silicium Oxide → SILICON DIOXIDE
Silico → COAL
SILICON also → DUSTS
Silicon Chloride → SILICON TETRACHLORIDE
SILICON DIOXIDE also → SPONGES
SILICONES also → EQUISETUM, GOLF BALLS, LIPSTICK, POLYURETHANE, PUTTY
Silicon Fluoride → SILICON TETRAFLUORIDE
SILICON TETRACHLORIDE
SILICON TETRAFLUORIDE
Silicote → DIMETHICONE (*with silicon dioxide*)
Silizaz → DIMETHICONE (*with silicon dioxide*)
SILK also → HYDROGEN CYANIDE
Silkweed → ASCLEPIAS
Silky Sophora → SOPHORA sp.
Silly Putty® → PUTTY also → BORIC ACID
SILO
"Silo-filler's Disease" → NITROGEN DIOXIDE
Silosan → PIRIMIPHOS METHYL
Silubin → BUFORMIN
Silvadene → SILVER SULFADIAZINE
Silvanol → LINDANE
SILVER also → ACETYLENE, SELENIUM, SOLDER
Silver Cyanide → SODIUM HYDROXIDE
SILVER IODIDE
Silver-Leafed Nightshade → SOLANUM sp.
Silver Liana → MASCAGNIA
Silverling → BACCHARIS sp.
SILVER NITRATE also → INKS, POTASSIUM PERMANGATE
SILVER PROTEIN, MILD
Silverstone → POLYTEF
SILVER SULFADIAZINE also → PROPYLENE GLYCOL
Silvery Horsebrush → TETRADYMA sp.
Silver Vitellin → SILVER PROTEIN, MILD
SILVEX
Silvisar → SODIUM CACODYLATE also → MONOSODIUM METHANEARSONATE, CACODYLIC ACID
Simatin → ETHOSUXIMIDE
Simax → CIODRIN
SIMAZINE
Simeon → PROSCILLARIDIN
Simeskellina → KHELLIN
Simethicone, activated → DIMETHICONE (*with silicon dioxide*)
Simmondsia chinensis → JOJOBA BEAN
Simovil → SIMVASTATIN
Simoxil → PENICILLINS – "NEWER and SYNTHETICS" > *Amoxicillin*
Simpalon → SYNEPHRINE
Simpamina → AMPHETAMINE
Simpamina-D → DEXTROAMPHETAMINE SULFATE
Simpatedrin → AMPHETAMINE
Simplene → EPINEPHRINE
Simplotan → TINIDAZOLE
Simulect → BASILIXIMAB
SIMVASTATIN also → CLOZAPINE, CYCLOSPORINE, ERYTHROMYCIN, GEMFIBROZIL, ITRACONAZOLE, MIBEFRADIL, NEFAZODONE, VERAPAMIL
Sinalbin → MUSTARD
Sinan → MEPHENESIN
Sinapis alba → MUSTARD
Sinapsis arvensis → MUSTARD
Sincomen → SPIRONOLACTONE
Sincoumar → ACENOCOUMAROL
Sindesvel → METHAQUALONE
Sindiatil → BUFORMIN
Sinequan → DOXEPIN
Sinerol → OXYMETAZOLINE
Sinesalin → BENDROFLUMETHIAZIDE
Sinestrol → DIETHYLSTILBESTROL
Sinetens → PRAZOSIN HYDROCHLORIDE
Sinevir → CYPROTERONE
Sinex → OXYMETAZOLINE
Singletary Pea → LATHYRUS sp.
Singoserp → SYROSINGOPINE
Singulair → MONTELUKAST
SINIGRIN also → BRUSSEL SPROUTS, CABBAGE, CAULIFLOWER, MUSTARD
Sinituho → PENTACHLOROPHENOL
Sinlestal → PROBUCOL
Sinnamin → APAZONE
Sinoflurol → TEGAFUR
Sinogan-Debil → METHOTRIMEPRAZINE
Sinomin → SULFAMETHOXAZOLE
Sinos → CYCLOPENTAMINE
Sinosid → PAROMOMYCIN
Sinox → DINITRO-σ-CRESOL
Sinox W → DINOSEB

Sintisone → PREDNISOLONE
Sintofolin → HEXESTROL
Sintolexyn → CEPHALEXIN
Sintomicetina → CHLORAMPHENICOL
Sintomycin → CHLORAMPHENICOL
Sintotrat → HYDROCORTISONE
Sintrom → ACENOCOUMAROL
Sinufed → PSEUDOEPHEDRINE
Sinvacor → SIMVASTATIN
Siogène → CHLORQUINALDOL
Sionon → SORBITOL
Siosteran → CHLORQUINALDOL
Siplarol → FLUPENTIXOL
Siptazin → CINNARIZINE
Siqualine → FLUPHENAZINE DECANOATE
Siqualone → FLUPHENAZINE DECANOATE
Siquil → TRIFLUPROMAZINE
Siragan → CHLOROQUINE
Sirdalud → TIZANIDINE
Sir Henry Morton Stanley → COFFEE
Siringina → SYROSINGOPINE
Sirlene → PROPYLENE GLYCOL
Sirolimus → RAPAMYCIN
Siros → ITRACONAZOLE
Sirtal → CARBAMAZEPINE
Siseptin → SISOMICIN
Sisobiotic → SISOMICIN
Sisolline → SISOMICIN
SISOMICIN
Sisomin → SISOMICIN
Sistan → METHAM SODIUM
Sisuril → HYDROFLUMETHIAZIDE
Sitosterol → POTATO
β-Sitosterol → BEETS
Sittamisiri → SUCROSE
SIUM SUAVE
Sivastin → SIMVASTATIN
Sivlor → LOVASTATIN
60 Minutes → FLECAINIDE
SJ 1977 → METHIXENE
SK 65 → PROPOXYPHENE
SK 331-A → XANTHINOL NIACINATE
ska Maria Pastora → SALVIA
SKATOLE
SK-Bamate → MEPROBAMATE
Skedule → CHLORMADINONE
Skelaxin → METAXALONE
Skelid → TILUDRONATE
Sk-Erythromycin → ERYTHROMYCIN STEARATE
SKF-trans-385 → TRANYLCYPROMINE
SKF 478 → DIPHENIDOL
SKF 478-A → DIPHENIDOL
SKF 1700-A → NYLIDRIN
SKF 2601-A → CHLORPROMAZINE
SKF 4657 → PROCHLORPERAZINE
SKF 5137 → DEXTROMORAMIDE
SKF 6574 → PHENAZOCINE HYDROBROMIDE
SKF 7988 → VIRGINIAMYCIN
SKF 8542 → TRIAMTERENE
SKF 12,141 → PENICILLINS – "NEWER and SYNTHETICS" > *Diphenicillin Sodium*
SKF 14,287 → IDOXURIDINE
SKF 14,336 → CLOMACRON
SKF 18,667 → POLOXALENE
SKF 20,716 → PROPERICIAZINE
SKF 23,880-A → CYCLOOCTYLAMINE
SKF 29,044 → PARBENDAZOLE
SKF 30,310 → OXIBENDAZOLE
SKF 33,134A → AMIODARONE
SKF 39,162 → AURANOFIN
SKF 40,383 → CARBUTEROL
SKF 41,558 → CEFAZOLIN SODIUM
SKF 62,698 → TICRYNAFEN
SKF 62,979 → ALBENDAZOLE
SKF 82,526-J → FENOLDOPAM MESYLATE
SKF 83,088 → CEFMETAZOLE
SKF 88,373 → CEFTIZOXIME
SKF 92,058 → METIAMIDE
SKF 92,334 → CIMETIDINE
SKF 100,168-A → IBOPAMINE
SKF 101,468 → ROPINIROLE
SKF 102,886 → HALOFANTRINE
SKF 104,864 → TOPOTECAN
Skikimmi → SHIKIMIC ACID
Skimmi → SHIKIMIC ACID also → ILLICIUM ANISATUM
Skinoren → AZELAIC ACID
"Skin-poppers" → HEROIN
Skiodan → METHIODAL SODIUM
Sklerolip → CLOFIBRATE
Skleromexe → CLOFIBRATE
Sklero-Tablinen → CLOFIBRATE
SK-Lygen → CHLORDIAZEPOXIDE
Skoke → PHYTOLACCA sp.
Skopolate → SCOPOLAMINE
Skota → SKUNK CABBAGE
SK-Penicillin G → PENICILLIN > *Potassium Benzylpenicillinate*
Skullcap → SCUTELLARIA LATERIFLORA
"Skunk" → CANNABIS
SKUNK CABBAGE → SYMPLOCARPUS also → CALCIUM OXALATE, VERATRUM sp.
Skunk Weed → SKUNK CABBAGE
SL-80.0750-23N → ZOLPIDEM TARTRATE
SLAFRAMINE also → RHIZOCTONIA LEGUMINICOLA

Slag Wool → MINERAL WOOL
Slaked Lime → CALCIUM HYDROXIDE
Slangkop → URGINEA sp.
Slathion → PARATHION
Sleep-Eze® → METHAPYRILENE
Sleepinal → METHAQUALONE
Sleeping Buddha → ESTAZOLAM
Sleepwell → METHAPYRILENE
Sleepy Grass → STIPA sp.
Slinkweed → GUTIERREZIA MICROCEPHALA
Slobber Factor → SLAFRAMINE also → RHIZOCTONIA LEGUMINICOLA
Slo-Bid → THEOPHYLLINE
Slo-Mag → MAGNESIUM CHLORIDE
Slo-Phyllin → THEOPHYLLINE
Sloprolol → PROPRANOLOL
Slow-K → POTASSIUM CHLORIDE
Slow-Mag® → MAGNESIUM CHLORIDE
Slow-Pren → OXPRENOLOL HYDROCHLORIDE
Smack → HEROIN
Smallhead Sneezeweed → HELENIUM sp.
Small Kalmia → KALMIA
Small Tobacco → NICOTIANA sp.
Smartweeds → POLYGONUM sp.
SMCO → *S*-METHYL CYSTEINE SULFOXIDE also → KALE
SMDC → METHAM SODIUM
S-mez → SULFAMETHAZINE
Smilagenin → SMILAX sp.
SMILAX sp.
Smilax aristolochiaefolia → SARSAPARILLA
Smite → SODIUM AZIDE also → POTASSIUM AZIDE
SmithKline → TICRYNAFEN
SmithKline Beecham → LOTRAFIBAN
Smith, Kline, & French Labs → TRIFLUOPERAZINE
Smithsonite → ZINC
Smoke → CANNABIS
Smoke Dust → PHENCYCLIDINE
Smoking/Smokers → CAFFEINE, CANNABIS, CYANIDES, NICOTINE and SALTS, PENTAZOCINE, PHENYLBUTAZONE, PROPOXYPHENE, TOBACCO
Smooth Puffer → LAGOCEPHALUS sp.
SMS 201-995 → OCTREOTIDE
Smut → USTILAGO
SN 307 → ONDANSETRON
SN 390 → QUINACRINE
SN 612 → ETHOXAZENE
SN 3115 → PLASMOCID
SN 6718 → CHLOROQUINE
SN 6771 → BIALAMICOL
SN 10,751 → AMODIAQUIN
SN 11,841 → AMSACRINE
SN 12,837 → CHLOROGUANIDE
SN 13,272 → PRIMAQUINE
SN 13,276 → PENTAQUINE
SN 34,615 → PROMECARB
SNAKE(S)
Snake Plant → SANSEVIERA sp.
SNAKEROOT, WHITE
Snake's Head → FRITILLARIA MELEAGRIS
Snakeweed → BROOMWEED also → GUTIERREZIA MICROCEPHALA, POLYGONUM sp.
Snakewort → POLYGONUM sp.
Snappers → AMYL NITRITE also → BUTYL NITRITE
SND 919 → PRAMIPEXOLE
Sneeze Powders → CONVALLARIA MAJALIS
Sneezeweed → CONVALLARIA MAJALIS
S. N. G. → NITROGLYCERIN
Snip → DIMETILAN
Sno Phenicol → CHLORAMPHENICOL
Snorts → PHENCYCLIDINE
Sno Tears → POLYVINYL ALCOHOL
"Snow" → COCAINE also → HEROIN
SNOWBERRY
Snowbirds → COCAINE
Snowdrop → GALANTHUS NIVALIS
Snow-On-The-Mountain → EUPHORBIA, *marginata*
SNP → PARATHION
SNUFF
"Snuff-taker's Lung" → PITUITARY
Soamin → SODIUM ARSANILATE
SOAP also → SODIUM PERBORATE
Soap Bark → QUILLAJA
Soap Berry → SAPINDUS sp.
Soapberry Tree → BALANITES sp.
Soapers → METHAQUALONE
Soapwort → BOUNCING BET
Sobelin → CLINDAMYCIN
Sobrero → NITROGLYCERIN
Sobril → OXAZEPAM
Socrates → CONIUM MACULATUM
Soda Ash → SODIUM CARBONATE
Soda Lye → SODIUM HYDROXIDE
Sodanton → PHENYTOIN
Sodar → DISODIUM METHANEARSONATE also → METHANEARSONIC ACID
Sodelut G → MEDROXYPROGESTERONE ACETATE
Sodestrin-H → ESTROGENS, CONJUGATED

SODIUM also → ALCOHOL, BENZYL; LITHIUM; PENICILLIN > *Benzylpenicillin, Sodium;* PICKLES, POTASSIUM, PREDNISONE, SAUERKRAUT, SODIUM BICARBONATE, SODIUM SULFATE
SODIUM ACETRIZOATE
Sodium-2-N-acetylsulfamyl-4,4´-diaminodiphenylsulfone → ACETOSULPHONE SODIUM
Sodium Acid Carbonate → SODIUM BENZOATE
Sodium Acid Phosphate → SODIUM PHOSPHATE, MONOBASIC
Sodium Acid Pyrophosphate → SAUSAGE
Sodium Acid Sulfite → SODIUM BISULFITE
SODIUM ALGINATE
Sodium Alum → ALUMINUM SODIUM SULFATE
Sodium Anilarsonate → SODIUM ARSANILATE
SODIUM ARACHIDONATE
SODIUM ARSANILATE
SODIUM ARSENATE
SODIUM ARSENITE also → PHOSPHORIC ACID
Sodium Aurothiomalate → GOLD SODIUM THIOMALATE
Sodium Aurothiosulfate → GOLD SODIUM THIOSULFATE
SODIUM AZIDE also → WINE
SODIUM BENZOATE also → CHEESES, DIAZEPAM, IBUPROFEN, KETOPROFEN, MARGARINE, MUSTARD, *N*-NITROSODIMETHYLAMINE, PROPOXYPHENE, TARTRAZINE
Sodium Biborate → BORAX
SODIUM BICARBONATE also → ACETALDEHYDE, AMPHETAMINE, ANTIPYRINE, CHLOROQUINE, DEXTROAMPHETAMINE, FENFLURAMINE, MECAMYLAMINE, MEPERIDINE, METHACYCLINE, METHENAMINE, NALIDIXIC ACID, NITROFURANTOIN, PHENOBARBITAL, PHENYLBUTAZONE, PRAZEPAM, PROCAINAMIDE, PROCAINE, PROPANTHELINE, QUINIDINE, QUININE, SALICYLIC ACID, TETRACYCLINE, THEOPHYLLINE
Sodium Bichromate → SODIUM DICHROMATE
Sodium Biphosphate → SODIUM PHOSPHATE, MONOBASIC also → PHOSPHATES
SODIUM BISULFITE also → CISPLATIN, ISOETHARINE, METOCLOPRAMIDE, MORPHINE, SULFUR DIOXIDE
Sodium Borate → BORAX
SODIUM BROMIDE also → PHENOBARBITAL
SODIUM CACODYLATE
SODIUM CARBONATE
SODIUM CARBOXYMETHYLCELLULOSE
Sodium CEZ → CEFAZOLIN SODIUM
SODIUM CHLORATE
SODIUM CHLORIDE also → LITHIUM, LITHIUM CHLORIDE, SILVER NITRATE
SODIUM CHROMATE also → SODIUM CHLORIDE
SODIUM CITRATE
Sodium Cromoglycate → CROMOLYN SODIUM
SODIUM CYANATE also → CYANATES
SODIUM CYANIDE
SODIUM CYCLAMATE also → LINCOMYCIN, TOLBUTAMIDE
SODIUM DEHYDROCHOLATE also → DEHYDROCHOLIC ACID
Sodium Diatrizoate → MERALLURIDE
SODIUM DICHROMATE
Sodium Dihydrogen Phosphate → SODIUM PHOSPHATE, MONOBASIC
Sodium Dimethylarsenate → SODIUM CACODYLATE
Sodium Dioctyl Sulfosuccinate → DOCUSATE SODIUM
Sodium Disulfide → LETTUCE
Sodium Dodecyl Sulfate → SODIUM LAURYL SULFATE
SODIUM EDETATE also → GRAPE
Sodium Edethamil → SODIUM EDETATE
Sodium EDTA → SODIUM EDETATE
Sodium Ethylmercurithiosalicylate → THIMEROSAL
SODIUM ETIDRONATE
SODIUM FLUORIDE also → HYDROFLUORIC ACID, ORANGE
Sodium Fluosilicate → SODIUM SILICOFLUORIDE
SODIUM GLUTAMATE also → ACETAZOLAMIDE, DILL, SHRIMP
Sodium Hexafluorosilicate → SODIUM SILICOFLUORIDE
Sodium Hexametaphosphate → SODIUM CHLORIDE
SODIUM HYALURONATE
Sodium Hydrate → SODIUM HYDROXIDE
Sodium Hydrogen Bis(2-propylpentanoate) → VALPROIC ACID
SODIUM HYDROXIDE also → ARSINE, HAIR, PRETZELS, SODIUM HYPOCHLORITE
Sodium γ-Hydroxybutyrate → SODIUM

Solanum tuberosum → POTATO
Solapsone → SOLASULFONE
Solaskil → LEVAMISOLE
Solasonine → SOLANUM sp.
Solasoldine → CESTRUM
Solaspin → CALCIUM ACETYLSALICYLATE
SOLASULFONE
Solatran → KETAZOLAM
Solatunine → SOLANINE
Solaxin → CHLORZOXAZONE
Soldep → TRICHLORFON
SOLDER
Soles → CANNABIS
Solestril → PROSCILLARIDIN
Solevar → NORETHANDROLONE
Solfocrisol → GOLD SODIUM THIOSULFATE
Solfoton → PHENOBARBITAL
Solganal → AUROTHIOGLUCOSE
Solganal B → AUROTHIOGLUCOSE
Solgol → NADOLOL
Solidago virgaurea → GOLDENROD
Solid Green → BRILLIANT GREEN
Soliphylline → CHOLINE THEOPHYLLINATE
Solis → DIAZEPAM
Soliwax → DOCUSATE SODIUM
Solocalm → PIROXICAM
Solodelf → TRIAMCINOLONE
Solomon's Seal → POLYGONATUM BIFLORUM
Solone → PREDNISOLONE
Solosin → THEOPHYLLINE
Soloxine → THYROXINE
Solozone → SODIUM PEROXIDE
Solprin → ACETYLSALICYLIC ACID also → CALCIUM ACETYLSALICYLATE
Solprina → CHLOROQUINE
Solpyron → ACETYLSALICYLIC ACID
Solubacter → TRICLOCARBAN
Solucort → PREDNISOLONE
Solu-Cortef → HYDROCORTISONE
Solu-Decortin-H → PREDNISOLONE
Solu-Delta-Cortef → PREDNISOLONE
Soluglaucit → DIETHYL-*p*-NITROPHENYLPHOSPAHTE
Solufilin → DYPHYLLINE
Solufyllin → DYPHYLLINE
Solu-Glyc → HYDROCORTISONE
Solu-Medrol → METHYLPREDNISOLONE
Solupen → PENICILLIN > *Potassium Benzylpenicillinate*
Solupred → PREDNISOLONE
Solu-Predalone → PREDNISOLONE
Solutrast → IOPAMIDOL
Solvan → DIPHENADIONE
Solvay Ash → SODIUM CARBONATE
Solvazinc → ZINC
Solvent Blue 8 → METHYLENE BLUE
Solvent Naphtha → NAPHTHA
Solvent Orange 3 → 2,4-DIAMINOAZOBENZENE
Solvent Yellow 2 → AURAMINE
Solvent Yellow 3 → σ-AMINOAZOTOLUENE
Solvezink → ZINC
Solvocin → TETRACYCLINE
Soma → CARISPRODOL
Soma → PHENCYCLIDINE
Somacton → GROWTH HORMONES
Somadril → CARISPRODOL
Somagerol → LORAZEPAM
Somagrebove → GROWTH HORMONES
Somalapor → GROWTH HORMONES
Somalgit → CARISPRODOL
SOMAN
Somatofalk → SOMATOSTATIN
Somato-gen → GROWTH HORMONES
Somatomax PM → SODIUM OXYBATE
Somatonorm → GROWTH HORMONES
SOMATOSTATIN also → INSULIN
Somatotropic Hormone → GROWTH HORMONES
Somatotropin → GROWTH HORMONES also → PIMOZIDE
Somatotropin Release Inhibiting Factor → SOMATOSTATIN
SOMATREM also → GROWTH HORMONES
Somavubove → GROWTH HORMONES
Somberol → METHAQUALONE
Sombrevin → PROPANIDID
Sombucaps → HEXOBARBITAL
Sombulex → HEXOBARBITAL
Somenopor → GROWTH HORMONES
Somese → TRIAZOLAM
Sometan → METHAM SODIUM
Sometribove → GROWTH HORMONES
Sometripor → GROWTH HORMONES
Somfasepor → GROWTH HORMONES
Somidobove → GROWTH HORMONES
Somilan → CHLORAL BETAINE
Sominat → DICHLORALANTIPYRINE
Somio → CHLORALOSE
Somipront → DIMETHYL SULFOXIDE
Somnafac → METHAQUALONE
Somnal → AMOBARBITAL
Somnalert → HEXOBARBITAL
Somnased → NITRAZEPAM
Somnatrol → ESTAZOLAM
Somnesin → MEPARFYNOL
Somnibel → NITRAZEPAM

Sommipron → APROBARBITAL
Somnite → NITRAZEPAM
Somnium → METHAQUALONE
Somnomed → METHAQUALONE
Somnopentyl → PENTOBARBITAL
Somnos → CHLORAL HYDRATE
Somnothane → HALOTHANE
Somnox → CHLORAL HYDRATE
Somnurol → BROMISOVALUM
Somonil → METHIDATHION
Somophyllin → AMINOPHYLLINE
Somophyllin T → THEOPHYLLINE
Sonacide → GLUTARALDEHYDE
Sonaform → CYCLOBARBITAL
Sonal → METHAQUALONE
Sonate → SODIUM ARSANILATE
Sonazine → CHLORPROMAZINE
Sone → PREDNISONE
Sonebon → NITRAZEPAM
Soneryl → BUTETHAL
Songar → TRIAZOLAM
Sonimen → CHLORDIAZEPOXIDE
Soni-Slo → ISOSORBIDE DINITRATE
Sonuctane → VINBARBITAL
SOOT
Sopangaminé → VITAMIN B_{15}
Sopaquin → CHLOROQUINE
Sopari → ARECA
Sopental → PENTOBARBITAL
Sopentyl → PENTOBARBITAL
Sopes → METHAQUALONE
Sophiamin → CHLORDIAZEPOXIDE
SOPHORA sp.
Sophorine → CYTISINE
Sopor → METHAQUALONE
Soprathion → PARATHION
Soprol → BISOPROLOL
Soprontin → ACEPROMAZINE
Sorbangil → ISOSORBIDE DINITRATE
SORBIC ACID
Sorbichew → ISOSORBIDE DINITRATE
Sorbidilat → ISOSORBIDE DINITRATE
Sorbid SA → ISOSORBIDE DINITRATE
Sorbilande → SORBITOL
Sorbilax → SORBITOL
SORBINIL
Sorbistat → SORBIC ACID also → POTASSIUM SORBATE
Sorbit → SORBITOL
SORBITAN TRIOLEATE
SORBITOL also → PHENOBARBITAL, SAUSAGE, SODIUM POLYSTYRENE SULFONATE, WINE
Sorbitrate → ISOSORBIDE DINITRATE
Sorbitur → SORBITOL
Sorbo → SORBITOL
Sorbonit → ISOSORBIDE DINITRATE
Sorbostyl → SORBITOL
Sorboxethenes → POLYSORBATES
Sordenac → CLOPENTHIXOL
Sordinal → CLOPENTHIXOL
SORGHASTRUM NUTANS
SORGHUM also → CYANATES, CYANIDES, DHURRIN, HYDROGEN CYANIDE, MILLET, NITRATES, NITRITES
Sorgo → SORGHUM
Soriatane → ACITRETIN
Soriflor → DIFLORASONE
Sormetal → CALCIUM DISODIUM EDETATE
Sorot → DEQUALINIUMS, *chloride*
Sorquad → ISOSORBIDE DINITRATE
Sorquetan → TINIDAZOLE
Sorrel → OXALIS sp.
Sorrels → RUMEX sp.
Sorreltree → ANDROMEDA sp.
SORSACA
Sorsaka → SORSACA
Sortis → ATORVASTATIN
Sorunex → HEXYLRESORCINOL
Sosegon → PENTAZOCINE
Sosol → SULFISOXAZOLE
Sostril → RANITIDINE
Sotacor → SOTALOL
Sotalex → SOTALOL
SOTALOL also → CLONIDINE, FLUOXETINE, MILK
SOTERENOL
Sotradecol → SODIUM TETRADECYL SULFATE
Souframine → PHENOTHIAZINE
Sour Dock → RUMEX sp.
Sour-grass → ARROWGRASS
Soursob → OXALIS sp.
Soursop → SORSACA
Sour Wood → ANDROMEDA sp.
South American Clump Palm → PALMS
South American Eggplant → SOLANUM sp.
South American Holly → ILEX sp.
SOUTHERN PRICKLY ASH
Sovcain → CINCHOPHEN
Sovcaine → DIBUCAINE
Soventol → BAMIPINE
Soverin → METHAQUALONE
Sowbread → CYCLAMEN
Sowell → MEPROBAMATE
Soxisol → SULFISOXAZOLE
Soxomide → SULFISOXAZOLE
Soxysympamine → METHAMPHETAMINE

SOYBEANS also → METAPROTERENOL, PEANUTS, THYROID
SOYBEAN OIL also → PROPOFOL
SP 54 → PENTOSAN POLYSULFATE
SP 732 → PROLINTANE
Spabucol → TRIMEBUTINE MALEATE
Spacolin → ALVERINE
Spametrin-M → METHYLERGONOVINE
Spandelle → POLYURETHANES
Spaderizine → CINNARIZINE
Span 85 → SORBITAN TRIOLEATE
Spanbolet → SULFAMETHAZINE
Spandex® → POLYURETHANES also → 2-MERCAPTOBENZOTHIAZOLE
Spanish Broom → SPARTIUM JUNCEUM
Spanish Fly → CANTHARIDES
Span-K → POTASSIUM CHLORIDE
Spanon → CHLORDIMEFORM
Spanor → DOXYCYCLINE
"Spans" → GRAMICIDIN
Spar → CALCIUM CARBONATE
Spara → SPARFLOXACIN
SPARFLOXACIN also → SUCRALFATE
Sparine → PROMAZINE
Sparsamycin A → TUBERCIDIN
Spartakon → LEVAMISOLE
SPARTEINE also → BAPTISIA, ERGONOVINE, SPARTIUM JUNCEUM, THERMOPSIS sp.
SPARTIUM JUNCEUM
Spartocin → SPARTEINE
Spasfon-Lyoc → PHLOROGLUCINOL
Spasmaverine → ALVERINE
Spasmocyclon → CYCLANDELATE
Spasmolyn → MEPHENESIN
Spasmo-Nit → PAPAVERINE
Spasmophen → OXYPHENONIUM BROMIDE
Spasmopriv → FENOVERINE
Spasuret → FLAVOXATE HCL
SPATHIPHYLLUM sp. also → LILLIES
Spatonin → DIETHYLCARBAMAZINE
SPE 2792 → ANISINDIONE
Spearworts → RANUNCULUS
"Special K" → KETAMINE
Special LA Coke → KETAMINE
Specifin → NALIDIXIC ACID
Speckled Jewels → IMPATIENTS sp.
Spectacillin → PENICILLINS – "NEWER and SYNTHETICS" > *Epicillin*
Spectam → SPECTINOMYCIN
Spectazole → ECONAZOLE NITRATE
SPECTINOMYCIN also → LITHIUM, NEOMYCIN
Spectogard → SPECTINOMYCIN
Spectracide → DIAZINON
Spectramedryn → MEDRYSONE
Spectramox → PENICILLINS – "NEWER and SYNTHETICS" > *Amoxicillin*
Spectra-Sorb → OXYBENZONE
Spectra-Sorb UV 24 → DIOXYBENZONE
Spectra-Sorb UV 284 → SULISOBENZONE
Spectrobid → PENICILLINS – "NEWER and SYNTHETICS" > *Bacampicillin*
Spectrum → CEFTAZIMIDES
Spedifen → IBUPROFEN
Speed → METHAMPHETAMINE
"Speed Balls" → HEROIN also → NITROGLYCERIN
Spenoxin → METHOCARBAMOL
SPERMACETI
Spermwax → SPERMACETI
Spewing Sickness → HELENIUM sp.
SPHAGNUM MOSS
Spheroidine → TETRODOTOXIN
Spheromycin → NOVOBIOCIN
Sphingomyelinase → SPIDERS
Spicebush → CALYCANTHUS FLORIDUS
Spider Lily → HYMENOCALLIS OCCIDENTALIS
Spider Pea → CASSIA
SPIDERS
Spikenard → ARALIA sp.
Spinacane → SQUALANE
SPINACH also → ANTHRAQUINONE, NITRATES, NITRITES, PHENCYCLIDINE, PHYTOLACCA sp., POTASSIUM NITRATE
Spinacia oleracea → SPINACH
Spindle Tree → EUONYMUS
Spindle Wood → EUONYMUS
Spineless Horsebrush → TATRADYMIA sp. also → TETRADYMA sp.
Spiny Dogfish → SQUALUS ACANTHIUS
SPIPERONE
Spiracine → PARSONSIA sp.
Spiraline → PARSONSIA sp.
Spiramin → TRANEXAMIC ACID
SPIRAMYCIN also → CARBOMYCIN
Spiretic → SPIRONOLACTONE
Spirin → SALICYLIC ACID
Spirit of Salt → HYDROCHLORIC ACID
Spirocid → ACETARSONE
Spirocort → BUDESONIDE
Spiroctan → SPIRONOLACTONE
Spirodecanone → SPIPERONE
Spiroderm → SPIRONOLACTONE
Spirofulvin → GRISEOFULVIN
Spirolone → SPIRONOLACTONE

STACHYS ARVENSIS
Stadadorm → AMOBARBITAL
Stadalax → BISACODYL
Stadol → BUTORPHANOL
Stafac → VIRGINIAMYCIN
Staff of Life → VITAMIN O
Stafoxil → PENICILLINS – "NEWER and SYNTHETICS" > *Floxacillin*
Stag Bush → VIBURNUM PRUNIFOLIUM
"Staggering Grain Toxicosis" → T-2 TOXIN
Stagger Weed → DICENTRA CANADENSIS
Staghorn Sumac → SUMAC, POISON
St. Anthony's Fire → BREAD
St. Bennet's Herb → CONIUM MACULATUM
St. George Disease → PIMELIA sp.
St. John's Bread → CAROB
St. John's Wort → HYPERICUM also → WARFARIN
St. John the Conqueror Root → JALAP
St. Lucia Grass → BRACHIARIA
Stalleril → THIORIDAZINE
Stam → PROPANIL
Stamine → PYRILAMINE
Stampede → PROPANIL
Stampen → PENICILLINS – "NEWER and SYNTHETICS" > *Dicloxacillin*
STAMPS and STAMP PADS
Stanazol → STANOZOLOL
Stangen → PYRILAMINE
Stanilo → SPECTINOMYCIN
STANLEYA PINNATA
Stanley Livingston → COFFEE
Stan-Mag → MAGNESIUM OXIDE
Stannochlor → STANNOUS CHLORIDE
STANNOUS CHLORIDE
STANNOUS FLUORIDE
STANOLONE
Stanomycetin → CHLORAMPHENICOL
STANOZOLOL also → WARFARIN
Stanquinate → IODOQUINOL
Stapenor → PENICILLINS – "NEWER and SYNTHETICS" > *Oxacillin Sodium*
Stapf → SORGHUM
Staphcil → PENICILLINS – "NEWER and SYNTHETICS" > *Floxacillin*
Staphcillin → PENICILLINS – "NEWER and SYNTHETICS" > *Methicillin Sodium*
Staphlipen → PENICILLINS – "NEWER and SYNTHETICS" > *Floxacillin*
Staphlex → PENICILLINS – "NEWER and SYNTHETICS" > *Floxacillin*
Staphlomycin → VIRGINIAMYCIN
Staphobristol-250 → PENICILLINS – "NEWER and SYNTHETICS" > *Cloxacillin/Sodium salt*
Staphybiotic → PENICILLINS – "NEWER and SYNTHETICS" > *Cloxacillin/Sodium salt*
Starazin → PROMAZINE
Star of Bethlehem → ORNITHOGALUM UMBELLATUM
STARCHES also → METHYLPHENIDATE
Starchwort → ARISAEMA
Stardust → COCAINE
STARFISHES
Star Grass → CYNODON
Staril → FOSFINOPRIL SODIUM
Starlicide → 3-CHLORO-*p*-TOLUIDINE
Staticin → ERYTHROMYCIN
Statimo → CARBAZOCHROME SALICYLATE
Statran → EMYLCAMATE
Stauffer 2061 → PEBULATE
Stauffer R-1303 → CARBOPHENOTHION
Stauroderm → FLURAZEPAM
STAVUDINE
Stayban → FLURBIPROFEN
Staycept → NONOXYNOL-9
Stazepine → CARBAMAZEPINE
Steapsin → LIPASE
STEARIC ACID
Stearyl Mercaptan → OCTADECANETHIOL
Steatoda paykulliana → SPIDERS
Steclin → TETRACYCLINE
Stecsolin → OXYTETRACYCLINE
STEEL
STEGANOTAENIA ARALIACEA
STEI-300 → HYPERICUM
Steladone → CHLORFENVINPHOS
Stelazine → TRIFLUOPERAZINE DIHYDROCHLORIDE
Stellamicina → ERYTHROMYCIN ESTOLATE
Stellarid → PROSCILLARIDIN
Stemetil → PROCHLORPERAZINE
Stemex → PARAMETHASONE ACETATE
STENBOLONE
Stenediol → METHANDRIOL
Stenolon → METHANDROSTENOLONE
Stenovasan → AMINOPHYLLINE
Stepanol WA 100 → SODIUM LAURYL SULFATE
STEPHANIA TETRANDA
Sterane → PREDNISOLONE
Sterax → DESONIDE
STERCULIA sp.
Sterculic Acid → OKRA also → MILK
STERIGMATOCYSTIN
Sterilon → CHLORHEXIDINE
Steripaque → BARIUM SULFATE
Sterlane → LAURYL OXYPROPYLAMINOBUTYRIC ACID

STERNO® also → ALCOHOL, ETHYL
Steroderm → DESONIDE
Sterogyl → VITAMIN D_2
Sterolone → PREDNISOLONE
Sterosan → CHLORQUINALDOL
Steroxin → CHLORQUINALDOL
Sterzac → TRICLOSAN
Stesolid → DIAZEPAM
Stevacin → OXYTETRACYCLINE
STH → GROWTH HORMONES
Stibanate → ANTIMONY SODIUM GLUCONATE
Stibine → ANTIMONY TRIOXIDE
Stibnite → ANTIMONY also → ANTIMONY TRISULFIDE
STIBOCAPTATE
Stibogluconate Sodium → ANTIMONY SODIUM GLUCONATE
STIBOPHEN
Stick → CANNABIS
Stick Button → BURDOCK
Stickerweed → SOLANUM sp.
Stickstofflost → MECHLORETHAMINE
Stickwort → AGRIMONIA EUPATORIA
Stiedex → DESOXIMETASONE
Stiemycin → ERYTHROMYCIN
Stiglyn → NEOSTIGMINE
Stigmasterol → POTATO
Stilalgin → MEPHENESIN
Stilamin → SOMATOSTATIN
STILBAMIDINE
STILBAZIUM IODIDE
Stilbestronate → DIETHYLSTILBESTROL
Stilbetin → DIETHYLSTILBESTROL
Stilboefral → DIETHYLSTILBESTROL
Stilboestroform → DIETHYLSTILBESTROL
Stilboestrol DP → DIETHYLSTILBESTROL
Stilbofax → DIETHYLSTILBESTROL
Stilkap → DIETHYLSTILBESTROL
Stillacor → DIGOXIN
Stillargol → SILVER PROTEIN, MILD
STILLINGIA sp.
Stillman's Bleach Cream → MERCURY, AMMONIATED
Stilnoct → ZOLPIDEM TARTRATE
Stilnox → ZOLPIDEM TARTRATE
Stilpalmitate → DIETHYLSTILBESTROL
Stilronate → DIETHYLSTILBESTROL
Stilny → NORDAZEPAM
Stimul → PEMOLINE
Stimulexin → DOXAPRAM
Stimu-LH → LH-RH
Stimulin → NIKETHAMIDE
Stimu-TSH → THYROTROPIN RELEASING HORMONE
Stinerval → PHENELZINE
Stingaree → STINGRAYS
Stinging Nettle → URTICA DIOICA also → HISTAMINE
STINGRAYS
Stinkblaar → DATURA, *strammonium*
Stink Bombs → HYDROGEN SULFIDE
Stinking Weed → CASSIA, *occidentalis*
Stinking Willie → SENECIO sp.
Stink Tree → ORIXA JAPONICA THUMB
Stinkweed → DATURA, *strammonium*; THLASPI ARVENSE
Stinkwood → ZIERIA ARBORESCENS
Stinkwort → INULA sp. also → DATURA
STIPA sp.
Stirofos → GARDONA
STIZLOBIUM DEERINGIANUM
STODDARD SOLVENT also → NAPHTHA
Stomolophus nomurai → JELLYFISH
Stomp → PENDIMETHALIN
STONE FISH
Stone Seed → LITHOSEPMUM RUDERALE
Stoplight → CALADIUM
Storage Disease → POLYVINYLPYRROLIDONE
STORAX also → PAPER
Storocain → OXETHAZINE
Stovarsolan → ACETARSONE
STOVES
Stoxil → IDOXURIDINE
STP → DIMETHOXYMETHYLAMPHETAMINE
STPP → SODIUM TRIPOLYPHOSPHATE
Strabolene → NANDROLONE
Straderm → FLUOCINONIDE
STRAMMONIUM also → BELLADONNA
Strathmore Weed → PIMELIA sp.
STRAWBERRY
Strawberry Aldehyde → ETHYL METHYLPHENYLGLYCIDATE
Strawberry Fields → LYSERGIC ACID DIETHYLAMIDE
Strawberry Plant → EUONYMUS, *americanus*
Strawberry Tomato → PHYSALIS sp.
STRELITZIA REGINA
Streptase → STREPTOKINASE
Streptocid album → SULFANILAMIDE
Streptodornase → STREPTOKINASE
Streptogramin → DALFOPRISTIN
STREPTOKINASE also → RETEPLASE
STREPTOMYCIN also → ALCURONIUM, AMOBARBITAL, CURARE, *d*-TUBOCURARINE CHLORIDE, DIMETHYL TUBOCURARINE IODIDE,

EPINEPHRINE, ETHACRYNIC ACID; ETHER, ETHYL; FRAMYCETIN, GALLAMINE TRIETHIODIDE, GENTAMICIN, HALOTHANE, KANAMYCIN, NEOMYCIN, NETILMYCIN, PAROMOMYCIN; PENICILLIN also → PENICILLIN > *Benzylpenicillin, Procaine*; POLYMIXIN B, PROCAINAMIDE, PROMETHAZINE, QUINIDINE, SODIUM BICARBONATE, SUCCINYLCHOLINE, THIACETAZONE, VANCOMYCIN
STREPTONIGRIN
Streptonivicin → NOVOBIOCIN
Streptonivocin → NOVOBIOCIN
STREPTOZOCIN also → 1-METHYL-1-NITROSOUREA
Streptozon → SULFAMIDOCHRYSOIDINE
Streptozotocin → STREPTOZOCIN
Streunex → LINDANE
Striatran → EMYLCAMATE
Strictylon → NAPHAZOLINE
String Bean → BEANS
STROBANE®
Strobane-T → TOXAPHENE
Strodival → OUABAIN
Stromba → STANOZOLOL
Stromectol → IVERMECTIN
Strongid → PYRANTEL
STRONTIUM also → CHEESES, COFFEE, FRUITS and FRUIT JUICES, TEA
STRONTIUM CHLORIDE
Strontium Hydrate → STRONTIUM HYDROXIDE
STRONTIUM HYDROXIDE
STRONTIUM SULFIDE
STROPHANTHIDIN also → PARQUETINA NIGRESCENS
STROPHANTHIN
Strophanthin K → STROPHANTHIN
STROPHANTHUS sp. also → BALANITES, GARCINIA sp.
STROPHARIA sp. also → MUSHROOMS
Strumacil → METHYLTHIOURACIL
Strumazol → METHIMAZOLE
Struvite → MAGNESIUM
STRYCHNINE also → ATROPINE, BRUCINE, IPECAC, STRYCHNOS
STRYCHNOS also → TELITOXICUM PERUVIANUM
STRYPHNODENDRON OBOVATUM
Stubble Berries → SOLANUM sp.
Stuff → CANNABIS
Stugeron → CINNARIZINE
Stutgeron → CINNARIZINE
Stutgin → CINNARIZINE
STX → SAXITOXIN
Stypticweed → CASSIA
Styquin → BUTAMISOLE
Styrax → STORAX
STYRENE
Styrol → STYRENE
Styrolene → STYRENE
SU 101 → LEFLUNOMIDE
SU 3088 → CHLORISONDAMINE
SU 3118 → SYROSINGOPINE
SU 4885 → METYRAPONE
SU 5864 → GUANETHIDINE
SU 8341 → CYCLOPENTHIAZIDE
SU 9064 → METOSERPATE
SU 10,568 → CLORTERMINE
SU 13,437 → NAFENOPIN
SU 21,524 → PIRPROFEN
Suacron → CARAZOLOL
Suanovil → SPIPERONE
Suavitil → BENACTYZINE
Subicard → PENTAERYTHRITOL TETRANITRATE
Subitex → DINOSEB
Sublimaze → FENTANYL
Subose → GLYHEXAMIDE
Sub-Quin → PROCAINAMIDE
Subtosan → POLYVINYLPYRROLIDONE
Sucaryl → SODIUM CYCLAMATE also → ANILINE
Sucaryl Calcium → CALCIUM CYCLAMATE
Succimal → ETHOSUXIMIDE
SUCCINYLCHOLINE also → APROTININ, BACITRACIN, BUNAMIDINE, CIMETIDINE, COLISTIN, CURARE, DEXPANTHENOL, DIAZEPAM, DICHLORVOS, DIGITALIS, DIHYDROSTREPTOMYCIN, DONEPEZIL, ECHOTHIOPHATE IODIDE; ETHER, ETHYL; FUROSEMIDE, GRAMICIDIN, HALOTHANE, HEXAFLUORENIUM BROMIDE, KANAMYCIN, LIDOCAINE, LITHIUM, LYSERGIC ACID DIETHYLAMIDE, MAGNESIUM SULFATE, MALATHION, METUREDEPA, MIVACURIUM, NEOMYCIN, NEOSTIGMINE, OXYTETRACYCLINE, PHENELZINE, PHYSOSTIGMINE, PROCAINE, PROPANIDID, PYRANTEL, PYRIDOSTIGMINE BROMIDE, QUINIDINE, QUINURONIUM, RONNEL, RUELENE, STREPTOMYCIN, TACRINE, THIOTEPA, TRICHLORFON
Succinyl Dichloride → SUCCINYLCHOLINE

MINERAL SEAL OIL, MONOCROTOPHOS, METHSUXIMIDE, METHYL CHLORIDE, METHYL CHLOROPHENOXY ACETIC ACID, METHYLENEDIOXYAMPHETAMINE, MONOSODIUM METHANEARSONATE, NAPHTHA, NAPROXEN, NIALAMIDE, NICOTINE, NITRAZEPAM, NITRIC ACID, NORTRIPTYLINE, OLEANDER, OPIPRAMOL, OXALIC ACID, OXANDROLONE, PAPAVERINE, PARATHION, PAROXETINE, PENTOBARBITAL, *p*-PHENYLDIAMINE, PHENYLPROPANOLAMINE, PHENYTOIN, PHOLEDRINE; PHOSPHORUS, RED; PINDOLOL, PINE, PHYLLANTHUS sp., PODOPHYLLUM, POTASSIUM, POTASSIUM CHLORATE, POTASSIUM CYANIDE, POTASSIUM DICHROMATE, PRACTOLOL, PRIMIDONE, PROBENECID, PROPOXYPHENE, PROPRANOLOL, PYRILAMINE, PYRIMINIL, QUINIDINE, QUININE, RESERPINE, RHODODENDRON, RICIN, RIFAMPIN, RILUZOLE, SALICYLAMIDE, SCOPOLAMINE, SODIUM ARSENITE, SODIUM CARBONATE, SODIUM CHLORATE, SODIUM CHLORIDE, SODIUM CYANIDE, SODIUM HYDROXIDE, SODIUM HYPOCHLORITE, SULFAMETHOXAZOLE, SULFURIC ACID, SULPHAPHENAZOLE, SULTHIAME, SUMATRIPTAN SUCCINATE, SUPROFEN; 1,1,2,2-TETRACHLOROETHANE; THALLIUM, THEOPHYLLINE, THIOPENTAL, THIOPROPAZATE, THIORIDAZINE, TIMOLOL, TOLBUTAMIDE, TRAMADOL, TRANYLCYPROMINE, TRIAZOLAM; 2,4,5-TRICHLOROPHENOXY ACETIC ACID; TRIETHYLENE GLYCOL, TRIHEXYPHENIDYL, VALERIAN sp., VERAPAMIL, VERATRUM, WARFARIN, XYLAZINE, ZINC, ZINC CHLORIDE, ZINC SULFATE

Suigonan → GONADOTROPINS, CHORIONIC

"Suilyuk Disease" → TRICHODESMA sp.

Sular → NISOLDIPINE

SULBACTAM also → AMLODIPINE, LANSOPRAZOLE, OMEPRAZOLE

Sulben → SULFABENZAMIDE

Sulcephalosporin → CEFSULODIN SODIUM

Sulcolon → SALICYLAZOSULFAPYRIDINE

Sulcon → SULFAQUINOXALINE

SULCONAZOLE NITRATE

Sulcosyn → SULCONAZOLE NITRATE

Sulcrate → SUCRALFATE

Sulduxine → SULFADIMETHOXINE

Suleo-M → MALATHION

Sulf-10 → SULFACETAMIDE SODIUM

SULFABENZAMIDE

Sulfabenzide → SULFABENZAMIDE

Sulfabid → SULFAPHENAZOLE

Sulfabon → SULFADIMETHOXINE

Sulfabrom → SULFABROMOMETHAZINE

SULFABROMOMETHAZINE

Sulfabutin → BUSULFAN

SULFACETAMIDE SODIUM

SULFACHLORPYRIDAZINE

Sulfacin → SULFISOXAZOLE

Sulfactin → DIMERCAPROL

SULFACYTINE

Sulfadiasulfone Sodium → ACETOSULPHONE SODIUM

SULFADIAZINE also → PHENYLPROPANOLAMINE

Sulfadimerazine → SULFAMETHAZINE

SULFADIMETHOXINE also → PHENOBARBITAL, PHENPROCOUMON, TOLBUTAMIDE, WARFARIN

Sulfadimezine → SULFAMETHAZINE

Sulfadimidine → SULFAMETHAZINE

Sulfadine → SULFAMETHAZINE

SULFADOXINE

Sulfaethidole → PENICILLIN also → PENICILLIN > *Penicillin V*; PENICILLINS – "NEWER and SYNTHETICS" > *Cloxacillin Sodium, Methicillin Sodium, Nafcillin Sodium, Oxacillin Sodium*

Sulfafloc → CELLULOSE

Sulfafurazol → SULFISOXAZOLE

SULFAGUANIDINE

Sulfaguine → SULFAGUANIDINE

Sulfaisomezol → SULFAMETHOXAZOLE

Sulfalar → SULFISOXAZOLE

SULFALENE

Sulfalex → SULFAMETHOXYPYRIDAZINE

SULFAMERAZINE also → SULFAMETHIAZINE

Sulfamerazinum → SULFAMERAZINE

SULFAMETER

SULFAMETHAZINE also → CAFFEINE, CHLORPROPAMIDE, CYCLOSPORINE, TRIMETHOPRIM, ZIDOVUDINE

SULFAMETHIZOLE also → CHLOROTHIAZIDE, METHENAMINE, PHENYTOIN, TOLBUTAMIDE, WARFARIN

SULFAMETHOXAZOLE also → AZATHIOPRINE, CHLORPROPAMIDE, DISULFIRAM, FOLIC ACID, METHOTREXATE, PHENYTOIN, PROPANTHELINE, QUINAPRIL, TEICOPLANIN, TRIMETHOPRIM, WARFARIN

Sulfamethoxydiazine → SULFAMETER

Sulfamethoxypyrazine → SULFALENE

SULFAMETHOXYPYRIDAZINE also → ETHYL BISCOUMACETATE; PENICILLIN also → PENICILLIN > *Penicillin V;* PENICILLINS – "NEWER and SYNTHETICS" > *Cloxacillin Sodium, Methicillin Sodium, Nafcillin Sodium, Oxacillin Sodium;* PHENYLBUTAZONE, SALICYLIC ACID, WARFARIN

Sulfametin → SULFAMETER

Sulfametizol → SULFAMETHIZOLE

Sulfametopyrazine → SULFALENE

Sulfametorine → SULFAMETER

Sulfametossipiridazina → SULFAMETHOXYPYRIDAZINE

Sulfametoxipirimidine → SULFAMETER

Sulfamezathine → SULFAMETHAZINE

Sulfamidinum → SULFAGUANIDINE

p-Sulfamidobenzoate → SACCHARIN

SULFAMIDOCHRYSOIDINE

Sulfamonomethoxine → SULFAMETER

Sulfamul → SULFATHIAZOLE

Sulfamylon → MAFENIDE

SULFANILAMIDE also → BUCLOSAMIDE, DIETHYLENE GLYCOL, MONENSIN, Preface

SULFANILIC ACID

Sulfanilylguanidine → SULFAGUANIDINE

SULFAPHENAZOLE also → CHLORPROPAMIDE, GLYBURIDE, PENICILLIN, PHENYTOIN, TOLBUTAMIDE, WARFARIN

Sulfapyelon → SULFAMETHIZOLE

Sulfapyridazine → SULFAMETHOXYPYRIDAZINE

SULFAPYRIDINE also → SULFATHIAZOLE

Sulfapyrimidine → SULFADIAZINE

Sulfa Q → SULFAQUINOXALINE

SULFAQUINOXALINE

Sulfarthrol → SULFUR

Sulfasalazine → SALICYLAZOSULFAPYRIDINE also → DIGOXIN, FOLIC ACID, SULFANILAMIDE

Sulfasan → BISETHYLXANTHOGEN

Sulfasol → SULFISOXAZOLE

Sulfa-Span → SULFAMETHAZINE

Sulfasuxidine → SUCCINYLSULFATHIAZOLE

Sulfasymazine → PENICILLIN also → PENICILLIN > *Penicillin V;* PENICILLINS – "NEWER and SYNTHETICS" > *Cloxacillin Sodium*

Sulfates → CABBAGE also → SODIUM SULFATE

Sulfathalidine → PHTHALYLSULFATHIAZOLE

SULFATHIAZOLE also → METHENAMINE, PROCAINE

Sulfavitina → SULFATHIAZOLE

Sulfazin → SULFISOXAZOLE

Sulfdurazin → SULFAMETHOXYPYRIDAZINE

Sulfentanil → SUFENTANIL

SULFINPYRAZONE also → ACETYLSALICYLIC ACID, CHLORTHALIDONE; PENICILLIN also → PENICILLIN > *Penicillin V;* PENICILLINS – "NEWER and SYNTHETICS" > *Cloxacillin Sodium;* PHENPROCOUMON, PROBENECID, SALICYLIC ACID, SALSALATE, SODIUM SALICYLATE, THEOPHYLLINE, WARFARIN

Sulfinylbismethane → DIMETHYL SULFOXIDE

Sulfinyl Chloride → THIONYL CHLORIDE

Sulfisomezole → SULFAMETHOXAZOLE

SULFISOXAZOLE also → CHLORPROPAMIDE, METHOTREXATE; PENICILLIN also → PENICILLIN > *Penicillin V;* PENICILLINS – "NEWER and SYNTHETICS" > *Cloxacillin Sodium;* PHENFORMIN, TOLBUTAMIDE, WARFARIN

Sulfites → BEER, CARAMEL, CIDER, DOPAMINE, GRAMICIDIN, POTATO, SHELLFISH, SHRIMP, SODIUM BISULFITE, SODIUM GLUTAMATE, SULFUR DIOXIDE

SULFOBROMOPHTHALEIN SODIUM also → PHENOBARBITAL, PHENYTOIN

Sulfoguenil → SULFAGUANIDINE

Sulfolex → SULFADIAZINE

Sulfona-Mae → DAPSONE

Sulfonamides → ACENOCOUMAROL, ACETAZOLAMIDE, ACETOHEXAMIDE; ALCOHOL, ETHYL; BISHYDROXYCOUMARIN, BUMETANIDE, CALCIUM CARBIMIDE, CELECOXIB, CHLOROPROCAINE, DIBUCAINE, GLIPIZIDE, INSULIN, MAGNESIUM OXIDE, METOCLOPRAMIDE, METOLAZONE, NALIDIXIC ACID, NITROFURANTOIN, PENICILLIN,

PHENOLSULFONPHTHALEIN, *p*-PHENYLENEDIAMINE, PHENYTOIN; POLISHES, SHOE; POLYTHIAZIDE, PROBENECID, PROCAINE, SHOES, SODIUM BICARBONATE, TETRACAINE HYDROCHLORIDE, TOLBUTAMIDE
Sulfone → LANSOPRAZOLE, OMEPRAZOLE
Sulfonmethane → HEMATOPORPHYRIN
Sulforaphane → RADISH
Sulformethoxine → SULFADOXINE
Sulformetoxine → SULFADOXINE
Sulforthodimethoxine → SULFADOXINE
Sulforthomidine → SULFADOXINE
SULFOTEP
Sulfothiorine → SODIUM THIOSULFATE
Sulfotrim → TRIMETHOPRIM also → SULFAMETHOXAZOLE
Sulfoxol → SULFISOXAZOLE
SULFOXONE
Sulftalyl → PHTHALYLSULFATHIAZOLE
SULFUR also → COAL, CUTTING OILS; FUELS, OIL; MOTOR OIL, SODIUM THIOSULFATE
Sulfur Chloride → SULFUR MONOCHLORIDE
SULFUR DIOXIDE also → *n*-BUTYL MERCAPTAN, CEMENT, COAL, PAPER, SHELLFISH, SODIUM METABISULFITE, SULFUR, SULFURIC ACID
Sulfur Fluoride → SULFUR HEXAFLUORIDE
SULFUR HEXAFLUORIDE
SULFURIC ACID also → CHLOROSULFONIC ACID, DIMETHYL SULFATE, GOLF BALLS, SODIUM BISULFITE, SULFUR DIOXIDE
Sulfuric Anhydride → SULFUR TRIOXIDE
Sulfuric Ether → ETHER, ETHYL
Sulfuric Oxychloride → SULFURYL CHLORIDE
SULFUR MONOCHLORIDE
Sulfur Mustard → MUSTARD GAS
Sulfurous Anhydride → SULFUR DIOXIDE
Sulfurous Oxide → SULFUR DIOXIDE
Sulfurous Oxychloride → THIONYL CHLORIDE
Sulfur Oxide → SULFUR DIOXIDE
Sulfur Phosphide → PHOSPHORUS PENTASULFIDE
SULFUR TETRAFLUORIDE
SULFUR TRIOXIDE
SULFURYL CHLORIDE
SULFURYL FLUORIDE
SULINDAC also → CYCLOSPORINE, DIMETHYL SULFOXIDE, FENBUFEN, LABETALOL, WARFARIN
Sulinol → SULINDAC
SULISOBENZONE
Sulitrene → SULFAMETHOXYPYRIDAZINE
Sulla → SULFAMETER
Sulmet → SULFAMETHAZINE
Sulmid → SULFANILAMIDE
Sulmycin → GENTAMICIN
Sulphadione → DAPSONE
Sulphafurazole → SULFASOXAZOLE
Sulphan Blue → BLUE VRS
Sulphasalazine → SALICYLAZOSULFAPYRIDINE
Sulphetrone → SOLASULFONE
Sulphonazine → SOLASULFONE
Sulphur → SULFUR
SULPHYDRILIC ACID
SULPIRIDE also → ALUMINUM HYDROXIDE
Sulpiril → SULPIRIDE
Sulpirine → DIPYRONE
Sulpitil → SULPIRIDE
Sulprim → SULFAMETHOXAZOLE
SULPROFOS
Sulprotin → SUPROFEN
Sulpyrin → DIPYRONE
Sulpyrine → DIPYRONE
Sulquin → SULFAQUINOXALINE
Sulredox → PENICILLAMINE
Sulreuma → SULINDAC
Sulsetrex Piperazine → ESTRONE
Sulsoxin → SULFISOXAZOLE
Sultamicillin → PENICILLINS – "NEWER and SYNTHETICS"
Sultân → PISTACIA sp.
Sultanol → ALBUTEROL
Sulten-10 → SULFACETAMIDE SODIUM
SULTHIAME also → PHENYTOIN
Sultirene → SULFAMETHOXYPYRIDAZINE
Sulxin → SULFADIMETHOXINE
Sulzol → SULFATHIAZOLE
SUM-3170 → LOXAPINE
Suma → CANNABIS
SUMAC, POISON also → MANGO, PENTADECYLCATECHOL, POISON IVY
Sumamed → AZITHROMYCIN
Sumatra Camphor → BORNEOL
SUMATRIPTAN SUCCINATE
Sumedine → SULFAMERAZINE
Sumetrolim → SULFAMETHOXAZOLE also → TRIMETHOPRIM
Sumetrolium → SULFAMETHOXAZOLE also → TRIMETHOPRIM
Sumial → PROPRANOLOL
Sumicidin → FENVALERATE
Sumifly → FENVALERATE

Sumipower → FENVALERATE
SUMITHION
Sumitomo → SUMITHION
Sumitox → MALATHION also → FENVALERATE
Summer Cypress → KOCHIA
Summer Poinsettia → EUPHORBIA, *heterophylla*
Summetrin → PAPAIN
Sumox → PENICILLINS – "NEWER and SYNTHETICS" > *Amoxicillin*
Sumycin → TETRACYCLINE
Sumyon → VALERIANA sp.
Sunbloc → CINOXATE
Sunbrella → *p*-AMINOBENZOIC ACID
Suncare → OCTYLDIMETHYL PABA
Suncide → PROPOXUR
Sundown → OCTYLDIMETHYL PABA
Sunette → ACESULFAME K
SUNFLOWER also → DATURA, MARIGOLD, SCLEROTINIA SCLEROTIORUM
Sunflower Daisy → WEDELIA sp.
Sunfural → TEGAFUR
Sungard → SULISOBENZONE
Sungoje → LASIOSIPHON sp.
Sungwoi → ERYTHROPHLEUM
Sunlight® → LEMON, TEA
Sunny Delight® → β-CAROTENE
Sunset Yellow FCF → FDC YELLOW #6
Sunshine → LYSERGIC ACID DIETHYLAMIDE
Sunshine Vitamin → VITAMIN D
Sun Spurge → EUPHORBIA
Sun Tea → TEA
Superanabolin → NANDROLONE
Super Arsonate → METHANEARSONIC ACID
Superbonder → CYANOACRYLATES
Super C → KETAMINE
Super-Caid → BROMADIOLONE
Super-EPA® → DOCOSAHEXAENOIC ACID
Super Glue → CYANOACRYLATES
Superil → DIETHYLCARBAMAZINE
Superinone → TYLOXAPOL
Super Joint → PHENCYCLIDINE
Superphosphate → FLUORIDES and FLUORINE
Superpyrin → ALOXIPRIN
Super-Rozol → BROMADIOLONE
Superseptyl → SULFAMETHAZINE
Super Three Cement → CYANOACRYLATES
Super Weed → PHENCYCLIDINE
Supinine → PYRROLIZIDINES
Suplexedil → FENOXEDIL
Supona → CHLORFENVINPHOS
Supotran → CHLORMEZANONE
Supprax → CEFIXIME
Supprelin → HISTRELIN
Suprachol → DEHYDROCHOLIC ACID, *sodium salt*
Supracide → METHIDATHION
Supracombin → SULFAMETHOXAZOLE
Supracor → TERFENADINE
Supracyclin → DOXYCYCLINE
Supraene → SQUALENE
Supral → PECILOCIN
Suprametil → METHYLPREDNISOLONE
Supramid → SULFAMETER
Supramycin → TETRACYCLINE
Suprane → DESFLURANE
Supranol → SUPROFEN
Suprapen → PENICILLINS – "NEWER and SYNTHETICS" > *Amoxicillin*
Supra-Puren → SPIRONOLACTONE
Suprarenalin → EPINEPHRINE
Suprarenin → EPINEPHRINE
Suprasec → LOPERAMIDE
Suprastin → CHLOROPYRAMINE
Supratonin → AMEZINIUM METHYL SULFATE
Suprefact → LH-RH
Supres → TRIMAZOSIN
Supressin → DOXAZOCIN MESYLATE
Suprilent → ISOXSUPRINE
Suprim → SULFAMETHOXAZOLE
Suprocil → SUPROFEN
SUPROFEN
Suprol → SUPROFEN
Supronal → SULFAMERAZINE
Suprotan → CHLORMEZANONE
Suracton → SPIRONOLACTONE
Sural → ETHAMBUTOL
Suralgan → TIAPROFENIC ACID
SURAMIN SODIUM
Surcopur → PROPANIL
Surem → NITRAZEPAM
Surfacaine → CYCLOMETHYCAINE
Surfathesin → CYCLOMETHYCAINE
Surfer → PHENCYCLIDINE
Surflan → ORYZALIN
Surgam → TIAPROFENIC ACID
Surgamic → TIAPROFENIC ACID
Surgamyl → TIAPROFENIC ACID
Surgeon General → AGENT ORANGE
Surgicel → CELLULOSE, *Oxidized*
Surgi-Cen → HEXACHLOROPHENE
Surika → FLUFENAMIC ACID
Surinam Balsam → BALSAM PERU
Surital → THIAMYLAL SODIUM
Surlid → ROXITHROMYCIN
SURMA also → LEAD, MASCARA

Surofene → HEXACHLOROPHENE
Sursumid → SULPIRIDE
Survector → AMINEPTINE
Sûs → GLYCYRRHIZA GLABRA
Susadrin → NITROGLYCERIN
Suscard → NITROGLYCERIN
Suscardia → ISOPROTERENOL
Su Seguro Carpidor → TRIFLURALIN
SUSHI also → FISH
Suspendol → ALLOPURINOL
Sus-phrine → EPINEPHRINE
Suspren → IBUPROFEN
Sustac → NITROGLYCERIN
Sustain → SULFAMETHAZINE
Sustaire → THEOPHYLLINE
Sustamycin → TETRACYCLINE
Sustane → BUTYLATED HYDROXYTOLUENE
Sustane 1-F → BUTYLATED HYDROXYANISOLE
Sustiva → EFAVIRENZ
Sustonit → NITROGLYCERIN
Sutan → 5-ETHYL DIISOBUTYLTHIOCARBAMATE
Suterberry → ZANTHOXYLUM sp.
SUTILAINS
Sütleğen → EUPHORBIA sp.
Sutoprofen → SUPROFEN
Suvren → CAPTODIAMINE
Suxamethonium Bromide → SUCCINYLCHOLINE
Suxamethonium Chloride → SUCCINYLCHOLINE
Sux-Cert → SUCCINYLCHOLINE
Suxilep → ETHOSUXIMIDE
Suximal → ETHOSUXIMIDE
Suxinutin → ETHOSUXIMIDE
Suzutolon → BETAHISTINE
S.V.C. → ACETARSONE
Swainsona → SWAINSONINE
SWAINSONIA sp. also → OXYTROPIS
SWAINSONINE also → ASTRAGALUS, OXYTROPIS
Swallow Wort → CHELIDONIUM MAJUS
Swamp Hellebore → VERATRUM sp.
Swamp Horsetail → EQUISETUM
Swamp Sumac → SUMAC, POISON
Swamp Turnip → ARISAEMA
Swat → BOMYL
Swedes → RAPE
Swedish Green → COPPER ARSENITE
Sweeta → ANILINE
Sweet Cane → CALAMUS
Sweet Cicely → OSMORHIZA CHILENSIS
Sweet Cinnamon → CALAMUS
Sweet Cumin → ANISE
Sweet Flag → CALAMUS
Sweet Grass → CALAMUS
Sweet Gum → STORAX
Sweet Myrtle → CALAMUS
Sweet Oil → OLIVES
Sweet One → ACESULFAME K
Sweet Pea → LATHYRUS sp.
SWEET POTATO also → CYANIDES
Sweet Root → CALAMUS
Sweet Rush → CALAMUS
Sweet-Scented Bedstraw → GALIUM sp.
Sweet Sedge → CALAMUS
Sweetshrub → CALYCANTHUS FLORIDUS
Sweet Woodruff → MELILOT
Swellhead → TETRADYMIA, TRIBULUS TERRESTRIS
Swimtrine → COPPER TRIETHANOLAMINE
Swiss Blue → METHYLENE BLUE
SWISS CHARD
Swiss Cheese Plant → MONSTERA sp.
Sydenham, Dr. Thomas → OPIUM
Sykose → SACCHARIN
Syloid → SILICON DIOXIDE
Sylvic Acid → ABIETIC ACID
Symbio → SULFADIMETHOXINE
Symmetral → AMANTADINE
Sympamin → DEXTROAMPHETAMINE SULFATE
Sympamine → AMPHETAMINE
Sympatedrine → AMPHETAMINE
Sympathin → NOREPINEPHRINE
Sympatholytin → *N*-(2-CHLOROETHYL)DIBENZYLAMINE
m-Sympatol → PHENYLEPHRINE
Symphoricarpus albus → SNOWBERRY
SYMPHYTINE also → COMFREY, PYRROLIZIDINES
Symphytum → COMFREY
Symplocarpus foetidus → SKUNK CABBAGE
Synacid → SODIUM HYALURONATE
Synaclyn → FLUNISOLIDE
Synacort → HYDROCORTISONE
Synacthen → TETRACOSACTIDE
SYNADENIUM GRANTII
Synadrin → PRENYLAMINE LACTATE
Synalar → FLUOCINOLONE ACETONIDE
Synamol → FLUOCINOLONE ACETONIDE
Synanceja sp. → STONE FISH
Synandone → FLUOCINOLONE ACETONIDE
Synanthic → OXFENDAZOLE
Synarel → NAFARELIN ACETATE
Synasteron → OXYMETHOLONE

Synastrin → SPARTEINE
Syncaine → PROCAINE
Syncelose → METHYL CELLULOSE
Synchrorin → MEGESTROL ACETATE
Syncillin → PENICILLINS – "NEWER and SYNTHETICS" > *Phenethicillin Potassium*
Syncl → CEPHALEXIN
Synclotin → CEPHALOTHIN
Syncortyl → DEOXYCORTICOSTERONE, *acetate*
Syncoumar → ACENOCOUMAROL
Syncurine → DECAMETHONIUM BROMIDE
Syndopa → L-DOPA
Syndrox → METHAMPHETAMINE
Synédil → SULPIRIDE
Synemol → FLUOCINOLONE ACETONIDE
Synephrin → SYNEPHRINE
SYNEPHRINE also → ORANGE
Synercid → DALFOPRISTIN
Synestrin → DIETHYLSTILBESTROL
Synestrol → DIENESTROL
Synflex → NAPROXEN
Syngacillin → PENICILLINS – "NEWER and SYNTHETICS" > *Cyclacillin*
Syngamix → DISOPHENOL
Syngenstrone → PROGESTERONE
Syngesteron(e) → PROGESTERONE
Syngestrotabs → ETHISTERONE
SYNHEXYL
Synklor → CHLORDANE
Synkonin → HYDROCODONE BITARTRATE
Synogil → NATAMYCIN
Synopen → CHLOROPYRAMINE
Synotic → FLUOCINOLONE ACETONIDE
Synpen → CHLOROPYRAMINE
Synpitan → OXYTOCIN
Synsac → FLUOCINOLONE ACETONIDE
Synsepal → PENICILLINS - "NEWER & SYNTHETICS" > *Amoxicillin*
Synstigmin → NEOSTIGMINE
Syntaris → FLUNISOLIDE
Syntarpen → PENICILLINS – "NEWER and SYNTHETICS" > *Cloxacillin/Sodium salt* also *Dicloxacillin*
Syntecort → PARAMETHASONE ACETATE
Syntestrine → DIETHYLSTILBESTROL
Syntetrin → ROLITETRACYCLINE
Syntex → SOYBEANS
Syntexan → DIMETHYL SULFOXIDE
Synthaderm → POLYURETHANES
Synthecilline → PENICILLINS – "NEWER and SYNTHETICS" > *Phenethicillin Potassium*
Synthenate → SYNEPHRINE
Synthepen → PENICILLINS – "NEWER and SYNTHETICS" > *Phenethicillin Potassium*
Synthepen-P → PENICILLINS – "NEWER and SYNTHETICS" > *Propicillin Potassium*
"Synthetic Heroin" → α-METHYL FENTANYL
Synthoestrin → DIETHYLSTILBESTROL
Synthomycetine → CHLORAMPHENICOL
Synthomycin → CHLORAMPHENICOL
Synthovo → HEXESTROL
Synthroid → THYROXINE
Syntocain → PROCAINE
Syntometrine → ERGONOVINE
Syntomycin → OXYTETRACYCLINE
Syntophylline → AMINOPHYLLINE
Syntrogène → HEXESTROL
Syprine → TRIENTINE
Syraprim → TRIMETHOPRIM
Syrian Rue → PEGANUM HARMALA
Syringa Berry → MELIA sp.
Syringopine → SYROSINGOPINE
SYROSINGOPINE
Syrup of Ipecac → IPECAC also → EMETINE
Syrupy Glucose → CORN SYRUP
Syscor → NISOLDIPINE
Systamex → OXFENDAZOLE
Systasol → DINOBUTON
Systemox → DEMETON
Systen → ESTRADIOL
Systodin → QUINIDINE
Systox → DEMETON
Systral → CHLORPHENOXAMINE
Sytam → SCHRADAN
Sytobex(-H) → VITAMIN B_{12}

T

T-2 TOXIN also → TRICHOTHECENE MYCOTOXINS, YELLOW RAIN
2,4,5-T → 2,4,5-TRICHLOROPHENOXYACETIC ACID
T_4 → THYROXINE
T_4 → CYCLONITE
T-113 → BUTONATE
T 1220 → PENICILLINS – "NEWER and SYNTHETICS" > *Piperacillin Sodium*
T 1258 → TEMAFLOXACIN
T 1824 → EVANS BLUE
T 2104 → TABUM
TAA → THIOACETAMIDE
Tab → MORPHINE
Taba → NICOTIANA sp.
Tabaco → TOBACCO
TABACU DI PISCADO
Tabalgin → ACETAMINOPHEN

Tabalon → IBUPROFEN
Tabasco → PIMENTO
Tabazur → NICOTINE and SALTS
Tabe → PIRENZEPINE HYDROCHLORIDE
Taberdog → DIAZINON
Tabergat → DIAZINON
TABERNAEMONTANA sp.
TABERNANTHE IBOGA also → IBOGAINE
Tabex → CYTISINE
Tabilin → PENICILLIN > *Potassium Benzylpenicillinate*
Tabloid → THIOGUANINE
Tabotamp → CELLULOSE, *Oxidized*
Tabouret des Champs → THLASPI ARVENSE
Tabrin → OFLOXACIN
TABUM
TAC → PHENCYCLIDINE
Tacaryl → METHDILAZINE
Tace → CHLOROTRIANISENE
Tachionin → TRICHLORMETHIAZIDE
Tachmalin → AJMALINE
Tachyrol → DIHYDROTACHYSTEROL
Tacitin → BENZOTAMINE
TACRINE also → CIMETIDINE, SUCCINYLCHOLINE
TACROLIMUS also → DIAMINOZIDE, DILTIAZEM, ERYTHROMYCIN, FLUCONAZOLE, ITRACONAZOLE, NEFAZODONE, NIFEDIPINE, PHENYTOIN, SIMVASTATIN
Tafil → ALPRAZOLAM
Tagagel → CIMETIDINE
Tagamet → CIMETIDINE
Tagathen → CHLOROTHEN
TAGETES MINUTA
Tag Fungicide → PHENYLMERCURIC ACETATE
Tag HL-331 → PHENYLMERCURIC ACETATE
Tahneeya → HALVA
TAI-284 → CLIDANAC
Taicelexin → CEPHALEXIN
Taiguic Acid → LAPACHOL
TAIPOXIN
Takanarumin → ALLOPURINOL
Takepron → LANSOPRAZOLE
Takesulin → CEFSULODIN SODIUM
Taktic → AMITRAZ
Takus → CERULETIDE
Taladren → ETHACRYNIC ACID
Talampicillin → PENICILLINS – "NEWER and SYNTHETICS"
Talat → PENICILLINS – "NEWER and SYNTHETICS" > *Talampicillin*
Talatrol → TROMETHAMINE
Talbutal → BUTALBITAL
Talc → MAGNESIUM SILICATE
Taleudron → PHTHALYLSULFATHIAZOLE
Talidine → PHTHALYLSULFATHIAZOLE
Talimol → THALIDOMIDE
Talinolol → VERAPAMIL
Talipexole → FLEROXACIN
Tall Mannagrass → GLYCERIA sp.
Tall Oil Resin → ROSIN
Tallow Tree → SAPIUM sp.
Talodex → FENTHION
Talofen → PROMAZINE
Talon → BRODIFACOUM
Talotren → THEOPHYLLINE
Taloxa → FELBAMATE
Talpen → PENICILLINS – "NEWER and SYNTHETICS" > *Talampicillin*
Talpicil → PENICILLINS – "NEWER and SYNTHETICS" > *Talampicillin*
Talsis → BISOXATIN ACTATE
Talucard → PROSCILLARIDIN
Talusin → PROSCILLARIDIN
Talwin → PENTAZOCINE
Tamara → CRATAEVA BENTHMII
TAMARIND
Tamaron → METHAMIDOPHOS
Tamas → BENZYDAMINE
Tamate → MEPROBAMATE
Tamban → SARDINELLA sp.
Tambocor → FLECAINIDE
Tamboril de campo → ENTEROLOBIUM
Tamed Iodine → IODOPHORS
Tamiflo → OSELTAMIVIR
Tamofen → TAMOXIFEN CITRATE
Tamoxasta → TAMOXIFEN CITRATE
TAMOXIFEN CITRATE also → AMINOGLUTETHIMIDE, EXEMESTANE, MIBEFRADIL, RIFAMPIN, WARFARIN
Tampons → CELLULOSE, SODIUM CARBOXYMETHYLCELLULOSE, SANITARY NAPKINS
TAMSULOSIN
TAMUS COMMUNIS
Tanacetum parthenium → FEVERFEW
Tanacetum vulgare → TANSY
Tanaclone → DYCLONINE
TANAECIUM sp.
Tanafol → CHLORMEZANONE
Tanakan → GINKGO BILOBA also → CHLOROQUINE
Tandacote → OXYPHENBUTAZINE
Tandearil → OXYPHENBUTAZINE
Tandix → INDAPAMIDE

Tang® → METHADONE HCL also → ORANGE
Tangantangan Oil → CASTOR OIL
Tangeretin → ORANGE
"Tango & Cash" → α-METHYL FENTANYL
Tanidil → PROMETHAZINE
Tankan → CHLOROQUINE
Tanner Grass → BRACHIARIA
TANNIC ACID also → BARIUM SULFATE, CHLORHEXIDINE, GALLIC ACID,TEA
Tannins → TEA
Tanston → MEFENAMIC ACID
TANSY also → TARAXACUM OFFICINALE
Tansy Mustard → DESCURAINIA PINNATA
Tansy Ragwort → SENECIO sp.
TANTALUM
Tan Tan → LEUCAENA sp.
Tantum → BENZYDAMINE
TAO → TROLEANDOMYCIN
Taoryl → CARAMIPHEN
TAP 144 → LEUPROLIDE
Tapar → ACETAMINOPHEN
Tapazole → METHIMAZOLE
Taperyva Hu → CASSIA, *occidentalis*
TAPIOCA
Taquidil → TOCAINIDE
Taractan → CHLORPROTHIXENE
Tarantula → SPIDERS
Tarasan → CHLORPROTHIXENE
Tarasyn → KETOROLAC
TARAXACUM OFFICINALE also → TEA
TARAXEIN
Tarbush → FLOURENSIA CERNUA
Tar Camphor → NAPHTHALENE
Tardigal → DIGITOXIN
Tardocillin → PENICILLIN > *Benzylpenicillin, Benzathine*
Tardomyocel → PENICILLIN > *Benzylpenicillin, Benzathine*
Tareg → VALSARTAN
"Tares" → FLOUR
Targocid → TEICOPLANIN
Targosid → TEICOPLANIN
Targretin → BEXAROTENE
Taricha granulosa → NEWT
Taricha torosa → TETRODOTOXIN
Taricotoxin → NEWT
Tarivid → OFLOXACIN
Taroctyl → CHLORPROMAZINE
Tarodyl → GLYCOPYRROLATE
Tarodyn → GLYCOPYRROLATE
Tarquinor → HALQUINOL
TARRAGON
Tars → LOLIUM TEMULENTUM
'Tar Smarts" → PITCH also → BENZO[α]PYRENE
Tartago → CASTOR BEAN
Tartarian Honeysuckle → LONICERA CAPRIFOLEUM
Tarter Emetic → ANTIMONY POTASSIUM TARTRATE
TARTRAZINE also → ACETYLSALICYLIC ACID, AMINOPYRINE, ANILINE, BENZOIC ACID, DEXTROAMPHETAMINE; ESTROGENS, CONJUGATED; GRAPE, HYDROXYUREA, IBUPROFEN, INDOMETHACIN, KETOPROFEN, MEFENAMIC ACID, MEGESTROL ACETATE, PARABENS, PHENYLBUTAZONE, PROPOXYPHENE, SECOBARBITAL SODIUM, SODIUM BENZOATE, STRAWBERRY, TETRACYCLINE
Tarweed → AMSINCKIA INTERMEDIA
Taseron → KETANSERIN
Tasis → BISOXATIN ACTATE
Task → DICHLORVOS
Tasmar → TOLCAPONE
Tasnon → PIPERAZINE
Tasso barbasso → VERBASCUM THAPSU
Tassu → EUPHORBIA, *dendroides*
TAT Ant Trap → THALLIUM
TATBA → TRIAMCINOLONE
Taterpex → CHLOROPROPHAM
TAT Home Guard Roach Poison → THALLIUM
TATRADYMIA sp.
Taumidrine → BAMIPINE
Taural → RANITIDINE
Tauredon → GOLD SODIUM THIOMALATE
TAURINE also → BACLOFEN
Tavanic → LEVOFLOXACIN
Tavegil → CLEMASTINE
Tavegyl → CLEMASTINE
Ta-Verm → PIPERAZINE
Tavist → CLEMASTINE
Tavor → LORAZEPAM
Taxilan → PERAZINE
TAXINE also → YEW
Taxiphyllin → MAGNOLIAS
Taxol → PACLITAXEL
TAXOTERE
Taxus sp. → YEW
Taxus brevifolia → PACLITAXEL
TAZAROTENE
Tazepam → OXAZEPAM
Tazicef → CEFTAZIMIDES
Tazidime → CEFTAZIMIDES

Tazobactam → PENICILLINS – "NEWER and SYNTHETICS" > *Piperacillin Sodium* also → PROBENECID
Tazorac → TAZAROTENE
TBC → *p-tert*-BUTYLCATECHOL
TBE → *sym*-TETRABROMOETHANE
Tb I-698 → THIACETAZONE
TBS → TRIBROMSALAN
TBZ → THIABENDAZOLE
3TC → LAMIVUDINE
TCA → TRICHLOROACETIC ACID
TCAB → 3,4-DICHLOROANILENE
TCAOB → 3,4-DICHLOROANILENE
T-Caps → CARBINOXAMINE MALEATE
TCBA → METHAZOLE
TCC → TRICLOCARBAN
TCDBD → DIOXIN
TCDD → DIOXIN
TCDS → TETRADIFON
TCE → TRICHLOROETHYLENE
T-Cell Growth Factor → ALDESLEUKIN also → INTERLEUKINS
TCGF → ALDESLEUKIN also → INTERLEUKINS
TCH → METHALLIBURE
TCP → TRITOLYL PHOSPHATE
2,4,5-TCP → 2,4,5-TRICHLOROPHENOL
TCSA → TETRACHLOROSALICYLANILIDE
TCV 116 → CANDESARTAN CILEXETIL
TD-480 → HEXAFLURATE
2,4 TDA → 2,4-TOLUENEDIAMINE
TDC #970 → ARSENAMIDE
TDE → ETHOGLUCID
TDE → 1,1-DICHLORO-2,2-BIS (*p*-CHLOROPHENYL)ETHANE
p,p´-TDE → 1,1-DICHLORO-2,2-BIS (*p*-CHLOROPHENYL)ETHANE
T. diffusa → TURNERA APHRODISIACA
TDI → TOLUENE-2,4-DIISOCYANATE
2,4-TDI → TOLUENE-2,4-DIISOCYANATE
TDP → COCARBOXYLASE
TE 031 → CLARITHROMYCIN
TEA also → DOXORUBICIN, ENDOSULFAN, HALOPERIDOL, KETOPROFEN, POLYSTYRENE, PROCARBAZINE, RUMEX sp., SENECIO sp., SHAVE GRASS, SODIUM, TANNIC ACID, WARFARIN
TEA also → ABRUS PRECATORIUS, ALOE, ALOYSIA, BARBASCO, BURDOCK, BUSH TEAS, CALAMUS, CATNIP, CEPHALANTHRUS, CESTRUM, CHAMOMILE, COCCULUS, COLTSFOOT, COMFREY, CONVALLARIA MAJALIS, COPPER CARBONATE, DATURA, DIGITALIS, DOGWOOD, DOXORUBICIN, DULCAMARIN, ECHIUM sp., ELDERBERRY, ENDOSULFAN, EUPATORIUM sp., EUPHORBIA, FLAX, FLUORIDES and FLUORINE, GERMANDER, GRAPHITE, HALOPERIDOL, HEDEOMA, HELIOTROPIUM sp., HOLLY, HYDRANGEA, HYPERICUM, ILEX sp., IRON, JUI, JUNIPER, KETOPROFEN, KOMBUCHA, LAGOCHILLUS, LARREA sp., LOBELIA, MARIGOLD, MELIA sp., MENTHOL; MISTLETOE, AMERICAN; MUSHROOMS, PAGAMEA MACROPHYLLA, PAPAVER sp., PAW-PAW, PAYLOOAH, PENNYROYAL, PEPPERMINT, PHORBOLMYRISTATE ACETATE, PHYTEUMA, PHYTOLACCA sp., POLYSTYRENE, PRASTERONE, PROCARBAZINE, PSEUDOEPHEDRINE, PYRROLIZIDINES, RETRORSINE, RHODODENDRON, RUMEX sp., SANGURA, SASSAFRAS, SCUTELLARIA LATFRIFLORA, SENECIO sp., SENNA, SHAVE GRASS, SMILAX, SODIUM, SOLANDRA sp. SORSACA, SYMPHYTINE, TANNIC ACID, TARAXACUM OFFICINALE, TILIA sp., TRIETHANOLAMINE, TRIETHYLAMINE, WARFARIN, XANTHORHIZA SIMPLICISSIMA
"Tea" → CANNABIS
TEA → TRIETHANOLAMINE, TRIETHYLAMINE
T.E.A. Chloride → TETRAETHYLAMMONIUM CHLORIDE
Tea-graver's Disease → TEA
Tearisol → CELLULOSE, *Oxidized*
TEA TREE
TEAK also → DESOXYLAPACHOL
Tebamin → POTASSIUM-*p*-AMINOSALICYLATE
Tebethion → THIACETAZONE
Tebloc → LOPERAMIDE
Tebonin → GINKGO BILOBA
Tebrazid → PYRAZINAMIDE
Tebron → HALOPROPANE
TECHNETIUM^{99m} also → MAGNESIUM SULFATE
Tecodin → OXYCODONE
Tecomin → LAPACHOL
Tecquinol → HYDROQUINONE
Tecramine → FLUFENAMIC ACID

Tecto → THIABENDAZOLE
Tectonia grandis → TEAK
Tedania ignis → SPONGE
"Teddies & Betties" → TRIPELENNAMINE
Tedelparin → DALTEPARIN
Tedion → TETRADIFON
Tedolan → ETODOLAC
TEDP → SULFOTEP
Teel → SESAME
Tefamin → AMINOPHYLLINE
Tefferol → IRON DEXTRINS
Tefilin → TETRACYCLINE
Teflon → POLYTEF
Teflox → TEMAFLOXACIN
Tefsiel C → TEGAFUR
TEG → TRIETHYLENE GLYCOL
Tega-Cetin → CHLORAMPHENICOL
TEGAFUR
Tega-Pyrone → DIPYRONE
Tegison → ETRETINATE
Tegopen → PENICILLINS – "NEWER and SYNTHETICS" > *Cloxacillin/Sodium salt*
Tegretal → CARBAMAZEPINE
Tegretol → CARBAMAZEPINE
Tëhara → PATINOA ICHTHYOTOXICA
Teichomycin A_2 → TEICOPLANIN
TEICOPLANIN
Tekodin → OXYCODONE
Tekresol → CRESOL, SAPONATED SOLUTION of
Tekwaisa → METHYL PARATHION
TEL → TETRAETHYL LEAD
Telazol → TILETAMINE
Teldane → TERFENADINE
Teldanex → TERFENADINE
Teldrin → CHLORPHENIRAMINE, *maleate*
Telebar → BARIUM SULFATE
Telemin → BISACODYL
TELEOCIDIN
Telepaque → IOPANOIC ACID
Teleprim → SULFAMETHOXAZOLE
Telesmin → CARBAMAZEPINE
Teletrast → IOPANOIC ACID
TELFAIRIA OCCIDENTALIS
Telfast → FEXOFENADINE
Telgin-G → CLEMASTINE
TELIOSTACHYA LANCEOLATA
TELITOXICUM PERUVIANUM
Tellurites → TELLURIUM
TELLURIUM
Tellurium Dioxide → TELLURIUM
Tellurium Fluoride → TELLURIUM HEXAFLUORIDE
TELLURIUM HEXAFLUORIDE
Tellurium Oxide → TELLURIUM
Tellurium Tetrafluoride → TELLURIUM
Telmin → MEBENDAZOLE
Telmintic → MEBENDAZOLE
TELMISARTAN
Telodrin → ISOBENZAN
Telone → CHLOROPICRIN
Telone II → 1,3-DICHLOROPROPENE
Telotrex → TETRACYCLINE
Telvar → MONURON
Telvol → PROMOXALONE
TEM → TRIETHYLENE MELAMINE
Temac → TEMAFLOXACIN
TEMAFLOXACIN
Temaril → TRIMEPRAZINE
Temasept → TRIBROMSALAN
TEMAZEPAM also → DISULFIRAM
Temefos → TEMEPHOS
TEMEPHOS
Temesta → LORAZEPAM
Temgesic → BUPRENORPHINE
Temik® → ALDICARB
Temlo → ACETAMINOPHEN
Temodal → TEMOZOLOMIDE
Temodar → TEMOZOLOMIDE
Temodol → TEMOZOLOMIDE
Temovate → CLOBETASOL
TEMOZOLOMIDE
Temposil → CALCIUM CARBIMIDE
Tempra → ACETAMINOPHEN
Temserin → TIMOLOL
Temuline → RYE GRASS
Tenacid → IMIPENEM
Tenaklene → PARAQUAT
Tenalet → CARTEOLOL HYDROCHLORIDE
Tenalin → CARTEOLOL HYDROCHLORIDE
Tenathan → BETHANIDINE
Tendex → PROPOXUR
Tendor → DEBRISOQUIN
Tenebrimycin → TOBRAMYCIN
TENECTEPLASE
Tenelid → GUANABENZ ACETATE
Tenemycin → TOBRAMYCIN
Tenesdol → DIPHENIDOL
Tenex → GUANFACINE
Teniathane → DICHLOROPHEN(E)
Teniatol → DICHLOROPHEN(E)
TENIPOSIDE
Tennecetin → NATAMYCIN
"Tennis Toe" → SHOES
Teno-basan → ATENOLOL
Tenoblock → ATENOLOL
Tenopt → TIMOLOL
Tenoran → CHLOROXURON

Tenormal → PEMPIDINE TARTRATE
Tenormin → ATENOLOL
Tenox BHA → BUTYLATED HYDROXYANISOLE
Tenox BHT → BUTYLATED HYDROXYTOLUENE
TENOXICAM also → RIFAMPIN
Tenpen → PENICILLIN > *Potassium Benzylpenicillinate*
Tensibar → BIETASERPINE
Tensilon → EDROPHONIUM
Tensinol → PEMPIDINE TARTRATE
Tensinyl → CHLORDIAZEPOXIDE
Tensobon → CAPTOPRIL
Tensofin → FLUPHENAZINE DECANOATE
Tensoprel → CAPTOPRIL
Tensopril → LISINOPRIL
Tensoral → PEMPIDINE TARTRATE
Tenso-Timelets → CLONIDINE
Tenuate → DIETHYLPROPION
Tenuate Dospan → DIETHYLPROPION
Tenyl → BETAHISTINE
Teofilcolina → CHOLINE THEOPHYLLINATE
Teofilina → THEOPHYLLINE
Teokolin → CHOLINE THEOPHYLLINATE
Teon → TEONANACATL
TEONANACATL also → PEYOTE
Teonova → THEOPHYLLINE
Teoptic → CARTEOLOL HYDROCHLORIDE
Teosona → THEOPHYLLINE
TEPA → TRIETHYLENE PHOSPHORAMIDE
Tepanil → DIETHYLPROPION
TEPHROSIA sp.
Tephrosin → TEPHROSIA sp.
Tepilta → OXETHAZINE
Tepogen → PENICILLINS – "NEWER and SYNTHETICS" > *Cloxacillin/Sodium salt*
TEPP → TETRAETHYL PYROPHOSPHATE
TEQUILA
Tequin → GETIFLOXACIN
Téralithe → LITHIUM CARBONATE
Teralutin → 17-α-HYDROXYPROGESTERONE
Teramine → CHLORPHENTERMINE
Terazol → TERCONAZOLE
TERAZOSIN
Terbasmin → TERBUTALINE SULFATE
Terbenol → NOSCAPINE
TERBINAFINE also → NORTRIPTYLINE, WARFARIN
TERBUFOS
Terbul → TERBUTALINE SULFATE
TERBUTALINE SULFATE also → PROPRANOLOL, SODIUM METABISULFITE, THEOPHYLLINE
TERCONAZOLE
Tercospor → TERCONAZOLE
Terdin → TERFENADINE
Terebinthinae → TURPENTINE
Terenac → MAZINDOL
Terenol → RESORANTEL
TERFENADINE also → CLARITHROMYCIN, DELAVIRDINE, DIRITHROMYCIN, DISULFIRAM, ERYTHROMYCIN, FLUCONAZOLE, FLUOXETINE, FLUVOXAMINE, GRAPEFRUIT, INDINAVIR, ITRACONAZOLE, KETOCONAZOLE, METRONIDAZOLE, MIBEFRADIL, MICONAZOLE NITRATE, NIFEDIPINE, RITONAVIR, SAQUINAVIR, TROGLITAZONE, TROLEANDOMYCIN, VERAPAMIL, ZILEUTON
Terfex → TERFENADINE
Terfluzine → TRIFLUOPERAZINE DIHYDROCHLORIDE
Tergitol 4 → SODIUM TETRADECYL SULFATE
Tergitol TP-9 → NONOXYNOL-9
Tergotol 15-S-9 → ALKYL-ω-HYDROXYPOLY(OXYETHYLENE)
Teriam → TRIAMTERENE
Teridax → IOPHENOXIC ACID
TERLIPRESSIN
Termil → CHLOROTHALONIL
TERMINALIA sp.
Terminator → DIAZINON
Ternadin → TERFENADINE
TERODILINE
Terolin → TERODILINE
Teronac → MAZINDOL
Terpate → PENTAERYTHRITOL TETRANITRATE
Terpene Polychlorinate → STROBANE®
Terpinen-4-ol → TEA TREE
Terpinyl Thiocyanoacetate → ISOBORNYL THIOCYANOACETATE
Terposen → RANITIDINE
Terra Cariosa → ROTTENSTONE
Terraclor → QUINTOZENE
Terracur P → FENSULFOTHION
Terrafungine → OXYTETRACYCLINE
Terraject → OXYTETRACYCLINE
Terralon-LA → OXYTETRACYCLINE
Terramycin → OXYTETRACYCLINE
Terrasytam → DIMEFOX
Terravenos → OXYTETRACYCLINE
Terretonin → ASPERGILLUS
Tersan → THIRAM
Tertran → IPRINDOLE

Tertroxin → LIOTHYRONINE
Teslac → TESTOLACTONE
Tesnol → PROPRANOLOL
Tespamin → THIOTEPA
Tessalon → BENZONATATE
Testa Ispaghula → PLANTAGO SEED
Testa-Scorbic → VITAMIN C
Testavol → VITAMIN A
Testoderm → TESTOSTERONE
TESTOLACTONE
Testolin → TESTOSTERONE
Testoral → FLUOXYMESTERONE
TESTOSTERONE also → ALCOHOL, ETHYL; CIMETIDINE, CHLORDANE, DDT, DIGOXIN, FLUCONAZOLE, GRAPEFRUIT, KETOCONAZOLE, MELATONIN, PHENOBARBITAL, PHENYLBUTAZONE, SPIRONOLACTONE, STANOLONE, THYROXINE, WARFARIN
Testred → METHYL TESTOSTERONE
Testro AQ → TESTOSTERONE
TETA → TRIENTINE
Tetan → TETRANITROMETHANE
Tetmosol → MONOSULFIRAM
Tetrabakat → TETRACYCLINE
TETRABENAZINE also → L-DOPA, NIALAMIDE
Tetrabid → TETRACYCLINE
Tetrablet → TETRACYCLINE
Tetrabon → TETRACYCLINE
sym-TETRABROMOETHANE
Tetrabromol Blue → BROMPHENOL BLUE
TETRABROMOSALICYLAMILIDE
TETRACAINE HYDROCHLORIDE also → ETHYL-*p*-AMINOBENZOATE, LIDOCAINE
Tetracap → TETRACHLOROETHYLENE
Tetracemate → DISODIUM EDETATE
Tetracemin → SODIUM EDETATE
Tetrachel → TETRACYCLINE
Tetrachlorethylene → TETRACHLOROETHYLENE
3,4,3′,4′-Tetrachloroazoxybenzene → METHAZOLE
1,1,2,2-TETRACHLORODIFLUORETHANE
Tetrachlorodiphenylethane → 1,1-DICHLORO-2,2-BIS(*p*-CHLOROPHENYL)ETHANE
1,1,2,2-TETRACHLOROETHANE
TETRACHLOROETHYLENE also → ALCOHOL, ETHYL; BARIUM, OLIVES
Tetrachloromethane → CARBON TETRACHLORIDE
TETRACHLOROSALICYLANILIDE also → BITHIONOL, DICHLOROPHEN
Tetrachlorvinphos → GARDONA
Tetracholoethene → TETRACHLOROETHYLENE
Tetracompren → TETRACYCLINE
TETRACOSACTIDE
Tetracosactrin → TETRACOSACTIDE
TETRACYCLINE also → ACENOCOUMAROL, ACETANALID, ALUMINUM, ALUMINUM HYDROXIDE, ANHYDROTETRACYCLINE, BARIUM, BENZYDAMINE, BISHYDROXYCOUMARIN, CALCIUM, CIMETIDINE, CITRIC ACID, DESMOPRESSIN, DOCUSATE SODIUM, FERROUS SULFATE, LITHIUM, MAGNESIUM, MAGNESIUM SILICATE, MAGNESIUM STEARATE, METHOXYFLURANE, METOCLOPRAMIDE, METRONIDAZOLE, MILK, NOVOBIOCIN, PECTIN, PENTOBARBITAL, PHENFORMIN, PHENYTOIN, PROCAINAMIDE, PROPANTHELINE, SODIUM BICARBONATE, STRONTIUM, SUCRALFATE, TARTRAZINE, THEOPHYLLINE, THIMEROSAL, TOLBUTAMIDE, *d*-TUBOCURARINE CHLORIDE, VITAMIN A, VITAMIN C, VITAMIN K, WARFARIN, ZINC
Tetracyclines → ACETYLCYSTEINE, BISMUTH, CALCIUM CARBIMIDE, CODEINE, CYCLOCOUMAROL, DESMOPRESSIN, INSULIN, IRON, LIPASE, PENICILLIN, POLYVINYLPYRROLIDONE, POTASSIUM, PROCAINAMIDE, STRONTIUM
Tetracyn → TETRACYCLINE
n-TETRADECANE
1-Tetradecanol → ALCOHOL, MYRISTYL
Tetradecin → TETRACYCLINE
Tetradecin Novum → TETRACYCLINE
Tetradecyl Alcohol → ALCOHOL, MYRISTYL
TETRADECYLTRIMETHYLAMMONIUM BROMIDE
TETRADIFON
Tetradon → FISH
Tetradonium Bromide → TETRADECYLTRIMETHYLAMMONIUM BROMIDE
TETRADYMA sp.
Tetradymol → TETRADYMA sp.
TETRAETHOXYSILANE
TETRAETHYLAMMONIUM CHLORIDE
TETRAETHYL LEAD → GASOLINE
Tetraethylplumbane → TETRAETHYL LEAD
TETRAETHYL PYROPHOSPHATE

Tetraethyl Silicate → ETHYL SILICATE
Tetraethylthiuram → MONOSULFIRAM
Tetrafenphos → TEMEPHOS
TETRAGONIA TETRAGONOIDES
Tetrahydroaminoacridine → TACRINE
Tetrahydrobenzenes → CYCLOHEXENES
Δ^9-TETRAHYDROCANNABINOL also → ANTIPYRINE, CANNABIS, DIAZEPAM, MELATONIN, PHENCYCLIDINE, TETRAHYDROZOLINE
TETRAHYDROFURAN
TETRAHYDROFURFURYL ALCOHOL
TETRAHYDROHARMINE also → AYAHUASCA
Tetrahydrolipostatin → ORLISTAT
Tetrahydrophenobarbital → CYCLOBARBITAL
py-Tetrahydroserpentine → AJMALICINE
Tetrahydroserpentine → AJMALICINE
Tetrahydrouroshiol → PENTADECYLCATECHOL
TETRAHYDROZOLINE
Tetraine → PROCAINE
3,5,3′,5′-Tetraiodothyroinine → THYROXINE
Tetrakap → TETRACYCLINE
TETRALIN
Tetralisal → LYMECYCLINE
Tetralite → NITRAMINE
Tetrallobarbital → BUTALBITAL
Tetralution → TETRACYCLINE
Tetralysal → LYMECYLINE
TETRAM®
Tetramavan → TETRACYCLINE
Tetramel → OXYTETRACYCLINE
Tetramethrin → PYRETHRUM, PYRETHRINS, and PYRETHROIDS
Tetramethylene Oxide → TETRAHYDROFURAN
TETRAMETHYL LEAD
N,N,N′,N′-Tetramethyl-3-(10H-phenothiazin-10-yl)-1,2-propane-diamine → AMINOPROMAZINE
TETRAMETHYL SUCCINONITRILE
Tetramethylthionine Chloride → METHYLENE BLUE
Tetramethylthiuram Disulfide → DISULFIRAM
Tetramide → MIANSERIN
TETRAMISOLE also → SUCCINYLCHOLINE
Tetramycin → TETRACYCLINE
Tetramyl → LYMECYCLINE
Tetran → OXYTETRACYCLINE
Tetran → POLYTEF
Tetranap → TETRALIN
TETRANITROMETHANE
Tetraolean → OLEANDOMYCIN
2,4,5,6-Tetraoxohexahydropyrimidine → ALLOXAN
Tetraphene → 1,2-BENZANTHRACENE
Tetrapon → THIRAM
TETRAPTERIS sp.
Tetra-Tablinen → OXYTETRACYCLINE
Tetraverin → ROLITETRACYCLINE
Tetraverine → TETRACYCLINE
Tetrazet → NEOMYCIN
Tetrex → TETRACYCLINE
Tetrim → ROLITETRACYCLINE
Tetrine → SODIUM EDETATE
Tetriv → ROLITETRACYCLINE
TETRODOTOXIN also → FISH, LEGOCEPHALUS, LEPISOSTEUS SPATULA, NEWT, SALAMANDERS
Tetrodoxin → TETRODOTOXIN
Tetron → TETRAETHYL PYROPHOSPHATE
Tetropil → TETRACHLOROETHYLENE
Tetrosol → TETRACYCLINE
Tetryl → NITRAMINE
Tetryzoline → TETRAHYDROZOLINE
Tetterwort → SANGUINARIA CANADENSIS
Teucrium chamaedys → GERMANDER, WILD
Tevcocin → CHLORAMPHENICOL
Tevcodyne → PHENYLBUTAZONE
Tevilon → POLYVINYL CHLORIDE
Texacort → HYDROCORTISONE
Texas Queen's Delight → STILLINGIA sp.
Texas Sarsaparilla → MENISPERMUM CANADENSE
Texas Wonder → CALADIUM
Texsolve S → NAPHTHA
TFM → TRIFLUOMEPRAZINE
T-2 Fusariotoxin → T-2 TOXIN
TGB → TIAGABINE
TH 100 → AMINOPHYLLINE
TH 152 → METAPROTERENOL
TH 1165a → FENOTEROL HYDROBROMIDE
TH 1314 → ETHIONAMIDE
TN 1321 → PROTIONAMIDE
TH 2180 → PROPANIDID
THA → TACRINE
Thacapsol → METHIMAZOLE
Thacapzol → METHIMAZOLE
Thalassin → SEA ANEMONES
Thalazole → PHTHALYLSULFATHIAZOLE
THALIDOMIDE
Thalitone → CHLORTHALIDONE
THALLIUM also → CEMENT, COAL, DIGITALIS, KALE, SELENIUM
Thalomid → THALIDOMIDE
THAM → TROMETHAMINE
Thanite → ISOBORNYL THIOCYANOACETATE

THC → Δ^9-TETRAHYDROCANNABINOL also → CANNABIS
Theal → DYPHYLLINE
THEBAINE also → OPIUM
"The Birds" → DOMOIC ACID
Thecodine → OXYCODONE
Theelin → ESTRONE
Theelol → ESTRIOL
Thefylan → DYPHYLLINE
The Garden → ANTIARIS TOXICARIA
The Hawk → LYSERGIC ACID DIETHYLAMIDE
Thein → CAFFEINE
Thekodin → OXYCODONE
Thelykinin → ESTRONE
THENALDINE TARTRATE
Thenclor → CHLOROTHEN
THENIUM CLOSYLATE
Thenyldiamine → ETHYLENEDIAMINE
Thenylene → METHAPYRILENE
Thenylpyramine → METHAPYRILENE
Theobid → THEOPHYLLINE
Theobroma cacao → CHOCOLATE
THEOBROMINE also → CACOA, CHOCOLATE, COCA, COCOA, DOXORUBICIN, ILEX sp., MILK, TEA
Theocap → THEOPHYLLINE
Theochron → THEOPHYLLINE
Theoclear → THEOPHYLLINE
Theodrox → AMINOPHYLLINE
Theo-Dur → THEOPHYLLINE
Theodyl → THEOPHYLLINE
Theograd → THEOPHYLLINE
Theolair → THEOPHYLLINE
Theolamine → AMINOPHYLLINE
Theolin → THEOPHYLLINE
Theolix → THEOPHYLLINE
Theolixir → THEOPHYLLINE
Theomin → AMINOPHYLLINE
Theon → THEOPHYLLINE
Theophyl → THEOPHYLLINE
Theophyldine → AMINOPHYLLINE
Theophyllamine → AMINOPHYLLINE
THEOPHYLLINE also → ACETAMINOPHEN, ADENOSINE, ALLOPURINOL, AMBUPHYLLINE, AMINOGLUTETHIMIDE, AMINOPHYLLINE, AMIODARONE, AZITHROMYCIN, BCG, CANNABIS, CARBONS, CETIRIZINE, CIMETIDINE, CIPROFLOXACIN, CLARITHROMYCIN, CLORAZEPATE DIPOTASSIUM, DILTIAZEM, ENOXACIN, ERYTHROMYCIN, ERYTHROMYCIN ETHYL SUCCINATE, ERYTHROMYCIN STEARATE, ERYTHROPOIETIN, ETHYLENEDIAMINE, FAMOTIDINE, FLEROXACIN, FLUCONAZOLE, FLUVOXAMINE, FUROSEMIDE, HYDROXYZINE, HYPERICUM PERFORATUM, ILEX sp., INTERFERONS, ISONIAZID, LIOTHYRONINE, LITHIUM, MAGNESIUM OXIDE, MERSALYL MEXILETINE, MINOCYCLINE, MORICIZINE, MORPHINE, NIMESULFIDE, NIFEDIPINE, NIZATIDINE, NORFLOXACIN, OMEPRAZOLE, PECTIN, PEFLOXACIN, PENTOXIFYLLINE, PHENOBARBITAL, PHENYTOIN, PROPAFENONE, PROPRANOLOL, PYRANTEL, RANITIDINE, RIFAMPIN, RITONAVIR, SODIUM BICARBONATE, SUCRALFATE, TACRINE, TEA, TERBUTALINE, TETRACYCLINE, TICLOPIDINE, TOBACCO, TOCAINIDE, TROLEANDOMYCIN, VANCOMYCIN, VERAPAMIL, VIDARABINE, VILOXAZINE, VITAMIN B_1, ZAFIRLUKAST, ZILEUTON
Theophylline Aminoisobutanol → AMBUPHYLLINE
Theophylline Ethylenediamine → AMINOPHYLLINE
Theoprastus → Preface
Théosalvose → THEOBROMINE
Theostat → THEOPHYLLINE
Theovent → THEOPHYLLINE
Theoxylline → CHOLINE THEOPHYLLINATE
Thephorin → PHENINDAMINE TARTRATE
The Pimp's Drug → COCAINE
Therabloat → POLOXALENE
Theraderm → BENZOYL PEROXIDE
Theralax → BISACODYL
Theralene → TRIMEPRAZINE
Therapas → BENZOYLPAS also → CALCIUM BENZOYLPAS
Therapav → PAPAVERINE
Therapin → XYLOMETAZOLINE
Theraplix → METHAZOLAMIDE
Theratuss → PIPAZETHATE
TherCys → BCG
Thermalon → THIAMBUTENE
THERMOACTINOMYCES VULGARIS also → SUCROSE
Thermofax → METHYL GALLATE
THERMOPOLYSPORA POLYSPORA
Thermopsine → THERMOPSIS sp.
THERMOPSIS sp.

Thermovyl → POLYVINYL CHLORIDE
Theruhistin → ISOTHIPENDYL
The Slavic Chronicles → CANNABIS
The Star Spangled Powder → COCAINE
"The Unfortunate Traveler" → GARLIC
THEVETIA sp.
Thevetin → THEVETIA sp.
THF → TETRAHYDROFURAN
Thiaben → THIABENDAZOLE
THIABENDAZOLE also → BENTONITE, GLUCOSE
Thiabenzole → THIABENDAZOLE
Thiacetarsamide → ARSENAMIDE
THIACETAZONE also → ISONIAZID
THIALBARBITAL
Thiamazole → METHIMAZOLE
THIAMBUTENE
Thiamcol → THIAMPHENICOL
Thiamine Diphosphate → COCARBOXYLASE
Thiamine Tetrahydrofurfuryl Disulfide → VITAMIN B_1
Thiaminium → VITAMIN B_1
Thiamin Propyl Disulfide → VITAMIN B_1
Thiamizide → DIAPAMIDE
Thiamol → VITAMIN B_1
THIAMPHENICOL
THIAMYLAL SODIUM also → EPINEPHRINE, NOREPINEPHRINE
Thiantan → DIETHAZINE
Thiaretic → HYDROCHLORTHIAZIDE
Thiazamide → SULFATHIAZOLE
Thiazenone → THIAZESIN
THIAZESIN
2-Thiazolamine → 2-AMINOTHIAZOLE
4-Thiazolidinecarboxylic Acid → TIMONACIC
THIAZOLSULFONE
Thiazosulfone → THIAZOLSULFONE
Thibenzole → THIABENDAZOLE
Thibone → THIACETAZONE
"Thick Ear" → TRIBULUS TERRESTRIS
Thickleaf Drymary → DRYMARIA sp., *pachyphylla*
"Thick Yellow Head" → TRIBULUS TERRESTRIS
Thidicur → SULFAMETHIZOLE
Thienamycin → CILASTATIN
Thienylic Acid → TICRYNAFEN
Thiergan → PROMETHAZINE
THIETHYLPERAZINE MALEATE
Thimecil → METHYLTHIOURACIL
Thimer → THIRAM
THIMEROSAL also → MERCURY, PHENYLMERCURIC NITRATE, TETRACYCLINE
Thimet → PHORATE
Thimul → ENDOSULFAN
THIOACETAMIDE
Thioacetazone → THIACETAZONE
Thiobarbitone → THIALBARBITAL
2,2′-Thiobis(4,6-dichlorophenol) → BITHIONOL
Thiobutyl Alcohol → *n*-BUTYL MERCAPTAN
Thiocarbamide → THIOUREA
Thiocarbazil → THIACETAZONE
THIOCARLIDE
Thiochrysine → GOLD SODIUM THIOSULFATE
THIOCOLCHICOSIDE
Thioctacid → THIOCTIC ACID
Thioctan → THIOCTIC ACID
THIOCTIC ACID
Thioctidase → THIOCTIC ACID
Thiocuran → SULFAMETHOXAZOLE
THIOCYANATES also → BRASSICA sp., CAULIFLOWER, DESCURAINIA PINNATA, KALE, LEAD, SODIUM NITROPRUSSIDE, TOBACCO
Thiocymetin → THIAMPHENICOL
Thiodan → ENDOSULFAN
Thiodiphenylamine → PHENOTHIAZINE
Thioethyl Alcohol → ETHANETHIOL
Thiofen → BITHIONOL
Thiofor → ENDOSULFAN
THIOGLYCOLIC ACID
THIOGUANINE also → ALLOPURINOL, CYTARABINE
THIOGUANOSINE
Thiola → TIOPRONIN
Thiomembumal → THIOPENTAL
Thiomerin Sodium → MERCAPTOMERIN SODIUM
Thiomersal → THIMEROSAL
Thiomersalate → THIMEROSAL
THIOMESTERONE
Thiomucase → HYALURONIDASE
Thionazin → ZINOPHOS
Thionembutal → THIOPENTAL
Thionex → ENDOSULFAN
Thionicol → THIAMPHENICOL
Thionylan → METHAPYRILENE
THIONYL CHLORIDE
Thioparamizone → THIACETAZONE
THIOPENTAL also → CHOLESTEROL, EPINEPHRINE, IMIPRAMINE, METHDILAZINE, METHOTRIMEPRAZINE, NOREPINEPHRINE, PROMAZINE, PROPIOMAZINE, PROTHIPENDYL, SCHRADAN, SUCCINYLCHOLINE, SULFISOXAZOLE

Thiopentalum → THIOPENTAL
Thiopentone → THIOPENTAL
THIOPERAZINE
Thiophenicol → THIAMPHENICOL
Thiophos → PARATHION
Thioplex → THIOTEPA
Thioproline → TIMONACIC
THIOPROPAZATE
Thioproperazine → THIOPERAZINE
THIORIDAZINE also → ALUMINUM HYDROXIDE, AMITRIPTYLINE, BROMOCRIPTINE, DEXTROAMPHETAMINE, DIPHENHYDRAMINE, IMIPRAMINE, LITHIUM, METRIZAMIDE, METUREDEPA, NALTREXONE, PAROXETINE, PHENYTOIN, PINDOLOL, QUETIPINE FUMARATE, QUINIDINE, RESERPINE, TOLBUTAMIDE, TRAZODONE HCL
Thiosalicylic Acid → THIMEROSAL
Thiosan → THIRAM
Thioseconal → THIAMYLAL SODIUM
THIOSEMICARBAZIDE
Thiosulfil → SULFAMETHIZOLE
Thiosulfurous Dichloride → SULFUR MONOCHLORIDE
Thiotan → DIETHAZINE
Thiotax → 2-MERCAPTOBENZOTHIAZOLE
THIOTEPA also → PANCURONIUM BROMIDE
Thiotepp → SULFOTEP
Thiotex → THIRAM
THIOTHIXENE
Thiothyron → METHYLTHIOURACIL
THIOURACIL also → MILK
THIOUREA also → WET SUITS
Thioxamyl → OXAMYL
Thioxolone → TIOXOLONE
THIP → AGARIN
Thipendyl → ISOTHIPENDYL
THIRAM also → DISULFIRAM, GLOVES, GOLF BALLS, 2-MERCAPTOBENZOTHIAZOLE, MONOSULFIRAM, RUBBER, SHOES
Thirame → THIRAM
Thirasan → THIRAM
Thitrol → 4(2-METHYL-4-CHLOROPHENOXY)BUTYRIC ACID
Thiurad → THIRAM
Thiuram → THIRAM
Thiuramyl → THIRAM
Thiuretic → HYRDOCHLOROTHIAZIDE
Thixokon → SODIUM ACETRIZOATE
THLASPI ARVENSE
Thom → KRATOM
Thoman → LEAD
Thomas Balsam → BALSAM TOLU
Thomas Nagle → GARLIC
THONZYLAMINE HYDROCHLORIDE
Thorazine → CHLORPROMAZINE
Thoria → THORIUM OXIDE
Thorium Dioxide → THORIUM OXIDE
THORIUM OXIDE
Thornapple → DATURA
Thoxan → ETOXADROL HCL
Thoxidil → MEPHENESIN
Threadleaf Groundsel → SENECIO sp.
THRELKELDIA PROCIFLORA
Threostat II → PROPYLTHIOURACIL
Thrombasal → PHENINDIONE
Thrombatin → PHENINDIONE
Thrombocid → PENTOSAN POLYSULFATE
Thrombocytin → SEROTONIN
Thrombohepin → HEPARIN
Thromboliquine → HEPARIN
Thrombolysin → FIBRINOLYSIN
Thrombophob → HEPARIN
Thrombosan → PHENINDIONE
Thrombran → TRAZODONE HCL
Throw Wort → LEONURUS CARDIACA
Thrush-XX → COPPER NAPHTHENATE
Thrusters → AMPHETAMINES
THS 839 → DENATONIUM BROMIDE
Thuja plicata → CEDAR, WESTERN RED
THUJA OCCIDENTALIS
Thuja Oil → THUJA OCCIDENTALIS
THUJONE also → ABSINTHIUM, APIOLE, OAKMOSS, TANSY, THUJA OCCIDENTALIS
THURFYL NICOTINATE
Thybon → LIOTHYRONINE
Thybromol → BROMTHYMOL BLUE
Thycapsol → METHIMAZOLE
THYLACHIUM AFRICANUM
Thylate → THIRAM
Thylogen → PYRILAMINE
Thylose → CELLULOSE
Thyme Camphor → THYMOL
THYMIDINE
THYMINE
THYMOL
THYMOXAMINE
Thypinone → THYROTROPIN RELEASING HORMONE
Thyradin → THYROID
Thyratrop → THYROTROPIN RELEASING HORMONE
Thyrefact → THYROTROPIN RELEASING HORMONE

Thyreostat → METHYLTHIOURACIL
Thyroblock → POTASSIUM IODIDE
Thyrocalcitonin → CALCITONIN
Thyrocrine → THYROID
Thyrogen → THYROTROPIN
Thyroglobulin(e) → THYROID
THYROID also → AMITRIPTYLINE, HAMBURGERS, IMIPRAMINE, KETAMINE, METHIMAZOLE, METHYL TESTOSTERONE, PHENTERMINE, POLYBROMINATED BIPHENYLS, PROPRANOLOL, SOYBEANS
Thyroidine → THYROID
Thyroid Stimulating Hormone → THYROTROPIN
Thyrojod → POTASSIUM IODIDE
Thyroliberin → THYROTROPIN
Thyroton → THYROTROPIN
Thyrotrophin → THYROTROPIN
Thyrotropic Hormone → THYROTROPIN
THYROTROPIN also → CHLOROQUINE
Thyrotropin Releasing Factor → THYROTROPIN RELEASING HORMONE
THYROTROPIN RELEASING HORMONE
Thyroxevan → THYROXINE
THYROXINE also → ACENOCOUMAROL, ACETONITRILE, AMIODARONE, AMPHETAMINE, CARBAMAZEPINE, CLOFIBRATE, FERROUS SULFATE, FUROSEMIDE, HALOFENATE, HEXOBARBITAL, LOVASTATIN, NALIDIXIC ACID, PHENOBARBITAL, PHENYTOIN, NITROFURANTOIN, PHENINDIONE, PREDNISONE, SALSALATE, SUCRALFATE, ZOXAZOLAMINE
Thytropar → THYROTROPIN
Tiazon → DAZOMET
Tibatin → CLOTRIMAZOLE
Tiabenzole → THIABENDAZOLE
TIAGABINE
TIAMULIN also → MONENSIN, NARASIN, SALINOMYCIN
Tiamutin → TIAMULIN
TIAPAMIL
TIAPROFENIC ACID
Tiazolidin → TIMONACIC
Tiberal → ORNIDAZOLE
Tibinide → ISONIAZID
Tibione → THIACETAZONE
Tibon → THIACETAZONE
Tibricol → NIFEDIPINE
Tibutol → ETHAMBUTOL
TIC → PHENCYCLIDINE
Ticar → PENICILLINS – "NEWER and SYNTHETICS" > *Ticarcillin Disodium*
Ticarcillin Disodium → PENICILLINS – "NEWER and SYNTHETICS" also → AMIKACIN, CLAVULANIC ACID, GENTAMICIN, TOBRAMYCIN
Ticarpen → PENICILLINS – "NEWER and SYNTHETICS" > *Ticarcillin Disodium*
Ticillin → PENICILLINS – "NEWER and SYNTHETICS" > *Ticarcillin Disodium*
Tick Bean → BEANS, Faba
Tickberry → LANTANA sp.
Ticlid → TICLOPIDINE HCL
Ticlobran → CLOFIBRATE
Ticlodix → TICLOPIDINE HCL
Ticlodone → TICLOPIDINE HCL
TICLOPIDINE HCL also → THEOPHYLLINE, WARFARIN
Ticlosin → TICLOPIDINE HCL
TICRYNAFEN also → WARFARIN
Tidemol → BUFORMIN
TIEMONIUM IODIDE
Tiempe → TRIMETHOPRIM
Tienam → IMIPENEM also → CILASTATIN
Tien Hua Fen → TRICHOSANTHIN
Tienilic Acid → TICRYNAFEN
Tienor → CLOTIAZEPAM
Tiesene → ZINEB
Tiezene → ZINEB
Tifomycine → CHLORAMPHENICOL
Tigacol → NICOTINYL ALCOHOL
Tigan → TRIMETHOBENZAMIDE
Tigason → ETRETINATE
Tigerfish → PTEROIS VOLITANS
Tiger Lily → LILIES
Tiger Tree → THEVETIA sp.
12-σ-Tigloyl-4-deoxyphorbol-13-isobutyrate → SYNADENIUM GRANTII
Tiguvon → FENTHION
Tiha → SYNADENIUM GRANTII
Tik-20 → DIAZINON
Tiklid → TICLOPIDINE
Tikofuran → FURAZOLIDONE
Tikosyn → DOFETILIDE
Tilade → NEDOCROMIL SODIUM
Tilarin → NEDOCROMIL SODIUM
Tilatil → TENOXICAM
Tilcarex → QUINTOZENE
Tilcotil → TENOXICAM
Tildiem → DILTIAZEM
Tildin → CARBROMAL
TILETAMINE
TILIA sp.

TILLITIA sp.
Tillman → PEBULATE
Tilmapor → CEFSULODIN SODIUM
TILMICOSIN
TILUDRONATE
Timacor → TIMOLOL
Timazin → FLUOROURACIL
Timbauba → ENTEROLOBIUM
Timber Milkvetch → MILKVETCH
Timelit → LOFEPRAMINE
Timentin → PENICILLINS – "NEWER and SYNTHETICS" > *Ticarcillin Disodium*
Timet → PHORATE
Timocort → HYDROCORTISONE
Timodyne → MEFEXAMIDE
TIMOLOL also → NIFEDIPINE, PAROXETINE, QUINIDINE
TIMONACIC
Timonil → CARBAMAZEPINE
Timoped → TOLNAFTATE
Timoptic → TIMOLOL
Timoptol → TIMOLOL
Timosin → CHLORDIAZEPOXIDE
TIMOTHY
Timothy Grass → RYE GRASS
Timovan → PROTHIPENDYL
Timpylate → DIAZINON
TIN also → STANNOUS CHLORIDE
Tinactin → TOLNAFTATE
Tinaderm → TOLNAFTATE
Tinarhinin → TETRAHYDROZOLINE
TIN ARSENATE and TIN ARSENITE
Tinavet → TOLNAFTATE
Tindanol → ETHANOLAMINE
Tin Dichloride → STANNOUS CHLORIDE
Tin Difluoride → STANNOUS FLUORIDE
Tindurin → PYRIMETHAMINE
TINIDAZOLE
Tinnevely Senna → SENNA
Tinostat → DIBUTYL TIN DILAURATE
Tinset → OXATOMIDE
Tintorane → WARFARIN
Tinuvin 327 → 2-(5-CHLORO-2H-BENZOTRIAZOL-2-YL)-4,6-BIS (1,1-DIMETHYLETHYL)-PHENOL
Tinya → EUPHORBIA
TINZAPARIN SODIUM
Tiobicina → THIACETAZONE
Tiocarlide → THIOCARLIDE
TIOCONAZOLE
Tioctan → THIOCTIC ACID
TIOFOSFAMID → THIOTEPA
Tiofosyl → THIOTEPA
Tiomesterone → THIOMESTERONE
Tio-Mid → ETHIONAMIDE
TIOPRONIN
Tiotilin → TIAMULIN
Tiotixene → THIOTHIXENE
Tiouracil → THIOURACIL
Tiovalon → TIXOCORTOL PIVALATE
TIOXOLONE
Tipitipi → ABRUS PRECATORIUS
Tipton Weed → HYPERICUM
Tirade → FENVALERATE
Tirampa → THIRAM
Tiratricol → TRIIODOTHYROACETIC ACID
Tirian → CHLOROGUANIDE
Tirilazad → NIMODIPINE, PHENYTOIN
TIROFIBAN
Tiroide → THYROID
Tiroidina → THYROID
Tirossina → THYROXINE
Tiroxina → THYROXINE
Tisercin → METHOTRIMEPRAZINE
TISSUE PLASMINOGEN ACTIVATOR also → RETEPLASE
TITANIUM DIOXIDE
Titanium Hydroxide → TITANIUM TETRACHLORIDE
Titanium Oxide → TITANIUM DIOXIDE
Titanium Oxychlorides → TITANIUM TETRACHLORIDE
Titanium Peroxide → TITANIUM DIOXIDE
TITANIUM TETRACHLORIDE
Titriplex III → DISODIUM EDETATE
Tityus → SCORPIONS
Tixair → ACETYLCYSTEINE
Tixantone → LUCANTHOE HYDROCHLORIDE
TIXOCORTOL PIVALATE
TIZANIDINE also → PHENYTOIN
TL 1578 → TABUM
TM 906 → TRIMEBUTINE MALEATE
TMA → TRIMELLITIC ANHYDRIDE also → TRIMETHYLAMINE
"TMA Flu" → TRIMELLITIC ANHYDRIDE
TMAN → TRIMELLITIC ANHYDRIDE
TMB → TRIMEBUTINE MALEATE
TML → TETRAMETHYL LEAD
TMP → TRIOXSALEN
TMPE → MESCALINE
TMPEA → MESCALINE
TMQ → TRIMETREXATE
TMSN → TETRAMETHYL SUCCINONITRILE
TMTD → THIRAM
TMZ → TEMOZOLOMIDE

TNFRiFc → ETANERCEPT
TNKase → TENECTEPLASE
TNM → TETRANITROMETHANE
TNR 01 → ETANERCEPT
TNT → 2,4,6-TRINITROTOLUENE
Toado → FISH
TOADS
TOBACCO also → ACROLEIN, CARBON MONOXIDE, CYANIDES, CYCLANDELATE, DDT, DIETHYLENE GLYCOL, FLUPHENAZINE, HALOPERIDOL, HYDROGEN CYANIDE, LEAD, NICOTIANA sp., *N*-NITROSONORNICOTINE, PERILLA FRUTESCENS, POLONIUM, PROPRANOLOL, PYRIDINE, RADIUM; SUMAC, POISON
Tobra → TOBRAMYCIN
Tobracin → TOBRAMYCIN
Tobradistin → TOBRAMYCIN
Tobralex → TOBRAMYCIN
Tobramaxin → TOBRAMYCIN
TOBRAMYCIN also → FUROSEMIDE, IBUPROFEN, MICONAZOLE NITRATE, NEOMYCIN; "PENICILLINS – "NEWER and SYNTHETICS" > *Piperacillin Sodium, Ticarcillin Disodium*; QUINIDINE, VANCOMYCIN
Tobrex → TOBRAMYCIN
TOCAINIDE also → PROPRANOLOL, THEOPHYLLINE
Toclase → CARBETAPENTANE CITRATE
Tocodilydrin → NYLIDRIN
Tocodrin → NYLIDRIN
α-Tocopherol → VITAMIN E
Tocosamine → SPARTEIN
ToDay → CEPHAPIRIN
Today Sponge® → DIOXANE also → 2,4-TOLUENEDIAMINE
Toeba Root → DERRIS ROOT
Toefa Root → DERRIS ROOT
Toe Negra → TELIOSTACHYA LANCEOLATA
Tofacine → TOFENACIN HCL
TOFENACIN HCL
TOFISOPAM
Tofisopam → TOFISOPAM
Tofosam → TOFISOPAM
Tofranil → IMIPRAMINE
TOFU
Togamycin → SPECTINOMYCIN
Togiren → ERYTHROMYCIN ESTOLATE
TOK → NITROFEN
"Toka" → TURKEY
Tokiocillin → PENICILLINS – "NEWER and SYNTHETICS" > *Ampicillin*
Tokiolexin → CEPHALEXIN
Tokocin → DIBEKACIN
Tokokin → ESTRONE
Toksobidin → TOXOGONIN
Tokuderm → BETAMETHASONE
Tolanase → TOLAZAMIDE
Tolansin → MEPHENESIN
Tolax → MEPHENESIN
TOLAZAMIDE also → KAOLIN, RIFAMPIN
Tolazolamide → TOLAZAMIDE
TOLAZOLINE HCL also → DOPAMINE, NOREPINEPHRINE
Tolazul → TOLONIUM CHLORIDE
Tolban → PROFLURALIN
Tolbugen → TOLBUTAMIDE
Tolbusal → TOLBUTAMIDE
TOLBUTAMIDE also → ACETYLSALICYLIC ACID; ALCOHOL, ETHYL; AMINOGLUTETHIMIDE, APAZONE, BISHYDROXYCOUMARIN, CHLORAMPHENICOL, CLOFIBRATE, ETHYL-*p*-AMINOBENZOATE, FUROSEMIDE, HEXESTROL, ISOCARBOXAZID, KAOLIN, KETOCONAZOLE, LEFLUNOMIDE, METHYLDOPA, OXYPHENBUTAZONE, PARALDEHYDE, PHENOBARBITAL, PHENYLBUTAZONE, PHENYRAMIDOL, PHENYTOIN, PROBENECID, PROPRANOLOL, RIFAMPIN, SERTRALINE, SODIUM SALICYLATE, SODIUM VALPROATE, SULFADIAZINE, SULFADIMETHOXINE, SULFAMETHOXYPYRIDAZINE, SULFANILAMIDE, SULPHAPHENAZOLE, SULFISOXAZOLE, TERBINAFINE, TETRACYCLINE, WARFARIN, ZAFIRLUKAST
Tolbutol → TOLBUTAMIDE
Tolbutone → TOLBUTAMIDE
TOLCAPONE also → L-DOPA
Tolcasone → TRICHLORMETHIAZIDE
Tolcil → MEPHENESIN
Tolectin → TOLMETIN
Toleron → FERROUS FUMARATE
Tolferain → FERROUS FUMARATE
Tolguacha → DATURA
Tolhart → MEPHENESIN
σ-TOLIDINE also → *N,N*-DI(ACETOACETYL)-σ-TOLIDINE
Tolifer → FERROUS FUMARATE
Toliman → CINNARIZINE
Tolinase → TOLAZAMIDE
Tolisartine → TOLPERISONE HCL

Tolit → 2,4,6-TRINITROTOLUENE
Tolmene → TOLMETIN
TOLMETIN also → WARFARIN
TOLNAFTATE also → MONOAMYLAMINE
Tolnate → PROTHIPENDYL
TOLONIUM CHLORIDE
Tolosate → MEPHENESIN
TOLOXATONE also → SUMATRIPTAN SUCCINATE
Toloxyn → MEPHENESIN
TOLPERISONE HCL
TOLRESTAT
Tolrestatin → TOLRESTAT
Tolseran → MEPHENESIN
Tolserol → MEPHENESIN
Tolseron → MEPHENESIN
TOLTERODINE TARTRATE
TOLUENE also → ALCOHOL, ETHYL; BUTANONE, DISTILLATE, GLUE, RUBBER
2,4-TOLUENEDIAMINE also → POLYURETHANE, SILICONES
TOLUENE-2,4-DIISOCYANATE also → POLYURETHANE
σ-Toluenesulfamine → σ-TOLUENE SULFONAMIDE
σ-TOLUENE SULFONAMIDE also → SACCHARIN
p-TOLUENE SULFONIC ACID
Toluene Trichloride → BENZOTRICHLORIDE
TOLUIDINE
Toluidine Blue O → TOLONIUM CHLORIDE
Toluina → TOLBUTAMIDE
Tolulexin → MEPHENESIN
Tolulox → MEPHENESIN
Toluol → TOLUENE
Toluquinone → MILLIPEDES
Tolu-safranine → SAFRANINES also → LIPSTICK
Tolvin → MIANSERIN
Tolvon → MIANSERIN
4-(σ-Tolyazo)-σ-toluidine → σ-AMINOAZOTOLUENE
Tolycar → CEFOTAXIME
p-Tolyl Acetate → *p*-CRESYL ACETATE
Tolyprin → APAZONE
Tolyspaz → MEPHENESIN
Tomabef → CEFOPERAZONE
Tomamas → LEDUM sp.
Tomarin → COUMAFURYL
Tomatidenol → POTATO
TOMATO also → LATEX, PEPPERS, SCOPOLAMINE, SODIUM GLUTAMATE
Tombran → TRAZODONE
Tomocalcin → CALCITONIN
Tomorin → COUMACHLOR
Tomoscar → MENOGARIL
Tonamil → THONZYLAMINE HYDROCHLORIDE
Tonco Bean → TONKA BEANS
Tonco-tonco → DATURA
Tonedron → METHAMPHETAMINE
Tonephin → VASOPRESSIN
Tonexol → TETRACAINE HYDROCHLORIDE
Tonibral → DEANOL
Tonic Water → QUININE
TONKA BEANS also → MELILOT
Tonka Bean Camphor → COUMARIN
Tonocard → TOCAINIDE
Tonocholin B → ACETYLCHOLINE
Tonoftal → TOLNAFTATE
Tonopres → DIHYDROERGOTAMINE METHANESULFONATE (MESYLATE)
Tonquin Bean → TONKA BEANS
Too-a-poo → BEANS, Winged
Toot → COCA
Toothache Tree → ZANTHOXYLUM sp.
Toothpick Ammi → AMMI VISNAGA
TOOTHPICKS
Topalgic → SUPROFEN also → TRAMADOL
Topamax → TOPIRAMATE
Topazone → FURAZOLIDONE
Topclip Parasol → CYPERMETHRIN
Top Form Wormer → THIABENDAZOLE
Tophol → CINCHOPHEN
Tophosan → CINCHOPHEN
Topicain → OXETHAZINE
Topicorte → DESOXIMETASONE
Topicycline → TETRACYCLINE
Topifug → DESONIDE
Topiglan → ALPROSTADIL
TOPIRAMATE also → PHENYTOIN, VALPROIC ACID
Topisolon → DESOXIMETASONE
Topitox → CHLOROPHACINONE
Topline → AMITRAZ
Topocaine → CYCLOMETHYCAINE
TOPOTECAN
Toprec → KETOPROFEN
Toprek → KETOPROFEN
Toprol-XL → METOPROLOL
"Tops & Bottoms" → TRIPELENNAMINE
Topsym → FLUOCINONIDE
Topsymin → FLUOCINONIDE
Topsyn → FLUOCINONIDE
Topsyne → FLUOCINONIDE
Toquilone → METHAQUALONE
Toradiur → TORSEMIDE
Toradol → KETOROLAC

Toraflon → METHAQUALONE
Torak → DIALIFOR
Torasemide → TORSEMIDE
Torate → BUTORPHANOL
Toratex → KETOROLAC
Torazina → CHLORPROMAZINE
Torbugesic → BUTORPHANOL
Torbutrol → BUTORPHANOL
Torch Oil → XYLITOL
Tordon → PICLORAM
Torecan → THIETHYLPERAZINE MALEATE
Torem → TORSEMIDE
TOREMIFENE CITRATE also → TAMOXIFEN CITRATE
Torental → PENTOXIFYLLINE
Toresten → THIETHYLPERAZINE MALEATE
Toricelocin → CEPHALOTHIN
Torinal → METHAQUALONE
Toriol → RANITIDINE
Torlamicina → ERYTHROMYCIN
Tormona → 2,4,5-TRICHLOROPHENOXYACETIC ACID
Tornalate → BITOLTEROL
Torrat → TRIMEPRANOL
TORSEMIDE
Toryn → CARAMIPHEN
Tosnone → CARBETAPENTANE CITRATE
Tossimex → BROMHEXINE
Tostram → ITRAMIN TOSYLATE
Tostrex → TESTOSTERONE
Tosufloxacin → ALUMINUM HYDROXIDE also → MILK
Totacef → CEFAZOLIN SODIUM
Totacillin → PENICILLINS – "NEWER and SYNTHETICS" > *Ampicillin*
Totalciclina → PENICILLINS – "NEWER and SYNTHETICS" > *Ampicillin*
Totapen → PENICILLINS – "NEWER and SYNTHETICS" > *Ampicillin*
Totazina → COLISTIN
Totifen → KETOTIFEN FUMARATE
Totomycin → TETRACYCLINE
Touch Me Not → IMPATIENTS sp.
Tournefortia gnapholodes → TABACU DI PISCADO
to-vo → XANTHOSOMA sp.
Toxakil → TOXAPHENE
TOXALBUMIN also → ABRIN
TOXAPHENE
Toxichlor → CHLORDANE
Toxicodendron diversilobum → POISON OAK
Toxicodendron radicans → POISON IVY
Toxicodendron vernix → SUMAC, POISON
TOXIFERINE also → ALCURONIUM
Toxilic Anhydride → MALEIC ANHYDRIDE
Toxinal → OXYTETRACYCLINE
Toxital® → PENTOBARBITAL
Toxivers → PIPERAZINE
Toxobin → POLYVINYLPYRROLIDONE
TOXOGONIN
Toxonin → TOXOGONIN
Toxylon pomiferum → OSAGE ORANGE
Toyomycin → CHROMOMYCIN A
TP 21 → THIORIDAZINE
2,4,5-TP → SILVEX
t-PA → TISSUE PLASMINOGEN ACTIVATOR
TPA → PHORBOL
TPA → TISSUE PLASMINOGEN ACTIVATOR
TPD → VITAMIN B_1
TPPA → THIOTEPA
TPP → COCARBOXYLASE
TPP → TRIPHENYLPHOSPHATE
TPS 23 → MESORIDAZINE
TR 495 → METHAQUALONE
Trabest → CLEMASTINE
TRACHEMENE GLAUCIFOLIA also → WILD CELERY and WILD PARSNIPS
Trachinus draco → GREAT WEEVER
TRACHYANDRA SALTII
Tracix → IMIPENEM
Tracosal → OXPRENOLOL HYDROCHLORIDE
Tracrium → ATRACURIUM
Tradon → PEMOLINE
Trafigal → TRIMETHOPRIM
TRAGACANTH GUM
TRAGIA sp.
Trakipeal → CHLORDIAZEPOXIDE
Tral → HEXOCYCLIUM
Tralgon → ACETAMINOPHEN
Tralin → HEXOCYCLIUM
Tramacin → TRIAMCINOLONE
TRAMADOL also → PHENPROCOUMON, SERTRALINE, WARFARIN
Tramal → TRAMADOL
Trametan → THIRAM
Tramisole → LEVAMISOLE
Trancolon → MEPENZOLATE BROMIDE
Trancopal → CHLORMEZANONE
Trancote → CHLORMEZANONE
Trandate → LABETALOL
Tranex → TRANEXAMIC ACID
TRANEXAMIC ACID
Tranexan → TRANEXAMIC ACID
Trangorex → AMIODARONE
Tranimal → DIAZEPAM

Tranimul → CHLORDIAZEPOXIDE
Trank → PHENCYCLIDINE
Trankimazin → ALPRAZOLAM
Trankvilan → MEPROBAMATE
Tranlisant → MEPROBAMATE
Tranpoise → MEPHENOXALONE
Tran-Q → HYDROXYZINE
Tranquase → DIAZEPAM
Tranquazine → PROMAZINE
Tranquilan → MEPROBAMATE
Tranquilax → MEDAZEPAM
Tranquiline → MEPROBAMATE
Tranquillin → BENACTYZINE
Tranquizine → HYDROXYZINE
Tranquo-Puren → DIAZEPAM
Tranquo-Tablinen → DIAZEPAM
Transamin → TRANEXAMIC ACID
Transanate → CHLORMEZANONE
Transbilix → MEGLUMINE IODIPAMIDE
Transcop → SCOPOLAMINE
Transcutal → 2-(2-ETHOXYETHOXY)ETHANOL
Transcycline → ROLITETRACYCLINE
Transderm → SCOPOLAMINE
Transderm-Nitro → NITROGLYCERIN
Transgo® → TRANSMISSION FLUIDS
Transiderm-Nitro → NITROGLYCERIN
TRANSMISSION FLUIDS
Transparente → MYOPORUM sp.
Transvaal Daisy → GERBERA
Transvaal Slangkop → URGINEA sp.
Trantan → SPECTINOMYCIN
Trantoin → NITROFURANTOIN
Tranvet → PROPIONYLPROMAZINE
Tranxene → CHLORAZEPATE
Tranxène → CLORAZEPATE DIPOTASSIUM
Tranxilène → CLORAZEPATE DIPOTASSIUM
TRANYLCYPROMINE also → ACENOCOUMAROL; ALCOHOL, ETHYL; AMPHETAMINE, AVOCADO, BEANS, CHEESES, CORN, DESIPRAMINE, DEXTROAMPHETAMINE, DIPHENHYDRAMINE, ETHYL BISCOUMACETATE, FLUOXETINE, HERRING, IMIPRAMINE, INSULIN, MEPERIDINE, METHAMPHETAMINE, METHIONINES, SERTRALINE, TOLCAPONE, TRIFLUOPERAZINE DIHYDROCHLORIDE, L-TRYPTOPHAN, VENLAFAXINE
Tranzine → PROMAZINE
Trapanal → THIOPENTAL
Trapex → METHYLISOTHIOCYANATE
Trasacor → OXPRENOLOL HYDROCHLORIDE
Trasamlon → TRANEXAMIC ACID
Trasicor → OXPRENOLOL HYDROCHLORIDE
Traslan → CHLORMADINONE
TRASTUZUMAB also → WARFARIN
Trasylol® → APROTININ
Tratul → CIMETIDINE
Traubenzucker → GLUCOSE
Traumacut → METHOCARBAMOL
Traumanase → BROMELAIN
Traumasept → POLYVINYLPYRROLIDONE IODINE COMPLEX
Traumatociclina → MECLOCYCLINES
Traumon Gel → ETOFENAMATE
Trausabun → MELITRACEN HCL
Travamin → PROTEIN HYDROLYSATES
Travamine → DIMENHYDRINATE
Travase → SUTILAINS
Traveler's Joy → CLEMATIS sp., *vitalla*
Travel-Gum → DIMENHYDRINATE
Travelin → DIMENHYDRINATE
Travelmin → DIMENHYDRINATE
Travert → INVERT SUGAR
Travin → BUSPIRONE
Trax-One → BROMADIOLONE
Trazinin → APROTININ
TRAZODONE HCL also → THIORIDAZINE, TRIFLUOPERAZINE DIHYDROCHLORIDE, WARFARIN
Trazolan → TRAZODONE HCL
Trecalmo → CLOTIAZEPAM
Trecator → ETHIONAMIDE
Trecul Queen's Delight → STILLINGIA sp.
Tree Asp → CATERPILLARS
Tree of Heaven → AILANTHUS ALTISSIMA
Tree of Life → THUJA OCCIDENTALIS
Tree Tobacco → NICOTIANA sp.
Trefanocide → TRIFLURALIN
Treficon → TRIFLURALIN
Treflan → TRIFLURALIN
Trèfle → CLOVERS
TREMA ASPERA
Tremaril → METHIXENE
Tremarit → METHIXENE
Trematone → TREMETOL
"Trembles" → EUPATORIUM sp., GOLDENROD; SNAKEROOT, WHITE
TREMETOL also → GOLDENROD; SNAKEROOT, WHITE
Tremin → TRIHEXYPHENIDYL
Tremonil → METHIXENE
Tremoquil → METHIXENE

TREMORINE also → L-DOPA, OXOTREMORINE
Tremorins → PENICILLIUM sp.
Tremuloidin → POPLAR
Trendar → IBUPROFEN
Trenimon → TRIAZIQUONE
Trentadil → BAMIFYLLINE
Trental → PENTOXIFYLLINE
Treomycetina → CHLORAMPHENICOL
Trepidan → PRAZEPAM
Trepidone → MEPHENOXALONE
Trepomycin → VIRGINIAMYCIN
Trescatyl → ETHIONAMIDE
Trescillin → PENICILLINS – "NEWER and SYNTHETICS" > *Propicillin Potassium*
Tresochin → CHLOROQUINE
Tresortil → METHOCARBAMOL
Trest → METHIXENE
Tresten → THIETHYLPERAZINE MALEATE
Tretamine → TRIETHYLENE MELAMINE
Trethylene → TRICHLOROETHYLENE
Tretinoin → VITAMIN A
Trevintix → PROTIONAMIDE
Trexan → NALTREXONE
TRF → THYROTROPIN RELEASING HORMONE also → INTERLEUKINS
TRH → THYROTROPIN RELEASING HORMONE
TRI → TRICHLOROETHYLENE
Tri-6 → LINDANE
Tri-Abrodil → SODIUM ACETRIZOATE
Triac → TRIIODOTHYROACETIC ACID
Triacana → TRIIODOTHYROACETIC ACID
TRIACETOXYANTHRACENE
Triacetyl-6-azauridine → AZARIBINE
Triacetyloleandomycin → TROLEANDOMYCIN also → THYMOL
Triacontanol → SORSACA
Triam → TRIAMCINOLONE
TRIAMCINOLONE
TRIAMTERENE also → HYDRALAZINE, HYDROCHLOROTHIAZIDE, INDOMETHACIN, INSULIN, LITHIUM, MECAMYLAMINE, POTASSIUM, QUINIDINE, VERATRUM sp.
TRIAMTOMA sp.
Trianel → TRIPARANOL
TRIANTHEMA sp.
Triantoin → MEPHENYTOIN
Triarylmethanes → COPY PAPER and COPY MACHINES
Triasox → THIABENDAZOLE
Triasporin → ITRACONAZOLE
Triasyn → ANILAZINE
Triatec → RAMIPRIL
Triatox → AMITRAZ
Triavil → AMITRIPTYLINE
TRIAZIQUONE
TRIAZOLAM also → ALUMINUM HYDROXIDE, CIMETIDINE, GRAPEFRUIT, ITRACONAZOLE, KETOCONAZOLE, NEFAZODONE, RANITIDINE
1,2,4-Triazole-3-amine → AMINOTRIAZOLE
Triazure® → AZARIBINE also → AZAURIDINE
Triazurol → CHLORAZANIL
Tri-Ban → PINDONE
Tribavirin → RIBAVIRIN
TRIBENOSIDE
Tribrissin → TRIMETHOPRIM
TRIBROMOETHANOL
Tribromomethane → BROMOFORM
TRIBROMSALAN also → DIBROMSALAN
Tribulosis → TRIBULUS TERRESTRIS
TRIBULUS TERRESTRIS
Triburon → TRICLOBISONIUM CHLORIDE
Tributon → 2,4,5 - TRICHLOROPHENOXYACETIC ACID
TRIBUTYL PHOSPHATE
S,S,S-TRI-*N*-BUTYL PHOSPHOROTRITHIOATE
Tricaine → COPPER
Tricalcium Arsenate → CALCIUM ARSENATE
Tricanix → TINIDAZOLE
Tricarbamix Z → ZIRAM
Trichazol → METRONIDAZOLE
TRICHILIA EMETICA
Trichlor → TRICHLOROETHYLENE
Trichloren → TRICHLOROETHYLENE
TRICHLORFON
TRICHLORMETHIAZIDE also → MAGNESIUM OXIDE, PHENTERMINE
TRICHLOROACETIC ACID also → WARFARIN
Trichloroacetyline → PHOSGENE
TRICHLOROBENZENES
Trichlorocarbanilide → ANILINE
3,4,4´-Trichlorocarbanilide → TRICLOCARBAN
1,1,1-TRICHLOROETHANE
Trichloroethanoic Acid → TRICHLORMETHIAZIDE
TRICHLOROETHANOL also → ALCOHOL, ETHYL; CHLORAL HYDRATE
Trichloroethene → TRICHLOROETHYLENE
TRICHLOROETHYLENE also → ALCOHOL, ETHYL; BARIUM, COFFEE, EPINEPHRINE, OXYGEN, PHOSGENE, SOYBEANS, WELDING

Trichloroethyl Phosphate → TRICLOFOS
TRICHLOROFLUOROMETHANE also → DICHLORODIFLUOROMETHANE
Trichlorohydrin → 1,2,3-TRICHLOROPROPANE
Trichloromethane → CHLOROFORM
Trichloromethiazide → TRICHLORMETHIAZIDE
TRICHLORONAPHTHALENE
Trichloronitromethane → CHLOROPICRIN
Trichlorphene → TRICHLORFON
2,4,5-TRICHLOROPHENOL
2,4,6-TRICHLOROPHENOL
2,4,5-TRICHLOROPHENOXYACETIC ACID also → AGENT ORANGE, DIOXIN
1,2,3-TRICHLOROPROPANE
TRICHLOROTRIFLUOROETHANE also → BARIUM
TRICHOCEREUS PACHANOI
Trichocide → METRONIDAZOLE
Tricho Cordes → METRONIDAZOLE
TRICHODESMA sp.
Trichodesmin → TRICHODESMA sp.
Tricho-Gynaedron → METRONIDAZOLE
TRICHOLOMA sp.
Trichomonex → IODOCHLORHYDROXYQUINOLINE
TRICHOMYCIN
Trichonat → TRICHOMYCIN
Trichopal → METRONIDAZOLE
Trichorad → 2-ACETAMIDO-5-NITROTHIAZOLE
Trichoral → 2-ACETAMIDO-5-NITROTHIAZOLE
Trichosanthes kirilowii → TRICHOSANTHIN
TRICHOSANTHIN
TRICHOTHECENE MYCOTOXINS also → YELLOW RAIN
Tricinolon → TRIAMCINOLONE
Tri-Clene → TRICHLOROETHYLENE
TRICLOBISONIUM CHLORIDE
TRICLOCARBAN
TRICLOFENOL
TRICLOFOS also → WARFARIN
Triclos → TRICLOFOS
TRICLOSAN
Triclox → PENICILLINS – "NEWER and SYNTHETICS" > *Cloxacillin/Sodium salt*
Tricocet → METRONIDAZOLE
Tricofuran → FURAZOLIDONE
Tricolam → TINIDAZOLE
Tricoloid → PROCYCLIDINE
Tricomicina → TRICHOMYCIN
Triconazole → TERCONAZOLE
Tricor → FENOFIBRATE
Tricortale → TRIAMCINOLONE
Tricortan → CORTICOTROPIN
Tricoxidil → MINOXIDIL
Tricresol → CRESOL
Tricresyl Phosphate → TRITOLYL PHOSPHATE
Tri-σ-cresyl Phosphate → TRI-σ-TOLYL PHOSPHATE
TRICROMYL
Tricuran → GALLAMINE TRIETHIODIDE
Tricyclamol → PROCYCLIDINE
Tricyclo[3.3.1.1^{3,7}]decan-1-amine → AMANTADINE
Tricyte → POLYSTYRENE
Tridesilon → DESONIDE
Tridestrin → ESTRIOL
TRIDIHEXETHYL CHLORIDE
Tridihexethyl Iodide → TRIDIHEXETHYL CHLORIDE
Tridihexethyls → PHENOBARBITAL
Tridil → NITROGLYCERIN
Tridione → TRIMETHADIONE
TRIEN → TRIENTINE
Tri-Endothol → ENDOTHALL
TRIENTINE also → ETHYLENEDIAMINE
Triester → NITROGLYCERIN
Tri-Ethane → 1,1,1-TRICHLOROETHANE
TRIETHANOLAMINE
TRIETHYLAMINE
TRIETHYLENEDIAMINE
TRIETHYLENE GLYCOL
Triethylene Glycol Diglycidyl Ether → ETHOGLUCID
TRIETHYLENE GLYCOL MONOMETHYL ETHER
TRIETHYLENE MELAMINE
TRIETHYLENE OLEATE POLYPEPTIDE CONDENSATE
TRIETHYLENE PHOSPHORAMIDE
Triethylenephosphoramide → THIOTEPA
Triethylenetetramine → TRIENTINE
TRIETHYL LEAD
TRIETHYL TIN
Triflucan → FLUCONAZOLE
Triflumen → TRICHLORMETHIAZIDE
TRIFLUOMEPRAZINE
TRIFLUOPERAZINE DIHYDROCHLORIDE also → ALUMINUM HYDROXIDE, CHEESES, METHYLDOPA, PARGYLINE, PREDNISONE, RESERPINE
TRIFLUOPERAZINE DIHYDROCHLORIDE
Trifluoperidol → TRIFLUPERIDOL

Trifluopromazine → TRIFLUPROMAZINE
Trifluoramine → NITROGEN FLUORIDE
Trifluorammonia → NITROGEN FLUORIDE
TRIFLUPERIDOL also → PHENINDIONE
TRIFLUPROMAZINE also → MEPROBAMATE
TRIFLURALIN also → ALACHLOR, *N*-NITROSODI-*n*-PROPYLAMINE
Triflurex → TRIFLURALIN
Trifolium sp. → CLOVERS
Trifolium alexandrinum → TRIPHOLIUM ALEXANDRINUM
Trifoside → DINITRO-σ-CRESOL
Triftazin → TRIFLUOPERAZINE DIHYDROCHLORIDE
Trigard → CYROMAZINE
Trigger → RANITIDINE
Tri-Globe → TRIMETHOPRIM
Triglochin sp. → ARROWGRASS
Triglycine → NITRILOTRIACETIC ACID
Triglycyllypressin → TERLIPRESSIN
Trigonella foenum graecum → FENUGREEK
Trigonelline → MIRABILIS JALAPA
Trigonyl → SULFAMETHOXAZOLE
Trigot → ERGOLOID MESYLATES
Trihalomethanes → CHLORINE
Triherbicide-CIPC → CHLOROPROPHAM
Trihexane → TRIHEXYPHENIDYL
TRIHEXYPHENIDYL also → DESIPRAMINE, DIPHENHYDRAMINE
Trihistan → CHLORCYCLIZINE
1,3,5-Trihydroxybenzene → PHLOROGLUCINOL
Trihydroxyestrin → ESTRIOL
Trihydroxypropane → GLYCERIN
Triiodomethane → IODOFORM
TRIIODOTHYROACETIC ACID
3,5,3´-Triiodothyronine → LIOTHYRONINE
Triiodothyroxine → PREDNISONE
Triiotrast → SODIUM ACETRIZOATE
Trikosterol → TRIPARANOL
Trilafon → PERPHENAZINE
Trilan → SULPIRIDE
Trilene → TRICHLOROETHYLENE
Trilifan → PERPHENAZINE
Triline → TRICHLOROETHYLENE
Trilisate → CHOLINE MAGNESIUM TRISALICYLATE
Trilit → 2,4,6-TRINITROTOLUENE
Trilobacin → PAXILLINE also → PAW-PAW
Triludan → TERFENADINE
Trilumbrin → PYRANTEL
Trim → TRIFLURALIN
Trimanyl → TRIMETHOPRIM
Trimar → TRICHLOROETHYLENE
Trimaton → METHAM SODIUM
TRIMAZOSIN
Tri-Me → METHYL TRITHION
TRIMEBUTINE MALEATE
Trimedat → TRIMEBUTINE MALEATE
Trimelarsan → MELARSONYL POTASSIUM
TRIMELLITIC ANHYDRIDE also → EPOXY RESINS
TRIMEPRANOL
TRIMEPRAZINE
Trimeprimine → TRIMIPRAMINE
Trimeproprimine → TRIMIPRAMINE
Trimesulf → SULFAMETHOXAZOLE
Trimetaphan → TRIMETHAPHAN CAMSYLATE
TRIMETHADIONE
TRIMETHAPHAN CAMSYLATE also → DIMETHYL TUBOCURARINE IODIDE, SUCCINYLCHOLINE
TRIMETHOBENZAMIDE
TRIMETHOPRIM also → AZATHIOPRINE, CYCLOSPORINE, DISULFIRAM, DOFETILIDE, FOLIC ACID, DIGOXIN, GLIPIZIDE, METHOTREXATE, PHENYTOIN, QUINAPRIL, SULFAMETHAZINE, TEICOPLANIN, ZIDOVUDINE
Trimethoxazine → TRIMETOZINE
TRIMETHOXYAMPHETAMINE also → MYRISTICIN
Trimethoxyphosphine → TRIMETHYL PHOSPHITE
TRIMETHYLAMINE also → BENZOYL PEROXIDE, CHOLINE, EGGS, HERRING, MERCURIALIS ANNUA, WHEAT
Trimethylaminoglycine → BETAINE
Trimethyl Aniline → PONCEAU 3R
1,3,5-TRIMETHYLBENZENE
Trimethylene → CYCLOPROPANE
TRIMETHYL PHOSPHITE
4,5´,8-Trimethylpsoralen → TRIOXSALEN
Trimetion → DIMETHOATE
Trimeton → PHENIRAMINE MALEATE
TRIMETOZINE
TRIMETREXATE
Trimexolone → RIMEXOLONE
TRIMIPRAMINE
Trimogal → TRIMETHOPRIM
Trimonase → TINIDAZOLE
Trimopan → TRIMETHOPRIM
Trimox → PENICILLINS – "NEWER and SYNTHETICS" > *Amoxicillin*
Trimysten → CLOTRIMAZOLE

Trimpex → TRIMETHOPRIM
Trinalgon → NITROGLYCERIN
Trinex → TRICHLORFON
Trinitrin → NITROGLYCERIN
1,3,5-TRINITROBENZENE
Trinitroglycerol → NITROGLYCERIN
2,4,6-Trinitrophenol → PICRIC ACID
Trinitrosan → NITROGLYCERIN
2,4,6-TRINITROTOLUENE also → TETRANITROMETHANE
Trinitrotoluol → 2,4,6-TRINITROTOLUENE
Trinoxol → 2,4-TRICHLOROPHENOXYACETIC ACID also → 2,4,5 TRICHLOROPHENOXYACETIC ACID
Trinuride → PHENETURIDE
Triocetin → TROLEANDOMYCIN
Trionine → LIOTHYRONINE
Triopac → SODIUM ACETRIZOATE
Triosil → SODIUM METRIZOATE
Triostat → LIOTHYRONINE
Triosul → SULFUR TRIOXIDE
Triothyrone → LIOTHYRONINE
Triovex → ESTRIOL
Trioxazine → TRIMETOZINE
Trioxone → 2,4,5 - TRICHLOROPHENOXYACETIC ACID also → AGENT ORANGE
TRIOXSALEN also → BENZYL SALICYLATE, HERACLEUM sp.
Trioxyethylrutin → TROXERUTIN
Trioxysalen → TRIOXSALEN
TRIPARANOL
Triparin → TRIPARANOL
Tripdiolide → TRIPTERYGIUM WILFORDII
TRIPELENNAMINE also → CHLORPROMAZINE, ETHYLENEDIAMINE, GUANETHIDINE, OPIUM, PENTAZOCINE
Triperidol → TRIFLUPERIDOL
Triphacyclin → TETRACYCLINE
Triphedinon → TRIHEXYPHENIDYL
Triphenidyl → TRIHEXYPHENIDYL
Triphenylmethanes → COPY PAPER and COPY MACHINES
TRIPHENYLPHOSPHATE
TRIPHOLIUM ALEXANDRINUM
Triphthasine → TRIFLUOPERAZINE DIHYDROCHLORIDE
Triphthazine → TRIFLUOPERAZINE DIHYDROCHLORIDE
Tripiperazine Dicitrate → PIPERAZINE
Triple Dye → GENTIAN VIOLET
Tripneustes → SEA URCHINS
Tripoli → ROTTENSTONE
Tripotassium Citrate → POTASSIUM CITRATE
Tripotassium Dicitrato Bismuthate → OMEPRAZOLE
Triprim → TRIMETHOPRIM
TRIPROLIDINE
Tripsina → TRYPSIN
TRIPTERYGIUM WILFORDII
Triptil → PROTRIPTYLINE
Triptisol → AMITRIPTYLINE
Triptizol → AMITRIPTYLINE
Triptolide → TRIPTERYGIUM WILFORDII
TRIPTORELIN
Triquine → QUINACRINE
TRIS → TROMETHAMINE
Trisamine → TROMETHAMINE
Tris-Amino → TROMETHAMINE
Trisaminol → TROMETHAMINE
Trisanil → TRIBROMSALAN
Tris BP → TRIS(2-3-DIBROMOPROPYL)PHOSPHATE
Tris Buffer → TROMETHAMINE
TRIS(2-3-DIBROMOPROPYL)PHOSPHATE
TRISETUM FLAVESCENS
TRIS(HYDROXYMETHYL)NITROMETHANE also → BRONOPOL
Tris Nitro → TRIS(HYDROXYMETHYL) NITROMETHANE
TRISODIUM EDETATE
Trisodium 1-(4-sulfo-1-naphthylazo)-2-naphtol-3,6-disulfonate → AMARANTH
Trisodium Citrate → SODIUM CITRATE
Trisodium σ-Phosphate → SODIUM PHOSPHATE, TRIBASIC
Trisodium Phosphonformate → FOSCARNET SODIUM
Trisodium Versenate → TRISODIUM EDETATE
Trisoralen → TRIOXSALEN
Tri-Sweet® → ASPARTAME
Tritace → RAMIPRIL
Triteren → TRIAMTERENE
Tritheon → 2-ACETAMIDO-5-NITROTHIAZOLE
Trithion → CARBOPHENOTHION
Triticum aestivum → WHEAT
Triticum sativum → WHEAT
Triticum vulgare → WHEAT
Tritisan → QUINTOZENE
Tritium → PAINT also → HYDROGEN
Tritoftorol → ZINEB
TRITOLYL PHOSPHATE also → SESAME
TRI-σ-TOLYL PHOSPHATE also → ALCOHOL, ETHYL; APIOLE, GASOLINE; GINGER, JAMAICAN; HYDRAULIC FLUIDS, RUBBER, SESAME, SHOES, WHISKEY

Triton A-20 → TYLOXAPOL
Triton WR-1339 → TYLOXAPOL
Trittico → TRAZODONE HCL
Triumbren → SODIUM ACETRIZOATE
Triurol → SODIUM ACETRIZOATE
Trivastal → PIRIBEDIL
Trivazol → METRONIDAZOLE
Trivetrin → TRIMETHOPRIM
Trivitan → VITAMIN D_3
Trizine → TRIMETHOPRIM
Trizma → TROMETHAMINE
Trobicin → SPECTINOMYCIN
Trochin → CHLOROQUINE
Trodax → NITROXYNIL
Trofan → L-TRYPTOPHAN
Trofurit → FUROSEMIDE
TROGLITAZONE also → CHLOESTYRAMINE RESIN, TERFENADINE, WARFARIN
Trolamine → TRIETHANOLAMINE
TROLEANDOMYCIN also → ALUMINUM HYDROXIDE, CISAPRIDE, ERGOTAMINE, ITRACONAZOLE, KETOCONAZOLE, METHYLPREDNISOLONE, METHYSERGIDE, OLEANDOMYCIN, TACROLIMUS, TERFENADINE, THEOPHYLLINE, TRIAZOLAM
Trolene → RONNEL
TROLNITRATE PHOSPHATE
Trolone → SULTHIAME
Tromasin → PAPAIN
Trombavar → SODIUM TETRADECYL SULFATE
Trombovar → SODIUM TETRADECYL SULFATE
Trometamol → TROMETHAMINE also → KETOROLAC
TROMETHAMINE
Tromethane → TROMETHAMINE
Tromexan → ETHYL BISCOUMACETATE
Tromexan Ethyl Acetate → ETHYL BISCOUMACETATE
Tropaeolin G → METANIL YELLOW
Tropaeolum majus → NASTURTIUM
Tropalin → TRIPARANOL
Tropax → OXYBUTYNIN CHLORIDE
Trophicardyl → INOSINE
Trophysan → PROTEIN HYDROLYSATES
Tropical Jellyfish → JELLYFISH
TROPICAMIDE
Tropic Snow → DIEFFENBACHIA sp.
TROPISETRON
Tropisetron → METOCLOPRAMIDE
Tropium → CHLORDIAZEPOXIDE
Tropotox → 4(2-METHYL-4-CHLOROPHENOXY)BUTYRIC ACID
Trosinone → ETHISTERONE
Trosyd → TIOCONAZOLE
Trosyl → TIOCONAZOLE
TROVAFLOXACIN
Trovan → ALATROFLOXACIN, TROVAFLOXACIN
TROXERUTIN
Trp → L-TRYPTOPHAN
Tru → PYRVINIUM PAMOATE
Truck Drivers → AMPHETAMINE
True® → MENTHOL
Trumpet Creeper → CAMPSIS RADICANS
Trumpet Flower → SOLANDRA sp.
Trusopt → DORZOLAMIDE
Truxal → CHLORPROTHIXENE
Truxaletten → CHLORPROTHIXENE
Trypafalvine → ACRIFLAVINE
Trypamidium → ISOMETAMIDIUM
TRYPAN BLUE
TRYPARSAMIDE
Tryparsone → TRYPARSAMIDE
Tryponarsyl → TRYPARSAMIDE
Trypothane → TRYPARSAMIDE
Trypoxyl → SODIUM ARSANILATE
TRYPSIN also → BEANS, SOYBEANS, TETRACYCLINE
Tryptacin → L-TRYPTOPHAN
TRYPTAMINE also → ANADENANTHERA, BROMOLYSERGIDE, CANARY GRASS, DIPLOPTERYS CABRERANA, PASSIFLORA sp.
Tryptan → L-TRYPTOPHAN
Tryptar → TRYPSIN
Tryptanol → AMITRIPTYLINE
Tryptil → PROTRIPTYLINE
Tryptizol → AMITRIPTYLINE
Tryptocalm → L-TRYPTOPHAN
L-TRYPTOPHAN also → CLOMIPRAMINE, *N,N*-DIMETHYLYTRYPTAMINE, FLUOXETINE, PHENYLALANINE, TRANYLCYPROMINE, SKATOLE
TRYPTOQUIVALINE and TRYPTOQUIVALONE
Trypure → TRYPSIN
T's and B's → PENTAZOCINE
T's and Blues → PENTAZOCINE also → TRIPELENNAMINE
Tschat → CATHA EDULIS
TSE KOO CHOY also → MERCUROUS CHLORIDE
TSH → THYROTROPIN
Tsiao → ZANTHOXYLUM sp.

Tsiklamid → ACETOHEXAMIDE
Tsiklodol → TRIHEXYPHENIDYL
Tsiklometiazid → CYCLOPENTHIAZIDE
Tsiklomitsin → TETRACYCLINE
TSN → TETRAMETHYL SUCCINONITRILE
TSP → SODIUM PHOSPHATE, TRIBASIC
TSPA → THIOTEPA
Tsudohmin → DICLOFENAC
TTE → TRICHLOROTRIFLUOROETHANE
TTMS → MONOSULFIRAM
TTX → TETRODOTOXIN
Tuads → THIRAM
Tuasol → TRIBROMSALAN
Tua Tua → FLAIRA
Tuazol → METHAQUALONE
Tuazolone → METHAQUALONE
Tubadil → *d*-TUBOCURARINE CHLORIDE
Tubarine → *d*-TUBOCURARINE CHLORIDE
Tuba Root → DERRIS ROOT
Tubazid → ISONIAZID
Tuberactinomycin B → VIOMYCIN
Tubercazon → THIACETAZONE
TUBERCIDIN
Tuberculin → FISH OILS, ISONIAZID
Tuberite → PROPHAM
Tuberose → POLYANTHES TUBEROSA
Tubex → CORTICOTROPIN
Tubilysin → ISONIAZID
d-TUBOCURARINE CHLORIDE also → BACITRACIN, CHLOROTHIAZIDE, CHLOROTHALIDONE, COLISTIN, CURARE, CYCLOTHIAZIDE, DIAZEPAM, ETHACRYNIC ACID, FUROSEMIDE, GALLAMINE TRIETHIODIDE HALOTHANE, HISTAMINE, KANAMYCIN, LIDOCAINE, LITHIUM, MAGNESIUM SULFATE, MANNITOL, METHYCLOTHIAZIDE, NEOMYCIN, PROCAINAMIDE, PROPRANOLOL, PYRIDOSTIGMINE BROMIDE, QUINETHAZONE, QUINIDINE, STREPTOMYCIN
Tu-cillin → PENICILLIN > *Benzylpenicillin, Procaine*
Tuclase → CARBETAPENTANE CITRATE
Tugon → TRICHLORFON
Tuki Phool → TARAXACUM OFFICINALE
"Tukki" → TURKEY
Tulalal → SANSEVIERA sp.
TULIP also → CALCIUM OXALATE
Tulipine → TULIP
Tulisan → THIRAM
Tullidora → COYOTILLO
Tumbaku → NICOTIANA sp.
Tumbleweed → KOCHIA
Tumfafiya → CALOTROPIN
Tumil-K → POTASSIUM GLUCONATE
rh Tumor Necrosis Factor Receptor → ETANERCEPT
Tums® → CALCIUM CARBONATE
Tuna → SODIUM GLUTAMATE
Tung Nut → TUNG TREE
Tungoil Tree → TUNG TREE
TUNG SHUEH
TUNGSTEN also → WINE
TUNGSTEN CARBIDE
TUNG TREE
Tunol → COD LIVER OIL
Tupa → SYNADENIUM GRANTII
TUR → CHLORMEQUAT CHLORIDE
Turada → SANSEVIERA sp.
Tural → SANSEVIERA sp.
Turec → BECLOBRATE
Turinal → ALLYLESTRANOL
Turisynchron → METHALLIBURE
Turk-E-San → ACETARSONE
TURKEY also → ACROLEIN
Turkey Bush → MYOPORUM sp.
Turkey Corn → CORYDALIS, DICENTRA CANADENSIS
Turkeyfish → PTEROIS VOLITANS
Turkey Mullein → EREMOCARPUS SETIGERUS
Turkey Pea → TEPHROSIA sp.
"Turkey X Disease" → PEANUTS
Turkish Tobacco → NICOTIANA sp.
Turloc → METUREDEPA
TURMERIC
Turnabouts → AMPHETAMINE
TURNERA APHRODISIACA
Turn Hoof → IVY
TURNIP also → PHENYL ISOCYANATE
Turnip Weed → RAPISTRUM RUGOSUM
Turn-On → ISOBUTYL NITRITE
TURPENTINE also → CAMPHOR, Δ^3-CARENE, CHLORINE, PINENE, PROMETHAZINE
Turpentine Weed → GUTIERREZIA MICROCEPHALA
TURTLES
Tururubi → LASIOSIPHON sp.
Tururibi → GNIDIA KRAUSSIANA
Tusscapine → NOSCAPINE
Tusseval → PIPERIDIONE
Tussilago farfara → COLTSFOOT
Tussionex® → PHENYLTOLOXAMINE also → HYDROCODONE BITARTRATE
TUT-7 → MENOGARIL

Titane → *sec*-BUTYLAMINE
TUTIN also → CORIARIA, HONEY
Tutu → CORIARIA
TV-485 → ETOFENAMATE
TV 1322 → ACEMETACIN
"*Tweens*" → POLYSORBATES also → GRAMICIDIN
Tweety Bird → METHYLENEDIOXYAMPHETAMINE
Two-grooved Milkvetch → MILKVETCH
Twsb → STIBOCAPTATE
Tx → OIL ORANGES
TYBAMATE
Tybatran → TYBAMATE
Tydantil → NIFURATEL
Tylan → TYLOSIN
Tylcalsin → CALCIUM ACETYLSALICYLATE
Tylciprine → TRANYLCYPROMINE
Tylenol® → ACETAMINOPHEN also → POTASSIUM CYANIDE
Tylinal → DIETHYLPROPION
TYLOSIN
TYLOXAPOL
Tyloxypal → TYLOXAPOL
TYLUS
Tylocine → TYLOSIN
Tylocrebrine → TYLOPHORA sp.
Tylon → TYLOSIN
TYLOPHORA sp.
Tylose → METHYL CELLULOSE
Tylose MGA → CELLULOSE, *Sodium Carboxymethylcellulose*
Tympanol → DIMETHICONE (*with silicon dioxide*)
Tymelit → LOFEPRAMINE
Type E Toxin → FISH
TYPEWRITER RIBBONS
TYRAMINE also → ACACIA; ALCOHOL, ETHYL; AMPHETAMINE, BEANS, BANANA, BEER, CAVIAR, CHEESES, CHOCOLATE, CREAM, FIG, FISH, ISOCARBOXAZID, LINEZOLID, LIVER, MECLOBEMIDE, MISO; MISTLETOE, AMERICAN; NIALAMIDE, ORANGE, PARGYLINE, PLEUROTIS sp., PASSIFLORA sp., PHENELZINE, SAUSAGE, TOLOXATONE, TOMATO, WINE, YEAST, YOGURT
Tyrimide → ISOPROPAMIDE IODIDE
Tyrocidine → TYROTHRICIN
Tyrosamine → TYRAMINE
Tyrosinase → BEANS
TYROTHRICIN also → GRAMICIDIN
Tyrylen → BUTAPERAZINE
Tyvid → ISONIAZID
Tyzanol → TETRAHYDROZOLINE
Tyzine → TETRAHYDROZOLINE
TZU 0460 → ROXATIDINE ACETATE

U

U 1363 → DIPHENADIONE
U 4527 → CYCLOHEXIMIDE
U 5965 → TETRACYCLINE
U 6013 → ISOFLUPREDONE
U 6324 → PHENYLTHIOUREA
U 6591 → NOVOBIOCIN
U 6987 → CARBUTAMIDE
U 7800 → FLUPREDNISOLONE
U 8344 → URACIL MUSTARD
U 8471 → MEDRYSONE
U 8839 → MEDROXYPROGESTERONE ACETATE
U 9889 → STREPTOZOCIN
U 10,136 → ALPROSTADIL
U 10,149 → LINCOMYCIN
U 10,858 → MINOXIDIL
U 10,997 → MILBOLERONE
U 11,555A → TRIETHYLAMINE
U 11,828 → NORMETHANDRONE
U 12,062 → DINOPROSTONE
U 14,583 → DINOPROST
U 14,743 → PORFIROMYCIN
U 17,835 → TOLAZAMIDE
U 18,409 → SPECTINOMYCIN
U 18,496 → 5-AZACYTIDINE
U 19,920 → CYTARABINE
U 21,251 → CLINDAMYCIN
U 22,550 → CALUSTERONE
U 24,973A → MELITRACEN HCL
U 26,225A → TRAMADOL
U 26,452 → GLYBURIDE
U 26,597A → COLESTIPOL
U 27,182 → FLURBIPROFEN
U 28,774 → KETAZOLAM
U 31,889 → ALPRAZOLAM
U 32,921 → CARBOPROST
U 32,921E → CARBOPROST
U 33,030 → TRIAZOLAM
U 36,059 → AMITRAZ
U 36,384 → CARBOPROST
U 41,123 → ADINAZOLAM
U 42,585E → LODOXAMIDE TROMETHAMINE
U 42,718 → LODOXAMIDE TROMETHAMINE
U 52,047 → MENOGARIL
U 53,217 → PROSTACYCLIN

U 100,766 → LINEZOLID
Ubab → WITHANIA SOMNIFERA
Ubas Tree → ANTIARIS TOXICARIA
UBIDECARENONE also → WARFARIN
Ubiquinone → UBIDECARENONE
Ubretid → HEXAMARIUM BROMIDE
Uburu Ocha → ANONA
UC 7744 → CARBARYL
UC 21,149 → ALDICARB
Ucaricide → GLUTARALDEHYDE
UCB 2543 → CARBETAPENTANE CITRATE
UCB 3983 → MESNA
UCB 4445 → BUCLIZINE
UCB 4492 → HYDROXYZINE
UCB 5062 → MECLIZINE
UDMH → 1,1-DIMETHYLHYDRAZINE
Udolac → DAPSONE
Udukaju → JATROPHA sp.
UFT → TEGAFUR
Ugba → CASTOR BEAN
Uguele → GLORIOSA
Ugurol → TRANEXAMIC ACID
Ujoviridin → INDOCYANINE GREEN
UK 738 → ETHYBENZTROPINE
UK 4271 → OXAMNIQUINE
UK 14,304-18 → BRIMONIDINE TARTRATE
UK 20,349 → TIOCONAZOLE
UK 33,274-27 → DOXAZOCIN MESYLATE
UK 48,340
UK 49,858 → FLUCONAZOLE
UK 92,480 → SILDENAFIL
UK 109,496 → VORICONAZOLE
UK 124,114 → SELAMECTIN
Ukapen → PENICILLINS – "NEWER and SYNTHETICS" > *Ampicillin*
Ukidan → UROKINASE
Ukwaju → TAMARIND
'ULBA → FENUGREEK
Ulcar → SUCRALFATE
Ulcedin → CIMETIDINE
Ulcerfen → CIMETIDINE
Ulcermin → SUCRALFATE
Ulcetrax → FAMOTIDINE
Ulcex → RANITIDINE
Ulcidine → CIMETIDINE
Ulcimet → CIMETIDINE
Ulcofalk → CIMETIDINE
Ulcogant → SUCRALFATE
Ulcolax → BISACODYL
Ulcomedina → CIMETIDINE
Ulcomet → CIMETIDINE
Ulcosan → PIRENZEPINE HYDROCHLORIDE
Ulcuforton → PIRENZEPINE HYDROCHLORIDE
Ulcus-Tablinen → CARBENOLOXONE
Uldumont → METHANTHELINE BROMIDE
Ulexine → CYTISINE
Ulfamid → FAMOTIDINE
Ulfaret → CEFSULODIN SODIUM
Ulfinol → FAMOTIDINE
Ulhys → CIMETIDINE
Ulmenide → CHENODEOXYCHOLIC ACID
Ulmus → ELM
Ultandren → FLUOXYMESTERONE
Ultidine → RANITIDINE
Ultrabion → PENICILLINS – "NEWER and SYNTHETICS" > *Ampicillin*
Ultracef → CEFADROXIL
Ultracide → METHIDATHION
Ultracillin → PENICILLINS – "NEWER and SYNTHETICS" > *Cyclacillin*
Ultracorten → PREDNISONE
Ultracortenol → PREDNISOLONE
Ultradine → POLYVINYLPYRROLIDONE IODINE COMPLEX
Ultradol → ETODOLAC
Ultrafur → FURALTADONE
Ultra-Lente → INSULIN
Ultram → TRAMADOL
Ultramarine Green → CHROMIC OXIDE
Ultran → PHENAGLYCODOL
Ultranol → VITAMIN D_3
Ultraparin → ENOXAPARIN
Ultrapen → PENICILLINS – "NEWER and SYNTHETICS" > *Propicillin Potassium*
Ultrase MT20 → PANCRELIPASE
Ultrasulfon → SULFADIMETHOXINE
Ultratiazol → PHTHALYLSULFATHIAZOLE
Ultrax → SULFAMETER
Ultroxim → CEFUROXIME
UM 792 → NALTREXONE
UM 952 → BUPRENORPHINE
Umatrope → GROWTH HORMONES
Umbellatine → BERBERINE
Umbelliferone → VANILLA
Umbethion → COUMAPHOS
Umbradil → IODOPYRACET
Umbrella Plant → PODOPHYLLUM
Umbrella Thorn → ACACIA
Umbrella Tree → MELIA sp.
UM DULUKWA → SOLANUM sp.
um Fude → CASTOR BEAN
Umine → PHENTERMINE
UML 491 → METHYSERGIDE
Umushegwe → CARISSA

Unacim → PENICILLINS – "NEWER and SYNTHETICS" > *Sultamicillin*
Unagen → DIPYRONE
Unakalm → KETAZOLAM
Unal → *p*-AMINOPHENOL
Unasyn → PENICILLINS – "NEWER and SYNTHETICS" > *Sultamicillin*
Unat → TORSEMIDE
γ-UNDECALACTONE
1-UNDECANEDITHIOL
Undecanone-2 → METHYL NONYL KETONE
UNDECYLENIC ACID
Undecylmercaptan → 1-UNDECANEDITHIOL
Unden → ESTRONE also → PROPOXUR
Unibaryt → BARIUM SULFATE
Unicin → TETRACYCLINE
Unicontin → THEOPHYLLINE
Unidasa → HYALURONIDASE
Unidigin → DIGITOXIN
Unidone → ANISINDIONE
Unifirst Corporation → TRICHLOROETHYLENE
Unifur → FURALTADONE
Unifyl → THEOPHYLLINE
Unihep → HEPARIN
Uniloc → ATENOLOL
Unimoll BB → BUTYL BENZYL PHTHALATE
Unimycetin → CHLORAMPHENICOL
Unimycin → OXYTETRACYCLINE
Union Carbide Corp. → NIAX® CATALYST ESN
Unipazole → METRONIDAZOLE
Unipen → PENICILLINS – "NEWER and SYNTHETICS" > *Nafcillin Sodium*
Uniphyl → THEOPHYLLINE
Uniphyllin → THEOPHYLLINE
Unipril → RAMIPRIL
Uniprofen → BENOXAPROFEN
Unipyranamide → PYRAZINAMIDE
Uniquin → LOMEFLOXACIN
Unisedil → DIAZEPAM
Unisept → CHLORHEXIDINE
Unisom → DOXYLAMINE SUCCINATE
Unisomnia → NITRAZEPAM
Unitane → TITANIUM DIOXIDE
Unitensin Tannate → CRYPTENAMINE TANNATE
Unithiol → DIMERCAPTOPROPANE SULPHONATE
Unitiol → DIMERCAPTOPROPANE SULPHONATE
Univasc → MOEXIPRIL
Univer → VERAPAMIL
Unizole → DIMETRIDAZOLE
Unkie → MORPHINE
Unosulf → SULFAMETHOXYPYRIDAZINE
UP 33-901 → CIBENZOLINE
UP 83 → NIFLUMIC ACID
UP 34,101 → PROPACETAMOL
Upas Tree → ANTIARIS TOXICARIA
Upcyclin → TETRACYCLINE
Upjohn → TOLAZAMIDE
Upland Cotton → COTTON
Uppers → AMPHETAMINE
Ups → AMPHETAMINE
Upstene → INDALPINE
Uracil-6-carbolyic Acid → OROTIC ACID
URACIL MUSTARD
Uradal → CARBROMAL
Uragon → BROMACIL
Uralgin → NALIDIXIC ACID
Uramustine → URACIL MUSTARD
Uranap → METHIONINES
Uranine → FLUORESCEIN
URANIUM also → HYDROFLUORIC ACID, RADON
Uranium Hexafluoride → HYDROFLUORIC ACID
Uranium Nitrate → URANYL NITRATE
Urantoin → NITROFURANTOIN
Uranyl Fluoride → HYDROFLUORIC ACID
URANYL NITRATE
URAPIDIL
Uraprene → URAPIDIL
Urari → CURARE
Urasorb DMO → OCTYLDIMETHYL PABA
Urbadan → CLOBAZAM
Urbanyl → CLOBAZAM
Urbason → METHYLPREDNISOLONE
Urbason-Solubile → METHYLPREDNISOLONE
Urbilat → MEPROBAMATE
Urbol → ALLOPURINOL
UREA also → SHARK
Ureaphil → UREA
URECHITES sp.
Urecholine → BETHANECHOL
UREDOFOS
Uregit → ETHACRYNIC ACID
p-Ureidobenzenearsonic Acid → CARBARSONE
Urem → IBUPROFEN
Ureophil → UREA
Urepearl → UREA
Urese → BENZTHIAZIDE
Uretham → URETHANE
Urethan → URETHANE
URETHANE also → BEER, BREAD, BROTIZOLAN, DIETHYL PYROCARBONATE, MEPROBAMATE, OLIVES, ORANGE, YOGURT

Uretren → TRIAMTERENE
Uretrim → TRIMETHOPRIM
Urex → FUROSEMIDE
Urex → METHENAMINE
Urfamicina → THIAMPHENICOL
Urfamycine → THIAMPHENICOL
Urgilan → PROSCILLARIDIN
URGINEA sp.
Urginea maritima → SQUILL
Uriben → NALIDIXIC ACID
URIC ACID also → CAULIFLOWER, CHLORAL HYDRATE, 6-CHLOROPURINE, CHLOROTHIAZIDE, CHLORTHALIDONE, GENTISIC ACID, HYDRALAZINE, MERCAPTOPURINE, METHYLDOPA, MOLYBDENUM, PEAS, PHENACEMIDE, SARDINES, SODIUM BICARBONATE, SUPROFEN
Uricemil → ALLOPURINOL
Uriclar → NALIDIXIC ACID
Uricovac → BENZBROMARONE
Uridinal → PHENAZOPYRIDINE
Urimeth → METHIONINES
Urinex → CHLOROTHIAZIDE
Urinorm → BENZBROMARONE
Urinox → OXOLINIC ACID
Uriodone → IODOPYRACET
Uripurinol → ALLOPURINOL
Urisal → SODIUM CITRATE
Urispas → FLAVOXATE HCL
Uritone → METHENAMINE
Uritrate → OXOLINIC ACID
Uritrol → NITROXOLINE
Urizept → NITROFURANTOIN
Urlea → BENDROFLUMETHIAZIDE
Uroalpha → THYMOXAMINE
Uro-Alvar → OXOLINIC ACID
Uro b → FENURON
Urobenyl → ALLOPURINOL
Uro-Carb → BENTHANECHOL
Urocaudal → TRIAMTERENE
Uro-Cedulamin → METHENAMINE
Urocit K → POTASSIUM CITRATE
Uro-Clamoxyl → PENICILLINS – "NEWER and SYNTHETICS" > *Amoxicillin*
Urocoli → NITROXOLINE
Urodiazin → HYDROCHLOROTHIAZIDE
Urodie → TERAZOSIN
Urodin → NITROFURANTOIN
Urodine → PHENAZOPYRIDINE
Urodixin → NALIDIXIN
UROKINASE also → INDOMETHACIN, PHENYLBUTAZONE
Urokon Sodium → SODIUM ACETRIZOATE
Urolene Blue → METHYLENE BLUE
Urolong → NITROFURANTOIN
Urolophus jamaicensis → STINGRAYS
Urolosin → TAMSULOSIN
Urolucosil → SULFAMETHIZOLE
Uroman → NALIDIXIC ACID
Uromandelin → METHENAMINE
Uromiro → IODAMIDE
Uromitexan → MESNA also → ETHANETHIOL
Uronal → BARBITAL
Uronamin → METHENAMINE
Uronase → UROKINASE
Uroneg → NALIDIXIC ACID
Uronorm → CINOXACIN
Uropan → NALIDIXIC ACID
Uropen → PENICILLINS – "NEWER and SYNTHETICS" > *Hetacillin*
Urophenil → THIAMPHENICOL
Uroplus → SULFAMETHOXAZOLE
Uro-Ripirin → EMEPRONIUM
Uroseptra → SULFAMETHOXAZOLE
Urosin → ALLOPURINOL
UROSTACHYS SAURURUS
Urosulfon → SULFACETAMIDE SODIUM
Uro-Tablinen → NITROFURANTOIN
Urotractan → METHENAMINE
Urotrate → OXOLINIC ACID
Urotropin → METHENAMINE
Urovist → MEGLUMINE DIATRIZOATE
Uroxacin → NORFLOXACIN
Urox B → BROMACIL
Uroxin → OXOLINIC ACID
Uroxol → OXOLINIC ACID
Urozide → HYDROCHLOROTHIAZIDE
Ursnon → FLUOROMETHOLONE
Ursol P → *p*-AMINOPHENOL
Urtias 100 → ALLOPURINOL
URTICA DIOICA
Urtosal → SALICYLAMIDE
Urupan → DEXPANTHENOL
Urushiol → ANACARDIACEA, POISON IVY; SUMAC, POISON
Urusonin → SPIRONOLACTONE
USAF EK 1569 → PHENYLTHIOUREA
Uskan → OXAZEPAM
Usnein → USNIC ACID
USNIC ACID also → PARMELIA MOLLIUSCULA
Usol Olive 6G → 4-CHLORO-σ-PHENYLENEDIAMINE
Usten → SALINOMYCIN
USTILAGO
Ustimon → HEXOBENDINE

Ustracide → METHIDATHION
Uta → STROPHANTHUS sp.
Utemerin → RITODRINE
Uteracon → OXYTOCIN
UtibiD → OXOLINIC ACID
Uticort → BETAMETHASONE
Utimox → PENICILLINS – "NEWER and SYNTHETICS" > *Amoxicillin*
Utinor → NORFLOXACIN
Utopar → RITODRINE
Utovlan → NORETHINDRONE
Utrogestan → PROGESTERONE
UV 9 → OXYBENZONE
Uval → SULISOBENZONE
Uvasorbs → BENZOPHENONES
Uvesterol-D → VITAMIN D_2
Uvinals → BENZOPHENONES
Uvinul M-40 → OXYBENZONE
Uvinul MS-40 → SULISOBENZONE
Uvistat → MEXENONE
Uzone → PHENYLBUTAZONE

V

V 18 → TETRADIFON
Vaccinium oxycoccus → CRANBERRY
Vacor → PYRIMINAL
Vadebex → NOSCAPINE
Vadosilan → ISOXSUPRINE
Vaflol → VITAMIN A
Vagamin → METHANTHELINE BROMIDE
Vagantin → METHANTHELINE BROMIDE
Vagestrol → DIETHYLSTILBESTROL
Vagidine → POLYVINYLPYRROLIDONE IODINE COMPLEX
Vagifem → ESTRADIOL
Vagilen → METRONIDAZOLE
Vagimid → METRONIDAZOLE
Vagistat → TIOCONAZOLE
Valaciclovir → VALACYCLOVIR
ValACV → VALACYCLOVIR
VALACYCLOVIR
Valadol → ACETAMINOPHEN
Valamin → ETHINAMATE
Valamina → FLUPHENAZINE DECANOATE
Valaxona → DIAZEPAM
Valbazen → ALBENDAZOLE
Valcote → VALPROIC ACID
Valcote → SODIUM VALPROATE
Valentinite → ANTIMONY TRIOXIDE
Valeramide OM → AMINOPENTAMIDE HYDROGEN SULFATE
Valerian → VALERIANA sp.
VALERIANA sp.
Valetan → DICLOFENAC
Valerone → DIISOBUTYL KETONE
Valexon → PHOXIM
VALINOMYCIN
Valip → TRIPARANOL
Valiquid → DIAZEPAM
Valisone → BETAMETHASONE
Valium → DIAZEPAM
Vallergan → TRIMEPRAZINE
Valmagen → CIMETIDINE
Valmid → ETHINAMATE
Valmidate → ETHINAMATE
VALNOCTAMIDE
Valodin → DIRITHROMYCIN
Valoid → CYCLIZINE
Valone → 2-ISOVALERYL-1,3-INDANEDIONE
Valpin → ANISOTROPINE METHYLBROMIDE
VALPROIC ACID also → ACETYLSALICYLIC ACID, CARBAMAZEPINE, CHLORAZEPATE, CHOLESTYRAMINE RESIN, ERYTHROMYCIN, ERYTHROPOIETIN, FELBAMATE, METHOSUXIMIDE, NAPROXEN, NORTRIPTYLINE, PHENOBARBITAL, PHENYTOIN, PRIMIDONE, ZIDOVUDINE
Valproïnezuur → VALPROIC ACID
Valrelease → DIAZEPAM
VALRUBICIN
VALSARTAN also → GRAPEFRUIT
VALSPODAR
Valstar → VALRUBICIN
Valsyn → FURALTADONE
Valtrex → VALACYCLOVIR
Valzin → DULCIN
VANADIUM also → SULFUR DIOXIDE
Vanadium Tetrachloride → VANADIUM
Vancenase → BECLOMETHASONE
Vancide → ZIRAM
Vancide 89 → CAPTAN
Vanclay → KAOLIN
Vancocin → VANCOMYCIN
Vancoled → VANCOMYCIN
VANCOMYCIN also → DALFOPRISTIN, FUROSEMIDE, METHOTREXATE, NIFEDIPINE, TEICOPLANIN
Vandal Root → VALERIANA sp.
Vandid → ETHAMIVAN
Vanectyl → TRIMEPRAZINE
Van Es' Walking Disease → SENECIO sp.
Vangard 45 → CAPTAN
Van Gogh, Vincent → ABSINTHIUM, DIGITALIS

van Helmont, Dr. → OPIUM
Vanicide TM-95 → THIRAM
VANILLA also → COUMARIN
Vanillal → ETHYL VANILLIN
Vanilla Pudding → PHENYTOIN
Vanillic Acid Diethylamide → ETHAMIVAN
Vanillic Diethylamide → ETHAMIVAN
Vanillin → VANILLA
Vanillylmandelic Acid → GLYCERYL GUAIACOLATE, METHOCARBAMOL, METHYLDOPA
Vaniqa → EFLORNITHINE
Van-Mox → PENICILLINS – "NEWER and SYNTHETICS" > *Amoxicillin*
Vanoxide → BENZOYL PEROXIDE
Vanoxin → DIGOXIN
Vanquin → PYRVINIUM PAMOATE
Vansil → OXAMNIQUINE
Vantyl → CINCHOPHEN
Vapam → METHAM SODIUM
Vapona → DICHLORVOS
Vaponefrin → EPINEPHRINE
Vapo-N-Iso → ISOPROTERENOL
Vaporin → NAPHTHA
Vaporole → ISOBUTYL NITRITE
Vapotone → TETRAETHYL PYROPHOSPHATE
Varagu Millet → PASPALUM sp.
Varaire → VERATRUM sp.
Varemoid → TROXERUTIN
Variaphylline LA → AMINOPHYLLINE
Variotin → PECILOCIN
Variton → DIPHEMANIL METHYLSULFATE
VARNISH
Varnish Naphtha → NAPHTHA
Varsol 1 → NAPHTHA
Vasal → PAPAVERINE
Vascardin → ISOSORBIDE DINITRATE
Vasiodone → IODOPYRACET
Vasodilan → ISOXSUPRINE
Vascor → BEPRIDIL
Vasculat → BAMETHAN
Vasculit → BAMETHAN
Vascunicol → BAMETHAN
Vaseline® → PETROLATUM also → PARAFFIN
Vaselinomas → PETROLATUM
Vasicine → PEGANUM HARMALA
Vaskulat → BAMETHAN
Vasocard → TERAZOSIN
Vasodiatol → PENTAERYTHRITOL TETRANITRATE
Vaso-Dilatan → TOLAZOLINE HCL
Vasodin → NICARDIPINE
Vasoglyn → NITROGLYCERIN
Vasokastan → AESCULUS sp.
Vasokellina → KHELLIN
Vasoklin → THYMOXAMINE
Vasolan → VERAPAMIL
Vasomed → TROLNITRATE PHOSPHATE
Vasomet → TERAZOSIN
Vasomotal → BETAHISTINE
Vasonase → NICARDIPINE
Vasoplex → ISOXSUPRINE
VASOPRESSIN also → ACETAMINOPHEN, CHLORPROPAMIDE, CLOFIBRATE, CYCLOPHOSPHAMIDE, GROWTH HORMONES, PHENFORMIN; PITUITARY, POSTERIOR
Vasoprine → ISOXSUPRINE
Vasorbate → ISOSORBIDE DINITRATE
Vasorome → OXANDROLONE
Vasospan → PAPAVERINE
Vasotec → ENALAPRIL
Vasotocin → VASOPRESSIN
Vasotran → ISOXSUPRINE
Vasotrate → ISOSORBIDE DINITRATE
Vasoxine → METHOXAMINE
Vasoxyl → METHOXAMINE
Vasten → PRAVASTATIN
Vastcillin → PENICILLINS – "NEWER and SYNTHETICS" > *Cyclacillin*
Vastribil → TROXERUTIN
Vasylox → METHOXAMINE
Vatracin → PENICILLINS – "NEWER and SYNTHETICS" > *Cyclacillin*
Vazofirin → PENTOXIFYLLINE
VC 1-13 → DICHLOFENTHION
VC9-104 → ETHOPROP
VC-13 → DICHLOFENTHION
V-cillin → PENICILLIN > *Penicillin V*
VCR → VINCRISTINE
VCS → METHAZOLE
VCS 506 → LEPTOPHOS
VDH → BISACODYL
Vectarion → ALMITRINE BISMESYLATE
Vectavir → PENCICLOVIR
Vectren → DIAPAMIDE
Vectren → ISOXICAM
VECURONIUM BROMIDE also → ATENOLOL, DANTROLENE, ESMOLOL, MIDAZOLAM, PROPRANOLOL
Veegun → PENICILLINS – "NEWER and SYNTHETICS" > *Amoxicillin, Ampicillin*
Veepa Oil → NIM OIL
Veesyn → PENICILLIN > *Penicillin V*
Vegaben → CHLORAMBEN
Vegetable Calomel → PODOPHYLLUM
Vegetable Mercury → PODOPHYLLUM
Vegetable Pepsin → PAPAIN

Vegfru → PHORATE
Vegiben → CHLORAMBEN
Vegolysen → HEXAMETHONIUM BISTRIUM
Vehem-Sandoz → TENIPOSIDE
Veinamitol → TROXERUTIN
Velacycline → ROLITETRACYCLINE
Velardon → PAPAIN
Vélar Fausse Giroflée → ERYSIMUM CHEIRANTHOIDES
Velban → VINBLASTINE
Velbe → VINBLASTINE
Vel 58-CS-11 → DICAMBA
Veldopa → L-DOPA
Velmol → DOCUSATE SODIUM
Velmonit → CIPROFLOXACIN
Velocef → CEPHADRINE
Velopural → HYDROCORTISONE
Velosef → CEPHADRINE
Velosulin → INSULIN
Velsicol-58-CS-11 → DICAMBA
Velsicol 104 → HEPTACHLOR
Velsicol 1068 → CHLORDANE
Velsicol Compound R → DICAMBA
Veltol Plus → ETHYL MALTOL
Veltrim → CLOTRIMAZOLE
Velvet Bean → STIZLOBIUM DEERINGIANUM
Velvetgrass → HOLCUS LANATUS also → CYANIDES
Velvet Plant → VERBASCUM THAPSUS
Vendarcin → OXYTETRACYCLINE
Venen → TRIPROLIDINE
Venetine → BARBERRY
Venetlin → ALBUTEROL
Venex → TRIBENOSIDE
Vengeance → BROMETHALIN
Veniten → TROXERUTIN
VENLAFAXINE also → FLUOXETINE, ISOCARBOXAZID, SERETRALINE
Venopex → CIMETIDINE
Venopirin → LYSINE ACETYLSALICYLATE
Venoruton → TROXERUTIN
Venostassin → AESCULUS sp.
Ventaire → PROTOKYLOL
Ventipulmin → CLENBUTEROL
Ventodisks → ALBUTEROL
Ventolin → ALBUTEROL
Ventox → ACRYLONITRILE
Ventramine → NIKETHAMIDE
Ventrazol → PENTYLENETETRAZOL
Ventussin → BENZONATATE
Vepisid → ETOPOSIDE
Veracillin → PENICILLINS – “NEWER and SYNTHETICS” > *Dicloxacillin*
Veracin → VERAPAMIL
Veracur → FORMALDEHYDE
Veradol → NAPROXEN
Veramax → MEDROXYPROGESTERONE ACETATE
Veramex → VERAPAMIL
Verano, Carmine → CAMPHOR
VERAPAMIL also → ACETYLSALICYLIC ACID; ALCOHOL, ETHYL; ATENOLOL, CEFTRIAXONE, CIMETIDINE, CLINDAMYCIN, CLONIDINE, CYCLOSPORINE, DIGITALIS, DIGOXIN, DISOPYRAMIDE, DOFETILIDE, ERYTHROMYCIN, FLECAINIDE, GRAPEFRUIT, LITHIUM, MIDAZOLAM, PENBUTOLOL SULFATE, PEPPER, PHENYTOIN, PRACTOLOL, PROPRANOLOL, QUINIDINE, RIFAMPIN, SIMVASTATIN, TACROLIMUS, THEOPHYLLINE, TIMOLOL, VINCRISTINE, VITAMIN D_3
Veraptin → VERAPAMIL
Veraseptyl → SILVER PROTEIN, MILD
VERATRALDEHYDE
VERATRAMINE
Veratran → CLOTIAZEPAM
Veratridine → VERATRINE
VERATRINE also → SABADILLA
VERATRUM sp . also → CYCLOPAMINE, DEXTROAMPHETAMINE, ETHACRYNIC ACID, PSEUDOEPHEDRINE, SABADILLA, SPIRONOLACTONE
Verax → BENZYDAMINE
VERBASCUM THAPSUS
VERBENA
VERBESINA ENCELIOIDES
Verbigen → CHLORAMBEN
Vercyte → PIPOBROMAN
Verdiana → TRIPARANOL
Verdisol → DICHLORVOS
Verdone → METHYL CHLOROPHENOXY ACETIC ACID
Verelan → VERAPAMIL
Verexamil → VERAPAMIL
Vergonil → HYDROFLUMETHIAZIDE
Verifax® → 4-PHENYL CATECHOL
Veritab → MECLIZINE
Veritol → PHOLEDRINE
Verladyn → DIHYDROERGOTAMINE METHANESULFONATE (MESYLATE)
Vermeerbos → GEIGERIA sp.
Vermeeric Acid → GEIGERIA sp.
Vermeerasiekte → GEIGERIA sp.
Vermicidin → MEBENDAZOLE

Vermicompren → PIPERAZINE
Vermillion → MERCURIC SULFIDE, RED
Vermirax → MEBENDAZOLE
Vermitiber → PYRVINIUM PAMOATE
Vermitin → NICLOSAMIDE
Vermitin → PHENOTHIAZINE
Vermizym → PAPAIN
VERMOUTH
Vermox → MEBENDAZOLE
VERNONIA sp.
Veroletten → BARBITAL
Veronal → BARBITAL
Verophene → PROMAZINE
Verospiron → SPIRONOLACTONE
Veroxil → INDAPAMIDE
VERRUCULOGEN also → PENICILLIUM sp.
Verruculotoxin → VERRUCULOGEN also → PENICILLIUM sp.
Verrugon → SALICYLIC ACID
Versapen → PENICILLINS – "NEWER and SYNTHETICS" > *Hetacillin*
Versatrex → PENICILLINS – "NEWER and SYNTHETICS" > *Hetacillin*
Versed → MIDAZOLAM
Versene 100 → SODIUM EDETATE
Versene CA → CALCIUM DISODIUM EDETATE
Vesicine → PEGANUM HARMALA
Versotrane → HYDRARGAPHEN
Verstran → PRAZEPAM
Versus → BENDAZAC
VERTEPORFIN
Vertigon → PROCHLORPERAZINE
Vertimec → ABAMECTIN
Vertine → PHENICARBAZIDE
Vertolan → SULFAMETHAZINE
Verton → CHLORMADINONE
Verucasep → GLUTARALDEHYDE
Vervain → VERBENA
Vesadin → SULFAMETHAZINE
Vesamin → SODIUM ACETRIZOATE
Vesanoid → VITAMIN A
Vesdil → RAMIPRIL
VESNARINONE
Vesnaroid → VITAMIN A
Vespéral → BARBITAL
Vesperone → BRALLOBARBITAL
Vespral → TRIFLUPROMAZINE
Vesprin → TRIFLUPROMAZINE
Vespula vulgaris → WASPS
Vestolit → POLYVINYL CHLORIDE
Vetacin → NETILMYCIN
Vetacortyl → METHYLPREDNISOLONE
Vetalar → KETAMINE
Vetalgina → LYSINE ACETYLSALICYLATE
Vetalog → TRIAMCINOLONE
Vetame → TRIFLUPROMAZINE
Vetamox → ACETAZOLAMIDE
Vetarsenobillon → NEOARSPHENAMINE
Vetavir → PENCICLOVIR
Vetbond → BUTYL CYANOACRYLATE
Vetch → VICIA
Vetcortenol → PREDNISOLONE
Vetcytine → CHLORAMPHENICOL
Vetdectin → MOXIDECTIN
Vetemac → RONNEL
Veteusan → CROTAMITON
Vetibenzamine → TRIPELENNAMINE
Veticol → CHLORAMPHENICOL
Vetidrex → HYDROCHLOROTHIAZIDE
Vetinol → NOSCAPINE
Vetisulid → SULFACHLORPYRIDAZINE
VETIVER
Vetivert → VETIVER
Vetiveryl → VETIVER
Vetocin → OXYTOCIN
Vetquamycin → TETRACYCLINE
Vetranquil → ACEPROMAZINE
Vetrazin → CYROMAZINE
Vetsin → SODIUM GLUTAMATE
Vet-Sorb → METHYL CELLULOSE
Vexol → RIMEXOLONE
Viadril → HYDROXYDIONE SODIUM
Viagra → SILDENAFIL
Vialibran → MEDIBAZINE
Vialidon → MEFENAMIC ACID
Vi-Alpha → VITAMIN A
Vianin → GENTIAN VIOLET
Vianol → BUTYLATED HYDROXYTOLUENE
Viansin → CHLORDIAZEPOXIDE
Viapta → METHENAMINE
Viarex → BECLOMETHASONE
Viarox → BECLOMETHASONE
Viasept → BISMUTH GLYCOL ARSANILATE
Vibazine → BUCLIZINE
Vibalt → VITAMIN B_{12}
Vibradox → DOXYCYCLINE
Vibramicina → DOXYCYCLINE
Vibramycin → DOXYCYCLINE
Vibraveineuse → DOXYCYCLINE
Vibrio sp. → SHELLFISH also → SHRIMP
Vibriomycin → DIHYDROTACHYSTEROL
VIBURNUM PRUNIFOLIUM
Vicard → TERAZOSIN
Vicelat → VITAMIN C
Viceton → CHLORAMPHENICOL
VICIA also → CYANIDES, HYDROGEN CYANIDE

Vicia faba → BEANS, Faba
Vicianin → SAMBUCUS sp.
Vicilan → VILOXAZINE
Vicine → BEANS, MOMORDICA CHARANTIA
Vicrom → CROMOLYN SODIUM
Victan → ETHYL LOFLAZEPATE
VIDARABINE also → ALLOPURINOL, THEOPHYLLINE
Videau → ENTOLOMA
Vi-De-3-hydrosol → VITAMIN D_3
Videx → DIDANOSINE
Vidine → POLYVINYLPYRROLIDONE IODINE COMPLEX
Vidopen → PENICILLINS – "NEWER and SYNTHETICS" > *Ampicillin*
Vidora → INDORAMIN
Vienna Green → CUPRIC ACETOARSENITE
VIGABATRIN also → PHENYTOIN
Vigantol → VITAMIN D_3
Vigne vierge → IVY, PARTHENOCISSUS QUINQUEFOLIA
Vigorsan → VITAMIN D_3
VIGUIERA ANNUA
Vikane → SULFURYL FLUORIDE
Villiaumite → SODIUM FLUORIDE
Vilona → RIBAVIRIN
VILOXAZINE also → CARBAMAZEPINE, THEOPHYLLINE
Vimicon → CYPROHEPTADINE
Vimocillin → PENICILLINS – "NEWER and SYNTHETICS" > *Hetacillin*
ViMycin → CHLORTETRACYCLINE
Vinactane → VIOMYCIN
Vinarol → POLYVINYL ALCOHOL
VINBARBITAL
VINBLASTINE also → MIBEFRADIL, ZIDOVUDINE
Vincaleukoblastine → VINBLASTINE
VINCA ROSEA
Vincent Van Gogh → ABSINTHIUM, DIGITALIS
Vincetoxin → CYNANCHUM
VINCLOZOLIN
VINCOFOS
Vincosid → VINCRISTINE
Vincrex → VINCRISTINE
VINCRISTINE also → ALLOPURINOL, ISONIAZID, ITRACONAZOLE, MIBEFRADIL, PROCARBAZINE, VERAPAMIL, ZIDOVUDINE
VINDESINE SULFATE
Vinegar → ACETIC ACID
Vinegar Naphtha → ETHYL ACETATE
Vine Maple → MENISPERMUM CANADENSE
Vinesthine → ETHER, VINYL
Vinethine → ETHER, VINYL
Vingard → VINCOFOS
Vinho de jurema → MIMOSA
Viniril → POLYVINYLPYRROLIDONE
Vinol → POLYVINYL ALCOHOL
VINORELBINE
Vinydan → ETHER, VINYL
VINYL ACETATE
Vinylacetonitrile → ALLYL CYANIDE
Vinylamine → ETHYLENEIMINE
γ-Vinyl-γ-aminobutyric Acid → VIGABATRIN
Vinylbenzene → STYRENE
VINYL BROMIDE
Vinyl Carbinol → ALCOHOL, ALLYL
VINYL CHLORIDE also → POLYVINYL CHLORIDE
Vinyl Cyanide → ACRYLONITRILE
VINYLCYCLOHEXENE DIOXIDE
Vinyl Ethanoate → VINYL ACETATE
Vinyl Ether → ETHER, VINYL
Vinylethylene → 1,3-BUTADIENE
VINYLIDENE CHLORIDE
VINYL TOLUENE
Vinyl Trichloride → 1,1,1-TRICHLOROETHANE
Viobamate → MEPROBAMATE
Viocid → GENTIAN VIOLET
Viocin → VIOMYCIN
Vio-D → VITAMIN D_2
Vioform → IODOCHLORHYDROXYQUINOLINE
Vioformio → IODOCHLORHYDROXYQUINOLINE
Viokase → PANCRELIPASE
Violeta de metilo → GENTIAN VIOLET
Violet 6B → FDC VIOLET #1
VIOMYCIN also → CAPREOMYCIN, POLYMIXIN B, VANCOMYCIN
Viophan → CINCHOPHEN
Viopsicol → CHLORDIAZEPOXIDE
Vio-Serpine → RESERPINE
Viosterol → VITAMIN D_2
Vioxx → ROFECOXIB
Viozene → RONNEL
Vipicil → PENICILLINS – "NEWER and SYNTHETICS" > *Cyclacillin*
Vipral → ACYCLOVIR
Viprynium → PYRVINIUM PAMOATE
Viquiera → HYDROGEN CYANIDE
Vira-A → VIDARABINE
Viracept → NELFINAVIR
Viractil → METHOTRIMEPRAZINE
Viraferon → INTERFERONS

Viramid → RIBAVIRIN
Viramune → NEVIRAPINE
Viraspray → HEXYLRESORCINOL
Virazid → RIBAVIRIN
Virazole → RIBAVIRIN
Virex → PROTEIN HYDROLYSATES
Virginicin → VIRGINIAMYCIN
Virginia Creeper → IVY
VIRGINIAMYCIN
Viridicatumtoxin → PENICILLIUM sp.
Virlix → CETIRIZINE
VIROLA sp. also → JUSTICA PECTORALIS var STENOPHYLLA
Virorax → ACYCLOVIR
Virosterone → TESTOSTERONE
Virudox → IDOXURIDINE
Viruxan → ISOPRINOSINE
Viruzona → METHISAZONE
Visabutina → OXYPHENBUTAZINE
Visammin → KHELLIN
Viscadan → KHELLIN
Viscarin → CARRAGEENAN
Visceralgine → TIEMONIUM IODIDE
Viscin → MISTLETOE, EUROPEAN or JAPANESE
Viscoleo → FLUPENTIXOL
Viscotoxins → MISTLETOE, EUROPEAN or JAPANESE
Viscum album → MISTLETOE, EUROPEAN or JAPANESE
Viscumin → MISTLETOE, EUROPEAN or JAPANESE
Visine → TETRAHYDROZOLINE
Visiren → OFLOXACIN
Visken → PINDOLOL
Visnagalin → KHELLIN
Visnagen → KHELLIN
Visotrast → SODIUM ACETRIZOATE
Vistagen → LEVOBUNOLOL
Vista-Methasone → BETAMETHASONE
Vistaril Pamoate → HYDROXYZINE
Vistaril Parenteral → HYDROXYZINE
Visu-beta → BETAMETHASONE
Visudyne → VERTEPORFIN
Vitacarn → LEVOCARNITINE
Vitacee → VITAMIN C
Vitacimin → VITAMIN C
Vitacin → VITAMIN C
VITAMIN A also → BEXAROTENE, CALCIUM, β-CAROTENE, CARROTS, CORTISONE, COUMAPHOS, MINERAL OIL, NEOMYCIN, SODIUM NITRATE, THIRAM, WATERCRESS, WHALES
Vitamin A Acid → VITAMIN A
VITAMIN B COMPLEX
VITAMIN B_1 also → BRACKEN FERN, CHLOROBUTANOL, DIGITALIS, FISH, MAGNESIUM TRISILICATE, METHIMAZOLE, OXYTETRACYCLINE, PRALIDOXIME, SODIUM METABISULFITE, SULFUR, SULFUR DIOXIDE, TANNIC ACID, TEA, WATERCRESS
VITAMIN B_2 also → CHLORTETRACYCLINE, GALACTOFLAVIN, MESTRANOL, OXYTETRACYCLINE, PROPANTHELINE, THALIDOMIDE
VITAMIN B_3 also → CORN, GLIPIZIDE, GLUCOSE, INSULIN, ITRACONAZOLE, LOVASTATIN, MECLIZINE, MILLET
Vitamin B_5 → THIRAM
VITAMIN B_6 also → CHLOROTRIANISENE, CYCLOSERINE, 4-DEOXYPYRIDOXINE, *L*-DOPA, DOPAMINE, DOXYLAMINE SUCCINATE, HOMOCYSTEINE, ISONIAZID, MESTRANOL, NITROFURANTOIN, OXYTETRACYCLINE, PENICILLAMINE, PHENYTOIN
VITAMIN B_{12} also → *p*-AMINOSALICYLIC ACID, CIMETIDINE, COLCHICINE, EGGS, ERYTHROPOIETIN, FOLIC ACID, MESTRANOL, METFORMIN, NEOMYCIN, NITROUS OXIDE, OMEPRAZOLE, POLYMIXIN B, POTASSIUM CHLORIDE, POTASSIUM CITRATE, RIFAMPIN, SODIUM CYANIDE, TATRACYCLINE, TOBACCO
Vitamin B_{12a} → VITAMIN B_{12}
VITAMIN B_{15} also → DICHLOROACETIC ACID, PEAS
Vitamin B_{17} → AMYGDALIN
Vitamin B_C → FOLIC ACID
Vitamin B_x → *p*-AMINOBENZOIC ACID
VITAMIN C also → ACETAMINOPHEN, ACETANILID, CALCIUM OXALATE, CAULIFLOWER, CHOLESTEROL, CHLOROBUTANOL, CHROMIUM PICOLINATE, CLOFIBRATE, CYSTINE, DEXTROAMPHETAMINE, FERROUS SULFATE, FLUPHENAZINE, IRON, LIPASE, MESTRANOL, MINERAL OIL, OXALIC ACID, PENTOBARBITAL, POTATO, PROPYLENE GLYCOL, SELENIUM, TETRACYCLINE, VITAMIN B_{12}, WARFARIN

VITAMIN D also → CALCIUM, CHOLESTYRAMINE RESIN, ISONIAZID, MEPHENYTOIN, MINERAL OIL, PHENTURIDE, PHENOBARBITAL, PHENYTOIN, PREDNISONE, PRIMIDONE, PROPYLENE GLYCOL
VITAMIN D_2
VITAMIN D_3
VITAMIN E also → β-CAROTENE, CHOLESTEROL, COD LIVER OIL, COTTONSEED OIL, CYSTINE, DIGITALIS, DIOXATHION, FATS, MINERAL OIL, OXYGEN, PEAS, SILVER, SODIUM METABISULFITE, TELLURIUM, WARFARIN
Vitamin G → VITAMIN B_2
VITAMIN H also → EGGS
Vitamin H_1 → *p*-AMINOBENZOIC ACID
VITAMIN K also → ACENOCOUMAROL, BROCCOLI, CHLORTETRACYCLINE, CHOLESTYRAMINE RESIN, COUMARIN, ETHYL BISCOUMACETATE, FISH, MINERAL OIL, NEOMYCIN, PENICILLIN, PHENOBARBITAL, PHYTIC ACID, QUINIDINE, SOYBEANS, SPIGELIA ANTHELMIA, STREPTOMYCIN, SULFAMETHAZINE, SULFAQUINOXALINE, SULFATHIAZOLE, SULFISOXAZOLE, TETRACYCLINE, WARFARIN
Vitamin M → FOLIC ACID
VITAMIN O
Vitamin PP → VITAMIN B_3
Vitamin R → METHYLPHENIDATE
VITAMIN U
Vitamist → VITAMIN C
Vitaneurin → VITAMIN B_1
Vitarubin → VITAMIN B_{12}
Vitascorbol → VITAMIN C
Vitaseptol → THIMEROSAL
Vitas U
Vitavax → CARBOXIN
Vitazechs → PYRIDOXAL-5-PHOSPHATE
Vitellin → LECITHIN
VITEX sp.
Vitis sp. → GRAPE
Vitispiranes → POTATO
Viton → LINDANE
Vitpex → VITAMIN A
Vitrasert → GANCICLOVIR
Vitravene → FORMIVERSEN SODIUM
Volonimat → TRIAMCINOLONE
Voltaren → DICLOFENAC
Voltarol → DICLOFENAC
Vomex A → DIMENHYDRINATE
"Vomiting Disease" → SNAKEROOT, WHITE
"Vomiting Gas" → CHLOROPICRIN
Vomiting Sickness → BLIGHIA SAPIDA, GEIGERIA sp.
Vomit Nut → NUX VOMICA
Vomitoxin → DEOXYNIVALENOL also → TRICHOTHECENE MYCOTOXINS
Vomitron → ONDANSETRON
Vonamycin Powder V → NEOMYCIN
Vondecaptan → CAPTAN
Vongory → CANNABIS
Vontil → THIOPERAZINE
Vontril → DIPHENIDOL
Vontrol → DIPHENIDOL
Voranil → CLORTERMINE
VORICONAZOLE
Vorlan → VINCLOZOLIN
Vorlex → METHYLISOTHIOCYANATE
Voveran → DICLOFENAC
VP-16-213 → ETOPOSIDE
VPM → METHAM SODIUM
Vucine → ETHACRIDINE
Vulcamicina → NOVOBIOCIN
Vulcamycin → NOVOBIOCIN
Vulkamycin → NOVOBIOCIN
Vitrocin → NITROFURAZONE
Vivactil → PROTRIPTYLINE
Vival → DIAZEPAM
Vivalan → VILOXAZINE
Vivarint → VILOXAZINE
Vivatec → LISINOPRIL
Vividrin → CROMOLYN SODIUM
Vividyl → NORTRIPTYLINE
Vivol → DIAZEPAM
VLB → VINBLASTINE
VM 26 → TENIPOSIDE
VM & P Naphtha → NAPHTHA
VOCHYSIA sp.
Vodol → MICONAZOLE NITRATE
Vogan → VITAMIN A
Vogan-Neu → VITAMIN A
Volatile Oil of Mustard → ALLYL ISOTHIOCYANATE
Volaton → PHOXIM
Voldal → DICLOFENAC
Volid → BRODIFACOUM
Volital → PEMOLINE
Volmax → ALBUTEROL
Volon → TRIAMCINOLONE
Vulkanox → 1,2-DIHYDRO-2,2,4-TRIMETHYLQUINOLINE
Vulklor → CHLORANIL
Vumon → TENIPOSIDE

VVPB 6453 → TRIMEPRANOL
VX also → SULFUR
VX 478 → AMPRENAVIR
Vydate → OXAMYL

W

W → L-TRYPTOPHAN
W 37 → BUFORMIN
W 090 → EBASTINE
W 483 → ETHOPROPAZINE
W 554 → FELBAMATE
W 1206 → FENCAMFAMINE
W 1544A → PHENELZINE
W 1760A → NAMOXYRATE
W 1962 → PROTHIPENDYL
W 3566 → QUINESTROL
W 4565 → OXOLINIC ACID
W 4744 → MECLOQUALONE
W 4869 → PREDNISOLONE
W 5975 → BETAMETHASONE
W 6421A → LEVOBUNOLOL
W 7000A → LEVOBUNOLOL
W 7320 → ALCLOFENAC
W 7618 → CHLOROQUINE
W 8495 → ISOXICAM
W 19,053 → ETIDOCAINE HCL
W 36,095 → TOCAINIDE
W 40,020 → PRAZEPAM
WAI-LING-SIN
Wai-munoh → NICOTIANA sp.
Wake → BEANS, Lima
Wakerobin → ARISAEMA
Wake-ups → AMPHETAMINE
Walconesin → MEPHENESIN
"Walkabout" → PYRROLIZIDINES
Walking Disease → AMSINKIA *Intermedia*
WALKING STICK
Wallace, Mike → FLECAINIDE
WALLPAPER
Wallpaper Poisoning → ARSENIC
WALNUT also → CYCLOPIAZONIC ACID, PEANUTS
Walter Reed 2721 → AMIFOSTINE
Wambunzila → ALCHORNEA CORDIFOLIA
Wampcocap → VITAMIN B_3
WANDERING JEW
Wanja → ADENIUM sp.
Wansar → DIPHENIDOL
Warabi → BRACKEN FERN
Waran → WARFARIN
Warbex → FAMPHUR
Warduzide → CHLOROTHIAZIDE

WARFARIN also → ACARBOSE, ACETAMINOPHEN, ACETYLSALICYLIC ACID; ALCOHOL, ETHYL; ALLOPURINOL, ALUMINUM HYDROXIDE, AMINOGLUTETHIMIDE, AMINOSALICYLIC ACID, AMIODARONE, AMOBARBITAL, ANTIPYRINE, APAZONE, AVOCADO, AZITHROMYCIN, BENZIODARONE, BENZO[α]PYRENE, BEZAFIBRATE, BROCCOLI, BUTABARBITAL, BUTETHAL, CARBAMAZEPINE, CEFAMANDOLE, CEFAZOLIN, CEFMETAZOLE, CHLORAL BETAINE, CHLORAL HYDRATE, CHLORAMPHENICOL, CHLORDIAZEPOXIDE, CHLORPROMAZINE, CHOLESTYRAMINE RESIN, CIMETIDINE, CINCHOFEN, CIPROFLOXACIN, CISAPRIDE, CLARITHROMYCIN, CLOFIBRATE, CLOVERS, CLOZAPINE, CORTICOTROPIN, COUMACHLOR, CYCLOSPORINE, DANAZOL, DEXTROTHYROXINE, DIAMINOZIDE, DIAZEPAM, DIAZOXIDE, DICHLORALANTIPYRINE, DICHLORALPHENAZONE, DIFLUNISAL, DIMETHICONE, DISULFIRAM, DOXYCYCLINE, ENOXACIN, ERYTHROMYCIN, ERYTHROMYCIN STEARATE, ETHACRYNIC ACID, ETHYCHLORVYNOL, ETHYLESTRENOL, ETOPOSIDE; FACTORS II, IX, and X; FAMOTIDINE, FELBAMATE, FENBUFEN, FENOFIBRATE, FENOPROFEN, FEVERFEW, FLOSEQUINAN, FLUCONAZOLE, 5-FLUOROURACIL, FLUOXETINE, FLUVASTATIN, FLUVOXAMINE, GASOLINE, GEMCITABINE, GEMFIBROZIL, GINKGO BILOBA, GINSENG, GLUCAGON, GLUTETHIMIDE,GRISEOFULVIN, HALOFENATE, HEPTABARBITAL, IBUPROFEN, ICE CREAM, INDOMETHACIN, INTERFERONS, ISOCARBOXAZID, ISONIAZID, ISOXICAM, ITRACONAZOLE, KANAMYCIN, KETOCONAZOLE, KETOPROFEN, LETTUCE, LEVAMISOLE, LEVOFLOXACIN, LINDANE, LOMEFLOXACIN, LORAZEPAM, LOVASTATIN, MAGNESIUM HYDROXIDE, MAGNESIUM OXIDE, MECLOFENAMIC ACID, MEFENAMIC ACID, MEPROBAMATE, MERCAPTOPURINE,

MESALAMINE, METHANDROSTENOLONE, METHOTREXATE, METHYLPHENIDATE, METHYL SALICYLATE, METHYL TESTOSTERONE, METOLAZONE, METRONIDAZOLE, MEXOLICAM, MICONAZOLE NITRATE, MINERAL OIL, MITOTANE, MORICIZINE, MOXALACTAM, NALIDIXIC ACID, NEOMYCIN, NIALAMIDE, NIMESULIDE, NIZATIDINE, NORFLOXACIN, NORTRIPTYLINE, OFLOXACIN, OMEPRAZOLE, OXYMETHOLONE, OXYPHENBUTAZONE, PAROXETINE, PENICILLIN; PENICILLINS – "NEWER and SYNTHETICS"; PERGOLIDE MESYLATE, PHENFORMIN, PHENOBARBITAL, PHENYLBUTAZONE, PHENYRAMIDOL, PHENYTOIN, PHYTIC ACID, PIRACETAM, PIROXICAM, POTATO, PRIMIDONE, PROCAINAMIDE, PROPAFENONE, PROPOFOL, PROPOXYPHENE, PROPYLTHIOURACIL, QUINIDINE, QUININE, RANITIDINE, RESERPINE, RIFAMPIN, RITINOVIR, ROFECOXIB, ROXITHROMYCIN, SALSALATE, SAQUINAVIR, SECOBARBITAL SODIUM, SERTRALINE, SIMVASTATIN, SODIUM, SODIUM BICARBONATE, SOYBEAN OIL, SPIGELIA ANTHELMIA, SPIRONOLACTONE, STANOZOLOL, STREPTOKINASE, STREPTOMYCIN, SUCRALFATE, SULFADIMETHOXINE, SULFAMETHIZOLE, SULFAMETHOXAZOLE, SULFAMETHOXYPYRIDAZINE, SULFAPHENAZOLE, SULFAQUINOXALINE, SULFINPYRAZONE, SULFISOXAZOLE; SULINDAC, TAMOXIFEN CITRATE, TEA, TERBINAFINE, TESTOSTERONE, TETRACYCLINE, THYROID, THYROXINE, TICLOPIDINE, TICRYNAFEN, TISSUE PLASMINOGEN ACTIVATOR, TOLBUTAMIDE, TOLMETIN, TOREMIFENE CITRATE, TRAMADOL, TRASTUZUMAB, TRAZODONE HCL, TRICHLOROACETIC ACID, TRICLOFOS, TRIMETHOPRIM, TROGLITAZONE, UBIDECARENONE; VITAMINS C, E, and K; ZAFIRLUKAST, ZILEUTON

Warficide → COUMAFURYL

Warfilone → WARFARIN

Warner-Lambert → PHENYTOIN

Wart Flower → CHELIDONIUM MAJUS

Wartweed → CHELIDONIUM MAJUS

Wartwort → CHELIDONIUM MAJUS

Warty Caltrop → KALLSTROEMIA

Waruzol → ASTEMIZOLE

WASABI

Washing Soda → SODIUM CARBONATE

WASPS also → DICLOFENAC

WATER also → POTASSIUM, RIBAVIRIN

WATER BOATMAN

WATERCRESS also → CHLORZOXAZONE

Water Dragon → CALLA LILY

Water Dropworts → OENANTHE sp.

Water Figwort → SCROPHULARIA sp.

Water Glass → SODIUM SILICATE

Water Hemlock → CICUTA sp.

Water Hemp → AMARANTHUS

Water Lovage → OENANTHE sp.

Water Moccasin → SNAKE(S)

Water Parsnip → SIUM SUAVE

Water-pepper → POLYGONUM sp.

Water Shamrock → BUCKBEAN

Water Smartweed → POLYGONUM sp.

Water Trefoil → BUCKBEAN

Wattle → ACACIA

Wavicide → GLUTARALDEHYDE

Wax Bean → BEANS

Waxberry → SNOWBERRY

Wax Flower → HOYA sp.

Wax Myrtle → TEA

Wax Plant → HOYA sp.

Waxsol → DOCUSATE SODIUM

WAY-CMA 676 → GEMTUZIMAB OZOGAMICIN

Waynecomycin → LINCOMYCIN

WBA 107 → DIFENACOUM

WBA 8119 → BRODIFACOUM

W.D.D. → IMIPRAMINE

WE 941 → BROTIZOLAN

WeatherBlock → BRODIFACOUM

Wedding Bell → MORNING GLORY

WEDELIA sp.

Wedges → LYSERGIC ACID DIETHYLAMIDE

Weed → CANNABIS

Weed 108 → MONOSODIUM METHANEARSONATE

Weedar → 2,4,5-TRICHLOROPHENOXYACETIC ACID

Weedazol → AMINOTRIAZOLE

Weed-E-Rad → MONOSODIUM METHANEARSONATE

Weedex → METHYL CHLOROPHENOXY ACETIC ACID

Weed Hoe → MONOSODIUM METHANEARSONATE also → METHANE ARSONIC ACID
Weedol → PARAQUAT
Weedone → 2,4-DICHLOROPHENOXYACETIC ACID
Weedone → PENTACHLOROPHENOL
Weedone → 2,4,5-TRICHLOROPHENOXYACETIC ACID
Weifacodeine → PHOLCODINE
Weir Vine → IPOMEA sp.
Weisspiessglanz → ANTIMONY TRIOXIDE
Welchol → COLESEVELAM
WELDING
Welfurin → NITROFURANTOIN
Wellbatrin → BUPROPION
Wellbutin → BUPROPION
Wellcome 33-A-74 → ATRACURIUM
Wellcome 248U → ACYCLOVIR
Wellcome prepn 47-83 → CYCLIZINE
Wellcoprim → TRIMETHOPRIM
Welldorm → DICHLORALANTIPYRINE
Wemid → ERYTHROMYCIN STEARATE
Welvic → POLYVINYL CHLORIDE
Werepe → MUCUNA sp.
Wescopen → PENICILLIN > *Potassium Benzylpenicillinate*
Wespuril → DICHLOROPHEN(E)
Westcort → HYDROCORTISONE
Western Bitterweed → HYMENOXYS sp.
Western Hellebore → VERATRUM sp.
Western Horsenettle → SOLANUM sp.
Western Labrador Tea → LEDUM sp.
Western Minniebush → MENZIESA FERRUGINEA
Western Water Hemlock → CICUTA sp.
Westrosol → TRICHLOROETHYLENE
Weteye Bombs → SARIN
WET SUITS
Wh 7286 → XYLAZINE
WHALES
WHEAT also → ETHYLMERCURIC CHLORIDE, FLOUR, FLUOROACETIC ACID, HEXACHLOROBENZENE, MERCURY, NITROGEN CHLORIDE, PHOSPHENE, POTATO, TETRACYCLINE
Whey Factor → OROTIC ACID
Whipcide → PHTHALOFYNE
WHISKEY also → ALPRAZOLAM, FERRIC CHLORIDE, NITROGLYCERIN, *N*-NITROSODIMETHYLAMINE, *n*-PROPYL ALCOHOL, SOLDER
White(s) → COCAINE
White Bole → KAOLIN
Whitebrush → ALOYSIA LYCIOIDES
White Camas → ZIGADENUS sp.
White Caustic → SODIUM HYDROXIDE
White Cedar → THUJA OCCIDENTALIS also → MELIA Sp.
"White Dope" → α-METHYL FENTANYL
White Flower → MIMOSA
White-Flowered Milkweed → EUPHORBIA, *corollata*
White Goosefoot → CHENOPODIUM
White Hellebore → VERATRUM sp.
White Helleborine → VERATRUM sp.
Whitehorn → HAWTHORN
White Lead → LEAD CARBONATE
White Lightning → LYSERGIC ACID DIETHYLAMIDE
White Mercuric Precipitate → MERCURY, AMMONIATED
White Mustard → MUSTARD
White Oil → MINERAL OIL
White Potato → POTATO
White Precipitate → MERCURY, AMMONIATED
White Rot → CENTALLA ASIATICA
White Sanicle → EUPATORIUM, *ogeratoides and urticelfolium*
White Sea Urchin → LYTECHINUS VARIEGATUS also → SEA URCHINS
White Snakeroot → EUPATORIUM, *rugosum also* → SNAKEROOT
WHITE SPIRIT
White Swallow-wort → CYNANCHUM
White Tar → NAPHTHALENE
White Vitriol → ZINC
White Water Lily → NUPHAR
Whitewort → POLYGONATUM BIFLORUM
Whore-house Tea → EPHEDRINE, *Ephedra*
"Who Watches the Watcher" → PHENYLCYCLOHEXENE
Wicopy → DIRCA PALUSTRIS
Widecillin → PENICILLINS – "NEWER and SYNTHETICS" > *Amoxicillin*
Wild Alum Root → GERANIUM
Wild Arum → CALLA LILY
Wild Balsam Apple → MOMORDICA CHARANTIA
Wild Barley → HORDEUM JUBATUM
Wild Black Cherry → CHERRY
Wild Calla → CALLA LILY
WILD CARROT also → CONIUM MACULATUM
WILD CELERY and WILD PARSNIPS

Wild Chamomile → CHAMOMILE
Wild Cherry → CHERRY also → CYANIDES
Wild Chervil → CHAEROPHYLLUM sp.
Wild Coleus → PERILLA FRUTESCENS
Wild Corn → VERATRUM sp.
Wild Cucumber → MOMORDICA CHARANTIA
Wild Custard Apple → ANONA
Wild Daisy → CHRYSANTHEMUM
Wilder Wein → IVY
Wild Garlic → ALLICIN
Wild Gentian → CHIRONIA PALUSTRIS subsp. TRANSVAALENSIS
Wild Geranium → GERANIUM
Wild Ginger →. ASARUM CANADENSE
Wild Gooseberry → NICANDRA PHYSALODES
Wild Heliotrope → HELETROPIUM sp.
Wild Indigo → BAPTISIA sp.
Wild Jasmin → CESTRUM
Wild Leek → ALLIUM TRICOCCUM
Wild Lemon → PODOPHYLLUM
Wild Licorice → ABRUS PRECATORIUS also → ABRIN
Wild Mustard → MUSTARD
Wild Onion → ALLIUM VALIDUM, A. CANADENSE, A. CERNEUM
Wild Opium → LETTUCE
Wild Parsley → WILD CARROT
Wild Parsnip → WILD CELERY and WILD PARSNIPS also → CICUTA sp., CONIUM MACULATUM, PARSNIPS, TRACHEMENE GLAUCIFOLIA
Wild Pepper → ARISAEMA
Wild Plum → PLUM
Wild Poinsettia → EUPHORBI, *heterophylla*
Wild Raisin → VIBURNUM PRUNIFOLIUM
Wild Rue → PEGANUM HARMALA
Wild Sage → LANTANA sp.
Wild Sunflower → DORONICUM sp.
Wild Tobacco → NICOTIANA sp. also → LOBELIA
Wild Turnip → ARISAEMA
Wild Water Lemon → PASSIFLORA
Wild Winter Pea → LATHYRUS sp.
Wild Yellow Barberton Daisy → GERBERA
Wild Yellow Plum → PLUM
Wiley, Dr. Harvey W. → BORAX
Wilkinite → BENTONITE
Willestrol → DIETHYLSTILBESTROL
Williams Grass → FESCUE, TALL
Willpo → PHENTERMINE
Willpower → PHENMETRAZINE
Wilprafen → JOSAMYCIN
Wilson's Leather Protector → ISOOCTANE also → PROPANE
Win 244 → CHLOROQUINE
Win 3046 → ISOETHARINE
Win 5063-2 → THIAMPHENICOL
Win 5494 → AMOTRIPHENE
Win 8077 → AMBENONIUM
Win 9154 → INOSITOL NIACINATE
Win 11,450 → BENORYLATE
Win 13,820 → BECANTHONE
Win 14,098 → PIMINODINE
Win 14,833 → STANOZOLOL
Win 17,757 → DANAZOL
Win 18,320 → NALIDIXIC ACID
Win 18,501-2 → OXYPERTINE
Win 20,228 → PENTAZOCINE
Win 22,005 → UROXINASE
Win 24,933 → HYCANTHONE
Win 32,784 → BITOLTEROL
Win 35-213 → ROSOXACIN
Win 39,103 → METRIZAMIDE
Win 39,424 → IOHEXOL
Win 40,680 → AMRINONE
Win 47,203 → MILRINONE
Wincoram → AMRINONE
Windflowers → ANEMONE sp.
Window Pane → LYSERGIC ACID DIETHYLAMIDE
WINE also → ALOE, CHEESES, CHOLESTIN, DOPAMINE, GRAPE, HISTAMINE, IRON, ISOCARBOXAZID, LIGUSTRUM VULGARE, NIALAMIDE, PHENELZINE, PHENOLPHTHALEIN, POTASSIUM METABISULFITE, PROCARBAZINE, PROPOXYPHENE, SCOPOLAMINE, SODIUM AZIDE, SODIUM BISULFITE, SULFUR DIOXIDE
Wine Fluorosis → FLUORIDES and FLUORINE
Winged Bean → BEANS, Winged
Win-Kinase → UROKINASE
Winobanin → DANAZOL
Winstrol → STANOZOLOL
Winter Cherry → SOLANUM sp.
Winter Cherry → PHYSALIS sp.
Wintermin → CHLORPROMAZINE
Wintersteiner's Compound F → CORTISONE
Winthrop Labs → PENTAZOCINE
Wintodon → BISMUTH GLYCOL ARSANILATE
Wintomylon → NALIDIXIC ACID
Winton's Disease → SENECIO sp.
Winuron → ROSOXACIN
Wire Grasses → ARISTIDA sp.

Wytensin → GUANABENZ ACETATE
Wytrion → TROLEANDOMYCIN
Wyvital → PENICILLINS – "NEWER and SYNTHETICS" > *Cyclacillin*

X

X → VITAMIN E
100X → 4-BROMO-2,5-DIMETHOXYAMPHETAMINE
X 537A → LASALOCID
X 1497 → PENICILLINS – "NEWER and SYNTHETICS" > *Methicillin Sodium*
X5079C → SARAMYCETINE
Xahl → CEPHALEXIN
Xamamina → DIMENHYDRINATE
Xametina → TRIMETHOBENZAMIDE
Xanax → ALPRAZOLAM
Xanef → ENALAPRIL
Xanor → ALPRAZOLAM
Xanteline → METHANTHELINE BROMIDE
XANTHINOL NIACINATE
Xanthinol Nicotinate → XANTHINOL NIACINATE
Xanthocepthalum microcephala → GUTIERREZIA MICROCEPHALA
XANTHOSOMA sp. also → CALADIUM
Xanthostrumarin → COCKLEBURS
Xanthotoxin → HERACLEUM sp. also → AMMI MAJUS, METHOXSALEN
Xanthum sp. → COCKLEBURS
Xanturat → ALLOPURINOL
Xavin → XANTHINOL NIACINATE
Xaxa → ACETYLSALICYLIC ACID
Xeloda → CAPECITABINE
Xenagol → PHENAZOCINE HYDROBROMIDE
Xenalon → SPIRONOLACTONE
Xanthines → LITHIUM
Xanthium → THEOPHYLLINE
Xanthocephalum sp. → BROOMWEED
XANTHOCILLIN
XANTHORHIZA SIMPLICISSIMA
XANTHORRHEA sp.
Xenar → NAPROXEN
Xenid → DICLOFENAC
Xenylamine → 4-AMINODIPHENYL
Xerac BP 5 → BENZOYL PEROXIDE
Xerac BP 10 → BENZOYL PEROXIDE
Xerene → MEPHENOXALONE
Xerumenex → TRIETHYLENE OLEATE POLYPEPTIDE CONDENSATE
Xilkagan → KALMIA
Ximaol → OCTAMOXIN
Ximos → CEFUROXIME
XIPAMIDE
Xitix → VITAMIN C
XL-7 → BITHIONOL
Xolamin → CLEMASTINE
Xopenex → LEVALBUTEROL
X-Otag → ORPHENADRINE
XTC → METHYLDIAMINE METHAMPHETAMINE
XU 62-320 → FLUVASTATIN
Xumbradil → IODOPYRACET
Xyduril → CLOFIBRATE
Xylan Hydrogen Sulfate → PENTOSAN POLYSULFATE
Xylapan → XYLAZINE
Xylasol → XYLAZINE
XYLAZINE also → HALOTHANE, MORPHINE
XYLENE also → ALCOHOL, ETHYL; DISTILLATE, GLUE
Xylite → XYLITOL
XYLITOL
Xyliton → XYLITOL
Xylocaine → LIDOCAINE
Xylocitin → LIDOCAINE
Xylol → XYLENE
Xylomed → XYLOSE
XYLOMETAZOLINE
Xylo-Mucine → CELLULOSE
Xylonest → PRILOCAINE
Xylopfan → XYLOSE
XYLOSE also → NEOMYCIN
d-Xylose → XYLOSE
Xylotocan → TOCAININDE
Xylotox → LIDOCAINE
Xymelin → XYLOMETAZOLINE
Xyranit → XYLITOL
XZ-450 → AZITHROMYCIN

Y

Y 3 → CHLOROPROPHAM
Y 6047 → CLOTIAZEPAM
Yacca → XANTHORRHEA sp.
Yadulan → CHLORTHIAZIDE
Yagé → AYAHUASCA
Yageine → HARMINE
Yahourt → YOGURT
Yaje → AYAHUASCA
Yakee → VIROLA sp.
Ya-Ko-Yoó → JUSTICA PECTORALIS var STENOPHYLLA
Yama → HYDROGEN CYANIDE

Yamacillin → PENICILLINS – "NEWER and SYNTHETICS" > *Talampicillin*
Yamanza → CLEMATIS sp.
Yamatetan → CEFOTETAN
Yams → DIOSCOREA also → CYANATES
Yangona → KAVA
Yaro → DATURA
Yarrow → ACHILLEA sp. also → ASTERS
Yatren → CHINIOFON
Yatrocin → NITROFURAZONE
Yaupon → HOLLY
Yautia → XANTHOSOMA sp.
YC 93 → NICARDIPINE
YEAST also → CHOLESTIN, HISTAMINE, SOLDER, YOGURT
Yellow 1 → 2-NAPHTHYLAMINE
YELLOW AB
Yellow Adder's Tongue → ERYTHRONIUM
Yellow Big Head Disease → PANICUM sp.
Yellow Brittlegrass → SETARIA sp.
Yellow Cedar → THUJA OCCIDENTALIS
Yellow Cross Shells → MUSTARD
Yellow Daisy → WEDELIA sp.
Yellow False Oat → TRISETUM FLAVESCENS
Yellow Foxtail → SETARIA sp.
Yellow-flowered Nightshade → URECHITES sp.
Yellow Heads → GNIDIA KRAUSSIANA
Yellow Henbane → NICOTIANA sp.
Yellow Indigo → BAPTISIA sp.
Yellow Jacket → HORNETS also → WASPS
"Yellow Jackets" → PENTOBARBITAL
Yellow Jasmine → GELSEMIUM SEMPERVIRENS
Yellow Jessamine → GELSEMIUM SEMPERVIRENS
Yellow Mercury Iodide → MERCUROUS IODIDE
Yellow Mustard → MUSTARD
Yellow Oat Grass → TRISETUM FLAVESCENS
YELLOW OB also → ORANGE
Yellow Oleander → THEVETIA
Yellow Orange S → FDC YELLOW #6
Yellow Parilla → MENISPERMUM CANADENSE
Yellow Phosphorus → FIRECRACKERS
Yellow Pond Lily → NUPHAR
Yellow Puccoon → HYDRASTIS CANADENSIS
YELLOW RAIN also → DEOXYNIVALENOL, 4-DEOXYSCIRPENOL, PENICILLIUM sp., PHOMA sp., T-2 TOXIN
Yellow Resin → ROSIN
Yellow Rice → PATULIN also → PENICILLIUM sp.
Yellow Rocket → BABAREA VULGARIS
Yellow Root → HYDRASTIS CANADENSIS also → XANTHORHIZA SIMPLICISSIMA
Yellows → LYSERGIC ACID DIETHYLAMIDE
Yellow Snowdrop → ERYTHRONIUM
Yellow Star Thistle → CENTAUREA sp.
Yellow Toadflax → LINARIA VULGARIS
Yellow Ultramarine → CALCIUM CHROMATE
Yellow Vetchling → LATHYRUS sp.
Yellow Water Lily → NUPHAR
Yellow Wax → BEESWAX
Yellow Weed → DUGALDIN also → HELENIUM sp.
Yellow Wood → TERMINALIA sp. also → ZANTHOXYLUM sp.
Yerba mate → ILEX sp.
Yerba Mora → SOLANUM sp.
"Yesterday, Today, & Tomorrow" → BRUNFELSIA
Yetyl → TOBACCO
YEW also → FORMIC ACID, TAXINE
YLANG YLANG OIL
YM 177 → CELECOXIB
YM 294 → INTERLEUKIN 11
YM 617 → TAMSULOSIN
YM 08054 → INDELOXAZINE
YM 09330 CEFOTETAN
YM 11,170 → FAMOTIDINE
YM 12,617 → TAMSULOSIN
YM 14,090 → INTERFERONS
Yobir → ALPRENOLOL
Yochinol → CHINIOFON
Yodoxin → IODOQUINOL
Yoghurt → YOGURT
YOGURT also → CIPROFLOXACIN, GALACTOSE, MILK, YEAST
Yohimbe → YOHIMBINE
YOHIMBINE also → THIOPENTAL
δ-Yohimbine → AJMALICINE
Yomesan → NICLOSAMIDE
Yopo → ANADENANTHERA also → COHOBA
Yoristen → FONAZINE MESYLATE
York Road Poison → GASTROLOBIUM sp.
Yorkshire Fog → HOLCUS LANATUS
Youteshu → THEOPHYLLINE
Ypenyl → DOPAN
Yperite → MUSTARD GAS
Yuan-Hau → YUANHAUCINE
YUANHAUCINE
Yuanhaudine → YUANHAUCINE
Yuanhaufine → YUANHAUCINE
Yuanhautine → YUANHAUCINE
YUCCA
Yu Cheng Incident → POLYCHLORINATED BIPHENYLS
Yuehehukene → MURRAYA PANICULATA
Yüksükotu → DIGITALIS
YUREMA also → MIMOSA

Z

ZIERIA ARBORESCENS
Zigacine → ZIGADENUS sp.
ZIGADENUS sp.
Zildasac → BENDAZAC
ZILEUTON also → PROPRANOLOL, THEOPHYLLINE, WARFARIN
Zimate → ZIRAM
Zimbabwe Pride → VERNONIA sp.
Zimeldine → ZIMELIDINE
ZIMELIDINE also → ALCOHOL, ETHYL
Zimmwaldite → MICA
Zimox → PENICILLINS – "NEWER and SYNTHETICS" > *Amoxicillin*
Zinacef → CEFUROXIME
Zinamide → PYRAZINAMIDE
ZINC also → ACETYLENE, CADMIUM, CIPROFLOXACIN, GALVANIZED METALS, GINGER ALE, GREPAFLOXACIN, LEMON, MILK, NITRATES, NITRITES, PAINT, PAPER, RUBBER, SESAME, THALLIUM, TIN
Zinc Acetate → ZINC
Zincaps → ZINC
Zinc Carbonate → ZINC
Zincate → ZINC
Zinc Chloride → ZINC
Zinc Chromate → ZINC
Zinc Dibenzyl Dithiocarbamate → ZINC also → RUBBER
Zinc Diethyldithiocarbamate → ZINC
Zinc Fluoborat → PERMANENT PRESS
Zinc Hydroxide → CORTICOTROPIN
Zinc Insulin → ZINC
Zincite → ZINC
Zincmate → ZIRAM
Zinc Nitrate → PERMANENT PRESS
Zinc Omadine → ZINC
Zincomed → ZINC
Zinc Oxide → ZINC
Zinc Phosphide → ZINC
Zinc Pyridinethione → ZINC
Zinc Pyrion → ZINC
Zinc Pyrithione → ZINC
"Zinc shakes" → WELDING also → ZINC
Zincspar → ZINC
Zinc Stearate → ZINC
Zinc Sulfate → ZINC also → TETRACYCLINE
Zinc Vitriol → ZINC
Zinc White → ZINC
Zinc-Tox → ZINC
ZINEB
Zinecard → RAZOXANE also → DEXRAZOXANE
Zingiber officinale → GINGER, JAMAICAN
Zinjabil → GINGER, JAMAICAN
Zinnat → CEFUROXIME
ZINNIA also → DATURA, SUNFLOWER
ZINOPHOS
Zinosan → ZINEB
Zinostatin → NEOCARZINOSTATIN
Zip Bond → CYANOACRYLATES
ZIPEROL
Zip Grip → CYANOACRYLATES
ZIRAM
Zirberk → ZIRAM
ZIRCONIUM
Zirconium Oxide → ZIRCONIUM
Zirconium Oxychloride → ZIRCONIUM
Ziride → ZIRAM
Zirtek → CETIRIZINE
Zithromax → AZITHROMYCIN
ZR 36,374 → ILOPROST
ZR 112,119 → ABECARNIL
ZL 101 → NIZATIDINE
ZM 204,636 → QUETIAPINE FUMARATE
Zitox → ZIRAM
Zitoxil → ZIPEROL
ZNA → DILL
ZnDMD → ZIRAM
Zoaquin → IODOQUINOL
Zocor → SIMVASTATIN
Zocord → SIMVASTATIN
Zofran → ONDANSETRON
Zolacef → CEFAZOLIN SODIUM
ZOLADEX
Zolamine → ETHYLENEDIAMINE
ZOLMITRIPTAN also → CIMETIDINE
Zoloft → SERTRALINE
Zolone → PHOSALONE
ZOLPIDEM TARTRATE also → RIFAMPIN
Zoltec → FLUCONAZOLE
Zolyse → CHYMOTRYPSIN
Zomax → ZOMEPIRAC
Zomaxin → ZOMEPIRAC
ZOMEPIRAC also → NAPROXEN
Zomig → ZOLMITRIPTAN
Zoniden → TIOCONAZOLE
Zonifur → NIHYDRAZONE
ZONISAMIDE also → PHENOBARBITAL
Zonite → SODIUM HYPPOCHLORITE
Zonometh → DEXAMETHASONE
Zoom → GUARANA
Zoomycetin → CHLORAMPHENICOL
Zophren → ONDANSETRON
Zopirac → ZOMEPIRAC
Zorac → TAZAROTENE
Zorane → METAXALONE
Zoroxin → NORFLOXACIN

Zostrix → CAPSAICIN
Zosyn → PENICILLINS – "NEWER and SYNTHETICS" > *Piperacillin Sodium*
Zothelone → QUINURONIUM
Zoton → LANSOPRAZOLE
Zovirax → ACYCLOVIR
Zoxamin → ZOXAZOLAMINE
ZOXAZOLAMINE also → CHLORDANE, PHENOBARBITAL, PHENYLBUTAZONE, THYROXINE
Zoxine → ZOXAZOLAMINE
ZP → ZINC
ZPT → ZINC
ZR 515 → METHOPRENE
Z Span → ZINC
Zuclomiphene → CLOMIPHENE CITRATE
Zuclopenthixol → CLOPENTHIXOL
Zumaril → ALCLOFENAC
Zumenon → ESTRADIOL
Zunden → PIROXICAM
Zurma → CASTOR BEAN
Zyban → BUPROPION
Zyderm → COLLAGEN
Zyflo → ZILEUTON
Zygadenus → ZIGADENUS sp.
Zygomycin → PAROMOMYCIN
ZYGOPHYLLUM sp.
Zyklolat → CYCLOPENTOLATE
Zyklon → HYDROGEN CYANIDE
Zyloprim → ALLOPURINOL
Zyloric → ALLOPURINOL
Zymafluor → SODIUM FLUORIDE
Zymino → PROTEIN HYDROLYSATES
Zymofren → APROTININ
Zyprexa → OLANZEPINE
Zyrlex → CETIRIZINE
Zyrtec → CETIRIZINE
Zythiol → MALATHION
Zyvox → LINEZOLID